# ENGLISH – GERMAN

# TECHNICAL AND ENGINEERING

# DICTIONARY

# ENGLISH – GERMAN
# TECHNICAL AND ENGINEERING
# DICTIONARY

by

## DR LOUIS DE VRIES

*formerly Professor, Iowa State University*

and

## THEO M. HERRMANN

Reprint 1972
Completely Revised and Enlarged
1967

## McGRAW-HILL BOOK COMPANY

NEW YORK        LONDON        SYDNEY        TORONTO

## OSCAR BRANDSTETTER VERLAG KG · WIESBADEN

# ENGLISH – GERMAN
# TECHNICAL AND ENGINEERING
# DICTIONARY

von

## Dr. Louis De Vries

*vormals Professor an der Iowa State University*

und

## Theo M. Herrmann

Nachdruck 1972
vollkommen überarbeitet und erheblich erweitert
1967

## McGRAW-HILL BOOK COMPANY

NEW YORK     LONDON     SYDNEY     TORONTO

## OSCAR BRANDSTETTER VERLAG KG · WIESBADEN

"I shall not think my employment
useless or ignoble if . . . my labors
afford light to the repositories of
science."

DR. SAMUEL JOHNSON

English-German Technical and Engineering Dictionary, Second Edition
Copyright © 1967 by the McGraw-Hill Publishing Company Limited,
Maidenhead, Berkshire, England.

ISBN 3 87097 043 X

Printed in the Federal Republic of Germany.

Alleinauslieferung und Vertrieb für Deutschland Oscar Brandstetter Verlag KG, Wiesbaden
Exclusive rights of sale in Germany Oscar Brandstetter Verlag KG, Wiesbaden

**16610**

*Dedicated to*
*Eva-Margaret-Louise*
*with*
*Gratitude and Regard*

# PREFACE

The rapidly progressing development in science and technology necessitated also the preparation of the English-German volume in order to meet the demands for a modern up-to-date dictionary. Not only were antiquated terms removed and new terms added resulting from the growth in scientific and technical knowledge, but the authors also endeavered to expand the vocabulary by a very considerable extent.

The authors desire herwith to express their sincere gratitude to the many contributors in the various branches of industry for their valuable terminologies. At the same time recognition is due to Mrs. De Vries for her untiring effort in the preparation of the first draft of the manuscript and thus contributed her ample share to the successful completion of this volume.

Since no work of this kind is ever complete, the authors are always grateful for further contributions and suggestions to make such a work more valuable and helpful to its users.

October 1967

LOUIS DE VRIES
THEO M. HERRMANN

# VORWORT

Die stürmisch fortschreitende Entwicklung von Wissenschaft und Technik machte die Bearbeitung auch des englisch-deutschen Bandes notwendig, um den Ansprüchen, die der Benutzer an ein modernes Wörterbuch stellt, gerecht zu werden. Hierbei wurden nicht nur veraltete Termini ausgemerzt und neue Begriffe, die in Zusammenhang mit dem Anwachsen des technischen und naturwissenschaftlichen Wissens entstanden, hinzugefügt, die Verfasser haben sich auch bemüht, durch straffe Selektion des Wortgutes die bekannte Vielfalt des in diesem Wörterbuch gebotenen Stoffes noch zu erweitern.

Wie auch bei der Überarbeitung des Gegenbandes gebührt hier den Firmen der verschiedensten Industriezweige Dank für die Bereicherung des ausgewählten Wortmaterials. Zugleich sei auch für die unermüdliche Arbeit von Frau De Vries gedankt, die bei der Vorbereitung des ersten Manuskriptentwurfes wertvolle Hilfe leistete und damit zum Gelingen dieses Werkes beitrug.

Da keine Arbeit dieser Art je vollkommen ist, sind die Autoren dankbar für Vorschläge, die helfen können, Lücken zu schließen oder Verbesserungen vorzunehmen.

Oktober 1967

LOUIS DE VRIES
THEO M. HERRMANN

# COLLABORATORS – MITARBEITER

This task could not have been accomplished without the assistance and co-operation of the men and women listed here, to whom the author is deeply indebted.

C. R. Addinall, Library Service Bureau, Merck and Company, Inc., Rahway, New Jersey.

Hilda J. Alseth, Librarian of College of Engineering, University of Illinois.

Clarence E. Bardsley, Engineer, War Department, United States Engineers Office, Pittsburgh.

Helen Basil, Librarian, Crane Company, Chicago.

J. Christian Bay, Librarian, The John Crerar Library, Chicago.

K. F. Beaton, Shell Oil Company, New York.

R. L. Bebb, Research Laboratory Firestone Tire and Rubber Company, Akron, Ohio.

Mildred Benton, U.S. Department of Agriculture, Washington, D.C.

C. V. Bertsche, Lawyers' and Merchants' Translation Bureau, New York.

Charles H. Brown, Librarian, Iowa State College, Ames, Iowa.

Henry Brutcher, Technical Translation Service, Altadena, California.

Marguerite Burnett, Librarian, Federal Reserve Bank, New York.

Phyllis H. Carter, Science Branch Librarian, Carnegie Institute of Technology, Pittsburgh.

Huber O. Croft, Mechanical Engineer, State University of Iowa, Iowa City, Iowa.

Lucile S. Davison, Research and Technical Development Department, United States Rubber Company, Passaic, New Jersey.

Jessie Dudley, Librarian, Dow Chemical Company, Midland, Michigan.

Laura Eales, Head Technology Department, Bridgeport Public Library, Bridgeport Connecticut.

H. B. Edwards, Waterways Experiment Station. Corps of Engineers, Vicksburg, Mississippi,

Maude Ellwood, Laurence, Woodhams & Mills, Kalamazoo, Michigan.

Elma T. Evans, Librarian, Curtiss-Wright Corporation, Buffalo, New York.

Fred L. Fehling, German Department, University of Iowa, Iowa City, Iowa.

Robert F. Ferguson, Hazel Atlas Glass Company, Zanesville, Ohio.

Anna F. Frey, Librarian, Western Precipitation Corporation, Los Angeles, California.

Marie S. Gaff, Librarian, E. I. DuPont De Nemours & Company, Wilmington, Delaware.

Elsie L. Garvin, Librarian, Eastman Kodak Company, Rochester, New York.

C. R. Hammond, Librarian, Naval Ordnance Laboratory, United States Naval Gun Factory, Washington, D.C.

Margaret Hazen, Associate Reference Librarian, Massachusetts Institute of Technology, Cambridge, Massachusetts.

Hercules Powder Company Research Staff, Wilmington, Delaware.

L. A. Huguemont, Research Laboratory, General Electric, Schenectady, New York.

Otto M. Joergensen, Shell Oil Company, Hamburg, Germany.

E. W. Kammer and Ruth H. Hooker, Naval Research Laboratory, Anacostia Station, Washington, D.C.

Viktor Kleinlercher, Vienna, Austria.

Dr. ing. Fritz E. H. Koch, Heidelberg, Germany.

Norbert Kohlmayr, Vienna, Austria.

Ludwig Köllmann, Jr., Minden-Westfalen, Germany.

Otto Konig, Ph. D., National Lead Company, Brooklyn, New York.

Ruth Kristoffersen, Reference Librarian, Iowa State College, Ames, Iowa.

Max Lanzendörfer, München-Pasing, Germany.

Marie Lugscheider, Librarian, Engineering Division, Ranger Aircraft Engines, Farmingdale, New York.

Paul Maegerle, Winterbach-Schorndorf, Germany.

Vivian J. MacDonald, Librarian, Aluminum Research Laboratories, Aluminum Company of America, New Kensington, Pennsylvania.

W. O. Maxwell, International Harvester Company, Chicago.

Mary E. McCullough, Koppers Company, Inc., Pittsburgh.

John Metschl, Director, Technical Information Service, Gulf Research and Development Company, Pittsburgh.

Robert Möller, Stuttgart-Cannstatt, Germany.

Helen J. Moss, Librarian, Yale University, School of Engineering, New Haven, Connecticut.

Hanna B. Muller, Librarian, The Museum of Modern Art, New York.

Wilbur Nelson, Research Professor of Aeronautics, University of Michigan, Ann Arbor, Michigan.

Natalie Nicholson, Librarian, Engineering School, Harvard University, Cambridge, Massachusetts.

Alice C. Olson and Dr. Mudge, International Nickel Company, New York.

Dr. Alexander Oppermann, Philadelphia.

Robert W. Orr, Director, Iowa State College Library, Ames, Iowa.

Lisa G. Otto, Librarian, Research Laboratory, Interchemical Corporation, New York.

Martha Peppel, Librarian, Lumus Company, New York.

Ethel C. Pierce, Librarian, Climax -Molybdenum Company, Detroit, Michigan.

F. A. Raven, David Taylor Model Basin, Carderock, Maryland, now at Massachusetts Institute of Technology, Cambridge, Massachusetts.

Dr. Raymond Schmidt, Frankfurt/Main, Germany.

B. R. Regen, Researcher and Translator, New York.

Richard Rimbach, Managing Editor of Instruments Publishing Company, Pittsburgh.

Margaret M. Rocq, Librarian, Standard Oil Company of California, San Francisco, California.

Herbert Rodeck, Armour Research Foundation, Technology Center, Chicago.

J. D. Ryder, Professor of Electrical Engineering, Iowa State College, Ames, Iowa.

Esther Schlundt, Reference Librarian, Purdue University, Lafayette, Indiana.

Else L. Schulze, Technical Librarian, Procter and Gamble Company, Ivorydale, Ohio.

Lewis L. Sell, Technical Translation Bureau, New York.

John M. Sharf, Armstrong Cork Company, Research Laboratory, Lancaster, Pennsylvania.

T. E. R. Singer, Research and Translation, New York.

Julian F. Smith, Technical Librarian, Institute of Textile Technology, Charlottesville, Virginia.

Leah E. Smith, Librarian, Bell Telephone Laboratories, New York.

Maurice H. Smith, Librarian, Institute of Aeronautical Science, New York.

Jane Ulrey, Librarian, Westinghouse Electric Company, Bloomfield, New Jersey.

Le Roy D. Weld, Professor of Physics, Coe College, Cedar Rapids, Iowa.

Evelyn Wimersberger, Head, Catalog Department, Iowa State College, Ames, Iowa.

Herbert E. Wolff, Wiesbaden, Germany.

Eleanor V. Wright, Librarian, Chrysler Corporation, Detroit, Michigan.

Max Wulfinghoff, Mechanical and Chemical Engineer, Cincinnati, Ohio.

# List of used abbreviations
## Aufstellung der verwandten Abkürzungen

| | | |
|---|---|---|
| acoust. | acoustics | Akustik |
| aerodyn. | aerodynamics | Aerodynamik |
| agr. | agriculture | Landwirtschaft |
| arch. | architecture | Baukunst |
| artil. | artillery | Artillerie |
| astron. | astronomy | Astronomie |
| atom. | atomics | Atomwissenschaft |
| aviat. | aviation | Luftfahrt |
| chem. | chemistry | Chemie |
| comm. | commerce | Handel |
| comput. | computer | Rechner |
| CRT | cathode ray tube | Kathodenröhre |
| cryst. | crystallography | Kristallografie |
| data proc. | data processing | Datenverarbeitung |
| electr. | electricity | Elektrizität |
| electron. | electronics | Elektronik |
| engin. | engineering | Ingenieurwesen |
| film | motion pictures | Filmwesen |
| geol. | geology | Geologie |
| geophys. | geophysics | Geophysik |
| g/m | guided missiles | Fernlenkwaffen |
| hydraul. | hydraulics | Hydraulik |
| mach. | machinery | Maschinenwesen |
| math. | mathematics | Mathematik |
| meas. | measure(ment) | Maß, Meßtechnik |
| mech. | mechanics | Mechanik |
| med. | medicine | Medizin |
| meteor. | meteorology | Wetterkunde |
| metall. | metallurgy | Metallurgie |
| mil. | military | Militärwesen |
| min. | mining | Bergbau |
| miner. | mineralogy | Mineralogie |
| nav. | navigation | Navigation |
| opt. | optics | Optik |
| paper mfg. | paper manufacturing | Papierherstellung |
| phono | phonography | Tonaufzeichnung |
| photo | photography | Fotografie |
| phys. | physics | Physik |
| p.p. | power plant | Triebwerk |
| print. | printing | Buchdruck, Druckwesen |
| rdo | radio | Funk, Radio |
| rdr | radar | Funkmeßwesen, Radar |
| r.r. | railroad | Eisenbahn |
| sec. rdr. | secondary radar | Sekundärradar |
| statist. | statistics | Statistik |
| tape rec. | tape recorder | Bandaufnahmegerät |
| | tape recording | Bandaufnahme |
| telegr. | telegraphy | Telegrafie, Tastfunk |
| teleph. | telephony | Telefonie, Sprechfunk |
| telet. | teletype | Fernschreibwesen |
| TV | television | Fernsehen |

# EXPLANATIONS

### 1. Arrangement of keywords

The principal keywords have been placed to the left for better emphasis. Since compound words make up the main part of a technical dictionary, these were grouped together for clarity. The principal keyword which stands in front is repeated by a tilde (sign of repetition) ~.

### 2. Alphabetical arrangement

The alphabetical order has been retained throughout without regard for the valuation of the individual term. If a keyword is simultaneously a verb, substantive or adjective (adverb), the verb comes first, followed by the substantive and the adjective, separated by a semicolon.

Example: **close, to ~**     abschließen, etc.

       **close**        Beschluß *m*, Schluß *m;* dicht, eng

Present participle and past participle of the preceding keyword **close, to ~** appear in alphabetical order as proper keywords. The same is true for adverbial forms of the adjective **close,** as e.g. **closely.**

### 3. Compound words

In addition to compounds which are made up in English by putting together two or more words, there are numerous compounds which are formed with prepositions such as **of, from, by, to, with.** In order to preserve the greatest possible clarity, the method of separating both types of compositions was employed. This is explained by the following example:

| | |
|---|---|
| Keyword (substantive and possible adjective) | **clamp** Aufspannfrosch *m*, Balkweger *m*, Briede *f*, Bügel *m*, etc. |
| Compounds of this keyword with prepositions | **clamp, ~ of bricks** Hag *m* **~ on dark slide** Anlegekassette *f* **~ for fine adjustment** Klemmung *f* der Feinverstellung, etc. |
| Compounds of this keyword without intermediary prepositions | **clamp, ~ bolt** Spannschraube *f* **~ bracket** Klemmhülse *f* **~ carrier** Halteklammer *f* **~ collar** Klemm-, Verschluß-ring *m* **~ cone** Einspannkonus *m*, etc. |

One should take notice that compounds with prepositions are arranged alphabetically not according to the above terms, but according to the following terms.

### 4. Spelling style

Variations in American and English spelling are not marked. American spelling was generally preferred.

### 5. Hyphen

If the keyword is composed of two words with a hyphen as for instance **cross-head** in the case of repetitions only one tilde ~ is used. The term is considered as keyword.

### 6. Illustrating references

Such references are given in two different versions. In the case of subject matters, the abbreviations listed on page 10 as for instance g/m, rdo, rdr, r.r., are used and placed after the German word. In order to limit the list of abbreviations, more directed illustrations were used in the text in plain language which (also in brackets) are coordinated with the German terms, as e.g. (stretched across flight deck of aircraft carrier) Zugriemen *m*, (in a curve) Knick *m*, (of a curve) Knickstelle *f*, etc.

## 7. Use of o r respectively o d e r

For reasons of saving space where equivalence of expressions permits, a repetition of entire words is avoided and the various possibilities of expressions are marked in English by **or** or in German by **oder.**

## 8. Abbreviations

In contrast to the first edition of this dictionary an appendix with abbreviations has been omitted since such an appendix within the limits of this volume on the one hand cannot claim the completeness and standardization demanded while on the other hand extensive volumes with Anglo-American/German abbreviations are already on the market.

# ERLÄUTERUNGEN

## 1. Stichwortanordnung

Die Hauptstichwörter sind zur besseren Hervorhebung nach links ausgerückt. Da zusammengesetzte Wörter den Hauptteil eines technischen Wörterbuches ausmachen, wurden diese übersichtlich in Gruppen zusammengefaßt. Durch die Tilde (Wiederholungszeichen) ∼ wird das Hauptstichwort, das am Anfang steht, wiederholt.

## 2. Alfabetische Anordnung

Die alfabetische Anordnung ist ohne Rücksicht auf die Wertigkeit des einzelnen Begriffes durchgehend eingehalten worden. Ist ein Stichwort gleichzeitig Verbum, Substantiv oder Adjektiv (Adverb), so steht das Verbum an erster Stelle, es folgen das Substantiv und das Adjektiv, getrennt durch ein Semikolon.

Beispiel: **close, to ∼**    abschließen, etc.

**close**    Beschluß *m*, Schluß *m;* dicht, eng

Partizip Präsens und Partizip Perfekt des vorangegangenen Stichwortes **close, to ∼** erscheinen in alfabetischer Anordnung als eigene Stichwörter. Dasselbe gilt auch für adverbiale Formen des Adjektivs **close,** wie z. B. **closely.**

## 3. Zusammengesetzte Wörter

Neben den Komposita, die im Englischen durch Zusammensetzung von zwei oder mehr Wörtern gebildet werden, gibt es eine Vielzahl von Wortzusammensetzungen, die mit Präpositionen, wie **of, from, by, to, with** gebildet werden. Um hier die größtmögliche Übersicht zu wahren, wurde der Weg der Trennung beider Arten von Komposita beschritten. Das nachfolgende Beispiel gibt darüber Aufschluß:

| | |
|---|---|
| Stichwort (Substantiv und evtl. Adjektiv) | **clamp** Aufspannfrosch *m*, Balkweger *m*, Briede *f*, Bügel *m* etc. |
| Zusammengesetzte Wörter dieses Stichwortes mit Präpositionen | **clamp, ∼ of bricks** Hag *m* **∼ on dark slide** Anlegekassette *f* **∼ for fine adjustment** Klemmung *f* der Feinverstellung etc. |
| Zusammengesetzte Wörter dieses Stichwortes ohne dazwischengesetzte Präpositionen | **clamp, ∼ bolt** Spannschraube *f* **∼ bracket** Klemmhülse *f* **∼ carrier** Halteklammer *f* **∼ collar** Klemm-, Verschluß-ring *m* **∼ cone** Einspannkonus *m* etc. |

Es ist zu beachten, daß zusammengesetzte Stichwörter mit Präpositionen nicht nach diesen sondern nach den folgenden Begriffen alfabetisch geordnet sind.

## 4. Schreibweise

Unterschiede in der amerikanischen und englischen Schreibweise sind nicht gekennzeichnet. Der amerikanischen Schreibweise wurde im allgemeinen der Vorzug gegeben.

## 5. Bindestrich

Besteht das Stichwort aus zwei Wörtern, die mit einem Bindestrich verbunden sind, z. B. **cross--head,** so wird bei Wiederholungen nur e i n e Tilde ∼ benutzt, d. h. der Terminus gilt als ein Stichwort.

## 6. Erläuternde Hinweise

Erläuternde Hinweise werden in zwei verschiedenen Versionen gegeben. Wo es sich um Sachgebiete handelt, werden die auf Seite 10 verzeichneten Abkürzungen, wie z. B. g/m, rdo, rdr, r.r., verwendet und grundsätzlich dem deutschen Wort nachgestellt. Um das Abkürzungsverzeichnis begrenzt zu halten, wurden mehr gezielte Erläuterungen im Klartext verwendet, die, ebenfalls in Klammern, den deutschen Begriffen beigeordnet sind, wie z. B. (stretched across flight deck of aircraft carrier) Zugriemen *m*, (in a curve) Knick *m*, (of a curve) Knickstelle *f*, etc.


## 7. Verwendung von o r bzw. o d e r

Aus Raumersparnisgründen wurde dort, wo es die Eindeutigkeit der Ausdrücke zuläßt, auf eine Wiederholung ganzer Wörter verzichtet und die verschiedenen Ausdrucksmöglichkeiten durch **or** im Englischen bzw. **oder** im Deutschen gekennzeichnet.


## 8. Abkürzungen

Im Gegensatz zur ersten Auflage dieses Wörterbuches wurde auf einen Anhang mit Abkürzungen verzichtet, da ein derartiges Unterfangen in dem hier gesteckten Rahmen einerseits den Anspruch auf Vollständigkeit und Standardisierung nicht erheben kann und andererseits bereits umfangreiche Bücher mit anglo-amerikanisch/deutschen Abkürzungen auf dem Markt sind.

# A

**A-1 on-off keying (CW)** A1 Telegrafie *f* (tonlos)
**A-2 tone keying (MCW)** A2 Telegrafie *f* (tönend)
**A-3 Telephony** A3 Telefonie *f*
**A-4 Facsimile** A4 Bildfunk *m*
**abac** Nomogramm *n*, Nomograf *m*, Rechentafel *f*
**abacus** Rechenbrett *n*, Säulendeckplatte *f*
**abaft** Achter *m*; hinterwärts, am Hinterteil *n*
**abampere** Abampere *n*
**abandon, to** ~ aufgeben, auflassen, (the plane) aussteigen, (an application, a patent) fallen lassen; preisgeben, überlassen, verlassen, verwerfen, verzichten
**abandoned workings** verlassene Baue *pl*
**abandonment** Aufgabe *f*, Patentanspruchsaufgabe *f* (Patentanmeldung), Preisgebung *f*, Verlassen *n*, Verzicht *m*, Verzichtleistung *f*
**abate, to** ~ (fire) abflauen, aufheben, sich beruhigen, nachlassen, (chem) sinken, (smoke) verzehren
**abated** herabgesetzt
**abatement** Abschlag *m*, Nachlassen *n*, Preisnachlaß *m*, Rabatt *m*, Ungültigmachen *n*, Vergünstigung *f*, Verminderung *f*
**abatis** Baumverhau *m*
**A battery,** A-Verstärker *m*, Heizbatterie *f*
**abaxial** achsenentfernt, außer Achse ~ **ray** außerachsialer Strahl *m*
**Abbe drawing cube** Abbesches Würfelchen *n*
**abbreviate, to** ~ abkürzen, kürzen, verkürzen
**abbreviated** abgekürzt, kurzgefaßt ~ **term** Stummelwort *n*
**abbreviation** Abbreviatur *f*, Abkürzung *f*, Kürzung *f*, Kurzwort *n*, Kurzzeichen *n*, Stummelwort *n* ~ **key** Abkürzungstaste *f* (typewriter)
**ABC telegraph** ABC-Telegraf *m*, Buchstabentelegraf *m*
**abcoulomb** Abcoulomb *n*
**abduct, to** ~ (heat) ableiten, abduzieren
**abeam** gegenüberstehend ~ **of** dwars
**aberration** Abbildungsfehler *m* (opt.), Aberration *f* (photo), Abirrung *f*, Ablenkung *f*, Abweichung *f*, (in lenses) Fehler *m* ~ **of form** Bildwölbung *f* (photo) ~ **of light** Abirrung *f* des Lichtes ~ **function** Fehlerfunktion *f*
**aberrationless** fehlerfrei
**abet, to** ~ (law) anreizen
**abetment** (law) Anreizung *f*
**abfarad** Abfarad *n*
**ability** Fähigkeit *f*, Können *n*, Qualifikation *f*, Vermögen *n*
**ability,** ~ **to earn a living** Erwerbsfähigkeit *f* ~ **to expand (or extend)** Ausdehnungsvermögen *n* ~ **to flow** Fließfähigkeit *f* ~ **to follow in the oscillation** Mitschwingungsvermögen *n* ~ **to hold a course** Fahrtrichtungsschaltung *f* ~ **to hover** Schwebe-fähigkeit *f* -vermögen *n* ~ **to lend itself to gluing** Leimfähigkeit *f* ~ **to maintain cutting power** Schneidhaltigkeit *f* ~ **to be modulated** Modulierbarkeit *f* ~ **to pay** Solvenz *f* ~ **to preserve keenness** Schneidhaltigkeit *f* ~ **to respond** Ansprechvermögen *n* ~ **to retain cutting power** Schneidhaltigkeit *f* ~ **to start** (motor) An-

drehvermögen *n* ~ **to supply** Abgabefähigkeit *f* ~ **to vibrate** Schwingfähigkeit *f* ~ **to withstand overstressing by harmonics** Oberwellenbelastbarkeit *f*
**ability test** Eignungsprüfung *f*
**abjudication** (of rights) Aberkennung *f*
**ablating** abschmelzbar, abschmelzend, verdampfend
**ablation** Ablation *f*, Abschmelzen *n*, Abtragung *f*, Fort-tragen *n*, -waschen *n*
**ablaze** lichterloh
**able** fähig, geschäftskundig, geschickt ~ **to compete** konkurrenzfähig ~ **to fly** flugfähig ~ **to pay** solvent ~ **to put up security** kautionsfähig ~ **to support load** tragfähig ~ **to vibrate** schwingfähig ~ **to work** arbeitsfähig
**ablution** Abwaschen *n*
**A-B method** A-B-Betrieb *m*
**Abney level** Spiegelneigungsmesser *m*
**abnormal** anormal, entartet, normwidrig, regelwidrig, ungewöhnlich, ungewollt ~ **condition** Mißstand *m* ~ **radiation** Funkstreuung *f*
**abnormality** Entartung *f*, Regelwidrigkeit *f*
**abnormally fast drive ( or run)** Rohgang *m*
**aboard** an Bord
**A-board** abgehende Plätze *pl*, Ausgangsplätze *pl* ~ **toll operation** A-B-Betrieb *m*, Betrieb *m* über A- und B-Plätze ~ **toll traffic** A- und B-Betrieb *m* im Fernverkehr
**abohm** Abohm *n*
**A-, B-operator** (local) A- oder B-Beamtin *f*
**abolish, to** ~ abschaffen, abtun, aufheben, außer Gebrauch *m* setzen, ungültig machen
**abolished, to be** ~ in Fortfall *m* kommen, in Wegfall *m* kommen
**abolition** Abschaffung *f*, Aufhebung *f*, Einziehung *f*, Ungültigmachen *n*
**aborted take-off** Startabbruch *m*
**abounding in springs** quellenreich
**about-face turn** Kehrtwendung *f*
**above** oben, oberhalb, über ~ **the average** über dem Durchschnitt *m* ~ **floor level** Überflur . . . ~ **grade** Überflur . . . ~ **ground** oberirdisch, über Tag *m* ~ **ground (or floor) hydrant** Überflurhydrant *m* ~ **-ground structures** Hochbauten *pl* ~ **-lying hydro-electrical power plant** obenliegendes Kraftwerk *n* ~ **platen device** Oberwalzenführung *f* ~ **-water tube** Überwasserrohr *n*
**abrade, to** ~ (ab)reiben, abschaben, abschleifen **to** ~ **with emery** abschmirgeln **to** ~ **(or rub) with pumice** abbimsen
**abraded lead particles** Bleispäne *pl*
**abrase, to** ~ abreiben, abtun, abschürfen
**abrasimeter** Scheuerprüfer *m*
**abrasion** Abnutzung *f*, Abreiben *n*, Abreibung *f*, Abrieb *m*, Abschaben *n*, Abschabung *f*, Abscheuerung *f*, Abschleifen *n*, Abtragung *f*, Verschleiß *m*
**abrasion,** ~ **-proof** abriebfest ~ **resistance** Abriebfestigkeit *f*, Scheuerfestigkeit *f* ~ **-resistant** verschleißfest ~ **strength** Abriebfestigkeit *f*, Kratzfestigkeit *f* ~ **test** Abreibungsprobe *f*,

Abreibversuch *m*, Rißhärteprobe *f*, Schleifversuch *m*, Verschleißprüfung *f* ~ **tester** Abreibeprüfmaschine *f*, Scheuerprüfer *m* ~ **testing machine** Abnutzungsmaschine *f*, Schleifmaschine *f* ~ **wear** Verschleiß *m* durch Schleifwirkung

**abrasive** Abreibungsmittel *n*, Abrieb *m*, Abschabung *f*, Blassand *m*, Gebläsesand *m*, Putzsand *m*, Reibmittel *n*, Schleifmaterial *n*, Schleifmittel *n*, Schmirgel *m*; abreibend, abschleifend, schmirgelartig

**abrasive,** ~ **action** Scheuerwirkung *f*, Schleifwirkung *f* ~ **and cutting-off machine** Trennschleifmaschine *f* ~ **base paper** Schleifmittelrohpapier *n* ~ **cloth** Schmirgelleinen *n* ~ **coated fabrics** Schleifgewebe *n* ~ **dust** Schleifstaub *m* ~ **effect** Scheuerwirkung *f* ~ **grain** Schleifkorn *n* ~ **hardness** Ritzhärte *f* ~**-hardness test** Ritz-(härte)probe *f* ~ **paper** Polierpapier *n*, Putzpapier *n*, Schleifpapier *n*, Schmirgelpapier *n* ~ **papers** Rostpapiere *pl* ~ **powder** Reibepulver *n*, Schmirgel *m* ~ **power** Schleifwirkung *f* ~ **raw materials** Schleifrohstoffe *pl* ~ **separator** Gebläsesandsichter *m* ~ **temper** Schleifgüte *f* ~**-wear tester** Abreibeprüfmaschine *f* ~ **wheel hardness tester** Härteprüfmaschine *f* für Schleifscheiben

**abreast** nebeneinander, in Dwarslinie *f* ~ **of** auf der Höhe von, querab

**abridge, to** ~ abkürzen, verkürzen

**abridgment** Abkürzung *f*, Auszug *m*, gekürzter Auszug *m*

**abrogation** Aberkennung *f*, Aufhebung *f*, Ausfallserscheinung *f*, Außerkraftsetzung *f*

**abrupt** abschüssig, jäh, kurz, plötzlich, schroff, steil ~ **change of cross section** sprungweise Querschnittsänderung *f* ~ **change of voltage** Spannungssprung *m* ~ **drop** Steilabfall *m* ~ **reversing** Umspringen *n* ~ **roll in the direction of flight** gerissene Rolle *f* ~ **sharp angle** scharfer Ansatz *m*

**abruptness** Plötzlichkeit *f*

**abscissa** Abszisse *f*, Auftragslinie *f* ~ **axis** X-Achse *f* ~ **of an image point** Bildabszisse *f* ~ **for maximum ordinate of trajectory** Gipfelentfernung *f* ~ **of point of cutoff** Brennschlußweite *f*

**absence** Abwesenheit *f*, Nichterscheinen *n* **in** ~ **of** mangels **in the** ~ **of** in Ermangelung *n* von **in the** ~ **of light** unter Lichtabschluß *m* ~ **of air** Luftabschluß *m* ~ **of any break of vision in a bifocal glass** Unmerklichkeit *f* des Bildsprunges bei einer Zweistärkenbrille ~ **of color** Farblosigkeit *f* ~ **of current** Stromlosigkeit *f* ~ **of danger** Gefahrlosigkeit *f* ~ **of distortion** Verzerrungsfreiheit *f* ~ **of disturbing radiations** Störfreiheit *f* ~ **of gases** Gasfreiheit *f* ~ **of haze in the image** Schleierfreiheit *f* des Bildes ~ **of twist** Drallfreiheit *f* ~ **of vibration** Vibrationsfreiheit *f*

**absent** abwesend **to be** ~ fehlen ~ **subscriber service** Fernsprech-kundendienst *m* oder -auftragsdienst *m*

**absentee** Abwesender *m* ~ **service** Auftragsdienst *m*

**absolute** absolut, unverschlüsselt, vollständig ~ **air filter** Feinstfilter *m* ~ **altimeter** absoluter Höhenmesser *m* ~ **altitude** absolute Höhe *f* ~ **angle of attack** absoluter Anstellwinkel *m* (aerodyn.) ~ **apparatus** absolut eichbares Meßgerät *n* ~ **block** absoluter Block *m*, unbedingter Halt *m*, unbedingte Raumfolge *f* (aerodyn.) ~ **ceiling** höchste Flughöhe *f*, höchste erflogene Gipfelhöhe *f*, Rechnungsgipfelhöhe *f*, Versuchsgipfelhöhe *f* ~ **deviation** Geschoßabweichung *f* ~ **direction** eindeutige Peilseite *f* ~ **efficiency** absoluter Nutzeffekt *m* ~ **electrometer** Spannungswaage *f* ~ **flying altitude** absolute Flughöhe *f* ~ **force** absolute Kraft *f* ~**-frequency meter** absoluter Frequenzmesser *m* ~ **height** Meereshöhe *f* ~ **humidity** absolute Feuchtigkeit *f*, Luftsättigung *f* ~ **pitch** absolutes Tongehör *n* ~ **pressure** absolute Spannung *f*, (rolling) absoluter Druck *m* ~ **quarantine** Isolierung *f* ~ **specific gravity** absolutes spezifisches Gewicht *n* ~ **speed** Fluggeschwindigkeit *f* über Grund ~ **stop signal** unbedingtes Haltesignal *n* ~ **system** (of Gauss) absolutes System *n* ~ **term** absolutes Glied *n* ~ **terms** Absolutberechnung *f* ~ **unit** absolute Einheit *f*, absolute Maßeinheit *f*, Höhe *f* ~ **value** absoluter Wert *m* ~ **zero** absoluter Nullpunkt *m*

**absolutely** durchaus, restlos ~ **accurate,** ~ **precise,** ~ **correct** haarscharf ~ **straight** schnurgerade

**absolve, to** ~ (from debts, etc) erlassen

**absorb, to** absorbieren, ansaugen, anziehen, auffangen, aufnehmen, aufsaugen, aufzehren, binden, einsaugen, (energy from an electron beam) entnehmen, (sound) dämpfen, (light) schwächen, verschlucken **to** ~ **heat** Wärme *f* aufnehmen **to** ~ **shocks** abfedern

**absorbable** absorbierbar, aufnahmefähig

**absorbability** Absorbierbarkeit *f*

**absorbance** Extinktion *f* ~ **index length of absorbing path** Extinktionsmodul *m*

**absorbed,** ~ **energy** verschluckte Energie *f* ~ **gas** absorbiertes Gas *n* ~ **substance** sorbierter Stoff *m* **to be** ~ angehen (Farbstoff)

**absorbency** Saugfähigkeit *f*

**absorbent** Absorptionsmittel *n*, Einsaugemittel *m*, Einziehungsmittel *n*, Schluckstoff *m*, sorbierender Stoff *m*

**absorbent** aufsaugend, einsaugend, saugfähig

**absorbent,** ~ **cotton** Watte *f* ~ **gauze treated with bismuth for burns** Brandbinde *f* ~ **oil** Absorptionsöl *n* ~ **paper** Bindenkrepp *m*, Fließpapier *n*, Löschpapier *n*, saugfähiges Papier *n*

**absorber** Absorptionsapparat *m*, Absorptionsmittel *n*, Aufnehmer *m* ~ **circuit** Absorptionskreis *m* ~ **tower** Absorptionsturm *m* ~ **tube** (or valve) Ballaströhre *f*, Absorptionsröhre *f*

**absorbing** absorbierend, aufnehmend

**absorbing,** ~ **apparatus** Absorptionsapparat *m* ~ **capacity of a lens** Absorptionswert *m* eines Glases ~ **circuit** Absorptionsstromkreis *m* ~ **device** Auffangvorrichtung *f* ~ **duct** Schalldämpfer *m* ~ **field** Auskoppelfeld *n* ~ **filter** Absorptionsfilter *m*, (of neutral or tinted glass) neutrales Blendglas *n* ~ **fluid** absorbierende Schicht *f* ~ **layer** absorbierende Schicht *f* ~ **material** absorbierende Schicht *f* ~ **medium** Absorptionsmittel *n* ~ **rod** absorbierender Stab *m* ~ **screen** Strahlenfilter *n*, Absorptionsschirm *m* ~ **sound** Schalldämpfungsmittel *n*

**absorption** (power) Absorbierbarkeit *f*, Absorp-

tion *f*, (of a gas) Aufnahme *f*, Aufsaugung *f*, Aufzehrung *f*, Aufziehen *n* (eines Farbstoffes), Bindung *f*, Dämpfung *f*, Einziehung *f*, Resorption *f*, Schluckung *f*, (of light) Verlust *m*, Verwärmung *f*
**absorption,** ~ **of energy** Kraftverbrauch *m*, Leistungsentnahme *f* ~ **of gas** Gasaufnahme *f* ~ **of gases** Gasaufzehrung *f* ~ **of heat** Wärmetönung *f* ~ **of iron** Eisenaufnahme *f* ~ **of power** Leistungsentnahme *f* ~ **of sulfur** Schwefelaufnahme *f* ~ **of water** Wasseraufsaugung *f*
**absorption,** ~ **bands** Absorptionsstreifen *m* ~ **bulb** Vorlage *f* ~ **capacity** Absorptionsvermögen *n*, Aufnahmevermögen *n*, Aufsaugevermögen *n*, Schluckfähigkeit *f* ~ **cell** Absorptionsküvette *f* ~ **circuit** Saugkreis *m* (rdr) ~ **coefficient** Absorptionskoeffizient *m*, (screening function) Abschirmkoeffizient *m* ~ **coloring** Einsaugungsfärbung *f* ~ **column** Absorptionsrohr *n* ~ **current** Absorptionsstrom *m*, Nachwirkungsstrom *m* ~ **curve** Absorptionskurve *f* ~ **discontinuity** Absorptionskante *f*, Absorptionssprung *m* ~ **dynamometer** Bremsdynamometer *n* ~ **flask** Aufnahmekolben *m* ~ **layer** Sumpf *m* ~ **limiting frequency** durch Absorption *f* begrenzte Frequenz ~ **loss** Schluckverlust *m* (acoust.) ~**-measuring method** absorptiometrische Methode *f* ~ **modulation (method)** Absorptionsmodulation *f* ~ **oil** Waschöl n ~ **plant** Absorptionsanlage *f* ~ **point** Aufnahmespitze *f* ~ **potential** Absorptionsanteil *m* des Potentials ~ **power** Aufnahmefähigkeit *f*, Aufsaugevermögen *n* ~ **rate** Absorptionsgrad *m* ~ **spectra of crystals** Absorptionskristallspektren *pl* ~ **spectra of x-rays** Absorptionsröntgenspektren *pl* ~ **spectrum** Absorptionsstreifen *m* ~ **tower** Absorptionsturm *m* ~ **tube** Absorptionsrohr *n*, Audiometer *n* ~**-type refrigeration machine (or unit)** Absorptionskälteanlage *f*, Absorptionskältemaschine *f* ~ **vessel of variable depth of fluid** Absorptionsgefäß *n* mit veränderlicher Schichtdicke ~ **wavemeter** Absorptionswellenmesser *m* ~ **wave trap** Wellenabsorptionssaugkreis *m*
**absorptive** absorptionsfähig, aufnahmefähig
**absorptive,** ~ **capacity** Saugfähigkeit *f*, Wasseraufnahmefähigkeit *f* ~ **height** Saughöhe *f* ~ **material** Schluckstoff *m* ~ **power** Absorptionskraft *f*, Absorptionsvermögen *n*
**absorptivity** Absorptionsvermögen *n*, Schluckbeiwert *m*
**abstain, to** ~ **from** unterlassen
**abstract, to** ~ abführen, ableiten, absondern, ausziehen, (energy from an electron beam) entnehmen, entziehen
**abstract** Abriß *m*, (of an article) Auszug *m*, Extrakt *m*, Übersicht *f*, Zusammenfassung *f*
**abstract,** ~ **of accounts** Rechnungsauszug *m* ~ **doctrine theorem** Lehrsatz *m* ~ **number** reine Zahl *f* (ohne Dimension) ~ **set** Studiodekor *m* ~ **voucher** Sammelbeleg *n*
**abstracting service** Referierdienst *m*
**abstraction** Entziehung *f* ~ **of water** Wasserentziehung *f*
**abukumalite** Abukumalith *n*
**abundance** Ausgiebigkeit *f*, Fülle *f*, Häufigkeit *f*, Reichhaltigkeit *f*, Reichtum *m*, Überfluß *m* ~ **of seams** Flözreichtum *m* **with** ~ **of lines** linienreich

**abundance,** ~ **anomalies** Häufigkeitsanomalien *pl* ~ **ratio** Häufigkeitsverhältnis *n*
**abundant** ausgiebig, ergiebig, reichhaltig, reichlich
**abuse** Mißbrauch *m*, vorschriftswidrige Behandlung *f*
**abut, to** ~ anliegen an, anstoßen, (against) auftreffen **to** ~ **upon** angrenzen
**abutment** Auflager *n*, Brückenendwiderlager *n*, Brückenwiderlager *n*, Drucklager *n*, Endauflager *n*, Endwiderlager *n*, Gegenlager *n*, Ortspfeiler *m*, Seitenschild *n*, Strebepfeiler *m* (einer Brücke), Wehrwange *f*, Widerlager *n*, Widerlagspfeiler *m* ~ **beam** Landstoß *m*, Stoßbalken *m*, Stoßbohle *f* ~ **boss** Anschlaghocker *m* ~ **pier** Widerlager *n* ~ **transom** Kämpfer *m*
**abutting** Verbrüstung *f* ~ **end** Stoßfläche *f* ~ **surface** Auflauffläche *f*
**abyss** Abgrund *m*
**a.c. A.C. a-c (alternating current)** Wechselstrom *m*
**a.c. A.C. a-c,** ~ **circuit** Wechselstromnetz *n* ~**-d.c.-set** Allstromgerät *n* ~ **dump** Wechselspannungsunterbrechung *f* ~ **erasing head** Wechselstromlöschkopf *m* ~ **motor** Wechselstrommotor *m* ~ **voltage** Wechselspannung *f* ~ **voltage on the grid** Gitterschwingung *f* ~ **voltmeter** Spannungsmesser *m* für Wechselspannung
**acacia** Akazie *f* ~ **oil** Akazienöl *n*
**acaroid,** ~ **gum (or resin)** Akaroidharz *n*
**accede, to** ~ beistimmen, genehmigen **to** ~ **to** einwilligen
**accedence** Beitritt *m*, Einwilligung *f*
**accelerate, to** ~ beschleunigen, hochfahren, in Schwung *m* bringen, schneller werden
**accelerated** beschleunigt ~ **aging test** Schnellalterungsversuch *m* ~ **combustion** beschleunigte Verbrennung *f* ~ **course** Schnellkurs *m* ~ **method** Abkürzungsverfahren *n* ~ **stop** Startlaufabbruch *m* ~ **stop distance available** Start-(lauf)abbruchstrecke *f*, verfügbare Start(lauf)-abbruchstrecke *f* ~ **take-off** Schnellstart *m* ~ **test** Kurzversuch *m*, Kurzzeitprüfung *f* ~ **turn** Kurve *f* mit Beschleunigung *f* (aviat.)
**accelerating,** ~ **agent** Abbindebeschleuniger *m* ~ **anode** Beschleunigungsanode *f* ~ **field** Beschleunigungsfeld *n* ~ **grid** Beschleunigungsgitter *n* ~ **pump** Beschleunigerpumpe *f*, Vergaser *m* ~ **relay** Fortschaltrelais *n* ~ **secondary-electron multiplier** Pendelvervielfacher *m* ~ **voltage** Beschleunigungsspannung *f*, Saugspannung *f*
**acceleration** Anlauf *m*, Anzugsmoment *n*, Anzugsvermögen *n*, Beschleunigung *f*, Erhöhung *f*, Geschwindigkeitssteigerung *f*, Heraufsetzung *f* (der Drehzahl), Steigerung *f*
**acceleration,** ~ **of the current** Strombeschleunigung *f* ~ **of a freely falling body** Fallbeschleunigung *f* ~ **due to gravity** Erdbeschleunigung *f*, Fallbeschleunigung *f*, Schwerebeschleunigung *f* ~ **of gravity** Gravitätsbeschleunigung *f* ~ **due to heaviness** Schwerebeschleunigung *f* (geophys.) ~ **along the path** Bahnbeschleunigung *f* ~ **by second gun anode** Nachbeschleunigung *f* ~ **due to weight** Gewichtsbeschleunigung *f*
**acceleration,** ~ **coil of the Gulstad relay** Beschleunigungswicklung *f* des Gulstadrelais ~

constant of a machine Anlaufzeitkonstante f ~
curve Beschleunigungskurve f ~ gauge Be-
schleunigungsfühler m ~ instrument Beschleu-
nigungsanzeiger m (aviat.) ~-measuring table
Beschleunigungsmeßtisch m ~ pickup Be-
schleunigungsaufnehmer m (or pickup) pendulum Beschleunigungspendel n ~ pump Beschleu-
nigungspumpe f
accelerative force Beschleunigungskraft f
accelerator Abbindebeschleuniger m, Beschleuniger m, Beschleunigungsmaschine f, Beschleunigungspedal n, Drosselklappe f, Förderer m,
Gashebel m, Regeltritt m, Schleuderhebel m
accelerator, ~ activity Beschleunigerwirkung f
~ control pedal Gasregulierungsfußhebel m ~
control rod Gasfußhebelstange f ~ electrode
Beschleunigungselektrode f, Nachbeschleunigungselektrode f ~ grid Beschleunigungsgitter
n (CRT), Geschwindigkeitsgitter n ~ pedal
Beschleunigungsfußhebel m, Gasfußhebel m,
Gasfußtritt m ~-pedal shaft Gasfußhebelwelle f
~ potential Beschleunigungsspannung f ~ shaft
Gasfußwelle f ~ spring Beschleunigungsfeder f,
Gasfußhebelfeder f ~ stall Strömungsabriß m
(durch Beschleunigen) ~ substance Reifungskörper m
accelerometer Beschleunigungsintegrationsgerät
n, Beschleunigungsmesser m, Beschleunigungs-
meßgerät n, Beschleunigungsmeter m
accent Ton m
accented letter Sonderbuchstabe m
accentuate, to ~ anheben, auszeichnen, betonen,
hervorheben to ~ contrasts die Kontrastwirkung f erhöhen
accentuation Anhebung f, Betonung f, Bevorzugung f
accentuator Anhebungsnetzwerk n
accept, to ~ abnehmen, annehmen, genehmigen,
hinnehmen to ~ the call abfragen to ~ a
message ein Telegramm n annehmen to ~ on
original conditions zu ursprünglichen Bedingungen pl annehmen to ~ something (unwanted
or unpleasant) in Kauf m nehmen
acceptability Abnahmemöglichkeit f, Eignung f,
Zulassungsfähigkeit f
acceptable annehmbar, einwandfrei to be ~ kon-
venieren
acceptance (of goods) Abnahme f, Akzept n,
Annahme f, Einnahme f, Übernahme f, Zusage f ~ of baggage Gepäckannahme f ~ of
material Materialabnahme f ~ of a record Anerkennung f eines Rekordes ~ of specification
test Abnahmelauf m
acceptance, ~ area Stabilitätsfläche f ~ center
Annahmestelle f ~ credit (confirmed irrevoc-
able credit) Rembourskredit m ~ factory test
Abnahmeprobe f, Abnahmeprüfung f ~ flight
Abnahmeflug m ~ gauge Abnahmelehre f ~
number Gutzahl f ~ receipt Übernahmeurkunde
f ~ report Abnahmebericht m ~ run of engine
Abnahmelauf m ~ test Abnahmemessung f,
Abnahmeversuch m, Übernahmeprobe f, Zu-
lassungsprüfung f
accepted anerkannt, gebilligt ~ and measured in
place an Ort m und Stelle f abgenommen und
gemessen ~ authority anerkannter Fachmann
m ~ bill Akzept n

accepting Abnehmen n ~ (or collecting) office
Annahmeamt n
acceptor durchlässiger Kreis m ~ circuit Bandfilter m (electron.) durchlässige Kette f, durchlässiger Kreis m, Saugkreis m, Serienresonanz-
kreis m, ~ pole Fangpol m
access Zugang m, Zutritt m ~ of bed load Zunahme f der Geschiebe ~ of steam Dampfzu-
tritt m
access, ~ canal Zufahrtskanal m ~ cover Zu-
gangsdeckel m ~ door Zugangsklappe f ~
hatch Einsteigluke f ~ plate Abdeckplatte f
(aviat.) ~ platform Aufstiegsbühne f ~ road
Zufahrtstraße f ~ shaft of the caisson Schachtrohr n ~ steelwork Bedienungskonstruktion f
(Leitern, Bühnen) ~ time (memory) Suchzeit f,
Zugangszeit f, (info proc.) Zugriffszeit f ~ way
Zufahrt f
accessibility Zugänglichkeit f ~ for inspection
and maintenance in flight Wartbarkeit f während des Fluges
accessible zugängig, zugänglich
accession Zustimmung f
accessories Apparatteile pl, Ergänzungsgeräte pl,
Hilfsmittel pl, Zubehör n, Zubehörteile pl ~ for
arms Waffenzubehörteile pl ~ of centrifugal
machines Zentrifugenzubehör n
accessories, ~ drive housing Geräteantriebs-
gehäuse n ~ industry Zubehörindustrie f
accessory Apparat m, Beteiligter m, Geräteteil
m, Zubehör n, Zubehörteil m, Zusatzeinrichtung f; hinzukommend, zusätzlich
accessory ~ agent Hilfsstoff m ~ apparatus
Nebenapparat m ~ compartment Apparateteil
n ~ defense Hindernis n ~ device Vorsatzgerät n, Zusatzeinrichtung f ~ drive Hilfsantrieb m, Hilfsgeräteantrieb m, Sonderantrieb
m ~ equipment Ausrüstungsgerät n, Hilfsvorrichtung f ~ housing Hilfsgerätegehäuse n, Zu-
behörgehäuse n ~ instrument Zusatzgerät n ~
material Hilfsstoff m ~ part Zubehörteil m ~
piece of equipment Ausrüstungsstück n ~ rack
Geräteträger m ~ section Hilfsgeräteträger m
accident Betriebsunfall m, Panne f, Unfall m,
Unglück n, Unglücksfall m, Unterbrechung f
accident, ~ insurance Unfallversicherung f ~
investigation Unfalluntersuchung f ~ prevention Unfallverhütung f ~ preventive Unfallver-
hütungsmaßregel f ~-proof unfallsicher ~s
regulations Unfallverhütungsvorschriften pl ~
report Unfallbericht m, Unfallmeldung f
~-report sheet Unfallmerkblatt n ~ signaling
system Unfallmeldeanlage f ~ statistics Unfall-
statistik f ~s that have occurred aufgetretene
Unfälle pl
accidental Akzidenzien pl, Versetzungszeichen
pl; beifällig, nebensächlich, zufällig
accidental, ~ contact zufällige Berührung f ~
dendriform exposure of film Verblitzen n des
Films ~ error Zufälligkeitsfehler m ~ ground
Erdschluß m ~ hit Zufallstreffer m ~ printing
Kopiereffekt m (magn. tape) ~ resumption of
work unvorhergesehenes Ingangsetzen n
accidentally durch Zufall m
accidentalness Zufälligkeit f
acclimatization Akklimatisierung f, Gewöhnung f
acclimatize, to ~ akklimatisieren, anpassen, einheimisch machen, gewöhnen

**acclimatized** akklimatisiert
**accolade** (shape of obstacle) Akkolade *f* (klammerförmige Gestalt eines Hindernisses)
**accomet accumulator** Akkometsammler *m* (rdo)
**accommodate, to** ~ akkommodieren, anpassen, aufnehmen, in Übereinstimmung bringen, unterbringen
**accommodating connection** Anpassungsschaltung *f*
**accommodation** Anpassung *f*, Gelaß *n*, Räumlichkeit *f*, Unterbringungsmöglichkeit *f*, Unterkunft *f* ~ **for an ultimate growth** Ausbaumöglichkeit *f*
**accommodation,** ~ **bill** Gefälligkeits-akzept *n*, -wechsel *m* ~ **ladder** Schiffsleiter *f* ~ **power** Anpassungsfähigkeit *f*
**accommodator** Akkomodationsapparat *m*
**accompaniment strings** Begleitsaiten *pl*
**accompany, to** ~ begleiten
**accompanying** begleitend, beifolgend, zugehörig
**accompanying,** ~ **action** Begleiterscheinung *f* ~ **bed** Begleitflöz *n* ~ **body** Begleitkörper *m*, Begleitstoff *m* ~ **illustration** nebenstehende Abbildung *f* ~ **metal** Begleitmetall *n* ~ **mineral** Begleitmineral *n* ~ **phenomenon** Begleiterscheinung *f* ~ **picture** (on data sheet) nebenstehendes Bild *n* ~ **substance** Begleitkörper *m*, Begleitstoff *m*
**accomplish, to** ~ ausführen, bewerkstelligen, durchführen, zuwegebringen
**accomplishment** Durchführung *f*, Leistung *f*, Vervollkommnung *f*, Vollendung *f*
**accomplishing** Vollzug *m*, erledigend
**accord** (music) Akkord *m*, Anklang *m*, Einklang *m*, Festlegung *f*, Übereinstimmung *f*, Vereinbarung *f*
**accordance** Abstimmung *f*, Einverständnis *n* **in** ~**, in** ~ **with** nach Maßgabe *f* von, **in** ~ **with illustration** bildmäßig **in** ~ **with instructions** auftragsgemäß **not in** ~ **with plans** außerplanmäßig
**according,** ~ **to contract** vertragsmäßig ~ **to a fixed pattern** schablonenartig, schablonenmäßig ~ **to fixed rates** tarifmäßig ~ **to instructions** vorschriftsgemäß, vorschriftsmäßig ~ **to law** gesetzmäßig ~ **to pattern standard regulation** probemäßig ~ **to plan** plangemäß, planmäßig ~ **to regulations** vorschriftsgemäß, vorschriftsmäßig ~ **to a rule** regelmäßig ~ **to sample** nach Muster *n* ~ **to sample standard regulation** probemäßig ~ **to scale** maßstäblich, laut Tarif *m* ~ **to a set pattern** schablonenartig, schablonenmäßig ~ **to size** größenordnungsmäßig ~ **to statute** satzungsgemäß
**accordion tube** Faltenschlauch *m*
**account, to** ~ **for** erklären, verbuchen, Rechenschaft ablegen
**account** Abrechnung *f*, Berechnung *f*, Bericht *m*, Konto *n*, Nachweisung *f*, Rechenschaft *f*, Rechnung *f* **on** ~ abschlägig, auf Abschlag *m* ~ **of the mine** Grubenrechnung *f*
**account,** ~ **holder** Konteninhaber *m* ~ **section** Rechnungsstelle *f*, Verbuchungsstelle *f*
**accountable** haftbar
**accountancy** Rechnungswesen *n*
**accountant** Rechnungsbeamter *m*, Rechnungsführer *m*, -prüfer *m*, -revisor *m*, Rendant *m*
**accounting** Abrechnung *f*, Buchung *f*, Verrech-

nung *f* ~ **book** Grubenregister *n* ~ **department** Kostenstelle *f*, Rechnungsdienst *m* ~ **machine** Buchhaltungsmaschine *f* ~ **office** Abrechnungsstelle *f* ~ **records** Buchhaltungsunterlagen *pl*
**accrescence** Anwuchs *m*
**accreting bank** anlandendes oder verlandetes Ufer *n*
**accretion** Anwuchs *m* ~ **along a bank** Anlandung *f* an einem Ufer ~ **of the blocking layer** Zuwachs *m* der Sperrschicht ~**s** Ansätze *pl*
**accrue, to** ~ auftreten, ergeben, entstehen
**accrued interest** aufgelaufene Zinsen *pl*
**accumulate, to** ~ akkumulieren, anfallen, anhäufen, anlagern, anlaufen, (sich) ansammeln, anschwellen, aufhäufen, aufspeichern, sich anreichern, zusammenhäufen
**accumulated,** ~ **interest** aufgelaufene Zinsen *pl* ~ **temperature** Temperatursumme *f*, Wärmesumme *f*
**accumulating,** ~ **speed** Sammelganggeschwindigkeit *f* ~ **stimulus** einschleichender Reiz *m*
**accumulation** Anhäufung *f*, Ansammlung *f*, Aufhäufung *f*, Aufspeicherung *f*, Häufung *f*, Speicherung *f*
**accumulation,** ~ **of amplitudes** Aufschaukeln *n* der Schwingungsanschläge ~ **of energy** Energieaufspeicherung *f* ~ **of gas** Gasansammlung *f* ~ **of heat** Wärmeansammlung *f*, Wärmestauung *f* ~ **of manufacturing tolerances** Abmaßsummierung *f* ~ **of space charges** Raumladungswolke *f* ~ **of talus** Schutthalde *f* ~ **of traffic** Verkehrsanhäufung *f* ~ **of unfilled orders** Auftragsrückstand *m* ~ **of ultrasonic waves** Überschallwellenanhäufung *f*
**accumulation electrode** Stau *m*, Speicherelektrode *f*
**accumulative,** ~ **limits** sich summierende Toleranzen ~ **material** akkumulierendes Material *n*
**accumulator** Akku *m*, Dampfdom *m*, Druckspeicher *m* (hydraul.), Hauptreservoir *n*, Kraftsammler *m*, Kraftspeicher *m*, Sammelbehälter *m*, Stromsammler *m*, Zwischenspeicher *m* ~ **with liquid electrolyte** Akkumulator mit Säurefüllung *f*
**accumulator,** ~ **acid** Akkumulatorensäure *f*, Füllsäure *f* ~ **bag** Speicherblase *f* ~ **battery** Akkumulatorenbatterie *f*, Sammlerbatterie *f*, Sekundärbatterie *f* ~ **box** Akkumulatorgefäß *n*, Akkumulatorkasten *m* ~ **case** Akkumulatorkasten *n* ~ **cell** Sammlerzelle *f*, Sekundärelement *n* ~ **cells** Akkumulatorenbatterie *f* ~ **charging apparatus** Akkumulatorenladeapparat *m* ~ **drive** Sammlerantrieb *m* ~ **effect** Stauwirkung *f* ~ **end celles** Ausgleichzellen *f* ~ **grid plate** Akkumulatorgitterplatte *f* ~ **jar** Akkumulatorgefäß *n*, Akkumulatorglas *n* ~ **lead plate** Akkumulatorbleiplatte *f* ~ **plate** Sammlerplatte *f* ~ **substation** Sammlerunterwerk *n* ~ **switsch** Batterieregulierschalter *m*, Batterieschalter *m* ~ **switsch regulator** Zellenschalter *m* ~ **tank** Akkumulationsbehälter *m*, Sammelbehälter *m* ~ **terminal** Akkumulatorklemme *f* ~ **tester** (instrument) Batterieprüfer *m* ~ **vessel** Sammlergefäß *n*
**accuracy** Fehlerfreiheit *f*, (clock) Ganggenauigkeit *f*, Genauigkeit *f*, Ordnung *f*, Präzision *f*, Richtigkeit *f*, aufgewandte Sorgfalt *f*, Steuerschärfe *f* ~ **of alignment** Richtschärfe *f* ~ **of**

**angle** Winkelrichtigkeit *f* ~ **of angle measurement** Winkelmeßgenauigkeit *f* ~ **of control** Regelgenauigkeit *f* ~ **of duplication** Kopiergenauigkeit *f* ~ **of fit** Paßgenauigkeit *f* ~ **to gauge** Maßhaltigkeit *f* ~ **of manufacture** Herstellungsgenauigkeit *f* ~ **of measurement** Meßgenauigkeit *f* ~ **of reading** Ablesegenauigkeit *f* ~ **of reproduction** Kopiergenauigkeit *f* ~ **of scale** Maßstabsgerechtigkeit *f* ~ **of shape** Formgenauigkeit *f* ~ **of sighting** Zielgenauigkeit *f* ~ **to size** Maßhaltigkeit *f* ~ **of the work diameter** Durchmessergenauigkeit *f*

**accuracy,** ~ **grade** Genauigkeitsgrad *m* ~ **landing** Genauigkeitslandung *f* ~ **tolerance** Formgenauigkeitstoleranz *f*

**accurate** fehlerfrei, haargenau, knapp, richtig, sorgfältig, zielgenau ~ **to gauge** lehrenhaltig ~ **to size of gauge** maßgerecht

**accurate,** ~ **alignment** genaue Ausrichtung *f* ~ **dimension** genaues Maß *n* ~ **duplication** Vervielfältigung *f* ~ **levelling** Feinhorizontierung *f* ~-**measuring instrument** Genauigkeitsmeßwerkzeug *n* ~ **reproduction** inhaltstreue Abbildung *f* ~ **scanning** Feinortung *f* (rdr) ~ **tuning** genaue Abstimmung *f*

**accurately** mit Genauigkeit *f*, genau ~ **controllable** fein regelbar ~ **defined** genau abgegrenzt

**accusation** Anklage *f*, Beanstandung *f*, Beschuldigung *f*

**accuse, to** ~ anklagen, beschuldigen, Beschwerde einleiten **to** ~ **of** bezichtigen

**accused** Beklagter *m*, Beschuldigter *m* ~ **defendant** Angeschuldigter *m*

**accuser** Ankläger *m*, Kläger *m*

**accustom, to** ~ (an)gewöhnen

**accustomed** gewohnt

**acenaphthene** Azenaphthen *n*

**acentric** azentrisch

**aceric acid** Ahornsäure *f*

**acerbity** Herbheit *f*

**acetamide** Azetamid *n*, Essigsäureamid *n*

**acetanilide** Azetanilid *n*

**acetate** Azetat *n* ~ **of** essigsauer ~ **of lead** Bleizucker *m*

**acetate,** ~ **base** Azetatträger *m* (d. Films) ~ **disk** Azetatplatte *f* ~ **record** Azetatplatte *f* ~ **spun rayon** Azetatzellwolle *f*

**acetic,** ~ **acid** Azetylsäure *f*, Essigsäure *f*, Grünspanessig *m* ~ **aldehyde** Azetaldehyd *n* ~ **anhydride** Essigsäureanhydrid *n* ~ **ether** Äthylazetat *n*, Essigäther *m*

**acetimeter** Azetimeter *n*, Essigprüfer *m*

**acetoacetic ester** Azetessigäther *m*

**acetone** Azeton *n*, Brenzessigäther *m*, Essiggeist *m* ~ **chloride** Chlorazetol *n*

**acetonitrile** Azetitril *n*

**acetophenone** Acetophenon *n*

**acetovanillone** Azetosyringon *n*

**acetyl** Azetyl *n* ~ **chloride** Azetylchlorid *n*, Essigsäurechlorid *n*

**acetylable** azetylierbar

**acetylate, to** ~ azetylieren

**acetylation** Azetylierung *f*

**acetylene** Äthin *n*, Azetylen *n*, Bogenlampe *f* ~ **brazing output** Azetylenhartlötausrüstung *f* (oxy-) ~ **cutter** Brennschneider *m* ~ **cutting plant** Azetylenschneideanlage *f* ~ **cylinder** Azetylenflasche *f* ~ **flare light** Azetylenleucht-

**feuer** *n* ~ **gas** Azetylengas *n* ~ **generator** Azetylenentwickler *m* ~ **lighting** Azetylenbefeuerung *f* ~ **pressure generator** Azetylenpreßgaserzeuger *m* ~ **purifying agent** Azetylenreinigungsmasse *f* ~ **regulator** Azetylendruckminderer *m* ~ **starter** Azetylenanlasser *m* ~ **welding** Azetylenschweißung *f*

**acetylized lignin** azeteliertes Lignin *n*

**achieve, to** ~ erzielen, zustande bringen

**achievement** Arbeitsleistung *f*, Erfolg *m*

**achromat** Achromat *m*

**achromatic** achromatisch, farbfehlerfrei, farblos ~ **color** unbunte Farbe *f* ~ **lens** Achromat *m*, scharfzeichnende Linse *f*, achromatisches Objektiv *n* ~ **locus** Weißgebiet *n* (im Farbtondiagramm) ~ **objective** Achromat *m* (photo) ~ **quartz fluorite lens** Quarzfluoritachromatlinse *f* ~ **quartz rock salt lens** Quarzflußspatachromatlinse *f*

**achromatism** Achromasie *f*, Achromatismus *m*, Farblosigkeit *f*

**achromatization** Abweichungskorrektur *f* (opt.), Achromasie *f*, Farbenkorrektion *f*

**achromatize, to** ~ achromatisieren

**achromatized** farbkorrigiert

**acicular** nadelförmig ~ **bismuth** Belonit *m* ~ **martensite** nadeliger Martensit *m*

**acid** Säure *f*; sauer ~ **of liquor** Säureflotte *f*

**acid,** ~ **and alkali proof plastic materials** säurebeständige Kunststoffe *pl* ~ **Bessemer steel** saurer Bessemer-Stahl *m* ~ **bottle** Säureflasche *f* ~ **brittleness** Beizbrüchigkeit *f*, Beizsprödigkeit *f* ~ **catalysed resin** säurehärtendes Harz *n* ~ **centrifuge** Säurezentrifuge *f* ~ **coloring matter** saurer Farbstoff *m* ~ **concentration** Säuredichte *f* ~ **container** Säurebehälter *m* ~ **content** Säuregehalt *m* ~ **corrosion** Säureanfressung *f*, Säureangriff *m* ~ **cure** Kaltvulkanisation *f* ~ **density** Säuredichte *f* ~ **developer** Säurenentwickler *m* ~ **egg** Druckbehälter *m*, Druckbirne *f* ~ **etching** Ätzung *f* mit Säure ~ **figure** Säurezahl *f* ~ **fixer** saures Fixiersalz *n* ~ **flotation method** Säureschwemmverfahren *n* ~ **flotation process** Säureschwemmverfahren *n* ~-**forming** säurebildend ~ **fume scrubber** Säurenebelwäscher *m* ~ **fumes (or gases)** Säuredämpfe *pl* ~ **and basic hearthconverting process** (sulfuric) Birnenprozeß *m* ~ **heat test** Erhitzungsversuch *m* mit Schwefelsäure ~ **heater** Säureerhitzer *m* ~ **hydroextractor** Nitrierschleuder *f*, Säurezentrifuge *f* ~ **ion** Säureion *n* ~-**laden** säurehaltig ~ **level** Säurestand *m* ~ **lined** sauer zugestellt (Ofen) ~-**lining** saures Futter *n* ~ **manufacture** Säureherstellung *f* ~ **mine water** saures Grubenwasser *n* ~ **number** Säurezahl *f* ~ **open-hearth furnace** saurer Martinofen *m* ~ **oxalate of potassium** Kaliumbioxalat *n* ~ **parting** Säurescheidung *f* ~ **plant** Sulfitanlage *f* ~ **plumping** Säureschwellung *f* ~ **pressure reservoir** Säuredruckbehälter *m*

**acidproof** säurebeständig, säurefest ~ **cast-iron** säurebeständiger Guß *m* ~ **grease** Säureschutzfett *n* ~ **lining** säurefeste Auskleidung *f* ~ **paint** säurefeste Farbe *f* ~ **tank** säurefestes Bassin *n* ~ **wire** Säuredraht *m*

**acid,** ~ **radical** Säurerest *m* ~ **reclaim** Säureregenerat *n* ~ **resistance** Säurebeständigkeit *f* ~-**resistant** säurebeständig ~-**resisting** säure-

beständig, säurefest ~ **seal paint** säurefeste
Auskleidung f ~ **settling drum** Säureabsatz-
behälter m ~ **siphon** Säureheber m ~ **sludge**
Säureabfall m, Säureharz n, Säureschlamm m
~**-soluble** säurelöslich ~ **steam-developing
method** Säuredampfentwicklungsverfahren n
~**-steeping bowl** Einweichbottich m (Säure)
~ **strength** Säuredichte f ~ **test** Säureprüfung f
~ **tower** Säureturm m ~ **treatment** Behandlung
f mit Säure ~ **vat** Einsäurebottich m ~**-washed**
säuregereinigt
**acidiferous** säurehaltig
**acidifiable base** säurefähige Base f
**acidification** Ansäuerung f, Sauermachen n,
Säuerung f
**acidify, to** ~ absäuern, ansäuern, sauermachen,
säuern
**acidifying** Ansäuerung f ~ **bath** (acid bath) Ab-
säuerungsbad n
**acidimeter** Säuremesser m
**acidity** Azidität f, Sauerkeit f, Säuerlichkeit f,
Säure f ~ **meter** Azidimeter n
**acidizing of wells** Erschließung f mit Säure (von
Ölgruben)
**acidless** säurefrei
**acidolytic degradation** azidolytischer Abbau m
**acidometer** Säuremesser m
**acidulate, to** ~ ansäuern
**acidulated water** angesäuertes Wasser n
**acidulating** Durchsäuerung f
**acidulation** Säuerung f
**acidulous** säuerlich ~ **spring** Sauerbrunnen m,
Sauerquelle f
**Ackerman steering** Achsschenkellenkung f
**acknowledge, to** ~ anerkennen, bestätigen (Meldung), erkennen, zugeben
**acknowledgment** Anerkennung f, Bekanntgeben
n, Bestätigung f, Empfangsbestätigung f ~ **of
receipt** Empfangsanzeige f ~ **signal** Quittungs-
zeichen n
**aclinic** aklinisch, nicht inklinierend ~ **lines**
Linien pl gleicher Spannung
**acme** Gipfel m, höchste Qualität f ~ **thread**
Trapezgewinde f
**aconitine** Akonitin n
**acorn** Kabelhalter m (naut) ~**-shaped part**
eichelförmig gestalteter Teil m ~ **tube** Eichel-
röhre f, Knopfröhre f ~ **valve** Knopfröhre f
**acoumeter** Akumeter m, Hörschärfenmesser n
**acouophony** Akufonie f
**acoustic** Gehör n; akustisch, tongebend
**acoustic,** ~ **absorptivity** Schallschluckung f ~
**action** Schallereignis n, (brought on microphone) Beaufschlagung f ~ **capacitance** akustische Kapazität f ~ **center of frequencies** Tonzentrum n ~ **chamber** Schallkammer f ~ **clarifier** Klangreiniger m ~ **compliance** akustische
Kapazität f oder Nachgiebigkeit f, akustischer
Blindwiderstand m (induktiv) ~ **correction**
Ausschalten n des Schallverzugs ~ **corrector**
(sound location) Verzugsrechner m ~ **delay-
line** akustisches Laufzeitglied n ~**-detecting
apparatus** Horchgerät n ~**direction finding**
Gehörpeilung f ~ **duct** Gehörgang m ~ **dust
pattern** Staubwelle f ~ **excitation** akustische
Erregung f ~ **feedback** akustischer Kurzschluß m, akustische Rückkoppelung f, Schallrückkoppelung f ~ **figure** Klangfigur f ~ **filter**

akustischer Filter m ~ **fixing of aircraft** Abhorchen n von Flugzeugen ~ **frequency** Schall-
frequenz f, Tonfrequenz f ~ **frequency branch**
akustischer Zweig m ~ **funnel** (of dictating
machine) Schallbecher m ~ **image** akustisches
Bild n ~ **impedance** Schallimpedanz f, Schall-
rückwirkung f, Schallwellenwiderstand m ~
**inertance** akustischer Blindwiderstand m,
akustische Trägheit f ~ **inlet** (of a microphone)
Einsprache f ~ **intonation** Klangeinsatz
m ~ **intrusion detector** akustisches Einbruchs-
signal n ~ **irradiation** Beschallung f ~ **measuring of altitude** Luftlotung f ~ **memory** akustischer Speicher m ~ **microphone** Schall-
mikrofon n ~ **orientation** akustische Ortung
f ~ **panel** Schallschutzplatte f ~ **pattern**
Klangbild n ~ **perception** akustische Empfindung f ~ **pickup** Schalldose f ~ **power** Schall-
leistung f ~ **pressure** Schalldruck m ~ **radiator**
Schallscheinwerfer m ~ **reactance** akustische
Gegenwirkung f, akustischer Blindwiderstand
m ~ **reflectivity** Schallreflexionskoeffizient m
~ **regeneration** Schallrückkoppelung f ~ **resistance** akustischer Widerstand m, Schallwir-
kungswiderstand m, (frictional) Schallrei-
bungswiderstand m ~ **screen** akustischer
Widerstand m ~ **sensibility** Ansprechempfindlichkeit f ~ **signal** hörbares Signal n, Läute-
signal n, Schallzeichen n ~ **shock** Knacken n,
Knackgeräusch n ~ **short circuit** akustischer
Kurzschluß m ~ **slow motion** Tonzeitdehnung f
~ **sounding gear** Ortungsgerät n ~ **source**
Resonanzschallquelle f ~ **stiffness** Schallhärte f
~ **telegraphy** Gehörtelegrafie f ~ **trans-
mittivity** Schalldurchlaßgrad m ~ **treatment**
(of a room) schalldämpfende Raumverklei-
dung f ~ **vault** Schallgewölbe n ~ **vibration**
Schallschwingung f ~ **wave** Schallwelle f
**acoustical** akustisch ~ **equivalent of inductance**
Induktivität f ~ **impedance** akustischer Scheinwiderstand m ~ **perception** Tonempfindung f,
Tonwahrnehmung f ~ **performance** Hörsamkeit f ~ **power** akustische Leistung f ~ **quantity**
akustische Größe f ~ **sensing** Schallempfang m
~ **value** akustische Größe f
**acoustically,** ~ **inactive (or inert)** schalltot ~ **live
room** Hallraum m
**acoustician** Schallingenieur m
**acoustics** Akustik f, Hörsamkeit f, Klangwirkung f, Schallehre f, Schalltechnik f
**acoustimeter** Geräuschmesser m
**acoustoelectric index** Verhältnis n eines Sende-
systems
**acquiesce, to** ~ beigeben, einverstanden sein
**to** ~ **in** beistimmen, sich abfinden mit, übereinstimmen
**acquiescence** Billigung f, Zustimmung f
**acquire, to** ~ erwerben, gewinnen **to** ~ **a license**
eine Lizenz f erwerben **to** ~ **the right to construct a dam** das Staurecht n erwerben **to** ~
**static electricity** elektrisch werden **to** ~ **a target**
ein Ziel n auf- oder erfassen
**acquisition** Erwerb m, Erwerbung f ~ **cost** Be-
schaffungskosten pl, Erwerbungskosten pl
~ **radar** Erfassungsradargerät n
**acreage** Anbaufläche f
**acribometer** Akribometer n
**acrid** beißend, herb, scharf

acridic (or acridinic) acid Akridinsäure *f*
acridine color (or dye) Akridinfarbstoff *m*
acrimonious bitter
acrobatic akrobatisch ~ figures Flugfiguren *pl*
~ flight Kunstflug *m*
acrolein Akrolein *n*
across durch, quer, querdurch, parallel zu, über,
über Eck *n* ~ the grain quer zur Faser
acrylate Acrylsäureester *m*
acrylic, ~ acid Akrylsäure *f* ~ plastics Akryl-
harzkunststoffe *pl*
act, to ~ arbeiten, eingreifen, einwirken, funktionieren, handeln, verhalten, wirken to ~
against entgegenwirken, zuwiderhandeln to ~
without authority eigenmächtig handeln to ~
for vertreten to ~ in opposite phases phasenverkehrt wirken to ~ in opposition zuwiderhandeln to ~ on one's own authority eigenmächtig handeln to ~ as a reducing agent reduzierend
wirken to ~ in reverse polarity phasenverkehrt
wirken to ~ in reverse rückwärtswirken to ~ as
a spring durchfedern to ~ together zusammenwirken
act Aktenstück *n*, Handlung *f*, Tat *f*, Vorgang *m*,
Werk *n* ~ of dispatching Aussendung *f* ~ of
God höhere Gewalt *f* ~ of hardening Hartmachen *n* ~ of location Anbringung *f* in einer
bestimmten Lage ~ performed in line of duty
Diensthandlung *f* ~ of printing Druckvorgang *m*
acting stellvertretend, wirkend ~ as a hygrofuge
feuchtigkeitsvertreibend
acting, ~ committee Aktionskomitee *n* ~ force
angreifende Kraft *f* ~ impedimentally upon
operation betriebshindernd ~ manager Betriebsführer *m*
actinic aktinisch, chemisch wirksam, lichtchemisch, (in radation) aktiv ~ effect Licht-
(ein)wirkung *f* ~ rays aktinische (Licht)-
strahlen *pl* ~ screen Leuchtschirm *m*
actinide elements Aktiniden *pl*
actinism Lichtempfindlichkeit *f*
actinium series Aktiniumzerfallsreihe *f*
actinogram Aktinogramm *n*
actinolite Aktinolith *m*, Strahleisen *n*, Strahl-
stein *m*
actinometer Aktinometer *n*, Strahlenmesser *m*,
Strahlenprüfer *m*, Strahlungsmesser *m*
actinon Aktiniumemanation *f*, Aktinon *n*
action Aktion *f*, Arbeitsweise *f*, (law) Bescheid
*m*, (of patent office) Beschluß *m*, Betonung *f*,
Betriebsverhalten *n*, (in film projection) Bewegung *f*, Einfluß *m*, Eingriff *m* (chem.), (of
gear) Eingriff *m*, Einwirkung *f*, Ereignis *n*, Erscheinung *f*, Funktion *f*, Gang *m*, Geschehnis *n*,
Handlung *f*, Lauf *m*, Rechtshandel *m*, Tat *f*,
Tätigkeit *f*, Verfügung *f*, Vorgang *m*, Werk *n*,
Wirken *n*, Wirkung *f*, (Zahnrad) wirkende
Kraft *f*
action, in ~ im Betrieb *m*, im Einsatz *m*, eingeschaltet out of ~ ausgeschaltet, außer Gefecht *n*
~ for annulment Nichtigkeitsklage *f* ~ of blast
Blaswirkung *f* ~ for cancellation Nichtigkeitsklage *f* ~ for damages Schadenersatzklage *f* ~ of
directional beam Richtwirkung *f* ~ for discontinuance (of infringement) Einstellungsklage
*f* ~ at a distance Fernbetrieb *m*, Fernwirkung *f*
~ of field through aperture lens Felddurchgriff
*m* (opt.) ~ of force Kraftwirkung *f* ~ of frost

Frostwirkung *f* ~ of heat Wärmewirkung *f* ~ for
infringement of patent rights Patentverletzungsklage *f* ~ of light Lichtwirkung *f* ~ of rust
Korrosion *f*, Rostangriff *m* ~ for stay (or
suspension) Einstellungsklage *f* ~ of wind
Windeinwirkung *f* ~ for withdrawal Rück-
nahmestreit *m*
action, ~ cycle Arbeitstakt *m* ~ line Arbeitslinie
*f* ~ magnitude Wirkungsgröße *f* ~ pending
altitude (for ballons) Wartehöhe *f* ~ period
Abtastzeit *f* ~ photograph Fotografie *f* zur
Veranschaulichung von Handgriffen und
Arbeitsvorgängen ~ principle Wirkungsprinzip
*n* ~ quantity Wirkungsgröße *f* ~ quantum
Wirkungsquantum *n* ~ sequence (aut. contr.)
Wirkungsablauf *m* ~ spot Abtastpunkt *m* ~
turbine Aktionsturbine *f*, Gleichdruckturbine *f*
~-variable Phasenintegral *n*, Wirkungsvariable
*f* ~ variable with time zeitlich veränderlicher
Vorgang *m*
actionable einklagbar, klagbar
actionometer Lichtstrahlenmesser *m*
activate, to ~ aktivieren, antreiben (Motor), aufstellen, beladen to ~ a surface with hydrogen
eine Oberfläche *f* mit Wasserstoff beladen
activated geladen, (in) Tätigkeit *f*
activated, ~ carbon Aktivkohle *f* ~ charcoal
Adsorptionskohle *f* ~ clay (or fuller's earth)
aktivierte Bleicherde *f* ~ molecule angeregtes
Molekül *n* ~ sludge belebter Schlamm *m*
~-sludge process Belebtschlammprozeß *m*
~ water aktiviertes Wasser *n*
activating, ~-agent Aktivierungsmittel *n* ~ impulse Betätigungsimpuls *m*
activation Aktivierung *f*, Empfindlichmachung *f*,
Formierung *f* ~ heat of atom and radical
reactions Aktivierungswärme *f* von Atom- und
Radialreaktionen ~ of a surface Oberflächenbeladung *f*
activator Aktivierungsmittel *n*, Beschleuniger *m*
active aktiv, emsig, geschäftig, lebhaft, regsam,
rührig, rüstig, tätig, wirksam ~ in lowering
surface tension kapillaraktiv
active, ~ admittance Wirkleitwert *m* ~ balance
of trade aktive Handelsbilanz *f* ~ bleaching
earth aktive Bleicherde *f* ~ circuit Nutzkreis *m*
~ component Wattkomponente *f*, Wirkkomponente *f*, Wirkspannungskomponente *f*, Wirkstromkomponente *f*, Wirkteil *m* ~ content
Wirkungsgehalt *m* ~ core aktive Zone *f*,
Reaktorkern *m* ~ current Verluststrom *m*,
Wattstrom *m*, Wirkstrom *m* ~ deposit radioaktiver Niederschlag *m* ~ earth pressure Erddruck *m*, Erdschuß *m* ~ energy meter Wirk-
leistungszähler *m*, Wirkverbrauchszähler *m*,
~ lattice Reaktorgitter *n* ~ line Abtastlinie *f*
(rdr) ~ line trace Zeilenhinlauf *m* ~ material
aktive Masse *f* ~ mixer Flachherdmischer *m*
~ paste aktive Masse *f* ~ polar surface wirksame Polfläche *f* ~ power Wirkleistung *f* ~ (or
real) power consumption Wirkleistungsver-
brauch *m* ~-power meter Wirkverbrauchs-
messer *m* ~ resistance Wirkwiderstand *m*
~ return loss of echo Echostromdämpfung *f*
~ rudder Bugruder *n* ~ singing point wirksamer
Schwingungseinsatzpunkt *m* ~ stroke Vorwärtsbewegung *f* (film) ~ surface wirksame
Oberfläche *f* ~ voltage Wirkspannung *f*

**activity** Aktivität *f*, Beschäftigungsgrad *m*, Betätigung *f*, Betrieb *m*, Bewegung *f*, Füllfaktor *m*, Tätigkeit *f*, Wirksamkeit *f* ~ **of crystals** Erregbarkeit *f* (electron.)

**activity,** ~ **(curve) graph** Aktogramm *n* ~ **distribution** Aktivitätsverteilung *f* ~ **limit** Aktivitätsverteilung *f* ~ **limit** Aktivitätsgrenze *f*

**actual** aktuell, tatsächlich, wirklich ~ **allowance** Istabmaß *n* ~ **cast iron** Gußmetall *n* ~ **cathode** reelle Kathode *f* ~ **clearance** Istspiel *n* ~ **conditions** Iststand *m* ~ **dimension** Istmaß *n* ~ **efficiency** Wirkleistung *f* ~ **finished size** Istgröße *f*, Istmaß *n* ~ **horsepower** effektive Pferdestärke *f*, verfügbare Pferdekraft *f* ~ **interference** Istübermaß *n* ~ **level** Augenblickspegel *m* ~ **line** natürliche Leitung *f* ~ **load** Nutzlast *f* ~ **metric horsepower** effektive Pferdestärke *f* ~ **movement of stern into the water** (launching) eigentlicher Ablauf *m* ~ **operating diagram** Wirkschema *n* ~ **operating valve clearance** Betriebsventilspiel *n* ~ **output** Nutzleistung *f* ~ **power** effektive Leistung *f* ~ **quantity** Istmenge *f* ~ **selsyn** Istgeber *m* ~ **size** Istmaß *n*, natürliche Größe *f* ~ **-size** naturgroß ~ **speed of fluid at center line of a closed duct** Meridiangeschwindigkeit *f* ~ **state of equipment** Istbestand *m* ~ **strength** Stärkebestand *m* ~ **time line** Istzeitlinie *f* ~ **value** (of controlled condition) Sollwert *m*, Istwert *m* (der Regelgröße) ~ **value gauge** Istwertanzeige *f* ~ **value indication** Istwertanzeige *f* ~ **velocity** wirkliche Geschwindigkeit *f* ~ **volume** Derbgehalt *m* ~ **working diagram** Wirkschema *n*

**actuality** Wirklichkeit *f*

**actuarial,** ~ **calculations** versicherungsmathematische Berechnungen *pl* ~ **expert** Sachverständiger *m* für Versicherungsmathematik *f* ~ **reserves** technische Reserven *pl* ~ **theory** Versicherungsmathematik *f*

**actuate, to** ~ ansprechen (electr.), antreiben, betätigen, erregen, wirken, in Bewegung *f* setzen, in Gang *m* bringen

**actuated** betätigt, gesteuert ~ **intermittently** absatzweise bewegt

**actuating** Ansprechen *n* (electr.) ~ **appliance** Betätigungsorgan *n* ~ **arm** Angriffschenkel *m* ~ **bar** Schaltwelle *f* ~ **cam** Stoßdaumen *m* ~ **cylinder** Arbeitszylinder *m*, Stellzylinder *m* ~ **lever** Antriebshebel *m*, Betätigungshebel *m*, Ladehebel *m* ~ **mechanism** Steuerbetätigung *f* ~ **ring** Haltering *f* ~ **rod** Regelstange *f* ~ **strut** Betätigungsstrebe *f* ~ **transfer function** Betätigungsübertragungsfunktion *f*

**actuation** Betätigung *f* ~ **by a system of impulses from magnets at the pulleys** Magnettrommelimpulsverfahren *n* ~ **of switsch** Schalterbetätigung *f*

**actuation shaft** Antriebswelle *f*

**actuator** Antrieb *m*, Betätigungsorgan *n*, Kraftschalter *m*, Stellmotor *m*, Versteller *m* ~ **disk** Einflußfläche *f* ~ **section** Stellgruppe *f* ~ **sleeve** Zylinderlaufbuchse *f* (hydraul.)

**acuity** Schärfe *f* ~ **of hearing** Gehörschärfe *f*, Hörschärfe *f* ~ **of image of camera** Auflösungsvermögen *n* ~ **of vision** Augenschärfe *f*

**acuity meter** Hörschärfemeßgerät *n*

**acute** akut, scharf, scharfsinnig, spitz, spitzig ~ **angle** spitzer Winkel *m* ~ **-angled** scharf-

winklig, spitzwinklig ~ **bisectrix** erste oder spitze Mittellinie *f* ~ **exposure** kurzzeitige Bestrahlung *f* ~ **triangle** spitzwinkliges Dreieck *n*

**acuteness** Scharfblick *m*, Schlagfertigkeit *f*

**acyclic** azyklisch

**adamantine** diamantartig ~ **drilling** Schrotbohren *n* ~ **luster** Diamantglanz *m* ~ **spar** Diamantspat *m*

**adamellite** Adamellit *m*

**adamsite** Adamsit *n*, Diphenyl-aminchlorarsin *n*, -chlorarsin *n*

**adapt, to** ~ adaptieren, anpassen, anschmiegen, aptieren, einpassen, herrichten

**adaptability** Anpassungsfähigkeit *f*, Brauchbarkeit *f*, Einsatzfähigkeit *f*, Verwendbarkeit *f*, Wendigkeit *f* ~ **to acceleration** Beschleunigungsverträglichkeit *f*

**adaptable** anpassungsfähig, gefügig, verwendbar

**adaptable-type wiper motor** Anpaßwischmotor *m*

**adaption** Angleichung *f*, Anpassung *f* ~ **to the dark** Dunkeladaption *f* ~ **to eliminate noise** Störanpassung *f* ~ **to terrain** Geländeanpassung *f*

**adapted** abgestimmt, angepaßt, bestimmt sein für, geeignet, zweckmäßig ~ **to circuit changes** umschaltbar ~ **for cross-country driving** geländefähig, geländegängig

**adapter** Allonge *f*, Amtsanschließer *m*, Anpaßstück *n*, Ansatz *m* (photo), Blattadapter *m*, Einschraubstück *n*, Führungsteil *m*, Hilfsträger *m*, Paßstück *n*, Retortvorstoß *m*, Spritzmulde *f*, Überwurfmutter *f*, Umspannfutter *n*, Verbindungsstöpsel *m*, Verbindungsstück *n*, Vorsatzgerät *n* (rdo), Vorstoß *m*, Zusatzgerät *n*, Zwischensatzstück *n*, Zwischensockel *m*, Zwischenstück *n*, Zwischenring *m* ~ **for fitting Armaturenstutzen** *m* ~ **to stand (or to tripod)** Stativaufsatz *m*

**adapter,** ~ **amplifier** Anpassungsverstärker *m*, Einschraubstück *n* ~ **(backplate)** Futterscheibe *f* ~ **bearing** Spannhülsenlager *n* ~ **bearing seal** Adapterlagerdichtung *f* ~ **bushes** Klemmbuchsen *pl* ~ **flange** Übergangsflansch *m* ~ **gauge** Vorrichtungsschmiegenlehre *f* ~ **nipple** Gewindestutzen *m* ~ **piece** Ansatzstück *n*, Ansatzstutzen *m* ~ **plate** Aufspannplatte *f* ~ **plug** Steckhülse *f* ~ **(reducer)** Übergangsstück *n* ~ **ring** Haltering *m* ~ **sleeve** Einsatzhülse *f*, Spannhülse *f* ~ **spring** Angleichfeder *f* ~ **thread** Mundlochgewinde *n* ~ **-transformer lamp** Reduktorlampe *f* ~ **unit** Anpassungsgerät *n*

**adapting** Anpassung *f*, Einpaß *m* ~ **piece** Fassonrohr *n*, Formstück *n*, Paßrohr *n*

**adaption of impedance** Scheinwiderstandsanpassung *f*

**adaptometer** Adaptometer *n*

**adaptor (see adapter)**

**Adcock,** ~ **antenna** Adcock-Antenne *f* ~ **range** Adcock-Vierkursfunkfeuer *n*

**add, to** ~ addieren, anfügen, anhängen, anschließen, (ein Rohr) aufsetzen, hinzugeben (chem.), hinzurechnen, nachsetzen, (up) summen, summieren, (one substance to another) versetzen, zugeben, zulaufen, zulegen, zusammenzählen, zuschlagen, zusetzen **to** ~ **on** anlagern **to** ~ **to** anreihen, sich anlagern, beifügen, beitragen, ergänzen **to** ~ **tubs** Förderwagen *m*

aufschieben **to ~ up** zusammenrechnen **to ~ vectorially** vektoriell addieren
**add-water cell** Füllelement *n* (electr.)
**added** zusätzlich **~ area for augmenting the tidal-air effect** Hinterlagerungsraum *m* **~ heat** zugeführte Wärme *f* **~ piece** Ansatz *m* **~ water quantity** Wasserzusatz *m*
**addendum** Ergänzung *f*, Nachtrag *m* **~ of the tooth** (in a gear wheel) Kopfhöhe *f* des Zahnes
**addendum, ~ angle of bevel gear** Kopfwinkel *m* des Kegelrades **~ circle** Kopfkreis *m* **~ envelope of the skew gear** Hüllfläche *f* des Schraubrades **~ line** Kopfkreis *m*
**adder** Addierer *m*, Addiereinrichtung *f* (data proc), Additionskreis *m*, Beimischer *m* (TV)
**adding** summierend **~ device** Addierwerk *n* **~ machine** Additions(rechen)maschine *f*, Rechenmaschine *f*
**addition** Anlagerung *f*, Annahme *f*, Aufzählung *f*, Beifügung *f*, Beimengung *f*, Beimischung *f*, Beisatz *m*, Erweiterungsbau *m*, Summation *f*, Zufuhr *f*, Zugabe *f*, Zusatz *m*, Zusatzstoff *m*, Zuschlag *m*, Zutat *f* **~ of heat** Wärmezufuhr *f* **~ of lime** Kalkzuschlag *m* **~ of limestone** Kalkzuschlag *m* **~ of nickel** Nickelzusatz *m* **~ of oil in saturation** Fettzusatz *m* bei der Saturation **~ of scrap** Schrottzugabe *f* **~ of a signal** Aufschaltung *f* (eines Signals) **~ to structure** Anbau *m* **~ of velocities** Geschwindigkeitsabhängigkeit *f*
**addition, ~ agent** Beimischungsstoff *m*, Zusatzeisen *n*, Zusatzmittel *n* **~ compound** Anlagerungsverbindung *f* **~ constant** Additionskonstante *f* **~ reaction** Anlagerungsreaktion *f*
**additional** beigefügt, extra, zusätzlich **~ air** Zusatzluft *f* **~ air nozzle** Zusatzluftdüse *f* **~ allowance** Zulage *f* **~ application** Zusatzanmeldung *f* (patent) **~ article** Zusatzartikel *m* **~ attachment** Zusatzvorrichtung *f* **~ bending strength** Zusatzknickkraft *f* **~ building** Nebengebäude *n* **~ brake resistance** Zusatzbremswiderstand *m* **~ breaker** Nachbrecher *m* **~ charge** Nachladung *f*, Nebenausgabe *f*, Zusatzladung *f* **~ charge for telephoning of telegram** Zusprechgebühr *f* **~ cost** Extrakosten *pl*, Mehrausgabe *f* **~ drag** Zusatzwiderstand *m* **~ evidence** Nebenbeweis *m* **~ expenditure** Mehraufwand *m* **~ expense** Nebenausgabe *f* **~ firing** Zusatzfeuerung *f* **~ freight** Frachtzuschlag *m*, Überfracht *f* **~ gun anode** Nachbeschleunigungselektrode *f* **~ heating** Rückheizung *f* **~ impulse** Zusatzheizung *f*, Zusatzimpuls *m* **~ income** Nebeneinkunft *f* **~ instrument** Zusatzeinrichtung *f* **~ iron** Zusatzeisen *n* **~ loss** Zusatzverlust *m* **~ multiple** Ansatzfeld *n* (teleph.) **~ passes** zusätzliches Durchlaufen *n* (durch die Maschine) **~ patent** Zusatzpatent *n* **~ path** Umwegleitung *f* **~ pay** Besoldungszulage *f*, Zulage *f* **~ payment** Nachzahlung *f* **~ plant** Nebenanlage *f* **~ plug-in** Zusatzeinschub *m* **~ postage** Strafporto *n* **~ pressure** Überdruck *m* **~ rack** Zusatzgestell *n* **~ resistance** zusätzlicher Widerstand *m*, Zusatzwiderstand *m* **~ scanning raster** wiederabtastender Raster *m* **~ scavenging air through auxiliary scavenging ports** Nachladung *f* **~ security** Nachsicherung *f* **~ service** besondere Leistung *f* **~ set** Zusatzaggregat *n* **~ stress** Zusatzspannung *f* **~ supply of power** Energie-

**nachfuhr** *f* **~ tank** Zusatzbehälter *m* **~ tare** Supertara *f* **~ tax** Nachsteuer *f* **~ team** Mehrgespann *n* **~ time** Zuschlagzeit *f* **~ twisting** zusätzlicher Drall *m*, zusätzliche Drehung *f* **~ units** Zusatzeinheiten *pl* **~ voltage** Zusatzspannung *f* **~ water** Zusatzwasser *n* **~ weight** Zusatzgewicht *n* **~ yoke** Mehrgespann *n*
**additive** Wirkstoff *m*; addierend, additiv, zusätzlich **~ compound** Anlagerungserzeugnis *n* (chem.) **~ (fuel)** Betriebsstoffzusatzmittel *n* **~ process** Zusatzverfahren *n* **~ property of plane rotations** Additivität *f* der Drehwinkel **~ reaction** Additionsreaktion *f*
**additivity** Additivität *f* **~ effect** Akkumulierungseffekt *m*
**additron** Additron *n*
**adress, to ~** anreden, ansprechen, mit einer Aufschrift versehen
**address** Adresse *f*, Anrede *f*, Anschrift *f*, Ansprache *f*, Aufschrift *f*, Denkschrift *f*, Ortsangabe *f*, Rede *f* **~ of the consignee** Versandanschrift *f* **~-read wire** Adressenlesedraht *m* **(public-) ~ system** Sprachverstärkungsanlage *f*
**addressed memory** volladressierter Speicher *m*
**addressee** Empfänger *m* (Adressat)
**addressing machine** Adressiermaschine *f*, Anschriftenmaschine *f* **~ stencil** Anschriftenmaschinenschablone *f*
**adduce, to ~** einblättern, erbringen
**adduct** Addukte *n*
**adduction** Beibringung *f*
**adept** geschickt, sachverständig
**adequate** angemessen, ausreichend, entsprechend **to be ~** hinreichen
**adhere, to ~** adhärieren, anbacken, anhaften, anhängen, ankleben, festhaften, haften, hängen, hangen, innehalten, **to ~ to** (directions) einhalten, kleben
**adherence** Adhäsion *f*, Anhaften *n*, Anhaftung *f*, Ankleben *n*, Haften *n*, Haftenbleiben *n*, Haftfähigkeit *f*, Haftheit *f*, enge Anschmiegung *f* **~ to the norm** Normentreue *f*
**adherent** anhaftend, klebend; Anhänger *m*
**adhering** Kleben *n*
**adhesion** Adhäsion *f*, Anhaftung *f*, Flächenanziehung *f*, Haftbarkeit *f*, Haften *n*, Haftfähigkeit *f*, Haftheit *f*, Haftigkeit *f*, Haftkraft *f*, Haftung *f*, Haftvermögen *n*, Kleben *n*, Lötung *f* **~ method** Abreißmethode *f* **~ resistance** Gleitwiderstand *m* **~ surface area** Klebfläche *f*
**adhesive** Haftmittel *n*, Klebemittel *n*, Kleber *m*, Klebstoff *m*; anhaftend, adhäsiv, gummiert, klebend, klebrig **~ capacity** Adhäsionskraft *f* **~ force** Anziehungskraft *f*, Klebkraft *f* **~ grease** Adhäsionsfett *n* **~ lacquer** Kleblack *m* **~ materials** Klebstoffe *pl* **~ pitch** Kabelwachs *n* **~ plaster** Heftpflaster *n* **~ power** Adhäsionskraft *f*, Adhäsionsvermögen *n*, Haftvermögen *n*, Klebekraft *f* **~ strength** Haftfestigkeit *f* **~ stress** Haftspannung *f* **~ substance** Leim *m* **~ tape** Klebband *n*, Klebestreifen *m*
**adhesiveness** Adhäsionsvermögen *n*, Klebrigkeit *f*, Haftvermögen *n*
**adhesives** Klebstoffkitte *pl*
**adiabatic** adiabatisch, wärmeundurchlässig **~ change of conditions** adiabatische Zustandsänderung *f* **~ compression** adiabatische Verdichtung *f* **~ compressor head** Gebläseförder-

höhe *f* ~ **curve** Adiabate *f* ~ **invariance** adiabatische Invarianz *f* ~ **lapse rate** adiabatische Maßveränderung *f*, adiabatischer Zustand *m* der Luft ~ **process of change** adiabatische Änderung *f* (thermodyn)
**adiactinic** aktinisch, undurchlässig ~ **radiation** aktinisch undurchlässige Strahlung *f*
**A-digit selector** erster Kennziffernwähler *m*
**adion** Adion *n*
**adipocere** Fettwachs *n*, Leichenwachs *n*
**adipose** fettig
**adit** Bergwerkseingang *m*, Stollen *m* ~ **drainage** Wasserstollen *m* ~ **end** Abbaustoß *m* (min.) ~ **level** Wasserlösungsstollen *m* ~ **window** Fluchtort *m* (min.)
**adjacent** angrenzend, anliegend, benachbart, naheliegend **to be** ~ naheliegen
**adjacent,** ~ **angle** Nebenwinkel *m* ~ **box** Anbaukasten *m* ~ **channel** Nachbarkanal *m* (rdo) ~ **channel interference** Nachbarkanalstörung *f* ~ **channel selectivity** Nahselektion *f* (Trennschärfe gegen benachbarten Kanal) ~ **column** Nebenkolonne *f* ~ **domains** Nachbargebiet *n* ~ **elements of the lattice plane** Netzebenenverband *m* ~ **land** anliegendes Grundstück *n* ~ **rock** Nebengestein *n* ~ **row** benachbarte Reihe *f* ~ **sector** Nebenabschnitt *m* ~ **side** Nebenseite *f* ~ **vision carrier** Nachbarbildträger *m* ~ **zone** Nachbarzone *f*
**adjective dyestuffs** adjektive Farbstoffe *pl*
**adjoin, to** ~ grenzen (an)
**adjoining** angelehnt, angrenzend, anliegend, anstoßend, benachbart, neben ~ **angle** Nebenwinkel *m* ~ **concession** Nachbarfeld *n* ~ **formation** Nachbargebilde *n* ~ **mountains** angrenzendes Gebirge *n* ~ **rail** Anschlagschiene *f* ~ **sector** Nebenabschnitt *m* ~ **sheet** Anschlußblatt *n* ~ **soil** angrenzendes Gebirge *n*
**adjoint** adjungiert
**adjournment** Aufschiebung *f*, Aufschub *m*, Aussetzung *f*, Vertagung *f*
**adjudge, to** ~ zuerkennen, zusprechen
**adjudicate** verdingen, verurteilen, zuerkennen, zuteilen
**adjunct** Attribut *n*, Bestandteil *m*, Zusatz *m*
**adjust, to** ~ abgleichen, abgreifen, abrichten, abschirmen, adjustieren, anhalten, anordnen, anpassen, arretieren, ausgleichen, ausrichten, berichtigen, bremsen, eichen, einpassen, einregeln, einregulieren, einrichten, einstellen, feinstellen, festhalten, festlegen, feststellen, herrichten, justieren, kompensieren, nachregulieren, nachstellen, orientieren, passen, regeln, regulieren, richten, sperren, stellen, umstellen, unterlegen, verstellen, visieren, wieder in Ordnung bringen, zurechtmachen, zusammenpassen, zielen auf, (sight) zurichten
**adjust, to** ~ (a dispute) schlichten, (lens) einmitten **to** ~ **the bearing** das Lager *n* nachstellen **to** ~ **levels** auf gleiches Niveau *n* einstellen **to** ~ **by measuring** bemessen **to** ~ **to neutral** neutral einstellen **to** ~ **neutrally** neutral einstellen **to** ~ **one frequency to another** eine Frequenz *f* nach einer anderen einstellen **to** ~ **the rudder in failure of an engine** gegensteuern **to** ~ **the sight** Visier *n* anstellen oder einstellen **to** ~ **to uniformity** auf Gleichheit *f* einstellen
**adjustable** adjustierbar, anstellbar, einstellbar,

justierbar, nachstellbar, regelbar, regulierbar, stellbar, veränderlich, verschiebbar, verstellbar
**adjustable,** ~ **from the cockpit** vom Führersitz *m* aus verstellbar ~ **on ground** am Boden *m* verstellbar
**adjustable,** ~ **air louver** Kühlerjalousie *f* ~ **arrangement** Verschiebemöglichkeit *f* ~ **ball-and-socket bracket drop hanger** Armhängelager *n* mit Kugelbewegung und in der Höhe verstellbarer Lagerachse ~ **ball-and-socket hanger** Hängelager *n* mit Kugelbewegung und in der Höhe verstellbarer Lagerachse ~ **base** Schraubenfuß *m* ~ **bearing** nachstellbares Lager *n* ~**-blade chucking reamer** Maschinenreibahle *f* mit aufgeschraubten Messern ~**-blade hand reamer** Handreibahle *f* mit verstellbaren Messern ~**-blade reamer** Reibahle *f* mit eingesetzten Messern ~**-blade shell reamer** Aufsteckreibahle *f* mit nachstellbaren Messern ~ **bolt** Zuganker *m* ~ **center for milling slight tapers** Höhenzenter *m* zum Fräsen schlanker konischer Gegenstände ~ **compass card** Stellrose *f* ~ **condenser** Anodenrückwirkung *f*, einstellbarer, regelbarer, variabler, veränderbarer oder veränderlicher Kondensator *m* ~ **cord suspended luminaries** Zugpendelleuchten *pl* ~ **crank brace** Nachstellwinkel *m* ~ **cutter** einsetzbare oder verstellbare Schneidezunge *f* ~ **depth gauge** verstellbare Tiefenlehre *f* ~ **diffusor** Verstelldiffusor *m* ~ **disc condenser** Drehkondensator *m* ~ **driver's seat** verstellbarer Führersitz *m* ~ **friction clamp** verstellbare Reibungssperre *f* ~ **gib** Nachstelleiste *f* ~ **height range** Höhenverstellbereich *m* ~ **holder** Einstellhülse *f* ~ **index** Merkzeiger *m* ~ **knurl holder** verstellbarer Rändelhalter *m* ~ **lever** Stellhebel *m* ~ **marker** Einstellmarke *f* ~ **pedal** nachstellbarer Fußhebel *m* ~ **pilot's seat** verstellbarer Führersitz *m* ~ **pin** Stellzapfen *m* ~**-pitch propeller** einstellbare Schraube *f* (am Boden) ~ **point** einstellbares Komma *n* (info proc.) ~ **pole support** Deichselausgleichvorrichtung *f* ~ **propeller** einstellbare oder verstellbare Luftschraube *f*, Einstellschraube *f* ~ **pullings-in frame** Spannstock *m* mit verstellbaren Füßen ~ **rake (angle)** Einstellwinkel *m* ~ **resistance** regelbarer Widerstand *m*, Regelwiderstand *m* ~ **resistor** Einstellwiderstand *m*, Regelwiderstand *m* ~ **slide** Schiebekontakt *m* ~ **spindle bearing** nachstellbares Spindellager *n* ~ **spring bolt** gefederter Zuganker *m* ~ **spring rear sight** Federschraubvisier *n* ~ **starter** Regelanlasser *m* ~ **stop** einstellbarer Anschlag *m* ~ **(tilting) stripper** hochkippbarer Abstreifer *m* ~ **table dogs** Stellvorrichtung *f* für den Tisch ~ **tail plane** im Fluge verstellbare Höhenflosse *f* ~ **tap** nachstellbarer Gewindebohrer *m* ~ **tool rest** Werkzeugauflage *f* ~ **transformer** Drehtransformator *m* ~ **turbojet unit** Turbotransformator *m* ~ **wire mesh resistor** Gittergleitwiderstand *m* ~ **wrench** Mutterstellschlüssel *m*, verschiebbarer Schraubenmutterschlüssel *m*
**adjustability** Anstellbarkeit *f*, Einstellbarkeit *f*, Nachstellbarkeit *f*, Regelbarkeit *f*, Regelfähigkeit *f*, Regulierfähigkeit *f*, Verstellbarkeit *f*
**adjusted** abgeschirmt, bemessen, eingestellt, geregelt, gerichtet ~ **to** zugeschnitten auf ~ **brake**

Friktionsautomatik *f* ~-**filament current** heizabgestimmter Strom *m*

**adjuster** Eichmeister *m*, Einsteller *m* ~ **board** Regulierbrett *n*, Verstellbrett *n* ~ **wheel** Ausgleichsrad *n*

**adjusting** Abschirmung *f*, Anpassung *f*, Einstelldrehung *f*, Richten *n*, Sperre *f*, (method of least squares) Ausgleichung *f*

**adjusting,** ~ **apparatus** Stellvorrichtung *f* ~ **arm** Regulierhebel *m* ~ **bar for range finder** Berichtigungslatte *f* ~ **bench** Justierstand *m* ~ **bracket** Verstellblock *m* ~ **clip** Spurplättchen *n* ~ **collar** Verstellring *m* ~ **(potentiometer) control** Einstellpotentiometer *m* ~ **(setting) control** Justierpotentiometer *n* ~ **correction** Adjustierung *f* ~ **device** Justier-einrichtung *f*, -vorrichtung *f*, Nachstellvorrichtung *f*, Regulierungsvorrichtung *f*, Stellvorrichtung *f*, Verstellvorrichtung *f* ~ **dog** Stellklaue *f* ~ **equipment** Anstellvorrichtung *f* ~ **figure** Justierzahl *f* (Normzahl) ~ **file** Genauigkeitsfeile *f* ~ **gauge** Einstellehre *f* ~ **gauge of a screwclamp shape** schraubzwingenähnliche Einstellehre *f* ~ **gear** Stellzeug *n* ~ **gib** Einstelleiste *f* ~ **hole** Einstellbohrung *f* ~ **key** Stellkeil *m*, Stellmutter *f*, Stellschlüssel *m* ~ **lathe for range finder** Berichtigungslatte *f* ~ **ledge** Verstelleiste *f* ~ **lever** Einstellhebel *m*, Zündverstellhebel *m* ~ **(or control or timing) lever fastening** Verstellhebelbefestigung *f* ~ **lever stop** Verstellhebelanschlag *m* ~ **linkage** Verstellgestänge *n* ~ **loop** Einstellbügel (telet.) *m* ~ **machine** Adjustagemaschine *f*, Richtereimaschine *f*, Zurichtereimaschine *f*, Zurichtungsmaschine *f* ~ **mandrel** Einstelldorn *m* ~ **mark** Paßmarke *f*, Einstellmarkierung *f* ~ **nut** Endstück *n* (Pol einer Leitung), Knebelmutter *f*, Reguliermutter *f*, Spannmutter *f*, Stellmutter *f* ~ **organs** Verstellglieder *pl* ~ **(or timing) path** Verstellweg *m* ~ **piece** Paßstück *n* ~ **pin** Einstellstift *m*, Paßstift *m* ~ **pivot** Einstellzapfen *m* ~ **point** Einschießpunkt *m* ~ **power** Verstellkraft *f* ~ **rail** Verstelleiste *f* ~ **reading on range finder** Berichtigungsmeßreihe *f* ~ **ring** Einstellring *m*, Stellring *m* ~ **rod** Ausgleichstange *f*, Einstellstange *f* (aviat.), Regelstange *f* ~ **roller** Einstellwalze *f* ~ **screw** Berichtigungsschraube *f*, Einstellschraube *f*, Einstellspindel *f*, Justierschraube *f*, Nachstellschraube *f*, Regulierschraube *f*, Spreizschraube *f*, Stellringschraube *f*, Stellschraube *f* ~ **screw of axlebox wedge** Achslagerstellkeilschraube *f* ~ **shop** Zurichterei *f* ~ **slide** Einstellschieber *m*, Schwächungsanker *m* ~ **slide block** Stellschlitten *m* ~ **slider** Kontaktschieber *m* ~ **spindle** Stellspindel *f* ~ **spring** Paßfeder *f* ~ **spring hanger (or link)** Federhängebolzen *m*, Federspannschraube *f* ~ **strip** Beigabe *f* ~ **strips** Einlegestücke *pl* ~ **tubbing** Aufbaugußring *m* ~ **tube** Einstelltubus *m* ~ **unit** Anpassungsgerät *n* (rdr) ~ **wedge** Anzugskeil *m*, Regulierkeil *m*, Stellkeil *m* ~ **wedges** Einlegestücke *pl* ~ **wheel** Stellrad *n* ~ **winding** Abgleichwicklung *f* ~ **worm** Nachstellschnecke *f* ~ **wrench** Einstellschlüssel *m*, Stellschlüssel *m*

**adjustment** Abstimmung *f*, Adjustage *f*, Angleichung *f*, Arretierung *f*, Ausgleich *m*, Ausgleichung *f*, Ausrichten *n*, Berichtigung *f*, Einblendung *f*, Einfügung *f*, Eingliederung *f*, Einregulierung *f*, Einstellung *f*, Einstellvorrichtung *f*, Engjustierung *f*, Justierung *f*, Nachstellung *f*, Rastereinstellung *f*, Regelbefehl *m*, Regelung *f*, Regulierung *f*, Rektifikation *f*, Richtigstellung *f*, Sollwertgeber *m*, Verbesserung *f*, Vermittlung *f*, Verstellung *f*, Zurichtung *f*

**adjustment, out of** ~ dejustiert, verstellt, nicht mehr richtig eingestellt ~ **for back rest on bed** Einstellung *f* für die Gegenlagerständer *m* auf dem Bett ~ **of contrast** Dynamikeinregelung *f* ~ **for definition** Scharfeinstellung *f* ~ **of error of closing** Ausgleichung *f* ~ **of the extrusion orifice** Düsenspaltverstellung *f* ~ **of the fixed-tail surfaces** Flossenverstellung *f* ~ **(repeater) of gain** Verstärkungsregelung *f* ~ **of the gap between the rolls** Walzenspaltverstellung *f* ~ **of height** Höheneinstellung *f* ~ **of the helical slide** Schrägzahnschieberverstellung *f* ~ **of ignition** Zündungseinstellung *f* ~ **of observations** Fehlerausgleichung *f* ~ **of offices** Gleichschaltung *f* ~ **of paper** Papiereinstellung *f* ~ **of position of brushes** Bürstenverschiebung *f* ~ **of rolls** Walzenanstellung *f*, Walzeneinstellung *f* ~ **of signal strength** Lautstärkeregulierung *f* ~ **of stroke** Hubverstellung *f* ~ **of tension in wires** Durchhangsregelung *f*, Regulieren *n* der Drähte ~ **of tools** Werkzeugverstellung *f* ~ **of top cutter** Obermesserverstellung *f* ~ **to value of darkness** Dunkelwertsteuerung *f* ~ **of vibration motion at rotary unit** Einstellung *f* der Querverreibung am Zylinderfarbwerk ~ **of volume range or contrast** Einregelung *f* der Dynamik

**adjustment,** ~ **aberration** Einstellfehler *m* ~ **bolt** Einstellbolzen *m* ~ **cock (or valve)** (for flames) Kleinsteller *m* ~ **cover** Einstelldeckel *m* ~ **feature** Einstellungsmerkmal *n* ~ **framing** Bildstrichverstellung *f* (TV) ~ **gear for range finder** Berichtigungsgerät *n* ~ **instructions** Fehlersuchtabelle *f* ~ **knob** Stellknopf *m* ~ **lever** Einstellhebel *m* (teletype) ~ **locking pin** Einstellsperrstift *m* ~ **machine** Adjustagemaschine *f*, Richtereimaschine *f*, Zurichtereimaschine *f* ~ **mark** Einstellmarke *f*, Justiermarke *f* ~ **needle** Einstellnadel *f*, Einstellzeiger *m* ~ **nut (or screw)** Einstellmutter *f* ~ **pin** Justierstift *m* ~-**scale correction** Adjustierung *f* ~ **screw** Justierschraube *f*, Gegenspindel *f* ~ **shaft** Drehspindel *f* ~ **wire** Ausgleichdraht *m*

**adjuvant** (substance) Hilfsstoff *m*

**admeasurement** Ausmessung *f*

**admeasuring apparatus** Ausmeßgerät *n*

**administer, to** ~ einflößen, eingeben, verwalten

**administration** Amtsführung *f*, Amtsverwaltung *f*, Anwendung *f*, Hauptverwaltung *f*, Leitung *f*, Verwaltung *f* ~ **building** Verwaltungsgebäude *n*

**administrative** administrativ, erteilend, verwaltend ~ **aircraft** Kurierflugzeug *n* ~ **area** Verwaltungsgebiet *n* ~ **authority** Verwaltungsbehörde *f* ~ **department** Amt *n* ~ **details** besondere Anordnungen *pl* ~ **district** Verwaltungsbezirk *m* ~ **function** Amt *n* ~ **office** Verwaltungsbüro *n* ~ **official** Verwaltungsbeamter *m* ~ **regulations** Verwaltungsvorschrift *f*

**administrator** Administrator *m*, Amtsverwalter *m*, Verwalter *m*

**admissibility** Zulässigkeit *f*

**admissible** zulässig ~ **character** zulässiges Zeichen *n* ~ **flow rates** zulässige Höchstgeschwindigkeiten *pl*
**admission** Admission *f*, Annahme *f*, Aufnahme *f*, Bekanntgeben *n*, Einlaß *m*, (to) Einlassen *n*, Eintritt *m*, Zufluß *m*, Zufuhr *f*, Zugeständnis *n*, Zulaß *m*, Zulassung *f*, Zutritt *m* ~ **of air** Luftzufuhr *f* ~ **of fluid to turbine blades** Beaufschlagung *f* ~ **of light** Lichtzufuhr *f* ~ **of manufacturing** Fabrikationsfreigabe *f*
**admission,** ~ **aperture** (engine) Einlaßschlitz *m* ~ **cam** Einlaßnocken *m*, (gas and oil engines) Einlaßscheibe *f* ~ **channel** Eintrittskanal *m* ~ **controller** Zuflußregler *m* ~ **edge** (slide valves) Einlaßkante *f* ~ **gear** Ansauggestänge *n*, Einlaßgestänge *n*, Einlaßsteuerung *f* ~ **line** Admissionslinie *f* ~ **pipe** Einlaßrohr *n*, Eintrittsrohr *n* ~ **port** Admissionskanal *m*, Dampfeinlaß *m*, Einlaßöffnung *f*, (engines) Einlaßschlitz *m* ~ **potential** Admissionsspannung *f*, Eintrittsspannung *f* ~ **pressure** Admissionsdruck *m*, Einlaßdruck *m*, Eintrittsdruck *m*, (of steam) Dampfeintrittsspannung *f* ~ **pressure regulator** Vordruckregler *m* ~ **relay** Berichtigungsrelais *n* ~ (part of the) **stroke** Füllungsstrecke *f* ~ **temperature** Zulauftemperatur *f* ~ **tension** Admissionsspannung *f* ~ **valve** Ansaugventil *n* ~**-valve cone** Einlaßkegel *m* ~**-valve rod** Einlaßventilstange *f* ~**-valve roller** Einlaßventilrolle *f*
**admit, to** ~ anerkennen, beaufschlagen, bekennen, eingestehen, einlassen, gestehen, werten, zuführen, zugeben, zugestehen, zulassen
**admittance** Admittanz (rdo), Leitwert *m*, Scheinleitwert *m*, Zutritt *m*, Wellenleitwert *m* **no** ~ Eintritt *m* verboten, verbotener Eingang *m*
**admitted** anerkannt, zugestanden
**admix, to** ~ beimengen, beimischen, strecken, zumischen
**admixed** zugemischt ~ **material** Zumischstoff *m*, Zusatzstoff *m* ~ **substance** zugemischter Stoff *m*
**admixture** Beimischung *f*, Beisatz *m*, Zumischstoff *m*, Zumischung *f*, Zusatz *m*, Zuschlag *m*
**admonition** Mahnung *f*
**adobe** Lehmstein *m*, Tonerde *f*
**adopt, to** ~ adoptieren, annehmen, sich aneignen, aufgreifen
**adopted** angenommen
**adoption** Annahme *f*, Einführung *f*
**adrift** lostreibend, treibend
**adroit** gewandt
**adroitness** Geschicklichkeit *f*
**adsorb, to** ~ (at points of attachment) auslagern
**adsorbate** adsorbierte Substanz *f*
**adsorbed layers** adsorbierte Schichten *pl*
**adsorbent** Adsorbens *m*, Einsaugemittel *m*
**adsorbing,** ~ **capacity** Adsorptionsfähigkeit *f* ~ **power** Adsorptionsvermögen *n* ~ **substance** Trägerstoff *m*
**adsorption** Ansaugen *n*, Flächenanziehung *f*, Haftung *f* ~ **atom** Adatom *n* ~ **layer** Adsorptionsschicht *f* ~ **property** Adsorptionsfähigkeit *f* ~ **space** Dicke *f* der adsorbierten Schicht
**adulterant** Verderbungsmittel *n* (chem.), Verfälschungsmittel *n*
**adulterate, to** ~ mischen, nachahmen, verderben, verfälschen, verschneiden, verunreinigen, zusetzen

**adulterated,** ~ **alcohol** Alkoholersatz *m* ~ **turpentine** Terpentinölersatz *m*
**adulteration** Fälschung *f*, Verfälschung *f*
**adumbrate, to** ~ abschatten
**ad valorem duty** Wertzoll *m*
**advance, to** ~ anrücken, ansteigen, anziehen, aufrücken, aufschlagen, befördern, ein- und ausfahren, erhöhen, fördern, fortschreiten, vordringen, voreilen, vorgehen, vorgehen, vorrücken, vorstellen, vorverlegen, vorwärts bringen, zunehmen, (money) vorschießen **to** ~ **a claim (or point)** Anspruch *m* geltend machen **to** ~ **an opinion** eine Ansicht *f* vorbringen
**advance** Anlauf *m*, Aufschlag *m*, Erhöhung *f*, Fortrückung *f*, Fortschreiten *n*, Fortschritt *m*, Verbesserung *f*, Vorauszahlung *f*, Vordringen *n*, Voreilen *n*, Vorgehen *n*, Vorrücken *n*, Vorschub *m*, Vorschuß *m*, Vorsprung *m*, (magnet) Vorstellen *n*, Vorstellung *f*, Vorstoß *m*, Vortrieb *m*, Vorwärtsbewegung *f*
**advance, in** ~ im voraus, vorläufig (notice) ~ **on the face** Abbaufortschreitung *f* ~ **in price of** . . . Preiserhöhung *f* von . . . ~ **in prices** Hausse *f* ~ **of scraper** Schabervorschub *m*
**advance,** ~ **attrition** vorzeitige Abnutzung *f* ~ **ball** Polierkugel *f* ~ **borehole** Vorbohrloch *n* ~ **cam** Anlaufkurve *f* ~ **gear** Vorschubgetriebe *n* ~ **heading** Ausrichtstrecke *f* (min.), Einschnittstollen *m* beim Tunnelbau, Richtstollen *m* ~ **ignition** Früheinspritzung *f* (Diesel), Frühzündungsdruck *m* ~ **mechanism** Vorschubmechanismus *m* ~ **notice** Vorbescheid *m* ~ **notification of incoming call** Vorbereitung *f* eines Ferngesprächs ~ **path** Voreilweg *m* ~ **post** Außenposten *m* ~ **pulse** Schiebeimpuls *m* ~ **speed** Vorschubgeschwindigkeit *f* ~ **time** Vorgabezeit *f*
**advanced** fortgeschritten, vorgeschoben, vorverlegt ~ **algebra** höhere Algebra *f* ~ **course** Fortbildungskurs *m* ~ **design** neuzeitliche Konstruktion *f* ~ **ignition** Frühzündung *f*, Vorzündung *f* ~ **presignal** vorverlegtes Vorsignal *n* ~ **sparking** vorgelegte Funkenlage *f* ~ **stall** vorgerückter überzogener Flug *m* ~ **training** Fortbildung *f*, Weiterbildung *f*
**advancement** Förderung *f*
**advancing** fortschreitend; Anziehen *n* (prices) ~ **and retreating blades** vor- und rücklaufende Blätter *pl* ~ **wave** fortschreitende Welle *f*
**advantage** Benutzung *f*, Gewinn *m*, Mehrgewicht *n*, Nutzen *m*, Nützlichkeit *f*, Vorteil *m*, Vorzug *m* ~ **factor** Flußfaktor *m*, Optimalbestrahlungsfaktor *m*
**advantageous** gewinnbringend, günstig, nützlich, vorteilhaft, zweckmäßig **to be** ~ Vorteil *m* bringen
**advection** Advektion *f*, Lufthorizontalbewegung *f* (meteor.) ~ **fog** Advektionsnebel *m*, Mischungsnebel *m*
**adventive** adventiv
**advent** Ankunft *f*
**adventure** Abenteuer *n*, gewagtes Unternehmen *n*
**adventurer** Bergeunternehmer *m*, Gewerke *m*
**adventurine feldspar** Aventurinfeldspat *m*
**adversary** Gegenpartei *f*, Gegner *m*
**adverse** normwidrig ~ **action** abschlägiger Bescheid *m* ~ **balance** Unterbilanz *f* ~ **decision**

abschlägiger Bescheid *m* ~ **weather** schlechtes Wetter *n*
**advertise, to** ~ ankündigen, anzeigen, bekanntmachen, inserieren, kundgeben, werben
**advertisement** Anzeige *f*, Inserat *n* ~ **broker (or canvasser)** Anzeigensammler *m* ~ **proof** Anzeigenabzug *m* ~ **typesetter** Anzeigensetzer *m*
**advertiser** Anzeiger *m*
**advertising** Reklame *f*, Werbung *f*, (for bids) Ausschreibung *f* ~ **airplane** Reklameflugzeug *n* ~ **art** Werbekunst *f* ~ **article** Werbeartikel *m*, Werbegegenstand *m* ~ **expense(s)** Insertionsgebühr *f*, Insertionskosten *pl*, Werbekosten *pl* ~ **flight** Reklameflug *m* ~ **novelty** Zugabeartikel *m* ~ **pillar (or billboard)** Anschlagsäule *f* ~ **poster** Werbeplakat *n* ~ **printed matter** Werbedrucksache *f* ~ **printed print** Werbedrucksache *f* ~ **signboard** Reklameschild *n* ~ **success** Werbeerfolg *m*
**advice** Avis *m*, Benachrichtigung *f*, Beratung *f*, Bericht *m*, Gutachten *n*, Meldung *f*, Rat *m*, Ratschlag *m*
**advisable** angezeigt, rätlich, zweckmäßig
**advise, to** ~ ankündigen, anraten, anweisen, avisieren, in Kenntnis setzen, raten
**adviser** Berater *m*, Ratgeber *m*
**advisory** beratend ~ **board** Beratungsstelle *f*, Verwaltungsrat *m* ~ **body** Beirat *m* ~ **deliberation** Beratung *f*
**advocate, to** ~ anraten, befürworten
**adz, to** ~ **and fix the chairs on the ties** die Schwellen *pl* einschneiden und aufstuhlen
**adz(e)** Axt *f*, Beil *n*, Breitbeil *n*, Dachsbeil *n*, Dechsel *m*, Haue *f*, Krummhaue *f*
**adze, to** ~ behauen **to** ~ **the ties** die Schwellen *pl* einblatten
**aeolian** äolisch ~ **bell** Äolsglocke *f* ~ **harp** Äolsharfe *f* ~ **note** Hiebton *m* ~ **rocks** äolisches Geröll *n* ~ **tone** Anblaston *f*
**aeolipile** Äolusball *m*
**aeolotropy** Äolotropie *f*
**aerate, to** ~ auflockern, durchlüften, lüften
**aerated** lufthaltig ~ **cement block** Gasbetonstein *m* ~ **plastics** Schaumstoff *m* ~ **spring** Sauerbrunnen *m*, Sauerquelle *f* ~ **water** kohlensaures, moussierendes Wasser *n*
**aeration** Auflockerung *f*, Belüftung *f*, Lüftung *f*
**aerating,** ~ **apparatus** Auflockerungsapparat *m* ~ **compartment** Durchlüftungskammer *f*
**aerator** Auflockerungsapparat *m*
**aerial** (see **antenna**) Antenne *f*; atmosphärisch, luftig, zur Luft gehörig, oberirdisch
**aerial,** ~ **arch** Luftsattel *m* ~ **billow** Luftwegwelle *f*, Luftwoge *f*, Windwoge *f* ~ **bomb** Fliegerbombe *f* ~ **cableway** Luftseilbahn *f* ~ **camera** Luftbildgerät *n*, Luftbild- *f*, Panoramakamera *f*, Reihenbildner *m* ~ **camouflage** Luftverschleierung *f* ~ **compass** Luftfahrtkompaß *m* ~ **conduit** Oberleitung *f* ~ **contact line** Fahrtdrahtoberleitung *f* ~ **craft** Flugmaschine *f* ~ **cutout** Freileitungssicherung *f* ~ **defense** Luftverteidigung *f* ~ **delivery** Lufttransport *m* ~ **delivery container** Abwurf-behälter *m*, -gondel *f*, Lufttransportbehälter *m* ~ **discharger** Überspannungsableiter *m* ~ **drum** Aufwickeltrommel *f* ~ **engineering** Luftbildtechnik *f* ~ **ferry** Schwebefähre *f* ~ **flashlight photograph** Blitzlichtluftbild *n* ~ **frog** Luftwegweiche *f* ~ **gas-**

**pressure sprinkler** Flugzeugzerstäubungsgerät *n* ~ **gas-spraying apparatus** Luftstromgerät *n* (aviat.) ~ **image** virtuelles Bild *n* ~ **line** oberirdische Leitung *f* ~ **-mapping camera** Luftbildmeßkamera *f* ~ **mapping lens** Meßobjektiv *n* ~ **mosaic** Luftbild-karte *f*, -skizze *f* ~ **navigation** Flugwesen *n*, Luftfahrtnavigation *f*, Luftortung *f* ~ **network** Freileitungsanlage *f*, Luftstreckennetz *n* ~ **night camera** Nachtluftbildgerät *n* ~ **observation** Luftbeobachtung *f* ~ **oscillator** Luftschallsender *m* ~ **photograph** Aufnahme *f* des Geländes, Luftaufnahme *f*, Luftbildaufnahme *f*, Senkrechtaufnahme *f* ~ **photograph plane** Luftbildflugzeug *n* ~ **photographer** Bildflieger *m* ~ **photographers** Flugzeugbildpersonal *n* ~ **photography** Luftbildwesen *n* ~ **-photography mosaic** Reihenbild *n* ~ **picture** Fliegeraufnahme *f* ~ **portable camera** Fliegerhandkamera *f* ~ **railway** Schwebebahn *f* ~ **ropeman** Seilbahnlader *m* ~ **ropeway** Drahtseilbahn *f*, Seilhängebahn *f* ~ **sounding line** Luftschallot *n*, Schallhöhenmesser *m* ~ **sport** Luftsport *m* ~ **spotting** Beobachtungsfliegen *n* ~ **survey** Luftvermessung *f*, Überwachung *f* durch Flugzeuge ~ **survey photograph** Luftmeßbild *n* ~ **target** Luftziel *n* ~ **tramway** Luftseilbahn *f* ~ **view** Flugzeugaufnahme *f*, Luftansicht *f* ~ **weight of trailing antenna** Belastungsgewicht *n* ~ **wire** Luftdraht *m*, Luftleitung *f*
**aerification** Luftbildung *f*
**aeriform** luftförmig
**aerobatic** kunstflugtauglich ~ **attitude** Kunstfluglage *f* ~ **contest** Kunstflugwettbewerb *m* ~ **display** Kunstflugvorführung *f* ~ **flight** Kunstflug *m* ~ **-flying trial** Kunstflugerprobung *f* ~ **maneuver** Kunstflugfigur *f* ~ **pilot** Kunstflieger *m* ~ **plane** Kunstflugzeug *n* ~ **qualities** Kunstflugeigenschaften *pl* ~ **sailplane** Segelkunstflugzeug *n* ~ **-training plane** Übungsflugzeug *n* für Kunstflug ~ **-training single seater** Kunstflugübungseinsitzer *m*
**aerobatics** Kunstfliegen *n*, Kunstflug *m* ~ **in formation** Kunstflug *m* im Verband
**aerocartograph** (machine plotting photogrammetric photograph) Aerokartograf *m*
**aeroconcrete** Schaumbeton *m*
**aerodrome** (also see **airdrome** and **airport**) Flugplatz *m* ~ **of departure** Ausgangsflugplatz *m* ~ **of destination** Zielflugplatz *m*
**aerodrome,** ~ **beacon** Flugplatzleuchtfeuer *n* ~ **control tower** Platzverkehrskontrollstelle *f* ~ **control van** Pistenwagen *m* ~ **elevation** Flugplatzhöhe *f* ~ **hazard beacon** Flugplatzwarnungsbake *f* ~ **identification sign** Flugplatzerkennungszeichen *n* ~ **lighting** Flugplatzbefeuerung *f* ~ **reference point** Flugplatzbezugspunkt *m* ~ **site** Flugplatzlage *f* ~ **traffic** Flugplatzverkehr *m* ~ **traffic circuit** Platzrunde *f* ~ **traffic zone** Flugplatzverkehrszone *f*
**aerodynamic** aerodynamisch, lufttechnisch, windschnittig ~ **balance** aerodynamischer Ausgleich *m*, Auswiegen *n*, Flächenanziehungsausgleich *m*, Innenausgleich *m* ~ **center** Brennpunkt *m*, Indifferenz-, Neutral-, Druck-punkt *m* (aviat.) ~ **drag** (on airplane) Luftwiderstand *m* ~ **efficiency** aerodynamischer Wirkungsgrad *m* ~ **fineness** aerodynamische Feinheit *f* ~ **flow research** Windkanalforschung *f* ~ **forces** dyna-

mische Luftkräfte *pl* ~ **forms** aerodynamische Linienführung *f* ~ **lift** Auftrieb *m* ~ **moment field** Rückstellfeld *n*, Luftmomentenfeld *n* ~ **qualities** aerodynamische Eigenschaften *pl* ~ **pressure** Flugstaudruck *m* ~ **resistance** Strömungswiderstand *m* ~ **seal** Spaltabdeckung *f* ~ **and inertia forces in turns** Gas- und Massendrehkraft *f* ~ **volume displacement** Luftverdrängung *f*

**aerodynamics** Aerodynamik *f*, Dynamik *f* luftförmiger Körper, Fluglehre *f*, Luftdrucklehre *f*

**aerodyne** Luftfahrzeug *n* schwerer als die Luft

**aeroembolism** Höhenkrankheit *f*

**aerofilm** Fliegerfilm *m* (photo)

**aerofoil** Flügel *m* (als Oberbegriff), Stromlinienflügel *m*, Tragfläche *f*

**aerogene gas** Aerogengas *n*

**aerogenous** aerogen

**aerograph** Druckluftspritzapparat *m*, Farbzerstäuber *m*, Spritzpistole *f*

**aerography** Aerografie *f*, Luftbeschreibung *f*

**aerohydraulics** Flughydraulik *f*

**aerolimnology** Aerolimnologie *f*

**aerolite** Aerolith *m*, Luftstein *m*

**aerological observatory** Luftwarte *f*

**aerology** Ärologie *f*, aeronautische Wetterkunde *f*, Luftkunde *f*, Luftlehre *f*

**aeromechanics** Aeromechanik *f*

**aeromedical** luftfahrtmedizinisch

**aeromedicine** Luftfahrtmedizin *f*

**aerometer** Aerometer *n*, Dichtemesser *m*, Dichtigkeits-, Flüssigkeitsdichte-, Schwere-messer *m*, Spindelzelle *f*

**aerometeorograph** Aerometeorograph *m*

**aero-mixture indicator** Gemischstärkemeßgerät *n*

**aeronaut** Luftfahrer *m*, Luftschiffer *m*

**aeronautical** aeronautisch, Luftfahrt ... ~ **board** Luftfahrtverwaltung *f* ~ **chart** Flugkarte *f*, Luftfahrerkarte *f* ~ **equipment standards** Luftfahrtgerätenorm *f* ~ **exhibition** Luftfahrtausstellung *f* ~ **ground light** Luftfahrtbodenfeuer *n* ~ **ground radio station** Bodenfunkstelle *f* ~ **information publication (AIP)** Luftfahrthandbuch *n* ~ **information service** Flugberatungsdienst *m* ~ **inspection** Bauaufsicht *f* ~ **light** Luftfahrtbodenfeuer *n* ~ **meteorological service** Flugwetterdienst *m* ~ **meteorology** Flugwetterkunde *f* ~ **mobile station** bewegliche Flugfunkstelle *f* ~ **radio** Flugfunk *m* ~ **radio service** Flugfunkdienst *m* ~ **research** Luftfahrtforschung *f* ~**research institute** Luftfahrtforschungsanstalt *f* ~ **station** Bodenfunkstelle *f* ~ **technique** Flugtechnik *f* ~ **telecommunication (service)** Flugfernmeldewesen *n* (-dienst *m*) ~ **weather service** Flugwetter-beratung *f*, -dienst *m*, Luftfahrtwetterdienst *m*

**aeronautics** Aeronautik *f*, Fluglehre *f*, Flugtechnik *f*, Flugwesen *n*, Luftfahrt *f*, Luftfahrtwesen *n*, Luftschiffahrtskunde *f*, Luftschiffkunst *f*

**aerophore** Atmungsapparat *m*, Beleuchtungsapparat *m*, Leitstrahlsender *m*

**aerophotogrammetric,** ~ **lens** Meßfliegerobjektiv *n* ~ **map** Karte *f* nach Luftbildaufnahmen ~ **plan** Plan *m* nach Luftaufnahmen ~ **survey** luftfotogrammetrische Aufnahme *f*

**aerophotographic** aerofotografisch, luftfotografisch ~ **apparatus** Luftbildaufnahmegerät *n* ~

**camera** Luftbildkamera *f* ~ **map** Luftbildkarte *f* ~ **mosaic (or plan)** Luftbildplan *m* ~ **section** Luftbildabteilung *f* ~ **sketch** Luftbildskizze *f*

**aeroplane** (also see **airplane**) Flugzeug *n* ~ **cross-spar** Flugzeugtraverse *f* ~ **flutter** Flugzeugflattern *n* ~ **spar** Flugzeugholm *m*

**aeropulverizer** Aeromühle *f*

**aerosiderite** Aerosiderit *m*, Lufteisenstein *m*

**aerosol** Aerosol *n* ~ **separating chamber** Aerosolabscheidekammer *f*

**aerostat** Aerostat *n*, Luftfahrzeug *n* leichter als die Luft

**aerostatic** aerostatisch

**aerostatics** Aerostatik *f*, Luftgleichgewichtslehre *f*

**aerostructure (of a flying boat)** Flugwerk *n* und Leitwerk *n* (eines Flugbootes)

**aerostrut** Fahrgestellstrebe *f*

**aerotechnics** Luftfahrttechnik *f*

**aerotopographic map** Luftbildplan *m*

**aerotow** Flugzeugschleppstart *m*

**aerotriangulation** Lufttriangulierung *f*

**aesthesiometer** Ästhesiometer *n*

**aethrioscope** Äthrioskop *n*

**affect, to** ~ angreifen, beeinflussen, befallen, einwirken, ergreifen, verändern

**affected by the air** luftempfindlich

**affection** Beeinflussung *f*

**affidavit** Aussage *f*, Ausweispapier *n*, eidesstattliche Erklärung *f*, eidliche Versicherung *f* ~ **of execution** eidesstattliche Erklärung *f*

**affiliate, to** ~ angliedern

**affiliated** nahestehend ~ **company** Zweiggesellschaft *f* ~ **concern** Zweiggeschäftsbetrieb *m*

**affiliation** Angliederung *f*

**affinability** Affinierbarkeit *f*

**affinable** affinierbar

**affination,** ~ **output** Affinationsausbeute *f* ~ **sugar** Affinade *f*

**affine, to** ~ affinieren

**affine,** ~ **geometric sentences** affine geometrische Aussagen *pl* ~ **transformation** affine Verzerrung *f*

**affining quality** Affinationswert *m*

**affinitive** affin

**affinity** Affinität *f*, Bindung *f*, Verbindungsfähigkeit *f*, Verbindungskraft *f*, Verwandschaft *f*, Ziehvermögen *n* ~ **for oxygen** Sauerstoffverwandschaft *f*

**affinors** Affinoren *pl*

**affirm, to** ~ behaupten

**affirmation** Aussage *f*, Wahrheitsbekräftigung *f* ~ **in lieu of oath** eidesstattliche Erklärung *f* oder Versicherung *f*

**affirmative, in the** ~ **(sense)** bejahenden Sinnes

**affix, to** ~ anhängen **to** ~ **a seal** besiegeln

**afloat** flott

**affluent** Nebenfluß *m*

**afflux** Zufluß *m* ~ **velocity** (of fluids) Anflußgeschwindigkeit *f*, Anströmungsgeschwindigkeit *f* (electron.)

**afford, to** ~ aufbringen, gewähren, eine Kraft *f* aufbringen, eine Last *f* aufbringen **to** ~ **assistance** Vorschub *m* leisten

**afforest, to** ~ bewalden

**afforestation** Aufforstung *f*

**A-fixture** A-förmiges Gestänge *n*

**A- (or H-)fixture line** Linie *f* mit Doppelgestänge

**afloat, to get ~** loskommen
**afocal ancillary lens system** afokales Vorsatzlinsensystem *n*
**afore, ~-mentioned** oben erwähnt, vorerwähnt **~ said** vorstehend
**A-frame** Auslegerbock *m*
**aft** Achter *m*; achteraus, achtern, hinterwärts **~ body section** Heckteil (d. Rakete) **~(er) deck** (small craft) Achterdeck *n* **~ end** Achterende *n* **~ frame assembly** hinterer Triebwerkabschnitt *m* **~ gate** Untertor *n*, (sluice) Niedertor *n*
**after nachher ~ accelerator** Nachbeschleunigungselektrode *f* **~ admission** Nachfüllung *f* **~ annealing** Nachtempern *n* **~ bake** Nachhärten *n* **~ blow** Nachblasen *n* **~ body** Achterschiff *n*, Rumpf *m* (g/m) **~-burner** Nachbrenner *n* **~-burning** Nachbrennen *n*, Nachverbrennung *f* **~ compensation** Nachkompensierung *f* **~ compressor** Nachschaltverdichter *m* **~ cooler** Nachkühler *m* **~ coupling** Nachkupplung *f* **~ damp** Nachschwaden *m*, böses Wetter *n* **~-deck** Heck *n* **~-effect** Nachwirkung *f* **~-effect function** Nachwirkungsfunktion *f* **~ fermentation** Nachgärung *f* **~-firing** klopfende Verbrennungen *pl* nach Abstellen der Zündung (Diesel) **~ flow** Nachfließen *n* (metall.), plastische Nachwirkung *f* **~-fractionating tower** Nachfraktionierturm *m* **~ frame** Achterspant *n* **~ (forward) gangway** achtere (vordere) Fallreep *n* **~ generation** Nachgasen *n* **~-glow** Bildnachleuchten *n*, Nachglühen *n*, Nachleuchten *n*, Phosphoreszens *f* **~-glow time** Nachleuchtdauer *f* **~-glow tube** Nachleuchtrohr *n* (electron.) **~ heat** Nachwärme *f* **~ hold** Achterraum *m* **~-image** Nachbild *n*, nachleuchtendes Bild *n* **~ peak** Achterpiek *f* **~-piece** (ship-building) Scheg *n* **~-portion of fuselage** Rumpfhinterteil *n* **~ product** Nacherzeugnis *n*, Nachprodukt *n* **~ product crystallizer** Nachproduktmaische *f* **~-retting of flax** Nachrotte *f* des Flachses **~-sales service** Kundendienst *m* **~ shock** Nachbeben *n* **~ shooting** (in quarries to reduce size of large pieces) Nachschießen *n* **to ~ shrink** nachschrumpfen **~ shrinkage** Nachschrumpfung *f* **~ stretching** Nachverstrecken *n* **~ treatment** Nachbehandlung *f*, Nachwaschen *n* **~ worker** (sugar mfg) Nachdrehschleuder *f* **~ working** Nachwirkung *f*
**agalite** Faserkalk *m*
**agalmatolite** Agalmatolith *m*, Bildstein *m*
**agario mineral** Bergmilch *f*
**agate** Achat *m* **~ ball** Achatkugel *f* **~ bearings** Achateinsätze *pl* **~ cup** Achathütchen *n* **~ mountings (or settings)** Achateinsätze *pl* **~ mortar** Achatmörser *m* **~ tip** Achatfuß *m*
**agathic acid** Agathendisäure *f*
**agatiferous** achathaltig
**age, to ~** ablagern, altern, aushärten (metall.)
**age** Alter *n*, Formation *f* (geol.), Zeitraum *m* **~ class** Geburtsjahrgang *m* **~ coating of lamp bulbs** Beschlagen *n* der Lampenbirnen **~ equation** Altersgleichung *f* **~ group** Altersgruppe *f* **~ limit** Altersgrenze *f* **~ resister** Alterungsschutzmittel *n*
**aged** abgelagert, gealtert
**age-harden, to ~** altern, anlassen, veredeln, vergüten (Leichtmetall)
**age-hardenable** vergütbar
**age-hardening** Altershärtung *f*, Alterungshär-

tung *f*, Anlassen *n*, Aushärtung *f*, Ausscheidungshärtung *f*, Veredeln *n*, Vergüten *n*, Vergütung *f*, Zeithärtung *f*
**agency** Agentur *f*, Mittel *n*, Niederlage *f*, Organ *n*, Vermittlung *f*
**agenda** zu erledigende Punkte *pl*, Tageseinteilung *f*, Tagesordnung *f*
**agent** Agens *n* (chem.), Agent *m*, Beauftragter *m*, Faktor *m*, Geschäftsträger *m*, Medium *n*, Mittel *n*, Vermittler *m* **~ for lacquered printing** Lackabdruckmittel *n* **~ of putrefaction** Fäulniserreger *m*
**agglomerate, to ~** agglomerieren, zusammen-backen, -ballen, -sintern
**agglomerate** Agglomerat *n*, Anhäufung *f*, Brekzie *f*, Sinter *m*, Sinterkuchen *m*, Sintererzeugnis *n* **~ cell** Brikettelement *n*
**agglomerated, ~ cake** Agglomerat *n*, Gesinter *n*, Sintererzeugnis *n* **~ sinter cake** Agglomeratkuchen *m*
**agglomerating, ~ method** Agglomerierverfahren *n* **~ plant** Agglomerieranlage *f*, Sinteranlage *f* **~ process** Agglomerierverfahren *n*
**agglomeration** Agglomeration *f*, Anhäufung *f*, Ballung *f*, Gesinter *n*, Sinterung *f*, Verschachtelung *f*, Zusammen-backen *n*, -ballen *n*, -ballung *f* **~ power** Klebekraft *f*
**agglutinant** Klebemittel *n*
**agglutinate, to ~** verkleistern, zusammenleimen
**agglutinate** Bindemittel *n*, Bindungsmittel *n*
**agglutination** Lötung *f*, Verkleisterung *f*
**aggradation** Aufschüttung *f*
**aggravate, to ~** bekräftigen, erschweren, verstärken
**aggravating condition** Erschwernis *f*
**aggravation** Bekräftigung *f*, Erschwerung *f*, Verstärkung *f*
**aggregate, to ~** anhäufen, ansammeln, zusammenhäufen
**aggregate** Aggregat *n*, Ansammlung *f*, Menge *f* (for concrete); Zuschlagstoff *m* gesamt **~ motion method** Wegsummenverfahren *n* (teletype) **~ motion principle** Additionsprinzip *n* (teletype) **~ motion translator** Summenübersetzer *m* (teletype) **~ peak level** Summenspitzenpegel *m* **~ recoil** Molekülrückstoß *m*
**aggregates** Dämmstoffe *pl* **~ of vacancies** Anhäufung *f* von Leerstellen
**aggregation** Aggregation *f*, Anhäufung *f*, Ansammlung *f*, Ganzheit *f*, Haufen *m*, Zusammenschluß *m*
**aggressive, ~ power** Angriffskraft *f* **~ tack** Trockenklebrigkeit *f*
**aggressiveness** Aggressivität *f*
**aggrieved** betroffen **~ party** Benachteiligter *m*, Geschädigter *m*
**agility** Gewandtheit *f*
**aging** Ablagerung *f*, Altern *n*, Alterung *f*, Veredelung *f* (Leichtmetall), Vergüten *n*, Vorreife *f* **~ of coal** Inkohlung *f* **~ at 100 degrees** Kochvergütung *f* **~ (at ordinary temperature)** Kaltvergütung *f* **~ at room temperature** Selbstalterung *f*
**aging, ~ effect** Alterungseinfluß *m* **~ hopper** Vorreifekasten *m* **~ operating** künstliche Alterung *f* **~ process** Alterungsverfahren *n*, Lagerprozeß *m* **~ temperature of metals** Alterungstemperatur *f* **~ test** Alterungsprobe *f*, Alterungsversuch *m* **~ time** Nachhärtungsfrist *f*

**agio** Agio *n*, Aufgeld *n*
**agitable** aufrührbar
**agitate, to** ~ aufrühren, aufwiegeln, bewegen, durchrühren, durchschütteln, erregen, hetzen, rühren, schütteln, umrühren, umschwenken **to ~ the transmitter** das Mikrofon schütteln
**agitated** stürmisch ~ **accumulator** kochender Akkumulator
**agitating** Rühren *n*, Rührung *f* ~ **arm** Schlagflügel *m* ~ **time** Rührzeit *f* ~ **vane** Rührflügel *m*
**agitation** Bewegung *f*, Erregung *f* ~ **of bath** Badbewegung *f* ~**-type flotation machine** Agitationsflotator *m*
**agitator** Agitationsmaschine *f*, Quirl *m*, Rührapparat *m*, Rührarm *m*, Rührer *m*, Rührmaschine *f*, Rührwerk *n*, Schütteleinrichtung *f*, Stocheisen *n*, Strudelrad *n* ~ **box** Agitationszelle *f* ~ **cell** Agitationszelle *f* ~ **shoe for wash mill** Schlämmschuh *m* ~ **tank** Rührtank *m* ~ **vessel** Rührkessel *m*
**agonic,** ~ **curve** Agone *f*, agonische Kurve *f* ~ **line** Agone *f*, agonische Kurve *f*, Nullisogone *f*
**agree, to** ~ eingehen, einigen, entsprechen, übereinstimmen, in Einklang *m* sein, einen Vertrag *m* schließen **to ~ to** beistimmen, einverstanden sein, einwilligen **to ~ upon** bedingen, verabreden, (sich) vereinbaren **to ~ with** sich vertragen mit, übereinkommen **to ~ on the chargeable duration of a call, to ~ on the total daily chargeable minutes** die Gesprächsdauer vereinbaren oder vergleichen **to ~ with an opinion** sich einer Ansicht *f* anschließen
**agreed** eingegangen, vereinbart
**agreement** Abkommen *n*, Abmachung *f*, Akkord *m*, Einigung *f*, Einklang *m*, Einvernehmen *n*, Einverständnis *n*, Festlegung *f*, Genehmigung *f*, Symbasis *f*, Übereinkommen *n*, Übereinkunft *f*, Übereinstimmung *f*, Vereinbarung *f*, Vereinigung *f*, Vergleich *m*, Vertrag *m*, Zustimmung *f* ~ **on compensation** Abfindungsvertrag *m* ~ **by piece** Akkord *m* nach Stückzahl ~ **for service** Arbeitsverhältnis *n*
**agricultural** landwirtschaftlich ~ **expert** Kulturtechniker *m* ~ **installation** landwirtschaftlicher Betrieb *m* ~ **work** Landarbeit *f*
**agriculture** Ackerbau *m*, Landwirtschaft *f*
**agriculturist** Landwirt *m*
**aground** gestrandet
**ahead** voraus, vorn, vornliegend, vorwärts ~ **position** Vorwärtsstellung *f* ~ **running** Vorwärtslauf *m* ~ **turbine** Vorwärtsturbine *f*
**Ahrent's siphon lead tap** Ahrentscher automatischer Bleistift
**Aich metal** Aichmetall *n*
**aid, to** ~ beistehen, beitragen, helfen
**aid** Anlage *f*, Beisitzer *m*, Beistand *m*, Helfer *m*, Unterstützung *f* **by the ~ of** an Hand von ~ **to vision** (research) Sehhilfsmittel *n*
**aided eye** bewaffnetes Auge *n*
**aikinite** Nadelerz *n*
**aileron** Hilfsflügel *m*, Querruder *n*, Quersteuer *n*, Verwindungsklappe *f* ~ **balance** Querruderausgleichteil *n*, Querruderentlastung *f* ~**-balancing tab** Querruderausgleichsklappe *f* ~**-boosting tab** Querruderausgleichsklappe *f* ~ **chord** Querrudertiefe *f* ~**-compensating tab** Querruderausgleichsklappe *f* ~**-connecting strut** Querruder-

verbindungsstrebe *f* ~ **cover(ing)** Querruderbespannung *f* ~ **crank** Querruderhebel *m* ~ **deflection** Querruderausschlag *m* ~ **horn** Querruderhebel *m* ~ **leading edge** Querrudervorderkante *f* ~ **lever** Querruderhebel *m* ~**-operating tube** Querruderstoßstange *f* ~ **pick-up** Fahrtgeber *m* ~ **positioning relay** (autopilot) Querlagenrelais *n* ~ **positioning signal** (autopilot) Querlagenbefehl *m* ~ **T-crank** Querruderkreuzgelenk *n* ~ **trimming** Querrudertrimmung *f* ~**-trim tab** Querrudertrimmklappe *f*
**aim, to** ~ anlegen, einstellen, richten, visieren, zielen **to ~ at** anrichten, anschneiden, anstreben, anzielen (einen Gegenstand), richten auf, zielen auf
**aim** Absicht *f*, Augenmerk *n*, Haltepunkt *m*, Richtlinie *f*, Richtung *f*, Ziel *n*, Zweck *m* ~ **verifier** Ziellinienprüfer *m*
**aimed, to** ~ **at** angestrebt ~ **shot** direkter Schuß *m*
**aiming** Richten *n* ~ **axis** Zielachse *f* ~ **circle** Bussolenrichtkreis *m*, Richtkreis *m*, Stellungsunterschied *m* ~**-circle collimator** Richtkreiskollimator *m* ~**-circle graduation** Richtkreiseinteilung *f* ~ **disk** Ziellöffel *m* ~ **group** Zieldreieck *n* ~ **mechanism** Richtgerät *n*, Richtmittel *n*, Zieleinrichtung *f* ~**-off allowance** (for height speed, lead, deflection) Vorhalt *m* ~ **point** Anlegepunkt *m*, Orientierungspunkt *m* (Luftbildwesen), Richtpunkt *m*, Zielpunkt *m* ~**-point azimuth reading** Richtpunktzahl *f* ~ **post** E-Latte *f*, Richt-stäbchen *n*, -stock *m* ~**-practice instrument** Richtübungsgerät *n*
**air, to** ~ abröschen, auslüften, (batteries) entlüften, (paper) lüften
**air** Atmosphäre *f*, Luft *f*, Wind *m* ~ **for (or of) combustion** Verbrennungsluft *f* ~**rarefied by heating** Heißluft *f* ~ **used for atomizing** Zerstäubungsluft *f*
**air,** ~ **absorption** Luftaufnahme *f* ~ **access** Luftzutritt *m* ~**accumulator chamber** Luftspeicher *m* ~ **activity** Flugbetrieb *m* ~ **adjustment** Korrekturluftdüse *f* (jet) ~ **admission** Luftzutritt *m* ~ **atomizer** Luftzerstäubungsapparat *m* ~ **axis** bahnfeste Achse (aerodyn.) ~ **baffle** Luftleitblech *n*, Prallblech *n* ~ **bag** Heizschlauch *m* ~ **bag injector** Schwimmsackinjektor *m* ~**-balancing breather** Luftdruckausgleichsleitung *f* ~ **base** Luftstützpunkt *m* ~ **bath** Luftbad *n* ~ **bearing** Luftlager *n* ~ **bearing gyro** luftgelagerter Kreisel *m* ~ **bell** Glasfehler *m*, Luftblase *f*, (when of irregular shape) Glasblase *f* ~ **belt** Windmantel *m* ~ **blast** Gebläse-luft *f*, -wind *m*, (gun) Gebläse *n* **to ~-blast** heißblasen
**air-blast,** ~ **breaker** Druckschalter *m* ~ **circuit breaker** Expansionsschalter *m* ~ **nozzle** Luftblasdüse *f* ~ **supply** Wind-zufluß *m*, -zufuhr *f*, -zuführung *f* ~ **switch** Druckgasschalter *m* (cross) ~ **switch** Druckluftschalter *m*, Preßgasschalter *m*, Preßluftschalter *m* ~ **temperature** Windtemperatur *f* ~ **tube** Blasrohr *n* ~ **velocity** Windgeschwindigkeit *f* ~ **wind** Blasluft *f*
**air,** ~ **blasting** Heißblasen *n* ~ **blasting period** Heißblasperiode *f* ~ **bleed** kleine Luftöffnung *f* ~ **bleeders** (motor) Bremsluftbohrung *pl* ~ **bleed hole** (motor) Bremsluftkanal *m* ~**-bleed plate** Luftdüsenplatte *f* ~**-blow** Heißblasen *n* ~ **blowing** Windfrischen *n* ~ **blow period** Heißblasperiode *f* ~ **boiling** Flimmern *n* der Luft

airborne durch die Luft, luftbefördert, Bord . . .
(in Zusammensetzungen) ~ **communication
jammer** Bordstörsender m ~ **particulates** luft-
transportierte Makroteilchen pl ~ **transmitter**
Bordsender m

air, ~ **bottle** Luftbehälter m ~ **box** Luftkammer
f, Unterdruckkammer f, Windkasten m
~ **brake** Bremsklappe f, Landeklappe f, Luftbremse f, Luftdämpfer m, (compressed-) Luft-
druckbremse f ~-**brake assembly** Druckluft-
bremsausrüstung f ~ **break contactor** Luft-
schütz m ~ **breakup** Zerlegung f der Luftkräfte
~ **breather** Luftansauger m ~-**breathing engine**
lauftansaugendes Triebwerk n ~-**brush** Luft-
bürste f ~ **bubble** Luftblase f, Lufteinschluß m
~ **buffer** Luftpuffer m, abgepuffertes Luftpolster n ~ **buoyancy** Luftauftrieb m ~ **cap**
Belüftungskappe f ~ **cap for fan spray** Flach-
strahlverteiler m (print) ~ **cap for round spray**
Rundstrahlverteiler m ~ **car** Autoflugzeug n
~ **carrier** Luftfahrzeug n ~ **cavity** Lufteinschluß m, eingeschlossene Luftblase f ~ **cell**
Luftspeicher m ~-**cell combustion system** Luft-
speicherdieselverfahren n ~-**cell engine** Luft-
speicherdieselmotor m ~-**cell-type engine** Luft-
speichermaschine f ~ **chamber** Luft-behälter m,
-kammer f, -kessel m, Wind-kammer f, -kessel
m, -mantel m, (of loudspeaker) Lautsprecher-
luftkammer f ~ **chamber of loud-speaker**
Trichtervorhof m ~-**chamber loud-speaker** Luft-
kammerlautsprecher m ~ **channel** Luft-kanal m,
-weg m, Wetterleitung f ~ **charge** Ladeluft f
~ **checker** Luftkammer f ~ **chimney** Ansaug-
schlot m ~ **chuck** Luftschlauchventil n, Preß-
luft-futter m, -spanneinrichtung f ~ **circuit** Luft-
kreislauf m, ~ **circuit breaker** Luftschalter m
~-**circulating oven** Luftumwälzungsofen m
~ **classifier** Luftklassifikator m ~ **cleaner** Luftreiniger m ~ **clutching** Preßluftsteuerung f ~ **cock**
Lufthahn m, Unterdruckkammer f ~(-**core**)-
**coil** Luftspule f ~ **column** Luftsäule f ~ **commerce** Luftverkehr m ~ **compensator** Luftausgleicher m ~ **composition** Luftbeschaffenheit f
~ **compression** Luftkompression f ~ **compressor** Drucklufterzeuger m, Luft-kompressions-
maschine f, -kompressor m, -presser m, luft-
verdichtende Maschine f ~-**compressor booster**
Ausladegebläse n ~-**compressor oil** Kompressoröl n ~-**compressor pump** Luftpumpe f ~ **condensor** Luft-verdichter m, -kondensator m,
Oberflächenkondensator m
air-**condition, to** ~ mit Klimaanlage f versehen
air-**conditioned** klimatisiert, luftgekühlt
air-**conditioning** Bewetterung f, (equipment)
Klimaanlage f, Klimatechnik f, Klimatisierung
f, Luft-trocknen n, -zurichtung f, -konditionie-
rung f, Raumbewetterung f ~ **apparatus** Luft-
reinigungsanlage f ~ **plant** Luftreinigungs- f,
Klima-anlage f
air, **conditioner** Klimaanlage f ~ **conditions** Luft-
verhältnisse pl ~ **conducting** luftleitend ~ **conduit** Luftleitung f, Wetterlutte f ~-**conduit-
adjusting apparatus** Luttenricht-, Wetterlutten-
richt-apparat m ~ **connecting piece** Luftan-
schlußstutzen m ~ **connection** Luftverbindung f
~ **consumption** Luftverbrauch m ~ **control drum**
Trommelschieber m ~ **control manifold and
header** Steuerluftverteiler m und Sammler m

~ **controls** Luft-steuer n, -steuerwerk n ~-**control service** Luftaufsichtsdienst m ~ **convention**
Luftabkommen n ~ **conveyer** Luftförderer m
~ **coolant** Kühlluft f ~ **cooled** luftgekühlt, wind-
gefrischt (steel) ~-**cooled engine** luftgekühlte
Maschine f, luftgekühlter Motor m ~-**cooled
intercooler** Luftzwischenkühler m ~-**cooled
transformer** luftgekühlter Transformator m
~-**cooled tube** luftgekühlte Röhre f ~ **cooler**
Luftkühler m ~ **cooling** Luftkühlung f ~ **core**
Luftkern m, eisenlos ~-**core choke** Luftdrossel
f, eisenfreie Drossel f ~-**core coil** eisenfreie
Drossel f ~-**core(d) coil** Luftkernspule f ~-**cored** (choke) **coil** eisenfreie Spule f ~-**cored frame**
(direction finder) Luftrahmen m ~-**cored solenoid** eisenloses Solenoid n, Luftspule f ~-**core
inductance coil** Induktanzspule f mit Luftkern
~-**core inductor** Luftkerninduktivität m ~-**core
magnet** eisenfreier Magnet m ~-**core transformer** Lufttransformator m ~-**core variometer**
Luftvariometer n
aircraft **Flugzeug** n, Luftfahrzeug n ~ **used for
acrobatic training** Kunstflugtyp m ~ **on skis**
Flugzeug n mit Schneekufen ~ **with open or
outrigger tail** Gitterschwanzflugzeug n
aircraft, ~ **accessories** Flugzeugzubehör n
~ **armament** Flugzeug-bestückung f, -bewaff-
nung f ~ **arrester net** Fangnetz n ~ **battery**
Bordbatterie f, Bordstromsammler n (electr.)
~ **brake parachute** Flugzeugbremsschirm m
~ **carpenter** Flugzeugtischler m ~ **carrier** Flug-
zeug-mutterschiff n, -träger m ~ **certificate**
Luftfahrzeug-bescheinigung f, -eintragungs-
schein m ~ **charges** Bordgebühr f ~ **construction** Flugzeugbau m ~ **electromechanic** Flug-
zeugelektromechaniker m ~ **equipment** Bord-
anlage f ~ **error** Bordablenkung f ~ **exhibition**
Flugzeugausstellung f ~ **fabric** Flugzeugbe-
spannstoff m ~ **factory** Flugzeugfabrik f
~ **frame** Flugzeugzelle f ~ **guidance (or guiding)**
Flugzeugleitung f ~ **industry** Luftfahrtindustrie
f ~-**instrument mechanic** Flugzeugfeinmechani-
ker m ~ **instruments** Flugzeuginstrumente pl
~ **intercom(munication)** Bord-gespräch n, -ver-
ständigung f (rdo) ~ **intercommunication
system** Bordsprechanlage f ~ **jack** Pendelbock
m ~ **leatherworker** Flugzeugsattler m ~ **log**
Luftfahrzeugbordbuch n ~ **material** Werkstoff
m für Luftfahrtzwecke ~ **mechanic** Bordmonteur m, Flugzeugwart m ~ **metalworker** Flug-
zeugmetaller m ~ **painter** Flugzeugmaler m
~ **parachute flare** Fallschirmleuchtbombe f
~ **performance** Leistung f des Flugzeuges
~ **plant** Flugzeugwerk n ~ **receiver** Flugzeug-
empfänger m ~ **refuelling** Flugzeugbetankung f
~ **repair** Flugzeugreparatur f ~ **salvage unit**
Flugzeugbergetrupp m ~ **sheetmetal worker**
Flugzeugklempner m ~ **station** Flugzeug-
station f, Luftfunkstelle f ~ **structural noise**
Zellenvibration f ~ **tinsmith worker** Flugzeugklempner m ~-**to-~-identification** Bord-zu-
Bord-Kennung f ~ **type instrument** Meßgerät n
für Bordzwecke ~ **works** Flugzeugwerk n
aircrew **Flugzeugmannschaft** f
air, ~ **cure (or vulcanization)** Vulkanisation f mit
direktem Dampf in der Luft ~ **cured article**
Heißluftvulkanisat m ~ **curing** Lufterhärtung f
(Betonstein) ~ **current** Luft-strom m, -strö-

mung *f*, -zug *m*, Wetter-strom *m*, -zug *m*, Wind--strom *m*, -strömung *f*, -zug m ~ **cushion** abgepuffertes Luftpolster *n*, Luft-kissen *n*, -polster *n* ~**-cushion cylinder** Luftpolsterzylinder *m* ~ **cushion pressure** Luftkissenbelastung ~ **cushioned** luftlagernd, luftgelagert ~**-cushioned freepiston-type engine** Freiflugkolbenmaschine *f* ~ **cylinder** Druckluft- *m*, Luft- *m*, Windzylinder *m* ~ **damper** Luft-dämpfer *m*, -schieber *m*, Windschieber *m* ~ **damping** Luftdämpfung *f* ~ **danger** Luftgefahr *f* ~ **data computer** Luftwerterechner *m* ~ **defense warning service** Luftschutzwarndienst *m* ~ **deflector** Luft-abweiser *m*, -leitblech *n* ~ **density** Luft-dichte *f*, -dichtigkeit *f*, -gewicht *n* ~**-density control valve** Luft-wichtenregelventil *n*, -wichtregelventil *n*, Vordrossel *f* ~ **depression** Luftunterdruck *m* ~ **dial gauge** Luftdruckmeßuhr *f* ~**-dielectric trimmer** Lufttrimmer *m* ~**-dielectric variable capacitor** Luftdrehkondensator *m* ~ **discharge** Luftentladung *f* ~ **discharge scroll** Luftaustrittsstutzen *m* ~ **display** Flug-veranstaltung *f*, -vorführung *f* ~**-distributing main** Hauptwindleitung *f* ~**-distributing system** Wetterführung *f* ~ **door** Wettertür *f* ~ **draft** Luftstrahl *m*, Betriebsluft *f* ~**-drain** Luftkanal *m* ~**-dried** geschwelkt ~**-dried brick** Lehmstein *m* ~**-dried wood** lufttrockenes Holz *n* ~**-drift triangulation** Winddreiecksrechnung *f* ~ **drill** Preßluftbohrer *m* ~**-drive** Ölförderung *f* mittels Druckluft **airdrome** (also see **aerodrome**) Flughafen *m*, Flugplatz *m*

**air**, ~**-dry** lufttrocken ~ **drying** Luft- *f*, Windtrocknung *f* ~ **drying apparatus** Lufttrockner *m* ~ **drying enamels** lufttrocknende Emaille ~ **drying plant** Windtrocknungsanlage *f* ~ **drying process** Windtrocknungsverfahren *n* ~ **duct** Entlüftungskanal *m*, Luftkanal *m*, Luftzug *m*, Lutte *f*, Rauchkanal *m*, Windsegel *n* ~ **ducts** Luftrohre *pl* (Rohrpost) ~ **eddy** Luftwirbel *m* ~ **eliminator** Luftabscheider *m* ~ **engineering** Lufttechnik *f* ~ **entering the carburetor** Vergaseransaugluft *f* ~ **entraining agent** Belüftungsmittel *n* ~ **entrainment in concrete** Lufteinschluß *m* in Beton ~ **escape** Entlüftungsventil *n*, Luftablaß *m* ~ **escape cage** Entlüftungseinsatz *m* ~ **evacuation valve breather** Entlüftungsklappe *f* ~ **exhaust** Luftaustritt *m* ~**-exhaust valve** Entlüftungsventil *n* ~ **exit hole** Luftaustrittsloch *n* ~ **extraction** Luftentnahme *f* ~ **fan** Luftbläser *m* ~ **fed drill** Druckluftbohrmaschine *f* ~**feed** Luftförderung *f* **airfield** Flugfeld *n*, Landeplatz *m*

**air**, ~ **filled** mit Luft *f* gefüllt ~ **filling chuck** Luftfüllschlauch-Mundstück *n* ~ **filter** Luft--filter *m*, -reiniger *m* ~ **filter in gas masks** Filtervorsatz *m* ~**-filter oil** Luftfilteröl *n* ~ **flap** Luftklappe *f* ~ **flask** Kompressortank *m* ~ **fleet** Luftflotte *f* ~**-flooding** Ölförderung *f* mittels Druckluft ~ **flow** Gebläsewind *m*, Luft-bewegung *f*, -strom *m*, -strömung *f*, Windströmung *f* ~**-flow diverter valve** Luftumleitventil *n* ~**-flow modulating valve** Luftdurchflußregelventil *n* ~**-flow path** Luftweg *m* ~**-flow separation** Abbrechen *n* der Luftlinien ~ **flue** Luft-kanal *m*, -weg *m* ~**-flux** Luftlinien *pl* ~**-flux coupling** Luftspaltkraftflußkopplung *f* ~ **fogging** Luftschleier *m* **airfoil** (also see **aerofoil**) Flügel *m*; Flügelkör-

per *m*, Hilfsflügel *m*, Stromlinienkörper *m*, Tragfläche *f*, Tragflügelprofil *n* ~ **for supersonic velocity** Überschallprofil *n* ~**-blade section** Profildicke *f* ~ **profile** Flügelprofil *n* ~ **section** Flügelquerschnitt *m*, Tragflächenprofil *n* ~ **wing** Tragflügel *m*

**air**, ~**force** Luft-kraft *f*, -waffe *f* ~ **fractionator** Lufttrenner *m* ~ **frame** Flugwerk *n*, Zelle *f* ~ **(-core) frame** Luftspule *f* ~**-frame industry** Zellenindustrie *f* ~ **frame side, on the** ~ zellenseitig ~ **freighter** Luftfrachter *m* ~ **friction** Luftreibung *f* ~ **friction damping** Luftdämpfung *f* ~ **funnel** Schnorchel *n* ~ **furnace** Flamm-, Wetter-, Wind-ofen *m* ~ **(radiation) furnace** Reverberierofen *m* ~**-furnace slag** Flammofenschlacke *f* ~ **gap** lufterfüllter Abstand *m*, Luft--spalt(e) *f*, -strecke *f*, -umhüllung *f*, (of horseshoe magnet) Maulweite *f*, (permanent magnet) Nutzraum *m*, Polschuhspaltbreite *f* ~ **gap choke** Luftspaltdrossel *f* ~**-gap lightning arrester** Luftblitzableiter *m* ~ **gap of magnet** Interferrikum *n* ~**-gap reactance coil** Funkdrosselspule *f* ~ **gap reactor** Drossel *f* mit Luftspalt ~**-gap transformer** Übertrager *m* mit unterteiltem Eisenkern ~ **gaps** Luftstöße *pl* ~ **gas** Luftgas *n* ~**-gas producer** Luftgaserzeuger *m* ~ **glow** Luftleuchten *n*, Nachthimmelslicht *n* ~ **goods** Luftgut *n* ~**-ground communication** Flugfunkverbindung *f* ~ **guide** Flugkursbuch *n* ~ **guiding sheet** Windführungsblech *n* ~ **gun** Blaspistole *f*, Luftgewehr *n*, Spritzpistole *f* ~ **hammer** Preßlufthammer *m* ~ **handling equipment** lufttechnische Anlagen *pl* ~**-hardened** luftvergütet ~ **hardening** Lufthärtung *f* **hardening steel** lufthärtender Stahl *m*, Luft-härtungstahl *m*, -härtestahl *m* ~ **heading** Parallelstrecke *f* (min.) ~ **heater** Heißlufterhitzer *m*, Luft-erhitzer *m*, -vorwärmer *m*, Winderhitzer *m* ~ **hoist** Drucklufthebezeug *n*, Preßluftaufzug *m* ~ **hole** Dunstloch *n*, Lichtloch *n*, Luftkanal *m*, Luftloch *n*, Windöffnung *f*, (of a charcoal pile) Zugloch *n*, Zug *m* (eines Kohlenmeilers) ~ **horn** Luftansaughutze *f*, Lufteintrittsstutzen *m* ~ **hose** Luftschlauch *m* ~ **humidifying plant** Luftbefeuchtungsanlage *f* ~ **humidity indicator** Hygrometer *n* ~ **hydraulic brake** Druckluftöldruckbremse *f* ~ **impact wrench** Preßluftschrauber *m* ~ **induction pipe** Luftsaugleitung *f* ~ **ingress** Lufteinströmung *f* ~ **injection** Drucklufteinspritzung *f*, Einblaseleitung *f*, Lufteinblasung *f* ~**-injection engine** Lufteinblasemaschine *f* ~**-injection pressure** Einführungsdruck *m* ~**-injection pressure-control valve** Einblasdruckregler *m* ~ **injector** Luftinjektor *m* **airing** Lüften *n*, Lüftung *f*, (of mines) Wetterwirtschaft *f* ~ **frame hanging room** Lufthänge *pl* ~ **plant** Be- und Entlüftungsanlage *f* ~ **process** Hängeverfahren *n* ~ **stage** Trockengestell *n* ~ **tube** Bewitterungsleitung *f* **air**, ~ **inlet** Luft-einlaß *m*, -einlaßöffnung *f*, -eintritt *m*, -zutritt *m* ~ **inlet branch** Lufteintrittsstutzen *m* ~ **inlet conduit** Zulaufkanal *m* ~ **inlet duct** Ansaugschacht *m* ~ **inlet flange** Luftleitungseinlaßflansch *m* ~ **inlet hole** Lufteinlaßbohrung *f* ~ **inlet opening** Lufteinströmöffnung *f* ~ **inlet screen** Lufteinlaufgitter *n* ~ **valve** Lufteinlaßventil *n* ~**-inrushes** Lufteinbrüche *pl* ~ **intake** Luft- ansaughutze *f*, -ansaugrohr *n*,

-ansaugstutzen *m*, -einlaß *m*, -einlaßrohr *n*, -eintritt *m*, -hutze *f*, -zuführungsrohr *n*, Schnorchel *n* ~ **intake casing** Einlaßstutzen *m* (jet) ~ **intake casing** (of rocket) Einlaufleitkranz *m* ~ **intake duct** Luftansaugkanal *m* ~ **intake guide vane** Einlaßleitschaufel *f* (jet) ~ **intake horn** Ansaugkrümmer *m* ~ **intake pipe** Luftsaugleitung *f* ~ **intake screen** Ansaugschutzort *m*, Lufteinlaufgitter *n*, Lufteintrittsgitter *n* ~ **intake slot** (of radiator) Kühlermaul *n* ~ **intake valve** Einlaßventil *n* für Luft ~ **interchanger** Luftaustauscher *m*, Luftmischer *m* ~ **interrupting switch** Lastschalter *m* ~ **jar squeezer** Druckluftpreßvorrichtung *f* ~**-jarring machine** Druckluftrüttler *m* ~**-jarring molding machine** Formmaschine *f* mit Druckluftrüttler ~ **jet** Blasschlitz *m*, Luftdüse *m*, Luftstrahl *m* ~**-jet pendulation** Luftstrompendelung *f* ~ **jolter** Rüttelformmaschine *f* für Druckluftbetrieb, (in molding) Preßluftrüttler *m* ~ **law** Luftrecht *n* ~ **layer** Luftschicht *f*, Windschicht *f*, Windstufe *f* ~ **layer of high humidity** wasserreiche Luftschicht *f* ~ **lead** Spill *n* ~ **legislation** Luftfahrtgesetzgebung *f* ~ **lens** Luftlinse *f* ~ **lift** Druckluftheber *m*, Luftheberanlage *f*, Preßluftförderung; Luftbrücke *f* ~ **lift agitator** Mischluftrührer *m* ~**-lift pump** Mammutpumpe *f*

**air-line** Belüftungsrohr *n*, Luftleitung *f*; Luft-linie *f*, -fahrtlinie *f*, -verkehrslinie *f*, -weg *m*, Windleitung *f* ~ **to the flood-light mast** Luftzuführung zum Lichtmast ~ **distance** Luftlinienentfernung *f*, Entfernung in der Luftlinie ~ **lubrication** Luftschmierungseinrichtung *f* ~ **operating company** Luftverkehrsgesellschaft *f* ~ **service** Fluglinendienst *m* ~ **system** Flugliniennetz *n*

**air**, ~ **liner** Verkehrsflugzeug *n* ~ **lines** Luftspektrallinien *pl* ~ **liquefaction** Luftverflüssigung *m* ~ **liquefier** Luftverflüssiger *m* ~ **liquefying plant** Luftverflüssigungsanlage *f* ~**-loaded accumulator** Druckluftakkumulator *m*, Druckölflasche *f* ~ **lock** Gasschleuse *f*, Luft-druckkammer *f*, -falle *f*, -sack *m*, -schleuse *f*, -tasche *f*, Vakuumschleuse *f*, Wetterschleuse *f* ~ **lock** (chamber) **of electron microscope** Mikroskopschleuse *f* ~**-lock device for plates** Einschleusvorrichtung *f* für Platten ~**(-core) loop** Luftspule *f* ~ **mail** Luftpost *f* ~ **mail letter** Luftpostbrief *m* ~ **mail pilot** Postflieger *m* ~ **mail postage** Luftpostgebühr *f* ~ **mail route** Luftpoststrecke *f* ~ **mail service** Luftpostdienst *m* ~ **mail stamp** Luftpostmarke *f* ~ **mail traffic** Luftpostverkehr *m* ~ **main** Luftleitung *f* ~ **man** Luftfahrer *m* ~ **manifold** Luftverteilerdecke *f* ~ **mass** Luftmasse *f* ~ **meter** Anemomesser *n*, Anemometer *n*, Luftmesser *m*, Windmesser *m* ~ **metering ring** Luftzumeßring *m* ~ **mile** Luftmeile *f* ~ **mixture** Mischluft *f* ~**-mixture setting** Luftdrosselstellung *f* ~ **moistener** Luftanfeuchter *m* ~**-moistening chamber** Befeuchtungskammer *f* ~ **molecule** Luftmolekül *n* ~ **monitor** Luftüberwachungsgerät *n* ~ **mortar** Luftmörtel *m* ~ **motor** Luftmotor *m* ~ **navigation** Luftfahrt *f* ~**-navigation obstacle** Luftfahrthindernis *n* ~ **nozzle** Luft-düse *f*, -trichter *m* ~ **objective** Flugziel *n* ~ **occlusion** Luftaufnahme *f*, Lufteinschluß *m* ~**-oil separator** Schmierstoffent-

schäumer *m* ~**-oil type shock strut** Öl-Luft-Federbein *n* ~ **opening vent** Luftabflußöffnung *f*, Luftabzugsöffnung *f* ~**-operated** preßluftbetätigt ~ **operated chuck** Druckluft-, Preßluft--futter *n* ~**-operated jarring machine** Rüttelformmaschine *f* für Druckluftbetrieb ~**-operated pump** Pumpe *f* mit Preßluftantrieb ~**-operated squeezing machine** Formmaschine *f* mit Druckluftpressung ~**-outlet conduit** Abluftkanal *m* ~ **outlet opening** Luftaustrittsöffnung *f* ~ **oven** luftgeheizter Ofen *m* ~ **passage** Flugkosten *pl*; Luft-durchgang *m*, -einlaß *m*, -kanal *m*, -schlitz *m* ~**-passage resistance** Luftdurchströmungswiderstand *m* ~**-passage ring** Luftüberleitungsring *m* ~**-passenger service** Personenluftverkehr *m* ~ **photogrammetry** Aerofotogrammetrie *f*, Luftbildmessung *f*, Luftfotogrammetrie *f* ~ **photograph** Luftfotografie *f* ~ **photographic apparatus** Luftbildgerät *n* ~ **photography** Aerofotogrammetrie *f*; Luft--bildaufnahme *f*, -bildmessung *f*, -fotografie *f* ~ **pilot nozzle** Luftschaltdüse *f* ~ **pipe** Entlüftungsleitung *f*, Luftleitung *f*, Luft-rohr *n*, -schlauch *m*, Zug-rohr *n*, -röhre *f*, Wetterlutte *f* (min.) ~ **pistol** Abblasepistole *f*

**airplane** (also see **aeroplane**) Flugzeug *n* **the ~ ascends or climbs** das Flugzeug steigt **the ~ takes off** das Flugzeug hebt ab ~ **with auxiliary fuel tanks** Flugzeug *n* mit zusätzlichen Kraftstoffbehältern ~ **of all-wood construction** Ganzholzflugzeug *n* ~ **for spraying insecticides** Streuflugzeug *n* für Insektenbekämpfung ~ **with suction slot** Absaugflugzeug *n* ~ **with sweepback** Flugzeug *n* in Pfeilform ~ **with variable lifting surface** Flugzeug *n* mit veränderlicher Flügelfläche

**airplane**, ~ **altitude relative to atmospheric density** Flughöhe *f* entsprechend der Luftdichte ~ **antenna** Flugzeugantenne *f* ~ **armament** Flugzeugbewaffnung *f* ~ **armor** Flugzeugpanzer *m* ~ **arrester** Flugzeugaufhalter *m* ~ **boresight axis** Flugzeugjustierachse *f* ~ **catapult cradle** Flugzeugschleuder *f* ~ **chock** Flugzeuglaufradkeil *m* ~ **control** Flugzeugsteuerung *f* ~ **crew** Flugzeugbesatzung *f* ~ **dope** Bespannungslack *m* ~ **engine** Flugmotor *m* ~ **fitted with skis** Schneeflugzeug *n* ~ **formation** Flugzeugverband *m* ~ **glider** Gleitflieger *m* ~ **heating** Flugzeugbeheizung *f* ~ **instruments** Flugzeuginstrumente *pl* ~ **linen** Flugzeugleinen *n* ~ **marking** Erkennungszeichen *n* ~ **markings** Hoheitsabzeichen *n* ~**-model construction** Flugmodellbau *m* ~**-model testing** Flugzeugmusterprüfung *f* ~ **motor** Flugmotor *m* ~ **oil** Flugzeugöl *n* ~ **passenger** Fluggast *m* ~ **performance** Flugzeugleistung *f* ~ **photograph** Flugzeugaufnahme *f* ~ **picture** Luftbildaufnahme *f* ~ **pilot** Flugzeugführer *m* ~ **piloting** Führung *f* des Flugzeuges ~ **port** Flugzeugpark *m* ~ **position finding** Flugzeugortung *f*

**airplane-radio**, ~ **apparatus** Flugzeugfunkgerät *n* ~ **equipment** Flugzeugfunkgerät *n* ~ **room** Bordfunkstelle *f* ~ **station** Bordfunkstelle *f*

**airplane**, ~ **receiving set** Bordempfangsgerät *n* ~ **recognition** Flugzeugerfassung *f* ~ **reserves** Flugzeugreserve *f* ~ **science** Flugzeugkunde *f* ~ **spotting** Richtungshören *n* ~ **structure** Flugzeugzellengerüst *n* ~**-testing station** Flugver-

suchsstation f ~ tire Flugzeugreifen m ~ towing
Flugschlepp n ~ traveling in contact with water
Wassern n ~-velocity slide rule Flugzeugge-
schwindigkeitslineal n ~ wing Tragflügel m
~ wireless set Flugzeugbordstation f
air, ~ pocket Luft-blase f, -einschluß m, -loch n,
-sack m, -tasche f, (descending gust) Fallbö f
~ pollution Luftverunreinigung f
airport (also see aerodrome and airdrome) Flughafen m, Flugplatz m ~ of departure Aus-
gangsflughafen m ~ of destination Bestim-
mungsflughafen m
airport, ~ administration Flughafenleitung f
~ authorities Flughafenleitung f ~ beacon
Flughafenleuchtfeuer n ~ border line Flughafengrenze f ~ boundary Flughafengrenze f
~-boundary light Flughafengrenzlicht n ~ buildings Flugplatzgebäude pl ~ design Flugplatz-
entwurf m ~ floodlight Flugplatzlicht n
~ fueling system Flughafenzapfanlage f
~ lighting Flugplatzbefeuerung f ~ lighting-
system Flugplatzbeleuchtungsanlage f ~ management Flughafenleitung f ~ subsoil Flugplatz-
untergrund m
air, ~-position Aufnahmeort m in der Luft
~-position indicator Lagebestimmungs-, Po-
sitionsbestimmungs-gerät n ~ preheating chamber Luftvorwärmkammer f
air pressure Luftdruck m, Luftspannung f, Windpressung f ~ at ground level Bodenluftdruck m
~ boiler Luftdruckkessel m ~ conveying Pfrop-
fenförderung f ~ drop across the cylinder
Druckabfall m am Zylinder ~ equalizing flap
Luftdruckausgleichshutze f ~ fuse Staudruck-
zünder m ~ gauge Druckbenzinuhr f, Luft-
druckmesser m ~ manifold Preßluftanschluß m
~ operated doors pneumatische Türbetätigung f
~ release valve Luftdruckverminderungsventil
n ~ relieve valve Luftabblasventil n ~ space
(loudspeaker) Luftkammer f ~ system Stau-
druckerzeugung f ~ tank Druckkessel m ~ test
Druckluftprüfung f ~ transportation Pfropfen-
förderung f ~ valve Preßluftventil n ~ valve
flap Druckluftventilklappe f (Rohrpost)
air, ~-proof luftdicht ~ pump Luftpumpe f,
Reifenluftständer m, (laboratory) Entlüftungs-
pumpe f, Glockenventilator m ~ pump receiver
Luftpumpenglocke f ~ pumping Preßluftpum-
pen n ~ purifier Luftreiniger m, Windfege f
~ pyrometer Luftpyrometer n ~ quake Lufter-
schütterung f ~-quenched steel lufthärtender
Stahl m ~ quenching Luftabschreckung f ~ radiator Luftkühler m ~ raft Floßboot n, Luft-
floß m, Schlauchboot n
air-raid, ~ alarm Flieger-alarm m, -warnung f,
Luft-alarm m, -warnung f ~ defense Luftschutz
m ~ precautionary measure Luftschutzmaß-
nahme f ~ protection Luftschutz m ~ protection
district Luftschutzrevier n ~ protection law
Luftschutzgesetz n, Luftschutzpflicht f ~ protection service Sicherheits- und Hilfsdienst m
~ shelter Luftschutzkeller m, Luftschutzraum
m ~ warden Alarmposten m ~ warning Flie-
geralarm m, Luftwarnung f
air, ~ receiver Bremskessel m, Druckluftbehälter m, Luftaufnehmer m, Lufthaube f ~ recirculating system Luftumwälzanlage f ~ record Luftfahrtrekord m ~ rectifier Luftrektifi-

kationsanlage f ~-reducing valve Luftdruck-
regler m ~-reduction method of smelting Röstreaktionsarbeit f ~-refining process Bessemer-
Verfahren n, Birnenprozeß m ~ refuelling Luftbetankung f ~ regeneration unit (rocket) Luft-
erneuerungsanlage f ~-regulating accessories
Armaturen pl für Luftdruckregulierung ~ release Luftabschneider m ~ release and drain
cock Entlüftungshahn m ~ relief Entlüftung f
~-relief valve Entlüftungsventil n ~ required
Luftbedarf m ~ requirement Luftbedarf m
~ requirement Luftbedarf m ~ reservoir Druckluftbehälter m, Windkessel m, Windsammler m,
~ resistance Luftwiderstand m ~ resisting luftbeständig ~ return Luftrückführung f ~ reversing valve Luftwechselklappe f ~ roll Luft-
walze f (der Papiermaschine) ~ route Flugstrecke f, Flugweg m, Luftstrecke f, Luftweg m
~ routes Streckennetz n ~ route map Strecken-
karte f ~ saddle Luftsattel m ~-sand mixture
Sandluftgemisch n ~ scavenging Luftbespülung
f ~ scape Luftbild n ~ scoop Luft-ansaughutze
f, -ansaugstutzen m, -fänger m, -hutze f,
-sackmaul n, -schippe f, -stutzen m, -zuführung
f, Stauraum m, Windfänger m ~ scoop pressures Druck m im Lufteintrittsstutzen ~-scoop
valve Luftsackventil n
airscrew Druckschraube f ~ boss Propellernabe
f ~ brake Luftschraubenbremse f ~ end Luftschraubenseite f ~ speed Schraubendrehzahl f
air, ~ seal Luftabschluß m ~-seasoned lufttrocken ~ seasoning (of wood) Lufttrocknung f
~ sensitivity Luftempfindlichkeit f ~ separation
Windsichtung f, (thickness) Luftabstand m
~-separation plant Lufttrennungsanlage f ~ separator Luftabscheider m, Windseparator m,
~ service Luftverbindung f ~-service airplane
Arbeitsflugzeug n ~ shaft Luftrohr n, Luft-
schacht m, Lüftungsschacht m, Wetterschacht
m ~ shed window Hallenfenster n
airship Luftschiff n (lenkbares) ~ carrier Luftschiffträger m ~ navigation Luftschiffahrt f
~ pilot Luftschiffführer m ~ port Luftschiff-
hafen m ~ service Luftschiffverkehr m ~ shed
Luftschiffhalle f ~ tender Luftschiffträger
m
air, ~ shutter Lufteinlaßschieber m ~ shut-off
valve Luftabsperrhahn m ~ sick luftkrank
~ sickness Luftkrankheit f sifter Wind-separa-
tor m, -sichter, -sortierer m ~ sifting Windsichtung f ~ silencer Beruhigungsvorrichtung f
~ slide Luftventil n ~ sluice Luftschleuse f
~ sounding Luftlotung f, Schallhöhenmes-
sung f
airspace Hohlschicht f, (of loudspeaker) Laut-
sprecherluftkammer f, Luftraum m ~ cable
Hohlraumkabel n, Lufthohlraumkabel n ~ in
grate Rostspalt m ~ paper Papierhohlraum m
~-paper-core cable Hohlraumkabel n, Papierhohlraumkabel n ~ ratio Luftporenanteil m
airspeed Eigengeschwindigkeit f (aviat.), Fluggeschwindigkeit f, Geschwindigkeit f gegenüber
umgebender Luft, Luftgeschwindigkeit f ~ indicator Fahrtgeber m, Fahrtmesser m, Luft-
geschwindigkeitsmesser m, Staudruckmesser m
~ indicator impact-pressure reading Staudruck-
messeranzeige f

air, ~ spout Kühllufteintrittsklappe *f* ~ spray Luftspritze *f* ~ **spring, with** ~ luftgefedert ~ **squeezer** Druckluftpreßformmaschine *f*, Formmaschine *f* mit Druckluftpressung, Preßformmaschine *f* für Luftdruckbetrieb ~ **squeezing** Luftpressung *f* ~-**squeezing core machine** Kernformmaschine *f* mit Druckluftpressung ~-**squeezing molding machine** Preßformmaschine *f* für Luftdruckbetrieb ~ **stain** Luftflecken *m* ~-**start, to** ~ luftanlassen ~ **starting equipment** Luftanlaßeinrichtung *f* ~ **station** Aufnahmeort *m* in der Luft ~ **stop ring** Ring *m* für Saugstange ~-**storage chamber** Luftspeicherkammer *f* ~-**storing process** Akkumulierungsverfahren *n* ~ **stove** mit Auspuffgas arbeitender Ansaugluftvorwärmer *m* ~ **strainer** Luftfilter *m* ~ **strangler** Luftdrossel *f*, Lufteinlaßklappe *f* (-schieber *m*) ~ **stratum of high humidity** wasserreiche Luftschicht *f* ~ **stream** Luftstrom *m* ~-**stream air flow** Fahrwind *m* ~-**suction device** Luftabsaugvorrichtung *f* ~-**suction line** Ansaugeluftleitung *f* ~-**suction valve** Ansaugeventil *n* ~ **supply** Luft-zufuhr *f*, -zuführung *f*, -zutritt *m*, Windlieferung *f*, Wind-zufluß *m*, -zufuhr *f*, -zuführung *f* ~ **supply duct** Windsaugschacht *m* (Gasmaschine) ~ **survey** Luftbildaufnahme *f*, Luftlandvermessung *f* ~-**survey camera** Luftbildmeßkamera *f* ~-**survey map** Karte *f* nach Luftbildaufnahmen ~-**survey plan** Plan *m* nach Luftaufnahmen ~ **suspension** Luftlager *n* ~ **suspension, with** ~ luftgefedert ~-**swept grinding plant** Luftstrommahlanlage *f* ~ **swirl** Luftwirbel *m* ~ **swivel** Luftzuführung *f* ~ **system** Luftsystem *n* ~ **tank (of brake)** Luftbehälter *m* ~ **target** Flugziel *n* ~-**temperature** Lufttemperatur *f*, Luftwärmegrad *m* ~ **tempering** Abschrecken *n* in Luft ~ **test with soap and water on welds** Abseifen *n* der Schweißnähte ~ **tester** Luftprüfer *m* ~ **thermometer** Luftthermometer *n* ~ ~ **throttle** Luftklappe *f* ~ **tight** lichtdicht, luftdicht, luftdicht gekapselt, winddicht ~ **tight joint** luftdichte Verbindung *f* ~ **tight seal** luftdichter Verschluß *m* ~ **tightness** Winddichtigkeit *f* ~-**time** Sendezeit *f* ~ **timetable** Luftkursbuch *n* ~-**to-air rocket** Luft-Luft-Rakete *f* ~-**to-ground** Boden-Boden ~-**to-ground radio telephony** Boden/Bord-Sprechverkehr *m* ~-**to-surface vessel** Abtastung der Meeresoberfläche vom Flugzeug aus ~ **tower** Luftturm *m*, Reifenluftständer *m* **air traffic** Flugbetrieb *m*, Flugverkehr *m*, Luftverkehr *m* ~ **agreement** Luftverkehrsabkommen *n* ~ **association** Luftverkehrsverband *m* ~ **control (ATC)** Flugsicherung *f* ~ **control boat** Flugsicherungsschiff *n* ~ **control clearance** Flugverkehrsfreigabe *f* ~ **control service** Flugsicherungskontrolldienst *m* ~ **controller** Flugverkehrslotse *m* ~ **law** Luftverkehrsgesetz *n* ~ **rule** Luftverkehrsvorschrift *f* air, ~ **train** Schleppzug *m* ~ **trajectory** Trajektorie *f* ~ **transmission path** Luftübertragungsweg *m* ~ **transmission ring** Luftüberleitungsring *m* ~-**transport company** Luftverkehrsgesellschaft *f* ~ **trap** Luftfalle *f* ~ **travel of thread** Luftstrecke *f* des Fadens ~-**travel ticket** Flugschein *m* ~ **traveler** Luftreisender *m* ~ **tube** Belüftungsrohr *n*, Heizschlauch *m*, Luftrohr *n*, (small) Luftröhrchen *n*, Luftschlauch *m* ~-**tube**

**radiator** Luftröhrenkühler *m* ~ **tubing mixing** Luftschlauchmischung *f* ~ **twirl** Luftwirbel *m* ~ **umbrella** Luftschirm *m* ~ **unworthiness** Luftuntüchtigkeit *f* ~ **uptake** Luftzug *m* ~ **uptake duct** Luftzug *m* ~ **uptake flue** Luftzug *m* ~ **valve** Luft-kanal *m*, -schieber *m*, -ventil *n*, Windklappe *f*, (intake) Luftöffnung *f* ~ **valve housing (valve casing)** Hahngehäuse *n* ~ **valve plug** Hahnküken ~ **valve of tank** Atmungsbehälter *m* für Tankgase ~ **vane** Windflügel *m* ~-**vapor eliminator separator** Kraftstoffentlüfter *m* ~ **velocity** Luftgeschwindigkeit *f* ~ **vent** Abzug *f*, Auslaßventil *n* für Luft, Entlüftungsstutzen *m*, Zugloch *n* ~ **vent baffle** Ölabweiser *m* (in der Entlüftung) ~-**vent cock** Entlüftungsventil *n* ~ **vent pipe** Entlüftungs-leitung *f*, -rohr *n* ~-**vent protector** Belüftungsklappe *f*, Schutzkappe *f* ~ **vessel** Windkessel *m* ~-**vessel end** Kesselboden *m* ~ **volume** Luftmenge *f* ~ **void** Luftpore *f* ~ **vortex** Luftdrehung *f* ~ **wall ionization chamber** Luftwändekammer *f*, Ionisationskammer *f* ~ **washer** Luftwascher *m*, Luftwaschkammer *f* **airway** Luftstraße *f*, Luftstrecke *f*, Luftverkehrslinie *f* ~ **bill** Luftfrachtbrief *m* ~ **lighting** Flugstreckenbefeuerung *f* ~ **man** Wetterstreckenhauer *m*, Wetterarbeiter *m* (min) ~ **marking** Streckenkennung *f* ~ **plotter** Luftwegzeichner *m* ~ **radio beacon** Luftstraßenfunkfeuer *m* ~ **weather forecast** Streckenwetterberatung *f* air, ~ **weight** Gewicht *n* der Luft ~ **wheel** Flugzeugreifen *m* ~ **wheel pile feeder** Luftwalzenstapelanleger *m* ~ **wing** Windflügel *m* ~ **wiper** Lufttrockner *m* ~ **worthiness** Lufttüchtigkeit *m* ~ **worthy** flugfähig, luftfähig **airy** luftig **aisle** Durchgang *m*, Gang *m* (innerhalb eines Lagers), (factory construction) Hallengang *m*, Seitenschiff *n* ~ **in a forest** Waldschneise *f* **(building) with one** ~ einschiffig **ajar** halboffen, (door) angelehnt **alabandite** Alabandin *f*, Braunsteinkies *m*, Glanzblende *f*, Manganblende *f*, Manganglanz *m* **alabaster** Alabaster *m* **alabastrite** Alabastergips *m* **alar septum** Seitenseptum *n* **alarm, to** ~ alarmieren, beunruhigen **alarm** Alarm *m*; Alarmapparat *m*; Besorgnis *f*, Bestürzung *f*, Beunruhigung *f*; Melder *m*, Warner *m*; Warnung *f*, Warnvorrichtung *f*, (electromagnetic) Wecker *m* ~ **and signal bell** Läutewerksglocke *f* ~ **apparatus** Alarmapparat *m*, Meldevorrichtung *f*, Signalapparat *m*, Warnvorrichtung *f* ~ **area** Warning *m* ~ **bell** Alarmglocke *f*, Alarmwecker *m* ~ **bell of line-counting machine** Zählglocke *f* des Zeilenzählers ~-**bell circuit** Läuteschaltung *f* ~ **check valve** Naßalarmventil *n* ~ **clock** Weckeruhr *f* ~ **contact making** Alarmkontaktgabe *f* ~ **cork** Knallkork *m* ~ **device** Alarmschalter *m*, Alarmvorrichtung *f* ~ **fuse** Alarmsicherung *f*, Schmelzsicherung *f* mit Signalgabe ~ **indication** Warnmeldung *f* ~ **lamp** Meldelampe *f* ~ **mechanism** Weckerwerk *n* ~ **plant** Alarmanlage *f* ~ **readiness** Alarmbereitschaft *f* **alarm-relay** Alarmrelais *n* ~ **of grid potential** Gitteralarmrelais *n* ~ **for "main lock-out chain**

faulty" Alarmrelais *n* für „Hauptsperrkette Gitteralarmrelais *n* ~ for "intermediate selector fails to operate" Alarmrelais *n* für „Verbindungswähler stellt nicht ein" (MK) ~ for "main lock-out chain faulty" Alarmrelais *n* für „Hauptsperrkette gestört" (MR) ~ for "marker permanently engaged" Alarmrelais *n* für „Markierer dauerbelegt" ~ for plate potential Anodenalarmrelais *n* (teleph.)

alarm, ~ ring Alarmring *f* ~ signal Alarmzeichen *n*, Meldezeichen *n* ~ station Alarmplatz *m* ~ (horn) stopping switch Alarmabschalter *m* ~ system Alarmanlage *f* ~ threshold Alarmschwelle *f* ~-type fuse Meldesicherung *f* ~ watch Taschenwecker *m* ~ whistle Lärmpfeife *f*

alaskaite Alaskait *m*

albedo Albedo *n*, Reflexionsfaktor *m* ~ limiting value Grenzalbedo *n*

albertite Albertit *m*

albite Albit *m*, Natronfeldspat *m* ~ twinning Albitzwilling *m*

albitite Albitit *m*

albolite künstliches Elfenbein *n*

albumen Albumin *n*, Eiweiß *n*

albumenized paper Albuminpapier *n*

albumin Eiweiß *n* ~ filter Eiweißfänger *m*

albuminometer Albuminometer *n*

albuminous eiweißartig

alchemy Alchemie *f*

alclad Alkladblech *n*, plattiertes Duralblech *n*, Duralumin *n* mit Außenschicht aus Reinaluminium; aluminiumplattiert

alcohol Alkohol *m*, B-stoff *m*, Spiritus *m*, Sprit *m*, Weingeist *m* ~ for engine operation Kraftsprit *m*

alcohol, ~-blended fuel Alkoholgemisch *n*, Alkoholtreibstoff *m* ~ blow torch Spirituslötlampe *f* ~ engine Spiritusmaschine *f* ~-nitric acid solution alkoholische Salpetersäure *f* ~ oxidation Alkoholoxydation *f* ~ pipe B-stoffrohr *n* ~ release valve B-stoffenttankungsventil *n* ~-soluble spritlöslich ~-soluble color spritlösliche Farbe *f* ~ transloading vehicle B-stoffempfangswagen *m* ~ vapor Spiritusdampf *m*

alcoholate of plants Alkoholauszug *m* aus Pflanzen

alcoholic alkoholhaltig, alkoholisch ~ ferment Alkoholgärungspilz *m*

alcoholmeter Alkoholometer *n*

alcoholysis Umesterung *f*

alcove Alkoven *m*

aldehyde Aldehyd *n*

aldehydic aldehydisch

alder wood Eisenholz *n*, Erlenholz *n*

aldol Aldol *n*

aldose Aldose *f*

A leg earthed a-Zweig *m* geerdet

a (b) leg a-(b) Draht *m*

a and b legs crossed a-gekreuzt und b-Zweig *m* gekreuzt

alembic Destillierkolben *m*

alemite fitting Fettpreßnippel *n*

alert Alarmbereitschaft *f*; aufmerksam, rüstig on the ~ alarmbereit to be on the ~ aufpassen ~ signal Vorwarnung *f*

alertness Aufmerksamkeit *f*

aleurometer Aleurometer *n*

alfalfa, ~ and grass drill Alfalfa- und Grasdrillmaschine *f* ~ teeth Alfalfazinken *pl*

alfenide Alfenid *n*

alga Alge *f*

algae Seetang *m*

algebra Algebra *f*

algebraic, ~ (al) algebraisch ~ equation algebraische Gleichung *f* ~ function algebraische Funktion *f* ~ surface algebraische Flächen *pl*

algebroid functions algebroide Funktionen *pl*

algorithm Algorithmus *m*

algraphy Algrafie *f*

alicyclic alizyklisch

alidade Alhidade *f*, Diopter *n*, Dioptrie *f*, (ruler) Diopterlineal *n*, (telescopic-sighting) Kippregel *f* ~ bubble Alhidadenlibelle *f* ~ circle Alhidadenkreis *m* ~ transporter Alhidadentransporteur *m*

alien Ausländer *m*, Fremder *m*; ausländisch, fremd ~ tones Fremdtöne *pl*

alienate, to ~ entfremden, übertragen, veräußern

alienation from proper use Zweckentfremdung *f*

alight, to ~ anwassern, aussteigen to ~ on water anwassern

alight brennend

alighting Landen *n* ~ on water Wasserung *f* ~ gear Fahrwerk *n* ~ speed Wasserungsgeschwindigkeit *f* (aviat.)

align, to ~ abfluchten, abgleichen, anreihen, anrichten, ausfluchten, ausgleichen, (sich) ausrichten, einfluchten, einreihen, einstellen, zum Fluchten *n* bringen, symmetrieren to ~ sights on anvisieren to ~ wheels Räder *pl* richten to ~ with in eine gerade Linie *f* bringen

aligned abgeglichen, ausgerichtet ~-grid tube Röhre *f* mit Elektronenbündelung ~ lugs gegenständige Augenlage *f* ~ nuclei orientierte Kerne *pl*

aligner Abstecker *m*

aligning, ~ of springs of a relay Ausrichten *n* der Federn eines Relais

aligning, ~ arbor Hilfsdorn *m* ~ board Fluchtholz *m* ~ condenser Nachstimmungskondensator *m* ~ device Richtgerät *n* ~ jig Ausrichtschablone *f* ~ pin Einstellstift *m*, Paßstift *m* ~ pole Latte *f*, Meßlatte *f* ~ structures by eye Flugzeugaufbauteile *pl* mit dem Auge abmessen ~ works Richtarbeiten *pl*

alignment Abfluchtung *f*, Abgleich *m*, Abgleichung *f*, Ausfluchtung *f*, Ausrichten *n*, Ausrichtung *f*, Axialität *f*, Einfluchtung *f*, Einreihung *f*, Einstellen *n*, Geradeführung *f*, gerade Linienstrecke *f*, Gleichrichtung *f*, Nacheichung *f*, Richtung *f*

alignment, out of ~ schlecht ausgerichtet to be in ~ fluchten in ~ with in gerader Linie *f* mit ~ of buildings Bauflucht *f* ~ of lights Richtfeuerlinie *f* ~ of nuclei Kernorientierung *f*

alignment, ~ bearing Deckpeilung *f* ~ chart Fluchtlinientafel *f* ~ coil Abgleichspule *f* ~ diagram Fluchtliniendiagramm *n* ~ error Ausrichtfehler *m* (acoust.) ~ nomogram Fluchtliniennomogramm *n* ~ scale Richtskala *m*, Zeilenhöhezeiger *m* ~ stake Richtstange *f* ~ table Fluchtlinientafel *f* ~ tool Einflüchter *m* ~ vestige Spur *f*

**a (b) limb** a-(b) Draht *m*
**aliovalent** andere Wertigkeit *f*
**aliquant** teilerfremd, ungleichteilend
**aliphatic** aliphatisch, fett ~ **hydrocarbon** Fettkohlenwasserstoff *m* ~ **series** Fettreihe *f*
**aliquot part** Bruchteil *m*
**alisonite** Kupferbleiglanz *m*
**alite, to** ~ alitieren
**alite** Alit *m*
**alited steel** alitierter Stahl *m*
**aliting** (a process of surface alloying iron with aluminum) Alitieren *n*
**alitize, to** ~ alitieren
**alive** lebend, lebendig, spannungsführend, unter Spannung befindlich, stromdurchflossen (electr.) **to be** ~ in Betrieb sein
**alizarin** Alizarin *n* ~ **dye** Alizarinfarbe *f* ~ **red** alizarinrot ~ **sodium monosulfonate** alizarinsulfosaures Natrium *n*
**alkalescence** Alkaleszenz *f*
**alkali** Alkali *n* ~ **as anonym to acid** Lauge *f* ~ **blue** alkaliblau ~ **carbonate** Alkalikarbonat *n* ~ **cartridge** Alkalipatrone *f* ~ **cellulose** Alkalizellulose *f* ~ **cyanide** Alkalizyanid *n*, Zyankali *n* ~ **halide film** Alkalischicht *f* ~ **metal** Alkalimetall *n* ~-**metal mercury trap** Falle *f* mit Alkalimetall ~-**metal photo-emissive cell** Alkalizelle *f* ~ **meter** Laugenmesser *m* ~-**proof** alkalibeständig, alkalifest, laugebeständig ~ **scale** (in radiators) Kesselsteinbildung *f* ~-**soluble** alkalilöslich ~ **solution** Alkalilauge *f* ~ **sulfides** Schwefelalkalien *pl* ~ **treatment** Behandlung *f* mit Lauge
**alkaline** alkalisch, alkalinisch, laugenartig ~ **accumulator** alkalischer Sammler *m*, alkalischer Akkumulator ~ **earth** Erdalkali *n*
**alkaline-earth** erdalkalisch ~ **group** Alkalimetallgruppe *f* ~ **metal** Erdalkalimetall *n* ~-**oxide lamp (or light)** (Aeo-)Aufzeichnungslampe *f*
**alkaline,** ~ **maturity** Alkalireife *f* ~-**metal group** Alkalimetallgruppe *f* ~ **mud** alkalischer Schlamm *m* ~ **photocell** Alkali-metallfotozelle *f*, -fotozelle *f* ~ **solution** Alkalilösung *f*, laugenartige Lösung *f*
**alkalinity** Alkaleszenz *f*, Alkalinität *f*, Alkalität *f* ~ **tester** Alkalitätsanzeiger *m*
**alkalinization** Alkalisierung *f*
**alkalize, to** ~ alkalisieren
**alkaloid** Alkaloid *n*
**alkoxylate** veräthern
**alkyd, resin lacquer** Alkydharzlack *m*
**alkydal lacquer** Alkydalspritlack *m*
**alkylation** Alkylierung *f*
**all** Gesamtheit *f*; ganz, sämtlich ~ **at once** auf einmal ~-**around performance** Leistung *f* in allen vorkommenden Betriebsfällen ~-**balsa model** Balsaflugmodell *n* ~-**black malleable cast iron** Schwarzguß *m* ~ **blank repeat key** Dauertaste *f* ~-**bottom sound** dumpfer Ton *m* ~-**busy circuit** Abschaltung *f* ~-**channel decoder** Mehrkanaldekodierer *m* ~-**clear signal** Alarmende *n*, Entwarnung *f*, Fertigsignal *n* (min.) ~-**clear signal light** Klarmeldelampe *f* ~-**cross perforating press** Perforierbreitpresse *f* ~-**electric interlocking** elektrische Verschlüsse *pl* (Stellwerk) ~-**electric supply** Vollnetzanschluß *m* ~-**glass construction** Allglasausfüh-

rung *f* ~-**glass syringes** Ganzglasspritzen *pl* ~-**interference eliminated** voll entstört ~-**level sample** Durchschnittsmuster *n* ~-**mains receiver** Allnetzgerät *n*, Allstromempfänger *m*, Empfänger *m* für Allstrombetrieb, Universalnetzempfänger *m*, Universalgleich- und Wechselstromempfänger *m*
**all-metal** Ganzmetall *n*, ganz aus Metall *n* ~ **airplane** Ganzmetallflugzeug *n* ~ **construction** Ganzmetall-bauart *f*, -bauweise *f* ~ **design** Ganzmetallbauweise *f* ~ **(projection) screen** Ganzmetallwand *f* ~ **tube base** Stahlröhrensockel *m*
**all-or-nothing relay** Hilfsrelais *n*
**all-out** allumfassend
**all-pass network** Allpaßnetzwerk *n*, Allwellenfilter *m*
**all-purpose,** ~ **adhesive** Alleskleber *m* ~ **type (or face)** Allzweckschrift *f*
**all,** ~-**ready signal light** Klarmeldelampe *f* ~-**relay selector** Relaiswähler *m* ~-**relay system** Relaissystem *n* ~ **rights reserved** alle Rechte *pl* vorbehalten
**all-round** universell, vielseitig, allen Zwecken dienend ~ **crane** Drehkran *m* ~ **fastness properties** Gesamtechtheitseigenschaften *pl* ~ **properties** Gesamteigenschaften *pl* ~ **view** Vollsicht *f* ~ **vision** Vollsicht *f* ~ **vision cab(in)** Vollsichtkanzel *f*
**all,** ~ **steel body** Ganzstahlkarosserie *f* ~-**steel casing** Ganzstahlgehäuse *n* ~-**top sound (or voice)** kreischender Ton *m* ~-**trunks-busy register** Gesamtbelegungszähler *m* ~-**up weight** aufgerüstetes Gewicht *n*, Gesamtfluggewicht *n* ~-**wave receiver** Allwellenempfänger *m* ~-**wave receiving set** Allwellenempfangsanlage *f* ~-**ways fuse** empfindlicher Kopfzünder *m* ~-**weather body** Allwetteraufbau *m*, Allwetterkarosserie *f* ~-**wheel drive** Allradantrieb *m* ~-**wheel tractor** Allradschlepper *m* ~-**wing airplane** Nurflügelflugzeug *n* ~-**wood construction** Ganzholzbauweise *f*
**allanite** Allanit *m*
**allantoin** Allantoin *n*
**allay, to** ~ mäßigen
**allelomorph** allelomorph
**allegation** (law) Vorbringen *n* ~**s** Angaben *pl*
**allege, to** ~ angeben, behaupten, unterstellen, vorbringen, vorgeben
**alleged anticipating matter** (patent law) Einspruchmaterial *n*
**allemontite** Antimonarsen *n*, Arsenikantimon *n*, Arsenikspießglanz *m*
**Allen,** ~ **head screw** Zylinderschraube *f* mit Innensechskant ~ **key** Vier-(Sechs-)Kantstiftschlüssel *m* ~ **screw** Innenvierkantschraube *f*, Zylinderschraube *f* mit Innensechskant ~ **set screw** Gewindestift *m* mit Innensechskant ~ **wrench** Inbus-Schlüssel *m*, Sechskantstiftschlüssel *m*
**alliance** Allianz *f*, Bündnis *n*, Vereinigung *f* ~ **of convenience** Zweckverband *m*
**alligation** Versetzung *f*, Mischungsrechnung *f*
**alligator** Alligator *m*, Luppenquetsche *f*, Steinhaue *f* ~ **clip** Froschklemme *f*, Krokodilklemme *f* ~ **cracking** netzartige Rißbildung *f* ~ **grab** Fangrachen *m*, Greifrachen *m*, Klauenfänger *m* ~ **grip wrench** Zahnschlüssel *m*, Zahn-

zange *f* ~ **shears** Alligatorschere *f*, Hebelschere *f* ~ **squeezer** Alligatorquetsche *f* ~ **wrench** Alligatorzange *f*
**alligatoring** Krokodilnarben *pl*, Rißbildung *f*
**allobar** Allobar *m*
**allocate, to** ~ aufnehmen (in eine Bearbeitungsvorrichtung), aufstellen, zuteilen
**allocation** Anordnung *f*, Anweisung *f*, Aufstellung *f*, Speichenverteilung *f* (info proc), Zuteilung *f* ~ **of frequencies** Wellenverteilung *f* ~ **of working expenses** Verwaltungskosten *pl* ~ **plan** Zellenplan *m* ~ **scheme** Belegungsplan *m*, Benutzungsplan *m*
**allochromatic** allochromatisch
**allomorphic** allomorph
**allot, to** ~ austeilen, bewilligen, überweisen, vorbestimmen, zuerkennen, zuerteilen, zumessen, zuteilen, zuweisen **to** ~ **to** beigeben
**allotted to** zugeteilt
**allotting switch** Überweisungswähler *m*
**allotment** Anweisung *f*, Auslösung *f*, Baulos *n*, Kontingent *n*, Parzelle *f*, terminmäßiger Abzug *m*, Überweisung *f*, Verteilung *f*, Zuordnung *f*, Zuteilung *f*, Zuweisung *f* ~ **of equipment** Gerätausstattung *f* ~ **for iron ore** Eisensteinfeld *n* ~ **of ore** Erzfeld *n* ~ **worked underground** Tiefbaubetrieb *m*
**allotriomorphic** allotriomorph ~ **crystal** Kristallit *m*
**allotropic** allotrop, allotropisch ~ **change** allotropische Modifikation *f*
**allotropism** Allotropie *f*
**allotropy** Allotropie *f*
**allotter** Anrufverteiler *m*, Rufordner *m*, Wählsucher *m*
**allow, to** ~ bewilligen, erlauben, einräumen, genehmigen, gestatten, gewähren, vergünstigen, vergüten, zugeben **to** ~ **for** berücksichtigen, in Betracht ziehen, Rücksicht nehmen auf **to** ~ **to ascend** (balloon) auflassen **to** ~ **to clarify** absitzen lassen **to** ~ **to deposit** absitzen lassen **to** ~ **discharge** eine Röhre *f* öffnen **to** ~ **discount** Rabatt *m* bewilligen oder geben **to** ~ **to drain** abtropfen lassen **to** ~ **to evaporate** abdampfen lassen **to** ~ **to fill** ziehen lassen (in (Bad)) **to** ~ **to react** einwirken lassen (chem.) **to** ~ **to settle** absetzen lassen **to** ~ **to shrink** krumpeln **to** ~ **to soak** einziehen lassen **to** ~ **time for payment** stunden
**allowable** gewährbar, zulässig ~ **error** zulässiger Fehler *m* ~ **production** bewilligte Erzeugung *f* ~ **variation** zulässige Abweichung *f* oder Toleranz *f* ~ **voltage** Spannungszulässigkeit *f* ~ **working stress** zulässige Betriebsspannung *f*
**allowance** Ablaß *m*, Abschlag *m*, ausgesetzte Summe *f*, Ausstattungssoll *n*, Bewilligung *f*, Berücksichtigung *f*, Einräumung *f*, Entschädigung *f*, Erlaß *m*, Erlaubnis *f*, Ermäßigung *f*, Gebührenermäßigung *f*, Gebührennachlaß *m*, Gehalt *n*, Genehmigung *f*, Preisnachlaß *f*, Spielraum *m*, Toleranz *f*, Vergünstigung *f*, Vergütung *f*, (for taxes) Zollnachlaß *m*, Zugabe *f*, Zulage *f*, zulässige Anzahl *f*, zulässige Abweichung *f* (teleph.), Zuschuß *m* ~ **for elevation** Druckausgleich *m* ~ **for the error of the day** Berücksichtigung *f* der Tageseinflüsse ~ **of letters patent** Erteilung *f* eines Patentes ~ **va-**

riation zulässige Abweichung *f* ~ **for waste** Überration *f*, Zuschlagration *f* ~ **in weight** Remedium *n* am Schrot
**alloy, to** ~ karatieren (gold or silver). legieren, mischen, verschmelzen, versetzen, zusetzen
**alloy** Hartlegierung *f*, Legierung *f*, Metallegierung *f*, Mischmetall *n*, Vermischung *f*, Versetzung *f*, Zusatz *m*, Zusatzmetall *n*; zweimetallisch ~ **that does not deteriorate with age** alterungsfreie Legierung *f* ~ **available on request** Antraglegierung *f* ~ **for cutting instruments (or tools)** Schneidlegierung *f*
**alloy,** ~ **cast iron** legiertes Gußeisen *n* ~ **die casting** Legierungsgußform *f* ~ **ingot** Legierungs--blöckchen *n*, -rohmassel *n* ~ **junction** Legierungsschicht *f* ~ **low in critical (or scare) materials** sparstoffarme Legierung *f* ~ **metal** Legierung *f* ~ **not restricted in use** Hauptlegierung *f* ~ **sheet** Legierungsblech *n* ~ **steel** Legierungs-, Sonder--stahl *m* ~-**treated steel** niedriglegierter Stahl *m* ~ **tool steel** legierter Werkzeugstahl *m* ~ **transistor** legierter Transistor *m*
**alloyage** Legierung *f*
**alloyed** legiert ~ **gold** Karatgold *n*
**alloying** Legieren *n*, Zulegieren *n* ~ **of gold (or of silver)** Karatierung *f*
**alloying,** ~ **addition** Legierungszusatz *m* ~ **constituent** Legierungsbestandteil *m* ~ **element** Legierungs-element *n*, -körper *m*, Zusatzelement *n* ~ **furnace** Legierungsofen *m* ~ **metal** Legierungszusatz *m* ~ **process** Legierungsprozeß *m* ~ **property** Legierbarkeit *f*
**allright position** Freilage *f*
**alluvial** Geröll *n*, goldhaltiges Geröll *n*; alluvial, angeschwemmt, angespült ~ **basin** Schwemmmulde *f* ~ **cone** Schuttkegel *m* ~ **deposit** angeschwemmte Lagerstätte *f* ~ **deposits** angeschwemmter Boden *m* ~ **gold** Waschgold *n*, in Anschwemmung vorkommendes Gold *n* ~ **land** Anlandung *f* ~ **ore** Seifenerz *n* ~ **stone** Schwemmstein *m*
**alluviation** Ablagerung *f*
**alluvion** Alluvion *n*, Versandung *f*
**alluvium** Ablagerung *f*, Alluvium *n*, angeschwemmter Boden *m*, angespültes Land *n*, Anschwemmung *f*, Anspülung *f*
**allyl,** ~ **bromide** Allylbromid *n* ~ **chloride** Chlorallyl *n* ~ **cyanide** Zyanalkyl *n* ~ **sulfide** Allylsulfid *n*
**allylice alcohol** Allylalkohol *m*
**allylisothiocyanate** Allylsenföl *n*
**almanac** Jahrbuch *n*, Kalender *m* ~ **of closing dates (or deadlines)** Terminkalender *m*
**almandine** Almandin *m*, Karfunkelstein *m*, roter Granat *m*
**almandite** Almandin *m*, Eisengranat *m*, Eisentongranat *m*, Karfunkelstein *m*
**almost** beinahe, fast, gleichsam ~ **"schlicht" mapping** fast schlichte Abbildung *f* (math.) ~ **vertical lode** seigerer Gang *m*
**alni magnetic flux** Alnimagnetfluß *m*
**aloe,** ~ **fiber** Aloefaser *f* ~ **rope** Aloeseil *n*
**aloft** empor, oben
**along** entlang, längs ~ **the lines of** ... nach Art des ... ~ **lines** leitungsgerichtet ~ **side** Bord an Bord, längsseits, nebenan, nebenstehend ~ **with** neben
**aloxite** Aloxit *m*, künstlicher Korund *m*

**alpace** Alpaka n
**alpax** Alpax m
**alpha, ~-beta Geiger tube** Alpha-Beta-Zählrohr n ~ **burst detector** Alphaexplosionsdetektor m ~ **change** Alpha-Emission f ~-**naphthol** Alphanaphtol n ~-**naphthylamine** Alphanaphthylamin n ~ **numeric** alphanumerisch ~ **particle** Alphateilchen n ~ **pulse counting insert** Alpha-Impulszählereinsatz m ~ **rays** Alphastrahlen pl ~ **topic** alphatopisch ~ **tron** Alphastrahlen pl ~ **tube** Zählrohr n für Alpha-Strahlen
**alphabet** Alphabet n ~ **perforating machine** Alphabetperforiermaschine f
**alphabetic order** alphabetische Reihenfolge f
**alphabetical** alphabetisch ~ **accounting machine** alphabetschreibende Tabelliermaschine f ~ **duplicating punch** Alphabetwiederholungslocher m ~ **interpreter** Alphabetlochschriftübersetzer m ~ **notation** Buchstabentonschrift f ~ **office code** Kennbuchstabe m ~ **punch** Alphabetlocher m ~ **verifier** Alphabetlochprüfer m
**alphatopic** alphatopisch
**alphatron** Alphatron n
**Alpine foothills** Alpenvorland n
**alpine railway** Gebirgsbahn f
**alstonite** Alstonit m
**altatite** Tellurblei n
**alter, to** ~ abändern, abwandeln, ändern, entstellen, umändern, variieren, (sich) verändern, verschieden sein, verstümmeln, wandeln, wechseln
**alteration** Abänderung f, Änderung f, Entstellung f, Umänderung f, Veränderung f, Wandel m ~ **of course sixteen points in succession** Kehrtschwenkung f (naut.) ~ **of shade** Nuancenveränderung f ~ **(or change) of shape (or design)** Gestaltsänderung f ~ **of stress** Wechselbeanspruchung f
**altered** umgeändert, umgearbeitet ~ **rock** metamorphisches Gestein n
**alternate, to** ~ abwechseln, abwechselnd etwas tun, alternieren, versetzen, wechseln, wechselweise aufeinanderfolgen
**alternate** (shift) ablösend, abwechselnd, alternierend, wechselseitig, wechselständig ~ **aerodrome** Ausweichflugplatz m, Ausweichhafen m ~ **angle** Wechselwinkel m ~-**channel interference** Störung f aus dem dritten Kanal ~ **ebb and flow of the tide** Tidesträmung f ~ **fuel** Ausweichkraftstoff m ~-**immersion test** Wechseltauchversuch m ~ **position** Ausweichstellung f, Wechselstellung f ~ **proportion** Wechselverhältnis n ~ **route** Ersatzweg m ~ **routing** Ausweichvermittlung f, Leitweglenkung f, Umwegschaltung f, Umwegsteuerung f ~ **shelling with highexplosive and chemical shells** Buntschießen n ~ **stress** Wechselfestigkeit f ~-**stress test** Hinund Herbiegeprobe f, Wechselfestigkeitsprüfung f ~ **target** Ausweichziel n ~ **tensionalcompressional strength** Zugdruckwechselfestigkeit f ~-**tooth construction** Wechselverzahnung f ~-**tooth cutter** Fräser m für Wechselverzahnung ~ **torsional strength** Verdrehungsdauerfestigkeit f, Verdrehwechselfestigkeit f ~ **two-way-communication** Zwischenhörbetrieb m ~ **two-way radio communication** Funkwechselsprechen n ~ **two-way radio traffic** Funkwech-

selverkehr m ~ **two-way traffic** Wechselverkehr m ~ **working kiln** Ofen m mit unterbrochenem Betrieb
**alternately** absatzweise, in Absätzen, umschichtig, wechselweise
**alternating** Wechseln n; abwechselnd, alternierend, wechselnd ~ **arc** Wechselstromlichtbogen m ~ **bending** Biegewechselbeanspruchung f ~ **bending strength** Biegewechselfestigkeit f ~ **bending stress** Wechselbiegespannung f ~ **bending test** Hin- und Herbiegeprobe f, Wechselbiegeversuch m ~ **bevel grind** wechselseitiger Schrägschliff m ~ **blinker** Wechselblinkleuchte f ~ **colored lights** Wechsel-, Mischfeuer n ~ **component of current** Wechselstromkomponente f ~ **component of voltage** Wechselspannungskomponente f ~ **current** Wechselstrom m
**alternating-current from mains** Netzwechselspannung f
**alternating-current, ~ arc** Wechselstrombogen m ~ **arc lamp** Wechselstrombogenlampe f ~ **arc welder** Wechselstromschweißumformer m ~ **alarm clock (or ringer)** Wechselstromwecker m ~ **alternation (or half wave)** Halbperiodenwechsel m ~ **ammeter** Wechselstrommesser m ~ **amplitude** Wechselamplitude f ~ **bell** Wechselstromwecker m ~ **(measuring) bridge** Wechselstrommeßbrücke f ~ **buzzer** Wechselstromwecker m ~ **calling relay** Rufwechselstromrelais n ~ **circuit** Wechselstromkreis m ~ **commutator motor** Wechselstromkommutatormotor m ~ **component** Wechselstromkomponente f ~ **conductivity** Wechselstromleitfähigkeit f ~ **continuous-current converter** Wechselstromgleichstromeinankerumformer m ~ **continuous dynamotor** Wechselstromgleichstromumformer m ~ **cycle** Wechselstromperiode f ~ **dialing** Einfrequenzfernwahl f, Wechselstromfernwahl f ~ **direct-current power mains** Allstromnetzteil n ~-**direct-current valve** Allstromröhre f ~ **dynamo** Wechselstrom-dynamo m, -dynamomaschine f ~ **engineering** Wechselstromtechnik f ~ **field** Wechselstromfeld n ~ **generator** Wechselstrom-dynamo m, -dynamomaschine f, -erzeuger m, -generator m ~ **heating** Wechselstromheizung f ~ **hum** Brodeln n, Brodelstörung f, Brummen n, Brummton m, Netzbrummen n, Netzton m, Wechselstromton m ~ **induction** Wechselstrominduktivität f ~ **interference** Wechselstrombeeinflussung f ~ **lighting circuit** Wechselstromlichtnetz n ~ **losses** Wechselstromverluste pl ~ **mains** Wechselstromnetz n ~ **mains receiver** Wechselstromempfänger m ~ **measurement** Wechselstrommessung f ~ **measuring set** Wechselstrommeßgerät n ~ **motor** Wechselstrommotor m ~ **multiple telegraphy** Wechselstrommehrfachtelegrafie f ~-**operated receiver** Wechselstromempfänger m, Empfänger m für Wechselstrom-Netzspeisung ~-**operated valve** Wechselstromröhre f ~ **potential-ratio method** Methode f nach dem Verhältnis der Wechselstrompotentiale ~ **power** Wechselstromleistung f ~ **pulsing** Wechselstromsignalisierung f ~ **receiver** Empfänger m für Wechselstrom-Netzanschluß, Wechselstromempfänger m ~ **rectifier** Wechselstromgleichrichter m ~ **relay** Wechselstromrelais n ~ **relay rectifier**

Wechselstromrelaisgleichrichter *m* ~ **resistance** Wechselstromwiderstand *m* ~ **selection** Wechselstromwahl *f* ~ **shunt motor** Wechselstromnebenschlußmotor *m* ~ **source** Wechselstromquelle *f* ~ **starter** Wechselstromanlasser *m* ~ **system** Wechselstromanlage *f* ~ **track circuit** Gleisstromkreis *m* für Wechselstrom ~ **traction** Wechselstromfahrbetrieb *m* ~ **transformer** Wechselstromtransformator *m* ~ **valve** Wechselstromröhre *f* ~ **voltage** Wechselspannung *f*

**alternating,** ~ **dew test** Wechselbetauungsversuch *m* ~ **direct stress testing machine** Zugdruckschwingungsprüfmaschine *f* ~ **discharge** alternierende Entladung, *f* ~ **effect** Wechselwirkung *f* ~ **field** Wechselfeld *n* ~ **flux** Wechselfluß *m* ~ **force** wechselwirkende Kraft *f* ~ **immersion test** Wechseltauchversuch *m* ~-**impact bending test** Wechselschlagbiegeversuch *m* ~-**impact machine** Dauerwechselschlagwerk *n* ~-**impact test** Dauerwechselschlagversuch *m*, Wechselschlagversuch *m* ~ **light supplementary apparatus** Wechsellicht-Zusatzeinrichtung *f* ~ **load deformation** Wechselverformung *f* ~ **loads** Wechselbelastung *f* ~ **magnetic field** magnetisches Wechselfeld *n* ~ **moment** Wechselmoment *n* ~ **motion** alternierende Bewegung *f* ~ **oblique picture** Pendelaufnahme *f* ~ **perforation** Zickzacklochung *f* ~ **potential** Wechselspannung *f* ~ **pressure** Wechseldruck *m* ~-**repeated-stress test** Dauerversuch *m* mit wechselnder Beanspruchung ~ **repetition of stress** Dauerwechselbeanspruchung *f* ~ **saw** Walzengatter *n* ~ **shock load** Wechseldruck *m* ~ **shutter** (in stereoscopy) Wechselblende *f* ~ **stress** Schwingungsbeanspruchung *f*, Wechselbeanspruchung *f*, Wechselspannung *f* ~ **stress amplitude** Spannungsausschlag *m* ~ **stress number** Lastwechselzahl *f* ~-**stress test** Wechselschlagversuch *m* ~ **talking currents** Sprachwechselströme *pl* ~ **tensionand compression-stress fatigue strength** Zugdruckschwingungsfestigkeit *f* ~ **thrust bearing** Wechseldrucklager *n* ~-**torsion test** Drehschwingversuch *m* ~ **quantitiy** Wechselgröße *f* ~ **voltage** Wechselspannung *f*

**alternation** Abwechslung *f*, Halbperiode *f*, Halbperiodendauer *f*, Halbwelle *f*, Wechsel *m*, wechselseitige Folge *f* ~ **of multiplicities** Vielfältigkeitsabwechslung *f* ~ **of polarity** Polaritätswechsel *m*, Polwechsel *m* ~ **of stress** Wechselspannung *f*

**alternation,** ~ **bending test machine** Hin- und Herbiegemaschine *f* ~ **law** Folgegesetz *n*

**alternative** Ausweg *m*, Entweder-oder-satz *m*, Wechselfall *m*; alternierend ~ **frequenzy** Ausweich-frequenz *f*, -welle *f* ~ **fuel engine** Motor *m* für verschiedene Kraftstoffe ~ **implacement** Wechselstellung *f* ~ **motor** Drehstrommotor *m* ~ **position** Wechselstand *m* ~ **route** Ersatzweg *m*, Hilfsweg *m* ~ **wave** Ausweichwelle *f* (rdr)

**alternatives** Alternativa *pl*

**alternator** Alternator *m*, Wechselstrom-dynamo *m*, -erzeuger *m*, -generator *m*, -maschine *f*, ~ **with rotating spark gap** Wechselstromgenerator *m* mit rotierender Funkenstrecke

**alternator,** ~ **armature** Wechselstromanker *m* ~ **disk set** Wechselstromgenerator *m* mit umlaufender Funkenstrecke ~ **magnet** Feldmagnet *m*

~ **selection panel** Umformerwahlschalter *m* (rdo) ~ **transmitter** Maschinensender *m*

**alti-electrograph** Alti-Elektrograf *m*

**altigraph** Höhenschreiber *m*

**altimeter** Entfernungsmeßgerät *n*, Höhenmesser *m*, Höhenzeiger *m* ~ **lag** Höhenmesser -verzögerung *f*, -widerstand *m* ~ **setting** Einstellung *f* des Höhenmessers

**altimetric,** ~ **device** Höhenmesser *n* ~ **dial (or disk)** Scheibenhöhenmesser *m* ~ **instrument** Höhenmesser *n*

**altimetry** Höhenmessung *f*, Höhenvermessung *f*

**altitude** Höhe *f* über Normal Null (NN), Meereshöhe *f* ~ **of bearing** Peilhöhe *f* ~ **at culmination** Kulminationshöhe *f* ~ **of lower rim (or limb)** Unterrandhöhe *f* ~ **of point of impact** Treffpunkthöhe *f* ~ **above (mean) sea-level** Höhe *f* über NN (Seehöhe), NN = Normal Null ~ **mittlere Seehöhe** ~ **of the sun** Sonnenstand *m* ~ **of target** Treffhöhe *f* ~ **of target at control point** Reglerpunkthöhenwinkel *m* ~ **of target at firing point** Abschußhöhe *f*

**altitude,** ~ **aptitude test** Höhentauglichkeitsprüfung *f* ~ **boost** Höhenlader *m* ~ **bracket** Höhenschicht *f* ~ **carburetor** Höhenvergaser *m* ~ **circle** Höhenkreis *m* ~ **control** Höhenregelung *f* ~ **control nozzle** Lagesteuerungsdüse *f* (der Rakete) ~ **correction** Höhenverbesserung *f* ~-**correction ruler** Erdkrümmungslineal *n* ~ **counter** Höhenzähler *m* ~ **depression** der Betriebshöhe entsprechender Unterdruck *m* ~ **depression chamber** Höhenunterdruckkammer *f* ~ **figure** Höhenzahl *f* ~ **flight** Höhenflug *m* ~ **gauge** (in central heating) Wasserhöhenanzeiger *m* ~ **graduation** Höhenskala *f* ~ **indication** Höhenangabe *f* ~ **lead** Höhenvorhalt *m* ~ **level** Höhenlage *f* ~ **marks** Höhenpunkte *pl* ~ **measurement** Höhenlotung *f* ~ **meteorological service** Höhenwetterdienst *m* ~ **meter** Höhenmesser *m* ~ **mixture control** Höhengemischregler *m* ~ **parallel** Höhengleiche *f* ~ **performance** Höhenleistung *f* ~-**plotting board** Höhenmeßplan *m* ~ **pressure chamber** Höhenprüf-anlage *f*, -kammer *f*, Unterdruckanlage *f* ~-**proved** höhenfest ~ **reading** Höhengabe *f* ~ **record** Höhenreihe *f* ~ **resistance** Höhenfestigkeit *f* ~ **scale** Höhenskala *f* ~ **setting screw** Höhenstellschraube *f* ~ **steps** Höhenstufen *pl* ~-**test chamber** Höhenkammer *f*, Höhenprüfstand *m*, Unterdruckkammer *f* ~-**tested** höhenfest ~ **throttle lever** (supercharger) Höhengashebel *m* ~ **value** (in vision) Höhenwert *m* ~ **zone** Höhenschicht *f* (Stratosphäre)

**alto-cumulus clouds** Altocumulus *m*, grobe Schäfchenwolken *pl*

**alto-stratus clouds** hohe Schichtbewölkung *f*

**alum** Alaun *n* ~ **mordant** (alumina mordant) Tonerdebeize *f* ~ **pickle** Alaunbrühe *f* ~ **shale** Alaunschiefer *m* ~ **slate** Alaunschiefer *m*, Vitriolschiefer *m* ~ **stone** Alaunstein *m* ~-**tanned** alaungar ~ **tanner (tawer)** Weißgerber *m* ~ **works** Alaunsiederei *f*

**alumed leather** alaungares Leder *n*

**alumina** Alaunerde *f*, Aluminiumoxyd *n*, Tonerde *f* ~ **content** Tonerdegehalt *m* ~ **cream** Tonerde *f* ~ **inclusion** Tonerdeeinschluß *m* ~ **lake** Tonerdelack *m*

aluminal silver Aluminiumsilber *n*
aluminate Aluminat *n* ~ of potash Kalitonerde *f*
~ liquor Aluminatlauge *f*
aluminiferous alaunhaltig, aluminiumhaltig, tonerdehaltig
aluminize, to ~ alitieren, aluminisieren, Aluminium *n* aufspritzen und einbrennen, veralmen, veraluminieren
aluminized screen Bildschirm *m* mit aufgedampfter Aluminiumhaut
aluminizing Alitieren *n*, Aufspritzen *n* von Aluminium
aluminothermic aluminothermisch
aluminothermy Aluminothermie *f*
aluminous alaunhaltig, tonerdereich, tonhaltig ~
abrasive Elektrokorund *m*, Kunstkorund *m* ~
amber Bernsteinalaun *m* ~ cement Schmelzzement *m* ~ flux Tonzuschlag *m* ~ limestone Alaunstein *m* ~ soap Alaunseife *f*
aluminox künstlicher Korund *m*
aluminium Aluminium *n*, Tonerdemetall *n* ~
acetate Aluminazetat *n*, Aluminiumazetat *n*, essigsaure Tonerde *f* ~ alloy Alulegierung *f*, Aluminiumlegierung *f* an ~ alloy Autogal A ~
ammonium sulfate Ammonalaun *m*, Ammoniakalaun *m* ~ backing Aluminiumeinlage *f* ~
bar Aluminiumbarren *m* ~-base alloy Legierung *f* auf Aluminiumgrundlage ~-bearing aluminiumhaltig ~ bisulfite Aluminiumbisulfit *n*
~ boride Aluminiumbor *n* ~ brass Aluminiumkupferzinklegierung *f*, Aluminiummessing *n*
~ casting Aluminiumguß *m* ~-cell rectifier Aluminiumzellengleichrichter *m* ~ chill casting Aluminiumkokillenguß *m*, Aluminiumschalenguß *m* ~ chloride chlorsaure Tonerde *f* ~ coat Aluminiumüberzug *m* ~ coating Metallisierung *f* ~ coherer Aluminiumfritter *m* ~ content Aluminiumgehalt *m*
aluminium-copper, ~-magnesium alloy Bondur *n*
~-magnesium alloys for aircraft construction Avionallegierungen *pl* ~-zinc alloy Aluminiumkupferzinklegierung *f*
aluminium, ~ cover Aluminiumbekleidung *f* ~
crank-case Aluminiumkurbelgehäuse *n* ~ cube Aluminiumwürfel *m* ~ dust Aluminiumstaub *m* ~ fillings Aluminiumfeilspäne *pl* ~ float Aluminiumschwimmer *m* ~ fluoride Fluoraluminium *n* ~ foil Alfol *n*, Aluminiumfolie *f* ~ foil strips Alubandkappen *pl* ~ housing Aluminiumgehäuse *n* ~ hydrate Aluminiumoxydhydrat *n* ~ (hydroxide) hydrate Tonerdehydrat *n* ~
hydroxide Aluminiumoxydhydrat *n* ~ ingot Aluminiumbarren *m* ~ iodate Aluminiumjodat *n* ~ iodide Aluminiumjodid *n* ~ kitchen sets (or utensils) Aluminiumkochgeschirr *n* ~
manufacture Aluminiumherstellung *f* ~ mounting Aluminiumfassung *f* ~ nitrate salpetersaure Tonerde *f* ~ ore Aluminiumerz *n* ~ oxide Aluminiumoxyd *n*, Fasertonerde *f* ~ paint Aluminiumfarbe *f* ~ pig Aluminiumbarren *m* ~
pipe Aluminium-rohr *n*, -röhre *f* ~ piston Aluminiumkolben *m* ~ plate Aluminiumgeschirr *n* ~ plating Alumetieren *n*, Alumetierung *f* ~
polish Aluminiumputzmittel *n* ~ potassium sulfate Aluminiumkaliumsulfat *n* ~ powder Aluminiumpulver *n* ~ rectifier Aluminiumgleichrichter *m* ~ reeds (accordion) Aluminiumtöne *pl* ~ rod Aluminiumstange *f* ~-rolling ingot

Aluminiumwalzbarren *m* ~ salt Tonerdesalz *n*
~ sand casting Aluminiumsandguß *m* ~ sheet Aluminiumblech *n* ~ shot Aluminiumgrieß *m*
~ silicofluoride Aluminiumfluorsilikat *n*, Kieselfluoraluminium *n* ~ sodium chloride Aluminiumnatriumchlorid *n* ~ sodium sulfate Natronalaun *m* ~ solder Aluminiumhartlot *n*
~ soldering Aluminiumlöten *n* ~ stearate stearinsaures Aluminium *n* ~ sulfate schwefelsaure Tonerde, *f* Tonerdesulfat *n* ~ sulfide Schwefelaluminium *n* ~ telluride Telluraluminium *n* ~ thiocyanate Aluminiumrhodanid *n*
~ tray Aluminiumschale *f* ~ trihydrate Aluminiumoxydhydrat *n* ~ tube Aluminium-rohr *n*, -röhre *f* ~ wire Aluminiumdraht *m* ~ works Aluminiumhütte *f*
alundum Alundum *n*, künstlicher Korund *m* ~
brick Alundumziegel *m*
alunite Alaunstein *m*, Alunit *m*
alunogen Alunogen *n*, Federalaun *n*
amadou Zündschwamm *m*
amalgam Amalgam *n*, Quecksilberlegierung *f* ~
pot retort Amalgamausbrenntopf *m* ~ press Amalgampresse *f* ~ trap (stamp milling) Amalgamfänger *m*
amalgamable amalgamierbar
amalgamate, to ~ amalgamieren, quicken, vermischen, verquicken, verschmelzen ~ claims zusammengelegte Berggerechtsame *pl*
amalgamater metal Quickmetall *n*
amalgamating, ~ liquid Amalgamierungsflüssigkeit *f* ~ mill Quickmühle *f* ~ pan Amalgamierpfanne *f* ~ plant Amalgamationsanlage *f* ~
solution Amalgambad *n*
amalgamation Amalgamierung *f*, Quickarbeit *f*, Vereinigung *f*, Verquickung *f*, Verschmelzung *f*, Zusammenfassung *f*, Zusammenschluß *m* ~ of similar concerns Zusammenlegung *f* gleichartiger Betriebe
amalgamator Amalgamator *m*, Amalgamiertisch *m*
amass, to ~ aufhäufen
amateur Bastler *m*, Liebhaber *m* ~ binding Liebhabereinband *m* ~ constructor Funk-, Radiobastler *m* ~ craftsman Bastler *m* ~ license Amateurlizenz *f* ~ photographer Amateurfotograf *m* ~ radio station Amateurfunkstelle *f* ~ wireless operator Funkamateur *m*
amatol Amatol *n*
amazonite Amazonenstein *m*, Smaragdspat *m*
amber Bernstein *m* ~ airway gelbe Luftstraße *f*
~gris graue Ambra *f* ~like bernsteinartig ~
oil Bernsteinöl *n* ~-seed oil Moschuskörneröl *n*
~ varnish Bernsteinlack *m*
amberoid Preßbernstein *m*
ambidextrous rechts und links zugleich
ambiency Umgebung *f*
ambient Außenwelt *f*, Umgebung *f*; umgebend
~conditions Umgebungsbedingungen *pl* ~
light Umgebungslicht *n* ~ noise Nebengeräusch *n* ~ surface Umfläche *f* ~ temperature Raumtemperatur *f*, Umgebungstemperatur *f*
ambiguity Mehrdeutigkeit *f*, Zweideutigkeit *f*
ambiguous doppelsinnig, vieldeutig, zweideutig
ambipolar ambipolar
ambit Kontur *f*, Umriß *m*
ambroin Ambroin *n*

**ambulance** Krankenwagen *m*, Sanitäts-auto *n*, -kraftwagen *m*
**ambulant** ambulant, umherziehend
**Ambursen dam** Ambursensperre *f*
**ameliorate, to** ~ verbessern
**amelioration** Melioration *f*
**amenability to receive polish** Polierfähigkeit *f*
**amend, to** ~ ändern, ausbessern, berichtigen, verbessern, (an application) umändern
**amended focal length** (rectifier) Ersatzbrennweite *f* (Entzerrungsgerät)
**amendment** Abänderung *f*, Änderung *f*, Berichtigung *f*, Eingabe *f*, Verbesserung *f*
**American,** ~ **standard scale for density** Baumé-standard *m* ~ **wire gauge** Amerikanische Drahtlehre *f*
**amianthus** Asbest *m*, Flachstein *m*, Strahleisen *n*
**amid acid** Amidsäure *f*
**amidate, to** ~ amidieren
**amide** Amid *n* ~ **chips** (sugar mfg) Amidschnitzel *pl*
**amido,** ~ **acid** Amidsäure *f* ~**carbonic acid** Amidokohlensäure *f* ~ **nitrogen** Amidstickstoff *m* ~**sulfuric acid** Amidosulfosäure *f*, (sulfamic acid) Amidoschwefelsäure *f*
**amidships** mitschiffs
**amine** Amin *n*
**amino,** ~ **acetic acid** Amidoessigsäure *f* ~ **acid** Amidosäure *f*, Aminosäure *f*, Aminsäure *f* ~**benzene** Amidobenzol *n*, Anilin *n* ~ **compound** Aminoverbindung *f* ~**glutaric acid** Aminoglutarsäure *f* ~ **group** (amino acids) Aminogruppe *f* ~**ketonic** aminoketonisch ~ **phenol** Amidophenol *n* ~ **sulfonic acid** Amidosulfosäure *f*
**aminography** Aminografie *f*
**ammeter** Ammeter *n*, Ampèremeter *n*, Strommesser *m*, Stromprüfer *m*, Stromzeiger *m* ~ **box** Ampèremeterkasten *m* ~ **selector switch** Ampèremeterumschalter *m* ~ **shunt** Ampèremeternebenwiderstand *m*
**ammonal** Ammonpulver *n*
**ammonia** Ammon *m*, Ammoniak *n*, Ammoniakgas *n* ~ **acetate** essigsaures Ammoniak *n* ~ **alum** Ammonalaun *m* ~ **bicarbonate** doppeltkohlensaures Ammoniak *n* ~ **carbonate** kohlensaures Ammoniak *n* ~ **compressor** Ammoniakverdichter *m* ~ **equipment** Ammoniakapparate *pl* ~ **lactate** milchsaures Ammoniak *n* ~ **leaching** Ammoniaklaugung *f* ~ **outlet valve** Ammoniakauslaßventil *n* ~**-oxygen mixture** Ammoniaksauerstoffgemisch *n* ~ **persulfate** überschwefelsaures Ammoniak *n* ~ **recovery** Ammoniakgewinnung *f* ~ **refrigerating coil** Ammoniakkühlschlange *f* ~ (washer) **scrubber** Ammoniakwascher *m* ~ **scrubbing** Ammoniakwäsche *f* ~ **separation** Ammoniakabscheidung *f* ~ **tartrate** weinsteinsaures Ammoniak *n* ~ **washing process** Ammoniakwäsche *f* ~ **water** Ammoniakwasser *n* ~ **works** Ammoniakanlage *f* ~ **yield** Ammoniakausbeute *f*
**ammoniacal** ammoniakalisch, ammoniakhaltig ~ **copper oxide** Kupferoxydammoniak *n* ~ **cuprous chloride** Ammoniakkupferchlorür *n* ~ **liquor** Ammoniakwasser *n*
**ammonium** Ammon *m*, Ammonium *n* ~ **acetate** essigsaures Ammoniak *n* ~ **arsenate** Ammoniumarseniat, arsensau(e)res Ammonium *n* ~

**benzoate** benzoesaures Ammonium *n* ~ **bicarbonate** doppeltkohlensaures Ammoniak *n* ~ **borate** borsaures Ammonium *n* ~ **butyrate** buttersaures Ammon oder Ammoniak *n* ~ **chlorate** chlorsaures Ammonium *n* ~ **chloride** Ammonchlorid *n*, Ammoniumchlorid *n*, Chlorammonium *n*, Salmiak *m*, Salmiaksalz *n*, salzsaures Ammoniak *n* ~ **chloride index** Spinnreife *f* ~ **chloroplatinate** Plastinsalmiak *m* ~ **chlorostannate** Zinnsalmiak *m* ~ **chromate** chromsaures Ammon *n* ~ **citrate** zitronensaures Ammon *n* ~ **cyanide** Zyanammonium *n* ~ **ferric alum** Eisenammonalaun *m* ~ **formate** ameisensaures Ammon, *n* Ammoniumformiat *n* ~ **hippurate** hippursaures Ammonium *n* ~ **hydrate** Ammoniumhydroxyd *n* ~ **hydrosulfide** Ammonhydrosulfid *n*, Ammoniumsulfhydrat *n* ~ **hydroxide** Ammoniak *n*, Ammoniumhydroxyd *n* ~ **imidosulfonate** Ammoniumimidosulfonat *n* ~ **iodide** Jodammon *n* ~ **lactate** milchsaures Ammoniak ~ **molybdate** molybdänsaures Ammon *n* ~ **nitrate** Ammoniumsalpeter *m*, salpetersaures Ammon(ium) *n* ~ **nitrate explosive** Ammonsalpeter *n* ~ **nitride** Stickstoffammonium *n* ~ **nitrite** salpetrigsaures Ammonium *n* ~ **oxalate** oxalsaures Ammon *n* ~ **perchlorate** überchlorsaures oder übersaures Ammon *n* ~ **persulfate** überschwefelsau-(e)res Ammon oder Ammoniak *n* ~ **phosphate** phosphorsaures Ammon *n* ~ **picrate** Ammoniumpikrat *n* ~ **selenate** selensaures Ammonium *n* ~ **selenide** Selenammonium *n* ~ **selenite** selenigsaures Ammonium *n* ~ **stannic chloride** Pinksalz *n*, Zinnammoniumchlorid *n*, Zinnchlorammonium *n* ~ **sulfate** Ammoniumsulfat *n*, schwefelsaures Ammon *n* oder Ammoniak *n* ~ **sulfide** Schwefelammon *n* ~ **sulfocyanate** Ammoniumrhodanid *n*, Ammoniumsulfozyanid *n*, Schwefelzyanammonium *n* ~ **tannate** gerbsaures Ammonium *n* ~ **tartrate** weinsau(e)res Ammon *n* oder Ammoniak *n* ~ **thiocyanate** Ammoniumrhodanid *n*, Ammoniumwolframat *n*, wolframsaures Ammonium *n* ~ **valerate** valeriansau(e)res Ammonium *n* ~ **vanadate** vanadinsaures Ammon *n*
**ammunition** Munition *f*
**amorphous** amorph, amorphisch, bildlos, formlos, gestaltlos, nichtkristallisch, strukturlos, unkristallinisch, unkristallisiert ~ **carbon** amorpher Kohlenstoff *m* ~ **wax** amorphes Paraffin *n*
**amortisseur winding** Dämpferwicklung *f*
**amortization** Abschreibung *f*, Amortisation *f*, Amortisierung *f*, Tilgung *f*
**amortize, to** ~ amortisieren, tilgen
**amount, to** ~ **to** betragen, sich belaufen auf
**amount** Ausmaß *n*, Bestand *m*, Betrag *m*, Ergebnis *n*, Menge *f*, Posten *m*, Zahl *f*
**amount,** ~ **of ascending current** Aufwindwert *m* ~ **of bank** Querneigungswert *m* ~ **of charge** Gebührenbetrag *m*, Höhe *f* der Gebühr ~ **of chips and swarf** Spannmenge *f* ~ **of compensation** Abfindungssumme *f* ~ **of contraction** Schrumpfmaß *n* ~ **of copper used** Kupferaufwand *m* ~ **of crown** Balligkeit *f* ~ **of deflection** Biegungspfeil *m*, Größe *f* der Ablenkung ~ **of deviation** Ablenkungsgröße *f*, Deviationswert *m* ~ **of energy** Energiebetrag *m* ~ **of flow** Gas-

stromstärke $f$ ~ of heat Wärme-betrag $m$, -menge $f$ ~ of iron Eisenmenge $f$ ~ of lead Vorhaltemaß $n$, Vorhaltswert $m$ ~ of loss Verlusthöhe $f$ ~ of metal in ores Metallgehalt $m$ ~ of modulation Modelungsgrad $m$ ~ of motion Bewegungsgröße $f$ ~ of ozone in the air Ozongehalt $m$ ~ of precipitation Niederschlagswert $m$ ~ of pressure applied Preßdruck $m$ ~ of radiation Strahlungsmenge $f$ ~ of rain Regenmenge $f$ ~ of resistance Widerstandshöhe $f$ ~ of the settling Senkungsmaß $n$ ~ of shift Größe $f$ der Ablenkung ~ of shrinkage Schrumpfmaß $n$ ~ of smoke nuclei in the air Rauchgehalt $m$ ~ of swell Auffederung $f$ ~ of thrust gain Rückgewinnungsgrad $m$ ~ of undersize Untermaß $n$ ~ of upward current Aufwindwert $m$ ~ of water held in rock Durchfeuchtung $f$

amount, ~ due der fällige Betrag $m$ ~ field Betragfeld $n$ ~ weighed out Einwägung $f$

ampangabeite Ampangabit $m$

ampelite (a black coloring, pigment) Ampelit $m$

amperage Ampèrestärke $f$, Ampèrezahl $f$, Stromstärke $f$, Stromwert $m$ ~ range Strommeßbereich $m$ ~ rating (of cable) Nennstrom $m$

ampere Ampère $n$ ~ balance Stromwaage $f$ ~ conductors Ampèredrähte $pl$

ampere-hour Ampèrestunde $f$ ~ capacity Ampèrezahl $f$ ~ efficiency Ampèrestundenwirkungsgrad $m$ ~ (or quantity) meter Ampèrestundenzähler $m$

ampere, ~ turn Ampèrewindung $f$ ~ turns Ampèrewindungszahl $f$ ~ wires Ampèredrähte $pl$

Ampère's rule Ampèresche Schwimmerregel $f$

amperite tube Spannungsreglerröhre $f$

amphibian Amphibie $f$, Amphibium $n$, Wasser- und Landflugzeug $n$ ~ flying boat Amphibiumflugboot $n$ ~ plane Amphibienflugzeug $n$, Wasserlandflugzeug $n$ ~ tank Amphibien-kampfwagen $m$, -panzerwagen $m$, -tank $m$, Schwimmerpanzerkampfwagen $m$, Wasserlandpanzerwagen $m$

amphibious amphibisch ~ truck Schwimmlastkraftwagen $m$

amphibole Amphibol $m$, Hornblende $f$, Pargasit $m$

amphibolite (hornblende schist) Amphibolit $n$

amphidromic, ~ tide Drehtide $f$ ~ point Amphidromiepunkt $m$

amphoterer Halbleiter $f$

amphoteric, to ~ amphotheren

amphoteric ion Zwitterion $n$

ample auskömmlich, genügend, reichlich, umfangreich of ~ size dimensioniert

amplidyne Amplidyne $f$, Querfeldverstärkermaschine $f$

amplification Entdämpfung $f$, Vergrößerung $f$, Verstärkerung $f$, Verstärkung $f$, (effect) Verstärkungswirkung $f$ ~ of gauge Spurerweiterung $f$ ~ in nepers Verstärkungsmaß $n$ ~ (or gain) of the photocurrent Verstärkung $f$ des Photostromes

amplification, ~ coefficient Verstärkungs-zahl $f$, -ziffer $f$ ~ constant Schutzwirkung $f$, Verstärkungs-faktor $m$, -konstante $f$, -zahl $f$ ~ curve Verstärkungskurve $f$ ~ curve of a transformer Transformatorkurve $f$ ~ effect (of the electron tube) Verstärkerwirkung $f$ ~ factor Schutzwirkung $f$, Verstärkungs-faktor $m$, -konstante $f$,

-zahl $f$, (through grip) Durchgriffsfaktor $m$ ~ range Verstärkungsbereich $m$ ~ stage Verstärkungsstufe $f$

amplified verstärkt

amplifier Lautverstärker $m$, Sendeverstärker $m$, Verstärker $m$, ~ for frame sawtooth time base Bildfrequenzverstärker $m$ ~ in stage, above input stage Nachverstärker $m$

amplifier, ~ bay Verstärker-bucht $f$, -bunker $m$ ~ bulb Verstärkerlampe $f$ ~ cascade Verstärkerkette $f$ ~ channel Verstärkernetz $n$ ~ exposure meter Belichtungsmesser $m$ ~ frame Verstärkergestell $n$ ~ installation Verstärkeranlage $f$ ~ knob Lauthörknopf $m$ ~ mounting Verstärkergestell $n$ ~ noise Verstärkergeräusch $n$ ~ panel Verstärkerbrett $n$ ~ permanently in circuit fest eingeschalteter Verstärker $m$ ~ plug Verstärkerstöpsel $m$ ~ plug-in Verstärkereinschub $m$ ~ rack Verstärker-bucht $f$, -bunker $m$, -brett $n$ ~ stage Verstärkungsstufe $f$ ~ station Verstärkeramt $n$ ~ transformer Verstärkertransformator $m$ ~ unit Verstärkereinschub $m$ ~ valve Verstärker-lampe $f$, -rohr $n$, -röhre $f$

amplify, to ~ (waves) aufschaukeln, verstärken

amplifying, ~ without distortion verzerrungsfreie Verstärkung $f$ ~ detector Audion-, Richt-verstärker $m$, Verstärkergleichrichter $m$ ~ gauge Dickenmesser $m$ ~ means Verstärker $m$ ~ relay station Verstärkeramt $n$ ~ stage Verstärkerkraftwagen $m$ (rdo) ~ tube Röhrenverstärker $m$, Verstärker-rohr $n$, -röhre $f$ ~-tube socket Verstärkerröhrenfassung $f$ ~ voltmeter Röhren-spannungsmesser $m$, -voltmeter $n$

amplitude Amplitude $f$, Ausmaß $n$, (as applied to oscillations) Ausschlag $m$, Betrag $m$, Ganghöhe $f$, Größe $f$, Scheitelwert $m$, Schwingweite $f$, Schwingungsausgleich $m$, Tidenhub $m$, Weite $f$, Wellenhöhe $f$

amplitude, with ~ fidelity amplitudengetreu ~ of aberration Größe $f$ der Abweichung ~ of accomodation Akkomodationsbreite $f$ ~ of beats Schwebungsamplitude $f$ ~ of deflection Ablenkungsamplitude $f$, größte Ablenkung $f$ ~ of flyback Sprunghöhe $f$ (TV) ~ of forced (or nonharmonic) vibrations Amplitude $f$ der erzwungenen Schwingungen ~ of oscillation Schwingungs-amplitude $f$, -ausschlag $m$, -bogen $m$, -höhe $f$, -weite $f$ ~ of pendulum swing Pendelausschlag $m$ ~ of return Sprunghöhe $f$ (TV) ~ of a symmetrical quantity Amplitude $f$ einer symmetrischen Wechselgröße ~ of the vibration Schwingungsausschlag $m$ ~ of vibration Schwingungsweite $f$ ~ of wave Wellenweite $f$

amplitude, ~ calibration unit Amplitudeneichvorrichtung $f$ ~ calibrator Amplitudeneichstufenwähler $m$ ~ characteristic Frequenz-gang $m$, -kennlinie $f$, -kurve $f$ ~ curve Amplitudenkurve $f$ ~ dependence Amplitudenabhängigkeit $f$ ~-discrimination selection Amplitudenselektion $f$ ~ discriminator Amplitudensieb $n$ ~ distortion Klirrverzerrung $f$, nichtlineare oder ungradlinige Verzerrung, (expressed in decibels) Spannungsabhängigkeit $f$ ~-distortion factor Amplitudenverzerrungsfaktor $m$ ~ envelope Amplitudenhüllkurve $f$ ~ factor Größen-, Scheitel-faktor $m$ ~ filter Amplitudenselektion $f$, Amplitudensieb $n$ ~ increase of a vibration Aufschaukeln $n$ der Schwingung $n$ ~ limiter

Abkapper *m* ~ **lopper** Amplitudenbegrenzer
*m* ~ **modulation** Amplituden-modulation *f*
(rdo), -modulierung *f* ~ **modulator** Amplituden-
modler *m*, -modlung *f* ~ **proportional to frequency** Frequenzamplitude *f* ~ **resistance**
Dämpfungsverzerrung *f* einer Leitung *f* ~ **response** Amplitudentreue *f* ~ **selector** Ampli-
tuden-selektion *f*, -sieb *n* ~ **separator** Ampli-
tuden-selection *f*, -sieb *n*, Synchronisier-ab-
trennung *f*, -aussiebung *f* ~**-to-time conversion**
Amplituden-zu-Zeit-Umsetzung *f*
**amply dimensioned** groß dimensioniert
**ampule** Ampulle *f*, Glaskirsche *f*
**amputation** Abnehmen *n*, Abschneiden *n*
**a m u** (atomic mass unit) atomare Maßeinheit *f*
**amygdaline** mandelartig
**amygdaloid** Mandelstein *m* ~ **(al)** mandelförmig
**amyl** Amyl *n* ~ **acetate** Essigsäureamylester *n* ~
**alcohol** Amylalkohol *m*, Amyloxydhydrat *n* ~
**chloride** Chloramyl *n* ~ **cyclohexane** Amyl-
zyclohexan *n* ~ **formiate** ameisensaures Amyl
*n* ~ **hydrate** Amyloxydhydrat *n* ~ **hydrosulfide**
Amylsulfhydrat *n* ~ **iodide** Amyljodid *n*
**amylene** Amylen *n*
**amylopectin** Amylopektin *n*
**anabatic** anabatisch, konvektiv ~ **wind** Aufwind
*m*
**anachromatic** anachromatisch
**anaerobic** anaerob
**anaglyph,** ~ **method** Anaglyphenverfahren *n* ~
**viewers** Anaglyphenbrillen *pl*
**anal fin** Kielflosse *f* (Flugzeug)
**analcite** Analcim *n*
**anallactic point** anallaktischer Punkt *m*
**analogous** ähnlich, analog, sinngemäß ~ **transistor** analoger Transistor *m*
**analogue,** ~ **computer** Analog-Computer oder
-Rechner *m*, Analogierechenmaschine *f* ~
**technique** Analogtechnik *f*
**analogy** Entsprechung *f*
**analysis** Analyse *f*, Analysis *f*, Auseinandersetzung *f*, Auswertung *f*, Bestimmung *f*, Betrachtung *f*, Messung *f*, Nachweis *m*, Zergli-
derung *f*, Zerlegung *f*, Zersetzungskunst *f*, Zusammensetzung *f*, Zusammenstellung *f*
**analysis,** ~ **by absorption** Absorptionsanalyse *f*
~ **by boiling (or by fractional) distillation** Siede-
analyse *f* ~ **and grade** Analyse *f* und Körnung *f*
~ **of mine gas** Grubengasanalyse *f* ~ **of path**
Bahnanalyse *f* ~ **of profit** Gewinnzerlegung *f* ~
**of the records obtained** Auswertung *f* der erhal-
tenen Versuchsergebnisse ~ **by residue** Rück-
standsanalyse *f* ~ **test** Befundaufnahme *f* (z. B.
eines zerlegten Motors)
**analyst** Analytiker *m*, Ausführender *m* einer
Analyse, Laborant *m*
**analytic (al)** analytisch
**analytical,** ~ **balance** chemische Waage *f*, Präzisionswaage *f* ~ **(chemical) balance** Analysen-
waage *f* ~ **geometry** höhere Geometrie *f* ~**-grade**
analysenrein ~ **solution** analytische Lösung *f*
~ **study of a balance sheet** Bilanzanalyse *f* ~
**test** Analysenprobe *f*
**analytically** analytisch, rechnerisch
**analyzable** analysierbar, scheidbar
**analyzation** Auflösung *f*
**analyze, to** ~ abbauen, analysieren, auflösen,
auswerten, bestimmen, scheiden, untersuchen,

zergliedern, zerlegen, zersetzen, (e. g., measurement results) auswerten
**analyzer** Abtasteinrichtung *f*, Analysator *m*,
Bildabtastvorrichtung *f*, (impage point) Bildpunktabtaster *m*, Bildzerlegungsvorrichtung *f*,
Meßzelle *f*, Prüfgerät *n*, Zergliederer *m*, Zerleger *m*
~ **tube adapter** Röhrenmeßsockel *m*
**analyzing** Abtasten *n*, Scheiden *n*, Scheidung *f*,
(of the picture) Bildabtastung *f* ~ **of picture**
Bildzerlegung *f*
**analyzing,** ~ **output counter with printed output**
Analysierzähldruckgerät *n* ~ **nicol** Analysator-
nikol *n*
**anamorphic optics** Entzerrungsoptik *f*
**anamorphosing lens** Zerrlinse *f*
**anamorphosis** Wandlungsbild *n*
**anamorphote lens** Anamorphotobjekt *n*
**anamorphotic,** ~ **ancillary lens system** anamor-
photisches Vorsatzlinsensystem *n* ~ **expansion**
anamorphotische Dehnung *f*
**anaphase** Anaphase *f*
**anastigmat** Anastigmat *n*
**anastigmatic** anastigmatisch ~ **folding magnifier**
anastigmatische Einschlaglupe *f*
**anastatic printing** anastatischer Druck *m*
**anatase** Anatas *m*
**anchor, to** ~ abspannen (einen Mast), ankern,
festlegen, verankern
**anchor** Ankerklotz *m*, (of penstocks) Fixpunkt
*m*, Querriegel *m*, Schließe *f*, Verankerung *f*
(eines Mastes) ~ **is at short stay** der Anker ist
Kurz Stag ~ **fouled by the stock** stockunklarer
Anker *m*
**anchor,** ~ **bearing** Ankerpeilung *f* ~ **bolt** Anker-
bolzen *m*, Fundamentanker *m*, Fundamentschraube *f*, Grundbolzen *m* der Verankerung,
(screw) Ankerschraube *f* ~ **cable** Ankertau *n*,
Verankerungskabel *n* ~ **capstan** Ankerlicht-
maschine *f* ~ **and capstan motor** (submarines)
Ankerspillmotor *m* ~ **chain** Teukette *f*, Verankerungskabel *n* ~**-chain stopper** Ketten-
stopper *m* ~ **clamp** Abspannklemme *f* ~ **compartment** Ankerraum *m* ~ **crown** Ankerkreuz *n*
~ **escapement** Anker-gang *m*, -gesperre *n*, -hem-
mung *f*, -unruhe *f*, ~ **fluke** Ankerflügel *m*,
Bohrlöffel *m* ~ **gear** Ankergeschirr *n* ~ **guard
plate** Schürze *f* ~ **hitch** Ankerstich *m* ~ **hook**
Ankerhaken *m*, Klammerhaken *m* ~ **ice**
Grundeis *n*, Schlammeis *n* ~ **light** Anker-laterne
*f*, -licht *n* ~ **line** Ankertau *n* ~ **pile** Strebe *f*, Zugpfahl *m* ~ **plate** Ankerplatte *f* ~ **press** Fang-
presse *f* ~ **rack stick** Ankerrödel *n* ~**-rope reel**
Ankerseilrolle *f* ~ **screw** Sicherungsschraube *f*
~ **shackle** Ankerschäkel *m* ~ **shank** Ankerrute
*f* ~ **sheeting** Verankerungswand *f* ~ **stake** Ankerpfahl *m* ~ **stock** Ankerbalken *m* ~ **stop**
Sicherungsanschlag *m* ~**-testing machine** Ankerprüfmaschine *f* ~**-testing machine with
slidding-weight balance** Ankerprüfmaschine *f*
mit Laufgewichtswaage ~ **tie** Verankerung *f*,
Verankerungsbalken *m* ~ **tie beam** Zugbalken
*m* ~ **wall** Ankerwall *m* ~ **watch** Ankerwache *f*
**anchorage** Anker-grund *m*, -platz *m*, -stelle *f*,
-stellung *f*, Befestigung *f*, Einsatzhafen *m*, Festpunkt *m* ~ **of the backstay of a suspension
bridge** Verankerung *f* eines Hängebrücken-
kabels ~ **basin** Liegestelle *f*

**anchored,** ~ **cathode spot** fixierter Kathodenfleck *m* ~ **mine** Ankermine *f* ~ **start** Fesselstart *m*

**anchoring** Abspannung *f* (eines Mastes), (operation) Ankern *n*, Verankerung *f* ~ **of the track** Gleisverankerung *f*

**anchoring,** ~ **bolt** Ankerbolzen *m* ~ **cable** Haltekabel *n* ~ **clips** Ankerlaschen *pl* ~ **depth** Einbindetiefe *f* ~ **formation** Ankerformation *f* ~ **picket** Ankerpfahl *m* ~**-point capacitor** Stützpunktkondensator *m* ~ **rod** Zuganker *m* ~ **rope** Abspannseil *n* ~ **wire** Abspanndraht *m*

**anchusin** Alkannarot *n*

**ancillary,** ~ **apparatus** Zusatzgerät *n* ~ **equipment** Zusatzausrüstung *f* ~ **heating surface** Nachschaltheizfläche *f* ~ **jack** Aushilfsklinke *f* ~ **labors** Nebenleistungen *pl* ~ **lens** Vorsatzlinse *f* (photo) ~**-lens system** Vorsatzsystem *n*, Zusatzeinrichtung *f*

**andalusite** Andalusit *m*, Hartspat *m*

**andersonite** Andersonit *m*

**andesite** Andesit *m*

**and modul** Und-Modul *n*

**andradite** Kalkeisensilikat *n*

**anechoic room** schalltoter Raum *m*

**anelastic** anelastisch

**anelasticity** Anelastizität *f*

**anelectrotonus** Anelektrotonus *m*

**anellate, to** ~ anellieren

**anemobiagraph** Anemobiagraf *m*

**anemoclinograph** Anemoklinograf *m*

**anemogram** Anemogramm *n*, Windgeschwindigkeitsaufzeichnung *f*, Windmeßkurve *f*, Windregistrierung *f*

**anemograph** Registrieranemometer *m*, Windradanemograf *m*, Windregistrierapparat *m*, Windschreiber *m*

**anemometer** Anemomesser *m*, Anemometer *n*, Luftströmungsmesser *m*, Windgeschwindigkeitsmesser *m*, Windmeßgerät *n*, Windstärkemesser *m* ~ **of windmill type** Flügelradwindmesser *m*

**anemometrically variable pitch** Windmühlenflügelverstellung *f*

**anemometry** Windmessung *f*

**anemoscope** Windfahne *f*, Windzeiger *m*

**anemostat** Luftregler *m* (Kabinenheizung im Flugzeug), Luftaustritt *m*

**aneroid** Dosenaneroid *n* ~ **altimeter** Aneroidluftdruckmesser *m*, Dosenhöhenmesser *m* ~ **barometer** Aneroidbarometer *n*, Dosenbarometer *n*, Federbarometer *n*, Federluftdruckmesser *m*, Kapselbarometer *n*, Kapselluftdruckmesser *m*; luftleere Barometerdose *f*, Wetterglas *n* ~**-liquid statoscope** Flüssigkeitsstatoskop *n*

**angle** Ecke *f*, Knie *n*, Knierohr *n*, Krümmung *f*, Winkel *m* **to** ~ **off** ecken ~ **of aberration** Aberrationswinkel *m* ~ **of action** Eingriffswinkel *m*, Wirkungswinkel *m* ~ **of action of gear** Eingriffslänge *f* ~ **of adjustment** Einstellwinkel *m*, Verstellwinkel *m* ~ **of admission (or inlet) of the steam** (turbine) Einlaßventilwinkel *m* ~ **of advance** Voreilwinkel *m* ~ **of approach** Anlaufwinkel *m*, Flugwinkel *m*, Überhang *m* ~ **of attack** Anblasewinkel *m* (aerodyn.), Anstellwinkel *m* (aerodyn.), Luftstoßwinkel *m* ~ **of attack recorder** Anstellwinkelschreiber *m* ~ **of avertence** Verschwenkungswinkel *m* ~ **in azi-**muth Azimutwinkel *m* ~ **of bank** Querneigungswinkel *m*, Winkel *m* in der Kurve, Winkel *m* der Schräglage, Krägungswinkel (aviat.) *m* ~ **of beam** Leitstrahlwinkel *m* ~ **of bending** (of test piece) Durchbiegungswinkel *m* ~ **of bomb release** Bombenabwurfwinkel *m* ~ **of brush load** Bürstenwinkel *m* ~ **at the center** Zentriwinkel *m* ~ **of two great circles** Kreiszweieck *n* ~ **in a circular segment** Peripheriewinkel *m* ~ **of clearance** Deckungswinkel *m* ~ **at climb** Steigungswinkel *m* ~ **of climb** Anstiegwinkel *m*, Steigwinkel *m* ~ **of collimation** Richtungswinkel *m* ~ **of cone of dispersion** Kegelwinkel *m* ~ **of connecting rod** Schubstangenausschlag *m* ~ **of contact** Eintritts-, Greif-, Rand-winkel *m* ~ **of convergence** Konvergenzwinkel *m* ~ **of crab** Kreuzungswinkel *m* ~ **of crank rotation** Kurbeldrehwinkel *m* ~ **between cranks** Kurbelversetzung *f* ~ **of crossing** Herzstückneigung *f*, Kreuzungsverhältnis *n* ~ **of cylinder setting** Zylinderwinkel *m* ~ **of declination** Deklinationswinkel *m*, Mißweisungswinkel *m* ~ **of declination of a star** Breite *f* eines Gestirns ~ **of deflection** Ablenk(ungs)winkel *m*, Richtungswinkel *m* ~ **of delation** Erweiterungswinkel *m*, Übertragungswinkel *m* ~ **of departure** Abgangsbahn-, Abstrahl-, Abschuß-, Erhebungs-, Schuß-winkel *m* (g/m) ~ **of depression** negativer Geländewinkel *m*, Tiefenwinkel *m* ~ **of descent** Abstiegwinkel *m*, Gleitwinkel *m* ~ **of deviation** Ablenkungswinkel *m*, Deviationswinkel *m*, Ausschlagwinkel *m* (des Schiffes) ~ **of dip** Inklinationswinkel *m* ~ **of direction** Richtungswinkel *m* ~ **of distribution** (of pressure) Verteilungswinkel *m* ~ **of dive** Flugneigungswinkel *m* ~ **of divergence** Divergenzwinkel *m* ~ **of down-wash** Abströmungswinkel *m* ~ **of drift** Versetzungswinkel *m*, Windwinkel *m* ~ **of eccentricity** Exzentrizitätswinkel *m* ~ **of elevation** Anschlag- *m*, Aufsatz-, Elevations-, Erhebungs-, Erhöhungs-, Höhen-, Richtungs-winkel *m*, Rohrerhöhung *f* ~ **of elevation below horizontal** Tiefenwinkel *m* ~ **of elevation dial** Aufsatztrommel *f* ~ **of elevation gear (or mechanism)** Aufsatzwinkeltrieb *m* ~ **of emergence** Ausfallwinkel *m*, Austrittswinkel *m* ~ **of emission** Ausstrahlungswinkel *m* ~ **of fall** Fallwinkel *m* ~ **for fastening** Befestigungswinkel *m* ~ **of field of vision** Gesichtswinkel *m* ~ **of firing** Schußwinkel *m* ~ **for first indication** Auffaßwinkel *m* (rdr) ~ **of flight** Flugwinkel *m* ~ **of fluctuation** Schwankungswinkel *m* ~ **of flute** Rillenwinkel *m* ~ **of friction** Ruhelagewinkel *m*, Ruhewinkel *m*, Schüttwinkel *m* ~ **of gear** Schiebewinkel *m* ~ **of glide** Gleitwinkel *m* ~ **of gradient** Erhebungswinkel *m*, Neigungswinkel *m* ~ **of grid retardation** Zündverzögerungswinkel *m* ~ **of grip** Umschlingungswinkel *m* ~ **of groove** Rillenwinkel *m* ~ **of helix** Schraubenlinienwinkel *m* ~ **of horizontal swing** Schwenkungswinkel *m*, Verschwenkungswinkel *m* ~ **of image** Bildwinkel *m* ~ **of impact** Aufschlagwinkel *m*, Auftreffwinkel *m*, Stoßwinkel *m* ~ **of impulse** Anfangswinkel *m* ~ **of incidence** Ansatzwinkel *m*, Anstellwinkel *m* (aerodyn.), Auftreffwinkel *m*, Einfallsrichtung *f* (accoust), Einfallswinkel *m*, Einstellwinkel *m*,

Glanzwinkel *m* ~ **of inclination** Fall-, Inklinations-, Kippungs-, Steigungs-winkel *m* ~ **of incline** Neigungswinkel *m* ~ **of (internal) friction** Reibungswinkel *m* ~ **of internal friction** Reibungs-zahl *f*, -ziffer *f* ~ **of intersection** Schnittwinkel *m*, Übergangswinkel *m* ~ **of joint** Stoßwinkel *m* ~ **of jump** Abgangsfehlerwinkel *m* (artil.) ~ **of lag** Nacheilungswinkel *m*, Verzögerungswinkel *m* ~ **of lead** Luv-, Steigungs-, Voreil-, Voreilungs-, Vorhalte-winkel *m* ~ **of lock** Einschlagwinkel *m* der Lenkung ~ **of mesh** Eingriffswinkel *m* ~ **of minimum elevation** Deckungswinkel *m* ~ **of nip** Greifwinkel *m* ~ **of opening** Kegelwinkel *m* ~ **of oscillation** Ausschubwinkel *m*, Schwingungswinkel *m* ~ **of parallax** Sehstreifen *m* ~ **at the periphery** Peripheriewinkel *m* ~ **of phase difference** Phasenverschiebungswinkel *m* ~ **of pintle traverse** Seitenrichtfeld *n* ~ **of pitch** Anstellwinkel *m*, Fallwinkel *m*, Längsneigungswinkel *m* (aviat.), (of an eccentric or cam) Aufkeilwinkel *m* ~ **of polarization** Polarisationswinkel *m* ~ **of position** Bahnwinkel *m*, Stellungswinkel *m* ~ **of principal incidence** Haupteinfallwinkel *m* ~ **formed by two radii** Radien-Winkel *m* ~ **of rake** Neigungswinkel *m* ~ **of reception** Erfassungswinkel *m* ~ **of reference for integration of acceleration** Geschwindigkeitsmeßwinkel *m* ~ **of reflection** Ausfall-, Aussprung-, Prall-, Reflexions-, Zurückwerfungs-winkel *m* ~ **of refraction** Brechungs-, Refraktions-winkel *m* ~ **of repose** Gleit-, Reibungs-, Ruhe-, Schütt-winkel *m*, ~ **of repose** of (the natural) **slope** Böschungswinkel *m* ~ **of ricochet** Abprallwinkel *m* ~ **of ricocheting** Prallwinkel *m* ~ **of rifling** Drallwinkel *m* ~ **of roll** Rollwinkel *m* (aviat.), Winkel *m* in der Kurve, Winkel *m* der Schräglage ~ **of roll of a ship** Schlingerwinkel *m* ~ **of rotation** Drehungswinkel *m*, Drehwinkel *m*, Umdrehungswinkel *m* ~ **of rudder** Zapfenwinkel *m* ~ **of screw thread** Flankenwinkel *m* des Schraubengewindes ~ **in a segment** Umfangswinkel *m* ~ **of shear** Scherungswinkel *m* ~ **of shearing deformation** Gleitung *f* ~ **of sides** lip Abrutsch-, Schiebe-winkel *m* ~ **of sight** Gesichts-, Positions-, Sicht-, Treffhöhen-winkel *m* ~ **of sight instrument** Deckungswinkelmesser *m* ~ **of sighting** Richtungswinkel *m*, Seitenrichtung *f* ~ **of a sine wave** Phasenwinkel *m* ~ **of site** Geländewinkel *m* ~ **of site instrument** Geländewinkelmesser *m* ~ **of site level** Geländewinkel-, Höhen-libelle *f* ~ **of site mechanism** Geländewinkelmeßvorrichtung *f* ~ **of skew** Schrägungswinkel *m* ~ **of slip** Winkelveränderliche *f* ~ **of slope** Böschungs-, Erhebungs-, Gelände-, Schrägungs-winkel *m* ~ **of squint** Schielwinkel *m* ~ **of stall** Höchstauftriebswinkel *m* (aviat.), Strömungsabreißwinkel *m* (aerodyn.) ~ **of strabism** Schielwinkel *m* ~ **of strike** Streichwinkel *m* ~ **of sweep** Pfeilwinkel *m* (aviat.) ~ **of sweepback** Pfeilwinkel *m* (aviat.) ~ **of swing** Kantungswinkel *m*, Schwingungswinkel *m* ~ **of tail setting** Schwanzflächenwinkel *m* ~ **of talus** Abdachungswinkel *m* ~ **of tangent** Elevationswinkel *m* ~ **of taper** Konizität *f* des Kegels ~ **of tilt** Kippungs-, Nadir-, Neigungs-winkel *m* ~ **of torque** Drehungs-, Torsions-,

Laufzeitwinkel *m* ~ **of traverse** Bestreichungs-, Verdrehungs-winkel *m* ~ **of transition time** Schwenkungs-winkel *m* ~ **of traverse correction** Querrichtwinkel *m* ~ **of twist** Drall-, Drehungs-, Torsions-, Verdrehungs-winkel *m* ~ **of upturn** Aufdrehwinkel *m* ~ **of view** Bildfeld *n*, Bild-, Blick-, Gesichts-winkel *m* ~ **of vision** Sehkegel *m* ~ **of wall friction** Wandreibungswinkel *m* ~ **of the wind** Luftwinkel *m* ~ **of wing setting** Flügeleinstellwinkel *m* ~ **of writing** Schriftlage *f* ~ **of yaw** Gierungs-, Seiten-, Wende-, Abtreib-winkel *m* (aviat.) ~ **between zero-chord and zero-lift incidence** Flügelschnittwinkel *m*

**angle,** ~**-axis rollers** geschränkte Röllchen *pl* ~ **balance** Winkelabgleich *m* ~ **bar** Eckschiene *f* ~**-belt drive** Winkelriementrieb *m* ~ **bisector** Winkelhalbierende *f* ~ **board** Schrägbrett *n* ~ **brace** Winkelband *n* ~ **bracket** Befestigungswinkel *m*, Wandarm *m*, Winkelarm *m*, Winkelkonsóle *f*, Winkelstütze *f* ~ **butt strap** Winkelverlaschung *f* ~ **check valve** Winkelventil *n* ~ **cock** Eckhahn *m*, Winkelhahn *m* ~ **crane** Winkelkran *m* ~**-cut papers** Diagonalpapiere *pl* ~ **cutter** Diagonalschneidemaschine *f*, Winkelfräser *m* ~**-cutting machine** Schrägschneidemaschine *f* **180-degree** ~ gestreckter Winkel *m* ~ **dependence** Winkelabhängigkeit *f* ~ **drilling** Winkelbohren *n* ~ **drive** Winkelantrieb *m*, Winkeltrieb *m* ~ **elbow** Winkelstück *n* ~ **extrusion head** Schrägspritzkopf *m* ~ **fade-in** Winkelaufblendung *f* ~ **fitting** Winkelbeschlag *m* ~ **flange** Winkelflantsch *m* ~**-forming machine** Eckenbiegemaschine *f* ~ **groove** Winkelkaliber *n* ~ **guide** Winkel *m* ~ **hinge** Winkelband *n* ~**-hinge sinking-in attachment** Winkelbandeinlaßapparat *m* ~ **iconal** Winkeleikonal *n* (opt.) ~**-indicating attachment** Winkelanzeigevorrichtung *f* ~ **iron** Eckband *n*, Eckeisen *n*, Winkeleisen *n* ~ **iron bending machine** Winkeleisenbiegemaschine *f* ~**-iron guide** Winkeleisenführung *f* ~**-iron hoop** Winkeleisenring *m* ~ **joint** Winkelgelenk *n*, Winkelstoß *m*, ~ **lever** Winkelhebel *m* ~ **lugs** Winkelgreifer *m*, Winkelsporen *pl* ~ **measurement** Winkelmaß *n* ~ **(measuring) grid** Winkelgitter *n* ~ **measuring inset circle** Winkelmeßteilkreis *m* ~ **measuring inset protractor** Winkelmeßeinsatz *m* ~**-measuring instrument** Winkelmaß *n*, Winkelmeßgerät *n* ~ **minute** Bogenminute *f* ~ **molding press** Winkelpresse *f* ~ **papers** Schrägpapiere *pl* (paper mfg.) ~ **pillow block** schräges Stehlager *n* ~ **pipe** Bogenrohr *n* ~ **planing** Schräghobeln *n* ~ **plate** Aufspannwinkel *m*, Winkelplatte *f*, Winkelstück *n* ~ **plumb** Winkellot *n* ~ **pole** Winkelstange *f* ~ **poles** Winkelgestänge *n* ~ **port** Eckpforte *f* ~ **prism** Winkelprisma *n* ~ **radii** Zentriwinkel *m* ~ **range** Ausschlagweg *m*, Winkelweg *m* ~ **ring** Winkelring *m* ~ **section** Winkelprofil *n* ~ **shape** Winkelprofil *n* ~**-shaped form** abgewinkelte Form *f* ~ **sheet iron** Winkelblech *n* ~ **side** Schenkel *m* eines Winkels ~ **space** Winkelraum *m* ~ **steel** Winkeleisen *n* ~ **steel bar** Winkelstahlschiene *f* ~ **stiffener** Gurtungs-, Verstärkungs-winkel *m* ~ **straggling** Zufallsstreuung *f* ~**-straightening machine** Winkeleisenrichtmaschine *f* ~ **stress of frame on reciprocation saws** Angelspannung *f* der

Gattersägen ~ **table** Winkeltisch *m* ~ **thermometer** Winkelthermometer *n* ~ **tool** Schrägstahl *m* ~ **tower** Winkelmast *m* ~ **tracking** Kontinuverfolgung *f* (rdr) ~ **union** Rohrkugelstütze *f* ~ **valve** Eckventil *n*, Winkelsperrventil *n* ~ **vertex** Scheitel *m* eines Winkels ~ **wheel** Winkelrad *n*
**angles** Winkeleisen *n*
**anglesite** Anglesit *m*, Bleiglas *n*, Bleisulfat *n*, Bleivitriol *m*, Bleivitriolspat *m*, Vitriol-bleierz *n*, -bleispat *m*
**Ångström unit** Ångströmeinheit *f*
**angular** eckig, scharfkantig, winkelförmig, winkelig, auf einen Winkel *m* bezüglich ~ **acceleration** Winkelbeschleunigung *f* ~ **acceleration of course** Kurswinkelbeschleunigung *f* ~ **adjustment** Schwenkbarkeit *f* ~ **adjustment of tilting table** Schrägstellbarkeit *f* des Werkzeugtisches ~ **advance** Winkelvoreilung *f* ~ **aperture** Öffnungswinkel *m* ~ **aperture of objective** Objektivöffnungswinkel *m* ~ **ballistic variable** Winkelgröße *f* ~ **bearing** Traglager *n* ~ **butt strap** Winkellasche *f* ~ **cam displacement** Nockenversetzung *f* ~ **-contact ball bearing** Ringschrägkugellager *n* ~ **-contact single row ball bearing** Ringschräglager *n* (selbsterhaltend, zweireihig) ~ **cut** Winkelschnitt *m* ~ **cutter** Lücken-, Winkel-fräser ~ **deflection** Winkel-ablenkung *f*, -ausschlag *m* ~ **dependence of scattering** Streuungswinkelabhängigkeit *f* ~ **deviation loss** Winkelabweichungsverlust *m* (acoust.) ~ **difference** Winkelunterschied *m* ~ **difference between present and future position** Visierwinkel *m* ~ **displacement** Ausschlag *m*, Winkelverlagerung *f*, Winkelverschiebung *f* ~ **displacement of control surface** Ruder-, Steuer--ausschlag *m* ~ **distance** Winkelabstand *m* ~ **distribution** Winkelverteilung *f* ~ **drilling head** Winkelbohrkopf *m* ~ **drive** Winkeltrieb *m* ~ **field of the lens** Bildwinkel *m* des Objektivs ~ **freedom** Schwenkbereich *m* ~ **frequency** Eck-, Kreis-, Winkel-frequenz *f*, Winkelgeschwindigkeit *f* ~ **function** Kreis-, Winkel-funktion *f* ~ **groove** Spitzbogenkaliber *n* ~ **guide** Winkelführung *f*, Winkelführung *f* ~ **height** Höhenwinkel *m*, Meßhöhenwinkel *m* ~ **height of target** Zielhöhenwinkel *m* ~ **height of target when gun is fired** Abschußhöhenwinkel *m* ~ **impulse** Flächenträgheitsmoment *m* ~ **kinetic energy** Rotationsenergie *f* ~ **meshing** Schrägeingriff *m* ~ **milling cutter** Lücken-, Winkelfräser *m* ~ **momentum** Drall *m*, Drehimpuls *m*, Winkeldrehmoment *m* ~ **momentum matrix elements** Drehimpulsmatrizenelemente *pl* ~ **motion** Drehung *f*, Winkelbewegung *f* ~ **movement** Lagenausschlag *m*, Nebenausschlag *m*, Winkelauswanderung *f*, Winkelbewegung *f* ~ **movement of control surface** Steuerausschlag *m* ~ **offset of sparks** Funkenversetzung *f* ~ **papers** Schnittpapiere *pl* ~ **point** Scheitelhöhe *f* ~ **position** Stellungswinkel *m*, Winkellage *f*, Winkelstellung *f* ~ **position signal** Lagewinkelkommando *n* ~ **program velocity** Umlenkgeschwindigkeit *f* ~ **rate** Auswanderungszeit *f* ~ **ratio** Winkelverhältnis *n* ~ **reamer** Winkelreibahle *f* ~ **resolving power** Schärfenwinkel *m*, Sehschärfegrenzwinkel *m* ~ **rules** (of square, triangular, or any other polygonal section)

Kantmaßstäbe *pl* ~ **strain** Winkelspannung *f* ~ **suppression cap (or plug)** Winkelentstörstecker *m* ~ **thread** scharfes oder scharfgängiges Gewinde *n*, Spitzengewinde *n* ~ **time** Auswanderungszeit *f* ~ **travel** Winkelauswanderung *f* ~ **turning** Winkeldrehung *f* ~ **velocity** Eckfrequenz *f*, Kreisfrequenz *f*, Wechselgeschwindigkeit *f*, Winkelbeschleunigung *f*, Winkelgeschwindigkeit *f* ~ **velocity of course** Kurswinkelgeschwindigkeit *f* ~ **velocity of the slipstream** Strahldrehgeschwindigkeit *f* ~ **velocity of spin** Trudeldrehgeschwindigkeit *f* ~ **velocity meter** Winkelgeschwindigkeitsgerät *m* ~ **wheel** Winkelrad *n*
**angularity** Eckigkeit *f*, Winkellage *f*, Winkelstellung *f*, Winkligkeit *f* ~ **tolerance** Neigungsfehlertoleranz *f*
**anharmonic,** ~ **constants** anharmonische Terme *pl* ~ **ratio** Doppelverhältnis *n* ~ **terms** anharmonische Terme *pl*
**anharmonicity** Anharmonizität *f*
**anhedral** positive V-Stellung *f* (aerodyn.)
**anholomonic** anholomon
**anhydrite** Karstenit *m*
**anhydrous** anhydrisch, entwässert, kristallwasserfrei, nichtwässerig, wasserfrei ~ **sodium carbonate** kalzinierte Soda *n*
**aniline** Anilin *n* ~ **black mordant** Anilinschwarzbeize *f* ~ **diiodide** Dijodanilin *n* ~ **dye** Anilin--farbe *f*, -farbstoff *m* ~ **formaldehyde** Anilinformaldehyd *n* ~ **hydrochloride** Anilinchlorhydrat *n*, Anilinhydrochlorid *n*, salzsaures Anilin ~ **oil** Anilin *n* ~ **salt** Anilinhydrochlorid *n*
**animal,** ~ **char** Knochenkohle *f* ~ **charcoal** Beinschwarz *n*, Tierkohle *f* ~ **gelatin** animalische oder tierische Leimung *f* ~ **glue** Tierleim *m* ~ **grease** tierisches Fett *n* ~ **oil** Tieröl *n* ~ **power** Göpelantrieb *m* ~ **size** Tierleim *m* ~ **sizing** tierische Leimung *f*
**animate, to** ~ **the fermentation** die Gärung beleben
**animated cartoon** Zeichentrickfilm *m*
**animating electrode** Ionisierungsgitter *n*
**animations** mechanische Trickvorrichtungen *pl* (film)
**anion** Sauerstoffion *n* ~ **vacancy** Anionfehlstelle *f*
**anionic** anionisch
**aniseikon** Fehlerscheibchensucher *m*
**anisic aldehyde** Anisaldehyd *n*
**anisole** Anisol *n*
**aniso-elasticity** Anisoelastizität *f*
**anisotropic** anisotrop ~ **consolidation** anisotrope Verdichtung *f* ~ **(shear) distortion** anisotrope Verzeichnung *f* ~ **solids** anisotrope Festkörper *pl*
**anisotropy** Anisotropie *f*
**ankerite** Ankerit *m*, Braunspat *m*
**annabergite** Nickelblüte *f*
**anneal, to** ~ abkühlen, anlassen, ausglühen, ausheizen, einbrennen, enthärten, entspannen, erhitzen, frischen, glühen, kühlen, nachlassen, normalisieren, tempern
**anneal** Vergütung *f*
**annealed** ausgeglüht, geglüht, weich geglüht ~ **sheet iron** ausgeglühtes Eisenblech *n* ~ **steel** vergüteter Stahl *m* ~ **(drawn) steel** angelassener Stahl *m* ~ **(grain) structure** Vergütungsge-

füge *n* ~ **wire** ausgeglühter Draht, Glühdraht *m*

**annealer** Ausglüher *m*

**annealing** Abatmen *n*, Anlassen *n*, Entspannung *f*, Frischen *n*, (operation) Glühbehandlung *f*, Glühen *n*, Glühung *f*, Nachglühen *n*, Tempern *n*, Temperung *f*, Vergütung *f*, Vorglühen *n* (full) ~ Ausglühen *n*

**annealing**, ~ **action** Glühfrischwirkung *f*, Temperwirkung *f* ~ **bath** Glühbad *n* ~ **box** Glühfrischkasten *m* ~ **chamber** Glühfrischkammer *f*, Glühraum *m* ~ **color** Anlauffarbe *f*, Anlaßfarbe *f* ~ **condition** Temperbedingung *f* ~ **curve** Anlaßkurve *f* ~ **cycle** Glühzeit *f* ~ **expansion** Glühausdehnung *f* ~ **furnace** Ausglüh-flammofen *m*, -ofen *m*, Glühofen *m*, Metalltemperofen *m*, Temperofen *m*, Vorglühofen *m* ~ **furnace for glass** Glastemperofen *m* ~ **heat** Glühfrischhitze *f* ~ **installation** Glühanlage *f* ~ **lacquer** Einbrennlack *m* ~ **method** Ausglüh-, Glüh-verfahren *n*, Glühweise *f* ~ **oven** Kühlofen *m*, Vorglühofen *m* ~ **plants** Glühanlagen *pl* ~ **point** Kühltemperatur *f* ~ **pot** Einsatz-, Glüh-topf *m*, Tempergefäß *n*, Tempertopf *m* ~ **practice** Glühbetrieb *m* ~ **process** Ausglühverfahren *n*, Glühfrischverfahren *n*, Glühprozeß *m*, Glühverfahren *n*, Temperverfahren *n* ~ **range** Temperungsbereich *m* ~ **temperature** Anlaßtemperatur *f*, Glühtemperatur *f*, Glühwärmegrad *m* ~ **time** Glühdauer *f*, Glühzeit *f*

**annerodite** Annerödit *m*

**annex, to** ~ anfügen, angliedern, anhängen, anschließen

**annex** Anbau *m*, Anhang *m*, Anlage *f*, Anschluß *m*, Beilage *f*, Nebenanlage *f*, Nebengebäude *n*

**annexation** Annexion *f*, Eingemeindung *f*, Einverleibung *f*

**annexed** angebaut, beifolgend ~ **building** Nebenbau *m*

**annihilation** Vernichtung *f* ~ **photon** Zerstrahlungsphoton *n* ~ **radiation** Vernichtungsstrahlung *f*, Zerstrahlung *f*

**annote, to** ~ mit Vermerk versehen

**annotation** Anmerkung *f* ~ **of litigation** Streitanmerkung *f*

**annotations** Wortangaben *pl*

**announce, to** ~ ankündigen, anmelden, ansagen, ausschreiben, bekanntgeben, bekanntmachen, kundgeben, melden, verkünden **to** ~ **a call** eine Verbindung *f* anbieten **announced** angekündigt, angemeldet, angezeigt

**announcement** Ankündigung *f*, Ansage *f*, Bekanntgabe *f*, Bekanntgeben *n*, Zeitungsanzeige *f* ~ **fee** Bekanntmachungsgebühr *f*

**announcer** Ansager *m*, Sprecher *m*

**annual** jährlich ~ **balance of account** Jahresabschluß *m* ~ **balance sheet** Jahresabschluß *m* ~ **costs** Jahresausgaben *pl* ~ **fee** Jahresgebühr *f* ~ **load curve** Belastungsgebirge *n* (electr.) ~ **motion motor** Jahresgangmotor *m* ~ **output** Jahreserzeugung *f* ~ **progress** Jahresgang *m* ~ **publication** Jahrgang *m* ~ **reimbursement** Jahresabrechnung *f* ~ **returns** jährliche statistische Aufstellung *f* ~ **ring** Jahresring *m* ~ **statement** Jahresabschluß *m* ~ **suscription rate** Jahresgebühr *f* ~ **tax** Jahresgebühr *f* ~ **variations** jährliche Schwankungen *pl* ~ **volume** Jahresmenge *f*

**annuity** Rente *f* ~ **bond** Rentenbrief *m* ~ **due** (patent) Taxe *f* fällig

**annul, to** ~ annullieren, aufheben, rückgängig machen, streichen **to** ~ **each other** einander aufheben

**annular** ringförmig ~ **ball bearing** Ringkugellager *n* ~ **barrel bearing** Ringtonnenlager *n* ~ **barrel vault** Ringgewölbe *n* ~ **bead** Ringwulst *f* ~ **beam** Ringstab *m* ~ **bedways** Ringführung *f* des Bettes ~ **bit** Bohrkrone *f*, Kronenbohrer *m*, Ringbohrer *m* ~ **boiler** Dampfmantel *m* ~ **borer** Kronenbohrer *m* ~ **burner** Rundbrenner *m* ~ **cavity** Kreisrinne *f* ~ **chamber** Ringkammer *f* ~ **clearance** Ringspalt *m* ~ **collar** Ringbund *m* ~ **concave mirror** ringförmiger Hohlspiegel *m* ~ **cone** Ringtrichterrichtungshörer *m* ~ **core** Ringkern *m* ~ **cutter** Ringbohrer *m* ~ **electrode** Ringelektrode *f* ~ **finned heat exchanger** ringförmiger Rippenheizkörper *m* ~ **float** Ringschwimmer *m* ~ **flow-area** Gebläseringfläche *f* ~ **fuel tank** Kraftstoffringbehälter *m* ~ **fused-quartz spiral** Schmelzraupe *f* ~ **gear wheel** Ringsteuerrad *n* ~ **gears** Hohlräder *pl* ~ **girder** Ringstab *m* ~ **groove** Ringnut *f* ~ **holder** Faltenhalter *m* ~ **jet discharge** Ringstrahler *m* (Schwebefahrzeug) ~ **kiln** Ringofen *m* ~ **lens** Ringlinse *f* ~ **live-steam main** Frischdampfringrohr *n* ~ **magnet** Ringmagnet *m* ~ **member** Ringstab *m* ~ **milling cutter** Ringfräser *m* ~ **opening** Eintrittskreisschlitz *m* ~ **recording head** Ringkopf *m* ~ **ring** Holzring *m*, Ringmutter *f* ~ **seat** Ringsitz *m* ~ **selfaligning ballbearing** Ringpendelkugellager *n* ~ **shape** Kreisring *f*, Ringform *f* ~ **sleeve** Ringmanschette *f* ~ **sliding contact** Ringschleifkontakt *m* ~ **slot** Kreisschlitz *m* ~ **spring** Ringfeder *f* ~-**spring shock absorber** Ringfederbein *n* ~ **steam ring** Ringleitung *f* ~ **stiffener** Aussteifungsring *m* ~ **tank bottom** Ringkesselboden *m* ~ **taperroller bearing** Ringkegellager *n*, Ringschräglager *n* ~ **tensile strength** Ringzugfestigkeit *f* ~ **thimble** Rundkausche *f* ~ **toothed wheel** Zahnrad *n* mit innerer Verzahnung ~-**type burner** Kranzbrenner *m* ~ **valve** Ringventil *n* ~ **vault** Ringgewölbe *n* ~ **water box** Ringsammler *m* ~ **wave** Ringwelle *f*

**annulating network** Zerrungskreis *m*

**annulled** gegenstandslos, nichtig

**annulment** Nichtigkeits-, Ungültigkeits-erklärung *f* ~ **of a patent** Erlöschen *n* eines Patentes, Patenterlöschung *f* ~ **department** Nichtigkeitsabteilung *f*

**annulus** Ringraum *m*, Ringspule *f*

**annunciator** Signalfallscheibe *f* ~ **board** Fallscheibenapparat *m*, (drop indicator) Fallscheibentafel *f*, Signallampentafel *f* ~ **disk** Fallscheibe *f* ~ **drop** Anrufklappe *f* ~ **light** Meldelampe *f* ~ **panel** Warnleuchttafel *f* ~ **wire** Klingeldraht *m*

**anodal** anodal ~ **light** Anodenlicht *n*

**anode** Anode *f*, positive Elektrode *f*, Sauerstoffpol *m*, Saugspannung *f*

**anode,** ~ **accumulator** Anodensammler *m* ~ **alternating-current conductance** Anodeninnenleitwert *m* ~ **alternating-current resistance** Anodeninnenwiderstand *m* ~ **aperture** Blendenloch *n* (CRT), Lochblende *f* ~ **battery** Anodenbatterie *f*, Netzanode *f* ~ **bend** Anoden-schwanz

*m*, -kennlinie *f* ~-bend detector Anodengleichrichter *m*, Richtverstärker *m* ~-bend rectification Anodengleichrichtung *f* ~ breakdown voltage Anodenzündspannung *f*, Durchschlag-
spannung *f* ~-cathode gap Anodenkathoden-
strecke *f* ~ characteristic Charakteristik *f* der
Röhre ~ circuit Anoden-kreis *m*, -stromkreis
*m* ~ clamp Anodenbügel *m* ~ collector Sammelanode *f* ~ compartment Anodenraum *m*
~ current Anodenstrom *m* ~ current density
anodische Stromdichte *f* ~ current grid voltage
Anodenstrom-Gitterspannungskurve *f* ~ curve
Anodenabsinkkurve *f* ~ density Anoden-dichte
*f*, -stromstärke *f* (spezifische Anodenstrom-
stärke) ~ detection Anodengleichrichtung *f* ~
detector Anodengleichrichter *m*, Richtverstärker *m* ~ diaphragm Anodenblende *f* ~ differential resistance Innenwiderstand *m* ~ feedback Anodenrückwirkung *f* ~ feed resistors
Anodenwiderstände *pl* ~ follower Kathodenbasisverstärker *m* ~ generator Anodenmaschine
*f* ~ grid-plate capacitance Anodengitterkapazi-
tanz *f* ~ hanger Anodenaufhänger *m*, Anodenbügel *m* ~ impedance Innenwiderstand *m* ~
input lead Anodenspannungszuführung *f* ~ input power zugeführte Anodenleistung *f* ~ keying
Anodentastung *f* ~ lug Anodenrohr *n* ~ modulation Anodentastung *f* ~ modulator Anoden-
tastgerät *n* (rdr) ~ mold Anodenform *f* ~ mud
Anodenschlamm *m* ~ partition Anodenblende
*f* ~ pin Anodenstift *m* ~ plate Anodenplatte *f*
~ potential Anodenspannung *f* ~ ray Kanalstrahl *m* ~ rays Anodenstrahlen *pl* ~ reactance
Anodenrückwirkung *f* ~ rectification Anodengleichrichtung *f* ~ remnants Anodenreste *pl*
~ resistance Anodenwiderstand *m*, (load)
Innenwiderstand *m* ~ scrap Anodenabfall *m*
~ screen Anodenschutznetz *n* ~-screening grid
Anodenschutzgitter *n*, Anodenschutznetz *n*
~ segment (of magnetron) Anodenschale *f*
~ sorting Anodenaussortierung *f* ~ spot Ano-
denfleck *m*, Fokus *m* ~ sputtering Anodenzer-
stäubung *f* ~ stopper Anodenschutzwiderstand
*m* ~ strap Anodenverbindungsstreifen *m* ~
supply Anodenspeisung *f* ~ supply voltage
Anodengleichspannung *f* ~ switching Anoden-
tastgerät *n* ~-to-filament circuit Anodenkreis
*m* ~ transformer Anodentrafo *m* ~ voltage
Anodenspannung *f*
anodic (or anodal) anodisch ~ area Stromaustrittszone *f* ~ coating Eloxalüberzug *m* ~
etching elektrolytisches Anätzen *n* ~ treatment
Eloxieren *n*
anodized eloxiert
anodizing Eloxierung *f* ~ operation Vered(e)-
lungsprozeß *m* ~ process Anodisierungsver-
fahren *n*
anoin Anoin *n*
anolyte Anodenflüssigkeit *f*, Anolyt *m*
anomalistic anomalistisch
anomalous anomal, normwidrig, unregelmäßig
~ cathode fall anomaler Kathodenfall *m* ~ propagation (of waves) anomale Wellenausbreitung *f*
anomaly Abweichung *f*, Anomalie *f*, Unregelmäßigkeit *f* ~ during the fermentation Gärungs-
erscheinung *f*
anorthite Anorthit *m*, Kalkfeldspat *m*

anoxic cell Anaerobe *f*
answer, to ~ antworten, (a call) beantworten,
erwidern, in die Leitung eintreten, sich in der
Leitung melden to ~ call at switchboard abfragen to ~ for Rechenschaft ablegen to ~ on
a circuit einen Ruf beantworten
answer Antwort *f*, Bescheid *m*, Erwiderung *f*,
Klagebeantwortung *f*, Lösung *f* ~ to telephone
call Anrufantwort *f* ~ back Namengeber *m*
~-back code Kennziffer *f* ~-back signal Rückmeldung *f* ~ lag Umschaltepause *f* ~ print Probekopie *f* (film)
answerable verantwortlich
answering Abfragen *n* ~ board Abfrageamt *n* ~
button (for trunk line) abgetastete Amtsleitung
*f*, Meldeleitung *f* ~ circuit Abfrageschaltung
*f* ~ equipment Abfrageapparat *m* ~ interval
Verzögerung *f* bei der Beantwortung ~ jack
Abfrageklinke *f*, Antwortklinke *f* ~ lever
Rückmeldehebel *m* ~ operator Abfragebeam-
tin *f* ~ (or outgoing) position Abfrageplatz *m*,
Teilnehmerplatz *m* ~ plug Abfrage-, Antwort-,
Melde-, Mithör-stöpsel *m* ~ signal Antworte-
zeichen *n*, Gegensignal *n* ~ a signal Rückmeldung *f* ~ station Gegenfunkstelle *f* ~ supervisory lamp Überwachungslampe *f* des rufenden
(verlangten) Teilnehmers ~ supervisory relay
Schlußzeichengebung *f* für den rufenden (verlangten) Teilnehmer ~ system Abfragesystem
*n* ~ wave Antwortwelle *f* ~ wire Abfrageschnur
*f*
antacid säurewidrig
antagonistic entgegengesetzt wirkend, gegenwirkend ~ ignition Abreißzünder *m* ~ spring
Abreißfeder *f* ~ spring housing Zuggehäuse *n*
antartic circumpolar current antarktischer zir-
kumpolarer Strom *m*
antecedent Vorderglied *n*, Vordersatz *m*, Vorgänger *m*, Vorgeschichte *f*; vorangehend, vorausgehend
antechamber Vorkammer *f*, Vorraum *m* ~ system of injection Vorkammereinspritzung *f*
antedate, to ~ vordatieren
antedated, vorausdatiert
antefilter Vorfilter *m*
anteisomeric acids Anteisosäuren *pl*
antenna (see also aerial) Antenne *f*, Meßdipol
*m* (rdo) ~ ammeter Antennenstromanzeiger
*m* ~ array Antennengebilde *n* ~ base cover
Antennenkappe *f* ~ binding post Antennenanschluß *m* ~ capacitance Antennenkapazität *f*
~ center-driven Antenne *f* mit Mittelanschluß
~ circuit Antennenkreis *m* ~ circuit-breaker
Überspannungsableiter *m* ~ coil Antennenspule *f* ~ condenser Antennen-, Verkürzungs-
-kondensator *m* ~ connection Antennenanschluß
*m* ~ cording Antennenlitze *f* ~ coupling Anten-
nenankoppelung *f* ~-coupling control Anten-
nenkoppelung *f* ~ cross talk Antennenüber-
sprechen *n* ~ current Antennenstrom *m* ~-cur-
rent indicator Antennenstromanzeiger *m* ~
dish Antennenreflektor *m* ~ duct Antennendurchführung *f* ~ effect Antenneneffekt *m* ~
equipment Antennengerät *n* ~ fairing Anten-
nenverkleidung *f* ~ fairlead Antennenschlacht
*m* ~ feed impedance Antenneneingangswider-
stand *m* ~ firing (of mines) Antennenzündung
*f* ~ ground plate Antennenerder *m* ~ helix

Antennenspule *f* ~ **house** Antennenabstimm-häuschen *n* ~ **impedance** Eingangswiderstand *m* ~ **inductance** Antenneninduktivität *f* ~-**inductance coil** Antennenverlängerungsspule *f* ~ **inductivity** Induktivität *f* der Antennenabstimmspule ~ **input** zugeführte Antennenleistung *f*, der Antenne *f* zugeführte Leistung *f* ~ **insulator** Antennenisolator *m* ~ **jack** Antennenanschlußbüchse *f*, Antennenbüchse *f* ~-**lead** Antennenzuleitung *f* ~ **lead(-in)** Antennen-ableitung *f*, -anführung *f*, -einführung *f* ~ **lead-in insulator** Antennendurchführung *f* ~ **load coil** Antennenverlängerungsspule *f* ~ **loading coil** Antennenverlängerungsspule *f* ~ **mast** Antennenmast *m*, Funkmast *m* ~-**matching unit** Antennenanpassungsgerät *n* (rdr) ~ **mount** Abtaster *m* ~ **outfit** Antenngerät *n* ~ **plug** Antennenstecker *m* ~ **power** Antennenleistung *f* ~-**proximity zone** Nahbereich *m* ~ **reel** Antennen-haspel *f*, -winde *f* ~ **repeat dial** Antennenanzeigeskala *f* ~ **resistance** Antennenwiderstand *m* ~ **resonance curve** Antennenresonanzkurve *f* ~ **rod** Antennenstab *m* ~ **series capacitor** Antennenverkürzungskondensator *m* ~ **spreader** Antennenstecker *m* ~ **support** Antennengerüst *m* ~ **suspension** Antennenaufhängung *f* ~ **switch** Antennenumschalter *m* ~ **switch with fuse** Antennenschalter *m* mit Sicherung ~ **system** Antennengebilde *n* ~ **terminal** Antennenklemme *f* ~ **tilt angle** Antennenneigungswinkel *m* ~ **trailing inductance coil** Antennenabstimmspule *f*, Antennendurchführung *f* ~ **tuning** Antennenabstimmung *f* ~ **tuning coil** Antennenabstimmspule *f* ~ **tuning condenser** Antennenabstimmkondensator *m* ~ **tuning house** Antennenabstimmhäuschen *n* ~ **tuning inductance** Antennenabstimmspule *f* ~ **tuning switch** Antennenwahlschalter *m* ~ **voltage** Antennenspannung *f* ~ **winch** Antennenhaspel *f* ~ **wire** Antennendraht *m* ~ **yard** Antennenquerstrebe *f*

**anterior** Vorderfront *f* ~ **curvature** Vorwölbung *f* ~ **end** Vorderrand *m*, Stirnrand *m* (geol.) ~ **stop** Vorderblende *f* ~ **surface of a lens** Deckglas *n* einer Linse

**anteroom** Vorraum *m*

**anthelion** Gegensonne *f*

**anthracene** Anthrazen *n* ~ **oil** Anthrazenöl *n*

**anthracite** Anthrazit *m* ~ **coal** Glanzkohle *f*, Kohlenblende *f* ~ **culm** Anthrazitgrus *m*, Anthrazitkulum *m* ~ **silt** Anthrazit-schlamm *m*, -schlammkohle *f*, Schlammanthrazit *m* ~ **slush** Anthrazit-schlamm *m*, -schlammkohle *f*, Schlammanthrazit *m*

**anthracitic** anthrazitisch ~ **coal** anthrazitische Kohle *f*

**anthracitization** Anthrazitbildung *f*

**anthraconite** Kohlenspat *m*, Stinkstein *m*

**anthranilic acid** Anthranilsäure *f*

**anthraquinone** Anthrachinon *n*

**anthygron armored wires and cables** Anthygronrohrdrähte *pl*

**anti-acid** säurewidrig

**anti-arcing screen** Rundfeuerschutz *m*, Schutzwand *f* gegen Überschläge

**antiballoon device** Ballon-abweiser *m*, -sperre *f*

**antibody** Gegengift *n*, Gegenstoff *m*, Immunkörper *m*

**antibonding orbital** lockernder Zustand *m*

**anti-breakage,** ~ **device** Kohlenbruchverhütung *f*, ~ **shovel** Senkschaufel *f*

**anticapacitance switch** kapazitätsarmer Schalter *m*

**anticapsize bag** Kenterschützbeutel *m*

**anticatalyst** Antikatalysator *m*, Paralysator *m*

**anticathode** (of X-ray tube) Antikathode *f*

**antichlorine** (sodium thiosulfate) Antichlor *n*

**anticipate, to** ~ erwarten, im voraus etwas tun, voraussehen, zuvorkommen **anticipated, as is** ~ voraussichtlich ~ **value** Erwartungswert *m*

**anticipation** Entgegenhaltung *f*, Erwartung *f*, (in prior art) Abweisungsmaterial *n*, (patents) Vorveröffentlichung *f* **in** ~ im Voraus ~ **mode** Vorwegnahme-Methode *f*

**anticipatory** vorwegnehmend, (patent law) vorausnehmend, (patents) neuheitsschädlich ~ **control** Vorsteuerung *f* ~ **disclosure** Entgegenhaltung *f* ~ **material** Abweisungsmaterial *n* ~ **references** entgegenstehendes Material *n*

**anticlinal** antiklinal ~ **axis** Faltenachse *f* ~ **bulges** Aufwölbung *f* ~ **flexure** Abbiegung *f*, Flexur *f*, Kniefalte *f*, Schichtensattel *m* (geol.) ~ **fold** Antiklinalfalte *f* ~ **formation** Sattelbildung *f* ~ **line** Faltenachse *f* ~ **ridge** Gebirgssattel *m* ~ **theory** Antiklinaltheorie *f*

**anticline** Antiklinale *f*, Gebirgssattel *m*

**anticlinorium** Antiklinorium *n*, Faltenbündel *n*

**anticlockwise** links, linksdrehend, entgegen dem Uhrzeigersinn, gegen den Uhrzeigersinn ~ **rotation** Linksdrehung *f*

**anti-coagulant stabilizer** Antikoagulationsmittel *n*

**anticoherer** Antikohärer *m*, Gegenfritter *m*

**anticoincidence circuit** Antikoinzidenzschaltung *f*

**anti-collision light** Warnblinker *m* (Flugzeug)

**anticorona,** ~ **collar** Sprühschutzwulst *f* ~ **ring** Sprühring *m*

**anticorrosion composition** Schutzanstrich *m*

**anticorrosive** rostfest, rosthindernd ~ **agent** Rostschutzmittel *n*

**anti-creep device** Leerlaufhemmung *f*

**anti-creepers** Wanderschutzklemmen *pl*

**anticrustator** Kesselsteinzerstörer *m*

**anticyclone** Antizyklon *m*, Gegenluftwirbel *f*, ~ **area** Hochdruckgebiet *n*

**anticyclonic,** ~ **(ally)** antizyklonal ~ **state** Antizyklonalstadium *n*

**anti-damping** Anfachung *f*

**antidazzle,** ~ **cap** Blendschutzklappe *f* ~ **device** Blendungsschutz *m* ~ **glasses** Blendungsschutzbrille *f* ~ **louver** Blendschutzklappe *f* ~ **screen** Blendungsschirm *m*

**anti-dazzling screen** Blendschutzscheibe *f*

**antidetonant** klopffest, Gegenklopfmittel *n*, Mittel *n* zur Erhöhung der Klopffestigkeit

**antidetonating fuel** klopffester Brennstoff *m*

**antidetonation compound** Gegenklopfmittel *n*, Klopfbremse *f*

**antidiffusing screen** Streustrahlenblende *f*

**antidiffusion screen** Strahlenfilter *n*

**antidim,** ~ **compound** Brillenglassalbe *f*, Klarsichtsalbe *f* ~ **eyepiece** (gas mask) Klarscheibe *f*

**antidisintegrant** abbauverhinderndes Mittel *n*

**antidistortion,** ~ **device** Entzerrer *m*, Entzerreranordnung *f*, Entzerrereinrichtung *f*, Entzerrungsanordnung *f* ~ **switch** Entzerrungsschalter *m*

antidive and rolling rudder Krängungsruder *n*
antidote Gegenstoff *m*
antidrag wire Gegenzugdraht *m* (aviat.), Holm-auskreuzung *f*
antidribble device Vorrichtung *f* zur Verhütung des Leckens
anti-drip nozzle nichttropfender Zapfhahn *m*
anti-echo circuit Absperrkreis *m*
antielectron Positron *n*
antifading, ~ antenna Schwundverminderungs-antenne *f* ~ device Schwundausgleich(er) *m*
antifatigue ermüdungsbeständig ~ means Spannungsentlastung *f*
antiferro-electricity Antiferroelektrizität *f*
antiflicker blade Beruhigungsflügel *m*, Zwischen-flügel *m*
antifluorite lattice Antiflußpatgitter *n*
antiflutter wire Hüllenverstärkungsdraht *m*
antifouling, ~ composition Bodenanstrich *m* ~ paint faulsichere Farbe *f*, Schutzanstrich *m* ~ painting Unterwasseranstrich *m* ~ preservative fäulnisverhütendes Mittel *n*
antifreeze Gefrierschutz *m*, (alcohol) Äthanöl *n* ~ agent Gefrierschutzmittel *m* ~ mixture Frost-schutz-mischung *f*, -mittel *n* ~ solution Frost-schutzmischung *f*, Gefrierschutzmittel *n*
anti-freezer Frostschutzmittel *n*
antifreezing kältebeständig ~ agent Frostschutz-mittel *n* ~ lubricant Frostschutzfett *n* ~ mix-ture kältebeständiges Gemisch *n* ~ preparation Frostschutzvorbereitung *f* ~ quality Kältebe-ständigkeit *f* ~ solution Frostschutzmittel *n*, Gefrierschutzlösung *f*
antifriction, ~ alloy Lagerlegierung *f* ~ bearing Antifriktionslager *n*, Wälzlager *n* ~ bearing grease Wälzlagerfett *n* ~ grease Heißachsen-schmiere *f* ~ metal Lagermetall *n*, Weißmetall *n* ~ properties Gleiteigenschaften *pl*
antifrost, ~ bite ointment Frostschutzsalbe *f* ~ screen Frostschutzscheibe *f*
antifroth oil Schaumöl *n*
antigas, ~ defense Gasabwehr *f* ~ paulin Gas-plane *f* ~-shelter curtain Gasschutzvorhang *m* ~ training Gasschutzübung *f*
antiglare, ~ (or dimming) cap Abblendhaube *f* ~ glasses Blendschutzgläser *pl* ~ goggles Blend-schutzbrille *f* ~ paint Blendschutzfarbe *f* ~ screen Blendschutzscheibe *f*, Blendungsschirm *m* ~ shield Blendschützer *m*, Blendschutzscheibe *f* ~ switch Abblendumschalter *m*
anti-growl resistance Entknurrungswiderstand *m*
antihalation, ~ backing lichthoffreie Schicht *f* ~ coating Lichthofschutzschicht *f* (photo) ~ dry plate solarisationsfreie Platte *f* (photo) ~ substance Lichthofschutzmittel *n*
antihalo, lichthoffrei ~ base Lichthofschutz *m* ~ means Lichthofschutzmittel *n*
anti-hum Dämpfer *m* ~ capacitor Entbrumm-kondensator *m* ~ device Entbrummer *m*
anti-icer Frostgraupeln *n*, Gefrierschützer *m* ~ device Vereisungsschutzgerät *n* ~ fluid Ver-eisungsschutzflüssigkeit *f* ~ solution Entei-sungsmittel *n*
anti-icing device Enteisungsvorrichtung *f*, Ver-eisungs-schutzanlage *f*, -schutzgerät *n*
anti, ~-induction cable induktionsfreies Kabel *n* ~-induction device Seiteninduktionsschutz *m*

~-induction network Nebensprechdämpfungs-netzwerk *n* ~-inductive arrangement, (or device) Induktionsschutz *m* ~-infectious infektionsver-hütend ~-interference condenser Entstörungs-kondensator *m* ~-interference device Entstör-gerät *n* ~-interference shrouding Funkschutz-ummantelung *f* ~-jam, to ~ entstören ~-jam-ming Störungsbeseitigung *f* (rdr)
antiknock (compound) Antiklopfmittel *n*; klopf-fest ~ agent Antiklopfmittel *n* ~ constituent Gegenklopfmittel *n* ~ fuel Antiklopfbrenn-stoff *m*, klopffester Betriebsstoff *m* oder Brenn-stoff *m* oder Kraftstoff *m* ~ fuel dope Anti-klopfzusatzmittel *n* ~ gasoline klopffestes Ben-zin *n* ~ value Klopffestigkeits-grad *m*, -wert *m*, Oktanwert *m*
antiknocking, ~ property, (quality or rating) Klopffestigkeit *f*
anti, ~-leak cement Dichtungsmittel *n* gegen Flüssigkeitsdurchtritt ~-lift wire Gegenhebe-draht *m* (aviat.), Gegenkabel *n* ~-logarithm Numerus *m* (Logarithmus) ~-microphonic tube federnde Röhrenfassung *f*, Röhrenfassung *f* ~-microphonic valve holder federnder Röhrenhal-ter *m*, Röhrenfassung *f* ~-mine defense Minen-abwehr *f*
antimonial antimonartig, antimonhaltig, anti-monisch, spießglanzartig ~ copper glance Ant-imonkupferglanz *m* ~ gray copper Antimon-fahlerz *n* ~ lead Antimonblei *n*, Hartblei *n* ~ silver Antimonsilber *n* ~ silver ore Weißsilber-erz *n*
antimonic antimonhaltig, antimonisch ~ acid Spießglanzsäure *f*
antimonide Antimonid *n*
antimonious antimonhaltig, antimonisch ~ acid antimonige Säure *f* ~ iodide Antimonjodür *n* ~ sulfide Antimonsulfür *n*
antimony Antimon *n*, Spießglanz *n* ~ ash Spieß-glanzasche *f* ~ bloom Antimonblüte *f* ~ chlo-ride Antimonchlorid *n*, Antimonchlorür *n*, Spießglanzbutter *f* ~ flux Sternschlacke *f* ~ glance Antimonglanz *m* ~ like antimonartig ~ metallurgy Antimonerzverhüttung *f* ~ ore Anti-monerz *n*, Pyrantimonit *m*, Spießglanzerz *n* ~-ore deposit Antimonerzvorkommen *n* ~ oxide Antimonoxyd *n*, Spießglanzoxyd *n*, Weisspieß-glanz *m*, -glanzerz *n* ~ oxychloride Antimonoxychlorür *n* ~ pentachloride Anti-monchlorid *n* ~ pentasulfide Antimonsulfür *n*, Goldschwefel *m* ~ plating Verantimonierung *f* ~ potassium tartrate Antimonkaliumtartrat *n*, Brechweinstein *m* ~ selenide Selenantimon *n* ~ silicate kieselsaures Antimonoxyd *n* ~ skim-mings Antimonabstich *m* ~ star Antimonstern *m* ~ sulfide Spießglanz *m* ~ trichloride Anti-monchlorür *n* ~ trioxide Antimonoxyd *n* ~ trisulfide Antimonsulfür *n*, Antimontrisulfid *n* ~ vermilion Kermes *m* ~ (white) trioxide Anti-monweiß *n*
antinode Bauch *m*, Bausch *m*, (loop) Bausch-schwingung *f*, Wellenbauch *m* ~ of oscillation Schwingungsbauch *m* ~ of stationary oscilla-tion Schwingungsschleife *f* ~ of stationary wave Schwingungsschleife *f* ~ of vibration Schwin-gungsbauch *m* ~ of a vibration Bauch *m* einer Schwingung
antinoise geräuschdämpfend

**anti-oxidant** Alterungsschutzmittel *n* (für Gummi), Alterungsstabilisator *m*, Antioxydationsmittel *n*
**anti-oxidation dope** Zusatzmittel *n* zur Verhütung der Oxydation
**anti, ~-parralax mirror** parallaxfreier Spiegel *m* **~-particle** Antiteilchen *n* **~-phase domains** gegenphasige Bereiche *pl* **~-pinking** (of fuel) klopffest
**anti-plug switch** nicht gesperrt
**antipodal** antipodisch
**antipode** Antipode *m*, Gegenteil *n*, Gegenfußpunkt *m*, Gegensatz *m*, optischer Antipode *m* (chem.)
**anti-pole** Gegenpol *m*
**anti-priming** Kesselschaumverhütung *f*
**anti-propeller end** von der Luftschraube abgewandte Seite *f* des Motors
**antipyretic** Antipyretikum *n*
**antipyrine** Antipyrin *n*
**antiquated, to become** ~ veralten
**anti-rattle** nichtklappernd ~ **roller** (for window) Fensterrolle *f*
**anti-, ~-reaction device** Rückkopplungssperre *f* **~-reflection coating** reflexmildernder Belag *m* (film) **~ -reflex coating** Antireflexbelag *m* (film)
**antiresonance** Antiresonanz *f*, Parallelresonanz *f*, Sperresonanz *f*, Sperrkreiswirkung *f* (rdo), Stromresonanz *f* ~ **band** Absperrbereich *m* ~ **circuit** Stromresonanzkreis *m* ~ **device (or apparatus)** Absperrwirkung *f* ~ **frequency** Eigenfrequenz *f* des Sperrkreises
**antiresonant** der Resonanz entgegenwirkend **to be** ~ der Resonanz entgegenwirken ~ **band** Sperrkreisbereich *m* ~ **circuit** Absperr-, Entkopplungs- *m*, Sperr-kreis *m* ~ **coil** Entzerrungsdrossel *f* ~ **frequency** Antiresonanzfrequenz *f*
**anti-roll device** Dämpfer *m* für die Rollbewegung
**anti-rolling tank** Schlingertank *m*
**antirot** fäulnis-hindernd, -widrig
**antirusting, ~ agent** Eisenschutzmittel *n* ~ **paint** nichtrostende Farbe *f*
**antiscale composition** Kesselsteinmittel *n*
**antiseptic** antiseptisches Mittel *n*; fäulnisbekämpfend
**anti-set-off, ~ device** Abschmutzvorrichtung *f* (print.) ~ **settling agent** Absetzverhinderungsmittel *n*
**antishimmy brace** Antiflatterbock *m* ~ **shackle** Antiflatterbock *m* ~ **suspension** Lenkstoßfang *m*
**anti-side-tone, ~ circuit** Dämpfungsschaltung *f* ~ **device** geräuschdämpfende oder rückhördämpfende Schaltung *f*, Schaltung *f* zur Verminderung der Nebengeräusche ~ **set** Sprechstelle *f* mit Rückhördämpfung ~ **telephone set** Fernsprecher *m* mit Schutzschaltung gegen Mikrophongeräusch
**antisinging, ~ device** Rückkopplungssperre *f* ~ **stability** Pfeifsicherheit *f*
**antiskid** rutschfest, rutschsicher ~ **chain** Gleitschutz *m* ~ **property** Griffigkeit *f* (Straße)
**antislip** gleitsicher ~ **cover** Trittschutz *m* ~ **finishing agent** Schiebefestmittel *n*
**anti-softener** (for rubber) Versteifer *m*
**anti-spark disc** Gegenfunk *m*

**antispraying insulator** Sprühschutzisolator *m*
**antistall, ~ gear** Sackflug-anzeiger *m*, -schutzgerät *n* ~ **indicator** Sackfluganzeiger *m*
**antistatic, ~ antenna** abgeschirmte Antenne *f*, antistatisches Tau *n* ~ **device** Entelektrikator *m*
**anti-Stokes lines** anti-Stokes'sche Linien *pl*
**antisubstance** Antikörper *m*
**antitarnish paper** Rostpapier *n*, rostschützendes Papier *n*, Silberpackpapier *n*
**antithesis** Gegensatz *m*, Gegenteil *n*
**antithetic** gegensätzlich
**anti-throwing agent** Haftmittel *n*
**antitoxin** Gegengift *n*, Schutzstoff *m*
**antitrades** Antipassat *m*, Gegenpassatwind *m*
**antitwilight** Gegendämmerung *f*
**antitype** Gegenbild *n*
**anti-typical** gegenbildlich
**anti-vibration, , ~ damper** Gummipuffer *m* ~ **foundation** Schwingungsfundament *n* ~ **means** Spannungsentlastung *f* ~ **mica** Dämpfungsglimmer *m* ~ **unit** Gummimetallelement *n*
**anti-wear agent in the oil** Verschleiß verhindernder Zusatz *m* im Öl
**antizodiacal light** Gegenschein *m*
**antizymotic** gärungshemmend
**antlers** Gestänge *n*
**anvil** Amboß *m*, Döpper *m*, Gegenzapfen *m*, Prallstock *m*, Reitel *m*, Stoßreitel *m*, Tastbolzen *m* ~ **with one arm** Galgenamboß *m* ~ **of a steam hammer** Schabotte *f* eines Dampfhammers
**anvil, ~ beak** Amboßhorn *n* ~ **bed** Amboßfutter *n* ~ **block** Amboßklotz *m*, Amboßstock *m*, Schabotte *f*, Schawatte *f*, Unteramboß *m* ~ **cloud** Amboßwolke *f* ~ **contact** Amboßkontakt *m* ~ **cutter** Blockmeißel *m* ~ **dross** Hammerschlacke *f* ~ **face** Amboßbahn *f*, Stoßfläche *f* ~ **horn** Amboßhorn *n* ~ **insertion piece** Amboßeinsatz *m* ~ **inset stake** Stöckel *m* ~ **pillar** Schießstock *m* ~ **stake** Amboßstöckel *m* ~ **stand** Amboßfutter *n* ~ **stock** Amboßfutter *n* ~'s **stock** Hammerstock *m* ~ **swage** Amboßgesenk *n* ~ **tool** Amboßgesenk *n*
**apatite** Apatit *m*, Phosphorit *m*
**aperiodic** aperiodisch, eigenschwingungsfrei, gegedämpft, grenzgedämpft, nichtabgestimmt, nichtperiodisch, nichtresonierend, unabgestimmt ~ **(ally)** unperiodisch
**aperiodic, ~ antenna** aperiodische Antenne *f* ~ **circuit** aperiodischer Stromkreis *m*, nichtschwingungsfähiger Kreis ~ **compass** aperiodischer oder gedämpfter Kompaß ~ **component of a short-circuit current** Gleichstromglied *n* des Kurzschlußstromes ~ **discharge** aperiodische Entladung *f* ~ **instrument** aperiodisch gedämpftes Instrument *n* ~ **measuring instrument** gedämpftes Meßinstrument *n* ~ **pendulum** aperiodisches Pendel *n* ~ **phenomenon** aperiodischer oder schwingungsfreier Vorgang *m* ~ **reading** gedämpfte Anzeige *f* ~ **resistance** Grenzwiderstand *m* ~ **vibration** Grenzschwingung *f* ~ **voltmeter** aperiodischer Spannungsmesser *m* ~ **wave** aperiodische Welle *f*
**aperiodicity** Aperiodizität *f*
**apertometer** Apertometer *n*, Aperturmeter *m*, Öffnungswinkelmesser *m*
**apertural, ~ defect** Öffnungsfehler *m* ~ **effect** Diffusionskreis *m*, Öffnungsfehler *m*, Schlitz-

effekt *m*, ~ **error** Öffnungsfehler *m* ~ **image** Öffnungsbild *n*

**aperture** (focal) Apertur *f*, Ausschnitt *m*, (in a plate) Belichtungsfenster *n*, Bildbühne *f*, Bildfenster (film), (stop) Blende *f*, Blendenausschnitt *m*, (of diaphragm or stop) Blendenöffnung *f*, Eintragsöffnung *f*, Filmprojektionsfenster *n*, Linsenöffnung *f*, Loch *n*, Lochweite *f*, Masche *f*, Mündung *f*, Öffnung *f*, Öffnungswinkel *m*, Schauloch *n*, Spalt *m* **to cut down** ~ ausblenden ~ **of the beam** Bündeldurchschnitt *m* ~ **of a belfry** Schalloch *n* ~ **of a bell** Schalloch *n* ~ **of a door** Türöffnung *f* ~ **of an optical panel** Spannwinkel *m* ~ **in scanning disk** Abtastöffnung *f* (TV)

**aperture,** ~ **aberration** Öffnungsfehler *m* ~ **angle** Öffnungswinkel *m* ~ **compensation** Öffnungsausgleich *m* ~ **corrector** Aperturblendenkorrektur *f*, Cosinus-Entzerrer *m* ~ **coupling** Blendenkupplung *f* ~ **diaphragm** Aperturblende *f* ~ **disk lens** Lochscheibenlinse *f* ~ **distortion** Bildpunktverzerrung *f* ~ **effect** Spalteffekt *m* ~ **guide** Fensterführung *f* (film), Filmbandführung *f* ~ **image** Öffnungsbildapertur *f* ~ **lens** Durchgrifflinse *f*, Fensterlinse *f* ~ **mask opening** Bildfensterausschnitt *m* ~ **plate** Gleitbahneinsatz *m*, Fensterplatte *f* (film), ~ **ratio** Öffnungsverhältnis *n*, relative Öffnung *f* ~ **setting wheel** Einstellrädchen *n* für Blende ~ **stop** Aperturblende *f*

**apertured** durchlöchert, löcherig ~ **anode** Diaphragmanode *f* ~ **disk** durchlochte Scheibe *f* ~ **disk anode** Lochscheibenanode *f* ~ **disk lens** Lochblendenlinse *f* ~ **electrode disk** (electron microscope) Lochelektrode *f* ~ **partition** Lochblende *f* ~ **stop** Diaphragmanode *f*, Spaltblende *f*

**apertureless** spaltfrei ~ **illumination** spaltfreie Beleuchtung *f*

**apex** Firstpunkt *m*, Scheitel *m*, Scheitelpunkt *m*, Spitze *f* ~ **of arch** Bogenscheitel *m* ~ **of the cone** Kegelspitze *f* ~ **of curve** Kurvengipfel *m* ~ **of lens** Linsenscheitel *m* ~ **of pin-core** Kegelspitze *f*

**apex,** ~ **angle** Positionswinkel *m* ~ **drive** Mittelpunktsspeisung *f* ~ **factor** Verstärkungsscheitelfaktor *m*

**aphelion** Sonnenferne *f*

**aphrite** Aphrit *m*, Schaumkalk *m*

**aphrizite** Aphrizit *m*

**aphthitalite** Glaserit *m*

**apice** Spitze *f* (eines Dreiecks)

**aplanat** Aplanat *n*

**aplanatic** aplanatisch, fehlerfrei

**aplanatism** Aplanasie *f*

**apochromatic** apochromatisch ~ **objective** Apochromat *m*

**apogee** Apogäum *m*

**apoglucic acid** Apogluzinsäure *f*

**apolar liquid** apolare Flüssigkeit *f*

**A pole** Spitzbock *m*

**apomecometer** Entfernungs-Höhenmesser *m*

**apophyllite** Fischaugenstein *m*

**A position** Abgangsplatz *m* (teleph), A-Platz *m*

**apostrophe** Auslassungszeichen *n*

**apothecaries' weight** Apotherkergewicht *n*

**apparatus** Anlage *f*, Apparat *m*, Apparatur *f*, Ausrüstung *f*, Einrichtung *f*, Gerät *n*, Gerät-

schaft *f*, Geschirr *n*, Instrument *n*, Meßgerät *n*, Vorrichtung *f*, Zubehör *n*

**apparatus,** ~ **for automatic measurement of cable lengths** Kabellängenmesser *m* ~ **for bending electrotypes** Galvano-Biegeapparat *m* ~ **for calibrating thermometers** Thermometereichapparat *m* ~ **for checking recorded items of information** Prüfgerät *n* für Nachrichtenaufzeichnung ~ **for the deaf** Hörgerät *n* ~ **for deforming elastically** Apparat *m* zum elastischen Deformieren ~ **for demonstration** Vorführungsapparat *m* ~ **for the determination of carbon in iron** Eisenkohlenstoffbestimmungsapparat *m* ~ **for direct drying by means of furnace gases** Feuertrockner *m* ~ **for distant transmission** Fernanzeigegerät *n* ~ **for drawing wires** Apparat *m* zum Anziehen von Drähten ~ **for free fall** Abfallapparat *m* ~ **for illuminating sight** Nachtvisierbeleuchtung *f* ~ **for increasing the symmetry** Symmetrierzusatz *m* (teleph.) ~ **for laying out sheets** Ablegeapparat *m* für Papierbogen ~ **for the levigation of emery** Schmirgelschlämmapparat *m* ~ **for long-range transmission** Fernübertragungsgerät *n* ~ **for measuring permeability** Diffusionsgerät *n* ~ **for rapid determination of ash** Aschenschnellbestimmer *m* ~ **for returning water of condensation** Kondenswasserrückspeiseanlage *f* ~ **for taking off gases** Gasentziehungsvorrichtung *f* ~ **for tearing off the shavings** Spänezerreißer *m* ~ **for testing elasticity of yarn** Garnelastizitätsmesser *m* ~ **for testing electric conductivity** Leitfähigkeitsprüfer *m* ~ **for testing the tearing strength of fabric** Zerreißgerät *n*

**apparatus,** ~ **compartment lower** Geräteschrank *m* ~ **connecting directly to switchboard** Amtszusatz *m* ~ **gamma** Apparategamma *n* (electron.), Apparaturgamma *n* ~ **glass** (stock) Geräteglas *n* ~ **lead** Apparatezuleitung *f* ~ **meter** Bestimmungsapparat *m* ~ **patent** Apparatepatent *n* ~ **plug** Gerätestecker *m*

**apparent** anscheinend, augenscheinlich, scheinbar **to be** ~ auftreten

**apparent,** ~ **attenuation** Scheindämpfung *f* ~ **component** Scheinkomponente *f*, Scheinwert *m* (electr.) ~ **density** Schuttgewicht *n* ~ **disc temperature** scheinbare Flächenhelligkeit *f* ~-**energy meter** Scheinleistungszähler *m* ~ **extract** scheinbarer Extraktgehalt *m* ~ **gravity** scheinbarer Extraktgehalt *m* ~ **heat of ignition by hot pellets** scheinbare Aktivierungswärme *f* bei Zündung durch heiße Kugeln ~ **horizon** natürlicher Horizont *m* ~ **impedance** Scheinwiderstand *m* ~ **output** Scheinleistung *f* ~ **power** Scheinleistung *f* ~-**power meter** Scheinleistungs-, Scheinverbrauchs-zähler *m* ~-**power transmission ratio** Scheinleistungsübersetzungsverhältnis *n* ~ **resistance** Impedanz *f*, scheinbarer oder wirksamer Widerstand *m*, Scheinwiderstand *m* ~-**resistance measurement** Scheinwiderstandsmessung *f* ~ **specific gravity** scheinbares spezifisches Gewicht *n* ~ **value** Scheinwert *m* (electr.)

**appeal, to** ~ Berufung einlegen (gegen), dringend bitten

**appeal** Appell *m*, Berufungsklage *f*, Beschwerde *f*, (law) Berufung *f* ~ **action** Beschwerdeverfahren *n*

**appealer** Berufungskläger *m*

**appealing** werbekräftig
**appeals** Rechtsmittel *pl*
**appear, to** ~ auftauchen, auftreten, aussehen, erscheinen, in Erscheinung treten, scheinen, sichtbar werden, vorkommen, zum Vorschein kommen **to** ~ **in an equation** in eine Gleichung eingehen
**appearance** Ansehen *n*, Anstrich *m*, Auftreten *n*, Augenschein *m*, Aussehen *n*, Erscheinung *f*, Gestaltung *f*, Gestellung *f*, Vorschein *m* ~ **of expansion** Treiberscheinung *f* ~ **of fracture** Bruchaussehen *n*, Bruchflächenaussehen *n* ~ **of the grain** Narbenbild *n* ~ **of sparks** Funkenbild *n*
**appearance potential** Entstehungs-, Erscheinungs-potential *n*, Erscheinungsspannung *f*, Ionisationsspannung *f*, Mindestvoltgeschwindigkeit *f*
**appelate, to have** ~ **jurisdiction** zuständig sein als Berufungsinstanz *f*
**appellant** beschwerdefähig; Berufungskläger *m*
**appellatory plaint** Beschwerdeschrift *f*
**appellor** Berufungskläger *m*
**append, to** ~ anhängen
**appendage** Anhängsel *n*
**appendix** Anhang *m*, Anlage *f*, Beilage *f*, Füllansatz *m*, Zusatz *m* ~ **folio** Anlageheft *n*
**apperceive, to** ~ heraushören
**appertain, to** ~ angehören
**applanation tonometer** Applanationstonometer *n*
**applause** Beifall *m* ~ **meter** Geräusch-spannungsmesser *m*, -spannungszeiger *m*
**apple,** ~ **coal** weiche Kohle *f* ~ **wood** Apfelbaumholz *n*
**appliance** Anlage *f*, Apparat *m*, Apparatur *f*, Ausrüstung *f*, Einrichtung *f*, Gerät *n*, Gerätschaft *f*, Mittel *n*, Vorrichtung *f* ~ **for hoisting** Hebewerkzeug *n* ~ **for lifting** Hebewerkzeug *n* ~ **for measuring insulation** Isolationsmeßgerät *n*
**appliance,** ~ **conversion** Umstellung *f* der Gasgeräte ~ **switch** Geräteanschlußkabelschalter *m*
**applicability** Anwendbarkeit *f*, Anwendungsmöglichkeit *f*, Eignung *f* Verwendungsfähigkeit *f*, Verwendungszweck *m*
**applicable** angebracht, anwendbar, verwendbar
**applicant** Anmelder *m*, Antragsteller *m*, Bewerber *m*, Gesuchsteller *m*, Nachsuchender *m*, Petent *m* ~ **for an examination** Prüfling *m*
**application** Anliegen *n*, (patents) Anmeldung *f*, Ansuchen *n*, Ansuchung *f*, Anwendung *f*, Aufbringen *n*, Auftragen *n* (Farbe), Ausführungsbeispiel *n*, Beanspruchung *f*, (for a position) Bewegung *f*, (patents) Eingabe *f*, Einsatz *m*, Gebrauch *m*, Gesuch *n*, Übertragung *f*, Umschlag *m*, Verwendung *f*
**application,** ~ **of action** Angriffspunkt *m* der Kraft ~ **of adhesive** Klebstoffauftrag *m* ~ **of complex quantities to alternating-current problems** komplexes Rechnen *n* mit Wechselstromgrößen ~ **of force** Angriffspunkt *m* der Kraft ~ **of a force** Aufbringen *n* einer Belastung ~ **of load** Belastung *f* ~ **of the models** Verwendung *f* der Modelle ~ **for patent for mining claims** Nutung *f* ~ **of power** Kraftangriff *m* ~ **of stress** Kraftangriff *m* ~ **of test signals** Zuführung *f* von Testsignalen ~ **for a situation** Stellengesuch *n*

**application,** ~ **blank** Antragsformular *n* ~ **form** Antragsformular *n* ~ **papers** Anmeldungsakten *pl* ~ **roll** Auftragwalze *f* ~ **rollers** Auftragerollen *pl*
**applicator** Bestrahlungstubus *m*, Elektrode *f*, Radiumkapsel *f*
**applied** angewandt, angewendet ~ **for** (patents) angemeldet ~ **at end of load** Endlast *f*
**applied,** ~ **art** Kunstgewerbe *n* of ~ **art** kunstgewerblich ~ **ignition** Fremdzündung *f* ~ **load on the surface** Auflast *f* ~ **magnetic field** anliegendes Magnetfeld *n* ~ **moment** Angriffsmoment *m* (mech. engin), Angriffspunkt *m* **shock** Stoßerregung *f* (acoust.), eingeprägter Stoß *m*
**apply, to** ~ anbringen, ansuchen, anwendbar sein, anwenden, applizieren, auflegen, auftragen, gebrauchen, nachsuchen, passen, verwenden, (around something) wenden **(by pressure)** umpressen, (paint) auftragen, **(to)** an die Leitung anlegen **to** ~ **bank** Querruder geben **to** ~ **a battery** eine Batterie *f* anlegen **to** ~ **brake** bremsen **to** ~ **by brush** aufpinseln **to** ~ **bushying potential** Besetztspannung anlegen **to** ~ **color** Farbe auftragen **to** ~ **emulsion** beschichten **to** ~ **a force** eine Kraft *f* aufbringen, eine Last *f* aufbringen **to** ~ **for letters patent** ein Patent anmelden **to** ~ **for a term of respite** Stundung verlangen **to** ~ **a tourniquet** abbinden **to** ~ **a voltage** eine Spannung *f* anlegen **to** ~ **a wash** schlichten **to** ~ **for** sich bewerben um, beantragen, einreichen **to** ~ **for a patent** ein Patent anmelden oder beantragen **to** ~ **to** angehen, ansetzen, gelten **to** ~ **for the grant of a concession** um Verleihung eines Feldes nachsuchen **to** ~ **a reflecting coating** verspiegeln **to** ~ **stresses** beanspruchen **to** ~ **the handles** das Geschirr garnieren
**applying** Aufbringen *n*, (a miner) Anlegen *n* eines Bergarbeiters ~ **gunite** Torkretverfahren *n*
**appoint, to** ~ anberaumen, anstellen, anweisen, ernennen
**appointment** Amt *n*, Anstellung *f*, Festsetzung *f*, Verabredung *f*, Vorladung *f* ~ **to a university lectureship** Habilitation *f* ~ **message** Festzeitgespräch *n*
**apportion, to** ~ bemessen, teilen, verteilen
**apportioning** Zuteilung *f* ~ **vessel** Dosiergefäß *n*
**apportionment** Kontingentierung *f*
**appraisable** berechenbar
**appraisal** Auswertung *f*, Schätzung *f*, Wertbestimmung *f*, Wertung *f* ~ **rating** Beurteilungsgrößen *pl*
**appraise, to** ~ abschätzen, taxieren, veranschlagen *m*
**appraiser** Taxator *m*
**appreciable** merkbar, merklich, namhaft, weitgehend **without** ~ **error** fast fehlerfrei
**appreciate, to** ~ abschätzen, anerkennen
**appreciation** Abschätzung *f*, Einschätzung *f*, Kursgewinn *m*, Wertschätzung *f* ~ **of difference in depth** Tiefenunterscheidung *f*
**appreciative** empfänglich
**apprehend, to** ~ empfinden, ergreifen, fassen, festnehmen
**apprentice** Lehrjunge *m*, Lehrling *m* ~ **hewer** Lehrhauer *m* ~ **seaman** Matrose *m* ~ **ship** Lehre *f*, Lehrzeit *f*
**apprise, to** ~ in Kenntnis setzen, notifizieren

**approach, to** ~ aneinander rücken, aufliegen (aviat.), angehen, sich annähern, (weather) aufziehen, im Anzug sein, heranmarschieren, herannahen, (bad weather) heraufkommen, kommen, sich nähern
**approach** (flight) Anflug *m*, Anmarschweg *m*, Annäherung *f*, (road) Annäherungsweg *m*, Anzug *m*, Bewegungsvorgang *m*, Inangriffnahme *f*, (e. g. to design) Lösung *f*, Näherung *f*, Vorgehen *n*, Zugang *m*, Zugangsweg *m* ~ **and run-off angle** Ein- und Auslaufwinkel *m*
**approach,** ~ **aids** Anflughilfen *pl* (aviat.) ~ **angle** Anflugwinkel *m* ~ **area** Anflugsektor *m* ~ **control** Anflugkontrolle *f* ~ **control office** Anflugkontrollstelle *f* ~ **corridor** Anflugschneise *f* ~ **disposition** Annäherungsaufstellung *f* ~ **embankment** Brückendamm *m* ~ **flow direction** (turbines) Anströmrichtung *f* ~ **hump** Zulaufberg *m* ~ **idling conditions** Anflugleerlaufzustand *m* (aviat.) ~ **infinity** unendlich werden ~ **ladder** Zugangsleiter *f* ~ **lane** Anflugschneise *f* ~ **light** Anflugfeuer *n*, Landefeuer *n* ~ **lighting** Anflugbefeuerung *f*, Grundlinienfeuer *n* ~ **locking** Anrücksperre *f* ~ **locking device** Annäherungsverschluß *m* ~ **path** Anflugbahn *f*, Anflugweg *m*, Schneise *f* ~ **procedure** Anflugmanöver *n* ~ **ramp** Zufahrtsrampe *f* ~ **road** Vormarschstraße *f*, Zufahrtsweg *m* ~ **roller conveyor** Auflaufrollgang *m* ~ **sector** Anflugschneise *f*, Einflugsektor *m* ~ **sequence** Anflugfelge *f* ~ **signal** Einflugzeichen *n* ~ **speed** Anfluggeschwindigkeit *f* ~ **surface** Anflugfläche *f*, Anflugebene *f* ~ **track** Anflugweg *m*, Zulaufstrecke *f* ~ **trench** Annäherungsgraben *m*, Laufgraben *m*, Zugangsgraben *m* ~ **way** Anflugweg *m* ~ **works** Annäherungswerke *pl* ~ **zone** Anflugsektor *m*
**approaching** annähernd ~ **car** anlaufender Wagen *m* ~ **course** Anflugskurs *m* ~ **passage** Zugangskanal *m* ~ **target** kommendes Ziel *n*
**approbation** Billigung *f*
**appropriate, to** ~ anweisen, sich aneignen, zuteilen, zuweisen
**appropriate** angemessen, einschlägig, entsprechend, fachmännisch, gehörig, passend, sachdienlich, sachgemäß, tauglich, zweckmäßig ~ **airworthiness requirement(s)** anzuwendende Lufttüchtigkeitsvorschrift(en) *pl*
**appropriated dividend** bewilligte (zugewiesene) Dividende *f*
**appropriation** Zuteilung *f*, Zuweisung *f*
**approval** Bewilligung *f*, Billigung *f*, Einwilligung *f*, Genehmigung *f*, Gutbefund *m*, (of an engine) amtliche Zulassung *f* eines Motors, Zustimmung *f*
**approve, to** ~ amtlich bestätigen, billigen, genehmigen, gutheißen
**approved** bewährt, gebilligt, genehmigt, gutbefunden, zugelassen ~**-type certificate** Musterprüfungsschein *m*
**approximate, to** ~ aufrunden (math.), sich nähern **to** ~ **a decimal** eine Dezimale *f* aufrunden
**approximate** angenähert, grob, ungefähr ~ **adjustement fire** grobes Einschießen *n* ~ **analysis** Rohanalyse *f* ~ **calculation** Näherungsrechnung *f* ~ **determination of a point's location** relative Ortsbestimmung *f* ~ **equation** Nähe-

rungsformel *f* ~ **estimate** ungefährer Anschlag *m* ~ **figure** Richtwert *m* ~ **formula** Näherungsformel *f* ~ **method** Näherungsverfahren *n* ~ **methods of analysis** Näherungsrechnung *f* ~ **quantity** Näherungsgröße *f* ~ **range of measurement** angenäherter Meßbereich *m* ~ **solution** Näherungslösung *f* ~ **value** Annäherungswert *m*, Näherungswert *m*
**approximately** annähernd, annäherungsweise, in groben Zügen
**approximation** Annäherung *f*, Annäherungswert *m*, Näherung *f*, Näherungsansatz *m*, Näherungslösung *f*, Näherungswert *m*, Vernachlässigung *f* **to a first** ~ in erster Annäherung *f*
**approximation,** ~ **equation** Näherungsgleichung *f* ~ **formula** Annäherungsformel *f*, Näherungsformel *f* ~ **method** Näherungsverfahren *n*
**appurtenances** Zubehör *n*
**apron** (rolling mill) Abdeckplatte *f*, Abstellfläche *f* (Flugplatz), Blechschutz *m* (auto), Fußsteig *m* (aviat.), Hallenvorplatz *m* (aviat.), Klappausleger *m*, Mutterplatte *f*, zementiertes Rollwerk *n*, Schloßplatte *f*, Schurz *m*, Schürze *f*, Seitenverankerung *f*, (paper mfg.) Siebleder *n*, Sohle *f*, Sohlenbefestigung *f*, (farm. mach.) endlose Transportkette *f*, Vorplatz *m* mit befestigtem Boden (aviat.), Vortuch *n*, neutrale Zone *f* (aviat.)
**apron,** ~ **of fertilizer spreader** automatischer Düngerzufuhrapparat *m* ~ **on a lathe** Räderplatte *f*
**apron,** ~ **area** Vorfeldbereich *m* (aviat.) ~ **block** Schutzblock *m* ~ **conveyor** Plattenförderbank *n*, Pfannentransporteur *m* (hinged) ~ **extension** Klappausleger *m* ~ **feeder** Bandaufgabe *f*, Plattenband *n*, ~ **hoist engine** Klappausleger-Einziehwerk *n* ~ **lathe** Räderplattendrehbank *f* ~ **margin** Vorfeldrandzone *f* ~ **nut** Mutterschloß *m* ~ **pan** Plattenbandsegment *n* ~ **shield** Unterschild *m*
**apse** Altarnische *f*
**apsides** Apsiden *pl*
**apt** angemessen, fähig, zweckdienlich
**aptitude** Eignung *f*, Geschick *n* ~ **for stereoscopic vision** Fähigkeit *f* stereoskopisch zu sehen ~ **test** Berufseignungsprüfung *f* ~**-testing laboratory** Personalprüfstelle *f*
**aqua,** ~ **fortis** Ätze *f*, Gelbbrennsäure *f*, Scheidewasser *n* ~ **regia** Goldscheide-, Königs-, Scheide-wasser *n* ~ **vitae** Lebenswasser *n*
**aquadag coating** Graphitbelag *m*
**aquarium reactor** Bassinreaktor *m*
**aquatic sport** Wassersport *m*
**aqueduct** Aquädukt *m*
**aqueous** wasserartig, wasserhaltig, wässerig ~ **ammonia** Salmiakgeist *m* ~ **corrosion** Feuchtigkeitskorrosion *f* ~ **homogenous reactor** homogener Wasserreaktor *m* ~ **solution** wässerige Lösung *f*
**araban** Araben *n*
**arabinic acid** Arabinsäure *f*
**arabinose** Arabinose *f*, Gummizucker *m*
**arable,** ~ **land** Ackerland *n*, Humuserde *f* ~ **soil** Ackerboden *m*, Ackerland *n*
**arabonic acid** Arabonsäure *f*
**arachis oil** Arachid *n* (Erdnußöl)
**arachnoid** spinnwebenartig
**aragonite** Aragonit *m*, Schalenkalk *m*, Schaumkalk *m*

**araldite** Araldit *n*
**arbiter** Schiedsrichter *m*
**arbitral commission** Schiedskommission *f*
**arabitration** schiedsrichterliches Verfahren *n*,
(judgment) Schiedsspruch *m*, Schlichtung *f* ~
**agreement** Schiedsvertrag *m* ~ **analysis** Schieds-
analyse *f*
**arbitrary** beliebig, eigenmächtig, eigenwillig,
nach freiem Ermessen, nach Gutdünken, willkürlich ~ **action** Willkür *f* ~ **correction** freie
Verbesserung *f* ~ **sequence computer** Sequenti-
ellrechner *m*, Sprunglauf *m*
**arbitrator** Schiedsrichter *m*, Schlichter *m* ~**'s
award** Schiedsgutachten *n*
**arbor** Achse *f*, Aufnahmedorn *m*, Aufsteckdorn
*m*, Aufsteckhalter *m*, Drehstift *m*, Hilfszapfen
*m* (beim opt. Teilkopf), Spindel *f*, Träger *m*,
Welle *f* ~ **for polishing balance wheels** Unruhe-
polierdrehstift *m*
**arbor,** ~ **press** Dornpresse *f* ~**-type lap** Läppdorn *m*
**arborescent** baumähnlich, dendritisch ~ **crystal**
Dendrit *m*
**arc** Bogen *m*, Bogenlampensender *m*, Bogenlinie *f*, Flammenbogen *m*, Kreisbogen *m*, Lichtbogen *m*
**arc, on the** ~ über dem Bogen *f* ~ **of amplitude
of the oscillation of a pendulum** Ausschlagbogen
*m* ~ **of belt contact** Umschlingungsbogen *m*
~ **of circle** Kreisbogen *m* ~ **with colored effect**
Effektlichtbogenlampe *f* ~ **of contact** Auflage-
bogen *m*, Eingriffbogen *m*, umspannter Bogen
*m* ~ **of fire** Bestreichungsbogen *m* ~ **of the
meridian** Meridianbogen *m* **to** ~ **over** überschlagen ~ **of rotation** Ausschlagbogen *m* ~ **of
time** Zeitbogen *m* ~ **of training** Bestreichungsbogen *m*
**arc,** ~ **adjuster** Schweißwiderstand *m* ~ **arrester**
Bogenableiter *m*, Bogenblitzableiter *m* ~ **back**
Bogenrückschlag *m*, Lichtbogenüberschlag *m*,
Rückzündung *f* ~ **baffle** Ablenkplatte *f*, Licht-
bogenleitfläche *f* im Extiron ~ **blow** Blaswirkung *f* ~ **bracket** Haltebügel *m* (Haltestift mit
Bügel) ~ **converter** Lichtbogengenerator *m*,
Lichtbogensender *m* ~ **core** Elektronenstraße
*f*, Lichtbogenkern *m* ~ **crater** Schweißblase *f*
~ (lamp) **current** Bogenstrom *m* ~ **cutting
machine** Lichtbogenschneidemaschine *f* ~ **discharge** Bogenentladung *f*, Funken- oder Licht-
und Bogenentladung *f*, Lichtbogenentladung *f*
~**-drop loss** Lichtbogenverlust *m* ~**-drop voltage**
Brennspannung *f*, Lichtbogenabfall *m* ~ **error**
Sehnenfehler *m* ~ **extinction (or extinguishing)**
Lichtbogenlöschung *f* ~ **flame** Bogenflamme *f*,
Lichtbogenaureole **(electric-)** ~ **furnace** Lichtbogenofen *m* ~ **generator** Lichtbogengenerator
*m*, Lichtbogenschwingungserzeuger *m* ~ **heating** Lichtbogenbeheizung *f*, Lichtbogenheizung
*f* ~**-heating apparatus** Gerät *n* mit Lichtbogenheizung ~ **ignition** Lichtbogenzündung *f* ~ **initiation** Bogenzündung *f* ~ **lamp** Bogenlampe *f*,
Lichtbogen *m* ~ **lamp carbon** Dochtkohlestoff
*m* ~**-lamp accessories** Bogenlampenzubehör *n*
~ **lamp with carbon electrodes** Effektbogenlampe *f* ~**-lamp mirror** Bogenlampenspiegel *m*
~**-lamp starter** Lampenanlasser *m* ~ **length of
an oscillation arch** Bogenlänge *f* ~ **light** Bogen-
licht *n* ~**-light source** Leuchtröhre *f* ~ **measure**

Bogenmaß *n* ~ **oscillation** Lichtbogenschwin-
gung *f* ~**-over** Überschlag *m* ~ **plot** Bogendia-
gramm *n* ~**-proof** lichtbogenbeständig ~ **quench
chamber** Lichtbogenlöschkammer *f* ~ **quenching
range** Löschbereich *m* ~ **rectifier** Lichtbogen-
gleichrichter *m* ~ **resistant** lichtbogenbeständig
~ **rocket motor** Lichtbogentriebwerk *n* ~ **root**
Bogenfußpunkt *m* ~ **shield** Lichtbogenhülle *f*
~ **shielding** Lichtbogenumhüllung *f* ~ **spectrum**
Bogenspektrum *n* ~ **stream** Lichtbogensaum *m*
~ **suppression coil** Erdungsspule *f* ~ **suppressor**
Schutzvorrichtung *f* gegen Funkenüberschlag
~ **terminal** Lichtbogenkopf *m* ~**-through** Durch-
zündung *f* ~ **transmitter** Lichtbogensender *m*
~**-type elevating mechanism** Zahnbogenrichtmaschine *f* ~ **welding** elektrische Schweißung *f*,
Bogen-, Lichtbogen-schweißung *f* ~**-welding
method** Lichtbogenschweißverfahren *n* ~**-weld-
ing seam** Lichtbogenschweißnaht *f* ~ **zone**
Lichtbogenbereich *m*
**arcade** Bogengang *m*
**arch, to** ~ wölben **to** (over) ~ einwölben
**arch** Bogen *m*, Brücke *f*, Fensterbogen *m*, Gewölbe *n*, Hängedecke *f*, Kuppe *f*, Oberbalken
*m*, Querbalken *m*, Wölbung *f* ~ **of a bridge**
Brückenbogen *m* ~ **of a chamber** Kammerge-
wölbe *n* ~ **for drainage pipes** Ablaufrohrbogen
*m* ~ **of a vault** Gewölbebogen *m*
**arch,** ~ **abutment** Gewölbewiderlager *n* ~ **action**
Gewölbewirkung *f* ~**-back** Rückzündung *f* ~
**bond** Bogenverband *m* ~ **brick** Bogenbaustein
*m*, Gewölbestein *m* ~ **centering** Gewölbeschalung
*f* ~ **core** Gewölbekern *m* ~ **dam** Bogenstau-
mauer *f* ~ **formation in sand** Glocke *f* ~ **formed
by scaffolding** Hängegewölbe *n* ~ **limb** Gewöl-
beschenkel *m* ~ **pressure** Gewölbedruck *m* ~
**punch** Henkellocheisen *n* ~ **roof** Kuppelgewölbe *n* ~ **span** Bogenspannweite *f* ~ **truss**
Bogenfachwerk *n* ~ **work** Gewölbebogen *m*
**arched** bogenförmig, gewölbt, kuppelartig ~
**back** (music) gewölbter Boden *m* ~ **bridge**
Bogenbrücke *f* ~ **buttres** Schwibbogen *m* ~
**connection** gewölbter Zulauf *m* ~ **crown** Gewölbe *n* ~ **deck beam** Spantbogenprofil *n*
~ **dome** Kuppelgewölbe *n* ~ **girder** Bogenträ-
ger *m* ~ **lock floor** gewölbte Schleusensohle *f*
~ **plate** Tonnenblech *n* ~ **roof** Deckengewölbe
*n*, Gewölbedach *n*, Tonnendach *n*, tonnenförmiges Dach *n* ~ **roof of a hearth** Herdge-
wölbe *n* ~ **shed** gewölbtes Fach *n* ~ **squall**
Gewitter-, Wolken-kragen *m*
**archeology** Altertumskunde *f*, Archäologie *f*
**archil-extract** Orseilleextrakt *m*
**Archimedean drill** Drehbohrer *m*
**Archimedes' axiom** Archimedisches Axiom *n*
**arching,** Aufwölbung *f* (géol.), Brückenbildung *f*,
Gewölbewirkung *f*, Wölbung *f* ~ **of sand** Verspannen *n*
**archipelago** Archipel *m*
**architect** Architekt *m*, Baumeister *m*
**architectonics** Bauwissenschaft *f*
**architectural,** ~ **acoustics** Bau-, Raum-akustik *f*
~ **design** Raumgestaltung *f* ~ **photography**
Gebäudeaufnahme *f*
**architecture** Architektur *f*, Baukunst *f*
**architrave** Architrav *m*
**archives** Archiv *n*
**arcing** Bogen-, Funken-bildung *f*, Funken-

überschlag *m*, Lichtbogenbildung *f*; bogenbildend (metal)
**arcing,** **~ back** Gleichrichterrückzündung *f* **~ chamber** Funkenkammer *f*, Lichtbogenkammer *f* **~ contact** Funkenentzieher *m*, Funkenlöscher *m* **~ ground** Flammenbogenerdschluß *m*, Lichtbogenerdung *f* **~-ground suppressor** Erdungsschalter *m* **~ ring** Schutzkorb *m* **~ step** Lichtbogenschritt *m*
**arcotron** Stabröhre *f*
**arcs in rare gases** Edelgasbogen *pl*
**arctic** arktisch **~ air** arktische Kaltluft *f* **~ circle** Polarkreis *m* **~ current** Polarstrom *m* **~ smoke** arktischer Nebel *m*
**arcuate** gewölbt **~ grid** Kreisbogengitter *n* **~ line** Bogenlinie *f*
**ardometer** Ardometer *n*, Gesamtstrahlungspyrometer *n*, Gluthitzemesser *m*
**are** (metric unit) Ar *n*
**area** Anbaufläche *f*, Areal *n*, Bereich *m*, Bezirk *m*, Fläche *f*, Flächenraum *m*, Gebiet *n*, Gebietsteil *m*, Gegend *f*, Geländeabschnitt *m*, Geländeteil *m*, (surface) Grundfläche *f*, Inhalt *m*, Quadratmaß *n*, Raum *m*, Raummenge *f*, Rayon *m*, Umkreis *m* (wing) **~** Flächeninhalt *m*
**area,** **~ of abrasion** Abrasionsfläche *f* **~ of application** Verwendungsgebiet *n* **~ of applicator at skin surface** Eintrittsfeld *n* **~ of ball imprint (or indentation)** Kalottenfläche *f* **~ of bearing** Auflagefläche *f* **~ of calms** Kalmen *pl* **~ of a circle** Kreis-inhalt *m*, -fläche *f* **~ of competency** Zuständigkeitsbereich *m* **~ of contact** Begrenzungs-, Berührungs-fläche *f* **~ of contact of wheel** Radauflagerfläche *f* **~ of consumption** Verbraucherbezirk *m* **~ of cup** Kalottenoberfläche *f* **~ of cut** Schnittfläche *f* **~ of cylinder** Walzenfläche *f* **~ of diagram** (engine indicator) Diagrammfläche *f* **~ of diffusion** Ausbreitungsbezirk *m* **~ between two dikes (or groins)** Bühnenfeld *n* **~ of fracture** Bruchfläche *f* **~ of hearth** Rostfläche *f* **~ of high pressure** Antizyklon *m* **~ of horizontal tail surfaces** Höhenleitwerkfläche *f* **~ of impact** Auftreffgelände *m* **~ of impact of the anticathode** Brennfleck *m* **~ of impression** Eindruckfläche *f* **~ of indentation** Eindruckfläche *f* **~ of internal-hollowed section** Hohlquerschnittsfläche *f* **~ of lateral dispersion** Breitenstreuung *f* **~ of measurement** Meßfeld *n* **~ under moment curve** Momentenfläche *f* **~ of low pressure** Zyklone *f* **~ of pressure** Druckfläche *f* **~ of the probe** Sondenausdehnung *f* **~ in which the product of a factory may be sold** Absatzkreis *m* **~ of rest** Auflagerfläche *f* **~ of roller bed** Rollfläche *f* **~ of section** Einheitsquerschnitt *m*, Querschnittsfläche *f* **~ of separation** Abreißgebiet *n* (aerodyn.), abgetrenntes Gebiet *n* **~ of shearing force** Querkraftfläche *f* **~ of spreading** Ausbreitungsbezirk *m* **~ of stabilizer** Höhenflossenfläche *f* **~ between successive positions** Zwischenfeld *n* **~ of supply** Verbrauchsgebiet *n* **~ of tail unit** Leitwerkfläche *f* **~ of vortex core** Wirbelkernfläche *f* **~ of water-passage** Wasserdurchflußfläche *f* **~ of waters** Seegebiet *n* **~ of waterway** benetzter Querschnitt *m* **~ of a windowpane** Feld *n* eines Glasfensters
**area,** **~ bearing sound track** vertonte Oberfläche *f* **~ code** Ortskennzahl *f* **~ conflagration**

Flächenbrand *m* **~ control** Bereichskontrolle *f* (aviat.), **~ covered** Meßfeld *n* **~ exposed to view** Sichtausschnitt *m* **~ fire** Flächenfeuer *n* **~ gassing** Flächenvergasung *f* **~ increment** Flächenzunahme *f*, Gebietszuwachs *m* **~ illumination** Raumleuchten *n* **~ integral** Flächenintegral *n* **~ load** Flächenlast *f* **~ monitor** Raumüberwachungsgerät *n* **~-perimeter ratio** Verhältnis *n* der Fläche zu ihrem Umfang **~-preserving mapping** inhaltstreue Abbildung *f* **~ quotient** Flächenquotient *m* **~ ratio** Flächenverhältnis *n* **~ sketch** Grundrißskizze *f* **~ to be surveyed** Vermessungsgebiet *n* **~ way** Lichtschacht *m* (vor Kellerfenster)
**areal** flächenförmig, flächenhaft, flächig **~ coil** flächenhafte Spule *f* **~ contact** flächenhafte Berührung *f* **~ density** Flächendichte *f* **~ path of current** flächenanziehungsartige Strombahn *f* **~ surveys** Flächenanziehungsaufnahmen *pl*
**areate, to** **~** mit Kohlensäure *f* sättigen
**arecoline bromide** Arecolinbromid *n*
**arena** Bühne *f*, Schauplatz *m*
**arenaceous** sandführend, sandig **~ shale** Sandschiefer *m*
**areometer** Aräometer *n*, Tauchwaage *f*
**areometric** aräometrisch
**areopycnometer** Aräopyknometer *n*
**argentan** Neusilber *n*
**argentiferous** silberführend, silberhaltig **~ gold** Silbergold *n* **~ sand** Silbersand *m* **~ tetrahedrite** Silberfahlerz *n*, Weißgültigerz *n*
**argentimeter** Argentometer *n*
**argentine** Aphrit *m* **~ mica** Silberglimmer *m*
**argentite** Glanzerz *n*, Glanzsilber *n*, Glaserz *n*, Schwefelsilber *n*, Silber-glanz *m*, -glas *n*, -glaserz *n*, -schwärze *f*
**argentometer** Silbermesser *m* (photo)
**argentosulfurous** silberschwefelig
**argillaceous** tonartig, tonhaltig, tonig **~ earth** Tonerde *f* **~ iron ore** Toneisenstein *m* **~ limestone** Tonkalk *m* **~ sand** Tonsand *m* **~ siderite** tonhaltiger Spateisenstein *m* **~ slate** Tonschiefer *m* **~ veined agate** Schichtling *m*
**argillite** Tonschiefer *m*
**argillocalcite** Tonkalk *m*
**argon tungsten-arc process** Argon-Wolframbogenverfahren *n*
**argonal rectifier** Argonalgleichrichter *m*
**argue, to** **~** diskutieren, erörtern, für und wider erörtern, streiten, vorbringen to **~ in support of an opposition** Einspruch unterstützen to **~ nonvalidity** die Gültigkeit eines Patentes anfechten
**argued statement of defense** Verteidigungsschrift *f*
**argument** Auseinandersetzung *f*, Ausführung *f*, Beweis *m*, Schlußfolgerung *f*, Streit *m*, Verhandlung *f* **~s and grounds for appeal** Berufungsbegründung *f*
**argumentation** Beweisführung *f*, (in support of an opposition) Begründung *f* eines Einspruches
**argus contact lever** Arguspendel *n*
**arid** dürr, trocken, wasserlos
**aridity** Dürre *f*
**arise, to** **~** entstehen, (sich) ergeben, erstehen, hervorkommen, (question) auftauchen
**arising,** **~ of sudden irregularity or unsteadiness** Umspringen *n* **~ out of** entstanden aus . . .
**arisings** Altmaterial *n*

**arithmetic** Arithmetik *f*, Rechenkunst *f*, Zahlenlehre *f*; rechnerisch ~ **element** Rechenwerk *n* ~ **mean** arithmetisches Mittel *n* ~ **operation** arithmetische Operation *f*, (data proc), Rechenoperation *f* ~ **progression** arithmetische Reihe *f* ~ **shift** Stellen(wert)verschiebung *f* ~ **unit** Rechenwerk *n* (info proc)

**arithmetical** arithmetisch ~ **and clerical accuracy** rechnerische Richtigkeit *f* ~ **element** Rechenwerk *n* ~ **error** Rechenfehler *m* ~ **operation** Rechenoperation *f* ~ **progression** arithmetische Reihe *f* ~ **shift** Stellenverschiebung *f*

**arithmetico-geometric mean series** arithmetisch-geometrische Reihe *f*

**arithmometer** Arithmometer *n*

**arkose** Kaolinsandstein *m*

**arm, to** ~ armieren, aufrüsten, bewaffnen, bewehren, rüsten, (Handgranate) scharfmachen, (Bombe) scharfstellen

**arm** Arm *m*, Ausleger *m*, Frosch *m*, Lineal *n*, Querträger *m*, Schenkel *m*, Steg *m*, Waffe *f*, Radarm *m* (mach.), (of shield) Stütze *f*, (of bridge or filter) Zweig *m* **with the -(s) free** freihändig

**arm,** ~ **of bracket** Konsolenarm *m* ~ **of the bridge** Brückenzweig *m* (electr. engin.) ~ **of the compasses (or dividers)** Zirkelschenkel *m* ~ **of a machine tool** Auslegerarm *m* ~ **of precision** Genauigkeitswaffe *f* ~ **of a river** Flußarm *m* ~ **of the sea** Meeresarm *m* ~ **for side lights** Laternenträger *m* ~ **of split trail** Holm *m*

**arm,** ~ **and arc protractor** Plansektor *m* ~ **and hand signal** Armzeichen *n* ~ **assembly** Schwenkarm *m* ~ **badge** Armspiegel *m* ~ **band** Armband *n*, Armbinde *f* ~**-binder handle** Armklemmhebel *m* ~ **brace** Scherenstrebe *f* ~ **braces** (harness) Armstützen *pl*, Gegenhalterstützen *pl*, Scheren *pl* ~**-clamp screw** Führungsarmklemmschraube *f* ~**-clamping nuts** Armklemmuttern *pl* ~ **cross** Armkreuz *n* ~ **dowel pin** Armprisonstift *m* ~ **extension bracket** Verlängerungsschiene *f* ~ **file** Armfeile *f* ~ **girdle** Armträger *m*, Armklemmstück *n* ~ **mooring** Hahnepot *m* ~ **pointer** Gradzeiger *m* ~ **rack** Armzahnstange *f* ~ **rest** Arm-auflage *f*, -lehne *f* ~**-rocker lever** Kipphebel *m* ~ **shaft** Armwelle *f* ~ **shunt** Abzweigung *f* ~ **signal** Armzeichen *n* ~ **strap** Halteschlaufe *f*, Halteriemen *m* ~**-swinging handle** Griff *m* zum Schwenken des Armes ~ **ways** Armführungs-prisma *n*, -wange *f* ~**-worm box** Armschneckenlager *n*

**armament** Armierung *f*, Aufrüstung *f*, Bestückung *f*, Bewaffnung *f*, Rüstung *f*, Rüstungsmaterial *n* ~ **industry** Rüstungsindustrie *f* ~ **plant** Rüstungswerk *n* ~ **works** Rüstungswerk

**armature** (dynamo) Anker *m*, Armatur *f*, Ausrüstung *f*, Beschlag *m*, Bewehrung *f*, Eichkatzenanker *m*, Gerät *n*, Läufer *m*, (turn) Leiterschleife *f*, Leuchte *f*, (of magnet) Magnetanker *m*, Metallbeschlag *m*, Relaisanker *m*, Rotor *m*, Zubehör *n*, Zubehörteil *m*

**armature,** ~ **of a permanent magnet** Anker *m* eines Dauermagnetes ~ **of a relay** Anker *m* eines Relais ~ **with salient pole** Polanker *m*

**armature,** ~ **air gap** Ankerluftspalt *m* ~ **bar** Ankerstab *m* ~ **bearing** Ankerlager *n* ~ **binding wire** Ankerbindedraht *m* ~ **bore** Ankerbohrung *f* ~ **bracket link** Ankereinstellhebel *m* ~ **coil**

Ankerspule *f*, Ankerwindung *f* ~ (-winding) **coil** Ankerwicklung *f* ~ **conductor** Ankerleiter *m* ~ **controls** Armatur *f* ~ **core** Ankerkern *m* ~ **core disk** Ankerblech *n* ~ **core with charging winding** Ankerkern *m* mit Ladewicklung ~ **core with ignition winding** Ankerkern *m* mit Zündwicklung ~ **core with lighting winding** Ankerkern *m* mit Lichtwicklung ~ **cross** Ankerstern *m* ~ **current** Anker-, Läufer-strom *m* ~ **diverter contactor** Ankerverteilerschutz *m* ~ **drive** Ankerantrieb *m* ~ **drum** Ankertrommel *f* ~ **end connections** Ankerklemmen *pl* ~ **excursion** Ankerumschlag *m* ~ **extension** Ankerfortsatz *m* ~ **field** Ankerfeld *n* ~ **flux** Ankerkraftfluß *m* ~ **gap** Ankerbohrung *f* ~ **guiding pin** Ankerführungsbolzen *m* ~**-head flange** vorderer Ankerflansch *m* ~ **hesitation** Schaltunsicherheit *f* ~**-holding screw** Ankerhalteschraube *f* ~ **hub** Rotornarbe (electr.) ~ **iron** Ankereisen *n* ~ **key** Ankerkeil *m* ~ **lamination** Ankerblech *n* ~ **leakage** Ankerstreuung *f* ~**-leakage flux** Ankerstreufluß *m* ~ **line of force** Ankerkraftlinie *f* ~ **pinion** Anker-ritzel *n*, -zahnrad *n* (electr.) ~ **plates** Ankerpaket *n* ~ **projection** (lug, shoulder) Ankervorsprünge *pl* ~ **reaction** Anker-gegenwirkung *f*, -rückwirkung *f* ~ **resistance** Ankerwiderstand *m* ~ **return spring** Ankerrückdruckfeder ~ **shaft** Ankerachse *f* (electr.) ~ **(-shaft) spindle** Ankerwelle *f* ~ **short** Lamellenschluß *m* ~ **shunt** Ankernebenschluß *m* ~ **slot** Ankernut *f* ~**-spider** Anker-büchse *f*, -körper *m*, -nabe *f*, -stern *m*, Läuferstern *m*, Rotorstern *m*, Ständerkörper *m* ~**-stray flux** Ankerstreufluß *m* ~ **stroke** Anker-hub *m*, -spiel *n* ~ **tooth** Ankerzahn *m* ~ **travel** Anker-bewegung *f*, -umschlag *m*, -weg *m* ~ **voltage** Ankerspannung *f*

**armed** bewaffnet, bewehrt, geladen; ~ **fuse** geschärfter Zünder *m*

**arming** Bewaffnung *f*, Rüsten *n* ~ **of poles** Ausrüstung *f* von Gestängen

**arming,** ~ **cable** Auslöseseil *n* (Fallschirm) ~ **cord** Auslöseleine *f* ~ **knob** Auslöseknopf *m* ~ **press** Deckelpressung *f* ~ **ring** Ringsprengkapsel *f* ~ **screw** Schärfspindel *f* ~ **solenoid** Schärfmagnetspule *f* ~ **time** Schärfzeit *f* ~ **wire** Schärfdraht *m*

**armmiter gear guard** (radial drill) Armkegelrad-Schutzdeckel *m*

**armor, to** ~ bewehren, panzern

**armor** Einbauten *pl*, Panzer *m*, Panzerung *f*; ~ **for electrical cables** Armierung *f*

**armor,** ~ **belt** Panzergürtel *m* ~**-bending press** Panzerbiegepresse *f* ~**-clad** gepanzert ~ **clamp** Bewehrungsschelle *f* ~**-piercing** panzer-brechend, -durchschlagend ~**-piercing ammunition** Panzermunition *f* ~**-piercing bullet** Stahlkerngeschoß *n* ~ **plate** Panzer-blech *n*, -platte *f*, -schale *f* ~**-plate planing machine** Plattenhobelmaschine *f* ~**-plate rolling mill** Panzerplattenwalzwerk *n* ~**-plated** armiert, gepanzert ~ **plating** Panzerung *f* ~ **protection** Panzerschutz *m* ~ **rods** Dämpferdrähte *pl* ~ **shield** Panzerschild *m* ~ **shielded** armiert ~ **steel** Panzerstahl *m* ~ **tipped** armiert

**armored** armiert, bewehrt, gepanzert, geschützt ~ **bakelite models** Bakelitmodelle *pl* mit Aluminiumbewehrung ~ **cable** armiertes oder bewehrtes oder geschütztes Kabel *n*, Armierungskabel n, Panzerdraht m, Panzerkabel n, Spiralleitung f ~ **car** Panzerkraftwagen m ~ **carriage** Panzerlafette f ~ **chassis** Panzerwanne f ~ **combat car** Panzerspähwagen m ~ **cruiser** Panzerkreuzer m ~ **cupola** Panzerkuppel f ~ **hose** armierter Schlauch m ~ **hull** Panzerwanne f ~ **skirting** Laufwerkpanzerung f ~ **tubular flooring** Zylinderstegdecke f ~ **valve** Panzerventil n ~ **vessel** Panzerschiff n ~ **war cable** Panzerfeldkabel n ~ **wire** Panzerdraht m ~ **wires** Rohrdrähte pl

**armorer** Büchsenmacher m

**armoring** Armierung f, Bewehrung f, Drahtverbindung f, Eiseneinlage f, (of a blast furnace) Verankerung f ~ **disc** Armierungsscheibe f ~ **machine** Kabelbewehrungsmaschine f ~ **wire** Bewehrungsdraht m, Schutzdraht m

**armorizing** Einsatzhärtung f

**armory** Kaserne f, Zeughaus n ~ **ordnance shop** Waffenmeisterei f

**arms** Bewaffnung f ~ **rack** Gewehrstütze f

**army** Armee f, Heer n

**aromatic** aromatisch ~ **fuel** Kraftstoff aus aromatischen Bestandteilen ~ **hydrocarbon** aromatischer Kohlenwasserstoff m ~ **oil** Duftöl n ~ **series** aromatische Reihe f

**arouse, to** ~ wachrufen

**arrange, to** ~ abmachen, akkordieren, anordnen, aufstellen, ausmachen, beiordnen, besprechen, einfluchten, einrichten, gestalten, gliedern, herrichten, ordnen, regeln, regulieren, richten, Streit beilegen, schlichten, übereinkommen **to** ~ **alphabetically** alphabetisch einreihen **to** ~ **clearly** übersichtlich gruppieren **to** ~ **in columns** einordnen **to** ~ **the cylinders in V-(form)** Zylinder gegeneinander geneigt anordnen **to** ~ **in** anreihen **to** ~ **in layers** schichten, einschichten **to** ~ **in lines** staffeln **to** ~ **like a pendulum** pendelnd lagern ~ **one above the other** übereinander anordnen **to** ~ **to oscillate** pendelnd lagern

**arranged** eingebaut ~ **in order** gegliedert ~ **side by side** nebeneinander angeordnet ~ **to swing (out) backward** nach hinten klappbar ~ **to turn (or swing) down** niederklappbar ~ **upright** aufrechtstehend

**arrangement** Abkommen n, Abmachung f, Anordnung f, Ansatz m, Aufbau m, Ausführung f, Disposition f, Durchbildung f, Einrichtung f, Einrüstung f, Gliederung f, Gruppierung f, Lagerung f, Ordnung f, Regelung f, Reihenfolge f, Übereinkommen n, Vereinbarung f, Vergleich m, Vorrichtung f

**arrangement,** ~ **of amplifier** (repeaters) Verstärkeranordnung f ~ **of cells** Zelleneinrichtung f ~ **for central code recording** Sammelmorseeinrichtung f ~ **of the change gears** Wechselradanordnung f ~ **of collimation point** Bildmarkenanordnung f ~ **for delivering the lids to the churns** Deckelauflegevorrichtung f ~ **of engines** Maschinenanordnung f ~ **s in force** geltende Abmachungen pl ~ **of ignition cable** Zündleitungsführung f ~ **for the interception of mischievous calls** Fangvorrichtung f ~ **in layers** Schichtung f ~ **of a mill train** Straßenanordnung f ~ **of molecules** Molekularanordnung f ~ **of pole attachments** Ausrüstung f von Gestängen ~ **of printing units** Druckwerkanord-

nung f ~ **of rolls** Walzenanordnung f ~ **of roll stands** Walzgerüstanordnung f ~ **of roll stands in separate lines** gestaffelte Straße f ~ **of roll stands in two separate lines** zweigestaffelte Straße f ~ **of rolling stands** Gerüstanordnung f ~ **of shafting** Wellenstranganordnung f ~ **of steps** Stufenmuster n ~ **in straight rows** reihenförmige Anordnung f ~ **to release switches** Abwerfeinrichtung f ~ **of the survey** Aufnahmeanordnung f ~ **of transmitter** Senderanordnung f ~ **of transverse frames** Ringanordnung f ~ **for underlining** Unterstreichvorrichtung f ~ **of valves** Ventilanordnung f ~ **of wires on pole lines** Gruppierung f von Leitungen ~ **of a workshop** Betriebsregelung f

**arranging in layers** Schichtung f

**array** Feld n (info proc.), Ordnung f, Reihe f ~ **of antennas** Antennenanordnung f (regelmäßiger Ausführung) ~ **theorem** Anordnungssatz m

**arrears** Rückstand m **in** ~ rückständig **to be in** ~ **in** Rückstand kommen **to be in** ~ **with** ins Stocken n geraten ~ **of interest** rückständige Zinsen pl

**arrest, to** ~ abfassen, abschalten, abstellen (mach.), anhalten, arretieren, aufhalten, bremsen, Einhalt tun, einrasten, einziehen, festhalten, festlegen, festnehmen, feststellen, hemmen, mit Beschlag belegen, sperren, zum Stillstand bringen, stillstehen, stocken **to** ~ **carbon drop** Entkohlung f aufhalten

**arrest** Abschaltung f, Anhalten n, Arretierung f, (of judgment) Aufschiebung f, Aussetzung f, Beschlagnahme f, Einhalt m, Hemmung f, (metallograph.) Haltepunkt m ~ **of circulation** Zirkulationshemmung f ~ **in development** Entwicklungsstillstand m

**arrestation point** Umwandlungspunkt m

**arrester** Ableiter m, Abschneider m, Stärkerklappe f ~ **case** Filtergehäuse n ~ **gear landing** Landung f mit Fangvorrichtung ~ **hook** (Lande)-Fanghaken m ~ **means** Rastenarretierung f

**arresting** Festlegemittel n, Feststellung f, Sperre f, Stillsetzung f ~ **device** Anschlagvorrichtung f ~ **gear** Abbremsvorrichtung f, Bremse f, Bremsstrecke f, Landebremsvorrichtung f ~ **hook** Fanghaken m, Landebremshaken m ~ **hook for winding post** Sperrnase f (photo) ~ **lever** Sperrhebel m ~ **switch** Verriegelungsschalter m

**arris** ausspringende Ecke f, Kaikante f ~ **cutter (or remover)** Gratabschneider m

**arrival** Ankunft f, Landung f ~ **curve** Empfangskurve f ~ **platform** Ankunftsbahnsteig m ~ **time** Eintreffzeit f ~ **track** Einfahrtgeleise n

**arrive, to** ~ ankommen, anlangen, einlaufen, eintreffen, kommen **to** ~ **at a solution** zu einer Lösung f gelangen

**arrived** eingegangen

**arriving current** Zustrom m

**arrow** Pfeil m ~ **engine** Dreireihenstandmotor m ~ **head** Exerzierzündspitze f, Pfeilspitze f ~ **-headed dimension line** in einen Pfeil m auslaufende Maßlinie f ~ **headed letters** Pfeilschrift f ~ **headed types** Pfeilschrift f ~ **indicating direction of current** Strömungspfeil m ~ **point** Exerzierzündspitze f ~ **-shaped** pfeilförmig

**~-shaped index** Pfeilmarke *f* **~-type** pfeilförmig
**~-type engine** W-Motor *m*
**arsenal** Arsenal *n*, Zeughaus *n*
**arsenate, ~ of** arsensau(e)r . . . **~ of copper** Holzkupfererz *m* **~ of lime** Pharmakolith *m*
**arsenic** Arsen *n*, Arsenik *n* **~ acid** Arsensäure *f*
**~ bisulfide** rotes Arsensulfid *n* **~ bloom** Arsenblüte *f*, weißes Arsenik **~-free** arsenfrei **~**
**hydride** Arsenwasserstoff *m* **~ iodide** Arsenjodid *n* **~ lime** Giftäscher *m* **~ metal** Arsenmetall *n* **~ ore** Arsenerz *n* **~ oxide** Arsenoxyd *n*
**~ pentasulfide** Arsenpentasulfid *n* **~ skimmings**
Arsenabstrich *m* **~ sulfate** Arsenikvitriol *n* **~**
**sulfide** Arsensulfid *n* **~ triiodide** Arsenjodid *n*,
Arsentryodid *n* **~ trioxide** Arsenoxyd *n*, Arsentrioxyd *n*, weißer Arsenik **~ trisulfide** gelbes
Arsensulfid *n*
**arsenical** arsenhaltig, arsenikalisch **~ cadmia**
Giftstein *m* **~ iron** Arsenikkies *m* **~ nickel** Arseniknickel *n* **~ (iron) pyrites** Arsenkies *m* **~**
**pyrites** Arsenikkies *m* **~ silver** Arseniksilber *n*
**arsenide** Arsenmetall *n*
**arsenides** arseniksaures Salz *n*
**arseniferous** arsenführend
**arsenious, ~ acid** arsenige Säure *f*, Arsentrioxyd
*n* **~ anhydride** Giftmehl *n* **~ sulfide** Arsentrisulfid *n*
**arsenite of** arsenigsauer
**arsenolite** Arsenblüte *f*, Arsenikkalk *m*, weißes
Arsenik *n*
**arsenopyrite** Arsenikkies *m*, Arsenkies *m*, Gift-
kies *m* (prismatischer), Mispickel *m*
**arsine** Arsin *n*
**arson** Brandstiftung *f*
**art** Fach *n*, Kunst *f*, Kunstfertigkeit *f*, Technik *f*
**art, ~ of analyzing** Zersetzungskunst *f* **~ of**
**construction** Baukunst *f* **~ of etching** Kaustik *f*
**art, ~ bronze** Kunstbronze *f* **~ casting** Kunst-
guß *m* **~(istic) foundry** Kunstgießerei *f* **~**
**leather** Kunstleder *n* **~ paper** Kunstdruckpapier
*n* **~ publishers** Kunstverlag *m* **~ term** Kunstausdruck *m*
**arterial, ~ highway** Autobahn *f* **~ road** Aus-
fallstraße *f*
**arteriography** Arteriografie *f*
**artesian casing** Filterbrunnenrohr *n* **~ well**
artesischer Brunnen *m*, Rohr-, Steig-brunnen
*m* **~-well pump** Rohrbrunnenpumpe *f* .
**artesian, ~ condition** artesische Bedingung *f* **~**
**head** artesische Druckhöhe *f*
**arthrography** Arthrografie *f*
**article** Abhandlung *f*, Ansatz *m*, Artikel *m*,
Aufsatz *m*, Erzeugnis *n*, Fabrikat *n*, Gegenstand
*m*, Ware *f* **~ of clothing** Bekleidungsstück *n*
**~ of equipment** Ausrüstungsgegenstand *n* **~**
**of export** Exportartikel *m* **~ of value** Wertgegenstand *m*
**article claim** Sachpatent-Anspruch *m*
**articles in bulk** Massenartikel *pl*
**Articles, The ~** Heuervertrag *m*
**articulate, to ~** anlenken, deutlich aussprechen,
gliedartig verbinden, gliedern
**articulate** artikuliert, deutlich, klar **~ speech**
Sprachlaute *pl*
**articulated** angelenkt (drehbar), deutlich, gegliedert, gelenkig, scharf hervortretend **~ arm**
Gelenkausleger *m* **~ connecting rod** Mittel-
Neben-pleuel *n* **~ drilling head** Gelenkspindel-

bohrkopf *m* **~ harrow** Gelenk-, Glieder-egge *f*
**~ jaw friction clutch** Gelenkspreizringkupplung
*f* **~ joint coupling** Gelenkkuppelung *f* **~ lever**
Knickhebel *m* **~ link gearing** (valve-gear drive)
Gelenkschwingensteuerung *f* **~ locomotive**
kurvenbewegliche Lokomotive *f* **~ locomotive**
**turntable** Lokomotivgelenkdrehscheibe *f* **~**
**mirror** Lenkspiegel *m* **~ pipe** Gelenkrohr *n*,
Gliederröhre *f* **~ rod** Anlenkpleuel *n*, Neben-
schubstange *f* **~ rod pin** Anlenkbolzen *m*
(für Nebenpleuel) **~ rotor system** Gelenkrotor-
system *n* **~ shaft** Gliederwelle *f* **~ spindle** Ge-
lenkspindel *f* **~ strut** Gelenkstiel *m* **~ system**
gelenkiges Blattaufhängungssystem *n* **~ train**
Gelenkzug *m* **~ voice** deutliche Sprache *f*
**articulating rod** Nebenpleuelstange *f*
**articulation** Artikulation *f*, Deutlichkeit *f*, Fügung *f*, Gelenk *n*, Gliederbau *m*, Gliederung *f*,
gute Verständlichkeit *f*, Klarheit *f*, Scharnier *n*,
Sprachdeutlichkeit *f*, Sprachverständlichkeit *f*,
Sprachverständigung *f*, Verständlichkeit *f* **~**
**and intelligibility** Verständlichkeit *f* **~ of letters**
Lautverständlichkeit *f* **~ for logatomes** Loga-
tomverständlichkeit *f*, Silbenverständlichkeit *f*
**articulation, ~equivalent** Verständlichkeitsäqui-
valent *n* **~ joint** Fugengelenk *n* **~ piece** Gelenkstück *n* **~ reduction** Minderung *f* der Verständlichkeit **~ reference equivalent** Bezugs-
artikulationsäquivalent *n*
**artifice** Kniff *m*, Kunstfertigkeit *f*, Kunstgriff
*m*, Trick *m*
**artificer** Feuerwerker *m*, Waffenmeister *m*
**artificial** erzwungen, künstlich, nachgemacht **~**
**age-hardening** Warmaushärtung *f* **~ aging** An-
laßhärtung *f*, Aushärten *n*, Kochvergütung *f*,
künstliche Alterung *f*, Warmaushärten *n*
(Leichtmetalle), Warmvergütung *f* **~ altitude**
künstliche (in der Höhenkammer erzeugte)
Höhe *f* **~ antenna** Ersatzantenne *f*, künstliche
Antenne *f* **~ (balancing) circuit (or line)** Ausgleichsleitung *f* **~ balancing line** Leitungsnach-
bildung *f* **~ basalt** Schmelzbasalt *m* **~ black signal** Testsignal *n* für Schwarz **~ blocking layer**
künstliche Sperrschicht *f* **~ cable** Kunstleitung
*f*, künstliches Kabel *n* **~ cement stone** Zement-
kunststein *m* **~ circuit** Kunstleitung *f*, künstliche Leitung *f* **~ coil** Kettenleiter *m* **~ cotton**
Kunstbaumwolle *f*, Zellstoffwatte *f* **~ cover**
künstliche Deckung *f* **~ currents** hervorgerufene Ströme *pl* **~ draft** künstlicher Zug *m* **~**
**earthquakes** künstliche Seismen *pl* **~ extension**
**line** Leitungsverlängerung *f* **~ fiber cellulose**
Kunstfaserzellstoff *m* **~ fog** künstlicher Nebel
*m* **~ foundation** künstliche Gründung *f* **~ gas**
künstliches Gas *n* **~ gum** Dextrin *n* **~ horizon**
Flieger-, Kreisel-horizont *m*, künstlicher Horizont *m* **~ horn** Kunsthorn *n* **~ infinity** künstliches Meßziel *n* **~ lake** Teich *m* **~ leather**
Kunstleder *n*, Lederersatz *m* **~ leather imitations-
papier *n* **~ leather base** Kunstlederrohstoff *n*
**~ light screen** Kunstlichtfilter *n* **~ line** Kettenleitung *f*, Kunstleitung *f*, künstliche Leitung *f*
**~ manganese dioxide** künstlicher Braunstein
*m* **~ marble** Alaungips *m*, Gipsmarmor *m* **~**
**parchment** Pergamentpapier *n* **~ product** Kunst-
produkt *n*, Kunststoff *m* **~ radio-activity**
künstliche Radioaktivität *f* **~ resin** Kunstharz
*n* **~ stone** Betonstein *m*, Kunststein *m* **~ target**

for **calibration** Justiergerät *n* ~ **white signal** Testsignal *n* für Weiß (TV) ~ **wood** Kunstholz *n* ~ **wool** Kunstwolle *f*
**artificially ventilated** fremdgelüftet
**artillery** Artillerie *f*
**artisan** Handwerker *m*
**artistic,** ~ **forged piece** Kunstschmiedestück *n* ~ **metal castings** Metallkunstguß *m*
**artist's,** ~ **colors** Öltubenfarben *pl* ~ **paint** Künstlerfarbe *f*
**asbestoid** asbestartig
**asbestos** Amiant *m*, Asbest *m*, Eisenamiant *m* ~ **apron** Asbestschürze *f* ~ **board** Asbestpappe *f* ~ **canvas** Asbestleinwand *f* ~ **cloth** Asbestgewebe *n*, Asbesttuch *n* ~ **clothing** Asbestschutzkleidung *f* ~ **cord** Asbestschnur *f* ~ **covered** asbestumsponnen ~ **fiber** Asbestfaserstoff *m* ~ **flour** Asbestmehl *n* ~ **gasket** Asbestdichtung *f* ~ **glove** Asbesthandschuh *m* ~ **joint** Asbestdichtung *f* ~ **lumber** Asbestholz *n* ~ **milk** Asbestaufschlemmung *f* ~ **packing** Asbest-dichtung *f*, -packung *f* ~ **pad** Asbestplatte *f* ~ **plate** Asbestplatte *f* ~-**protected** asbestgeschützt ~ **quarry** Asbestgrube *f* ~ **rope** Asbestkordel *f* ~ **screen** Asbestschirm *m* ~ **slate** Asbestschiefer *m* ~-**slate board** Asbestschieferbrett *n* ~ **suit** Asbestanzug *m* ~ **transom** Asbestschieferbrett *n* ~ **washer** Asbestdichtung *f* ~ **wire gauze** Asbestdrahtnetz *n* ~ **wood** Asbestholz *n* ~ **wool** Asbestflocken *pl*
**asbolite** Erdkobalt *m*, Kobaltmanganerz *n*, Rußkobalt *m*
**ascend, to** ~ ansteigen, aufsteigen, auftreiben, emporkommen, emporsteigen, ersteigen
**ascendency** Auftrieb *m*
**ascender** Oberlänge *f*
**ascending** ansteigend, aufsteigend ~ **air current** Aufwind *m* ~ **branch of trajectory** aufsteigender Ast *m* ~ **cloud** Aufgleitwolke *f* ~ **convection current** Wärmeaufwind *m* ~ **conveyor** Steigförderer *m* ~ **current** aufwärtsgerichteter (Luft-) Strom *m* ~ **flank** Anstiegflanke *f* ~ **letter** Oberlänge *f* ~ **pipe** Steigerohr *n* ~ **power** steigende Potenz *f* ~ **reduction** Zurückführung *f* auf eine höhere Benennung (math) ~ **step** Firste *f*, Firstenstoß *m* ~ **tube** Steigerohr *n*
**ascension** Ansteigung *f*, Aufgang *m*, Aufsteigen *n* ~ **place** Aufstiegplatz *m*
**ascent** Ansteigen *n*, Anstieg *m*, Auffahrt *f*, Aufgang *m*, Aufstieg *m*, Aufwärtsbewegung *f*, Aufwärtshub *m*, Ausfahren *n* (der Bergleute), Steigen *n*, Steigflug *m* (aviat.), Steigung *f* ~ **into the stratosphere** Stratosphärenaufstieg *m* ~ **on gradients** Bergfahrt *f*
**ascertain, to** ~ ausfindigmachen, eruieren, ermitteln, feststellen, konstatieren, vergewissern **to** ~ **by measuring** bemessen **to** ~ **mean value** mitteln
**ascertainable** feststellbar
**ascertained,** ~ **attenuation** Planungsdämpfung *f* ~ **efficiency** gemessene Wärmeausnutzung *f*
**ascertaining the visual acuity** Sehschärfenbestimmung *f*
**ascorbic acid** Ascorbinsäure *f*
**ascribe, to** ~ zurückführen, zuschreiben **to** ~ **to** zumessen
**asdic** Horchgerät *n*, Sonar *m*, Unterwasserortungsgerät *n*

**aseptic** aseptisch
**ash, to** ~ äschern, veraschen
**ash** Asche *f*; Eschenholz *n*
**ash,** ~ **apparatus** Aschenbestimmer *m* ~ **conveyor** Aschenförderanlage *f* ~ **damper** Aschenfallklappe *f* ~ **determination** Aschengehaltsbestimmung *f* ~ **door** Aschenfallklappe *f*, Bodenklappe *f* ~ **dump** Aschhaufen *m* ~ **formation** Schlackenbildung *f* ~ **forming** aschebildend, schlackebildend; Schlackbilden *n* ~ **free** aschenfrei ~ **fusing point** Aschenschmelzpunkt *m*, Aschestromregler *m* ~ **gray** aschgrau ~ **handling equipment** Aschentransportanlage *f* ~ **handling plant** Aschenförderanlage *f* ~ **hoist** Aschheißmaschine *f* ~ **hole** Feuergrube *f* ~ **pan** Aschenschüssel *f* ~ **percentage** Aschengehalt *m* ~ **pit** Aschenfall *m*, Aschengrube *f*, Aschenraum *m*, Feuergrube *f*, Löschgrube *f* ~ **plate** Aschenzacken *m*, Hinterzacken *m* ~ **removal** Ascheziehen *n*, Entaschung *f* ~-**shade** aschfarben ~ **shoot** Aschschütte *f* ~ **tray** Aschenbecher *m* ~ **tree** Esche *f* ~ **valve** Aschenfallklappe *f* ~ **washing** Aschenwäsche *f*
**ashes boiler** Aschensieder *m*
**ashing** Braschen *pl*, Polieren *n*, Veraschung *f*
**ashlar** (dressed) Deckquader *n*, Hausstein *m* ~ **facing** Quader-mauerwerk *n*, -verblendung *f*, -verkleidung *f* ~ **masonry** Bruchsteinmauerwerk *n* ~ **stone** Bruchstein *m*
**ashlaring** Dachverschalung *f*, Quadermauer *f*
**ashless** aschenfrei ~ **filter** aschenfreier Filter *m* ~ **paper** (filter) aschenfreies Papier *n*
**ashore** an Land, ans Ufer
**ashy** fahl
**ask, to** ~ bitten, fordern, nachfragen **to** ~ **for** anfragen, erbitten
**asp wood pulp** Espenholzschliff *m*
**aspect** Anblick *m*, Ansehen *n*, äußere Erscheinung *f*, Gesichtspunkt *m*, Lage *f*, Standpunkt *m* ~ **of signal** Signalbild *n* (r.r.) ~ **control relays** Begriffsteller *m* (r.r.) ~ **ratio** Aspektverhältnis *f* (aviat.) Bildformat *n*, Bildhöhe und Zeilenlänge *f*, Bildkantenverhältnis *n*, Flächenverhältnis *n*, Flügelstreckung *f*, (picture) Formfaktor *m*, Längenverhältnis *n*, Seitenverhältnis *n*, Streckenverhältnis *n*, Streckung *f*
**asphaline** Asphalin *n*
**asphalt, to** ~ asphaltieren
**asphalt** Asphalt *m*, Asphaltstein *m*, Bergpech *n*, Bergteer *m*, Erdharz *n*, Erdpech *n*
**asphalt,** ~ **block** Asphalt-klotz *m*, -platte *f* ~ **bound** Mischmakadam *m* ~ **cement** Asphaltkitt *m* ~ **coat** Asphaltüberzug *m* ~ **coating** Bitumenanstrich *m* ~ **floor** Asphaltestrich *m* ~ **floor(ing)** Asphaltfußboden *m* ~ **intaglio printing ink** Asphalttiefdruckfarbe *f* ~ **pavement** Asphaltpflaster *n* ~ **process** Askandruck *m* ~ **putty** Asphaltkitt *m* ~ **(-paved) road** Asphaltstraße *f* ~ **rock** Pechgang *m* ~ **surfacing** Deckasphaltschicht *f*
**asphaltene** Asphalten *n*
**asphaltic** asphalthaltig, asphaltisch, erdharzig, erdpechhaltig ~ **base** Rohparaffin *m* ~ **hydrocarbon** asphaltischer Kohlenwasserstoff *m* ~ **mastic** Asphaltguß *m* ~ **plaster** Asphalt-, Bitumen-bewurf *m*
**asphaltum** Mineralpech *n*

aspherical nichtsphärisch
asphyxiate, to ~ ersticken
asphyxiating erstickend ~ gas erstickender
Kampfstoff m, Giftgas n, Stickgas n
asphyxiation Erstickung f
aspirant Anwärter m
aspirate, to ~ ansaugen, aufsaugen, saugen
aspirate Hauch-laut m, -zeichen n ~ air Ansaugluft f ~ sound Hauchlaut m
aspirated and pressure charged mit und ohne
Aufladung f
aspirating hose Mundschlauch m
aspiration Ansaugen n, (sucking-in) Streben n,
Einsaugen f ~ vakuum Ansaugvakuum n
aspirator Luftsauger m, Saug-apparat m,
-flasche f, Sauger m, Windsemaphor m
ASR radar (aerodrome surveillance radar) Flug-
platz-Rundsuch-Radar n
assail, to ~ anfechten
assault, to ~ angreifen, anstürmen, tätlich beleidigen
assault Angriff m, Ansturm m, Sturm m, Vorstoß m
assay, to ~ analysieren, probieren, prüfen
assay Auswertung f, Probe f, Versuch m ~ of
buddled ore Waschprobe f ~ of copper ore
Kupfererzprobe f ~ of washed ore Waschprobe f
assay, ~ apparatus Probegerätschaft f ~ balance
Probierwaage f ~ button Probierkorn n ~
crucible Probier-tiegel m, -tüte f ~ furnace Ka-
pellen-, Probier-ofen m ~ laboratory Probier-
laboratorium n ~ spoon Probierlöffel m ~
weight Probiergewicht f
assayer Anrichter m, Probierer m, Prüfer m ~'s
tongs Probierkluft f, Probierzange f
assaying Probierkunst f ~ by the cupel Probe f
durch Abtreiben ~ vessel Probiergefäß n
assemblage Apparateaufbau m, Aufbau m,
Scharen n, Zusammenbau m, Zusammenstellung f ~ with key piece Schurzwerk n ~ on
roof of street car isolierter Stromabnehmer-
träger m
assemble, to ~ abbinden, aufbauen, aufmontieren, aufrüsten, aufstellen, (sich) scharen,
sammeln, zusammenbauen, zusammenberufen,
(aus Teilen) zusammensetzen, to ~ a piping
eine Rohrleitung f verlegen to ~ butt on butt
anpfropfen
assembled zusammengesetzt ~ by bolts zu-
sammengebolzt
assembled, ~ ball bearing fertig montiertes Kugellager n ~ condition Einbauzustand m ~
position Gebrauchslage f
assembler Monteur m
assembling, ~ bolt Bolzenschrauben m zum
Montieren ~ department Montagehalle f ~
fixture Lehre f zum Zusammenbau ~ jig Zusammenbaulehre f ~ pin Verbindungszapfen
m ~ in place Montage f an Ort und Stelle ~
plant Montagehalle f ~ point Füllort m ~
room Montagehalle f, Montageraum m, Rüsthalle f ~ shop Montierhalle f ~ space Montageraum m ~ station Rangierbahnhof m ~ tool
Montagewerkzeug n ~ work Zusammenstell-
arbeit f
assembly Aufbau m (aus mehreren Teilen), Aufstellung f, Apparatesatz m, Appell m, Bau-

gruppe f, Einbau m, Einsatz m (mech.), Errichtung f, Gesamtheit f, Montage f, Mon-
tagegruppe f, Montierung f, Rüstung f, Satz m,
Verbindung f, Versammlung f, Zusammenbau m, Zusammenstellung f
assembly, ~ of conduction electrons Gesamtheit
f der Leistungselektronen new ~ of an engine
Erstzusammenbau m eines Motors ~ of individual ferries fahrgliederweiser Einbau m ~
of lenses Linsensatz m ~ of prefabricated machine parts Baukastensystem n ~ of unit parts
Baukasteneinheit f
assembly, ~ adhesive Montagekleber m ~ bench
Montagewerktisch m ~ clearance Einbauspiel
n ~ conveyer Zusammenbauband n ~ core Re-
aktorkernanordnung f ~ department Mon-
tageabteilung f ~ drawing Zusammenstellungszeichnung f ~ equipment Zusammenbauvorrichtung f ~ guide claw Führungsklaue f ~ jig
Zusammenbauvorrichtung f ~ language Block-
sprache f ~ line Fließband n für Zusammenbau, Fördertisch m, Montagebahn f, Montage-
straße f ~ line manufacture Fließband-
brikation f am laufenden Band ~ line production Fließfertigung f, Reihenfertigung f ~
line work Fließarbeit f ~ plant Montagewerk
n, Sammlungswerkstatt f ~ point Montage-
platz m, Sammelplatz m, Versammlungsplatz
m ~ position Bereitstellungsplatz m ~ rig Zu-
sammenbaugroßvorrichtung f ~ schemes Mon-
tagepläne pl ~ screw Befestigungs-, Verbin-
dungs-schraube f ~ shop Montagewerkstatt
f ~ stand Montage-gestell n, -block m ~ view
Gesamtansicht f ~ welding jig Zusammen-
schweißvorrichtung f ~ work Zusammenbau-
arbeit f
assent, to ~ to beistimmen
assent Zustimmung f
assert, to ~ aussagen, behaupten, beteuern
assertion Angabe f, Beteuerung f, Billigung f,
Geltendmachen n (Anspruch)
assess, to ~ abschätzen, einschätzen, (charges)
festsetzen to ~ an amount (eine) Summe f ermitteln, to ~ a charge eine Gebühr f festsetzen
to ~ damage Entschädigung f festsetzen to ~
damages (or indemnity) Schadenersatz m festsetzen
assessable steuerpflichtig
assessed value Nennwert m
assessing of charges Gebührenermittlung f
assessment Abschätzung f, Einschätzung f,
Steuer f, Umlage f, Wertermittlung f, Wertung
f ~ system Umlageverfahren n ~ system for
annuities Rentendeckungsverfahren n
assessor Assessor m, Beisitzer m, Gerichtsbeisitzer m
asset Aktivwert m ~ depreciation account Ver-
mögensabschreibungskonto n ~ investment
Aktivposten m ~ item Aktivposten m ~
register Anlageverzeichnis n, Objektkartei f
~ revaluation Neubewertung f von Aktivposten ~ transfers Umschreibung f von Anlage-
werten
assets Aktive f, Aktivposten m, Bereitstellungen
pl ~ of salvage value Anlagegegenstände pl
asseverate, to ~ zusichern
assiduous ausdauernd, emsig, fleißig, unver-
drossen

**assign, to** ~ abtreten, anberaumen, beiordnen, bevollmächtigen, eingliedern, ernennen, übertragen, (to) zedieren, (to) zuordnen, zuteilen, zuweisen, (a representative) bestellen **to** ~ **a reason for** motivieren **to** ~ **a task** beauftragen **to** ~ **a trunk** Nummer *f* der Verbindungsleitung ansagen oder bezeichnen **to** ~ **a value to** bewerten **to** ~ **a value** (to) einen Wert *m* beilegen

**assign** Rechtsnachfolger *m* ~ **pushbutton** Zielzuweisungsschalter *m*

**assignation** Abtretung *f*, Anweisung *f*, Bevollmächtigung *f*, Ernennung *f*, Zuweisung *f*

**assigned, to be** ~ **to a circuit (or a position)** eine Leitung *f* oder einen Schrank *m* bedienen ~ **couple** Kräftepaardichte *f* ~ **force** Kraftdichte *f* ~ **frequency** Kenn- *f*, Soll-frequenz *f*, Vorselektion *f*, Vorwahl *f*, Vorwählbarkeit *f*, zugeteilte (Soll)frequenz *f* ~ **wave length** Antwortwelle *f*

**assignee** Bevollmächtigter *m*, Rechtsnachfolger *m*

**assignment** Abtretung *f*, Bestellung *f*, Bevollmächtigung *f*, Eingliederung *f*, Einweisung *f*, Übertragung *f*, Übertragungsurkunde *f*, Zuordnung *f*, Zuteilung *f*, Zuweisung *f*, (to) Zession *f* ~ **and assumption agreement** Abtretungs- und Übernahmevertrag *m* ~ **of hours** Dienstplan *m*, Stundenplan *m* ~ **of an international circuit** Benutzung *f* oder Verwendung *f* einer zwischenstaatlichen Leitung ~ **of a mission** Auftragserteilung *f* ~ **of ownership** Besitzeinweisung *f*

**assignment,** ~ **form** Anweiseschein *m* ~ **key** Dienstleitungs-, Überweisungs-taste *f*, Zuteilschalter *m*, Zuteiltaste *f* ~ **method** Verteilungsmethode *f*

**assignor** Abtretender *m*, Anweisender *m*, Übertragender *m*, Zedent *m*

**assimilable** angleichungsfähig, assimilierbar

**assimilate, to** ~ angleichen, assimilieren, ausgleichen

**assist, to** ~ aushelfen, beistehen, helfen, mitwirken, raten, unterstützen

**assistance** Beistand *m*, Hilfe *f*, Hilfeleistung *f*, Hilfskraft *f*, Mitarbeit *f*, Unterstützung *f*, Vorschub *m*

**assistant** Assistent *m*, Begleiter *m*, Gehilfe *m*, Helfer *m*, Mitarbeiter *m* ~ **for copying** Kopierer *m* (photo)

**assistant,** ~ **driver** Begleiter *m*, Beifahrer *m* ~ **holding bar relay** Brückenhilfsrelais *r* ~ **mechanic** Hilfsmonteur *m* ~ **mine surveyor** Markscheidersteiger *m* ~ **starter** Hilfsstarter *m*

**assisted** unselbständig ~ **discharge** unselbständige Entladung *f* ~ **take-off** Abflug *m* mit Starthilfe

**assisting feed roll** Speisezylinder *m*

**associate, to** ~ assoziieren, hinzufügen, sich hinzugesellen, verbinden

**associate** Begleiter *m*, Gefährte *m*, Mitarbeiter *m* ~ **judge** Gerichtsbeisitzer *m* ~ **power of attorney** Untervollmacht *f*

**associated** angelagert, angeschlossen, entsprechend zugeordnet, verbunden, vereinigt, ~ **liquid** angelagerte Flüssigkeit *f* ~ **path** Hilfsleiter *m* ~ **wave** Begleitwelle *f*

**association** Bindung *f*, Gesellschaft *f*, Genossenschaft *f*, Konsortium *n*, Verband *m*, Verein *m*, Vereinigung *f*, Zusammenstellung *f* ~ **of German engineers** VDI (Verein Deutscher Ingenieure) ~ **of ideas** Begriffsgesellung *f* ~ **by relation** Beziehungsassoziation *f*

**assonance** Ähnlichkeit *f*, Assonanz *f*, Übereinstimmung *f*

**assort, to** ~ sondern

**assortment** Sortiment *n*, Sortierung *f*

**assumed,** ~ **loads** Belastungsannahmen *pl* ~ **value** Annahmewert *m*

**assumption** Annahme *f* ~ **of congruence** Kongruenzannahme *f* ~ **of load** Belastungsannahme *f* ~ **of proportionality** Proportionalitätsannahme *f*

**assure, to** assekurieren, versichern

**astable** astabil ~ **multivibrator** astabile Kippschaltung *f*, astabiler Multivibrator *m*

**astatic** astatisch, richtungsunempfindlich ~ **coil** astatische oder selbstschirmende Spule *f* ~ **condition** Astasie *f* ~ **couple** astatisches Nadelpaar *n* ~ **load** astatische Belastung *f* ~ **microphone** nichtgerichtetes oder nichtrichtungsempfindliches Mikrofon *n* ~ **pair** astatisches Nadelpaar *n* ~ **system** astatisches System *n*

**astaticism** astatischer Zustand *m*

**astatics** Astasie *f*

**astaticty** Astasierung *f*

**astatine** Astatin *n*

**astatization** Astasierung *f*

**astatize, to** ~ astasieren, astatisch machen

**astatizing** Astatisierung *f*

**asterisk** Stern *m*, Sternchen *n* ~ **protection device** Sternsymbol-Schutzeinrichtung *f*

**asterism** Asterismus *m*

**astern** achteraus, achtern, hinten, rückwärts ~ **course** achterliche Kurslinie *f* ~ **gear** Rückwärtssteuerung *f* ~ **guide** Rückwärtsgleitschiene *f* ~ **light** Heck-lampe *f*, -leuchte *f*, -licht *n* ~ **running of the propeller** Rückwärtslauf *m* der Schraube ~ **signal** Hecklichtzeichen *n*

**asteroid** Asteroid *m*

**asthenosphere** Asthenosphäre *f*

**astigmatic** astigmatisch ~ **image defect error** astigmatischer Bildfehler ~

**astigmatism** Astigmatismus *m*, Punktlosigkeit *f*, Randunschärfe *f* ~ **control** Randschärfen--korrektur *f*, -regler *m*

**astralon** Astralon *n*

**astray** vom rechten Weg ab

**astride** rittlings

**astrigent** adstringierend, zusammenziehend ~ **effect** Adstringenz *f*

**astro,** ~ **camera** Astrokamera *f* ~ **compass** Astrokompaß *m* ~ **dynamics** Astrodynamik *f*

**astrogation** Astronavigation *f*, Raumnavigation *f*

**astrographic telescope** Himmelsfernrohr *n*

**astroids** Astroiden *pl*

**astrolabe** Sternhöhenmesser *m*, Winkelmesser *m*

**astrometer** Astrofotometer *n*

**astronautical craft** Weltraumschiff *n*

**astronautics** Astronautik *f*, Raumflug *m*, Weltraumschiffahrt *f*

astronomer Sternforscher *m*
astronomic telescope Himmelsfernrohr *n*
astronomical sternkundlich ~ chart Himmelskarte *f* ~ fixing of position astronomische
Ortsbestimmung *f* (finding) ~ telescope Stern-
rohr *n*
astronomy Sternkunde *f*
astrophotography Astrofotografie *f*
astrophysical astrophysikalisch
astrophysicist Astrophysiker *m*
A-switchboard Abgangsplätze *pl*, A-Platzschrank
*m*, A-Schrank *m*
asymeter Asymeter *m*
asymmetric asymmetrisch, nichtsymmetrisch ~
conductance richtungsabhängiges Leitvermögen
*n* ~ sideband transmission Restseitenbandver-
fahren *n* (electr.)
asymmetrical asymmetrisch, ungleichförmig, ungleichmäßig, unsymmetrisch
asymmetrical, ~ characteristic unsymmetrische
Kennlinie *f* ~ compressive stress ausmittige
Druckbelastung *f* ~ conductance asymmetrische Leitfähigkeit *f* (cryst.) ~ conductivity richtungsabhängige Leitfähigkeit *f* ~
heterostatic circuit asymmetrisch-heterosta-
tische Schaltung *f* ~ loading Querlastigkeit *f*
~ modulation Trägeramplitudenabweichung *f*
~ resistance gerichteter oder richtungsab-
hängiger Widerstand *m*
asymmetry Abgleichfehler *m*, Asymmetrie *f*,
Mißverhältnis *n*, Ungleichförmigkeit *f*, Unsymmetrie *f*
asymptote Asymptote *f*; asymptotisch ~ correction chart Hyperbelplan *m*
asymptotic(al) asymptotisch ~ behavior Asymp-
totik *f* ~ directions Asymptotenrichtungen *pl*
expression asymptotische Gleichung *f* ~ focal
length Asymptoten-Brennweite *f* ~ image
formation Asymptoten-Abbildung *f* ~ lines
Asymptotenlinien *pl* ~ value Asymptotenwert
*m*
asynchronous asynchron ~ alternator Asynchron-
generator *m* ~ computer Asynchronrechner *m*
~ discharger asynchrone Scheibenfunkenstrecke *f* ~ motor Asynchronmotor *m* ~ remote-
controlled power station Asynchronspeicher-
werk *n* ~ starting motor Asynchronanwurf-
motor *m*
assume, to ~ annehmen, für wahr annehmen,
gelten lassen, unterstellen, voraussetzen to ~
as existing vorgeben to ~ duty den Dienst *m*
antreten to ~ significance auswirken
assuming Übernahme *f*; angenommen ~ a pasty
state Teigigwerden *n*
assumption Annahme *f*, Ansatz *m*, Unterstellung
*f*, Vermutung *f*, Voraussetzung *f*
assurance Beteuerung *f*, Sicherheit *f*, Versicherung *f*
assure, to ~ gewährleisten, versichern, zusichern
assured Versicherungsnehmer *m*; zuversichtlich
assurer Versicherer *m*
ataxia Ataxie *f*, Unregelmäßigkeit *f*
athermal athermisch
athermanous atherman, wärmedurchlässig ~
to infrared rays undurchlässig für Ultrarot *n*
athodyd Lorinmaschine *f*
athwart dwars quer, querab ~ sea Dwarssee *f*

~ ship dwarsschiffs ~ ship force Querschiffskraft *f* ~ the stream querstroms
atlas Atlas *m*
atmidometer Dunstmesser *m*
atmometer Verdunstungs-gefäß *n*, -messer *m*
atmosphere Atmosphäre *f*, Dunsthülle *f*, Dunst-
kreis *m*, Luft *f* Lufthülle *f*, Luftkreis *m*, Wetter *n*
atmospheric(al) atmosphärisch ~ absorption
atmosphärische Absorption *f* ~ action Luft-
einwirkung *f* ~ agencies Atmosphärilien *pl* ~
condenser Berieselungskondensator *m* ~ condition Luftzustand *m* ~ conditions Wetterlage *f*,
Wetterverhältnisse *pl* ~ corrosion Wetter-
korrosion *f* ~ density Luftdichte *f* ~ disturbance atmosphärischer Störpegel *m* ~ duct
atmosphärische Leitschicht *f*, Troposphärenkanal *m* ~ dynamo Atmosphärendynamo *n* ~
electric current circuit luftelektrischer Stromkreis *m* ~ electricity Luft-, Wolken-elektri-
zität *f* ~ excess pressure Atmosphärenüberdruck *m* (atü) ~ exposure Witterungseinfluß *m*
induction atmosphärischer Ansaugvorgang *m*
~ influences Atmosphärilien *pl* ~ layer atmosphärische Schicht *f*, Luftschicht *f* ~ moisture
Luftfeuchtigkeit *f* ~ nitrogen Luftstickstoff *m*
~ noise atmosphärische Störung *f* ~ oxygen
Luftsauerstoff *m* ~ pollution Staubgehalt *m*,
Verunreinigung *f* der Luft ~ pressure Atmosphäre, Atmosphären-, Luft-, Stau-druck *m* ~
pressure region Luftdruckbereich *m* ~ rerum
atmosphärische Redestillation *f* ~ temperature
Umgebungslufttemperatur *f* ~ tension Atmo-
sphärendruck *m* ~ transmittance atmosphärische Wärmedurchlässigkeit *f* ~ water Niederschlagswasser *n* ~ wave Höhen-, Raum-welle *f*
atmospherics atmosphärische Störung *f*, Gewitterstörung *f*, luftelektrische Störungen *pl*,
Luftstörungen *pl*, Störbefreiung *f*
ATO (assisted take-off) Starthilfe *f*
atoll Lagunenriff *n*
atom Atom *m*, Grundteilchen *n*, Urteilchen *n* ~
chipping Atomumwandlung *f* ~ fission Atom-
-kernspaltung *f*, -zertrümmerung *f* ~ form
factor Atomformfaktor *m*, Strukturfaktor *m*
~ shell Atomhülle *f* ~ smasher Atomauflöser
*m* ~ smashing Atomzertrümmerung *f* ~ splitting
Atomzertrümmerung *f*
atomic atomar, atomartig, atomhaltig, atomisch
~(al) atomistisch ~ arrangement Atomord-
nung *f* ~ boiler Reaktor *m* ~ charge Elementarquantum *n* ~ combining power atombindende Kraft *f* ~ deterrent force Atomstreit-
macht *f* ~ displacement Atomverrückung *f*,
Atomverschiebung *f* ~ domain atomare Anordnung *f* ~ force constants Kopplungspara-
meter *n* ~ hull Atomrumpf *m* ~ hydrogen arc
welding Arcatomschweißung *f* ~-hydrogen arc
welding atomare Wasserstoffschweißung *f* (Was-
sergasschweißung) ~ lattice Stromgitter *n* ~
level Atomhüllenniveau *n* ~ nucleus Atomkern
*m* ~ number Atom-nummer *f*, -ordnungszahl
*f*, -zahl *f*, -ziffer *f*, Kernladungszahl *f*, Stellenzahl *f* ~ orbit Valenzzustand *m* ~ photoelectric
effect Fotoionisation *f* ~ pile Atommeiler *m*
~ ray method Atomstrahlmethode *f* ~ scattering factor Atomformfaktor *m* ~ structure
Raumgitter *n* ~ structure of diamond Diamant-

gerüst *n* ~ **structure factor** Atomformfaktor *m*
~ **theory** Atomlehre *f* ~ **torso** Atomrumpf *m*
~ **union** Atomverbindung *f* ~ **weight** Atom-⸗
gewicht *n*
**atomically dispersed** atomdispers
**atomicity** Atomigkeit *f*
**atomism** Atomismus *m*
**atomizable** zerstäubbar
**atomization** Zerstäubung *f* ~ **jet** Verneblerdüse
*f*
**atomize, to** ~ (a coat or film) aufstäuben, in
kleinste Teilchen aufspalten, (under high
pressure) vernebeln, verpuffen, zerstäuben **to**
~ **as a fan-shaped spray** fächerförmig zerstäuben
**atomized** feingepulvert, gespritzt
**atomizer** Nebelapparat *m*, Zerstäuber *m*, Zer-
stäubungsapparat *m* ~ **cone** Zerstäuberkegel
*m*, Zerstäuberschirm *m* ~ **nozzle** Zerstäuberdüse *f* ~ **pipe** Zerstäuberrohr *n* ~ **plate** Düsenplatte *f* ~ **valve** Einspritzventil *n*
**atomizing,** ~ **agent** Zerstäubungsmittel *n* ~
**burner** Zerstäuberbrenner *m* ~ **carburetor** Ein-
spritzvergaser *m* ~ **drier** Zerstäubungstrockner
*m* ~ **valve** Zerstäuberventil *n*
**atonal interval** atonales Intervall *n*
**A-tone of tuning fork** Kammerton *m*
**atrophy, to** ~ einschrumpfen, schwinden, ver-
kümmern
**atrophy** Schwund *m*, Schwunderscheinung *f*
**attach, to** ~ anbauen, anhängen, anheften, anfügen, angliedern, anmachen, anschließen,
anstecken, (sheet) aufziehen, beimessen, mit
Beschlag belegen, einrichten, festmachen,
hängen, hangen, heften, kuppeln, unterstellen,
zuteilen **to** ~ **to** anbringen beigeben **to** ~ **to**
**a series** anreihen **to** ~ **onto** anarbeiten
**attach cam** Andrückkurve *f*
**attachable** anfügbar, anknüpfbar, aufsteckbar
~ **accessories** Aufsteckzubehör *n* ~ **base plate**
Zulegeplatte *f* ~ **camera** Ansatzkamera *f* ~
**mechanical stage** aufsetzbarer Kreuztisch *m*,
Objektführer *m* (Stereo-Mikroskop) ~-**tape**
**blinker lamp** Anbaublinkleuchte *f*
**attached** beifolgend ~ **to** befestigt, gebunden
(binden), kommandiert zu, zugeteilt **to be** ~
**to** anhaftend ~ **to nucleus** (temporarily) kern-
gebunden
**attached,** ~ **device** Vorsatzgerät *n* ~ **flow** anliegende Strömung *f* (aerodyn.) ~ **governor** an-
gebauter Regler *m* ~ **oscillating direction indicator** Anbaupendelwinker *m* ~ **part** Ansetzstück *n* ~ **piece** Ansetzstück *n* ~ **pier** Pfeiler-
vorlage *f* ~ **thermometer to protect a metal
surface** attachiertes (oder eingefügtes) Thermometer *n*
**attaching,** ~ **lug** Befestigungsknagge *f* ~ **pad**
Anbauflansch *m*, Auflagefläche *f* ~ **parts** An-
bauteile *pl* ~ **plug** Anschlußstecker *m* ~ **(or
connecting) tube** Ansatzrohr *n*
**attachment** Anbringung *f*, Ansatz *m*, Aufhängung *f*, Aufsatz *m*, Befestigung *f*, Beschlagnahme *f*, Fassung *f*, Gehänge *n*, Kupplung *f*,
Nebenvorrichtung *f*, Unterstellung *f*, Zubehör
*n*, Zubehörteil *m*, Zusatzgerät *n*, zusätzliche
Bearbeitungsvorrichtung *f*
**attachment,** ~ **of the control surface** Ruder-
befestigung *f* ~ **of the cylinder** Zylinderbefesti-

gung *f* ~ **for dividing prime numbers** Primz-
zahlenteileinrichtung *f* ~ **for shifting** Zusatz-
drehkondensator *m* ~ **to the transmitting and
receiving unit** Simultanzusatz *m* ~ **of wing to
body** Rumpfanschluß *m*
**attachment,** ~ **collar** Befestigungslasche *f* ~
**feed-trip dog and lever** Anschlag *m* und Hebel
*m* für die selbsttätige Auslösung des Rund-
tischselbstganges ~ **frequency** Einfangfrequenz
*f* ~ **hook** Aufhängehaken *m* ~ **link** Befestigungsglied *n* ~ **objective** Vorsatzobjektiv *n* ~ **plate**
Hängelasche *f* ~ **prism** Vorsatzprisma *n* ~ **rail**
Befestigungsschiene *f* ~ **ring** Aufschraubring
*m* ~ **srew** Anzugs-, Druck-, Klemm-schraube *f*
~ **strap** Klemmschelle *f*
**attachments** Vorrichtung *f* ~ **to cutter** Bezug *m*
für Schneideapparat ~ **to tailpiece of otter**
Bezug *m* für Otterschwanzstück (nav.)
**attack, to** ~ abweisen, anfressen, angreifen,
anstürmen, zerfressen
**attack** Angriff *m*, Angriffsgefecht *n*, Anhall *m*,
Einflüsse *pl*, Inangriffnahme *f* ~ **by acid** Säureangriff *m* ~ **by chemical action** chemischer
Angriff *m* ~ **by corrocion** Korrosionsangriff *m*
~ **by the current** Stromangriff *m* ~ **of fluid
about an airfoil** Beaufschlagung *f*
**attackable** angreifbar
**attackability** Angreifbarkeit *f*
**attacked** angegriffen ~ **by acids** von Säure *f*
angegriffen
**attacking** Erprobung *f* ~ **force** (origin of force)
Adhäsionskraft *f* ~ **power** Angriffskraft *f*
**attain, to** ~ erreichen, erzielen, gewinnen
**attainable** erreichbar
**attained** zurückgelegt ~ **accuracy** erreichte Genauigkeit *f*
**attainment** Erzielung *f*
**attemperator** Bottichschwimmer *m*
**attempt, to** ~ probieren, versuchen
**attempted,** ~ **rescue** Rettungsversuch *m* ~ **salvage**
Bergungsversuch *m*
**attend, to** ~ bedienen, beiwohnen, eine Versammlung *f* besuchen **to** ~ **to** sich befassen mit,
betreuen, nachsehen, pflegen, warten
**attendance** Bedienung *f*, Wartung *f* ~ **card** An-
wesenheitskarte *f*
**attendant** Abnehmer *m*, Aufwärter *m*, Bedienungsmann *m*, Gehilfe *m*, Wärter *m* ~
**of a pile driver** Ramm-Meister *m*
**attendant,** ~ **action** Begleiterscheinung *f* ~
**circumstance** Begleitumstand *m* ~ **path** Fuß-
pfad *m* (r.r.) ~ **phenomenon** Begleiterscheinung
*f*, Nebenerscheinung *f* ~**'s set** Bedienungs-
station *f*, Hauptstelle *f* ~ **telephone** Bedienungs-
fernsprecher *m*
**attended** besetzt
**attention** Acht *f*, Achtung *f*, Aufmerksamkeit *f*,
Augenmerk *n*, Wartung *f* **to the** ~ **of** zur Kennt-
nisnahme ~ **of** zu Händen (on letters)
**attention signal** Achtungssignal *n*
**attentive** achtsam, aufmerksam, sorgfältig
**attentiveness** Aufmerksamkeit *f*
**attenuant** Verdünnungsmittel *n*
**attenuate, to** ~ abdämpfen (electr.), abflächen,
abschwächen, dämpfen, gedämpft werden,
schwächen
**attenuated** gedämpft ~ **tooth** verschwächter
Zahn *m*

**attenuating,** ~ **band** Dämpfungsbereich *n* ~ **sound** Schalldämpfungsmittel *n*
**attenuation** Abklingen *n*, Ausdehnung *f*, Bedämpfung *f*, Dampf *m*, Dämpfung *f*, räumliche Dämpfung *f*, Drosselwirkung *f* (electr.), Flachen *n*, Geräuschbekämpfung *f*, Schwächung *f*, Schalltilgung *f*, Verdünnung *f*, Vergärung *f*, Verkleinerungsgrad *m*
**attenuation** ~ **of high frequencies** Absenkung *f* in den Höhen ~ **in the pass range** Lochdämpfung *f* ~ **of pulsations (or vibrations)** Stoßdämpfung *f* ~ **of sound** Abfall *m* der Schallintensität ~ **of first sound** Abfall *m* der Wärmewellenintensität erster Art ~ **in the space period** Ruhedämpfung *f* (teletype) ~ **in suppressed band** Sperrdämpfung *f* ~ **of ultrasonic beam** Dämpfung *f* eines Ultraschallstrahls ~ **per unit length** Dämpfungskonstante *f*, Dämpfung *f* je Längeneinheit
**attenuation,** ~ **band** Sperrbereich *m* ~ **characteristic of equalization** Enzerrerkurve *f* ~ **characteristics** Dämpfungsverlauf *m* ~ **coefficient** Dämpfungs-konstante *f*, -ziffer *f*, Schwächungskoeffizient *m* ~ **coil** Abflachungsdrossel *f* ~ **compensator** Entzerrer *m* ~ **compensator series type** Längsentzerrer *m* ~ **constant** Dämpfungs-entzerrung *f*, -exponent *n*, Dämpfung *f* je Längeneinheit, Dämpfungs-konstante *f*, -maß *n*, spezifische Dämpfung *f*, Übertragungsmaß *n*, Verlustkonstante *f* ~ **curve** Dämpfungskurve *f*, -verlauf *m* ~ **device** Dämpfungseinrichtung *f* ~ **distortion** Dämpfungs-, Frequenz--verzerrung *f* ~ **equalization** Dämpfungs-ausgleich *m*, -entzerrung *f* ~ **equalizer** Dämpfungs--ausgleicher *m*, -entzerrer *m* ~ **equalizer series type** Längsentzerrer *m* **equalizer-shunt type** Querentzerrer *m* ~ **equivalent** Dämpfungsmaß *n* ~ **factor** Dämpfungszahl *f*, Übertragungsmaß *n* ~ **frequency curve** Dämpfungsfrequenzkurve *f* ~ **layer** Sperrschichtfotoeffekt *m* ~ **length** Gesamtdämpfung *f* ~ **measurement** Dämpfungsmessung *f* ~ **measuring device** Dämpfungsmesser *m*, Dekrementmesser *m*, Dekremeter *n* ~ **mesh** Dämpfungsglied *n* ~ **network** Dämpfungsglied *n* ~ **peak** Dämpfungspol *m* ~ **ratio** Dämpfungs-grad *m*, -faktor *m*, -maß *n* ~ **shunt** Schwächungsnebenschluß *m* ~ **standard** Dämpfungsmaß *n* ~ **switch** Dämpfungsschalter *m* (electr.)
**attenuator** Abschwächer *m*, Dämpfungsglied *n*, Dämpfungswiderstand *m*, Eichleitung *f*, (sound recording and reproduction) Entzerrer, Lautstärkeregler *m*, Schwächungsglied *n*, Verlängerungsleitung *f* ~ **pad** Dämpfungsglied *n* (electr.)
**attest, to** ~ beglaubigen, besagen, bescheinigen
**attestation** Vereidigung *f*
**attested copy** beglaubigte Abschrift *f*
**attic** Boden *m*, Dachgeschoß *n* **in the** ~ am Boden *m* ~ **chamber** Dachstube *f*
**attitude** Fluglage *f*, Haltung *f*, Standpunkt *m*, Stellung *f*, Verhalten *n* ~ **of flight** Fluglage *f* ~ **of target** Zielhöhe *f* ~ **indicator** Fluglageanzeiger *m*, Kreiselhorizont *m*
**attle** taubes Gestein *n*
**attorney** Anwalt *m*, Beauftragter *m*, Bevollmächtigter *m*, Rechtsanwalt *m*, Sachverwalter *m* ~ **for applicant** (patents) Vollmacht *f*

**attract, to** ~ anziehen
**attracted** gezogen ~ **disk electrometer** absolutes Kelvin-Elektrometer *n* ~ **position** Anzugstellung *f* (of a relay)
**attracting and resetting poles** Anzugs- und Rückzugspol *m* (Magnet)
**attraction** Anziehen *n*, Anziehung *f*, Anziehungskraft *f*, Zugkraft *f*
**attractive** ansprechend, anzeihend, reizend ~ **force (or power)** Anziehungskraft *f*
**attribute, to** ~ beilegen, beimessen, zumessen, zurückführen, zuschreiben
**attribute** Eigenschaft *f*
**attrition** Abbau *m*, Abnutzung *f*, Abreibung *f*, Abtragung *f*, mechanischer Abrieb *m*, Schleifwirkung *f*, Verschleiß *m*, Zerreibung *f*, Zerstampfung *f* ~ **product** Ausmahlprodukt *n* ~ **quality** Abnutzbarkeit *f* ~ **rate** Ausfallquote *f*
**A-tube** Kernrohr *n*
**auburn** nußbraun
**auction, to** ~ **under compulsion** zwangsweise versteigern ~ **sale** Versteigerung *f*
**auctioneer** Versteigerer *m*
**audibility** Hörbarkeit *f*, Lautstärke *f*, Vernehmbarkeit *f*, Wahrnehmbarkeit *f* ~ **factor** Hörbarkeitsfaktor *m* ~ **meter** Lautstärkemesser *m* ~ **network** Hörbarkeitslautstärkeregler *m* ~ **test** Lautstärkemessung *f* ~ **threshold** Hörschwelle *f*
**audible** akustisch, hörbar, hörfrequent, vernehmlich ~ **and visible** optischakustisch ~ **and visible warning device** optische akustische Warnanlage *f*
**audible,** ~ **busy signal** Besetztton *m* ~ **detection** hörbare (sichtbare) Anzeigung *f* ~ **event** Schallereignis *n* ~ **frequency** Hör-, Ton-frequenz *f* ~ **range** akustischer Bereich *m* ~ **reception** Hörempfang *m* ~ **ringing signal** Freizeichen *n* (rückwärtiges Zeichen für den abgehenden Ruf) ~ **signal** hörbarer Anruf *m*, hörbares Rufzeichen *n* ~ **signal device** akustische Signalvorrichtung *f* ~ **test** akustische Besetztprüfung *f*, Knackprüfung *f*, Summerprüfung *f*
**audibly** lautbar
**audience** Auditorium *n*, Zuhörer *m*, Zuhörerschaft *f*, Zuschauer *m*
**audio** Gehör *n* ~ **action accompanying television** Begleiten *n* zum Fernsehbild *n* ~ **aid** Hör--gerät *n*, -hilfe *f* ~ **amplification** Niederfrequenz-, Tonfrequenz-verstärkung *f* ~ **amplifier** Niederfrequenzverstärker *m*, Tonfrequenzverstärker *m* ~ **and video film broadcast** Tonrundfunk *m* ~ **band** Tonfrequenzband *n* ~ **carrier** Tonträger *m* **choke** Niederfrequenzdrossel *f* ~ **circuit** Hörfrequenzkreis *m* ~ **control desk** Tonregietisch *m* ~ **control engineer** Toningenieur *m* ~ **detector** zweiter Detektor *m*, Zwischenfrequenzgleichrichter *m* ~ **filter** Tonsieb *n*
**audio-frequency** Hör-, Nieder-Tongeber *m* ~ **circuit** Summerschaltung *f* ~ **section** Summerteil *n*
**audio,** ~ **mixer** Tonmischpult *n* ~-**oscillator** Niederfrequenzoszillator *m*, Schwebungssummer *m*, Tonfrequenzgenerator *m* (TG) ~ **peak limiter** Tonfrequenz-Amplitudenbegrenzer *m* ~ **perspective in sound reproduction (and**

**acoustic)** räumliche Tonwirkung f ~ **range** Hörfrequenzbereich m, Hörgrenze f ~ **reactor** Niederfrequenzdrossel f ~ **receiving device** Tonempfangseinrichtung f ~ **sensation** Hör-, Schall-empfindung f ~ **sensation area** Empfindungsgrenze f des Ohres, Hörfläche f ~ **sensation curve** Ohrenkurve f ~ **sensation scale** Gehörempfindungsskala f ~ **sensitivity** Hörempfindlichkeit f ~ **signal** Gehöranzeige f ~ **stage** Tonkreis m ~ **stage amplifier** Audionstufe f (electr., rdr) ~ **tone** Tonfrequenz f ~ **traffic** Sprechverkehr m ~ **transmitter** Tonsender m ~ **trap** Tonsperrkreis m ~**wave potential** Tonwellenspannung f

**audiogram** Audiogramm n
**audiology** Audiologie f
**audiometer** Gehörmesser m, Hörschärfenmesser n
**audiometry** Gehörprüfung f, Gehörschärfenmessung f
**audion** Audion n, Dreielektroden-rohr n, -röhre f, Elektronenröhre f, Empfangsgleichrichterröhre f, Gleichrichterröhre f (Empfangsgleichrichter), Pliotron n
**audion, ~ amplifier** Audionverstärkerröhre f ~ **circuit** Audionschaltung f ~ **detector** Audiondetektorröhre f, Röhrendetektor m ~ **receiver** Audionempfänger m
**audit, to** ~ revidieren
**auditing** Revision f
**auditor** Hauptbuchhalter m, Rechnungsbeamter m, Rendant m, Revisor m, Zuhörer m
**auditorium** Zuschauerraum m ~ **receiving set** Saalempfänger m
**auditory** Gehör n ~ **canal** Gehörgang m ~ **direction finding** Hörpeilung f ~ **location** Abhorchen n von Flugzeugen ~ **masking** Gehörmaskierung f ~ **perspective** Raumwirkung f, Schall-, Sprach-, Sprech-, Ton-frequenz f ~ **amplification** Niederfrequenz-, Tonfrequenz--verstärkung f ~ **amplification stage** Niederfrequenzverstärkungsstufe f ~ **amplifier** Niederfrequenz-, Tonfrequenz-verstärker m ~ **chopper** Tonfrequenzzerhacker m ~ **circuit** Niederfrequenzschaltung f ~ **control** Tonfrequenz-Fernsteuerung f ~ **current** Sprechstrom m ~ **generator** Tongenerator m, Tonfrequenzgenerator m (rdo) ~ **modulated transmitter** modulationsgesteuerter Sender m ~ **oscillator** Tongenerator m ~ **peak limiter** Tonfrequenz-Spitzenbegrenzer m ~ **powerline carrier control** Rundsteuertechnik f ~ **reception** Niederfrequenztonselektion f ~ **rediffusion** Leitungsrundfunk m ~ **section** Niederfrequenzteil m ~ **stage** Niederfrequenzstufe f ~ **telegraphy** Tonfrequenztelegrafie f ~ **transformer** Niederfrequenztransformator m, Niederfrequenzübertrager m, Sprechtransformator m ~ **tube** Niederfrequenzröhre f
**auger** Bohrer m, Bohrschappe f, Erdbohrer m, Lochbohrer m, Löffelbohrer m, Schneckenbohrer m, Schneckengang m, Schöpfbohrer m, Schwellenbohrer m, Ventilbohrer m, Vorbohrer m, (pole) Nagelbohrer m, (shell) Schappenbohrer m ~ **bit** Bohreisen n, Bohrschappe f, Bohrspitze f, Erdbohrer m, Holzbohrer m, Schappe f (offene), Schappen-, Schlangen-, Schrauben-bohrer m ~ **bit with advance cutter**

Lochbohrer m mit Vorschneider ~ **boring** Handdrehbohrung f, Schappenbohren n ~ **edge** Schappenkante f ~ **hole** Bohrloch n ~ **nose shell** Kratzer m ~ **stem guides** Bohrführungsstücke pl ~ **transition** Auger-Übergang m ~ **type bit** Schneckenbohrer m ~ **yield** Auger-Ausbeute f
**augite** Augit m
**augitic** augitartig
**augment, to** ~ vermehren, verstärken
**augmentation** Steigerung f, Vermehrung f ~ **control** Dämpfungsregler m ~ **distance** Extrapolationsabstand m
**aural** Gehör n; akustisch ~ **and visual audience** Schauhörerschaft f ~ **bearing** Gehörpeilung f ~ **carrier** Tonsignal n (TV) ~ **comparison** Hörvergleichsmessung f ~ **dazzle** (or **dazzling**) akustische Blendung f ~ **determination of direction** Richtungshören n ~ **homing** Gehörzielflug m ~ **masking** Tonmaskierung f (acoust.) ~ **null** akustisches Minimum n (Peilung), Auralnull f, Hörminimum n, Radiostille f ~ **null direction-finder** akustischer Peiler m ~ **radio range** akustische Bake f, akustischer Leitstrahlsender m ~ **reception** Hörempfang m ~ **sensation scale** Gehörempfindungsskala f ~ **sensitivy** Lautstärkeempfindung f ~ **threshold** Hörschwelle f
**auranine** Auranin n
**aurate** Goldoxydsalz n
**aureola** Hof m
**aureole** Heiligenschein m, Strahlenkrone f
**auric, ~ acid** Goldsäure f ~ **compound** Auri-, Goldoxyd-verbindung f ~ **salt** Goldoxydsalz n ~ **sulfide** Goldsulfid n
**aurichalcite** Messingblüte f
**auriferous** goldführend, goldhaltig, güldig, ~ **alluvium** Goldseife f ~ **earth** Golderde f ~ **pyrites** Goldkies m ~ **quartz** Goldquarz m ~ **sinter** Goldsinter m
**aurin** Pararosolsäure f
**auripigment** gelbe Arsenblende f, gelbes Arsensulfid n
**aurora, ~ australis** Südlicht n ~ **borealis** Nordlicht n
**aurothiosulfuric acid** Goldthioschwefelsäure f
**aurous, ~ chloride** Aurochlorid n ~ **compound** Goldoxydverbindung f ~ **cyanide** Goldzyanür n ~ **hydroxide** Goldhydroxydul n ~ **iodide** Goldjodür n ~ **oxide** Goldoxydul n ~ **sulfide** Goldsulfür n
**auscultation** Auskultation f, Auskultieren n
**austempering** Austemperung f, Zwischenstufenvergütung f
**austenite** Austenit n ~ **antirust steel** austenitischer rostfreier Stahl m
**austenitic** austenitisch ~ **manganese steel** Manganhartstahl m ~ **stainless steel** austenitischer rostfreier Stahl m ~ **steel** austenitischer Stahl m
**austerity** Einschränkungsmaßnahme f
**Austin-Cohen transmission formula** Austin-Cohensche Formel f
**Australian ironwood** australisches Eisenholz n
**autarchy** Autarkie f
**authentic** authentisch, quellenmäßig, zuverlässig ~ **copy** beglaubigte Abschrift f
**authenticate, to** ~ beurkunden

**authentication** Beglaubigung *f* ~ **modes** authentische Tonarten *pl*
**authenticity** Echtheit *f*, Zuverläsiigkeit *f*
**author** Autor *m*, Schriftsteller *m*, Urheber *m*, Verfasser *m* ~ **'s proof** Verfasserverbesserung *f*
**authoritative** maßgebend ~ **quotation** Belegstelle *f*
**authority** Amt *n*, Ansehen *n*, Autorität *f*, Befehlsbefugnis *f*, Befugnis *f*, Behörde *f*, Berechtigung *f*, Dienstgewalt *f*, Dienststelle *f*, Ermächtigung *f*, Genehmigung *f*, Instanz *f*, Kapazität *f*, Macht *f*, Sachverständiger *m*, Vollmacht *f*, Vorgesetztenverhältnis *n*, Zeugnis *n* ~ **to prospect** Schürfbefugnis *f* ~ **reference** Beweisstelle *f*
**authorization** Berechtigung *f*, Bevollmächtigung *f*, Ermächtigung *f*, Genehmigung *f*, Mandat *n* ~ **procedure** Genehmigungsverfahren *n*
**authorize, to** ~ berechtigen, bevollmächtigen, erlauben, ermächtigen, genehmigen
**authorized** befugt, behördlich zugelassen, berechtigt, bevollmächtigt, dienstlich, zuständig ~ **to sign** zeichnungsberechtigt
**authorized,** ~ **agent** Bevollmächtigter *m* ~ **person** Berufener *m* ~ **share capital** bewilligtes registriertes Aktienkapital *m* ~ **strength and equipment** Soll-stand *m*, -stärke *f*
**auto,** ~ **adaption** Selbstanpassung *f* ~ **alarm** selbsttätiges Seenotalarmgerät *n* ~ **barotropy** Autobarotrop *n* ~ **bias** automatische Vorspannung *f*
**autocapacity,** ~ **coupled** durch gemeinsame Kapazität *f* gekoppelt ~ **coupling** Kupplung *f* durch gemeinsame Kapazität
**auto,** ~ **catalysis** Autokatalyse *f* ~ **centering chuck** selbstzentrierendes Spannfutter *n* ~ **changer** Plattenwechsler *m*
**autochthonous** autochthon
**auto,** ~ **clave** Autoklav *m*, Dämpfer *m* ~ **clave method** Autoklavenverfahren *n* ~ **collimating telescope** Autokollimations-Fernrohr *n* ~ **collimator** Autokollimations-Fernrohr *n* ~ **control** Haltezeichengeber *m*, Dienstzeichengeber *m* (teleph.) ~ **correlation function** Autokorrelationsfunktion *f* ~ **correlogram** Autokorrelogramm *n* ~ **dial** Zieltaster *m* ~ **diffusion** Selbstdiffusion *f* ~ **drop switch** Selbstabwurfschalter *m*
**autodyne** Autodyn *n*, Schwingaudion *n*, selbstschwingende Mischröhre *f*, selbsterregender Schwingungskreis *m*, selbstschwingendes Audion *n*, Selbstüberlagerer *m*, Selbstüberlagerungsschaltung *f*
**autodyne,** ~ **beat receiver** Schwingaudionempfänger *m* ~ **frequency meter** Schwingaudionwellenmesser *m* ~ **oscillator** Autodynempfänger *m* ~ **receiver** Autodynempfänger *m*, Schwingaudionempfänger *m*, Überlagerungsempfänger mit Selbsterregung ~ **reception** Autodyn-, Schwingaudion-, Selbstüberlagerungs-empfang *m*, Überlagerungsempfang *m* mit Selbsterregung ~ **wave meter** Schwingaudionwellenmesser *m*
**auto,** ~**-electronic emission** Autoelektronenemission *f*, Feldelektronenemission *f* ~**-focus enlarger** Vergrößerungsapparat *m* mit selbsttätiger Einstellung ~**-frettage** Autofrettage *n*, Kaltrecken *n*, Selbstschrumpfung *f* ~**-genic soldering** Selbstlötung *f*

**autogenous** autogen, durch sich selbst entstanden ~ **cutting** autogenes Schneiden *n* ~ **cutting shop** Schneidwerk *n* für autogenes Schneiden ~ **ignition temperature** Selbstentzündungstemperatur *f* ~ **welding** autogene Schweißung *f*, Gasschmelzschweißung *f* ~ **welding connection** Schweißanschluß *m* ~ **welding by fusion** Schmelzschweißung *f* ~ **welding plate** Autogenschweißplatte *f* ~ **welding by pressure** Preßschweißung *f* ~ **welding technique** Autogentechnik *f*
**autogenously,** ~ **welded** autogen geschweißt ~ **welded steel-tube fuselage** Rumpf *m* aus autogen geschweißten Stahlröhren
**autogiro** Flug-, Hub-, Trag-schrauber *m*, Windmühlenflugzeug *n*
**autographic,** ~ **color** Autografiefarbe *f* ~ **printing paper** Autografiepapier *n* ~ **recorder** selbstschreibende Meßvorrichtung *f*, Schaubildzeichner *m* ~ **recording apparatus** Diagrammapparat *m*
**autographical** autografiert ~ **printing** Autografie *f*
**auto,** ~**-heterodyne** Autoheterodyn *n*, Schwingaudion *n*, Selbstüberlagerer *m* ~**-ignition** Selbstzündung *f* ~**-inductive** autoinduktiv ~**-inductive coupling** Kopplung *f* durch gemeinsame Induktivität ~**-inductively coupled** autoinduktiv oder durch gemeinsame Induktivität *f* gekoppelt ~**-ionization** Selbstionisation *f*
**autolysis** Autolyse *f*
**automate, to** ~ automatisieren
**automated** automatisiert
**automatic** automatisch, maschinenmäßig, schablonenmäßig, selbsttätig, selbstregelnd ~ **account card feed** Kontenautomat *m* ~ **aerial camera** Reihenbildgerät *n* ~ **air-speed control** Eigengeschwindigkeitsgeber *m* ~ **arc-welding machine** Lichtbogenschweißautomat *m* ~ **assembly machine** Montageautomat *m* ~ **back bias** automatische Rückwärtsregelung *f* ~ **background control (ABC)** selbsttätige Helligkeitsregelung *f* (TV), Kontrastautomatik *f* ~ **band-width selection** Trennschärfenregelung *f* ~ **bar machine** Stangenautomat *m* ~ **bass compensation (a. b. c.)** automatische Baßanhebung *f* ~ **bass control** gehörrichtige Lautstärkeregelung *f* ~ **bevel gear planer** automatische Kegelrandhobelmaschine *f* ~ **biasing of grid** automatische Gittervorspannung *f* ~ **black-level control** Schwarzwertautomatik *f* (TV) ~ **branch exchange** Automatennebenamt *n* ~ **break** selbsttätige Unterbrechung *f* ~ **breaking timing device** Abrißselbstversteller *m* ~ **breech mechanism** selbsttätiger Verschluß *m* ~ **brightness control (ABC)** selbsttätige Helligkeitsregelung *f* ~ **camera** Reihenbildner *m* ~ **carriage** automatische Vorschubeinrichtung *f*, Zeilenautomat *m* ~ **case shift** automatischer Zeichenwechsel *m* ~ **cash-dispenser** Münzgeber *m* ~ **cathead** automatische Winde *f* ~ **centering** Zwangseinmittung *f* ~ **central office** Selbstanschlußamt *n* ~ **check(ing)** Selbstprüfung *f* (data proc.) ~ **chromatogramm scanner** Radiopapierchromatograf *m* ~ **chuck lathe** Futterautomat *m* ~ **circuit break** Selbstausschalter *m*, Selbstunterbrecher *m* ~ **circuit-breaker circuit** Selbstunterbrecherschal-

tung f ~ **circulation** selbsttätiger Umlauf m ~
**clearing** selbsttätige Schlußzeichengebung f,
selbsttätiges Schlußzeichen n ~ **coding** automatische Programmfertigung f ~ **connection**
Wählvermittlung f ~ **contact breaker** selbsttätiger Unterbrecher m ~ **continuous-service
plant** selbsttätiges Laufkraftwerk n ~ **control**
Automatisierung f, Eigensteuerung f, Selbst-
regelung f, Selbststeuer n, Selbststeuerung f
~ **control angle (timing angle)** Selbstverstell-
winkel m ~ **control device** Selbststeuergerät n
~ **control equipment** selbsttätige Regelvorrichtung f ~ **control system** Selbststeueranlage f
~ **control technology** Regelungstechnik f ~
**core-breaking tool** automatischer Kernbohrer
m ~ **coupling** selbstspannende Kupplung f ~
**crankshaft turning lathe** Kurbelwellendreh-
automat m ~ **cross-feed** Quervorschub m ~
**cutoff device** (for soldering irons etc.) Absteller m ~ **cut-out** Selbstunterbrecher m, Sicherungsautomat m ~ **cycle control** Ablaufauto-
matik f, Programmsteuerung f ~ **cycle operation** Programmschaltwerk n ~ **dead reckoning**
automatische Koppelnavigation f ~ **delivery**
Selbstausleger m ~ **delivery apparatus** Verkaufsautomat m ~ **deviation measuring** Test-
maschine f (g/m) ~ **device for lighting of stair-
cases** Treppenbeleuchtungsautomat m ~ **die
sinking machine** (Gesenk-) Kopierfräsautomat
m ~ **direction finder (ADF)** Kreiselkompaß m,
selbstanzeigender Peiler m ~ **discharge** Selbst-
entleerung f ~ **discharge hopper** Selbstentlader
m, Selbstentleerer m ~ **discharging device**
Selbstentleerungsvorrichtung f ~ **disconnection** Selbstauslösung f ~ **disconnector** Absteller m ~ **disengaging motion** Selbstaus-
rücker m ~ **distance control** Entfernungsnach-
lauf m (rdr) ~ **double-casting machine** Doppel-
komplettgießmaschine f ~ **doubling arrangement** Selbstdoublierung f ~ **drive** Selbstantrieb
m ~ **engine control** Kommandogerät n ~ **error
correction** automatische Fehlerkorrektur f ~
**exchange** Automatenamt n, Wählvermittlung
f ~ **extension** automatisches Ausfahren n ~
**fastening device for tip trough** selbsttätige Mul-
denfeststellung f ~ **feed** automatischer Vorschub m, Selbstgang m des Vorschubes, (of
machine tool) Druckschaltung f ~ **floating
pump** selbsttätige Schwimmerpumpe f ~ **foil**
Schaltfolie f (tape rec.) ~ **fore-end** Abreißvor-
derschaft f ~ **free-fall bit** Freifallbohrer m
(selbsttätiger) ~ **freight lift** Last f ohne Füh-
rerbegleitung ~ **fuse setter** Zünderstellmaschine
f ~ **gain control** automatische Lautstärkeregelung f, automatischer Schwundausgleich m,
automatische Schwundregelung f, selbsttätiger
Lautstärkenausgleich m, selbsttätige Verstärkungsregelung f ~ **gain regulation** selbsttätige Verstärkungsregelung f ~ **gas apparatus**
Gasautomat m ~ **gas lighter** Gasselbstanzün-
der m ~ **gear miller** Räderfräsautomat m ~
**grab** Greifkübel m ~ **grab bucket** Selbstgreifer
m ~ **grid** automatisch vorgespanntes Gitter n
~ **grid bias** automatische Gittervorspannung f
~ **grinder for cutter** Fräserschärfautomat m ~
**high speed voltage regulator** Schnellregler m
~ **hobbing machine** Abwälzfräsautomat m ~
**holding device** selbsttätige Besetzthaltung f ~

**hunting** freie Wahl f ~ **ignition** Selbstzündung
f ~ **ignition timing** Zündzeitpunktselbstver-
stellung f ~ **incarrier noise suppression** Stummabstimmung f ~ **insertion of repeaters with
operator dialing** selbsttätige Einschaltung f
von Verstärkern durch Fernwahl der Beamtin
~ **interlock** Selbstsperrung f ~ **interrupter**
Selbstunterbrecher m ~ **isolating valve** automatisches Abschlußventil n ~ **jig lathe** automatische Profildrehbank f ~ **knot-hole pegging
machine** Astausflickautomat m ~ **lathe** Mehr-
spindelstangenautomat m ~ **lead** Vorschub-
räder pl ~ **lift indicator** Füllwaage f ~ **lifting
of a tool on return stroke** selbsttätiges Abheben
n des Werkzeuges beim Rückwärtsgange ~
**light-figure selector** Leichtwertautomatik f ~
**limiter** automatischer Lautstärkebegrenzer m
~ **line** Leitungsautomat m ~ **line counter** Zeilenzähler m ~ **line feed** automatischer Zeilenvorschub m (teletype) ~ **load brake** selbsttätige
Lastbremse f ~ **lock-on** automatische Aufschaltung f (rdr), automatische Erfassung f ~
**loop compensation** Schleifenausgleich m ~
**lubrication** selbsttätige Schmierung f ~ **machine**
Automat m ~ **meshing and demeshing device**
Vorrichtung f zum selbsttätigen Ein- und Ausrücken von einem Getriebezahnrad ~ **message
accounting (AMA)** Gebührenerfassungsein-
richtung f, Zetteldruckverfahren n ~ **miller**
Fräsautomat m ~ **Morse system** Schnellmorse-
system n ~ **multiple bender** Mehrfachbiegeauto-
mat m ~ **multiple cutter finishing lathe** automatische Vielstahlfertigdrehbank f ~ **multiple
riveter** Mehrfachnietautomat m ~ **navigator**
Kurskoppler m ~ **noise-dive stop** Sturzflug-
anschlag m ~ **noise leveller** automatischer
Störbegrenzer m ~ **noise limiter** Stärke-,
Rausch-begrenzer m ~ **numbering** automatische
Numerierung f ~ **nutmilling machine** Muttern-
fräsautomat m ~ **oil-pressure lubrication** automatischer Ölumlauf m ~ **onsetting machine**
mechanische Aufschubvorrichtung f ~ **operation** Selbstgang m ~ **order-wire distributor**
selbsttätiger Dienstleitungsverteiler m ~ **overload control** automatische Überlastregelung f
~ **phase control (APC)** automatische Phasenregelung f (TV) ~ **piling machine** Rüsselapparat
m, ~ **pilot** automatischer Pilot m, automatische
(Kurs)steuerung f, Kreisellagengeber m, Robot-
pilot m, Selbststeuergerät n ~ **pistol** Maschi-
nen-, Selbstlade-pistole f ~ **pitch control** Auf-
bäumregler m (aviat.) ~ **pitching machine** Pech-
automat m ~ **plug board** Schnellschalter m ~
**power feed** zwangsläufiger selbsttätiger Vorschub m ~ **printing machine** Kopierautomat m
(photo) ~ **profiling machine** Kopierfräsautomat m ~ **programming** automatische Programmfertigung f ~ **propeller** Luftschrauben-
automatik f ~ **pull-out** automatische Abfangvorrichtung f ~ **punch** automatischer Locher
m ~ **punching machine** Stanzautomat m ~
**radio monitor** automatischer Sendermonitor
m ~ **reclosing under short-circuit conditions**
Kurzschlußfortschaltung f ~ **recording** selbst-
registrierend ~ **regulation** Selbstregulierung f
~ **regulator** gesteuertes Regelventil n ~ **release**
Selbstauslösung f ~ **removal** Abrißautomatik
f ~ **resetting of direction finder to zero** Nach-

laufeinstellung f ~ **retransmission** selbsttätige Weitergabe f ~ **return switch for direction indicator** Winkerrückstellschalter m ~ **ribbon reverse** automatische Bandumschaltung f ~ **rifle** Selbstladegewehr n ~ **rocker** Schaukelkurvette f ~ **rod magazine** selbsttätige Stangenzuführung f ~ **route selection** selbsttätige Leitweglenkung f ~ **routine test equipment** Einrichtung f für regelmäßige Prüfungen ~ **routing** Lochstreifenübertragung f (teletype) ~ **safety brake** Federspeicherbremse f ~ **screw machine** Automat m, Schraubenautomat m, Schraubendrehbank f ~ **selling machine** Selbstverkäufer m ~ **sensitivity (or volume control)** automatische Empfindlichkeitsregelung f, Regelautomatik f ~ **sequence call** Kettengespräch n ~ **sequence manufacture** automatisierte Serienfertigung f ~ **sequencing** Programmsteuerung f ~ **short circuit reclosing** Kurzschlußfortschaltung f ~ **short-circuit valve** Durchlaufwechselventil n ~ **shutter** automatischer Verschluß m ~ **speed control** Einrichtung f für selbsttätige Beschleunigung ~ **spotter** automatischer Treffanzeiger m ~ **sprinkler** Feuerlöschsprengapparat m ~ **stamping press** Stanzautomat m ~ **starter** Anlasser m, Selbstanlasser m, Starter m ~ **starting device** selbsttätige Anwerfvorrichtung f ~ **steel** Automatenstahl m (mit kurzbrechendem Span) ~ **stock feed** Materialvorschub m ~ **stop** Anschlag m für die selbsttätige Vorschubauslösung ~ **stop device** Fahrsperre f, Anschlagvorrichtung f für den Revolverschlitten ~ **substation** selbsttätiges Unterwerk n ~ **switch** Selbstschalter m, Umschalterrelais n ~**-switch gear** Selbstschalterkasten m ~ **synchronizer** Synchronantrieb m, Synchronservomechanismus m ~ **system** Wählsystem n ~ **table traverse** Tischselbstgang m ~ **tape relay** Lochstreifenweitergabe f ~ **tapping machine** Gewindebohrautomat m ~ **tare unit** automatische Tariereinheit f ~ **telegraph** Maschinen-, Reihen-, Schnell-telegraf m ~ **telephone exchange** Selbstanschlußamt n ~ **telephone room** Vermittlungswählersaal m ~ **telephone system** Selbstanschluß-, Wähler-, Zuteilungs-system n ~ **telephony** Selbstanschlußwesen n ~ **telescope** Automatenfernrohr n ~ **three-knife trimmer** Dreimesserautomat m ~ **time announcer** selbsttätiger Zeitansager m ~ **timer device** Selbstversteller m ~ **tipper** Selbst-entlader m, -entleerer m ~ **-tipping car** Selbst-entlader m, -entleerer m ~ **toll dialing** Selbstfernwahl f ~ **tone compensation** automatische Klangkorrektion f ~ **towing apparatus** Autoschlepp m ~ **tracking** automatische Zielverfolgung f ~ **traction controller** selbsttätige Zugsteuerung f ~ **trailer coupling** selbsttätige Anhängerkupplung f ~ **transmitter** Maschinen-sender m, -geber m ~ **tray elevator** Kippaufzug m ~ **trip** Selbstauslösung f ~ **tripping** automatische Abschaltung f oder Unterbrechung f ~ **tripping action** Selbstauslösung f ~ **trouble signal** Fehlerglocke f ~ **tuck bar** automatisch einstellbarer Fangteil m ~ **tuning indication** Scharfeinstellung f ~ **tuning means** (sharp) Automatik f ~ **twistdrill grinder** Spiralbohrerschärfautomat m ~ **ventilating louvre** Flatterjalousie f ~ **volume contractor** Dynamikpresser

m ~ **volume control** automatische Lautstärkeregelung f, automatischer Pegelregler m oder Schwundausgleich m, automatische Schwundregelung f, Fadingausgleich m, Schwundausgleichschaltung f, selbsttätiger Lautstärkenausgleich m oder Pegelregler oder Schwundregler, Verstärkungsregelung f ~ **volume control amplifier** automatischer Lautstärkeregler m, Regelverstärker m ~**volume control circuit** automatischer Schwundregler m ~ **volume control means** Fading-, Schwund-automatik f ~ **volume control mixer hexode** Fadingmischhexode f ~ **volume control potential** Regelspannung f ~ **volume control system** automatischer Lautstärkeregler m oder Schwundregler m ~ **volume expander** Dynamikdehner m ~ **washer** Sprudelwascher m ~ **water-level regulator** Wasserstandsregler m ~ **weapon** Selbstlader m ~ **weld** Maschinenschweißung f ~ **welding machine** Schweißautomat m ~ **welding timer** Schweißtakter m ~ **winding** Selbstaufzug m ~ **wire feed** Drahtzuführungsautomat m ~ **working under control of the originating operator** von der Abgangsbeamtin gesteuerter Wählbetrieb m

**automatically** schablonenartig ~ **acting** automatisch wirkend, selbsttätig wirksam ~ **advanced ignition** Zündanlage f mit automatischer Zündzeitpunktverstellung ~ **opening parachute** automatischer Fallschirm m ~ **operated coupling** automatische Schaltkupplung f

**automatics with charging valve** Automatik f mit Auffüllventil

**automaton** Automation f, Automatisierung f

**automaton** Automat m

**automobile** Kraftfahrzeug n, Kraftwagen m ~ **with inside drive** Innenlenker m

**automobile, ~ accident** Kraftfahrunfall m ~ **body** Karosserie f, Kraftfahrzeugaufbau m ~ **body sheet iron** Karosserieblech n ~ **body sheet steel** Automobilblech n ~ **engine** Automobilmotor m ~ **exhibition** Kraftfahrschau f ~ **fault-finder** Autostörsuchgerät n ~ **highway** Autobahn f ~ **show** Kraftfahrschau f ~ **steam engine** Fahrzeugdampfmaschine f ~ **tractor** Zugwagen m

**automolite** Automolit m

**automorphic** idiomorph

**automorphism of a configuration** Automorphismus m einer Konfiguration

**automotive** kraftfahrttechnisch ~ **engine** Fahrzeugmotor m ~ **grease** Kraftwagenfett n ~ **rating** Leistung f (bei Fahrzeugmotoren)

**autonomous** autonom, selbständig

**autonomy** Selbstverwaltung f

**auto, ~ oxidation** Selbstoxydation f ~ **parallax finder** Sucher m mit automatischer Parallaxenkorrektur ~ **photoelectric** selbstfotoelektrisch

**autopilot** Aufrichtmotor m, Dämpfungsregler m, Fahrtgeber m, Lotgeber m, Rudergetriebe n, Selbststeueranlage f ~ **control** Kursgeber m

**auto, ~ pneumatic circuit breaker** Druckgasschalter, m, Druckluftschalter m ~ **protective tube** Strahlenschutzröhre f ~ **punch-impact hardness tester** Schlaghärteprüfer m ~ **radar plot** Elektronenkartenabbildung f ~ **radiograph** Autoradiogramm n ~ **repeater** Richtungsspannungserzeuger m

**autorotation** Autorotation *f*, Eigendrehung *f* ~ **entry** Übergang *m* in die Autorotation
**auto,** ~ **scaler** automatisches Zählwerk *n* ~ **self-excitation** direkte Selbsterregung *f*, innere Mitkopplung *f* ~**sonic** schallgesteuert, tonautomatisch ~ **spotter** automatischer Treffanzeiger *m* ~ **stellung** Eigensignal-Stellung *f* ~ **stereogram** Autostereogramm *n* ~ **stop** automatische Abschaltung *f* ~**-switch** Selbstschalter *m* ~ **switch rack** Zuteilungsgestell *n* ~**-synchronous motor** selbstumlaufender Synchronmotor *m* ~**-towed start** Autoschleppstart *m* ~ **transductor** Auto-, Start-transduktor *m* ~ **transformer** Auto-, Einspulen-, Spar-transformator *m*, Sparwandler *m* ~ **transmitter** Lochstreifensender *m* ~**type** Rasterbild *n* ~**-type printing ink** Autotypiefarbe *f* ~**typy** Autotypie *f*, Rasterätzung *f* ~ **vac system** Unterdruckförderung *f*
**autoxidator** Autoxydator *m*
**autunite** Autunit *m*
**auxanometer** Auxanometer *n*
**auxiliary** Aushelf . . ., Aushilfe *f*, Hilfs . . ., Zusatz *m*; behelfend, behelfs . . ., zusätzlich
**auxiliary,** ~ **adit** Hilfsstollen *m* ~ **agent** Hilfsstoff *m* ~ **aiming mark (or point)** Hilfsziel *n* ~ **air** Hilfsluft *f* ~ **air compressor** Hilfsluftverdichter *m* ~ **air distribution** Servoluftumsteuerung *f* ~ **airfoil stabilizer** Düsenflügel *m* ~ **angle** Hilfswinkel *m* ~ **anode** Hilfsanode *f* ~ **antenna** Behelfsantenne *f*, Hilfsantenne *f* ~ **apparatus** Hilfsapparat *m*, Hilfseinrichtung *f*, Hilfsvorrichtung *f*, Nebenapparat *m*, Rüstsatz *m* ~ **appliance** Hilfseinrichtung *f* ~ **attachment** Zusatzeinrichtung *f*, zusätzliche Einrichtung *f* oder Vorrichtung *f* ~ **ballast tank** Hilfsballasttank *m* ~ **base** Hilfsbasis *f* ~ **battery** Verstärkungsbatterie *f* ~ **bearing** Hilfspeilung *f* ~ **binder** Überlage *f* ~ **bus bars** Hilfssammelschienen *pl* ~ **bushing relay for automatic message accounting** Belegungshilfsrelais *n* bei Gebührenerfassung ~ **calculating line** Rechenhilfslinie *f* ~ **call-back relay** Rückfragehilfsrelais *n* ~ **call transfer relay** Umlegungshilfsrelais *n* ~ **cam** Beiläuferexzenter *m* ~ **carrier** Hilfsträger *m* ~ **central office** Hilfsknotenamt *m* (teleph.) ~ **circuit** Hilfsschaltung *f* (electr.), Hilfsstromkreis *m* (teleph.) ~ **clock** Nebenuhr *f* ~**(repeater) compass** Tochterkompaß *m* ~ **compensation set** Zusatzausgleichsaggregat *n* ~ **compressor** Zusatzverdichter *m* ~ **conduit** Hilfsleitung *f* ~ **connection rod** Nebenpleuel *n* ~ **consideration** Hilfsbetrachtung *f* ~ **construction part** Nebenbauteil *m* ~ **contact** Hilfskontakt *m* ~ **contours** Hilfskurven *pl* ~ **control pressure line** Notsteuerleitung *f* ~ **control surface** Hilfsruder *n* (aviat.) ~ **corrector relay** Hilfskorrektionsrelais *n* ~ **crab** Hilfskatze *f* ~ **dam** Hilfsdamm *m*, Vorsperre *f* ~ **datum** Hilfsabbildung *f* (Meßkammer) ~ **department** Hilfsbetrieb *m* ~ **detail** Nebenarbeit *f* ~ **device** Hilfseinrichtung *f*, Hilfsvorrichtung *f* ~ **direction-finder station** Peilnebenstelle *f* ~ **director** Kommandohilfsgerät *n* ~ **drain** Flügelort *m* ~ **drive** Nebenantrieb *m*, (jet) Abzweiggetriebe *n* ~ **drive gear** Geräteantriebpritzel *m* ~ **driven units** durch Hilfsantriebe angetriebene Geräte *pl* ~ **(testing) electrode** Hilfselektrode *f* ~ **embank-**

**ment** Hilfsdamm *m* ~ **engine** Auxiliarmaschine *f*, Servomotor *m* ~ **equipment** Zusatzeinrichtung *f* ~ **exchange** Nebenstelle *f* ~ **exciting coil** Hilfserregerwicklung *f* ~ **fault relay** Störungshilfsrelais *n* ~ **field** Hilfsfeld *n* ~ **field railroad** Eisenbahnzubringerlinie *f* ~ **float** Hilfsschwimmer *m* (Flugzeug) ~ **force** Hilfskraft *f* ~ **frequency** Hilfsfrequenz *f* ~ **front sight** Hilfskorn *n* ~ **furnace for finishing glassware** Einbrennofen *m* ~ **gasoline tank** Hilfsbenzinbehälter *m* ~ **gear** Geländegang *m* ~ **gearbox** Geräteträger *m* ~ **grate** Hilfsrost *m*, Notrost *m* ~ **grid** Hilfsgitter *n* ~ **(control) grid** Hilfsnetz *n* ~ **gripper** Vorgreifer *m* ~ **ground direction-finding station** Hilfspeilstelle *f* ~ **ground target** Erdvergleichsziel *n* ~ **heater** Zusatzheizung *f* ~ **hoist** Hilfshubwindwerk *n* ~ **hoisting (gear) tackle** Hilfshubwerk *n* ~ **holding bar switching relay** Brückenschaltehilfsrelais *n* ~ **horizon camera** Horizontzusatzkamera *f* ~**ignition** Hilfszündung *f* ~ **ignition device** Hilfszündvorrichtung *f* ~ **ignition lead** Reservezündung *f* ~ **implement** Hilfsvorrichtung *f* ~ **jack** Aushilfsklinke *f*, Hilfsklinke *f*, Wiederholungsklinke *f* ~ **jet** Zusatzdüse *f* ~ **jet carburetor** Hilfsdüsenvergaser *m* ~ **key strip transfer relay** Weichenhilfsrelais *n* ~ **lamp** Wiederholungslampe *f* ~ **landing field** Ausweichflugplatz *m*, Hilfslandeplatz *m* ~ **lift of crane** Hilfshubwerk *n* ~ **line** Hilfslinie *f* ~ **lines** Aushilfsadern *pl* (electr.), Mehrfachanschlüsse *pl* ~ **lock** Hilfsklinke *f* ~ **machine** Hilfsmaschine *f*, Zusatzmaschine *f* ~ **manhole** Hilfsbrunnen *m* ~ **map grids** Ergänzungsplanquadrat *n* ~ **master clock** Unterhauptuhr *f* ~ **material** Hilfsstoff *m* ~ **means** Hilfsmittel *n* ~ **mechanism** Hilfseinrichtung *f* ~ **member** Nebenbauteil *m* ~ **modulation set** Modulationssatz *m* ~ **motor for adjusting** Servomotor *m* zum Adjustieren ~ **motor for turning** Servomotor *m* zum Drehen ~ **mount** Hilfslafette *f* ~ **oscillator radio** Hilfsoszillator *m* ~ **outlet** Hilfsableitung *f* ~ **parachute** Ausziehfallschirm *m*, Hilfsschirm *m* ~ **personnel** Hilfskraft *f* ~ **pipe for supplying air in blast furnaces** Notform *f* ~ **piston** Hilfskolben *m* ~ **plane** Hilfsebene *f* ~ **plug-in units** Hilfseinschübe *pl* ~ **point** Collinscher Hilfspunkt *m* ~ **pole** Hilfspol *m* ~ **position** Hilfsplatz *m* ~ **power station** Hilfsunterwerk *n* ~ **power unit (APU)** Hilfstriebwerk *n* ~ **predictor** Kommandohilfsgerät *n* ~ **pump** Hilfspumpe *f* ~ **push-button relay** Tastenhilfsrelais *n* ~ **quantity** Hilfsgröße *f* ~ **radio apparatus** Funkzusatzgerät *n* ~ **range** Gruppengetriebe *n* ~ **record** Hilfsabbildung *f* (Meßkammer) ~ **relay** Hilfsrelais *n* ~ **rider** Beifahrer *m* ~ **ring transfer relay** Rufweiterschaltungshilfsrelais *n* ~ **road** Nebenstraße *f* ~ **roller** Andrückwalze *f* ~ **roller assembly bracket** Schwenkarm *m* ~ **roller bar** Profilrohr *n* ~ **roller system** Andruckvorrichtung *f* ~ **sailplane** Segelflugzeug *n* mit Hilfsmotor ~ **scale** Hilfsmaßstab *m* ~ **scavenging** Zusatzspülung *f* ~ **scavenging port** Nachladeschlitz *m* ~ **seat** Begleitsitz *m* (next to driver) ~ **service supply** Eigenbedarfsversorgung *f* (Kraftwerk) ~ **shaft** Hilfswelle *f*, Nebenwelle *f* ~ **shutter** Nebenverschluß *m* ~ **sight** Kollimator *m* ~

**signal** Schlußzeichenwecker *m* ~ **signal with synchronizing signals (BAS signal)** Hilfs-BAS-Signal *n* ~ **skid** Hilfsschiff '*n*, Notsporn *m* (aviat.) ~ **skid car** Vorspannschleppwagen *m* ~ **slat for movable surfaces** Umlenkflügel *m* ~ **slide valve** Hilfsschieber *m* ~ **spar** Hilfsholm *m*, Hilfsträger *m* (aviat.) ~ **spark** Steuerfunken *m* ~ **spark gab** Hilfsfunkenstelle *f* ~ **store** Hilfsspeicher *n*, Zusatzspeicher *m* (data proc.) ~ **strainer** Hilfsknotenfänger *m*, Katzenfang *m* ~ **suction lead** Hilfslenzleitung *f* ~ **supply (system)** Hilfsnetz *n* ~ **switch** Hilfskontakt *m* (auf Schütz) ~ **tank** Hilfs-, Zusatz-behälter *m* **taphole** Notstich *m* ~ **target** Hilfsziel *n* ~ **telephone line** Ausweichvermittlung *f* ~ **theorem** Hilfssatz *m* ~ **thread** Überlage *f* ~ **tool(s)** Hilfsgerät *n* ~ **track** Hilfskette *f* ~ **track vehicle** Hilfskettenfahrzeug *n* ~ **transmission** Gruppen-, Zusatz-getriebe *n* ~ **transmitter** Hilfssender *m* ~ **trolley** Hilfslaufkatze *f* ~ **unit on board** Hilfsaggregat *n* für Schiffe ~ **valves** Notschütz *n* ~ **variable** Hilfsvariable *f* ~ **variables** Hilfsgrößen *pl* (autom. contr.) ~ **vessel** Begleitschiff *n*, Hilfsschiff *n* ~ **voltage** Hilfsspannung *f* ~ **winding** Hilfswickelung *f* ~ **wing** Hilfsflügel *m* **wing flap** Zusatzflügel *m* ~ **wing slot** Hilfsflügelschlitz *m*

**auxiometer** Auxometer *n*

**auxochromous group** auxochrome Gruppe *f* (chem.)

**avail, to** ~ benutzen, in Anspruch nehmen

**available** anwendbar, erhältlich, nutzbar, ständig, verfügbar, verwendbar, vorhanden, vorrätig, zur Verfügung **to be** ~ bereitstehen **to make** ~ herausbringen (Modell) ~ **for assignment (or commitment or employment)** einsatzfähig

**available,** ~ **buoyancy (or lift)** Nutzhubkraft *f* ~ **code group** Verfügungssignale *pl* ~ **energy** Gütezahl *f* (magnet) ~ **head** verfügbares Gefälle *n* ~ **line** nutzbarer Teil *m* der Bildzeile (facsimile) ~ **noise power** verfügbare Rauschleistung *f* ~ **range of setpoint adjustment** Sollwertbereich *m* ~ **variable group** Signalverfügung *f* (signal codes) ~ **width of a bore** (nutzbarer) Bohrlochquerschnitt *m*

**availability** Erreichbarkeit *f*, Verfügbarkeit *f* (teleph.)

**avalanche** Lawine *f* ~ **of earth** Grundlawine *f* ~ **of electrons** Elektronenlawine *f* ~ **of sand and stones** Mure *f*

**avalanche,** ~ **baffle works** Lawinenschutzmauer *f* ~ **breakdown** Lawinendurchschlag *m* ~ **ionisation** totale Ionisierung *f* im Geigerzählrohr ~ **wind** Lawinenwind *m*

**avalent** nullwertig

**avenue** Anfahrt *f*, Baumgang *m*, Baumallee *f*

**average, to** ~ havarieren, den Durchschnitt bilden (or nehmen), mitteln **to (take the)** ~ das Mittel bilden

**average** Durchschnitt *m*, Havarie *f*, Mittel *n*, Mittelwert *m*; durchschnittlich, mittler **on an** ~ durchschnittlich **on the** ~ durchgängig ~ **out gradients to** Stufen *pl* ausmitteln ~ **of mean deviation** mittlere (quadratische, normale) Abweichung *f*

**average,** ~ **absolute pulse amplitude** mittlere absolute Impulsamplitude *f* ~ **age** Durch-

schnittsalter *n* ~ **analysis** Durchschnittsbestimmung *f* ~ **calculating operation** gemittelte Operationszeit *f* ~ **capacity** Durchschnittsleistung *f* ~ **consumption** Durchschnittsverbrauch *m* ~ **cost** Durchschnittskosten *pl* ~ **damage** gewöhnliche Havarie *f* ~ **degree of polymerization** Durchschnittspolymerisationsgrad *m* ~ **density** mittlere Dichte *f* ~ **excitation energy** mittlere Anregungsenergie *f* ~ **eye (standard eye for photometry)** mittleres normales Auge *n* ~ **force** Durchschnittskraft *f* ~ **illumination** Mittelbildhelligkeit *f*, mittlere Helligkeit *f* (TV) Ruhebelichtung *f*, Ruhelicht *n* ~ **information content per symbol** mittlerer Informationsbelag *m* (einer Nachrichtenquelle) ~ **information rate per time** mittlerer Informationsfluß *m* (einer Nachrichtenquelle) ~ **installation** Aggregat *n* ~ **life** mittlere Lebensdauer *f* ~ **load** Durchschnittsbelastung *f* ~ **measure** Durchschnittsmaß *n* ~ **noise factor** Bandrauschzahl *f* ~ **number** Durchschnittstiter *m*, Durchschnittszahl *f* ~ **output** Durchschnittsausbringung *f*, Durchschnittsleistung *f* ~ **performance** Durchschnittsleistung *f* ~ **picture level (APL)** mittlere Bildhelligkeit *f* (TV) ~ **position action** Pendelregelung *f* ~ **power** mittlere Leistung *f*, Durchschnittsleistung *f* ~ **power output** mittlere Ausgangsleistung *f* ~ **pulse power** mittlere Impulsleistung (rdr) ~ **quality** Durchschnittsqualität *f* ~ **sample** Durchschnitts-muster *n*, -probe *f* ~ **shading component** Grundhelligkeit *f* ~ **shading value** Mittellicht *n* ~ **shop** Durchschnittsbetrieb *m* (Mittelbetrieb) ~ **speed** mittlere oder durchschnittliche Geschwindigkeit *f*, Durchschnittsgeschwindigkeit *f* ~ **speed in revolutions per minute** Mittelwert *m* in Umdrehungen pro Minute ~ **thickness of armor** Durchschnittspanzerstärke *f* ~ **ticket time** durchschnittliche Gesprächszeit *f* ~**transinformation rate per time** mittlerer Transinformationsfluß *m* (einer Nachrichtenverbindung) ~ **value** arithmetischer Mittelwert *m*, Durchschnittswert *m* ~ **value indicator** Mittelwertzeiger *m* ~ **velocity** mittlere Geschwindigkeit *f* ~ **voltage** mittlere Elektrodenspannung *f*

**averaged motion** ausgeglichene Bewegung *f*

**averaging** durchschnittlich betragend, Mittelbildung *f*

**avert, to** ~ ablenken, abwenden

**averted** abgewandt

**avertence** Verschwenkung *f*

**aviation** Aviatik *f*, Fliegerei *f*, Flugwesen *n*, Luftfahrt *f*, Luftfahrtwesen *n*, Flugschiffahrt *f*

**aviation,** ~ **accessories** Luftfahrtbedarf *m* ~ **authorities** Luftfahrtbehörden *pl* ~ **board** Luftamt *n* ~ **convention** Luftverkehrsabkommen *n* ~ **exhibition** Luftfahrtausstellung *f* ~ **industry** Flugzeugbau *m*, Flugzeugindustrie *f* ~ **meeting** Flugveranstaltung *f* ~ **pioneer** Flugpionier *m* ~ **rally** Sternflug *m* ~ **record** Flugrekord *m* ~ **sign** Luftfahrtkennzeichen *n* ~ **symbol** Luftfahrtkennzeichen *n* ~ **weather service** Luftfahrtwetterdienst *m* ~ **weather station** Flugwetterwarte *f*

**aviator** Flieger *m* ~ **outfit** Fliegerausrüstung *f*

**aviatrix** Fliegerin *f*

avidity Avidität *f*
avigation Luftfahrtnavigation *f*
avocation Nebenbeschäftigung *f*
avoid, to ~ abwenden, ausschalten, ausweichen, eliminieren, sich entziehen, umgehen, vermeiden, (einer Sache) aus dem Wege gehen **to ~ a remedy of law** Rechtsmittel *n* aufheben
avoidance Vermeidung *f* ~ **of mistakes** Verständigung *f*
avoirdupois weight Handelsgewicht *n*
awake, to wachrufen
award, to ~ zuerkennen, zuerteilen, zuteilen **to ~ damages** eine Entschädigung zuerkennen **to ~ a medal** prämiieren **to ~ a price** prämiieren
award Ehrensold *n* **the ~ of contract** die Zuteilung des Vertrages ~ **to the inventer** Erfinderzuwendung *f*
aware, to be ~ of unterrichtet sein
awash in überflutetem Zustand *m*
awkward linkisch, unbeholfen, ungeschickt ~ **moment** kritischer Augenblick *m* ~ **pieces** unförmige Stücke *pl*
awl Ahle *f*, Pfriem *m*
awning Markise *f*, Sonnenzelt *n*, Wagenplane *f*
a w u (atomic weight unit) atomare Gewichtseinheit *f*
ax, axe Axt *f*, Beil *n*, Hacke *f*, Schellhammer ~ **of the image** Bildachsen *pl*
axes arranges at an angle achswinklig
axial achsial, achsig, achsrecht, axial, mittig
axial, ~ and cross adjustment Längs- und Querverschiebung *f* ~ **angle** Achsenwinkel *m* ~ **armature relay** Achsenankerrelais *n* Relais *n* mit Achsenlagerung ~ **back-pressure turbine** Axialgegendruckturbine *f* ~ **ball bearing** Axialkugellager *n* ~ **bearings** Längslager *n* ~ **canal** Längskanal *m* ~ **cavity** Trichterlunker *m* ~ **-clearance tolerance** Axialkluft *f* ~ **compression** Knickbeanspruchung *f* ~ **compressor** Axialverdichter *m* ~ **conduct of fire** Gleichlaufverfahren *n* ~ **direction** Achsenrichtung *f* ~ **dispersion** Längenstreuung *f* ~ **feed method** Axialverfahren *n* ~ **float** Axialluft *f*, Axialspiel *n* ~ **flow** axialsymmetrische Strömung *f* ~ **flow compressor** Axialkompressor *m*, Axialverdichter *m* ~ **-flow fan** Axialventilator *m*, Schraubenlüfter *m*, Torpedolüfter *m* ~ **flow impulse turbine** Gleichdruckaxialturbine *f* ~ **flow radiator** Axialkühler *m* ~ **-flow turbine** Axialturbine *f* ~ **-flow-type turbine** Axialturbine *f* ~ **force** Achskraft *f*, Axialkompressor *m*, Axialkraft *f*, Normalkraft *f*, Normalspannung *f* ~ **grooved ball bearing** Axialrillen-Kugellager *n* ~ **inducer** Voraxialrad *n* ~ **inducer stage** Achsialvorstufe *f* ~ **lead style resistor** Widerstand *m* mit achsialen Anschlüssen ~ **load** Achsialbeanspruchung *f*, Achsialdruck *m* ~ **magnification** Tiefenvergrößerung *f* ~ **movement** Axialverschiebung *f* ~ **play** Axialspiel *n*, Endspiel *n* ~ **pressure** Achsenlängs-druck *m*, -schub *m*, Axialdruck *m*, Axialschub *m*, Seitenschub *m* ~ **pump** Schraubenpumpe *f* ~ **ratio** Achsenverhältnis *n* ~ **spacing of turns** (pitch) Ganghöhe *f* ~ **sponginess** sekundärer Lunker *m* ~ **spring** Achsialfeder *f* ~ **stress** Axialbeanspruchung *f*, Normalspannung *f* ~ **superchargers** Axialader *pl* (aviat.) ~ **sup-**

port of a pin-type insulator Klöppel *m* ~ **tension** Axialzug *m* ~ **thrust** Achsenlängs-druck *m*, -schub *m*, Axialdruck *m*, Axialschub *m*, Seitenschub *m* ~ **trolley** Rollenstromabnehmer *m* für Fahrdraht in Gleismitte ~ **tube** Achsenrohr *n* ~ **velocity** Axialgeschwindigkeit *f* ~ **wiring** Axial-, Längs-verspannung *f*
axially, ~ **-flexible screw conveyor** axialelastischer Schneckenförderer *m* ~ **symmetric** axialsymmetrisch, rotationssymmetrisch ~ **symmetrical** rotationssymmetrisch ~ **symmetrical load** drehsymmetrische Belastung *f*
axinite Axinit *m*
axiom Axiom *n*, Grundsatz *m*, Hauptsatz *m* (math.) ~ **of congruence** Kongruenzaxiom *n* ~ **of inaccessibility** Unerreichbarkeitsaxiom *n* ~ **of parallels** Parallelenaxiom *n*
axiomatic concept axiomatischer Begriff *m*
axiomatics Axiomatik *f*
axiometer Axiometer *m*
axioms, ~ **of order** Anordnungsaxiome *pl* ~ **of plane geometry** ebene Axiome *pl* ~ **of solid geometry** räumliche Axiome *pl*
axis Achse *f*, Angel *f*, Hauptlinie *f*, Koordinatenachse *f*, Mittellinie *f*, Richtlinie *f*
axis, ~ **of the abscissas** X-Achse *f* ~ **of aircraft** Flugzeugachse *f* ~ **of the anticline** Sattelachse *f* ~ **of the arch** Sattelachse *f* ~ **of the bore** Seelenachse *f* ~ **of buoyancy** Schwimmachse *f* ~ **of camera** Achse der Bildkamera ~ **of centers** Spitzenachse *f* ~ **of the channel center** Achse *f* der Fahrrinne ~ **of collimation** Zielachse *f* ~ **of cone** Kegelachse *f* ~ **of coordinates** Koordinationsachse *f* ~ **of curvature** Krümmungsachse *f* ~ **of eddy** Wirbelachse *f* ~ **of flotation** Schwimmachse *f* ~ **of groove** neutrale Kaliberlinie *f* ~ **of a groove** neutrale Linie *f* ~ **of grooves** Walzlinie *f* (of rolls) ~ **of a hinge** Angelzapfen *m* einer Türangel, Gickel *m* ~ **of incidence** Einfallslot *n* ~ **of instantaneous rotation** augenblickliche Rotationsachse *f*, Momentachse *f* ~ **of ordinates** Ordinatenachse *f* ~ **of the ordinates** Y-Achse *f* ~ **of oscillation** Schwingungsachse *f* ~ **of pressure** Drucklinie *f* ~ **of reference** Bezugsachse *f* ~ **of refraction** Brechungsachse *f* ~ **of revolution** Drehungsachse *f*, Spindel *f* ~ **of roll** Walzenachse *f* ~ **of rotation** Dreh-, Drehungs-, Impuls-, Kreisel-, Rotations-, Schwenk-, Umdrehungs-achse *f* ~ **of sight** Zielachse *f* ~ **of signal communication** Nachrichtenachse *f* ~ **of similitude of circles and spheres** Ähnlichkeitsachsen *pl* von Kreisen und Kugeln ~ **of the spindle** Spindelachse *f* ~ **of swing** Verkantungsachse *f* ~ **of supply** Nachschubstraße *f* ~ **of symmetry** Spiegelachse *f*, Symmetrieachse *f* ~ **of thrust** Stoßachse *f* ~ **of tilt** Bildhauptwaagerechte *f*, Haupthorizontale *f* (photo), Hauptwaagerechte *f*, Kippachse *f* ~ **of traverse correction** Querrichtachse *f* ~ **of the trough** Muldenachse *f* ~ **of turn** Wendeachse *f* ~ **of vision** Gesichtsachse *f* ~ **of vortex** Wirbelachse *f* ~ **of a waveguide** Hohlleiterachse *f*
axis, ~ **inclination error** Achsneigungsfehler *m* ~ **indicator** Achsenabzeiger ~ **intercept** Achsenabschnitt *m* ~ **parallel** achsenparallel ~ **parallel to the path** Bahnachse *f* ~ **test disk for astigmatic eyes** Achsensehprüfscheibe *f*

für astigmatische Augen *f* ~ **through center of
gravity** Schwerpunktachse *f*
**axle** Achse *f*, Kropfachswelle *f*, Spindel *f*, Welle
*f* ~ **and-bar-suspended motor** Tatzenmotor *m*
~ **coupled with dynamo shaft** mit einer Dyna-
mowelle gekuppelte Achse *f* ~ **of pulley** Blocknagel *m*
**axle,** ~ **base** Achsenabstand *m* ~ **beam** Wellenbalken *m* ~ **bearing** Achslager *n*, Achslagerung
*f*, Kupplung *f* ~**-bearing step** Achslagerschale
*f* ~ **bearings** Rollenachslager *n* (r. r.) ~ **bore**
Achsloch *n*
**axle-box** Achsengehäuse *n*, Achsenlager *n*, Lagerbüchse *f*, Lagerschale *f* ~ **adjusting wedge**
Achslagerstellkeil *m* ~ **body** Achskiste *f* ~
**cover** Achsbüchsendeckel *m* ~ **grease** Fett *n*
zur Schmierung des Achsenlagers ~ **guide**
Achsenbüchsenführung *f* ~ **keep** Achsenlager-
unterkasten *m* ~ **liner** Achslagergleitplatte ~
**oil** Achsenschmieröl *n* ~ **yoke** Achsbüchsjoch
*n*
**axle,** ~ **brace** Achsverstrebung *f* ~ **bracket** Ach-
senbock *m*, Achsträger *m* ~ **camber** Achssturz
*m* ~ **cap** Achsenkappe *f* ~ **cap recorder** Achs-
kappenzähler *m* ~ **case (or casing)** Achsgehäuse *n* ~ **center** Mittelachse *f* ~ **center piece**
Achsmittelstück *n* ~**-changing device** Achs-
wechselvorrichtung *f* (r. r.) ~ **clip** Achsband *n*
~**-clip holder** Achsbandhalter *m* ~ **collar** Achsbund *m*, (wheel) Stoßscheibe *f* ~ **coupling plate**
Achszwinge *f* ~ **cutting-off and centering machine** Achsenabstech- und Zentrierbank *f* ~
**dip** Achssturz *m* ~ **disk** (wheel) Nabenscheibe *f*
~ **drive** Achsantrieb *m* ~**-drive bevel wheel**
Tellerrad *n* ~ **drive pinion** Achsantriebsritzel *n*
~ **driving shaft gasket** Achswellendichtung *f* ~
**end** Achsende *n* ~**-end pivot** Achszapfen *m* ~
**flange (disk)** Achsscheibe *f* ~ **friction** Achsen-
reibung *f* ~ **grease** Achsfett *n*, Achsenschmiere
*f*, Wagenfett *n* ~ **guard** Achshalter *m* ~ **guard
strut** Achsgabelstrebe *f* ~ **head** Achsenkopf
*m* ~ **housing** Achsgehäuse *n* ~ **(housing) tube**
Achsrohr *n* ~**-inclination difference** Hubhöhe *f*
~ **journal** Achs-hals *m*, -kopf *m*, -schenkel *m*,
-stummel *m*, -zapfen *m*, Lagerhals *m* ~ **journal
steering knuckle** Achshals *m* ~ **load** Achsbelastung *f*, Achsdruck *m*, Achsenlast *f* ~
**lowering** Achssenken *n* ~ **middle lathe** Achs-
mitteldrehbank *f* ~ **mounted motor** Tatzen-
lagermotor *m* ~ **mounting** Achslagerung *f* ~
**neck** Achsenhals *m*, Achsenzapfen *m*, Achsschenkel *m* ~ **nut** Achsauflage *f*, Achsmutter *f*
~ **pin** Achsennagel *m*, Achsnagel *m*, Lünse *f* ~
**pivot pin** Längszapfen *m* ~**-pivot steering** Achsschenkellenkung *f* ~ **plate** Achsscheibe *f* ~
**play** Achsspiel *n* ~ **plus two** (shrunk-on) **wheels**
Radsatz *m* ~ **pointing system** Achsenrichtverfahren *n* (g/m) ~ **pressure** Achsdruck *m* ~
**ring** Buchsring *m*, (wagon) Achsring *m* ~**-rough-
ing lathe** Achsschenkelschruppdrehbank *f* ~
**seat** Achssitz *m* ~ **shaft** Differenzialseiten-
welle *f*, Mittelachse *f*, Radnabenzapfen *m*
Steckachse *f*, Wellenbalken *m*, Wellenbaum *m*
~**shift(ing)** Achsverlagerung *f* ~ **sleeve** Achsgehäuse *n* ~ **slide** Achslagerführung *f* ~ **spindle**
Achsschenkel *m* ~ **spindle bolt** Achsschenkelbolzen *m* ~**-spindle-bolt assembly** Achsschenkelträger *m* ~**-spindle-bolt thrust bearing** Achs-

schenkeldrucklager *n* ~ **spring** Achsfeder *f* ~
**springing** Achsenfederung *f* ~ **stay** Achsab-
strebung *f* ~**streamlining** Achsverkleidung *f*
~ **support** Achsenstütze *f* ~ **suspension** Achs-
federung *f* ~ **tee** Achsen-T-Stück *n* (Bremse)
~ **tree** Flügelweite *f*, Kronenwelle *f*, Wellenbaum *m* ~ **tree bed** Achsfutter *n* (of wagon)
~ **tree bed bolster** Achsschemel *m* ~ **tree box**
Achsbüchse *f* ~ **tree of a plow** Pflughaupt *n*
~ **tree stay** Achsstütze *f* ~ **washer** Achshals *m*
~ **weight** Achsbelastung *f* ~ **wheel** Achsenrad
*n* ~ **working machine** Achsenbearbeitungs-
maschine *f* ~ **yoke** Achskabel
**axoid of a rigid motion** Polfläche *f* einer Bewegung
**axometer** Axometer *n*
**axometric** axometrisch
**Ayrton shut** Ayrtonscher Nebenschluß *m*, Mehr-
fachnebenwiderstand *m*
**Az-El (azimuth-elevation)scope** Azimut-Höhen-
bildschirm *m* (rdr)
**azelaic acid** Azelainsäure *f*
**azeotropic** azeotropisch
**azide** Azid *n*
**azimuth** Azimut *m*, Kompaßzahl *f*, rechtweisende Peilung *f*, Richtungswinkel *m*, Schei-
telbogen *m*, Scheitelkreis *m*, Seitenwinkel
*m*
**azimuth,** ~ **of course** Kurswinkel *m* ~ **of the
plate perpendicular** Azimut *m* der Aufnahmerichtung ~ **of target** Zielseitenwinkel *m*, (gun)
Kurswinkel *m*
**azimuth,** ~ **angle** Azimutwinkel *m*, Höhenwinkel *m*, Horizontalwinkel *m* ~ **apparatus**
Peilapparat *m*, Peilgerät *n* ~ **card** Kursrose *f*
(nav.) ~ **change** Teilringänderung *f* ~ **circle**
Teilscheibe *f* ~ **compass** Peilkompaß *m* ~
**cursor** Azimutlinie *f* ~ **disc** Richtkreis *m* ~
**distance** Zenithdistanz *f* ~ **drum** Teiltrommel *f*
~ **field pattern** Horizontaldiagramm *n* ~ **finder**
Azimutzeiger *m* ~ **gyro** Azimutkreisel *m* ~
**indicating goniometer** Azimutpeilvorrichtung
*f* (rdr) ~ **and elevation indicator** Geber *m* (rdr)
~ **instrument** Peilscheibe *f* ~ **knob** Triebscheibe *f* zur Seitenrichtung ~ **micrometer** Teilring *m*, Teiltrommel *f* ~ **reading** Nadelzahl *f*,
Teilringzahl *f* ~ **receiver** Seitenempfänger *m* ~
**resolution** Horizontalwinkelauflösung *f* (rdr)
~ **ring** Abtriftring *m* ~ **scale** Seitenteilkreis
*m* ~ **scope** Seitenrichtrohr *n* ~ **setter seat** Richt-
sitz *m* ~ **spread** Azimutstreuung *f* ~ **tube'** Sei-
tenpeilrohre *f* ~ **value** Azimutwert *m* (rdr),
Breitenwert *m* (of vision)
**azimuthal** azimutabhängig, azimutal ~ **accuracy**
Genauigkeit *f* der Seitenbündlung ~ **angle**
azimutale Abweichung *f* ~ **beam angle** Öffnungswinkel *m* ~ **illumination** Auflicht *n* ~
**lever** Richtungslineal *n* ~ **mode** Azimutal-
schwingung *f* ~ **projection** Azimutalprojection
*f*
**azine group** Azingruppe *f*
**azo,** ~ **compound** Azokörper *m* ~ **cyclic** azo-
zyklisch ~ **derivative** Azokörper *m*
**azotometer** Stickstoffmesser *m*
**azure** hochblau ~ **color** Lasurfarbe *f*
**azurite** Bergblau *n*, Berglasur *f*, blauer Malachit
*m*, Kupferblau *n*, Kupferlasur *f*, Lasurit *m*
~ **blue** Schmelzblau *n*

# B

**babbit, to** ~ ausgießen (Lager)
**babbit** Lagermetall *n*, Lagerweißmetall *n* ~ **bearing** Verbundlager *n* ~ **bushing** Weißmetallagerschale *f* ~**-lined bearing** Lager *n* mit Weißmetallfutter ~**-lined bushing** Leerlaufbuchse *f* mit Weißmetallfutter ~ **lining** Weißmetallfutter *n* ~ **(metal) lining** Weißmetallausguß *m* ~**-metal** Metall-Legierung *f*, Lagermetall *n* (Zinn, Kupfer, Blei, Antimon), Weißmetall *n* ~ **sleeve bearing** Verbundgußgleitlager *n*
**babbited,** ~ **bearing** Lager *n* mit Weißmetallfutter, Weißmetallager *n* ~ **bushing** Leerlaufbuchse *f* mit Weißmetallfutter
**babbitting fixture (or jig)** Umgießvorrichtung *f*
**babble** Diaphonie *f*, Gemurmel *n*
**babbling** Murmeln *n* (acoust.)
**baby,** ~ **Bessemer steel plant** Kleinbessemerei *f* ~ **keg-light** Stufenlinsen-Kleinscheinwerfer *m* ~ **press** Vorpresse *f* (paper mfg.) ~ **unit** Vorpreßanlage *f* (print.)
**bacillary structure** faserige Struktur *f*, Faserstruktur *f*
**bacillus** Keim *m*
**back, to** ~ hinterlegen, umbiegen, (of wind) krimpfen, widerdrücken **to** ~ **off** hinterdrehen, hinterschleifen **to** ~ **out** herausspritzen, hochwirbeln, (of iron in a mold) kochen **to** ~ **up** hinterlegen, rückwärts heranfahren, zurückfahren **to** ~ **off a screw** eine Schraube etwas lösen **to** ~ **the shell** das Galvano hintergießen (print.) **to** ~**-titrate** rücktitrieren **to** ~**tune** einpfeifen (rdr)
**back** Buckel *m*, Dach *n*, Firste *f*, Rücken *m*, Rückseite *f*, Schirm *m*, äußere Wölbung *f* (rdr) ~ **of a blast furnace** Hinterknobben *m* eines Hochofens ~ **of a book** Buchrücken *m* ~ **of a fabric** Stoffrückseite *f* ~ **of floodgate wall** Schleusenrückwand *f* ~ **of a gallery** Firste *f* eines Stollens (min.) ~ **of inclined face** Rückenflanke *f* ~ **of a seat** Rück(en)lehne *f* ~ **of tooth** Zahnrücken *m*
**back,** ~ **ampere turns** Gegenamperewindungen *pl* ~**-arc** Rückzündung *f* bei Gasgleichrichtern ~ **axle** Hinterachse *f* ~ **balance** Gegengewicht *n* im Bremsberg ~ **band** Gabeltragriemen *m* ~ **beam** (of ILS) rückwärtiger Leitstrahl *m* ~**-bearing** Wegflugpeilung *f* ~ **bias** Rückwärtsregelspannung *f* ~**-bone** Hauptstütze *f*, Rückgrat *n* ~**-bone casing** Turbinenträger *m* (Triebwerk) ~ **borehole** Firstenborloch *n* ~ **boring attachment** Hinterbohreinrichtung *f* ~ **brake** Gegen-, Hinter-bremse *f* ~ **bridge wall** Fuchsbrücke *f* ~ **calipers** Taster *m* mit Zahnbogen ~ **center** Panne *f*, Reitnagel *m* (of lathe) ~ **clearance** Rückenspiel *n* ~ **cloth** (back grey) Zwischenläufer *m* (Mit-, Unter-läufer) ~ **component** Hinterglied *n* (einer Linse) ~ **conductance** Sperrleitwert *m* ~ **cone of bevel gear** Ergänzungskegel *m* des Kegelrades ~ **connections** Rückanschluß *m* ~**-coupled** rückgekoppelt ~**-coupling condition** Rückkopplungsbedin-

gung *f* ~ **crank pumping** kombiniertes Pumpen *n* ~ **current** Rückströmung *f* ~ **cushion** Rückenpolster *n* ~ **diffusion** Rückdiffusion *f* ~ **dike** Achterdeich *m*, Hinterdamm *m* ~ **discharge** Rückenentladung *f* ~ **door** Schlupftür *f* ~ **drilling** Hinterbohren *n* ~**-drilling attachment** Hinterbohreinrichtung *f* ~ **drop** Kulisse *f* ~ **ear** Nebenschleife *f* ~ **echo** Hinterbündelecho *n*, Rückzipfelecho *n*
**backed,** ~ **foil** kaschierte Folie *f* ~**-off cutter** Fräser *m* mit hinterdrehten Schneidezähnen, hinterdrehter Fräser *m* ~ **stamper** verstärkte Matrize *f* (acoust.)
**back,** ~ **edge** Hinterkante *f*, Schneidrücken *m* ~**-effect cell** Hinterwandzelle *f* ~ **electromotive force** gegenelektromotorische Kraft *f* ~ **elevation** Hinterfront *f* ~ **emission** Gegenemission *f* ~ **end cover for air pump** Boden *m* zur Luftpumpe ~ **end of a shell** Granatspiegel *m* ~**-end plate** Hinterboden *m* ~ **face** Rückwand *f* ~ **facing** Hinterendplandrehen *n* ~ **facing attachment** Innendrehvorrichtung *f* ~ **fairing** Rückenverkleidung *f* ~ **fall weir** Kropf *m* ~ **feed thread** Rückfördergewinde *n* ~ **fill** Auffüllung *f*
**backfill, to** ~ anschütten
**back,** ~ **filling** Auffüllung *f*, Baugrube *f*, Hinterfüllung *f*, Verdichten *n*, Versatz *m* ~ **filling machine** Rakel-appretiermaschine *f* (textiles), -maschine *f*, -stärkemaschine *f* ~ **filling starcher** Rakel-maschine *f*, -stärkemaschine *f*, -appretiermaschine *f* (textiles) ~ **finish** Linksappretur *f*, Rückseitenappretur *f*
**backfire, to** ~ rückknallen, rückschlagen, zurückknallen
**backfire** Ausgangsklemme *f*, Auspuffknallen *n*, entnehmbare Klemme *f*, Fehlzündung *f*, Frühzündung *f*, Frühzündung *f* geht nach hinten los, Gegenfeuer *n*, Rückfeuerung *f*, Rückprall *m*, Rückzündung *f* ~ **into carburetor** Zurückschlagen *n* im Vergaser ~ **grid** Flammenrückschlagsieb *n*, gitterförmige Flammenrückschlagsicherung *f*
**backfiring** Flammenrückschlag *m*, Gleichrichterrückzündung *f*
**back,** ~**-flame hearth** Pultofen *m* ~ **flash** Fehlzündung *f*, Rückschlag *m* ~ **flow** Rückstrom *m* ~ **flow pipe** Rückstromrohr *n* ~ **flush** Rückspülung *f* ~ **focal legth** hintere Brennweite *f*, Schnittweite *f* ~ **focal plane** hintere Brennebene *f* ~ **force** gegenelektromotorische Kraft *f* ~ **fork** (motorcycle) Hintergabel *f* ~ **fork bridge** Hinterradgabelverbindung *f* ~ **frame of a rig** Hinterteil *n* eines Bohrgerüstes
**back-gear** Radvorgelege *n*, Trommelvorgelege *n*, Übersetzungszahnrad *n*, Vorgelege *n*, Vorgelegewelle *f* ~ **case** Vorgelegeschutzkappe *f* ~ **case guard** Vorgelegeräderhaube *f* ~ **cover guard** Vorgelegeräderschutzkappe *f* ~ **lever** Vorgelegeeinrückhebel *m*, vorgelegter Einrückhebel *m* ~ **mechanism** Spindelvorgelege *n* ~ **pull pin** Rädervorgelegeziehkeil *m* ~ **sliding**

pinion and stem gear Mitnehmerkupplung *f*,
Schieberäder *pl* zur Vorgelegeeeinrückung
back, **~-geared motor** Motor *m* mit Zahnradvorgelege **~ gearing** Zahnradübersetzung *f*,
Zahnradvorgelege **~ gearing arrangement**
Vorgelege *n* **~-glance** (unobstructed view)
Umblick *m*
background Fond *m*, Grund *m*, Grundhelligkeit *f*, (noise) Grundton *m*, Hintergrund *m*,
Hinterkleiden *n* (photo), Nulleffekt *m*, Rauschen *n*, Untergrund *m* ~ **(brightness) control**
Grundhelligkeitsregler *m* **~ collimating mark**
(or point) Bildhintergrundmarke *f*, Hinter-
grundmarke *f* ~ **control** Helligkeitsregler *m*
(rdr) ~ **count** Nulleffekt *m* ~ **counting rate**
Hintergrundzählrate *pl*, Nulleffektimpulse *pl*
~ **counts** Hintergrundzählstöße *pl*, Nulleffektimpulse *pl* ~ **effect** Hintergrundeffekt *m* ~ **effect measurements** Nulleffektmessungen *pl*
~ **eradication** Schleierentfernung *f* ~ **film** Hin-
tergrundtrickbild *n* ~ **grains** Kornuntergrund
*m* ~ **illumination** Gegenbeleuchtung *f*, Mittelbildhelligkeit *f* ~ **level** Grundniveau *n* ~ **noise**
Eigenrauschen *n*, Grundgeräusch *n*, Laufge-
räusche *pl* (tape rec.), Störgeräusch *n*, Wärmerauschen *n* ~ **noise level** Grundgeräuschpegel *m*
~ **noise of radio** Funkenrauschen *n* ~ **pattern**
Flächenmuster *n* ~ **plate** Hintergrundtrick-
bild *n* ~ **projection** Hintergrundprojektion *f*
~ **radiation** Hintergrunds-, Nulleffekt-strah-
lung *f* ~ **rate** Nulleffektrate *f* ~ **scattering** Un-
tergrundstreuung *f* ~ **stability** Nullpunktsta-
bilität *f*
back, ~ **guide roller** Gegenführungsrolle *f* ~ **hand
welding** Nachlinksschweißung *f*, Rückwärtsschweißung *f* ~ **haul cable** Leerseil *n*, Rückhol-
seil *n* ~ **heating** Überhitzung *f*, Rückheizung *f*
~ **illumination** Umfeldbeleuchtung *f*, Hinter-
grundbeleuchtung *f* ~ **induction** Gegeninduktion *f* ~ **iron** Abziehmesser *n*, Putzmesser *n*
~ **joiner** (bookbinding) Einhänger *m* ~ **kick**
Rückentladung *f* (rdo) ~ **knotter** Knotenfänger *m*
backlash Endspiel *n*, Gleichrichterrückzündung
*f*, Hubverlust *m*, leerer Gang *m*, Leergang *m*,
positiver Gitterstrom *m*, Rückzündung *f*,
Schlupf *m*, Spiel *n*, Spielraum *m*, toter Gang *m*,
totes Spiel *n*, unvollständige Gleichrichtung *f*,
(apart from play) Gegenschlag *m*, (of screw
thread) Flankenspiel *n* no ~ frei von totem
Gang ~ **of gears** Spiel *n* der Getrieberäder
~ **of a screw** toter Gang *m* einer Schraube
~ **of teeth** Zahnflankenspiel *n* ~ **of valve**
negativer (Ionen)gitterstrom *m*
backlash, ~ **error** Losefehler *m* (aut. contr.)
~**-free reversal** spielfreies Umsteuern *n* ~ **potential (or voltage)** Gegenspannung *f* (rdo)
backlast Flankenspiel *n*
back, ~ **lens** Hinterlinse *f* ~ **lighting** Hintergrund-
beleuchtung *f*, Rückzündung *f* ~ **lobe** (of
antenna) Hinter-keule *f*, -zipfel *m*, -lappen *m*,
Rückzipfel *m* ~ **log** Arbeitsrückstand *m*, Rückhalt *m* ~**-log of needs** Nachholbedarf *m* ~ **lot**
Ateliergelände *n* (film) ~ **magnetization** Ge-
genmagnetisierung *f*
back-off, ~ **angle** Freiwinkel *m* (b. Räumnadel),
Rückenwinkel *m* ~ **cam** Abdrückkurve *f*
back, ~ **orders** zurückgestellte Aufträge *pl* ~ **pad**

Kammkissen *n*, Rückenpolster *n* ~ **part** Hintergestell *n*, Hinterteil *n* ~**-pedaling brake**
Rücktrittbremse *f* ~ **pick (or shot)** Wiederkehr *f* des Schützenschlages ~ **plan** Hinteransicht *f* ~ **plate** Hinterboden *m*, Hinterwand
*f*
backplate Hinterzacken *m*, Rückplatte *f*, Ver-
schlußleiste *f* ~ **of a firebox** Feuerbuchsrück-
wand *f* ~ **catch (or latch lock)** Zurrstück *n*
~ **pin** Verschlußleistenachse *f*
back, ~ **pointer** Ausfallbündelzeiger *m* ~ **pressure** (of exhaust) Auspuffwiderstand *m*, Aus-
tritt(dampf)spannung *f*, Gegendruck *m*, (apart
from play) Gegenschlag *m*, Rückdruck *m*,
Rückprall *m*
back-pressure, ~ **bottle filler** Gegendruckfla-
schenfüllapparat *m* ~ **engine** Gegendruckma-
schine *f* ~ **pass-off turbine** Gegendruckent-
nahmeturbine *f* ~ **racking apparatus** Abfüllapparat *m* mit Gegendruckventil *n* ~ **roller**
Gegendruckwalze *f* ~ **spring** Gegendruckfeder
*f* ~ **turbine** Gegendruckdampfturbine *f* ~ **valve**
Rückschlagventil *n*
back, ~ **projection** Durchprojektion *f* ~**-pro-
jection equipment** Rückprojektionsanlage *f*
~ **prop** Krücke *f* ~ **pull** Bremszug *m* ~ **puppet**
Hinterdocke *f* ~ **rake** Neigungswinkel *m* einer
Schneidekante ~ **reaction** Rückreaktion *f*
~**-reflection photogram** Rückstrahldiagramm *n*
~**-reflection process** Rückstrahlverfahren *n*
~ **release** Rückauslösung *f*, rückwärtige Auflösung *f* ~ **resistance** Sperrwiderstand *m*
back-rest Anschlaglineal *n* beim Steigungsprü-
fer, Führungsbrille *f*, Hintersupport *m*,
Lünette *f*, Rück(en)lehne *f*, Rückenschutz *m*,
Stützlager *n* ~ **for swing tool** Gegenführung *f*
~ **for turret** Gegenführung *f* (Revolver) ~
**clamp bolt** Gegenlagerständer-Feststellschraube
*f* ~ **gib screw** Gegenlagerleistenschraube *f*
back, ~ **ring** Rückruf *m* ~ **run stop bar** Haltefal-
len *pl* ~ **saw** Bogensäge *f*
backscatter Rückwärtsstreuung *f* (rdr) ~ **factor**
Rückstreuungsfaktor *m* ~ **gauging** Rückstrahl-
messung *f* ~ **measuring chamber** Rückstrahl-
meßkammer *f* ~ **process** Rückstrahlverfahren *n*
back, ~**-scattered electrons** rückwärts gesteuerte
Elektronen *pl* ~**-scattering** Rückstreuung *f*
~**-scattering coefficient** Reflexionsfaktor *m*
(rdr) ~ **seat** Rücksitz *m* ~ **shaft scroll** (textiles)
Mandause *f* ~**-shifting of cone belt** Konusrie-
menrückführung *f* ~ **shifting of wind** Linksablenkung *f* (aviat.) ~ **shifting of wind counterclockwise** regelmäßiger Windwechsel *m* ~
**shunt keying** Tastung *f* mit Trägerunterdrük-
kung *n* ~ **side** Hinterseite *f*, Rückseite *f* ~**-sided**
windschief (wood) ~**-sided timber** windschiefes
Holz *n* ~ **sight** Diopterkimme *f*, Rückblick *m*,
Zielstachel *m* ~ **sight of view-finder** Korn *n*,
Sucherkorn *m* ~ **sight socket** Aufsatzhülse *f*
~ **signal** Rücksignal *n* to ~ **slide** zurücksacken
~**-slope angle** Hinterschliffwinkel *m*, (of cutting
tool) Hinterschleifwinkel *m* ~ **space** Bandrück-
setzen *n* (tape rec.) ~**-space key** Rückstelltaste *f* ~**-space pawl** Rückstellklinke *f* ~ **spacer**
Rücktaste *f*, Rückschalttaste *f* ~**-spacer key**
Rücktaste *f* ~**-spacing control** Korrektur-,
Rückhol-, Wiederhol-taste *f* (tape rec.) ~**-spa-
cing key** Rückzugtaste *f* ~**-spring lock** Bastard-

schloß *n* ~ **squab** Rückenpolsterung *f* ~ **square** Anschlagwinkel *n*, Winkelanschlag *m* ~- **squaring attachment** Rückabbrichtvorrichtung *f* ~ **stand** Hintergestell *n* ~ **stay** Ankerplatte *f*, Backstag *m*, Bergstütze *f*, Verstrebung *f* ~ **stay rope** Rückhaltskabel *n* ~ **steam** Gegendampf *m* ~ **stone** Rückstein *m* ~ **stop** Anschlag *m*, rückwärtiger Anschlag *m*, Anschlagstück *n*, Rückanschlag *m*, Rücklaufsperre *f*, Ruheschiene *f*, (target) Geschoßfang *m* ~ **strap** Rückengurt *m* (Fallschirm), Verbindungsriemen *m* ~ **stream** Gegendampf *m* ~ **streaming** Rückströmung *f* ~ **stripping machine** Einfaß- und Fälzelmaschine *f* ~ **stroke** Rück-bewegung *f*, -gang *m*, -hub *m*, -prall *m*, -schlag *m*, -stoß *m*, -wärtsbewegung *f* ~-**swept** nach hinten verjüngt ~-**swept wing** pfeilförmiger Flügel *m* (aviat.) ~ **taper** Ablaufhahn *m* (tooth), Hinterschneidung *f* ~ **tenon saw** Zapfensäge *f* ~ **tension** Bremszug *m* ~ **tire** Hinterradreifen *m* ~-**titration** Rücktitrieren *n* ~-**to**-~ **connection** (of rectifier elements) Kreuzschaltung *f* ~-**to**-~ **display** Kehrbild *n* (rdr) ~-**to**-~ **method** Rückarbeitsverfahren *n* ~-**to**-~ **pip matching** Gegenschriftanzeige *f* (rdr) ~ **torque** Rückkehrmoment *n* (rdo) ~ **tools** Filet *n* ~ **twist** Gegendrall *m* ~-**type parachute** Rückenfallschirm *m*, Rückenkissenfallschirm *m*

**back-up**, ~ **safety device** Notschutzhilfsgerät *n* ~ **protection** Reserve-, Überlagerungs-schutz *m* ~ **ring** Stützring *m* ~ **roller** Andruckrolle *f* ~ **stock** Reservebestand *m*

**back**, ~ **view** Hinteransicht *f* ~ **wall** Rückwand *f* ~-**wall barrier (blocking) layer cell** Hinterwandsperrschichtzelle *f* ~-**wall rectifier** Hinterwandgleichrichter *m*

**backward** rückläufig, rückliegend, rückständig, rückwärts ~-**acting regulator** Rückwärtsregelung *f* ~-**and-forward bending test** Hin- und Herbiegeprobe *f* ~-**and-forward movement** Hin- und Herbewegung *f*, Rück- und Vorwärtsgang *m* ~-**bent vane** rückwärts gekrümmte Schaufel *f* ~ **current** Sperrstrom *m*, (in rectifier) Rückstrom *f* ~ **difference** absteigende Differenz *f* ~ **differential equation** Rückwärts-Differentialgleichung *f* ~ **erosion** rückschreitende Erosion *f* ~ **glance** Rückblick *m* ~ **movement** Rücklauf *m* ~ **propagation** Rückwärtsausbreitung *f* ~ **push** Rückschlag *m*, Rückstoß *m* ~ **recall signal** Rückwärtsrückruf *m* ~ **regulation** Rückwärtsregelung *f* ~ **resistance** Sperrwiderstand *m* ~ **signalling** Rückwärtssignal *n*, Rückwärtssignalgabe *f*, Rückwärtszeichengebung *f* ~ **stagger** Rückwärtsstaffelung *f* ~ **stop dog** Rücklaufanschlag *m* ~ **transfer admittance** Rückwärts-Scheinleitwert *m* ~ **vision** Rückwärtssicht *f* ~-**wave oscillator** Rückwärtswellenoszillator *m*

**back**, ~ **warmer** Rückenwärmer *m* ~ **warp** Unterkette *f* ~ **wash** Rückstau *m*, zurückgeworfene Wellenbewegung *f*, zurücklaufende Strömung *f* ~ **washer** Plätter *m*

**backwater** Abwasser *n*, Stauwasser *n* ~ **curve** Stau-kurve *f*, -linie *f* ~ **effect** Pfeilerstau *m* (due to bridge piers) ~ **surge** Schwall *m*

**back**, ~ **wave** Zwischenzeichenwelle *f* ~ **wing** Hinterflügel *m*

**backer block** Matrizenuntersatz *m*

**backing** Armkreuzringe *pl*, Bodendrähte *pl*, Gegenhalt *m*, Hintergrunddekor *n* (film), Hinterkleiden *n*, Hinterstampfung *f*, Linksablenkung *f*, Lichthofschutzschicht *f* (film), Rückhalt *m*, Träger *m*, Unterlage *f*, versteifende Ausfütterung *f*, Zurückdrehen *n* (meteor.) ~ **anchor** Kattanker *m* ~ **bearings** Stützrollen *pl* ~ **board** Abpreßbrett *n* ~ **circuit** Kompensationskreis *m* ~ **counterclockwise** Ablenkung *f* nach links ~ **device** Gegenhalter *m* ~ **frame** Anlegrahmen *m* ~ **light** Rückfahrlaterne *f* ~ **material** Stahlunterlage *f* ~ **metal** Hintergießmetall *n* ~-**off** Hinterdrehung *f* ~-**off cam** Abschlagfortschaltkurve *f* ~-**off clearance** Hinterschliff *m* ~-**off device** Hinterdrehvorrichtung *f* ~-**off lathe** Hinterdrehbank *f* ~-**off speed** Abschlagdrehzahl *f* ~-**off time** Abschlagszeit *f* ~-**off and winding-on** Abschlagen *n* und Aufwinden *n* ~-**out** Düsenende *n* ~-**out punch** Durchtreiber *m* ~ **paper** Rohpapier *n*, Stereotypiepapier *n* ~ **pump** Vorvakuumpumpe *f*, (for vaccum) Vorpumpe *f* ~ **roll** Stützwalze *f* ~ **run** Kappnaht *f* ~ **sand** Füllsand *m*, Haufensand *m* ~ **storage** Ergänzungsspeicher *m* ~ **strip** Schweißunterlage *f* ~-**up roll adjustment** Stützrolleneinstellung *f* ~ **wires** Bodendrähte *pl*, (paper) Armkreuzringe *pl*

**backless** rückenfrei

**bacteria suction filters** Bakterienfilternutschen *pl*

**bacterial warfare** Bakterienkrieg *m*

**bactericidal** keimtötend

**bacterizide effect** bakterizide Wirkung *f*

**bacteriologist** Bakteriologe *m*

**bacteriosis** Schleimfäule *f*

**bad**, ~ **condition** Schadhaftigkeit *f* ~ **debts** Zahlungsausfälle *pl* ~ **spot** Bandfehlstelle *f* (tape rec.) ~-**visibility landing aid** Schlechtwetterlandeanlage *f* ~ **weather** schlechtes Wetter *n*, Unwetter *n* ~-**weather landing** Schlechtwetterlandung *f* ~ **weather line** Schlechtwetterstrecke *f* ~ **wool** Ausschußwolle *f*

**badge** Abzeichen *n*, Orden *m*, Rangabzeichen *n*

**badger**, ~-**hair pencil (or brush)** Dachshaarpinsel *m* ~ **hole** Dachsbau *m*

**badigeon** Stuckmörtel *m*

**badlands** Ödland *n*

**badly damaged** schwer beschädigt

**baffle, to** ~ drosseln, stauen, mit Leitblech versehen

**baffle** Ablenkblech *n*, Diaphragm *n*, Lautsprechertruhe *f*, Leitfläche *f*, Leitwand *f*, Lenkwand *f*, Mischblech *n*, Ölfänger *m*, Prellplatte *f*, Schallbrett *n*, Schallschirm *m*, Schalltilgungsmittel *n* (film), Schlingerwand *f* (Flugmotoren) Sperre *f*, Staukörper *m*, Stoßbalken *m*, Streulichtblende *f*, Tauchwand *f*, Umlenkblech *n*, Unterbrecherklappe *f*, Wand *f*, Wandstoffbekleidung *f*, Widerstandskörper *m*, Zugdämpfer *m*, Zunge *f*, Zwischenwand *f*, (loudspeaker) Scheidewand *f*, (partition or board) Schallwand *f*, (plate) Prellblech *n*, (plate, in furnace) Leitblech *n*, (of the separator) Prallwand *f*

**baffle**, ~ **blanket** Schallschluckhülle *f* ~ **board** Schutzbrett *n*, Spritzbrett *n* (paper mfg.) ~ **chamber** Blendengehäuse *n* ~ **cleaner** Tropfenfänger *m* (f. Gasmaschine) ~ **cloth** Bespannung *f* ~ **conductivity** Luftdurchlässigkeit *f* der Leit-

bleche ~ converter Trennplattenumformer *m*
(microwaves) ~ deflector Formblech *n* ~ effect
Stauwirkung *f* ~ grid Staugitter *n* ~ head
Schwallwand *f* (Tankw.) ~ locations Lage *f* der
Trennwände ~ piers Energievernichtungseinrichtung *f* ~ plate Ablenker *m*, Prall-blech *n*,
-platte *f*, -schirm *m*, Prellplatte *f*, Schlingerwand *f*, Spritzfangblech *n*, Stauscheibe *f*,
Trennungsplatte *f*, versetzte Platte *f*, Verteilungsplatte *f* ~ plates with protecting bricks
Rohrwandstein *m* (Kessel) ~ pulley Pralltrom-
mel *f* ~ purifier Stoßreiniger *m* ~ ring Fingerring *m*, Ringleitblech *n* ~ screw Prallschraube *f*
~ separator Stoßreiniger *m* ~ sheet (in a magnetron) Fangblech *n* ~ tightness enges Anliegen *n*
der Kühlluftleitbleche ~ tripod Schlingerstativ *n*
**baffler** Hebel *m*, Nocke *f*
**bag, to** ~ einsacken, sacken, sich bauschen
**bag** Beutel *m*, Blase *f*, Sack *m*, Tasche *f* ~ and
portfolio leather Aktenmappenleder *n* ~-brush-
ing machine Sackbürstemaschine *f* ~ conveyer
Sackförderer *m* ~ electrode Beutelelektrode *f*
~ filter Beutel-, Sack-filter *n* ~ lining Beutel-
futter *n* ~ making machine Taschenherstel-
lungsmaschine *f* ~ paper Beutelpapier *n* ~ piler
Sackstapler *m* ~ sealing Beutelschweißen *n*
~ sealing machine Beutelverschließmaschine *f*
**bagasse** Bagasse *f*, Zuckerrohrtrestor *m* ~ carrier Siruptransporteur *m*, Trebertransporteur
*m* ~ cellulose Bagassezellstoff *m* ~ fiber Zuk-
kerrohrfaser *f* ~ lignin Bagasselignin *n*
**bagatelle** Bagatelle *f*, Kleinigkeit *f*, Lapalie *f*
**baggage** Gepäck *n* ~ car Gepäck-, Pack-wagen
*m* ~ checking Gepäckaufgabe *f* ~ compartment
Gepäckabteil *n*, Gepäckraum *m* ~ convoy Ge-
päcktroß *m* ~ hold Gepäckraum *m* ~ pouch
Packtasche *f* ~ rack Gepäck-brett *n*, -galerie *f*,
-halter *m*
**bagger** Einsacker *m*
**bagging** Einsacken *n* ~ down Lagerung *f* ~ gear
Setzkastenspülung *f* ~ hopper Einsacktrichter
*m* ~ platform Einsackplattform *f* ~ scale Absackwaage *f* ~ spout attachment Sackfüllervor-
richtung *f*
**baggot** paketierte Luppenstäbe *pl*
**baggy** wellig, sackartig
**bail, to** ~ löffeln, schlämmen, schöpfen to ~ out
abspringen, mit dem Fallschirm abspringen,
(water) ausschöpfen
**bail** Bügel *m*, Gehänge *n*; Bürgschaft *f*, Henkel
*m*, Gewähr *f*, Kaution *f* ~ of patent stopper
Flaschenbügel *m* ~ brake (of a winding
machine) Rückschlagbremse *f*
**bailer** Bohrlöffel *m*, Klappsonde *f*, Löffel *m*,
Ölschöpfer *m*, Schlammbüchse *f*, Schlammlöffel *m*, Schöpfer *m*, Schöpflöffel *m*, Schöpfpolle *f*
~ boring Büchsbohren *n* ~ fishing hook Büch-
senfanghaken *m*
**bailing** Löffeln *n* ~ auger Schöpfbohrgerät *n* ~
bucket Schöpfeimer *m* ~ cylinder Löffelzylinder
*m* ~ drum Schlammkran *m*, Schöpfbaspel *f* ~
gear Löffelvorrichtung *f* ~-out Abspringen *n*
(aviat.)
**bailment** Novation *f*
**bailor** Gewährsmann *m*
**bailsman** Gewährsmann *m*
**bain-marie** Wärmeschrank *m* (in Großküche)

**bait, to** ~ hetzen
**bait** Buttern *n*, Fangmittel *n*
**bakable** ausheizbar
**bake, to** ~ ausheizen, backen, brennen, sintern,
trocknen to ~ on anbacken to ~ out ausheizen
to ~ together festbacken to ~ superficially anbacken
**baked** getrocknet ~ crust Schmelzrinde *f* (eines
Meteoriten) ~ dirt gestampfter Lehmboden *m*
~-out ausgeheizt ~ wall Füllmauer *f*
**bakelite** Bakelit *n* ~-bonded Kunstharzbindung *f*
~-bonded grinding bakelitgebundene Schleifscheibe *f*
**baker** Bäcker *m*, Trockenkammer *f*
**baking** Festbacken *n*, Sintern *n*, Sinterung *f*,
(of insulating coils) Ausbacken *n*, (resins)
Hitzehärtung *f*; backend
**baking,** ~ appliance Bratgerät *n* ~ coal backende
Sinterkohle *f* ~ current Formierungsstrom *m*
~ enamel Einbrennlack *m* ~ finish Einbrenn-
farbe *f* ~ hood Backhaube *f* ~ machine Back-
maschine *f* ~-out temperature Ausheiztem-
peratur *f* ~ oven Backofen *m*, Trocken-kammer *f*, -ofen *m* ~ press Kochpresse *f* ~ process
Brennprozeß *m* ~ temperature Brenntemperatur *f* ~ trough Backtrog *m*
**balance, to** ~ abgleichen, abstimmen, abwägen,
ausbalancieren, ausgleichen, auslasten, auswichten, balancieren, begleichen, einspielen,
entlasten (Ventil), nachbilden, saldieren, symmetrieren, wägen, zum Ausgleich bringen, ins
Gleichgewicht bringen, im Gleichgewicht erhalten, im Gleichgewicht sein, das Gleichgewicht herstellen (circuits) to ~ the airplane in
construction auswiegen to ~ dynamically dynamisch auswuchten to ~ out ausbalancieren,
ausgleichen, auskoppeln, aussteuern, auswiegen, auswuchten, entkoppeln, neutralisieren
to ~ up einspielen, gegenseitig abwägen
**balance** Abgleich *m*, Abgleichung *f*, Aufrech-
nung *f*, Ausgleich *m*, Ausgleichung *f*, Auswichten *n*, Auswuchtung *f*, Bestand *m*, Differenz *f*,
Gleichgewicht *n*, Nachbilden *n*, Rest *m*, Saldo
*m*, Waage *f*, Wippe *f*
**balance,** ~ of an airplane Flugzeuggleichgewicht
*n* ~ of the control surface Ruderentlastung *f*
~ of flow Strömungsausgleich *m* ~ on hand
vorhandener Bestand *m* ~ of moment of momentum Drehimpulsbilanz *f* ~ of moments
Momenten-ausgleich *m*, -gleichgewicht *n* ~ of
payments in clearing Zahlungsausgleich *m* ~ of
profit Ertrags-, Gewinn-saldo *m* (wet) ~ of
propeller after lubrication angeglichene Luftschraube *f* nach der Schmierung ~ of slide
Stößelausbalancierung *f* ~ by spring compensation Federausgleich *m* ~ of weight Gewichtsausgleich *m*, Gewichtsausgleichung *f*
**balance,** ~ air entry Bremslufteintritt *m* (motor)
~ arm Waagebalken *m* ~ attenuation Fehlerdämpfung *f* ~ beam Schwunghebel *m*, Waagebalken *m* ~ bridge of Poncelet Ponceletbrücke *f*
~ cable Ausgleichseil *n* ~ cancelled Saldo *m*
gestrichen ~ car Ausgleichwagen *m* ~ card
Bestandskarte *f* ~ carried forward Saldovortrag
*m* ~ condenser Symmetrierkondensator *m* ~
condition Gleichgewichtsfall *m* ~ counter Sub-
traktionszähler *m* ~ counter positions saldieren-
de Zählerstellen *pl* ~ deficiency Nachbildungs-

fehler *m* ~ **equipoise** Balance *f* ~ **error** Abgleichfehler *m*, Ausgleichungsfehler *m* ~ **gear** (governor) Ausgleichwerk *n* ~ **hole** Ausgleich-bohrung *f*, -loch *n* ~ **indicator** Ausgleichs-anzeiger *m*, -stromanzeiger *m*, Lastigkeitswaage *f* ~ **knife-edge** Waageschneide *f* ~ **length** Abgleichlänge *f* ~ **lever** Schwengel *m* ~ **material** Auswuchtstück *n* ~ **method** Abgleichverfahren *n*, Nullmethode *f* ~ **network** Nachbildung *f* ~ **nut** Balanciermutter *f* ~ **pan** Waageschale *f* ~ **piece** Auswuchtstück *n*, Wägestück *n* ~ **pipe** Ausgleichsrohr *n* ~ **plate** Entlastungsplatte *f* ~ **point** Abgreifpunkt *m* ~ **poiser** Schwingkölbchen *n* ~ **range** Abgleichsbereich *m* ~ **relief valve** Druckausgleichsventil *n* ~ **resistance** Ausgleichswiderstand *m* ~ **resistor** Nachbildwiderstand *m* ~ **rod** Balancierlatte *f* ~ **room** Waagezimmer *n* ~ **rope** Unterseil *n* ~ **selection** Saldenauswahl *f* ~ **selector** Saldensteuerapparat *m* ~ **sheet** Abrechnungsnachweis *m*, Rechnungsabschluß *m* ~ **sheet items** Bilanzposten *pl* ~ **spring** Federwaage *f*, Unruhefeder *f* ~ **tab** Ausgleichsfläche *f*, Ausgleichsruder *n* ~ **tank** Ausgleichstopf *m* ~ **test pit** Schleudergrube *f* ~ **tester** Nachbildungsprüfer *m* ~-**testing set** Nachbildungssatz *m* (teleph.) ~ **trough** Balanciertrog *m* ~ **turning tool** Schälstahlhalter *m* ~-**unbalance transformer** Symmetriertopf *m* ~ **valve** Gleichgewichtsventil *n* ~ **weight** Balanciermutter *f*, Belastungsgewicht *n*, Fall-, Gegen-, Schwung-gewicht *n* ~-**weight vibration absorber** als Schwingungsdämpfer wirkendes pendelndes Gegengewicht *n* ~ **weights of the crankshaft** Gegengewichte *pl* der Kurbelwelle ~ **weights of needle valve** Gegengewichte *pl* des Nadelventils ~ **wheel** Hemmungs-, Schwung-, Steig-rad *n*, Unruhe *f* ~-**wheel arbor** Unruhewelle *f* ~ **window** Schwingflügel *m* (Fenster) **balanced** abgeglichen, ausgeglichen, ausgewuchtet, druckentlastet, vorgesteuert, (ailerons) entlastet, symmetrisch ~ **to ground** erdsymmetrisch **balanced,** ~ **amplifier** ausgeglichener Verstärker *m*, Gegentaktverstärker *m* ~ **angle** Winkelabgleich *m* ~ **antenna** entkoppelte oder vom Sender entkuppelte Antenne *f* ~ **armature** ausbalancierter Anker *m*, vierpoliges Ankersystem *n* (acoust.) ~-**armature loud-speaker** Vierpollautsprecher *m* ~-**armature magnetic loud-speaker** entlasteter magnetischer Lautsprecher *m* ~ **attenuation** Symmetriedämpfung *f* ~ **attitude of airplane** Gleichgewichtsflugstand *m* ~ **bridge** Brückenabgleich *m* ~-**bridge network** Kreuzspulgerät *n* ~ **centrifugal weights (or masses)** ausgewuchtete Schleudermassen *pl* ~ **circuit** symmetrischer Kreis *m* ~ **condition** Gleichgewichtszustand *m*, Symmetrie *f* ~ **currents** symmetrische Ströme *pl* ~ **differential transformer** Ausgleichstransformator *m*, Ausgleichsübertrager *m* ~ **feeder** Paralleldrahtleitung *f* ~ **filter** Ausgleichfilter *n* ~-**frame cultivator** Hackmaschine *f* mit balanciertem Rahmen ~ **gate** Gleichgewichts-, Riegel-, Ständer-tor *n*, Wanne *f* mit Gegengewichtsausgleich ~ **guard** Hängemeißel *m* ~ **lever** Hebelstange *f* ~ **line** erdsymmetrische Leitung *f* ~ **load** symmetrische oder ausgleichene Belastung (of a triphase system) ~ **loop (antenna)** abgleichene Rahmenantenne *f* ~ **mixer**

symmetrische Mischstufe *f* ~ **modulator** abgeglichener Modulator *m* ~ **motor** ausgewuchteter Motor *m* ~ **peel** Chargierschwengel *m* ~ **phases** gleichbelastete Phasen *pl* ~ **piston** Kolben *m* mit Druckausgleich ~ **piston valve** entlasteter Kolbenschieber *m* ~ **polyphase system** symmetrisches Mehrphasensystem *n* ~ **position** Gleichgewichts-, Ruhe-lage *f* ~ **pressure** Gleichdruck *m* ~-**pressure method** Druckausgleichverfahren *n* (Indikatoreichung) ~-**pressure torch** Gleichdruckbrenner *m* ~ **quadripole** symmetrischer Vierpol *m* ~ **receiver** ausbalancierter Empfänger *m* ~ **receiver antenna** vom Sender entkoppelte Empfangsantenne *f* ~ **relief valve** rückdruckfreies Überdruckventil *n* ~ **roll** ausbalancierte Walze *f* ~ **rudder** Balanceruder *n* ~ **sense windings** symmetrische Lesewicklung *f* ~ **spool valve** rückdruckfreies Steuerschiebeventil *n* ~ **surface** entlastetes Ruder *n* ~ **three-winding transformer** Ausgleichs-transformator *m*, -übertrager *m* ~ **transformer** Ausgleichstransformator *m* ~ **transmission line** ausgleichene Übertragungsleitung *f* ~ **twin feeder** symmetrische Zweidraht-Speiseleitung *f* ~-**two-terminal-pair network** ausgeglichener Vierpol *m* ~ **valve** entlasteter Schieber *m*, entlastetes Ventil *n* ~ **voltages** erdsymmetrische Spannungen *pl* ~ **wicket gate** Klapptor *n* im Gleichgewicht ~ **wire-circuit** symmetrischer Kreis *m*
**balancer** Schwinge *f*, Schwinghebel *m*
**balancing** Abgleich *m*, Abgleichen *n*, Abgleichung *f*, Ausbalancierung *f*, Ausgleich *m*, Ausgleichen *n*, Ausgleichung *f*, Auswuchten *n*, Auswuchtung *f*, Schütteln *n*, Zentrierung *f*
**balancing,** ~ **of accounts** Gegenrechnung *f*, Rechnungsabschluß *m* ~ **of books (or accounts)** Kassenabschluß *m* ~ **of a circuit** Ausgleichen *n* oder Nachbildung *f* einer Leitung ~ **of circuits** Leitungsabgleichverfahren *n*, Symmetrierung *f* (teleph.) ~ **by condensers** Kondensatorabgleich *m* (teleph.) ~ **of the deflecting voltage** Symmetrierung *f* der Ablenkspannung ~ **for end thrust** Entlastung *f* vom Seitenschub ~ **of the masses** Massenausgleich *m* ~ **of operators' loads** Ausgleich *m* in der Belastung der einzelnen Plätze ~ **of the propeller** Auswuchten *n* der Luftschraube ~ **of a telescope for motion in altitude** Balancierung *f* eines Fernrohres in Höhenbewegung
**balancing,** ~ **action** Pufferwirkung *f* ~ **antenna** Abstimmungsantenne *f* ~ **apparatus** Nullinstrument *n* ~ **arrangement** Abgleichvorrichtung *f* ~ **blade** Zwischenflügel *m* (film) ~ **booster** Ausgleichzusatzmaschine *f* ~ **calculation** Ausgleichsrechnung *f* ~ **capacitor** Neutralisationskondensator *m* ~ **capacity** Ausgleichskapazität *f*, -kondensator *m*, Gegengewicht *n* ~ **cells** Ausgleichzellen *pl* ~ **circuit** Abgleichvorrichtung *f*, Nachbildung *f* ~ **condenser** Neutralisationskondensator *m* ~ **current** Ausgleichsstrom *m* ~ **disc** Entlastungsscheibe *f* (Pumpe) ~ **equipment** Puffergerät *n* ~ **fixture** Auswuchtvorrichtung *f* ~ **force** Gleichgewichtskraft *f* ~ **forward of hinge point** Nasenausgleich *m* ~ **lever** Balancier *m* ~ **machine** Auswuchtmaschine *f* ~ **mandrel** Auswuchtdorn *m* ~ **mass** Auswuchtmasse *f* ~ **method** Kompensations-

verfahren n ~ **network** Ausgleichs-, Kunst-
-leitung f, künstliche Leitung f, Leitungsnach-
bildung f, Entzerrungskreis m (zum Hervorheben einiger Frequenzbereiche) ~ **nut** Aus-
wuchtmutter f ~-out Ausbalancierung f, Aus-
koppelung f, Entkoppelung f, Neutralisierung f
~-out of feed-back Rückkoppelungsunterdrük-
kung f ~-out of jamming Entkoppelung f von
Störern ~ **pipe** Ausgleichsleitung f ~ **pipe**
**connection** Anschlußausgleichsleitung f ~
**piston** Gegendruckkolben m ~ **piston of turbine**
Balancierkolben m ~ **potentiometer** Abgleichpotentiometer n ~ **repeating coil** Nachbildungs-
überträger m ~ **resistance** Abgleich-, Ausgleichswiderstand m ~ **resistor** Symmetriewi-
derstand m ~ **side** Nachbildungsseite f ~ **speed**
Ausgleichsgeschwindigkeit f ~ **stand** Auswucht-
stand m, Gleichgewichtswaage f ~ **surface**
Ausgleichfläche f (aviat.) ~ **tab** Entlastungs-
ruder n ~ **tabulator** Kontrolltabulator m ~
**tester** Abgleichprüfer m ~ **time** Ausgleichzeit f
~ **torque** Ausgleichdrehmoment m ~ **transfor-**
**mer** Ausgleichumformer m ~ **two-way repea-**
**ters** Abgleichschaltung f für Verstärker ~ **valve**
Ausgleichventil n ~ **weight** Stabilitätsgewicht n
~ **wire** Balancierdraht m (electr.)
**balast** Betonzuschlagstoff m
**balata** Balata f ~ **belt** Balatariemen m ~ **belting**
Balatariemenleder n
**balcony** Altan m, Balkon m
**bale, to** ~ ballen, emballieren
**bale** Ballen m, Bündel n, Pack n, Stückgut n ~
**breaker** Ballenbrecher m ~ **goods** Güter pl in
Ballen ~ **knife** Ballenmesser n ~ **opener** Ballen-
öffner m ~ **splitting press** Maschine f zum Zerschneiden von Gummiballen ~ **straw** Preß-
stroh n ~ **tie maker** Ballenbindeapparat m
**baled** zu Paketen gepreßt ~ **scrap** Schrott-
paket n
**baler** Wasserschöpfer m
**baling** Ballen n ~ **band** Ballenband n ~ **board**
Ballenbrett n ~ **box** Füllkasten m ~ **hoop** Bal-
lenreifen m ~ **machine** Ballenpackmaschine f,
Emballiermaschine f ~ **press** Ballenpresse f,
Pack-, Paketier-presse f ~ **wire** Bindedraht m
**balk** Auskeilen n, Balken m, Fehler m, Knüppelrost m, Längsträger m, Mittel-, Streck-, Trag-,
Zug-balken m
**balked landing** Durchstart m (aviat.)
**balky engine** Motor m mit Fehlzündungen
**ball, to** ~ **up** knäueln, zusammenballen
**ball** Ball m, Deul m, Knäuel n, Külbchen n,
Kugel f, Libelle f (Wendezeiger), Pendellinse f,
Posten m, Wolf m
**ball,** ~ **for antifrying protection** Prasselschutzkugel f ~ **of clay** Tonballen m ~ **of fire** Feuerkugel f ~ **of massecuite** Füllmasseknoten m ~
**of slurry** Schlammnest n ~ **of steel** Lotte f ~ **of**
**twine** Bindfadenknäuel n ~**s of wrought iron**
Puddelluppe f
**ball to be forced into the material being tested**
Eindringkörper m
**ball,** ~**-and-line float** Tiefschwimmer m ~**-and-**
**race mill** Horizontalkugelmühle f (roulette)
~**-and-roller-bearing** Kugel- und Tonnenlager n
~**-and-roller-bearing elements** Walzlagerteile pl
~ **and socket** Kugelzapfen m ~**-and-socket**
**bearing** Lager n mit Kugelgelenk ~**-and-socket**

**gear shifting** Kugelschaltung f ~**-and-socket**
**hinge** Kugellagergelenk n ~**-and-socket joint**
Gelenklager n, Knochengelenk n, Kugel-an-
schluß m, -gelenk n, -kuppelung f, -scharnier n,
Universalgelenk n ~ **and spigot** Vor- und Rück-
sprungverbindung f ~ **adapter piece** Kugelstück
n ~ **annulus** Kugellaufring m ~ **bearing** Kugel f,
Kugellager n, Kugellagerung f, Kugel- und
Rollenlager n, Wälzlager n
**ball-bearing,** ~**-bracket drop hanger** Kugelkon-
solhängelager n ~ **bushing** Kugellagerleerlauf-
buchse f ~ **cage** Kugellagerbuchse f ~ **cup**
Kugellagergehäuse n ~ **drop hanger** Kugel-
hängelager n ~ **grease** Fett n für Kugellager ~
**inner race (or ring)** Kugellagerinnenring m ~
**mill** Kugellagermühle f ~ **mounting** Kugeltrag-
lager n ~ **oil** Schmieröl n für Kugellager ~
**outer race** Kugellageraußenring m ~ **pillow**
**block** Kugelstehlager n, Stehkugellager n ~
**pivot** Kugellagerzapfen m ~ **post hanger** Kugel-
säulenkonsollager n ~ **race** Kugellager-fassung
f, -laufring m, -ring m ~ **retainer** Kugellagerringkäfig m ~ **run out** ausgelaufenes Kugellager n ~ **set** Kugellagersatz m ~ **sleeve** Kugel-
lagerbüchse f ~ **slewing gear** Kugeldrehverbin-
dung f (crane) ~ **spacer** Kugellagerringkäfig m
~ **steel** Kugellagerstahl m ~ **stud** Bundbolzen
m ~ **tester** Kulatest m ~ **type spindle** Kugella-
gerspindel f ~ **wallbracket hanger** Kugelwand-
konsollager n ~ **way in track** Kugelführung f
~ **worn out** ausgelaufenes Kugellager n
**ball,** ~ **blasting** Kugelstrahlen n ~ **box** Kugel-
büchse f ~ **buoy** Kugel-boje f, -tonne f ~
**burnishing** Glanz-, Kugel-polieren n ~ **cage**
Kugelkäfig m ~ **cap** Kugelschnellverschluß m
~ **carriage** Kugelwagen m ~ **centering** Kugel-
zentrierung f ~ **charge** Kugelladung f ~ **check**
**nozzle** Kugelverschlußdrüse f ~ **clay** Bindeton
m ~ **clip** Klauen-, Krallen-feder f ~ **closure**
Kugelverschluß m ~ **cloud** Wolkenballen m
(meteor.) ~ **cock** Schwimmerhahn m, Schwimm-
ventil n ~ **connecting branch** Kugelstutzen m ~
**connection** Kugelverschraubung f ~ **contact**
**tip** Meß-, Tast-kugel f ~ **controlled distribution**
Kugelsteuerung f ~ **crusher (or mill)** Ballen-
mühle f ~ **cumulus** zusammengeballte Wolkenmasse f ~ **cup** Kugel-pfanne f, -schale f
**(indentation machine with)** ~ **drop hammer**
Kugelfallhammer m ~ **end** kugelförmiges Endstück n, Kugelkopf m ~ **(end) pin** Kugelzapfen
m ~**-ended strut** Kugelendstück n ~ **float**
Kugelschwimmer m ~**-form condenser** Kugelkühler m ~ **gap** Kugelfunkenstrecke f ~ **gauge**
Kugellehre f ~ **gauge cock** Kugelprobehahn m
~**-gauging device** Kugelmeßvorrichtung f ~
**grip** Ballengriff m ~ **grip screw** Griffschraube f
~ **groove** (Lauf)rille f ~ **guide** Kugelführung f
~ **hammer** Ballhammer m ~ **hand grenade**
Kugelhandgranate f ~ **handle** Kugelgriff m ~
**hardness** Kugeleindruck m ~**-hardness test** Kugeldruck-härteuntersuchung f, -probe f, Kugelprobe f ~ **head** Kugelkopf m ~ **housing** Kugel-
gehäuse n ~ **impact hardness tester** Kugel-
schlaghärteprüfer m ~ **impression** Eindruckkalotte f, Kugeleindruck m ~**-impression test**
Kugeldruck-härteuntersuchung f, -probe f,
Kugelprobe f ~ **imprint** Kugeleindruck m ~ **in-**
**clinometer** Kugelneigungsmesser m ~ **indenta-**

tion Kugeleindruck *m* ~-indentation test Kugel-druckhärteuntersuchung *f*, Kugeldruckprobe *f*, Kugelprobe *f* ~-indentation testing apparatus Kugeldruckhärteprüfer *m* ~ iron Luppeneisen *n*

**ball-joint** Kugellagerverbindung *f*, Kugelver-schraubung *f* ~ case Kugelgelenkgehäuse *n* ~ housing Kugelgelenkgehäuse *n* ~ steering Kugelkreislauflenkung *f*

**ball,** ~ knob Kugelknopf *m* ~ lever Kugelhebel *m* ~ lightning Kugelblitz *m* ~-like kugelähn-lich ~ load Kugelladung *f* ~ lock Kugelverblok-kung *f*, Kugelverschluß *m* ~ lump Luppe *f* ~ mandrel Warmlochdorn *m* (in piercing mills, reelers) ~ measuring (gauging) attachment Ku-gelmeßvorrichtung *f* ~ measuring stand Kugel-meßhalter *m* ~ measuring table Kugelmeßvor-richtung *f* ~ mill Kugel-, Schwing-mühle *f* ~ mill for wet grinding Naßkugelmühle *f* ~ mill with screen Siebkugelmühle *f* ~ mount Kugel-blende *f*, -lafette *f* ~ notch Kugelraste *f* ~ nut Kugel-mutter *f* ~ oiler Kugelöler *m* ~ pane iron Balleisen *n* ~ peen Kugelfinne *f* ~-peen hammer Hammer *m* mit Kugelfinne, Kugelhammer *m* ~ pin Kugelbolzen *m* ~ pin assembly Kugel-bolzengruppe *f* ~ pivot Kugelgelenk *n* ~ point Kugel-kuppe *f*, -spitze *f* ~-pointed contact tip Meßhütchen *n* mit Kugelfläche ~-point(ed) pen Kugel-schreiber *m*, -spitzfeder *f* ~-point setscrew Druckschraube *f* mit Kugelkuppe ~ press Luppenhammer *m* ~ pressure Kugeldruck *m* ~-pressure hardness Kugeldruckhärte *f* ~-pressure test Kugel-druckhärteuntersuchung *f*, -druckprobe *f*, -probe *f* ~ pressure (thrust) test Kugeleindruckversuch *m* ~ pump Kugel-ventilpumpe *f* ~ race Kugel-käfig *m*, -kammer *f*, -korb *m*, -lagerring *m*, -laufbahn *f*, -reihe *f*, -ring *m*, -spur *f*, Leitring *m*, Rollkugelkranz *m*, (of bearing) Laufring *m*, (thrust) Druckkugel-lager *n* ~-race adjuster ring Kuppelspurlager *n* ~ race bearing Laufringkugellager *n* ~ race ring Kugellaufring *m* ~ reception Ball-empfang *m* (electr.) ~ release Ballaufnahme-auslöser *m* ~ resolver Kugelkoordinatenwand-ler *m* ~ retainer Kugelkäfig *m* ~ retainer ring Ringkäfig *m* ~ retaining valve Kugelrück-schlagventil *n* ~ seat Kugelsitz *m* ~ shape Kugelform *f*, Kugelgestalt *f* ~-shaped kugel-förmig ~ shield galvanometer Kugelpanzer-galvanometer *n* ~ shooting hardening Kugel-strahlen *n* ~ socket Kugel-pfanne *f*, -schale *f* ~ spacer Ringkäfig *m* ~ spark gap Kugel-funkenstrecke *f* ~ spinner Kugelwirbel *m* ~ squeezer Luppenquetsche *f* ~ stage Kugel-tisch *m* (micros) ~-stage microscope Kugel-mikroskop *n* ~ steel Kugelstahl *m* ~ stud Kugelzapfen *m* ~ support Kugelhalter *m* (flanged) ~ T Kugel-T-Stück *n* ~ test Eindruck-verfahren *n*, Härteprobe *f* ~ tester Fallhärte-prüfer *m* ~ thrust Druckkugel *f*, Kugeldruck *m* ~-thrust bearing Enddruckkugellager *n*, Kugel-drucklager *n*, Wechsellager *n* ~-thrust hardness Kugeldruckhärte *f* ~-thrust-hardness tester Kugeldruckhärteprüfer *m* ~ (thrust) hardness tester Kugeldruck-maschine *f*, -presse *f* ~ thrust test Kugel-druckhärteuntersuchung *f*, -druckprobe *f*, -probe *f* ~-tipped contact feeler Meßkugel *f*, Meßschnabel *m* mit Kugeln

~-top attachment Destillieraufsatz *m* ~ turner Kugeldrehapparat ~-turning rest Kugeldreh-support *m* ~-turret Bodenkanzel *f* ~-type base Kugelfuß *m* ~-type cell Kugelkalotte *f* ~ (check) valve Kugelventil *n* ~ valve Kegel-ventil *n*, Kugelverschluß *m*, Ventilkugel *f* ~-valve-type lubricator Kugelverschlußöler *m* ~ variometer Kugelvariometer *n* ~ weights Masse *f* des Pendels

**ballast, to** ~ schottern to ~ a road eine Straße beschottern

**ballast** Ballast *m*, Lastladung *f*, Schotter *m*, Schotterbett *n* ~ for cruising Fahrtballast *m* ~ to throw overboard Manövrierballast *m*

**ballast,** ~ bed Schotterschüttung *f* ~ concrete Schotterbeton *m* ~ digging Schotterbruch *m* ~ dredger Baggerprahm *m* ~ elevator Schotter-becherwerk *n* ~ hammer Schotterschlegel *m* ~ lamp Stromregelröhre *f*, Widerstandslampe *f* (im elektrischen Kreis) ~ pit Kiesgrube *f* ~ rake Kiesharke *f*, -rechen *m* ~ resistance Ballastwiderstand *m*, selbstregelnder Vorwider-stand *m* ~ resistor Ballastwiderstand *m*, Barretter *m*, Stromregelgröße *f*, selbstregelnder Vorwiderstand *m* ~ sack Schleppsack *m* ~ screener Schottertrommel *f* ~ tank Ballasttank *m*, Tauchtank *m* ~-tank flood valve Ballasttank-flutventil *n* ~ tank vent valve Ballasttankaus-laßventil *n* ~ tube Stromregelröhre *f*, Wider-standslampe *f* ~ wagon Erdwagen *m* ~ weights Ballastklötze *pl* (U-Boot)

**ballasted up** ausgewogen

**ballasting** Kiesschüttung *f*, Beschotterung *f*, (pumps for adjusting ballast) Ballastbetrieb *m* ~ resistance Ausgleichs-, Beruhigungs-wider-stand *m*, Widerstandsballast *m* ~ tool Schotter-werkzeug *n*

**balling** Luppen *n*, Luppenmachen *n*, Wickeln *n* ~ up Zusammenballung *f* ~ furnace Schweiß-ofen *m* ~ property Ballungsfähigkeit *f* ~ spindle Ballingspindel *n*

**ballistic** ballistisch ~ air density ballistisches Luftgewicht *n* ~ cam Kurvenkörper *m* ~ coefficient Querschnittsbelastung *f*, Wurfzahl *f* ~ correction for initial condition of the bore (or of the moment) Gebrauchsstufe *f* ~ curve Geschoßbahn *f* ~ density and wind factors ballistische Tageseinflüsse *pl* ~ density and wind factors in time-of-flight seconds Baltasekun-den *pl* ~ director Kommandogerät *n* ~ effi-ciency äußerer Wirkungsgrad *m* ~ galvano-meter Stoßgalvanometer *n* ~ indication un-gedämpfte Anzeige *f* (of instrument) ~ measuring instrument ungedämpftes Meß-instrument *n* ~ pendulum Stoßwaage *f* ~ read-ing ungedämpfte Anzeige *f* (of instrument) ~ sum of all interior factors Gebrauchsstufe *f* ~ test Abnahmebeschuß *m* ~ values Bahnwerte *pl* ~ variables Bahngrößen *pl* ~ velocity Geschoß-geschwindigkeit *f*

**ballistics** Ballistik *f*, Lehre *f* vom Schuß, Schieß-lehre *f*, Schießwesen *n*, Wurflehre *f*

**ballistite** Ballistit *f*

**ballonet** Ballonett *n*, Gaszelle *f*, Luftsack *m* ~ envelope (or diaphragm) Luftsackhülle *f*

**balloon** Luftballon *m*, (flask) Ballon *m* ~ basket Ballonkorb *m* ~-borne transmitter Ballonsender *m* ~ cable Ballongurt *m* ~ catch Begrenzungs-

ring *m* (Haspel) ~ **conductor** Ballonader *f* ~
**cover** Ballonhülle *f* ~ **crew** Ballonmannschaft *f*
~ **descent** Ballonabstieg *m* ~ **envelope** Ballonhülle *f* ~ **fabric** Ballonstoff *m*, Hüllenstoff *m*
~ **flask** Kugelflasche *f* ~ **inflated with hydrogen**
Wasserstoffballon *m* ~ **inflated with illuminating
gas** Leuchtgasballon *m* ~ **net** Ballonnetz *n* ~
**pilot** Ballonführer *m*, Freiballonführer *m* ~
**rudder** Steuersack *m* ~ **silk** Ballonseide *f* ~
**sonde** Ballonsonde *f* ~ **statoscope** Ballonvario-
meter *m* ~ **survey** Ballonaufnahme *f* ~ **tail
valve** Füllansatzventil *n* ~ **take-off site** Ballon-
aufstiegplatz *m* ~ **theodolite** Ballontheodolit *m*
~ **tire** Ballonreifen *m*, Überballonreifen *m* ~
**winch** Ballonwinde *f*
**ballute** (of spacecraft) Ballonschirm *m*
**balmy** lau
**balsam,** ~ **fir** Balsamtanne *f* ~ **poplar** Balsam-
pappel *f*
**balsa wood** Balsa *n*, Balsaholz *n*, Floßbaumholz
*n*
**baltimorite** Baltimorit *m*
**balun** Symmetrier-glied *n* oder -übertrager *m* ~
**transformer** Spulentransformator *m*
**baluster** Geländerpfosten *m*, Traille *f*
**balustrade** Brustlehne *f*, Geländer *n*
**Bamag method (or** ~**-shaft process)** Bamag-
schachtverfahren *n*
**bamboo** Bambus *m*, Bambusholz *n*
**banana,** ~ **jack** Bananensteckerbuchse *f* ~ **pin**
Bananenstecker *m* ~ **plug** Bananenstecker *m* ~
**plug connection** Bananensteckerverbindung *f*
**Banbury mixer** Innenmischer *m* (mit Stempel)
**Banca tin** Bankazinn *n*
**band, to** ~ verankern (of a blast furnace)
**band** Bandage *f*, Bereich *m*, Bergemittel *n*, Binde
*f*, Gürtel *m*, Haftblei *n*, Heftschnur *f*, Leiste *f*,
Riemen *m*, Schar *f*, Streifen *m*, Zeile *f*, (range
of frequencies) Radiokonstante *f*, (ring) Band *n*
~s Bandstahl *m*, Haftblei *n* (glass mfg)
**band,** ~ **of armature** Ankerbandage *f*, Umspan-
nung *f* ~ **of contacts** Kontaktband *n* ~ **between
cutoff points** Lochweite *f* ~ **of crystal** Kristall-
streifen *m* ~ **of rifle** Schaftring *m* ~s **of ferrite**
Ferritstreifen *pl* ~s **of the spectrum** Kantensystem *n* der Bande ~s **of straight lines** Scharen
*pl* von Geraden (Kurven)
**band,** ~ **adjustment** Bandbreitenregelung *f* ~
**amplifier** Bandverstärker *m* ~ **antenna** Bandantenne *f* ~ **armored cable** Bandpanzerleitung *f*
~ **articulation** Bandverständlichkeit *f* ~ **brake**
Bandbremse *f* ~ **builder** Ringklebemaschine *f* ~
**carrier** Bandpost *f* ~ **center** (spectroscope)
Nullinie *f* ~ **chain** Bandkette *f*, Gallsche Kette *f*
~ **conveyer** Bandförderer *m*, Bandpost *f*, Förderband *n*, Gurt-förderband *n*, -förderer *m*
~ **edge** Bandenkante *f*, Bandenkopf *m* (spectral
analysis) ~ **elevator** Gurtbecherwerk *n* ~ **elimination filter** Bandsperre *f*, Bandenausfilterung
*f* (acoust.) ~ **equipment** Beschienung *f* ~ **expanding clutch** Bandkupplung *f* ~ **(pass)filter**
Bandfilter *n* ~-**filter circuits** Bandfilterkreise *pl*
~-**follow-up switch** Frequenzbandfolgeschalter
*m* ~ **galvanometer** Bändchengalvanometer *n* ~
**head** Bandenkopf *m* ~ **iron**
Bandeisen *n*, Bandsäge *f* ~ **iron hoop** Band-
eisenreif *m* ~ **microphone** Bandmikrofon *n*
~ **model** Bändermodell *n*

**band-pass** Bandendurchlässigkeit *f*, Banddurch-
laßbereich *m*, Bandpaß *m*, Durchlaßbreite *f*,
Empfangsbandbreite *f*, Lochweite *f*, Wellenbereich *m*, (of a filter) Lochbreite *f* eines Frequenzsiebes ~ **ampflifier** Bandfilterverstärker
*m* ~ **coupled** bandfiltergekoppelt ~ **filter** Band-
paßfilter *n*, Filtersiebkette *f*, Frequenzbereich-
filter *n*, Frequenzfilter *n*, Siebkette *f* ~ **filter
coil** Bandfilterspule *f* ~ **width of a filter** Durchlaßbereich *m* (rdo)
**band,** ~ **pressure level** Schalldruckpegel *m*
(acoust.) ~ **progression** Bandenzug *m* ~ **proportional** Proportionalitätsband *n* ~ **pulley** Riemenscheibe *f* ~-**range radio** Bereichband *m*
~-**rejection filter** Bandsperre *f* ~ **rope** Bandseil
*n*, Flach- und Bandseil *n* ~ **saw** Bandsäge *f*
~ **saw blade** Bandsägeblatt *n* ~ **selector** Bandfilter *n*, Wellenschalter *m* ~ **selector switch**
Bandbreitenschalter *m* (electron.) ~ **shape**
Bandform *f* ~ **shaped** bandförmig ~ **spacing**
Streifenabstand *m* ~ **spectrum** Banden-, Strei-
fen-spektrum *n* ~ **spread** Bandbreitenregelung *f*,
Bandspreizung *f* (rdo) ~ **spreading** Banddeh-
nung *f* (rdo) ~ **steel** Bandstahl *m* ~ **steel for
pens** Schreibfederbandstahl *m* ~ **stop** Bandsperre *f* (rdo) ~ **stop filter** Bandsperrfilter *n*
~ **string** Heftschnur *f* ~ **switch** Bereichs-,
Frequenzband-, Grobstufen-, Wellen-schalter
*m* ~ **tension** Spannung *f* des Bandes ~ **turning**
Ringdrehen *n* ~ **voltage** Bandströmung *f* ~
**wheel** Bandscheibe *f* ~-**wheel-powered pump**
Scheibenantriebpumpstation *f* ~ **width** Bandbreite *f*, Bandweite *f* (rdo), Durchlaßbreite *f*,
Frequenzabstand *m*, Lochbereich *m*, Lochbreite *f*, Lochweite *f*, Tonumfang *m*, (of a filter)
Frequenzdurchlässigkeit *f* ~ **width of a filter**
Lochbreite *f* eines Frequenzsiebes ~ **width at
50 per cent down** Halbwertbreite *f*
**band-width,** ~ **adjuster** Bandbreiteeinstellung *f*
~ **control** Bandbreiteregler *m* ~ **curve** Ab-
stimm-, Selektivitäts-kurve *f* ~ **method** Band-
breitenverfahren *n*
**band wiper arm** Bandwischhebel *m*
**bandage, to** ~ bandagieren
**bandage** Bandage *f*, Binde *f*, Bindenkrepp *m*,
Verband *m*, Verbandsstoff *m*, Verbandzeug *n*,
Wundverband *m* ~ **scissors** Verbandsschere *f*
**banded** bandstreifig, gestreift, lagenförmig, streifig, streifig verwachsen ~ **bituminous coal**
Streifenkohle *f* ~ **filter** Streifenfilter *m* ~ **ogive
bullet** Randkegelgeschoß *n* ~ **structure** Bän-
derung *f*, Streifengefüge *n*, Zeilenstruktur *f*
**banderole** Meßfähnchen *n*, Verschlußstreifen *m*
**bang** schallender Schlag *m*
**bang open, to** ~ aufknallen
**bang-bang control** Zweipunktregelung *f*, Zweipunktsteuerung *f*
**banging** Bumsen *n* (acoust.)
**banister** Geländersäule *f*, Treppengeländer *n*
**banjo** Hinterachstrichter *m* ~ **bolt** Hohlschraube
*f*, Winkelrohranschlußstutzen *m* ~ **connection**
Schwenkanschluß *m*, ringförmiges Rohranschlußstück *n* ~-**type axle** durchgehende Achse
*f* ~ **union** schwenkbarer Rohranschluß *m*
**bank, to** ~ aufhäufen, auslenken (g/m), kränken,
kurven, Kurve machen, sich in die Kurve
legen (aviat.), (a blast furnace or ship) dämpfen, (fire) aufdämmen, (of a hearth bottom)

stampfen **to ~ up** anschütten, aufstampfen **to ~ out** auf Halde bringen **to ~ a plane** in die Kurve legen **to go into a steep ~** auf den Flügel stellen **to ~ transformers** Transformatoren gruppieren

**bank** Bank *f*, Bankgeschäft *n*, Böschung *f*, Damm *m*, Deich *m*, Federnpaket *n*, Flußdamm *m*, Kurvenlage *f*, Pier *m*, Querlage *f* (aviat.), Querneigung *f*, Reihe *f* gleichartiger Teile (mech.), Schweifgestell *n*, Strosse *f*, Ufer *n*, Uferböschung *f*, Untiefe *f*, Wendung *f*, (of contacts) Kontaktbank *f*, (of contacts, keys, etc.) Reihe *f*, (of issue) Notenbank *f*, (selector) Kontaktbank *f* **increased ~** (in a curve) Überhöhung *f*

**bank, ~ of clouds** Wolkenwand *f* **~ of contacts** Kontaktband *n*, Kontaktkranz *m* **~ of cylinders** Zylinderreihe *f* **~ of ditch** Grabenböschung *f* **~ of keys** Klaviatur *f*, Schalterreihe *f*, Schlüsselreihe *f*, Tastenreihe *f* **~ of lamps** Lampenaggregat *n*, Lampenfeld *n*, Mehrfachstrahler *m* **~ of lights** Lampenaggregat *n* **~ of nozzle** Düsenstock *m* (Klimaanlage) **~ of retorts** Retorten-bank *f*, -reihe *f* **~ of stationary contacts** Wählersegment *n* **~ of transformers** Transformatorenbank *f*

**bank, ~-and-pitch indicator** Quer- und Längsneigungsanzeiger *m* **~-and-turn indicator** Querneigungsmesser *m*, Steueranzeiger *m* **~-and--wiper switch** Hebdrehwähler *m* **~ attitude** Querlage *f*, Querneigung *f*, Schräglage *f* (aviat.) **~ bill** Kassenanweisung *f*, Kassenschein *m* **~ business** Bankgeschäft *n* **~ cable** Kontaktleitungsvielfachkabel *n*, Vielfachkabel *n*, Zuteilungsvielfachkabel *n* **~ cashing** Bankinkasso *n* **~ chain** Gelenkkette *f* **~ collection** Bankinkasso *n* **~ commission** Bankprovision *f* **~ contact** Bankkontakt *m* **~ current accounts** Girokonten *pl* **~ defenses** Dammschutz *m*, Uferbefestigung *f* **~-deposit account book** Girogenenbuch *n* **~ engine** Schiebelokomotive *f* **~ error** Querlagenfehler *m* (aviat.) **~ eye (mouth) of a shaft** Hängebank *f* **~ full** Bordvoll *m* (soil mech) **~ furnace** Batterieofen *m* **~ hook** Förderhaken *m* **~ indicator** Schräglageanzeiger *m* **~ line** Uferlinie *f* (of a stream) **~man** Abnehmer *m* **~ multiple** Vielfachfeld *n*, Zuteilungsvielfach *n* **~ note** Kassen-anweisung *f*, -schein *m* **~ paper** Bankpostpapier *n* **~ post paper** Hartpostpapier *n* **~ protection** Uferschutz *m* **~ rating** Gruppenleistung *f* **~ reclamation** Abbau *m* (von Förderhalden) **~ slope system** Böschungssystem *n* **~-to-~cabling** Kontaktverkabelung *f* **~ wire** Kontaktdraht *m* **~ wires** Kontaktleitungsvielfachverdrahtung *f*, Vielfachfelddrähte *pl* **~ wiring** Vielfachverdrahtung *f*

**banked, ~ coil** Mehrlagenspule *f* **~ turn** kurze Kehrtwendung *f* **~ winding** Reihenwicklung *f*, Stufenwicklung *f*, verschachtelte Wicklung *f*

**banker** Wechsler *m* **~'s buying rate** Geldkurs *m*

**banking** Bankung *f*, Geldhandel *m*, Hängebankarbeiten *pl*, Kurvenflug *m*, Schräglage *f*, Stampfen *n*, (a blast furnace) Dämpfen *n* **~ of plane in a curve** Schräglage *f* des Flugzeuges in der Kurve **~-up of a river** Aufstau *m* eines Flusses

**banking, ~ angle** Hängewinkel *m* **~ business** Wechselstube *f* **~ face** Anschlagfläche *f* **~**

**indicator** Querneigungsanzeiger *m* **~ turn** Querneigung *f* **~-up soil** Dammerde *f*

**bankruptcy** Konkurs *m*

**banksman** Abzieher *m* (am Schacht) **(bottom-) ~** Anschläger *m*

**banner** Hauptzeile *f* (Presse) **~ cloud** Wolkentuch *n*

**banquette** Auftrittstufe *f*

**bantam tube** Miniaturröhre *f*

**bar, to ~** ausschließen, sperren, verriegeln, versperren

**bar** Balken *m*, Bar *n* (atmosphärische Luftdruckeinheit), Barren *m*, Bengel *m* (print.), Bügel *m*, Lamelle *f*, Leiste *f*, Lumme *f*, Metalleiste *f*, Pole *f*, Profilstange *f*, Querhaupt *n*, Querstück *n*, Querträger *m*, Riegel *m*, Schenke *f*, Schiene *f*, Schlagbaum *m*, Splint *m*, SSR-Anzeige *f* (sec.rdr.), Stab *m*, Stange *f*, Stangeneisen *n*, Strebe *f*, Streifen *m*, Traverse *f*, Untiefe *f*, Vollstange *f*, Walzader *f*, Zain *m*, (sheet) Breiteisen *n*, (high sea bed parallel with the shore) Barre *f*, (of molding box) Schore *f*, (roller of breaking waves) Brandungsgürtel *m* **to be a ~ to** ein Hinderungsgrund *m* sein (law) **~ to practicing one's profession** Untersagung *f* der Berufsausübung

**bar, ~ of flat section** Flachstange *f* **~ of fraction** Bruchstrich *m* **~ of a harbor** Hafenbaum *m*, Hafenschlengel *m* **~ of hexagonal section** Sechskantstange *f* **~ of rectangular section** Vierkantstange *f* **~ of round section** Rundstange *f* **~ of a printer's form** Schließnagel *m* **~ for sheet guides** Welle *f* zur Leitung **~ for sheet steadier** Stange *f* zum Bogenstreicher **~ of soap** Seifenriegel *m* **~ for wash-up blade** Schiene *f* für Messingrakel

**bar, ~-actuating vibrator roller** Schaltstange *f* zur Hebwalze **~-and-dot raster** Linien- und Punktgitter *n* **~-arranging device** Stabordner *m* **~ automatic** Stangenautomat *n* **~-bending device** Stabbiegevorrichtung *f* **~-buffer spring** Schienenpufferfeder *f* **~ capstan** Handspill *n* **~-chamfering machine** Stangenspitzmaschine *f* **~-coal cutter** Stangenschrämmaschine *f* **~ code** Strichcode *m* **~ construction** Stabkonstruktion *f* **~ copper** Stangenkupfer *n* **~ cutter** Stabeisenschere *f*, Stabschere *f* **~ expansion (or extension)** Stabdehnung *f* **~ feed** Stangenvorschub *m* **~ filter** Stabfilter *m* **~ folding** Fälzen *n* **~ generator** Balkengenerator *m* (TV) **~-graph** Strichdiagramm *n* **~ grate** Gitterrost *m*, Stabrost *m* **~-guide bush** Werkstofführungsring *m* **~ head** Knebelkopf *m* **~ iron** Rund-, Stab-eisen *n* **~-iron bundling machine** Stabeisenbündelmaschine *f* **~-iron rolling train** Stabstraße *f* **~-iron twisting machine** Stabeisenverwindemaschine *f* **~ lead** Stangenblei *n* **~ list** Eisenliste *f* (schedule) **~ loaded for pressure column** auf Druck belasteter Stab *m* **~ magnet** Magnetstab *m*, Stabmagnet *m* (rod) **~ material** Stabmaterial *n* **~ mill** Stabwalzwerk *n* **~ mill products** Stabeisen *n* **~ mining** Kieswäsche *f* **~ pattern** Balkenmuster *n* (TV), Strichraster *m* (TV) **~ pin** Kettensteg *m* **~ pivot axis** Stangendrehachse *f* **~ remnant** Stangenrest *m* **~ reversing lever** Steuerungshebel *m* **~-rolling mill** Stabeisenwalzwerk *n*, Stabwalzwerk *n* **~-rolling train** Stabeisenstraße *f* **~ scale** Maßstableiste *f*

~ scan Abtastebene f ~ section Stabeisenprofil n, Stabprofil n ~ shape Stab-form f, -profil n ~-shaped stabförmig ~ (iron) shears Stabeisenschere f ~ sight Stangenvisier n ~ silver Barrensilber ~ soap Stangenseife f ~ solder Stangenlötzinn n ~ spring Stabfeder f ~ steel Stangenstahl m, (round) Stabstahl m ~ stiffness Stabsteifigkeit f ~ stock Stabmaterial n ~ stock for automatics Maschinenstahl m ~ stock lathe Drehbank f für Stangenarbeit ~ store(yard) Stangenlager n ~ stress Stabkraft f ~ stripping knife Stangenschälmesser n ~ timer Stempelholz n ~ tin Stangenzinn n ~ turning mechanism Kippantrieb m ~ turret lathe Stangenrevolverdrehbank f ~-type transformer Schienenstromwandler m ~ wear Stangen-abnützung f, -verschleiß m ~ wimble Riegelbohrer m ~ winding Stabwicklung f ~ work Stangenarbeit f ~-wound armature Anker m mit Stabwicklung, Stabanker m ~ zinc Barrenzink n

baratte Baratte f (sulfiding)

barb Widerhaken m ~-type radiator Bartkühler m

barbed, ~ fork Widerhakengabel f ~ hook Fischhaken m ~-wire Stacheldraht m

barbed-wire, ~ barrier (or entanglement) Drahtverhau m ~ fence Stacheldrahtzaun m

barbette Barbette f, Geschützbank f

barble, to ~ verwürfeln

bare bloß, blank (Draht), entblößt, kahl, nackt, ohne Zubehör, unbewachsen (Gelände) to ~ abisolieren, entblößen to ~ the wire Draht abisolieren

bare, ~ cable blankes Kabel n oder Seil n ~ conductor blanker Draht m oder Leiter m ~ electrode blanke Elektrode f, Blankelektrode f, nackte Elektrode f ~ engine Motor m ohne Verkleidung oder ohne Zubehör ~ place Kahlstelle f ~ reactor Reaktor m ohne Reflektor ~ spot durchgescheuerte Stelle f ~ wire blanker oder nackter Draht m ~ wire electrode Blachstab m

bareness Nacktheit f

bargain, to ~ handeln

bargain Handel m, Vertrag m ~ by the job (or the lump) Generalgedinge n ~ master Gedingenehmer m

bargaining Ortsgedinge n

barge Baggerschute f, Kahn m, Kipprahm m, Lastkahn m, Leichter m, Leichterschiff n, Lichter m, Ponton m, Prahm m, Schaluppe f, Schute f a ~ is horse-drawn ein Schiff n wird getreidelt ~ is laid up ein Schiff n ist aufgelegt (not working) ~ is poled along ein Schiff n wird gestakt (by a barge pole) ~ course Fußschicht f

barite Bariumsulfat n, Baryt m, Schwererde f

barium Barium n, Baryum n ~ azide Baryumazid n ~ carbonate kohlensau(e)res Barium n, Witherit m ~ cyanate zyansaures Baryt n ~ dioxide Bariumhyperoxyd n, Bariumsuperoxyd n ~ ferrocyanide Ferrozyanbarium n ~ hydrosulfite Bariumthiosulfat n ~ hydroxide Ätzbaryt m, Bariumhydrat n, Bariumhydroxyd n, Bariumoxydhydrat n, Barythydrat n ~ hydroxide solution Barytlauge f ~ nitrate Baryumnitrat n ~ oxide Bariumoxyd n, Baryt m, Baryterde f ~ peroxide Bariumhyperoxyd n, Bariumperoxyd n, Bariumsuperoxyd n, Ba-

ryumsuperoxyd n ~ phosphate phosphorsaures Barium ~ platinocyanide Bariumplatinzyanür n ~ sulfate Bariumsulfat n, Barytweiß n, Permanentweiß n ~ sulfate Schwerspat m (an artificial) ~ sulfide Bariumsulfid n ~ sulfite Bariumsulfit n ~ sulfocyanate Bariumsulfozyanid n ~ thiocyanate Bariumrhodanid n ~ thiosulfate Bariumthiosulfat n ~ tungstate Bariumwolframat n ~ yellow Barytgelb n

bark, to ~ ablohen, entrinden, schälen to ~ off abrinden

bark Barke f, Baumrinde f, Borke f, Bork(en)holz n, Holzschwarte f, Rinde f ~ breaking machine Lohbrechmaschine f ~ cutter Rindenschneidmaschine f ~ tanning Lohgerben n ~ wood Borke f, Bork(en)holz n

barked, ~ fracture splitterige Bruchfläche f ~ wood Schälholz n

barker Rindenschälmaschine f, Schäler m, Schälmaschine f ~ machine Putzmaschine f

Barkhausen, ~ effect (or jump) Barkhausen Sprung m ~-Kurz circuit Barkhausen-Kurz--Schaltung f ~-Kurz oscillations Barkhausen--Kurz-Schwingungen pl, B-K-Schwingungen pl, elektrischer Elektronentanz m ~-phon Barkhausen-Phon n

barking Abborken n ~ machine Putzmaschine f, Rindenschälmaschine f, Schäler m, Schälmaschine f

barkometer Brühenmesser m

barley Gerste f ~ cleaner and separator Gerstenputz- und Sortiermaschine f ~ coal Anthrazitart f

barm Gest f

barn Scheune f

barnacle Entenmuschel f, Schülpe f

baroclinic current barokline Strömung f

barocyclonometer Barozyklonometer n

barogram Barogramm n

barograph Barograf m, Höhenschreiber m, Luftdruckschreiber m, registrierendes Barometer n, Schreibbarometer n

barographic chart Barografenblatt n

barometer Barometer n, Dosenaneroid n, Schweremesser m (aneroid) ~ measuring atmospheric pressure at airdrome not reduced to sea level Feinhöhenmesser m ~ pressure Barometerdruck m ~ reading Barometerstand m ~ scale Barometerskale f ~ tube Barometerröhre f

barometric barometrisch ~ altitude Druckhöhe f (meteor.) ~ altitude transducer Druckhöhengeber m ~ bellows barometrische Druckdose f ~ chart Isobaren-, Luftdruck-karte f ~ check Barometerkorrektur f (direction finder) ~ column Barometersäule f ~ height Barometerhöhe f ~ leg pipe Fallwasserrohr n ~ maximum Antizyklon m ~ pressure Außenluftdruck m, barometrischer Druck m, Barometerstand m, Luftdruck m, Quecksilbersäule f ~ pressure gradient barometrische Luftdrucksteigung f ~ switch Barometerrelais n (rdr) ~ tail pipe Fallwasserrohr n ~ thermometer Siede-, Hypso--thermometer n, Siedewärmegradmesser m ~ time release unit barometrisches Auslöse-Zeitwerk n ~ tube Fallrohr n, Fallwasserrohr n

barometry Luftdruckmessung f

baroresistor Höhenausgleichwiderstand m

**baroscope** Baroskop *n*
**barotropic current** barotrope Strömung *f*
**barrack** Baracke *f* ~ **camp** Barackenlager *n* ~
**construction** Barackenbau *m* ~ **encampment**
Barackenlager *n*
**barracks** Truppenunterkunft *f*, (permanent)
Kaserne *f* ~ **yard** Kasernenhof *m*
**barrage** Abdämmung *f*, Abschließen *n*, Feuersperre *f*, Sperrfeuer *n*, Stauwerk *n*, Talsperre *f*,
Trommelfeuer *n*, Wehr *f*, (of a river arm) Absperren *n* ~ **balloon** Sperrballon *m* ~ **cable**
Seilsperre *f*, Sperrseil *n* ~ **density** Sperrdichte
*f* ~ **height** Sperrhöhe *f* ~ **interval** Sperrab-
stand *m* ~ **jammer** Breitbandstörsender *m* ~
**jamming** Sperrstörung *f* (rdr), breitbandiges
Stören *n* ~ **kite** Sperrdrachen *m*
**barrandite** Aluminiumeisenphosphat *n* (wasserhaltiges)
**barred** gesperrt ~ **zone** Sperrgebiet *n*
**barrel** Büchse *f*, Faß *n*, Gebinde *n*, Lägel *n*,
Laufbuchse *f*, Pumpenstiefel *m*, Rohr *n*, Tonne
*f*, Trommel *f*, Vergaserlufttrichter *m*, Zylinderbüchse *f*, (of gun) Lauf *m* ~ **of built-up
gun** Mantelringrohr *n* ~ **of lens (or of objective)** Fassung *f* eines Objektives ~ **of a ram**
Kolbenführung *f* ~ **of a roll** Walzenballen *m*,
Walzballen *m*, Walzenkörper *m*, Walzkörper
*m* ~ **of turnbuckle** Spannschloßhülse *f* ~ **of
well winch** Brunnentrommel *f*
**barrel,** ~ **aileron roll** Rolle *f* (aviat.) ~ **arch**
Tonnengewölbe *n* ~ **blank** Laufrohling *m*
~ **brazer** Lauflöter *m* ~ **bung** Faßspund *m* ~
**buoy** Faßboje *f*, Faßtonne *f* ~ **burnishing**
Kugelpolieren *n* ~ **burst** Rohr-krepierer *m*,
-zerspringer *m* ~ **cleaner** Rohr-reiniger *m*,
-wischer *m* ~ **compass** Trommelkompaß *m*
~ **controller** Schaltwalzenanlasser *m* ~ **converter** liegender Konverter *m*, Trommelkon-
verter *m* ~ **cooling** Rohrkühlung *f* ~ **cooling
system** Laufkühlung *f* ~ **cradle** Faßkrippe *f*
~ **dash** Faßtrichter *m* ~ **dent-removing tool**
Rohrausbeuldorn *m* ~ **drainer** Faßentleerungsvorrichtung *f* ~ **elevator** Faßelevator *m* ~
**enlargement** Ausbrennung *f* ~ **erosion** Ausbrennung *f* ~ **fender** Faßaufschlagfender *m*
~ **fine borer** Laufbohrer *m* ~ **float** Tonnen-
-fähre *f*, -floß *n* ~ **forger** Rohrschmied *m*
~**-head plate** Rohrstirnwand *f* ~ **hook** Lauf-
klaue *f* ~ **hoop** Faßreifen *m* ~ **housing** Trommel-
gehäuse *n* ~ **howel** Krummhaue *f* ~ **gauge**
Seelenmesser *m* ~ **jacket** Lauf-, Rohr-mantel
*m* ~ **key** Hohlschlüssel *m* ~ **length** Rohrlänge *f* ~ **liner** Futterrohr *n* ~**-locking gear**
Laufverriegelung *f* ~**-locking lever** Rohrhalte-
hebel *m* ~**-locking ring** Laufsitzring *m* ~**-
locking stud** Gewehrkuppelungsstück *n* ~ **lug**
Laufhaken *m* ~ **mill** Topf-, Trommel-mühle
*f* ~ **mixer with staggered baffles** Freifallmischer *m* ~ **muzzle** Rohrmündung *f* ~ **nut** Spannschloßmutter *f* ~ **oil saver** Ölfangzylinder *m*
~ **packing** Laufsitzring *m* ~ **polishing** Kugel-
trommelpolieren *n* ~ **position** Rohrlage *f* ~
**process** Tonnenverfahren *n* ~ **protector** Lauf-
schützer *m*, Rohrschoner *m* ~ **radiator** Faß-,
Trommel-kühler *m* ~ **recoil** Rohrrücklauf *m*
~ **reflector** Laufseelenprüfer *m*, Rohrspiegel
*m* ~ **ring** Laufring *m* ~ **roll** Faßrolle *f*, Überschlag *m* über den Flügel ~ **roof** Tonnendach

*n*, tonnenförmiges Dach *n* ~ **setter** Laufmon-
teur *m* ~**-shaped** kegelig, tonnenförmig, walzenförmig ~**-shaped distortion** tonnenförmige
Verzeichnung *f* ~**-shaped roller bearing** Ton-
nen-lager *n*, -rollenlager *n* ~ **sheets** Faßbleche
*pl* ~ **shell** Trommel-mantel *m*, -wand *f* ~
**shutter** Trommelblende *f* (film) ~ **slide** Rohr-
schlitten *m* (artil.) ~ **spring** (watch) Zugfeder *f*
~ **stop** Faßschienenklotz *m* ~ **support** Rohr-
stütze *f* ~ **switch** Walzenschalter *m* ~ **throttle**
Rundschieber *m* ~ **track** Faßschienen *pl*
~**-type engine** Trommelmotor *m* ~**-type motor**
Axialzylindermotor *m* ~**-type rotary engine**
Trommelumlaufmotor *m* ~**-type stationary
engine** Trommelstandmotor *m* ~ **vaulting**
Tonnen-flechtwerk *n*, -gewölbe *n* ~ **wheel**
Walzenrad *n*
**barreled road** gewölbte Straße *f*
**barreling** Drall *m*, Drehung *f* ~ **station** Faß-
füllstelle *f*
**barren** dürr, erzleer, fruchtlos, kahl, taub, unfruchtbar ~ **ground** taubes Gestein *n* ~ **rock**
Abraum *m* ~ **sand** unergiebiger Sand *m* ~
**track** Taubfeld *n*
**barretter** Barretter *m*, Eisenwasserstoffwiderstand *m*, Stromregulator *m* ~ **electromotive
force** Eisenwasserstoffwiderstand *m*
**barricade, to** ~ verrammeln, zusperren
**barricade** Barrikade *f*, Hindernis *n*, Sperre *f*,
Straßen- und Wagensperre *f*, Verriegelung *f*
**barricading** Sperrung *f*
**barrier** Abbrandstreifen *m* (film), Abdeckstreifen *m*, Aufschüttung *f*, Barriere *f*, Damm
*m*, Fangvorrichtung *f* (on runway), Potential-
schwelle *f*, Potentialwall *m*, Schranke *f*, Schutzwand *f*, Schwelle *f*, Spaltungswall *m*, Sperre *f*,
Staukörper *m*, Steg *m* (electr.), Trennlinie *f*
(film), Umgitterung *f*, Zugschranke *f* ~ **of
lattice** Gitterschranke *f* ~ **without railings**
gitterloses Gelände *n* ~ **of railroad** Eisenbahnwegschranke *f* ~ **between record grooves**
Schallplattensteg *m* ~ **with a rod** Stangen-
schranke *f*
**barrier,** ~ **containing fixed space charges** Raum-
ladungsrandschicht *f* ~ **diffusion method**
Trennwanddiffusionsmethode *f* ~**-film cell**
Halbleiterfotozelle *f* ~**-film rectifier** Sperrgleichrichter *m* ~ **fortification** Sperrbefestigung
*f* ~ **frequency** Grenzfrequenz *f* ~ **gate** Gattertor *n* ~ **grid** Schutzgitter *n* ~ **grid mosaic** Bild-
anode *f* ~ **grid tube** Sperrgitterröhre *f* ~ **guard**
Schutzerdungsgestell *n* ~ **height** Höhe *f* des
Potentialwalls
**barrier-layer** Halbleiterschicht *f* ~ **capacitance**
Kapazität *f* einer Sperrschicht ~ **cell** Gleich-
richter-, Sperrschicht-zelle *f* ~ **photocell**
Halbleiter-, Sperrschicht-fotozelle *f* ~ **photoelectric effect** Sperrschichtfotoeffekt *m* ~
**rectifier** Sperrschichtgleichrichter *m*
**barrier,** ~ **penetration factor** Gamow-Faktor *m*
~ **pillar** Sicherheitspfeiler *m* ~ **plane front-
-wall cell** Vorderwandzelle *f* ~**-plane photocell**
Halbleiterfotozelle *f* ~**-plane rear-wall cell**
Hinterwandzelle *f* ~ **reef** Damm-, Wall-riff
*n* ~ **resistance** Randschicht-, Vor-widerstand
*m* ~**-type cell** Sperrschichtzelle *f*, Sperrzelle *f*
~ **voltage** Randspannung *f*
**barring** Joch *n* ~ **gear** Drehvorrichtung *f*

barrow Förderwagen *m*, Lastkarre *f*, Trage *f*
(wheel) ~ Schiebekarren *m* ~ charging Begichtung *f* von Hand, Karrenbegichtung *f* ~ man
Fördermann *m*, Karrenläufer *m* ~ tram För-
derstrang *m* ~ way Förderstrecke *f*, Laufbohle
*f* ~ wheel Karrenrad *n*
bars Blätter *pl*, Führung *f*, Stabmaterial *n*,
Stabstahl *m*
bartack Stopfnaht *f*
barter Tausch *m* ~ system Kompensationsgeschäft *n*
bartering Umtausch *m*
barycentric baryzentrisch
barye Bar *n* (the centimeter-gram-second absolute unit of pressure)
barymetry Luftschweremessung *f*
baryta Bariumoxyd *n*, Baryt *m*, Baryterde *f* ~
paper Barytflußspatpapier *n*
barytes Permanentweiß *n* ~ lake Spatlack *m*
~-reduction Spatverschnitt *m*
barytic barytartig
barytrons schwere Elektronen *pl*
bas-relief flacherhaben
bas-signal Bild-, Auftast- und Synchronsignal *n*
(TV)
basal, ~ bristle Basalborste *f* ~ metabolism
Grundumsatz *m* ~ plane (or surface) Basis-
fläche *f* (cryst.), Grundfläche *f*
basalt Basalt *m* ~ wall Basaltverwerfung *f*
basaltic rock Basaltfelsen *m*
bascule, ~ bridge Klappenbrücke *f* ~ mount
Pendelaufhängung *f*
base, to ~ basieren, fußen to ~ on sich stützen auf
base Anhaltspunkt *m*, Anlage *f*, Anschlag *m*,
Aufnahmeabstand *m*, Base *f* (chem.), Basis *f*,
Bett *n*, Boden *m*, Bodenfläche *f*, Fundament *m*,
Fuß *m*, Fußpunkt *m*, Grund *m*, Grund-bau *m*,
-fläche *f*, -lage *f*, -platte *f*, -sohle *f*, -stoff *m*,
Klotz *m*, Postament *n*, Quetschfuß *m*, Richtpunkt *m*, Sockel *f* (Röhre), Sohle *f*, Stoßfläche
*f*, Träger *m*, Unter-bau *m*, -fläche *f*, -gestell *n*,
-lage *f*, -lagsteg *m* (print.), -satz *n*, -teil *m*, Ver-
kehrsstützpunkt *m*, (board) Grundbrett *n*,
(line) Grundlinie *f*, (plate) Sockel *m*, (of pavement) Packlage *f*; unedel
base, ~ of ascent Aufsteigstelle *f* ~ of calcula-
case Hülsenboden *m* ~ of clouds Wolkenuntergrenze *f* ~ of a column Fußgestell *n* einer
Säule ~ of control stick Steuerknüppelfuß *m*
tion Berechnungsgrundlage *f* ~ of a cartridge
*m* ~ of a crystal Kristallnfläche *f* ~ of
engine bed Lagerfuß *m* ~ of equilibrium Gleichgewichtsboden *m* ~ with external contacts
Außenkontaktsockel *m* ~ of foundation Fun-
damentabsatz *m* ~ of frame Gestellfuß *m* ~
of the girder Trag-, Träger-fuß *m* ~ of hearth
Herdsohle *f* ~ of a lamp Röhrenfuß *m* ~ of
a mixture Grundbestandteil *m* eines Gemisches ~ of nozzle Düsenbohrung *f* ~ of
operations Operationsbasis *f* ~ of pole Mast-
sockel *m* ~ of the rear sight Visierfuß *m*,
Visiersattel *m* ~ of rim Felgenboden *m* ~
of shell Bodenkammer *f*, Geschoßboden *m*
~ of slope Böschungsfuß *m* ~ of stand Stativ-
standfläche *f* ~ of structure Bauwerksohle *f*
~ of a tube Röhrenfuß *m* ~ of a valve Röhrenfuß *m* ~ of verification Hilfsstandlinie *f* ~
of vertical stabilizer Steuerflossenwurzel *f*

base, ~ adjustment Basiseinstellung *f* ~ air
field Aufmarschflugplatz *m* ~ airport Flug-
zeugstützpunkt *m* ~ angle (gun) Grundrich-
tungswinkel *m* ~ angles in an isosceles triangle
Basiswinkel *m* im gleichschenkligen Dreieck
~ band Basisband *n*, Grundfrequenzband *n*
~ board Boden-, Objektiv-brett *n*, (dabs strip)
Scheuerleiste *f* ~ board-extensible camera Lauf-
bodenkamera *f* ~ branch Basiszweig *m* ~
bushing Grundbüchse *f* ~-cap screw Sockel-
gewinde *n* ~ carriage Basiswagen *m*, Standlinienschlitten *m* ~ casting Amboß-klotz *m*,
-lager *n*, Klotz *m*, Schabotte *f*, Schwatte *f* ~
centered basiszentriert ~ chamber oil under-
guard Untergehäuseölschutz *m* ~ charge
Bodenkammer-, Grund-ladung *f* (explosives)
~ circle Grundkreis *m* ~-circle pitch Grund-
kenntniskreisteilung *f* ~ cone Grundfaden *m*
~-connection diagram Sockelschaltung *f* ~
construction Unterbauweise *f* ~ copper (in
copper oxide rectifier) Mutterkupfer *n* ~
~ course Unterbau *m* ~ cover Abschlußplätt-
chen *n* ~ covering Sockelmantel *m* ~ current
Basiskreis *m* ~ curve Grundkenntniskurve *f*
~ delay-action fuse Bodenabstandzünder *m*
~ depth Hülsentiefe *f* ~ detonator Bodenzünder *m* ~ disc unterer Teller *m* (Motoraufhängung) ~ electrode Grundplatte *f* (Selengleichrichter) ~ end station Langbasisanlage
*f*, Meßstation *f* ~ exchange capacity Basenaus-
tauschvermögen *n* ~ exchange process Basen-
austauschverfahren *n* ~ failure Bruch am Fuße
einer Rutschung ~ former Basenbildner *m*
~ forming basenbildend ~ frame Untergestell
*n* ~ friction Reibung *f* auf Fundamentsohle ~
harbor Hauptliegehafen *m* ~ impact hit Boden-
treffer *m* ~ lacquer coat Grundlack *m* ~ lead
Basisklemme *f* ~ leg Queranflugteil *m* der
Platzrunde (aviat.) ~ length Basislänge *f* (of
the mount of a photogrammetric or surveying
camera) ~ lighting Grundlinienbefeuerung *f*
(aviat.)
baseline, ~ arm Basis *f*, Basislineal *n*, Basislinie *f* (rdr), Grundlinie *f*, Richtungslinie *f*,
Standlinie *f* (rdo), Zeitbasis *f* (CRT) ~ for
map reference Stoßlinie *f* ~ method of measurement Standlinienverfahren *n* ~ noises Stör-
rauschen *n* ~ triangle Standliniendreieck *n*
base, ~ load operation Grundlastbetrieb *m* ~
load of production Grundlast *f* der Förderleistung *f* ~-load power station Grundkraftwerk
*n* ~-load station Ausgleichwerk *n* ~ map
Grundkarte *f* ~ material Grundmaterial *n* ~
materials Ausgangsmaterialien *pl* ~ measuring
Basismessung *f* ~-measuring apparatus Basis-
meßapparat *m* ~-measuring subtense bar
Basismeßlatte *f* ~ meridian Grundmeridian *m*
~ metal Grund-material *n*, -metall *n*, -werk-
stoff *m*, unedles Metall *n* ~ metallic rectifier
Zellengleichrichter *m* ~ notation Stellenwertschreibung *f* ~ oil ungemischter Ausgangs-
schmierstoff *m* ~ operation Strichfahren *m*
(of power plants) ~ pan Bodenwanne *f* ~ pay
Grundgehalt *n* ~ percussion fuse Bodenzünder
*m* ~ photographic Fotorohpapier *n* ~ pin
Sockelstift *m* ~-pin contacts Röhrenkontakte
*pl* ~ plane Grundhobel *m*
base-plate Abdeckblech *n*, Abstützplatte *f*,

Aufstellplatte *f*, Bodenplatte *f*, Bodenstück *n*, Fundamentplatte *f*, Grundplatte *f*, Lagerplatte *f*, Schlenplatte *f*, Schwelle *f*, Sohlplatte *f*, Unterlage *f*, Unterlagsplatte *f*, (camera) Fußplatte *f* ~ **deflection** Spiegelauslenkung *f* ~ **for frog** Herzstückunterlage *f* (r.r.) ~ **latch** Bodenstücksperre *f* ~ **socket** Kugelpfanne *f* **base,** ~ **point** Hauptrichtungs-, Träger-punkt *m* ~ **portion** Unterteil *m* ~ **pressure** Sohldruck *m* ~ **price** Grundpreis *m* ~ **printing** Basendruck *m* ~ **prism** Endprisma *n* ~ **projection** Basisprojektion *f* ~ **rail** Sockelleiste *f* ~ **range finder** Basisentfernungsmesser *m* ~ **rate (pay)** Grundlohn *m* ~ **ratio** Basisverhältnis *n* (aviat.) ~ **reflector window** Ausblicköffnung *f* ~ **region** Basiszone *f* ~ **register** Höhe *f* der Fußringrille, Indexregister *n* ~ **ring** Fundamentring *m*, Fußring *m*, Fußwinkel *m*, Sockelring *m* ~ **ring of gun** Horizontierungsring *m* ~ **ruler** Basislineal *n* ~ **section** Kernladung *f* ~ **sheet** Deck-, Grund-blatt *n* ~ **sheet spacer** Unterlage *f* ~ **shoe** Anker-, Wurzel-platte *f* ~ **signal** (combined main and shunting signal) Grundsignal *n* ~ **sill** Grundschwelle *f* (for maintaining elevation of channel bed) ~ **slab** Grund-, Unterlags--platte *f* ~ **speed** Grunddrehzahl *f* (motor) ~ **spiral angle** Grundschrägungswinkel *m* ~ **spray** Bodenkammerschrapnell *n* ~ **stone** Auflagerstein *m* ~ **support** Lagerbock *m*, Standholm *m* ~ **surge** Basiswolke *f* ~ **thread** Grundfaden *m* ~ **tin** Halbzinn *n* ~ **triangle** Grunddreieck *n* ~**-tray** Bodenblech *n*
**based, to be** ~ **on** beruhen auf ~ **upon** gegründet auf
**baseless** grundlos
**basement** Erdgeschoß *n*, Grundbau *m*, Keller *m*, Kellergeschoß *n*, unter Bodenhöhe liegendes Erdgeschoß *n*, Untergeschoß *n*
**basic** basisch, fundamental, maßgebend, grundsätzlich (in facts)
**basic,** ~ **aircraft** Flugzeug *n* im Grundgewichtszustand oder im Leergewichtszustand ~ **airspeed** Fahrtausgangswert *m* ~ **Bessemer pig** Thomasroheisen *n* (iron) ~ **Bessemer process** Thomasverfahren *n* ~ **Bessemer steel** Thomas--eisen *n*, -stahl *m* ~ **cellulose** Ausgangszellulose *f* (pulp mfg) ~ **change gears** Grundwechselräder *pl* ~ **channel** Vorkanal *m* ~ **circuit** Grundschaltung *f* ~**-circuit diagram** Grundschaltbild *n*, Prinzipschaltung *f* (showing underlying principle) ~ **color** Grundfarbe *f* ~ **concept** Grundbegriff *m* ~ **consideration** Grundforderung *f* ~ **converter** basischer Konverter *m* ~ **converter steel** Thomas-eisen *n*, -flußeisen *n*, -flußstahl *m*, -stahl *m* ~ **data** Grundzahlen *pl* ~ **design** Grundausführung *f*, Grundentwurf *m* ~ **element of a unit** Teileinheit *f* ~ **engine type** konstruktiv grundlegendes Motormuster *n* ~ **equation** Grundgleichung *f* ~ **equipment** Grundausrüstung *f* ~ **figure** Grundfigur *f* ~ **form** Grundform *f* ~ **formants** Hauptformanten *pl* ~ **gasoline** Ausgangskraftstoff *m* ~ **group** Vorgruppe *f* ~ **hole** Einheitsbohrung *f* ~ **hue** Grundfarbe *f* ~ **identity** grundlegende Identität *f* ~ **index** Grundgewichtsindex *m* ~ **interference suppression** Grundentstörung *f* ~ **knowledge** Grundkenntnis *f* ~ **lattice** Grundgitter *n* ~ **laying point** Festlegepunkt *m* ~

**lined** basisch zugestellt ~**-lined converter** Thomasbirne *f* (Bessemer) ~ **load** basische Last *f*, bleibende Belastung *f*, Einheitsbelastung *f*, vorhandene Last *f* ~ **load per day** Kampftagesrate *f* ~ **materials** Ausgangswerkstoff *m* ~ **metal** Grundmetall *n* ~ **model** Grundausführung *f* ~ **moment** Moment *m* des Grundgewichts ~ **multivibrator** Urmultivibrator *m* ~ **network** Grundnachbildung *f* ~ **noise** Eigenrauschen *n* ~ **pattern** Grundprinzip *n* ~ **pig iron** Phosphoreisen *n* ~ **pitch** Grundsteigung *f* ~ **preparation used as a starting point** Ausgangspräparat *n* (chem.) ~ **principle** Grund-regel *f*, -gesetz *n*, Leitsatz *m* ~ **rate** Grund-gebühr *f*, -ziffer *f* ~ **rate of premium** Grundprämie *f* ~ **region** Grundgebiet *n* ~ **repeater unit** Einheitsverstärker *m* ~ **requirements of a gear tooth system** Verzahnungsgesetz *n* ~ **research** Grundlagenforschung *f* ~ **rules of tooth engagement** Verzahnungsgesetz *n* ~ **science** Grundlagenforschung *f* ~ **shaft** Einheitswelle *f* ~ **shape of form** Ausgangsform *f* ~ **solution of lead acetate** Bleiessig *m* ~ **speaking pair** Sprechadern *pl* ~ **steel** basisches Siemens-Martin-Material *n* ~ **steelworks** Thomasstahlwerk *n* ~ **stocks of raw materials** Rohstoffgrundlage *f* ~ **sulfate white lead** Sulfobleiweiß *n* ~ **super-group** Vor-Übergruppe *f* ~ **timing circuit** Taktgeber *m* ~ **timing frequency** Grundfrequenz *f* ~ **tool form** Werkzeuggrundform *f* ~ **training** Grundausbildung *f* ~ **trait** Grundzug *m* ~ **unit** Grundeinheit *f* ~ **value** Grundwert *m* ~ **weight** Grundgewicht *n* ~ **weld** Grundschweißstelle *f* ~ **wiring** Grundschaltung *f* ~ **work-hardening** Grundverfestigung *f*
**basicity** Basität *f*, Basizität *f*
**basifier** Basenbildner *m*
**basifying** basenbildend
**basilica** Basilika *f*
**basin** Ausgußschale *f*, Bassin *n*, Becken *n*, Behälter *m*, Einguß *m*, Flachbecken *n*, Hafen *m*, Hafenbecken *n*, Hafendock *n*, Kessel *m*, Mulde *f*, Napf *m*, Schale *f*, Tank *m*, Zuflußgebiet *n*, (structural) Baugrube *f* ~ **into which the dry dock opens** Dockvorhafen *m* ~ **for a pump storage station** Becken *n* für ein Pumpspeicherwerk ~ **for locking (or saving), water** Sparbecken *n* ~ **where percolation takes place** Versickerungsbecken *n* ~ **provided for the shipping from ice during the winter** Winterhafen *m* ~**-shaped valley** Talmulde *f* ~ **sill** Schwelle *f* im Tosbecken ~**-type** muldenförmig
**basis** Basis *f*, Fundament *n*, Grund *m*, Grundfläche *f*, Grundlage *f*, Grundstock *m*, Ständer *m*, Tragschicht *f*, Unterlage *f* **on the** ~ **of** an Hand von, auf Grund ~ **for allocation** Verteilungsschlüssel *m* ~ **of comparison** Vergleichs-grundlage *f*, -unterlage *f* ~ **for rectification** Entzerrungsunterlage *f* ~ **of valuation** Richtwert *m*
**basis amount** Steuermeßbetrag *m* (einheitlicher)
**basket** Erzmaß *n*, Korb *m*, Zementierschirm *m*, (cockpit) Gondel *f* ~ **of a capital** Glocke *f* eines Kapitells
**basket,** ~ **coil** Korbspule *f* ~ **filled with gravel** Senkkorb *m* ~ **frame** Kimmweiden *pl* ~ **frame (or framework)** Korbgerippe *n* ~ **rim** Korbzarge *f* ~ **ring** Korbring *m* ~ **screen** Korbfilter *m*

~-shape grate Korbrost *m* ~ shield Schutzkorb *m* ~ suspension Korbstelleinen *pl* (Ballon) ~ toggle Korbknebel *m* ~(-type) coil Korbbodenspule *f* ~ weave Leinwandbindung *f*
~-weave coil Korbboden-, Korb-spule *f* ~
winding Korbwicklung *f* ~-wound coil Korb-
deckelspule *f*
basketry Korbflechterei *f*
basonic axis Muldenlinie *f* eines Flözes
basophil granule basophiles Körnchen *n*
bass Baß *m*, Brandschiefer *m*, Tieftonbereich *m*
~ bar Baßbalken *m* (string instrument) ~
boosting Baßanhebung *f* ~ box Baßkasten *m*
(musical instruments) ~ compensation Baßanhebung *f* ~ compensator Baßentzerrer *m*
~ control Baßregler *m* ~-cut filter Baßfilter *n*
~ drum große Trommel *f*, Pauke *f* ~ lift Anhebung *f* der Bässe ~ loudspeaker Baßlaut-
sprecher *m* ~ keys Brummkasten *m* ~ note
tiefer Ton *m* ~ notes Tiefen *pl* ~ sound tiefer
Ton *m* ~ speaker Tieftonlautsprecher *m* ~
voice Baß *m* ~ wood amerikanische Linde *f*,
Bast *m*, Lindenholz *n*
basset Ausbiß *m*, Ausgehende *n*
bassoon siphon Fagottwasserabguß *m*
bassy condition Tiefenhervorhebung *f*
bast Bast *m* ~ fiber Bastfaser *f* ~ packaging
tape (or band) Bastband *n*
bastard, ~ cut Mittelhieb *m* ~ file Bastardfeile
*f*, Vorfeile *f*
bastion Bastei *f*, Bastion *f*
bastite Bastit *m*
bat Hieb *m*, Signalkelle *f* ~ bolt Bolzen *m* mit
flachem Verlängerungsstück
batch Brand *m*, Fritte *f*, Füllung *f* (Speisung),
Gebäck *n*, Glassatz *m*, Haufen *m*, Masse *f*,
Mischtrommel *f*, Partie *f*, Posten *m*, Postenverfahren *n* (distill. proc), Satz *m*, Schicht *f*,
(glass) Gemenge *n*, (viscose, glass) Charge *f*
~ of brick (or of tile) Ziegelbrand *m*
batch, ~ carbonation Einzelsaturation *f* ~
charge Speisung *f* ~ counter Stückzähler *m*
~ mixer Chargenmischer *m* ~ operation diskontinuierliche Anlage *f*, Satzbetrieb *m* ~ process
Chargenbetrieb *m*, Postenverfahren *n* ~
processing Vorgruppierung *f* ~ production
Reihenproduktion *f* ~ sizes Chargengröße *f*
~ still Retorte *f* ~ turret for tire fabrics Ma-
terialturm *m* für Reifengewebe ~ working
Reihenbeförderung *f*
batches, in ~ partienweise, schubweise
batching Schichten *n* (in Lagen) ~-off Abnehmen
*n* der Mischung ~ bracket Träger *m* für Auf-
wickelwalzen ~ machine for the decatizer for
batching of goods with wrapper Dekatierwickel-
maschine *f* zum Aufwickeln der Stoffe mit
Zwischenläufertuch ~ oil Batschöl *n* ~ roller
Aufwickel-, Holz-rolle *f*
batchwise Bereitung *f* durch Postenverfahren
bate Beize *f*
bath Bad *n*, Küvette *f* the ~ has become impoverished das Bad ist zu metallarm
bath, ~ circulation Flottenumwälzung *f* ~ level
Badspiegel *m* ~-patenting Tauchpatentieren *n*
(metall.) ~ potential Badspannung *f* (electrolysis) ~ ratio Flottenverhältnis *m* ~ reaction
Badreaktion *f* ~ sample Badprobe *f* ~ solution
Badflüssigkeit *f* ~ surface Badspiegel *m* ~

travel Badstrecke *f* (of thread) ~ tub gunner's
pit Bodenwanne *f* ~ voltage Badspannung *f*
bathe, to ~ baden
bath-extract, to ~ baden (chromieren)
bathometer Tiefenmesser *m*
bathymeter Tiefenseemesser
bathymetric charts Tiefenkarten *pl*
bathyscaphe Tauchboot *n*
bathysphere Tauchkugel *f*
batswing burner Fächer-, Fischschwanz-, Fle-
dermaus-, Schlitz-, Schnitt-brenner *m*
batten, to ~ down verschalken to ~ the hatches
die Luken schalken
batten Bohle *f*, Latte *f*, Leiste *f*, Richtscheit
*n* ~ and space bulkhead Gitter-, Latten-,
Traljen-schott *n* ~-type seat Lattensitz *m*
batter, to ~ aushämmern, böschen, zerschlagen
to ~ out böschen
batter Anlauf *m*, Ausbauchung *f*, Böschung *f*,
Böschungsanlage *f*, Neigung (der Mauer) *f*,
Verjüngung *f* ~ of dam (or of weir) Böschungsneigung *f* ~ gauge Lattenprofillehre *f* ~ level
Bergwaage *f*, Neigungsmesser *m*
battered abgeböscht ~ pile geneigter Pfahl *m*
~ wall Schrägwand *f*
battering wall ausgebauchte Mauer *f*, Schrägwand *f*
battery Batterie *f*, Element *n*, Grundteil *m*,
Zelle *f*
battery, ~ of boilers Kesselgruppe *f* ~ of bottle
cells *m* Flaschenbatterie *f* ~ of cams Nocken-
satz *m* ~ of cylinders Flaschenrost *m* (bottles)
~ of filter wells Filterbrunnenaggregat *n* ~
of fuel cocks Ventilbatterie *f* ~ with high
starting capacity startfeste Batterie *f* ~ of
holes gleichzeitiger Abschuß *m* mehrerer Bohrlöcher (blasting) ~ of Leyden jars Leydener
Flaschenbatterie *f* ~ of light construction
Bordbatterie *f* ~ of magnets Magnetmagazin
*n*, zusammengesetzter Magnet *m* ~ of oil
tanks Öltankbatterie *f* ~ of primary cells Ele-
mentenbatterie *f*
battery, ~ accumulator Akkumulator *m* ~
attendant Batteriewärter *m* ~ boiler Etagenkessel *m* ~ booster Anlaßmagnet *m*, Batterie-
zusatzmaschine *f* ~ box Batteriekasten *m*,
Batterieschrank *m*, Elementbecher *m*, Elektrospeicher *m* ~ bus Batteriesammelschiene *f*
~ car Sammlerkraftwagen *m*, Triebwagen-
lokomotive *f* ~ carbon Batteriekohle *f* ~ case
Batterieschrank *m* ~ cell Batterie-element *n*,
-zelle *f* ~ charger Batterieladesatz *f*, Ladeeinrichtung *f*, Lademaschine *f*, Ladesatz *m*
~-charger rectifier Batterieladegleichrichter *m*
battery-charging Batterieklemme *f*, Laden *n* der
Batterie, Ladung *f* des Akkumulators ~ set
Batterieladesatz *m* ~ station Ladestelle *f* ~
unit Lademaschinensatz *m*
battery, ~ clip Element-, Batterie-klemme *f* ~
commutator Batteriewechsel *m* ~ compartment Akkumulatorraum *m*, Akkuraum *m*,
(submarines) Akkumulatorenraum *m* ~ container Batteriekasten ~ cover Batterieab-
deckung *f* ~ current Batteriestrom *m* ~
cutout switch Batterieabschalter *m* ~ dies
Pochstempel *m* ~ discharge Batterieentladung
*f*, Erschöpfung *f* der Batterie ~ displacement
Stellungsunterschied *m* ~ electrode Batterie-

elektrode *f* ~ **eliminator** Ersatzgerät *n* für Batterie ~-**filling accessories** Einfüllarmaturen *pl* ~ **frame** Batteriegestell *n* ~ **front** Batteriefront *f* ~ **gauge** Batterie-galvanometer *n*, -prüfer *m* ~ **heating** Batterieheizung *f* ~ **ignition** Batteriezündung *f* ~ **indicator** Batteriezähler *m* ~ **jack** Batterieklinke *f* ~ **jar** Batterie-, Element-glas *n* ~ **knife** Batterieschaber *m* ~ **lamp** Batterielampe *f* ~ **lead** Batterie-leitung *f*, -zuführung *f*, -zuleitung *f* ~ **life** Spieldauer *f* ~ **location plot** Batteriebild *n* ~ **loop** Batterie-Erdschleife *f* ~ **manning table** Batterietafel *f* ~ **mid-point** Batteriemitte *f* ~ **mud** Elementschlamm *m* ~-**operated** batteriegespeist ~-**operated receiver** Batterieempfänger *m* ~ **pad** Batterieunterlage *f* ~ **pan** Batteriepfanne *f* (sugar contrating) ~ **plate terminal** Batterieanschlußplatte *f* ~ **pole** Batteriepol *m* ~ **post** Pochsäule *f* ~ **pricker** Anodenstecker *m* ~ **receiver** Batterieempfänger *m* ~ **resistance** Batteriewiderstand *m* ~ **ringing** Batterieanruf *m* ~-**ringing telephone** Fernsprecher *m* mit Batterieanruf *m* ~-**ringing** (magneto-)**telephone station** Fernsprecher *m* für Batterie-(Induktor-)Anruf ~ **room** Batterie-, Sammler-raum *m* ~ **shoe** Pochschuh *m* (min.) ~ **stand** Batteriegestell *n* ~ **stop** Sammlerkontakt *m* ~ **sump jar** Batteriesäureauffanggefäß *n*

**battery-supply**, ~ **bridge** Speisebrücke *f* ~ **circuit** Speisebrücke *f* ~ **circuit noise** Restbrumm *m*, Stromversorgungsgeräusch *n* ~ **coil** Speisespule *f*

**battery**, ~ **supplying plate current to vacuum tubes** Anodenbatterie *f* ~ **switch** Batterieschalter *m* ~ **(cell) switch** Zellenschalter *m* ~-**switch conductor** Zellenschalterleitung *f* ~ **syringe** Batterieheber *m* ~ **terminal connection** Batterieklemme *f* (clip) ~ **terminals** Batteriepol *m* ~ **tester** Batterie-, Element-prüfer *m* ~ **vent valve** Batterieentlüftungsventil *n* ~ **voltage** Batteriespannung *f* ~ **voltmeter** Elementprüfer *m* ~ **wire** Batteriedraht *m* ~'**s zero point** Batterienullpunkt *m*

**battle** Kampf *m*, Schlacht *f* ~ **airplane** Kampfflugzeug *n* ~ **cruiser** Schlachtkreuzer *m* ~ **fleet** Schlachtflotte *f* ~ **ship** Linien-, Schlacht-schiff *n*

**battled wall** gezinnelte Mauer *f*

**baud** Baud *n*

**Baudot**, ~ **system** Baudotsystem *n* ~ **telegraph** Baudottelegraf *m*

**Baumann sulfur print** Baumannsche Schwefelprobe *f*

**Baumé spindle** Bauméspindel *f*

**Bauschinger multiplying lever extensometer** Bauschinger Rollenapparat *m*

.**bauxite** Bauxit *m*

**bay** Abteilung *f* eines Gestells, Brückenstrecke *f* Bucht *f*, Einbuchtung *f*, Fach *n*, Fachwerkfeld *m*, Feld *n*, Gerippeabteil *m*, Gestell (Verstärkergestell), Hallenfeld *n*, Hallenschliff *m*, Joch *n*, Jochfeld *f*, Meerbusen *m*, Täschchen *n*, Wehröffnung *f*, (of trench) Schlag *m*, (small gulf) Bai *f*, (of registers or selectors) Wählerbucht *f* ~ **of amplifiers** Bunker *m* für Verstärker ~ **of a bridge** Brücken-bogen *m*, -feld *n* ~ **of a door** Türnische *f* ~ **of racks** Zählergestell *n* ~ **of registers** Zählerbucht *f* ~ **of a**

**shaft** Schachtfeld *n*, Schachtverzug *m* ~ **of a sluice** Haupt einer Schleuse ~ **of spar** Holmfeld *n* ~ **of a window** Fensternische *f*

**bay**, ~-**berry wax** Myrtenwachs *n* ~ **horse** Fuchs *m* ~ **section** Brückenfahrzeug *n* ~ **window** Erker *m* ~ **work** Riegelwand *f*

**bayleyite** Bayleyit *m*

**bayonet** Seitengewehr *n* ~ **base** Bajonettsockel *m* ~ **catch (or clutch)** Bajonettkuppelung *f* ~ **guard** Parierstange *f* ~ **hilt** Seitengewehrgriff *m* ~ **joint** Bajonett-verbindung *f*, -verschluß *m*, Renkverband *m* ~ **knot** Faustriemen *m* ~ **lock** Bajonettverschluß *m* ~ **mounting** Bajonettscheibenbefestigung *f* ~ **plate** Bajonettscheibe *f* ~ **socket** Stiftsockel *m* ~ **stacks** Bayonett-Auspuffrohrsystem *n* ~ **stud** Seitengewehrhalter *m* ~-**type fitting** Bajonettanschluß *m* ~ **unit** Bajonettverschluß *m* (of lamps)

**B-battery** Anodebatterie *f*, Anodenbatterie *f*, B-Batterie *f*, Zündakkumulator *m* ~ **eliminator** Ersatzgerät *n* für Anodenbatterie, Netzanode *f*

**B-board** ankommende Plätze *pl*, Eingangsplätze *pl*

**B-digit selector** zweiter Kennzifferwähler *m*

**be, to** ~ **on** eingeschaltet sein

**beach, to** ~ auf den Strand laufen lassen

**beach** Gestade *n*, Strand *m*, Ufer *n* ~ **deposit** Uferwall *m* ~ **gravel** Strandkies *m* (an der Seeküste) ~ **sand** Silbersand *m*

**beaching carriage** Schwimmerwagen *m* (aviat.) ~ **gear** Wasserflugzeugtransportwagen *m*

**beacon, to** ~ markieren, signalisieren, vermarken

**beacon** Bake *f*, Fanal *n*, Feuer *n*, Funkbake *m*, Funkfeuer *n*, Leuchte *f*, Leuchtfeuer *n*, Leuchtturm *m*, Strahlenwerfer *m* ~ **antenna** Bakenantenne *f* ~ **buoy** Baken-boje *f*, -tonne *f* ~ **course** Bakenkurs *m* ~ **light** Blickschulung(s)signal *n* ~ **list** Leuchtfeuerverzeichnis *n* ~ **register** Leuchtfeuerverzeichnis *n* ~ **station** Richtungssender *m*, Strahlungswerfer *m*

**beaconing** Bebakung *f*

**bead, to** ~ bördeln, börteln, falzen, rändeln, sickern, umbördeln (a tube), umfalzen **to** ~ **over** umbördeln

**bead** Bördelrand *m*, Kugel *f*, Pastille *f*, Perle *f* (Glasstäbchen), Perlstab *m*, Randwulst *f*, Schweißraupe *f*, Sicke *f*, Siedesteinchen *n*, Wulst *m*, (sight) Korn *n*, (in welding) Raupe *f* ~ **on tire** Laufdeckenwulst *m*, Reifenwulst *m*, Schweißraupe *f*

**bead**, ~ **aperture** Kornöffnung *f* ~ **capacitor** Perlkondensator *m* ~-**clad conductor (or lead)** Litze *f* mit Perlen ~ **covering** Wulstkernbelag *m* ~-**covering machine** Umwicklungs-, Wulstkern-maschine *f* ~ **drive** Bortelantrieb *m* ~-**edged tire** Wulstreifen *m* ~ **flipping** Umwickeln *n* des Wulstkerns ~ **hammer** Perlhammer *m* ~ **lightning** Perlschnurblitz *m* ~ **test** Bördelversuch *m* ~ **thermistor** Perlthermistor *m* ~ **transistor** Perltransistor *m* ~ **tube** Perlrohr *n* ~ **winding machine** Maschine *f* zum Herstellen der Wulstringe

**beaded** mit Glasperlen isoliert ~ **bar** Wulststab *m* ~ **edge** Bördel-, Wulst-rand *m* ~-**edge**

**tire** Reifen *m* mit Wulst ~ **iron** Hospeneisen *n* ~ **lightning** Perlschnurblitz *m* ~-**rim tubular capacitor** Wulstrohrkondensator *m* ~ **screen** Perl- oder Silberwand *f* ~ **tire** Wulstreifen *m*
**beadedless tire** Reifen *m* ohne Wulst
**beader bit (or knife)** Fassonhobelmesser *m*
**beading** Sicken *n*, Wulst *m* ~ **device** Wulstvorrichtung *f* ~ **die** Bördelmatrize *f* ~ **hand tool** Rohrstemmeisen *n* ~ **machine** Bördelmaschine *f*, Sickenmaschine *f* ~ **press** Falzmaschine *f*
**beadless tire** Reifen *m* mit geradem Wulst
**beak** Ausguß *m*, Helm-rohr *n*, -schnabel *m*, Nase *f*, Schnabel *m*, (of an anvil) Hörnchen *n* ~ **head** Gallion *n* ~ **iron** Bankhorn *n*, Galgenamboß *m*, Schlagstöckchen *n*, Sperrhorn *n*
**beaker** Becher *m*, Becherglas *n*, Kochbecher *m*, Ringschale *f* ~ **cover** Uhrglas *n* ~ **flask** Becherglaskolben *m*
**beam, to** ~ **up** aufbäumen, **to** ~ **the warp** die Kette bäumen
**beam** Achse *f*, Balancier *m*, Balken *m*, Baum *m*, Bootsbreite *f*, Bündel *n* (opt.), Feuer *n* (rdo), Keule *f*, Leitlinie *f*, Leitstrahl m (nav.), Profilträger *m*, Querriegel *m*, Richtstrahl *m*, Schiene *f*, Schwengel *m*, Schwinge *f*, Spindel *f*, Strahl *m*, Strahlung *f*, Strahlungsbüschel *m*, Tragebaum *m*, Träger *m*, Trägerschwelle *f*, Unterzug *m*, Welle *f*, (iconoscope) Abtaststrahl *m*, (light) Büschel *m*, (of light) Strahlen-bündel *n*, -büschel *n*, (of ship) Weite *f*, (support) Deckbalken *m*, (swing) Schwinghebel *m* **on her** ~ **ends** auf der Seite liegend ~ **of balance** Waagebalken *m* ~ **of a boat** Schiffsbreite *f* ~ **of corrugated sheet metal** Wellblechholm *m* ~ **of electrons and ions** Wellenbündel *n* ~ **of float** Schwimmerbreite *f* ~ **of a flying boat** Körperlänge *f* ~ **of light** Lichtstrahl *m*, Parallelstrahlenbündel *n* ~ **of nonuniform section** Knotenstrahl *m* ~ **of parallel light** paralleles Lichtstrahlenbüschel *n* ~ **of rays** Strahlbündel *n*, Strahlen-bündel *n*, -büschel *n* ~ **of a vessel** Schiffsbreite *f*
**beam,** ~ **angle** Öffnungswinkel *m*, Strahlrichtung *f* ~ **antenna** Baken-, Richtstrahl-antenne *f*, Richtstrahler *m* ~ **arm** Baumarm *m* ~ **array** Richtantennennetz *n* ~ **balance** Balkenwaage *f* ~ **bearing** Kettbaumlager *n* ~ **bearing support** Baumlagerung *f* ~ **bending press** Balkenbiegepresse *f* ~ **capstrip** Holmgurt *m* ~ **catcher** Ablenkplatte *f*, Strahlabfänger *m* ~ **chuck** Baummitnehmer *m* ~ **compass** Stangenzirkel *m* ~-**control apparatus** Leitstrahlgerät *n* ~-**control pilotage** Leitstrahlführung *f* ~ **cross section** Büschel-, Strahl-querschnitt *m* ~ **current** Strahlstrom *m* ~ **cutoff** Strahlsperrung *f* ~ **cutter** Stanzpresse *f* ~ **deflection** Strahlablenkung *f* ~ **deflection tube** Elektronenstrahl-, Quersteuer-röhre *f* ~-**deflector switch** Abblendschalter *m* ~ **deviation** Strahlablenkung *f* ~ **direction indicator** Strahlungsrichtungsanzeiger *m* ~ **eclipse** Strahlsperrung *f* ~ **emission** Richtsendung *f* ~-**emitting** Strahlenformung *f* ~ **engine** Balancier(dampf)maschine *f* ~ **focus** Strahlschärfe *f* ~ **forming** Strahlenformung *f* ~ **girder** Profileisenträger *m* ~-**guided curvature** Führungskrümmung *f* ~ **hanger** Gestängehänger *m* ~ **hole** Bestrahlungskanal *m*, Strahlröhre *f*, Strahlenkanal *m* ~ **interfero-**

**meter** Strahleninterferometer *n* ~ **joist** Träger *m* ~ **loading** Balkenbelastung *f*, Bootsbreitenbelastung *f* ~ **lobe switching** Keulenumtastung *f* ~ **movement** Strahlbewegung *f* ~ **perveance** Strahlperveanz *f* ~ **positioning** Strahleinstellung *f* (TV) ~-**power tetrode** Endtetrode *f* (Elektronenbündelung) ~-**power tube** Bündelendröhre *f* ~-**power valve** Endröhre *f* mit Elektronenbündelung ~ **pump** Balkenpumpe *f* ~-**receiving station** Richtempfangsanlage *f* ~-**reception detachment** Richtempfängertrupp *m* ~ **reflection** Strahlenzurückwerfung *f* ~ **restrained at one end** eingespannter Balken *m* ~-**reversing lens** Strahlumkehroptik *f* (TV) ~-**rider guidance** Leitstrahlenkung *f* (Rakete) ~-**rolling mill** Trägerwalzwerk *n* ~ **scale** Hebelwaage *f* ~ **sea** Dwarssee *f* ~-**shaped pointer** Balkenzeiger *m* ~ **sharpening** Strahlschärfung *f* ~ **splicing** Stoßfugenüberlaschung *f* ~ **splitter** Lichtverteiler *m* (film), Strahlenteiler *m*, (densitometer) halbversilberter Spiegel *m* ~ **splitting** Strahlspaltung *f* ~ **spreader** Spreizschwelle *f* ~-**spring indicator** Stabfederindikator *m* ~ **station** drahtloser Richtweiser *m*, Einstrahlfunkstelle *f*, Leitstrahlsender *m*, Peilstelle *f*, Relaissendeanlage *f*, Richtfunkfeuer *n*, Richtungssender *m*, Strahlungswerfer *m* ~ **suppression** Strahlaustastung *f* (TV) ~ **survival** Strahlerhaltung *f* ~ **switching** Leitstrahlumschaltung *f*, Keulenumtastung *f* ~ **switch-off point (or valve)** Abschaltwert *m* ~ **system** Einstrahlsystem *n* ~ **test** Balkenprobe *f* ~ **track** Fahrbahn *f*, Schienenstrang *m* ~ **transmission** Richtsendung *f* ~ **transmitter** Einstrahlsender *m*, Strahl-sender *m*, -sendestelle *f* ~ **trolley track** Laufkatzenfahrschiene *f* ~ **tube** Elektronrichtstrahler *m*, Laufzeitröhre *f*, Triftröhre *f*, Röhre *f* mit Elektronenbündelung, (torpedo) Breitseitrohr *n* ~ **tube working with two fields** Zweifeldröhre *f* ~-**type automatic punching machine** Auslegerstanzmaschine *f* ~-**type retarder** Balkengleisbremse *f* ~-**type riveting machine** Auslegernietmaschine *f* ~ **voltage** Beschleunigungsspannung *f* ~ **web** Holmsteg *m* ~ **well** Schwengelbohrloch *n* ~ **width** Öffnungswinkel *m*, Strahlbreite *f*
**beamed,** ~ **short-wave transmitter** Richtstrahler *m* ~ **transmission** gerichtete Ausstrahlung *f* (electr.)
**beaming** Bündelung *f*, Konzentration *f*, Wellenbündelung *f*, (textiles) Falten *n* ~ **device** Bündelungsanlage *f* ~ **frame** Bäumstuhl *f* ~ **machine** Bäummaschine *f* ~-**machine minder** Maschinenbäumer *m*
**beams** Balkenwerk *n*, Gebälk *n* ~ **of carriers** Trägerbündel *n*
**beamwise bending** Blattbiegung *f* senkrecht zur Drehebene
**bean,** ~-**and-pea attachment** Bohnen- und Erbsendreschvorrichtung *f*, Zusatzgerät *n* für Bohnen und Erbsen (Drusch) ~ **aphis** schwarzer Befall *m* ~ **ore** Bohnerz *n* ~ **planter** Bohnenpflanzmaschine *f*
**bear, to** ~ anpeilen, aushalten, Bezug haben auf, drücken, ertragen, sich erstrecken, peilen, tragen, eine bestimmte Richtung haben **to** ~ **down** wuchten **to** ~ **on** stützen **to** ~ **up** anluven **to** ~ **upon** einwirken auf **to** ~ **in mind**

gedenken to ~ on shore nach Land *n* zu halten to ~ testimony zeugen to ~ witness Zeugnis *n* ablegen

**bear** Bodensau *f*, Härtling *m*, Ofenbär *m* ~ **trap** Bärenfalle *f* ~-**trap dam** Dachwehr *f*

**beard** Bart *m* ~-**type radiator** Bartkühler *m*

**bearer** Kabelhalter, Träger *m*, Trägerstütze *f*, Unterzug *m*, (windmill) Auflageknagge *f* ~ **bar** Tragstange *f* ~ **feet of crankcase** Aufhängepratzen *m* am Kurbelgehäuse ~ **wire** Aufhängungsdraht *m*

**bearing** Anpeilung *f* (aviat.), Anschlag *m*, Auflager *n*, Auftreten *n*, ausgepeilte Richtung *f* (nav.), Azimut *m*, Drehzapfen *m*, Drücken *n*, Einfluß *m*, Ertragen *n*, Funkpeilung *f*, Kompaßkurs *m*, Kompaßzahl *f*, Lager (mach.), Lagerbüchse *f*, Lagerstelle *f*, Lagerung *f*, Lauflager *n*, Lochleibung *f*, Ortsbestimmung *f* (rdo), Peilung *f*, Richtung *f*, Richtungswinkel *m*, Schnitt *m*, Seitenwinkel *m*, Ständer *m*, Standortpeilung *f*, Stauchung *f*, Streichen *n* einer Schicht, Streichwinkel *m*, Stütze *f*, Tragfläche *f*, Unterlager *n*, (direction finder) Lage *f*, (of a drilling engine) Bohrkopf *m*, (of a shaft) Wellenlager *n*, (step) Spurlager *n* the ~ **runs hot** das Lager läuft (sich) warm a ~ **has wiped** ein Lager hat geschmiert the ~ **seizes** das Lager brennt fest

**bearing** ~ **of a beam** freitragende Länge *f* eines Balkens ~ **of bedplate** Grundplattenlager *n* ~ **for bogie pin** Drehzapfenlager *n* ~ **of the centrifugal** Halslager der Schleuder (sugar mfg) ~ **of the coast** Verlauf der Küste ~ **on column** Ständergleitfläche *f* ~ **of the countercrank** Gegenkurbellager *n* ~ **for distributor roller** Lager *n* für Übertragwalze ~ **for horizontal shaft** Führungsbolzen *m* ~ **of roller** Achse *f* der Rolle ~ **of spring buckle** Federbundzapfen *m* ~ **by stars** Gestirnpeilung *f* ~ **by the sun** Sonnenpeilung *f*

**bearing,** ~ **accurary** Peilgenauigkeit *f* ~ **alloy** Lagerlegierung *f* ~ **area** Auflagerfläche *f*, Gründungssohle *f* ~ **arrangement** Peilvorrichtung *f* ~ **axle** Tragachse *f* ~ **azimuth** Seite *f* (rdo) ~ **ball** Lagerkugel *f* ~ **base** Lagerfuß *m* ~ **beam** Lagerbalken *m* ~ **bearing plate of radio direction finder** Funkpeilscheibe *f* ~ **block** Kämpferstein *m*, Lagerbock *m*, Lagerstuhl *m*, Stehbock *m* ~ **body** Lagerkörper *m* ~ **box** Lager-gehäuse *n*, -kasten *m*, -körper *m*, -schale *f* ~-**box lining** Lagerschalenausguß *m* ~ **bracket** Gleit-, Lager-bock *m*, Lager--bügel *m*, -schild *n*, -stütze *f* ~ **bracket fixture** Lagerbockverschraubung *f* ~ **bracket for spring bar** Federwiderlager *n* ~ **brass** Lagerschale *f* ~ **bronze** Rotguß *m* ~ **bush for starter** Lagerbuchse *f* für Anlasser ~ **bushing** Lagerschale *f* ~ **cage** Lager-hülse *f*, -käfig *m*, -korb *m* ~ **cap** Lager-bügel *m*, -deckel *m*, -kappe *f* ~ **capacity** Trag-fähigkeit *f*, -kraft *f* ~ **caps** Lagerdeckelschalen *pl* ~ **car** Rostbalken *m* ~ **carrier** Lagerbalken *m* ~ **carrying shaft** Schaltspindellager *n* ~ **casing** Lagergehäuse *n* ~ **channel** Doppelpeilung *f* (rdo) ~ **chart** Peiltabelle *f*, Ortungskarte *f* (rdo) ~ **clamp** Lagerflansch *m* ~ **clearance** Lager-luft *f*, -spiel *n* ~ **collar** Lagerkragen *m* ~ **column** Rahmensäule *f* ~ **compass** Peilkompaß *m*

~ **condition(s)** Auflagerbedingung *f* (arch., statics) ~ **cone** Lagerinnenring *m* ~ **cover** Lagerdeckel *m* ~ **cup** Lageraußenring *m* ~ **depth** Tragtiefe *f* ~ **design** Lagerbauart *f* ~ **displacement** Peilverlagerung *f* ~ **disk** Steuerwinkelscheibe *f* ~ **end pressure** Lagerkantenpressung *f* ~ **error** Peil-fehler *m*, -verlagerung *f* ~ **face of thread** Tragfläche *f* der Gewindeflanke ~ **factor** Zuschlagfaktor *m* (für Lager) ~ **field** Peilfeld *n* ~ **finder** Peileinrichtung *f* (für Sehrohre) ~ **frame** Lagerraum *m* ~ **friction** Lagerreibung *f* ~ **groove** Schlitzkanal *m* ~ **hinge** Auflagergelenk *n* ~ **house** Lagerkörper *m* ~ **housing** Lagergehäuse *n* ~ **indicator** Peilungsanzeiger *m* ~-**indicator means** Standanzeiger *m* ~ **jewel** Lagerstein *m* ~ **jewel holder** Trägerleiste *f* ~ **land** hervorstehende tragende Fläche *f* (mech.) ~ **life calculations** Lebensdauerberechnungen *pl* ~ **load** Lagerbelastung *f* ~ **measurement** Richtungsmessung *f* ~ **metal** Babbitt-, Lager-metall *n* ~ **meter** Richtungsanzeiger *m* ~ **method** Peilungsart *f* ~ **neck** Lagerhals *m* ~ **object** Peilziel *n* ~ **oil** Maschinenlageröl *n* ~ **pedestal** Lagerbock *m* ~ **perpendicular to the earth** Erdlot *n* (aerial photo) ~ **pile** Druckpfahl *m*, stehender Pfahl *m*, Stützpfeiler *m*, Tragpfahl *m* ~-**pile foundation** stehende Pfahlgründung *f* ~ **pin** Lager-bolzen *m*, -zapfen *m* ~ **plate** Grundplatte *f*, Kursscheibe *f*, Lagerschild *n*, Peilvorrichtung *f*, Visiervorrichtung *f* ~ **plate of direction finder** Peilaufsatz *m* ~ **plate of radio direction finder** Funkpeilscheibe *f* ~ **play** Lagerspiel *n* ~ **point** Richtungspunkt *m* ~ **post** Trägerstift *m*, Pinnenträger *m* ~ **potentiometer** Peilungspotentiometer *n* ~ **pressure** Anpreß-, Auflage-, Auflager-, Lager-, Leibungs-, Lochleibungs-, Stauch-druck *m*, Stützkraft *f*, (distribution) Lastaufnahme *f* ~ **pressure on supports** Auflagerreaktion *f* ~ **projector** Peilungsprojektor *m* ~ **pulley** Tragrolle *f* ~ **rail** Tragschiene *f* ~ **reaction** Auflagerdruck *m*, Stützknagge *f* ~ **reaction force** Lagerstützkraft *f* ~ **resolution** Azimutauflösung *f* (nav.) ~ **retainer** Anschlagscheibe *f* ~ **retention bolt** Lagerfußschraube *f* ~ **ring** Trag-, Gleit-ring *m* ~ **roller** Lagerrolle *f* ~ **seat** Lagerschale *f* ~ **sense** Peilseite *f* ~-**sense switch** Peilseitenschalter *m* ~ **shell** Lager-gehäuse *n*, -schale *f* ~ **shift** Peilverlagerung *f* ~ **side cover** Lagerabschlußdeckel *m* ~ **sleeve** Lager-hülse *f*, -korb *m* ~ **socket** Spurplatte *f* (railway turntables) Drehpfanne *f* ~ **spacer** Distanzbüchse *f* ~ **spring** Trag-, Stütz--feder *f* ~ **spring for railroad carriages** Eisenbahntragfeder *f* ~ **station** Richtungsempfangsstation *f* ~ **strength** Tragfähigkeit *f* ~ **stress of a rivet** Stauchspannung *f* ~ **string roll** Peilfadenaufroller *m* ~ **strip** Auflagerleiste *f* ~ **support** Lagergestell *n*, Lagerruhe *f*, Lagerstützschale *f*, (in turbine and jet engines) Lagerstern *m* ~ **surface** Anlagekante *f*, Arbeitsfläche *f*, Auflagefläche *f*, Auflager *n*, Lagerauflage-, Lager-, Lagerober-, Lauf-fläche *f*, Stützpunkt *m* ~ **surfaces of doors (or of windows)** Anschläge *pl* ~-**surface area** Druckfläche *f* ~ **table** Auflagefläche *f*, Peiltisch *m* ~ **temperature gauge** Lagertemperaturmeßgerät *n* ~ **testing machine** Lagerprüfmaschine *f* ~

**tolerance** Lagerluft *f* (mech.) ~ **transmission unit** Peilungsanzeiger *m* ~ **tube** Seitensichtrohr *n* ~ **wall** Lagerwand *f* ~ **yoke** Lagergabel *f*

**bearings** Achsenlager *n*, Tragweite *f* ~ **determined from the airplane** Eigenpeilung *f* (aviat.)

**beat, to** ~ abklopfen, klopfen, mahlen (paper mgf), pochen, rühren, schlagen, stauchen, zerfasern, (a way or path) bahnen **to** ~ **back** zurückschlagen **to** ~ **down** feststampfen, rammen **to** ~ **in** einschlagen (nail) **to** ~ **into** hereinschlagen (the ground) **to** ~ **off** losschlagen **to** ~ **out** ausbeulen, ausklopfen, ausschmieden, ausschweifen **to** ~ **through** hindurchschlagen **to** ~ **down the earth** die Erde rammen **to** ~ **the form** Schwärze *f* auftragen **to** ~ **away the ground** auffahren **to** ~ **the ink** Schwärze *f* auftragen **to** ~ **out iron** das Eisen abbreiten oder flachschmieden **to** ~ **the lathe (or the slay)** die Lade anschlagen **to** ~ **to windward** anluven, aufkreuzen

**beat** Anschlagtaste *f*, Ausschlag *m*, Ausschlag *m* eines Zeigers, Interferenz *f*, Klappensignal *n*, Klatsche *f*, Mahlarbeit *f*, Pulsation *f*, Runde *f*, Schlag *m*, Schwebung *f*, Schwebungsfrequenz *f*, Schwebungssummer *m*, Wellenzug *m*, (of pulses) Takt *m* **in** ~ im Takt *m* ~**s of audible frequency** Schwebungen *pl* von Hörfrequenz ~ **of a scythe** Dengelbahn *f*

**beat,** ~ **amplitude** Schlagamplitude *f* ~ **cycle** Schwebungsperiode *f* ~ **frequency** Pfeifen *n* (TV), Pfiff *m* (TV), Schwebungsfrequenz *f* ~ **frequency oscillation** Einpfeifen *n* ~ **frequency oscillator (BFO)** Frequenzschwingungssteuerung *f*, Schwebungssummer *m* ~ **indicator** Schwebungsanzeiger *m* ~ **method** Heterodyne *f*, Schwebungs-methode *f*, -verfahren *n* ~ **note** Interferenz-, Schwebungs-, Stoß-, Überlagerungs-ton *m* ~**-note method** Suchtonverfahren *n* ~**-note pitch** Schwebungstonhöhe *f* ~ **picture** Klatschenbild *n* ~**-ray plaque** Flachträger *m* ~ **receiver** Heterodyne-, Interferenz-, Schwebungs-, Überlagerungs-empfänger *m* ~ **reception** Heterodyn-, Interferenz-, Schwebungs-, Überlagerungs-empfang *m* ~ **regulation** Anschlagsregelung *f* ~ **tone** Interferenz-, Schwebungs-ton *m* ~**-tone oscillator** Schwebungssummer *m*

**beaten,** ~ **cobwork** Plisseebau *m* ~ **path** Trampel-pfad *m*, -weg *m* ~ **piece** gekumpeltes Stück *n* ~ **silver** Blattsilber *n* ~**-stuff tester** Mahlungsgradprüfer *m* ~ **track** Trampel-pfad *m*, -weg *m* ~ **zone** bestrichener Raum, bestrichenes Gelände *n*

**beater** Feinzeugholländer *m*, Flügel *m* eines Rührwerkes, Ganzzeugholländer *m*, Holländer *m*, Klöpfel *m*, Krücke *f*, Mahlholländer *m*, Pantschmaschine *f*, Prätschmaschine *f*, Rührkrücke *f*, Schaufel *f*, Schaumschläger *m*, Schlägel *m*, Schläger *m*, Schlagmaschine *f*, Stampfer *m*, Stoffmühle *f*, Stopfhacke *f*, Walkhammer *m*, Walzenmühle *f*, (weaving) Flackmaschine *f* ~ **of a centrifugal machine** Flügel *m* einer Zentrifugalsichtmaschine ~ **with screw motion** Schraubenführungholländer *m*

**beater,** ~ **bar** Messer *n* am Holländer ~ **blade** Schlägermesser *n* ~ **button** Batteurknopf *m* ~ **drum** Leistentrommel *f* ~ **house** Holländersaal

*m* ~**man** Holländermüller *m* ~ **mill** Schläger-*f*, Schlag-, Schlagkreuz-, Schlaggrad-mühle *f* ~ **movement** Schlägerantrieb *m* (film) ~ **plate** Grundwerk *n* ~ **room** Holländersaal *m* ~ **shaft** Schlägerwelle *f* ~ **stock** Holländermasse *f*

**beating** Rührung *f*, Schlagen *n*, Schwebung *f* ~ **in of cable** Vorbereitung *f* des Kabelendes zum Überziehen des Ziehschlauchs ~ **of the waves** Wellenschlag *m*

**beating,** ~ **arm** Schlagleiste *f* ~ **board** Sprungbrett *n* ~ **brush** Abziehbürste *f* (Korrektur), Abklopfbürste *f* (print.) ~ **current** Schwebungsstrom *m* ~ **device** Schlagvorrichtung *f* ~ **effect** Schwebungs-, Interferenz-vorgang *m* ~ **engine** Feinzeug-, Ganzzeug-holländer *m*, (pulps, etc.) Breimühle *f* ~ **machine** Klopfstuhl *m* ~ **mill** Stampf-, Stoß-kalander *m* ~ **opener** Klopfwolf *m* (textiles) ~ **process** Schwebungsvorgang *m* ~ **shoe** Pochschuh *m* ~ **stone** Schlagstein *m* ~ **vat** Schlageküpe *f*

**Beaufort,** ~ **notation** Beauforts Wetterskala *f* ~ **scale** Beaufortskala *f* ~ **scale of wind force** Beauforts Windstärkeskala *f* ~ **scale of wind intensities** Beaufortsche Windskala *f* ~**'s scale** Windstärkenskala *f* von Beaufort

**beaver dam** Biberwehr *f*

**becket** Führungsschlaufe *f*, Knebelstropp *m*

**becking mill** Aufweitewalzwerk *n*

**become, to** ~ werden, stehen **to** ~ **bankrupt** fallieren **to** ~ **choked** verstopfen **to** ~ **clogged** verstopfen **to** ~ **coated** beschlagen, überziehen **to** ~ **cold** kalt gehen **to** ~ **covered** überziehen **to** ~ **covered with rust** überrosten **to** ~ **darker** nachdunkeln **to** ~ **dim (or dull)** anlaufen **to** ~ **effective** auswirken **to** ~ **even** verflachen **to** ~ **fixed by rust** anrosten **to** ~ **foggy** sich verschmieren **to** ~ **fouled** sich verschmieren **to** ~ **free** frei werden **to** ~ **hazy** sich verschmieren **to** ~ **indistinct** verschwimmen **to** ~ **insolvent** fallieren **to** ~ **jammed** sich klemmen **to** ~ **known** herauskommen **to** ~ **less steep** flach werden **to** ~ **lost** sich verfliegen, Orientierung *f* verlieren **to** ~ **muddy** anschlämmen **to** ~ **perceptible** auswirken **to** ~ **popular** Anklang finden **to** ~ **richer** in sich anreichern **to** ~ **scrap** entfallen **to** ~ **sick** erkranken **to** ~ **smaller** kleiner werden **to** ~ **smeared** sich verschmieren **to** ~ **spring hard** federhart werden **to** ~ **viscous** verdicken **to** ~ **weaker** abflauen **to** ~ **worse** verschlechtern **to** ~ **wrecked** zerstört werden **to** ~ **yellow** vergilben

**becoming,** ~ **effective** tatsächlich zur Wirkung kommend ~ **hot** Heißwerden *n* (chem.) ~ **quiet** Ruhigwerden *n*

**Becquerel,** ~ **cell** Elektrolytzelle *f* ~ **effect** Becquereleffekt *m*

**becquerelite** Becquerelit *m*

**bed, to** ~ auflagern, betten, festmachen (mach.), einschaben (surfaces or brasses) **to** ~ **up** umstampfen

**bed** Ablagerung *f*, Bank *f*, Bett *n*, Bettung *f*, Fundament *n*, Gefüge *n*, Geleise *n* (r.r.), Gestell *n*, Horizont *m*, Lagerstätte *f*, Schicht *f*, Unterlage *f*, (~ding) Lage *f*, Lager *n*, (river) Sohle *f* **in** ~ lagerförmig (geol.) ~(ding) **of boiler** Kessellager *n* ~ **of broken stone** Schotterbett *n* ~ **of clay** Ton-schicht *f*, -unterlage *f* ~ **of grains** Graupenbett *n* ~ **of ore** flaches Erztrumm $n$ ~ of potash salts Kalisalzlager $n$ ~ of rails Schienenbett $n$ ~ of river Flußsohle $f$ ~ of road metal Schotterbett $n$ ~ of supporting means Träger $m$

**bed,** ~ **bolt** Lager-, Träger-bolzen $m$ ~ **box** Kastenfuß $m$ ~**-box leg** Drehbankfuß $m$ ~ **bracket** Bettkonsol $n$ ~ **coke** Bett-, Füll-koks $m$ ~ **drive gear** Antriebsrad $n$ für Karren ~ **drive pinion** Rollrad $n$ ~ **elevation** Niveau $n$ der Sohle, Sohlenhöhe $f$ ~ **forming** bettbildend (discharge) ~**-forming discharge** bettbildender Durchfluß $m$ ~ **frame** Bandtraggerüst $n$ ~ **gap** Bettkröpfung $f$ ~ **guideways** Bettführungsbahnen $pl$ ~ **joint** Längsfuge $f$ ~**lam** Durcheinander $n$ ~**like jointing** Bankung $f$ (geol.) ~ **load** Geschiebe $n$ ~**-load transport** Geschiebe-führung $f$, -fracht $f$ (solids discharge)

**bedplate** Amboßklotz $m$, Amboßlager $n$, Auflagerplatte $f$, Bettung $f$, Bodenplatte $f$, Chabotte $f$, Fundament $m$, Fundamentrahmen $m$, Fußlasche $f$, Grundwerk $n$, Klotz $m$, Maschinengestell $n$, Messerblock $m$ (paper mfg), Platine $f$, Schabotte $f$, Schawatte $f$, Schwelle $f$, Sohlenplatte $f$, Sohlplatte $f$, Stuhlplatte $f$, Zacke $f$, Zacken $m$ ~ **of a rag engine** Platte $f$ des Holländers ~ **with crosshead guides** Gabelbalken $m$ (Dampfmaschine) ~ **beam** Grundlager $n$ ~ **box** Grundwerkkasten $m$ ~ **knife** Holländergrundwerkmesser $m$ ~ **miller** Bettfräsemaschine $f$ ~**-type joint** Fußlaschenverbindung $f$

**bed,** ~ **plug dowel** eingebetteter Dübel $m$ ~ **position** Lagerstelle $f$ (eines Maschinenteils) ~**rock** Felsboden $m$, festes Gebirge oder Gestein, Mutterfels $m$, Untergrund $n$ ~ **size** Fundamentgröße $f$

**Bedson** (continuous) **rod mill** Bedson Drahtstraße $f$

**bed,** ~ **spring antenna** Matratzenantenne $f$ ~**stone** Bodenstein $m$ ~**way** Bett-, Führungs-bahn $f$ ~**way cross-section** Führungsbahnquerschnitt $m$ ~ **ways** Bettbahn $f$, Bettführungsbahnen $pl$

**bedded** bankig, gebankt ~**-in mold** Form $f$ ganz im Boden ~ **vein** Lagergang $m$

**bedding** Auflagefläche $f$, Bettung $f$, Bettzeug $n$, Einbettung $f$, Lagerung $f$, Schichtung $f$, (of machines) Verbleiben $n$, (Fußboden) Verlegen $n$ ~**in molding** Formen $pl$ halb im Boden ~**-in process** Einlaufvorgang $m$ ~ **of pipes** Verlegung $f$ von Rohren

**bedding,** ~ **concrete** Unterbeton $m$ ~ **pile** Stützpfeiler $m$ ~ **plane** Schicht-, Schichtungs-fläche $f$ ~ **planes** Schichtfugen $pl$ ~ **value** Bettungsziffer $f$

**bedew, to** ~ betauen

**bee** Biene $f$ ~ **glue** Bienenharz $m$, Kitt-, Kleb-, Vor-wachs $n$ ~**'s wax** Bienenwachs $n$

**beech** Buche $f$ ~ **charcoal** Buchenkohle $f$ ~ **(nut) oil** Bucheckernöl $n$ ~**-tar pitch** Buchenpech $n$ ~**wood** Buchenholz $n$

**beehive** Bienenkorb $m$ ~ **coke oven** Bienenkorbkoksofen $m$ ~ **furnace** Bienenkorbofen $m$ ~ **neon lamp** Bienenkorblampe $f$ ~ **(coke) oven** Bienenkorbofen $m$ ~ **shelf** durchlöcherter Bügel $m$ (gas drying)

**beep** Testbandsignal $n$ (acoust)

**beeswing bran** Flugkleie $f$

**beet,** ~**-and-bean drill** Rüben- und Bohnendrill-

maschine $f$ ~ **attachment** Rübensäeapparat $m$ ~ **bin** Rübenlager $n$ ~ **dumping** Rübenentladen $n$ ~ **flume** Rübenschwemme $f$ ~**-leaf catcher** Rübenkrautfänger $m$ ~ **pricker** Gribbel $m$ ~ **puller** Rübenheber $m$ ~ **pulp** ausgelaugte Rübenschnitzel $n$ ~ **quota** Rübengrundlieferungsrecht $n$ ~ **rasp** Rübenreihe $f$ ~ **sapogenin** Rübenharzsäure $f$ ~ **scab** Rübenschrot $n$ ~**-seed ball** Rübensamenknäuel $n$ ~**-seed disinfectant (or seed dressing)** Rübensamenbeizmittel $n$ ~ **seeder** Rübensäemaschine $f$ ~ **sickness** Rübenmüdigkeit $f$ ~ **slice** Rübenschnitzel $n$ ~ **tails** Schwänze $pl$ (sugar mgf) ~ **tiredness** Rübenmüdigkeit $f$ ~ **tool** Rübengerät $n$ ~ **washer** Quirlwäsche $f$ ~ **washer with revolving arm agitators** Rübenquirlwäsche $f$ ~ **wheel** Rübenhubrad $n$

**beetle** Handstampfe $f$ (mach.), Holzschlägel $m$, Käfer $m$, Schlägel $m$, (instrument) Erdstampfe $f$, (textiles) Stoßkalander $m$

**beetling engine (or mill)** Stampfappretur $f$, Stampfkalander $m$

**befall, to** ~ zustoßen

**before,** ~ **(after) use** vor (nach) Benutzung $f$ ~ **hand** vorher

**begin, to** ~ anfangen, anschneiden, beginnen, entstehen **to** ~ **with** vorab, von vornherein **to** ~ **to melt** anschmelzen **to** ~ **a mine** einschlagen **to** ~ **to rust** anrosten **to** ~ **to spin** anspinnen

**beginner** Anfänger $m$

**beginning** Anfang $m$, Antritt $m$, Beginn $m$, Ursprung $m$ **for the** ~ vorderhand **from the** ~ von Grund auf ~ **at mid-load** mit einer halben Spule $f$ beginnend ~ **at midsection** mit einem halben Spulenfeld $n$ beginnend

**behalf, on** ~ **of** auf Veranlassung von

**behave, to** ~ sich benehmen, sich betragen, verhalten

**behaved wood** imprägniertes Holz $n$

**behavior** Aufführung $f$, Benehmen $n$, Betragen $n$, Betriebsverhalten $n$, Haltung $f$, Verhalten $n$

**behaviorism** Verhaltungsweise $f$

**behavioristic** extrospektiv

**behenic acid** Bensäure $f$

**Behm,** ~ **depth indicator** Schallot $n$ ~ **detonating cartridge** Behm-Lotpatrone $f$, Knallpatrone $f$ ~ **ear lead** Behm-Ohrlot $n$ ~ **echo depth sounder** Schallot $n$ ~ **echosounding machine** Behm-Ohrlot $n$ ~ **period meter** Behm-Zeitmesser $m$, Kurzzeitmesser $m$

**beige** naturfarben

**bel** Bel $n$

**belay, to** ~ belegen, festmachen

**belaying,** ~ **cleat** Belegklampe $f$ ~ **pin** Karvel-, Kovillen-, Beleg-nagel $f$ (naut)

**B-electrode tube** Pliotron $n$

**belemnite** Belemnit $m$, Fingerstein $m$, Strahlkeil $m$

**belfry** Glocken-stube $f$, -stuhl $m$

**Belgian,** ~ **furnace** belgischer Ofen $m$ ~ **looping mill** belgische Drahtstraße $f$ ~ **rod mill** belgische Drahtstraße $f$ ~ **wire mill** Wechselduo $n$

**B-eliminator** Ersatzgerät $n$ für Anodenbatterie, Netzanode $f$, Netzanschlußgerät $n$

**belite** Belit $m$

**belix angle** Schrägungswinkel $m$

**bell** Blase $f$, Dach $n$, Glocke $f$, Klingel $f$, Klöp-

pel *m*, Konus *m* (TV), Läutewerk *n*, Muffenende *n* (teleph.), Schelle *f*, Signal *n*, Ziehtrichter *m*, (in butt welding of tubes) Manschette *f*, (call) Wecker *m*, (welding tubes)
Trichter *m*, (wind instrument) Schallbecher
*m* ~ **of a capital** Glocke *f* eines Kapitells
**bell, ~-and-hopper** Gasverschluß *m* **~-and-hopper
arrangement** Gichtverschluß *m*, Parry-Trichter
*m* **~-and-spigot joint** Muffenverbindung *f* ~
**and spigot (pipe)** übereinandergreifendes Rohr
*n* **~-and-spigot pipe** Rohr *n* mit Vor- und Rück-
sprung-Verbindung ~ **buoy** Glocken-boje *f*,
-tonne *f* **~-cage** Glockenstuhl *m* ~ **circuit** Klin-
gelleitung *f*, Weckerstromkreis *m* ~ **compass**
Glocken-, Kuppel-kompaß *m* ~ **cover** Schutzglocke *f* **~-crank** winkelhebelartig ~ **crank** Knie-,
Wirbel-, Winkel-hebel *m* (lever) **~-crank drive**
Winkelantrieb *n* **~-crank handle** Bremswinkel
*m* **~-crank lever** krummer Hebel *m* ~ **crusher**
Glockenmühle *f* ~ **dome** Glockenschale *f* ~
**function lever comb** Klingelkamm *m* ~ **funnel**
Glockentrichter *m* ~ **gong** Glockenschale *f*
~ **hammer** Glockenklöppel *m*, Schlaghebel *m*
~ **hoist** Gicht-, Glocken-winde *f* ~ **housing**
Laterne *f* (Motor) ~ **insulator** Glockenisolator
*m* **~ item counter** Glockenzählwerk *n* (adding
mach.) ~ **jar** Campane *f*, Glas-glocke *f*,
-ballon *m*, Glocke *f*, Tauchglocke *f* ~ **jar filter**
Glockenfilter *n* ~ **latch bar** Glockenverriege-
lungsstab *m* ~ **metal** Bronze *f*, Glocken-bronze
*f*, -erz *n*, -gut *n*, -metall *n*, -speise *f* **~mouth**
Gleichrichter *m*, (petroleum) Auswalzen *n*,
Einlauftrompete *f* (Schalltrichter) **~-mouth
duct** Leitung *f* mit glockenförmigem Ende *f*
~ **mouthed** glockenförmig, glockenförmig auf-
geweitet **~mouthed fishing socket** Fangglocke
*f* **~mouthed socket** Glockenfänger *m* **~-mouth-
ing** Abrundung *f* (Walzkaliber) ~ **operating
gear** Gicht-, Gichtglocken-, Glocken-winde
*f* **~-operating lever** Glockenauslösehebel *m*
~ **process** Glockenverfahren *n* ~ **pull** Notleine
*f* ~ **push** Klingeltaster *m* ~ **reset bar** Glocken-
rückstellstab *m* **~-ringing apparatus** Glocken-
läutemaschine *f* **~-ringing machine** Läutemaschine *f* ~ **screw** Fangglocke *f* ~ **seal** Glok-
kenverschluß *m* (water) ~ **set** Wecker *m*
(rdo)
**bell-shaped** glockenförmig ~ **crusher** Glockenmühle *f* ~ **curve** Glockenkurve *f* ~ **impulse
sector** gezahnte Glocke *f* ~ **magnet** Glocken-
magnet *n* ~ **part of trumpet** Schalltrichter *m*
~ **top** Glocke *f* ~ **valve** Kronenventil *m*
**bell, ~ signal** Glocken-signal *n*, -zeichen *n*,
Klingelzeichen *n* ~ **socket** Rohrfangtüte *f* ~
**stop** Weckerausschalter *m* ~ **striker** Glockenklöppel *m* ~ **stroke** Glockenschlag *m* ~ **swipe**
Glockenschwengel *m* ~ **taper turning** Konus
*m* (der Bildröhre) ~ **tower** Glockenturm *m* ~
**transformer** Klingeltransformator *m* ~ **trap**
Glockenverschluß *m* (sanitary) **~-type auto-
clave press** Glockenheizkessel *m* **~-type
distributing gear** Gichtglocke *f*, Glocke *f* (blast
furnace) **~-type intermediate housing** Glocken-
zwischengehäuse *n* **~-type wash tower** Glocken-
wäscher *m* ~ **valve** Glockenventil *n* ~ **water
closet** Glockenwasserverschluß *m* ~ **winch**
Gichtglockenwinde *f* ~ **wire** Klingeldraht *m*
~ **work** glockenartiger Abbau *m*

**Bell's dephosphorizing (or pigwashing) process**
Bells Roheisenentphospherungsverfahren *n*
**bellied** bauchig
**Bellini-Tosi system** Bellini-Tosi System *n*,
Kreuzrahmenpeilverfahren *n*
**bellow, to** ~ brüllen
**bellows** Ausblaseapparat *m* (print.), Ausdeh-
nungsmanschette *f*, Balg (photo), Barometerdose *f*, Blasbalg *m*, Blasebalg *m*, Dehnungs-
ausgleicher *m*, Druckdose *f*, Faltenbalg *m*,
Federbalg *m*, Federungsblech *n*, Federungsblock *m*
**bellows, ~-and-strap arrangement** Faltenbalg *m*
~ **blowpipe** Lötbrenner *m* **~-body camera** Blase-
balgkamera *f* ~ **camera** Balgkamera *f* ~
**chamber** Aneroidkammer *f* ~ **dust extractor**
Kastenausbläser *m* ~ **extension** (of camera)
Balgauszug *f* ~ **folds** Balgfalten *pl* **~-framed
door** Falttor *n* ~ **gauge** Aneroidmanometer *n*
~ **joint** Balgdichtung *f* (sealing) ~ **ring** Balg-
linse *f* ~ **support** Balgenstütze *f*
**belly, to** ~ aufblähen **to** ~ **out** ausbauchen
**belly** Bauch *m*, Bauchseite *f*, Fuchs *m*, Kohlensack *m* (Hochofen), Leib *m*, (string instrument)
Decke *f*
**belly, ~ band** Bauchgurt *m* (of safety belt)
**~-band pad** Kammkissengurt *m* ~ **helve** Brust-
hammer *m* ~ **landing** Bauchlandung *f* (aviat.),
Plattfuß *m* ~ **load** Rumpfaußenlast *f*, Unter-
last *f* **~-pad strap** Kammkissengurt *m* ~ **pipe**
Rüssel *m* ~ **radiator** Bauchkühler *m* ~ **tank**
Abwurfbehälter *m*, Rumpfaußentank *m* ~
**turret** Bodenkanzel *f*
**bellying** Ausbauchung *f*
**below** unter, unterhalb, unten
**below, ~ grade** Tiefschnitt *m*, Unterflur *m*
~ **grade construction** Tiefbauarbeiten *pl* ~
**ground** untertags, unter Tage ~ **ground (floor)
hydrant** Unterflurhydrant *m* **~-mentioned**
nachstehend erwähnt ~ **track level** Tiefschnitt
*m* ~ **zero** unter Null
**belt** Gurt *m*, Gurtband *n*, Gürtel *m*, Koppel *n*,
Leibgurt *m*, Leibriemen *m*, Riemen *m*, Transmission *f*, (conveyer) Band *n*, (painting) Ab-
setzstreifen *m*, (cable) Polster *n* ~ **for conical
drum drive** Konusriemen *m* ~ **composed of
several layers of material** mehrfacher Riemen
*m* ~ **of vorticity** Wirbelband *n*
**belt, ~ appliance** Riemenzubehör *n* ~ **armor**
Gürtelpanzer *m* ~ **arrangement** Bandanordnung
*f* ~ **band** Gurtband *n* ~ **beam** Bandträger *m*
~ **canal** Ringkanal *m* ~ **carrier** Bandförderer
*m* ~ **carrier system** Fließbandanlage *f* ~ **clamp**
Riemenspanner *m* ~ **conveyer** Band-förderer
*m*, -transporteur *m*, Förderband *n*, Förder-
bandstapler *m*, Gurt-förderband *n*, -förderer
*m*, Gurttransporteur *m* ~ **conveyor flight** Band-
straße *f* **by the ~ conveyor system** am laufenden
Band *n* ~ **conveyor weighing** Förderbandwaage
*f* ~ **coupler** Riemenverbinder *m* ~ **coupling**
Bandkupplung *f* ~ **cross section** Riemenquer-
schnitt *m* ~ **dressing** Riemen-appretur *f*, -fett
*n* ~ **drive** Pesen-, Riemen-antrieb *m*, Riemen-
trieb *m*, Riemenübertragung *f* ~ **driven** mit
Riemenantrieb ~ **drop hammer** Wickelhammer
*m* ~ **drum** Gurttrommel *f* ~ **dynamometer**
Transmissionsdynamometer *m* ~ **elevator** Gurt-
elevator *m* ~ **fastener** Riemen-band *n*, -ver-

binder *m*, -verschluß *m*, -klammer *f* (hook)
~ **fastener screw** Riemenschraube *f* ~ **feed**
Gurt-ladung *f*, -zufuhr *f* ~-**feed guide** Gurthebel *m* ~-**feed lever** Patronenträger-, Zu-
bringer-hebel *m* ~ **feed pawl** Gurt-schieber *m*,
-zuführungskralle *f* ~ **feed slide** Gurtzufüh-
rungsschlitten *m* ~ **feeder** Bandaufgabe *f*
~-**filling machine** Gurt-füller *m*, -füllmaschine *f*
~ **frame** Bandgestell *n* (Förderanlage) ~
**gearing** Riemenvorgelege *n* ~ **grease** Treib-
riemenfett *f* ~ **grinder** Bandschleifmaschine *f*
~ **guard** Riemenverdeck *n* ~ **guide** Riemen-
-führung *f*, -gabel *f*, -leiter *m* ~ **guide pulley**
Riemenleitrolle *f* ~ **highway** Umgehungsstraße *f* ~-**holding pawl** Gurthebel *m* ~ **house**
Riemenhaus *n* ~ **idler** Bandrolle *f* ~ **joint**
Riemenverbindung *f* ~ **knife (splitting) machine**
Bandmesser(spalt)maschine *f* ~ **lacer** Riemenverbinder *m* ~ **lever shifter** Riemenhebel-
schalter *m* ~-**loading machine** Gurtfüller *m*
~ **loop** Koppelschlaufe *f* ~ **mounter** Riemenaufleger *m* ~ **pawl** Zuführer *m* ~ **ply** Bandein-
lage *f* ~**power cane mill** Rohrmühle *f* mit Kraftantrieb ~ **profile** Riemenprofil *n* ~ **pull** Riemenzug *m* ~ **pull regulator** Gurtzugregler *m* ~
**pulley** Gurtscheibe *f*, Bandscheibennabe *f*,
Riemenrad *n*, Riemenscheibe *f*, Scheibe *f* für
Riemenbetrieb ~ **punch** Riemen-locher *m*,
-lochzange *f* ~ **railroad** Gürtelbahn *f* ~ **ring**
Gürtelring *m* ~ **sander** Bandschleifer *m*, Bandschleifmaschine *f* (Sandpapier für Holzverar-
beitung) ~ **saw** Bandsäge *f* ~**scanner** Trommelabtaster *m* (TV) ~ **scraper** Bandabstreifer *m*
~ **seat** Gurtsitz *m* ~ **separator** Band-scheider
*m*, -separator ~ **shifter** Ausrückhebel *m*,
Riemen-absteller *m*, -ausrücker *m*, -führer *m*,
Treibriemenrückvorrichtung *f* ~ **shifter bar**
Ausrückerwelle *f* ~-**shifter fingers** Riemen-
ausrückgabel *f* ~ **shifting** Riemenverschiebung
*f* ~-**shifting device** Riemen-ausrückvorrichtung
*f*, -schaltvorrichtung *f* ~-**shifting finger** Riemen-
gabel *f* ~-**shifting fingers** Riemenausrücker-
gabel *f* ~ **slip** Riemenrutsch *m* ~ **slip control
unit** Bandschlupfkontrollgerät *n* ~ **speed** Gurt-
geschwindigkeit *f* ~ **strap** Gurtband *n* ~ **stress**
Riemenbeanspruchung *f* ~ **stretcher** Riemenspanner *m* ~ **suspension** Gurtaufhängung *f*
~ **tension** Riemenzug *m* ~ **tension adjustment**
Riemenspannverstellung *f* ~ **tensioning device**
Riemenspanner *m* ~ **tightener** Riemenbefesti-
ger *m* ~ **trailing idler** Bandleitrolle *f* ~ **transmission** Riemen-antrieb *m*, -transmission *f*,
-übertragung *f* ~ **tunnel** Riemenhaus *n* ~-**type
capstan** Gurtabzug *m* ~-**type dehydrator** Bandtrockner *m* ~-**type ticket carrier** Bandpost *f*,
Förderband *n*, laufendes Band *n* ~ **width** Rie-
menbreite *f* ~ **wrapper** Bandumführungsrie-
men *m*
**belted** gegürtet ~ **insulation cable** Gürtelkabel
*n*
**belting** Riemenleder *n* ~ **butt** Riemenkernstück
*n* ~-**in run** Durchdrehlauf *m* (Fremdantrieb)
~ **leather** Transmissionsleder *n*
**bench** Bank *f*, Berme *f*, Bezugspunkt *m*, Feil-
bank *f* (mach.), Maschinenbett *n*, auf Füßen
ruhendes Maschinenbett *n*, (work) Werkbank *f*
~ **for cleaning castings** Gußputztisch *m*
**bench**, ~ **axe** Zimmeraxt *f* ~ **center** Grundplat-

te *f* ~ **centering device** Bankspitzenapparat
*m* ~ **dogs** Hobelbankhaken *pl* ~-**drill stand**
Tischbohrständer *m* ~ **filing machine** Werk-
bankfeilmaschine *f* ~ **holdfast** Bankhaken *m* ~
**jolter** Vibriertisch *m* ~ **lathe** kleine Drehbank
*f*, Tischdrehbank *f* ~ **machine** Bankmaschine
*f* ~ **mark** Bodenpunkt *m*, Denkmal *n*, Fixpunkt *m*, Höhenmarke *f*, Lattenpunkt *m*,
Merkpunkt *m*, Niveaufixpunkt *m*, trigonometrischer Punkt *m* ~ **micrometer** Stand-
schraublehre *f* ~ **miller** Werkbankfräsmaschine
*f* ~ **milling machine** Tischfräsmaschine *f* (für
Holzbearbeitung) ~ **model** Tischgerät *n* ~
**molding work** Bankformerei *f* ~ **photometer**
Bankfotometer *n* ~ **punch** Bankdurchschlag
*m* ~ **r. p. m.** Standdrehzahl *f* ~ **scale** Prüf-
meßbank *f* ~ **screw cutting machine** Tischge-
windeschneidmaschine *f* ~ **screw vise** Bank-
schraube *f* ~ **shears** Bock-, Stock-schere *f* ~
**stock** Handvorrat *m* ~ **test** Laborversuch *m*,
Prüfstandversuch *m*, Werkstattprüfung *f*
~-**type drilling machine** Werkbankbohrma-
schine *f* ~ **vise** Bankschraubstock *m*
**bend, to** ~ abfassen, abkanten, abwinkeln, anspannen, ausknicken, beugen, biegen, durchbiegen, eine Feder *f* anspannen, krümmen,
krumm werden, neigen, verbiegen, (a ray)
abknicken **to** ~ **back** (a ray) zurückwerfen
**to** ~ **in** einbiegen, einknicken **to** ~ **off (or aside
or downward)** abbiegen **to** ~ **out** ausbiegen **to**
~ **over** falten, umbiegen **to** ~ **sharp** ausknicken
**to** ~ **up** hochbiegen **to** ~ **up(ward)** aufbiegen
**to** ~ **back twisted ends of wires** die Würgestelle
umbiegen **to** ~ **cables** einstecken **to** ~ **on edge**
hochkantbiegen **to** ~ **the edges of a sheet** börteln **to** ~ **the points** (of nail) verzwicken **to**
~ **at right angles** kröpfen
**bend** Bogen *m*, Bogenrohr *m*, Bug *m*, Bügel *m*,
Knick *m*, Knierohr *n*, Krümmer *m*, Krümmung
*f*, Kurve *f*, (as axle) Kröpfung *f*, (of a curve)
Knie *m*, (break possible) Biegung *f*, (pipe)
Rohrkrümmer *m*, gekrümmtes Rohr *n* ~ **of
characteristic** Kennlinienknick *m*
**bend,** ~ **allowance** Biegungs-einräumung *f*,
-zuschuß *m* ~ **coupling** Koppelbogen *m* ~-**over
test** Faltversuch *m* ~ **pipe** Krummrohr *n*,
Rohrkrümmer *m* ~ **pulley** Knicktrommel *f*
~ **radius** Krümmungsradius *m* ~ **test** Biege-
probe *f*, Faltversuch *m* ~ **wood shaper** Tischfräsmaschine *f*
**bendable** biegsam, verbiegbar
**bender** Biegeapparat *m* ~ **crystal** Biegekristall
*m* ~'**s board** Buchbinderpappe *f*
**bending** Abbiegung *f*, Abfassen *n*, Ablenkung *f*,
Beugen *n*, Beugung *f*, Biegebeanspruchung *f*,
Biegen *n*, Biegung *f*, Durchbiegung *f*, Faltung
*f*, Krümmen *n*, Krummwerden *n*, Verbiegung
*f*, (at a sharp angle) Ausknickung *f* ~ **at angles**
Abkröpfung *f* ~ **of beams** Balkenbiegung *f* ~
**of the bearing springs** Durchbiegen *n* der Trag-
federn ~ **of a chimney** Schiefführung *f* eines
Schornsteins ~ **in two directions** Doppelbie-
gung *f*
**bending,** ~ **angle** Biegewinkel *m* ~ **apparatus**
Biegeapparat *m* ~-**back point** Umkehrpunkt *m*
~ **beam** Biegewange *f* ~ **block** Biegeblock *m*
~ **brittle point** Kältebiegeschlagfestigkeit *f* ~
**change strength** Biegewechselfestigkeit *f* ~

coil Krümmungsspule $f$ ~ couple Biegemoment $n$ ~ crack Biegungsriß $m$ ~ cradle Biegegehänge $n$ ~ cylinder Biegezylinder $m$ ~ device Biegevorrichtung $f$ ~ die Biege-form (Bif) $f$, -stempel $m$, ~-fatigue range (or strength) Biegungsschwingungsfestigkeit $f$ ~ fixture Biegevorrichtung $f$ ~ (flexure) strain Biegeformung $f$ ~ force Biegekraft $f$ ~ force constant Kraftkonstante $f$ ~-impact test Biegeschlagversuch $m$ ~ jaw Biegebacke $f$ ~ limit Biegegrenze $f$ (beim Biegeversuch) ~ load Biege-beanspruchung $f$, -last $f$ ~ machine Biege-, Langfalzund Zudrückmaschine $f$, Biege-walze $f$, -maschine $f$ (iron) ~ mandrel Biegedorn $m$ ~ mode Biegungsart $f$ ~ moment Biegemoment $m$, Biegungsmoment $m$ ~ oscillation Biegeschwingung $f$ ~ pliers Biegezange $f$ ~ point Biegestelle $f$ ~ press Abkantmaschine $f$, Biegepresse $f$, (bending plates for shipbuilding) Biegewalze $f$, ~ pressure Biegedruck $m$ ~ property Biegefestigkeit $f$ ~ radius Biegungshalbmesser $m$ ~ ram Biegestempel $m$ ~ resilience Biegespannung $f$, Biegungsspannung $f$ ~ resistance Biegesteifigkeit $f$ ~ roll Biegewalze $f$ ~ rolls Biegewalze $f$ ( bending plates for shipbuilding) ~ stiffness Biegungssteifigkeit $f$ ~ strain Biege-beanspruchung $f$, -druck $m$, -spannung $f$, Biegungsspannung $f$, (load) Biegebelastung $f$ ~ (flexure) strain Biegeformung $f$ ~ strength Biegefestigkeit $f$, Biegungsfestigkeit $f$ ~ strength under alternating loading Biegwechselfestigkeit $f$ ~ stress Biege-beanspruchung $f$, -spannung $f$, Biegungs-beanspruchung $f$, -kraft $f$, -spannung $f$ ~ stress by wind pressure Beanspruchung $f$ auf Biegung durch den Winddruck ~-stress durability Dauerbiegefestigkeit $f$ ~ table Biegetisch $m$ ~ test Biege-probe $f$, -prüfung $f$ ~ test in tempered state Härtungsbiegeprobe $f$ ~-test machine Prüfmaschine $f$ für Faltversuche ~ tests on built-up welded joints Aufschweißbiegeversuch $m$ ~-test shackle Biegevorrichtung $f$, Einspannkopf $m$ für Biegeversuche ~-test specimen Biegeprobekörper $m$ ~ tongs Biegezange $f$ ~ value Biegezahl $f$ ~ vibration Biegungs-, Knick-schwingung $f$ ~ vibration testing machine Biegeschwingungsmaschine $f$
**bends** Höhenkrankheit $f$, Tiefdruckluftkrankheit $f$ (aviat.)
**beneficiation** Aufbereitung $f$
**benevolent fund** Unterstützungskasse $f$
**benic acid** Bensäure $f$
**Bennerfelt (independent direct-arc) furnace** Bennerfelt-Ofen $m$
**bent** Binder $m$, Bock $m$, Knie $n$; gebogen (biegen), gekröpft, krumm, verbogen (tube), (tube, pipe) abgeknickt ~ in a circle kreisförmig gebogen
**bent**, ~ antenna geknickte Antenne $f$ ~ course geknickter Radiokurs $m$ ~ crank Bogenhebel $m$ ~ down geknickt ~ (slaters') hammers Haubrücken $pl$ ~ lenses durchgebogene Gläser $pl$ ~ lever Kniehebel $m$ ~-lever embossing press Kniehebelprägepresse ~ link gekröpftes Glied $n$ ~ parting tool gebogener Geißfuß $m$ ~ pipe gebogenes oder gekrümmtes Rohr $n$, Bogen-, Krumm-, Schenkel-rohr $n$ ~ plate gekantetes Blech $n$ ~ pointer geknickter Zeiger

$m$ ~-shank tapper tap Maschinenmuttergewindebohrer $m$ mit langem gebogenem Schaft, Mutternbohrer $m$ mit (langem) gebogenem Schaft ~ shovel Boden-scharre $f$, -schaufel $f$, Erdschaber $m$ ~ side tool gebogener Seitenstahl $n$ ~ stave for repairs Flickdaube $f$ ~ swage Verkröpfgesenk $n$ ~ thermometer Winkelthermometer $n$ ~ threading tool gekröpfter Gewindestahl $m$ ~ tube Schenkelrohr $n$, Schlangenrohrvorwärmer $m$ ~-tube boiler Steilrohrkessel $m$ ~-up bar Schrägeisen $n$ ~-up rod aufgebogener Stab $m$
**bentonite** Bentonit $n$
**benzal chloride** Benzalchlorid $n$
**benzaldehyde** Benzaldehyd $n$, Benzoylwasserstoff $m$, Bittermandelöl $n$
**benzanilide** Benzalinid $n$
**benzene** Benzol $n$, Steinkohlenteerbenzin $n$ ~ used as rubber solvent Gummilösungsbenzin $n$ ~ benzene, ~ hydrocarbon Benzolkohlenwasserstoff $m$ ~ nucleus Benzolkern $m$ ~ recovery plant Benzolwäsche $f$ ~ scrubber (or washer) Benzolwäscher $m$ ~ series Benzolreihe $f$ ~ sulfonic acid Benzolsulfonsäure $f$ ~ sulfonylchloride Benzolsulfochlorid $n$ ~ wash oil Benzolwaschöl $n$
**benzidine** Benzidin $n$
**benzil** Benzil $n$
**benzine** Benzin $n$, Erdölbenzin $n$, Petroleumbenzin $n$ ~ separator Benzinabscheider $m$
**benzoate** benzoesauer
**benzoated** benzoiniert
**benzoic**, ~ acid Phenylameisensäure $f$ ~ ether Benzoeäther $m$
**benzoin** Benzoin $n$, (gum) Benzoe $n$
**benzoinated** benzoiniert
**benzoinitrile** Benzonitril $n$
**benzol** Benzene (Steinkohlenbenzol), Phenylwasserstoff $m$, Steinkohlenteerbenzin $n$ ~ coal naphtha Benzol $n$ ~ mixture Benzolgemisch $n$ ~ recovery Benzolgewinnung $f$ ~-recuperation plant Benzolgewinnungsanlage $f$ ~ still Benzolabtreiber $m$
**benzoper acid** Benzopersäure $f$
**benzophenone** Benzophenon $n$
**benzoquinone** Benzochinon $n$
**benzoyl**, ~ bromide Benzoylbromid $n$ ~ chloride Benzoylchlorid $n$ ~ hydride Benzoylwasserstoff $m$
**benzoylate, to** ~ benzoylieren
**benzyl** Benzyl $n$ ~ acetate essigsaures Benzyl $n$ ~amine Benzylamin $n$ ~ chloride Benzylidenchlorid $n$, Chlorbenzyl $n$ ~ cyanide Benzylzyanid $n$
**bepartition angle** Bipartitionswinkel $m$
**beret parachute** Pilzfallschirm $m$
**Berger-type smoke agent mixture** Bergermischung $f$
**Berges drive** Bergestrieb $m$
**Berlin blue** Berlinerblau $n$
**berm** Auftritt $m$, Berme $f$, Erdaufwurf $m$, inneres Bankett $n$ ~ planted with reeds bepflanzte Berme $f$
**bernotar** Bernotar
**Bernoulli's theorem** Bernoulli's Lehrsatz $m$
**berth, to** ~ festmachen
**berth** Anlegestelle $f$, Back $f$, Bett $n$, Helling $f$, Liegeplatz $m$ (auf Zügen), Schlafkoje $f$, (dredg-

ed), Sohle *f*, (for ship) Ankergrund *m*, (place for mooring) Anlegeplatz *m*, (place for moorings) Festmacheplatz *m* ~ **for fuel-oil bunkering** Tankanlage *f* ~ **mole** Hellingmole *f*

**berthing maneuver** Einbringemanöver *n* (of airship)

**berthierite** Eisenantimonerz *n*, Eisenantimonglanz *m*

**Betrand-Thiel process** Bertrand-Thiel Verfahren *n*

**beryl** Beryll *m*

**beryllia** Beryllerde *f*

**beryllium** Beryllium *n* ~ **hydroxide** Beryllhydrat *n* ~ **oxide** Beryllerde *f*

**besmear, to** ~ bemalen

**besom** Besen *m*

**besprinkle, to** ~ beträufeln

**Bessel,** ~**'s function** Besselsche Funktion *f* ~ **function of the first kind** Bessel-Funktion *f* erster Art

**Bessemer,** ~ **converter** Bessemer-Birne *f* ~ **heat** Bessemer-schmelze *f* ~ **installation** Bessemer-Anlage *f* ~ **mill** Bessemer-Anlage *f* ~ **operation plant** Bessemer-Anlage, -Betrieb *m* ~ **plant** Bessemer-Anlage *f*, Konverterwerk *n* ~ **practice** Konverterbetrieb *m* ~ **process** Birnenprozeß *m* ~ **(converting) process** Bessemer-Verfahren *n* ~ **(converter) steel** Bessemer-Stahl *m* ~ **work plant** Bessemer-Anlage, -Betrieb *m*

**Bessemerizing** Verblasen *n* in flüssigem Zustande

**best** Bestwert *m*; optimal, optimum **at** ~ **bestenfalls in the** ~ **way** bestens

**best,** ~ **economy curve** Schaulinie *f* größtmöglicher Wirtschaftlichkeit ~ **performance** Bestleistung *f* ~ **quality** höchste Qualität *f* ~ **quality coal** Bestmelierte *f* (Kohlensorte) ~ **references** erstklassige Referenzen *pl* ~ **turn** wendigste Kurve *f*

**bestow, to** ~ erteilen, verleihen

**beta iron** Beta-Eisen *n*

**betafite** Betafit *m*

**betaine** Betain *n*

**betanaphthol benzoate** Benzonaphtha *n*

**beta-ray** Betastrahl *m* ~ **plaque** Plattenträger *m* ~ **protection screen** Betastrahlenschutzscheibe *f* ~ **protection-table top** Betastrahlenschutztischaufsatz *m* ~ **screen** Betaschirm *m* ~**-spectra** Betaspektron *pl* ~ **spectrometer** Betastrahlspektrometer *m*

**betatopic** betatopisch

**betatron** Vielfachbeschleuniger *m*

**betterment** Besserung *f*

**better-than-average** überdurchschnittlich

**between,** ~ **decks** Zwischendeck *n* ~**-lens diaphragm** Mittelblende *f* ~**-lens shutter** Objektiv-, Zwischenlinsen-verschluß *m* ~**-the--lens shutter** Zentralverschluß *m*

**bevatron** Bevatron *n*

**bevel, to** ~ abflachen, abkanten, abschrägen, ausschärfen, schmiegen, schweifen, mit der Schmiege messen, (metals) abfassen **to** ~**-edge** facettieren

**bevel** Anschnitt *m*, Böschung *f*, Fase *f*, Gehre *f*, Gehrmaß *n*, Gehrung *f*, Kegel *m*, Schmiege *f*, Schräge *f*, Schrägmaß *n*, Stellwinkel *m*, Winkellineal *n* ~**(ing)** Abflachung *f*, Abschrägung

*f*; schräg ~ **of valve seat** Ventilsitzabschrägung *f* **bevel,** ~ **cant** abgeschrägte Kante *f* ~ **crown wheel** Kegeltellerrad *n* ~ **cut** Schrägschnitt *m* ~ **cutting** Anschneiden *n* von Stemmkanten ~ **differential** Kegelraddifferential ~ **drive gear** Antriebskegelrad *n* ~**-driving pinion** Kegelantriebsritzel *m* ~**-edged lens** facettierte Linse *f* ~**-epicyclic-type reduction gear** Kegelradumlaufgetriebe *n* ~ **friction wheel** Kegelreibrad *n* ~ **gauge** Schrägmaß *n* ~ **gear** getriebenes Kegelrad *n*, Kegel-getriebe *n*, -rad *n*, -zahnrad *n*, konisches Getriebe *n*, konische Verzahnung *f*, Ritzel *n*, Stirn- und Kegelräder *pl*, schiefe Verzahnung *f* ~ **gear with conical pinion** Kegelradausgleichsgetriebe *n* ~ **gear with involute-tooth spiral** Evolventenzahnkegelrad *n*

**bevel-gear,** ~ **alignment** Kegelradmontage *f* ~ **blank** Kegelabstand *m* ~ **case** Kegelradgehäuse *n* ~ **cutter** Kegelrad-fräser *m*, -hobelstahl *m* ~ **drive** Kegelradantrieb *m* ~ **drive jet** Kegelradgetriebe *n* ~ **generator** Kegelradhobler *m* ~ **planer** Kegelradhobelmaschine *f* ~ **rear--axle drive** Kegelradhinterachsantrieb *m* ~ **testing fixture** Kegeltriebprüfvorrichtung *f* ~**-tooth system** Kegelradverzahnung *f* ~ **wheel** Kegelrad *n*

**bevel,** ~ **gearing** Kegelrädergetriebe *n*, Kegelradgetriebe *n* ~ **helical reduction gear** Kegelschneckengetriebe *n* ~ **joint** schräge Verbindung *f* ~ **pinion** Antriebskegelrad *n*, konisches Zahnrad, *n* treibendes Kegelrad *n* ~**-pinion ball bearing** Kegelradkugellager *n* ~ **pinion drive shaft** Kegelradritzelantriebswelle *f* ~ **planer** Geradefacettehobel *m* ~**-pointed screw** Schraube *f* mit Kegelansatz ~ **protractor** Universalwinkelmesser *m* ~ **rule** Stellwinkel *m* ~ **side gear** Achskegelrad *n* ~ **spur gear** Kegel-rad *n*, -stirnradgetriebe *n* ~ **squares (hexagon)** Sechseckwinkel *pl* ~ **tool** Stahl *m* für schräge Flächen ~**top transfer chain** dachförmige Transportkette *f* ~ **turning device** Kegeldreheinrichtung *f* ~ **wheel** konisches Rad *n*, Winkelrad *n* ~**-wheel drive** Kegeltrieb *m* ~**-wheel reversing wheel** Kegelradwendegetriebe *n*

**beveled** abgefast, abgekantet, abgeschrägt, schräg, verjüngt ~**casing** Kegelröhren *pl* ~ **clearing cam** (innen) abgeschrägter Fangteil *m* eines Hebers (Fangklappe) ~ **cutter** Kegelfräser *m* ~ **driving wheel** Kegeltreibrad *n* ~**-edge** winkelrandig ~ **glass** Facettenglas *f* ~ **mirror glass** gefeldertes Spiegelglas *n* ~ **rule** Lineal *n* mit abgeschrägter Kante ~ **track section** Trapezjoch *n*

**beveler** Hobel *m*

**beveling** Abfasen *n*, Schrägabschneiden *n* ~ **cut** Backenschmiede *f* ~ **machine** Zuschärfmaschine *f* ~ **plane** Schräghobel *m*

**beware, to** ~ sich in acht nehmen ~! Obacht!

**beyond** außerhalb, dahinter, über ~**horizon communication** Überreichweiteverbindung *f* ~ **range** außer Bereich

**bezel** Beobachtungsfenster *n*, Frontring *m*, Schrägkante *f*, Zuschärfungsfläche *f*, zugeschärfte Kante *f*, (in casing or panel or radio apparatus) Fenster *n*, (in radio set panel) Guckloch *n*

**B-H curve** Magnetisierungskurve *f*

**bias, to ~** einseitig einstellen, vormagnetisieren, vorspannen (electr.), Vorspannung erteilen **to ~ out** durch eine Gegenkraft *f* ausgleichen, (by counterforce) ausgleichen **to ~ a tube** Röhre *f* einstellen
**bias** Defekt *m*, einseitige Vorspannung *f*, einseitige Wirkung *f*, Gegenkraft *f*, Nullabgleichung *f* (gyro), Verzerrung *f* (telegr.), Vormagnetisierung *f*, Vormagnetisierungsstrom *m* (acoust.), Vorspannung *f*, Überwiegen n; einseitig wirkend, schief, schief geschnitten **~ of a circuit** Differenz-spannung *f*, -strom
**bias, ~ angle** Schneidewinkel *m* des Gewebes **~ bindings** Schrägbänder *pl* **~ box** Vorspannungsgerät *n* (TV) **~ cell** Gitterbatterie *f* **~ cutter (or cutting hole)** Diagonalschneidemaschine *f* **~ distortion** einseitige Verzerrung *f* (telegr.) **~ driver** Vormagnetisierungstreiber *m* **~ generator** Generator *m* für die Gittervorspannung **~ lighting** Grundniveau *n* der Beleuchtung *f* **~ magnetization** Vormagnetisierung *f* **~ meter** Verzerrungsmeßgerät *n* **~ potential** Ruhespannung *f* **~ potentiometer** Polarpotentiometer *n* **~ rectification** Anodengleichrichtung *f* **~-reducing potential** Schaltspannung *f* **~ supply** Netzanschlußgerät *n* **~ telegraph distortion** asymmetrische Verzerrung *f* **~ voltage** Vorspannung *f* (rectifier) **~ winding** Polarisations-, Vormagnetisierungs-wicklung *f*
**biased** einseitig überwiegend, vorgespannt **~ fabric** Diagonalstoff m **~ magneto bell** Wechselstromwecker *m* mit Ankerumlegefeder
**biasing, ~ current** Vormagnetisierungsstrom *m* **~-grid voltage** Gitterverschiebungsspannung *f*, Gittervorspannung *f* **~ illumination** Vorlicht *n* **~ potential** Sperr-, Vor-spannung *f* **~ resistor** Gittervorwiderstand *m* **~ spring** Ankerumlegefeder *f* (alarm clock) **~ stop** Reintonblende f **~ voltage** Pendelrückkoppelungsspannung *f*, (grid bias) Gittervorspannungsverlagerung *f*
**biatomic** doppelatomig
**biax magnetic element** zweiachsiges magnetisches Element *n*
**biaxial** zweiachsig **~ crystal** zweiachsiger Kristall *m* **~ mica** weißer Glimmer *m*
**biaxiality** Zweiachsigkeit *f*
**bibasic** bibasisch (zweibasisch)
**Bibby coupling** Bibby-Kuppelung *f*
**bibcock** Ausflußhahn *m*
**bibliographic list** Schrifttumverzeichnis *n*
**bibliography** Schrift-material *n*, -tum n
**bibliometer** Löschpapierprüfer *m*
**bicarbonate of** doppelkohlensauer
**bi-characteristic** Bicharakteristik *f*
**bichloride** Bichlorid n
**bichromate** Bichromat *n* **~ of potash** doppeltchromsaures Kali *n* **~ cell** Bichromatgefäß *n*, Chromsäureelement *n* **~-emulsion reproduction process** Bichromatkolloidkopierprozeß *m* **~ titration** Bichromattitration *f*
**bichromated gum process** Gimmichromverfahren *n*
**bickern** Schlagstöckchen *n*
**biconical, ~ antenna** Doppelkonusantenne *f* **~ horn** Doppelkonustrichter *m* (acoust.)
**biconvex, ~ airfoil profile** bikonvexer Flügelschnitt *m* **~ airfoil section** bikonvexes Profil *n*

**~ lens** Bikonvex- (opt.), Zwiebel-linse *f* **~ or lenticular profile** Linsenprofil *n*
**bicurve section (or profile)** Wellenprofil *n*
**bicycle** Fahrrad *n*, Rad *n* **~ crane** Velozipedkran *m* **~ frame** Fahradgestell *n* **~ frontfork blades** Vorderradscheiben *pl* **~ tread** Fahradlauffläche *f* **~ tubing** Fahrradröhre *f* **~-valve tubing** Ventilschlauch *m*, dünner Gummischlauch *m*
**bicyclic** bizyklisch
**bid, to ~** auffordern, bieten
**bid** Angebot *n* **~ price** Geldkurs *m*
**bidding combination** Arbeitsgemeinschaft *f*
**bidirectional, ~ antenna** Doppelkonusantenne *f*, zweiseitige Richtantenne *f* **~ pulses** Zweirichtungsimpulse *pl*
**bifilar** bifilar, doppel-adrig, -fädig, -fäsig, -gängig, zwei-adrig, -fädig **~ cable** doppeladriges oder zweiadriges Kabel *n* **~ oscillograph** Schleifenoszillograf *m* **~ resistor** Bifilarwiderstand *m* **~ suspension** Bifilar-, Doppelfaden-aufhängung *f* **~ winding** Bifilardraht-, Bifilar-, Zweidraht-wicklung *f*, bifilare Wicklung *f* **~ wire** Bifilardraht *m*
**bifocal** mit zwei Brennpunkten **~ glasses** Zweistärkenbrille *f*
**bifuel system** Zweikraftstoffverfahren *n*
**bifurcate, to ~** (sich) gabeln; doppelgängig, gabelartig
**bi-furcated** gegabelt, zweizackig **~ chute** Hosenschnurre *f*, Zweigrutsche *f* **~ pipe** Gabelrohr *n* **~ rivet** Gabelniete *f* **~ smoke nozzle (or hood)** Fuchsanschlußhosenstücke *pl* **~ tube** Gabelrohr *n* **~ tubular rivet** Gabelrohrniete *f*
**bifurcating cable** Abzweigkabel *n*
**bifurcation** Ablenkung *f*, Abzweigung *f*, Gabelteilung *f*, Gabelung *f*, (foundry ladle) Gabel *f*, **~ of current** Stromspaltung *f* (stream)
**bifurcation, ~ buoy** Spaltungstonne *f* **~ signal** Spaltungszeichen *n* **~ station** Gabelamt *n*
**big** groß **~ bell** Unterglocke *f* **~ end** Pleuelfuß *m*, (of connecting rod) Schubstangenkopf *m* **~ end of master rod** Kurbelwellenende *n* des Hauptpleuels **~-end bearing** Kurbellager *n* des Hauptpleuels, Pleuelstangenlager *n* **~ spot** große Soffitte f
**bight** Bucht *f*, Einbuchtung *f*, Schleife *f*
**bigrid valve** Doppelgitterröhre *f*
**bilamellar** zweiplattig
**bilateral** doppelseitig, zweiseitig **~(ly)** beidseitig **~-area track** Zweizackenschrift *f* (film) **~ characteristic** achtförmige Richtcharakteristik *f* **~ drive** doppelseitiger Antrieb *m* **~ track** Zweizackenschrift *f* (sound film) **~ track recording with double-vane shutter** Doppelzackenschrift *f* mit Abdeckung **~ transducer** bilateraler Wandler *m* (acoust.) **~ variable-area sound track** Verschiebezackenspur *f*
**bilge** Bilge *f*, Bucht *f*, Kielraum *m*, Kimm *m*, Kimme *f* **~ block** Kimmschlitten *m* **~ harping** Kimmsente *f* **~ keel** Schlingerkiel *m* **~ pump** Bilgenwasser-, Kiel-, Lenz-pumpe *f* **~ strake** Kimmungsplanke *f* **~ water** Schlagwasser *n* **~ well** Bilgebrunnen *m*
**bilinear scalar** bilinear skalar
**bilk, to ~** prellen
**bill** Affiche *f*, Ankerspitze *f*, Kostenrechnung *f*, Plakat *n*, Rechnung *f*, Schnabel *m*, Zeche *f*, Zettel *m*

bill, ~ of charges Klageschrift *f* ~ of deposit Depositenschein *m* ~ of entry Einfuhrschein *m*, Eingangsmanifest *n*, Einklarierung *f*, Zolleinfuhrschein *m* ~ of exchange Devise *f*, Tratte *f*, Wechsel *m* ~ of font Schriftzettel *m* ~ of health Gesundheitspaß *m* ~ of indictment Anklageschrift *f* ~ of lading Begleitpapier *n*, Frachtbrief *m*, Konnossement *n*, Ladungs-, Verlade-schein *m* ~ of materials Materialliste *f* ~ of sale Verkaufsurkunde *f* ~ of tonnage Meßbrief *m*

bill, ~ board antenna Antennenwand *f*, Querstrahler *m* ~ broker Wechselagent *m* ~ discount Wechseldiskont *m* ~ distributor Zettelverteiler *m* ~ drawn in favor of Wechsel *m* ausgestellt zu Gunsten von ~ feed device Rechnungsvorschub *m* ~ hook Hippe *f* ~ paper Wechselformularpapier *n*, Werttitelpapier *n*

billet, to ~ einquartieren, unterbringen to ~ out ausquartieren

billet Barren *m*, Bolzen *m*, Deul *m*, Ortsquartier *n*, Ortsunterkunft *f*, Platine *f*, Stahlknüppel *m*, Stelle *f*, Unterbringung *f*, Unterkunft *f*, Zaggel *m*, (metal) Knüppel *m*, Zaggel m ~-boring machine Knüppelbohrmaschine *f* ~ charging crane Blockchargierkran *m* ~ end Knüppelende *n* ~ mill Knüppel-walze *f*, -walzwerk *n* ~ overhauling Knüppelputzerei *f* ~ press Blockpresse *f* ~-reheating furnace Knüppelwärmofen *m* ~ roll Knüppel-, Vor-walze *f* ~ roll stand Knüppelgerüst *n* ~-rolling train Knüppelstraße *f* ~ shears Knüppelschere *f* ~ size ingot Knüppel *m* ~ yard Knüppellager *n*

billeting Belegung *f*

billiard-cue chalk Billardkreide *f*

billietite Billietit *m*

billing machine Fakturiermaschine *f*

billion electron volts Giga-Elektronenvolt *n*

billy Vorspinnmaschine *f* (wool mfg)

bimensal period Bimester *n*

bimetal Bimetall *n*, Verbundguß *m* ~ plate plattiertes Blech *n* ~ spring (bi-spring) Bifeder *f* ~ strip Bimetallbügel *m*

bimetalactionometer Bimetalaktionometer *n*

bimetallic bi-, zwei-metallisch, aus zwei Metallen bestehend ~ strip Bimetallstreifen *m* ~ wire Bimetall-, Doppelmetall-draht *m*

bimetallical contact springs Thermobimetallkontaktfedern *pl*

bimetallism Doppelwährung *f*

bimonthly period Bimester *n*

bimorph crystal Doppelplatten-, Zweielement--kristall *m*

bin Behälter *m*, Behältnis *n*, Bunker *m*, Einwurftrichter *m*, Fülltrichter *m*, Lagerfach *n*, Sammelrumpf *m*, Sieb *n*, Tasche *f*, Vorratsbunker *m* ~ effect Silowirkung *f* ~ storage Lagerung *f* in Fächern

binant Binant *m*

binaries Doppelstern *m*

binary binär, zweigliedrig, zweistoff ~ alloy Zweistofflegierung *f*, binäre Legierung *n* ~ cell (Speicher)binärzelle *f*, binäres Speicherelement *n* ~ code Binärkodewort *n* ~ code decimal notation binärgesetzte Dezimalschreibweise *f* ~ collision Zweierstoß m ~ component (star) Doppelsternkomponente *f*, Zweifachverbindung *f*, binäre Verbindung *f* ~ counter Binär-

zähler *m* ~ digit Binärziffer *f*, Zweierschritt *m* ~ digital computer binäre Digitalrechenmaschine *f* ~ lens Zwillingslinse *f* ~ mixture Zweistoffgemisch *n* ~ notation binäre Darstellungsweise *f* ~ number Dualzahl *f* ~ scaler Zweifachuntersetzer *m* ~ system Zweistoffsystem *n*, binäres System *n* ~-to-decimal conversion Binärdezimalkonvertierung *f*

binaural beidohrig, binaural, doppelhörig ~ audition zweiohriges Hören *n* ~ balance Mitteneindruck *m* ~ effect Raumtoneffekt *m* ~ hearing zweiohriges Hören *n* ~ method in sound locating Binauralverfahren *n* ~ sound--location method Zweiohrverfahren *n*

bind, to ~ abbinden, backen, binden, docken, einbinden, einfassen, verknüpfen, (machines) fressen, (motor) zum Eingriff kommen to ~ in boards kartonieren to ~ off ketteln

bind, ~ risers Bindesteiger *pl* ~-up Einschnürung *f*

binder Abspannung *f*, Band *n*, Binde-draht *m*, -maschine *f*, -material *n*, -mittel *n*, Bindungsmittel *n*, Drahtheftklammer *f*, Feststellhebel *m*, Garbenbinder *m*, Klemmhebel *m*, Trägermetall *n*, Verbinder *m* ~ cover Binderdecke *f* ~ frame Riemenleiter *m* ~ plug Klemmbolzen *m*

bindheimite Bleiniere *f*

binding Bandage *f*, Binde *f*, Bindung *f*, Bund *m*, Einband *m*, Fressen *n*, Verband *m*, Verkittung *f*; backend not ~ freibleibend ~ by means of cardan (or shaft) Gelenkwellenverbindung *f*

binding, ~ agent Binde-, Bindungs-mittel *n* ~ attachment Binde-apparat *m*, -vorrichtung *f* ~ bolt Fuß-, Sockel-schraube *f* ~ bond Verbindungsbahn *f* ~ border Einfassungsborte *f* ~ cloth Einbandleinen *n*, Leinwandbindung *f* ~ constituent Trägermetall *n* ~ cord Bindeleine *f* ~ edge of the street Einheftkante *f* des Blattes ~ energy Bildungs-, Bindungs-energie *f* ~ force Klebkraft *f* ~-in-wire Bindedraht *m* ~ joist Binderbalken *m*

binding, ~ leaf Liegekamm *m* ~ material Bindemittel *n*, Bindungsmittel *n*, Klebstoff *m*, Trägermetall *n* ~ means Bindemittel *n*, Bindungsmittel *n* ~ nut Gegenmutter *f* ~-off machine Kettelmaschine *f* ~ (or interlacing) point Bindepunkt *m* ~ post Anschlußklemme *f*, Drahthalter *m*, Drahtklemme *f*, Klemme *f*, Klemmschraube *f*, Polklemme (electr.), Verbindungsklemme *f* ~-post voltage Klemmenspannung *f* ~ power Binde-, Klebe-, Kleb-kraft *f* ~ power property (or power quality) Bindevermögen *n*, Back-, Binde-, Kleb-fähigkeit *f* ~ process Bindverfahren *n* ~ quality Bindefähigkeit *f* ~ rafter Füllsparren *m* ~ ring Heftring *m* ~ rivet Heftniet *n* ~ screw Anschlußschraube *f*, Kabelklemme *f*, Klemme *f*, Klemmschraube *f*, Polklemme *f* des Feldelements, Stellringschraube *f* ~ stone Binder *m* ~ strap Abdichtungslinie *f* (gas mask) ~ strength Verbandfestigkeit *f* ~ strips Einfaßleisten *pl* ~ tape Bindeband *n* ~ thread Bindfaden *m*, Spagat *m* ~ threads Liage *f* ~ twine Abbindegarn *n* ~ wire Binde-, Wickel-draht *m*

bing tale Bleierzgedinge *n*

binnacle Kompaß-haus, *n* -häuschen n

binocular Doppelfernglas *n*, Doppelglas *n*, (glasses) Feldstecher m, (micros.) für gleichzeitiges Sehen *n* mit beiden Augen; beidäugig, binokular, zweiäugig ~ **coil** Binokularspule *f* ~ **field of view** beidäugiges Blickfeld *n* ~ **mat** Schlüssellochmaske *f* (film) ~ **observation telescope** Doppelbeobachtungsfernrohr *n* ~ **parallax difference** Konvergenzwinkelunterschied *m* ~ **stand magnifier** Lupenmikroskop *n* ~ **telescope** Doppelfernrohr *n* ~ **telescopic magnifiers** Doppelfernrohrlupe *f* ~ **tube** Doppeltubus *m* ~ **vision** zweiäugiges Sehen *n*

**binoculars** Fernglas *n*

**binode** Binode *f*, Doppelknoten *m*, Verbundröhre *f*

**binomial** Binom *n*, zweigliedrig (math.), aus zwei Gliedern bestehender Ausdruck *m* (math.) ~ **corner** Binomial-Winkelsystem *n* ~ **distributing** Bernouillische Gleichung *f* ~ **series** Binomialreihe *f* ~ **theorem** binomischer Lehrsatz *m* ~ **twist** Binomialtorder *m*

**binormal** binormal

**biochemical** biochemisch

**biochemistry** Biochemie *f*

**biotite** Magnesiaglimmer *m*, schwarzer Glimmer *m*

**Biot-Savart law** Biot-Savart'sches Gesetz *n*, Durchflutungsgesetz *n*

**bipack method** Zweipackverfahren *n*

**bipartite** zweiteilig

**bipartition** Zweiteilung *f*

**biphase** zweiphasig ~ **connection** Zweiphasenschalter *m* ~ **equilibrium** zweiphasiges Gleichgewicht *n*

**biplane** Doppel-, Zwei-decker *m*, Zweistieler *m* ~ **effect** gegenseitiger Flächeneinfluß *m* ~ **filament system** Doppelwendelsystem *n* ~ **formula** Doppeldeckerbauart *f* ~ **fuselage** Rumpfzweidecker *m* ~ **interferences** Doppeldeckerstörungen *pl* ~ **model** Doppeldeckerflugmodell *n*

**biplug** Doppel-, Zwillings-stecker *m*

**bipod** Gabelstütze *f*, Zweibein *n*

**bipolar** doppelpolig, zweipolig ~ **coordinate** Bipolarkoordinate *f* ~ **dynamo** zweipolige Dynamomaschine *f* ~ **generator** bipolarer Impulsgenerator *m* ~ **receiver** doppelpoliger Hörer *m* ~ **stretcher** Impulsdehnungsstufe *f*

**bipolarity** Bipolarität *f*

**bipotential equation** Bipotentialgleichung *f*

**biprism** Zwillingsprisma *n*

**bi-propellant** Doppeltreibstoff *m* (g/m)

**biquadratic equation** Gleichung *f* vierten Grades

**biquartz** Doppelplatten *pl*, Doppelplattenkristall *m*

**biquinary notation** biquinäre Schreibweise *f*

**birch** Birke *f* **of** ~ aus Birke(nholz) ~ **plywood** Birkensperrholz *n* ~ **wood** Birkenholz *n*

**bird,** ~'**s-eye view** Vogelperspektive *f*, Vogelschau *f* ~ **flight** Ruder-, Vogel-flug *m* ~ **organ** Dompfaffenorgel *f* ~ **warbler** Zwitscherpfeife *f* ~ **whistle** Vogelpfeife *f*

**birdie** Zwitschern *n* (electr.)

**birefractive** doppeltbrechend

**birefringence** Doppelbrechung *f*, Betrag *m* der Doppelbrechung

**birefringent** doppeltbrechend

**Birkengang furnace** Birkengang-Ofen

**Birmingham sheet-and hoop-iron standard gauge** Birmingham Normallehre *f* für Bleche und Bandeisen

**birth-and-death process** Erneuerungsprozeß *m*

**biscuit** Hartbrot *n*, Zwieback *m*; Metallschwamm *m*, reduziertes Metall *n* ~**-shaped combustion chamber** flacher Verbrennungsraum *m*

**bisect, to** ~ halben, halbieren, hälften

**bisecting line of an angle** Winkelhalbierungslinie *f*

**bisection** Halbierung *f*, Hälftung *f*, Zweiteilung *f*

**bisector** Bisektrix *f*, Halbierungslinie *f*

**bisectrix** Halbierende *f*, Halbierungslinie *f* (line)

**bisignal zone** Doppelsignalzone *f* (nav.)

**bisilicate** Bisilikat *n*

**bismite** Wismut(h)blüte *f*, Wismutocker *m*

**bismuth** Wismut(h) *n*, *m* ~ **alloy** Wismutlegierung *f* ~ **arsenate** Wismut(h)arseniat *n* ~ **blende** Wismut(h)blende *f* ~ **bromide** Bromwismut *m* ~ **chloride** Chlorwismut *n* ~ **compound** Wismutverbindung *f* ~ **content** Wismutgehalt *m* ~ **iodide** Wismutjodig *n* ~ **litharge** Wismutglätte *f* ~ **nitrate** salpetersaures Wismut(oxyd) *n* ~ **ocher** Wismut(h)blüte *f*, Wismutocker *m* ~ **ore** Wismuterz *n* ~ **oxide** Wismutglätte *f*; Wismutoxyd *n* ~ **oxychloride** Wismutoxychlorid *n* ~ **oxyiodide** Wismutoxyjodid *n*, Wismutsubjodid *n* ~ **oxynitrate** Wismutniederschlag *m* ~ **phenolate** Phenolwismut *m*, *n* ~ **precipitate** Wismutniederschlag *m* ~ **refining** Wismutraffination *f* ~ **salt** Wismutsalz *n* ~ **silicate** Eulytin *m*, Kieselwismut *m* ~ **solder** Wismutlot *n* ~ **spiral** Wismutspirale *f* ~ **subbenzoate** Wismutsubbenzoat *n* ~ **subcarbonate** Wismutsubkarbonat *n* ~ **subgallate** Wismutsubgallat *n* ~ **sulfide** Schwefelwismut *n*, Wismutglanz *m* ~ **trichloride** Chlorwismut *n* ~ **trioxide** Wismutoxyd *n* ~ **white** Wismutweiß *n*

**bismuthic acid** Wismutsäure *f*

**bismuthiferous** wismuthaltig

**bismuthinite** Wismutglanz *m*

**bismutite** Wismutit *m*, Wismutspat *m*

**bistable** bistabil ~ **control** Zweipunktsteuerung *f* ~ **multivibrator** Flip- und Flop-Generator *m*, Multivibrator *m* ~ **trigger circuit** bistabile Triggerschaltung *f*

**bister** Bisterbraun *n* ~ **shade** bisterfarbiger Ton *m*

**bistoury** Bistouri *m*

**bisulfite** doppelschwefelsauer ~ **waste liquor** Bisulfitlauge *f* ~**-zinc vat** Bisulfitzinkstaubküpe *f*

**bit** Backe *f*, Backenmeißel *m*, Bart *m*, Bit *n*, Bohreinsatz *m*, Bohrschneide *f*, Einsatzbohrer *m*, Hobeleisen *n*, Maul *n*, Meißel *m*, Plättchen *n*, Schneideneinsatz *m*, Schnipsel *n*, Schnipselchen *n*, Stückchen *n*, (harness) Gebiß *n*, (polishing or finishing) Senkkiste *f* ~ **with bottom flush** Spülbohrer *m* ~ **for hand brace** Holzbohrer *m* (wood tools) für Bohrkurbel ~**s for handbrace** Bohrer *m* für Bohrkurbel ~**s of paper** Papierschnitzel *n* ~ **with parallel cutting edges** Parallelschneidenmeißel *m* ~ **of vibration** Schwingungsabschnitt *m*

**bit,** ~ **bar** Ballen *m* ~ **brace** Bohrwinde *f* ~ **charging** Ausbauarbeiten *pl* ~ **density** Binärzifferndichte *f* ~ **drill** Bohrer *m* ~ **gauge** Meißellehre *f*, Meißelschablone *f* ~ **head** Meißelkopf *m* ~ **hook** Meißelfanghaken *m* ~ **rods** Meißelgestänge *n* ~ **stock** Kurbelbohrer *m* ~**-stock drill** Faustleier-, Leier-bohrer *m*

**bitangent** doppeltberührend
**bitartrate** zweifachweinsauer
**bite, to** ~ angreifen (Feile, Schneidewerkzeug), beißen, erfassen, fassen, greifen **to** ~ **catch** festhaken
**bite** Bissen *m*, Stich *m*
**biter** Beißer *m*
**biting** scharf
**bitone horn** Doppeltonhorn *n*
**bitt, to** ~ einen Törn um die Beting nehmen, Beting *f*, Kreuzpoller *m*
**bitter** bitter
**Bitter bands (or patterns)** Bitter'sche Streifen *pl* (powder)
**bitterness** Herbheit *f*
**bitterwood** Bitterholz *n*
**bitumastic** bituminöser Mastix *m*
**bitumen** Bergpech *n*, Bitumen *n*, Erdharz *n*, Erdpech *n* ~**-bound road bases** bitumengebundener Straßenunterbau *m* ~ **coat** Bitumenanstrich *m*
**bituminiferous** asphalthaltig, bitumenhaltig
**bituminization** Bituminierung *f*
**bituminous** bitumig, bituminös, erdharzig, erdpechartig, fett, pechhaltig ~ **coal** backende Kohle *f*, Back-, Fett-, Steinkohle *f*, pechhaltige Kohle *f* ~**-coal carbonization** Steinkohlenentgasung *f* ~**-coal gas** Steinkohlengas *n* ~**-coal tar** Steinkohlenteer *m* ~ **coat** Bitumenanstrich *m* ~**-concrete facing** Verkleidung *f* von Bitumenbeton ~ **earth** Bitumen *n* ~ **insulation** bitumenartige Isolation *f* ~ **lignite** ölreiche Braunkohle *f* ~ **marl** Stückmergel *m* ~ **material** bituminöser Stoff *n* ~ **pitch** Asphaltpech *n* ~ **rock** pechhaltiges Gestein *n* ~ **sand** pechhaltiger Sand *m* ~ **seal** Bitumendichtung *f* ~ **shale** Brand-, Kohlen-schiefer *m*, pechhaltiger Schiefer *m* ~ **slate** Porzellanschiefer *m* ~ **wood** Lignit *m*
**bivalence** Zweiwertigkeit *f*
**bivalent** zweiwertig ~ **radical** zweiwertiges Radikal *n*
**bivalves** Zweischaler *pl*
**bivariant** divariant, zweifachfrei ~ **equilibrium** devariantes oder zweifachfreies Gleichgewicht *n*
**bivector** Bivektor *m*
**black, to** ~ schwärzen
**black** Moorkohle *f*, Ruß *m*; schwarz, übergar ~**-after-black** Schwarz *n* hinter Weiß (TV)
**black-and-white** Helldunkel *n* ~ **channel** Schwarz-Weiß-Kanal *m* (TV) ~ **picture** Schwarzweißbild *n* (TV) ~ **picture transmission** Schwarzweißbildsendung *f* ~ **reception** Schwarz-Weiß-Empfang *m* (TV)
**black,** ~ **amber** Fuchsambra *m* ~ **annealing** erste Glühung *f* ~ **aphis** schwarzer Befall *m* ~ **ash** Schwarzschmelze *f* ~**band** Blackband *m*, kohlehaltiger Spateisenstein *m*, Kohleneisenstein *m*, Schwarzstreif *m*, schwarzer Balken *m* ~**band ironstone** Blackband *m*, Kohleneisenstein *m* ~**board** Schreib-, Wand-tafel *f* ~**board chalk** Schulkreide *f* (sticks) ~ **body** schwarzer Körper oder Strahler *m*, (phys.) ~**-body radiation** Hohlraumstrahlung *f*, schwarze Strahlung *f* ~**-body temperature** Schwarztemperatur *f* ~ **bolt** roher oder unbearbeiteter Bolzen *m* ~ **charcoal** Schwarzkohle *f* ~ **coal** Schwarzkohle *f* ~ **copper** Schwarzkupfer *n* ~**damp** Gemisch *n* von Kohlendioxyd und Stickstoff ~ **dot**

schwarze Peilpunktmarke *f* ~ **dot mark (or pointer) to read bearings** (on dials) schwarze Anzeigemarke *f* ~ **dye** Schwärze *f*
**blacken, to** ~ schlichten, schwärzen
**blackening** Schwarzfärbung *f*, Schwärzung *f*, Gießerschwärze *f*
**blacker-than-black region** Ultraschwarzpegel *m* (TV)
**black,** ~ **finish** schwarze Lackierung *f* ~ **finishing** brünieren ~ **fly** schwarzer Befall *m* ~ **frost** Barfrost *m* ~ **fuel peat** Pechtorf *m* ~ **glossy fuel peat** Lebertorf *m* ~**-heart malleable iron** Schwarzkern-, Schwarzkerntemper-guß *m*
**blacking** Gießerschwärze *f*, Schlichte *f*, Schwärzen *n*, Wichse *f*, (wash) Schwärze *f* ~ **brush** Schwarzquast *m* ~ **leather** Wichsleder *n* ~ **method** Schwärzungsmethode *f* ~ **swab sprayer** Schwärzeverteiler *m*
**black,** ~ **ink** Druckerschwärze *f* ~ **iron mica** Schwarzstein *m* ~ **iron ore** Psilomelan *n* ~ **iron oxide** Eisenmohr *m* ~ **iron plate** Schwarzblech *n* ~**-iron plate** schwarzes Blech *n* ~ **iron sand** sandiger Magneteisenstein *m*
**blackish** schwärzlich
**black,** ~ **jack** Formhammer *f* ~ **japan** Asphaltlack *m* ~ **lake** Lackschwarz *n* ~ **lead** Grafit-, Ofen-schwärze *f*, Pottlot *n*, Reißblei *n* ~**-lead crucible** Grafittiegel *m* ~**-lead lubrication** Grafitschmierung *f* ~**-lead powder** Eisenschwärze *f* ~**-lead spar** Schwarzbleierz *n* ~**-lead wash** Grafitschwärze *f* ~**-leading brush** Grafitierbürste *f* ~ **level** Schwarzpegel *m* ~ **level control** Schwarzsteuerung *f* ~**-level picture signal** Oberstrich *m* (TV) ~**-level value** Schwarzwert *m* (TV) ~**-light source** ultraviolette Lichtquelle *f* ~ **lignite** schwarzer Lignit *m* ~**-line paper** Eisengallus-, Heliografie-, Negrografie-, Schwarzpositiv-papier *n* ~ **liquor** Ablauge *f*, Natronablauge *f* ~ **liquor(s)** schwarze Lauge *f* ~ **listener** Schwarzhörer *m* ~ **mark** schwarze Peilpunktmarke *f* (on scale) ~ **market** Schleichhandel *m* ~ **mercuric sulfide** Quecksilbermohr *m* ~ **mica** schwarzer Glimmer *m*
**blackness** Schwärze *f*, Trübung *f*, Undurchsichtigkeit *f*
**black,** ~ **oil** Dunkelöl *n* ~ **oil circulation** Schwarzölumlauf *m* ~**-oil circulation furnace** Feuerung *f* der Zirkulationsanlage für dunkles Öl
**blackout, to** ~ abblenden, austasten (by blanking signal), verdunkeln
**blackout** Austastung *f*, Leuchtverbot *n*, Verdunk(e)lung *f*, ~ **of flyback** Rücklaufverdunkelung *f* ~ **blind** Verdunklungsrollo *n* ~ **curtain** Verdunkelungsvorhang *m* ~ **effect** Betäubung *f* ~ **level** Austastpegel *m* (TV) ~ **marker** Dunkelpunkt *m* (rdr) ~ **measure** Verdunk(e)lungsmaßnahme *f* ~ **pulse** Austastgemisch *n* (TV), Sperrimpuls *m* ~ **signal** Austastsignal *n* (TV)
**black,** ~ **paper for covers** schwarzes Papier *n* ~ **peak** Maximum *n* an Schwarz (TV) ~**-pot attendant** Ölkesselwärter *m* ~ **powder** Naßbrand-, Schwarz-pulver *n* ~ **powder pellets** Sprengsalpeter *m* ~ **printing** Schwarzdruck *m* ~ **radiation** schwarze Strahlung *f* ~**(-body) radiation** Schwarzstrahl *m* ~ **recording** Schwarzwertempfang *m* ~ **red** beginnende Glut *f* ~ **rubber tubing** Paragummischlauch *m*

~ **salt** Ochras *m*, Pottaschefluß *m* ~ **sand**
Altsand *m*, Haufensand *m* ~**(ing) sand**
schwarzer Sand *m* ~ **scope** Dunkelschirm *m*
(facsim) ~ **screen** Sonnenblende *f* ~ **screen
television set** Fernseher *m* mit Grauscheibe
~ **sheet** Schwarzblech *n* ~ **sheet steel** schwarzes
Blech *n* (ordinary) ~**-short** schwarzbrüchig ~
**shot** fehlerhafte Metallisierung *f* ~ **signal**
Schwarzwert *m* ~ **slag** schwarze Schlacke *f*
**blacksmith** Beschlag-, Fahnen-, Grob-, Hammer-, Huf-schmied *m*, Schmied *m* ~**'s chisel**
Blockmeißel *m* ~**'s forge** Schmiedeherd *m*
~**'s punch** Schmiededurchschlag *m* ~**'s slag**
Schmiedeschlacke *f* ~**'s tongs** Schmiedezange
*f* ~**'s tool** Schmiedegerät *n*
**black,** ~ **spar** Bleischwärze *f* ~ **spotter** Entstördiode *f* (TV) ~ **squall** Donner-, Gewitter-bö
*f* ~**strap** schwarzes Öl *n*, Restmolasse *f* ~**strap
molasses** Endmelasse *f* ~ **substitute** brauner
Faktis *m* ~**(-body) temperature** schwarze
Temperatur *f* ~**-to-white amplitude** Schwarzweißsprung *m* ~ **top paving** Asphaltbelag *m* ~
**transmission** Schwarzwertübertragung *f* ~
**varnish** Lackschwarz *n* ~**(level) voltage**
Schwarzspannung *f*
**Blackwall hitch** Hakenschlag *m*
**blackwash, to** ~ schlichten, schwärzen
**black,** ~ **wash** Gießerschwärze *f*, Kohlenschichte
*f*, Schlichte *f*, Schwärze *f* ~**-wash sprayer**
Grafitierapparat *m* ~**-white control** Helldunkel-,
Zweipunkt-steuerung *f* ~**-white range** Schwarzweißsprung *m* ~ **wire** unverzinkter Draht *m*
**blacks** schwarzes Papier *n*
**bladder** Blase *f*
**blade** Blatt *n*, Blattfeder *f*, Flosse *f*, Flügel *m*,
Halm *m*, Klinge *f*, Messer *n*, Schar *f*, Schenkel
*m*, Zunge *f*, (compressor) Schieber *m*, (of
cutter) Messerklinge *f*, (of knife) Messerklinge
*f*, (of knife switch) Blatt *n*, (gas turbine) Austrittsschaufel *f*, (oar) Blatt *n*, (shovel) Schaufelblatt *n*, (turbine) Schaufel *f* ~ **of cloth shears**
Tuchscherblatt *n* ~ **(disk) of the diaphragm**
Blendenscheibe *f* ~ **for loom** Schlagnase *f* für
Webstühle ~ **of an oar** Ruderblatt *n* ~ **of
shears** Scherblatt *n* ~**s** Blätter *pl* (knife) ~**s of
a turbine** Beschaufelung *f*
**blade,** ~ **angle** Anstell-, Blatt-, Schaufelwinkel *m* ~ **area** Blattfläche *f* (of propeller)
~ **carrier** Schaufelträger *m* ~ **channel** Schaufelkanal *m* ~ **chord** Blattiefe *f* ~ **chuck** Messerbeilage *f* ~ **control of lifting rotors** Windflügelverstellung *f* ~ **drag brace** Blattwiderstandsstrebe *f* ~ **drum** Messertrommel *f* ~ **eddy**
Blattwirbel *m* ~ **edge** Schraubenblattkante *f*
~ **edge sheating** Blattkantenbeschlag *m* ~ **edge
tipping** Blattkantenbeschlag *m* ~ **element**
Blattelement *n* ~ **end** Deckscheibe *f* ~ **face**
Schaufelbrust *f* ~ **fold** Hauerfalz *m* ~ **folding
cradle** Blattstütze *f* (Hubschrauber) ~ **form**
Blattform *f* ~ **grip** Blattadapter *m*, -stück *n* ~
**holder** Messerhalter *m* ~**-hub unit** Blattnabeneinheit *f* ~ **lag damper** Schwenkgelenkdämpfer
*m* (Hubschrauber) ~ **leading edge** Blattvorderkante *f* ~ **leaving angle** Laufschaufelaustrittswinkel *m* ~ **lobe** Schaufellappen *m* ~ **loss
factor** Laufschaufelverlustbeiwert *m* ~ **machine**
Planierschaufel *f* ~ **overlap** Überschneidung *f*
der Blattkreise *f* (Hubschrauber) ~ **peripheral**

**velocity** Blattspitzenumfangsgeschwindigkeit *f*
~ **pitch** Blattanstellung *f* (Hubschrauber),
Blattanstellwinkel *m*, Schaufelteilung *f* ~
**pitch horn** Blattverstellhorn *n* ~ **position** Blattstellung *f* ~ **profile** Blattquerschnitt *m* ~ **rim**
Laufkranz *m* (turbine) ~ **ring** Schaufelring *m*
(Triebwerk) ~ **ring carrier** Leitschaufelträger *m*
~ **rod** Nebelpleuel *n* ~ **rod of a forked-rod and
blade-rod assembly** inneres Pleuel *n* eines
Gabelpleuels ~ **root** Blattwurzel *f*, Wurzel *f*
des Luftschraubenflügels ~ **root bearings**
Blattwurzellager *n* ~ **root end fitting** Blattwurzelanschlußstück *n* ~ **root socket** Blattwurzelbeschlag *m* oder -futter *n* (Hubschrauber) ~ **section** Blattquerschnitt *m* ~ **setting**
Schaufelanstellung *f* ~**-shank cuff** Blattfußmanschette *f* (Luftschraube) ~ **shank cuff**
Blattwurzelverkleidung *f* ~ **shredder** Flügelzerfaserer *m* ~ **spacing** Schaufelzeilung *f* ~ **taper**
Schaufelverjüngung *f* ~ **thickness** Blattstärke *f*
~ **tip** Blattspitze *f* ~**-tip eddy** Blattspitzenwirbel *m* ~**-tip metal sheathing** Blattspitzenbeschlag *m* ~**-tip vortex** Blattspitzenwirbel *m*
~ **tipping** Blattspitzenbeschlag *m* ~ **tips of
propeller** Blattenden *pl* ~ **torque balancing**
Blattauswuchten *n* (Hubschrauber) ~ **tracking**
Blattspurprüfung *f* (Hubschrauber) ~ **twist**
Blattverdrehung *f*, Blatt-, Schaufel-verwindung *f* ~**-type connecting rod** Mittelpleuel *n*
~ **vortex** Blattwirbel *m* ~ **wheel** Eimerrad *n*
(turbine) ~ **width** Schneidenbreite *f* ~**-width
ratio of propeller** Blattbreitenverhältnis *n*
**bladed** flügelig, (lathlike) plattig ~ **wheel** beschaufeltes Rad *n* (turbine)
**blades** Blätter *pl* (knife)
**blading** Beschaufelung *f*, Schaufelung *f*
**blanc fixe** Bariumsulfat *n*, Barytweiß *n*
**blanch, to** ~ weißsieden
**blancher** Ausglüher *m*
**blanchimeter** Blanchimeter
**blank, to** ~ abblenden, anschlagen (print.),
blankschlagen, stanzen **to** ~ **off** abschließen
(aviat.) **to** ~ **out** auslöschen, ausstanzen **to** ~
**spots in television** abdunkeln
**blank** Blank *n*, Durchschuß *m*, Leerstelle *f*,
Lücke *f*, Münzplatte *f*, Niete *f*, Null *f*, Radkörper *m*, Roh-stück *n*, -teil *m*, Schema *n*,
Schrötling *m*, Stanzstück *n*, Vordruckblatt *n*,
(after pressure application or molding) Preßling *m*, (for blasting caps) Näpfchen *n*, (metall.)
Blech-ausschnitt *m*, -ausstoß *m*, (paper) leeres
Blatt, (value) Blindwert *m*; blanko, blind, frei,
leer, unbedruckt, unbeschriftet, unbeschrieben,
weiß **in** ~ **condition (or form)** unbearbeitet ~
**in the map (or in the survey)** Lücke *f* in der
Karte oder Aufnahme ~ **to be tooled (or
machined)** Werkstück *n* ~ **already machined**
vorgearbeitetes Werkstück *n*
**blank,** ~ **bar** Bildsynchronisierlücke *f* (between
successive frames) ~ **blow** Leerschlag *m* (Bohrhammer) ~ **cartridge** Platzpatrone *f* ~ **casing
bit** Verrohrungsbohrkrone *f* ~ **charge** Manöverkartusche *f* ~ **column detector** Leerspaltensucher *m* ~ **cover** Blendanstrich-, Blind-,
Voll-deckel *m* ~ **cutter bit** Einsatzmesser *n* (ungeschliffen für Stahlhalter) ~ **determination**
Blindprobe *f*, Blindversuch *m* ~ **end** Vollboden
*m* ~ **experiment** Leerversuch *m* ~ **film** Roh-

film *m* ~ **flange** Blindflansch *m*, blinder Flansch *m*, Blindscheibe *f*, Vorschweißflansch *m* ~ **form** Blankoformular *n* ~ **groove** ummodulierte Rille *f* (phono) ~ **hardness test** Blindhärteprüfung *f* ~ **head** Vollboden *m* ~ **holder** Blechhalter *m* ~ **instruction** Leerbefehl *m* (data proc) ~ **key** Abstands-, Blank-taste *f* ~ **letter of attorney** Blankovollmacht *f* ~ **line** Unterschlagszeile *f* (print.) ~-**out signal** Austastsignal *n* ~ **page of a book** erste Seite. *f* eines Buches ~ **paper** Schöndruck *m* ~ **(sheet of) paper** leeres Blatt *n* ~ **piece (or part)** Rohling *m* ~ **pipe** hermetisches Rohr *n* (well) ~ **round for removing a projectile stuck in the bore** Aushilfskartusche *f* ~ **sheet** Mönchsbogen *m* (print.), blinder Bogen *m* ~ **side** Blankseite *f* (paper) ~ **slug** Blindgußzeile *f* ~ **swage** Blindpräge *f* ~ **test** blinder Versuch *m*, Blindversuch *m* ~ **test sample** Blindprobe *f* ~ **titration** Blindtitration *f* ~ **window** blindes Fenster *n*

**blanked,** ~ **out** gestanzt ~ **picture (video) signal** Bildsignal *n* mit Abtastung ~ **video signal** videofrequentes BA-Signal *n*

**blanker** Vorschmiedegesenk *n*

**blanket, to** ~ abschirmen, einnebeln, überdecken

**blanket** Decke *f*, Druckfilz *m* (print), Drucktuch *n* (print.) Schlafdecke *f*, Schurz *m*, Wandstoffbekleidung *f*, ~ **printing machine** Deckendruckmaschine *f* ~ **sluice** Planherd *m*

**blanketing** Abdeckung *f*, Austastung *f*, Überstrahlung *f* ~ **of picture signal** Untergehen *n* der Bildspannung ~ **of signals** Signalüberdeckung *f*

**banketing,** ~ **curve** Hüllkurve *f* ~ **frequency** Abdeckungsfrequenz *f*

**blanking** Austastung *f*, Dunkelsteuerung *f* (CRT), Dunkeltastung *f* ~ **and operating clamp circuits** Tast- und Klemmschaltungen *pl* ~ **and punching tools** Schnitte und Stanzen *pl* ~ **cutter** Blankpräge *f* ~ **device** Austastgerät *n* ~ **die** Blankpräge *f* ~ **dies** Besteckstanzen *pl*, Stanzwerkzeug *n* ~ **gate** Austast-Torimpuls *m* ~ **impulse** Wegtastimpuls *m* ~ **level** Austastpegel *m*, Dunkelsteuerwert *m*, Dunkeltastwert *m* (CRT) ~ **machine** Aushaumaschine *f* ~ **means** Strahlabblender *m* ~ **plate** Sperrscheibe *f* ~ **plug** Verschluß-, Blindstopfen *m* ~ **pulse** Austastimpuls *m* ~ **signal (or impulse)** Austastzeichen *n* ~ **sweep unit** Sperrkippsender *m* ~ **time** Sperrzeit *f* ~ **tool** Blechstanz-, Formgebe-, Stanz-werkzeug *n* ~ **tool steel** Stanzstahl *m* ~ **zone** Austastbereich *m*, Dunkelsteuerbereich *m* (CRT)

**blankness** Leere *f*

**blanks** Vormaterial *n*

**blast, to** ~ abblasen (of castings), abstrahlen, anblasen, blasen, erblasen, schießen, sprengen, zersprengen **to** ~ **in** einblasen **to put on the** ~ anblasen

**blast** Explosionsdruckwelle *f*, (machine) Gebläse *n*, Luftstoß *m*, Luftstrom *m* mit hoher Geschwindigkeit, Schuß *m*, Sprengen *n*, Wind *m*, Windstoß *m*, Zerknallstoß *m* **the** ~ **is on** das Gebläse arbeitet

**blast,** ~ **action** Strahlwirkung *f* ~ **air** Gebläseluft *f* ~ **apparatus** Blasapparat *m*, Gebläse-

werk *n*, Windbläser *m* ~ **bay** Splitterboxe *f* ~ **box** Wind-kasten *m*, -kessel *m* ~-**box cover** Windkastenplatte *f* ~ **cabinet** Blasraum *m*, Gebläsekasten *m* ~ **chamber** Blasraum *m* ~ **connection** Düsenstock *m* ~-**control valve** Luftventil *n* ~ **cooling furnace** Zugkühlofen *m* ~ **cupola** Gebläsekupolofen *m* ~ **cylinder** Gebläsezylinder *m* ~ **deflector** Strahlabweiser *m* ~ **drying** Windtrocknung *f* ~ **effects** Staubaufwirbelung *f* (Mondlandung) ~ **engine** Gebläsemaschine *f* ~ **fence** Düsenstrahlschutzwand *f*, Strahlabweiser *m*

**blast-furnace** Blas-, Blau-, Floß-, Gebläse-, Gebläseschacht-, Hoch-, Krumm-, Schacht-, Verblase-ofen *m* ~ **without appendices** Hochofenrumpf *m* ~ **with bucket hoist** Hochofen *m* mit Kübelbegichtung ~ **with closed front** Hochofen *m* mit geschlossener Brust ~ **of medium height** Halbhochofen *m* ~ **with open front** Hochofen *m* mit offener Brust ~ **with skip hoist** Hochofen *m* mit Kippgefäßbegichtung

**blast-furnace,** ~ **bear** Hochofensau *f* ~ **blowing engine** Hochofengebläse *n* ~ **blowing plant** Hochofengebläseanlage *f* ~ **bosh** Hochofenrast *f* ~ **burden** Hochofenmöller *m*, Möller *m* ~ **bustle pipe** Hochofenwindring *m* ~ **casting** Hochofenabstich *m* ~ **(-slag) cement** Hochofen-, Hütten-zement *m* ~ **charge** Hochofenbeschickung *f* ~ **charging** Hochofen-begichtung *f*, -beschickung *f* ~ **charging equipment** Hochofenbegichtungsanlage *f* ~ **coke** Hochofen-, Hütten-koks *m* ~ **crucible** Hochofengestell *n* ~ **design** Schachtofenbau *m* ~ **elevator** Hochofenaufzug *m* ~ **engineer** Hochöfner *m* ~ **fittings** Hochofenarmatur *f* ~ **foreman** Hochofenmeister *m* ~ **frame** Hochofengerüst *n* ~ **gas** Gicht-, Hochofen-, Schwach-gas *n* ~-**gas cleaner** Gichtgasreiniger *m* ~ **gas-driven blowing engine** Hochofengasgebläse *n* ~-**gas engine** Gichtgasmaschine *f* ~ **gas engine** Hochofengasmaschine *f* ~ **gas main** Gichtgasleitung *f* ~ **gas-purifying plant** Hochofengasreinigungsanlage *f* ~ **gas valve** Hochofengasschieber *m* ~ **gun** Stichlochstopfmaschine *f* ~ **hearth** Hochofengestell *n* ~ **hoist** Hochofenaufzug *m* ~ **hopper** Hochofengicht *f* ~ **jacket** Hochofenpanzer *m* ~ **lines** Hochofenprofil *n* ~ **lining** Hochofenfutter *n* ~ **lintel plate** Hochofentragring *m* ~ **lump slag** Hochofenstückschlacke *f* ~ **man** Hochofenmann *m*, Hüttenmann *m* ~ **manager** Hochofenleiter *m* ~ **mantle** Hochofentragring *m* ~-**melting operation (or practice)** Hochofenführung *f* ~ **method** Röstreduktionsarbeit *f* ~ **mixer** Roheisenmischer *m* ~ **mouth** Hochofengicht *f* ~ **operator** Hochofenleiter *m*, Hochöfner *m* ~ **operation** Hochofenbetrieb *m* ~ **output** Hochofenerzeugung *f* ~ **plant** Hochofen-anlage *f*, -werk *n*, Hütte *f* ~ **practice** Hochofenbetrieb *m* ~ **process** Hochofenverfahren *n* ~ **ring** Hochofentragring *m* ~ **salamander** Hochofensau *f* ~ **settler** Hochofenvorherd *m* ~ **shaft** Hochofen-schacht *m*, -umkleidung *f* ~ **slag** Hochofenschlacke *f* ~ **slag cement** Hochofenschlackenzement *m* ~ **smelting** Schachtofen-arbeit *f*, -schmelzen *n* ~ **smelting operation** Hochofenvorgang *m* ~ **sow** Hochofensau *f* ~ **stack** Hochofenschacht *m* ~-**stack casing**

Hochofenschachtpanzer *m* ~ **steam blower** Hochofendampfgebläse *n* ~ **throat** Hochofengicht *f* ~ **top** Hochofengicht *f* ~ **tuyère** Hochofenwindform *f* ~ **well** Hochofengestell *n* ~ **works** Hüttenwerk *n*

**blast,** ~ **gate** Anlaßtor *n*, Auslaßtor *n*, Windabsperrschieber *m*, Windschieber *m* ~ **gauge** Winddruckmesser *m*, Windmesser *m* ~ **gun** Blasdüse *f*, Strahlrohr *n* ~ **head** Strahlkopf *m* (Sandstrahlgebläse) ~ **heating** Winderhitzung *f* ~**hole** Blase *f* im Metall, Blaseloch *n* ~ **inlet** Windeinströmungsöffnung *f* ~ **lamp** Blas-, Einbrenn-, Gebläse-lampe *f* ~-**lamp work** Lampenarbeit *f* ~ **loading** Kraft *f* der Expansionswelle *f* ~ **main** Windleitung *f* ~ **meter** Gebläsemesser *m* ~ **nozzle** Blasdüse *f*, Blasmundstück *n*, Strahldüse *f* ~-**nozzle orifice** Blasdüsenöffnung *f* ~-**nozzle tip** Blasdüsenkopf *m* ~ **opening** Blasöffnung *f* ~ **passage** Winddurchgang *m* ~ **pipe** Blase-, Blas-rohr *n*, Windleitungs-, Wind-, Windzuführungs-rohr *n*, Windleitung *f* ~-**pipe leading** Gebläseleitung *f* ~-**pipe line** Windrohrleitung *f* ~ **pressure** Detonationsdruck *m*, Explosions-, Gebläse-, Luft-druck *m*, Ventilatorpressung *f*, Winddruck *m*, Windpressung *f* ~-**pressure meter** Winddruckmesser *m* ~-**pressure tank** Druckluft-, Druckwind-, Wind-kessel *m* ~ **roasting** Verblaseröstung *f* ~ **room** Blashaus *n* ~ **room with plain grated floor** Putzhaus *n* mit Bodenrostfläche ~ **sand** Gebläsesand *m* ~ **stone** Windstein *m* ~ **superheater** Luftüberhitzer *m* ~ **supply** Windlieferung *f* ~ **tests** Ansprengversuche *pl* ~ **tuyère** Wind-düse *f*, -form *f* ~ **valve** Windventil *n* ~ **velocity** Luftgeschwindigkeit *f* ~-**volume meter** Windmengemesser *m* ~ **wall** Splitterwall *m* ~ **wave** Druck-, Expansions-welle *f* ~ **whistle** Ton *m*

**blaster** Sprenger *m*

**blasting** Bohr- und Schießarbeit *f*, Schießarbeit *f*, Sprengarbeit *f*, Sprengen *n*, Sprengung *f*, (of microphone) Übersteuerung *f* ~ **by means of a mine** Minensprengung *f* ~ **by means of torpedo** Torpedieren *n* (petroleum) ~ **away of a shelf of rocks** Abraumsprengung *f*

**blasting,** ~ **agent** Sprengmittel *n* ~ **bomb** Sprengbombe *f* ~ **cap** Hütchen *n*, Sprengkapsel *f* ~ **cartridge** Bohr-, Spreng-patrone *f* ~ **charge** Bohrladung *f*, Granatfüllung *f*, Sprengkörper *m* (fixed) ~ **charge** Sprengbüchse *f* ~-**charge box** Ladungskasten *m* ~-**charge container** Ladungs-büchse *f*, -gefäß *n* ~ **charges in series** Reihenladung *f* ~ **company** Sprenggesellschaft *f* ~ **detachment** Sprengtrupp *m* ~ **detonating machine** Zündmaschine *f* ~ **equipment** Bestrahlungsanlage *f* ~ **gelatin** Spreng-gelatine *f*, -gummi *m* ~ **grit** Gebläsekies *m* ~ **ignition cable** Sprengkabel *n* ~-**machine exploder** Glühzündapparat *m* ~ **magnet** Flammblaser *m* ~ **method** Verblaseverfahren *n* ~ **needle** Schießnadel *f* ~ **operation** Blasarbeit *f* ~ **plan** Sprengplan *m* ~ **powder** Sprengpulver *n* ~ **process** Sprengvorgang *m* ~ **sand** Blas-, Gebläse-sand *m* ~ **tools** Bohr- und Schießzeug *n*

**Blau(water) gas** Blauwassergas *n*

**blaze, to** ~ aufflammen **to** ~ **off steel** Stahl abbrennen

**blaze** Flamme *f*, Lichtschein *m*

**blazing** lichterloh

**bleach, to** ~ blanchieren, bleichen, entfärben

**bleach** Bleiche *f*, Bleichflüssigkeit *f*, Bleichmittel *n*, Chlorkalklösung *f*, Chlorlauge *f* ~ **effect** Bleicheffekt *m* ~ **style** Bleichartikel *m*

**bleached oil** gebleichtes Öl *n*

**bleacher's helper** Bleichmittel *n*; Bleichergehilfe *m*

**bleaching, fast to** ~ bleichecht

**bleaching** Ausbleichen *n*, Bleiche *f*, Bleicherei *f*, Bleichung *f*, Entfärbung *f*, Weißgrad *m* ~ **ability of a fabric to resist** ~ Bleich-, Schweiß-echtheit *f* **ability of a fabric to withstand** ~ Lichtechtheit *f* **ability of a fabric to resist (or withstand)** ~ Kochechtheit *f* ~ **with calcium hypochlorite** Chlorkalkbleiche *f* ~ **of cellulose** Bleichen *n* des Zellstoffes

**bleaching,** ~ **agent** Bleich-, Entfärbungsmittel *n* ~ **apparatus** Bleichapparat *m* ~ **bath** Bleichbad *n* ~ **chest** Bleichbottich *m* ~ **earth(s)** Bleicherde *f* ~ **engine** Bleichholländer *m* ~ **liquid** Chlorkalklösung *f* ~ **liquor** Bleichflüssigkeit *f* ~ **mordant** Bleichbeize *f* ~-**out process** Ausbleichverfahren *n* ~ **plant** Bleichanlage *f* ~ **powder** Bleich-kalk *m*, -pulver *n*, Clorkalk *m* ~ **process** Bleichverfahren *n* ~ **resistance** Waschechtheit *f* ~ **soda** Bleichsoda *f* ~ **stain** Bleichfleck *m*

**bleary** verschwommen, (of a picture) soßig ~ **appearance** verschwommenes Aussehen *n* ~ **zone** Verwaschungszone *f*

**bled** (a pamphlet cut down too much) zu stark beschnitten ~ **off** abfallend (Bilder) ~ **steam** Abzapfdampf *m*, angezapfter Dampf *m*

**bleed, to** ~ (paper) abbluten, ablassen, (a pipe) abzapfen, entleeren, entlüften, entnehmen, aus kleinen Öffnungen ausfließen, umleiten **to** ~ **the brake** die Bremse *f* lüften

**bleed** Ausblühen *n*, Wasserausströmen *n* (min.) ~ **air** Abzapfluft *f* (vom Triebwerk) ~ **air system** Abzapfluftanlage *f* ~ **fuel** Überlaufkraftstoff *m* ~ **hole** kleine Luftöffnung *f* ~ **line** Steuerleitung *f* ~ **opening** Steuerleitungsanschluß *m* ~ **port** Abzapf-, Entlüftungs-anschluß *m* ~ **post** Abzapf-, Entlüftungs-anschluß *m* ~ **valve** Entlüftungs-, Ablaß-ventil *n*

**bleeder** (drain valve) Ablaßventil *n*, Anzapf *m*, Musterhahn *m*, (petroleum) Probehahn *m* ~ **air-control piston** Bremsluftsteuerkolben *m* ~ **assembly** Entlüftungsschraube *f* ~ **chain** Belastungswiderstandskette *f*, Spannungsteiler *m* ~ **current** Anzapf-dampf *m*, -strom *m* ~ **pipe to let off gas** Gasabflußrohr *n* ~ **resistor** Ableit(ungs)-, Belastungs-, Vorbelastungs-widerstand *m* ~ **steam** Anzapfdampf *m* ~ **throttle** Entlüftungsdrossel *f* ~ **turbine** Anzapf-, Entnahme-turbine *f* ~ **valve** Anzapf-, Gichtgas-rohr-ventil *n* ~ **well** Abzapfbrunnen *m*

**bleeding** Ablassen *n*, Bluten *n*, Blutung *f*, Entnehmen *n*, Farbensaum *m* (film), (of dyes) Abfärben *n*, (steam) Anzapfen *n* ~ **of the prints** Fließen *n* der Drucke

**blemish** Fehler *m* (in wood), Flecken *m*, Makel *m*

**blemishes** Stichflecken *pl*

**blend, to** ~ anmengen, anpassen, durchmengen, durchmischen, einmischen, ineinander übergehen lassen, melieren, mengen, mischen, vermengen, verschmelzen, verschneiden, verschwimmen, zusammenlaufen, to ~ the controls die Ruderbetätigung koordinieren

**blend** Mischung *f*, Verschnitt *m*

**blende, ~ concentrate** Blendekonzentrat *n* **~-roasting furnace** Blenderöstofen *m*

**blended** meliert verschnitten **~ altitude rate** zugemischte Vertikalgeschwindigkeit *f* (aviat.)

**blender** Mischer *m*

**blending** Mischung *f*, Übergang *m*, Vermengung *f*, Verschmelzung *f*, (as of fuels) planmäßige Vermischung *f* **~ and mixing** Überblender *m* **~ with the ground** Geländeanpassung *f*

**blending, ~ agent** Verschnitt-, Zusatz-mittel *n* **~ equipment (or device)** Mischapparatur *f* **~ ratio** Mischungsverhältnis *n* **~ value** Mischungs-wertigkeitsziffer *f* bei Oktanzahlbestimmung **~ varnish** Zusatzlack *m*

**blick, to ~** blicken

**blick** Aufblick *m*

**blimp** Kleinluftschiff *n*, kleines unstarres Luftschiff *f*, Kamerahülle *f*

**blimped camera** umhüllte Kamera *f*

**blind, to ~** abdecken (aviat.), blenden, verblenden, verschließen **to ~ flange** blind flanschen

**blind** Markise *f*; blind **~s** Deckwerk *n* (fort) **~ for a window** Fensterblende *f*

**blind, ~ alley** Sackgasse *f* **~ anchor** einarmiger Hafenanker *m* **~ anchorage** Blindverankerung *f* **~ angle** toter Winkel *m* **~ beam** Blindbaum *m* **~ calcualtor** unsichtbar schreibende Rechenmaschine *f* **~ coal** Kohlenblende *f* **~ cover** Blinddeckel *m* (Blenddeckel) **~ (blocking) embossing** Blindprägung *f* **~ end bore** Sackloch *n* **~-end cylinder** geschlossener Zylinder *m* **~ flange** Blindflansch *m*, Verschlußflantsche *f*, Vollflantsch *m*

**blind-flying** Blind-fliegen *n*, -flug *m*, Instrumenten-fliegen *n*, -flug *m* **~ bank** Blindflug-kurve *f* **~ curve** Blindflugkurve *f* **~ equipment** Blindflug-gerät *n*, -gerätausrüstung *f* **~ hood** Blindflughaube *f* **~ instruments** Blindflug-instrumente *pl* **~ school** Blindflugschule *f* **~ training plane** Blindflugschulflugzeug *n*

**blind, ~fold operation of crank wheels** Blindkurbeln *n* **~ frame** Blendrahmen *m* **~ holder** Blindkassette *f* **~ hole** Blindbohrung *f*, Sackloch *n* **~ landing** Nebel-, Blind-landung *f* (aviat.) **~-landing equipment** Funklandegerät *n* **~ pass** blindes Kaliber *n* **~ pit** Stapelschacht *m* **~ plug** Blind-stopfen *m*, -stöpsel *f* **~ printing plate** Blindplatte *f* **~ riser** verlorener Kopf *m* **~ road** Sackgasse *f* **~ sector** Schattensektor *m* **~ shaft** Blindschacht *m* **~ shell** Blindgänger *m*, Versager *m* **~ space** sichttoter Raum *m* **~ spot** blinde Stelle *f* im Funkempfang, Empfangsloch *n*, Schattenstelle *f*, tote Zone *f*, toter Schußwinkel *m* (aviat.), Totlage *f*, Totpunkt *m* **~ supervision** Überwachung *f* ohne Schlußzeichen vom angerufenen Teilnehmer **~ tapping** Sackgewinde *n* **~ track** totes Gleis *n* **~ traffic** Blindverkehr *m* **~ transmission** Blindübermittlung *f* **~ turn** Blindflugkurve *f* **~ wall** Blendfassade *f*, blinde Mauer *f* **~ window** Blendfenster *n* **~ zone** Schattenzone *f*

**blindate** Klappblende *f*

**blinder for head** Supportfeststellbolzen *m*

**blinding** Blendung *f*

**B-line** Adressenregister *n*

**blink, to ~** blinken

**blink** Bügel *m* **~ lamp** Blinkleuchte *f* **~ microscope** Blinkmikroskop *n*

**blinker** Blendrahmen *m* **~ apparatus** Blinkgerät *n*, Lichtsprechgerät *n* **~ arrangement** Blinkschalung *f* **~ beacon** Blinkfeuer *n*, Blinkgerät *n* **~ device** Blinkeinrichtung *f* **~-flow indicator** Durchflußblinker *m* **~ lamp** Blinklampe *f* **~ light** Blinkfeuer *n*, Winker *m* **~ message** Blinkmeldung *f* **~ method** Blinkverfahren *n* **~ post** Zwischenstelle *f* **~ signal** Blinkzeichen *n* **~ signaling** Blinkerei *f* **~ signaling in Morse code** Morsefeuer *n* **~ unit** Blinkgeber *m*

**blinkers** Blendleder *pl*

**blinking stop** Morseblende *f*

**blip** Ablenkzeichen *n*, Echozeichen *n* (rdr) **~-scan ratio** Anzeigewahrscheinlichkeit *f* (rdr)

**blister** Blase *f* **to get ~s** blasig werden Schnallen *pl* (paper mfg.) **~ in a casting** Gußblase *f* **~ in the glass** Glasblase *f*

**blister, ~ agent** ätzender Kampfstoff *m* **~ copper** Blasen-, Rohr-kupfer *n* **~ copper slag** Schwarzkupferschlacke *f* **~ dent** Einbeulung *f* **~ formation** Blasenbildung *f* **~-forming** hautätzend **~ gas** ätzender Kampfstoff *m* **~ resistance** Beulsteifigkeit *f* **~ steel** Brenn-, Zement-stahl *m* **~(ed) steel** Blasenstahl *m*

**blistered** blasig **~ casting** blasiger Guß *m*

**blistering** blasenziehend; (in welding) Ausblühung *f*; (paint, metal, photography) Blasenbildung *f*

**blistery** blasig

**blizzard** Blizzard *m*, Schneegestöber *n*, Schneesturm *m*

**bloat** aufgebläht

**bloating clay** Blähton *m*

**blob, ~ density** Klecksdichte *f* **~ structure** Klecksigkeit *f*

**Bloch band** Energieband *n*

**block, to ~** (chem.) abbinden, abriegeln, aufklotzen (print.), blocken, sichern, sperren, verblocken, verhacken, verhauen, verriegeln, versetzen, versperren, verstopfen **to ~ blank** austasten **to ~ off** abschirmen **to ~ out timber** Holz *n* schneiden **to ~ up** blockieren verbauen, zustellen

**block** Bildstock *m* (Druckerei), Block *m* (data proc.), Block *m* (Holz), Druckstock *m*, Flasche *f* (mech.), Flasche *f* des Flaschenzugs, Flaschenzuges *m*, Kästchenblock *m*, Klischee, *n*, Kloben *m*, Klotz *m*, Rollenkloben *m*, Rollenzug *m*, Sperrer *m*, Stein *m*, Widerlager *n*, (iron) Eisenblock *m*, (of masonry) Mauerblock *m*, (of purchase) Block *m*, (coke-furnace) Batterie *f* **~(ing)** Sperre *f* **set of gauge ~s** Endmaßsatz *m* **~ and tackle** Flaschenzug *m*, Flaschenzugleine *f*, Rollenzug *m* **~ of ice** Eisberg *m*, Eisscholle *f* **~ of information** Wörterblock *m*, Wortgruppe *f* (data proc.) **~ on a railroad track** Blockwerk *n*

**block, ~ access** Blockübertragung *f* **~ antenna** Gemeinschaftsantenne *f* **~ apparatus** Blockeinrichtung *f* **~ armature** Blockarmatur *f* **~ barrier** Blockwall *m* **~ battery** Blockbatterie *f* **~ brake** Backen-, Klotz-bremse *f* **~-carring truck** Rollwagen *m* **~ caving** Blockbruchbau *m* **~ chain** Block-, Hebezug-kette *f* **~ chains** Roll-, Umlauf-kette *f* **~-chain sprocket** Block-, Hebe-kettenrad *n* ~-chain-sprocket hob Blockketten-radwalzfräser *m* ~ **coefficient** Völligkeitsgrad *m* ~ **condenser** Abschlußkondensator *m* ~(ing) **condenser** Blockkondensator *m* ~ **condenser for signaling purposes** Rufsperrkondensator *m* (teleph.) ~ **conduit** Formstückkanal *m* ~ **cutter** Druckformenschneider *m* ~ **diagram** Ablaufplan *m*, Blockschaltbild *n*, Blockzeichnung *f*, Flußplan *m*, Übersichtsschema *n* ~ **faulting** Blockverwerfung *f* ~ **gap** Blockabstand *m* ~ **gauge** Klischeehöhenmesser *m*, Vorrichtungsblocklehre *f* ~ **house** Blockhaus *n* ~ **instrument case** Blockschrank *m* ~ **lava** Block- oder Schollenlava *f* ~ **letters** Blockschrift *f*, gotische Schrift *f* ~ **lifting** Aufhängung *f* der Blöcke ~ **link** Innenglied *n* ~ **lock** Blockschloß *n* (signal gear) ~**maker** Abklatscher *m* (print.) ~ **operations** Blockbetrieb *m* ~-**outs** Ausklinkungen *pl* ~ **pin** Blocknagel *m* ~ **post** Zugfolgestelle *f* ~ **print(ing)** Blockdruck *m* ~-**printing machine** Modelldruckmaschine *f* ~ **proof** Klischeeabzug *m* ~ **proofing press** Autotypieandruckpresse *f* ~ **pulley** Hängebankseil-, Vollriemen-scheibe *f* ~ **rack** Klischeeregal *n* ~ **reception** Blockempfang *m* ~ **safe** blocksicher ~ **schematic diagram** Blockschaltbild *n* ~ **section** Blockabschnitt *m* (r.r.) ~ **shackle** Flaschenzugbügelhaken *m* ~ **signal** Blocksignal *n*, Haltezeichen *n* ~ **signal box** Blockhütte *f* (r.r.) ~ **signaling apparatus** Mastsignal *n* ~ **signals** Blockschrift *f* ~ **slicing machine** Blockschneidemaschine *f* ~ **sort** Gruppensortierung *f* ~ **stamp** Blockstanze *f* (for cutting out lids, etc) ~ **station** Blockstelle *f* (r.r.)

**block-system** Blockfeld *n*, Blocksystem *n* (r.r.), Raumfolge *f*, Zugfolge *f* ~ **with track normally closed** Streckenblock *m* mit Grundstellung der Signale auf Halt ~ **with track normally open** Streckenblock *m* mit Grundstellung der Signale auf freie Fahrt ~ **plant** Blockanlage *f* ~ **relay** Blockrelais *n*

**block,** ~ **telephone** Blockfernsprecher *m* ~ **terminal** Endverzweiger *m* ~ **time** Blockzeit *f* ~ **tin** Blockzinn *n* ~-**to**-~ Blockzeit *f* (aviat.) ~ **tool** Kastenwerkzeug *n* ~ **tool box** Blockstichelhaus *n* ~ **toolholder** Kastenstahlhalter *m* ~ **transport crane** Blockkran *m* ~ **trench** Sackgraben *m* ~ **type** Blockschrift *f* ~ **valve** Abteilventil *n* ~ **wall** Blockwall *m* ~ **wire-stitching machine** Blockdrahtheftmaschine *f* ~**work** Blocklagen *pl* ~ **yard** Werkplatz *m*

**blockade, to** ~ **a port** einen Hafen blockieren

**blockade** Blockade *f*, Verkehrssperre *f* ~ **of the continent** Kontinentalsperre *f* ~ **area** Blockadé-gebiet *n* ~ **zone** Blockadegebiet *n*, Sperrgebiet *n*

**blockading sea areas** Sperrung *f* von Seegebieten

**blockage** Verstopfung *f* ~ **effect** Sperreffekt *m*

**blocked** geblockt, gesperrt ~ **grid keying** Gittersperrspannungstastung *f* ~ **impedance** blockierte Impedanz *f* (acoust.) ~ **semiwave** Sperrhalbwelle *f*

**blocker** Sperrer *m*

**blockette** Teilblock *m*

**blocking** Abriegelung *f*, Blocken *n*, Blockierung *f*, Festklemmen *n*, Sperrblockierung *f*, Sperrung *f*, Verstopfung *f*, (of steel beams) Ausklinkung *f* ~ **of chutes** Schurrenverstopfung *f* ~ **of the**

receiver Zustopfen *n* des Empfängers

**blocking,** ~ **action** Sperrwirkung *f* ~ **brake** Haltebremse *f* ~ **capacitor** Sperrkondensator *m* ~ **characteristics** Sperrkennlinie *f* ~ **condenser** Sperrkondensator *m* ~ **course** Blöckenreihe *f* (auf einem Gesims) ~ **current** Blockstrom *m* ~-**current circuit** Blockierungsstromkreis *m* ~ **down** Vorläppen *n* ~ **fortification** Sperrbefestigung *f* ~ **gear** Entlastungs-, Feststell-vorrichtung *f* ~ **impulse** Sperrimpuls *m* ~ **layer** Halbleiter-, Sperr-schicht *f* ~ **layer photocell** Halbleiterfotozelle *f* ~ **lever** Arretier-, Arretierungs-hebel *m* (general) ~ **light stop** Sperrblende *f* ~ **medium** Absorptionsmittel *n* ~ **oscillator** Sperr-kippsender *m*, -schwinger *m* ~-**out** Abdecken *n* ~-**out ink** Korrekturtusche *f* ~ **pawl** Sperrhaken *m* ~ **period** positive Sperrzeit *f*, Zündwinkel *m* ~ **piece** Futterholz *n* ~ **pin** Arretierstift *m* ~ **ratio** Sperrverhältnis *n* ~ **rectifier** Sperrgleichrichter *m* ~ **signal** Blockstation *f* ~ **slide** Feststellschieber *m* ~ **spring** Sperrfeder *f* ~ **switch** Blockierschalter *m* ~ **time** Sperrzeit *f* ~ **unit** Gerätsblock *m* ~ **value** Sperrwert *m* ~ **voltage** Sperrspannung *f*

**blocky** kompakt (coke)

**blomstrandite** Betafit *m*, Blomstrandit *m*

**blood** Blut *n* ~ **clot** Blutgerinsel *n* ~ **charcoal** Blutkohle *f* ~ **coagulation** Blutgerinnung *f* ~ **count** Blutkörperchenzählung *f* ~ **counter** Blutkörperchenzähler *m* ~-**counter chamber** Blutkörperchenzählkammer *f* ~ **fluid** Blutflüssigkeit *f* ~**group** Blutgruppe *f* ~ **molasses** Blutmelasse *f* ~ **plasma** Blutflüssigkeit *f* ~ **poisoning** Blutvergiftung *f* ~ **rain** Blutregen *m* ~ **seasoning** Blutglanz *m* ~ **serum** Blutflüssigkeit *f* ~-**soaked** blutgetränkt ~**stone** Blutstein *m*, Jaspis *m* mit roten Flecken ~ **transfusion** Bluttransfusion *f* ~ **type** Blutgruppe *f* ~ **vessel** Blutgefäß *n*

**bloom, to** ~ auswalzen, vorblocken, vorhämmern, vorstrecken, vorwalzen, die Luppen auswalzen

**bloom** Beschlag *m*, Bildweichheit *f*, Dackel *m*, Deul *m*, Eisenblock *m*, (on varnish) Hauch *m*, Hof *m* (TV), (ore) Luppe *f*, Schein *m* (der Mineralöle), Schimmer *m*, Schrei *m*, Schürbel *m*, Schwefelausschlag *m*, Stahlblock *m*, Überstrahlung *f*, Verfärbung *f*, Vorblock *m*, vorgewalzter Block *m*, Walzblock *m*, Wolf *m* ~ **and slab yard** Blocklager *n* ~ **of oil** Abglanz *m* von Öl ~ **of steel** Lotte *f*

**bloom,** ~ **buggies** Blockwagen *pl* ~ **crop end** Blockende *n* ~ **end** Blockende *n* ~ **reheating furnace** Blockwärmeofen *m* ~ **roll** Vorwalze *f* ~ **shears** Blockschere *f* ~ **steel** Luppenstahl *m* ~ **wagon** Luppenwagen *m* (metall.)

**bloomer** Block-gerüst *n*, -walze *f*; SSR-Anzeige *f*

**bloomery** Luppen-feuer *n*, -frischhütte *f*; Rennfeuer *n*, -herd *m* ~ **fire** Luppenfrischfeuer *n*, Renn-feuer *n*, -herd *m* ~ **forge (or hearth)** Renn-feuer *n*, -herd *m* ~ **iron** Herdfrischeisen *n*, Rennstahl *m* ~ **process** Rennverfahren *n*

**blooming,** Anlaufen *n* (Farbe), Aufhellung *f* schwarzer Stellen, Ausblühen *n*, Blühen *n*, Leuchtfleckaufweitung *f* (rdr), Vorstrecken *n*, Vorwalzen *n*, Überstrahlung *f*, Wolkigwerden *n*, (defect in maroons) Bronzieren *n* ~ **condition** flau ~ **mill** Blockwalzwerk *n*, Brammenwalze *f*, Grobwalzwerk *n*, Vorstrecke *f*, Vorwalzwerk *n* ~-mill housing Blockwalzenständer *m* ~-mill stand Block-gerüst *n*, -walzgerüst *n* ~-mill train Block-straße *f*, -strecke *f* ~ pass Block-kaliber *n*, -walzkaliber *n*, Vorstich *m*, Vorstreckkaliber *n* ~ roll Blockwalze *f*, Vorblockwalze *f*, Vorstreckwalze *f*, Zängwalze *f* ~-roll train Blockwalzstraße *f* ~ stand Vorgerüst *n* ~ strand Vorstrecke *f*, Vorwalzstrecke *f* ~train Vorstraße *f*

**blooper** Klebepresse *f*

**blooping,** ~ **elimination** Geräuschbeseitigung *f* ~ **patch (or splice)** Tonklebestelle *f*

**blossom** Düppelblüte *f* (rdr)

**blot, to** ~ auslöschen beklecksen, löschen, unsauber abziehen (print.) **to** ~ **out** tilgen, auspinseln

**blot** Flecken *m*, Sau *f*, Verschluß *m* ~ **test** Fehlerprobe *f*

**blotch** Pustel *f* ~ **print** Deckerdruck *m* ~ **prints** großflächige Drucke *pl* ~ **roller** Bodenwalze *f* (print.)

**blotching** Kraterbildung *f*

**blotted** klecksig

**blotter** Fließblatt *n*, Löscher *m*

**blotting,** ~ **board** Lochkarton *m* ~ **pad** Schreibunterlage *f* ~ **paper** Abschmutz-, Fließ, Lösch-papier *n*

**blow,, to** ~ abblasen, abschmelzen, ansprechen, aufblähen, blasen, (Sicherung) durchbrennen, durchschmelzen, erblasen, verblasen, wehen, (out) ausblasen, (tanks) lenzen **to** ~ **against** anblasen, anströmen **to** ~ **away** (air or wind) abströmen **to** ~ **back** rückschlagen **to** ~ **by** durchblasen (piston rings) **to** ~ **cold** kalt erblasen, mit Kaltwind *m* blasen **to** ~ **down** abblasen, ablassen, ausblasen, herunterblasen, herunterfrischen **to** ~ **fuse** Sicherung *f* durchbrennen **to** ~ **hot** heißblasen, mit Heißwind *m* blasen **to** ~ **in** (in a blast furnace) anblasen **to** ~ **in(to)** einblasen **to** ~ **(or dust) off** abblasen **to** ~ **out** auslöschen, erlöschen, platzen, (furnace) kaltlegen, (mining) ausbrennen, auskochen **to** ~ **shut** zublasen **to** ~ **through** durchblasen **to** ~ **toward** zuströmen **to** ~ **up** aufblasen, aufkochen, aufpumpen, auftreiben, heißblasen, sprengen, zersprengen, vergrößern (photo) **to** ~ **with cold air** mit Kaltwind *m* blasen **to** ~ **with hot air** mit Heißwind *m* blasen **to** ~ **in the furnace** Hochofen *m* anblasen **to** ~ **off with gunpowder** mit Pulver *n* absprengen **to** ~ **off steam** Dampf *m* abblasen, auslassen

**blow** Anschlag *m*, Blasevorgang *m*, Durchschmelzung *f*, Hieb *m*, Schlag *m*, Schmelzgang *m*, Stoß *m* ~ **by** Durchblasen *n* ~ **of the hammer** Hammerschlag *m*

**blow,** ~ **back** Gasdruck *m* ~-**back-operated weapon** Gasdrucklader *m* ~ **bending test** Schlagbiegeprobe *f* ~ **case** Druckfaß *n* ~**down piping** Abblaseleitung *f*, Rohrleitung *f* zur raschen Entleerung ~**down pit** Ablaßgrube *f* ~ **down tank** Abblasetank *m*, Reservoir *n* für rasche Entleerung ~ **furnace** Blaseofen *m* ~**gun** Abblasepistole *f* ~**hole** Blase *f*, Blasenhohlraum *m*, Blasenraum *m*, Fehlstelle *f*, Gaspore *f*, Innenlunker *m*, Lunker *m*, Pfeife *f*, schwarze Stelle *f*, (in casting) Gasblase *f*, (metal) Gas-

einschluß *m*, (in a cast piece) Luftblase *f* ~**hole in a casting** Gußblase *f* ~**hole formation** Lunkerung *f* ~**hole segregate** Gasblasensteigerung *f* ~**holes below the surface** Randblase *f* (in an ingot) ~-**impact machine** Fallwerk *n* ~(-pipe) **lamp** Lötlampe *f* ~ **mold** Blasform *f* ~ **molding** Blasformverfahren *n* ~ **molding machine** Blasformmaschine *f* ~ **molding unit** Blasformanlage *f*

**blowoff** Probehahn *m* ~ **cock** Ablaß-, Ausblase--hahn *m*, Luftventil *n*, Schlammhahn *m* ~ **device** Abblasevorrichtung *f* ~ **line** Abblaseleitung *f* ~ **outlet** Ausblasöffnung *f* ~ **pipe** Ausblaseleitung *f* ~ **plug** Ablaßzapfen *m* ~ **pressure** Drucküberschuß *m* ~ **tank** Entspannungsgefäß *n* ~ **valve** Abblase-hahn *m*, -ventil *n*, Ablaßventil *n*, Probehahn *m*, Sicherheits-, Überdruck-, Windabblase-ventil *n* ~ **valve (or drain plug)** Entleerungsschraube *f* mit Überdruckventil

**blowout** Ausbruch *m*, Durchschmelzung *f*, Durchschlag *m* (electr.), Funklösung *f*, (blast furnace) Niederblasen *n*, Platzen *n*, Reifenpanne *f* ~ **coil** Blas-, Durchschmelz-spule *f* ~ **distributor** Ausblaseverteiler *m* (submarine) ~ **gun** Ausblasepistole *f* ~ **patch** Mantelmanschette *f* (tire), Zerreißscheibe *f* ~ **port** Ausblasöffnung *f* (rocket) ~ **preventer** Ausbruchsventil *n* ~ **valve** Ausblase-hahn *m*, -ventil *n*, Ausbruchsventil *n*

**blowpipe** Auftreibeisen *n*, Blaserohr *n*, Blasrohr *n*, Brenner *m*, Entlüftungsrohr *n*, Gebläsebrenner *m*, Gebläselampe *f*, Lötrohr *n*, Pustrohr *n*, Rüssel *m*, Schneidbrenner *m*, Schweißbrenner *m* ~ **action** Stichflammenwirkung *f* ~ **bead** Schmelzperle *f* ~ **flame** Lötflamme *f* ~ **nipple** Lötrohrspitze *f* ~ **(test) piece** Lötrohrprobe *f* ~ **solderer** Rohrlöter *m* ~ **test** Lötprobe *f*

**blow,** ~ **pit** Ausblasebutte *f*, Diffuseur *m*, Waschbehälter *m* ~ **test** Blas-, Pust-probe *f* ~ **test on wet filter paper** Spritzprobe *f* (Farbstoffprüfung) ~-**through valve** Abgas-, Dampfabblas-ventil *n* ~**torch** kleine Lötlampe *f*, Pustlampe *f* ~-**up tank** Aufkocher *m*, Probehahn *m*, Schnarch-, Schnarr-ventil *n* ~-**up test** Prüfung *f* unter Druck

**blower** Blaseeinrichtung *f*, Blasemeister (Bessemer), Bläser *m*, Druckluftventilator *m*, Fächer *m*, Gebläse *n*, Kompressor *m*, Lader *m*, Lüfter *m*, Lüftungsmaschine *f*, (blasender) Ventilator *m*, Vorverdichter *m*, Wind-bläser *m*, -gebläse *n*, -maschine *f*; (in converter operation) Schmelzer *m*, (engine) Gebläsemaschine *f* ~ **for the drier** Trocknergebläse *n*

**blower,** ~ **agitator** Rührflügel *m* ~ **bar** Bläser *m* ~ **base** Lüftersockel *m* ~ **casing** Gebläse-, Lader-gehäuse *n* ~ **control** Lüfterüberwachung *f* ~**-cooled** zwanggekühlt ~**-cooled engine** durch Gebläse gekühlter Motor *m* ~ **drive** Ladergetriebe *n* ~**-engine house** Gebläsemaschinenhaus *n* ~ **engineer** Gebläsemaschinenwärter *m* ~ **eye** Ladereinlaß *m* ~ **fan** Gebläsewerk *n* ~ **fault relay** Lüfterstörungsrelais *n* ~ **fault release** Lüfterstörungsentriegelung *f* ~**-fed engine** Kompressor-, Lader-motor *m* ~ **motor** Gebläse-, Vorverdichter-motor *m* ~ **nozzle** Durchblasdüse *f* ~ **panel** Lüfterfeld *n* ~ **pipe** Gebläse-, Zubläser-rohr *n* ~-**pressure pipe** Gebläsedruckstutzen *m* ~-**pressure pipe butt** Vorverdichterdruckstutzen *m* ~ **ratio** Laderübersetzung *f* ~-**regulating man** Gebläsebediener *m* ~ **rim** Laderringkanal *m* ~**rim air temperature** Ladelufttemperatur *f* ~ **separator** Gebläseseparator *m* (grain) ~ **shaft** Gebläse-, Vorverdichter-welle *f* ~ **spiral** Laderspirale *f* ~ **spreader** Blasmaschine *f* ~ **stream** Umströmung *f* (Windtunnel) ~ **test plant** Blasprüfstand *m* ~-**throttle slide valve** Gebläsedrosselschieber *m* ~-**type supercharger** Gebläsevorverdichter *m* ~ **vane** Gebläseflügel *m*

**blowers actuated by water power** Wasserstrahlgebläse *n*

**blowing** Abschmelzen *n*, Ansprechen *n*, Aufblasen *n* Druckbrennen *n*, Durchschmelzen *n*, Entfaltung *f*, Frischen *n*, Frischung *f*, Verblasen *n*, ~ **in** Einblasen *n* ~ **out** Ausblasen *n* ~ **up** Heißblasen *n* ~ **of a fuse** Abschmelzen *n* einer Sicherung

**blowing,** ~ **agent** Blähmittel *n* ~ **cylinder** Gebläsezylinder *m* ~ **device** Einblasevorrichtung *f* (for charging shaft furnaces) ~-**down distributor** Ausblaseverteiler *m* (submarine) ~ **effect** Blaswirkung *f* ~ **engine** Gebläsemaschine *f* ~-**engine house** Gebläsemaschinenhaus *n* ~ **engine with slide valve** Schiebergebläse *n* ~ **engine worked by blast-furnace gas** Hochofengasgebläsemaschine *f* ~ **fan** Blasventilator *m* ~ **furnace** Blasofen *m* ~ **head** Blaskopf *m* ~ **hot** Heißblasen *n* ~-**in practice** Anblasetechnik *f* (blast furnace) ~ **iron** Glasmacherpfeife *f* ~ **magnet** Blasmagnet *m*, Flammblaser *m* ~-**off fuzz** Faserflug *m* ~-**off pressure** Abblasedruck *m* ~ **operation** Windfrischvorgang *m* ~-**out** Ausblasen *n* ~ **plant** Gebläseanlage *f* ~ **plug** blasende Kerze *f* ~ **point** Abschmelzstromstärke *f* ~ **position** Blasestellung *f* ~ **pressure** Preßluftdruck *m* ~ **process** Blaseverfahren *n* ~ **tuyère** Windform *f* ~ **valve** Gebläseventil *n* ~ **wedge** Sprengkeil *m*

**blown** blasig, geblasen, (fuse) durchgebrannt ~ **free** freigeblasen ~-**off** weggeblasen

**blown,** ~ **glass** geblasenes Glas *n* ~-**out hole** Bohrlochpfeife *f* ~-**out shot** Ausbläser *m*, Pfeife *f* (blasting) ~ **primer** Zündhütchenausfall *m* ~ **sand** Flugsand *m* ~ **steel** Frischstahl *m*

**blowy** blasig

**blue, to** ~ bläuen: **to** ~-**glow** glimmen, Glimmlicht *n* zeigen

**blue** blau ~ **of the sky** Himmelsblau *n*, blaues Himmelslicht *n* ~ **adder** Blaubeimischer *m* (TV) ~ **brittle** blaubrüchig ~ **brittleness** Blaubruch *m*, Blaubrüchigkeit *f*, Bläue *f* ~ **electron gun** Blaustrahlsystem *n* (TV) ~ **filter** Blau-filter *n*, -scheibe *f* ~ **fracture test** Blaubruchversuch *m* ~-(**water)gas** Koksgas *n* ~ **glow** Glimmlicht *n* ~ **haze** Glimmlicht *n* ~ **heat of iron** Blauwärme *f* ~-**hot** blauglühend ~ **iron earth** Blau-eisenerde *f*, -eisenerz *n*, -eisenspat *m*, Eisen-blau *n*, -blauerde *f*, -blauspat *m* ~ **lias** Blaulias *n* ~ **light** Fackelfeuer *n* ~ **light mirror** Blauspiegel *m* ~ **metal** Zwischenstein *m* (of copper) ~**oil** Blauöl *n* ~ **pencil** Blaustift *m* ~ **powder** Poussiere *f*, Schmelz *m* (metal)

**blueprint, to** ~ eine Blaupause herstellen

**blueprint** Blau-, Licht-, Plan-pause *f*, Bauplan *m* ~ **drawing** Blaudruck *m* ~ **lamp** Lichtpaus-

lampe *f* ~ **paper** Blaunegativ-, Eisenblau- *n*, Lichtpaus-papier *n*

**blue,** ~ **printer** Lichtpausgerät *n*, Blaudruckplatte *f* ~ **printing linen** Lichtpausleinen *n* ~ **record** Blauauszug *m* ~ **rot** Blaufäule *f* ~-**sensitive** blauempfindlich ~ **short** blaubrüchig ~ **shortness** Blaubruch *m*, Blaubrüchigkeit *f* ~ **sky** blauer Himmel *m* ~ **stone** Blaukugel *f*, Zwischenstein *m* ~ **tint** Blaustich *m* ~ **verditer** Kupferblau *n* ~ **vitriol** Blaustein *m*, Kupfervitriol *n* ~-**writing oscillograph** Blauschriftoszillograf *m*

**blued,** ~ **sugar** geblauter Zucker *m* ~ **surface resulting from tempering** blaue Anlaßfarbe *f*

**blueing** Blauanlaufen *n*, Bläue *f*, Bläuen *n* ~ **machine for tempering iron pieces** Blaumaschine *f* zum Anlassen von Eisenteilen

**blueness** Bläue *f*

**bluff** (Felsen)klippe *f*

**bluffing wheel** Drehsandstein *m*

**bluish** bläulich **with** ~ **haze** blaustichig ~ **tinge** Bläuen *n* **with** ~ **tint** blaustichig

**blunder, to** ~ pfuschen, verfahren

**blunder** Fehler *m*, Fehlgriff *m*, Mißgriff *m*, Schnitzer *m*

**blunderer** Stümper *m*

**blunt, to** ~ abstumpfen **to** ~ **glass** Glas *n* blind machen **to** ~ **the points** verzwicken (of nails)

**blunt** stumpf ~ **angle** verbrochene Ecke *f* ~-**cornered** stumpfeckig ~-**cornered rectangle (or square)** Kreisviereck *n* ~ **edge** angenutzte Schneide *f* ~-**edged** stumpfkantig ~ **ended** abgestumpft ~ **ended switch** Schleppwechsel *m* ~ **file** Stumpffeile *f* ~ **seam** Stumpfnaht *f* ~ **stub switch** Schleppwechsel *m* ~ **tool** stumpfes Werkzeug *n*

**blunted** abgestumpft

**blunting of threads** Abrunden *n* von Gewinden

**blur, to** ~ beflecken, verschwimmen, verwaschen, verwischen; unscharf machen, verschmieren (TV), verschwimmen (TV)

**blurred** unscharf, verschwommen ~ **appearance** verschwommenes Aussehen *n* (of image) ~ **image** verschwommenes Bild *n* ~ **voice** verschwommene Sprache *f* ~ **zone** Verwaschungszone *f*

**blurring** Verschwimmung *f*, Verwischung *f* (film); (of minimum) Trübung *f*, (in X ray) Verharzung *f*, Verschmierung *f* ~ **of precision** Funktrübung *f*

**bluring,** ~ **circle** Unschärfering *m*, Zerstreuungskreis *m*, Zerstreuungsscheibchen *n* ~ **circles of eye** Augenzerstreuungsbilder *pl*, Zerstreuungsbilder *pl* des Auges ~ **factor** Klirrfaktor *m* ~ **potential** Schwimmspannung *f*

**blush, to** ~ anlaufen (steel)

**blushing** (of lacquer) Trübung *f*, (of lacquer or varnish) Weißwerden *n*

**board, to** ~ beschalen, besteigen, entern, verschalen, verzimmern, mit Brettern verschlagen **to go on** ~ besteigen **to** ~ **the floor** den Fußboden *m* dielen **to** ~ (grain) levantieren **to** ~ **up** bedielen, verschlagen

**board** Abbaustrecke *f*, Amt *n*, Ansteckpfahl *m*, Ausschuß *m*, Beköstigung *f*, Bohle *f*, Bord *m* (ship, plane etc), Brett *n*, Pappe *f*, Planke *f*, Quader *m*, Tafel *f* **on** ~ **ship** auf Schiffen ~**s** Pappdeckel *m* ~**s suspended between two posts**

Schwebestrich *m* ~ for composed types Satzbrett *n* ~ of directors Aufsichtsrat *m* ~ of health Gesundheitsamt *n* ~ for instruments Instrumentplanchette *f* ~ for meters Zählerplatte *f* ~ of mining company Bergrat *m* ~ of the plane table Mensel-, Meßtisch-platte *f* ~ with pot eyes Ösenbrett *n* ~ for sorts Handmatrizenbrett *n*

board, ~-and-pillar system (or work) Pfeilerbau *m* ~ bell Schrankwecker *m* ~ drop hammer Brettfall-, Stangenreib-hammer *m* ~ effect Bordeffekt *m* ~-felt Pappenfilz *m* ~ fire extinguisher Bordfeuerlöscher *m* ~ glazed (or glazing) Deckelsatinage *f* (paper mfg) ~ indicator Klappenschrank *m* ~-lift drop hammers Brettfallhämmer *pl* ~ machine Rundsiebpappenmaschine *f* (paper mfg) ~ mechanic Bordmonteur *m* ~ meeting Verwaltungsratversammlung *f* ~ mold Pappenform *f* ~ partition Schalwand *f* ~ rack Brettregal *n* ~ signal Scheibensignal *n* ~ television apparatus Tableaufernseher *m* ~ walk Bretterbahn *f*, Graben-, Holz-, Latten-rost *m*

boarded abgedeckt ~ ceiling getäfelte Decke *f* ~ floor Dielung *f* ~ floor of a turntable Abdeckung *f* einer Drehscheibe

boarder lights Oberlichter *pl*

boarding Bodenbelag *m*, Bretterverkleidung *f*, Dielen *n*, Dielung *f*, Schwartenbrett *n*, Täfelung *f*, Verkleidung *f*, Verschalung *f*, Verzimmerung *f* ~ of a roof Bretterschalung *f* eines Daches

boarding, ~ fence Bretterzaun *m* ~ grapnel Enterhaken *m* ~ joist Dielenbalken *m*, Dielenlager *n*, Polsterholz *n* ~ (or crippling) machine Krispelmaschine *f* ~ ladder Aufsteigleiter *f*

boasting rohe Behausung *f*

boat Barke *f*, Boot *n*, Schiff *n* a ~ is moving astern ein Schiff fährt rückwärts ~ by ~ bootsweise ~ is alongside Schiff *n* liegt an der Kaje ~'s awning Bootsonnensegel *n* ~ going upstream Schiff *n* auf der Bergfahrt

boat, ~ boom Backspiere *f* ~builder Bootsbauer *m* ~-building workers Bootswerft *f* ~'s crew Bootsbesatzung *f* ~ deck Bootsdeck *n* ~ ferry Kahnfähre *f* ~'s grating Bootsgräting *f* ~ hoist Bootsheißmaschine *f* ~ hook Bootshaken *m* ~ slings Bootshißstropp *n*, Bootsstropp *m* ~'s tackle Bootstakel *n* ~ watch Bootsposten *m*

boatswain Bootsmann *m* ~'s chair Luftkabelfahrstuhl *m* ~'s store Bootsmannshellegatt *n*

bob Gewicht *n* des Pendels, Klöppelkugel *f*, Pendellinse *f*, Perpendikel-ende *n*, -gewicht *n*, Senkkörper *m*

bobbin Ablauf-, Auflauf-haspel *m*, Aufwinderöhre *f*, Bandseiltrommel *f*, Bobine *f*, Fadenhänger *m*, Flachseiltrommel *f*, Gurttrommel *f*, kleine Spule *f*, kleine Wicklung *f*, Spule *f*, Spulenkasten *m*, Spulenkörper *m*

bobbin, ~ core Spulenkern *m* ~ doffer Abzieherin *f* (textiles) ~ frame Spulenrahmen *m* ~ height Spulenlänge *f* ~ net Tüll *m* ~ overall diameter mittlere Spulenweite *f* ~ rail Spulwagenführung *f* ~ spring Drahtspulenfeder *f* ~ target Klappscheine *f*

bobbing Verschwindscheibe *f*, Schwankung *f*

bockform plateholder aufklappbare Kassette *f*

bode plot Bodediagramm *n*

bodenbenderite Bodenbenderit *m*

bodied oil Dicköl *n*

bodily körperlich ~ injury Körperverletzung *f* ~ waves Raumwellen *pl*

bodkin Ahle *f*, Pfriem *m*, Schnürnadel *f*

body, to ~ eindicken (drying oils)

body Apparatkörper *m*, Ballen *m*, Boot *n* (aviat.), Hauptteil *m*, Kasko *n*, Körper *m*, Leib *m*, Rumpf *m*, Rumpfwerk *n*, (ceramics) Scherben *pl*, (of converter) Mittelstück *n*, (of a gas producer) Schacht *m*, (speech book) Hauptstück *n*

body ~ of air Luftkörper *m* ~ of binoculars Fernrohrkörper *m* ~ with bushes Gehäuse *n* mit Büchsen ~ of carburetor Vergasergehäuse *n* ~ of the compass card (or rose) Rosenkörper *m* ~ of connecting rod Pleuelstangenschaft *m* ~ of a cupola Kupolofenschacht *m* ~ of frog Weichenkörper *m* (electr.) ~ of a furnace Ofenmassiv *n* ~ of fuse Satzstück *n*, Zünderkörper *m* ~ of a letter Schriftkegel *m* ~ of miners Bergknappschaft *f* ~ of paint Deckungskraft *f* ~ of a paint Farbkörper *m* ~ of the plow Pflugkörper *m* ~ of projectile Geschoßhülle *f* ~ of a pump Pumpenstiefel *m* ~ of a reverberatory furnace Arbeitsraum *m* des Flammofens, Herdraum *m* des Flammofens ~ of revolution Dreh-, Rotations-, Umdrehungs-körper *m* ~ of roll Walzballen *m*, Walzenbund *m* ~ of rotation Rotations-, Umdrehungs-körper *m* ~ of shell Geschoß-hülle *f*, -körper *m* ~ of telescope Fernrohrkörper *m* ~ of the X-ray tube Röntgenrohrkörper *m*

body, ~ angle of a cone Kegelöffnungswinkel *m* (math.) ~ apron Bleigummischürze *f* ~ axis flugzeugfeste Achse *f*, Körperachse *f* ~ board Kunstdruckkarton *m* ~ bolster Kastenwiege *f* ~ capacitance Handkapazität *f* ~-capacity effect Handkapazität *f* ~-centered kubischzentriert, raumzentriert (cryst.) ~-centered cubic lattice raumzentriertes kubisches Gitter *n* (cryst.) ~-centered cubic metals kubischraumzentrierte Metalle *pl* ~ chamber Wohnkammer *f* (geol.) ~ color Deckfarbe *f* ~ cone Ganzpolkegel *m* ~ conformity Anpassung *f* an den menschlichen Körper ~ contact Körperschluß *m* ~ corporate Körperschaft *f* ~ counterbore Halssenker *m* ~ design Aufbau *m* der Maschine ~-flanging device Zargenbördeleinrichtung *f* ~-forming machine Kastenbiegepresse *f*, Zargen-rundmaschine *f*, -biegemaschine *f* ~ fount Grundschrift *f* ~ frame Aufbaurahmen *m* ~ framework Kastengerippe *n* ~ gauge Ballenmesser *m* ~-heat Eigenwärme *f* ~ heavier than water sanker Körper *m* ~ length Manteltiefe *f* ~ panel Rumpffeld *n* ~ paper Roh-, Träger-, Grund-papier *n* (of condensers) ~ presenting rotation symmetry rotationssymmetrischer Körper *m* ~ radiator Rumpfkühler *m* ~ resistance Form-, Körper-, Rumpf-widerstand *m* ~ sand Füll-, Haufen-sand *m* ~ section radiography Röntgenschichtverfahren *n* ~ shutter release Gehäuseauslöser *m* (photo) ~ tank Rumpftank *m* ~ transverse member Rumpfquerstrebe *f* ~ tube Objektivtubus *m* ~ varnish Schleiflack *m* ~-wing interference gegenseitige Beeinflussung *f* von Tragflügel und Rumpf

**bodying** Eindickung *f* (of oil)
**Boetius furnace** Böetius-Feuerung *f*
**bog, to** ~ absacken **to** ~ **down** sich festlaufen
**bog** Moor *n*, Morast *m*, Schwindgrube *f*,
Senkgrube *f* ~-**iron ore** Eisenschörl *m*, Eisen-
sumpferz *n*, Ortstein *m*, Rasen-eisenerz *n*,
-eisenstein *m*, -erz *n*, Sumpferz *n*, Sumpfeisen-
stein *m*, Wiesenerz *n* ~ **lime** Seekreide *f* ~ **ore**
Bohnerz *n*, Morasterz *n* ~-**ore purifier** Rasen-
erzreiniger *m* ~ **peat** Moortorf *m*
**boggy** versumpft, sumpfig ~ **soil** Schlammboden
*m*
**Boghead coal** Boghead-Kohle *f*
**bogie** Drehgestell *n* (r.r.), Fahrwerk *n*, (pivoted)
Drehschemel *m* ~ **brake** Drehgestellbremse *f* ~
**crane** Rollkran *m* ~-**frame** Radgestell *n* ~
**girder** Fahrwerksträger *m* ~ **wheel** Ketten-
laufrad *n*, Laufrad *n*, Tragrolle *f* ~-**wheel**
**base** Drehgestellachsenstand *m* ~-**wheel sus-**
**pension frame** Rollenwagen *m*
**bogus** nachgemacht, schwindelhaft, unecht ~
**board** Schrenz- und Speltpappe *f*
**Bohemian (red) garnet** böhmischer Granat *m*
**Bohr atom** Bohrsches Atommodell *n*
**boil, to** ~ abkochen, ansieden, aufkochen auskochen, brodeln, entschalen (Seide), kochen,
sieden, wellen **to** ~ **away** versieden **to** ~ **blank**
blank kochen **to** ~ **blank and crystallize in**
**tanks** in Kasten verkochen **to** ~ **briskly** durchkochen **to** ~ **down** eindampfen, eindicken, einengen, einkochen, herunterfrischen, verkochen
**to** ~ **off** abkochen, auskochen, fortkochen
(chem.), verkochen, (sugar, silk, etc) fertig-
kochen **to** ~ **out** abbrühen, abdämpfen, auskochen **to** ~ **over** überkochen, übersieden **to**
~ **thoroughly** durchkochen **to** ~ **up** aufkochen,
ebullieren **to** ~ **to afterproduct** auf Nachprodukt verkochen **to** ~ **to grain** auf Korn verkochen **to** ~ (linseed oil) **upon litharge** Firnis
kochen **to** ~ **a strike of sugar** einen Sud
stramm abkochen **to** ~ **stringproof** auf Faden
verkochen (sugar mfg), auf Kasten verkochen
**boil** Frischreaktion *f* (metall.), Geschwür *n*,
Rohfrischperiode *f* ~ **adjuvant** Siedeerleich-
terer *m* ~-**away loss** Verdampfungsverlust *m*
(bei siedender Flüssigkeit) ~-**proof** kochfest
**boiled** abgekocht, gekocht, sacht (silk) ~
(linseed) **oil** Leinölfirnis *m*
**boiler** Dampfkessel *m*, Flammenrohrkessel *m*,
Heißwasserspeicher *m*, Heizkessel *m*, Kessel
*m*, Kocher *m*, Kochkessel *m*, Küpe *f*, Pfanne *f*,
Wasserkessel *m*, Warmwasserbereiter *m* ~
**with bottom end flanged outward** ausgehals-
ter Kesselboden *m* ~ **for central heating**
Zentralheizungskessel *m* ~ **that exploits**
**energy contained in waste gases** Abhitzkessel
*m* ~ **with vertical large tubes** weitrohriger
Steilrohrkessel *m*
**boiler,** ~ **acceptance** Kesselabnahme *f* ~
**accessories** Kesselzubehör *n* ~ **anti-incrustant**
**composition** Kesselsteinverhütungsmittel *n* ~-
**anti-scaling composition** Kesselsteinverhütungsmittel *n* ~ **auger** Kesselbohrer *m* ~ **barrel** Langkessel *m* ~ **blow-down set** Kesselentlaugungs-
einrichtung *f* ~ **bottom** Kesselboden *m* ~
**brace** Verankerung *f* eines Kessels ~ **casing**
Dampfkessel-, Kessel-bekleidung *f*, Kesselmantel *m* ~-**cleansing compound** Kesselstein-

lösemittel *n* ~ **coal** Kesselkohle *f* ~-**coaling**
**plant** Kesselbekohlungsanlage *f* ~ **control panel**
Kesseltafel *f* ~ **drum** Kesseltrommel *f* ~
**efficiency** Kesselwirkungsgrad *m* ~ **end** Kesselboden *m* ~ **end flanged inward** eingehalster
Boden *m* ~ **envelope** Kesselmantel *m* ~
**factory** Kesselfabrik *f* ~-**feed** (to) kesselspei-
sen; Kesselspeisung *f* ~-**feed pump** Dampfkes-
selspeisepumpe *f* ~-**feed water** Kesselspeisewasser *n* ~-**feed-water meter** Kesselspeisewas-
sermesser *m* ~ **feeder** Kesselspeisepumpe *f* ~
**firing** Dampfkessel-, Kessel-feuerung *f* ~
**fittings** Kessel-armatur *f*, -ausrüstung *f* ~ **flue**
Fuchskanal *m*, Kesselflammrohr *n* ~ **flue gas**
Kesselgas *n* ~ **flue pass** Kesselzug *m* ~ **fuel**
Kesselgut *n* ~ **furnace** Kesselfeuerung *f* ~ **gas**
Rauchgas *n* ~ **grate** Kesselrost *m* ~ **head** Kesselboden *m* ~ **heads** Behälterböden *pl* ~-
**heating surface** Kesselheizfläche *f* ~ **holder**
Dampfkesselfüße *pl* ~**house** Kesselhaus *n*
~**house operation** Kesselhausbetrieb *m* ~
**horse-power** Kesselpferdestärke *f* ~ **jacket** Kes-
selverkleidung *f* ~ **level** Kesselsohle *f* ~ **maker**
Kesselbauingenieur *m*, Kesselschmied *m* ~
**operation** Kesselbetrieb *m* ~ **pipe** Siederohr *n*
~ **plant** Dampfkessel-, Kessel-anlage *f* ~ **plate**
Dampfkessel-, Grob-, Kessel-blech *n*, Kes-
selplatte *f* ~ **pressure** Kessel-druck *m*, -span-
nung *f* ~ **priming** Kesselschäumen *n* ~ **putty**
Schwarzkitt *m* ~ **room** Kesselraum *m* ~ **scale**
Kesselniederschlag *m* ~-**scale deposit** Kesselsteinablagerung *f* ~-**scale formation** Kesselsteinbildung *f* ~ **scales** Pfannenstein *m* ~-**scal-**
**ing hammer** Kesselhammer *m* ~-**scaling tool**
Kesselabklopfwerkzeug *n* ~ **setting** Kessel-
lagerung *f* ~ **shell** Kesselwand *f*, Langkessel *m*
~ **shop** Kessel-bauanstalt *f*, -bauwerkstatt *f* ~
**sludge** Kessellauge *f* ~ **and smoke tube working**
**machine** Siede- und Rauchrohrbearbeitungs-
maschine *f* ~-**supervising apparatus** Kessel-
überwachungsapparat *m* ~ **support** Kesselstuhl
*m* ~ **switchboard** Kesselwarte *f* ~ **tap** Kesselge-
windebohrer *m* ~ **test** Kesselprüfung *f* ~
**tester** Kesselprobierpumpe *f* ~ **tube** Dampfkes-
selsiederohr *n*, Kesselrohr *n* ~ **type** Kessel-
bauart *f* ~ **wall** Kesselmantel *m* ~ **waste heat**
Kesselabwärme *f* ~-**water feed pump** Kessel-
speisewasserpumpe *f* ~-**water treatment** Kes-
selwasserreinigung *f*
**boiling** Aufwallen *n*, Kochen *n*, Wallung *f*, (of
a melting bath) Schäumen *n*, (noise) Brodeln
*n*; kochend ~-**out** Abbrühen *n*, Abdämpfen *n*
~ **over** Übersieden *n*, (foam) Überschäumung
*f* ~ **of afterproduct** Nachproduktverkochung *f*
~ **to string proof** Fadenkochen *n* ~**s** Schlacke
*f*, vor dem Kochen abgestochene Puddel-
schlacke *f*
**boiling,** ~ **agent** Kochlösung *f* ~ **bath** Sudbad *n*
~ **burner** Kronenbrenner *m* (chem.) ~ **cooling**
Siedekühlung *f* ~ **course** Siedeverlauf *m* ~ **and**
**crabbing machine** Brennbock *m* ~ **curve** Siedeverlauf *m* ~ (-point) **curve** Siedekurve *f* ~
**drum** Schlammkasten *m* ~ **flask** Koch-,
Siede-kolben *m* ~ **noise** kochendes Geräusch *n*
~ **period** Kochperiode *f* ~ **point** Kochpunkt *m*,
Siedebeginn *m*, Siedepunkt *m* ~-**point barome-**
**ter** Hypsothermometer *n* ~-**point curve** Siede-
linie *f* ~-**point thermometer** Siedethermometer

n, Siedewärmegradmesser m ~ **process** Versud m ~ **progress** Siedeverlauf m ~ **range** Siedegrenze f ~ **rate** Siedeverlauf m ~ **stage** Rohfrischperiode f ~ **temperature** Siedegrenze f ~ **test** Kochprobe f ~ **tub** Schlammkasten m ~**-tube caulking machine** Siederohrdichtmaschine f ~ **vessel** Aufkochgefäß n

**bold** auffällig, dreist, hervortretend, kühn, verwegen **in ~-face type** fett gedruckt **~-face type** Fettdruck m

**boll** die Gewindebacke m

**bollard** Böller m, Poller m, Schiffshalter m, Schiffspfahl m

**bollards** Dalben pl, Dallen m, Duckdalben pl

**bolograph** Bolometerdiagramm n

**bolometer** Bolometer n ~ **detection** Bolometer-Strahlungsmessung f ~ **mount** Bolometerabschluß m

**bolometric(al)** bolometrisch

**bolster** Matrizenrahmen m, Schwartenbrett n, Wiege f (Fahrzeug), (textiles) Achsenträger m ~ **plate** Aufspann-, Frosch-platte f

**bolt, to** ~ anbolzen, beuteln, durchgehen, durchwerfen, festschrauben, schließen, verbolzen, verriegeln, verschrauben, vorstrecken, (breechblock) abriegeln, (rolling-mill work) vorwalzen, (watchworks) schnell ablaufen **to** ~ **to** anschrauben **to** ~ **together** zusammenschrauben

**bolt** Bolzen m, Bolzen m mit Vorlegescheibe und Mutter, Dorn m, Falle f, Gewehrschloß n, Mutterschraube f, Riegel m, Schar f, Schloß n, Schraube f, Schraube f mit Mutter, Wirbel m, (screw) Schraubenbolzen m ~ **for anchorage in masonry** Steinschraube f ~ **of cross-tube** Hilfsachsenbolzen m ~ **of flange** Flanschschraube f ~ **with head** Kopfschraube f ~ **of a hinge** Dorn m eines Scharnierbandes ~ **for lever** Hebelbolzen m ~ **and nut articles** Bolzenmaterial n ~ **with thread** Gewindebolzen m ~ **upright** kerzengerade

**bolt, ~-aligning device** Riemenfluchtgerät n ~ **base** Schloßfuß m ~ **beater** Bolzenschläger m ~ **body** Verschlußstück n ~ **catch** Kammerfang m **~-catch piece** Kammerfangstück n ~ **chisel** Kreuzmeißel m ~ **circle** Schraubenlochkreis m **~(hole)circle** Lochkreis m ~ **clipper** Bolzenabschneider m **~-crownbeater** Bolzenkronenschläger m ~ **cutter** Schraubenbolzenabschneider m, Schraubenschneidemaschine f ~ **deduction** Bolzenschwächung f ~ **diameter** Bolzendurchmesser m ~ **die** Preßbacke f, Schneidbacken m ~ **guide** Patronenführungsleiste f ~ **handle** Kammergriff m, Kammerstengel m, Schloß-, Verschluß-hebel m ~**head** Schraubenkopf m ~ **header** Nageleisen n **~-hole circle** Schraubenlochkreis m ~ **holes in boss** Nabenbolzenbohrung f ~ **hook** Hakenschraube f ~ **housing** Schloßgehäuse n ~ **keeper** Zuhaltung f ~ **knob** Kammerknopf m ~ **lever** Spannhebel m ~ **link pinion** Kettenkurbel f ~ **lug** Kammerwarze f ~**-milling machine** Bolzenschaftfräsemaschine f ~ **nab** Schließhaken m ~ **nut** Bolzenmutter f ~ **and nut** Mutterschraube f ~ **pin** Schließe f ~**-pitch circle** Schraubenlochkreis m ~ **position** Abriegelungsfront f ~ **screwing** Bolzenverschraubung f ~ **(pipe) screwing machine** Gewindeschneidemaschine f ~ **slide** Kammerbahn f

~ **slot** Aufspannschlitz m ~ **spring** Zugfeder f ~ **staple** Schließhaken m ~ **stock** Schrauben--eisen n, -material n ~ **thread cutting machine** Bolzengewindeschneidemaschine f ~ **valve** Bolzenventil n ~ **washer** Bolzenscheibe f ~ **wire** Schraubendraht m ~ **wrench** Bolzenschlüssel m

**bolted** verschraubt ~ **connection** Schraubverbindung f, geschraubte Verbindung f (Stahlbau) ~ **flanges** Flanschverschraubung f ~ **joint** Schraubenverbindung f, Verschraubung f ~ **manhole cover** mit Bolzen befestigter Mannlochdeckel m ~ **pipe joint** Rohrverschraubung f ~ **steel chain** Stahlbolzenkette f ~ **tenon** vernagelter Zapfen m ~ **union** Bolzenverbindung f

**bolter** Beutelsieb n, Mühlbeutel m, Schoßrübe f, Sieb n, Stockrübe f, Verbolzer m

**bolting** Bolzenverschraubung f, Sieben n, Siebung f, Verbolzung f, Verschraubung f, Vorstrecken n, schnelles Ablaufen n (eines Uhrwerkes) ~ **cloth** Müllergaze f, Siebtuch n ~ **machine** Sichtmaschine f ~ **mill** Beutelmaschine f ~**-resistant beet** schosserwiderstandsfähige Rübe f ~ **roll** Vorsteckwalze f ~ **sieve** Beutelsieb n ~ **silk** Beutelgaze f

**boltless** nietlos

**bomb, to** ~ mit Bomben belegen

**bomb** Bombe f ~ **bay** Bombenschacht m ~**-bay doors** Bombenklappe f ~**-bay release gear** Schachtauslöser m ~ **body** Bombenhülle f (aerial bomb) ~ **calorimeter** Explosionskalorimeter n, kalorimetrische Bombe f ~ **carrier** Bombengehänge n ~ **cell** Bombenmagazin n, Lastenraum m ~ **chute** Bombenwurfschacht m, Schachtzelle f ~**-chute trough** Bombenschüttkasten m ~**-control mechanism** Bombenauslösungsvorrichtung f ~ **container** Bombenmagazin n ~ **detonator** Bombenzünder m ~ **doors** Bombenklappe f ~ **dropping** Bombenwurf m ~ **fuse** Bombenzünder m ~ **head** Bombenkopf m ~ **hit** Bomben-einschlag m, -treffer m ~**-nose** Bombenkopf m ~**-proof** bomben-fest, -sicher, granatsicher ~**-proof shelter** bombensichere Deckung f ~ **rack** Abwurfschacht m, Aufhängevorrichtung f für Bomben, Bombenabwurfeinrichtung f, Bombenanhangvorrichtung, Bombenaufhängevorrichtung f, Bombengehänge n, Bombenlagerungsgestell n (aviat.) Bombenmagazin n (aviat.) ~ **release** Bombenauslösung f, Bombenlösevorrichtung f (aviat.)

**bomb-release, ~ control** Abwurfschalter m, Auslösevorrichtung f ~ **gear** Bombenabwurfvorrichtung, Bombenabzugvorrichtung f ~ **handle** Bombenhebel m ~ **lever** Bomben-abzughebel m, -wurfhebel m, Wurfhebel m ~ **mechanism** Abwurf-gerät n, -vorrichtung f, Bombenauslösungsvorrichtung f, Bombenabwurfgerät n (aviat.) ~ **moment** Auslösezeitpunkt m ~ **point** Auslösepunkt m ~ **slip** Bombenschloß n ~ **switch** Bombenknopf m ~ **unit** Abwurfgerät n

**bomb, ~resistant** bombenfest ~ **sight** Abwurfsehrohr n, Bombenabwurfzielgerät n, Bomben--richtgerät n, -zielapparat m, -visier n, Zielfernrohr n ~ **splinter** Bombensplitter m ~ **stowage** Bombenhalterung f ~ **suspension** Bombenaufhängung f ~ **trajectory** Bomben-fallkurve f, -flugbahn f, Wurfbahn f, Wurfparabel f

**bombard, to** ~ beschießen, bombardieren, (with electrons, etc) beschießen

**bombarded particle** getroffenes Teilchen *n*
**bombarding particle** Geschoßteilchen *n*
**bombardment** Aufprall *m*, Beschießung *f*, Beschuß *m*, Bombardement *n*, Bombardierung *f*, Prall *m* ~ **by ions** Ionenprall *m* ~ **force of electrons** Elektronenwucht *f* ~**-group wedge** Gruppenkeil *m* ~ **test** Beschußprobe *f* ~ **velocity** Einfallgeschwindigkeit *f*
**bomber** Bombenflugzeug *n*
**bombing** Bombardierung *f*, Bombenwurf *m* ~ **angle** Vorhaltewinkel *m* ~ **calculator** Abwurfrechengerät *n* ~ **machine** Bombardierungsflugzeug *n* ~ **plane** Bombenflugzeug *n* ~ **range** Wurfweite *f* ~ **target** Bombenziel *n*
**bonanze ore body** reicher Erzkörper *m*
**bond, to** ~ abbinden, binden, durchtränken, haften, metallisieren (electr.), verlaschen **to** ~ **together** zusammenkitten
**bond** Aneinanderlagerung *f*, Band *n*, Bindemittel *n*, Binder *m*, Bindung *f*, Bindungsmittel *n*, Dichtung *f*, Garantieschein *m*, Kaution *f*, Lasche *f*, Metalleinsatz *m* (Bremse); Revers *m*, Schuldschein *m*; Verbindung *f*, Verbindungsstelle *f*, Verbindungsstück *n*, Verkittung *f*, (masonry) Verband *m* **in** ~ unter Zollverschluß *m* ~ **of an armature** Spannband *n*
**bond,** ~ **direction** Valenzrichtung *f* ~ **energy** Bildungsenergie *f* ~ **moment** Bindungsmoment *n* ~ **orbital** Bindungsbahn *f* ~ **paper** Feinpapier *n* ~ **strength** Haftvermögen *n* (Beton an Eisen) ~ **stress** Haftspannung *f* ~ **type** Bindungsart *f*
**bondage** Kette *f*
**bonded** aufgeklebt ~ **masonry** verbundenes Mauerwerk *n* ~ **plants** Betriebsanstalten *pl* (Zoll) ~ **seal** Verbundstoffdichtung *f* ~ **strain gauge** aufgeklebter Widerstandsdehnungsmeßstreifen *m* ~ **warehouse** Transitlager *n*, Zolllagerhaus *n*
**Bonder** Binder *m* (registered trade mark)
**bonderize, to** ~ bondern
**bonderized** gebondert, mit Bonder-Chemikalien *pl* behandelt ~ **steel** gebonderter Stahl *m*
**bonderizing** Bondern *n* ~ **machine** Bondermaschine *f*
**bonding** Abbinden *n*, Masseverbindung *f*, Verkettung *f*, Verkittung *f* ~ **with grinding wheels** Bindung *f* bei Schleifscheiben
**bonding,** ~ **agent** Bindemittel *n*, Bindungsmittel *n* ~ **capacity** Bindevermögen *n* ~**-capacity property (or quality) strength** Bindefähigkeit *f* ~ **clay** Bindeton *m* ~ **compound** Klebemasse *f* ~ **electron** Valenzelektron *n* ~ **jumper** Masseverbinder *m*, Kurzschlußbügel *m* ~ **lead** Masseverbinder *m* ~ **lug** Masse-Anschlußlasche *f* ~ **material** Imprägnierungsmittel *n* ~ **orbital** gemeinsame Elektronenbahn *f* ~ **power** Bindefähigkeit *f* ~ **properties** Verbundwirkung *f* ~ **ribbon** Bleimantelverbinder *m* (teleph.) ~ **strength** Bindungsstärke *f*, Leimfestigkeit *f*, Verbindungskraft *f* ~ **strip** Bleimantelverbinder *m* ~ **tie** Bindestab *m*
**bonds** Sicherheitsleistung *f*
**bone, to** ~ klöppeln
**bone** Bein *n*, Knochen *m*, (made of) knöchern ~ **ash** Knochenasche *f* ~ **black** Beinschwarz *n*, Knochen-kohle *f*, -schwarz *n*, Spodium *n* ~ **charcoal** Beinschwarz *n*, Spodium *n* ~ **coal** Knochenkohle *f* ~**-coal steaming apparatus**

Knochenkohledämpfer *m* ~**-conduction receiver** Knochenleitungshörer *m* ~ **drill** Trepanatorium *n* (med.) ~**-dry** staubtrocken ~ **dust** Knochenmehl *m* ~ **fracture** Knochenbruch *m* ~ **glass** Milchglas *n* ~ **glue** Knochenleim *m* ~**-grease-extracting apparatus** Knochenentfettungsvorrichtung *f* ~ **key** Beintaste *f* ~ **lace** geklöppelte Kante *f* ~ **meal** Knochenmehl *m* ~ **oil** Knochenöl *n* ~ **saw** Beinsäge *f*
**boning** Nivellieren *n* ~ **rod** Visiertafel *f*
**bonnet** Deckel *m*, Kappe *f*, Haube *f*, Schutzkorb *m* ~ **fastener** Motorhaubenhalter *m* ~ **latch** Haubenverschluß *m*
**bonus** Bearbeitungszugabe *f*, Bearbeitungszuschlag *m*, Gehaltszulage *f* ~ **in kind** Naturalbonus *m*
**bony** knochig
**book, to** ~ anmelden, anschreiben, buchen **to** ~ **a toll call** ein Gespräch *n* anmelden
**book** Buch *n* **by** ~ **post** unter Kreuzband *n* ~ **in boards** Pappband *m* ~ **of patterns** Musterbuch *n* ~ **of samples** Musterbuch *n*
**book,** ~**-back bending device** Rückenbrechvorrichtung *f* ~ **backing** Einbandrücken *m* ~**binding** Buchbinderei *f* ~ **capacitor** Klappkondensator *m* ~**case** Bücher-gestell *n*, -schrank *m* ~ **clamp** Hobelzwinge *f* ~**keeper** Buch-führer *m*, -halter *m* ~**keeping transfer** Umbuchung *f* ~ **lining paper** Vorsatzpapier *n* ~**mark** Lesezeichen *n* ~ **paper** Werkdruckpapier *n* ~**-printing establishment** Werkdruckerei *f* ~**seller** Buchhändler *m* ~**shelf** Bücherbrett *n*, Büchergestell *n* ~**shop** Buchhandlung *f* ~ **stitching** Buchheften *n* ~ **trimming machine** Buchbeschneidemaschine *f*
**booked call** Daueranmeldung *f*
**booking** Anmeldung *f*, Aufzeichnung *f* der Gesprächsanmeldungen, Verbuchung *f* ~ **of a call** Gesprächsanmelden *n*
**booklet** Broschüre *f* ~ **stitched sideways** seitlich geheftete Broschüre *f*
**boom, to** ~ brausen, dröhnen
**boom** Backspiere *f*, geschäftlicher Aufschwung *m*, Hausse *f* (Börse), Hochkonjunktur *f*; Ladebaum *m*, Spiere *f*; Tankerrohr *n* (Luftbetankung), Leitwerkträger *m*, (crane) Ausleger *m* ~ **of a harbor** Hafen-baum *m*, -schlengel *m* ~ **of a truss** Gurtung *f* eines Brückenträgers
**boom,** ~ **effect** Kellerton *m* ~ **elevator** Auslegerhebewerk *n* ~ **nozzle** Tankerrohrmundstück *n* ~ **stand** Auslegerstativ *n* ~ **swing** Schwenkbereich *m* (Bagger)
**booming** Dröhnen *n* ~ **start** Anlaßknall *m*
**boomy** hohl ~ **reproduction** hohle Wiedergabe *f* ~ **sound** hohler Ton *m*
**boost, to** ~ aufladen, verstärken (battery)
**boost** Anhebung *f*, Erhöhung *f*, Lader *m*, Ladedruck *m*, Steigerung *f* ~ **compressor** Zusatzverdichter *m* ~ **control** Ladedruckregler *m* **threestage automatic** ~ **control** selbsttätige, dreistufige Ladedruckregelung *f* ~ **device** Aufladevorrichtung *f* ~ **gauge** Ladedruckmesser *m* ~ **intercooling** Ladeluftkühlung *f* ~ **pickoff** Ladedruckanzapfung *f* ~ **pressure** Ladedruck *m* ~**-pressure control** Ladedruckregelung *f* ~ **pressure gauge** Ladedruckmesser *m* ~ **pump** Förderpumpe *f* ~ **ratio** Ladedruck *m* ~ **regulator** Ladedruckregler *m* ~ **reversal** Ladedruck-

umkehr f ~ **temperature** Ladelufttemperatur f
**booster** Aufladegebläse n, Hilfsmotor m, Nachschaltverdichter m, Puffersatz m (electr.),
Servomotor m, Startvorsatz m, Übertragungsladung f (Zünder), Zündverstärker m, Zusatz-
dynamo m, Zusatzgleichrichter m, Zusatzmaschine f, (battery) Spannungserhöher m
~ **to reduce volts** Gegenschaltmaschine f
**booster, ~ action of intake manifold** druckstau-
ende Wirkung f (Druckstauwirkung) der Ein-
strömleitung ~ **aggregate** Zusatzaggregat n
(electr.) ~ **amplifier** Zusatzverstärker m, NF-
Zwischenverstärker m ~ **battery** Zusatzbatterie
f ~-**battery metering** Gesprächszählung f durch
Zuschalten einer Zählspannung, Zählung f
durch Spannungserhöhung ~ **brake** Vakuum-
Servo-Bremse f ~ **charge** Anfeuerungssatz m,
Eingangszündung f, Initiale f, Sprengkapsel f
~ **coil** Anlaßzünd-, Summeranlaß-spule f ~
**diode** Schaltdiode f ~ **drive** Ladergetriebe n ~
**fan** Sonderventilator m ~ **foil** Verstärkerfolie f
~ **ignition** Summeranlaßzündung f ~ **light**
Aufheller m (film) ~ **magneto** Anlaßmagnet-
zünder m ~ **motor** Startraketenmotor m ~ **oil**
**pump** Hilfsöl-, Öldruckverstärker-pumpe f ~
**potential** Zusatzspannung f ~-**power supply**
Impulszusatz m ~ **pump** Förder-, Drucksteige-
rungs-, Treibdampf-, Überdruck-pumpe f ~
**pump fuel discharge line** Förderpumpenaus-
gangsleitung f ~ **pump fuel pressure** Förder-
pumpendruck m ~ **radiator** Zusatz-, Hilfs-küh-
ler m ~ **relay** Hilfsrelais n ~ **rocket** Startrakete f
**set** Zusatz m ~ **switch** Zusatzschalter m ~
**transformer** Saugtransformator m ~-**type oil**
**diffusion pump** Öltreibdampfpumpe f ~ **voltage**
Treiberspannung f
**boosting** zusätzlich ~ **battery** Pufferbatterie f ~
**charge** Teil-, Zwischen-ladung f ~ **device** An-
schubvorrichtung f ~ **transformer** Zusatztransformator m ~ **voltage** Zusatzspannung f
**boot** Schuh m, Stiefel m, Werkzeugverschluß m
~ **jack** Klinkenfanggabel f ~**leg** Kniestück n
**booth** abgetrennter Verschlag m, Verschlag m,
(at exposition) Stand m
**bootstrap** Bandbefehl m ~ **circuit** Bootstrap-
Schaltung f
**boracite** Borazit m, Boraxspat m
**boral** Boral n ~ **shield** Boralschirm m
**boborate** Borat n; borsauer ~ **of soda** Borax m
**borax** Borax m, Natriumtetraborat n ~ **box**
Lötbüchse f
**bordage** Bohlenbelag m
**border, to** ~ besäumen, bördeln, börteln, einfassen, krämpen, naheliegen, rändeln, rändern, säumen, umbördeln **to** ~ **on** angrenzen
**to** ~ **upon** streifen
**border** Bord m, Einfassung f, Gradstein m,
Kante f, Kimme f, Leiste f, Markscheide f,
Rabatte f, Rand m, Randverzierung f, Saum m,
Umrandung f, Zarge f, ~ **(case)** Kastenrand m
~ **(edging)** Bordüre f ~**(-line)** Grenzfall m ~
**of the compass card (or rose)** Rosenrand m ~ **of**
**a door panel** Fries-, Füllungs-glieder pl ~ **of the**
**fingernail** Fingernagellimbus m ~ **of map**
Kartenrand m
**border, ~ fortification** Grenzbefestigung f ~-
**line between dark and light areas** Helldunkel-
grenze f ~-**line cases** Grenzfälle f ~-**line ray**

Grenzstrahl m ~ **matrix** Ornamentmatrize f
~ **region** Grenzgebiet n ~ **rule** Einfassungslinie
f ~ **rules of blocks** Klischeerandlinien pl ~
**strip** Umrandung f ~ **tile** Ortziegel m
**bordering** Ränderung f ~ **on** angrenzend (math.)
~ **machine** Bördel-, Rändel-maschine f ~
**tool** Bördeleisen f
**Bordoni regulating transformer** Bühnenlicht-
Wechselstromsteller m
**bore, to** ~ abbohren, absinken, abteufen, anbohren, ausdrehen, bohren, durchbohren,
schlagen, teufen, (re)~ aufbohren **to** ~ **out**
ausbohren **to** ~ **out the stay bolts** die Stehbolzen pl abbohren
**bore** Ausbohrung f, Ausbrennung f, Ausdrehung
f, Bohrung f, Bohrloch n, Hub m, innerer
Zylinderdurchmesser m, Laufinneres n, lichte
Weite f, Rohrseele f, Rohrwand f, (cylinder)
Bohrung f, (gun) Seele f, (of gun) Laufseele f,
(phys. geog.) Seebeben n ~ **of boss** Naben-
bohrung f ~ **of cylinder** Hub m des Zylinders
~ **of the gun** Rohrseele f ~ **of a gun barrel**
Seele f eines Geschützrohres ~ **of hub** Naben-
bohrung f ~ **of spindle** Spindelbohrung f
**bore, ~ bit** Bohrmeißel m ~ **brush** Rohrbürste f
~ **contact tip** Meßhütchen n mit Kegelfläche ~
**frame** Bohr-gerüst n, -gestell n ~ **gauge** Kaliber-
ring m, Kaliberzylinder m, Laufbohrungsmes-
ser m ~ **grinding spindle** Aufbohrspindel f
**borehole** Bohrloch n, Bohrung f ~ **lining** Bohrrohr n ~ **logging method** Bohrlochuntersu-
chungsmethode f ~ **made from galleries** innere
Bohrung f ~ **pump** Abteuf-, Bohrloch-pumpe f,
Ventilbohrer m ~ **spacing** Bohrlochdistanz f ~
**tube** Bohrrohr n
**bore, ~-inspection telescope** Innensehrohr n ~
**lever** Bohrhebel m ~ **rods** Bohrgestänge n
~-**running fit** Kulissen-, Lauf-bohrung f
~-**safe** rohrsicher ~-**safe fuse** rohrsicherer
Zünder m, Rohrsicherheit f ~ **searcher** Rohr-
wandbeobachtungsgerät n ~ **sight** Justiergerät
n ~ **shaper cutter** Scheibenformfräser m ~
**sight line** Justierlinie f ~ **sighting** Justierung f
~ **slide fit** Gleit-, Rutsch-bohrung f ~ **surface**
Seelenwand f
**bored** durchgebohrt ~ **pile** Bohrpfahl m ~ **well**
Bohrbrunnen m
**borer** Bohrer m, Bohrmaschine f, Holzwurm m
~ **with circular bit** Kreisausheber m
**boric, ~ acid** Borax-, Bor-säure f ~ **acid oint-**
**ment** Borsalbe f ~ **anhydride** Borsäureanhydrid
n ~ **oxide** Bortrioxyd n
**borine** Bor n
**boring** Abbohrung f, Anbohrung f, Aufbohren
n, Ausbohrarbeit f, Bohren n, Bohrloch n,
Bohrung f, Lochung f, Span m, (cored holes)
Bohren n ~s Bohrklein n, Bohrmehl n, Bohrspäne pl, Drehspäne pl, Gußspäne f
**boring, ~ apparatus** Bohr-anlage f, -apparat m
~ **bar** Ausbohrspindel f, Bohr-, Loch-stange f
~ **bar receptacle** Bohrstangenaufnahme f ~
**bar top slide** Bohrstangen-Obersupport m ~
**bit** Bohrerschneide f ~ **blade** Bohrstichel m ~
**block** Bohrscheibe f, Messerkopf m ~ **brace**
Bohrkrücke f, Bohrwinder f, Brustleier f ~
**clamp** Bohrbündel n ~ **cock** Anbohrhahn m ~
**contractor** Bohrunternehmer m ~ **cutter** Bohrstichel m ~ **dust** Bohrmehl n ~ **frame** Bohrbank

*f* ~ **head** Bohrkopf *m*, Bohrreitstock *m* ~ **head pilot bar** Führungsstange *f* zum Bohrkopf ~ **head shank** Bohrkopfschaft *m* ~ **hose** Bohrschlauch *m* ~ **jig** Bohrvorrichtung *f* ~ **journal** Bohrbericht *m*, Bohrrapport *m*, Bohrtagesbericht *m* ~ **kernel** Bohrkern *m* ~ **lathe** Ausbohrdrehbank *f* ~ **log** Bohrprotokoll *n*, Schichtenverzeichnis *n* ~ **machine** Ausbohr-, Bohr-maschine *f* ~ **mill** Bohrer *m*, Bohrmaschine *f*, Bohrwerk *n* ~ **mud** Bohrtrübe *f* ~ **pipe box** Anbohrrohrschelle *f* ~ **plant** Bohr-anlage *f*, -apparat *m* ~ **platform** Bohrtafel *f* ~ **quill** Bohrpinole *f* ~ **rake** Bohrrechen *m* ~ **rig** Anbohrapparat *m* ~ **rod** Bohr-kolben *m*, -stock *m* ~ **saddle** Bohr-bock *m*, -kern *m* ~ **silt** Bohrschmand *m* ~ **slide** Bohrschlitten *m* ~ **sludge** Bohrtrübe *f* ~ **socket** Bohrfutter *m* ~ **sounding** Peilen *n* (by pole) ~ **template** Bohrlehre *f* ~ **test** Bohrversuch *m* ~ **tool** Anbohrwerkzeug *n*, Ausbohrstahl *m* (inside), Ausdrehstahl *m*, Bohrstahl *m*, Bohrwerkzeug *n* ~**-tool holder** Bohrstahl-, Bohrstangen-halter *m* ~**-tool snout** Bohrstangenhalter *m* ~**-tool support** Bohrgerätefuß *m* ~ **tools** Bohrgestänge *n* ~ **tower** Bohrturm *m* ~ **trestle** Bohrgerüst *n* ~ **tripod** Dreibein *n* ~ **tube** Bohrröhre *f* ~ **and turning mill** Bohr- und Drehwerk *n* ~ **valve** Anbohrschieber *m*

**bornite** Bornit *m*
**boro, ~benzoic acid** Borbenzoesäure *f* ~**-bromide** Borbromid *n* ~**-butane** Borbutan *n* ~**calcite** Borokalzit *n* ~**ethane** Boräthan *n*
**boron** Bor *n*, Borium *n* ~ **carbide** Borkarbid *n* ~ **counter tube** Bortrifluorid-Zählrohr *n* ~ **fluorid** Boriumfluorid *n* ~ **nitride** Borstickstoff *m* ~ **steel** Borstahl *m* ~ **sulfide** Borsulfid *n*
**borotungstic acid** Borwolframsäure *f*
**borough** Stadtteil *m*
**borrow, to** ~ aufnehmen, borgen, entlehnen, leihen
**borrow, ~ area** Entnahmestelle *f* ~ **material** Aushubmaterial *n* ~ **pit** Entnahmestelle *f* ~ **pit cross section** Untergrundquerschnitt
**borrower** Entleiher *m*
**bort** Bort *m*
**boryl potassium tartrate** Boraxweinstein *m*
**bosh** Rast *f* (of a furnace) ~ **angle** Rastwinkel *m* ~ **band** Rastankerring *m* ~ **brickwork** Rastmauerwerk *n* ~ **casing** Rastpanzer *m* ~ **cooling box** Rastkühlkasten *m* ~ **cooling plate** Rastkühlkasten *m* ~ **plate** Rastplatte *f* ~**-plate box** Rastkühlkasten *m*
**boson** Boson *n*
**boss** Anguß *m*, Auge *n*, Gesenke *n*, Gewindeauge *n*, Nabe *f*; Vorgesetzter *m*; Vorsprung *m*, Warze *f* ~ **of drum** Trommelnabe *f* ~ **for treadle holder** Fußtritthalter *m*
**boss, ~ buffer** Anschlag *m* (mech.) ~ **hammer** Boßhammer *m* ~ **limitation** Anschlag *m*
**bossed cam wheel** Verteilerzahnrad *n*
**bossing** Nabensenken *n*
**botch, to** ~ verpfuschen
**both, to** ~ **sides** doppelseitig **on** ~ **sides** beidseitig
**both, ~ halves of the girder** beide Riegelhälften *pl* ~ **posts** beide Stiele ~ **sides of** beiderseits ~ **sloped girders** beide Schrägstäbe *pl* ~**way** doppeltgerichtet ~**way junction** doppeltgerichtete Verbindungsleitung *f*

**botryoidal** traubig, traubenförmig ~ **blende** Schalenblende *f*
**bott** Lehmstopfen *m*
**bottle, to** ~ auf Flaschen füllen, abfüllen in Flaschen, sich einschnüren (tensile test), zusammenschnüren
**bottle** Flasche *f* ~**-bottom molder** Flaschenbodenformer *m* ~ **breakage** Füllrohr *n* ~ **capsule** Flaschenkapsel *f* ~**-casting machine** Flaschengießmaschine *f* ~**-cleaning machine** Flaschenreinigungsmaschine *f* ~ **closures** Verschlußhütchen *n* ~ **collar** Wulst *m* der Flaschenmündung ~ **contents** Flascheninhalt *m* ~ **conveyor** Flaschenzugtransporteur *m* ~**-corking machine** Flaschenkorkmaschine *f* ~**-draining truck** Auslaufgestell *n* für Flaschen, Flaschenabtropf-, Flaschenauslauf-gestell *n* ~ **emptier** Flaschenausgießer *m* ~ **feed** Zuführung *f* der Flaschen ~ **filling machine** Flaschenfüllmaschine *f* ~ **gourd** Flaschenkürbis *m* ~ **guiding cam** Flaschenführungskurve *f* ~ **jack (lifting)** Flaschenwinde *f* ~ **kiln** Kuppelofen *m* ~ **lift** Flaschenhub *m* ~**-molding press** Flaschenpreßmaschine *f*
**bottleneck** Engpaß *m*, Flaschenhals *m*, Gleishals *m* ~ **drawing end** Rohrangel *f* ~ **ingot mold** Flaschenhaltergurt *m*
**bottle, ~nose whale blubber** Döglingstran *m* ~**-ring maker** Flaschenmundstückformer *m* ~ **shutter** Flaschenverschluß *m* ~**-soaking apparatus** Flascheneinweichapparat *m* ~ **stopper** Flaschenverschluß *m* ~ **stopper with lever** Flaschenhebelverschluß *m* ~ **valve** Füllstutzflaschenventil *n* ~ **wrapper** Flaschenhülse *f*
**bottled, ~ gas** Gas *n* in Stahlflaschen ~ **motor-fuel gas** Speichergas *n* (e. g. propane, butane)
**bottling** Abziehen *n*, Querzusammenziehung *f*, Zusammenschnürung *f* ~ **apparatus** Abziehapparat *m* (brewing) ~ **plant** Abfüllanlage *f* (brewing) ~ **room** Flaschenabfüllraum *m*
**bottom** Basis *f*, Boden *m*, Fuß *m*, Rückstand *m*, Sohle *f*, unteres Ende *n*, Unterteil *n*, Unterwasserschiff *n* (of a ship up to the water line) **from the** ~ **toward the top** von unten nach oben
**bottom, ~ of the bed** Sohle *f* des Flözes ~ **of bore** Seelenboden *m* ~ **of borehole** Bohr-ort *m*, -sohle *f* ~ **of cell** Küvettenboden *m* ~ **of crankcase** Kurbelgehäuseunterteil *m* ~ **of ditch** Grabensohle *f* ~ **of feeder** Oberkanalsohle *f* ~ **of headrace feeder** Obergrabensohle *f* ~ **of an ingot** Blockfuß *m* ~ **of manhole** Schachtsohle *f* ~ **of parachute pack** Rückenschale *f* ~ **of the sea** Meeresgrund *m* ~ **of shaft** Schachtsohle *f* ~ **of shell** Gehäuseboden *m* ~ **of terminals** Klemmenunterseite *f* ~ **of trough** Muldentiefstes *n* ~ **of the valley** Talsohle *f*
**bottom, ~ aileron** Unterflügelquerruder *n* ~ **apron plate** Unterschutz *m* (Auto) ~ **assembly** Behälterwanne *f* (power plant) ~ **bench of a seam** Unterbank *f* eines Flözes ~ **bend** unteres Knie *n* ~ **blade** Untermesser *m* ~ **block** Unter-kloben *m*, -flasche *f* (of a flask lift), -gehänge *n* ~ **blow valve** Bodenventil *n* ~**-blown converter** Konverter *m* mit von unten eintretendem Wind ~ **board** Aufstampfboden *m*, Boden *m*, Fußbrett *n*, Platinenbrett *n*, Leerboden *m* (in molding) ~ **box** Unterkasten *m*, (drag) Formunterteil *m* ~ **bracket bearings**

**and parts** Tretlager *n* und Tretlagerteile ~ **bush**
Grundbüchse *f* ~ **camber** Druckseite *f* (of
wing), Unterflächenwölbung *f* ~ **captain** Grubenaufseher *m* **to** ~-**cast** steigend abgießen oder
gießen, steigend gegossen ~ **casting** Gießen *n*
im Gespann, steigender Guß *m* ~-**charging**
**machine** Bodeneinsetzmaschine *f* ~ **clack**
Saugklappe *f* ~ **cleanout** Bodenablaß *m* ~
**clearance** Fußspiel *n*, ~ **coal** Grundkohle *f*
~ **coat** erster Auftrag *m* (Lack) ~ **contact**
Bodenkontakt *m*, Herdanschluß *m* ~ **contour**
**line** Tiefenlinie *f* ~ **cover** Bodendeckplatte *f* ~
**covering** Bodenbelag *m* ~ **crosscut** Sohlenquer-
schlag *m* ~ **current** Bodenstrahlung *f*, Grundströmung *f* der Flut, Sohlenströmung *f* ~-**cutter**
**molding machine** Unterfräsmaschine *f* ~
**cylinder casting** Zylinderboden *m* ~ **dead**
**center** Totpunkt *m*, unterer Totpunkt *m* ~
**die** Untergesenk *n* ~ **discharge** Bodenentleerung *f*, untere Entleerung *f*, unterer Strahl
*m* ~-**discharge car** Bodenentleerer *m* ~-**dis-**
**charge conduit** Grundablaß *m* ~ **dishes** Boden-
wölbplatten *pl* ~-**door skip** Kübel *m* mit beweglichem Boden (for filling caissons) ~
**drive** unterer Antrieb *m* ~-**drying kiln** Bodenbrennofen *m* ~-**dump bucket** Kübel *m* mit
beweglichem Boden ~-**dump scow** Klappschute
*f* ~ **dump truck** Bodenentleererlastwagen *m* ~
**dumping bucket** Kübel *m* mit Bodenentleerer
~ **edge** Unterkante *f* ~ **electrode** Bodenelektrode *f* ~ **end** kaltes Ende *n* ~ **felt** Unterfilz *m* ~
**fermentation** Untergärung *f* ~-**fermentation**
**yeast** Unterhefe *f* ~-**fermented** untergärig ~
**filler** Anhänger *m* ~ **fire** Sohlenfeuer *n* ~ **flange**
Bodenflansch *f*, Fußflansch *m*, Gurtungsblech
*n*, Unterflansch *m*, Untergurt *m* (aviat.) ~ **flap**
**bucket** Klappkübel *m* ~ **flue** Sohlkanal *m* ~
**fuller** Ballgesenk *n*, Ballhammerunterteil *n*,
runder Setzhammer *m* ~ **gate** Bodenklappe *f*
~ **girder** unterer Riegel *m*, Unterrahmen *m*,
Unterrahmenstück *n* ~ **grid** Bodenrost (Kühlschrank) ~-**half mold** Unterform *f* ~ **heated**
bodenbeheizt ~ **heating** Boden-, Herd-behei-
zung *f* ~-**heating furnace** Bodenbrennofen *m*
~ **hole** Grundloch *n* ~-**hole packer** Sohlen-
stopfbüchse *f* ~-**hole pressure** Sohlendruck *m*
~-**hole-sample packer** Bodenmusterapparat *m*
~ **iron** Schaleneisen *n* ~ **jackcar** Bodeneinsatzwagen *m* (Konverter) ~ **layer of a bed** Unterbank *f* eines Flözes ~ **level** untere Grundstrecke
*f* (min.) ~ **longeron** Rumpfunterholm *m* ~
**margin gauge** Randmaßstab *m* ~ **material**
schweres Material ~ **measure** liegendste
Bank *f* (min.) ~ **outlet sluice** Grundablaßschütz
*n* ~ **paste** Bodenteig *m* ~ **piece** Unterstück *n* ~
**pillar** Grundpfeiler *m* ~ **plate** Boden *m*,
Boden-blech *n*, -platte *f*, Gespann *n*, Fundament *m*, Sohlen-, Sohl-platte *f* **to** ~-**pour** steigend abgießen, steigend gießen ~-**pour ladle**
Gießpfanne *f* mit Stopfenausguß, Stopfenpfanne *f* ~ **pouring** steigender Guß *m* ~
**pouring of multiple molds** Gespannguß *m* ~
**pressure** Sohl-, Unter-druck *m* ~ **product**
Sumpfprodukt *n* ~ **ray** unterer Strahl *m* ~
**ring** Deckring *m* ~ **rings of compass bowl**
Kesselbodenring *m* ~ **rods** Untergestänge *f* ~
**roll** Unterwalze *f* ~ **sealing** Sohlendichtung *f*
~ **section of a gas producer** Generatorunter-

teil *n* ~ **sediments** Bodensatz *m* ~**s sediments**
**and water** Bodensatz *m* und Wasser ~ **settlings**
Bodensatz *m* ~ **sheave** Unterflasche *f* (for a
flask lift) ~ **side** Unterseite *f* ~ **sill** Grundsohle *f*
~ **slide** Unterschieber *m* ~ **slide rest** Unter-
support *m* ~ **sprocket** Nachwickelrolle *f*,
untere Transportrolle *f* (film) ~ **stitching**
Bodenheftung *f* ~ **strip** Schleifleiste *f* ~ **surface**
Bodenfläche *f*, untere Begrenzungsfläche *f*,
Unterseite *f* ~ **swage** Untergesenk *n*, Unterteil *m* eines Gesenks ~ **swedge** hohler Setzhammer *m* ~-**tap crane ladle** Kranstopfen-
pfanne *f* ~-**tap ladle** Stopfenpfanne *f* ~ **tearer**
Bodenreißer *m* ~ **tool** Abschrot *m* ~ **valve** Bodenventil *n* ~-**valve poppet** Bodenventilkegel
*m* ~ **water** Liegend-, Unter-wasser *n* ~ **wing**
**spar** Unterflügelholm *m* ~ **yeast** untergärige
Hefe *f*

**bottomed** söhlig
**bottoming** Decken *n*, Grundbau *m*, Schüttung *f*,
Sperrung *f*, Vorgrundierung *f* ~ **brush** Grun-
dierbürste *f* ~ **die** Fertigschneideisen *n* ~
**paste** Grundmasse *f* (Lack) ~ **tap** Gewindebohrer *m* für Grundlöcher oder Sacklöcher,
Grundbohrer *m*, Nachschneider *m*
**bottomry** Bodmerei *f*
**bottoms** Bodenprodukt *n* (Destilliersäule); Faßgeläger *n*
**bottstick** Ton-propfen *m*, -stopfen *m*
**boucherization** Boucherisierung *f*, Tränkung *f*
mit Kupfervitriol
**boucherize, to** ~ boucherisieren, mit Kupfervitriol tränken
**bought scrap** Kauf-bruch *m*, -schrott *m*
**boulder** Felsblock *m*, Geröll *n* ~ **clay** Geschiebe-
-lehm *m*, -mergel *m* ~ **flint** Flintstein *m* ~
**wall** Geschiebewall *m*
**boulders** Felsgeröll *n*, Geschiebeblöcke *pl* ~ **of**
**quartzite** Findlingsquarzit *m*
**bounce, to** ~ anschlagen, aufprallen, prallen,
prellen
**bounce** Springen *n*, Sprung *m*, (in sound reproduction) Brillanz *f* ~ **plate** Prallplatte *f* ~
**rocket** Schlagrakete *f*
**bouncing** Tauchen *n* ~-**ball synchronization**
Nachsynchronisierung *f* ~ **movement while**
**landing on water** Landung *f* in Stampfbewegung ~ **pin** Schlagstift *m*, Springstabindikator
*m*, Springzeiger *m*
**bound, to** ~ abgrenzen, begrenzen, prallen
**bound** Aufprall *m*, Einzelsprung *m*, Grenze *f*,
Satz *m*, Schranke *f* **by** ~**s** sprungweise, gebunden, unfrei, verpflichtet ~ **off** abgebunden ~
**charge** gebundene Ladung *f* ~ **electricity**
gebundene Elektrizität *f* ~ **vortex** gebundener
Wirbel *f*
**boundaries of improved river channels** Streich-
linien *pl* der Flußufer
**boundary** Bauabteilung *f*, Begrenzung *f*, Gemarkung *f*, Grenze *f*, Markscheide *f*, Rand *m*,
Trennungslinie *f*, Ziel *n* **to mark the** ~ abgrenzen ~ **of groove** Kaliberbegrenzung *f* ~
**of a profile (or section)** Profilbegrenzung *f* ~
**between units** Nahtstelle *f* (position)
**boundary,** ~ **angle** Randwinkel *m* ~ **beacon** Be-
grenzungsbake *f* ~ **break-away separation**
Grenzschicht(e)ablösung *f* ~ **collocation** Rand-
kollokation *f* ~ **condition** Randbedingung *f*

~ **day marking** Umgrenzungstagesmarkierung f ~ **demarcation** Markscheidung f ~ **effect** Sperrschicht f ~ **equation** Grenzgleichung f ~ **field** Randfeld n ~ **film** äußerste Schicht f, Grenzfilm m, Grenzschicht f ~ **illumination searchlight** Begrenzungsleuchten n ~ **layer** Grenzfläche f, Grenzschicht f ~ **layer blowing** Grenzschichtabblasung f ~ **layer control** Grenzschicht-absaugung f, -steuerung f (aerodyn.) ~ **layer influence** Grenzschichtbeeinflussung f ~ **layer separation** Grenzschichtabspaltung f ~ **layer velocity** Grenzschichtgeschwindigkeit f ~ **light** Begrenzungs-, Grenz-Platzumrandungs-, Umrandungs-licht n ~ **lighting** Randbefeuerung f ~ **lights** Umgrenzungsfeuer pl (Flughafen) ~ **line** Begrenzungs-, Grenz-, Scheide- f, Stoß-, Umgrenzungs-linie f ~ **marker** Begrenzungsgerät n, Umrandungs-zeichen n, -zeiger m, Grenzstein m (stone) ~ **markers** Umgrenzungsmarker pl (Flugh.) ~ **marking** Begrenzungsreiter m ~ **path** Rennsteig m, Rennstieg m ~ **point** Wandpunkt m ~ **point lemma** Randpunktsatz m ~ **pole** Grenzpfahl m ~ **post** Markpfahl m ~ **problem** Randwertaufgabe f ~ **sea-air** Grenzschicht f Luft/Meer ~ **stone** Markstein m ~ **surface** Grenzfläche f, Sprungschicht f (between two liquids) ~ **value** Grenz-, Rand-wert m ~ **-value problem** Grenzwertproblem n

**bounded** beschränkt

**boundless** grenzenlos, maßlos, schrankenlos, unbeschränkt

**bounds** Berandung f by ~ gebunden, sprungweise, unfrei, verpflichtet

**bounty** Exportvergütung f, Prämie f ~ **on exportation** Ausfuhrprämie f

**bourdon,** ~ **drive** Röhrenfedermeßwerk n ~ **gauge** Röhrenfederlehre f ~ **manometer** Bourdon-Manometer n ~ **pressure gauge** Bourdon--Druckmesser m ~ **tube** Bourdon-Röhre f, Bourdonsche Röhre f, Schwingsche Röhre f ~ **-tube pressure gauge** Rohrfederdruckmesser m

**bournonite** Bleifahlerz n, Bouronit m, Schwarzspießglanzerz n, Spießglanzbleierz n

**bourn-out** Ausschöpfung f

**bourrelet** Führungs-, Zentrier-wulst f

**bourrette yarn** Bourrettegarn n

**bouse, to** ~ auftaljen

**boven body** Bovenstumpen m

**bow, to** ~ neigen, streichen (a string), verbiegen

**bow** Bogen m, Bogenlineal n, Bug m (nav.), Bügel m (Stromabnehmer) Krümmung f, Kurve f, Kurvenlineal n, Schleife f, (of a sinker) Befestigungsbügel m **at the** ~ vorn **on the** ~ vor dem Bug ~ **of the boat** Bootsbug m ~ **of case** Gehäusebügel m

**bow,** ~ **anchor** Buganker m ~ **armor** Bugpanzer m ~ **buoyancy tank** Bugauftriebtank m ~ **cap** Bugkappe f ~ **cockpit** Bugkanzel f (aviat.) ~ **compartment** Vorderkaffe f ~ **compasses** Null-, Nullen-zirkel m ~ **contakt** Bügelkontakt m (electr.) ~ **file** Raumfeile f ~ **frog** Bogenfrosch m (string instrument) ~ **girder** gebogener Träger m ~ **heaviness** Buglastigkeit f (aviat.) ~ **-heavy** buglastig, bugschwer, kopflastig ~ **line** Bulin f, Laufknoten m, Vorleine f ~ **line on a bight** doppelter Fahlstich m ~ **-line-knot**

Pfahlstich m, Packerknoten m (Fallschirm) ~ **member** Bogenlehre f (arch.) ~ **pen** Zirkelfeder f ~ **pencil** Zirkelstift m ~ **radiator** Brustkühler m ~ **saw** Bogen-, Bügel-, Schwanz-, Schweif--säge f ~ **-saw frame** Bügelgatter n ~ **spacer** Meßzirkel m ~ **spring** Federbügel m ~ **string** Bogenschnur f ~ **-type elctrode** Bügelelektrode f ~ **wave** Kopf-, Front-, Stirn-welle f ~ **-wing skid** Schutzbügel m

**Bowden** Bowden-Kabel n, Bowden-Steuerzug m ~ **control cable in flexible steel conduit** Bowden-Zug m ~ **wire** Bowden-Zug m, Duzdraht m, Stößeldraht m ~ **-wire control** Stoßdrahtsteuerung f ~ **-wire release** Drahtauslösung f (Verschluß)

**bowed,** ~ **roll** gekrümmte Walze f ~ **spring washer** Federscheibe f ~ **weft** Fadenverschiebung f (Gewebe)

**bower anchor chain** Ankerkette f

**bowing under load** Durchbiegung f

**bowl, to** ~ kegeln, rollen lassen, schieben

**bowl** Bassin m, Becher m, Napf m, Satte f, Schale f, Tank m, Tiegel m ~ **of the calender** Kalanderwalze f ~ **classifier** Klassierer m ~ **gudgeon** Kugelzapfen m ~ **mill** Griffinmühle f ~ **overshot** Hülse f des "Overshot" ~ **padding mangle** Walzenfoulard m ~ **paper** Kalanderwalzenpapier n ~ **-type classifier** Klassifikator m mit Vorbehälter

**bows** Bug m (seaplane)

**bowspring valve cup** Bogenfederventilkapsel f

**Bowsprit** Bugspriet n

**box, to** ~ ausbüchsen, ausbuchsen, bekleiden, emballieren **to** ~ **the compass** die Kompaßpunkte pl der Reihe nach aufzählen (aviat.) **to** ~ **a wheel** ein Rad n ausbüchsen oder ausbuchsen

**box** Behältnis n, Bohrspindel f, Buchsbaum m, Büchse f, Dose f, Etui n, Futteral n, Kasten m, Kiste f, Koffer m, Muffe f, Schachtel f, Umhüllung f, (foundry) Formrahmen m (print.) Kapsel f ~ **for casting** Formkasten m ~ **of fire door** Feuerzarge f ~ **of the gyro(scope)** Kreiselgehäuse n ~ **a mold** Gießkasten m ~ **(of stamping machine)** Trog m ~ **for tool block** Blockstichhaus n ~ **with a two-segment electrometer system** Binantenschachtel f ~ **of a vehicle** Bock m

**box,** ~ **-annealed** kastengeglüht ~ **-annealed sheet** kastengeglühtes Blech n ~ **annealing** Kisten-, Kasten-glühung f ~ **annealing furnace** Haubenglühofen m ~ **barrow** Kastenwagen m ~ **beam** Kastenträger m ~ **bill** Gerät n ~ **blank** Kartonagenzuschnitt m ~ **board** Karton m, Kistenbrett n, Schachtelpappe f ~ **borer** Kastenbohrer m ~ **bubble** Düsenlibelle f ~ **camera** Büchsen-, Kasten-kamera f ~ **cannister** Schüttkasten m ~ **car** bedeckter oder geschlossener Eisenbahnwagen m ~ **car detector** Impulsspitzendetektor m ~ **carburizing** Pulverzementieren n ~ **cart** Bock-, Kasten-wagen m, Kastenkarre f ~ **casting** Kastenguß m ~ **charging crane** Muldenchargierkran m ~ **chuck** Zweibackenfutter n ~ **clamp** Klemme f, Schraubstock m ~ **classifier** Spitzkasten m ~ **column** Kastensäule f ~ **compass** Bussole f ~ **compound** Muffenausgußmasse f ~ **container** geschlossener Behälter m ~ **cover** Kastendeckel m

*m* ~ **drain** Sinkkasten *m* ~ **enamels** Kartonagenglacépapier *n* ~-**end wrench** Sechskantsteckschlüssel *m* (hexagon) ~ **equipment** Kofferapparat *m* (für Bildaufnahme) ~ **formation** Rottenformation *f* ~ **frame** Kastengestell *n*, Schüttkasten *m* ~-**frame construction** Kastenkonstruktion *f* ~ **girder** Kastenträger *m* ~ **groove** Kastenkaliber *n*, geschlossenes Kaliber *n* ~**head** Dosen-, Kabel-endverschluß *m* ~ **horizon** Dosenhorizont *m* ~ **iron** Plättglocke *f* ~ **key** Aufsteckschlüssel *m* ~-**kite** Kastendrachen *m* ~-**kite tail** Kastensteuerung *f* ~ **level** Dosenlibelle *f* ~ **loom** Drehwebstuhl *m* (textiles) ~ **magazine** Kastenmagazin *n* ~ **mold** Kastenform *f*, Kastengußform *f* ~ **motion** Schützenwechsel *m* ~ **mount** parallelepipedische Fassung *f* ~ **number** Chiffreadresse *f* ~ **number advertisement** chiffrierte Anzeige *f* ~' **nut** Überwurfmutter *f* ~ **part** Formkastenhälfte *f* ~ **pass** Quadrat-, Flach-kaliber *n*, geschlossenes Kaliber *n*, (in rolling) Kastenkaliber *n* ~ **pin** Kastenstift *m* ~ **plate** Kastenplatte *f* ~ **prizer** Wickelpresser *m* (tobacco) ~ **receiver** Dosenfernhörer *m* ~ **relay** Dosenrelais *n* ~ **screw** Schraubenbüchse *f* ~ **searching** Kastensuche *f* (rdr) ~ **section** Kastenprofil *n* ~-**section-type jib** Blechausleger *m* ~-**shear apparatus** Schergerät(buchse) *f* ~ **spanner** Steck-, Sechskantsteck-, Stock-schlüssel *m* ~ **spar** Holmkasten *m*, Kastenholm *m* (aviat.), Kastenspiere *f* ~ **staple** Schließkappe *f* ~ **stitch** viereckige Vernähung *f* ~ **stud** Kern-böckchen *n*, -stütze *f* ~ **table** Form-, Kastentisch *m* ~ **tail unit** Kastenleitwerk *n*, Schwanzwelle *f* ~ **thread** Innengewinde *n*, Muffe *f*, weibliches Gewinde *n* ~ **timber** Holzgeripppe *n* ~ **timbering** Schalenholzzimmerung *f* ~-**toe cement** Kappensteife *f* ~ **tool** Kastenwerkzeug *n* (Drehstahl) ~ **trail** Kastenlafette *f* ~ **turbine wheel** Kastenrad *n* ~-**type apparatus** Plattenapparat *m* ~-**type contruction** Hohlkastenbauweise *f* ~-**type cooler** Kistenkühler *m* ~-**type design** kastenförmig ~-**type hand truck** Kastenwagen *m* ~-**type impeller** Laderrad *m* mit kastenförmig geschlossenen Schaufelkanälen ~-**type leg** Drehbankfuß *m* ~-**type pattern** Kastenmodell *n* ~ **wall** Kastenwand *f* ~ **winch** Kastenwinde *f* ~ **wood** Buchsbaumholz *n* ~ **wrench** Ring-, Steck-schlüssel *m*
**boxed** abgedeckt ~ **catch** Schließkappe *f* ~ **spindle** Gehäusespindel *f*
**boxing** Bettungskoffer *m*, Einschließen *n*, Kistenverpackung *f*, Verschalungsmaterial *n* ~ **of the tiles** Verkiesen *n* der Schwellen (r.r.)
**boycott** Acht *f*, Boykott *m*
**Boyle's law** Boylesches Gesetz *n*
**B position** B-Platz *m*, B-Platzschrank *m*
**brace, to** ~ abfangen, abspannen, abstreben (aviat.), abstützen, aussteifen, ausstreben, klammern, verklammern, verspannen, versteifen, verstreben, (a frame) absprießen **to** ~ **back** backbrassen **to** ~ **full** abbrassen
**brace** Aufhängeriemen *m*, Band *n*, Bohrleier *f*, Brasse *f*, Brustbohrer *m*, Brustleier *f*, Doppelbügel *m*, Faustleier *f*, Haubenstütze *f*, Klammer *f*, Stag *m*, Steife *f*, Stützbalken *m*, Verbindungsklammer *f*, (cross or push) Strebe *f* ~ **with key** Krückel *m*

**brace,** ~ **drill** Leierbohrer *m* ~ **head** Bohrkrückel *m*, Krückelstock *m* ~ **head on top of the bore rod** Krückelstück *n* ~ **pipe** Querversteifungsrohr *n* ~ **plate** Bindeblech *n* ~ **rod** Stützstange *f* (Seitenflosse) ~ **spoon bits for brush blocks** Bürstenhals-Löffelbohrer *m* ~ **strut** Abfangsstrebe *f* (d. Fahrwerks), Spannstrebe *f*
**braced,** ~-**beam construction** Gitterwerk *n* ~-**beam construction equipped with diagonal web members** Diagonalversteifung *f* ~ **girder** Hängewerk *n*, Parallelträger *m*, Steifrahmen *m* ~ **link** Verbindungsstrebe *f* ~ **tail unit** abgestrebtes Leitwerk *n* ~ **wing** verstrebter Flügel *m*
**braceless** unverspannt, verspannungslos ~ **plane** verspannungsloses Flugzeug *n*
**bracelet compass** Armbandkompaß *m*
**braces** unterer Querriegel *m*; zusammenfassende Klammern *pl* (print.)
**brachistochrone** Brachystochrone *f*
**bracing** Auskreuzung *f* (aviat.), Verband *m* (Tragwerk), Verspannung *f*, Verspreizung *f*, Versteifung *f*, Verstrebung *f*, ~ **cable** Spannkabel *n* ~ **cable ditch** Holzverschalung *f* zum Absteifen eines Kabelgrabens ~ **clamp for lead** Leitungsabfangschelle *f* ~ **member** Rastenspreize *f* (piano) ~ **members** Verbände *pl* (of steel construction) ~ **rib** Versteifungsrippe *f* ~ **strut** Abfangstiel *m* ~ **tube** Tragrohr *n* ~ **wire** Spanndraht *m*, Spannseil *n*, Verspannungsdraht *m*, Verspannungskabel *n* ~ **wire eyelet** Spannseilöse *f* ~ **wire fitting** Verspannungsbeschlag *m*
**bracket, to** ~ eingabeln, einklammern, einordnen
**bracket** Arm *m*, Armleuchter *m*, Auskragung *f*, Ausleger *m*, Befestigungsarm *m*, Bildstock *m*, Bratzen *m*, Bügel *m*, Eingabelung *f*, Federhand *f*, Gabelung *f*, Galgen *m*, Gegentor *n*, Knagge *f*, Kragarm *m*, Krage *f*, Kragträger *m*, Lagerblech *n*, Schelle *f*, Ständer *m*, Stehlager *n*, Stütze *f*, Stützpunkt *m*, Tischen *n*, Träger *m*, Tragkonsole *f*, (of pump) Pumpenträger *m*, Wandstärkeverlaufträger *m*
**bracket,** ~ **for bell crank** Gabelbock *m* ~ **for chain adjuster** Halter *m* für Kettenspanner ~ **for kerb** Konsole *f* für Spurbegrenzung ~ **of microscope** Mikroskopbügel *m* ~ **similar in shape to that of a snap gauge for measuring cylindrical work** rachenlehrenähnlicher Bügel *m* ~ **for spherical compensators** Kugelträger *m*
**bracket,** ~ **assembly** Lagerblock *m* ~ **bearing** Konsollager *n* ~ **bearing for transmission drives** Transmissionslager *n* ~ **carrier** Federbock *m* ~ **drop hanger** Armhängelager *n*, Konsolhängelager *n* ~ **elevating screw** Auftriebspindel *f* ~ **fire** Gabelschießen *n* ~ **jib crane** Schwenkkran *m* ~ **pin** Konsole *f* ~ **plate** Befestigungsplatte *f* ~ **pole** Auslegermast *m* ~ **supporting jet** rudder Strahlruderhaltebock *m* ~ **table** Konsoltisch *m* ~-**type board** (hinged panels) Flügelschalttafel *f*
**bracketing** Eingabelung *f*, Gabel-bilden *n*, -bildung *f* ~ **a beam** Strahlfliegen *n* (aviat.) ~ **method of adjustment of fire** Gabel-bilden *n*, -bildung *f*
**brackets** eckige Klammern *pl* (print.); Rahmenimpulse *pl* (sec. rdr.)

**brackish water** Brackwasser *n*
**brad** Bodenspieker *m*, Fußbodennagel *m*, Spiekernagel *m*
**bradawl** Vorstecker *m*
**bradenhead** Rohrkopf *m*
**Bradfield insulator** Bradfield-Isolator *m*
**Bradford breaker** Trommelsiebbrecher *m*
**Bragg scattering** Braggsche Streuung *f*
**bragite** Bragit *m*
**braid, to** ~ flechten, klöppeln, umflechten, umklöppeln, umspinnen **to ~ around** beflechten
**braid** Besatz *m*, Kolbenstickerei *f*, Paspel *m*, Zopf *m*, schraubenförmige Wicklung *f* ~(ing) Litze *f* ~ **of hair** Haarflechte *f* ~-**covered cable** Kabel *n* mit Umklöppelung
**braided** besponnen, geflochten, geklöppelt ~ **cable** umklöppeltes Kabel *n* ~ **connector** Litzenverbinder *m* ~ **cord** geschlagene Leine *f* ~ **hose** Schlauch *m* mit Geflechteinlagen ~ **metal covering (wire, cable)** Drahtumklöppelung *f* ~ **silk cord** geflochtene Seidenschnur *f* ~ **wire** umklöppelter oder umsponnener Draht *m*
**braider** Klöppel-, Umflecht-maschine *f*
**braiding** Umklöppelung *f*, Umspinnung *f* ~ **of cotton** Baumwollumklöppelung *f* ~ **with spun yarn** Gespinstumflechtung *f* ~ **machine** Docken-, Flecht-, Umklöppel-maschine *f*
**brail, to** ~ **up** aufgeien
**brain,** ~ **voltage** Gehirnspannung *f* ~ **unit** elektronisches Steuergerät *n*
**braise** Koksstaub *m*
**brake, to** ~ abbremsen, anhalten, arretieren, bremsen, festhalten, feststellen, sperren **to ~ excessively** überbremsen
**brake** Abkantbank *f*, Abschwächer *m*, Blechbiegemaschine *f*, Breche *f*, Bremse *f*, Bremsvorrichtung *f*, Fahrbremse *f*, Hebelstange *f*, Hemmschuh *m*, Klotz *m*, Pumpenschwengel *m*, Umbruch *m* (in a drive or gallery) ~ **for airplane engine** Bremsklatsche *f* ~ **and** ~-**lifting magnets** Brems- und Bremsluftmagnete *pl* ~ **on the transmission shaft** Vorgelegebremse *f* **the** ~ **slips** die Bremse rutscht
**brake,** ~ **action** Angriffwinkel *m*, Hemmung *f* ~ **adjuster** Bremsnachstellvorrichtung *f* ~ **adjusting bearing** Bremsregulierlager *n* ~ **adjusting screw** Bremsnachstellschraube *f* ~ **adjustment** Bremseinstellung *f* ~ **anchor pin** Bremsträger *m* ~ **anchor plate** Bremsbock *m* ~ **angle** Bremswinkel *m* ~-**applying handle** Bremsgriff *m* ~ **arm** Bremsarm *m* ~-**articulation piece** Bremsgelenkstück *n* ~-**away coupling** Abreißkupplung *f* ~ **backing plate** Bremsdeckplatte *f* ~ **band** Brems-band *n*, -ring *m*, Reibungsband *n* ~ **band of cotton straps** Bremsband *n* aus Baumwollgurten ~-**band lining** Bremsbandbelag *m*, Bremsbelag ~-**band voltage stress** Bandspannung *f* ~ **bar** Bremszugstange *f* ~ **block** Bremsklotz *m*, Bremsblock *m* (d. Schlauchbackenbremse) ~ **body** Brech-, Einsatz-körper *m* ~ **box** Bremsbock *m* ~ **bracket** Bremsträger *m* ~ **braking position** Bremsstellung *f* ~ **bushing** Bremsbüchse *f* ~ **cable** Brems-draht *m*, -kabel *n*, -seil *n*, Zugriemen *m* (stretched across flight deck of aircraft carrier) ~ **cam** Brems-daumen *m*, -nocke *f* ~ **camshaft** Bremswelle *f* ~ **chain** Bremskette *f* ~ **cheek**

**Bremsbacke** *f* ~ **clevice** Bremsstapel *m* ~ **clutch** Bremsbacke *f* ~ **collar** Bremsband *n* ~-**compensating lever** Bremsausgleichhebel *m* ~-**compensating shaft** Bremsausgleichwelle *f* ~ **contactor** Bremsschutz *m* (zur Gangschaltung) ~ **control** Brems-antrieb *m*, -betätigung *f*, -steuerung *f* ~ **coupling** Bremskuppelung *f* ~ **crank** Bremskurbel *f* ~ **cross** Bremsausgleich *m* ~ **cross shaft** Bremsquelle *f* ~ **disc** Bremsscheibe *f* ~ **distance** Bremsbahn *f* ~ **drum** Bremstrommel *f* ~ **drum surface** Bremsfläche *f* ~ **dynamometer** Bremsdynamometer *n* ~ **effect** Bremswirkung *f* ~ **energizer** Bremsverstärker *m* ~ **equalization** Bremsausgleich *m* ~ **equalizer** Bremsausgleicher *m* ~ **equipment** Bremsausrüstung *f* ~ **fastening flange** Bremsbefestigungsring *f* ~-**field oscillator** Bremsaudion *n* ~-**field tube** Bremsröhre *f* ~-**field valve** Bremsröhre *f* ~ **flap device** Bremsklappenapparatur *f* (g/m) ~ **flap tube** Bremsröhre *f* ~ **fluid** Bremsflüssigkeit *f* (hydraul.) ~ **fluid reservoir** Bremsölbehälter *m* ~ **foot pump** Bremsfußpumpe *f* ~ **gear** Bremsschaltung *f* ~ **guard** Bremsschutz *m* ~ **guide** Bremsführung *f* ~ **hand lever** Bremshandhebel *m* ~ **handle** Handgriff *m* der Bremse ~ **hanger** Bremsgehänge *n* ~**head** Brems-backe *f*, -block *m*, -schuh *m* ~ **horsepower** Brems-pferdekraft *f*, -pferdekraftstärke *f*, -leistung *f*, effektive Leistung *f*, effektive Pferdestärke *f*, Effektivleistung *f*, gebremste Pferdestärke *f*, nutzbare Leistung *f*, Nutzleistung *f* ~ **hose** Bremsschlauch *m* ~ **housing** Bremsgehäuse *n* ~ **knock** Bremsstoß *m* ~ **lag** Bremsverlustzeit *f* ~**lever** Bremshebel *m*, Hebelgriff *m* ~ **lever arm** Bremshebelarm *m* ~-**lever connecting rod** Bremszugstange *f* ~-**lever guide** Bremshebelführung *f* ~ **lever side plate** Bremshebelwange *f* ~-**lifting magnet** Bremslüftmagnet *m* ~ **line** Bremsleitung *f* ~ **liner (or lining)** Brems-belag *m*, -futter *n* ~(-**band**) **lining** Bremsband *n* ~ **lining disc** Reibscheibe *f* ~ **load** Bremsbelastung *f* ~ **lock** Bremsverschluß *m* ~ **magnet** Bremsmagnet *m*, Kurzzeitmesserbremse *f* ~**man** Bremser *m* ~ **masks** Bremsschilde *pl* ~ **mean effective pressure** effektiver oder nutzbarer Mitteldruck *m*, mittlerer Wirkungsdruck *m* ~ **meter** Bremslängenmesser *m* ~ **mounting plate** Brems-anbauflansch *m*, -träger *m* ~-**operating lever** Bremsbetätigungshebel *m* ~ **operation** Bremsbetätigung *f* ~ **pedal** Bremsbackenpedal *n*, -fußhebel *m*, -pedal *n*, -tritt *m*, Fußbremshebel *m* ~ **pedal gear** Bremsfußhebelübersetzung *f* ~ **pedal pressure** Bremspedaldruck *m* ~ **piston** Bremskolben *m* ~ **power** Bremsleistung *f*, Fangkraft *f* ~ **pressure** Brems-druck *m*, -kraft *f* ~-**pressure lead** Bremsleitung *f* ~ **pressure plate** Bremsscheibe *f* ~ **propeller** Bremsluftschraube *f* (aviat.) ~ **puck** Bremsbelag *m* ~ **pull rod** Bremslasche *f* ~ **pulley** Bremsscheibe *f* ~ **ratchet** Bremssperre *f* ~ **reducing valve (reducer)** Bremskraftminderer *m* ~ **release cylinder** Bremslösezylinder *m* ~ **release device** Bremslösevorrichtung *f* ~ **release solenoid** Magnetbremslüfter *m* ~ **ring** Bremsring *m* ~ **rod** Brems-leiste *f*, -stange *f*, -zugschere *f* ~-**rod linkage** Bremsgestänge *n* ~ **running in an oil bath** Ölbremse *f* ~ **screw**

Bremsbolzen *m* ~ **shaft** Brakenstange *f*, Brems-welle *f* ~ **shoe** Brems-backe *f*, -block *m*, -klotz *m*, -schuh *m* ~ **shoe carrier** Bremsbackenträger *m* ~ **shoe check spring** Bremsbackenrückzug-feder *f* ~ **shoe clearance** Bremsbackenspiel *n* ~ **shoe holder** Bremsbackenhalter *m*, -klaue *f* ~ **shoe lining** Bremsbackenbelag *m* ~ **shoe mounting bolts** Bremsbackenlagerbolzen *pl* ~ **shoe principle** Bremsschuhprinzip *n* ~ **shoe return spring** Bremsbackenrückholfeder *f* ~ **shoe stoplight** Bremsbackenlicht *f* ~ **shoes** Bremsglieder *pl*, Schleifsohlen *pl* ~ **skid** Splitterbremse *f* ~ **socket** Bremsbüchse *f* ~ **solenoid** Bremslüfter *m*, Kuppelspule *f* ~ **spider** Bremsspinne *f* ~ **(release) spring** Bremsfeder *f* ~ **spring pliers** Bremsfederzangen *pl* ~ **strap** Bremsband *n*, Zugriemen *m* ~**-support bearing** Bremsstützlager *n* ~ **tension** Bremsspannung *f* ~ **test** Bremsbackenprobe *f* ~**-testing stand** Bremsstand *m* ~ **torque** Bremsmoment *m* (of a meter) ~**-torque bench with rocking plate** Pendelrahmenbremsstand *m* ~ **triangle** Brems-dreieck *n* ~ **truss bar (or beam)** Bremsdrei-eck *n* ~ **tube** Bremsrohrwelle *f* ~ **unit** Brems-körper *m* ~ **valve** Bremsventil *n* ~ **wear adjuster** Bremsnachstellung *f* ~ **wheel** Brems-, Hemm-rad *n* ~ **wire** Bremsseil *n*

**braked,** ~ **run** gebremster Auslauf *m* (aviat.) ~ **wheel** Bremsbackenrad *n*

**braking** Bremsung *f*, Verzögerung *f* ~ **by short-circuiting armature** Ankerkurzschlußbremsung *f*

**braking,** ~ **action** Bremswirkung *f* ~ **area** Brems-fläche *f* ~ **axle** Bremsachse *f* ~ **bar** Bremsknebel *m* ~ **club** Bremsknüppel *m* ~ **couple** Brems-moment *m* (of a continuously rotating instru-ment) ~ **decline** Bremsgesenk *n* ~ **device** Bremsbacken-, Hemm-vorrichtung *f* ~ **effect** Brems-arbeit *f*, -wirkung *f* ~ **energy** Brems-arbeit *f* ~ **fan** Bremsschirm *m* ~ **gear** Brems-werk *n* ~ **incline** Brems-berg *m*, -hang *m* ~ **parachute** Bremsfallschirm *m* ~ **period** Brems-dauer *f* ~ **power** Bremsvermögen *n* ~ **power limiting device** Bremskraftbegrenzer *m* ~ **power transmission** Bremskraftübertragung *f* ~ **resi-stance of friction** Bremswiderstand *m* ~ **roller** Bremswalze *f* ~ **roller lever** Bremsrollenhebel *m* ~ **system** Bremssystem *n* ~ **torque** Brems-drehmoment *m* ~ **trace** Bremsspur *f*

**branch, to** ~ verzweigen **to** ~ **off** abbiegen, abschwenken, abzweigen, bespulen **to** ~ **a current** einen Strom *m* abzweigen

**branch** Abzweig *m*, Abzweigstelle *f*, Abzweigung *f*, Ast *m* (science, tree), Ausläufer *m*, Branche *f*, Fach *n*, Gewerbe *n*, Glied *n*, Nebenarm *m*, Niederlage *f*, Teilkörper *m*, Weiche *f*, Zweig *m*, Zweigbahn *f*, (of manufacture) Bereich *m*, (info proc.) Verzweigung *f*, (of subject) Teil-gebiet *n*

**branch,** ~ **of a band** Bandenzweig *m* ~ **of a business organization** Zweiggeschäftsbetrieb *m* ~ **of a curve** Kurvenast *m* ~ **of industry** Er-werbs-, Industrie-, Nahrungs-zweig *m* ~ **of lightning discharge** Nebenstrahl *m* ~ **of manufacture** Betriebszweig *m* ~ **of ore** Erz-trumm *m* ~ **of a river** Flußarm *m* ~ **of service** Dienstzweig *m* ~ **of trade** Erwerbs-, Nahrungs--zweig *m* ~ **of tree** Baumast *m*

**branch,** ~ **box** Abzweigdose *f* ~ **cable** Stich-,

Zweig-kabel *n* ~ **camp** Neben-, Zweig-lager *n* ~ **circuit** Ableitung *f*, Abzweigleitung *f*, Zweigstromkreis *m* ~ **circuit distribution center** Leitungsverteiler *m* ~ **connection** Gabel *f* ~ **culvert** Stichkanal *m* ~ **current** Abzweig-strom *m*, Teilstrom *m*, Zweigstrom *m* ~ **(ed) current** verzweigter Strom *m* ~ **depot** Zweig--lager *n*, -park *m* ~ **establishment** Filiale *f*, Zweiganstalt *f* ~ **exchange** Unter-, Zweig-amt *n* ~ **executive office** Außenstelle *f* ~ **factory** Zweigwerk *n* ~ **feeder line** Anschlußlinie *f* (aviat.) ~ **gallery** Abzweigstollen *m* (min.), Seitenschlag *m*, Zweig-stollen *m*, -strecke *f* ~ **instruction** Sprungbefehl *m* ~ **joint** Abzweig-punkt *m* ~ **knot** Astknoten *m* ~ **lamp** Trum-lampe *f* ~ **line** Anschluß-bahn *f*, -leitung *f*, Gleisabzweigung *f*, Gleisanschluß *m*, Neben-bahn *f*, Nebenlinie *f* (r.r.) Verzweigungsschnitt *m*, Zweig-bahn *f*, -leitung *f*, -linie *f* ~ **(pipe-)-line** Abzweigungsrohr *n*, -röhre *f* ~**-off** Ab-zweigstelle *f* (r.r.) ~**-off point** Abzweigpunkt *m* ~ **office** Unterzentrale *f* (teleph.) ~ **order** Abzweigbefehl *m* ~ **park** Zweigpark *m* ~ **piece** Abzweigstück *n*, -stutzen *m* ~ **pipe** Ab-zweigrohr *n*, Abzweigung *f*, Nebenrohr *n*, Zweig-leitung *f*, -rohr *n* ~ **pipe line** Seitenlei-tung *f* ~ **piping** Zweigrohrleitung *f* ~**(ing) point** Verzweigungspunkt *m* ~ **point** Windungs-, Knoten-, Verzweigungs-punkt *m* (electr.) ~**-stream line** Verzweigungsstromlinie *f* ~ **road** Neben-straße *f*, -weg *m* ~ **rod** Kolben-stange *f*, Querarm *m* ~ **spur** Zweigleitung *f* ~ **strip** Verteilerstreifen *m* ~ **T** Abzweig-stück *n*, -stutzen *m* ~**-T (of a pipe)** Abzweigmuffe *f* ~**-T of a pipe** Rohrabzweigstück *n* ~ **terminal** Abzweigklemme *f* ~**-terminal line** Sackbahn *f* ~ **tube** Nebenrohr *n* ~**-type singularity** Ver-zweigung *f* ~ **wood** Astholz *n* ~ **works** Zweig-fabrik *f*

**branched** gegabelt, verzweigt ~ **circuit** ver-zweigter Stromkreis *m* ~ **currents** verzweigte Ströme *pl* ~ **laminar model** laminares Modell *n* mit Verzweigungen ~**-off** abgezweigt ~ **resonant circuit** Drosselkreis *m*, Parallelreso-nanzkreis *m*

**branching** Verästelung *f*, Verzweigung *f*, (off) Abzweigung *f*; doppelgängig ~ **off** Ablenkung *f* ~ **of current** Stromspaltung *f* (river) ~ **of currents** Stromverzweigung *f* ~ **of the fuselage** Rumpfgabelung *f* ~ **of the pipe** Rohrver-zweigung *f* ~ **of a river** Flußspaltung *f*

**branching,** ~ **dendrites** Dendritenverästelung *f* ~ **fraction** Verzweigungsanteil *m* ~ **jack** Ab-zweig-, Parallelimpedanz-klinke *f* ~ **line** Gabel-linie *f* ~ **link** Verzweigungsglied *n* ~**-off of conductor** Leitungsabzweigung *f* ~**-off point** Abzweigungsstelle *f* ~**-off station** Knotensta-tion *f* ~ **point** Abzweigpunkt *m* ~ **probability** Abbruchs-, Verzweigungs-wahrscheinlichkeit *f* ~ **ratio** Verzweigungsverhältnis *n* ~ **rule** Ver-zweigungsregel *f* ~ **terminal** Abzweigklemme *f*

**branchless** astfrei, astrein

**branchy** astreich

**brand, to** ~ anmerken, einbrennen, kennzeich-nen, markieren, signieren, mit einem Brand-zeichen versehen

**brand** Bezeichnung *f*, Brandzeichen *n*, Machart *f*, Marke *f*, Markenbezeichnung *f*, Signum *n*,

Sorte *f*, Stempel *m* ~ **of steel** Stahlmarke *f* ~ **lens** Markenglas *n* ~**-new** funkelnagelneu
**branded oil** Markenöl *n*
**branding iron** Brandeisen   *n*, Brenneisen *n*, Brennstempel *m*, Signierwerkzeug *n*, Thermokauter *m*
**brannerite** Brannerit *m*
**branning** Bundbleiche *f* ~ **machine** Kleiabreibemaschine *f*
**branny fibrous stock** Übergangsgries *m*
**brasmoscope** Brasmoskop *n*
**brasque, to** ~ auskleiden
**brass** Gelbkupfer *n*, Messing *n*, Lagerbüchse *f* (mach.), Lagerschale *f* (mach.) **(yellow)** ~ Gelbguß *m* **like** ~ messingartig ~ **of connecting rod** Lagerschale *f* der Pleuelstange
**brass,** ~ **binding post** Messingklemme *f* ~ **border** Messingeinfassung *f* ~ **brace** Messingklammer *f* ~ **bushes** Pfannenlager *n* ~ **bushing** Messinglager *n* ~ **cartridge case** Messingkartusche *f* ~ **case** Messinggehäuse *n* ~**-cased** mit Messing *n* verkleidet ~ **casing** Messingfassung *f* ~ **castings** Messingguß *m* ~ **curves** Messingbogenregletten *pl* ~ **cylinder** Messingzylinder *m* ~ **die-casting** Messingspritzguß *m* ~ **die pressing** Messingpreßstück *n* ~**-dipping plant** Gelbbrennanlage *f* ~ **disk** Rotgußscheibe *f* ~ **doctor** Messingrakel *n* ~ **embossing plate** Messingprägeplatte *f* ~ **engraving** Messingätzung *f* ~ **finishing machine** Messingschlichtmaschine *f* ~ **foil** Messing-blatt *n*, -folie *f* ~**founder** Gelbgießer *m* ~ **foundry** Bronze-, Gelb-, Messing-gießerei *f* ~ **gauze** Messingnetz *n* ~ **jaw socket** Messingbacke *f* ~ **knuckles** Schlagring *m* ~ **mounting** Messingfassung *f* ~ **nut** Messingmutter *f* ~ **peg** Messingstöpsel *m* ~ **pinion** Messingritzel *n* ~**-plate** vermessingen ~ **plate** Messingblech *n* ~ **plating** Vermessingung *f* ~ **powder** Streuglanz *m* ~ **reglet** Messinglinie *f* ~ **relief embossing plate** Messing-Reliefprägeplatte *f* ~ **rim** Messingreifen *m* ~ **rules** Messinglinien *pl* ~ **screw** Messingschraube *f* ~ **scum** Messing-abstrich *m*, -abzug *m* ~ **section** Messingprofil *n* ~ **shaving** Messingspan *m* ~ **sleeve** Muffe *f* von Messing ~ **solder** Messing-lot *n*, -schlaglot *n* ~ **spaces** Messingausschluß *m* ~ **tape** Messingband *n* ~**-taped** mit Messingband *n* umwickelt ~ **taping** Messingbandumlappung *f* ~ **terminal** Messingklemme *f* ~ **tubing** Messingrohr *n* ~ **turning** Messingspan *m* ~ **types** Messingschrift *f* ~**ware** Messingware *f* ~ **wind instrument** Blechblasinstrument *n*, Messingblasinstrument *n* ~ **wire** Messingdraht *m* ~ **wire for wood screws** Messingholzschraubendrähte *pl*
**brassard** Armband *n*, Armbinde *f*
**brasses** Rotgußschalen *pl*
**brassic acid** Brassinsäure *f*
**brassy** messingartig
**brattice** Schachtscheider *m* (min.) ~ **way** Firstenstrecke *f*
**Braun tube** Braunsche Röhre *f*, Elektronen-, Kathoden-strahlröhre *f*
**braunite** Braunit *m*, Hartmanganerz *n*
**brayer** Farbläufer *m*, Keule *f* (print.), Reiber *m*
**braze, to** ~ auflöten, hartlöten, löten
**braze** Aschenstaub *m*, Lötstelle *f* ~**-on flange** Lötflansch *m*
**brazed** aufgelötet, geschweißt, hartgelötet ~

**flange** Auflötflantsch *m* ~ **joint** Lötfuge *f* ~**-on flange** aufgelöteter Flansch *m* ~ **pipe** gelötetes Rohr *m*
**brazen erzen** ~ **dish** Justiermaß *n*
**brazier** Wärmeofen *m* ~ **head screw** Flachrundschraube *f* ~**-type rivet** Blechniete *f*
**braziery** Rotgießerei *f*
**brazilwood paper** Braunholzpapier *n*
**brazing** Hartlötung *f*, Lötstelle *f*, Lötung *f* (by hard solder) ~ **of duplicate work** Massenlötung *f*
**brazing,** ~ **copper** Kupferlot *n* ~ **flux** Hartlötmittel *n* ~ **metal** Messing-lot *n*, -schlaglot *n* ~ **method** Lötverfahren *n* ~ **mixture** Hartlot *n* ~ **process** Hartlötverfahren   *n* ~ **spelter** kupferzinklegiertes Hartlot *n*, Messinghartlot *n* ~ **tongs** Löt-zwinge (-zange) *f* ~ **torch** Lötbrenner *m*
**breach, to** ~ eine Bresche schlagen, schießen
**breach** Bresche *f*, Bruch *m* ~ **of confidence** Vertrauensbruch *m* ~ **of contract** Vertragsbruch *m* ~ **in a dike** Bruchstelle *f* im Deich ~ **of a dike** Deichbruch *m* ~ **in an embankment** Bruchstelle *f* im Deich ~ **of trust** Untreue *f*
**breaching point** Sprengstelle *f*
**breadth** Breite *f*, Breitseite *f*, Weite *f* ~ **of the day** lichte Breite *f* ~ **of line** Linienbreite *f* ~ **of summit** Kronenbreite *f* ~ **of tooth** Zahnbreite *f* ~ **of top** Kronenbreite *f* ~ **of variable-area sound track** Zackentonspurbreite *f*
**breadth gauge** Breitenmaß *n*
**break, to** ~ abfallen, ausschalten, brechen, degradieren, durchreißen, durchschlagen, entsperren, entzweigehen, knicken, öffnen, reißen, springen, unterbrechen, zerbrechen, zerreißen, zerspringen, zuwiderhandeln; (contact) abschalten, (through) durchbrechen, (of sea) branden, (up) aufschließen
**break, to** ~ **across** entzweibrechen **to** ~ **away** abbrechen, abkippen (aviat.) **to** ~ **camp** Lager abbrechen **to** ~ **coarsely** vorbrechen (of ore) **to** ~ **contact** ausschalten, einen Kontakt aufheben **to** ~ **down (or away)** abreißen **to** ~ **down** abbrechen, demontieren, durchschlagen, niederbrechen, niederschlagen, herunterblokken, Panne erleiden, spalten, strecken, vorstrecken, vorwalzen, zubruchgehen, zusammenbrechen **to** ~ **down by** aufschlüsseln nach **to** ~ **in** anlernen, einarbeiten, einbrechen, gewöhnen **to** ~ **into** eindringen, einsteigen (a house) **to** ~ **off** abbrechen, ausbrechen, losbrechen **to** ~ **open** aufbrechen, aufschlagen platzen, sprengen **to** ~ **out** ausbrechen, ebullieren, losbrechen **to** ~ **part way** aufreißen **to** ~ **up** abbauen, abspalten, abwracken, aufbrechen, auflockern, auseinandergehen, auseinanderreißen, demontieren, kleiner machen (coal, stone, etc), losbrechen, zerkleinern, zersetzen
**break, to** ~ **off in bits** abbröckeln **to** ~ **up a bridge** abbrücken **to** ~ **down from bulk to unit containers** umschlagen **to** ~ **a circuit** einen Stromkreis *m* öffnen **to** ~ **a connection** eine Verbindung *f* trennen **to** ~ **a connection** eine Verbindung *f* trennen **to** ~ **a contact** einen Kontakt *m* schließen **to** ~ **contract** vertragsbrüchig werden **to** ~ **the flax** braken **to** ~ **a joint** eine Verbindung *f* abschrauben **to** ~ **up the lump** gar aufbrechen (metal) **to** ~ **in**

pieces zerschlagen to ~ into pieces entzwei-
schlagen to ~ to pieces zersplittern to ~ the
oil seal das Losbrechmoment (des Motors)
beim Anlassen überwinden to ~ a record
einen Rekord m schlagen to ~ the stall den
Strömungsabriß beenden to ~ in zwo entzwei-
schlagen
break Anriß m, Ausbruch m, Bruch m, Drahtbruch m (teleph. line), Durchbiegung f, Einbruch m, Gedankenstrich m, Haltepunkt m,
Spalte f, Speichenmesser m, Sprung f, Unterbrecher m, Unterbrechung f, Verwerfung f, (in
a curve) Knick m, Knickpunkt m, (of a curve)
Knickstelle f, Kontaktabstand m, Schaltstrecke
f ~-and-make Unterbrechung f und Schließung
f ~ in an axle Achsenbruch m ~ of magnetic
flux elektromagnetischer Abriß m ~ of magnetic separation (of isotopes) elektromagnetischer
Abriß m ~ of the shaft in the interior of the
boring Gestängeausreißer m ~ of the types
Anguß m der gegossenen Typen ~ of vision
Bildsprung f ~ (or shift) of vision Bildsprung
m ~ in a wall Blinde f im Mauerwerk
break arrangement Pausenregelung f
break-away Unterbrechung f ~ of dislocation
Auflösung f einer Versetzung ~ of flow Abreißen n der Strömung ~ signal generator
Andrehsignalgenerator m ~ torque abbrechen-
des Drehmoment m
break, ~-before-make contact Folgekontakt m
(Öffnen vor Schließen) ~-contact Öffner m
(electr.), Ruhe-, Trenn-, Unterbrechungs-kon-
takt m
breakdown Ausfall m, Betriebsstörung f, Bruch
m, Dauerbruch m, Defekt m, Durchbruch m,
Durchschlagen n, Niederbrechung f, Panne f,
Störung f, Umbruch der Linie, Versagen n,
Zerwalzen n, Zündung f, Zusammenbruch m,
(electr.) Durchschlag m, Durchschlagskraft f
breakdown, ~ by . . . Aufteilung nach . . . ~ in
amorphous media Durchschlag m amorpher
Substanzen ~ of cost(s) Kostenaufgliederung f
~ of insulation Fehler m (motors) ~ of a
roll Walzenbruch m ~ by surface conduction
Überschlag m (of electr. insulation)
breakdown, ~ action Baskülverschluß m ~ bench
Durchschlagmeßtisch m ~ condition Durchschlagsbedingung f ~ couple Abfallmoment n
(sudden stoppage or stalling of polyphase
motors) ~ die Vorstein m ~ diode Zenerdiode f
~ division Aufteilung f ~ due to overvoltage
Überspannungsdurchschlag m ~ due to thermal instability Wärmedurchschlag m ~ field-
strength Durchschlagsfeldstärke f ~ frame
Basküle f ~ lorry Abschleppwagen m ~ moment Abfallmoment n ~ point Glimmspannung f, (sudden stoppage or stalling of polyphase motors) Abfallmoment n ~ (resistance)
potential Durchschlagsfestigkeit f ~ potential
Einsatzspannung f ~ rating Durchschlagsfestigkeit f ~ signal Abbrechungssignal n ~ stand
Streckgerüst n ~ strength Durchschlagsfestigkeit f, Spannungssicherheit f ~ test Bruchprobe
f, Durchschlagprobe f (insulation) ~ tester
Spannungsprüfer m ~ torque Abfallmoment n,
Kippmoment n ~ truck Abrüstwagen m ~
voltage Durchschlagsfestigkeit f, Durch-
schlags-, Überschlag-, Zünd-spannung f ~

voltage in air Luftdurchschlagspannung f ~
voltage of the creep distance Gleitfunkenspan-
nung f ~ wagon Hilfsgerätewagen m
breakdowns Vorprodukte pl (Walzwerke)
break, ~ frequency Frequenz f im Kennlinienknick ~ gap Abriß m ~ impulse Öffnungsim-
puls m ~-in Pausenzwischenruf m ~-in procedure
Dazwischensprechen n ~ jack Trenn-, Unterbrechungs-klinke f ~ key Unterbrechungstaste f ~ line kurze
Linie f ~ link Trennbügel m ~-off Durchbruch
m, Durchschlag m ~-off cross heading Durchhieb m ~-off diagramm Reißdiagramm n ~-off
point Strömungsabreißpunkt m ~-off region
Reißgebiet n ~-out block Abschraubblock m
~-out plate Meißelabschraubvorrichtung f
~-out tongs Abschraubzange f, ~ Haltezeichen
n (print.) ~ point Durchbruch m, Verzweigungspunkt m, Zwischenstop m (info proc.) ~ pulse
Öffnungsstoß m ~ relay Unterbrechungs-
relais n ~ signaling system Pausensignalanlage f
~ spark Unterbrechungsfunke m ~staff of a
smith's bellows Blasebalgschwengel m ~
-through Durchbruch m ~-to-make ratio Impulsverhältnis n ~water Wellenbrecher m
breakable brechbar, zerbrechlich
breakage Brechen n, Bruch m ~ of a roll Walzenbruch m ~ allowance zulässiger Bruch m ~
load Bruchlast f
breaker Aufbrecher m, Brecher m, Brechmaschine f, Brechtopf m (Walzwerk), Protektorein-
lage f, Schlagmühle f, Vorkarde f, Wellenbrecher m, (circuit) Schulter m, (textiles) Vor-
kratze f ~s Brandung f, Brechmesser m, Sturzsee f ~ with mantle of armored steel plates
Panzerbrecher m
breaker, ~ arm Unterbrecherhebel m ~ block
Block-blei n, -brecher m ~ bolt Zerreißbolzen
m ~ cam Unterbrechernocken m ~ card
Grobkrempel m, Schrubbelmaschine f, Vorkrempel m ~cord strip Cordgewebe n ~ current
Öffnungsstrom m ~ jaw Brechbacke f ~ plant
Brechhaus n ~ plate Prallfläche f (of hammer
mill) ~ plow Brech- oder Rodepflug m ~ point
Abbrennbürste f, Brechpunkt m (electr.),
Unterbrecherspitze f, Störgrenzpegel m ~ point
gauge Unterbrecherkontaktlehre f ~ stack
Feuchtglättwerk n (paper mfg.), feuchtes
Glättwerk n (paper mach.) ~ strip Panzerlage f
~-tripped signal Schalterfallmeldung f ~-type
Brechertyp m
breaking Ausschalten n, Brechung f, Bruchstelle
f, Gewinnung f, Grobzerkleinerung f, Knickung
f, Reißerscheinung f, Springen n, Trennung f,
Zerbrechen n, Zerreißen n ~ of the arc Abreißen
n des Lichtbogens ~ up of an avalanche at the
starting point Lawinenverbauung f ~ a code
Entschlüsselung f ~ of a dike Deichbruch m
~ of the drill stem Gestängebruch m ~ of an
embankment Deichbruch m ~ up of ice by
explosives Eissprengen n ~ of local calls for toll
calls through clearing Fernamtstrennung f ~
up of a mole Molenbruch m ~ up of a pier
Molenbruch m ~ the priming Säugung f
unterbrechen ~ into small pieces Zerstückelung f
breaking, ~ band Reißband n ~ capacity Aus-
schalt-leistung f, -vermögen n, Nennausschalt-

leistung f ~ **capacity of a switch** zulässiger Ausschaltstrom m ~ **card** Vorkarde f ~ **coal** Abkohlen n ~ **cone** Brechkegel m ~ **cord on parachute** Abreißschnur f ~ **current** Öffnungsinduktionsstrom m ~ **delay** Abfallverzögerung f
**breaking-down** Abwerfen n (rock), Durchschlagen n, Vorkaliberwalzen n, Vorstrecken n, Zerfall m, Zubruchgehen n, (into components) (Niederbrechung f) ~ **of the suction column** Abreißen n der Saugsäule
**breaking-down, ~ mill** Grobstrecke f, Vorwalzwerk n ~ **mill press** Blockbrechpresse f ~ **mill train** Grob-eisenstraße f, -straße f ~ **pass** Schnellvorwalzkaliber n, Streckkaliber n, Vorstich m, Vorstreckkaliber n ~ **roll** Schnellvorstreck-, Streck-, Vorkaliber-, Vorstreck-walze f ~ **stand** Vorgerüst n ~ **strand** Vorstrecke f ~ **temperature** Abbautemperatur f ~ **test** Durchschlagsversuch m ~ **train** Vorstraße f
**breaking, ~ drop effect** Wasserfalleffekt m ~ **elongation** Bruchdehnung f ~ **engine** Halbstoffholländer m ~ **factor** Bruchfaktor m ~ **hammer** Anschlagfäustel n ~**-in** Anlegemanöver n, Einbruch m; Hochtrainieren n, Hochzüchten n (armor) ~**-in shot** Auflockerungssprengung f, Einbruchschluß m ~ **knife** Schlagschneide f ~ **length** Bruchlast f, Reißkilometer m, Reißlänge f (synthetic fibers) ~ **lever** Unterbrecherhebel m ~ **limit** Berstdruck m, Bruchgrenze f, Zerreißgrenze f ~ **load** Bruchbelastung f, Bruch-festigkeit f, -last f, Knicklast f, Reißlast f, Zerreißbelastung f ~ **load test** Belastungsprobe f bis zum Bruch ~ **machine** Schlagwerk n ~**-off** Abbruch m ~**-out** Losbrechen n ~**-out point** Abzweigstelle f ~ **piece** Brechblock m (des Walzwerks) ~ **plant** Brechachse f, Zerkleinerungsanlage f ~ **point** Bruchpunkt m (chem.) Festigkeitsgrenze f, Sollbruchstelle f, Zerreißgrenze f, Störgrenzpegel m ~ **resilience** Knickspannung f ~ **ring** Brechring m ~ **shaft** Brech-achse f, -spindel f (des Walzwerks) ~ **speed** Reißgeschwindigkeit f ~ **step** Außertrittfallen n ~ **strain** Bruch-grenze f, -last f, -spannung f, -widerstand m, Reißfestigkeit f, Reißlast f, Reißwiderstand m, Zugfestigkeit f ~ **strength** Bruchfestigkeit f, Bruchwiderstand m, Knickfestigkeit f, Knickwiderstandsfähigkeit f, Reißfestigkeit f, Reißwiderstand m, Tragfestigkeit f, Trennungsfestigkeit f, Zerbrechungsfestigkeit f, Zugfestigkeit f ~ **stress** Bruchbeanspruchung f, Bruchfestigkeit f, Bruchspannung f, Knickspannung f, Zerreißbelastung f ~ **tension** Bruchdehnung f ~ **test** Bruchprobe f, Knickversuch m, Zerreißprobe f ~**-up** Abwracken n, Aufbrechen n, Losbrechen n, Zermürbung f ~ **weight** Bruchlast f
**breakwater** Fangbuhne f, Rißbank f, Schirmwerk n, Wellenbrecher m
**breast** Brust f, Feldesbreite f ~ **of slipway** Vorhelling f
**breast, ~ beam** Brustbaum m (naut) ~ **boards** Pfändung f ~ **borer** Faustleier f ~ **box** Stoffauflauf m, Stoffeinlauf m (paper mfg.), Verteilungskasten m ~ **drill** Bohrdreher m, Bohrwinder m, Brustbohrer m, Brustleier f, Faustleierbohrer m ~ **harness** Sielengeschirr n ~ **height** Fensterbrüstung f ~ **lead** Taucherherz n

~ **line** Querleine f ~ **plate** Bohrgerüst n, Brust-panzer m, -schild m, -platte f (for drilling), -blatt n (harness) ~**plate harness** Vorderzeug n ~**plate microphone** Brustmikrofon n ~**plate transmitter** Brustmikrophon n ~ **roll** Brust-walze f, -baum m (textiles) ~ **roller** Friktionsrolle f (Walzwerk), Stufenrolle f, Vortrommel f ~ **rollers** Ständerrollen pl ~**rope** Dwarsfeste f ~ **side of work** Feldesbreite f ~ **strap** Blatt n (harness) ~**summer** Oberschwelle f ~ **support** Bruststütze f ~ **water wheel** Kropfrad n, mittelschlächtiges Wasserrad n ~ **wheel** Kropfrad n ~**work** Brustwehr f, Brustwerk n, Fensterbrüstung f, Schulterwehr f, Umwehrung f
**breasting, ~ of the beater** Sattel m des Holländers, Holländerkropf m ~ **backfall** Berg m oder Sattel m eines Papierholländers (paper mfg.)
**breath** Atem m, Hauch m ~ **filter layer** Mundschicht f ~ **protection sheet (or shield)** Hauchschutz m (bei Spaltlampengeräten)
**breathe, to ~** atmen, ein- oder ausatmen
**breathed consonant** stimmloser Konsonant m
**breather** Atmungsventil n, Druckausgleichsöffnung f, Entlüfter m, Entlüftungsöffnung f, Entlüftungsstutzen m, Entlüftungsvorrichtung f, Luft-, Saug-, Schnarch-, Schnüffel-ventil n ~ **line** Entlüftungsleitung f ~ **pipe** Entlüftungsleitung f ~ **reserve tank** Atem-Nachfüllbehälter m ~ **roof** Atmungsdach n (tank)
**breathing** Atemtätigkeit f, Atmen n, Atmung f, Druckablassen n (releasing pressure in crankcase); atmend ~ **of crankcase** Atmung f des Kurbelgehäuses ~ **of microphone** Atmen n des Kohlemikrofons, Mikrofonatmen n
**breathing, ~ apparatus** Atemgerät n, Atmungs-apparat m, -gerät n ~ **apparatus for high-altitude flying** Höhen-atmer m, -atmungsgerät n ~ **apparatus for stratosphere flying** Einrichtung f für Höhenatmung ~ **appliance** Atmungsgerät n ~ **bag** Atemsack m ~ **characteristics of a motor** Gaswechseleigenschaften pl des Motors ~ **mast** Luftmast m, Schnorchel n ~ **microphone** atmendes Mikrofon n ~ **resistance** Atemwiderstand m (gas mask) ~ **tube** Atem-, Atmungs-schlauch m (oxygen apparatus) ~ **ventilator of tank** Atmungsbehälter m für Tankgase
**breccia** Breccie f, Brekzie f, Bresche f, Brockengestein n, Schuttbreccie f, Trümmergestein n ~ **structure** Brecciengefüge n
**breech** Lade-raum m, -öffnung f (of gun), Verschluß m (mechanism) ~ **bar** Verschlußschiene f
**breechblock** Fallblock m, Federkolben m, Geschütz-, Gewehr-verschluß m, Schieber m, Stoßboden m, Verschluß m, Verschlußblock m, -keil m, -schraube f ~ **carrier** Verschlußträger m ~ **cover** Verschlußkappe f, -überzug m ~ **handle** Verschlußhebel m ~ **mechanism** Blockverschluß m ~**-mechanism part** Verschlußteil m ~ **screw** Gewehrverschluß-schraube f ~ **slide** Verriegelungsschlitten m
**breech, ~ body** Verschlußkörper m ~ **bolster** Richtkissen n ~**-bolt recess** Lager n für den Verschlußbolzen ~ **bore gauge** Kaliber-ring m, -zylinder m ~ **buffer plug** Verschlußpufferdorn

*m* ~ **carrier** Verschlußtrage *f* ~ **casing** Schloß-schützer *m*, Verschlußmantel *m* ~**-closing spring** Schließfeder *f* ~ **door** Boden-verschluß *m*, -klappe *f* (torpedo) ~**loader** Hinterlader *m* ~ **lock** Ladeklappe *f* ~**locking crank** Ver-schlußkurbel *f* ~ **mechanism** Geschützver-schluß *m*, Ladevorrichtung *f*, Zylinderver-schluß *m* ~**-mechanism connector arm** Rück-holer *m* ~**-mechanism crank** Öffnerkurbel *f* ~**-mechanism lever** Öffnerhebel *m* ~ **piece** Laufmundstück *n* ~ **pin** Schwanzschraube *f* ~ **plate** Bodenstück *n*, Schloßplatte *f* ~ **rail** Ver-schlußschiene *f* ~ **rear plate** Verschlußplatte *f* ~ **recess** Keilloch *n* ~**-recess surface** Keil-lochfläche *f* ~ **ring** Verschlußstück *n* ~ **starting mechanism** Patronenanlaßvorrichtung *f*
**breeches,** ~ **buoy** Hosenboje *f* ~ **chute** Hosen-rutsche *f*
**breeching** Hinterzeug *n*, Schornsteinfuchs *m*
**breed, to** ~ erzeugen, ziehen, züchten
**breed** Gattung *f*
**breeding** Kultur *f*, Zucht *f* ~ **apparatus** Zuchtap-parat *m* ~ **gain** Brutgewinn *m*
**breeze** Abrieb *m*, Brise *f*; Grus *m*, kleinstückiger Koks *m*, Koksgrus *m*; Kühlte *f*, Lösche *f*, Wind *m* **(coke)** ~ Kokslösche *f* ~ **brick** Aschestein *m* ~ **bunker** Abfallbunker *m* ~ **coal** Feinkohle *f*
**B-register** Adressenregister *n*
**Brescian steel** Persanerstahl *m*
**breve** Kürzezeichen *n*
**brevier** Antiqua *f*, Petit *f*
**brevity** Abkürzung *f*, Kürze *f* ~ **code** Kürzungs-kode *m*
**brew** Gebräu *n* ~ **kettle** Braukessel *m*
**brewer** Brauer *m* ~**s' grain molasses** Biertreber-melasse *f* ~**'s grains** Treber *pl* ~**'s pitch** Brauer-pech *n* ~**'s sugar** Brauzucker *m* ~**'s vat** Faß *n*
**brewery** Brauerei *f* ~ **stirring device** Brauerei-rührwerk *n*
**brewing** Bierbrauen *n*, Brauen *n*, Gebräu *n*, Sud *m* ~ **cup** Braubottich *m* ~ **house** Sudhaus *n* ~ **pan** Braupfanne *f* ~ **tun** Braubottich *m*
**Brewster angle** Polarisationswinkel *m*
**bribe, to** ~ bestechen
**bribe** Bestechung *f*
**bribery** Bestechung *f*
**brick, to** ~ einmauern, mauern **to** ~ **up** auf-mauern, zumauern
**brick** Barn-, Bau-, Brand-, Mauer-, Ziegel-, Ziegelback-stein *m*, gebrannter Stein *m*, Häufigkeitsprogramm *n*, Ziegel *m*, (building) Blauziegel *m* ~ **and slab machine** Ziegel- und Kachelherstellungsmaschine *f* ~ **arch** Form-stein *m*, Gewölbebackstein *m* ~ **baking drum** Steinhärtekessel *m* ~**bat** Stein-, Ziegel-Brocken *m* ~ **burner** Ziegelbrenner *m* ~ **burning** Ziegel-brand *m*, -brennen *n* ~ **channel** gemauer-ter Kanal *m* ~ **chisels** Chariereisen *pl* ~ **clay** Ziegel-erde *f*, -ton *m* ~ **course** Ziegelschicht *f* ~ **course laid on edge** Rollschicht *f* ~ **foundation** Ziegelsteinfundament *n* ~ **kiln** Backsteinofen *m*, Ziegelei *f*, Ziegeleiofen *m*, Ziegelhütte *f*, Ziegelofen *m* ~**layer** Maurer *m*, Ziegelmauer *f* ~**laying** Ziegelarbeit *f* ~ **lining** Ausmauerung *f* ~ **maker** Ziegel-brenner *m*, -streicher *m* ~**mak-ing press** Steimetz-, Ziegel-presse *f* ~**-on-edge paving** Ziegelhochkantpflaster *n* ~**-pavement**

**construction** Klinkerpflasterbau *m* ~ **pit** gemauerter Schacht *m* ~ **press** Ziegelpresse *f* ~**-red** ziegelrot ~ **road** Klinkerstraße *f* ~ **screen** Backsteinraster *m* ~ **steam works** Dampf-ziegelei *f* ~ **stone** Putzstein *m* (fire) ~ **wall** Backsteinmauer *f*, Ziegelmauer *f* ~ **work** Aus-mauerung *f*, Backsteinmauerwerk *n*, Gemäuer *n*, Mauerwerk *n*, Ziegelmauerwerk *n* ~ **work casing** Ziegelsteinverkleidung *f* ~**work dam** Mauerdamm *m* ~**work lining** Futtermauer *f* ~**works** Ziegelbrennerei *f*, Ziegelei *f*
**bricking** Mauerung *f*
**bridge, to** ~ aufbrücken, in Brücke schalten **to** ~ **across** in Brücke schalten, überbrücken
**bridge** Brücke *f*, Meßbrücke *f*, Steg *m*, Über-laden *n*, (of a ship) Deck *n* **(pilot)** ~ Komman-dobrücke *f* **to be in** ~ in Brücke liegen (teleph.) ~ **above upper lock gate** Oberhauptbrücke *f* ~ **for alternating current** Wechselstrombrücke *f* ~ **of airproof cases** Kastenbrücke *f* ~ **of boiler furnace** Feuer-bock *m*, -brücke *f* ~ **of collapsible boats** Faltbootbrücke *f* ~**s on edge** hochge-kantete Stege *pl* ~ **for loading coal** Kohlenver-ladebrücke *f* ~ **of nose** Nasensteg *m* ~ **with plug contacts** Stöpselmeßbrücke *f* ~ **for stocking coal** Kohlenverladebrücke *f* ~ **for supporting pipe** Rohrbrücke *f* ~ **of valve** Ventilbügel *m* ~ **with a slewing jib crane running on top** Ver-ladebrücke *f* mit Drehkran
**bridge,** ~ **amplifier** Brückenverstärker *m* ~ **apex** Brückenscheitel *m*, Scheitel *m* der Brücke ~ **arch** Brückengewölbe *n* ~ **arm** Brückenarm *m* ~ **arrangement** Brückenschaltung *f* ~ **balance** Brücken-ausgleich *m*, -gleichgewicht *n* ~ **balance point** Nulldurchgang *m* ~ **balancing current** Brückenausgleichstrom *m* ~ **boom** Brückenträger *m* ~ **branch** Brückenzweig *m* ~**builder** Brückenbauer *m* ~ **building** Brücken-bau *m*, -schlag *m* ~ **cam** Schloßbrücke *f* ~ **capacity** Verkehrslast *f* ~ **causeway** Brücken-bahn *f* ~ **center line** Brückenachse *f* ~ **circuit** Brücken-schaltung *f*, -anordnung *f* (arrange-ment) ~ **coils** Brückenarm *m* ~ **commander** Brückenkommandant *m* (engin.) ~ **connection** Brückenschaltung *f* ~ **connector** Überbrük-kungsklemme *f* ~ **construction** Brückenbau *m* ~**-construction party** Brückentrupp *m* ~ **crane** Brückenkran *m*, Hochbahnkran *m*, Verlade-brücke *f* ~ **crossing** Brückenüberführung *f* ~ **design** Brückenbau *m* ~ **designer** Brücken-bauer *m* ~ **diagram** Brückenskizze *f* ~ **die** Stegdorn *m* ~ **drive** Brückenantrieb *m* ~ **duplex connection** Brückengegensprechschaltung *f* ~ **duplex system** Brückengegensprechsystem *n* ~ **exit** Brückenausgang *m* ~ **feedback** Brücken-rückkopplung *f* ~ **girder** Brückenträger *m* ~ **guide** Bandöse *f* ~ **head** Brückenkopf *m* **head line** Brückenkopflinie *f* ~**head postion** Brük-kenkopfstellung *f* ~ **joint** Brückenstoß *m* (rails) ~ **key** Brückenschlüssel *m* (electr.) ~ **line** Brückenlinie *f* (engin.) ~ **load** Brückenlast *f* ~ **measurement** Brückenmessung *f* ~ **megger** Brückenisolationsmesser *m*, Brückenmegger *m* ~ **member** Brückenteil *n* ~ **modulator** Ring-modulator *m* ~ **network** Vierpolkreuzglied *n* ~ **part** Brückenteil *m* ~ **piece** Einsatzbrücke *f* ~ **pile** Brückenpfeiler *m*, Grundpfahl *m* ~ **planks** Brückenbelag *m* ~ **plate** Verladebrücke *f* ~

**raft** Brückenfloß n ~ **rail** Brückschiene f ~ **railing** Brückengeländer n ~ **rectifier** Brückengleichrichter m ~ **relay** Brückenrelais n ~ **resistance** Brückenwiderstand m ~ **roadway** Brückenfahrbahn f ~ **site** Brückenstelle f ~ **span** Brückenspannung f ~ **spot weld** Laschenpunktschweißung f ~ **substructure** Brückenüberbau m ~ **superstructure** Brückenüberbau m, Fachwerkträger m ~ **test** Brückenmessung f ~ **transformer** Ausgleichübertrager m ~ **transition** Reihenparallelschaltung f mit Brückenschaltung ~ **type unloading installation** Verladebrücke f ~ **wall** Brückenkörper m (**front**) ~ **wall** Feuerbrücke f ~ **wall tube** Rohr n an der Feuerbrücke ~**wall tubes** Brückenmauerrohre pl ~ **wire** Brückendraht m ~**wire cap** Brückenglühzünder m (explosives)

**bridged,** ~ **across** in Brücke geschaltet ~ **ringer** Rufumsetzer m (zwischen Leitungsadern) ~ **T network** überbrücktes T-Glied n

**bridging** Brückenschlag m, Schwartenbrett n, Überbrückung f, Zusammenbacken n ~ **capacitor** Quer-, Überbrückungs-kondensator m ~ **coil** Abzweig-, Parallel-spule f; Quer-rolle f, -spule f ~ **condenser** Parallel-, Quer-, Überbrückungs-kondensator m ~ **connection** Überbrückungsschaltung f ~ **equipment** Brückengerät n ~ **jack** Parallelklinke f ~ **joist** Dielenlager n, Polsterholz n ~ **switch** Überbrückungstaste f

**bridle, to** ~ zügeln

**bridle** Federbride f, Zaum m, Zäumung f, Zaumzeug n, Zügel m ~ **cable** Gummiabschlußkabel n ~ **mooring** Kettengehänge n ~ **path** Reitweg m ~ **ring** Zügelring m ~ **rod** Gegenbalancier m ~ **wire** Bandhaken m, Endpeitsche f, Gummiabschlußkabel n, Hakenband n, Überführungsdraht m

**brief, to** ~ informieren

**brief** Kurznachricht f, Schriftsatz m, Schriftstück n, (foolscap paper) Kanzleipapier n; bündig, knapp, kurz, kurzgefaßt ~ **for appeal** Berufung f ~**case** Akten-mappe f, -tasche f

**briefing** Befehlausgabe f, Berichterstattung f, Einsatzbesprechung f ~ **of flight crews** Flugbesprechung f ~ **room** Aufklärungs-, Flugbesprechungs-raum m

**bright** glänzend, heiter, hell, klar, lebhaft, leuchtend, licht, stark beheizt, weißglühend, (metal) blank ~ **and turned bolts and screws** blanke und gedrehte Schrauben pl ~ **adaptation** Hellanpassung f ~**annealed** blankgeglüht ~ **annealing** Blankglühen n ~ **annealing furnace** Blankglühofen m ~ **blue** hochblau ~**burn, to** blankbrennen ~**crystalline fracture** grobkörnige Bruchfläche f

**bright-dark,** ~ **change-over switch** Helldunkelschalter m ~ **field condenser** Helldunkelfeldkondensator m ~ **field slider** Helldunkelfeldschieber m

**bright-dim relay** Abdunkelrelais m

**bright-drawn** blankgezogen ~ **material** blank gezogenes Material m ~ **hexagon steel bar** blank gezogener Sechskantstahl m ~ **sections** blanke Profile pl

**bright,** ~ **emitter** Reinkathode f ~**emitting cathode** Hochtemperaturkathode f ~ **enamel** gebürsteter Glacé m ~ **enamels** Hochglanzpa-

piere pl ~ **field diaphragm** Hellfeldblende f ~ **field image** Hellfeldbild n ~**field observation** Hellfeldbeobachtung f ~**finished** glanzdekatiert ~**finished sheet** hochglanzpoliertes Blech n ~ **golden** goldhell ~**gray** hellgrau ~ **green** hochgrün ~ **ground illumination** Hellfeldbeleuchtung f (micros) ~**level value** Weißwert m (TV) ~ **light** Fernlicht n ~ **lights** Spitzlichter pl ~ **luster** Hochglanz m ~ **machined bolt** bearbeiteter Bolzen m ~ **orange** Gelbglut f ~ **polished carbon steel** Silberstahl m ~ **red** hochrot, kirschrot ~**red heat** Hellrot-glühhitze f, -glut f ~ **red-hot** hellrotglühend ~ **rim** Lichtrand m ~ **(cold-) rolled** blankgewalzt ~ **search** Hellsuchen n ~ **side** Lichtseite f ~ **steel** Blankstahl m ~ **stock** Brightstock m (oil), unvollendetes Zylinderöl n ~ **valve** hochbeheizte Röhre f ~ **vermillion** hochrot ~ **wire** blanker Draht m,

**brighten, to** ~ abblicken, aufblicken, blicken, erheitern, erhellen, (color) auffrischen **to** ~ **up** aufhellen

**brightener** Abklärer m (dyeing)

**brightening** Aufhellung f ~ **with acid** Säureavivage f ~ **with fat** Fettavivage f ~ **power** Aufhellvermögen n ~ **pulse** Helligkeitsimpuls m

**brightness** Beleuchtungsstärke f, Flächenhelle f, Glanz m, Helligkeit f, Leucht-dichte f, -stärke f ~ **of background** Grundhelligkeit f ~ **of the crosslines** Fadenhelligkeit f (opt.) ~ **of image** Bildhelligkeit f ~ **of spot** Fleckhelligkeit f ~ **of the sun** Sonnenhelligkeit f ~ **of a surface** Leuchtdichte f

**brightness,** ~ **brilliance** Helligkeit f ~ **control** Helligkeits-regelung f, -regler f ~ **density of screen** Bildwandleuchtdichte f ~ **difference** Leuchtdichtekontrast m ~ **distribution** Helligkeitsverteilung f ~ **fluctuation** Helligkeitsschwankung f ~**intensity change** Helligkeitsveränderung f ~ **intensity measurement** Helligkeitsmessung f ~ **knob** Bildhelligkeitsregler m ~ **level** Leuchtdichtewert m ~ **meter** Belichtungsmesser m ~ **setting** Helligkeitseinstellung f ~ **temperature** Leuchttemperatur f ~ **tester** Weißgradmesser m ~ **value** Helligkeitswert m

**Bright's bells** Doppelläufer m

**brilliance** Brillanz f, Flächenhelle f, Helligkeit f (rdr), Hochglanz m, Kontrast m (TV), Wiedergabenatürlichkeit f ~ **of picture** Bildhelligkeit f ~ **of sound reproduction** Wiedergabebrillanz f

**brilliance,** ~ **control** Helldunkelsteuerung f, Helligkeitsregler m (rdr), Kontrastregler m (TV) ~ **modulation of control electrode** Aufhellungsorgan n ~ **wave** Helligkeitswelle f (electron.)

**brilliancy** Glanz m, Helligkeit f

**brilliant** glänzend, hell ~ **black** Glanzschwarz n ~ **bronze** Glanzbronze f ~ **green** Brillantgrün n ~ **lisle** Glanzflor m ~ **polish** Hochglanzpolitur f ~ **starch** Stärkeglanz m ~ **starch for linen** Wäscheglanzstärke f ~ **varnish** Glanzlack m ~ **white** Brillantweiß n

**brim, to** ~ rändeln

**brim** Kimme f, Kranz m, Rand m ~**stone impression** Schwefelabdruck m

**brine** Lauge f, Salz-lauge f, -lösung f, -sohle f, -wasser n, Seewasser n; Sole f, Sulze f, Sülze f ~ **cooler** Solekühler m ~ **ditch** Solkanal m ~

**gauge** Solwaage *f* ~ leaching Salzlaugung *f* ~ outlet Soleaustritt *m* ~ **pump for rapid pickling** Lakepumpe *f* zum Schnellpökeln ~ **salt** Quell-, Sol-salz *n* ~ **socket** Sollöffel *m* ~ **spring** Solquelle *f*
**brineing** Versalzen *n*
**brinell, to** ~ brinellieren
**Brinell,** ~ **ball(-hardness) test** Brinell-Kugeldruckprobe *f* ~ **coefficient of hardness (or hardness number)** Brinellsche Härtezahl *f*, Brinell-Zahl *f* ~ **hardness** Brinell-Härte *f* (hardness test for metals) ~ **hardness test** Brinell-Härteprobe *f*, Brinell-Meßverfahren *n*, Kugelhärteprobe *f* ~ **hardness-test method** Eindruckverfahren *n* ~ **hardness tester** Brinell-Härteprüfer *m* ~ **test** Brinell-Probe *f* ~ **unit** Brinell-Einheit *f* (metal)
**bring, to** ~ **bringen to** ~ **about** bewirken, herbeiführen, zustandebringen **to** ~ **action (against)** klagen, klagbar werden **to** ~ **back** nachholen, zurückbringen **to** ~ **down** herabsetzen, zum Absturz bringen (aviat.) **to** ~ **down the roof** zu Bruche bauen (min.) **to** ~ **forth** hervorbringen **to** ~ **home** einbringen **to** ~ **in** einbringen, einholen, zur Produktion bringen **to** ~ **judiciary (or legal) action** gerichtlich belangen **to** ~ **out** ausbringen, herausbringen **to** ~ **suit (against)** klagbar werden **to** ~ **to** (circuit point) hinführen **to** ~ **together** ansprengen (Endmasse) **to** ~ **up** anfahren, heranschaffen, heranziehen, herbeischaffen, nachziehen, vorbringen, vorführen, zuführen **to** ~ **up to** einstellen auf
**bring, to** ~ **an action** erheben **to** ~ **into agreement** nachsteuern **to** ~ **to coincidence** zur Deckung bringen **to** ~ **complaints to notice** Beanstandungen *pl* geltend machen **to** ~ **into continuity (or coincidence)** in Koinzidenz bringen **to** ~ **into line** ausstraken **to** ~ **to the nearest round figure** aufrunden **to** ~ **into phase** gleich aufsetzen, (in) gleich aufsetzen **to** ~ **into position** in Stellung bringen **to** ~ **a reaction to completion** abreagieren **to** ~ **to speed** anfahren **to** ~ **up steam** anlassen **to** ~ **into step** synchronisieren **to** ~ **on stream** ingangsetzen (petroleum) **to** ~ **into view** sichtbar machen **to** ~ **to white heat** weißglühen
**bringer** Zubringer *m*
**bringing,** ~ **forth** Hervorbringung *f* ~ **forward** Heranführung *f* ~ **down lumps** Stückfall *m* (min.) ~ **in the rotor of the turbogenerator** Einfahren *n* des Turbogeneratorläufers ~**-up** Vorbringen *n* ~**-up of fuel** Kraftstoffnachschub *m*
**brinneling** Reibkorrosion *f*
**briquette, to** ~ brikettieren, paketieren, pressen
**briquette** Brikett *n*, Formling *m*, Paket *n*, Preßkohle *f*, Preßkuchen *m*, Preßling *m*, Ziegel *m* ~ **manufacture** Brikettherstellung *f* ~ **mold** Preßkuchenform *f* ~ **plant** Brikettier-anlage *f*, -anstalt *f* ~ **press** Brikettpresse *f* ~ **sample** Preßkuchenprobe *f*
**briquetted coal** Preßkohle *f*
**briquetting** Brikettierung *f*, Paketierung *f* ~ **of pulp** Schnitzelbrikettierung *f* ~ **asphalt** Pech *n* für die Preßkohlenindustrie ~ **plant** Brikettfabrik *f*, Brikettier-anlage *f*, -anstalt *f*, Brikettierungsanlage *f* ~ **press** Brikettpresse *f* ~ **process** Brikettierverfahren *n* ~ **property** Brikettierfähigkeit *f*

**brisant** zerbrechend ~ **explosives** brisante Sprengstoffe *pl*
**brisk** lebhaft, schwunghaft, stark
**bristle** Borste *f* ~**-finishing works** Borstenzurichterei *f*
**britannia metal** Babbittmetall *n*
**British** englisch ~ **Association Standard Screw Thread** Thuryuhrschraubengewinde *n* ~ **mile** englische Meile *f* ~ **Standard Institution (BSI)** Normenausschuß *m* ~ **Thermal Unit (BTU)** Britische Wärmeeinheit (1 BTU = 252 Kalorien)
**brittle** anbrüchig, brechbar, bröck(e)lig, brüchig, faul, hart-, heiß-, kurz-brüchig, morsch(ig), mürbe, rösch, springhart, spröde, sprödig, zerbrechlich ~ **when red-hot (or red-short)** rotbrüchig
**brittle,** ~ **failure** Sprödbruch *m* ~ **fracture** Sprödbruch *m* ~ **fractures of steel** Brucharten *pl* des Stahles ~ **iron** sprödes Eisen *n* ~ **lacquer** Reißlack *m* ~ **metal** Rotmessing *n* ~ **point** Bruchpunkt *m* (of glass) ~ **silver ore** Rösch-erz *n*, -gewächs *n*
**brittleness** Brüchigkeit *f*, Faulbruch *m* (metal), Faulbrüchigkeit *f*, Sprödigkeit *f*, Zerbrechlichkeit *f* ~ **owing to caustic lye treatment** Laugensprödigkeit *f* ~ **when hot** Warmsprödigkeit *f*
**broach, to** ~ aufreiben, ausräumen, nutenziehen, räumen, ziehen (holes)
**broach** Ahle *f*, Glätträumnadel *f*, Räum-ahle *f*, -nadel *f*, -stahl *m*, -werkzeug *n*, Reibahle *f*, Zieh-dorn *m*, -stange *f* **to** ~ **a hole** Loch auftreiben
**broach,** ~ **collecting grid** Dornsammelrost *m* ~ **elevator** Räumnadelheber *m* ~ **guiding tray** Räumnadelführung *f* ~ **life** Räumnadelstandzeit *f* ~ **post** Helmstänge *f* ~ **travel** Räumhub *m* (der Räummaschine)
**broaching** Räumen *n*, (cask) Anstich *m* ~ **of a groove (or keyway)** Räumen *n* einer Nut
**broaching,** ~ **bit** Erweiterungs-, Nach-bohrer *m* ~ **cutter** Räumnadel *f* ~ **fixture** Räumvorrichtung *f* ~ **machine** Loch-, Räum-, Räumnadelzieh-maschine *f* ~ **press** Räumpresse *f* ~ **stroke** Räumhub *m* (der Räummaschine) ~ **tap** Anstichhahn *m* ~ **tool grinding machine** Räumnadelschleifmaschine *f* ~ **tool sharpening machine** Räumnadelschärfmaschine *f*
**broad** breit, weit, roh (rdo)
**broad** Lampenaggregat *n* (film) ~ **ax** Breitbeil *n*
**broadband,** ~ **amplifier** Breitbandverstärker *m* (rdr) ~ **antenna** Breitbandantenne *f* ~ **balun** Breitband-Symmetrie-Überträger *m* ~ **carrier system** Breitbandträgerfrequenzsystem *n* ~ **circuit** Breitbandkreis *m* ~ **conveyor** Breitbandübertragung *f* ~ **filter** Bandfilter *m* oder Siebgebilde *n* oder Siebkette *f* von großer Lochweite ~**-receiver** Breitbandempfänger *m*
**broad,** ~ **beam** Großfeldabsorption *f*, breiter Strahl *m* ~ **beam head lamp** Breitstrahler *m*
**broadcast, to** ~ durch Rundfunk übertragen oder verbreiten, rundfunken, senden
**broadcast** Rundfunk *m*, Sendung *f* ~ **amplifier** Rundfunkverstärker *m* ~ **apparatus** Rundfunkgerät *n* ~ **listener** Radiohörer *m* ~ **receiver** Rundfunk-, Rundspruch-empfänger *m* ~ **reception** Rundfunkempfang *m* ~ **relay** Draht-, Drahtrund-funk *m* ~ **seeder** Breit-saatsäe-

maschine *f*, -säemaschine *f* ~ **system** Rundschreibsystem *n* ~ **telegraphy** Rundschreibeanlage *f* ~ **transmission** Rundfunkübertragung *f* ~ **transmitter** Rundfunksender *m* ~ **wave** Rundfunkwelle *f*
**broadcaster** Rundfunkteilnehmer *m*
**broadcasting** Aufführung *f*, Rundspruch *m*, Rundfunk *m* ~ **in clear** Klartextfunken *n* ~ **at dictation speed** Schreibfunk *m* ~ **engineering** Rundfunktechnik *f* ~ **range** Hörbereich *m* ~ **receiver circuit** Rundfunkempfängerkreis *m* ~ **station** Funkhaus *n*, Funksender *m*, Rundfunk-sender *m*, -station *f*, -stelle *f*, Sender *m*, Sendestelle *f* ~ **studio** Rundfunksenderaum *m* ~ **-studio performance** Sendespiel *n*
**broad,** ~ **channel** Einziehung *f*, Halskehle *f* ~ **chisel** Sticheisen *n* ~ **dimension** breite Seite

lisiermaschine *f* ~ **-faced steel gear** Kammwalze *f* ~ **flanged beams** Breitflanschträger *m* ~ **iron wedge** Plötz *m* ~ **lath** getrennte Latte *f* ~ **-nose tool** Breitschneidestahl *m* zum Vorstechen ~ **pendant** Doppelständer *m* ~ **pick** Flachhaue *f* ~ **pulse** breiter Impuls *m*, Impuls *m* mit langer Laufzeit ~ **sheet** Atlantenformat *n* ~ **sheet** Inplano *n*, Querformat *n*
**broadside** Anschlagzettel *m*, Atlantenformat *n*, Breitseite *f*, Breitstrahler *m* (film), Inplano *n*, Patent *n*, Querformat *n*, Salve *f* ~ **array** Antennenanordnung *f* für Querstrahlung, Dipolebene *f* ~ **on** breitseits ~ **port** Breitseitpforte *f* ~ **sea** Dwarssee *f* ~ **slip** Breitseitestapel *m*
**broad,** ~ **specification** weitgehende Beschreibung *f* ~ **strip mill** Breitbandwalzwerk *n* ~ **tuning** rohe oder unscharfe Abstimmung *f* ~ **wedge formation** Breitkeil *m*
**broadwise** der Breite nach
**broaden, to** ~ abflachen, erbreiten, erweitern, verbreitern, verflachen
**broadening** Breiten *n*, Breitung *f*, Verbreiterung *f* ~ **of line (or strip)** Linienverbreiterung *f* (TV)
**broadly tuned** unscharf abgestimmt (rdo)
**brocaded gauze** Nadelstuhlgewebe *n*
**brocading** Broschieren *n*
**brochantite** Brockantit *m*
**brochure** Druckheft *n*
**broiling product** Rösterzeugnis *n*
**broke** Ausschuß *m*, Fabrikationsabfälle *pl*
**broken** durchbrochen, durchschnitten, unterbrochen, (sky) fast bedeckt, (line) strichliert, strichpunktiert, (sky) wolkig; Ausschuß *m* (paper mfg.), Fabrikationsabfälle *pl* ~ **away** weggebrochen ~ **down** umgebrochen, aufgeschlüsselt ~ **in** eingelaufen (Motor)
**broken,** ~ **brick** Ziegelbrocken *m* ~ **castings** Gußbruch *m* ~ **clouds** unterbrochene Wolken *pl* ~ **coal** Brechkohle *f* ~ **-contact lens** ungekittete zusammengesprengte Linse *f*, zusammengesprengtes Objektiv *n* ~ **-contact objective** zusammengesprengtes Objektiv *n* ~ **contact prisms** (closely fitted together) aneinandergesprengte Prismen *pl* ~ **-contact prisms** zusammengesprengte Prismen *pl* ~ **core** gebrochener Kern *m* ~ **corner** abgestoßene Kante *f* ~ **glass** Glas-brocken *m*, -bruch *m*, -scherben *pl* ~ **grain** zersprengte Narben *pl* ~ **ground** unebenes Gelände *n* ~ **hardening** unterbrochene

Härtung *f* ~ **line** gestrichelte Linie *f*, Knicklinie *f* ~ **-line curve (or graph)** gerissene Kurve *f* ~ **matter** gequirlter Satz *m* ~ **mold** gebrochene Form *f* ~ **ore** Erzklein *m* ~ **outlines** zerlegte Formen *pl* ~ **shade** gebrochene Farbe *f* ~ **slope** gebrochene Böschung *f* ~ **stone** Bruchstein *m*, grober Kies *m*, Kleinschlag *m*, Schotter *m*, Splitt *m*, Steinschlag *m* ~ **-stones manufacturing plant** durchgeschlagener Probestab *m* ~ **type** abgefallene Schrift *f* ~ **wire** Drahtbruch *m*
**broker** Makler *m*, Unterhändler *m*
**brokerage** Maklerkosten *pl* ~ **(fee)** Unterhändlergebühr *f*
**brokes** Ausschuß *m*, Fabrikationsabfälle *pl*
**bromate** Bromat *n*, bromsauer
**brombenzyl cyanide** Brombenzylzyanid *n*
**bromcresol purple** Bromkresolpurpur *m*
**bromide** Bromid *n* ~ **paper** Bromidpapier *n*, Bromsilber-papier *n* (photo), -gelatinepapier *n*
**bromine** Brom *n* ~ **fluoride** Bromfluor *n* ~ **hydrate** Bromhydrat *n* ~ **-phenol blue** Bromphenolblau *n* (chem.)
**bromlite** Alstonig *m*
**bromoacetyl bromide** Bromacetylbromid *n*
**bromobenzoic acid** Brombenzoesäure *f*
**bromocyanide process** Bromozyanidprozess *m*
**bromocyanogen** Bromzyan *n*
**bromoil printing paper** Bromöldruckpapier *m*
**bromonaphthalene** Bromnaphthalin *n*
**bromostannic acid** Zinnbromwasserstoffsäure *f*
**bromyrite** Bromspat *m*
**bronchography** Bronchografie *f*
**bronchophony** Röhrenstimme *f*
**bronze, to** ~ bronzieren, brünieren
**bronce** Bronze *f*, Rotguß *m* ~ **and silver printing** Argentindruck *m*
**bronze,** ~ **bearing bush** Bronzelagerschale *f* ~ **blocks** Bronzeklötze *pl* ~ **bush** Rotgußlagerschale *f* ~ **bushed bearings** Bronzebüchsen *pl* ~ **-casing** Bronzegehäuse *n* ~ **chaser** Bronzeziselör *m* ~ **coating** Bronzebezug *m* ~ **color** Erzfarbe *f* ~ **foundry** Bronzegießerei *f* ~ **hinge** Bronzegelenk *n* ~ **joint** Bronzegelenk *n* ~ **-lead alloy** Bleibronze *f* ~ **lustre** Bronzeglanz *m* ~ **metal box** Rotgußbüchse *f* ~ **pigment** Bronzefarbe *f* ~ **-pigmented lacquer** Bronzelack *m*, -mischlack *m*, -tinktur *f* ~ **-pigmented varnish** Bronze-mischlack *m*, -tinktur *f* ~ **plating** Bronzebezug *m* ~ **printing varnish** Bronzedruckfirnis *m* ~ **screw wire** Bronzeschraubendraht ~ **sleeve** Bronzebezug *m* ~ **slide** Gleitbahn *f* von Bronze ~ **-statue foundry** Bildgießerei *f* ~ **type** Bronzeschrift *f* ~ **wire** Bronzedraht *m* ~ **work** Bronzearbeit *f*
**bronzing,** ~ **lacquer** Bronzelack *m* ~ **machine** Abstaubmaschine *f* ~ **medium** Bronzeunterdruckfarbe *f* ~ **pickle** Brünierbeize *f* ~ **varnish** Bronzetinktur *f*
**bronzite** Bronzitfels *m*
**bronzy** bronzierend ~ **appearance** Bronzierungserscheinung *f* ~ **sheen** bronziger Schimmer *m*
**brooch** Agraffe *f*
**brook** Bach *m*
**broom** Besen *m* ~ **handle** Besenstiel *m*
**brought to** beigedreht
**brow** einfallende Ortsstrecke *f* ~ **band** Stirnriemen *m*

**brown, to** ~ brünieren; braun
**Brown and Sharp wire gauge** Amerikanische Drahtlehre *f*
**brown,** ~ **coal** Berg-, Braun-, Fett-kohle *f* ~ **coal for low-temperature retort process** Schwelkohle *f* ~**-coal briquette** Braunkohlenbrikett *n*, ~**-coal firing** Braunkohlenfeuerung *f* ~**-coal gas** Braunkohlengas *n* ~**-coal grit** Braunkohlensandstein *m* ~ **hematite** Braun-erz *n*, -eisenerz *n*, -eisenstein *m* ~ **iron ore** Braun-eisenerz *n*, -eisenstein *m*, -erz *n*, Limonit *m* ~ **magnesite** Magnesiumeisen *n* ~ **mechanical pulp** Braun--schliff *m*, -holzstoff *m* ~ **ore** Blauerz *n*, Braun--eisen *n*, -eisenerz *n*, -eisenstein *m* ~ **paper (wrappings)** Braunholzpapier *n* ~**-print paper** Pauspapier *n* ~ **print paper** Sepiapapier *n* ~ **spar** Braun-kalk *m*, -spat *m* ~**-stone** Grau--braunstein *m*, -mangan *n* ~**-type mine ignition** Antennenzündung *f*
**browner** Braunmacher *m*
**Brownian motion (or movement)** Brownsche Bewegung *f*
**browning,** ~ **antirust coating (on rifle-barrel)** Brünierung *f* ~ **(or black burn) furnace** Schwarzbrennofen *m*
**brownish** bräunlich, braunstichig
**Brownmillerite cement** Brownmillerit-Zement *m*
**browns** Braunholzpapier *n*
**brucite** Brucit *m*
**bruise, to** ~ schrotten, verbeulen, verschrot(t)en, zermahlen, zerquetschen, zerstoßen
**bruise** Beule *f*, Quetschung *f*
**bruising mill** Schrotmühle *f*
**Brunswick green** Braunschweiger Grün *n*
**brush, to** ~ anstreichen, bestreichen mit, bürsten, entlangstreifen, pinseln, sich in Büscheln entladen, (in chilling) bestreichen **to** ~ **against** streifen, vorbei-streichen, -streifen **to** ~ **off** abbürsten, abkehren, plamotieren, (sugar) abhacken **to** ~ **on** aufstreichen **to** ~ **out** aufschlagen (paper)
**brush** Bürste *f*, Bürstenarm *m*, Gesträpp *n*, Kontaktarm *m*, Kontaktarmträger *m*, Quast *m*, Reisholz *n*, Schleifbürste *f* (electr.), Strahlungsbüschel *m* (electr.), Wedel *m*, (gear) Bürstenträger *m*, (carbon) Bürste *f*, (iconoscope) Abtaststrahl *m*, (painting) Pinsel *m* **(contact)** ~ Kontaktbürste *f*
**brush,** ~ **in permanent contact** dauernd aufliegende Bürste *f* ~ **of radiations** Strahlen-bündel *n*, -büschel *n* ~ **of rays** Strahlbündel *n*, Strahlen-bündel *n*, -büschel *n* ~ **for spark(ing) plugs** Zündkerzenbürste *f* ~ **for washing** Waschpinsel *m*
**brush,** ~ **bolter** Mehlbürstmaschine *f* ~**-breaker plow** Buschrodepflug *m* ~ **carriage** Bürsten--rahmen *m*, -träger *m* ~ **coat** Bürstenstrich *m* ~ **collar** Bürstenträger *m* ~ **compare check** Bürstenkontrolle *f* ~ **contact pressure** Bürstenauflagedruck *m* ~ **contrivance** Bürstenvorrichtung *f* ~ **damper** Bürstenfeuchter *m* ~ **detacher** Bürstendetascheur *m* ~ **discharge** Büschelentladung *f*, elektrische Glimmentladung *f*, Korone *f*, Sprühen *n*, Sprühentladung *f*, St. Elmsfeuer *n*, (between glow discharge and sparking) Bürstenentladung *f* ~**-discharge loss** Sprühverlust *m* ~ **displacement** Bürstenverschiebungswinkel *m* ~ **enamel** gebürstetes Pa-

pier *n* ~ **fitter** Bürsteneinzieher *m* ~ **flue** Rohrbürste *f* ~ **form of contact** Bürstenkontakt *m* ~ **form of discharge** Bürstenentladung *f* ~ **friction loss** Bürstenreibungsverlust *m* (electr.) ~ **gear** Bürstenarm *m* ~ **holder** Bürstenhalter *m* (electr.), Kontaktstückhalter *m* ~**-holder key** Bürstenschlüssel *m* ~**-holder rod** Bürstenstift *m* ~ **hook** Faschinenmesser *n* ~**-lifting device** Bürstenabhebevorrichtung *f* ~ **light** Büschel-, Glimm-licht *n* ~ **mark** Paßzeichen *n* ~ **matrix** Matrizen-karton *m*, -papier *n* ~ **opening** Bürstenfenster *n* ~**out** Feinmahlen *n* ~ **out (mater) matrix** Handschlagmater *n* ~ **pickling** Streichbeize *f* ~ **position** Bürstenstellung *f* ~ **printing** Bürstdruck *m* ~ **proof** Bürstenabzug *m* ~ **rack** Bürstenträgergestell *n* ~**-raising gear** Bürstenabhebevorrichtung *f* ~ **ring** Bürstenträger *m* ~ **rocker** Bürstenarm *m* (straight arm) ~ **rod** Bürsten-arm *m*, -träger *m*, -welle *f*, Kontaktarmträger *m*, Kontaktschlitten *m* ~ **roll** Putzbürste *f* ~ **rollers** Bürstenwalzer *m* (pulp and paper) ~ **selector** Bürstenwähler *m* ~ **shaft** Bürsten-arm *m*, -armspindel *f* ~ **shift** Bürstenverschiebung *f* ~ **shifting** Bürstenverstellung *f* ~**-shifting motor** Verstellmotor *m* ~ **spreader** Bürstenstreichmaschine *f* ~ **spring** Kontaktarm *m* (für eine Sprechleitung), Kontaktbürste *f*, Schleiffeder *f* ~ **support** Bürstenträger *m* ~ **tube** Rohrbürste *f* ~ **weir** Buschwehr *n* ~ **wire** Bürstendraht *m* ~**wood** Hecke *f*, Reisig *n* ~**wood borer** Bürstenholzbohrer *m* ~**wood cooling stack** Reisiggradierwerk *n* ~**work revetments** Spreutlagen *pl* ~ **yoke** Bürstenjoch *n* (electr.)
**brushable** streichfähig
**brusher** Striegler *m* (paper mfg.)
**brushes** Folienbürsten *pl*
**brushing** bestreichbar; Ausstreichen *n*, Bürsten *n*, Feinmahlen *n*, Nachreißen *n* (des Gesteins) ~ **coal** Abkohlen *n* ~ **device** Bürstvorrichtung *f* ~ **filler** Streichspachtel *f* ~ **lacquer** Streichlack *m* ~**-off** Abhacken *n* ~**-off device** Abstreichbürste *f* ~**-on color** Streichfarbe *f* ~ **worm** Bürstenschnecke *f*
**brute force (or method)** Orientierung *f* der Kerne durch starke Felder bei sehr tiefer Temperatur (atom.) ~ **jamming** Massivstörung *f* (rdr)
**B switchboard,** B-Platz *m*, Eingangsplätze *pl*
**B-tube** B-Speicherröhre *f* ~ **(B-box)** Indexregister *n* (info proc)
**bubble, to** ~ brodeln **to** ~ **forth** hervor-quellen, -sprudeln **to** ~ **(or gush) over** übersprudeln **to** ~ **through** durchperlen, einblasen **to** ~ **up** aufsteigen
**bubble** Blase *f*, Bläschen *n*, Glaskanzel *f*; Libelle *f*, Querlibelle *f*, Schaumblase *f* ~ **of the level** Libellenblase *f* ~ **of water** Wasserblase *f*
**bubble,** ~ **bombing sight** Libellenlotgerät *n* ~ **cap** Fraktionierbodenglocke *f* ~ **chamber** Blasenkammer *f* ~ **collapse** Blasenzusammensturz *m* (cavitation) ~ **column** Schaumsäule *f* ~ **formation** Blasenbildung *f* ~ **gauge** Gasblasenströmungsmesser *m* ~**-hole** Blähpore *f* ~ **level** Libelle *f* ~ **nucleation by ions** Blasenbildung *f* an Ionen ~ **plate** Glockenboden *m* ~ **separator** Blase(n)abscheider *m* ~ **sextant** Blasen-, Libellen-, Luftblasen-sextant *m* ~

test Blase-, Blas-, Pust-probe *f ~* **tower** Frak-
tionierturm *m ~* **tray** Glockenboden *m*,
Sprudelplatte *f*
**bubbling** Aufwallung *f*, Brodel *m ~* **up** Aufquellen *n*, Aufwallen *n ~*-**type electrode** Durch-
perlungselektrode *f ~* **washer** Durchflußwa-
scher *m*
**bubbly** blasig
**Buchholz protective relay** Buchholz-Relais *n*
**Büchi supercharging** Büchi-Aufladung *f*
**buck, to ~** beuchen, bocken, brechen, entgegen-
arbeiten, gegenarbeiten **to ~ against** anlaufen
(a potential) **to ~ up in front** aufbäumen, vorn
aufkippen (aviat.)
**buck** Beuche *f ~* **dowel** Bockdalben *m ~* **ram**
Steifleinen *n ~* **riveting** Gegenhaltnietung *f*
*~* **saw** Handspannsäge *f ~* **shot** Roller *m*,
Schrot *m ~* **skin** Hirschleder *n ~* **stay** Anker-
säule *f ~* **stay for condenser box** Strebe *f* für
Kondensator, Strebe *f* für Kühlerwanne
*~* **wheat coal** Klein-, Nußgrus-kohle *f ~* **wheat**
**flour** Heide-mehl *n*, -kornmehl *n ~* **battery**
Zusatzbatterie *f*
**bucker** Erzpocher *m*
**bucket** Auffänger (elektrode); Baggereimer *m*,
Becher *m*, Bütte *f*, Eimer *m*; Erzmaß *n*; Förderkorbkübel *m*, Kastengehänge *n*, Kübel *m*,
Ladetrichter *m*, Mulde *f*, Schachtfördergefäß *n*,
Schöpfzelle *f*, (turbine) Laufschaufel *f* **by the ~**
**eimerweise ~ with bottom discharge** Kastengehänge *n* mit Bodenentleerung *~* **of overflow**
**weir** Sturzbett *n ~*s **of a turbine** Beschaufelung *f*
**bucket, ~ attachment** Becherbefestigung *f ~*
**attemperator** Eisschwimmer *m ~* **base** Schau-
felfuß *m ~* **belt** Paternoster *n ~* **car** Kübelwagen *m ~* **carriage** Laufwagen *m ~*-**carriage**
**track** Schrägaufzugbahn *f ~* **chain** Eimerkette *f*
*~* **chain conveyor** Eimerkettenförderer *m ~*
**channel** Eimerrinne *f ~* **charging** Kübelbegichtung *f ~*-**charging gear** Begichtungsanlage *f* mit Kübel, Kübelbegichtungs-
anlage *f ~* **conveyer** Bechertransporteur *m*,
Becherwerk *n*, Exkavator *m ~* **conveyer with**
**return run** Becherwerk *n* mit Rückführung
*~*-**conveyer dredge (or excavator)** Eimerketten-
bagger *m ~* **depth** Schaufeltiefe *f ~* **discharge**
Kübelentleerung *f ~* **dredge** Eimer-, Löffel-
-bagger *m ~* **elevator** Becher-transporteur *m*,
-werk *n*, Eimerwerk *n*, Kettenelevator *m*,
Kübelaufzug *m*, Schöpfwerk *n*, Umlaufzug *m*
*~* **elevator with return run** Becherwerk *n* mit
Rückführung *~*-**elevator boot** Becherwerks-
kopf *m ~* **filling** Kübelbegichtung *f ~* **grab**
Greiferkatze *f ~* **gutter** Eimerrinne *f ~* **handling**
**crane** Kübel-, Mulden-transportkran *m ~* **han-**
**ger** Kastengehänge *n ~* **ladder excavator** Eimer-
kettenbagger *m ~* **land dredge** Eimertrocken-
bagger *m ~* **leaving angle** Laufschaufelaustrittswinkel *m ~* **lift** Eimerkunst *f*, Saugsatz *m*
in einem Pumpenschacht *~***like** kübelartig
*~* **loss factor** Laufschaufelverlustbeiwert *m*
*~* **ring** Schaufelkranz *m* (gas turbine) *~* **service**
Kübelbetrieb *m ~* **suspension rod** Hängewerk *n*,
Kübelstange *f ~* **tender** Eimerkettenleiter *f*
*~***trencher** Grabenbagger *m* (mit Eimern) *~*
**trough** Schaufelmulde *f ~* **twist** Schaufelver-
windung *f ~*-**type machine** Becherwerk *n ~*-
**valve piston of a lift pump** Hubpumpenkolben *m*

*~* **wheel** Schaufel-, Schöpf-, Zellen-, Eimer-
(turbine), Becher-rad *n ~* **wheel excavator**
Schaufelradbagger *m ~* **wheel loader** Schau-
felradlader *m ~* **wheel pump** Schaufelradpumpe
*f ~* **wheel reclaimer** Schaufelradrücklader *m*
**bucking, ~ circuit** Kompensationskreis *m ~*
**cloth** Laugentuch *n ~* **coil** Gegenwindung *f*,
Kompensations-, Zitter-spule *f ~* **iron** Erz-
pocheisen *n ~* **lye** Beuchlauge *f ~* **tub** Beuchfaß *n*
**buckle, to ~** ausbeulen, ausknicken, krümmen,
schnallen, sich krumm ziehen, sich krümmen,
stauchen, verbiegen, verziehen, zuschnallen,
(column) knicken **to ~ on** festschnallen **to ~**
**out** ausbeulen
**buckle** Falte *f* (Blech), Federbride *f*, Feder-
bund *m*, Federkasten *m*, Koppelschloß *n*,
Schnalle *f*, Spange *f*; gebogen, gekrümmt,
krumm, verbogen, wellig *~* **plate** Federplatte *f*
**buckled, ~ girth** Schnürsattelgurt *m ~* **wheel** ver-
wundenes Rad *n*
**bucklers** dünne Ketten *pl*
**buckling** Atmen *n* (film), Ausbeulen *n*, Ausknickung *f*, Knickerscheinung *f*, Knickung *f*,
Krümmen *n*, Krummwerden *n*, Stauchung *f*,
Überfluß *m*, Verbeulung *f*, Verbiegen *n*, Verbiegung *f*, Verziehen *n*, Verziehung *f*, Wölbung
*f*, (of plates) Ausbeulung *f*, geometrische
Flußwölbung *f ~* **out of the plane of the girder**
Ausknicken *n* aus Trägerebene
**buckling, ~ coefficient** Beulbeiwert *m ~* **error**
Knickfehler *m ~* **load** Knick-beanspruchung *f*,
-kraft *f*, -last *f ~* **mill** Knickwalzwerk *n ~* **re-**
**silience** Knickspannung *f ~* **resistance** Beul-
festigkeit *f*, Knicksicherheit *f ~* **resistant** beu-
lensteif *~* **strength** Beul-, Bieg-festigkeit *f*,
Knickwiderstandsfähigkeit *f*, (of metal) Knickfestigkeit *f ~* **stress** Knick-beanspruchung *f*,
spannung *f ~* **test** Beulenmessung *f*, Knickversuch *m*
**Bucky screen** Streustrahlenblende *f*
**bud, to ~** okulieren
**budding** angehend *~* **knife** Okulier-, Pfropf-
-messer *n*
**buddle, to ~** schlämmen
**buddle** Kehrherd *m*
**buddling** Schlämmung *f*
**budget** Etat *m*, Haushalt *m*, Haushaltsplan *m*
*~* **item** Etatposten *m ~* **year** Rechnungsjahr *n*
**buff, to ~** glänzen, glanzschleifen, lederfarbig,
polieren, schwabbeln
**buff** Schwabbelscheibe *f ~* **shade** Chamoiston *m*
**buffalo skin** Büffelleder *n*
**buffer, to ~** (Stoß) abdämpfen, puffern (a
solution)
**buffer** Auffangvorrichtung *f*, Betätigungswinkel
*m*, Bremse *f*, Buffer *m*, Dämpfungsfähigkeit *f*,
Federanschlagpuffer *m*, Luftpufferstoff *m*,
Luftvorholer *m*, Oderschaltung *f*, Poliermaschine *f*, Polierscheibe *f*, Prellbock *m*, Puffer *m*,
Pufferkreis *m*, Pufferspeicher *m*, Rohrrück-
laufbremse *f*, Scheuerschutz *m* (Fallschirm),
Stoßfänger *m*, Stoßkissen *n*, Trennverstärker
*m*, Zwischenkreis *m* (tuning)
**buffer, ~ action** Pufferwirkung *f ~* **amplifier**
Trennstufe *f* (electr.) *~* **bar** Auffangstange *f*,
Stoßvorrichtung *f ~* **base plate seating** Puffer-
grundplatte *f ~* **base spring seating** Puffer-
grundplatte *f ~* **battery** Ergänzungs-, Puffer-

-batterie f ~ **bearing** Pufferlager n (artil.) ~
**block** Abstützbock m, Pufferklotz m (mach.)
~ **bolt** Pufferbolzen m ~ **brake** Rohrbremse f
~ **case** Puffergehäuse f ~**-casing press** Hülsen-
pufferpresse f ~ **circuit** Pufferschaltung f
~ **contact** Amboß-, Puffer-kontakt m ~ **crossbeam** Pufferquerriegel m ~ **disk** Pufferteller m
~ **dynamo** Pufferdynamo m ~ **effect** Pufferwirkung f ~ **fluid** Bremsflüssigkeit f ~ **gas**
Schutzgas n ~ **gear** Pufferungsteil m ~ **isolator**
Trennstufe f ~ **load** Pufferstand m ~ **nut** Puffer-
mutter f ~ **nut spanner** Puffermutterschlüssel m
~ **oil chamber** Bremsölbehälter m ~ **plunger**
Puffer-schaft m, -stange f ~ **resistor** Puffer-
widerstand m ~ **shank** Pufferstange f ~ **sleeve**
Pufferhülse f ~ **solution** Pufferlösung f ~ **spring**
Buffer-, Puffer-, Stoß-feder f ~ **stage** Pufferstufe f, Sperrkreis m, Trennstufe f (electr.)
~ **stop** Prell-träger m, -bock m ~ **stops** Gleis-
abschluß m (terminal) ~ **storage** Ausgangs-,
Eingangs-speicher m ~ **store** Pufferspeicher m
(info proc) ~ **tube** Trenn-, Isolier-röhre f ~
**tuning** Zwischenkreisabstimmung ~ **wall** Gor-
dungswand f
**buffered solution** gepufferte Lösung f
**buffering** Pufferung (chem.)
**buffeting** Flatterschwingung f, fortwährende
Schwingung f, Schütteln n (eines Luftfahrzeuges)
**buffing** Aufrauhen n, Frimmeln n, Schleifen n,
Schwabbeln n ~ **brush** Borstenscheibe f ~ **cylinder** Dolierwalze f ~ **lathe** Schwabbelma-
schine f ~ **load** Pufferdruck m, Rangierstoß m
~ **machine** Aufrauh-, Schwabbel-maschine f
~ **slicker** Abnarb-, Dolier-eisen n ~ **wheel**
Bimsmaschine f, Filzpolierscheibe f, Glanz-,
Schwabbel-scheibe f ~ **work** Schwabbelar-
beit f
**bug** Gerätfehler m, Radarmarke f, Störung f,
zeitweises Versagen n
**buggi,** ~ **ladle** Pfanne f mit Wagen ~ **man** Fördermann m
**bugle** Flügel-, Signal-horn n
**build, to** ~ anlegen, aufführen, bauen, erbauen,
konstruieren **to** ~ **in** einbauen **to** ~ **up** aufschweißen, ausfachen, bilden, einschwingen,
errichten, verbauen, sich ansammeln **to** ~ **up
a bed** Schweißraupe f auftragen **to** ~ **a dam**
dämmen **to** ~ **in front** verbauen **to** ~ **a landing**
Landung f bauen **to** ~ **slovenly** fluchtlos bauen
**builder** Baumeister m, Bautechniker m, Erbauer m, Konstrukteur m, Versatzarbeiter m
~**'s hoist** Bauwinde f
**building** Bau m, Bauwerk n, Gebäude n, Haus n
~ **and loan society** Baugenossenschaft f ~ **in
day work** Regiearbeit f ~ **of dikes and dams
across a valley** Talsperrenbau m ~ **in series**
Serienbau m
**building, (public)** ~ **authorities** Baubehörde f,
Baupolizei f ~ **bag** Arbeitsschlauch m ~ **berth
for large vessels** Großhelling f ~ **block** Baustein m ~ **block system** Baukastensystem n ~
~ **boards** Baupappe f ~ **code (by law)** baupoli-
zeiliche Vorschrift f ~ **construction** Hochbau m
~ **construction (or industry)** Bauwesen n ~ **contractor** Bau-meister m, -unternehmer m ~
**cradle** Helligtisch m ~ **ground** Baugrund m
~ **hatch** Montageluke f ~ **joiner** Bautischler m

~ **line** Bauhorizont m, Bau-, Flucht-linie f
~ **lumber** Bauholz n ~ **material** Baumaterial n
~ **mortar** Baumörtel m ~ **office** Baubüro n
~ **obstacle** Verbauung f
**building-out,** ~ **capacitor** Abgleich-, Ergänzungs-
-kondensator m ~ **circuit** Kondensatoraus-
gleichschaltung f ~ **network** Ergänzungsnetz-
werk m ~ **resistance** Ausgleichswiderstand m
~ **section** Anpassungsglied n, Spulenfelder-
gänzung f, zusätzliche Nachbildung, Zusatz-
nachbildung f (of transmission line)
**building,** ~ **paper** Baupappe f ~ **plan** Bauzeichnung f ~ **program** Bauprogramm n ~ **project**
Bauvorhaben n ~ **purpose** Bauzweck m ~ **regulations** Bauvorschrift f ~ **restriction zone
around airports** Bauschutzbereich m ~ **sand**
Bausand m ~ **site** Bau-platz m, -stelle f ~**-site
topography** Bauteilterrain n ~ **slip** Helgen m
~ **specimen** Baumuster n ~ **stone** Bau-, Mauer-
-stein m ~ **structure** fliegende Anlage f (with
equipment) ~ **trade** Bau-fach n, -gewerbe n,
-handwerk n ~ **trade joinery** Bautischlerei f
~**-up** Aufbau m aus mehreren Teilen, Aufhäufung f, Einregelungszeit f, Zusammenbau m
~**-up of acoustic energy** Anhallen n ~**-up of
tires** Radkranzschweißung f ~**-up of tone**
Tonaufbau m ~**-up of waste dumps** Halden-
hochschüttung f ~**-up of wheel tires** Bandagen-
schweißung f ~**-up current** Einschwingstrom m
~**-up period** Anfachzeit f, Regelgeschwindig-
keit f (Vorhaltezeit) ~**-up (process)** Einschwingen n ~**-up process** Aufschaukelvor-
gang m, Einschwungvorgang m ~**-up time** Aufbauzeit f, Aufschaukelzeit f, Einschwing-dauer
f, -zeit f ~**-up transient oscillation process** Einschwingvorgang m ~**-up welding** Auftrag-
schweißung ~ **yard** Bau-hof m, -platz m,
-stelle f
**build-up** Anhäufung ~ **and decay distortion**
Verzerrung f durch Ein- und Ausschwingen
~ **or increase of amplification** Aufschaukeln n
(of ascillations) ~ **of picture** Bildaufbau m
(line by line) ~ **and vent valve** Druckaufbau-
und Entlüftungsventil n
**built,** ~ **with brick** gemauert ~ **of concrete** betoniert ~ **beam with keys** verdübelter Träger m
**built-in** eingebaut ~ **antenna** eingebaute Antenne
f ~ **blinker light** Einbaublinkleuchte f ~ **camshaft** eigene Nockenwelle f ~ **coverless switch**
Einbauschalter m ~ **direction** Einbauwinker m
~ **drive** eingebauter Antrieb m ~ **dryers** nachgeschaltete Trockeneinrichtung f ~ **gauge head**
Einbaumeßkopf m ~ **motors** Einbaumotoren pl
~ **movements** Einbau- und Einsteckwerke pl
~ **rear (reversing) light** Einbaurückfahr-
leuchte f ~ **stop lamp** Einbaubremslaterne f
~ **tumbler switch** eingebauter Kippschalter m
~ **unit** Einbauaggregat n ~ **wall-box frame**
Mauerkasten m
**built,** ~**-on motors** Anbaumotoren pl ~ **onto**
angebaut
**built-up** zusammengebaut ~ **and semibuilt
crankshaft** gebaute und halbgebaute Kurbelwelle f ~ **area** bebautes Gelände n ~ **barrel**
Ringrohr n, festes Seelenrohr n ~ **beam** gewachsener oder zusammengesetzter Balken m
~ **coil** zusammengesetzte Schlange f ~ **connection** Durchgangsverbindung f ~ **crankshaft**

zerlegbare Kurbelwelle *f* ~ **crossing with base frog** Schienenherzstück *n* ~ **crossing with base plate** Schienenherzstück *n* ~ **cylinder** aus Kopf und Laufbuchse zusammengesetzter Zylinder *m* ~ **edge** Aufbauschneide *f* (Drehstahl) ~ **girder** gebauter Balken *m* oder Träger *m*, gewachsener Balken *m* ~ **ground area** bebaute Fläche *f* ~ **gun** Ringkanone *f* ~ **lens** Sprossenlinse *f* ~ **mica** Mikanit *n* ~ **period** Anschwingzeit *f* ~ **rib** Fachwerkrippe *f* ~ **rotary switch** Paketschalter *m* ~ **welding** Auftragsschweißung *f* ~ **wing** Fachwerkflügel *m* ~ **wooden tube** Holzbandrohr *n*

**bulb** Ballon *m*, Cuvette *f*, Kolben *m*, Kugel *f*, kugelförmiges Gefäß *n*, Küvette *f*, Trog *m*, Wulst *m*, (glass) Glasballon *m*, Glasbirne *f*; Glühlampe *f* (electr) ~ **of a tube** Röhrenkolben *m* ~ **of the X-ray tube** Röntgenrohrkörper *m*

**bulb,** ~ **angle** Winkelwulst *f* ~ **angle iron** Wulstwinkeleisen *n* ~ **barometer** Gefäßbarometer *n* ~ **blackening** Kolbenschwärzung *f* ~ **blowing machine** Kolbenblasmaschine *f* ~ **connection (or terminal)** Glühlampenanschluß *m* ~ **cutting machine** Kolbenschmelzmaschine *f* ~ **efficiency** Gefäßwirkungsgrad *m* ~ **exposure** Ballaufnahme *f* ~ **filling** Füßchenfüllung *f*

**bulbhead rail** Doppelkopfschiene *f*

**bulb,** ~ **horn** Ballhupe *f* ~ **iron** Wulsteisen *n* ~ **mold** Kolbenform *f* ~ **pile** Pfahl *m* mit Fußverbreiterung oder mit zwiebelförmiger Spitze ~ **remover** Kolbenentferner *m* ~ **ring** Kolbenwulst *m* ~ **seating** Glühlampenträger *m* ~ **section** Flachwulstprofil *n* ~**-shaped** kolben-, zwiebel-förmig ~ **socket** Fassung *f* ~ **thermometer** Kugelthermometer *n* ~ **tray** Kolbenbrett *n* ~**-type combustion chamber** kugelförmiger Brennraum *m* ~**-type condenser** Kugelkühler *m* ~ **washing machine** Kolbenspülmaschine *f*

**bulbed tube** Kugelrohr *n*

**bulbous** knollig ~ **bow** Bugwulst *f* ~ **pipes** ausgebuchtete Rohre *pl* ~ **shape** Kugelgestalt *f* ~ **tubes** ausgebuchtete Rohre *pl*

**bulbus** Augapfel *m* (electron.)

**bulge, to** ~ anschwellen, aufweiten, (sich) ausbauchen, ausbeulen, ausweiten, bossieren, einbeulen, sich bauschen, stauchen to ~ **out** aufblähen, ausbauchen, verdicken (of tubes)

**bulge** Ansatz *m*, Anschwellen *n*, Anschwellung *f*, Ausbauchung *f*, Ausbuchtung *f*, Beule *f*, Buckel *m*, Wellenbauch *m*, Wulst *m*, (in curve) Erhöhung *f* ~**s** Außendruckkörper *pl* ~**-out** Aufweitung *f* ~ **of a curve** Buckel *m* einer Kurve ~ **of the earth** Erdkrümmung *f* ~**-testing apparatus** Einbeulapparat *m* ~ **of a vibration** Bauch *m* einer Schwingung

**bulged** ballig gedreht (Riemenscheibe), ramponiert ~**-in** eingedrückt ~ **tube** eingedrücktes Rohr *n* (boilers)

**bulging** Ausbauchung *f*, Ausbuchtung *f* (metall.), Verbeulung *f* ~ **out** Ausbauchung *f* ~ **of lead blocks used for testing explosives** Bleiblockausbauchung *f*

**bulging,** ~ **device** Druckvorrichtung *f* ~ **test** Anschwell-, Aufweite-probe *f*, Einbeulungsprüfung *f*, Einbeulversuch *m*, Stauchprobe *f*, Stauchversuch *m*, (tubes) Ausschwellprobe *f*,

(tube) Expandierprobe *f* ~**-test machine** Prüfmaschine *f* für Einbeulversuche ~**-test specimen** Stauchprobe *f*

**bulgy** bauchig

**bulk, to** ~ lose aufschütten

**bulk** Haupt-masse *f*, -menge *f*, Klumpen *m*, Kompression *f*, lose Ladung *f*, Masse *f*, Raumbedarf *m*, Umfang *m* **in** ~ unverpackt, in losen Haufen ~ **(base) of semiconductor** Halbleiterinneres *n*

**bulk,** ~ **cargo** gestürzte Ladung *f* ~**-carrying conveyances** Beförderungsmittel *pl* ~ **clearance** Bauchfreiheit *f* ~ **container** Packgefäß *n* ~ **conveyance** Massenförderung *f* ~ **delivery** Großlieferung *f* ~ **density** Schüttgewicht *n* ~ **depot** Tanklager *m* ~ **eraser** Löschspule *f* (tape rec) ~ **factor** Füllkonstante *f*, Schüttvolumen *n* ~ **freight** Sperrgut *n*, sperriges Gut *n* ~ **generator** Innenrohrgenerator *m* ~ **goods** Massengüter *pl*, Schüttgut *n* ~ **handling equipment** Massengut-Umschlagsanlagen *pl* ~ **handling installation** Verladeanlage *f*

**bulkhead** Boden *m*, Bohlwand *f*, Damm *m*, Kesselboden *m*, Querspant *m*, Querwand *f*, Rumpfspant *m*, Scheidewand *f*, Schott *n*, Spant *n*, Spantring *m*, Spreizspant *m*, Spundwand *f*, Strömungsschott *m* (g/m), wasserdichte Schiffswand *f*, Zwischenwand *f* ~ **door** Schottüre *f* ~ **partition** Schottwand *f*

**bulk,** ~ **heading** Spundung *f* ~ **liquid** Flüssigkeitsmasse *f* ~ **material** Schüttgut *n* ~ **meter** Mengenzähler *m* ~ **modulus** Elastizitätsmodul *n* für Druck, reziproker Wert *m* der Kompressabilität ~ **paper** Dickdruckpapiere *pl* ~**-reduction point** Eisenbahntankstelle *f*, Umschlagstelle *f* ~ **resistance** Bahnwiderstand *m* ~ **sample** zusammengegossenes Muster *n* ~ **specific gravity** Raumgewicht *n* (ohne Wasser) ~ **stocks** Bestände *pl* an unverpackten Waren ~ **storage** lose Aufbewahrung *f* ~ **tariff** Bausch-, Pauschalgebühren-tarif *m* ~ **technical** großtechnisch ~ **test** Massenprobe *f* ~ **toroid** Mehrfachringspule *f* ~ **travel voucher** Flugscheinheft *n* ~ **unloader** Verladebrücke *f* ~ **viscosity** Kompressionsviskosität *f* ~ **weight** Schüttgewicht *n*

**bulky** massig, sperrig, unhandlich, voluminös ~ **color** Raumfarbe *f* ~ **paper** dickgriffiges Papier *n* ~ **part** Großteil *n* ~ **scrap** sperriger Schrott *m* ~ **test piece** sperrige Probe *f*

**bull** Haussier *m* ~ **bit** flacher Bohrer *m* ~ **chisels** Straßenaufreißer *m*

**bulldog** Bulldogg *m*, Saigerschlacke *f* (metal) ~ Dörner-, Puddel-schlacke *f* ~ **spear** Rohrkrebsfänger *m* ~ **type slip socket** Keilfangglocke *f*

**bulldozer** Bär *m* (Diesel), Biegemaschine *f*, Bulldozer *m*, horizontale Biegepresse *f*, Planierraupe *f*, Stauchpresse *f* ~ **attachment** Räumvorrichtung *f*

**bull,** ~ **frog** Ochsenfrosch *m*

**bullgear train** Triebräderwerk *n*

**bullhead,** ~ **pass** Flachbahnkaliber *n* ~ **rail** Doppelkopfschiene *f*

**bull,** ~ **ladle** Transportpfanne *f* ~ **nose tool** Schruppstahl *m* mit Halbmondspitze oder mit Rundschneide ~ **plate** Kolbenhals *m* ~ **points** Straßenaufreißer *m* ~ **ring** Mahl-ring *m*, -teiler *m* ~ **rope** Aufzugseil *n*, endloses Kabel *n*, rundes Hanfantriebskabel *n* ~ **switch** Hauptschalter *m* (film) ~ **wheel** Bohrseil-trommel *f*, -winde *f*, Hauptrad *n*, Kehrrad *n*, Transmissionsscheibe *f* ~**-wheel drive power** Kehrradantrieb *m* ~**-wheel flanges** Felgenkante *f* ~**-wheel shaft** Tischantriebsradwelle *f*

**bullet** Geschoß *n*, Kugel *f*; Schubdüsennadel *f* ~ **cylinder** Geschoßkanal *m* ~**-deflector plate** Abgleitschiene *f* ~ **extractor** Geschoßzieher *m* ~ **hole** Geschoßsplittersonde *f*, Schußöffnung *f* ~ **slowdown** Geschoßbahnabfall *m* ~**-trephine** Kugelbohrer *m* ~ **tube** Geschoßkanal *m*

**bulletproof** beschuß-, kugel-, schuß-sicher, kugelfest ~ **glass** kugelsicheres oder schußsicheres Glas *n* ~ **glass mount** Linsenlafette *f* ~ **plating** gewehrschußsicheres Blech *n* ~ **self-sealing tank** selbstabdichtender, beschußsicherer Behälter *m* ~ **vest** Panzerweste *f*

**bulletin** Ankündigung *f*, Anschlag *m*, Heft *n* ~ **board** Anschlagtafel *f*

**bullion** Gold- oder Silberbarren *m* ~ **scale** Kornwaage *f*

**bull's,** ~ **eye** Katzenauge *n*, Kauche *f*, Schwarze *n* einer Schießscheibe, Zentrum *n*, Zielpunkt *m* ~ **discharge** Auslauf *m* mit Schauglas ~ **glass** Butzenscheibe *f*, Mondglas *n* ~ **lens** Beleuchtungs-, Butzenscheiben-linse *f*

**bulwark** Bollwerk *n*, Schanzkleid *n*

**bump, to** ~ glucksen, stoßen (kochende Flüssigkeit), verbeulen

**bump** Beule *f*, Bö *f*, Bums *m* (tape rec), Höcker *m*, Prellstoß *m*, Rückprall *m*, Steigbö *f* (aviat.), Vorsprung *m*

**bumper** Abweiser *m*, Dämpfer *m*, Prellbock *m*, Puffer *m*, Stoß-fänger *m*, -schiene *f* (Auto), -stange *f* ~ **bag** Landepuffer *m*, Stoßkissen *n* ~ **bar holder** Stoßstangenträger *m* ~ **cross stay** Stoßquerträger *m* (Auto) ~ **pad** Stoßpolster *n* ~ **plate feeder** Stoßaufgabevorrichtung *f* ~ **rod** Stoßbügel *m* ~ **spud** Zungenbohrer *m*

**bumpiness** Bockigkeit *f*, Böengefühl *n*, Böigkeit *f*

**bumping** Stoßen *n*, stoßweises Sieden *n* ~ **bag** Landungspuffer *m*, Stoßsack *m* (aviat.) ~ **hammer** Stampfhammer *m* ~ **mechanism** Prellvorrichtung *f* ~ **post** Prell-bock *m*, -pfahl *m* ~ **table** Rütteltisch *m*, Schüttel-, Stoß-herd *m*

**bumpless transfer** stoßfreie Umschaltung *f* (autom. contr.)

**bumpy** böig, bockig (meteor.), holperig ~ **air** böige Luft *f* ~ **road** holp(e)riger Weg *m* ~ **weather** böiges Wetter *n*

**bunch, to** ~ verhacken, verhauen, versetzen, zusammenballen, zusammenschalten **to** ~ **up** aufhocken **to** ~ **(together)** bündeln

**bunch** Bund *m*, Bündel *n*, Büschel *m*, Butz *m*, Butze *f*, Butzen *m*, Haufen *m*, Nest *n*, Niere *f*

**bunched,** ~ **circuit** Simultanleitung *f* ~ **conductors** Leiterbündel *n*

**buncher** Bündeler *m*, Bündelmaschine *f*, Eingangsresonator *m* ~ **space** Einkoppelstrecke *f*, Steuerraum *m*

**bunches** Erzklumpen *pl* ~ **of circuits** Leitungsbündel *n* ~ **of conductors** Leiterbündel *n* ~ **of incoming** (outgoing) **trunks** ankommendes (abgehendes) Leitungsbündel *n* ~ **of ore** Putzen *m* ~ **of rays** Strahlenbündel *n* ~ **in the glass** Glasblase *f*

**bunching** Paketbildung *f*, Zusammen-ballung *f*, -ballen *n* (of electrons in groups), (of electrons in Klystron) Bündelung *f* ~ **of cable conductors for testing** Prüf-ende *n*, -stumpf *m* ~ **of charges** (or **of electrons**) Ladungszusammenballung *f* ~ **of ions** Ionenpaketierung *f*

**bunching,** ~ **angle** Laufwinkel *m* im Triftraum ~ **attachment** Bündelapparat *m* ~ **machine** Würgemaschine *f* ~ **parameter** Ballungsmaß *n* ~ **range of electrons** Überholungsgebiet *n* der Elektronen ~ **strip** Parallelklinkenstreifen *m*

**bundle, to** ~ bündeln

**bundle** Bündel *n*, Gebinde *n*, Pack *n*, Ring *m*; Stückgut *n*; (skeins) Docke *f* ~ **of documents** Aktenstoß *m* ~ **of file of papers** Faszikel *m* ~ **of iron bars** Gespann *n* ~ **laminations** Blechpaket *n* ~ **of radiations** (or **rays**) Strahlen-bündel *n*, -büschel *n* ~ **of trunks** Leitungsbündel *n* ~ **of water tubes** Siederohrbündel *n*

**bundle,** ~ **conductor** (of powerline) Bündelleiter *m* ~ **loader** Garbenauflader *m*

**bundled wood** Bündelholz *n*

**bundles of fascines** Buschwürste *pl*, Senkfaschine *f*

**bundling** Bündelung *f*

**bung, to** ~ einspunden, spunden, verspunden

**bung** Kapselstoß *m*, Spund *m* ~ **board** Spundbohle *f* ~**-cleaning apparatus** Spundputzapparat *m* ~ **flap** Spundlappen *m*

**bunged** gespundet

**bungee** Bungee (aviat.), federndes Element *n*, elastisches Bauelement *n* ~ **hole** Anstich *m*, Faß-, Spund-, Zapf-loch *n*

**bunging apparatus** Spundapparat *m*

**bungle, to** ~ pfuschen, verpfuschen

**bungled** gepfuscht

**bungler** Pfuscher *m*, Stümper *m*

**bungling work** Machwerk *n*

**bunk** Koje *f*

**bunker, to** ~ bunkern

**bunker** Behälter *m*, Bunker *m*; Eingangshohlraum *m*, Sammelrumpf *m*, Vorratstasche *f*, (coal) Kohlenbunker *m* ~ **belts** Bunkerbänder *pl* ~ **coal** Bunkerkohle *f* ~ **gate** Bunkerverschluß *m*, Entnahmescheibe *f* ~ **position** Bunkerstellung *f* ~ **ventilation** Bunker-entlüftung *f*, -lüftung *f*

**bunkered** gebunkert

**Bunsen,** ~ **burner** Bunsen-Brenner *m* ~ **cell** Bunsen-Element *n* ~ **flame** Bunsen-Flamme *f* ~ **photometer** Bunsen-Fotometer *m*, Fettfleckfotometer *m*

**Buntens funnel pipette** Buntensche Lufthalle *f*

**bunting** Fahnentuch *n* ~ **iron** Glasbläserrohr *n*

**buoy, to** ~ abbaken, ausbojen (a channel), **to** ~ **up** auftreiben

**buoy** Boje *f*, Seemarke *f*, Seetonne *f*, Tonne *f*, (for seaplane) Ankerflott *n* ~ **lantern** Bojenlaterne *f* ~ **rope** Bojentau *n* ~ **store** Tonnenhof *m* ~ **tender** Seezeichendampfer *m*

**buoyancy** Auftrieb *m*, dynamischer Auftrieb *m*, Hubkraft *f*, Schwebefähigkeit *f*, Schwimm-auftrieb *m*, -fähigkeit *f*, -kraft *f*, Tragfähigkeit *f* ~ **chamber** Schwimmer *m* ~ **coefficient** Auftriebsbeiwert *m* ~ **proceeding process** Schwimmaufbereitungsverfahren *n* ~ **tank** Auftriebtank *m*

**buoyant** aufwärtsstrebend, diastasereich, schwimmfähig ~ **force** Auftriebskraft f ~- **force method** Auftriebsmethode f (hydrometer) ~ **gas** Trag-, Füll-gas n (balloon) ~ **lift** statischer Auftrieb m ~ **moored mine** Auftriebmine f
**buoying,** ~ **agent** Auftriebmittel n ~ **force** Auftriebskraft f
**bur, to** ~ auszupfen
**bur** Bart m, Grat m ~ **ore** Klettenerz n ~ **trimmer** Gratputzer m
**burble,** ~ **angle** kritischer Anstellwinkel m ~ **fence** Wirbelwulst f (des Ballonschirms) ~ **point** Gefahrenmoment m, Grenzschicht(e)-ablösungs-, Spaltungs-, Übergangs-punkt m
**burbling** Ablösung f (aerondyn.) ~ **point** Abreißungspunkt m, Unstetigkeitsstelle f ~ **region** Ablösungsgebiet n ~ **stage of flow** Zustand m der abgerissenen Strömung (aerodyn.) **zone of** ~ Abreißgebiet n
**burden, to** ~ aufladen, begichten, belasten, bepacken, beschweren
**burden** Belastung f, Beschickung f, Beschickungsgut n, Beschickungsmaterial n, Bürde f, Gattierung f, Gicht f, Last f, Möller m, Satz m, Schmelzgut n, Tragfähigkeit f, Traglast f, Vorgabe f (min.) ~ **of proof** Beweislast f
**burden,** ~ **balance** Gattierungswaage f ~ **calculation** Möllerberechnung f ~ **charge** Möllergicht f ~**-charging carriage** Möllerwagen m ~ **material** Möller m ~ **squeezer** Luppenmühle f ~ **yield** Möllerausbringungen f
**burdening** Belastung f, Begichtung f, Beschickung f, Beschwerung f, Möllerberechnung f
**burdock** Klette f
**bureau** Amt n, Büro n
**bureau,** ~ **of construction** Konstruktionsbüro n ~ **for material testing** Materialprüfungsamt n ~ **of standards** Eichamt n, Eichbehörde f ~ **of yards and docks** Schiffswerftenbüro n
**burette** Bürette f, Maßnahme f, Maßregel f, Maßrohr n, Meßrohr f, Tropfröhre f ~ **clamp** Bürettenklemme f ~ **float** Bürettenschwimmer m ~ **pincer** Meßröhrenklemme f ~ **stand** Bürettengestell n ~ **tip** Bürettenausflußspitze f ~ **valve** Bürettenhahn m
**burgee** Windrüssel m
**burglar,** ~ **alarm** Einbrecherglocke f, Einbruchmelder m ~**-alarm bell** Einbruchwecker m ~ **proof window** einbruchsicheres Fenster n
**burglary** Einbruch m
**burial** Beerdigung f, Begräbnis n, Bestattung f ~ **ground** Friedhof m
**buried,** ~ **antenna** eingegrabene Antenne f, Boden-, Erd-antenne f ~ **cable** Erdkabel n, versenktes Kabel n ~ **engine** versenkter Motor m ~ **pipework** erdverlegte Rohrleitung f ~ **wiring** Leitungsverlegung f unter Putz, Unterputz-Leitungsverlegung f
**burin** Grabstichel m
**burl, to** ~ belesen, noppen
**burlap** Packleinwand f
**burled sheet** Noppenblech n
**burler** Nopper m
**burless cut** gratloser Schnitt m
**burling,** ~ **iron** Noppeisen n ~ **machine** Zupfmaschine f
**burn, to** ~ (from below) abbrennen, abschneiden (cutting by oxyhydrogen) brennen, feuern,

flammen, überrösten, verbrennen, verfeuern, verglimmen **to** ~ (coke) backen **to** ~ **away** wegschmelzen **to** ~ **down** vollständig verbrennen **to** ~ **in** einbrennen **to** ~ **in a pile** im Reaktor brennen **to** ~ **off** wegbrennen **to** ~ **off (or down)** abbrennen **to** ~ **on** löten **to** ~ **out** ausbrennen, ausglühen, durchbrennen (Spulen), verbrennen, zundern **to** ~ **to** (ashes) festbrennen **to** ~ **to storage cell plates** verschmelzen mit Sammlerplatten **to** ~ **to waste** vernichten **to** ~ **together** zusammenbrennen **to** ~ **through** durchbrennen (Spulen) **to** ~ **up** aufbrennen, verglühen (spacecraft at re-entry)
**burn, to** ~ **black** schwarzbrennen **to** ~ **nuclear fuel** Kernbrennstoff m zum reagieren bringen **to** ~ **slightly** besengen **to** ~ **slowly** schwelen
**burn** Brandwunde f ~**-back** (in welding) Rückbrand m ~ **button** (of a door) Türwirbel m ~**-off** Abschmelzgeschwindigkeit f ~**-out** Ausbrennen n (electr.), Brennschluß m, Durchbrennen n, Durchschmelzung f, Flammabriß m ~**-out speed** Brennschlußgeschwindigkeit f ~**-up** effektiver Abbrand m, Verbrauchsverhältnis n, Verbrauch m durch Neutronenspaltung
**burnable** brennbar, verbrennlich
**burned,** ~ (off) verbrannt ~**-in spot** Einbrennstelle f ~**-off** vollständig gar
**burned,** ~ **clay** gebrannter Ton m ~ **gas** Verbrennungsgas m ~ **sponge** Schwammkohle f
**burner** Brenner m, Einspritzdüse f, Kohlenbrenner m, (charcoal) Köhler m ~ **with nonvariable head** Einzelbrenner m
**burner,** ~ **arrangement** Brenneranordnung f ~ **capacity** Brennerleistung f ~ **characteristic** Brennerkenngröße f ~ **cone** Mischkegel m ~**-cup** Verbrennungstopf m ~ **efficiency** Brennerleistung f ~ **mouth** Brennerdüse f ~ **nozzle** Brenner-düse f, -mund m, -mundstück f ~ **nozzle assembly** Düsenstock m ~ **orifice** Brenneraustrittsöffnung f ~ **pipe** Brenner-, Misch-rohr n ~ **plate** Zerstäuberlochplatte f ~ **pliers** Gasbrennerzange f ~ **rocket motor** Heizbehälter m ~ **system** Brenneranlage f ~ **tip** Brenner-kopf m, -mund m, -mundstück n, -spitze f ~ **tube** Brennerrohr n ~ **unit** Brenneranlage f ~ **wall tube** Rohr n der Brennerwand
**burnettising** Brünierung f (von Holz)
**burnettize, to** ~ mit Zinkchlorid n tränken
**burning** brennend; Brand m, Brennen n, Verbrennung f, Verfeuerung f ~**-off** Abbrand m (electr.), plötzliche Verbrennung f ~ **out** Durchbrennen n, Verbrennen n ~ **through** Durchbrennen n ~ **of commutator** Kommutatorbrand m ~ **of the contacts** Kontaktabbrände pl
**burning,** ~ **agent** Verbrennungshilfsstoff m ~ **brand** Brennstempel m ~ **composition ring** Satzring m ~ **furnace** langsamgehender Ofen m ~ **glass** Brennglas n ~ **hour** Brennstunde f ~ **kiln** Brennofen m ~ **oil** Brennöl f, Leuchtpetroleum m ~**-out energy** Ausbrennenergie f ~ **oven** Brennofen m ~ **period** Brenndauer f (of an incandescent lamp) ~ **point** Brenn-, Entzündungs-, Zünd-punkt m ~ **quality** Brennprobe f ~ **rate** Brenngeschwindigkeit f (rocket) ~ **reamer** Rohranfräser m ~ **temperature** Brenntemperatur f ~ **test** Brenn-probe f, -versuch m ~ **voltage** Brennspannung f

**burnish, to** ~ bräunen, brünieren, glänzen, glanzschleifen, hochpolieren, polieren, preßpolieren, satinieren **to ~ extra-bright** hochglanzpolieren
**burnish gold** Glanzgold n
**burnished, ~ chroming** Glanzverchromung f ~ **gilding** Wasservergoldung f ~ **gilding of metals** Vergoldung f mit Blattgold ~ **gold** Glanzgold n ~ **silver** Glanzsilber n
**burnisher** Filzstock m, Gerbeisen n, Glätt-bein n, -heft n, -knochen m, -stahl m, -stange f, -zahn m, Heißsatiniermaschine f (photo), Satiniermaschine f
**burnishing** Glättung f, Polieren n, Prägepolieren n ~ **shell** Glätt-, Kalibrier-hülse f ~ **steel** Glätteisen n ~ **stick** Putzholz n ~ **surface** Polierfläche f (tape rec.) ~ **tool** Polierwerkzeug n
**burnt** gebrannt, übergar, überhitzt, verbrannt; faulbrüchig (metall.)
**burnt, ~ brick** Backstein m ~ **dolomite** Sinterdolomit m ~ **gases** Abzugsgas n, Feuergase pl **the ~ gases escape** die Abgase pl entweichen ~ **gypsum** Deckkalk m, Gyps-kalk m, -mehl n ~**-in lacquered** brandlackiert ~ **insulation** eingefressene Isolation f ~ **iron** Brandeisen n ~ **lime** Fettkalk m, gebrannter Kalk m ~ **out** durchgebrannt (fuse), ausgelaufen (bearing) ~ **sand** eingebrannter Sand m ~ **segment** eingefressenes oder verbranntes Segment n ~ **spots** Brandstellen pl ~ **steel** verbrannter Stahl m
**burr, to** ~ abgraten, krempeln, schnarren, aufkarden (raise nap)
**burr** Grat m, Bohrgrat m, Lochputzen n, Walzbart m, Preßnaht f, (spinnerei) Grat m, (of the hole) Grat m, Walzgrat m ~**-picking machine** Wollzupfmaschine f ~ **removing device** Entgrater m
**burred edges of barrel rifling** Felderquetschung f
**burring** Auswicken n, Entkletten f ~ **action** Entklettung f ~ **attachment** Abgrateinrichtung f ~ **machine** Abbürstemaschine f, Abgratungspresse f ~ **press** Abgratpresse f ~ **reamer** Entgrat-, Rohr-fräser m
**burrow, to** ~ eingraben, aushölen
**burrow** Erdgrube f, Grubenhalde f (mine), Halde f
**burry condition** Schneidnarbe f
**burst, to** ~ bersten, brechen, krepieren, platzen, reißen, springen, zerplatzen, zersprengen, zerspringen **to ~ forth** herausbrechen **to ~ off** abbersten **to ~ off (or out)** losplatzen **to ~ open** aufklaffen, aufplatzen, einsprengen, sprengen
**burst** Entladungsstoß m, Hilfsträgergleichlaufimpuls m (TV), Schauer m, Stoß m (of cosmic rays); gesprengt ~ **in the barrel** Rohrzerscheller m ~ **in the bore** Rohr-detonierer m, -krepierer m, -zerscheller m ~**s of cosmic rays** Höhenstrahlen-schauer m, -stöße pl ~ **of cosmic rays** Raumstrahlenstoß m ~ **of fire** Feuerschlag m, Feuerstoß m, Reihenfeuer n
**burst, ~ cloud** Sprengwolke f ~ **due to frost** Frostsprengung f ~ **effect of projectile** Geschoßwirkung f ~ **factor** Berstzahl f ~ **pulse** Durchbruchstoß m (rdr) ~ **pulse corona** einmaliger Koronadurchbruch m ~ **range** Sprengpunktentfernung f ~ **smoke** Sprengwolke f ~ **test** Aufplatzungsprobe f ~ **tire** geplatzter Reifen m

**burster** Burster m ~ **course** Zerschellerschicht f ~ **tube** Kammerhülsenrohr n
**bursting** Bersten n, Platzen n, Springen n; springend ~ **of a dike** Dammbruch m ~ **charge** Ausstoß-, Spreng-ladung f ~ **charge of shell** Granatfüllung f ~ **diaphragm** Platzmembran(e) f ~ **disk** Bruchplatten pl ~ **layer** Zerschellerschicht f ~ **limit** Berstdruck m ~ **plate** Einreißplatte f (hydraul.) ~ **pressure** Druckbeständigkeit f ~ **projectile** Explosivgeschoß n ~ **strain** Berstdruckprüfung f ~ **strength** Berst-druck-prüfung f, -festigkeit f, -druck m (of paper, etc) ~ **stress** Sprengspannung f ~ **test** Zerplatzprobe f ~**-test apparatus** Zerplatzprobenvorrichtung f
**bury, to** ~ begraben, eingraben, vergraben, in die Erde verlegen **to ~ alive** verschütten
**burying, ~ of a cable** Kabelverlegung f ~ **depth** Eingrabetiefe f
**bus** Omnibus m; Sammelschiene f, Schiene f, Vielfachleitung f
**busbar** Sammelschiene f, Schiene f, Strom-, Stromzuführungs-, Zuleitungs-schiene f ~**-chamber** Sammelschienenkasten m, Stromkreisunterbrecher m ~**-clamp** Doppelkammer f, Flansch m für Sammelschienen ~**-connection** Schienenverbinder m (electr.) ~**-supports** Lager pl für Sammelschienen ~**voltmeter** Sammelschienenvoltmeter n
**bus, ~ isolator** Schienentrenner m ~ **transfer equipment** Sammelschienenübertragungsausrüstung f
**bush, to** ~ ausbüchsen, ausfüttern, füttern, verbüchsen, (bearing) ausbuchsen ~ **for roller carriage** Büchse f zum Walzenstuhl **to ~ with** ausfüttern mit ~ **with cast-in liner** ausgegossene Lagerschale f
**bush** Buchse f, Büchse f, Hag m, Lagerschale f, Laufbuchse f, Kugellagerbuchse f, Muffe f, Rohrstutzen m, Steckbuchse f, Steckerbuchse f, Strauch m, Überwurf m (mech.), Zapfenlager m (bearing) ~**(ing)** Lagerbüchse f, (in sheave of bar) Bestrüppung f ~ **of jack** Klinkenhülse f ~ **with steel casting bronze** mit Bronzebüchsen pl versehen
**bush, ~ bearing** (for flywheel) Büchse f ~ **chains** Büchsenketten pl ~ **hammer** Boßhammer m ~ **hammers** Charierhämmer pl ~ **harrow** Buschegge f ~ **mold** gewachsener Boden m ~**(ing) rind** Schale f ~ **wrench** Ausschraubwerkzeug n für Spundringe
**bushed, ~ hole** ausgetuchtes Loch n ~ **inlet end** Einlaßlagerende n
**bushel** Scheffel m ~ **measure** Scheffelmaß n
**busheled iron** Schweißeisen n
**bushing** Absatzstück n, Austuchung f, Bohrbüchse f, Buchsring m, Buchse f, Büchse f, Durchführung f, Einführungsisolator m, Einsatz m (Tiefbohranlage), Futterstück n, Garnierung f, Hohlbolzen m, Hülse f, Lager n, Lagerpfanne f, Laufbuchse f, Leerlaufbuchse f, Muffe f, Radbüchse f, Reif m, Rohrdurchführung f, Futter n (metall.), (shell of) Lagerschale f ~ **with flange** Leerlaufbuchse f mit Bund ~ **with grease-pocket feed** Leerlaufbuchse f mit Fettschmierung
**bushing, ~ chain** Stiftkette f ~ **current transformer** Durchführungswandler m, Stützer-

stromwandler *m* ~ **guide** Führungsbuchse *f* ~
**insulator** Wanddurchführungsisolator *m* ~
**potential device** Potentialvorrichtungsisolator
*m* ~ **transformer** Durchsteckwandler *m* ~**-type**
**current transformer** Durchführungstromwand-
ler *m*
**bushes** Gebüsch *n*
**business** Angelegenheit *f*, Arbeit *f*, Beschäftigung *f*, Geschäft *n*, Geschäftslage *f* **to do** ~ Ge-
schäfte machen ~ **boom** Konjunktur *f* ~ **cycle**
Wechsellage *f* ~ **end of a tool** angreifendes
Ende *n* eines Werkzeuges ~ **hours** Dienstschicht *f*, -stunden *f* ~ **practices** Geschäftsge-
bräuche *pl* ~ **telephone** Geschäftsanschluß *m*
(subscriber's telephone on business premises)
**busk** Blankscheit *n*
**bustle pipe** Ringleitung *f*, Windring *m* (of blast
furnace)
**busy, to** ~ als besetzt kennzeichnen, besetztmachen, Besetztzeichen belegen, sperren, einen
Wähler belegen, einen Wähler sperren **to** ~ **a**
**selector** einen Wähler *m* belegen (sperren) **to** ~
**oneself, to be** ~ sich beschäftigen
**busy** belegt, besetzt, emsig, geschäftig, rührig
~ **back jack** Besetztklinke *f* ~**-back signal** Besetztzeichen *n* (teleph.) ~ **condition** Besetzt-
stellung *f*, -zustand *m* ~ **earth** Besetzterde *f*
~ **hour** Hauptverkehrsstunde *f* ~ **hours** verkehrsstarke Zeit *f* ~ **junction** besetzte Verbindungsleitung *f* ~ **lamp** Besetztlampe *f* ~ **line**
besetzte Leitung *f* ~ **period** Hauptverkehrszeit *f*, -stunde *f*, Hauptgesprächszeit *f* (teleph.)
~ (**engaged**) **position** Besetztzustand *m* ~ **relay**
Prüfrelais *n* ~**(-test) relay** Besetztrelais *n* ~
**response** Besetztmeldung *f* ~ **signal** Besetzt-
-meldung *f*, -zeichen *n* ~**-signal light** Besetzt-
schauzeichen *n* ~ **test** Besetzt-prüfen *n*, -prü-
fung *f* ~ **test with tone signal** Besetztprüfung *f*
mit Summerton ~ **tone** Besetztton *m* ~ (**back**)
**tone** Besetztton *m* ~ **trunk** besetzte Verbindungsleitung *f* ~ **wire** Sperrleitung *f*
**busying** Belegen *n*, Belegung *f* ~ **potential** Besetztspannung *f* ~ **relay** Belegungsrelais *n*
**but, to** ~ **wood** holzen
**butadiene** Butadien *n* ~ **natrium** Buna ~ **rubber**
Butadienkautschuk *m*
**butane** Butan *n* ~ **oxidation** Butanoxydation *f*
~ **recovery** Butanwiedergewinnung *f*
**butanol** Butanol *n* ~**-acetone ferment** Butanol-
azetonvergärung *f*
**but-board bearing** Nebenlager *n*
**butt, to** ~ anschlagen, anstoßen, stoßen,
stumpf aneinanderfügen **to** ~ **against** angrenzen, anliegen an **to** ~**-joint** stumpf aneinanderfügen **to** ~**-solder** stumpf löten **to** ~**-weld**
stumpf schweißen
**butt** Grenzstrecke *f*, Grifftück *n*, Kolben *m*,
Kugelfang *m*, Stammende *n*, unteres Ende *n*,
(hinge) Einsetzband *n* ~ **of rifle** Gewehrkolben
*m* ~ **end** Stumpf *m* ~**-ended spoke** Speiche *f*
mit verdicktem Ende ~ **hinge** Fisch-, Schar-
nier-band *n*, Stumpfkolben *m* ~ **joint** Blech-
stoß *m*, Dwarsnaht *f*, Hirnfuge *f*, Laschennietung *f*, Stoß-fuge *f*, -naht *f*, -verbindung *f*,
stumpfe Verbindung *f*, Stumpf-stoß *m*, -ver-
bindung *f*, winkelrechter Stoß *m*, (of rifle)
Stoß *m* ~**-joint riveting** Laschennietung *f*
~ **jointed** stumpf angefügt ~**-mortising units**

Stemmeinheiten *pl* ~**-muff coupling** Muffenkupplung *f* ~ **pier** Eckpfeiler *m* ~ **plate** Kol-
ben-, Stoß-, Blattverstärkungs-platte *f* ~**-**
**pressure inlet** Gebläsedruckstutzen *m* ~ **ring**
Anschlagring *m* ~**-seam welding** Stumpfnaht-
schweißung *f* (**flat**) ~ **strap** Flachlasche *f*
~ **strap** Lasche *f*, Rahmeneinlage *f*, Stoß-blech
*n*, -lasche *f*, -platte *f* ~ **stroke** Kolbenhieb *m*
~ **swivel plate** Klammerfuß *m* ~**-treated** mit
zubereitetem Stangenende *n* ~**-treated pole**
Stange *f* mit zubereitetem Ende ~ **weld**
Schweißnahtstoß *m*, Stumpfnaht *f* ~**-welded**
stumpfgeschweißt ~**-welded pipe** stumpfge-
schweißtes Rohr *n* ~**-welded tube** stumpfge-
schweißtes Rohr *n* ~ **welder** Stumpfschweiß-
maschine *f* ~ **welding** Stoß-lotung *f*, -ver-
schweißung *f*, Stumpf-schweißen *n*, -schwei-
ßung *f* ~**-welding machine** Stumpfschweiß-
maschine *f* ~**-welding process** Stumpfschweiß-
verfahren *n*
**butter,** ~ **paper** fettdichtes Papier *n* ~ **valve**
Butterhahn *m*
**butterfly** Schmetterling *m*, Weichzeichner *m*
(film) ~ **bolt** Flügelschraube *f* ~ **burner** Schmet-
terlingsbrenner *m* ~ **circuit** Schmetterlings-
kreis *m* ~ **cock** Flügelhahn *m* ~ **indicator**
Sternschauzeichen *n* ~ **nut** Flügel-, Flügel-
schrauben-, Schmetterlings-mutter *f* ~ **screw**
Flügelschraube *f* ~ **valve** Drehschütz *m*, Dros-
selklappenventil *n*
**butting,** ~ **of teeth** Auftreffen *n* von Zahn auf
Zahn ~ **against** Anstoßen *n* ~ **angle** Stoßwinkel
*m* ~ **face** Anlauffläche *f* ~ **ring** Anlauf-ring *m*,
-scheibe *f*
**button, to** ~ **on** aufknöpfen **to** ~ **up** zuknöpfen
**button** Blocktaste *f*, kleine Kapsel *f*, Knopf *m*,
Pastille *f*, Rastenknopf *m*, Regulus *m*, Wirbel
*m* (**assaying**) ~ Korn *n* ~ **for silencing local bell**
Weckertaste *f*
**button,** ~ **blank** Knopfrohling *m* ~ **braker**
Knopfmaschine *f* (textile finishing) ~ **catcher**
Sandfang *m* (paper mfg) ~ **die** Nuß *f* (of tools)
~ **head** Halbrundkopf *m* ~**-head bolt** Halbrundschraube *f* ~**-head rivet** Halbrundniet *m*,
halbversenktes Niet *n* (Kesselniet) ~**-head screw**
Halbrundkopfschraube *f* ~ **hole** Knopfloch *n*
~ **knob** Bedienungsknopf *m* ~ **loop** Anknöpf-
lasche *f* ~ **scale** Kornwaage *f* ~ **stick** Knopf-
gabel *f* ~ **transmitter** Kapselmikrofon *n* ~ **trap**
Sandfang *m* (paper mfg) ~ **tube** Knopfröhre *f*
~**-type blowcock** Druckknopfventil *n*
**buttons pilot lamp** Tastenkontrollampe *f*
**buttress, to** ~ abstützen (Pfeiler)
**buttress** Gegen-pfeiler *m*, -stütze *f*, Strebenpfei-
ler *m*, Wirkmesser *n* (blacksmith), (flying)
Strebebogen *m* ~ **dam** aufgelöste Staumauer *f*
~ **thread** Säge-gewinde *n*, -profilgewinde *n*
~(**-type**) **thread** Trapezgewinde *n*
**butty system** Geding*e* *n* (min.)
**butylaldehyde** Butylaldehyd *n*
**butyl,** ~ **chloralhydrate** Crotonchloralhydrat *n*
~ **compounds** Butylverbindungen *f*
**butylcyclohexane** Butylcyclohexan *n*
**butylene** Butylen *n*
**butyraldehyde** Butyraldehyd *n* ~ **peroxide**
Butyraldehydperoxyd *n*
**butyrate** Butyrat *n*; buttersauer
**butyric acid** Buttersäure *f*, Buttersäureäthylester *n*

**buy, to** ~ beziehen, einkaufen, kaufen **to** ~ **up** auskaufen **to** ~ **on installment plan** auf Abzahlung kaufen **to** ~ **wholesale** aufkaufen

**buyer** Abnehmer *m*, Einkäufer *m*, Käufer *m*, Verbraucher *m*

**buying** Ankauf *m* ~ **power** Kaufkraft *f* ~ **price** Kaufpreis *m*

**buzz, to** ~ brausen, schnarren, summen, überfegen ~ **(ing)** Summen *n* ~ **track** Brumm-spur *f*, -streifen *m*, Prüfstreifen *m* (film); Summ-spur *f*, -streifen *m*

**buzzer** Hupe *f*, Schelle *f*, Schnarre *f*, Schnarrsummer *m*, schwingender Unterbrecher *m*, Selbstunterbrecher *m*, Summer *m*, Unterbrecher *m*, Wecker *m*, (alarm) Schnarrwecker *m* ~ **for busy signal (or tone)** Summer *m* für Besetztzeichen ~ **for dialing tone** Summer *m* für Amtszeichen ~ **for ringing tone** Summer *m* für frei

**buzzer,** ~**-driven circuit** durch Summer erregter Kreis *m* ~**-driven wavemeter** Wellenmesser *m* mit Summererregung ~ **excitation** Summererregung *f* ~ **generator** Summergenerator *m* ~ **(striking) horn** Summeraufschlaghorn *n* ~**-indicator drop** Weckerfallklappe *f* ~ **interrupter** Summerunterbrecher *m* ~ **key** Summerknopf *m* ~ **lead** Summerleitung *f* ~ **pratice set** Klopfer *m* ~ **relay** Selbstunterbrecher-, Summer-relais *n* ~ **set** Summerzusatz *m* ~ **signal** Brumm-, Summer-zeichen *n* ~**-type starting ignition** Summeranlaßzündung *f* ~**-type transformer** Pendelumformer *m* ~ **visual indicator** Summerschauzeichen *n* ~ **voltage** Summerspannung *f* ~ **wave meter** Summerwellenmesser *m*

**buzzing,** ~ **of nozzle** Schnarren *n* der Düse ~ **noise** Brummen *n* ~ **tone** Summerton *m* ~ **vibration of nozzle** Schnarrschwingung *f* der Düse

**B vertical** Vertikalkanal B

**B wire** b-Ader *f*, Ader *f* zum Stöpselhals

**BX cable** Bandpanzerleitung *f*

**BY** die Leitung ist besetzt

**by-padded air** Falschluft *f*

**by-pass, to** ~ ableiten, aussparen, überbrücken, überströmen, umgehen

**by-pass** Abkürzung *f*, Ableitung *f*, Ausweichstelle *f*, Beiweg *m*, Entlastungskanal *m*, Entwässerungskanal *m*, Nebenauslaß *m*, Nebenleitung *f*, Nebenstrang *m*, Nebenverkehrsweg *m*, Nebenweg *m*, Parallelweg *m*, separater

Gang *m*, Überstromkanal *m*, Überstromweg *m*, Umbruch *m*, Umfahrung *f*, Umfahrungsstrecke *f*, Umgang *m*, Umlaufkanal *m*, Umleitung *f*, Umwegeleitung *f* (electr.), Verzweigung *f*

**by-pass,** ~ **air flow** Sekundärluftstrom *m* (jet) ~ **anode** Nullanode *f* ~ **apparatus** Umlaufapparat *m* ~ **area** Überholstelle *f* (Flughafen) ~ **automatic-telephone system** Kreislaufselbstanschlußsystem *n* ~ **cable** Anschlußkabel *n* ~ **capacitor** Quer-, Vorbeischleifungs-, Überbrückungs-kondensator *m* ~ **circuit** Umgehungsschaltung *f* ~ **clack valve** Umgangsklappe *f* ~ **condensater** Umgehungs-, Parallelkapazität *f* ~ **condenser** Ableit-, Nebenschluß-, Parallel-, Quer-, Überbrückungs-kondensator *m* ~ **connection** Umgehungsschaltung *f* ~ **door** Mantelstromleitklappe *f* ~ **duct** Sekundärluft-Ringspalt *m* ~ **engine** Mantelstromtriebwerk *n* ~ **filter** Nebenschlußfilter *m* ~ **light signal** Lichthupe *f* ~ **line** Umgehungsleitung *f* ~ **lubricating oil filter** Nebenstromölfilter *n* ~ **opening** Freilauf *m* ~ **pipe** Umführungsleitung *f* ~ **ratio** Durchsatzverhältnis *n* (rocket) ~ **regulator** Nebenwegregler *m* ~ **(road)** Entlastungs-, Umgehungs-straße *f* ~ **system** Kreislauf-, Umgehungs-system *n* ~ **tracks** Ausweichsgleis *n* ~ **transformer** Durchgangstransformer *m* ~ **turbine** Anzapfturbine *f* ~ **valve** Abschalt-, Anzapf-, Entlastungs-, Luftumgangs-, Rücklauf-, Umgehungs-, Umlauf-ventil *n* ~ **valve control** Überströmventilregulierung *f*

**by-passing** Umgehung *f*, untere Umgehung *f*

**bypath** Parallelweg *m*

**by-pit** Nebenschacht *m*

**by-product** Abfall *m*, Abfall-erzeugnis *n*, -produkt *n*; Neben-erzeugnis *n*, -produkt *n*, verwertbarer Abfallstoff *m* ~ **coke** Hütten-, Zechen-koks *m* ~ **coke oven** Destillationsofen *m* ~ **coke-oven gas** Koksgas *n* ~ **coke-oven plant (or practice)** Destillationskokerei *f* ~ **coking** Destillationskokerei *f* ~ **coking plant (or practice)** Nebenproduktkokerei *f* ~ **gas** Destillationsgas *n* ~ **oven** Kammerofen *m* ~ **plant** Gewinnungs-, Nebenprodukt-anlage *f* ~ **recovery** Nebenproduktgewinnung *f* ~**-recovery plant** Nebenproduktgewinnungsanlage *f* ~ **tar** Absatzteer *m* (coal tar)

**by-road** Querweg *m*

**by-wash** Entlastungs-, Entwässerungs-kanal *m*, Umlaufgraben *m*

# C

**cab** Fahrerhaus *n*, Droschke *f*, Führerstand *m* (Lokomotive), Taxi *n* ~ **over engine** Stirnsitz *m* ~ **illumination** Fahrerhausleuchte *f* ~ **lining** Innenverkleidung *f* ~-**tire cable** Gummischlauchleitung *f*
**cabane** Baldachin *m*, Spann-bock *m*, -turm *m* ~ **bracing** Rahmenspannturm *m*, Spannturmverspannung *f*, Tragrohr *n* ~ **strut** Baldachinstrebe *f*
**cabaric root** Haselwurz *f*
**cabin** geschlossener Führersitz *m*, Kabine *f*, Kabinenraum *m*, Kajüte *f*, Kammer *f* ~ **for safety lamps** Lampenstube *f* ~ **on ship** Koje *f*
**cabin,** ~ **altimeter** Kabinendruck(höhen)messer *m* ~ **altitude** Kabinendruck *m* ~ **boy** Schiffsjunge *m* ~ **cruiser** Kabinenflugzeug *n* ~ **door** Kabinentür *f* ~ **heating** Kabinenheizung *f* ~-**heating jet** Kabinenheizsprüher *m* ~ **height** Höhe *f* der Kabine ~ **hood** Kabinendach *n* ~ **leakage tester** Kabinendruckprüfgerät *n* ~ **length** Länge *f* der Kabine ~ **plane** Kabinenflugzeug *n* ~ **roof** Kabinendach *n* ~ **vent switch** Kabinenbelüftungsschalter *m* ~ **wall** Kabinenwand *f* ~ **window** Kabinenfenster *n*
**cabinet** Gehäuse *n*, Holzgehäuse *n*, Schrank *m*, Zelle *f* ~ **file** Münz-, Wälz-feile *f* ~ **framework** Schrankgestell *n* ~ **lamp** Schrankleuchte *f* ~ **leg** Kastenfuß *m* ~ **maker** Tischler *m*, Schreiner *m* ~**making** Kunsttischlerei *f*, Tischlerei *f* ~ **speaker** Flächenlautsprecher *m* ~ **stand** Aufbaustand *m* ~ **work** Kunsttischlerei *f*
**cable, to** ~ kabeln, verkabeln, verseilen
**cable** Draht-schloß *n*, -seil *n*, Festmachetau *n*, Kabel *n*, Kabelseil *n*, Leine *f*, Leitung *f*, Leitungsschnur *f*, Windanker *m*, (conductor) Seil *n*, Tau n ~ **and conduit (operated) switch** Bowdenzug-, Seilzug-schalter *m* ~ **for electric surveys** Kabel(seile) für elektrisches Schürfen ~ **for four-wire working** Vierdrahtkabel *n* ~ **in gutta-percha covering** Guttaperchakabel *n* ~ **in space** Raumseil *n* ~ **for twin-band telephony** Zweibandkabel *n* ~ **for weak currents** Fernmeldekabel *n*
**cable,** ~ **acorn** Kabelschoner *m* (naut.) ~ **address** Draht-anschrift *f*, -wort *n* ~ **amplifier** Kabelverstärker *m* ~ **anchorage** Anker *m* des Seiles ~ **apron** Seilschürze *f* ~ **armoring** Kabelschoner *m* ~**armoring machine** Kabelbewehrungsmaschine *f* ~ **arrangement** Anschlußschema *n* (photo) ~ **assembly** Kabelgruppe *f* ~-**assignment record** Benutzungsplan *m*, Kabelbesetzungsbuch *n* ~ **attenuation** Kabeldämpfung *f* ~ **backstay** Pardun *n*, Pardune *f*, ~ **balancing** Kabel-, Kabelader-ausgleich *m* ~ **balancing method** Kabelausgleichsverfahren *n* ~-**balancing network** Kabelnachbildung *f* ~ **beacon** Kabelbake *f* ~ **bearer** Kabel-aufhänger *m*, -stütze *f*, -träger *m* ~ **block** Drahtseil-hindernis *n*, -kloben *m* ~ **boring** Seilbohren *n* ~ **box** Endverschluß *m* ~ **bracket** Kabel-stütze *f*, -träger *m* ~ **braiding** Kabelumklöppelung *f* ~ **brake** Seilbremse *f* ~ **branching** Kabelverzweigung *f*

~ **break** Kabel-bruch *m*, -unterbrechung *f* ~ **bucket** Drahtseilkübel *m* ~ **buoy** Kabel-boje *f*, -tonne *f* ~ **bushing** Kabeltülle *f*, Ringkausche *f* ~ **cap** Anschlußkappe *f* ~ **car** Berg- und Talbahn *f*, Luftkabelfahrstuhl *m* ~ **carrier** Kabeltrage *f* ~ **cellar** Kabelkeller *m* ~ **chamber** Abzweigkasten *m* ~ **channel** Kabelrinne *f* ~ **chute** Aufstiegkanal *m*, Kabelschacht *m* ~ **circuit** Kabelkreis *m* ~-**circuit diagram** Kabelliste *f* ~ **circuit shell** Zwischenstecker *m* (elektr.) ~ **clamp** Drahtseilklemme *f*, Kabel-halter *m*, -klemme *f*, Leitungsklemme *f* ~ **cleat** Leitungshalter *m* ~ **clip** Kabel-klammer *f*, -schuh *m* ~ **code** Kabelschrift *f*, Seekabelalphabet *n* ~ **code direct printer** Direktschreiber *m* ~ **compound** Kabel-ausgußmasse *f*, -schmiere *f* ~ **conductor** Kabel-ader *f*, -leitung *f*, Litze *f* ~-**conductor splicing** Verbinden *n* der Kabeladern ~ **conduit** Kabel-kanal *m*, -rohr *n*, -strang *m*, Zuleitungskanal *m* ~ **connecting box** Kabelverbindungsmuffe *f* ~ **connecting clamp** Polklemme *f* ~ **connection** Kabelanschluß *m*, Seilverbindung *f* ~-**connection box (or case)** Kabelverzweiger *m* ~ **connector** Kabelanschluß-brücke *f*, -stück *n*, Leitungsverbinder *m* ~ **control** Seilsteuerung *f* ~ **control box** Seilzugspannkasten *m* ~-**control set** Kabelüberwachungsgerät *n* ~-**cord fabrics** Kordgewebe *n* ~ **core** Kabel-ader *f*, -kern *m*, -seele *f*, -seil *n* ~-**core conductor** Kabel-innenleiter *m*, -seele *f* ~ **cover-tile** Kabelabdeckstein *m* ~ **covering** Kabel-hülle *f*, -umhüllung *f* ~-**covering wire** Kabeldeckdraht *m* ~ **crane** Kabelkran *m* ~ **current transformer** Kabelstromwandler *m* ~ **curtain** Seilvorhang *m* ~ **cutter** Sprenggreifer *m* ~ **departure** Kabelabgang *m* ~ **desiccant** Kabelwachs *n* ~ **detector** Kabelsuchgerät *n* ~ **disc** Kabelscheibe *f* ~ **disconnect adjuster** Schnelltrennspannschloß *n* ~-**distributing box** Kabelverteiler *m* ~ **distributing cabinet** Kabelverteilerschrank *m* ~ **distributing plug** Abzweigmuffe *f* ~-**distributing system** Kabelverteilungssystem *n*
**cable-distribution,** ~ **box** Kabelüberführungs-endverschluß *m*, -kasten *m*, Kabel-, Linien-verzweiger *m* ~ **case** Kabelverzweiger *m* ~ **head** Kabelendverzweiger *m* ~ **plug** Abschluß-, Kabelabzweig-, Kabelverzweigungs-, Verzweigungs-muffe *f* ~ **point** (junction between cable system and overhand wires) Kabelaufführung *f* ~ **pole** Kabelüberführungssäule *f*
**cable,** ~ **drawing-in box** Kabeleinziehkasten *m* ~ **drawing-in implement** Kabeleinziehgerät *n* ~ **dredger** Kabelbagger *m* ~ **drill (or drilling)** Seil-bohren *n*, -bohrung *f* ~ **drive** Drahtseiltrieb *m*, Seilantrieb *m* ~ **drum** Kabelhaspel *m*, Kabel-, Seil-, Winden-trommel *f* ~ **duct** Kabel-kanal *m*, -rohr *n*, -strang *m* ~-**duct clip (or holder)** Kabelhalter *m*, Kabelschelle *f* ~ **dummy section** Kabelschrank *m* ~ **elevator** Kabelwinde *f* ~ **elongation** Seilstreckung *f* ~ **end** Anschlußkappe *f* ~-**end bell** Gehäuse *n* für Kabelendstück ~ **entry** Kabeleinführung *f*

~ extension arm einseitiger Querträger *m* für Luftkabel ~ eye Kabel-öse *f*, -schuh *m* ~ failure Kabelbruch *m* ~ fairing Seilverkleidung *f* ~ fan ausgeformtes Kabel *n*, Kabelzopf *m* ~ fault Fehlerort *m* an unterirdischen Kabeln, Kabelfehler *m* ~ ferry Zugfähre *f* ~ fill Kabelauslastung *f* ~ filling yarn Kabelfüllgarn *n* ~ fitting Taubeschlag *m* ~ fittings Kabelgarnituren *pl* ~ force Seilkraft *f* ~ form ausgeformtes Kabel *n*, Form *f* der Verkabelung, Kabelform *f*, -zopf *m* ~ gland Kabeleinführungshals *m* ~ grease Kabelgleitfett *n* ~ grip Einziehstrumpf *m*, Kabel-einziehstrumpf *m*, -schuh *m*, Ziehstrumpf *m* ~ grip split Nachziehschlauch *m* ~ guard Absperrgestell *n* ~ guides Zügeführung *f* ~ hanger Kabelaufhänger *m* ~ head Abschlußmuffe *f*, End-, Kabelend-verschluß *m* ~ hoist Drahtseilflaschenzug *m* ~ hoisting gear Seilhubeinrichtung *f* ~ holder Leitungshalter *m* ~ hook Kabelträger *m* ~ house Kabelhaus *n* ~ hut Kabel-häuschen *n*, -hütte *f*, Untersuchungshäuschen *n* ~-impregnating tank Kabeltränkkessel *m* ~ inlet Kabeleinführung *f* ~ insulating material Kabelisoliermasse *f* ~ interruption Kabelunterbrechung *f* ~ joining chamber kleiner Kabelbrunnen *m* ~ joint Drahtseilschloß *n*, Kabelabzweig *m*, Kabellötstelle *f*, Kabelspleißung *f*, Lötstelle *f*, Seilschloß *n*, Verbindungsstelle *f* ~-joint box Abzweigkabel *n*, Kabelabzweigkasten *m*, Kabellötbrunnen *m* ~ jointing Kabellötung *f* ~-jointing detachment Kabellöttrupp *m* ~-junction box Kabelverteiler *m* ~ key Kabeltaste *f* ~-laid rope Kabelseil *n*, Kabelschlag *m*, kabelweise geschlagenes Tau *n* ~ lashing Drahtbund *m* ~ lay Kabelschritt *m*, Verseilung *f* ~ lay up Drall *m* ~laying Kabelverlegung *f* ~-laying detachment Kabelbautrupp *m* ~-laying trolley car or truck Verlegewagen *m* ~ laying winches Kabeleinziehwinden *pl* ~ layout Kabelplan *m* ~'s length Kabellänge *f* ~ letter Kabelbrief *m* ~ lifting Kabelhochführung *f* ~ line Kabel-ader *f*, -leitung *f*, -linie *f*, Seilzug *m*, verkabelte Leitung *f* ~ locker Kettenlast *f* ~ loom Kabelbaum *m* ~-machine driver Drahtverseiler *m* ~ make-up Kabelaufbau *m* ~-making machine Kabelmaschine *f* ~ manhole Abzweigkasten *m* ~ manifold (for ignition wires) Kabelrohr *n* ~ manufacture Kabelherstellung *f* ~ marker Kabelmerkstein *m* ~ matching gear Kabelanpaßgerät *n* ~ Morse code Kabelalphabet *n* ~s mortgaging act Kabelpfandgesetz *n* ~ mounting Kabelhalterung *f* (Stufenschalter) ~ network Kabelnetz *n* ~ oil Kabelöl *n* ~-operated seilbetätigt ~ outlet Kabelausführung *f* ~ pair Kabeladernpaar *n* ~ paraffin Kabelwachs *n* ~ passage Kabeldurchlaß *m* ~-paying-out machine Kabelauslegemaschine *f* ~ pipe Kabelrohr *n* ~ pit Kabelgrube *f* ~ plant Kabel-anlage *f*, -netz *n* ~ plant without distribution boxes Kabeladernetz *n* ~ plow Seilpflugmaschine *f* ~ plug Renkstecker *m* ~ plug-type connector Leitungssteckverbinder *m* ~-powered seilbetätigt ~ project Kabelplanung *f* ~ protected against interference induktionsgeschütztes Kabel *n*, Kabel *m* mit induktionsgeschützten Leitungen ~-protection pipe Kabelschutzrohr *n* ~ pully Seil-führungsrolle *f*,

-rolle *f*, -scheibe *f* ~ puncture Fehlerort *m* an unterirdischen Kabeln ~ quadrant Seilscheibe *f* ~ rack Kabel-gerüst *n*, -gestell *n*, -rost *m*, -traggerüst *n* ~ railroad Seilbahn *f* ~ railway Drahtseilbahn *f*, Kabelbahn *f* ~ reel Kabel-haspel *m*, -trommel *f* ~ reel car Kabelverlegewagen *m* ~-reel trailer Anhänger *m* für Kabeltransport, Kabelwagen *m* ~ release (gear) Seilauslöservorrichtung *f* ~-repair ship Kabelreparaturschiff *n* ~ repeater Kabelverstärker *m* ~ restorer Kabelreparaturschiff *n* ~ ring Tragring *m* ~ rings (for antenna) Kabeltragringe *pl* ~ roller Seilführungs-rolle *f*, -scheibe *f* ~ rope Seil *n* ~ run Kabelweg *m* ~-scaffolding detachment Kabelgerüsttrupp *m* ~ screening Kabelabschirmung *f* ~ sealing ends Kabelstutzen *pl* ~ shackle Kabelschloß *n* ~shaft Kabelhochführungsschacht *m* ~ sheathing Kabel-mantel *m*, -umhüllung *f* ~ sheave Kabelrolle *f* ~ shelf Kabel-brett *n*, -gestell *n*, -rost *m*, -träger *m* ~ shield Kanalmundstück *n* ~ shielding conduit Kabelschirmleitung *f* ~ ship Kabelschiff *m* ~ shoe Kabelschuh *m* ~ side Kabelseite *f* ~ slack Kabellose *f* ~ sleeve Kabel-muffe *f*, -verbinder *m* ~ socket Endverschluß *m*, Kabel-einführung *f*, -schuh *m* ~ solderer Kabellöter *m* ~ sounding Kabelsonde *f* ~ spacer Kabel-führungsplatte *f*, -abstandsschelle *f* ~ splice Kabel-spleißung *f*, -verbindung *f*, Seilspleiß *m* ~ splitting machine Kabelauftrennmaschine *f* ~ stage Kabelgatt *n* ~ steamer Kabeldampfer *m* ~ stock Kabelqualität *f* ~ strand Kabel-ader *f*, -litze *f* ~-stranding machine Kabel-umflechtmaschine *f*, -verseilmaschine *f* ~ stub Anschlußpeitsche *f* ~ support Kabel-stütze *f*, -traggerüst *n* ~-support rack Kabelendgestell *n* ~ supports Drahtträger *m* ~ suspender Tragband *n* für Luftkabel, Kabeltragringe *pl* ~ suspension Kabelaufhängung *f*, Kabelhalter *m* ~ system Kabel-anlage *f*, -netz *n* ~ tank Kabeltank *m* ~ tap Kabelabzweig *m* ~ tape Kabelarmierung *f* ~ tensiometer Seilzugmesser *m* ~ tension diagram Seilspannungsschaubild *n* ~ tension meter Seilzugmesser *m* ~ terminal Kabel-abschluß *m*, -endverschluß *m*, -endverzweiger *m*, -öse *f*, -schuhanschluß *m*, Verschlußkopf *m* ~-terminal box Kabel-klemmkasten *m*, -muffe *f* ~ terminal rack Kabelanschlußgestell *n* ~-terminal screw Kabelklemmschraube *f* ~ termination Kabelabschluß *m*, Trennendverschluß *m* ~ terminator Gehäuse *n* für Kabelendstück ~ test with compressed air Kabeldichtigkeitsprüfung *f* mittels Druckluft ~-testing car Kabelmeßkarren *m* ~-testing equipment Kabelprüfeinrichtung *f* ~-testing instrument Kabelmeßgerät *n* ~-testing machine Kabelprüfmaschine *f* ~ thimble Kabelschleife *f*, Seilbuchse *f* ~ tile Kabelstein *m* ~ toll Fernleitungskabel *n* ~ (drilling) tools Seilbohrgerät *n* ~ tower Kabelturm *m* ~ trailer Kabeltrommelanhänger *m* ~ transmission Übertragung *f* auf Kabeln ~-traveling crab Seillaufkatze *f* ~ tree Kabelbaum *m* ~ trench Kabelgraben *m* ~ trolley Kabelwagen *m* ~ trough Kabelkasten *m* ~ trough(ing) Kabelrinne *f* ~ trough for bridges (or tunnels) Kabelkasten *m* an Brücken oder in Tunneln ~ trunk Kabelbaum *m* ~

**tube** Falzrohr *n* ~-**tuning unit** Kabelabgleich-
kasten *m* (rdr) ~ **turning** Kabelschrank *m* ~
**turning section** Ansatzschrank *m* (teleph.) ~
**vault** Kabelrost *m* ~ **wax** Kabelwachs *n* ~**way**
Kabelkanal *m*, Schwebebahn *f*, Seillaufbahn *f*,
Seilbahn *f* ~**way bridge crane** Brückenkabel-
kran *m* ~**way carriage** Hängewagen *m* ~ **wheel**
Seilrad *n* ~ **winch** Kabelwinde *f*, Seilwinde *f*
~ **wire** Kabelader *f*, Kabeldraht *m* ~-**wire rope**
Seil *n* ~ **work(s)** Kabelwerk *n*
**cabling** Ausstäbung *f*, Verdrahtung *f*, Verkabelung *f* ~ **system** Verkabelungssystem *n*
**cacodylic acid** Kakodylsäure *f*
**cacoxenite** Kakoxen *n*
**cadastral** Katastermesser *m* ~ **survey** Kataster-
aufnahme *f*, Landvermessung *f*
**cadastre** Kataster *m*
**cade oil** Kaddig-, Kade-öl *n*
**cadence, to** ~ taktgeben
**cadence** Takt *m*, Tonschluß *m* ~ **signal** Taktzei-
chen *n* ~ **tapper** Taktgeber *m*, (keying) Tasten
*pl*
**cadmia** Galmei *m*, Ofengalmei *m*, Zinkofen-
bruch *m*, Zinkschwamm *m*
**cadmiated iron** kadmiertes Eisen *n*
**cadmium** Kadming *n*, Kadmium *n* ~ **arsenite**
Kadmiumarsenit *n* ~ **hydrate** Kadmium-
(oxyd)hydrat *n* ~ **plating** Kadmieren *n*,
Kadmiumüberzug *m* ~-**plating plant** Ver-
kadmiumierungsanlage *f* ~ **storage cell** Kad-
miumsammler *m* ~ **strip** Kadmiumband *n* ~
**sulfocyanate** Kadmiumrhodanid *n* ~ **vapor
lamp** Kadmiumdampflampe *f*
**cadre** Kader *m*, Stamm *m*, Stammpersonal *n*
**cage** Aufzug *m*, Behältnis *n*, Biete *f* (press),
Fördergerippe (min.), Fördergestell *n*, Gehäuse *n*, Käfig *m*, Lagerkapsel *f*, Laterne *f*,
Scheibenkäfig *m*, (pits, mines) Fahr-, Förder-
-korb *m*, Förder-kasten *m*, -schale *f*, (drawing)
Korb *m*, (roller) Rollenkäfig *m* ~ **of delivery
valve** (motor) Druckventileinsatz *m* ~ **antenna** Doppelkegel-, Doppelkonus-, Käfig-,
Konus-, Reusen-antenne *f* ~ **bar** Käfigstab
*m* ~ **coil** Käfigspule *f* ~ **dipole** Käfigdipol *m*
~ **guide** Förderschacht *m*, Spurlatte *f* ~ **mill**
Desintegrator *m* ~ **oil box** Etagenpresse *f* ~
**oil plate press** Etagenpresse *f* ~ **oil press**
Etagenpresse *f* ~ **press** Seiherpresse *f* ~ **rotor**
Käfigläufer *m* ~ **seat** Aufsetzvorrichtung *f*
~-**uncage lever** Feststellhebel *m* ~ **winding
plant** Gestellförderanlage *f*
**cageable,** ~ **gyro** feststellbarer Kreisel *m* (Kompaß) ~ **yaw roll gyro** feststellbarer Kurs-
querneigungskreisel *m*
**caged,** ~ **switch** Rastschalter *m* ~ **valve** hängendes Ventil *m*
**cagelike** käfigförmig
**cager** Anschläger *m*, Fesselung *f* (g/m), Fesse-
lungshebel *m* (g/m)
**cages** Schnäbelung *pl*
**caging** Fessel ... (gyro), Feststell ... ~ **alignment** Justierung *f* der Fesselung ~ **coil** Fesse-
lungsspule *f* (g/m) ~ **knob** Feststellknopf *m*
~ **relay** Feststellrelais *n* ~ **switch** Feststell-
schalter *m*
**cagged ingot** vorgewalzter Block *m*
**cairngorm** Rauchquarz *m*
**caisson** Dockverschlußsponton *n*, Druckluft-

kasten *m*, Hinterwagen *m*, Kasten *m*, Senk-
schacht *m*, Wanne *f* (Bau), (navy) Docktor *n*,
(compressed-air) Senkkasten *m*, (ship)
Schwimmtor *n* ~ **of a building slip** Helling-
ponton *m* ~ **chamber** Pontonkammer *f* ~
**lock gate** Rollponton *n*, Schiebespule *f*
**cake, to** ~ backen, ballen, festbacken (Kohle),
greifen, sintern, zusammenbacken, (paint)
stocken, **to** ~ **together** (by heat or pressure)
festbacken **to** ~ **upon** anbacken
**cake** Kuchen *m*, Schrei *m* (metall.) ~ **of clinker**
Schlackenkuchen *m* ~**s of ore** Erzplatten *pl*
~ **of rose copper** Kupferrosette *f* ~ **of rubber**
Kautschukblock *m* ~ **of slag** Schlackenkuchen
*m*
**cake,** ~ **ore** Kuchenerz *n* ~ **space** Kuchenraum
*m* ~ **surface** Kuchenoberfläche *f* ~ **thickness**
Kuchenstärke *f* ~-**top strewing-machine** Streu-
selmaschine *f* ~-**wax** Wachsplatte *f*
**caked** zusammengesintert ~ **with carbon** Öl-
kohleansatz *m* aufweisend *m* ~ **sugar** klumpiger
Zucker *m*
**caking** Festbacken *n*, Sinterung *f*, Backen *n*
(coal or sugar) ~ **coal** backende Kohle *f*,
Back-, Fett-kohle *f* ~ **floor** Greifhaufen *m*
~ **property** Back-, Ballungs-fähigkeit *f* ~
**quality** Backfähigkeit *f*
**calamine** Galmei *m*
**calamite** Kalamit *m*
**calamus oil** Kalmusöl *n*
**calan** Calan *m*
**calandria** Röhrenheizkörper *m* ~ **pan** Calandria-
Kochapparat *m* ~-**type pan** Röhrenvakuum *n*
**calc,** ~**silicate rocks** Kalksilikatgesteine *pl*
~-**sinter** Kalksinter *m* (Travertin) ~-**spar** Kalk-
spat *m*
**calcar arch** Frittofen *m*
**calcareous** kalk-artig, -haltig, kalkig ~ **clay**
Kalkmergel *m* ~ **deposit** Kalkablagerung *f* ~
**earth** Kalkerde *f* ~ **gravel** Kalkkies *m* ~-**shale**
**quarry** Kalkschieferbruch *m* ~ **spar** isländischer
Kristall *m* ~ **tufa** Tuffstein *m*
**calcic** kalziumhaltig
**calcify, to** ~ verkalken
**calcimeter** Kalkmesser *m*
**calcination** Abschwefelung *f*, Brand *m*, Brennen
*n*, Röstung *f* ~ **in clamps** Röstung *f* in Stadeln
~ **of gypsum** Gipsbrennerei *f* ~ **in heaps** Röstung *f* in Haufen
**calcination,** ~ **assay** Röstprobe *f* ~ **process**
Röstprozeß *m*, ~ **test** Röstprobe *f*
**calcine, to** ~ abbrennen, abschwelen, abschwefeln, ausglühen, brennen, einäschern, glühen,
kalzinieren, rösten, zubrennen
**calcine** Schwefelkiesabbrand *m*
**calcined** gebrannt, geröstet ~ **borax** gebrannter
Borax *m* ~ **dolomitic** gebrannter Dolomit *m*
~ **gold** Goldkalk *m* ~ **iron** Eisenkalk *m* ~
**lime** gebrannter Kalk *m* ~ **magnesia** gebrannte
Magnesia *f* ~ **magnesite** gebrannter Magnesit
*m* ~ **ore** Rösterz *n* ~ **product** Röst-erzeugnis *n*,
-produkt *n*
**calciner** Röster *m*, Röstofen *m*
**calcining** Brennen *m*, Glühen *n*, Rösten *n* ~ **at
white heat** Weißbrennen *n*
**calcining,** ~ **apparatus** Röstapparat *m* ~ **drum**
Rösttrommel *f* ~ **furnace (or kiln)** Brenn-,
Kalzinier-, Röst-ofen *m* ~ **method** Röstver-

fahren *n* ~ **operation** Röstvorgang *m* ~ **plant** Röstanlage *f* ~ **process** Brennprozeß *m* ~ **product** Rösterzeugnis *n* ~ **stall** Röststadel *m* ~ **test** Ansiedescherben *m*
**calciothorite** Kalziumthorit *m*
**calcite** Atlas-, Kalk-spat *m*, Kalkstein *m*, Kalzit *m*
**calcium** Kalzium *n* ~ **acetate** essigsaurer Kalk *m* ~ **antimony phosphate** Kalziumantimonphosphat *n* ~ **borate** Boraxkalk *m*, Kalkborat *n* ~ **carbide slag** Kalziumkarbidschlacke *f* ~ **carbonate** isländischer Doppelspat *m*, kohlensau(e)rer Kalk *m* ~ **chloride** Chlorkalk *m*, Chlorkalzium *n*, Kalziumchlorid *n* ~ **cyanamide** Kalkstickstoff *m* ~ **ferrocyanide** Ferrozyankalzium *n* ~ **fluoride** Fluorkalzium *n*, Flußspat *m*, flußsau(e)rer Kalk *m* ~ **glycerophosphate** Kalziumglyzerinphosphat *n* ~ **hydrate** Kalziumhydroxyd *n* ~ **hydrosulfide** Kalziumsulfhydrat *n* ~ **hydroxide** gelöschter Kalk *m*, Kalkhydrat *n*, Kalziumhydroxyd *n* ~ **hypochloride** unterchlorsaurer Kalk *m* ~ **hypochlorite** Chlorkalk *m*, unterchlorigsaurer Kalk *m* ~ **lactate** milchsaurer Kalk *m* ~ **nitrate** Kalksalpeter *m*, salpetersaurer Kalk *m* ~ **oxalate** oxalsaurer Kalk *m* ~ **oxide** ätzender oder gebrannter Kalk *m*, Kalkerde *f*, Kalziumoxyd *n* ~ **phenolate** Phenolkalzium *n* ~ **phosphate** Kalziumphosphat *n* ~ **phosphide** Phosphorkalzium *n* ~ **potassium sulfate** Kalziumkaliumsulfat *n* ~ **sodium sulfate** Kalziumnatriumsulfat *n* ~ **sulfate** Gips *m*, Kalksulfat *n*, schwefelsaurer Kalk *m* ~ **(lime) sulfite** schwefligsaurer Kalk *m* ~ **sulfocyanate** Kalziumrhodanid *n* ~ **tungstate** Kalziumwolframat *n*, wolframsaures Kalzium *n*
**calculable** berechenbar
**calculagraph** Kalkulagraf *m*, Zeitschreiber *m*, Zeitstempel *m* (automatic)
**calculate, to** ~ ausrechnen, berechnen, errechnen, kalkulieren, rechnen, veranschlagen, (again) durchrechnen **to** ~ **a charge** gattieren **to** ~ **the volume of a solid** die Masse kubizieren
**calculated** gegißt, theoretisch, vorausberechnet ~ **analysis** kalkulierte oder errechnete Analyse *f* ~ **ceiling** errechnete oder theoretische Gipfelhöhe *f* ~ **matter** Trockensubstanz *f* ~ **speed** Rechnungsgeschwindigkeit *f* ~ **value** Tafelwert *m*
**calculating** Vorrechnung *f* ~ **disk** Rechenscheibe *f* ~ **instrument** Rechenapparat *m* ~ **machine** Rechenmaschine *f* ~ **punch** elektronisches oder lochendes Rechengerät *n* ~ **roughly** bei überschlagender Rechnung, überschlägig berechnet ~ **rule** Rechenschieber *m* ~ **ruler** Artillerierechenschieber *m*
**calculation** Abrechnung *f*, Ausrechnung *f*, Berechnung *f*, Bezifferung *f*, Errechnung *f*, Kalkül *n*, Kalkulation *f*, Kostenrechnung *f*, Nachrechnung *f*, Rechnung *f*, Voranschlag *m*, Zählung *f* ~ **involving fractions** Bruchrechnung *f* **to be far out in** ~s sich in der Rechnung irren
**calculation,** ~ **of brakes** Bremsenberechnung *f* ~ **of cost** Kostenberechnung *f* ~ **of current consumption** Strombedarfsberechnung *f* ~ **of fluxing** Schlackenberechnung *f* ~ **of lattice energy** Gitterberechnung *f* ~ **of mixture** Gattierungsberechnung *f* ~ **of the number of**

**selectors** Wählerberechnung *f* ~ **of production** Vorkalkulation *f* ~ **of production cost** Selbstkostenberechnung *f* ~ **of stability** Stabilitätsberechnung *f* ~ **of strength** Festigkeitsberechnung *f* ~ **of trim** Trimmberechnung *f*
**calculator** Berechner *m*, Rechner *m*, digitaler Rechenautomat *m*, digitale Rechenmaschine *f*, Vorrechner *m*
**calculograph** Gesprächszeitmesser *m*
**calculus** Rechnen *n* ~ **of differences** Differenzrechnung *f* ~ **of variations** Variationsrechnung *f*
**caldron** Heiz-, Koch-kessel *m*
**caledonite** Kaledonit *n*
**calender, to** ~ kalandern, kalandrieren, mangeln, satinieren
**calender** Glander *f*, Glätt-presse *f*, -werk *n*, Kalander *m*, Mangel *f*, Satinier-maschine *f*, -werk *n*, Trockenglattwerk *n*, Walzenglättwerk *n* Walzenmangel *f*, Walzwerk *n* ~ **of dates due** Fristkalender *m*
**calender,** ~ **beam** Mangelbaumstuhl *m* ~ **coater** Auftragskalander *m* ~ **color** Oberflächenfärbung *f* (paper mfg) ~ **cut** Quetschfalte *f* ~ **cuts** Satinierfalten *pl* ~ **finish** Walzendekatur *f* ~ **grain** Kalandereffekt *m* ~ **grease** Kalanderfett *n* ~**man** Kalanderführer *m* ~ **operator** Kalandrierer *m* ~ **pressure** Klemmdruck *m* der Kalanderwalzen *m* ~ **roll** Kalanderwalze *f* ~**-roll paper** Walzenpapier *n* ~ **roller** Abzugswalze *f* ~ **run** Kalandrieren *n* ~ **train** Kalanderreihe *f* ~ **spots** Kalanderschaden *m* ~ **year** bürgerliches Jahr *n*
**calenderer** Roller *m*
**calendering** Kalandrieren *n*, Kalandrierung *f* ~ **effect** Mangeleffekt *m* ~ **machine** Glander *f* ~ **unit** Kalandrierwerk *n*
**caleometer** Caleometer *n*
**calf** Wade *f* (textiles) ~ **binding** Franz-, Ganzleder-band *m* ~**-bound volume** Ganzlederband *m* ~ **paper** Kalbslederpapier *n* ~ **wheel** Flaschenzug-, Förder-trommel *f*
**caliber** Bohrung *f*, Kernmaß *n*, Laufweite *f*, Lehre *f*, Rohrweite *f*, Seelen-durchmesser *m*, -weite *f*, Stichmaß *n*, (of firearms) Kaliber *n* ~ **gauge** Kaliber-maß *n*, -messer *m* ~ **length** Kaliberlänge *f*
**calibered** kalibrig
**calibrate, to** ~ abgleichen, ablehren, auswägen, eichen, kalibrieren, leistungsmäßig durchmessen (a motor)
**calibrated** geeicht, kalibriert, (gauge) abgelehrt ~ **airspeed (CAS)** berichtigte Eigengeschwindigkeit *f* oder Fahrt *f* ~ **altitude** berichtigte, angezeigte Höhe *f* ~ **fan brake** geeichte Bremsluftschraube *f* ~ **foil** Eichfolie *f* ~ **instrument** Etalonapparat *m* ~ **leak** kalibriertes Kapillarrohr *n* ~ **nozzle** Meßdüse *f* ~ **orifice** Staurand *m* (hydr) ~ **phase changer** Meßkette *f* ~ **slide-wire** eingeteilter Schleifdraht *m* ~ **value** Einstellwert *m* ~ **weir for measuring discharge** Meßwehr *n*
**calibrating,** ~ **apparatus** Eich-gerät *n*, -instrument *n* ~ **coil** Eichspule *f* ~ **equipment** Eichvorrichtung *f* ~ **head** Kalibrierkopf *m* ~ **pipette** Kalibrierpipette *f* ~ **resistance** Eichwiderstand *m* ~ **roll** Kalibrierwalze *f* ~ **room for meters** Zählereichsaal *m* ~ **tool** Kalibrier-

werkzeug n ~-transmitter test oscillator Meßsender m
**calibration** Abmessung f, Ausliterung f (Tankwagen), Eichen n, Eichung f, Funkbeschickung
f, Grad-einteilung f, -teilung f, Kalibrierung f,
Stärkemessung f, Teilung f ~ of direction finder
Funkbeschickungsaufnahme f ~ of testing
instrument Eichung f eines Meßgerätes ~ of
**Venturi** Venturiquerschnitt m
**calibration, ~ absorption curve** Eichabsorp-
tionskurve f ~ accuracy Eichgenauigkeit f ~
adjustment Frequenzkorrektur f ~ button Kon-
trollindexknopf m ~ cable Eichkabel n ~
capacitor Eichkondensator m ~ chart Eich-
tafel f (rdr) ~ check Eichenkontrolle f (electron.)
~ circle Eichzirkel m ~ circuit Eichkreis m ~
condenser Eich-, Eichungs-kondensator m ~
correction Ausschalten n der Grundstufe
~-current regulator Eichstromregler m ~
curve Eich-, Meß-kurve f ~ curves Eichunter-
lage f (rdr) ~ instrument Eich-gerät n, -instru-
ment n ~ marker Eichmarke f ~ meters Eich-
zähler m ~ post Eichpfahl m ~ office Eichamt
n ~ samples Eichproben pl ~ shunt Eichneben-
schluß m ~ test Eich-lauf m, -versuch m,
Schub- und Leistungsprüfung f (motor) ~
unit Markteil m ~ variations Eichabwanderun-
gen pl ~ voltage Eichspannung f ~-voltage
regulator Eichspannungsregler m
**calibrator** Eicher m, Graduer m
**calico** Buchbinderkattun n, Kaliko m ~ printer
Zeugdrucker m
**calific value** Heizwert m
**californium** Californium n
**caliper, to** ~ abtastern, lehren, mit Tastlehre
messen to ~ a hole against standard eine Bohrung ablehnen
**caliper** (innerer) Durchmesser m, Holzmeß-
gerät n, Tastvorrichtung f ~ compasses Taster-
zirkel ~ compresses Kalibriertaster m ~ gauge
Rachen-, Schrauben-, Taster-, Tast-lehre f,
Zylinderstickmaß n ~ leg Tasterschenkel m ~
**scale** Schublehre f
**calipers** Dickenmesser m, Greifzirkel m, Innen-,
Krumm-, Taster-, Tast-zirkel m, Taster m ~
with regulating screw Taster m mit Feinstellschraube ~ with jaws to measure thread gauge
**capacitively** Fadendicketaster m
**calite** Calit n
**calk, to** ~ abdichten, ausblatten, dicht machen,
dichten, durchzeichnen, Fugen verstopfen, kal-
fatern, pausen, stemmen, verstemmen, verstreichen
**calk, ~ weld** Dichtungsschweißung f, Stemm-
naht f ~ welding Kalfaterung f durch Schweißung
**calked joint** Stemmfuge f
**calking** Abdichtung f, Dichtungsstoff m, Kal-
fatern n, Kalfaterung f, Pause f, Pauszeichnung
f, Stemmen n, Verstemmen n, Verstemmung f
~ chisel Dichtmeißel m, (for use on metal edges
proper) Fülleisen n ~ edge Stemmkante f ~
hammer Meißel-, Stemm-hammer m ~ iron
Dichteisen n ~ machine for boiler pipes Siederohrdichtmaschine f ~ mallet Dicht-, Kalfater-
-hammer m ~ paper Pauspapier n ~ putty
Abdeckpaste f ~ ring Packring m ~ tool
Stemmsetze f ~ wire Stemmdraht m

**call, to** ~ ansprechen als, errufen, nennen,
rufen to ~ again wieder rufen, den Anruf m
wiederholen to ~ at (port) anlaufen to ~ away
abberufen to ~ for erfordern, (someone) herbeirufen to ~ for funds Kapital n anfordern oder
einfordern to ~ in einfordern, einziehen to ~
in question bezweifeln to ~ in a repeater station
ein Amt n in die Leitung eintreten lassen to
~ on a person bei jemandem vorsprechen to ~
selectively wahlweise rufen to ~ to life wachrufen tu ~ up anläuten, anklingeln, anrufen,
aufrufen
**call** Abfragung f (teleph.), Abruf m (comm.)
Anruf m, Aufruf m, Berufung f, Gespräch n,
Ruf m, Zuruf m at or on ~ auf Abruf m ~
booked by pre-arrangement Daueranmeldung
f ~ filed for later completion zurückgestellte
Verbindung f ~ on hand unerledigter Anruf
m ~ liable to a charge gebührenpflichtiger Anruf m ~ with request for charges Gespräch n
mit Gebührenansage f
**call, ~ accumulation** Gesprächsspeicherung f ~
allotter Anrufordner m ~ announcer Nummern-
anzeigeeinrichtung f ~-back and cut-in relay
Rückfrage- und Aufschalterrelais n ~-back
device Namengeber m (telet.) ~ bell Anruf-
-glocke f, -wecker n, Lockklingel f, Weckanruf
m ~ circuit Dienstleitung f ~-circuit key
Dienstleitungstaste f ~-circuit method Dienstleitungsbetrieb m ~ count Stichzählung f
~-counter key Leistungszähltaste f ~-distribu-
ting system Anrufverteilungssystem n, Ver-
teilersystem n ~ distribution Verteilung f
der Anrufe ~ distributor Anrufverteiler m
~ finder Anrufsucher m ~ fill Nutzungs-,
Wirkungs-grad m ~ identifier Anruffeststeller
m (teleph.) ~ indicator Anrufanzeiger m,
Lampenfeld n für Rufanzeige, Nummernanzeiger m ~-indicator position Arbeitsplatz
m mit Rufanzeige ~ letter Rufzeichen n
~-meter Gesprächszähler m ~ metering Gebührenerfassung f ~-order ticket Gesprächsblatt f ~ rate indicator Gebührenanzeiger m ~
receiving switch Anrufverteiler m ~ relay An-
rufrelais n ~ relay for cable working Anruf-
relais n für Kabelbetrieb m ~ sender Nummern-
geber m ~ sign Rufzeichen n (rdo) ~ signal
Anruf-, Peil-zeichen n, Weckzeichenweiter-
gabe f ~-signal apparatus Anrufapparat m
~ station Sprechstelle f ~ transfer relay Um-
legungsrelais n ~ wire Dienst-, Sprech-leitung
f
**called, ~ for** angefordert to be ~ heißen to be
for oral argument (or hearing) zur mündlichen
Verhandlung gelangen
**called, ~ destination** Zielamt n (teleph.) ~
extension gerufener Teilnehmer m ~ line
angerufene Leitung f ~ party angerufener oder
verlangter Teilnehmer m ~ party's reply relay
Teilnehmermelderelais n ~ station Gegenfunkstelle f, Gegenstelle f (rdo) ~ subscriber verlangter Teilnehmer m ~ (wanted, required)
subscriber angerufener Teilnehmer m
**caller** anrufender Teilnehmer m to be ~ besetzt
sein durch eigenen Anruf
**Callier factor** (of print contrast) Callier-Faktor
m
**calligraphy** Schönschreiben n

**calling** Amt *n*, Anrufen *n*, Rufen *n* ~ **code** Ruf-schlüssel *m*, -zeichen *n* ~ **concentrator** Anrufschrank *m* (teleph.) ~ **condition** Anrufanreiz *m* ~ **cord** Ruf-, Verbindungs-schnur *f* ~ **device** Anrufeinrichtung *f* für Übertragungen, Anrufer *m* ~ **frequency** Ruffrequenz *f* ~ **harbor** Nothafen *m* ~ **indicator** Anrufklappe *f* ~-**in lamp** Anruflampe *f* ~ **jack** Abfrageklinke *f* ~ **lamp** Anrufglüh-, Anruf-, Ruf-lampe *f* ~-**lamp strip** Anruflampenstreifen *m* (teleph.) ~ **magneto** Anrufinduktor *m* ~ **number barring circuit** Rufnummernsperre *f* ~ **party** anrufender oder rufender Teilnehmer *m* ~-**party release** Auslösung *f* einer Verbindung durch Einhängen des rufenden Teilnehmers, Vorwärtsauslösung *f* ~ **periods** Anrufzeiten *pl* ~ **pilot** Ruflampe *f* ~ **plug** Ruf-, Verbindungs-stöpsel *m* ~ **rate** Gesprächsdichte *f* ~ **relay** Rufrelais *n* ~ **signal** Anrufsignal *n* ~ **station** (an)rufende Station oder Stelle *f* ~-**subscriber release** Auslösung *f* beim Einhängen des Hörers durch den Anrufenden ~ **wave** Anruf-, Ruf-welle *f*
**callitype** Kallitypie *f*
**callityping** Kallitypie *f*
**callous** abgestumpft, hornig
**Callow** (traveling belt) **screen** Callow-Sieb *n*
**calm, to** ~ beruhigen, sich legen **to become** ~ sich beruhigen
**calm** Flaute *f*, Regungslosigkeit *f*, Ruhe *f*, Windstille *f* ~ **of cancer** Windstille *f* des Krebses ~ **of capricorn** Windstille *f* des Widders
**calm,** ~ **air** ruhige Luft *f* ~ **belt** Doldrums *n*, Kalmengürtel *m* ~ **region** Beruhigungsstrecke *f* ~ **zone** Kalmen-gürtel *m*, -zone *f*
**calmness** Geräuschlosigkeit *f*
**calomel** Calomel *n*, Hornquecksilber *n*, Kalomel *n*, Merkurhornerz *n*
**calorescence** Kaloreszenz *f*
**caloric** Wärmestoff *m*; kalorisch ~ **unit** Wärmeeinheit *f*
**calorie** Kalorie *f*
**calorific** Heizkraft *f*; kalorisch, Wärme erzeugend ~ **balance** Wärmebilanz *f* ~ **capacity** Wärmekapazität *f* ~ **efficiency** Wärme-leistung *f*, -wirkungsgrad *m* ~ **intensity** Heizeffekt *m*, pyrometrischer Wärmeeffekt *m*, Verbrennungstemperatur *f*, Wärmeintensität *f* ~ **power** Erwärmungskraft *f*, Heizwert *m* ~ **net** ~ **power** unterer Heizwert *m* ~ **requirement** Wärmebedarf *m* ~ **value** absoluter Heizeffekt *m*, Brenn-, Heiz-wert *m*, Heizwertzahl *f*, Wärmewert *m* ~ **value of gas** Gasheizwertzahl *f*
**calorimeter** Heizwertmesser *m*, Kalorimeter *n*, Kühlwasseranzeiger *m*, Wärmemesser *m*
**calorimetric** (al) kalorimetrisch ~ **bomb** kalorimetrische Bombe *f*
**calorimetry** Wärmemengen-, Wärme-messung *f*
**caloriscope** Kaloriskop *n*
**calorize, to** ~ kalorisieren
**calorized steel** alitierter Stahl *m*
**calorizing** Kalorisation *f*, Kalorisieren *n*, Kalorisierung *f*
**calotte** Kalotte *f*, Kugelabschnitt *m*, Glockenlager (r.r.)
**calutron** Calutron *n*
**calyx** Brockenfänger *m*, Kelch *m*
**cam** Daumen *m*, Daumennocke *f*, Däumling

*m*, Exzenter *m*, Frosch *m*, Hebedaumen *m*, Herzscheibe *f*, Hocker *m*, Kamm *m*, Knagge *f*, Kurvenkörper *m* (math.), Lenkfinger *m*, Mitnehmer *m*, Mitnehmerklaue *f* (beim Fahrwerk), Nadelheber *m*, Nase *f*, Nocke *f*, Nocken *m*, Nockenbahn *f*, Nockenscheibe *f*, Scheibennocke *f*, Steuerung *f*, Tatze *f*, (wheel) Daumenscheibe *f* ~ **and lever steering mechanism** Schraubengang-, Spindel-lenkung *f* ~ **for brush** Exzenterscheibe *f* ~ **for cut-out** Auslösenase *f* ~ **for guide rail** Kurvenscheibe *f* zum Versenken der Greifer *f* ~ **for sucker bar swing** Exzenter *m* für Saugerschwingung ~ **of breech-mechanism lever** Öffnerhebelnocken *m* ~ **of lifting cog** Wellendaumen *m*
**cam,** ~ **assembly** Hubscheibe *f* (hydraul.), Schrägscheibe *f* ~ **bearing** Kugellaufring *m* ~ **box** Daumenkasten *m*, Nockengehäuse *n* ~ **bracket** Kurvenstück *n* ~ **case** Nockengehäuse *n* ~ **contact** Nockenkontakt *m* ~ **contour** Nockenform *f* ~ **control** Nockensteuerung *f* ~ **design** Nockenkonstruktion *f* ~ **disk** Kurven-bahn *f*, -scheibe *f*; Nocken-schale *f*, -scheibe *f*, Schlagrad *n* ~ **drive** Kurventrieb *m*, Nockenantrieb *m* ~ **drum** Kurven-, Nocken-trommel *f* ~ **engine** Nockenmotor *m* ~ **examination** Nokkenuntersuchung *f* ~ **follower** Exzenterrolle *f*, Nockenstößel *m*, Ventilstößel *m* ~-**follower guide** Stößelführung *f* ~ **gear** Daumenantrieb *m*, Daumensteuerung *f*, Exzenterrad *n*, Kurvenscheibengetriebe *n*, Nockensteuerung *f* ~ **gearing** Kammgetriebe *n*, Steuerung *f* durch unrunde Scheiben ~ **grinding machine** Steuerwellenexzenterschleifmaschine *f* ~ **groove in breechblock** Führungsleiste *f* ~ **head** Nockenkopf *m* ~ **key** Zapfenschlüssel *m* für die Herzscheibe ~ **layout** Kurvenblatt *n* ~ **lever** Nocken-, Wälz-hebel *m* ~-**lever steering** Wälzhebelsteuerung *f* ~ **lever template** Kurvenhebelschablone *f* ~ **lobe** Nockenbuckel *m* ~ **milling machine** Steuerwellenexzenterfräsmaschine *f* ~ **movement** Herzexzenterbewegung *f* ~ **ope-movement** Herzexzenterbewegung *f* ~ **operated switch** Nockenschalter *m* ~ **operating grippers** Greiferkurve *f* ~ **plate** Hubscheibe *f*, Kurvenbahn *f*, Kurven-, Nocken-scheibe *f* ~ **position** Nockenstellung *f* ~ **profile** Nokkenprofil *n* ~ **retaining ring** Schloßteilhaltering *m* ~ **ring** Nockentrommel *f* ~ **roller** Kurvenroller *f* ~ **segment** Kurvenstück *n* zur Ventilklappe
**camshaft** Daumen-, Nocken-, Steuer-welle *f* ~ **bearing** Nocken-, Steuer-wellenlager *n* ~ **casing** Nocken-, Steuer-wellengehäuse *n* ~ **control** Steuerung *f* (camera housing) ~ **controller** Nockensteuerschalter *m* ~ **cycle** Steuerwellenkreis *m* ~ **drive** Nockenwellen-, Steuerwellen-antrieb *m* ~ **driven magneto** Steuerwellenmagnetzünder *m* ~ **gear** Nockenschaltwerk *n* ~ **gear key** Nockenwellengetriebeschlüssel *m* ~ **gear wheel** Nockenwellenrad *n* ~ **gears** Nockenwellengetriebe *n* ~ **grinding attachment** Nockenwellenschleifeinrichtung *f* ~ **housing** Nockenwellengehäuse *n*, Steuerwellengehäuse *n* ~ **housing cover** Steuerwellengehäusedeckel *m* ~ **return oil line** Steuerwellenablaufölleitung *f* ~ **roller** Nockenwellenrolle *f* ~ **speed** Steuerwellendrehzahl *f* ~

**tester** Nockenwellenprüfgerät *n* ~ **timing gear** Nockenwellenantriebsrad *n*
**cam,** ~ **sheet** Kurvenblatt *n* ~ **sleeve** Nockenbuchse *f* ~ **slot** Kurven-, Verriegelungs--schlitz *m* ~ **spindle** Nockenwelle *f* ~ **springs** Nockenkontakt *m* ~ **steering** Nockensteuerung *f* ~ **stick** Nockenstange *f* ~ **stud** Mitnehmerschraube *f* ~ **switch** Nockenschalter *m* ~ **template** Kurvenschablone *f* ~-**throttle control** Kurventrieb *m* ~ **track** Nockenlaufbahn *f* ~ **trunnion** Hubscheibenlagerzapfen *m* (hydraul.), Schrägscheibenlagerzapfen *m* ~ **turning lathe** Nockendrehbank *f* ~-**type drive** Nockengetriebe *n* ~-**type engine** Kurvenscheibenmotor *m* ~-**wheel of the strainer** Schlagrad *n* des Knotenfängers
**camaïen** Monochromie *f*
**camber, to** ~ ausbauchen, biegen, umbiegen, wölben
**camber** Bombierung *f*, Flügelwölbung *f*, Krümmung *f*, Pfeilhöhe *f*, Radsturz *m*, Sturz *m* der Kraftwagenräder, Tiefe *f* der Flächenkurve, Überhöhung *f*, Wölbung *f*, (iron rails) senkrechte Krümmung *f* ~ **of an arch** Pfeil *m* ~ **of rolls** Balligkeit *f* der Walzen ~ **of spring** Pfeilhöhe *f* oder Sprengung *f* der Feder ~ **of wheels** Radsturz *m*
**camber,** ~-**and-pivot inclination** Nachlauf *m* ~ **angle** Sturzwinkel *m* ~-**changing flap** Wölbungsklappe *f* ~ **position** Krümmungsrücklage *f* ~ **ratio** Wölbungsverhältnis *n*
**cambered** gesprengt ~ **airfoil** gewölbtes Tragflügelprofil *n* ~ **axle** gestürzte Achse *f* ~ **insert** Schubdüseneinsatz *m* ~ **mirror glass** gewölbtes Spiegelglas *n* ~ **profile** gewölbtes Profil *n* ~ **road** gewölbte Straße *f* ~ **roll** ballig ~ **spring** gesprengte Feder *f* ~ **wing** gewölbter Flügel *m* (airplane)
**cambering** Umkröpfung *f* ~ **of cranked axle** Achskröpfung *f* ~ **of spring** Krümmung *f* der Feder
**cambrian** kambrisch (Kambrium)
**camel** Hebemittel *n*, Kamel *n* ~'**s hair belting** Kamelhaarriemen *m*
**camera** Aufnahme-apparat *m*, -gerät *n*, Bildkamera *f*, Kamera *f*, Kammer *f*, Lichtbildgerät *n*
**camera,** ~ **of projection** Bildwurf-, Projektions--kamera *f* ~ **for retarded-action picture projection work** Zeitlupenaufnahmekamera *f* ~ **for single photographs** Einfach-, Einzel-kam ~ **for slow-motion picture projection work** Zeitlupenaufnahmekamera *f* ~ **on stand (or on tripod)** Stativkamera *f* ~ **for stop-motion micrography** Mikroraffer *m* ~ **for strip photograph on vertical axis** senkrechte Reihenkamera *f* ~ **for taking pictures** Bildfänger *m*, Laufbild-fänger *m*, -kammer *f* ~ **for time--lapse motion micrography** *m* Mikroraffer *m*
**camera,** ~ **access door** Zugangsklappe *f* im Kameraraum (aviat.) ~ **adapter** Zwischenhalterung *f* ~ **alignment** Bündelzentrierung *f* ~ **axis** Kamera-Achse *f* ~ **bellow** Kamerabalg *m* ~ **body** Kameragehäuse *n* ~ **case** Kameragehäuse *n* ~ **casette** Kamerakassette *f* ~ **chain** Kamerazug *m* (TV) ~ **compartment** Kameraraum *m* (aviat.) ~ **coverage** Kamerabereich *m* ~ **dolly** Fahrstativ *n*, Kamerafahrgestell *n* ~

**door** Klappe *f* zur Kamera ~ **drive-shaft pin** Kameraanschlußzapfen *m* ~ **extension** Kameraauszug *m* ~ **feed mechanism** Kameralaufwerk *n* ~-**film-feed mechanism** Filmkameralaufwerk *n* ~ **gun** Zielbildkamera *f* ~ **hood** Kameralichtklappe *f* ~ **line** Schärfenfeldgrenze *f* ~ **lucida** Camera clara *f* ~ **magazine** Kassette *f* ~ **main circuit** Hauptstromkreis *m* (bei Meßkamera) ~ **mount** Kameraaufhängung *f* ~ **mounting** Kameragestell *n* ~ **obscura** Dunkelkammer *f* ~ **plastica** Doppel-bildwerfer *m*, -projektor *m* ~ **position** Aufnahmeort *m* ~ **screen** Mattscheibe *f* ~ **station** Aufnahme-ort *m*, -standpunkt *m*, Standort *m*, Standpunkt *m* (Erdbildaufnahme), Stationspunkt *m* ~ **technician** Aufnahmetechniker *m* ~ **trip circuit** Kameraauslösestromkreis *m* ~ **truck** Kamerafahrgestell *n* ~ **tube** Abtaströhre *f* (TV), Kamera-, Bildfänger-röhre *f* ~ **window** Kamerafenster *n*
**camlock mounting** Camlockbefestigung *f*
**camming arrangement** Steuerkurveneinrichtung *f*
**camouflage, to** ~ maskieren, tarnen, verdecken, verkleiden, verschleiern
**camouflage** Blendanstrich *m*, Maske *f*, Maskierung *f*, Sichtschutz *m*, Tarnung *f* ~ **covering** Tarn-bezug *m*, -decke *f* ~ **effect of shadows cast by a body** Schlagschatten *m* ~ **matting** Tarngeflecht *n* ~ **net** Tarnnetz *n* ~ **netting** Tarn-geflecht *n*, -netz *n* ~ **paint** Sichtschutz-, Tarn-anstrich *m* ~ **radar used in jamming** Funkmeßtäuschungsgerät *n* ~ **screen** Tarndecke *f* ~ **strip** Tarnband *n* ~ **wire netting** Drahtnetz *n* zum Tarnen
**camouflaged** getarnt, verdeckt
**camouflaging** Tarnen *n*, Verschleierung *f*
**camouflet** Quetsch-ladung *f*, -mine *f*
**camp,** Lager *n*, Lagerplatz *m* ~**bed** Feldbett *n* ~ **kettle** Feldkessel *m* ~ **sheathing** Bollwerk *n* ~ **site** Lager-ort *m*, -stelle *f*
**campaign** Aktion *f*, Reise *f* (Hochofen)
**camphine** Kampfin *n*
**camphor** Kampfer *m* ~ **oil** Kampferöl *n*
**camphorated monobromide** Kampfermonobromid *n*
**camphoric,** ~ **acid** Kampfersäure *f* ~ **anhydride** Kampfersäureanhydrid *n*
**can, to** ~ in Büchsen konservieren
**can** Blech-gefäß *n*, -schachtel *f*, Büchse *f*, Dose *f*, Kanister *m*, Kanne *f*, Konservenbüchse *f*, Spulentopf *m*, Topf *m*, (of a coil) magnetischer Schirm *m* ~ **and churn-washing machine** Kannenwascherbau *m* ~ **anode** Topfanode *f* ~ **carrier** Kannenträger *m* ~ **coiler** Abzugsdrehwerk *n* ~ **drier** Zylindertrockner *m* ~ **hook** Löschhaken *m* ~ **oiling** Kannenschmierung *f* ~ **opener** Büchsenöffner *m* ~ **tenter** Kannenwärter *m* (textiles) ~ **turntable** Kannentellerrahmen *m*
**Canada balsam** Kanadabalsam *m*
**canal** Fleet *n*, Fluß *m*, Gracht *f*, Kanal *m*, Werkgraben *m*, Wasser-lauf *m*, -straße *f* ~ **in a cut** Kanal *m* im Einschnitt ~ **that crosses a water divide** Scheitelkanal *m* ~ **with filter bed** Sickerkanal *m* ~ **on embankment** Kanal *m*
**canal,** ~ **aqueduct** Kanalbrücke *f* (**lining of)** ~ **bed** Kanalbett *n* ~ **boat** Pinasse *f* ~-**cleaning**

**outfit** Kanalreinigungsgerät *n* ~ **embankment** Kanaldamm *m* ~ **lock** Kanalschleuse *f* ~ **piling** Kanalspundwand *f* ~ **ray** Kanalstrahl *m* ~**-raydischarge ion source** Kanalstrahlionenquelle *f* ~**-ray discharges** Kanalstrahlentladungen *pl* ~**-ray tube** Kanalstrahlröhre *f* ~ **rays** Anoden-, Kanal-strahlen *pl*, positive Strahlen *pl* ~ **slide valve** Kanalschieber *m* ~ **tug** Kanalschlepper *m* ~ **water level** Kanalwaage *f*
**canalization** Entwässerungsanlage *f*, Kanalisierung *f*
**canalize, to** ~ kanalisieren
**canard** Vorderschwanz-, Enten-flugzeug *n*, kopfgesteuertes Flugzeug *n* ~ **model** Entenflugmodell *n* ~**-type aircraft** Entenmodell *n*
**canary medium** goldgelber Stoff *m* (photo)
**cancel, to** ~ abbestellen, annullieren, auflösen, aufheben, auslöschen, ausstreichen, durchstreichen, einander aufheben, entwerten, für ungültig erklären, kassieren, löschen, rückgängig machen, stornieren, streichen, tilgen **to** ~ **a call** eine Gesprächsanmeldung *f* streichen (lassen) **to** ~ **out** sich wegheben (a factor), sich fortheben ~ **key** Löschtaste *f*
**canceled** gegenstandslos, storniert ~ **call** gestrichene Gesprächsanmeldung *f*
**cancellated structure** Netzgefüge *n*
**cancellation** Abbestellung *f*, Absage *f* (eines Auftrages), Annullierung *f*, Aufhebung *f*, Auflösung *f*, Kündigung *f*, Löschung *f*, Nichtigkeitserklärung *f*, Streichen *n*, Streichung *f*, Unterdrückung *f* ~ **of a call** Streichung *f* einer Gesprächsanmeldung ~ **of a subscriber's contract** Kündigung *f* eines Fernsprechanschlusses ~ **coil** Abgleich-, Kompensations--spule *f* (g/m)
**cancelling,** ~ **button** Abstellknopf *m* ~ **key** Rücknahmetaste *f*
**candelabrum** Kandelaber *m*
**candidacy** Anwartschaft *f*
**candidate** Anwärter *m*, Bewerber *m*
**candle** Candeler *m*, Kerze *f*, Licht *n* ~ **(power)** Kerze *f* ~ **flame** Kerzenflamme *f* ~**-holder** Handleuchter *m* ~**-lamp holder** Kerzenfassung *f* ~ **power** Kerzen-lichtstärke, -stärke *f*, Leucht-, Licht-stärke *f* ~**-power characteristic** Kerzenstärkencharakteristik *f* ~**-power curve** Leuchtkraftkurve *f* ~**-power test** Feststellung *f* der Lichtstärke ~ **shape low voltage lamp** Kerzenkleinlampe *f* ~**stick** Handleuchter *m*, Leuchter *m*
**cane** Rohr *n* (Pflanze), Rohrholz *n*, Rohrstock *m*, Schilf *n*, Schilfrohr *n*, Stock *m* ~**-and-sorghum mill** Rohr- und Rohrhirsemühle *f* ~ **culture** Rohranbau *m* ~ **seat** Rohrsitz *m* ~ **sugar** Rohrzucker *m* ~ **weave** Rohrgeflecht *n*
**canister** Blech-emballage *f*, -kanne *f*, Büchse *f*, Kanister *m*, Kanne *f* ~ **filter** Büchsenfilter, *m* ~ **(mask)** Filterbüchse *f*
**cankerous growth in wood** Holzkrebs *m*
**canned** eingehülst ~ **coil** Käfigspule *f*, Spulentopf *m* ~ **food** Büchsenkonserve *f*
**cannel coal** langflammige Kohle *f*, Fackel-, Kannel-kohle *f*
**cannelure** Hülsengliederung *f*
**cannibalization** Ausschlachtung *f*
**cannibalize, to** ~ ausschlachten (von Fahrzeugen)

**canning** Einhülsen *n* ~ **technology** Einmachtechnik
**cannister grain** Treibsatz *m* (g/m)
**cannon** Geschütz *n*, Kanone *f* ~ **barrel** Kanonenrohr *n* ~**-boring lathe** Geschützrohrdrehbank *f* ~ **engine** Kanonenmotor *m* ~ **plug** Sammelstecker *m* ~ **tube shield** laufförmige Haube *f*
**cannula** Kanüle *f*
**canoe** Kahn *m*, Kanu *n*, Paddelboot *n*
**canon pinion** Viertelrohr *n*
**canonical,** ~ **dissection** kanonische Zerschneidung *f* ~ **ensemble** kanonische Gesamtheit *f* ~ **equation** kanonische Gleichung *f*
**canopy** Baldachin *m*, Kabinenhaube *f*, Kappe *f* (Fallschirm), Schutzdach *n* ~ **of cockpit** Führersitzverkleidung *f*
**canopy,** ~ **bag** innerer Verpackungssack *m* (Fallschirm) ~ **defrosting** Sichtscheibenenteisung *f* ~ **latching mechanism** Kabinendachverriegelungsvorrichtung *f* ~ **liftbar** Kabinendachhebestange *f* ~ **remover** Kabinendachabwerfer *m*
**cant, to** ~ abkanten, abschrägen, umkanten, verkanten, (timber) abfasen **to** ~ **up** umkanten
**cant** Querneigung *f*, schiefer Radstand *m*, Schräge *f*, Verkantung *f* ~ **of rails** Schienenüberhöhung *f*
**cant,** ~ **chisel** Kantbeitel *m* ~ **file** Barettfeile *f* ~ **firmer chisel** Kantbeitel *m* ~ **hook** Bedienungshaken *m* ~ **level** Radstand-, Verkantungs-libelle *f* ~ **timber** abgefastes oder abgekantetes Holz *n* (Balken)
**canted** abgefast, abgekantet, abgeschrägt ~ **bulkhead** Schrägspant *m* (Flugzeug) ~ **timber** abgefastes Holz *n*, abgekanteter Balken *m*
**canteen** Feldflasche *f*, Flasche *f*, Kantine *f*
**canter** Handgalopp *m*
**cantilever, to** ~ auskragen
**cantilever** Auskragung *f*, Ausladung *f* (Brücke), Ausleger *m*, Auslegekran *m*, Freiträger *m*, Konsole *f*, Kragarm *m*, Sparrenkopf *m*, Überhang *m*; (airplane construction) freigespannt, (construction type) freitragend ~ **arm** Kragarm *m*, Reichweite *f* ~ **beam** Auslegerbalken *m*, Gerber-, Frei-, Konsol-, Krag-träger *m*, Kragbalken *m* ~**-beam bridge** Kragträgerbrücke *f* ~ **bridge** Auslegerbrücke *f* ~ **condition(s)** Auflagerbedingung *f* (arch., statics) ~ **crane** Auslegerbockkran *m* ~ **extension** Kragarm *m* ~ **frame** freitragender Rahmen *m* ~ **gantry crane** Kantileverkran *m* ~ **landing gear** Einbeinfahrgestell *n* ~ **load acting at the post** Kragarmlast *m* am Stiel ~ **monoplane** freitragender Eindecker *m* ~ **retaining wall** Winkelstützmauer *f* ~ **roof** Ausleger-, Kragen-dach *n* ~ **shaft** freitragende Welle *f* ~ **sheet piles** auskragende Spundwand *f* ~ **spar** Gerberträger *m* ~ **spring** Auslegerfeder *f* ~ **structure** Gerberträger *m* ~ **tower** Fischbauchmast *m* ~ **trembler** Konsolenticker *m* ~**-type tower** (with double-tapered mast) Kreuzteilungsmast *m* ~ **undercarriage** freitragendes Fahrgestell *n* ~ **wing** freitragender Flügel *m*
**cantilevered** freischwebend
**canting** Verkanten *n* ~ **of the shaft** Ecken *n* der Welle ~ **while moving** Ecken *n* beim Fahren ~ **burner** Eckenbrenner *m*
**cantle** Hinterzwiesel *m*

**canvas** Baumwoll-gewebe *n*, -stramin *m*, Bedachungsstoff *m*, Bindertuch *n*, Drilch *m*, Hanfleinwand *f*, Leinen *m*, Leinenstoff *m*, Leinwand *f*, Plache *f*, Segeltuch *n*, Segeltuchstoff *m*, Verpackungsplane *f*, Zeltleinwand *f* ~ **and duck** Segeltuchgewebe *n* ~ **for embroidery** Papierstramin *m*

**canvas,** ~ **air conduit** Tuchlutte *f* ~ **bag** Jutesack *m* ~ **belt** Baumwollriemen *m* ~ **bucket** Wassersack *m* ~ **clothing** Drilchanzug *m* ~ **cover** Segeltuchhülle *f* ~ **drain** Stofflutte *f* ~ **hood** Plane *f* ~ **note** Leinenpost *f* ~ **pipe** Stofflutte *f* ~ **tarpaulin** Segeltuchhülle *f* ~ **tire** Leinwandreifen *m* ~ **tool roll** Rolltasche *f* aus Stoff für Werkzeug ~ **water bucket** Wassertragesack *m* ~ **wrap** Segeltuchhülle *f*

**canvassing** Erhebung *f*

**canyon** geschlossener Raum *m*

**caoutchouc** Gummi *m* (*n*), Gummielastikum *n*, Kautschuk *m*, Weichgummi *m*

**cap, to** ~ abdecken, mit einer Kappe versehen **to** ~ **the piles** die Pfähle beholmen **to** ~ **a well** (eine) Quelle oder ein Bohrloch versiegeln, zuschütten

**cap** Abschlußkappe *f*, Anschlußkontakt *m* an der Spitze des Glasballons (rdo), Aufsatz *m*, Deckel *m*, Deckschwelle *f*, Haube *f*, Kalotte *f*, Kappe *f*, Kapsel *f*, Kopfkappe *f*, Kronholz *n*, Kronschwelle *f*, Muschel *f*, Mütze *f*, Nase *f*, Ohrmuschel *f*, Pfahlkopf *m*, Schutzkappe *f*, Sockel *m* (Röhre), (oil drilling) Hut *m*, (inserted over pipe at point where latter penetrates roofs to prevent entrance of water) Rohrständerabdichtung *f* ~ **for shaft bearing** Kurbelwellenlager *n* ~ **on bed of river** Grundschwelle *f* ~ **of fountain pen** Füllfederhalterkappe *f* ~ **of projectile** Geschoßkappe *f* ~ **of tank** Tankverschluß

**cap,** ~-**and-pin suspension insulator** Kappenisolator *m* ~ **applier** Kappenaufleger *m* ~ **band** Besatzstreifen *m* ~ **board** Kappe *f* ~ **chamber** Zünderglocke *f* ~ **closure** Verschlußkappe *f* ~ **crimper** Würgeverbindungszange *f* ~ **filler** Sockelfüllgerät *n* ~ **fitting device** Kappenaufleger *m* ~ **frame** Deckelrahmen *m* ~ **head** Rohrkopf *m* ~ **holder (or spring holder)** Deckelhalter *m* (Federträger *m*) ~ **jet** Manteldüse *f* ~ **nut** Abschluß-, Hut-, Kapsel-, Überwurf-mutter *f* ~ **packing** Kapsellliderung *f* ~ **peak** Mützenschirm *m* ~ **piece** Pfannendeckel *m* ~ **pile** Pfahlhaube *f* ~ **plate** Firsteisen *n* ~ **pot** bedeckter Hafen *m* ~ **screw** Deckelschraube *f*, Hutmutter *f*, Kopfschraube *f*, Metallschraube *f*, Überwurfmutter *f* ~-**shaped diaphragm** Kalottenmembran *f* ~-**shaped object** Kalotte *f* ~ **sleeve** Sockelbuchse *f* ~ **strip** Flügelrippenkappe *f*, Formrippe *f*, Nasen-, Stirn-leiste *f* ~ **stripe** Besatzstreifen *m* ~-**type gasket** Hutmanschette *f* ~-**type joint** Kalottenverbindung f (für Schüttelrutschen)

**capability** Befähigung *f*, Fähigkeit *f*, Leistungsfähigkeit *f* ~ **of being cupped** Tiefziehfähigkeit *f* ~ **of being fashioned (formed, kneaded, molded)** Bildsamkeit *f* (by pressure and/or heat application) ~ **of being replaced** Ersatzfähigkeit *f* ~ **of condensation** Kondensierbarkeit *f* ~ **of resistance** Widerstandsfähigkeit *f* ~ **to attack** Angriffsfähigkeit *f*

**capable** fähig

**capable,** ~ **of** befähigt ~ **of assuming a load** lastbar ~ **of baking (or caking)** backfähig

**capable,** ~ **of being assayed** prüfbar ~ **of being assimilated** assimilierbar ~ **of being braked** bremsbar ~ **of being cast** gieß-bar, -fähig ~ **of being cemented** härtbar ~ **of being detached** zerlegbar ~ **of being diluted** verdünnbar ~ **of being dropped** ausklappbar (on hinges) ~ **of being hardened** härtbar ~ **of being lowered** herabsenkbar ~ **of being mixed** versetzbar ~ **of being moistened** benetzbar ~ **of being moved** vorschlagbar ~ **of being opened** aufschließbar ~ **of being polished** polierbar ~ **of being poured** gießbar ~ **of being raised** (on hinges) aufklappbar ~ **of being rent** zerreißbar ~ **of being screwed on** aufschraubbar ~ **of being swung open** aufklappbar ~ **of being swung out** (on hinges) ausklappbar ~ **of being taken apart** auseinandernehmbar ~ **of being tested** prüfbar ~ **of being thrown out of gear** auskuppelbar ~ **of being torn** zerreißbar ~ **of being treated** versetzbar ~ **of being turned out (or over)** umkehrbar ~ **of being wetted** benetzbar ~ **of being wrought** schmiedbar

**capable,** ~ **of changing form** formänderungsfähig ~ **of cross-country travel** geländegängig ~ **of development** bildungsfähig ~ **of dilution** verschnittfähig ~ **of dripping (or forming drops)** tropfbar ~ **of expanding** aufspreizbar ~ **of explosion** explosibel ~ **to fly** flugfähig ~ **of movement** bewegungsfähig ~ **of oscillation** schwingfähig ~ **of (high) production** leistungsfähig ~ **of reaction** reaktionsfähig ~ **of reduction** reduktionsfähig ~ **of saturation** sättigungsfähig ~ **of spreading** bestreichbar ~ **of supporting** tragkräftig ~ **of up swelling** quellfähig

**capacitance** Belastbarkeit *f*, Feder (electr.), Kapazitanz *f*, Kapazität *f*, Kapazitätswiderstand *m*, Kapazitiv *n*, kapazitive Reaktanz *f*, kapazitiver Blindwiderstand *f*, kapazitiver Widerstand *f*, Kondensanz *f* ~ **of a condenser** Kapazität *f* eines Kondensators ~ **of the diaphragm** Membrankapazität *f* ~ **for shock protection** Berührungsschutzkapazität *f* ~ **per unit length** Kapazität *f* je Längeneinheit ~ **of wiring** Schaltkapazität

**capacitance,** ~ **altimeter** Kapazitanzhöhenmesser *m* ~ **beam switching** kapazitive Leitstrahldrehung *f* ~ **box with plugs** Stöpselkondensator *m* ~ **bridge** Kapazitätsmeßbrücke *f* ~ **bushing** Blindwiderstandsisolator *m* ~ **current** Verdrängungs-, Verschiebungs-strom *m* ~ **deviations** Kapazitätsunterschiede *pl* ~ **strain gauge** kapazitiver Dehnungsmeßstreifen *m*

**capacitative** kapazitiv ~ **voltage divider** kapazitiver Spannungsteiler *m*

**capacitive,** ~ **coupling** elektrische oder kapazitive Kopplung *f* ~ **feedback** kapazitive Rückkupplung *f* ~ **leaks** kapazitiver Verlust *m* ~ **reactance** kapazitiver Blindwiderstand *m* oder Widerstand *m*, Kondensanz *f*, Widerdruck *m* ~ **resistance** Kapazitätswiderstand *m* ~ **sawteeth generator** kapazitiver Sägezahngenerator *m*

**capacitivity** Dielektrizitätskonstante *f*

**capacitor** Kondensator *m*, Mehrfachkondensator *m* ~ **for indoor location** Innenraumausfüh-

rung *f* (Kondensator) ~ **for shock protection**
Berührungsschutzkondensator *m*
**capacitor,** ~ **bank** Kondensatorbatterie *f*
~ **contactor** Kondensatorschütz *m* ~ **discharge
ignition unit** Kondensatorzündgerät *n* ~ **fuse**
Kondensatorzünder *m* ~ **input** Kondensator-
eingang *m* ~**-input filter** Filter mit kapaziti-
vem Eingang ~ **loud-speaker** elektrostatischer
Lautsprecher *m* ~ **pickup** Kondensatorton-
abnehmer *m* ~ **speaker** elektrostatischer Lautsprecher *m*
**capacitron** Kapazitron *n*
**capacity** Absorbierbarkeit *f*, Arbeitsbereich *m*,
Aufnahmefähigkeit *f*, Ausbringen *n*, Ausbringung *f*, Belastungsgrenze *f*, Fähigkeit *f*,
Fassung *f*, Fassungs-raum *m*, -vermögen *n*,
Fesselung *f*, Fördermenge *f*, Füllung *f*, Gehalt
*m*, Inhalt *m*, Kapazität *f*, Ladungsfähigkeit,
Ladungsvermögen *n*, Leistung *f*, Leistungsfähigkeit *f*, Qualifikation *f*, Raumgehalt *m*,
Tragkraft *f*, Vermögen *n* **(carrying)** ~ Fassungsvermögen *n*, (maximale) Stellenzahl *f*,
Zahlbereich *m* (info proc.)
**capacity,** ~ **for being cast** Gießfähigkeit *f* ~ **for
being drawn** Ziehfähigkeit *f* ~ **for being poured**
Gieß-barkeit *f*, -fähigkeit *f* ~ **for being rolled**
Walzbarkeit *f* ~ **for change** Umwandlungs-
fähigkeit *f* ~ **for deformation** Formänderungsvermögen *n* ~ **for doing work** Arbeitsfähigkeit
*f* ~ **for forming foam froth** Schaumbildungsvermögen *n* ~ **for forming slag** Verschlackungsfähigkeit *f* ~ **for performing work** Arbeitsvermögen *n* ~ **for resistance** Widerstandsfähigkeit
*f* ~ **for work** Arbeitsvermögen *n*
**capacity,** ~ **of bottle** Flascheninhalt *m* ~ **of compression champer** Verdichtungsrauminhalt *m* ~
**of dissolution** Auflösungsvermögen *n* ~ **of
dye-stuffs to flocculate out** Ausflockbarkeit *f*
der Farbstoffe *m* ~ **grab** Greiferinhalt *m* ~
**of a line** Leistungskapazität *f* ~ **of memory**
Speicherkapazität *f* ~ **of a power station when
completed** Ausbauleistung *f* eines Kraftwerkes
~ **of solution** Auflösungsvermögen *n* ~ **of a
standard-gauge railroad** Leistungsfähigkeit *f*
einer Vollbahnstrecke ~ **of a transmitter** Senderleistung *f* (rdo)
**capacity to withstand stresses (or take loads)**
Beanspruchbarkeit *f*
**capacity,** ~ **balance** Kapazitäts-ausgleich *m*,
-symmetrie *f* ~ **balancing** Querausgleich *m*
~ **balancing method** Kapazitätsausgleichsverfahren *n* ~ **bridge** Kapazitätsmeßbrücke *f* ~
**coefficient** Lieferzahl *f* ~**-coupled** kapazitiv
gekoppelt ~ **coupling** elektrische Kopplung *f*,
Kapazitätskopplung *f*, kapazitive Kopplung *f*
~ **current heating** kapazitive Hochfrequenz-
erhitzung *f* ~ **decade radio** Kapazitätsdekade
*f* ~ **factors** Kapazitätsfaktoren *pl* ~ **fuse** Hochleistungssicherung *f* ~ **independent of frequency**
frequenzunabhängige Kapazität *f* ~ **load** kapazitive Belastung *f*, Spitzenbelastung *f* ~ **operated
relay** kapazitiv verstimmtes Alarmgerät *n* ~
**range** Meßbereich *m* ~ **rating** Leistungsbe-
rechnung *f* ~ **reactance** Kapazitanz *f*, kapazitiver Blindwiderstand *m*, Kapazitätsreaktanz
*f*, kapazitive Reaktanz *f*, kapazitiver Blindwiderstand *m* ~ **resistance** blinder Widerstand
*m* ~ **susceptance** kapazitive Suszeptanz *f*,

kapazitiver Leitwert *m* ~ **test** Kapazitäts-
-messung *f*, -probe *f* ~ **unbalance** Kapazitäts-
-ungleichheit *f*, -unsymmetrie *f* ~**-unbalance
meter** Kopplungsmesser *m*
**cape** Kap *n*, Pelerine *f*, Umhang *m*, (geogr.)
Landspitze *f* ~ **chisel** Kreuzmeißel *m*, Meißel
*m*
**capel** Hornstein *m*
**capillarity** Haarröhrchenkraft *f*, Haarkraft *f*,
Kapillarität *f*, Oberflächenspannung *f* (Oberflächenenergie *f*)
**capillary** Leuchtkörper *m*; haar-artig, -förmig,
kapillar ~ **action** Kapillarkraft *f* ~ **attraction**
Haar-kraft *f*, -rohranziehung *f* -röhrchen-
wirkung *f* ~ **bracket** Röhrenhalter *m* (für Geiß-
lerröhre) ~ **combustion tube** Verbrennungs-
kapillare *f* ~ **crystal** Trichit *m* ~ **current** Kapil-
laritätsstrom *m* ~ **duct** Haarkanal *m* ~ **electrometer** Kapillarelektrometer *m* ~ **fissure** Haarriß *m* ~ **force** Kapillarkraft *f* ~**-gravity waves**
Kapillar-Schwerewellen *pl* ~ **native copper**
Haarkupfer *n* ~ **pressure** Kapillardruck *m* ~
**red oxide of copper** Chalkotrichit *m*, Kupferblüte *f* ~ **rise** kapillarer Aufstieg *m*, kapillare
Steighöhe *f*, Kapillaritätsansteigung *f*, Saughöhe *f* ~ **silver** Haarsilber *n* ~ **siphon** Heber-
haarrohr *n* ~ **structure** adrige oder faserige
Struktur *f* ~ **test** Kapillaritätsversuch *m* ~
**tube** Haar-rohr *n*, -röhrchen *n*, Kapillare *f*,
Kapillarrohr *n*, Röhrchen *n* ~ **tube of thermometer** Thermometer-röhre *f*, -säule *f* ~ **waves**
Kapillarwellen *pl* ~ **wire** Haardraht *m*
**capital** Fond *n*, großer Buchstabe *m*, Kapital *n*,
Kapitell *n*, Majuskel *f*, Säulen-knauf *m*,
-knopf *m*, -kopf *m*, Titelbuchstabe *m* ~ **and
revenue expenditure** Kapital- und Kostenauf-
wand *m* ~ **account** Kapitalkonto *n* ~ **assets**
Vermögen *n* ~ **construction** Anlageerstellung
*f* ~ **cost of installation** Gesamtanlagekosten
*pl* ~ **invested** Anlagekapital *n* ~ **letter** Groß-
buchstabe *m* ~ **value (or assets)** Vermögens-
werte *pl*
**capitalization** Kapitalisierung *f*
**capitalize, to** ~ kapitalisieren
**capitalized value of property** Ertragswert *m*
**capitals** Buchstaben *pl*
**caponier** Eskarpenkaponniere *f*, Grabenstreiche
*f*
**capped,** ~ **box** Kappenschachtel *f* ~ **joint** über-
einandergelgte Fuge *f* ~ **nut** Kugelmutter *f* ~
**roof** Haubendach *n* ~ **shell** Kappengeschoß *n*
**capper** Verschließer *m*
**capping** Abdeckung *f*, Bedeckung *f*, Deckschwelle *f*, Kron-holz *n*, -schwelle *f*; Mauer-
-abdeckung *f*, -deckplatte *f*, -kranz *m* (of a
wall) ~ **brick** Deck-, Kappen-ziegel *m* ~ **head**
Verschließkopf *m* ~ **heads** Verschlußstellen *pl*
~ **machine** Kapp-, Verschließ-maschine *f* ~
**piece** Deckschwelle *f*, Jochpfette *f* ~ **plate**
Blattsitz *n*, Wandrahmen *m* ~ **unit** Ver-
schließaggregat *n*
**capricious coupling** Streukoppelung *f*
**caproic acid** Capronsäure *f*
**caprylic acid** Caprylsäure *f*
**caps** Aufsatzbretter *pl*, Hängestützen *pl* (min.)
**capsize, to** ~ kentern, zum Kentern bringen,
sich überschlagen, umschlagen
**capsizing** Kentern *n* ~ **gradient** Kippneigung *f*

**capstan, to** ~ spillen
**capstan** Ankerwinde *f*, Ankerspill *n*, Auflaufhaspel *m*, Bandantriebsachse *f* (tape rec.), Erdwinde *f*, Haspel *f*, Kabestan *n*, Knebelgriff *m*, Revolverschlitten *m*, Seilrangierwinde *f*, Spill *n*, Tonrolle *f*, Verholspill *n*, stehende Winde *f*, Winde *f* ~ **of lathe** Drehknopf *m*
**capstan,** ~ **bar** Spake *f* ~ **drum** Spillwindentrommel *f* ~ **handle** Drehkreuz *n* ~**head screw** Knebelschraube *f* ~ **idler** Andruckrolle *f* (tape rec.) ~ **installation** Spillanlage *f* ~ **lathe** Revolverdrehbank *f*, Revolverbank *f* mit Zwischenschlitten ~ **rope** Spielseil *n* ~ **screw** Kreuzlochschraube *f* ~ **spike** Stellstift *m*
**capstone** Abdeckstein *m*
**capsular** kapselartig
**capsular element** Dosenmembran *f*
**capsule** Druckdose *f*, Kapsel *f*, Dose *f*, Würfel *m*, (explosives) Deckhütchen *n* ~ **aneroid** Aneroidbarometer *n*, Dosenaneroid *n*, Kapselbarometer *n*, Kapselluftdruckmesser *m*, Wetterglas *n* ~ **clay** Kapselton *m* ~ **control** Druckdosensteuerung *f* ~**-controlled** dosengesteuert (aviat.) ~ **diaphragm draught gauge** Dosenmembranzugmesser *m* ~ **fragments** Kapselscherben *pl* ~ **landing indicator** Kapsellandungsmesser *m* ~**-type dynamometer** Meßdose *f*
**captance** Kaptanz *f*
**caption** Überschrift *f* ~ **overlay** Schrifteinblendung *f*
**captivate, to** ~ einfangen, fangen, fesseln
**captive,** ~ **balloon** Fesselballon *m* ~ **kite** Fesseldrachen *m* ~ **kite balloon** Drachenfesselballon *m* ~ **nut** unverlierbare Mutter *f* ~ **plane** Fesselflugzeug *n*
**capture, to** ~ abfangen, aufbringen, einbringen, einfangen, einnehmen, erbeuten, erobern, fangen, kapern, nehmen, (electrons) auffangen
**capture** Einnahme *f*, Gefangennahme *f* ~ **of spring** Quellfassung *f*
**capture,** ~ **area** Absorptionsfläche *f* ~ **coefficient** Einfangfaktor *m* ~ **cross section** Einfang-, Wirkungs-querschnitt *m* ~ **efficiency** Einfangausbeute *f* ~ **ratio** Einfangverhältnis *n* ~ **spot** Fangstelle *f* ~**-to-fission** Einfang *m* zur Spaltung
**captured electrons** klebende Elektronen *pl*
**capturing of electrons** Kleben *n* der Elektronen
**car** Fahrzeug *n*; Förder-, Gleis-wagen *m*; Gondel *f*, Lastkarre *f*, Maschine *f*, Wagen *m* ~ **and locomotive works** Lokomotivfabrik *f* ~ **for transporting long iron bars** Langeisenwagen *n*
**car,** ~ **axle** Wagenachse *f* ~ **bay** Gondelfeld *n* ~ **body** Wagen-kasten *m*, -aufbau *m* ~ **builder** Wagenbauer *m* ~ **casting** Gießen *n* mittels Gießwagen ~**-catching device** Wagenfänger *m* ~ **circulations** Wagenumläufe *pl* ~ **commander** Wagenführer *m* ~**-construction part** Waggonbauteil *m* ~ **door** Schlag *m*, Wagenschlag *m* ~ **dumper** Wagen-kippenbrücke *f*, -kipper *m* ~ **dumping** Waggonentladung *f* mit Kipper ~ **frame** Lauf-, Wagen-gestell *n* ~ **haulage** Rangierwinde *f* ~ **headlight** Wagenlaterne *f* ~**-heater switch** Wagenheizerschalter *m* ~**-license fee** Kraftwagensteuer *f* ~**-line** horizontaler Gurt *m* (r.r.) ~**-load** Lieferung *f*, Wagenladung *f* ~ **loading** Waggonverladung

*f* ~ **oil** Wagenöl *n* ~ **panel** Gondelfeld *n* ~ **radio receiver** Autodynempfänger *m* ~**shed** Wagen-halle *f*, -schuppen *m* ~ **shunter** Verschieber *m* ~ **spotter** Rangierspill *n* ~ **spring** Tragfeder *f* ~ **suspension cable** Gondelseil *n* ~ **suspension ring** Gondelring *m* ~ **tipper** Waggonkipper *m* ~**-transfer platform** Wagenschiebebühne *f* ~ **truck** Wagen-gestell *n*, -untergestell *n* ~ **turntable** Waggondrehscheibe *f* ~**-type mixer** fahrbares Mischpult *n* ~ **unloading** Waggonausladung *f*
**caramel cooker** Karamelkocher *m*
**carat** Karat *n*; karätig
**caravan** Karawanenwagen *m*
**carbamic acid** Amidokohlen-, Carbamin-säure *f*
**carbamyl chloride** Chlorkohlensäureamid *n*
**carbanilic acid** Carbanilsäure *f*
**carbide** Karbid *n*, Karbur *n*, Kohlenstoffmetall *n*; karbidisch ~ **band (or streak)** Karbidzeile *f* ~ **bearing** Sintermetallager *n* ~ **capacity** Karbidfüllung *f* ~ **carbon** Karbidkohle *f* ~**-clad machine parts** hartmetallgepanzerte Maschinenteile *pl* ~ **drum** Karbidtrommel *f* ~ **furnace** Karbidofen *m* ~ **industry** Karbidindustrie *f* ~ **lamella** Karbidlamelle *f* ~ **lamp** Karbidlampe *f* ~ **metal** Hartmetall *n* ~ **powder** Karbidstaub *m* ~ **separation** Karbidseigerung *f* (metall.) ~ **slag** Karbidschlacke *f* ~ **sludge** Karbidschlamm *m* ~**-to-water gas generator** Karbideinwurfentwickler *m* ~**-to-water generator** Einwurf-apparat *m*, -entwickler *m*
**carbide-tipped,** hartmetallbestückt ~ **drill** *m* Hartmetallbohrkrone *f* ~ **saw blade** Hartmetallsägeblatt *n* (mit eingesetzten Hartmetallzänen) ~ **tool** Hartmetalldrehstahl *m*, hartmetallbestücktes Werkzeug *n*, Werkzeug *n* mit Hartmetallschneide *n* ~ **turning tool** Drehmeißel *m* mit Hartmetallschneiden *m*
**carbine** Karabiner *m*
**carbohydrate** Kohle(n)hydrat *n*
**carbolated lime** Karbolkalk *m*
**carbolic acid** Karbol *n*, Karbolsäure *f*, Phenol *n*
**carbolineum** Karbolineum *n*
**carbon** Kohle *f*, Kohlenspitze *f* (electr.), Kohlenstoff *m*, Kohlezylinder *m* (electr.), (electric arc light) Bogenlampenkohle *f*, (of an arc lamp) Kohlenstift *m* ~ **alloy** Kohlenstofflegierung *f* ~ **arc** Kohlelicht-, Kohlen-bogen *m* ~**-arc welding** Kohlelichtbogenschweißung *f* ~**back microphone** Postmikrofon *n* ~**-bag electrode** Kohlenbeutelelektrode *f* ~**-bag transmitter** Kohlenbeutelmikrofon *n* ~ **bar** Kohlenstab *m* ~ **battery** Kohlenbatterie *f* ~ **black** Druckerschwärze *f*, Ruß *m* ~ **(gas) black** Gasruß *m* ~ **block** Kohleklotz *m*, Kohlenstoffstein *m* ~**-block protector** Kohlenblitzableiter *m* ~ **block protector** Riffelkohlenblitzableiter *m*, Spannungsbegrenzer *m* ~ **break switch** Kohlenschalter *m* (electr.) ~ **brick** Kohlen-stoffstein *m*, -ziegel *m* ~ **brush** Kohle *f*, Kohlenbürste *f*, Schleifkohle *f* (electr.) ~ **chloride** Chlorkohlenstoff *m* ~ **compound** Kohlenstoffverbindung *f* ~ **concentration** Kohlenstoffsättigung *f* ~ **consumption** Kohlenverbrauch *m* ~ **content** Kohlen-gehalt *m*, -stoffgehalt *m* ~ **copy** Durchschlag *m*, Kopie *f* ~**-copy pad** Durchschreibeblock *m* ~ **deposit** Rußablage-

rung f ~ **diaphragm** Kohlenmembran f ~
**diapositive** Pigmentdiapositiv n ~ **dioxide**
Kohlen-dioxyd n, -säure f, -säureanhydrid n,
-säuregas n, Stickgas n ~ **dioxide flask** Kohlen-
säureflasche f ~ **dioxide motor** Kohlensäure-
motor m ~ **disulfide** Kohlensulfid n, Schwefel-
kohlenstoff m ~ **duplicate system** Durchschrei-
beverfahren n ~ **dust** Kohlengrieß m, Kohlenpulver n ~**-dust transmitter** Kohlenstaubmikro-
fon n ~ **electrode** Kohlenelektrode f ~ **feed**
Kohlennachschub m ~**-feeding mechanism of**
**searchlight** Lampenregelwerk n ~ **filament**
Kohle(n)faden m ~**-filament lamp** Kohle(n)-
fadenlampe f ~ **filler in liquidoxygen blasting**
**charges** Kohlenstoffträger m ~**-film potentiome-**
**ter** Kohleschichtpotentiometer n ~**-free** kohlefrei ~ **grain** Kohlenkorn n ~ **granule** Kohle(n)-
korn n ~**-granule chamber** Kohlenkörner-
kammer f ~**-granule microphone** Kohlenkörnermikrofon n ~**-granule noise** Körner-
rauschen n (microphone) ~**-granule thrust-**
**measuring device** Kohlendruckverfahren n
~**-granule transmitter** Kohlenkörnermikrofon
n, Kohlenpulvermikrofon n ~ **granules** Koh-
lenkörner pl ~ **holder** Kohlenhalter m ~ **isotope**
**ratio** Kohlenstoffisotopenhäufigkeit f ~ **light-**
**ning arrester** Kohle(n)blitzableiter m ~ **light-**
**ning rod protector** Kondensatorblitzableiter
m ~**like** kohlen-ähnlich, -artig ~ **microphone**
Kohle(n)mikrofon n ~ **monoxide** Kohlen-
monoxyd n, Koksgas n, Kohlenoxyd n ~
**monoxide combustion** Kohlenoxydverbrennung
f ~ **noise** Mikrofongeräusch n, (of microphone)
Kohlenkorngeräusch n ~ **oxysulfide** Kohlen-
oxysulfid n ~ **paper** Durchschlag-, Durch-
schreib-, Kohle-, Pigment-papier n **pencil** Zeichenkohle f ~ **penetration** Kohlenstoffdurch-
dringung f ~ **period** Aufkohlungsperiode f ~
**pickup** Kohlenstoffaufnahme f ~ **pile** Kohlen-
drucksäule f ~ **pole** Kohlepol m ~**-powder**
**microphone (transmitter)** Kohlenpulvermikrofon n ~ **protector** Kohleblitzableiter m ~
**ratio** Kohlenstoffgehaltverhältnis n ~ **raiser** die
Kohlenstoffaufnahme förderndes Mittel n ~
**recorder** Rußschreiber m ~ **refuse** C-Berge pl
~ **residue** Kohlenrückstand m ~ **residue of oil**
Ölkoks m ~**-resistance microphone** Kohle-
widerstandmikrofon n ~ **resistor** Kohlenwi-
derstand m ~ **ribbon feed device** Papierband-
führung f ~ **rod** Kohle f, Kohle-stab m, -puppe
f (of dry cell) ~ **seating** Kohlenauflage f ~
**silicide** Kohlenstoffsilizium n ~ **steel** Kohlenstoffstahl m, unlegierter Stahl m ~**-stick**
**microphone** Stabmikrofon n ~ **suboxide** Koh-
lensuboxyd n ~ **terminal** Kohlepol m ~
**tetrachloride** Chlorkohlenstoff m, Kohlen-
stofftetrachlorid n, Tetrachlorkohlenstoff m
~ **tetrafluoride** Kohlenstofftetrafluorid n, Tetra-
fluorkohlenstoff m ~ **tool steel** Kohlenstoffwerk-
zeugstahl m, unlegierter Werkzeugstahl m ~
**transmitter** Kohle(e)mikrofon n, ~ **wipper**
Schleifkohle (electr.) ~**-zinc battery** Kohle-Zinksammler m ~**-zinc cell** Kohle-Zinkelement n
**carbonaceous** kohlen-ähnlich, -artig, -haltig,
-stoffhaltig, kohlig, Kohlenstoff enthaltend,
Kohlenstoff abgebend ~ **material** Kohle f,
Kohlungsmittel n ~ **ramming mass** Kohlenstoff-
stampfmasse f

**carbonate, to** ~ saturieren mit Kohlendioxyd,
kohlensau(e)res Salz n, Karbonat n
**carbonate,** ~ **of** kohlensauer ~ **of lead** kohlensaures Bleioxyd n ~ **of lime** isländischer Kristall
m, Schlemmkreide f ~ **of silver** Grausilber n
**carbonation** Kohlensäuresaturation f, Saturation
f mit Kohlendioxyd ~ **juice** Schlammsaft m
~**-juice return** Schlammsaftrücknahme f ~
**tank** Saturationsapparat m für Kohlendioxyd
**carbonator** Saturationsapparat m für Kohlendioxyd
**carbonic** kohlensauer, Kohlenstoff enthaltend
~ **acid** Kohlendioxyd n, Luftsäure f ~ **acid gas**
Kohlensäure f ~ **acid level** Kohlensäurespiegel
m ~ **acid starter** Kohlensäureanlasser m
**carboniferous** karbonisch, kohle(n)haltig, Kohlenstoff enthaltend, kohlenstoffhaltig ~ **forma-**
**tion** Karbon n ~ **limestone** Kohlen-, Kulm-
-kalk m ~ **rock** Steinkohlengebirge n ~
**sandstone** Kohlensandstein m
**carbonite** Anthrazit n
**carbonization** Durchkohlung f, Entgasung f, Inkohlung f, Karbonisation f, Kohlung f, Trok-
kendestillation f, Verkohlen n, Verkohlung f,
Verkokung f, (byproducts preparation) Destillation f ~ **of bituminous coal** Steinkohlendestil-
lation f ~ **of lignite** Braunkohlenschwelung f
~ **of solid fuels** Brennstoffschwelung, Schwelung f von Brennstoffen ~ **of wood** Holzverkohlung f
**carbonization,** ~ **gas** Schwelgas n ~ **involving**
**gas recirculation** Spülgasschwelung f ~ **plant**
Kokerei f, Kokereianlage f, Steinkohlendestil-
lation f ~ **process** Schwelvorgang m ~ **retort**
Verkokungsretorte f ~ **zone** Schwelzone f
**carbonize, to** ~ karbonisieren, kohlen, verkohlen, verkoken, (superficially, slightly, or
partially) ankohlen, (at low temperature)
schwelen **to** ~ **under vacuum** abschwelen **to**
~ **under vacuum at a low temperature** verschwe-
len
**carbonized** aufgekohlt, gar ~ **fuel** verkohlter
Brennstoff m ~ **lignite** Braunkohlenkoks m
~ **residue** Restkohle f
**carbonizing,** ~ **agent** Kohlungsmittel n ~
**chamber** Brennkammer f ~ **flame** reduzierende
Flamme f ~ **oven** Verkohlungsofen m ~ **period**
Garungsdauer f ~ **plant** Karbonisier-anlage f,
-anstalt f
**carbonyl** Karbonyl n ~ **chloride** Chlorkohlenoxyd n, Kohlenoxydchlorid n, Phosgen n
**carborundum** Karborund n, Siliziumkarbid n
~ **detector** Karborunddetektor m ~ **saw** Fu-
genschneidgerät n ~ **wheel** Karborundum-
scheibe f
**carboxylic acid** Karbonsäure f
**carboy** Ballon m, Glasballon m, Korbflasche f,
Schwefelsäureballon m, Transportflasche f
**(acid)** ~ Säureballon m ~**-emptying apparatus**
Ballonentleerungsapparat m
**carburan** Karburan m
**carburated fuel** Frischgas n
**carburet, to** ~ karburieren
**carbureted, hydrogen gas** Kohlenwasserstoffgas
n ~ **water gas** karburiertes Wassergas n
**carbureting oil** Karburieröl n
**carburetion** Karburation f, Karburierung f, Ver-
gaseranordnung f, Vergasung f

**carburetor** Düse *f*, Karburator *m*, Schamotteüberhitzer *m*, Schwimmer *m*, Verdampfapparat
*m*, Verdampfer *m*, Vergaser *m* ~ **adapter** Ver-
gaserzwischenstück *n* ~ **adjustment** Regulierung *f* des Vergasers, Vergaser-einstellung *f*,
-regelung *f* ~ **air control** Vergaserluftkontrolle
*f* ~-**air heater** Ansaugluftvorwärmer *m*, Ver-
gaser-luftheizer *m*, -luftvorwärmer *m* ~ **air
scoop** Vergaser-luftfänger *m*, -luftschippe *f* ~
**air temperature** Vergasereintrittslufttemperatur
*f* ~ **anti-icer** Vergaserfrostschutzpumpe *f* ~
**body** Vergasergehäuse *n* ~ **capacity** Vergaserleistung *f* ~ **choke** Mischrohr *n* ~ **control** Vergasergestänge *n* ~ **engine** Vergaser-maschine
*f*, -motor *m* ~ **filter** Vergaserfilter *m* ~ **float**
Kipp-, Vergaser-schwimmer *m* ~ **float spindle**
Schwimmernadel *f* ~ **gasket** Vergaserdichtung
*f* ~ **heater** Heizvergaser *m* ~ **hot-air door** Ver-
gaserwarmluftklappe *f* ~ **hot spot** Vergaser-
vorwärmung *f* ~ **icing** Vergaser-eisbildung *f*,
-vereisung *f* ~ **induction system** Vergaserluft-
einlaß *m* ~ **intake screen** Korbsieb *n* ~ **jacket**
Vergaserheizmantel *m* ~ **jet** Vergaserdüse *f*
~ **linkage** Vergasergestänge *n* ~ **metering**
Düsenausflußleistung *f* des Vergasers ~
**needle** Schwimmernadel *f* ~ **needle valve** Kraft-
stoffnadelventil *n* ~ **primer** Tipper *m* des Vergasers ~ **scoop ram** Vergaseransaugluftstutzen
*m* ~ **throat** Vergasertrichter *m* ~ **throttle** Ver-
gaserdrossel *f* ~ **throttle bore** Vergaserlufttrichter *m*
**carburization** Aufkohlung *f*, Karburierung *f*,
Kohlung *f*, Einsatzhärtung *f* durch Aufkohlung, Zementierung *f*
**carburize, to** ~ aufkohlen, einsetzen, kohlen,
karburieren, zementieren
**carburized iron** gekohltes Eisen *n*
**carburizer** Aufkohlungsmittel *n*, Einsatzmittel *n*,
Kohlungsmittel *n*, Zementpulver *n*
**carburizing** Einsatzhärten *n*, Karbonisieren *n*
~ **addition** kohlender Zusatz *m* ~ **agent** Einsatzpulver *n*, Härtesalz *n*, Kohlungsmittel *n*
**(re)~agent** Aufkohlungsmittel *n* ~ **material**
Einsatzpulver *n*, Härtesalz *n* ~ **pot** Einsatztopf *m*, Zementier-kasten *m*, -kiste *f* ~
**property** Einsatzhärtbarkeit *f* ~ **steel** Einsatzstahl *m*
**carcass** Leiche *f*, Kadaver *m*, Karkasse *f* **of a ship**
Gerüst *n*, Rohbau *m* ~ **of a tire** Leinwand-
körper *m* (Reifenunterbau) ~-**utilization plant**
Kadaververwertungsanlage *f*
**carcel, ~ burner** Karcel-brenner *m*, -lampe *f*
~ **lamp** Karzel (Brenner) *m*
**carcinotron** Karzinotron *m* (Rückwärtswellen-
generator *m*)
**card, to** ~ krempeln, rauhen, (textiles) kardieren
**card** Diagrammblatt *m*, Karte *f*, Krempel *f*,
Zettel *m*, (weaving) Kratze *f* ~ **carrier** Rosenträger *m* ~ **clothier** Krempelbeziecher *m* ~
**compass** Kompaß *m* mit Kompaßrose ~ **cutter**
Karten-, Muster-schläger *m* ~ **cycle total
transfer** Summenübertragung *f* im Kartengang ~ **feed** Kartenzuführung *f* ~ **feeding**
Kartenvorschub *m* ~ **field** Lochkartenfeld *n*
~ **file** Kartei *f*, Kartothek *f* ~ **fitting** (textiles)
Kardenbeschlag *m* ~ **holder** Kartenhalter *m* ~
**index** Kartei *f*, Zettelkatalog *m* ~ **post** Rosensäule *f* ~ **printing installation** Kartendrucker

*m* ~-**programmed** auf Karten festgelegtes
Programm *n* ~-**proof punch** Lochprüfer *m* ~
**punch** Kartenlocher *m* ~ **reader** Kartenabtaster
*m* ~ **rebound** Kartenrückprall *m* ~ **reconditioner**
Kartenbügler *m* ~ **record system** Karteisystem
*n* ~ **reproducer** Kartendoppler *m* ~ **run** Kartendurchlauf *m* ~ **support** Rosenträger *n* ~-
**to-tape converter** kartengesteuerter Streifenlocher *m*
**cardan** Kardan *m*; kardanisch ~ **axes** Kardan-
achsen *pl* ~ **axle** Treibwelle *f* ~ **gear** Kardan-
getriebe *n* ~ **joint** Kardan-gelenk *n*, -kuppelung
*f* ~ **ring** Kreuzgelenk-, Kardan-ring *m* (gyro
compass) ~ **shaft** Kardanwelle *f* ~ **shaft brake**
Getriebsgehäusebremse *f*, Motorbremse *f* ~
**tube** Gelenkrohr *n* ~ **wave** Gelenk-, Kreuz-
-welle *f*
**cardanic** kardanisch ~ **suspension** Kardanaufhängung *f*; kardanisch aufgehängt; kardanische Aufhängung *f*
**cardboard** geklebter Karton, Karton *m*, Pappe *f*,
Pappendeckel *m* ~ **for folding boxes** Faltkarton
*m*
**cardboard, ~ articles** Kartonagewaren *pl* ~-
**bending machine** Pappenbiegemaschine *f* ~
**box** Kartonage *f*, Pappkarton *m* ~ **calenders**
Pappensatiniermaschine *f* ~ **container** Papp-
schutzhülle *f* ~ **ferrule** Papphülse *f* ~ **glazing
rolls** Pappsatinierwalzwerk *n* ~ **lining** Pappen-
aufzug *m* ~ **lining machine** Pappenbeklebe-
maschine *f* ~ **machine** Kartonmaschine *f* ~
**slide** Bodenleitzunge *f* ~ **top** Pappscheibe *f*
~ **tube cutting machine** Papphülsenabstech-
maschine *f* ~ **washer** Pappscheibe *f* ~ **winder**
Garnaufkarter *m*
**carded-cotton, ~ dyeing** Wattefärberei *f* ~-**wool
spinning mill** Streichgarnspinnerei *f* ~ **wool
yarn** Streichgarn *n* ~-**worsted yarn** Sayet(te)-
garn *n* (~ **worsted**) **yarn** Streichgarn *n*,
Streichengarn *n*
**cardigan, ~ arrangement** Fangeinrichtung *f* ~
**cam lever** Fanghebel *m*
**cardinal, ~ number** Grundzahl *f*, Hauptzahl *f* ~
**point** ausgezeichneter Punkt *m* (opt.), Haupt-
strick *m*, Himmelsgegend *f*, Himmelsrichtung
*f*, (of a lens) Grundpunkt *m*, Linsengrund-
punkt *m* ~ **points** ausgezeichnete Punkte *pl*
(opt.), Kardinal-punkte *pl*, -striche *pl*
**carding bench** Krämpelbank *f*
**cardioid** Herzkurve *f* ~ **characteristic** Herzcharakteristik *f*, Kardioide *f* ~ **curve (or diagram)**
Kardioidenkennlinie *f* (rdo) ~ **directive
diagram** Kardioide *f* ~ **formation** Kardioiden-
bildung *f* ~ **microphone** Mikrofon *n* mit Herz-
form ~-**shaped path** Herzbahn *f*
**carditioner** Kartenbügler *m*
**care** Acht *f*, Achtsamkeit *f*, Aufsicht *f*, Fürsor-
ge *f*, Pflege *f*, Sorge *f*, Sorgfalt *f*, Vorsicht
*f*, Wartung *f*, Adresse *f* **(in)** ~ **of**, c/o per Adresse *f* ~ **of the body** Körperpflege *f* ~ **of furnace
lining** Schonung *f* des Ofenfutters
**careen, to** ~ **a ship** einem Schiff Schlagseite geben
**careening** Kielholen *n* ~ **basin** Werfthafenbecken
*n*
**career** Karriere *f*, Laufbahn *f*, Lebenslauf *m*
**careful** behutsam, pfleglich, sorgfältig, vorsichtig
~ **treatment** Schonung *f*

carefully besorgt
**carefulness** Sorgfalt *f*
**careless** unachtsam, unvorsichtig ~ **cutting of gates** falsche Anschnitte *pl*
**carelessness** Fahrlässigkeit *f*, Flüchtigkeit *f*, Nachlässigkeit *f* Sorglosigkeit *f*, Unachtsamkeit *f*
**cargo** Fracht *f*, Ladung *f*, Lieferung *f*, Schiffsladung *f* the ~ **has shifted** die Ladung ist übergangen
**cargo,** ~ **aircraft** Lastenflugzeug *n* ~ **airplane** Lastflugzeug *n* ~ **boat** Last-boot *n*, -dampfer *m* ~ **carrier** Frachtbeförderungsbehälter *m* (Hubschrauber) ~**-carrying glider** Lasten--segelflugzeug *n*, ~**segler** *m* ~ **chute** Lastenfallschirm *m* ~ **compartment** Frachtraum *m* ~ **crane** Stückgutkran *m* ~ **distribution** Frachtverteilung *f* ~ **flying boat** Frachtflugboot *n* ~**-handling gear and facilities** Umschlageinrichtung *f* ~ **hold** Lastraum *m* ~**-intake certificate** Ladebericht *m* ~ **out-take certificate** Ausladebericht *m* ~ **parachute** Lastfallschirm *m* ~ **scooter** Lastenroller *m* ~ **steamer** Frachtdampfer *m* ~ **storage** Frachtlagerung *f* ~ **train** Lastenzug *m* ~**-transport aircraft (or airplane)** Lastflugzeug *n* ~ **truck** Lastkraftwagen *m* ~ **winch** Schiffsladewinde *f*
**carillon** Glockenspiel *n*
**Carius furnace** Wasserbadschießofen *m*
**carius tubes** Bombenrohre *pl*
**carminative** windschutztreibend
**carmine** Karmin *m*
**carnalite** Karnallit *m*
**carnauba wax** Karnaubawachs *n*
**carnaubic acid** Karnaubasäure *f*
**carnelian** Karneol *m*
**Carnot cycle** Carnot-Prozeß *m*, Carnotscher Kreisprozeß *m*
**carnotite** Carnotit *m*
**Caron process** Caron-Verfahren *n*
**carotene** Karotin *n*
**carpenter** Schreiner *m*, Tischler *m*, Zimmerer *m*, Zimmermann *m* ~'**s bench** Hobelbank *f* ~'**s brace** Brustleier *f* ~'**s journeyman** Zimmergeselle *m* ~'**s (chalk) line** Trassierleine *f* ~'**s marking gauge** Reißmaß *n* ~'**s square** Winkelmaß *n* ~'**s workshop** Tischler-, Zimmer-werkstätte *f*
**carpentry** Zimmerei *f*, Zimmerhandwerk *n*
**carpet** Teppich *m* ~ **felt** Teppichunterlagspapier *n* ~ **knife** Fadenschneider *m* ~ **loom** Teppichstuhl *m*
**carpolite** Fruchtversteinerung *f*
**carriage** Bettschlitten *m*, Bremsgestell *n*, Haltung *f*; Fuhr-lohn *m*, -werk *n*; Gefährt *n*, Karre *f*, Karren *f*, Konsolschlitten *m*, Kutsche *f*, Laufwerk *n*, Langdreh-, Normal-schlitten *m*, Papierwagen *m*, Schlitten *m*, Wagen *m* **(tool)** ~ Support *m* **by** ~ per Achse *f* ~ **for radiation source** Strahlerwagen *m* ~ **for removal** Fahrgestell *n* ~ **of a boring machine** Bohrschlitten *m* **(gun)** ~ **with overhead shield** Schirmlafette *f*
**carriage,** ~ **advance** Wagenvorschub *m* ~ **auger** Wagenbohrer *m* ~ **blacksmith** Wagenschmied *m* ~ **body** Karosserie *f* ~ **bolt** Holzversenkschraube *f*, Wagenbolzen *m* ~ **builder** Karossier *m*, Wagenbauer *m* ~ **building** Wagenschlosserei *f* ~ **cam** im Schlitten montierte

Schloßteile *pl* ~ **cask** Fuhrfaß *n* ~ **control** Vorschubkontrolle *f* ~ **door** Schlag *m* ~ **entrance** Torweg *m* ~ **fee** Abfuhrgebühr *f* ~ **feed** Transportwagenschaltwerk *n* ~ **feed lever** Hauptschlittenvorschubhebel *m* ~ **fork** Schlittengabel *f* ~ **frame** Lafettenrahmen *m* (artil.) ~ **gauge** Meßrahmen *m* ~ **grease** Wagenschmiere *f* ~ **guide** Bettschlittenführung *f* ~ **handle** Schlittenkurbel *f* ~ **lantern** Wagenlaterne *f* ~ **lever** (typewriter) Wagenheber *m* ~ **lock lever** Wagenfeststeller *m* ~ **movement** Wagenbewegung *f* ~ **paid** frachtfrei ~ **panels** Hartpappe *f*, Karosseriepappe *f*, Panneaubekleidungspapier *n* ~ **position indicator** Wagenstellungsanzeiger *m* ~ **position repeating cam disc** Wagenquittungskurve *f* ~ **rails** Wagenführung *f* ~ **receding motion** Wagenrückgang *m* ~**-release key** (typewriter) Wagenauslösetaste *f* ~ **return** Papierschlittenrückführung *f*, Rückführung *f* des Papierschlittens, Wagenrücklauf *m* ~ **return lock bar** Auslöser *m* (telet.) ~ **roll retainer** Wagenkugelkäfig *m* ~ **saddle** Längsschlitten *m* ~ **stroke** Wagenhub *m* ~ **support rails** Wagenführung *f* ~ **travel** Schlittenverschiebung *f* ~ **way** Damm *m* der Straße ~**-way cover** befahrbare Schachtabdeckung *f* ~ **wing** Führungsansatz *m*
**carrik bitt** Beting *f*, Spillbeting *m*
**carried,** ~ **away by tide** durch den Strom *m* abgetrieben (Schiff) ~**-forward call** Vortagsanmeldung *f* ~ **on a pole line** an einer Stangenlinie *f* geführt
**carrier** Brems(belag)träger *m*, Fördermaschine *f*, Mitnehmer *m*, Rahmen *m*, Sammler *m* (print.), Schlitten *m*, Spediteur *m*, Transport *m*, Transportmittel *n*, Transporteur *m*, Übertrager *m* (chem.), Unterlage *f*, (current) Trägerstrom *m* **(screen)** ~ Schirmträger *m* ~ **(pneumatic dispatch)** ~ Rohrpostbüchse *f* ~ **amplifier** Trägerverstärker *m* ~ **amplitude** (in telemetric or teletransmission work) Aussteuerungsgrad *m* ~ **amplitude regulation** Trägeramplitudenänderung *f* ~ **arm of leaf spring** Feder-hand *f*, -stuhl *m* ~ **bar** Lauf-, Trag-stange *f* ~**-based plane** Trägerflugzeug *n* ~ **beat** Trägerschwebung *f* ~ **bed** Trägergestein *n* ~ **bolt** Mitnehmerbolzen *m* ~**-borne aircraft** Decklandeflugzeug *n* ~ **bracket** Haltestück *n* ~ **changer** Schlittenwechsler *m* ~ **channel** Trägerstrom--kanal *m*, -weg *m* ~ **circuit** Drahtfunkleitung *f*, Trägerfrequenz-leitung *f*, -stromkreis *m* ~ **current** Träger-strom *m*, -welle *f*, leitungsgerichteter Strom *m* ~ **current control** Trägersteuerungssystem *n* ~ **current multiple telegraphy** Wechselstrommehrfachtelegrafie *f* ~ **current radiotelephony** leitungsgerichtete Hochfrequenztelefonie *f* ~ **current suppressor** Trägersperröhre *f* ~ **current telegraphy** Trägerstromtelegrafie *f* ~ **current telephony** Trägerstromfernsprechen *n* ~ **frame** Halterrahmen *m* (Tragrahmen *m*)
**carrier-frequency** Trägerfrequenz *f* ~ **principle** Trägerfrequenzverfahren *n* ~ **pulse** impulsmodulierte Trägerschwingung *f* ~ **rating** Trägerfrequenzseite *f* ~ **shifting** Trägerfrequenz--umtastung *f*, -verschiebung *f* ~ **voltage** Trägerfrequenzspannung *f* ~ **wave train** Trägerfrequenzwellenzug *m*

carrier, ~ gating Trägerwegtastung *f* ~ generator Trägergenerator *m* ~ hinge pin Konsolen-
scharnierbolzenstift *m* ~ hub Mitnehmernabe *f*
~ interruption (in synchronizing work) Trägerwegtastung *f* ~ isolating choke coil trägersper-
rende Drossel *f* ~ leak Trägerrest *m* ~ level
Trägerpegel *m* ~ line Drahtfunkleitung *f*,
Trägerleitung *f* (teleph.) ~ line section Doppel-
aderleitungsabschnitt *m* ~ link Trägerfrequenz-
verbindung *f* ~ noise Trägerrauschen *n* ~ noise
level Trägerrauschpegel *m* ~ oscillation Trägerschwingung *f* ~ peg Mitnehmerstift *m*
~ pigeon Brieftaube *f* ~ pin Haltestift *m* ~ plane
Flugzeug *n* eines Flugzeugträgers ~ plate Mitnehmer *m* (in axle assembly), Trägerplatte *f*
~ plate cover Mitnehmerscheibe *f* ~ power
Trägerleistung *f* ~ power output rating Träger-
ausgangsleistung *f* ~ repeater Trägerstromver-
stärker *m* ~ roller Transportwalze *f* ~ running
time Fahrzeit *f* (Rohrpost) ~ sentence Binde-
satz *m* (Ankündigungssatz) ~ shift Träger-
frequenz-verschiebung *f*, -umtastung *f* ~ stud
with arc-shaped bracket Haltestift *m* mit Bügel
~ support Träger *m* ~ supports Stützträger *pl*
im Hochvakuumofen ~ suppression Trägerwegtastung *f*, Unterdrückung *f* der Trägerfrequenz, (in quiescent carrier telephony) Trä-
gerunterdrückung *f* ~ tap choke coil Drossel *f*
mit Leitungsabzweigung ~ telegraphy Leitungsübertragung *f* ~ telemetering Trägerfre-
quenzfernmessung *f* ~ telephone Trägerstromfernsprechen *n* ~ telephone system Träger-
strom-fernsprechanlage *f*, -fernsprechung *f* ~
telephony Leitungsübertragung *f*, Trägerfre-
quenztelefonie *f* ~ terminal Trägerfrequenz-
gerät *n* ~ terminal circuit Drahtfunkendschal-
tung *f* ~-to-noise ratio Träger-Rausch-Verhält-
nis *n* ~ transmission Drahtfunk *m*, Hochfrequenzverbindung *f*, Übertragung *f* mit Trägerströmen (über Leitungen) ~ tubing Tragerohr *n*
~ vector Trägervektor *m* ~ vehicle Trägerfahr-
zeug (g/m) ~ voltage Trägerspannung *f* ~ wave
Trägerwelle *f* ~ wave amplifier Trägerfrequenz-
meßverstärker *m* ~-wave telegraphy Träger-
frequenztelegrafie *f* ~ wave train Trägerwellen-
zug *m* ~ yoke Aufhängungsdraht *m* ~ yoke
Gleitschienenjoch *n*
carriers geladene Teilchen *pl*
carrousel Kreistransporteur *m*
carry, to ~ leiten, mittragen, tragen, überbringen, übertragen, (threads) einziehen, (current)
führen to ~ along mitschleppen to ~ away abtragen, abdecken, fortführen (roofs) to ~ back
to the beginning auf Koordinatenursprung *m*
umrechnen to ~ forward mitreißen to ~ further
anknüpfen an etwas to ~ into zuführen to ~
off (or away) abführen to ~ off ableiten, aufräumen, fortleiten to ~ off heat Wärme *f* ableiten to ~ on treiben, fortsetzen to ~ on
development weiterentwickeln to ~ on solid
frames in fester Stuhlung *f* lagern to ~ out
austragen, ausüben, leisten, vollstrecken, (test)
ausführen to ~ out acrobatics Kunstflug *m* ausführen to ~ out an attack einen Angriff *m*
durchführen to ~ over mitreißen, stornieren,
(distillation) hinüberreißen, überschreiben to ~
through durchführen, durchsetzen
carry, to ~ current on the line die Leitung mit

Strom beschicken to ~ a current Strom *m* führen to ~ the traffic den Verkehr *m* abwickeln
carry Übertrag *m* (info proc.) ~ contact Zehner-
übertragkontakt *m* ~ power Tragfähigkeit *f*
carrying tragend ~ along of tube oscillation Mitnahme *f* der Röhrenschwingung ~ by means of
barrels Tonnenabfuhrsystem *n* ~ into effect (or
practice) Ausübung *f* (eines Gedankens) ~-out
Befolgung *f* ~-over Überführung *f*
carrying, ~ agent Träger *m* ~ apparatus Transportgerät *n* ~ axle Lauf-, Trag-achse *f* ~ basket
Transportkorb *m* ~ bolt Stützschraube *f* ~
capacity Belastbarkeit *f*, Ladegewicht *n* eines
Wagens, Leistung *f*; Trag-kraft *f*, -last *f*, -ver-
mögen *n*, -fähigkeit *f* ~ case Bereitschaftstasche *f*, Handkoffer *m*, Tragkasten *m* ~ fork
Mitnehmergabel *f* ~ frame Traggerüst *n* ~-gate
Hauptstrecke *f* (petroleum) ~ liquid over Flüs-
sigkeitsverschleppung *f* ~ member Haltezapfen
*m* (Sammelbezeichnung für Stativteile) ~ sling
Tragbügel *m* ~ spindle Tragspindel *f* ~ strap
Trag-band *n*, -riemen *m*, Trage-gurt *m*, -riemen
*m* ~ trade Zwischenhandel *m* ~ wire Tragdraht
*m*
cart, to ~ einkarren to ~ the ground die Erde
abkarren
cart Flurfördermittel *n*, Förderwagen *m*, Fuhre
*f*, Fuhrwerk *n*, Grubenwagen *m*, Karre *f*, Karren *m*, Laufkarren *m*, Lore *f*, Transportwagen
*m*, Wagen *m* (push) ~ Schiebtruhe *f* ~ for
tools Gerätetransportwagen *m*
cart, ~ load Fuder *n*, Fuhre *f* ~ stove Kärcher-
ofen *m* ~ traffic Fuhrgewerbe *n* ~-type mixer
fahrbares Mischpult *n* ~-type radio station
fahrbare Funkstation *f*, Karrenstation *f* ~
wright Wagner *m* ~ wheel Karren-, Wagen-rad
*n*, Kreiskorn *n* (a type of antiaircraft sight) ~
winder Aufkarter *m* (weaving)
cartage Fuhrlohn *m*, Zustellungsgebühr *f* ~ expenses Anfahrkosten *pl*, Transportkosten *pl*
cartel Arbeitsgemeinschaft *f*, Kartell *n* ~ contract Gewerkschaftsvertrag *m* ~ ship Austauschschiff *n*
Cartesian, ~ curve kartesische Kurve *f* ~ surface
kartesische Fläche *f*
cartilage Knorpel *m*
cartographer Kartenzeichner *m*, Kartograf *m*,
Tabellenzeichner *m*
cartographic kartografisch
carton Karton *m*, Pappdeckel *m*, Pappe *f*, Papp-
schachtel *f*, Zellstoffpapier *f*
cartoning, ~ machine (or cartoner) Kartonher-
stellungsmaschine *f* ~ system Kartonier-
system *n*
cartridge Geschoßhülle *f*, Sicherungspatrone *f*
~ carrier Patronenträger *m* ~ case Geschoß-
hülse *f*, Hülse *f*, Hülsenkartusche *f*, Kartusche
*f*, Kartusch-, Patronen-hülse *f* ~-case rim
Hülsenrand *m* ~-case ward Kartuschvorlage *f*
~ chamber Patronenlager *n* ~ clip Ladestreifen *m* ~ container Patronenraum *m* ~ cylinder
Patronenschleuse *f*, Revolvertrommel *f* ~
drum Ladetrommel *f*, Panzer-, Patronen-
-trommel *f* ~ extractor Entladestock *m* ~
flange Hülsenkopf *m* ~ fuse Patronen-, Stöp-
sel-sicherung *f* ~ heater Heizpatrone *f* ~
packer Patronenpackmaschine *f* ~ paper Kar-
dus-, Karton-papier *n*, Kartuschpappe *f*, Li-

nienpapier *n* ~ **papers** Patronenpapier *n* ~-**powered tool** Bolzenschießgerät *n* ~ **starter** Anlaßpistole *f*, Patronenanlasser *m* ~ **wad** Patronenfilzpfropfen *m*
**carve, to** ~ ausarbeiten, aushauen, ausschrämen, einschneiden, schnitzen, schnitzeln, verschrämen, ziselieren
**carved** verschrämt ~ **face** verschrämter Stoß *m* (min.) ~-**type poster** Schneideschriftplakat *n* ~ **weir** durchbrochenes Wehr *n* ~ **wood** geschnitztes Holz *n*
**carvel-built boat** Karvelboot *n*
**carver** Schnitzer *m*
**carving** Ausschram *m*, Schnitzarbeit *f* ~ **knife** Schnitz-, Vorlege-messer *n* ~ **knife and fork** Tranchierbesteck *n*
**cascade, to** ~ wasserfallartig herunterfallen
**cascade** Dampf-sprühe *f*, -sprühregen *m*, Gitter *n*, Kaskade *f*, Schaufelgitter *n* **to (join in)** ~ in Kaskade *f* schalten **in** ~ in Reihe *f* ~ **of separating units** Trennungsstifenkaskade *f*
**cascade,** ~ **amplifier** Kaskadenverstärker *m* ~ **bomb** Kaskadenbombe *f* ~ **circuit** Kaskadenschaltung *f* ~ **connection** Kaskaden-, Stufen--schaltung *f* ~ **control** Folgesteuerung *f* ~ **converter** Kaskadenumformer *m* ~ **detection** Kaskadennachweis *m* ~ **particles** Multiplikationsschauer *m* ~ **principle** Kletterprinzip *n* ~ **printer** Mehrfachkopiermaschine *f* (film) ~ **process** Kaskadenverfahren *n* ~ **rectifier circuit** Kaskadenschaltung *f* ~ **shower** Kaskadenschauer *m* ~-**type screen** Doppelschirm *m* ~ **voltage transformer** Kaskadenspannungswandler *m*
**cascaded carry** Kaskadenübertrag *m*
**cascading connection** Hintereinanderschaltung *f* (motor)
**case, to** ~ abfüttern, ausfüttern, auskleiden, einhängen, einstecken (in ein Gehäuse), emballieren, ummanteln, verkleiden, verrohren, verschalen **to** ~ **off sheets** die Schoten abfieren (naut.) **to** ~ **glass** Glas *n* überfangen
**case** Angelegenheit *f*, Behältnis *n*, Büchse *f*, Etui *n*, Fall *m*, Formrahmen *m*, Futter *n*, Futteral *n*, Gehäuse *n*, äußere Hülle *f*, Kasten *m*, Kiste *f*, Mantel *m*, Rollkolben *m*, Sache *f*, Schachtel *f*, Scheide *f*, Schicht *f*, Schrank *m*, Umstand *m*, (metall.) Randzone *f* (legal) ~ Rechtsfall *m* **in** ~ falls **in** ~ **of** im Falle **in** ~ **of need** nötigenfalls **as the** ~ **may be** beziehungsweise **the** ~ **in question (or under review)** der vorliegende Fall *m* ~ **at issue** Streitfall *m*
**case,** ~ **of avertence** Schwenkungsfall *m* ~ **of convergence** Konvergenzfall *m* ~ **of emergency** Bedarfsfall *m* ~ **of horizontal swing** Schwenkungsfall *m* ~ **of loss** Schadenfall *m* ~ **of need** Notfall *m* ~ **of resonance** Resonanzfall *m* ~ **of stress** Spannungszustand *m* ~ **of wax cloth** Wachstuchfutteral *n* ~ **of a window** Fensterfutter *n* ~ **of yarn** Garnkiste *f*
**case,** ~ **annealing** Einsatzglühung *f* ~ **blower** Kapselgebläse *n* ~ **board** Kistenbrett *n* ~-**bonded charge** gehäuseverbundener Treibsatz *m* ~ **drain** Gehäuseablaß *m* (hydraul.) ~ **fittings** Etuibeschläge *pl*
**case-harden, to** ~ anstählen, einsatzhärten, einsetzen, härten, zementieren, im Einsatz härten

**case-hardenability** Einsatz-fähigkeit *f*, -härtbarkeit
**case-hardened** einsatzgehärtet, gehärtet, im Einsatz gehärtet, schalenhart ~ **crossing** Hartguß--herzstück *n*, -kreuzung *f* (r.r.) ~ **frog** Hartgußherzstück *n*, Hartgußkreuzung *f* (r.r.) ~ **roller** glasharte Walze *f* ~ **steel** Schalenguß-, Zement-stahl *m*
**case-hardener** Einsatzhärter *m*
**case-hardening** Einsatz-härten *n*, -härtung *f*, Einsetzen *n* ~ **by carburization** Einsatzhärtung *f* durch Zementieren ~ **bath** Härtebad *n* ~ **box** Einsatzkasten *m*, Zementier-kasten *m*, -kiste *f* ~ **compound** Einsatz-mittel *n*, -pulver *n* ~ **furnace** Einsatz-, Härte-, Zementierungs-ofen *m* ~ **material** Einsatzmaterial *n* ~ **pot** Einsatztopf *m* ~ **powder** Einsatzpulver *n* ~ **property** Einsatzhärtbarkeit *f* ~ **quality** Einsatzhärtbarkeit *f* ~ **steel** Einsatzstahl *m* ~ **temperature** Einsatztemperatur *f* ~ **work in a cyanide bath** Zementation *f* im Salzbad
**case,** ~ **noise** Reibungsgeräusch *n* ~-**rotated free-rotor gyro** lagerfreier Kreisel *m* mit rotierendem Gehäuse *m* ~ **shift** Zeichenwechsel *m* ~ **spring** Gehäusefeder *f* (Uhr)
**cased,** ~ **borehole** verrohrtes Bohrloch *n* ~-**butt coupling** Muffenkupplung *f* ~ **catch** Schließkappe *f* ~ **floor** Friesfußboden *m*, Halbparkett *n* ~-**in bore** verrohrte Bohrung *f* ~-**in film guide** gekapselte Filmführung *f* ~ **panel** eingestemmte Füllung *f* ~ **sash** Aufziehfenster *n* ~ **well** Rohrbrunnen *m*
**casein** Kasein *n* ~ **fiber** Kaseinwolle *f* ~ **glue** Kaseinleim *m* ~ **paint** Kaseinfarbe *f*
**casemate** Kasematte *f* ~ **of a window** Fensterflügel *m* ~ **bore** Kastenbohrung *f* ~ **drill** Kastenbohrer *m* ~ **staple** Fensterkrampe *f* ~ **window** Flügelfenster *n*
**casement** Futterrahmen *m* ~ **fastener** Vorreiber *m* (Fenster)
**cash, to** ~ einziehen, bar **to** ~ **a bill of exchange (or draft)** einen Wechsel auslösen
**cash** Bargeld *n*, Geld *n*, Kasse *f* ~ **in hand** verfügbares Geld *n*, Bestand *m* ~ **on delivery** gegen Nachnahme *f*, Nachnahme *f*, Zahlung *f* gegen Lieferung, zahlbar bei Lieferung ~-**on-delivery service** Nachnahmeinkasso *n* ~-**on-hand journal (or record)** Geldbestandsbuch *n*
**cash,** ~ **balance** Kassen-bestand *m*, -abschluß *m* ~ **book** Ausgabenbuch *n* ~ **box** Kassette *f*, Münzbehälter *m* ~ **liabilities** Anforderungsverpflichtungen *pl* ~ **money** bares Geld *n* ~ **order** Kassen-anweisung *f*, -schein *m* ~ **prize** Geldpreis *m* ~ **register** Kontroll-, Registrier-, Zahl--kasse *f* ~ **register for sales slips** Zahlkassenquittungsdrucker *m* ~ **register with key action** Tastenregistrierkasse *f* ~ **register with lever action** Hebelregistrierkasse *f* ~ **settlement** Geldabfindung *f*
**cashew tree** Nierenbaum *m*
**cashier, to** ~ kassieren
**cashier** Kassierer *m* ~'s **office** Kasse *f*
**cashing,** ~ **of a bill** Wechseleinlösung *f* ~ **up** Kassenabschluß *m*
**casing** Bekleidung *f*, Bohrrohr *n*, Einfassung *f*, Einkapselung *f*, Futterrohr *n*, Gehäuse *n*, Hülle *f*, Hülse *f*, Kapsel *f*, Kapselung *f*, Mantel *m*, Mantelrohr *n*, Panzer *m*, Panzerung *f*,

Reifenmantel *m*, Schalung *f*, Schutzhülle *f*, Taschenbügel *m*, Tübbings *pl*, Umhüllung *f*, Umkleidung *f*, Ummantelung *f*, Verkleidung *f*, Verrohrung *f*, Verschalung *f*, Wickelung *f*; (tubing) Ausfütterung mit Röhren, (of converter floor) Ring *m* (**outer**) ~ Gehäuse *n* ~ **for goods to be cooled** Kühlgutbehälter *m* ~ **for latch mechanism** Schloßgehäuse *n* ~ **for overrunning clutch** Freilaufschale *f* ~ **for traveling crab** Laufkatzengehäuse *n* ~ **of the centrifugal** Mantel *m* der Schleuder ~ **of a pipe** Rohrbekleidung *f* ~ **of the turntable** Trog *m* der Drehscheibe ~ **with bellmouthed collars** Bohrröhre *f* mit tonnenförmigen Muffen ~ **with bulging collars** Bohrröhre *f* mit stark gewölbten Muffen ~ **with collars** Muffenverrohrung *f* ~ **with inserted joints** eingezogene Röhre *f*, Muffenbohrröhre *f*

**casing**, ~ **bailer** Verrohrungsreinigungsbüchse *f* ~ **bowl** Rohrfangglocke *f* ~ **buffers** Hülsenpuffer *pl* ~ **cap** Bohrdeckel *m* ~ **clamp** Rohrschelle *f* ~ **clamp wedge** Verrohrungshaltekeil *m* ~ **clamps with wedges** Kopfzugstück *n* ~ **clevis** Rohreinhängebügel *m* ~ **cover** Gehäusedeckel *m* ~ **cutter** Inrohrschneider *m*, Rohrschneider *m* ~ **disk** Verrohrungsdeckelscheibe *f* ~ **elevator** Rohrelevator *m* ~ **fitting** Verbindungsstift *m* für Bohrrohr ~ **flange** Futterrohrflansch *m* ~ **float** Rohrschwimmer *m* ~ **float collar** Muffe *f* mit Ventil ~-**float shoe** Ventilschuh *m* ~ **head** Bohrkopf *m*, Kappe *f*, Rohrkopf *m*, Verrohrungskopf *m* ~-**head gas** Erdgas *n*, nasses Erdgas *n* ~-**head gasoline** Bohrrohr-benzin *n*, -gasolin *n* ~ **hook** Futterrohr-, Rohrförder-haken *m*, Rohrschaltung *f* ~ **jack** Futterrohrheber *m* ~ **joint** Futtersperrverbindung *f* ~ **knife** Futterrohrspalter *m*, Inrohr-, Rohr-schneider *m* ~ **lifter** Rohrheber *m* ~ **lug** Gehäuseauge *n* ~ **machine** Packmaschine *f* ~ **paper** Tauenpapier *n* ~ (**lining**) **paper** Kistenausschlagpapier *n* ~ **pipe** Bohr-, Futter-rohr *n* ~ **pressure** Bohrrohrdruck *m* ~ **pressure screws** Verrohrungsheber *m* ~ **ripper** Rohrschlitzer *m* ~ **shell** Gehäusewandung *f* ~ **shoe** Rohrschuh *m* ~ **slip** Abfangkeil *m* ~ **spear** Backen-, Krebs-fänger *m* ~ **spider** Rohrkeilkranz *m* ~ **splitter** Rohrschlitzer *m* ~-**spud straightener** Birne *f*, Bohrbirne *f* ~ **straightener** Rohrstrecker *m* ~ **string** Rohrstrang *m* ~ **support** Rohrständer *m* ~ **swage** Rohrtreibbirne *f* ~ **tester** Wassersperrungsprüfer *m* ~ **vane** Gehäuseschaufel *f* ~ **wagon** Rohrwagen *m*

**cask** Faß *n* ~ **to be returned** Leihgebinde *n*

**cask**, ~ **bridge** Tonnenbrücke *f* ~ **bush** Spundbüchse *f* ~ **clasp** Faßspange *f* ~-**clasp holder** Faßspangenhalter *m* ~ **cleaning-off machine** Abhobelmaschine *f* für Fässer ~-**croze cutting (or -crozing) machine** Faßkrösemaschine *f* ~-**decanting apparatus** Faßabfüllapparat *m* ~ **deposit** Faßgeläger *n*, Geläger *n* ~-**emptying apparatus** Faßabfüllapparat *m* ~-**groove (or -grooving) cutting machine** Faßkrösemaschine *f* ~ **hoop** Faß-band *n*, -reifen *m* ~-**illuminating apparatus** Ausleuchtapparat *m* für Fässer ~ **planing machine** Abhobelmaschine *f* für Fässer ~ **raising apparatus** Faßaufsatzform *f* ~ **rivet** Faßniete *f* ~ **sediment** Faßgeläger *n* ~ **setting up apparatus** Faßaufsatzform *f* ~ **stave** Faßstab *m*

**casket** Kassette *f*, Transportbehälter *m*

**casks** Fastage *f*, Fustage *f*, Leergut *n*

**Cassegrain reflector** Cassegrain-Spiegelteleskop *n*

**Cassel,** ~ **furnace** Kasseler Ofen *m* ~ **yellow** Kasseler Gelb *n*

**casse paper** Abfall *m*, Ausschuß *m*

**casserole** Kasserole *f*

**cassiterite** Bergzinn *n*, Kassiterit *m*, Zinnerz *n*, Zinngraupen *pl*, Zinn-oxyd *n*, -stein *m*

**cast, to** ~ abformen (aviat.), abgießen, gießen, sich krumm ziehen, vergießen, werfen **to** ~ **adrift** loslassen **to** ~ **around** herumgießen, (something) umgießen **to** ~ **from the bottom** mit dem Steigrohr *n* gießen **to** ~ **from the top** fallend (ab)gießen **to** ~ **in open sand** im Herd gießen **to** ~ **off** loswerfen, (copy) abschätzen **to** ~ **off from a buoy** von einer Boje *f* loswerfen **to** ~ **on** angießen **to** ~ **on end** stehend abgießen oder gießen **to** ~ **up** nach oben gießen

**cast, to** ~ **badly** vergießen **to** ~ **cold** kaltvergießen **to** ~ **horizontal** waagerecht gießen **to** ~ **integral** eingießen **to** ~ **loose** loslassen **to** ~ **sideways** krumm werden **to** ~ **solid** massiv gießen **to** ~ **upright** aufrecht gießen

**cast** Abdruck *m*, Abguß *m*, Besetzung *f* (film), Formstück *n*, Gegensonne *f*, Gipsverband *m*, Guß *m*, Polytypie *f*, Vieldruck *m*; aufgegossen, gegossen, (into) vergossen ~ **in dry mold** Trockenguß *m* ~ **in pairs** paarweise gegossen ~ **of bloom** (of oil) Abglanz *m*, Widerschein *m* ~ **of the lead** Lotwurf *m* ~ **on an object** Eigenschatten *m*

**cast,** ~ **aluminium** Aluminiumguß *m*, Gußaluminium *n* ~ **articles** Gußwaren *pl* ~ **basalt** Schmelzbasalt *m* ~ **brass** Gußmessing *n*, Messingguß *m* ~ **bronze** Gußbronze *f* ~ **bulkhead fitting** Gußleuchte *f* ~ **casing** Gußkasten *m* ~ **concrete** Gußbeton *m*, ~ **copper** Kupferguß *m* ~ **housing** Gußgehäuse *m* ~-**in concrete current-limiting reactor** Betonseilspule *f* ~-**in place pile** Ortspfahl *m* ~-**insitu concrete** an Ort und Stelle gegossener Beton *m* ~ **integral** angegossen, eingegossen, in einem Stück *n* gegossen

**cast-iron** Eisenguß *m*, Grauguß *m*, Guß *m*, Gußeisen *n*, schmiedbarer Guß *m*; gußeisern ~ **base** Gußeisensockel *m*, gußeiserner Fuß *m* ~ **bearing** Gußlager *n* ~ **bowl** Gußbad *n* in ~ **box** gußgekapselt ~ **brazing** Graugußlöten *n* ~ **core** Gußboden *m* ~ **cylinder** Gußzylinder *m* ~ **economizer** Gußeisenvorwärmer *m* ~ **frame** Gußbock *m* ~ **mixture** Gußgattierung *f* ~ **mold** Blockform *f*, Kokille *f* ~ **paling** (single bar) Gitterstange *f* ~ **pipe** gußeiserne Leitung *f*, gußeisernes Rohr *n*, Gußeisen-, Guß-rohr *n* ~ **roll** Guß-, Weich-walze *f* ~ **roller** gußeiserne Rolle *f* ~ **scrap** Gußschrott *m* ~ **shoe** gußeiserner Pfahlschuh *m* ~ **sleeve** Gußeisenmuffe *f* ~ **splinters** Gußspäne *pl* ~ **strainer** gußeiserner Durchschlag *m* ~ **strength** Gußfestigkeit *f* ~ **structure** Gußstruktur *f* ~ **tappet** Gußeisenstößel *m* ~ **treatment** Gußeisenbearbeitung *f* ~ **tubbing** gußeiserne Cuvelage *f* ~ **welding** Gußschweißung *f* ~ **welding without preheating** Gußeisenkaltschweißung *f* ~ **welding with pre- and postheating** Gußeisenwarmschweißung *f* ~ **working** Gußeisenbearbeitung *f*

cast, ~ lead Gußblei n ~-lens blank Ausguß-
scheibe f ~ line Summenlinie f ~ metal Guß m,
Gußmetall n, Metallguß m ~ metal carbide
gegossenes Metallkarbid n ~-metal case Gußgehäuse n ~-metal housing Gußgehäuse n ~-
metal scrap Gußbruch m ~-off burr Abschlag-
rad n (weaving) ~-off position Abschlagstel-
lung f ~-on angegossen ~-on mount Aufguß m
(print.) ~ piece Gußstück n ~ plate Gußplatte f
~ raffinade gegossene Raffinade f ~ resin Edelkunstharz n ~ seam Gußnaht f ~-skein farm
wagon Hofwagen m mit verstärkten Achsen
~ slab Bramme f ~ steel Gußstahl m; Stahl-
-formguß m, -guß m; Tiegelstahl m
cast-steel gußstählern ~ barrel Gußstahllauf m
~ body Körper m von Gußstahl ~ bullet Guß-
stahlkugel f ~ crankcase Stahlgußgehäuse n
~ crossing Gußstahlherzstück n ~ crucible
Gußstahltiegel m ~ drum Stahlgußtrommel f
~ frog Gußstahlherzstück n ~ ingot Gußstahl-
block m ~ manufacture Gußstahlfabrikation f
~ parts Flußeisengußwaren pl ~ plate Gußstahlblech n ~ products Stahlguß m ~ reducing
shackles Stahlgußreduzierschäkel m ~ ring
Gußstahlring m ~ roll Stahlgußwalze f ~ stiffening ring Eckring m aus Gußstahl ~ tubing
gußstählerne Cuvelage f ~ wire Gußstahldraht
m
cast, ~ structure Gußgefüge n ~ tooth gegossener
Zahn m ~ washer gußeiserne Unterlagsscheibe
f ~ weld Gießschweißung f ~-welding powder
Gußschweißpulver n ~ zinc Gußzink m
castability Gießbarkeit f, Vergießbarkeit f
castable gußfähig, vergießbar
castellate, to ~ bezinnen
castellated, ~ nut Kronenmutter f ~ yoke Joch-
körper m mit Längsnuten
castellations Kronierung f
caster Gießer m, Lenkrad n, schwenkbare
Rolle f, Vorlauf m ~ action Schwenkung f
~(ing) tail wheel schwenkbares Spornrad n
~ wheel Lenkrolle f, Lenkrad n
casting Abguß m, Formling m, gegossener Rohling m, Gießen n, Gießerei f, Guß m, Gußeisen n, -teil m, -stück n; Vergießen n ~ takes
place every ... hours alle ... Stunden wird
zum Abstich geschritten ~ from the cupola
Kupolofenguß m ~ in chill gegen Kokille gießen ~ in chills Kapselguß m ~ in crucibles
Tiegel-gießerei f, -guß m ~ of iron pots Topf-
gießerei f ~ of malleable (cast) iron Glühstahlguß m ~ of rollers Walzengießen n ~ of shadows Schattenwurf m ~ of works of art Kunstgießerei f ~ under suction Sauguß m ~ with
blowholes blasiger Guß m
casting, ~ alloys Gußlegierungen pl ~ bay Gieß-
grube f ~ bed Gießbett n ~ box Formkasten m,
Flasche f (in founding), Gußkasten m ~ breaker Gußbrechmaschine f ~ brush Gußputzbürste f ~ buggy Gießwagen m ~ cart Guß-
karren m ~ channel Gußrinne f ~ chill insertion Gießkokilleneinsatz m ~ cleansing hammer
Gußputzhammer m ~ crane Guß-, Gießerei-
kran m ~ crucible Schöpfhaspel f ~ device for
bushes Lagerschalengießvorrichtung f ~ drum
Gießtrommel f ~ equipment Gießeinrichtung f,
Gießvorrichtung f ~ factor Zuschlagstücke pl
für Gußstücke ~-floor level Hüttenflur m ~

gate Gußtrichter m ~ git Gußtrichter m ~ gutter Gußgerinne n ~ house Gieß-halle f, -haus n,
-hütte f ~ implements Gießanlage f ~ ladle
Füllöffel m; Gabelpfanne f, Gieß-kelle f,
-pfanne f, Pfanne f ~ line Gießstrecke f ~
loam Formlehm m (pig) ~ machine Gießma-
schine f ~ matrix Gußmatritze f, Gußmutter f
~metal Gieß-, Guß-metall n ~-model draft An-
zugußmodell n (power plant engin.) ~ mold
Gieß-form f, -kasten m, -lade f ~ molded in
flask Kastenguß m ~ net Wurfnetz n ~ nozzle
Gießmuschel f ~ operation Gießarbeit f ~
pattern Gußmodell n ~ pig iron Gießereiro-
eisen n ~ pit Gieß-grube f, -platz m, -stelle f
~-pit crane Gießgrubenkran m ~-pit refractories Stahlwerksverschleißmaterial n ~ plate
Gespannplatte f, Gießtafel f ~ platform Gießbühne f ~ procedure Gießverfahren n ~ process
Gießverfahren n ~ properties Gießfähigkeit f
~ resin Gießharz n ~ shovel Wurfschaufel f
~ square Gießwinkel m ~ store Gußlager n
~ stores Gußmagazin n ~ strain Gußspannung
f ~ strand Gießband n ~ stress Gußspannung f
~ temperature Gieß-, Vergieß-temperatur f
~ wax Gußwachs n ~ weight of a low-frequency induction furnace Abgußgewicht n eines
Niederfrequenzinduktionsofens ~ wheel Gießtrommel f
castings Gußkörper m pl, Gußwaren pl ~ for
machinery building Maschinenbauguß m ~ of
ductile cast iron Gußeisen n mit Kugelgraphit
~ to be cleaned Putzgut n (founding) ~ cleaner
(or scraper) Gußschleifer m
castle Burg f, Schloß n ~like burgartig ~ nut
Kronenmutter f ~ nut without turret Kronenmutter f ohne Hals
castor Fußrolle f, Laufrolle f
castor wheel Deichsel-rolle f, -träger m
casual connection Wirkungszusammenhang m
casualty Ausfall m, Verlust m to be a ~ ausfallen ~ power supply equipment Notstromanlage f
cat, to ~ katten
cat Katze f, (navy) Katt f ~ davit Kattdavit m
~'s-eye Katzenauge n, Rückstrahler m ~ head
Haspel f, Katzenkopf m (Zwischenfutter)
~-head man Spillarbeiter m ~ walk Holzrost
m, Kriechgang m; Lauf-boden m, -gang m;
Steg m ~ whisker Pinselelektrode f (rdo),
Spiralfeder f (eines Kristalldetektors) ~-whisker
detector Bürsten-, Pinsel-detektor m
catabatic katabatisch
catacaustic reflektierende Brennfläche f, Re-
flexionsbrennfläche f, rückwerfende Brennfläche f, Rückstrahlbrennfläche f ~ surface
reflektierende oder rückstrahlende Brennfläche f
catadioptric katadioptrisch
Catalan, ~ direct process katalonisches Renn-
feuer n ~ forge Katalanschmiede f ~-forge
process katalonisches Rennfeuer n ~ furnace
Luppenfeuer n
catalogue, to ~ durchmustern
catalogue Broschüre f, Kartei f, Kartothek f,
Katalog m, Liste f, Mappe f, Verzeichnis n
~ number Bestellnummer f
catalysis Aufspalt m, Katalyse f
catalyst Beschleuniger m, chemischer Beschleuniger m, Katalysator m, Kontakt m, Kontakt-
-mittel n, -substanz f

**catalytic** katalytisch ~ **agent** Katalysator *m*
**catalyzer** Beschleuniger *m*, chemischer Beschleuniger *m*, Katalysator *m*
**catapult, to** ~ abschleudern (aviat.), katapultieren **to** ~ **a glider** ein Gleitflugzeug *n* in die Luft schleudern
**catapult** Katapult *m*, Schleuder *f*, Schleuder- -anlage *f*, -bahn *f*; Schleudersitzkanone *f*, Schleuderstartvorrichtung *f*; Start-hilfe *f*, -schleuder *f* **(launching)** ~ Startkatapult *n*
**catapult,** ~ **airplane** Schleuderflugzeug *n* ~ **angle** Katapultwinkel *m* ~ **equipment** Katapultiereinrichtung *f* ~ **fittings** Katapultierbeschläge *pl* ~ **launching** Katapult-, Gummiseil-start *m* ~ **launching gear** Schleudervorrichtung *f* ~ **mechanism** Startvorrichtung *f* (aviat.) ~ **plane** Katapultflugzeug *n* ~ **rail** Gleitstange *f* ~ **seat** Schleudersitz *m* ~ **ship** Katapultschiff *n* ~ **ship for postal aircraft** Flugstützpunkt *m* ~ **take-off** Katapult-, Schleuder-start *m*
**catapultable** katapultfähig
**cataract** Katarakt *m*, Stromschnelle *f* ~ **cylinder** Bremszylinder *m*
**catch, to** ~ abfangen, angreifen, anhaken, eingreifen, einklinken, einrasten, einschnappen, fangen, sammeln, sich klemmen, (electrons) auffangen, (energy from an electron beam) entnehmen **to** ~ **up with** einholen, überholen **to** ~ **with iron bands** anhaspen **to** ~ **fire** anfangen zu brennen, Feuer fangen
**catch** Anschlag *m*, Arretierhebel *m*, Arretierung *f*, Eingriff *m*, Falle *f*, Fallenachse *f*, Fangeisen *n*, Fischfang *m*, Frosch *m*, Geisselfuß *m*, Gesperr *n*, Haken *m*, Klaue *f*, Klinke *f*, Mitnehmer *m*, Nase *f*, Öhr *n*, Schließe *f*, Schließhaken *m*, Schnalle *f*, Sperrhaken *m*, Sperrklinke *f*, Sperrstift *m*, Spritzer *m*, (lock) Hakenklinke *f* **(oil-drilling)** ~ Glückshaken *m* ~ **for pulley block** Flaschenzugradverschluß *m*
**catch,** ~ **band** Fangband *m* ~ **basin** Sammel-, Senk-loch *n*, Straßeneinlauf *m* ~ **bin** Ausgleichbunker *m* ~ **bolt** Mitnehmerbolzen *m* ~ **crop** Ackervorbau *m*, Fangpflanze *f* ~ **device for elevator cages** Förderkorbfangvorrichtung *f* ~-**fitted clip** Nasenklemmplatte *f* ~ **hook** Fanghaken *m* ~ **pan** Schmutzfänger *m* (für Kabelschächte) ~ **pit** Klärbecken *m*, Schlammsack *m*, Wasserabzugsgraben *m* ~ **ring** Mitnehmerring *m* ~ **rod** Fallstange *f* ~ **spring** Einschnappfeder *f* ~-**thread device** (textiles) Fadenfangvorrichtung *f* ~-**up of electrons** Überholungsgebiet *n* der Elektronen ~-**water drain** Sickerkanal *m* ~ **word (or by-word)** Begriff *m* ~ **wrench** Fanggabel *f*
**catchall** Saftabscheider *m*, (tool machine) Auffangtrog *m*
**catcher** Anschlag *m*, Auskoppelraum *m*, Fänger *m*, Greifkorb *m*, Hinterwalzer *m*, Rohrgreifer *m*, Schnapper *m*, Senke *f*, (in beam tube) Auffanganode *f*, Auffänger *m*, Entnahmeelektrode *f* ~ **electrode** Sammelelektrode *f* ~ **foil** Absorptions-, Fang-folie *f* ~ **plate** Fängerteller *m* ~'s **side** Austrittsseite *f* (metall.) ~ **space** Sammelraum *m* (in klystron), Auskopplungsraum *m*
**catches** Aufsatz-knaggen, -stützen *pl*
**catching** Abfassen *n* ~ **device** Auffangvorrichtung *f* ~ **diode** Klemmdiode *f* ~ **groove** Fang-

**rille** *f* ~ **piece** Fangfrosch *m* ~-**rod elevator** Förderstuhl *m*, Gestängefanggerät *n* ~ **width** Einfallwerte *pl*
**catchment,** ~ **area (or watershed)** Abflußgebiet *n* ~ **basin** Sammel-gebiet *n*, -trichter *m* ~ **basin of a torrent** Oberlauf *m* eines Wildbaches ~ **drainage area** Einzugsgebiet *n*
**catechol tannin** Brenzkatechingerbstoff *m*
**categorically** nachdrücklich
**category** Begriffs-fach *n*, -klasse *f*, Gruppe *f*, Kategorie *f* ~ **rating** Artberechtigung *f* (aviat.)
**catelectrotonus** Katelektrotonus *m*
**catenary** Fahrleitung *f*, Kettenlinie *f*, Korbbogen *m*
**catenary,** ~ **construction** Kettenfahrleitung *f* ~ **(curve)** Kettenlinie *f* ~ **hanger** Hängedraht *m* ~ **linkage** Kettenbindung *f* ~ **suspension** Entlastungsseil *n*, Kettenfahrleitung *f*
**catenoid** Katenoid *n*, Kettenfläche *f*
**catenoidal** kettenlinienförmig
**catenoidal** kettenlinienförmig
**cater-corner** diagonal, übereck
**caterpillar** Gleiskettengerät *n*, Raupe *f* ~ **band** Radgürtel *m* ~ **chassis** Raupenfahrgestell *n* ~ **drive** Gleiskettenantrieb *m* ~ **gate** Raupenschütze *f* ~ **prime mover** Gleiskettenzugmaschine *f* ~ **track** Gleiskette *f*, Raupe *f* ~-**tracked** raupengängig ~ **traction** Raupenzug *m* ~ **tractor** Gleisketten-schlepper *m*, -zugmaschine *f*; Ketten-, Raupen-schlepper *m* ~-**type landing gear** Raupenfahrwerk *n*
**catgut** Katgut *n*
**cathead** Winde *f* ~ **of crane** Kranschnabel *m* ~ **stopper** Kattstopper *m*
**cathedral** Kathedrale *f*; negative V-Stellung *f* (der Tragfläche) ~ **angle** Kathedralenwinkel *m*
**cathetometer** Kathetometer *n*
**cathetus** Kathete *f*
**cathode** Heizfaden *m*, Kathode *f*, negative Elektrode *f* ~ **anchor** Kathodenhalterung *f* ~ **arrester** Kathodenableiter *m* ~ **assembly** Kathodenwickel *m* ~ **beam** Brennpunkt *m*, Elektronen-bündel *n*, -strahl *m*, Kathodenstrahlbündel *n* ~ **bearing a coat (or layer) of oxide** Schichtkathode *f* ~ **bias** Kathodenvorspannung *f* ~ **border** Kathodensaum *m* ~ **conductance** Kathodenleitwert *m* ~-**connecting strip** Kathodenbändchen *n* ~-**coupled circuit** Anodenbasisverstärker *m* ~ **current** Emissions-, Entladungs-, Faden-, Kathoden-, Raumlade-strom *m* ~ **dark space** Kathodendunkelraum *m* ~ **deposit** Kathodenniederschlag *m* ~-**drop lightning arrester** Kathodenfallableiter *m* ~-**drop space** Fallraum *m* ~ **fall** Kathodenfall *m* ~ **fall space** Fallraum *m* ~ **filament** Kathodenfaden *m* ~ **flicker effect** Kathodenfackelerscheinung *f* ~ **follower** Kathoden-verstärker *m*, -folger *m*, -folgestufe *f*, -schaltung *f* ~ **follower output** Kathodenauskopplung *f* ~ **follower type phase splitter** Phasenumkehrschaltung *f* ~ **glow** Kathodenlicht *n*, Kathodenlichthaut *f* ~-**glow lamp** Kathodenglimmlicht *n* ~-**glow tube** Kathodolumineszenzlampe *f* ~ **grid** Brems-, Fang-gitter *n* ~-**heater tube** indirekt geheizte Röhre *f* ~-**heating time** Anheizzeit *f*, Kathodenanheizzeit *f* ~ **hook** Kathodenhaken *m* ~ **hum** Kathodenrauschen *n* ~ **influx** Ionenstrom *m* ~ **jump of potential** Kathodensprung *m* ~ **keying** Ka-

thodentastung *f* ~ **lead** Kathodenzuleitung *f* ~ **lens** Vorsammellinse *f* ~ **line** Kathodenlinie *f* ~ **loop** Kathodenhaken *m* ~ **luminescence** Kathodenlumineszenz *f* ~ **output** Kathoden-ausgang *m*

**cathode-ray** Elektronenbündel *n*; Elektronen-, Kathoden-strahl *m* ~ **beam** Kathodenstrahl-bündel *n* ~ **beam-intensity modulation** Kathodenstrahlintensitätskontrolle *f* ~ **direction finder (CRDF)** Sichtpeilgerät *n* ~ **image** Glüh-fadenbild *n* ~ **oscillograph** Elektronenstrahl-oszillograf *m*, Kathodenstrahl *m*, Kathoden-strahloszillograf *m*, Kathoden-, Kathoden-strahl-oszillograf *m* ~ **pencil** Kathodenstrahl-bündel *n* ~ **tube (CRT)** Braunsche Röhre *f*, Elektronenstrahlröhre *f*, Elektronrichtstrahler *m*, Entfernungsfeinmeßröhre *f*, Kathoden-strahlröhre *f*; optische Aktivität *f*, Wieder-gaberöhre *f*, Strahlquelle *f* (electron micro-scope), Tubus *m* ~ **tube display** Elektronenbild *n* ~ **tuning-indicator tube** Elektronenstrahlröhre *f*

**cathode,** ~ **reactivation (or rejuvenation)** Katho-denneuaktivierung *f* ~ **resistor biasing** Katho-denvorspannung *f* mit Widerstand ~ **screen** Ab-schirmkappe *f* ~ **sleeve** Kathodenröhrchen *n* ~ **space** Kathodenraum *m* ~ **spot** Fokus *m*, Quecksilberkathode *f* ~ **spot of mercury arc** Quecksilberlichtbogenkathodenfleck *m* ~ **sput-tering** Kathodenzerstäubung *f* ~ **surface** Ka-thodenfläche *f* ~ **terminal** Kathodenanschluß *m* ~ **tube** Kathodenröhre *f* ~-**tube concentration coil** Vorsammelspule *f*

**cathodic** kathodisch ~ **area** Stromeintrittszone *f* ~ **bombardment** Elektronenaufprall *m* ~ **dis-charge** kathodische Entladung *f* ~ **disintegra-tion** Kathodenzerstäubung *f* ~ **evaporation** Ka-thodenzerstäubung *f* ~ **flux** Ionenstrom *m*

**cathodophone** Kathodophon *m*, membranloses Mikrofon *n*

**catholyte** Kathodenflüssigkeit *f*, Katholyt *m*

**cation** Kation *n*, positives Ion *n*

**catkin valve** Catkin-röhre *f*

**catline,** ~ **guard** Spielseilschutz *m* ~ **sheave** Spielseilrolle *f*

**catoptric** katoptrisch

**cattle,** ~ **guard** Schutzdraht *m* ~-**stunning appa-ratus** Schlachtviehbetäubungsapparat *m*

**caul** Polierblech *n*

**caulk, to** ~ einstemmen **to** ~ **circularly** ring-verstemmen ~ **welding** Verlöten *n*

**caulket rivet** verstemmtes Niet *n*

**caulking,** ~ **iron** Schöreisen *n* ~ **machine for pipes** Rohrdichtmaschine *f*

**causal** ursächlich ~ **connection nexus** Kausal-nexus *m* (photq)

**causality** Ursächlichkeit *f* ~ **requirements** Kau-salitätsforderungen *pl* ~ **violation** Kausalitäts-verletzung *f*

**causation** Kausalität *f*, Ursächlichkeit *f*

**causative** ursächlich

**cause, to** ~ bewirken, erzeugen, machen, ver-ursachen (damage) anrichten **to** ~ **to chap** auf-schrunden **to** ~ **to crack** aufschrunden **to** ~ **to float in cement** einzementieren **to** ~ **to flow** be-rieseln **to** ~ **to be in resonance with** in Resonanz bringen **to** ~ **to trickel over** berieseln **to** ~ **the roof of a mine to fall in or to fall down** den Bruch niedergehen lassen

**cause, to** ~ **breakdown** stören **to** ~ **damage** Scha-den anrichten **to** ~ **friction** reiben **to** ~ **inter-ference** stören **to** ~ **pain** quälen **to** ~ **uneasiness** beunruhigen **to** ~ **vaporization of getter** Getter *m* abschießen

**cause** Angelegenheit *f*, Grund *m*, Rechtsstreit *m*, Schuld *f*, Ursache *f*, Veranlassung *f* (legal) ~ **Rechtsfall** *m* ~ **of the accident** Unfallursache *f* ~ **of action** Klagegrund *m* **to** ~ **(ultra-) sound waves to act (or impinge) upon** Beschallung *f*

**cause way** Chaussee *f*, Damm-, Kunst-straße *f*

**caustic** Ätzmittel *n*, Beize *f*, Beiz-, Haft-mittel *n*; Kaustik *f* (opt.); ätzend, beißend, beizend, kaustisch, scharf ~ **alkali** Ätzkali *n* ~ **embritt-lement** interkristalline Korrosion *f* ~ **lime** Ätzkalk *m*, ätzender oder gebrannter Kalk *m* ~ **pencil of copper sulfate** Kupferstift *m* ~ **pot** Laugenkessel *m* ~ **potash** Ätzkali *n*, Kalium-hydroxyd *n* ~ **potash (lye) solution** Ätzkali-lauge *f* ~ **powder** Ätzpulver *n* ~ **power** Beiz-kraft *f* ~ **salt** Ätzsalz *n* ~ **soda** Ätznatron *n*, Natriumhydroxyd *n*, Natronlauge *f*, Sodastein *m* ~ **soda solution** Natronätzlauge *f* ~-**(lye) solution** Ätzlauge *f* ~ **stick** Ätzstift *m* ~ **sur-face** brechende Brennfläche *f* (opt.), Brenn-fläche *f* ~ **tip** Kaustikspitze *f* ~ **washing** Lau-genwaschanlage *f*

**causticity** Ätzkraft *f*, Kaustizität *f*

**causticize, to** ~ kaustizieren

**caustics** Kaustikflächen *pl*

**cauterization** Abbeizen *n*, Zerfressung *f*

**cauterize, to** ~ abbeizen, ätzen, ausätzen, bei-zen, einbrennen, zerfressen, zubrennen **to** ~ **slightly (or superficially)** anätzen

**cautery** Brennstift *m*, Kauter *m* ~ **loop (or snare)** Glühschlinge *f*

**caution** Achtung *f*, Androhung *f*, Umsicht *f*, Vorsicht *f*, Warnungstafel *f* ~ **board** War-nungsschild *n* ~ **signal** Vor-, Warn-signal *n*

**cautionary letter** Warnungsschreiben *n*

**cautious** behutsam, vorsichtig

**cave, to** ~ aushöhlen, nachfallen **to** ~ **in** ein-stürzen

**cave** Höhle *f* ~-**in** Einsturz *m* ~ **roof** Höhlen-dach *n*

**cavern** unterirdischer Hohlraum *m*

**cavernous lode** offener Gang *m*

**cavetto** Hohlkehle *f*

**caving** Verschlämmung *f* ~ **coal** Hereinbre-chendes m (min.) ~ **formations** nachfallende Gebirgsschichten *pl* ~ **material** nachfallendes Gut *n* ~ **rock** nachfälliges Gebirge *n* ~ **system** Bruchbau *m*, Nachfall *m*

**cavitation** Blasenbildung *f*, Hohlsog *m* (phys.), Kavitation *f*, (in supersonic waves) Hohl-raumbildung *f*, (in metals) Lunkerung *f*, (of propellers, pumps, etc.) Ablösung *f* ~ **of gases** Gasblasenbildung *f*

**cavities** Bruchfeld *n*, Hohlraumresonatoren *pl*, Löcher *pl*

**cavity** Drusenraum *m*, Groblunker *m*, Grube *f*, Höhle *f*, Hohlheit *f*, Höhlung *f*, Hohlraum *m*, Lunker *m*, Vertiefung *f*, (in a casting) Saug-höhle *f* ~-**coupled filter** hohlraumgekoppeltes Filter *n* ~ **design** Hohlraumresonatoren *pl* ~ **drag** Kavitätswiderstand *m* ~-**frequency me-ter** Hohlraumfrequenzmesser *m* ~ **growth** Hohlraumwachstum *m* ~ **magnetron** Hohlraum-, Rad-, Vielfach-magnetron n ~ **parameters** Hohlraumabmessungen pl ~ **radiation** Hohlraumstrahlung f, Schwarzstrahl m ~ **resonance** Höhlungsresonanz f ~ **resonant point** Hohlraumresonanzpunkt m ~ **resonator** Hohlraumresonator m (electron.), Schwingtopf m ~ **resonator magnetron** Resotank m ~ **temperature** schwarze Temperatur f, Schwarztemperatur f ~ **troide** Scheibenröhre f ~ **wall** Hohl-mauer f, -wand f

**cawk** unreiner Baryt m

**C battery** Gitter-batterie f, -vorspannung f, -widerstand m

**C bias** Gittervorspannung f

**CB/dial line** ZB/W-Leitung f

**CB-line** ZB-Anschluß m

**C clamp** C-förmige Schraubzwinge f

**C-digit selector** dritter Kennzifferwähler m

**C display** C-Schirm m (rdr)

**cease, to** ~ aufhören, abbrechen, eingehen, einstellen, enden, endigen, in Fortfall kommen, in Wegfall kommen **to** ~ **to exist** erlöschen **cease, to** ~ **brightening** abblicken **to** ~ **fire** das Schießen abbrechen, Feuer n einstellen **to** ~ **germinating** auskeimen **to** ~ **glowing** abglühen **to** ~ **steaming** ausdämpfen **to** ~ **work** Arbeit f niederlegen

**cedar** Zeder f ~ **oil** Zedernholzöl n

**cede, to** ~ abtreten, nachgeben, sich geben, übertragen

**ceiling** Bewölkungshöhe f, Decke f, Gipfelhöhe f (aviat.), Hauptwolkenuntergrenze f (meteor.), Wegerung f, Wolkenhöhe f ~ **of caisson working chamber** Decke f der Arbeitskammer ~ **of timbers** Balkendecke f ~ **with bays** Felder-Kassetten-decke f ~ **with one engine out** Gipfelhöhe f bei Ausfall eines Motors ~ **with one motor stopped** Gipfelhöhe f mit einem stehenden Motor

**ceiling,** ~ **altitude** Flughöhe f ~ **balloon** Flughöhenballon m ~ **brush** Deckenbürste f ~ **crab** Deckenlaufkran m ~ **duct** Deckendurchführung f ~ **fan** Deckenfächer m ~ **height** Flughöhe f ~ **-height indicator** Flughöhenanzeiger m ~ **-lamp holder** Deckenfassung f ~ **light** Deckenleuchte f, Flughöhenstrahl m, Wolkenscheinwerfer m ~ **light projector** Wolkenhöhenmesser m ~ **panel heating** Deckenheizung f ~ **plaster(ing)** Deckenputz m ~ **price** Höchstpreis m ~ **projector** Wolkenhöhenmeßscheinwerfer m ~ **reflector** Deckenlichtspiegler m ~ **sink water trap** Deckensinkkasten m ~ **speed** Gipfelgeschwindigkeit f (aviat.) ~ **spreader** Deckenverteiler m ~ **sprinkler** versenkter Sprinkler m ~ **type coil** Deckenkühlsystem n ~ **unlimited** unbegrenzte Wolkenhöhe f ~ **ventilator** Deckenventilator m ~ **void** Deckenhohlraum m ~ **voltage of an exciter** höchste Erregungsspannung f ~ **zero** Wolkenhöhennullpunkt m

**ceilometer** Wolkenhöhenmeßgerät n

**celadonite** Grünerde f

**celanese** Glanzstoff m

**celestial** himmlisch ~ **body** Himmelskörper m ~ **chart** Himmelskarte f ~ **globe** Himmelsglobus m, -wölbung f ~ **navigation** astronomische Navigation oder Ortung f, Himmelsortung f ~ **phenomenon** Himmelserscheinung f ~ **pole** Himmelspol m ~ **sign** Himmelszeichen

n ~ **signs** Himmelssymbole pl ~ **sphere** Himmels-kugel f, -sphäre f

**celite** Celit m

**cell** Bad n, Dose f, Element n (electr.), Flügelzelle f, Gaszelle f, Grundteil m, Hohlraum m, Kette f, Schacht m, Speicherzelle f, Tragwerk n (aviat.), Zelle f, (photo-tube) Küvette f, (photometry) Strahler m (battery) ~ Gegenzelle f (electrolysis) ~ Badkasten m (regulating) ~ Schaltzelle f **to set up a** ~ ein Element n ansetzen ~ **for field telegraph** Feldelement n ~ **with air depolarizer** Element n mit Luft als Depolarisator

**cell,** ~ **activity** Zelltätigkeit f ~ **box** Zellengefäß n ~ **capacitance** Meßzellenkapazität f ~ **carbons (battery rods)** Elementenkohlen pl ~ **casing** Zellenmantel m ~ **cavity** Porenraum m ~ **changer** Probenwechsler m ~**-cluster model** Zellen-Cluster-Modell n ~ **container** Elementbecher m ~ **contents** Zelleninhalt m ~**-faced piston ring** Kolbenring m mit Phosphidnetzwerk auf der tragenden Fläche ~**-group arrangement** Bäderanordnung f (electrolysis) ~ **holder** Kammerhalter m (opt.) ~ **inspection lamp** Untersäurelampe f ~**-less bucket wheel** kammerloses Schaufelrad n ~**-like** zell-ähnlich, zellenartig' ~ **membrane** Zellwand f ~ **pit furnace** Zellentiefofen m ~ **pitch** Zellpech n ~ **potential** Akkumulator-, Bad-spannung f ~ **reverser** Küvettenwechsler m ~ **seat** Küvettenhalter m (Sitz für Küvetten) ~ **space** Porenraum m ~ **stand** Küvettenständer m ~ **structure** Zellenaufbau m ~ **switch** Zellenschalter m ~**-type bucket wheel** Kammerrad n ~ **voltage** Akkumulator-, Bad-, Zellen-spannung f ~ **volume** Gitterzellenvolumen n ~ **wall** Zellenwand f, Zellwand f ~**-wall-destroying** zellwandzerstörend

**cellar** Hauskeller m, Keller m ~ **floor** Kellerfußboden m ~ **sink-water trap** Kellersinkkasten m

**cellaring** Unterkellerung f

**celled** zellig

**cellemite** Zellonemaillit n

**celler, to** ~ einkellern

**celloid** zellähnlich

**celloidine paper** Zelloidinpapier n

**cellone** Zellon n, Zellonlack m

**cellophane** Glashaut f, Zell-glas n, -haut f, Zellophan n, Zellulosefolie f

**cellucotton** Papier-, Zellstoff-watte f

**cellular** zellähnlich, zellenartig, zellförmig, zellig ~ **air filters** Zellenluftfilter pl ~ **arrragement** Zelleneinbau m ~ **block** Zellenblock m ~ **board** Zell-papier n, -pappe f ~ **girder** Fuhrwerkträger m ~ **pasteboard** Holzpappe f ~**-plastic insulation** Schaumstoffisolierung f ~ **pyrites** Zellenkies m ~ **radiator** Zellenkühler m ~**-rubber articles** Zellengummiartikel m ~ **structure** Netz-, Netzwerk-struktur f, zellenartige Struktur f, Zellenstruktur f ~ **substance** Zellmasse f ~ **switchboard** gekapselte Schaltanlage f ~ **tissue** Zellengewebe n ~ **tube** Holzröhre f ~**-type mooring pier** Haltepfahl m ~**-type structure** netzwerkartiges Gefüge n ~ **wheel** Kastenrad n

**cellule** Flügelzelle f, Gaszelle f, Zelle f; Trag-zelle f, -werk n (aviat.)

**celluloid** Kunsthorn *n*, Zellhorn *n*, Zelluloid *n* ~ **disk** Zelluloidtafel *f* ~ **facing** Zelluloidauskleidung *f* ~ **frames** Zellhornbrille *f* ~ **showcard** Zelluloidplakat *n* ~ **template** Klarzelle *f* (photo) ~ **wheel** Zelluloidrad *n*
**cellulose** Holzfaserstoff *m*, Lignose *f*, Zellstoff *m*, Zellulose *f* ~ **from coniferous trees** Nadelholzzellstoff *m*
**cellulose,** ~ **acetate dope, lacquer** Zellonlack *m*, Zelluloseazetat *n* ~ **acetate rayon** Zelluloseazetatseide *f* ~ **asbestos** Zellstoffasbest *m* ~ **derivate** Zellpack *m* ~ **face** Blankseite *f* (film) ~ **factory** Zellulosefabrik *f* ~ **finish** Zelluloselack *m* ~ **lacquer** Zelluloselack *m* ~ **nitrate disk** Lackfolie *f* ~ **nitrate rayon** Zellulosenitratseide *f* ~ **number** Zellulosezahl *f* ~ **paper** Zellulosepapier *n* ~ **paper with satin finish** geglättetes Zellulosepapier *n* ~ **side** Blankseite *f* (film) ~ **solution in cuprammonium hydroxide** Kuoxam *n* ~ **wall** Zellulosewandung *f* ~ **xanthogenate** Zellulosexanthogenat *f*
**cembra pine** Arobe *f*
**cement, to** ~ ausfugen, aufkohlen, backen, binden, einsetzen, einzementieren, fluhen, kitten, kleben, leimen, verkitten, verkleben, verleimen, zementieren **to** ~ **in** (place or position) aufkitten, einkitten **to** ~ **together** zusammenkitten **to** ~ **up** zukitten
**cement** Aufkohlungsmittel *n*, Bindemittel *n*, Bindungsmittel *n*, Glasverkitter *m*, Kitt *m*, Klebemittel *n*, Zement *m*, Zementmörtel *m* (mortar), Zwischenmasse *f* (petrog.), ~ **bag** Zementsack *m* ~ **bar** Zementstahlstab *m* ~ **bunker** Zementsilo *m* ~ **clinker** Zement-klinker *m*, -schlacke *f* ~ **concrete** Zementbeton *m* ~ **consisting of Portland cement and finely ground blastfurnace slag** Hochofenzement *m* ~ **conveyer** Zementförderband *n* ~ **copper** Zementkupfer *n* ~ **duct** Kabelformstück *n* aus Zement ~ **facing** Zementbewurf *m* ~ **factory** Zementbrennerei *f* ~ **flag** Zementfliese *f* ~ **float collar** Zementiermuffenventil *n* ~ **float shoe** Zementierschuh *m* ~ **floor** Zementestrich *m* ~ **foundation** Zementsockel *m* ~ **glue** Zementleim *m* ~ **grout** Zementbrei *m* ~ **iron** Zementeisen *n* ~ **kiln practice** Zementofenpraxis *f* ~ **laying-on machine** Zementauftragemaschine *f* ~**-lime concrete** Zementkalkbeton *m* ~**-lime mortar** Zementkalkmörtel *m* ~**-mill drive** Zementmühlenantrieb *m* ~**-mill plant** Zementmahlanlage *f* ~ **mortar** Wassermörtel *m* ~ **paste** Zementbrei *m* ~ **pat** Zementkuchen *m* ~ **plant** Zementwerk *n* ~ **plaster** Zement-bewurf *m*, -putz *m*, -verputz *m* ~ **production** Zementdarstellung *f* ~ **rendering** Zementverputz *m* ~ **retainer** Zementausbringer *f*, Zementierpacker *m* ~ **scales** Zementwaage *f* ~ **shed** Zementschuppen *m* ~ **silver** Zementsilber *n* ~ **steel** Blasenstahl *m* ~ **stone** Kalksteinniere *f* ~ **stone quarry** Zementmergelgrube *f* ~ **testing** Zementprüfung *f* ~ **tile** Zementfliese *f* ~**-trass concrete** Zementtraßbeton *m* ~ **work** Zementieren *n* ~**-works installation** Zementwerksanlage *f* ~**-works outfit** Zementfabrikeinrichtung *f*
**cementation** Aufkohlung *f*, Einsatzhärten *n*, Oberflächenkohlung *f*, Zementation *f*, Zementdarstellung *f*, Zementierung *f*, Zementierverfahren *n* ~ **carbon** Zementkohle *f* ~ **furnace**

Einsatz-, Zementier-ofen *m* ~ **process** Zementation *f*, Zementierverfahren *n*
**cemented** zementiert ~ **armor** Eisenpanzer *m* ~ **bifocal** Aufkittlinse *f* ~ **carbide** Hartmetall *n*, Sinterkarbid *n* ~ **carbide product** gesintertes Hartmetall *n* ~**-carbide tip** Hartmetallplättchen *n* ~ **carbide (cutting) tool** Hartmetallwerkzeug *n* ~ **hard carbide** Hartmetall *n* ~ **lens** eingekittete, gekittete oder verkittete Linse *f* ~ **metal carbide** gesintertes Metallkarbid *n* ~ **steel** Zementstahl *m* ~ **zone** Einsatz *m*
**cementing** Dichtung *f*, Verkittung *f*, Zementierung *f*, Zementieren *n*; backend ~ **by jet** Zementieren *n* im Spritzverfahren
**cementing,** ~ **agent** Bindemittel *n*, Dichtungsmittel *n*, Trägermetall *n*, Zementiermittel *n* ~ **apparatus** Farbenspritzapparat *m* ~ **box** Glühtopf *m*, Zementier-kasten *m*, -kiste *f* ~ **carbon** Härtungskohle *f* ~ **head** Zementkopf *m* ~ **job** Zementierverfahren *n* ~ **lens** Aufkittlinse *f* ~ **material** Bindemittel *n*, Dichtungsmaterial *n*, Dichtungsstoff *m*, Trägermetall *n*, (for cementation of iron) Glühmittel *n* ~ **medium** Dichtungs-, Kohlungs-mittel *n* ~ **medium cement** Einsatzmittel *n* ~ **pot** Einsatztopf *m* ~ **powder** Härte-, Zement-pulver *n* ~ **process** Zementierverfahren *n* ~ **property** Backfähigkeit *f*
**cementite, to** ~ aufkohlen
**cementite** Austenit *n*, Eisenkarbid *n*, Härtungskohle *f*, Zementit *m* ~ **carbon** härtender Kohlenstoff *m* ~ **disintegration** Zementitzerfall *m* ~ **spheroidizing** Zementierzusammenballung *f*
**cementlike** zementartig
**census** Volkszählung *f*
**center, to** ~ ankernen, ankörnen, auf die Mitte einstellen, auslasten, auswichten, einmitten, mitten, mittig einstellen, zentrieren **to** ~ **punch** ankörnen
**center** Achse *f*, Bogenlehre *f*, Einlage *f*, Knotenpunkt *m*, Mitte *f*, Mittelpunkt *m*, Zentrale *f*, Zentrum *n* ~(**ing**) Lehrbogen *m* ~ **to** ~ von Achse zu Achse **to bring to the** ~ **of its run** einspielen
**center,** ~ **of action of the atmosphere** Aktionszentrum *n* in der Atmosphäre ~ **of anticyclone** Hochdruckzentrum *n* ~ **of area** Flächenmittelpunkt *m* ~ **of attack** Angriffsmittelpunkt *m* ~ **of bar** Stabkern *m* ~ **of bore** Bohrlochachse *f* ~ **of borehole** Achsmitte *f* des Bohrloches ~ **of buoyancy** Auftriebs-mittelpunkt *m*, -schwerpunkt *m*, Deplacementschwerpunkt *m*, Formschwerpunkt *m* (aviat.) ~ **of colineation** Sammelpunkt *m* ~ **of crystallization** Kristallisations-kern *m*, -zentrum *n* ~ **of curvature** Krümmungsmittelpunkt *m* ~ **of cyclone** Tiefdruckzentrum *n* ~ **of the depression** Depressionszentrum *n* ~ **of dispersion** mittlerer Treffpunkt *m* ~ **of displacement** Auftriebs-, Verdrängungs- -mittelpunkt *m* ~ **of disturbance** Störstelle *f* ~ **of dividing head** Teilkopfmitte *f* ~ **of the earth** Erdmittelpunkt *m* ~ **of figure** Körpermittelpunkt *m* ~ **of the front** Frontmitte *f* ~ **of gravity** Baryzentrum *n*, Gleichgewichtspunkt *m*, Schwerkraftzentrum *n*, Schwerpunkt *m* **(position of or at)** ~ **of gravity** Schwerpunktlage *f* ~ **of gravity of airplane** Flugzeugschwerpunkt *m* ~ **of gravity of a surface** Flächen-

schwerpunkt *m* ~-of-gravity system (or -of-mass system) Schwerpunktsystem *n* ~ of growth Wachstumszentrum *n* ~ of gyration Drehpunkt *m* ~ of high barometric pressure Hochdruckzentrum *n* ~ of the hurricane Orkanmitte *f* ~ of image Abbildungs-, Bild-mittelpunkt *m* ~ of impact mittlerer Treffpunkt *m* ~ of indentation Eindruckmitte *f* ~ of inertia Beharrungs-, Massen-punkt *m*, Massenzentum *n*, Trägheitsmittelpunkt *m* ~ of key Mittelschiene *f* der Taste ~ of the lens Objektivmitte *f* ~ of lift Auftriebsmittelpunkt *m* ~ of low Tiefdruckzentrum *n* ~ of low barometric pressure Tiefdruckzentrum *n* ~ of the low Depressionszentrum *n* ~ of mass Beharrungs-, Massen--punkt *m*, Massenzentrum *n*, Massemittelpunkt *m*, Schwerpunkt *m* ~ of mass law Schwerpunktssatz *m* (position of or at) ~ of mass Schwerpunktlage *f* ~ of mass system Schwerpunkt-Koordinatensystem *n* ~ of motion Drehpunkt ~ of optical image (or picture or plate) optischer Bildmittelpunkt *n* ~ of origin Entstehungszentrum *n* ~ of oscillation Schwingungs-mittelpunkt *m*, -zentrum *n* ~ of percussion Stoß-mittelpunkt *m*, -punkt *m*, -zentrum *n* ~ of picture Bildmittelpunkt *m* ~ of plate Bildmittelpunkt *m* ~ of pressure Angriffs-, Druck-, Widerstands-mittelpunkt *m* ~ of (aerodynamic) pressure Luftkraftangriffspunkt *m* ~ of pressure of fixed tail surface Flossendruckpunkt *m* ~-of-pressure movement Druckpunktverlegung *f* ~ of projection Projektionszentrum *n* ~ of resistance Widerstandszentrum *n* ~ of revolution Drehpunkt *m* ~ of rotation Dreh-achse *f*, -pol *m*, -punkt *m*, krystallografischer Drehpunkt *m* ~ of sintered cylinder Spaltstoffzentrum *n* ~ of span Feldmitte *f* ~ of suspension Schwingungszentrum *n* ~ of target pip Dunkelpunkt *m* ~ of a telegraph system Drehkreuzachse *f* ~ of thrust Antriebs-mittelpunkt *m*, -zentrum *n*, Druckzentrum *n*, Vortriebsmittelpunkt *m* ~ of the thunderstorm Gewitterherd *m* ~ of the tire contact Radauflagepunkt *m* ~ of tire impact Auflagepunkt *m*, Radauflegepunkt *m* ~ of traffic Verkehrsknotenpunkt *m* ~ of tropical cyclone Mitte *f* des Sturmes
center, ~ adapter sleeve Hülse *f* für Körnerspitze ~ angle Spitzenwinkel *m* ~ bar Mittel--leiter *m*, -schenkel *m* ~-base impedance Fußpunktwiderstand *m* ~ beam Mittelträger *m* ~ bearing Mittellager *n* ~ bit Anbohrer *m*, Löffelbohrer *m*, Mittelstück *n*, Zentrumbohrer *m* ~ board Mittelschwert *n* ~ bore Zentrierbohrung *f* ~ broad Zentrumschärfe *f* ~ cable filler Beilaufzwickel *m* ~ carrier mittlerer Bremsbelagträger *m* ~ casting crane Gießgrubenkran *m* ~ clearance Spitzenweite *f* ~ cock bit Zapfenbohrer *m* ~-coupled loop zentrale Kopplungsschleife *f* ~ cradle Spitzenbock *m* ~ cradle circle Spitzenbockteilkreis *m* ~-disc roll Scheibenwalze *f* ~ distance Achsabstand *m*, Achsen-abstand *m*, -entfernung *f*; Mittenentfernung *f*, Spitzenentfernung *f* ~ drill Anbohrer *m*, Zentrierbohrer *m* ~ drilling Längsbohren *n* ~ driver Mitnahmekörner *m* ~ driver attachment Spitzenmitnehmereinrichtung *f* ~-driving-shaft gear Mittelwellenantriebsrad *n*

~ electrode Mittelelektrode *f*, Zündstift *m* (spark plug) ~ extend Aufweitung *f* der Bildmitte ~ frame Mittelteil *m* der Zelle ~ frequency Mitten-, Ruheträger-frequenz *f* ~ grinder Spitzenschleifapparat *m* ~ head Zentrierwinkel *m* ~ height Spitzenhöhe *f* ~ hole Anbohrung *f* ~ housing Mittelgehäuse *n* ~ iron Zentriereisen *n*, Schwengellager *n* ~ keelson Mittelkielschwein *n* ~ key Austreiber *m* ~ lathe Spitzendrehbank *f* ~ leg Mittelschenkel *m* ~ length Mittenmaß *n* ~ limb Mittelschenkel *m* center-line Mitte *f*, Mittellinie *f*, neutrale Linie *f*, Skelettlinie *f* (Profil) ~ of excavator Baggerachse *f* ~ of the frame Rahmensymmetrieachse *f* ~ of groove neutrale Kaliberlinie *f* ~ of rivet holes Wurzellinie *f* ~ of spindle Spindelmitte *f*
center-line, ~ lights Mittellinienfeuer *n* (Flugplatz) ~ marking Achsenmarke *f* ~ movement Mittenachsenbewegung *f* ~ tip tank Flügelspitzentank *m* ~ wire Mittelfaden *m*
center, ~ loaded antenna mittenbelastete Antenne *f* ~ mark Ankörnung *f*, Körner *m*, Körnermarke *f* ~ marker Markierdorn *m* ~ offset bezogene Außermittigkeit *f*, Mittenversetzung *f* ~ piece Mittelstück *n* ~ pin (of compass) Pinne *f* ~ pin of compass Kompaßspinne *f* ~ pin socket Drehpfanne *f* ~-pin support Pinnenträger *m*, Trägerstift *m* ~ plane section of wing Flügelmittelteil *m* ~ plate Drehgestellzapfenlager *n*, Mittelplatte *f* ~ plate for bogie pin Drehzapfenlager *n* ~ plug Keiltreiber *m* ~ point Mitte *f*, Zentrierspitze *f* ~ pole Deichsel *f* (wagon) ~-pole prop Wagenstütze *f* ~ portion of ingot Blockmitte *f* ~ punch Ankörner *m*, Körner *m*, Locheisen *n*, Mittelpunktsucher *m* ~ punch mark Körnerschlag *m* ~-punching apparatus Mittelpunktsucher *m* ~ ram Mittelstempel *m* ~ rap Mittelanzapfung *f* ~ ray Mittelstrahl *m* ~ reamer Spitzensenker *m* ~ repeat Mittelrapport *m* ~ rest Brille *f* oder Lünette *f* ~ rests Spitzenböcke *pl* ~ ring Zentrierring *m* ~ roller Mittelwalze *f* ~ section Mittelflügel *m*, (of wing) Baldachin *m* ~ section strut Baldachin-holm *m*, -strebe *f*; Spannturmstrebe *f* ~ selvedge Mittelleiste *f* ~ shield Mittelschild *n* ~ sill Zentralträger *m* (r. r.) ~ size (or dimension) Mittenmaß *n* ~ square Zentrierwinkel *m* ~ steering column Standrohr *n* (Auto) ~ stop Mittenanschlag *m* ~ stub Mittelzapfen *m*, waagerechter Mittelstab *m* ~ support Königstuhl *m* ~ tap Mittelabgriff *m* ~ tap connection Mittelanzapfung *f* ~-tap key modulator circuit Tastschaltung *f* ~-tap key modulator scheme Mittelpunkttastschaltung *f* ~-tapped winding (with mid-point tap) in der Mitte angezapfte Wicklung *f* ~ tapping Mittelanzapfung *f* ~ tapping point Mitte *f* der Differentialspule, mittlere Anzapfstelle *f* ~ through plate Mittelkielplatte *f* ~-to-center distance Achsen-, Mitten-abstand *m* ~-to-center spacing Schwerpunktabstand *m* ~ tooth Mittelzinken *m* ~ triangulation Mittelpunkttriangulation *f* ~ tubular chassis Mittelrohrrahmen *m* ~ web Seele *f* ~-web wheel Scheibenrad *n* ~-web-wheel rolling mill Scheibenradwalzwerk *n* ~ wheel (watchmaking) großes Bodenrad *n* ~ wire Seelendraht *m* ~ zero mittlerer Nullpunkt *m* ~-zero

**instrument** Instrument *n* mit Nullpunkt in der Mitte
**centered** gemittet, mittenrichtig, zentriert, zentrisch eingestellt ~ **face** Flächenzentrierung *f*
**centering** Auswichten *n*, Einmitten *n*, Lehrbogen *m*, Lehrbrett *n*, Lehre *f*, (of arch) Lehrgerüst *n*, Mittung *f*, Zentrieren *n*, Zentrierung *f* ~ **of the bubble of water level** Einspielen(lassen) *n* der Libellenblase ~ **the cutter lip by grinding** Mittigschleifen *n* der Brustfläche
**centering,** ~ **apparatus** Ankerapparat *m* ~ **attachment for circular nibbling works** Zentriervorrichtung *f* für Rundschnitte ~ **chuck** Zentrierfutter *n* ~ **cone** Klem-, Zentrier-kegel *m* ~ **control** Bild-einstellung *f*, -nachstellung *f* (TV) ~ **device** Zentriervorrichtung *f* ~ **diaphragm** Zentrierblende *f* ~ **disk** Mittungsschneide *f* ~ **error** Einspielfehler *m* ~ **lens with ruled cross** Zentrierglas *n* mit Strichkreuz ~ **machine** Zentrierapparat *m* ~ **pin** Mittungs-, Zentrier--stift *m* ~ **rim** (loudspeaker) Zentrierrand *m* ~ **ring** Zentrierring *m* ~ **rod** starres Lot *n* ~ **screw** Mittungsstift *m*, Zentrierschraube *f* ~ **screw chuck** Zentrierschraubenfutter *m* ~ **shoulder** Zentrierrand *m* ~ **spigot** Zentriereinpaß *m* ~ **spring** Zentrierfeder *f* ~ **tapers for inside diameters** Innenzentrierkegel *pl* ~ **telescope** Zentrierfernrohr *n* ~ **tool** Zentrierstahl *m* ~ **tulip** Zentriertulpe *f*
**centerless,** ~ (**circular**) **grinder** spitzenlose Rundschleifmaschine *f* ~ **grinding machine** spitzenlose Schleifmaschine *f* ~ **thread grinding** spitzenloses Gewindeschleifen *n*
**centesimal** zentesimal ~ **balance** Zentesimalwaage *f* ~ **circle graduation** zentesimale Kreisteilung *f* ~ **graduation** Neugradteilung *f* ~ **weighing machine** Zentesimalwaage *f*
**centigrade** (scale) Celsiusskala *f*; hundertgradig
**centigram** Zentigramm *n*
**centiliter** Zentiliter *m*
**centimeter** Zentimeter *m* ~ **cube** Zentimeterwürfel *m* ~ **division** Zentimeterfelderteilung *f* **25-~ lathe** Drehbank von 25 cm Spitzenhöhe ~ **wave** Zwerg-, Zentimeter-welle *f* ~ **wave length** Zentimeterwelle *f*
**centinormal** hundertstelnormal
**central** mittig, zentral, zentrisch, ~ **adjusting** Zentralanstellung *f* ~ **angle** Winkelbogen *m* ~ **angular beasing** Zentralspurlager *n* ~ **barrier** Kernpotentialberg *m* ~ **battery** gemeinsame Batterie *f* ~**-battery signaling** selbsttätige Schlußzeichengebung *f* ~**-battery signaling exchange** Handamt *n* mit selbsttätiger Schlußzeichengebung ~ **beacon** Zentralbake *f* ~ **bit** Kolbenmeißel *m* ~ **blower** gemeinsames Gebläse *n* ~ **body** Mittelstück *n* ~ **brick** Mittelstein *m* ~ **buffer-coupler** Zentralpuffer *m* (r. r.) ~ **capacitor method** Zentralkompensation *f* (power-factor correction) ~ **channel** (of a charcoal pile) Quandelschacht *m* ~ **clock station** Uhrenzentrale *f* ~ **committee** Hauptausschuß *m* ~ **compartment** Geräteraum *m* ~**-compartment shell** Geräteraumgerippe *n* ~ **computer** eigentliche Rechenmaschine *f* ~ **condensation plant** Zentralkondensationsanlage *f* ~ **conductor** Mittelleiter *m* ~ **contact rail** Stromschiene *f* in Gleismitte ~**-contact trough**

Mittelkontaktgefäß *n* ~ **definition** Mittenschärfe *f* ~ **directionfinder station** Peil-hauptstelle *f*, -leitstelle *f* ~ **engine** Mittelmotor *m* ~ **engine nacelle** Zentralmotorgondel *f* ~ **equation** Zentralgleichung *f* ~ **exchange** Endamt *n*, Endanstalt *f* (teleph.) ~ **feed hole of the perforated tape** Führungsloch *n* in der Mitte des Stanzstreifens ~ **field** (of vision) Mittelfeld *n* ~ **fixed shaft** Zentralwelle *f* ~ **float** Mittelschwimmer *m* ~**-float plane** Mittelschwimmerflugzeug *n* ~ **focusing** Mitteltrieb *m* ~ **gap** Leerstelle *f*, zentrales Loch *n* ~ **gas offtake** Zentralabzugsrohr *n* ~ **gas-producter plant** Zentralgeneratoranlage *f* ~ **gate** Eingußrohr *n* ~ **girder** Mittelbalken *m* ~ **gravity** Schwerkraftzentrum *n* ~ **guide** Zentralführung *f* ~ **handling line** Reglerleine *f* ~ **heating** Sammelheizung *f*, Zentralheizung *f* ~**-heating fittings** Zentralheizungsarmaturen *pl* ~**-heating from piece** Zentralheizungsformstück *n* ~**-heating plant** Zentralheizungsanlage *f* ~ **hub bore** Nabenkegelbohrung *f* ~ **hub unit** Anlage *f* in kreisförmiger Bauart ~ **indication** Zentralstrahlanzeige *f* ~ **lubrication** Zentralschmierung *f* ~ **monitoring position** Überwachungszentrale *f* ~ **office** Amt *n*, Fernsprechamt *n*, Fernsprechvermittlung *f*, Schaltanlage *f*, Vermittlungs--amt *n*, -stelle *f*, Zentralabteilung *f*, Zentrale *f* ~**-office maintenance man** Amtsmechaniker *m* ~ **office of measurement** Meßzentrale *f* ~**-office trunk (P.A.B.X.)** Amtsleitung *f* ~ **part of the fuselage** Rumpfmittelstück *n* ~ **perspective** Zentralperspektive *f* ~ **pier** Mittelpfeiler *m* ~ **pin wheel** zentrales Stiftrad *n* (film) ~ **pipe** Trichterrohr *n* (founding) ~ **pivot** Königszapfen *m* ~ **plane of the lens** Mittelebene *f* des Objektivs ~ **plant** Zentralanlage *f* ~ **point** Mittelpunkt *m*, Vermittlung *f* (Rohrpost) ~ **portion** Blockkern *m* ~ **position** Mittellage *f* ~ **power take-off** zentraler Außenabtrieb *m* ~ **pressure** Mittendruck *m* ~ **projection** Zentralprojektion *f* ~ **quad** Kernplattenvierer *m* ~ **quad of cable** Kernvierer *m* ~ **radio office** Betriebszentrale *f* ~ **rail** Mittelschiene *f* ~ **ray** Zentralstrahl *m* ~ **row** Mittelreihe *f* ~ **runner** Mittelrenner *m*, Trichterrohr *n* ~ **screw** Zentralspindel *f* ~ **shutter** Zentralverschluß *m* ~ **skid** Zentralkufe *f* ~ **spindle** Zentralspindel *f* ~ **spindle-type grinding mill** Pendelmühle *f* ~ **stake** Quandelschacht *m* ~ **station** Zentral--anlage *f*, -station *f* ~ (**or main**) **steam power plant** Dampfkraftzentrale *f* ~ **stem** Zentralspindel *f* ~ **supercharger** gemeinsames Gebläse *f* ~ **tapping** Mittenanzapfung *f* (electr.) ~ **tightening spindle** Zentralspindelanpressung *f* ~ **traffic office** Betriebszentrale *f* ~ **tube (or shell)** Kammerhülse *f* ~**-tube charge** Kammerhülsenladung *f* ~ **unit** Zentralgerät *n* ~ (**hot-**) **water supply** Zentralwasserversorgung *f* ~ **waterworks** Zentralwasserstation *f* ~ **web** Mittelrippe *f* ~ **wheel** Mittelrad *n* ~ **wiring** Amtverkabelung *f*
**centralization** Zentralisierung *f*, Zusammen-fassung *f*, -legung *f*
**centralize, to** ~ in einem Mittelpunkt *m* vereinigen, zentralisieren, zusammenlegen **to** ~ **in one place** an einem Punkt *m* vereinigen
**centralized,** ~ **control** Zentralbedienung *f* ~

lubrication Sammelschmierung f ~ preparation plant Zentralaufbereitungsanlage f ~ ventilation Zentralwetterführung f
centrally regulated zentralsteuerbar
centraradian (one hundredth radian) Centraradian n
centrate, to ~ in die Mitte stellen (einstellen), auf Mitte
centre see center
centrifiner Stockaufschläger m
centrifugal Schleuder f, Schleuderguß m (casting); vom Mittelpunkt fortstrebend, zentrifugal ~ with nonperforated basket sieblose Schleuder f ~ with safety closing device Schleuder f mit Sicherheitsverschluß ~ with steaming apparatus Schleuder f mit Dampfdeckvorrichtung ~ with top discharge Schleuder f mit oberer Entleerung
centrifugal, ~ acceleration Flieh-beschleunigung f, -kraftbeschleunigung f ~ action Schleudern n, Zentrifugalwirkung f ~ air pump Schleudergebläse n ~ arming device (or fuse) Fliehbacke f ~ beater Zentrifugalstoffmühle f ~ belt Schleuderband n ~ (air) blower Schleudergebläse n ~ blower Umlaufgebläse n, Zentrifugalgebläse n ~ brake Geschwindigkeitsbremse f ~ cast Schleudergußstück n, Zentrifugal-gießen n, -guß m ~ casting method Zentrifugalgießverfahren n ~ casting process Schleuderguß-, Zentrifugalgieß-verfahren n ~ chamber Wirbelkammer f ~ cleaner Schleuderwascher m ~ clutch Fliehgewichtskuppelung f, Fliehkraftkuppelung f ~ compressor Kreiselverdichter m, Schleuder-kompressor m, -lader m, -verdichter m ~ concentrator Schnellaufverdichter m ~ condense pump Kondensatkreiselpumpe f ~ contactor Fliehkraftschalter m ~ coupling Zentrifugalkuppelung f ~ dressing machine Zentrifugalsichtmaschine f ~ drum Schleudertrommel f ~ exhauster Zentrifugalexhaustor m ~ fan Schleuderlüfter m ~ flier Schleuderflügel m ~ flyball Fliehkraftregler m ~ force Abstrebekraft f, Entfernungskraft f, Flieh-gewicht n, -kraft f, Schleuderkraft f, Schwungkraft f, Zentrifugalkraft f ~ fuse Fliehkraftzünder m ~ governor Fliehkraftregler m, Zentrifugal-regler m, -regulator m ~ grinder Pendelmühle f ~ impeller Kreiselpumpe f, Kreiselverdichter m, Zentrifugalstufe f des Verdichters ~ ingot mold Schleudergußform f ~ load test Schleuderprüfung f ~ lubrication Schleuderschmierung f ~ machine Schleudermaschine f ~-machine bottom Zentrifugenboden m ~ mill Schleudermühle f ~ moment of the plane of buoyancy Zentrifugalmoment n der Schwimmebene ~ motion Zentrifugalbewegung f ~ oil extractor Spänezentrifuge f ~ oil foam remover Entschäumungsschleuder f ~-pendulum-type tachometer Fliehpendeldrehzahlmesser m ~ power Schwungkraft f ~ pump Kreisel-, Schleuder-, Zentrifugal-pumpe f ~ regulator Schwungkugelregler m ~ rotor Kreiselladerrad n ~ running Schleuderablauf m ~ screen Schleudersieb n ~ separator Schleuderapparat m, Zentrifugalscheider m ~ sifter Zentrifugalsichter m ~ sirup Ablaufsirup m ~ speed Schleuderdrehzahl f ~ stand Schleuderstand m ~ strainer Zentrifugalsor-

tierer m ~ supercharger Kreisel-gebläse n, -lader m, -vorverdichter m, Schnellaufverdichter m ~ switch Fliehkraftschalter m ~ (automatic) time control Fliehkraftverstellung f ~ timer (or governor) Fliehkraftversteller m ~-type speedometer Fliehkraftgeschwindigkeitsmesser m ~-type supercharger Fliehkraftgebläse n, Schleuderlader m ~ washer Zentrifugalwascher m ~ water pump Kreiselwasserpumpe f ~ wedge block Fliehkraftbacke f ~ weight Fliehgewicht n ~ weight carrier Fliehgewichtsträger m ~ weight deflection (or deviation) Fliehgewichtsausschlag m ~-weight-drive type of fuse mechanism Fliehgewichtsantrieb m ~ whirler Schwungmaschine f
centrifugally, ~ cast (iron) Schleuderguß m ~ cast pipe Schleudergußrohr n
centrifuge, to ~ abschleudern, ausschleudern, schleudern, zentrifugieren
centrifuge Ausschleuder-, Ausschwingmaschine f, Schleuder f, Zentrifuge f ~ for washing sugar Deckschleuder m
centrifuging Schleuderung f
centripetal zentripetal, zum Mittelpunkt m hinstrebend ~ force Anstrebe-, Zentripetal-, Zustrebe-kraft f ~ motion Zentripetalbewegung f
centrobaric schwerzentrisch
centrode Walzbahn f
centroid Beharrungs-, Flächenschwer-punkt m, Flächenschwerpunkt m eines Querschnittes, Massen-punkt m, -zentrum n ~ axis Schwerachse f ~ diagram Energiestufendiagramm n
centroidal axis Schwerlinie f
centrosymmetrical zentralsymmetrisch, mittensymmetrisch
cephalon Kopfschild n
ceramet Keramik-Metallgemisch n
ceramic keramisch ~ cooling coil Tonkühlschlange f ~ disk Keramikplatte f ~ hood-shaped fixed capacitor condenser Hütchenkondensator m ~ insulation keramische Isolation f ~ insulator of plug keramischer Isolator m der Zündkerze ~ insulator of spark plug keramischer Steinisolator m der Zündkerze ~ junction capacitor keramischer Sperrschichtkondensator m ~ liner keramische Feuerhaut f ~ plug Steinkerze f ~ tools Keramikwerkzeuge pl ~ transfer picture keramisches Abziehbild n
ceramics Keramik f
cerargyrite Hornerz n, Kerat n, Silberhornerz n, Silberspat m
cerate Wachspflaster n ~ paste Wachspaste f
ceraunograph Ceraunograf m
ceresin Zeresin n
ceric zerig ~ acid Wachssäure f ~ chloride Cerichlorid n ~ compound Ceritverbindung f ~ oxide Cerioxyd n
cerine Zerin m
cerite Cererz n, Cerinstein m ~ earth Zeriterde f
cerium Cer n, Cermetall n, Zer n, Zermetall n ~ carbonate kohlensau(e)res Ceroxydul n ~ content Cergehalt m ~ earth Cererde f ~ fluoride Cerfluorid n ~ metal Cermetall n ~ ore Cererz n ~ oxide Ceroxyd n ~ protoxide Ceroxydul n
cerotic acid Cerotinsäure f
cerous compund Ceroverbindung f ~ salt Cerosalz n

certificate Attest *n*, Ausweis *m*, Diplom *n*,
Gutachten *n*, Gutschrift *f*, Protokoll *n*, Schein
*m*, Zeugnis *n*
certificate, ~ of accuracy Prüfungsattest *n* ~ of
airworthiness Lufttüchtigkeitszeugnis *n* ~ of
appointment Bestellungsurkunde *f* ~ of competency Befähigungs-zeugnis *n*, -nachweis *m*
~ of conduct Führungszeugnis *n* ~ of correction Eichtabelle *f* ~ of navigation Navigations-
zeugnis *n* ~ of origin Ursprungszeugnis *n* ~ of
performance Leistungsabzeichen *n* ~ of permission to work a mine Mutungsschein *m* ~ of
quality Gütezeugnis *n* ~ of registration Ein-
tragungsurkunde *f* ~ of registry Beilbrief *m*,
Eintragungsbescheinigung *f* ~ of renewal Er-
neuerungsschein *m* ~ of service Dienstbeschei-
nigung *f* ~ of tonnage Meßbrief *m* ~ of warranty
Garantieschein *m*
certificate share Anteilschein *m*
certification Beglaubigung *f*
certified eingetragen ~ copy beglaubigte Abschrift *f* ~ copy of application Prioritätsbeleg *m*
~ engineer Diplomingénieur *m* ~ statement beglaubigte Angabe *f*
certify, to ~ beglaubigen, bescheinigen, bezeu-
gen to ~ as airworthy die Lufttüchtigkeit bescheinigen
cerussite Bleikarbonat *n*, Bleispat *m*, Cerussit *m*,
Kohlenbleispat *m*, Weißblei *n*
cesiated bedeckt mit Zäsium *n*
cesium Zäsium *n* ~ alum Zäsiumalaun *m* ~ cell
Zäsiumzelle *f* ~ photoelectric cell Zäsiumfoto-
zelle *f*
cessation Aufhören *n*, Aussetzen *n*, Einstellung
*f*, Stillstand *m*, Unterbrechung *f*
cessing Inseln *pl* im Film
cession Abtretung *f*
cess pipe Abtrittschlot *m*
cesspool Abortgrube *f*, Abzugskanal *m*, Abzugsschleuse *f*, Sammelloch *n*, Senkgrube *f* ~ of a
pit Schachtsumpf *m* ~ system Abfuhrsystem *n*
cestral-tube Kegelstrahlröhre *f* (Kestral-Röhre)
cetane, ~ number Cetan-wert *m*, -zahl *f* ~ number rating Cetanzahl *f*
cetene Cetylen *n* ~ number Cetenzahl *f*
cetyl Cetyl *n*
cetyl acid Cetylsäure *f*
Ceylon (graphite) plumbago Ceylongraphit *m*
ceylonite Zeylonit *m*
C-frame Sondenbügel *m*
chad tape Stanzstreifen *m*
chadless, ~ perforation Schuppenbildung *f* ~ tape
Prägestreifen *m*, teilgelochtes Band *n*
chads Schnitzel *n*
chafe, to ~ abscheuern, (through) durchscheu-
ern, frottieren, schaben, scheuern, sich reiben,
wund reiben
chafe, ~ marks Scheuerstellen *pl* ~ rod Scheuer-
-bock *m*, -pfahl *m*
chafed wund
chafer Käfer *m*, Wulstschutzstreifen *m*
chaff Düppel *m* (rdr), Häcksel *m* ~ burst Düppelstoß *m* ~ cloud Düppelstraße *f* ~ cutter Fut-
terlade *f*, Häckselbank *f*, Häcksellade *f* ~ dispenser Düppelabwurfvorrichtung *f* ~ dropping
Düppelung *f* ~ element Einzeldüppel *m* ~
stream Düppelstrom *m*
chaffer Wärmröhre *f*

chafing Schamfielen *n*, Scheuern *n* ~ mark
Schabstelle *f* ~ patch Scheuerpflaster *n*,
Schutzstreifen *m* ~ resistance Reibfestigkeit *f*
chain, to ~ fesseln, ketten, mit einer Kette messen to ~-dot strichpunktieren
chain Gliederkette *f*, Kette *f* ~ (system) Kettenleiter *m* ~ of bubbles (filtration) Blasenkette *f*
~ of contacts Kontaktkette *f* ~ for counter-
weighing the crossrail Kette *f* zum Ausbalan-
zieren des Querbalkens ~ of egg insulators
Eierkette *f* ~ of fortifications Festungsreihe *f*
~ of hills Höhenzug *m*, Hügelkette *f* ~ of
triangulation Dreieckskette *f* ~ to prevent expansion of store casks Faßspannkette *f* ~ with
spike hooks Kette *f* mit Einschlagkeilen
chain, ~ adjusterKetteneinsteller *m* ~-adjusting
screw Kettenspannschraube *f* ~ agents Ban-
denträger *m* ~ attachment link Kettenbefesti-
gungsglied *n* ~ belt endlose Kette *f*, Glieder-
keilriemen *m* ~ block Kettenflaschenzug *m*
~ blocks spurgeared Differentialzeuge *pl* mit
gradliniger Zahnung ~ branching Kettenver-
zweigung *f* ~-branching probability Kettenver-
zweigungswahrscheinlichkeit *f* ~ breaking Ket-
tenabbruch *m* ~ (suspension) bridge Ketten-
brücke *f* ~ broadcast station Gemeinschafts-
sender *m* ~ bushing Kettenhülse *f* ~ cable Glie-
derkabel *n*, Kabelkette *f* ~ carrier roller Mit-
nehmerwalze *f* ~ case Kettenkasten *m* ~ (relay)
circuit Kettenschaltung *f* ~ coal cutter Ketten-
schrämmaschine *f* ~ control Kettensteuerung *f*
~ conveyer Kettenförderer *m*, Kratzwinde *f*
~ coupler Kettenendbolzen *m* ~ coupler link
Kettenverbindungsglied *n* ~ cutter (endless
band of cutting teeth) Fräserkette *f* ~ decay
Kettenumwandlung *f* ~ delivery Wendbogen-
einrichtung *f* ~ discharge Kettenabnahme *f*
~ disintegration Kettenzerfall *m* ~-dotted line
strichpunktierte Linie *f* ~ dredger Exkavator *m*
~ drive Ketten-antrieb *m*, -getriebe *n*, -radan-
trieb *m*, -trieb *m*, Laufkatze *f* ~-drive propelling
mechanism Fahrwerkkettentrieb *m* ~ drum
Ketten-stern *m*, -trommel *f* ~ dummy coupling
Kettenleerkuppelung *f* ~ elevator Kettenbecher-
werk *n* ~ feeder Kettenbeschicker *m* ~ fission
yield Spaltproduktausbeute *f* ~ follower Kettenzieher *m* ~ friction Kettenreibung *f* ~-gated
analyzer Brückendetektor *m* ~ gearing Ketten-
übersetzung *f* ~ grapplers Kettenanker *m* ~
grate Kettenrost *m* ~ grate firing Kettenrost-
feuerung *f* ~-grate stoker feed Wanderrostbeschickung *f* ~-grate stokers Gitterfeuerung *f*
~ gripper bar Greiferbrücke *f* ~ gripper delivery Kettengreiferauslage *f* ~ guard Kettenschutz *m* ~ guide Kettenführung *f* ~ guide pully
Ketten-führungsrolle *f*, -leitrolle *f* ~ guide wall
Kettenführungswand *f* ~ haulage Kettenför-
derung *f* (min.) ~ hawse Kettenklüse *f* ~ hoist
Haspelwinde *f*, Kettenkran *m*, Kettenzug *m*
~ hoisting gear Kettenhubeinrichtung *f* ~
housing Kettenhaube *f* ~ induction Kettenein-
leitung *f* ~ installationKetteneinbau *m* ~ insulator Kettenisolator *m* ~ insulators for high
tension Hochspannungsisolatorenketten *pl* ~
jack Kettenwinde *f* ~ leader Kettenzieher *m*
~ lever Ketten-, Kuppluns-hebel *m* ~ lift Ket-
tenaufzug *m* ~ line Querrippe *f* ~-line formation Kettenreihe *f* (aviat.) ~ lines Steglinien *pl*

(paper mfg.) ~ **link** Ketten-glied *n*, -öse *f*, -schake *f* ~ **link pin** Kettengliedbolzen *m* ~ **load** Kettenlast *f* ~ **locker** Kettenkasten *m* ~ **loop** Seilschlinge *f* ~ **lug eye** Kettenschenkelauge *n* ~ **man** Kettenzieher *m* ~ **mark** Querrippe *f* ~ **marks** Steglinien *pl* (paper mfg.) ~ **mesh mat** Kettenmatte *f* ~ **molecule** Kettenmolekül *n* ~ **mortiser** Kettenfräsmaschine *f* ~ **mortising machine** Kettenstemmaschine *f* ~**-nose pliers** Flachzange *f* mit langer Maul- und Nadelspitze ~ **path** Kettenlauf *m* ~ **pattern** Kettenschema *n* ~ **pipe cutter** Gliederrohrschneider *m* ~ **pipe tongs** Kettenrohrzangen *pl* ~ **pipe vise** Kettenrohrschraubstock *m* ~**-pipe wrench** Kettenrohrzange *f* ~ **pitch** Kettenleitung *f* ~ **pits** Kettenschächte *pl* ~ **pull** Kettenzug *m* ~ **pulley** Kettenrolle *f* ~ **pulley wheel** Haspelrad *n* ~ **pump** Kettenpumpe *f*, Paternosterwerk *n* ~ **reaction** Kettenreaktion *f* ~ **reciprocal action** Kettenwechselwirkung *f* ~ **rivet** Kettenniete *f* ~ **riveting** Parallelnietung *f* ~ **road** Kettenbahn *f* ~ **roller** Ketten-laufrolle *f*, -rolle *f* ~ **roller bearing** Kettenrollenlager *n* ~ **runner** Streckenläufer *m* ~ **saw** Glieder-, Ketten-säge *f* ~ **sheave** Kettenrollen-, Rollen-scheibe *f* ~ **side bar** Ketten-lasche *f*, -platte *f*, -seitensteg *m* ~ **stitch** Kettel-, Ketten-stich *m* ~ **strand** Kettenstrang *m* ~**-stretching device** Kettenspanner *m* ~ **stripper** Kettenabweiser *m* ~ **surveying** Kettenmessung *f* ~**-switch-apparatus** Kettenschaltapparat *m* ~ **stud** Kettenbolzen *m* ~ **tackle block** Kettenflaschenzug *m* ~ **tension adjuster** Kettenspanner *m* ~ **tension device** Kettenspannvorrichtung *f* ~**-testing machine** Kettenprüfmaschine *f* ~ **thermal explosion** Kettenwärmexplosion *f* ~ **tightener** Kettenspanner *m* ~ **toggle** Katzenkopf *m* ~ **tongs** Kettenzange *f* ~ **tool** Kettenaufleger *m* ~ **tractor** Kettenschlepper *m* ~ **tramway** Kettenbahn *f* ~ **transmission** Kettenübertragung *f* ~ **travel** Kettenlauf *m* ~ **tread** Kettenlauffläche *f* ~ **trompe** Paternoster-, Ketten-gebläse *n* ~ **twisting** Andrehen *n* oder Anknoten *n* der Ketten, (textiles) Kettenanknoten *n* ~**-type bucket conveyor (or elevator)** Becherwerk *n* mit Kettenführung, ~**-type traveling grate** Kettenrost *m* ~ **wheel** Ketten-nuß *f*, -rad *n*, -rolle *f* ~**-wheel gear** Kettenradvorgelege *n* ~**-wheel shaft** Kettenradwelle *f* ~ **winch** Kettenwinde *f* ~ **windlass** Kettenspill *n* ~ **wire** Nähdraht *m* (paper mfg.), Querdraht *m*

**chainer** Anschläger *m*
**chaining** Messen *n* (mit der Kette)
**chainlike** kettenartig
**chainomatic balance** Kettenwaage *f*
**chair** Schachtfördergefäß *n*, Stuhl *m*, (of a roof pole) Stangenschuh *m*; Vorsitz *m* ~ **plate** Stuhlplatte *f*
**chairman** Aufsichtsratvorsitzender *m* (of a board of directors), Obmann *m*, Vorsitzender *m* ~ **of the board** Vorstand *m* einer Gesellschaft
**chalcedony** Chalcedon *m*, Sardonyx *m*
**chalcocite** Chalkosin *m*, Kupferglanz *m*
**chalcolite** Chalkolith *m*
**chalcopyrite** Chalcopyrit *m*, Eisenkupferkies *m*, Gelbkupfererz *n*, Kupferkies *m*, Kupferpyrit *m*
**chalcostibite** Kupferantimonglanz *m*
**chalcotrichite** Chalkotrichit *m*, Kupferblüte *f*

**chalice** Kelch *m*, Meßkelch *m*
**chalk, to** ~ kreiden, weiß werden **to** ~ **off** abzeichnen
**chalk** Kreide *f* ~ **bench** Abkalktisch *m* ~ **dressing** Kreidebereitung *f* ~ **fog** Kalkschleier *m* ~ **line** Zimmerschnur *f* ~ **marl** Kreidemergel *m* ~ **overlay paper** Kreidepapier *n* ~ **reduction** Kreideverschnitt *m* ~ **test for cracks** Kalkmilchprobe *f* zur Risseprüfung (protective coating)
**chalking** Abkreiden *n*, Kreiden *n* ~ **a line** Abschnürung *f*
**chalky** kreide-artig, -haltig ~ **clay** Mergelkalk *m* ~ **screen** Kalkschleier *m* ~ **soil** Wiesenkalk *m*
**challenge, to** ~ abfragen (sec. rdr.); ablehnen, aufnehmen, anrufen, beanstanden **to** ~ **validity of a patent** die Patentgültigkeit anfechten
**challenge** Abfragung *f* (sec. rdr.), Anruf *m*, Erkennungszeichen *n* (mil.), ~**-radar** Abfrageradar *n* ~ **trophy** Wanderpreis *m*
**challenger** Abfragesender *m* (sec. rdr.)
**chalybeate** eisenähnlich, eisenartig ~ **spring** Stahlquelle *f*
**chamber, to** ~ kesseln (von Bohrlöchern)
**chamber** Abbaustrecke *f*, Ausströmraum *m*, Bassin *n*, Gehäuse *n*, Kammer *f*, (of gun) Laderaum *m* ~ **of cock** Hahnsitz *m* ~ **for collecting deposits** Niederschlagskammer *f* ~ **for tail gate** Nische *f* ~ **of commerce** Handelskammer *f* ~ **of ore** Erzbutze *f* ~ **of a pump** Pumpenstiefel *m*
**chamber,** ~ **acid** Kammersäure *f* ~ **body** Kammerkörper *m* (rocket) ~ **coefficient** Hallzahl *f* (acoust.) ~ **current** Kammerstrom *m* ~ **depth micrometer** Kammertiefenmesser *m* ~ **drying oven** Kammertrockner *m* ~ **envelope** Kesselmantel *m* ~ **filter press** Kammerfilterpresse *f* ~ **filter press without leaching with aeration** Kammerfilterpresse *f* mit einfacher Auslaugung ~ **furnace** Kammerofen *m* ~ **kiln** Kammerofen *m* ~ **mount** Kammerhalter *m* (Beobachtungskammer) ~ **oven** Kammerofen *m* ~ **pressure** Ofendruck *m* ~ **resonator** Hohlraumresonator *m* ~ **wall** Kesselmantel *m*
**chambering** (explosives) Kammerschießen *n* ~ **shots** Kesselschüsse *pl*
**chameleon mineral** Chamäleon *n*
**chamfer, to** ~ abecken, abfassen, abkanten, abrunden (Zähne), abschärfen, abschrägen, auskehlen, bekanten, böschen, einkehlen, kannelieren, schwächen, schweifen, verjüngen, mit Werkzeug senken **to** ~ **the ends of the staves** abkimmen
**chamfer** Abrundung *f*, Abschnitt *m*, abgeschrägte Kante *f*, Auskehlung *f*, Fase *f*, Kantenbrechen *n*, Rief(e)lung *f*, Schrägkante *f*, Schratte *f*, Zuschärfung *f* ~ **of shaft** Wellenabsatz *m* ~ **of a tap** Gewindebohransschnitt *m*
**chamfer,** ~ **angle of the thread** Gewindeauslaufwinkel *m* ~ **tool** kegeliger Senker *m*
**chamfered** abgekantet, abgeschrägt, ausgekehlt, schräg ~ **edge** abgeschrägte Kante *f*, abgestoßene Kante *f* ~ **drilling bit** ausgekehlter Spatenmeißel *m* ~ **joint** abgefaste Fuge *f*
**chamfering** Abfasen *n*, Abschrägung *f*, Anfasen *n*, Böschung *f*, Kannelierung *f* ~ **drill** Senker *m* ~ **machine** Zahnkantenfräsmaschine *f* ~ **machine for pasteboard** Abschärfmaschine *f* für Pappe, Pappenabschärfmaschine *f*

**chamois, to** ~ sämischgerben
**chamois** sämisches Leder f ~-**dressed leather**
sämischgares Leder f ~ **leather** Gems-, Fen-
ster-, Putz-, Sämisch-, Wild-leder n ~ **tannage**
Sämischgerbung f
**chamoisite ore** Chamosit m
**chamotte** gebrannter Ton m
**champfer, to** ~ riefen
**champion lode** Hauptgang m
**chance, to** ~ verfallen
**chance** Aussicht f, Möglichkeit f, Zufälligkeit f
~ **conveyance** Fuhrgelegenheit f ~ **hit** Zufallstreffer m ~ **machine** Zufallsmaschine f ~ **solution** Zufallslösung f
**chancel** Altarraum m, Kanzelle f ~ **screen** Git-
terbrüstung f
**chancellery** Kanzlei f
**chandelier** Armleuchter m, Kandelaber m
**chandelle** Chandelle f, hochgezogene Kehre f
**change, to** ~ (sich) ändern, abändern, abwandeln, abwechseln, auswechseln, einwechseln,
(of the wind) herumdrehen, übergehen, umändern, umbauen, umgestalten, umrechnen,
umschichten, umschlagen, umsetzen, um-
stecken, umsteigen (Zug), (of attitude) umstellen, umwandeln, (sich) verändern, verwandeln, wechseln **to** ~ **color** verfärben, verschießen **to** ~ **colors** schillern **to** ~ **direction** umsteuern **to** ~ **from** umschalten **to** ~ **the fuse**
die Sicherung auswechseln **to** ~ **guard** Posten
ablösen **to** ~ **to higher gear** aufwärtsschalten
**to** ~ **to lower gear** abwärtsschalten **to** ~ **over**
schalten, umrüsten, umschalten, umsetzen,
umstellen, umsteuern **to** ~ **over from first to
second speed** vom ersten auf den zweiten Gang
m übergehen **to** ~ **over into** übergehen **to** ~ **at
random** beliebig auswechseln **to** ~ **suddenly**
sich sprunghaft ändern
**change** Abänderung f, Abwechslung f, Änderung
f, Umänderung f, Umformung f, Umschlag m,
Umsetzung f, Umwandlung f, Veränderung f,
Verwandlung f, Wandel m, Wechsel m, Wechseln n
**change,** ~ **of air** Luftwechsel m ~ **of altitude**
Höhenänderung f ~ **of blades** Messerwechsel m
~ **in brightness** Leuchtdichteänderung f ~ **of
brightness** Helligkeitswechsel m ~ **in capacity**
Kapazitätsänderung f ~ **in charge** Umladung f
~ **of color** Farbenänderung f, Farben-ver-
änderung f, -wechsel m; Farbumschlag m ~ **of
colors** Labradorisieren n (min.) ~ **of condition**
Zustandsänderung f ~ **of connections for receiving** Umschaltung f auf Empfang ~ **of connections for transmitting** Umschaltung f auf
Senden ~ **of control** Steuerwechsel m (aviat.)
~ **of course** Kursänderung f ~ **of cross section**
Querschnittänderung f ~ **of deviation** Ablen-
kungs-, Deviations-änderung f ~ **in dimension**
Maßänderung f, Umgrößerung f ~ **in direction**
Richtungswechsel m ~ **of direction** Richtungsänderung f ~ **in distance** Abstandsänderung f
~ **of feed** Vorschubänderung f ~ **of filter cloth**
Auswechseln n der Filtertücher ~ **in flap angle**
Klappenverstellung f ~ **in flow stress** Fließ-
spannungssprung m ~ **of focusing** Umfokus-
sierung f ~ **of form** Deformation f, Gestaltsveränderung f ~ **of gauge** Spurveränderung f
~ **of geomagnetic** erdmagnetisches Feld n ~ **of**

**gradient** Gefällbruch m ~ **of height** Höhenänderung f ~ **of impulse** Impulsveränderung f
~ **in instruction** (in connection with a call) Änderung f einer Gesprächsanmeldung ~ **from
laminar to turbulent flow** Abreißen n (aerodyn.)
~ **of latitude** Breitenänderung f ~ **of level** Höhenunterschied m ~ **of load** Lastwechsel m
~ **in load distribution** Lastigkeitsänderung f
~ **in measure** Maßveränderung f ~ **of measure**
Formatveränderung f ~ **of momentum** Impulsveränderung f ~ **of place name** Ortsumbe-
nennung f ~ **of polarity** Polwechsel m ~ **in
position** Lagenänderung f ~ **of position** Lageänderung f, Lage-, Stellungs-wechsel m, Verlegen n ~ **in potential** Potentialsprung m
(electr.) ~ **in pressure** Druckveränderung f,
(plus to minus) Druckwechsel m ~ **of pressure** Druckänderung f ~ **of register** (plus
to minus) Druckwechsel m, Kassenleergang m
~ **in rotation** Drehungsänderung f ~ **in scale**
Maßveränderung f ~ **in section** Querschnitts-
übergang m ~ **of sectional area of piping** Quer-
schnittsveränderung f der Rohrleitung ~ **in
self-inductance** Selbstinduktionsänderung f ~
**in shade** Schattenveränderung f ~ **of shade**
Tonverschiebung f ~ **in shading values** Schattenveränderung f ~ **of shape** Formänderung f,
Gestaltsveränderung f ~ **in size** Maßänderung
f, Umgrößerung f ~ **of slags** Schlackenwech-
sel m ~ **of solubility** Löslichkeitsveränderung f
~ **in sound impression** Klangbildveränderung f
~ **in speed** Gangänderung f ~ **of speed** Gang-,
Geschwindigkeits-wechsel m ~ **of state** Zustandsänderung f ~ **of state of aggregate**
Aggregatzustandsänderung f ~ **of strain** Span-
nungsveränderung f (metall.) ~ **of strengthening** Wechselverfestigung f ~ **in structure** Struk-
turwandel m ~ **of target** Zielwechsel m ~ **of
target speed** Änderungsgeschwindigkeit f ~ **in
temperature** Temperaturänderung f ~ **of temperature** Temperaturveränderung f, Wärme-
-änderung f, -wechsel m ~ **of tendency** Stimmungswechsel m (stock exchange) ~ **of the
tide** Rückkehr f der Ebbe und Flut, Widerzeit f
~ **of tone** Stimmungswechsel m (stock exchange) ~ **of volume** Raumänderung f ~ **in
voltage** Spannungsänderung f ~ **in volume**
Raumveränderung f, Volumenveränderung f
~ **of water level** Wasserstandsänderung f ~ **of
weave** Bindungswechsel m ~ **of wind** Zurückspringen n des Windes
**change,** ~ **cock** Wechselhahn m ~ **coincidence**
Zufallskoinzidenz f ~ **collar** Wechselmuffe f
~ **finger** Wechselfinger m
**change-gear** Getriebeläufer m, Wechselrad n,
Wechselrädergetriebe n, Zahnrädergetriebe n,
Zahnradwechsel m ~ **box** Wechselgetriebe-
kasten m ~ **bracket** Schere f, Stelleisen n ~
**cover** Wechselradverdeck n ~ **handle** Wechsel-
hebel m ~ **quadrant** Wechselräderschere f ~
**ratios** Wechselradverhältnisse pl
**change-gears of indexing head** Teilkopfwechsel-
räder pl
**change-lever** Umschalthebel m
**change-over** Überblendung f (film), Übergang m,
Übertritt m, Umschaltung f, Umstellbarkeit f,
Umstellung f ~ **bar magnet** Umschaltestan-
genmagnet m ~ **button** Wendetaste f ~ **clutch**

Wechselkupplung *f* ~ **cock** Umschalthahn *m* ~ **condenser** Wechselkondensator *m* ~ **contact** Ruhe- und Arbeitskontakt *m* ~ **contact unit** Wechselkontakt *m* ~ **contactor** Umschaltschütz *m* ~ **cue** Überblendungszeichen *n* (film) ~ **device** Übergangseinrichtung *f* ~ **gear** Schaltgetriebe *n* ~ **line** Überströmleitung *f* ~ **mark** Überblendmarke *f* ~ **marker generator** Überblendmarkengeber *m* ~ **mechanism** Umstellvorrichtung *f* ~ **panel** Umschaltfeld *n* ~ **plate** Umleitplatte *f* ~ **process** Umschaltvorgang *m* ~ **relay** Umschaltrelais *n* ~ **switch** Polwechsler *m*, Polwender *m*, Umschalter *m* ~ **switch contact plate** Umschalterkontaktplatte *f* (Scheinwerfer) ~ **system** Umschaltanlage *f* ~ **time** Umschaltzeit *f* ~ **valve** Umschalt-, Wechsel--ventil *n*

change, ~ **parts** auswechselbare Werkzeuge *pl* ~-**piece** Wechselstück *n* ~ **point** Haltepunkt *m*, Umschlagpunkt *m* ~ **ratio** Umrechnungsfaktor *m* ~ **rod** Wechselstange *f* ~ **round** Umschwenken *n*

change-speed Umschaltung *f* ~ **fork** Wechselschiene *f* ~ **gear** Geschwindigkeits-umstellvorrichtung *f*, -wechselgetriebe *n*, Wechselgetriebe *n*, Zahnradwechselgetriebe *n* ~ **gear type** Wechselradtyp *m* ~ **governor** Verstellregler *m* ~ **lever** Umschalthebel *m* ~ **lever ball** Getriebearmkugel *f* ~ **mechanism** Umschaltvorrichtung *f* ~ **motor** Motor *m* mit mehreren Drehzahlstufen

change, ~-**tank** mixer Mischer *m* mit Wechselbehälter ~ **tape** Mutationsband *n* ~-**tune** switch Wellenumschalter *m* ~-**valve** box Wechselventilkasten *m* ~-**wheel** gear Gangwechsel *m* (speed)

changeable schillernd, veränderlich, wechselbar, wechselnd ~ **luster** Schillerglanz *m*

changed umgeändert, umgebaut

changes, ~ **in range per unit time** Entfernungsänderung *f* per Zeiteinheit ~ **of rate about the pitch axis** Nickbeschleunigung *f* ~ **of rate about the roll axis** Rollbeschleunigung *f* ~ **of rate about the yaw axis** Gierbeschleunigung *f*

changing veränderlich, wechselnd; Übergang *m* ~ **of barrels** Rohrwechsel *m* ~ **the bit** Umsetzen *n* des Bohrgerätes ~ **the doffer pinion** Abzugswechsel *m* ~ **on the left-hand side** Linksanschlag *m* ~ **of shifts** Schichtwechsel *m* ~ **of wind direction** Herumholen *n* des Windes

changing, ~ **box** Wechselkasten *m* ~ **collar** Wechselring *m* ~ **color** wechselfarbig ~ **current** Aufladestrom *m* ~ **device** Wechselvorrichtung *f* ~ **information** Informationsübermittlung *f* ~ **magazine** Wechsel-kassette *f*, -magazin *n* (film) ~ **note** Wechselnote *f* ~ **operation** Wechselvorgang *m* ~ **over** Umschalten *n*, Umschaltung *f* ~ **ring** Wechselring *m* ~ **zero position** veränderliche Nullstellung *f*

channel, to ~ aushöhlen, auskehlen, ausreifen, einkehlen, führen, kannelieren, kehlen, riefen, riffeln **to cut a** ~ ausbogen

channel Absatzweg *m*, Abzucht *f* (hydr.), Anschnitt *m*, Arm *m*, Ausflußkanal *m*, Ausflutung *f*, Auskehlung *f*, Aussparung *f*, Fahrwasser (navig.), Falz *m*, (of fairway) Flußbett *n*, Flutgang *m*, Furche *f*, Gefluder *n*, Gerinne *n*, Hohlkehle *f*, Kanal *m*, (ship building) Keep *f*,

Kehle *f*, Kette *f*, Nachrichtenkanal *m*, Pfeife *f*, Riefe *f*, Rinne *f* (aviat.), Rinnsal *n*, Rüste *f*, Schram *m*, Schratte *f*, Seegatt *n*, Spur *f*, Stromrinne *f*, Umlaufkiste *f*, U-Profil *n*, Wassergraben *m*, Wasserlauf *m*, Weg *m*

channel, ~ **for beams** Balkenpfalz *f* ~ **of communication** Nachrichtenmittel *n*, Nachrichtenverbindung *f* ~ **for the eye** Öhrfurche *f* (needle) ~ **of a river** Talweg *m* eines Flusses **a** ~ **is scoured out** eine Stromrinne *f* vertieft sich ~ **of spillway** headrace Obergraben *m* ~ **with a strong current** Schußgerinne *n*

channel, ~ **area** Kanalgebiet *n* ~ **bank** Kanalumsetzer *m* ~ **bar** U-Eisen *n* ~ **brick** Kanalziegel *m* ~ **capacity** Kanalkapazität *f* (eines Nachrichtenkanals) ~ **chisel** Kanalmeißel *m* ~ **control** Einzelregler *m* ~ **designation** Kanalbezeichnung *f* ~ **displacement** Kanalverschiebung *f* ~ **filter** Kanalfilter *n* ~ **formed by breakers** Brandungskehle *f* ~ **iron** U-Eisen *n* ~ **letter** Kanalbuchstabe *m* (rdo) ~ **limit** Lochgrenze *f* ~ **loading** Kanalbelastung *f* (rdo) ~ **marker** Rinnenmarker *m* (aviat.) ~ **mixing** loudspeaker Mischlautsprecher *m* ~ **modulator** Kanalumsetzer *m* ~ **number** Kanalnummer *f* ~ **passage** (between the banks) Fahrrinne *f* ~ **patch** U-Flicken *m* ~ **position** Kanallage *f* ~ **radius** Kanalradius *m* ~ **rib** Fachwerkrippe *f* ~ **sample** Schlitzprobe *f* ~ **section** Doppel-T-Profil *n*, U-Profil *n*, U-Querschnitt *m* ~ **selected basic length** Rinnengrundlänge *f* (aviat.) ~ **selector** Verkehrsartenschalter *m*, Kanalwähler *m* ~ **supergroup** Kanalübergruppe *f* ~ **test tone level** Knalmeßpegel *m* ~ **threshold** Kanalschwelle *f* ~ **traffic** Revierfahrt *f* ~ **transducer** elektroakustische Kette *f* ~-**translating** equipment Kanalmodulatordemodulator *m* ~-**type (or core-type) induction furnace** Rinnenofen *m* (Induktionsofen) ~ **width** Kanalbreite *f* (rdo), Lochgrenze *f*, Rinnenbreite *f* (aviat.)

channelled, ~ **plate** Riffelblech *n* ~ **punch** kannelierter Stampfer *m* ~ **spectrum** Platteninterferenzspektrum *n*

channelling Kanaleffekt *m*, Mehrfachausnutzung *f* (rdo) ~ **corrosion** langadrige Rostanfressung *f* ~ **effect** Kanalverlust *m* ~ **effect factor** Kanalverlustfaktor *m*

channellized transmitter Sender *m* mit Simultankanälen

chaos Wirrnis *f*

chap, to ~ aufreißen

chap Maul *n* eines Schraubstocks

chapel Kapelle *f*

Chapelet furnace Chapeletofen *m*

chaplet Kern-böckchen *n*, -nagel *m*, -stütze *f* ~ **hinge** Paternosterband *n*

chapped gerissen, narbenbrüchig

chapping knife Kabelmesser *n*

chapter Kapitel *n*

char, to ~ ankohlen, brennen, kohlen, schmoren, verkohlen, verkoken, zu Kohle brennen

char, ~ **felling** Kahlschlag *m* ~ **filter** Knochenkohlefilter *m* ~-**revivifying kiln** Glühofen *m* zur Wiederbelebung der Knochenkohle

character Art *f*, Beschaffenheit *f*, Charakter *m*, Eigenschaft *f*, Gepräge *n*, Schriftzeichen *n*, Verlauf *m*, Wesen *n*, Zeichen *n* (info proc.) ~ **of grain** Kornbeschaffenheit *f* ~ **of soil at**

surface Oberflächenbeschaffenheit *f* des Bodens ~ of the wind Windeigentümlichkeit *f*
**character,** ~ **allocation to keyboard** Tastenfeld *n* (Zuordnung der Zeichen) ~ **counter** Zeichenzählvorrichtung *f* ~ **recognition** Schriftzeichenerkennung *f* ~ **set** Alphabet *n*, Zeichenvorrat *m* (comput.)
**characteristic** Besonderheit *f*, Bestimmungsstück *n*, Eigenschaft *f*, Eigentümlichkeit *f*, Grundzug *m*, Kennlinie *f*, Kennmerkmal *n*, Kennwert *m*, Kennzeichen *n*, Kurve *f*, Mal *n*, Merkmal *n*, Signalgruppe *f* (rdr), Zug *m*; arteigen, ausgeprägt, bezeichnend, charakteristisch, eigen, eigenartig, eigentümlich, kennzeichnend ~ **of action** Wirkungsart *f* ~ **of circuit control** Auslösecharakteristik *f* ~ **(line) of injection** Spritzkennlinie *f* ~ **of a thermionic valve** Kennlinie *f* der Elektronenröhre, Röhren-charakteristik *f*, -kennlinie *f*
**characteristic,** ~ **angular momentum** Eigendrehimpuls *m* ~ **curve** Charakteristik *f*, Gütegradschaubild *n*, Kennlinie *f*, Kennwertlinie *f* ~ **curve of attenuation** Dämpfungskennlinie *f* ~ **curve of iron filament ballast lamps** Kennlinie *f* von Eisenwiderständen ~ **curves** Kennkurven *pl* ~ **data** Kenndaten *pl*, Kennzeichnungen *pl* ~ **distortion** Apparatverzerrung *f*, Einschwingverzerrung *f* (telegr.), Regelverzerrung *f* ~ **drooping in relation to load** lastabhängig geneigte Kennlinie *f* ~ **energy** Eigenenergie *f* ~ **equation** Stammgleichung *f* ~ **factor** Kennzahl *f* ~ **feature** Haupt-, Unterscheidungs-merkmal *n* ~ **film curve** Schwärzungskurve *f* ~ **frequencies of a sound** Formanten *pl* eines Klanges ~ **frequency** Eigenfrequenz *f* ~ **frequency band** Funkkanal *m* ~ **function** Eigenfunktion ~ **horn sound** Trichterklang *m* ~ **impedance** Feldwiderstand *m* einer Hohlleiterwelle, Wellen-, Kenndämpfungs-, Kenn-, Wellen-widerstand *m* ~ **lines** Kennlinien *pl* ~ **magnitude** Kenngröße *f* ~ **number** Eigenwert *m* ~ **property** Charakteristik *f* ~ **quantity** Kenngröße *f* ~ **radiation** Eigenstrahlung *f* ~ **resistance** Wellenwiderstand *m* ~ **signal** Kennsignal *n* ~ **straightening** Kennlinienkippung *f* ~ **structure** Wesengefüge *n* ~ **time** Sollzeit *f* ~ **value** Eigenwert *m* ~ **variable** Kennwert *m* ~ **velocity of flow** kennzeichnende Geschwindigkeit *f*
**characteristics** Charaktereigenschaft *f*, Daten *pl*, Gegebenheiten *pl*, (numerical or physical) Kennziffer *f* ~ **of combustion** Verbrennungsverlauf *m*
**characterization** Charakterisierung *f*, Kenntlich-machen *n*, -machung *f*, Kennzeichnung *f*
**characterize, to** ~ kennzeichnen
**characterized (by)** gekennzeichnet (durch)
**characters** Buchstaben *pl*
**charcoal** Holzkohle *f*, Kohle *f*, Kohlholz *n* ~ **from piles** Meilerkohle *f*
**charcoal,** ~ **bed** Löschboden *m* ~ **black** Holzkohlenmehl-, Holzkohlen-schwärze *f* ~ **blast furnace** Holzkohlenhochofen *m* ~ **bottom** Löschboden *m* ~ **brazier** Holzkohlen-, Kohlen--pfanne *f*, Lötofen *m*, Löttopf *m* ~ **breeze** Holzkohlenlösche *f* ~ **breeze mixed with clay** Holzkohlengestübbe *n* ~ **burning** Holzverkohlung *f*, Köhlerei *f*, Meiler *m* ~ **burning in long piles**

Haufenverkohlung *f* ~ **burning in pits** Grubenverkohlung *f* ~ **crayon** Reißkohle *f* ~ **dust** Holzkohlenpulver *n*, Lösche *f* ~ **dust mixed with clay** Holzkohlengestübbe *n* ~ **filter** Kohlenfilter *m* ~ **fines** Holzkohlenpulver *n* ~ **fining process** Löscharbeit *f* ~ **fire** Holzkohlenfeuer *n* ~ **furnace** Frisch-feuer *n*, -herd *m* ~ **heap** Holzkohlenmeiler *m* ~ **hearth** Holzkohlen-, Frisch-feuer *n*, Frischherd *m*, Lösch-, Rennherd *m* ~**-hearth iron** Frischfeuereisen *n* ~**-hearth steel** Herd-, Holzkohlen--frischstahl *m* ~ **iron** Frischfeuereisen *n*, Holzkohlen(roh)eisen *n* ~**-iron sheet** Holzkohlenblech *n* (blasting) ~**-mixture container** Kohlenstoffträger *m* (blasting) ~ **pencil** Zeichenkohle *f* ~ **pig** Holzkohlenroheisen *n* ~ **pig iron** Holzkohlenroheisen *n* ~ **pile** Holzkohlen-, Kohlen-meiler *m*, Meiler *m* ~ **powder** Holzkohlenmehl *n*, Holzkohlenpulver *n* ~ **process** Löscharbeit *f* ~ **retort** Holzkohlenretorte *f*
**chare, to** ~ scharrieren **to** ~ **an ashlar** einen Stein *m* scharrieren
**charge, to** ~ anfüllen, anstürmen, aufgeben, aufgichten (metall.), (a battery) aufladen, (an account) aufrechnen, aufschütten, auftragen, aufwerfen, beauftragen, begichten, beladen, belasten, beschicken, berechnen, (law) beschuldigen, chargieren, (magazines) durchladen, einfüllen, einschütten, einsetzen, erheben, laden, schöpfen, schütten, speisen, unterstellen, verladen, zugeben; in Anrechnung bringen, mit einer Gebühr belegen **to** ~ **to the account** anrechnen (Konto belasten) **to** ~ **an accumulator** einen Akkumulator *m* laden **to** ~ **a battery** die Batterie laden (aufladen) **to** ~ **in a bill** ansetzen **to** ~ **extra** nachfordern **to** ~ **the furnace** gichten **to** ~ **the magazine** Platten *pl* in Kassetten einlegen (photo) **to** ~ **out** weiterbelasten **to** ~ **with** bezichtigen **to** ~ **to revenue** auf budgetierte Kosten *pl* verbuchen **to** ~ **a surface with hydrogen** eine Oberfläche *f* mit Wasserstoff beladen **to** ~ **one, two, three fees** einfach, doppelt, dreifach berechnen
**charge** Abgabe *f*, Anklage *f*, Ansatz *m*, Anweisung *f*, Aufladung *f*, Belastung *f*, Berechnung *f*, Beschickung *f*, Beschickungsgut *n*, Beschickungsmaterial *n*, Brand *m*, Charge *f*, (furnace) Durchsatz *m*, Einlage *f*, (of a furnace) Einsatz *m*, Einsatzwerkstoff *m*, Eintrag *m*, Füllung *f* (Speisung), Gattierung *f*, Gicht *f*, Glassatz *m*, Kochereintrag *m*, Lade *f*, Ladung *f*, Last *f*, Ofeneinsatz *m* (metall.), Satz *m*, Treibsatz *m* (Rakete) ~ **per batch** Einsatz *m* pro Partie ~ **of a chamber** Kammerladung *f* ~ **of a conductor** Ladung *f* eines Leiters ~ **of electricity** Elektrizitätsladung *f* ~ **of glass** Glasladung *f* ~ **of mixture** Ladegemisch *n* ~ **on nucleus** Kernladungszahl *f* ~ **of pig** Heize *f* (smelting) ~ **upon a reality** dingliche Belastung *f* ~ **of secondary cell, of storage cell** Ladung eines Sammlers ~ **of space grid** Gitterraumladung *f* ~ **on nucleus** Kernladungszahl *f*
**charge,** ~ **accumulation** Speicherwirkung *f*, Ladungsspeicherung *f* ~ **carrier** Ladungsträger *m* ~ **cloud** Ladungswolke *f* ~ **conjugation** Ladungskonjugation *f* ~**-current density** Stromladungsdichte *f* ~**-current potential** Vierer-

strompotential *n* ~ **density** Ladedichte *f* (explosives) **~-density modulation** Dichtemodulation *f*, Ladungsdichtesteuerung *f* ~ **dissipation** Ladungsableitung *f* ~ **distortion** Ladungsverformung *f* ~ **eliminator** Ableiter *m* ~ **end** Eintragende *n* ~ **equalization** Ladungsausgleich *m* ~ **equation** Ladungsgleichung *f* ~- **exchange cross section** Unladungsquerschnitt *m* ~-**exchange operator** Austausch-, Ladungsaustausch-operator *m* ~-**exchange phenomenon** Ladungsaustauscherscheinung *f* ~ **factor** Ladegrad *m* ~ **frequency** Ladungsfrequenz *f* ~ **image** Ladungsbild *n* ~-**independent** ladungsunabhängig ~ **invariance** Ladungsunveränderlichkeit *f* ~-**mass ratio** spezifische Ladung *f*, Verhältnis *n* von Ladungs zu Masse ~ **multiplet** Ladungsmultiplett *n* ~-**out sheet** Entnahmebeleg *m* ~ **pattern** Ladungsbild *n* ~ **renormalization** Ladungsrenormierung *f* ~ **shape** Treibsatzprofil *n* ~ **spin** Ladungsspin *m* ~ **storage** Ladungsspeicherung *f* ~-**storage tube** Bild-, Ladungs-speicherröhre *f* ~ **switch** Zellenschalter *m* ~-**to-tap time** Schmelzzeit *f* ~ **transfer** Ladungsübertragung *f* ~ **trust** Betreuung *f* ~-**weight ratio of projectile** Ladungsverhältnis *n*
**chargeable** gebührenpflichtig, verantwortlich ~ **call** gebührenpflichtiges Gespräch *n* ~ **time** gebührenpflichtige Gesprächsdauer *f* ~-**time indicator** Gesprächszeitmesser *m* ~-**time lamp** Zeitlampe *f*
**charged** aufgeladen, berechnet, geladen, stromführend ~ **carrier** Ladungsträger *m* ~ **condition** geladener Zustand *m* (of storage cells) ~ **minutes** Gebührenminuten *pl* ~ **paper** beschwertes Papier *n* ~ **particle** Elektrizitätsträger *m* ~ **particle carriers** geladene Teilchen *pl* ~ **wire** spannungsführender Draht *m*
**charger** Aufgeber *m*, Batterieladegerät *n*, Füller *m*, Gichtmann *m*, Lade-schale *f*, -streifen *m* (for cartridges) ~ **indicator** Ladungsschieber *m* ~-**reader** Lade- und Meßgerät *n*
**charges** Ausgaben *pl*, Kosten *pl*, Spesen *pl* ~ **for conveyance** Zustellungsgebühr *f*
**charging** Aufladen *n*, Beschickung *f*, Einschleusen *n*, Einschüttung *f*, Einsetzen *n*, Füllen *n*, Füllrumpf *m*, Gichtung *f*, Laden *n*, Schütten *n* ~ **of accumulator** Sammlerladung *f* ~ **of a condenser** Laden *n* eines Kondensators ~ **by distance** Gebührenabmessung *f* nach der Entfernung ~ **of fuel** Brennstoffschüttung *f* ~ **of a surface** Oberflächenbeladung *f* ~ **by time** Gebührenbemessung *f* nach der Dauer ~ **of the working cylinder** Ladung *f* des Arbeitszylinders
**charging, ~ apparatus** Schüttvorrichtung *f* ~ **appliance** Beschickungs-, Chargier-vorrichtung *f* ~ **area** Schüttfläche *f* ~ **arrangement** Begichtungseinrichtung *f* ~ **bar** Schwengel *m* ~ **bay of a glass tank** Einlegevorbau *m* ~ **belt** Aufladeband *n* ~ **blower** Aufladegebläse *n* ~ **(switch)board** Ladeschalttafel *f* ~ **box** Beschickungs-, Lade-mulde *f*, Mulde *f* ~ **bucket** Einsetzmulde *f*, Füllgefäß *n*, Gichtkübel *m* ~ **capacitor** Ladekondensator *m* ~ **car** Begichtungs-, Gattierungs-wagen *m* ~ **carriage** Fördergefäß *n* ~ **chamber** Ladeschrank *m* ~ **characteristic curve (or line)** Ladekennlinie *f* ~ **chute** Verladerutsche *f* ~ **circuit** Ladestromkreis *m* ~

**coefficient** Ladekoeffizient *m* ~ **commutator** Ladeumschalter *m* ~ **condenser** Ladekondensator *m* ~ **cone** Aufgabetrichter *m* ~ **control box** Laderegulierkasten *m* ~ **converter** Lade-umformer *m* ~ **crane** Beschickungskran *m*, Chargier-, Einsatz-kran *m* ~ **current** Ladestrom *m* ~ **current control lamp** Ladelampe *f* ~- **current impulse** Ladestromstoß *m* ~ **curve** Ladekurve *f* ~ **deck** Abzugsbühne *f* ~ **device** Begichtungs-, Beschickungs-, Einsetz-vorrichtung *f* (rolling mill) ~ **device for elevators** Becherwerkaufgabevorrichtung *f* ~ **diagram** Fülldiagramm *n* ~-**independent** ladungs-unabhängig ~ **door** Arbeitstür *f*, Beschickungsöffnung *f*, Einsatz-, Einsetz-tür *f*, Füllklappe *f*, Gichteinwurfsöffnung *f* (of blast furnace), Gichtöffnung *f* ~-**door sill** Einsatztürschwelle *f* ~ **elevator** Aufgabebecherwerk *n* ~ **end** schickungsseite *f* ~ **equipment** Beschickungsanlage *f*, Ladeeinrichtung *f* ~ **face** Ladefläche *f*, Ladungsseite *f* ~ **facilities** Ladevorrichtung *f* ~ **facility** Aufgabevorrichtung *f* ~ **flap** Füllklappe *f* ~ **floor** Begichtungs-, Chargier-, Gicht-bühne *f*, Kocherboden *m*, Ofenbühne *f*, Setzboden *m* ~ **gauge** Gichtmaß *n* ~ **gear** Begichtungsanlage *f* ~ **generator** Lade-dynamo *m*, -maschine *f* ~ **grid** Einwurfrost *m* (für Kohle) ~ **head** Füllkopf *m* ~ **hole** Füll-, Gicht-öffnung *f* ~ **hopper** Füll-schacht *m*, -trichter *m*, -vorrichtung *f* ~ **hose** Abfüllschlauch *m* ~ **indicator lamp** Ladekontrollampe *f* ~ **installation** Begichtungsanlage *f* ~ **machine** Ausdrück-, Einsetz-, Füll-maschine *f*, Ladedynamo *m*, Lademaschine *f*, Stromerzeugungsanlage *f* ~-**machine operator** Chargierer *m* ~ **man** Gichtmann *m* ~ **material** Einsatzgut *n* ~ **mechanism** Aufgabevorrichtung *f* ~ **opening** Beschickungs-, Einschütt-, Einwurf-öffnung *f* ~ **operation** Begichtung *f* ~ **panel** Ladeschalttafel *f* ~ **peel** Schwengel *m* ~ **period** Ansaugehub *m* (gas engine) ~ **platform** Aufgabesohle *f*; Beschickungs-, Chargier-, Lade-bühne *f*; Gicht-boden *f*, -brücke *f*, -bühne *f*, -ebene *f* (of blast furnace) ~ **plug** Ladestöpsel *m* ~ **port** Füllöffnung *f* ~ **pressure** Aufladeluftdruck *m* ~ **process** Aufladeverfahren *n* ~ **pump** Ladepumpe *f* ~ **rate** Stromaufnahme *f* ~ **rectifier** Ladegleichrichter *m* ~ **resistance** Ladewiderstand *m* ~ **scales** Gattierungswaage *f* ~ **scoop** Ballastschaufel *f* ~ **set** Lade-satz *m*, -aggregat *n* (electr.) ~ **side** Einsetzseite *f* ~ **socket** Einfüllstutzen *m* ~ **space** Stellfläche *f* ~ **spout** Eingußschnauze *f* ~ **station** Ladestation *f* ~ **stock** Einsatzprodukt *n* ~ **stove** Füllofen *m* ~ **surface** Beschickungsoberfläche *f* **(battery)**-~ **switch** Ladeschalter *m* ~ **time** Chargierzeit *f*, Einsatzdauer *f*, Ladezeit *f* ~ **tower** Gichtturm *m* ~ **truck** Füllwagen *m* ~ **tube** Laderöhre *f* ~ **unit** Begichtungsanlage *f* ~ **valve** Einzugs-, Füll-ventil *n*, Schleuse *f* ~ **voltage** Ladespannung *f* ~ **winding** Ladewicklung *f* ~ **zone** Begichtungszone *f* (Kupolofen), Setzraum *m*
**charing chisel** Scharriereisen *n*
**chariot** Katze *f*, Schlitten *m* (telegr.)
**charlock-destroying powder** Hederichvernichtungspulver *n*
**Charpy, ~ impact-testing machine** Charpyscher Pendelhammer *m* ~ **notch** Charpy Rundkerbe *f*, Rundkerb *m*

charred verkohlt, verschmort ~ bone Knochenkohle f ~ horn Hornkohle f ~ leather Leder-
kohle f ~ molasses slop Melasseschlempekohle
f ~ slop Schlempekohle f ~ sponge Schwammkohle f
charring Ankohlen n, Brennen n, Verkohlen n,
Verkohlung f ~ in pits Grubenverkohlung f ~
plant Verkohlungsanlage f
chart, to ~ festlegen
chart Karte f, Paßkarte f, Plan m, Registrier-
streifen m, Tabelle f, Tafel f, Übersicht f ~ of
equal variations Isogenenkarte f ~ of a recording
instrument Diagramm n
chart, ~ bearing drum Diagrammtrommel f ~
equipment Kartenausrüstung f ~ glass Skalen-
glasscheibe f ~ graduation Skaleneinteilung f
~ paper Karten-, Lankartendruck-, Plandruck-
Registrier-papier n ~ paper feed Papiervorschub m ~ roll Meß-, Papier-streifenrolle f ~
room Karten-haus n, -raum m ~ sheet Kartenblatt n ~ speed Papiervorschub m
charted radio bearing unverbesserte oder abgelesene Funkseitenpeilung f
charter, to ~ heuern
charter Freibrief m, Satzung f ~ by weight Verfrachtung f nach Gewicht ~ master Gruben-
pächter m ~ party Charterpartie f
chartered course Sollkurs m
charterer Befrachter m
charts for the projection perimeter Gesichtsvor-
drucke pl zum Projektionsperimeter
chase, to ~ auskreuzen, ausmeißeln, Gewindestrehlen, treiben, ziselieren to ~ a screw thread
ein Gewinde n nachschneiden
chase Fassung f (Zeichnung), Rahmen m ~ lock
Rahmenschloß n ~ printing Schließrahmen m
~ ring Prägering m
chased work Bunzen, Ziselier-arbeit f
chaser Geleitboot n, Geleitschiff n (navy), (for
screw threads) Gewindestahl m, Nachschneider
m, Strehler m (Gewindestahl), Ziseleur m ~
tooth Kammzahn m
chases without quoins Schließrahmen pl ohne
Schließzeuge
chasing Nachschneiden n, Treiben n (techn.),
Ziselierung f ~ of screw thread tap bottoming
Gewindenachschnitt m
chasing, ~ dial Gewinde-uhr f, -schneideanzei-
ger m ~ hammer Teller-, Treib-hammer m ~
star Gewindestern m ~ tool Gewinde-strähler
m, -strehler m, Nachschneider m, Strehler m,
Strehlwerkzeug n, Treib-form f, -werkzeug n
chasm Spalte f
chassis Chassis n, Fahrwerk n, Fahrgestell n,
Gestell n, Grundplatte f, (of loudspeaker) Haltevorrichtung f, Wagenuntergestell n ~ carrying capacity Fahrgestelltragfähigkeit f ~ connection Chassisanschluß m ~ dynamometer
Fabrikprüfstand m ~ earth (or ground) Fahr-
gestellmasse f ~ frame Fahrwerkrahmen m ~
frame of loudspeaker Lautsprecherkorb m ~
longitudinal Rahmenlängsträger m ~ side
member Rahmenlängsträger m ~ truck Laufgestellrahmen m
chatoyant changierend, schillernd
chats Mittelprodukt n (min.)
chatter, to ~ erschüttern, klappern, prellen,
rattern, (of machines) schlagen, zittern

chatter Erschütterung f, (of a key contact)
Schlag m ~(ing) Klappern n ~ of contacts prellende Kontaktöffnung f ~ mark Rattermarke f
Chatterton's compound Chattertonmasse f
chauffeur Fahrer m, Führer m, Kraftfahrer m,
Wagenführer m
check, to ~ (baggage) abfertigen, abhalten, absperren, ankreuzen (aviat.), auffieren, aufhalten, (the motion) bremsen, Einhalt tun, einhalten, hemmen, hintanhalten, kontrollieren,
kraquelieren (painting), mit einem gegebenen
Werte übereinstimmen, (the calibration) nacheichen, nachmessen, (computations) nachprüfen, nachrechnen, nachsehen, nachzählen, proben, prüfen, reißen (vom Anstrich), revidieren,
Schach bieten, schricken, sichern, überprüfen,
überwachen, vergleichen, zurücktreiben to ~
alignment die Ausrichtung prüfen to ~ back
rückversichern to ~ a bore eine Bohrung f
messen to ~ the dimension nachmessen to ~
for freedom auf leichten Gang m prüfen to ~
the frequency die Frequenz prüfen to ~ a message
ein Telegramm n prüfen to ~ meter readings
Meßskalen nachprüfen to ~ the roundness and
parallelism auf Zylindrizität f prüfen to ~ a
shaft for truth eine Welle f auf Rundlauf prüfen to ~ the soldered connection die Lötstelle
prüfen to ~ tank dips Tankpeilungen nachprüfen to ~ the time of day die Zeit vergleichen to
~ up untersuchen
check Arretierfeder f, Carreau n, Eindämmung f, (for arresting tool or springs) Fangbügel m, Formkasten m, Hemmung f, Karo
n, Kassenanweisung f, Kassenschein m, Kontrolle f (of a message), Lücke f, Nachun-
tersuchung f, Riß m, Scheck m, (holding back)
Schlappe f, Überwachung f, Unterbrechung f,
Zettel m ~(ing) Kontrolle f, Nacheichung f,
Prüfung f ~ and restrictor valve Rückschlag-
drosselventil n ~ and vent valve Rückschlag-
und Entlüftungsventil n ~ of the head Schließ-
knie n ~ of shears Scherblatt n
check, ~ analysis Kontrollanalyse f ~-back
indicator Antwortgeber m ~-back position
indicator Rückmelder m, Spiegelfeld n ~ bearing of fix Kontrollpeilung f ~ block Scheiben-
klampe f ~ bolt Arretierbolzen m ~ book
Scheckheft n ~ cable Abfangseil n (aviat.),
Fangschlaufe f, Grenzstropp m, Lenkkabel n
~ calculation Rechenprobe f ~ calibration
Nacheichung f ~ column method Zeilensum-
menprobe f ~-control number Kontrollnummer f ~ cracks Schrumpfrisse pl ~ determination
Gegenprobe f, Kontrollbestimmung f ~ digit
Kontroll-, Prüf-ziffer f ~ draft Aschenschieber
m ~ flight Meß-- oder Einweisungsflug m ~
gauge Prüf-lehrdorn m, -maß n ~ group Kon-
trollgruppe f ~ key Kontrolltaste f ~ lever
Sperrhebel m ~ line Kreuzleine f ~ list Klar-,
Kontroll-, Überwachungs-liste f ~ mark
Prüfzeichen n ~ number Kohlen-, Kontroll-,
Prüf-nummer f ~ nut Gegen-, Konter-mutter f
~-out Funktionsprüfung f ~ pad Nummern-
block m ~ paper Sicherheitspapier n ~ pattern
(diamond or dice pattern) Würfelmuster n ~
pilot Prüfpilot m ~ plate Anlaufscheibe f ~
plug Prüfschraube f ~ point Kontroll-, Ver-
gleichungs-punkt m ~ program Prüfprogramm

*n* ~ **rail** Gegenschiene *f*, Zwangsschiene *f*
~ **rein** Kreuzleine *f* ~ **relay** Prüfrelais *n* ~ **ring**
Ansatz-, Anschlag-ring *m* ~ **rope** Hemmseil *n*
~ **routine** Prüfprogramm *n* ~ **sample** Kontroll-
muster *n* ~ **screw** Halteschraube *f* ~ **sheet** Befundbogen *m* ~ **spring** Schrankfeder *f* ~ **strap**
Arretierriemen *m* (textiles) ~ **table** Verteiler *m*
~ **test** Gegenprobe *f* ~ **tone** Kontrollton *m*
~-**up** Durchsicht *f*, Nach-prüfung *f*, -unter-
suchung *f* ~-**up experiment** Parallelversuch *m*
~-**up transmitter** Kontrollsender *m* ~ **valve** Ab-
sperr-, Klappen-, Regulier-ventil *n*, Rück-
schlagklappe *f*, Rückschlagklappen-, Rück-
schlag-, Rückström-, Sperr-ventil *n* ~-**valve**
**suction pump** Membranpumpe *f*
**checked** angehakt, angekreuzt, gesperrt, kariert ~ **manoeuvre** parierte Flugbewegung *f* um
die Querachse ~ **pitching manoeuvre** parierte
Flugbewegung *f* um die Querachse
**checker, to** ~ karieren, würfeln
**checker** Aufnahmebeamter *m*, Aufseher *m*, Fadenprüfer *m*, Gitter *n*, Gitterwerksraum *m*,
Gitterwerksstein *m*, Prüfbeamter *m*, Prüfer *m*,
Prüfungsbeamter *m* ~ **boardlike** schachbrettförmig ~ **board pattern** Schachbrettmuster *n*
~ **board staff** Felderlatte *f* ~-**brick** Fachwerks-,
Gitter-, Gitterwerks-, Kammer-stein *m* ~-
**brick superheater** Schamotteüberhitzer *m* ~
**brickwork** Gittermauerwerk *n* ~ **chamber** Git-
terkammer *f*, -werksraum *m*, Kammer *f* ~
**firebrick** Kammerstein *m* ~ **flue** Gitterdurchgang *m* ~ **opening** Gitterdurchgang *m* ~ **pas-**
**sage** Fachwerk-schacht *m*, -kanal *m*, Gitterdurchgang *m*, Gitterwerksschacht *m* ~ **work**
Fachwerk *n*, Fachwerkskörper *m*, Gitterwerk
*n*, Kammerfüllung *f*, Kammermauerwerk *n*,
Würfelwerk *n* (of gas producer) ~ **work of**
**wooden slats** Holzhordenkonstruktion *f* ~
**work grillage** Fachwerkrost *m* ~ **work stack**
Fachwerkschacht *m*
**checkered** gewürfelt, geschacht, kariert, würfelig ~ **goods** gewürfeltes Zeug *n* ~ **jaw** geriffelte
Brechbacke *f* ~ **(metal) plate (or sheet)** Flur-
platte *f*, Riffelblech *n* ~ **wire** Riffeldraht *m*
**checkering** Fachwerkmauerung *f*
**checking** Angießen *n*, Besetztprüfung *f*, Nachprüfung *f*, (slight crisscross cracking) Netz-
adern *pl*, Prüfen *n*, Verzögerung *f* ~ **against**
**prior art** Entgegenhaltung *f* ~ **of baggage** Gepäckaufgabe *f*
**checking,** ~ **arrangement** Eichvorrichtung *f* ~
**case** Meßkoffer *m* (rdo) ~ **circuit** Kontrollstromkreis *m* ~-**code time** Kodeprüfzeit *f* ~
**device** Kontrollvorrichtung *f* ~ **equipment** Kontrollanlage *f* ~ **jib** Einstell-, Prüf-lehre *f* ~ **lever**
Arretierhebel *m* ~ **measurement** Kontrollindex-
maß *n* ~ **plane** Hemmungsebene *f* ~ **record**
Prüfnachweis *m* ~ **system** Kontrollsystem *n*
**cheek** Backe *f*, (gun) Lafettenwand *f*, Wand *f*,
Wange *f*, (in molding) Zwischenkasten *m* ~
**brake** Doppelbackenbremse *f* ~ **piece** Wangen-
stück *n* ~ **stone of gutter** Gossenbordstein *m*
~ **strap** Backenstück *n*
**cheese** Käse *m* ~ **antenna** Segmentantenne *f* ~
**head** runder Kopf *m*, Rundkopf *m* ~ **head**
**with two holes** Kreuzlochkopf *m* ~-**head screw**
Zylinder-, Linsen-schraube *f* ~-**head selftapping**
**screw** Zylinderblechschraube *f* ~-**headed screw**

Zylinderkopfschraube *f* ~-**headed shoulder**
**screw** Linsenschraube *f* mit Ansatz ~ **pile**
Kreuzpulsäule *f*
**chelate,** ~ **combination** Scherenbindung *f* ~
**compound** metallorganische Verbindung *f*
**chelation** Scherenbindung *f*
**cheleutite** Wismutkobaltkies *m*
**chellac bonded textolite** Textolit *n* mit Schell-
lacküberzug
**chemical** chemisch, chemikalisch ~ **air warfare**
aerochemischer oder luftchemischer Krieg *m*
~ **affinity** chemische Verwandtschaft *f* ~ **agents**
chemische Kampfmittel *pl* ~ **agents that re-**
**main in suspension in the air** Schwebstoffe *pl*
~ **analysis** Untersuchung *f* ~ **battalion** Nebel-
werferabteilung *f* ~ **binding effect** Einfluß *m*
der chemischen Bindung ~ **bomb** chemische
Bombe *f*, Gas-, Kampfstoff-bombe *f* ~ **bond**
chemische Bindung *f* ~ **combination** chemische
Verbindung *f* ~ **composition** chemische Zusammensetzung *f* ~ **compound** chemische Verbindung *f* ~ **constitution** chemischer Aufbau *m*
~ **cotton** reine Zellulose *f* ~ **defense** Gasschutz
*m* ~ **dosimeter** chemischer Dosimeter *m* ~
**dressing agents** Appreturzusatzmittel *n* ~
**durability** (of glass) Angreifbarkeit *f* ~ **effic-**
**iency** Umsetzungs-, Wirkungs-grad *m* ~ **en-**
**gineer** technischer Chemiker *m* ~ **equation**
Verbindungsgleichung *f* ~ **filter layer** (diatomite) Diatomitschicht *f* ~ **fog** Farbschleier *m*
(photo) ~ **formula using dashes, lines** Strich-
formulierung *f* ~-**fuel motor** chemischer Kraftstoffmotor *m* ~ **fumes** Kampfstoffschwaden *m*
~ **fuze** chemischer Zünder *m* ~ **impurity**
Fremdatom *n* ~ **interpretation** chemische
Deutung *f* ~ **method of soil consolidation**
Joosten(-sches) Verfahren *n* ~ **plant** Chemika-
lienfabrik *f* ~ **process** chemischer Vorgang *m*,
Chemismus *m* ~ **processing** chemische Bearbeitung *f* ~ **property** Eigenschaft *f* ~ **pulp**
**card** Zellstoffkarton *m* ~ **reprocessing** chemische Aufarbeitung *f* ~ **resistance** chemische
Beständigkeit *f* ~ **retransformation** chemische
Rückwandlung *f* ~ **rocket projector** schweres
Wurfgerät *n* ~ **scales** chemische Waage *f* ~
**separation (of isotopes)** chemische (Isotopen-)
Trennung *f* ~-**spray apparatus** Zerstäubergerät
*n* ~ **straw pulp** Strohzellstoff *m* ~ **symbol** che-
misches Zeichen *n* ~ **test paper impregnated**
**with potassium iodide** Jodkaliumpapier *n* ~
**tracer** chemischer Indikator *m*
**chemically,** ~ **combined (or fixed)** chemisch gebunden ~ **pure (or clean)** chemischrein ~ **re-**
**sistant glass** gegen chemischen Angriff widerstandsfähiges Glas *n*
**chemicals** Chemikalien *pl*
**chemiluminescence** Chemilumineszenz *f*
**chemism** Chemismus *m*
**chemist** Chemiker *m*, Laborant *m*
**chemistry** Chemie *f* ~ **of salts** Halochemie *f*
**chemoluminescence continuum** Chemolumines-
zens-Kontinuum *n*
**chemotechnical** chemotechnisch
**cheque** Scheck *m* ~-**book** Scheckbuch *n*
**cheralite** Cheralit *m*
**cherry** Kugelfräser *m* ~ **coal** weiche, nicht bak-
kende Kohle *f* ~ **picker** geschlitzte Fangbüchse
*f* ~ **red** kirschrot ~-**red heat** Kirschglut *f*,

Kirschrot-glut *f*, -glühhitze *f* ~ **wood** Kirsch-
baumholz *n*
**cherrying attachment** Tiefensteuereinrichtung *f*
**chert** Hornstein *m*
**cherts** Kieselschiefer *m*
**cherty** hornsteinartig ~ **limestone** Kieselkalk-
stein *m*
**chess** Schach *n* ~-**board** Schachbrett *n* ~-**board**
**layout** schachbrettartige Aufstellung *f*
**chest** Bütte *f*, Kasten *m*, Kiste *f*, Lade *f* ~ **of**
**drawers** Schubkasten *m*
**chest,** ~ **bellows** Kastengebläse *n* ~ **blowing**
**machine** Kastengebläse *n* ~-**pack parachute**
Brustfallschirm *m* ~ **strap** Brustgurt *m* (parachute) ~ **type telephone signal** Brustfernsprecher *m* ~ **wheel** Kastenrad *n*
**chestnut** Kastanie *f* ~ **wood** Kastanienholz *n*
**chevkinite** Tschevkinit *m*
**chiastolite** Chiastolith *m*, Hohlspat *m*, Kreuzstein *m*
**chick** Klampe *f*, Küken *n*
**chicken** Huhn *n* ~ **wire** Hühnerdraht *m* ~-**wire-**
**glass** Zelldrahtglas *n*
**chicle** gummiartige Substanz *f*
**chief** Chef *m*, Haupt *n*, Vorgesetzter *m*; hauptsächlich, oberst ~ **beam** Dachbinderbalken *m*
~ **central office** Knotenamt *n* ~ **constituent**
Hauptbestandteil *m* ~ **designer** Chefkonstruk-
teur *m* ~ **electrician** Oberbeleuchter *m* ~ **en-**
**gineer** leitender Ingenieur *m*, Oberingenieur *m*
~ **(or principal) frame** Scherspant *m* ~ **in-**
**spector of mines** Berghauptmann *m* ~ **inter-**
**ference zone** Störnebel *m* ~ **matter** Hauptsache
*f* ~ **offender** Hauptschuldiger *m* ~ **operator**
Aufsichts-beamter *m*, -beamtin *f*, Betriebsmeister *m*, Meister *m*, Vorarbeiter *m* ~ **ope-**
**rator's desk** Aufsichts-platz *m*, -tisch *m* ~ **pilot**
Chefpilot *m* ~ **rafter** Füllsparren *m* ~ **stoker**
**chiefly** hauptsächlich, vorwiegend
Oberheizer *m*
**Chile,** ~ **mill** Kollergang *m* ~ **niter (or salpeter)**
Chilisalpeter *m*
**Chilean mill** chilenische Mühle *f*
**chill, to** ~ abkühlen, abschrecken, erkälten,
erkalten, glashart machen (metall.), Kühlen,
in Kokillen gießen, in der Kokille vergießen
**to** ~-**cast** hartgießen **to** ~ **iron** das Roheisen
abschrecken
**chill** Form *f*, Frost *m*, Gußeisenform *f* (mold),
Kühleisen *n*, Schalengußform *f*, Schreckplatte *f*
~ **casting** Hartguß *m*, Hartgußstück *n*, Kokillenguß *m*, Kokillengußstück *n*, Schalenguß
*m* ~ **form** eiserne Form *f* ~ **mold** Kapsel *f*
(foundry), Kokillenform *f*, Kühlkokille *f* ~
**room** Kältelagerraum *m*
**chilled** schalenhart, in Kokille gegossen, tief-
gekühlt ~ **bottom** gehärteter Pflugkörper *m*
~-**cast iron** Hartgußkörper *m* ~-**cast-iron**
**shell** Hartgußgranate *f* ~ **core** Schalengußkern
*m* ~ **fire bar** Hartgußroststab *m* ~ **goods** Kühl-
gut *n* ~ **(cast) iron** Hartguß *m* ~-**iron runner**
Hartgußläufer *m* ~ **margin** Abkühlungsfläche *f*
~ **milk** Kaltmilch *f* ~ **plow** Pflug *m* mit Hartgußkörper und Schar (Schwingpflug) ~ **roll**
Hartgußwalze *f*, Hartwalze *f* ~ **roll iron** Walzenguß *m* ~ **roller** glasharte Walze *f* ~ **rolls**
geriffelte Walzen *pl* ~ **shot** Hartgußgeschoß *n*
~ **work** Kapselguß *m*

**chilling** Abkühlung *f*, Abschrecken *n*, Abschrek-
kung *f* ~ **effect** Abschreck-, Kühl-wirkung *f*
~ **device** Kühlvorrichtung *f* ~ **form** Abschreckform *f* ~ **point** Kristallisationsbeginn *m* (chem.)
**chilly** naßkalt, naßkühl
**chime** Gargel *f*, Läutewerk *n*
**chimera** Chimäre *f*
**chimney** Blasehals *m* (eines Ventilators), Dige-
storium *n*, Esse *f*, Kamin *m*, Quandelschacht *m*,
Rauchfang *m*, Schlot *m*, Schornstein *m* ~ **of**
**a building** Hausschornstein *m*
**chimney,** ~ **base** Schornsteinsockel *m* ~ **breech-**
**ing** Kaminabzug *m* ~ **cowl** Schornsteinaufsatz
*m* ~ **damper** Schornstein-, Zug-schieber *m* ~
**draft** Essen-, Kamin-, Schornstein-zug *m* ~
**effect** Schornsteinwirkung *f* ~ **fender** Kamin-
gitter *n* ~ **flue** Anzugs-, Rauch-kanal *m*, Rauch-
rohr *n* ~ **hood** Kaminhaube *f*, Rauchfang-,
Rauch-mantel *m*, Schurz *m* ~ **intake at base**
Fuchs *m* ~ **jambs** Kamin-einfassung *f*, -ge-
wände *n* ~ **mantel** Schurz *m* ~ **mantle** Rauch-
fang-, Rauch-mantel *m* ~ **spring** Kaminschnäp-
per *m* ~ **stack** Essenschaft *m* ~ **stalk (or shaft)**
Dampfesse *f* ~ **sweeper** Essenkehrer *m*, Feuer-
rüpel *m* ~ **valve** Essen-, Kamin-ventil *n*
**chin** Kinn *n* ~ **curb socket (or groove)** Kinnket-
tengrube *f* ~ **strap** Kinn-riemen *m*, -stößel *m*,
Sturmriemen *m* ~ **support** Kinnstütze *f* ~ **tug**
Kinnstößel *m*
**china** Porzellan *n* ~-**clay** Kaolin *n* ~-**clay wash-**
**ing** Kaolinschlämmerei *f*
**China,** ~ **grass** Nessel *f* ~ **kaolin** Chinaton *m*
**chine** (seaplane) Bodenquerschnitt *m*, Kimm *m*,
Kimme *f*, Kimmkante *f*, Kimmträger *m*
**Chinese blast furnace** chinesischer Gebläseofen
*m*
**chink** metallisches Klingen *n*, Sprung *m*, Spaltöffnung *f* (in clarinet) ~ **between mouthpiece**
**and reed** Mundstücköffnung *f*
**chinked (wood)** gerissen
**chinking** Klirren *n*
**chinkolobwite** Chinkolobwit *m*
**chinook** Chinookwind *m*, Föhn *m*
**chintz paper** Kattundruckpapier *n*
**chip, to** ~ abrauhen, behauen, behobeln, knippen, meißeln, schnitzeln, schnitzen, schroten,
zerspanen **to** ~ **off** abblättern, abklopfen, abplatzen **to** ~ **out (or off) with the chisel** ab-
stemmen mit dem Meißel
**chip** Fragment *n*, Hobelspan *m*, Schnipsel *n*,
Schnipselchen *n*, Span *m*, Splitter *m*, Stückchen *n* ~ **ability** Zerspanbarkeit *f* ~ **ax** Schlicht-
beil *n* ~ **basket** Spankorb *m* ~-**board** Maschi-
nengraukarton *m*, Schrenz- und Speltpappe *f*,
Spanholzplatte *f* ~-**board press** Spanplatten-
presse *f* ~ **box** Schnitzelkasten *m* ~ **breaker**
Span-brecher *m*, -brechernute *f*, -formernute *f*
~ **breaker flute** Spanbrechernut *m* ~ **breaker**
**(or breaking) grooves** Spanbrechernuten *pl*
~ **breaker (or breaking) steps** Spanbrecher-
stufen *pl* ~ **briquetting press** Spänebrikettie-
rungspresse *f* ~ **clearance** Spänefall *m*, Spanlücke *f* **(suitable)** ~ **chute** Spänefang *m* ~ **cross**
**section (or profile)** Spannpreßquerschnitt *m*
~ **deflector** Spanlenker *m* ~ **detaching** spanabhebende Arbeit *f* ~ **exhaust system** Spanab-
saugung *f* ~ **handling installation** Spanbeför-
derung *f* ~ **loft** Schnitzelspeicher *m* ~ **pan**

Späneschale ~ **pressure** Spanndruck *m* ~**-proof** nicht abblätternd ~ **removel** Spanabnahme *f*, spanabhebende Bearbeitung *f*, Spanentleerung *f* ~ **screen** Schnitzelsortierer *m*, Schüttelsortierer *m*, Sortiertrommel *f* für Hackspäne (paper mfg.) ~ **separator** (or **centrifuge**) Späneschleuder *f* ~ **space** Spanlücke *f*, Spanraum *m* ~ **space** (or **clearance**) Spänedurchlaß *m* ~ **tray** Spanfangschale *f*, Spänepfanne *f*, Spänekasten *m* ~ **width** Spanbreite *f* ~ **wood** Spanholz *n*
**chipper** Gußputzer *m*, Hacker *m*, Hackmaschine *f*, Putzmeißel *m* ~ **tube** Schwellenröhre *f*
**chipping** Abmeißeln *n* ~ **chisel** Flachmeißel *m* ~ **hammer** Meißelhammer *m* ~ **knife** Kabelmesser *n* ~ **work** Stemmarbeiten *pl*
**chipping** Abfall *n*, Brocken *m*, Kleinschlag *m*, Splitt *m* ~ **of stone** Steinbrocken *pl*
**chips** Bohrspäne *pl*, Späne *pl*, Stoffabfall *m* ~ **of wood** Holzabfälle *pl* ~ (or **shavings-**) **clarifying washer** Klärspänewaschmaschine *f*
**chirality** Chiralität *f*
**chirps** Trillern *n*, Zirpen *n*
**chisel, to** ~ meißeln, schroten, stemmen **to** ~ **off** abmeißeln
**chisel** Meißel *m*, Stecheisen *n*, Stemmeisen *n*, Treibwerkzeug *n* (**calking**) ~ Stemmeißel *m* (**chipping**) ~ Beitel *m* ~ **joint** Meißelspitz *f* (**ripping**) ~ Stechbeitel *m* (**scrap**) ~ Schrotmeißel *m* ~ **shank** Bohrereinsteckende *n* ~ **steel** Meißelstahl *m* ~ **work** Ziselierarbeit *f*
**chiselled toothing** gehauene Verzahnung *f*
**chiseling,** ~ **out** Ausstemmung *f* ~ **work** Stemmarbeiten *pl*
**chit, to** ~ aufschießen, keimen
**chloanthite** Chloanthit *m*, Weißnickel-erz *n*, -kies *m*
**chloracetic acid** Chloressigsäure *f*
**chloracetophenone** Chlorazetophenon *n* ~ **solution** Chlorazetophenonchlorpikrinlösung *f*
**chloramine** Chloramin *n*
**chlorate** Chlorat *n* ~ **discharge** Chloratätze *f* ~ **explosive** Chloratsprengmittel *n*
**chlorauric acid** Goldchlorwasserstoffsäure *f*
**chlorhydrate** Chlorhydrat *n*
**chlorid,** ~ **acid** Bleichsäure *f*, Chlorsäure *f* ~ **acid anhydride** Chlorsäureanhydrid *n*
**chloride** Chlorid *n*, Chlorür *n* ~ **of arsenic** Chlorarsenik *n* ~ **of chrome mordant** Chlorchrombeize *f* ~ **of lime** Bleich-kalk *m,* -mittel *n*, -pulver *n*; Chlor-kalk *m*, -lauge *f* ~ **of silver** Hornerz *n* ~ **of soda** Chlornatron *n* ~ **of zinc** Zinkbutter *f*
**chloride,** ~ **cake** Chloridschlacke *f* ~ **solution** Chloridlauge *f* ~ **storage cell** Chloridsammler *m* ~ **volatilizing process** Chloridverdampfungsprozeß *m*
**chloridizable** chlorierungsfähig
**chloridization** Chlorierung *f*
**chloridizing roasting** chlorierendes Rösten *n*
**chlorinate, to** ~ chloren, chlorieren
**chlorinated rubber** Chlorkautschuk *m*
**chlorinating,** ~ **agent** Chlorierungsmittel *n* ~ **roasting** chlorierendes Rösten *n*
**chlorination** Chloration *f*, Chlorierung *f*, Verchlorung *f* ~ **of waste water** Abwasserchlorung *f*, Chlorung *f* von Abwasser ~ **barrel** Chlorationstrommel *f*
**chlorine** Bleichsäure *f* ~ (**gas**) Chlor *n*, Chlor-

gas *n*, Chlorin *n* ~ **liquefying plant** Chlorverflüssigungsanlage *f* ~ **monoxide** Chlormonoxyd *n* ~**-oxygen-hydrogen explosion** Chlorknallgasexplosion *f* ~**-peroxide bleach** Chlorsauerstoffbleiche *f* ~ **trifluoride** Chlortrifluorid *n*
**chlorinity** Chlorgehalt *n*
**chlorite** Chlorit *m* ~ **slate** Chloritschiefer *m*
**chloritic** chloritisch
**chloroamyl** Chloramyl *n*
**chloroauric acid** Aurichlorwasserstoffsäure *f*
**chlorobenzene** Chlorbenzol *n*
**chlorobenzoic acid** Chlorbenzoesäure *f*
**chlorocarbonic acid** Chlorkohlensäure *f*
**chlorochromic acid** Chlorchromsäure *f*
**chloro-dihydrophenarsazine** Chlorodihydrophenarsazin *n*
**chloroethane** Chloräthyl *n*
**chlorometer** Chlormesser *m*
**chloropalladic acid** Palladiumchlorwasserstoff *m*
**chloropalladous acid** Palladochlorwasserstoffsäure *f*
**chlorophyll** Blättergrün *n*, Blattgrün *n*
**chloropicrin** Chlorpikrin *n*
**chloroplatinic acid** Chlorplatinsäure *f*, Platinchlorwasserstoffsäure *f*
**chlorostannic acid** Stannichlorwasserstoffsäure *f*, Zinnchlorwasserstoffsäure *f*
**chlorostannous acid** Stannochlorwasserstoffsäure *f*
**chlorosulfonic acid** Chlorsulfonsäure *f*
**chlorous** chlorig ~ **acid** Chlorigsäure *f*, Chlorürsäure *f*
**chock** Bock *m*, Bremskeil *m*, Einbaustück *n*, Hemmkeil *m*, Holzpfeiler *m*, Staukeil *m*, Zapfenlager *n* (of roller) **to** ~ **a wheel** ein Rad *n* hemmen ~ **balk** Knaggenbalken *m* ~ **block** Bremsklotz *m* (aviat.), Abbremsklotz *m* ~ **release** Holzpfeilerfreiträger *m*
**chocks** Bremsbackenklötze *pl*, Bremsklötze *pl*
**choice** Auswahl *f*, Wahl *f*; trefflich, wahlweise ~ **of frequency** Frequenzwahl *f*
**choir** Chor *m* ~ **loft** Empore *f*
**choke, to** ~ abdrosseln, abwürgen, drosseln, ersticken, stauen, verstopfen **to** ~ **out** vollständig abdrosseln, unterdrücken **to** ~ **teeth of a file** eine Feile *f* verschmieren
**choke** Abflachungsdrossel *f*, Drehfeldspule *f*, Drossel *f*, Drosselspule *f*, Gegendruck *m* bei Ölbohrungen, Luftklappe *f* des Vergasers, Stärkerklappe *f*, Starterklappe *f*, Vergaserlufttrichter *m*, Würgebohrung *f*, Zubruchgehen *n* ~ **arm** Regulierhebel *m* ~ **bore** Drosselbohrung *f* ~ **circuit** Sperrkreis *m* ~ **coil** Schutzdrossel *f* ~ **condenser** Abflachungskondensator *m*, Überspannungsschutzdrossel *f* ~**-coupled amplifier** Drossel-spulenverstärker *m*, -verstärker *m* ~ **coupling** Drosselkopplung *f* ~ **damp** Ferch *m*, Nachschwaden *m*, Stickwetter *pl*, böses Wetter *n*, ~ **flange** Drosselflansch *m* (im Hohlleiter) ~ **input** Drosseleingang *m* ~ **joint** Drosselverbindung *f*, induktive Hohlleiterkopplung *f* ~ **modulation** Anodenspannungsmodulation *f*, Heising-Modulation *f*, Parallelröhrenmodulation *f* ~ **piston** Kurzschlußschieber *m* ~ **plunger** Kurzschlußschieber *m* ~**-transformer** Drosseltransformator *m* ~ **tube** Drosselrohr *n*, Luft-düse *f*, -trichter *m* ~ **tube stop-screw** Lufttrichterhalteschraube *f* ~ **valve** Vordrossel *f*

choked abgedrosselt, verstopft **to become ~** sich zusetzen **~ fire** Schmauchfeuer n **~ flange** Drosselflansch m **~ pump** unklare Pumpe f

choker valve Drosselventil n

choking Einschnürung f, Streckenschlag m (in a rolling mill), Verstopfung f; stickig **~ coil for higher harmonics** Oberwellendrossel f **~ effect** Drosselwirkung f **~ effect on the flow** Drosselwirkung f, Drosselung f des Kühlluftstromes **~-off** Unterbindung f **~-out** Unterdrückung f **~ tube** Drosselröhre f **~-up** Verschlämmung f **~-up with sand** Versandung f

cholestanol Cholestanol n

cholesterin Cholesterin n

choline Cholin n

choose, to **~** aussuchen, (aus)wählen

choosing Bemessung f

chop, to **~** hacken, zerhacken **to ~ off** abhacken, abschroten **to ~ out** verhacken, verhauen, versetzen **to ~ up** einhacken **to ~ to pieces** zerstückeln; Schnittlänge f (paper mfg.) **~ hammer** Stielhammer m

chopped zerhackt **~ filled plastic** Schnitzelmasse f **~ light** Wechsellicht n **~ straw** Häcksel m, n

chopper Drehschalter m, Futterklinge f, Hacker m, Hackmaschine f, Schnitzelmesser n, Unterbrecher m, Zerhacker m **~ amplifier** Verstärker m mit Eingangsunterbrecher **~ bar** Schreibstange f **~-bar controller** Fallbügelregler m (elektr. Meßgeräte) **~ bar recorder** Fallbügelschreiber m **~ bar recording** Punktschreibung f **~ diaphragm** Schwingblende f **~ disk** Lochscheibenunterbrecher m **~ excitation** Zerhackererreger m **~-type monitoring** Fallbügelabtastung f (aut. contr.) **~ wheel** Lichtsirene f

chopping Kabelung f **~ axe** Hackaxt f **~ bench** Schnitzelbank f **~ bit** Meißel f **~ blade** Futterklinge f **~ knife** Futterklinge f, Hackmesser n **~ machine** Häckelschneider m, Abschlagemaschine f (match mfg.) **~ sea** Kabbelsee f

choppy abgebrochen (air wave), gekräuselt, kabbelig (sea) **~ sea** unruhige See f mit kurzen Brechern

chord Gurt m (of truss or girder), Gurtung f, Leine f, Profilsehne f, Saite f, Schnur f, Sehne f (geom.), Seil n, Strang m, Tiefe f (of wing) **~ of combination of wings** Gesamtflügeltiefe f **~ at a plane** Flächensehne f (aviat.) **~ at root** Flügeltiefe f an der Wurzel **~ at tip** Flügeltiefe f an der Spitze **~ of a truss** Gurtung f eines Brückenträgers

chord, **~ buzzer** Saitensummer m **~ depth of control surface** Rudertiefe f **~ distribution** Tiefenverteilung f **~ incidence** Anstellwinkel m **~ length** Profiltiefe f, Länge f der Flügelsehne **~ line** Flügel-, Profil-sehne f **~ member** Gurtstab m **~ plate** Gurtplatte f **~ taper** Tiefenverjüngung f **~-type armature** Sehnenanker m (Zündspule) **~ winding** gesehnte Wicklung f **~-wise** längsaxial, in Richtung der Längsachse **~-wise bending** Blattverbiegung f in Drehebene

chordal, **~ cut** Ader-, Sehnen-, Tangential--schnitt m **~ section** Ader-, Flader-, Sehnen-, Tangential-schnitt m **~ surface** Sehnenebene f

chorded, **~ cooling tube** Rohrhärte f **~ winding** Sehnenwicklung f (electr.)

chords Glasschlieren n

chore leichte Arbeit f

Christmas tree antenna Tannenbaum-, Weihnachtsbaum-antenne f

chroma Farbstärke f, Farbton m (TV) **~ coder** Buntsignalumsetzer m (TV) **~ control** Farbtonregler m (TV) **~-saturation** Farbensättigung f (TV)

chromate Chromat n, Chromsalz n **~ of potassium** Chromkali n

chromatic chromatisch **~ aberration** chromatische Aberration f oder Abweichung f, chromatischer Fehler m, Farbabweichung f, Farbenfehler m, Farbenzerstreuung f (opt.), Farbfehler m **~ bleeding** chromatischer Schaum m **~ color** bunte Farbe f **~ defect** Farb(en)fehler m **~ dispersion** Fächerung f **~ identity** Farbengleichheit f **~ pitch pipe** chromatischer Tonangeber m **~ purity** Farbenreinheit f **~ selection** Farbauszug m **~ sensation** Farbenempfindung f **~ sensitivity** Farbenempfindlichkeit f **~ soft bleeding, soft focus** chromatischer Schaum m **~ spectrum** Farbenspektrum n **~ thermometer** Meßbleistift m

chromaticities Farbwerte pl

chromaticity Farbton m (TV), Farbtonempfindlichkeit f des Auges (TV) **~ scale** Farbenskala f **~ scale system** Farbmaßsystem n

chromatics Farbenlehre f

chromatid Chromatid n **~ break** Chromatidenteilung f

chromatin Chromatin n

chromatism Farbenzerstreuung f

chromatogram scanner Radiopapierchromatograf m

chromatograph strip Chromatografiestreifen m

chromatographic (or electrophoresic) paper strips Chromatografie- oder Elektrophorese-Papierstreifen m

chromatometer Farbenmesser m

chromatometrics Farbenmessung f

chromatometry Farbenmessung f

chromatopsia test Farbsehprüfung f

chrome, to **~ after dyeing** nachchromieren **to ~-plate** verchromen

chrome Chrom n, Parisergelb n **~ alum** Chromalaun n **~-alumen solution** Chromeiweißkopierlösung f **~ black** Chromschwarz n **~ gelatin** Chromgallerte f **~ green** Chromgrün n **~ imitation board** Chromoersatzkarton m **~ iron ore** Chromeisenstein m, Chromit m **~ molybdenum steel** Chrommolybdänstahl m **~ nickel** Chromnickel n **~-nickel steel** Chromnickelstahl m, Nickelchromstahl m **~ ore** Chromerz n **~ plated piston ring** Chromring m **~ plating** Verchromung f **~ red** Chromrot n **~ steel** Chromstahl m **~ tan liquor** Chromierungsbrühe f **~-tungsten steel** Chromwolframstahl m **~ yellow** Chromgelb n

chromic chromhaltig **~ acid** Chromsäure f **~ acid cell** Chromsäureelement n **~ anhydride** Chromsäureanhydrid n **~ carbide** Chromkarbid n **~ chloride** Chlorchrom n **~ chromate** Chromchromat n **~ compound** Chromverbindung f **~ cyanide** Chromzyanid n **~ fluoride** Chromfluorid n **~ hydroxide** Chromhydroxyd n, Chromooxydhydrat n **~ iodide** Chromjodid n **~ iron** Chromeisen n, Ferrochrom n **~ molybdate** Chrommolybdat n **~ oxalate** Chromo-

oxalat *n* ~ **oxide** Chromooxyd *n* ~ **potassium cyanide** Chromizyankalium *n* ~ **potassium sulfate** Chromkaliumsulfat *n* ~ **salt** Chromisalz *n*, Chromoxydsalz *n* ~ **spinel** Pikotit *m*
**chromiferous** chromhaltig
**chrominance** Buntheit *f* (TV) ~ **information** Buntinformation *f* (TV) ~ **signal** Buntsignal *n* ~ **signal elimination** Buntsignalsperre *f* ~ **subcarrier** Bunthilfsträger *m*
**chromit** Ferrochrom *n*
**chromite** Chrom-eisen *n*, -eisenstein *m*, -erz *n*, Chromit *m*, Eisenchrom *n*
**chromium** Chrom *n* ~ **acetate** Chromazetat *n* ~ **carbide** Chromkarbid *n* ~ **chloride** Chlorchrom *n*, ~ **chromate** Chromchromat *n* ~ **cyanide** Chromzyanid *n* ~ **fluoride** Chromfluorid *n*, Fluorchrom *n* ~ **hydroxide** Chromhydroxyd *n*, Chromooxydhydrat *n* ~ **molybdate** Chrommolybdat *n* ~ **oxalate** Chromoxalat *n* ~ **oxide** Chrom-grün *n*, -oxyd *n* ~ **oxychloride** Chromooxychlorid *n*, Chromylchlorid *n* ~ **oxyfluoride** Chromoxyfluorid *n* **to** ~-**plate** verchromen ~ **plating** Verchromung *f* ~-**plating plant** Verchromungsanlage *f* ~ **potassium sulfate** Chromalaun *n* ~ **salt** Chromsalz *n* ~ **silicon steel** Chromsiliziumstahl *m* ~ **steel** Chromstahl *m*
**chromize, to** ~ inchromieren
**chromo,** ~**board** Chromokarton *m* ~ **blotting board** Chromolöschkarton *m* ~ **paper** Kunstdruckpapier *n*
**chromogenic** farbenerzeugend
**chromolithography** Farbensteindruck *m* (print.)
**chromonitric acid** Chromsalpetersäure *f*
**chromoscope** Farbfernsehbildröhre *f*
**chromosphere** Chromosphäre *f*
**chromosulfuric acid** Chromschwefelsäure *f*
**chromotypography** Buntdruck *m*
**chromous,** ~ **chloride** Chromchlorür *n*, Chromchlorid *n* ~ **compound** Chromoverbindung *f*, Chromoxydulverbindung *f* ~ **fluoride** Chromfluorür *n* ~ **hydroxide** Chromhydroxydul *n* ~ **oxide** Chromoxydul *n*
**chromyl chloride** Chromooxychlorid *n*, Chromylchlorid *n*
**chronaximeter** Chronaximeter *n*
**chronic exposure** dauernde Strahlungseinwirkung *f*
**chronoelectrical recording** chronoelektrische Registrierung *f*
**chronograph** Chronograf *m*, registrierender Zeitmesser *m*, Zeit-blinker *m*, -schreiber *m*
**chronological** zeitgeordnet, zeitlich, zeitlich aufeinanderfolgend ~ **sequence** zeitliche Reihenfolge *f*
**chronology** Zeitrechnung *f*
**chronometer** Chronometer *m*, Zeit-messer *m*, -meßgerät *n* ~ **escapement** freie Hemmung *f* ~ **test calibration** Uhrprüfung *f*
**chronometric** zeitmeßkundlich
**chronopher** Zeitzeichengeber *m*
**chronoscope** registrierender Zeitmesser *m*, Zeitmeßgerät *n*
**chronotachometer** Stichdrehzähler *m*
**chrysoberyl** Chrysoberyll *m*, Goldberyll *m*
**chrysocolla** Kupfergrün *n*, Malachitkiesel *m* ~ **ore** Chrysokollerz *n*
**chrysoidine** Chrysoidin *n*

**chuck, to** ~ einspannen, spannen
**chuck** Aufspannfutter *n*, Bohrmaschinenfutter *n*, Drehbankfutter *n*, Einspannvorrichtung *f*, Futter *n*, Klemme *f*, Klemmkonus *m*, Planscheibe *f*, Spannfutter *n*, Spannschraube *f*, Spannvorrichtung *f*, Spannwerkzeug *n*, Verschluß *m* ~ **and bar work** Futter- und Stangenarbeit *f* ~ **for circular grinding** Aufspannfutter *n* zum Rundschleifen
**chuck,** ~ **capacity** Spannbereich *m* ~ **collet** Spundfutter *n* (Drechslerbuchse) ~ **fingers** Spannzange *f* ~ **jaw** Einspann-, Futter-backe *f* ~ **jaws** Einspannbacken *pl*, Spannfutter *n* ~ **lathe** Spindeldrehbank *f* ~ **operation** Spannfutterbetätigung *f* ~ **plates** Futterplatten *pl* ~ **table** Spannfuttertisch *m*
**chucking** Ein-, Fest-spannen *n*, Spannen *n* ~ **by compressed air** Preßluftfestspannung *f* ~ **capacity** Spanndurchmesser *m* ~ **device** Einspannvorrichtung *f* ~ **equipment** Kraftspanneinrichtung *n*, Spannmittel *pl* ~ **mechanism** Spanngetriebe *n* ~ **reamer** Maschinenreibahle *f* ~ **tool** Spannzeug *n* ~ **wedge** Spannkeil *m* ~ **work** Spannarbeiten *pl*
**chug, to** ~ puffen
**chugging** unregelmäßige Verbrennung *f*
**chunky** in Klumpen
**church** Kirche *f* ~ **tower** Kirchturm *m* ~**yard** Friedhof *m*, Kirchhof *m*
**churn, to** ~ schütteln, eine Flüssigkeit *f* durchschütteln
**churn** Butterfaß *n*, Erdbohrer *m*, Kneter *m* ~ **drill** Seil-bohren *n*, -bohrung *f*
**churning,** ~ **of water** (as by a baffled pump) Durcheinanderwirbeln *n* des Wassers ~ **machine** Kirnvorrichtung *f*
**chute, to** ~ rutschen
**chute** Absturzrinne *f*, Bremsschirm *m* (aviat.), Rinne *f*, Rutsche *f*, Schacht *m*, Schleife *f*, Schurre *f*, Schütte *f*, Schüttrinne *f*, Sturz *m*, Trog *m*, Verschlag *m* ~ (**of film**) Gleitbahn *f* ~ **for sack conveyance** Sackwendelrutsche *f* (**tip**) ~ Sturzrinne *f*
**chute,** ~ **adjustment** Rutscheneinstellung *f* ~ **boot of a sounding rocket** Fallschirmbehälter *m* ~ **separator** Rutschabscheider *m* ~ **trap** Schüttklappe *f*
**chutes and ducts** Rinnen und Leitungen *pl*
**cigar,** ~ **bander** Zigarrenbändchenumleger *m* ~ **cutter** Zigarrenabschneider *m* ~ **lighter** Zigarrenanzünder *m*
**cigarette,** ~ **lighter** Feuerzeug *n* ~-**machine ribbon** Zigarettenmaschinenband *n* ~-**tube machine** Zigarettenhülsenmaschine *f*
**cilia forceps (or tweezers)** Zilienpinzette *f*
**cinch strap** Gurtstrippe *f*
**cinchona bark** Chinarinde *f*
**cinchonine hydrochloride** Cinchoninhydrochlorid *n*
**cinder** Asche *f*, Koks *m*, Schlacke *f*, Sinter *m*, Skorie *f*, Stückschlacke *f*, Zinder *m*, Zunder *m* ~ **box** Schlackenkasten *m* ~ **coal** durch Intrusion veränderte Kohle *f* ~ **collector** Schlacken-sammelgefäß *n*, -sammler *m* ~ **dump** Schlackenhalde *f* ~ **fall** Aschenfall *m*, Schlakkengang *m* ~ **frame** Funken-rost *m*, -sieb *n* ~ **funnel** Aschentrichter *m* ~ **heat** Abschweiß-, Entzünderungs-wärme *f* ~ **inclusion** Schlacken-

einschluß *m* ~ **iron** Schlacken-eisen *n*, -spieß *m* ~ **notch** Schlacken-form *f*, -loch *n*, -öffnung *f*, -stich *m*, -stichloch *n* ~ **outlet** Aschenabzug *m* ~ **paste** Schlichte *f* ~ **pig** Schlackenroheisen *n* (metall.) ~ **pit** Aschengrube *f*, Schlacken-gang *m*, -grube *f* ~ **pit man** Pitsreiniger *m* ~ **plate** Schlackenzacken *m* ~ **pocket** Schlacken--sammelgefäß *n*, -sammler *m* ~**-quenching trough** Lösche *f* ~ **spout** Schlackenrinne *f* ~ **stone** Schlackenstein *m* ~ **tap** Schlacken-loch *n*, -öffnung *f*, -stich *m*, -stichloch *n* ~ **yard** Schlakkenhalde *f*

**cinders** Braschen *pl*

**cindery** schlackig

**cinema** Kino *n* ~ **base-line apparatus** Kinobasisgerät *n*

**cinematic spotter (or spotting) device** Kinoflekker *m*

**cinematograph** Bildfänger *m*, Filmaufnahmekammer *f*, Laufbild-fänger *m*, -kammer *f*, -werfer *m* ~ **picture projector** Bildwerfer *m*

**cinematographic,** ~ **action** Bewegungsvorgang *m* ~ **flight** Reihenbildflug *m*

**cinematography** Kinematografie *f*

**cinnabar** Merkurblende *f*, Zinnober *m* ~ **green** Zinnobergrün *f* ~ **ore** Zinnober-erde *f*, -erz *n* ~ **scarlet** Zinnoberscharlach *n*

**cinnamaldehyde** Zimtaldehyd *n*

**cinnamic acid** Cinnamylsäure *f*

**cinnamon** Zimt *m* ~ **stone** Hessonit *m*

**cinnamylic acid** Cinnamylsäure *f*

**cinophot** Belichtungsmesser *m*

**cipher, to** ~ chiffrieren, schlüsseln, verschlüsseln, verziffern

**cipher** Chiffre *f*, Chiffreschrift *f*, Initialen *pl* als Wasserzeichen (paper mfg.), Null *f*, Nummer *f*, Schlüssel *m*, Schlüsselbuchstabe *m*, Ziffer *f* ~ **code** Chiffre *f*, Chiffernschlüssel *m* ~**-code typewriter** Geheimschreibmaschine *f* ~ **device** Handschlüsselvorrichtung *f* ~ **disk** Verschlüsselungsscheibe *f* ~ **key** Geheimwort *n* ~ **machine** Schlüsselmaschine *f* ~ **roll for counters** Zahlenrolle *f* für Zählwerke ~ **text** Geheimtext *m* ~ **writing** Chiffrier-, Geheim-schrift *f*

**ciphered message** Chiffretelegramm *n*

**ciphering,** ~ **code** Chiffreschlüssel *m* ~ **machine** Chiffriermaschine *f* ~ **message** Chiffriertext *m*

**ciphony** verschlüsselter Sprechverkehr *m*

**circle, to** ~ kreisen, umkreisen **to** ~ **out** ausblenden

**circle** Drehungskreis *m*, Kreis *m*, Kreislinie *f*, Ringel *m*, Ronde *f*, Rondell *n*, Umfang *m*, Zirkel *m* 1/400 of a ~ Neugrad *m*

**circle,** ~ **of altitude** Höhenzirkel *m* ~ **of confusion** Fehlerscheibchen *n*, Zerstreuungskreis *m* ~ **of curvature** Krümmungskreis *m* ~ **of diffusion** Unschärfering *m*, Zerstreuungskreis *m* ~ **of glass surrounding a stone** Hof *m* ~ **and helix** Kreis *m* und Spirale *f* ~ **of the horizon** Gesichtskreis *m* ~ **of latitude** Ost-Westlinie *f* ~ **of least aberration** Kreis *m* der geringsten Aberration ~ **of least confusion** Kreis *m* der geringsten Unschärfe ~ **of least diffusion** Zerstreuungsscheibchen *n* ~ **of longitude** Längenkreis *m* ~ **of unit radius** Einheitskreis *m* ~ **of vision** Sehkreis *m*

**circle,** ~ **bearing** Kreispeilung *f* ~ **bulb** Projektionsbirne *f* ~**-cutting machine** Kreisschere *f*

~ **diagram** Kreis-, Orts-diagramm *n* ~ **disk** Wälzzylinder *m* ~**-dot mode** Zirkel-Punkt-Methode *f* ~ **graduation** Kreisstellung *f* ~**-in** Abblenden *n* (film) ~ **marker** Kreis-, Rund--anzeiger *m* ~ **minute** Kreisminute *f* ~**-out** Aufblenden *n* (film) ~ **protractor** Vollkreistransporteur *m* ~ **reading point** Kreisablesestelle *f* ~ **segment** Kreisabschnitt *m* ~ **voltage** Kreisspannung *f*

**circles,** ~ **in any position** Kreise *pl* in beliebiger Lage ~ **diagram** Stromlaufschema *n* (rdo)

**circling,** ~ **guidance lights** Platzrundenführungsfeuer *pl* (aviat.) ~ **mark** kreisende Marke *f* ~ **target** kreisendes Ziel *n*

**circlip** Seeger-, Spreng-ring *m*

**circuit, to come into** ~ eingeschaltet werden

**circuit** Anlage *f*, Fernmeldeverbindung *f*, Gleichrichter *m* oder Röhrenanordnung *f*, Kette *f* (electr.), Kreis *m*, Kreislauf *m*, Leitung *f*, Platzwechsel *m*, Röhrenschaltung *f*, Schaltglied *n*, Schaltung *f*, Schaltungsanordnung *f*, Schlängelung *f*, Strom *m*, Stromkreis *m*, Umfang *m*, Umlaufkreis *m* **in** ~ angeschlossen **out of** ~ stromlos **over a closed** ~ in geschlossener Bahn **f to do a** ~ eine Platzrunde *f* fliegen **via** ~ Transitstromkreis *m* ~ **with (or without) busy lamp** Leitung *f* mit (oder ohne) Besetztlampe ~ **in good order** einwandfreie Leitung *f* ~ **of holding coil** Haltestromkreis *m* ~ **of the iconoscope** Schaltung *f* des Ikonoskops ~ **in which magnetization does not proceed at uniform rate** Bremsfeldschaltung *f* the ~ **is regular** die Leitung ist normal geschaltet **the** ~ **is singing** die Leitung pfeift ~ **for sound-and flash-ranging station** Meßleitung *f*

**circuit,** ~ **analog** Schaltungsanalogon *n* ~ **angle** Phasenwinkel *m* (der Stromrichterschaltung) ~ **arrangement scheme** Schaltung *f* ~ **blocking** Sperrkreiskoppelung *f* ~ **break** Ausschaltung *f* ~ **breaker** Ausschalter *m*, Leistungsschalter *m*, Schaltschütz *m*, Schutzschalter *m*, Selbstschalter *m*, selbsttätiger Unterbrecher *m*, Sicherungsautomat *m*, Spannungssicherung *f*, Stromkreisunterbrecher *m*, Stromunterbrecher *m*, Trennschalter *m*, Überlastschalter *m*, Unterbrecher *m* ~**-breaker capacity** zulässiger Ausschaltstrom *m* ~**-breaker oil** Schalteröl *n* ~ **breaking** Unterbrechungsvermögen *n* ~**-breaking capacity** Abschalt-. Schalter-leistung *f* ~ **breaking key** Ausschalttaste *f* ~ **busy hour** Hauptverkehrsstunde *f* für eine Leitung oder Leitungsgruppe ~ **capacitance** Kreis-, Schalt--kapazität *f* ~ **carrying audiofrequency (or voice) current** Sprechstromkreis *m* ~ **closer** Einschalter *m* ~**-closing lever** Stromschlußhebel *m* ~**-closing position** Stromschließstellung *f* ~ **component** Schaltelement *n* ~ **configuration** Schaltung *f* ~ **connection** Schaltverbindung *f* ~ **constant** Leitungskonstante *f* ~ **crossed with another** Leitung *f* in Berührung mit einer anderen ~ **design** Schaltungsaufbau *m* ~ **detail** Teilstromlauf *m* ~ **detector** Stromprüfer *m*

**circuit-diagram** Leitungsschema *n*, Schalt-aderbild *n*, -anordnung *f*, -bild *n*, -schema *n*, -skizze *f*, Schaltungszeichnung *f*, Stromlauf *m*, Stromlauf-plan *m*, -skizze *f*, -zeichnung *f* ~ **for hoist** Hubschaltung *f* ~ **for travel** Fahrschaltung *f*

**circuit,** ~ **discipline** Verkehrsdisziplin *f* (rdo) ~

**driver** Schwingungsgenerator *m* ~ **efficiency** Kreiswirkungsgrad *m* ~ **element** Grundschaltung *f*, Schalt-aderelement *n*, -element *n*, -mittel *n* ~ **element to influence the transit time** Laufzeitglied *n* ~ **gap admittance** Spaltleitwert *m* eines Kreises ~-**group busy hour** Hauptverkehrsstunde *f* für eine Leitung oder Leitungsgruppe ~ **holding time** Belegungsdauer *f* (rdo) ~ **identification** Ausklingeln *n* der Adern, Leitungskennung *f* ~ **interrupter** Lastschalter *m* ~ **iron** Schlußeisen *n* ~ **label** Schaltbild *n* (im Gerät) ~-**layout record** Leitungsübersicht *f* ~ **log** Betriebsbuch *n* ~ **looped back and forth** Schleifenschaltung *f* ~ **means** Schaltmittel *n* ~ **means causing fast rise of saw-tooth potential** Entladeschaltung *f* ~ **means causing slow rise of saw-tooth voltage** Ladeschaltung *f* ~ **means designed to form synchronizing signals** Anpassungsschaltung *f* ~ **message number** Laufnummer *f* ~ **meter** Geräuschspannungsmesser *m* ~ **model** Prüf-einrichtung *f*, -gestell *n*, -schaltung *f*, Versuchsschaltung *f* ~ **noise** Leitungsgeräusch *n*, Widerstandsrauschen *n* ~ **noise level** Kreisrauschpegel *m* ~ **number** Leitungsnummer *f* ~-**opening capacity** Abschalt-, Schalter-leistung *f* ~-**opening spark** Unterbrechungsfunke *m* ~ **organization designed to economize plate current** Sparschaltung *f* ~ **out of order** gestörte Leitung *f* ~ **parameter** Schaltelement *n* ~ **plan** Leitungsplan *m* ~ **points** Anschlüsse *pl* (zum Prüfen) ~ **scheme** Schaltung *f* ~ **studies** Schaltungslehre *f* ~ **switch** Gegentaktschaltung *f* ~ **terminal** Stromklemme *f* ~ **termination** Leitungsabschluß *m* ~ **tester** Leitungsprüfer *m* ~ **time** Leitungszeit *f* ~ **usage** Belegungsdauer *f* in vom Hundert ~-**usage record** Beobachtungs-, Überwachungs-blatt *n* ~ **wiring** Stromlaufbahn *f* ~ **worked on up-and-down basis** Leitung *f* für Verkehr in beiden Richtungen

**circuital,** ~ **integral** Rotationsintegral *n* ~ **relations** Umlaufrelationen *pl*

**circuitation** Rotationsintegral *n* eines Vektors

**circular** Umdruck *m*, Zirkular *n*; kreisförmig, kreisrund, ringförmig, rund, umlaufend, völlig ~ **air grid** Rosettenschieber *m* ~ **antenna** Ringantenne *f* ~ **aperture** Kreislochblende *f* ~ **approach** kreisförmige Annäherung *f* ~ **arc (or curve)** Kreisbogen *m* ~ **ballbearing** Spurkugellager *n* ~ **bead** Rundnaht *f* (welding) ~ **bell** Dosenwecker *m* ~ **box-type loom** Drehladewebstuhl *m* ~ **brush** Bürstenscheibe *f* ~ **bubble** Düsenlibelle *f* ~ **calcining kiln** Röstschachtofen *m* ~ **cam arm** Heber *m* ~ **cardboard cutter** Pappkreisschere *f* ~ **case** Runddgehäuse *n* ~ **channel** gekrümmtes Gerinne *n*, kreisförmiges Gerinne *n*, Kropfgerinne *f* ~ **chart** Kreisblatt *n* ~-**checked wood** kernschäliges Holz *n* ~-**coil filament** Wendelkreis *m* ~ **compass** Kreisbussole *f* ~ **cone** Kreiskegel *m* ~ **convex concentrator** Rundherd *m* ~ **conveyor** Umlaufförderband *n* ~ **course** Rundbahn *f* ~ **cowl** Motorring-, Ring-haube *f* ~ **cowl with controllable gills** Ringhaube *f* mit Regelklappen ~ **cross section** kreisrunder oder runder Querschnitt *m*, Kreisquerschnitt *m* ~ **current** Kreisstrom *m* ~ **currents** Kreisströmung *f* ~ **cutter** Rundschneidemaschine *f* ~ **cutting device** Kreisschneidevor-

richtung *f* ~ **cutting guide** Kreisschneidemaschine *f*, Rundführung *f* ~ **cylinder** Kreiszylinder *m* ~-**cylinder failure surface** kreiszylindrische Gleitfläche *f* ~ **deckle edge knife** Büttenrandkreismesser *n* ~ **diaphragm** Kreisblende *f* ~ **disc** Kreisscheibe *f* ~ **discs** umlaufende Trockenteller *pl* ~ **disc with a hole** Kreislochplatte *f* ~ **disc tool** Formscheibenstahl *m* ~ **dividing machine** Kreisteilmaschine *f* ~-**electric wave** elektrische Kreiswelle *f* ~ **fabric** Rundware *f* ~ **face grinding machine** Kolbenringschleifmaschine *f* ~ **feed** Kreisvorschub *m* ~ **file** Raspelfeile *f* ~ **flight** Kreisflug *m* ~ **fluctuation movement** (of spin) Zitterbewegung *f* ~ **flux** Kreisfluß *m* ~ **form** Kreisschablone *f* ~ **forming tool** Rundformstahl *m* ~ **front sight** (anti-aircraft sight for machine guns) Kreiskorn *n* ~-**front-sight frame** Kreiskornrahmen *m* ~-**front-sight support** Kreiskorn *n* ~ **function** Kreisfunktion *f*, Winkelfunktion *f* ~ **furnace** Rundofen *m* ~ **gills** Flachshechelmaschine *f*, (textiles) Glatthechel *m* ~ **grinding machine** Rundschleifmaschine *f* ~ **groove** Ringnut *f* (ringförmige Nut) ~ **grooved nut** Rillenmutter *f* ~ **grooving saw** Kreisfalzsäge *f*, Nutkreissäge *f* ~ **guide** Rundführung *f* ~ **heat deflector** Kühlscheibe *f* ~ **involute** Kreisevolvente *f* ~ **kiln** Rundkupfer *n* ~ **knife** Kreismesser *n* ~ **knitting machine** Rundstuhl *m* ~ **line** Kreis-linie *f*, -zweieck *n* ~ **load area** Kreisbelastungsfläche *f* ~ **loci** Kreisdiagramm *n* ~ **magnetic wave** zylindersymmetrische TM-Welle *f* ~ **measure** Bogenmaß *n* ~ **motion** Achsendrehung *f*, Kreisbewegung *f* ~ **movement** Kreisbahnbewegung *f* ~ **nut** Lochmutter *f* ~ **parameter** Kreisparameter *n* ~ **passage** Rundkanal *m* ~ **path** Kreisbahn *f* ~ **perforation** kreisrunde Löcher *pl* ~ **pitch** Umfangteilung *f*, Zahnkreisteilung *f*, (of turbine) Schaufelabstand *m* ~ **planform** kreisförmiger Grundriß *m* ~ **plotting protractor** Vollkreistransporteur *m* ~ **polarizer** Zirkularpolarisator *m* ~ **polarization** zirkulare Polarisation *f* ~ **profiling** Rundkopieren *n* ~ **profiling apparatus** Zylinderkurvenfräseapparat *m* ~ **protractor** Vollkreisgradmesser *m* ~ **rack** Drehring *m* ~ **radiation pattern** Rundstrahler *m* ~ **rail** Schienenkranz *m* ~ **rail-road** Gürtelbahn *f* ~ **ring** Kreisring *m* ~ **ring in the plane** ebener Kreisring *m* ~ **ring sector** Kreisringstück *n* ~ **saw** Kreissäge *f* ~-**saw bench** Tischkreissäge *f* ~ **scale** Kreis-einteilung *f*, -skala *f*, -teilung *f* ~ **scan(ning)** Kreisabtastung *f* (rdr) ~ **screen** Dreh-, Kreis-raster *m* ~-**seam weld** Rundnahtschweißung *f* ~ **section** Kreisschnitt *m* ~ **sector** Kreissektor *m* ~ **segment** Kreisabschnitt *m* ~ **shape** Ringform *f*, Rundprofil *n* ~ **shears** Rollschermesser *n* ~ **shelves** umlaufende Trockenteller *pl* ~ **shot** Ringschlitz *m* ~ **slide valve** Rundschieber *m* ~ **slitting knife** Tellermesser *n* ~-**slitting saw** Nutkreissäge *f* ~ **slot** Ringnut *f* ~ **space width** Zahnstärke *f* im Teilkreis ~ **spirit level** Dosenlibelle *f* ~ **spot** kreisförmiger Punkt *m* ~ **spring** Ringfeder *f* ~ **spring steel flange clamp** Federstahlspannband *n* ~ **stage** Rundtisch *m* (opt.) ~ **stiffener** Aussteifungsring *m* ~ **string border machine** Rundleistenmaschine *f* ~ **subsidence** Kesselbruch *m* (geol.) ~ **surface** Kreisfläche *f* ~ **swivel table** Rund-

schwenktisch *m* ~ **table (or rest)** Rundsupport
*m* ~ **thickener** Rundeindicker *m* ~ **thunder-**
**storm** Ringgewitter *n* ~ **time-base tube** Polarkoordinatenröhre *f* ~ **tooth thickness** Zahnstärke *f* im Rollkreis ~ **trace** kreisförmige Zeitbasis *f* (rdr) ~ **track** Laufring *m*, Schienenkranz *m* ~ **track mount** Lafettenkranz *m* ~
**trembler** Gleichstromdosenwecker *m* ~ **trough**
Kreisrinne *f* ~ **twist** Rundschlag *m* ~ **vortex**
**(or annular) eddy producer** Wirbelring *m* ~
**wave** Kreiswelle *f* (electr.) ~**-wedge densitome-**
**ter** Keilschwärzungsmesser *m*, Kreiskeilschwär-
zungsmesser *m* ~ **wire** runder Leiter *m* ~ **work**
Schlauch *m* (weaving)
**circularity** Kreisform *f* ~ **tolerance** Unrund-
heitstoleranz *f*
**circularly,** ~ **arranged cage antenna** Kreisgrup-
penreusenantenne *f* ~ **polarized antenna** zirkular polarisierte Antenne *f* ~ **polarized light**
zirkular polarisiertes Licht *n* ~ **unstable** kreis-
labil
**circulate, to** ~ fließen, kreisen, laufen, treiben,
umlaufen, umlaufen lassen (oil), verkünden,
zirkulieren **to** ~ **around** umströmen **to** ~
**through** durchstreichen
**circulating** kreisend, umlaufend
**circulating** Umwälzung *f* ~ **conveyor** Umlauf-
förderer *m* ~ **forced lubrication** Druckumlaufschmierung *f*, Zwangsumlaufschmierung *f* ~
**fuel** umlaufender Spaltstoff *m* ~ **library** Leih-
-bibliothek *f*, -bücherei *f* ~ **lubricating oil**
Umlaufschmieröl *n* ~ **map** Verkehrskarte *f*
~ **memory** Umlauf-speicher *m*, -register *n* ~ **oil**
Umlauf- *n*, Zirkulations-öl *n* ~ **oil pump** Zir-
kulationsölpumpe *f* ~ **passage** Zirkulations-
steg *m* ~ **pipe** Umlaufleitung *f* ~ **pump** Um-
lauf-, Umwälz-, Zirkulations-pumpe *f* ~ **pump**
**for cooling water** Kühlwasserumlaufpumpe *f*
~ **register** Umlauf-speicher *m*, -register *n* ~
**splash lubrication** Spritzumlaufschmierung *f*
~ **store** Umlaufspeicher *m* ~ **tank** Kreislauf-
behälter *m* ~ **water** Rückkühlwasser *n* ~-
**water inlet** Kühlwassereintritt *m* ~**-water outlet**
Kühlwasseraustritt *m*
**circulation** Auflage *f* (print.), Durchflutung *f*
(magnetic potential), Durchzug *m*, Fließen *n*,
Fluß *m*, Kreislauf *m* (of fluid), Kreisumlauf *m*,
Rundlaufen *n*, Spülung *f*, Strömung *f*, Umlauf *m*, Umlaufintegral *n* (of a vector), Zir-
kulation *f* **in** ~ kursfähig, in Umlauf *m* **to be**
**in** ~ kursieren ~ **of bank notes** Notenumlauf *m*
~ **of dye liquor** Flottenbewegung *f* ~ **of ions**
Ionenrundlauf *m* ~ **of money** Geldumlauf *m*
**circulation,** ~ **constant** Umlaufgröße *f* ~ **evapo-**
**rator** Umlaufverdampfer *m* ~ **heating** Umlauf-
heizung *f* ~ **layer** Umlaufschicht *f* ~ **loop** Kreislauf *m*, Umlaufleitung *f* ~ **map** Wegekarte *f* ~
**oil pump** Umlauflötpumpe *f* ~ **pipe** Umlauf-
rohr *n* ~ **regulating valve** Umwälzregelventil *n*
~ **shaft** Fahrschacht *m* ~**-system lubrication**
Umlaufschmierung *f* ~ **theorem** Zirkulationssatz *m* ~ **theory** Zirkulationstheorie *f*
**circulator** Zirkulator *m*
**circulatory disturbance** Kreislaufstörung *f*
**circumcirculate, to** ~ umfließen, umspülen
**circumference** Kreisumfang *m*, Peripherie *f*,
Ringumfang *m*, Umfang *m*, Umkreis *m* ~ **of**
**cam** Scheibenumfang *m* ~ **of protection** Schutz-

umfang *m* ~ **of pulley** Scheibenumfang *m* ~ **of**
**pulley rim** Riemenscheibenumfang *m* ~ **of**
**wheel** Radumfang *m*
**circumferential,** ~ **circle** Umfangkreis *m* ~
**contour turning** Umrißdrehen *n* ~ **force** Um-
fang(s)kraft *f* ~ **groove** Rundfuge *f*, umlaufende
Kerbe *f* ~ **highway** Umgehungsstraße *f* ~ **joint**
Ringdichtung *f* ~ **line** Umgrenzungslinie *f* ~
**oscillation** Umfangsschwingung *f* ~ **pitch (of**
**gear)** Stirnteilung *f* ~ **reinforcement** Umschlie-
ßungsbügel *m*, Umwehrung *f* ~ **resistance** Um-
fangswiderstand *m* ~ **runner** Felgenläufer *m*
~ **seam** Rundnaht *f* ~ **speed** Umfangsgeschwindigkeit *f* ~ **stress** Ringspannung *f* ~
**surface** Umfläche *f* ~ **tensile strength** Ringzugfestigkeit *f* (Rohr) ~ **tension** Umfangsspannung *f* ~ **vibration** Umfangsschwingung *f* ~
**weld** Schrägnaht *f* ~ **welding joint** Rund-
schweißnaht *f* ~ **wire** Umfangskabel *f*
**circumferentially enclosed propeller** ummantelte
Schraube *f*
**circumferentor** Gradbogen *m*
**circumnavigate, to** ~ umgehen, umfliegen
**circumscribe, to** ~ abgrenzen, begrenzen, umschreiben
**circumscribed circle of section** Profilumkreis *m*
**circumscription** Umschreiben *n*, Umsicht *f*
**circumstance** Umstand *m*, Verhältnis *n*, Zustand
*m*
**circumstantial** ausführlich, umständlich ~ **evi-**
**dence** Indizienbeweis *m*
**circumvallation line** Zirkumsvallationslinie *f*
**circumvent, to** ~ umgehen
**cirro,** ~**-cumulus** Zirrokumulus *m* ~**-cumulus**
**clouds** Lämmerwolken *pl*, Schäfchenwolken *pl*
~**-stratus** Zirrostratus *m* ~**-stratus cloud**
Schleierwolke *f*
**cirrus** Zirrus *m* ~ **cloud** Cirrus-, Feder-, Zirrus-
-wolke *f*
**cistern** Bassin *n*, Sammelbrunnen *m*, Wasserbehälter *m*, Zisterne *f* ~ **barometer** Gefäß-baro-
meter *n*, -luftdruckmesser *m* ~ **car** Zisternen-
wagen *m*
**citadel** Zitadelle *f*
**citation** Entgegenhaltung *f*, Vorladung *f*, Zitat *n*
**cite, to** ~ anführen, vorladen
**citric acid** Zitronensäure *f*
**citronella oil** Zitronenöl *n*
**city** Stadt *f* ~ **area** Stadtgebiet *n* ~ **council** Stadt-
rat *m* ~ **district** Geschäftsgegend *f* ~ **gas**
Leuchtgas *n* ~ **map** Stadtplan *m* ~ **planning**
Städtebau *m* ~ **water** Wasserleitungswasser *n*
~ **water supply** städtische Wasserleitung *f*
**civic** bürgerlich
**civil** bürgerlich, zivil ~ **air-line operating com-**
**pany** Luftverkehrsunternehmen *n* ~ **aviation**
Verkehrsluftfahrt *f*, Zivilflugwesen *n* ~ **day**
bürgerliche Zeit *f* ~ **engineer** Bau-, Zivil-inge-
nieur *m* ~ **engineering** Bauingenieurwesen *n*
~ **right** bürgerliches Ehrenrecht *n* ~ **time** mittlere Sonnenzeit *f* ~ **twilight** bürgerliche Dämmerung *f*
**civilian population** Zivilbevölkerung *f*
**clack** Ventilklappe *f* ~ **bailer** Klappenventil-
büchse *f* ~ **box** Klappenbohrer *m* ~ **concrete**
**box** Klappenbetonbüchse *f* ~ **seat** Ventilsitz *m*
~ **valve** Klappen-büchse *f*, -ventil *n*, Rückhaltklappe *f*

**clad, ~ metal** Verbundguß *m* **~ sheet (or plate)** plattiertes Blech *n* **~ vessel** abgeschirmter Behälter *m*

**cladding** metallischer Überzug *m* **~ material** Auflagemetall *n* **~ process** Plattierverfahren *n*

**claim, to ~** anfordern, in Anspruch nehmen, beanspruchen, requirieren, verlangen

**claim** Anforderung *f*, Anrecht *n*, Beanspruchung *f*, Behauptung *f*, Bergwerksanspruch *m*, Bezugsrecht *n*, Forderung *f*, Inanspruchnahme *f*, Mutung *f*, Patentanspruch *m*, Recht *n*, Verlangen *n* (law) **~ for indemnification** Schadenersatzanspruch *m* **~ of replacement** Ersatzanspruch *m* **~ based on a bill of exchange** Wechselforderung *f* **~ (patent) title** Anspruch *m*

**claimant** Anspruchsteller *m*, Muter *m*, Patentanmelder *m*

**claimholder** Muter *m*

**claims respecting the wages of a ship's crew** Heuerrückstand *m*

**clamber, to ~** klettern

**clammy** feuchtkalt, klamm

**clamp, to ~** aufspannen, befestigen, einmieten, einspannen, festklammern, festklemmen, festlegen, feststellen, klemmen, sich einklemmen, spannen, verklammern, zurren **to ~ (to)** anklammern (an) **to ~ on** anschellen **to ~ securely** festzurren **to ~ tightly** festspannen

**clamp** Aufspannfrosch *m*, Balkweger *m*, Bride *f*, Bügel *m*, Drahtseilschloß *n*, Einspannkopf *m*, Fangkluppe *f*, Formkastenpresse *f*, Greifbacke *f*, Greifbaken *m*, Haft *f*, Haken *m*, Halter *m*, Klammer *f*, Klampe *f*, Klemme *f*, Klemmhülse *f* (d. Radbremse), Klemmplatte *f*, Krampe *f*, Miete *f*, Muffe *f*, Röhrenhalter *m*, Röststadel *m*, Schelle *f*, Spannblech *n*, Stadel *f*, Zange *f*, Ziehband *n*, Zurrung *f*, Zwinge *f*, **(screw) ~** Schraubzwinge *f* **~ of bricks** Hag *m* **~ for fine adjustment** Klemmung *f* der Feinverstellung **~-on dark slide** Anlegekassette *f* **~ for downfeed** Klemmschraube *f* für die Vertikalbewegung

**clamp, ~ bolt** Spannschraube *f* **~ bracket** Klemmhülse *f* **~ carrier** Halteklammer *f* (Tragklammer) **~ collar** Klemm-, Verschluß-ring *m* **~ cone** Einspannkonus *m* **~ coupling** Klemm-, Schalen-kupplung *f* **~ diode** Schwarzsteuer-, Klemm-diode *f* **~ dog** Spannkloben *m* **~ fit** Klemmsitz *m* **~ fixing** Klammerbefestigung *f* **~ foot** Klemmfuß *m* **~ gate valve** Durchgangsschieber *m* mit Bügel **~ handle** Knebelgriff *m* **~-hub wheel** Rad *n* mit eingepreßter Nabe **~ iron for walls** Mauerklammer *f* **~ knob** Klemmknopf *m* **~(ing) lever** Klemmhebel *m* **~ locking device** Spannbügelverschluß *m* **~ plate** Klemmschuh *m* **~ ring** Klemmring *m* **~ rot** Mietenfäule *f* **~ screw** Bündelschraube *f* **~(ing) screw** Preßschraube *f* **~ (or clamping) slide** Klemmschlitten *m* **~ stirrup** Spannbügel *m* (Kraftstoffpumpe) **~ strap** Klemmbügel *m*, Spannband *n* **~ welding machine** Einspannschweißvorrichtung *f*

**clamped** eingeklemmt **~ (together)** durch einen Bügel *m* verbunden **~ cantilever** eingespannter Freiträger *m* **~ floor** Friesfußboden *m* **~ length** Klemmlänge *f*

**clamper** Klemmschaltung *f* **~ circuit** Haltestromkreis *m*, Klemmschaltung *f*

**clamping** Befestigung *f*, Einspannen *n*, Einspannung *f*, Festspannen *n*, getastete Schwarzsteuerung *f*, Klemmen *n*, Klemmung *f*, Regelhaltung *f*, Spannen *n* **~ action** Spannvorgang *m* **~ angle** Aufspannwinkel *m* **~ band** Zurrgurt *m* **~ band lever** Spannbandhebel *m* **~ bar** Druckbalken *m*, Preßleiste *f* **~ bolt** Anzieh-, Befestigungs-bolzen *m*, Befestigungsschraube *f* Druckbolzen *m*, Halteschraube *f*, Klappenstellschraube *f*, Klemmbolzen *m*, Supportbefestigungsschraube *f*, Tischfeststellschraube *f*, Verbindungsschraube *f* **~ bush** Spannbüchse *f* **~ capacity** Schließleistung *f* **~ chuck** Einspannklaue *f*, Einspannkopf *m*, Klemmfutter *n*, (for bench lathes) Futterplatten *pl* **~ circuit** Blokkierschaltung *f*, Haltestromkreis *m*, Kappschaltung *f* **~ claw (cable)** Klemmzange *f* **~ collar** Schelle *f* **~ collar and handle** Schelle *f* mit Handgriff **~ cone** Spannkegel *m* **~ cover plate** Spannlasche *f* **~ cross** Zuspannkreuz *n* **~ device** Aufspann-, Einspann-, Festspann-vorrichtung *f*, Spanneinrichtung *f*, Spannschiene *f*, Spannzeug *n*, Verriegelung *f*, Zurrung *f*, Klemmvorrichtung *f* **~ eye** Klemmauge *n* **~ face** Befestigungsfläche *f* **~ flange** Spannflansch *m* **~ frame** Druckrahmen *m* **~ girder** Balken *m* zum Festspannen **~ grooves** Aufspann-schlitze *pl* oder -nuten *pl* **~ handle** Klemmkopf *m* **~ head** Einspannkopf *m* **~ holder** Einspannhalter *m* **~ housing** Einspanngehäuse *n* **~ jaw** Klemm-, Spann-backe *f* **~ knob** Arretierknopf *m* **~ lever** Klemmhebel *m*, Reitstockfeststellhebel *m*, Spannschieber *m* **~ mechanism** Festklemmvorrichtung *f* **~ notch** Klemmnute *f* **~ nut** Halte-, Überwurf-mutter *f* **~ piece** Klemmstück *n* **~ place** Einspannstelle *f* **~ plate** Aufspannflansch *m*, Befestigungsflansch *m*, Klemmen-, Klemm-, Spann-platte *f* **~ pliers** Würgeverbindungszange *f* **~ power** Spannkraft *f* **~ rail** Aufspannschiene *f* **~ ring** Eisenzwinge *f*, Klemmring *m*, Schelle *f*, Spannring *m* **~-ring handle** Klemmringhebel *m* **~ ring on dial sight** Klemmplatte *f* **~ screw** Gegenmutter *f*, Klemm-, Preß-schraube *f* **~-screw handle on dial sight** Klemmvorrichtung *f* **~ sheet** Klemmblech *m* **~ shoe** Spannpratze *f* **~ sleeve** Klemm-hülse *f*, -muffe *f* **~ slots** Aufspannschlitze *pl* oder -nuten *pl* **~ spindle** Zugspannspindel *f* **~ spring** Federklammer *f* **~ stand** Befestigungsbock *m* **~ strip** Halteleiste *f* **~ stud** Spannbolzen *m* **~ support** Aufspannblock *m* **~ surface of slide** Stößelspannfläche *f* **~ tool** Aufspann-, Einspann-, Spann-werkzeug *n* **~ washer** Zwischenscheibe *f* **~ work** Spannarbeiten *pl*

**clamshell dredge** Greifbagger *m*

**clandestine radio transmitter** Geheimsender *m*

**clang** Klang *m*

**clanging** Klirren *n*

**clank** Kettengeklirr *n*

**clap, to ~** klappen

**clapboard** Dachschindel *f*, Synchronklappe *f*

**clapper, to ~** klappen

**clapper** Glockenschwengel *m*, Klatsche *f* (film), Klöppel *m*, Synchronklappe *f* (film) **~ block** Stichelhalter *m* **~ block pin** Stichelhalterstift *m* **~ box** Klappen-führung *f*, -halter *m* **~ rod** Klöppelstange *f* **~ stick** Klöppelstange *f* **~ valve** Schlagventil *n*

**clappers** Klappensignal *n*, Klatsche *f* ~ **to mark sound and picture track for synchronization** Synchronisierklappe *f*

**clarification** Abklärung *f*, Abläuterung *f*, Aufklärung *f*, Klärung *f*, Läuterung *f* ~ **of solution** Laugenreinigung *f* ~ **of wash water** Waschwasserklärung *f* ~ **(or filter) plant** Kläranlage *f*

**clarifier** Klärapparat *m*, Klärmittel *n*, Klärungsmittel *n*, Läuterpfanne *f*, Nachwaschsetzmaschine *f*

**clarify, to** ~ abhellen, abklären, abläutern, aufhellen, aufklären, klar machen, klären, läutern

**clarifying** Defäkation *f*, Defekation *f*, Läutern *n*, Schönen *n*, Scheiden *n*, Scheidung *f* ~ **of wine** Weinklärung *f* ~ **agent** Klärungsmittel *n* ~ **basin** Kläranlage *f* ~ **filter** Klärfilter *n* ~ **plant** Kläranlage *f* ~ **rake** Klärrechen *m* ~ **tank** Klärbehälter *m*

**clarity** Klarheit *f*, Klärung *f* **(functional)** ~ Übersichtlichkeit *f* (of a control panel)

**clarkeite** Clarkeit *m*

**clash, to** ~ aufeinandertreffen (Gegensätze), in Widerspruch stehen

**clash point of light valve** maximale Aussteuerung *f* des Lichtstrahles (in sound recording)

**clashing** Klirren *n*

**clasp, to** ~ einhaken, klammern, umschlingen

**clasp** Agraffe *f*, Haft *f*, Häkchen *n*, Haken *m*, Kettel *f*, Klammer *f*, Krampe *f*, Spange *f*, (lock) Überwurf *m*, Verschluß *m*, (of lock) Zunge *f* ~ **handle** Fallhebel *m*

**class** Art *f*, (of fabricated products) Baureihe *f*, Gattung *f*, Gruppe *f*, Güte *f*, (educ.) Jahrgang *m*, Klasse *f*, Lehrgruppe *f*, Ordnung *f*, Qualität *f*, Sorte *f*

**class,** ~ **of accuracy** Klasse *f* (bei Meßgeräten) ~ **of aircraft** Luftfahrzeugklasse *f* ~ **of circuit for control purposes** Rang *m* (Wichtigkeitsgrad) einer Leitung ~ **of equipment** Gerätklasse *f* ~ **of fit** Gütegrad *m* einer Passung, Sitzart *f* ~ **of ore** Erzgattung *f* ~ **of pig iron** Roheisengattung *f* ~ **of record** Rekordklasse *f* ~ **of sand** Sandsorte *f* ~ **of surface** Flächenklasse *f* ~ **of vessel** Fahrzeugklasse *f* ~ **of worker** Arbeiterkategorie *f* ~ **of xanthenium (or pyrenium) dyes made by dehydrogenation** Dehydrenium-Farbstoffe *pl*

**class,** ~ **A (B) amplifier** A-(B-) Verstärker *m* ~ **B modulation** B-Modulation *f*, Gegentakt-B-Modulation *f* ~ **C amplifier radio** C-Verstärker *m* ~ **division** Klasseneinteilung *f* ~ **interval** Klassengröße *f* ~ **microscope** Kursmikroskop *n* ~ **name** Gattungsname *m*

**classed** eingeteilt

**classes,** ~ **of aerodromes** Arten *pl* der Flugplätze ~ **of mappings** Abbildungsklassen *pl* ~ **of rock filling** Kategorien *pl* der Bruchsteine

**classic** mustergültig

**classical** formal, klassisch ~ **mechanics** Grundgleichungen *pl*

**classification** Anordnung *f*, Beurteilung *f*, Eingliederung *f*, Eingruppierung *f*, Einstufung *f*, Einteilung *f*, Fachordnung *f*, Klassen-einteilung *f*, -ordnung *f*, Klassierung *f*, Klassifikation *f*, Klassifizierung *f*, Sichtwirkung *f*, Zusammenstellung *f* ~ **of equipment** Stoffgliederung *f* ~ **of ore** Erzbezeichnung *f* ~ **of pig iron** Roheiseneinteilung *f* ~ **by points** Punktwertung *f* ~ **of runnings** Ablauftrennung *f* ~ **according to**

**size** Größenklasse *f* ~ **of waves** Welleneinteilung *f*

**classification,** ~ **code** Klassifizierungskode *m* ~ **formula** Wertungsformal *f* ~ **keyboard** Fachwählertastatur *f* ~ **mill** Klassifikationsanlage *f* ~ **track** Abstellgeleise *n* (r.r.), Richtungsgleis *n* (r.r.) ~ **unit** Klassiereinschub *m* ~ **yard** Rangiergelände *n*

**classified** eingeteilt, geheim, klassifiziert

**classifier** Klassifikator *m*, Klassierapparat *m*

**classify, to** ~ abfachen, auslesen, eingliedern, einordnen, einreihen, einstufen, einteilen, Geheimhaltung verfügen, gliedern, klassieren, klassifizieren, sichten, sondern, sorten, sortieren **to** ~ **a charge** gattieren

**classifying** Klassieren *n*, Siebung *f* ~ **drum** Klassiertrommel *f* ~ **grate** Klassierrost *m* ~ **plant** Klassieranlage *f* ~ **screen** Klassiersieb *n*

**clastic** klastisch (rock) ~ **rocks** Trümmer *pl*

**clatter, to** ~ klirren, rasseln

**clatter** Gerassel *n*

**clause** Bestimmung *f*, besondere Bedingung *f*, Klausel *f*, Zusatz *m* ~ **of amnesty** Amnestieklausel *f*

**clausthalite** Klaustalit *m*, Kobalt-bleiglanz *m*, -bleierz *n*, Selenblei *n*

**claw, to** ~ **off** abarbeiten

**claw** Greifer *m* (film), Greifdorneisen *n* (e.g. for lumbering), Haken *m*, Klaue *f*, Kralle *f*, Mitnehmerklaue *f*, Nagelheber *m*, Pratze *f*, Schar *f*, Schere *f*, Tatze *f* **(pull-down)** ~ Gegentaktaufzeichnung *f*, Greifer *m* (film) ~ **for casting off** Abwerfklaue *f* ~ **of the clutch** Kupplungsklaue *f*

**claw,** ~**-baffle combustion chamber** Faltenbrennkammer *f* ~ **beam** Pratzenstempel *m* ~ **belt fastener** Riemenkralle *f* ~ **bit** Klauenbohrer *m*, Ohrenschneide *f* ~ **clutch** Klauenkuppelung *f* ~ **coupling** Antriebklaue *f*, Kleu(en)kuppelung *f* ~ **crane** Pratzenkran *m* ~**-feed system** Greifersystem *n* ~ **hammer** Hammer *m* mit gespaltener Finne, Klauenhammer *m* ~ **hatchet** Klauenbeil *n* **in-and-out** ~ Bildgreifer *m* (of threading mechanism) ~**-in spindle** Festspannschraube *f* ~ **movement** Greiferantrieb *m* (film) ~ **offset** Klauenversetzung *f* ~**-pole stationary-field synchronous generator** Klauenpolgenerator *m* ~ **slipping** Greiferschlupf *m* (film) ~ **supply** Greifernachschub *m* (film) ~**-(gear-)switch collar** Klauenschaltmuffe *f* ~ **wrench** Nagelheber *m*

**clay, to** ~ verletten **to** ~ **(a wall)** kleiben

**clay** Bergletten *pl*, Bleicherde *f*, Klei *m*, Lehm *m*, Mergel *m*, Schlick *m*, Tegel *m*, Ton *m* ~ **ballots** Bälle *pl* (ceramics) ~ **band** (clay ironstone) Eisenerzlehm *m*, Toneisenstein *m*, tonhaltiger Spateisenstein *m* ~**-bearing** tonhaltig ~ **bed** Tonlager ~ **binder** Tonbinder *m* ~ **body** Tonsubstanz *f* ~ **bond** Bindeton *m*, Tonbindemittel *n* ~**-bond fire clay** Tonerdeschamotte *f* ~**-bond silica brick** Quarzstein *m* ~ **brick** Lehmziegel *m*, Tonziegel *m*, Ziegelstein *m* ~ **conduit** Formstück *n* (Tonform), Tonformstück *n*, Tonrohr *n* ~ **cone** Tonkegel *m* ~ **content** Tongehalt *m* ~ **core** Tonkern *m* ~ **crucible** Tontiegel *m* ~ **cutter** Tonschneider *m* ~ **deposit** Tongrube *f*, Tonlager *n* ~**-digging spade** Tonstechspaten *n* ~**-filled fissure** Schmerkluft *f* ~

**filling** Dichtungsmaterial *n*, Füllungsmaterial *n*, Lehmfüllung *f* ~ **gall** Tongalle *f* ~ **grouting** Toninjektion *f* ~ **gun** Stichlochstopfmaschine *f* ~ **industry** Tonindustrie *f* ~ **ironstone** Eisenton *m*, Toneisenstein *m* ~ **ironstone deposit** Toneisensteinlager *n* ~ **lens** Tonlinse *f* ~ **like** tonartig ~ **lute** Tonkitt *m* ~ **maker** Tonmischer *m* ~ **mixed with silver** Silberletten *m* ~ **mortar** Feuerkitt *m* (for setting firebricks) ~ **pigeon** Tontaube *f* ~ **pipe** Steinzeugrohr *n* ~ **pit** Lehmgrube *f*, Tongrube *f* ~ **plug** Lehmstopfen *m*, Tonpropfen *m*, Tonstopfen *m* ~ **refractories** Schamottegut *n* ~ **retort** Tonmuffel *f* ~ **rolling mill** Tonwalzwerk *n* ~ **schist** Schieferton *m*, Tonschiefer *m* ~ **shale** Schieferletten *m*, tonhaltiger Schiefer ~ **slate** Schieferton *m*, Tonschiefer *m* ~ **sludge** Aufschlämmung *f* (von Ton) ~ **soil** Kleisode *f*, Lehmboden *m*, Tonboden *m* ~ **tile** Steinzeug *n*, Tonziegel *m* ~ **treatment** Bleicherdebehandlung *f* ~ **vessel** Tongefäß *n* ~ **ware** Steinzeug *n* ~ **wash** Tonschlämme *f* ~ **works** Tonwerk *n*

**clayed sugar** gedeckter Zucker *m*

**clayey** lehmig, lettig, tonartig, tonig ~ **marl** Tonmergel *m* ~ **sand** tonhaltiger Sand *m* ~ **soil** tonige Erde *f*

**claying** Decken *n* ~ **of a borehole** Verletten *n* eines Bohrloches ~ **apparatus** Deckapparat *m* ~ **bar** Trockenbohrer *m* (min.)

**clayish** lettig, tonig

**clead, to** ~ mit Brettern abdecken

**clean** blank, rein, sauber

**clean, to** ~ abwischen, aufräumen, (den Kessel) ausschaben, (das Feuer) ausschüren, auswaschen, läutern, polieren, putzen, reinigen, säubern, scheuern, spülen, waschen **to** ~ **off** abputzen, abstäuben, ausblasen **to** ~ **out** ausfegen, auskehren, ausputzen **to** ~ **up** abtreiben, aufsäubern, austreiben, entgasen, saubermachen **to scrape and** ~ **brightly** blank machen (Draht) **to** ~ **off burrs** den Grat entfernen **to** ~ **(se) casks** faßschwanken **to** ~ **contacts** Kontakte *pl* reinigen **to** ~ **(or gin) the cotton** die Baumwolle egrenieren **to** ~ **(or wash) the form** entschwärzen **to** ~ **a planed surface of wood** das Holz abschlichten

**clean,** ~**-air receiver** Reinluftraum *m* ~ **break** funkenfreie Unterbrechung *f* ~ **coal** Reinkohle *f* ~ **configuration** Flugzeug *f* ohne Außenbordlasten ~ **copy** Reinschrift *f* ~ **cut edges** scharf begrenzte Umrisse *pl* ~ **exhaust** rauchfreier Auspuff *m* ~ **lines** klare Linienführung *f* ~ **machine proof** Ausfallbogen *m* ~**-oil** Frischöl *n* ~**-oil circulation cracking process** Reinöl-kreislauf *m*, -zirkulation *f* ~ **proof** Reindruck *m*, Revisionsbogen *m* ~ **reactor** kalter Reaktor *m* ~ **sheet** Aufhängebogen *m* (print.) ~**-up** Aufräumung *f*, plötzliche Gasaufzehrung *f* ~**-water overflow** Klarwasserüberlauf *m*

**cleaned,** ~ **by sandblasting** mit dem Sandstrahlgebläse *n* gereinigt ~**-up bottom** ausgeräumte Ofensohle *f*

**cleaner** Abstreifer *m*, Kratzer *m*, Nachreiniger *m*, Reiniger *m*, Reinigungsmittel *n*, Spatel *m* ~ **for machinery** Entstauber *m* ~ **cap** Filtergehäusedeckel *m* ~ **element** Filterpatrone *f*

**cleaning** Abblasen *n* (Eisen), Aufbereitung *f*, Nullpunktschärfung *f* (Peilung), Putzen *n*, Reinigen *n*, Reinigung *f*, Säuberei *f*, Säuberung *f*, Verputzen *n* ~ **of blast-furnace gas** Gichtgasreinigung *f* ~ **of castings** Gußputzerei *f* (Gußputzen) ~ **of coal** Kohlenaufbereitung *f* ~ **at several densities** Mehrwichtesortierung *f*

**cleaning,** ~ **action** Putzwirkung *f*, Reinigungseffekt *m* ~**-and-dressing-operation** Putzarbeit *f* ~ **apparatus** Reinigungsapparat *m* ~ **barrel** Scheuerfaß *n*, Scheuertrommel *f* ~ **bench** Putztisch *m* ~ **brush** Putzleiste *f* ~ **capacity** Putzleistung *f* ~ **compound** Putzmaterial *n* ~ **department** Putzerei *f* ~ **device** Reinigungswert *m* ~ **door** Einsteigöffnung *f* ~ **efficiency** Putzleistung *f* ~ **equipment** Reinigungsanlage *f* ~ **hole** Reinigungs-luke *f*, -öffnung *f* ~ **knife** Abzieh-, Putz-messer *n*, Reinigungsschaber *m* (beim Spaltfilter) ~ **materials** Putzzeug *n* ~ **milling cutter** Reinigungsfräser *m* ~ **oil** Putzöl *n* ~ **patch (or rag)** Reinigungsdocht *m* ~ **pit** Putz-, Reinigungs-grube *f* ~ **plant** Reinigungsanlage *f* ~ **port** Reinigungsklappe *f* ~ **position** Putzstand *m* ~ **practice** Putztechnik *f* ~ **process** Putzverfahren *n*, Wasch-prozeß *m*, -vorgang *m* ~ **rag** Putzlappen *m* ~ **rod** Gewehr-, Reinigungs-, Wisch-stock *m* ~ **room** Putzerei *f*, Putzhaus *n*, (foundry) Gußputzerei *f* ~**-room equipment** Gußputzanlage *f* ~ **sand** Putzsand *m* ~ **shop** Putzerei *f* ~ **staff** Rohrreiniger *m* ~ **table** Putztisch *m* ~ **tool for the nozzle** Düsenreiniger *m* ~ **vent** Reinigungsöffnung *f* ~ **waste** Flachswerg *n*, Putzwerg *n*, Putzwolle *f* ~ **wires** Reinigungsdrähte *pl*

**cleanliness** Reinheit *f*, Reinlichkeit *f*, Sauberkeit *f*

**cleanness** aerodynamisch günstige Ausbildung *f* (aerodyn.)

**cleanout** Ablaßöffnung *f* ~ **door** Reinigungs-, Schlacken-, Schlack-tür *f* ~ **flange and plug** Bodenablaßflansch *m* und Stopfen *m* ~ **hole** Reinigungsöffnung *f*

**cleanse, to** ~ abputzen, abspritzen, abspulen, abwaschen, säubern **to** ~ **and lime** schwöden (hides)

**cleanser** Ausräummaschine *f*, Putzmittel *n*

**cleansing** Decke *f* (sugar) ~ **of ores** Erzaufbereitung *f* ~ **cellar** Abfüll-keller *m*, -raum *m* ~ **agent** Reinigungsmittel *n* ~ **gas** Spülgas *n*

**clear, to** ~ abhellen, abraumen, abräumen, abschwächen, aufräumen, aufrollen, aufwältigen, (furnace) ausbrechen, (a way or path) bahnen, (Fehler) beheben, (faults) beseitigen, bloßlegen, durchfahren, einhängen, entblocken, freigeben, klar halten, läutern, (forest) lichten, löchen (data proc.), räumen, (land) roden, saldieren, sich aufklären, schummeln, verzollen, vorbeikommen **to** ~ **away** forträumen **to** ~ **in** einklarieren **to** ~ **out** eine Verbindung *f* auflösen, leeren, ausräumen **to** ~ **up** aufhellen, aufklären, erhellen, klären, sich aufklären, sich aufhellen (weather) **to** ~ **up a mine** aufwältigen **to** ~ **of rubbish** entrümpeln

**clear, to** ~ **an adit** einen Stollen *m* aufräumen **to** ~ **the attic** den Boden *m* entrümpeln **to** ~ **cellar** auskellern **to** ~ **a connection** eine Verbindung *f* aufheben **to** ~ **the counter** den Zähler *m* löschen **to** ~ **the debris** das taube Gestein aufräumen **to** ~ **decks** klar machen (navy)

to ~ the engine freibrennen to ~ a fault eine Störung *f* beseitigen to ~ items Posten abtragen to ~ the line freie Fahrt geben, freigeben to ~ pools from mud Teiche ausschlämmen to ~ the rock das taube Gestein aufräumen to ~ before writing löschen vor dem Schreiben
clear Lichte *n*; anschaulich, aufgearbeitet, außer Eingriff (teleph.), betriebsfähig, blank, deutlich, durchsichtig, eindeutig, entblockt (r.r.), farblos, frei, glashell, glasklar, heiter (weather), hell, kenntlich, klar, lauter, licht, rein, sichtig, stromfähig (current or wire), übersichtlich, unbesetzt, unzweideutig, wolkenlos in the ~ licht ~ (of, from) räumlich frei ~ out Auflösung *f*
clear, ~ aperture of a lens lichte Linsenweite *f* ~ charging space inside cross bearers freie Ladefläche *f* zwischen den Drehgestellen ~ cut scharf geschnitten ~ cutting Abtrieb *m* (forest) ~ felling Kahlschlag *m* ~ filtrate durchsichtiges Filtrat *n* ~ fish oil Helltran *m* ~-gas plant Klargasanlage *f* ~-glass adapter Klarglasscheibenhalter *m* ~-glass globe Klarglasglocke *f* ~ glass lens (or window) glasklares Fenster *n* ~-glass screen Blank-, Klarglaseinstell-scheibe *f* ~ height lichte Höhe *f* ~ ice glasartiges Vereisen *n*, Klareis *n* ~ ice glaze glasartige Vereisung *f* ~ lacquer farbloser Lack *m* ~ light tint hellklare Farbe *f* ~ liquor Klärsel *n* (sugar) ~ opening Durchflußquerschnitt *m* ~-out relay Rückführ-, Auslöse-relais *n* ~ pilot burner weiße Lampe *f* ~ portion helle Stelle *f* ~ pronunciation deutliche Aussprache *f* ~ selvedges Leistenreinheit *f* ~ signal Abfahr- (r.r.), Freigabe-, Frei-signal *n* ~ sky heiterer Himmel *m* ~ test Prüfung *f* auf Betriebsfähigkeit ~ text Klartext *m* ~-text teleprinter Klarfernschreibmaschine *f* ~ varnish Klarlack *m* ~ voice deutliche oder reine Sprache *f* ~ water Freiwasser *n* (civil eng.) ~ way Freifläche *f* (aviat.) ~ weather helles Wetter *n* ~ white hellweiß ~ width Lichtweite *f*
clearance Abstand *m*, Anfahrmaß *n*, Anstellwinkel *m*, Ausdehnungsspiel *n*, ausgeglichteter Raum *m*, Ausklarieren *n* (ship), Ausnehmung *f*, Aussparung *f*, freier Raum *m*, Helligkeit *f* (eines Bildes), lichter Raum *m*, Lichtmaß *n*, Raum *m*, schädlicher Raum *m* (dead space), Schnüffelspiel *n*, Spalt *m*, Spaltweite *f*, Spiel *n*, Spielraum *m*, Vertrieb *m*, Vorrollung *f* (of goods), Zollbelastung *f*, Zwischenraum *m* (piston) ~ Kolbenspielraum *m* ~ (radially) between car wheel and guard Abstand *m* zwischen Radoberkante und Kotflügel ~ between collars Kaliberöffnung *f* ~ between flywheel rim and pinion Ritzelabstand *m* (Anlasser) ~ of freight Güterabfertigung *f* ~ of injection pin Spritzzapfenspiel *n* ~ of motion Bewegungsspiel *n* ~ between outer race and ball bearing Gehäuseabstand *m* ~ of piston Kolbenspiel *n* ~ of pole lines Abstand *m* der Leitungen vom Erdboden ~ of push rods Stoßstellenabstand *m* ~ above the road level Bodenfreiheit *f* ~ at root Spitzenspiel *n* ~ for take-off Startfreigabe *f*
clearance, ~ angle Ansatz-, Hinterschleif-, Hinterschliff-, Frei-, Schleif-winkel *m* (Drehstahl) ~ angle of turning steel Anstellwinkel *m* des Drehmeißels ~ area Spaltfläche *f* ~ bearing

Gleitlager *n* ~-chit circular Laufzettel *m* ~ diagram Durchgangsprofil *n* ~ drawing Spielraumzeichnung *f* ~ fit Bewegungssitz *m*, Spielpassung *f* ~ gauge Umgrenzung *f* des lichten Raumes, Umgrenzungslehre *f* ~ hole Durchsteckbohrung *f* ~ hole for cuttings Schneidezahnloch *n* ~ indicator rod Begrenzungsstab *m* ~ inward Einklarierung *f* ~ limitation Abstand-, Lade-begrenzung *f*, Ladeprofil *n* ~ line Umgrenzungslinie *f* ~-loading gauge (or limit) Ladelehre *f* ~ loss Spaltverlust *m* ~ outward Ausklarierung *f* ~ paper Klarierungsschein *m* ~ plane Hindernisbezugsebene *f* (aviat.) ~ play Schnüffelspiel *n* ~ sale Ausverkauf *m*, Räumungsverkauf *m* ~ space Geschoßanlage *f*, Verdichtungsraum *m* ~ time Abschaltzeit *f* ~ volume Tot-, Verdichtungs-raum *m*, Zylinderhöhe *f*
clearer Putzvorrichtung *f* ~ plate Reinigerplättchen *n* ~ roller Putz- und Fangwalze *f*
clearing Aufhebung *f*, Aufheiterung *f*, Auflösung *f*, Aufräumarbeit *f*, Beseitigung *f*, Defäkation *f*, Defekation *f*, Einhängen *n*, Freigabe *f*, Klärung *f*, Läuterung *f*, Lichtung *f*, Säuberung *f*, Scheidung *f*, Schlußzeichengabe *f*, Schlußzeichengebung *f*, Schummeln *n*, Unterbrechung *f*, Verrechnung *f*
clearing, ~ of a fault Störungsbeseitigung *f* ~ of ice Eisabgang *m* ~ of obstacles Beseitigung *f* von Sperren ~ of the right of way Räumen *n* der Trasse ~ the river bed Verlegung *f* eines Flußbettes ~ a section (of track) Entblockung *f* ~ of silo Siloentleerung *f*
clearing, ~ agent Klärmittel *n*, Klärungsmittel *n* ~ agreement with foreign country Verrechnungsabkommen *n* ~ attachment Ausfaserapparat *m* ~ basin Klär-sumpf *m*, -teich *m* ~ bath Klärbad *n* ~ block Freigabeblock *m* ~ button Schlußtaste *f* (teleph.) ~ current Schlußzeichenstrom *m* ~ cylinder Läutertrommel *f* ~ device Freigabeorgan *n* ~ dollar Verrechnungsdollar *m* ~ field Reinigungsfeld *n* ~ gauge Durchgangsprofil *n* ~ house Abrechnungsstelle *f* ~-house business Giroverkehr *m* ~ indicator Schlußzeichen *n* ~ iron Formstecher *m* ~ key Löschtaste *f* ~ lamp Schlußlampe *f* ~-lamp repeating Schlußzeichenübertragung *f* ~ lever Löschhebel *m* (bei Rechenmaschinen) ~ liquor Kläre *f* ~-out drops Schlußklappen *pl* ~-out of a shaft Aufmachen *n* eines Schachtes ~-out signal Schlußzeichen *n* ~ pan Läuterpfanne *f* ~ pulse Abschaltestromstoß *m* ~ pump Lenkpumpe *f* ~ rail lever Löschschienenhebel *m* ~ relay Auslöse-, Schlußzeichen-relais *n* ~ rod Entladestange *f* (firearms) ~ section Abrückabschnitt *m* (electr.) ~-signal Schlußzeichen *n* ~-signal arrangement Schlußzeicheneinrichtung *f* ~-signal lamp Schlußzeichenlampe *f* ~ signalizing Schlußzeichengabe *f* ~ sump Klär-sumpf *m*, -teich *m* ~ thickness Aufhellungsdicke *f* ~ time Entionisierungszeit *f* ~-up Aufklärung *f*, Aufräumung *f* ~ vat Läuterbottich *m*
clearly defined klar bestimmt
clearness Betriebsfähigkeit *f*, Deutlichkeit *f*, Durchsichtigkeit *f*, Helligkeit *f*, Klarheit *f*, Reinheit *f*, Schärfe *f*, Sicht *f*, Sichtigkeit *f*, ~ of sound Wiedergabebrillanz *f*

cleat, to ~ anklammern, befestigen mit Klammern, festklampen, festkrampen
cleat Anguß m, Belegnadel f, Halteleiste f (zur Sicherung durch Niederhalten), Keil m, Klampe f, Klemmplatte f, Kreuzholz n, Leiste f, Stützblock m, Tragkonsole f ~ insulator Isolierklemme f, Klemmisolator m ~ work Leistenarbeit f
cleating Befestigung f
cleats Taquet n (on a glass-cutting table)
cleavability Spaltbarkeit f
cleavable spaltbar
cleavage Abspaltung f, Aufspaltung f, Ausweichen n (of rock upon blasting), Blätterbruch m, falsche Schieferung f, Schieferung f, Spaltung f, Transversalschieferung f ~ of fat Fettspaltung f
cleavage, ~ brittleness interkristalline Brüchigkeit f, Spaltbrüchigkeit f ~ crack Spaltriß m ~ crystal Spaltungskristall m ~ face Spaltebene f, Spaltfläche f, Spaltungsfläche f ~ fissure Spaltriß ~ loss Spaltverlust m ~ plane Gleit-, Grenz-, Spalt-fläche f, Spaltebene f, Spaltflächenzeichnung f, Spaltungsebene f, Spaltungsfläche f, Trennungsfläche f ~ product Spaltkörper m, Spaltungsprodukt n ~ resistance Trennwiderstand m ~ strength Spaltfestigkeit f ~ surface Spaltfläche f ~ test Spaltversuch m ~ traces Spaltflächen pl (cryst.)
cleave, to ~ abspalten, klöben, schlitzen, spalten, spleißen, zerspalten to ~ hard rocks with quoins ketzern
cleaver Spaltkeil m
cleaving Blätterbruch m ~ chisel Spaltmeißel m
cleft Bruch m, Schlitz m, Spalte f; gerissen, klüftig
clench, to ~ einhaken, einklinken, fest zusammenpressen to ~ together zusammenschlagen
clenched stapling Klammerheftung f
clenching (winding) screw Knebelschraube f
Clerget, ~ divisor Clerget-Divisor m ~ method Clerget-Verfahren n
clerical error Schreibfehler m
clerk Beamter m, Expedient m, Kontorist m (male), Kontoristin f (female), Schreiber m
Cletrac steering mechanism Cletrac-Lenkgetriebe n
clever gescheit, geschickt, klug, sinnvoll
cleverness Geschicklichkeit f, Gewandtheit f
clevis Bügel m, Gabelkopf m, Haken m am Förderseil ~ bolt Gabelkopf-, Schäkel-bolzen m ~ hook Einhängebügel m ~ insulator Schäkelisolator m ~ joke end Gabelkopf m ~ pin Gabelkopfbolzen m ~ plate Festhalteplatte f ~ stud Stangenkopfgewinde n
clew, to ~ down herunterholen to ~ up aufgeien
cliché Bildstock m, Druckstock m, Klischee n
click, to ~ anschlagen, einschnappen, klappen, knacken, knipsen, prellen (key), ticken
click Knack m, Knacken n, Knackgeräusch n, Schaltklinke f, Schlag m, Sperrhaken m, Sperrkegel m, Sperrklinke f, Sperrstift m, Türklopfer m ~ gauge Tickmanometer n ~ safety Schlagstücksicherung f ~ spring Einklink-, Sperr-feder f ~-stop arrangement Rastvorrichtung f, Tastvorrichtung f (rdo) ~-stop device Rastvorrichtung f ~ stop indicator Rüstenschauzeiger m ~ suppressed entstört ~ suppressor Hörschutz m ~ work Gesperr n

clicking Anschlag m des Klopfers ~ disturbance Knackstörung f
client Abnehmer m, Auftraggeber m, Kunde m
clientele(s) Kundschaft f
cliff Felsen m, Klippe f, Uferabbruch m
clifted (or clifty) felsig, gespalten
climate Klima n
climatic klimatisch ~ belt Klimagürtel m ~ change Klimaänderung f
climatology Klimalehre f, Klimatologie f
climax, to ~ steigern
climax Gipfel m, Gipfelpunkt m
climb, to ~ aufsteigen, besteigen, ersteigen, klettern, steigen to ~ a ramp Steigung f nehmen to ~ up ausfahren, hinaufklettern to ~ vertically senkrecht steigen
climb Anstieg m, Emporsteigen n, Steigen n, Steigflug m ~-and-ceiling factor Steigzahl f ~ angle Kippwinkel m ~ characteristics Steigflugeigenschaften pl ~ curve Steig-kurve f, -linie f ~ indicator Steigmesser m ~ miller Gleichlauffräsmaschine f ~ milling Gleichfräsen n ~-out Steigflug m ~ phase Steigflugphase f ~ power Steigflugleistung f, Steigleistung f ~ rating Steigleistung f ~ take-off power Startleistung f im Steigflug (helicopter)
climbers Klettereisen n, Kletterschuhe pl (electr.), Steigeisen pl
climbing Fahren n (min.), Steigen n ~ ability Kletter-, Steig-fähigkeit f, Steigvermögen n ~ angle Aufstieg-, Steigungs-winkel m ~ capacity Steigfähigkeit f ~ coefficient Anstiegsbeiwert m ~ distance Steigstrecke f ~ film evaporator Kletterverdampfer m ~ flight Aufwärts-, Steig-flug m ~ iron Kletter-, Steig-eisen n ~ ladder Aufstiegleiter m ~ path Aufstiegbahn f (rocket) ~ plant Schlinggewächs n ~ power Steig-fähigkeit f, -kraft f ~ program Umlenkprogramm n (g/m) ~ quality Steigeigenschaft f ~ roller Abfangwalze f ~ (turn) roller Kletterwalze f ~ shaft Aufstieg-, Steig-schacht m ~ shoes Kletterschuhe pl ~ speed Steiggeschwindigkeit f ~ turn gezogene Kurve f, Steigkurve f, steigende Umdrehung f ~ turntable Kletterdrehscheibe f
clinch, to ~ nageln, stauchen (a rivet) to ~ cables einstecken
clinched and riveted niet- und nagelfest
clincher Haspe f ~ rim Wulstfelge f (wheel) ~ tire Wulstreifen m
clinching stroke Nietschlag m
cling, to ~ anhaften, ankleben, hängen, hangen, sich anhängen, sich anklammern to ~ to haften, sich ankrallen, umschlingen to ~ tenaciously anklammern
clinic Klinik f
clinical klinisch
clink, to ~ klirren
clink Haspe f
clinker, to ~ ausräumen, ausrosten, ausschlacken (clean out), brennen, festbacken, schlacken, sintern, verschlacken, ziehen
clinker Klinker m, Schlacke f, Schlacken-klotz m, -klumpen m, Sinterschlacke f ~ brick Klinker m ~-built geklinkert ~ disposal Aschenaustragung f ~-quenching trough Lösche f ~ remover Schlackenaustrager m ~ scavenger Entschlacker m

**clinkering** Aschenverflüssigung *f*, Backen *n*, Schlackenziehen *n*, Sintern *n*, Sinterung *f*, Verschlackung *f* ~ **coal** schlackende Kohle *f* ~ **door** Schlackentür *f*, Schlacktür *f* ~ **spear** Stocheisen *n*

**clinkery** schlackig

**clinking** Rissigwerden *n* (in castings) ~ **test** Klinkprobe *f*

**clinkstone** Klingstein *m*

**clinoclasite** Strahlerz *n*

**clinometer** Böschungswaage *f*, Fallwinkel-, Gefälle-, Gefälls-, Geländewinkel-messer *m*, Klinometer *n*, Libelle *f*, Neigungswinkelmesser *m*, Winkelmesser *m*, Winkelwasserwaage *f* ~ **sight** Libellenaufsatz *m* (artil.)

**Clinton ore** Clinton-Erz *n*

**clip, to** ~ abkappen, abschneiden, abschroten, beschneiden, knipsen, scheren, stützen, verkürzen, verstümmeln **to** ~ **on** mit Klammer befestigen

**clip** Anschlußschelle *f* (aviat.), Bandschelle *f* (belt), Bogen *m*, Bügel *m*, Drahtschloß *n*, Drahtseilschloß *n*, Geschoßheber *m*, Halter *m*, Klammer *f*, Klemme *f*, Klemmer *m*, Klemmplatte *f*, Krampe *f*, Lasche *f*, Patronenrahmen *m*, Reifen *m*, Schelle *f*, Schneppe *f*, Spange *f*, Vorschuh *m*, Ziehband *n* (spring) ~ Federklemme *f* (wire-rope) ~ Seilklemme *f* ~ **for air tubes** Rohrschelle *f* ~ **for fixing light rails** Gleisklemme *f* ~ **for stentering** Spannkluppe *f*

**clip,** ~ **binder** Klammerhefter *m* ~ **bolt** Hakenschraube *f* ~ (collar) **brake** Halsringbremse *f*, Ringbremse *f* ~ **chain** Kluppenkette *f* ~ **connector** Verbindungsklammer *f* ~ **position** Befestigungsstreifen *m* ~ **screen** Schüttelsieb *n* ~ -**spring switch** Federschalter *m*

**clipped,** ~ **dot** spitzer Morsepunkt *m* oder Punkt *m* ~ **dots** spitze Morsepunkte *pl*

**clipper** Amplitudenabschneider *m*, Begrenzerkreis *m*, Kapper *m*, Karabinerhaken *m*, Synchronisierimpuls-abtrennung *f*, -aussiebung *f*, Schnellsegler *m* (navy), Spitzenbegrenzschaltung *f*, Taktgeber *m* (data proc.) ~ **circuit** Amplitudenbegrenzer *m* ~ **diode** Klipperdiode *f* ~ **limiter** Klipper *m* mit Ansprechschwelle ~ **stage** Impulsabtrennstufe *f* ~ **tube** Begrenzerröhre *f*

**clippers** Beißzange *f*, Schneidzange *f*

**clipping** Beschneidung *f* (loss of initial or final speech sounds), Klemmung *f*, Papierschnitzel *n*, Silbenabschneidung *f*, Wortabschneidung *f*, Wortverstümmelung *f*, Wörterverstümmelung *f* (of words) ~ **of amplitude crests (or peaks) by limiter means** Abschneiden *n* der Amplitudenspitzen ~ **of speech** Abschneiden *n* von Silben und Worten

**clipping,** ~ **circuit** Begrenzerkreis *m* ~ **time** Zeitkonstante *f* des Begrenzers

**clippings** Abschnitte *pl*, Blechabfälle *pl*, Münzkrätze *f* ~ **waste** Abgang *m*

**clips for monkey** Auslösehaken *m* (trip gear)

**cliseometer** Beckenneigungsmesser *m*

**clive** Rohrdichtungskegel *m*, Rohreinhängebügel *m*

**cloaca** Abzugsschleuse *f*

**clock** Taktgeber *m* (data proc.), Uhr *f*, Zeitmesser *m*, Zeitgeber *m* (data proc.), Zwickel *m* ~ **case** Uhrgehäuse *n* ~ -**dial target direction-**

**indicator** Zwölfuhrzeiger *m* ~ **frequency** Takt-, Zeitgeber-frequenz *f* ~ **governor** (with vanes) Windfang *m* ~ **hour figure of the vector group** Schaltgruppennumerierung *f* nach dem Zifferblatt ~ **installation** Uhrenanlage *f* ~ **meter** Pendelzähler *m* ~ **outlet** Uhröffnung *f* ~ **pivot** Uhrzapfen *m* ~ **pulse** Taktimpuls *m* ~ **signal** Uhrzeitzeichen *n* ~ **spring** Uhrfeder *f* ~ -**spring steel** Uhrfederstahl *m* ~ **steel** Sperrkegelstahl *m* ~ **substation** Uhrenunterzentrale *f* ~ **time** Uhrzeit *f* ~ **track** Taktspur *f* (tape rec.) ~ **winding** Uhraufzug *m*

**clockwise** rechtsdrehend, rechtsgängig, rechtsläufig, regelmäßig, in der Richtung oder im Sinne des Uhrzeigers, im Uhrzeigersinn ~ **and anti-clockwise running** Rechts- und Linkslauf *m* ~ **direction** Uhrzeigersinn *m* ~ **rifling** Rechtsdrall *m* ~ -**rotating system** Rechtssystem *n* ~ **rotation** Drehung *f* im Uhrzeigersinn, Rechtsdrehung *f*, Rechtslauf *m*

**clockwork** Federspannmotor *m* (motor), Räderwerk *n*, Uhrwerk *n* (mechanism) ~ **action** Uhrwerkantrieb *m* ~ **arc lamp** Uhrwerkbogenlampe *f* ~ **delay fuse** Uhrwerkverzugszünder *m* ~ (motor) **drive** Federkraftantrieb *m* ~ **feed** Uhrwerknachschub *m* ~ **fuse** Uhrwerkzünder *m* ~ **time switch** Uhrwerkzeitschalter *m* ~ **train** Feder-kraftantrieb *m*, -zug *m*

**clod** Klumpen *m* ~ **crusher** Ackerschleife *f*, Erdschollenquetsche *f*, Schollen-brecher *m*, -zerteiler *m*

**clog, to** ~ festsetzen, verstopfen, verstopft werden, zusetzen

**clog** Hemmschuh *m*

**clogged** verschlämmt, verstopft ~ **relief valve** Druckentlastungsventil *n*

**clogging** Festsetzung *f*, Verkleisterung *f*, Verschlämmung *f*, Verstopfung *f* ~ **of nozzle** Düsenverschmutzung *f*

**clophene capacitor** Clophenkondensator *m*

**close, to** ~ abschließen, geschlossen werden, klappen (e. g. lid of a chest), kurzschließen, schließen, sperren, verschließen, zumachen, zuschließen, zustellen **to** ~ **in** sich nähern **to** ~ **a circuit** einen Stromkreis *m* herstellen oder schließen **to** ~ (or make) **contact** Kontakt schließen **to** ~ **up** anschließen, aufrücken **to** ~ **up** (open) **the contacts** Kontakte enger (weiter) stellen **to** ~ **by heating** zubrennen **to** ~ **by melting** zuschmelzen **to** ~ **ranks** aufschließen **to** ~ **up the structure** das Gefüge auflockern **to** ~ **a switch** einen Schalter oder Stromkreis schließen **to** ~ **the throttle** Gas ganz wegnehmen, Motor abdrosseln

**close** Beschluß *m*, Schluß *m*; dicht bei, dicht, eng, kompakt, nahe, nah, klamm ~ **to** nahebei, in der Nähe, neben ~ (together) gedrängt, gedrungen ~ -**annealed** kastengeglüht ~ -**annealed sheet** kastengeglühtes Blech *n* ~ **annealing** Glühen *n* in geschlossenen Behältern, Kasten-, Kisten-glühung *f* ~ **approximation** gute Annäherung *f* ~ -**burning coal** Glüh-, Mager-kohle *f* ~ -**by** dicht bei ~ **contact** enge Anschmiegung *f* ~ -**coupled body** engsitziger Aufbau *m* ~ **coupling** feste Koppelung *f*, ~ **effect** Nahwirkung *f* ~ **field** Nahfeld *n* ~ -**filtered** durch einen engmaschigen Filter *m* gedrückt ~ **fit** Festsitz *m*, Edelpassung *f* ~ **fit**

of thread Gewindepassung *f* (fein) **~-fitting** knapp **~ goods** dichte Ware *f* (textiles) **~- grain wood** engjähriges Holz *n* **~-grained** fein- körnig, kleinluckig **~-grained iron** Feineisen *n* **~-grained pig iron** feinkörniges Roheisen *n* **~-grained texture** dichtes Gefüge *n* **~-grained wood** Derbholz *n* **~ grains** geschlossene Narben *pl* **~-in-fall-out** primärer radioaktiver Nieder- schlag *m* **~-in protection security** Nahsicherung *f* **~ matter** enggehaltener Satz *m* (print.) **~- medium shot** Halbtotale *f* **~-mesh control network** engmaschiges Festpunktnetz *n* **~- meshed** engmaschig **~-order construction** ge- schlossener Bau *m* **~-packed** dichtgepackt, dicht gepackt **~-packed structure** dichteste Kugelpackung *f* **~ pig** feinkörniges Roheisen *n* **~ pursuit** scharfes Nachdrängen *n* **~-range conveyer** Nahförderer *m* **~-range direction finding** Nahpeilung *f* **~ sand** feiner oder un- durchlässiger Sand *m* **~ scanning** Feinab- tastung *f* (rdr) **~-setting of the rolls** enge Walzen- stellung *f* **~-spaced rotary beam antenna** Dreh- richtantenne *f* mit kleinem Elementenabstand **~ spacing of fins** dichtstehende Rippenan- ordnung *f* **~ talking** Nahbesprechung *f* **~- talking microphone** Nahbesprechungsmikrofon *f* **~-texture** dicht (paper) **~ texture uniform throughout** gleichmäßig dichtes Gefüge *n* **~-up focusing attachment** Naheinstellgerät *n* **~-up view** Nah-, Groß-aufnahme *f* **~ warning ring** Nahwarnring *m* **closed** abgeschlossen, abgesperrt, eingefedert (as a compensator jack), eingeschaltet, ge- schlossen, gesperrt, verschlossen **~by a resistance** durch einen Widerstand abgeschlos- sen

**closed, ~ antenna** geschlossene Antenne *f* **~ armor** geschlossene Armierung *f* oder Be- wehrung *f* **~ basin** Flutbecken *n* **~ bearing** Stecklager *n* **~ box-type impeller** geschlossenes Laufrad *n* **~ circuit** geschlossener Kreis *m* oder Stromkreis *m*, Kreisleitung *f*, Ruhe- stromkreis *m*

**closed-circuit, ~ alarm system** Alarmeinrich- tung *f* mit Ruhestrom **~ arrangement** Ring- schaltung *f* **~ classification** Klassifikation *f* im geschlossenen Kreislauf **~ connection** Ruhe- stromschaltung *f* **~ cooling** Rückkühlung *f* **~ current principle** Ruhestromprinzip *n* **~ pipe line** Ringleitung *f* (of blast-furnace hot-air line) **~ proof** ruhestromsicher **~ signalling** Ruhe- strombetrieb *m* **~ system** Ringleitungssystem *n*, Ruhestromsystem *n* **~ television system** interne Fernsehanlage *f* **~ ventilation** Um- laufkühlung *f* **~ working** Ruhestrombetrieb *m*

**closed, ~ circuits** geschlossene Polygon *n* **~ coil** geschlossene Dampfschlange *f* **~ core** eisengeschlossen, geschlossener Eisenweg *m* oder Kern *m* **~-core transformer** Transforma- tor *m* mit geschlossenem, magnetischem Kreis **~ crystallizer** geschlossene Maische *f* **~ cup** geschlossener Tiegel *m* **~ cycle** Kreislauf *m* **~-end cylinder** einseitig geschlossener Zylinder *m* **~-end impedance** Kurzschluß-impedanz *f*, -widerstand *m* **~ fermenter** geschlossenes Gär- gefäß *n* **~ flash point** Flammpunkt *m* im ge- schlossenen Tiegel **~ fore part (or front)** geschlossene Brust *f* **~ furnace** Gefäßofen *m*

**~ impeller** geschlossenes Laufrad *n* **~-in production** eingeschlossene Produktion *f* **~ line** abgeschlossene Leitung *f* **~ line of position** Anschluß *m* **~ linkage** Kraftschluß *m* **~ loop** geschlossener Kreislauf *m*; geschlossene Schlei- fe *f* **~-loop effect** Rahmen-effekt *m*, -wirkung *f* **~- loop frequency response** Frequenzgang *m* der Regelung **~-loop oscillations** Regel- schwingungen *pl* **~-loop servo system** ge- schlossener Regelkreis *m* **~ (open) magne- tic circuit** geschlossener (offener) Eisen- kreis *m* **~ mold** geschlossene Form *f* **~ nozzle** geschlossene Düse *f* **~ oscillatory circuit** geschlossener Schwingungskreis *m* **~ pass** geschlossenes Kaliber *n* **~ path** ge- schlossener Weg *m* **~ phase** abgeschlossene Phase *f* **~ port** verbotener Hafen *m* **~ position** „Ein"-Stellung *f* (of switch), Einschaltstellung *f* **~ pressure** Lagedruck *m* **~ sand mold** ge- schlossene Gußform *f* **~ shell** abgeschlossene Schale *f* **~ shop** betriebsfremde Program- mierung *f* **~ steam** indirekter Dampf *m* **~ stub** Kurzschlußblindschwanz *m* **~ subroutine** ge- schlossenes Teilprogramm *n* **~ support** Ohren- lager *n* (mach.) **~ tester** geschlossenes Gefäß *n* (flash point) **~-top roll housing** Kappen-, Rahmen-ständer *m* **~ transaction file** Kartei *f* abgeschlossener Vorgänge **~ truck** geschlos- sener Wagen *m* **~ tube** Bajonettrohr *n* **~ tuyère** geschlossene Form *f* **~ upon itself** in sich geschlossen **~ wagon** geschlossener Wagen *m* **~ well** abgeschlossene Sonde *f*

**closely, ~ set warp** dicht gestellte Kette *f* **~ spaced** dichtstehend

**closeness** Dichte *f*, Dichtheit *f*, Dichtigkeit *f*, Enge *f*, Gedrängtheit *f*, Genauigkeit *f* **~ of grain** Korndichte *f* **~ in winding** Windschutz- dichte *f*

**closer** Ankerstein *m* (brick), Halbstein *m* (half brick), Spannkopf *m* (einer Spannzange) **(cir- cuit) ~** Schalter *m* **~ roll** Patrizenwalze *f*

**closet** Abortanlage *f*, Klosett *n* **~ hopper** Abort- trichter *m*

**closing** Abschließen *n*, Absperrung *f*, Absperren *n* (of a river arm) **~ of an arm of the sea** Ab- schließung *f* eines Meeresarms **~ of circuit** Stromschluß *m* **~ of a circuit** Schließung *f* eines Stromkreises **~ a ring** Ringschließung *f* **~ of valve** Ventilschluß *m*

**closing, ~ apparatus** Abstellvorrichtung *f*, Ver- schluß *m* **~ button** Druckknopfschalter *m* zum Schließen **~ cap head** Zünderkappe *f* **~ circuit** Einschaltkreis *m* **~-clasp** Überfall *m* (of lock) **~ contact** Schließkontakt *m* (electr.) **~ cylinder pin** Schließzylinderzapfen *m* **~ date** Termin *m* **~ device** Abschluß-, Schließ-vorrichtung *f*, Taschenbügel *m*, Verschluß *m* **~ device for bottles** Flaschenverschluß *m* **~ die** Schließnippel *m* **~ drum** Schließtrommel *f* **~ flap** Abschluß- klappe *f* **~ gear of the tongs** Zangenschließ- werk *n* **~ hasp** Überfall *m* (of lock) **~ head** Schließ-kopf *m*, -vorrichtung *f* **~-in arrange- ment** Abschlußvorrichtung *f* **~ indicating lamp** Signallampe *f* beim Einschalten **~ line** Schluß- linie *f* **~ loop** Verschlußschlaufe *f* (parachute) **~ magnet** Sperrmagnet *m* **~ motor** Schließmotor *m* **~ needle** Absperrnadel *f* **~ piston** Absperr- kolben *n* **~ pulley** Schließrolle *f* **~ push-**

**button** Einschaltknopf *m* ~ **shape** Formschluß *m* (metall.) ~ **speed** Schließgeschwindigkeit *m* ~ **stock** Endbestand *m* ~ **stone** Schlußstein *m* ~ **time** Schließzeit *f*, Sperrzeit *f* ~-**up** Stillegung *f* ~ **valve** Abschlußventil *n* ~ **window of a thermocouple** Abschlußfenster *n* eines Thermoelements

**closure** Abschluß *m*, Schließbeschlag *m*, Schließen *n*, Schließung *f*, Verschluß *m*, Verschlußdeckel *m*

**clot** Flocken *pl*, Rührhaken *m*

**cloth** Gewebe *m*, Kaliko *m*, Leinen *n*, Leinwand *f*, Segeltuch *n*, Stoff *m*, Tuch *n*, Zeug *n* ~ **accumulator** Warenspeicher *m* ~ **air conduit** Tuchlutte *f* ~ **bag** Stockbeutel *m* ~ **beam** Zeugbaum *m* (weaving) ~ **binding** Leineneinband *m* ~ **buff** Tuchscheibe *f* ~-**centered paper** Faden-, Gewebe-papier *n*, Papyrolin *n* ~ **contact** Warenanstrich *m* ~ **covering** Stoffbespannung *f*, Stoff-überzug *m*, -umkleidung *f* ~ **cutter** Zuschneider *m* ~ **cutting** Zeugzuschneiden *n* ~-**disk cover** Stoffscheibenbelag *m* ~-**embossing plant** Gaufrieranstalt *f* ~-**faced paper** Papyrolin *n* ~ **filter** Filter-, Kolier-, Seiher-tuch *n*, Tucheinsatz *m* ~-**finishing establishment** Appreturanstalt *f* ~-**finishing machine** Preßkalander *m* ~-**folding machine** Gewebelegemaschine *f* ~-**lined envelope** Leinenumschlag *m* ~ **mop** Lappenscheibe *f* ~-**piece glazing** Glanzieren *n* von Geweben *f* ~ **pressing** Gewebepressen *n* ~ **print** Leinwandpause *f* ~ **printing** Zeugdruck *m* ~-**printing factory** Zeugdruckerei *f* ~-**printing plant** Gaufrieranstalt *f* ~ **prover** Weberglas *n* ~-**screen dust arrester** Entstaubungsanlage *f* mit Staubfilter aus Stoff, trockener Staubsauger *m* ~ **shears** Tuchschere *f* ~ **strainer** Tuchvorfilter *n* ~ **tracing** Leinwandzeichnung *f* ~ **wheel polishing** Schwabbelpolieren *n* ~ **width** Warenbreite *f* ~ **winding** Stoffverwickelung *f*

**clothe, to** ~ auskleiden, bekleiden

**clothes,** ~ **closet** Kleiderablage *f* ~-**line winder** Wäscheleinenwickler *m*

**clothing** Bekleidung *f*, Kleidung *f* ~ **lined with ebonite** Hartgummiauskleidung *f*

**clotted** geronnen, klumpig ~ **dirt** Schmutzknoten *m*

**clotty** klumpig

**cloud, to** ~ moirieren, verhüllen, (sound) verdecken

**cloud** Wolke *f*, Schattenfleck *m* ~ **and collision warning radar** Nebel- und Antikollisionsradar *m* ~ **and pour test** Bestimmung *f* des Trübepunktes und des Erstarrungspunktes ~ **of electrons** Elektronen-schwarm *m*, -wolke *f* ~ **of gas** Gaswolke *f* ~ **of gas (or smoke), produced by blasts (or explosions)** Schwaden *pl* ~ **of haze** Dunstwolke *f*

**cloud,** ~ **altimeter** Wolkenhöhenmesser *m* ~ **bank** Wolken-bank *f*, -feld *n* ~ **banner** Wolkenstreifen *m* ~ **base** Wolkenuntergrenze *f* ~ **base ceiling** Wolkenuntergrenze *f* ~ **breaking procedure** Durchstoßverfahren *n* (aviat.) ~ **burst** Wolkenbruch *m*, Platzregen *m* ~**burst hardening** Kugelstrahlen *n*, Kaltverfestigung *f* ~ **cap** Wolkenkappe *f* ~ **ceilometer** Wolkenhöhenmesser *m* ~ **chamber** Nebelkamera *f*, Nebel-, Wilson-kammer *f* ~ **chamber expansion** Nebelkammerexpansion *f* ~ **chamber photo-**

**graphy** Nebelkammeraufnahme *f* ~ **column** Rauchsäule *f* ~ **cover** Wolkendecke *f* ~-**density measurement apparatus** Wolkendichtemesser *m* ~ **dome** Wolkenkuppe *f* ~ **electricity** Wolkenelektrizität *f* ~ **flight** Wolkenflug *m* ~ **formation** Wolkenbildung *f* ~ **forms** Wolkenformen *pl* ~ **gas** Nebelgas *m* ~ **head** Wolkenkuppe *f* ~ **height** Wolkenhöhe *f* ~ **layer** Wolkenschicht *f* ~ **limit** Nebelbildungsgrenzwert *m* ~ **map** Bewölkungs-, Wolken-karte *f* ~ **mass** Wolkenmasse *f* ~ **measurement** Wolkenmessung *f* ~ **mirror** Wolkenspiegel *m* ~ **motion** Wolkenzug *f* ~ **point** Kristallisationsbeginn *m* (chem.), Trübepunkt *m* ~ **pulse** Raumladungsimpuls *m* ~ **rake** Wolkenrechen *m* ~ **reflector** Wolkenspiegel *m* ~ **searchlight** Wolkenscheinwerfer *m* ~ **shadow** Wolkenschatten *m* ~ **soaring** Wolkensegeln *n* ~ **test** Trübepunkt *m* ~ **top** Wolkenkuppe *f* ~ **track** Nebelspur *f* ~-**track interpretation** Nebelspurauswertung *f* ~ **train** Wolkenzug *f*

**cloudiness** Bewölkung *f*, Trübe *f*, Trübung *f*, Umwölkung *f*

**clouding** Gehörmaskierung *f*, (of glass) Trübung *f*, Verdeckung *f*, Verhüllen *n*, Wolkenbildung *f* ~ **of a lens** Linsentrübung *f*

**cloudless** wolkenlos

**cloudy** bedeckt, bewölkt, trüb, trübe, (sky), wolkig **to become** ~ trüb werden (Flüssigkeit), ~ **layer** Nebel-, Wolken-schicht *f*

**clough** arch Freiarche *f*

**clout nail** Absatzstift *m* (boot)

**clove,** ~ **hitch** Mastwurf *m*, Webeleinstee *m* ~ **oil** Nelkenöl *n*

**clover** Klee *m*, Kleeblatt *n*, Kleie *f* ~ **dodder** Kleeseide *f* ~-**leaf antenna** Kleeblattantenne *f* ~-**leaf arrangement (or connection)** Kleeblattumleitung *f* ~-**leaf body** Kleeblattaufbau *m* ~-**leaf intersection** Kleeblattkreuzung *f* ~-**leaf (or flyover) junction** Kleeblattkreuzung *f*

**club** Keule *f*, Knüppel *m*, Kolben *m* ~ **ribbon** Vereinsband *n* ~-**type insulator** Knüppelisolator *m*

**clue** Anhaltspunkt *m*, Leitfaden *m*, Schlüssel *m*, Vorläufer *m*

**clump block** Klampblock *m*

**clumsiness** Eckigkeit *f*, Schwerfälligkeit *f*

**clumsy** holperig, plump, unbeholfen, ungeschickt

**cluster, to** ~ bündeln (von Raketentriebwerken)

**cluster** Büschel *m* ~ **of crystals** Kristalldruse *f* ~ **of plane surfaces** Ebenenbüschel *n* Fallschirmtraube *f*, Gruppe *f*, Nest *n*, Tröpfchen *n* (cryst.)

**cluster,** ~ **attachment** Gruppenstanze *f* ~ **gear** Doppelrücklaufrad *n*, Stufenzahnrad *n* ~ **mill** Mehrwalzwerke *pl*

**clustered** gebündelt ~ **sugar** Zuckerknoten *m* ~ **teeth** (of a saw) Gruppenzahnung *f*

**clustering** Schwarmbildung *f*

**clusters** Molekülkomplexe *pl*

**clutch, to** ~ einkuppeln, kuppeln, schalten

**clutch** Ausrückkupplung *f*, ausrückbare oder schaltbare Kupplung, Haken *m*, Klaue *f*, Kupplung *f*, Kupplungskreuz *n*, Reibkegel *m*, Schaltkuppelung *f* ~ **(coupling)** Ein- und Ausrückkuppelung *f* ~ **for starting handle** Andrehkurbelklaue *f* ~ **for take-down** Freilauf *m* für Warenabzug

**clutch,** ~ **adjustment** Kupplungsnachstellung

*f* (radial) ~ **bearing** Kupp(e)lungslager *n* ~
**brake spring** Kupplungsbremsfeder *f* ~ **cam**
Kupplungsdaumen *m* ~ **closer** Kupp(e)lungs-
ausrückmuffe *f* ~ **collar** Kupplungsmuffe *f* ~
**cone** Kupplungskegel *m* ~ **coupling** schaltbare
Kupplung *f* ~ **coupling box** Ausrückklaue *f*
~-**coupling sleeve** Kupplungsmuffe *f* ~ **disengaging shaft** Kupplungsausrückwelle *f* ~ **disk**
Kupplungslamelle *f* ~-**disk lining** Kupplungs-
scheibenbelag *m* ~ **dog** Antriebsklaue *f*, Kupplungsklaue *f* ~ **drive** Kupplungsantrieb *m* ~-
**drive collar** Kupplungsring *m* für den Fräsdorn
~-**driven impulse** Kupplungsschnapper *m* ~
**facing** Kupplungs-belag *m*, -scheibe *f* ~ **facing
spring** Feder *f* unter dem Kupplungsbelag ~
**fork** Kupplungsgabel *f* ~ **guard** Kupplungs-
schutz *m* ~ **half** Kupplungshälfte *f* ~ **housing**
Freilaufbuchse *f*, Kupplungsgehäuse *n* ~ **hub**
Kupplungsnabe *f* ~ **interlock** Kupplungsverrie-
gelung *f* ~ **key** Schlüsselbolzen *m* ~ **leather**
Kupplungsleder *n* ~ **lever** Ausrück-, Kupp-
lungs-, Schalt-hebel *m* ~-**lever shifter** Kupp-
lungshebelschalter *m* ~ **linkage** Kupplungsgestänge *n* ~ **magnet** Arbeits-, Kupplungs-magnet
*m* ~ **member** Kupplungsteil *m* ~ **operating
device** Kupplungsgestänge *n* ~ **operation**
Kupplungsbedienung *f* ~ **operator** Kupplungs-
-ausrücker *m*, -schalter *m* ~ **pedal** Kupplungs-
-hebel *m*, -fußhebel *m*, -pedal *n* ~-**pedal lever**
Kupplungspedalhebel *m* ~ **pedal shaft** Kupp-
lungsfußhebelwelle *f* ~ **plate** Kupplungsplatte *f*
~ **pressure plate** Kupplungsdruckplatte *f* ~-
**release sleeve** Kupplungsausrückmuffe *f* ~
**ring** Kreuzring *m*, Kupplungs-rad *n*, -ring *m* ~
**rod** Klauenstange *f* ~ **rods** Kupplungsgestänge
*n* ~ **shaft** Antriebswelle *f*, Kupplungs-welle *f*,
-spindel *f*, Mitnehmerwelle *f* ~ **shifter** Klauen-
kupplungs-, Kupplungs-ausrücker *m*, Kupp-
lungsschalter *m* ~-**shifter stand** Kupplungs-
schalterbock *m* ~ **sleeve** Schiebemuffe *f* ~
**sleeve guide** Kupplungsmuffenführung *f* ~
**slide** Kupplungsschieber *m* ~ **spider** Kupplungs-
armkreuz *n* ~ **spindle** Kupplungswelle *f*, Wickel-
teller *m* (tape rec.) ~ **spreader** Kupplungsexpan-
sionsring *m* ~ **spring** Kupplungsfeder *f* ~
**sprocket wheel** Kupplungskettenrad *n* ~ **stop**
Kupplungsbremse *f* ~ **stop disc** Anschlag-
lamelle *f* ~ **taper** Kuppelkonus *m* ~ **throwover
(or yoke)** Kupplungsgabel *f* ~ **weight** Kupp-
lungsfliehgewicht *n* ~ **wheel** Kupplungsrad *n*
~-**withdrawal lever** Kupplungsausrückhebel *m*
~-**withdrawal shaft** Kupplungsauslösewelle *f*
**clutching** Einschalten *n* ~ **device** Kupplungsein-
richtung *f*
**clutter** Störfleck *m* (rdr), Störungszeichen *n* (rdr)
~ **diagram** Festzeichenbild *n* ~ **elimination**
Festzeichenunterdrückung *f* (rdr) ~ **gating**
Festzeichenaustastung *f*
**C network** C-Glied *n*
**coacervation** Flockenbildung *f*
**coach** Einpauker *m*, Wagen *m* ~-**built** karossiert
~ **house** Wagenremise *f* ~-**man** Kutscher *m*
~-**man's hatchets** Stellmacherbeile *pl* ~ **screw**
Stellmacherschraube *f*, Holzschraube *f* mit
Vierkantkopf, Vierkantkopf *m* ~ **screw spanner**
Vierkantschlüssel *m* ~ **wrench** englischer
Schraubenschlüssel *m*
**coaggregation** Tröpfchen *n*

**coagulable** gefrierbar, koagulierbar ~ **albumin**
gerinnbares Eiweiß *n*
**coagulant** Koagulierungsmittel *n*
**coagulate, to** ~ erstarren, gerinnen, koagulieren,
verdicken, zusammenlaufen
**coagulated** geronnen
**coagulating,** ~ **agent** Koagulierungsmittel *n* ~
**bath** Fällbad *n*
**coagulation** Ausflockung *f*, Festwerden *n*, Flok-
kenbildung *f*, Gerinnen *n*, Gerinnung *f*, Koagu-
lierung *f*, (of colloids and smoke particles)
Zusammenballen *n* ~ **point** Erstarrungspunkt
*m*
**coagulator** Gerinnstoff *m*
**coagulometer** Koagulationsmesser *m*
**coagulum** Gerinnsel *n*, Koagel *n*
**coal, to** aufschütten **to** ~ **a ship** ein Schiff *n*
bekohlen
**coal** Kohle *f* ~ **to be washed** Waschkohle *f*
**coal,** ~-**allotting piston** Kohlenzuteilkolben *m*
~ **area** Steinkohlenvorkommen *n* ~ **ashes**
Steinkohlenasche *f* ~ **auger** Schrämmaschine *f*
~ **bag** Kohlensack *m* ~ **barrow** Kohlentrans-
portkarren *m* ~ **basin** Kohlenbecken *n* ~
**basket** Füllkorb *m* ~-**bearing** kohlig ~ **bed**
Kohlen-flöz *n*, -schicht *f* ~ **bin** Kohlenbunker *m*
~-**black** kohlschwarz ~ **blending** Kohlen-
mischung *f* ~-**blending plant** Mischanlage *f* ~
**brick** Kohlenbrikett *n* ~ **briquette** Kohlen-
preßstein *m*, Steinkohlenbrikett *n* ~-**bunker
space** Kohlenladeraum *m* ~ **bunkering port**
Bekohlungshafen *m* ~ **bunkering wharf** Bekohlungshafen *m* ~ **burning** Kohlenfeuerung *f*
~ **business** Kohlenhandel *m* ~ **by-product
industries** Industrie *f* der Kohlen- und Teer-
nebenprodukte ~-**car dumper** Kohlenwagen-
wipper *m* ~ **carbonization** Kohlenentgasung *f*,
Steinkohlenschwelung *f* ~ **carbonization coking**
Steinkohlenverkokung *f* ~-**carbonizing plant**
Kokerei *f* ~-**carbonizing practice** Kokereitechnik *f* ~-**carbonizing process** Kokereiprozeß
*m* ~ **charge** Kohlen-einsatz *m*, -füllung *f* ~
**charging** Kohlenschüttung *f* ~ **cinders** Stein-
kohlenschlacke *f* ~ **consumed at the pit** Selbst-
verbrauch *m* an Kohle ~-**conversion industry**
Umwandlungstechnik *f* ~ **crusher** Kohlen-
mühle *f* ~ **crushing** Kohlenzerkleinerung *f* ~
**damper** Kohlenwischer *m* ~ **deposit** Kohlen-
-lager *n*, -lagerstätte *f*, -vorkommen *n*; Stein-
kohlen-lager *n*, -vorkommen *n* ~ **deposits that
can be extracted** gewinnbares Kohlenvorkom-
men *n* ~ **distributor** Telleraufgabe *f* ~ **district**
Kohlengebiet *n*, Kohlenrevier *n* ~ **dock** Kohlenhafen *m* ~ **dressing** Kohlenaufbereitung *f* ~
**drier** Kohlentrockner *m* ~ **dust** Feinkohle *f*,
Kohlenmehl *n*, Kohlenstaub *m*, Lösche *f*,
Schlackenstaub *m*, Staubkohle *f*, Steinkohlen-
staub *m*
**coal-dust,** ~ **breeze** Gestübbe *n* (metall.) ~
**collecting worm** Staubaufgabe *f* ~ **engine** Koh-
lenstaubmotor *m* ~ **facing** Kohlenschichte *f* ~
**fired** staubbeheizt ~ **firing** Brennstaubfeuerung
*f*, Kohlenstaubfeuerung *f*, Staubfeuerung *f* ~
**mill** Kohlenstaubmühle *f* ~ **particle** Kohlen-
teilchen *n* ~ **preparation plant** Staubaufberei-
tungsanlage *f*
**coal,** ~ **engineering** Kohlentechnik *f* ~ **feeder**
Redler *m* ~ **feeding** Kohlenschüttung *f* ~ **field**

Becken *n*, Kohlen-becken *n*, -revier *n* ~-fired
kohlebeheizt ~-fired boiler Kohlenkessel *m*
~ firing Kohlen-befeuerung *f*, -feuerung *f* ~
flame Kohlenflamme *f* ~ flotation Kohlenwäsche *f* ~ gas Kohlen-, Kokerei-, Leucht-,
Steinkohlen-gas *n* ~-gas manufacture Steinkohlengaserzeugung *f* ~ grab Greiferkübel *m*,
Kohlengreifer *m* ~-grate firing Kohlenrost-
feuerung *f* ~-handling car Kohlentransport-
karren *m*, Kohlenwagen *m* ~-handling crane
Kohlenverladekran *m* ~-handling equipment
Kohlentransportanlage *f*, Kohlenverladeanlage
*f* ~-handling machinery Kohlenverladeanlage
*f* ~ heap Kohlenmeiler *m* ~ hewer (overhand
stooping) Steigortshauer *m* ~like kohlen-ähnlich, -artig ~loader Kohlenlader *m* ~ lorry car
Kohlenförderwagen *m* ~ lump Kohlenstück *n*
~ measure Karbon *n*, Kohlenflöz *n* ~-measure
ironstone Kohleneisenstein *m* ~ merchant
Kohlenhändler *m* ~-milling plant Kohlen-,
Kohlenstaub-mahlanlage *f* ~ mine Kohlenbergwerk *n*, Kohlengrube *f*, Steinkohlenbergwerk *n* ~ miner Kohlenhauer *m* ~ mining
Kohlen-abbau *m*, -bergbau *m* ~-mining
district Kohlenrevier *n* ~-mixing installation
Kohlenmischanlage *f* ~ mud Kohlenschlamm
*m* ~ oil Kohlenteeröl *n* ~ outcrop Kohlenaus-
biß *m* ~ output Kohlenförderung *f* ~ paste
(mixture of coal and oil) Kohlenbrei *m* ~ pick
hammer Abbauhammer *m* ~ picker Abbauhammer *m* ~ picking belt Kohlenklaubriemen *m* ~
pile Kohlen-haufen *m*, -meiler *m* ~ pit Kohlengrube *f*, Zeche *f* ~ preparation Kohlenaufbereitung *f* ~-preparation plant Kohlenaufbe-
reitungsanlage *f* ~ pulverizer Kohlenstaubmühle *f* ~-pulverizing plant Kohlen-mahlanlage
*f*, -müllerei *f*, -staubmahlanlage *f*, -staubmüllerei
*f* ~ region Kohlengebiet *n*, Steinkohlenvorkommen *n* ~ remover Abkohler *m* ~ research Koh-
lenforschung *f* ~ scoop Kohlenlöffel *m* ~
seam Kohlen-flöz *n*, -schicht *f*, Steinkohlen-
schicht *f* ~ separation Kohlenaufbereitung *f* ~
shed Kohlenschuppen *m* ~ shovel Kohlenschau-
fel *f* ~ slack Feinkohle *f*, Kohlengestübbe *n*,
Kohlenlösche *f* ~-sprinkler hose Kohlenzer-
stäubungsschlauch *m* ~ stack Kohlenbansen
*m* ~ storage Kohlenlager *n* ~ storage place
Kohlenmagazin *n* ~ supplies of recent formation junge Kohlenvorkommen *pl* ~ tar Koh-
len-, Steinkohlen-teer *m* ~(gas) tar Gasteer *m*
~-tar creosote Steinkohlenkreosot *n* ~-tar
naphta Solventnaphta *n* ~-tar oil Steinkohlen-öl
*n*, -teeröl *n*, Teeröl *n* ~-tar pitch Steinkohlen-
-pech *n*, -teerpech *n* ~ tip Kohlenkippe *f* ~
tipper Kohlenkipper *m* ~-unloading crane Koh-
lenentladekran *m* ~ vein Kohlenflöz *n* ~
washer Kohlenwäscher *m* ~ washing Kohlenwäsche *f* ~-washing plant Kohlenwäsche *f* ~
wharf Kohlenhafen *m* ~ yard Kohlen-lager *n*,
-platz *m*

coalesce, to ~ ineinanderfließen, koaleszieren,
verschmelzen, verwachsen, zusammenballen,
zusammenfließen

coalesced filaments verklebte Einzelfäden *pl*

coalescence Koaleszenz *f*, Verschmelzung *f*,
metallografische Zusammenballung *f*

coalescing Zusammenballung *f* ~ property Ballungsfähigkeit *f* (cryst.)

coaling, ~ bed Warmbett *n* ~ door Feuertür *f*,
Feuerungstür *f* ~ equipment Kohlenbeschik-
kungsvorrichtung *f* ~ installation Bekohlungsanlage *f* ~ plant Bekohlungsanlage *f* ~ station
Bekohlungs-, Kohlen-station *f* ~ table Warm-
bett *n*

coalite Halbkoks *m*

coalition Angliederung *f*, Koalition *f*

coaly kohlig

coarse grob, grob gepreßt, plump, rauh, rösch,
(ore) arm ~ adjustment Grobeinstellung *f*,
Steileinstellung *f* (Propeller), (in microscope)
Triebschraube *f* ~-adjustment switch Grob-
stufenschalter *m* ~-and-fine-adjustment (of
microscope) Zahn und Triebbewegung *f* ~
beam Grobkennung *f* (nav.) ~ coal jig Grob-
kornmaschine *f* ~ control Grobeinstellung *f* ~
copper metal erster Kupferstein *m* ~ copper
slag Schwarzkupferschlacke *f* to get ~ copper
Kupfer *n* schwarz machen ~ crush grobzer-
kleinern ~ crusher Vorzerkleinerungsmühle *f*
~ crushing Grobzerkleinerung *f*, Vorbrechen *n*
~-crushing grinder Vorzerkleinerungsmühle *f*
~-crushing rolls Grobwalzwerk *n* ~-crystal-
line grobkristallin ~ denier grober Titer *m*
~ dust particle Staubkorn *n* ~ error voltage
bridge Zehnereinstellbrücke *f* (Tacan) ~ feed
starker Vorschub *m* ~-filamentous grobfädig
~ file Grob-, Schrubb-, Stroh-feile *f* ~ filter
Groböl-, Roh-, Vor-filter *n*, Rohsieb *n*, ~-fine
action Grob-Fein-Regelung *f* ~ fit Grobspan-
nung *f* ~ grain grobes Korn *n*, Grobkorn *n*
~-grain cement Grobmörtel *m* ~-grain concrete Grobmörtel *m* ~-grain grinding wheel
grobkörnige Schleifscheibe *f* ~-grain iron
Grobkorneisen *n* ~-grain powder großkörniges
Pulver *n* ~-grain structure grobes Gefüge *n*
~-grain washer Grobkornsetzmaschine *f* ~-
grained grob-haarig, -faserig, -kristallin, -nar-
big, -körnig (film) ~-grained fracture grobkörnige Bruchfläche *f* ~ granulation Grob-
bruch *m* ~ gravel Grobschotter *m* ~ grid weitmaschiges Gitter *n* to ~-grind grobschleifen ~
ground grob gemahlen; grobe Mahlung *f*
~-ground corn Schrot *m* ~-ground grist Grob-
schrot *n* ~ hackle Abzugshechel *f* ~ indicator
Grobzeiger *m* ~ levelling Grobhorizontierung
*f* ~ meal Schrotmehl *n* ~ mesh(ed) grobmaschig ~ metal slag Rohsteinschlacke *f* ~
motion Grobverstellung *f* ~ ore Groberz *n*,
Stückerz *n* ~-ore furnace Groberzofen *m* ~
pitch große Steigung *f*, Steilstellung *f* (Propeller) ~ pitch thread grobgängiges Gewinde *n*
~ plaster Berapp *m*, Bewurf *m*, Rauhwerk *n*,
Spritzbewurf *m* ~ pointer dial Grobskala *f* ~-
pored grobporig ~ position circle Zielkreis *m*
~ pottery Töpfergut *n* ~ product Grobkorn *n*
~ pumping time Grobpumpzeit *f* ~ radioloca-
tion Grobortung *f* (rdr) ~ rake Grobrechen *m*
~ rasp Bleiraspel *f* ~ reduction Grobzerkleinerung *f* ~ reduction impact crusher Prallbrecher
*m* ~ regulating rods Grobregelstäbe *pl* ~
resistor Hochfahrwiderstand *m* ~ rod Grob-
regelstab *m* ~ sand Flußkies *m*, Sandgrieß *m*
~ scan Teilabtastung *f* ~ scanning Grobab-
tastung *f* (rdr) ~ screen photograph Grob-
rasteraufnahme *f* ~ sea gravel Fluß-kies *m*,
-schotter *m* ~ sea sand Flußschotter *m* ~

setting Grobeinstellen *n* ~ **sieve** Grobsieb *n*
~ **sight** Grobvisier *n* ~**-sized carbide** Grob-
karbid *n* ~ **slurry** Grobschlamm *m* ~ **stage**
Grobstufe *f* ~ **structure** Grobgefüge *n*, Großstruktur *f* ~ **structure analysis** Grobstrukturanalyse *f* ~ **thread** Grobgewinde *n*, steil-
gängiges Gewinde *n* ~**-threaded** grobdrahtig,
grobfädig ~ **threading attachment** Steilge-
windeschneideinrichtung *f* ~**-to-medium crushing** Vorzerkleinerung *f* ~**-toothed sidemill-
ing cutter** Scheibenfräser *m* mit Grobver-
zahnung ~ **tuning** Grobabstimmung *f*, grobe
Abstimmung, Grobstimmung *f*, rohe Abstimmung *f* ~ **tuning drive** Grobtrieb *m* ~
**vacuum** Grobvakuum ~ **voltage bridge** Zehner-
Einstellbrücke *f* (Tacan)
**coarsely,** ~ **crystalline** grobkristallinisch ~ **cutout** geschroppt, geschruppt ~ **dispersed** grob-
dispers ~ **ground corn** Schrott *m* ~ **ground
dried beets** Rübenschrot *n* ~ **perforated** grob-
gelocht ~ **powered** grob gepreßt ~ **ringed**
grobjährig
**coarsen, to** ~ vergröbern
**coarseness** Grobheit *f*, Grobkörnigkeit *f*, Körnung *f*, Rohheit *f*
**coarsening** Vergröberung *f* ~ **of grain** Kornver-
gröberung *f*
**coast** Gestade *n*, Küste *f* ~ **beacon** Küstenleucht-
feuer *n* ~ **geodetic survey** Küstenvermessung *f*
~ **guard** Küstenwache *f* ~ **line** Küstenlinie *f*
~**(al) radio station** Küstenfunkstelle *f* ~ **refraction** Küstenbrechung *f* (rdo) ~ **station**
Küstenfunkstelle *f* ~ **survey** Küstenaufnahme
*f* ~**-survey camera** Küstenaufnahmekamera *f*
~ **town** Seestadt *f* ~ **wind** Küstenwind *m*
**coastal,** ~ **beach belt** Nehrung *f* ~ **chart** Küstenkarte *f* ~ **communication system** Küstennach-
richtennetz *n* ~ **defense** Küstenwehr *f* ~ **direction-finding station** Küstenpeilstelle *f* ~ **engineering** Küstenbauten *pl* ~ **fortification** Küstenbe-
festigung *f* ~ **light** Küstenfeuer *n* ~ **line**
Küstenlinie *f* ~ **patrol** Küstenwacht *f* ~
**refraction** Küstenbrechung *f* (rdr) ~ **sector**
Küstenabschnitt *m* ~ **service** Küstendienst *m*
~ **stream** Küsten-, Ufer-strömung *f*
**coaster** Küstenfahrer *m*, Untersatz *m* ~ **brake**
Freilaufbremse *f*
**coasting** Freilauf *m* ~ **curve** Auslaufkurve *f* ~
**flight** Flug *m* nach Brennschluß (g/m) ~ **port**
Küstenhafen *m* ~ **trade** Küstenfahrt *f*
**coat, to** ~ anstreichen, aufbringen, auftragen,
auskleiden, begießen (film), bekleiden, (a film
with emulsion) beschichten, bestreichen, einhüllen, einkleiden (chem.), (film) gießen, putzen, streichen, überstreichen, überziehen, (with
electrolytically deposited copper) mit einem
galvanischen Überzug versehen, umkleiden,
umwickeln, (a lens) vergüten, (with paint)
verkleiden **to** ~ **over** verstreichen **to** ~ **with
aluminium** allitieren, aluminisieren **to** ~ **with
asphalt** asphaltieren **to** ~ **the concrete** den
Beton verkleiden **to** ~ **with graphite** graphitieren **to** ~ **with nickel** vernickeln **to** ~ **with
paper** kaschieren **to** ~ **with small broken stone**
aufschottern **to** ~ **with steel** verstählen **to** ~
**by vapor** aufdämpfen
**coat** Aufstrich *m*, Auftrag *m* (of paint), Belag *m*,
Belegung *f* (of condenser), Decke *f*, Farben-

überzug *m*, Formmantel *m* (foundry), (of
paint) Lage *f*, Putz *m*, Rock *m*, Überzug *m*,
Schicht *f* ~ **of clay** Tonbeschlag *m* ~ **of color**
Farbanstrich *m* ~ **of flatting varnish** Schleif-
lacküberzug *m* ~ **of ice** Eishülle *f* ~ **of paint**
Anstrich *m*, Farbanstrich *m* ~**s of paint** Farb-
schutzüberzug *m* ~ **of snow** Schneemantel *m*
**coat,** ~ **bag** Mantelsack *m* ~ **film** Schutzschicht
*f* ~ **stitcher** Fantasiewirker *m*
**coated** geschichtet, getaucht (electrode), überzogen, umhüllt, ~ **cathode** geschichtete
Kathode *f* ~ **electrode** getauchte Elektrode *f*,
Mantel-, Tauch-elektrode *f*, ummantelte oder
umwickelte Elektrode *f* ~ **fabric** Lackgewebe
*n* ~ **film** gegossener Film *m* ~ **objective** vergütetes Objektiv *n* ~ **paper** gestrichenes Papier
*n* ~ **rod** Mantelelektrode *f* ~ **smokeless powder**
Kernpulver *n*, rauchloses Pulver *n* ~ **tape**
Zweischichtband *n* (tape rec.)
**coating** (painting) Anstrich *m*, Aufdampfen *n*,
Auflage *f*, Auskleidung *f*, Bedampfung *f*, Beschlag *m*, Beschichtung *f*, Futter *n*, Haut *f*,
Lage *f*, lichtempfindliche Schicht *f*, Schicht *f*,
Streichen, Strich *m*, Überstreichen *n*, Überzug *m*, Umhüllung *f*, Umkleidung *f*, Um-
wandung *f*, Verputz *m*, (of paint) Putz *m*,
~ **with aluminium** Aufspritzen *n* von Aluminium ~ **of blocks** Blockhalde *f* ~ **with broken
rock** Aufschotterung *f* ~ **with broken stones**
Aufschütten *n* des Steinschlags ~ **of a lens**
eine Linse vergüten ~ **of tar paint** Kohlenteeranstrich *m*
**coating,** ~ **activation** Überzugaktivierung *f* ~
**agent for molds** Einstreichmittel *n* für Formen
~ **appliance** Streichvorrichtung *f* ~ **calenders**
Beschichtungskalander *m* ~ **colors** Zuricht-
deckfarben *pl* ~ **composition** Anstrichstoffe *pl*
~ **compound** Überzugsmischung *f* ~ **drum**
Kandiertrommel *f* ~ **head** Beschichtungskopf
*m* (film) **a** ~ **imitating a metal or other surfaces**
Imitationslack *m* ~ **lacquer** Decklack *m* ~
**lacquer for buttons** Knopfüberzugslack *m* ~
**machine** Streichmaschine *f* (emulsion) ~
**machine for stained paper** Buntpapiermaschine
*f* ~ **materials** Beschichtungsstoffe *pl* (photo) ~
**paper** Streichkarton *m*, Streich-, Streichroh-
-papier *n* ~ **plant** Aufdampfanlage *f* ~ **plant
for paper** Streichanlage *f* für Papier ~ **properties** Deckkraft *f* ~ **thickness** Auftragsschicht *f*
**coatings** Belegung *f*
**coaxial** gleichachsig, koaxial, konzentrisch,
mittig ~ **antenna** Koaxialantenne *f* ~ **cable**
koaxiales Kabel *n*, konzentrische Leitung *f*,
konzentrisches Kabel *n* ~ **cable with silkthread
supported central conductor** Fadenkabel *n* ~
**coil** koaxiale Spule *f* ~ **connector** konzentrisches Verbindungsstück *n* ~ **contrarotating
propellers** Doppelpropeller *pl* ~ **line** Hohlrohrleitung *f*, konzentrisches Kabel *n*, konzentrische Leitung *f* ~ **pair** Koaxialleitung *f* ~
**transmission line** Koaxialleitung *f*
**coaxiality tolerance** Toleranz *f* der Achsen-
gleichheitsabweichungen
**cob stacker** Maiskolben-aufschichter *m*, -elevator *m*
**cobalt** Kobalt *m* ~ **arsenate** Kobaltarseniat *n* ~
**bloom** Kobalt-blume *f*, -blüte *f*, -schlag *m*
~**(ic) bromide** Kobaltbromid *n* ~ **chloride**

Chlorkobalt *m* ~ **green** Kobaltgrün *n* ~ **plating** Verkobaltung *f* ~ **pyrites** Kobaltkies *m*
**cobaltic,** ~ **compound** Kobaltverbindung *f* ~ **oxide** Kobaltgrün *n*
**cobaltiferous** kobalthaltig
**cobaltite** Glanzkobalt *m*, Kobaltglanz *m*
**cobaltocobaltic oxide** Kobaltoxyduloxyd *n*
**cobaltous,** ~ **bromide** Kobaltbromür *n* ~ **chloride** Kobaltchlorür *n* ~ **oxide** Kobaltoxydul *n*
**cobber** Schneider *m*, Schleuder *f*, Separator *m*, Sichter *m*, Trenner *m*
**cobbing** Scheiden *n*
**cobble, to** ~ stauchen
**cobble** Brocken *m* ~ **coal** Füllkohle *f*
**cobbler** Schuster *m* ~'s **wax** Schuhmacherwachs *n*
**cobbles** Festläufer *m* (Walzwerk), Würfelkohle *f*
**cobblestone** Feldstein *m*, Kopfsteinpflaster *n*
**cobbling** Stauchung *f*
**cobs** glockenähnliche Bildverzerrung *f*
**cobweblike** spinnwebenartig
**coccolite** körniger Augit *m*
**cock, to** ~ spannen **to** ~ **for firing** entsichern **to** ~ **a gun (or rifle)** den Hahn am Gewehr spannen
**cock** Absperrglied *n*, Absperrorgan *n*, Absperrteil *m*, Absperrvorrichtung *f*, Hahn *m*, Hahnventil *n*, Kran *m* **(stop)** ~ Absperrhahn *m* ~ **with long pipe to strain off store casks** Abseihhahn *m*
**cock,** ~ **adjustment** Hahneinstellung *f* ~ **bagging** Filter-sack *m*, -schlauch *m* ~**chafer** Maikäfer *m* ~ **nail** Hahnschraube *f* ~ **pin** Hahnschraube *f* ~ **plug contact pressure air** Hahnkükenanpreßluft *f* ~ **spring** Winkelfeder *f* ~ **stopper** Stöpselhahn *m* ~ **system** Hahnbatterie *f* ~ **wrench** Hahnschlüssel *m*
**cocked (firearms)** gespannt
**cockersprag** Abbaustempel *m*
**cocking** Spannen *n* (von Waffen) ~ **and firing arrangement** Spannabzug *m* ~ **of piston** Kippen *n* des Kolbens ~ **of the piston** Verkanten *n*
**cocking,** ~**apparatus** Durchladeeinrichtung *f* ~ **bolt** Spannstück *n* ~ **cable** Spannseil *n* ~ **cam** Spannfalle *f* ~ **cylinder** Spannzylinder *m* ~ **handle** Abzugshebel *m* ~ **lever** Spann-hebel *m*, -stück *n* ~ **lever pin** Bolzen *m* zum Spannstück, Hahnbolzen *m* ~**-lever shaft** Spannwelle *f* ~ **pawl** Spann-Nase *f* ~ **piece** Schlößchen *n* ~ **screw** Spannschraube *f* ~ **slide** Spannschieber *m* ~ **spring** Spann-, Widerspann--feder *f* ~ **spring catch** Widerspannabzug *m*
**cockle, to** ~ fälteln
**cockling,** ~ **of the paper** Rumpeln *n* des Papiers, Wellig-liegen *n*, -werden *n* ~ **sea** Kabbelsee *f*
**cockpit** Führersitz *m*, Kabine *f*, Kanzel *f*, Sitzraum *m* ~ **compass** Rumpfkompaß *m* ~ **cover** Schiebe-, Schutz-haube *f* ~ **cowling** Führersitzhaube *f* ~ **cutout (or cutaway)** Rumpfausschnitt *m* ~ **d. c. bus** Kabinengleichstromsammelschiene *f* ~ **facing** Kanzelverkleidung *f* ~ **flood light** Kabinenflutleuchte *f* ~ **heater** Sitzraumwärmer *m* ~ **heating** Führerraumbeheizung *f* ~ **roof** Führerraumüberdachung *f* ~ **rudder handle** Pedalverstellgriff *m* ~ **spotlight** Kabinenhandleuchte *f* ~ **starter** Bordanlasser *m*
**cockscomb pyrites** Kammkies *m*

**cock-up** hochstehend
**coco fiber** Kokosfaser *f*
**coconut,** ~ **fat** Rest *m* der höheren Alkohole aus Loryl ~ **matting** Kokosmatte *f*
**cocoon** Kokon *m*
**co-cumulative energy spectrum** Energie-Summenspektrum *n*
**co-current flow** parallele Strömung *f*
**code, to** ~ chiffrieren, schlüsseln, verschlüsseln, verziffern
**code** Alphabet *n*, Bestimmungen *pl*, Code *m*, Geheimschrift *f*, Gesetzbuch *n*, Kode *m*, Kodex *m*, Kurzsignal *n*, Schlüssel *m*, Sprache *f*, Stichwort *n*, Tarntafel *f*, (telegram) Telegrammschlüssel *m*, Vorschrift *f* **(office)** ~ Kennziffer *f* **in** ~ verschlüsselt ~ **of performance** Leistungsvorschriften *pl* ~ **of signals** Signalbuch *n*
**code,** ~ **bar** Zuteilungsschiene *f* ~ **beacon** Kodeleuchtturm *m* ~ **book** Kodebuch *n*, Satzbuch *n*, Schlüsselheft *n* ~ **call** Deckanruf *m* ~ **character** Kode-, Telegrafier-zeichen *n* ~ **check** Kodeprüfung *f* ~ **combination** Strombild *n* ~**-contracting method** Koderaffung *f* ~ **conversion** Umkodierung *f* ~ **converter** Umkodierer *m* ~ **delay** Kode-Impulsabstand *m* ~ **element** Kode-, Zeichen-element *n* ~ **garbling** Kodeverwirrung *f* (SSR) ~ **group** Buchstabengruppe *f*, Funkabkürzung *f*, Schlüsselgruppe *f* ~ **key** Chiffrierschlüssel *m* ~ **language** Geheimsprache *f*, verabredete Sprache *f* ~ **letter** Kennbuchstabe *m* ~ **letters** Rufzeichen *pl* ~ **message** Schlüsselmeldung *f* ~ **name** Deckname *m* ~ **note** sendereigener Ton *m* ~ **number** Anforderungszeichen *n*, Kennzahl *f*, Tarnzahl *f* ~ **panel** Tuchzeichen *n* ~ **plug** Schlüsselstecker *m* ~ **punching** Schlüssellochung *f* ~ **reception** Telegrafieempfang *m* ~ **restrictor** Sperrmitläufer *m* (telegr.) ~ **ringing** Ruf *m* mit verabredeten Zeichen, wahlweiser Ruf *m* nach einem Rufschlüssel ~ **selector** Berzirkswähler *m*, Gruppenwähler *m* in sechsstelligen Netzen, Netzgruppenwähler *m* ~ **sheet** Schlüsselblatt *n* ~ **signal** Flaggensignal *n*, Funkrufzeichen *n*, Kennungssignal *n*, Peilzeichen *n* ~ **switch** Amtsnamenwähler *m*, I Gruppenwähler *m* in sechsstelligen Netzen ~ **table** Tarntafel *f* ~ **time** Aufgabezeit *f* ~ **translation** Entschlüsselung *f* ~ **word** Deckname *m*
**coded** verschlüsselt ~ **decimal** kodiertes Dezimal *n* ~**-decimal digit** kodierte Dezimalziffer *f* ~ **program** kodiertes Programm *n* ~ **pulse train method** Impulstelegrammethode *f* (Fernwirktechnik)
**codeine** Kodain *n*
**coder** Coder *n*, Kodiergerät *n*, Verschlüßler *m*
**co-determination** Mitbestimmung *f*
**codify, to** ~ kodifizieren
**coding** Chiffrierung *f*, Kodieren *n*, Schlüsseln *n*, Verschlüsselung *f* **(en)**~ Schlüsselung *f* ~ **delay** Kodierverzögerung *f*, willkürliche Zeitverzögerung *f* ~ **disk** Kodierscheibe *f* ~ **line** (Kodierte) Befehlszeile *f* ~ **pulse multiple** Impulsverschlüsselung *f*, Mehrfachimpulskode *n* ~ **truck** Schlüsselkraftwagen *m*
**codress message** Funkspruch *m* mit verschlüsselter Anschrift
**co-driver** Beifahrer *m*

**coefficient** Ableitung *f*, Beiwert *m*, (numerical) Beizahl *f*, (from experiment or experience) Erfahrungszahl *f*, Faktor *m*, Festwert *m*, Füllfaktor *m*, Hilfswert *m*, Kennziffer *f*, Koeffizient *m*, Wertezahl *f*, Zahl *f*, Ziffer *f*
**coefficient, ~ of absorption** Absorptionsgrad *n* **~ of buckling** Knickzahl *f* **~ of compressibility** Steifezahl *f*, Zusammendrückungs-zahl *f*, (-modul) **~ of contraction** Koeffizient *m* der Zusammenziehung **~ of consolidation** Verfestigungsziffer *f* **~ of coupling** (of two circuits) Kopplungsfaktor *m* **~ of cubical expansion** Raumausdehnungszahl *f* **~ of cyclic irregularity** Ungleichförmigkeitsgrad *m* **~ of deflection (or deviation)** Ablenkungszahl *f* **~ of dialtion** Ausdehnungskoeffizient *m* **~ of differential tones** Differenztonfaktor *m* **~ of difficulty** Schwierigkeitsgrad *m* **~ of diffusion** Streukoeffizient *m*, totaler Streufaktor *m* **~ of dirtying** Verschmutzungsfaktor *m* **~ of discharge** Ausfluß-, Fließ-koeffizient *m* **~ of disintegration** Zerfallziffer *f* **~ of dispersion** Zerstreuungskoeffizient *m* **~ of dissociation** Dissoziationsgrad *m* **~ of drag** Mitführungskoeffizient *m* **~ of earth pressure at rest** Ruhedruckziffer *f* **~ of elasticity** Dehnsteife *f*, Elastizitäts-beiwert *m*, -modul *m*, -zahl *f*, -ziffer *f*, Quetschungszahl *f* **~ of elongation** Dehnungszahl *f* **~ of evaporation** Verdampfungsziffer *f* **~ of excellence** Gütegrad *m* **~ of expansion** Ausdehnungs-beiwert *m*, -koeffizient *m*, -zahl *f*, Raumdehnungszahl *f* **~ of (longitudinal) expansion** Längsdehnungszahl *f* **~ of fineness** Völligkeitsgrad *m* **~ of fluctuation** Ungleichförmigkeitsgrad *m* **~ of force normal to chord** Pfeilkrafttreibwert *m* **~ of force parallel to (or along) chord** Sehnenkraftbeiwert *m* **~ of form ballistics** Formweise *f* **~ of friction** Reibungs-beiwert *m*, -koeffizient *m*, -wert *m*, -zahl *f*, -ziffer *f* **~ of ground friction** Bodenreibungs-zahl *f*, -ziffer *f* **~ of hardness** Härte-wert *m*, -zahl *f*, Koeffizient der Festigkeit **~ of harmonic distortion** Klirrfaktor *m* **~ of heat expansion** Wärmedehnzahl *f* **~ of heat transfer** Wärmeübergangswert *m* **~ of heat transmission** Wärmedurchgangszahl *f* **~ of hysteresis** Hysteresekoeffizient *m* **~ of impact** Stoßzahl *f* **~ of induction** Erregungszahl *f* **~ of irregularity** Beiwert *m* der Ungleichmäßigkeit **~ of lateral flow** Quertriebsbeiwert *m* **~ of lift** Auftriebs-erzeugung *f*, -koeffizient *f* (aviat.) **~ of linear absorption** totaler linearer Absorptionskoeffizient *m* **~ of linear expansion** Längenausdehnungszahl *f*, linearer Ausdehnungskoeffizient *m* **~ of load factor** Belastungsfaktor *m* (des Kraftwerkes) **~ of magnification** Vergrößerungszahl *f* **~ of mutual inductance** Gegeninduktivitätskoeffizient *m* **~ of mutual induction** Koeffizient *m* der gegenseitigen Induktion **~ of negative feedback** Gegenkopplungsfaktor *m* **~ of non-linear distortion** (of a valve) Klirrfaktor *m* einer Verstärkerröhre **~ of occupation** Wirkungsgrad *m* **~ of performance** Nutzeffekt *m* **~ of permeability** Durchlässigkeits-beiwert *m*, -koeffizient *m* **~ of recombination** Wiedervereinigungskoeffizient *m* **~ of reduction** Abnahme-, Reduktions-koeffizient *m* **~ of refraction** Refraktionskoeffizient

*m* **~ of resistance** Festigkeits-koeffizient *m*, -zahl *f*, -ziffer *f*; Widerstandszahl *f* **~ of restitution** Rückkehrkoeffizient *m* **~ of rigidity** Steifigkeitszahl *f* **~ of roughness** Rauhigkeitsbeiwert *m* **~ of safety** Sicherheitszahl *f* **~ of self-inductance** Koeffizient *m* der Selbstinduktion, Selbstinduktivitätskoeffizient *m* **~ of self-induction** Selbstinduktionskoeffizient *m*, Selbstinduktivität *f*, Selbstinduktivitätskoeffizient *m* **~ of shear** Gleitmaß *n* **~ of sliding** Gleitreibungsziffer *f* **~ of sphere** (photometry) Kugelfaktor *m* **~ of subgrade reaction** Bettungsziffer *f* **~ of swelling** Beiwert *m* der Schwellung, Schwellbeiwert *m* **~ of thermal conductivity** Wärmeleitzahl *f* **~ of thermal expansion** Ausdehnungsziffer *f*, Wärmeausdehnungs-koeffizient *m*, -zahl *f* **~ of transfer constant** Fortpflanzungskonstante *f* **~ of turbidity** Trübungskoeffizient *m* **~ of uniformity** Gleichförmigkeits-zahl *f*, -wert *m* **~ of utilization factor** Belastungsfaktor *m* (des Kraftwerkes) **~ of variation** prozentuale quadratische Streuung *f* **~ of viscosity** Zähigkeitsbeiwert *m* **~ of visibility** Sichtbarkeitsbeiwert *n* **~ of volume change** spezifischer Porenwasserverlust *m* **~ of warp** Verwindungszahl *f* **~ of wear** Abnutzungsgröße *f*
**coefficient, ~ data** Richtwert *m* **~ potentiometer** Koeffizientenpotentiometer *n*
**coelostat** Zölostat *m*
**coercimeter** Koerzitivkraftmesser *m*
**coercive** in Schranken *pl* haltend **~ force** Koerzitiv-, Rückhalts-kraft *f* **~ measures** Zwangsverfahren *n*
**coercivity** Koerzitivkraft *f*
**coexist, to ~** gleichzeitig vorhanden sein, koexistieren
**coffer** Füllmauer *f*, Koffer *m*, Unterfaß *n* **~ of ceiling pan** Kassette *f*
**cofferdam** Fangdamm *m*, Hilfsdamm *m*, Kastendamm *m* **~ with double sheeting** Kastenfangedamm *m*
**coffered ceiling** Felder-, Kassetten-decke *f*
**coffering** wasserdichte Ausmauerung *f* eines Schachtes
**cofferwork of loam earth** Piseebau *m*
**coflexure** Durchbiegung *f* (einer Linse) **~ bending** Flächenkrümmung *f*
**cog, to ~** aufkämmen, auswalzen, blocken, herunterblocken, verzahnen, vorblocken, vorstrecken, vorwalzen, Zähne *pl* in ein Zahnrad einsetzen **to ~ down** strecken, herunterwalzen, vorwalzen
**cog** Berg(e)versatzpfeiler *m*, Daumen *m*, eingesetzter Zahn (beim Rad), Einfädelschlitz *m* (tape rec.), Holzpfeiler *m*, Knagge *f*, Nase *f*, Radzahn *m*, Versatzpfeiler *m* (min.), Walzblock *m*, Zahn *m* **~ of a wheel** Radzahn *m*
**cog, ~ depth** Zahntiefe *f* **~ rack** Zahnstange *f* **~ railway** Zahnradbahn *f* **~ roller** Zahnzylinder *m* **~ tool** Kammstahl *m*
**cog-wheel** Daumen-, Kamm-, Kronen-, Kron-, Pol-rad *n*, Rad *m* mit stumpfen Zähnen, Sperrad *n*, Zahnrad *n* **~escapement** Zahnradauslösung *f* **~ gearing** Zapfenrädergetriebe *n* **~ locomotive** Lokomotive *f* einer Zahnradbahn **~ slow motion** Zahnradfeinbewegung *f* **~ transmission** Zahnradübertragung *f*

**cogged** gezahnt, (of an ingot) ausgewalzt ~ **cylinder** Kammwalze f ~ **fork** Zahngabel f ~ **ingot** Vorblock m, vorgewalzter Block m
**cogger** Zwischenwalzer m
**cogging** Verkämmung f, Verzahnung f, Vorstrecken n, Vorwalzen n ~ **of face (or surface)** Planverzahnung f
**cogging,** ~-**down pass** Schnellvorwalz-, Streck--kaliber n ~-**down roll** Schnellvorstreck-, Streck-walze f ~ **joint** Überkämmung f, Verkämmung f ~ **mill** Block-, Vor-walzwerk n ~-**mill train** Block-strecke f, -straße f ~ **pass** Blockkaliber n, Blockwalzkaliber n, Vorstich m, Vorstreckkaliber n ~ **roll** Block-, Vorblock-, Vorstreck-walze f ~-**roll train** Blockwalzstraße f ~ **stand** Blockgerüst n ~ **strand** Vorwalzstrecke f, (of rolls) Vorstrecke f ~ **train** Vorstraße f
**cohere, to** ~ fritten, zusammenhaften
**coherence** Fritterung f, Frittung f, Kohärenz, Zusammenhang m ~ **length of wave trains** Wellenzugkohärenzlänge f, Kohärenzlänge f der Wellenzüge ~ **range** Mitnahmebereich m
**coherent** zusammenhängend ~ **oscillator** durchstimmbarer Oszillator m, Kohärenzoszillator m ~ **pulse operation** Impulskohärenzverfahren f
**coherer** Fritterempfänger m, Frittröhre f, Kohärer m, Wellenanzeiger m, Wellendetektor m ~ **protector** Frittsicherung f ~ **resistance** Frittwiderstand m ~ **terminal** Fritterklemme f ~-**type acoustic shock reducer** Frittersicherung f
**cohesion** Bindekraft f, Haltekraft f, Kohäsion f, Zusammenhaftvermögen n, Zusammenhalt m, Zusammenhang m ~ **strength** Kohäsionsfestigkeit f
**cohesionless** nicht bindig, kohäsionslos
**cohesive** bindig, zähe ~ **energy** Bindungsenergie f ~ **force** Kohäsionskraft f ~ **soil** bindiger Boden m
**cohesiveness** Klebefähigkeit f, Kohäsionsvermögen n
**cohobate, to** ~ nochmals destillieren
**coil, to** aufrollen, (cable) aufschließen, aufspulen, aufwickeln, bespulen, pupinisieren, umwickeln, wickeln, winden, **to** ~ **up** aufrollen, spulen
**coil** Gewinde n, Kehrwendel n, Magnetspule f, Rahmen m, Ring m, Rolle f, Schlange f, Schlangenrohr n, Spirale f, Spule f, aufgeklärtes Tau n, Wendel f, Wicklung f, Windung f, Zündspule f ~ **of cable** Kabelring m ~ **of a distributed winding** Einzelspule f ~ **in opposite direction** gegenläufige Windung f (electr.) ~ **of pipe** Rohrschlange f ~ **of rope** Seilwindung f ~ **of the spring** Federwindung f ~ **of a winding connected to a commutator** Wicklungselement n ~ **of wire** Drahtring m, Drahtrolle f ~ **for hot wire** (blinker unit) Wendel f des Hitzedrahtes (Blinkgeber)
**coil,** ~ **antenna** Spulenantenne f ~ **assembly** Spulensatz m ~ **axis** Spulenachse f ~ **build-down** Bundabbau m ~ **build-up** Bundaufbau m ~(-**loaded) cable** Spulenkabel n ~ **can** Spulenabschirmtopf m ~ **capacity** Eigenkapazität f der Spule, Spulenkapazität f ~ **clutch** Federbandkupplung f ~ **condenser** Schlangenkühler m ~ **core material** Spulenkernmaterial n ~

**coupler** Spulenkoppler m ~ **current supply** Spulenstromversorgung f ~ **drag** Zwischenstück n für Schlangenrohr ~ **end** Wickelkopf m (electr.) ~ **form** Spulen-, Wickel-körper m, Wicklungshalter m ~ **former** Spulenrahmen m ~ **galvanometer** Spulengalvanometer n ~ **holder** Spulenhalter m ~ **ignition** Zündspulenzündung f ~-**ignition system** Batteriezündung f, Spulenzündanlage f, Zündspulensystem n ~ **jacket** Umkapselung f der Spulen ~-**loaded** pupinisiert, spulenbelastet ~-**loaded cable** Pupinkabel n, mit Spulen belastetes Kabel n ~-**loaded circuit** bespulte Leitung f, Pupinleitung f ~ **loading** Bespulung f, Längsspulung f, Pupinverfahren n, Spulenbelastung f ~ **neutralization** induktive Neutralisation f ~ **piece** Spulenstück n ~ **rack** Spulengestell n ~ **resistance** Spulenwiderstand m ~ **section** Spulenabschnitt m (load) ~ **spacing** Spulen-abstand m, -entfernung f ~ **spring** Band-, Schnecken-, Schrauben-, Spiral-feder f ~-**spring connector** Spiralfederklemme f (telegr.) ~ **support** Wicklungshalter m ~ **switch mechanism** Spulenrevolver m ~ **tap** Spulen-ableitung f, -abzweig m, -zapfung f ~ **testing** HF-Vakuumprüfung f ~-**type evaporator** Rohrschlangenverdampfer m ~ **vacuum** Vakuumapparat m mit Heizschlangen ~ **vacuum machine** Schlangenvakuumapparat m ~ **width** Spulenweite f ~ **winder** Wickler m ~ **winding** Spulen-, Zweit-wicklung f ~ **winding tester** Windschutzschlußprüfer m
**coiled** geknäuelt, gewunden ~ **bundle** Ring m ~ **casing** Spiralgehäuse n ~-**coil filament** Doppelwendelglühdraht m ~-**disk wheel** Wickelrad n ~ **exponential horn** aufgewundener Exponentialtrichter m ~ **exponential loud-speaker** aufgewundener Exponentialtrichter m ~ **horn** aufgewundener Lautsprechertrichter m, aufgewundenes Lautsprecherhorn n, Falterhorn n ~ **rope** aufgeschlossenes Tau n ~ **spring** Keil-, Schnecken-feder f ~-**up filament lamp** Wendeldrahtlampe f ~-**up paper ribbon** Papierschlange f ~ **wire** Spiral-, Wendel-draht m ~-**wire gong** (of coincollector stations) Klangfeder f
**coiler with collapsible drum** Haspel f mit reduzierbarem Durchmesser
**coiler,** ~ **drum** Wickeltrommel f ~ **plate** Drehteller m ~ **stand** Drehtopfständer m ~ **tension rolling mill** Ziehwalzwerk n
**coiling** Aufschließen n, Aufspulen n, Bandablage f, Wicklung f ~ **length (of drum)** Aufwickellänge f ~ **machine** Wickelmaschine f
**coils** Bobinenpapier n, eisengekapselte Spulen pl
**coin, to** ~ ausmünzen, ausprägen, prägen **to** ~ **money** münzen
**coin** Geld n, Geldstück n, Münze f ~ **box** Kassiervorrichtung f ~ (**collecting**) **box** Münzbehälter m ~-**box circuit** Automatenleitung f ~ **collector** Kassiervorrichtung f ~-**collector telephone station** Fernsprechamtautomat m ~-**in-the slot electricity meter** Elektrizitätsselbstverkäufer m ~ **plate** Münzplatte f, Schrötling m ~ **receptacle** Münzbehälter m ~ **relay** (coin collect or return) Kassierrelais n ~ **slot** Geld-, Münz-einwurf m
**coinage** Gepräge n, Münz-kunst f, -prägung f, -wesen n ~ **alloy** Münzlegierung f

**coincide, to** ~ übereinstimmen, (with) zusammenfallen, zusammentreffen **to** ~ **with** sich decken mit (geom.)

**coincidence** Deckung *f*, Gleichzeitigkeit *f*, Koinzidenz *f*, Zusammenfallen *n*, Zusammentreffen *n* ~ **amplifier** Koinzidenzverstärker *m* ~ **circuit** Koinzidenzschaltung *f* ~ **connection** Koinzidenzschaltung *f* ~ **detector** Koinzidenzdetektor *m* ~ **error** Abkommfehler *m* ~ **gating circuit** Koinzidenzdurchlaßschaltung *f* ~ **range finder** Koinzidenzentfernungsmesser *m*, Meßsucher *m*, Schnittbildentfernungsmesser *m* ~ **setting** Koinzidenzeinstellung *f* ~ **stage** Koinzidenzstufe *f* ~ **telemeter** Koinzidenzentfernungsmesser *m* ~ **tube** Koinzidenzröhre *f*

**coincident** zusammenfallend, gleichzeitig

**coined** geprägt ~ **word for measurements of dielectric properties** Dielektrometrie *f*

**coiner** Münzschläger *m* ~'s **stamp** Münzschwengel *m*

**coining** Prägung *f* ~ **of money** Münzen *n* ~ **die** Münzenprägematrize *f*, Prägestanze *f* ~ **ferrule** Prägring *m* ~ **hammer** Münzschlaghammer *m* ~ **machine** Prägemaschine *f* ~ **press** Druckwerk *n*, Münzprägemaschine *f* ~ **press for gold** Goldprägepresse *f*

**coir** Kokosfaser *f* ~ **rope** Basttau *n*

**coke, to** ~ garen, verkohlen, verkoken

**coke** Coke *m*, Koks *m* ~ **ash** Koksasche *f* ~ **barrow** Kokskarren *m* ~ **bed** Koksbett *n* ~ **bin** Koksfüllrumpf *m* ~ **blast furnace** Kokshochofen *m* ~ **breeze** Abfallkoks *m*, (from lignite, tar) Grude *f*, Gruskoks *m*, Koksabrieb *m*, Koksgestübbe *n*, Koksklein *n* ~ **brick** Koksstein *m* ~ **bucket** Kokskübel *m* ~**-bucket-handling crane** Kokskübelkran *m* ~ **bunker** Koksbunker *m* ~ **burden** Koksgicht *f* ~ **burning** Koksbrennen *n*, Verkokung *f* ~ **charge** Koksgicht *f* ~**-charging car** Koksbegichtungswagen *m* ~ **cinder** Kokslösche *f* ~**-combustibility** Koksverbrennlichkeit *f* ~ **crusher roll** Koksbrechwalze *f* ~**-discharge side** (of a coke oven) Kokslöschseite *f* ~ **discharger** Koksausdrückmaschine *f* ~**-discharging machine** Koks-ausdrückmaschine *f*, -aussstoßvorrichtung *f* ~ **drawer** Kokszieher *m* ~**-drawing machine** Koksziehvorrichtung *f* ~ **dross** Koksklein *n* ~ **dust** Koks-abrieb *m*, -grus *m*, -lösche *f*, -mehl *n*, Koks-staub *m* ~ **filter** Koksfilter *m* ~ **fines** Koksgrus *m* ~ **fire** Koksfeuer *n* ~**-fired** koksgefeuert ~ **firing** Koks-beheizung *f*, -feuerung *f* ~ **fork** Koksgabel *f* ~ **formation** Koksentstehung *f* ~ **furnace** Koksofen *m* ~ **grab** Koksgreifer *m* ~ **guide** Koksführungsschild *n* ~ **hopper** Koksfüllrumpf *m* ~ **knocker** Kasselsteinabklopfer *m*, Rohrputzturbine *f* ~ **leveler** Planiermaschine *f* ~ **lump** Koksbrocken *m* ~ **manufacture** Koksbereitung *f*, Kokserzeugung *f* ~ **oven** Kokerei-, Koks-ofen *m*

**coke-oven,** ~ **coke** Hüttenkoks *m* ~ **construction** Koksofenbau *m* ~ **discharge side** Koksseite *f* ~ **gas** Destillationsgas *n*, Kokereigas *n* ~ **installation** Koksofenanlage *f* ~ **operator** Koksöfner *m* ~ **oven with a long inclined retort** Schrägeretortenofen *m* ~ **plant** Kokerei *f*, Koksofenanlage *f*, Koks-werk *n*, -zeche *f* ~**-plant operation** Kokereiwesen *n*

**coke,** ~ **packing** Koksfüllung *f* ~ **per charge**

Satzkoks *m* ~ **pusher** Ausdrück-, Ausstoß-maschine *f*; Koks-ausdrückmaschine *f*, -aussstoßvorrichtung *f* ~ **quality** Koksgüte *f* ~ **quenching** Kokslöschung *f* ~**-quenching car** Kokslöschwagen *m* ~**-quenching station** Kokslöschturm *m* ~**-quenching tower** Kokslöschturm *m* ~ **ratio** Kokssatz *m* ~ **recovered** Fallkoks *m* ~ **residue** Rohkoks *m* ~ **scrubber** Koks-skrubber *m*, -wascher *m* ~**-storage bin** Koks-bunker *m*, -vorratstasche *f* ~**-storage pocket** Koksvorratstasche *f* ~ **tar** Zechenteer *m* ~ **test** Verkokungszahl *f* ~**-watering car** Koksabwurframpe *f* ~ **yield** Koks-ausbeute *f*, -ausbringen *n*

**coked** gar ~ **residue** Koksrückstand *m*

**coking** Backen *n*, Entgasung *f*, Garung *f*, Kokerei *f*, Koksbrennen *n*, Kokung *f*, Verkohlung *f*, Verkokung *f* ~ **in ovens** Ofenverkokung *f* ~ **in ridges** Haufenverkohlung *f* ~ **of lignite** (or **peats, etc.**) Braunkohlenvergasung *f* ~ **of the oil** Ölkohlebildung *f* ~ **of pit coal** Verkokung *f* von Steinkohle

**coking,** ~ **capacity** Backfähigkeit *f* ~ **chamber** Koks-, Verkokungs-kammer *f* ~ **coal** Fett-, Koks-kohle *f* ~ **index** Backfähigkeit *f* ~ **mass** Kokskuchen *m* ~ **nuts** Fettnußkohle *f* ~ **plant** Kokerei *f*, Koks-werk *n*, -zeche *f*, Verkokungsanlage *f* ~ **practice** Kokerei *f*, Kokereiwesen *n* ~ **process** Kokereiprozeß *m*, Verkohlungs-, Verkokungs-vorgang *m* ~ **property** (or **quality**) Back-fähigkeit *f*, -vermögen *n* ~ **test** Verkokungsprobe *f*, (of coal) Rohanalyse *f* ~ **time** Verkokungszeit *f* ~ **under pressure** Druckverkokung *f* ~ **works** Verkokungsanstalt *f*

**col** Gebirgspaß *m*, Sattel *m* zwischen zwei Antizyklonen

**colander** Waschsieb *n*

**colation** Durchseihen *n*

**co-latitude** Polabstand *m*

**colature** Durchseihen *n*

**Colby furnace** Colby- Ofen *m*

**colcothar** Eisen-, Glanz-, Polier-rot *n*

**cold** kalt **to** ~**-blast** kalt erblasen **to** ~**-cast** kaltgießen **to** ~**-draw** kaltziehen **to** ~**-forge** kalthämmern **to** ~**-form** kaltverformen **to** ~**-hammer** hartschlagen, kalthämmern **to** ~**-press** kaltpressen **to** ~**-roll** kaltwalzen **to** ~**-saw** kalt sägen **to** ~**-straighten** kaltrichten **to** ~**-strain** kaltstrecken **to** ~**-work** kalt bearbeiten, kalt-strecken, -verformen

**cold** Frost *m*, Kälte *f* ~ **adhesive putty for attaching demolition charges** Kaltklebekitt *m* ~ **age-hardening** Kaltaushärtung *f* ~ **aging** Kaltaushärtung *f* ~ **air** Kaltluft *f* ~**-air blast** Kaltwind *m* ~**-air port** Kühlluftbohrloch *n* ~ **application** (in road construction) Kalteinbau *m* ~ **area** aktivitätsfreier Raum *m* ~ **bend** Kaltbiegung *f* ~ **-bend expressed in degrees of a circle** (fatigue-testing unit) Biegewinkel *m* ~**-bend test** Kaltbiegeversuch *m* ~ **bending** Kalt-biegen *n*, -biegung *f* ~**-bending test** Kaltbiegeprobe *f* ~ **blast** Kalt-gebläse *n*, -wind *m*

**cold-blast,** ~ **furnace** Kaltgebläseofen *m* ~ **inlet** Kaltwindeingang *m* ~ **main** Kaltwindleitung *f* ~ **pig iron** kalterblasenes Roheisen *n* ~ **sliding valve** Kaltwindschieber *m*

**cold,** ~ **blowing** Kaltblasen *n* ~ **blue vat** Vitriolküpe *f* ~ **boiler** Vakuumkocher *m* ~ **bonding** Kaltkleben *n* ~ **brittleness** Kaltsprödigkeit *f* ~-**cast** kaltvergossen ~ **cathode** kalte Kathode *f* ~ **cetane rating** Cetanzahlbestimmung *f* beim Start aus dem kalten Zustand (Anlaßverfahren) ~ **charge** fester Einsatz *m* ~ **chisel** Eisen-, Hart-, Kalt-, Kaltschrott-, Metall-meißel *m*, Nuteisen *n* ~ **clearance** Kaltspiel *n* ~ **color hue** kalter Farbton *m* ~ **compressing** Kaltdrücken *n* ~ **crack** Kalt-schweiße *f*, -schweißstelle *f* ~ **cure (or curing)** Kaltvulkanisation *f* ~ **deformation** Kaltverformung *f* ~-**die** Kaltmatrize *f* ~-**die block** Kaltmatrize *f* ~ **drawing** Autofrettage *n*, Kalt-recken *n*, -zug *m* ~- **drawing bench** Kaltziehbank *f* ~-**drawing die** Kaltziehmatrize *f* ~-**drawn** blankgezogen, kalt geschmiedet, kaltgezogen ~ **due to evaporation** Verdunstungskälte *f* ~ **emission** Feldelektronenemission *f* ~ **emission of electrons** Autoelektronenemission *f* ~ **end** kalte Lötstelle *f*, Kalt-lötstelle *f*, -verbindung *f* ~ **extrusion process** Kaltstrangpressvorgang *m* ~ **flaw** Kalt--schweiße *f*, -schweißstelle *f* ~ **flow** Fließvermögen *n* in der Kälte ~-**forging die** Preßdorn *m* ~-**formed** kaltverformt ~ **forming** Kaltverformung *f* ~-**forming property** Kaltverformbarkeit *f* ~ **front** Einbruchs-, Kalt-front *f* ~-**front squall** Front-, Linien-böe *f* ~-**galvanized** verzinkt ~ **galvanizing** Verzinkerei *f* ~ **glue** Kaltleim *m* ~ **greenhouse** Kalthaus *n* ~-**ground** (pulp) Kaltschliff *m* ~-**hammered** federhart ~-**hammering** Kaltschmieden *n* ~ **hobbing** Einsenken *n* (Gesenkfertigung) ~-**iron** hartgeschlagenes Eisen *m* ~ **iron** kalter Einsatz *m* ~ **junction** kalte Lötstelle *f*, Kaltlötestelle *f*, Kaltverbindung *f* ~ **laboratory** aktivitätsfreies Laboratorium *n* ~ **lap** Kalt-schweiße *f*, -schweißstelle *f* ~ **metal** kalter Einsatz *m* ~-**mix surfacings of roads** Kalteinbaubelag *m* ~-**predefecated juice** kalt vorgeschiedener Saft *m* ~-**press method** Kaltpreßverfahren *n* ~-**pressed bolt (or screw)** preßblanke Schraube *f* ~ **pressing** Kalt-drücken *n*, -pressen *n* ~ **protective** Kälteschutzmittel *n* ~ **reduction mill** Kaltwalzwerk *n* ~ **resisting** kältebeständig ~ **riveting** Kaltnietung *f* ~-**roll neck grease** Fett *n* für Kaltwalzen ~-**rolled** kaltgewalzt ~ **rolled strip** Kaltband *n* ~-**rolling** Kaltwalzen *n* ~-**rolling mill** Kalt-walzerei *f*, -walzwerk *n* ~-**rolling practice** Kaltwalzerei *f* ~-**rolling property** Kaltwalzbarkeit *f* ~ **saw for cutting of risers** Trichterkaltsäge *f* ~ **sensitivity** Empfindlichkeit *f* gegen Kaltverformung ~ **set** Schrothammer *m* ~-**setting** kaltabbindend ~-**setting adhesive** Kaltkleber *m* ~ **settling** Kaltabsetzen *n* ~ **shaping** Kaltverformung *f* ~-**short** kaltbrüchig ~-**shortness** Kalt-bruch *m*, -brüchigkeit *f*, -sprödigkeit *f* ~ **shot (or shut)** Kalt-schweiße *f*, -schweißstelle *f* ~ **shuts** kalte Einschüsse *pl* ~ **stamping nut barr** Kaltpreßmuttereisen *n* ~-**starting** Kaltstart *m* ~-**starting unit** Kaltstartgerät *n* ~ **storage** Kühl-haus *n*, -werk *n* ~-**storage meat** Gefrierfleisch *n* ~-**storage room** Kühlraum *m* ~ **store** Kühlhalle *f* ~ **straining** Kalt-härtung *f*, -strecken *n* ~ **stretching** Kaltstrecken *n* ~ **strip mill** Kaltwalzwerk *n* ~-**swaging machine** Kaltgesenkdrückmaschine *f* ~ **tempering** Kalt-

aushärtung *f* ~ **test** Kälteprobe *f* ~-**test thermometer** Kältethermometer *n* ~-**thread-rolling machine** Kaltgewindewalze *f* ~ **trap** Kondensations-, Kühl-falle *f* ~ **treatment** Kältebehandlung *f* ~ **twisting** Kaltverdrehung *f* ~ **upsetting die** Kaltstauchmatrize *f* ~ **upsetting machine** Kaltvorstauchmaschine *f* ~ **varnish** Kaltlack *m* ~ **wave** Kältewelle *f* ~ **welding** Kaltschweißung *f* ~ **work** Kaltrecken *n* ~ **work hardening** Stauchhärtung *f* ~ **work-hardening** Kaltverfestigung *f* ~-**work sensitivity** Empfindlichkeit *f* gegen Kaltverformung ~ **workability** Kaltverarbeitungsfähigkeit *f* ~-**worked** autofrettiert, kaltgereckt, kaltverformt ~ **working** Kaltbearbeitung *f*, kalte Verformung *f*, kalter Gang *m*, Kalt-strecken *n*, -verarbeitung *f*, -verformung *f*, (of blast furnace) Rohgang *m* ~-**working process** Kaltrecken *n*

**coldness** Kälte *f*

**coleopter** Ringflügler *m*, Koleopter *m*

**collaborate, to** ~ mitarbeiten, zusammenarbeiten, zusammenwirken

**collaboration** Mitarbeit *f*, Mitwirkung *f*, Zusammenarbeit *f* ~ **check list** Mitzeichnungsliste *f*

**collaborator** Mitarbeiter *m*

**collagen** Leimgewebe *n*

**collapse, to** ~ einstürzen, knicken, niederbrechen, schlappmachen, umklappen, zerfallen, zusammenbrechen, zusammenfallen, zusammenlegen, zusammenschrumpfen, zusammensinken, zusammenstürzen

**collapse** Einfallen *n*, Einsturz *m*, Knickung *f*, Sturz *m*, Zusammenbruch *m*, Zusammenfallen *n* ~ **of casing** Rohrpressung *f* ~ **of a mole** Molenbruch *m* ~ **of a pier** Molenbruch *m* ~ **of wall** Kippen *n* in der Mauer

**collapsed**, ~ **casing** Rohrverwerfung *f* ~ **ribbon-shaped cross section** geklappter Querschnitt *m*

**collapsible** auseinandernehmbar, zerlegbar, zusammenklappbar, zusammenlegbar ~ **bit** ausziehbarer Meißel *m*, Flügelbohrer *m* ~ (rubber) **boat** Faltboot *n* ~ **die** Klappgesenk *n* ~ **dinghy** Faltboot *n* ~ **drum** Klapptrommel *f* ~ **heat coil** Hitzrolle *f* mit Gleitstift ~ **hyperbolic paraboloid** bewegliches hyperbolisches Paraboloid *n* ~ **hyperboloid of one sheet** einschaliges, bewegliches Hyperboloid *n* ~ **mast** Steckmast *m* ~ **model of a general ellipsoid consisting of circular sections** bewegliches Kreisschnittmodell *n* des dreiachsigen Ellipsoids ~ **radio mast** Steckmast *m* ~ **reel** Fallhaspel *f* ~ **tap** Gewindeschneidekopf *m* (Innengewinde)

**collapsing** Eindrückung *f*, (tires) Abplattung *f* ~ **field** zusammenbrechendes Feld *n* ~ **length** Knicklänge *f*

**collar, to** ~ krümmen, ringeln

**collar** Beilegering *m* (print.), Bohrlochöffnung *f*, Bund *m*, Hals *m*, Halsband *n*, Hubbegrenzeranschlag *m*, Hülse *f*, Kaliberrand *m*, Kragen *m*, Lagerschale *f*, Manschette *f*, Rand *m* (math.), Reif *m*, Ring *m*, Ringlager *n*, Rosette *f*, Schlußring *m*, Schulterscheibe *f*, Umschlag *m*, Verbindungsmuffe *f* des Whipstockes, Walzrand *m*, Walzring *m*, weiblicher Muffenteil *m*, Wirbel *m*, Wulst *m*, Zapfenlager *n*, Zwinge *f* (top) ~ Anschlußmuffe *f* (Tiefbohrung) ~ **of a roll** Walzenrand *m* ~ **on slide valve** Schieberbund *m*

collar, ~ beam Hainbalken *m* ~ bearing Hals-, Axial-, Längs-lager *n* ~ buster Muffenbrecher *m* ~ coupling Muffenschloß *n* ~ curve Halsbandkurve *f* ~ end bearing Bundlager *n*, Lager *n* mit Bundringen ~-head nut Flachbundmutter *f* ~ insignia Kragenspiegel *m* ~- joint casing Muffenbohrröhre *f* ~ journal Kammzapfen *m* ~ nut Achs-, Bund-mutter *f* ~ patch Kragen-patte *f*, -spiegel *m* ~ (barrel) ring Klauenring *m* ~ screw Bundschraube *f* ~ socket Muffenhülse *f*, (petroleum) Halsdose *f* ~ space Manschettenraum *m* ~ step bearing Ringspurlager *n* ~ strap Halsband *n*, Unterlitze *f*, Verankerungsbügel *m* ~ stud Bundbolzen *m* ~ tab Kragenpatte *f* ~ tongues Kragendrehzange *f* ~ with thread Gewindemuffe *f* ~ without thread Muffenhülse *f* ~-type box Halsschachtel *f*
collate, to ~ kollationieren, mischen (data proc.), vergleichen, zusammenfassen
collated telegram Telegramm *n* mit Vergleichung
collateral Nebensicherheit *f*; nebenherlaufend ~ circumstance Nebenumstand *m* ~ contact Doppelkontakt *m* ~ factor Nebenumstand *m* ~ patent Nebenpatent *n* ~ series zugehörige Zerfallsreihen *pl*
collation Auswertung *f*
collator Mischer *m* (data proc.), Ordner *m*
collect, to ~ annehmen, ansammeln, auffangen, beitreiben, einholen, einziehen, fangen, sammeln, sich ansammeln, zusammentragen to ~ money einkassieren to ~ pay beziehen to ~ terms ordnen
collect, ~ chuck Vorderendfutter *n* ~ message Rimessengespräch *n*
collected gesammelt, versammelt, zusammengestellt
collected condensate pump Sammelkondensatpumpe *f*
collecting Abfangen *n*, Sammeln *n* ~ anode Auffang-, Fang-anode *f* ~ bar Sammelschiene *f* ~ chamber Sammelraum *m* ~ cone (wind tunnel) Auffangtrichter *m* ~ cylinder Sammelflasche *f* (bei Kälteanlagen) ~ diaphragm Fangschirm *m* ~ drum Auflaufhaspel *m* ~ dump Sammelort *m* ~ electrode Niederschlags-, Sammel-elektrode *f* ~ flask Sammelflasche *f* ~ flue Sammelkanal *m* ~ funnel Fang-, Sammel-trichter *m* ~ launder Auffangrinne *f* (Rundherd) ~ main Hydraulik *f*, Sammelkanal *m*, Vorlage *f* ~ mirror Fangspiegel *m* ~ mouth Auffangöffnung *f* ~ office Annahmestelle *f*, Telegrammannahmestelle *f* ~ pan Fangschale *f* ~ pipe Sammel-drain *m*, -leitung *f*, -rohr *n* ~ plate Abscheideplatte *f* ~ point Sammelstelle *f*, Saugspitze *f* (gas) ~ pump Tropfpumpe *f* ~ reservoir Auffangschale *f* ~ ring Sammelring *m*, Schleifring *m* ~ station Sammelplatz *m* ~ tank Sammel-becken *n*, -kasten *m* ~ tray Auffangbehälter *m* ~ trough Sammelmulde *f* ~ unit Bergungskommando *n* ~ vessel Auffang-, Sammel-gefäß *n*
collection Annahme *f*, Ansammlung *f*, Auswahl *f*, Beitreibung *f*, (money) Einziehung *f*, Kollekte *f*, Sammlung *f*, Schatz *m* ~ in plant Betriebsappell *m* ~ pocket Fangraum *m*
collective gemeinschaftlich, kollektiv, zusam-

mengefaßt ~ address group Sammeladreßgruppe *f* ~ advertising Gemeinschaftswerbung *f* ~ agreement Tarifvertrag *m* ~ call sign Sammelrufzeichen *n* ~ call signal Sammelrufzeichen *n* ~ consignment Sammelladung *f* ~ and cyclic control of rotor blades gleichsinnige und periodische Blattwinkelsteuerung *f* ~ denomination Sammelbezeichnung *f* ~ electron Bändermodell *n* ~ element Sammelsystem *n* ~ flight order Sammelauftrag *m* ~ (or focusing) lens Sammellinse *f* ~ marks Kollektivmarken *pl* ~ number Folgenummern *pl* ~ performance Gemeinschaftsleistung *f* ~ pitch gleichsinnige Blattwinkelverteilung *f*, gemeinsame Blattverstellung *f* (helicopter) ~ protection Sammelschutz *m* ~ report Sammelmeldung *f* ~ shipment Sammelladung *f* ~ system Sammelsystem *n* ~ telegram Sammeltelegramm *n* ~ weather radiogram Wettersammelfunkspruch *m* ~ wiring Sammelverlegung *f*
collectively sämtlich
collector Anode *f* einer Kathodenstrahlröhre (TV), (electrode) Auffängerplatte *f*, Bildfänger *m* (rdo), Empfänger *m*, (ring) Kollektor *m*, (ring) Kommutator *m*, Sammler *m*, Sammelrohr *n*, Sammeltank *m*, Schleifring *m*, Stromsammler *m*, Stromwender *m* ~ in open circuit leerlaufender Kollektor *m*
collector, ~ anode Sammelanode *f* ~ bar Kollektorlamelle *f* ~ bow Stromabnehmerbügel *m* ~ brush Abnehmer-, Kollektor-bürste *f* ~ dissipation (of multiplier) Auffängerplattenverlustleistung *f* ~ electrode Abnahme-, Absaug-elektrode *f*, Auffänger *m*, Auffängerkäfig *m*, Sammelelektrode *f*, Suchelektrode *f* ~ junction Kollektor-randschicht *f*, -übergang *m* ~ line Leckleitung *f* ~ ring Bürstensammelring *m*, Kollektorring *m*, Laderspirale *f*, ringförmige Sammelleitung *f*, Sammelring *m*, Schleifer *m* ~ segment Kollektorlamelle *f*, Kommutatorbürste *f*, Kommutatorlamelle *f* ~ shoe (sliding contact) Schleifschuh *m* ~- shoe gear Schleifstück *n*, Stromabnehmer *m* für (oberirdische) Stromschiene *f* ~ sleeve Kollektorbuchse *f* ~ space (in klystron) Sammelraum *m* ~ tank Leckbehälter *m* ~ type generator Schleifringläufer *m*
college Hochschule *f*, Lehranstalt *f*
collet Klemmhülse *f*, Backenlager *m*, Futter(gehäuse) *n* (mach.), Klemmring *m*, Konushülse *f*, Kragen *m*, Schneideisenkapsel *f*, Schraubhülse *f*, Spannpatrone *f*, Spannzange *f* ~ for taper shanks Konushülse *f* für kegelige Schäfte
collet, ~ attachment Spannzangeneinrichtung *f* ~ chuck Futtering *f*, Handhebelspannung *f*, Spann-, Zangenspann-futter *n* ~ chuck attachment Patronenspanneinrichtung *f* ~ chuck head Zangenspannkopf *m* ~ chucking Zangenspannung *f* ~ chucking attachment Spannpatroneneinrichtung *f* ~ finger Spannfinger *m* ~ head Zangenkopf *m*
collide, to ~ anstoßen, kollidieren, schricken, zusammenstoßen
collier Bergmann *m*, Kohlenbergmann *m*; Kohlen-dampfer *m*, -schiff *n*
colliery Hüttenzeche *f*, Kohlenbergwerk *n*, Kohlengrube *f*, Kohlenzeche *f*, Steinkohlenbergwerk *n*, Zeche *f* ~ fan Wettermaschine *f* ~ power station Zechenkraftwerk *n*

**colligative property** konzentrationsabhängige Eigenschaft f
**collimated beam** paralleler Strahl m
**collimating, ~ axes** Bildkoordinatensystem n ~ **axis** Rahmenachse f ~ **lens** Kollektivlinse f ~ **mark** Kollimationslinie f, Meß-, Rahmen-, Raum-, Ziel-marke f ~ **pinhole** kollimierendes Nadelöhr n ~ **point** Bild-feldmarke f, -marke f, Meßpunkt m ~ **ray** Zielstrahl m ~ **shield** Bleiabschirmung f ~ **staff** Ziellatte f
**collimation** Gesichtslinie f, Kollimation f ~ **line** Absehlinie f ~ **mast** Justiermast m ~ **plane** Kollimationsebene f
**collimator** Kollimator m, Richtglas n, Richtkreis m, Spaltrohr n ~ **for night firing** Nachtzieleinrichtung f ~ **bridge** Vorsatzrohr n ~ **shield** Bleiabschirmung f ~ **telescope** Kollimatorfernrohr n
**collinear antenna** Kollinearantenne f
**collineations** Kollineationen pl
**collision** Kollision f, Stoß m, Zusammenprall m, Zusammenstoß m, Zusammentreffen n ~ **on the ground** Zusammenstoß m am Boden
**collision, ~ bulkhead** Kollisionsschott n ~ **coefficient** Rückkehrkoeffizient m, Stoßzahl f ~ **cross section for capture** Stoßquerschnitt m für Einfang ~ **damping** (of spectral lines) Spektrallinienverbreiterung f durch Ionen- und Atomstoß ~ **deactivation probability** Stoßentaktivierungswahrscheinlichkeit f ~ **excitation** Stoßanregung f (eines Gases) ~ **force of electrons** Elektronenwucht f ~ **hazard** Zusammenstoßgefahr f ~ **transition probability** Stoßübergangswahrscheinlichkeit f
**collisional multiplication of carriers** Stoßvervielfachung f der Träger
**collocation** Kollokation f
**collodion** Colloxylin n, Klebäther m, Kollodium n ~ **cotton** Kollodiumfasermasse f ~ **layer** Kollodiumschicht f
**collodionize, to ~** kollodionieren, mit Kollodium behandeln
**colloid** Kolloid n ~ **of antimony trisulfide** Antimontrisulfidsol n ~ **chemistry** Kolloidchemie f
**colloidal** gallertartig, gelartig, kolloidal ~ **cellulose** Zellstoffschleim m ~ **coal** Fließkohle f ~ **electrolyte** Mizelle f ~ **fuel** Fließkohle f ~ **lubricants** colloidale Öle pl ~ **matter** Kolloid-(al)-stoff m, -substanz f ~ **salt** Mizelle f ~ **substance** Kolloid(al)-stoff m, -substanz f
**collotype** Farbenlichtdruck m
**colmation** Kolmation f
**colocynth** Koloquinte f
**cologarithm** negativer Logarithmus
**colon** Doppelpunkt m
**colonial, ~ produce** Kolonialwaren pl ~ **trade** Kolonialhandel m
**colonist** Ansiedler m, Kolonist m
**colonization** Besiedlung f, Siedlung f
**colonize, to ~** ansiedeln, besiedeln
**colony** Ansiedlung f, Kolonie f, Niederlassung f
**colophonite** Pechgranat m
**colophony** Geigenharz n, Kolophon(ium) n
**color, to ~** abfärben, färben, kolorieren, streichen, (by coat of clay) angießen **to ~ print** farbdrucken
**color** Anstrichfarbe f, Farbe f, Farbton m,

Farbstoff m, Färbung f, Tönung f ~ **to fit in** Eindruck-, Paß-farbe f **a ~ like metal** Metallfarbe f ~ **of paint** Anstrichfarbe f
**color, ~ agglutinant** Farbenbindemittel n ~ **balancing** (of color film) Farb-, Farben-ausgleich m ~ **band** Farbstreifen m ~ **bar generator** Farbbalkengenerator m (TV) ~ **bar pattern** Farbbalkenmuster n (auf dem Bildschirm für Meßzwecke) ~ **bar signal** Farbbalkensignal n (TV) ~**-bearer guidon** Fahnenträger m ~**-blind** farbenblind ~ **blindness** Farbenuntüchtigkeit f ~ **block** Farbdruckstock m ~ **box** Tuschkasten m ~ **box (or trough)** Chassis n (Foulard) ~ **box frame** Chassisrahmen m ~ **break-up** Farbenzerlegung f (TV) ~ **burst** Hilfsträgergleichlaufpuls m (TV), Schock m ~ **camouflage** Farbtarnung f ~ **change** Farben-, Farb-umschlag m ~ **chart** Farbatlas m, Farbentafel f, Farbenskala f, Farbmaßsystem n ~ **chemistry** Farbenchemie f ~ **coat** Farbdeckschicht f ~ **code** Farben-kode f, -kennzeichnung f (cable), Farbschlüssel m ~ **code chart** Kennfarbentafel f ~ **coding** Farbkennzeichnung f, (of conductors) Bezeichnung f ~ **comparator** Farbvergleicher m ~ **comparing** Farbenvergleich m ~ **comparison** Farbangleichung f ~ **component** Farbkomponente f, (images) Farbauszüge pl ~ **contamination** Farbwertverschiebung f (TV) ~ **control** Buntspannungsregler m (TV) ~**-corrected** farbkorrigiert ~ **cycle system** Farbringsystem n ~ **defect** Farbenfehler m ~ **demodulator** Farbdemodulator m (TV) ~ **determination** Farbbestimmung f ~ **devoid of hue** unbunte Farbe f ~ **difference** Farbunterschied m ~ **discharge effect** Buntätzeffekt m ~ **discharge paste** Buntätzfarbe f ~ **discrimination** Farbenunterscheidung f ~**-discrimination faculty** Farbenunterscheidungsvermögen n ~ **disk** Farbenscheibe f ~ **dispersion** Farbenzerstreuung f ~**-distinguishing cone** (of eye) farbenempfindliches Zäpfchen n ~**-distinguishing cones** (of the eye) farbenempfindliche Zäpfchen pl ~ **doctor** Farbabstreich-rakel f, -messer n ~ **embossing** Farbraster m ~ **fidelity** Farbtreue f (film), Farbentreue f, Farbwiedergabetreue f (TV) ~ **field** Farbteilbild n (TV) ~ **filter** Farb-, Licht-filter n ~ **finish** (of a machine) Anstrich m ~ **flicker** Farbenflimmern n (TV) ~ **foil** Farbfolie f ~ **fringe** Farbensaum m, Farbsaum m, farbiger Saum m ~ **fringing** Farbrandeffekt m, Randaufbruch m (TV) ~ **gradation** Farben-, Farb-stufe f ~ **grid** Farbgitter n (TV) ~ **grinder** Farbenreiber m ~**-grinding machine** Farbenreibmaschine f ~**-grinding mill** Farbreibmaschine f ~ **guide** Paßkreuz n ~ **indicator** Farbenanzeiger m ~ **information** (luminance plus chrominance) Farbinformation f (TV) ~ **intensity** (in heat-treating) Farbunterschied m ~ **killer** Farbsperre f (TV) ~ **kinescope** Chromoskop m, Farbfernsehbildröhre f ~ **level** Farbhöhe f (TV) ~ **light signal** Farblichtsignale pl (r.r.) ~ **lithography** Farbenlithographie f ~ **maker** Farbenanmacher m ~ **making table** Farbtisch m ~ **manufactory** Farbenfabrik f ~ **matching** Farbangleichung f, Farbenvergleich m ~ **measurement** Farbmessung f ~ **measurer** Farbenmeßapparat m ~**-measuring apparatus**

Farbmeßapparat *m* ~ **mixer** Farbenkreisel *m*, Farbmischer *m* (opt.) ~-**mixing machine** Farbmischmaschine *f* ~ **modulator** Farbmodulator *m* (TV) ~ **overload** Farbübersättigung *f* (TV) ~ **pencil** Markierstift *m* ~ **perception** Farbempfindung *f*, Farbensehen *n*, Farbenwahrnehmung *f* ~ **photograph** Farbenfilmaufnahme *f* ~ **photography** Heliochromie *f* ~ **picture screen** Farbschirm *m* ~ **picture tube** Chromoskop *m*, Farbfernsehbildröhre *f* ~ **plate** Farbenplatte *f* ~ **print** Farbkopie *f* ~ **printing** Bunt-, Farben-druck *m* ~ **problem** Farbenplatte *f* ~ **process etcher** Farbätzer *m* ~ **producing** farbenerzeugend ~ **pyramid** Farbenoktaeder *m* ~ **record** Farbauszug *m* ~ **refraction** Farbenbrechung *f* ~ **registration** Farbenüberdeckung *f* ~ **removal apparatus** Bleichapparat *m* ~ **rendition** Farbwiedergabe *f* (TV) ~ **reproduction** Farb(en)wiedergabe *f* ~ **response** spektrale Empfindlichkeit(sverteilung) *f*, spektrale lichtelektrische Ausbeuteverteilung *f* ~ **response of photocell** spektrale Empfindlichkeit *f* ~-**response curve** Charakteristik *f* der spektralen Empfindlichkeit ~ **retention** Farbtonbeständigkeit *f* ~ **retouching** Farbretusche *f* ~ **reversion** Farbumschlag *m* ~ **ribbon reverse** Farbbandumschaltung *f* ~ **sample** Farbmuster *n* ~ **sampling frequency** Farbschaltfrequenz *f* (TV) ~ **sampling rate** Farbschaltfrequenz *f* (TV) ~ **scale** Farbatlas *m*, Farben-lehre *f*, -skala *f* ~ **scheme** Farbenfolge *f* ~ **schlieren method** Farbschlierenverfahren *n* ~ **screen** Farbscheibe *f* ~-**screen photography** Farbrasteraufnahme *f* ~-**screen plate** Farbrasterplatte *f* ~ **sensation** Farbempfindung *f* ~-**sensitive** farbenempfindlich ~ **sensitivity** spektrale Empfindlichkeit(sverteilung) *f*, spektrale, lichtelektrische Ausbeuteverteilung *f* ~-**sensitivity curve** Charakteristik *f* der spektralen Empfindlichkeit ~ **sensitivity of eye** Augenfarbempfindlichkeit *f* ~ **sensitivity of the eye** Farbempfindlichkeit *f* des Auges, Farbempfindlichkeitsverteilung *f* ~ **separation** Farbauszug *m* ~ **sequence** Farbenordnung *f* ~ **shade** Farbschattierung *f* ~ **shop** Farbküche *f* ~ **solids** Farbkörper *m* ~ **spectrum** Farbenspektrum *n*, Farbkreis *m* ~ **square** Farbenviereck *n* ~ **standard** Farbnorm *f* ~ **strainer** Färbesieb *n* ~ **straining** Seihen *n* der Farbe ~ **surface** Farbfläche *f* ~ **switching procedure** Färbungsschaltverfahren *n* ~ **sync (or synchronism)** Farbgleichlauf *m* (TV) ~ **television** farbiges Fernsehen *n*, Farbfernsehen *n* ~ **temperature** Farbtemperatur *f* ~ **tinting** (color shading) Farbenabtötung *f* ~ **tolerance** Farbtoleranz *f* ~ **transition** Farbübergang *m* (TV) ~ **transmission** Farbübertragung *f* ~ **transparency viewer** Farbfilmbetrachtungsgerät *n* ~ **triangle** Farbendreieck *n* ~ **triple** Farbdrilling *m* (TV) ~ **tube** Chronoskop *n*, Farbfernsehbildröhre *f* ~ **value** Farbwert *m* ~ **vision** Farben-sehen *n*, -tüchtigkeit *f*, -wahrnehmung *f* ~-**vision deficiency** Farbenfehlsichtigkeit *f* ~ **wheel** Farbenkreisel *m*, Farbmischer *m*
**colorable** färbbar
**coloration** Färbung *f*
**colored** farbig, gefärbt ~ **board** Buntpappe *f* ~

**clay** Farberde *f* ~ **cloth** bunter Lappen *m* ~ **crayon** Farbkreide *f* ~ **discharge** Buntätze *f* ~ **discharge print** Buntätzdruck *m* ~ **disk** Farbenschattierung *f* ~ **filter** Farbfilter *m*, Farbglas *m* ~ **first printed resist** Vordruckbuntreserve *f* ~ **fringe** farbiger Saum *m* ~ **glass** Farb-, Farben-glas *n* ~ **glass revolver** Farbglasrevolver *m* ~ **image** Farbenbild *n* ~ **impression** Buntdruck *m* ~ **map sketch** Kroki *n* ~ **overprint resist** Überdruckbuntreserve *f* ~ **pencil** Bunt-, Farb-stift *m* ~ **plate** Farbenkunstdruck *m* ~ **quartz** Bergfluß *m* ~ **resist** Buntreserve *f* ~ **screen** Lichtfilter *m* ~ **throughout** durchgefärbt ~ **tracer serving** farbige Zähl-ader *f* ~ **wax pencil** Fettstift *m*
**colorimeter** Farbenmaß *n*, Farbenmeßapparat *m*, Farbenmesser *m*, Farbmeßapparat *m*, Farbmesser *m*, Kolorimeter *n*
**colorimetric** kolorimetrisch
**colorimetry** Farbenmessung *f*, Farbmessung *f*, Kolorimetrie *f*
**coloring** Anstrich *m*, Färbung *f*, färbend ~ **agent** Färbemittel *n*, Färbungsmittel *n* ~ **earth** Farberde *f* ~ **matter** Farbstoff *m*, Pigment *n* ~ **power** Färbevermögen *n*, Färbkraft *f* ~ **substance** Farbstoff *m*
**colorless** farblos ~ **glass** Weißglas *n*
**colors** Fahne *f* ~ **for sighting purposes** Blendfarbstoffe *pl* **having few** ~ farbarm
**colour** see **color**
**Colpitts**, ~ **circuit** Dreipunktschaltung *f*, Röhrenschaltung *f* mit kapazitiver Rückkoppelung ~ **oscillator** Dreieckpunktkreis *m*, Dreipunktkreisschwingungserzeuger *m* ~ **oscillator circuit** Colpitts-Schaltung *f*
**colter** Pflug-eisen *n*, -messer *n*, Voreisen *n* ~ **and jointer** Kolter *m* und Vorschneider *m*
**columbate** Niobat *n*
**columbic acid** Niobsäure *f*
**columbium** Niob *n*
**column** Gestell *n*, Gruppe *f*, Halbseite *f* (print.), Kolonne *f*, Kolumne *f*, Maschinenständer *m*, Pilaster *m*, Pfeiler *m*, Pfosten *m*, Reihe *f*, Rubrik *f*, Säule *f*, senkrechte Reihe *f*, Spalte *f*, Ständer *m*, Stativsäule *f*, Stelle *f* (data proc.), Stellenwert *m*, Stiel *m*, Stütze *f* ~ **for control lever linkage** Schaltsäule *f* ~ **of flames** Feuersäule *f* ~ **of a miller** Fräßständer *m* ~ **of vehicles** Fahrzeugkolonne *f* ~ **of water** Wassersäule *f* ~ **of water and spray** Fontäne *f*
**column**, ~ **base** Säulenfuß *m* ~ **cap** Säulen-kappe *f*, -knopf *m*, Verschlußstück *n* ~ **correction** Fadenkorrektion *f* ~ **driving miters** Kegelräder *pl* für die Mittenwelle ~ **galley** Kolumnenschiff *n* ~-**guided** Säulenbauart *f* ~-**guided slewing crane** Drehkran *m* in Säulenbauart ~-**guiding tools** Säulenführwerkzeuge *pl* ~ **head** Kolumnentitel *m*, Säulenkopf *m* ~ **jacket** Mantelrohr *n* ~ **jib crane** Säulendrehkran *m* ~ **mount** Säulengestell *n* ~ **screw press** Säulenfriktionsspindelpresse *f* ~ **sleeve** Führungssäule *f* ~ **speaker** Schallzeile *f* ~ **splits** Absplitthebel *m* ~ **still** Abtreib-, Destillier-säule *f* ~ **support** Trägermaterial *n* ~ **switching** Säulenumschaltung *f* ~ **traffic** Kolonnenverkehr *m* ~ **type** Säulenbauart *f* ~-**type chromatography** Säulenchromatographie *f* ~-**type drilling machine** Säulen-, Ständer-bohrmaschine *f* ~

**width** Kolumnenbreite f ~ **with spherical end bearings** Pendelsäule f
**columnal** Stielglied n
**columnar** säulenartig, säulenförmig ~ **anthracite** stengliger Anthrazit m ~ **argillaceous red iron ore** Nagelerz n ~ **coal** Stangenkohle f ~ **ionization** Ionisationssäule f ~ **structure** Säulen-struktur f, -form f
**colza oil** Kletten-, Rüb-öl n
**comagmatic** petrografisch
**comb, to** ~ kämmen
**comb** (textiles) Aushacker m, Hechel f, Kamm m, Meßharke f, Rechen m, Schaftnase f, Wellenkamm m ~ **for beaming** (textiles) Faltenkamm m ~ **of butt** Kolbennase f ~ **condenser** Wabenkondensator m ~ **nephoscope** Wolkenrechen m ~ **pliers** Hechlerzange f
**combating** Bekämpfung f ~ **of noise** Lärm-abwehr f, -bekämpfung f
**combers (or combs) for flax** Flachshechler m
**combing** Kämmen n (of textiles), Kämmling m
**combination** Bindung f, Kombination f, Verbindung f, Vereinigung f, Verknüpfung f, Zusammen-schluß m, -setzung f, -stellung f ~ **of movements** zusammengesetzte Deckoperationen pl (geol.)
**combination, ~ baking steam oven** Verbunddampfbackofen m ~ **bar** Zuteilungsschiene f ~ **bevel** Doppelschmiege f ~ **circular saw blade** Kreissägeblatt n mit Gruppenverzahnung (Holz) ~ **cistern and siphon barometer** Gefäßheberbarometer n ~ **coefficient** Vereinigungskoeffizient m ~ **coke-oven and blast-furnace gas** Mischgas n ~ **compressor** Kombinationsverdichter m ~ **connector** Ortsfernleitungswähler m ~ **energies** Bindungsenergien pl ~ **exposure** Kombinationsaufnahme f ~ **fishing socket** Universalfangglocke f ~ **frequency** Kombinations-, Schwebungs-frequenz f ~ **fuse** Doppelzünder m (und Zeitzünder), Einheitszünder m, Mehrfachzünder m ~ **handset** Mikrotelefon n ~ **jarring and squeezing machine** kombinierte Rüttel- und Preßformmaschine f ~ **kneader and mixer with tilting-type trough body** Flügelknetmaschine f mit Wipptrog ~ **lever** (motion gear) Voreilhebel m ~ **lock** Alphabet-, Kombinations-, Vexier-, Geheim-schloß n ~ **mill** Doppelrohrmühle f, halbkontinuierliches Walzwerk n, Verbundmühle f ~ **monkey wrench** Hakenschlüssel m ~ **oven** Verbundofen m ~ **pliers** Kombinations-, Kombi-, Mehrfach-zange f ~ **principle** Kombinationsprogramm n ~ **protractor** Kombinationswinkelmesser m ~ **punching and riveting machine** kombinierte Loch- und Nietmaschine f ~ **radial** Tragstützlager n ~ **reciprocating engine and jet** Motorluftstrahltriebwerk n ~ **rig** kombinierte Bohranlage f ~ **rods** Einheitsgestänge n ~ **rolling-mill practice** kombiniertes Walzverfahren n ~ **rule** Assurelinie f (print.), Kombinationslinie f (print.) ~ **sand pump** Universalsandpumpe f ~ **scale** Verbundskala f ~ **shade** Mischfärbung f ~ **side rake and tedder** kombinierter seitwärts ablegender Heurechen m und Wender m ~ **socket** Keil-, Universal--fänger m ~ **sucker bar** Saugstange f ~ **stop** (organ) Kombinationsregister n ~ **suit** Überanzug m ~ **thill and pole** (wagon) kombinierte

Ein- und Zweispannerdeichsel f ~ **(al) tone** Kombinationston m ~ **tone distortion** Intermodulationsverzerrung f ~ **turning** Verbunddrehen n ~ **-type valve filler** Ventilfüller-Kombinat m ~ **of water gas and distillation gas** Doppelgas n ~ **wheel-track drive** Rädergleiskettenantrieb m ~ **wheel-track vehicle** Hilfskettenfahrzeug n ~ **wrench** englischer Schraubenschlüssel m, Kombischlüssel m
**combinatorial, ~ analysis** Kombinatorik f ~ **circuit** Schaltnetz n ~ **topology** kombinatorische Topologie f
**combine, to** ~ binden, kombinieren, mischen, verbinden, vereinigen, zusammenfassen, zusammensetzen **to ~ with** verknüpfen
**combine** Interessengemeinschaft f, Kombinat n, kombinierte Mäh- und Dreschmaschine f (Mähdrescher) ~ **circuit** Vierersprechkreis m ~ **comb** Übersetzerrechen m ~ **disk** Übersetzerscheibe f ~ **harvester** Mähdrescher m
**combined** gemischt, geschlossen, kombiniert, zusammengesetzt ~ **with** gebunden ~ **accounting unit** Sammelkasse f ~ **action** Summenwirkung f ~ **antenna** Gemeinschaftsantenne f ~ **Bessemer and Martin processes** Duplexverfahren n ~ **blinker and parking lamp** Blinkrelaisstadtleuchte f ~ **blinker and stop lamp** Blindbremsleuchte f ~ **boring and milling machine** Bohr- und Fräswerk n ~ **bright and dark field illumination** Mischfeldbeleuchtung f ~ **cable** gemischtpaariges Kabel n ~ **carbon** gebundener Kohlenstoff m ~ **choke and condenser suppressor** Drosselkondensator m ~ **chrominance signal** kombiniertes Buntsignal n (TV) ~ **circuit** Phantom-, Vierer-kreis m ~ **coupling** gemischte Kopplung f ~ **dynamo** Pendellichtanlaßbatteriezünder m ~ **effect** Summenwirkung f ~ **felt-tube and cell filter** Filzrohrzellenfilter n ~ **felt-tube and cloth-sack filter** Filzrohraschfilter n ~ **felt-tube and radial-type filter** Filzrohrsternfilter m ~ **flash and tail lamp** Blinkrelaisschlußleuchte f ~ **flying and landing strut** Fang-strebe f, -stiel m ~ **ignition and starting switch** Zündanlaßschalter m ~ **lens** Gesamtobjektiv n ~ **lever** Hebeblock m ~ **line and recording operation** Meldefernplatzbetrieb m ~ **-line and recording service** beschleunigter Fernverkehr m ~ **line and recording toll central office** Schnellamt n ~ **listening and speaking key** Sprech- und Mithörschalter m ~ **lister** kombinierter Dammkulturpflug m und Maisdrill m, kombinierte Reihensä- und Zudeckmaschine f ~ **local and toll operation** Verfahren n mit einfachem Anruf ~ **local and toll selector** Wähler m für Orts- und Fernverkehr ~ **nosepiece and propeller hub** Flanschnabe f ~ **oscillator detector** Mischstufe f ~ **output** Gesamtleistung f ~ **parts** miteinander verbundene Teile pl ~ **pieces of coal and shale** verwachsene Kohle f ~ **quartz plate** Kombinationsquarzplatte f (Ehringhaus) ~ **ram and leveling machine** Planierstößel m ~ **resistance** kombinierter Widerstand m ~ **rotation and reflection** Drehspiegelung f (cryst.) ~ **seed and fertilizer drill** Sä- und Düngerstreumaschine f ~ **shrinkage-and-caving method** Etagen- und Firstenbruchbau (vereinigter) m ~ **stone breaker and roller mill** Steinbrecher-

walzenmühle *f* ~ **stop** Bremsblinkschlußleuchte
*f*, Bremsschlußnummernlaterne *f* ~ **stress**
Summenspannung *f* ~ **with sulfuric acid**
schwefelsauer ~ **sweep rake and stacker** kombinierter Heuschlepprechen *m* mit Hochwinde
und Abladevorrichtung, kombinierter Heu-
schubrechen und Heustapler *m* ~ **transmitter
and receiver** Senderempfänger *m* ~ **trunk and
local exchanges** vereinigtes Fern- und Ortsamt
*n* ~ **water** gebundenes Wasser *n*
**combiner** Kombinator *m*, Zusammenschaltein-
richtung *f* ~ **wheel** Kombinatorscheibe *f*
**combining,** ~ **ability** Verbindungsfähigkeit *f* ~
**attachment** Zinkenfräsgerät *n* (Heimelektro-
werkzeug) ~ **circuit** Stammkreis *m* ~ **equivalent**
Verbindungsgewicht *n* ~ **nozzle** Fangdüse *f*,
(injectors) Druckdüse *f* ~ **power** Verbindungs-
-fähigkeit *f*, -kraft *f* ~ **tool** Zinkenfräser *m* ~
**transformer** Viererabzweigübertrager *m* ~ **volume** Verbindungsvolumen *n* ~ **wheel** Streich-
rad *n* ~ **weight** Verbindungsgewicht *n*
**combustibility** Abbrennbarkeit *f*, Brennbarkeit *f*,
Verbrennbarkeit *f*, Verbrennlichkeit *f*
**combustible** brennbar, brennfähig, entzündbar,
feuergefährlich, verbrennbar, verbrennlich,
zündbar ~ **composition** Anfeuerung *f* ~ **constituent** brennbarer Bestandteil *m* ~ **gases**
Brenngase *pl* ~ **mixture** Kraftstoffgemisch *n*
~ **pit** Ausbrennschacht *m*
**combustibles** Brennbares *n*
**combustion** Ausbrand *m*, Brand *m*, Stoffwechselvorgang *m*, Verbrennung *f* ~ **by explosion** Ex-
plosivverbrennung *f*
**combustion,** ~ **accessories** Feuerungen *pl* ~
**arch** Zündgewölbe *n* ~ **boat** Glüh-, Verbren-
nungs-schiffchen *n* ~ **bomb** Explosionskalorimeter *n* ~ **capsule** Glühschälchen *n* ~ **chamber**
Brenn-kammer *f*, -raum *m*, -schacht *m*, Feuer-
-büchse *f*, -kasten *m*, -raum *m*; Heizraum *m*,
Rauchkammer *f*; Verbrennungs-kammer *f*,
-raum *m*, -rohr *n*, -schacht *m* **slab-shaped** ~
**chamber** flacher Verbrennungsraum *m* ~-
**chamber liner** (jet) Brennkammereinsatz *m* ~
**condition** Verbrennungsverhältnis *n* ~ **cowl**
Düsenmantel *m*, (jet) Brennrohr *n* ~ **cup** Glühschale *f* ~ **curve** Verbrennungslinie *f*
**combustion-cutoff,** ~ **frequency** Brennschluß-
frequenz *f* ~ **integral** Brennschluß-integral *n*,
-kommando *n* ~ **point** Brennschlußpunkt *m* ~
**signal** Brennschlußsignal *n* ~ **signal site** Brenn-
schlußstellung *f* (g/m) ~ **system** Brennschluß-
anlage *f* (g/m)
**combustion,** ~ **degree** Ausbrennengrad *m* ~ **drag**
Brennwiderstand *m* ~ **efficiency** Ausbrennengrad *m*, Verbrennungsleistung *f* ~ **engine**
Brennermotor *m* **(internal)** ~ **engine** Ver-
brennungsmaschine *f* ~ **formula** Verbrennungs-
formel *f* ~ **furnace** Verbrennungsofen *m* ~
**gas** Verbrennungsgas *n* ~ **gases** Feuergase *pl*
~ **ignition liner** Zündflammrohr *n* ~ **method**
Einspritzverfahren *n* ~ **mist** Brennstoffnebel *m*
~ **nozzle** Verbrennungsdüse *f* ~ **period** Brenn-
-periode *f*, -zeit *f* ~ **pressure** Verbrennungsdruck *m* ~ **process** Brennverlauf *m* ~ **space**
Verbrennungsraum *m* ~ **starter** Verbrennungs-
anlasser *m* ~ **system** Einspritzverfahren *n* ~
**test stand** Brennstand *m* ~ **tube** Verbrennungs-
rohr *n* ~ **turbine** Brennkraftturbine *f* ~ **tuyère**

Verbrennungsdüse *f* ~ **unit** Heizbehälter *m* ~
**unit with ring-gap mixing nozzle** Ringspalt-
mischdüsenofen *m* ~ **zone** Brenn-, Verbren-
nungs-zone *f*
**combustive process** Brennprozeß *m*
**come, to** ~ kommen, ins Ziel gehen **to** ~ **about**
zustande kommen **to** ~ **in** eingehen, (waves)
einfallen **to** ~ **out** herauskommen, hinaus-
kommen **to** ~ **together** aneinander rücken **to**
~ **up** heraufkommen **to** ~ **up (to)** entsprechen,
genügen **to** ~ **to an agreement** sich verständi-
gen, vereinbaren **to** ~ **to the aid of** eingreifen **to** ~ **on aim** ins Ziel gehen **to** ~ **out of
center** unrund werden **to** ~ **in on the beam**
Eigenpeilen *n* auf Funkfeuer **to** ~ **in on a
circuit** in eine Leitung eintreten **to** ~ **up
to daylight** zu Tage ausstreichen **to** ~ **over
foggy** daaken **to** ~ **up to the grass** ausblühen
**to** ~ **to hand** eingehen **to** ~ **in to land** zur
Landung ansetzen **to** ~ **to light** zutage kommen oder treten **to** ~ **near(er)** aneinander
rücken **to** ~ **to pass** sich ereignen **to** ~ **in
phase (or in step)** in Tritt kommen **to** ~ **within
the requirements** den Anforderungen entsprechen **to** ~ **to rest** anschlagen, zum Stillstand
kommen **to** ~ **to terms** einigen, übereinkommen **to** ~ **true** zutreffen **to** ~ **out of a well**
aus dem Bohrloch herauskommen
**comet** Haarstern *m*, Komet *m*, Schweifstern *m*
~ **finder** Kometensucher *m* ~ **projector** Kome-
tenprojektor *m*
**cometary orbits** Kometenbahnen *pl*
**comfort** Annehmlichkeit *f*, Bequemlichkeit *f*,
Wohlbehagen *n* ~ **air conditioning** Komfort-
klimatisierung *f* ~ **factor** Behaglichkeitswert *m*
~ **zone** Behaglichkeitszone *f*, Beharrlichkeits-
grenze *f* (Klimaanlage)
**coming into force** Inkrafttreten *n*
**co-mingling** Vermischung *f*
**comma** Beistrich *m*, Komma *n*
**command, to** ~ befehligen, führen, kommandieren ~ **capsule** Führungskapsel *f* ~ **guidance**
Leitstrahlsteuerung *f* (Rakete) ~ **rate** Signal-
größe *f* ~ **rate** Steueränderungsgeschwindig-
keit *f* (gyro) ~ **variable** Führungsgröße *f* (aut.
contr.)
**commandeer, to** ~ beitreiben
**commence, to** ~ anfangen, beginnen, einleiten,
eintreten
**commencement** Anfang *m* ~ **of an act or a law**
Inkrafttreten *n* ~ **of delivery (or feeding)** För-
derbeginn *m*
**commend, to** ~ empfehlen
**commendation** Anerkennungsschreiben *n*, Belohnung *f*
**commendable** ratsam
**commensurable** mit gleichem Maße meßbar,
kommensurabel, verhältnismäßig
**commensurate, to** ~ auf ein gemeinsames Maß
bringen, anpassen
**comment** Anmerkung *f*, Auslegung *f*, Bemerkung
*f*, Erklärung *f*
**commentation** Randglosse *f*
**commerce** Geschäft *n*, Handel *m*, Verkehr *m*
**commercial** betriebsmäßig, (of efficiencies) einträglich, gangbar, geschäftlich, handelsgängig,
kommerzial, kommerziell, marktgängig, wirtschaftlich ~ **air line (company)** Luftverkehrsge-

sellschaft f ~ **airplane** Handelsflugzeug n ~ **airport** Verkehrsflughafen m ~ **airship** Verkehrsluftschiff n ~ **alternating currents** technische Wechselströme pl ~ **aluminium** Handelsaluminium n ~ **apparatus** Industriegeräte pl ~ **artist** Grafiker m ~ **aviation** Verkehrsflugwesen n, Verkehrsluftfahrt f ~ **aviator** Verkehrsflieger m ~ **beet** Fabrikrübe f ~ **benzene** Handelsbenzol n a ~ **brand of zinc** Garantiezink n ~ **car** Geschäftswagen m, Handelsautomobil n ~ **cast iron (or casting)** Handelsguß m ~ **center** Handels-, Wirtschafts-zentrum n ~ **clerk** Handelsangestellter m ~ **continuous service (C.C.S.)** Dauerbetrieb m ~ **current supply** Netzanschluß m ~ **directory** Handelsadreßbuch n ~ **district** Geschäftsgegend f ~ **document** Geschäftspapier n ~ **efficiency** Ausnutzungsfaktor m, Gesamtwirkungsgrad m, wirtschaftlicher Nutzen m ~ **enterprise** Handelsunternehmen n ~ **experience** kaufmännische Erfahrung f ~ **flight** Streckenflug m ~ **flying boat** Verkehrsflugboot n ~ **forms** Geschäftsformulare pl ~ **frequency** technische Frequenz f ~ **grade** Handelsqualität f; handelsüblich ~ **harbor** Handelshafen m ~ **intercourse** Handelsverkehr m ~ **iron** technisches Eisen n ~ **iron (or steel)** Handelseisen n ~ **item** handelsüblicher Artikel m ~ **law** Handelsrecht n ~ **lead** Handelsblei n ~ **length** handelsübliche Länge f ~ **limit** handelsübliche Toleranz f ~ **litharge** Kaufglätte f ~ **material** technischer Werkstoff m ~ **office** Börsenamt n ~ **paper** Geschäftspapier n ~ **plane** Verkehrsflugzeug n ~ **port** Handelshafen m ~ **power** Netzspannung f ~ **power reconnection** Netzrückschaltung f ~ **product** Handelsprodukt n ~ **quality** Handelsqualität f, handelsübliche Qualität f ~ **receiver** Verkehrsempfänger m ~ **seaplane** Wasserverkehrsflugzeug n ~ **seaport** Handelsseehafen m ~ **section (of bank)** Handelsabteilung f ~ **sheets** Handelsbleche pl ~ **sign** Firmenschild n ~ **size** handelsüblich ~ **speed** Verkehrsgeschwindigkeit f ~ **status** Geschäftslage f ~ **steel** Handelsstahl m ~ **switchboard** Börsenamt n ~ **traveler** Handlungsreisender m ~ **zinc** Handelszink n ~ **zinc oxide** technisches Zinkoxyd n
**commercially available** im Handel erhältlich
**comminute, to** ~ pulverisieren, pulvern, vermahlen, zerkleinern
**comminuted** feingepulvert, pulverisiert ~ **fracture** Stückbruch m, Zerschmetterungsbruch m
**comminution** Ausmahlung f, Feinzerkleinern n, Vermahlung f, (by abrasion) Zerreibung f, (of ores) Zerreiben n, Zerstäubung f
**commisssion, to** ~ beauftragen
**commission** Auftrag m, Ausschuß m, Indiensthaltung f, Provision f, (crime) Begehung f, (navy) Indienststellung f **to get out of** ~ (motor) ausfallen ~ **of weights and standards** Eichrat m
**commission,** ~ **agent** kommissionierter Vertreter m ~ **business** Zwischenhandel m ~ **statement** Vergütungsabrechnungsbeleg m
**commissioned** patentiert
**commissioner** Beauftragter m, Generalbevollmächtigter m, Kommissar m
**commissioning** Indienststellen n, Indienststellung f

**commit, to** ~ anvertrauen, (an act) begehen **to** ~ **a breach of trust** im Dienst untreu sein
**committee** Ausschuß m, Beirat m ~ **of action** Aktionskomitee n ~ **on units and formulas** Ausschuß m für Einheiten und Formeln
**commodities** Gebrauchsgüter pl
**commodity** Gut n, Ware f
**common** allgemein, allgemein üblich, gemein, gemeinsam, gemeinschaftlich, gewöhnlich, üblich **in** ~ gemeinsam
**common,** ~ **accuracy** normale Genauigkeit f ~ **base** gemeinsame Grundplatte f ~ **base connection** Basisgrundschaltung f ~ **battery** gemeinsame Batterie f, Zentralbatterie f ~ **-battery central office** Fernsprechvermittlungsamt n mit Zentralbatteriebetrieb ~ **-battery circuit** ZB-Leitung f ~ **-battery exchange area** Netz n mit Zentralbatteriebetrieb ~ **-battery subscriber** Zentralbatterieteilnehmer m ~ **-battery supply** Zentralbatteriespeisung f ~ **-battery telephone set** Fernsprechapparat m mit Zentralbatterie ~ **-battery telephone station** Teilnehmersprechstelle f ~ **-battery working** Zentralbatteriebetrieb m ~ **bit** Spitzbohrer m ~ **blue vat** Vitriolküpe f ~ **bolter** Sichtzeug n ~ **brown coal** gemeine Braunkohle f ~ **carriage bolt** Schloßschraube f ~ **chord** Dreiklang m ~ **cold** Erkältung f ~ **collector circuit** Kollektorgrundschaltung f ~ **connection** gemeinsame Verbindung f ~ **corundum** Diamantspat m ~ **denominator** gemeinsamer Nenner m, General-, Haupt-nenner m ~ **divisor** gemeinschaftlicher Teiler m ~ **drive** gemeinsamer Antrieb m ~ **emitter circuit** Emittergrundschaltung f ~ **factor** gemeinsamer Faktor m ~ **fennel oil** Bitterfenchelöl n ~ **-frequency broadcasting** Gleichwellenrundfunk m ~ **fuse composition** rascher Satz m ~ **-impedance coupling** galvanische Kopplung f ~ **information carrier** einheitlicher Datenträger m ~ **item** Standardoder Normteil n ~ **law** Gewohnheits-, Land-recht n ~ **leg** gemeinsamer Pol m (Netzwerktheorie) ~ **line** Knotenlinie f ~ **logarithm** dekadischer Logarithmus m, Dezimallogarithmus m ~ **logarithms** dekadische Logarithmen pl ~ **main** Sammelleitung f ~ **mica** Silberglimmer m ~ **mortar** Baumörtel m ~ **multiple** gemeinsames Bündel n ~ **path** gemeinsamer Weg m ~ **point** Ausgangspunkt m ~ **practice** Gewohnheit f ~ **resistance** gemeinsamer Widerstand m ~ **return** gemeinsame Rückleitung f ~ **resin** Harzpech n ~ **salt** Kochsalz n ~ **signaling battery** Signalbatterie f ~ **signaling paths** gemeinsame Wege pl zur Übermittlung der Zeichen ~ **stamped tin with an alloy of lead** Pfundzinn n ~ **truss** Freigebinde n ~ **-user channel** Kanal m für gemeinsamen Gebrauch ~ **wealth** Gemeinwesen n ~ **windlass** Hornhaspel m ~ **workman** Lohnarbeiter m
**commutative ionization** lawinenartige Ionisation f
**communal reception** Blockempfang m
**communicable** ansteckend
**communicate, to** ~ kommutieren, mitteilen, in Verbindung stehen, in Verkehr stehen, verkehren
**communicating** kommunizierend
**communication** Anschluß m an das Netz, Fort-

pflanzung *f*, Meldung *f*, Mitteilung *f*, Nachricht *f*, Verbindung *f*, Verbindungsmittel *n*, Verkehr *m*, Verständigung *f* ~ **with aircraft** Bodenbordverkehr *m* ~ **of motion** Fortpflanzung *f* der Bewegung

**communication,** ~ **apparatus** Fuge *f*, Funkgerät *n* ~ **art** Schwachstrom-, Fernmelde-technik *f* ~ **band** Übertragungsband *n* ~ **cable** Schwachstromkabel *n* ~ **center** Fernmeldezentrale *f* ~ **channel** Übertragungskanal *m* ~ **circuit** Fernmeldeleitung *f* ~-**construction squad** Fernsprechanschlußtrupp *m* ~ **cord** Notleine *f*, (alarm in railroad cars) Alarmhebel *n* ~ **engineering** Nachrichtentechnik *f* ~ **equipment** Fernmeldegerät *n* ~ **facility** Fernmelde-anlage *f*, -einrichtung *f* ~ **failure** Verbindungsausfall *m* ~ **line** Fernmeldeleitung *f* ~-**line repeater** Leitungsverstärker *m* ~ **net** Nachrichtennetz *n* ~ **plane** Verbindungsflugzeug *n* ~ **road (or route)** Verbindungsstraße *f* ~ **satellite** Fernmeldesatellit *m* ~ **section** Nachrichtenabteilung *f* ~ **station** Fernmeldeanlage *f* ~ **system** Verständigungsanlage *f* ~ **trench** Ableitungs-, Annäherungs-, Korridor-, Verbindungs-, Verkehrs-graben *m*

**communications** Nachrichtenwesen *n* ~ **activity** Nachrichtenverkehr *m* ~ **cable** Nachrichten-, Schwachstrom-kabel *n* ~ **plane** Nachrichtenflugzeug *n* ~ **relay satellite** Fernmeldesatellit *m* ~ **switching panel** Fernmeldeschaltplatte *f* ~ **transmitter** Nachrichtensender *m*

**communicator** Fernmelder *m*

**communiqué** Bericht *m*

**community** Gemeinde *f*, Gemeinschaft *f*, Gemeinwesen *n* ~ **of interests** Interessengemeinschaft *f* ~ **automatic exchange** Landfernsprechnetz *n* für Selbstanschlußbetrieb ~ **dial service** Wählbetrieb *m* auf dem Lande

**commutable** umschaltbar

**commutate, to** ~ umschalten

**commutated,** ~-**antenna direction-finder** Umschaltantennenpeiler *m* ~ **current** kommutierter Strom *m*

**commutating** Umschalten *n* ~ **machine** Führermaschine *f* ~ **pole** Wendepol *m* ~ **winding** Wendefeldwicklung *f*

**commutation** Ablösung *f*, Kommutation *f*, Kommutierung *f*, Stromwendung *f*, Umschaltung *f* ~ **of quarters** Wohnungsgeldzuschuß *m*

**commutation,** ~ **curve** Umschaltkurve *f* ~ **formula** Vertauschungsformel *f* ~ **ripple** Kommutierungswelle *f* ~ **rule** (magnetism) Austauschrelation *f* ~ **ticket** Monatskarte *f*

**commutator** Kollektor *m*, Kommutator *m*, Schalter *m*, Stromwechsler *m*, (with segments) Stromwender *m*, Unterbrecher *m*, Unterbrecherscheibe *f*, Zahnradunterbrecher *m* ~ **bar** Kollektorstab *m* ~ **(end) bearing** Kollektorlager *n* ~ **break** Kommutatorunterbrecher *m* ~ **brush** Kollektor-, Kommutator-, Zündverteiler-bürste *f* ~ **bush** Kollektorbüchse *f* ~ **interrupter** Kommutatorunterbrecher *m* ~ **meter** Magnetmotorzähler *m* ~ **motor** Kommutatormotor *m* ~ **noise** Kollektorgeräusch *n* ~ **pitch** Stromwenderschritt *m* ~ **returning device** Kollektornachdrehvorrichtung *f* ~ **ripple** Kommutierungswelle *f* ~ **riser** Stromwenderfahne *f* ~ **segment** Kollektor-lamelle

*f*, -**stab** *m*, Kommutatorlamelle *f*, Schleifsegment *n* ~ **shaft** Stromwenderwelle *f*, Welle *f* des Stromwenders ~ **sleeve** Kollektorbuchse *f*, Kommutatorbuchse *f*, Stromwendernabe *f* ~ **switch** Fahrtrichtungsschalter *m* ~ **turning device** Abdrehvorrichtung *f* für den Kollektor ~ **turning-off device** Kollektorabdrehvorrichtung *f*

**commuting observable** kommutierende Observable *f*

**compact, to** ~ dichten, verdichten, zusammendrücken **to** ~ **with roller** abwalzen

**compact** bündig, derb, dicht, fest, festgelagert, (of grains or of machines) gedrängt, gedrungen, geschlossen, kompakt, vollwandig ~ **in shape** kompakte Form *f*

**compact** Preßling *m*, Übereinkommen *n*, Verbundstoff *m* (metall.) ~ **arrangement** geschlossene Ständerbauart *f* ~ **construction (or design)** gedrängte Bauart *f* ~ **galena** Bleischweif *m* ~-**grained** feinkörnig, feinkristallin(isch) ~ **gypsum** Alabaster *m*, körniger Gips *m* ~ **material** Verbundstoff *m* ~ **sand** fester Sand *m* ~ **tip for contacts** Sondermetallauflage *f* für Schaltstücke

**compacting** Dichtung *f* ~ **equipment** Verdichtungsgerät *n*

**compaction,** ~ **method** Verdichtungs-methode *f*, -weise *f* ~ **pile** Verdichtungspfahl *m*

**compactness** Bündigkeit *f*, Dichte *f*, Dichtheit *f*, Dichtigkeit *f*, Festigkeit *f*, Gedrängtheit *f*, gedrungener Bau *m* ~ **of construction** Geschlossenheit *f* des Aufbaus

**compactor** Straßenfrosch *m* ~ **rolls** Verdichterwalzen *pl*

**companding** Pressung-Dehnung *f* der Dynamik

**compandor** Kompandor *m*, Presser-Dehner *m* ~ **action** Kontrasthebung *f*

**companion** Begleiter *m*, Gefährte *m*, Genosse *m*, Gespann *n* (print.) ~ **dimensions** Anschlußmaße *pl* ~ **engine** symmetrisch entsprechender Motor *m* auf der anderen Seite des Rumpfes ~ **flange** Gegenflansch *m* ~**ship** Genossenschaft *f* ~**way** Kajütentreppe *f* ~-**work switch** Mitlaufwähler *m*

**company** Firma *f*, Genossenschaft *f*, Gesellschaft *f*, Handelsgesellschaft *f*, Kompanie *f* ~-**operated depot** Lager *n* in eigener Verwaltung

**comparability** Vergleichbarkeit *f*

**comparable** vergleichbar

**comparative** vergleichbar, vergleichend ~ **degree** Mittelgrad *m* ~ **dyeing** Vergleichsausfärbung *f* ~ **experiment** Vergleichsversuch *m* ~ **figure** Vergleichszahl *f* ~ **measuring of thickness** Dickenvergleichsmessung *f* ~ **performance** vergleichsweise Leistung *f* ~ **quality** Güteverhältnis *n* ~ **speech test** vergleichender Sprechversuch *m* ~ **table** Vergleichstabelle *f* ~ **test** vergleichender Versuch *m*

**comparatively** vergleichsmäßig

**comparator** Gleichheitsprüfer *m*, Vergleicher *m* (info proc.) ~ **circuit organization** Abwägeschaltung *f*

**compare, to** ~ gegenüberstellen, gleichsetzen, vergleichen

**compared** gegeneinander gestellt, gegenübergestellt ~ **with** gegen, gegenüber **to be** ~ **with** gegenüberstehen

**comparing device for use on the bill feed** Ver-

gleichseinrichtung *f* an der Rechnungsführung
**comparison** Nebeneinanderstellung *f*, Vergleich
*m*, Vergleichung *f* **by way of** ~ zu Vergleichs-
zwecken *pl* ~ **of two mutual inductances**
Induktionsmeßwaage *f*
**comparison,** ~ **basis** Vergleichsbasis *f* ~ **device**
**for drawings** Zeichnungsvergleicheinrichtung *f*
~ **electrode** Bezugselektrode *f* ~ **lamp** Ver-
gleichslampe *f* ~ **line** Vergleichslinie *f* ~
**measuring** Vergleichsmessung *f* ~ **method** Ver-
gleichs-methode *f*, -verfahren *n* ~ **prism** Ver-
gleichsprisma *n* ~ **solution** Vergleichslösung *f*
~ **spark drawer** Meßfunkenzieher *m* ~ **spark**
**gap** Meßfunkenstrecke *f* ~ **theorems** Vergleichs-
sätze *pl*
**compartment** abgeteilter Raum *m*, Abteil *n*,
Abteilung *f*, Coupé *n*, Fach *n*, Feld *n*, Geländeabschnitt *m*, Kabine *f*, Kaffe *f*, Kammer *f*,
Ladebezirk *m* (Schwerpunktrechnung), Parzelle *f*, Schacht *m*, Schott *n*, Unterteilung *f*,
Verschlag *m*, Zelle *f* ~ **of a pit** Schachtabteilung *f* ~ **of a set in a guided missile** Betriebsfeld
*n* ~ **of a shaft** Schachtabteilung *f*
**compartment,** ~ **air exhaust** Raumabluft *f* ~ **air**
**supply** Raumzuluft *f* ~ **drier** Trockenschrank
*m* ~ **kiln** Kammerofen *m* ~**-type dryer** Trockenkammer *f* ~**-type grate stocker** Zonenwander-
rost *m* ~**(emergency) ventilator** Raum(not)-
lüfter *m* (im U-Boot)
**compartmentalization** Abschottung *f* (naut.)
**compass** Bussole *f*, Kompaß *m*, Passer *m*, Peilrahmen *m*, Spitzenzirkel *m*, Spitzzirkel *m*, Umfang *m*, Zirkel *m* **(naval)** ~ Seekompaß *m* ~ **on**
**board of airplane** Nahkompaß *m* ~ **with**
**detachable legs** Einsatzzirkel *m* ~ **with shifting**
**points** Steckzirkel *m* ~ **with wing** Zirkel *m* mit
Führungsbügel
**compass,** ~ **accessories** Kompaßzubehör *n* ~
**azimuth** Kompaßkurswinkel *m* ~ **bearing**
Kompaßpeilung *f*, Kompaßrichtung *f* ~ **bowl**
Kompaß-büchse *f*, -kessel *m* ~ **box** Kompaß-
büchse *f* ~ **brick** Krummziegel *m* ~ **card** Kompaßrose *f*, Rose *f*, Rosenblatt *n*, Rosenkarte *f*
~ **case** Bussolengehäuse *n*, Kompaßkessel *m* ~
**ceiling** Tonnendecke *f* ~ **circle** Bussolenkreis
*m* ~ **compensation** Kompaßausgleichung *f* ~
**correction card** Deviationstabelle *f* ~ **corrector**
Kompaßkompensator *m* ~ **course** Kompaß-,
Steuer-kurs *m* ~ **course made good** gesegelter
Kompaßkurs *m* ~ **declination** Kompaß-ab-
weichung *f*, -mißweisung *f* ~ **deviation** Funkfehlweisung *f*, Kompaßlenkung *f* ~ **error**
Kompaß-ablenkung *f*, -fehler *m* ~ **float** Kom-
paßschwimmer *m* ~ **gimbals** kardanische Aufhängung *f* ~ **heading** Kompaßsteuerkurs *m*
~ **housing** Bussolengehäuse *n* ~ **key** Zirkel-
schlüssel *m* ~ **liquid** Kompaßflüssigkeit *f* ~
**method of laying guns** Nadelverfahren *n* ~
**needle** Kompaßnadel *f* ~ **north** Kompaßnorden
*m* ~ **oak timber** Eichenkrummholz *n* ~ **pedestal**
Kompaßsäule *f* ~ **platform** Peildeck *n* ~ **points**
Kompaßpunkte *pl* ~ **reading** Kompaßzahl *f* ~
**ring** Bussolenring *m* ~ **rose** Kompaßrose *f*,
Rosenblatt *n*, Rosenkarte *f*, Rosenträger *m* ~
**saw** Stichsäge *f* ~ **scale** Kompaßrose *f* ~ **sense**
Kompaßseite *f* ~ **swinging** Kompensierung *f* ~
**traverse** Bussolenzug *m* ~ **variation** Abweichung *f* des Kompasses

**compasses** Scharnierzirkel *m* ~ **with shifting**
**points** Reißzirkel *m*
**compatibility** Vereinbarkeit *f* (TV), Verträglichkeit *f* ~ **with drying oils on cooking** Verkoch-
barkeit *f* (Lack) ~ **conditions** Kompatibilitäts-,
Integrabilitäts-bedingungen *pl* ~ **problem** Ma-
terialauswahlproblem *n*
**compatible** vereinbart, verträglich **to be** ~ zu-
sammenbestehen ~ **color television** kompatib-
les Farbfernsehverfahren *n*
**compendious** gedrängt, kurzgefaßt
**compendium** Auszug *m*
**compensate, to** ~ abgleichen angleichen, aufheben, ausbalancieren (electr.), auslenken
(g/m), auswuchten, entzerren, ersetzen, vergüten **to** ~ **for** ausgleichen, ausschalten,
kompensieren, neutralisieren **to** ~ **for damage**
einen Schaden ersetzen **to** ~ **in output circuit**
ins Gleichgewicht bringen
**compensated** ausgeglichen ~ **continuous-wave**
**transmission** Senden *n* durch Verstimmung,
ungedämpftes Senden *n* mit Verstimmung ~
**current transformer** Stromwandler *m* mit Hilfswicklung ~ **motor** Motor *m* mit Kompensationswicklung ~ **semi-conductor** neutralisierter
Halbleiter *m* ~ **volume control** kompensierte
Lautstärkeregelung *f* ~ **wattmeter** Leistungsmesser *m* mit Hilfswicklung
**compensating** Kompensieren *n*, Vergüten *n* ~
**action** Abgleichsignal *n* ~ **allowance** Ausf-
gleichszulage *f* ~ **bar** Kompensierungsstab *m*
~ **beam** Ausgleichhebel *m* ~ **bend** (steam pipes)
Ausdehnschleife *f* ~ **body** Ausgleichskörper *m*
~ **cam** Korrekturscheibe *f* ~ **cell** Kompen-
sationsküvette *f* ~ **chamber** Ausgleichskammer
*f* ~ **(trickle)-charge** Erhaltungsladung *f* ~ **cir-**
**cuit** Ausgleich-leitung *f*, -stromkreis *m*, Ent-
zerrer-, Kompensations-schaltung *f* ~ **compu-**
**tation** Ausgleichsrechnung *f* ~ **course** Kompen-
sierungskurs *m* ~ **current** Ausgleichsstrom *m*,
Kompensationsstrom *m* ~ **device for pressure**
Druckausgleicheinrichtung *f* ~ **disc** Ausgleichsscheibe *f* ~ **factor** Ausgleichszahl *f* ~ **field**
Kompensationsfeld *n* ~ **filter** Ausgleichsfilter
*n* ~ **flow** Ausgleichströmung *f* ~ **fore-and-aft**
**magnet** Ausgleichungslängsmagnet *m*, Kom-
pensationslängsmagnet *m* ~ **gear** Ausgleichsrad *n* ~ **jet** Kompensator-, Korrektur-düse *f*,
(carburetor) Ausgleich-düse *f* ~ **layer** Aus-
gleichbogen *m* ~ **lead** Ausgleichleitung *f* ~
**lead wire** Kompensations-draht *m*, -leitung *f*
~ **lever** Ausgleichhebel *m* ~ **loop** (steam pipes)
Ausdehnschleife *f* ~ **loops** Kompensations-
schleifen *pl* ~ **magnet** Ausgleichungsmagnet *m*
~ **measure** Ergänzungsmaßnahme *f* ~ **method**
Kompensationsverfahren *n* ~ **network** Dämp-
fungs-ausgleicher *m*, -entzerrer *m*, Kompensa-
tionsschaltung *f* ~ **piston** Temperaturaus-
gleichskolben *m* ~ **plane eyepiece** Kompens-
planokular *n* ~ **planimeter** Kompensationspla-
nimeter *n* ~ **potentiometer circuit** Kompensa-
torschaltung *f* ~ **-pressure measuring apparatus**
Kompensationsdruckmeßordnung *f* ~ **pro-**
**perties** Kompensationswirkung *f* ~ **reactor**
Kompensationsdrosselspule *f* ~ **repeater** entzerrender Verstärker *m* ~ **resistance** Abgleich-,
Ausgleichs-widerstand *m*, Ayrtonscher Nebenschluß *m*, Berührungs-, Kompensations-, Vor-

-widerstand *m* ~ **resistor** Ausgleichswiderstand *m* ~ **rod** Kompensierungsstab *m* ~ **roller** Ausgleichsrolle *f* ~ **speed** Ausgleichsgeschwindigkeit *f* ~ **spring** Ausgleichfeder *f* ~ **surface** Ausgleichflosse *f* ~ **suspension gear** Ausgleichsgehänge *n* ~ **switch** Kompensationsschalter *m* ~ **tank** Ausgleichbehälter *m* ~ **thwartship (magnet)** Ausgleichungsquermagnet *m* ~ **time** Ausgleichzeit *f* ~ **transverse magnet** Kompensationsquermagnet *m* ~ **truck** Ausgleichswagen *m* (Kran) ~ **voltage** Ausgleichsspannung *f* ~ **washer** Ausgleichscheibe *f* ~ **wave** Verstimmungswelle *f* ~ **weight** Ausgleichsgewicht *n* ~ **winding** Kompensationswicklung *f*

**compensation** Abfindung *f* (financial), Abstandsgeld *n*, Ausbalancierung *f* (electr.), Ausgleich *m*, Ausgleichen *n*, Ausgleichung *f*, Beschickung *f* (direction finding), Entschädigung *f*, Ersatz *m*, Kompensation *f*, Kompensieren *n*, Kompensierung *f*, Konoidierung *f*, Schadenersatz *m*, Schadloshaltung *f*, Vergütung *f*

**compensation, ~ of capacity deviations** Längsausgleich *m* (teleph.) ~ **of the compass** Kompensieren *n* des Kompasses ~ **for compass deviation** Kompensation *f* des Magnetkompasses ~ **of the control surface** Ruderausgleichung *f* ~ **of the elevator** Höhenruderausgleich *m* ~ **of errors** Fehlerausgleichung *f* ~ **for expenses** Aufwandsentschädigung *f* ~ **for play** Spielausgleich *m* ~ **for property damage** Sachschadenersatz *m* ~ **of reactive current** Blindstromkompensation *f* ~ **for wear** Abnutzungsausgleich *m*, Verschleißausgleich *m*

**compensation, ~ arrangement** Kompensationsvorrichtung *f* ~ **chamber** kompensierte Ionisationskammer *f* ~ **circuit** Ausgleichsleitung *f* ~ **control** Umpumpbegrenzer *m* ~ **device** Kompensationsvorrichtung *f* ~ **error** Deckungsfehler *m* ~ **lever** Abgriffbügel *m* (rdr) ~ **means** (cam, etc. in direction finding) Berichtigungsgerät *n* ~ **measurement** Ausgleichsmessung *f* ~ **pendulum** Schichtungspendel *n* ~ **pipe** Gaspendelleitung *f* ~ **regulator** Kompensationseinsteller *m* ~ **resistance** Ausgleichwiderstand *m* ~ **set** Ausgleichaggregat *n* ~ **theorem** Ausgleichungslehrsatz *m* ~ **wave** Verstimmungs-, Zwischenzeichen-welle *f* ~ **winding** Ausgleichswicklung *f*

**compensator** Ausgleicher *m*, Ausgleichs-gefäß *n*, -vorrichtung *f*, Blindleistungsmaschine *f*, Geschwindigkeitsregler *m* (film), Kompensator *m*, Kompensiereinrichtung *f* (Kompaß), Korrektor *m*, Phasenschieber *m*, Regulator *m* ~ **to accentuate and deaccentuate (bands, frequencies etc.)** Entzerrer *m* zum Anheben und Senken

**compensator, ~ cam** Funkbeschickungskurvenscheibe *f* ~ **jack** Ausgleichstrebe *f* ~ **lever** Spannhebel *m* ~ **rail** Kompensatorschiene *f* ~ **sleeve** Kompensatorhülse *f* (hydraul.) ~ **spool** Steuerkolben *m* (hydraul.)

**compensatory leads** Kompensationsleitungen *pl*
**compensograph** Kompensograf *m*
**compete, to** ~ an einem Wettbewerb teilnehmen, konkurrieren **able to** ~ konkurrenzfähig
**competence** Bewährung *f*, Rechtszuständigkeit *f*, Sachkenntnis *f*, Sachkunde *f*
**competency of a court** Gerichtsstand *m*

**competent** befugt, fachkundig, fachmännisch, fähig, maßgebend, zuständig
**competing process** konkurrierende Prozesse *pl*
**competition** Konkurrenz *f*, Wettbewerb *m* **(starting)** ~ Anlaßwettbewerb *m* ~ **event** Wettkampf *m* ~ **model** Wettbewerbsmodell *n*
**competitive** konkurrenzfähig ~ **procurement procedure** öffentliches Ausschreibungsverfahren *n*
**competitiveness** Konkurrenzfähigkeit *f*
**competitor** Bewerber *m*, Konkurrent *m*, Mitbewerber *m*, Wettbewerbsteilnehmer *m*, Wettkämpfer *m*
**compilation** Zusammen-fassung *f*, -stellung *f*
**compile, to** ~ zusammenstellen **to** ~ **results** auswerten
**compiling routine** erzeugendes Programm *n* (info proc.)
**complain, to** ~ Beschwerde einleiten, beschweren, klagen, reklamieren, sich beschweren
**complainant** Kläger *m*
**complaint** Anklage *f*, Anklageschrift *f*, (law) Antrag *m*, Beanstandung *f*, Beschwerde *f*, Klage *f*, Reklamation *f* ~ **for discontinuance (or stay or suspension)** Einstellungklage *f* ~ **desk** Beschwerdestelle *f* ~ **regulations** Beschwerdeordnung *f* ~ **section** Beschwerdestelle *f*
**complement** Bemannung *f*, Besatzung *f*, Bestückung *f*, Ergänzung *f*, Ergänzungswinkel *m*, Vervollständigung *f*, Zutat *f* ~ **of angle** Winkelkomplement *n*
**complemental series** Ergänzungsreihe *f*
**complementarity** Komplementarität *f*
**complementary** ergänzend ~ **angle** Ergänzungs-, Komplement-winkel *m* ~ **color** Ergänzungs-, Gegen-, Komplementär-, Neben-farbe *f* ~ **evaporator** Nachverdampfer *m* ~ **modulus** Komplement *n* ~ **recording** Ausgleichentzerrung *f* ~ **rectifier** Hilfsgleichrichter *m* ~ **set** Komplementärmenge *f* ~ **shades** gegenseitig sich hebende Farben *pl* ~ **spur gear** Ergänzungsstirnrad *n*
**complete, to** ~ ausbauen, durcharbeiten, endbearbeiten, ergänzen, erweitern, fertigstellen, nachtragen, vervollständigen, vollenden **to** ~ **a call** eine Gesprächsverbindung herstellen **to** ~ **a circuit** einen Stromkreis schließen **to** ~ **drying** (of a lacquer) durchtrocknen **to** ~ **(or fill out) a form** ein Formular ausfüllen **to** ~ **printing** ausdrucken **to** ~ **a run after landing** auslaufen (aviat.) **to** ~ **to schedule** planmäßig abschließen
**complete** ganz, gesamt, geschlossen, total, voll, vollkommen, vollständig, vollwertig, vollzählig ~ **airfoil** Gesamtstrake *f* ~ **analysis** Vollanalyse *f* ~ **arrangement of penetrometer** Gesamtanordnung *f* der Sonde ~ **assembly** Ganzmontage *f* ~ **blower (or compressor)** Vollgebläse *n* ~ **blowpipe** Lötrohrbesteck *n* ~ **break-down** Außerbetriebsetzung *f* (mach.) ~ **carry** vollständiger Übertrag *m* ~ **circle** Vollkreis *m* ~ **circuit** Hin- und Rückleitung *f* (electr.) ~ **cleaning** Hauptreinigung *f* ~ **combustion** vollständige oder vollkommene Verbrennung ~ **expectation of life** mittlere Lebensdauer *f* ~ **extension** Endausbau *f* ~ **fading** synchrone Schwunderscheinung *f* ~ **field order** Gesamtbefehl *m* ~ **form** geschlossene Form *f* ~ **framing**

of a window Fenstergerähme n ~ **ignition** Zünddurchschlag m ~ **impulse** vollständiger Stromstoß m ~ **installation** Volltriebwerk n ~ **keyboard** Volltastatur f ~ **loco-fireboxes** fertige Feuerbüchsen pl ~ **miscibility** vollständige Mischbarkeit f ~ **modulation** vollständige Aussteuerung f ~ **multiple** Gesamtvielfachfeld n ~ **n-gon in space** vollständiges räumliches n-Eck n ~ **operation** Volloperation f ~ **overhaul** vollständige Überholung f ~ **plants** komplette Anlagen pl ~ **pulse** vollständiger Stromstoß m ~ **radiation** Hohlraumstrahlung f ~ **round** vollständig kompensierter Schuß m ~ **screening** Vollentstörung f ~ **series of mixed crystals** Mischkristallreihen pl ~ **splice at one section** Gesamtstoß m ~ **sterilisation** Keimfreiheit f ~ **stroke of plunger (or piston)** Kolbenvollhub m ~ **sweep** (mine sweeping) Durchlauf m ~ **turn** (of screw) volle Umdrehung f ~ **unit** geschlossenes Aggregat n ~ **volley** Vollsalve f

**completed** ergänzt ~ **material discharge** Fertigstoffauslauf m ~ **shell** besetzte Schale f

**completely** restlos, weitgehend

**completing** Komplettierung f, Vervollständigung f

**completion** (of a position) Ausbau m, Ergänzung f, Erweiterung f, Fertigstellung f, Komplettierung f, Vervollständigung f, Vollendung f ~ **in time** fristgemäße Fertigstellung f ~ **of call** Gesprächsschluß m ~ **on schedule** fristgemäße Fertigstellung f ~ **of titration** Abtitrierungspunkt m

**completion,** ~ **data** Fertigstellungsdaten pl ~ **reports** Fertigstellungsberichte pl ~ **test** Lückentest m

**complex** komplex, verwickelt, unübersehbar, zusammengesetzt

**complex** Mehrstoff m ~ **of approach** Begegnungskomplex m ~ **of figures** Zahlenkomplex m

**complex,** ~ **admittance** komplexer Scheinleitwert m ~ **alloy** Mehrstofflegierung f ~ **alloy steels** Komplexstähle pl ~ **attenuation constant** Fortpflanzungsgröße f ~ **combination** Komplexverbindung f ~ **compound** Komplexverbindung f ~ **conjugate** komplexkonjugiert ~ **display** zusammengesetztes Bild n ~ **domain** komplexes Gebiet (math.), Unterbereich m ~ **forming substance** Komplexformer m ~ **function** nichtreelle Funktion f ~ **harmonic wave** zusammengesetzte Schwingung f ~ **musical sound** zusammengesetzter Klang m ~ **(harmonic) oscillation** zusammengesetzte Schwingung f ~ **metallic salt** Metallkomplexsalz n ~ **ore** Komplexerz n ~ **phenomenon** Komplexerscheinung f ~ **plane** komplexe Zahlenebene f (Gaußebene) ~ **-ration tracing receiver** Ortskurvenschreiber m ~ **salt** Komplexsalz n ~**-screw motion** komplexschraubenartige Bewegung f ~ **sine wave** zusammengesetzte Sinuswelle f ~ **steel** Stahl m, der neben Kohlenstoff drei oder mehr Legierungsbestandteile enthält ~ **tone** Kombinationston m (acoust.) ~ **(harmonic) wave** zusammengesetzte Schwingung f

**complexes of organic acid with inorganic salts** Ansolvosäuren pl

**complexing agent** Komplexbildner m

**compliance** Einwilligung f, Federung f, Federungswiderstand m, (of mechanical filter) Filterfederungswiderstand m, Folgeleistung f, Nachgiebigkeit f

**complicate, to** ~ erschweren

**complicated** beschwerlich, kompliziert, schwierig, umständlich, verwickelt, zusammengesetzt ~ **setting** komplizierter Satz m

**complication** Komplikation f, Kompliziertheit f, Verflechtung f, Verwicklung f

**complications** Weiterungen pl

**comply, to** ~ (with) erfüllen **to** ~ **with** einwilligen, entsprechen, Folge leisten, nachkommen

**compole field** Wendefeld n

**component** anteilig, zusammensetzend, (part) integrierend

**component** Anteil m, Komponente f, Teil m, Teileinheit f, Zubehörteil m ~ **of compressive force** Druckkomponente f ~ **out of phase** phasenungleiche Komponente f ~ **of speed** Teilgeschwindigkeit f ~ **of stability** Stabilitäts-anteil m, -beitrag m ~ **of structure** Aufbaukomponente f ~ **of target travel during time of flight of projectile** Auswanderungsstrecke f

**component,** ~ **balance** Komponentenwaage f ~ **batteries** Großbatterie f ~ **battery** Einzelbatterie f ~ **board** Steckkarte f ~ **coil** Teilspule f ~ **current** Teilstrom m ~ **drill** abnehmbarer Bohrer m ~ **field** Feldkomponente f ~ **force** Seiten-, Teil-kraft f ~ **frequency** Frequenzkomponente f, Teilfrequenz f ~ **line** Stammkreis m ~ **machinery** dazugehörige Maschinenanlage f ~ **magnification** Einzelvergrößerung f ~ **metal** Zusatzmetall n ~ **oven assembly** Thermostat m ~ **part** Bestand-, Einzel-teil n, eingepaßter Teil m ~ **prism** Teilprisma n ~ **scale** Komponentenwaage f ~ **voltage** Spannungskomponente f, Teilspannung f

**components** Einbauten pl, Inhaltsstoffe pl, Musterwerkstücke pl ~ **of charge** Einsatzverhältnisse pl

**compose, to** ~ abfassen, setzen, verfassen, zusammensetzen

**composed** gefaßt, zusammengesetzt **to be** ~ **of** bestehen ~ **TV signal** Impulsgemisch n

**composing,** ~ **galley** Schiffchen n (print.) ~ **machine** Setzmaschine f ~ **rack** Kastenregal n ~ **room** Setzerei f ~ **stick** Korrigierwinkel-, Winkel-haken m (print.) ~ **table** Setztisch m ~ **tool** Setzwerkzeug n

**composite** zusammengesetzt ~ **action** Verbund m ~ **aircraft** Doppel-, Huckepack-, Zwillings-flugzeug n ~ **beam** Verbundträger m ~ **cable** gemischtpaariges Kabel n, geschichtetes Kabel n, Kabel n mit verschiedenen Aderstärken ~ **candle** Stearinkerze f ~ **circuit** Simultanleitung f, zusammengesetzte Leitung f ~**-coil wattmeter** Ausgleichwattmeter. n ~ **compound lens** Satzobjektiv n ~ **connection** Doppelbetriebsschaltung f ~ **construction** Gemischtbau m, Verbundkonstruktion f ~ **controlling voltage** kombinierte Steuerspannung f ~ **crystal** Doppelplatten pl ~ **electromotive force** zusammengesetzte elektromotorische Kraft f ~ **field** zusammengesetztes Feld n ~ **float** Tiefschwimmer m ~ **formation** Mischbildung f ~ **lattice**

**beam** Verbundfachwerkträger *m* **~-loaded**
viererpupinisiert **~-loaded cable** viererpupi-
pinisiertes Kabel *n* ~ **loading** Viererpupinisierung *f* ~ **magnifier** zusammengesetzte Lupe
*f* ~ **mold** Mehrfachform *f* ~ **objective** mehrlinsiges Objektiv *n* ~ **picture signal** Signalgemisch
*n* ~ **resistor** Massewiderstand *m* ~ **(through)**
**ringing** (Durch)rufen *n* in Simultanschaltung
~ **sample** zusammengesetztes Muster *n* ~ **set**
Einrichtung *f* für Unterlagerungstelegrafie,
Endstelle *f* einer Simultanleitung, Simultaneinrichtung *f* (telegr.) ~ **shot** Kombinationsaufnahme *f* ~ **signal** Bildaustastsynchronsignal
*n* (TV), BAS-Signal *n*, Signalgemisch *n*,
Videosignal *n* ~ **signaling system** Erdsystem *n*
~ **speaker** Lautsprecherkombination *f* ~ **syn-**
**chronization signal** totales Synchronsignal *n* ~
**television signal** komplexes Fernsehsignal *n* ~
**wave filter** Kombinationsfilter *n* (acoust.),
Überlagerung *f* von Wellen ~ **weld** Dicht- und
Festnaht *f* ~ **wheel** Metallfiberrad *n* ~ **window**
Verbundfenster *n* ~ **wire** unterteilter Draht *m*
~ **working** Simultanbetrieb *m*
**composited circuit** Simultan-leitung *f*, -ver-
bindung *f*
**composition** Aufbau *m*, Aufmachung *f*, Aufsatz
*m*, Ausstattung *f*, Beschaffenheit *f*, Gemisch *n*,
Glassatz *m*, Legierung *f*, (of light beams) Licht-
verteilung *f*, Masse *f*, Satz *m*, Schriftsatz *m*,
Verbindung *f*, Zusammenfassung *f*, Zusammensetzung *f*
**composition,** ~ **of the air** Luftzusammensetzung
*f* ~ **of colors** Farbenstellung *f* ~ **of conductor**
Leiteraufbau *m* ~ **of forces** Kräftekomposition
*f*, Kraftzusammensetzung *f* ~ **of gas** Gaszusam-
mensetzung *f* ~ **of the gases and smoke result-**
**ing from an explosion** Schwadenzusammen-
setzung *f* ~ **of the ground** Bodenbeschaffenheit *f* ~ **of matter** Stoffverbindung *f* ~ **of packets**
Paket-, Stück-satz *m* ~ **for powder** Pulversatz
*m* ~ **of slips** Paket-, Stück-satz *m* ~ **of the soil**
Bodenbeschaffenheit *f* ~ **by weight** Gewichts-
zusammensetzung *f*
**composition,** ~ **caster** Gießsetzmaschine *f* ~
**formula** empirische Formel *f* ~ **head** Komposi-
tionskopf *m* ~ **limits** Zusammensetzungsgrenze
*f* ~ **metal** Kompositionsmetall *n* ~ **part** Satz-
säule *f* (g/m) ~ **plane** Verwachsungsfläche *f*
(geol.) ~ **resistor** Massewiderstand *m* ~ **rider**
**roller** Massereibwalze *f* ~ **roller** Massewalze *f*
~ **stock** Masse *f* ~ **surface** Verwachsungsfläche
*f* (cryst.)
**compositor** Schriftsetzer *m* ~ **and printer** Schwei-
zerdegen *m*
**composure** Fassung *f*
**compo-type bush** Compobüchse *f*
**compound, to** ~ kompoundieren, legieren,
mischen, versetzen, zusammensetzen
**compound** Bindung *f*, Einzäunung *f*, Gemisch *n*,
Masse *f*, Mischung *f*, Mittel *n*, Verbindung *f*,
Zusammensetzung *f*; aus mehreren Stoffen zusammengesetzt, mehrfach zusammengesetzt ~
**action** Summenwirkung *f* ~ **arrangement**
Tandem-, Verbund-ordnung *f* ~ **beam** Verbundbalken *m* ~ **bow** Verbundbügel *m* ~
**casing** Kompoundverrohrung *f* ~ **casting** Verbundguß *m* ~ **catenary construction** Kettenfahrleitung *f* mit Hilfstragedraht ~ **characteris-**

**tic** Doppelschlußverhalten *n* ~ **circuit** zusammengesetzte Leitung *f* ~ **control action** Sum-
mierungsregelung *f* ~ **core** Zwischenkern *m* ~
**course** Koppel-, Zwischen-kurs *m* ~ **cupola**
Kupolofen *m* mit erweitertem Herd ~ **curvature**
**(or curve)** dreidimensionale Krümmung *f* ~
**curve** Korbbogen *m* ~ **cycle** Mehrfachkreislauf
*m* ~ **dyestuff** Kombinierfarbstoff *m* ~ **dynamo**
Doppelschlußmaschine *f*, Dynamo *m* mit gemischter Wicklung ~**(-wound) dynamo** Verbunddynamo *m* ~ **effect** Verbundwirkung *f*
~ **engine** Verbundmaschine *f* ~ **excitation**
Doppelschluß-, Verbund-erregung *f* ~ **gear**
**handle** Übersetzungsräderhebel *m* ~ **glass**
Verbundglas *n* ~ **gyrostatic level** Mehrfachkreiselneigungsmesser *m* ~ **ingot** Verbund-
block *m* ~ **interest** Zinseszins *m* ~ **lever** Differentialhebel *m* ~ **lever arrangement** Hebelsy-
stem *n* ~**-lever testing machine** Prüfmaschine *f*
mit Differentialhebel ~ **lift** Verbundheber *m*
~ **magnet** Lamellenmagnet *m*, Magnetbündel *n*,
magnetisches Magazin *n*, zusammengesetzter
Magnet *m* ~ **metal tool rest** Metallkreuzschlitten *m* ~ **micrometermotion table** Koordinaten-
meßtisch *m* ~ **mineral oil** leicht gefettetes
Mineralöl *n* ~ **(ingot) mold** Verbundkokille *f*
~ **motion** zusammengesetzte Bewegung *f* ~
**(-wound) motor** Kompound-, Verbund-motor
*m* ~ **nucleus** Zwischenkern *m* ~ **oil** gefettetes
Öl *n* ~ **parachute** Stufenfallschirm *m* ~**-plug**
**header** Rücklaufbüchse *f* mit Schraubbügel
~ **regulator** Kompoundierungseinsteller *m* ~
**rest** Oberschieber *m* ~ **(slide) rest** Kreuzsupport
*m* ~ **rest hand wheel** Kreuzsupporthandrad *n* ~
~ **shutter** Kompoundverschluß *m* ~ **signal**
zusammengesetztes Zeichen *n* ~ **slide** Kreuzschlitten *m* ~ **slide motion** Kreuztischverschie-
bung *f* ~ **slide rest** Kreuzsupport *m* (Drehbank) ~ **slides** Kreuzschaltung *f* ~ **spring**
Verbundfeder *f* ~ **spring hammer** Verbund-
federhammer *m* ~ **steam blowing engine** Ver-
bunddampfgebläsemaschine *f* ~ **steam engine**
Verbunddampfmaschine *f* ~ **steel** Verbund-
stahl *m* ~ **swivel base** Doppeldrehteil *m* ~
**table** Kreuzsupporttisch *m* ~ **table motion**
**range** Kreuztischbewegungsbereich *m* ~ **tube**
Verbundröhre *f* ~ **twin** Kristallzwilling *m* ~
**two-cylinder engine** Zwillingsverbundmaschine
*f* ~ **winding** Doppel-, Feld-wicklung *f*, gemischte
Wicklung *f*, Verbundwicklung *f* ~**-wound**
**generator** Kompound-, Doppelschluß-dyna-
momaschine *f*, Verbunddynamo *m* ~**-wound**
**motor** Kompound-, Doppelschluß-, Verbund-
-motor *m*
**compounded** mit Isoliermasse *f* getränkt ~
**color** Mischfarbe *f* ~ **(loaded) latex** gefüllter
Latex *m* ~ **oil** Kompoundöl *n*
**compounding** Versatz *m*, Versetzung *f* ~ **of**
**differences of synchronism** Auflaufen *n* der
Gleichlauffehler
**compregnate, to** ~ komprimieren
**compregnated laminated wood** Kunstharzpreß-
holz *n*
**comprehend, to** ~ erfassen, fassen
**comprehended** gefaßt
**comprehension** Zusammenfassung *f*
**comprehensive** umfassend, vielseitig
**compress, to** ~ drücken, eindrücken, einengen,

einpressen, einschnüren, knicken, komprimieren, pressen, stauchen, verdichten, zusammendrücken, zusammenpressen
**compressed** eingeengt, (shock absorber) eingefedert, komprimiert, verdichtet ~ **air** Druckluft *f*, komprimierte Luft *f*, Preßluft *f*
**compressed-air,** ~ **accumulator** Druckluftakkumulator *m* ~ **apparatus** Entstauber *m* ~ **atomizer** Druckluftzerstäuber *m* ~ **attachment for chuck** Preßluftspanneinrichtung *f* ~ **battery charging line** Bordbatteriefülleitung *f* ~ **blower** Druckluftbläser *m* ~ **bottle** Druckluftflasche *f* ~ **brake** Druckluftbremse *f* ~ **brake cylinder** Bremsluftbehälter *m* ~ **brake hoses** Druckluftbremsschläuche *pl* ~ **caisson** Luftdrucksenkkasten *m* ~ **chamber** Druckluftraum *m* ~ **cleaner** Blaspistole *f* ~ **cock** Druckluftübernahme *f* ~ **condenser** Druckluftkondensator *m* ~ **conditioning** Zusatzbelüftung *f* ~ **connection** Druckluftanschluß *m* ~ **consumption** Preßluftverbrauch *m* ~ **container** Druckluftbehälter *m* ~ **control piston** Druckluftsteuerkolben *m* ~ **control relay** Druckluftkontrollrelais *n* ~ **conveyer** Druckluftförderer *m* ~ **counterrecoil mechanism** Luftvorholer *m* ~ **current** Preßluftstrom *m* ~ **cylinder** Druckluftzylinder *m*, Preßluft-flasche *f*, -zylinder *m* ~ **distributor** Druckluftverteilerstutzen *m* ~ **drive** Preßluftabtrieb *m* ~**-driven model** Preßluftmotormodell *n* ~ **engine** Preßluftmotor *m* ~ **equilibrator** Luftausgleicher *m* ~ **gun** Preßlufthammer *m* ~ **hammer** Druckluft-, Luftdruck-hammer *m* ~ **hoist** Preßlufthebezeug *n* ~ **hose** Druckluftschlauch *m* ~**-impulse catapult** preßluftgetriebener Katapult *m* ~ **jet** Druckluftstrahl(gerät) *n*, Preßluftstrahl *m* ~ **landing leg** Luftfederstrebe *f* ~ **landing strut** Luftfederbein *n* ~ **lift pump** Druckluftheberpumpe *f*, Heberpumpe *f* mit Druckluft ~ **line** Druck-, Druckluft-, Preßluft-leitung *f* ~ **loudspeaker** Preßluftlautsprecher *m* ~ **motor** Druckluftmotor *m* ~**-operated turnover machine** Wendelplattenformmaschine *f* für Druckluftbetrieb ~ **panel** pneumatisches Pult *n* ~ **pipe** Druck(luft)leitung *f* ~ **pipe-connection valve** Druckluftanschlußventil *n* ~ **piston** Preßluftkolben *m* ~ **plant (or installation)** Druckluftanlage *f* ~ **process** Luftdruckverfahren *n* ~ **rammer** Preßluftstampfer *m* ~ **receiver for starting** Druckluftanlaßgefäß *n* ~ **recuperator** Luftvorholer *m* ~ **service** Druckluftbetrieb *m* ~ **shock absorber** Luftstoßdämpfer *m* ~ **spray apparatus** Druckluftstrahlgerät *n* ~ **starter** Druckluftanlasser *m*, pneumatischer Anlasser *m*, Preßluftanlasser *m* ~ **starting device** Druckluftanlaßvorrichtung *f* ~ **storage tank** Vorratsluftbehälter *m* ~ **stream** Druckluftstrom *m* ~ **supply** Druckluft-, Preßluft-zuführung *f* ~ **tank** Preßluftbehälter *m*, Windsammler *m* ~ **tank for brake** Druckluftbremsbehälter *m* ~ **tool** Druckluftwerkzeug *n* ~ **trailer brake coupling** Druckluft-Anhängerbremsenschluß *m* ~ **tube** Druckluftschlauch *m* ~ **tunnel** Überdruckwindkanal *m* ~ **valve** Druckluftstutzen *m* ~ **vibrator** Druckluftvibrator *m*
**compressed,** ~ **asphalt** Stampfasphalt *m* ~ **concrete** Druckbeton *m* ~**-concrete articles** Zementgußwaren *pl* ~ **cone connection** Klemm-

kegelverbindung *f* ~ **drug** gepreßtes Medikament *n* ~ **fuel** Preßkohle *f* ~ **gas** komprimiertes Gas *n* ~**-gas condenser** Preßgaskondensator *f* ~**-gas nozzle** Druckgasdüse *f* (g/m) ~**-iron-core coil** Massekernspule *f* ~**-iron dust core** Eisenpulverkern *m* ~**-iron powder core** Massekern *m* ~ **oxygen** Hochdrucksauerstoff *m* ~**-powder charge** Pulverpreßling *m* ~**-powder core** Eisenstaub-, Masse-kern *m* ~ **powder core** gepreßter Eisenpulverkern *m* ~ **region** Druckgebiet *n* ~ **rubble** Preßgipsplatte *f* ~**-sheet packing** Schichtpackung *f* ~ **springs** Federn *pl* mit Vorspannung ~ **steam** komprimierter Dampf *m* ~**-steel shell** Preßstahlgranate *f* ~ **-vapor system** Dampfkompressionssystem *n* ~ **wadding** Verbandwatte *f* ~**-water reservoir** Druckwasserbehälter *m* ~ **yeast** Preßhefe *f*
**compressibility** Kompressibilität *f*, Komprimierbarkeit *f*, Verdichtbarkeit *f*, Zusammendrückbarkeit *f* ~ **barrier** Kompressibilitätsgrenze *f* ~ **effect** Kompressibilitätseinfluß *n* ~ **error** Kompressibilitätsfehler *m* ~ **factor** Preßbarkeitsfaktor *m*
**compressible** verdichtbar, zusammendrückbar, zusammenpreßbar ~ **flow** kompressible Strömung *f* ~ **fluid jet** kompressibler Strahl *m* ~ **soil** sich setzender Boden *m*
**compressing** Verdichtung *f* ~ **collar** Anpreßbund *m* ~ **engine** Kompressionsmaschine *f* ~ **pump** Komprimierpumpe *f* ~ **ring** Kompressionsring *m* ~ **spring for carbon brush** Bürstenfeder *f* für Kohlebürste
**compression** Druck *m*, Druck-amplitude *f*, -festigkeit *f*, Dynamikeinebnung *f*, (of film) Einschnürung *f*, Kompression *f*, Pressung *f*, Stauchdruck *m*, Stauchung *f*, Verdichtung *f*, Zusammendrückung *f* ~ **(stroke)** Verdichten *n* **(air)** ~ Luftverdichtung *f* **in** ~ auf Druck *m* beansprucht ~ **of air particles** (in sound wave) Zusammenpressung *f* der Luftteilchen ~ **of band** Bandeinengung *f* ~ **of film** Filmeinschnürung *f* ~ **of ground (or soil)** Bodendruckpressung *f* ~ **of the root** Drückung *f* der Gewindekernausrundung ~ **and tension test** Druckversuch *m* ~ **of vapor by its own pressure** Brückenkompression *f* ~ **of volume** Dynamikverengung *f*
**compression,** ~ **apparatus** Druckvorrichtung *f* ~ **arrangement** Druckvorrichtung *f* ~ **booster** Drucksteigerer *m* (Zündkerzenprüfgerät) ~ **buckling** Schlagbiegeprüfung *f* ~ **chamber** Kompressionsraum *m* ~ **chord** Druckgurt *m* ~ **cock** Verdichtungshahn *m* ~ **condenser** Quetschkondensator *m* ~ **conduction** Quetschleitung *f* ~ **cone** Kompressionstubus *m* ~ **coupling** Klemm-, Schalen-kupplung *f* ~ **curve** Verdichtungskurve *f* ~ **cycle** Verdichtungstakt *m* ~ **direction** Stauchrichtung *f* ~ **effect** Kompressionswirkung *f* ~**-endurance (or fatigue) test** Druckermüdungsversuch *m* ~ **flange** Druckgurt *m* ~ **force** Druckkraftfestigkeit *f* ~ **gauge** Kompressometer *n* ~**-gauge cock** Druckwasserhahn *m* ~ **grease gun** Druckschmierbüchse *f* ~ **heat** Kompressionswärme *f* ~ **ignition** Selbst-, Verdichtungs-, (engine) Eigen-zündung *f* ~**-ignition engine** Motor *m* mit Eigenzündung oder Verdichtungszündung ~**-ignition motor** Eigenzündungsmotor *m* ~ **joint** Klemmverbindung *f* ~ **knob** Kompres-

sionsknopf *m* ~ **leg** Federbein *n* ~ **line** Verdichtungskurve *f* ~ **mark** Druckstelle *f* ~-**measuring spring** Druckmeßfeder *f* ~ **member**
Druck-glied *n*, -gurt *m*, -stab *m*, Gegenstrebe
*f*, Innenstrebe *f*, Innenstiel *m* ~ **mold** Preßform
*f* ~ **molded plastics** Formpreßstoffe *pl* ~
**molding** Formpressen *n* ~ **oil cup** Preßöler *m*
~-**oil-cup lubrication** Preßölschmierung *f* ~
**packing ring** Verdichtungsring *m* ~ **plate**
Preßplatte *f* ~ **point** oberer Totpunkt *m* des
Verdichtungshubes ~ **pressure** Kompressions-,
Verdichtungsdruck *m* **(final)** ~ **pressure** Ver-
dichtungsenddruck *m* ~ **pump** Kompres-
sionspumpe *f*, Ladepumpe *f* ~ **ratio** Druckver-
hältnis *n*, Kompressionssatz *m*, Kompressions-
verhältnis *n*, Verdichtungs-grad *m*, -verhältnis
*n*, -zahl *f* ~ **refrigeration** Druckkühlung *f*,
Kühlung *f* durch Kompression ~ **release**
Kompressionsnocken *m*, Verdichtungsminderer *m*, Verdichtungsnocken *m* ~ **relief** Ver-
dichtungsverminderung *f*, Verminderung *f* der
Verdichtung ~ **relief cock** Verdichtungsminderer *m* ~ **resistance** Kompressionswiderstand
*m* ~ **rib** Hauptflügelrippe *f* ~ **ring** Kolben-
-dichtring *m*, -gasring *m* ~ **screw** Justierschraube *f* ~ **set** Verdichtungsverformung *f* ~
**shock** Verdichtungsstoß *m* ~ **sleeve** Klemmring *f* ~ **space** Verdichtungsraum *m* ~ **spring**
Druckfeder *f*, Vorspannung *f* ~ **springs**
Druckfedern *pl* ~ **standardizing box** Druckmeßdose *f* ~ **strain** Druckverformung *f* ~
**strength** Druckkraftfestigkeit *f* ~ **stress** Druck-
-beanspruchung *f*, -belastung *f* ~-**stress-defor-
mation diagram** Spannungsdruckdiagramm *n*
~-**stress tester** Druckkraftprüfer *m* ~ **stroke**
Druckhub *m*, Kompressions-hub *m*, -schub *m*,
-takt *m*, Verdichtungshub *m* ~ **strut** Federbein
*n*, Federstrebe *f*, Gegenstrebe *f*, Querstück *n*,
Querträger *f* ~ **tap** Kompressions-, Zisch-hahn
*m* ~ **test** Druckprobe *f*, Stauchversuch *m* ~-
**test machine** Prüfmaschine *f* für Druckver-
suche ~-**test specimen** Druckprobe *f*, Druck-
probekörper *m* ~-**testing machine** Druck-
prüfungsmaschine *f* ~ **trajectory** Drucktrajek-
torie *f* ~ **trimmer** Quetschtrimmer *m* ~ **vacuum gauge** Kompressionsvakuummeter *m* ~
**valve** Verdichtungswelle *f*, Zischhahn *m* ~
**volume** toter Zylinderraum *m*, Verdichtungs-
raum *m* ~ **wave** Verdichtungswelle *f*, (explosion)
Verdichtungsstoßwelle *f* ~ **yield point** Quetschgrenze *f*
**compressional**, ~ **members** Druckstäbe *pl* ~
**relaxation** Druckrelaxation *f* ~ **vibrations**
Druckschwingungen *pl* ~ **wave** Druckwelle *f*
(acoust.)
**compressionless** kompressionslos, kompressorlos
**compressive**, ~ **amplitude** Druckamplitude *f* ~
**cleaving** Druckspaltung *f* ~ **force** Druckkraft *f*
~ **impact stress** Schlagdruckbeanspruchung *f*
~ **load** Druckbelastung *f*, Quetschung *f* ~-**load
application** Druck-beanspruchung *f*, -belastung
*f* ~ **protection spiral** Druckschutzspirale *f* ~
**resilience strain** Druckspannung *f* ~ **strain**
negative Spannung *f*, Stauchung *f* ~ **strength**
Druck-festigkeit *f*, -widerstand *m* ~ **stress**
Druck-kraft *f*, -spannung *f*, negative Spannung
*f* ~ **twinformation** Druckzwillingsbildung *f* ~

**wave** Kompressionswelle *f* ~ **yield strength**
Druckstreckgrenze *f*
**compressometer** Kompressionsmesser *m*
**compressor** Gebläse *n*, Kompresser *m*, Kompressor *m*, Kompressorverstärker *m*, Lader *m*,
Nachschalteverdichter *m*, Verdichter *m*, Vorverdichter *m*, Zusammendrücker *m*, (in volume-
range control) Dynamikpresser *m* **(air)** ~
Luft-gebläse *n*, -verdichter *m* ~ **(in volume-range
control)** Presser *m* ~ **adiabatic work head
characteristic** Ladeförderhöhenlinie *f* ~ **bearing jet** Verdichterlager *n* ~ **blades** Verdichter-
schaufeln *pl* ~ **casing** Kompressorgehäuse *n* ~
**casing jet** Verdichtergehäuse *n* ~ **characteristics**
Verdichterkennfeld *n* ~ **contactor** Kompressor-
schütz *m* ~ **curves** Kompressorenkennlinien *pl*
(Druckkapazität) ~ **delivery pipe** (jet) Kom-
pressorleitrohr *n* ~ **diffusor** Gebläsegitter *n* ~
**discharge pressure** Gebläseenddruck *m* ~ **drum**
Kompressortrommel *f* ~ **efficiency** Gebläse-
wirkungsgrad *m* ~ **exit** Verdichterende *n* ~
**housing jet** Verdichtergehäuse *n* ~ **impeller** Geb-
läseläufer *m* ~ **inducer rotor** Ladevorsatz-
läufer *m* ~ **inlet** Gebläseeintritt *m* ~ **inlet
temperature** Verdichtereintrittstemperatur *f* ~
**inlet valve** Saugventil *n* ~ **intercooler** Ladeluftkühler *m* ~ **oil** Kompressoröl *n* ~ **plant**
Kompressoranlage *f* ~ **pulsation limit** Pumpgrenze *f* ~ **rear frame** Verdichteraustrittge-
häuse *n* ~ **rotor** Verdichterläufer *m* (Triebwerk)
~ **rotor impeller** Ladelaufrad *n* ~ **scroll** (jet)
Verdichterspirale *f* ~ **stall** Strömungsabriß *m*
im Verdichter ~ **stator** Gebläsegitter *n*, Verdichtergehäuse *n* ~ **throttle (or throttling)
valve** Verdichterabsperrventil *n* ~ **valve** Kom-
pressorventil *n*
**comprise, to** ~ enthalten, umfassen
**comprising limiter tube and network** Frequenzweiche *f*
**compromise, to** ~ ausgleichen, gefährden,
(code, etc.) bloßstellen
**compromise** Abgeltung *f*, Ausgleich *m*, Bloß-
stellung *f*, Übereinkommen *n*, Vergleich *m* ~
**cross** Mittelding *n* ~ **network** angenäherte
(mittlere) Nachbildung *f*
**compromised** gütlich, geordnet
**comptometer** Rechenmaschine *f*
**comptometry** Maschinenrechnen *n*
**Compton, ~ effect** Compton-Effekt *m* ~ **electron**
Rückstoßelektron *n*
**comptroller** Leiter *m* der Buchhaltungsabteilung
*m*, Schichtmeister *m*
**compulsive force** Zwangskraft *f*
**compulsory** obligatorisch ~ **centering** Zwangszentrierung *f* ~ **payment for damages** Regreß-
pflicht *f* ~ **pilotage** Lotsenzwang *m* ~ **registration** Eintragungszwang *m* ~ **sale by auction**
Zwangsversteigerung *f* ~ **work** Pflichtarbeit *f*
~ **working** (of a patent) Ausführungszwang *m*
**computable** berechenbar, zählbar
**computation** Abschätzung *f*, Ausrechnung *f*,
Auswertung *f*, Berechnung *f*, Errechnung *f*,
Rechnung *f*, Überschlag *m*, Zählung *f* **by** ~
rechnerisch ~ **of an adjustment** Fehleraus-
gleichsrechnung *f* ~ **of coordinates** Koordina-
tenberechnung *f* ~ **of mean trajectory** Mittel-
bahnberechnung *f* ~ **of missile oscillation**
Überlagerungsrechnung *f* ~ **of quantities**

Massenberechnung *f* ~ **of spring** Federberechnung *f*
**computation,** ~ **table** Rechentafel *f*
**compute, to** ~ ausrechnen, berechnen, errechnen, kalkulieren, rechnen, überschlagen, veranschlagen, zusammenrechnen
**computed** berechnet ~ **blasting charge** Planladung *f* ~ **(demolition) charge** Planladung *f* ~ **high water** Berechnungshochwasser *n*
**computer** Kalkulator *m*, Rechen-blatt *n*, -gerät *n*, -scheibe *f*, Rechner *m* ~ **for flight data** Betriebsdatenschieber *m* ~ **for guidance into line of sight** Einlenkrechner *m* ~ **with pointer** Zählwerk *n* mit Zeiger
**computer,** ~ **code** Maschinenkode *m* ~ **control unit** Befehlswerk *n* (data proc.) ~ **instruction** Maschinenbefehl *m* ~ **operation** Programmablauf *m*, Rechnerbetrieb *m*
**computing,** ~ **of charges** Gebührenermittlung *f* ~ **center** Rechenstelle *f* ~ **inception** Berechnung *f* des Eintretens ~ **interval** Rechenzeit *f* ~ **machine** Rechenmaschine *f* ~ **recording comparator** Komparator *m* ~ **shaft** Rechenwelle *f* ~ **slide rule for transfer of sound-and flash-ranging data to control chart** Auswertelineal *n* ~ **station** (sound and flash ranging) Auswertestelle *f* ~ **tables** Rechenzettel *m*
**computor** Rechner *m*, programmgesteuerter Rechenautomat *m*
**concatenation** Verkettung *f* ~ **connection** (of motors) Hintereinander-, Kaskaden-schaltung *f*
**concave** (mirror) hohl, hohlgeschliffen, konkav, rundhohl ~ **bank** Konkave *f* eines Flusses, einbuchtendes Ufer *n* ~ **brick** Krummziegel *m* ~ **cathode** Hohlkathode *f* ~~ **lens** bikonkave Linse *f* ~ **cutter** Fräser *m* für halbkreisförmige Profile ~ **filled weld** Hohlkehlschweißung *f* ~ **flank** hohle Flanke *f* ~ **glass** Hohlglas *n* ~ **grating** Hohl-, Konkav-gitter *n* ~ **ground** hohlgeschliffen ~ **lens** Hohllinse *f*, Zerstreuungslinse *f* ~ **milling cutter** Hohlfräser *m* ~ **mirror** Sammelspiegel *m* ~ **mold** (e. g. in structural metal) Hohlkehle *f* ~ **(or convex) radii** vertiefte oder erhabene Radien *pl* ~ **(mirror) reflector** Hohlspiegel *m* ~ **saws** Konkavsägen *pl* (Tellersägen) ~ **shaping attachment** Konkavhobelvorrichtung *f* ~ **slope of river bend** Gleithang *m* ~ **sound** Hohlsonde *f* ~ **tapered** hohlkehlig ~ **tile** Nonne *f* ~ **washer** Hohlscheibe *f* ~ **weld** leichte Schweißnaht *f*
**concavity** Hohlrundung *f*, Konkavität *f*
**concavo,** ~-**concave** doppelthohl ~-**convex** hohlerhaben ~-**convex lens** konkavkonvexe Linse *f*
**conceal, to** ~ maskieren, überdecken, verbergen, verdecken, verheimlichen, verstecken
**concealed** gedeckt, unauffällig, versteckt ~ **battery** gedeckte oder verdeckte Batterie *f* ~ **camera** Geheimkamera *f* ~ **camp** Deckungslager *n* ~ **space** verdeckter Hohlraum *m* ~ **wiring** verdeckte Leitungsführung *f*, Unterputzleitung *f*
**concealing power** Deckkraft *n*, Farbdeckfähigkeit *f*
**concealment** Deckung *f*, Hinterhalt *m*, Verschleierung *f*, Verschweigung *f* ~ **by smoke** Vernebelung *f*
**concede, to** ~ konzessionieren, vergünstigen, zugestehen

**conceivable** absehbar
**concentrate, to** ~ anreichern, armtreiben, aufbereiten, (light) bündeln, eindicken, einengen, einkochen, entwässern, konzentrieren, verdichten, verdicken, sammeln, verkochen, verstärken, zusammenlegen **to** ~ **(by evaporation)** eindampfen **to** ~ **(metals)** spuren ~ **of ore** Erzschlich *m*
**concentrate** Konzentrat *n*, Schlich *m*, Schliech *n* ~ **bin** Konzentratbehälter *m* ~ **discharge** Konzentrataustrag *m* ~ **pulp** Konzentrattrübe *f*
**concentrated** begrenzt, entwässert, geballt, geschlossen, konzentriert, punktförmig verteilt, rein, verdichtet, verstärkt, zusammengefaßt ~ **ammonia liquor (or water)** verdichtetes Ammoniakwasser *n* ~ **attack** konzentrierter Angriff *m* ~ **capacity** punktförmige Kapazität *f*, punktförmig verteilte Kapazität *f* oder Ladung *f* ~ **charge** geballte Ladung *f* ~-**filament (lamp) bulb** Dunkelfeldbirne *f* ~ **force** Einzelkraft *f* ~ **inductance** (for electrical inertia) punktförmig verteilte Induktivität *f* ~ **light** gerichtetes Licht *n* ~ **load** Einzelkraft *f*, (on beam) Einzellast *f*, Punktbelastung *f* ~ **matt** (of fine metal) Spurstein *m* ~ **metal** Spurstein *m* ~ **nuclear fuel** konzentrierter Kernbrennstoff *m* ~ **operation** Betriebszusammenfassung *f* ~ **tin** Zinnschlich *m* ~ **trunks** in einem Zentralschrank vereinigte Leitungen *pl*
**concentrating** Bündelung *f* ~ **coil** Konzentrationsspule *f* ~ **cup** Fokussierungseinrichtung *f* ~ **switchboard** Zentralschrank *m* ~ **table** Anruftisch *m* (teleph.)
**concentration** Anhäufung *f*, Anreicherung *f*, Bündelung *f*, (of rays) Einblendung *f*, Entwässerung *f*, Gehalt *m*, Grädigkeit *f* (chem.), Heranführung *f*, Konzentration *f*, Konzentrieren *n*, Konzentrierung *f*, Lösungsdichte *f*, Sättigung *f*, Stärke *f*, Verdichtung *f*, Verdickung *f*, Versammlung *f*, Zusammenlegung *f*, Zusammenziehung *f* **at the maximum of** ~ höchstprozentig (chem.) ~ **of circuits** Zusammenlegung *f* der Leitungen ~ **of the feeding liquor** Nachsatzkonzentration *f* ~ **by freezing** Einengung *f* durch Gefrieren ~ **of incoming calls** Anrufzusammenfassung *f* (teleph.) ~ **of the light rays** Bündelung *f* des Lichts ~ **of space charges** Raumladungswolke *f*
**concentration,** ~ **carrier** Konzentrationsträger *m* (metall.) ~ **(focusing) coil** (cathode-ray oscillograph) Sammelspule *f* ~ **cup** Kathodenbecher *m* ~ **factor** Anreicherungsfaktor *m*, (soils) Ordnungszahl *f* ~ **lens** Konzentrationslinse *f* ~ **line** Sammeldienstleitung *f* ~ **potential** Konzentrationsspannung *f* ~ **ratio** Konzentrations-verhältnis *n* ~ **ring** Korbring *m* ~ **slag** Spurschlacke *f* ~ **switchboard** Zentralanrufschrank *m* ~ **table** Aufbereitungsherd *m* ~ **transport** Aufmarschtransport *m*
**concentrator** Anreicherungsapparat *m*, Eindicker *m*, Eindickzylinder *m*, Entwässerungszylinder *m*, Reiniger *m*, Zentralanrufschrank *m*, Zentralschrank *m*, Zentralumschalter *m* ~ **panel** Klinkenfeld *n*
**concentric** achsparallel, besitzend, gleichmittig, konzentrisch, konzentriert, mittig, zentrisch ~**(al)** konzentrisch ~ **cable** koaxiales oder konzentrisches Kabel *n*, konzentrische Leitung *f*

**~ cable with silk-thread supported central conductor** Fadenkabel *n* **~ circles** konzentrische Kreise *pl* **~ drive and reduction gear** Untersetzungsgetriebe *n* mit konzentrischen An- und Abtriebswellen **~ enlargement** zentrale Erweiterung *f* **~ feeder (antenna)** konzentrische Speiseleitung *f* **~ float** Ring-, Zentral-schwimmer *m* **~ groove** Auslauf-, End-rille *f* (phono) **~ light diffuser** Lichtverteilungsschirm *m* **~ line** Hohlrohrleitung *f* **~ pipe line** konzentrisches Kabel *n* **~-pipe (or tube) transmission line** konzentrische Speiseleitung *f* **~ row atomizer** Kreiszeilendüse *f* **~ shaft** Konzenterbolzen *m* **~ windings** Transformationsröhrenwicklung *f*
**concentrically** konzentrisch
**concentricity** Mittenrichtigkeit *f*, Konzentrizität *f*, Rundheit *f* **~ check** Rundlaufprüfung *f* **~ error** Rundlauffehler *m* **~ gauge** Rundlaufmeßuhr *f* **~ tolerance** Toleranz *f* der Mittigkeitsabweichungen
**concept** Auffassung *f* **~(ion)** Begriff *m*
**conception** Auffassung *f*, Vorstellung *f*
**conceptually** begrifflich
**concepts** Möglichkeiten *pl*
**concern** Betreff *m*; Konzern *m*
**concerning** betreffend, bezüglich, hinsichtlich **~ actual use in service** betrieblich **~ profession** gewerblich **~ threshold** schwellig **~ trade** gewerblich
**concerted** aufeinander abgestimmt
**concertina** Drahtrolle *f* **~ roll** Drahtwalzenhindernis *n*
**concession** Bewilligung *f*, Einräumung *f*, Konzession *f*, Vergünstigung *f* **~ of mine** Grubenmaß *n*, Verleihung *f* **~ syndicate** Mutungsgemeinschaft *f*
**concessions, to make** **~** konzessionieren
**conchoid** Muschellinie *f*
**conchoidal** muschelig **~ fracture** muscheliger Bruch *m*, muschelige Bruchfläche *f* **~ fracture of porcelain** muscheliger Bruch *m* des Porzellans
**conciliation board** Gütestelle *f*
**concise** bündig, knapp, kurzgefaßt
**conciseness** Kürze *f*
**conclude, to** **~** abschließen, beschließen, enden, schließen
**conclusion** Abschluß *m*, Beschluß *m*, Ende *n*, Folge *f*, Folgerung *f*, Nachwort *n*, Rückschluß *m*, Schluß *m*, Schlußfolgerung *f* **~ of business** Geschäftsabschluß *m*
**conclusive** beweiskräftig **~ force** Beweiskraft *f*
**conclusiveness** Beweiskraft *f*
**concomitant** auftretend **~ symptom** Begleitsymptom *n*
**concordant** zusammenstimmend
**concrete, to** **~** betonieren, einbetonieren, fluhen, fritten, konsolidieren, konstruktiv, verfestigen, aus Beton bestehend
**concrete** Beton *m*, Fluh *f*, Grundmörtel *m*, Gußmörtel *m* **~ and hollow tile slab** Steindecke *f* **~ arch** Betongewölbe *n* **~ area** Betonfläche *f* **~ base** Betonbettung *f* **~ beam** Betonbalken *m* **~ bed** Beton-bettung *f*, -fundament *n* **~ block** Beton-block *m*, -stein *m*, Zementformstück *n* **~-block conduit** Zementformstückkanal *m* **~ bomb** Betonbombe *f* **~ box** Betonkasten *m* **~**

**break** Betonbruch *m* **~ breaker** Weghobel *m* **~ bridge** Betonbrücke *f* **~ buggy** Betonkipper *m* **~ building** Betonbau *m* **~ bunker** Betonbunker *m* **~ casting** Guß-mörtel-stück *n*, -teil *m* **~ column** Betonpfeiler *m* **~ composing the body of a wall** Beton *m* der Hintermauerung **~ construction** Betonbau *m* **~ core wall** Kernmauer *f* von Beton **~ counterweight** Gegengewicht *n* von Beton **~ cover** Beton-decke *f*, -deckplatte *f* **~ covering** Betondeckung *f* **~ deformation pick-up** Betondeformationsgeber *m* **~ dugout** Betonunterstand *m* **~ elevator** Turm *m* zum Heben des Betons **~ emplacement** Bunkerstand *m* **~ evidence** Sachbeweis *m* **~ floor** Betondecke *f*, Beton-fußboden *m*, -sohle *f*, Zementfußboden *m* **~ fortification** Betonwerk *n* **~ foundation** Beton-bettung *f*, -fundament *n*, -sockel *m* **~ foundation block sunk in drilled holes** aufgelöste Fundamente *pl* **~ fracture** Betonbruch *m* **~ girder** Betonbalken *m* **~ hollow brick** Hohlblockstein *m* **~ jacket** Betonummantelung *f* **~ juice** trockener Saft *m* **~ mass** Betonmasse *f* **~ mixer** Beton-maschine *f*, -mischer *m*, -mischmaschine *f*, -mischwerk *n* **~ obstacle** Betonsperre *f* **~ pillar** Betonsäule *f* **~ pillbox** Betonbunker *m* **~ pipe** Beton-, Zement-rohr *n* **~ plant** Anlage *f* zum Mischen des Betons **~ plates** Betonstirnplatten *pl* **~ pole** Beton-, Schleuder-, Zement-mast *m* **~ press** Betonpresse *f* **~ pumice block** Schwemmstein *m* **~ pump** Betonpumpe *f* **~ reactor** Betonspule *f* **~ reinforcing wire** Betonarmierdraht *m* **~ road** Betonstraße *f* **~ road finisher** Straßendeckenfertiger *m* **~ road surface padded down** gedichtete Betonstraßendecke *f* **~ road surface true to profile** profilgerechte Betonstraßendecke *f* **~ road surface without undulations** wellenfreie Betonstraßendecke *f* **~ rounds** Moniereisen *pl* **~ shelter** Beton-deckung *f*, -unterstand *m* **~ shield** Beton-abschirmung *f*, -panzer *m* **~ slab** Beton-balken *m*, -platte *f* **~ slab paving** Betonplattenbelag *m* **~ strip** Betonbahn *f* **~ structure** Betonbauwerk *n* **~ structures** Betonbauten *pl* **~ sufficiently wet to flow** Gußbeton *m* **~ support** Beton-gestell *n*, -säule *f* **~ surface pavement** Straßenoberbeton *m* **~ (slab) topping** Aufbeton *m* **~ tower** Schleudermast *m* **~ treatment** Betonverarbeitung *f* **~ turret** Betonturm *m* **~ weir** Betonwehr *n* **~ wharf** Lade-brücke *f*, -steg *m* **~ work** Betonwerk *n*
**concreting** Betonieren *n*
**concretion** Betonieren *n*, Konkretion *f*
**concretionary** konkretionär **~ horizon** verfestigte Schicht *f*
**concur, to** **~** zusammenlaufen
**concurrence** Gleichzeitigkeit *f*
**concurrent** **~ axes** sich schneidende Achsen *pl* **~ centrifuge** Durchstromzentrifuge *f* **~ forces** Kräfte *pl* die in einem Punkt angreifen
**concurrently** gleichzeitig **~ tuned jammer** abstimmbarer Störsender *m*
**concussion** Erschütterung *f*, Zerknallstoß *m* **~ of the brain** Gehirnerschütterung *f* **~ spring** Federdämpfer *m*
**condemn, to** **~** verwerfen
**condemned** als nicht gebrauchsfähig eingestuft
**condensability** Verdichtbarkeit *f*
**condensable** kondensierbar, verdichtbar

condensance Kondensanz *f*, kapazitive Reaktanz *f*, kapazitiver Blindwiderstand *m*
condensate Kondensat *n*, Niederschlagswasser *n* ~ collector Kondensatsammelbehälter *m* ~ pressure reducer Kondensatentspanner *m* ~ return system Kondensatrückführungssystem *n* ~ run-off Schwitzwasserablaß *m* ~ separator Kondenstopf *m*
condensated kondensiert
condensation Kondensation *f*, Kondensierung *f*, Niederschlag *m*, Verdichtung *f*, Verflüssigung *f* ~ of air particles (in sound wave) Zusammenpressung *f* der Luftteilchen ~ by contact Oberflächenkondensation *f*
condensation, ~-accumulator vessel Kondensatsammelgefäß *n* ~ apparatus Kondensationsapparat *m* ~ basis Kondensationsbasis *f* ~ column Kondensationssäule *f* ~ discharge Kondensatablauf *m* ~ groove Schwitzwasserrinne *f* ~ level Kondensations-basis *f*, -punkt *m* ~ losses Abkühlungsverlust *m* ~ machine Kondensationsmaschine *f* ~ operation Kondensationsbetrieb *m* ~ point temperature Taupunkttemperatur *f* ~ product Kondensat *n*, Kondensationsprodukt *n* ~ runoff Kondensatablauf *m* ~ tank Kondensatbehälter *m* ~ trap Kondensatsammelgefäß *n*
condense, to ~ (sich) anhäufen, ausscheiden, dichten, eindicken, kondensieren, (vapor) niederschlagen, sich niederschlagen, tauen, (gases) verflüssigen, verdichten, verkürzen, zusammenpressen to ~ water vapor Wasserdampf *m* ausscheiden
condensed kondensiert, verdichtet ~ diagram überhöhtes Diagramm *n* ~-steam pump Kondensatpumpe *f* ~ type Engschrift *f* ~ water Schwitzwasser *n* ~-water discharge Kondenswasseraustritt *m*
condenser Ansammlungsapparat *m* (electr.), Kondensationsapparat *m*, Kondensator *m*, Kondensor *m*, Kühlapparat *m*, Kühler *m*, Verdichter *m*, Verdichtungsapparat *m*, Verflüssiger *m*, Vorlage *f* ~ primary ~ Luftkühler *m* ~ of reflux type Kondensator *m* nach dem Gegenstromverfahren *n* ~ with spiral condensing tube Kühler *m* mit Schlangenrohr ~ with spiral tube sealed to body Kühler *m* mit eingeschmolzenem Schlangenrohr
condenser, ~ aperture Kollektoröffnung *f* ~ aperture setting ring Kondensorstellring *m* ~ armature Kondensatorbelegung *f* ~-balancing method Kapazitätsausgleich *m* durch Zusatzkondensatoren, Kapazitätsausgleichsverfahren *n* durch Zusatzkondensatoren ~ bank Kondensatorenbatterie *f* ~ box Kondensatorenkasten *m*, Kondensatorwanne *f*, Kühlerkiste *f* ~ case Bechergehäuse *n* ~ charge Kondensatoraufladungen *pl* ~ charge and discharge Kondensatorumladung *f* ~ circuit Kondensatorkreis *m* ~ coating Kondensatorbelegung *f* ~ coil Kondensorspule *f*, Kühlschlange *f* ~ coupling, Kapazitätskuppelung *f*, elektrische oder kapazitive Kopplung *f* ~ covering Kondensatorbelag *m* ~ discharge Kipp-, Kondensator-entladung *f* ~ electroscope Kondensatorelektroskop *n* ~ element Sammelsystem *m* ~ enlarger Vergrößerungsapparat *m* mit selbsttätiger Einstellung (photo) ~ ferrule Konden-

satorohrring *m* ~ front lens Kondensoroberteil *m* (Vorderlinse) ~ interference suppression Kondensatorentstörung *f* ~ iris Kollektorblende *f* ~ jacket Kühlermantel *m* ~ leads Kondensatordurchführung *f* ~ leathers Nitschelleder *n* ~ lens Beleuchtungs-, Kondensator-, Kondensor-, Sammel-linse *f* ~ lightning arrester Kondensatorblitzschutzvorrichtung *f* ~ lining Kondensatorbelag *m* ~ load kapazitive Belastung *f* ~ loud-speaker elektrostatischer Lautsprecher *m* ~ meter Kondensatorelektrometer *n* ~ microphone elektrostatisches Mikrofon *n*, Kondensatormikrofon *n*, kapazitives oder statisches Mikrofon *n* ~-microphone circuit Kondensatormikrofonschaltung *f* ~ motion head Kondensortrieb *m* ~ motor Verflüssigermotor *m* ~ nose Vorlagemündung *f* ~ plate Kondensatorplatte *f* ~ plates Belegungen *pl* ~ pot Kondensatorenkasten *m* ~ reel Kondensatorwickel *m* ~ reel for switching device Anschlußrolle *f* des Anschaltgerätes ~ relay system Lichtleitsystem *n* ~ scaling tool Kondensatorschaber *m* ~ siren Kondensatorsirene *f* ~ sliding sleeve Kondensor-schiebhülse *f*, -schiebrohr *n* ~ speaker elektrostatischer Lautsprecher *m* ~ spindle Kondensatorachse *f* ~ stop Kollektorblende *f* ~ system Kondensationsanlage *f*, Sammelsystem *n* ~ tape Kondensatorenband *n* ~ telephone Kondensatortelefon *n* ~ transmitter Kondensatormikrofon *n*, kapazitives Mikrofon *n* ~ tube Kondensationsröhre *f*, Kondensator-, Kondens-, Kühl-rohr *n* ~-type direct-current multiplier Vervielfältigungskondensator *m* ~-type bush (or terminal) Kondensatorklemme *f* ~ water Brüden-, Kondensations-, Kondens-, Oberflächen-wasser *n* ~ winding Kondensatorwickel *m*
condensing Influenz *f* ~ of a text Zusammenziehung *f* eines Textes
condensing, ~ apparatus Kondensationsapparat *m* ~ coil Kühlschlange *f* ~ jacket Kühlmantel *m* ~ lens Beleuchtungslinse *f* ~ pipe Verdichtungsrohr *n* ~ pump Druckpumpe *f* ~ reheat turbine Kondensationsturbine *f* ~ trap Kondensatscheider *m* ~ tube Dampfablaßschlauch *m* ~ turboalternator set Kondensations-Turbosatz *m* ~ water Rückkühlwasser *n*
condensive reactance Kapazitanz *f*, kapazitive Reaktanz *f*, kapazitiver (Blind)widerstand *m*
condition, to ~ angewöhnen, bearbeiten, bedingen, konditionieren, trocknen
condition Bedienung *f*, Bedingung *f*, Beschaffenheit *f*, Eigenschaft *f*, Erfordernis *n*, Gestaltung *f*, Lage *f*, Stand *m*, Verfassung *f*, Verhältnis *n*, Voraussetzung *f*, Wesen *n*, Zustand *m*
condition, ~ of acceptance Übernahmebedingung *f* ~ of affairs Sachverhalt *m* ~ of auction Versteigerungsbedingung *f* ~ of contest Wettbewerbsausschreibung *f* ~ of continuity Raumbedingung *f* ~ of enveloping Einhüllbedingung *f* ~ of equilibrium Gleichgewichtslage *f* ~ of flow Strömungszustand *m* ~ of intersection Schnittlinienbedingung *f* ~ of linearity Linearitätsbedingung *f* ~ of material at time of supply Anlieferungszustand *m* ~ of nourishment Nahrungsbedingung *f* ~ of practice Betriebsbedingungen *pl* ~ of sale Lieferungsbedingung *f* ~ as supplied Lieferungszustand *m* ~ for self-

**oxcillation** Anfachungsbedingung *f* ~ **for shock waves** Stoßwellenbedingung *f* ~ **as supplied** Lieferungszustand *m* ~ **of support** Auflagerbedingung *f* ~ **for symmetry** Symmetriebedingung *f* ~ **of testing** Prüfbedingung *f*
**condition,** ~ **code** Brauchbarkeitskode *m* ~ **condemned** für unbrauchbar erklärt
**conditional,** bedingt ~ **backpoint instruction** bedingter Stopbefehl *m* ~ **backpoint instrument** bedingter Stopbefehl *m* ~ **block** bedingte Raumfolge *f* ~ **branch** bedingter Sprung *m* ~ **breakpoint instruction** bedingter Haltebefehl *m* ~ **instability** Feuchtlabilität *f* ~ **instruction** bedingter Befehl *m* ~ **interlock** bedingte Verschlußeinrichtung *f* ~ **jump** bedingter Sprung *m* ~ **observation** bedingte Beobachtung *f* ~ **transfer (of control)** bedingter Sprung *m*
**conditioned by** beruhend auf, gegründet auf
**conditioner** Konditionierapparat *m*
**conditioning** Bearbeitung *f*, Konditionierung *f* ~ **agent** Befeuchtungsmittel *n* (zum Konditionieren) ~ **bobbin** (textiles) Befeuchtungsspule *f* ~ **device** Vorrichtung *f* zur Herstellung bestimmter Versuchsbedingungen ~ **machine** Egalisier-, Garnanfeucht-maschine *f* ~ **period** Nachhärtefrist *f*
**conditions** Bestimmungen *pl*, Gegebenheiten *pl* ~ **of contract** Vertragsbedingungen *f pl* ~ **of cutting edges** Schneidenzustand *m* ~ **of delivery** Bezugsbedingungen *pl* ~ **of installation (or of fitting)** Einbauverhältnis *n* ~ **for mounting (or installation)** Einbaubedingungen *pl* ~ **of payment** Zahlungsbedingungen *f pl* ~ **of shear** Schubbedingungen *pl* ~ **for stress** Spannungszustand *m* ~ **under way** Fahrteigenschaften *pl*
**conducive** beitragend, dienlich, förderlich
**conduct, to** ~ abführen, betreiben, dirigieren, durchführen, durchleiten, fortleiten, führen, geleiten, leiten, vornehmen, zuführen to ~ **an acceptance test** einen Abnahmeversuch *m* ausführen to ~ **away** ableiten to ~ **in** einleiten to ~ **over** überleiten to ~ **past** (something) vorbeiführen to ~ **research** forschen to ~ **a survey** eine Vermessung *f* durchführen
**conduct** Betragen *n*, Führung *f* ~ **of the furnace operation** Schmelzführung *f* ~ **of heat** Schmelzführung *f* ~ **of the melting operation** Schmelzführung *f*
**conductance** Anleitung *f*, Konduktanz *f*, Leitfähigkeit *f*, Leitung *f*, Leitvermögen *n*, Leitwerk *n*, Leitwert *m*, Stufenleiter *m*, Wirkleitwert *m* ~ **(of valve)** Durchgangswiderstand *m* ~ **bridge** Konduktanzmeßbrücke *f* ~ **characteristic** Leitwertcharakteristik *f* ~ **curve** Leitwertlinie *f* ~ **factor** Leiterzahl *f* ~ **per unit length** Wirkleitwert *m* pro Längeneinheit
**conductibility** Leitungs-fähigkeit *f*, -vermögen *n*
**conductimeter** Leitfähigkeitsmesser *m*
**conducting** leitfähig ~ **aluminum** Leitaluminium *n* ~ **arc** Auslader *m* ~ **bit** Zuführungsbohrer *m* ~ **cable** Zuleitungskabel *n* ~ **direction** Durchlaßrichtung *f* ~ **filament** Leuchtfaden *m* ~ **finish** leitender Überzug *m* ~**-hearth furnace** herdbeheizter Ofen *m* ~ **a high voltage** hochspannungsführend ~ **layer** leitende Schicht *f* ~ **main** Umlaufleitung *f* ~ **material** Leitungsmaterial *n* ~ **period** Brennzeit *f*, Stromflußwinkel *m* ~ **piece** Leitstück *n* ~ **power** Leitungs-

fähigkeit *f* ~ **rod** Leitstange *f*, Zuleitung *f* ~ **salt** Leitsalz *n* ~ **sphere** leitende Kugel *f* ~ **tube** Leitungsrohr *n* ~ **wire** Drahtleitung *f*, Leitungs-, Strom-draht *m*
**conduction** Fortleitung *f*, Führung *f*, Leitung *f*, Leitungsfähigkeit *f*, Übertragung *f* ~ **by excess electrons** (in-type conduction) Überschußhalbleitung *f* ~ **of heat** Wärme-durchgang *m*, -fortleitung *f*, -leitungsfähigkeit *f* ~ **by holes** Defekthalbleitung *f*
**conduction,** ~ **alternating current** Leitungswechselstrom *m* ~ **band** Leitungsband *n* ~ **current** Leit-, Leitungs-strom *m* ~ **deafness** Leitungstaubheit *f* ~ **drier** Trockner *m* mit indirekter Beheizung ~ **electron** kernfernes Elektron *n*, Leitungselektron *n* ~ **loss** Leitungstaubheit *f* ~ **path** Leitungsbahn *f* ~ **properties** (of monovalent metals) Leitungseigenschaften *pl* ~ **resistance** Leitwiderstand *m* ~ **stratum** leitende Schicht *f*
**conductive** direkt, galvanisch, konduktiv, leitend, leitfähig, leitungsfähig ~ **component** Wirkteil *m* ~ **coupling** direkte oder galvanische Kopplung *f* ~ **pen (or pencil)** Graphitstift *m* ~ **ripple pickup** galvanische Brummschleife *f* ~ **rubber hose** Schlauchleitung *f* (Gummi) ~ **zone** leitendes Gebiet *n*
**conductively connected** galvanisch verbunden
**conductivity** Leitfähigkeit *f*, Leitungs-fähigkeit *f*, -koeffizient *m*, -vermögen *n*, Leitvermögen *n*, spezifischer Leitwert *m* **(specific)** ~ spezifische Leitfähigkeit *f*, spezifisches Leitvermögen *n* ~ **by holes** Defektleitung *f* (transistor)
**conductivity,** ~ **cell** Leitfähigkeitsgefäß *n* ~ **counters** Leitfähigkeitszähler *pl* ~ **indicator** Leitfähigkeitsmeßbrücke *f* ~ **modulation** Leitfähigkeitsbeeinflussung *f* ~ **recorder** Leitfähigkeitsschreiber *m* ~ **standard** Leitfähigkeitsnormal *n* ~ **tester** Leitfähigkeitsprüfer *m*
**conductometer** Leitungsmesser *m*, Leitwertmesser *m*
**conductor** Ableiter *m* (electr.), Ableitungsmittel *n*, Ader *f*, Bohrlochschutzrohr *n*, Bohrtaucher *m*, Draht *m*, Elektrizitätsleiter *m*, Führer *m*, Führungsrohr *n*, Führungsschiene *f*, Konduktor *m*, Leiter *m*, Leitung *f*, Leitungsdraht *m*, Schaffner *m*, Stromleiter *m* (current), Zugführer *m* (r.r.), Zuleitung *f* (electr.) ~ **of a cable** Kabelader *m* ~ **for moist places** Feuchtraumleitung *f* ~ **with negative resistance** mit negativem Widerstand *m* behafteter Leiter *m*
**conductor,** ~ **bar** Zuführungsschiene *f* ~ **constant** Leitungskonstante *f* ~ **copper** Leitungskupfer *n* ~ **core** Leitungsseele *f* ~ **diameter** Leiter-durchmesser *m*, -stärke *f* ~ **element** Leiterstück *n* ~ **fitting** Kabelfassung *f* ~ **lead** Leitungsführung *f* ~ **part** Leiterteil *m* ~**-rail ramp** Auflaufschiene *f* ~ **resistance** Leiterwiderstand *m* ~ **string** Standrohr *n* ~ **structure (or system)** Leiter-gebilde *n*, -system *n* ~ **thickness** Stromleiterdicke *f*
**conduit** Abzucht *f*, Führung *f*, Kanal *m*, Leitkanal *m*, Leiter *m*, Leitung *f*, Rohr *n*, Röhre *f*, Röhrenzug *m*, Rohrleitung *f*, Schütte *f*, Stollen *m*, Umlauf *m* ~ **for detonation cord** Sprengrohr *n* **for lead-in cables** Einführungskanal *m* für Hauseinführungen (teleph.)
**conduit,** ~ **box** (lead-in wires) Abzweigdose *f*

~ **cable** Röhrenkabel *n* ~ **coupling** Übergangs-
formstück *n* ~ **diameters** Rohrweiten *pl* ~
**lining** Stollenpanzerung *f* ~ **plate** (lead-in
wires) Abzweigdose *f*, Leitblech *n* ~ **support**
Kabelunterstützung *f*, Rohrlager *n* ~ **wire**
Rohrdraht *m* ~ **wiring** Verlegung *f* in Röhren
**cone, to** ~ aufkegeln (Kegelverfahren), pressen
**cone** (of rays) Bündel *n*, Düsenkegel *m*, Kegel *m*,
kegelförmige Messerwalze *f*, Konus *m*, Konuslautsprecher *m*, konische Kreuzspule *f*, Kuppe
*f*, Spitzkegel *m*, Trichter *m*, Trombe *f*, Tubus
*m*, Zapfen *m*
**cone,** ~ **with bent point** Diaskleralkegel *m* ~ **of**
**blast** Sprengkegel *m* ~ **of burst** Sprengkegel *m*
~ **of debris** Aufschüttungskegel *m* ~ **of deli-**
**very valve** Druckventilkegel *m* ~ **of die** Kegel
*m* der Schärfeinlage ~ **of dispersion** Feuer-,
Geschoß-garbe *f*, Streukegel *m*, Streuungs-garbe
*f*, -kegel *m* ~ **of fire** Feuergarbe *f* with ~ **heads**
**(or lost heads)** tiefversenkt und breitgestaucht
~ **of light** Beleuchtungs-, Licht-kegel *m* ~ **of**
**nulls** Null-konus *m*, -kegel *m* ~ **of range** Schuß-
garbe *f* ~ **of rays** Lichtkegel *m*, Strahlen-bündel *n*, -büschel *n*, -kegel *m* **of** ~ **revolution**
Rotationskegel *m* ~ **of rupture** Zerreißke-
gel *m* ~ **of second order** Kegel *m* zweiter
Ordnung ~ **of shearing stress** Schubspannungs-
kegel *m* ~ **of silence** Schweige-, Null-kegel *m*
(beim Funkfeuer) ~ **of silence marker** Null-
kegelfunkfeuer *n* ~ **of starting-air valve** An-
laßventilkegel *m* ~ **of view (or vision)** Sehkegel
*m*
**cone,** ~ **angle** Konuswinkel *m* ~ **angle of valve**
Ventilsitzabschrägung *f* ~ **'s antenna** Kegel-
antenne *f* ~'s **apex to apex** Kegelspitzen *pl*
zueinander ~'s **base to base** Kegelspitzen *pl*
voneinander ~ **bearing** Kegellager *n* ~-**bed**
Kegelrand *m* ~ **belt** Keilriemen *m* ~ **block**
Ausrichtbeil *n* ~ **(friction) brake** Kegelbremse
*f* ~ **breaker** Einweichtrommel *f*, Knetmühle *f*,
Papierzerfaserer *m* ~ **clamp** Klemmkegel *m* ~
**clutch** Kegel-, Kegelreibungs-kupplung *f* ~
**core** Konushülse *f* ~ **coupling** Kegelkupplung *f*
~ **crusher** Kegelbrecher *m* ~ **diaphragm** Konusmembran *f* ~ **end** Ansatzspitze *f* ~ **engine**
Kegelstoffmühle *f* ~ **friction clutch** Kegel-,
Konus-kupplung *f* ~ **granulator** Kegelgranu-
lator *m* ~**head** Kegelkopf *m* ~ **hub** Konusnabe
*f* ~ **impeller** Kegelschnellrührer *m* ~ **impeller**
**mixer** Kegelkreiselmischer *m* ~ **impression (or**
**imprint)** Kegeleindruck *m* ~ **loud-speaker**
Konuslautsprecher *m* ~ **mandrel** Drehdorn *m*
mit Konusscheibe ~ **mill** Glockenmühle *f* ~
**non-return valve** Kegelrückschlagventil *n* ~
**nut** Konusmutter *f* ~ **penetration test** Ein-
dringungsversuch *m* ~ **penetrometer** Kegelein-
dringungsapparat *m* ~-**point setscrew** Druckschraube *f* mit Spitze ~ **pulley** kegeliges Antriebsrad *n*, Kegelriemenscheibe *f*, Kegel-
scheibe *f*, Stufen-rad *n*, -scheibe *f* ~-**pulley drive**
Antrieb *m* durch Stufenscheibe, Kegel(trom-
mel)trieb *m* ~ **ratched adjustment clutch** Kegel-
klingenkupplung *f* ~ **screen** Trichtersieb *n* ~-
**shaped** kegelförmig ~-**shaped loaf sugar** Hut-
zucker *m* ~-**shaped rudder** Kegelsteuer *n* ~-**sha-**
**ped shell** Kegelmantel *m* ~ **speaker** Trichterlautsprecher *m* ~ **stop** Klemmkegel *m* ~ **stop**
**collar** Feststellkegel *m* ~ **test** Kegel(druck)-

probe *f* ~-**thrust test** Kegeldruckprobe *f* ~
**valve** Kegelventil *n* ~ **winder** (weaving) Weft-
garnspuler *m*
**coned brake joint** Kegelbremsgelenk *n*
**confer, to** ~ erteilen, konferieren, vergleichen,
übertragen **to** ~ **power of attorney** Vollmacht *f*
erteilen
**conference** Besprechung *f*, Rücksprache *f*,
Tagung *f* ~ **call** Rund-, Sammel-gespräch *n*,
Sammelverbindung *f* ~ **call installation** Rund-
gesprächseinrichtung *f* ~ **call between sub-**
**scribers' installations without special apparatus**
Sammelgespräch *n* mit gewöhnlichen Sprech-
stellenapparaten ~ **call with loudspeakers**
Sammelgespräch *n* mit Lautsprechern ~-**cal-**
**ling equipment** Konferenzschaltung *f*, Rund-
gesprächs-, Sammelgesprächs-einrichtung *f* ~
**circuit** Rundgesprächsverbindung *f* ~ **(call)**
**circuit** Sammelgesprächsverbindung *f* ~ **com-**
**munication** Konferenzschaltung *f* ~ **(call)**
**connection** Sammelgesprächsverbindung *f* ~
**jack** Konferenzklinke *f* ~ **report** Bespre-
chungsbericht *m* ~ **room** Besprechungsraum *m*
**confidence interval** Vertrauensbereich *m*
**confidential** geheim, vertraulich ~ **matter** Ver-
trauenssache *f*
**configuration** Anordnung *f*, Konfiguration *f*,
Stellung *f* von Gegenständen zueinander, (e. g.
of molecules) Konstellation *f*, Zustandsform *f*
~ **of coast** Küstengestaltung *f* ~ **of flow**
Stromlinienbild *n* ~**s of flow** Strömungsgebilde
*n* ~ **of ground** Bodengestaltung *f*, Geländegestalt *f* ~ **in the plane** ebene Konfiguration *f* ~
**of saddle point** Sattelpunktskonfiguration *f*
~ **in space** räumliche Konfiguration *f* ~ **of**
**terrain** Geländegestaltung *f* ~ **interaction**
Wechselwirkung *f* zwischen Elektronenanordnungen
**confine, to** ~ absperren, beschränken, internieren, (in a borehole) einschließen **to** ~ **by dikes**
eindämmen
**confined** beschränkt, eingeengt, eng, gehalten
~-**compression test** Druckversuch *m* mit behinderter Seitenausdehnung ~ **space liable to**
**contain explosive mixtures** explosionsgefähr-
licher Betriebsraum *m*
**confines** Bezirk *m*, Grenze *f*, Umfang *m*
**confining,** ~ **field** magnetisches Plasmabegren-
zungsfeld *n* ~ **liquid** (as an oil, for use with
water) Absperrflüssigkeit *f*
**confirm, to** ~ beglaubigen, bekräftigen, bestätigen, erhärten, festigen, feststellen **to** ~ **by**
**oath** beeidigen **to** ~ **in writing** verbriefen
**confirmation** Beglaubigung *f*, Bestätigung *f*,
Feststellung *f*, Firmung *f*, Konfirmation *f* ~
**of appeal** Berufungsbegründung *f* ~ **by oath**
Beeidigung *f*
**confirmatory** bestätigend ~ **decision** Bestäti-
gungsurteil *f* ~ **order** Befehlsbestätigung *f*
**confiscate, to** ~ beschlagnahmen, einziehen
**confiscation** Beschlagnahme *f*
**conflagration** Feuer *n*
**confluence** Kontaktstelle *f*, Nebenfluß *m*, Zu-
sammenfluß *m* ~ **of two watercourses** Zusam-
menfluß *m* zweier Wasserläufe
**confluent zone** Kontaktzone *f*
**confocal surfaces** konfokale Flächen *pl*
**conform, to** ~ anpassen, prüfen, richten, in

Übereinstimmung befinden, in Übereinstimmung bringen **to ~ to** entsprechen, genügen **to ~ to a curve** sich in einer Kurve *f* bewegen, (of railway track) einer Kurve *f* folgen **to ~ to a straight line** sich einer Geraden *f* anschmiegen

**conformability** Paßfähigkeit *f*

**conformable** konkordant **~ to** angemessen **~ representation** konforme Abbildung *f*

**conformal** formengleich, winkeltreu **~ mapping** fast schlichte oder konforme Abbildung **~ representation** konforme Abbildung *f* **~ transformation** konforme Abbildung *f*

**conformity** Einverständnis *n*, Konkordanz *f*, Lagerung *f*, gleichförmige Lagerung *f*, Übereinkommen *n*, Übereinstimmung *f*, Winkeltreue *f* **in ~ with** entsprechend **to be in ~ with the load** der Belastung *f* entsprechen

**confront, to ~** gegenüberstellen

**confrontation** Gegenüberstellung *f*

**confronted (with)** gegeneinander gestellt

**confuse, to ~** irreführen, verwechseln, verwirren, verzwicken

**confusing lights** verwirrende Lichtquellen *pl* (aviat.)

**confusion** Bestürzung *f*, Durcheinander *n*, Verwechselung *f*, Verwirrung *f* **~ of ideas** Begriffsverwirrung *f* **~ reflector** Verwirrungsreflektor *m* **~ region** Verwaschungsgebiet *n* (rdo nav.) **~ zone** Verwaschungszone *f* (TV)

**congeal, to ~** einfrieren, (fat) erstarren, gefrieren, gerinnen, koagulieren, zusammenfrieren

**congealable** gefrierbar

**congealed** festgeworden

**congealing** Erstarrung *f* **~ apparatus** Gießapparat *m* **~ apparatus for gelatin plates** Gelatineplattenkühlapparat *m* **~ point** Erstarrungs-, Gefrier-punkt *m* **~ temperature** Erstarrungstemperatur *f*

**congelation** Gefrieren *n*

**congenial** gleichgestimmt

**congested** überfüllt, verstopft **~ area** Ballungsgebiet *n*

**congestion** Anhäufung *f*, Überfüllung *f*, Verstopfung *f* **~ meter** Überlastungszähler *m*

**conglomerate, to ~** ballen, konglobieren, konglomerieren, zusammen-backen, -ballen

**conglomerate** Konglomerat *n*, Nagelfluh *f*, Puddingstein *m* (geol.), Trümmergestein *n* **~ grain** Viellinge *pl* **~ structure** Konglomergefüge *n*

**conglomeration** Zusammenballen *n*, Zusammenballung *f*

**Congo, ~ red** Kongorot *n* **~ yellow** Kongogelb *n*

**congruence** Kongruenz *f*, (geom.) **~ theorem** Kongruenzsatz *m*

**congruency** Kongruenz *f*

**congruent** kongruent **~-band system** Gleichlageverfahren *n* **~ melting point** Kongruenzschmelzpunkt *m* **~ triangle** kongruentes Dreieck *n* **~ triangles** kongruente Dreiecke *pl*

**congruity** Winkeltreue *f*

**congruous** zusammenstimmend

**conic** konisch, verjüngt **~ diaphragm** Kegelblende *f* **~ pressure grease head** Kegelwulstschmierkopf *m* **~ projection** Linsenperspektive *f* **~ section** Kegelschnitt *m*

**conical** kegelförmig, kegelig, konisch, verjüngt **~ bearing** Spitzenlager *n* **~ bore in boss** Nabenkonusbohrung *f* **~ buddle** Kegelherd *m* (min.) **~ buoy** Spitztonne *f* **~ cheese** konische Kreuzspule *f* **~ clutch** Kegelklauenkupplung *f* **~ cock** Kegelhahn *m* **~ coil** konische Spule *f* **~ cone claw** Kegelklauenkupplung *f* **~ countersunk hole** kegelige Senkung *f* **~ coupling** Kegelklauenkupplung *f* **~ crusher** Glockenmühle *f* **~ diaphragm** Konusmembran *f* **~ dipole** Ganzpolkegel *m* **~ drum** Kegeltrommel *f* **~ feeder** Gießtrichter *m* **~ friction drum** kegelige Reibtrommel *f* oder Reibwalze *f* **~ friction roller** kegelige Reibtrommel *f* oder Reibwalze *f* **~ grinding** Konischschleifen *n* **~ head** Spitzkopf *m* **~ horn** konisches Horn *n*, Konustrichter *m* **~-indentation hardness** Kegeldruckhärte *f* **~ lubricator** Kegelwulstschmierkopf *m* **~ magnetic shield** Abschirmkonus *m* (TV) **~ member of discharge** Düsennadel *f* **~ mill** kegelförmige Kugelmühle *f* **~ nipple** Kegelwulstschmierkopf *m* **~ pen** Kegelfeder *f* **~ pendulum** Kegelpendel *n* **~ penetrometer** Kegeldrucksonde *f* **~ pin (or pivot)** Spitzzapfen *m* **~ projection** Kegelprojektion *f* **~ ring** Keilring *m* **~ roll inside conical refiner** kegelförmige Messerwalze *f* **~-roller bearing** Kegelrollenlager *n* **~ scan** konische Abtastung *f* **~ seat** Kegel-, Krater-sitz *m* **~ settler** Verdickungstrichter *m* **~ shaft end** Kegelwellenstumpf *m* **~ shape** Kegelgestalt *f* **~ sleeve** Leinendurchführung *f* **~ speed pulley** Stufenrad *m* **~ spiral spring** Kegelfeder *f* **~ surface** Kegelfläche *f* **~ top mark** Kegeltoppzeichen *n* **~ wheel** Kegelrad *n* **~ (taper) winding** kegelige (oder konische) Wicklung *f* **~-wing valve** Kegelventil *n*

**conically, ~ distributed blow** Kegelströmung *f* **~ shaped wooden bobbin** konischer Holzdorn *m* **~ threaded** konisch ausgedreht

**conicalness** Kegelform *f*, Konizität *f*

**conicity** Konizität *f*

**conics** Lehre *f* von den Kegelschnitten

**coniferous forest (or woods)** Nadelwald *n*

**coniform** kegelförmig

**coning** Konizität *f*, Konuswinkelbildung *f* **~ and quatering** (sampling) Kegelverfahren *n*

**conjecture** Verrückung *f*

**conjoined rule of three** Kettenregel *f*

**conjugate, to ~** konjugieren, paarweise verbinden

**conjugate, ~ (picture)** Koppelbild *m* **~ attenuation constant** konjugiert-komplexe Dämpfung *f* **~ axis** Nebenachse *f* **~ complex** konjugiert komplex **~-complex quantities** konjugiert komplexe Größen *pl* **~ distance** konjugierte Schnittweite *f* **~ functions** konjugierte Funktionen *pl* **~ impedance** konjugiert-komplexer Scheinwiderstand *m* eines Vierpols **~ impedances** konjugiert-komplexe Wechselstromwiderstände *pl* **~ intercept** konjugierte Schnittweite *f* **~ line** konjugierte Gerade *f* **~ phase constant** konjugiert-komplexes Winkelmaß *n* **~ transfer coefficient** konjugiertkomplexes Übertragungsmaß *n* **~ transfer constant** konjugiert komplexes Übertragungsmaß *n* **~ variables** konjugierte Größen *pl*

**conjunct motion** stufenweise Bewegung *f*

conjunction Verbindung *f*, Vereinigung *f*, ~ of successive photographs Anschluß aufeinanderfolgender Aufnahmen, Folgebildanschluß *m* conjunctive address group Verbundadreßgruppe *f* conjuncture Konjunktur *f* conn, to ~ a ship Ruderkommando *n* geben connate water fossiles Wasser *n* connect, to ~ angliedern, anhängen, ankuppeln, anschalten, anschließen, anspannen, anstücken, einschalten, ketten, koppeln, kuppeln, schalten, mit Schaltaderdraht verbinden, verbinden, in Verbindung bringen, in Verbindung setzen, verknüpfen, zuschalten to ~ in circuit again wiedereinschalten to ~ to earth (or ground) mit Erde verbinden to ~ in multiple vielfach schalten to ~ in multiple arc gemischt schalten to ~ in opposition gegeneinander schalten to ~ at the outlet side nachschalten to ~ in parallel nebeneinander schalten, parallel schalten to ~ in series hintereinanderschalten, in Reihe schalten, vorschalten to ~ in series multiple gemischt schalten to ~ at will wahlweise einschalten

connect, to ~ differentially gegeneinander schalten to ~ electrically to the metal structure of aircraft abbinden to ~ like a loop zur Schleife schalten to ~ through durchschalten, durchverbinden to ~ two coils in series zwei Spulen gleichsinnig in Reihe schalten to ~ wrongly mißleiten

connect, ~ key Verbindungstaste *f* ~ load Anschlußleistung *f*

connected angeschlossen, angekoppelt, gekuppelt, geschaltet, verbunden to be ~ eingeschaltet sein (electr.), in Verbindung stehen ~ with each other untereinander verbunden ~ in identical phase gleichphasig angeschlossen ~ in multiple vielfachgeschaltet ~ in opposition gegeneinander geschaltet

connected, ~ across a resistance über einen Widerstand *m* geschaltet ~ control eingeschaltete Regelung *f* ~ differentially gegeneinander geschaltet ~ load Anschlußwert *m* ~ value Anschlußwert *m*

connecting Schalten *n*, Verbinden *n* ~ adapter Anschlußstutzen *m* ~ band Verbindungsschiene *f* ~ bar Bleileiste *f*, Traverse *f*, Welle *f* ~ block Anschluß-klemmbrett *n*, -schiene *f*, Klemmleiste *f* ~ bolt Anschluß-, Verbindungs-bolzen *m* ~ box Verbindungsmuffe *f* ~ branch Einsatzstutzen *m* ~ bridge Schaltbrücke *f* ~ bush Schalt-buchse *f*, -hülse *f* ~ cable Schaltkabel *n* ~ casing Verbindungshülse *f* ~ casing of interfering device with neutral Nullung *f* eines Störers ~ chain Bindekette *f* ~ circuit Verbindungssatz *m* (teleph.), Verbindungsstromkreis *m* ~ culvert Verbindungskanal *m* ~device Amtsanschließer *m*, Anschaltvorrichtung *f*, Schaltglied *n*, Verbindungsorgan *n* ~ ear Anschlußöse *f* ~ file Anschlußmann *m*, Fühlungshalter *m*, Verbindungsmann *m* ~ flex Leitungsadern *pl* ~ frame Anschlußspant *n* ~ gear Lenksteuerung *f*, Vorgelege *n* ~ hose Schlauchbinder *m* ~ intermediary Netzanschlußgerät *n* ~ jack Vermittlungsklinke *f* ~ lead wire Verbindungskabel *n* ~ lever Leiste *f* (Klemmleiste) ~ line Verbindungsbahn *f*, Verbindungslinie *f* ~ a line into a telephone

circuit Einschleifen *n* einer Leitung ~ lines of equal grid-net error Isogrive *f* ~ link Verbindungs-glied *n*, -lasche *f*, Zwischenglied *n* ~ link of chain Kettenverbindungsglied *n* ~ link with end piece Zugstange *f* mit Stangenkopf *f* ~ link guide Kulissenführung *f* ~ longeron Trennholm *m* ~ lug Verbindungsstift *m* ~ means Verbindungsklemme *f* ~ member Bindeglied *n* ~ nut Laschen-, Überwurf-mutter *f* ~ panel Steckanordnung *f* ~ part Bindeglied *n* ~ passage Verbindungskanal *m* ~ piece Ansatzstück *n*, Anschlußstück *n*, Stutzen *m*, Verbindungs-stück *n*, -stutzen *m* ~ pin Haltebolzen *m* ~ pipe Anschlußrohr *n*, Verbindungs-leitung *f*, -rohr *n*, -röhre *f* ~ pipe line Anschlußrohrleitung *f* ~ piping Anschlußrohrleitung *f* ~ plate Anschlußplatte *f*, Verbindungsblech *n* ~ platform Übergangsbrücke *f* ~ plug Verbindungsstöpsel *m*, (ground) Stoßstecker *m* ~ point Verbindungspunkt *m* ~ rack Verbindungsgestell *n* ~ rail Anschlußschiene *f* ~ railroad Eisenbahnverbindung *f* ~ relay Durchschalterelais *n* ~ ring Schleif-, Verbindungs--ring *m*

connecting-rod Kurbelstange *f*, Lenkstange *f*, Pleuel *n*, Pleuelstange *f*, Schubstange *f*, Treibstange *f*, Zuganker *m* ~ for conveying troughs Förderrinnenschubstange *f* ~ assembly Kurbeltrieb *m* ~ bearing Pleuel-lager *n*, -stangenlager *n* ~ bearing-bush Pleuellagerschale *f* ~ big-end bearing Pleuelstangenendlager *n* ~ blade Pleuelschaft *m* ~ bush Pleuelbüchse *f* ~ bushing Pleuelstangenlager *n* ~ cap Kolbenstangendeckel *m* ~ head Schubstangenkopf *m* ~ inspection plate Pleuelstangenschauplatte *f* ~ joint Pleuelstangenverbindung *f* ~ motion Schubstangenbewegung *f* ~ scooping nose Pleuelschöpfnase *f* (Luftpresser) ~ set Anschlußgerät *n* ~ strap Pleuelstangenbügel *m* ~ system with corrected dead-center angles Ausgleichspleuel *n*, Pleuel *n* für Zündmomentausgleich

connecting, ~ screw Anschlußschraube *f* ~ section Zwischenstück *n* ~ set Anschlußgerät *n* ~ shackle Verbindungsschäkel *m* ~ shaft Transmissionswelle *f* ~ sleeve Kabelverbinder *m*, Verbindungs-hülse *f*, -muffe *f* ~ socket Anschlußstutzen *m* ~ strap Pol-brücke *f*, -leiste *f* ~ strip Durchverbindungsstreifen *m*, Klemmleiste *f* ~ surface Verbindungsfläche *f* ~ tag Sockelstift *m* ~ terminal Anschlußklemme *f* ~ thread Bind-, Binde-faden *m* ~ through extension Durchschaltung *f* ~ through relay Durchschalterrelais *n* ~ traverse Stoßschwelle *f* ~ tube Ansatzrohr *n*, Verbindungs-rohr *n*, -röhre *f* ~up Beschaltung *f* ~-up circuit anschließender Stromkreis *m* ~ wire Ableitungsdraht *m* ~ worm Verbindungsschnecke *f*

connection Anschaltung *f*, Anschluß *m*, Anschlußleitung *f* (electr.), (gas) Anschlußstück *n*, Bezeichnung *f*, Beziehung *f*, Einrückung *f*, Gespräch *n*, (of cells) Poldraht *m*, Schaltung *f*, Stutzen *m*, Verband *m*, Verbindung *f*, Verhältnis *n*, Verknüpfung *f*, Zusammenhang *m*, Zwischenstück *f*

connection, ~ of corners Eckverband *m* ~ to feed rack Schubstange *f* für die Schaltzahnstange ~ in groups Gruppenverbindung *f* of ignition armature Zündankeranschluß *m* ~ of incandescent lamps Glühlampenschaltung *f* ~ of interfering device case with neutral wire Störernullung *f* ~ at joint Stoßverbindung *f* ~ to the lampholders Fassungstragarm *m* ~ by means of cardan shafts Verbindung *f* durch Gelenkwellen ~s for measurement Meßschaltung *f* ~ for party-line central-office equipment Leitungswählerschaltung *f* für Zweiganschlüsse ~ of polyphase circuits Mehrphasenschaltung *f* ~ for pressure gauge Anschluß *m* für Manometerleitung ~ of a recorder Schreiberanschluß *m* ~s to safety lock Verbindungsstange *f* zur Sicherung ~ to supply system Netzanschluß *m* ~ for trailer hose Kuppelkopf *m* (für Anhängerschlauch) ~ for welding and fusion burners Anschluß *m* für Schweiß- und Schneidbrenner

connection, ~ and butt straps Anschluß *m* und Stoßdeckung *f* ~ angle Befestigungswinkel *m* ~ assembly Verbindungsleitung *f* ~ banjo Schwenkstutzen *m* ~ block Anschlußleiste *f* (electr.) ~ box Anschlußkasten *m*, Klemm-dose *f*, -kasten *m*, Verbindungsdose *f* ~ branch Anschlußstutzen *m* ~ cable (or cord) Anschlußschnur *f* ~ diagram Verbindungsschema *n* ~ fitting Anschlußnippel *n* ~ jack Anschlußklinke *f* (electr.) ~ key Stromschlußhebeltaste *f* ~ lever Verbindungshebel *m* ~ nut Überwurfmutter *f* ~ piece Anschlußstutzen *m* ~ plate Klemmenplatte *f*, Knotenblech *n* ~ point Verbindungspunkt *m* ~ rose Anschlußdose *f* ~ shaft Verbindungswelle *f* ~ sleeve Stützenhülse *f* ~ strength Kupplungskraft *f* ~ strip Anschlußstreifen *m*; Klemm-leiste *f*, streifen *m*; Löt-, Lötösen-, Verbindungs-streifen *m* ~ stud Sechskantstiftschlüssel *m* ~ switch Gegentaktschaltung *f* ~ terminal Anschlußklemme *f* ~ terminal board Verbindungsklemmbrett *n* ~ time Ausführungszeit *f* (einer Verbindung) ~ time lag relay Einschaltverzögerungsrelais *n* ~ wire Verbindungsdraht *m*

connectivity Zusammenhangszahl *f*
connector Anschluß-teil *m*, -stecker *m*, Leitungswähler *m*, Linienwähler *m*, Steckkontakt *m*, Steckvorrichtung *f*, Triebzahn *m*, Verbinder *m*, Verbindungs-klemme *f*, -stecker *m*, -stück *n*, (battery) Klemme *f* ~ block Anschlußklemme *f* ~ box Abzweigdose *f* ~ fitting Kugelfassung *f* ~ link Steckglied *n*, Verbindungsstück *n* ~ receptacle Anschlußdose *f*, (flush mounting) Flansch-, Kabel-, Steck-dose *f* ~ set Anschlußgerät *n* ~ snap Karabinerhaken *m* ~ socket Steckbüchse *f*, Steckerhülse *f* ~ strip Buchsenleiste *f* (female)
connotation Merkmal *n*, Nebenbedeutung *f*
conodes Konoden *pl*
conoid Konoide *f*; glockenförmig
conoidal konoidisch
co-normal Konormale *f*
conoscope Kristallachsenmesser *m*
conoscopic konoskopisch
conscious error realisierter Fehler *m*
conscript, to ~ ausheben, einziehen
conscription Aufgebot *n*, Aushebung *f*, Einziehung *f*
consecutive aufeinanderfolgend, nachfolgend

~ number Folgenummer *f*, laufende Nummer *f*, Laufnummer *f* ~ computer Sequentiellrechengerät *n* in ~ order in der Nummernfolge *f* ~ order of start Startfolge *f* ~ (or consequent) reaction Folgereaktion *f* ~ scanning einfacher Raster *m* (ohne Zeilensprung)
consecutively hintereinander, nacheinander
consensus Übereinstimmung *f*
consent, to ~ zugeben to ~ to einwilligen
consent Bewilligung *f*, Billigung *f*, Einwilligung *f*, Genehmigung *f*, Zusage *f*, Zustimmung *f* ~ judgment Anerkennungsurteil *n*
consequence Auswirkung *f*, Ergebnis *n*, Konsequenz *f*
consequent folgend, folgerecht ~ of a ratio Hinterglied *n* eines Verhältnisses ~ point Folgepunkt *m* ~ pole Folge-pol *m*, -punkt *m*, zusätzlicher Pol *m*
consequential condition Folgeverbindung *f* (math.)
conservation Bewahrung *f*, Erhaltung *f*, Haltbarmachung *f*, Konservierung *f* ~ of areas Flächenerhaltung *f* ~ of energy Energiesatz *m* ~ of the mass Erhaltung des Masse ~ of momentum Erhaltung *f* des Impulses ~ theorem Erhaltungssatz *m*
conservative erhaltend, unter Berücksichtigung *f* ausreichender Sicherheit ~ design reichliche Bemessung *f* von Konstruktionsteilen ~ estimate vorsichtige Abschätzung *f* ~ flux quellenfreier Fluß *m*
conserve, to ~ bewahren, haltbar machen, konservieren, schonen
consider, to ~ berücksichtigen, betrachten, einrechnen, erwägen, wägen
considerable ansehnlich, bedeutend, beträchtlich, erheblich, namhaft, stark ~ load variation Belastungsstoß *m*
consideration Acht *f*, Bedenken *n*, Bedeutung *f*, Berücksichtigung *f*, Betracht *m*, Betrachtung *f*, Entschädigung *f*, Erwägung *f*, Prüfung *f*, Rücksicht *f*, Überlegung *f* ~ of similitude Ähnlichkeitsbetrachtung *f*
consign, to ~ to übertragen
consignee Empfänger *m*
consignment Anlieferung *f*, Kommissionslager *n*, Konsignation *f*, Lieferposten *f* ~ of valuables Wertsendung *f* ~ note Frachtbrief *m* ~ number Leitungszahl *f*
consignor Absender *m* ~ code Versandstellenkode *m*
consist, to ~ in (of) bestehen in (aus)
consistence Konsistenz *f*
consistency Bestand *m*, Dicke *f*, Dickflüssigkeit *f*, Festigkeit *f*, (textiles) Flottenverhältnis *n*, Folgerichtigkeit *f*, Konsequenz *f*, Konsistenz *f*, Zusammenhalt *m*, Zusammenhang *m* (of pulp) ~ of measuring values Reproduzierbarkeit *f* von Meßwerten, Übereinstimmung *f* ~ of reading Regelmäßigkeit *f* der Ablesung
consistency, ~ check Eignungsprüfung *f* ~ factor Festigkeitszahl *f* ~ index Konsistenzzahl *f* ~ variables Konsistenzparameter *n*
consistent bleibend, dicht, konsequent, konsistent, übereinstimmend, verträglich ~ with sound engineering technisch einwandfrei ~ grease Konsistentenfett *n* ~ lubricant konsistentes Fett ~ pattern (of hits) geschlossenes Trefferbild *n*

consistently lückenlos

consisting, ~ (of) bestehend ~ of four parts vierteilig

consistometer Konsistenzmesser m

console Bedienkasten m, Konsole f, Kontrollpult n, Kontrolltisch m, Kragstück n (arch.), Säulenhängelager n, Sichtgerät n (rdr) ~ flood light Konsolenflutleuchte f ~ receiver Standempfänger m (rdr)

consolidate binden, festigen, festmachen, konsolidieren, setzen, stampfen, verdichten, verfestigen, zusammendrücken

consolidated verdichtet being ~ (of sediment) zusammensitzen n ~ floor befestigte Sohle f ~ profit and loss account konsolidierte Gewinn- und Verlustrechnung f

consolidation Ausbau m, Befestigung f, Dichtung f, Konsolidierung f, Sackmaß n, Sinterung f, Verbindung f, Verdichtung f, Vereinigung f, Verfestigung f, Verschmelzung f, Zusammenschluß m ~ of the subsoil (by weight) Bodenpressung f ~ settlement Setzung f durch Verdichtung ~ test Druckversuch m mit behinderter Seitenausdehnung, Zusammendrückungsversuch m

consolidometer Odeometer n, Zusammendrückungsapparat m

consolute gegenmischbar, gegenseitig mischbar, unbeschränkt mischbar

consonance Einklang m, Gleichklang m

consonant Konsonant m, Mitlaut m; gleichtönend ~ articulation Konsonantendeutlichkeit f

conspicuous auffindbar, deutlich sichtbar, in die Augen fallend, kenntlich to be ~ auffallend to become ~ hervortreten

conspiracy Verabredung f (law)

conspire, to ~ verabreden (law)

constancy Beständigkeit f, Gleichheit f, Gleichmäßigkeit f, Konstanz f ~ of frequency Frequenzkonstanz f ~ of the interfacial angle Winkelkonstanz f ~ of performance Betriebssicherheit f ~ of pitch Tonkonstanz f ~ of speed Tourenkonstanz f ~ of visual size Sehgrößenkonstanz f ~ of volume Raumbeständigkeit f ~ of zero setting Nullpunktkonstanz f

constant anhaltend, beharrlich, beständig, bleibend, dauernd, einwellig, fortlaufend, fortwährend, gleichbleibend, gleichmäßig, konstant, ruhend, standhaft, ständig, stet, stetig, unveränderlich ~ of shape formbeständig

constant Festwert m (math.), Festzahl f, Konstante f, Unveränderliche f ~ of anharmonicity Anharmonizitätskonstante f ~ of inertia Trägheitskonstante f, ~ of (measuring) instrument Instrumentkonstante f ~ of integration Integrationskonstante f ~ of a meter Zählerkonstante f ~s of the plate Orientierungselemente pl ~ of proportionality Proportionalitätskoeffizient m, -konstante f ~ of radioactive transformation Zerfallskonstante f ~ of retardation Verzögerungskonstante f ~ in time zeitlich gleichbleibend ~ per unit length Grundwert m

constant, ~ bearing course Kollisionskurs m (g/m) ~ braking retardation konstante Bremsverzögerung f ~ circulation oiling Umlaufschmierung f

constant-current, ~ charge Laden n mit gleichbleibendem Strom ~ modulation Anodenspannungsmodulation f, Heising-Modulation f, Parallelröhrenmodulation f ~ source Konstantstromquelle f ~ system System n mit gleichbleibender Stromstärke ~ transformer Transformator m für gleichbleibenden Strom

constant, ~ diameter gleichdick ~ direction error Halbkreisfehler m, konstanter Richtungsfehler m ~ drip arrangement ständiges Nachtropfverfahren n ~ entropy unveränderliche Entropie f ~ field Gleichfeld n ~ filtration temperature temperaturkonstante Filtration f ~ frequency variable dot keying Zeitmodulation f ~ glow (glowtube) potential Brennspannung f ~ head gleichbleibender Wasserdruck m ~ impact Einheitsstoß m ~ K-filter Konstant-K-Filter m ~ level carburetor Vergaser m mit gleichbleibendem Stand ~ light Gleichlicht n ~ loading Gleichbeanspruchung f ~ loop Schleifenberührung f ~ magnetic field magnetisches Gleichfeld n ~ mesh gear Getriebe n mit ständigem Eingriff ~ mesh transmission Aphongetriebe n ~ missile angle festgehaltene Lage f ~ parallel flow gleichförmige Schichtenströmung f ~ potential supply Gleichspannungsquelle f ~ power (or force) konstante Kraft f ~ pressure Gleichdruck m

constant-pressure, ~ bomb Gleichdruckbombe f ~ boost Gleichdruckaufladung f ~ combustion Gleichdruckverbrennung f, Verpuffungsverfahren n ~ combustion engine Gleichdruckmaschine f, Gleichdruckmotor m ~ cycle Diesel-, Gleichdruck-prozeß m, Gleichdruckverfahren n ~ gas turbine Gleichdruckgasturbine f ~ gasturbine plant Gleichdruckgasturbinenanlage f ~ heat accumulator Gleichdruckspeicher m ~ line Gleichdrucklinie f ~ method Gleichdruckverfahren m ~ system Gleichdruckanlage f,

constant, ~ rate of loading gleichmäßige Laststeigerung f ~ resistance network Glied n mit konstantem Widerstand ~ speed gleichförmige Geschwindigkeit f, unveränderliche Umlaufzahl f

constant-speed, ~ airscrew Luftschraube f gleichbleibender Drehzahl ~ control (unit) Regler m für gleichbleibende Drehzahl ~ follow up mechanism Isodromrückführung f ~ governor Drehzahlregler m für Verstellschraube gleichbleibender Drehzahl ~ motor Motor m mit gleichbleibender Geschwindigkeit ~ propeller Luftschraube f mit gleichbleibender Geschwindigkeit ~ pulley Einscheibe f ~ scanning Abtastung f mit gleichbleibender Geschwindigkeit

constant, ~ supervision laufende Überwachung f ~ temperature bath Bad n bei steter (unveränderlicher) Temperatur ~ voltage charge Laden n mit gleichbleibender Spannung ~ voltage generator Gleichspannungsgenerator m ~ voltage modulation Reihenröhren-, Vorröhren-modulation f ~ volume combustion Gleichraumverbrennung f ~ volume cycle Gleichraumprozeß m ~ volume line Linie f gleichen Rauminhalts ~ volume process (or system) Gleichraumverfahren n ~ volume test Versuch m auf der Baustelle bei gleichblei-

bendem Volumen ~ **weight** Gewichtskonstanz
*f* ~ **wind** gleichmäßiger Wind *m*
**constantan** Konstantan *n*
**constantly** fortwährend, konstant, stets ~ **active
bell** Fortschellklingel *f*
**constellation** Sternbild *n*
**constituent** Anteil *m*, Bestandteil *m*, Gemengteil
*m*, Komponente *f* (Farbe); zusammensetzend
~ **of a (chemical) combination** Komponente *f*
einer Verbindung ~ **particle** Materieteilchen *n*
**constitute, to** ~ aufstellen, ausmachen, begründen, bilden
**constituting a bar to the grant of a patent** pa-
tenthindernd
**constitution** Aufbau *m*, Körperbeschaffenheit *f*,
Staatsverfassung *f*, Verfassung *f* ~ **diagram**
Zustandsdiagramm *n*
**constitutional,** ~ **curve** Zustandskurve *f* ~ **diagram** Zustandsschaubild *n* ~ **field** Zustandsfeld *n* ~ **formula** Konstitutionsformel *f*
**constitutive,** ~ **equation** Materialgleichung *f*
~ **properties** konstitutive Eigenschaften *pl*
**constrained** unfrei, zwangsläufig ~ **current
operation** Betrieb *m* mit erzwungener Erregung
~ **motion** erzwungene Bewegung *f* ~ **oscillation**
erzwungene oder fremderregte Schwingung *f*
~ **vibrations** erzwungene Schwingungen *pl*
**constraint** Zwang *m* **by** ~ zwangsläufig ~
**hypersurface** Zwangsbedingungenhyperfläche *f*
**constraints** Zwangsbedingungen *pl*
**constrict, to** ~ anwürgen, verengen, verjüngen,
zusammenschnüren
**constriction** Einschnürung *f*, Verengung *f*, Verjüngung *f*, Zusammenschnürung *f*, Zusammenziehung *f* ~ **of an extended dislocation** Einschnürung *f* in eine aufgespaltene Versetzung
~ **of volume** Dynamikverengung *f* ~ **energy**
Einschnürungsenergie *f* ~ **model** Einschnü-
rungsmodell *n* ~ **ratio** Klemmung *f*
**construct, to** ~ anlegen, ausführen, bauen, bilden, darstellen, entwerfen, erbauen, errichten,
gestalten, konstruieren **to** ~ **in open cutting** im
offenen Schacht ausschlitzen **to** ~ **on similar
lines** gleichbauen
**constructed diameter** Nenndurchmesser *m*
**construction** Anlage *f*, Aufbau *m*, Aufführen *n*,
Ausbau *m*, Ausbildung *f*, Ausführung *f*, Bau
*m*, Bauart *f*, Bauen *n*, Baukonstruktion *f*,
Bauwerk *n*, (fabric) Bindung *f*, Darstellung *f*,
Erbauung *f*, Errichtung *f*, Gestalt *f*, Gestaltung *f*, Herstellung *f*, Konstruktion *f*, Zusammensetzung *f* **in** ~ im Bau begriffen **new** ~
Neubau *m* **under** ~ im Bau *m* **type of** ~ Bauweise *f*, Bauform *f*, Baumuster *n*
**construction,** ~ **of an airdrome** Anlegen *n* eines
Flugplatzes ~ **using climbing rollers** Abfang-
verfahren *n* ~ **of fieldworks** Schanzarbeit *f*
~ **of the funicular (link) polygon** Seileckkon-
struktion *f* ~ **under license** Lizenzbau *m* (aviat.)
Nachbau *m* ~ **of a position** Stellungsbau *m*
~ **of slopes** Dammschüttungsarbeiten *pl* ~ **of
a work** Bauausführung *f*
**construction,** ~ **assembly** Einrichtung *f* ~ **box
principle** Bauprinzip *n* ~ **data** Konstruktions-
unterlagen *pl* ~ **detachment** Arbeitertrupp *m*,
Bauabteilung *f* ~ **drawing** Bauzeichnung *f*,
Konstruktionszeichnung *f* ~ **engineering** Bau-
praxis *f*, Bauwesen *n* ~ **float** Einbaufähre *f* ~

**gang** Baukolonne *f*, Bautrupp *m* ~ **jig** Anbau-
vorrichtung *f* ~ **joint** Baufuge *f* ~ **license**
Baulizenz *f* ~ **line** Mittellinie *f*, neutrale Linie
*f* ~ **line of groove** Kaliberlinie *f* ~ **lines** Profil
*n*, Profillinie *f* ~ **lines of brickwork** (of blast
furnace) Steinzeichnung *f* ~ **material** Bauzeug *n* ~ **materials** Werkstoff *m* ~ **office** Bau-
amt *n* ~ **plan** Bauentwurf *m*, Bauplan *m* ~
**platform** Arbeitsbühne *f* ~ **principle** Konstruktionsprinzip *n* ~ **process** konstruktive
Tätigkeit *f* ~ **purpose** Bauzweck *m* ~ **requirements** Bauvorschrift *f* ~ **schedule** Bauprogramm
*n* ~ **set** Baueinheit *f* ~ **site** Bau-platz *m*, -stelle
*f* ~ **specification** Bauvorschrift *f* ~ **speed** ver-
anschlagte Fluggeschwindigkeit *f* ~ **squad**
Bautrupp *m* ~ **staff** Baustab *m* ~ **stage** Bau-
stufe *f* ~ **superintendent** Bauführer *m* ~ **supervision** Bauaufsicht *f* ~ **system** Bauweise *f* ~
**trestle** Gerüstbrücke *f* ~ **unit** Bau-kolonne *f*,
-teil *m*, -trupp *m*, -zug *m* ~ **work** Bauarbeit *f*
~ **worker** Bauarbeiter *m*
**constructional** baulich, konstruktionstechnisch,
konstruktiv ~ **asymmetry** Schränkfehler ~
**change** Umkonstruktion *f* ~ **competition** Bau-
wettbewerb *m* (Modelle) ~ **detail** Konstruk-
tionseinzelheit *f* ~ **details of circuit wiring**
schaltungstechnischer Aufbau *m* ~ **dimension**
Anbaumaß *n* ~ **drawing** Ausführungszeichnung
*f* ~ **engineering** Maschinenbau *m* ~ **feature**
Baumerkmal *n* ~ **rights** Baurechte *pl* ~ **units**
Bauelemente *pl*
**constructive** aufbauend, konstruktiv
**constructor** Bauer *m*, Erbauer *m*, Gestalter *m*,
Konstrukteur *m*
**construe, to** ~ auslaugen; auslegen
**consult, to** ~ zu Rate ziehen
**consultation** Beratung *f*, Rücksprache *f* ~ **call**
Rückfragegespräch *n*
**consulting,** ~ **engineer** beratender Ingenieur *m*
~ **office** Beratungsstelle *f* ~ **service** Beratungs-
dienst *m*
**consumable,** ~ **article** Verbrauchsgegenstand *n*
~ **electrode** abschmelzende Elektrode *f*,
Schweißelektrode *f* ~ **load** Betriebs-, Ver-
brauchs-last *f* ~ **stocks** Verbrauchsbestand *m*
**consumation of error** Fehlerabgleichsmethode *f*
**consume, to** ~ abfressen, abzehren, aufbrauchen,
aufzehren, verarbeiten, verbrauchen, verzehren
**consumed** verbraucht **to be** ~ **by fire** verbrennen
**consumer** Abnehmer *m*, Konsument *m*, Verbraucher *m* ~ **'s goods** Verzehrungsgegenstand
*m* ~ **goods industry** Konsumgüterindustrie *f*
**consuming** Verarbeitung *f* ~ **of rf-energy** Vernichtung *f* (oder Abfackeln *n*) überschüssigen
Gases von HF-Energie ~ **branch** Verbrauchs-
güterindustrie *f* ~ **device** Verbraucher *m* ~
**industry** Verbraucherindustrie *f* ~ **power** Aufnahmefähigkeit *f* ~ **works** Verbraucherwerk
*n*
**consummate, to** ~ besiegeln
**consumption** Abbrand *m*, Abzehrung *f*, Aufwand *m*, Aufzehrung *f*, Bedarf *m*, Konsum *m*,
Leitungsverbrauch *m*, Verbrauch *m* ~ **of coke**
Koksverbrauch *m* ~ **of current** Stromentnahme
*f* ~ **of electrodes** Elektrodenabbrand *m* ~
**of materials** Materialverbrauch *m*

consumption, ~ characteristic Verbrauchskennlinie f ~ curve Verbrauchskurve f ~ decrease Verbrauchsverringerung f ~ figures Verbrauchsziffern pl ~ goods Verzehrungsgegenstand m ~ indicator Verbrauchsmesser m ~ pipe Verbrauchsleitung f ~ rate Verbrauchsquote f ~ tax Verbrauchssteuer f ~ test Verbrauchs-prüfung m, -wettbewerb m ~ transmitter Verbrauchsgeber m ~ type item Verbrauchsartikel m

consumptiongraph Verbrauchskurve f

contact Abzug m, Anschluß m, Berührung f, Eingriff m, Fühlung f, Kontakt m, Kontaktstück n, Leitungsberührung f, Schaltstück n, Schließer m, Sockelstift m (block), Verbindung f

contact, to be in ~ berühren to make ~ with abtasten ~ of back part key lever Ruhekontakt m ~ to earth Erdschluß m ~ of knife edges Schneidenberührung f in ~ with the liquid flüssigkeitsberührt ~ with wire Drahtberührung f

contact, ~ adhesive Haftkleber m ~ adjustment Kontaktverstellung f ~ agent Kontaktmittel n ~ altimeter Kontakthöhenmesser m ~ angle Kontaktwinkel m ~ arc Berührungsbogen ~ arc welding Kontaktlichtbogenschweißung f ~ area gedrückte Fläche f ~ arm Abreißhebel m, Kontaktarm m, Meßbügel m ~ arm hanger Meßbügelträger m ~ ball Kontaktkugel f ~ band grinding method Kontaktbandschleifverfahren n ~ bank Kontakt-bank f, -feld n, -reihe f, -satz m ~ bead Kontaktperle f ~ between lines Leitungsberührung f ~ black Kontaktruß m ~ blade Meßschneide f ~ blade assembly Kontaktfedersatz m ~ blade holder Schneidenauflage f ~ blades equipment Schneideneinrichtung f ~ block Taststein m ~ bounce Kontaktprellung f (Relais) ~ bow Schleifbügel m ~ box Übergangsdose f ~ breaker Kontakt-, Strom-unterbrecher m, Unterbrecher m

contact-breaker, ~ cam Kontaktnocken m ~ cam sleeve Unterbrechernockenhülse f ~ cover Kurzschluß-, Unterbrecher-deckel m ~ fastening screw Unterbrecherbefestigungsschraube f ~ lever Unterbrecherhebel m ~ opening Unterbrecherkontaktöffnung f ~ plate Unterbrecherplatte f ~ point Unterbrecherkontakt m ~ pull type spring Unterbrecherzugfeder f ~ shoulder Unterbrechernocken m ~ spanner Unterbrecherschlüssel m

contact, ~ breaking catch Abreißklinke f ~ breaking device Abreißvorrichtung f ~ breaking spark Abreißfunken m ~ bridge Schalterbrücke f ~ bridge piece Kontaktbügel m ~ bush Kontaktbüchse f ~ carriage Kontaktschlitten m, Kontaktrahmen m ~ chatter Kontaktprellen n, Prellen n der Kontakte ~ clamp Kontaktklemme f ~ clearance Kontakt-abstand m, -weite f ~ clip Kontaktschelle f (electr.) ~ closing Schluß m ~ closing period (or time) Kontaktschließdauer f ~ comb Kontaktkamm m ~ controllable from one point von einem Punkt aus regelbare Anstrichverstellung f ~ correction error Endmaßmessung f ~ corrosion Berührungskorrosion f ~ curve Wälzkurve f ~ deposit Kontakt-gang m, -lagerstätte f ~ detector Fest-, Kontakt-detektor m, Kon-

taktgleichrichter m ~ device Kontaktwerk n ~ discontinuity Kontaktunstetigkeit f ~ disk Kontaktscheibe f ~ effect Kontaktwirkung f ~ electricity Berührungselektrizität f ~ ends Schaltenden pl (der Wicklung) ~ (sur)face Berührungs-, Kontakt-fläche f ~ fault Kontaktfehler m ~ file Kontaktfeile f ~ filtration Kontaktfiltration f ~ finger Kontaktfinger m ~ fire Berührungszündung f ~ flight Flug m mit Bodensicht, Kontaktflug m ~ flying Schauzeichenkurssteuerung f ~ follow Folgeweg m ~ gap Kontaktabstand m ~ generator Berührungssystementwickler m ~ goniometer Anlegegoniometer n ~ half ring Kontakthalbring m ~ heating (heated clothing or pads) Berührungsheizung f ~ indicator Kontaktanzeiger m ~ irradiation equipment Nahbestrahlungsgeräte pl ~ jaw Kontakt, Messerkontakt-, Spann-backe f ~ jaws Einspannbacken pl ~ knob Wechselweibchen n ~ lamination Kontaktlamelle f ~ length gedrückte Länge f, Kontakthebel m ~ lenses Haftgläser pl, Kontakt-linsen pl, -schalen pl ~ lever Fühlhebel m ~ light Kontaktlicht n, Landebahnfeuer n ~ (or set-down) light Aufsetzlicht n ~ line Fahrleitung f ~ line voltage Fahrdrahtspannung f ~ between lines Leitungsberührung f ~ load Berührungsbelastung f ~ lode Kontaktgang m ~ loss Übergangsverlust m ~ maker Kontaktgeber m, Stromschließer m ~ making Kontaktgebung f ~ making clock (CMC) Kontaktuhr f ~ mechanism Kontakt m, Kontaktwerk n ~ metamorphism Kontaktmetamorphose f ~ microphone Kontaktmikrofon n ~ migration (or creeping) Kontaktwanderung f ~ mine Kontakt-, Stoß-, Tret-mine f ~ mission Fühlunghalten n ~ modulated amplifier Verstärker m mit Eingangsunterbrecher ~ murmer Kontaktrauschen n ~ navigation Flug m nach Bodensicht ~ noise Kratzgeräusch n ~ opening rate Kontaktöffnungsgeschwindigkeit f ~ panel Kontaktfeld n ~ path Kontaktstrecke f ~ pattern Tragbild n ~ piercing point Stechspitze f ~ pin Kontaktstift m ~ plane Berührungsfläche f, Fühlunghalter m ~ plate Brücke f des Stöpsels ~ plug Kontaktstöpsel m ~ point Kontaktspitze f, Punktschweißelektrode f, Schließ-, Stoß-stelle f, (airplane landing) Ansatzpunkt m ~ potential Berührungs-, Kontakt-spannung f, Kontaktpotential n ~ potential barrier Kontaktpotentialwall m ~ potential difference Kontaktspannung f ~ potential series Spannungsreihe f ~ pressure Anpreß-, Anpressungs-, Auflage-druck m, Flächenpressung f, Kontaktdruck m ~ pressure adjustment Vorrichtung f zur Verminderung der Meßkraft (Optimeter) ~ pressure ring Kontaktdruckfeder f ~ print Kontaktabdruck m (photo) ~ printer Schlitzkopiermaschine f (photo) ~ printing Kopieren n ~ process Kontaktverfahren n ~ proof kontaktsicher ~ rail Stromschiene f ~ rating Schaltleistung f ~ rectifier Drehvariometer n, Kontaktdetektor m, Sperrschichtgleichrichter m ~ reflection Tragbild n ~ resistance Kontakt-, Übergangs-widerstand m ~ resistances Siebwiderstände pl ~ rock Kontaktgestein n ~

**roller** Führungsrolle *f*, Rollenelektrode *f* ~ **rule** Anlagemaßstab *m* ~ **safety condenser** Berührungsschutzkondensator *m* ~ **safety device** Berührungsschutz *m* ~ **scanning** Kontaktabtastung *f* ~ **screw** Kontakt-, Meß-, Stromschluß-schraube *f* ~ **segment** Kontaktscheibe *f*, -stück *n* ~ **series** Spannungsreihe *f* ~ **sheet** Anschlußblatt *n* ~ **shoulder** Unterbrechernocken *m* ~ **side of the belt** Laufseite *f* des Riemens ~ **spacing** Kontaktabstand *m* ~ **sphere** Kontaktkugel *f* ~ **spring** federnder Kontakt *m*, Kontaktfeder *f*, Schichtquelle *f*, Unterbrecherfeder *f* ~ **spring assembly** Kontaktfedersatz *m* ~ **spud** Kontaktklotz *m*, Schaltstück *n* ~ **strip** Anschlußleiste *f* ~ **stud** Kontaktkopf *m*, kurzer Kontaktstift *m*, Schalterkontakt *m*, (of an electric controller) Kontaktknopf *m* ~ **substance** Kontakt-mittel *n*, -substanz *f* ~ **support (or carrier)** Kontaktträger *m* ~ **surface** Auflage-, Grenz-, Lauf-fläche *f* ~ **surface of the journal** Zapfenanlagefläche *f* ~ **switch** Kontakt-stück *n*, -umschalter *m* ~ **terminal** Kontaktklemme *f* ~ **tip** Meß-bolzen *m*, -hütchen *n* ~ **treadle of mercury** Quecksilberschienenkontakt *m* ~ **trigger** Druckstück *n* ~ **trolley** Kontaktrolle *f* ~ **trough** Kontaktgefäß *n* ~ **tube** Einschaltröhre *f*, Stromzuführungsdüse *f* (in welding) ~ **unit** Kontaktsatz *m* ~ **vacuum gauge** Kontaktmanometer *m* ~ **wear** Kontaktabnutzung *f* ~ **welder** Schweißvorrichtung *f* für Kontakte ~ **wheel** Kontaktscheibe *f* (für Bandschleifer) ~ **wiper** Kontaktfinger *m* ~ **wire** Fahrdraht *m* (antenna) ~ **wire insulator** Fahrdrahtisolator *m*

**contacted oil** mit Bleicherde behandeltes Öl *n*
**contacting** Kontaktbehandlung *f*
**contactor** Einschalter *m*, Fernschalter *m*, Hüpfer(schalter) *m*, Kontakt *m*, Kontakteinrichtung *f*, Kontaktgeber *m*, pneumatischer Schalter *m*, Schalter *m*, Schaltschütz *m*, Schütz *n*, Tastrelais *n* ~ **control** Automatenschalterregelung *f* ~ **equipment** Schützensteuerung *f* ~ **interlocking** Schützverriegelung *f* ~ **lever** Fühlhebel *m* ~ **rack** Schützengerüst *n* ~ **release** Schützenriegelung *f* ~ **timer** Schaltzeitrelais *n*
**contain, to** ~ binden, enthalten, fassen, fesseln, hinhalten
**container** Auffangkasten *m*, Aufnehmer *m*, Behälter *m*, Behältnis *n*, Dose *f*, Filtereinsatz *m*, Gefäß *n*, Kanister *m*, Preßtopf *m*, Rezipient *m*, Topf *m*, Transportbehälter *m* ~ **for adding dye-stuffe** Farbstoffzusatzgefäß *n* ~ **for the discs** Tellerablage *f* ~ **for dusting material** Streugutbehälter *m* ~ **for liquids** Flüssigkeitsbehälter *m* ~ **with a narrow neck** Blechflasche *f* ~ **with source** Präparathalter *m* mit Strahler ~ **of a tube** Röhrenkolben *m* ~ **of a valve** Röhrenkolben *m* ~ **for water heating** Wasserblase *f*
**container, ~ cover** Behälterdeckel *m* ~ **frame** Trommelrahmen *m* ~ **skid** Behälterkufe *f*
**containers** Fustage *f* ~ **accounts** Gebindebewegungsmeldung *f*
**containing** haltig, hältig, inhaltlich ~ **acid** säurehaltig ~ **aluminum** aluminiumhaltig ~ **arsenic** arsenhaltig, arsenikalisch ~ **asphalt** asphalt-

haltig, erdpechhaltig ~ **bismuth** wismuthhaltig ~ **fusel** fuselhaltig ~ **iron** eisenhaltig ~ **lime** kalkhaltig ~ **lithium** lithiumhaltig ~ **magnesium** magnesiumhaltig ~ **marl** mergelhaltig ~ **mercury** quecksilberhaltig ~ **molybdenum** molybdänhaltig ~ **mud** schlammhaltig ~ **no metal** unhaltig ~ **no oxygen** sauerstoffrei ~ **oil** ölhaltig ~ **oxide** oxydhaltig ~ **paint mist** farbnebelhaltig ~ **paraffin** paraffinhaltig ~ **platinum** platinhaltig ~ **selenium** selenhaltig ~ **size** schlichtehaltig ~ **slime (or sludge)** schlammhaltig ~ **steam** dampfhaltig ~ **sulfur** schwefelhaltig ~ **tellurium** tellurhaltig ~ **tar** teerhaltig ~ **tin** zinnhaltig ~ **tin oxide** zinnoxydhaltig ~ **tungsten** wolframhaltig ~ **wall** Umfassungsmauer *f* ~ **water** wasser-haltend, -haltig ~ **zinc** zinkhaltig
**containment, ~ valve** Hauptabschlußventil *n* ~ **time** Begrenzungszeit *f*
**contaminable** ansteckbar
**contaminant** Verunreinigung *f*
**contaminate, to** ~ (painting) in Berührung bringen mit, schädlich machen, verfärben (TV), verschmutzen, verseuchen, verunreinigen
**contaminated** schmutzig ~ **by dross** verkrätzt
**contamination** Vergiften, Vergiftung *f*, Verschmutzung *f*, Verseuchen *n*, Verseuchung *f*, Verunreinigung *f* ~ **of area** Geländevergiftung *f* ~ **meter** Verseuchungsmeßgerät *n*
**contemporaneous** gleichzeitig
**contemporary** Zeitgenosse *m*; modern, zeitgenössisch
**content** Ausfüllung *f*, Inhalt *m* (data proc.), Rauminhalt *m* (cubic) ~ **of nuclei** Kerngehalt *m* ~ **by volume** Raumgehalt *m* ~(s) Gehalt *m*, Inhalt *m*
**contention** Streit-gegenstand *m*, -punkt *m*
**contentiously pleaded** streitig verhandelt
**contents** Ausdehnung *f*, Fassungsraum *m*, Inhalt *m* (data proc.), Raummenge *f* ~ **(table of)** Sachregister *n* ~ **for pressure sprayers** Druckbestäuberspritzmittel *n* ~ **gauge** Inhaltsmesser *m* für Behälter, Vorratsmesser *m* ~ **indicator** Vorratsanzeiger *m* ~ **transmitter** Vorratsgeber *m*
**contest, to** ~ bestreiten
**contest** Wettbewerb *m* ~ **proceedings** Streitverfahren *n* ~ **rules** Wettbewerbsbestimmung *f*
**contestable** anfechtbar, bestreitbar
**contestant** Wettbewerber *m* ~ **to a suit** Streiter *m*
**contested, ~ action** Anfechtungsprozeß *m* ~ **case** Streitfrage *f*, Streitpunkt *m*
**context** Zusammenhang *m*
**contiguity** Berührung *f*, Nähe *f*
**contiguous** angrenzend, anstoßend, benachbart ~ **to be** ~ naheliegen
**continent** Erdteil *m*, Festland *n*
**continental** binnenländisch, kontinental ~ **drift** Kontinentalverschiebung *f* ~ **horsepower** metrische Pferdestärke *f* ~ **icecap** Binneneis *n* ~ **letter** Sonderbuchstabe *m* ~ **margins** Kontinentalränder *pl* ~ **shelf** Kontinentalschelf *n*
**contingencies** Möglichkeiten *pl*, unvorhergesehene Ausgaben *pl*
**contingency spares** Vorratsteile *pl* für besonderen Bedarf
**contingent** Kontingent *n*; eventuell ~ **complex** komplex, konjugiert ~ **liabilities** Bürgschafts-

und Haftungsverhältnisse *pl* ~ **reversion** bedingter Übergang *m*
**contingently** eventuell
**continual** andauernd, ständig
**continuance** Bestand *m*, Dauer *f* ~ **of a patent** Patentdauer *f*
**continuation** Auslauf *m*, Fortsetzung *f* ~ **principle** Fortsetzungsprinzip *n*
**continue, to** ~ anhalten, dauern, fortbestehen, fortführen, fortsetzen, währen
**continue process** Kontinuverfahren *n*
**continued,** ~ **emergency power** erhöhte Dauerleistung *f* für Notfälle ~ **fraction** Kettenbruch *m* ~ **fraction representation** Kettenbruchdarstellung *f* ~ **interaction** fortgesetzte Beeinflussung *f*
**continuing computation** fortlaufende Rechnung *f*
**continuity** Beständigkeit *f*, Folge *f*, Fortdauer *f*, Koinzidenz *f*, Raumzusammenhang *m*, Stete *f*, Stetigkeit *f* ~ **of a curve** stetiger Verlauf *m* einer Linie ~ **of operation** Dauerbetrieb *m*
**continuity,** ~ **check** Durchgangsprüfung *f* ~ **concept** Stetigkeitsbegriff *m* ~ **control** Einsatzkontrolle *f* ~ **equation** Kontinuitätsgleichung *f*, Raum-bedingung *f*, -gleichung *f* ~ **preserving contact** Folge-, Schlepp-kontakt *m* ~ **suite** Abwicklungsgruppe *f* ~ **test** Prüfung *f* auf Stromfähigkeit, Stromdurchgangsprüfung *f*, Stromfähigkeitsprüfung *f* ~ **tester** Durchgangsprüfgerät *n*, Stromschlußprüfer *m*, Windungsschlußprüfer *m*
**continuous** anhaltend, berührungslos, beständig, dauernd, durchgehend, durchlaufend, fortdauernd, fortlaufend, fortwährend, kontinuierlich, laufend, nachfolgend, pausenlos, stet, stetig, (waves) tönend, stufenlos, ununterbrochen, zusammenhängend ~ **absorptions spectra** Absorptionskontinua *pl* ~ **air venting** Dauerentlüftung *f* ~ **alkaline earth spectra** Erdalkalikontinua *pl* ~ **assembling line** Taktverfahren *n* ~ **attention method** ständige Bereitschaft *f* ~ **beam** Durchlaufbalken *m*, durchlaufender Balken *m* oder Träger *m* ~ **beam of light** Dauerlicht *n* ~ **belt conveyer** laufendes Band *n* ~ **bucket elevator** Becherwerk *n* mit Rückführung ~ **cadmium spectra** Cadmiumkontinua *pl* ~ **casting** Stranggießen *n* ~ **center sill** durchgehender Zentralträger *m* ~ **charge** Dauerbelastung *f* ~ **charging (or loading)** Dauerladung *f* ~ **checking (or check)** laufende Kontrolle *f* ~ **chip** Fließ-, Längs-span *m* ~ **combustion** Dauerbrand *m* ~ **controller** stetiger Regler *m* ~ **conveyer** Stetigförderer *m* ~ **copying** Durchlaufkopieren *n* ~ **counter** Sammelzähler *m*
**continuous-current** Gleichstrom *m*, Umformer *m* ~ **alternating current converter** Gleichstromwechselstromeinankerumformer *m* ~ **circuit** Gleichstromleitung *f* ~ **field** Gleichstromfeld *n* ~ **line** Gleichstromleitung *f* ~ **rotary-current converter** Gleichstromdrehstromankerumformer *m* ~ **supply** Gleichstromspeisung *f* ~ **wire** Gleichstromleitung *f*
**continuous,** ~ **curve** kontinuierliche Kurve ~ **dash signal** Dauerstrich *m* ~ **dimming** Dauerabblendung *f* ~ **distillation** stetige Destillation *f* ~ **drawbar** durchgehende Zugstange *f* (r.r.)

~ **duty** ununterbrochener Dienst *m* ~ **duty temperature rise** Dauererwärmung *f* ~ **dyeing** Weiterfärben *n* ~ **electrode** Dauerelektrode *f* ~ **electromotive force** Gleichspannung *f* ~ **electronic spectra** Elektronenkontinua *pl* ~ **emission spectrum** Emissionskontinuum *n* ~ **excitation of waves (or point where oscillating begins)** Anfachung *f* ~ **existence** Fortbestand *m* ~ **feed** ununterbrochener gleichmäßiger Vorschub *m* ~ **feed of film** stetiger Bildwechsel *m* ~ **feed facing head** Plandrehkopf *m* mit selbsttätigem Vorschub ~ **film** Endlosfilm *m* ~ **film printer** Durchlaufkopiermaschine *f* ~ **filtration** Durchflußfiltration *f* ~ **finishing machine** Duffmaschine *f* ~ **finishing mill** kontinuierliche Fertigstraße *f* ~ **fire** Dauer-, Lagenfeuer *n* ~ **flow** Dauerstrom *m*, Störungskontinuum *n* ~ **flow calorimeter** Junkerskalorimeter *n* ~ **flow heater** Durchlauferhitzer *m* ~ **flow system** Dauerflußanlage *f* ~ **forms (endless office forms)** Endlosformulare *pl*, Endlosformat *m* ~ **furnace** Kanalofen *m* ~ **gauging optimeter** Optimeter *m* für laufende Messung von Band- und Drahtmaterial ~ **heating furnace** Durchlauf-, Roll-ofen *m*, kontinuierlicher Wärmeofen *m* ~ **hunting** ununterbrochener Wählerlauf *m* ~ **hydrogen** Wasserstoffkontinua *pl* ~ **immersion test** Dauertauchversuch *m* ~ **interference** Dauerstörung *f* ~ **inventory** laufende Inventur *f* ~ **job card** Sammelstückkarte *f* ~ **kiln** Ringofen *m* ~ **kneading machine** kontinuierlicher Kneter *m* ~ **lehr** Kanalkühlofen *m* ~ **line-recording instrument** Linienschreiber *m* ~ **load (or stress)** Dauer-betrieb *m*, -beanspruchung *f* ~ **load method** Krarupverfahren *n* ~ **loadability** Dauerbelastbarkeit *f* ~ **loading** Krarupumspinnung *f*, punktförmig verteilte Induktivität *f*, stetige Belastung *f*, verteilte Induktion *f* ~ **loading of cable** Krarupisierung *f* ~ **loopingrod mill train** kontinuierliche Staffelstraße *f* ~ **lubrication bearing** Dauerschmierlager *n* ~ **machine** Dauerschlagwerk *n* ~ **mapping** stetige oder topologische Abbildung *f* ~ **mercury spectrum** Quecksilberkontinuum *n* ~ **mill train** kontinuierliche Straße *f* ~ **mixer (continuous kneader mixer)** Fließmischer *m* ~ **motion of film** stetiger Bildwechsel *m* ~ **note** Dauerton *m* ~ **open width washer** Kontinübreitwaschmaschine *f* ~ **operation** Dauerbetrieb *m*, Fließ-arbeit *f*, -betrieb *m* ~ **oscillation** ungedämpfte Schwingung *f* ~ **output** Dauerleistung *f* ~ **oxygen spectra** Sauerstoffkontinua *pl* ~ **parapet** durchlaufender Erdaufwurf *m* in ~ **phase** fein verteilt ~ **photocathode** zusammenhängende Fotokathode *f* ~ **pipe stove** Röhrenwinderhitzungsapparat *m* ~ **pointing traversing wheel** Folgezeigerantrieb *m* ~ **power** Dauerleistung *f* ~ **process** kontinuierliches Verfahren *n*, Kreisprozeß *m* ~ **processing machine** Durchlauf-Entwicklungsmaschine *f* (film) ~ **production** Fließarbeit *f* ~ **radiation** Bremsstrahlung *f* ~ **radio alert** Funkdauerbereitschaft *f* ~ **rare gas spectra** Edelgaskontinua *f* ~ **rating** Dauerbetrieb *m* ~ **recombination spectra** Rekombinationskontinua *pl* ~ **reheating furnace** kontinuierlicher Wärmeofen *m* ~ **ring** Meßbügel *m* ~ **roasting furnace**

Schuttröstofen *m* ~ **rod mill train** kontinuierliche Drahtstraße *f* ~ **rolling train** Doppelduostraße *f*, Doppelzweiwalzenstraße *f* ~ **rope** durchlaufendes Seil *n* ~ **rotary contact printer** Durchlaufkontaktdruckmaschine *f* (film) ~ **run of a curve** stetiger Verlauf *m* einer Linie ~ **run test stand for dynamos** Dauerlaufprüfbank *f* für Lichtmaschinen ~ **running condenser (or capacitor)** Dauerlaufkondensator *m* ~ **running winding** Dauerlaufwickel *m* ~ **sampling** ununterbrochene Probeentnahme *f* ~ **screw feature** endlose Feinbewegung *f* ~ **sequence** fortlaufende Reihe *f* ~ **series** fortlaufende Reihe *f* ~ **series of lines** Linienzug *m* ~ **service** Dauerbetrieb *m*, Dauerbetrieb *m* mit gleichbleibender Belastung ~ **shaft** durchgehende Welle *f* ~ **shifting device** Fortschalteeinrichtung *f* ~ **signal** Dauerton *m* ~ **signal transmitter** Dauerstrichsender *m* ~ **slab** durchlaufende Platte *f* ~ **slow motion** endlose Feinbewegung (optics) ~ **sound** Dauerton *m* ~ **spar** durchgehender Holm *m* (aviat.) ~ **spectra** Fotodissoziationskontinua *pl* ~ **spectrum** Kontinuum *n*, Streckenspektrum *n* ~ **speed cone** Kegel(trommel)trieb *m* ~ **steamer** Kontinüdämpfer *m* ~ **strip annealing** Banddurchziehglühen *n* ~ **strip camera** Bildstreifenkamera *f* ~ **strip furnace** Banddurchziehofen *m* ~ **stroke** Dauerschlag *m* ~ **switching device** Fortschalteeinrichtung *f* ~ **switching operation** Fortschaltung *f* ~ **tension** Gleichspannung *f* ~ **test** Dauer-abbremsung *f*, -prüfung *f*, -schlagprobe *f* ~ **three-phase rotary converter** Gleichstrom-Drehstrom-Einankerumformer *m* ~ **tone** Dauerton *m* ~ **tone signaling** Dauerstromwahl *f* ~ **torque** Dauer(dreh)moment *n* ~ **twist** Drehkreuz *n* (electron.) ~ **twisting** einfacher Drall *m* ~ **type annealing furnace** Glühtunnelofen *m* ~ **type furnace** Tunnelofen *m* ~ **value control** Festwertregelung *f* ~ **variation** gleitende Variation *f* ~ **voltage** Gleichspannung *f* ~ **wave (CW)** kontinuierliche Welle *f*, ungedämpfte Schwingungen *pl*, ungedämpfte Welle *f*

**continuous-wave,** ~ **circuit** Tastfunkverbindung *f* ~ **jammer** Dauerstrichstörsender *m* ~ **radar** Dauerstrichradar *n* ~ **signal** Dauerstrich *m*, ungedämpftes Zeichen *n* ~ **telegraphy** Funktelegrafie *f*, ungedämpfte Wellenfunktelegrafie *f* ~**transmitter** ungedämpfter Sender *m*, Zwischenkreissender *m*

**continous,** ~ **waves** Gleichwellen *pl*, ungedämpfte oder unmodulierte Wellen *pl* ~ **weld** durchlaufende Naht *f*, durchgehende Schweißung *f* ~ **welding** Dauerschweißbetrieb *m*, Nahtschweißung *f*, Rollennahtschweißung *f* ~ **wing** durchlaufender oder freitragender Flügel *m* ~ **wire entanglement** durchlaufendes Drahthindernis *n* ~ **work** Kontinübetrieb *m* (Färberei) ~ **working** Dauerbetrieb *m*, laufender Arbeitsgang *m* ~ **X-rays** kontinuierliche Röntgenstrahlen *pl*

**continuously,** ~ **adjustable** dauernd einstellbar ~ **adjustable inductor** regelbare Induktivität *f* ~ **connected battery connected to a discharge circuit** Batterie *f* in Dauerladeschaltung ~ **distributed** gleichmäßig verteilt ~ **distributed capacity** stetig verteilte Kapazität *f* ~ **distributed inductance** gleichmäßig oder stetig verteilte

Induktivität *f* ~ **distributed load** gleichförmige Belastung *f* ~ **distributed loading** gleichförmige Ladung *f* ~ **indicating** anzeigend, fortlaufend ~ **loaded** stetig belastet, ununterbrochen belastet, (cable) krarupisiert ~ **loaded cable** Krarupkabel *n*, stetig belastetes Kabel *n* ~ **loaded circuit** Krarupleitung *f*, stetig belastete Leitung *f* ~ **loaded conductor** Krarup-ader *f*, -leiter *m* ~ **loaded line** gleichmäßig belastete Leitung *f* ~ **ringing alarm** Fortschellwecker *m* ~ **running** durchlaufender Betrieb *m* ~ **sounding bell** Rasselglocke *f* ~ **variable** stufenlos regulierbar ~ **variable drive** stufenlos regelbares Getriebe *n* ~ **variable drive unit** kontinuierlicher Steuervorsatz *m* ~ **variable wave length** kontinuierlich bestreichbarer Wellenbereich *m*

**continuum theory** Kontinuitätstheorie *f*

**contort, to** ~ krümmen, verzerren

**contortion** Krümmung *f*, Verdrehung *f*, Verzerrung *f*

**contour** Außenlinie *f*, Einhüllende *f*, Fasson *f*, Form *f*, Gestalt *f*, Höhenlinie *f*, Höhenschicht *f*, Hülle *f*, Kontur *f*, Leitkurve *f* (of cam compensator), Schichtlinie *f*, Umfang *m*, Umgrenzungslinie *f*, Umkreis *m*, Umriß *m* ~ **and elevation plot** Schichtlinienplan *m* ~ **of groove** Kaliberform *f* ~ **of a groove** Rillenprofil *n* ~ **of sea bed** Tiefenkurve *f* ~**s of sea bed** Isobaten *pl* ~ **of wings** Tragflächenumriß *m*

**contour,** ~ **boring** Umrißbohren *n* ~ **gauge** Profillehre *f* ~ **integral** Umlaufintegral *n* ~ **interval** Schichtenabstand *m* ~ **level** Schichtebene *f* ~ **line** Höhenschicht *f*, Höhenschichtlinie *f*, Horizontalkurve, Profilbegrenzung *f*, Schichtenlinie *f*, Umhüllungslinie *f* ~ **lines of terrain** Isophyse *f*, Schichtlinien *pl* des Geländes ~ **map** Höhenkurvenkarte *f*, Höhenschichtenkarte *f*, Konturkarte *f*, Schichtlinienkarte *f*, Schichtlinienplan *m*, Umrißkarte *f* ~ **mapping** Konturdarstellung *f* ~ **oscillation** Querschwingung *f* (cryst.) ~ **plan** Höhenschichten-, Schichtlinien-karte *f* ~ **plane** Schichtebene *f* ~ **radius** Abrundungsradius *m* ~ **shape of pattern** Modellform *f* ~ **surface** Umfläche *f* ~ **vibration** Kristallquerschwingung *f*, Querschwingung *f*

**contouring** Schichtlinienzeichnung *f*

**contourogram** Schaubild *n* der Oberflächenbeschaffenheit

**contourograph** Gerät *n* zur Aufzeichnung der Oberflächenbeschaffenheit

**contract, to** akkordieren, beschränken, eingehen, einlaufen, einschrumpfen, einspringen, einziehen, kontrahieren, einen Vertrag schließen, schrumpfen, schwinden, soggen (metall.), verengen, verengern, (a tensor) verjüngen, verkürzen, verschrumpfen, zusammenschrumpfen, zusammenziehen **to** ~ **for** bedingen **to** ~ **in area** einschnüren

**contract** Abkommen *n*, Abmachung *f*, Abschluß *m*, Akkord *m*, Auftrag *m*, Fernsprechanschluß *m*, Kaufvertrag *m*, Kontrakt *m*, Submission *f* (stipulated), Teilnehmervertrag *m*, Übereinkommen *n*, Vertrag *m* (for public works) **by** ~ kontraktlich **one who has work done by** ~ Auftraggeber *m* ~ **of agreement** Vertragsabschluß *m* ~ **between employer and employee** Dienstvertrag *m* ~ **of employment**

Arbeitsvertrag *m* ~ **for service** Arbeitsverhältnis *n*, Dienstvertrag *m*

**contract,** ~ **agreement** Vertragsabschluß *m* ~ **carriers** Vertragsspediteur *m* ~ **delay penalty** Konventionalstrafe *f* ~ **expiration** Vertragsablauf *m* ~ **penalty** Vertragsstrafe *f* ~ **provisions** Vertragsbedingungen *pl* ~ **termination** Vertragsauflösung *f* ~ **value** Pflichtwert *m* ~ **work** Akkordarbeit *f*, Gedingearbeit *f*

**contractant** Vertragspartei *f*

**contracted,** ~ **discharge** kontrahierte Entladung *f* ~ **Morse figures (or signals)** abgekürzte Morsezahlen *pl* (Morsezeichen) ~ **word coinage** Kurzwort *n*

**contracter** Kompressor *m*, Kompressorverstärker *m*

**contractible** zusammenziehbar

**contractile force** Kontraktionskraft *f*

**contracting,** ~ **combine** Arbeitsgemeinschaft *f* ~ **firm** Bauunternehmung *f* ~ **nozzle** Leitvorrichtung *f* mit abnehmendem Querschnitt ~ **party** Kontrahent *m* ~ **state** Vertragsstaat *m*

**contraction** Einsprung *m* (Gewebe), Kontraktion *f*, Schrumpfung *f*, Schwinden *n*, Schwindmaßverkürzung *f*, Schwindung *f*, Verengung *f*, Verjüngung *f*, Verkürzung *f*, Zusammenziehung *f* ~ **of area** Einschnürung *f*, Querzusammenziehung *f* ~ **of cooling rivet** Eingehen *n* der Niete bei Abkühlung ~ **of cross-sectional area** Querschnittszusammenziehung *f* ~ **in length** Längszusammenziehung *f* ~ **of a seam** Verdrückung *f* eines Flözes ~ **of volume** Volumkontraktion *f*

**contraction,** ~ **allowance** Schwindungszugabe *f* ~ **cavity** Lunker *m* ~ **coefficient** Ausflußeinschnitt *m* ~ **crack** Schrumpfriß *m* ~ **joint** Schwindfuge *f* ~ **measure** Schrumpfmaß *m* ~ **joint** Schwindfuge *f* ~ **measure** Schrumpfmaß *n* ~ **rule** Schwindmaß *n*, Schwindmaßstab *m* ~ **shrinkage** Schrumpfung *f* ~ **strain** Schrumpfspannung *f*, Schrumpfungsriß *m*, Schwindspannung *f* ~ **tension** Schrumpfspannung *f*

**contractor** Auftragnehmer *m*, Ausführer *m*, Hauptgedingenehmer *m*, Kontrahent *m*, Lieferant *m*, Lieferer *m*, Submittent *m*, Unternehmer *m*, Vertragsschließender *m* ~ **of wells** Bohrunternehmer *m*

**contracts placed** Kontraktabschlüsse *pl*

**contractual** kontraktlich, vertraglich ~ **obligation** Vertragsverbindlichkeit *f* ~ **repair shop** Vertragswerkstatt *f*

**contractually bound (or obligated)** vertraglich verpflichtet

**contradiction** Einrede *f*, Verneinung *f*, Widerspruch *m*

**contradictory** unvereinbar, widersprechend

**contradistinction** Gegensatz *m*

**contraflow turbine** Gegenflußturbine *f*

**contraption** Aggregat *n*, Apparat *m*

**contraries** störende Begleitstoffe *pl* (paper mfg.)

**contrarotating** gegenläufig ~ **propeller** Gegenlaufschraube *f*

**contrary** Gegenteil *n*; entgegengesetzt, widersprechend ~ **to agreement** vertragswidrig in ~ **direction** in entgegesetztem Sinne, widersinnig ~ **to law** rechtswidrig, widerrechtlich ~ **to order** ordnungswidrig ~ **to regulations** unvorschriftsmäßig ~ **to rule** regelwidrig in ~

**sense** widersinnig **of** ~ **sense** gegenläufig ~ **to** im Gegensatz zu

**contrast, to** ~ von etwas abstechen, einen Gegensatz bilden, entgegensetzen, entgegenstellen, gegenüberstellen, gegenüber etwas hervortreten lassen

**contrast** Abstechung *f*, Abstich *m*, Gegensatz *m*, Gradation *f*, Kontrast *m*, Steilheit *f* (film), Vergleich *m* **in** ~ **with** im Gegensatz zu ~ **of colors** Gegenfarbe *f* ~ **of film** Filmsteilheit *f*

**contrast,** ~ **amplifier** Kontrastverstärker *m* ~ **border** Kontrastrand *m* ~ **characteristic** Dynamiklinie *f* ~ **control** Bildverstärkungsregelung *f*, Kontrastregler *m* ~ **equalizer (gamma correction)** Gradationausgleich *m* ~ **expansion** Dynamiksteigerung *f* ~ **filter** Kontrastfilter *n* ~ **photometer with cubical cavity** Kontrastfotometer *n* mit Würfel ~ **picture** kontrastreiches Bild *n* ~ **range** Helligkeitssprung *m*, Kontrastumfang *m* ~ **ratio** Kontrastverhältnis *n* (TV) ~ **ratio between lowest and highest intensities of notes (or passages)** Dynamik *f* ~ **reduction** Kontrastherabsetzung *f* ~ **regulator** Dynamikregler *m* ~ **rendition** Kontrastwiedergabe *f* ~ **screen** Kontrastfilter *n* ~ **sensibility** Kontrast-empfindlichkeit *f*, -empfindung *f* ~ **sensivity** Augen-, Kontrast-empfindlichkeit *f*, Kontrastempfindung *f* ~ **staining** Kontrastfärbung *f* ~ **threshold** Kontrastschwelle *f*

**contrasted** abgehoben, entgegengesetzt

**contrasting** Vergleichung *f* ~ **sample** Gegenprobe *f*

**contrasty,** ~ **image** hartes Bild *n* ~ **picture** hartes Bild *n*, detailreiches Bild *n*

**contravariant** kontravariant

**contravene, to** ~ verletzen, zuwiderhandeln

**contravention** Übertretung *f*, Zuwiderhandlung *f*

**contribute, to** ~ beitragen

**contributing (drainage) area** Beitragsfläche *f*

**contribution** Kontribution *f*, Zuschuß *m* ~**s by mines** Bergwerksabgaben *pl*

**contributor** Mitarbeiter *m*

**contributory, to be a** ~ **determinant** mitbestimmen ~ **infringement of patent** mittelbare Patentverletzung *f*

**contrivance** Apparat *m*, Apparatur *f*, Erfindung *f*, Plan *m*, Stellage *f*, Vorrichtung *f* ~ **for extraction** Auszugmittel *n*

**contrive, to** ~ dichten, entwerfen, erfinden, planen

**control, to** ~ ansteuern, aussteuern, beaufsichtigen, beeinflussen, besprechen, bestätigen, bremsen, Gewalt haben über, kontrollieren, mitlesen, modulieren, nachmessen, nachprüfen, nachsehen, prüfen, regeln, regulieren, schalten, steuern, überwachen **to** ~ **the field** das Feld behaupten

**control** Antrieb *m*, Arretierfeder *f*, Aufsicht *f*, Aussteuerung *f*, Beaufsichtigung *f*, Bedien(ungs)organ *n*, Beeinflussung *f*, Betätigung *f*, Bewirtschaftung *f*, Bremse *f*, Festpunktnetz *n*, Getriebe *n*, Kontrolle *f*, Leitung *f*, Leitwerk *n*, Modulation *f*, Nachprüfung *f*, Potentiometer *n*, Prüfung *f*, Regelhahn *m*, Regelung *f*, Regulierung *f*, Schaltung *f*, Steuereinrichtung *f*, Steuerorgan *n*, Tastung *f*, Überwachung *f*, Vermessung *f* (geol.)

**control,** ~ **of combustion** Regelung *f* der Ver-

brennung ~ of excitation Erregerkreistastung
f ~ of the intersection of the planes Schnitt-
linien-bedingung f, -steuerung f ~ and load
governing valve Steuer- und Lastregelventil n
~ in which magnetization does not proceed at
uniform rate Bremsfeldschaltung f ~ of
operation Betriebskontrolle f ~ by pilot beam
Leitstrahlführung f ~ of position Stellungs-
regelung f ~ of principal distance Bildweiten-
steuerung f (Entzerrungsgerät) ~ of public
opinion Menschenführung f ~ and rationing
Bewirtschaftung f ~ of the spot brightness
Steuerung f der Punkthelligkeit ~ of torrents
Wildbachverbauung f ~ of (angular) vane
position Stellungszuordnung f ~ of volume
range Dynamikregelung f
control, ~ accuracy Regelgenauigkeit f ~ action
Regel-verhalten n, -verlauf m, -vorgang m ~
action transmitter Kommandostelle f ~ (remote
control) ~ actuator Antriebswelle f ~ amplifier
Mischgerät n, Steuerverstärker m ~ and
instrument panel Instrumentenbrett n ~ aneroid
barometrische Reglerdose f (for engine), Reglerdose f ~ appliance Steuervorrichtung f ~
arrangement Steueranlage f ~ bar Steuerlineal
n ~ battery Steuerbatterie f ~ battery subscriber Zentralbatterieteilnehmer m ~ beam
Steuerwippe f ~ bell crank Schaltwippe f (Türbetätigung) ~ board Überwachungstafel f ~
board for turbine plant Turbinenüberwachungs-
tafel f ~ box Bediengerät n, Schaltkasten m ~
brake Regelbremse f ~ break Gruppenunter-
brechung f ~ breakout force Initialkraft f
(der Steuerorgane) ~ brush Kontrollkohlenstift m ~ bushing Regulierbüchse f ~ button
Drehknopf ~ cab Führerhaus n ~ cabin Füh-
rer-kanzel f, -raum m, -stand m ~ cable
Steuer-kabel n, -seil n, -zug m
control-cable, ~ coupling Steuerstromkupplung
f, Steuerzugkupplung f ~ guide Seilführungs-
büchse f, Steuerzugführung f ~ guiding bush
Seilführungsbuchse f
control, ~ cam Steuernocken m ~ capstan
Steuerwinde f ~ capsule chamber Reglerdosen-
kammer f ~ car Führergondel f ~ carbon
Kontrollkohlenstift m ~ center Warnzentrale
f ~ chain Steuerkette f ~ chair Steuersessel m
~ chamber Steuerkammer f ~ change Grup-
penwechsel m ~ channel Steuerungsweg m
(administr.) ~ character Steuerzeichen n ~
characteristic Steuerungseigenschaft f ~ chart
Steuerungsdiagramm n ~ circuit Anreizkreis
m, Regelschaltung f, Steuerkreis m, Steuer-
stromkreis m, Überwachungsstromkreis m ~
circuitry Steuerschaltung f ~ clamping device
Feststellvorrichtung f ~ clearance (or play)
Steuerspiel n ~ clock Steckuhr f ~ cock Kontrollhahn m ~ coil Richtspule f ~ column
Knüppel m (aviat.), ausbaubarer Steuerungs-
knüppel m, Steuerknüppel m, Steuersäule f
~ column central Knüppel m in Nullage ~
connection Bremsschaltung f ~ console Be-
dienungskonsole f, Regulierpult n, Steuer-
konsole f ~ constant Rückstellungskonstante f
~ contact Regelkontakt m ~ contour Kon-
trollinie f ~ counter Befehlszähler m (data
proc.) ~ crank Steuerkurbel f ~ cubicle Ab-
hörkabine f ~ cup Kontrollnapf m (Schmier-

pumpe) ~ current (teleph.) Kuppelstrom m,
Überwachungsstrom m ~ cylinder Steuer-
zylinder m ~ dam Stauwehr n ~ damper fuse
Schmelzsicherung f (für Dämpfungsregler)
~ deflection Steuerausschlag m ~ desk Fern-
lenk-, Regler-, Schalt-pult n ~ determination
Kontrollbestimmung f ~ device Bedienungs-
einrichtung f, Kontrollapparat m, Regeleinrichtung f, Regelung f (Regeleinrichtung),
Regelvorrichtung f ~ dial Einstellskala f ~
differential Steuerdifferentialdrehmelder m ~
distance Schaltweg m ~ effect Steuerwirkung
f ~ effectiveness Regelgüte f ~ electrode
Lichtsteuerungselektrode f (einer Braunschen
Röhre), Steuer-, Steuerungs-elektrode f, Wehneltzylinder m ~ element Schaltelement n ~
engineering Regelungstechnik f ~ equipment
Regelvorrichtung f, Schaltausrüstung f ~
experiment Gegenversuch m ~ faces Steuer-
kanten pl ~ field Steuerfeld n ~ filter post
(CFP) Filterposten m ~ fin Steuerflügel m
~ fittings Steuerbeschläge pl ~ force Steuer-
kraft f ~ force indicator Seitensteuerdruck-
anzeiger m ~ force recorder Steuerkraft-
schreiber m ~ frequency Leitfrequenz f ~
gauge Prüflauflibelle f ~ gear Kontrollanlage
f, Schaltung f, Stellvorrichtung f, Steuer-
apparat m, Steuerorgan n, Überwachungsvor-
richtung f, Verstellgetriebe n ~ generator
Steuergenerator m ~ gondola Führergondel
f ~ grid Raumladungs-, Steuer-, Steuerungs-
-gitter n ~ grid potential Raumladegitter-
spannung f ~ grip Drehgriff m ~ handle
Lenker m ~ head Bedienungsknopf m ~ head
locking nut Steuerungsgegenmutter f ~ heavy
steuerlastig ~ house Zentrale f für die Bedienung der Maschinen ~ impulse Verstel-
lungsimpulslänge f ~ installation Regelwerk
n ~ instrument Mitleseapparat m, Über-
wachungsgerät n ~ keyer Zeichengeber m
~ knob Bedienungsgriff m, Bedienungsknopf
m, Betätigungsknopf m ~ laboratory Kontroll-
laboratorium n ~ lamp Kontrolleuchte f ~
lamp for high beam Fernlichtanzeigerleuchte
f ~ level Kontrollibelle f ~ lever Bedienungshebel m, Bedienhebel m, Führungshebel m, Lenker m, Schalthebel m, Steuerhebel m, Steuerungshebel m, (jet) Be-
diengestängehebel m ~ lever die Schalthebel-
preßform f ~ lever knob Bedienhebelknopf m,
Kugelknopf m ~ lever system Bediengestänge
n ~ lifting cam Steuerhubrad n (Ölpumpe) ~
limitation Steuerbegrenzung f ~ line Reglerlinie f, Steuerzug m ~ line from mains Netz-
steuerleitung f ~ line pressure Steuerleitungs-
druck m ~ linkage Bedienungs-, Steuer-
-gestänge n ~ lock solenoid Entriegelungsmagnet m ~ loop Regelkreis m ~ magnet
Richtmagnet n ~ measures Steuerabläufe pl
~ mechanism Betätigungsvorrichtung f, Schaltwerk n, Steuerleitung f, Steuerung f, Steuerungsgetriebe n ~ member Steuerorgan n ~ method
of fuel oil pump Regulierschema n der Brennstoffpumpe ~ motion components Regelbau-
gruppen pl ~ move (or manipulation) Bedienungshandgriff m ~ nozzle Kontrollhahn m
~ number Kontrollnummer f ~ oscillation
Regelschwingung f ~ overbalance Überaus-

gleich *m* der Steuerung ~ **overlay** Bildschirmvorsatz *m* (rdr) ~ **panel** Bedientafel *f*, Bedienungs-feld *n*, -pult *n*, Schaltbrett *n* ~ **panel lighting** Schalttafelbeleuchtung *f* ~ **pass** Regelzug *m* ~ **pawl** Steuerklinke *f* ~ **pedal** Regelfußhebel *m*, Seitensteuerpedal *m* ~ **period** Steuerzeit *f* ~ **piston** Steuerkolben *m* ~ **plunger** Hemmstange *f*, Vorlaufdorn *m* ~ **point** Aufgabesollwert *m*, Istwert *m* der Regelgröße, Meßstelle *f*, Orientierungspunkt *m*, Reglerpunkt *m*, Tonsteuerstelle *f* ~ **point net** Festpunktnetz *n* ~ **portion** Regelteil *m* ~ **position indicator** Steuerlagenanzeiger *m* ~ **position recorder** Ruderausschlagschreiber *m* ~ **potential** Regelspannung *f* ~ **precision** Regelungsgenauigkeit *f* ~ **pressure** Steuerdruck *m* ~ **printer** Kontrolldrucker *m*, Kontrollempfänger *m*, Mitlesedrucker *m* ~ **program** Steuerprogramm *n* (data proc.) ~ **pulley (or roller)** Umlenkscheibe *f* ~ **pulpit** Meßhaus *n*; Regelimpuls *m* ~ **pulse** Steuerimpuls *m* ~ **quantity** Regelgröße *f* ~ **range** Ruderausschlagbereich *f* ~ **ratio** Steuerfaktor *m* ~ **reaction time** Steuerzeit *f* ~ **receiver** Steuerempfänger *m* ~ **recorder** Mitleserekorder *m* ~ **register** Befehls-, Kontroll-register *n* ~ **regulation** Hebereinstellung *f* ~ **relay** Kontroll-, Steuer--relais *n* ~ **repeater** Steuerempfänger *m* ~ **resistance characteristics** Regelwiderstandskennlinie *f* ~ **resolution** Ansprechschwelle *f* ~ **response** Ansprechen *n* auf Steuerausschlag ~ **rheostat** Regulierwiderstand *m* ~ **ring** Mitlesering *m* ~ **rod** Regelstab *m* ~ **rod ball joint** Kugel *f* für Regelstange ~ **rod calibration** Eichung *f* der Kontrollstäbe ~ **rods** Steuergestänge *n* ~ **room** Befehls-, Kommando-, Kontroll-raum *m*, Meßwarte *f*, Regelraum *m*, Regierraum *m*, Schaltwarte *f*, Zentrale *f* ~ **roster** Überwachungsliste *f* ~ **sample** Kontrollprobe *f* ~ **section** Kontrollamt *n* (teleph.) ~ **sequence** Befehlsfolge *f*, (info proc.) Steuerungsablauf *m* ~ **set** Steuermaschinensatz *m* ~ **shaft** Schalt-, Steuer-, Verstell-welle *f* (aviat.) ~ **signal** Kommando *n* ~ **sleeve** Regelhülse *f*, Schaltmuffe *f* ~ **slip** Kontrollstreifen *m* ~ **spool valve** Steuerschieber *m* ~ **spring** Gegen-, Spring-feder *f* (Steuerfeder) ~ **stage** Steuerstufe *f* ~ **stand (or post)** Steuerstand *m* ~ **stand assembly** Steuerknüppelbock *m* ~ **station** Kommandostand *m*, Kontroll-, Steuerungs--stelle *f* ~ **stick** ausbaubarer Steuerknüppel *m*, Lenk-, Steuer-knüppel *m* ~ **stick cradle** Steuerknüppelschwenkrohr *n* ~ **strip** Führungsband *n* (of printer), Kontrollbildreihe *f*
**control-surface** Leitfläche *f*, Ruder *n*, Ruderfläche *f*, Steuerfläche *f*, Steuerorgan *n*, Steuerruderlager *n* ~ **axis** Ruderachse *f* ~ **hinge**

~ **loads** Leitwerkbelastung *f* ~ **moment** Rudermoment *n* ~ **travel** Steuerflächenausschlag *m*
**control**, ~ **switch** Bedienungsschalter *m* (electr.), Bremsschalter *m*, Steuerschalter *m*, Steuerwähler *m* ~ **switch gear (or mechanism)** Schaltwerk *n* ~ **synchro** Steuer-Drehmelder *m* ~ **system** Anschaltvorrichtung *f*, Steuerwerk *n* ~ **system layout** Steuerungsschema *n* ~ **tag** Laufzettel *m* ~ **test** Prüfmessung *f* ~ **time coefficient** Stellzeitkoeffizient *m* ~ **time switch**

Steuerzeitschalter *m* ~ **tower** Aufsichtsturm *m*, Befehlsgebäude *n*, Befehlsturm *m*, Kontrollturm *m* (aviat.), Stellwerk *n* ~ **track** Kontrollspur *f* (film) ~ **transformer** Steuerempfänger *m* ~ **transformer instrument** Meßwandler *m* ~ **transmitter** Steuergeber *m*, Steuersender *m* ~ **(or regulating) travel** Regelweg *m* ~ **tube** Regelröhre *f* ~ **turn-and-bank indicator** Steuerwendezeiger *m* ~ **unit** Bediengerät *n*, Bedienungsgerät *n* (rdo), Befehlsstelle *f* (airport), Leitwerk *n* (info proc.), Regler *m*, Steuerwerk *n* (info proc.) ~ **valve** Kontrollventil *n*, Steuerröhre *f*, Steuerventil *n*, Strömungsregler *m* ~ **valve with stroke preselection** Hubvorwähler *m* ~ **valve cap** Steuerventilklappe *f* ~ **value** Einstellwert *m* ~ **variables** Regelungsgröße *f* ~ **vent for exhaust gas** Abgasregelklappe *f* ~ **voltage** Steuerspannung *f*, (potential) Steuerpotential *n* ~ **voltage amplifier** Regelspannungsverstärkerröhre *f* ~ **watch (or clock)** Prüfuhr *f* ~ **wave** Steuerwelle *f* (time) ~ **wheel** Kontroll-, Steuer-rad *n* ~ **wheel center plate (or disk)** Steuerradseilscheibe *f* ~ **winding** Regel-, Steuer-wicklung *f* ~ **wire (or cable) duct** Steuerzugkanal *m* ~ **word** Leitzelle *f*, Steuerwort *n* (info proc.)
**controllability** Kontrollierbarkeit *f*, Lenkbarkeit *f*, Regelbarkeit *f*, Steuerfähigkeit *f*
**controllable** erfaßbar, kontrollierbar, lenkbar, regelbar, regulierbar, steuerbar, verstellbar ~ **airscrew** Verstellschraube *f* ~ **check-valve** absperrbares Rückschlagventil *n* ~ **cowling** verstellbare Triebwerksverkleidung *f* ~ **flapped cowl** Haube *f* mit verstellbaren Luftabflußklappen ~ **gill** Luftführungsregelklappe *f*, Spreizklappe *f* ~ **pitch propeller** verstellbare Luftschraube *f*, Verstellluftschraube *f* ~ **ventilation** regelbare Belüftung *f*
**controlled** gelenkt, gesteuert ~ **by a crystal** Kristallgitter *n* ~ **aerodrome** Flugplatz *m* mit Verkehrskontrolle ~ **build-up** systematische Changierung *f* (textiles) ~ **carrier modulation** Hapug-Modulation *f* ~ **communication** Leitverkehr *m* ~ **condition** Regelzustand *m* ~ **economy** gelenkte Wirtschaft *f*, Zwangswirtschaft *f* ~ **flight** Flug *m* unter Verkehrskontrolle ~ **ignition** künstliche Zündung *f* ~ **item** Vorbehaltsartikel *m* ~ **map** genau bemessene Karte *f* ~ **mine** Beobachtungs-, Grund-mine *f* ~ **mosaic** Bildplan *m* ~ **plate conductance** Steilheit *f* ~ **plate transconductance** gegenseitige Leitfähigkeit *f* ~ **rocket** gesteuerte Rakete *f* ~ **sender** fremdgesteuerter Sender *m* ~ **spillway** gesteuerter Überlauf *m* ~ **spin** freiwilliges Trudeln *n* ~ **sweep oscillations** unselbständige Kippschwingungen *pl* ~ **system** gesteuertes System *n*, Regelstrecke *f* ~ **valve** gesteuerte Klappe *f* ~ **(angular) vane position** Lagezuordnung *f* ~ **variable** Regelgröße *f* ~ **weir** gesteuerter Überlauf *m* ~ **zone entry hubs** gesteuerte Impulseingänge *pl*
**controller** Bediener *m*, Fahrschalter *m*, Kontrolleur *m*, Kontrollinstrument *n*, Regelgerät *n*, Regler *m*, Regulator *m*, Revisor *m*, Schaltapparat *m*, Schalter *m*, Schichtmeister *m*, Steuerapparat *m*, Steuerschaltung *m*, Umleiter *m*, Wächteruhr *f* ~ **for mining railway** Grubenbahnfahrschalter *m*

controller, ~ action Regelverhalten n ~ chain
Reglerkette f ~ contact Fahrschalterkontakt
m ~ cylinder Schaltwalze f ~ cylinder bush
Schaltwalzenbüchse f (Lenkschloß) ~ handle
Schalthebel m (des Fahrschalters) ~ output
pressure Steuerdruck m ~ resistance Fahr-
schalterwiderstand m ~ shutter Regelbacken m
~ system Direktorsystem n
controlling, ~ action Steuervorgang m ~ apparatus Schaltapparat m, Schalter m ~ apparatus for sugar boiling Verkochungskontroll-
apparat m ~ cabin Zentrale f für die Bedienung
der Maschinen ~ circuit maßgebende oder
wichtige Leitung f ~ couple Richtmoment n
~ device Kontrolleinrichtung f, Steuerung f
~ dimensions Hauptabmessungen pl ~ distributor korrigierender Verteiler m ~ double
ground conditions in timegraded or distance-
dependent protective systems Doppelerdschluß-
erfassung f beim widerstandsabhängigen Zeit-
staffelschutz ~ equipment Regelanlagen pl ~
exchange betriebsführende Anstalt f ~ factor
bestimmende Größe f ~ force Richtkraft f ~
instrument Schaltapparat m, Schalter m ~
magnet Richtmagnet m ~ mechanism Kon-
trollmechanismus m, Schaltgetriebe n, Schaltwerk n ~ office betriebsführende Anstalt f
~ operator betriebsführende Beamtin f ~
pressure gauge Kontrollindexmanometer n ~
spring Rückführfeder f ~ strip Schablonen-
streifen m, Schaltband n ~ surface Leitwerks-
fläche f ~ surfaces Leitwerk n ~ torque Ein-
stellmoment n ~ valve Regelventil n
controls Bedienteile pl, Bedienungsgestänge n,
Bewirtschaftungsvorschriften pl, Steuerwerk
n (Flugzeug), Steuerung und Bedienungsein-
richtung f (Flugzeug) ~ pedals and levers
Bedienungsgriffe pl
controversial matter Streitsache f
controversy Diskussion f, Polemik f, Streit m
(wissenschaftlicher), Streitfall m, Streitfrage f,
Streitpunkt m
contusion Quetschung f
convalescent Heilanstalt f ~ leave Erholungs-
urlaub m
convected mitbewegt, mitgeschleppt
convection Fortpflanzung f, Konvektion f, Konvektionsstrom m, Strahlung f, Übertragung f,
Wärmeübertragung f ~ cap Verteilungskappe
f ~ current Konvektionsstrom m (electron.) ~
current modulation Strommodulation f ~ drier
Trockner m mit direkter Beheizung ~ heat Wärmemitführung f ~ heat surface feuerberührte
Heizfläche f ~ section Konvektionszone f
convective auf Fortpflanzung beruhend, konvektiv ~ flow Konvektionsströmung f
convectron Konvektron n
convene, to ~ einberufen
convenience Bequemlichkeit f, Schicklichkeit f
convenient angemessen, bequem, geeignet, gemächlich, handlich, passend, schicklich, zweckdienlich
convening authority Gerichtsherr m
convention Abkommen n, Abmachung f, Übereinkommen n, Versammlung f, Vertrag m
~ concerning the protection of cables Kabel-
schutzvertrag m ~ date (patent) Prioritäts-
datum n ~ year Prioritätsjahr n

conventional herkömmlich, üblich ~ bracing
Normalspannung f ~ (or normal) distance of
distinct vision konventionelle Sehweite f ~
milling Gegenlauffräsen n ~ sign Karten-
signatur f
converge, to ~ konvergieren, zusammenlaufen,
zuspitzen
convergence Konvergenz f ~ in mean Konvergenz f im Mittel ~ angle gauge Winkelmesser
m für konvergierende Winkel ~ distance Brennweite f ~ mode Konvergenztyp m
convergency Konvergenz f
convergent konvergent, konvergierend, ver-
engend ~ beams konvergente Bündel pl ~
lens Kollektiv n, Sammellinse f ~ series konvergente Reihe f
convergently beginning series Anfangskonver-
gente f
converging verjüngt, zusammenstrahlend ~
and diverging nozzle doppeltrichterförmiges
Rohr n ~ lens Sammel-glas n, -linse f ~ rays
konvergente Lichtstrahlen pl ~ series konvergente oder zusammenlaufende Reihen pl
(math.)
conversant with beschlagen
conversation Besprechung f, Gespräch n, Unterhaltung f, Unterhaltungssprache f, Unterredung f the ~ is distorted die Sprache ist
verzerrt ~ by subscription Abonnementsge-
spräch n ~ unit Gesprächseinheit f
converse Gegenteil n, Umkehrung f; umgekehrt
~ of the principle Umkehrung f des Prinzips
conversion Überführung f, Umänderung f, Umbau m, Umcodierung f, Umformung f, Umkehrung f, Umsatz m, Umsetzung f, Umwandlung f, Umwertung f, Unterschlagung f
(law), Verwandlung f ~ of austenite Austenit-
umwandlung (metall.) f ~ into discs (or rosettes)
Scheibenreißen n (metall.) ~ of the course
Kursverwandler m ~ of iron into steel by
carbonization Stahlkohlen n ~ of paper Pa-
pierveredelung f ~ into steel Stählung f ~ of
yield Umformung m des Ausstoßes ~ of weight
Gewichtsumrechnung f
conversion, ~ basis Umrechnungsbasis (Geld) f
~ coefficient Umwandlungskoeffizient m ~
conductance Mischsteilheit f, Überlagerungssteilheit f ~ detector erster Detektor m ~
efficiency Wirkungsgrad m ~ engine Einbaumotor m ~ factor Konversions-, Umformungs-,
Umrechnungs-, Umschwungs-, Umwandlungs-
-faktor m ~ gain Mischverstärkung f, Über-
lagerungsverstärkung f ~ gain ratio Misch-
verstärkungsgrad m ~ loss Misch-, Umsetzungs-
-verlust m ~ plant Austauschanlage f ~ process
Umwandlungsvorgang m ~ rate Umrechnungsfaktor m ~ ratio Umsetzungsverhältnis n ~
style Umwandlungsartikel m ~ table Um-
rechnungstabelle f, Umrechnungstafel f, Um-
wandlungstabelle f ~ transconductance Mischsteilheit f, Überlagerungssteilheit f ~ transducer Frequenzwandler m ~ transition Kon-
versionsübergang n ~ voltage gain Spannungsverstärkung f
convert, to ~ überführen, umbauen, umformen,
umrechnen, umsetzen, umstellen, umwandeln,
verarbeiten, verwandeln, zementieren to ~
from manual to automatic working zum Wahl-

betrieb *m* überleiten (teleph.) **to ~ into steel**
verstählen
**converted** umgeändert, umgearbeitet, umgebaut, umgerechnet **to be ~ into** übergehen,
sich umwandeln in (chem.)
**converted, ~ counter** Zählerumkehrung *f* **~**
**energy** umgesetzte Energie *f* **~ steel** Blasen-,
Zement-stahl *m*
**converter** Anrichter *m*, Ausgangstrommel *f*,
Birne *f*, Drehtransformator *m*, Frischbirne *f*,
Frischer *m*, Gleichrichter *m*, Konverter *m*,
Konverterbirne *f*, Mischstufe *f*, Richter *m*,
Umcodierer *m*, Umwandler *m*, Verarbeiter *m*,
Verdampfer *m* (flüss. Sauerstoff), Vorsatzgerät
*n*, Wandler *m* (TV), Windfrischapparat *m*,
(rotary) Umformer *m*, (valve-type grid-con-
trolled) Stromrichter *m*
**converter, ~ air box** Konverterwindkasten *m* **~**
**belly** Konverterbauch *m* **~ blast box** Konverterwindkasten *m* **~ blower set** Umformer-
lüftersatz *m* **~ body** Konverterbauch *m* **~**
**bottom** Konverterboden *m* **~ charge** Kon-
vertereinsatz *m* **~ charging platform** Kon-
verterbühne *f* **~ circuit** Umformerstromkreis
*m* **~ efficiency** Umwandlerausbeute *f* **~ house**
Konverterhalle *f* **~ iron** Flußeisen *n* **~ lining**
Konverter-auskleidung *f*, -futter n **~ mill**
Konverterwerk *n* **~ mouth** Konverter-helm *m*,
-mündung *f*, -öffnung *f* **~ nose** Konverter-hals
*m*, -schnauze *f* **~ plant** Konverteranlage *f* **~**
**plate method** Konverterplatten-Methode *f* **~**
**platform** Konverterkanzel *f* **~ practice** Konverterbetrieb *m* **~ pulpit** Konverterkanzel *f*
**~ ring trunnion** Konverterringzapfen *m* **~**
**section** Umwertetrupp *m* **~ set** Umform-
aggregat *n* **~ shaped** birnenförmig **~ shop**
Konverterwerk *n* **~ shutter** Wechselverschluß
*m* **~ steel** Konverterstahl *m* **~ switchboard**
Umformerschalttafel *f* **~ trunnion** Konverter-
zapfen *m* **~ trunnion ring** Konverterring *m* **~**
**tube** Mischröhre *f*, Übersetzungsrohr *n*, Umkehrröhre *f* **~ unit** Umformersatz *m* **~ wind**
**box** Konverterwindkasten *m*
**convertibility** Umsetzbarkeit (Banknote), Umwandelbarkeit *f*
**convertible** überführbar, umsetzbar, umstellbar, verwandelbar **~ aircraft** Verwandlungs-
flugzeug *n* **~ bond** Wandelschuldverschrei-
bung *f* **~ condenser** Satzkondensor *m* **~ lens**
Satzobjektiv *n* **~ set** Objektivsatz *m*, Satz-
objektive *pl* (Doppelprotar) **~ track-wheeled**
**vehicle** Vielradfahrzeug *n* **~ wheel-track drive**
Rädergleisketten-, Räderraupen-antrieb *m*
**converting** Bearbeitung *f* (print.), Frischen *n*,
Verblasen in flüssigen Zustand, Windfrischen *n*, Zementdarstellung *f* **~ into an asso-**
**ciation** Vergesellschaftung *f* **~ apparatus** Um-
setzergerät *n* **~ furnace** Gefäßofen *m* **~ gear**
Umwandlungsrad *n* **~ pot** Frischungs-prozeß
*m*, -verfahren *n*, Zementierkiste *f* **~ vessel**
Frischbirne *f*
**convertiplane** Verwandlungshubschrauber *m*
**convex** erhaben, gewölbt, hochrund, konvex,
runderhaben **~ bank** konvexes oder vorspringendes Ufer *n* **~ body** konvexer Körper
*m* **~ curvature** konvexe Krümmung *f* **~ cutter**
Fräser *m* für halbkreisförmige Profile **~ fillet**
**weld** überwölbte Kehlnaht *f* **~ glass** gebogenes

Glas *n* **~ lens** Brennglas *n*, Konvexlinse *f* **~**
**milling cutter** Halbkreisfräser *m* (nach außen
gewölbt) **~ seal** Linsendichtung *f* **~ slope**
gewölbte Böschung *f* **~ (or curved) surface of**
**a cone** Kegelmantel *m* **~ ~ weld** volle Schweißnaht *f*
**convexity** Ausbuchtung *f*, Bauchung *f*, Kon-
vexität *f*
**convexly turned** ballig gedreht
**convexo-convex** doppelthochrund
**convey, to ~** abtransportieren, abtreten, befördern, fördern, fortführen, fortleiten, führen,
leiten, transportieren, überbringen, überführen, übertragen, überweisen, verfahren, zuführen **to ~ through** durchführen
**conveyance** Beförderung *f*, Förderung *f*, Fortleitung *f*, Transport *m*, Überführung *f*, Übertragung *f*, Verkehrsgewerbe *n*, Verkehrsmittel
*n*, Vorschub *m*, Wanderung *f*, Zuführung *f* **~**
**of material** Gütertransport *m* **~ of possession**
Besitzeinweisung *f* **~ of slimes** Schlammför-
derung *f*
**conveyer** Conveyor *m*, Förderapparat *m*, Förderer *m*, Fördergerät *n*, Förderstand *m*,
Ladeband *n*, laufendes Band *n*, Paternoster
*m*, Transporteur *m*, Zubringer *m* **~ and**
**lifting machine** Förder- und Hubgerät *n* **~**
**apparatus** Transportgerät *n* **~ band felt** Trans-
portfilz *m* **~ belt** Förder-bahn *f*, -band *n*,
-riemen *m*, laufendes Band *n*, Tragband *n*,
Transportband *n*, Transportriemen *m* **~-belt**
**manufacture** am laufenden Band *n* fabrizieren
**~ belting** Förderbänder *pl*, Transportband *n*
**~ bucket** Förder-gefäß *n*, -korbkübel *m* **~**
**bundle carrier** Garbenträger *m* **~ chain** Transportkette *f* **~ chute** Förder-rinne *f*, -rutsche *f*
**~ drive** Transporteurantrieb *m* **~ drying ma-**
**chine** Bandtrockner *m* **~ fork** Transportgabel
*f* **~ (pipe) line** Förderleitung *f* **~-line pro-**
**duction** Fließbandfertigung *f*, Fließfertigung
*f* **~ machinery** Conveyoranlage *f* **~ mechanism**
Transportmechanismus **~ ~ plant** Transport-
anlage *f* **~ roller** Förder-rolle *f*, -walze *f*
**~ screw** Förder-, Transport-schnecke *f* **~**
**shoot** Förderrutsche *f* **~ system** Fertigungs-
straße *f*, Förderanlage *f* **~ track** Förderbahn *f*
**~ trough** Förderrinne *f* **~ type elevator** Pater-
nosteraufzug *m* **~ vessel** Fördergefäß *n* **~**
**worm** Förderschnecke *f*
**conveyers** Stetigförderer *m*
**conveying, ~ agent** Förderorgan *n* **~ bridge**
**for coal open working** Abraumförderbrücke
*f* **~ bucket** Transportbecher *m* **~ by (com-**
**pressed) air** pneumatische Förderung *f* **~**
**capacity** Förderleistung *f* **~ capsule** Förder-
büchse *f* **~ (machinery) chain** Förderkette *f*
**~ device** Fördermittel *n* **~ implement** Transportgerät *n* **~ installations** Fördereinrich-
tungen *pl* **~ machinery** Beförderungsmittel *n*,
Förderanlage *f*, Fördervorrichtung *f*, Transportanlage *f* **~ main** Förderleitung *f* **~ plant**
Förderanlage *f* **~ plant for stone powder** Ge-
steinsstaubförderanlage *f* **~ problem** Transportproblem *n* **~ purpose** Transportzweck *m*
**~ route** Fördergutstrom *m* **~ speed** Fördergeschwindigkeit *f* **~ system** Fördersystem *n*,
Förderungssystem *n*, Transportsystem *n* **~**
**track** Förderstrecke *f* **~ trough** Kastenrinne

*f* ~ **tube (or tubing)** Förder-rohr *n*, -röhre *f*
**conveyor:** see conveyer
**convincing** beweiskräftig
**convoluted** geknäuelt
**convolution** Faltung *f*, Umwindung *f*, Windung
*f* ~ **of the winding** Umgang *m* der Wicklung
(electr.) ~ **sum** Faltungssumme *f* (math.) ~
**theorem** Faltungssatz *m*
**convoy, to** ~ begleiten, geleiten
**convoy** Begleitung *f*, Begleitzug *m*, Geleit *n*,
Geleitfahrzeug *n*, Geleitzug *m*, Konvoi *m*,
Schiffsgeleit *n*, Transport *m* ~ **guard** Geleit-
schutz *m* ~ **plane** Begleitflugzeug *n* ~ **protection** Geleitzugschutz *m*
**cook, to** ~ abkochen, kochen *m*
**cooked** überentwickelt (film)
**cooker** Kocher *m*
**cooking** Kochen *n* ~ **appliances** Kochgerät *n*
~ **range** Kochherd *m* ~ **utensils** Koch-gerät *n*,
-geschirr *n* ~ **vat** Kochkessel *m*
**cool, to** ~ erkalten, erkälten, kühlen, (off) abglühen, (off or down) abkühlen **to** ~ **down**
sich abkühlen, herunterkühlen, verkühlen **to**
~ **down suddenly** abschrecken **to** ~**-hammer**
hartschlagen
**cool** frisch, kalt, kühl, matt (iron) ~**-hammered
iron** hartgeschlagenes Eisen *n* ~ **running of an
engine** Betriebstemperatur *f*, Motorbetrieb *m*
bei niedriger Betriebstemperatur
**coolant** Kühlflüssigkeit *f*, Kühlmittel *n*, Kühlöl
*n*, Kühlstoff *m*, Kühlungssystem *n* ~ **for milling** Fräskühlmittel *n*
**coolant,** ~ **cabinet** Klimaschrank *m* ~ **circuit**
Kühlkreislauf *m* ~ **circulation pump** Kühlwasserumlaufpumpe *f* ~ **drains** Kühlmittel-
abfluß *m* ~ **equipment** Kühlwasser-, Naßdreheinrichtung *f* ~ **exit temperature** Kühlstoff-
austrittstemperatur *f* ~ **gutter** Kühlmittelrinne
*f* ~ **jacket** Kühlstoffraum *m* ~ **line** Kühlmit-
telleitung *f* ~ **loop** Kühlungskreis *m* ~ **tank**
Kühlmittelbehälter *m* ~ **tank reservoir** Kühl-
stoffbehälter *m* ~ **temperature gauge** Kühlstoff-
temperaturmesser *m*
**cooled,** ~ **by forced air coils** bewegte Kühlung
*f* ~ **anode transmitting valve (CAT)** Senderöhre
*f* mit Anodenkühlung
**cooler** Kühlapparat *m*, Kühle *f*, Kühler *m*,
Kühlschiff *n*, Kühlstock *m*, Kühlwerk *n* ~
**area** Kühlerfläche *f* ~ **box** Kühlwanne *f* ~ **by
stages** Zonenkühler *m* ~ **crystallizer** Kühlmai-
sche *f* ~ **housing** Nischenkasten *m* ~ **surface**
Kühlerfläche *f*
**coolidge tube** Glühkathodenröhre *f*, Röntgenröhre *f* mit Heizkathode
**cooling** Abkühlung *f*, Kühlung *f* ~ **by air** Luftkühlung *f* ~ **by evaporation** Verdampfungs-
kühlung *f* ~ **by expansion** Entspannungsküh-
lung *f* ~ **by heat sink** Kontaktkühlung *f* (Transistor) ~ **by means of circulating water** Wasser-
umlaufkühlung *f* ~ **by outspread surface** Flä-
chenkühlung *f*
**cooling,** ~ **agent** Abkühlmittel *n*, Kälteträger
*m*, Kühlmittel *n* ~ **air** Kühlluft *f*
**cooling-air,** ~ **baffle** Kühlluftleitblech *n* ~
**blower** Kühlluftgebläse *n* ~ **control valve**
Kühlluftregelventil *n* ~ **deflector** Kühlluftleitblech *n* ~ **duct** Kühlluftabführung *f* ~ **exit**
Kühlaustrittsöffnung *f* ~ **flow** Kühlluftstrom

*m* ~ **impeller** Kühlluftgebläserad *n* ~ **jacket**
Kühlluftmantel *m* ~ **slot** Kühlluftschlitz *m*
~ **tap** Kühlluftspalt *m*
**cooling,** ~ **apparatus** Kühlapparat *m* ~ **area**
Kühlfläche *f* ~ **baffle** Kühlband *n*, Kühlluftleitblech *n*, Luftleitblech *n* ~ **basin** Kühl-
becken *n* ~ **bath** Kühlbad *n* ~ **bed** Kühlbett
*n*, Morgankühlbett *n*, Warmlager *n* ~ **block**
Kühlerblock *m* ~ **box** Kühlkasten *m* ~ **brine**
Kühlsole *f* ~ **cap** Kühlhaube *f* ~ **capacity**
Kühlleistung *f* ~ **cell** Kühlzelle *f* ~ **chamber**
Kühl-kammer *f*, -raum *m* ~ **coil** Kühlschlange
*f* ~ **column** Kühlturm *m* ~ **cone** Kühlblech *n*
~ **conveyor** Kühlband *n* ~ **cooper** Kühlschiff
*n* ~ **crack** Schrumpfriß *m* ~ **crystallizer
with stirring device** Kühlrührmaische *f* ~ **curve**
Abkühlungskurve *f* ~ **device** Kühlvorrichtung *f*
~ **disk** Kühlplatte *f* ~ **down** Abkühlung *f* ~
**down period** Kaltblaseperiode *f* ~ **down station**
Abkühlungsplatz *m* ~ **drag** aerodynamischer
Widerstandsanteil *m*, der zur Kühlung dienende
aerodynamische Widerstandsanteil *m* ~ **due
to expansion** Entspannungsabkühlung *f* ~
**effect** Kühlwirkung *f* ~ **fan** Kühl-gebläse *n*,
-ventilator *m* ~ **fan increasing gear** Lüfterüber-
setzungsgetriebe *n* ~ **fin** Kühl-fahne *f*, -lamelle
*f*, -rippe *f*, Lamelle *f* ~ **fin area of cylinder**
Rippenfläche *f* ~ **flange** Kühl-flansch *m*,
-lamelle *f*, -rippe *f* ~ **flaps** Kühlluftklappen *pl*
~ **floor** Kühle *f*, Kühlstock *m* ~ **fluid** Kühlflüssigkeit *f* ~ **fold** Kühlsicke *f* ~ **gills** Kühlluftklappen *pl* ~ **guide vane** Kühlluftleitblech *n*
~ **insert (or baffle)** Kühleinsatz *m* ~ **installation
for turning** Naßdreheinrichtung *f* ~ **jacket** Kühlermantel *m*, Kühlmantel *n* ~ **jacket cover**
Kühlwasseraumdeckel *m* ~ **jet nozzle** Kolben-
ölspritzdüse *f* ~ **jig** Kühlvorrichtung *f* ~
**liquid** Kühlflüssigkeit *f* ~ **load** Kältebedarf *m*
~ **medium** Abkühlmittel *n*, Kühlmittel *n* ~ **method** Kühlungs-art *f*, -methode *f*, -system *n*
~ **oil** Kühlöl *n* ~ **opening** Kühlöffnung *f* ~ **pan**
Kühl-pfanne *f*, -schiff *n* ~ **passage** Kühlgang *m*
(g/m) ~ **period** Abkühlungszeit *f*, Kaltschüren *n*
~ **pipe** Kühl-kanal *m*, -leitung *f* ~ **plant for
low temperature** Tiefkühlanlage *f* ~ **plate** Kühlplatte *f* ~ **plate box** Kapelle *f*, Kühlnische *f*
~ **pond** Kühlwasserteich *m* ~ **power** Kühlleistung *f* ~ **power loss** für Kühlzwecke ange-
wendeter Leistungsanteil *m*, Kühlverlustleistung
*f* ~ **rate** Abkühlungsgeschwindigkeit *f* ~ **reservoir** Kühlteich *m* ~ **rib** Kühllamelle *f*, Kühlrippe *f* ~ **ring** Kühlring *m* ~ **roller** Kühlwalze *f*
~ **section** Kühlerelement *n* ~ **sieve** Kühlsieb *n*
~ **speed** Abkühlungsgeschwindigkeit *f* (metall.)
~ **strain** Abkühlungsspannung *f* ~ **surface** Abkühlungsfläche *f*, Kühlfläche *f*, Kühlober-
fläche *f* ~ **suspension gearing** Kühlgehänge *n*
~ **system** Kühl-anlage *f*, -system *n*; Kühlungs-
-art *f*, -methode *f*, -system *n* ~ **table** Kühl-bett
*n*, -tisch *m*, Warmlager *n* ~ **tank** Kühlgefäß *n*,
-mantel *m*, -schiff *n* ~ **tower** Gradierwerk *n*,
Kühl-turm *m*, -werk *n* ~ **train** Kühlzug *m* ~
**trap** Kühlfalle *f* ~ **tray** Kühlschiff *n* ~ **tube**
Kühlerrohr *n*, Kühl-rohr *n*, -wanne *f* ~ **vane**
Kühl-flügel *m*, -rippe *f* ~ **vat** Abkühlfaß *f*
~ **vessel with absorbing surface** Kühlgefäß *n*
mit Aufsaugefläche ~**-water** Kühlwasser *n*
**cooling-water,** ~ **circulation** Kühlwasserkreis-

lauf *m*, Kühlwasserumlauf *m* ~ **connection**
Kühlwasseranschluß *m* ~ **consumption** Kühl-
wasserverbrauch *m* ~ **discharge** Kühlwasser-
ablauf *m* ~ **discharge pipe** Kühlwasserableitung
*f* ~ **drain cock** Kühlwasserablaßhahn *m* ~ **feed
pipe (or supply)** Kühlwasserzufluß *m*, -zulauf *m*,
-zuleitung *f* ~ **filling neck** Kühlwassereinfüll-
stutzen *m* ~ **flange** Kühlwasseranschlußstutzen
*m* ~ **inlet** Kühlwassereintritt *m* ~ **line** Kühlwasser-rohr *n*, -rohrleitung *f* ~ **manifold** Kühl-
wasserverteilerrohr *n* ~ **outlet** Kühlwasseraustritt *m* ~ **pipe** Kühlwasser-rohr *n*, -rohrleitung
*f* ~ **pipe line** Kühlwasserleitung *f* ~ **preheater**
Kühlwasservorwärmer *m* ~ **pump** Kühlwasserpumpe *f* ~ **safety switch** Kühlwasserkontroll-
schalter *m* ~ **sight glass** Kühlwasserschauglas *n*
~ **supply** Kühlwasserbeschaffung *f* ~ **tank**
Kühlwasserbehälter *m*, Kühlwasserkasten *m*
~ **thermostat** Kühlwasserthermostat *m*
**coop** Bottich *m*, Kaue *f*, Kufe *m*
**cooper** Böttcher *m*, Büttner *m*, Faßbinder *m*,
Küfer *m* ~'s **bent shank draw knives** Küfer-
krummesser *pl* ~'s **cleaning knife** Küferputz-
messer *n* ~'s **drawing knife** Küfermesser *n*
~'s **drawing scraper** Küferzugschaber *m* ~'s
**heading knife** Küferschroppmesser *n* ~'s **hooks**
Faßhaken *pl* ~'s **plane** Fügblock *m*, Fügebank
*f* ~'s **round scraper** Küferschaber *m*
**cooperate, to** ~ beitragen, mitwirken, zusammenarbeiten, zusammenwirken
**cooperating contact (or terminal)** Gegenkontakt
*m*
**co-operating unit** Gespannschaft *f*
**cooperation** Gemeinschaft *f*, Gemeinschaftsarbeit *f*, Kooperation *f*, Mitarbeit *f*, Mitwirkung *f*, Zusammenarbeit *f* ~ **data link** auf Zusammenarbeit beruhender Werteaustausch *m*
**cooperative** Genossenschaft *f*; gemeinsam ~
**(research) aid council** Notgemeinschaft *f* ~
**effort** Gemeinschaftsarbeit *f* ~ **society (or
store)** Konsumverein *m*
**coordinate, to** ~ ausrichten, beiordnen, einheitlich leiten, gleichschalten
**coordinate** Beiwert *m*, Koordinate *f* ~ **in space**
Raumkoordinate *f* ~ **axes** Achsenkreuz *n* ~ **(or
curve) chart** Netztafel *f* ~ **lattice** Koordinaten-
raster *m* ~ **map grid** Gitternetz *n* ~ **measuring
apparatus** Koordinatenmeßgerät *n* ~ **paper**
Koordinatenpapier *n* ~ **potentiometer** Koordi-
nationskompensator *m* ~ **reading scale** Transversalmaßstab *m* ~ **reporting network** Ko-
ordinationsmeldenetz *n* ~ **scale** Planzeiger *m* ~ **scheme** Koordinatennetz *n* ~ **square**
Planquadrat *n* ~ **system** Koordinatensystem *n* ~ **system in plotting machines** Ma-
schinenkoordinatensystem *n* ~ **table drilling
machine** Koordinatenbohrmaschine *f* ~ **table
grinding machine** Koordinatenschleifmaschine *f*
~ **tolerance** Koordinaten-Maßangaben *pl*
**coordinated** beigeordnet *f* ~ **turn** Normalkurve *f*,
scheinlotrichtige Kurve *f*
**coordinates** Koordinaten *pl* ~ **of the image** Bild-
koordinaten *pl*
**coordination** Abstimmung *f*, Beeinflussung
(power and telephone line) *f*, Gleichschaltung *f*,
Vereinheitlichung *f*, Zuordnung *f*, Zusammenspiel *n* ~ **of new methods** gegenseitiges Einarbeiten *n* ~ **lattice** Koordinations-gitter *n*

~ **number** Koordinationszahl *f* ~ **shells** Ko-
ordinationsschalen *pl*
**coordinatograph** Koordinatograf *m*
**co-oscillational tide** Mitschwingungsgezeit *f*
**co-owner** Miteigentümer *m*
**cop, to** ~ kannettieren, fangen
**cop** Auflaufhaspel *m*, Aufwinderöhre *f*, Einschußspule *f* (textiles), Eintragsspule *f* (textiles), Garnkötzer *m*, Kannette *f* (textiles) ~
**base** Kopsbasis *f* ~ **building cam disc** Fort-
schaltkurve *f* ~ **building motion** Fortschalt-
einrichtung *f*
**copal** Kopalharz *n* ~ **dull varnish** Kopalmatt-
lack *m* ~ **lacquer** Kopallack *m* ~ **resin** Kopalharz *n* ~ **spirit varnish** Spiritus-Kopallack *m*
~ **varnish** Kopal-firnis *m*, -lack *m*
**copalite** Kopalin *m*
**cope, to** ~ ausklinken, verdingen **to** ~ **with** be-
herrschen, Rechnung tragen
**cope** Oberform *f* ~ **chisel** Nuteisen *n* ~ **cophasal**
gleichphasig, kophas, in Phase mit ~ **flask**
Oberkasten *m* ~ **pattern** obere Modellhälfte *f*
~ **state** Phasengleichheit *f*
**copiable** kopierfähig
**copied** abschriftlich
**copilot** Kopilot *m*, zweiter Flugzeugführer *m*
**coping** Abdeckstein *m*, Abdeckung *f*, (of steel
beams) Ausklinkung *f*, Deckplatte *f*, Deckquader *m*, First *m*, Mauerabdeckung *f*, Mauer-
deckplatte *f*, Verschalung *f* ~ **of a roof** First-
-haube *f*, -linie *f*
**coping**, ~ **attachment** Ausklinkvorrichtung *f*
~ **level of lock** Schleusenplattform *f* ~ **machine**
Ausklinkmaschine *f* ~ **piece (or plate)** Blatt-
stück *n*, Wandrahmen *m* ~ **saw** Bogensäge *f*
(Holzsäge) ~ **stone** Abdeckplatte *f*
**coplanar** in einer Ebene mit, in gleicher Ebene
liegend ~ **line** Komplanare *f*
**copped strands** Stapelglasseide *f*
**copper, to** ~ kupfern, verkupfern **to** ~-**plate**
kupferplattieren, verkupfern
**copper** Kessel *m*, Küpe *f*, Kupfer *n*, Pfanne *f*
**(of)** ~ kupfern ~ **and bronze** Buntmetalle *pl*
~ **and iron pyrites** Kupfereisenkies *m* ~ **obtained by melting waste copper** Krätzkupfer *n*
**copper**, ~ **acatate** essigsaures Kupferoxyd *n* ~
**acetylide** Kupferazetylen *n* ~ **alloy** Kupfer-
legierung *f* ~ **ammoniate** Kupferammoniat *n*
~ **arsenate** Kupferarseniat *n* ~ **arsenite** arsenig-
sau(e)res Kupfer *n*, Kupferarsenit *n* ~ **asbestos**
Kupferasbest *m* ~ **asbestos gasket** Kupfer-
asbestdichtung *f* ~ **ashes** Kupferasche *f* ~
**assay** Kupferprobe *f* ~ **band** Kupferband *n*
~ **bar** Kupferbarre *f* ~ **bath** Kupferbad *n*
~ **bearing** kupferhaltig, kupferführend ~
**bearing alloy** Lagerlegierung *f* mit Kupfergehalt ~ **bearing steel** Stahl *m* mit Kupferbei-
mengung *f* ~ **billets** Rundkupfer *n* ~ **bit** Lötkolben *m* ~ **bit with an edge** Hammerlötkolben
*m* ~ **blast furnace** Kupfer-hochofen *m*, schachtofen *m* ~ **block** Kupferklotz *m* ~ **block**
**lightning rod (or** ~-**block protector)** Kupfer-
und Glimmerblitzableiter *m* ~ **bloom** Kupfer-
-blüte *f*, -federerz *n* ~ **borate** borsaures Kupfer
*n* ~ **bus** Sammelschiene *f* aus Kupfer ~ **cable**
Kupferstill *m* **to get** ~ **cakes** rosettieren ~ **calx**
Kupferkalk *m* ~ **(ed) carbon** Galvanokohle *f*
~ **carbonate** Bergblau *n*, kohlensau(e)res Kup-

fer *n*, Kupferkarbonat *n* ~ **chloride** salzsaures Kupferoxyd *n* ~ **citrate** Kupferzitrat *n* ~ **clad steel wire** Bimetalldraht *m* ~ **clad wire** verkupferter Draht *m* ~ **coat** Kupferüberzug *m* ~ **coating** Kupferüberzug *m* ~ **coil** Kupferschlange *f* ~ **collar** Kupfer-mantel *m*, -ring *m* ~ **collar relay** Kupferring-, Verzögerungs-relais *n* ~ **color** Kupfer-farbe *f*, -rot *n* ~ **concentrate** Kupferkonzentrat *n* ~ **conductivity standard** Kupferleitfähigkeitsnormal *n*, Leitfähigkeitsnormal *n* des Kupfers ~ **cone** Kupferkonus *m* ~ **content** Kupfergehalt *m* ~ **converting plant** Kupferbessemerei *f* ~ **cyanide** Kupferzyanid n ~ **cylinder** Kupfertrommel *f* ~ **damping** Kupferdämpfung *f* ~ **deposit** Kupferniederschlag *m* ~ **die casting alloy** Kupferspritzgußlegierung *f* ~ **dishes** Kupferkapseln *pl* ~ **efficiency** Kupferwirkungsgrad *m* ~ **embossing** Kupferprägerei *f* ~ **factor** Wicklungsfaktor *m* ~ **filings** Kupferfeilicht *n* ~ **finery** Kupferfrischofen *m* ~ **fitting** Kupferbeschlag *m* ~ **foil** Blattkupfer *n*, Kupfer-blatt *n*, -blech *n* ~ **formate** Kupferformat *n* ~ **foundry** Kupfergießerei *f* ~ **frame** Kupferbügel *m* ~ **fumes** Kupferrauch *m* ~ **fur** Pfannenstein *m* ~ **furnace** Kupfergarherd *m* ~ **gasket** Kupfer-dichtring *m*, -dichtung *f* ~ **gauze** Kupfer-drahtnetz *n*, -gaze *f*, -gewebe *n* ~ **gauze brush** Kupfergewebebürste *f* ~ **glance** Kupferglanz *m*, -glas *n* ~ **granules** Kupfergranalien *pl* ~ **green** Kupfergrün *n* ~ **hammer** Kupferhammer *m* ~ **head** Kupfermantel *m*, Verzögerungsring *m* ~ **heating coils** Kupferheizschlangen *pl* ~ **horn ore** Kupferhornerz *n* ~ **hydride** Kupferwasserstoff *m* ~ **hydroxide** Kupfer-hydrat *n*, -hydroxyd *n* ~ **ingot** Kupfer-barre *f*, -blöckchen *n* ~ **iodide** Kupferjodid *n* ~ **iron sulfate** Kupfereisenvitriol *n* ~ **jacket** Kupfermantel *m* ~ **jacketed** mit einem Kupfermantel *m* versehen ~ **jacketed relay** Kupfermantelrelais *n* ~ **jaw socket** Kupferbacke *f* ~ **jointing sleeve** Arldsche Kupfer-röhre *f*, -hülse *f*, -verbindungshülse *f* ~ **leaching** Kupferlaugung *f* ~ **lead alloy** Kupferblei *n* ~ **linoleate** leinölsaures Kupfer *n* ~ **loss** Kupferverlust *m* ~ **(Joule) losses with direct current** Gleichstromverluste *pl* ~ **magnesium alloy with aluminium and iron** Novokonstant *n* ~ **master alloys** Kupfervorlegierungen *pl* ~ **matte** Kupferstein *m* ~ **mesh** Kupfer-drahtgeflecht *n*, -gewebe *n* ~ **metal** Kupferstein *m* ~ **mine** Kupferbergwerk *n* ~ **nickel** Arseniknickel *n* ~ **nickel alloy** Kupfernickel *n* ~ **nickel (blastfurnace) matte** Kupfernickelrohstein *m* ~ **nickel (converter) matte** Kupfernickelfeinstein *m* ~ **nitrate** salpetersaures Kupferoxyd *n* ~ **number** Kupferzahl *f* ~ **ore** Kupfererz *n* ~ **oxide** Kupfer-kalk *m*, -oxyd *n* ~ **oxide cell** Kupferoxydulzelle *f*, Kupronelement *n* ~ **oxide rectifier** Kupferoxydgleichrichter *m* ~ **packing** Kupferdichtung *f* ~ **phosphate** Liberthenit *m* ~ **phosphide** Phosphorkupfer *n* ~ **pipe** Kupferrohr *n* ~ **plant** Kupferwerk *n* ~ **plate** verkupfert; Kupfer-blech *n*, -plättchen *n*, -platte *f* ~ **plate engraver** Kupferstecher *m* ~ **plate engraving** Kupferstich *m* ~ **-plate for photogravure** Kupferätzplatte *f* ~ **plate printing** Kupfer-druck *m*, -tiefdruck *m*, -verfahren *n* ~ **plate printing machine** Plattendruckmaschine

*f* ~ **plate rectifier** Plattengleichrichter *m* ~ **-plated** unterkupfert, verkupfert ~ **plating** Kupferüberzug *m*, Verkupferung *f* ~ **pole** Kupferpol *m* ~ **potassium chlorate** Kupferkaliumchlorat *n* ~ **potassium chloride** Kupferkaliumchlorid *n* ~ **precipate** Kupferniederschlag *m* ~ **punt** Scheuerprahm *m* ~ **pyrites** Chalcopyrit *m*, Kupfer-kies *m*, -sinter *m* ~ **rain** Sprüh-, Streu-kupfer *n* ~ **rectifier** Trockengleichrichter *m* ~ **red** Kupferrot *n* ~ **refinery** Kupferraffinerie *f* ~ **refining** Kupfer-garmachen *n*, -raffination *f*, -reinigung *f* ~ **refining hearth** Rosettierherd *m* ~ **refining plant** Kupferraffinerie *f* ~ **regulus** Kupferkönig *m* ~ **resinate** harzsaures Kupfer *n* ~ **rivet** Kupferniet *n* ~ **rolling mill** Kupferwalzwerk *n* ~ **rust** Kupferrost *m* ~ **salicylate** Kupfersalizylat *n* ~ **scale** Kupferasche *f*, Kupfer-glühspan *m*, -hammerschlag *m*, -schlag *m*, -sinter *m* ~ **scrap** Bruchkupfer *n*, Kupferschrott *m* ~ **scum** Kupferschaum *m* ~ **segments** Lamellenkupfer *m* ~ **selenide** Selenkupfer *n* ~ **setting** Kupferheftung *f* ~ **sheath** Kupfermantel *m* ~ **sheathing** Kupferbeschlag *m* ~ **sheet** Kupferblech *n* ~ **silicide** Siliziumkupfer *n*, ~ **slag** Kupfergarschlacke *f* ~ **sleeve** Kupferrohr *n* ~ **sleeve joint** Kupferröhrenverbindung *f* ~ **slug** Verzögerungsring *m* ~ **smeltery** Kupferhütte *f* ~ **smelting** Kupferverhüttung *f* ~ **smelting plant** Kupferhütte *f* ~ **smith** Kupferschmied *m* ~ **smoke** Kupferrauch *m* ~ **solder** Kupferlot *n* ~ **soldering lug** kupferner Kabelschuh *m* zum Anlöten ~ **solution** Kupferlauge *f*, Kupferlösung *f* ~ **sponge** Kupferschwamm *m*, Schwammkupfer *n* ~ **steel** kupferhaltiger Stahl *m* ~ **stranded wire** Kupferlitze *f* ~ **strip** Kupferstreifen *m* ~ **sulfate** Kupfervitriol *n* ~ **sulfide** Kupferglas *n*, Kupferschwefel *m*, Kupfersulfid *n* ~ **tape** Kupferband *n* ~ **thiocyanate** Kupferrhodanid *n* ~ **tinsel** Kupferlahn *m* ~ **tube** Kupfermantel *m*, Kupferrohr *n* ~ **turnings** Kupferdrehspäne *pl* ~ **value** Kupferzahl *f* ~ **wash roll** Siebreinigungswalze *f* ~ **weight** Kupfergewicht *n* ~ **wire** Kupferdraht *m*, Kupferleitungsdraht *m* ~ **zinc brazing mixture (or spelter)** Schlaglot *n* ~ **zinc cell** Kupferelement *n*, Kupferzinkelement *n*

**copperas** Kupferwasser *n*

**coppered** gekupfert, verkupfert ~ **carbon** Kupferkohle *f* ~ **relay** Kupfermantel-, Verzögerungs-relais *n* ~ **steel** kupferhaltiger Stahl *m* ~ **steel plates** gekupferte Stahlbleche *pl* ~ **wire** verkupferter Draht *m*

**coppery** kupfrig

**coppice** Gebüsch *n*

**coprecipitation** Mitfällung *f*

**copse** Gehölz *n*

**copy, to** ~ abbilden, abschreiben, abziehen, (by printing or impression) Abzüge machen, aufnehmen (messages), ausfertigen, doppeln, kopieren, mitschreiben, nachahmen, nachbilden, nachmachen, nachzeichnen, schablonieren **to** ~ **off** abzeichnen **to** ~ **in slips** in Fahnen abziehen **to** ~ **through** durchkopieren

**copy** Abbild *n*, Abbildung *f*, Abdruck *m*, Abschrift *f*, Abzug *m*, Ausfertigung *f*, Exemplar *n*, Kopie *f*, Nachbild *n*, Nachbildung *f*, Pause *f*, Vorbild *n* ~ **of contract** Vertragsabschrift *f*

~given to the author Belegexemplar n
copy, ~ board Projektionstisch m ~ cutter Manuskriptverteiler m ~ hold Erbpacht f
copyholder Modelltisch m, (for typewriters) Konzepthalter m ~ elevating handwheel Handrad n für Höhenverstellung des Modelltisches ~ horizontal adjustment Querbewegung f (Modelltisch) ~ screw bearing Schablonenbockspindellager n ~ vertical adjustment Senkrechtbewegung f (Modelltisch)
copy, ~ map Umdruckkarte f ~ milling Kopierfräsen n ~ milling machine Kopierfräsmaschine f ~ paper Durchschlagpapier n ~ planing Kopierhobeln n ~ proof Bürstenabzug m, Fahne f (print.) ~ turning Kopierdrehen n, Nachformdrehen n ~ writer Schriftleiter m
copying Abschreiben n, Abschreibung f, Vervielfältigung f ~ apparatus Lichtpaus-, Vervielfältigungs-apparat m ~ arc lamp Lichtpausbogenlampe f ~ attachment Kopiereinrichtung f ~ boards Kopierlöschkarton m ~ calipers Kopierzirkel m ~ device Kopiereinrichtung f ~ equipment Reproduktionsgerät n ~ ink Kopiertinte f ~ lathe Schablonendrehbank f ~ lathe design Drehbankbau m ~ lead (for pencils) Kopiermine m ~ machine Fassonier-, Nachform-maschine f ~ mechanism Kopiereinrichtung f ~ paper Kopierpapier n ~ papers Blaupausenpapier n ~ paste Hektografenmasse f ~ pencil Kopierstift m ~ press Kopierpresse f ~ pressure Nachformkräfte pl ~ process Kopierverfahren n (photo) ~ property Kopierfähigkeit f ~ rolls Kopierrollenpapier n ~ shaper Kopierhobelmaschine f ~ slide Kopierschieber m ~ system Nachformsystem n ~ telegraph Kopiertelegraf m ~ tool Kopierdrehstahl m ~ work Kopierarbeit f
copyist Kopist m
copyright Nachdrucksrecht n, Verlagsrecht n, Urheberrecht n ~ in broadcasting Funkurheberrecht n ~ mark Urheberschutzvermerk m ~ protection Verlagsrechtsschutz m ~ reserved alle Rechte vorbehalten
copyrighted name gesetzlich geschützter Name m
coracite Corazit m
coral Koralle f, Korallentier n ~ red Persisch(es) Rot n ~ reef Korallenriff n ~ rock Korallenfelsen m
coralline Korallin n
corbel, to ~ auskragen
corbel Kragstück n ~ of a capitel Glocke f eines Kapitells ~ under a window jamb Fensterkonsole f ~ tree Kraftbalken m
corbeled vorgekragt
corbelling Auskragung f
cord, to ~ zuschnüren
cord Bindfaden m, Faden m, Klafter n, Knoten m (in glass), Kordelung f, Leine f, Litze f, Saite f, Schnur f (flexible), Strang m, Strick m having ~s streifig ~ of hand desk telephone Handapparatschnur f ~ and plug Steckerschnur f
cord, ~ amoring Schnurschutz m ~ barrel Schnurtrommel f ~ carrier Seilbahn f ~ circuit Schnurpaar n, Schnurstromkreis m, Verbindungsschnur f (teleph.)
cord-circuit-repeater Schnurverstärker m ~ central(station) Schnurverstärkeramt n ~ fuse rack

Schnurverstärkerversicherungsstelle f ~ plant Schnurverstärkeranlage f ~ switchboard Schnurverstärkerschrank m ~ trouble Schnurverstärkerstörung f
cord, ~ connection clip Schnuranschlußklemme f ~ covering Schnurbespinnung f ~ (endless) drive Schnurlaufantrieb m ~ fastener Schnurbefestigung f, -klemme f ~ fuse detonation Leitfeuerzündmittel n ~ grip Schnuranschlußklemme f ~ junction ring Seilring m ~ lead Leitungslitze f ~ line Schnurleitung f ~ packing Schnurpackung f ~ pair Stöpselschnur f ~ pendant Schnurpendel m/n ~ plug Schnurstecker m ~ protecting means Schnurschutz m ~ protecting wire helix Schnurschutzspirale f ~ pulley Schnur-rolle f, -wirtel m ~ repairing center Schnurwerkstatt f ~ shake test Schnurprüfung f durch Schütteln ~ terminal Anschlußvorrichtungen pl der Schnur, Kabelschuh m, Schnuröse f, Schnurstecker m ~ terminal strip Abzweigklemmbrett n ~ test Schnurprüfung f ~ testing jack Schnurprüfklinke f ~ (cable)-tire Kordreifen m ~ type fuse Leitfeuerzündmittel n ~ type fuse detonation Leitfeuerzündung f ~ weight Schnurgewicht n
cordage Leinenwerk n, Metergedinge n, Takelung f ~ making machine Tauchwerkherstellungsmaschine f ~ oil Seilöl n ~ reel Taurolle f
cordeau detonierende Zündschnur f
cordierite Dichroit m
cording Schnürung f ~ quires Bindebücher pl (paper mfg.), Schutzbogen m (äußere Lagen)
cordite Kordit n
cordless schnurlos ~ private branch exchange schnurloser Umschalter m ~ switchboard schnurloser Klappenschrank m
cordlike strickförmig
cordon, to ~ off absperren
cordon Gürtel m, Kordon m (of a wire), Litze f
cords Schlieren pl (im Glas)
corduroy Cordsamt m ~ road Knüppel-damm m, -holzweg m, -weg m, Prügelweg m ~ weave Cordbindung f
cordway conveyors Seilpostanlagen pl
core, to ~ entmischen, kernbohren, seigern
core Ader f (electr.), Dorn m, Eisenpaket n, (of casting) Gußkern m, Innenschicht f (acoust.), Kegel m, kegelförmige Messerwalze f, Kern m, Mark n, (dust) Massekern m, (of ions) Rumpf m, (of transformer) Säule f, Schacht m, Seele f, (cable) Seilseele f, Wulst m, Zentralstück n
core, ~ of a cable Ader f, Leiter m, Seele f eines Kabels ~ of a coil Spulenkern m ~ of cotton Garnrolle f ~ of depression Depressionskern n ~ of a magnet Kern m ~ of projectile Geschoßkern m ~ of a rosted ore Herz n eines gerösteten Erzes ~ of spool Spulen-kern m, -achse f ~ of the vortex Wirbelzentrum n ~ of weld Schweißbutzen m
core, ~ baking Kern-trocknen n, -trocknung f ~ baking equipment Kerntrockenanlage f ~ baking oven Kerntrocken-kammer f, -schrank m ~ bar Dorn m, Kegelmesser m ~ barrel Kernrohr n ~ barrel coupling tap Kernröhrenmuffenhänger m ~ binder Kernbindemittel n ~ bit Bohrkrone f, Kernbohrer m ~ blast machine Kernblasmaschine f ~ blower Kernblasma-

schine f ~ **boring crown** Meißelkrone f ~ **box** Kernbüchse f (metall.), **Kernkasten** m ~ **box shelf** Kernkastenregal n ~ **(or cleaning) brush** Reinigungsbürste f ~ **cable** adrige Leitung f ~ **cast** Kernguß m ~ **catcher** Kernfänger m ~ **coil** Kernspule f ~ **conveyor** Kernrollband n ~ **deposit** Kernablage f ~ **diameter** Kerndurchmesser m ~ **diameter of screw** Kerndurchmesser m der Schraube oder des Schraubengewindes ~ die Kern-büchse f, -formplatte f, -patrone f ~ **disk** Kernplattenscheibe f ~ **drill** Kernkronen-, Löffel-bohrer m ~ **drill rig** Kernbohrausrüstung ~ **drilling** Kernbohrung f ~ **drilling head** Kernbohrkopf m ~ **drying** Kerntrocknung f ~ **ducts** Abzüge pl (electr.) ~ **electron group** Elektronenrumpf m ~ **extractor** Bohrkernzieher m ~ **former** Wikkelkörper m ~ **hardness** Kernhärte f (bei eingesetzten Teilen) ~ **hole** Putzloch n ~ **hole cap** Kernlochdeckel m ~ **hole cover** Kernlochverschluß m ~ **hole pilot** Kernlochzapfen m ~ **image** Pufferspeicher m ~ **induction furnace** kernloser Induktionsofen m, Induktionstiegelofen m ~ **iron** Kerneisen n ~ **isomerism** Rumpfisomerie f ~ **jarring machine** Rüttler m für Kerne ~ **jolter** Kernrüttler m ~ **knock out** Kernausdrücker m ~ **lamination** Kernblech n ~ **laying-up machine** Leiterverseilmaschine f ~ **lifter** Kernheber m ~ **lifter ring** Kernringheber m ~ **loss** Kernverlust m ~ **losses** Eisenverluste pl, Kernplattenverluste pl ~ **machine** Ausstoßformmaschine f, Kernausstoßmaschine f ~ **magnet measuring element** Kernmagnetmeßwerk n ~ **making bench** Kernmacherstich n ~ **making department** Kernmacherei f ~ **making equipment** Kernformeinrichtung f ~ **mark** Kernmarke f ~ **material** Kernwerkstoff m ~ **materials** Kernbaustoffe pl ~ **memory** Kernspeicher m (info proc.) ~ **method** Kernschichtmethode f ~ **mold** Kernform f ~ **molding** Kernformen n ~ **molding machine** Kernformmaschine f ~ **nail** Kernnagel m ~ **oil** Kernöl n ~ **oven** Kernofen m ~ **oven plant** Kernofenanlage f ~ **oven truck** Kerntrockenwagen m ~ **part** Kernstück n ~ **pin** Kernstift m ~ **plate** Kern-, Kernform-platte f ~ **plug** Kernpfropfen m ~ **print** Kern-auge f, -loch n, -marke f, (foundry) Gießstöpsel m ~ **pump** Kernabsaugpumpe f ~ **radiator** Blockkühler m ~ **receiving plate (or tray)** Kernanlageplatte f ~ **receiving sleeve** Kernrohr n (Zündspule) ~ **recovery pulse** Kernrückstellimpuls m ~ **refining** Kernrückfeinen n (metall.) ~ **rollover machine** Kernformmaschine f mit Wendeeinrichtung ~ **sample** Kernprobe f ~ **sand** Kernsand m ~ **sand mixer** Kernsandmischmaschine f ~ **setter** Kerneinleger m ~ **shell** Bohrkernhülle f ~ **shop** Kernmacherei f ~ **socket** Kernhülse f ~ **spindle** Kernspindel f ~ **store** Kernspeicher m (info proc.) ~ **strength** Kernfestigkeit f ~ **tank** Reaktorgefäß n ~ **template** Kernschablone f ~ **thread** Kern-, Seelen-faden m ~ **transformer** Kernplattentransformator m, Kernstrafo m ~ **tray** Kernblech n ~ **turning lathe** Kerndrehbank f ~ **turnover draw machine** Kernformmaschine f mit Handabhebung ~ **type induction furnace** Induktionsrinnofen m ~ **type transformer** Kerntransformator m ~ **wall** Kernmauer f ~

**wire-straightening machine** Kerndrahtrichtmaschine f ~ **zone** Kernzone f
**cored** ausgehölt ~ **carbon** Dochtkohle f ~ **casting** Kernguß m ~ **electrode** Seelenelektrode f ~ **hole** Kernloch n, mittels Kern im Gußstück erzeugter Hohlraum m, Schacht m, vorgegossenes Loch n (precast hole) ~ **interval (or distance)** Kernstrecke f ~ **solder** Kolophoniumzinn n ~ **wire** Seelendraht m ~ **work** Hohl-, Kern-guß m
**coreless** eisenlos, kernlos, ohne Kern
**coreometer** Pupillenweitmesser m
**coring** Entmischung f
**coriolis force** Corioliskraft f
**cork, to** ~ korken, pfropfen, verkorken **to** ~ **over** überpantoffeln
**cork** Absperr-glied n, -organ n, -teil m, -vorrichtung f; Flaschenverschluß m, Kork m, Pfropfen m ~ **board** Levantierholz n ~ **brick** Korkstein m ~ **chips** Korkabfälle pl ~ **chute** Korkfallrinne f, Korkzuführungskanal m ~ **disk for crown bottle caps** Kronenkorkscheiben pl ~ **gasket** Korkdichtung f ~ **hopper** Kronenkorkbehälter m ~ **molds** Korkrohrbekleidung f ~ **pad** Korkklotz m ~ **pressure spring** Korkenanpreßfeder f
**corkscrew** korkzieherförmig gewunden; Korkenzieher m, Korkzieher m ~ **antenna** Schraubenantenne f ~ **field** schraubenförmiges Feld n ~ **rule** Korkziehergesetz n
**cork, ~ slab** Korkplatte f ~ **sole** Korkeinlegesohle f ~ **sorting device** Korkensortierwerk n ~ **stirrer** Korkrührwerk m ~ **stopper** Korkstopfen m ~ **tip** Korkmundstück m ~ **tympan** Korkaufzug m ~ **wirer** Drahtverschnürer m
**corker** Korker m ~ **base** Korkerfuß m
**corking apparatus** Verkorkvorrichtung f
**corky** korkig
**corn, to** ~ körnen
**corn** Getreide n, Mais m ~ **and grain milling** Flockenwalze f, Getreidemahlen n ~ **and pea hopper** Mais- und Erbsensamenbehälter m ~ **binder** Maisbinder m ~ **cleaner** Kornschwinge f ~ **cultivator** Maishackmaschine f ~ **drill** Maisdrillmaschine f ~ **husker** Maisenthülser m ~ **husker and shredder** Maisenthülser und Zerschneider m ~ **husker and silo filler** Maisenthülser und Elevator m ~ **lister** Dammkulturpflug und Maisdrill m, Reihensäe- und Zudeckmaschine f ~ **machine** Maismaschine f ~ **milling machine** Getreidespitzmaschine f ~ **planter** Maispflanzmaschine f ~ **sheller** Mais-, Maiskolben-schäler m
**corneal microscope** Hornhautmikroskop n f
**corned leather** genarbtes Leder n
**cornelian-wood handle** Hartriegelwerkzeugheft n
**corneous** hornig ~ **silver** Goldhornerz n
**corner, to** ~ aufkaufen **to** ~ **on** ecken
**corner** ausspringende Ecke f, Ecke f, Eckstück n, Spitze f, Winkel m ~ **of cube** Würfeleck n ~ **of the negative** Formatecke f
**corner, ~ antenna** Winkelreflektorantenne f ~ **arch** Gegenpfeiler m ~ **brace** Eckanker m ~ **bracket** Eckstück n ~ **(or border) clamp** Ecklammer f ~ **(or angle) connection** Eckanschluß m ~ **cramp** Scheinecke f ~ **cut** Eckenschliff m ~ **cutting** Abschattierung f (TV) ~ **cutting machine** Eckenausstoßmaschine f ~ **detail**

Eckenschärfe *f* ~ **fillet** Eckleiste *f* ~ **flange** Eckgurt *m* ~ **joint** Ecknaht *f*, Winkelstoß *m* ~ **joint welding** Eckschweißung *f* ~ **mark** Eckwasserzeichen *n* ~ **mark(s)** Eckenmarkierung *f* ~ **mounting machine** Eckenverbindemaschine *f* ~ **notch** einspringender Winkel *m* ~ **ornament** Eckenverzierung *f* ~ **piece of frame** Rahmeneckbeschlag *m* ~ **pillar** Eckpfeiler *m* ~ **plate** Eckband *n* ~ **post** Eckbalken *m* ~ **protection** Kantenschutz *m* ~ **punching machine** Eckenstanzmaschine *f* ~ **radiusing** Kantenrunden *n* ~ **reflector antenna** Raumdipol (rdr), Winkelreflektor *m* ~ **ridge** Eckfirst *m* ~ **sealing paper** Eckenschließpapier *m* ~ **seat** Eckplatz *m* ~ **shelf** Eckbrett *n* ~ **stake** Ortspfahl *m* ~ **staples** Eckenhaftklammer *pl* ~ **stitching** Eckenheftung *f* ~ **stone** Eck-, Grund-, Kopf-stein *m* ~ **tool** Eckstahl *m* ~ **truss** Eckaussteifung *f* (Kastenträger) ~ **working machine** Eckenbearbeitungsmaschine *f* (print.)
**cornered** eckig, unrund
**cornering machine** Eckenrundstoßmaschine *f*
**cornet** Kornett *n*
**cornice** Gesims *n*, Kranz *m*, Kranz-gesims *n*, -leiste *f*, -reif *m*; Mauerbrüstung *f*, Randleiste *f*, Sims *m* ~ **flashing** Simsabdeckung *f*
**cornish boiler** Einflammenrohrkessel *m*
**corollary** Folgerung *f*, Folgesatz *m* (math.), Zusatzannahme *f* ~ **assumption** Zusatzannahme *f*
**corona** Entladungsstoß *m*, Glimmen *n*, Glimmentladung *f*, Hof *m*, Korona *f*, Koronenerscheinung *f*, Kranzleiste *f*, Lichtkranz *m*, Sprühen *n*, Sprühentladung *f*, stille Entladung *f*, Strahlenkrone *f* ~ **ball** (of whip antenna) Prasselschutzkugel *f* ~ **brushing** Sprühentladung *f* ~ **discharge** Korona *f*, Sprühen *n*, Sprühentladung *f*, St. Elmsfeuer *n*, strahlartige Entladung *f* ~ **effect** Korona-erscheinung *f*, -effekt *m* ~ **loss** Koronaverlust *m* ~ **losses** Glimmverluste *pl*
**coronagraph** Koronagraf *m*
**coronal discharge** Koronaentladung *f*
**coronet** Hufkrone *f*
**coronium line** Coroniumlinie *f*
**co-rotational** mitrotierend ~ **frame** mitrotierendes Bezugssystem *n* ~ **time flux** mitrotierender Fluß *m*
**corporation** Gesellschaft *f*, Gesellschaft *f* mit beschränkter Haftung, Gilde *f*, Innung *f*, Körperschaft *f*
**corpse** Leiche *f*
**corpuscle** Atom *n*, Körperchen *n*, Massenteilchen *n*, Stoffteilchen *n*, Teilchen *n*
**corpuscular** atomistisch ~ **radiation** Korpuskularstrahlung *f* ~ **theory** Wärmestofftheorie *f*
**corral, to** ~ einfrieden
**corral** Einfriedung *f*
**correct, to** ~ ändern, angleichen, anpassen, ausschalten, berichtigen, einem Zustand anpassen, (jam or stoppage) beheben, (a distortion) entzerren, korrigieren, richtigstellen, verbessern **to** ~ **a bearing** eine Peilung verbessern
**correct** fehlerfrei, regelrecht, richtig, schicklich, wahr **in** ~ **phase relationship** richtigphasig **at or in** ~ **scale** maßstabgerecht **in** ~ **scale** maßstabgerecht ~ **to two decimal places** genau auf zwei Dezimalstellen

**correct,** ~ **angle of lead** Sollvorhalt *m* **bearing (or line) for approach** Landeachse *f* ~ **left-to-right** seitenrichtig ~ **preset (or rated) frequency** Sollfrequenz *f* ~ **tone filter** tonrichtiger Filter (opt.)
**corrected** berichtigt, gerichtet, umgerechnet, verbessert ~ **to** reduziert auf ~ **airspeed** Fahrtausgangswert *m* ~ **altitude** richtiggestellte Höhe *f* ~ **elevation** Aufsatzerhöhung *f* ~ **eye** berichtigtes Auge *n* ~ **line** Korrekturzeile *f* ~ **north** rechtweisend Nord ~ **radio bearing** beschickte oder verbesserte Funkseitenpeilung *f*, korrigierte (korrektierte) Funkpeilung *f* ~ **range** Aufsatzentfernung *f* ~ **white** Auffärbung *f* (paper), aufgefärbtes Weiß *n*, Aufhellung *f*, nunanciertes Weiß *n*
**correcting** Nachjustierung *f* ~ **of distortion** Entzerrung *f* ~ **brush** Korrektionsbürste *f* ~ **cam** Korrektionsdaumen *m* ~ **circuit** Entzerrerschaltung *f* ~ **currents** Gleichlauf-stöße *pl*, -ströme *pl*, -zeichen *pl* ~ **device** Entzerrer *m*, Korrektionseinrichtung *f* ~ **displacement** Korrekturverschiebung *f* ~ **distributor** korrigierender Verteiler *m* ~ **element** Stellglied *n* ~ **field** Richtfeld *n* ~ **impulse** Gleichlaufstromstoß *m*, Korrekturimpuls *m* ~ **lens** berichtigendes Glas *n* ~ **magnet** Gleichlaufmagnet *m* ~ **network** Ausgleichsnetzwerk *n*, Entzerrerkette *f* ~ **relay** Gleichlaufrelais *n* ~ **repeater** entzerrender Verstärker *m* ~ **ring** Gleichlaufring *m* ~ **screw** Korrektionsschraube *f* ~ **segment** Gleichlaufsegment *n* ~ **unit** Stellwerk *n* ~ **voltage** Korrekturspannung *f* ~ **wheel** Korrektionsrad *n*
**correction** Änderung *f*, Berichtigung *f*, (in direction finding work) Beschickung *f*, Einstellung *f*, Entdämpfung *f*, (of aerial photographs) Entzerrung *f*, Korrektion *f*, Korrektur *f*, Reglerkorrektur *f*, Rektifikation *f*, Richtigstellung *f*, Verbesserung *f* ~ **by addition** Zusatzberichtigung *f* ~ **of amplitudes** Amplitudenentzerrung *f* ~ **for barrel corrosion** Gebrauchsstufe *f* ~ **for deviation of directionfinder bearing** Funkbeschickung *f* ~ **for displacement** Richtkreiskorrektur *f* ~ **of distortion** Berichtigung *f* der Verzerrung ~ **for drift** Seitenverbesserung *f* ~ **for exterior ballistic factors** Tagesverbesserung *f* ~ **of geometrical distortions** Entzerrung *f* von Geometriefehlern (TV) ~ **for (of) gravity** Schwereberichtigung *f* ~ **for interior and exterior ballistic factors** Ausschalten *n* der besonderen und Witterungseinflüsse ~ **of non-linear distortions** Linearisierung *f* ~ **of nonlinear harmonic distortion** Klirrdämpfung *f* ~ **for powder temperature** Gebrauchsstufe *f* ~ **of stream** Strahlberichtigung *f* ~ **for temperature** Wärmegradberichtigung *f* ~ **for velocity error** Ausschalten *n* der Grundstufe ~ **for weight of the projectile** Gebrauchsstufe *f*
**correction,** ~ **adjustment drum** Berichtigungswalze *f* ~ **collar** Korrektionsring *m* ~ **device** Korrektionseinrichtung *f* ~ **factor** Berichtigungs-faktor *m*, -wert *m*; Korrektionsfaktor *m*, Korrektionsglied *n* ~ **key** Verbesserungstaste *f* ~ **lens** Korrektionssystem *n* ~ **mark** Berichtigungsmarke *f* ~ **method** Berichtigungsverfahren *n* ~ **potentiometer** Korrekturpotentiometer *m* ~ **scale on sight** Reglerteilung *f* ~ **signal** Korrekturkommando *n* (g/m) ~

**tables** Rechenzettel *m* ~ **term** Korrekturglied *n* ~ **value** Beiwert *m* (Peilung), Berichtigungswert *m* (to compensate bearing errors) ~ **wedge** Berichtigungswalze *f* ~ **wheel** Korrektionsrad *n*
**corrective** Verbesserungsmittel *n*; berichtigend ~ **action** Änderung *f* der Stellgröße ~ **agent** (corrigent) Korrektionsmittel *n* ~ **circuit** Korrektionsschaltung *f* ~ **error** Ausgleichsfehler *m* ~ **factor** Berichtigungsbeiwert *m*, Korrekturfaktor *m* ~ **impulse** Gleichlaufimpuls *m* ~ **network** Ausgleichsnetzwerk *n*, Entzerrer *m*, Entzerrerkette *f* ~ **screen** Tuschiersieb *n* ~ **term** Korrektionsglied *n*
**correctly,** ~ **aligned** fluchtrecht ~ **trimmed** richtiglastig
**correctness** Richtigkeit *f* ~ **of the angles** Winkeltreue *f*
**corrector** Brennlängenschieber *m*, Korrektor *m*, Regler *m*, Schieber *m* ~ **and adjuster** Entfernungsmeßkontrollanlage *f* ~ **adjustment** Korrektoreneinstellung *f* ~ **circuit** Ausgleichskreis *m* ~ **relay** Gleichlaufrelais *n* ~ **vane** Abgleich-Beiregel-platte *f* ~ **wheel** Korrektionsrad *n*
**correlate, to** ~ in (wechselseitige) Beziehung setzen
**correlation** Beziehung *f*, Korrespondenz *f*, Übereinstimmung *f*, Verbindung *f*, Wechselbeziehung *f*, Zuordnung *f*, (of things) Zusammengehörigkeit *f* ~ **factor** Korrelationskoeffizient *m* ~ **function** Einflußfunktion *f* ~ **matrix** Korrelationsmatrix *m* ~ **sample** Ringversuchsmuster *n*
**correlative** Korrelat *n*
**correspond, to** ~ entsprechen, einen Briefwechsel führen, korrespondieren (mit), in Verkehr stehen, übereinstimmen
**correspondence** Briefwechsel *m*, Korrespondenz *f*, Schriftverkehr *m*, Schriftwechsel *m*, Übereinstimmung *f*, Verschmelzung *f* der Halbbilder bei stereoskopischer Betrachtung (photo) ~ **course** Fernkurs *m* ~ **principle** Korrespondenzprinzip *n*
**correspondent** Berichter *m*, Korrespondent *m*; gleichnamig (math.)
**corresponding** diesbezüglich, übereinstimmend, entsprechend (to) ~ **account** Gegenkonto *n* ~ **grades** Gegenqualitäten *pl* ~ **train** Anschlußzug *m*
**corridor** Gang *m*, Schneise *f*
**corrode, to** ~ abätzen, abfressen, abrosten, abzehren, anätzen, auffressen, angreifen, anrosten, ätzen, beizen, durchfressen, einfressen, fressen, korrodieren, rosten, verrosten, verrotten, zerfressen, zernagen. zerstören
**corroded** abgefressen, verrostet
**corrodibility** Korrodierbarkeit *f*, Korrosions-, Rost-empfindlichkeit *f*
**corrodible** ätzbar, korrodierbar, korrosionsempfindlich, rostempfindlich
**corroding** Rosten *n*; beizend, korrosionsfördernd ~ **agent** Korrosionsbildner *m*, korrosionsförderndes Mittel *n*, Rostbildner *m*, Rostmittel *n* ~ **agents (or media)** Angriffsmittel *pl* ~ **medium** korrosionsförderndes Mittel *n*, Rostmittel *n*
**corrosible** ätzbar
**corrosion** Abnutzung *f*, Abzehrung *f*, Anfraß *m*, Anfressung *f*, Anrostung *f*, Ätzen *n*, Ausna-

gung *f*, Ausspülung *f*, Beizen *n*, chemischer Angriff *m*, Durchfressen *n*, Einfressung *f*, Korrossion *f*, Rostanfressung *f*, Rostbildung *f*, Rosten *m*, Rostfraß *m*, Rostung *f*, Verrostung *f*, Verschleiß *m*, Zerfressen *n*, Zerfressung *f*, Zerstörung *f* ~ **of the lead sheathing** Kabelmantelkorrosion *f* ~ **by niter** Salpeterfraß *m*
**corrosion,** ~ **allowance** Korrosionsgrenze *f* ~ **cell** Korrosionselement *n* ~ **creep test depending on time** Korrosionszeitstandversuch *m* ~ **fatigue fracture** Korrosionsdauerbruch *m* ~ **fatigue strength** Korrosionsdauerfestigkeit *f* ~ **inhibition** Korrosionsverhinderung *f* ~ **inhibitor** Korrosionsschutzmittel *n* ~ **phenomenon** Korrosionserscheinung *f* ~ **pit** Korrosions-, Rost-narbe *f* ~ **preventative** Korrosionsschutzmittel *n* ~ **preventing** korrosionshindernd ~ **preventing paint** Korrosionsschutzanstrich *m* ~ **prevention grease** Korrosionsschutzfett *n* ~ **proof** korrosionsbeständig ~ **proofing treatment** Korrosionsbehandlung *f* ~ **resistance** Korrosions-festigkeit *f*, -widerstand *m*, Rostfestigkeit *f* ~ **resistance of a metal** Ätzungswiderstand *m* eines Metalls ~ **resistant** korrosionsbeständig, rostsicher ~ **resisting** korrosionsfest, rostbeständig ~ **resisting quality** Korrosionsbeständigkeit *f* ~ **resisting steel** korrosionsfreier Stahl *m* ~ **test** Ätz-, Ätzungsversuch *m*, Korrosionsprüfung *f*, Korrosionsversuch *m*, Rostungsversuch *m*
**corrosive** Abbeizmittel *n*, Beize *f*, Beizmittel *n*, Korrosionsbildner *m*; ätzend, beizend, korrodierbar, korrodierend, scharf, zerfressend ~ **action** Korrosionsvorgang *m*, Rostangriff *m* ~ **agent** Korrosionsbildner *m* ~ **attack** Rostangriff *m* ~ **effect** Korrosionswirkung *f* ~ **environment** korrosionsförderndes Medium *n* ~ **gas** Ätzgas *n* ~ **power** Ätzkraft *f*, Beizkraft *f* ~ **stick** Ätzstift *m* ~ **sublimate** Ätzsublimat *n*, Quecksilberschlorid *n* ~ **substance** Ätzmittel *n*
**corrosiveness** Korrosionswirkung *f*
**corrugate, to** ~ riffeln, mit Sicken versehen, wellen ~ **opening** Wellenbohrung *f*
**corrugated** gerieft, geriffelt, gerippt, gewellt, wellenförmig ~ **barometric cell** gewellte Barometerdose *f* ~ **board** Well-papier *f*, -pappe *f* ~ **bottom** Wellblechboden *m* ~ **brick** Wellstein *m* ~ **curve** wellige Kurve *f* ~ **flue boiler** Wellrohrkessel *m* ~ **friction socket** Rohrreibungsglocke *f* ~ **glass** gestreiftes Glas *n*, Riffelglas *n* ~ **hose** Faltenschlauch *m* ~ **iron buildings** Wellblechbauten *pl* ~ **iron cover** Wellblechdecke *f* ~ **lead sheeting** Riffelblei *n* ~ **metal tube** Metallwellblechrohr *n* (Entstörung) ~ **metal wing covering** Wellblechbekleidung *f* ~ **paper** Well-karton *m*, -papier *n*, -pappe *f* ~ **plate condenser** Wellplattenkondensator *m* ~ **pressure box (or capsule)** gewellte Barometerdose *f* ~ **sheet duralumin** Duralwellblech *n* ~ **sheet iron** Wellblech *n* ~ **sheet metal** Riffel-, Runzel-blech *n* ~ **sheet-metal covering** Wellblechbeplankung *f* ~ **sheet rolling mill** Wellblechwalzwerk *n* ~ **sheet steel** Wellblech *n* ~ **spring washer** Wellenfederring. *m* ~ **steel** geschweißter Stahl *m* ~ **suspension means** (of loudspeaker cone) gewellter Einspannrand *m* ~ **tip** gewellter Griff *m* ~ **tube** Faltenschlauch *m*, Wellenrohr *n*, Wellrohr *n* ~ **tube of tombac** Tombak-Wellrohr

*n* ~ **tubing** Wellrohr *n* ~ **tubing seal** Wellrohrdichtung *f*
**corrugating,** ~ **press** Wellblechpresse *f* ~ **roll** Wellblechwalze *f* ~ **rolling mill** Wellblechwalzwerk *n*
**corrugation** Riffelung *f*, Rippe *f*, Sicke *f*, Wellung *f* ~ **test** Welligkeitsprüfung *f* ~ **vane** Verstärkungsprofil *n*
**corrugations on rails** Wellenbildung *f* auf den Schienen
**cortical,** ~ **layer** Rindenschicht *f* ~ **stimulator** elektronischer Kortexstimulator *m*
**corumdum** Korund *m*
**cosecant** Kosekans *m*, Kosekante *f* ~ **squared antenna** Cosec-Quadrat-Antenne *f* ~ **squared beam** Kosekanzbündel *n* (rdr) ~ **squared pattern** Cosec-Quadrat-Charakteristik *f*
**coseparation** Mitfällung *f*
**coset** Nebenklasse *f* (math.)
**cosine** Kosinus *m* ~ **emission law** Lambertgesetz *n* ~ **law** Kosinusgesetz *n* ~ **line** Kosinuslinie *f* ~ **series** Kosinusreihe *f* ~ **term** Kosausdruck *m*, Kosinusglied *n*
**cosletizing** Rost- und Korrosions-Schutzbehandlung *f* für Stahl
**cosmic,** ~ **radiation** kosmische Strahlung *f*, kosmische Ultrastrahlung *f* ~ **ray radiation** Höhen-, Raum-strahlung *f* ~ **ray shower** kosmischer Schauer *m* ~ **ray track** Höhenstrahlenschauer *m* ~ **rays** Höhenstrahlen *pl*, kosmische Strahlen *pl*, kosmische Ultrastrahlen *pl*, kosmische Ultrastrahlung *f*, Weltraumstrahlen *pl*
**cosmogony** Kosmogonie *f*
**cosmography** Weltbeschreibung *f*
**cosmology** Kosmologie *f*
**cosmonautics** Kosmonautik *f*, Raumflugwesen *n*
**cosmotron** Kosmotron *n*, Protonen-Synchroton *n*
**cost, to** ~ kosten **to** ~ **individually** einzeln in Ansatz bringen
**cost** Kosten *pl*, Preis *m*, Unkosten *pl*
**cost,** ~ **of advertisement** Insertions-gebühr *f*, -kosten *pl* ~ **of attendance** Bedienungskosten *pl*, Wartungskosten *pl* ~ **of circuit** Schaltaufwand *m* ~ **of construction** Anlagekosten *pl*, Bauaufwand *m*, Baukosten *pl*, Herstellungskosten *pl* ~ **of conversion** Umbaukosten *f* ~ **of current** Stromkosten *pl* ~ **of erection** Aufstellungskosten *pl* ~ **of installation** Anlage-, Aufstellungs-kosten *pl* ~ **of interest and amortization of capitel** Abschreibung *f* und Verzinsung *f* des Anlagekapitals ~ **of land** Grundpreis *m* ~ **of maintenance** Anlagekosten *pl* ~ **to manufacture** Selbstkostenpreis *m* ~ **of operation** Bedienungskosten *pl* ~ **of proceedings** Prozeßkosten *pl* ~ **of productions** Gestehungs-, Herstellungs-kosten *pl*, Macherlohn *m*, Selbstkosten *pl* ~ **of rebuilding** Neubaukosten *pl* ~ **of telegram** Telegrammspesen *pl* ~ **of transmission of energy** Energieleistungskosten *pl* ~ **of upkeep** Instandhaltungs-, Unterhaltungs-kosten *pl*
**cost,** ~ **accounting department** Selbstkostenberechnungsabteilung *f* ~ **analysis** Zeitstudienwesen *n* ~ **breakdown** Kostenaufgliederung *f* ~ **category** Wertstufe *f* ~ **per unit of heat** spezifischer Wärmepreis *m* ~ **price** Ankaufs-, Einstands-, Gestehungs-, Kosten-, Selbstkosten-

-preis *m* ~ **reduction** Einsparung *f* ~ **sheet** Kostenaufstellung *f* ~ **study** Berechnung *f* ~ **system** Selbstkostenrechnung *f* ~ **vouchers (or records)** Kostenbelege *pl*
**costed stores** Magazinmaterialien *pl* mit Wert
**costing** Nachrechnung *f*, Vorrechnung *f*
**costs** Ausgaben *pl*, Spesen *pl* ~ **of housing** Unterbringungskosten *pl*
**co-subscribe, to** ~ mithalten
**Cosyns stage** Cosynsscher Kreuztisch *m*
**cotangent** Kotangens *m*, Kotangente *f*
**cote** Köte *f*
**cotonize, to** ~ cottonisieren
**cottage steamer** Kastendämpfer *m*
**cotter, to** ~ verklammern, versplinten
**cotter** Dorn *m*, Fangkeil *m*, Keil *m*, konischer Keil *m*, Querkeil *m*, Querriegel *m*, Schließe *f*, Schließkeil *m*, Splint *m*, Vorstecker *m*, (pin) Vorsteckkeil *m* ~ **bolt** Bolzen *m* mit Splint ~ **case** Keilgehäuse *n* ~ **extractors** Splintzieher *pl* ~ **file** dickflache Feile *f* ~ **hole drill** Keillochbohrer *m* ~ **key** Schlüsselkeil *m* ~ **lock** Splintverschluß *m* ~ **mill** Langfräser *m* ~ **pin** Bolzen *m* mit Kopf und Splintloch, Lösekeil *m*, Schließbolzen *m*, Sicherungsvorstecker *m*, Vorstecker *m*, Vorsteckstift *m* ~ **pin detachable roller chain** versplintete Nietbolzenkette *f* ~ **pin extractor** Splintenzieher *m* ~ **pin hole** Splintloch *n* ~ **pin wire** Splintdraht *m* ~ **retention** Querkeilbefestigung *f* ~ **screw** Keilschraube *f* ~ **slot** Keilloch *n* ~ **splint pin** Vorstecksplint *m*
**cottered,** ~ **chain** Kette *f* mit versplinteten Bolzen ~ **joint** Querkeilverbindung *f* ~ **pin** versplinteter Bolzen *m*
**cotton** Baumwolle *f*, (in laquer terminology) Wolle *f*, baumwollen ~ **asbestos** Asbestwolle *f* ~ **bale breaker picker** Baumwollballenöffner *m* ~ **batting** Watte *f* ~ **belt** Baumwolltuchriemen *m* ~ **belt(ing)** Baumwollriemen *m* ~ **binder** farbiger Kennfaden *m* ~ **braided covering** Baumwollumklöppelung *f* ~ **brake lining** Baumwollbremsband *n* ~ **card** Baumwollkratze *f* ~ **carding process** Baumwollkarden *n* ~ **cleaning waste** Putzbaumwolle *f* ~ **cord** Baumwolleine *f* ~ **and corn drill** Baumwoll- und Maisdrillmaschine *f* ~ **and corn lister** Baumwoll- und Maisanhäufler *m*, Dammkulturpflug *m* mit Baumwoll- und Maisdrill ~ **covered cable** Baumwollkabel *n* (silk- and) ~ **covered cable** Baumwoll(seiden)kabel *n* ~ **covered wire** Baumwolldraht *m*, mit Baumwolle umsponnener Draht *m* ~ **covering** Baumwollumspinnung *f* ~ **deviler** Baumwollwolfer *m* ~ **drawing frame tenter** Baumwollstrecker *m* ~ **dressing installation** Baumwollaufbereitungsanlage *f* ~ **drill** Baumwoll-drillich *m*, -drell *m*, Drell *m* ~ **duck** Baumwollgewebe *n*, Bindertuch *n* ~ **fabric** Baumwollgewebe *n* ~ **fiber** Baumwoll-faser *f*, -hartgewebe *f* ~ **filter** Wattefilter *n* ~ **filter cloth** Baumwollfiltertuch *n* ~ **finisher minder** Baumwollvollender *m* ~ **gauze** baumwollene Gaze *f* ~ **gin** Baumwoll-entkörnungsmaschine *f*, -kratze *f*; Egreniermaschine *f* ~ **hard-waste-breaker tenter** Baumwollreißer *m* ~ **hard-waste breaking** Baumwollabfallreißerei *f* ~ **hosiery** Baumwollstrickerei *f* ~ **lap-machine minder** Baumwollausbreiter *m* ~ **lapper** Baumwollausbreiter *m* ~ **line** Baumwoll-

leine f ~ **linters** Baumwollabfall m ~ **linters pulp** Baumwollzellstoff m ~ **machine** Baumwollmaschine f ~ **oil** Baumwollsaatöl n ~ **packing** Baumwolldichtung f ~ **picker tenter** Baumwollrupfer m ~ **planter** Baumwolldrillmaschine f, Baumwollpflanzer m ~ **plug** Wattebausch m ~ **reel valve** Garnrollenröhre f ~ **reeler** Weifer m ~ **rope** Baumwollseil n ~ **roving frame tenter** Baumwollvorgarnspinner m ~ **seed oil** Baumwoll-saatöl n, -samenöl n ~ **shaker** Baumwollschlagmaschinearbeiter m ~ **speeder tenter** Baumwollfeinstrecker m ~ **steamer** Baumwolldämpfer m ~ **tape saturated with tannin** tanningetränktes Band n ~ **texture** Baumwollgewebe n ~ **thread** Baumwollfaden m ~ **threader** Baumwollüberspinner m ~ **twine** Baumwollgarn n ~ **twister** Baumwollzwirner m ~ **waste** Baumwoll-abgang m, -abfall m, Putzwolle f, Werg n ~-**waste hand** Baumwollabfallsortierer m ~ **willower** Baumwollwolfer m ~ **winding** Baumwollhaspelei f ~ **wood** Baumwollholz n, kanadisches Pappelholz n, Rohbaumwolle f ~-**wool ball** Wattekugel f ~-**wool probe** Watteträger m ~ **yarn** Baumwollgarn n
**couch, to** ~ gautschen (paper mfg.) **to** ~ **paper sheets** Papier n gautschen ~ **stool** Büttenstuhl m
**coucher** Abgautschapparat m, Gautscher m
**"cough" button** Räuspertaste f
**coulisse** Kulisse f
**coulomb** Coulomb n ~ **meter** Coulombmesser m, Coulombzähler m ~'**s balance** Coulombsche Waage f ~'**s barrier** Coulombscher Potentialwall m ~'**s law** Coulombsches Gesetz
**Coulombian repulsion** Coulombabstoßung f
**counsel** Anwalt m, Beratung f, Rat m, Rechtsbeistand m, Sachwalter m
**count, to** ~ durchmustern, rechnen, zählen **to** ~ **off** abzählen ~ **from right to left (or counterclockwise)** linksläufig zählen **to** ~ **up** buchen
**count** (silk and art silk) Denier n, (of yarn) Feinheitsnummer f, (of cosmic rays) Stoß m, Zählrate f, Zählung f ~ **card** Zählkarte f ~-**down** Antwortbakenausbeute f ~ **printer** Zählbetragsdrucker m ~ **rate** Impulsfrequenzmessung f ~ **totals** Zählbeträge pl
**countable** abzählbar
**counted,** ~ **amount** Zählbetrag m ~ **amount printer** Zählbetragsdrucker m ~ **cut-off wall** Abdichtungsschleier m
**counter,** ~ entgegenhandeln **to** ~ **draw** einen Gegenabdruck m machen oder herstellen (print.) **to** ~ **stain** nacheinanderfärben **to** ~ **veneer** gegenfurnieren
**counter** Annahmeschalter m, Gillung f, Heck n, Ladentisch m, Punze f, Schalter m, Schaltertisch m, Tisch m, Zähleinrichtung f, Zähler m (Durchfluß), Zählwerk n, Ziffernrolle f; gegen ~ **with click action** Zählwerk n mit springenden Ziffern ~ **of strokes** Hubzähler m
**counter,** ~-**air current** Luftschleier m ~ **ampere turns** Gegenamperewindungen pl
**counteract, to** ~ entgegenarbeiten, entgegenwirken, gegenarbeiten, gegenwirken, neutralisieren, zuwiderhandeln
**counteracting,** ~ **force** Gegenkraft f ~ **stirring mechanism** gegeneinanderarbeitendes Rührwerk n ~ **winding** Differentialwicklung f

**counteraction** Gegenwirkung f ~ **plaintiff** Widerklänger m
**counterbalance, to** ~ abstimmen, aufwiegen, ausbalancieren, ausgleichen, auswägen, auswiegen, auswuchten, das Gegengewicht halten
**counterbalance** Bremsgewicht n, Entlastungsvorrichtung f, Gegenausgleich m, Gegengewicht n, Gegengewichtsbalancier n, Massenausgleich m, Seitenhalter m ~ **control** Saldosteuerung f ~ **cylinder** Ausgleichszylinder m ~ **weight** Ausgleichgewicht n
**counter-balanced, to be** ~ **by** ausgeglichen werden durch
**counter-balancing spring** Ausgleichsfeder f
**counter,** ~ **bill** Gegenwechsel m ~ **blade** Vorschneidemesser n ~ **blade stone remover** Steinfängervorlage f ~ **block** Gegenblock m ~ **bolt** Gegenvorstecker m ~-**boost pressure** Auflagergegendruck m
**counterbore, to** ~ mit Werkzeug n senken
**counterbore** Halssenker m, Krauskopf m, Senker m, Senkwerkzeug n, Versenker m
**counter,** ~ **boring** Herstellung f einer zylindrischen oder kegeligen Vertiefung für den Schraubenkopf, Versenken n ~ **boring tool** Abflachmesser n, Zapfensenker m ~ **brace** Gegenschräge f, -diagonale f, -druckspreize f, -strebe f ~ **brake** Gegenbremse f ~ **calibrating room** Zählereichsaal m ~ **calibration** Zählrohreichung f ~ **cell** gegengeschaltete Sammlerzelle f oder Zelle f ~ **chain wheel** Gegenkettenrad n ~ **charge** Gegenklage f ~ **circuit** Zählschaltung f ~ **claim** Gegen-forderung f, -rechnung f, -schuld f; Rückanspruch m
**counterclockwise** dem Uhrzeiger entgegengesetzt, im Gegensinn zum Uhrzeiger, linksdrehend, linksgängig, linksläufig ~ **direction** Gegenzeigersinn m ~ **rifling** Linksdrall m ~ **rotating** linksdrehend ~ **rotation** Links-drehung f, -lauf m
**counter,** ~ **connected** gegengeschaltet ~ **connection** Gegenschaltung f ~ **control** Zählerkontrolle f ~ **controls** Summenkontrolle f ~-**coupling half** Gegenkupplungshälfte f ~ **cover** Zählwerkdeckel m ~-**counter measures** Schutzmaßnahmen pl
**countercurrent** Gegen-lauf m, -strom m, -strömung f; gegenläufig ~ **boiler** Gegenstromkessel m ~ **condenser of the multipass type** Elementenbündelverflüssiger m ~ **condenser with water basin (or water catcher)** Gegenstromkondensator m mit Wasserfänger ~ **interchangers** Gegenstromaustauscher m ~ **juice mixer** Gegenstromvorwärmer m ~ **(or rotary) pan mixer** Gegenstromtellermischer m ~ **pipe cooler** Gegenstromröhrenkühlgerät n ~ **principle** Gegenstromprinzip m ~ **process** Gegenstromverfahren n ~ **rain-type condenser** Gegenstromregenkondensator m
**counter,** ~ **deadtime** Zähler-Totzeit f ~ **dial** Zählwerkzifferblatt n ~ **die** Patrize f ~ **draft** Gegenwechsel m ~ **drawing** Pause f, Pauszeichnung f ~ **effect** Gegenwirkung f ~ **efficiency** Zählerausbeute f ~ **electrode** Gegenelektrode f ~ **electromotive force** entgegengesetzte Gegenkraft f, gegenelektromotorische Kraft f, Gegenzelle f ~ **entry** Zählereingang m ~ **excavation** Gegenort n

**counterfaller** Gegenschläger *m* (textiles) ~ **motion** Gegenwinderbewegung *f* ~ **shaft** Gegenwinderwelle *f*
**counterfeit, to** ~ nachahmen, nachbilden, nachmachen
**counterfeit** Nachdruck *m*; nachgeahmt
**counter-feiting** Fälschung *f*
**counter,** ~ **filter** Nachfilter *m* ~ **flange** Gegenflansch *f* ~ **floor** Blindboden *m* ~ **flow** Gegenstrom *m*, -strömung *f* ~**-flow heat exchanger** Gegenstromwärmeaustauscher *m* ~**-flow principle** Gegenstromprinzip *n* ~**-flowing injection** Gegenstromeinspritzung *f* ~ **foil** Kontrollschein *m* ~ **fort** Gegenpfeiler *m* (hydraul.), Pfeiler *m*, Strebepfeiler *m*, Verstärkungspfeiler *m* ~ **gangway** Gegenort *m* ~ **gas amplification** Gasverstärkung *f* des Zählrohres ~ **head** Begleitstrecke *f*
**counterhold, to** ~ gegenhalten
**counter,** ~ **indicator** Tourenzähler *m* ~ **knife** Gegenmesser *n* ~ **lever** Gegendruckhebel *m* ~ **lifetime** Zählerlebensdauer *f* ~ **light** Gegenlicht *n* ~ **list** Zählerschreibung *f* ~ **list exit** Zählerausgang *m* für Posten ~ **lode** übersetzender oder anscharrender Gang *m* ~ **luffing** Wippbewegung *f*
**countermand, to** ~ abbefehlen, abbestellen, (an order) absagen (Bestellung), einen Gegenbefehl geben
**counter,** ~ **measure** Gegenmaß-nahme *f*, -regel *f* ~ **mechanism** Distanzvorrichtung *f* ~ **motion** Gegenbewegung *f* ~ **move** Gegenzug *m* ~ **movement** Gegenbewegung *f* ~ **nut** Kontermutter *f* ~ **obligation** Gegenverpflichtung *f* ~ **offer** Gegen-angebot *n*, -vorschlag *m* ~ **oil** Öl *n* für Maßapparate ~ **operating voltage** Zählerbetriebsspannung *f* ~ **order** Abbestellung *f*, Gegenauftrag *m*, Gegenbefehl *m* ~ **overshooting** Überanregung *f* eines Zählrohres ~ **overvoltage** Zählerüberspannung *f* ~ **pad** Gegenkufe *f* ~ **part** Abdruck *m*, Gegenbild *n*, Gegenstück *n* ~ **plateau** Zählerplateau *n* ~ **poise** Ausgleichskapazität *f*, Beschwerungs-rohr *n*, -stück *n*, Bremsgewicht *n*, Gegengewicht *n*, Waagegegengewicht *n*, Wippe *f* ~**poise connector** Gegengewichtsanschluß *m* ~ **positions** direktsaldierende Zählerstellen *pl*
**counterpressure** Gegendruck *m*, Widerdruck *m* ~ **chamber** Gegendruckkammer *f* ~ **filling apparatus** Gegendruckfüllapparat *m* ~ **racker** Abfüllapparat *m* mit Gegendruck, Gegendruckabfüllapparat *m* ~ **racking apparatus** Gegendruckfaßfüllapparat *m* ~ **valve** Gegendruckventil *n*
**counterprint** Gegendruck *m*
**counter,** ~ **problem** Zählrohrproblem *n* ~ **profile** Gegenprofil *n* ~ **proof** Gegenabdruck *m* (print.), Gegenabzug *m* ~**-proposal** Gegenantrag *m* ~ **punch** Gegen-punze *f*, -punzen *m*, -stanze *f*, -stempel *m* ~ **range** Anlaufgebiet *n*
**counter-recoil,** ~ **buffer** Vorlaufhemmstange *f* ~ **cylinder** Vorholzylinder *m* ~**-cylinder support** Vorholerstütze *f* ~ **cylinder yoke** Vorholerlager *n* ~ **mechanism** Vorholer *m* ~ **mechanism bushing** Vorholerlager *n* ~ **mine** Gegen-mine *f*, -stollen *m* ~ **motion** Vorlaufbewegung *f* ~ **piston** Vorholerkolben *m* ~ **spring** Rücklauffeder *f*

**counter,** ~ **recovery time** Totzeit *f*, Zählrohrerholungszeit *f* ~ **reel** Rollenpapier *n*, Sekarerollen *pl* (paper mfg.) ~ **re-ignition** Zählrohrwiederzündung *f* ~ **remittance** Gegenrimesse *f* ~ **resolving time** Zählrohrauslösungszeit *f* ~ **revolving airscrews** gegenläufige Luftschrauben *pl* ~ **ring** Gegenring *m* ~**-rotating drive** Gegensinngetriebe *n* ~ **rotation** Antirotation *f*, Gegen-drehung *f*, -lauf *m* ~**-rotation propeller (or counter screw)** Gegenschraube *f* ~ **scrap** Gegenböschung *f*, Konterescarpe *f* ~ **screw** Unterschraube *f*
**countershaft** Gegen-, Mittel-, Neben-welle *f*, Vorgelege *n*, Vorgelegewelle *f* ~ **for ceiling suspension** Deckenvorgelege *n* ~ **axle** Blindachse *f*, Blindwelle *f* ~ **bearing** Vorgelegelager *n* ~ **gear** Gegenwellengetriebe *n*, Vorgelegerad *n* ~ **pulley** Gegenstufenscheibe *f* ~ **wheel** Vorgelegerad *n*
**counter-shock** Widerstoß *m*
**countersign, to** ~ gegenzeichnen
**countersign** Erkennungswort *n*, Losungswort *n*, Parole *f*
**countersignature** Gegenzeichnung *f* (Unterschrift)
**countersink, to** ~ ankörnen, ansenken (Werkstücke, Bolzen), senken, versenken
**countersink** Bohrfräser *m*, Einschliff *m*, kegeliger Senker *m*, Kranzkopf *m*, Krauskopf *m*, Senker *m*, Senkwinkel *m*, Spitzsenker *m*, Versenkbohrer *m*, Versenker *m*, Versenkung *f* ~ **for center punch holes** Körnerlochsenker *m* ~ **bolt** Versenkschraube *f*
**countersinking** Versenkarbeit *f*, Versenkung *f* ~ **for switch handle** Senkung *f* für Schaltgriff ~ **bit** Ankörnbohrer *m*
**counter,** ~ **slope** Gegenböschung *f*, -hang *m* ~ **sluice** Gegenschleuse *f* ~ **socket** Gegentrichter *m* ~ **spring** Gegendruckfeder *f*, Gegenfeder *f* ~ **starting potential** Zählerzündspannung *f* ~ **station** Gegenstation *f* ~ **stay** Gegenstütze *f* ~ **steam** Kontredampf *m* ~ **steam brake** Gegendampfbremse *f* ~ **storage** Speicherzähler *m* ~ **stream principle** Gegenstromprinzip *m* ~ **suit** Gegenklage *f*, Widerklage *f*
**countersunk** versenkt; Zapfen *m* mit viereckigem Loch ~ **angle** Sehwinkel *m* ~ **bolt** Bolzen *m* mit versenktem Kopf ~ **carriage bolt** Senkschraube *f* mit Vierkant ~ **collar** Senkbund *m* ~ **head** versenkter Kopf *m* ~ **head grooves pin (type C)** Senkkerbnagel *m* ~ **head rivet** Kopfversenkniete *f*, Senkkopfniete *f* ~ **head screw** Senkschraube *f* ~ **hole** kegelige Aussenkung *f* ~ **nut** Senkmutter *f* ~ **oval head screw** Linsensenkschraube *f* ~ **rivet** Senkkopfniete *f*, Versenkniet *n*, versenktes Niet *n* ~ **rivet with shallow button head** Linsenniet *n* ~ **rivet head** Nietsenkkopf *m* ~ **riveting** Senknietung *f*, Versenknietung *f* ~ **screw** eingelassene Schraube *f*, Senkschraube *f*, versenkte Schraube *f* ~ **screw head** versenkter Schraubenkopf *m* ~ **wire nail** Senkstift *m*
**counter,** ~ **support** Gegenständer *m* ~ **thrust** Gegenstoß *m* ~ **time lag** Zählerverzögerung *f* ~**-torque control** Gegenstrombremsung *f* ~ **total exit** Summenwerkausgang *m* ~ **trade** Antipassat *m*, Gegenpassatwind *m* ~ **tube** Lichtzähler *m*, Zählrohr *n* ~**-tube channel**

Zählgerätekanal *m* ~ **twilight** Gegendämmerung *f* ~ **vailing duty** Ausgleichszoll *m* ~ **value** Gegenwert *m* ~ **voltage conductor** Gegenspannungsdraht *m* ~ **wall (counter-mure)** Gegenmauer *f* ~ **wedge** Gegenkeil *m* ~ **wheel** Zählwerkrad *n*, Zählrad *n*
**counterweight** Ausgleichsgewicht *n*, Belastungsgewicht *n*, Beschwerungs-rohr *n*, -stück *n*, Gegengewicht *n*, Kontragewicht *n*, Spanngewicht *n* ~ **of a crane (or derrick)** Auslegergegengewicht *n* ~ **for left side of crossrail** linksseitiges Gegengewicht *n* des Querbalkens ~ **for rods** Gestängegegengewicht *n* ~ **for roller carriage** Gewicht *n* für Walzenstuhl
**counterweight,** ~ **bolt** Gegengewichtsbolzen *m* ~ **boom** Gegengewichtsausleger *m* ~ **brake** Wurfhebelbremse *f* ~ **carrier catch** Gegengewichtsnase *f* ~ **chain** Gegengewichtskette *f* ~ **fitting** Schnurzugpendel *m, n* ~ **float** Schwimmer *m* des Gegengewichts ~ **(or counterpoise) loop** Gegengewichtsschleife *f* ~ **suspension fitting** Glühlampenpendel *n* (electr.)
**counter,** ~ **wheel** Zählrad *n*, Zählwerkrad *n* ~ **wire** Gegenkabel *n* (aviat.)
**counters for test values** Meßwertspeicher *m*
**counting** Zählung *f* ~ **of pulses** Impulszählung *f* ~ **apparatus for miner's trucks** Förderwagenzähleinrichtung *f* ~ **chain of relays** Zählkette *f* ~ **chamber** Zählkammer *f* ~ **channel** Zählkanal *m* ~ **circuit** Zählerkreis *m*, Zählschaltung *f* ~ **decade** Zähldekade *f* ~ **device** Zähl-einrichtung *f*, -werk *n* ~ **disc** Zählscheibe *f* ~ **gas** Zählgas *n* ~ **house** Zechenhaus *n* ~ **individual pulses** Einzelimpulszählung *f* ~ **ionization chamber** Ionisationskammer *f* ~ **loss** Totzeitkorrektur *f* ~ **mechanism** Zählvorrichtung *f* ~ **rate** Zähl-frequenz *f*, -geschwindigkeit *f*, -rate *f* ~ **rate characteristic** Zählratecharakteristik *f* ~ **rate meter** Zählgeschwindigkeitsmesser *m* ~ **reel** Zählweise *f* ~ **relay** Relais *n* für Stromstoßzählung ~ **roll** Zählwalze *f* ~ **train** Zählwerk *n* ~ **tube** Zählröhre *f* ~ **units** Zähleinschübe *pl* ~ **weighers** Zählwaagen *pl*
**countless** unzählig
**country** Feld *n*, Gebirge *n*, Land *n*, Landkreis *m* ~**(side)** Gelände *n* ~ **of destination** Bestimmungsland *n* ~ **of origin** Heimatstaat *m*, Ursprungsland *n* ~ **of publication** Erscheinungsland *n*
**country,** ~ **adjacent to the coast** Küstenstrich *m* ~ **fog** Landnebel *m* ~ **rock** Nebengestein *n*
**coupé** geschlossener Zweisitzer *m*
**couple, to** ~ anhängen, ankuppeln, aufpropfen, einkuppeln, einrücken, koppeln, kraftschlüssige Verbindung herstellen, kuppeln, schalten, verankern, verbinden **to** ~ **back** rückkoppeln **to** ~ **at the outlet side** nachschalten **to** ~ **poles** Telegrafenstangen verkuppeln **to** ~ **up with** ankuppeln an
**couple** äußere Lötstelle *f*, Drall *m*, Koppel *n*, Kräftemoment *n*, Paar *n*, Zweiheit *f* ~ **forces** Kräftepaar *n* ~ **wire** Drahtschenkel *m*
**coupled** gekoppelt, gekuppelt ~ **axle** Kuppelachse *f* ~ **boundary** gekoppelte Grenzfläche *f* ~ **brushes** Bürstenpaar *n* ~ **camera** Koppelkammer *f* ~ **circuit chain** aus gekoppelten Kreisen bestehender Kettenleiter *m*, Gebilde *n* aus gekoppelten Kreisen ~ **circuits** gekoppelte

Kreise *pl* ~ **effect** Kopplungseffekt *m* ~ **oscillatory circuits** gekoppelte Schwingungskreise *pl* ~ **poles** Doppelgestänge *n* ~ **rigidly** phasenstarr verkoppelt ~ **wheel** Kuppelrad *n* ~ **wheel with lateral play** verschiebbares Kuppelrad *n* ~ **wheel set** Kuppelradsatz *m*
**coupler** Gerätesteckvorrichtung *f*, Koppler *m*, Kopplungsspule *f*, Kuppler *m*, Kupplung *f*, Wagenkuppler *m* ~ **on pliers** Zangenring *m*
**coupler,** ~ **jaw** Kuppelklaue *f* ~**-key recess** Rohrhalter *m* (Verbindung mit der Rücklaufbremse) ~ **knuckle** Ankuppelungsgelenk *n* ~ **link** Kupplungslied *n*, (of a chain) Verschlußglied *n* ~ **lock** Arretierstift *m*, Kupplungssperre *f* ~ **plug** Gerätestecker *m* ~ **plug and socket connection** Gerätesteckvorrichtung *f* ~ **recess** Lagerbock *m* ~ **shim** Kupplungsbeilegescheibe *f* ~ **socket** Geräteanschlußkabelsteckdose *f*
**couplet** Antennenpaar *n*
**coupling** Ankopplung *f* (rdo), Einrücken *n*, Keilschloß *n* (einer Kuppelung), Kopplung *f*, Kupplung *f*, Muffe *f*, Platzzusammenschaltung *f* (mit Platzschalter), Schlauchverbindung *f*, Verschraubung *f*, Vorschaltung *f* ~ **(shaft)** Wellenkupplung *f* ~ **for abutting shafts** Kupplung *f* zur Verbindung zweier Wellenenden ~ **of injection timing device** Spritzverstellerkupplung *f* ~ **of two oscillatory circuits** Kopplung *f* von Schwingungskreisen ~ **through pin** Mitnehmerstift *m* ~ **with plain ends** gerade Muffe *f* mit zweifach konischem Gewinde ~ **of short range** Kopplung *f* kurzer Reichweite ~ **with recessed ends** doppelt-konische Muffe *f* mit Einschnitten ~ **for tube conduit** Gewindemuffe *f* ~ **and uncoupling hoses** Anschließen *n* und Abnehmen *n* der Schläuche
**coupling,** ~ **adapter** Mitnehmerstück *n* ~ **arrangement** Kupplungsvorrichtung *f* ~ **box** Abzweigkasten *m*, Kupplungsmuffe *f* ~ **bushing** Kupplungsbüchse *f* ~ **cage** Fangkorb *m* ~ **cap** Kupp(e)lungsmutter *f* ~ **capable of being disengaged** ausrückbare Kupplung *f* ~ **capacity** Kopplungskapazität *f* ~ **cases** Kopplungsmöglichkeiten *pl* ~ **chain** Mitnehmerkette *f* ~ **changer** Kopplungswechsler *m* ~ **circuit** Ankoppelkreis *m* (electr.) ~ **coefficient** Kopplungskoeffizient *m*, Kopplungsziffer *f* ~ **coil** Koppelspule *f*, Kopplungsspule *f*, Kuppelspule *f* ~ **condenser** Kopplungskondensator *m* ~ **control** Kopplungsregler *m* ~ **crosshead pin** (in Diesel) Gelenkteil *m* ~ **cross piece** Mitnehmerquerstück *n* ~ **device** Kuppelvorrichtung *f* ~ **dog** Kupplungshahn *m* ~ **element** Kopplungslied *n* ~ **factor** Kopplungsfaktor *m*, Kopplungskoeffizient *m*, Kopplungsrichtwert *m*, Kopplungsziffer *f* ~ **flange** Anschlußflansch *m*, Einbaustutzen *m* ~ **gear** Einrückvorrichtung *f* ~ **hook** Kuppelhaken *m* ~ **hose** Kuppelschlauch *m* ~**-hysteresis effect** Ziehen *n*, Zieherscheinung *f* ~ **key** Platzumschalter *m* ~ **lever** Kupplungshebel *m* ~ **lining** Kupplungsbelag *n* ~ **link** Befestigungsglied *n*, Verbindungsglied *n* ~ **loop** Kopplungsschleife *f* ~ **member** Befestigungsglied *n*, Kopplungsglied *n* ~ **members** Koppelelemente *pl* ~ **mouth** Kupplungsmaul *n* ~ **nut** Spannschloß *n*, Überwurfmutter *f*, Verbindungsmutter *f* ~ **part** Mitnehmerteil *n* ~ **piece** Ansatz-, Kuppel-stück *n* ~ **pin** Kupp-

lungsbolzen *m*, Verbindungsstück *n*, Verschlußbolzen *m* ~ **pin with nut lock** Gewindebolzenverschluß *m* ~ **pipe between turbine and condenser** Abdampfstutzen *m* ~ **plug** Verbindungsstecker *m* ~ **probe** Kopplungssonde *f* ~ **reducer** Reduktionsklemme *f* ~ **relay** Kuppelrelais *n* ~ **resistance** Kopplungswiderstand *m* ~ **right and left** gerade Muffe *f* mit Rechts- und Linksgewinde ~ **rod** Kuppel-, Kupplungs-stange *f* ~ **shaft** Kuppel-, Kupplungs-welle *f* ~ **shaft end** Kuppelwellenstumpf *m* ~ **sleeve** Kupplungshülse *f* ~ **spindle** Kuppelspindel *f* ~ **spindle thread for railroad work** Eisenbahnkupplungsspindelgewinde *n* ~ **supply line** Kupplungsspeiseleitung *f* ~ **tappet** Auslöseknagge *f* ~ **terminal** Kupplungsklemme *f* ~ **transformer** Kopplungs-transformator *m*, -übertrager *m* ~ **tube** Kopplungsröhre *f* ~ **way** Kupplungsschlitz *m* ~ **wheel** Kupplungsrad *n* ~ **winding** Schubwicklung *f* ~ **wrench** Kupplungsschlüssel *m*

**coupon** (on castings) Anguß *m*, Gutschein *m*, Kupon *m*, Zinsschein *m*

**courier** Kurier *m*

**course** Bahn *f*, (of a ship) Fahrt *f*, Flugrichtung *f*, Gang *m*, Geschoßbahn *f*, Kurs *m*, (of stone) Lage *f*, Lauf *m*, Laufbahn *f*, Rennbahn *f*, Rennstrecke *f*, Richtung *f*, (bricks) Schicht *f*, Streichen *n*, Verlauf *m*, Weg *m*, Zug *m*, (to aim at) Zielkurs *m* of ~ selbstverständlich

**course,** ~ **of action** Aktionsverlauf *m*, Gefechtsverlauf *m* ~ **at arrival** Ankunftskurs *m* ~ **away from airport** ablaufender Kurs *m* ~ **away from beacon** ablaufender Zielkurs *m* ~ **toward beacon** anlaufender Zielkurs *m* ~ **of beam** Leitrichtung *f*, Strahlengang *m* ~ **of blocks** Blockreihe *f*, Lagen der Blöcke ~ **of bricks** Backsteinschicht *f*, Ziegelsteinschicht *f* ~ **of business** Geschäftsgang *m* ~ **of calculation** Rechengang *m* ~ **of combustion** Verbrennungsablauf *m* ~ **of construction** Baukurs *m* ~ **(or behavior) of the control surface** Ruderlauf *m* ~ **of cure** (in vulcanizing rubber) Programm *n* ~ **of the current** Stromverlauf *m* ~ **of curve** Kurvenverlauf *m* ~ **at departure** Abgangskurs *m* ~ **of developement** Entwicklungsgang *m* ~ **and distance calculator (or computer)** Kurs- und Entfernungsschätzer *m* ~ **of events** Hergang *m* ~ **of evaporation** Verdunstungsverlauf *m* ~ **of exchange** Kurszettel *m* ~ **of flight** Flugweg *m* ~ **(or curve) of forces** Kräfteverlauf *m* ~ **of headers** Binderschicht *f*, Kopfschicht *f*, (masonry) Kopfstückenschicht *f*, Streckschicht *f* ~ **of instruction** Ausbildungskursus *m*, Bildungsgang *m*, Kursus *m*, Lehrgang *m*, Lehrkursus *m* ~ **of law** Rechtsweg *m* ~ **of (university) lectures** Kolleg *n* ~ **of life** Lebenslauf *m* ~ **of manufacture** Arbeitsgang *m*, laufende Fabrikation *f* ~ **of ore** Erztrumm *m* ~ **of the proceedings** Verlauf *m* des Verfahrens ~ **of production** Produktionsgang *m* ~ **of rays** Strahlverlauf *m* ~ **of reaction** Reaktionsverlauf *m* ~ **of a river** Lauf *m* eines Flusses ~ **of roofing felt** Bahn *f* (Dachpappe) ~ **of the ship** Schiffsweg *m* ~ **of stretchers** Läuferschicht *f* ~ **of thread** Gewindegang *m*

**course,** ~ **angle** Lagewinkel *m* ~ **angle lead** Kurswinkelvorhalt *m* ~ **arrow** Kurspfeil *m* ~ **bearing** Kurs-peilung *f*, -richtung *f* ~ **cal**culator Kursrechner *m* ~ **change or shift** Kurswechsel *m* ~ **compass** Richtgeber *m* ~ **computer** Kurs-rechengerät *n*, -rechner *m* ~ **correction** Kursrichtigstellung *f* ~ **error** Fahrtfehler *m*, Kursfehler *m* ~ **figure** Kurszahl *f* ~ **gyro battery** Bordbatterie *f* ~ **heading in wind** Windkurs *m* ~**-indicating beacon** Bake *f* für Kursanzeige ~**-indicating radio beacon** Richtungsanzeiger *m* ~ **indicator** Kursanzeiger *m*, Zielgerät *n* ~ **laid straight endwise** Läuferschicht *f* ~ **light** Kursrichtungslicht *n* ~ **made good** Flugrichtung *f* über Grund ~ **mark** Kurszahl *f* ~ **motor** (autopilot) Kursmotor *m* ~ **pointer** Kurszeiger *m* ~ **recorder** (aviat.) Kursschreiber *m* ~**-setting switch** Richtungsgeber *m* ~ **sight** Kursvisier *n* ~ **and speed calculator** Dreieckrechner *m* ~**-steering apparatus** Kurssteueranlage *f* ~ **time signal** Kurszeitzeichen *n* ~ **triangle** Kursdreieck *n* ~ **(or speed) triangle** Fahrtdreieck *n*

**courses of blocks** Blocklagen *pl*

**court** Instanz *f*, (glass-roofed) Lichthof *m* **out of** ~ außergerichtlich ~ **of appeals** Beschwerdegericht *n*, höhere Instanz *f* ~ **of arbitration** Schiedsgericht *n* ~ **of common pleas** Zivilgerichtshof *m* ~ **of competent jurisdiction** zuständiger Gerichtshof *m* ~ **of honor** Ehrenrat *m* ~ **of inquiry** Untersuchungskommission *f* ~ **(or jury) of honor** Ehrengericht *n* ~ **of justice** Gerichtshof *m*

**court yard** Hof *m*

**covalence** Netzbindung *f*

**covalent,** ~ **bond (or bonding)** kovalente Bindung *f*

**covariance** Kovarianz *f*

**covariant** Kovariante *f* ~ **constant** kovariant, konstant ~ **derivative** kovariante Ableitung *f*

**cove** Bucht *f*, Eckleiste *f* für Decken-, Stuck-kehle *f*

**coved,** ~ **edges** Böschungskante *f* ~ **skirting** Kehlsockel *m*

**covellite** Covellit *m*, Kupferindigo *m*

**cover, to** ~ abdecken, auftragen, auskleiden, ausschlagen, (with paint, insulation) bekleiden, belegen, beschirmen, beschlagen, bestreichen, (with) beziehen, decken, einhüllen, einkapseln, maskieren, sich bedecken, überdecken, überfangen, überspinnen, überziehen, umdecken, ummanteln, umspinnen, umwickeln, verdecken, verkleben, verkleiden, verschalen, zudecken, (a distance) zurücklegen

**cover, to** ~ **over** überstülpen **to** ~ **up** eindecken, zudecken **to** ~ **with** beschütten ~ **the back** den Rücken decken **to** ~ **with boards** bebohlen, bedielen **to** ~ **with small broken stone** aufschottern **to** ~ **with chesses** bebohlen **to** ~ **its costs** freibauen **to** ~ **a distance** eine Strecke zurücklegen **to** ~ **with drawings** vollzeichnen **to** ~ **with ice** vereisen **to** ~ **with metal (or wood)** beplanken (aviat.) **to** ~ **oneself** with sich dekken mit **to** ~**-print** aufsetzen **to** ~ **by pouring** übergießen **to** ~ **with rime** bereifen **to** ~ **a risk** ein Risiko decken **to** ~ **subsequently** nachdecken **to** ~ **with tapestry** tapezieren **to** ~ **wire** den Draht umspinnen, (with cotton etc) bespinnen, (with cotton) bewickeln **to** ~ **wire with braid** den Draht umklöppeln

**cover** Abdeckung *f*, Belag *m*, Decke *f*, Deckel *m*,

Deckplatte *f*, Deckung *f*, Einband *m*, (binding) Einbanddecke *f*, Formbelag *m* (paper mfg.), Futteral *n*, Haut *f*, Hülle *f*, Kapsel *f*, Kastendeckel *m*, Klappe *f*, Kuvert *n*, Plane *f*, Schirmdach *n*, Schlußdeckel *m*, Schutz *m*, Schutzdecke *f*, Schutzdeckel *m*, Schutzhaube *f*, Schutzmantel *m*, Überdeckung *f*, Überzug *m*, Umschlag *m*, Verkleidung *f*, Verkleidungsdeckel *m*, Verschlußdeckel *m* ~ **and bells** (of blast furnace) doppelter Gichtverschluß *m* ~ **of shaft tubbing** Cuvelagedeckel *m* ~ **of slide valve** Deckfläche *f*

**cover,** ~ **assembly** Behälterdeckel *m* ~ **beading** Deckelwulst *m* ~ **belt holding roller** Deckbandrolle *f* ~ **binding** Schalenholzzimmerung *f* ~ **bolt** Deckelschraube *f* ~ **buffer** Deckelanschlag *m* ~ **card** Umschlagkarton *m* ~ **convoy** Bedeckung *f* ~ **design** Umschlagzeichnung *f* ~ **edge** Deckelrand *m* ~ **equipment** Deckelausrüstung *f* ~ **film** Schutzschicht *f* ~ **glass** Deckglas *n*, Instrumentenglas *n*, (of microscope) Deckgläschen *n* ~ **glass cowl** Abschlußglashaube *f* (bei Scheinwerfern) ~ **glass gauge** Deckglastaster *m* ~ **hook** Schrankhaken *m* ~ **joint** Deckelverschraubung *f* ~ **latch** Deckelsperre *f* ~ **lock of the centrifuge** Deckelverriegelung *f* der Schleuder ~ (**manhole**) Abdeckplatte *f* ~ **name** Deckname *m* ~ **nut** Gewindedeckel *m*, Hutmutter *f* ~ **pivot** Deckelzapfen *m* ~ **plate** Abdeck-platte *f*, -scheibe *f*, Abschlußdeckel *m*, Gurtplatte *f*, Kopfplatte *f*, Lasche *f*, Schieber *m*, Stoßblech *n* ~-**plate catch** Deckelriegel *m* ~-**plate latch** Deckriegel *m* ~-**plate lock** Deckelsperre *f* ~-**print effect** Überdruckeffekt *m* ~-**print style** Überdruckartikel *m* ~-**printing** (overprinting) Überdrucken *n* ~-**printing unit** Umschlagwerk *n* ~ **rail** Deckschiene *f* ~ **ratio** Schattenverhältnis *n* ~ **remover** Abhebe-, Abschöpf-löffel *m* ~-**ring** (or **washer**) Abdeckring *m* ~ **screw** Deckelschraube *f* ~ **sheet** Deck-blatt *n*, -blech *n*, Umschlagseite *f* ~ **slab** Schachtabdeckplatte *f* ~ **slide** Bodenklappe *f* ~ **slides** Deckplättchen *n* ~ **slip** Haubenhalter *m* ~-**slip thickness** Deckglasdicke *f* ~ **strap** Deckelriemen *m* ~ **strip** Abschlußkappe *f* ~ **strip of tail-plane tip** Flossenendkappe *f* ~ **stud** Deckelstiftschraube *f* ~ **trench** Deckungsgraben *m* ~ **tube** Verkleidungsrohr *n* ~ **valve** Deckelschieber *m* ~ **washer** Deckscheibe *f*

**coverage** Belegung *f*, Bewältigung *f*, Deckfähigkeit *f*, Deckkraft *f*, Erfassungsbereich *m* (rdr), luftbildmäßige Erfassung *f*, Reichweite *f* ~ **of a frequency band** Überstreichung *f* eines Frequenzbereichs

**coverage,** ~ **diagram** Erfassungsdiagramm *n*, Umriß *m* des abgetasteten Gebietes (rdr) ~ **factor** Bedeckung *f* des Netzes, Netzbedeckung *f*

**coveralls** Kombination *f*

**covered** abgedeckt, ausgelegt, bedeckt, gedeckt, überzogen, ummantelt, umsponnen, umwickelt, zurückgelegt, (wire) besponnen ~ **by** angelehnt ~ **in all round** allseitig geschlossen ~ **with boils** beulig ~ **with foliage** bewachsen ~ **with insulating varnish** mit Isolierlack *m* überzogen ~ **with leaves** belaubt ~ **with metal** blechbeschlagen, metallbeplankt **to become** ~

**with moisture** mit Feuchtigkeit *f* beschlagen ~ **with stressed smooth metal skin** glattblechbeplankt ~-**in terminal connection** verdeckter Klemmenanschluß *m* (electr.)

**covered,** ~ **crucible** bedeckter Hafen *m* ~ **electrode** Mantelelektrode *f*, ummantelte oder umwickelte Elektrode *f* ~ **foil detector** eingehüllter Detektor *m* ~ **lights** gedeckte Lichter *pl* ~ **line** versenkte Linie *f* ~ **market** Markthalle *f* ~ **position** gedeckte Stellung oder Stellung *f* ~ **quay** Kaifläche *f* ~ (**water**) **reservoir at or below ground level** Wasserkeller *m* ~ **shade** gedeckter Farbton *m* ~ **slip** gedeckte Helling *f* ~ **tail booth** verkleidete Heckzelle *f* ~ **terrain** bedecktes Gelände *n* ~ **wire** isolierter oder umsponnener Draht *m*

**covering** Abdeckung *f*, Anstrich *m*, Aufnahme *f*, Außenhaut *f* (aviat.), Bedeckung *f*, Bekleidung *f* (wing), Belag *m*, Belegung *f*, Beplankung *f*, Bespannung *f* (aviat.), Bezug *m*, Decke *f*, Decken *n*, Deckung *f*, Futter *n*, Mantel *m*, Isolierung *f*, Schutzdecke *f*, Überfang *m*, Überziehen *f*, Überzug *m*, Umhüllung *f*, Umkleidung *f*, Umschlag *m*, Umspinnung *f*, Umwicklung *f*, Verdeckung *f*, Verkleidung *f*, Verschalung *f*

**covering,** ~ **of bars** Einbettungstiefe *f* ~ **for cockpit** verstellbare Bespannung *f* eines Führersitzes ~ **with fibers** Beflockung *f* (elektrostatische) ~ **of the fuselage** (or **body**) Rumpfbekleidung *f* ~ **with insulating material** Abdecken *n* mit Isolierstoff ~ **with metal** (or **wood**) Beplankung *f* ~ **for ruling rubbers rollers** Liniergummiwalzenbezug *m* ~ **of step** Stufenbelag *m* ~ **of two sheets** zweiblättrige Überlagerungsfläche *f* ~ **for walls** Wandbekleidungsstoff *m*

**covering,** ~ **bath** Deckbad *n* ~ **capacity** Deckfähigkeit *f* ~ **chain** Zuschleifkette *f* ~ **clamp** Abdeckschelle *f* ~ **coat** deckender Anstrich *m* ~ **damping roll swanskin and sleeves** Wischwalzen-Molton und -Schläuche *pl* ~ **device** Abdeckvorrichtung *f* ~ **disc** Überfangscheibe *f* (für Begrenzungsleuchte) ~ **feuer** Deckungsfeuer *n*, Feuerschutz *m* ~ **frame** Eindeckrahmen *m* ~ **layer of body tissue** Körperoberfläche *f* ~ **letter** Begleitschreiben *n* ~ **liquor** Deck-kläre *f*, -klarsel *n* ~ **machine** Beplankungsgroßvorrichtung *f*, Umspinnmaschine *f* ~ **material** Bespannungsstoff *m*, Deckmittel *n*, Dickmittel *n* ~ **motion** Deckbewegung *f* ~ **operation** (X-ray practice) Deckoperation *f* ~ **panel** Verkleidungsblech *n* ~ **party** Bedeckungsabteilung *f*, Begleitkommando *n* ~ **plate** Belagplatte *f* ~ **position** Aufnahmestellung *f*, Zwischenfeld *n* ~ **power** (of paint) Farbdeckfähigkeit *f*, (for paints) Ausgiebigkeit *f*, Deckkraft *n* ~ **rail** (or **border**) Abdeckleiste *f* ~ **sheet** Belagbogen *m* ~ **shop** Bespannerei *f* ~ **sirup** Decksirup *m* ~ **skin** Behäutung *f* ~ **strip** Deckenleiste *f*, Eckkappe *f*, Hautblechstreifen *m* ~ **surface** Überlagerungsfläche *f* ~ **terminal** Abdeckklemme *f* ~ **tube** Mantelrohr *n* ~ **warp** Deckkette *f*

**covibrant,** ~ **conducter** mitschwingender Leiter *m* ~ **string** mitschwingende Saite *f*

**covibrate, to** ~ mitschwingen

**covibration** Mit-, Resonanz-schwingung *f*

coving Hohlkehle *f*

covolume Kovolumen *n*, Kovolum *n*

cowcatcher Gitter-, Schienen-räumer *m*

cowl, to ~ verkleiden

cowl Aufsatz *m*, Haubenverkleidung *f*, Kapuze *f*, Windhaube *f* ~ **fairing** stromlinienförmiger Übergang *m* von der Haube zum Rumpf ~ **flap** Kühlluftregelklappe *f*, Spreizklappe *f* an der Motorhaube ~**-mounting support** Halterung *f* für Motorhaube ~ **side-panels** Windführungs-seitenbleche *pl* (Auto) ~ **skirt** zylindrischer Teil *m* der Haubenverkleidung ~ **well** ring-förmige Vertiefung *f* am Umfang der Haubenverkleidung

cowled kappenförmig

cowling Aufsetzen *n*, F-Haube *f*, Verkleidung *f*, Verschalung *f* **(engine)** ~ Motor-haube *f*, -haubenblech *n* ~ **clip** Kappenverschluß *m* ~ **flaps** Belüftungsklappen *pl* ~ **ring** Strömungshaube *f*

co-worker Fachgenosse *m*, Mitarbeiter *m*

Cowper, ~**-Coles** (cold-galvanizing) **method** Cowper-Coles-Methode *f* ~ **stove** Cowper *m* ~ **stove chimney** Cowperesse *f* ~ **stoves** Winderhitzer *m*

coxswain Steuermann *m*

crab, to ~ krabben, dem Wind entgegendrehen, (of airplane) schieben

crab Hebezug *m* (mach.), Krabbe *f*, Winde *f* ~ **apple** Härtling *m* ~ **fitted with driver's stand** Führerstandlaufkatze *f* ~ **winch** Krüppelspillwinde *f*

crabbling windschiefer Flug *m*, krabbenartiges Fliegen *n* ~ **of an airplane** Flugzeugschieben *n* ~ **machine** Einbrennmaschine *f*

crack, to ~ aufknacken, aufreißen, aufspringen, ausspringen, bersten, knacken, (of loudspeaker) knarren, knattern, knicken, knistern, krachen, kracken, platzen, reißen, (painting) Risse bilden, rissig werden, ritzen, schlitzen, spalten, springen, zerspringen **to ~ off** abbersten, abspringen **to ~** (oil) abspalten

crack Ablösung *f*, Anriß *m*, (failure) Bruchriß *m*, Bruchstelle *f*, Fehlstelle *f*, Kluft *f*, Knack *m*, Knick *m*, Riß *m*, Ritz *m*, Ritze *f*, Spalt *m*, Spalte *f*, Sprung *m*, Windriß *m* ~ **from contraction** Kontraktionsspalte *f* ~ **in the cooling jacket** Kühlmantelriß *m* ~ **in pipes** Rohrbruch *m* ~ **formed during hardening** Hartborst *m*, Härteriß *m*

crack, ~ **detector** Anrißsucher *m* ~**-extension-force** Rißausweitungskraft *f* ~ **lashing** Rödelbund *m* ~**s per pass** Koeffizient *m* der Transformation durch Kracken ~**-proof** reißfest ~ **propagation** Rißausbreitung *f* ~ **resistance** Rißfestigkeit *f* ~ **silencer** Knallschutzgerät *n* ~ **strength** Rißfestigkeit *f*

cracked (ein)geknickt, gesprungen (springen), klüftig, rissig, (metal or glass) gerissen (reißen) ~**-carbon resistance** aufgedampfter Kohlewiderstand *m* ~ **gasoline** gekracktes Benzin, Krackbenzin *n*, Spaltbenzin *n* ~ **insulator** gesprungener Isolator *m* ~ **oil** Aufspaltöl *n* ~ **place** Rissigkeit *f* ~ **wood** kernschäliges Holz *n*

cracking Anriß *m*, Aufspalten *n*, Knacken *n*, Knattern *n*, Kracken *n*, Kracking *n*, Rißbildung *f*, Rupfen *n*, Spaltung *f*, Springen *n*, Sprungbildung *f*, Zerknicken *n*, Zersetzung *f*,

(oil) Abspaltung *f* ~ **on hardening** Härterißbildung *f* ~ **on welding** schweißrissig

cracking, ~ **charge** Erschütterungsladung *f* ~ **furnace** Feuerung *f* für die Toppingsanlage oder Krackanlage ~ **limit** Berstdruck *m* ~**-off ring** Abspringring *m* ~ **process** Krackprozeß *m*, Krackverfahren *n* ~ **stock** Rohstoff *m* für Kracking ~ **unit** Spaltanlage *f*

crackle, to ~ knacken, knarren, knistern, krachen, (painting) kraquelieren, verpuffen

crackle Prasseln *n* ~ **finish** Eisblumenlackierung *f*, Reißlack *m* ~ **laquer** Reißlack *m*

crackling Knistern *n*, Kratzgeräusch *n*, Nebengeräusch *n*, Verpuffung *f* ~ **(noise)** Knallgeräusche *pl* ~ **sound** (of tin) Schreien *n* ~ **sound of tin** Kreischen *n* des Zinns

cracks, ~ **in cage** Käfigbruch *m* (Kugellager) ~ **in the direction of the grain** Spiegelkluft *f*

cracky brüchig

cradle Ablaufgerüst *n*, Biegebalken *m*, Gabel *f* des Tischfernsprechers, Gestell *n* (mech.), Lagerstuhl *m*, Lattenkäfig *m*, Mulde *f*, (artil.) Rohrwiege *f*, Rostkatze *f* (metall.), Schienenstuhl *m*, Schlitten *m*, Stachelschwein *n*, Telefongabel *f*, Waffenlagerung *f*, Wälzkörper *m*, Wiege *f* ~ **of a penstock** Sockel *m*

cradle, ~ **carrier** Gabelträger *m* ~ **crane** Pratzenkran *m* ~ **dynamometer** Pendelwaage *f* ~ **fixation** (pan mounting) Wannenbefestigung *f* ~ **frame** Pendelrahmen *m*, Wiegenträger *m* ~ **guard** Erdungsbügel *m*, Schutznetz *n* ~ **guide rails** Wiegengleitbahn *f* ~ **hammer drill** Schiffchenbohrhammer *m* ~ **joint** Richtgelenk *n* ~ **lashing (or lock)** Wiegenzurrung *f* ~ **lock frame** Zurrbrücke *f* ~ **lock handle** Zurrgriff *m* ~ **relay** Kammrelais *n* ~ **rest** Gabelstütze *f* (teleph.) ~ **roll motion scale** Wälzlängenskala *f* ~ **support** Wiegenträger *m* ~ **switch** Gabelumschalter *m* ~ **tip wagon** Wiegentipper *m* ~ **trough** Wiegentrog *m* ~ **wall** Lafettenwand *f*

cradled electric dynamometer elektrische Pendelwaage *f*

craft (marine, land or air) Fahrzeug *n*, Gewerbe *n*, Kunstfertigkeit *f*

craftmanship Werkmannsarbeit *f*

craftsman Gewerbetreibender *m*, Handwerker *m*

crag Felswand *f*, Klippe *f*

craggy felsig

cram, to ~ mästen, vollstopfen

cramming Anfüllung *f*

cramp, to ~ ankrampen, beengen, klammern, klemmen

cramp Klammer *f*, Zwinge *f* ~ **iron** Haspe *f*, Krammeisen *n*, Krampe *f*, Kropfeisen *n* ~ **iron for fastening the jambstones on the wall** Gewändanker *m* ~ **iron for wood** Bankeisen *n* ~**-iron machine** Stollenherstellungsmaschine *f*

cramped beengt, eingeengt ~ **working space** beengter Arbeitsraum *m*

crane Aufzug *m*, Hebelade *f*, Heh (Heißvorrichtung), Höhenwinde *f*, Kran *m*, Kranich *m*, Kranwinde *f*, Ladebaum *m*, Laufkatze *f* **(overhead)** ~ **for pig bed** Gießbettkran *m*

crane, ~ **apparatus** Krananlage *f* ~ **arm of lattice type** gitterförmiger Ausleger *m* ~ **balks** Fahrbahnträger *m* ~ **beam (or boom)** Kranausleger *m* ~ **bridge** Kranbrücke *f* ~ **cable** Kranseil *n* ~ **carriage** Laufkatze *f* ~ **chain** Krankette

f ~ **construction** Kran-bau *m*, -konstruktion *f*
~ **column (or post)** Kranstütze *f* ~ **design**
Krankonstruktion *f* ~ **driver** Laufkranführer
*m* ~ **erection** Kranmontage *f* ~ **fitter** Kran-
monteur *m* ~ **fly** Schnake *f* ~ **frame** Krangerüst *ŋ* ~ **hook** Förderhaken *m* ~ **installation**
Krananlage *f* ~ **jib** Kran-arm *m*, -ausleger *m*
~ **ladle** Gießpfanne *f* mit Kran-gehänge, -gieß-
pfanne *f*, -pfanne *f* ~-**ladle bail** Kranpfannen-
gehänge *n* ~ **ladle with handtipping gear** Krangießpfanne *f* mit Getriebekippvorrichtung ~
**load** Kranlast *f* ~ **magnet** Kranmagnet *m* ~
**operator** Kranführer *m* ~ **operator's cabin**
(suspended and traveling on rail) Führersitz-
katze *f* ~ **pontoon** Kranprahm *m* ~ **rail** Kranschiene *f* ~ **rope** Kranseil *n* ~ **runway** Kran-
bahn *f* ~-**travel gearing** Kranfahrwerk *n* ~
**travel unit** Kranfahrwerksantrieb *m* ~ **traveling crab** Kranlaufkatze *f* ~ **traveling on
rails** Auslegerlaufkatze *f* ~ **trolley** Krankatze *f*
~ **truck** Kranwagen *m* ~-**truck ladle** Gießpfanne *f* mit Gehänge und Wagen ~-**type
motor driven trolley hoist** Laufwinde *f* mit
elektrischem Hub und Fahrwerk
**craniometer** Kraniometer *n*
**crank, to** ~ andrehen, ankurbeln, anlassen, (an
engine) anwerfen, kröpfen **to ~ an engine** den
Motor *m* durchdrehen **to ~ the engine** die
Maschine *f* andrehen, den Motor *m* anlassen
**crank** Drehling *m*, Gleitstein *m* einer Kulisse,
Handkurbel *f*, Kröpfung *f*, Krummachse *f*,
Kurbel *f*, (of a shaft) Kurbelkröpfung *f*, Leit-
arm *m*, Ruderhebel *m*, (of car) Vordreh-
schleuder *f*, Winkelhebel *m*, Zieharm *m*,
Zwinge *f* ~ **of breech mechanism** Schubkurbel *f*
~ **and flywheel pump** Kurbelgetriebeschwung-
radpumpe *f* ~ **for racklever** Stange *f* für Ver-
satzhebel ~ **of a shaft** Wellenkröpfung *f* ~ **of
main shaft** Bohrkurbel *f*
**crank,** ~ **angle** Kurbel-stellung *f*, -winkel *m* ~
~ **arm** Kurbelarm *m* ~ **arrangement** Kurbel-
-einrichtung *f*, -gestänge *n* ~ **assembly** Kurbeltrieb *m* ~ **axle** Kropfachswelle *f*, Kurbelachse *f*
~ **back balance** Kurbelwellengegengewicht *n*
~ **bar** Kurbelstange *f* ~ **bed lowering lever** doppelarmiger Hebel *m* ~ **block** Kurbelstein *m* ~
**box** Drehdose *f* ~ **brace** Faustleier *f* ~ **cam**
Kurbelscheibe *f*
**crankcase** Kurbelgehäuse *n*, Kurbelkasten *m*
(Kurbelkammer *f* ) ~ **breather** Kurbelge-
häuseentlüfter *m* ~ **breather pipe** Kurbelge-
häuseentlüftung *f* ~ **cover** Kurbelgehäusedek-
kel *m* ~ **dilution** Kurbelgehäuseschmierölver-
dünnung *f* ~ **explosion** Kurbelkastenexplosion
*f* ~ **front cover** Steuerräderdeckel *m* ~ **front
section** vorderer Kurbelgehäuseteil *m* ~ **gallery**
Kurbelgehäusebohrung *f* ~ **lower half** Ge-
häuseunterteil *n* ~ **main** ~ **section** mittlerer
Kurbelgehäuseteil *m* beim Doppelsternmotor
~ **oil** Kurbelgehäuse-, Kurbelkasten-öl *n* ~
**scavenging** Kurbelkastenspülung *f* ~ **section**
Kurbelgehäuseteil *m* ~ **sump** Ölmulde *f* ~
**supercharger** Kurbelkastenvorverdichter *m* ~
**upper half** Gehäuseoberteil *n*
**crank,** ~ **center** Kurbelzentrum *n* ~ **chamber**
Kurbel-gehäuse *n*, -kammer *f*, -kasten *m*,
-wanne *f* ~-**check rounding** Kurbelwangenausrundung *f* ~ **claw** Kurbelklaue *f* ~ **contact lever**

Winkeltasthebel *m* ~ **disk** Kurbel-blatt *ŋ*,
-scheibe *f* ~ **disk provided with an opening** ausgespartes Kurbelblatt *n* ~ **drawing press** Kur-
belziehpresse *f* ~ **drive** Kurbel-antrieb *m*,
-getriebe *n* ~-**drive motor** Motor *m* für Stan-
genantrieb ~ **end** Kurbelkopf *m* ~ **fulling mill**
Druckwalze *f* ~ **gear** Kurbel-getriebe *n*,
-trieb *m* ~ **guard** Kurbel-gehäuse *n*, -schutz-
haube *f*, Schutzhaube *f* ~ **guide** Kurbelschleife *f* ~ **handle** Handkurbel *f*, Kurbelgriff
*m*, Schloßkurbel *f* ~ **handle for changing feeds**
Handkurbel *f* für den Vorschubwechsel ~
**handle for raising tool block** Supportkurbel *f*
~ **handle tube** Kurbelgriffrohr *n* ~ **handle with
ball** Kugelhandkurbel *f* ~ **latch** Kurbelriegel *m*
~ **lever** Kurbelschiene *f* ~-**operated plate
shears** Kurbeltafelschere *f* ~-**operated window
pane** Kurbelfensterscheibe *f* ~ **pendulum** Win-
kelpendel *m* ~ **pin** Kropfhals *m*, Kurbel-griff
*m*, -wellenhubzapfen *m*, -zapfen *m*; Lauf-,
Treib-, Well-zapfen *m* ~ **pin bearing** Kurbel-
zapfenlager *n* ~ **pin clamping screw** Kurbel-
wellenklemmbolzen *m* ~ **pit** Kurbelwanne *f*
~ **point** Brech-, Winkel-punkt *m* ~ **press** Knie-
hebel-, Kurbel-presse *f* ~ **pressure** Kurbel-
druck *m* ~ **race** Kurbeltrog *m* ~ **racing** Schleu-
derkurbel *f* ~ **rod eyelet** Kurbelöse *f* ~ **rod
with roller** Schaltbolzen *m* mit Auflaufrolle
**crankshaft** Drillbohrer *m*, gekröpfte Welle *f*,
Kropfachswelle *f*, Kurbelachse *f*, Kurbelwelle
*f* ~ **assembly** Kurbeltriebwerk *n* ~ **bearing**
Kurbel-, Kurbelwellen-lager *n* ~ **bearing cover**
Kurbelwellenlagerdeckel *m* ~ **cover** Kurbel-
lochdeckel *m* ~ **cover plate** Kurbelwellen-
schutz *m* ~ **driven magneto** Kurbelwellen-
magnetzünder *m* ~ **end** Kurbelwellenstumpf *m*
~ **extension** Kurbelwellenverlängerung *f* ~
**feather** Kurbelwellenfederkeil *m* ~ **grinder**
Kurbelwellenschleifmaschine *f* ~ **housing** Mo-
torgehäuse *n* ~ **journal** Kurbelwellenende *n*
~ **key** Kurbelwellenfederkeil *m* ~ **main bearing**
Kurbelwellenhauptlager *n* ~ **pinion** Kurbelwel-
lenritzel *n* ~ **(belt) pulley** Kurbelwellenriemen-
scheibe *f* ~ **revolutions per minute** Kurbelwel-
lendrehzahl *f* ~ **rotational speed** Kurbelwellen-
drehzahl *f* ~ **seal** Kurbelwellenabdichtung *f*
~ **thrust bearing** Kurbelwellenpaßlager *m*
~ **vibration** Kurbelwellenschwingung *f* ~-**vibration damper** Kurbelwellenschwingungs-
dämpfer *m*
**crank,** ~ **sheave** Kurbelscheibe *f* ~ **slide oscillator** Kurbelschleife *f* ~ **slotting machine** Kur-
belstoßbohrmaschine *f* ~ **socket** Andrehklaue *f*
~ **starter** Kurbel-andrehvorrichtung *f*, -anlas-
ser *m*, Kurbelwellenanlasser *m* ~ **throw** Kurbelkröpfung *f* ~ **turning lathe** Kurbeldrehbank
*f* ~-**type swivel gun mount** Schwenkarmlafette *f*
~-**type telescopic mast** Kurbelmast *m* ~-**type
wiper** Kurbelwischer *m* ~ **web** Kurbel-arm *m*,
-blatt *n*, -schenkel *m*, -wange *f* ~ **wheel** Steuer-
rad *n* ~-**wheel drive** Kurbelradantrieb *m* ~
**windlass** Kurbelhaspel *f*
**cranked** gebogen, gekröpft, geschweift, mit einer
Kurbel versehen ~ **(or dropped) axle** gekröpfte
Achse *f* ~ **fishplate** Übergangslasche *f* (r.r.)
~ **flat iron** gekröpftes Flacheisen *n* ~ **handle**
Aufsteckkurbel *f* ~ **peg** Winkelstecker *m* ~
**portion of shaft** Kurbelkröpfung *f* ~ **punch** ge-

kröpfter Durchschlag *m* ~ **roughing tool** gekröpfter Schruppstahl *m* ~ **scissors** (for trimming wicks) Winkelschere *f* ~ **shaft** gekröpfte Welle *f* ~ **valve** Schlangenventil *n* ~ **wing** Knickflügel *m*

**cranking,** ~ **back** Zurückkurbeln *n* ~ **starter** Durchdrehanlasser *m*

**crankless engine** kurbelwellenloser Motor *m*, Taumelscheibenmotor *m*

**crash, to** ~ abstürzen (aviat.), krachen, stürzen **to** ~ **an airplane** das Flugzeug zu Bruch bringen

**crash** Absturz *m* (aviat.), Bruch *m*, Bruchlandung *f*, Krach *m*, Sturz *m* ~ **ability** Bruchlandeeignung *f* ~ **dive** Kopfsturz *m*, Schnelltauchen *n* ~ **equipment** Bergungsgerät *n* ~ **fire** Aufschlagbrand *m* ~ **frame** Aufpralldämpfungsrahmen *m* ~**-hat** Sturzhelm *m* ~ **helmet** Sturzhelm *m* ~**-landing** Bruchlandung *f* ~ **padding** Schutzpolsterung *f* ~ **proof tank** bruchsicherer Behälter *m* ~ **rescue** Rettung *f* ~ **rescue service** Rettungsdienst *m* ~ **switch** Schockschalter *m* ~ **tender** Tanklöschfahrzeug *n*

**crate** Kasten *m*, Kiste *f*, Kistenverschlag *m*, Kratte *f*, Lattenkiste *f*, Lattenverschlag *m*, Packkorb *m*, Stachelschwein *n*, Verschlag *m*

**crater** Kessel *m*, Krater *m*, Lichtspritze *f*, Trichter *m* ~ **area** Kraterfläche *f* ~ **aureols** Krateraureolen *pl* ~**-card** Kratersichtscheibe *f* ~ **cathode** Punktkathode *f* ~ **charge** Trichterladung *f* ~ **eliminator** Regelautomat *m* (eines Schweißgenerators) ~ **formation** Aufrauhung *f* des Brennfleckes, Kraterbildung *f* ~ **lake** Maar *n* ~ **lamp** Lichtspritze *f* ~ **neon lamp** Punktglimmlampe *f* ~ **point lamp** Lichtspritze *f* Lichtspritzelampe *f* ~**-shaped** punktförmig ~(**-point) tube** Lichtspritzeröhre *f*

**cratering** Auskolkung *f*

**crating** Lattenbeschlag *m*

**crave** Sprung *m* (plastics)

**crawl, to** ~ kriechen, schleichen, wandern

**crawl** Wanderungsgeschwindigkeit *f* (TV)

**crawler,** ~ **crane** Gleisketten-, Laufketten-, Raupen-kran *m* ~ **mounted conveyor** Raupenfördergerät *n* ~ **track assembly** Raupenfahrwerk *n* ~ **tractor** Raupenschlepper *m* ~ **truck** Raupe *f* ~ **undercarriage** Raupenunterbau *m*

**crawling** mit einer Unterfrequenz in Tritt kommen, Krabbeln *n*, Schleichen *n* ~ **speed** Kriechgang *m*

**crayon** Griffel *m*, Reißkohle *f*, Stift *m*, Zeichenstift *m* ~**-holder** Zeichenfeder *f*

**crazed** haarrissig

**creak, to** ~ knarren

**cream** Rahm *m*, Sahne *f* ~ **of tartar** doppeltweinsaures Kali ~**-colored** falb, kremfarben ~ **cooler** Rahmkühler *m* ~ **separator** Entrahmungsmaschine *f*, Milchentrahmer *m*, Milchseparator *m*, Milchzentrifuge *f* ~ **storage** Rahmkammer *f*

**creamery** Butterei *f*

**creaming agent** Aufrahmungsmittel *n*

**crease, to** ~ falten, runzeln, sieken, zerknittern, zerknüllen **to** ~ **the cover** den Umschlag *m* nuten

**crease** Bruch *m*, Eselsohr *n*, Falte *f*, Kniff *m*, Sieke *f* ~ **test-resistance** Knitterprobe *f* ~ **times** Abfallzeit *f*

**creased bend** Faltenrohrbogen *m*

**creasing** Faltenwerfen *n* ~ **of the paper** Knittern *n* des Papiers, Wellig-liegen *n*, -werden *n* ~ **and scoring lines** Rill- und Ritzlinien *pl*

**creasing,** ~ **block** Gegenstanze *f* ~ **die** Siekeneisen *n* ~ **hammer** Siekenhammer *m* ~ **resistance** Knitterfestigkeit *f* ~ **stake** Aufsatzamboß *m* zum Biegen von Werkstücken ~ **stakes** Siekenstöcke *pl* ~ **tool** Siekenzug *m*

**create, to** ~ begründen, erzeugen, hervorbringen, schöpfen, zeugen **to** ~ **a magnetic field** ein Magnetfeld *n* aufbauen

**creatine** Kreatinin *n*

**creation** Hervorbringung *f*, Schöpfung *f* ~ **of residual stresses** Selbstverspannung *f* ~ **operator** Erzeugungsoperator *m* ~ **rate** Paarbildungsgrad *m*

**creative** schöpferisch ~ **conception** Erfindergeist *m*

**credentials** Beglaubigungsschreiben *n*

**credit, to** ~ glauben, kreditieren, trauen, (Ansehen, Kredit) verschaffen

**credit** Ansehen *n*, Autorität *f*, Glaubwürdigkeit *f*, Haben *n*, Kredit *m* ~ **for bills** Wechselkredit *m*

**credit,** ~ **account** Verrechnungskonto *n* ~ **control** Kreditüberwachung *f* ~ **item** Aktivposten *m* ~ **lines** Vorspann *m* (film) ~ **memorandum** Gutschriftsanzeige *f* ~ **side** Habenseite *f* ~ **symbol** Habenzeichen *n* ~ **title** Vorspann *m* (film) ~ **transaction** Termingeschäft *n*

**crediting** Gutschrift *f* ~ **of flight time** Anrechnung *f* von Flugzeit

**creditor** Gläubiger *m* ~**'s** Habenseite *f*

**credits** Vorspann *m* (film)

**creek** Bucht *f*

**creel** Aufsteckgatter *n* (bei der Spinnmaschine), Lieferwerk *n* ~ **frame** Aufsteckrahmen *m* ~ **plate** Gattertisch *m*

**creep, to** ~ abgleiten, kriechen, schleichen, schliefen, wandern **to** ~ **through** durchkriechen, durchschlüpfen **to** ~ **up** anpirschen

**creep** Aufquellen *n*, Aufquellen *n* des Liegenden (min.), Bewegung *f* des Liegenden (min.), Bodensenkung *f*, Gleiten *n*, Kriech *n*, Gekriech *n*, Schuttkriechen *n*, Sohlenauftrieb *m*, Sohlendruck *m* (min.), Wanderungsgeschwindigkeit *f* (TV), Zählerleerlauf *m* ~ **of material** Materialwanderung *f* ~ **of the rails** Wandern *n* der Schienen ~ **of track** Verschieben *n* des Geleises in der Längsrichtung

**creep,** ~ **behavior** Kriechverhalten *n* ~ **buckling** plastisches Kriechen *n* ~ **distance** Kriechweg *m*, Kriechstrecke *f* (electr.) ~ **limit** Kriechfestigkeit *f*, Standfestigkeitsgrenze *f* ~ (**resistance) limit** Dauerstandfestigkeit *f* ~ (**leakage) paths** Isolationswege *pl* ~ **rate** Dehngeschwindigkeit *f*, Kriechgeschwindigkeit *f* (metall.) ~ **strain** Kriechdehnung *f* ~ **strength** Dauerfestigkeit *f*, Dauerstandfestigkeit *f*, Kriechgrenze *f*, Standfestigkeit *f* ~ **strength depending on time** Zeitstandfestigkeit *f* ~ **test** Dauerstandversuch *m*, Standfestigkeitsprüfung *f* ~ **testing machine** Kriechprüfmaschine *f* ~ **wash** Gekriech *n*

**creepage,** ~ **path** Kriechweg *m* (electr.) ~**-proof** kriechstromfest ~ **spark** Gleitfunken *m* ~ **spark gap** Gleitfunkenstrecke *f*

**creeper** Dregg-anker *m*, -haken *m*, Kettenbahn *f*

(min.), Schlinggewächs *n* ~ **blade** Kratzer-schaufel *f* ~ **lattice** Füllungslattentuch *n*
**creeping** Kriechen *n*, langsame Vertikalbewegung *f* (film), Wanderung *f*, (in rolling) Nacheilung *f* ~ **of the rails** Schienenwandern *n* ~ **current** Kriechstrom *m* ~ **distance** Kriechstrecke *f* ~ **film** Steigfähigkeit *f* ~ **speed** Feingang *m* (Elektrozug) ~ **strength** Kriechfestigkeit *f* ~ **title** Fahrtitel (film) ~ **wheat grass** Spitzgras *n*
**crematorium** Feuerbestattungsanstalt *f*
**crematory** Einäscherungsofen *m*
**Cremona diagram** Kräfteplan *m*
**crenated** gekerbt
**crenelate, to** ~ krenelieren
**crenelated** zinnenförmig ~ **poles** Pole *pl* mit Nuten
**creosote, to** ~ mit Kreosot *n* oder mit Teeröl *n* tränken
**creosote** Kreosot *n* ~ **oil** Kreosotöl *n*
**creosoted pole** mit Kreosot getränkte Stange *f*, geteerte Stange *f*
**creosoting** Kreosottränkung *f*
**crepe** Flor *m*, Krepp *m* ~ **rubber** Schaumgummi *n*
**crepitate, to** ~ knacken, knistern **to (de)~** verpuffen
**crepitation, (de)~** Verpuffung *f*
**crepuscular rays** Dämmerungsstrahlen *pl*
**crescendo pedal** (organ) Rollschweller *m*
**crescent** Kreiszweieck *n*, (musical instrument) Schellenbaum *m* ~-**shaped** mondförmig ~ **wrench** verstellbarer Schraubenschlüssel *m*
**cresol** Kresol *n*, Kresylsäure *f*
**cresotic acid** Kresotinsäure *f*
**cresset** Stocklaterne *f* .
**crest** Amplitude *f*, Berg *m* (einer Kurve), Bergkamm *m*, Firstkamm *m*, Grat *m*, Höhepunkt *m*, Höhen-rippe *f*, -rücken *m*, Krone *f*, Kuppe *f*, (value) Scheitelwert *m*, Schwelle *f*, Spitze *f* (eines Gewindes), Spitzenspiel (eines Zahnrades), Wellen-berg *m*, -kamm *m* ~ **of the dune** Dünenkamm *m* ~ **of a wave** Wellengipfel *m* ~ **of the weir** Fachbaum *m*, Wehrkrone *f*
**crest,** ~ **angle** Scheitelweg *m* ~ **clearance** Spitzenspiel *n*, Zahnkopfabrundung *f* ~ **factor** Scheitelfaktor *m* ~ **gate** Überlaufverschluß *m* ~ **indicator** Höchstwertanzeiger *m*, Maximumwertzeiger *m* ~ **length** Kammlänge *f* ~ **load** Belastungsspitze *f* ~ **meter** Scheitelmesser *m* ~ **points** Höhenpunkte *pl* ~ **power** Höchstleistung *f* ~ **template** Spitzenlehre *f* ~ **tile** Kaminziegel *m* ~ **value** Größt-, Scheitel-, Spitzen--wert *m* ~ **voltage** Gipfel-, Scheitel-, Spitzen--spannung *f* ~ **voltmeter** Impulsmesser *m*
**crestal area** Scheitelgebiet *n*
**cresylic acid** Cresylsäure *f*, Kresolsäure *f*, Kresylsäure *f*
**cretaceous** kreidehaltig ~ **marly limestone** Pläner *m* ~ **system** Kreidesystem *n* (geol.)
**cretonne** Doppelschirting *m*
**crevasse** Durchbruch *m*, (of a resonance or curve) Einsattelung *f*, Einsenkung *f*, Kluft *f*, Senke *f*, Zerklüftung *f* ~ **of resonance curve** Resonanzkurven-einsattelung *f*, -senke *f* ~ **curve** Resonanzkurve *f* des Quarzkristalls
**crevice** Innenriß *m*, Riß *m*, Ritz *m*, Ritze *f*, Spalt *m*

**creviced wood** kernschäliges Holz *n*
**crew** Belegschaft *f*, Bemannung *f*, Besatzung *f*, Führungsmannschaft *f*, Gefolgschaft *f*, Mannschaft *f*, Personal *n* ~ **change** Personalwechsel *m* ~ **member** Besatzungsmitglied *n* ~'**s nacelle** Besatzungsrumpf *m* ~'**s quarters** Mannschaftsraum *m*
**crib, to** ~ verbauen, verzimmern
**crib** Kranz *m* aus Holz oder Eisen, Krippe *f*, Senkkiste *f*, Werkzeugausgabe *f* ~-**biter** Krippensetzer *m* ~ **tubbing** Cuvelagezimmerung *f*, wasserdichter Ausbau *m* ~ **work** Senkkiste *f* ~ **work of wood** Holzgeflecht *n*
**cribber** Krippensetzer *m*
**crimp, to** ~ anwürgen, kräuseln, krimpfen, umbiegen **to** ~ **over** umfalzen
**crimp** Bruch *m*, Krause *f*, Kräuselung *f*, Würgung *f* ~ **in cartridge case** Würgerille *f*
**crimper** Randschleifer *m*
**crimping** Kröpfung *f* ~ **iron** Walkeisen *n* ~ **machine for wire netting** Krippmaschine *f* für die Drahtweberei ~ **press** Börtelpresse *f*
**crimson** hochroter Farbstoff *m*, Karmesin *n*
**crincle (or ripple) lacquer** Reißlack *m*
**cringle** Kausche *f*, Lägel *n*
**crinkle, to** ~ fälten, kräuseln, krumpeln
**crinkle** Kräuselkrankheit *f* ~ **finish** Kräusellack *m*, Schrumpflack *m*
**crinkled paper** Knitter-, Kräusel-papier *n*
**crinkling** Knickung *f*, Kräuschung *f* ~ **test** Knickerversuch *m*
**cripple, to** ~ verkrüppeln, zum Krüppel machen
**cripple leap-frog test** Teildurchprüfung *f*
**crippler** Krispelholz *n*
**crippling,** ~ **load** Knicklast *f* ~ **resilience** Knickspannung *f* ~ **stress** Knickspannung *f* ~ **test** Knickversuch *m*
**crisp, to** ~ kräuseln
**crisp image (or picture)** hartes Bild *n*
**crispening** Konturenbetonung *f*
**crisping** Kräuseln *n*
**criss-cross, to** ~ kreuz und quer laufen
**criss-cross** gitterartige Störung *f*
**criss-crossed winding** Kreuzwicklung *f*
**cristaline phases** kristalline Phasen *pl*
**cristobalite lattice** Cristobalitgitter *n*
**crit** kritische Bedingung *f*, kritische Masse *f*
**criteria programmed** programmierte Kriterien *pl*
**criterion** Anhaltspunkt *m*, Aufschluß *m*, Beurteilungsmerkmal *n*, Hauptmerkmal *n*, Kennmerkmal *n*, Kennzeichen *n*, Kriterion *n*, Kriterium *n*, Maßstab *m*, Merkmal *n* ~ **of climb** Flug-, Steig-zahl *f* ~ **of control effectiveness** Gütekriterium *n* ~ **of convergence** Konvergenzkriterium *n* ~ **of degeneracy** Entartungskriterium *n* ~ **for judgment** Beurteilungsmaßstab *m*
**critical** bedenklich, gefährdet, kritisch
**critical,** ~ **adjustment** kritische Einstellung *f* ~ **aircraft** kritisches Luftfahrzeug *n* ~ **altitude** Gleichdruckhöhe *f*, kritische Höhe, Volldruckhöhe *f*, Vollgashöhe *f* ~ **angle** Berechnungsgrenzwinkel *m*, Gefahrwinkel *m*, (Abstrahl-) Grenzwinkel *m* ~ **angle of attack where greatest lift is produced** kritischer Anstellwinkel *m* ~ **angle of a refractometer** Refraktometergrenzwinkel *m* ~ **angle of visual acuity** Sehschärfegrenzwinkel *m* ~ **anode voltage** kritische Ano-

denspannung f ~ **assembly** kritische Anordnung f ~ **bending speed** biegekritische Drehzahl f ~ **build-up resistance** kritischer Widerstand m für die (Selbst-) Erregung ~ **build-up speed** kritische Drehzahl f für die (Selbst-) Erregung ~ **case** Grenzfall m ~ **closing speed** kritische Gebrauchsgeschwindigkeit f ~ **components** komplizierte Einzelteile pl ~ **compression ratio** kritischer Druck m ~ **condition** Grenzbedingung f ~ **coupling** kritische oder optimale Kupplung f ~ **cruising altitude** Dauerleistungsgleichdruckhöhe f ~ **current** Grenzstromstärke f ~ **damping** kritische Dämpfung f ~ **density** orthobare Dichte f ~ **dimension** kritische Abmessung f ~ **engine** kritischer Motor m ~ **flicker frequency** Flimmer-, Verschmelzungs-frequenz f ~ **flicker fusion** Verschmelzungsfrequenz f ~ **focusing** kritische Einstellung f ~ **frequency** kritische Frequenz f, Durchdringungs-, fotoelektrische Schwellen-, Schwellen-frequenz f ~ **frequency of a filter** Grenzfrequenz f eines Filters ~ **grid current** kritischer Gitterstrom f ~ **grid voltage** Gitterzündspannung f, kritische Gitterspannung f ~ **head** kritische Druckhöhe f ~ **interval** Haltepunktsdauer f, Haltezeit f, kritischer Intervall m ~ **item** Engpaßartikel m ~ **limit** Abfallgrenze f ~ **load** Grenzbelastung f ~ **material** Sparwerkstoff m ~ **metal** Sparmetall n ~ **moment** kritischer Augenblick m, Wendepunkt m ~ **no-flicker frequency** Verschmelzungs-, Beruhigungs-frequenz f (film) ~ **opening speed** kritische Öffnungsgeschwindigkeit f ~ **point** Gefahrpunkt m, Höhepunkt m, kritischer Punkt m, Springpunkt m, Wechselpunkt m ~ **(thermal) point** Haltepunkt m, Umwandlungspunkt m ~-**point determination** Haltepunktbestimmung f ~ **potential** Löschspannung f ~ **pressure** kritischer Druck m ~ **range** Haltepunktsdauer f, Haltezeit f, kritischer Intervall m ~ **rate of mass transport** kritisches Verhältnis n des Massentransports ~ **resistance** Grenzwiderstand m, kritischer Widerstand m ~ **setting** kritische Einstellung f ~ **shear stress** Initialschubspannung f (cryst.), Schubfestigkeit f ~ **silence** Tonminimum n ~ **size** kritische Größe f ~ **speed** Durchsackgeschwindigkeit f, Gefahrmoment m, kritische Geschwindigkeit f ~ **speed of rotation** kritische Drehzahl f ~ **state** Notstand m ~ **value** Grenzwert m, kritischer Wert m ~ **velocity** Grenzgeschwindigkeit f ~-**velocity ratio** Machsche Zahl f ~ **voltage curve** Kippspannungsdiagramm m ~-**voltage parabola** Magnetfeldröhrenkennlinie f ~ **wave length** Grenzwellenlänge f, kritische Wellenlänge f ~ **zone** Schutzgürtel m
**criticality** Kritikalität f
**critically sharp (defined)** strichscharf
**criticism** Referat n
**criticize, to** ~ rezensieren, zensieren
**Crocco's eddy theorem** Wirbelsatz m
**crockery** Geschirr n, Steingut n ~ **(ware)** Tonwaren pl
**crocodile** Krokodil n, Ziehbank f ~ **shears** Alligator-, Hebel-schere f ~ **squeezer** Alligatorquetsche f
**crocoite** Krokoit m, roter Bleispan m, Rotbleierz n, -spat m

**crocus** Englischrot n, Polierrot n, (polishing powder) englisches Rot n, Pariser Rot n ~ **of antimony** Metallsafran m ~ **cloth** Crokustuch n
**crook, to** ~ krümmen
**crook** Häkchen n, Stimmbogen m (music) ~ **stick** Drahtgabel f (for laying or hoisting wires)
**crooked** krumm, schief, ungerade, verbogen ~ **crowbar** Spitzzange f ~ **hole** krummes Bohrloch n ~ **pincher** Geißfuß m ~ **tiller** gebogene Ruderpinne f ~ **wharf** Ankerbühne f
**Crookes, ~ space** Crookes'scher Dunkelraum m ~ **tube** Crookes'sche Röhre f
**crooking** Krummwerden n
**crop, to** ~ abmähen, abschöpfen, knöpfen, scheren, schopfen **to** ~ **out** ausstreichen (geol.) **to** ~ **up** auftreten
**crop** Ernte f; Kropf m, Scheideerz n ~ **of crystals** Kristallanschluß m **(inferior)** ~ **chute** Abfallrutsche f ~ **coal** minderwertige Kohle f am Ausgehenden ~ **conveyer** Schrottförderer m ~ **cutoff** Klinkensteuerung f ~ **end** verlorener Kopf m, (of an ingot) abgeschopfter Lunkerkopf m, (scrap) Knüppelende n ~ **estimate** Ernteschätzung f ~ **failure** Mißernte f ~ **restriction** Ernteeinschränkung f ~ **rotation** Fruchtfolge f, -wechsel m ~ **shears** Abschöpfschere f
**cropped** gekröpft ~-**up** eingestellt
**cropping** Schur f ~ **bench** Abstechdrehbank f ~ **machine** Tuchschermaschine f ~ **out** anstehend ~ **shears** Abschöpfschere f
**cross, to** ~ durchkreuzen, durchsetzen, kreuzen, queren, schränken, überfahren, überqueren, überschreiten, übersetzen, übersteigen, verschalten, verschränken, vertauschen **to** ~ **a boundary** übergreifen **to** ~-**connect** kreuzen (wires) **to** ~-**connect lines** Leitungen pl am Verteiler schalten **to** ~-**cut** ablängen, abschroten **to** ~-**hatch** kreuzschraffieren **to** ~-**form** plankopieren **to** ~-**link** vernetzen **to** ~ **off** abstreichen (auf Liste); verschlagen **to** ~ **out** ausbeißen (oil), zu Tage ausgehen (geol.) durchstreichen, streichen **to** ~ **(or strike) out** wegstreichen **to** ~ **out a thing** einen Strich durch etwas machen **to** ~-**profile** plankopieren **to** ~ **two wires** zwei Leitungen (durch)kreuzen **to** ~ **up** auftreten
**cross** quer, zuwiderlaufend
**cross** Berührung f, Drahtverwicklung f (teleph.), Drahtkreuzung f, Kreuzmaß n (print.), Kreuzstück n, Kreuz n, Kreuzzeichen n, Leitungsberührung f, Leitungsschluß m, Querverbindung f, T-förmiges Rohrstück n, Umschaltung f **(flanged)** ~ Kreuzstück n ~ **and vertical feed handle** Handgriff m für die selbsttätige Quer- und Vertikalbewegung
**cross, ~ adding** Querkontrolle f ~ **adjustment** Querverschiebung f (of electron microscope) ~-**advance-contact** Plan-vor-Kontakt m ~ **anchoring** Querstütze f ~ **appeal** Anschlußberufung f ~ **arm** Quer-arm m, -träger m, -zweig m ~ **arm brace** Schrägstütze f ~-**axle under-carriage** Fahrgestell n mit durchgehender Achse
**crossbar** Ausleger m, Brücke f, Bündel n, Einlaßgalgen m, Joch n, Knebel m, Kreuzarm m, Lumme f, Quer-balken m, -haupt n, -holz n, -kluft f, -leiste f, -riegel m, -schiene f, -stange f, -steg m, -strebe f, -stück n, -träger m; Riegelholz

*n*, Sprosse *f*, Wandriegel *m* ~ **of a gin** Riegel *m* eines Hebezeugs ~ **of the manhole** Mannlochbügel *m*
**crossbar,** ~ **distribution panel** Kreuzschienenverteiler *m* ~ **exchange** Vermittlung *f* mit Koordinatenschaltersystem ~ **generator** Fernsehgittergeber *m* ~ **light** Querbalkenfeuer *n* ~ **micrometer** Kreuzfadenmikrometer *n* ~ **principle** Kreuzschienenprinzip *n* ~ **switch** Koordinatenwähler *m*
**cross-bared end** Gitterkopfwand *f*
**crossbars** Gatter *n* ~ **of a winch** Haspelkreuz *n* ~
**crossbeam** Ausleger *m*, Bügel *m*, Ducht *f*, Dwarsbalken *m*, Holm *m*, Querbalken *m*, Querhaupt *n*, Querstück *n*, Querträger *m*, Sandstrake *f* (dry docks), Schwertbalken *m*, Strebe *f* ~ **of a cask** Faßriegel *m* ~ **turret head** Revolverquerbalkensupport *m*
**cross,** ~ **bearer** Rost-abträger *m*, -balken *m* ~ **bearing** Anschnitt *m*, Kreuzpeilung *f* (rdo) ~ **beater mill** Schlagkreuzmühle *f* ~**-bedding** Kreuzschichtung *f* ~ **belt** Quertransportband *n*, Schulterriemen *m* ~ **belt separator** Kreuzbandscheider *m* ~ **bill** Gegenwechsel *m* ~ **bit** Kreuzbohrer *m*, -meißel *m* (Bohrgerät *n*) ~ **blocking** Kreuz-, Quer-hacke *f* ~ **board** Pfeilendurchhieb *m* ~ **bolt** Querbolzen *m*, Riegelanschlagbolzen *m* ~ **bond** Kreuzverband *m* ~ **bonding** Querbindung *f* ~**-bore** Querbohren *n* ~ **brace** Mittelriegel *m* ~ **brace (diagonal)** Kreuzverstrebung *f* ~**-braced** ausgekreuzt ~**-bracing** Auskreuzung *f*, Kreuzverband *m*, Kreuzverspannung *f*, Querverstrebung *f* (arch.) ~ **bracing wire** Verspannungsdiagonale *f* ~ **branch** Querkluft *f* ~**-breaking strength** Durchbiegungsfestigkeit *f* ~**-breaking test** Biegeversuch *m*, Querbiegeversuch *m* ~ **breeding** Kreuzung *f* ~ **bubble** Kreuzlibelle *f* ~ **bunker** Querschiffsbunker *m* ~ **carrying capacity** Rahmentragfähigkeit *f* ~ **center** Kreuz *n* ~ **chisel** Breiteisen *n* ~ **claim** Gegenklage *f* ~**-claim plaintiff** Widerkläger *m* ~ **coat** Kreuzgang *m* (Lackierung) ~ **coil** Kreuzspule *f* ~**-coil antenna** Kreuzantenne *f*, Kreuzrahmenantenne *f*, zwei gekreuzte Rahmenantennen *pl* ~ **composition** Quersatz *m* ~ **compound arrangement** Zweiwellenanordnung *f* ~ **conduction** Kreuzleitung *f* ~**-connecting** Kreuzen *n* ~**-connecting block** Querverbindungsfeld *n* ~**-connecting board** Zwischenverteiler *m* ~**-connecting terminal** Kabelverzweiger *m* ~**-connecting terminals** Überführungskasten *m* ~**-connecting wire** Schaltader *f*, Schaltaderdraht *m* ~ **connection** Gegenschalten *n*, Kreuzverbindung *f*, Querverbindung *f*, Schaltdrahtverbindung *f*, (of wires) Kreuzung ~**-connection box** Abzweigkasten *m* ~**-connection field** Zwischenverteiler *m* ~**-control beam tube** Quersteuerröhre *f* ~ **conveyor** Querbandförderer *m* ~**-cord (in ring)** Verbindungsschnur *f* ~**-counter flow pass** Kreuzstromweg *m* (Wärmeaustauscher)
**cross-country** querfeldein ~ **car** Geländekraftfahrzeug *n* ~ **driving** Geländefahren *n* ~ **field path** Feldweg *m* ~ **flight** Überlandflug *m* ~ **higway** Autobahn *f* **having** ~ **mobility** geländegängig ~ **operation** Geländegängigkeit *f* ~ **passenger car** geländegängiger Personenkraftwagen *m* ~ **road** Querweg *m* ~ **truck** gelände-

gängiger Lastkraftwagen *m* ~ **vehicle** Geländefahrzeug *n*
**cross,** ~ **coupling** Übersprechkopplung *f* ~ **coupling error** Fehler *m* durch nicht winkelgerechten Einbau des Kreisels *m*, Kreuzkopplung *f* des Kreisels ~ **course** durchsetzender Gang *m*, Quergang *m* ~ **culvert** Stichkanal *m* ~ **current** Kreuzstrom *m*, Querströmung *f*
**crosscut** Hirnschnitt *m*, Kreuzhieb *m*, Ortsquerschlag *m*, Querschlag *m*, Querschnitt *m* ~ **and centering machine** Abläng- und Zentriermaschine *f* ~ **end** Hirnfläche *f* ~ **frame saw** Gatterquersäge *f* ~ **ingot** Flachblock *m* ~ **saw** Abkürzsäge *f*, Fuchsschwanz *m*, Fuchsschwanzsäge *f*, Hirnsäge *f*, Quersäge *f*, Schrotsäge *f*, (for logs) Steifsäge *f*, Trecksäge *f*, Trummsäge *f*, Zugsäge *f* ~ **sawing machine** Fuchsschwanzsägemaschine *f* ~ **teeth** (saw) Zahnschränkung *f* ~ **wood** Hirn-, Stirn-holz *n*
**cross,** ~ **cutter** Querschneidemaschine *f*, Querschneider *m* ~**-cutting** Kreuzschnitt *m* (film) ~ **cutting shears** Unterteilschere *f* ~ **demand** Gegenforderung *f* ~ **direction** Querrichtung *f* ~**-disc-type impulse coupling** Kreuzscheibenschnapper *m* ~ **drift** Qurdrift *f* ~**-drilled spigot** Splintzapfen *m* ~**-drilling attachment** Querbohreinrichtung *f* ~ **drum** Quertrommel *f* ~**-dyed style** Überfärbeartikel *m* ~ **entry** Gegenposten *m* ~**-face** Überblenden *n* ~ **fading** Umblenden *f* (TV) ~ **fall** Quergefälle *n* (of the water), Querneigung *f*
**cross-feed** Planzug *n*, Quervorschub *m* (Drehbank) ~ **dial** Planvorschubteilscheibe *f* ~ **dog** Quervorschubklinke *f* ~ **drive** Quervorschubantrieb *m* ~ **gear** Planzugrad *n* ~ **handle** Planzugkurbel *f*, Quervorschubkurbel *f* ~ **handywheel** Handrad *n* für den Quervorschub ~ **pinion** Planzugritzel *f*, Querspindel *f* für den rechten und linken Schlitten, Quervorschubspindel *f* ~ **serew**, Planvorschubspindel *f*, Planzugspindel *f* ~ **stop** Queranschlag *m*
**cross,** ~ **field** Querfeld *n*, Transversalfeld *n* ~ **filament** Querfaden *m* ~ **fire** Beeinflussung *f* (Induktion) durch Telegrafie, Kreuzfeuer *n* ~ **fissure** Querkluft *f* ~**-fissure vein** Querspaltengang *m* ~ **fitting** Rohrkreuz *n* ~**-flow type heat exchanger** Gegenstrom-Wärmeaustauscher *m* ~ **flux** Streufluß *m* ~ **footer** Querrechner *m* ~ **force** Querkraft *f* ~ **forceps** Kreuzpinzette *f* ~ **frame** Querrahmen *m* ~ **frisket** Quergreifer *m* ~ **frog** Herzstück *n* einer Kreuzung (r. r.) ~ **gallery** Kreuz *n* (min.) ~**-grain** quer zur Faserrichtung *f* ~ **grain** Querfaser *f*, Querfurnier *n*, Stirnseite *f*, Wimmer *m* (in wood) ~ **grained fiber** quergeschliffene Faser *f* ~**-grained fibers** Querschliffholzstoff *m* ~ **grating** Kreuzgitter *n* ~ **grinder** Querschleifer *m* ~**-hair diopter** Fadendiopter *m* ~**-hair plate** Stichkreuzplatte *f* ~ **hairs** Achsenrichtverfahren *n*, Faden-, Haarkreuz *n* ~ **handle** Handkreuz *n* ~**-hatch pattern** Schachbrettmuster *n* (TV) ~**-hatched** gescheckt, kariert, kreuzweise schraffiert ~ **hatching** Kreuzschraffierung *f*
**cross-head** Gleitbacke *f*, Kreuzkopf *m* (jet), Querhaupt *n*, Schräg- oder Querspritzkopf *m*, Zugstangenkopf *m* ~ **bearing** Kreuzkopfzapfenlager *n* ~ **engine** Kreuzkopfmotor *m* ~ **guide** Gleitplatte *f* (mach.), Kreuzkopffüh-

rung f ~ key Rohrhalter m (Verbindung mit der Rücklaufbremse) ~ pin Gabelzapfen m, Kreuzkopfzapfen m ~ shoe Gleitschuh m ~-type engine Kreuzkopfmaschine f
cross, ~ heading Abbaustrecke f, Querschlag m ~ holder Querhalter m ~ hole Kreuzloch n ~-hole nut Kreuzlochmutter f ~ ignition tube Umzündstutzen m ~-induction Nebensprechen n ~-intersecting canal Stichkanal ~-jet channel Querströmungskanal m ~ jig Quersetzmaschine f ~ joint Kreuz-gelenkkuppelung f, -verbindung f, -verzapfung f ~-jointing Auskreuzen n key holding pin Querkeilhaltebolzen m ~ lamination Holzfüllstück n ~ letter Kabelbuchstabe m (bei dem positive und negative Stromstöße abwechseln) ~ level Quer-libelle f, -neigungsmesser m ~-level bubble Verkantungslibelle f ~-level drive Verkantungstrieb m ~ leveling Einspielen n der Libelle ~-leveling mechanism Radstandtrieb m ~ levels Kreuzlibellen pl ~ lever Gestängekreuz n, Kunstkreuz n (min.) ~ licensing gegenseitige Lizenzerteilung f ~ line Bruchstrich m, Querleitung f, Strichkreuz n ~-line center Kreuzungspunkt m ~-line center on intersection Kreuzungsmittelpunkt m ~-line eyepiece Strichkreuzokular n ~ line micrometer Netzmikrometer n ~-line screen Kreuzlinienraster m ~ lines Achsenrichtverfahren n, Fadenkreuz n ~ link Kreuzgelenkkuppelung f, Zapfenkreuz n ~ linking Querverbindung f ~ lode Quergang m ~ loop Kreuzschlaufe f ~-loop antenna Kreuzdrehrahmenantenne f ~ magnetization Quermagnetisierung f ~-magnetized (or -magnetic) quermagnetisiert ~ mark Kreuzmarke f ~ member Quer-stück n, -träger m ~ modulation Kreuzmodelung f, -modulation f, Quermodulation f ~ motion Planbewegung f ~-mouthed borer (or chisel) Kronenbohrer n ~ movement Planvorschub m ~-neutralization Gegentaktneutralisation f
crossover Bündelknoten m (rdr), Drahtkreuzung f, Einschnürungspunkt m (in electron optics), Gleisverbindung f (r.r.), Kreuzungspunkt m, Kreuzungsweiche f, Querhaupt n, Querstück n, Querträger m, Schnittpunkt m (beam), Strebe f, Traverse f ~ of electrons Elektronenüberkreuzungsstelle f ~ of frequencies Frequenzschnitt m (sound film)
crossover, ~ aperture Einschnürungsblende f ~ brushing machine Querbürstmaschine f ~ clamp Kreuzungsklemme f ~ conveyor Querbandförderer m ~ duct Querkanal m ~ frequency Überschneidungsfrequenz f ~ network Aufteilungsfilter n (sound production), Frequenzweiche f ~ point Kreuzungspunkt m ~ road Verbindungsgleis n ~ T Doppel-T-Stück n ~ spiral Überleitrille f (phono) ~ tube Umzündstutzen m ~ valve Umschaltventil n
cross, ~ part of a press Querholz n einer Presse ~ peen Kreuzfinne f ~-peen hammer Hammer m mit Kreuzfinne, Querschlaghammer m ~-peen sledge Kreuzschlaghammer m ~-perforated tape Querlochstreifen m ~ pickup Nebensprechen n ~-piece Ausleger m, Flächenspiere f (aviat.), Hirnleiste f, Kreuzstück n (flanged), Querhaupt n, Querstück n, Querträger m, Querverband m, Steg m ~ piece of

framing of a lock gate Riegel m einer Schleusenzarge ~ pin Mitnehmerstift m ~ plate Querblech n ~ plugout connection Ausnahmequerverbindung f (signal) ~ point Kreuzungspunkt m ~ pointer Kreuzzeiger m ~-pole force Querschiffskraft f ~ product direktes Produkt n ~ rafter Sparrenwechsel m ~ rail Quer-führung f, -haupt n ~-rail slide Querbalkensupport m ~ range wind Querwind m ~ ratio Doppelverhältnis n ~ reference gegenseitiger Hinweis m, Kreuzverweisung f ~-reference card Hinweiskarte f ~ resistance Querwiderstand m (of network) ~ return plan-zurück f ~ road Kreuzweg m, Querstraße f, Querweg m, Wegkreuzung f ~ roads Straßenkreuzung f ~ roller bit Kreuzrollenmeißel m ~ roller rock bits Kreuzrollen-Felsmeißel m ~ row Querreihe f ~ rule under heading Kopfabschlußlinie f ~-run of the fabric Querbahn f ~ scale Quermaßstab m ~ scavenging Querspülung f, Querstromspülung f ~ screw(ing) joint Kreuzverschraubung f ~ seam Kreuznaht f
cross-section Drahtquerschnitt m, Durchschnitt m, Grundriß m, Querprofil n, Querriß m, Querschnitt m, Trennschnitt m ~ of a bar Stabquerschnitt m ~ at bottom of thread Kernquerschnitt m (bei Schrauben) ~ of a furnace Ofenquerschnitt m ~ of a girder Trägerquerschnitt m ~ of hole Lochquerschnitt m ~ of mouth of nozzle Endquerschnitt m ~ of passage Durchflußquerschnitt m ~ of shaft Schaftquerschnitt m (bei rohen Schrauben) ~ of steam discharge Abflußquerschnitt m ~ of suction pipe Saugrohrquerschnitt m ~ of total inflow Gesamteinströmquerschnitt m ~ of weld Nahtquerschnitt m ~ of the (conducting) wire Leitungsquerschnitt m
cross-section, ~ drawing Durchschnittszeichnung f ~ paper Koordinatenpapier n ~ work Querprofilaufnahmen pl
cross-sectional, ~ area Querdurchschnitt m, Querschnitt m, Querschnittsfläche f ~ area of conductor Leiterquerschnitt m ~ area of contraction Einschnürungsquerschnitt m ~ area of inlet Einströmungsquerschnitt m ~ area of nozzle aperture (or orifice) Durchtrittsquerschnitt m ~ area of passage Durchgangsquerschnitt m ~ area ratio Querschnittsverhältnis n ~ area of undistorted portion of field Einzugsgebiet n (direction-finding frame) ~ area of winding Wicklungsquerschnitt m ~ contour Querschnittsverlauf m ~ dimension Querabmessung f ~ drawing Schnittzeichnung f ~ efficiency (or production) Leistungsquerschnitt m ~ polish (or specimen) Querschliff m ~ view Quer-riß m, -schnittansicht f
cross, ~ shaft Fingerhebelwelle f, Querwelle f, Seitenvorgelege n ~-shaped kreuzförmig ~ shearing machine Transversalschermaschine f ~ sleeve Kreuzmuffe f
cross-slide Kreuz-schlitten m, -support m, Planschieber m, Querschlitten m, Quersupport m (on a lathe), Quersupportführung f, Unterschieber m ~ adjustment screw Schraube f für Querschlittenverstellung ~ arrangement (or attachment) Kreuztischeinrichtung f ~ feed lever Querschlittenvorschubhebel m ~ handwheel Querschlittenhandrad n ~ rest Plan-

**schlitten** *m*, Quersupport *m*, Quersupportführung *f* ~ **vernier scale** Noniusmaßstab *m* am
Querschlitten
**cross,** ~ **slides** Planzugführung *f* ~ **sliding quer-**
verschiebbar ~ **sliding turntable** Drehscheibe *f*
mit Querbewegung ~ **slip** Quergleitung *f*
~**-slip line** Quergleitlinie *f* ~ **slope landing**
Landung *f* quer zum Hang ~ **slot** Kreuzschlitz
*m* ~ **slotted** quergeschlitzt ~**-slotted screw**
Kreuzschlitzschraube *f* ~ **spider** Fadenkreuz *n*
~ **spiderline** Fadenkreuz *n* ~ **springer** Gratbogen *m* ~ **staff** Kreuzmaß *n* ~**-staff head**
Kreuzscheibe *f*, Winkelkopf *m* ~**-stitch** Kreuz-
stich *m* ~**-stone** Kreuzstein *m* ~ **stop** Queranschlag *m* ~ **string** Querschnur *f* (textiles) ~
**strip** Querschiene *f* ~ **striped** quergestreift ~
**stripes** Diagonalstreifen *pl* ~**-strung** kreuz-
saitig ~ **strut** Querstiel *m* ~ **struts** Strebenkreuz
*n* ~ **stud** Kreuzstrebe *f* ~ **switch** Kreuzweiche *f*
~ **(-over) system** Kreuzungssystem *n* ~ **tail**
**(butt, strap, or stud)** Gabelstück *n* der Pleuelstange
**crosstalk, to** ~ übersprechen; Kopiereffekt *m*
(tape rec.), Kreuzkopplung *f*, Nebensprechen *n*
**(near-end)** ~ **attenuation** Nebensprechdämpfung *f* ~ **circuit** Nebensprechkopplung *f* ~
**coupling** Nebensprechkopplung *f*, Neben-
sprechverlust *m* ~ **current** Nebensprechstrom
*m* ~ **error** Schieffehlerkomponente *f* (rdr) ~
**measurement** Nebensprechmessung *f* ~ **meas-**
**uring set** Nebensprechmesser *m* ~ **meter** Neben-
sprechdämpfungs-messer *m*, -zeiger *m*, Nebensprechmesser *m* ~ **path** Mitsprechkopplung *f*,
Nebensprechkopplung *f*, Nebensprechweg *m*,
Übersprechkopplung *f*, Übersprechweg *m*
~**-proof** nebensprechfrei ~ **transmission equi-**
**valent** Nebensprechdämpfung *f*, Übersprechdämpfung *f* ~ **volume** Nebensprechpegel *m*
**cross,** ~ **thread** Kreuzgewinde *n* ~ **tice** Bügel *m*,
Quer-binde *f*, -halter *m*, -haupt *n*, -schwelle *f*,
-stück *n*, -träger *m*; Schwelle *f*, Spurstange *f*,
Strebe *f* ~ **tie rod** Queranker *m*, Traverse *f* ~
**total** Querkontrolle *f* ~ **travel** Planweg *m*,
Querbewegung *f* ~ **traversing** Katzfahren *n* ~
**tree** Dwarssaling *f* ~ **tube** Querrohrträger *m*
(of chassis) ~**-tube boiler** Quer-siedekessel *m*,
-sieder *m* ~**-type folded dipole** Kreuzfaltdipol *n*
~**-type disc** Kreuzscheibe *f* ~**-type driving disc**
Mitnehmerkreuzscheibe *f* ~**-type screw fitting**
Kreuzverschraubung *f* ~ **wall** Querwand *f* ~
**way** kreuzweise ~ **way of the grain** Hirnseite *f*
~ **web** Fadenkreuz *n* ~ **wheel** Malteserkreuz *n*
~ **wind** Dwarswind *m*, Seitenwind *m* ~ **wind**
**component** Querwind-anteil *m*, -komponente *f*
~**-wind force** Querkraft *f*, Seitenwindkraft *f*
~**-wind landing** Landung *f* mit Seitenwind
~**-wind landing gear** Schiebefahrwerk *n*,
Seitenwindfahrwerk *n* ~**-wind take-off** Start *m*
mit Seitenwind ~**-wind undercarriage** Schiebe-
fahrgestell *n* ~ **winding** Kreuzwickelbildung *f*
~**-winged guided missile** Kreuzflügler *m* ~ **wire**
Fadenkreuz *n* ~ **wire (or diagonal) bracing**
Seilauskreuzung *f* ~**-wire meter** Fadenkreuz-
messer *m* ~**-wire micrometer** Fadenkreuzmikro-
meter *n* ~**-wire sight** Fadenkreuzdiopter *n* ~ **wires**
**(or hairs)** Stichkreuz *n* ~**-wise** kreuzweise, über
Eck, schränkweise ~ **working** Querbau *m*
**crossed** gekreuzt, geschränkt ~ **arms paddle**

**agitator** Kreuzbalkenrührer *m* ~ **belt** gekreuz-
ter Riemen *m* ~ **check** Verrechnungsscheck *m*
~**-coil antenna** Kreuzrahmen *m*, Kreuzrahmenantenne *f* ~**-coil device** Kreuzspulgerät
*n* ~**-coil measuring instrument** Kreuzspulmeß-
gerät *n* ~ **pair** gekreuzte Doppelader *f* ~
**spherical** gekreuzt sphärisch ~ **threads** Fadenkreuz *n* ~**-timber lashing** Kreuzbund *m*
**crossing** Durchquerung *f*, Furt *f*, Herzstück *n*,
Kreuzen *n*, Kreuzstelle *f*, Kreuzung *f*, Kreuzungspunkt *m*, Linienkreuzung *f*, Platzwechsel
*m*, Querträger *m*, Schränkung *f*, Straßenkreuzung *f*, Überführung *f*, Übergang *m*, Übergreifen *n*, Überquerung *f* (electr.), Übersetzen
*n*, Überwinden *n*, Verschränkung *f*, Vertauschung *f*, Wetterkreuz *n* ~ **of battery terminals**
Kreuzung *f* der Batteriepole ~ **of ditches** Grabendurchfahrt *f* ~ **(of a river by a canal) on the**
**level** Niveaukreuzung *f* ~ **of paths** Leitungs-
kreuzung *f* ~ **over the railroad** Eisenbahnüber-
gang *m* ~ **of the rolls** Schranken *pl* der Walzen
~ **of trenches** Grabendurchfahrt *f* ~ **with**
**wheelflange ramp** Herzstück *n* mit Flanschen
~ **of wires** Ader-, Leitungs-kreuzung *f*
**crossing,** ~ **gate** Bahnschranke *f* ~ **(stream-)** ~
**means** Übergangsmittel *n* ~**-off** Streichen *n*
~**-over** Überkreuzung *f* ~**-over clamp** Kreuz-
klemme *f* ~ **reamer** Nachschneidekreuz *n* ~
**site** Übersetzstelle *f* ~ **station** Übergangsbahn-
hof *m* ~ **track** Kreuzgleis *n* ~ **valve** Kreuzventil
*n* ~ **warp** Drehkette *f*
**crotch** Gabelholz *n*, Verzweigung *f* (Schnur)
**crotched patenthesis** eckige Klammer *f*
**croton chloralhydrate** Crotonchloralhydrat *n*
**Crotonic acid** Crotonsäure *f*
**crouch, to** ~ hocken
**croup** Kruppe *f* ~ **pad** Rückenkissen *n*
**crow** Bieger *m*, Krähe *f* ~**'s-foot** Gänsefuß *m*
(aviat.), Glückshaken *m*, Hahnepot *m*, schmiedeeiserner Vierspitz *m* (mil. obstacle) ~**'s nest**
Krähennest *n*, Spatzenhaus *n*
**crowbar** Brecheisen *n*, Brechklaue *f*, Brechstange *f*, Geißfuß *m*, Gleishebebaum *m*, Hebebaum *m*, Hebeeisen *n*, Hebeleisen *n*, Krücke *f*,
Kuhfuß *m*, Stemmeisen *n*, Ziegenfuß *f* ~ **with**
**splint end** Brecheisen *n* mit Kuhfuß
**crowd, to** ~ drängen, zulaufen
**crowd** Haufen *m*, Menge *f*, Menschenmenge *f*,
Schar *f*, Zudrang *m* ~ **noise** Saalgeräusch *n*
**crowded** zusammengedrängt
**crowding** Vorschub *m* (Bagger)
**crowfly distance** Luftlinienentfernung *f*
**crowfoot** standfestes Stativ *n*, Hahnpot *m* (naut.)
**crown, to** ~ ballig drehen
**crown** Bekrönung *f*, Kopf *m*, Kreuz *n*, Krone *f*,
obenliegendes Querjoch *n* einer Werkzeugmaschine, Scheitel *m*, Überhöhung *f*, Verteilerring *m* ~ **of arch** Bogenscheitel *m* ~ **of the**
**arch** Bekrönung *f* ~ **of the road** Scheitelpunkt
*m* ~ **of roadbed** Bahnkrone (r.r.) ~ **of a sluice**
Haupt *n* einer Schleuse ~ **of thorns** Dornen-
krone *f* (electron., rdr)
**crown,** ~ **bit** Kronenbohrer *m* ~ **block** feste
Flasche *f*, Turmflaschenzug *m*, Turmrolle *f* ~
**burner** Kranzbrenner *m* ~ **circle** Kopfkreis *m*
(in a gear wheel) ~ **circle diameter** Kopfkreis-
durchmesser *m* ~ **cork bottle** Kronenkork(en)-
Flasche *f* ~ **cork sealing machine** Kronenkork-

verschließmaschine *f* ~ **corker** Kronenkork-verschließmaschine *f* ~ **drill** Kronenbohrer *m* ~-**face pulley** ballig gedrehte Riemenscheibe *f* ~ **face rim** ballig gedrehte Kranzfläche *f* ~ **gear** Kronenrad *n*, Kronrad *n*, Planrad *n*, Zahnscheibe *f* ~ **glass** Mondglas *n*, Solinglas *n* ~-**grinding wheel** Topfschleifscheibe *f* ~ **height** Pfeilhöhe *f* ~ **hinge** Scheitelgelenk *n* ~ **iron** Grenzeisen *n* ~ **knot** Hahnepot *m* auf einem Knoten, Kreuzknoten *m*, Schildknoten *m* ~ **lens** Kron(glas)-linse *f* ~ **line** Kopfkreis *m* (in a gear wheel) ~ **nut** Kronenmutter *f* ~ **piece** Kopfstück *n* ~-**plate riveting apparatus** Deckennietapparat *m* ~ **port** Oberhafen *m* ~ **pulley** Bohrrolle *f*, Turmscheibe *f* ~ **rim** balliger Kranz *m* ~ **saw** Ringsäge *f* ~ **sheave** Bohrrolle *f* ~ **thickness** Scheiteldicke *f* ~ **top** Kronenaufsatz *m* ~ **wheel** Bolzenzahnkranz *m*, Kegeltellerrad *n*, Kronenrad *n*, Kronrad *n*, Planrad *n*, Planradverzahnung *f*, Zahnradunterbrecher *m*, Zahnscheibe *f*

**crowned** ballig ~ **tooth bevel gears** Balligverzahnung *f*

**crowning** Balligkeit *f*, Wölbung *f* ~ **attachment** Einrichtung *f* für das Balligverfahren ~ **mechanism** Balligeinrichtung *f* ~ **tool** Wölbwerkzeug *n*

**croze** Gargel *f* (the groove itself), Kröse *f* ~ **iron** Kröseisen *n*

**crozer** Gargelkamm *m*

**crucial** ausschlaggebend

**crucible** Gestell *n*, Schmelztopf *m*, Tiegel *m*, Tute *f*, Tüte *f* ~ **with muffler** bedeckter Hafen *m* (glassblowing)

**crucible**, ~ **bottom** Rückstein *m* ~ **cast steel** Tiegelflußstahl *m*, Tiegelgußstahl *m* ~ **charge** Tiegeleinsatz *m* ~ **coking test** Tiegelverkokung *f* ~ **drier** Tiegeltrockner *m* ~ **edge** Tiegelrand *m* ~ **furnace** Gefäß-, Kessel-, Pfannen-, Tiegelofen *m* ~ **iron** schmiedbarer Guß *m* ~ **lining** Tiegelfutter *n* ~ **maker** Tiegelbrenner *m* ~ **melting plant** Tiegelschmelzerei *f* ~ **melting practice** Tiegelschmelzbetrieb *m* ~ **melting process** Tiegelschmelzverfahren *n* ~ **mold** Tiegel(hohl)form *f* ~ **molding** Tiegelformerei *f* ~ **oven** Tiegelbrennofen *m* ~ **(process) practice** Tiegelbetrieb *m* ~ **rim** Tiegelrand *m* ~ **shank** Gieß-, Tiegel-gabel *f* ~ **steel** Gußstahl *m*, schmiedbarer Guß *m*, Tiegelgußstahl *m*, Tiegelstahl *m* ~ **steel foundry** Tiegelschmelzhütte *f* ~-**steelfurnace plant** Tiegelschmelzstahlhütte *f* ~ **steel melting** Tiegelschmelzerei *f* ~ **test** Verkorkungsprobe *f* ~ **tongs** Bauch-, Gieß-, Tiegelzange *f*

**cruciform** kreuzförmig ~ **fin** kreuzförmige Flosse *f* (Rakete) ~ **girder** Kreuzträger *m*, Ringkreuz (aviat.) ~ **twins** Durchkreuzungszwillinge *pl* (cryst.)

**crude** grob, roh, unbearbeitet, unrein ~ **antimony** Rohantimon *n* ~ **asphalt** Asphaltgestein *n*, Asphaltstein *m* ~ **bauxite** Rohbauxit *m* ~ **benzene** Benzolvorerzeugnis *n* ~ **benzol** Rohbenzol *n* ~ **bismuth** Rohwismut *n* ~ **bottom** Bodensatz *m* nach der Destillation von Rohöl ~ **brass** Rohmessing *n* ~ **brine** Rohsole *f* ~ **copper** Rohkupfer *n* ~ **cresol** Rohkresol *n* ~ **felt** Rohfilzpappe *f* ~ **fractionation** Vorzerlegung *f* (chem.) ~ **gas** Rohgas *n* ~-**gas cleaning**

Gichtgasreinigung *f* ~-**gas-cleaning plant** Gichtgasreinigungsanlage *f* ~ **gas burner** Rohgasbrenner *m* ~-**gas machine** Gichtgasmaschine *f* ~-**gas main** Gichtgasleitung *f* ~ **glass** Rohglas *n* ~ **glycerin** Rohglycerin *n* ~ **helium** Rohhelium *n* ~ **iron** Roheisen *n* ~ **lanoline** Rohlanolin *n* ~ **lead** Rohblei *n* ~ **lead bullion** Werkblei *n* ~ **lignite** Rohbraunkohle *f* a ~ **lump of different metals** Gans *f* ~ **naphtha** Roh-naptha *n*, -öl *n*, -petroleum *n* ~ **naphtalene** Rohnaphthalin *n* ~ **oil** Erdöl *n*, Naphtha *n* & *f*, Rohöl *n*, Rohpetroleum *n* ~-**oil burner** Rohölbrenner *m* ~ **oil engine** Rohölmotor *m* ~ **ore** Roherz *n* ~ **paraffin** Gatsch *m* ~ **petroleum** Erdöl *n*, Roherdöl *n*, Rohnaphtha *n*, Rohöl *n*, Rohpetroleum *n* ~ **potash** Ochras *m*, Pottaschefluß *m* ~ **potassium salt** Kalirohsalz *n* ~ **scale** Rohparaffin *n* in Schuppenform ~ **scale wax** Filterkuchen *m* ~ **silk** Ekrüseide *f* ~ **silver** Rohsilber *n* ~ **stabilization** Rohölstabilisierung *f* ~ **state** Rohzustand *m* ~ **sulfur** Rohschwefel *m* ~ **tar** Urteer *m* ~ **turpentine** Terpentinfirnis *f* ~ **zinc** Rohzink *n*

**cruise, to** ~ kreuzen (navy), reisen

**cruise** Marschflug *m*, Ozeanfahrt *f*, Reiseflug *m* ~ **ceiling** Reisefluggipfelhöhe *f* ~ **engine** Marschtriebwerk *n* ~ **flight** Reiseflug *m* ~ **thrust** Schub *m* im Reiseflug

**cruiser** Aufklärungsschiff *n*, Flottenkreuzer *m*, Kreuzer *m* ~ **with flight deck** Flugdeckkreuzer *m* ~ **car** Streifenwagen *m* der Polizei

**cruising** Kreuzen *n*, Reise *f* ~ **altitude** Fahr-, Flug-höhe *f* ~ **boost** Ladedruck *m* bei Reiseleistung ~-**climb limit** Leistungsgrenze *f* im Steigflug für Dauerleistung ~ **comsumption** Verbrauch *m* bei Reisegeschwindigkeit ~ **fuel consumption** Kraftstoffverbrauch *m* bei Reiseflug ~ **fuel tank** Reichweitenbehälter *m* ~ **horsepower output** Pferdekraftnutzwert *m* beim Fliegen ~ **hover height** Schwebeflughöhe *f* ~ **level** Flughöhe *f*, Reiseflughöhe *f* ~ **maneuver** Fahrübung *f* ~ **manifold pressure** Ladedruck *m* bei Reiseleistung ~ **output** Reiseleistung *f* ~ **performance** Reiseflugleistung *f* ~ **power** Dauerleistung *f*, Reisedauerleistung *f*, Reiseflugleistung *f* ~ **practice** Fahrübung *f* ~ **radius** Aktionsbereich *m*, Fahrbereich *m* ~ **range** Flugbereich *n* ~ **speed** Marschgeschwindigkeit *f*, ökonomische Fahrt *f*, Reisegeschwindigkeit *f* ~ **turbine** Vorwärtsturbine

**crumble** abbröckeln, abkröseln, bröckeln, kräuseln (painting), krümeln, losbröckeln, verkrümeln, zerbröckeln, zerfallen, zerknittern, zerknüllen, zerstäuben, zusammenbrechen, zusammensinken, zusammenziehen (painting)

**crumbling** Abbröckeln *n*, Abbröckelung *f*, Ausbröckelung *f*, Mauerfraß *m*, Zerkrümeln *f* ~ **iron** Kröseleisen *n* ~ **machine** Knittermaschine *f*

**crumbly** bröck(e)lig, krümelig, zerbröckelnd, zerreiblich

**crumped** gekröpft

**crumple, to** ~ krumpeln, verdrücken, zerdrükken, zerknittern, zerknüllen

**crumpled** knitt(e)rig, schrumpfig

**crumpling strength** Knitterfestigkeit *f*

**chrunch, to** ~ zerknistern

**chrunching** Knirschen *n* (of sand)

**crush, to** ~ aufschließen (cool), auspressen, aus-

rotten, brechen, eindrücken, mahlen, pochen, quetschen, verbeulen, verdrücken, vermahlen, walzen, zerdrücken, zerkleinern, zerknittern, zerknüllen, zermahlen, zerquetschen, zerstampfen, zerstückeln, zusammendrücken, to ~ **coarsely** schroten to ~ **gently** schonend zerkleinern to ~ **ore** Erz *n* pochen
**crush load** Überbelastung *f*
**crushable leg** Bruchelement *n*
**crushed** eingedrückt, gebrochen, gedrückt, verdrückt ~ **coal** Brechkohle *f* ~ **coke** Brechkoks *m*, Koksmehl *n* ~ **electrodes** Elektrodenmehl *m* ~ **gravel** Brechkies *m* ~ **limestone** Kalkschotter *m*, Kalksteinschotter *m* ~ **oven coke** Brechkoks *m* ~ **rock** Schotter *m*, Steinschlag *m* ~ **rock wrapped in wire mesh** Drahtschotter *m* (for rivercontrol work) ~ **sugar** Knoppern *pl*, Lompenzucker *m*, Lumpenzucker *m* ~ **sugar cane** Bagasse *f*
**crusher** Brecher *m*, Brechwerk *n*, Erzquetschmaschine *f*, Mühle *f*, Quetsche *f*, Schlagmühle *f*, Schroter *m*, Steinklopfer *m*, Vorbrecher *m*, Zerkleinerungsmaschine *f* ~ **for olives** Olivenquetsche *f*
**crusher,** ~ **housing** Brechermantel *m* ~ **index** Stauchwert *m* ~ **jaw** Brechbacke *f* ~ **gauge** Gespannungsmesser *m*, Stauchzylinder *m* ~ **gauges** Druckdosen *pl* ~ **roll** Läufer *m* ~ **run** Brechgut *n* ~ **worm** Brechschnecke *f*
**crushers** Druckdosen *pl*
**crushing** Brechen *f*, Chiffonieren *n*, Grobmahlung *f*, Mahlung *f*, Pochen *n*, Quetschen *n*, Stauchen *n*, Verdrücken *n* (paper mfg.), Vermahlung *f*, Walzen *n*, Zerdrücken *n*, Zerkleinerung *f*, Zermalmung *f*, Zerquetschen *n*, Zerschlagen *n*, Zertrümmerung *f* **(preliminary)** ~ Vorzerkleinerung *f* ~ **and pulverizing plant** Zerkleinerungsanlage *f*
**crushing,** ~ **cylinder** Quetschwalze *f* ~ **device** Zerkleinerungsvorrichtung *f* ~ **duty** Mahlleistung *f* ~ **effect** Stauchwirkung *f* ~ **gauge** Brechmesser *m* ~ **hard materials** Hartzerkleinerung *f* ~ **head** Brechkegel *m* ~ **limit** Quetschgrenze *f* ~ **machine** Brecher *m*, Brechmaschine *f*, Zerkleinerungsmaschine *f*, Zerreibwolf *m* ~ **mill** Brechwalzwerk *n*, Grobzerkleinerungsmühle *f*, Mahlanlage *f*, Mahlwerk *n*, Quetsch-(walz)werk *n*, Schlagmühle *f*, Walzenbrecher *m*, Zerkleinerungsmühle *f* ~ **mouth** Brechmaul *n* ~ **operations** Knäppern *n* ~ **plant** Brechanlage *f*, Brecheranlage *f*, Brechhaus *n*, Brechwalzwerk *n* ~ **plate** Stampfplatte *f* ~ **practice** Mahltechnik *f* ~ **ring** Brechring *m* ~ **roll** Zerkleinerungswalze *f* ~**-roll shell** Walzenring *m* ~ **rollers** Rollenmühle *f* ~ **rolls** Walzwerk *n* **(malt-)** ~ **room** Schroterei *f* ~ **strain** Druckspannung *f* ~ **strength** Druckfestigkeit *f*, Scheitendruckfestigkeit *f* (Rohr) ~ **strength of a cube** Würfelfestigkeit *f* ~ **stress** Druck-beanspruchung *f*, -kraft *f*, -spannung *f*; Zerdrükkungsspannung *f* ~ **surface** Mahlfläche *f* ~ **test** Druckprobe *f*, Druckversuch *m*, Stauchversuch *m* ~**-test specimen** Druckprobe *f* ~ **unit** Zerkleinerungsanlage *f*
**crust, to** ~ bekrusten, verkrusten
**crust** Kruste *f*, Schanze *f*, Schaum *m*, Überzug *m* ~ **of cobalt** Kobaltbeschlag *m* ~ **of weathered material** Verwitterungskruste *f*

**crustal deformation** Krustenverformung *f*
**crusted** schalig
**crusty** krustig
**crutch** Dolle *f*, Krücke *f*, Krückwerk *n*, Stollkrücke *f* ~ **mounting** Bock *m* (engine)
**crutcher** Forke *f* (metall.), Seifenmischer *m*
**crux** (of a matter) Kern *m* ~ **of a decision** Entscheidungskern *m*
**cryogenic,** ~ **engineering** Tiefsttemperaturtechnik *f* ~ **gyro** kryogener oder magnetischer Kreisel *m*
**cryolite** Eisstein *m*, Kryolith *m* ~ **glass** Spatglas *n*
**cryomagnetic** kryomagnetisch
**cryometer** Gefrierpunktmesser *m*
**cryoscope** Gefrierpunktmesser *m*
**cryoscopics** Gefrierpunktlehre *f*
**cryostat** Kälteregler *m*, Kryostat *n*
**cryotron** bespulter Widerstand *m*
**cryptanalysis** Geheimanalytik *f*, Geheimschrift -analyse *f*, -entzifferung *f*, Schlüsselauswertung *f*
**crypto,** ~ **crystalline** kryptokristallin ~ **equipment** Schlüsselgerät *n* ~ **gram** verschlüsselter Spruch *m* ~ **grammic** verschlüsselt ~ **graphing** Schlüsselung *f* ~ **graphy** Geheimschrift *f*, Verschlüsselung *f* ~ **meter** Deckfähigkeitsmesser *m*, Deckkraftmesser *m*, Kryptometer *n* ~ **volcanic earthquake** Intrusionsbeben *n*
**crystal** Korn *n*, Kristall *m*, Quarz *m* ~**s** Leuchtergehänge *n* (chandelier), Glasbehang *m* ~ **agent** Kristallisator *m* ~ **(structure) analysis** Kristallstrukturuntersuchung *f* ~ **angle** Kristallecke *f* ~ **axis** Kristallachse *f* ~ **boundary** Korngrenze *f*, Kristallgrenzlinie *f* ~ **center** Kristallkern *m* ~ **chip** Kristallsplitter *m* ~ **collimator** Kollimator *m*, Hilfsfernrohr *n* ~ **combination** Kristallverbindung *f* ~ **complexes** Viellinge *pl* ~ **control** Kristall-, Quarz-steuerung *f* ~**-control means** Kristalltaktgeber *m* ~**-control receiver** Kristalldetektorempfänger *m* ~ **control stage** Kristallsteuerstufe *f* ~**-controlled** kristallgesteuert, quarzgesteuert ~**-controlled drive unit** Quarzsteuervorsatz *m* ~**-controlled transmitter** kristallgesteuerter Sender *m* ~**-controlled tube generator** quarzgesteuerter Röhrengenerator *m* ~ **counter** Kristallzähler *m* ~ **cup** Kristallpfanne *f* ~ **cut** Kristallschnitt *m* ~ **cutter** Kristallschreiber *m* (phono) ~ **(structural) defect** Kristallbaufehler *m* ~ **detector** Drehrariometer *n*, Festdetektor *m*, Kontaktgleichrichter *m*, Kristalldetektor *m*, Röhrendetektor *m* ~ **diamagnetism** Kristalldiamagnetismus *m* ~ **diode** Gleichrichterelement *n*, Kristallrichtröhre *f* ~ **domains** Kristallbereiche *pl* ~ **drive** Kristallsteuerung *f* ~ **drive unit** Quarzsteuervorsatz *m* ~ **edge** Kristallkante *f* ~ **effects** Einfluß *m* des Kristallgitters ~ **elevation** Kristallhub *m* ~ **face** Kristall-ebene *f*, -fläche *f* ~ **filter** Kristallfilter *m* ~ **flexural vibration** Kristallbiegungsschwingung *f* ~ **frequencychanger efficiency** Wirkungsgrad *m* eines Kristallmodulators ~ **goniometer with one (two) circle(s)** ein- oder zwei-kreisiges Kristallgoniometer *n* ~ **goniometry** Kristallwinkelmessung *f* ~ **grain** Kristall-korn *n*, -splitter *m* ~ **grating** Kristallgitter *n* ~ **growing** Kristallziehen *n* ~ **growth** Kristallwachstum *n*

**~ holder** Kristallfassung *f* (piezoelectric), Kristallhalterung *f*, Quarzbehälter *m* **~ (oscillator or resonator) holder** Fassung *f* für Kristalle **~ hyperfine structure** Kristallhyperfeinstruktur *f* **~(line) ice** Destillateis *n* **~ junction line** Kristallgrenzlinie *f* **~ lacquer** Kristall-Lack *m* **~ lattice** Kristall-, Raum-gitter *n* **~ lattice imperfection** Kristallbaufehler *m* **~-lattice plane** Netzebene *f* **~-lattice vibration** Kristallgitterschwingung *f* **~-like** kristallartig **~ marking** Spaltflächenzeichnung *f* **~ master oscillator** Quarzsteuersender *m* **~ microphone** Kristallmikrofon *n*, piezoelektrisches Mikrofon **~ monitor** Kristalltaktgeber *m* **~ mount** Fassung *f* für Kristalle, Kristallfassung *f*, Kristallhalterung *f* **~ mounting** Quarzfassung *f* **~ nucleus** Keimkristall *m*, Kristallkeim *m* **~ orientation** Kristallorientierung *f* **~ oscillator** Schwingkristall *m* **~ pickup** Kristallschalldose *f* **~ plane** Kristall-ebene *f*, -fläche *f* **~ plate** Quarzscheibe *f* **~ prism** Kristallkörper *m* **~ pulling** Kristallziehverfahren *n* **~ pulling machine** Kristallziehofen *m* **~ receiver** Detektorenempfänger, Kristallfernhörer *m* **~ rectifier** Kontakt-, Kristall-gleichrichter *m* **~ ridges** Translationsstreifung *f* **~ rotation** Kristalldrehung *f* **~ scale** Kristallgitterskala *f* **~ sceleton** Kristallgerippe *n* **~ scintillator** Kristallzintilator *m* **~ sludge** Kristallbrei *m* **~ spectrograph** Kristall(gitter)spektrograf *m* **crystal-stabilized, ~ drive (oscillator) stage** quarzgesteuerte Steuerstufe *f* **~ master stage** fremdgesteuerte Steuerstufe *f* **~ oscillator stage** fremdgesteuerte, quarzgesteuerte Steuerstufe *f* **~ superheterodyne receiver** Zwischenfrequenzempfänger *m* mit Kristallsteuerung **~ transmitter** kristallstabilisierter Sender *f* **crystal, ~ stabilizer** Kristalltaktgeber *m* **~ stabilizing** Kristallsteuerung *f* **~ stage** Quarzstufe *f* **~ standard** Quarznormal *m* **~ structure** Kristallaufbau *m*, Kristallbau *m* **~ suspension** Quarzhalterung *f* **~ terms** Kristalltermen *pl* **~ tetrode mixer** Punkttetrode *f* **~ tuning fork** Kristallstimmgabel *f* **~ turret** Quarztrommel *f* **~ unit** Kristallgerät *n* **~ vibration** Kristallschwingung *f* **~ video rectifier** Bildkristallgleichrichter *m* **~ wafer** Kristallscheibchen *n* **~ weakness** kristallinische Schwäche *f*

**crystalliform** kristallartig, kristallförmig, kristallin

**crystalline** kristallartig, kristallhell, kristallin, kristallinisch, kristallografisch **~ aggregate** Kristallaggregat *n* **~ aggregation** Kristallhaufwerk *n* **~ force** Kristallisationsvermögen *n* **~ form** Kristall-bau *m*, -form *f* **~ fracture** körniger Bruch *m* (iron) **~ grain** Kristallit *m*, Kristallkorn *n* **~ group** Kristallgebilde *n* **~ growth** Kornwachstum *n*, Kristall-anschluß *m*, -ausbildung *f*, -bildung *f*; Kristallisationswachstum, Kristallwachstum *n* **~ iron** körniges Eisen *n* **~ lens** Kristallinse *f* **~ material** kristallines Material *n* **~ matter** Kristallgebilde *n* **~ modification** Umkristallisation *f* **~ particle** Kristallteilchen *n* **~ pig iron** Spangeleisen *n* **~ refinement** Kornverfeinerung *f* **~ silver** Kristallsilber *n* **~ solid solution** Mischkristall *n* **~ structure** Kristall-bau *m*, -gebilde *n*, -struktur *f* **~ transformation** Umkristallisation *f* **~ wax** kristallisiertes Paraffin *n*

**crystallinity** Kristallinität *f*

**crystallite** Dendrit *m*, Kristallgerippe *n*, Kristallit *m* **~ growth** Kristallwachstum *n*

**crystallizability** Kristallisationsfähigkeit *f*

**crystallizable** kristallisationsfähig, kristallisierbar

**crystallization** Anschuß *m*, Kornbildung *f*, Kristallanschuß *m*, Kristallbildung *f*, Kristallisation *f*, Kristallisierung *f*, Kristallkernbildung *f* **~ in tanks** Kastenarbeit *f* **~ works** Pattinsonieranstalt *f*

**crystallize, to ~** anschießen, ansetzen, kristallisieren, kräuseln (of tung oil films), **to ~ out** auskristallisieren, herauskristallisieren, soggen

**crystallized, ~ cinnabar** Zinnoberspat *m* **~ tin plate** Perlmutterblech *n*

**crystallizer**, Anschlußfaß *n*, Kristallisationsapparat *m*, Kristallisator *m*, Kristallisierapparat *m*, Sudmaische *f* **~ with upper worm drive** Maische *f* mit oberem Schneckenradantrieb *f* **~ pan** Kochmaische *f*

**crystallizing** Auskristallisieren *n* **~ and curing** Zuckerhausarbeit *f* **~ dish** Kristallisierschale *f* **~ pan** Kristallisiergefäß *n* **~ process** Kristallisationsvorgang *m* **~ vessel** Kristallisations-, Kristallisier-gefäß *n*

**crystallogram** Kristallogramm *n*

**crystallographer** Kristallforscher *m*

**crystallographic** kristallografisch **~ axis** Kristallachse *f* **~-axis orientation** Kristallachsenrichtung *f* **~ classes** kristallografische Klassen *pl* **~ data** kristallografische Daten *pl* **~ growth** Kristallanschuß *m* **~ plane** Symmetrieachse *f*, -ebene *f*

**crystallography** Kristallkunde *f*, Kristallehre *f*, Kristallografie *f*

**crystalloid** kristalloid

**crystalloidal** kristalloid

**crystalloluminescence** Kristallumineszenz *f*

**crystallometer** Kristallometer *n*

**CTC (centralized train control) interlocking** Zentralstellwerk *n*

**cubage** umbauter Raum *m*

**cubanite** Weißkupfererz *n*

**cubature** Inhaltsbestimmung *f* eines Körpers, Kubikberechnung *f*

**cube, to ~** kubieren

**cube** Kubikzahl *f*, Kubus *m*, Quadratblock *m*, Würfel *m* **~ gambier** Würfelgambir *n* **~ insert** Würfeleinsatz *m* (Kühlschrank) **~ photometer** Integrationswürfel *m* **~ root** Kubikwurzel *f*, dritte Wurzel *f* **~ root law** Kubikwurzelgesetz *n* **~-shaped briquette** Würfelbrikett *n* **~-shaped coil** Würfelspule *f* **~-type contrast photometer** Kontrastfotometer *n* mit Würfel

**cubic** würfelförmig, würfelig **~ (al)** kubisch **~ body-centered metals** kubischraumzentrierte Metalle *pl* (cryst.) **~ capacity** Hubraum *m*, Kubikinhalt *m* **~ capacity of cylinder** Zylindervolumen *n* **~ centimeter** Kubikzentimeter *m* **~ closed packed structure** kubisch dichteste Kugelpackung *f* **~ contents** Kubikinhalt *m* **~ curve** Kurve *f* dritter Ordnung **~ dilatation** Volumdilatation *f* **~ face centered** kubischflächenzentriert **~ foot** Kubikfuß *m* **~ inch** Kubikzoll *m* **~ lattice** kubisches Gitter *n* **~ measure** Körpermaß *n*, Kubikmaß *n* **~ measurement** Körpermaß *n* **~ meter** Fest-, Kubik-, Raum-meter *m* ~ **number** Kubikzahl *f* ~ **point group** kubische Punktgruppe *f* ~ **root** Kubikwurzel *f* ~ **roots** Einheitswurzeln *pl* ~ **screw** Würfelschraube *f* ~ **space lattice** kubisches Raumgitter *n* ~ **strength** Würfelfestigkeit *f* ~ **summit** Würfeleck *n* (cryst.) ~ **texture** Würfeltextur *f*

**cubical** würfelförmig ~ **antenna** Würfelantenne *f* ~ **contents** körperlicher Inhalt *m* ~ **lattice** Würfelgitter *n* ~ **measure** Körpermaß *n*

**cubicle** Kabine *f* (film, TV), Schaltzelle *f*, Schlafraum *m*, Zelle *f*

**cubing** Rauminhaltmessung *f* ~ **formula** Kubierungsformel *f*

**cubo-octahedron** Kubooktaeder *n*

**cudgel** Knebel *m*, Prügel *m*

**cue** Stichwort *n*, Übertragungsbefehl *m*

**cuff** Ärmelaufschlag *m*, Aufschlag *m*, Manschette *f*, Stulpe *f* ~ **patch** Ärmelpatte *f* ~ **valve** Schieberventil *n*

**cull, to** ~ auszupfen, belesen, kletten, noppen **to** ~ **ore** das Erz klauben

**cullage** Abfallware *f*

**cullet** Glas-brocken *m*, -bruch *m*, -scherben *pl*

**culm** Anthrazitart *f* (kleine), Grus *m*, Kulm *m*, Steinkohlenklein *n*

**culmiferous** anthrazithaltig

**culminate, to** ~ seinen Höhepunkt erreichen

**culminating point** Höhepunkt *m*

**culmination** Gipfelung *f* ~ **point** Gipfelpunkt *m*

**culpable** schuldig

**culprit** Täter *m*

**cultivate, to** ~ anbauen, bauen, bebauen, behauen, bestellen (a field), bewirtschaften (soil), kultivieren, pflegen, ziehen, züchten

**cultivated,** ~ **fields** bebaute Felder *pl* ~ **rubber** Plantagenkautschuk *m*

**cultivating lands** Urbarmachung *f* von Land

**cultivation** Ackerbau *m*, Anbau *m*, Bestellung *f*, Bodenbearbeitung *f*, Bodenbestellung *f*, Kultivierung *f*, Kultur *f* ~ **of nature** Naturschutz *m*

**cultivator** Grubber *m*, Krümmer *m*, Kultivator *m* ~ **tooth** Kultivatorzinke *f*

**culture** Anbau *m*, Kultur *f*, Kulturlandmarken *pl* (man-made landmarks)

**cultured** gebildet

**culvert** Abzugskanal *m*, Aquädukt *m*, Bachdurchlaß *m*, Durchlaß *m*, Einlaufkanal *m*, Entwässerungsstollen *m*, Kanaldüker *m*, Rinnstein *m*, Siel *n*, Umlauf *m*, Wasser-ablaß *m*, -ableitung *f*, -abzug *m* ~ **convergence** Einlauf *m* ~ **covered with a slab** Plattendurchlaß *m* ~ **siphon** Drücker *m*

**culverts arranged in the floor** Grundläufe *pl*

**cumalic acid** Cumalinsäure *f*

**cumarinic acid** Cumarinsäure *f*

**cumbersome** beschwerlich, schwerfällig, viel Platz beanspruchend

**cumene** Cumol *n*, Kumol *n*

**cumic acid** Cuminsäure *f*

**cuminic acid** Cuminsäure *f*

**cumol** Cumol *n*

**cumulate, to** ~ sich anhäufen

**cumulation** Anhäufung *f*, Kumulation *f*

**cumulative** addierend, additiv, anhäufend, kumulativ ~ **(differential) compound excitation** Mitverbunderregung *f* ~ **compound motor** Schlupfmotor *m* ~ **curve** Summenkurve *f* ~ **dose** Gesamtdosis *f* ~ **frequency** Summenhäufigkeit *f*

~ **frequency curve** Faserzahldiagramm *n* ~ **grid detection** Gitterstromgleichrichtung *f* ~ **rectification** Audiongleichrichtung *f* ~ **tolerances** Toleranzketten *pl*

**cumuliform** haufenwolkenartig, kumulusförmig

**cumulo,** ~-**nimbus** Kumulonimbus *m* ~-**nimbus clouds** Gewitterwolken *pl* ~-**stratus (clouds)** Rollkumulus *m*, Wulstkernkumulus *m*

**cumulus** Kumulus *m* ~ **cloud** Haufenwolke *f*, Kumulus *m*, Kumuluswolke *f* ~ **formation** Kumulusbildung *f*

**cuneal** keilförmig ~ **line** Keilstrich *m*

**cuneate** keilförmiger Kristallit *m*

**cuneiform** keilförmig

**cup, to** ~ austiefen, tiefziehen

**cup** Becher *m*, Kalotte *f*, Kugelabschnitt *m*, Napf *m*, Näpfchen *n*, Pokal *m*, Rondell *n*, Schale *f* (oil), Tasse *f* ~ **of cartridge** Hütchen *n* ~ **of the eyepiece** Augenmuschel *f* ~ **for grease lubrication** Büchse *f* für Fettschmierung ~ **of the rhumb card** Kompaßpfanne *f*

**cup,** ~-**and-cone arrangement** Gasverschluß *m* (furnace), Gichtverschluß *m* ~-**and-cone bearing** Kegelkugellager *n* ~-**and-cone charger** Doppelrichterapparat *m* (blast-furnace) ~ **anemometer** Schalenkreuzwindmesser *m* ~-**anemometer type of speed indicator** Schalenkreuzfahrtmesser *m* ~ **blade** Topfschaufel *f* ~ **board** Schrein *m*, Vorrats-, Küchen-, Teller-schrank *m*, Schrank *m* ~ **bolt** Pfannenbolzen *m* ~ **brush** Topfbürste *f* ~ **collar** Schmierfänger *m* ~ **discharger** Schießbecher *m* (of rifle grenade) ~ **dish** Hohlschale *f* ~ **electrometer** Schalenelektrometer *n* ~-**fluted** kalottengeriffelt ~ **grease** Staufferfett *n* ~-**grinding wheel** Topfschleifscheibe *f* ~ **head (oval or mushroom head)** Flachrundkopf *m* ~ **leather** getriebenes Leder *n* ~ **notch** Becherausschnitt *m* ~ **point** Ringschneide *f* ~ **shake** Ringklaft *f* (in wood) ~-**shaped** topfförmig, becherförmig ~-**shaped bulb** Becherlampe *f* ~-**shaped cam disc** Kurvenscheibe *f* ~ **shaped cutter** Topffräser *m* ~-**shaped die** Aufsatzhammer *m* ~-**shaped gong** Kelchglocke *f* ~-**shaped hammer** Schellhammer *m* ~-**shaped part** Kugelschale *f* ~ **socket** Pfannenkappe *f* ~ **spring** Tellerfeder *f* ~ **standard** Pfannenbolzen *m* ~ **test** Löffel-, Schröpf-probe *f* ~-**type anemometer** Schalenwindstärkemesser *m* ~-**type condenser (or capacitor)** Rundbecherkondensator *m* ~-**type current meter** Schalenkreuzstrommesser *m* ~-**type guiding condenser (or capacitor)** Becherführungskondensator *m* ~-**type housing** Rundbechergehäuse *n* ~ **valve** Kronenventil *n* ~ **vane** Schalenflügel *m* ~ **wheel** Becher-, Topf-scheibe *f* (emery) ~ **wheel** Schmiergeltasse *f*

**cupel, to** ~ kapellieren, kupellieren, treiben

**cupel** Glühschale *f*, Kapelle *f*, Kupelle *f*, Probescherbe *f*, -scherben *m*, Scheidekapelle *f*, Scherbe *f*, Test *m*, Treibscherben *m* ~ **ashes** Treibasche *f* ~ **furnace** Abtreibeofen *m* ~ **holder** Kapellenträger *m* ~ **mold** Kapellenform *f* ~ **pyrometer** Legierungspyrometer *n* ~ **test** Kapellenprobe *f* ~ **tongs** Kapellenkluft *f*

**cupeling** Treiben *n* ~ **furnace** Abtreibe-, Kapellen-, Treib-ofen *m*

**cupellate, to** ~ abtreiben

**cupellation** Kapellenprobe *f*, Kupellation *f*,

Kupellieren *n*, Probe *f* durch Abtreiben, Treibprozeß *m* ~ **loss** Kapellenraub *m*
**cupola** Beobachtungskuppel (used for observation) *f*, Beobachtungsturm *m*, Kuppe *f*, Kuppel *f*, Maschinengewehrstand *m*, Panzerturm *m*, Turmaufsatz *m* ~ **blower** Kupolofengebläse *n* ~ **charging** Kupolofenbegichtung *f* ~ **charging crane** Kupolofenbegichtungskatze *f* ~ **forehearth** beheizter Kupolofenvorherd *m* ~ **furnace** Kapellen-, Kupol-, Kuppel-, Schachtofen *m* ~ **hearth** Kupolofenherd *m* ~ **installation** Kupolofenanlage *f* ~ **keeper** Kupolofenbediener *m* ~ **leg** Kupolofentragsäule *f* ~ **lining** Kupolofenausfütterung *f* ~ **loss** Abbrand *m* ~ **malleable iron** Kupolofentemperguß *m* ~ **mantle** Kupolofenmantel *m* ~ **melting** Kupolofenschmelzen *n* ~ **metal** Kupolofeneisen *n* ~ **mixture** Kupolofengattierung *f* ~ **operation** Kupolofenbetrieb *m* ~ **practice** Kupolofenbetrieb *m*, Kupolofenführung *f* ~ **process** Kupolofenverfahren *n* ~ **receiver (or settler)** Kupolofenvorherd *m* ~ **shaft** Kupolofenschacht *m* ~**-shaped** kuppelförmig ~ **shell** Kupolofenmantel *m* ~ **slag** Kupolofenschlacke *f* ~ **spout** Kupolofenrinne *f* ~ **stack** Kupolofenschacht *m* ~ **tenter** Kupolofenschmelzer *m* ~**-type gun mount** Kuppellafette *f*
**cuppability test** Tiefziehprobe *f* (specimen)
**cupped** ausgehöhlt ~ **gripping (or biting) point** Ringschneide *f*
**cupping** Reißkegelbildung *f*, Tiefung *f*, Windriß *m* (of wood) ~**(or dishing) depth** Kümpelungstiefe *f* ~**-ductility value** Tiefungswert *m* ~ **glass** Schröpfkopf *m* ~ **machine** Blechprüfapparat *m* ~ **method (or process)** Tiefziehverfahren *n* ~ **test** Tiefungsprobe *f*, Tiefziehprobe *f*, Treibprobe *f* ~ **test machine** Tiefziehprüfmaschine *f* ~ **value** Tiefungswert *m*
**cuppy wire** von innen längsrissiger Draht *m*
**cuprammonium** Cuproxam *n* ~ **rayon** Glanzstoff *m*, Kupferkunstseide *f* ~ **silk** Kupferseide *f* ~ **solution** Kuoxam *n*
**cupreous** kupferig ~ **vitriol** kupferhaltiges Vitriol *m*
**cupric**, ~ **acid** Kupfersäure *f* ~ **arsenite** Kupferarsenit *n* ~ **chloride** Cupri-, Kupfer-, Kuprichlorid *n*, salzsaures Kupferoxyd *n* ~ **citrate** Kupferzitrat *n* ~ **compound** Cupri-, Kupferoxyd-verbindung *f* ~ **cyanide** Kupferzyanid *n* ~ **ferrocyanide** Ferrocyankupfer *n* ~ **formate** Kupferformiat *n* ~ **hydroxide** Kupfer-hydroxyd *n*, -oxydhydrat *n* ~ **iodide** Kupferjodid *n* ~ **oxide** Cuprioxyd *n*, Kupferoxyd *n* ~ **salicylate** Kupfersalizylat *n* ~ **salt** Kupferoxydsalz *n* ~ **silicate** Kupferoxydsilikat *n* ~ **sulphide** Kupfersulfid *n* ~ **thiocyanate** Kupferrhodanid *n*, Rhodankupfer *n*
**cupriferous** kupferführend, kupferhaltig, kupferig
**cuprite** Cuprit *m*, Kupferblüte *f*, Kupferglas *n*, Kupferrot *n*
**cuproautunite** Kupferautunit *m*
**cupromanganese** Kupfermangan *n*, Kupromangan *n*, Mangankupfer *n*
**cupron cell** Kupronelement *n*
**cupronickel jacket** Nickelkupfermantel *m*
**cuprosilicon** Siliziumkupfer *n*
**cuprous**, ~ **bromide** Kupferbromür *n* ~ **chloride**

Cuprochlorid *n*, Kupferchlorür *n*, Kuprochlorid *n*, salzsaures Kupferoxydul *n* ~ **compound** Kupferoxydulverbindung *f* ~ **cyanide** Kupferzyanür *n* ~ **halide** Kupferhalogenid *n* ~ **hydroxide** Kupferhydroxydul *n*, Kupferoxydulhydrat *n* ~ **iodide** Cuprojodid *n*, Kupferjodür *n* ~ **oxide** Cuprooxyd *n*, Kuprooxyd *n*, Kupferoxydul *n* ~ **oxide cell** Kupferoxydulzelle *f*, Kupronelement *n* ~ **oxide rectifier** Cuproxtrockengleichrichter *m*, Trockengleichrichter *m* ~ **pickling hath** Kuprodekopierbad *n* ~ **salt** Kuprosalz *n* ~ **silicate** Kupferoxydulsilikat *n* ~ **sulphide** Kupfersulfür *n*, Schwefelkupferoxydul *n* ~ **thiocyanate** Kupferrhodanür *n*
**cuprozippeite** Kupferzippeit *m*
**curative agent** Vulkanisationsmittel *n*
**curb, to** ~ dämmen, hemmen
**curb** Bordschwelle *f*, Bordstein *m*, Drosselung *f*, Einfassung *f*, Fassung *f* (general mechanical), Kranz *m* aus Holz oder Eisen, Schutzwinkel *m*, Springquelle *f* ~ **of a well** Brunnen-brüstung *f*, -einfassung *f*
**curb**, ~ **beam** Brückenschwelle *f* ~ **bit (bridle)** Kandare *f* ~**-bit set** Stangengebiß *n* ~ **chain** Kinn, Sperr-kette *f* ~ **rafter** Obersparren *m* ~ **rein** Kandarenzügel *m* ~ **roof** gebrochenes Dach *n* ~ **stone** Abweisstein *m*, Bordschwelle *f*, Bordstein *m*, Gradstein *m*, Prellstein *m*, Randstein *m*, Rinnstein *m*, Straßenkante *f*
**curbed signaling** Curbsenden *n*
**curbing** Curbsenden *n*
**curcuma** Curcuma *f*
**curd, to** ~ eindicken, gerinnen
**curd soap** Kernseife *f*
**curdle, to** ~ gerinnen, verdicken
**curdled** geronnen
**curds** Gerinnsel *n*, käsiger Niederschlag *m*
**curdy precipitation** käsiger Niederschlag *m*
**cure, to** ~ aushärten, einsalzen, gesundmachen, heilen, präparieren (painting), räuchern, schleudern, trocknen
**cure** Abhilfe *f*, Heilbehandlung *f*, Kur *f*, Übervulkanisation *f* ~ **in molds** Muldenheizung *f*
**cured weight** Salzgewicht *n*
**curfew** Verkehrssperre *f*
**curie**, ~ **therapy** Radiumbestrahlung *f* ~ **unit quantity of radon** Radiumemanationseinheit *f*
**Curie point** Curie'scher Punkt *m*
**curing** Nachbehandlung *f*, Schleuderung *f* (sugar mfg.) ~ **of defects (or errors)** Fehlerbeseitigung *f* ~ **cellar** Pökel-raum *m* (-keller) ~ **period** Abbindezeit *f* ~ **plant** Räucherei *f* ~ **range** Vulkanisationsbereich *m* ~ **rate** Vulkanisationsgeschwindigkeit *f* ~ **temperature** Schubtemperatur *f*
**curite** Lebererz *n*
**curl, to** ~ kräuseln, krümmen, ringeln, rollen (of film), Schlingen bilden
**curl** Kringel *m*, Quirl (Maß der Wirbelgröße), Rotation *f* eines Vektors, Schlinge *f* im Draht, Wirbel *m*
**curled** geringelt ~ **birchwood** Birkenmaser *f* ~ **exponential horn** aufgewundener Exponentialtrichter *m* ~ **exponential loud-speaker** aufgewundener Trichter *m* ~ **horn (loud-speaker)** Faltenhorn *n* ~**-over (or rolled) spring eye** gerolltes Federauge *n* ~ **selvedges** umgeschlagene Leisten *pl*

**curling** Kräuseln *n* (of or in paper) Einrollen *n* ~ **cut** Schälanschnitt *m* ~ **paper** Papilottenpapier *n* ~ **selvedges** sich rollende Leisten *pl*
**curly top (disease)** Kräuselkrankheit *pl*
**currency** Gebräuchlichkeit *f*, Gültigkeit *f*, Münze *f*, Währung *f* ~**-conversion table** Währungsumrechnungstabelle *f*
**current** Fluß *m*, Lauf *m* Strom *m*, Strömung *f*, Zug *m*, **in** ~ kursfähig **up** ~ aufsteigender Luftstrom *m*
**current,** ~ **of air** Luftzug *m* ~ **on breaking** Abschaltstromstärke *f* ~ **of breath** Luftgleichstrom *m* (in sound recording) ~ **due to the concentration gradient** Diffusionsstrom *m* ~ **on contact** Schaltstrom *m* ~ **to earth** Erdgeschoßstrom *m* ~ **of dielectric convection** dielektrischer Verschiebungsstrom *m* ~ **of electromagnetic induction** Strominduktion *f* ~ **of holes** Löscherstrom *m* ~ **of hydrogen** Wasserstoffstrom *m* ~ **of residual charge** Nachwirkungsstrom *m* ~ **of a river** Tiefenlinie *f* ~ **at the terminals** Klemmstrom *m* ~ **of water** Wasserstrom *m*
**current,** ~ **account** Girokonto *n*, Kontokorrent *n* ~ **account book** Saldokonto *n* ~ **alternator** Polwechsler *m* ~ **amplification** Stromverstärkung *f* ~ **amplitude** Stromamplitude *f* ~ **anchor** Stromanker *m* ~**-anchor line** Stromankerlinie *f* ~ **antinode** Strombauch *m* ~ **attenuation** Stromdämpfung *f* ~ **balance** Stromwippe *f* (electr.), Stromwaage *f* (weigher) ~ **beat** Stromschwebung *f* ~ **bedding** Diagonalschichtung *f* ~ **capacity** Stromleistung *f* ~ **carrier** Stromträger *m*
**current-carrying** stromdurchflossen, stromführend ~ **capacity** Hochbelastbarkeit *f*, Strombelastbarkeit *f* (of conductor) ~ **conductor** stromdurchflossener Leiter *m* ~ **line** beschickte Leitung *f* ~ **lug** stromführende Fahne *f* ~ **rail** Stromzuführungsschiene *f* ~ **spring** Stromzuführungsfeder *f* ~ **wire** stromführender Draht *m*
**current,** ~ **change** Stromänderung *f* ~ **changer** Gleichrichter *m* ~ **characteristics** Stromwert *m* ~ **charge** Strombelag *m* (antenna) ~ **chart** Strom-, Strömungs-karte *f* ~ **circuit** Stromkreis *m* ~ **coil** Stromspule *f* ~**-collecting brush** Stromabnahmebürste *f* ~**-collecting rail** Stromschiene *f* ~ **collector** Stromabnehmer *m* ~ **collector arm** Stromabnehmerbügel *m* ~ **collector line** Schleifleitung *f* ~ **component** Stromkomponente *f* ~ **computation** Stromberechnung *f* ~ **connecting bar** Stromschiene *f* ~ **consumption** Strom-aufnahme *f*, -verbrauch *m* ~ **coverage** Strombelag *m* ~ **converter** Stromumformer *m* ~**-conveying** stromführend ~ **curve** Stromkurve *f* ~ **data** Strom-, Strömungs-angabe *f* ~ **density** Belastung *f*, Stromdichte *f* ~ **density on the object** Strahlungslast *f* (electron microscopy) ~ **detector** Galvanoskop *n*, Stromdetektor *m* ~ **displacement** Stromverdrängung *f* ~**-displacement motor** Stromverdrängungsmotor *m* ~ **distribution** Strom-belag *m*, -verlauf *m* ~ **distribution noise** Stromverteilungsrauschen *n* ~ **distributor** Stromverteiler *m* ~ **divider** Stromweiche *f* ~ **(or power) drain** Stromentnahme *f* ~ **driver** Schreibstromstufe *f* ~ **echo** Leitungsecho *n* ~ **electricity** Galvanismus *m* ~ **equation** Stromgleichung *f* ~ **event** Tagesereignis *n* ~

**expenses** laufende Kosten *pl* (not running expenses) ~ **feedback** Stromgegenkopplung *f* ~**-flow resistance** Durchgangswiderstand *m* ~ **flow reversing switch** Stromrichtungsumschalter *m* ~ **flowing from grid to screen** Steuergitterstrom *m* ~ **flowing in potential circuit** Spannungsstrom *m* ~ **fluctuation** Energieänderung *f*, Stromschwankung *f* ~ **follower** Gitterbasisschaltung *f* ~ **forcing** Felderhöhung *f* ~ **generated by handoperated dynamo** Handdrehdynamostromquelle *f* ~**-illumination characteristic** Stromlichtstärkecharakteristik *f* ~ **impulse** Stromimpuls *m*, Stromstoß *m*, Telegrafierstromstoß *m* ~**-impulse contact** Stromstoßkontakt *m* ~ **indicator** Stromanzeiger *m* ~ **intensity** Stromstärke *f* ~ **intensity at make** Einschaltstromstärke *f* ~ **lead-in (lead off)** Stromdurchführung *f* ~ **leakage** Stromübergang *m* ~ **less** stromlos ~ **limiter** Strombegrenzer *m* ~**-limiting coil** Begrenzungsdrossel *f* ~ **limiting device** Strombegrenzer *m* ~**-limiting fuse** Grobsicherung *f* ~**-limiting reactor** Strombegrenzungsspule *f* ~ **limiting resistor** Strombegrenzungswiderstand *m* ~ **limiting switch** Strombegrenzungsschalter *m* ~ **load** Strombelastung *f* ~ **loop** Strombauch *m* ~ **loss** Stromverlust *m* ~ **measurement** Strommessung *f* ~**-measuring instrument** Strommeßgerät *n* ~ **meter** hydrometrischer Flügel *m*, Stromdurchflußmesser *m*, Strömungsmesser *m*, Stromzähler *m* ~ **(-displacement) meter** Verdrängungsmesser *m* ~ **meter held in position by means of a rod** Stangenflügel *m* ~**-meter measurement** Flügelmessung *f* (hydraul.) ~ **multiplication** Stromvervielfachung *f* ~ **node** Strom-knoten *m*, -nullstelle *f* ~ **noise** Stromrauschen *n* ~ **oscillation** Stromschwingung *f* ~ **pack** Stromversorgungsteil *m* ~ **path** Strom-bahn *f*, -faden *m*, -verlauf *m*, -weg *m* ~ **phase** Stromphase *f* ~ **potential** Stromspannung *f* ~ **price** Kurs *m* ~ **pulsation** Stromstoß *m* ~ **pulse** Stromstoß *m* ~ **rate** Kurs *m* ~ **rate of exchange** Tageskurs *m* ~ **ratio (or gain)** Stromverstärkungsfaktor *m* ~ **recorder** Stromschreiber *m* ~ **rectifier** Stromgleichrichter *m* ~ **refraction** Strombrechung *f* ~ **regulation** Stromregulierung *f* ~ **regulator** Stromregulator *m* ~ **relay** Stromwächter *m* ~ **reserve** Kommutator *m* ~ **reversal** Stromwechsel *m* ~**-reversal bridge** Stromwendebrücke *f* ~ **reverser** Polwender *m* ~**-reversing key** Stromwender *m* ~ **reverter** Stromumkehrer *m* ~ **ripple** Kommutierungswelle *f*, Stromwelle *f* ~ **rise** Einschaltung *f*, Stromanstieg *m* ~ **rush** plötzlicher Stromstoß *m* ~ **sensitivity** Stromempfindlichkeit *f* ~ **sequence** Stromfolge *f* ~ **sheet** flächenanziehungsartige Strombahn *f*, Spulenseite *f* ~**-sheet formula** Stromflächenformel *f* ~ **stabilizing** Stromstabilisierung *f* ~ **stay** Stromleine *f* ~**-storage cell** Heizsammler *m* ~ **supply** Strom-lieferung *f*, -versorgung *f*, -zuführung *f* ~ **supply to extension telephones** Nebenstellenspeisung *f* ~ **supply from public mains** Netzstrom *m* ~**-supply brush** Stromzuführungsbürste *f* ~ **supply cable** Stromzuleitungskabel *n* ~ **supply loss** sekundäre Dämpfung *f* ~ **surge** Einschaltüberströme *pl*, Stromstoß *m*, Telegrafierstromstoß *m* ~ **take-off device** Stromabnehmer *m* ~ **telephone-influence factor** Fernsprechform-

faktor *m* des Stromes ~ **test** Prüfung *f* der Speisestromstärken ~ **transformer** Stromstoß-umformer *m*, Stromtransformator *m*, Strom-umformer *m*, Stromwandler *m* ~ **transmitting** Stromgeben *n* ~ **(or flow) triangle** Stromdrei-eck *n* ~ **tube** Stromfaden *m* ~**-utilization device** Stromverbraucher *m* ~ **value** Stromwert *m* ~ **variation** Stromänderung *f*, Stromschwankung *f* ~ **vector** Stromvektor *m*, Stromzeiger *m* ~ **velocity** Stromgeschwindigkeit *f* ~**-voltage characteristic** (of the two-electrode valve) Ent-ladekurve *f*, Entladungscharakteristik *f* einer Diode, Stromspannungskennlinie *f*, Strom-spannungskurve *f* ~ **wave** Stromwelle *f* ~**-wave-length characteristic** (of a photocell) Charakteri-stik *f* der spektralen Empfindlichkeit ~ **winding** Stromwicklung *f* ~ **yield** Stromausbeute *f* ~ **zero point** Stromnullpunkt *m*
**curriculum** Lebenslauf *m*, Lehrplan *m*
**curried leather** zugerichtetes Leder *n*
**currier's,** ~ **black** Lederschwärze *f* ~ **knife** Falz-eisen *n*
**curry, to** ~ gerben
**curry comb** Kardätsche *f*, Mähnenkamm *m*, Stiegel *m*
**cursor** Peilanzeiger *m*, Reiter *m*, Schiebekontakt *m*, Schieber *m*, Seitenwinkelanzeiger *m*, Stell-ring *m* ~ **lines** Markierungslinien *pl*
**cursory** flüchtig, oberflächlich
**curtail, to** ~ beschneiden, schmälern, stützen, verkürzen
**curtailing** Schmälerung *f*
**curtain, to** ~ vorhängen
**curtain** Antennenanordnung *f*, Fenstervorhang *m*, Gardine *f*, Herdmauer *f*, Kernmauer *f*, Läufer *m*, Neutronenfänger *m*, (cloth) Stoff-vorhang *m*, (von Farbe) Vorhang *m* ~ **of aurora** Nordlichtdraperie *f* ~ **of signals** Signalwand *f*
**curtain,** ~ **antenna** Fächerantenne *f* ~ **array** Dipolebenenanordnung *f*, Dipolwand *f* (d. Antenne) ~ **clasp** Vorhanghalter *m* ~**-drawing device** Garderobezugvorrichtung *f* ~ **fade-in** Vorhängeaufblendung *f* ~**-fading shutter** Vor-hangblende *f* ~**-lace machine** Garderobeweb-stuhl *m* ~ **pole** Vitragenstab *m* ~**-type shutter** (radiator) Rollvorhang *m* ~ **wall** Abschirm-mauer *f*
**curtains** Dichtungsschirm *m*
**curtate** verkürzte horizontale Linie *f*
**curvature** Abrundung *f*, Biegung *f*, Bogen *m*, Einbiegung *f*, Falte *f*, Krümmung *f*, Kurven-form *f*, Richtfaktor *m* (einer Verstärkerröhre), Wölbung *f*
**curvature** ~ **of beams** Balkenkrümmung *f* ~ **of the brush face** Griffläche *f* ~ **of the character-istic** Krümmung *f* der Charakteristik ~ **of a curve in space** Krümmung *f* einer Raumkurve ~ **of the direction** Richtungskrümmung *f* ~ **of the earth** Erdkrümmung *f* ~ **of equipotential surfaces of gravity** Krümmung *f* der equipo-tentialen Flächen des Schwerkraftfeldes ~ **of field (or image)** Bildwölbung *f* ~ **of image field** Bildkrümmung *f*, Bildfeldwölbung *f* ~ **of a line** Streckenkrümmung *f* ~ **at peak** Scheitelwöl-bung *f* ~ **of a plane curve** Krümmung *f* einer ebenen Kurve ~ **of a road** Wegkrümmung *f* ~ **of stream-line** Stromlinienkrümmung *f* ~ **of a surface** Krümmungsmaß *n* einer Fläche ~ **of**

**track** Gleiskrümmung *f*
**curvature,** ~ **bevel** Winkelpasser *m* ~ **term** Krüm-mungsfunktion *f*
**curve, to** ~ abbiegen, biegen, bombieren, krüm-men, kurven, schweifen, wölben
**curve** Biegung *f*, Bogen *m*, Diagramm *n*, Gleis-kurve *f*, Krümmung *f*, Kurve *f*, Kurvenlineal *n*, Linie *f*, Linienzug *m* (math.), Schaulinie *f*, Schleife *f*, Verlauf *m* **(stepped** ~) Treppenkurve *f*
**curve,** ~ **of buoyancy** Auftriebskurve *f* ~ **of condition** Zustandskurve *f* ~ **of deflection** Bie-gungslinie *f* ~ **of discharge** Entladungskurve *f* ~ **of displacement by load** Lastenmaßstab *m* ~ **of distortion** Verzeichnungskurve *f* ~ **of equal pressure** Isobare *f* ~ **of error** Fehlerlinie *f* ~ **of flight path** Bewegungslinie *f* ~ **of gyroscopic moment** Kreiselmomentlinie *f* ~ **of large radius** flache Kurve *f* **(polar)** ~ **of light distribution** Lichtverteilungskurve *f* ~ **of line** Gleiskrüm-mung *f* ~ **of lowering of groundwater level** Ab-senkungskurve *f* ~ **of neutral stability** Indiffe-renzkurve *f* ~ **of normal magnetization** Kom-mutierungskurve *f* ~ **of plunger (or piston) travel** Kolbenweglinie *f* ~ **of projection** Wurf-linie *f* ~ **of response** Schwellwertkurve *f* ~ **in space** Raumkurve *f* ~ **of spectrum** Energiever-teilungskurve *f*, Spektralkurve *f* ~ **of spur gear** Teilbahn *f* des Stirnrades ~ **of state** Zustands-kurve *f* ~ **of the stern** Heckkurve *f* ~ **of switch** Weichenbogen *m* ~ **of the trajectory** ballistische Kurve *f*, Gestalt *f* der Flugbahn
**curve,** ~ **compensated** drehgeschützt (electron.) ~**-cutting machine** Kurvenschere *f* ~ **cutting shears** Kurvenscheren *pl* ~**-drawing recorder** Kurven-, Linien-schreiber *m* ~ **factor** Aus-bauchungsfaktor *m* ~ **fitting** Kurvenanpassen *n* ~ **following a thirdpower law** kubische Kurve *f* ~ **gauge** Kurvenlehre *f* ~ **light** Kurvenschein-werfer *m* ~ **negotiating characteristic** Kurven-gängigkeit *f* ~ **path** Kurvenzug *m* ~ **pen** Kurvenziehfeder *f* ~ **ruler** Kurvenlineal *n* ~ **section** Kurvenjoch *n* ~ **shape** Kurvenform *f* ~ **sheet** Kurven-blatt *n*, -tafel *f* ~ **template** Kur-venlineal *n* ~ **tracer** Kurvenschreiber *m* ~ **trend** Krümmungsverlauf *m*
**curved** bogenförmig, gebogen, gekrümmt, ge-schweift, kreisförmig gebogen, krumm, krumm-linig, krummschalig, ungrad ~ **in a parabola** parabolisch gekrümmt
**curved,** ~ **aileron tip former** Querruderrand-bogen *m* ~**-angle segment** Winkelbogenseg-ment *n* ~ **boundary** krummliniger Rand *m* ~ **downstream floor** gekrümmtes Sturzbett *n* ~ **electrode** Sichelelektrode *f* ~ **form** geschweifte Form *f* ~ **front line** Frontbogen *m* ~ **head** ge-wölbter Boden *m* ~ **line** Bogenlinie *f*, elastische Linie *f*, krummer Strang *m* (r.r.) ~ **piece** Kur-venstück *n* ~ **piece of metal** Bügel *m* ~ **pipe** Bogenrohr *n*, gebogenes oder gekrümmtes Rohr *n* ~ **rocker** Rollenbahn *f* ~ **roller conveyor** Kurvenrollgang *m* ~ **runner plate** gewölbte Filmbahn *f* ~ **sail** (windmill) gewölbter Flügel ~ **scale** Bogenskala *f* ~ **scissors** (for wick trim-ming) Dochtschere *f* ~ **section** Kurvenstück *n* ~ **sector** Kurvenbahn *f* ~ **shapes** gebogene Zu-schnitte *pl* ~ **slide** Wendelrutsche *f* ~ **slot** Kurvenschlitz *m* ~ **spaces** gekrümmte Räume

*pl* ~ **spur track** Gleiskurve *f* ~ **surface** gekrümmte Oberfläche *f* ~ **swage** Verkröpfgesenk *n* ~ **teeth** Bogenverzahnung *f* ~ **track** krummer Strang *m* (r.r.), Kurvenbahn *f* ~ **well** gebogene Bohrung *f* ~ **wing tip** Abschlußbogen *m* des Flügels (aviat.)

**curvilinear** gekrümmt, krummlinig ~ **distortion** Feldkrümmungsverzerrung *f* ~ **flight** Kurvenflug *m* ~ **polygon** krummliniges Vieleck *n* ~ **saw** Frettsäge *f*

**curvimeter** Kurvenmesser *m*

**curving** Schweifung *f* ~ **of spring** Krümmung *f* der Feder

**curvometer** Krümmungslinie *f*

**cushion, to** ~ (explosive) abfedern, polstern **to** ~ **with steam** Gegendampf geben

**cushion** Buffer *m*, Kissen *n*, Polster *n*, Puffer *m*, Reibzeug *n* (mach.) ~ **cam** Führungsteil *m* unterhalb der Senkerspitze ~ **lubrication** Kissenschmierung *f* ~**-shaped distortion** kissenförmige Verzeichnung *f* ~ **sheet** Patentgummi *n* ~ **shot** Hohlraumschießen *n* ~ **tire** Hochelastikreifen *m* ~**-tire equipment** Elastikbereifung *f*

**cushioned** federnd ~ **blasting** Hohlraumschießen *n* ~ **seat** Polstersitz *m* ~ **socket** federnde Röhrenfassung *f* ~ **tube holder (or socket)** federnde Röhrenfassung *f*, federnder Röhrenhalter *m* ~ **valve holder** federnde Röhrenfassung *f*

**cushioning** Dämpfung *f*, Einlagerung *f*, Prellvorrichtung *f*, Prellung *f*, (in explosives) Pufferung *f* ~ **box** Prellbuchse *f* ~ **cylinder** Luftpolster *n*, Pufferzylinder *m* ~ **spring** Fangfeder *f*

**cusp** Kuppe *f*, Rückkehrpunkt *m* (rdr, math.), Wendepolpunkt *m*, Wendepunkt *m* ~ **of a curve** Rückkehrpunkt *m* einer Kurve ~ **of the first kind** Hellebardenspitze *f* ~ **of pressure wave** Druckwellenkopf *m* ~ **of second kind** Schnabelspitze *f* ~ **point** Umkehrpunkt *m*

**cusped cavity** spitz auslaufender Hohlraum *m*

**custodian** Bewacher *m*, Hauswart *m*

**custody** Bewachung *f* Gewahrsam *n* ~ **deposit** Verwahrungsdepot *n*

**custom** Brauch *m*, Gebrauch *m*, Gewohnheit *f* ~ **of craftsmen** Handwerksbrauch *m* ~ **house** Zollamt *n* ~ **house document** Zollpapier *n* ~ **house duty** Zollabgabe *f* ~**-made** der üblichen Art ~**make, to** ~ auf Bestellung *f* machen ~ **warehouse** Zollspeicher *m*

**customary** althergebracht, gebräuchlich, üblich ~ **in a country** landesüblich ~ **in a place** ortsüblich ~ **type** der üblichen Art

**customer** Abnehmer *m*, Auftraggeber *m*, Besteller *m*, Käufer *m*, Kunde *m*, Mandant *m*, Verbraucher *m* ~ **complaint** Reklamation *f* ~ **service** Kundendienst *m*

**customs** Zoll *m* ~ **airport** Zollflughafen *m* ~ **clearance** zollamtliche Abfertigung *f* ~ **declaration** Abfertigungsschein *m* ~ **duty** Einfuhrzoll *m* ~ **entry** zollamtliche Abfertigung *f* ~ **examination** Zolluntersuchung *f* ~**-free trade zone** Freihandelszone *f* ~ **frontier** Zollgrenze *f* ~ **launch** Zollkreuzer *m* ~ **law** Zollgesetz *n* ~ **official** Zollbeamter *m* ~ **port** Zollhafen *m* ~ **seal** Zollverschluß *m* ~ **service** Zollwesen *n* ~ **tariff rate** Zollsatz *m* ~ **wharf** Zollmole *f*

**cut, to** ~ anschneiden, beschneiden, (in circles) fräsen, (slag or cinders) fressen, (in filing) hauen, (a rope) kappen, knippen, lösen (kalt), mähen, scheren, (wood) schlagen, schleifen, schneiden, schnitzeln, schnitzen, spanabheben, stechen, stoßen, trennen, vermindern, verringern, verschneiden

**cut, to** ~ **across** überschneiden, durchschneiden **to** ~ **across the grain** querschneiden **to** ~ **across the ground** das Gebirge durchörtern **to** ~ **away with the chisel** wegmeißeln **to** ~ **a cable** ein ein Kabel *n* schneiden **to** ~ **in circuit** vorschalten **to** ~ **a connection** eine Verbindung *f* trennen **to** ~ **crops** schopfen **to** ~ **by degrees** abstoßen, abstufen **to** ~ **down** fällen **to** ~ **down the trees** abholzen **to** ~ **fine** genau berechnen **to** ~ **groove** Nut oder Nute stoßen **to** ~ **in** einhacken, (valve) einschleifen, einschalten, einstechen **to** ~ **in (in two-motion selector)** eindrehen **to** ~ **in circuit again** wiedereinschalten **to** ~ **keyway** Nut(e) stoßen **to** ~ **a keyway in a shaft** eine Welle *f* nutzen **to** ~ **to length** ablängen **to** ~ **lengthwise** der Länge nach schneiden **to** ~ **lines of force** Kraftlinien schneiden **to** ~ **a lode** anfahren **to** ~ **the nick** einhobeln **to** ~ **new ground** ein neues Feld aufhauen **to** ~ **off** (light) abdunkeln, abhacken, abhauen, abriegeln, abschalten, abschneiden, abspannen (steam) absperren, abstechen, abtrennen, ausschalten, austasten, beschneiden, verriegeln **to** ~ **off a bed** (geol.) ein Flöz *n* abdämmen **to** ~ **off a subscriber** einen Teilnehmer *m* trennen **to** ~ **off supplies** Zufuhr *f* abschneiden **to** ~ **open** aufschneiden **to** ~ **operating time** Bearbeitungszeit *f* verringern **to** ~ **out** abschalten (relay or release gear) ansprechen, ausblenden (tape rec.), (by autogenous method) ausbrennen, ausschalten, ausschneiden, herausschneiden, wegfressen, außer Strom *m* setzen **to** ~ **out of action** abstellen **to** ~ **out the circuit** außer Strom *m* setzen **to** ~ **out resistance** Widerstand *m* ausschalten **to** ~ **into pieces** stückeln **to** ~ **to size** zuschneiden **to** ~ **smaller** beschneiden, abstützen **to** ~ **from solid** aus dem Vollen herausarbeiten **to** ~ **steep down** (building) steil abbösschen **to** ~ **steps** abtreppen **to** ~ **through** durchfressen, durchschalten, durchschneiden, durchstechen, durchverbinden **to** ~ **trenches** (to ore veins) schrämen **to** ~ **in two** entzweischneiden **to** ~ **up** einhacken, durchschneiden, zerschneiden **to** ~ **up (or in)** einschneiden **to** ~ **up timber** Holz schneiden

**cut** beschnitten, geschliffen, geschnitten ~ **across (the grain)** quergeschnitten ~ **gear** geschnittene Zähne *pl* ~ **glass** facettiertes oder geschliffenes Glas *n* ~ **goods** geschnittene Ware *pl* ~ **nose of the pier** Kopf *m* des Strompfeilers ~**-off** abgeschnitten, abgeschaltet, getrennt ~ **open** aufgeschnitten ~ **thread** geschnittenes Gewinde *n* ~ **tooth** gefräster oder geschnittener Zahn *m*

**cut** Abbildung *f*, Abhieb *m*, Abschnitt *m* (in earthwork) Abtrag *m*, Bildstock *m*, Durchstich *m*, Einschnitt *m*, Fasson *f*, Hau *m*, Hieb *m*, Krecke *f*, scharfe Überblendung (film), Scheibe *f* (of a gem) Schliff *m*, Schnitt *m*, Schnitte *f*, Schrott *m*, Streifen *m*

**cut,** ~ **of file** Feilenhieb *m* ~**-and-cover shelter** Unterschlupf *m* ~**-and-fill stoping** Firstenbau

*m* mit Versatz **~-and-try method** Ausprobieren *n*, empirisches Ermittlungsverfahren *n* **~-away for ejected cartridge cases** Durchtrittsöffnung *f* **~-away model** Schnittmodell *n* **~-away portion** Ausschnitt *m* **~ diamond** geschliffener Diamant *m* **~ film** Schnittfilm *m* **~ hard-stone products** Hartsteinschleifereierzeugnisse *pl* **~-image range finder** Schnittbildentfernungsmesser *m*
**cut-in** Vorschaltung *f* **~-continuous-wave transmission** rein ungedämpftes Senden *n* **~ current rush** Einschaltstoß *m* **~ marginal zones** eingezogene Marginalien *pl* **~ notch** Einschaltrastung *f* **~ pressure** Einschaltdruck *m* **~ relay** Einschalttrelais *n* **~ relay for local marker** Aufschalterelais *n* für Hausmarkierer **~ scene** Einschnitt *m* (film) **~ speed** Einschaltdrehzahl *f* **~ tension** Einschaltspannung *f* **~ wages** Lohnabbau *m*
**cutlass** Hirschfänger *m*
**cutler** Messerschmied *m*
**cutler's workshop** Messerschmiede *f*
**cutlery** Besteck *n*, Messerschmiedewaren *pl*, Messerwaren *pl* **~ paper** rostfreies Papier *n* **~ steels** Schneidewerkzeugstähle *m pl*
**cutline** Form *f*
**cutocellulose** Bastzellulose *f*
**cutoff** (rest) Abstechsupport *m*, Abschneiden *n*, Abschnittlänge *f*, Anschneidegesenk *n*, Sperrpunkt *m* (Röhre), Absperrung *f*, Abtrennung *f*, Durchstich *m*, (spectrum) Einschnitt *m*, Füllung *f*, (steam) Füllungsabschluß *m*, (point) Grenze *f*, Grenzfrequenz *f* in Hertz, (vacuum pump) Hahn *m*, Verschluß *m*, Zylinderfüllung *f* **~ of tube** Röhrensperrung *f*
**cutoff, ~ angle** abgestumpfter Winkel *m*, Brennschlußwinkel *m* **~ angular velocity** Grenzkreisfrequenz *f* **~ attenuator** Hohlrohrabschwächer *m* **~ bias** Gittersperre *f* **~ biasing potential** verriegelnde Vorspannung *f* **~ boundary** Lochgrenze *f* **~ button** Trenntaste *f* **~ cam** Abstechkurve *f* **~ canal** Stichkanal *m* **~ choke** Kippdrossel *f* **~ circular saw** Abkürzkreissägemaschine *f* **~ coefficient** Abschaltbeiwert *m*, Einstellwert *m* **~ constant** Abschaltfunktion *f* **~ contact** Brennschlußkontakt *m* **~ control** Brennschlußvermessung *f* **~ coordinates (or parameters)** Meßkoordinaten *pl* **~ corner** abgeschnittene Ecke *f* **~ coupling** Ausdehnungskupplung *f*, lösbare Kupp(e)lung *f* **~ distance** Abschirmeffektradius *m* **~ end** Abstichseite *f* **~ energy** Grenzenergie *f* **~ evaluation receiver** Brennschlußempfängerwagen *m* (g/m), Empfängerwagen *m* **~ frequency** Abschneidegrenzfrequenz *f*, Grenzfrequenz *f* (acoust.); (in terms of radians per second) Grenzkreisfrequenz *f*, Grenzfrequenz *f*, kritische Frequenz *f*, Trennfrequenz *f* **~ frequency of a filter** Grenzfrequenz *f* eines Filters **~ input** Abschaltleistung **~ jack** Trenn-, Überführungs-, Unterbrechungs-klinke *f* **~ key** Löschtaste *f*, Trennschalter *m*, Trenntaste *f*, Unterbrechungstaste *f* **~ lens** Verzögerungslinse *f* **~ levels** Grenzpegel *m* **~ lever** Stophebel *m* **~ magnet (or stopping solenoid or magneto)** Abstellmagnet *m* **~ modulation** Ausblendsteuerung *f* **~ parabola** Magnetfeldröhrenkennlinie *f* **~ part** Abfallstück *n* **~ period** Abschaltdauer *f* (film), Dunkelpause *f* **~ piston** Stopzug *m* **~ point** Ab-

trennungs-, Drossel-, Grenz-punkt *m*, Füllungsschluß *m* (steam), Sperrpunkt *m* (Röhre) **~ potential** (of tyratron) Löschspannung *f* **~ pressure** Abschaltdruck *m* **~ radius** Abschneideradius *m* **~ region** Reißgebiet *n* **~ relay** Abschalt-, Abtrenn-, Trenn-relais *n*, Relais *n* für Amtsanlassung **~ rod** Sicherheitsstab *m*, Stopgestänge *n* **~ sector field** abgehacktes Sektorfeld *n* **~ signal** Brennschlußkommando *n* (g/m), Schubabschaltung *f* (g/m); Trennzeichen *n* (teleph.) **~ signal receiver** Funkkommandoempfänger *m* **~-signal transmission** Brennschlußgabe *f* (g/m) **~ signal transmitter** Kommandosender *m* **~-signal transmitter output stage** Kommandosenderendstufe *f* **~ speed** Abschaltgeschwindigkeit *f* **~ spring** Ausschaltfeder *f* **~ stage** Abschaltstufe *f* **~ steam** abgesperrter Dampf *m* **~ switch** Stromunterbrecher *m* **~ temperature** Abschalttemperatur *f* **~ theorem** Abschneidesatz *m* **~ trench** Abdichtungsgraben *m*, Spundwand *f* **~ tuning condenser** Verstimmungskondensator *m* **~ valve** Abschaltklappe *f*, Abschlußventil *n*, (with expansion) Expansionsventil *n* **~ velocity** Brennschlußgeschwindigkeit *f* (g/m), Sollgeschwindigkeit *f* **~ voltage** Einsatz-, Unterdrückungs-spannung *f* **~ wall** Abdichtungssporn *m* (-mauer), Herdmauer *f*, Spundwand *f* **~ walls** Schutz *m* gegen Unterläufigkeit oder Unterspülung **~ wavelength** Abschneidewellenlänge *f*, Grenzwellenlänge *f*, kritische Wellenlänge *f* **~ wheel** Trennschleifmaschine *f*
**cutoffs** Schutz *m* gegen Unterspülung *f*, Schutz *m* gegen Unterwaschung
**cut-on filter** Kantenfilter *n*
**cutout** (exhaust) Auspuffklappe *f*, Ausschalter *m*, Ausschnitt *m*, Aussparung *f*, Schalter *m*, Schmelzsicherung *f*, Sicherung *f*, Sicherungskleinautomat *m*, Stromunterbrecher *m*, Unterbrecher *m* **~ button** Zündstromausschalter *m* **~ fuse** Abschmelzdraht *m*, Stromsicherung *f* **~ key** Abschalttaste *f* **~ lever** Freiauspuffhebel *m* **~ motor** Klinkmotor *m* **~ plunger** Taststange *f* **~ relay** Abschaltrelais *n* **~ section** Sichtausschnitt *m* **~ sector** Sichtausschnittbogen *m* **~ sheet** Modellierbogen *m* **~ switch** Trennschalter *m* **~ valve** Absperrventil *n*
**cutover** Inbetriebnahme *f* (eines Amtes, einer Teilnehmerleitung), schnelles Umschalten *n*, Überschneidung *f* (tape rec.), Überspringen *n* (tape rec.) Umschaltung *f* (teleph.) **~ of a central office** Überleitung *f* eines Amtes
**cut, ~ plane cutters** Lochhobeleisen *pl* **~ polished specimen** Anschliff *m* **~ pressure** Spanndruck *m* **~-set network** Netzwerk *n* mit wählbarer Trennschaltung **~ splice** Buchtspleiß *m* **~ thread** geschnittenes Gewinde *n* **~-through relay** Anschlußrelais *n*, Durchschalterrelais *n* **~ washer** geschnittene Unterlagscheibe *f* **~ water** Gallion *n*, Knie *n* **~ water of a bridge** Pfeilerhaupt *n* **~ wood** Schnittholz *n* **~ work** Einschnittarbeit *f* **~ work printing** Illustrationsdruck *m* **~ worm** Erdraupe *f*
**cutter** Abschneidemaschine *f*, Abschneider *m* (print.), Abschrot *m*, Fräsmesser *n*, Hauer *m*, Keil *m*, Kerbhauer *m* (min.), Kutter *m*, Messer *n*, Plattenführungsschnitt *m*, Rührwerk *n*, Schere *f*, Schneidestichel *m*, Schneidewerk *n*,

Schneiddose *f* (phono), Schneidmaschine *f*, Schneidstahl *m*, (tooth) Schneidwerkzeug *n*, Schnitzelmaschine *f*, Schnitzer *m*, Schrämer *m*, Setzeisen *n*, Stichel *m*, Zentrumbohrer *m*, Zwickzange *f* **(wire)** ~ Vorschneider *m* ~ **of the borer** Bohrschneide *f* ~ **for fluting reamers** Nutenfräser *m* für Reibahlen ~ **for gear wheels** Zahnfräser *m* ~ **for milling half circles** Halbkreisformfräser *m* ~ **for milling-machine and boring heads** Messer *n* für Fräs- und Bohrköpfe ~ **for tooth profile** Zahnformfräser *m* **cutter,** ~ **angle** Schneidewinkel *m* ~ **arbor** Fräsdorn *m* ~ **bar** Bohrwelle *f*, (mower) Fingerbalken *m* ~ **bar of a boring machine spindle** Bohrspindel *f* ~ **bit ground to form** Einsatzstahl *m* (geschl. f. Stahlhalter) ~ **bits** Schrammkronen *pl* ~ **blade** Fräsermesser *n* ~ **blades** Bohrstichel *m* ~ **block** Bohrkolben *m*, Fräskopf *m*, Messerwalze *f* ~ **carriage** Schneidschlitten *m* ~**-chain excavator** Fräsbagger *m* ~ **clamp** Messerklemme *f* ~**-clearance gauge** Fräserschleiflehre *f* ~ **diameter** Fräsdurchmesser *m* ~ **disc** Messerscheibe *f* ~ **drum** Messertrommel *f* ~ **head** Bohrkopf *m*, Bohrkrone *f*, Fräskopf *m*, Messerkopf *m*, Rollenkrone *f*, Werkzeugkopf *m* ~ **holder** Stahlhalter *m* ~**-holder support** Messerblock *m* ~ **lip aligning gauge** Einstellfinger zum Ausrichten *n* der Schneidbrust ~ **material** Schneidenwerkstoff *m* ~ **pick** Messerpicke *f* ~**-relieving attachment** Fräshinterdrehapparat *m* ~ **spindle** Drehstahl-, Fräs-spindel *f*, Messerwelle *f* ~ **spindle assembly** Frässpindellager *n* ~ **spindle sleeve** Frässpindelhülse *f* ~ **steel spindle** Frässtahlspindel *f* ~ **testing micrometer gauge for milling** Fräslehre *f* ~ **tooth** Schneidzahn *m* ~ **velocity** Fräsdrehzahl *f*
**cutters of special length** Sonderlängenstähle *pl*
**cutting** Anschneiden *n*, Einschnittarbeit *f*, Hausarbeit *f*, Durchstich *m*, (steel) zerspanende Fertigung, Fräsen *n*, (of oils, sugars) Gerinnen *n*, Hau *m*, Rohrsteckling *m*, Verringerung *f*, (of wood) Schlag *m*, Schlagen *n*, Schleifen *n*, Schneiden *n*, Schnitt *m*, Schram *m*, Span *m*, scharfe Überblendung *f* (film); schneidend, schneidig, spanabhebend **(under)** ~ Schrämarbeit *f* ~ **above track level** Hochschnitt *m* (min.) ~ **below track level** Tiefschnitt *m* ~ **of blowers** Anhauen *n* von Wetterbläsern ~ **the caulking edges** Anschneiden *n* von Stemmkanten ~ **out of the circuit** Ausfall *m* ~ **of coal** Kohlenhauen *n* ~ **and economy of time** Zeitersparnis *f* ~ **and editing** Schnitt *m* (film) ~ **in of an exchange** Überleitung *f* eines Amtes ~ **of feeders** Anhauen *n* von Wetterbläsern ~ **of operations** Abstechvorgang *m* ~ **of prices** Preisherabsetzung *f* ~ **out by reversing** ruckweises Ausschalten *n* ~ **of a seam** Abbau *m* eines Flözes ~ **off the steam** Dampfabsperrung *f* ~ **of a vein** Abbau *m* eines Flözes
**cutting,** ~ **ability** Schneidfähigkeit *f* ~ **across** Durchörterung *f*, (petroleum) Durchschlag *m* ~ **action** Schneidvorgang *m*, Schnittwirkung *f* ~ **action of slag** Schlackenangriff *m* ~ **agent** Verschnittmittel *n* ~ **alloy** Schneidmetall *n* ~ **alloy tip** Hartmetallbestückung *f* ~**-and bending tools** Schnitt- und Biegewerkzeuge *pl* ~**-and drawing tools** Schnitt- und Ziehwerkzeuge *pl* ~**-and stamping tools** Schnitt- und Prägewerkzeuge *pl* ~ **angle** Keilwinkel *m*

(dig-in or drag-angle). Schneid-, Schneide-, Schnitt-winkel *m* ~ **angle gauge** Schnittwinkellehre *f* ~ **away** Aussparung *f* ~ **bar** Ausbohrspindel *f* ~ **blade** Abdeckflügel *m* (film), Abschneidmesser *n*, Flügelblende *f* (film). ~ **blast** Wind *m* mit starker Pressung ~ **capacity** Schneidfähigkeit *f*, Schneidevermögen *n*, Schnittleistung *f* ~ **circle** Schneidkreis *m* ~ **compound** Kühlmittel *n* ~ **cradle** Schneidlade *f* ~ **crowned teeth** Balligverzahnung *f* ~ **cycle** Hobeln *n* ~ **cylinder** Messerzylinder *m* ~ **depth** Spandicke *f* ~ **die** Schneideisen *n*, Schnittmatrize *f* ~ **die making** Schnittbau *m* ~ **die plate** Stanz-, Stempel-matrize *f* ~ **direction** Schneidrichtung *f* ~ **disc** Schneidscheibe *f* ~ **edge** Schneid-, Schnitt-kante *f*, Schneidwinkel *m*, Werkzeugschneide *f* ~ **edge of bit** Meißelschneide *f* ~ **edge at bottom of well casing** Brunnenkranz *m* ~ **edge of the caisson** Schneide *f* des Senkkastens ~**-edge weld** Schneidbrennerkante *f* ~ **efficiency** Schnittleistung *f* ~ **flame** Schneidflamme *f*, schneidende Flamme *f* ~ **forceps** Scherenzange *f* ~ **grapnel** Schneidanker *m* ~ **groove** Fräsrille *f* ~ **hardness** Schneidhärte *f* ~ **head** Schneidestichel *m*, Schneidhülse *f*, Schneidkopf *m* ~ **height** Reichhöhe *f* ~ **line** Abschnittlinie *f* (print.), Schnittlinie *f* ~ **load** Schnittbelastung *f* ~ **lubricant** Schmiermittel *n* für die spangebende Bearbeitung ~ **machine** Abschneide-, Scher-, Schrämm-, Tuchschermaschine *f* ~ **machining** Spanabnahme *f*
**cutting-off** Abhacken *n*, Abriegelung *f*, Abschneiden *n*, Abschnürung *f*, Abstechen *n*, Trennschleifen *n* ~ **of piles** Pfahlabschneiden *n* ~ **the seed stalk** Abstoßen *n* des Samenstengels
**cutting-off,** ~ **apparatus** Verdunk(e)lungseinrichtung *f* ~ **burner** Schweißbrenner *m* ~ **cam** Abstechkurve *f* ~ **frame saw** Abschneideschweifsäge *f* ~ **lathe** Abstechdrehbank *f* ~ **machine** Abstechmaschine *f* ~ **pliers** Schneidezange *f* ~ **separation** Abstich *m* ~ **tool** Abschneidestahl *m*, Abstechstahl, Abstechstahlhalter *m*, Einstechstahl *m*, Schneidstahl *m*
**cutting,** ~ **oil** Bohrfett *n*, Schneidöl *n* ~ **operation** Zerspannungsvorgang *m* ~**-out of a bend (or of a curve)** Abschneiden *n* einer Krümmung ~**-out patterns** Schneidschablonen *pl* ~ **outfit** Schneidausrüstung *f* ~ **pattern** Beschneideschablone *f* ~ **pattern template** Beschneidschablone *f* ~ **plant** Schneidanlage *f* ~ **pliers** Beiß-, Schneide-, Zwick-zange *f* ~ **position** Schnittstellung *f* ~ **power** Schneidevermögen *n*, Schneidfähigkeit *f*, Schnittkraft *f*, Schnittleistung *f*, Spanleistung *f* **ability to maintain (or retain)** ~ **power** Schneidhaltigkeit *f* ~ **press** Schnittpresse *f* ~ **pressure** Arbeits-, Schneid-, Schnitt-druck *m*, Schnittkraft *f* ~ **procedure** Schnittverfahren *n* ~ **process** Schneidvorgang *m* ~ **property** Schneidfähigkeit *f*, Zerspanungseigenschaft *f* ~ **punch** Abschneidestempel *m* ~ **quality** Schneidfähigkeit *f* ~ **range** Gewindebereich *m* ~ **(or shearing) resistance** Schnittwiderstand *m* ~ **scar** Schneidnarbe *f* ~ **shaping** spanabhebende Verformung *f* ~ **speed** Schneid-, Schnitt-geschwindigkeit *f* ~ **stencil** Zuschneideschablone *f* ~ **stroke** Schnitthub *m* (e. Stößels) ~ **stylus** Schneidstichel *m* (phono) ~ **surface** Schneidfläche *f* ~ **table** Abschneidetisch *m*

~ **(the) teeth** Verzahnen *n* ~ **teeth of octoid form** Oktoidenverzahnung *f* ~**-through** Durchschaltung *f* ~**-through extension** Durchschaltung *f* ~ **tip** Schneid-düse *f*, -mundstück *n* ~ **tool** Drehmesser *n*, Drehstahl *m*, Schneidstahl *m*, Schneidwerkzeug *n*, Schnittwerkzeug *n*, Schreibstift *m*, spanabhebendes Werkzeug *n*, Stanzwerkzeug *n*, Zerspanungswerkzeug *n*, spanabhebendes Werkzeug *n* ~ **(-off) tool post** Abstechstahlhalter *m* ~ **tools** spanabhebende Werkzeuge *pl*, Zerspanungswerkzeuge *pl* ~ **torch** Schneidbrenner *m* ~**-torch flame** Schneidbrennerflamme *f* ~**-torch tip** Schneidbrenner-mundstück *n*, -spitze *f* ~ **value** Schneidfähigkeit *f* ~ **width** Schnittbreite *f* ~ **wire** Abschneidedraht *m* ~ **worm gears** Schneckenzahnrad *n*

**cuttings** Bohrmehl *n*, Schneidspäne *pl*, Schnitte *pl*, Schnitzel *m pl* ~ **of a boring** Bohrschmant *m* ~ **shoot** Spanfall *m*

**cuttle, to** ~ umtafeln ~ **motion** Tafler *m*

**cuttler** Breitfalter *m*, Legevorrichtung *f*

**cutwater** Vordersteven *m*

**cuvette** Küvette *f*

**CW (continuous wave)** ungedämpfte Welle *f*

**C washer** Vorsteckscheibe *f*

**C wire** Ader *f* zum Stöpselkörper, c-Ader *f*, c-Leitung *f*, Ader *f*

**cyanacetic** zyanessigsauer ~ **acid** Zyanessigsäure *f*

**cyanamide** Zyanamid *n*

**cyanate** Zyanat *n*, zyansauer

**cyanic acid** Zyansäure *f*

**cyanidation** Zyanid-prozeß *m*, -verfahren *n*, Zyan-laugerei *f*, -laugung *f*

**cyanide, to** ~ zementieren ~ **(prussiate)** blausaures Salz *n* (Zyanid), Zyanid *n*, Zyansalz *n* ~ **of copper** Zyankupfer *n* ~ **of potash** Zyankali(um) *n*

**cyanide,** ~ **bath** Zyanbad *n*, Zyansalzbad *n* ~**-bath hardening** Zyansalzbadhärtung *f* ~**-free** zyanidfrei ~ **hardening** Zyanhärtung *f* ~ **poisoning** Zyanvergiftung *f* ~ **process** Zyanidprozeß *m*, -verfahren *n*

**cyaniding** Einsatzhärten *n*, Zementierung *f*, Zyaneinsatzhärtung *f*, Zyanhärtung *f*, Zyanlaugerei *f*, -laugung *f*, Zyansalzbadhärtung *f*

**cyanite** blätteriges Beryll, Disthen *m*, Zyanit *m*

**cyanization** Zyanisierung *f*

**cyanize, to** ~ zyanisieren

**cyanizing** Zyaneinsatzhärtung *f* ~ **works** Zyanisierwerk *n*

**cyanocarbonic acid** Zyankohlensäure *f*

**cyanogen** Kohlenstickstoff *m*, Zyan *n*, Zyangas *n*, Zyanogen *n* ~ **bromide** Bromzyan *n*, Ce-stoff *m*, Zyanbromid *n* ~ **chloride** Clorzyan *n* ~ **compound** Zyanverbindung *f* ~ **discrepancy** Zyan-Diskrepanz *f* ~ **gas** Zyangas *n*, Zyanogen *n* ~ **iodide** Zyanjodid *n* ~ **iodide etching** Jodzyanätzung *f* ~ **iodide reducer** Jodzyanabschwächer *m* ~ **mixtures** Zyan-Säuerstoff-(Stickstoff-) Gemische *pl* ~ **poisoning** Zyanvergiftung *f* ~ **salt** Zyansalz *n* ~ **sludge** Zyanschlamm *m*

**cyanometer** Himmelblaumesser *m*

**cyanonitride** Zyanstickstoff *m*

**cyanoplatinic acid** Platinizyanwasserstoffsäure *f*

**cyanotoluene** Zyantoluol *n*

**cyanotrichite** Kupfersamterz *n*

**cyanuric acid** Trizyansäure *f*, Zyanürsäure *f*

**cybernetics** Kybernetik *f*

**cycle, to** ~ periodisch wiederkehren, radfahren, sich regelmäßig wiederholen

**cycle (of operations)** Arbeitsspiel *n*, Arbeitstakt *m*, Frequenz *f*, Gang *m*, Hertz *n* · (electr.), Kartengang *m* (Lochkarte), Kreiseinwelligkeit *f*, Kreislauf *m*, Kreisprozeß *m*, Kreisumlauf *m*, Periode *f* pro Sekunde, Periodendauer *f*, Periodenspannung *f*, Programm *n*, Ring *m*, Schleifendurchlauf *m* (data proc.), Schwingung *f*, Schwingungszeit *f*, Takt *m*, Turnus *m*, Umlauf *m*, Wechselzahl *f*, Zyklus *m* ~ **for anodizing** Vered(e)lungszeit *f* ~ **of operation** Ablaufphase *f* ~ **per second** Stromwechsel *m* pro Sekunde ~ **per second** Hertz-Einheit *f* ~ **of solar activity** Sonnentätigkeitsperiode *f* ~ **of stress** Belastungswiederholung *f*, Lastwechsel *m* ~ **of wind** Förderspiel *n* ~ **of work** Arbeitsfolge *f*

**cycle,** ~ **checking** laufende Prüfung *f* ~ **count** Schleifenzählung *f* (data proc.) ~ **criterion** Umlaufzahl *f* ~**-driving gear** Fahrradantrieb *m* ~ **fittings** Fahrradbestandteile *pl* ~ **handle bar** Fahrradlenkstange *f* ~ **index** Schleifenindex *m* (data proc.), Spielzahlanzeiger *m* ~ **lighting set (or system)** Radlichtanlage *f* ~ **period** Wiederkehrzeit *f* ~ **rate counter** Periodenzähler *m* ~ **ratio** Lastenspielzahlverhältnis *n* ~ **reset** Rückstellung *f* ~ **store** Umlaufspeicher *m* ~ **time** Wechseldauer *f* (Probenwechsler) ~ **tread crank** Fahrradtretkurbel *f*

**cycles per second (cps, c/s)** Schwingungen *pl* pro Sekunde, Hertz

**cyclic** ringförmig, zyklisch ~ **admittance** Drehfeldadmittanz *f* ~ **blade pitch** periodische Blattverstellung *f* ~ **compound** Ringverbindung *f* ~ **constant** Umlaufgröße *f* ~ **control** periodische Blattsteuerung *f* ~ **control column** Steuersäule *f* (Hubschr.) ~ **coordinate** zyklische Koordinate *f* ~ **course** Kreislauf *m* ~ **fatigue loading** Dauerwechselbeanspruchung *f* ~ **flow** Kreis-, Zirkulations-strömung *f* ~ **frequency** Kreisfrequenz *f* ~ **impedance** Drehfeldimpedanz *f* ~ **load** Lastwechsel *m* ~ **loading** (in materials testing) Wechselbeanspruchung *f* ~ **operation** periodischer Vorgang *m* ~ **permutation (or variation)** zyklische Vertauschung *f* ~ **pitch** periodische Blattwinkelverstellung *f* ~ **process** Kreis-prozeß *m*, -vorgang *m* ~ **shift** zyklische Verschiebung *f* ~ **stick** Steuerknüppel *m* (Hubschrauber) ~ **storage** periodischer Speicher *m* ~ **store** Umlaufspeicher *m*

**cyclical** periodisch ~ **binary code** zyklisch-binärer Kode *m* ~ **hydrocarbons** zyklische Kohlenwasserstoffe *pl* ~ **process temperature** Kreisprozeßtemperatur *f* ~ **tensile stress test** Zugversuch *m* mit pulsierender Beanspruchung ~ **variation of stress** Spannungswechsel *m*

**cyclically unsaturated** ringungesättigt

**cyclids** Zykliden *pl*

**cycling** Pendelung *f* ~ **movement** Pendelbewegung *f* ~ **time** Zykluszeit *f*

**cyclo,** ~ **genesis** Entwicklung *f* eines Zyklons (Wirbelsturm) ~ **graph** Zyklograf *m* ~ **gyro** Radflügelflugzeug *n*, Umlaufflugzeug *n* ~ **hexadiene** Zyclohexadien *n* ~ **hexane** Zyclohexan *n* ~ **hexanol** Zyklohexanol *n* ~ **hexanone** Anon *n* ~ **hexene** Zyclohexen *n*

**cycloid** Radlinie *f*, Zykloid *n* (in stroboscopic aberration)
**cycloidal** zykloidisch ~ **height** Zycloidhöhe *f* ~ **height of a magnetron** Magnetronzycloidhöhe *f* ~ **path** (of electrons) Roll-bahn *f*, -kreis *m* ~ **tooth system** Zyklodenverzahnung *f* ~ **toothed gear** Zahnrad *n* mit Zykloidenverzahnung
**cycloids** Zykloiden *pl*
**cyclometer** Umlaufzähler *m*
**cyclometry** Kreismessung *f*
**cyclone** Späneabschneider *m*, Wirbelsturm *m*, Zyklon *m* ~ **baffles** Wirbelleitwerk *n* ~ **circuit** Zyklonkreislauf *m* ~ **clarifier** Klärzyklon *n* ~ **impeller** Hoesch-Rührer *m* ~ **impeller mixer** Ekato-Korbkreiselmischer *m* ~ **low** Tiefdruckwirbel *m* ~ **separator** Zyklonenschneider *m* ~ **thickener** Eindickzyklon *m*
**cyclonic** zyklonal ~ **thunderstorm** Wirbelgewitter *n*
**cyclopentadiene** Zyklopentadien *n*
**cyclopentane** Zyklopentan *n*
**cyclopentene** Zyklopenten *n*
**cyclopropane** Zyklopropan *n*
**cyclorama lights** Horizontleuchten *pl* (theater)
**cyclostyle paper** Stenzilpapier *n*
**cyclotron** Zyklotron *n* ~ **and inverted forms there of** Elektronenturbine *f* ~ **orbits** Zyklotronumläufe *pl* ~ **tube** Vielfachbeschleuniger *m*
**cylinder** (for gas) Flasche *f*, Hohlwalze *f*, Laufbuchse *f*, Probierröhrchen *n*, Rolle *f*, (of pumps) Stiefel *m*, Trommel *f*, Walze *f*, Zylinder *m*, of cathode-ray oscillograph) Zylinderelektrode *f* ~ **for the hydraulic cross traverse** Längshydraulikzylinder *m* ~ **with pleat distributing device** Faltenverteilungswalze *f* ~ **of a printing machine** Presseur *m* ~ **of the rag engine** Holländerwalze *f* ~ **of revolution** Rotationszylinder *m* ~ **of sinking pit** Senkzylinder *m*
**cylinder,** ~ **bank** Flanschengruppe *f* ~ **barrel** Zylinder-laufbüchse *f*, -mantel *m* ~**-barrel sealing surface** Laufbuchsendichtfläche *f* ~ **base** Zylinderflansch *m* ~**-base temperature** Zylinderflanschtemperatur *f* ~ **bearer** Schmitzleiste *f* ~ **block** Gehäuse-, Zylinder-block *m* ~ **blowhole** Gasflaschenpore *f* ~ **bolt** Zylinderbolzen *m* ~ **bore** innerer Zylinderdurchmesser *m*, Zylinder-bohrung *f*, -lauffläche *f* ~ **boring apparatus** Zylinderbohrapparat *m* ~ **boring machine** Zylinderbohrmaschine *f* ~ **boring mill** Zylinderbohrwerk *n* ~ **bottom walve** Zylinderbodenventil *n* ~ **cam** Kurventrommel *f* ~ **capacity** Zylinder-inhalt *m*, -volumen *n* ~**-cased air filter** Gehäuse-Luftfilter *m* ~ **casting** Zylinderguß *m* ~ **catch** Auffanggabel *f* ~ **charge** Zylinderladung *f* ~ **cleading (or lagging)** Zylinderverkleidung *f* ~ **clearance** schädlicher Raum *m* ~ **clearer roller** Zylinderputzwalze *f* ~ **clothing** Garnitur *f* der Trommel ~ **coking** Verkokung *f* in Retorten ~ **cooling fin** Zylinderkühlrippe *f* ~ **cover** Siebmantel *m*, Walzenüberzug *m*, Zylinderdeckel *m* ~ **diameter** Zylinder-durchmesser *m*, -wandung *f*, -weite *f* ~ **diaphragm** Zylinderblende *f* ~ **dipole antenna** Rohrdipolantenne *f* ~ **displacement** Zylinderinhalt *m* ~ **draining machine** Rundsiebentwässerungsmaschine *f* ~ **dressing clamp** Aufzugsklappe *f* ~**-dried paper** maschinengetrocknetes Papier *n* ~ **drier** Zylindertrockner *n*

~ **drying machine** Dampftrockenmaschine *f* ~ **engraving works** Walzengravieranstalt *f* ~ **envelope** Gasflaschen-, Zylinder-mantel *m* ~ **escapement** Zylinderhemmung *f* ~ **fairing** Zylinderhaube *f* ~ **finish** Walzendekatur *f* ~ **fittings** Zylinderausrüstung *f* ~ **foot** Zylindergestell *n* ~ **fulling machine** Walkzylinder *m* ~ **gap** Zylinderluke *f* ~ **gas** Flaschen-azetylen *n*, -gas *n* ~ **gas bottling** Flaschengasabfüllung *f* ~ **gate** Zylinderventil *n* ~ **gauge** Kaliberzylinder *m* ~ **gear** Zylinderrad *n* ~ **gear rack** Zylinderzahnstange *f* ~ **grinder** Zylinderschleifmaschine *f* ~ **grinding machine** Walzenabschleifmaschine *f* ~ **head** Kesselboden *m*, Zylinderdeckel *m*, Zylinderkopf *m* ~**-head cover** Zylinderkopfhaube *f* ~**-head gasket** Zylinderkopfdichtung *f* ~**-head oil drain** Ölrücklaufleitung *f* der Kipphebelschmierung ~ **hold-down bolt** Zylinderbefestigungsschraube *f* ~ **holding down bolt** Befestigungsschraube *f* für Zylinder ~ **holding stud** Zylinderhaltebolzen *m* ~ **housing** Zylindergehäuse *n* ~ **indexing**, Trommelschaltung *f* ~ **interval** Zylinderluke *f* ~ **iron** Zylinderguß *m* ~ **jacket** Dampfhemd *n*, Zylinder-mantel *m*, -wandung *f* ~ **liner** eingesetzte Zylinderlaufbüchse *f*, Einsatzzylinder *m*, Zylinder-auskleidung *f*, -büchse *f*, -einsatz *m*, -futter *n*, -hülle *f*, -laufbüchse *f* ~ **lubrication** Zylinderschmierung *f* ~ **mill** Walzenwalke *f* ~ **moistening hose** Feuchtwalzenschlauch *m* ~ **oil** Zylinderöl *m* ~ **output** Zylinderleistung *f* ~ **packing gauge** Aufzugslehre *f* ~ **pad** Zylinderauflagefläche *f* ~ **paper machine** Rundsiebmaschine *f* ~ **piston** Zylinderkolben *m* ~ **pocket** Gasflaschenpore *f* ~ **press** (textiles) Muldenpresse *f* ~ **pressure** Zylinderinnendruck *m* ~**-pressure gauge** Inhaltsmanometer *n* ~**-pressure regulator** Flaschenschlußventil *n* ~ **printing** Walzendruck *m* ~**-protecting device** Walzenschutzvorrichtung *f* ~ **pump** Kolbenbohrpumpe *f* ~ **sanding machine** Walzenschleifmaschine *f* ~ **saw** Trommelsäge *f* ~ **sheet** Margebogen *m* ~ **side frame** Zylinderlagerbock *m* ~ **singeing machine** Zylindersenge *f* ~ **sizing machine** Zylinderschlichtmaschine *f* ~ **stiffening piece** Zylinderverstrebung *f* ~ **stock solution** Zylinderöllösung *f* ~ **stud** Stiftschraube *f* am Kurbelgehäuse zur Zylinderbefestigung ~ **stuffing box** Zylinderstopfbüchse *f* ~ **support** Walzenstuhl *m* (print.) ~ **support bearing** Zylinderauflager *n* ~ **trip** Druckabsteller *m* (print.) ~ **valve** Flaschenschlußventil *n*, Zylinderventil *n* ~ **volume** Zylinder-inhalt *m*, -volumen *n* ~ **wall** Zylinderwandung *f* ~ **weld** Gasflaschenmantel *m* ~ **welding** Gasflaschenschweißverfahren *n* ~ **worn out of true** unrunder Zylinder *m*
**cylindric,** ~ **coil** Zylinderspule *f* ~ **nozzle** Walzendüse *f* ~ **optical apparatus** Gürtelleuchte *f* ~**-prismatic breech mechanism** Zylinderdrehverschluß *m*
**cylindrical** tonnenförmig, walzenförmig, walzig, zylindrisch ~ **ball bearing** Ringzylinderlager *m* ~ **billet** Rundknüppel *m* ~ **bit** Zylinderbohrer *m* ~ **bloom** Rundblock *m* ~ **body** zylindrischer Teil *m* ~ **boiler** Walzenkessel *m* ~ **bottle** Rollflasche *f* ~ **brush** Rohrbürste *f* ~ **cam** Mantelkurve *f* ~ **cathode** Hohl-, Rund-kathode *f* ~ **coil** zylindrische Spule *f* ~ **coordinate** Zylinder-

koordinate $f$ ~ counter chamber zylindrischer Reaktor $m$ ~ cross-staff with divided circle Winkeltrommel $f$ ~ die for thread rolling Gewindewalzrolle $f$ ~ drainer Trommelknotenfänger $m$ ~ dressing machine Mehlbürstmaschine $f$ ~ electrode Zylinderelektrode $f$ ~ fit Rundpassung $f$ ~ function Zylinderfunktion $f$ ~ gauges Grenzlehrdorn $m$ ~ gearing Umfangsverzahnung $f$ ~ grid Gitterhülse $f$ (Filter) ~ grinding Rundschleifen $n$ (plain) ~ grinding Wellenschliff $m$ ~ handle Walzengriff $m$ ~ harmonics Zylinderfunktion $f$ ~ head screw Zylinderschraube $f$ ~ magnetic separator Walzenmagnetsichter $m$ ~ mask-carrying canister Tragbüchse $f$ ~ milling fulling machine Zylinderwalke $f$ ~ mount Walzenblende $f$ ~ padlock Bolzenriegel $m$ ~ parabolic antenna Zylinderparabol-Antenne $f$ ~ part of a boiler Langkessel $m$ ~ part of propellant tanks Behälterschuß $m$ (g/m) ~ pin Zylinderstift $m$ ~ plug gauge Kaliberdorn $m$, Lehrbolzen $m$ ~ projection Merkatorkartenprojektion $f$ ~ ring armature Zylinderringanker $m$ ~ roller thrust bearing Scheibenzylinderlager $n$ ~ screw gear zylindrisches Schraubrad $n$ ~-shaped core

Zylinderkern $m$ ~-shaped powder charge Zylinderladung $f$ ~-shaped screen Sieb-trommel $f$, -zylinder $m$ ~ shell Hülse $f$ (Hohlzylinder) ~ slide valve Rohrschieber $m$ ~ sliding surface kreiszylindrische Gleitfläche $f$ ~ slot antenna Rohrschlitzantenne $f$ ~ sluice Zylinderventil $n$ ~ spar Rundholm $m$ ~ tank zylindrischer Tank $m$ ~ tap Zylinderschneidschraube $f$ ~ top mark Zylindertoppzeichen $n$ ~ trickling apparatus Zylinderberieselungsapparat $m$ ~ two-die thread rolling machine Gewindewalzmaschine $f$ mit zwei Walzrollen ~ valve hohes Zylinderventil $n$, -schütz $n$ ~-wedge breech mechanism Rundkeilverschluß $m$ ~ winding Röhrenwicklung $f$ ~ wire dünner Draht $m$ ~ yoke Konzentrierspule $f$

cymograph Kymografion $n$
cymometer Wellenmesser $m$
cymoskope Wellenanzeiger $m$, Wellendetektor $m$
cypher code Zifferschrift $f$
cypress wood Zypressenholz $n$
cyrtolite Cyrtolith $m$
cyrtometer Cyrtometer $n$
cytidine phosphate Zytidinphosphat $n$
cytoplasm Zellplasma $n$

# D

**dab, to** ~ abklatschen, abtupfen, betupfen, klischieren, mit der flachen Hand leicht schlagen
**dab** Klecks *m*
**dabbed drawing** gewischte Zeichnung *f*
**dabber** Abklatscher *m*, Filzwalze *f*
**dabbing machine** Abklatschmaschine *f*
**dacryocystography** Dakryozystografie *f*
**dagger** Dolch *m*, Kreuz(zeichen *n* (print.)
**daily** täglich ~ **allowance** Tagegeld *n* ~ **check** tägliche Vergleichung *f* ~ **definitive** Definitivhauptbuch *n* ~ **load curve** Tagesbelastungskurve *f* ~ **march of temperature** täglicher Temperaturwechsel *m* (meteor.) ~ **output (or production)** Tagesleistung *f* ~ **rate** Tages-rate *f*, -satz *m* ~ **report** Tagesmeldung *f* ~ **scale** Tagessatz *m* ~ **schedule** Tagesdienstplan *m* ~ **strength report** Iststärkenachweisung *f* ~ **supply tank for fuel oil** Brennstofftagesbehälter *m* ~ **task** Tagesdurchlauf *m* ~ **wages** Tagelohn *m*
**dairy** Meierei *f*, Molkerei *f*
**Dalton's theory** Daltonsches Gesetz *n*
**dam, to** ~ abdämmen, anhalten, arretieren, bremsen, dämmen, festhalten, feststellen, hemmen, sperren, verdämmen **to** ~ **in** eindämmen **to** ~ **up** anstauen, dämmen, eindämmen, stauen **to** ~ **up the course of a stream** einen Fluß *m* verdämmen **to** ~ **up the water** das Wasser *n* anstauen
**dam** Abdämmung *f*, Absperrdamm *m*, Damm *m*, Deich *m*, Fangbuhne *f*, Flußdamm *m*, Klappenwehr *f*, Notverschluß *m*, Pier *m*, Schlange *f*, Schütze *f*, Sperrdamm *m*, (construction) Stauanlage *f*, Staumauer *f*, Stauwerk *n*, Talsperre *f*, Verdämmung *f*, Wall *m*, Wallstein *m*, Wehr *n*, Wehrkörper *m* ~ **on bed of river** Grund-schwelle *f*, -wehr *n* ~ **with frames and needles** Nadelwehr *n* ~ **of a harbor** Mole *f* ~ **across a valley** Talsperre *f*
**dam,** ~ **beam seal** Dammbalkenverschluß *m* ~ **crest** Krone *f* der Talsperre *f* ~ **embankment** Dammkörper *m* ~ **plate** Wallplatte *f* ~ **stone** (of blast furnace) Wallsteinplatte *f* ~ **stone** (of blast furnace) Wallstein *m* ~ **timbers built into head race** Oberwasserdammbalken *m* ~ **works** Spreutlagen *pl*
**damage, to** ~ beschädigen, schaden, Schaden *m* zufügen, verderben
**damage** Beschädigung *f*, Einbuße *f*, (to ship of cargo) Havarie *f*, Leckschaden *m*, Nachteil *m*, Schade *m*, Schaden *m*, Schiffunfall *m*, Verlust *m*
**damage,** ~ **to a cargo** Seeschaden *m* ~ **to cultivated fields** Flurschaden *m* ~ **done by hail** Hagelschlag *m*, Hagelsturm *m* ~ **due to mining operations** Bergschaden *m* ~ **to a ship** Seeschaden *m* ~ **in transit** Transportschaden *m*
**damage,** ~ **assessment** Schadenschätzung *f* ~ **clearance** Instandsetzungsdienst *m* ~ **criteria** Schadenmaßstab *m* ~ **rate in fatigue** Ermüdungsverlust *m*
**damaged** beschädigt, morsch(ig), ramponiert, schadhaft **to be** ~ Schaden nehmen ~ **at sea** seebeschädigt, (ship) havariert

**damaged,** ~ **goods** Ausschußware *f*, beschädigte Ware *f* ~ **plane** Bruch *m*
**damages** Entschädigung *f*, Schadenersatz *m* ~ **for personal loss** Personenschadenersatz *m*
**damascene, to** ~ damaszieren
**damascening** Damast *m*, Damaszierung *f*
**Damascus steel** Damaststahl *m*, Damaszenerstahl *m*
**damask, to** ~ damassieren
**damask** Damast *m*, Damaszierung *f* (Stahl) ~ **steel** Damaststahl *m*, Damaszenerstahl *m*
**dammar (or dammer) resin** Kauri-, Stein-harz *n*
**damming** Abdämmung *f*, Dämmung *f* ~ **detail** Abdammtrupp *m* ~ **effect** Stauwirkung *f* ~ **-off a spring** Eindämmen *n* einer Quelle *f* ~ **position** Staulage *f* ~ **skirt** Stauschild *n* (an Walzen) ~ **squad** Abdammtrupp *m* ~ **-up** Anhäufung *f*, Anstauung *f*, Einschleusung *f*, Stauung *f* ~ **-up of water by any means** Anstau *m*
**damp, to** ~ abdampfen, (of charcoal) abdämpfen, abflächen, befeuchten, (out) dämpfen, feuchten, (paper) matrisieren, schäften, verlangsamen (aviat.) **to** ~ **out** vollständig dämpfen, unterdrücken, zum Verschwinden *n* bringen
**damp** Dunst *m*; feucht, naß, stickig ~ **air** feuchte Luft *f* ~ **cold** naßkalt ~ **concrete mass** erdfeuchte Betonmasse *f* ~ **finish** feuchte Appretur *f* ~ **fog** Nebelschwaden *pl* ~ **room** feuchter Raum *m* ~ **storage tests** Feuchtlagerversuche *pl*
**damped** gedämpft **to be** ~ gedämpft werden ~ **balance** gedämpfte Waage *f* ~ **current** Bremsstrom *m* ~ **oscillation** gedämpfte Schwingung *f* oder Welle *f* ~ **oscillations** gedämpfte Schwingungen *pl* ~ **periodic element** gedämpft schwingendes Organ *n* ~ **pointer action** Zeigerdämpfung *f* (electr. Meßgeräte) ~ **sine vibration** (asymptotic wave, vibration, oscillation) abklingende Kraft *f* ~ **sine wave** gedämpfte Sinusschwingung *f* ~ **wave** gedämpfte Welle *f*
**dampen, to** ~ anfeuchten, (superficially) annetzen, anwässern, benetzen, (silence) dämpfen, feuchten
**damper** Anfeuchter *m*, Aschenschieber *m*, Dämpfer *m*, Dämpfungsdiode *f*, Dämpfungsregler *m*, Dämpfungsvorrichtung *f*, Drossel *f*, Drosselklappe *f*, Klappe *f*, Lüftungsschieber *m*, Ofenklappe *f* (butterfly) Rauchklappe *f*, Schalldämpfer *m*, Widerstandskörper *m*, Windabsperrschieber *m*, Zugklappe *f*, Zugregler *m*
**damper,** ~ **cutout switch** Dämpferkanalschalter *m* ~ **device** Schwingungsdämpfer *m* ~ **plate** Luftschieber *m* ~ **ring** Dämpferring *m* ~ **rod** (piano) Abhebestange *f* ~ **spindle** Klappenspindel *f* ~ **valve** Dämpfungsventil *n* ~ **weight** Ausgleichgewicht *n* ~ **winding** Dämpferwicklung *f*
**dampers** Anfeuchtmaschine *f*, Feuchtglätte *f*, Feuchtpresse *f*
**damping** Abdämpfung *f*, Abklingen *n*, Bedämpfung *f*, Echounterdrückung *f* (acoust.), Dämpfung *f*, räumliche oder zeitliche Dämpfung *f*, Flachen *n*, Streuung *f*, Stromdämpfung *f*

**damping,** ~ **of grid circuit** Bekämpfung *f* des Gitterkreises *m* ~ **(out) of oscillations** Schwingungsdämpfung *f* ~ **of pulsations** Stoßdämpfung *f* ~ **of reverberations** Echounterdrückung *f* ~ **(out) of vibrations** Schwingungs-, Stoßdämpfung *f*

**damping,** ~ **action** Drosselwirkung *f* ~ **bolt** Schraubenbolzen *m* durch den eine Dämpfungswirkung erzielt wird ~ **capacity** Dämpfungsfähigkeit *f*, Rückdehnung *f*, Werkstoffdämpfung *f* ~ **chamber** Dämpferkammer *f* ~ **coefficient** Dämpfungskonstante *f*, Luftdämpfungsbeiwert *m* ~ **coefficient (or constant)** Dämpfungsfaktor *m* ~ **(or decay) coefficient** Dämpfungsziffer *f* ~ **(or amortisseur) coil** Dämpfungsspule *f* ~ **constant** Abklingkonstante *f*, Dämpfungsexponent *m* ~ **constant (or factor)** Dämpfungsziffer *f* ~ **constant of a cable** Verlustkonstante *f* ~ **couple** (of a deflecting instrument) Dämpfungsmoment *n* ~ **cup** Dämpfungsbehälter *m* ~ **curve** (electr.) Dämpfungskennlinie *f* ~ **de-attenuation** Dämpfungsreduktion *f* ~ **decrement** Dämpfung *f*, Dekrement *n*, Schwingungsdekrement *n* ~ **device** Dämpfungs-einrichtung *f*, -vorrichtung *f* ~ **diode** Dämpfungsdiode *f*, Zellendiode *f* (TV) ~ **down** Dämpfen *n* ~ **element** Dämpfungsglied *n* ~ **factor** Dämpfungszahl *f* ~ **(or attenuation) factor** Dämpfungsfaktor *m* ~ **factor (or coefficient)** Dämpfungskonstante *f* ~ **force** Dämpfungskraft *f* ~ **gain** Dämpfungsreduktion *f* ~ **gyro** Dämpfungskreisel *m* ~ **gyro regulator** Dämpfungsregler *m* ~ **lug** Dämpfungsfahne *f* ~ **meter** Dämpfungsmesser *m* ~ **mica** Dämpfungsglimmer *m* ~ **plate** Dämpfungsklotz *m* ~ **pole** Dämpfungspol *m* ~ **power** Dämpfungsfähigkeit *f* ~ **power (or property)** Dämpfungsvermögen *n* ~ **property** Dämpfungsfähigkeit *f*, Vibrationsaufnahme *f* ~ **pulley** Beruhigungsrolle *f* ~ **ratio** Dämpfungsgrad *m* ~ **reduction** Dämpfungs-reduktion *f*, -verminderung *f*, Entdämpfung *f* ~ **regeneration** Dämpfungsreduktion *f* ~ **resistance** Bremswiderstand *m* (electr.), Dampf-, Dämpfungs-widerstand *m* ~ **resistor** Dämpfungswiderstand *m* ~ **resulting in signal power (or in useful radiation)** Nutzdämpfung *f* ~ **roller** Anfeuchtwalze *f* ~**rolls** Anfeuchtmaschine *f*, Feuchtglätte *f*, Feuchtpresse *f* ~ **solution** Feuchtwasser *n* ~ **spring** Dämpferfeder *f* ~ **strip** Dämpfungsstreifen *m* ~ **surface** Dämpfungsfläche *f* (aviat.) ~ **time** Einstellzeit *f* eines Galvanometers *m* ~ **time constant** Dämpfungszeitkonstante *f* ~ **torque** Dämpfungsmoment *n* ~ **tube** Dämpfungsröhre *f* (in Zeilenablenkkreis *m*) ~ **winding** Dämpferwicklung *f* ~ **(or amortisseur) winding** Dämpfungswicklung *f*

**dampness** Feuchte *f*

**dampproof** feuchtigkeits-beständig, -dicht, -sicher ~ **and non-metallic sheathed wires** Feuchtraum- und Mantelleitungen *pl* ~ **course** wasserdichte isolierende Schicht *f* ~ **installation cable** Feuchtraumleitung *f* ~ **membrane** Dichtungshaut *f*

**dampproofing** Abdichtung *f*

**dandy roll** Entwässerungswalze *f*, (paper) Druckwalze *f*, Siebwalze *f* (paper mfg.) ~ **roller** Vordruckwalze *f* (Egoutteur)

**danger** Achtung *f*, Gefahr *f*, Gefährdung *f*, Gefährlichkeit *f*, Not *f*, Warnungstafel *f*

**danger,** ~ **of becoming dusty** Verstaubungsgefahr *f* ~ **of breaking** Bruchgefahr *f* ~ **of burning** Verbrennungsgefahr *f* ~ **of feed-back** Rückkuppelungsgefahr *f* ~ **of fire** Brandgefahr *f* ~ **of fouling** Verschmutzungsgefahr ~ **of frost action** Frostgefahr *f* ~ **of gas** Gasgefahr *f* ~ **of gliding** Rutsch-gefahr *f*, -gefährlichkeit *f* ~ **of ice formation (or of icing)** Vereisungsgefahr *f* ~ **of interception** Abhörgefahr *f* (rdo) ~ **to life** Lebensgefahr ~ **of lightning** Blitzgefahr *f* ~ **of pole reversal** Umpolgefahr *f* ~ **of radiation** Strahlungsgefährdung *f* ~ **of rusting** Rostgefahr *f* ~ **of slagging** Verschlackungsgefahr *f* ~ **of slipping** Rutschgefährlichkeit *f*

**danger,** ~ **alarm system** Gefahrmeldeanlage *f* ~ **area** Gefahren-raum *m*, -gebiet *n* (aviat.) ~ **arrow** Blitzpfeil *m*, Hochspannungspfeil *m* ~ **coefficient** Reaktivitätskoeffizient *m* ~ **flag** Warnflagge *f* ~ **light** Signallaterne *f* ~ **point** Gefahrenpunkt *m* ~ **report** Gefahrenmeldung *f* ~ **signal** Gefahrensignal *n*, Gefahrzeichen *n*, Notsignal *n*, Warnungsschild *n* ~ **zone** bestrichener Raum *m*, Gefahrengebiet *n*, Visierbereich *n*, Warngebiet *n*

**dangerous** gefährlich, gefahrvoll ~ **hill** gefährlicher Hügel *m*, gefährliche Steigung *f* ~ **lights** gefährdende Lichtquellen *pl* (aviat.) ~ **materials class** (e. g. that comprising inflammable liquids) Gefahrenklasse *f*

**dangle, to** ~ baumeln

**Daniell's cell** Daniell-Element *n*

**Danly,** ~ **brake** Danlybremse *f* ~ **clutch** Danlykupplung *f*

**dappled** scheckig

**Darby's recarburization** Darbysche Kohlung *f*

**dark** Dunkelheit *f*; dunkel, finster ~ **adaptation** Dunkel-adaptation *f*, -anpassung *f* ~ **adapted eye** dunkel adaptiertes Auge *n* ~ **brown cod oil** Dunkeltran *m* ~**brown laminated subbituminous coal** Papierkohle *f* ~ **burn** Ermüdung *f* eines Leuchtstoffes ~ **cell** Dunkelzelle *f* ~**-colored** dunkelfarbig ~ **conduction** Dunkelstrom *m* ~ **contrast method** Dunkelfeldverfahren *n* ~ **current** Dunkelstrom *m* (photoelectric cell) ~ **discharge** Dunkelentladung *f* ~ **field** Dunkelfeld *n* ~ **field equipment** Dunkelfeldeinrichtung *f* ~ **field illumination** Dunkelfeldbeleuchtung *f* ~ **field observation** Dunkelfeldbeobachtung *f* ~ **ground** Dunkelfeld *n* ~ **ground (or stop)** Dunkelfeldblende *f* ~ **ground illumination attachment** Dunkelfeldbeleuchtungseinrichtung *f* ~ **interval** Dunkelzeit *f* ~ **lantern** Blendlaterne *f* ~ **period** Dunkel-pause *f*, -zeit *f* ~ **picture portions** Tiefen *pl* ~ **positions** Dunkelstellungen *pl* (geol.) ~ **pre-sparking current** Vorstrom *m* ~ **red** (hot metal) dunkelrot ~**-red hot bulb** dunkelglühende Glühhaube *f* ~ **resistance** Dunkelwiderstand *m* ~ **room** Dunkelkammer *f*, lichtdichte Kassette *f* ~**-room focusing screen** Dunkelkammerscheibe *f* ~ **room pan impermeable to light** lichtdichte Kassette *f* ~ **search of searchlights** Dunkelsuche *f* **searching instrument** Dunkelsuchgerät *n* ~ **slide** Kassette *f* ~ **slide for camera insertion** Einlage *f* ~ **slide carriage** Kassettenschlitten *m* ~ **slide carrier** Kassettenrahmen *m* ~ **slide cash box** Kassettenhalter *m* ~ **space** Dunkelraum *m* ~ **space around the cathode** Kathodendunkelraum *m* ~ **spot** Dunkelpunkt *m* ~ **titan ore pigment** Titanmennige *f* ~**-trace screen** Dunkelschriftschirm *m* ~**-trace tube sciatron** Blau-, Dunkel-schriftröhre *f*

**darken, to** ~ abdunkeln, auslöschen, sich bräunen, dunkel werden, dunkeln, erlöschen, zum Erlöschen bringen, (lamp) löschen, (subsequently) nachdunkeln, (dull) trüben, schwärzen, verdunkeln **to** ~ **ship** abblenden

**darkening** Auslöschung *f*, Erlöschen *n*, (complete) Verdunk(e)lung *f*, Verfinsterung *f* ~ **agent** Abdunklungsmittel *n* ~ **plant** Verdunk(e)lungsanlage *f*

**darkness** Dunkel *n*, Dunkelheit *f*, Finsternis *f*

**darn, to** ~ stopfen

**darning needle** Stopfnadel *f*

**D'Arsonval galvanometer** D'Arsonval Galvanometer *n*, Drehspulgalvanometer *n*

**dart, to** ~ schießen

**dart** Pfeil *m* ~ **bit** Pfeilbohrer *m*

**dash, to** ~ heftig stoßen, schlagen, schleudern, übergießen **to** ~ **dot** strichpunktieren **to** ~ **to pieces** zerrütten

**dash** Balken *m*, (in a sentence) Gedankenstrich *m*, (in Zeichnungen) gestrichelte Linie *f*, Morsestrich *m*, Schießen *n*, Schlag *m*, Strich *m*, kurzer Strich *m*, plötzlicher Zusammenstoß *m* ~**-and-dot line** strichpunktierte Linie *f* **dashboard** Armaturenbrett *n*, Instrumentenbrett *n*, Spritzbrett *n*, Spritzwand *f*, Stirnwand *f* (Auto) ~ **clock** Borduhr *f* (aviat.) ~ **illumination** Schaltbrettbeleuchtung *f* ~ **leather** Spritzleder *n* ~ **light or lamp** Schaltbrettleuchte *f* ~**-mounted fan** Spritzbrettkühler *m*

**dash,** ~**-controlled** vom Armaturenbrett *n* aus regelbar ~ **counting** Strichzählung *f* (teleph.) ~ **curve** Strichkurve *f* ~**-dot mode** Strich-Punkt-Verfahren *n* ~**-dot relay** Zeitrelais *n* mit Bremszylinder *m* ~**-dotted** strichpunktiert ~**-dotted line** strichpunktierte Linie *f* ~ **lamp** Spritzbrettlampe *f* ~ **lighting** Instrumentenbrettbeleuchtung *f* ~**-line curve** gestrichelte Schaulinie *f* ~**-lined** strichliniert ~ **number** Unterbezeichnung *f* (aviat.) ~ **panel** Trennwand *f*

**dashpot** Bremszylinder *m*, Buffer *m*, Dämpfer *m*, Dämpfungsvorrichtung *f*, Luftkissen *n*, Luftpuffer *m*, Ölbremse *f*, Puffer *m*, (mit Luftpolster oder Öl arbeitender) Stoßdämpfer *m*, Stoßfang *m*, Stoßfänger *m* ~ **lever** Bremshebel *m* (telet.) ~ **relay** Relais *n* mit Bremszylinder ~ **type shock absorber** hydraulischer Stoßdämpfer *m* ~ **type timer** Zeitrelais *n* mit Öldämpfung

**dash,** ~ **relief valve** Dampfpufferregelventil *n* ~ **tachometer** Ferndrehzahlmesser *m* ~ **(or wash) wheel** Waschrad *n*

**dashed,** ~ **curve** gestrichelte Kurve *f* ~ **line** gestrichelte Linie *f*

**dasymeter** Dampfdichtemesser *m*, Gasdichtemesser *m*

**data** Angaben *pl*, Aufschluß *m*, Befunde *pl*, Daten *pl*, Meß- und Versuchswerte *pl*, Unterlagen *f pl* ~ **of box measurements** Kolliliste *f*

**data,** ~ **acquisition** Datenerfassung *f* ~ **book** Schießkladde *f* ~ **cable** Datenkabel *n* ~ **card** Betriebsdatentafel *f* ~ **circuit** Datenübertragungsanlage *f* ~ **collection** Datensammlung *f*

~ **computer** Kommandogerät *n* ~ **distribution** Datenverteilung *f* ~ **dump** Informationsverlust *m* ~ **flow** Meßwertübertragung *f* ~ **flow chart** (or **diagram**) Datenflußplan *m* ~ **handling system** Datenverarbeitungssystem *n* ~ **link** Datenübertragungsverbindung *f* ~ **logger** Datenspeicher *m*, Meßwertdrucker *m* ~ **obtained during flight** erflogene Meßwerte *f* ~ **printer** Zählbetrags-, Zahlen-drucker *m* ~ **processing** Datenverarbeitung *f* ~ **processing equipment** Meßwertverarbeitungsanlage *f* ~ **processor** Daten-verarbeitungsgerät *n*, -verarbeiter *m* ~ **reduction** Datenreduktion *f* ~ **rejected** ausgeschlossene Werte *pl* ~ **report** Datenbericht *m* ~ **section** Datenteil *m* ~ **sheets** Datenlisten *pl* ~ **smoothing** Werteglättung *f* ~ **subject to change without notice** Änderungen *pl* vorbehalten ~ **transmission** Daten-, Kommando-übertragung *f* ~**-transmission lines for searchlights** Richtwerteverbindung *f* ~ **transmitter** Übertragungsgerät *n*

**date, to** ~ datieren **to** ~ **back** retrodatieren **to become out of** ~ veralten

**date** Datum *n*, **set** ~ **for a case** Termin *m* zur Verhandlung einer Sache anberaumen

**date,** ~ **of delivery** Lieferfrist *f* ~ **of execution** (or **of fabrication**) Ausführungsdatum *n* ~ **of maturity** Fälligkeitsdatum *n*, Fälligkeitstermin *m*, Verfalltag *m* ~ **of payment** Bezahlungsdatum *n*, Zahlungstermin *m* ~ **of receipt of tenders** Versteigerungstermin *m* ~ **of reduction to practice** Ausführungsdatum *n* ~ **of service** Zustellungsdatum *n* ~ **of valuation** Bewertungsstichtag *m*

**date,** ~ **block** Kalenderblock *m* ~ **book** Notizkalender *m* ~**-indicating watch** Datumanzeigeuhr *f* ~ **(-limit) line** Datumgrenze *f* ~ **required** Liefertermin *m* ~ **stamp** Datumstempel *m* ~**-time group** Datum-Zeit-Gruppe *f*

**dated item** befristet verwendbarer Artikel *m*, Fristartikel *m*

**dating** Altersmessung *f* ~ **back** Rückdatierung *f*

**datum** Angabe *f*, Bezug *m*, Festpunkt *m*, gegebene Größe *f*, Grundlage *f*, Höhe *f* (über Bezugspunkt), Kartennull *n*, Meßwert *m* ~ **altitude** Bezugshöhe *f* ~ **dimension** Bezugs-, Grund-maß *n* ~ **error** Meßwertfehler *m* ~ **height** Bezugshöhe *f* ~ **level** Bezugs-ebene *f*, -fläche *f*, -horizont *m* (soil mech.), Grundebene *f* ~ **line** Ausgangslinie *f*, Bezugshorizont *m*, Grundlinie *f*, Horizont *m*, Horizontale *f*, Normallinie *f*, Standlinie *f* ~ **performance** Bezugsleistung *f*, Luftfahrzeugleistung *f* bezogen auf ... ~ **plane** Bezugs-ebene *f*, -fläche *f*, Grundebene *f* ~ **point** Anhalts-, Bezugs-punkt *m* ~ **surface** Bezugs-ebene *f*, -fläche *f*

**daub, to** ~ beklecksen, verschmieren, versudeln

**daub** Deckenstrich *m* (beim Lederlackieren), Tonkleister *m*

**dauber** Faßpicher *m* ~ **of paper** Papiersudler *m*

**daubing** Transchmiere *f*, Verschmierung *f*

**daughter,** ~ **activities** erzeugte Aktivität *f* ~ **atom** Folgeelement *n* ~ **instrument** Tochtergerät *n* ~ **products** Folgeprodukte *pl*

**davidite** Davidit *m*

**davit** Bootsdavit *m*, Davit *m*, Jütte *f*, Schiffskran *m* ~ **guy** Davitbackstag *m*

**dawn** Dämmerung *f*, Morgen *m*, Morgen-dämmerung *f*, -grauen *n*

**day** Tag *m* **at** ~ zutage **for** ~s tagelang ~ **of sett-lement** Gütetermin *m* ~**-and-night transfer key** Nachtschalter *m* ~ **blindness** Tagblindheit *f* ~ **break** Tagesanbruch *m* ~ **flight** Tagflug *m*

**daylight** Etagenhöhe *f*, Tageslicht *n* ~ **changing** Tageslichtwechselung *f* ~ **exposure** Tageslichtaufnahme *f* ~ **factor** Tageslichtfaktor *m* ~ **lamp** Tageslichtlampe *f* ~ **loading magazine** Tageslichtkassette *f* (film) ~ **opening** Öffnungsweite *f*, lichte Öffnungsweite *f* ~ **photograph** Tagesaufnahme *f* ~ **saving time** Sommerzeit *f* ~ **unloading** Tageslichtentnahme *f*

**day,** ~ **marking** Tages-kennzeichen *n*, -markierung *f* ~ **load** Tagesbelastung *f* ~ **operations** Tagesflugbetrieb *m* ~ **parker** Dauerparker *m* ~ **range** Tagesreichweite *f* ~ **rate** Tagesgebühr *f*, volle Gebühr *f* ~ **shift** Tagesschicht *f* ~ **ticket** Tageskarte *f* ~**-to-**~ **variations** tägliche Schwankungen *pl* ~ **wave** Tageswelle *f* ~ **wind** Tageswind *m* ~ **work** Stundenlohnarbeiten *pl*

**day's,** ~ **run** Etmal *n* ~ **work** Tagesarbeit *f*

**dazzle, to** ~ blenden

**dazzle** (of eye) Blendung *f* ~ **factor** Blendungsfaktor *m* ~ **paint** Blendanstrich *m* ~**-painted** bunt bemalt

**dazzler** Blender *m*

**dazzling** blendend

**DC, d.c., d–c** (**direct current**) Gleichstrom *m*

**DC/AC converter** Wechselrichter *m*

**DC,** ~ **amplifier** Gleichspannungsverstärker *m*, Gleichstromverstärker *m* ~ **arc** Gleichstromlichtbogen *m* ~ **balancer** Ausgleichsmaschine *f* ~ **bias** (**cores**) Gleichstromvormagnetisierung *f* ~ **centring** Gleichstromzentrierung *f* ~ **clamp diode** Niveau-, Schwarzsteuer-diode *f* ~ **component** Gleichstrom-anteil *m*, -komponente *f*, Nullkomponente *f* ~ **conductance** Gleichstromleitwert *m* ~ **control grid current** Steuergittergleichstrom *m* ~ **coupling** galvanische Kopplung *f* ~ **current transformer** Gleichstromwandler *m* ~ **dialling pulses** Gleichstromwählimpulse *pl* ~ **distribution** Gleichstromverteilung *f* ~ **dump** Informationsverlust *m* ~ **erasing head** Gleichstromlöschkopf *m* ~ **form factor** Mittel-, Effektiv-wert *m* ~ **machine** Gleichstrommaschine *f* ~ **magnetic biasing** Gleichstromvormagnetisierung *f* ~ **mains** Gleichstromnetz *n* ~ **pulse** Gleichstromstoßabgabe *f* ~ **read and write zero** Vormagnetisierungs-Nullaufzeichnung *f* und Lesen *n* ~ **reinsertion** Schwarzsteuerung *f*, Wiederherstellung *f* des Schwarzpegels ~ **relay** Gleichstromrelais *n* ~ **restoration** Wiederherstellung *f* des Schwarzpegels ~ **restorer** Gleichstromzuschaltung *f*, Nullpunktfesthalteschaltung *f* ~ **screen grid current** Schirmgittergleichstrom *m* ~ **tachogenerator** Gleichstromtachogenerator *m* ~ **traction** Gleichstrombahnbetrieb *m* ~ **voltage service conditions** Gleichspannungsbeanspruchung *f*

**deaccentuation** Absenkung *f* (of frequencies)

**deaccentuator** Vorentzerrungsglied *n*

**deacidify, to** ~ entsäuern

**deacidized tar** entsäuerter Raffinationsabfall *m*

**deacidizing** Entsäuerung *f* ~ **vessel** Entsäuerungstrog *m*

**deactivate, to** ~ aberregen, (Radium) entaktivieren, inert machen, unwirksam machen

**deactivation** Desaktivierung *f*, Inaktivierung *f*

**dead, to** ~**-burn** totbrennen **to** ~**-melt** abstehen lassen **to** ~**-press** (explosives) totpressen **to** ~**-roast** totrösten

**dead** beruhigt, erloschen, duff, dürr, festgelagert, (of materials, points) falsch, gestorben, glanzlos, leblos, leer, matt, ruhig, spannungslos (electr.), stromlos, taub (Gestein), tot ~ **abatis** Baumsperre *f* ~ **angle** toter Winkel *m* ~ **axle** feststehende Achse *f*, Laufachse *f* ~ (**or stationary**) **axle** feststehende Achse *f* ~ **band** Unempfindlichkeitsbereich *m*, unwirksamer Bereich *m* ~ **battery** entladene Batterie *f*

**deadbeat** aperiodisch-ausschwingend, aperiodische Dämpfung *f* (aviat.), eigenschwingungsfrei (phys.), nicht periodisch, unperiodisch ~ **compass** aperiodischer Kompaß *m* ~ **discharge** aperiodische Entladung *f* ~ **escapement** schleifende Hemmung *f* ~ **galvanometer** aperiodisches Galvanometer *n* ~ **indication** Standanzeige *f* ~ **measurement** stationäre Messung *f* ~ **measuring instrument** aperiodisches oder gedämpftes Meßinstrument *n* ~ **reading** gedämpfte Anzeige *f*, Standanzeige *f*

**dead,** ~ **body** Leiche *f* ~ **bolt** Absteller *m* ~**-burned** totgebrannt ~**-burned dolomite** totgebrannter Dolomit *m* ~**-burning** Totbrennen *n* ~**-burnt limestone** totgebrannter Kalkstein *m* ~ **calm** Flaute *f* ~ **cards** inaktive Karten *pl* ~ **center** feststehende Reitstockspitze *f*, genaue Mitte *f*, Körnerspitze *f*, Reitstockspitze *f*, ruhende Spitze *f*, Ruhepunkt *m* (engin.), toter Punkt *m*, tote oder feste Spitze *f*, Totlage *f*, Totpunkt *m* ~**-center ignition** Totpunktzündung *f* ~**-center indicator** Totpunktanzeiger *m* ~**-center position** Totpunktlage *f* ~**-center pulley** Antriebsscheibe *f* für tote Spitzen ~ **channel** Altwasser *n* ~ **cone** toter Trichter *m* ~ **corner** toter Winkel *m* ~ **end** blindes Ende *n*, Gleisstumpf *m* (r.r.), Schalltote *f* (acoust.), totes Gleis *n*, tote Windung *f* ~**-end binding** Endbund *m* (electr.) ~**-end clamp** Abspannklemme *f* ~**-end loggers' road** Holzweg *m* ~**-end nut** Verschlußmutter *f* bei Rohrleitungen ~**-end siding** Sackgleis *n* ~**-end tie** Abspannbindung *f* der Leitung, Abspannbund *m* Abspannung *f* ~**-end tower** Abspannmast *m* ~**-end turns** tote Windungen *pl* ~**-ending** Abspannung *f* ~ **engine** ausgefallener oder gestorbener Motor *m* ~ **eye** Bullauge *n*, Jungfer *f*, Kausche *f* ~**-eye ring** Kauschring *m* ~**-eye socket** Kausche *f* ~ **face** Blendfassade *f* ~ **files** abgelegte Akten *pl* ~ **flat** matt, stumpf ~ **floor** Blindboden *m* ~ **fold** tote Falte *f* ~ **freight** Leerfracht *f* ~ **gas** Abzugsgas *n* ~ **gilding** Mattgoldung *f* ~ **gold** Mattgold *n* ~ **ground** satter Erdschluß *m*, toter Raum *m* ~ **grounding** vollständiger Erdschluß *m* ~ **head** Anguß *m*, Gießkopf *m*, Massel *f*, Saugmassel *f*, verlorener Kopf *m* ~ **hole** blindes Loch *n* ~ **knife** feststehendes Messer *n* ~ **landing** Landung *f* mit stehendem Triebwerk *n* ~ **level** ebene Fläche *f* ~ **light** Lichtpforte *f* ~ **lime** verwitterter oder abgestandener oder abgestorbener Kalk *m*

**deadline** außer Betrieb befindliche Leitung *f*, endloses Seil *n*, Engpaß *m*, Druckende *n* (print.), Frist *f*, Grenzlinie *f*, Indifferenzlinie *f* (electr.), letzter Zeitpunkt *m* (Termin), Seilschlinge *f*, Stichtag *m*, Termin *m*, tote Leitung *f*

**dead load** Belastung *f*, Belastung *f* durch Eigengewicht, Gewichtsbelastung *f*, ruhende Belastung *f*, ruhende oder ständige Last *f*, statische Belastung *f*, Totlast *f*
**deadlock, to ~** auf den toten Punkt kommen, steckenbleiben, sich festfahren
**dead, ~-lock** Einriegelschloß *m*; toter Punkt *m* **~ lode** tauber Gang *m* **~ low water** hohle Ebbe *f* **~ man** (anchor block) Ankerblock *m*, Behelfsanker *m* **~-man's-handle** Sicherheits-einrichtung *f*, -fahrschaltung *f* **~ matter** Ablegesatz *m* **~-melting** Abstehenlassen *n* **~ mold** verlorene Gießform *f* **~ mold casting** Formguß *m* **~ movement (or motion)** Frei-, Leer-hub *m* **~ number** unbenützte Nummer *f* **~-number tone** Summerzeichen *n* zur Anzeige unbenützter Leitung **~ path** (thermometer) toter Gang *m* **~ plate** (of grates) Schür-, Vorstell-platte *f* **~ plug ring** isolierter Stöpselring *m* **~ position** Totlage *f* **~ reckoning** Besteck *n*, Besteckführung *f*, (position by) gegißtes Besteck *n*, gegißte Besteckrechnung *f*, Gissung *f*, Koppeln *n* (aviat.), Koppelkurs *m*, Schiffsortung *f* **~-reckoning position** Koppelort *m* **~ ring** fester verklebter Kolbenring *m* **~ rock** Quergestein *n* **~-rolled rubber** übermastizierter Gummi *m* **~ roller bed** nicht betriebsfähiger Rollgang *m* **~ room** echototer Raum *m* (acoust.) **~ run(ning)** toter Gang *m* **~ sector** Befestigungsstreifen *m* **~ short** (circuit) vollständiger Kurzschluß *m* **~ smooth** spiegelglatt **~-smooth file** Fein-, Doppel-schlichtfeile *f* **~ soft steel** extraweicher oder niedriggekohlter Stahl *m* **~ space** Schwebungslücke *f*, toter oder feuerleerer oder gedeckter Raum *m*, toter Winkel *m* **~ spot** abgedeckte oder abgeschirmte Stelle *f*, Auslöschzone *f*, Empfangslücke *f*, Funkschatten *m*, Schwingloch *n* (oscillator), tote Zone *f* (Wellenausbreitung *f*), toter Punkt *m* (acoust.) **~ start** Start *m* aus der Ruhelage, stehender Start *m* **~ steam** Abdampf *m* **~-stick landing** Landung *f* ohne Motor **~ (or surplus) stocks** tote Bestände *pl* **~ stop gripping device** Sperrfangvorrichtung *f* (Aufzug *m*) **~ stop method** Nullpunktmethode *f* **~-stroke hammer** Federhammer *m* **~ studio** nachhallfreies Studio *n* **~ time** Abklingzeit *f*, Ladeverzug *m*, Ladeverzugszeit *f*, Nachwirkzeit *f*, Verzugszeit *f*, Totzeit *f* **~-time correction** Totzeitkorrektur *f* **~ travel** toter Gang *m*, Totlauf *m* **~-true** haargenau **~ water** Kiel-, Totwasser *n*; Vorstrom *m* **~ weight** Belastungsgewicht *n*, Eigengewicht *n*, Gewichtslast *f*, (safety valves and presses) Gewichtsbelastung *f*; Leergewicht *n*, Tara *f*, totes Gewicht *n*, Verpackungsgewicht *n* **~-weight Brinell machine** Brinell-Presse *f* mit Hebelwaage **~-weight gauge** Kolbenlehre *f* **~-weight scale** Ladungsskala *f* **~-weight valve** Schwergewichtsventil *n* **~ white** mattweiß **~ window** blindes Fenster *n* **~ wire** stromloser Leiter *m* **~ wood** abgestorbenes Holz *n*, Reisig *n*, Totholz *n* **~ works** (mine) Grubenvorrichtungs- und Ausrichtungsarbeiten *pl* **~ zone** Schlierenzone *f*, (of ribbon microphone) totes Feld *n*, stille Zone *f* (rdo reception), tote Zone *f*
**deaden, to ~** abdämpfen, abstumpfen, anschärfen, ausschärfen, dämpfen, mattieren (colors), quälen, töten

**deadener** (for stopping vibration of wire) Schalldämpfer *m*
**deadening** Schalltilgung *f*
**deadly** tötlich
**deads** Abfall *m*, Berge *pl*, Berg(e)klein *n*, taubes Gestein *n*
**de-aerating** Entgasung *f* **~ plant** Entgasungsanlage *f* (bei Speisewasseraufbereitung *f*)
**de-aeration** Entgasung *f*
**de-aerator** Entgaser *m* (Speisewasser *n*), Entschäumungsschleuder *f*, Luftabscheider *m* **~ tank** Entlüftungsbehälter *m*
**deaf** taub **~ aid** Hörhilfe *f* **~ and dumb** taubstumm **~ ear** taubes Ohr *n*
**deafen, to ~** betäuben
**deafener** Dämpfer *m*
**deafness** Taubheit *f*
**deal, to ~** austeilen, handeln, Handel treiben, verfahren **to ~ with** vornehmen **to ~ with discrepancies** Abweichungen *pl* behandeln **to ~ descriptively with** eingehend behandeln (in einer Abhandlung)
**deal** Abkommen *n*, Abschluß *m*, Brett *n*, Diele *f*, Fichtenholz *n*, Föhrenholz *n*; Menge *f*, Planke *f*, Portion *f*, Transaktion *f* **~ board** Diele *f* **~ wood** Tannenholz *n*
**de-alcoholize, to ~** entgeisten
**dealer** Händler *m* **~ delivering materials** Lieferant *m*
**dealing, ~ in stocks** Effektenhandel *m*
**dear** kostspielig, teuer
**dearness** (of provisions) Teuerung *f*
**de-asphaltizing** Entasfaltierung *f*
**death rays** Todesstrahlen *pl*
**deathnium** Rekombinationsstelle *f*
**de-attenuation** Entdämpfung *f*, Dämpfungsverminderung *f*
**debacle** Beginn *m* des Eisgangs *m*
**debark, to ~** ausschiffen
**debarkation** Ausbootung *f*, Ausschiffung *f*
**debasement** Verschlechterung *f*
**debate, to ~** erörtern
**debate** Aussprache *f*, Erörterung *f*
**debeader** Wulstschneidemaschine *f*
**debenture** Pfandbrief *m*, Schuldschein *m*, Verschreibung *f*
**debenzolation** Benzolwäsche *f*
**debenzolize, to ~** Benzol *n* abtreiben
**debilitate, to ~** entkräftigen
**debit, to ~** in Anrechnung *f* bringen, belasten, ein Konto *n* belasten, zur Last *f* schreiben, auf der Sollseite *f* verbuchen **~ and credit** Soll *n* und Haben *n* **~ side** Soll *n* **~ voucher** Einnahmebeleg *m*
**deblooming** Bleichen *n*, Entscheinen *n* **~ color** Entscheinungsfarbe *f*
**de-blooper** Schallfilmlochvorrichtung *f*
**deblurring** Enttrübung *f*
**debossing** Tiefprägen *n*
**debouch, to ~** debouchieren, hervorbrechen
**debris** Bergeversatz *m*, Faulschlamm *m*, Rollstücke *pl*, Steinschutt *m*, Trümmer *pl*, Überbleibsel *n* **~ cone** Schuttkegel *m*
**debrominate, to ~** entbromen
**debt** Schuld *f*
**debtor** Schuldner *m* **~ and creditor** Soll *n* und Haben *n*
**debug** ausprüfen

**debugging** Störbeseitigung *f*
**debunched** dekonzentriert
**debunching** (in klystron) Entbündeln *n*
**deburring** Entgraten *n*
**debutanizer** Entbutanisierungsapparat *m*
**Debye-Scherrer ring or circle** (powder pattern) Debye-Scherrer-Ring *m*
**decade** Dekade *f*, Zehnergruppe *f* ~ **attenuator box** dekadischer Abschwächer *m* ~ **capacitance box** dekadischer Kondensator *m* ~ **counter tube** dekadische Zählröhre *f*, Dezimalzählrohr *n* ~ **inductance box** dekadische Induktivität *f* ~ **resistance box** Dekadenwiderstand *m* ~ **resistor** Dekadenwiderstand *m* ~ **scaler** Dekadenzähleinrichtung *f*, dezimaler Untersetzer *m* ~ **tube** Kathodenstrahlzählröhre *f*
**decadence** Verfall *m*
**decadent**, ~ **oscillation** gedämpfte Schwingung *f* ~ **wave** gedämpfte Welle *f*
**decadic resistor** Widerstandsatz *m*
**decagon** Zehneck *n*
**decagram** Dekagramm *n*
**decahedron** Zehnflächner *m*
**decahydronaphthalene** Decahydronaphthalin *n*
**decal** Abzieh-bild *n*, -papier *n*
**decalage** Flügelschränkung *f*, Schränkung *f* der Flügel gegeneinander
**decalcify, to** ~ entkalken
**decalcomania picture** Abziehbild *n*
**decalescence** Abschreckung *f*, Dekaleszenz *f*, (point) Haltepunkt *m*, sprunghafte Temperaturabnahme *f*
**decameter** Dekameter *m n*
**decametric waves** Dekameterwellen *pl* (HF-Bereich *m*)
**decanning** Enthülsen *n*
**decant, to** ~ abdekantieren, abfüllen, abgießen, abklären, abresten, abschlemmen, abseihen, absetzen lassen, abtreiben, abziehen, dekantieren, umfüllen
**decantation** Abgießung *f* ~ **apparatus** Dekantierapparat *m*
**decanted** abgefüllt
**decanter** Dekantiergefäß *n*, Klärflasche *f*, Karaffe *f*
**decanting**, ~ **apparatus** Abfüllapparat *m*, Schlemmapparat *m* ~ **bottle** Abklärflasche *f* ~ **centrifuge** Dekantierzentrifuge *f* ~ **cylinder** Dekantierzylinder *m* ~ **device** Umfüllvorrichtung *f* ~ **glass** Dekantierglas *n* ~ **jar** Dekantiertopf *m* ~ **machine** Schlemmaschine *f* ~ **plan for canalization** Kanalisationskläranlage *f*
**decapitation** Enthauptung *f*
**decarbonization** Entkohlung *f*, Kohlenstoffentziehung *f* ~ **plant** Entkarbonisierungsanlage *f*
**decarbonize, to** ~ dekarbonisieren, entkohlen, von Ölkohle *f* befreien
**decarbonizer** Ölkohleentferner *m*
**decarburization** Frischen *n*, Kohlenstoffentziehung *f*
**decarburize, to** ~ dekarbonisieren, entkohlen, frischen
**decare** Dekare *f*
**decatize, to** ~ dekatieren **to** ~ **with dry steam** trockendekatieren
**decatizing**, ~ **machine** Kesseldekatiervorrichtung *f* ~ **mark** Dekatierfalte *f*
**decatron** Dekatron *n*

**decay, to** ~ abklingen, auf Null *f* abnehmen, ausschwingen, dämpfen, entarten, ersticken, faulen, modern, stocken, verderben, verfallen, verfaulen, verklingen, verrotten, verwesen, vollständig abfallen, zerfallen, zerstört werden **to** ~ **in the open air** vermodern
**decay** Abklingen *n*, Ausschwingen *n*, Entarten *f*, Fäulnis *f*, Stoffzerfall *m*, Verfall *m*, Verwesung *f*, Verwitterung *f*, vollständiger Abfall *m*, Zerfall *m*, Zersetzung *f*, Zerstörung *f* ~ **of current** Abfall *m* des Stromes *m* (Zeitkonstante *f*)
**decay**, ~ **branch** Zerfallzweig *m* ~ **chain** Zerfallsreihe *f* ~ **characteristics** Nachleuchtcharakteristik *f* ~ **coefficient** Dämpfungsfaktor *m*, Zerfallskonstante *f* ~ **constante** Abklingkonstante *f* ~ **current** Ausschwingstrom *m* (of transients) ~ **curve** Zerfallkurve *f* ~ **electron resulting from disintegration** (mesotron) Zerfallelektron *n* ~ **factor** Abklingfaktor *m* (acoust.) ~ **formula** Zerfallsformel *f* ~ **modulation** Zerfallmodul *m* ~ **modulus** mittlere Lebensdauer *f* ~ **nonlinear distortion** Ausschwingungsverzerrung *f* ~ **period** Nachwirkzeit *f* ~ **process** Ausschwing-, Ausschwung-vorgang *m* ~ **rate** Zerfallsgeschwindigkeit *f* ~ **scheme** Zerfallsschema *n* ~ **time** Abfall-, Abkling-, Nachleucht-, Nachwirk-, Zerfall-zeit *f* ~-**time of the fluorescent screen** Nachleuchtdauer *f* des Fluoreszenzschirmes *m* ~ **train** abklingender Kurvenzug *m*
**decayed** eingegangen, verfault
**decaying** abfallend, baufällig, faul, hinfällig ~ **current** abklingender Strom *m*
**deceased** gestorben, verstorben
**deceit** Betrug *m*, Falschheit *f*; falsch
**deceitful** hinterlistig, täuschend, trügerisch
**deceive, to** ~ täuschen
**deceive station** Scheinfunkstelle *f*
**decelerate, to** ~ abbremsen, bremsen, retardieren
**decelerating electrode** Brems-, Verlangsamungs-, Verzögerungs-elektrode *f*
**deceleration** Abnehmen *n* der Geschwindigkeit, Bremswirkung *f*, Bremsung *f*, negative Beschleunigung *f*, Stau *m*, Verlangsamen *n*, Verminderung *f* (der Drehzahl *f*), Verzögerung *f* ~ **of electrons** Bremsung *f* der Elektronen, Elektronenbremsung *f*
**deceleration**, ~ **curve** Auslaufkurve *f* ~ **parachute** Bremsfallschirm *m* ~ **radiation** Bremsstrahlung *f* ~ **test** Auslaufversuch *m* (balloon)
**decelerator** Hemmungskörper *m*
**decelerometer** Verzögerungsmesser *m*
**decentering** Dezentrierung *f*
**decentralize, to** ~ dezentralisieren, zerlegen
**decentration** Dezentrierung *f*
**deception** Täuschen *n*, Täuschung *f* ~ **signal traffic** Verschleierungsverkehr *m*
**deceptive** trügerisch
**decess** Freidrehung *f*
**dechlorinate, to** ~ entchloren
**dechroming bath** Entchromungsbad *n*
**decibel** Dezibel *n*, Phone *f* (acoust.)
**decibelmeter** Dezibelmeter *n*, Pegelmesser *m*
**decide, to** ~ den Ausschlag *m* geben, austragen, beschließen, bestimmen, entscheiden, sich entschließen
**decided** ausgeprägt, ausgesprochen, beschlossen, bestimmt, entschieden, entschlossen
**deciding** maßgebend

**deciduous forest (or wood)** Laubwald *m*
**decigram** Dezigramm *n*
**deciliter** Deziliter *n*
**decilog** Dezilog *m*
**decimal** Dezimalstelle *f* ~ **base** Dezimalsystem *n*
~ **count** Dezimaltiter *n* ~ **digit** Dezimalziffer *f*
~ **fraction** Dezimalbruch *m* ~ **gauge** Dezimallehre *f* ~ **index system** Dezimalklassifikation *f*
~ **measure** Dezimalmaß *n* ~ **notation** dezimale
Schreibweise *f* ~ **number** Dezimalzahl *f* ~
**number system** dezimales Zahlensystem *n* ~
**numbering** dekadische Bezifferung *f* ~ **place**
Dezimalstelle *f*, Zehnerstelle *f* ~ **point** Dezimalkomma *n* ~ **point alignment** Kommaausrichtung *f* ~ **power** Zehnerpotenz *f* ~ **resistance**
Dekadenwiderstand *m* ~ **rheostat** Dekadenrheostat *m* ~ **scales** Dezimalwaage *f* ~ **system**
Dezimal-, Zehner-system *n* ~ **target** Zehnerringscheibe *f* ~**-to-binary conversion** Dezimalbinärkonvertierung *f*
**decimate, to** ~ dezimieren, stark schwächen
**decimeter** Dezimeter *n* ~ **apparatus** Dezimetergerät *n* ~ **circle (or circuit)** Dezikreis *m* ~ **range**
Dezimetergebiet *n* ~ **wave lengths** Dezimeterwellen *pl* (rdr)
**decimetric waves** Dezimeterwellen *pl*
**decineper** Dezineper *n*
**decinormal** zehntelnormal
**decipher, to** ~ beschiffrieren, dechiffrieren, entschlüsseln, entziffern
**deciphering** Entschlüsselung *f*, Entzifferung *f*
**decision** Beschluß *m*, Entscheid *m*, Entscheidung
*f*, Entschluß *m*, Entschlußfassung *f*, Gerichtsbeschluß *m*, Rechtsspruch *m*, Urteil *n*, Verfügung *f*
**decision,** ~ **content** Entscheidungsgehalt *n*
(comput.) ~ **element** Entscheidungs-element *n*,
-gerät *n*, Verknüpfungsglied *n*
**decisive** ausschlaggebend, durchschlagend, maßgebend ~ **factor** Ausschlag *m*, ausschlaggebender Faktor *m* ~ **point** Kernpunkt *m*
**decisiveness** entscheidende Eigenschaft *f*, Entschiedenheit *f*
**deck** Etage *f* (Förderkorb *m*), (supported on
beams or bearers) Rost *m*, Tragfläche *f*, (in
switch) Kontaktbahn *f* ~ **of a bridge** Bedielung
*f* ~ **of fuselage** Rumpfoberseite *f*
**deck,** ~ **armor** Deckpanzer *m* ~ **auxiliaries**
Deckhilfsmaschinen *pl* ~ **cargo** Deckladung *f*
~ **gear** Deckausrüstung *f* ~ **hands** Deckmannschaft *f* ~ **house** Bohlenbeleg *m*, Deckhaus *n*
~ **layouts** Ausführung *f* der Glockenböden *pl*
~ **machine** Etagenmaschine *f* ~ **plate (or washer)**
Scheibe *f* ~ **plate girder bridge** vollwandige
Trägerbrücke *f* ~ **stopper** Taustopper *m* ~
**tackle** Gien *n*
**decked** gedeckt
**decking** Bodenbelag *m*, Deck *n* ~ **carriage** Einstoßschlitten *m* ~ **plant** Aufstoßvorrichtung *f*
~ **ram** Druckstempel *m*, Förderkorbaufsatzstößel *m*
**deckle** Deckelrahmen *m* ~ **edge** Büttenrand *m*
~**-edged paper** Büttenrandpapier *n* ~ **frame**
Auflauf-, Deckel-rahmen *m* ~ **strap** Formdeckel *m*
**decks cleared for action** klar Schiff
**declarant** Aussagender *m*
**declaration** Angabe *f*, Aussage *f*, Erklärung *f*,

eidliche Versicherung *f* ~ **of assignment** Abtretungserklärung *f* ~ **of cancellation (or invalidity)** Kraftloserklärung *f* ~ **of property (or
net worth)** Vermögenserklärung *f*
**declarator** Vereinbarungssymbol *n* (info proc.)
**declaratory,** ~ **action** Feststellungsklage *f* ~
**judgement** Feststellungsurteil *n*
**declare, to** ~ angeben, aussagen, aussprechen,
behaupten, erklären **to** ~ **a rate of bonus (or
dividend)** Dividendensatz *m* deklarieren
**declared,** ~ **infected** für verseucht erklären ~
**temperature** festgesetzte Temperatur *f* ~ **value**
Wertangabe *f*
**declassify, to** ~ die Geheimhaltung *f* aufheben
**declinable** abwandelbar
**declinate** abgebogen
**declination** Abfall *m*, Abweichung *f*, (of the
compass) Deklination *f* ~ **of radio direction
finder** Bordablenkung *f* des Funkpeilers *m*
**declination,** ~ **chart** Mißweisungskarte *f* ~
**circle** Deklinationskreis *m* ~ **compass** Deklinatorium *n* ~ **gear** Deklinationsgetriebe *n* ~
**map** Mißweisungskarte *f* ~ **tides** Deklinationstiden *pl*
**declinator** Richtkreisbussole *f*
**decline, to** ~ abebnen, abfallen, ablehnen, abnehmen, absinken, abwandeln, abweichen, beanstanden, fallen, nachlassen, sinken, verfallen,
verweigern, weigern **to** ~ **to grant a concession**
eine Mutung *f* abweisen
**decline** Abfall *m*, Abnahme *f*, Fall *m*, Niedergang *m*, Rückgang *m*, Verfall *m*, Verminderung
*f*, Zurückgehen *n* ~ **in activity** Abschwächung *f*
~ **of output** (engine power) Leistungsabnahme
*f* ~ **period of control potential** Ausregelzeit *f*
~ **in potential** Potentialabnahme *f*
**declined** abgeebbt
**declinometer** Deklinationsmesser *m*
**declivity** Abdachung *f*, Abschüssigkeit *f* (landscape), Gehänge *n*, Neigung *f* ~ **of a roof** Abdachung *f* eines Daches *n*
**declutch, to** ~ ausrücken, entkuppeln
**declutching** Entkuppelung *f* ~ **level** Entkuppelungshebel *m* ~ **member** Entkuppelungsglied *n*
**decoct, to** ~ auskochen
**decocting medium** Abkochmittel *n*
**decoction** Abkochung *f*, Absud *m* ~ **method**
Dekoktionsverfahren *n*
**decode, to** ~ dechiffrieren, dekodieren, entschlüsseln, entziffern
**decoder** Dekodierer *m*, Entmischer *m* (telet.),
Entschlüssler *m*, Übersetzer *m*, Umrechner *m*
**decoding** Entschlüsselung *f*, Entzifferung *f*, Umrechnung *f* ~ **cabinet** Demodulationseinrichtung *f* ~ **circuit** Dekodier-, Entschlüsselschaltung *f* ~ **matrix** Dekodiermatrix *m*
**decohere, to** ~ entfritten
**decoherence** Entfrittung *f*, Frittung *f*
**decoherer** Dekohärer *m*
**de-coil** abwickeln, haspeln
**decoiler** Abzugshaspel *f*
**decolor, to** ~ abfärben
**decolorant** Entfärbungsmittel *n*
**decoloration** Verfärbung *f*
**decolored** abgefärbt
**decolorization** Entfärbung *f*
**decolorize, to** ~ entfärben, verfärben
**decolorizing** Bleichen *n* ~ **agent** Entfärbungs-

mittel *n* ~ **carbon** Entfärbungskohle *f* ~ **power** Entfärbungsvermögen *n*
**decomposability** Zersetzbarkeit *f*
**decomposable** aufschließbar, zerlegbar, zersetzbar
**decompose, to** ~ abbauen, auflösen, aufschließen, entmischen, faulen, scheiden, trennen, zergliedern, zerfallen, zerlegen, zersetzen **to** ~ **through long boiling** verkochen, durch Kochen zerstören
**decomposed** morsch(ig), zersetzt ~ **residuum** Verwitterungsrückstand *m* ~ **rock** verwittertes Gestein *n* ~ **trap** Tuffwacke *f*
**decomposer** Zersetzer *m*
**decomposing** Scheiden *n*, Scheidung *f* ~ **agent** Zersetzer *m* ~ **vat** durchgehende Küpe *f*
**decomposition** Abbau *m*, Auflösung *f*, Aufschließung *f*, Aufschluß *m*, Entmischung *f*, Spaltung *f*, Trennung *f*, Umsetzung *f*, Stoffzerfall *m*, Zerfall *m*, Zerlegung *f*, Zersetzung *f* ~ **through long boiling** Verkochungserscheinung *f* ~ **in a chemical change** Umsetzungsgeschwindigkeit *f* ~ **of light** Auflösung *f* des Lichtes *n*, Lichtzerlegung *f* ~ **to partial fractions** Partialbruchzerlegung *f* ~ **of steam** Dampfzersetzung *f*
**decomposition,** ~ **cell** Zersetzungszelle *f* ~ **chamber** Zersetzungskammer *f* ~ **process** Zerfallprozeß *m*, Zersetzungsvorgang *m* ~ **product** Zersetzungsprodukt *n* ~ **voltage** Zersetzungsspannung *f*
**decompression,** ~ **cam** Dekompressionsnocken *n* ~ **chamber** Höhenkammer *f* ~ **lever** Entkompressions-, Verdichtungsverminderungs-hebel *m*
**decompressor** Dekompressionshebel *m*
**decontaminate, to** ~ entgasen, entgiften
**decontaminating agent** Entgiftungsmittel *n*
**decontamination** Entgiftung *f*, Entseuchen *n*, Entseuchung *f* ~ **of equipment** Geräteentgiftung *f*
**decontamination,** ~ **agent** Entgiftungsstoff *m* ~ **apparatus** Entgiftungsgerät *n* ~ **(sprinkling) drum** Streutrommel *f* ~ **instrument** Pulverzerstäuber *m* ~ **plow** Entgiftungspflug *m* ~ **point** Entgiftungsplatz *m* ~ **service** Entgiftungsdienst *m* ~ **squad** Entgiftungstrupp *m* ~ **sprinkling drum** Entgiftungstrommel *f* ~ **truck** Entgiftungskraftwagen *m*
**decontrol** Kontrollaufhebung *f*
**decorate, to** ~ verzieren
**decorating lehr** Einbrennmuffel *f*
**decoration** Ausschmückung *f*, Auszeichnung *f*, Verzierung *f* ~ **of glass and crystal** Glasveredelung *f*
**decorative,** ~ **coatings** Schmucküberzüge *pl* ~ **frame** Zierrahmen *m* ~ **printing** Buntdruck *m* ~ **slat** Zierleiste *f* ~ **sleeve** Zierhülse *f*
**decorticate, to** ~ borken, entrinden
**decortication** Entbasten *n* der Stengel
**decorticator** Schälmaschine *f*
**decouple, to** ~ (to suppress or lessen feedback) auskoppeln, entkoppeln
**decoupling** Entkoppelung *f* ~ **of feedback** Rückkoppelungsunterdrückung *f* ~ **condenser** Entkoppelungskondensator *f* ~ **network** Entkoppler *m*, Entkopplungsschaltung *f* ~ **resistance** Entkoppelungswiderstand *m*
**decoy, to** ~ heranlocken

**decoy** Lock-schiff *n*, -ziel *n*; Scheinanlage *f*, Täusch-reflektor *m*, -ziel *n* ~ **for low-flying aircraft** Tiefffliegerfalle *f* ~ **aerodrome** Scheinflugplatz *m* ~ **echo** Köderecho *n*, Radarboldecho *n* ~ **missile** Köderflugkörper *m* ~ **transmitter** Scheinsender *m*
**decrease, to** ~ abfallen, (wind) abflauen, abnehmen, einziehen, erniedrigen, fallen, herabsetzen, kleiner werden, reduzieren, sinken, verkleinern, vermindern, verringern **to** ~ **the efficiency** den Wirkungsgrad herabsetzen **to** ~ **the power** die Spannung *f* verringern **to** ~ **the sensitivity of a relay** ein Relais *n* weniger empfindlich einstellen
**decrease** Abfall *m*, Abnahme *f*, Ermäßigung *f*, (speed) Herabsetzung *f*, Minderung *f*, Verkleinerung *f*, Verminderung *f*, Verringerung *f* ~ **time** Abfallzeit *f*
**decrease,** ~ **of brightness** Abnahme *f* der Helligkeit oder Schärfe gegen den Rand des Bildes ~ **in concentration** Konzentrationsverlust *m* ~ **in cross-sectional area** Querschnittsverringerung *f* ~ **of current** Stromabnahme *f* ~ **in energy** Energieabnahme *f* ~ **in fuel consumption** Verbrauchsverringerung *f* ~ **in intensity** Intensitätsrückgang *m* ~ **of lift** Auftriebsverminderung *f* ~ **of light** Lichtabfall *m* ~ **of potential** Potentialabfall *m* ~ **of pressure** Druckverminderung *f* ~ **in price** Preisabschlag *m* ~ **of receipts** Mindereinnahme *f* ~ **of solubility** Löslichkeitsverminderung *f* ~ **of spot with intensity increase** Einstellfehler *m* ~ **in strength** Verschwächung *f* ~ **in temperature** Wärmeabnahme *f* ~ **of thrust** Schubrückgang *m* ~ **of velocity** Geschwindigkeitsabnahme *f* ~ **in volume** Volumenverminderung *f* ~ **in weight** Gewichtsabnahme *f*
**decreasing** Abnehmen *n*, Kleinerwerden *n*; abnehmend ~ **current** abfallender Strom *m* ~ **gear ratio** Untersetzungsverhältnis *n* ~ **pitch** abnehmende Steigung *f*, progressive Schnecke *f* ~ **speed** abfallende Drehzahl *f* ~ **stroke** abnehmende Changierung *f*
**decree, to** ~ beschließen, verfügen, verordnen
**decree** Beschluß *m*, Erlaß *m*, Gesetzvorschrift *f*, Verfügung *f* ~ **of the court** Gerichtsbeschluß *m*
**decrement** Abnahme *f*; Dämpfungs-verhältnis *n*, -wert *m*; Dekrement *n* ~ **of damping** Dämpfungsdekrement *m* ~ **table** Ausscheidetafel *f*
**decremental hardening** Teilhärtung *f*
**decremeter** Dämpfungsmesser *m*, Dekrementmesser *m*, Dekremeter *n*
**decrepit** ausgemergelt
**decrepitate, to** ~ zerknistern
**decrepitation water** Verknisterungswasser *n*
**decrypt, to** ~ entschlüsseln
**decurrent** abfließend
**decussation** Faserkreuzung *f*
**dedendum** Zahnfuß *m* ~ **of a tooth** Fußhöhe *f* (Zahnrad *n*) ~ **angle** Fußkegelwinkel *m* ~ **angle of bevel gear** Fußwinkel *m* des Kegelrades *n* ~ **circle** Fußkreis *m* ~ **line** Fußkreis *m*
**dedication ceremony** Einweihungsfeier *f*
**deduce, to** ~ ableiten, folgern, herleiten
**deduct, to** ~ ablassen, abrechnen, in Abzug bringen
**deductable** abzugsfähig
**deduction** Ableitung *f*, Ablesen *n*, Abschlag *m*,

Abtrag *m*, Abziehen *n*, (pay) Abzug *m*, Folgerung *f*, Herleitung *f*, Lohnabzug *m*, Nachlaß *m*, Schlußfolge *f*, Subtraktion *f*, Vergünstigung *f*, Verminderung *f* ~ **of rivet holes** Nietabzug *m* ~ **from salary** Gehaltsabzug *m*
**dedusting, ~ plant** Entstäubungsanlage *f* ~ **screen** Entstäubungssieb *n*
**dee** D-Elektrode *f* ~ **lines** D-Elektrodenträger *pl*
**deed** Aktenstück *n*, Tat *f*, Werk *n*
**deed, ~ of assignment** Abtretungsurkunde *f*, Übertragungsurkunde *f*, Zessionsurkunde *f* ~ ~ **of concession** Mutungsurkunde *f* ~ **of conveyance** Abtretungs-, Überschreibungs-, Übertragungs-urkunde *f* ~ **of transfer** Übertragungserklärung *f*, Zessionsurkunde *f*
**deem, to ~ adjudge** befinden
**de-emphasis** Deakzentuierung *f* (acoust.), Deemphase *f*, Entzerrung *f*, Nachentzerrung *f* (TV), Höhenebnung *f*, Rückentzerrung *f*
**de-emphasizing of high frequencies** Benachteiligung *f* der hohen Frequenzen
**de-energization** Aberregung *f* ~ **of a relay** Relaisaberregung *f*
**de-energize, to ~** aberregen, entmagnetisieren, entregen, inert machen, ein Relais *n* zum Abfall bringen, außer Strom *m* setzen
**de-energized** stromlos **to be ~** abfallen
**de-energizing resistor** Entregerwiderstand *m*
**deep, to ~-draw** tiefziehen
**deep** Tiefe *f* (Gewässer); (color) dunkel, satt, tief ~ **adit** Grund-, Wasserlösungs-stollen *m* ~ **beam** wandartiger Träger *m* ~ **cut** Tiefschnitt *m* ~-**dimension picture** Durchzeichnung *f* (TV) ~ **dip** Untertauchung *f* ~-**dished boiler head** tiefgewölbter Boden *m* ~ **drawing** (aluminium mfg.) Kümpeln *n*, Tiefziehen *n* ~-**drawing cold rolled steel** Tiefziehstahl *m* ~-**drawing purposes** Tiefziehzwecke *pl* ~-**drawing quality** Tiefziehqualität *f* ~-**drawing steel** Tiefziehzweck *m* ~-**drawing test** Tiefungsversuch *m* ~-**etch test** Tiefätzprobe *f* ~ **etching** Tiefätzung *f* ~-**fold sacks** Tieffaltsäcke *pl* ~ **foundation** Tiefgründung *f* ~-**freezing compartment** Tiefkühlfach *n* ~ **gap** tiefgehende Ausladung *f* ~-**going** tiefgehend ~-**groove ball bearing** Ring-, Scheibenrillenlager *n* ~-**groove ball journals** Radialrillenkugellager *n* ~-**groove-type radial ball bearing** Radiaxlager *n*, Tiefbettkugellager *n* ~ **hole boring** Tiefbohren *n* ~ **hollow** Kessel *m* ~ **level** (oil drilling) Grundstrecke *f*, Hang *m* (min.) ~ **level sluice** Tiefbaugrube *f* ~ **mine working** Tiefbaubetrieb *m* ~ **mining** Tiefbaubetrieb *m* ~ **pile** Hochstapel *m* ~ **pump** Bohrpumpe *f* ~-**race ball bearing** Tiefbettkugellager *n* ~ **red** tiefrot ~-**row ball bearing** Hochschulterkugellager *n* ~-**sea** Tiefsee *f* ~-**sea lead** Tiefseelot *n* ~-**sea navigation** Hochseeschiffahrt *f* ~-**sea sounding** Tiefseelotung *f* ~-**seated** tiefliegend ~-**seated decomposition** Tiefenzersetzung *f* ~-**seated** (plutonic) **rock** Tiefengestein *n* ~ **shade** satter Farbton *m* ~ **sluice gate** Tiefschütz(e) *m*, *f* ~ **sounding apparatus** Tiefsondiergerät *n* ~ **through holes** Durchgangslöcher *pl* ~ **trough conveying** Tieftrogbandförderung *f* ~ **water** Hochsee *f*, Tiefenwasser *n*, Tiefsee *f* ~ **water circulation** Tiefenzirkulation *f* ~-**webbed** hochstegig (Träger) ~ **well** Tiefbrunnen *m* ~-**well drilling plant** Aufschlußbohrung *f* einer

Tiefbohranlage, Tiefbohranlage *f* ~ **well pump** Tiefpumpe *f* ~ **workings** Tiefbau *m* ~ **X-ray therapy** Röntgentiefentherapie *f*
**deepen, to ~** abdunkeln (print.), abteufen, teufen, (color) verdunkeln, vertiefen
**deepening** Abteufen *n*, Luftdruckvertiefung *f* (aviat.), Tieferwerden *n*, Vertiefung *f* ~ **of the channel** Vertiefung *f* des Fahrwassers ~ **hammer** Tiefhammer *m*
**deeply, ~ involved in debt** überschuldet ~ **penetrating** tiefgehend
**dees** Duanten *pl*
**de-excitation** Abregung *f*
**deface, to ~** besudeln
**default, to ~** unterlassen
**default** Nichterfüllung *f*, Nichterscheinen *n* **in ~ of** in Ermanglung *f* von, mangels ~ **of payment** Nichtzahlung ~ **summons** Zahlungsbefehl *m*
**defaulting subscriber** Gebührenschuldner *m*, zahlungsunfähiger Teilnehmer *m*
**defeat, to ~** besiegen, schlagen
**defeat** Niederlage *f*
**defecant** (sugar mfg.) Saftreinigungsmittel *n*
**defecate, to ~** klären, reinigen, scheiden
**defecated** (sugar mfg.) geklärte Lösung *f* ~ **juice** geschiedener Saft *m*
**defecation, ~ with milk of lime** nasse Scheidung *f* ~ **lime** Scheidekalk *m* ~ **scum** Scheide-kalk *m*, -schlamm *m* ~ **tank** Scheidepfanne *f*
**defecator** Scheidepfanne *f* ~ **for dry lime** Trokkenscheidepfanne *f*
**defect** Fehlen *n*, Fehler *m*, Fehlerscheinung *f* (phenomenon), Fehlstelle *f*, Fleck *m*, Flecken *m*, Gebrechen *n*, Makel *m*, Mangel *m*, Narbe *f*, Stoffehler *m*, Teilungsfehler *m* ~ **in a casting** Gußblase *f* ~ **of image** Abbildungsfehler *m* (opt.) ~ **of sight** Fehlsichtigkeit *f* ~ **in timber** Holzfehler *m* ~ **of vision** Sehfehler *m*
**defect, ~ conduction** Defekthalb-, Ersatz-, Störstellenhalb-leitung *f* ~ **generation** Fehlstellenerzeugung *f* ~ **mobilities** Störstellenbeweglichkeit *f*
**defection** Mangel *m*, Scheidung *f*
**defective** defekt, fehlerhaft, lückenhaft, mangelhaft, schadhaft, unvollkommen ~ **contact** Wackelkontakt *m* ~ **cord** fehlerhafte Schnur *f* ~ **focus** Fehlanpassung *f* ~ **ion** Lockerion *n* ~ **joint** Fehlverbindung *f* ~ **ringing** mangelhafter Ruf *m* ~ **trunk** gestörte Verbindungsleitung *f* (teleph.) ~ **(or unsound) weld** Fehlschweißung *f*
**defectively, ~ sighted** fehlsichtig ~ **sighted eye** fehlsichtiges Auge *n*
**defectiveness** Mangelhaftigkeit *f*
**defects in metals** Fehlordnung *f* (chemische) in Metallen
**defend, to ~** schützen, in Schutz nehmen, verteidigen
**defended objective** Schutzobjekt *n*
**defense** Abwehr *f*, Einrede *f*, Einwendung *f*, (plea) Klagebeantwortung *f*, Rechtfertigung *f*, Schutz *m*, Verteidigung *f*, Wehr *f* ~ **belt** Schutzgürtel *m* ~ **boom** Backspiere *f* ~ **means** Abwehrmittel *n*
**defenses** Dammschutz *m*
**defensive** Abwehr *f*, Defensive *f* ~ **measure** Abwehrmaßnahme *f* ~ **weapon** Verteidigungswaffe *f*
**defer, to ~** aufschieben, aussetzen, einstellen, verschieben, zurückstellen

**deferment** Aussetzung *f*, Aufschub *m*, Gestellungsaufschub *m*, Zurückstellung *f*

**deferred, ~ action alarm** verzögerte Zeichengebung *f* ~ **call** zurückgestellte Verbindung *f* ~ **charge** Abgrenzungsposten *m*, Nachleistung *f* ~ **items** zurückgestellte Posten *pl* ~ **rate** ermäßigte Gebühr *f* ~ **share** Genußschein *m* ~ **telegram** Telegramm *n* zu ermäßigter Gebühr, zurückgestelltes Telegramm *n*

**defervescence** Entfieberung *f*

**deficiency** Ausfall *m*, Defizit *n*, Fehler *m*, Fehlstelle *f* (cryst.), Lücke *f*, Mangel *m*, Mangelleiter *pl*, Manko *n*, (in stock pile) Minusbestand *m*, Unterschluß *m* ~ **in correction** Korrektionsmangel *m* ~ **in iron** Eisenmangel *m* ~ **of materials** Stoffmangel *m* ~ **in receipts** Mindereinnahme *f* ~ **of sight** Fehlsichtigkeit *f* ~ **shortage** Ergänzungsbedarf *m*

**deficient** mangelhaft **to be ~ in** ermangeln ~ **in gas** gasarm ~ **in metal** metallarm ~ **for welding** hadrig

**deficit** Ausfall *m*, Manko *n*, Unterbilanz *f*

**deficity, ~ electrons** Löcher *pl* ~ **(or p-type) conduction** Mangel- oder p-leitung *f*

**defiladed** gedeckt, verdeckt

**defile, to** ~ besudeln

**defile** Bergenge *f*, Defilee *n*, Defilement *n*, Enge *f*, (on maps) Engpaß *m*, gedeckter Hang *m*, Hohlweg *m*, Klamm *f*, Tobel *m*

**definable** bestimmbar

**define, to** ~ (limits) abgrenzen, begrenzen, bestimmen, erklären, festsetzen, markieren

**defined** abgegrenzt, ausgeprägt, benannt (math.), umrissen

**defining, ~ aperture** Anodenblende *f* (of TV tube) ~ **equation** Bestimmungsgleichung *f* ~ **quantity** Bestimmungsgröße *f*

**definite** abgegrenzt, bestimmt, sicher ~ **integral** bestimmtes Integral *n* **on a ~ plan in accordance with theoretical principles** gesetzmäßig ~ **time element** begrenzt verzögerte Auslösung *f*

**definitely** bestimmt, klar

**definiteness** Bestimmtheit *f*

**definition** Abgrenzung *f*, Auflösungsvermögen *n* (Objektiv *n*), Begrenzung *f*, Begriffsbestimmung *f*, Begriffsfestlegung *f*, Begriffsfestsetzung *f*, Bestimmung *f*, Bildqualität *f* (film), Deutlichkeit *f*, Erklärung *f*, Festsetzung *f*, (telegr. signal) Güte *f*, Punktstärke *f* (rdr), Scharfabbildung *f*, Strichschärfe *f*, Umgrenzung *f*

**definition, ~ at center** Zentralschärfe *f* ~ **of contours** Konturenschärfe *f* ~ **in depth** Schärfentiefe *f*, Tiefenschärfe *f* ~ **of (or at) the edges** Eckenschärfe *f* ~ **of the field emission** Auflösung *f* (TV) **(degree or)** ~ **of image** Rasterfeinheit *f* ~ **in line direction** Längsschärfe *f* ~ **of lines** Linienschärfe *f* ~ **of minimum beam** Minimumstrahldefinition *f* **(degree or)** ~ **of picture** Bildfeinheit *f*, Feinheit *f*, Rasterung *f* ~ **in picture direction** Querschärfe *f*

**definition, ~ chart** Probebild *n* (TV) ~ **control** Scharfabbildungssteuerung *f* ~ **equation** Definitionsgleichung *f*

**deflagrate, to** ~ abbrennen, aufflammen, deflagrieren, explosionsartig verbrennen, verbrennen, verpuffen

**deflagrating, ~ jar** Abbrennglocke *f* ~ **spoon** Abbrennlöffel *m*

**deflagration** rasches Abbrennen *n*, Verpuffung *f*

**deflasher** Abgratvorrichtung *f*

**deflate, to** ~ (tire) ablassen, ausblasen, Luft *f* ausblasen, entleeren

**deflating sleeve** Entleerungsschlauch *m*

**deflation** Windabtragung *f* ~ **sleeve** Entleerungsloch *n* ~ **time** Entleerungszeit *f*

**deflator** Entlüftungsdorn *m*

**deflect, to** ~ abbiegen, ablenken, den Lichtstrahl *m* ablenken, abweichen, ausbiegen, auslenken (g/m), ausschlagen, beugen, biegen, durchbiegen, umbiegen **to ~ backward** nach rückwärts ausschlagen **to ~ the magnetic needle** die Magnetnadel *f* ablenken

**deflect defocussing** Defokussierung *f* infolge Ablenkung *f*

**deflectability** Ablenkbarkeit *f* ~ **of electron beam** Strahlhebelarm *m*

**deflectable** durchbiegungsfähig

**deflected** abgelenkt, ausgeschlagen ~ **by nucleus** kerngestreut ~ **well** abgewichene Bohrung *f*

**deflecting, ~ from true path** Auslenkung *f* ~ **baffle** Lautsprecherleitwand *f* ~ **bar** (in transmission gear) Ablenkstange *f* ~ **cam** Abstreichdaumen *m*, Ausklinknocken *m* ~ **coil** Ablenkspule *f* ~ **couple** ablenkendes Drehmoment *n* ~ **current** Ablenkstrom *m* ~ **devices of double slip points** Zungenvorrichtung *f* für doppelte Kreuzungsweiche ~ **effect** Ablenkungseffekt *m* ~ **electrode** Ablenkungselektrode *f* ~ **field** Ablenkfeld *n* ~ **force** rücktreibende Kraft *f* ~ **inductance** Ablenkinduktivität *f* ~ **labyrinth** Lautsprecherleitfläche *f* ~ **louvers** Ableitungsöffnung *f* ~ **piece** Ablenkschuh *m* ~ **magnet** Drehspule *f*, Magnetnadel *f* ~ **magnetic field** Ablenkungsmagnetfeld *n* ~ **member** Ablenkelement *n* ~ **plate** Ablenkungs-kondensator *m*, -elektrode *f*; Ablenk-, Prall-platte *f* ~ **plates of cathode-ray tube** Plattenpaar *n* der Braunschen Röhre ~ **pulley** Umlenkscheibe *f* ~ **roller** Umlenkrolle *f* ~ **system** Ablenksystem *n* ~ **tools** Ablenk-, Ausweich-werkzeuge *pl* ~ **torque** Meßmoment *n* ~ **tube** (cathode-ray oscilloscope) Ablenk-rohr *n*, -röhre *f* ~ **vane** Lautsprecher-leitfläche *f*, -leitwand *f* ~ **voltage** Ablenkspannung *f* ~ **voltage control** Steuerung *f* der Ablenkspannungen ~ **yoke** Ablenkjoch *n*

**deflection** Abbiegung *f*, Ablenkung *f*, Abweichung *f*, Anschlag *m*, Ausschlag *m* (Zeigerinstrument *n*), Ausschlagen *n*, (spring) Biegung *f*, Biegungsausschlag *m*, (in magnetron) Bremsung *f* der Elektronen, Deviation *f*, Durchbiegen *n*, Durchbiegung *f*, Durchfederung *f*, Einbiegung *f*, Nachgeben *n*, Pfeilhöhe *f*, Ruderausschlag *m*, (angle of) Seitenrichtung *f*, Umlenkung *f*, Verschiebung *f*, Zeigerstellung *f*

**deflection, ~ of a beam** Balkenbiegung *f* ~ **of a chord** Schwingungsausschlag *m* einer Saite *f* ~ **of the control stick** Knüppelausschlag *m* ~ **from course by wind** Abdrängung *f* ~ **in both directions** Ausschlag *m* nach beiden Seiten *f pl* ~ **in either direction** doppelseitiger Zeigerausschlag *m* ~ **in one direction** einseitiger Ausschlag *m* ~ **of electrons** (in magnetron) Elektronenbremsung *f* ~ **of flap** Klappenausschlag *m* ~ **of a girder** Biegungspfeil *m* ~ **of an indicator** Ausschlag *m* eines Anzeigegerätes *n* ~ **to the left** Linksablenkung *f* ~ **of lines of force**

Kraftlinienablenkung *f* ~ **of the needle** Nadelausschlag *m* ~ **of the oscillation** Schwingungsausschlag *m* ~ **of the plumb line** Ablenkung *f*
des Lotes *n* ~ **of pointer** Zeigerausschlag *m* ~ **to
the right** Rechtsablenkung *f* ~ **of spar** Holm-
durchbiegung *f* ~ **from true path** Auslenkung *f*
~ **of the vertical** Lotschwankung *f* ~ **of the
vibration** Schwingungsausschlag *m* ~ **of the
wind** Windablenkung *f*
**deflection,** ~ **aberration** Ablenkfehler *m* ~ **amplifier** Ablenkverstärker *m* ~ **apparatus** Bildablenkgerät *n* ~ **axis** Flugzeughochachse *f* ~
~ **balance** Wegvergleich *m* ~ **board** Seitenricht-
schieber *m* ~ **chamber** (in electron microscope)
Ablenkkammer *f* ~ **circuit** Ablenkschaltung *f*
(TV) ~ **coefficient** Ablenkungskoeffizient *m* ~
**coil** Ablenkspule *f* ~ **component** Seitenablenk-
komponente *f* ~ **condenser** Ablenkplatte *f*, Ab-
lenkungskondensator *m* ~ **correction** Schieber-
verbesserung *f*, Seitenänderung *f* ~ **curve**
Biege-, Verformungs-linie *f* ~ **defocusing** Ab-
lenkdefokussierung *f*, Defokussierung *f* bei
Ablenkung ~ **device** Ablenkvorrichtung *f* ~
**diagram** Biegeplan *m* (Statik) ~ **difference**
Feuervereinigungswinkel *m* ~ **dispersion** Quer-
streuung *f* ~ **distortion** Ablenkverzeichnung *f*
~ **error** Seitenabweichung *f* ~ **factor** Ausschlag-
faktor *m*, Reziprokwert *m* der Ablenkempfindlichkeit ~ **field** Ablenkfeld *n* ~ **force** Ab-
lenkungskraft *f* ~-**free surface** verzugsfreie
Fläche *f* ~ **gyro** Hochachsenkreisel *m* ~ **hardness** Auslenkhärte *f* ~ **indicator** Ablenkstrich *m*
(rdr), Durchbiegungsmesser *m* ~ **lead** Seitenvorhalt *m*, Seitenwinkelvorhalt *m* ~ **member**
Ablenkorgan *n* ~ **method** Ausschlagmethode *f*
~ **moment** Umlenkungsmoment *m* (optical) ~
**multiplier** Ablenkungsvervielfacher *m* ~ **plate**
Ablenkplatte *f* (CRT), Ablenkungskondensa-
tor *m*, Zeitplatte *f* ~ **polarity** Ablenkpolarität *f*
~ **potentials** symmetrische Ablenkspannungen
*pl* ~ **potentiometer** Ausschlagkompensator *m* ~
**prism** Einblickprisma *n* ~ **property** Federung *f*
~ **rate** Durchbiegungsmaß *n* (Werkstoffe) ~
**reading** Festlegezahl *f* ~ **registering** Ablenk-
registrierung *f* ~ **sensitivity** (reciprocal of
deflection factor) Ablenkungsempfindlichkeit *f*
~ **setting** Klappenausschlag *m* ~ **sheave** Draht-
seilumlenkrolle *f* ~ **space** (in electron microscope) Ablenkkammer *f* ~ **tensor** Beugungssen-
sor *f* ~ **test** Durchbiegeversuch *m* ~ **test by
pressure on the hub** Nabendurchbiegeprobe *f* ~
**unit** Bildablenkgerät *n* ~ **voltage** Ablenkspannung *f* ~ **weir** Ablenkungswehr *n* ~ **yoke** Ablenkjoch *n* (TV), Ablenkspule *f*
**deflectometer** Ablenkungs-, Biegepfeil-, Bie-
gungs-, Durchbiegungs-messer *m*
**deflector** Ablenker *m*, Ablenkfläche *f*, Ablenk-
klappe *f* (Flugzeug *n*), Deflektor *m*, Leitfläche *f*,
Prallblech *n*, Prallschirm *m*, Rohrpostweiche *f*,
Sprühteller *m*, Wasserverteilungsteller *m*,
Weiche *f* ~ **of piston** Kolbenablenker *m*
**deflector,** ~ **bag** Hülsenfänger *m* ~ **bar** Abweiser *m* ~ **coil** Ablenkspule *f* ~ **field** Ablenkfeld
*n* ~ **means** Ablenkgerät *n* ~ **oil ring** Spritzscheibe *f* ~ **piston** Nasen-, Nasenhauben-kol-
ben *m* ~ **plate** Ablenkplatte *f*, Ablenkungskon-
densator *m*, Blendblech *n*, Schutzplatte *f* ~ **roll**
Umlenkrolle *f* ~ **system** Ablenkeinheit *f* ~

(wave) **trap** Abschneider *m*
**deflocculate, to** ~ entflocken
**defloculated,** ~ **graphite** entflockter Graphit
*m*
**deflocculation** Entschuppen *n*, Flockung *f* ~
**agent** Dispergierungs-, Dispersions-mittel *n*
**defluent** abfließend
**defoamer** Antischaummittel *n*
**defocus-dash mode** Defokus-Strich-Verfahren *n*
**defocused** dekonzentriert
**defocusing** Defokussierung *f*, Entbündeln *n*
**defogging** Sichtscheibenkondensschutz *m*
**defoliate, to** ~ abblatten, losblättern
**defoliation** Abblatten *n*
**deforest, to** ~ abholzen
**deform, to** ~ deformieren, entstellen, die Gestalt *f* verändern, umformen, verbiegen, ver-
formen, verunstalten, verzeichnen, verzerren,
sich werfen
**deformability** Deformierbarkeit *f* (of nuclei),
Formänderungsfähigkeit *f*, Formveränderungs-
vermögen *n*, Umformbarkeit *f*, Verformbarkeit *f*, Verformungsvermögen *n*
**deformable** verformbar
**deformation** Abgleitung *f*, Änderung *f*, Ausdehnung *f*, Deformierung *f*, Entstellung *f*, Gestaltsveränderung *f*, Formveränderung *f*,
Quetschung *f*, Umformung *f*, Verbiegung *f*,
Verbildung *f*, Verformung *f*, Verformungsfähigkeit *f*, Verwerfung *f*, Verzerrung *f* ~ **of
areas** Flächendilatation *f* ~ **(working) of contacts** Kontaktverformung *f* ~ **in extension**
Längenänderung *f* ~ **due to lag** Nachwirkungs-
deformation *f* ~ **of a surface into itself** Deformation *f* einer Fläche in sich
**deformation,** ~ **energy** Formänderungsarbeit *f* ~
**gradient** Deformationsgradient *m* ~ **point** Erweichungspunkt *m* ~ **resistance** Formände-
rungswiderstand *m*
**deformed** mißgestaltet, verformt, verwachsen,
windschief ~ **bar** Spezialeisen *n* ~ **bars** Rippen-
eisen *pl* ~ **nuclei** abgeplattete Kerne *pl* ~ **surface** deformierte Fläche *f* ~ **wave** verzerrte
Welle *f*
**deforming** Umformung *f* ~ **strain** deformierende
oder verformende Spannung *f*
**deformity** Mißbildung *f*
**defray, to** ~ (Kosten) bestreiten
**defrost, to** ~ (refrigerators) abtauen
**defrost control** Abtausteuerung *f*
**defroster** Entfroster *m* ~ **blower** Entfrosterge-
bläse *n* ~ **hose** Entfrosterschlauch *m* ~ **nozzle**
Entfrosterdüse *f*
**defrosting** Enteisung *f* ~ **device for trolley wires**
Abtaueinrichtung *f* ~ **pump** Frostschutzpumpe
*f* ~ **shields** Frostschutzscheiben *pl*
**defruiter** Gerät *n* zur Unterdrückung nicht-
synchroner Antworten (sec. rdr.)
**defuel, to** ~ enttanken
**defuelling facility** Enttankeinrichtung *f*
**defuse, to** ~ entschärfen
**degarbling** Entwirrung *f* (sec. rdr.)
**degas, to** ~ entgasen, entgiften
**degasification** Entgasung *f*
**degasifying plant** Entgasungsanlage *f*
**degasing** Entgasungsmittel *n*, Gasaustreibung *f*
**degassed solution** entgaste Lösung *f*
**degasser** Entgasungsanlage *f*

**degassing** Entgiftung $f$ **~ agent** Entgiftungsmittel **~ apparatus** Entgiftungsgerät $n$ **~ service** Entgiftungsdienst $m$

**degate, to ~** den Anguß $m$ entfernen

**degauss, to ~** degaussieren, entmagnetisieren

**degaussing cable** Entmagnetisierungskabel $n$

**degeneracy** Entartung $f$

**degenerate, to ~** abarten, ausarten, entarten

**degenerate** entartet **~ electron gas** entartetes Elektronengas $n$ **~ electron state** entarteter Elektronenzustand $m$ **~ form** Kummerform $f$ **~ levels** entartete Terme $pl$

**degeneration** Degenerierung $f$, Entartung $f$, Gegenkopplung $f$

**degenerative, ~ circuit** Schaltung $f$ mit Gegenkopplung **~ (or negative) feedback** Gegenkopplung $f$, negative Rückführung $f$

**degerminator** Entkeimungsapparat $m$

**degradation** Abbau $m$, Abtragung $f$, Degradation $f$, Energieverlust $m$, Herabsetzung $f$, Rangverlust $m$, stufenförmige Verwitterung $f$ **~ of bands** Bandabschattierung $f$

**degrade, to ~** herabsetzen

**degras** Degras $n$, Wollfett $n$

**degrease, to ~** abfetten, abschäumen, degraissieren, entfetten

**degreasing** Entfetten $n$, Entfettung $f$ **~ agent** Entfettungsmittel $n$ **~ composition** Entfettungsmasse $f$ **~ unit** Entfettungsanlage $f$

**degree** Abstufung $f$, Grad $n$, Maß $n$, Ordnung $f$, Stufe $f$, Titel $m$ **nth ~** nte Ordnung $f$

**degree, ~ of absorption** Aufzehrgrad $m$ **~ of accuracy** Genauigkeitsgrad $m$ **~ of achromatic correction** Achromasie $f$ **~ of acidity** Säuregrad $m$ **~ of adjustment** Verstellwinkel $m$ **~ of admission** Füllgrad $m$ **~ of amplification** Verstärkungsgrad $m$ **~ of angle** Winkelgrad $m$ **~ of angular measure** Kreisgrad $m$ **~ of approach** Annäherungsgrad $m$ **~ of an arc** Kreisgrad $m$ **~ of attenuation** Regelfaktor $m$ **~ of bank** Querneigungsgrad $m$ **~ of beating** Mahlungsgrad $m$ **~ of bending** Durchbiegungsgrad $m$ **~ of a circle** Kreisgrad $m$ **~ of cleaning** Reinigungsgrad $m$ **~ of cleanliness** Sauberkeitsgrad $m$ **~ of climb** Steigungsgrad $m$ **~ of cloudiness** Bewölkungs-grad $m$, -größe $f$ **~ of coloring** Farbtönung $f$ **~ of compaction** Lagerungsdichte $f$ **~ of compression** Verdichtungs-grad $m$, -verhältnis $n$ **~ of concentration** Sättigungs-, Stärke-grad $m$ **~ of condensation** Verdichtungsgrad $m$ **~ of consolidation** Verfestigungsgrad $m$ **~ of consumption** Aufzehrgrad $m$ **~ of contraction** Schwindungsgrad $m$ **~ of contrast regulation** Regelgrad $m$ **~ of coverage** Bedeckungsgrad $m$ **~ of damping** Dämpfungsgrad $m$ **~ of deformation** Verformungsvermögen $n$ **~ of density** Dichtegrad $m$ **~ of dilution** Verdünnungsgrad $m$ **~ of disintegration** Auflösungsgrad $m$ **~ of dissociation** Dissoziationsgrad $m$ **~ of distinctness** Schärfe $f$ **~ of distribution** Verteilungsgrad $m$ **~ of division** Zerteilungsgrad $m$ **~ of draft** Verzugsmöglichkeit $f$ **~ of dryness** Trocknungsgrad $m$ **~ of dynamic-range regulation** Regelgrad $m$ **~ of elasticity** Eleastizitätsgrad $m$ **~ of error** Genauigkeitsmaß $n$ **~ of fermentation** Vergärungsgrad $m$ **~ of filling of bands** Besetzungsgrad $m$ der Banden $pl$, Bandenbesetzungsgrad $m$ **~ of fineness** Feinheits-

grad $m$ **~ of firmness** Festigkeitsgrad $m$ **~ of fluidity viscosity** zerstäubte Flüssigkeit $f$ **~ of formation** Bildungsgrad $m$ **~ of freedom** Freiheitsgrad $m$ **~ of freeness** (pulp) Schopperriegler $m$ **~ of hardness** Härtegrad $m$ **~ of heat** Hitze-, Wärme-grad $m$ **~ of immersion** Tauchtiefe $f$ **~ of importance** Dringlichkeitsstufe $f$ **~ of inflation** Blähungsgrad $m$ **~ of inhibition** Blähungsgrad $m$ **~ of irregularity** Ungleichförmigkeitsgrad $m$ **~ of latitude** Breitengrad $m$ **~ of longitude** Längengrad $m$ **~ of modulation** Aussteuerungskoeffizient $m$, Beeinflussungsfaktor $m$, Modulations-grad $m$, -ziffer $f$ **~ of modulation factor** Aussteuerungsgrad $m$ **~ of necessity** Dringlichkeitsstufe $f$ **~ of nitration** Nitrierungsgrad $m$ **~ of order** Ordnungsgrad $m$ **~ of overcompaction** Überdichtungsgrad $m$ **~ of oxidation** Oxydationsstufe $f$ **~ of permeability** Durchlässigkeitsgrad $m$ **~ of porosity** Porositäts-, Undichtigkeits-grad $m$ **~ of priority** Bewertungsklasse $f$ **~ of purity** Reinheitsgrad $m$ **~ of radio noise suppression** Entstörgrad $m$ **~ of double refraction** Stärke $f$ der Doppelbrechung **~ of remolding** Grad $m$ der Störung **~ of reset** Rückstellgrad $m$ **~ of saturation** Sättigungsgrad $m$ **~ of sensitivity** Empfindlichkeitsgrad $m$ **~ of settling** Sackmaß $n$ **~ of shrinkage** Schwindmaß $n$, Schwindungsgrad $m$ **~ of slipperiness** Schlüpfrigkeitsgrad $m$ **~ of stability** Stabilitäts-grad $m$, -maß $n$, -wert $m$ **~ of strength** Stärkegrad $m$ **~ of suitability for hardening** Härteeignungseigenschaft $f$ **~ of supercharging** Überverdichtungsgrad $m$ **~ of tension** Spannungsgrad $m$ **~ of tuning** Abstimmungsgrad $n$ **~ of turbidity** Trübungsgrad $m$ **~ of turbulence** Verwirbelung $f$ **~ of turn** Drehungsgrad $m$ **~ of uniformity** Gleichförmigkeitsgrad $m$ **~ of vacuum** Güte $f$ des Vakuums $n$ **~ of value** Bewertungsklasse $f$ **~ of viscosity** Viskositätsstufe $f$ **~ of wear** Verschleißmaß $n$

**degree, ~ Baumé** Baumégrad $m$ **~ Brix** Bricgrad $m$ **~ centigrade** Celsiusgrad $m$ **~ Engler** Englergrad $m$ (EG) **~ Kelvin** absolute Temperatur $f$ **~ sign** Gradzeichen $n$

**degrees, by ~** stufenweise **to (or in) some ~** einigermaßen **~ from cross-fuse setting** Grad vom Kreuz **180 ~ out of phase** gegenfasig

**degum, to ~** degummieren (Seide)

**dehumidification** Luftentfeuchtung $f$, Trocknung $f$ (Fallschirm $m$)

**dehumidifier** Luftentfeuchtungsanlage $f$, Trockenvorlage $f$

**dehumidify, to ~** entfeuchten

**dehydrate, to ~** dehydratisieren, dehydrieren, enthydratisieren, entwässern, kristallwasserfrei machen, Wasser $n$ entziehen

**dehydrated** entwässert, wasserfrei **~ coal-tar** wasserfreier (Stein-) Kohlenteer $m$

**dehydrating, ~ agent** Wasserentziehungsmittel $n$ **~ apparatus** Entwässerungsapparat $m$ **~ breather** Luftentfeuchter $m$ **~ chamber** Entfeuchterkammer $f$ **~ columns** Entwässerungskolonnen $pl$

**dehydration** Dehydratisierung $f$, Dehydrierung $f$, Entwässerung $f$, Wasserentziehung $f$ **~ of hydrocarbons by H and D atoms** Dehydrierung $f$ von Kohlenwasserstoffen durch H- und D-Atome **~ of oil** Öltrocknen $n$ **~ (or squeezing) effect** Entfeuchtung $f$

**dehydrator** Trockenapparat *m*, Verdampfer *m*
**dehydro-cholesterol** Dehydrocholesterin *n*
**dehydrodinaphthylenediamine** Dinylin *n*
**dehydrogenate, to** ~ dehydrieren
**dehydrogenation** Dehydrierung *f*
**dehydrogenizing catalysis** Dehydrogenisierungs-
katalyse *f*
**de-ice, to** ~ enteisen
**de-icer** Eisschutz *m*, Enteiser *m*, Enteisungsgerät
*n*, Enteisungsvorrichtung *f* ~ **boots** Gummi-
schlauchenteiser *pl*
**de-icing** Enteisung *f* ~ **of aircraft wings** Flügel-
enteisung *f* ~ **of carburetor** Vergaserenteisung *f*
**de-icing,** ~ **device for trolley wires** Abtaueinrichtung *f* ~ **equipment** Enteisungsanlage *f* ~ **fluid**
Enteisungsflüssigkeit *f* ~ **pump** Enteiserpumpe
*f* ~ **system** Enteisungsanlage *f*
**de-inking solution** Entfärbelösung *f*
**de-ionization** Entionisierung *f* ~ **impulse** Lösch-
impuls *m* ~ **rate** Entionisierungsgeschwindig-
keit *f* ~ **time** Entionisierungszeit *f*
**de-ionize, to** ~ entionisieren
**deject, to** ~ herabstimmen
**delaminate, to** ~ ableimen, entleimen
**delay, to** ~ aufschieben, hinhalten, hinziehen,
säumen, verlegen, verschleppen, verzögern,
zögern
**delay** Anstand *m*, Aufschub *m*, Betriebslaufzeit
*f*, Frist *f*, Fristung *f*, Stillstandsfrist *f*, Unterbrechung *f*, Verzögerung *f*, Verzug *m*, Wartezeit *f*, Zaudern *n*, Zeitverlust *m* ~ **in delivery**
Lieferungsverzögerung *f* ~ **of obligation** Lei-
stungsverzug *m* ~ **in performance** Leistungs-
verzug *m* ~ **in production** Betriebsunterbrechung
*f* ~ **in starting** Anzugsverzögerung *f* ~ **of train**
Verspätung *f* eines Zuges *m*
**delay,** ~ **acceleration** Nachbeschleunigung *f* ~
**action** Verzögerung *f* ~**-action cap** Zeitzünder
*m* ~**-action control relay** Steuerverzögerungsrelais *n* ~**-action fuse** Exerzierkopfzünder *m*,
Zünder *m* mit Verzögerung ~ **(ed)-action fuse**
Verzögerungszünder *m* ~**-action stopping relay**
Abstellverzögerungsrelais *n* ~ **arming** verzögertes Schärfen *n* ~ **bias** Verzögerungsspannung *f*
~ **basis operation** Betrieb *m* mit Vorbereitung
(Wartezeit) ~ **cable** Verzögerungsleitung *f* ~
**circuit** Verzögerungsschaltung *f* ~ **composition**
(explosives) Verzögerungssatz *m* ~ **corrector**
Vorhaltrechner *m* ~ **demodulator** Verzögerungs-
demodulator *m* ~ **distortion** Laufzeitstörung *f*,
Phasenverzerrung *f* ~ **element** Verzögerungs-
element *n*, -glied *n* ~ **equalizer** Phasenentzerrer
*m* ~ **equalizing** Laufzeitkorrektur *f*, Phasenentzerrung *f* ~ **equalizing cone** Laufzeitausgleich-
kegel *m* ~ **frequency distortion** Laufzeit/Fre-
quenzverzerrung *f* ~ **fuse** Aufschlagzünder *m*
mit Verzögerung, Verzugszünder *m* ~ **igniter**
Zeitzünder *m* ~ **line** Laufzeit-glied *n*, -kette *f*;
Verzögerungsleitung *f*, Zeitlinie *f* ~**-line**
**register** Verzögerungsleitungsregister *n* ~**-line**
**storage** Speicher *m* mit einer Verzögerungsleitung ~**-lock discriminator** Laufzeitfolger *m*
(rdr) ~ **network** (lump constant) Laufzeit-,
Verzögerungs-kette *f* ~ **pellet** Verzögerungssatz *m* ~ **period** Totzeit *f* ~ **switch** Verzögerungs-
schalter *m* ~**-time** Verzugszeit *f* ~**-time action**
Auslösezeit *f* ~**-time characteristic** Laufzeit-
charakteristik *f* ~**-time distortion** Laufzeitver-

zerrung *f* ~**-time variation** Laufzeitverlauf *m*
~ **timer** Zeitverzögerungsteil *m* ~**-transmission**
**time** Betriebslaufzeit *f* ~ **voltage** Verzögerungsspannung *f* ~**-weighting term** Laufzeitkorrektur *f*, Phasenentzerrung *f* ~ **working** Betrieb *m*
mit Vorbereitung
**delayed** befristet, verspätet, verzögert ~ **action**
Arbeiten *n* mit Verzögerung
**delayed-action,** ~ **alarm** verzögerte Zeichengebung *f* ~ **bomb** Knallkoffer *m* ~ **fuse** Innen-,
Kopf-zünder *m* ~ **release** Verlaufwerk *n* ~
**switch** Schleppkupplung *f* ~ **timing** Zeitlupen-
tempo *n* ~ **voltage** Verzögerungsspannung *f*
**delayed,** ~ **alpha particles** verzögerte Alphateilchen *pl* ~ **arming** verzögertes Schärfen *n* ~
**assessment** Nachverbrennung *f* ~**-automatic**
**gain control** verzögerte Regelung *f* ~ **automatic**
**volume control** verzögerte automatische Lautstärkeregelung *f*, verzögerter Schwundausgleich *m* ~**-blanking signal** verzögertes Austastsignal *n* ~ **call** Vormerkgespräch *n* ~**-coinci-**
**dence counting** Zählung *f* bei verzögerter Koinzidenz *f* ~ **critical** verzögert-kritisch ~ **fading**
**compensation** verzögerter Schwundausgleich *m*
~ **feed (cards)** verzögerte Kartenzuführung *f*
~**-fission neutrons** verzögerte Neutronen *pl* ~
**jump** Absprung *m* mit verzögerter Öffnung ~
**neutron emitter** Emitter *m* von verzögerten Neutronen ~ **opening** verzögertes Öffnen *n* ~**-pulse**
**interval** Verlustzeit *f* ~**-pulse tripping cam**
Leerlaufnocke *f* ~**-release** abfallverzögert ~
**scanning** verzögerte Abtastung *f* ~ **stopping**
**control** Anhalteverzögerung *f* ~ **sweep** verzögerte Ablenkung *f* (rdr), verzögerte Abtastung *f* oder Zeitablenkung *f* ~**-time electronic**
**flash** Langzeitelektronenblitz *m* ~ **time fuse**
Langzeitzündung *f* ~ **train** verspäteter Zug *m*
**delayer** Verzögerungsmittel *n*
**delaying** hinhaltend ~ **action** hinhaltender
Widerstand *m* ~ **position** Aufnahme *f*, Auf-
nahme-, Widerstands-stellung *f*
**Delco ignition system** Delco-Zündung *f*
**delcredere** Delcredere *n*
**deleaded solution** entbleite Lösung *f*
**delegate, to** ~ übertragen, zuteilen
**delegate** Abgeordneter *m*, Beauftragter *m*
**delete, to** ~ auslöschen, tilgen
**deletion** Streichung *f*, Tilgung *f*
**delftware** Delfter Porzellan *n*
**deliberate, to** ~ überlegen, verhandeln
**deliberate** absichtlich, bedacht, vorsätzlich ~
**fire** ruhiges Feuer *n* ~ **transmission of false**
**radio messages** Funktäuschung *f*
**deliberately** mit Vorsatz, wissentlich
**deliberation** Beratung *f*, Überlegung *f*
**delicacy** Empfindlichkeit *f* ~ **of regulatory device**
Empfindlichkeit *f* der Regulierung
**delicate** bedenklich, empfindlich, fein, heikel,
heikelig, kritisch ~ **adjustment** kritische Einstellung *f* ~ **balance** Hebelwaage *f* ~ **casting**
verwickelt gestaltetes Gußstück *f* ~ **focusing**
**(or setting)** kritische Einstellung *f* ~ **gradation**
Zartheit *f* (der Abtönung)
**delicately,** ~ **sensitive** feinfühlig ~ **stepped** feinstufig ~ **stepped type** feinstufige Ausführung *f* ~
**delime, to** ~ entkalken
**deliming agent** Entkalkungsmittel *n*
**delimit, to** ~ abgrenzen

**delimitation** Abgrenzung *f*, Begrenzung *f*, Grenzscheide *f*, Markierung *f*, Vermarkung *f*
**delimited** zweckgebunden
**delimiter** Begrenzer *m* (info proc.), Begrenzungssymbol *n* (info proc.)
**delineate, to** ~ abgrenzen, durch Linien festlegen, schreiben, skizzieren, zeichnen
**delineation** Zeichnung *f* ~ **of an objective** Objektkennung *f* ~ **on screen** Beschreibung *f* von Bildern auf dem Schirm
**deliquescent** zerfließend, zerschmelzend
**deliquesque, to** ~ schmelzen, zerfließen
**deliquesquence** Schmelz-flüssigkeit *f*, -produkt *n*, Zerfließung *f* (chem.)
**deliver, to** ~ abgeben, ablenden, abliefern, aufgeben, austragen, befördern, beliefern, einliefern, erlösen, liefern, zuführen, zustellen **to** ~ **at the surface** ausladen **to** ~ **up** ausliefern **to** ~ **current** den Strom abgeben **to** ~ **current to** mit Strom beliefern **to** ~ **(telegrams)** bestellen **to** ~ **a thing to a person** jemandem etwas zustellen **to** ~ **subsequently** nachliefern
**deliverable** bestellbar (mail)
**delivered** gewonnen (Schub)
**delivering,** ~ **bowl** Abzugswalze *f* ~ **liveroller table** Abfuhrrollgang *m* ~ **office** Bestellanstalt *f* ~ **point** Lieferstelle *f* ~ **roller** Zugwalze *f*
**delivery** Abgabe *f*, Ablage *f* (print.), Ablaß *m*, Ablieferung *f*, Anlieferung *f*, Aufgabe *f*, Ausgabe *f*, Ausguß *m*, Auslaß *m*, Ausleger *m* (sheet), Aushändigung *f*, Auflieferung *f*, Auslieferung *f*, Beförderung *f*, Bestellung *f*, Fördermenge *f*, Förderung *f*, Lieferung *f*, Übergabe *f*, Zuführung *f*, Zustellung *f*
**delivery,** ~ **to addressee** Zustellung *f* ~ **of current** Stromabgabe *f* ~ **of force** Kraftabgabe *f* ~ **to foreign markets** Auslandslieferung *f* ~ **(or supply) of gas** Gasversorgung *f* ~ **by installments** Sukzessivlieferung *f* ~ **of messages** Telegrammbestellung *f* ~ **of plant in final working order** Übergabe *f* (einer Anlage) ~ **of a pump** Förderleistung *f* einer Pumpe ~ **on sale** Feilhalten *n* ~ **of telegrams** Telegrammzustellung *f*
**delivery,** ~ **air** Förderluft *f* ~ **apparatus** Auslegeapparat *m* ~ **area** Liefergebiet *n* ~ **belt** Zubringerband *n* ~ **belt conveyer** Zubringerförderband *n* ~ **bill** Lieferschein *m* ~ **board (or table)** Ablegetisch *m* ~ **box** Ablegekasten *m* ~ **car** Förderwagen *m* ~ **charges** Lieferungskosten *pl* ~ **chute** Aufgaberutsche *f*, Ausschüttrinne *f* ~ **cock** Ablaßhahn *m* ~ **cylinder** Ausführzylinder *m* (print.) ~ **date** Beförderungsfrist *f*, Liefertermin *m* ~ **device** Ausschleusungsvorrichtung *f* (Förderanl.) ~ **disc** Arbeitshubscheibe *f* (Schmierpumpe) ~ **efficiency of blower** Förderleistung *f* des Gebläses ~ **end** Auslauf-ende *n*, -seite *f* Austragsende *n*, Austrittseite *f* ~ **expense** Bestellgebühr *f* ~ **fan** Ausstoßventilator *m* ~ **field** Auskoppelfeld *n* ~ **form** Lieferschein *m* ~ **gas** Förderungsgas *n* ~ **gate** Auslauf *m* (aus dem Bunker), Rechenausleger *m* ~ **gripper** Auswerfgreifer *m* ~ **group** Adressatenkurzbezeichnung *f* ~ **guide** Ausführung *f* ~ **guides** Auslaßführungen ~ **head** Drucksäule *f*, Förderhöhe *f* ~ **hood** Auswurfhaube *f* ~ **hose** Ablauf-, Druck-, Zapfschlauch *m* ~ **indicator** Lieferanzeiger *m* ~ **lattice** Ausgangslattentisch *m* ~ **mechanism**

Zapfer *m* ~ **meter** Durchlaufmesser *m* ~ **order** Bezugsschein *m* ~ **output** Druckleistung *f* ~ **period** Bestellzeit *f* ~ **pipe** Ableitung *f*, Abzugs-rohr *n*, -röhre *f*; Ausguß-röhre *f*, -rohr *n*; Auslaßrohr *n*, Ausströmungsrohr *n*, Druckleitung *f*, Druckrohr *n*, Entleerungsleitung *f*, Rücklaufrohr *n*, Zuleitung *f* ~ **pipe connection** Druckrohranschluß *m* ~ **plate** Zuführungsteller *m* ~ **platform** Abzugshängebank *f* ~ **plug** Abschlußzapfen *m* ~ **point of a conveyor** Auswurf *m* ~ **port** Entlade-, Liefer-hafen *m* ~ **pressure** Enddruck *m* (Luftkompressor) ~ **rate** Liefergrad *m* ~ **reel** Aufrollungswalze *f* ~ **requirement** Lieferungsbedingung *f* ~ **roadway** Ladestraße *f* ~ **roll** Ausgangswalze *f*, Eingangszylinder *m* ~ **roller** Ablaufwalze *f*, Ablieferungswalze *f* ~ **roller conveyor** Ablaufrollgang *m* ~ **roller drive** Lieferwerkantrieb *m* ~ **screw** Ablaßschraube *f* ~ **section** Ausgang *m* ~ **shaft** Auslegerechenwelle *f* ~ **side** Auslauf-, Austritt-seite *f*, Schutzblech *n* zur Ablage ~ **side frame** Auslegerseitenteil *m* ~ **side of a pump** Druckseite *f* der Pumpe, Förderseite *f* ~ **side of rolls** Walzenaustritt *m* ~ **sheet** (of hopper) Abfuhrplatte *f* ~ **space** Auslaufraum *m*, Ausströmraum *m*, Diffusor *m* ~ **specification** Lieferungsbedingung *f*, Liefervorschrift *f* ~ **speed** Austrittsgeschwindigkeit *f* ~ **spool** (wire-winding machines) Vorrats-rolle *f*, -spule *f* ~ **station** Abnahmestelle *f* ~ **support bracket** Auslegerdrehwand *f* ~ **system** Lieferwerk *n* ~ **table** Auslegetisch *m* ~ **tape** Ausführband *n* ~ **tape-guide** Ausgangsbandleitung *f* (print.) ~ **thickness** Austrittsdicke *f* ~ **truck** Geschäftswagen *m*, Lieferwagen *m* ~ **tube** Abgangsrohr *n*, Entbindungsrohr *n* ~ **unit** Fallschirmlast *f* ~ **(discharge) unit** Abgabevorrichtung *f* ~ **valve** Abflußventil *n*, Aufgabeschieber *m*, Ausguß-, Auslaß-, Druck-ventil *n* ~ **valve chamber** Auslaßgehäuse *n* ~**-valve spring** Druckventilfeder *f* ~ **wagon** Förder-, Liefer-wagen *m* ~ **zone** Abnahmezone *f*
**deload, to** ~ entspulen
**delorenzite** Delorenzit *m*
**delta** Deltasymbol *n* ~ **circuit** Dreieck-glied *n*, -schaltung *f* ~**-connected** in Dreieckschaltung *f* ~ **connection** Deltaschaltung *f*, Dreieckschaltung *f* ~ **function derivative** Ableitung *f* der Deltafunktion ~**-matched impedance antenna** symmetrische Anzapfantenne ~**-matching** Deltaanpassung *f* ~**-matching transformer** Deltaanpassungsübertrager *m* ~ **metal** Deltametall *n* ~ **network** Dreieckschaltung *f* ~ **noise** Deltarauschen *f* ~**-star switch** Dreiecksternschalter *m* ~ **voltage** Dreieckspannung *f* ~ **wing** Delta-, Dreieck-flügel *m*
**deluge** Flut *f* ~ **system** Überflutungsanlage *f* ~ **valve** Regenwandventil *n*
**delusion** Irreführung *f*, Täuschung *f*
**delusions,** ~ **of interpretation** Beziehungswahn *m* ~ **of reference** Beziehungswahn *m*
**deluster, to** ~ mattmachen, mattieren
**delusterant** Mattierungsmittel *n*
**delustered** mattgeschliffen
**delustering agent** Mattierungsmittel *n*
**deluxe model** Luxusausführung *f*
**demagnetization** Entmagnetisieren *n*, Entmagnetisierung *f* ~ **cable** Entmagnetisierungskabel *n*

demagnetize, to ~ entmagnetisieren
demagnetized sample entmagnetisierter Probe-
körper *m*
demagnetizer Entmagnetisierer *m*
demagnetizing entmagnetisierend ~ factor Ent-
magnetisierungsfaktor ~ turns gegenmagneti-
sierende Windungen *pl*
demand, to ~ anfordern, beanspruchen, beantragen, einfordern, erfordern, erfragen, fordern, requirieren, verlangen
demand Anforderung *f*, Anspruch *m*, Aufforderung *f*, (upon) Beanspruchung *f*, Bedarf *m*,
Ersuchen *n*, Erfordernis *n*, Forderung *f*, Kauf-
lust *f*, Nachfrage *f*, Zumutung *f*
demand, in ~ gesucht to be in great ~ großen
Zulauf haben ~ for concession without previous
discovery blinde Mutung *f* to ~ voting trust
control Geschäftsaufsicht beantragen.
demand, ~ attachment Maximumanzeiger *m* ~
flow Abfluß *m* ~ indicator Rufanzeiger *m*
(electr.) ~ meter Höchstverbrauchszähler *m* ~
regulator Bedarfsregler *m* ~ totalizing relay
Bedarfssammelrelais *n*
demanded verlangt
demanding no maintenance wartungsfrei
demands Inanspruchnahme *f* (load) ~ of reality
Realforderung *f*
demarcation Abgrenzung *f*, Abmarkung *f*,
Grenzberichtigung *f*, -bestimmung *f*, (line of)
Grenz-linie *f*, -regelung *f*, -scheidung *f*
demarcate, to ~ abgrenzen
dematerialization Dematerialisation *f*
demeanor Auftreten *n*, Benehmen *n*, Betragen *n*
demerit Nachteil *m* / rating negative Bewertung *f*
demeshing Außereingriffkommen *n* (von Zahnrädern) ~ gears (Zahnräder) Ausrücken *n*
demijohn Ballon *m*, Glasballon *m*, Korbflasche *f*,
Schwefelsäureballon *m*
demineralization Entmineralisierung *f*
demiss, to ~ chartern ohne Besatzung
demist, to ~ entfeuchten
demodulate, to ~ demodulieren, entmodeln,
gleichrichtern
demodulation Demodulation *f*, Empfangsgleichrichtung *f*, Entmodelung *f*, Entmodulierung *f*,
Gleichrichtung *f* ~ of speech Sprachwiedergabe *f* ~ effect Modulationsunterdrückung *f*
demodulator Demodulator *m*, Detektor *m*, Ent-
modeler *m*, Gleichrichter *m*
demolish, to ~ abbauen, einreißen, niederreißen
umstürzen, zerstören
demolishing Abbau *m* ~ point Abbauspitze *f*
(Steinbrecher) ~ work Abbrucharbeit *f*
demolition Abbruch *m*, Demolierung *f*, Verheerung *f* ~ of a building Abtragen *n* eines Bau-
werkes *n* ~ charge geballte Ladung *f*, Zerstör-
ladung *f* ~ equipment Zündgerät *n* ~ party (or
squad) Räumungstrupp *m* ~ vehicle Spreng-
ladungsträger *m*
demonstrate, to ~ beweisen, demonstrieren,
nachweisen, (anschaulich) vorführen
demonstration Beweisführung *f*, Darlegung *f*,
Demonstration *f*, Erweis *m*, Vorführung *f*,
Vorstellung *f* ~ call Propagandagespräch *n*,
Werbegespräch *n* ~ car Vorführwagen *m*
~ chart Unterrichtstafel *f* ~ experiment Schauversuch *m* ~ flight Vorführungsflug *m* ~
material Anschauungsmittel *n* ~ stand for

blinkers Blinkaufsteller *m*
demount, to ~ abbauen, abbrechen, ausbauen
demountable abnehmbar ~ magazine Steckspule
*f* ~ reel Steckspule *f* ~ rim abnehmbare Felge *f*
~ tube (or valve) zerlegbare Röhre *f*, Senderöhre *f* an laufender Pumpe
demounting Abbau *m*, Demontage *f*
demulsibility test Entemulgierversuch *m*
demulsify, to ~ entemulgieren
demulsifying Emulsionsbrechen *n*
demurrage Liegegeld *n*, Verzugskosten *pl*
demurrer Einspruch *m*, Einwand *m*, Einwendung
*f*, Rechtseinwand *m*
den Grundwerkkasten *m* (paper mfg.)
denaturant Denaturierungsmittel *n*, Vergällungs-
mittel *n*
denaturation Denaturierung *f*
denature, to ~ denaturieren, ungenießbar machen, vergällen
denatured, ~ alcohol Vergällungsholzgeist *m* ~
sugar vergällter Zucker *m* ~ with toluene
toluolvergällt
denaturize, to ~ denaturieren
denaturizing agent Denaturierungsmittel *n*
dendriform baumähnlich ~ exposure of film
Filmverblitzung *f*
dendrite Baumstein *m*, Dendrit *m*, Tannenbaum-
kristall *m*
dendritic baumförmig, dendritisch ~ crystal
Dendrit *m* ~ structure Primärgefüge *n*
dendrometer Baummesser *m*
denial Aberkennung *f*, Verneinung *f*, Verweigerung *f* ~ for nonpayment Sperre *f* eines Anschlusses (teleph) ~ under oath Abschwörung *f*
denier Denier *m* (0,05 gr, Gewichtseinheit *f*),
Einzeltiter *m*, Titer *m* ~ variation Titer-ab-
weichung *f*, -welle *f*
denim Drillich *m*
denitrating plant Denitrieranlage *f*
denitration Denitrierung *f*
denomination Benennung *f*
denominator Nenner *m*
denotation Andeutung *f*, Bezeichnung *f*, Entsprechung *f*
denote, to ~ bedeuten, bezeichnen
denounce, to ~ anzeigen
dense blasenfrei, derb, dicht, kompakt ~ lignite
gemeine Braunkohle *f* ~ liquid Schwerflüssig-
keit *f* ~ medium Schwertrübe *f* ~-medium
material Schwerstoff *m* ~-medium screen Trübe-
sieb *n* ~ negative dichtes Negativ *n* ~-neutral
(tint) glass neutralschwarzes Farbglas *n*
densener Kokilleneinlage *f*, Kühleisen *n*
densified impregnated (laminated) wood Preß-
schichtholz *n*
densifier Trübeeindrücker *m*
densify, to ~ dichten
densimeter Aräometer *n*, Densimeter *n*, Dichtemesser *m*, Dichtigkeitsmesser *m*, Luftdichte-
messer *m*, Senkwaage *f*
densimetric method densimetrische Methode *f*
densitometer Densitometer *n*, Schwärzungsmesser *m*
densitometry fotografische Dichtemessung *f*
density (of electrons in beam) Belegung *f*, Deckkraft *f* (photo), Densität *f*, Dichte *f*, Dichtheit *f*,
Dichtigkeit *f*, Dicke *f*, Feste *f*, Festigkeit *f*,
spezifisches Gewicht *n*, Raumgewicht *n*,

Schwärzung *f* (photo), Schwärzungsdichte *f* (film), Wichte *f*

**density,** ~ **of the air** Luftdichte *f* ~ **of bursts** Sprengpunktdichte *f* ~ **of charge** Ladungsdichte *f* ~ **of the cloth** Warendichte *f* ~ **of coverage** Verschleierungsfähigkeit *f* ~ **of electron current** Elektronenstromdichte *f* ~ **of electrons** Elektronenbelegung *f* ~ **of energy** Energiedichte *f* ~ **of the total electromagnetic energy** räumliche Energiedichte *f* ~ **of exhaust smoke** Abgastrübung *f* ~ **of fire** Feuergewicht *n* ~ **of flux changes** Flußwechseldichte *f* (tape rec.) ~ **of a gas** Gasdichte *f* ~ **of gas in the air** Luftanreicherung *f* ~ **of gas obstacle** Belegungsstärke *f* ~ **of ionozation** Ionenkonzentration *f* ~ **of lines of force** Kraftliniendichte *f* ~ **of loading** Ladedichte *f* ~ **of a packing of circles** Dichte *f* einer Kreislagerung ~ **of separation** Trennungsdichte *f* ~ **of snow** Schneedichte *f* ~ **of specimen** Probendichte *f* ~ **by surface** Flächendichte *f* (electr.) ~ **of traffic** Verkehrsdichte *f*

**density,** ~ **altitude** Cina-Atmosphärenhöhe *f*, Luftdichtenhöhe *f* ~ **anisotropy** Dichteanisotropie *f* ~ **control grid** Dichtereguliergitter *n* ~ **detector gauge** Dichtefühler *m* ~ **determination** Dichtebestimmung *f* ~ **distribution** Dichtefeld *n* ~ **field** Dichtefeld *n* ~ **fluctuations** Dichteschwankungen *pl* ~ **graduation** Schwärzungsabstufung *f* ~ **grid** Dichtereguliergitter *n* ~ **height** Höhe *f* entsprechend der Normalatmosphäre ~ **matrix** Dichtematrix *m* ~ **meter** Densitometer *n* ~ **modulation** Dichtemodulation *f* ~ **operator** Dichteoperator *m* ~ **range** Schwärzungsbereich *n*, ~ **recorder** Wichtewage *f* ~ **reference scale** Schwärzungsskala *f* ~ **regulating grid** Dichtegitter *n* ~ **regulator** Wichteregler *m* ~ **regulator grid** Dichtereguliergitter *n* ~ **scale** Dichteskala *f*, Schwärzungsumfang *m* ~ **tracks** Intensitätsschrift *f* ~ **transition curves** Dichteübergangskurven *pl* ~ **unit** Luftdichtegerät *n* ~ **value** Tönung *f* ~ **values** Helligkeitswerte *pl* ~ **variation** Dichteunterschied *m*

**densometer** Luftdurchlässigkeitsprüfer *m*

**dent, to** ~ einbeulen, einkeilen, verbeulen, zahnen

**dent** Eindruck *m* (durch Schlag oder Stoß erzeugt), Vorsprung *m* ~ **depression** Delle *f*

**dentate, to** ~ auszacken

**dentated line** gezackte, ausgezackte oder ausgezahnte Linie *f*

**dented** ausgezackt, eingebeult ~ **chisel** Gradierbeitelchen *n* ~ **line** (aus)gehackte oder gezackte Linie *f* ~ **spring leaf** Federblatt *n* mit Längsnut und Rippe

**denticulate, to** ~ zähneln; gezahnt

**denticulation** Verzahnung *f*

**denting** Einkerbung *f* ~ **stress** Beulspannung *f*

**denture** Gebiß *n* ~ **spring** Gebißfeder *f*

**denudation** Abtragung *f*

**denude, to** ~ entblößen

**denuded area** Kahlschlagfläche *f*

**denumerable** abzählbar

**denunciation** Anzeige *f*, (of a treaty) Kündigung *f*

**deny, to** ~ ableugnen, absprechen, verweigern

**deodorant** Desinfektionsmittel *n*, desodorierendes Mittel *n*, Geruchbekämpfungsstoff *m*, Thoriumfaden *m*

**deodorization** Desoderierung *f*, Entrüchelung *f*, Entstänkerung *f*, Geruchlosmachung *f*

**deodorize, to** ~ desodorieren, von Gerüchen befreien

**deodorizer** desodorierendes Mittel *n*, Rauchverzehrer *m*

**deodorizing** Entrüchelung *f*, Entstänkerung *f*, Geruchlosmachung *f* ~ **cartridge** Entstänkerungspatrone *f*

**de-oil, to** ~ ausfetten, entölen (Kälteanlage)

**deoxidant** Desoxidationsmittel *n*

**deoxidation** Deoxidation *f*, Desoxidation *f* ~ **period** Feinungsperiode *f*

**deoxidize, to** ~ desoxidieren

**deoxidized,** ~ **indigo** Indigoweiß *n* ~ **temper-hardening automatic screw machine steel** beruhigter Vergütungsautomatenstahl *m*

**deoxidizer** Beruhigungssubstanz *f*, Desoxidationsmittel *n*

**deoxidizing,** ~ **action** Desoxidationswirkung *f* ~ **additive** Beruhigungszuschlag *m* ~ **agent** Desoxidations-, Reduktions-mittel *n* ~ **flux** Desoxidationsschlacke *f* ~ **slag** Desoxidations-, Reduktions-schlacke *f*

**dep (deperdussin) control** Deperdussin Steuerung *f*

**depart, to** ~ abgehen, sich abkehren, abreisen, abrücken, abweichen, aufbrechen ausrücken

**departing flight** Abflug *m*

**department** Abteilung *f*, Beruf *m*, Dienstzweig *m*, Fach *n*, Gewerbe *n*, Gruppe *f*, Ministerium ~ **of development** Entwicklungsabteilung *f* ~ **of knowledge** Wissensgebiet *n* ~ **head** Gruppenleiter *m* ~ **store** Kauf-, Waren-haus *n*

**departmental allocation** Geschäfts-einteilung *f*, -verteilung *f*

**departmentalize, to** ~ abteilen

**departure** Abfahrt *f*, Abflug *m*, Abgang *m*, Abkehr *f*, Abmarsch *m*, Abreise *f*, Abschied *m*, (from orders) Abweichung *f*, Aufbruch *m*, Ausfahrt *f*, Ausgang *m*, Ausgehen *n*, (von einem Punkte) Auszug *m*, Start *m* ~ **point** Aufbruchsort *m* ~ **station** Abgangsamt *n* ~ **time** Aufbruchs-, Start-zeit *f* ~ **track** Ausfahrgleis *n*

**depend, to** ~ herabhängen to ~ **on** abhängen, sich richten nach, sich stützen auf to ~ **upon** abhängig sein von

**dependability** **test** Zuverlässigkeits-probe *f*, -prüfung *f*

**dependable** verläßlich, zuverlässig

**dependence** Abhängigkeit *f* ~ **on length** Längenabhängigkeit *f* ~ **on pressure of compressibility** Druckkoeffizient *m* der Kompressibilität

**dependency,** ~ **of attenuation on frequency** Dämpfungsverlauf *m* ~ **on frequency** Frequenzabhängigkeit *f* ~ **allowance for children** Kinderzulage *f* ~ **benefits** Familienunterhalt *m*

**dependent** Unterhaltsberechtigter *m*; abhängig, unselbständig to be ~ **on** abhängen sein von, angewiesen sein auf ~ **upon direction** richtungsempfindlich ~ **on frequency** frequenzabhängig ~ **on the gain** verstärkungsabhängig ~ **on load** belastungsabhängig ~ **sweep oscillations** unselbständige Kippschwingungen *pl* ~ **variable** Abhängige *n*, abhängige Veränderliche *f*

**depending,** ~ **on the voltage** spannungsabhängig

**dephase, to** ~ außer Phase bringen, in der Phase verschieben

**dephased** außer Phase, falschphasig, phasenverschoben, versetzt ~ **condition** Phasenverschiebung *f*, verschobene Phase *f* ~ **current** phasenverschobener Strom *m*
**dephasing** Verschiebung *f*
**dephlegmation** Teilkondensation *f*
**dephlegmator** Dephlegmator *m*
**dephosphorize, to** ~ entphosphorieren
**dephosphorization** Entphosphorung *f*
**depict, to** ~ abbilden, darstellen
**depilation** Enthaarung *f*
**deplane, to** ~ (aus dem Flugzeug) aussteigen
**deplaning point** Absetz-ort *m*, -platz *m*
**deplated finish** mit Sandstrahl behandelt und nachher elektrisch oxidiert
**deplete, to** ~ ausbeuten
**depleted** erschöpft, geleert, geräumt ~ **material** abgereichertes Material *n*
**depleting** erschöpfend
**depletion** Abreicherung *f*, Ausbeutung *f*, Entleerung *f*, Erschöpfung *f* ~ **of particles** Abwanderung *f* von Teilchen ~ **layer** Sperrschicht *f*
**deploy, to** ~ auseinanderziehen, betätigen, entfalten, öffnen (Bremsschirm)
**deployment** Aufmarsch *m*, Entwicklung *f*, Öffnung *f*, Streckung *f* (parachute) ~ **of production (or efficiency)** Leistungsentfaltung *f* ~ **bag** Bremsschirmpackhülle *f* ~ **shock** Streckstoß *m* (parachute) ~ **speed** Streckdauer *f* (parachute)
**depolarization** Depolarisation *f*
**depolarize, to** ~ depolarisieren
**depolarizer** Depolarisator *m*
**depolymerization** Entpolymerisation *f*
**deponent** Aussagender *m*
**deport, to** ~ deportieren, verbannen
**deportation** Ausweisung *f*, Deportation *f*, Fortschaffung *f*
**deposed** abgesetzt
**deposit, to** ~ ablagern, abschneiden, absetzen, sich absetzen, absitzen, ansetzen, auflegen, aufschwemmen, aufschweißen, aufschütten, ausfallen, ausscheiden, einschießen, hinterlegen, lagern, niederfallen, niederlegen, niederschlagen, setzen **to** ~ **the earth** Erde aufschütten **to** ~ **by evaporation** aufdämpfen **to** ~ **hard chronium** hartverchromen **to** ~ **mud** anschlämmen **to** ~ **sediment** sedimentieren
**deposit** Abgelagerte *n*, Ablagerung *f*, Abscheidung *f*, Absatz *m*, Angeld *n*, Ansatz *m*, Anzahlung *f*, Aufschwemmung *f*, Ausfällen *n*, Ausscheidung *f*, Belegung *f*, Beschlag *m*, Depositum *n*, Einlage *f*, Flöz *n*, Kontoeinlage *f*, Lager *n*, Lagerstätte *f*, Niedergesunkenes *n*, Niederlage *f*, Niederschlag *m*, Pfand *n*, Satz *m*, Schmelzgut *n*, Schweißgut *n*, Vorschuß *m*
**deposit, ~ on boiler surface** Kesselniederschlag *m* ~ **from water** Anschwemmung *f* ~ **of coke** Kokansatz *m* ~ **of cooling water** Niederschlag *m* des Kühlwassers ~ **of dry oil** Ölkruste *f* ~ **of potash salts** Kalisalzlager *m*
**deposit, ~ account** Depositenkonto *n* ~ **attack** Belagkorrosion *f* ~ **book** Bankbuch *n* ~ **business** Lombardgeschäft *n* ~ **contract** Verwahrungsvertrag *m* ~ **department** Depositenkasse *f* ~ **ground** Bodenablagerung *f* ~ **receipt** Depositenschein *m* ~ **tip** Bodenablagerung *f*
**deposited** abgesetzt, abgelagert (geol.) **to be** ~

**niederfallen** ~ **from solutions** aus Lösungen kondensiert ~ **carbon resistor** Kohleschichtwiderstand *m*
**depositing** Lagern *n*, Niederschlag *f* ~ **grid** Ablagerost *m*
**deposition** Absatz *m*, Abscheidung *f*, Absetzen *n*, Absetzung *f*, Aussage *f*, Erklärung *f*, Niederlegung *f*, Niederschlag *m*, Niederschlagung *f*; Tatbestandaufnahme *f* ~ **of ashes** Aschenablagerung *f* ~ **of the fuel** Kraftstoffausbreitung *f* ~ **of gravel** Kiesablagerung *f* ~ **of mud** Schlammablagerung *f* ~ **of sand** Sandablagerung *f* ~ **of soil from flowing water** Kolmation *f*
**deposition, ~ area** Niederschlagsfläche *f* ~ **tube** Absitzrohr *n*
**depositor** Hinterleger *m*
**deposits** Sinkstoffablagerung *f* ~ **of natural gas** Erdgasvorkommen *pl*
**depot** Ablage *f*, Depot *n*, Lager *n*, Lagerplatz *m*, Lagerschuppen *m*, Magazin *n*, Niederlage *f*, Park *m*, Proviantamt *n*, Stapelplatz *m*, Station *f*, Warenlager *n* ~ **administration** Magazin-, Material-verwaltung *f* ~ **ship** Beischiff *n*, schwimmender Flugstützpunkt *m*
**depreciate, to** ~ entwerten, sinken (im Wert)
**depreciate value** Tragewert *m*
**depreciation** Abschreibung *f*, Entwertung *f*, Herabsetzung *f* des Wertes, (in value) Minderwert *m*, Wertminderung *f*
**depress, to** ~ anschlagen, herabdrücken, niederbrechen, senken **to** ~ **a button (or a key)** eine Taste *f* oder einen Hebel *m* niederdrücken
**depressable** senkbar
**depressed** aufgelöst, gedrückt, niedergeschlagen ~ **arch** abgeflachter Bogen *m* ~ **floor** vertiefte Sohle *f*
**depression** Auslösung *f*, barometrische Tiefe *f*, Bodensenkung *f*, Depression *f*, Drücken *n*, Druckrückgang *m*, Druckverminderung *f*, Einsenkung *f*, Erdfall *m*, Höhenrichtung *f* (unter der Horizontalen), Kerb *m*, Kerbe *f*, Kessel *m*, Mulde *f*, Niederdrücken *n*, Niederung *f*, Senke *f*, Senkel *m*, Senkung *f*, Tiefdruck *m*, Tiefdruckgebiet *n*, Tiefe *f*, Tiefstand *m* (econ), Unterdruck *m*, Vertiefung *f* ~ **of ground** Bodensenkung *f* ~ **of a key** Anschlag *m* einer Taste ~ **on a slope** Hangmulde *f*
**depression, ~ angle** Senkungs-, Neigungs-winkel *m* ~ **box** Unterdruckkammer *f* ~ **clouds** Depressionsgewölk *n* ~ **-control gear** Hubbegrenzung *f* nach unten ~ **line** Muldelinie *f* ~ **piston** Rückstoßkolben *m* ~ **-position finder** Azimutentfernungsmesser *m*, Erdkrümmungslineal *n* ~ **telemeter** Depressionsentfernungsmesser *m*, Entfernungsmesser *m* mit senkrechter Basis ~ **washed out of river bed by current** Auskolkung *f*
**depressurization valve** Entlüftungsventil *n* (Außentank)
**deprivation** Aberkennung *f*
**deprive, to** ~ **of** entziehen
**deprocessing** entkonservieren
**depropanizer** Entpropanisierer *m*
**deproteinization** Aufspalten *n* von Eiweißsubstanzen
**depth, in** ~ treppenweise
**depth** (of a pit) Förderteufe *f*, Höhe *f* (Trägerprofil), Mächtigkeit *f*, Pegel *m*, Teufe *f*, Tiefe *f*, Tiefenlinie *f* **(practical)** ~ nutzbare Tiefe *f*

**depth,** ~ **of aileron** Querrudertiefe *f* ~ **of bore** Bohrtiefe *f* ~ **of borehole** Bohrlochtiefe *f* ~ **of boring** Bohrtiefe *f* ~ **of case** Einsatztiefe *f*, Härtetiefe *f* ~ **of cavity** Lochtiefe *f* ~ **of charge** Füllungshöhe *f* ~ **of chill** Härtetiefe *f* ~ **of chilling** Abschrecktiefe *f* ~ **of counterbore** Senktiefe *f* ~ **of crack** Rißtiefe *f* ~ **of curvature** Durchbiegung *f* einer Linse ~ **of cut** Einschnitttiefe *f*, Schnitthöhe *f*, Schnittiefe *f*, Spandicke *f*, Spantiefe *f* ~ **of drawing** Ziehtiefe *f* ~ **of embedment** Einbettungstiefe *f* ~ **of engagement of gears** Eingriffstiefe *f* ~ **of field scale** Abbildungstiefenskala *f*, Tiefenschärfenrechner *m* ~ **of fields** Fokustiefe *f*, Schärfentiefe *f* (focal), Sehtiefe *f*, Tiefenbereich *m*, Tiefenschärfe *f* (film) ~ **of fin** Rippenhöhe *f* ~ **of fire** Feuertiefe *f* ~ **of flat from periphery** Scheitelhöhe *f* ~ **of focus** Fokus-, Schärfen-, Seh-tiefe *f*, Tiefe *f* (of eye), Tiefenschärfe *f* ~ **of fusion** Einschmelztiefe *f* ~ **of gap** Ausladung *f* ~ **of groove** Kalibertiefe *f* ~ **of hardening zone** Härtetiefe *f* ~ **of hold** Raumtiefe *f* ~ **of horizontal tail unit** Höhenleitwerktiefe *f* ~ **of immersion** Eintauchtiefe *f*, Eintauchtiefe *f* des Thermometers (of the thermometer), Tauchtiefe *f* (of a ship) ~ **of impression** Eindringungstiefe *f* ~ **of in-feed** Einstechtiefe *f* ~ **below insertion point** Tiefe *f* unter Ansatzpunkt ~ **of keyway** Nuttiefe *f* ~ **of a letter** Schriftkegel *m* ~ **of modulation** Aussteuerungskoeffizient *m*, Beeinflussungsfaktor *m*, Modulations-grad *m*, -tiefe *f*, -ziffer *f* ~ **of modulation factor** Aussteuerungsgrad *m* ~ **of the navigable channel** nutzbare Fahrtiefe *f* ~ **of nitration** Nitriertiefe *f* ~ **of packing** Dichtungstiefe *f* ~ **of page** Seitenhöhe *f* ~ **of penetration** Eindringungstiefe *f* ~ **of penetration of burning (or fusion)** Einbrandtiefe *f* ~ **of pole hole** Einstelltiefe *f* (für Stangen) ~ **of profile** Profiltiefe *f* ~ **of quenching** Abschrecktiefe *f* ~ **of rainfall** Niederschlags-, Regen-höhe *f* ~ **of recess** Gesenktiefe *f* ~ **of set** Bunzentiefe *f* ~ **of shade** Dunkelstufe *f* (Farbe) ~ **of the shed** Sprunghöhe *f* ~ **of the ship's hold** Hohl *n* des Schiffsraumes ~ **of still water** Wehrhöhe *f* ~ **of thread** Gewindetiefe *f* ~ **of throat** Ausladung *f* ~ **for tonnage** Vermessungstiefe *f* ~ **of tooth** Zahnhöhe *f* ~ **of tooth below pitch line** Zahnfußtiefe *f* ~ **of trench** Bettungs-, Graben-tiefe *f* ~ **vertex** Scheiteltiefe *f* ~ **of vision** Schärfentiefe *f* ~ **of volume flow** Bandströmung *f* ~ **of water** Wasser-höhe *f*, -tiefe *f* ~ **of wear** Abnutzungstiefe *f*

**depth,** ~ **adjustment** Tiefeneinstellung *f* ~ **bomb** Unterwasserbombe *f* ~ **contour** Tiefen-kurve *f*, -linie *f* ~ **contours** Isobaten *pl* ~ **control of mine** Tiefensteller *m* (navig) ~ **data** Tiefenangabe *f* ~ **determination of flaws** Fehlerscheibchentiefenbestimmung *f*, Tiefenbestimmung *f* ~ **dislocation** Tiefenverlagerung *f* ~ **distributed** tiefgegliedert ~ **dose** Tiefendosis *f* ~ **effect** throwing power Tiefenwirkung *f* ~ **error** Abstandsfehler *m* ~ **feed control** Einstechvorschubregelung *f* ~ **feed motion** Tiefenvorschub *m* ~ **feed scale** Einstechtiefenskala *f* ~ **finding** Tiefenmessung *f* ~ **float** Tiefschwimmer *m* ~ **gauge** Tiefen-meßapparat *m*, -lehre *f*, -messer *m*, -meßvorrichtung *f*, -taster *m* ~ **hardening** Durchhärtung *f* ~ **indicator** Behm-Lot *n*,

Teufen-, Tiefen-anzeiger *m*, Tiefenmesser *m* ~ **measuring** Meßtiefe *f* ~ **measuring applicance** Tiefenmeßgerät *n* ~ **parallax** Tiefenparallaxe *f* ~ **perception** Tiefenlokalisation *f* ~ **regulator** Tiefeneinstellvorrichtung *f*, Tiefenruder *n* ~ **rudder** Höhensteuer *n* ~ **slide** Tiefenschlitten *m* ~ **sounder** Behm-Lot *n* ~ **test** Tiefenschürfung *f* ~ **washer** Anlaufring *m*
**depthing tool** Eingriffzirkel *m*
**depulp, to** ~ entpülpen
**deputy** Aufseher *m*, Beauftragter *m*, Ersatzmann *m*, Fahrhauer *m*, Stellvertreter *m*; stellvertretend ~ **of sinkers** Drittelführer *m* ~ **managing director** stellvertretender Geschäftsführer *m*
**derail, to** ~ entgleisen
**derailment** Entgleisung *f*
**derangement** Verrückung *f* ~ **of masses** Massenstörung *f*
**derated** abgedrosselt (aviat.)
**derating curves** Leistungsverlustkurven *pl*
**dereflection** Entspiegelung *f*
**derelict** (treibendes) Wrack *n*; herrenlos
**derelicts** Seetriften *pl*
**deresinated rubber** entharzter Kautschuk *m*
**derivable** ableitbar
**derivate, to** ~ ableiten; abgeleiteter Körper *m* (chem.)
**derivation** Ablenkung *f*, Abweichung *f*, Herleitung *f*, Stromableitung *f* ~ **to the skin** Hautableitung *f* ~ **branch line** Abzweigung *f* (electr.)
**derivative** Abbauprodukt *n*, Abgeleitete *f*, Abkömmling *m*, Ableitung *f*, Ableitungsmittel *n*, Derivat *n*, Differentialquotient *m*, Erzeugnis *n*, abhängige Größe *f* ~ **of function** abgeleitete Funktion *f* ~ **with respect to time** zeitlicher Differentialquotient *m*
**derivative,** ~ **action** D-Einfluß *m* (in controller), differenzierender Einfluß *m*, Vorhaltwirkung *f* ~ **action time** Vorhaltzeit *f* (aut. contr) ~ **control** Differentialregelung *f* ~ **term** abgeleiteter Ausdruck *m* ~ **trend with respect to time** zeitlicher Verlauf *m*
**derivator** Steilheits-, Tangenten-messer *m*
**derive, to** ~ abstammen, derivieren, herleiten, verzweigen **to** ~ **from** ableiten, entnehmen, stammen aus
**derived** abgeleitet **to be** ~ abstammen ~ **from** beruhend auf
**derived,** ~ **circuit** Abzweigstromkreis *m* ~ **function** abgeleitete Funktion *f* ~ **image in contradistinction to original image** abgeleitetes Bild *n* zum Unterschied vom Urbild (Kontaktkopien) ~ **quantity** abgeleitete Größe *f* ~ **resistance** Zweigwiderstand *m* ~ **set** Ableitung *f* einer Menge *f* ~ **unit** abgeleitete Einheit *f*, (of measure) abgeleitete Maßeinheit *f*
**de-riveting** Nietkopfschneiden *n*
**dermal resistance** Hautwiderstand *m*
**derrick** Ausleger *m*, Auslegerwippkran *m*, Bohrgerüst *n*, (drilling) Bohrturm *m*, schwenkbarer Halbportalkran *m*, Knappe *m* (min.), Kran *m*, Kranbaum *m*, Ladebaum *m*, Lademast *m*, Turm *m* ~**-car crane** Waggonkran *m* ~ **cornice** Turmkranz *m* ~ **dredger** Säulenschwenkbagger *m* ~ **floor** Arbeitsbühne *f*, Turmboden *m* (horizontal) ~ **girt** Turmverbindung *f* ~ **guy** Geibaum *m* ~ **man** Turmsteiger *m* ~ **platform**

Bohrbühne *f* ~ **pole** Baumast *m*, Hilfsgestänge *n* (zum Aufrichten schwerer Gestänge) ~ **pulley** Turmrolle *f* ~ **rigging** Derrickabspannung *f* ~-**style antenna mast** Gittermast *m* ~ **supported by guys** Derrickkran *m* mit Ankerseilen ~ **tension cable** Turmspannseil *n* ~ **taveling on rails** Auslegerlaufkatze *f* ~-**type crane** Auslegerdrehscheibenkran *m* ~-**type revolving crane** Auslegerkran *m* ~-**wagon crane** Wagenkran *m* ~ **wheel** Steuerrad *n*

**derricking controller** Hubfahr-, Hubsteuer-schalter *m*

**desaccharification** Entzuckerung *f*

**desand, to** ~ entsanden

**descale, to** ~ abklopfen (Kessel), dekapieren, entkrusten, entzundern

**descaling** Entzunderung *f*

**descant** Diskant *m*

**descend, to** ~ absinken, abstammen, absteigen, abwärtsbewegen, (scales) ausschlagen, (into a mine) einfahren, fallen, herabfahren, herabsinken, heruntergehen, herunterkommen, niedergehen, sinken **to** ~ **from** stammen aus **to** ~ **by parachute** mit dem Fallschirm *m* abspringen **to** ~ **into a shaft** einen Schacht *m* befahren **to** ~ **on water** anwassern

**descendent branch of trajectory** absteigender Ast *m* der Flugbahn

**descending** Niedergehen *n*; absteigend ~ **air current** Abwind *m* ~ **gallery** Fallende *n* ~ **letters** geschwänzte Schrift *f*

**descent** Abfall *m*, Abstieg *m*, Abwärtsbewegung *f*, (into a mine) Anfahrt *f*, (into a mine) Einfahrt *f*, (inclination) Gefälle *n*, Neigung *f*, Niedergang *m*, Niedergehen (des Flugzeuges), Sinkflug *m*, Talfahrt *f* ~ **of the charges of a blast furnace** Niedersinken *n* der Gichten eines Hochofens ~ **into a mine** Grubenfahrt *f* ~ **by stages** Fahrstuhllandung *f*

**descent,** ~ **phase** absteigender Ast *m* (d. Flugbahn), Sinkflugphase *f* ~ **plate** Kropf *m*, Sattel *m* (paper mfg.)

**describe, to** ~ beschreiben, darstellen **to** ~ **a circle** einen Kreisbogen *m* schlagen

**described,** ~ **above** vorgenannt ~ **below** nachstehend

**description** Beschreibung *f*, Darstellung *f*, Erfindungsbeschreibung *f*, (patents) Oberbegriff *m*, Schilderung *f*, Umschreiben *n* (math.) ~ **of coast** Küstenbeschreibung *f*

**descriptive** beschreibend ~ **geometry** darstellende Geometrie *f*, Stereographie *f*

**desensitize, to** ~ phlegmatisieren, unempfindlich machen

**desensitizing dye** desensibilisierende Farbe *f*

**desert, to** ~ ausreißen, sich verfranzen, verlassen (aviat.)

**desert** Wüste *f* ~ **belt** Wüstengebiet *n* (geol.) ~ **sand** Wüstensand *m* ~ **storm** Wüstensturm *m* ~ **test** Wüstenuntersuchung *f*

**deserve, to** ~ Anspruch haben auf, verdienen

**deserving (of) credit** kreditwürdig

**desiccant (material)** austrocknendes Mittel *n*, Entfeuchtermaterial *n*

**desiccate, to** ~ austrocknen, dörren, trocknen

**desiccated** wasserfrei ~ **cell** Füllelement *n* (electr.)

**desiccating** Trocknung *f* ~ **apparatus** Entwässe-

rungsapparat *m* ~ **cylinder** Trockenzylinder *m* ~ **device** Dörrapparat *m* ~ **machine for wood pulp** Zelluloseentwässerungsmaschine *f*

**desiccation** Austrocknung *f*, Vertrocknung *f*, Wasserentziehung *f*

**desiccative** Trockenmittel *n*

**desiccator** Austrocknen *n*, Druckluftanlage *f*, Exsiccator *m*, Trockenapparat *m*, Trockner *m* ~ **plate** Exsiccatoreinsatz *m*

**design, to** ~ anordnen, aufbauen, auslegen, bauen, berechnen, durchbilden, entwerfen, erbauen, gestalten, konstruieren, konzipieren, (weaving) das Muster absetzen, patronieren, planen, zeichnen **to** ~ **an installation** eine Anlage entwerfen **to** ~ **a model** ein Modell entwerfen

**design** Abriß *m*, Anordnung *f*, Aufbau *m*, Aufmachung *f*, konstruktiver Aufbau *m*, Aufriß *m*, Ausführung *f*, Bauart *f*, Bauwerk *n*, Berechnung *f*, Bild *n*, Dessin *n*, Dimensionierung *f*, konstruktive Durchbildung *f*, Entwurf *m*, Fasson *f*, Formgestaltung *f*, Gestalt *f*, Gestaltung *f*, Grundriß *m*, Konstruktion *f*, Modell *n*, Muster *n*, Musterzeichnung *f*, Plan *m*, Planung *f*, Riß *m*, Vorsatz *m*, Vorzeichnung *f*, Zeichnung *f*

**design,** ~ **with border** abgepaßtes Muster *n* ~ **of circuit** Berechnung *f* der Schaltung ~ **and construction** Gestaltung *f* und Bauausführung *f* ~ **of dams** Berechnung *f* der Talsperren ~ **of grooves** Kaliberkonstruktion *f* ~ **of grooves for sectional iron** Formeisenkalibrierung *f* ~ **of pass** Kaliberform *f* ~ **of passes** Kaliberkonstruktion *f* ~ **of roll passes** Walzkalibrierung *f* (step) ~ **of a selfsupporting overhead cable** Stufenmuster *n* eines selbsttragenden Luftkabels ~ **in full size** Musterriß *m* ~ **for stress uniformity** auf gleichmäßige Beanspruchung konstruieren

**design,** ~ **airspeed** Bemessungsfluggeschwindigkeit *f* ~ **amount** bei der Konstruktion zugrunde gelegter Sollwert *m* ~ **basic requirements** grundlegende Bauanforderungen *pl* ~ **capacity** Soll-Leistung *f* ~ **elements** Entwurfsbestandteile *pl* ~ **engineering** Konstrukteurtätigkeit *f* konstruktive Tätigkeit (aviat.) ~ **factor** Bauzahl *f* ~ **features** bauliche Kennzeichen *pl*, Baumerkmale *pl*, Gestaltungsmerkmale *pl* ~ **flow** Ausbauwassermenge *f* (hydraulic power station) ~ **formula** Bauformel *f* ~ **head** Konstruktionsfallhöhe *f* ~ **landing weight** Bemessungslandegewicht *n* (aviat.) ~ **load** rechnerische Belastung *f*, Bemessungsbelastung *f* ~ **office** Konstruktionsbüro *n* ~ **manoevring speed** Bemessungsgeschwindigkeit *f*, Ruderbetätigung *f* (aviat.) ~ **maximum weight** Bemessungshöchstgewicht *n* (aviat.) ~ **minimum weight** Bemessungs-Kleinstgewicht *n* ~ **paper** Linien-, Muster-, Tupf-papier *n* ~ **parameter** Entwurfparameter *m* ~ **patent** Gebrauchsmuster *n* ~ **(or starting) point** Auslegungspunkt *m* ~ **power** berechnete Leistung *f* ~ **practice** Konstruktionspraxis *f* ~ **pressure** Konzessionsdruck *m* ~ **problem** Entwurfsfrage *f* (engin.) ~ **roller** Musterwalze *f* ~ **sheet** Konstruktionsblatt *n* ~ **strength** Gestaltsfestigkeit *f* ~ **stress** zulässige Beanspruchung *f* oder Spannung *f* ~ **table** Bemessungstafel *f* ~ **take-off weight** Bemessungsstartgewicht *n* ~ **taxying weight** Bemessungsrollgewicht *n* (aviat.) ~ **wheel load** Bemessungsradlast *f* ~ **wing area** Bezugsflügelfläche *f* (aerodyn.)

**designate, to** ~ anmerken, beschriften, bezeichnen, kennzeichnen, markieren *m*, signieren

**designated,** ~ **fire zone** bezeichnete Feuergefahrenzone *f*, brandgefährdete Zone *f* ~ **representative** namhaft gemachter Vertreter *m*

**designation** Ansprache *f*, Beschriftung *f*, Bestimmung *f*, Bezeichnung *f*, Designation *f*, Entwerfen *n*, Konstruieren *f*, Marke *f*, Markierung *f*, Signum *n* ~ **of a listant material used in aircraft construction** Homogenholz *n* ~ **of quantity** Mengenbezeichnung *f*

**designation,** ~ **card** Bezeichnungsschild *n* ~ **code (or digit)** Kennziffer *f* ~ **strip** Bezeichnungsstreifen *m*

**designed** ausgebildet, berechnet ~ **for stress uniformity** auf gleichmäßige Beanspruchung *f* hin konstruiert ~ **to swing in front** nach vorn klappbar ~ **height** Sollhöhe *f* ~ **speed** berechnete Geschwindigkeit *f*

**designer** Erbauer *m*, Gestalter *m*, Konstrukteur *m*, Konstruktionszeichner *m*, Musterausnehmer *m*, Musteraussetzer *m*, Zeichner *m* **(structural)** ~ Statiker *m*

**designing** Durchbildung *f* ~ **of construction lines** Profilgestaltung *f* ~ **of grooved rolls** Walzkalibrierung *f* ~ **of section** Profilieren *n*, Formgebung *f* ~ **machine** Schablonen-(stech)maschine *f*

**desiliconization** Entsilizierung *f*

**desiliconize, to** ~ entsilizieren

**desilting** Entsandung *f*

**desilver, to** ~ entsilbern

**desilvering** Entsilberung *f* ~ **operation** Entsilberungsarbeit *f*

**desilverization** Entsilberung *f* ~ **plant** Entsilberungsanlage *f*

**desilverize, to** ~ entsilbern

**desintegration** Zerfall *m*

**desired** gewünscht ~ **frequency** Nutzfrequenz *f* ~ **temperature** Solltemperatur *f* ~ **trajectory** Sollflugbahn *f* ~ **value** Aufgabenwert *m* (aut. contr.), Sollwert *m* ~ **value (or set-point) adjuster** Sollwertsteller *m* ~ **value of controlled variable** vorgegebener Wert *m* ~ **wedge** angestrebte Keilbreite *f*

**desist, to** ~ **from** Abstand nehmen von

**desisting** Abstand *m*

**desize, to** ~ (fabrics) entschlichten

**desizing bath** Entschlichtungsbad *n*

**desk** Pult *n* ~ **stand** Pultgestell *n* ~ **stand telephone set** Fernsprechapparat *m* mit festem Mikrofon ~-**telephone set** Tischgehäuse *n* ~ **(stand) telephone set** Säulentischfernsprecher *m* ~ **tray** Briefablegekasten *m*

**deslag, to** ~ entschlacken

**deslagging hammer** Schlackenhammer *m*

**Deslandres diagram** (of spectral-band system) Kantenschema *n*

**deslime, to** ~ entschleimen

**desliming** Entschleimung *n* ~ **screen** Entschlämmungssieb *n*

**desolate** wüst

**desorption** Desorption *f*

**de-spin mechanism** Drallbremseinrichtung *f*

**dessiccant** Exsiccatorfüllmittel *n*

**dessiccator** Austrockner *m*

**Dessinier rolling mill** Dessinierwalzwerk *n*

**dessuade, to** ~ entraten

**destaticize, to** ~ elektrostatische Ladungen entfernen

**destaticizer** Entlader *m*

**destination** Bestimmung *f*, Bestimmungsort *m*, Ziel *n* ~ **bearing** Zielpeilung *f* (aviat.) ~ **dialling** Zielwahl *f* (teleph.) ~ **key** Zieltaste *f* (ZT) ~ **screen** Transparant *m* (f. Richtungsanzeiger) ~ **sign** Richtungsschild *n* ~ **station** Bestimmungsort *m*

**destitute** mittellos

**destroy, to** ~ aufreiben, tilgen, zerstören **to** ~ **by caustic** ausätzen **to** ~ **bacteria** entkeimen

**destroyed** abgetötet

**destroyer** Zerstörer *m* (navy) ~ **column** Zerstörersäule *f* (rdr)

**destructibility** Zerstörbarkeit *f*

**destructible** zerstörbar

**destruction** (of kinetic energy) Beruhigung *f*, Niederkämpfung *f*, Unbrauchbarmachen *n*, Verheerung *f*, Vernichtung *f*, Vertilgung *f*, Zerfall *m*, Zerstörung *f*, Zertrümmerung *f* ~ **in form of pits** lochfraßähnliche Zerstörung *f* ~ **of a wall** Mauereinsturz *m* ~ **limit** Zerreißgrenze *f* ~ **test** Dauerlauf *m* (mach.)

**destructive** zerstörend ~ **blow** vernichtender Schlag *m* ~ **distillation** Entgasung *f*, Trockendestillation *f*, Verkohlen *n*, Verkohlung *f*, Zersetzungsdestillation *f* ~ **distillation with steam** Dampfschwelung *f*, Zerstörungsfeuer *n* ~ **lightning** Schadenblitz *m* ~ **path** Zerstörungsstreifen *m* ~ **readout** destruktives Lesen *n* ~ **static test** statische Untersuchung *f* bis zum Brechpunkt

**desugarize, to** ~ absüßen

**desugarizing** Entzuckerung *f*

**desulfurization** Abschwefelung *f*, Entschwefelung *f*, Schwefelentfernung *f*

**desulfurize, to** ~ abschwefeln, entschwefeln

**desulfurizer** Entschwefelungsmittel *n*

**desulfurizing** Entschwefelung *f*, Schwefelentfernung *f* ~ **agent** Entschwefelungsmittel *m*

**desuperheat control valve** Dampftemperaturregelventil *n*

**desuperheater** Heißdampfkühler *m*

**desuperheating** Kühlung *f* des Heißdampfes

**detach, to** ~ (ab)kommandieren, ablösen (aviat.), abnehmen, abschalten, absondern, abstellen, ausscheiden, detachieren, lösen, loslösen, lostrennen **to** ~ **from a holder (or file)** abheften **to** ~ **in leaves (or in scales)** losblättern

**detachable** abnehmbar, abschraubbar, ansteckbar, ausbaubar, lösbar, zerlegbar ~ **ball journal bearing** Schulterkugellager *n* ~ **bottom** Losboden *m* ~ **chain** zerlegbare Kette *f* ~-**chain sprocket wheel** Kettenrad *n* für zerlegbare Kette ~ **cheek pad** Austeckbacke *f* ~ **circular table** aufsetzbarer Rundtisch *m* ~ **coil** Steckspule *f* ~ **connection** lösbare Verbindung *f* ~ **dual controls** ausbaufähige Doppelsteuerung *f* ~ **engine cowling** abnehmbare Triebwerksverkleidung *f* ~ **engine mounting** abnehmbarer Motorvorbau *m*, abnehmbares Triebwerksgerüst *n* ~ **magazine** Ansteckmagazin *n* ~ **panels** abnehmbare Verkleidungsbleche *pl* ~

**rim** abnehmbare Felge f **~ swiveling copying unit** schwenkbare Aufsatzkopiereinheit f **~ trailing section** Endkasten m **~ undercarriage** abwerfbares Fahrgestell n **~ wing** abnehmbarer Flügel m

**detached** abgesetzt, abgesondert, einzeln, freistehend

**detaching** Auslösung f, (waterways) Abschälung f **~ bell** Abziehglocke f **~ device** Abziehvorrichtung f **~ hook** Auslösehaken m. **~ nut** Abziehmutter f

**detachment** Abteilung f, Kommando n, Lösung f, Staffel f, Untergruppe f **~ of a cutting** Spanabhebung f

**detail, to ~** abkommandieren, ausscheiden, heranziehen

**detail** Einzelheit f, Kleinigkeit f, Trupp m, Truppe f (mil.), Truppenabteilung f **~ of image** Feinheit f **~ of picture** Bildfeinheit f, Feinheit f

**detail, ~ contrast ratio** Einzelheiten-Kontrastverhältnis n **~ draftsman** Zeichner m für Einzelzeichnungen **~ drawing** Detaillieren n, Einzel-, Teil-zeichnung f **~ rendition** Detailwidergabe f (TV) **~ sketch** Ausschnittskizze f

**detailed** ausführlich, eingehend **~ balancing** Gegenteil n von Stoßionisation **~ calculation** Durchrechnung f **~ description** ausführliche Beschreibung f **~ inspection** Stückprüfung f **~ specification** weitgehende Beschreibung f **~ ~ statement** Ausführung f, die ausführliche Rechnung f

**detailer** Konstrukteur m

**details** Angaben pl, Ausführungen pl

**detain, to ~** anhalten, arretieren, aufhalten, behindern, bremsen, festhalten, feststellen, hemmen, internieren, sperren

**detar, to ~** entteeren

**detarer** Teerabscheider m, Teerscheider m

**detaring** Entteerung f, Teerabscheidung f

**detartarizer** Wasserreinigungsapparat m

**detect, to ~** anzeigen, aufdecken, auffangen, aufnehmen, beobachten, entdecken, erfassen, erkennen, feststellen, gleichrichten, nachweisen **to ~ by feel** ertasten **to ~ by smelling** riechen **to ~ by touch** ertasten

**detectability** Trennbarkeit f

**detectable** nachweisbar

**detecting, ~ action** Detektorwirkung f, Gleichrichterwirkung f **~ efficiency** Nachweisempfindlichkeit f **~ element** Fühler m **~ tube (or valve)** Audion n

**detection** Anzeigung f, Aufnahme f, Beobachtung f, Demodulation f, Empfangsgleichrichtung f, Entdeckung f, Entmodelung f, Entmodulierung f, Erfassung (rdr), Gleichrichtung f, Nachweis m, Nachweisung f, Ortung f **~ of an object (or target)** Zielaufsuchen n (rdr)

**detection, ~ coefficient** Gleichrichterwirkung f **~ efficiency** Nachweisempfindlichkeit f **~ range** Auffassungsreichweite f, Entdeckungsreichweite f, Erfassungsentfernung f **~ system** Warnanlage f

**detectophone** Lauschmikrofon n

**detector** Anrichter m, Anzeiger m, Demodulator m, Detektor m, Empfangsvermittler m, Fühlgerät n, Gegentaktaudion n, Gemeßröhre f, Nachweisinstrument n, Stromanzeiger m,

Sucher m, Vermittler m, Wechselwinkel m **~ of motion** Bewegungsempfänger m **~ for use against submarines** Abhörapparat m

**detector, ~ action** Detektor-, Gleichrichter-wirkung f **~ aperture** Detektorblende f **~ balanced bias** Radarstörbeseitigung f, störpegelgesteuerte Gleichrichterschaltung f **~ bar** (railway signaling) Fühlschiene f **~ circuit** Detektorkreis m **~ circuit tuning** Auditionskreisabstimmung f **~ coefficient** Gleichrichterwirkungsgrad m **~ mixer tube** Mischröhre f **~ operating current** Detektorarbeitsspannung f **~ phone circuit** Detektorfernhörerkreis m **~ receiver** Detektorenempfänger m (rdo) **~ receiving set** Detektorgerät n **~ track section** Isolierschiene f **~ tube (or valve)** Audion-, Detektor-, Gleichrichter-röhre f

**detent** Anschlag m, Arretierung f, Auslösung f, Hilfszug m für Anlasser, Klinke f, Rastklinke f, Rastnase f, Sperr-kegel m, -klinke f, -zahn m **~ ball** Arretierungskugel f **~ gear of an indicator** Anhaltevorrichtung f eines Indikators **~ lever** Auslösehebel m, Expansionshebel m, Expansionsstange f **~ pin** Festhaltestift m **~ plate** Anschlagscheibe f **~ spring** Arretierfeder f

**detention** Einbehaltung f **~ for investigation** Untersuchungshaft f **~ of pay** Besoldungseinbehaltung f

**deter, to ~** abschrecken,

**detergent** Putzmittel n, Reinigungsmittel n, Zusatzmittel n zum Schmieröl zur Verhütung der Ölschlammabsonderung

**deteriorate, to ~** abnutzen, entarten, schlechter werden, verfallen, verschlechtern, in der Wirkung f zurückgehen

**deterioration** (Gummi) Alterung f, eingetretene Mängel pl, Verfall m, Verschlechterung f, Zersetzung f **~ of the discussion** Scheitern n der Verhandlung **~ of emission** Emissionsrückgang m

**determinable** erfaßbar, feststellbar

**determinant** Bestimmungsgröße f (math.), Determinante f **~ of the line equations** Gleichungsdeterminante f für Leitungen **~ theorem** Determinantensatz m

**determinate equation** bestimmte Gleichung f

**determination** Ableitung f, Ausschluß m, Bestimmung f, Entschlossenheit f, Ermittlung f, Festlegung f, Festsetzung f, Feststellung f, Prüfung f **~ procedure** Bestimmungsverfahren n

**determination, ~ of age** Altersbestimmung f **~ of azimuth (or bearing)** Winkelpeilung f **~ of the center of rotation** Ermittlung f des Drehpunktes **~ of the character of ore** Erzbewertung f **~ of charges** Gebührenbemessung f **~ of course** Fahrtbestimmung f **~ of efficiency by total losses** Gesamtverlustverfahren n, Verlustverfahren n **~ of height** Höhenbestimmung f **~ of iron** Eisenbestimmung f **~ of the limen** Schwellenbestimmung f **~ of the line by French method** Ermittlung f der Schadenlinie French **~ of mass** Massenermittlung f **~ of output** Leistungsbestimmung f **~ of position** Lagebestimmung f **~ of shrinkage** Schwindmaßbestimmung f **~ of sign** Vorzeichenbestimmung f **~ of size** Bemessung f **~ of the stresses** Nachweis m der Spannungen **~ of time** Zeitbestimmung f **~ of yield** Ertragsfeststellung f **~ of the yield** Etatfestsetzung f

**determinative** ausschlaggebend ~ **factor** maßgebender Faktor *m*

**determine, to** ~ ausmachen, ausmitteln, beschließen, bestimmen, ermitteln, festlegen, festsetzen, feststellen **to** ~ **a bearing** einen Winkel *m* messen **to** ~ **a course** koppeln **to** ~ **the rate** die Gebühr *f* festsetzen **to** ~ **the time** die (Uhr)zeit *f* feststellen

**determined,** bemessen, bestimmt, entschieden, entschlossen

**determining** maßgebend ~ **borehole** Grenzbohrloch *n* ~ **factor** ausschlaggebender Faktor *m*, Richtwert *m*

**determinism** Zwangslauflehre *f*

**deterrent powder** oberflächenbehandelnder Puder *m*

**detin, to** ~ entzinnen

**detinning,** Entzinnung *f*

**detonate, to** ~ bersten, detonieren, explodieren, klopfen, knallen, knattern, krepieren, platzen, verpuffen, zerknallen

**detonating** detonierend ~ **canal** Zündkanal *m* ~ **cap** Sprengkapsel *f* ~ **characteristics** Klopfwert *m* (Kraftstoff) ~ **charge** Aufladen *n*, Zündladung *f* ~ **charge boost** Aufladung *f* ~ **column** detonierende Strecke *f* ~ **compositum** Knallsatz *m* ~ **cord** Knallzündschnur *f* ~ **explosive** Brisanzsprengstoff *m* ~ **fuse** Detonationskapsel *f* ~ **gas** Wasserstoffknallgas *n* ~ **primer** Explosionszünder *m* ~ **rod** Schlagstift *m* ~ **slab** Zerschellerschicht *f* ~ **tube** Detonationskapsel *f* ~ **wire** Zünderdraht *m*

**detonation** Detonation *f*, Explosion *f*, Klopfen *n*, Knall *m*, Verpuffung *f*, Zündung *f* ~ **agent** Zündmittel *n* ~ **inducers** Explosionszentren *pl* ~ **limits** Detonationsgrenzen *pl* ~ **photographs** Detonationsaufnahmen *pl* ~ **pressures** Detonationsdrucke *pl* ~ **test** Prüfung *f* des Klopfverhaltens (eines Motors) ~ **velocity** Detonationsgeschwindigkeit *f* ~ **wave** Detonations-, Knall-welle *f*

**detonator** Kapsel *f*, Knallkörper *m*, Knallsignal *n*, Knallzündmittel *n*, Sprengkapsel *f*, Sprengkapselzünder *m*, Zünder *m*, Zündkasten *m* (electr.), Zündung *f* ~ **anvil** Amboß *m* der Zündglocke ~ **apparatus** Zündergerät *n* ~ **cable** Zündkabel *n* ~ **cap** Zündkapsel *f* ~ **casing** Zünderhalter *m*, Zündladungskapsel *f* ~ **charge** Eingangszündung *f*, Hauptladung *f*, Perkussionsladung *f*, Zündladungskörper *m* ~ **composition** Zündsatz *m* ~ **housing** Zündladungskapsel *f* ~ **percussion pellet of fuse** Zündbolzen *m* ~ **plunger casing** Zündachse *f* ~-**safe** detonationssicher ~ **slide** Detonatorschieber *m* ~ **tube** Sprengkapsel *f*

**detour, to** ~ umfahren

**detour** Umgehung *f*, Umweg *m*

**detouring** Umgehung *f*

**detract, to** ~ abziehen

**detraction of armature** Ankerabfall *m*

**detrain, to** ~ ausladen

**detraining** Ausladung *f* ~ **of troops** Truppenausladung *f* ~ **station** Ausladebahnhof *m*

**detriment** Benachteiligung *f*, Nachteil *m*

**detrimental** schädlich (rdo)

**detrital** aus Geröll *n* bestehend ~ **deposit** Trümmerlagerstätte *f* ~ **sand** Geröllsand *m*

**detritus** Felsgeröll *n*, Geschiebe *n*, Gesteins-

schutt *m*, Trümmermasse *f*

**detruck, to** ~ ausladen

**detrucking,** ~ **area** Ausladegebiet *n* ~ **point** Auslade-platz *m*, -stelle *f*

**detunability** Verstimmbarkeit *f*

**detune, to** ~ aus der Resonanz *f* bringen, verstimmen

**detuned** verstimmt

**detuning** Verstimmen *n*, Verstimmung *f* ~ **of resonant circuit** Schwingkreisverstimmung *f* ~ **width** Störbreite *f*

**deuterated target** die zu beschließende Deuteriumsprobe *f*

**deuteride** Deutrid *n*

**devaluation** Abwertung *f*

**devastate, to** ~ verheeren, verwüsten

**devastation** Verheerung *f*, Verwüstung *f*

**develling** Reißprozeß *m*

**develop, to** ~ abwickeln, aufschließen (mine), ausbauen, ausbilden, ausgestalten, durchbilden, entfalten, (parachute) entwickeln, hervorbringen **to** ~ **a green tone on storing** vergrünen

**developable** abwickelbar, entwickelbar ~ **surface** abwickelbare Fläche *f*

**developed** ausgereift ~ **armament probable error** Treffwahrscheinlichkeit *f* der Waffe ~ **blade area of propeller** abgewickelte Flügelfläche *f* ~ **projection** Abwicklung *f* ~ **surface** Flächenprojektion *f* ~ **view of plunger** Kolbenabwicklung *f*

**developer** Entwickler *m* (film), Entwicklungsflüssigkeit *f*, Rufer *m*

**developing** Entwickeln *n* (film) ~ **by airpassage** Luftgangentwicklung *f* ~ **of piping** Lunkelbildung *f*

**developing,** ~ **agent** Entwicklungsflüssigkeit *f* ~ **bath** Entwicklungsbad *n* ~ **equipment** Entwicklungsgeräte *pl* ~ **hanger** Filmrahmen *m* ~ **ink** Entwicklungsfarbe *f* ~ **liquid** Entwicklungsflüssigkeit *f* ~ **padding machine** Entwicklungsfoulard *m* ~ **paper** Entwicklungspapier *n* ~ **rack** Entwicklungsrahmen *m* ~ **solution** Entwicklerlösung *f* (photo) ~ **speed** Entwicklungsgeschwindigkeit *f* ~ **tank** Entwicklerbottich *m* ~ **time** Hervorrufungszeit *f* ~ **tray** Entwicklerschale *f*

**development** Abwälzung *f*, Abwicklung *f* (geom.), Ausbildung *f*, Bildung *f*, Entfaltung *f* (Fallschirm), Entstehung *f*, Entwicklung *f*, Entwicklungsstadium *n*, Verlauf *m*, Weiterentwicklung *f* ~ **of conical section** Kegelabwicklung *f* ~ **of cracks** Entstehung *f* von Rissen *pl* ~ **of crystal** Kristall-ausbildung *f*, -bildung *f* ~ **of a cylinder** Abwicklung *f* eines Zylindermantels ~ **of edge** (cartography) Abwicklung *f* des Randes ~ **of heat** Wärmeentwicklung *f*

**development,** ~ **factor** Bildspurzeit *f*, Entwicklungsfaktor *m* ~ **shock** Entfaltungsstop *m* (Fallschirm) ~ **solution** Reihenentwicklungslösung *f* ~ **study** Ausbauplan *m* ~ **(al) work** Entwicklungsarbeit *f*

**deviability** Ablenkbarkeit *f*

**deviate, to** ~ abirren, (a ray) abknicken, ablenken, abweichen, (rays) abzweigen, auseinanderweichen, umleiten, verschwenken **to** ~ **from course** vom Fahrweg *m* abkommen

**deviated well** abgewichene Bohrung *f*

**deviating** abweichend ~ **mirror** Umlenkspiegel *m* ~ **prism** Umlenkprisma *n* ~ **roller** Umlenkrolle *f*

**deviation** Abkommen *n*, Ablenkung *f*, Abmaß *n*, Abtrieb *m*, Abtrift *f*, Abweichung *f*, Auseinanderweichen *n*, Ausschlag *m*, Deviation *f*, Linksweiche *f*, Regelunterschied *m*, (of pencil or beam) Richtungsfehler *m*, Schwenkung *f*, Sprengweite *f* (artil.), Streuung *f*, Umleitung *f*
**deviation, ~ from course** Kursabweichung *f*, Versetzung *f* **~ from the desired value** Regelabweichung *f* **~ of frequency** Frequenzabweichung *f* **~ from the index value** Abweichung *f* vom Sollwert **~ of the plumb line** Ablenkung *f* des Lotes **~ from right course** Versegelung *f* **~ in slope** Steilheitsabweichung *f* **~ from trajectory** Seitenabweichung *f* **~ from true bearing** Abweichung *f* von der wahren Richtung **~ from vertical** Scheinlotschwankung *f* **~ of wind** Abweichung *f* des Windes **~ from zero** Nullabweichung *f*
**deviation, ~ amplifier** Ablenkungsverstärker *m* **~ bridge** Vergleichsbrücke *f* **~ card** Abweichungskarte *f* (aviat.) **~ check** Ablenkungsprüfung *f*, Deviations-kontrolle *f*, -nachprüfung *f* **~ clock** Abweichungs-, Deviations-tafel *f* **~ control** Ausregelung *f* der Abweichung **~ curve** Ablenkungs-, Deviations-linie *f* **~ due to diurnal (or seasonal) factors** Peilstrahlwegablenkung *f* **~ figure** Deviationswert *m* **~ flag** Flugwegablagezeiger *m* **~ indicator** Abweichungsanzeiger *m* **~ line** Ablenkungs-, Deviations-linie *f* **~ logbook** Kompaßtagebuch *n* **~ mark** Deviationspfahl *m* **~ meter** Abweichungsanzeiger *m* **~ pole (or post)** Deviationspfahl *m* **~ prism** Ablenkungsprisma *n* **~ ratio** Hubverhältnis *n*, Modulationsindex *m* **~ recorder** Abweichungsanzeiger *m* **~ roller** Ablenkungsrolle *f* **~ sensitivity** Ablenkungsempfindlichkeit *f* **~ stress** Spannungs-unterschied *m*, -abweichung *f* **~ table** Ablenkungstafel *f* **~ variable** Abweichungsgröße *f*
**device** Apparat *m*, Apparatur *f*, Ausrüstung *f*, Erdachtes *n*, Erfindung *f*, Gerät *n*, Instrument *n*, Meßgerät *n*, Mittel *n*, Plan *m*, Vorrichtung *f* **~ used for tilting (or tipping)** Kanter *m*
**device, ~ for adjusting** Arretiervorrichtung *f*, Festlegemittel *n*, Hemme *f*, Rast *f* **~ for adjusting lock** Sperrvorrichtung *f* **~ for arresting** Arretiervorrichtung *f*, Hemme *f*, Rast *f* **~ for carrying centering** Gerüsthalter *m* **~ for controlling gear** Arretiervorrichtung *f* **~ for the drawing of plans of elevation** Aufrißzeichenvorrichtung *f* **~ for flattening rectifier currents** Glättungseinrichtung *f* **~ for holding down (a clamp or jack)** Niederhalter *m* **~ for hollow grinding** Hohlkehlenschleifvorrichtung *f* **~ for instantaneous high rate discharge** Stoßbelastungsgerät *n* **~ for locking** Arretiervorrichtung *f*, Hemme *f*, Rast *f* **~ for long range initation of detonation** Fernzündgerät *n* **~ for lubricating axles** Bohr *n* zum Schmieren der Achsen **~ for maintaining tautness (or constant pressure)** Prallhaltevorrichtung *f* **~ for narrowing** (textiles) Deckvorrichtung *f* **~ for protecting colored ornaments** Farblichtorgel *f* **~ for room protection** Raumschutzgerät *n* **~ for setting** Arretiervorrichtung *f*, Hemme *f* **~ for smoothing rectifier currents** Glättungseinrichtung *f* **~ for stopping** Arretiervorrichtung *f*, Festlegemittel *n*, Hemme *f*, Rast *f* **~ for stopping the card**

**(or rose)** Rosen-abhebevorrichtung *f*, -arretiervorrichtung *f* **~ for stopping lock** Sperrvorrichtung *f* **~ for writing tables** Tabellenschreibeinrichtung *f*
**devil's claw** Kropfeisen *n* (naut.), Teufelsklaue *f*
**deviler** (cotton) Wolfer *m*
**devise, to ~** dichten, entwerfen, erfinden
**devised** kontruiert
**devitalize, to ~** inert machen
**devitrification** (glass) Versteinung *f*
**devitrified glass** entglastes Glas *n*
**devitrify, to ~** entglasen
**devoid, ~ of electrodes** elektrodenlos **~ of physically interfering substances** materienlos (electron.) **~ of sense** gegensinnig **~ of sense of logic** widersinnig
**devour, to ~** verschlingen
**devulcanization** Entvulkanisation *f*
**devulcanizing** Entvulkanisation *f*
**dew, to ~** betauen
**dew** Tau *m* **~ cap** Taukappe *f* **~ drop** Tautropfen *m* **~ point** Entparaffierung *f*, Taupunkt *m* **~ point thimble** Taupunktfühler *m* **~ retting** Tauröste *f*
**Dewar vessel** Dewar-Gefäß *n*
**dewater, to ~** entwässern
**dewatering** Entwässerung *f* **~ plant** Pumpanlage *f* **~ tower** Entwässerungsturm *m*
**dewax, to ~** entwachsen, paraffinieren
**dewaxing** Entparaffinierung *f*
**deweylite** Eisengymnit *m*, Eisenhymnit *m*
**dewindite** Dewindit *m*
**dewretting** Landrotte *f*, Luftröste *f*
**dexterity** Fertigkeit *f*, Geschicklichkeit *f*, Gewandtheit *f*
**dexterous** geübt
**dextrin** Dextrin *n*
**dextrinized potato flour** Kartoffelwalzmehl *n*
**dextrogyrate** rechtsdrehend **~ quartz** Rechtsquarz *m*
**dextropolarization** Rechtspolarisation *f*
**dextropropagating surge (or wave)** rechtslaufende Welle *f*
**dextrorotatory** rechtsdrehend
**dextrorotation** Rechtsdrehung *f*
**dextrose** Traubenzucker *m*
**dezinc, to ~** entzinken
**dezincing** Entzinkung *f*
**DF** (~ direction finding) Peilen *n*, Peilwesen *n*
**dhow** Dhau *f*
**diabase** Diabas *m*
**diabetometer** Diabetometer *n*
**diacathode ray** Kanalstrahl *m*
**diacaustic** brechene Brennfläche *f*, Brechungsbrennfläche *f*
**diacetic acid** Diessigsäure *f*
**diacetone alcohol** Diazetonalkohol *m*
**diacetyl** Diacetyl *n*, Diacetyldioxim *n* (dioxime) **~ dioxime** Dimethyglyoxim *n*
**diaclase** Bruchstelle *f*
**diadochite** Phosphoreisensinter *m*
**diaeresis** Trema *n*
**diagenesis** Diagenese *f*
**diagnostic routine** Diagnoseprogramm *n*, Fehlersuchprogramm *n* (info proc.)
**diagonal** Diagonale *f*, über Eck *n*, Gegendiagonale *f*, Schräge *f*, Winkellinie *f*; diagonal, quer, schräg **~ arch** Gratbogen *m* **~ bar** Schrägeisen

*n* ~ **bogie coupling** Diagonalkupplung *f* ~ **bond** (masonry) Festungsverband *m* ~ **brace** Diagonalstrebe *f* ~ **bracing** Diagonalverband *m*, Spannkreuz *n* ~ **bracing wire** Kreuzkabel *n*, Spanndrahtdiagonale *f* ~ **cross brace** Kreuzverstrebung *f* ~ **cut** Schrägschnitt *m* ~ **cutting pliers** Seitenschneider *m* ~ **direction** Quere *f* ~ **eyepiece** Steilsichtprisma *n* ~ **fabric** Diagonalstoff *m* ~ **fault** Diagonalverwerfung *f* ~ **layer** Diagonalschlag *m* ~ **member** Diagonalstab *m*, -strebe *f* ~ **members** Strebenkreuz *n* ~ **pitch** Schrägteilung *f* ~ **pull** Schrägzug *m* ~ **reinforcement** Diagonalbewehrung *f* ~ **relationship** (in periodic system) Diagonalbeziehungen *pl* ~ **rib** Gratbogen *m* ~ **rod** Schrägstab *m* ~ **rolling** Schrägwalzen *n* ~ **rolling process** Schrägwalzverfahren *n* ~ **stay** Kreuzspreize *f* ~ **strut** Diagonalstrebe *f* ~ **ties** Kreuzzangen *pl* ~ **trussing** Querstrebe *f* ~ **weave** Diagonalbindung *f*

**diagonalization** Diagonalisieren *n*

**diagonally across** schrägüber

**diagram** Abbildung *f*, Bild *n*, Diagramm *n*, Figur *f*, Illustration *f*, Kurve *f*, Kurvenbild *n*, Kurvendarstellung *f*, Plan *m*, Rechenschema *n*, Schaltung *f*, Schaubild *n*, Schaulinie *f*, Schautafel *f*, (circuit) Schema *n*, Tafel *f*, Zeichnung *f*

**diagram,** ~ **of connections** Leitungsschema *n*, Schaltbild *n* ~ **of connections for railroad stations** Blockschaltplan *m* ~ **of crank effort** Drehkraftlinie *f* ~ **for effect of range of stress on fatigue strength** Dauerfestigkeitsschaubild *n* ~ **of errors** grafische Fehlertafel *f* ~ **of forces** Kraftschema *n* ~ **of heights** Höhendiagramm *n*

**diagram,** ~ **lines** Serienlinien *pl* ~ **paper** Indikatorpapier *n* ~ **recorder** Diagrammapparat *m* ~ **sheet** Kurvenblatt *n* ~ **space pattern** Richtungscharakteristik *f* ~ **tachometer** Kurventachymeter *n*

**diagrammatic** bildlich, grafisch, schaubildlich, zeichnerisch ~ **drawing** Grafikon *n*, Schemazeichnung *f* (aircr.) ~ **illustration** Blockzeichnung *f* ~ **plan (or view)** Schemabild *n* ~ **representation** grafische oder schematische Darstellung *f* ~ **section** Prinzipschnitt *m* ~ **sketch** Prinzipskizze *f*

**diagrammatical** kurvenmäßig

**diagraph** Storchschnabel *m*

**diahochite** Diadochit *m*

**dial, to** ~ mit dem Kompaß aufnehmen, markscheiden, die Nummer wählen, die Nummernscheibe aufziehen oder drehen, wählen **to** ~ **through** durchwählen **to** ~ **(or set up) a number** einstellen **to** ~ **unit's digit** einerwählen

**dial** Anzeiger *m*, Bussole *f*, Drehscheibe *f*, Einstellscheibe *f*, Nummernschalter *m*, Nummernscheibe *f*, Scheibe *f*, Skala *f*, Skalenblatt *n*, (graduated) Skalenscheibe *f*, Teilkreis *m*, Teilkreis *m*, Teilung *f*, Triebscheibe *f*, Trommel *f*, Wähler *m*, Wählscheibe *f*, Wählvermittlung *f*, Zeigerplatte *f* ~ **of direction finder** Ablesescheibe *f* (Peilen)

**dial,** ~ **apparatus system** Zeigerapparatsystem *n* ~ **box** Nummernscheibenkästchen *n* ~ **card** Windrose *f* ~ **central office** Vermittlungsamt *n* mit Wahlbetrieb *m*, Wahlamt *n*, Wähleramt *n* ~ **depth gauge** Meßuhrtiefenlehre *f* ~ **disk** Griffplatte *f* ~ **drive** Rippscheibenantrieb *m*

~ **exchange area** Netz *n* mit Wahlbetrieb *n* ~ **face** Skala *f*, Zifferblattvorderseite *f* ~ **feed** Revolvertisch *m* ~ **gauge** Anzeigegerät, Fühlhebel *m*, Meßuhr *f*, Zeigerinstrument *n* ~ **gauge stop** Meßuhranschlag *m* ~ **gauge tracer** Meßuhr *f* ~ **indicator bracket** Fühluhrhalter *m* ~ **indicator stop** Fühlhebelanschlag *m* ~ **glass** Schutzglas *n* für Zifferblatt ~ **housing** Zählwerkgehäuse *n* ~ **hum** Freizeichen *n*, Summton *m* ~ **indicator** Trommelzeiger *m* ~ **intercommunicating system** Wählkästchen *n* ~ **key** Nummernscheibenkontakt *m* ~ **knob** Dreh-, Zeiger-knopf *m* ~ **light resonance indicator** Leuchtzeiger *m* ~ **line** Wählanschluß *m* ~ **mechanism** Anzeigemechanismus *m* ~ **micrometer** Uhrschraublehre *f* ~ **needle** Rippnadel *f* ~ **pencil wheel** Fingerscheibe *f* (teleph.) ~ **plate** Nummernscheibe *f* ~ **pointer** Trommelzeiger *m* ~ **prefixes** Kennzahlen *pl* (teleph.) ~ **private branch exchange** Privatnebenstellenanlage *f* mit Wahlbetrieb ~ **pulse** Wählimpuls *m* ~ **pulse spring** Nummernschalterkontaktfeder *f* ~ **pulses supervising relay** Wahlbegleitrelais *n* ~ **scale** Ringskala *f* ~ **scoring machine** Zeigersynchronisator *m* ~ **service** Wählerbetrieb *m*, Wählbetrieb *m* ~ **sight** Einblick *m*, Rundblickfernrohr *n*, Visierfernrohr *n*, Zeigervisier *n*, Zeigerzieleinrichtung *f*, Winkelzielfernrohr *n* ~ **speed indicator** Nummernschalterprüfgerät *n* ~ **switch** Finger-, Nummern-scheibe *f*, Nummernschalter *m*, Zuteilungsscheibe *f* ~ **system** automatisches System *n*, Wählsystem *n* ~ **system of cordless B position** B-Platz *m* mit Wahlzusatz *m* ~ **system equipment (or installation)** Wahleinrichtung *f* ~ **system tandem operation** Fernsprechbetrieb *m* mit selbsttätiger Durchgangsvermittlung *f* ~ **telephone set** Fernsprechapparat *m* für Wahlbetrieb *m* ~ **telephone system** Fernsprechnetz *n* mit Wahlbetrieb ~ **templet scale** Strichplattenteilkreis *m* ~ **test** Skalenprüfung *f* ~ **tester** Prüfeinrichtung *f* für Nummernscheiben ~ **thermometer** Zeigerthermometer *n* ~ **thickness indicator** Meßuhrdickenmesser *m* ~ **toll circuit** Fernwahlleitung *f* ~ **(ing) tone** Amtszeichen *n* ~ **tone** Wählton *m*, Amtsfreizeichen *n* ~ **unit** Fernschalt(zusatz)gerät *n*

**dialing** Einstellung *f* einer Nummer, Fernwahl *f* mit Nummernscheibe oder Tastensatz, Wählen *n* ~ **of units** Einerwahl *f*

**dialing,** ~ **A position** A-Platz *m* mit Wahlzusatz ~ **device** Impulsgeber *m* ~ **impulse** (supervisory control) Anreizkontakt *m*, Wählimpuls *m*, Wählstromstoß *m* ~ **jack** Wählklinke *f* ~ **key** Wählschalter *m* ~ **out** Fernwahl *f* ~ **tone** Amtsfreizeichen *n*, Amtssummerzeichen *n*, Aufforderungssignal, Signalspule *f* ~ **trunk** Einstellweg *m*

**dialkylperoxide** Dialkylperoxyd *n*

**diallage rock** Gabbro *m*

**dialyze, to** ~ dialysieren

**dializer** Dialysator *m*

**dialogite** Dialogit *m*

**dialogue recording** Sprachaufnahme *f*, Sprechaufnahme *f*

**dialyser** Dialysierapparat *m*

**dialysis** Dialyse *f*

**dialystic** dialysisch

**diamagnetic** diamagnetisch ~ **body** diamagnetischer Körper *m* ~ **substance** diamagnetischer Stoff *m*

**diamagnetism** Diamagnetismus *m*

**diameter** Durchmesser *m*, Durchschnitt *m*, Kreisdurchmesser *m*, Stärke *f* **with a narrow internal** ~ englumig ~ **and gauged length of test bars** Probestababmessung *f* ~ **class** Stärkeklasse *f*

**diameter,** ~ **of the baffle** Blendemesser *m* ~ **of ball** Kugeldurchmesser *m* ~ **of ball impression** Kugeleindruckmesser *m* ~ **of a bearing** Bohrung *f* eines Lagers ~ **of the bolt with circle** (or **of the bolthole circle**) Lochkreisdurchmesser *m* ~ **of bore** Bohrung *f*, Kaliber *n*, Lochdurchmesser *m* ~ **of chamfer circle** Spiegeldurchmesser *m* ~ **of chuck work** Drehdurchmesser *m* für Futterarbeit ~ **of coil** Spulendurchmesser *m* ~ **of core** Aderstärke *f* ~ **of diaphragm** Blendendurchmesser *m* (opt.) ~ **of faceplate** Planscheibendurchmesser *m* ~ **of furnace throat** Gichtweite *f* ~ **of gear cut** Raddurchmesser *m* ~ **of gear ring addendum circle** Zahnkranzkopfkreisdurchmesser *m* ~ **of hole** Lochweite *f* ~ **of impression** Eindruckdurchmesser *m* ~ **of lug** Fahnendurchmesser *m* (Kollektor) ~ **of orbit** Bahndurchmesser *m* ~ **of projectile** Geschoßdurchmesser *m* ~ **of root circle** Fußkreisdurchmesser *m* ~ **of shank** Schaftdurchmesser *m* ~ **of thread** Gewindestärke *f* ~ **of wire** Draht-dicke *f*, -durchmesser *m*, -stärke *f* ~ **of the workpiece** Werkstückdurchmesser *m*

**diametral** diametral ~ **pitch** Durchmesserteilung *f*, Modulteilung *f* ~ **pitch threads** Diametral-Pitch-Gewinde *n* ~ **voltage** Durchmesserspannung *f*

**diametrical** diametral, entgegengesetzt, gegenüber(liegend) ~ **clearance** Radialspiel *n* ~ **clearance measuring instruments** Radialluftmeßgerät *n* ~ **pitch** Teilkreisdurchmesser *m* ~ ~ **voltage** größte verkettete Spannung *f*

**diametrically opposite** diametral, entgegengesetzt, gegenüber(liegend)

**diamine** Diamin *m*

**diaminobenzene** Diamidobenzol *n*

**diaminotoluene** Diamidotoluol *n*

**diamond** Diamant *m*, Raute *f*, Spießkant *f* ~ **annular bits** Diamantbohrkronen *pl* ~ **antenna** Rhombusantenne *f* ~ **bit** Diamantkrone *f* ~ **bort** Diamant-bort *m*, -abfall *m* ~ **chips** Diamantsplitter *m* ~ **cone** Diamantkegel *m* ~ **core boring bit** Diamantkernbohrer *m* ~ **core heads** Kernbohrkrone *f* mit Diamanten besetzt ~ **crossing** Doppelherzstück *n*, Kreuzungsstück *n*, Spitzwinkelkreuzung *f* ~ **cut** Rautenschnitt *m* ~ **cutter** Diamantschleifer *m* ~ **die** Ziehdiamant *m* ~ **drawing dies** Diamantziehsteine *pl* ~ **dresser** Abdrehdiamant *m* ~ **drill** Diamantbohrer *m* ~ **dust** Diamantine *f* aus Glas ~ **fabric** Atlastrikot *m* ~ **fine-boring** Diamantfeinstbohrwerke *pl* ~ **fishing crown** Diamantfangkrone *f* ~ **grinding wheel** Diamantschleifscheibe *f* ~ **ink** Glasätztinte *f* ~ **mat** Rautenmatte *f* ~ **mortar** Diamantmörser *m* ~ **pass** Rautenkaliber *n*, (in rolling) Rhombus *m*, Spießkantkaliber *n* ~ **point bit** Spitzmeißel *m* ~ **point tool** Diamantspitzstahl *m* ~ **pointed cutting tool** eckiggeschliffener Drehstahl *m* ~

**pyramid hardness** Diamantpyramidenhärte *f* ~ **reamer** Diamanterweiterungsbohrer *m* ~ **rock-drill crown** Diamantbohrkrone *f* ~ **saw** Diamantsäge *f* ~ **shaped** rautenförmig ~ **shaped antenna** Rhombusantenne *f* ~ **speroconical penetrator** Diamantprüfspitze *f* ~ **structure** Diamantgitter *n* ~ **tipped wheel** Schleifscheibenabrichtdiamant *m* ~ **toolholder** Diamanthalter *m* zum Abdrehen der Schleifscheibe ~ **trueing device for grinding wheels** Schleifscheibenabrichtvorrichtung *f* ~ **turning tools** Diamantdrehstrahl *m* ~ **winding** Gleichspulenwicklung *f* ~ **wire drawing** Diamantzug *m* von Drähten ~ **work** Netzverband *m* ~

**diamonding** Werfen *n*

**dianisidine** Dianisidin *n*

**diaper** Diapyr *n*, Durchspießungsfalte *f*

**diapered** gemustert

**diaphane** Diaphanie *f*

**diaphanic,** ~ **base paper** Buntglasrohpapier *n* ~ **paper** Buntglaspapier *n*

**diaphanometer** Diaphanometer *n*, Lichtdurchlässigkeitsprüfer *m*

**diaphanoscope** Diaphanoskop *n*, Durchleuchtungsgerät *n*

**diaphanous** durchleuchtend, durchscheinend, transparent

**diaphany** Transparentpapierdruck *m*

**diaphone** Diafon *n*, Heuler *m*

**diaphragm, to** ~ abblenden, abdecken **to** ~ **out** ausblenden

**diaphragm** Blende *f*, Diaphragm *n*, Dose *f*, Düsenplatte *f*, Fensterblende *f*, Lautsprecherdose *f*, Leitapparat *m* (Turbine), Lichtblende *f*, Lochblende *f*, Luftschwinger *m*, Mehrlochdüse *f*, Membran(e) *f*, Membrandose *f*, Öffnungsblende *f*, federnde Platte *f*, Querschotte *f*, Querwand *f*, Rasterblende *f*, Schallblech *n*, poröse Scheidewand *f*, Schwingplättchen *n*, Schwingwand *f*, Stauscheibe *f*, federnde Trennscheibe *f*, Trennungsplatte *f*, Zwischenboden *m*, Zwischenwand *f*

**diaphragm** ~ **adjusting ring** Blendeneinstellring *m* ~ **assembly** Trennscheibe *f* (Triebwerk *n*) ~ **atomizer** Lochplattenzerstäuber *m* ~ **cap** Membrandeckel *m* ~ **carrier** Leitradträger *m* ~ **case** Federplattengehäuse *f* ~ **chamber** Ausgleichskammer *f*, (motor) Dosenkammer *f* ~ **with circular corrugations** Riffelmembrane *f* ~ **displacement** Membranauslenkung *f* ~ **gauge** Plattenfederlehre *f* ~ **governor** Membranregler *m* ~ **holder** Blendeneinsatz *m* ~ **housing** Membranblock *m* ~ **leaf** Blendenlamelle *f* ~ **leaf guide ring** Lamellenführungsring *m* ~ **membrane** Schwingblatt *n* ~ **method** (for pencil modulation) Blendenverfahren *n* ~ **moving like a piston** als Ganzes schwingende Membran(e) *f* ~ **oscillation** Membranschwingung *f* ~ **photometer** Blendfotometer *m* ~ **piston** Membrankolben *m* ~ **plane** Blendenebene *f* ~ **plate** Stauscheibe *f* ~ **plate of the crankcase** Kurbelgehäusezwischenwand *f* ~ **pressure gauge** Plattenfedermanometer *n* ~ **pump** Membranpumpe *f* ~ **rays** Strahlabblender *m* ~ **ring** Abdeckblende *f* ~ **ring mode filter** Ringblendenfilter *n* ~ **securing screw** Membranhalteschraube *f* ~ **setting ring** Irisblendeinstellung *f* ~ **spring** Membranfeder *f* ~ **suction pump**

Diamant-, Membran-saugpumpe f ~ support Blendenträger m ~ turret Blendenrevolver m ~ type rate-of-climb indicator Stauscheibenvariometer n ~ valve Membranventil n

**diaphragmless microphone** membranloses Mikrofon n

**diapositive** Diapositiv n, Durchsichtbild n, Glasbild n

**diaprojection** Durchprojektion f

**diary** Tagebuch n

**diaschistic rock** Schizolith m

**diascleral lamp** Diaskleralleuchte f

**diascope** Diaskop n

**diaspore** Diaspor n

**diastafor** Diastofor n

**diastase** Diastase f

**diastatic** diastatisch

**diastimeter** Distanzmesser m

**diastolic shock** Klappenstoß m

**diastrophism** Diastrophismus m

**diathermancy** Diathermansie f, Durchlässigkeit f für ultrarote Strahlen, Wärmedurchlässigkeit f

**diathermic** wärmedurchlässig

**diathermy** Diathermie f ~ knife Hochfrequenzkauter m

**diatom ooze** Diatomschlamm m

**diatomaceous** Filtergur f ~ earth Bergmehl n, Diatomeen-erde f, -pelit m, Fetton m, Kieselgur f

**diatomic** zweiatomig

**diatomite** Kieselgur f ~ layer Diatomitschicht f

**diazo coating** Diazolbelag m

**diazobenzene** Diazobenzol n

**diazonium compound** Diazoniumverbindung f

**diazotizable** diazotierbar

**diazotozation** Diazotierung f

**diazotizing** Diazotierung f

**dibasic** zweibasisch

**dibble, to** ~ dibbeln

**dibbling machine** Dibbelmaschine f

**dibenzyl** Dibenzyl n

**dibromethane** Äthylenbormid n, Dibromäthan n

**dibromobenzene** Dibrombenzol n

**dibutylether** Dibutyläther n

**dicalcium silicate** Bikalziumsilikat n, doppeltkieselsaurer Kalk m

**dichlorethylene** Azetylendichlorid n, Dichloräthylen n

**dichloride** Doppelchlorid n

**dichlorobenzene** Dichlorbenzol n

**dichloroethyl sulfide** Dichloräthylsulfid n

**dichloromethane** Dichlormethan n

**dichotomic** dichotomisch

**dichroic fog** dichroitischer oder zweifarbiger Schleier m

**dichroism** Dichroismus m

**dichroite** Dichroit m

**dichromatic fog** dichriotischer oder zweifarbiger Schleier m

**dictaphone** Diktafon n ~ reception Speicherempfang m (Schnelltelegraf)

**dictate, to** ~ diktieren

**dictating,** ~ machine Diktiergerät n ~ tube Sprechschlauch m

**dictograph** Diktiermaschine f

**diderichite** Diderichit m

**Didot point system** Didotpunktmaß n

**didymium** Didyn n

**die, to** ~ erlöschen, sterben to ~ away abklingen, ausschwingen, dämpfen, auf Null abnehmen, verhallen to ~-cast spritzen, eine Form f pressen to ~-cast threads Gewinde pl spritzen to ~ ~ down abflauen, abklingen, ausschwingen, gedämpft werden to ~ out abklingen, auslaufen, ausschwingen, erblassen, (of waves) verklingen

**die** Backe f, Blockschnitt m, Form f, Gesenk n, Gesenkform f, (screw) Gewindebacken m, Gußform f, Lochstempel m, (bottom) Matrize f, Matrizenuntersatz m (print.), Mönch m, Patrone f, Prägeplatte f (print.), Prallstück n, Preß-platte f, -stempel m, -werkzeug n, Schneidbacken m, Schneideeisen n, Schnitt m (mach.), Spritzgußform f, Stanze f, Stanzstempel m, Unterstanze f, Unterstempel m, Würfel m, (drawing) Ziehstein m, (tube mfg.) Ziehtrichter m

**die,** ~ adapter Haltevorrichtung f ~ angle Ziehwinkel m ~ base Düsenkörper m ~ block Stanzblock m ~ block swage Gesenk n ~ box mold Matrize f ~ brushing Prägepolieren n ~ carrier plate Aufspannplatte f

**die-cast** Gußgesenk n, Spritzguß m ~ condenser Spritzgußkondensator m ~ metal Druckguß m, gespritztes Metall n, Spritzgußmetall n ~ mount Spritzgußfassung f ~ steel Schalengußstahl m

**die-casting** Schalen-, Spritz-, Stempel-guß m ~ alloy Spritzlegierung f ~ die steel Spritzgußformstahl m ~ machine Spritzgußmaschine f ~ practice Spritzgußtechnik f ~ pressure (in founding) Spritzdruck m ~ process Spritzgußverfahren n ~ research Druckgußforschung f

**die,** ~ cavity Gesenkvertiefung f ~ change Steinauswechslung f ~ chaser Schneideisengewindestrehler m ~ coupling Fang-glocke f, -luppe f ~ cutting (or printing) Stanzen n ~ cutting plate Stanzblech n ~ engraving Gaufriergravierung f ~ filling machine Matrizenfeilmaschine f ~ forged point of frog im Gesenk n geschmiedete Herzstückspitze f ~ forging Gesenkschmiederei f ~ grid Isolierrost m ~ guide Führungsbacke f ~ hammer Nummernschlägel m ~ head Mundringhalter m, Schneidkopf m, Setzkopf m, Werkzeugaufnehmer m ~ head chaser Gewindeschneidkopfbacken pl ~ hob Schneideisengewindebohrer m, Strehlbohrer m für Schneideisen ~ hob for cutting open dies Backengewindebohrer m ~ holder Schneideeisenhalter m, Tasse f ~-maker Werkzeugmacher m ~-making Schnittbau m ~ mark Stempeleinschlag m ~ marks Matrizenmarken pl ~ matrix technique Matrizentechnik f ~ milling Gesenkfräserei f ~ mold Spritzgußform f ~ nipple Fangnippel m, Gewindefänger m ~ plate Gewinde-, Gewindeschneide-eisen n, (cutting) Stanzmatrize f, Zieheisen n, Ziehring m ~ presser part Gesenkpreßteil m ~ reservoir Materialkammer f ~ restriction Stauscheibe f ~ ring Matrizenhalter m, Mundring m ~ shaper Kopierstoßmaschine f ~ shaping Gesenkhoblerei f ~ shop Gesenkmacherei f, Präge f ~ sinker Gesenkfräsmaschine f ~ sinking Gesenkfräsen n ~ sinking cutter Gesenkfräser m ~ sinking machine Gesenkkopiermaschine f ~ sinking mill Nachformfräsmaschine f ~ slotting machine Räummaschine f ~ space

**lights** Pressenbeleuchtung *f* ~ **stamp** Prägepresse *f*, Stempelschaft *m* ~ **stamping** Stahlstichdruck *m* ~ **steel** Gesenkstahl *m*
**die-stock** Gasrohrschneidekluppe *f*, Gewindeschneideeisen *n*, Gewindeschneidkluppe *f*, Schneideisenhalter *m* ~ **chaser** Schneideisenstrehler *m* ~ **die** Schneideisen *n*
**die,** ~ **tooling shop** Gesenkschlosserei *f* ~ **tracing and finishing shop** Gesenkmacherei *f* ~ **wiping paper** Wischpapier *n* für Prägestempel ~ **work** Gesenkarbeit *f*
**dielectric** Dielektrikum *n*, Isolierzwischenlage *f*, Nichtleiter *m* ~ **(al)** dielektrisch ~ **absorption** dielektrische Absorption *f* ~ **antenna** dielektrische Antenne *f* ~ **break-down** Spannungsdurchschlag *m* ~ **capacity** Kapazität *f* ~ **coefficient** Dielektrizitätskonstante *f* ~ **conductance** dielektrische Leitfähigkeit *f* ~ **constant** Dielektrizitätskonstante *f*, elektrische Durchlässigkeit *f*, spezifische induktive Kapazität *f* ~ **current** Verdrängungsstrom *m* ~ **displacement** dielektrische Verschiebung *f* ~ **hysteresis** dielektrische Hysterese *f*, dielektrische Nachwirkung *f* ~ **leakance** dielektrische Ableitung *f* ~ **losses** dielektrische Verluste *pl* ~ **path** dielektrische Leitung *f* ~ **polarization** dielektrische Polarisation *f* ~ **radiator** dielektrischer Strahler *m* ~ **resistance** Isolationswiderstand *m* ~ **rigidity** Durchschlagsfestigkeit *f*, dielektrische Festigkeit *f* ~ **rod antenna** Stielstrahlantenne *f* ~ **rod-radiator** dielektrischer Stabstrahler *m* ~ **strength** Durchschlagsfestigkeit *f*, Durchschlagkraft *f*, dielektrische Festigkeit *f*, Spannungsfestigkeit *f*, dielektrische Widerstandsfähigkeit *f* ~ **test voltage** Prüfspannung *f* ~ **vane phase shifter** dielektrischer Phasenschieber *m* ~ **viscosity** dielektrische Nachwirkung *f* ~ **wave** Rohrwelle *f* ~ **waveguide** dielektrischer Wellenleiter *m* ~ **wire** dielektrischer Leiter *m*
**dielectrometer** Dekameter *n*
**dielectronic capture** dielektronischer Einfang *m*
**dies** Rohrgewinde *n* ~ **for die casting** Spritzmatrize *f* ~ **for screw-stock** Schneideisen *n* für Gewindekluppen
**Diesel,** ~ **aero engine** Dieselflugmotor *m* ~ **air compressor** Dieselluftverdichter *m* ~ **combustion** Dieselverbrennung *f* ~ **convertible ignition unit** Dieselumstellzünder *m* ~ **cycle** Dieselprozeß *m* ~ **d. c. generating set** Dieselgleichstromaggregat *n* ~ **electric drive** Dieselelektrischer Antrieb *m* ~ **engine** Dieselmotor *m*
**Diesel-engine,** ~ **with air cell and controlled turbulence** Dieselmotor *m* mit Luftwirbelkammer ~ **with divided combustion** Wirbelkammermotor *m* ~ **with precombustion chamber** Vorkammermotor *m* ~ **with solid injection** kompressorloser Dieselmotor *m*, Strahlzerstäubungsmaschine *f*
**Diesel-engine,** ~ **crane** Dieselmotorkran *m* ~ **fuel** Dieseltriebstoff *m* ~ **motor** Dieselflugmotor *m* ~ **oil** Dieselmotorenbrennstoff *m* ~ **practice** Dieselmotorenbetrieb *m*
**Diesel,** ~ **fuel** Dieseltriebstoff *m*, Gasöl *n* ~ **fuel distributor pump** Kraftstoffeinspritzpumpe *f* ~ **fuel injector** Zerstäuberdüse *f* ~ **index number** Dieselzahl *f* ~ **indices** Dieselindizes *pl* ~ **lubricating oil** Dieselmotorenschmieröl *n* ~ **oil** Diesel-, Treib-öl *n* ~ **principle** Dieselverfahren

*n* ~ **rail coach** Dieseltriebwagen *m* ~ **rail crane** Dieseldrehkran *m* ~ **reefer** Dieselkühllastwagen *m* ~ **tractor** Dieselschlepper *m* ~ **truck** Diesellastwagen *m* ~**-type aircraft engine** Flugzeugschwerölmotor *m*
**diethylamine** Diäthylamin *n*
**diethylaniline** Diäthylanilin *n*
**diethylclyclohexane** Diäthylclyclohexan *n*
**diethylether** Diäthyläther *m*
**diethylhexane** Diäthylhexan *n*
**diethylmalonic ester** diäthylmalonsaures Äthyl *n*
**differ, to** ~ abweichen, sich nicht decken, differieren, unterscheiden, verschieden sein **to** ~ **from** abliegen **to** ~ **fundamentally** sich grundlegend unterscheiden **to** ~ **in phase** ungleiche Phase *f* haben
**difference** Abstand *m*, Abweichung *f*, Differenz *f*, Irrung *f*, abweichendes Merkmal *n*, Unstimmigkeit *f*, Unterschied *m*, (angle) Verlustwinkel *m*, Verschiedenheit *f*, Zerwürfnis *n*
**difference,** ~ **in altitude (or elevation)** Höhenunterschied *m* ~ **in brightness** Helligkeitsunterschied *m* ~ **of a condenser** dielektrischer Verlustwinkel *m* ~ **between dead and observed reckoning** Besteckversetzung *f* ~ **in depth of modulation** Differenz *m*, Modulationstiefe *f* ~ **in dimension** Meßdifferenz *f* ~ **in focal adjustment** Einstellungsdifferenz *f* (opt.) ~ **of focus** Fokusdifferenz *f* ~ **in the height of liquid levels** (in turbines) Spiegelhöhenunterschied *m* ~ **of height level** Höhenunterschied *m* ~ **in illumination** Helligkeitsunterschied *m* (photo) ~ **in intensity** Helligkeitsunterschied *m* ~ **in latitude** Breitengradunterschied *m* ~ **in level of the lines of vision** Höhenschielen *n* ~ **in longitude** Längengradunterschied ~ **of opinion** Meinungsverschiedenheit *f* ~ **in (or of) phase** Phasendifferenz *f*, Phasenverschiebung *f* ~ **of phases** Phasenunterschied *m* ~ **of potential** Spannungsunterschied *m* ~ **in scale** Maßstabs-differenz *f*, -unterschied *m* ~ **of shade** Nuancenunterschied *m* ~ **in sound intensity** Lautstärkenunterschied *m* ~ **in structure** Strukturverschiedenheit *f* ~ **of summation frequencies** Summendifferenz *f* ~ **in temperature** Temperaturunterschied *m*, Wärmeunterschied *m* ~ **of thread pitch from the real size** Abweichung *f* der Gewindesteigung *f* vom Sollmaß *n* ~ **of time** Zeitunterschied *m* ~ **of vergence** Neigungsdifferenz *f*
**difference,** ~ **amplifier** Eingangsdifferenzverstärker *m* ~ **channel** Differenzkanal *m* ~ **frequency** Differenzfrequenz *f* ~ **method** Differenzverfahren *n* ~ **number** Neutronenüberschuß *m* ~ **scheme method** Differenzschemaverfahren *n* ~ **tone** Differenzton *m*
**different** abweichend, ander, anders, ungleich, verschieden **to be** ~ verschieden sein **of** ~ **ages** von ungleichem Alter **of** ~ **height (length or magnitude)** verschieden groß **of** ~ **sign** von verschiedenem Vorzeichen *n* ~ **kinds of needles** Rippnadelsorten *pl* ~ **types** unterschiedliche Sorten *pl*
**differentiable** differenzierbar, unterscheidbar
**differential** Ableitung *f*, Differential *n*; veränderlich ~ **across metering orifice** Wirkdruck *m* ~ **action** Differenzial-bewegung *f*, -wirkung *f* ~ **aileron** Differentialquerruder *n* ~ **aileron control** Querrudersteuerung *f* ~ **alarm** Diffe-

rentialmelder *m* ~ **amplifier** Differentialverstärker *m* ~ **analyzer** Differentialgleichungslöser *m* ~ **arrangement** Differentialschaltung *f* ~ **axledrive bevel gear** Differentialantriebskegelrad ~ **balance** Differenztonfaktorwage *f* ~ ~ **bellows** Differenzdruckdose *f* ~ **bevel gear** Ausgleichskegelrad *n* ~ **booster** Zusatzmaschine *f* mit Differentialerregung ~ **brake** Differentialbremse *f* ~ **bridge** Differentialbrücke *f* ~ **cage** Ausgleichsgetriebegehäuse *n* ~ **calculus** Differenz-, Differential-rechnung *f* ~ **capacitor** Differentialkondensator *m* ~ **carrier** Differentiallagerkasten *m* ~ **casing** Differentialgehäuse *n* ~ **chain block** Differentialflaschenzug *m* ~ **circuit** Gegenschaltung *f* ~ **coefficient** Differentialquotient *m* ~ **coefficient of the ‚nth' order** Differentialquotient *m* „n"ter Ordnung ~ **coil** Betätigungsspule *f* ~ **compound winding** Antikompoundwickelung *f*, Gegenverbundwickelung *f* ~ **compound-wound dynamo** Gegenverbunddynamo *m* ~ **compound-wound motor** Gegenverbundmotor *m* ~ **condenser** Differentialkondensator *m* ~ **connection** Differential-, Gegen-schaltung *f* ~ **contraction** Differentialschrumpfung *f* ~ **control** Differenzdruckregler *m* ~ **counting chart** Differentialzähltafel *f*, Zähltafel *f* ~ **cross section** differentieller Wirkungsquerschnitt *m* ~ **current reading** Differenzstromanzeige *f* ~ **deviation** Differenzabweichunganzeiger *m* ~ **displacement** differentielle Versetzung *f* ~ **distortion** Verzeichnungsunterschied *m* ~ **draft gauge** Differentialzugmesser *m* ~ **drive** Differentialantrieb *m* ~ **driving gear** Ausgleichsgetriebe *n* ~ **duplex system** Differentialgegensprechsystem *n* ~ **earth ringing** Durchrufen *n* in Simultanschaltung mit Erdrückleitung ~ **equation** Differentialgleichung *f* ~ **exitation** Gegenkompoundierung *f*, Gegenverbunderregung *f* ~ **feed** Differentialvorschub *m*, Differentialvortrieb *m* ~ **feed controlling device** Differentialnachlaßvorrichtung *f* ~ **fly frame** Differentialfleier *m* ~ **frequency** Differenzfrequenz *f* ~ **gain control** selektiver Abschwächer *m*, Differentialverstärkungsregelung *f* ~ **galvanometer** Differentialgalvanomesser *m* ~ **gap (sensitivity limits)** Ansprechgrenzen *pl* ~ **gauge** Differentialmanometer *n* ~ **gauge equipment** Differenzmeßanlage *f* ~ **gear** Ausgleichsgetrieberad *n*, Gegengetriebe *n* ~ **gear (ing)** Differential-, Ausgleichs-getriebe *f* ~ **gear casing** Ausgleichgehäuse *n* ~ **gear with self-locking device** selbstsperrendes Ausgleichgetriebe *f* ~ **gearsetting mechanism** Ausgleichsperre *f* ~ **generator** Differentialgeber *m* ~ **head** Druckgefälle *n* ~ **heat of dilution** Differentialverdunstungswärme *f* ~ **housing** Differentialgehäuse *n* ~ **linkage** Differentialgestänge *n* ~ **lock** Ausgleichsgetriebesperre *f*, Ausgleichssperre *f* ~ **lock lever** Ausgleichsgetriebesperrhebel *m* ~ **manometer** Differential-, Fein-druckmesser *m*, Differentialmanometer *n* ~ **method** Differentialverfahren *n* ~ **nut** Differenzmutter *f* ~ **operation** Differentialbestätigung *f* ~ **pinion** Differentialritzel *m* ~ **pinion shaft** Ausgleichsachse *f* ~ **piston** Differentialkolben *m*, Stufenkolben *m* ~ **pressure** Differentialdruck *m*, Differenzdruck *m*, Druckunterschied *m*, Wirkdruck *m* ~ **pressure pro-**

ducer Wirkdruckgeber *m* ~ **pressure switch** Differentialdruckschalter *m* ~ **pressure transmission pipes** Wirkdruckleitung *f* ~ **pressure transmitter** Wirkdruckwandler *m* ~ **pulley** Differentialrolle *f* ~ **pulleys** Differentialflaschenzug *m* ~ **pump** Stufenkolbenpumpe *f* ~ **pupilloscope** Differentialpupilloskop *n* ~ **quantity** Differential *n*, unendlich kleine Größe *f* ~ **quotient** Differentialquotient *m* ~ **reactance** Differentialdrosselung *f* ~ **reactor** Differentialdossel *f* ~ **receiver** Differentialempfänger *m* ~ **recovery rate** differentielle Erholungsrate *f* ~ **reduction** Differentialreduktion *f* ~ **regulator** Differentialregelwerk *n*, Hauptstromwerk *n* ~ **relay** Differentialrelais *n* ~ **repeater** Differentialempfänger *m* ~ **repeating coil** Differentialtransformator *m* ~ **resistance** innerer Widerstand *m* ~ **settlement** Setzungsunterschied *m* ~ **shaft** Ausgleichs-, Differential-welle *f* ~ **sheave** Differentialrolle *f* ~ **side gear** Hinterachswellenrad *n* ~ **slide wire** Schleifdraht *m* der Meßbrücke ~ **spider** Differentialkreuz *n* ~ **spider pinion** Ausgleichskegelrad *n* ~ **star piece** Sternausgleichsstück *n* ~ **steam calorimeter** Differentialkalorimeter *n* ~ **susceptibility and permeability** differentielle Suszeptibilität *f* und Permeabilität *f* ~ **synchro** Differentialdrehmelder *m* ~ **synchro transmitter** Differential-Drehfeldgeber *m* ~ **tariff** Staffeltarif *m* ~ **thermometer** Metallthermometer *n* ~ **tone** Differenzton *m* ~ **transformer** Ausgleichtransformator *m*, Brückenüberträger *m*, Differentialtransformator *m*, Differentialüberträger *m* (dreispuliger) ~ **transmitter** Differential-geber *m*, -sender *m* ~ **turbine** Gegenlaufmotor *m* ~ **walking beam** Umsetzhebel *m* (der Stabilisatorsteuerung) ~ **wheel** Differentialrad *n* ~ **wheel cylinder** Stufenradbremszylinder *m* ~ **winding** Differentialwicklung *f* ~ **winding motion** Differentialaufwindebewegung *f* (textiles) an einer Spinnmaschine *f*

**differentially wound** differentialgewickelt
**differentiate, to** ~ ableiten (math.), (with respect to) differenzieren, unterscheiden
**differentiated dike rock** Spaltungsgestein *n*
**differentiating** ~ **amplifier** Differenzierverstärker *m* ~ **circuit** Differentialschaltung *f* differenzierendes Netzwerk *n*
**differentiation** Ableitung *f*, Differenzierung *f* ~ **due to gravity sinking of crystals** gravitative Kristallisationsdifferentiation *f*
**differentiator** Differenziergerät *n*, Differenzierschaltung *f* ~ **circuit** Differenzierschaltung *f*
**differently tuned** verschieden abgestimmt
**difficult** beschwerlich, schwer, schwierig ~ to **dissolve** schwerlöslich ~ **to fuse (or melt)** schwer schmelzbar ~ **to please** schwer zufriedenzustellen ~ **to sell**, schwer verkäuflich ~ **to volatilize** schwerflüchtig ~ **terrain** schwieriges Gelände *n*
**difficulties** Weiterungen *pl* ~ **in levelling** Egalisierschwierigkeiten *pl*
**difficultly**, ~ **fusible** heißgradig, schwerschmelzbar, strengflüssig ~ **meltable** hochschmelzend ~ **permeable** schwerdurchlässig
**difficulty** Beschwerde *f*, Beschwerlichkeit *f*, Hindernis *n*, Schwierigkeit *f* ~ **with washers** Dichtungsschwierigkeit *f*

**diffract, to** ~ ablenken, beugen, biegen, brechen, (light ray) zerlegen
**diffracting** beugsam
**diffraction** Abbeugung *f*, Ablenkung *f* (opt.), Beugen *n*, Beugung *f*, Diffraktion *f*, Zerstreuung *f* ~ **at grazing incidence** streifende Beugung *f* ~ **of light** Beugung *f* des Lichts ~ **of rays** Strahlenbrechung *f* ~ **of sound** Schalldruckverstauung *f*
**diffraction,** ~ **angle** Beugungswinkel *m* ~ **beam** Beugungsbündel *n* ~ **diagram** Elastogramm *n* ~ **disc** Beugungsscheibchen *n* ~ **fringes** Beugungsfransen *pl*, Beugungsstreifen *pl* ~ **grating** Beugungsgitter *n* ~ **grating spectroscope** Gitterspektroskop *n* ~ **hand-grating spectroscope** Gitterhandspektroskop *n* ~ **image** Beugungsbild *n* ~ **lines** Beugungsstreifen *pl* (opt.) ~ **pattern** Beugungs-bild *n*, -figur *f* ~ **picture** Beugungsbild *n* ~ **polarization** Dityndallismus *m*, Doppelbeugung *f*, Pseudododichroismus *m* ~ **scattering** Berechnungsstreuung *f* ~ **screen** Beugungsgitter *n* ~ **sensitivity** Ablenkungsempfindlichkeit *f* ~ **spectrum** Gitterspektrum *n* ~ **stripping** Beugungsstripping *n*
**diffractive properties** Lichtabbeugung *f*
**diffusate** Diffusat *n*
**diffuse, to** ~ ausbreiten, ausschütten, diffundieren, durchwandern, einwandern, hindurchwandern (Gase), verteilen, wandern, zerstreuen **to** ~ **away** wegdiffundieren **to** ~ **into** hineindiffundieren, -wandern **to** ~ **into each other** ineinanderlösen **to** ~ **out** herausdiffundieren **to** ~ **to the sides** nach den Seiten *pl* streuen
**diffuse** zerstreut ~ **bounderies** unscharfe Kanten *pl* ~ **light** zerstreutes Licht *n* ~ **porous** zerstreutporig ~ **reflection** zerstreute oder diffuse Reflektion *f* ~ **reflection device** Remissionsansatz *m* ~ **reflection factor** gestreute Reflektion *f* ~ **scattering** Streustrahlung *f* ~ **spot** Schwärzungsfleck *m* ~ **transmission density** Durchlässigkeitsfaktor *m*
**diffused** diffus, durchdiffundiert, unscharf ~ **air tank** Druckbecken *n* ~ **junction** Diffusionsschicht *f* ~ **layer** eingewanderte Schicht *f* (Transistor) ~ **reflection of light** diffuse Lichtreflexion *f* ~ **reflection of sound** diffuse Schallflexion *f*
**diffuseness** Unschärfe *f*
**diffuser** Austreuung *f*, Diffuseur *m*, Diffusor *m* (Triebwerk *n*), Ladeleitrad *n*, Leitapparat *n* beim Kreiselgebläse *n*, Leitrad *n*, Leitvorrichtung *f* (jet), Lichtstreuungskörper *m*, Lufttrichter *m*, Zerstäuber *m* ~ **bottom** Diffusorboden *m* ~ **chamber** Leitkammer *f* ~ **deflection vane** Diffusorleitschaufel *f* ~ **neckpiece** Diffusoreinsatz *m* ~ **scrim** diffusierendes Gewebe *n* ~ **section** Laderleitapparat *m* ~ **thermometer** Diffusorthermometer *n* ~ **top** Diffusorhaube *f* ~ **types of shade** Diffusionsreflektor *m* (phys.) ~ **vane** Diffusor-, Zerstreuer-schaufel *f*
**diffusibility** Diffusionsvermögen *n*, Diffusor *m*
**diffusible** diffusionsfähig
**diffusing,** ~ **away** Fortdiffundierung *f* ~ **groove** Streurille *f* (Scheinwerfer *m*) ~ **lens** Streuscheibe *f*
**diffusion** Ausbreitung *f*, Ausstreuung *f*, Diffusion *f*, Diffusor *m*, Durchdringung *f*, Durchfahrt *f*, Durchfließung *f*, Einwanderung *f*, Lichtstreu-

ung *f*, Wanderung *f*, Wetteraustausch *m* (min.), Zerstreuung *f* ~ **of a pair of voids** Diffusion *f* eines Leerstellenpaares ~ **of the point image** Bildpunktverlagerung *f* ~ **by reflection** gestreute Reflexion *f* ~ **of sound** Schallverteilung *f* ~ **by transmission** gestreute Durchlassung *f* ~ **in velocity tensor** Diffusionstensor *m* im Geschwindigkeitsraum
**diffusion,** ~ **apparatus** Diffusionsgerät *n* ~ **apparatus with bottom discharge** Diffusionsapparat *m* mit unterer Entleerung ~ **apparatus with hydraulic for closing bottom manholes** Diffusionsapparat *m* mit unterem hydraulischem Mannlichverschluß ~ **barrier** Diffusionswand *f* ~ **capacitance** Diffusionskapazität *f* ~ **cell** Diffuseur *m*, Diffusionsgefäß *n* ~ **cloud chamber** Diffusionsnebelkammer *f* ~ **coefficient** Diffusionskonstante *f* ~ **cone** Konusmembran *f* ~ **constant** Diffusionskonstante *f* ~ **cross section** Diffusionsquerschnitt *m* ~ **current** Diffusionsstromdichte *f* ~ **effect** Streuung *f* ~ **equation** Diffusionsgleichung *f* ~ **flow** Diffusionsstromdichte *f* ~ **halo (or halation)** Diffusionslichthof *m* ~ **heat** Diffusionswärme *f* (gas) ~ **juice preheater** Diffusionssaftvorwärmer *m* ~ **kernel** Diffusionsintegralkern *m* ~ **pulp water** Ablaufwasser *n* der Diffusion ~ **pump** Diffusionspumpe *f* ~ **resistance** Ausbreitungswiderstand *m* ~ **theory of rectifier** Gleichrichter-Diffusionstheorie *f* ~ **vane** Diffusorschaufel *f* (jet) ~ **waste water** Diffusionswässer *pl*
**diffusional jog** Diffusionssprung *m*
**diffusionless process** Überklappvorgang *m*
**diffusivity** Diffusionsvermögen *n*, Temperaturleitzahl *f*
**dig, to** ~ **(out)** ausgraben, (trench) ausheben, baggern, einstechen, graben, roden, schachten, schanzen, schürfen, stechen **to** ~ **in** einbauen, sich eingraben, einschanzen, untergraben **to** ~ **out** ausbaggern **to** ~ **up** umgraben **to** ~ **a ditch** einen Graben *m* ziehen **to** ~ **up the earth** den Boden *m* aufgraben **to** ~ **peat** Torf *m* stechen
**digest, to** ~ aufschließen, digerieren, einweichen, versetzen **to** ~ **with water** einsumpfen, faulen
**digest** Schrifttumsabriß *m*, Übersicht *f*
**digester** Autoklav *m*, Dampftopf *m*, Digestor *m*, Entvulkanisationskessel *m*, Kocher *m*, Zellstoffkocher *m*
**digesting** Versetzung *f* ~ **flask** Digestionskolben *m*
**digestion** Aufschließung *f*, Aufschluß *m* ~ **of emulsion** Reifung *f* ~ **bottle** Digerierflasche *f* ~ **flask** Druckkolben *m* ~ **nucleus** Reifkeim *m*
**digger** Bergmann *m*, Gräber *m*, Löffelbagger *m*, Rodemaschine *f*
**digging** Erdarbeit *f*, Graben *m*, Schurf *m* ~ **of coal** Kohlengewinnung *f*
**digging,** ~ **bars** Straßenaufreißer *m* ~ **depth** (dredging) Grabtiefe *f* ~ **force at bucket lip** Umfangskraft *f* (Bagger *m*) ~ **resistance** Grabwiderstand *m* ~ **spoon** Schaufel *f* zum Ausheben des Erdreichs aus dem Stangenloch
**diggings** Minenbezirk *m*
**digit** Gruppe *f*, Stufe *f* (automatic teleph.), Wählstufe *f*, Ziffer *f* ~ **absorber** Impulsvernichter *m* ~ **absorbing selector** Apparat *m* zur Stromstoßunterdrückung, Nummernschlucker *m*, Stromstoßunterdrücker *m* ~ **absoption**

Unterdrückung *f* von Stromstößen, Unterdrückung *f* von Wählimpulsen ~ **conductor** Stellenleiter *m* ~ **control of features** Ziffernkontrolle *f* der Einrichtungen ~ **driver** Ziffernschreiber *m* ~ **field limit** Stellenfeldgrenze *f* ~ **key strip transfer key** Wählumschaltetaste (WU) *f* ~ **key strip transfer relay** Weichenrelais *n* ~ **line** Ziffernleitung *f* ~ **pulse** Stellenimpuls *m* ~ **punching** Normal-, Zahlen-lochung *f* ~ **selector** Impulsbuchsen *pl* ~ **sense channel** Stellenlese-, Stellenrichtungs-kanal *m*

**digital** ziffernmäßig; Taste *f* (piano) ~ **code** Ziffernkode *f* ~ **coding** Digitaldarstellung *f* ~ **computer** Digitalrechner *m*, numerische Rechenmaschine *f* ~ **counter** Ziffernzählwerk *n* ~ **display** Digitalanzeige *f* ~ **readout** ziffernmäßige Ablesung *f* ~ **representation** digitale Darstellung *f* ~ **signal** numerisches Signal *n* ~ **storage element** Ziffernspeicherelement *n* ~ **time and frequency meter** Zeit- und Frequenzmesser *m* ~ **transducer** Digitalempfänger *m* ~ **unit** Recheneinheit *f*

**digitalized data** digitalisierte Daten *pl*

**digitize, to** ~ digital darstellen, digitalisieren, umsetzen

**digitonine** Digitonin *n*

**digits, of many** ~ vielstellig

**diglycol tubular powder** Diglykolröhrenpulver *n*

**diglycolnitrate flake powder** Diglykolnitratblättchenpulver *n*

**digression** Abschweifung *f*, Umschweif *m*

**dihedral** Flächenwinkel *m*, V-Form *f*, (angle) V-Stellung *f*; dihedrisch, geschränkt, V-förmig (aviat.), zweiflächig ~ **angle** dihedricher Winkel *m*, Dihedralwinkel *m*, V-Form *f*, Öffnungswinkel *m* (of nav. lights) ~ **classes** Diederklassen *pl* ~ **leading edge** V-Stellung *f* der Flügelvorderkante

**dihedron** ... Dieder ....

**dihydric** dihydratisch

**dihydroxybenzene** Dioxylbenzol *n*

**dihydroxyl-benzene** Hydrochinon *n*

**dihydroxynaphthalene** Dioxynaphthalin *n*

**dihydroxyquinone** Dioxychinolin *n*

**dihydroxytoluene** Dioxytoluol *n*

**dike, to** ~ **(or dam) in** eindeichen

**dike** Buhne *f*, Buhnenwurzel *f*, Damm *m*, Deich *m*, Fangbuhne *f*, Flußdamm *m*, Gang *m*, Gesteinsgang *m*, Hafendamm *m*, Landanschluß *m* der Buhne, Pier *m*, Schlange *f*, Sommerdeich *m*, Wall *m* ~ **with chambers** Kammerdeich *m* ~ **of wattled piles** Schlickfänger *m* ~ **directly alongside the waterway** Schardeich *m* ~ **rocks** Ganggesteine *n*

**diking** Dämmung *f*

**dilapidated** baufällig

**dilatability** Dehnbarkeit *f*, Dehnungsfähigkeit *f*

**dilatable** ausdehnbar, dehnbar

**dilatancy** Verfestigung *f*, (of sand) Volumvergrößerung *f*

**dilatation** Dilatation *f* ~ **of length** Längendehnung *f*

**dilate, to** ~ ausdehnen, sich auslassen, ausweiten, erweitern

**dilated,** ~ **pupil** erweiterte Pupille *f* ~ **seam** schwebendes Flöz *n*

**dilation** Ausdehnung *f*, Dehnung *f*, Volumenmehrung *f* ~ **in melting** Schmelzausdehnung *f*

**dilational wave** Dehnungswelle *f*

**dilative (or dilatable) soil** dehnbare Böden *pl*

**dilatometer** Ausdehnungs-, Dehnbarkeits-, Dehnungs-messer *m*, Dilatometer *n*

**dilatory** säumig ~ **plea** Fristgesuch *n*

**diligent** arbeitsam, fleißig

**diluent** Streckmittel *n*, Streckungsmittel *n*, Verdünnungskörper *m*, Verdünnungsmittel *n*, Verschnittmittel *n*

**dilute, to** ~ schwächen, strecken, verdünnen, verflüssigen, versetzen, verwässern **to** ~ **with water** wässern

**dilute** dünn, dünnflüssig ~ **medium** Dünntrübe *f*

**diluted** verdünnt ~ **ammoniacal water** verdünntes Gaswasser *n* ~ **chromium alums** eingebettetes Chromaleun *n* ~ **crystals** eingebettete Kristalle *pl* ~ **soluble oil** Bohrwasser *n* ~ **solution** verdünnte Lösung *f* (chem.)

**diluting** Versetzung *f* ~ **medium** Verdünnungsmittel *n*

**dilution** Übergang *m*, Verdünnung *f* ~ **value** Löslichkeitszahl *f*, Verschnittfähigkeit *f* (of lacquers, etc.)

**diluvial** diluvial, diluvisch ~ **ore** Wascherz *n*

**dim, to** ~ abblenden, (light) abdunkeln, schwächen; blaß, finster, matt, trübe **to** ~ **head lamps** Scheinwerfer *pl* abblenden **to** ~ **one's headlights** abblenden **to** ~ **searchlights** Scheinwerfer *pl* abblenden

**dim,** ~ blaß, finster, matt, trübe ~ **light** Abblendlicht *n* ~**-out cap** Abblendkappe *f* ~ **red** dunkelrot ~ **red heat** Dunkelrotglut *f*

**dimension, to** ~ bemessen, dimensionieren

**dimension** Abmaß *n*, Abmessung *f*, Ausdehnung *f*, Ausmaß *n*, Bemessung *f*, Dimension *f*, Größe *f*, Maß *n*, Maßangabe *f*, Sollmaß *n*, Umfang *m* ~ **of the design** Ausgabemaß *n* ~ **constant** Dimensionskonstante *f* ~ **error** Maßfehler *m* ~ **line** Maßlinie *f*

**dimensional** die Abmessung betreffend, dimensional, größenmäßig, maßdimensional ~ **analysis** Ähnlichkeits-betrachtung *f*, -prinzip *n*, Dimensionalbetrachtungen *pl* ~ **case** eindimensionaler Fall *m* ~ **constant** Maßfaktor *m* ~ **difference** Maßabweichung *f* ~ **discrepancy** Maßabweichung *f* ~ **drawing** Maßzeichnung *f* ~ **equation** Dimensionsgleichung *f* ~ **factor** Maßfaktor *m* ~ **instability** Wachsen *n* ~ **outline** Maßskizze *f* ~ **perturbation** dimensionale Störung *f* ~ **stability** Maßbeständigkeit *f* ~ **standard** Abmessungsnorm *f* ~ **unit** Maßeinheit *f* ~ **variable** benannte Größe *f*

**dimensionally stable** maßhaltig

**dimensioned** dimensioniert ~ **drawing** Maßzeichnung *f* ~ **sketch** Maßskizze *f* ~ **top spring** bemessene Feder *f*

**dimensioning** Bemessung *f*, Dimensionierung *f*, Maßgebung *f*

**dimensionless variable** dimensionslose Veränderliche *f*

**dimensions** Raummaß *n*

**dimerics** Dimeren *n* (chem.)

**dimers** Dimeren *n* (chem.)

**dimethylamidobenzol** Diäthylanilin *n*

**dimethylamine hydrochloride** Dimethylaminchlorhydrat *n*, -hydrochlorid *n*

**dimethylaminoazobenzene** Dimethylamidoazobenzol *n*

**dimethylaniline** Dimethylanilin *n*
**dimethylated** dimethyliert
**dimethyl,** ~ **benzene** Dimethylbenzol *n* ~ **glyoxime** Dimethylglyoxim *n* ~ **sulphate** Dimethylsulfat *n*
**dimetric** dimetrisch
**diminish, to** ~ abnehmen, abschwächen, einschneiden, herabmindern, kleiner werden, mindern, schwächen, verjüngen, verkleinern, verringern **to** ~ **in size** abtragen **to** ~ **a wall** eine Mauer *f* verschwächen
**diminished interval** vermindertes Intervall *n*
**diminishing,** ~ **factor** Verkleinerungsfaktor *m* ~ **plank** Verjüngungsplanke *f* ~ **tariff** Staffeltarif *m* ~ **value** Wertminderung *f*
**diminution** Abfallen *n*, Abnahme *f*, Kleinerwerden *n*, Minderung *f*, Schmälerung *f*, Verkleinerung *f*, Verminderung *f*, Verringerung *f* ~ **of air** Luftabnahme *f* ~ **of density** Dichtigkeitsverlust *m* ~ **of pressure** Spannabfall *m* ~ **in size** Abtragung *f*
**dimmer** Abblendregler *m*, Blendglas *n* (opt.), (theater) Bühnenlichtregulator *m*, Dunkelpunktstufe *f*, Dunkelschalter *m*, Mattscheibe *f* (electr.), Schwächungsvorrichtung *f*, Strahlablender *m* ~ **board** Bühnenlichtstellwerk *n* ~ **coil** Regulierspule *f* ~ **control** Abblendregler *m*, Bühnenstellwerk *n*, Helligkeitsregler *m* ~ **switch** Abblendschalter *m*, Bühnenlichtregulator *m*, Dunkelschalter *m*
**dimming** Abblendung *f* ~ **action** Abblendvorbang *m* ~ **adjustment** Abblendeinstellung *f* ~ **cap** Abblendekapotte *f* ~ **effect** Blendleistung *f* ~ **filament** Abblendfaden *m* ~ **process** Abblendvorgang *m* ~ **resistor** Abdunkelwiderstand *m* ~ **ring** Blindring *m* ~ **sheet** Abblendblech *n* ~ **switch** Abdunkelschalter *m*, Dunkelpunktstufe *f*, Helldunkelschalter *m*, Verdunkler *m*
**dimness** Feste *f*, Finsternis *f*
**dim-nickel-plated** matt vernickelt
**dimorphism** Dimorphie *f*, Doppelgestaltung *f*, Zweispaltigkeit *f*
**dimorphous** dimorph
**dimple, to** ~ versenken, warzen (metall.), ziehsenken
**dimple** Grübchen *n* (in der Wange), Vertiefung *f*
**dimpled** angesenkt
**dinaphthyl** Dinaphthyl *n*
**dinas** Ganister *m*
**Dinas,** ~ **brick** Dinasstein *m* ~ **clay** Dinaston *m*
**dinghy** Beiboot *n*, Dingi *n*, Dingy *n*, kleines Ruderboot *m*, Schlauchboot *n* ~ **release** Schlauchbootauslösung *f* ~ **release lever** Schlauchbootabwurfhebel *m*
**dinging hammer** Ausbeulhammer *m*
**dinitroglycerin explosive** Dinitroglyzerinsprengstoff *m*
**dinking die** leicht abschneidende Stanze *f*
**dint** Beule *f*
**diode** Diode *f*, Gleichrichter *m*, Richtröhre *f*, Ventilröhre *f* mit zwei Elektroden, Zweielektrodenröhre *f* ~ **circuit** Diodenschaltung *f* (electr.) ~ **date** Diodendurchlaßbereich *m* ~ **detection** Dioden-, Zweipol-gleichrichtung *f* ~ **detector** Diodengleichrichter *m* ~ **load resistance** Diodenaußenwiderstand *m* ~ **matrix** Diodenmatrix *f* ~ **path** Diodenstrecke *f* ~

**probe-type-voltmeter** Tastvoltmeter *n* ~ **rectification** Diodengleichrichtung *f*, Zweipolgleichrichtung *f* ~ **triode** Binode *f*, Binodröhre *f* ~ **tube** Zweipolröhre *f* ~ **valve** Ventilröhre *f* mit zwei Elektroden
**diophase** Kupfersmaragd *m*
**diopter** Brechkrafteinheit *f* (opt.), Brechungseinheit *f*, Diopter *n*, Dioptrie *f*, Lichtbrechkraft *f* einer Linse, Meterlinse *f*, Seh-schlitz, -spalt *m*, Visierlineal *n* ~ **focusing** Okulareinstellung *f* ~ **focusing mount** Dioptrieneinstellung *f* ~ **hair** Diopterfaden *m* ~ **scale** Dioptrienteilung *f* ~ **slit** Diopterschlitz *m* ~ **thread** Diopterfaden *m*
**dioptometer** Dioptometer *n*
**dioptric** dioptrisch ~ **(al)** lichtbrechend ~ **device** Durchblickvorrichtung *f* ~ **lens** diotrische Linse *f*
**dioptrics** Dioptrik *f*, Lichtbrechungslehre *f*
**dioptrometer** Dioptrometer *n*
**dioptry** Brechungseinheit *f*
**diorite** Diorit *m*, Grünstein *m*
**dioxane addition** Dioxanzusatz *m*
**dioxide** Dioxyd *n*
**dip, to** ~ abbeizen, abbrennen, betupfen, durchhängen, einfallen (min.), einsenken, eintauchen, eintränken, schöpfen, tauchen, tunken **to** ~ **brass** Messing *n* abbrennen **to** ~ **a flag** die Flagge dippen **to** ~ **into** (einsenken) hineintauchen **to** ~ **lights** abblenden **to** ~ **nose down** abkippen **to** ~ **up** aufschöpfen
**dip** Abdachung *f* (Gelände *n*), magnetische Deklination *f*, (horizon) Depressionswinkel *m*, (of line wire) Durchbiegung *f*, Einfallen *n*, Einfallen *n* der Schichten, (of a resonance or other curve) Einsattelung *f*, Einsenkung *f*, Fallen *n*, Inklination *f*, Kimme *f*, (of lines) Leitungshang *m*, gezogenes Licht *n*, Neigung *f*, Resonanzkurve *f*, -einsattelung *f*, -senke *f*, Senke *f*, Senkung *f*, Tauchen *n*, Tauchung *f* ~ **up** ~ der Fallrichtung *f* entgegen ~ **in a curve** Delle *f* ~ **of the horizon** Kimmtiefe *f* ~ **of the paddle wheel** Schaufelradtauchung *f*
**dip,** ~ **brazing** Löten *n* im Bade, Tauchlöten *n* ~ **candle** gezogene Kerze *f* ~ **cavity** Tauchhöhlung *f* ~ **cell** Tauchzelle *f* ~ **circle** Neigungskreis *m* ~ **coat** getauchter Überzug *m* ~ **dyeing** Tauchfärbung *f* ~ **dyeing machine** Tauchfärbemachine *f* ~ **error** Kimmfehler *m* ~ **gauge** Winkelhaken *m* für Durchgangsprüfung ~ **gettering** Tauchgettern *n* ~ **hardening** Tauchhärten *n* ~ **lubrication** Tauchölschmierung *f* ~ **molding** Heißtauchen *n*, Tauchen *n* ~ **needle** Inklatorium *n*, Magnetometer *n*, Neigungsnadel *f* ~ **orbit** (of electrons) Tauchbahn *f* ~ **pipe** Tauchrohr *n*, (pickup or scoop) Fangrohr *n* ~ **rod** Meßstange *f*, Peilstab *m* (Tankanlage) ~ **roller** (immersion roller) Eintauchwalze *f* ~ **shooting** Neigungsschießen *n* ~ **slope** Neigung *f* eines Ganges ~ **soldering** Tauchlötung *f* ~ **stick** Peilstab *m* (Tankanlage) ~ **stock** Peilstab *m* ~ **tank** Tauchbehälter *m* ~ **tube** Peilrohr *n* ~ **varnishing apparatus** Lackiertauchapparat *m*
**dipentene** Dipenten *n*
**diphase** zweiphasig ~ **equilibrium** zweiphasiges Gleichgewicht *n*
**diphenic acid** Diphensäure *f*

diphenyl, ~ amine Diphenylamin *n* ~ chlorarsine Chlorsinkkampfstoff *m* ~ chloroarsine Diphenyl-aminchlorarsin *n*, -chlorarsin *n* ~ cyanarsine Zyanchlorarsinkkampfstoff *m* ~ ketone Benzophenon *n* ~ tin chloride Zinndiphenylchlorid *n*
diphosgene Perstoff *m*
diphthong Doppellaut *m*, Doppelvokal *m*
diplet Dublett *n*
diplex, ~ operation Diplexbetrieb *m*, Doppelsprechbetrieb *m* ~ radio transmission Diplexfunkübertragung *f* ~ reception Diplexempfang *m* ~ system Diplexsystem *n*, Doppelsprechsatz *m* ~ telegraph Diplextelegraf *m* ~ telegraphy Doppelschreiben *n* ~ transmission Diplexübertragung *f*
diplexer Diplexer *m*, Doppelzweckantenne *f*
diploid cell diploide Zelle *f*
diploma Diplom *n*
diplopia Doppelsichtigkeit *f*
dipole Antennendipol *n*, Antennenpaar *n*, Dipol *m*, Doppelpol (rdo), gerader Oszillator *m* ~ adjustment Dipoleinstellung *f* ~ antenna Dipolantenne *f*, Direktive *f*, Stabantenne *f* ~ approximation Dipolnäherung *f* ~ array Dipolgruppe *f*, -zeile *f* ~ connector Dipolanschlußklemme *f* ~ dipole broadening Dipol-Dipol-(Linien)-Verbreitung *f* ~ dipole interaction Dipol-Dipol-Wechselwirkung *f* ~ layer Dipolschicht *f* ~ moment dipolares Moment *n*, Dipolmoment *n* ~ pair Dipolschleifenpaar *n* ~ path Zweipolstrecke *f* ~ sections Schleifenteile *pl* ~ singularity Dipolsingularität *f* ~ source covering Dipolquellenbelegung *f* ~ sources Dipolquellen *pl* ~ surface density Dipolflächendichte *f* ~ tube Zweipolstrecke *f*
dipped, ~ candle gezogenes Licht *n* ~ electrode getauchte Elektrode *f*
dipper Büttgeselle *m*, Eintaucher *m*, Former *m*, Schöpfer *m* ~ arm Löffelstiel *m* ~ relay Dippelrelais *n* ~ slide Löffelklappe *f*
dipping Abbeizen *n*, Eintauchen *n* ~ of candles Lichtziehen *n* ~ of liquids in vessels Abstechen von Flüssigkeiten *pl* in Gefäßen
dipping, ~ colorimeter Eintauchkolorimeter *m* ~ conveyer Tauchförderer *m* ~ end Eintauchende *n* ~ filler Tauchspachtel *f* ~ fluid Tauchflüssigkeit *f* ~ frame Senkküpe *f* ~ lamp Tauchlampe *f* ~ matches into the inflammable compound Betupfen *n* der Zündhölzchen ~ motion (ship) Tauchbewegung *f* ~ needle Abweichungs-, Inklinations-nadel *f* ~ plant Tauchanlage *f* ~ process Tauchverfahren *n* ~ pyrometer Eintauchpyrometer *n* ~ refractometer Eintauchrefraktometer *n* ~ red Dochtspieß *m* ~ switch Schalter *m* zur Tauchlampe *f* ~ varnish Tauchlack *n*
dipple, to ~ abhauen
direct, to ~ anweisen, beauftragen, führen, hinrichten, hinweisen, leiten, lenken, richten to ~ past (something) vorbeiführen to ~ to zielen auf to ~ upon einführen
direct alsbaldig, direkt, explizit, gerade ~ (ly) unmittelbar ~ acting pump direkt angetriebene Pumpe *f*, direktwirkende Pumpe *f* ~ action direkte Wirkung *f*, Vorwärtsgang *m* ~ arc furnace unmittelbarer Lichtbogenofen *m* ~ arc heating direkte Lichtbogenerhitzung *f* ~ arc

nonconducting-hearth furnace unmittelbarer Lichtbogenofen *m* ohne Herdbeheizung ~ axis Längsfeld-, Pol-achse *f* ~ axis component of the voltage Längsspannung *f* ~ axis subtransient electromotive force subtransitorische Längs-EMK *f* ~ axis synchronous impedance synchrone Längsimpedanz *f* ~ axis transient electromotive force transitorische Längs-EMK *f* ~ call unmittelbare Verbindung *f* ~ capacitance to ground Teilerdkapazität *f* ~ circuit connection unmittelbare Verbindung *f* ~ communication Direktsprechen *n* ~ compass Sichtkompaß *m* ~ component of grid voltage Gitterwiderstand *m* ~ connected unmittelbar gekuppelt ~ connected motor-driven pump Pumpenaggregat *n* ~ connection Querverbindung *f* ~ connection circuit unmittelbare Verbindung *f* ~ control Direktsteuerung *f*, Vorwärtsregelung *f* ~ control autogiro Autogiro *n* mit Direktsteuerung ~ coupled unmittelbar angetrieben ~ coupled amplifier Gleichstromverstärker *m* ~ coupling direkte oder galvanische Kopplung *f* ~ cost direkte Kosten *pl* ~ current (Anoden)gleichstrom *m*
direct-current, (D. C.) ~ air flow Gleichluftstrom *m* ~ alternating-current converter (or inverter) Gleichstromwechselstromumformer *m* ~-alternating-current inverter Wechselrichter *m* ~ ammeter Gleichstrom-messer *m*, -meßinstrument *n* ~ amplification Gleichstromverstärkung *f* ~ amplifier Gleichstromverstärker *m* ~ amplifier stage Gleichstromverstärkerstufe *f* ~ arc lamp Gleichstrombogenlampe *f* ~ arc welder Gleichstromschweißumformer *m* ~ arcwelding converter Gleichstrom-Lichtbogen-Schweißumformer *m* ~ arcwelding generator Gleichstrom-Lichtbogen-Schweißgenerator *m* ~ balancer Ausgleichsmaschine *f* ~ bell Gleichstromwecker *m* ~ bridge Gleichstrommeßbrücke *f* ~ charging test method Gleichstromkapazitätsmessung *f* ~ charging unit Gleichstromladeanlage *f* ~ circuit Gleichstrombereich *m*, Gleichstromleitung *f* ~ commutator Gleichstromkommutator *m* ~ commutator meter Magnetmotorzähler *m* (mit Stromwender *m*) ~ component Gleichstromkomponente *f*, Grundhelligkeit *f* ~ component of grid voltage Gittervorspannung *f* ~ component reaches screen by way synchronizing pulses Durchschlagen *n* der Gleichstromkomponente *f* ~ compound generator Gleichstromdoppelschlußgenerator *m* ~ control voltage Steuergleichspannung *f* ~ controlled coil gleichstromvormagnetisierte Spule *f* ~ controlled three-legged coil gleichstromvormagnetisierte Spule *f* ~ converter Gleichstromtransformator *m*, Gleichstromumformer *m*, Gleichumrichter *m* ~ dialing Gleichstromfernwahl ~ distribution system Gleichstromnetz *n* ~ driving motor Gleichstromantriebsmotor *m* ~ dynamo Gleichdynamo *f*, Gleichstromdynamo *f*, Gleichstrom-erzeuger *m*, -generator *m*, -lichtmaschine *f* ~ electric energy Gleichstromenergie *f* ~ electric plant Gleichstromanlage *f* ~ electrical machine Gleichstrommaschine *f* ~ exitation Gleichstromerregung *f* ~ feed Speisegleichstrom *m* ~ galvanometer Gleichstromgalvanometer *m* ~ generator Gleichstrom-erzeuger *m*,

-dynamo *f*, -generator *m* ~ **grid current** Gittergleichstrom *m* ~ **line** Gleichstromleitung *f* ~ **magneto** Gleichstrommagnetapparat *m* ~ **mains** Gleichstrombordnetz *n* ~ **(power) mains** Gleichstromnetz *n* ~ **measurement** Gleichstrommessung *f* ~ **meter** Gleichstrommeßgerät *n*, Gleichstromzähler *m* ~ **motor** Gleichstrom-elektromotor *m*, -motor *m* ~ **network** Gleichstromnetz *n* ~ **operated receiver** Empfänger *m* für Gleichstrom, Empfänger *m* für Gleichstrom-Netzanschluß, Gleichstromempfänger *m* ~ **operated valve** Gleichstromröhre *f* ~ **output** Gleichstromentnahme *f* ~ **polarization** Gleichstromvormagnetisierung *f* ~ **potential** Gleichstrompotential *n* ~ **potential of rectifier** Relaisspannung *f* ~ **power** Gleichstromleistung *f* ~ **pulse** Gleichstromstoßabgabe *f* ~ **receiver** Empfänger *m* für Gleichstrom, Empfänger *m* für Gleichstrom-Netzanschluß ~ **(mains) receiver** Gleichstromempfänger *m* ~ **relay** Gleichstromrelais *n* ~ **resistance** Gleichstromwiderstand *m*, Ohmscher Widerstand *m* ~ **restorer** Gleichstromzuschaltung *f* ~ **ripple** Kräuselung *f* des Gleichstroms ~ **saturable coil** gleichstromvormagnetisierte Spule *f* ~ **series dynamo** Gleichstromreihenschlußmaschine *f* ~ **series motor** Gleichstromreihenschlußmotor *m* ~ **shunt motor (or dynamo)** Gleichstromnebenanschlußmaschine *f* ~ **shunt wound motor** Gleichstromnebenschluß-generator, -motor *m* ~ **signal** Gleichstromzeichen *n* ~ **signaling method** Gleichstromzeichengabe *f* ~ **source** Gleichstromquelle *f* ~ **supply** Gleichspannungsquelle *f*, Gleichstromnetz *n*, Gleichstromspeisung *f* ~ **supply voltage** Speisegleichspannung *f* ~ **telegraphy** Gleichstromtelegrafie *f* ~ **telegraphy over composited circuits** Unterlagerungstelegrafie *f* ~ **terminal** Gleichstrompol *m* ~ **testing voltage (DCTV)** Prüfspannung *f* ~ **three-wire plant** Gleichstromdreileiteranlage *f* ~ **traction** Gleichstromfahrbetrieb *m* ~ **transmission** Gleichstromübertragung *f* ~ **two-wire plant** Gleichstromzweileiteranlage *f* ~ **voltage potential** Gleichspannung *f*, Gleichspannungspotential *n* ~ **wire** Gleichstromleitung *f* ~ **working voltage (DCWV)** Nennspannung *f* (capacitor)

**direct,** ~ **cycle** (p. p.) direkter Kreislauf *m* ~ **dialling** direkte Wahl *f* ~ **(or local) direction finder** Eigenpeiler *m* ~ **drive** direkter Antrieb *m*, unmittelbarer Gang *m* ~ **drive motor** Achsmotor *m* ~ **driven blower** Gemischverteiler *m*, unmittelbar angetriebener Lader *m* ~ **driven dynamo** direkt gekuppelte Dynamomaschine *f* ~ **driven propeller** direkt angetriebene Luftschraube *f* ~ **drying** Feuertrocknung *f* ~ **earth capacitance** Teilerdkapazität *f* ~ **electric connection** galvanische Verbindung *f* ~ **engine** Motor *m* ohne Untersetzungsgetriebe ~ **exchange line** Hauptanschlußleitung *f* ~ **expenditure** direkte Kosten *pl* ~ **extraction** (with iron) Rennarbeit *f* ~ **extrusion** Warmstangenpressen *n* ~ **feed of the antenna** direkte Antennenspeisung *f* ~ **feeding mill** Einblasemühle *f* ~ **fire kiln** Rauchdarre *f* ~ **fired oil heater** direkt-beheizter Ölerhitzer *m* ~ **flow of air** Gleichluftstrom *m* ~ **fuel injection** Direkteinspritzung *f* ~ **function of** .. lineare Funktion *f* von .. ~ **heating arc**

**furnace** Lichtbogenwiderstandofen *m* ~ **hit** Treffer *m*, Volltreffer *m* ~ **illumination** Auflichtbeleuchtung *f*, gerade Beleuchtung *f* ~ **impact** gerader Stoß *m* ~ **impulse automatic telephone system** Selbstanschlußsystem *n* mit unmittelbarer Stromstoßgebung ~ **impulses** unmittelbare Stromstoßgabe *f* ~ **impulsing** unmittelbare Stromstoßgabe *f* ~ **indexing** Direktteilverfahren *n* ~ **injection** direkte Einspritzung *f*, Zylindereinspritzung *f* ~ **injection molding** angußloses Spritzen *n* ~ **inker** unmittelbar in die Leitung geschalteter Farbschreiber *m* ~ **inlet valve** unmittelbares Einlaßventil *n* ~ **interelectrode capacitance** Teilkapazität *f* ~ **international circuit** durchgehende Leitung *f* ~ **labor cost** Einzellohnkosten *pl* ~ **lateral light** direktes Licht *n* ~ **laying** direktes Richten *n* ~ **light electron microscope** Auflichtelektronenmikroskop *n* ~ **lighting** direkte Beleuchtung *f* ~ **line** Amtsleitung *f* ~ **line away from transmitter** achterliche Kurslinie *f* ~ **measurement** Direktmessung *f* ~ **measurement of length** echte Längenmessung *f* ~ **mechanical sound recording** akustische Schallaufnahme *f* ~ **method** direktes Verfahren *n* ~ **mirage** Luftspiegelung *f* nach oben ~ **mode** Direktverfahren *n* (nav.) ~ **motor drive** Einzelantrieb *m* ~ **pickup and viewing** direktes Fernsehen *n* ~ **playback** direktes Abspielen *n* ~ **pressure in pursuit** Nachdrängen *n* ~ **printer** Direktschreiber *m* ~ **printing** Direktdruck *m*, Drucktelegraf *m*, Fernschreiber *m* ~ **process** (iron) Rennarbeit *f*, (for production of wrought iron) Rennverfahren *n* ~ **process malleable iron** Renneisen *n* ~ **process slag** Rennfeuerschlacke *f*, Rennschlacke *f* in ~ **proportion** geradlinig ~ **pulses** unmittelbare Stromstoßgabe *f* ~ **pulsing system** System *n* mit direkten Stromstößen oder mit unmittelbarer Stromstoßgabe ~ **radiator loudspeaker** (without horn) trichterloser Lautsprecher *m* ~ **ratio** gerades Verhältnis *n* ~ **ray** Bodenstrahl *m* ~ **reading** Ablesung *f*, Direktablesung *f*, unmittelbare Ablesung *f*, direkt zeigend ~

**direct-reading,** ~ **compass** Nahablesekompaß *m*, Nahkompaß *m* ~ **direction finder** direkt anzeigender oder selbstanzeigender Peiler *m*, Sichtpeiler *m* ~ **frequency dial** Skala *f* mit direkter Frequenzablesung ~ **galvanometer** Galvanometer *n* mit unmittelbarer Ablesung ~ **instrument** Instrument *f* mit unmittelbarer Ablesung ~ **measuring instrument** Skalenmeßgerät *n* ~ **(transmission) measuring set** Pegelzeiger *m* ~ **radiator thermometer** Nahthermometer *n* ~ **tachometer** Nahdrehzahlmesser *m*

**direct,** ~ **reception** Suchempfang *m* ~ **record working** (teleph.) Meldeamtsbetrieb *m* ~ **recording** Direktschreibung *f* ~ **revolution counter** Nahdrehzahlmesser *m* ~ **rolling pressure** direkter Walzdruck *m* ~-**route group** Querleitungsbündel *n* ~-**routing system** System *n* mit unmittelbar gesteuerter Wahl ~ **sales** direkter Verkauf *m* ~ **selection** unmittelbare Wahl *f* ~ **self-contained drive** riemenloser Einzelantrieb *m* ~ **sound reproduction from negatives** Abhören *n* der Negative ~ **stress** Normal-beanspruchung *f*, -spannung *f* ~ **support** unmittelbare Unterstützung *f* ~ **take-off** Senkrechtstart *m* ~ **take-off and landing autogiro** Autogiro *n* für Senkrecht-start und -landung f ~ **tensile strength** Kopfzugfestigkeit f ~ **train** durchgehender Zug m ~ **transmission** direkte Ablesung f, Handgeben n ~ **tripping circuit control** unmittelbare Auslösung f ~ **trunking** Abfragebetrieb m ~ **trunking operation** Anrufbetrieb m ~ **view finder** Durchsichtssucher m (photo) ~ **view storage cathode-ray tube** Direktsichtspeicherröhre f ~ **view tube** Direktsichtbildröhre f ~ **viewing** Aufsichtsbetrachtung f, unmittelbare Betrachtung f (TV) **~-vision prism** geradsichtiges Prisma n **~-vision receiver** Direkt-Sicht-Empfänger m **~-vision spectroscope** Geradsichtspektroskop n **~-vision view finder** Mattscheibenbeobachtungsgerät n ~ **voltage** Gleichspannung f **~-voltage component** Gleichspannungskomponente f **~-voltage im⁻ pulse** Gleichstromstoß m ~ **wave** Bodenwelle f, direkte Welle f, direkter Strahl m, Oberflächenwelle f **~-wire circuit** Einfachleitung f ~ **X-ray** Röntgenstrukturanalyse f

**directed** gerichtet **to be ~ (at)** einseitig gerichtet **~ net** geleitetes Netz n ~ **ray of light** Schlaglicht n

**directing** Leiten n, (through selectors) Steuerung f ~ **beam** Richtstrahler m ~ **couple** Rückstellung f ~ **force** Direktions-, Richt-kraft f ~ **gun** Grundgeschütz n ~ **horn** Bündelungstrichter m ~ **line** Richtlinie f ~ **magnet** Richtmagnet n ~ **means** Weiche f (Rohrpost) ~ **moment** Richtmoment n ~ **pulse** Steuerstromstoß m ~ **staff** Absteckstange f

**direction** Anordnung f, Anweisung f, Einweisung f, Hinweis m, Leitung f, Nachweis m, Orientierung f, Ortsangabe f, Richtlinie f, Richtung f, Streichen n, Streichen n einer Schicht, Umsteuern n, Verhaltungsmaßregel f, Vorschrift f, Weisung f

**direction, ~ of air flow** Anblas-, Anström-richtung f ~ **of the apparent perpendicular** Scheinlotrichtung f (geophysics) ~ **of application of force** Schlagrichtung f ~ **of approach** Anflug-, Annäherungs-richtung f ~ **of arrow** Pfeilrichtung f ~ **of attack** Angriffsrichtung f ~ **of beam** Abstrahlrichtung f (rdr) ~ **in building** Baukurs m ~ **of circulation** Strömungsrichtung f ~ **of course** Laufrichtung f ~ **of current** Stromrichtung f ~ **of the current** Strömungsrichtung f ~ **of current passage** Stromdurchlaßrichtung f ~ **of cut** Schnittrichtung f ~ **of declination** Ausweichrichtung f ~ **of deflection** Ausweichrichtung f, Biegungspfeil m ~ **of dip** Inklinationsrichtung f ~ **for dissolving** Lösungsvorschrift f ~ **of drift** Treibrichtung f (aviat.) ~ **of drive** Antriebsrichtung f ~ **for evidence** Beweisbeschluß m ~ **of exposure** Sektor m der dominierenden Winde oder Wellen ~ **of fire** Feuerleitung f, -richtung f, Schußrichtung f ~ **of flame path** Flammenführung f ~ **of flames** Flammenrichtung f ~ **of flight** Bahnrichtung f, Flugbahnrichtung f, Flugrichtung f ~ **of flight over the ground** Flugrichtung f über Grund ~ **of flow** Fließ-, Strömungs-richtung f ~ **of flute** Spannutenrichtung f ~ **of a force** Kraftrichtung f ~ **of the force** Kraftsinn m ~ **of full dip of the mineral deposit** Fallinie f der Lagerstätte ~ **of gas flow** Gasführung f ~ **of**

**grain** Faserrichtung f ~ **of the gyro axis** Achsenrichtung f ~ **of incidence** Einfallsrichtung f ~ **of induction** Induktionsrichtung f ~ **of inspection** Einblickrichtung f ~ **of journey** Fahrtrichtung f ~ **of landing** Landerichtung f ~ **of lift** Auftriebsrichtung f ~ **of light** Ausstrahlungsrichtung f ~ **of the line** Linienführung f ~ **of lines of force** Kraftlinienrichtung f ~ **of march** Marschrichtung f ~ **of migration** Wanderungsrichtung f (of ions), Wanderungssinn m ~ **of motion** Bahn-, Marsch-richtung f ~ **of object to be photographed** Aufnahmerichtung f ~ **of polarization** Polarisationsrichtung f ~ **of the position line** Standlinienrichtung f ~ **of principal curvature** Hauptkrümmungsrichtung f ~ **of propagation** Ausbreitungsrichtung f, Fortpflanzungsrichtung f, (waves etc.) Fortschreitungsrichtung f ~ **of reflection** Ausfallsrichtung f ~ **of regard** Betrachtungs-, Blick-richtung f ~ **of relative wind** Anströmungsrichtung f ~ **of relative wind for zero lift** Anblasrichtung f verschwindenden Auftriebs, Flugwindrichtung f, verschwindenden Auftriebes ~ **of river flux** Flußrichtung f ~ **of rolling** Walzrichtung f ~ **of rotation** Dreh-richtung f, -sinn m; Rotationsrichtung f, -sinn m, Umdrehungsrichtung f, Umlaufrichtung f ~ **of scanning** Abtastrichtung f, Zerlegungssinn m ~ **of screwing** Schraubungssinn m ~ **of sight** Blickrichtung f ~ **of slip** Gleitrichtung f (cryst.) ~ **of stress** Spannungsrichtung f ~ **of swinging motion** Schwenkrichtung f ~ **of thrust** Schubrichtung f ~ **of track** Bahnrichtung f ~ **of transmission** Durchlaßrichtung f ~ **of transverse power feed** Plandrehrichtung f ~ **of travel** Warenlaufrichtung f ~ **of vector** Richtung f des Vektors, Vektorrichtung f ~ **of view** Blickrichtung f ~ **of vision** Visierrichtung f ~ **of winding** Wickelsinn m, Windschutzrichtung f

**direction, ~ arm** Richtungslineal n ~ **beam** Raumstrahler m ~ **bearing** Peilrichtung f ~ **changing lever** Umsteuerhebel m ~ **cone** Aufnahmetubus m ~ **controlled** richtungsabhängig ~ **cosine** Richtungskosinus m (math.) ~ **coupler** Richtkoppler m ~ **distortion** Richtungsverzerrung f ~ **distribution** Richtungsverteilung f ~ **error** Richtungsfehler m

**direction-finder** Flugrichtungsgerät n, Funkpeiler m, Peilempfänger m, Peiler m, Peilfunkeinrichtung f, Peilgerät n, Richtungsanzeiger m, Richtungsfinder m, Zielfluggerät n (rdo, nav.) ~ **accessory set** Peilzusatz m ~ **adapter** Peilvorsatz m (rdr) ~ **antenna** Peilantenne f ~ **coil** Peilspule f ~ **compass** Peilkompaß m ~ **deviation** Funk-, Richtungs-bestimmungsfehler m (rdo nav.) ~ **dial** Peilrose f ~ **frame** Peilrahmen m ~ **installation** Peilanlage f ~ **log** Funkplatte f ~ **loop** Peilrahmen m ~ **operator** Peilfunker m ~ **post** Peilstelle f ~ **receiver** Suchgerätempfänger m ~ **repeater compass** Funkpeiltochterkompaß m ~ **signal** Peilzeichen n ~ **station** Funkpeilstelle f, Peilfunkbetriebsstelle f, Peilhaus n, Peilstelle f ~ **on to target** Einpeilung f eines Zieles ~ **truck** Peilkraftwagen m ~ **unit** Peilanlage f ~ **wireless station** Bodenpeilstelle f

**direction-finding** Peil . . ., Peilen n, Peilung f, Peilwesen n, Richtungs-bestimmung f, -ermittlung f; Standortsbestimmung f ~ **control sta-**

tion Peilleitstelle *f* ~ **equipment** Peilgerät *n* ~ **error** Fehlweisung *f*, Peilfehler *m* ~ **facility** Peilanlage *f* ~ **plant** Peilfunkanlage *f* ~ **range** Peilreichweite *f* ~ **receiver** Peilfunkempfänger *m* ~ **sense switch** Peilseiteschalter *m* ~ **station** Funkpeilstelle *f*, Peilfunkstelle *f* ~ **warning set** Warnpeilgerät *n* ~ **wireless service** Bodenfunkdienst *m*

**direction,** ~ **indicator** Fahrtrichtungsanzeiger *m* (Winker, Blinker), Folgezeiger *m* ~ **indicator arm** Winkerarm *m* ~ **indicator switch** Winkerschalter *m* ~ **lamp** Richtungslampe *f* ~ **light** Richtungslicht *n* ~ **motion** Zugrichtung *f* ~ **receiving station** Richtempfangsstation *f* ~ **(or heading) relative to the air** Eigenrichtung *f* ~ **slide** Richtungsschlitten *m* ~ **and starting switch** Fahrtrichtungsanlaßschalter *m* ~ **tag** Paketanhänger *m* ~ **uncertainty** Richtungsunschärfe *f*

**directional** gerichtet, (in signal transmission) wahlweise ~ **accuracy** Richtgenauigkeit *f* ~ **action** Richtungseffekt *m* ~ **antenna** Richtantenne *f*, -strahler *m* ~ .**(n)avigation** Seitennavigation *f*, -ortung *f* ~ **beacon** Ansteuerungsfeuer *n* ~ **beam** Peilfunk *m* ~ **breakdown** Richtungsdurchschlag *m* ~ **characteristic** Richt(ungs)-charakteristik *f* ~ **coincidence** Richtungskoinzidenz *f* ~ **control** Kurssteuerung *f* ~ **control valve** Drehschieber *m* ~ **correlation** Richtungskorrelation *f* ~ **counter** Richtungsmessung *f*, Richtzähler *m* ~ **coupler** Richtkoppler *m* ~ **derivative** Richtungsabhängigkeit *f* ~ **diagram** Peilkurve *f*, Richtcharakteristik *f*, Richtkennlinie *f* ~ **difference** Richtungsunterschied *m* ~ **distortion** Richtungsfehler *m* (TV) ~ **drilling** dirigiertes Bohren *n*, Richtungsbohren *n* ~ **effect** (in film processing) Laufrichtungskoeffizient *m*, Richteffekt *m* (uni) ~ **effect** Richtwirkung *f* ~ **effect of waves** Wellenbündelung *f* ~ **error** Funkfehlweisung *f* ~ **error of the tooth flank** Flankenrichtungsfehler *m* ~ **errors** Nachteffekt *m* ~ **finder** Richtungsaufsucher *m* ~ **gain** Richtungsverstärkungsfaktor *m* ~ **gyro** Kurskreisel *m* ~ **gyro control (unit)** Kursgeberanlage *f*, Kurskreiselführgerät *n* ~ **gyroscope** Gyrorektor *m*, Kurskreisel *m*, Kurszentrale *f*, Richtungskreisel *m* ~ **homing** gerichtete Eigenpeilung *f* (rdo nav.) ~ **instability** Richtungsinstabilität *f* ~ **isolator** Richtungsisolator *m* ~ **loop antenna** Peilrahmen *m* ~ **loudspeaker** gerichteter Lautsprecher *m*, Richtlautsprecher *m*, Richtstrahler *m* ~ **marker** Richtungspfeil *m* ~ **microphone** gerichtetes Mikrofon *n*, Relais-, Richt-, Schall-mikrofon *n* ~ **navigation** Richtungsnavigation *f* ~ **oscillation** Flatter *m* ~ **oscilloscope** Richtröhre *f* ~ **pattern** Richtdiagramm *n* ~ **phase changer** Richtungsphasenschieber *m* ~ **position of antenna** Antennenstellung *f* ~ **potential** gerichtete Empfangsspannung *f* ~ **pressure** einseitiger Druck *m* ~ **property** richtende Eigenschaft *f*, Richtfähigkeit ~ **quantity** gerichtete Größe *f*, Richtgröße *f* ~ **quantization** Raum-, Richtungs-quantelung *f* ~ **radiation** Längsstrahlung *f* ~ **radio** Funkpeilgerät *n*, Peilfunk *m* ~ **radio beacon** Richtfunkfeuer *n* ~ **radio bearing** Funkseitenpeilung *f* ~ **radio link** Richtfunkstrecke *f* ~ **radio range**

Funkpeilreichweite *f* ~ **radio transmitter** Funkpeiler *m* ~ **radio unit** Fernrichtungsanlage *f* ~ **receiver** Peil-, Richt-empfänger *m* ~ **receiving set** Richtempfänger *m* ~ **reception** gerichteter Empfang *m* **with** ~ **response** richtungsempfindlich ~ **response pattern** Richtcharakteristik *f* ~ **sensitivity** Richtungsempfindlichkeit *f* ~ **signal** gerichtete Empfangsspannung *f* ~ **signal potential** Peilspannung *f* ~ **single-lever control of table feed** Einhebelschaltung *f* für Tischvorschübe ~ **stability** Richtungs-festigkeit *f*, -stabilität *f*, Stabilitätsrichtung *f* (aviat.) ~ **stress** Richtspannung *f* ~ **stability** Seitenstabilität *f* ~ **tap** Richtkoppler *m* ~ **transmission** Richtsenden *n* ~ **transmitter** Peilsender *m*, Richtsender *m*, gerichtete Sendestation *n* ~ **trim** Seitentrimmung *f* ~ **wireless telegraphy** gerichtete Funktelegrafie *f*

**directionless** richtungslos

**directions** Betriebsanweisung *f* ~ **for installing** Einbauvorschrift *f* ~ **for tracking trouble** Wegleitung *f* ~ **for use** Anwendungsvorschrift *f*, Bedienungsvorschrift *f*, Gebrauchsanweisung *f*, Gebrauchsvorschrift *f*

**directive** Direktive *f*, Richtlinie *f*; gerichtet, richtfähig ~ **antenna** Antenne *f* mit seitlichem Minimum, Richtantenne *f* ~ **antenna effect** Bündelung *f* ~ **beacon installation** Sammelstrahlanlage *f* ~ **beacon station** Richtfunkfeuer *n* ~ **characteristic** Richtungscharakteristik *f* ~ **diagram** Richtcharakteristik *f* ~ **effect** Richteffekt *m* ~ **force** Direktions-, Richt-kraft *f* ~ **gain** Antennenverstärkung *f*, Bündelungsgewinn *m* ~ **pattern** Richtdiagramm *n* ~ **pickup** Richtkoppler *m* ~ **power** (antenna) Richtvermögen *n* ~ **radio beacon** Richtfunkbake *f*, Richtfunkfeuer *n* ~ **reception** gerichteter Empfang *m*, Richtempfang *m* ~ **response pattern** Richtcharakteristik *f* ~ **screen** Perlenwand *f* ~ **surface** Oberflächengerichtet ~ **tap** Richtkoppler *m* ~ **transmission** gerichtete Übertragung *f* ~ **transmitter station** Richtfunkfeuer *n*

**directiveness** Richtbarkeit *f*

**directivity** Bündelung *f*, Richtfähigkeit *f* (electron.), Richtungsvermögen *n*, Richtungsverteilung *f* (rdo), Richtvermögen *n*, Richtwirkung *f* ~ **angle** Hauptstrahlwinkel *m* in der Vertikalebene ~ **factor** Bündelungsfaktor *m*, Richtvermögenverhältnis *n* ~ **index** Bündelungsmaß *n*, Richtungsverstärkungsfaktor *m* ~ **pattern** Richtdiagramm *n*

**directly** gleich ~ **controlled selector switch** direkteinstellbarer Wähler *m* ~ **driven** nicht übersetzt ~ **heated cathode** direkt geheizte Glühkathode *f* ~ **heated valve** direkt geheizte Röhre *f* ~ **operated control surfaces** direkt betätigte Ruder *pl* ~ **reversible Diesel engine** direkt umsteuerbarer Dieselmotor *m*

**director** Direktor *m* (Antenne), Dirigeur *m*, Feuerleitgerät *n*, Leitender *m*, Leiter *m*, Regisseur *m*, Register *n* (teleph.), Richtkreis *m*, Richtungsgeber *m*, Umleiter *m* ~ **of the board** Aufsichtsratmitglied *n* ~ **of photography** Bildmeister *m* ~ **of the plant** Betriebsdirektor *m* ~ **of play** Spielleiter *m*

**director,** ~**'s aide** Leitungsgehilfe *m* ~ **foretop** Vormarsstand *m* ~ **gradient** Direktorgradient *m* ~ **lamp** Richtlampe *f* ~ **meter** Umrechnerzählwerk *n* ~ **reading** Richtkreiszahl *f* ~ **selector** Umleitungswähler *m* ~ **system** Direktorsystem *n*, Umleitersystem *n*, Zentralrichtanlage *f* (nav.) ~ **telescope** Richtkreisaufsatzrohr *n*

**directorate** Bauaufsicht *f*, Vorstand *m*

**directory** Adreßbuch *n* ~ **of personnel** Betriebsunterlage *f* ~ **number** Rufnummer *f* ~ **number of subscriber having more than one line** Sammelnummer *f*

**directrices of conics** Leitlinien *pl* der Kegelschnitte

**directrix** Direktrix *f*, Leitkurve *f*, Leitlinie *f*, Mantellinie *f*, Richtlinie *f*, Richtungskurve *f*

**diresorcinolphthalein** Fluoreszein *n*

**dirigibility** Lenkbarkeit *f*

**dirigible** lenkbares Luftschiff *n*, Lenkluftschiff *n*; lenkbar ~ **balloon** Lenkballon *m*

**dirk** Dolch *m*

**dirt** Dreck *m*, Mist *m*, Schlamm *m*, Schmutz *m*, Versatzberg *m* ~ **basin** Schmutzfangschale *f* ~**-belt** Schmutzband *n* ~ **cage** (centrifuge) Flügeleinsatz *m* ~ **carrier** Waschbergeabzieher *m* ~ **chamber** Staubkeller *m* ~ **fender for reversible harrow** Schmutzfänger *m* für wendbare Egge ~ **moving** Erdbewegungen *pl* ~ **noise** Schmutzgeräusch *n* ~ **road** Erdstraße *f*, Fahrweg *m*, Feldweg *m* ~ **sample** Schmutzprobe *f* ~ **seal** Staubschutz *m* ~ **sediment** Schmutzanfall *m* ~ **shield** Schutzschild *n* ~**-solving agent** Schmutzlösungsmittel *n* ~ **tara** Putzprozente *pl* ~ **trap** Schmutzfänger *m*

**dirty, to** ~ besudeln, beschmutzen

**dirty** schmierig, schmutzig ~ **conditions** Verschmutzung *f* ~ **ends** Schmutzfäden *pl*

**disability** (law) Hemmungsgrund *m*, Unfähigkeit *f* ~ **insurance** Invaliditätsversicherung *f* ~ **rate** Invaliditätswahrscheinlichkeit *f*

**disable, to** ~ kampfunfähig machen, unwirksam machen, verletzen, versehren

**disable time** Ruderblockierungszeit *f* (g/m)

**disaccharate** Disacharat *n*

**disaccharid** Biose *f*, Disacharid *n*

**disacidified tar** entsäuerter Raffinationsabfall *m*

**disadjustment** Dejustierung *f*

**disadvantage** Nachteil *m*, Übelstand *m*, Unzuträglichkeit *f* ~ **factor** Absenkungsfaktor *m*

**disadvantageous** nachteilig, unvorteilhaft

**disaggregating** Quälen *n*, Totmahlen *n*

**disaggregation** Zerlegung *f* in Bestandteile

**disagree, to** ~ nicht übereinstimmen, in Widerspruch stehen

**disagreeable** unangenehm **of** ~ **smell** übelriechend

**disagreeing** unstimmig

**disagreemant** Unstimmigkeit *f*

**disalignment** Lageänderung *f*, Verschiebung *f*

**disallow, to** ~ zurückweisen

**disallowance** Zurückweisung *f*

**disallowing action** (patent) abschlägiger Bescheid *m*

**disappear, to** ~ ausgehen, sich erschöpfen, schwinden, verlorengehen, verschwinden, sich verstecken **to** ~ **from sight** außer Sicht kommen

**disappearance** Auflösung *f*, Schwund *m*, Schwunderscheinung *f*, Verschwinden *n* ~ **in (course of) time** zeitlicher Ablauf *m*

**disappearing** verschwindend ~ **antenna** Versenk-

antenne *f* ~**-filament pyrometer** Glühfadenpyrometer *n* ~ **gun mount** Verschwindeturm *m* ~ **point** Auflösbarkeitsgrenze *f* ~ **target** Verschwindscheibe *f* ~ **turret** Versenkdrehturm *m*

**disapproval** Mißbilligung *f*

**disapprove, to** ~ mißbilligen

**disapproved** nicht genehmigt

**disarm, to** ~ abrüsten, entschärfen, entwaffnen, unschädlich machen

**disarmament** Abrüstung *f*

**disassemble, to** ~ abbauen, abmontieren, abrüsten, ausbauen, auseinandernehmen, demontieren, zerlegen

**disassembled** abgerüstet

**disassembling** Abbau *m*

**disassembly** Ausbau *m*, Demontage *f* ~ **into individual ferries** fahrgliederweiser Ausbau *m*

**disassociation** Loslösung *f*, Trennung *f*

**disavow, to** ~ widerrufen

**disavowal** Verneinung *f*

**disband, to** ~ auflösen, entlassen

**disbanded** aufgelöst

**disbanding** Auflösung *f*

**disbarked wood** geschältes Holz *n*

**disburse, to** ~ verausgaben

**disbursement** Auslage *f*, Geldausgabe *f* ~ **return** Auszahlungsnachweis *m*

**disc: see also disk**

**disc** Scheibe *f*; scheibenförmig ~ **of the rotary spark gap** Funkenzieherscheibe *f*

**disc,** ~ **area** Dreh-, Propeller-kreisfläche *f* ~ **bit** Scheibenmeißel *m* ~ **cam** Kurvenscheibe *f* ~ **carrier** Lamellenträger *m* (Kupplung) ~ **flywheel** Scheibenschwungrad *n* ~ **fuse** Lötplättchen *n* ~ **harrow** Diskusegge *f* ~ **mill** Scheibenmühle *f* ~ **piercer** Scheibenlochwalzwerk *n* ~ **sanders** Scheibenschleifmaschine *f* ~**-seal diode** Scheibendiode *f* ~**-seal triode** Scheibentriode *f* ~**-seal tube** Höchstfrequenzröhre *f*, Scheibenröhre *f* ~**-seal valve** Scheibenröhre *f* ~ **serration** Schaufelvernutung *f* ~ **signal** Scheibensignal *n*

**discard, to** ~ (the top of an ingot) abschöpfen, ausrangieren, ausscheiden, außer Gebrauch setzen, verwerfen, weglegen **to** ~ **the top** (of an ingot) schopfen

**discard** Ausschuß *m*, Entfall *m*, Gekrätz *n* ~ **solution** abgestoßene Lauge *f*, Endlauge *f*

**discarded** abgesetzt

**discarding** Abscheidung *f*

**discern, to** ~ unterscheiden

**discernibility** Wahrnehmbarkeit *f*

**discernible** auffindbar, unterscheidbar

**discharge, to** ~ ablassen, abkehren, abladen, (naut.) abmustern, ausfließen, ausladen, auslassen, auslaufen, ausmustern, auspuffen, (the centrifugal) ausräumen, ausstoßen, ausströmen, ausströmen lassen, ausstürzen, austragen, (into) einmünden, (electr. battery) entladen, sich entladen, entlassen, entlasten, entleeren, löschen (Schiff), münden, ziehen **to** ~ **a boiler under pressure** ausblasen **to** ~ **a debt** eine Schuld ablösen **to** ~ **a digester under pressure** abblasen, ausblasen **to** ~ **(or fire) a gun** abfeuern **to** ~ **into** sich in etwas ergießen **to** ~ **a kiln** eine Darre abräumen **to** ~ **material** auswerfen **to** ~ **with a wheelbarrow** auskarren

**discharge** Abfall *m*, Abfluß *m*, Abführung *f*, Abladung *f*, Ablaß *m*, Abströmung *f*, Abzug *m*,

Ätzbeize f (print.), (of water) Ausfluß m, Ausguß m, Ausgußkasten m, Auslaß m, Auslaßventil n, Auslauf m, Ausstoß m, Ausströmung f, Ausströmung f unter Druck, Austragung f, Austritt m, Auswurf m, Bleichen n, Durchbruch m (bei Geiger-Müller-Zähler), Entladung f, (figure) Entladungsgebilde n, Entlassung f, Entleerung f, Erguß m, Fördermenge f, Leistungsfähigkeit f, Löschung f, (of a gun) Lösung f, Menge f, Rückströmung f, Wasserfracht f ~ of dredged material Löschen n, Verklappen n des Baggerguts ~ of exhaust gases Abgasung f ~ of a gun Abschuß m ~ of printing Ätzdruck m ~ from stamping trough Austrag m aus dem Pochtrog ~ of water Wasserauslauf m, Wasserverbrauch m
**discharge,** ~ **afterglow** Entladungsnachglimmen n ~ **apparatus** (torpedo) Ausstoßvorrichtung f ~ **belt** Abwurfband n ~ **blasts** Minenschießen n ~ **box** Austragkasten m ~ **capacity** Entladefähigkeit f ~ **center** Entlassungsstelle f ~ **channel** Ablaufgerinne n ~ **characteristic** Entladungscharakteristik f ~ **chute** Ablaufschurre f, Ausflußrinne f, Auslaufrutsche f, Austragschurre f ~ **chute of an ore-handling bridge crane** Abwurfschurre f ~ **circuit** Anoden-, Entladungs-kreis m ~ **cock** Abfluß-, Ablaß-hahn m ~ **coefficient** Abfuhrziffer f, Ausflußzahl f, Durchflußkoeffizient m ~ **connection** Ausgußstutzen m, Entladungsverbindung f ~ **control** Momentausrückung f ~ **conveyor** Abtransporteur m ~ **culvert** Abflußkanal m ~ **current** Anoden-, Entlade-, Entladungs-, Kriech-strom m ~ **curve** Abflußmengen-, Entlade-kurve f ~ **delay** Entladungsverzug m ~ **depth (or head)** Abflußhöhe f ~ **device** Entladevorrichtung f ~ **dome for vents** Entlüftungsdom m ~ **door** Entleerungsklappe f ~ **duct** Durchflußrohr n ~ **end** Abwurfende n, Auslaufende n, Ausstoßseite f, Austragsende n ~ **fittings** (torpedo) Ausstoßeinrichtung f ~ **flap** Auslaßklappe f ~ **frequency** Entladungsfrequenz f, Frequenz f der Durchflüsse ~ **funnel** Auslauf-, Ausström-, Schütt-trichter m ~ **gap** Entladungsstrecke f ~ **gate** Ausfluß-, Entleerungs-öffnung f ~ **head** Druckhöhe f, Förderdruck m ~ **header** Luftauslaßrohr n ~ **hole** Entlastungsloch n ~ **hopper** Schütt-Trichter m ~ **jet** Ausflußstrahl m ~ **key** Entlade-, Entladungs-schalter m, Steckschlüssel m für Ablaß ~ **knee** Auslaßellbogen m, Krümmer m ~ **lamp** Entladungslampe f ~ **lamp indicator** Glimmlampenindikator m ~ **lever** Entladungshebel m ~ **limit** Entladegrenzspannung f ~ **line** Entlastungsleitung f ~ **loss** Ausström-, Austritt-verlust m ~ **nozzle** Ausström-, Ausstoß-, Austritts-düse f, Austragsöffnung f, Blasdüse f, Einspritzdüse f, (jet) Schub-, Spritz-, Strahl-düse f, Zapfhahn m ~ **(pressure) nozzle** Druckdüse f ~ **opening** Ausflußloch n, Abfluß-, Ablaß-, Abzugs-, Ausfall-, Ausströmungs-, Austrags-, Durchfluß-öffnung f, Durchlaß m ~ **orifice** Ausstoßöffnung f ~ **outlet** Entleerungsöffnung f ~ **paste** Ätzfarbe f ~ **path** Entladungsweg m, Nebenweganode f ~ **pattern** Ätzmuster n ~ **pipe** Abflußleitung f, Abflußrohr n, Ablaßrohr n, Ablaufrohr n, Abrohr n, Abzugsrohr n, Abzugsröhre f, Abzugskanal m, Ausfluß-, Ausgangs-, Auslauf-, Aus-

stoß-, Austrag-, Auswurf-, Damm-, Druck-rohr n, Druckrohrleitung f, Schüttrohr n, Zapfrohr n ~ **(drain) pipe** Abfallrohr n ~ **point** Austrittsöffnung f ~ **port** Löschungshafen m ~ **potential** Entladungspotential n, Entladungsspannung f ~ **pressure** (pumps) Förderdruck m ~ **pressure line** (submarine) Ausdruckleitung f ~ **printing** Ätzbeizdruck m, Enlevagedruck m ~ **procedure (or process)** Entladevorgang m ~ **projection** Wurfweite f ~ **pulley** Abwurftrommel f ~ **pulse** Entladungsstoß m ~ **rate** Ausflußmenge f, Entladefähigkeit f ~ **rate of current** Entladestromstärke f ~ **register** Einzelzähler m, Einzelzählwerk n ~ **(gas-) relay** Entladungsrelais n ~ **resistance** Entladungswiderstand m ~ **resisting** ätz-beständig ~ **resistor** Entladungswiderstand m ~ **screen** Austragssieb n ~ **service valve** druckseitiges Ventil n an der Kühlmaschine ~ **setting** Brechspaltweite f, Spalteinstellung f ~ **side** Ausstoßseite f, Druckseite f (Pumpe) ~ **slope** Entladungssteilheit f ~ **sluice** Ausfallschütze f ~ **space** Entladungs-raum m, -strecke f ~ **spark gap** Entladefunkenstrecke f ~ **spout** Ablaßrinne f, Abwurfhaube f ~ **star** Ausschubstern m ~ **stop valve** Druckabsperrventil n ~ **stroke** Druckhub m ~ **style** Ätzartikel m ~ **surge** Entladungsstoß m ~ **switch** Zellenschalter m ~ **temperature** Druckrohrtemperatur f ~ **time** Ausflußzeit f ~ **tip** Auslaufspitze f ~ **tube** Entladungs-gefäß n, -rohr n, -röhre f ~**-tube buzzer** Glimmröhrensummer m ~ **valve** Abfluß-, Ablaß-, Ausfluß-ventil n, Ausstoßhahn m, Druckventil n, (under pressure) Förderventil n ~ **velocity** Ausfluß-, Ausströmungs-, (of turbines) Austritts-geschwindigkeit f ~ **vents** Austrittschlitze pl ~ **vessel** Entladungsgefäß n ~ **voltage** Abgabe-, Anoden-, Entlade-spannung f ~ **water** Abflußwasser n
**dischargeability** Ätzbarkeit f
**discharged** abgesetzt, ausgekippt
**discharger** Austragsvorrichtung f, Entlader m, Entladevorrichtung f, Funkenstrecke f, Funkenzieher m
**discharges** Durchflüsse pl
**discharging** Ausladung f, Ausräumen n, Ausschießen n der Diffuseure, Ausschleusen n, Entladen n ~ **of the delivery pipe** Entlastung f der Druckleitung
**discharging,** ~ **basin** Brunnensumpf m ~ **carrier** Entladungsträger m ~ **circuit** Entladungsstromkreis m ~ **culvert in a dike** Pumpsiel m ~ **device** Ablaßvorrichtung f ~ **excavated material** Bodenkippen n ~ **expenses** Entladekosten f ~ **funnel** Ablauftrichter m ~ **hole** Ausmündung f, Gußsteinauslauf m ~ **pipe** Blasrohr n ~ **spout** Ausflußrinne f ~ **spring** Entlastungsfeder f ~ **station** Entladestation f ~ **tray** Abgaberost m ~ **trough** Brunnensumpf m ~ **vault** Laubungsbogen m
**disclaim, to** ~ verzichten
**disclaimer** Verzicht m, Verzichtleistung f
**disclose, to** ~ aufdecken, aufschließen, enthüllen, erschließen
**disclosure** Aufschluß m, Beschreibung f (patent)
**discoid** scheibenförmig
**discolor, to** ~ abfärben

**discoloration** Farbenveränderung *f*, Farbumschlag *m*, Mißfärbung *f*, Verfärbung *f*, Vergällung *f* ~ **factor** (of oil) Entfärbungszahl *f*
**discolored** mißfarbig ~ **(or faded) shade** verschossener Farbton *m*
**discomfort** Unbehagen *n*
**discomposition effect** Wigner-Effekt *m*
**discone antenna** Scheibenkonusantenne *f*
**disconnect, to** ~ abhängen, abkuppeln, abschalten, abstellen, auskuppeln, auslösen, ausrücken, ausschalten, entkuppeln, freigeben, einen Stöpsel herausziehen, (eine Leitung) isolieren, lösen, loskuppeln, die Verbindung lösen oder trennen, unterbrechen **to** ~ **from binding post** abklemmen
**disconnect** Anschluß *m*, Entkupplung *f*, Trennstelle *f*, Verschraubung *f* ~ **key** Trenntaste *f* ~ **plug** Trennstecker *m* ~ **relay** Trennrelais *n* (TR) ~ **signal** Schlußzeichen *n*
**disconnectable** auskuppelbar, ausrückbar ~ **coupling** ausrückbare Kupplung *f* ~ **dual control** auskuppelbare Doppelsteuerung *f*
**disconnected** abgeschaltet, ausgeschaltet sein, getrennt
**disconnectible** abschaltbar, ausschaltbar
**disconnecting** Ausschalten *n*, Trennen *n* ~ **circuit** ausschaltender Stromkreis *m* ~ **device** Abschaltgerät *n* ~ **fork** Ausrückergabel *f* ~ **jack** Trennklinke *f* ~ **key knob** Trenntaste *f* ~ **lamp** Schlußlampe *f* ~ **point** Trennstelle *f* ~ **spring** Ausschlußfeder *f* ~ **strip** Trennstreifen *m* (teleph.) ~ **switch** Ausschalter *m*, Trennschalter *m*
**disconnection** Abhängigkeit *f*, Abhängung *f*, Abschalten *n*, Abschaltung *f*, Auskupp(e)lung *f*, Ausrückung *f*, Ausschalten *n*, Ausschaltung *f*, Ausschluß *m*, Entkopplung *f*, Entkuppelung *f*, Fernsprechsperre *f*, Leitungsunterbrechung *f*, Trennen *n*, Trennung *f*, Unterbrechung *f* ~ **of the blast** Windabstellung *f* ~ **by reversing** ruckweises Ausschalten *n*
**disconnection,** ~ **fault** Störung *f* durch Drahtbruch ~ **key** Abschalttaste *f* ~ **noise** Unterbrechungsgeräusch *n* ~ **switch** Trennschalter *m*
**disconnector** Trennschalter *m*
**discontinuance** Abbruch *m*, Aussetzung *f*, Unterbrechung *f* ~ **of oscillations** Schwingungsabreißen *n*
**discontinue, to** ~ abbrechen, absetzen, aufgeben, aufheben, aufhören, aussetzen, einstellen, unterbrechen, unterbrochen sein, unterbrochen werden
**discontinued wave** gedämpfte Welle *f*
**discontinuity** Diskontinuität *f*, Störstelle *f*, Ungleichförmigkeit *f*, Ungleichmäßigkeit *f*, Unterbrechung *f*, Unstetigkeit *f* ~ **conditions** Unstetigkeitsbedingungen *pl* ~ **effect** Abreißeffekt *m* ~ **interaction** Unstetigkeitswechselwirkung *f* ~ **point** Abreißungs-, Trennungs-punkt *m* (aerodyn.) ~ **region** Reißgebiet *n* ~ **surface** Unstetigkeitsfläche *f* ~ **types** Unstetigkeitstypen *pl*
**discontinuous** aussetzend, diskontinuierlich, ungleichförmig, ungleichmäßig, unstet, unstetig, unterbrochen ~ **and crystallographic groups of motions** diskontinuierliche und kristallografische Bewegungsgruppen *pl* ~ **action** unstetige Wirkung *f* ~ **film feed (or movement)** bildweise

Filmschaltung *f* ~ **groups of motions** diskontinuierliche Bewegungsgruppen *pl* ~ **movement** schaltweise Bewegung *f* ~ **running** intermittierender Betrieb *m* ~ **solution** unstetige Lösung *f* ~ **space charges** diskrete Raumladungen *pl* ~ **waves** gedämpfte Wellen *pl*
**discontinuously** absatzweise
**discord** dissonanter Akkord *m*, Mißklang *m*
**discordancy** Abweichung *f*
**discordant** unstimmig
**discount, to** ~ diskontieren, vergüten
**discount** Ablaß *m*, Abschlag *m*, Erlaß *m*, Ermäßigung *f*, Gebührennachlaß *m*, Minderbewertung *f*, Nachlaß *m*, Preisabzug *m*, Rabatt *m*, Skonto *n*, Vergütung *f* ~ **house** Wechselbank *f*
**discoupling** Entkopplung *f*
**discourse** Abhandlung *f*, Ausführung *f*
**discover, to** ~ aufdecken, auffinden, ausfindigmachen, ausmachen, ausrichten, entdecken, erschließen, erschroten, erschürfen, feststellen, nachweisen **to** ~ **a mine** ein Bergwerk fündig machen
**discoverer** Entdecker *m*, Erfinder *m*
**discovery** Aufschluß *m*, Aufsuchen *n*, Entdeckung *f*, Ermittelung *f*, Fund *m* ~ **of a mineral** Fündigkeit (min.) ~ **shaft** Fundgrube *f*
**discrepancy** Ungenauigkeit *f*, Unstimmigkeit *f*, Verschiedenheit *f*, Widerspruch *m* (math.) ~ **switch** Quittungsschalter *m*
**discrepant values** auseinandergehende Werte *pl*
**discrete** abgesondert ~ **band spectrum** Viellinienspektrum *n* ~ **quantity** kleine Quantität *f* ~ **sentence intelligibility** Satzverständlichkeit *f* (acoust.) ~ **word intelligibility** Wortverständlichkeit *f* (acoust.)
**discretion** Belieben *n*, Ermessen *n*, Fingerspitzengefühl *n*, Klugheit *f*, Umsicht *f*, Willkür *f*
**discretionary** beliebig
**discriminable** unterscheidbar
**discriminant of an algebra** Diskriminante *f* einer Algebra
**discriminate, to** ~ absondern, unterscheiden
**discriminating,** ~ **relay** Berichtigungsrelais *n* Differentialrelais *n*, Selektivschutz *m* ~ **selector** Mitlaufwerk *n* ~ **switch** Einwegschalter *m* ~ **tone** Unterscheidungston *m* (teleph.)
**discrimination** Auflösung *f* (rdr), Auflösungsvermögen *n*, Unterdrückung *f* (von Schwingungen), Unterscheidung *f* ~ **experiment** Wahlversuch *m*
**discriminator** Diskriminator *m*, Frequenzdetektor *m*, Trennschärfe *f*
**discuss, to** ~ absprechen, auswerten, besprechen, diskutieren, erörtern
**discussable** erörterbar
**discussion** Auseinandersetzung *f*, Aussprache *f*, Besprechung *f*, Diskussion *f*, Erörterung *f*, Gespräch *n*, Meinungsaustausch *m*, Verhandlung *f*
**disembark, to** ~ ausbooten, ausschiffen, (aus dem Flugzeug) aussteigen
**disembowel, to** ~ Innenteile entfernen
**disengage, to** ~ abkuppeln, abschalten, sich absetzen, ausklinken, (a clutch) auskuppeln, auslösen, ausschalten, ausrücken, (motor) außer Eingriff kommen, entkuppeln, entwickeln, sich entziehen, lösen, loslösen **to** ~ **the safety arm** entsichern

disengage, ~ limiter Beschleunigungsbegrenzer ~ warning light circuit Auskuppelwarnleuchtenschaltung *f*
disengageable abschaltbar, ausklinkbar
disengaged ausgeklinkt, ausgekuppelt, außer Eingriff, Leitung frei, los, unbesetzt ~ jack freie oder unbesetzte Klinke *f* ~ line freie Leitung *f* ~ position Ausrückstellung *f*
disengagement Absetzbewegung *f*, Auskupp(e)-lung *f*, Ausrückung *f*, Entkuppelung *f*, Freigabe ~ of heat Wärmeentwicklung *f*
disengaging Ausrücken *n*, Entrastung *f*; auskuppelbar ~ bar Ausrückschiene *f* ~ clutch Entkuppelung *f* ~ coupling ausrückbare oder lösbare Kupplung *f* ~ device Ausrückvorrichtung *f* ~ fork Ausrückgabel *f* ~ gear Auskupplungsvorrichtung *f*, Ausrücker *m*, Ausrückvorrichtung *f* ~ lever Ausrückhebel *m* ~ rod Ausrückstange *f* ~ shaft Ausrückwelle *f* ~ strap Ausrückbügel *m*
disentangle, to ~ auseinanderwickeln, entwirren
disentanglement Entflechtung *f*
disequilibrium Übergewicht *n*, Ungleichgewicht *n*
disgorge, to ~ abspritzen
disgorging plant Degorgieranlage *f*
disguise, to ~ bemänteln, verkleiden, unkenntlich machen
dish, to ~ pressen, tiefziehen, vertiefen
dish Fänger *m*, Parabolreflektor *m*, Satte *f*, Schale *f*, Schüssel *f*, Teller *m* ~ of confussion Brennfleckausdehnung *f* ~ with perforated bottom Siebschale *f*
dish, ~ cloth Wischtuch *n* ~ development Schalenentwicklung *f* (photo) ~ drainer Trockengestell *n* ~ dryer Geschirrtrockner *m* ~ insulator for antenna supporting Tellerisolator *m* ~ radius Wölbungsradius *m* ~ reflector Parabolreflektor *m* ~-shaped schalen-, schüssel-förmig ~-shaped base (foot or press) Tellerfuß *m*
disharmonious verstimmt
dished abgerundet, hohlgeschliffen, schalenförmig ~ bulkhead Kugelschott *n* ~ electrode Schalenelektrode *f* ~ end gewölbtes Ende *n* (Rohr) ~ ends gekümpelte Böden *pl* ~ head gewölbter Boden *m* ~ iron plate getriebenes Eisenblech *n* ~ laminae kugelförmig gewölbte Lamellen *pl*
dishing Kümpelarbeit *f* ~ machine Bombiermaschine *f*
disinclined, to be ~ sich abneigen
disincorporate, to ~ ausgliedern
disinfect, to ~ beizen, beräuchern, desinfizieren, entgiften, entkeimen, entseuchen, entwesen
disinfectant Beiz-, Desinfektion-, Räuchermittel *n*
disinfected keimfrei
disinfecting Entwesen *n* ~ agent Entseuchungsmittel *n* ~ chamber Desinfektionskammer *f*, Entseuchungsschrank *m* ~ steep Tauchbeize *f* (Pelz)
disinfection Beizung *f*, Desinfektion *f*, Entkeimung *f*, Entseuchen *n*, Entseuchung *f*, Entsuchungsgerät *n*, Entwesung *f* von Ungeziefer
disinflate, to ~ die Luft entweichen lassen
disintegrability Materiezertrümmerbarkeit *f*, Zertrümmerbarkeit *f*
disintegrable verwitterungsfähig, zertrümmerbar ~ substance zertrümmerbares Material *n*

disintegrant Abbaumittel *n*
disintegrate, to ~ abbauen, auflockern, auflösen, in seine Bestandteile auflösen, aufschließen, auseinanderfallen, entmischen, verwittern, zerfallen, zerkleinern, (slag) zerrieseln, zersetzen, zerstäuben, zerstückeln, zusammenfallen
disintegrated zersetzt
disintegrating ~ belt Zerfallgurt *m* ~ cathodes Aufdampfkathoden *pl* (electron.) ~ charge Nipolit-Preßstück *n* ~ slag Zerfallschlacke *f*
disintegration Abbau *m*, Abfall *m*, Abtragung *f*, Auflockerung *f*, Auflösung *f*, Auflösung *f* in die Bestandteile, Auseinanderstellung *f* der Bestandteile, Desaggregation *f*, Entmischung *f*, Umwandlung *f*, Zerfall *m*, Zerlegung *f*, Zersetzung *f*, Zertrümmerung *f* ~ of atom Atomzerfall *m* ~ in a chemical change Umsetzungsgeschwindigkeit *f* ~ of coal layers Abbau *m* von Kohleschichten
desintegration, ~ constant Umwandlungs-, Zerfalls-konstante *f* ~ electron Umwandlungselektron *n* ~ energy Zerfallsenergie *f* ~ rate Zerfalls-geschwindigkeit *f*, -verhältnis *n* ~ time Zerfallzeit *f* ~ voltage kritischer Spannungsabfall *m*
disintegrator Brechmaschine *f*, Desintegrator *m*, Knotenbrecher *m*, Pulverisiermaschine *f*, Raspelmaschine *f*, Schlagmühle *f*, Schleudermühle *f*, Wäscher *m*, Zerfaserer *m*, Zerkleinerer *m*
disk Fiberscheibe *f*, Kreisscheibe *f*, Lamelle *f*, Läufer *m*, Nockenbahn *f*, Platte *f*, Ronde *f*, Rondell *f*, Scheibe *f*, Teller *m* ~ for clocks Uhrplatte *f* ~ with lobes Daumenscheibe *f*, Exzenter *m*, Hebedaumen *m* ~ of pig iron Blattel *f*, Plattel *f* ~ with spiral perforations Spirallochscheibe *f*
disk, ~ anode Telleranode *f* ~ antenna Scheibenantenne *f* ~ area (of propeller) Drehkreisfläche *f*, Rotorkreisfläche *f* ~ armature Flachring *m*, Scheiben-anker *m* ~ attenuator Scheibenabschwächer *m* ~ attrition mill Kolloplex-, Scheiben-, Stift-mühle *f* ~ barker Messerschälmaschine *f*, Scheibenschäler *m* (paper mfg.) ~ bearing Scheibenlager *n* ~ beater Scheibenholländer *m* ~ bit Scheibenmeißel *m* ~ brake Scheibenbremse *f* ~ buffer Stoßpuffer *m* ~ capacitor Scheibenkondensator *m* ~ cell Scheibenküvette *f* ~ chart recorder Kreisblattschreiber *m* ~ chuck Scheibenfutter *n* (multiple) ~ clutch Lamellenkupplung *f*, Scheibenkupplung *f* ~ coil Flachspule *f*, Scheibenwicklung *f* ~ condenser Plattenkondensator *m* ~ cooler Scheibenkühler *m* ~ corn planter Scheibenmaisdrill *m* ~ coupling Flanschen-, Scheibenkupplung *f* ~ cover Scheibenschützer *m* ~ crusher Teller-, Scheiben-mühle *f* ~ cultivator machine Scheibenhackmaschine *f* ~ diaphragm Blendscheibe *f* ~ discharger rotierende oder umlaufende Funkenstrecke *f*, Scheibenelektrometer *m*, Scheibenfunkenstrecke *f* ~ drier (rotary shelf drier) Tellertrockner *m* ~ drill Scheibendrillmaschine *f* ~ drive key Bremsscheiben-Mitnehmerkeil *m* ~ electrode Tellerelektrode *f* ~ electrometer Scheibenelektrometer *n* ~ feeder Abstreich-teller *m*, -tisch *m* ~ filter Scheibenfilter *n* ~ furrow opener Scheibenfurchenöffner *m* ~ gap transmitter Plattenfunkenstrecke *f* ~ grinding machine Abricht-

schleifmaschine *f* ~ **harrow** Scheibenegge *f*, Tellerscheibenegge *f* ~ **hiller** Scheibenhäufler *m* ~ **honing machine** Scheibenhackmaschine *f* ~ **impeller mixer** Scheibenkreiselmischer *m* ~ **indicator** Scheibenzähler *m* (g/m), Scheibenzeigerwerk *n* ~ **insulator** Scheibenisolator *m* ~ **insulator for antenna supporting** Tellerisolator *m* ~ **landing wheel** Scheibenlaufrad *n* ~ **lever** Scheibenhebel *m* ~ **loaded** Wellenleiter *m* mit Lochblenden ~ **loading** Rotorbelastung *f* ~ **marker** Markierungsscheibe *f* ~ **marker attachment** Scheibenmarkierungsapparat *m* ~ **mill** Scheibenmühle *f* ~**-piercing process** Scheibenwalzverfahren *n* ~ **pile** Scheibenpfahl *m* ~ **piston** Scheibenkolben *m* ~ **piston blower** Kreiskolbengebläse *n* ~ **planter** Scheibenpflanzmaschine *f* ~ **player** Plattenspieler *m* ~ **plow** Scheibenpflug *m* ~ **ray** Scheibenstrahl *m* ~ **record** Schallplatte *f* ~ **recorder** Plattenschneider *m* ~ **rim (or circumference)** Felgenscheibe *f* ~ **ring** Tellerring *m* ~ **sander** Tellerschleifer *m* ~ **screentype filter** Siebscheibenfilter *m* ~ **seal** Ringeinschmelzung *f*, Scheibeneinschmelzung *f* ~**-sealed triode** Scheibentriode *f* ~ **shaped** scheibenförmig ~**-shaped cathode** Tellerkathode *f* ~**-shaped photocarrier** Bildträgerscheibe *f* ~**-shaped winding** Scheibenwicklung *f* ~ **shutter** Scheibenblende *f* ~ **signal** Scheibensignal *n* ~ **spacing** Blendenabstand *m* ~**-spinning machine** Tellerspinnmaschine *f* ~ **spring** Scheiben-, Teller-feder *f* ~ **system** Scheibensystem *n* ~ **trimmer** Scheibentrimmer *m*

**disk-type,** ~ **alternator** Scheibenankergenerator *m* ~ **gramophone** Plattenspieler *m* ~ **grinding wheel** Tellerschleifscheibe *f* ~ **horn** Tellerhorn *n* ~ **jack** Tellerwinde *f* ~ **milling cutter** Frässcheibe *f* ~ **phonograph** Plattenspieler *m* ~ **reamer** Scheibenreibahle *f* ~ **resistor** Scheibenwiderstand *m*

**disk,** ~ **valve** Platten-, Scheiben-, Teller-ventil *n* ~ **washer** Tellerscheibe *f* ~ **weeder** Scheibenmesser *n* für Unkraut ~ **wheel** Scheibenrad *n* ~ **winding** Scheibenwicklung *f* ~ **wobble** Plattenschlag *m*

**dislocate, to** ~ ausrenken, verdrängen, verrenken

**dislocated** versetzt

**dislocation** Dislokation *f*, Lagerungsstörung *f*, Verlagerung *f*, Verrückung *f*, Verschiebung *f*, Versetzung *f*, Verwerfung *f* (min.) ~ **of strata** Schichtenstörung *f* ~ **line** Versetzüngslinie *f* ~ **network** Versetzungsnetzwerk *n* ~ **strain** Versetzungsformänderung *f*

**dislocking bracket** Verriegelungsbügel *m*

**dislodge, to** ~ von seinem Platz entfernen, verdrängen, vertreiben **to** ~ **casting scale** Gußhaut *f* ablösen

**dislodgment** Verrückung *f*                          =

**dismantle, to** ~ abbauen, abbrechen, abmontieren, abrahmen, abrüsten, abtakeln, abwracken, ausbauen, auseinandernehmen, ausschalen, demontieren **to** ~ **a bridge** abbrücken **to** ~ **the engine** den Motor auseinandernehmen

**dismantled** zerlegbar

**dismantling** Abbau *m*, Abmontierung *f*, Abwrackung *f*, Ausbau *m*, Auseinanderbau *m*, (a gun) Auslegen *n*, Demontage *f*, Rückbau *m*

~ **of a line** Abbruch *m* der Linie ~ **charges** Abbruchkosten *pl* ~ **point** Trennstelle *f*

**dismasted** entmastet

**dismember, to** ~ auseinandernehmen

**dismiss, to** ~ abbauen, abkehren, ablehnen, ablohnen, abtreten, (a suit) abweisen, entheben, entlassen, kassieren, zurückweisen **to** ~ (legal action) **with costs** kostenpflichtig abweisen

**dismissal** Abbau *m*, Amtsenthebung *f*, Dienstentlassung *f*, Entlassung *f*, Entsetzung *f*, Fortschaffung *f*, Zurückweisung *f* ~ **of claim** Klageabweisung *f*

**dismissed** abgesetzt

**dismount, to** ~ abmontieren, abnehmen, abrüsten, auseinandernehmen, absitzen, absteigen, (aus dem Führersitz) aussteigen, demontieren, niederreißen, zerlegen

**dismountable,** ~ **antenna** Steckmast *m* ~ **mast** ausfahrbarer oder umlegbarer Mast *m*

**dismounted** zerlegbar

**disorder** Störung *f*, Unordnung *f*, Verwirrung *f* ~ **correlations** Fehlordnungskorrelationen *pl* ~**-order transitions** Unordnungs-Ordnungs-Umwandlungen *pl* ~ **pressure** Entropiedruck *m* ~ **scattering** Fehlordnungsstreuung *f*

**disordered** fehlgeordnet (cryst.), ungeordnet ~ **arrangement** ungeordneter Zustand *m* ~ **motion** ungeordnete Bewegung *f* ~ **scattering** ungeordnete Streuung *f* ~ **state** ungeordneter Zustand *m*

**disorderly** ordnungswidrig

**disorientation** Entorientierung *f*

**disparage, to** ~ herabsetzen

**disparagement** Herabsetzung *f*

**disparity** Abweichung *f*, Mißverhältnis *n*, Unstimmigkeit *f*, Verschiedenheit *f*

**dispatch, to** ~ abfertigen, (goods) abgehen lassen, absenden, ausschicken, aussenden, entsenden, Arbeiten schnell erledigen, expedieren, fertigen, versenden

**dispatch** Abfertigung *f*, Abfertigungsstelle *f*, Absendung *f*, Beförderung *f*, Depesche *f*, Inmarschsetzung *f*, Versand *m* ~ **of freight** Güterabfertigung *f*

**dispatch,** ~ **boat** Aviso *m* ~ **carrier** Nachrichtenvermittler *m* ~ **facilities** Versandbetrieb *m* ~ ~ **jar** Versandgefäß *n* ~ **(or shipping) note** Versandanzeige *f* ~ **station** Versandstation *f*

**dispatcher** Abfertiger *m*, Absender *m*, Expedient *m*, Flugdienstberater *m*, Versender *m*

**dispatching,** ~ **of luggage** Gepäckabfertigung *f* ~ **point** Absendestelle *f*

**dispel, to** ~ vertreiben

**dispensable** entbehrlich

**dispensary** Offizin *f*

**dispense, to** ~ ausschenken, austeilen **to** ~ **by tap** verzapfen **to** ~ **with** absehen (von), entbehren; ohne etwas fertig werden

**dispenser** Abwickelhaspel *f*, Düppelabwurfvorrichtung *f* (rdr), Meßbecher *m* ~ **cathode** Vorratskathode *f* ~ **unit** Distributeurmaschine *f*

**dispensing** Düppelabwurf *m* (rdr) ~ **pump** Zapfsäule *f* ~ **rate** Düppelabwurfdichte *f*

**disperge, to** ~ dispergieren

**dispersal,** ~ **of manufacturing facilities** Fertigungsverlagerung *f* ~ **area** Abstellplatz *m* ~ **effect** Entleerungseffekt *m*

**disperse, to** ~ auseinandertreiben, dispergieren, türmen, verbreitern, versprengen, verteilen, zersprengen, zerstreuen, zerteilen **to ~ light** Licht zerstreuen
**disperse** dispers ~ **system** Dispersoid *n*
**dispersed** dekonzentriert, dispers, dispergiert, verstreut
**dispersing,** ~ **agent** Dispergierstoff *m* ~ **effect** Dispergierfähigkeit *f* ~ **lens** Streuungslinse *f*, zerstreuende Linse *f*, Zerstreuungslinse *f* ~ **ring** Verteilerring *m*
**dispersion** Dispersion *f*, Dispersität *f*, Streubereich *m*, Streuung *f*, Teilheit *f*, (of pigments) Verteilung *f*, Zerlegung *f*, Zerstreuung *f*, Zerteilungsgrad *m* ~ **in depth** (ballistics) Längenstreuung *f* ~ **of hardness value** Härtestreuung *f* ~ **of light** Lichtzerlegung *f* ~ **of a light source** Streuung *f* einer Lichtquelle
**dispersion,** ~ **area** Streuungsfläche *f* ~ **binder** Bindemitteldispersion *f* ~ **caused by fuse differences** Zünderstreuung *f* ~ **coefficient** Streufaktor *m*, Streuzahl *f*, totaler Streufaktor *m* ~ **current** Ausbreitungsstrom *m* ~ **diagram** Streuungsbild *n* ~ **error of gun** Streuung *f* des Geschützes ~ **hardening** Ausscheidungshärtung *f* ~ **lens** Streulinse *f* ~ **medium** Dispersionsmittel *n* ~ **pattern** Bodentreffbild *n* ~ **prism** Dispersions-, Streuungs-prisma *n* ~ **rate** Dispersionsgrad *m* ~ **terms** Dispersionsterme *pl*
**dispersive** dispers ~ **power** Dispersions-, Zerstreuungs-vermögen *n*
**dispersoid analysis** Dispersoidanalyse *f*
**displace, to** ~ absetzen, entsetzen, verdrängen, verlagern, verlegen, verschieben, versetzen, verstellen, verwerfen **to ~ the liquid** die Flüssigkeit verdrängen
**displaceability** Verschiebbarkeit *f*
**displaceable** verschiebbar ~ **clamp lever** Klemmverschiebehebel *m* ~ **tension (or voltage)** verschiebbare Spannung *f*
**displaced** verschoben, versetzt ~ **in phase** fasenverschoben ~ **phase** verschobene Fase *f* ~ **threshold** versetzte Schwelle *f*
**displacement** Förderleistung *f*, Gittervorspannungsverlagerung *f*, Saugleistung *f*, Überdeckung *f*, Unschärfe *f*, Verdrang *m*, Verdrängung *f*, Verlagerung *f*, Verrückung *f*, Verschiebung *f*, Versetzung *f*, Verstellung *f*, Wanderung *f*, Wechsel *m*, Weg *m*, (of engine) Zylinderinhalt *m*
**displacement,** ~ **of the bands** Streifenverschiebung *f* ~ **of the center of gravity** Schwerpunktsverlegung *f*, Schwerpunktswanderung *f* ~ **of the center of mass** Schwerpunktsverschiebung *f* ~ **of the center of pressure** Druckpunktwanderung *f* ~ **of the coast line** Strandverschiebung *f* ~ **of cylinder** Hubraum *m* ~ **in elevation** Höhenauswanderung *f* ~ **of metal** Metallverschiebung *f* ~ **in null point** Nullpunktverschiebung *f* ~ **of phase** Ausschlag *m* ~ **in potential** Potentialverschiebung *f* ~ **by slipping** Gleitverschiebung *f* ~ **of soil** Bodenverdrängung *f* ~ **of a sound** Schallausschlag *m* ~ **of state** Zustandsverschiebung *f* ~ **of the support** Auflagerverschiebung *f* ~ **of water** Wasserverdrängung *f*
**displacement,** ~ **adjustment** Längsbewegung *f* ~ **antiresonance** Gegenresonanzabweichung *f* ~ **bottle** Hubflasche *f* ~ **characteristic** Förder-

leistungscharakteristik *f* ~ **compressor** Kapselgebläse *n*, -lader *m*, Verdrängungslader *m* ~ **constant** Verschiebungskonstante *f* ~ **current** Verdrängungsstrom *m* ~ **curve** Förderleistungskurve *f* ~ **derivative** Verrückungsableitung *f* ~ **device** Verschiebungseinrichtung *f* ~ **force** Stellkraft *f* ~ **gradient** Verschiebungsgradient *m* ~ **gyro** Lagekreisel *m* ~ **kernel** Verschiebungsintegralkern *m* ~ **law** Verschiebungssatz *m* ~ **pickup** Wegaufnehmer *m* ~ **potential** Verrückungspotential *n* ~ **resonance** Resonanzabweichung *f* ~ **speed** Verrückungsgeschwindigkeit *f* ~ **supercharger** Drehkolbenlader *m*, Kolbenlader *m*, Lader der Verdrängerbauart ~ **thickness** Verdrängungsdicke *f* ~**-type pumps** Verdrängerpumpen *pl*
**displacing** Versetzung *f* ~ **action** schiebende Wirkung *f* ~ **device** Verfahreinrichtung *f*
**display, to** ~ offen ausbreiten, darstellen, entfalten, (importance) hervorheben, zur Schau stellen, vorführen, vorlegen **to ~ goods for sale** Waren *pl* zum Verkauf aufstellen
**display** Aufwand *m*, Auslage *f*, Ausstellung *f*, Bild *n* (rdr), Darstellung *f* (math.), Vorstellung *f*
**display,** ~ **booth** Ausstellungsstand *m* ~ **cabinet** Vitrine *f* ~ **carton** Schaupackung *f* ~ **circuits** Bilderanzeigeschaltung *f* ~ **console** Bildschirmträger *m* (rdr) ~ **drum** Trommelspeicher *m* ~ **film** Vorführungsfilm *m* ~ **glass** Schauglas *n* ~ **racks** Schauständer *pl* ~ **scale** Bildschirmmaßstab *m* (rdr) ~ **scope** Bildschirm *m* ~ **selector** Bildwähler *m* ~ **stand** Schaufenstergestell *n* ~ **time** Anzeigedauer *f* ~ **tube** Musterglas *n*, Schauglas *n* ~ **type** Auszeichnungsschrift *f* ~ **unit** Anzeigevorrichtung *f*, Sichtgerät *n* (rdr)
**displayed beam width** dargestellte Strahlbreite *f*
**disposable,** ~ **fund** flüssige Mittel *pl* ~ **load** ausnutzbare Ladefähigkeit *f*, Nutzlast *f*, Zuladung *f*
**disposal, at one's** ~ verfügbar **to have at one's** ~ verfügen über **at the** ~ **of** zur Verfügung **at** ~ **for special duty** zur besonderen Verfügung
**disposal** Absetzen *n*, Anordnung *f*, Beseitigung *f*, Disposition *f*, Verfügung *f* **right of** ~ **by the owner of the soil** Verfügungsrecht *n* des Grundeigentümers ~ **of waste products** Ausschaltung *f* wertloser Erzeugnisse ~ **of waste water** Abwasserbeseitigung *f*
**disposal,** ~ **certificate** Aussonderungsnachweisung *f* ~ **ground** Abfuhrgelände *n* ~ **site** Abfuhrgelände *n* ~ **well** Wegschaffungsgrube *f*
**dispose, to** ~ anordnen, verfügen **to ~ in depth** nach der Tiefe gliedern **to ~ of** abtun, absetzen
**disposed in depth** tief gegliedert
**disposition** Ansatz *m*, Aufstellung *f*, Ausführung *f*, Gemüt *n*, Stimmung *f*, Veranlagung *f*, Verfügung *f*, Zug *m* ~ **in depth** Tiefengliederung *f* ~ **(or texture) of fibers** Lagerung *f* der Fasern ~ **of guns** (navy) Geschützaufstellung *f* ~ **of lights** Verteilung *f* der Feuer ~ **of property** Verfügung *f* über Material ~ **statement** Lagebericht *m*
**dispositive clause** Sollvorschrift *f*
**dispossess, to** ~ aberkennen, enteignen
**dispossession** Aberkennung *f*, Besitzentziehung *f*
**disproof** Widerlegung *f*
**disproportion** Mißverhältnis *n*

**disproportionate** unverhältnismäßig, verhältniswidrig

**disprove, to** ~ als falsch zurückweisen

**dispute, to** ~ anfechten

**dispute** Streit *m*, Streitfall *m*, Streitfrage *f*, Streitpunkt *m*, Streitsache *f*, Wortstreit *m*, Zank *m*, Zerwürfnis *n*

**disputed** strittig

**disqualification** Ausschließung *f*, Untauglichkeitserklärung *f*

**disqualified** nichtberechtigt **to be** ~ die Befähigung einbüßen

**disqualify, to** ~ ausschließen, disqualifizieren

**disregard, to** ~ absehen (von), außer Acht lassen, geringschätzen, vernachlässigen, verwerfen, unberücksichtigt lassen

**disregard** Zurücksetzung *f*

**disrupt, to** ~ auseinanderdrängen, auseinanderreißen, durchschlagen, reißen, zerreißen

**disruption** Auseinanderbersten *n*, Bersten *n*, Durchschlagen *n*, Moleküldissoziation *f*, Zerreißen *n* ~ **of the construction schedule** Unterbrechung *f* der Bauprogramme ~ **of insulations** Isolationsdurchschlag *m* ~ **spark** Entladungsfunke(n) *m*

**disruptive** disruptiv ~ **discharge** Durchbruch *m*, Durchbruchsentladung *f*, Durchschlag *m* (electr.), Funkenentladung *f*, Überschlag *m* elektrischer Funken ~ **jammer** durchschlagender Störsender *m* ~ **rigidity** Durchschlagsfestigkeit *f* ~ **strength** Durchschlagsfestigkeit *f*, dielektrische Festigkeit *f* ~ **voltage** Durchschlagsspannung *f*, Überschlagsspannung *f*

**disruptor,** ~ **flap** Störklappe *f* ~ **(wire)** Stördraht *m*

**dissect, to** ~ halbieren, zergliedern, zerlegen

**dissected country** durchschnittenes Gelände *n*

**dissecting,** ~ **darkfield condenser** Präparierdunkelfeldkondensor *m* ~ **lens** Präparierlupe *f* ~ **microscope** Präpariermikroskop *m* ~ **set** Präparierbesteck *n* ~ **work** Präparierarbeit *f*

**dissection** Zerlegung *f*

**dissector** Zerleger *m*, (of Farnsworth) Bildzerleger *m* ~ **aperture** Zerlegblende *f* ~ **tube** Bildaufnahmeröhre *f*

**disseminate, to** ~ aussäen, aussenden (rdo), ausstreuen, einsprengen, einsprenkeln, verbreiten, verteilen **to** ~ **information** Mitteilungen bekanntgeben

**disseminated** eingesprengt, zerstreut

**dissemination** Aussendung *f*, Ausstreuung *f*, Verteilung *f*

**dissent, to** ~ anderer Meinung sein

**dissimilar** unähnlich, ungleich, verschieden, verschiedenartig ~ **terms** Ungleichung *f* (math.)

**dissimilarity** Ungleichartigkeit *f*, Verschiedenheit *f*

**dissimulated electricity** gebundene Elektrizität *f*

**dissipate, to** ~ (heat, etc.) ableiten, (energy) aufbrauchen, auflösen, aufzehren, einrühren, (fog) steigen, verbrauchen, vergeuden, verteilen, (energy) verzehren, zerstreuen

**dissipated** verbraucht

**dissipating** sich auflösend

**dissipation** Aufzehrung *f*, Dämpfung *f*, Energieverbrauch *m*, Vergeudung *f*, (of potential) Verlust *m*, Verlustzeit *f*, (of energy) Verteilung *f*, Verzehrung *f*, Zerstreuung *f*

**dissipation,** ~ **of energy** Energie-vernichtung *f*, -verzehrung *f*, -zerstreuung *f*, Kraftverbrauch *m* ~ **of energy of fall** Vernichtung *f* der Energie des Gefälles ~ **of fire** Feuerzersplitterung *f* ~ **of forces** Kraftzersplitterung *f*, Zersplitterung *f* ~ **of heat** Wärme-abfuhr *f*, -ableitung *f* ~ **of setting heat** Abgeben *n* der Abbindungswärme

**dissipation,** ~ **factor** Verlustfaktor *m* ~ **law** Wiedervereinigungsgesetz *n* ~ **limit** Belastungsgrenze *f* (Elektronenröhren) ~ **loss** Verlustleistung *f*

**dissipationless line** verlustlose Leitung *f*

**dissipative** verlustbehaftet, verzehrend ~ **component** Wirkkomponente *f* ~ **impedance** Wirkkomponente *f* des Scheinwiderstandes ~ **linear** dissipativ linear ~ **part** dissipativer Anteil *m* ~ **resistance** Dämpfungswiderstand *m*, reiner Ohmscher Widerstand *m*, Verlustwiderstand *m*, Wirkwiderstand *m* ~ **stress** dissipativer Spannungsanteil *m* ~ **system** Verteilungssystem *n*

**dissipativity** Dissipativität *f*

**dissociable** dissoziierbar

**dissociate, to** ~ absondern, abtrennen, auflösen, dissoziieren, entmischen, spalten, trennen, zerfallen, zerlegen, zersetzen

**dissociated** getrennt

**dissociation** Abscheidung *f*, Abtrennung *f*, Auflösung *f*, Aufspaltung *f*, Ausscheidung *f*, Dissoziation *f*, Dissoziierung *f*, Entmischung *f*, Spaltung *f*, Zerfall *m*, Zerlegung *f*, Zersetzung *f* ~ **of impurity centers** Störstellendissoziation *f*

**dissociation,** ~ **chamber** Zersetzungskammer *f* ~ **coefficient** Auflösungsbeiwert *m* ~ **constants** Dissoziationskonstanten *pl* ~ **continuum** Dissoziationskontinuum *n* ~ **equation** Dissoziationsgleichung *f* ~ **equilibrium** Dissoziationsgleichgewicht *n* ~ **field effect** Dissoziationseffekt *m* ~ **isotherm** Dissoziationsisotherme *f* ~ **limit** Zerfallgrenze *f* ~ **power** Dissoziationsvermögen *n* ~ **process** Dissoziationsvorgang *m* ~ **property** Zersetzbarkeit *f*

**dissociative** dissoziativ ~ **capture** dissoziativer Einfang *m*

**dissoluble** lösbar

**dissolution** Auflösung *f*, Lösung *f*

**dissolve, to** ~ abblenden, aufgehen, aufheben, auflösen, aufschließen, lösen, überblenden, zerfließen, zerlassen **to** ~ **in** aufblenden (film) **to** ~ **in each other** ineinander lösen **to** ~ **out** auslösen **to** ~ **and reprecipitate** umfallen

**dissolved** aufgelöst, gelöst ~ **acetylene cylinder gas** Dissousgas *n* ~ **sugar** aufgelöster Zucker *m*

**dissolver** Auflösebehälter *m* ~ **chests** Alaunauflöser *m*

**dissolving** Auflösen *n*; sich auflösend, lösend ~ **action** Lösewirkung *f* ~ **capacity** Lösungsfähigkeit *f*, -vermögen *n* ~ **chest** Auflösebehälter *m* ~ **chests** Alaunauflöser *m* ~ **intermediary** Lösungsvermittler *m* ~ **power** (of telescopes), photographic, and shortwave telescope) Auflösungsvermögen *n*; Lösungs-fähigkeit *f*, -vermögen *n* ~ **shutter** Überblendungsblende *f* ~ **vat** Laugenfaß *n*

**dissonance** Diskordanz *f*, Dissonanz *f*

**dissonant** diskordant, mißklingend, verstimmt

**dissuade, to** ~ abraten

**dissymmetric** unsymmetrisch ~ **condition** Schiefe *f*

**dissymmetrical, ~ network** unsymmetrisches Netzwerk *n* **~ transducer** asymmetrischer Wandler *m* (acoust.)
**dissymmetry** Asymmetrie *f*, Spiegelungleichheit *f*, Unsymmetrie *f*
**dissyntonized** verstimmt
**distaff** Rocken *m*
**distance** Abstand *m*, Distanz *f*, Einstellentfernung *f* (focusing), Entfernung *f*, Strecke *f*, Tiefenabstand *m*, Wegstrecke *f*, Weite *f*, Zwischenraum *m*, Zwischenstück *n* **~ along an arc** Bogenabstand *m* **~ apart on centers** Mittenabstand *m*
**distance, ~ of acceleration** Beschleunigungsstrecke *f* **~ in air** Luftstrecke *f* **~ between axes** Achsabstand *m* **~ in azimuth traveled by target during time of flight** Hauptauswanderungsstrecke *f* **~ between back lens and image** Schnittweite *f* **~ between bearings** Lagerabstand *m* **~ in broken line** Entfernung *f* in gebrochener Linie **~ from center of drill spindle to column** Bohrspindelausladung *f* **~ of center of gravity from mouth to venturi** Luftangriffabstand *m* **~ between center lines** Mittenabstand *m* **~ from center of top to center of bottom openings in a radiator section** Nabenabstand *m* **~ between centers** (Förderband) Achsabstand *m*, Mittelabstand *m*, Spitzenabstand *m*, Spitzenweite *f*, (Getriebe) Zentralenabstand *m* **~ between the centers of rotation** Drehpunktabstand *m* **~ between centers of successive rulings** Gittermittenabstand *m* **~ between clamps** Einspannlänge *f* **~ over a closed circuit** Entfernung *f* in geschlossener Bahn **~ between compression (or main) ribs** Hauptrippenabstand *m* **~ between corners** Eckenabstand *m* **~ from the earth** Erdferne *f* **~ from the edge** Abstand vom Rand **~ from electron source to lens** Gegenstandsweite *f* **~ between false ribs** Hilfsrippenabstand *m* **~ between frequencies** Frequenzabstand *m* **~ to go** Entfernung *f* zum Zielort **~ between grid wires** Stegabstand *m* **~ of the horizon** Kimmabstand *m* **~ of intersection of rays passing through a lens** Schnittweite *f* von durch eine Linse hindurchgehenden Strahlen **~ of lattice mesh** Netzabstand *m* **~ between layers** Lagenabstand *m* **~ between legs** Stützenentfernung *f* **~ between lines** Zeilenabstand *m* **~ of the object** Bildabstand *m* **~ from object to lens (or from origin)** Startweite *f* **~ of observation** Wahrnehmungsabstand *m* **~ between points** Spitzenentfernung *f* **~ between points (or nodes) of support** Knotenabstand *m* **~ from present to future position (or from reference point to future position)** Gesamtauswanderungsstrecke *f* **~ between principal planes (or points)** Hauptebenenabstand *m* **~ between ribs** Rippenentfernung *f* **~ from rivet center to outside of angle** Streichmaß *n*, Wurzelmaß *n* **~ between screen and lens** Bildweite *f* **~ between ships** Schiffabstand *m* **~ between spindle centers** Spindelmittenentfernung *f* **~ in straight line** Entfernung *f* in gerader Linie **~ between successive rulings** (in diffraction grating) Mittenabstand *m* **~ between supports** Auflagerentfernung *f*, Spannbreite *f* **~ of the target** Zielentfernung *f* **~ from tip to tip of an airfoil** gesamte Breite *f* **~ between two block stations** Blockabstand *m*

**~ between wings** Deckabstand *m* (aviat.)
**distance, ~ action** (by selsyntype motor) Fernsteuerung *f* **~-beam lamp** Weitstrahler *m* **~-between** Zwischenbreite *f* **~ bolt** Distanzschraube *f* **~ calculator** Kursrechner *m* **~ chart** Entfernungstabelle *f* **~-circles** Entfernungskreise *pl* (geodätische) **~ collar** Distanzrohr *n* **~ condition** Abstandsbedingung *f* (photo) **~ control** Abstandssteuerung *f* **~-controlled** fernbetätigt **~-controlled synchronization** selbständiges Synchronisieren *n* (TV) **~ covered** durchlaufene oder zurückgelegte Strecke *f* **~ covered between two bearings** Versegelung *f* **~-difference measurement** Ablagemessung *f* (rdr) **~ error** Abstandsfehler *m*, Entfernungsfehler *m* (rdr) **~ flight** Streckenflug *m* **~ flown** Flugstrecke *f* **~ focusing device** Entfernungseinstellung *f* **~ fog** Entfernungsschleier *m* **~ gauge** Abstandsmesser *m* **~ gear** (torpedo) Distanzeinstellungsapparat *m* **~ gone** zurückgelegte Entfernung *f* **~ indication** Entfernungsanzeige *f* **~ indicator** Distanzzeiger *m* **~ interrogating signal** Entfernungsabfragesignal *n* **~ mark** Entfernungs-Meßmarke *f*, Markierkreis *m* (rdr) **~ measurement** Entfernungsmessung *f* **~ measuring equipment (DME)** DME-Gerät *n*, Entfernungsmeßgerät *n* **~ measuring instrument** Streckenmeßgerät *n* **~-measuring theodolite** Streckenmeßtheodolit *m* **~ meter** Entfernungsanzeiger *m* **~ piece** Abstand-büchse *f*, -fuß *m*, -hülse *f*; Abstandssäule *f* (rdr), Abstandsstück *n*, Distanzstück *n*, Einsatzstück *n*, Entfernungsbrücke *f* (photo), Klebstift *m*, Laterne *f*, Zwischenscheibe *f* **~ pin** Distanzbolzen *m* **~ portion** Fernteil *m* **~ protection** (for powerline) Streckenschutz *m* **~ pulse** Entfernungsmeßimpuls *m* **~ range** Reichweite *f* **~ ratio** Abstandsverhältnis *n* **~ relay** Distanzrelais *n* **~ release** Fernauslöser *m* **~ reply pulse** Entfernungsantwortimpuls *m* **~ revolution counter** Ferndrehzahlmesser *m* **~ ring** Abstandsring *m* **~ run** zurückgelegte Entfernung *f* **~ sailed** durchlaufene Strecke *f* **~ scale** Entfernungsskala *f* **~ separating peaks** Wellenbergglochweite *f* **~ setting mark** Einstellmarke *f* für Entfernung **~ shot** Fernaufnahme *f* **~ signal** Fernsignal *n* **~ staff** Distanzlatte *f* **~ thermometer** Fernthermometer *n* **~ traffic** Fernbetrieb *m* **~ traversed** Laufstrecke *f* **~ washer** Abstandsring *m*, Distanzscheibe *f*
**distant** abstehend, fern **~ collision** Fernstoß *m* **~ control** Fern-bedienung *f*, -schaltung *f*, -steuerung *f* **~ early warning (DEW)** Frühwarnung *f* (rdr) **~ earthquake** Fernbeben *n* **~ effect** Fernwirkung *f* **~ exchange** Gegenamt *n* **~ fog** Fernendunst *m* **~ object** Ferngegenstand *m*, Fernobjekt *n*, **~ office** Gegenamt *n* **~ office in the circuit** Gegenamt *n* (teleph,) **~-operated non-selfrunning time-base method** fremdgesteuerte unselbständige Kippmethode *f* **~-operated time-base method** fremdgesteuerte Kippschwingmethode *f* **~ past** Vergangenheit *f* **~-patrol plane** Fernaufklärer *m* **~ point** Fernpunkt *m* (Stereoaufnahmen) **~ pursuit** überholende Verfolgung *f* **~ radio shielding** Fernentstörung *f* **~-reading pyrometer** Fernthermometer *n* **~-reading-speed indicator** Fernfahrtanzeiger *m* (aviat.) **~-reading tachometer** Fern-

drehzahlmesser *m* ~ **reception** Fernempfang *m*
~ **station** fernes Amt *n* ~**-station reception** Fernempfang *m* ~ **thunderstorm** Ferngewitter *n* ~
**time recorder** Zeitfernzähler *m* ~ **transmission**
Fernanzeiger *m* ~ **view** Fernsicht *f* ~ **visibility**
Fernsicht *f* ~ **zone** Fernzone *f*

**distemper** üble Witterung *f*, Tempera-, Wasserfarbe *f*

**distensibility** Dehnfähigkeit *f*

**distention** Ausdehnung *f*, Aufgeblasenheit *f*

**disthene** blättriger Beryll *m*, Cyanit *m*, Disthen *m*

**distill, to** ~ abdestillieren, abziehen, destillieren, entgasen, entschwelen, schwelen, übersieden, umsieden **to** ~ **out** herausdestillieren **to** ~ **with steam** mit Wasserdampf abblasen **to** ~ **under vacuum** abschwellen **to** ~ **under vacuum at a low temperature** verschwellen

**distillable** destillierbar

**distillate** Destillat *n*, Übersud *m*, Umsud *m* ~
**engine** Spiritusmotor *m*

**distillation** Abziehen *n*, Brennen *n*, Destillation *f*, Destillierung *f*, Entgasungserzeugnis *n*, (fractional) Siedetrennung *f*, Übersieden *n*, Vergasen *n*, (dry) Verkohlen *n* ~ **of bituminous coal** Steinkohlen-destillation *f*, -entgasung *f* ~ **of coal at low temperature** Tiefverkokung *f* ~ **by descent** abwärtsgehende Destillation *f* ~ **in retorts** Retortenvergasung *f* ~ **under vacuum** Verschwelung *f*

**distillation,** ~ **characteristic** Siedeverhalten *n* ~ **column** Destilliersäule *f* ~ **curve** Verdampfungskurve *f* ~ **flask** Siedekolben *m* ~ **loss** Verlust *m* bei der Destillation ~ **overlap** Destillatüberlappung *f* ~ **process** Muffel-prozeß *m*, -verfahren *n* ~ **product of petroleum** Erdöldestillat *n* ~ **range** Destillationsbereich *m* ~ **residue** Destillationsrückstand *m* ~ **test** Destillationsprobe *f* ~ **tube** Destillationsrohr *n* ~ **zone** Vorwärmungszone *f*

**distilled** destilliert ~ **beverage** abgezogenes Getränk *n*

**distiller** Destillateur *m*, Destillierapparat *m*
~**'s grain molasses** Brennereitrebermelasse *f*
~**' grains** Treber *pl* ~**'s wash** Schlempe *f*

**distillery** (grain-alcohol) Branntweinbrennerei *f*, Brennerei *f*, Spiritusbrennerei *f* ~ **attemperator** Brennereikühler *m*

**distilling,** ~ **of lignite** (peats, etc) Braunkohlenvergasung *f* ~ **apparatus** Destillierapparat *m* ~ **filter** Destillierfilter *m* ~ **flask** Destillationskolben *m* ~ **furnace** Destillationsofen *m*, Destillierofen *m* ~ **plant** Destillationsanlage *f*, Destillierapparat *m* ~ **retort** Destillierkolben *m* ~ **tube** Destillationsrohr *n* ~ **unit** Brüdenpumpe *f* ~ **vessel** Destillationsgefäß *n*, Destillierblase *f*, Destilliergefäß *n*

**distinct** anschaulich, ausgeprägt, ausgezeichnet, deutlich, kenntlich, klar, übersichtlich ~ **image** scharfes Bild *n*

**distinction** Auszeichnung *f*, Gepräge *n*, Unterscheidung *f*

**distinctive** auffallend, unterscheidend ~ **mark** Wahrzeichen *n* ~ **marking** Unterscheidungsmarkierung *f* ~ **sign** Wahrzeichen *n* ~ **signal** Kennungssignal *n*

**distinctness** Klarheit *f* ~ **of image** Bildhelligkeit *f*

**distinguish, to** ~ auszeichnen, erkennen, kennzeichnen, unterscheiden, zur Unterscheidung

**to** ~ **oneself** sich auszeichnen

**distinguishable** erkennbar, kennbar, unterscheidbar

**distinguished** ausgeprägt, ausgezeichnet

**distinguishing,** ~ **characteristic** Unterscheidungsmerkmal *n* ~ **difference** kennzeichnender Unterschied *m* ~ **feature** Unterscheidungsmerkmal *n* ~ **mark** Bezeichnungsnagel *m* (Leitung), Kennzeichen *n*, Unterscheidungsabzeichen *n* ~ **marks** Kollizeichen *n* ~ **signal** Unterscheidungssignal *n*

**distort, to** ~ abwinkeln, deformieren, entstellen, krumm werden, sich krumm ziehen, verbiegen, verdrehen, verwerfen, verzerren, verzeichnen, (opt.) verziehen, sich verziehen, verzwicken, sich werfen, (steel) sich ziehen

**distorted** verzerrt, verzogen ~ **grain effect** Runzelkorn *n* (photo) ~ **grid** verzerrtes Netz *n* ~ **picture** Zerrbild *n* ~ **pile** gestörter Reaktor *m* ~ **sound** Mißton *m* ~ **waveform** verzerrte Wellenform *f* ~ **wave method** Störwellenmethode *f*

**distorting,** ~ **duct** Verzerrungsleitung *f* ~ **lens** Zerrlinse *f*

**distortion** Deformation *f*, Fälschung *f*, Fehler *m*, Formänderung *f*, (on hardening) Härtungsverzug *m*, Hinderung *f*, (factor) Klirrfaktor *m*, Schmeißen *n*, Übersprecherscheinung *f*, Verdrehung *f*, Verformung *f*, Verkehrung *f*, Verkrümmung *f*, Verwindung *f*, Verzerrung *f*, Verzeichnung *f* (film, foto), Verzerrung *f* (rdr), Verziehen *n*, Verzug *m*, Werfen *n*

**distortion,** ~ **of amplitude** Dämpfungsverzerrung *f* ~ **of bearing** Funkfehlweisung *f*, Peilfehler *m*, Peilstrahlwegablenkung *f* ~ **of bearing on site** Bordablenkung *f* ~ **of frequency** Dämpfungsverzerrung *f* ~ **of grain** Kornverzerrung *f* ~ **of orientation** anisotropische Verzeichnung *f* ~ **of profile** Profilverzerrung *f* ~ **of scanning pattern** Rasterverzerrung *f* ~ **of television image** Bildplastik *f* ~ **of tone** Dämpfungsverzerrung *f*

**distortion,** ~ **bridge** Klirrfaktormeßbrücke *f* ~ **constant** Verzerrungskonstante *f* ~**-correcting device** Entzerrer-anordnung *f*, -einrichtung *f*, Entzerrungseinrichtung *f* ~**-corrector** compensator Entzerrungsanordnung *f* ~**-corrector device** Entzerrungsanordnung *f* ~ **due to curvature** Feldkrümmungsverzerrung *f* ~ **due to feedback** Rückkuppelungsverzerrung *f* ~ **due to field curvature** Verzerrung *f* durch Feldkrümmung ~ **due to hardening** Härteverzug *m* ~ **due to lack of flatness** Feldkrümmungsverzerrung *f* ~ **error** Verzerrungsfehler *m* ~ **factor** Klirrfaktor *m* ~**-factor meter** Klirrfaktormesser *m* ~**-free** verwindungsfrei, verzeichnungsfrei ~**-free eyepiece** verzerrungsfreies Okular *n* ~ **measurement** Verformungs-, Verzerrungsmessung *f* ~**-measuring system** Verzerrungsmesser *m* ~ **phase** Störfase *f* ~ **power** Verzerrungsleistung *f* ~ **spring** Verdrehungsfeder *f* ~ **tensor** Verzerrungstensor *m* ~ **test** Beulenmessung *f* ~ **tolerance** zulässige Zeichenverzerrung *f* ~**-transmission impairment** Minderung *f* der Übertragungsgüte durch Frequenzbandbegrenzung

**distortional** verzerrend

**distortionless** verzerrungsfrei ~ **circuit** verzerrungsfreie Leitung *f* ~ **microphone** verzerrungsfreies Mikrofon *n*

**distract, to** ~ ablenken
**distraction** Ablenkung *f*
**distrain, to** ~ auspfänden
**distraint** Beschlagnahme *f*
**distress, to** ~ bedrängen, quälen
**distress** Notlage *f*, Notstand *m* ~ **at sea** Seenot *f*
~ **call (or signal)** Not-ruf *m*, -signal *n*, -zeichen *n*, -zeichenanruf *m* ~ **frequency** Notfrequenz *f* ~ **traffic** Notverkehr *m*
**distribute, to** ~ ablegen (print.), aufstellen, aufteilen, ausgeben, (color) ausstreichen, austeilen, austragen, einteilen, schütten, verteilen **to** ~ **a composition** einen Satz ablegen (print.) **to** ~ **the letters of a form** eine Form ablegen
**distributed** gegliedert, verteilt ~ **in depth** nach der Tiefe gestaffelt, tief gegliedert ~ **in lumps** punktförmig verteilt
**distributed,** ~ **amplifier** Kettenverstärker *m* ~ **capacitance** verteilte Kapazität *f* ~ **capacitance of a circuit** Selbstkapazität *f* ~ **capacity** Kapazitätsbelag *m*, verteilte Kapazität *f* ~ **charge** gestreckte Ladung *f* ~ **constant** Leitungskonstante *f* ~ **inductance** punktförmig verteilte Induktivität *f* ~ **parameter** verteilter Parameter *m* ~ **roughness** flächenhafte Rauhigkeiten *pl* ~ **winding** verteilte Wicklung *f*
**distributing** Schütten *n* ~ **agency** Verteilungsstelle *f* ~ **amplifier** Verteilerverstärker *m* ~ **apparatus for types** Ablegeapparat *m* für Lettern ~ **arm** Verteilungsarm *m* ~ **bar** Ablegerstange *f*, Verteilschiene *f* (electr.) ~ **board** Verteilungs-tafel *f*, -tisch *m* ~ **box** Lampenabzweigekasten *m*, Schieberkasten *m* (electr.) ~ **brush** Verstreichbürste *f* ~ **bus bar** Verteilungsschiene *f* ~ **cabinet** Verteilungsschrank *m* ~ **cable** Aufteilungs-, Teilnehmer-kabel *n* ~ **car** Ablegewagen *m* ~ **casing** Verteilungsröhre *f* ~ **chute** Schüttelrinne *f*, Verteilungsschurre *f* ~ **cylinder** Nacktwalze *f*, Reibzylinder *m* ~ **device** Verteiltrichter *m* ~ **eccentric** Steuerungsexzenter *m* ~ **facilities** Verteilereinrichtung *f* ~ **frame** Umschaltegestell *n*, Verteiler *m* ~ **frame wire** Z-Draht *m* ~ **fuse** Verteilungssicherung *f* ~ **gear** Steuerung *f* ~ **lens bar** Verteilungsschiene *f* ~ **line** bewegliche Leitung *f* ~ **pipe** Verteilungsröhre *f* ~ **point** Abzweigpunkt *m*, Ausgabestelle *f*, Park *m*, Speisepunkt *m*, Verteilerstelle *f*, Verteilungspunkt *m* ~ **pole** Abgangsgestänge *n*, Überführungs-säule *f*, -stange *f*, Übergangsgestänge *n* (electr.); Verteilungsmast *m* ~ **print (of film)** Verleihkopie *f* ~ **rod** Verteilungsstab *m* ~ **roll** Lochwalze *f* ~ **slide valve** Grund-, Steuer-schieber *m* ~ **sub-station** Abzweigunterstation *f*; Verteilerwerk *n* ~ **switch** Linienwähler *m* ~ **switchboard** Verteilerschrank *m* ~ **system** Verteilungssystem *n* ~ **tower** Verteilungsturm *m* ~ **valve** zylindrischer Drehschieber *m*, Verteilungsschieber *m* ~ **-valve motion** Kulissensteuerung *f* ~ **ware-house** Ausgabemagazin *n*
**distribution** Aufstellung *f*, Aufteilung *f*, Ausgabe *f*, Gemischverteilung *f*, Gliederung *f*, Streuung *f*, (current) Stromverteilung *f*, Verbreitung *f*, Verteilung *f*, Vertrieb *m*
**distribution,** ~ **of aerodynamic forces** Luftkraftverlauf *m* ~ **in angle** Winkelverteilung *f* ~ **of apportionment** (profit reserve) Verteilung *f* der Gewinnrücklage ~ **of blast** Windverteilung *f*

~ **of brightness** Leuchtdichteverteilung *f* ~ **of charge density** Ladungsdichtenverteilung *f* ~ **of circulation** Umlaufverteilung *f* ~ **of density** Dichteverteilung *f* ~ **in depth** Tiefengliederung *f* ~ **of energy** Geschwindigkeitsverteilung *f* ~ **of exposure** Lichtausbreitung *f* ~ **of film** Filmvertrieb *m* ~ **of forces** Kraftverteilung *f* ~ **of ink** Farbenverreibung *f* ~ **of load** Lastangriff *m* ~ **of messages** Telegrammverteilung *f* ~ **of orders** Auftragserteilung *f*, Befehlsausgabe *f* ~ **of precipitation** Niederschlagsverteilung *f* ~ **of rainfall** Niederschlagsverteilung *f* ~ **of requisition** Kopiegabe *f* bei Bedarfsanforderungen ~ **of steam** Dampfverteilung *f* ~ **of temperature** Wärmeverteilung *f* ~ **of tools** Werkzeugausgabe *f* ~ **of traffic** Verkehrs-bewegung *f*, -gestaltung *f* ~ **of velocities** Geschwindigkeitsverteilung *f* ~ **of vortices** Wirbelverteilung *f* ~ **in width** Breitengliederung *f*
**distribution,** ~ **arm** Verteilungsarm *m* ~ **board** Schalttafel *f* ~ **box** Verteilerkasten *m*, Verteilungskasten *m*, Verzweiger *m* ~ **cable** Verteilungskabel *n* ~ **case** Verteilungskasten *m* ~ **coefficient** Teilungskoeffizient *m*, Verteilungskonstante *f* ~ **company** Verteilungsgesellschaft *f* ~ **conduit** Verteilerleitung *f* ~ **constant** Verteilungskonstante *f* ~ **cylinder** Reibzylinder *m* ~ **duct** Verteilerleitung *f* ~ **factor** Verteilungsfaktor *m* ~ **fluctuation** Verteilungsrauschen *n* ~ **frame** Verteiler *m*, Verzweiger *m* ~ **function** Verteilungsfunktion *f* ~ **gear** Verteilungsanlage *f* ~ **gear controlling the starting air** Anlaßluftsteuerung *f* ~ **grid** Verteilungsgitter *n* ~ **header** Verteilstück (-Rohr) *n* ~ **key** Verteilerschlüssel *m* ~ **law** Verteilungs-gesetz *n*, -satz *m* ~ **license** Verteilungs-erlaubnis *f*, -lizenz *f* ~ **list** Anschriftenübermittlung *f* ~ **manifold assembly** Verteiler *m* ~ **medium** Ausgabemittel *n* ~ **modulus** Verteilungsmodul *n* ~ **network** Drahtfunknetz *n*, Verteilungsnetz *n* ~ **noise** Verteilungsrauschen *n* ~ **panel** Schalttafel *f* ~ **plate** Verteilungsplatte *f* ~ **point** Weitergabestelle *f* ~ **ratio** Verteilungskoeffizient *m* ~ **reinforcement** Verteilungseisen *n* ~ **rule** Ablegespan *m* ~ **switchboard** Verteilerschalttafel *f* ~ **system** Stromsystem *n* (electr.), Verteilsystem *n* ~ **terminal** Endverzweiger *m*, Unterverteiler *m*, Verteilkasten *m* ~ **transformer** Speisetransformator *m* ~ **valve** Steuerventil *n*
**distributive law** Distributivgesetz *n*
**distributor** Ableger *m* (print.), Aufgabetrichter *m*, Austeiler *m*, Kollektor *m*, Leitapparat *m*, Lieferer *m*, Steuerungseinrichtung *f*, Verteiler *m*, Verteilrichter *m*, Verteilsstelle *f*, Zerteiler *m*, (ignition) Zündverteiler *m* ~ **with jumping sparks** Verteiler *m* mit springenden Funken
**distributor,** ~ **arm** Verteilerfinger *m* ~ **board** Verteilertafel *f* ~ **box** Verteiler-, Zwischen-kasten *m* ~ **brush** Verteilerbürste *f* ~ **bucket** Gichtkübel *m* ~ **cam** Verteilerzahnrad *n* ~ **cap** Verteilerdose *f* ~ **cap cover** Verteilerdeckel *m* ~ **carriage** Ablegeschlitten *m* ~ **cogwheel** Verteilerzahnrad *n* ~ **column (or tower)** Verteilerturm *m* ~ **cover** Verteilerdeckel *m* (Motor) ~ **disk** Scheibe *f* des Verteilers ~ **disk with connection** Stützverteilerscheibe *f* ~ **drive** Ablegerantrieb *m* ~ **driving spindle** Verteilerantriebsspindel *f* ~ **duct** Kabel-graben *m*, -kanal

*m* ~ **face** Verteilerscheibe *f* ~ **finger** Verteilerbürste *f* ~ **gear** Verteilergetriebe *n* ~ **head** Batteriezündverteilerkopf *m*, (engine) Verteilerkopf *m*, Verteilerscheibe *f* ~ **link** Verteilerleiste *f* ~ **main** Verteilleitung *f* ~ **pipe** Verteilerrohr *n* ~ **piston** Rückführschieber *m* (jet), Verteilerkolben *m* ~ **plate** Verteilerplatte *f*, Verteilerscheibe *f*, Verteilteller *m* ~ **plateau** Verteilerscheibe *f* ~ **ring** Verteilerring *m* ~ **roller journals** Verreibwalzen *pl* ~ **rotor** Verteiler-läufer *m*, -laufstück *n* ~ **segment** Verteilersegment *n* ~ **shaft** Steuerwelle *f* (Schmierpumpen), Verteiler-achse *f*, -welle *f* ~ **sleeve bearing** Verteilergleitlager *n* ~ **stand** Verteilerbock *m* ~ **system** Verteileranlage *f* ~ **terminal** Verteilerklemme *f* ~ **track** Verteilergleitbahn *f*
**district** Bezirk *m*, Distrikt *m*, Gau *m*, Gebiet *n*, Gebietsteil *n*, Gegend *f*, Kreis *m*, Rayon *m*, Revier *n*, Strich *m*, Zone *f* ~ **exchange** (main) Bezirksamt *n* ~ **heating system** Fernheizung *f*
**disturb, to** ~ (sediment) aufrühren, beunruhigen, stören **to** ~ **existing (previous) drainage conditions** die Vorflut stören
**disturbance** Begleiteffekt *m*, Beschwerung *f*, Beunruhigung *f*, Störgröße *f*, (noise) Störung *f*, Unruhe *f*, Verrückung *f* ~ **of beds** Lagerungsstörung *f* ~ **of directional force** Richtströmung *f* ~ **of retention** Merkstörung *f* ~ **of traffic** Verkehrs-stockung *f*, -störung *f*
**disturbance,** ~ **calculation** Störungsrechnung *f* ~ **current** Störstrom *m* ~ **-fall indicator** Störungsfallkappe *f* ~ **function** Störfunktion *f* ~ **impulse** Störimpuls *m* ~ **network** Störungsnetz *n* ~ **potential** Störpotential *f* ~ **variable** Störgröße *f* ~ **variable feed forward** Störgrößenaufschalschaltung *f* ( (aut. contr.) ~ **velocity** Nebenbewegung *f*
**disturbed** gestört, durch Störung *f* außer Betrieb gesetzt ~ **area** Störgebiet *n* ~ **-one output** gestörtes Einersignal *n* ~ **-zero output** gestörtes Nullsignal *n*
**disturber** Störer *m*
**disturbing** störend ~ **current** Störstrom *m* ~ **effect** Störwirkung *f* ~ **moment** Störungsmoment *n* ~ **pulse** Störimpuls *m* ~ **station** Störer *m* ~**-unbalance-measuring set** Geräuschunsymmetriemesser *m* ~ **wave** Störschwingung *f*
**disulphamic acid** Disulfaminsäure *f*
**disulphonic acid** Disulfosäure *f*
**disuse** Nichtgebrauch *m*
**ditch, to** ~ ditschen, graben, notwassern
**ditch** Drosselspalt *m* (Hohlleiter), Gerinne *n*, Graben *m*, Kanal *m*, Krecke *f*, Rinne *f*; Wall-, Wasser-, Werk-graben *m* ~ **canal** Sackkanal *m* ~ **conveyor** Grabenband *n* ~ **-crossing ability** Überschreitfähigkeit *f* ~ **digging** Grabenherstellung *f* ~ **-digging machine** Grabenbagger *m* ~ **jumper** Grabenspringer *m* ~ **profile** Grabenprofil *n* ~ **sample** Spülprobe *f*
**ditcher** Grabenaushebemaschine *f*, Tieflöffelbagger *m*
**ditchers** Becherwerke *pl*
**ditching** Notlandung *f* (auf dem Wasser), Notwasserung *f* ~ **plow** Grabenflug *m*
**dithionic acid** Dithionsäure *f*
**dithyramb** Dithyrambe *f*
**ditungstic acid** Diwolframsäure *f*
**diurnal** ganztägig, täglich ~ **heating** tägliche

Erwärmung *f* ~ **motion gear** Tagesgang *m* ~ **movement** Tagesgang *m* ~ **parallax** tägliche Parallaxe *f* ~ **variation** Tagesänderung *f*, Tagesgang *m* ~ **variations** tägliche Schwankungen *pl*
**divalent** zweiwertig
**divariant** divariant, zweifachfrei ~ **equilibrium** zweifachfreies Gleichgewicht *n*
**dive, to** ~ stechen, Sturzflug *m* ausführen oder machen, tauchen, versenken
**dive** Absturz *m*, Fall *m*, Stechflug *m*, steiler Abflug *m*, Sturzflug *m* ~ **angle** Bahnneigungswinkel *m* ~ **flap** Stechflugklappe *f*, Sturzflugbremse *f*
**diver** Taucher *m*
**diverge, to** ~ abweichen, auseinanderfahren, auseinandergehen, auseinanderweichen, sich nicht decken, zerstreuen, **to** ~ **from** divergieren
**divergence** Abweichung *f*, Auseinandergehen *n*, Auseinanderlaufen *n*, Auseinanderweichen *n*, Divergenz *f*, Streuung *f* **case of** ~ Divergenzfall *m* ~ **of lines of latitude** Breitenausdehnung *f*
**divergence,** ~ **difference** Feuerverteilungswinkel *f* ~ **loss** Divergenzverlust *m*
**divergent** abweichend, auseinandergehend, auseinanderlaufen, auseinanderstrebend, divergent, divergierend ~ **beam light** Fächerleuchte *f* ~ **lens** Zerstreuungslinse *f* ~ **nozzle** Diffusordüse *f* ~ **values** auseinandergehende Werte *pl*
**diverging** divergent, divergierend, zerstreuend ~ **attack** exzentrischer Angriff *m* ~ **concavoconvex lens** konvexkonkave Linse *f* ~ **cone** Abflußrohr *n* ~ **lens** zerstreuendes Glas *n*, Streuungslinse *f* ~ **meniscus** konvexkonkave Linse *f* ~ **series** auseinanderlaufende Reihen *pl* (math.) ~ **wave** divergierende Welle *f*
**diverse** mehrere, verschieden, vielgestaltig
**diversified** unterschiedlich
**diversion** Ablenkung *f*, Diversion *f*, Nebenangriff *m*, Umleitung *f* ~ **of the line** Umlegung *f* der Leitung ~ **of a river channel** Streckung *f* eines Flußbettes
**diversion,** ~ **channel** Umflugkanal *m*, Umlauf *m* ~ **dam** Buhne *f* ~ **maneuver** Ablenkungsmanöver *n* ~ **means** (in apparatus) Ableitung *f* ~ **structure** Leitwerk *n* ~ **tunnel** Ableitungs-, Umlauf-stollen *m*, Umleitungstunnel *m* ~ **valve** Verteilerventil *n*
**diversity** Mannigfaltigkeit *f*, Ungleichheit *f*, Verschiedenheit *f* ~ **of colors** Vielfarbigkeit *f*
**diversity,** ~ **factor** Gleichzeitigkeitsfaktor *m*, Verschiedenheitsfaktor *m* ~ **receiver** Diversity-Empfänger *m* ~ **reception** Mehrfachempfang *m* ~**-reception method** Mehrfachempfangsverfahren *n* ~ **system** Mehrfachempfangssystem *n*
**divert, to** ~ abkehren, ableiten, ablenken, abwenden, abziehen, umleiten
**diverted transmission** Umtrieb *m* (Riemenumführung *m*)
**diverter** Ablenkgerät *n*, Abweiser *m* ~ **guide** Weiche *f* (Walzwerk), ~ **segregator valve** Umleit-Trennventil *n* ~ **segregator valve system** Kaltstartanlage *f* (Hubschr.)
**divide, to** ~ abteilen, dividieren, einteilen, scheiden, teilen, trennen, unterteilen, mit einer Teilung *f* versehen, verteilen, zerfallen, zerteilen **to** ~ **by boundaries** abgrenzen **to** ~ **into commercial lengths** auf handelsübliche Längen *pl* zerteilen **to** ~ **up** stückeln

divide Scheide *f*, Wasserscheide *f* ~-by-two circuit Frequenzhalbierschaltung *f*
divided (up) abgeteilt, eingeteilt, gebrochen, geteilt, getrennt, zerteilt ~ arc Limbusteilstrich
*m* ~ attention Vielfeldaufmerksamkeit *f* ~
axle gebrochene oder geteilte Achse *f* ~ bonus
Gewinnanteil *m* ~ circle Kreisleitung *f*, Teilkreis *m* ~ circuit verzweigter Stromkreis *m*,
Zweigstromkreis *m* ~ coil verteilte Spule *f* ~
combustion chamber unterteilter Brennraum *m*
~ combustion chamber engines Kammermoto-
ren *pl* ~ combustion engine Zweibrennraum-
motor *m* ~ control surfaces geteiltes Seitenleitwerk *n* ~ image range-finder inverted telemeter
Teilbildentfernungsmesser *m* ~ ironcore unterteilter Eisenkern *m* ~ plate crystal Doppelplattenkristall *m* ~ ruler Lineal *n* mit Teilung *f*
~ stuffing-box gland zweiteilige Stopfbüchsenbrille *f* ~ trough kneader Doppelmulden-
kneter *m*
dividend Dividende *f*, Teilungszahl *f* ~ mine
Ausbeutezeche *f* ~ warrant Dividendenschein
*m*
divider Außenteiler *m*, Einstrich *m*, Spitzenzirkel *m*, Teiler *m*, Verteiler *m*, Verzweiger *m*
~ circuit Frequenzteilerschaltung *f* ~ roller
Scheibenteilwalze *f* ~ stage Teilerstufe *f*
dividers Greifzirkel *m*, Handzirkel *m*, Kreis-
reißer *m*, Querholz *n*, Stechzirkel *m*, Teilzirkel
*m*, Zirkel *m*
dividing Scheidung *f*, Teilung *f*, Trennung *f*,
Nachbearbeitung *f* der Skalenteilung ~ attachment Teilaufsatz *m* ~ box Abzweig-kasten *m*,
-muffe *f*, Endverzweiger *m* ~ device Teilvorrichtung *f* ~ device for calculating punch Divi-
sionseinrichtung *f* für Rechenlocher ~ disk
Teil-platte *f*, -scheibe *f* ~ engine Teilmaschine *f*
~ factor Teilungszahl *f* ~ filter Frequenzweiche *f*, Trennfilter *m* ~ gear Teilgetriebe *n*
~ head Teilkopf *m* ~ head for spur gears Stirn-
räderteilkopf *m* ~ head for toothed wheels
Zahnräderteilkopf *m* ~ head axis Teilachse *f*
~ line Teilstrich *m*, Trennungslinie *f* ~ machine
Teilmaschine *f* ~ mechanism Teileinrichtung *f*
~ network Frequenzteilerschaltung *f*, Frequenzweiche *f* (acoust.), Trenn-Netzwerk *n* ~
rod Stange *f* mit Einstellungen ~ rule Abtei-
lungslinie *f* ~ screw Meßschraube *f* ~ spindle
Teilspindel *f* ~ stages Teilungsstufen *pl* ~ strip
Trennschiene *f* ~ surface Trennfläche *f* ~ wall
Trennbrett *n*, Trennspant *m*, Trenn(ungs)wand
*f*, Trennungswerk *n* ~ taster Teilzirkel *m* ~
template Teilschablone *f* ~ worm wheel Teil-
schneckenrad *n*
diving Abwärtsgleiten *n*, Tauchen *n* ~ apparatus
Tauchgerät *n* ~ bell Taucherglocke *f* ~-bell
gear Tauchergerät *n* ~-bell suit Taucheranzug
*m* ~ course Sturzflug *m*; Taucherlehrgang *m*
~ depth Tauchtiefe *f* ~ plane Tauchfläche *f* ~
position Sturzlage *f* ~ speed Sturzfluggeschwin-
digkeit *f* ~ start stürzender Start *m* ~ test
Sturzflugerprobung *f* ~ trial Sturzflugerprobung *f* ~ turn gedrückte Kurve *f*, Sturzflug-
kurve *f*
divining rod Wünschelrute *f*
divisibility Teilbarkeit *f*
divisible einteilbar, teilbar, unterteilbar
division Abteil *n*, Abteilung *f*, Abtrennung *f*,

(of patent application) Abzweigung *f*, Aufspaltung *f*, Bahnstrecke *f*, Division *f*, Einteilung *f*, Feld *n*, Gliederung *f*, Gradeinteilung *f*,
Intervall *n*, Skala *f*, Spaltung *f*, Teil *m*, Teilstrich *m*, Teilung *f*, Verteilung *f*
division, ~ of area Flächeneinteilung *f* ~ of the
circle Sechzigerteilung *f* ~ of the compass card
(or rose) Roseneinteilung *f* ~ into degrees Abstufung *f* ~ of the discharge Abflußmengenver-
teilung *f* ~ of five Fünferteiler *m* ~ of a pit
Schachtabteilung *f* ~ of scale Skalateilung *f* ~
of a shaft Schachtabteilung *f*, Schachttrumm *n*
~ of spar Holzteilung *f* ~ in squares Felderein-
teilung *f* ~ into square network (or grid) Qua-
dratnetzeinteilung *f* ~ for training Ausbil-
dungsabteilung *f*
division, ~ grate Teilungsrechen *m* ~ manager
Abteilungsleiter *m* ~ mark Strichmarke *f* ~
plane Trennungsebene *f* ~ plate Abdeckblech
*n*, Verteilungsplatte *f* ~ scale Strichskala *f* ~
sign Teilungszeichen *n* ~ testing device Teil-
kopfprüfgerät *n* ~ wall Scheidewand *f*, Trennwand *f*
divisional application abgezweigte Anmeldung *f*
divisor Divisor *m*, Teiler *m*
dizziness Schwindel *m*
dizzy schwindlig
do, to ~ bewerkstelligen, fertigen, handhaben,
leisten, (business) tätigen, tun, verrichten
docimastic dokimastisch
dock, to ~ anlegen, docken
dock Binnenhafen *m*, Dock *n*, Innenbecken *n*,
Luftschiffhalle *f* (wet) ~ (alongside a canal)
Hafenbecken *n* ~ chief Dockmeister *m* ~ conveyor Kaiband *n* ~ crane Dockkran *m* ~ dues
Dockgebühren *pl* ~ harbor Dockhafen *m* ~
installation Dockanlage *f* ~ pump Dockpumpe
*f* ~ warehouse Hafenlagerhaus *n* ~ warrant
Ladeschein *m* ~ yard Dock *n*, Schiffswerft *f*,
Werft *f*
dockage Dockgebühren *pl*
docket Begleitschein *m*, Rolle *f*, Sache *f* ~ number Aktenzeichen *n*, Rollennummer *f* ~ tab
rider Aktenschwanz *m*
docking Rendezvous *n* (g/m) ~ collar Anschluß-
trichter *m* (g/m) ~ gear (airship) Haltegestell
*n* ~ trolley (airship) Landegestell *n*
doctor Abstreichmesser *m* (mach.), Färbezylin-
der *m*, Messer *n* (paper mfg.), Schienen *pl*
(paper mfg.), Streichmesser *n* (mach.) ~
arrangement Rakelführung *f* ~ blade Abstreif-
messer *m*, Schaber *m* (mach.) ~ blade carrier
Rakelträger *m* ~ knife Kratzeisen *n*, Rakel *m*
~ roll Schaberwalze *f*, Schaberklingenwalze *f*
~ rule Rakellineal *n* ~ streak Rakelstreifen *m*
~ test (petroleum) Versuch *m* mit Plombit ~
treatment (petroleum) Plombitbehandlung *f*
doctored solution Entschwefelung *f*
doctrine Doktrin *f*, Lehre *f* ~ of equilibrium
Gleichgewichtslehre *f* ~ of phases Phasenlehre *f*
document Akte *f*, Aktenstück *n*, Arbeitsunterlage *f*, Ausweispapier *n*, Beweisstück *n*, Schriftstück *n*, Urkunde *f* ~ conveying the grant Verleihungsurkunde *f* ~ cover Aktendeckel *m* ~
input Eingang *m* an Belegen ~ paper Urkunden-
papier *n* ~ proof dokumentenecht (Kugel-
schreiber) ~ recorders Dokumentationsgeräte
*pl*

**documentary** urkundlich ~ **evidence** Urkunden-beweis *m* ~ **record** (of patents) Erteilungsakten *pl* ~ **reel** Dokumentarfilm *m* ~ **verification** Beurkundung *f*
**documentation** belegmäßige Bearbeitung *f*, Dokumentation *f*, Schrifttum *n*
**dodecagon** Zwölfeck *n*
**dodecahedral** dodekaedrisch ~ **slip** Dodekaeder-gleitung *f*
**dodecahedron** Dodekaeder *n*, Zwölfflach *n*
**dodge, to** ~ abschatten (photo), elektronisch aufhellen (photo), umgehen
**doff, to** ~ auswechseln
**doff** (textiles) Wechsel *m*
**doffer** Filet *n*, (textiles) Filettrommel *f*
**doffing,** ~ **motion** Unterwindeeinrichtung *f* ~ (waste) **roller** Anspinn-, Umspinn-walze *f*
**dog** Anschlag *m* (mech.), Aufspannfrosch *m*, Bauklammer *f*, Daumen *m*, Frosch *m*, (e. g. for lumbering) Greifdorneisen *n*, Hund *m*, Kamm *m*, Klaue *f*, Klinke *f*, Kloben *m*, Knagge *f*, Kropfeisen *n*, Nocke *f*, Spannkloben *m*, Sperrzahn *m*, Ziehzange *f* (mech.) ~ **and chain (or belt)** Ziehzeug *n* (min.)
**dog,** ~ **arm** Mitnehmerhebel *m* ~ **clutch** Antriebklaue *f*, Klauenkuppelung *f* ~ **clutch constant mesh gearbox** Allklauengetriebe *n* ~ **clutch member** Schaltklaue *f* ~-**controlled** nockenbetätigt ~ **coupling** Klauenkupplung *f*, Klaukupplung *f* ~ **curve** Hundekurve *f* ~'**s ear** Eselsohr *n* ~ **hair** (textiles) falsche Haare *pl* ~-**headed spike** Hakenstift *m* ~ **hook** Gaffel *f*, Klammerhaken *m* ~ **leg** Einlaufrillenfehler *m* (phono) ~ **leg chisel** Grundeisen *n* ~ **movement** Schlägerantrieb *m* (film) ~ **nail** Schloßnagel *m* ~ **plate** Mitnehmerplatte *f* ~-**point** (set) **screw** Druckschraube *f* mit am Ende angedrehtem Druckzapfen ~ **ring** Klauen-Dichtring *m* ~ **tooth course** Wolfszahnschicht *f* ~ **wood** Hart-riegelholz *n*
**dogger steel** Doggerstahl *m*
**dogma** Lehre *f*
**doldrums** Doldrums *n*, Kalmen *pl*, Kalmengür-tel *m*, Kalmenzone *f*, Stillen *pl*, Zone *f* der Windstillen
**dolerite** Dolerit *n*
**doll** Puppe *f*; Weichenpfahl *m*
**dolly** Gerät *n* (min.), fahrbarer Montagebock *m*, fahrbares Montagegestell *n*, Nietpfanne *f*, Preßstempel *m*, aufgesetzter Rammklotz *m*, Transportwagen *m*, Widerlager *n* ~ **bar** Nietendöpper *m* ~-**in** Vorfahren *n* der Kamera, Kameravorschub *m* ~ **mixer** fahrbares Misch-pult *n* ~-**out** Zurückfahren *n* der Kamera ~ **truck** Transportwagen *m* mit Rungen ~ **tub** Rührfaß *m*
**dolomite** Bitter-kalk *m*, -kalkspat *m*, -spat *m*, Braunspat *m*, Dolomit *m*, Dolomitspat *m*, Rauchkalk *m* ~ **bottom** Dolomitherd *m* ~ **brick** Dolomitstein *m* ~-**calcining kiln** Dolo-mitbrennofen *m* ~ **cupola furnace** Dolomit-brennofen *m* ~ **kiln** Dolomitbrennofen *m* ~ **lining** Dolomitzustellung *f* ~ **powder** Dolomit-mehl *n*
**dolomitic** dolomithaltig ~ **lime** Dolomitkalk *m*
**dolphin** (logging) Pfahlbündel *n*
**dolphins** Dalben *pl*, Dallen *pl*, Duckdalben *pl*
**domain** Bereich *m* ~ **of convergence** Konver-

genzbereich *m* ~ **of definition** Definitionsbe-reich *m* ~ **of dependence** Abhängigkeitszone *f* ~ **of integrity** Integritätsbereich *n* ~ **of multi-pliers** Multiplikatorenbereich *m*
**domain,** ~ **boundaries** Domänengrenzflächen *pl* ~ **collocation** Gebietskollokation *f* ~ **size** Domänengröße *f* ~ **wall** Kristallbereichswand *f* ~-**wall switching** Blockwandverschiebung *f*
**dome** Aufsatz *m*, Dach *n*, Dom *m*, geologischer Dom *m*, Gewölbe *n*, Haube *f*, Helmdecke *f*, Holländerhaube *f*, Kappe *f*, Kuppe *f*, Kuppel *f*, Kuppelhalle *f*, Pilz *m*, Rondell *n* ~ **of the mill** Mühlenhaube *f*
**dome,** ~-**base angle ring** Domwinkelring *m* ~ **bottom** Domboden *m* ~ **construction** Kuppel-bau *m* ~ **gas** Domgas *n* ~ **head counter-sunk screw** Linsensenkkopfschraube *f* ~ **illumination** Kuppelbeleuchtung *f* ~ **kiln** Kuppelofen *m* ~ **light** Deckenlicht *n* ~ **lighting** Kuppelbeleuch-tung *f* ~ **nut** Haubenmutter *f* ~ **pressure** Domdruck *m* ~ **reflector** Tiefstrahler *m* ~ **ring** Kuppelkranz *m* ~ **seating** Domsitz *m* ~-**shaped** gewölbt ~-**shaped combustion chamber** halb-kugelförmiger Verbrennungsraum *m* ~-**shaped roof** Gewölbedach *n* ~-**shaped top** Bergkuppe *f* ~-**tipped fillister head** Linsenkopf *m*
**domed** gewölbt ~ **cover** Behälterhaube *f* (Metall-haube *f*) ~ **cover glass** Blendennippel *m*
**domestic** einheimisch, heimisch, inländisch ~ **air service** Inland-(Flug)Linienverkehr *m* ~ **appliances** Haushaltsgeräte *pl* ~ **cheesecloth** Nesseltuch *n* ~ **coke** Kleinkoks *m* ~ **consump-tion** Inlandverbrauch *m* ~ **electric bell** Haus-wecker *m* ~ **fuel** Hausbrand *m* ~ **lighting plant** Hauszentrale *f* ~ **lubricating oil** Schmieröl *n* für den häuslichen Gebrauch ~ **plant** Haus-haltanlage *f* ~ **price** Inlandpreis *m* ~ **service** Dienstverhältnis *n* ~ **telegraph circuit** Tele-grafenleitung *f* für den inneren Verkehr ~ **telephone circuit** Fernsprechleitung *f* für den inneren Verkehr ~ **traffic** Inlandverkehr *m* ~ **volume** Zimmerlautstärke *f*
**domeykite** Arsenikkupfer *n*
**domical vault** Kuppelgewölbe *n*
**domicile** Aufenthaltsort *m*, Wohnsitz *m*
**dominant** Dominante *f* (mus.) ~ **mode** Grund- oder Haupttyp *m* (Wellenleiter) ~ **wave** Grund-, Haupt-welle *f*
**dominate, to** ~ niederhalten
**dominating** beherrschend
**donation** Geschenk *n*, Schenkung *f*
**done** bewerkstelligt; gar
**donkey** Esel *m* ~ **boiler** Hilfskessel *m* ~ **engine** Hilfsmaschine *f* ~ **pump** Dampfspeisepumpe *f*
**donor** Elektronenspender *m* ~ **impurity** Dona-torverunreinigung *f* ~ **level** Donatorenniveau *n*
**do-nothing instruction** Leerbefehl *m* (data proc.)
**donut** Flußverstärker *m*, Ringröhre *f*
**donutron** Doppelkäfigmagnetron *n*
**door** Pforte *f*, Tor *n*, Tür *f* ~ **with two leaves** Flügeltür *f*
**door,** ~ **actuating cylinder** Klappenarbeitszylin-der *m* ~ **air tank** Türluftbehälter *m* ~ **arch** Türbogen *m* ~ **bay** Türjoch *n* ~-**bolting in-stallation** Türverriegelungsanlage *f* ~ **boy** Tür-zieher *m* ~ **case** Tür-einfassung *f*, -futter *n*, Tür-verkleidung *f* ~ **check** Türpuffer *m* ~ **checkrod** Türspanner *f* ~ **close pressure post**

Klappeneinfahranschluß *m* ~ **closer (or lock)** Türschließer *m* ~ **contact interrupter** Türkontaktschalter *m* ~ **cushion** Türkissen *n* ~ **fastening** Türverschluß *m* ~ **fitting** Türbeschlag *m* ~ **frame** Türrahmen *m* ~ **framing** Rahmenstuhl *m* ~ **guide** Türführung *f* ~ **gunnet** Türband *n* ~ **handle** Tür-drücker *m*, -griff *m* ~ **hinge** Tür-, Klappen-scharnier *n* ~ **jamb** Türpfosten *m* ~ **joint** Türfuge *f* ~ **keeper's cell** Pförtnergemach *n* ~ **knob transformer** Hohlleiter-, Koaxial- -übergang *m* ~ **knob tube** Kohlrabiröhre *f* ~ **knobturning machine** Türknopfdrehbank *f* ~ **latch** Türverschluß *m* ~ **lining** Türfüllung *f* ~ **lintel** Türkappe *f* ~ **loop** Türband *n* ~ **mat** Fußmatte *f* ~ **mounting** Türbeschlag *m* ~ **opener** Türöffner *m* ~ **panel** Türverkleidung *f* ~ **pillar** Türsäule *f* ~ **plate** Vorsatzplatte *f* ~ **post** Türpfosten *m* ~ **push** Türkontakt *m* ~ **rail** Türsteg *m* ~ **scraper** Fußkratzen *n*, Kotkratze *f* ~ **sealing** Türdichtung *f* ~ **spring** Türfeder *f* ~ **step** Türpuffer ~ **switch** Türkontaktschalter *m*, Türschalter *m* ~ **valve** Türventil *n* ~ **way** Torweg *m*

**dope, to** ~ lackieren, überstreichen

**dope** Calcium-Silizium *n*, Firnis *m*, Gußmasse *f*, Isoliermasse *f*, Klebelack *m*, Lack *m*, (paint) Lackierung *f*, (dynamite mfg.) Zumischpulver *n* ~ **for aviation gasoline** Betriebsstoffzusatzmittel *n* ~**-spray section shop** Spritzlackiererei *f*

**doped** dotiert ~ **envelope** Firnishülse *f* ~ **gasoline** Bleibenzin *n* ~ **junction** dotierte Schicht *f*

**doper** Anlaßkraftstoffeinspritzvorrichtung *f*

**doping** (of semiconductors) Dotierung *f*, Tränkung *f*, (plywood) Trennmittel *n*, Zusatzmittel *n* (plywood) ~ **compensation** Dotierungsausgleich *m* ~ **system** Anlaßeinspritzsystem *n*

**Doppler,** ~ **broadening** Doppler-Breite *f* ~ **effect** Doppler Effekt *m* ~ **radar** Doppler-Radar *n* ~ **shift** Doppler-effekt *m*, -verschiebung *f*

**dormant** eingefroren, schlafend (geol.) ~ **bolt** versenkte Schraube *f*

**dormer** Lagerschwelle *f* des Fußbodens ~ **of a floor** Unterzug *m* eines Fußbodens ~ **window** Dachfenster *n*, Giebelfenster *n*, Luke *f*, Ochsenauge *n*

**dorsal,** ~ **antenna** Rückenflossenantenne *f* ~ **cup** Dorsalkapsel *f* ~ **fin** Rückenflosse *f* (aircraft), Rückenleitfläche *f* ~ **ridge** Rückenflosse *f* ~ **surface of the foot** Fußrücken *m*

**dosage** Dosierung *f*, (concrete) Mengenverhältnis *n* ~ **grid** Dosierungsgitter *n* ~ **rate** Dosisleistung *f*

**dosaged** dosiert

**dosaging valve** Dosierventil *n*

**dose, to** ~ abwiegen

**dose** Dosis *f*, Gabe *f* ~ **of an isotope** Isotopendosis *f*

**dose,** ~ **determination** Zumessung *f* ~**-effect curve** Dosiswirkungskurve *f* ~**-effect relation** Dosiswirkungskurve *f* ~ **fractionation** Dosisfraktionierung *f* ~ **meter** Dosismeßgerät *n* ~ **protraction** Dosisprotrahierung *f* ~ **rate** Strahlungsintensität *f* ~ **rate meter** Dosisleistungsmesser *m* ~ **surface** Oberflächendosis *f* ~ **value chamber** Dosisleistungskammer *f* ~ **value indication** Dosisleistungsanzeige *f* ~ **value measurement** Dosisleistungsmeßgerät *n* ~

**dosed** dosiert

**doser** Geber *m*

**dosimeter** Dosimesser *m*, Dosimeter *m*

**dosing** Zumessen *n* ~ **machine** Dosiermaschine *f*, Mengeneinteilungsmaschine *f* ~ **nozzle** Zumeßdüse *f* ~ **pump** Dosierpumpe *f*

**dosser** Kiepe *f*, Tragkorb *m*

**dot, to** ~ punkten, punktieren, tüpfeln

**dot** Morsepunkt *m*, Punkt *m*, Punktmarke *f* ~ **of light** Lichtpunkt *m*

**dot,** ~**-bar generator** Punkte-Balken-Generator *m* (TV) ~ **curve** Punktkurve *f* ~ **cycle** Schrittperiode *f*, Signalisierungszyklus *m* ~**-dash line** Strichpunktlinie *f* ~**-dash mode** Punkt-Strich-Verfahren *n* ~ **frequency** Telegrafierfrequenz *f*, Telegrafiergrundfrequenz *f* ~ **generator** Punktgenerator *m* (TV) ~ **halo** Punkthof *m* ~ **interlace** Punktsprung *m* (TV) ~ **interlacing** Punktsprungverfahren *n* (TV), Punktverflechtung *f*, Zwischenpunktabtastung *f* ~**-keying** Punkttastung *f* ~ **mark** Punktmarke *f* ~ **printing** Punktdrucken *n* ~ **product (or vector)** inneres Produkt *n* ~ **record** Punktkurve *f* ~ **sequential system** Punktfolgefarbenverfahren *n*

**dotted** (e. g. line on graph) gestrichelt, punktiert ~ **line** gestrichelte oder punktierte oder gepunktete Linie *f*, Strichlinie *f* ~**-line** Punktschreiber *m* ~**-line curve (or graph)** gerissene Kurve *f* ~ **rule** Punktlinie *f*

**double, to** ~ doppeln, doublieren, duplieren, falten, verdoppeln **to** ~ **back** umbiegen **to** ~ **print** einkopieren **to** ~ **V-groove** (welding) ausixen

**double** Schleier *m*; doppelt, zweifach ~ **acting** doppeltwirkend

**double-acting,** ~ **cutting machine** Doppelknippmaschine *f* ~ **cylinder** doppelwirkender Zylinder *m* ~ **engine** doppelwirkende Maschine *f* ~ **gear pump** doppelte Zahnradpumpe *f* ~ **hoist** Doppelwinde *f* ~ **hydraulic cylinder** Zweiweg-Arbeitszylinder *m* ~ **pump** doppelwirkende Pumpe *f*

**double,** ~ **action** zweiseitiges Arbeiten *n* ~**-action harp** Doppelpedalharfe *f* ~**-action pliers** Hebelvorschneider *m* ~**-action trigger** Wiederspannabzug *m* ~**-albumenized paper** brillantes Albuminpapier *n* ~**-amplification circuit** Reflexschaltung *f* ~**-amplitude-peak** Spitze-zu-Spitze-Amplitude *f* ~**-angle cutter** Winkelfräser *m* mit doppelter Schräge ~**-angle fishplate** Doppelwinkellasche *f* ~ **angle milling cutter** Prismenfräser *m* ~**-angle shears** Winkeleisenscheren *pl* ~ **antenna array** Doppelspiegel *m* ~ **antimony fluoride** Doppelantimonfluorid *n* ~**-aperture accelerator lens** (in cathode-ray tube) Beschleunigungslinse *f* aus zwei Lochelektroden ~**-aperture accelerator lens** Beschleunigungslinse *f* aus zwei Lochelektroden ~**-aperture lens** Immersions-linse *f*, -objektiv *n* ~**-apron entanglement** Drahtzaun *m*, Flandernzaun *m* ~**-apron fence** Drahtzaun *m*, Flandernzaun *m* ~ **arm kneader** Doppelarmkneter *m* ~**-armature mit zwei Ankern *pl* ~**-armature relay** Doppelankerrelais *n* ~**-armed** doppelarmig ~**-ball non-return (or check) valve** Doppelkugelrückschlagventil *n* ~ **bank** Arbeit *f* (mit zwei Dritteln) ~**-bank radial engine** Doppelsternmotor *m* ~**-banking** doppelruderig ~**-barreled** doppelläufig ~**-barreled pointer** Doppelbalkenzeiger *m* ~ **battery** Doppelbatterie *f* ~ **battery switch** Doppelzellenschalter *m* ~-**bay** Zweistieler *m* ~-**beam cathode ray tube** Zweistrahlröhre *f* ~-**beam trolley system** Zweischienenhängebahn *f* ~ **beam tube** Doppelstrahlröhre *f* **with a** ~ **bearing** doppeltgelagert ~ **bell-and-hopper arrangement** Doppelparry *m*, Doppelverschluß *m*, doppelter Gichtverschluß *m* ~ **bell jar plant** Doppelglockenanlage *f* ~ **bench** Doppelwerkbank *f* ~ **bend** Doppelbogen *m* ~ **bend test** Doppelfaltversuch *m* ~-**beta decay** doppelter Betazerfall *m* ~-**beta disintegration** doppelter Betazerfall *m* ~ **bevel ball bearing** Doppelschrägkugellager *n* ~-**bevel joint** V-Stoß *m* ~-**bevel ring** Doppelkegelring *m* ~-**bifilar gravimeter** Doppelbifilargravimeter *n* ~ **bit** Zwischenbohrer *m* ~-**blade** doppelschauflig ~ **blip** Doppelzeichen *n* (rdr) ~ **block** Flaschenzug *m* ~-**block condenser** (ocean cable) Doppelblockkondensator *m* ~-**blow cold upsetting machine** Doppeldruckkaltpresse *f* ~ **boiler** Wasserbad *n* ~ **bollard** Poller *m* ~ **bond** Doppelbindung *f* ~-**bond rule** Doppelbindungsregel *f* ~ **bordered flange pipe** Doppelbördelflanschrohr *n* ~ **bottom** Doppelboden *m* ~-**bottom compartment** Doppelbodenzelle *f* ~ **brace** Zangenverbindung *f* (electr.) ~ **bracket** Doppelausleger *m* ~-**break-and-make relay** Relais *n* mit zwei Wechselkontakten ~-**break contact** Doppeltrennkontakt *m* ~-**break jack** Doppelunterbrechungsklinke *f* ~-**break relay** Doppelunterbrechungsrelais *n* ~-**break switch** zweipoliger Schalter *m* ~ **brewing plant** Doppelsudwerk *m* ~ **bridge** (Kelvin) Doppelbrücke *f* ~-**bridge two-way repeater** Doppelbrückenverstärker *m* ~-**broken chippings** Edelsplitt *m* ~ **bucket** (turbine) Doppelschaufel *f* ~ **bus bars** Doppelsammelschienen *pl* ~-**butt strap joint** Laschenverbindung *f* ~-**button carbon microphone** Doppelkapselmikrofon *n* ~-**button transmitter** Doppelkapselmikrofon *n*, Doppelmikrofon *n*, Druckzugmikrofon *n* ~ **cage motor** Doppelnutmotor *m* ~ **cage motor engine** Doppelkäfigmotor *m* ~ **cage rotor** Doppelkäfigläufer *m* ~ **camber of spring** doppelte Federkrümmung *f* ~ (**or stereo**) **camera** Doppelkamera *f* ~-**canneled** zweiballig ~ **carbon-arc lamp** Doppelbogenlampe *f* ~ **card tenter** (textiles) Doppelkrempelführer *m* ~-**case shift** einfache Umschaltung *f* ~ **casing** Doppelröhre *f* ~ **catenary construction** Kettenfahrleitung *f* mit doppeltem Tragseil ~ **cell switch** Doppelzellenschalter *m* ~ **centering machine** doppelseitige Zentriermaschine *f* ~ **chain-stich sewing machine** Doppelkettenstichnähmaschine *f* ~ **chair** Kreuzungsstuhl *m* ~-**chamber-type tunnel kiln** Doppelkammerofen *m* ~-**channel** Zweikanal *m* ~-**channel duplex** Zweikanal-Duplexbetrieb *m* ~-**channel simplex** Zweikanal-Simplexbetrieb *m* ~ **chloride** Doppelchlorid *n* ~-**circuit receiver** Zweikreisempfänger *m* ~-**circuit receiving set** Sekundärhärte *f* ~ **circuit reception** Sekundärempfang *m* ~ **circulation** zweiseitige Flottezirkulation *f* ~-**clad vessel** doppelt umkleideter Behälter *m* ~ **claw** Doppelgreifer *m* (film) ~ **closed shell nuclei** Kerne *pl* mit doppelt abgeschlossenen Schalen ~ **closure** Doppelparry *m*, Doppelverschluß *m* ~ **clove hitch**

Schifferknoten *m* ~-**clutch** Zwischenkupplung *f* ~-**coated film** Doppelschichtfilm *m*, mehrfachbeschichteter Film *m* ~-**coated film stock** doppelbeschichteter Film *m* ~ **column** (navy) Doppelkiellinie *f* ~-**column press** Zweisäulenpresse *f* ~-**column type** Zweiständerbauart *f* ~-**column vertical boring mill** Zweiständerkarusselldrehbank *f* ~ **commutation** doppelpolige Umschaltung *f* ~ **commutator** Doppelumschalter *m* ~-**commutator motor** Doppelkollektormotor *m* ~ **compensating lever** Doppelschwinge *f* ~-**compound** (**slide**) **rest** Doppelkreuzsupport *n* ~ **concave lens** bikonkave Linse *f* ~ **condenser** Doppelverflüssiger *m* ~ **conductor** Doppelleitung *f* ~-**conductor cord** zweiadrige Schnur *f* ~ **cone** Doppelkegel *m*

**double-cone,** ~ **antenna** Doppelkegel-, Doppelkonus-antenne *f* ~ **blender** Trommelmischer *m* ~ **impeller** Doppelkegelkreiselrührer *m* ~ **impeller mixer** Doppelkegelkreiselmischer *m* ~ **mixer** Doppelkegeltrommelmischer *m* ~ **piston** (**or plunger**) **valve** Doppelkegelkolbenventil *n*

**double,** ~ **connecting piece** Doppelstutzen *m* ~ **connection** Doppelunterbrechungsverbindung *f* ~ **contact** Doppelkontakt *m* ~-**contact bulb** Doppelfadenglühlampe *f* ~-**contact separately mounted regulator** Zweikontaktwegbauregler *m* ~ **control tube** Doppelsteuerröhre *f* ~-**core barrel** Doppelkernrohr *m* ~-**core barrel drill** Doppelkernbohrer *m* ~ **core drilling equipment** Doppelkernbohrgarnitur *f* ~-**cord operation** Zweischnurbetrieb *m* (teleph.) ~-**cord switchboard** Zweischnurklappenschrank *m* ~-**cord system** Zweischnursystem *n* ~-**cotton-covered** doppelt umsponnen mit Baumwolle *f* ~-**crank drawing presses** Breitziehpressen *pl* ~-**cross coil antenna** Doppelkreuzrahmenantenne *f* with ~ **cross section** doppelschnittig ~-**cross stratagem** Gegenvertrauensspiel *n* ~-**crystal spectrometer** Zweikristallspektrometer *n* ~ **cup** Doppelglockenisolator *m* ~ **cup-shaped gasket** Doppeltopfmanschette *f* ~ **current** Doppelstrom *m*

**double-current,** ~ **key** Doppelstromtaste *f* ~ **operation** Doppelstrombetrieb *m* ~ **signalling** Doppelstromsystem *n* ~ **telegraph signal** Doppelstromtelegrafierzeichen *n* ~ **translation** Doppelstromübertragung *f* ~ **working** Doppelstrombetrieb *m*

**double,** ~ **curvature** räumliche Krümmung *f* ~-**cut** zweihiebig ~ **cutout** Doppelausschalter *m* ~ **cutter** Zweischneider *m* ~-**cutting bit** zweischneidiger Bohrer *m*, (two spiral grooves) Doppelspiralbohrer *m* ~-**cutting drill** zweischneidiger Bohrer *m* ~-**cutting shears** Doppelscheren *pl* ~ **cyanide** Zyandoppelsalz *n* ~ **cylinder** beidseitig offener Zylinder *m* ~ **cylinder channel** Doppelwalzenkanal *m* ~ **cylinder hay loader** Doppelzylinderheuauflader *m* ~-**cylinder steam dynamo** Zweizylinderdampfdynamo *m* ~ **dark slide** Doppelschlittenkassette *f* ~-**deck cage** Zweietagenkorb *m* ~-**deck tool** Aufbauwerkzeug *n* ~ **decomposition** Umsetzung *f* ~-**detection receiver** Superheterodynempfänger *m* ~-**ender** (file) Löffellanzette *f* ~-**entry compressor** zweiseitiger Kompressor *m* ~ **envelopment** Doppelumfassung *f* ~-**expansion engine** zweifache Expansionsmaschine *f*

~ **exposure** doppelte Belichtung *f*, Doppelbelichtung *f* ~**-extrastrong pipe** extrastarkes Doppelrohr *n* ~**-eye cable** Kabelnachziehschlauch *m*, Kabelziehstrumpf *m* (mit zwei Schlaufen) ~**-face hammer** Fäustel *n* ~**-face twill** beidrechter Köper *m* ~**-faced** doppelseitig ~**-faced pattern plate** doppelseitige Modellplatte *f* ~ **feedback** mehrfache Rückkopplung *f* ~ **fermentation process** Doppelgärverfahren *n* ~ **file** Doppelreihe *f* ~ **fillet flanged butt weld** Stirnkehlnahtprobe *f* ~ **fillet weld** Halsnaht *f* ~**-film projector** Zweibandprojektor *m* ~**-finder telescope** Doppelsuchfernrohr *n* ~ **fire gas producer** Doppelfeuergaserzeuger *m* ~**-fired continuous furnace** Zweizonenstoßofen *m* ~ **flanged** zweiflan(t)schig ~**-flanged butt joint** Bördelstoß *m* ~**-flanged butt weld** Bördelschweißung *f* ~ **flanged seam** Bördelnaht *f* ~ **flexible Doppelschnur** *f* (electr.) ~ **flexible cord** Zweileiterschnur *f* ~ **flow** Doppelfluß *m* ~**-flue boiler** Zweiflammrohrkessel *m* ~**-fluid cell** Zweiflüssigkeitselement *n* ~**-focus tube** Doppelfokusröhre *f* ~ **fold number** Doppelfalznummer *f* ~ **folder** Doppelfalzapparat *f* ~ **folding magnifier** Doppeleinschlaglupe *f* ~ **folds** Doppelfalzzahl *f* (film) ~ **frequency meter** Doppelfrequenzmesser *m* ~ **frequency shift keying** Zweikanalfrequenzumtastung *f* ~ **frog** Doppelherzstück *n*, Kreuzungsstück *n* ~ **fulling machine with two main pairs of rollers** Doppelzylinderwalze *f* mit zwei Hauptwalzpaaren ~ **furnace** Doppelpuddelofen *m* ~ **gain controller** Doppelschwächungswiderstand *m* ~ **gap ferrite core** Doppelspalt-Ferritkern *m* ~ **garnet** Kreuzband *n* ~ **gate** Doppelschütz *n* ~**-gauged variable capacitor (or condenser)** Doppeldrehkondensator *m* ~ **gear pump** Doppelzahnradpumpe *f* ~**-gear train** Doppelübersetzungsgetriebe *n* ~ **gear wheel** Doppelzahnrad *n* ~**-glass mirror** Doppelglasspiegel *m* ~ **goniometer** Doppelwinkelmesser *m* ~ **governor** Doppelpendel *n* ~ **graduated scale** doppelseitig kalibrierte Skala *f* ~ **grate** Doppelrost *m* ~ **grating method** Doppelgitterverfahren *n* ~**-grid lamp** Zweigitterröhre *f* ~**-grid tube** Doppelgitterröhre *f* ~ **grinding** Doppelmahlen *n* ~ **-groove insulator** Trennisolator *m* ~ **guide crane** Doppellenkerkran *m* ~ **half hitch** Schifferknoten *m* ~**-handed borer** zweimännischer Bohrer *m* ~ **hardening** Doppelhärtung *f* ~ **head box wrench** Doppelringschlüssel *m* ~ **-head telephone (or receiver)** Kopfdoppelfernhörer *m* ~**-head wrench** doppelmäuliger Schraubenschlüssel *m*, Doppelschraubenschlüssel *m* ~**-headed chaplet** Wandstärkenkernstütze *f* ~ **headed rail** Pilzschiene *f* ~**-helical** doppelschneckenförmig ~ **helical gear** Pfeilrad *n*, Rad *n* mit Winkelzähnen, Winkelverzahnungsgetriebe *n*, Winkelzahngetriebe *n* ~ **helical gearing** Getriebe *n* mit Winkelverzahnung ~ **helical rotor-type air turbine** Pfeilradmotor *m* ~ **helical teeth** Winkelverzahnung *f* ~ **helical tooth** Winkelzahn *m* ~ **helical tooth gear** Stirnrad *n* mit Winkelverzahnung ~ **highpass filter** Doppelspulenkette *f* ~ **hoe** Gabelhaue *f* ~ **hoes** Doppelhaue *f* ~ **hoisting** zweitrümmige Förderung *f* ~ **holed capstan head** Kreuzlochkopf *m* ~ **hook hoist block** Zwillingsflaschenzug *m* ~

**hook sluice** Hakenschütze *f* ~ **hopper feeder** Doppelkastenspeiser *m* ~ **hull** Doppelhülle *f* ~**-hump curve** zweispitzige Kurve *f* ~**-hump resonance curve** zweihöckrige oder zweispitzige oder zweiwellige Resonanzkurve *f* ~ **ignition** Doppelzünder *m*, Zweifunkendoppelzündung *f* ~ **ignition contact breaker** Zweifachzündunterbrecher *m* ~ **image** Doppelbild *n*, Echo *n* ~ **(ghost)image** Spiegelbild *n* ~**-image device** Doppelbildeinrichtung *f* ~**-image micro-meter** Doppelbildmikrometer *m* ~**-image range finder** Doppelbildentfernungsmesser *n*, Mischbildentfernungsmesser *m* ~**-impact association** Zweierstoßassoziation *f* ~ **impression** Dublieren *n* ~**-impulse method** Doppelimpulsverfahren *n* ~ **incline** Ablauf-berg *m*, -rücken *m* ~**-inclined fire grate** Sattelrost *m* ~ **insulator** Kreuzungsisolator *m* ~ **intermediate pinion gear** Zwischenradwelle *f* ~ **ionisation chamber** Doppelionisationskammer *f* ~ **iron for cooper's plane** Doppeleisen *n* für Fügblöcke ~ **irradiation experiment** Versuch *m* mit zweifacher HF-Einstrahlung ~ **jack** Doppelfalzklinke *f* ~ **jacket** Doppelmantel *m* ~**-jawed brake** Doppelbackenbremse *f* ~**-jet carburetor** Doppeldüsenvergaser *m* ~ **jig** Doppeljigger *m* ~ **joint** Doppelgelenk *n* ~ **key** Nachschlüssel *m* ~ **knife** Doppelmesser *n* ~ **ladder** Bockleiter *f* ~**-layer belt** Doppelriemen *m* ~**-layer lens** Doppelschichtlinse *f* ~ **lead** Doppelleitung *f* ~ **lead screw** doppelgängige Schnecke *f* ~ **length number** Zahl *f* doppelter Wortlänge ~ **lens** Doppel-linse *f*, -objektiv *n* ~**-lever contact breaker** Doppelunterbrecher *m* ~**-lever switch** Doppelhebelschalter *m*, Doppelkurbelumschalter *m*, doppelpoliger Hebelumschalter *m*, zweipoliger Hebelschalter *m* ~ **lever transmission** Doppelhebelübersetzung *f* ~ **life** doppelte Haltbarkeit *f* ~**-lift cam (with two lobes)** zweistufiger Nocken *m* ~ **limit** Grenzlehrdorn *m* ~ **limiter** Auftastimpulskreis *m*, Doppelbegrenzer *m* ~ **line** Doppellinie *f* ~**-line ruling pen** Doppelziehfeder *f* ~ **link** Doppelbandgelenk *n*, Doppelkrampe *f* ~**-link level luffing crane** Doppellenker-Wippdrehkran *m* ~ **lock carriage** zweisystemiger Schlitten *m* ~ **lock stitch** Doppelsteppstich *m* ~ **locks (twin)** Zwillingsschleuse *f* ~ **loop** Doppelschleife *f* ~**-loop knot** Doppelknoten *m* ~**-loop winding with parallel connections** Doppelschleifenwicklung *f* mit Parallelverbindungen (electr.) ~**-magazine composing machine** Doppelmagazinsetzmaschine *f* ~ **magnifier** Doppellupe *f* ~**-make contact** Doppelschließ-, Doppelarbeits-kontakt *m* ~**-make (break) relay** Relais *n* mit zwei Schließ-(Trenn-) kontakten ~**-make relay** Doppelrelais *m* ~ **-meaning** doppelsinnig, zweideutig ~ **measurement of charge** Doppelkraftmessung *f* ~ **modulation** Doppelmodulation *f* ~ **motion** gegenläufige Bewegung *f* ~**-motion agitator** Doppelrührer *m* ~**-motor airplane** Doppelrührer *m* ~**-motor airplane** Zweimotorenflugzeug *n* ~ **mounting (of gun)** Doppellafette *f* ~ **muletwist** geschleiftes Garn *n* ~ **multiplex insulator** Zweifachapparat *m* ~ **-needle telegraph** Doppelnadeltelegraf *m* ~ **nozzle** Zweifachdüse *f* ~ **open-end spanner** Doppelschraubenschlüssel *m* ~ **output pentode** Doppelpentoden-Endröhre *f* ~**-oxygen hose-**

cutting torch Dreischlauchbrenner *m* ~ **pawls**
Doppelsperrklinke *f* ~**-peak resonance curve**
zweihöckrige oder zweispitzige oder zweiwellige Resonanzkurve *f* ~**-peaked curve** zweispitzige Kurve *f* ~ **pendulum** Doppelpendel *n*
~ **penta prism** Doppelpentagonprisma *n* ~
**petticoat** Doppelglocke *f* ~**-phantom circuit**
Achterkreis *m*, Achterleitung *f* ~ **pica** Textschrift *f* ~ **pick** Doppelpicke *f*, Doppelspitz-,
Doppelstopf-hacke *f* ~ **pile** doppelflorig ~
**-pipe condenser** Doppelrohrverflüssiger *m* ~
**pipe stock** Duplexschneidkluppe *f* ~**-piston
Diesel engine** Doppelkolben-Dieselmotor *m*
~**-piston engine** Boxermotor *m*, Doppelkolbenmotor *m*, Gegenkolbenmotor *m* ~**-pit plant**
Doppelschachtanlage *f* ~**-pitch system** Ver-
bundteilungssystem *n* ~**-plate sounder** Klopfer
*m* mit zwei Klangscheiben ~ **plug** Doppel-
-stecker *m*, -stöpsel *m* ~ **plug and jack** Stecker
*m* mit Büchse ~**-plugged cord** Schnur *f* mit zwei
Steckern ~**-ply belt** zweifacher Riemen *m* ~
**pneumatic tires** Doppelpneu *m* ~**-point interpolation in space** Doppelpunkteinschaltung *f*
im Raum ~ **point thread chaser** Gewindestahl
*m* (als Gabelstahl) ~**-point threading chaser**
Gewindegabelstahl *m* ~**-pointed** zweizackig ~
**-pointed dowel (or wire nail)** Verbandstift *m*
~**-pointed pen** Parallelschreibfeder *f* ~**-pointed
pick** Doppelkeilhaue *f* ~**-polar** doppelpolig ~
**pole** Doppelgestänge *n*, Doppelmast *m*, (wide
spread) Kuppelstange *f* ~**-pole** doppel-, zwei-
-polig ~ **(pin) pole** Doppelstütze *f* ~**-pole
alternator** Doppelpolwechsler *m* (rdo) ~ **pole
armature** zweipoliges Ankersystem *n* (loud-
speaker) ~**-pole change-over switch** zweipoliger
Umschalter *m* ~**-pole doublethrow relay** Anlaßrelais *n*, zweipoliges Umschaltrelais *n* ~
**-pole ignition** Doppelpolzündung *f* ~**-pole receiver** doppelpolter Fernhörer *m* ~**-pole
switch** doppelpoliger oder zweipoliger Schalter
*m* ~**-pontoon seaplane** Zweischwimmerflugzeug
*n* ~ **precision** doppelte Genauigkeit *f* ~ **preselection** doppelte Vorwahl *f* ~ **press** Doppel-
presse *f* ~**-pressure regulation** Zweidruckrege-
lung *f* ~ **prism for angles of ninety degrees**
Prismenkreuz *n* ~ **projection** Doppelbildwurf
*m*, Doppelprojektion *f* ~ **projector** Doppel-
-bildwerfer *m*, -projektor *m* ~**-pronged** zwei-
zinkig ~**-pronged fork** zweizinkige Gabel *f* ~
**-pronged tiller** Gabelheft *n* ~ **propelling rake**
Doppelkurbelrechen *m* ~ **protractor** Doppelwinkelmesser *m* ~**-puddle furnace** Doppelpuddelofen *m* ~ **pully block** zweischeibiger Block
*m* ~ **punch** Doppellochung *f* ~ **punch and blank
column detection** Doppel- und Fehllochaus-
suchvorrichtung *f* ~ **pupil view finder** Doppel-
okular-Bildsucher *m* ~ **quartz wedge** Doppel-
quarzkeil *m* ~ **rack** Doppelregal *n*, doppelte
Zahnstange *f* ~**-rack motion** Doppelzahnstan-
genbewegung *f* ~ **ram's-horn** Widderkopf *m*
~ **reactor** Doppeldrossel *f* ~ **reamer** Doppel-
rundbohrer *m* ~ **reception** Doppelempfang *m*
~ **rectifier circuit organization** Doppelwegschaltung *f* ~ **reduction final drive** Doppelantrieb *m* ~**-reduction gear** Doppelübersetzungsgetriebe *n* ~**-refined iron** doppelt geschweißtes
Eisen *n* ~**-refined steel** Doppelraffinierstahl *m*,
Feinstahl *m* ~**-refracting crystal** doppelbre-

chendei Kristall *m* ~ **refraction** Doppelbrechung *f* ~ **regulating switch** Doppelzellenschalter *m* ~**-relay repeater** Zweirohrverstär-
ker *m* ~ **resistance box** gleichläufiger Doppel-
kurbelwiderstand *m* ~**-return** zweikehrig ~
**reversing mill** Umkehrduo *n* ~**-rib frame** Fang-
stuhl *m* ~**-rib loom** Raschel *f* ~**-rib warp goods**
Fangkettenware *f* ~ **ridge** Ablaufrücken *m*
~**-ring sight** Doppelringvisier *n* ~**-ring type
furnace** Zweirinnenofen *m* ~**-riveting** Doppel-
nietung *f* ~ **roasting** Doppelröstung *f* ~**-roll
crusher** Doppelwalzenmühle *f* ~ **roller clutch**
Doppelrollenkupplung *f* ~ **roller separator**
Doppelwalzenscheider *m* ~ **round bit** Doppel-
rundbohrer *m* ~ **rove** Doppeltgespinst *n* ~**-row**
zweireihig ~**-row ball bearing** Doppelkugel-
lager *n*, zweireihiges Kugellager *n* ~**-row radial
engine** Doppelsternmotor *m* ~**-row riveting**
zweireihige Nietung *f* ~**-row staggered radial
engine** Zweisternmotor *m* ~**-row steep-angle
ball bearing** Ringschräglager *n* ~**-row velocity
wheel** Geschwindigkeitsrad *n* ~ **rule** Doppellinie *f* ~ **salt** Doppelsalz *n* ~**-sawtooth potential**
doppelsägeförmige Spannung *f* ~**-scale** Doppel-
skala *f* ~**-scal(l)ippers (or S compass)** Dop-
pel-S-Taster *m* ~ **scattering experiments** Dop-
pelstreuversuche *pl* ~ **screw mixer** Doppel-
schneckenmischer *m* ~ **screw thread** doppelte
Gewindesteigung *f* ~**-screw vise** Stufenscheibenantrieb *m* ~ **seals** Doppellötungen *pl* ~
**seat** Doppelsitz *m* ~**-seat valve** Doppelsitzventil *n* ~**-seated** doppelsitzig ~**-seated sleeve
valve** Rohrdoppelsitzventil *n* ~ **section wing**
Doppelflügel *m* (aviat.) ~ **seizure** Doppelbe-
legung *f* (teleph.) ~ **set of brushes** Doppel-
bürstensatz *m* ~ **shaking screen** Doppelstoß-
schwingsieb *n* **in** ~ **shear** zweischnittig ~ **shear**
schnittige Scherung *f* ~**-shear riveting** zweischnittige Nietung *f* ~ **shearing** zweischnittige
Abscherung *f* ~**-sheave pulley** Zweirillenscheibe
*f* ~ **shed** Doppelglocke *f* ~**-shed insulator**
Doppelglockenisolator *m* ~**-shed porcelain insulator** Porzellandoppelglocke *f* ~**-sheer riveting** zweischnittige Nietung *f* ~**-shoe brake** Doppelbackenbremse *f* ~ **(or two)-shoe-type brake**
Zweibackenbremse *f* ~**-shovel plow** Doppel-
schaufelpflug *m* ~ **shroud knot** doppelter Wand-
knoten *m* ~**-side-band reception** Zweiseiten-
bandempfang *m* ~ **side-band transmission** Zwei-
seitenbandsenden *f* ~**-side-band transmitter**
Zweiseitenbandsender *m* ~**-sided** doppelseitig
~**-sided dial indicator** Uhr *f* mit doppelseitigem
Zifferblatt ~**-sided eccentric press** Doppelstän-
derexzenterpresse *f* ~**-sided goods** doppelflächi-
ge Waren *pl* ~**-sided pattern plate** zweiseitiges
Modell *n*, Reliefplatte *f* ~**-sided rack** doppelseitiger Rahmen *m* ~ **sieve** Doppelplanrätter *m*
~ **signals** Doppelzeichen *n* ~**-silk covered** mit
Seide *f* doppelt umsponnen, zweifach mit Seide
*f* umsponnen ~ **sinks** Doppelspülsteine *pl* ~
**sintering** Doppelröstung *f* ~ **sizing of fabrics**
zweiseitiges Appretieren *n* von Geweben ~
**skid** Zwillingskufe *f* ~ **skip charging** Begichtung
*f* mit zwei Aufzügen ~**-sleeve valve engine**
Doppelschiebermotor *m* ~ **slide** Doppelschie-
ber *m* ~ **slide crank chain** Doppelführungskette
*f* ~ **slider coil** Spule *f* mit zwei Gleitkontakten
~ **sliding door** Doppelflügeltür *f* ~ **slit method**

Doppelblendenmethode *f* ~ **sluice** Doppelschütz *n* ~ **sluice gate** Doppelschütze *f* ~ **-sluice weir** Doppelschützenwehr *n* ~ **snap roll** schnelle Doppelrolle (aviat.), schnelles Doppelüberschlagen *n* ~ **socket** Zwillingssockel *m*, Überschiebemuffe *f* ~ **spanner** Doppelschraubenschlüssel *m* ~ **spar** Doppelspat *m* ~ **spindle** Doppelspindel *f* ~**-spindle cylinder boring machine** zweispindelige Zylinderbohrmaschine *f* ~**-spiral bit** (two spiral grooves) Doppelspiralbohrer *m* ~ **spoke** Doppelspeiche *f* ~**-spool relay** zweispuliges Relais *n* **with** ~ **spools** zweispulig ~**-spot tuning** Zweipunkteabstimmung *f* ~ **squirrel cage motor** Spezialnutläufermotor *m* ~ **squirrel-cage winding** Doppelkäfigwicklung *f* ~**-stage compressor** zweistufiger Kompressor *m* ~**-stage reducing valve** Doppeldruckreduzierventil *n* ~**-stage system** Zweikreiser *m* ~ **stand** Doppelstativ *n* ~**-stand rolling mill** zweigerüstiges Walzwerk *n* ~ **standard planing machine** Zweiständerhobelmaschine *f* ~**-standard plate sheat** Zweiständerblechtafelschere *f* ~ **star** Doppelstern *m* ~ **star bit** Doppelspatenmeißel *m* ~ **stich** Doppelmasche *f* ~**-strand chain** doppelstränige oder zweistränige Kette *f* ~**-strap web joint** doppelte Stegverlaschung *f* ~**-stream amplifier** Doppelstrahlröhre *f* ~ **striations** Doppelschichten *pl* ~**-stroke bell** Doppel-läutewerk *n*, -schlagwecker *m* ~**-stroke (cycle)engine** Zweitaktmaschine *f* ~ **stroke time** Doppelhubzeit *f* ~**-suction** doppelflutig ~ **super effect** Doppelüberlagerungseffekt *m* ~ **super-heterodyne reception** Doppelsuperempfang *m* ~ **surface curvature** Schraubenlinie *f* ~ **suspension** Bifilaraufhängung *f* **(flanged)** ~ **-sweep T** doppeltes oder zweiseitiges Krümmer-T-Stück *n*, geschweiftes T-Stück *n* ~ **-swing jaw crusher** Doppelschwingenbrecher *m* ~ **switch** Doppelweiche *f* ~**-tandem engine** Doppelreihenmaschine *f* ~**-tapered muff coupling** Doppelkegelkupplung *f* ~**-tapered tower** Fischbauchmast *m* ~ **tariff** Doppeltarif *m* ~**-tariff meter** Doppeltarifzähler *m* ~ **taxation** Doppelbesteuerung *f* ~**-tee joist** I-Träger *m* ~**-telegraph set** Zweifachtelegraf *m* ~ **telescope** Doppelfernrohr *n* ~ **tensor** Doppeltenser *m* ~ **tensor equations** Doppeltensorgleichungen *pl* ~ **tensor field** Doppeltensorfeld *n* ~ **terminal** Doppelklemme *f* ~ **testbox** Verdopplerprüfkästchen *n* ~**-T girder** Doppel-T-Träger *m* ~**-T gland** Doppel-T-Eisenträger *m* ~ **thread** doppelgängiges, doppeltes oder zweigängiges Gewinde *n* ~**-threaded screw** Doppelgewindeschraube *f* ~ **three phase** Saugdrosselschaltung *f* ~ **three-throw** doppelte Drillingspreßpumpe *f* ~**-throat Venturi tube** Doppel-saugrohr *n*, -venturirohr *n* ~**-throw** doppelt gekröpft, Kröpfungspaar *n* ~**-throw disconnecting switch** Trennumschalter *m* ~**-throw knife switch** Doppelmesserschalter *m* ~**-throw pins** Kurbelzapfenpaar *n* ~**-throw pump** Zweikurbelpumpe *f* ~**-throw single-pole switch** Nockenschalter *m* ~**-throw switch** Hebelumschalter *m*, Umschalter *m*, zweipoliger Umschalter *m*, Wechselschalter *m* ~ **thrust bearing** Doppeldrucklager *n* ~**-timed** zweiseitig ~ **tires** Doppelbereifung *f*, Zwillingsbereifung *f* ~**-T iron** Doppel-T-Eisen *n*, I-Eisen *n* ~ **toggle system** Kniehebelsystem *n*

~**-tone color** Doppeltonfarbe *f* (Lack) ~**-tone effect** Doppeltoneffekt *m* (Lack) ~**-tone printing** Doppeltondruck *m* ~**-tongued flute** Doppelzungenpfeife *f* ~ **tool post (or toolholder)** doppelter Support *m* ~**-track** Doppelgeleise *n*; Doppelspur *f* (tape rec.), zweigleisig ~**-track bed** Zweibahnenbett *n* ~**-track conveying** doppeltrümmige Förderung *f* ~**-track hoist bridge** doppeltrümmiger Schrägaufzug *m* ~**-track line** Doppelbahn *f* ~**-track railroad** Doppelgeleisebahn *f* ~**-track railway** zweigleisige Bahn *f* ~**-track skip hoist** doppeltrümmiger Schrägaufzug *m* ~ **tracked** zweispurig, zweitrümmig **with** ~ **tracks** doppelgleisig ~**-T rail** Doppel-T-Schiene *f* ~**-transmission** Doppelsenden *n* ~ **tread** doppelte Lauffläche *f* ~ **tree** Doppelschwengel *m*, Ortscheit *n*, Zweispannwaage *f* ~ **triode** Doppeltriode *f*, Zweifachdiode *f* ~**-T-supporting frame** Doppel T-Trägerrahmen *m* ~**-tuned amplifier** doppelt abgestimmter Verstärker *m*, Zweikreisverstärker *m* ~ **turnable type of player** Tellergerät *n* ~ **twist** zweikettig ~**-twisting machine** Doppeldrahtzwirnmaschine *f* ~ **two-high mill** Doppelduo-Walzwerke *pl* ~ **two-housing mill with roughing train** Doppelduo *n* mit Vorgerüst ~**-type edge** Zweifachschneide *f* ~**-type excitating winding** Doppelerregerwicklung *f* ~**-type regulator (or governor)** Doppelregler *m* ~**-U butt joint** Doppel-U-Stoß *m* ~ **U, J, or S insulator spindle** U-, J-, S- förmige Doppelstütze *f* ~ **union** Doppelstutzen *m* ~**-V-antenna** Doppel-V-Antenne *f* ~**-V-engine** Fächermotor *m* ~**-vent valve** Schlangenventil *n* ~ **Venturi tube** Doppel-saugrohr *n*, -venturirohr *n* ~ **V-groove** X-Stoßnaht *f* ~**-V (butt)joint** X-Stoß *m* ~**-voids** Doppelleerstellen *pl* ~**-voltage generator** Generator *m* für zwei Spannungen ~ **voltmeter** Doppelvoltmeter *m* ~ **wall** Doppelwand *f* ~**-wall box** Herdeisen *n* ~**-walled** doppelwandig, doppelunterbrechungswandig ~ **warp** zweikettig ~**-wave rectification** Gleichrichtung *f* beider Halbwellen ~**-webbed** zweistegig ~ **wedge** Doppelkeil *m* (Flügelprofil), Keilpaar *n* ~**-wedge saccharimeter** Doppelkeilpolarisationsapparat *m* ~ **weighing** Doppelwägung *f* ~**-wet tool grinder** Werkzeugdoppelnaßschleifständer *m* ~ **wheel grinder** Doppelschleifständer *m* ~**-wide** doppelt breit ~ **winch** Doppelwinde *f* ~ **wind channel (or tunnel)** Doppelwindkanal *m* ~ **winder** (textiles) Dupliervwickelmaschine *f* ~ **winding** bifilare Wicklung *f* ~**-winding reactor** Doppelwicklungsspule *f* ~**-wire** doppeldrähtig ~**-wound** bifilar, bifilargewickelt, mit zwei Wicklungen *pl* ~**-wound generator** Doppelschlußgenerator *m* ~**-wound relay** Relais *n* mit zwei Wicklungen ~**-wound resistor** Bifilarwiderstand *m* ~**-wound transformer** Übertrager *m* mit zwei Wicklungen ~ **wrench** Doppelschraubenschlüssel *m* ~ **wringing post** Doppelpfahl *m* ~ **x-ray flash tube** Doppelröntgenblitzröhre *f* ~ **yoke** Poljoch *n* ~**-yoke lever** Doppelgabelhebel *m* ~**-zone producer** Doppelfeuergenerator *m*

**doubled yarn** gezwirntes Garn *n*
**doubler** Dopplungsstück *n*, (textiles) Dublierer *m*, Verdoppler *m* ~ **stage** Verdopplerstufe *f* (rdo) ~ **and twister** Doppelzwirner *m*

**doublet** Antennendipol *n*, Dipol *m*, Doppellinie *f* (opt.), Doppellinse *f*, Doppelquelle *f*, Doppel-V-Antenne *f*, Dublette *f* ~ **antenna** Dipolantenne *f* ~ **differences** Dublettabstände *pl* ~ **separation** (spectrum) Dublettenaufspaltung *f* ~ **splitting** Dublettaufspaltung *f*

**doubling** Doppeln *n*, Faltung *f*, Verdopplung *f* ~ **of frequency** Frequenzverdoppelung *f* ~ **of hare** Hakenschlag *m* ~ **of the image** (twin tube) Verdopplung *f* des Bildes

**doubling,** ~ **device working with oblique rollers** Schrägwalzendoublierung *f* ~ **effect** Echoeffekt *m* ~ **machine** Biegemaschine *f* ~ **process** Dublierverfahren *n* ~ **rod** Doublierstange *f* ~ **test** Faltversuch *m* ~ **time** Verdopplungszeit *f*

**doubly,** ~ **fed polyphase shunt commutator motor** ständergespeister (statorgespeister) Mehrphasennebenschlußmotor *m* ~ **refracting** doppelbrechend ~ **supported raft** Doppelfähre *f*

**douche, to** ~ berieseln

**douche** Brause *f*, Dusche *f*

**dough** Teig *m* **to** ~ **in** einmaischen, einsteigen ~ **development** Teigbereitung *f* ~ **mill** Gummikneter *m*

**doughing in** Einsteigen *n*

**doughnut** Flußverstärker *m*, Ringröhre *f* ~ **disk plate** Ringplatte *f* ~ **ring** Ringprellplatte *f* ~-**shaped** wulstförmig ~ **tire** Überballonreifen *m* ~ **tube** Betatron *n*, Elektronenschleuder *f*

**doughy** teigig

**dovetail, to** ~ (ein)schwalben, spunden, verzinken, zinken

**dovetail** Schwalbenschwanz *m*, Zinke *f* ~ **cutter** Schwalbenschwanzfräser *m* ~ **dowel** (doppelter) schwalbenschwanzförmiger Dübel *m* ~ **guide** Schwalbenschwanzführung *f* ~ **guides** Führungsschwänze *pl* ~ **indent** schwalbenschwanzförmiger Zahn *m*, Zusammenzinken *n* ~ **joint** Schwalbenschwanzverbindung *f* ~ **key** Schwalbenschwanzkeil *m* ~ **plane** Grathobel *m*, (for grooves) Federhobel *f* ~ **(clamp) plate** Klemmschiene *f* ~ **seat** schwalbenschwanzförmige Nut *f* ~ **slide** Objektivschlittenwechselvorrichtung *f* ~ **templet** Einschiebschablone *f*

**dovetailed** schwalbenschwanzförmig ~ **groove** Schwalbenschwanznute *f* ~ **ring** Ringschwalbe *f* ~ **twill** ineinandergeschobener Köper *m*

**dovetailing** Einschwalben *n*, Schwalbenschwanzverzapfung *f* ~ **machine** Zinkenschneidmaschine *f* ~ **plane** Zinkenhobel *m*

**dowel, to** ~ dübeln, verdübeln **to** ~-**in** eindübeln

**dowel** Diebel *m*, Döbel *m*, Dübel *m*, Haltezapfen *m*, Holzpflock *m*, Keil *m*, Lochstift *m*, hölzerner Nagel *m* oder Stift *m* ~ **bush** Führungsbuchse *f* ~ **joint** verdübelte Verbindung *f* ~ **pin** Diebel *m*, Döbel *n*, Dübel *m*, Paßstift *m*, Prisonstift *m*, Stehbolzen *m*, Zylinder-zapfen *m*, -stift *m* ~ **pin plate** Paßstiftplatte *f*

**doweled** gedübelt ~ **sheet templet** aufgestiftete Blechschablone *f*

**doweling** Verdübelung *f* ~ **machines** Spitzdübelautomaten *pl*

**Dowlais mill** Doppelduo-Walzwerke *pl*

**dowlas** Lederleinwand *f*

**down, to** ~ herunterlassen, niederlegen **to** ~ **the gear** das Landungsgestell *n* herunterlassen (aviat.)

**down** abwärts, nieder, unten ~ **of a mine** einziehender Schacht *m* ~ **by the stern** achterlastig, (navy) steuerlastig

**down,** ~ **cast** Einziehstrom *m* ~ **comer** Gichtgasabzugsrohr *n* ~ **comer tube** Fallrohr *n* ~ **coming wave** Höhen-, Raum-welle *f* ~ **current** Abstrom *m* (aerodyn.), Abwind *m*, herunterwehender Wind *m* ~ **cut miller** Gleichlauffräse *f* ~ **cut milling** Fräsen *n* im Gleichlauf, Gleichlauffräsen *n* ~ **dip** Fallrichtung *f*

**downdraft** nach unten gerichtete Strömung *f*, Unterwind *m* ~ **carburetor** Fallstromvergaser *m*, Niederzugvergaser *m* ~-**type furnace** Ofen *m* mit niedergehender Flamme

**down,** ~ **feed** Tiefenvorschub *m*, Vertikalbewegung *f* (engin.) ~-**feed adjustment** Einstellung *f* der Vertikalbewegung ~-**feed screw** Vertikalschaltspindel *n* ~ **flow** Abstrom *m* ~ **gate** Einguß-kanal *m*, -lauf *m*, -trichter *m*; Gießtrichter *m*, Lauf *m*, Trichter-einlauf *m*, -lauf *m*, -zulauf *m* ~ **gust** Fallbö *f*, Luftlochfallbö *f* ~ **gust of wind** Fallwind *m* ~ **hand weld** Schweißung *f* von oben ~ **haul** Niederholer *m*

**downhill** bergab ~ **grade** abschüssige Bahnstrecke *f* (r.r.) ~ **run** (skiing) Abfahrt *f*

**down,** ~ **inclined gate** einfallende Strecke *f* (min) ~ **lead** Ableitung *f*, Antennenableitung *f*, (of an antenna) Hochführung *f*, Speiseleitung *f* ~ **leads** (of an antenna) Herabführung *f* ~ **line** Ausfahrleitung *f* ~ **lock** Ausfahrverriegelung *f* (Fahrwerk) ~ **load** Flugzeuggewicht *n* in Abwind (aviat.) ~ **payment** Anzahlung *f*, Handgeld *n* ~ **pipe** Abfall-, Fall-, Regenfall-rohr *n* ~-**pointing hole** Seelenbohrloch *n* ~-**position indicator** Knickstrebenschalter *m* ~ **pour** Platzregen *m*, Schauer *m* ~ **rule** (under heading) Längslinie *f* ~ **shaft** Einziehschacht *m* ~ **spout** Abfallrohr *n* ~ **stop** Ausfahrbegrenzungsanschlag *m* (Fahrwerk)

**downstream** Talfahrt *f*, Talseite *f*, Unterwasserseite *f*; flußabwärts, luftseitig, stromab, stromabwärts, unterhalb ~ **anchor** Windanker *m* ~ **anchor cable** Windanker-leine *f*, -tau *n* ~ **apron** Sohle *f* der unteren Haltung, luftseitige Sohlenbefestigung *f*, Sturzbett *n* ~ **back of dam (or weir)** Wehrrücken *m* ~ **batter** luftseitige Verjüngung *f* ~ **cofferdam** unterer Fangdamm *m* ~ **end of pier** luftseitiges Ende *n* des Pfeilers ~ **face** Luftseite *f* ~ **face of dam (or weir)** Wehrrücken *m* ~ **facing of weir (or dam)** Abfallmauer *f* ~ **floor** Sturzbett *n* ~ **floor with steps** treppenförmiges Sturzbett *n* ~ **line of anchors** Windankerlinie *f* ~ **pressure** Ausgangsdruck *m* ~ **profile on center** luftseitiges Profil *n* der Mittellinie ~ **radius of crest** luftseitiger Halbmesser *m* der Krone ~ **slope** Landabdachung *f*

**downstroke** Abwärtsbewegung *f*, Niedergang *m*, Niederschlag *m* ~ **of knife** Messerniedergang *m* ~ **of slide** Stößelniedergang *m* ~ **edge** (of an impulse) Abstrich *m* ~ **press** Oberkolbenpresse *f*

**down,** ~ **take flue** absteigender Zug *m* ~ **throw** Absenkung *f*, gesunkener Flügel *m* einer Verwerfung (aviat.) ~ **tilt** Abwärtsneigung *f* ~ **time** Abschaltzeit *f*, Außerbetriebs-, Ausfall-, Steh-, Tot-zeit *f* (bei Maschinen) ~-**trench** Rücklaufgraben *m*

**downward** abwärts, niederwärts ~ **air current** Abwärtsböe *f* ~ **borehole** Sohlenbohrloch *n* ~ **component** Abkomponente *f* ~ **conveying** Abwärtslüftung *f* ~ **current** Abstrom *m* (aerodyn.) ~ **flight** Abstieg *m* ~ **force** Sinkkraft *f* ~ **gust** Abwärtsböe *f* ~ **journey** Niederfahrt *f* ~ **modulation** subtraktive Modulation *f* ~ **motion** Abwärtsbewegung *f*, Herab-, Senk-bewegung *f* ~ **movement** Abwärtsbewegung *f* **in a** ~ **position** abwärts gerichtet ~ **pressure** Abstrich *m*, (stylus in record player) Auflagedruck *m* ~ **strike (or stroke)** Abwärtsgang *m*, Abwärtshub *m* ~ **tendency** fallende Tendenz *f* ~ **travel** Rückfahrt *f* ~ **trend** Hang *m* nach unten ~ **welding** Schweißung *f* mit Abwärtsführung

**down warping** (of strata) Einwölbung *f*

**downwash** Abstrom *m*, Abwind *m* (aerodyn.), Flügelabstrom, Luftabzug *m*, Luftstromneigung *f* (aviat.) ~ **and wake condition** Abwindverhältnis *n* ~ **angle** Abwindwinkel *m* ~ **corcection** Abwindberichtigung *f* ~ **effect** Abwindwirkung *f*

**downwind** Rückenwind *m*, in Windrichtung *f*, herunterwehender Wind *m*; windabwärts ~ **area** Abwindgebiet *n* ~ **landing** Windlandung *f*, Landung *f* mit Rückenwind ~ **leg** Mitwindteil *m* (traffic circuit)

**downy,** ~ **feather** Flaumfeder *f* ~ **mildew** Kräuselkrankheit *f*, falscher Mehltau *m* ~ **plume** Flaumfeder *f*

**dowse, to** ~ (a beam) abblenden

**dowser** Feuerschutzschirm *m* (film); Wünschelrutengänger *m*

**dowsing rod** Wünschelrute *f*

**Dowson,** ~ **gas** Dowsongas *n* ~**-gas process** Mischgasprozeß *m*

**Dowty wheel** kniebeweglicher Radtypus *m* am Fahrgestell (aviat.)

**dozer tractor** Planierschlepper *m*

**dozzle metal** verlorener Kopf *m*

**drab** gelblichgrau

**draconitic** drakonitisch

**draff** Treber *pl* ~**-drying apparatus** Trebertrockenapparat *m* ~ **press** Trübpresse *f*

**draft, to** ~ ausarbeiten, ausheben, entwerfen, patronieren **to** ~ **(or draw) up** (a document) abfassen

**draft** Abnahme *f*, Abriß *m*, Abzug *m*, Akzept *n*, Anweisung *f*, Anzug *f*, Aufriß *m*, (of chimney) Auftrieb *m*, Aushebung *f*, Druck *m*, (wire drawing) Durchmesserabnahme *f*, (wire drawing) Durchmesserverminderung *f*, Durchzug *m*, Entwurf *m*, Entwurfzeichnung *f*, Fadeneinzug *m*, Geschirreinzug *m*, (boiler) Kesselzug *m*, Konizität *f*, Konzept *n*, (of a lamp) Luftzug *m*, Passage *f*, Querschnittsabnahme *f*, Querschnittsverminderung *f*, Querschnittsverringerung *f*, (of drawings) Riß *m*, Schafteinzug *m*, Skizze *f*, (of ships) Tiefgang *m*, Tratte *f*, Verjüngung *f*, Wassertiefe *f*, Wechsel *m*, Ziehung *f*, Zug *m*, Zugluft *f* ~ **of air** Verwehung *f* ~ **of water** (seaplane) Wasserverdrängung *f*

**draft,** ~ **action** Zugwirkung *f* ~ **agreement** Vertragsentwurf *m* ~ **animal** Zugtier *n* ~ **carburetor** Steigstromvergaser *m* ~ **change gear** Verzugswechsel *m* ~ **change shaft** Verzugswechselwelle *f* ~ **change wheel** Verzugswechselgetriebe *n* ~ **chimney** Feuerzug *m* ~ **control** Zugregu-

liervorrichtung *f* ~ **cylinder** Abzugszylinder *m* ~ **difference indicator** Tiefgangmesser *m* ~ **diverter** Strömungssicherung *f* ~ **efficiency of animals** Zugleistung *f* ~ **equipment** Zuganlage *f* ~ **flue** Abzugsschacht *m* ~ **form** Wechselformular *n* ~ **furnace** Zugofen *m* ~ **gauge** Unterdruckmesser *m*, Zugmesser *m*, Zugmesserlehre *f* ~ **gear** Verzugsgetriebe *n* ~ **hole** Fuchsöffnung *f*, Zugloch *n*, Zugöffnung *f* ~ **horse** Zugpferd *n* ~ **horses** Bespannung *f* ~ **intensifier** Zugverstärker *m* ~ **loss** Zugverlust *m*

**draftman's** Detailkonstrukteur *m*, Gestalter *m*, Konstruktionszeichner *m*, Musteraussetzer *m*, Musterzeichner *m*, Zeichner *m* ~ **right-angled triangle** Zeichenwinkel *m*

**draft,** ~ **mark** Gedingezeichen *n* ~ **meter** Gebläsemesser *m* ~ **preventer** Fensterdichter *m* ~ **registrant** Dienstpflichtiger *m* ~ **regulator** Zugregler *m* ~ **stage** Druckstufe *f* ~ **system** Streckwerksabdeckung *f* ~ **tube** Saugrohr *n* ~ **undergrate** Unterwind *m* ~ **underground** Zug *m* unter Tage

**drafting** Stärkeabnahme *f* (Walzwerk) ~ **ink** Linierfarbe *f* ~ **machine** Zeichenmaschine *f* ~ **office** Entwurfswerkstätte *f* ~ **paper** Linien-, Patronen-papier *n* ~ **practices** Zeichnungsmethoden *pl* ~ **room** Zeichenbüro *n* ~ **system** Streckenwerkssystem *n*

**drag, to** ~ dreggen, mitschleppen, nachgeben (naut.), reißen, schleifen, schleppen, zausen, zerren, ziehen, nach sich ziehen **to** ~ **along** mitreißen, mitziehen **to** ~ **in** hineinziehen **to** ~ **off** abschleppen **to** ~ **on** hinziehen **to** ~ **over** überheben **to** ~ **for a submarine cable** ein Seekabel *n* suchen **to** ~ **through** durchziehen, hindurchziehen **to** ~ **a wing** einen Flügel *m* hängen lassen

**drag** Ausräumer *m* für Bohrlöcher, Dregganker *m*, untere Formhälfte *f*, Hemmzeug *n*, Luftwiderstand *m*, Radschuh *m*, Rücktrieb *m* (aviat.), Strömungswiderstand *m*, (flask) Unterkasten *m*, Widerstand *m* ~ **anchor** Treibanker *m* ~ **antenna** Schleppantenne *f* ~ **axis** Widerstandachse *f* ~ **balance (or scale)** Luftwiderstandwaage *f* ~ **bar** Kupplungsstange *f*, Zugeisen *n* ~ **belt** Kratzband *n* ~ **bit** Blatt-, Spaten-meißel *m* ~ **bolt** Kuppelbolzen *m* ~ **brace** Knickstrebe *f* (Bugfahrwerk), Widerstandsstrebe *f* **(external)** ~ **bracing** Stirnverspannung *f* ~ **bucket** Schöpfbecher *m* ~ **chain** Hemm-, Schlepp-, Schurf-, Zug-kette *f* ~ **chain conveyor** Tragkettenförderer *m* ~ **chute** Landebremsschirm *m* ~ **classifier** Kratzerklassifikator *m* ~ **coefficient** Luftwiderstandsbeiwert *m*, Widerstands-beiwert *m*, -koeffizient *m* ~ **component** Komponente *f* in der Bewegungsrichtung ~ **cowling** Saughaube *f* ~ **factor** Luftwiderstandsbeiwert *m* ~ **feed** Schleppeinleger *m* ~ **flask** Oberkasten *m* ~ **flow** Hauptfluß *m* ~ **forces** Widerstandskräfte *pl* ~ **form** Formwiderstand *m* (aerodyn.) ~ **formula** Widerstandsformel *f* ~ **gyro** Schleppkreisel *m* ~ **hinge** Schwenkgelenk *n* (Hubschrauber) ~ **hook** Kuppel-, Zieh-haken *m* ~ **increase** Widerstandsanstieg *m* ~ **lever** Schlepphebel *m* ~ **lift rotation** Gleitzahl *f* ~ **line** Kranleine-, Kranschürf-, Schürfkübel-, Schleppkübel-bagger *m* ~ **line equipment** Greifereinrichtung *f* ~ **link** Lenkschubstange *f*, Schleppglied *n*, Schwenkgelenk *n* (Hubschrauber) ~ -link conveyer Redler *m* ~ loading Mitschleppkraft *f* ~ measurements Luftunterstandsmessungen *pl* ~ minimum Widerstandminimum *n* ~ net Schleppnetz *n* ~ parachute Bremsfallschirm *m* ~ plate Hemmblech *n* ~ pointer Schleppzeiger *m* ~ power of efficiency Schleppleistung *f* ~ pressure Mitschleppdruck ~ producing surface Widerstandsfläche *f* (Fallschirm) ~ product Widerstandsfläche *f* ~ rakes Handschleprechen *m* ~ range Mitnahmebereich *m* (rdo) ~ reducing luftschnittig ~ reduction Widerstandsverminderung *f* ~ rod Lenkschubstange *f* ~ rope Handhabungstau *n*, (gun) Langtau *n*, Schleppseil *n* ~ sack Schleppsack *m* ~ shoe Hemmschuh *m* ~ spring Kupplungsstangen-, Schrank-, Schlepp-, Tür-feder *m* ~ strut Abstands-, Knick-strebe *f* ~ switch Schleppschalter *m* (Wagenheizer) ~ twist Ausräumschnecke *f* ~-type tachometer Wirbelstromtachometer *n* ~ washer Zugöse *f* ~ wheel Bremse *f* ~ when taxiing Rollwiderstand *m* ~ wire Fangkabel *n* (aviat.), Holmauskreuzung *f*, Stirnseil *n*

**dragger-out** Paketierofenmann *m*

**dragging** Mitbewegung *f*, Zerrung *f*, Ziehen *n* ~ of a balloon Schleiffahrt *f* ~ in transversal direction Querschlepper *m*

**dragging,** ~ bucket Schleppschaufel *f* ~ coefficient Mitführungskoeffizient *m* ~ section schleifender Teil *m*

**dragon** Drache *m* ~'s-teeth obstruction Höckerhindernis *n*

**drain, to** ~ abfließen, abfließen lassen, ablassen, ablaufen, abseihen, abtropfen, abwässern, ausleeren, ausschleudern, austrocknen, dränen, (off) dränieren, entwässern **to** ~ **a bed (or seam)** ein Flöz *n* lösen **to** ~ **a cutting** einen Einschnitt *m* trockenlegen **to** ~ **off** abläutern, ableiten, absaugen, abziehen **to** ~ **the oil** das Öl *n* ablassen **to** ~ **a pit** eine Schacht *f* trocken legen **to** ~ **under pressure** ¶eerdrücken **to** ~ **swamps** entsumpfen

**drain** Abfluß *m*, Abflußleckleitung *f*, Abflußleitung *f*, Ablaß *m*, Ablauf *m*, Abzucht *f*, Abzug *m*, Ausfluß *m*, Ausguß *m*, Auslauf *m*, Drän *m*, Dränröhre *f*, Gassenrinne *f*, Gerinne *n*, Gosse *f*, Graben *m*, Kanal *m*, (in trench) Künette *f*, Rinne *f*, Schwindgrube *f*, Senkgrube *f*, Straßenrinne *f*, Wasserablaß *m* ~ **for engine exhaust box** Entwässerungsauspuff *m* ~ **and restrictor fitting** Ablaß *m* mit Doppelstück ~ **in a town** Binnentief *n*

**drain,** ~ **assembly conduit** Ablaßsammlungsleitung *f* ~ **board** Ablaufbrett *n* ~ **box** Ausgußkasten *m* ~ **cock** Abfluß-, Ablaß-, Auslauf-, Entleerungs-, Entwässerungs-, Leerlaß-hahn *m* ~ **device** Entleerungsvorrichtung *f* ~ **ditch** (or channel) Abflußkanal *m* ~ **hole** Wasserloch *n* ~ **(age) hose** Ablaßschlauch *m* ~ **metal** Gerinnstücke *pl* ~ **net** Drännetz *n* ~ **pad** Saugeinlage *f* (Flanschdichtung) ~ **pan** Tropfblech *n* ~ **passage** Abfluß-, Ablauf-kanal *m* ~ **pipe** Ablaßrohr *n*, Ablaufleitung *f*, Absaugleitung *f*, Abzugrohr *n*, Abzugsröhre *f*, Blechrinne *f*, Dränröhre *f*, Entwässerungsrohr *n*, Rinnstein *m*, Sickerrohr *n* ~ **pipeline** Entwässerungsleitung *f* ~ **pit** Gesenk *n* ~ **plug** Ablaßstopfen *m*,

Leerschraube *f* ~ **plug for oil sump** Ölablaßpfropfen *m* ~-rinsing apparatus Kanalspüler *m* ~ **spade** Drainierspalten *m* ~ **spout** Entnahmestutzen *m* ~ **table** Abtropf-kasten *m*, -tisch *m* ~ **tank** Wassertopf *m* ~ **tap** Ablaßhahn *m* ~ **tile** Rinnenziegel *m* ~ **trap** Geruchverschluß *m* ~ **valve** Abflußventil *n*, Ablaßhahn *m*, Abzapfhahn *m*, Entleerungsventil *n*, Entweichungsventil *n*, Lenzschieber *m*

**drainage** Abfluß *m*, Abwässern *n*, Ausschöpfung *f*, Austrocknung *f*, Drainage *f*, Drainieren *n*, Dränage *f*, Dränung *f*, Entwässerung *f*, Melioration *f*, Rohrleitungsentwässerung *f*, Wasserableitung *f*, Wasserhaltung *f* ~ **by compressed air** Drainage *f* mittels Preßluftinjektion ~ **by ditches (or gutters)** Grabenentwässerung *f* ~ **of an excavation** Ausschöpfen *n* einer Baugrube ~ **in the open** Trockenlegung *f* in offner Baugrube

**drainage,** ~ **appliance** Entwässerer *m*, Entwässerungsvorrichtung *f* ~ **area** Auslauffläche *f*, Flußgebiet *n* ~ **band** Entwässerungssieb *n* ~ **blanket** Entwässerungsteppich *m* ~ **board** Tropfbrett *n* ~ **channel** Sammelgraben *m* ~ **coil** Erdungs-, Saug-drossel *f* (electr.) ~ **culvert** Sickerkanal *m* ~ **ditch** Abflußgraben *m*, Abflußrinne *f*, Abzugsgraben *m*, Draingraben *m*, Entwässerungsgraben *m*, Sammelgraben *m*, Sickerwasser *n*, Zuggraben *m* ~ **elevator** Entwässerungsbecherwerk *n* ~ **fitting** Muffenrohrstück *n* ~ **gallery** Wasserlösungsstollen *m* ~ **horn** Esel *m* ~ **level** Wasser-lösungssohle *f*, -stollen *m* ~ **path** Entwässerungsweg *m* ~ **pit** Sickerloch *n* ~ **plant** Pumpanlage *f* ~ **pump** Entwässerungs-, Lenz-pumpe *f* ~ **screen** Entwässerungssieb *n* ~ **sieve** endloses Sieb *n* ~ **station** Pumpstation *f* ~ **system** Abwässerungs-, Entwässerungs-anlage *f* ~ **time** Leerzeit *f* ~ **tube** Drän *m*, Leckleitung *f* ~ **tunnel** Entwässerungsstollen *m* ~ **valve** Entwässerungsschleuse *f* ~ **well** Schwindgrube *f* ~ **wells** Entwässerungsbrunnen *m*

**drained** entwässert ~ **weight** Feuchtegewicht *n*

**drainer** Absetzbehälter *m*, Abtropfschale *f*, Stoffkasten *m* ~ **man** Bleichergehilfe *m*

**draining** Austrocknung *f*, Drainage *f*, Drainieren *n*, Dränage *f*, Entwässerung *f*, Trockenlegung *f*, Wasserablaß *m* ~ **by means of a funnel** Abnutschen *n* ~ **a spring** Ausschöpfen *n* einer Baugrube

**draining,** ~ **board** Abtropfbrett *n* ~ **channel** Ableitungskanal *m* ~ **chest** Absetzbehälter *m*, (paper mfg.) Entwässerungskasten *m*, Stoffkasten *m* ~ **device** Entleerungsgerät *n* ~ **dish** Abtropfschale *f* ~ **felt** Entwässerungsfilz *n* ~ **hose** Ablaufvorrichtung *f* (für Waschlauge) ~ **level** Wasserlösungssohle *f* ~ **machine** Ausschleudermaschine *f*, Schöpfmaschine *f* ~-off Abläuterung *f* ~ **pan** (oil pan) Ablaufwanne *f* ~ **plant** Vorflutanlage *f* ~ **pump** Entleerpumpe *f* ~ **rack (or stand)** Abtropfgestell *n* ~ **stand** Abtropfständer *m* ~ **tank** Absetzbehälter *m*, Stoffkasten (paper mfg.) *m* ~ **tower** Abtropfturm *m* ~ **transformer** Saugtransformator *m* ~ **well** Senkgrube *f*

**drape, to** ~ behängen, drapieren

**drape forming** Streckformverfahren *n*

**draper** Tuchhändler *m* ~ **s' caps** Kleiderstoffpackpapier *n*

**drapery** Drapierung *f*, Faltenwurf *m*; Tuchge-schäft *n*; Vorhang *m*
**draping** Stoffbedeckung *f*, Vorhang *m*, Wand-stoffbekleidung *f*
**drastic** starkwirkend
**draught** Gebräu *n*, Tauchtiefe *f* **on ~** vom Faß *n* **~ gauge** Flutometer *n* **~ machines for soil-tilling** Bodenbearbeitungszuggerät *n* **~ mark(s)** Ahming *f*
**draw, to ~** abheben, abseihen, anlassen (Ver-gütung), anreißen, anziehen, aufnehmen, auf-schleppen, aufzeichnen, ausheben, fassen, för-dern, (in spinning) laminieren, Lichtbogen ziehen, mitziehen, nachglühen, reißen, (liquid etc.) schöpfen, strecken, verbrauchen, ver-güten, vorreißen, zapfen, zeichnen **to ~ atten-tion to** erinnern **to ~ back** zurückziehen **to ~ a bill** einen Wechsel *m* ziehen **to ~ in a cable** ein Kabel *n* einziehen **to ~ off clinker** ent-schlacken **to ~ conclusions** Schlußfolgerungen *pl* ziehen **to ~ off the crankshaft** von der Kur-belwelle *f* abziehen **to ~ a curtain over** zu-hängen **to ~ a diagram** hinzeichnen **to ~ down while cold** kalt strecken **to ~ away downwards** nach unten abziehen **to ~ off gas** Gas *n* ab-saugen **to ~ in by hammering** einhämmern **to ~ in** ansaugen, einsaugen, einzeichnen, ein-ziehen, hineinziehen **to ~ by lots** auslosen **to ~ a map** auswerten, kartieren **to ~ money (from the bank)** Geld *n* abheben **to ~ near** aufziehen, heranziehen, im Anzug *m* sein **to ~ off** ab-dekantieren, abführen, abfüllen, ablassen, ab-leiten, absaugen, abstreichen, abzapfen, ab-ziehen, ansaugen, ausschlacken, (masonry) mit dem Lineal *n* abziehen **to ~ off slag** ab-schlacken, entschlacken, Schlacke *f* abziehen **to ~ out** aufreißen, ausheben, ausräumen, aus-schmieden, ausziehen, herausziehen, hervor-locken, zainen, ziehen **to ~ out and flatten steel** den Stahl *m* schienen **to ~ a pattern** das Modell *n* herausziehen **to ~ a perpendicular line** ein Lot *n* fällen **to ~ rivets** abnieten **to ~ to scale** maß-stäblich zeichnen **to ~ by sight** nach dem Augenmaß *n* zeichnen **to ~ to size** auf Maß *n* ziehen **to ~ up standards** Normen *pl* aufstellen **to ~ a straight line** eine gerade Linie *f* bilden **to ~ by suction** aspirieren **to ~ to** hinziehen **to ~ together** zusammenziehen, zuziehen **to ~ through** hindurchziehen **to ~ through trans-parent paper** durchzeichnen **to ~ a tube (or pipe)** ein Rohr *n* ziehen **to ~ up** aufsetzen, auf-stellen, aufziehen, ausfertigen, entwerfen, (a letter) formulieren, (with a windlass) haspeln, (molding) verfassen **to ~ (the pattern) very slow at the start** anlüften **to ~ water** Tiefgang *m* haben
**draw** Abhebung *f*, Abhub *m*, Aushub *m*, Hub *m*, Längsspiel *n*, (in casting) Nachsatz *m*, Sicke *f* **~ and buffer devices** Zug- und Stoßvorrichtung *f* **~ of the eyepiece** Okularauszug *m*
**drawback** Mißstand *m*, Überstand *m* **~ ram** Rückdrückkolben *m* **~ spring** Arretierfeder *f*
**drawbar** Anbauschiene *f*, Anhängerschiene *f*, Zugeisen *n*, Zugstange *f*, Zugwiderstand *m* **~ pull** Zugleistung *f*
**drawbeam** Haspelwalze *f*, Haspelwelle *f* **~ of a well** Brunnenschwengel *m*
**draw, ~ bench** Ziehbank *f* **~ bench for wire**

**drawing** Drahtzug *m* **~ bridge** Aufzug-, Klapp-, Zug-brücke *f* **~ broach** Räumnadel *f* zum Ziehen **~ cards** versetzte Diagramme *pl* **~ chuck** Konusfutterstück *n* **~ cord** Zieh-, Zug-schnur *f* **~-cutting** ziehender Schnitt *m* **~ cylinder** Abhebezylinder *m* **~ die** die Ziehform *f*
**drawdown** Absenkung *f* **~ wave** Entnehmesunk *m*
**draw, ~ former** Ziehbiegemaschine *f* **~ gate** Abzugschieber *m*, Schleusen-schütz *m*, -schütze *f* **~ gear** Zugvorrichtung *f* **~ head** Ziehwerk *n* **~ hook** Zughaken *m* **~-hook guide** Zughaken-führung *f* (r.r.)
**draw-in, ~ bolt** Zugschraube *f* **~ coil** Einzugs-wicklung *f* **~ collet** Spannzange *f* **~ collet chuck** Spannpatronenfutter *n* **~ (or clamping) nut** Festspannmutter *f* **~ rod for arbor** Anzugdorn *m* zum Fräsdorn **~ spindle** Spannseele *f* **~ torque** Einziehdrehmoment *n* **~ tube** Spannrohr *n* **~ type magnetic switch** Einzugmagnetschalter *m* **~ type three-jaw clutch (or chuck)** Zugdrei-backenfutter *n*
**draw, ~ key** Ziehkern *m* **~ knife** Abzieh-, Gerade-, Schnitt-messer *n*; Zieheisen *n*, Zieh-messer *n*, Zugmesser *f* (mus. instr.), Zugkontakt *m* **~ knob** Registerknopf *m* (mus. instr.), Zugkontakt *m* **~ lever** Zug-hebel *m* **~ machine (for patterns)** Abhebeform-maschine *f*
**drawoff, ~ cock** Ablaßhahn *m* **~ nut** Abdrück-mutter *f* **~ pan** Aufnahmeboden *m* **~ plate** Abzugsboden *m* **~ roll** Abzugswelle *f*
**draw, ~-out radiator** ausziehbarer oder einzieh-barer Kühler *m* **~-out-type circuit breaker** Aus-ziehausschalter *m* **~-out type of equipment** herausziehbare Geräteeinheit *f* **~ pin** Aufzieh-stift *m* **~ piston** Abhebekolben *m* **~ plate** Zieh-eisen *n* **~ punch** Zugstempel *m* **~ rod** Zugstange *f* **~ sheet** Grundblatt *n* **~ slate** Nachfallpacken *m* **~ spindle** Zugspindel *f* **~ spool** Bohrhaspel *m* **~ spring** Zugfeder *f* **~ stop (of organ)** Registerzug *m* **~ string** Zugband *n* **~ tongs** Kniehebelklemme *f* **~ tube (of microscope)** Ausziehtubus *m*, Zug *m* **~ twister** Streckzwirn-maschine *f* **~-type broaching** Nutenziehma-schine *f* **~ vice** Froschklemme *f* **~ well** Zieh-brunnen *m* **~ winch** Zughaspel *f* **~ wire** Ein-ziehdraht *m* **~ works** Bohrwinde *f*, Hebewerk *n*
**drawer** Fach *n*, Gefach *n*, Lade *f*, Schubkasten *m*, Schublade *f*, Zeichner *m*, Zieher *m* (metall.) **~ of the case** Schrankfach *n* **~ grid** Schub-ladenrost *m* **~ section** Schubladenfach *n* **~-type core-baking oven** Regulierkerntrockenofen *m* **~-type core oven** Kerntrockenofen *m* mit aus-ziehbaren Trockenzellen
**drawing** Abbildung *f*, Anlage *f*, (air cooling) Anlassen *n*, Aufzeichnung *f*, Ausziehen *n*, Recken *n* des Drahtes, Riß *m* Rückglühung *f*, Skizze *f*, Strecke *f* (metall.), Streckziehen *n*, (dyes) Verziehen *f*, Zeichnung *f*, Zeichnungs-bild *n*, Ziehen *n*, Ziehung *f*, Zug *m* **~ of current by grid** Gitterstromaufnahme *f* **~ of idle (or reactive or wattless) current** Blindstromauf-nahme *f* **~ of minerals** Förderung *f* **~ together of lines** Paarbildung *f* **~ in water colors** ge-tuschte Zeichnung *f* **~ of wire** Grobzug *m*
**drawing, ~ ability** Streckbarkeit *f* **~ apparatus** Zeichenapparat *m* **~ arm** Zeichenarm *m* **~ bars** Stangenzug *m* **~ bench for thick wires** Grobzug *m* **~ block** Zieh-eisen *n*, -platte *f*, -ring *m* **~**

**board** Planchette *f*, Reißbrett *n*, Zeichenbrett *n*, Zeichentisch *m* **~-bush mandrel** Ziehbuchsendorn *m* **~ cage** Förderkorb *m* **~ chalk** Zeichenkreide *f* **~ charcoal** Zeichenkohle *f* **~ compass** Stechzirkel *m* **~ compasses** Reißzirkel *m* **~ compound** Ablaßmittel *n*, Ziehfett *n* **~ crack** Spannungsriß *m* **~ cylindrically** Zylindrischziehen *n* **~ device** Zeichenvorrichtung *f*, Ziehvorrichtung *f* **~ die** Einziehwerkzeug *n*, Zieheisen *n*, -kern *m*, -matrize *f*, -ring *m*, -scheibe *f*, -werkzeug *n* **~-die hole** Ziehloch *n* **~ dies for cartridge shells** Einziehmatrizen *pl* für Patronenhülsen **~ effect** Ablaßwirkung *f* **~ engine** Göpel *m* **~ frame** (for wire) Drahtleier *f*, Förder-gerippe *n*, -gestell *n*, -schale *f*; Vorspinnmaschine *f* **~ frames** (textiles) Durchzug *m* **~ furnace** Ablaßofen *m* **~ gap** Ziehspalt *m* **~ grease** Ziehfett *n* **~ grid** Zeichennetz *n* **~ groove** Ziehriefe *f* **~ hook** Ziehhaken *m* **~-in** (air) Ansaugen *n*, Einziehen *n*, (textiles) Fadeneinziehen *n* **~-in frame** Garnaufwindemaschine *f* **~-in hook** Einziehhaken *m* **~-in wire** Ziehdraht *m* **~ incorporated** Aktenzeichnung *f* **~ ink** Zeichentusche *f* **~ installation** Zieherei *f* **~ instrument** Zeichengerät *n* (case of) **~ instruments** Reißzeug *n* **~ jaw** Ziehbacke *f* **~ knife** Geradeisen *n* **~ machine** Ziehvorrichtung *f* **~ mandrel** Ziehdorn *m* **~ material** Ziehgut *n* **~ materials** Zeichenmaterial *n* **~ means** Zugmittel *n* **~ mill** (for wires) Drahtziehwerk *n* **~ model** Entwurf *m* **~ number** Zeichnungsnummer *f* **~-off** Abzug *m* **~-off bung** Abfüllspund *m* **~-off (or rack) machine** Abfüllmaschine *f* **~-off of steam** Dampfentnahme *f* **~ oil** Ablaßöl *n* **~-out** Ziehvorgang *m* **~-out device** Ziehangel *f* **~ paper** Zeichenpapier *n* **~ pass** Schnellvorwalz-, Streck-, Zieh-kaliber *n* **~ pattern** Zeichenvorlage *f* **~ pen** Reiß-, Zieh-feder *f* **~ pencil** Zeichenstift *m* **~ pin** Heftzwecke *f* **~ plane** Zeichenebene *f* **~ plant** Ziehanlage *f* **~ point** Reiß-nadel *f*, -spitze *f* **~ power** Ziehkraft *f* **~ press** Tiefziehpresse *f* **~ process** Streckziehverfahren *n*, Ziehvorgang *m* **~ quality** Ziehfähigkeit *f* **~ roller** Streckwalze *f* **~ room** Zeichensaal *m*, Zieherei *f* **~ rope** Manntausendseil *n* **~ rule** Reißschiene *f*, Zeichenlineal *n* **~ scraper** Zugschaber *m* **~ scratch** Ziehriefe *f* **~ sheets** Ziehbleche *pl* **~ size (or dimension)** Zeichnungsmaß *n* **~ stand** Zeichengestell *n* **~ table** Zeichentisch *m* **~ tear** Einziehfalte *f* (metal) **~ temperature** Ablaßtemperatur *f* **~-through** Durchzug *m* **~ tongs** Kniehebelklemme *f* **~ tool** Zieh-ring *m*, -werkzeug *n* **~-up** Ausfertigung *f*, Ausstellung *f*, Einzug *m*, Heranziehung *f* **~ wire** Drahtzug m **~-work installation** Ziehwerksanlage *f*

**drawn** bezogen, gezeichnet **~ apart** auseinandergezogen **~ cold** gezogen **~ container** gezogene Dose *f* **~ filament lamp** Drahtlampe *f* **~ glass** gezogenes Glas *n* **~ grain** gezogene Narben *pl* **~ hexagon brass bar** gezogenes Sechskantmessing *n* **~-in** eingezogen **~-in antenna** eingefahrene Antenne *f* **~-in system** Verlegung *f* in Isolierrohr **~ iron wires** gezogene Flußeisendrähte *pl* **~ metal part** Ziehteil *m* **~-off** abgesaugt, (mixture) ausgesaugt **~ (out)** gezogen **~-out lip** Ausgußschnauze *f* **(metal) ~-out slag streak** Längszeile *f* **~ piece** Ziehteil *m* **~ steel**

gezogener Stahl *m* **~ steel wires** gezogene Stahldrähte *pl* **~ taper** Wachsstock *m* **~ tube** gezogene Röhre *f* **~-up by a notary** notariell **~ wire** gezogener Draht *m*

**dray, to ~** baggern

**dray** Förderkasten *m*

**dreadful** schrecklich

**dredge, to ~** ausbaggern, baggern

**dredge** Bagger *m*, Baggermaschine *f*, Schleppsack *m* **~-bucket teeth** Erzgreifermesser *n*

**dredged material** Baggergut *n*

**dredger** Bagger *m*, Schlammschaufel *f*, Schwimmbagger *m* **~ for clearing dry earth** Erdbagger *m* **~ bucket** Baggerkasten *m* **~ drum** Turas *m* **~ joint pin** Baggerbolzen *m* **~ transporting the ballast** Prahmbagger *m*

**dredges** Becherwerke *pl*

**dredging** Baggerung *f* **~ of a canal** Ausbaggerung *f* eines Kanals

**dredging, ~ bit** Baggerbohrer *m* **~ chain** Baggerkette *f* **~ craft** Naßbagger *m* **~-machine parts** Baggerteile *pl* **~ roll** Turas *m* **~ service** Baggerbetrieb *m* **~ tumbler** Eimertrommel *f*, Kettentreibscheibe *f*, Turas *m*

**dregs of tar** Teersatz *m*

**drench, to ~** abbrausen

**drench pit** Beizbütte *f*

**drencher** Regenwandbrause *f*

**dress, to ~** abdrehen, ablehren, abrauhren, abrichten, abrunden, abziehen, anrichten, anziehen, (ore) aufarbeiten, ausrichten, bearbeiten, behauen, beizen, dressieren, düngen, fertigmachen (finish type), gerben, gerade machen, nacharbeiten, nachbearbeiten, nachwalzen, putzen, rauhen, richten, schlichten, (filter) überziehen, wiederzurichten, zubereiten, zurechtschneiden, zurichten **to ~ and cover** durchdecken **to ~ ship** Toppflaggen *pl* setzen **to ~ the filterpress** Filterpresse *f* überziehen **to ~ left or right** sich nach links oder rechts ausrichten **to ~ the points** (print.) Punktur *f* zurichten **to ~ a quarrystone** einen Bruchstein *m* bossieren **to ~ a timber with the twibil** einen Balken *m* mit der Queraxt abputzen **to ~ trim** beputzen

**dressed** verputzt **~ leather** zugerichtetes Leder *n* **~ ore** aufbereitetes Erz *n*, Schlich *m* **~ plank (or board)** gehobelte Bohle *f* (Planke) **~ products** aufbereitetes Gut *n* **~ rock** Rundhöcker *m* **~ stone** Bruchstein *m*, Steinmetzquader *m* **~-stone facing** Quaderverkleidung *f* **~-stone masonry** Mazerwerk *n* von behauten Steinen **~ tin** Zinnschlich *m*

**dresser** Abziehvorrichtung (für Schleifscheibe), (kitchen) Anrichte *f*, Anrichter *m*, Aufbereiter *m*

**dressing** Appretieren *n*, Aufbereitung *f*, Ausrichten *n*, Bandage *f*, Bearbeiten *n*, (of tools) Nachbearbeitung *f*, Putzen *n*, Verbandsstoff *m*, Verbandszeug *n*, (building) Verblendung *f*, Verkleidung *f*, Wiederzurichtung *f*, Wundverband *m*, Zubereitung *f*, Zurichtung *f* **~ of ores** Erzaufbereitung *f* **~ of paper** Appretur *f* **~ with pigment finishes** Deckfarbenzurichtung *f* **~ of the raw beryl** Aufschluß *m* des Rohberylls **~ with slabs** Betäfelung *f* mit Platten, Plattenverblendung *f* **~ with tables** Plattenverblendung *f*

**dressing, ~ bench** Beschneidebank *f* **~ board** Auslegepappe *f* **~ brush** Schlichtbürste *f* **~ capacity** Abrichtleistung *f* **~ chisel** Balleneisen *n*, Grobmeisel *m* **~ equipment** Verbandmittel *n* **~ expenses** Aufbereitungskosten *pl* **~ floor** Aufbereitung *f*, Wäsche *f* (min.) **~ frame** Nachwalzgerüst *n* **~ glue** Appreturleim *m* **~ loss** Waschverlust *m* **~ machine** Appretur-, Leim-, Sicht-maschine *f* **~ mull** Verbandmull *m* **~ plant** Aufbereitungsanlage *f* **~ procedure** Richtvorgang *m* **~ shop** Putzerei *f*, Zurichterei *f* **~ table** Aufbereitungsherd *m*, Putztisch *m* **~ tool** Abdrehwerkzeug *n*, Abrichter *m*, Abrichtwerkzeug *n*, Richtwerkzeug *n*

**dribble, to ~** tröpfeln

**dribble oil** Tropföl *n*

**dried** getrocknet **~ defecation scum** Trockenschlamm *m* **~-pulp elevator** Trockenschnitzelelevator *m* **~-pulp screw conveyer** Trockenschnitzelschnecke *f* **~ sugar-beet cossettes** getrocknete vollwertige Zuckerrübenschnitzel *pl* **~ sugar-beet leaves** Trockenblatt *n*

**drier** Aufhänger *m*, Eindampfschale *f*; Trocken-apparat *m*, -maschine *f*, -mittel *n*, -schrank *m*, -stoff *m*; Trockner *m*, Verdampfer *m* **~ with agitator (agitator drier)** Rührwerkstrockner *m* **~ with cross-shaped internal-distribution system** Trockenapparat *m* mit Kreuzeinbau **~ with recycle of air** Umlufttrockner *m* **~ with single roller** Einwalzentrockner *m*

**drier, ~ coil** Nachverdampfungsschlange *f* **~ furnace** Trockenofen *m* **~ oil** Trockenöl *n* **~ section** Trockenpartie *f* (paper mfg.) **~ tube** (of drier) Trommelrohr *n*

**drift, to ~** absacken, abtreiben, aufreiben, (a hole) ausdornen, dahintreiben, durchtreiben, (of airplane) schieben, treiben, triften, vertreiben

**drift** Abdrängung *f*, Abtreiben *n*, Abtrieb *m*, Abtrift *f* (nav.), Abweichung *f*, Anschwemmung *f*, Aushaueisen *n*, Ausstoßwerkzeug *n*, Auswanderung *f* (gyro), Dorn *m*, Driftströmung *f*, (of electrons) Fluß *m*, Galerie *f*, Gang *m*, Locheisen *n*, Lochhammer *m*, Lochpfeife *f*, Lochräumer *m*, Lochstanze *f*, (in discharge) Mitnehmen *n*, Nageltreiber *m*, Richtung *f*, Schubbewegung *f*, Seitenabweichung *f*, Sohlenstrecke *f*, Stemmer *m*, Stollen *m*, Strecke *f* (min.), Treibstoff *m*, Trieb *m*, Trift *f*, Verschiebung *f*, Versetzung *f*, Vertreiben *n*, Zehrungsstempel *m* **~ of an airplane** Flugzeugversetzung *f* **~ of an arch** Seitenschub *m* eines Bogens **~ of frequency** Frequenz-abweichung *f*, -schwankung *f* **~ of strata** Einfallen *n* (Streichen) der Schichten

**drift, ~ angle** Abtriftwinkel *m* (aviat.), Luv *n*, Luv-, Seitentrift-, Vorhalte-winkel *m* **~ bar** Abdriftstab *m*, Abtriebhebel *m* (aviat.) **~ bolt (or pin)** Treibbolzen *m* **~ compass** Abtriftkompaß *m* **~ controls** Dralluftruder *n* **~ conveyor** Streckenband *n* **~ correction** Abdrängungsberichtigung *f*, Abdriftverbesserung *f*, Auswanderungskorrektur *f* (gyro), Drallverbesserung *f* **~ correction angle** Luvwinkel *m* **~ current** Driftströmung *f* **~ due to rotation of projectile** Drallabweichung *f* **~ due to wind** Seitentrift *m* **~ energy** Driftenergie *f* **~ fence** Ablagerungszaun *m* **~ hammer** Dornhammer

*m*, Durchschlaghammer *m* **~ ice** Eisgang *m*, Treibeis *n* **~ indicator** Abdrängungsmesser *m*, Abdriftanzeiger *m*, Abtriftanzeiger *m*, Abtriftmesser *m* **~ key** Austreibkeil *m* **~ landing** Schiebelandung *f* **~ landing gear** Schiebefahrwerk *n* **~ meter** Abdrift-messer *m*, -visier *n* **~ mine** Stollenbergbau *m* **~ mining** Stollenbetrieb *m*, Streckenbetrieb *m* **~ mobility** Driftgeschwindigkeit *f* **~ pin** Dorn *m* **~ punch** Dornhammer *m*, Durchtreiber *m*, Handdurchschläger *m*, Treiber *m* **~ recorder** Abtriftschreiber *m* (aviat.) **~ sail** Schleppsegel *n* (aviat.) **~ sand** Flug-, Schwemm-sand *m* **~ sight** Rücktriftzielgerät *n* **~ space** Laufraum *m* **~ speed** mittlere Geschwindigkeit *f* **~ test** Aufweiteprobe *f*, Lochprobe *f* **~ trail** räumliche Rücktrift *f* **~ transistor** Drifttransistor *m* **~ tube** Laufzeit-, Quersteuer-, Trift-röhre *f* **~ tube working with two fields** Zweifeldröhre *f* **~ tunnel** Laufraumelektrode *f* **~ velocity** Driftgeschwindigkeit *f*, Schiebegeschwindigkeit *f* **~ way** Richtstollen *m* **~ wave** Kompressionswelle *f* **~ wood** Flöß-, Treib-holz *n* **~ work** Hereintreibarbeit *f*

**drifted deposit** zusammengetriebenes Lager *n*

**drifter** Hammerbohrmaschine *f*, Vorrichtungsarbeiter *m*

**drifting** (punching or spreading) Dornen *n*, Hereintreibarbeit *f*, Streckenbetrieb *m*, Querschlag *m* (min.) **~ force** Abdrängkraft *f*, Driftkraft *f* **~ ice** Eisgang *m* **~ sand** Fließ-, Schwemm-, Treib-sand *m* **~ snow** Stöberschnee *m* **~ test** Aufdornprobe *f*

**drill, to ~** abbohren, abrichten, anbohren, anlernen, ausbilden, aussäen, bohren, drillen, durchbohren, üben **to ~ finished holes** nachbohren **to ~ in** in eine petroleumhaltige Schicht *f* geraten **to ~ out** aufbohren, ausbohren **to ~ up** aufbohren

**drill** Ausbildung *f*, Bohregge *f*, Bohrer *m*, Bohrkrone *f*, Bohrmaschine *f*, Bohrung *f*, Drillbohrer *m*, (cloth) Drillich *m*, Drillmaschine *f*, Einsatzbohrer *m*, Erdbohrer *m*, Furche *f*, Pech *n*, Säemaschine *f*, Schientuch *n*, Spiralbohrer *m*, Übung *f* **~ with ferrule** Rollenbohrer *m* **~ (ing)** (a new hole) **with twist drill** Bohren *n*

**drill, ~ bit** Bohrer *m*, Bohrerspitze *f*, Bohrmeißel *m*, Drillbohrereinsatz *m*, Spitzbohrer *m* **~ bit with cross-type cutting edge** Kreuzschneider *m* **~-bit grab** Drillbohrfänger *m* **~ blower** Bohrgebläse *n* **~ bow** Bohr-, Drell-, Fiedel-bogen *m* **~ breakage** Spiralbohrerbruch *m* **~ chain and swivel** Bohrkette *f* mit Wirbel **~ charger** Lademeister *m* **~ chips** Bohrgut *m* **~ chuck** Bohr-futter *m*, -kopf *m* **~ club** Bohrkeule *f* **~ collars** Schwerstangen *pl* **~ core** Bohrkern *m* **~ coupling** Muffe *f* für Bohrstangen **~ cradle** Bohrwagen *m* **~ drift** Austreiber *m* **~ gauge** Spiralbohrerlehre *f* **~ gauge plate** Bohrerlehre *f* **~ grinder** Spiralbohrerschleifmaschine *f* **~ ground** Exerzierplatz *m*, Truppenübungsplatz *m* **~-grubber plow** Grubbersämaschine *f* **~ gun** Handbohrmaschine *f* **~-hardness-testing machine** Bohrmaschine *f* zur Prüfung der Bearbeitungsfähigkeit von Guß **~ harrow** Drill(en)-, Furchen-egge *f* **~ head** Bohrkopf *m* **~ head sliding** Bohrschlitten *m* **~ holder** Spiralbohrerhalter *m* **~ hole** Bohrloch *n*, Bohrung *f* ~ **hole for grouting** Loch *n* zum Einspritzen ~ **jig** Drillmaschinenvorrichtung *f* ~**-jig bush** Bohrbüchse *f* für Bohrvorrichtung ~ **kit** Bohrausrüstung *f* ~ **machine** Drillmaschine *f* ~**-marked center** angebohrter Radiusmittelpunkt *m* ~ **pipe** Bohrgestänge *n* (einschließlich Bohrstange), Bohrrohr *n*, Gestängerohr *n*, Hohlgestänge *n* ~ **pipe elevator** Gestängefahrstuhl *m* ~**-pipe flush joint** Innenmuffe *f* (beim Bohrgestänge) ~ **pipe protectors** Gestängeschützer *m* ~**-pipe slips** Bohrrohrabfangkeil *m*, Gestängeabfangkeile *pl* ~ **pipe thread** Gestängerohrgewinde *n* ~ **pipe threads** Bohrgestänge-Gewinde *n* ~ **plow** Drillsäemaschine *f* ~ **point** Bohrerspitze *f*, Drillbohrer *m* ~ **point (or grinding) gauge** Spiralbohrerschleiflehre *f* ~ **poles** Bohr-, Dreh-gestänge *n* ~ **press** Bohrbank *f*, Bohrmaschine *f* ~ **pressure** Bohrdruck *m* ~ **rig** Anbohrapparat *m* ~ **rod** Bohrstange *f* ~ **rod with welded-on delivery worm** Bohrstange *f* mit aufgeschweißtem Förderwedel ~ **sets** Bohrersatz *m* ~ **sleeve** Bohrerhülse *f* ~ **socket** Bohr-futterkegel *m*, -halter *m* ~ **spacing** Drillweite *f* ~ **speeder** Schnellbohreinrichtung *f* ~**-spindle guide** Bohrspindelführung *f* ~ **steel** Bohrstahl *m* ~ **stem** Meißelschaft *m*, Schwerstange *f* ~**-stem bushing** Mitnehmereinsatz *m* ~**-stem test** Rohrlochprüfung *f* mit Gestänge ~ **tang** Bohrerzapfen *m* ~ **template** Bohrschablone *f* ~ **test** Bohr--probe *f*, -versuch *m* ~ **thrust** Bohrdruck *m* ~ **tube** Hohlgestänge *n* ~ **upsetting machine** Bohrerstauchmaschine *f*

**drilled** ausgebohrt ~ **head bolt** Schraube *f* mit Sicherungsloch im Kopf ~ **holes** Siederohrlöcher *pl* ~ **jig** Bohrlehre *f* ~ **well** Röhrenbrunnen *m*

**driller** Bohr-führer *m*, -hauer *m*, -maschine *f*, -meister *m* ~ **mine** Bohrmine *f*

**drilling** Aufbohren *n*, Bohrarbeit *f*, Bohrspäne *pl*, Bohrung *f* ~ **alongside** Umbohrung *f* ~ **of a borehole** Brunnenabsenkung *f* ~ **and counterboring fixture** Bohr- und Senkvorrichtung *f* ~ **by dredging** Baggerbohrung *f* ~ **in** eine Schicht *f* anbohren ~ **for lightness** Erleichterungsbohrung *f* ~ **of a well** Brunnenabsenkung *f*

**drilling,** ~ **apparatus** Bohr-anlage *f*, -apparat *m* ~ **bench** Bohrtisch *m* ~ **bit** Spatenmeißel *m* ~ **bridge** Bohrbrücke *f* ~ **capacity** Bohrleistung *f* ~ **center** Bohrungsmittenabstand *m* ~ **clamp** Seilklemme *f* ~ **contractor** Bohrunternehmer *m* ~ **control** Bohrdruckkontrolle *f* ~ **crew** Bohrmannschaft *f* ~ **crown** Bohrkrone *f* ~ **deep** Tiefbohren *n* ~ **device** Anbohrapparat *m*, Bohrvorrichtung *f* ~ **ferrule** Bohrzwinge *f* ~ **foreman** Oberbohrmeister *m*, Turmmeister *m* ~ **hammer** Treibfäustel *m* ~ **hoist** (oil drilling) Haspelwinde *f* ~ **implements** Bohrgezähe *n* ~ **jib** Bohrschablone *f* ~ **jig** Bohrvorrichtung *f* ~ **line** Bohrseil *n* ~ **machine** Bohrmaschine *f* ~ **mud** Bohr-schlamm *m*, -schmand *m* ~ **operation** Bohrarbeit *f* ~**-out misfires** Anbohrer *m* von Versagern ~ **outfit** Bohrgarnitur *f* ~ **output** Bohrwerk *n* (Tiefbohranl.) ~ **pipe** Bohrstange *f* ~ **rig** Bohr-anlage *f*, -apparat *m*, -gestell *n*, -kran *m* ~**-rig frame** Bohrkrangerüst *n* ~ **rope** Bohrseil *n* ~ **shaft** Bohr-strang *m*, -welle *f* ~ **site** Bohr-punkt *m* (Tiefbohrg.),

-stelle *f* ~ **spindle gear wheel** Bohrspindelrad *n* ~ **spoon** Bohrlöffel *m* ~ **string** Bohrstrang *m* ~ **support** Bohrstütze *f* ~ **tailstock** Bohrreitstock *m* ~ **test** Bohrversuch *m* ~**-tool joint** Bohrgerätekuppelung *f* ~ **tools** Bohrgezähe *pl* ~ **weight indicator** Drillometer *m* ~ **work** Bohrarbeit *f*

**drillings** Bohrklein *n*, Bohrmehl *n*, Drehspäne *pl*

**drillometer** (petroleum) Belastungsanzeiger *m*, Bohrdruckmesser *m*

**D-ring** dreieckiger Ring *n*

**drink, to** ~ trinken

**drink** Getränk *n*

**drinkable** trinkbar

**drinking,** ~ **bowl** Tränkbecken *n* ~ **cup** Trinkbecher *m* ~ **set** Trinkgerät *n* ~ **straw** Trinkhalm *m* ~ **water** Trinkwasser *n* ~**-water line** Trinkwasserleitung *f* ~**-water supply** Trinkwasserversorgung *f*

**drip, to** ~ abtröpfeln, herabtropfen (chem.), träufeln **to** ~ **down** heruntertröpfeln **to** ~ **off** abtropfen **to** ~ **out** aussträpfeln **to** ~ **subsequently** nachtropfen

**drip** Nachtropfen *n*, Tröpfel *n*, Tropfen *m* ~**-board** Abtropfer *m* ~ **catcher** Tropfenfänger *m* ~ **cock** Schnüffelventil *n*, Tropfhahn *m* ~ **cup** Tropf-becher *m*, -zylinder *m* ~ **edge** Abtropfkante *f* ~ **feed** Tropfölschmierung *f* ~ **flap** Tropftuch *n*, Regentraufe *f* ~ **gauge** Tropfenmesser *m*, Tropfglas *n* ~ **gutter** Tropfrinne *f* ~ **nozzle** Tropfdüse *f* ~ **oil** Tropföl *n* ~**-oil pipe** Leckkraftstoffleitung *f* ~ **oiler** (lubricator) Tropföler *m* ~**-oiling bearing** Tropföllager *n* ~ **pan** Abtropf-blech *n*, -schale *f*, Auffangschale *f*, Ölfänger *m*, Tropfbehälter *m*, Tropfblech *n*, Tropfschale *f*, Tropfwanne *f* ~**-pan grate** Tropfwannenrost *m* ~ **proof** tropfwasser-geschützt, -dicht ~ **ring** Abtropfring *m*, Tropfring *m* ~ **statoscope** Tropfenstatoskop *n* ~ **stone** Filterstein *m*, Kranzleiste *f*, Tropfstein *m* ~ **table** Abtropftisch *m* ~ **tray** Tropfschale *f*, Verdunstrinne *f* ~**-tray shelf** Tropfschalentraggitter *n* ~ **tube** Tropfenfänger *m*, Tropfrohr *n* ~ **valve** Schnüffelventil *n* ~ **water receiver** Tropfwasserbehälter *m*

**dripless** nicht tropfen

**dripping** Tropfen *n*; nachtropfen ~ **board** Abtropfbrett *n* ~ **edge** Abtropfkante *f* ~ **plant** Rieselanlage *f* ~ **rack** Abtropfbrücke *f* ~ **(or dropping) vessel** Tropfgefäß *n* ~ **wet** tropfennaß

**drive, to** ~ eine Strecke *f* auffahren, (in) eintreiben, fahren, führen, lenken, (a motor) mitnehmen, (piles) rammen, treiben, vortreiben **to** ~ **back** zurücktreiben **to** ~ **beyond cutoff** übersteuern **to** ~ **into a corner** in die Enge treiben **to** ~ **down** nieder-stoßen, -treiben **to** ~ **galleries** erlängen (von Strecken) (min.) **to** ~ **a gallery on** einen Stollen *m* vortreiben **to** ~ **a gallery to the hade of a seam** einen Abhau *m* machen (min.), eine einfallende Strecke *f* treiben **to** ~ **a heating through the ground** das Gebirge *n* durchörtern **to** ~ **high** hochtreiben **to** ~ **home** festschlagen **to** ~ **in** (piles) einrammen, (nail) einschlagen **to** ~ **a level** auffahren (min.) **to** ~ **the lines** die Zeilen *pl* enger machen (print.) **to** ~ **off** abdestillieren, abtreiben, austreiben **to** ~ **on** antreiben, auftreiben, forttreiben **to ~ out** (by pressure) abdrücken, ablassen, ausbringen (print.) ausfahren, austreiben **to ~ the propeller** Luftschraube *f* antreiben **to ~ a rivet** Niete *f* eintreiben **to ~ off the road** vom Fahrzeug *n* abkommen **to ~ through** durchtreiben **to ~ together** zusammentreiben **to ~ up** anfahren

**drive** Anfahrt *f*, (machine) Antrieb *m*, Betriebsart *f*, Durchsatz *m*, Fahrt *f*, Gelenkrohr *n*, Getriebe *n*, Transmission *f*, Trieb *m* **~ of the carriage** Wagenantrieb *m* **~ by compressed air** Preßluftantrieb *m* **~ of draft system** Streckwerksantrieb *m* **~ of the ignition distributor** Antrieb *m* des Zündverteiler **~ by rope** Seiltrieb *m* **~ of selector** Wählerantrieb *m* **~ by stages** stufenweises Rammen *n*

**drive, ~ appliance** Rohrpreßvorrichtung *f* **~ assembly** Laufwerk *n* **~ bearing** Antriebslager *n* **~ bushing** Antriebsstutzen *m*, Mitnehmereinsatz *m* **~ chain** Präzisionsrollenkette *f*, Treib-, Trieb-kette *f* **~ characteristic** Aussteuerungskennlinie *f* (electr. tube) **~ compartment** Fahrerraum *m* **~ coupling** Antriebskopf *m* **~-current** Schreibstrom *m* **~ disc** Innenlamelle *f* **~ field** Schreibfeld *n* **~ fit** Treibsitz *m* **~ flange** Antriebsflansch *m* **~ frequency** Steuerfrequenz *f* **~ (wheel) gear** Antriebsrad *n*, (transmission) Antriebszahnrad *n* **~-in counter** Autoschalter *m* (Bank) **~-in restaurant** Rasthaus *n* **~ key** Mitnehmer *m* **~ line** Schreibleitung *f* **~ magnet** Antriebsmagnet *m* **~ mechanism** Triebwerk *n* **~ member** Trieborgan *n* **~ mission** Fahrauftrag *m* **~ oscillator** Steuer--generator *m*, -röhre *f* **~ pattern** Antriebsbildfehler *m* (facsimile) **~ pin** Mitnehmerstift *m* (acoust.) **~ pin hole** Mitnehmerloch *n* (phono) **~ pinion** Triebel *n* **~ pipe** Rammrohr *n* **~ plate** Antriebsflansch *m* **~ power** Antriebsleistung *f* **~ pulley** Antriebsriemenscheibe *f*, Antriebsscheibe *f*, Riemenscheibe *f*, Triebscheibe *f* **~ pulley support** Antriebsscheibenkonsole *f* **~ pulse** Steuerimpuls *m* **~ range** Aussteuerungs--bereich *m*, -intervall *n*, Steuerungsbereich *n* **~ roller** Antriebswalze *f* **~ screw** Schlagschraube *f*, Triebschraube *f* **~ shaft** Achsen-, Antriebs-, Gelenk-, Kardan-, Steuer-, Trieb--welle *f* **~-shaft housing** Gelenkwellenrohr *n*, Kardanstützrohr *n* **~ shoe** Bohrschuh *m* **~ side** Antriebsseite *f* **~ slack check** Prüfung *f* des toten Ganges **~ spline** verzahnte Antriebswelle *f* **~ spring** Transportfeder *f* **~ sprocket** Antriebskettenrad *n* **~ stage** Steuerstufe *f* **~ test** Fahrversuch *m* **~ tube** Steuerröhre *f* **~ unit** Antriebsaggregat *n* **~ way in-movement** Einfahrt *f* **~ wheel** Einzahnscheibe *f* (film), Treibrad *n* **~ winch** Antriebswinde *f* **~ winding** Steuerwicklung *f* **~ windings** Schreibwicklungen *pl*

**driven** angetrieben, betrieben, geschlagen, getrieben **~ by helical gears** schraubenradangetrieben

**driven, ~ disc** Außenlamelle *f* **~ element** gespeistes Element *n* (antenna), aktiver Strahler *m* **~ end of a belt** schlaffes Trumm *n* **~ end of shaft** Abtrieb *m* **~ gear** getriebenes Zahnrad *n* **~ multivibrator** fremdgesteuerter Multivibrator *m* **~ plate** Schwungscheibe *f* **~ pulley** Gegenscheibe *f* **~ rivet** geschlagenes Niet *n* **~ sender**

fremdgesteuerter Sender *m* **~ shaft** Abtriebswelle *f*, getriebene Welle *f* **~ side of a belt** schlaffes Trumm *n*

**driver** Antreiber *m*, Antriebszapfen *m*, Besan *m*, Fahrer *m*, Führer *m*, Kraftfahrer *m*, Kraftwagenführer *m*, Kutscher *m*, Läufer *m*, abtreibender Maschinenteil *m*, Mitnehmer *m*, Ritzel *n*, Treiber *m*, Treiberstufe *f*, Treibmittel *n*, Wagenführer *m* **~ module** Treiber-modul *n*, -glied *n* **~ nose** Mitnehmerbolzen *m* **~ plate** Mitnehmerblech *n* **~ qualification record** Fahrbefähigungsnachweis *m* **~ rod** Stößel *m*, Stößer *m* **~ section (or stage)** Treib-, Steuer-stufe *f* **~ system** Treibersystem *n* **~ tube** Treiberröhre *f* **~-tube transformer** Treiberröhretransformator *m*

**driver's, ~ brake valve** Führerbremsventil *n* **~ cab** Führerhaus *n* **~ compartment** Fahrerhaus *n*, Führerstand *m* **~ hatch** Fahrerluke *f* **~ license** Führerschein *m* **~ periscope** Fahroptik *f* **~ seat** Führersitz *m* **~ visor** Fahrersehklappe *f* (tank)

**driving** Ausfahren *n*, Führung *f*, Niederhauen *n* (min.), Schlagen *n* **~ by accumulators** Akkumulatorenantrieb *m* **~ of rivets** Nietarbeit *f* **~ a well point** Brunnenabsenkung *f*

**driving, ~ arm** Griffarm *m* (mach.), Rast *f* **~ arrangement** Antriebsvorrichtung *f* **~ axle** Antriebsachse *f*, drehende Achse *f*, Treib-, Trieb-achse *f* **~ ball** Treibkugel *f* (Hammer) **~ band** Band *n*, Führungsband *n*, (on ammunition or gun) Führungsring *m*, (shell) Granatring *m*, Schnurantrieb *m* **~ bar** Triebstange *f* **~ belt** Antriebsriemen *m*, Transmissions-, Treib-riemen *m* **~ blade** Mitnehmerflügel *m* **~ bogie** Triebdrehgestell *n* **~ bridge** Mitnehmerbrücke *f* **~ bush** Antriebs-, Mitnehmer-buchse *f* **~ cam** Antriebsnocken *m* **~ cap** Bohrrohrkopf *m*, (on pile) Schlaghaube *f* **~ chain** Antriebs-, Treib-kette *f* **~ chain wheel** Antriebskettenrad *n* **~ circuit** Antriebsstromkreis *m*, äußerer Steuerkreis *m* **~ clutch** Antriebskupplung *f* **~ collar** Nabe *f* **~-cone pulley** Antriebsstufenscheibe *f* **~ cord** Antriebsschnüre *pl* **~ crank** Treibkurbel *f* **~ cushion** Dämpf-, Ramm--kissen *n* **~ direct-current motor** Antriebsgleichstrommotor *m* **~ disk** Mitnehmerscheibe *f* **~ dog** Anlasser-, Antriebs-klaue *f*, Drehherz *n* **~ drill** Fahrausbildung *f*, Fahrübung *f* **~ edge of the lands in bore of gun** Führungskante *f* der Felder in der Geschützseele **~ element** Antriebselement *n*, System *n* eines Zählers **~ end (of belt)** straffes Trumm *n* **~ engine** Antriebsmaschine *f* **~ fit** (petroleum) Fest-, Preß-, Ruhe-, Treib-sitz *m* **~ flange** Mitnehmerflansch *m* **~ force** Antriebskraft *f*, Trieb *m*, Triebkraft *f* **~ gallery** Streckenvortrieb *m* **~ gear** Antriebrad *n*, Laufwerk *n*, Trieb *m*, Triebrad *n*, Triebzeug *n* **with ~ gear above (below)** mit oberem (unterem) Antrieb *m* **~ gearing** Triebwerk *n* **~ gears** Antriebswerk *n* **~ head of casing** Futterrohrrammhaupt *m* **~ impedance** Antriebswiderstand *m* **~-in (of a nail)** Einschlag *m* **~-in mandrel** Eintreibdorn *m* **~ instructor** Fahrlehrer *m* **~ jaw** Antriebsklaue *f* **~ jet** Treibstrahl *m* **~-key coupling** Treibkeilverbindung *f* **~-key-type transmission** Ziehkeilgetriebe *n* **~ knob** Antriebsknopf *m* **~ lesson**

Fahrübung *f* ~ **lever** Antriebs-, Treib-hebel *m*
~ **lug (or tongue)** Mitnehmerzunge *f* ~ **magnet**
Antriebs-, Kraft-, Schalt-magnet *m* ~ **means**
Triebmittel *n* ~ **mechanism** Antriebs-, Trieb-
-werk *n* ~ **member** Antriebsteil *n*, Trieborgan
*n* ~ **motor** Antriebsmotor *m*, Triebmittel *n*,
Triebmotor *m* ~ **order** Fahrauftrag *m* ~-**out**
Überschreitung *f* (print.), weiter Satz *m*
~ **part** Triebwerksteil *m* ~ **pawl** Stoßklinke
*f* ~ **pedestal** Antriebsständer *m* ~ **pin** An-
triebsbolzen *m*, Treibzapfen *m* ~ **pinion**
Antriebsritzel *n*, Triebzahnrad *n* ~ **piston**
Treibkolben *m* ~ **plate** Anpreßplatte *f* ~ **platform** Rammebene *f* ~-**point admittance** Ein-
gangsadmittanz *f* ~-**point impedance** Eingangs-
scheinwiderstand *m* (acoust.) ~ **potential** (of
photocell) Saugspannung *f* ~ **power** Antriebs-,
Trieb-kraft *f* ~ **pulley** Antriebriemscheibe *f*,
Antriebsscheibe *f*, Scheibenschwungrad *n*,
Treibrolle *f*, Treibscheibe *f* ~ **pulley of beater**
Schlägerantriebsscheibe *f* ~ **pulse** Treiberim-
puls *m* ~ **quality** Fahreigenschaft *f* ~ **rack**
Zahnstange *f* ~ **rack for delivery carriage**
Seitenteil *n* für Presse ~ **range** Aktionsbereich
*m* ~ **record** Rammprotokoll *f* ~ **resistance**
Rammwiderstand *m* ~ **ring** Mitnehmerring *f*
~ **rod** Antriebs-, Pleuel-, Schub-stange *f* ~
**roller** Transport-, Treib-rolle *f* ~ **rope** Antriebsseil *n* ~ **safety** Fahrsicherheit *f* ~ **section**
Antriebselement *n* ~ **shaft** Antriebs-achse *f*,
-stange, -welle *f*, antreibende Welle *f*, Transmissionswelle *f*, Treibscheibe *f*, Verbindungswelle *f* ~-**shaft bushing** Antriebswellenbüchse *f*
~-**shaft coupling** Antriebswellenkuppelung *f*
~ **shield** Vortriebsschild *m* ~ **shocks (or vibration)** Fahrterschütterung *f* ~ **side** Antriebsseite
*f* ~ **snow** Schneegestöber *n* ~ **socket** Einramm-
glocke *f* ~ **speed (or r.p.m.)** Antriebsdrehzahl
*f* ~ **spindle** Antriebsspindel *f*, Bandantriebsachse *f* (tape rec.) ~ **sprocket** Antriebkettenrad *n*, Ketten-
triebrad *n*, Triebrad *n* ~ **stage** Vorstufe *f* ~
**starting pillar** Antriebsäule *f* ~ **step pulley** Antriebsstufenscheibe *f* ~ **string** Antriebsseil *n* ~
**surface** Mitnehmerfläche *f* ~ **system** Triebwerk
*n* ~ **tang** Mitnehmerlappenführung *f* ~ **time**
Fahrzeit *f* ~ **torque** Antriebsdrehmoment *n* ~
**trailer** Steuerwagen *m* ~ **(or power) transmission medium** Zugkraftglied *n* ~ **tube hole**
Antriebswelle *f* ~ **turn** gedrückte Kurve *f* ~
**type obstacle picket** Hindernisschlagpfahl *m* ~
**weight** Antriebs-, Uhr-gewicht *n* ~ **weight of a
pile driver** Bärgewicht *n* ~ **wheel** Triebrad *n* ~
**wheel for fliers** Fleierantriebsrädchen *n* ~
**wheel shaft** Tischantriebsradwelle *f* ~ **worm**
Antriebsschnecke *f*
**drizzle, to** ~ sprühen
**drizzle** Nieseln *n*, Nieselregen *m*, feiner Regen
*m*, Sprühregen *m*
**drizzling rain** Staubregen *m*
**drogue** Bremsschirm *m* (f. Schleudersitz), Fangtrichter *m*, Schleppziel *n*, Stabilisierungsschirm
*m*, Steuerschirm *m*, Stufenfallschirm *m*, Treibanker *m*, Wasseranker *m*, geschleppte Zielscheibe *f* ~ **chute** Hilfsschirm *m* ~ **gun** Hilfs-
schirmpatrone *f* (g/m) ~ **parachute** Bremsfallschirm *m*
**drometer** Drometer *n*

**drone, to** ~ dröhnen
**drone** Dröhnen *n*, Zielflugzeug *n*
**droogmansite** Droogmansit *m*
**droop, to** ~ abfallen, durchhängen, (of a curve)
sinken
**droop** Abfall *m*, Senkung *f*, Statik *f*, Wölbung *f*
~ **of muzzle of gun** Ausbuchtung *f* der Mündung oder Lippe ~ **or (permanently drooping)
voltage characteristic** (governor) dauernde Ungleichförmigkeit *f*
**drooped ailerons** herunterhängende Hilfsflügel
*pl*, Wanderquerruder *pl*
**drooping**, ~ **characteristic** abnehmende oder
fallende Charakteristik *f* ~ **characteristic line**
Neig-, Neigungs-kennlinie *f*
**drop, to** ~ abfallen, abnehmen, absenken, abwerfen, (a leaf) aufklappen, ausfahren, einfallen, (of wind) einlullen, fallen, (procedure)
fallen lassen, fällen, (a table leaf) herunterklappen, sinken, tröpfeln, verzichten **to** ~
**anchor** den Anker *m* fallen lassen **to** ~ **bombs**
Bomben *pl* werfen **to** ~ **the cage one level** den
Förderkorb *m* um eine Etage senken **to** ~ **down**
umklappen, vollständig abfallen **to** ~ **forge**
gesenkschmieden, im Gesenk *n* schmieden **to**
~ **in** eintröpfeln **to** ~ **off** abtropfen, vollständig
abfallen **to** ~ **out** (of relay) abfallen, ausfallen,
ausscheiden, entfallen, herausfallen **to** ~ **a
perpendicular** ein Lot *n* errichten oder fällen,
Senkkörper *m* fällen **to** ~ **the pilot** den Lotsen
*m* absetzen **to** ~ **stamp** schmieden im Gesenk *n*
**drop** Abfall *m*, Abfallen *n*, Abnahme *f*, Absturz
*m*, Ausfall (in current), Dämpfung *f*, Fall *m*,
(stage) Fallbühne *f*, Fallhöhe *f*, Fallklappe *f*
(teleph.), Gefälle *n*, Riemenfallhammer *m*,
(teleph.) Schauzeichen *n* (teleph.), Sinken *n*, Tropfen *m*
(water), Verlust *m*
**drop**, ~ **by** ~ tropfenweise ~ **of potential**
Potentialgefälle *n* ~ **in pressure** Druckabfall *m*
~ **of runway** Gefällstrecke *f* ~ **of a stamp (play)**
Spiel *n* eines Pochstempels *m* ~ **of starting fuel**
Zündtropfen *m* ~ **in temperature** Temperatur-
abfall *m*, Wärmegefälle *n*, Wärmeunterschied
*m* ~ **of water** Wassertropfen *m*
**drop**, ~ **action of pile driver** Auslösehaken *m*,
Auslöseschere *f* der Kunstramme *f* ~ **analysis**
Tüpfelanalyse *f* ~ **arm** Lenkstockhebel *m* ~
**arm shaft** Lenkhebelwelle *f* ~ **bar** Marken-
stange *f* ~ **barrel** Kipplauf *m* ~ **base rim** Tiefbettfelge *f* ~ **belly tank** Abwurfbehälter *m* ~
**block type breechblock** (or block breech mechanism) Fallblockverschluß *m* ~ **bottom** Bodenklappe *f*, Bodenplatte *f* ~ **bottom bucket**
Klappkübel *m* ~ **bottom cupola** Kupolofen *m*
mit Bodenklappe *f* ~ **box sley (weaving)** Wech-
sellade *f* ~ **bucket hoist** Hängekübelaufzug *m*
~ **casing** Fallröhren *pl* ~ **center rim** Tiefbettfelge *f* ~ **center wheel** Rad *n* mit Tiefbettfelge
~ **charge** Tropfenladung *f* ~ **chisel** Kuttermeißel *m* ~ **cock** Tropfenhahn *m* ~ **collector**
Tropfenkollektor *m* ~ **conduit** Tropfleitung *f*
~ **cord** Schalter *m* mit Schnur ~ **corn planter**
Tropfpflanzmaschine ~ **counter** Tropfenzähler
*m* ~ **distribution** (of a curve) Senkverteilung *f*
~ **door** Klapptür *f* ~ **door girder** Bodenklappen-
träger *m* ~ **drill** Horstsäemaschine *f* ~ **electrode**
Tropfenkollektor *m* ~ **fingers** Bogenhalter *m*
~ **forge** Gesenkschmiede *f* ~ **forge die** Schmie-

degesenk *n* ~ **forged** mit dem Fallhammer *m* geformt ~ **forging** Fallschmieden *n*, Gesenkschmieden *n*, Gesenkschmiederei *f*, Gesenkschmiedestück *n*, Warmpressen *n* ~ **forging block** Gesenkblock *m* ~ **gate sluice** Fallstütze *f* ~ **hammer** Fall-, Gesenk-, Gleis-, Rahmen--hammer *m*, Schmiedepresse *f* ~ **hammer die** Warmschmiedegesenk *n* ~ **hammer ram** Fallbär *m* ~ **hammer test** Fallhammerversuch *m* ~ **hammer-type pile driver** Zugramme *f* ~ **hanger** Hängerlader *n* ~ **hanger bearing box** Hängelagerschale *f* ~ **hanger frame** Hängebock *m*, Hängelager-bock *m*, -körper *m* ~ **hanger shoe** Hängelagerfuß *m* ~ **height** Fallhöhe *f* ~**-in** Störsignal *n* (tape rec.) ~**-in pin** Raststift *m*, Tastenfalle *f* ~**-in tape loading slot** Bandeinlegeschlitz *m* ~ **indicator** Anrufzeichen *n*, Fallklappe *f*, Klappenschrank *m*, Tropfenzeiger *m* ~ **indicator relay** Fallklappenrelais *n* ~ **indicator shutter** Anrufklappe *f* ~ **keel** Attrappengewicht *n*, Kielschwert *n* ~ **leaf** herunterklappbarer Tischflügel *m* ~ **light** Hängelicht *n* ~ **message** Abwurfmeldung *f* ~ **method** Tüpfelmethode *f* ~**-out** Abfallwert *m*, Beschichtungsloch *n* (tape rec.), Signalauffall *m* (tape rec.) ~ **panel** (in flat slab) Unterlagsplatte *f* ~ **pin earthing switch** Fallstifterdungsschalter *m* ~ **pin switch** Fallstiftschalter *m* ~ **plate** Tüpfelplatte *f* ~ **plate-type molding machine** Absenkformmaschine *f* ~ **point** Abwurfstelle *f* ~ **pouch** Melde-Abwurfhülle *f* ~ **press** Riemenfallhammer *m* ~ **(forge) press** Ziehpresse *f* ~ **reaction** Reaktion *f* beim Fall *m* ~ **scene** Zwischenaktvorhang *m* ~ **shaft** Einfall-, Senk--schacht *m* ~ **shape** Tropfenform *f* (streamline) ~ **shape general lighting lamps** Allgebrauchslampen *pl* in Tropfenform ~ **shot** Weichschrot *n* ~ **shutter** Fallscheibe *f*, Fallschieberverschluß *m*, Fallverschluß *m*, Guillotineverschluß *m* (photo), Klappblende *f* ~ **side** (of cart or truck) umlegbare Wand *f* ~ **side car** Gleiswagen *m* mit abklappbarer Längswand ~ **signal** Fallklappe *f* (teleph.) ~ **stamping** Preßling *m* ~ **staple** Blindschacht *m* ~ **starter** Fallanlasser *m* ~ **structure** Absturzbauwerk *n* (hydraul.) ~ **system of record changing** Fallsystemplattenwechsel *m* ~ **tank** Abwurfbehälter *m*, abwerfbarer Außentank *m* ~ **tap** Tropfenhahn *m* ~ **test** Abwurfversuch *m*, Fallhammerversuch *m*, Fallprobe *f*, Schlagprobe *f*, Schlagversuch *m*, Tropf- oder Tüpfelprobe *f* ~ **testing machine** Fallwerk *n* ~ **tin** Tropfzinn *n* ~ **type of sprag** Fallstütze *f* ~ **type switchboard** Klappenschrank *m* ~ **unit** Fallvorrichtung *f* (g/m) ~ **valve** hängendes Ventil *n* ~ **valve gear** Freifallsteuerung *f* ~ **weight** Fallkugel *f* ~ **-weight method** Tropfgewichts-, Tröpfchen--methode *f* ~ **weight service** Fallkugelbetrieb *m* ~ **weight test** Fallprobe *f* ~ **window** herablaßbares Fenster *n* ~ **wire** Einführungsdraht *m*, Sprechstelleneinführung *f* ~ **wires** (textiles) Falldrähte *pl* ~ **wise** tropfenweise ~ **work** Fallwerk *n* ~ **(or trip) worm** Fallschnecke *f* ~ **zinc** Tropfzink *n* ~ **zone** Absetzplatz *m* (von Fallschirmtruppen)

**droplet** Tröpfchen *n* ~ **growth** Tropfenwachstum *n* ~ **spectra** Tröpfchenspektra *n*

**droppable** abwerfbar ~ **starting assistance** ab-

werfbare Starthilfe *f* ~ **undercarriage** abwerfbares Fahrgestell *n* (aviat.)

**dropped down on hinges** vorschlagbar

**dropper** Ausläufer *m*, Eguttör *m*, Tropfenzähler *m*

**dropping** (bombs) Abwurf *m*, Bohrgerät *n*, (punch) Durchgehen *n*, Fällung *f*, Senkung *f* ~ **a message** Meldeabwurf *m* (aviat.) ~ **a stick of bombs** Reihenabwurf *m* ~ **of supplies from aircraft** Nachschubabwurf *m* ~ **of the voltage** Absinken *n* der Spannung

**dropping**, ~ **board** Abtropfbrett *n* ~ **bottle** Tropfflasche *f* ~ **characteristic** fallende Charakteristik *f* ~ **electrode** Tropfenelektrode *f* ~ **flank** Ablaufkurve *f* (aviat.) ~ **flask** Tropffläschchen *n* ~ **funnel** Tropftrichter *m* ~ **glass** Tropf-glas *n*, -röhre *f* ~ **mercury electrode** Quecksilbertropfelektrode *f* ~**-out** Lecken *n* ~ **piece** Fallstück *n* ~ **pipette** Tropfpipette *f* ~ **piston** Fallkolben *m* ~ **point** Tropfpunkt *m* ~ **range** Abwurfplatz *m* ~ **resistor** Vorwiderstand *m* ~ **time** Tropfungszeit *f* ~ **tube** Tropfröhre *f* ~ **way** Abwärtshub *m* (aviat.)

**droppings** Losung *f*

**drosometer** Taumesser *m*

**dross, to** ~ abheben, abziehen, entschlickern, schlickern

**dross** Abschaum *m*, Abstrich *m*, Abzug *m*, (iron) Braschen *pl*, Gekrätz *n*, Geschür *n*, Glätte *f*, Grus *m*, (of coal coke) Gruskohle *f*, Hartling *m*, Hüttenafter *n*, Nußgruskohle *f*, Schaum *m*, Schlacke *f*, Schlicker *m*, Skorie *f*, Trübe *f*, Unreinigkeit *f* in einer Flüssigkeit *f* ~ **of liquation** Krätzschlacke *f*

**dross**, ~ **dilution of ore** Gangart *f*, Ganggestein *n* ~ **plant** (or **works**) Abzugswerke *pl* ~ **sieving apparatus** Metallscheinsiebapparat *m*

**drosser** Schaumabheber *m*

**drosses** Gekrätzaschen *pl*

**drossing** Abziehen *n*, Schaumabheben *n*, Schlickarneit *f* ~ **operation** Schaumabheben *n*

**drossy** schlacken-ähnlich, -artig, schlackig

**drought** Dürre *f*, Trockenheit *f*

**drown, to** ~ ertrinken, (pumps) ersaufen, überfluten **to** ~ **(a noise)** übertönen

**drowned** ersoffen (min.), versoffen (min.), verwässert

**drowing** Übertönen *n*

**drum** Bandseiltrommel *f*, Baratte *f*, Benzinkanister *m*, Faß *n*, Fäßchen *n*, Flachseiltrommel *f*, Fördertrommel *f*, Kanister *m*, Rollkolben *m*, Schaltwalze *f*, Spule *f*, Trommel *f*, Walze *f* ~ **of a capital** Glocke *f* eines Kapitells ~ **with scanning apertures** Trommelblende *f* ~ **for steaming** Küpe *f* ~ **of winch** Windetrommel *f*

**drum**, ~ **armature** Anker *m* mit Trommelwicklung, Trommelanker *m* ~ **band brake** Trommelbandbremse *f* ~ **barker** Holzentrindungstrommel *f* ~ **barrage** Trommelwehr *n* ~ **barrow** Drahttrage *f* ~ **bearing** Trommellager *n* ~ **block** Trommelblock *m* ~ **cam** Steuertrommel *f* ~ **can** Benzinkanister *m* ~ **chamber** Trommelraum *m* ~ **charging valve** Flanschenventil *n* ~ **chart recorder** Trommelschreiber *m* ~ **cobber** Naß-, Walzen-separator *m* ~ **compass** Trommelkompaß *m* ~ **controller** Stellschalter *m*, Steuerwalze *f*, Walzenregler *m* ~ **dam** Trom-

melwehr *n* ~ **dial** Zugnummernschalter *m*
(teleph.) ~ **diaphragm** Trommelblende *f* ~
**drier** Walzentrockner *m* ~ **drier with cross-shaped internal distribution system** Trommeltrockner *m* mit Kreuzeinbau ~ **drier with dip tank** Tauchwalzentrockner *m* ~ **dumper** Trommelkippanlage *f* ~ **dyeing** Faßfärbung *f* ~ **edge** Trommelrand *m* ~ **feed** Trommelmagazinzuführung *f* ~ **feed spindle** Trommelvorschubspindel *f* ~ **governor** Flachregler *m* ~ **groove (or spiral)** Trommelrille *f*
**drumhead** Trommelfell *n*
**drum,** ~ **hoist** Hubwindwerk *n* ~ **index** Trommelzeiger *m* ~ **indicator** Trommelanzeiger *m* ~ **information assembler and dispatcher (DIAD)** Rufnummernspeicher und -geber *m* ~ **length** (of a cable) Fabrikationslänge *f*, Trommellänge *f* ~ **lens** Gürtellinse *f*, diotrische Linse *f* (film) ~ **lightning arrester** Walzenblitzableiter *m* ~ **lime** Faßäscher *m* ~ **memory** Trommelspeicher *m* ~ **mill** Trommelmühle *f* ~ **milling machine** Trommelfräsmaschine *f* ~ **mixer** zylindrische Mischtonne *f* ~ **motor** Zahnmotor *m* ~ **plate** Bremsabdeckblech *n*, Trommelblech *n* ~ **process** Trommelverfahren *n* ~ **radiator** Faß-, Trommel-kühler *m* ~ **receiver** Walzenempfänger *m* ~ **roll** Trommelwirbel *m* ~ **rotor** Trommelläufer *m* ~ **scanner** Trommelabtaster *m*, Spiegelkranz *m* ~ **screen** Siebtrommel *f* ~ **separator** Trommelsichter *m* ~ **shaft (or axis)** Trommel-achse *f*, -welle *f* ~ **shaped recorder** Trommelschreiber *m* ~ **shaped scale** Trommelskala (Autosuper) ~ **sheets filter** Trommelschichtenfilter *m* ~ **shutter** Trommelblende *f* ~ **sink float separator** Trommelsinkscheider *m* ~ **skin** (parchment) Trommelfell *n* ~ **speed** Trommeldrehzahl *f*, Trommelgeschwindigkeit *f* ~ **spider** Trommelstern *m* ~ **starter** Walzenanlasser *m* ~ **store** Trommelspeicher *m* (info proc.) ~ **stuffing** Faßschmiere *f* ~ **switch** Rasten-, Trommel-, Walzen-schalter *m* ~ **tanned** faßgar ~ **turret** Trommelrevolverkopf *m*
**drum-type,** ~ **air filter** Kesselluftfilter *n* ~ **alternator** Trommelankergenerator *m* ~ **cam** Trommelkurve *f* ~ **controller** Steuerwalze *f* ~ **dial** Trommelskala *f* (Autosuper) ~ **milling machine** Trommelfräsmaschine *f* ~ **starters and regulating starters** Anlaß- und Regelwalzen *pl* ~ **starting switch** Anlaßwalzenschalter *m*
**drum,** ~ **washer** Trommelwascher *m* ~ **wheel** Trommelrad *n* ~ **winding** Trommelwicklung *f* ~ **winding with diametral pitch** Durchmesserwicklung *f* ~ **winding with fractional (or shortened) pitch** Sehnenwicklung *f*
**drumminess** Dumpfheit *f*
**drummy** (speech) dumpf
**drumstick** Schlägel *m*, Trommelschlägel *m*
**drunken thread** schwimmendes Gewinde *n*
**druse** Druse *f*
**drused** nierenförmig
**drusy** drusig, nierenförmig ~ **cavity** Druse *f*, Drusenraum *m*
**dry, to** ~ abdarren, abtrocknen, auftrocknen, darren, dörren, trocknen, verdorren **to** ~ **in the air** anrauschen **to** ~ **cyanide** karbonitrieren **to** ~ **so that dust will not cling** staubtrocken **to** ~ **galvanize** sherardisieren **to** ~**-grind** trokkenzerkleinern **to** ~**-hard** (paper mfg.) über-

trocknen **to** ~ **at a high temperature** scharf trocknen **to** ~ **with a mop** dweilen, schwabbern **to** ~ **out** austrocknen **to** ~ **refine** trockenzerkleinern **to** ~ **a second time** nachtrocknen **to** ~ **without tackiness** klebfrei trocknen **to** ~ **thoroughly** durchtrocknen **to** ~ **up** auf-, aus-, ein--trocknen, verdorren **to** ~ **wood** (in a stove) Holz *n* dörren
**dry** Badekaue; dürr, niederschlagsarm, regenarm, trocken, (copper) übergar, wasserfrei ~ **adabatic lapse rate** trockene adiabatische Abweichungsrate *f* (aviat.) ~ **air plant** Trockenluftanlage *f* ~ **balance** trockenes Gleichgewicht *n* ~ **binder** Trockenbinder *m* ~ **blast plant** Windtrocknungsanlage *f* ~ **blast process** Windtrocknungsverfahren *n* ~ **bleaching** Trockenbleiche *f* ~ **bond strength** Trockenbindefestigkeit *f* ~ **bottom gas producer** Festrostgaserzeuger *m*, Gaserzeuger *m* mit fixem Rost ~ **box** Schutzkammer *f* ~ **bulb thermometer** trockenes Thermometer *n*, trockener Wärmegradmesser *m*, Trockenthermometer *n* ~ **bulk density** Trockenraumgewicht *n* ~ **cell** Füllelement *n*, Kohle-Zinksammler *m*, (battery) Trockenelement *n* ~ **cell battery** Anodentrockenbatterie *f* ~ **cleaner** Trockenreiniger *m* ~ **coal** Trockenkohle *f* ~ **compass rose** Trockenrose *f* ~ **compression-type machine** Trockenkompressionsmaschine *f* ~ **content** Trockengehalt *m* ~ **copper** übergares Kupfer *n* ~ **core cable** Hohlraumkabel *n*, Kabel *n* mit Luftraumisolierung *f*, Papierhohlraumkabel *n*, Papierkabel *n* ~ **crushing** Trockenmahlen *n* ~ **crushing roll** Trockenwalze *f* ~ **deposition** ungefährlicher Niederschlag *m* ~ **dial water meter** Trockenläufer *m* (Wasserzähler *m*) ~ **disc joint** Hardyscheibe *f*, Trockengelenk *n* ~ **distillation** Entgasung *f*, Trockendestillation *f*, Trockenentgasung *f*, Verkohlung *f*, Zersetzungsdestillation *f* ~ **distillation product** Schwelprodukt *n* ~ **dock** Trockendock *n* ~ **dock and locks** Dockschleuse *f* ~ **drilling** Trockenbohrung *f* ~ **dross** trockner Schaum *m* ~ **dust exhauster** trockner Staubsauger *m* ~ **electrolytic condenser** Trokkenkondensator *m* ~ **elongation** Trockendehnung *f* ~ **end of the paper machine** Trockenpartie *f* ~ **extract** Trockenanzug *m* (chem.) ~ **finishing machine** Trockenappreturmaschine *f* ~ **gas** Trockengas *n* ~ **gas meter** trockene Gasuhr *f* ~ **goods elevator** Trockenelevator *m* ~ **grinding** Trocken-mahlen *n*, -schliff *m*, -vermahlung *f* ~ **grinding machine** Trockenschleifmaschine *f* ~ **grinding paper** Trockenschliffpapier *n* ~ **grinding process** Trockenmahlverfahren *n* ~ **gum** Trockengummi *n* ~ **harbor** Strandhafen *m* ~ **hole blinde** (trockene) Bohrung *f* ~ **house** Trockenkammer *f* ~ **ice** feste Kohlensäure *f* ~ **joint** kalte oder schlechte Lötstelle *f* ~ **liner** Trockenzylinderlaufbüchse *f* ~ **liner-type cylinder** Zylinder *m* mit nicht benetzter Außenwand der Laufbuchse ~ **mat** Matrize *f* ~ **matter** Trockensubstanz *f* ~ **measure** Hohlmaß *n* ~ **mechanical wood pulp** trockener Holzschliff *m* ~ **meter** Hohlmaß *n* ~ **method granulation** Luftgranulation *f* ~ **mix of concrete requiring tamping** Stampfbeton *m* ~ **mounting** Trockenaufziehverfahren *n* ~ **net weight** Trockenleergewicht *n* ~ **objective** Trok-

kenlinse f ~ **open caisson** Senkbrunnen m ~ **operation** Trockenlauf m ~ **packing** Versatz m ~ **pavement** (laid without mortar) Trockenpflaster n ~ **pendant sprinkler** hängender Trokkensprinkler m ~ **phase** Trockenstadium n ~ **photovoltaic cell** Halbleiter-, Sperrschicht-fotozelle f ~ **pipe (sprinkler) system** Sprinklertrockenanlage f ~ **pipe (alarm) valve** Trockenalarmventil n ~ **pipe (alarm) valve set** Trockenalarmventilstation f ~ **plate** Trockenplatte f ~ **plate clutch** Trockenscheibenkupplung f ~ **plate rectifier** Trockengleichrichter m ~ **point engraver** Kaltnadelradierer m ~ **polishing** Trokkenpolieren n ~ **posting method** Trockenaufziehverfahren n ~ **pressed briquette** Trockenbriket n ~ **proofing paper** Abzugspapier n ~ **puddling** Trockenpuddeln n ~ **quencher gas** trockenes Löschgas n ~ **rectifier** Sperrschichtgleichrichter m ~ **rectifying cell** Trockengleichrichtzelle f ~ **reed relay** Relais n mit Schutzgaskontakten pl ~ **rendering** Ausschmelzen n (Fett) ~ **residue** Trockenrückstand m ~ **rot** Trockenfäule f ~ **rotted wood** brandiges Holz n ~ **rubble** Trockenmauerwerk n von Bruchstein ~ **rubble ashler** Trockenmauerwerk n von oder in behauenen Bruchsteinen ~ **run** Leerlaufprüfung f ~ **sand** ausgeglühter oder fetter Formsand m, Formmasse f, Masse f, (foundry) ausgeglühter Sand m, Streusand m, Trockensand m ~ **sand castings** Trockenguß ~ **sand mold** gebrannte Form f, Masseform f, Massenform f, Trockengußform f ~ **scrubbing** Trokkenwäsche f ~ **seed dressing** Trockenbeheizung f ~ **separation** trockene Scheidung f ~ **separator** Trockenscheider m ~ **shaver** elektrischer Rasierer m ~ **slag** kurze oder zähflüssige Schlacke f ~ **spell** Trockenperiode f ~ **spray and wet spray equipment** Puder- und Spritzeinrichtung f ~ **steam** Abdampf m, Trockenbohrung f, trockener oder überhitzter Dampf m ~ **steaming** Trockendekatur f ~ **steel mold** gebrannte Stahlgußform f ~ **stone form wall** Schalung(smauer) f von trockenen Steinen ~ **stone pitching** Stein-packung f, -satz m ~ **stone revetment** Bruchsteinschüttung f ~ **storing** Trockenaufspeichern n ~ **strength** Trockenfestigkeit f ~ **suction lift** Saughöhe f bei leerer Pumpe und Leitung ~ **sump** trockener Ölfluß m, Trockensumpf m ~ **sump lubrication** Trockensumpfschmierung f ~ **tack** Trockenklebrigkeit f ~ **treatment** trockenes Verfahren n ~ **tube (or pipe)** Trockenrohr n ~ **vat** Trockenbütte f ~ **wall** Steinpackung f ~ **walling** Bergemauer f ~**-way process** trockene Aufbereitung f ~ **weight** Leer-, Trocken-gewicht n ~ **weight of soil** Bodentrockengewicht n

**drying** Austrocknen n, Austrocknung f, Trocknung f ~ **of blast** Windtrocknung f ~ **by blowing** Trockenblasen n ~ **by circulating air** Zirkulationstrocknung f ~ **by graduated temperatures** Stufentrocknen n ~ **by heat** Heißtrocknung f ~ **by rays** Strahlentrocknung f ~ **(up)** Vertrocknung f

**drying**, ~ **accelerator** Trocknungsbeschleuniger m ~ **agent** Trocken-mittel n, -substanz f ~ **apparatus** Entwässerungs-, Trocken-apparat m ~ **apparatus with evaporation** Eindampfgerät n ~ **assembly** Trockeneinrichtung f ~ **bays** Trok-

kenfeld n ~ **cabinet** Trockenraum m ~ **chamber** Darrhaus n, Trockenraum m, Trockenschrank m, Trockner m ~ **closet** Trockenschrank m ~ **conveyer** Trockenförderer m ~ **cup with evaporation** Eindampfgerät n ~ **cylinder** Dampftrommel f, Trockenzylinder m ~ **cylinder in front of the machine** Vortrockenzylinder m ~ **cylinder developing process** Trockentrommelentwicklungsverfahren n ~ **device for rotors** Ankertrocknungsvorrichtung f ~ **drum** Trokkentrommel f ~ **effect** Trockenwirkung f ~ **equipment** Trockenanlage f ~ **fabrics** Entnässen n von Geweben ~ **frame** Trocken-gerüst n, -rahmen m ~ **furnace** Trockenofen m ~ **hearth** Trockenherd m ~ **horse** Winddocke f ~ **installation** Trockenanlage f ~ **kiln** Darrbühne f, Darre f, Sau f, Trocken-darre f, -ofen m ~ **loft** Trockenmansarde f ~ **machine** Trockenmaschine f ~ **medium** Trocknungsmittel n ~ **oil** Trockenfirnis m ~**-out** Abdämpfen n, Trocknen n ~ **oven** Dörrofen m, Trockenschrank m ~ **paper** Kopierlöschpapier n ~ **place** Aufhängeboden m, Darrhaus m, Trokkenboden m ~ **plant** Trockenanlage f, Trocknerei f, Trocknungsanlage f ~ **press** Auswindemaschine f ~ **process** Trockenverfahren n ~ **oven** Darrofen m; Trocken-kammer f, -ofen m, -schrank m ~ **rack** (tile) Trocken-gerüst n, -gestell n, -ständer m ~ **rolls** Auswindemaschine f ~ **room** Aufhängeboden m, Darre f, Darrkammer f, Trocken-boden m, -kammer f ~ **shed** Trockengerüst n ~ **sheet** Trockenblech n ~ **space** Trockenraum ~ **stove** Schwellendörrofen m, Trockenofen m ~ **substance** Trockenstoff m ~ **temperature** Trockentemperatur f ~ **through** Durchtrocknen n (Farbe) ~ **time** Trockenzeit f ~ **tower** Trockenturm ~ **tunnel** Trockendarre f, Trockenheit f

**dual** doppelt, dyadisch, zweifach ~ **of a tensor** dualer Tensor m

**dual**, ~ **access** Doppelzugriff m ~ **airpressure gauge** Doppeldruckluftmesser m ~ **amplification** gleichzeitige Hoch- und Niederfrequenzverstärkung f, Reflexverstärkung f ~ **beam cathode ray tube** Zweistrahl-Kathodenstrahlröhre f ~ **capacitance** Doppelkapazität f ~ **carburetor** Zwillingsvergaser m ~ **card** Verbundkarte f ~ **check valve** Doppel-, Zwillings--rückschlagventil n ~ **component** Dualkomponente f ~ **control(s)** Doppel-kontrolle f, -steuer n, -steuervorrichtung f; Zweifachsteuerung f ~ **control flying** Flug m am Doppelsteuer ~ **control heptode** doppelgesteuerte Heptode f ~ **cycle jet propulsion unit** Zweikreisstrahlenantrieb m ~ **directing means** (for routing purposes) Doppelweiche f ~ **disk brake** Zweischeibenbremse f ~ **drilling system** Doppelbohrsystem n ~ **drive** Doppelantrieb m ~ **element gear pump** Zwillingszahnradpumpe f ~ **feed tape carriage** doppelter Lochbandvorschub m ~ **fins and rudders** doppeltes Seitenleitwerk n ~ **float valve** Zwillingsschwimmerventil n ~ **function** doppelter Zweck m ~ **ignition** Doppelzünder m ~ **ignition system** Doppelzündungssystem n ~ **ignition magneto** Zwillingsmagnetzünder m ~ **instruction** Ausbildung f am Doppelsteuer ~ **ion** Zwitterion n ~ **lateral coil** Wabenspule f ~ **lattice** Mischgitter n ~ **pot**

**transducer** Doppelpotentiometer-Geber *m* ~ **potentiometer** Doppelpotentiometer *n* ~ **pre-reducer** Dualvoruntersetzer *m* ~ **printing** zweifache Schreibung *f* ~ **punch card** Verbundkarte *f* ~ **purpose** Mehrzweck *m* ~ **purpose illumination** Zweizweckbeleuchtung *f* ~ **ram vertical surface** Doppelaußenräummaschine *f* ~ **receiver** Reflexempfänger *m* ~ **reception** Röhrenempfang *m* mit gleichzeitiger Hoch- und Niederfrequenzverstärkung *f* ~ **role work** Doppelgängeraufnahme *f* ~ **shut-off valve** Zwillingsabspurventil *n* ~ **speed synchro** Fein-Grob-Drehmelder *m* ~ **strand winding** bifilare Wicklung *f*

**dualsyn** Drehmeldemotor *m* (gyro)

**dual,** ~ **take-up** Doppelauflauf *m* ~ **tandem design** Doppeltandemsatz *m* ~ **tandem wheel** Tandemdoppelrad *n* ~ **tire** Doppel-, Zwillings-reifen *m* ~ **tires** Doppelbereifung *f* ~ **track recorder** Doppelspurtonbandgerät *n* ~ **valve** Mehrfachröhre *f* ~ **valve spool** Vierwegesteuerschieber *m* ~ **wheel undercarriage** Doppelradfahrgestell *n*

**duality** Zweiheit *f* ~ **principle** (in the plane) Dualitätsprinzip *n*

**dually invariant configuration** dual-invariante Konfiguration *f*

**duant electrometer** Duantenelektrometer *m*

**duants** (cyclotron) Duanten *pl*

**dub, to** ~ nachsynchronisieren (film), (recording) schneiden, synchronisieren (film)

**dub** Kopie *f* (tape rec.)

**dubbing** Lederfett *n*, Nachsynchronisieren *n*, Nachvertonung *f*, Rückspielen *n*, Tonmischung *f* (acoust.), Überspielen *n* (tape rec.) ~ **sound** Tonmischung *f* ~ **studio** Synchronstudio *n*

**duck, to** ~ sich ducken

**duck** Bindetuch *n* (Riemen), Ente *f*, Segeltuch *n* ~ **board** Knüppel-, Latten-rost *m*, Rost *m*, Spurtafel *f* ~ **foot** Flachschneidehackmesser *m*, (agr. nach.) Gänsefußschar *f* ~ **foot electrode** Entenfußelektrode *f* ~ **kill sprinkler** Schmelzlotsprinkler *m* ~ **nose bit** halbelliptischer Löffelbohrer *m*

**duct** Durchführung *f*, Gang *m*, Dukt *m*, Kabelrohrstrang *m*, Kanal *m*, Kanalzug *m*, Kühlluftführung *f*, Leitkanal *m*, Leitung *f*, Rohr *n*, Röhre *f*, (wave prop.) Leitschicht *f*, Stollen *m* (for pipe lines), Wellenleiter *m* (electr.) ~ **that gives tidal air effect** Vorlagerungsrohr *n*

**duct,** ~ **area** Querschnittfläche *f* der Kühlluftführung *f* ~ **artificial line** Nachbildung *f* (telegr.) ~ **bank** Kabelkanal *m* ~ **blade** Duktorlineal *n* ~ **bolt** Durchführungsbolzen *m* (electr.) ~ **cleaner** Rohrbürste *f* ~ **cleaning tool** Rohrbürste *f* ~ **cooling** Kühlung *f* durch Düsenkühler ~ **edge shield** Kabeleinziehtülle *f*, Kabelverlegung *f*, Schleifbogen *m* ~ **entrance** Kabelzugöffnung *f* ~ **plug** Abdichtstöpsel *m* (für Kabelkanäle) ~ **rod** Schiebegestänge *n* ~ **rodding** Einführen *n* in einen Kanalzug ~ **roller** Farbkastenwalze *f* ~ **route** Kabelkanalführung *f* ~ **run** Kabelkanalsystem *n* ~ **sealing** Abdichten des Kanalzuges ~ **switch** Duktorschaltung *f* ~ **type fan** Luttenventilator *m* ~ **ventilation** Luttenwetterung *f* ~ **width** Kanalbreite *f* ~ **work** Kabelkanal *m*, Leitungen *pl*

**ducted** ummantelt ~ **cooling system** Düsenkühlung *f* ~ **fan** Düsenfächer *m*, Mantelschraube *f* (Flugzeug) ~ **fan jet engine** Zweikreisturbinenluftstrahlwerk *n*, ZTL-Triebwerk *n* ~ **fan powerplant** Zweistromtriebwerk *n* ~ **radiator** Düsenkühler *m*

**ductile** ausdehnbar, biegsam, biegungsfähig, bildsam, dehnbar, ducktil, geschmeidig, nachgiebig, spannbar, streckbar, zähe, ziehbar ~ **fracture** Formungsbruch *m*

**ductilimeter** Dehnbarkeitsmesser *m*, Dehnungsmesser *m*

**ductility** Ausziehbarkeit *f*, Biegsamkeit *f*, Dehnbarkeit *f*, Dehnungsvermögen *n*, (test) Ducktilität *f*, Fälligkeitstermin *m*, Formänderungsfähigkeit *f*, Formbarkeit *f*, Geschmeidigkeit *f*, Kaltbildsamkeit *f*, Spannbarkeit *f*, Streckbarkeit *f*, Verformungsfähigkeit *f*, Zähigkeit *f*, Ziehbarkeit *f*, Ziehgrad *m* ~ **testing machine** Blechprüfapparat *m* ~ **value** Ziehwert *m*

**ductilize, to** ~ biegsam machen, (wire) dehnbar machen

**ductilized filament** biegsamer Glühfaden *m*

**ducting of the air** Kühlluftführung *f* in Kanälen

**ductor** (textiles) Abstreichmesser *m* ~ **key** Farbklotz *m* ~ **knife screw** Farbmesserschraube *f* ~ **movement** Farbgebung *f* (print.) ~ **roller** Duktor *m*

**dud** Blindgänger *m*, Bodenkrepierer *m*

**Duddell arc** sprechende Bogenlampe *f*

**due** angemessen, fällig, gebührend, gehörig, schuldig ~ **to** beruhen auf, herrührend ~ **to official order** aus dienstlicher Veranlassung *f*

**due,** ~ **bill** Gutschein *m* ~ **date** Fälligkeitsdatum *n*, Verfalltag *m* **with** ~ **regard for** mit Rücksicht *f* auf

**dues** Abgabe *f*, Gebühr *f*

**duff** Feinkohle *f*

**duffel bag** Bekleidungs-, Kleider-sack *m*

**duffer** Stümper *m*

**dufrenite** Dufrenit *m*, Grün-eisenerde *f*, -eisenstein *m*, Kraurit *m*

**dugout** Dachsbau *m*, Stollen *m*, Unterschlupf *m*, Unterstand *m*

**dug,** ~ **peat** Handtorf *m* ~ **well** Schachtbrunnen *m*

**dulcifiant** Süßstoff *m*

**dulcifying material** Zymbal *n*

**dulcin** Dulzin *n*

**dull, to** ~ abstumpfen, anschärfen, ausschärfen, mattieren, stumpf machen

**dull** abgestumpft, düster, duff, dumm, dumpf, erloschen, glanzlos, (sound or voice) hohl, leblos, matt, schwach beheizt, still, stumpf, träge, trübe

**dull,** ~ **black** mattschwarz ~ **blue** mattblau ~ **chrome-plated** matt verchromt ~ **clear varnish** Hartmattlack *m* ~ **coal** Mattkohle *f* ~-**edged** ungesäumt ~-**edged timber** wahnkantiges Holz *n* ~ **emitter** Kaltkathode *f* ~ **emitter tube** Oxydkathodenröhre *f* ~ **emitter valve** Röhre *f* mit dunkelrotglühendem Faden ~ **emitting cathode** Sparkathode *f* ~ **finish** Mattglanz *m* ~ **finish sprint** Mattabzug *m* ~ **(or weak) luster** schwacher Glanz *m* ~ **pickling** Mattbrennen *n* ~ **red** dunkelrot, dunkle Rotglut *f*, mattrot ~ **red heat** Dunkelrotglut *f* ~ **red hot** dunkelrotglühend ~ **season** Sauregurkenzeit *f* ~ **sound**

hohler Ton *m* ~ **sound due to blooping patch** Klebestellengeräusch *n* ~ **(emitting) valve** schwach beheizte Röhre *f* ~ **white** mattweiß
**dulled** abgestumpft, mattiert ~ **by the addition of a pigment** pigmentmattiert ~ **(or beveled) edge** Wahnkante *f*
**dulling** Dulling *n*, Mattierung *f*, Stumpfwerden *n* (Werkzeug) ~ **action** Abstumpfwirkung *f*
**dullness** Dumpfheit *f*, Glanzlosigkeit *f*, Stille *f*, Stumpfheit *f* ~ **of business** Geschäftsstille *f* ~ **of hearing** Schwerhörigkeit *f*
**dumb,** ~ **antenna** verstimmte Antenne *f* ~ **barge** Schleppkahn *m* ~**-bell** Hantel *f*, Stäbchenform *f* ~**-bell slot** hantelförmiger Schlitz *m* ~ **contact** Leerlaufkontakt *m* ~ **iron** Federhand *f* (vorne und hinten) ~ **piece** unganze Platte *f*
**dumbo** Abtasten *n* der Meeresoberfläche vom Flugzeug aus
**dumies** Summationsindices *pl*
**dummy** belanglos (info proc.), blind, falsch, fingiert, künstlich
**dummy** (parachute) Attrappe *f*, (of a book) Blindband *m*, Blindmuster *n*, Ersatz *m*, Formatbuch *n* (print.), Puppe *f*, Probeband *m* (print.) ~ **airport** Scheinflughafen *m* ~ **airscrew** Bremsluftschraube *f* ~ **antenna** Antennenersatzstromkreis *m*, Ersatzantenne *f*, künstliche Antenne *f* ~ **block** Preß-, Vorlege-scheibe *f* ~ **diffuser** Scheindiffusor *m* ~ **drift** Blindort *n* ~ **engine** Motorattrappe *f* ~ **figure** Strohpuppe *f* ~ **fuse** Blindsicherung *f* ~ **harbor** Scheinhafen *m* ~ **hole** Blindloch *n* (tape rec.) ~ **instruction** Scheinbefehl *m* ~ **jack** Blindbuchse *f* ~ **letter** Füllbuchstabe *m* ~ **line** Blindleitung *f* ~ **load** künstliche Antenne *f* ~ **lug** freie Lötöse *f* ~ **obstacle** Scheinsperre *f* ~ **packing** Blindortversatz *m* ~ **pass** totes Kaliber *n*, (rolling mill) Blind-stecker *m*, -stich *m* ~ **payments** Scheinzahlungen *pl* ~ **piece** Blindstück *n* (Pumpe) ~ **piston** (turbine) Ausgleichkolben *m*, Blind-kerze *f*, -stopfen *m*, -stöpsel *m*, Druckausgleichkolben *m*, (steam turbine) Gegendruckkolben *m* ~ **plunger** Blindkolben *m* der Ölschmierpumpe ~ **port** Scheinhafen *m* ~ **rivet** Füll-, Heft-niet *n* ~ **road** Blindstrecke *f* ~ **roll** Schleppwalze *f* ~ **section** Ansatzschrank *m* (teleph.), Kabelschrank *m* ~ **sheet** Tragblech *n* ~ **signal** Scheinfunk *m* ~ **socket** Blindbuchse *f*, Steckdosenhalteblech *n* ~ **spire** tote Windung *f* ~ **statement** Leeranweisung *f* (info proc.) ~ **ticket** Blind-, Zahl-zettel *m* ~ **turn** tote Windung *f* ~ **works** Scheinanlagen *pl*
**dumontite** Dumontit *n*
**dump, to** ~ abkippen, abladen, abstürzen, ausstürzen, entladen, entleeren, fallen lassen, herunterwerfen, stürzen, umkippen **to** ~ **out** auskippen, ausstürzen **to** ~ **overboard** über Bord werfen
**dump** Ablage *f*, Depot *n*, Kippe *f*, Lager *n*, Lagerplatz *m*, Magazin *n*, Park *m*, Schutthaufen *m*, Spannungsunterbrechung *f*, Stapelplatz *m*, Sturzlager *n* ~ **above track level** Hochhalde *f* ~ **bailer** Zementierschlammbüchse *f* ~ **barrow** Kippkarren *m*, Muldenkippwagen *m* ~ **body** (truck) Kippkastenkarosserie *f* ~ **body trailer** Kippanhänger *m* ~ **bucket** Kippkastengehänge *n*, Kippkübel *m* ~ **car** Abwurfwagen *m*, Kipp-lore *f*, -wagen *m* ~ **car tipping to**

**either side** zweiseitiger Kippwagen *m* ~ **cart** Entlader *m*, Kippkarre *f*, Lore *f* ~ **check** Abwurfkontrolle *f* ~ **crank** Abwurfkurbel *f* ~ **goods** Rutschgut *n* ~ **heap** Schutthaufen *m* ~ **house** (rubbish) Kohlenbunker *m* ~ **iron** Federarm *m* ~ **site** Schuttabladeplatz *m* ~ **slag** Haldenschlacke *f* ~ **transporter** Lagerplatzbrücke *f* ~ **truck** Kippwagen *m*, Muldenkippwagen *m* ~ **truck for transporting barren rock** Abraumabsetzmaschine *f* ~ **type coal barrow** Kohlankippkarre *f* ~ **type hopper truck** Muldenkippwagen *m* ~ **type lift truck** Hubkippwagen *m* ~ **valve** Abladehahn *m*, Bodenventil *n*, (Schnell)-Ablaßventil *n*
**dumped material** Haldenmaterial *n*
**dumper** (blast furnace) Gichtmann *m*, Muldenkippwagen *m*
**dumping** Abladen *n*, Abstürzen *n*, Ausschüttung *f*, Entleerung *f* (Greifer), Kippen *n*, Schnellablaß *m*, Umkippen *n* ~ **into excavation** Innenkippe *f* ~ **and sacking scale** Ausschütt- und Absackwaage *f*
**dumping,** ~ **bridge** Sturzbrücke *f* ~ **cage** Förderkorb *m* ~ **excavated material** Bodenkippen *n* ~ **grate** Aschfallklappe *f* ~ **grid** Einwurfrost *m* ~ **ground** Haldensturz *m*, Sturzplatz *m* ~ **height** Ausschütt-, Lade-, Sturz-höhe *f* ~ **jettisoning of fuel** Schnellentleerung *f* des Brennstoffs *m* (aviat.) ~ **kiln** Klapphorde *f* ~ **level** Höhenlinie *f* an der das Abstürzen stattfindet, Kippstrosse *f* ~ **mechanism** Kippvorrichtung *f* ~ **platform** Kippbühne *f* ~ **point** Abwurfstelle *f* ~ **radius** Ausschüttweite *f* ~ **tip** Hochkippe *f* ~ **type tank** Leerlaßbehälter *m* ~ **waggon** Selbstentladewagen *m* ~ **width** Ladeweite *f* (Bagger)
**dun, to** ~ Zahlung *f* anfordern
**dune** Düne *f* ~ **protection** Dünenbau *m*
**dung** Dreck *m*, Dung *m*, Losung *f* ~ **water** Jauche *f*
**dunking,** ~ **circuit** Schwarzsteuerkreis *m* ~ **period** Schwarzsteuerzeit *f*
**dunnage** Unterlegbohlen *m*
**dunning,** ~ **letter** Mahnbrief *m* ~ **proceedings** Mahnverfahren *n*
**duo,** ~ **circuit** Duoschaltung *f* ~ **clad metal** plattiertes Blech *n* ~ **cone loudspeaker** Doppelkonuslautsprecher *m* ~ **control airplane** Doppelsteuerflugzeug *n* ~ **decimal** zwölfteilig ~ **decimal notation** duodezimale Schreibweise *f*
**duodiode** Doppeldiode *f*, Doppelrichtröhre *f*, Doppelzweipolröhre *f*, Duodide *f* ~ **pentode** Bidipentode *f* ~ **rectifier** Zweifachdiodengleichrichter *m*, Zweifachzweipolröhre *f* ~ **tetrode with suppressor grid** Bidipentode *f* ~ **tetrode tube** Doppelzweipolvierpolröhre *f* ~ **triode** Biditriode *f* ~ **triode tube** Doppelzweipoldreipolröhre *f* ~ **valve** Zweifachdiodengleichrichter *f*, Zweifachzweipolröhre *f*
**duo filter** Duofilter *m* (Zweifach-Kraftstoff-Filter)
**duolateral,** ~ **coil** Wabenspule *f* ~ **winding** Wabenwicklung *f*
**duo,** ~ **rolls** Zwillingswalzen *pl* ~ **service** doppelte Bestätigung *f* ~**-tricenary notation** duotrizinäre Schreibweise *f* ~ **triode** Doppeltriode *f*
**dupe, to** ~ anführen

dupe Duplikat *n* (film) ~ **negative** Duplikat-
negativ *n* ~ **positive** Duplikatpositiv *n*
duping Doppelung *f* (film) ~ **process** Dupver-
fahren *n* (film)
**Dupin's,** ~ **cyclids** Dupinische Zyklen *pl* ~
**theorem on triply orthogonal system of surfaces**
Duplinischer Satz *m* über dreifach orthogonale
Flächensysteme *pl*
duplet Dublett *n*
**duplex, to** ~ duplex betreiben, duplizieren, gegensprechen (teleph.)
**duplex** Duplexverbindung *f*, viererverseiltes Kabel *n*, doppelt ~ **artificial circuit** Duplexnach-
bildung *f* ~ **balance** Brückengleichgewicht *n*,
Duplexabgleich *m* ~ **beating engine** Doppel-
holländer *m* ~ **blasting ignition cable** Doppelsprengkabel *n* ~ **board** Duplexkarton *m* und
-pappe *f* ~ **cable** Kabel *n* mit Viererverseilung,
viererverseiltes Kabel *n* ~ **carburetor** Doppel-
vergaser *m* ~ **circuit** Duplexleitung *f*; Gegen-
sprech-leitung *f*, -schaltung *f*; Viererkreis *m* ~
**clutch** Zweiwegkupplung *f* ~ **coils** Brückenarm
*m* ~ **conductor lead wire** Zweileiterkabel *n* ~
**conductor wire** Doppelleitung *f* ~ **connected**
gegengeschaltet ~ **connection** Gegenschaltung
*f*, Gegensprechschaltung *f* ~ **dusting equipment**
Duplexentstäubungseinrichtung *f* ~ **frame**
(motorcycle) Doppelrohrrahmen *m* ~ **governor**
Doppelregler *m* ~ **halftone** Duplexautotypie *f*
~ **halftone photography** Duplexrasteraufnahme
*f* ~ **installation** Gegensprecheinrichtung (teleph.) ~ **key** Gegensprechtaster *m* ~ **mirror**
Hilfsspiegel *m* ~ **nozzle** Zweimengendüse *f*,
Zwillingsdüse *f* ~ **operation** Duplexbetrieb *m*,
Duplexverkehr *m*, Gegensprechbetrieb *m*,
Gegenverkehr *m*, Viererbetrieb *m*, Zwischenhörbetrieb *m* ~ **operation on a single frequency**
Gegenverkehr *m* auf gleicher Frequenz ~
**paper** doppelseitiges oder zweifarbiges Papier
*n* ~ **piston rings** Duplexkolbenringe *pl* ~
**plunger lubricator** Dampfschmierapparat *m* ~
**pressure gauge** Doppeldruckmesser *m* ~ **printing machine** Doppeldruckmaschine *f* ~ **process**
Duplexverfahren *n*, Gegensprechen *n* ~ **pump**
Duplexpumpe *f*, doppelte Pumpe *f* ~ **radio
traffic** Funkwechselverkehr *m* ~ **repeater** Dop-
pelstromgegensprechübertragung *f*, Gegen-
sprechübertragung *f*, Zweiwegeverstärker *m* ~
**rings** Zweierschalen *pl* ~ **set** Gegensprechsatz
*m*, Gegensprechschaltsatz *m* ~ **shaking table**
Doppelplanrätter *m* ~ **shot** Doppelaufnahme *f*
~ **slip** Doppelgleitung *f* ~ **sparking** Doppel-
funke *m* ~ **star connection** Doppelsternschal-
tung *f* ~ **steam-force pump** Duplexdampfpumpe
*f* ~ **steel** Duplexmetall *n* ~ **structure** Doppel-
gefüge *n* ~ **switch** Doppelschalter *m* ~ **system**
Doppelbetrieb *m*, Gegensprechsystem *n*, Zwei-
fachbetrieb *m* (teleph.) ~ **system of writing**
Gegenschreiben *n* ~ **telegraphy** Gegensprechen
*n*, Gegensprechtelegrafie *f* ~ **telephone connection** Doppelsprechschaltung *f* ~ **(or double-
deck) tool holder** Aufbaustahlhalter *m* ~ **traffic**
doppelseitiger oder doppelgerichteter Verkehr
*m* ~ **valve** Doppelventil *n* ~ **variable-area track**
Abdeckdoppelzackenspur *f*, Doppelzackenschrift *f* mit Abdeckung *f* ~ **(or two) way**
**traffic** Gegensprechverkehr *m* ~ **working** Dop-
pelverkehr *m*, Duplexbetrieb *m*, Duplexverkehr

*m*, Gegenschreibbetrieb *m* (telegr.), Gegenverkehr *m*
**duplexer** Duplexer *m*, Sende/Empfangsweiche *f*
(rdr), Simultanantenne *f*
**duplexing** Gegensprechen *n*, Verbundbetrieb *m*
mit zwei verschiedenen Öfen *pl* ~ **process** Ver-
bundverfahren *n*
**duplicate, to** ~ doppeln, doppelt ausführen,
dubbeln, dubben, duplizieren, nachahmen,
pausen, verdoppeln
**duplicate** Duplikat *n*, Kopie *f*, Pauszeichnung *f*,
Zweitschrift *f* **in** ~ abschriftlich, in zweifacher
Ausfertigung *f* oder Ausführung *f* ~ **casting**
Massenguß *m* ~ **circuitry** doppelte Schaltkreise
*pl* ~ **copy** Durchschrift *f* ~ **determination** Kontrollbestimmung *f*, Parallelversuch *m* ~ **jack**
Parallelimpedanzklinke *f* ~ **lens** Ersatzglas *n*
~ **pad** Durchschreibeblock *m* ~ **part** Austauschstück *n* ~ **piece** Reihen-, Serien-teil *m* ~
**production** Reihenherstellung *f* ~ **sample** Kontrollprobe *f*
**duplicating** Doubeln *n* ~ **apparatus** Vervielfältigungsapparat ~ **book** Durchschreibebuch *n*
~ **device** Wiederholungseinrichtung *f* ~ **emulsion** Dupemulsion *f* ~ **ink** Vervielfältigungsfarbe *f* ~ **milling machine** Kopier-, Nachform-
-fräsmaschine *f* ~ **negative** Dupnegativ *n* ~
**paper** Saugpostpapier *n* ~ **positive stock** Dup-
positivfilm *m* ~ **rack** Kartenbell *n* (Haupt-
kartenbett)
**duplication** Duplizierung *f*, Verdoppelung *f* ~
**of effort** Doppelarbeit *f* ~ **check** Duplikatver-
gleich *m*, Zwillingskontrolle *f* (info proc.)
**duplicator** Vervielfältiger *m*
**dup-negative** Dup-Negativ *n*
**duprite** Rotkupfererz *n*
**durability** Bestand *m*, Beständigkeit *f*, Bewährung *f*, Dauerhaftigkeit *f*, Lebensdauer *f* ~ **in
wear** Gebrauchstüchtigkeit *f*
**durable** beständig, dauerhaft, haltbar, nachhaltig, stark ~ **dotfastener** Druckknopf *m*
**durain** Mattkohle *f*
**duralumin** Dural *n*, Duralumin *n*, Ganzmetall *n*
~ **covering** Duralbeplankung *f* ~ **framework
of a wing** Flügelduralumingerüst *n* ~ **fuselage**
Duraluminrumpf *m* ~ **rib** Duraluminrippe *f* ~
**rivet** Duralniete *f* ~ **section** Duraluminprofil *n*
~ **spar** Duralholm *m* ~ **tube** Duralrohr *n*
**duration** Bestand *m*, Dauer *f*, Verweilzeit *f*,
Zeitdauer *f*, Zeitperiode *f*
**duration,** ~ **of approach flight** Anflugzeit *f* ~ **of
blow** Schlagdauer *f* ~ **of boiling (or brewing)**
Sudzeit *f* ~ **of burning** Brenndauer *f* ~ **of a call**
Gesprächsdauer *f* ~ **of collision** Stoßdauer *f*
~ **of connection** (operation) Einschaltdauer *f*
~ **of cut** Schnittdauer *f* ~ **of the dip** Zugdauer *f*
~ **of effect** Wirkungsdauer *f* ~ **of exposure** Be-
lichtungsdauer *f*, Belichtungszeit *f*, Expositionszeit *f* ~ **of filling** Füllungsdauer *f* ~ **of heat**
Chargendauer *f*, Schmelzdauer *f* ~ **of image
spot** Bildpunktdauer *f* ~ **of impulse** Impulsdauer *f* ~ **of injection** Einspritzdauer *f* ~ **of lead
application** Einschaltdauer *f* ~ **of load application** Belastungsdauer *f* ~ **of message in minutes**
Gesprächsminuten *pl* ~ **of passage** Durchzugs-
dauer *f* ~ **of pouring** Gießdauer *f* ~ **of the
principle pressure** (of a shock wave) Haupt-
druckdauer *f* ~ **of shock** Stoßdauer *f* ~ **of the**

**stability of a disperse system** Aufrahmzeit *f* **~ of storage** Aufbewahrungsdauer *f* **~ of sunshine** Sonnenscheindauer *f* **~ of test** Versuchsdauer *f* **~ of tool edge** Standfestigkeit *f* der Werkzeugschneide *f* **~ of transmission** Lauf-, Übertragungs-zeit *f* **~ of voltage application** Einschaltdauer *f*

**duration, ~ flight** Dauerflug *m* **~ process** (in phosphorescence decay) Dauerprozeß *m*

**durometer** Härteprüfer *m*, Härteprüfgerät *n*

**dusk** Dämmerung *f*, Dunkel *n*, Zwielicht *n*

**dust, to ~** (a coat or film) aufstäuben, ausstäuben, bestäuben, einpudern, einstauben, stäuben, streuen **to ~ off** abstäuben, abputzen, ausblasen, entstauben

**dust** Abtrieb *m*, Kehricht *m*, Mehl *n*, Nebel *m*, Staub *m* **~ and spark chamber** Staub- und Funkenkammer *f* (Kupolofen) **~ blower** Staubbläser *m* **~ box** Putzkasten *m* **~ bunker** Staubbunker *m* **~ cap** Schutz-, Staub-deckel *m*; Staubkapsel *f*, Staubschutzsieb *n* (Hupe) **~ carryover** Mitreißen *n* von Staub **~ cart** Müllwagen *m* **~ catcher** Schlacken-kammer *f*, -kasten *m*; Staub-abscheider *m*, -behälter *m*, -kammer *f*, -sack *m* **~ chamber** Flugstaubkammer *f*, Staubfänger *m*, Staubsammelraum *m* **~ cloth** Staubtuch *n* **~ coal** staubförmige Kohle *f* **~ coke** Koksabrieb *m* **~ collecting** Staubabscheidung *f* **~ collecting appliance** Entstaubungsvorrichtung *f* **~ collecting chamber** Entstaubungskammer *f*, Staubkammer *f* **~ collecting equipment** Entstaubungsanlage *f* **~ collection** Entstaubung *f*, Staubabsaugung *f* **~ collector** Staub-abscheider *f*, -fänger *m*, -sammler *m*, -sack *m* **~ color** Staubfarbe *f* **~ content of the air** Staubgehalt *m* **~ core** Eisenstaub-, Staub-, Pulver-kern *m* **~ counter** Staub-meß-gerät *n*, -zähler *m* **~ cover** Staub-decke *f*, -kappe *f* (oil cloth) Wachstuchhaube *f* **~ culm** Lösche *f* **~ deposit** Staubablagerung *f* **~ devil** Staubwirbel *m*, Staubwirbelwind *m* **~ duct** Staubabzug *m* **~ excluder** Vorschußkappe *f* **~ exhaust** Entstaubung *f*, Staubsaugung *f* **~ exhauster** Staubsauger *m* **~ exhausting device** Staubsaugevorrichtung *f* **~ extraction capacity** Abscheideleistung *f* **~ extraction plant** Entstaubungsanlage *f* **~ extractor** Ausblaseapparat *m* **~ fan** Staubventilator *m* **~ figures** Staubfiguren *pl* **~ filter** Staubfilter *m* **~ filtering mask** Staubmaske *f* **~ formation** Staubbildung *f* **~-free** staubfrei **~ fuel-fired** brennstaubbefeuert **~ guard** Staubschutz *m* **~ haze** Staubnebel *m* **~ hood** Staub-haube *f*, -kappe *f*, Verdeckhülle *f* **~ kit** Bestäubungsvorrichtung *f* **~-laden** staub-erfüllt, -geschwängert, -haltig **~-laden air** Staubluft *f* **~ laden air receiver** Staubluftraum *m*, staubhaltige Luft *f* **~ loss** Verstäubungsverlust *m* **~ mask** Staubschutzmaske *f* **~ mass** Staubmasse *f* **~ mixture** Staubgemisch *m* **~ noise** Staubrauschen *n* (of film) **~ nuisance** Staub-belästigung *f*, -plage *f* **~ ore** Mulm *m* **~ pan** Kehrschaufel *f* **~ plug** Abschlußklappe *f* **~ pocket** Staubsack *m* **~-poor** staubarm

**dust-proof** staubdicht, staubsicher **~ cover** staubdichte Kappe *f*, Staubschutzabdeckung *f* **~ housing** staubdicht gekapselte Verschalung *f*

**dust, ~ protecting bellows** Staubschutzbalg *m* **~**

**protecting envelope** Staubschutzhülle *f* **~ reclaiming mill** Staubseparationsanlage *f* **~ removal** Entstaubung *f*, Staubabsaugung *f* **~ removal with recirculation** (of clean air) Entstaubung *f* mit Luftrückführung *f* **~ seal** Staubdichtung *f* **~ seal washer** Staubdichtscheibe *f* **~ separating appliance** Entstaubungsvorrichtung *f* **~ separation** Entstaubung *f*, Staubabscheidung *f*, Staubseparation *f* **~ separation method** Staubabscheideverfahren *n* **~ separator** Staubabscheider *m* **~ shield** Staubschild *n* **~ sludge** Staubschlamm *m* **~ squall** Staubbö *f* **~ supply** Staubzufuhr *f* **~ tight** staubdicht **~ trap** Staubfänger *m* **~ trapping agent** Staubbindemittel *n* **~ washer** Staubschutzblech *n* (Lager) **~ whirlwind** Staubwirbelwind *m*

**duster** (tan liquor) Abtränkbrühe *f*, Haderndrescher *m*, Lappen *m*, Wischtuch *n*, Staubanzug *m*, staubstreuendes Flugzeug *n*, Versteckfarbe *f* **~ airplane** staubstreuendes Flugzeug *n*

**dusting** Einstreumaterial *n*, Entstauben *n*, Staubbildung *f*, Stäuben *n*, Streuen *f*, Verstäuben *n* **~ apparatus** Bestäubungsapparat *m* **~ arrangement** Streuvorrichtung *f* **~ band** Abstaubband *n* **~ box** Einstaubkasten *m* **~ brush** Staubpinsel *m* **~ loss** Stäubungsverlust *m* **~ machine** Puderapparat *m* **~ material** Streugut *n* **~ willey** Staubzyklop *m*

**dustless** staubfrei

**dustlike** staubartig, pulverartig, pulverig

**dusty** staubartig, staubförmig, staubig

**Dutch, ~ bond** Kreuzverband *m* **~ clinker** Klinker *m* **~ foil (or gold)** Metallgold *n* **~ gold** Rauschgold *n* **~ man** Spülbohrkopf *m* **~ metal** Schaumgold *n* **~ paper** Büttenrandpapier *n*, Holländischbütten *n* **~ roll** horizontales Rollen *n*, Windfahnenbewegung *f* (aviat.) **~ scoop** Schwungschaufel *f* (hydraul.) **~ tongs** Froschklemme *f*

**dutiable** zollpflichtig

**duties** Aufgabenbereich *n*

**duty** Abgabe *f*, Aufgabe *f*, Beanspruchung *f*, Berufspflicht *f*, Dienst *m*, Dienstleistung *f*, Förderabgabe *f*, Gebühr *f*, Leistung *f*, Maut *f*, Pflicht *f*, Steuer *f*, Verpflichtung *f*, Verwendungszweck *m*, Zoll *m* **in the line of ~** in Ausübung *f* des Dienstes **to be on ~** im Dienst to be **on ~** Dienst *m* haben **to be on temporary ~** auf zeitweiligem Dienst *m* sein **~ on board** Dienst *m* an Bord *n* (nav.) **~ of a machine** Betrieb *m* **~ on malt** Malzaufschlag *m*

**duty, ~ afloat** Dienst *m* an Bord **~ battery** Überwachungsbatterie *f* **~ chart** Dienstplan *m* **~ cycle** Arbeits-spiel *n*, -zyklus *m*, Einschaltdauer *f*, Impulsfaktor *m*, Spieldauer *f*, Tast-grad *m*, -verhältnis *n* **~ cycle meter (or cyclometer)** Tastmeßgerät *n* **~ drawback** Vered(e)lungsverkehr *m* **~ factor** Arbeitsphase *f*, Tastverhältnis *n* (Impulstechnik) **~-free** zollfrei **~ paid** versteuert **~ roster** Betriebsunterlage *f*, Dienst-eid *m*, -plan *m*, -tafel *f* **~ section** Rufabteilung *f* **~ steamboat** Dampfpinasse *f* vom Dienst *m* **~ unpaid** unversteuert, unverzollt

**D voltage** D-Spannung *f*

**dwarf, ~ pine** Latschenkiefer *f* **~ wave** Zwergwelle *f*

**dwell, to** ~ sich aufhalten, verweilen ~ **of a cam** Zone *f* der Nockenbahn bei welcher keine Änderung der Stößellage erfolgt

**dwindle, to** ~ abnehmen, sich auskeilen, zerrinnen, zusammenschrumpfen

**dwindled** angeebbt

**dwindling** Schwund *m*, Schwunderscheinung *f* ~ **away** Auskeilung *f*

**dyad** zweiteiliges Element *n*, Tensor *m*, Zweiheit *f*

**dye, to** ~ abfärben, färben **to** ~ **again** nacheinanderfärben **to** ~ **fast** echtfärben **to** ~ **first (or ground)** vordecken **to** ~ **to pattern** nach Nuance *f* färben **to** ~ **shades for mixtures** Melangen *pl* färben **to** ~ **in the size** in der Schlichte *f* färben

**dye** Farbe *f*, Färbemittel *n*, Farbstoff *m*, Färbung *f* ~ **absorption (or affinity)** Anfärbbarkeit *f*, Anfärbevermögen *n* ~ **band** Färbeband *n* ~ **bath** Farbbad *n* ~ **coated film** Farbenschichtfilm *m* ~ **experiment** Farbfadenversuch *m* **any** ~ **forming solid particles on the fabric** Körperfarbe *f* ~ **line paper** Lichtpausrohpapier *n* ~ **liquor** Farbflotte *f* ~ **number** Farbzahl *f* ~ **parchment yellow** das Pergament *n* gilben ~ **sensitization** Farbenempfindlichkeit *f*

**dyestuff** Farbstoff *m* ~ **for lakemaking** Lackfarbstoff *m* ~ **grinding machine** Farbenreibmaschine *f* ~ **mixing machine** Farbenmischmaschine *f* ~ **tube** Färbehülse *f* ~ **vat** Färbekessel *m*, Küpe *f* ~ **wood** Farbholz *n* ~ **works** Färberei *f*

**dyed** gefärbt ~ **in the grain** waschecht

**dyeing** Einfärbung *f*, Färberei *f*, Färbung *f*; färbend ~ **and color vats** Färbküpen *pl* ~ **in the froth** Schaumfärberei *f* ~ **to pattern (or to shade)** Nachmusterfärben *n*

**dyeing,** ~ **difference** Farb-absatz *m*, -unterschied *m* ~ **inside out** Linksfärben *n* ~ **pad** Färbefoulard *n* ~ **paddle** Färbehaspel *f* ~ **plant** Einfärbungsanlage *f* ~ **power** Färbekraft *f*

**dyer** Färber *m* ~ **'s spirit** Zinnkomposition *f*

**dying,** ~ **down time** Ausschwingzeit *f* ~**-out** Ausschwingen *n*, Abklingen *n* (sound) ~**-out constant** Ausschwingungskonstante *f* (transient) ~**-out oscillation** abklingende Schwingung *f* ~**-out process** Ausschwingungsvorgang *m* ~**-out time** (sound) Abklingdauer *f*, Ausschwingdauer *f*, Abklingzeit *f* ~**-out transient** Ausschwingvorgang *m*

**dynameter** Dynameter *n*, Vergrößerungsmesser *m*

**dynamic** dynamisch ~ **air pressure** Stau *m* ~ **balance** dynamisches Gleichgewicht *n* ~ **balancing of the upper roll** Auswuchtung *f* der Oberwalze *f* ~ **ball-indentation hardness** Kugelfallhärte *f* ~ **ball-indentation test** Kugelfallprobe *f* ~ **characteristic** Arbeitskennlinie *f*, Arbeitskurve *f* ~ **coercive force** dynamische Koerzitivkraft *f* ~ **compression stress** Schlagdruckbeanspruchung *f* ~ **compression test** Schlagdruckversuch *m* ~ **compressor** Dynamikverzerrer *m* ~ **condenser electrometer** Schwingkondensator-Meßverstärker *m* ~ **cooling** dynamische Abkühlung *f* ~ **cooling and heating** dynamische Kühlung *f* und Erwärmung *f* ~ **counterweight** Fliehkraftpendel *n*, pendelndes Gegengewicht *n* ~ **demonstrator** Demonstrationsdiagramm *n* ~ **diaphragm electrometer**

Meßverstärker *m* mit schwingender Membrane *f* ~ **elasticity** Elastodynamik *n* ~ **endurance test** Dauerversuch *m* mit pulsierender Beanspruchung ~ **equilibrium** dynamisches Gleichgewicht *n* ~ **expander** Dynamikentzerrer *m*, Wuchtsteigerer *m* ~ **flow chart (or diagram)** Datenflußplan *m* ~ **friction torque** Rutschmoment *n* ~ **hardness test** Einhiebverfahren *n* ~ **hardness testing machine** Schlaghärteprüfer *m* ~ **head** dynamisches Druckgefälle *n* ~ **impedance** Impedanz *f* bei Parallelresonanz ~ **indentation test** Sprunghärte *f* ~ **indicator** Staubdruckmesser *m* ~ **lift** dynamischer Auftrieb *m* ~**-lift theory** Drachentheorie *f* ~ **load** dynamische Belastung *f* ~ **loading spring** Schwinglastfeder *f* ~ **loud-speaker** (elektro)dynamischer Lautsprecher *m*, Tauchspulenlautsprecher *m* ~ **luminous sensitivity** Wechsellichtempfindlichkeit *f* ~ **memory** dynamischer Speicher *m* ~ **plate resistance** Anodenwiderstand *m* im Betriebszustand ~ **pressure** Erzeugungsdruck *m*, Flugstau *m*, Staudruck *m* ~ **pressure curve** Staudruckkurve *f* ~ **pressure distribution** Staudruckverteilung *f* ~ **pressure nozzle** Staudruckanschluß *m* ~ **pressure ratio** Staudruckverhältnis *n* ~ **pressure regulation** Staudruckregelung *f* ~ **pressure tendency meter** Staudrucktendenzmesser *m* ~ **range** Lautstärkenbereich *m* (der Stimme), Lautstärkeumfang *m* ~ **range of sound** Klangintensitätsbereich *m* ~ **range control means** Dynamikregler *m* ~ **range expansion** Dynamiksteigerung *f* ~ **reproducer** elektrodynamischer Tonabnehmer *m* (phono) ~ **sensitivity** dynamische Empfindlichkeit *f* ~ **sequential control** Programmablaufänderung *f* (data proc.) ~ **soaring** dynamischer Segelflug *m* ~ **spring deflection** Stoßweg *m* ~ **storage** dynamischer Speicher *m*, dynamische Speicherung *f* ~ **straight line** Arbeitsgerade *f* ~ **strength** dynamische Festigkeit *f*, Schwingungsfestigkeit *f* ~ **stress** Schlagbeanspruchung *f* ~ **stresses** dynamische Beanspruchung *f*, Zerspannungsleistung *f* ~ **subroutine** dynamisches Unterprogramm *n* ~ **tensile test** Schlagzerreißversuch *m* ~ **test** Dauerversuch *m*, Fallprobe *f*, dynamische Prüfung *f*, Schlag-probe *f*, -versuch *m* ~ **thrust** dynamischer Antrieb *m* (Stoß) ~ **torsional balancer** dynamischer Dämpfer *m* ~ **volume range** Dynamik *f*

**dynamical** dynamisch ~ **equation** Bewegungsgleichung *f* ~ **equilibrium** dynamisches Gleichgewicht *n* ~ **friction** dynamische Reibung *f* ~ **matrix** Kopplungsmatrix *f*

**dynamically,** ~ **balanced** dynamisch ausgewuchtet ~ **balanced rotating masses** gewuchtete rotierende Körper *pl*

**dynamicizer** Parallel-Serien-Konvertor *m*

**dynamics** Bewegungslehre *f*, Dynamik *f* ~ **of gases** Gasdynamik *f* ~ **of rotation** Rotationsdynamik *f*

**dynamism** Dynamik *f*, Lautstärkeverhältnis *n*

**dynamite, to** ~ auseinander sprengen

**dynamite** Dynamit *n*, Sprengstoff *m*

**dynamo** Dynamomaschine *f*, Generator *m*, Lichtmaschine *f*, Stromerzeuger *m* ~ **armature** Dynamoanker *m* ~ **battery ignition unit** Lichtbatteriezünder *m* ~ **bracket** Lichtmaschinen-

halterung *f* ~ **carrier** Lichtmaschinenträger *m*
~ **collector brush** Dynamoschleiffeder *f* ~ **drive**
Dynamoantrieb *m* ~**-electric amplifier** dyna-
moelektrischer Verstärker *m* ~**-electric machine** Maschine *f* mit Elektromagnet *m* ~ **fitter**
Dynamomonteur *m* ~ **gear** Dynamorad *n* ~
**laminations** Dynamoblech *n* ~ **lighting** Maschi-
nenbeleuchtung *f* ~ **magnet** Feldmagnet *m* ~
**magneto** Lichtmagnetzünder *m* ~ **oil** Dynamo-
öl ~ **pinion shaft** Dynamoritzelwelle *f* ~ **pulley**
Dynamoscheibe *f* ~ **room** E-Werk *n* ~ **sheet
iron** Dynamoblech *n* ~ **strap** Dynamoriemen *m*
~**-strap-end-clip** Dynamoriemenverschluß *m* ~
**(or lighting) winding** Lichtwicklung *f* ~ **yoke**
Lichtmaschinengehäuse *f*
**dynamometer** Dynamometer *n*, Kraftmesser *m*,
Leistungsmesser *m* ~ **hub** Leistungsmeßnabe *f*
~ **test stand** Brems-, Leistungsmeß-stand
*m*

**dynamometric** dynamometrisch ~ **dynamo**
Bremsdynamo *n*, Pendeldynamo *m*
**dynamometrical brake** Leistungsbremse *f*
**dynamotor** Einankerumformer *m*, Umformer
*m*

rotierender Umformer *m* mit gemeinsamem
Rotor
**dynatron** Dynatron *n* ~ **characteristic** Dynatron-
wirkung *f* ~ **method** Dynatronart *f* ~ **oscillation**
Habann-, Dynatron-schwingung *f* ~ **oscillator**
Dynatron-generator *m*, -oszillator *m*, -summer
*m*; Habannröhre *f*, Röhre *f* negativen Widerstandes ~ **tube with negative resistance** Verstärker *m* mit negativem Widerstand *m* ~**-type
of tube with negative resistance** Negativwider-
standsröhre *f*
**dynatrons** schwere Elektronen *pl*
**dyne** Dyn *n*, Dyne *f* ~ **per square centimeter**
Mikrobar *n*
**dynode** Parallelelektrode *f*, Sekundäremissionskathode *f*, Vervielfältigungsoberfläche *f* ~
**structure** Dynodenstruktur *f*
**dyotron** Dyotron *n*
**dyscrasite** Aluminiumsilber *n*, Antimonsilber *n*,
Diskrasit *m*, Silberspießglanz *m*, Spießglanz-
silber *n*
**dysprosium** Dysprosium
**dyssymetric distortion** unsymmetrische Verzerrung *f*

# E

eagle Adler m ~ stone Adlerstein m
eagre Seebeben n
ear Ähre f, Ansatz m, Fahrdrahtklemme f
Henkel m, Ohr n, Öhr n, Ohrmuschel f, Öse f,
Zipfel m by ~ nach dem Gehör n ~ of a cord
Seilöhr n, Tauöse f
ear, ~ deafening ohrenbetäubend ~ defender
Ohrenklappe f ~ drum Trommelfell n ~ flap
Ohrmuschel f ~-marked stocks Sperrbestand
m ~ muff Ohrenkappe f ~-phone Hörer m,
Kopfhörer m ~-piece Hörer-kapsel f, -muschel
f; Ohrmuschel f ~ plug Gehörschützer m
~-response characteristic Ohrempfindlichkeits-
kurve f ~ spindle Ährenspindel f ~ trumpet
Hörrohr n ~ tube Hörschlauch m
earliest separation products Erstausscheidungen
pl
early beizeiten, früh, (of motions or time) frühzeitig ~ attention Vorsorge f ~ bolter Früh-
schosser m ~ gate former (or generator) Früh-
impulsgenerator m ~ gated coincidence tube
unverzögerte Koinzidenzröhre f ~ ignition
Frühzündung f ~ value Anfangswert m ~
warning Vorwarnung f ~-warning radar set
Frühwarnradar n
earn, to ~ erwerben, gewinnen, verdienen to ~
a salary Gehalt n beziehen
earned income Arbeitseinkommen n
earning Erwerb m ~ capacity Rentabilität f
earnings Arbeitslohn m, Bruttoertrag m, Verdienst m
earth, to ~ erden to (put to) ~ an Erde f legen
earth Boden m, Erdball m, Erde f, Netzerde f ~
antenna Bodenantenne f ~ arrester Erdableiter
m, Erdverbindungskurzschließer m ~ auger
Erdbohrer m ~ axis erdfeste Achse f (aerodyn.)
~'s axis Erdachse f ~ bank Erd-, Schütt-damm
m ~ bar Erdungsschiene f ~ bodily tides Erd-
gezeiten pl ~ borer Erdbohrer m ~ brace Unterriegel m ~ capacity Erdkapazität f, Kapazität f
gegen Erde ~ circuit Erd-leitung f, -schleife f
~ clip Erdschelle f ~-colored fahl ~ conductivity Leitfähigkeit f des Erdbodens, spezifische
Leitfähigkeit f ~-connected geerdet ~ connection Erd-anschluß m, -bau m, -leitung f, -ver-
bindung f; Massenanschluß m ~ (or ground)
connection Maße pl (electr.) ~ covered heap
Meiler m ~'s crust Erdkruste f ~ current Erdstrom m ~-current equalizer Erdstromaus-
gleicher m ~-current prospecting Erdstrom-
schürfung f ~ currents natürliche Erdströme pl
~ dam Erd-, Stau-damm m; Staumauer m ~
dam with cellular core wall Zellkerndamm m ~
dike Staudamm m ~ direction finder Erdpeil-
gerät n ~ disturbance Erdstörung f ~'s equator
Erdäquator m, Erdgleicher m ~ fault Erdfehler
m ~ fill Erdaufschüttung f ~-fill dam gewalzter
Erddamm m, Staudamm m ~ flattening approximation Erdabplattungsnäherung f ~ flax
Asbest m ~-free erdfrei (electr.) ~ glaze Erd-
glasur f ~'s gravitational field Erdfeld n der
Schwerkraft ~ grid stage Gitterbasisstufe f

~s horizontal field Horizontalkomponente f
des Erdmagnetfeldes ~ induction Erdinduktion
f ~ inductor Erdinduktor m, Induktionsmeßgerät n für das magnetische Erdfeld ~-inductor
compass Erdinduktionskompaß m ~ lead
Erdungsleitung f
earth-leakage, ~ circuit breaker Fehlerstrom-
schutzschalter m ~ detector Erdsucher m ~ indicator Erdschlußanzeiger m ~ meter Erd-
schlußmesser m ~ monitor Isolationswächter m
earth, ~ loop Erdfehlerschleife f (teleph.) ~
magnetic erdmagnetisch ~'s magnetic field
erdmagnetisches Feld n ~ mass Erdmasse f ~
materials Erdstoffe pl ~ metal Erdmetall n ~
movement Erdschuß m ~ moving Erdbewegungen pl ~'s natural resources Bodenschätze pl ~
noise Erdgeräusch n ~'s oblateness Erdabplat-
tung f ~'s orbit Erdbahn f ~ parapet (in front
of concrete wall) Erdvorlage f ~-phantom circuit Viererkreis m mit Erdrückleitung ~ plate
Erdelektrode f, Erder m, Erdplatte f, plattenförmige Massenelektrode f (einer Zündkerze)
~ pole Erdpol m ~ potential Erdpotential n ~
pressure (active) Bodendruck m, Erddruck m,
(passive) Erdwiderstand m ~ pressure at rest
Ruhedruck m
earth-quake Beben n, seismische Bewegung f,
Erdbeben n ~ due to folding Faltungsbeben n
~ factor Erdbebenfaktor m
earth, ~ rammer Erdstampfe f ~ rate Erddrehung f ~'s rate signal Erddrehungs-Korrek-
tursignal n ~ resistance meter Erdwiderstands-
messer m ~ return Erd(rück)leitung f ~-return
automatic system Erdsystem n ~-return circuit
Einfachleitung f, Stromkreis m mit Erdrückleitung ~-return system System n mit Erdrück-
leiter ~ roof Abraum m (geol.) ~ screen (ungeerdetes) Gegengewicht n ~ screw Erd-
schraube f, Schraubenfuß m ~'s shadow Erd-
schatten m ~ shake due to tunneling Tunnelbe-
ben n ~ short-circuit Masseschluß m ~ shovel
Lochschaufel f ~ slip Bergrutsch m ~-space
service Erde/Weltraum-Funkdienst m ~ spring
Erdungsfeder f ~ station Erde-Funkstelle f ~
stay Fußanker m ~ strip Erdungsstreifen m
~'s surface Erdboden m ~ telegraphy (using
the ground as a conductor) Erd-funken n,
-telegrafie f ~ terminal Erdklemme f ~ terminal
arrester Erdfunkenstrecke f ~ tester Erdlei-
tungsprüfer m, Erdungsmesser m, Meßgerät n
zum Messen des Erdwiderstandes ~ traverse
Unterriegel m ~'s vertical field Vertikalintensität f ~ wall Erdwall m ~ wax Ceresin m ~ wire
Erd-draht m, -leitung f, -seil n ~ wire connector
Erdseilklemme f ~ wire for poles Stangenerd-
leitung f ~ wireless telegraphy Erdfunkerei f
earthwork(s) Bankett n, Bodenbewegung f, Erd-
bau m, Erdarbeiten pl, Feldschanze f, Unterbau m ~ engineering Unterbau m
earth-worm Erdraupe f
earthed geerdet, an Erde f liegend ~ battery
geerdete Batterie f ~ cradling geerdetes Schutz-

(erdungs)netz n ~ grid Brems-, Fang-gitter n
~-grid stage Gitterbasisstufe f ~ neutral conductor geerdeter Nulleiter m ~ neutral system
System n mit geerdetem Mittel- oder Sternpunkt
~ phantom circuit Simultanleitung f ~ rail
Masseschiene f ~ shield geerdeter Schutzring m
earthen, to ~ irden
earthen, ~ block irdenes Formstück n ~ bowl of
pipe Tonpfeifenkopf m
earthenware irdenes Geschirr n, Steingut n, Tongut n, Tongutwaren f, Tonwaren pl, Töpferware f ~ block Formstück n (Tonform) ~-block
conduit Formstückkanal m ~ dish Steingut-
schale f ~ duct Steingutrohr n ~ manufacturing
plant Steingutfabrikeinrichtung f ~ pipe irdenes
Rohr n, Tonrohr n
earthing Erden n, Erdschluß m (electr.), Erdung
f, Massenanschluß ~ angle Erdungswinkel m ~
brush Erdungskohle f ~ cable Erdkabel n ~
circuit Erdverbindung f ~ clamp Erdschelle f ~
connection Erdanschluß m ~ contact Erdung f
~ devices Erdungsvorrichtungen pl ~ insulator
Erdungstrenner m ~ isolating switch Trenner-
dungsschalter m ~ key Erdungstaste f ~ percentage Erdungsgrad m ~ pin Erdungsstift m
~ resistance Erdungswiderstand m ~ screw
Erdungsschraube f ~ strap Masseband n ~
switch Erdungsschalter m ~ terminal Erd-
leitungs-, Körperschluß-klemme f ~ wire
Körperschlußkabel n ~-wire ground cable
Kurzschlußkabel n
earthy erdartig, erdig, mulmig ~ asphalt Erd-
asphalt m ~ brown coal erdige oder mulmige
Braunkohle f ~ brown hematite Brauneisenmulm m ~ calcium carbonate Bergmilch f ~
cerussite Bleierde f ~ coal Erdkohle f ~ cobalt
Erdkobalt m ~ color Erdfarbe f ~ dilution of
ore Gangart f, Ganggestein n ~ fluor Flußerde
f ~ gypsum Gipserde f, Mehlgips m ~ magnetite
Eisenschwärze f ~ mineral Erdmineral n ~ ore
Mulm m ~ vivianite Eisenblauerde f
ease, to ~ abspannen, entlasten, erleichtern, lindern, mildern to ~ away auffieren, fieren to ~
down herunterfieren, heruntergehen (mit dem
Druck beim Füllen), sacken lassen to ~ off
abschrecken
ease Leichtigkeit f ~ of brushing (for lacquers)
Streichfähigkeit f ~ of control leichte Regelbarkeit f ~ of handling leichte Steuerbarkeit f ~ of
maintenance leichte Wartung f
easel Staffelei f
easement allmähliche Kurve f; Nutz-, Nutzungs-
recht n; Übergangskurve f
easily leicht ~ accessible griffbereit, leicht zugänglich ~ combustible leicht entzündlich oder
verbrennbar ~ dismantled leichte Zerlegbar-
keit f ~ followed up übersichtlich ~ fusible
leicht-flüssig, -schmelzbar, -schmelzlich,
schnellflüssig ~ fusible glazing leichtflüssige
Glasur f ~ liquefiable leichtflüssig ~ machined
leicht zu bearbeiten ~ meltable leicht-schmelz-
bar, -schmelzlich ~ shifted gear leicht zu schal-
tendes Getriebe n ~ slacking lignite coal erdige
Braunkohle f ~ soluble leichtlöslich ~ tilted
hochkippbar ~ understandable übersichtlich ~
volatilized leichtflüchtig
easing, ~ of a bend Abflachung f einer Krümmung ~ the break Lüften n der Bremse

east Ost m, Osten m, Ostpunkt m ~-iron pulley
Graugußriemenscheibe f ~ ward ostwärts
~-west line Ost-Westlinie f ~ wing Ostflügel m
easterly östlich
eastern östlich
East India tin Brenkas n
easy bequem, einfach, gemächlich, leicht of ~
access leicht zugänglich of ~ action (or motion)
leichtgängig ~ to cleave (or split) leichtspaltig
~ to find auffindbar ~ to fly leicht zu fliegen ~
to locate leicht auffindbar ~ to operate leicht
bedienbar ~ to read gut leserlich
easy, ~ axis Vorzugsachse f ~-axis direction
Vorzugsrichtung f ~ control Bedienungskom-
fort m ~ fit Gleitpassung f ~ flow leichtfließend
~ gradient sanfter Abhang m, schwache Steigung f ~ handling große Handlichkeit f ~ observation Bedienungserleichterung f ~-out
Gewindezieher m ~ running fit leichter Laufsitz m ~ solving schnell lösend ~-to-operate
device leicht zu bedienende Vorrichtung f
~-to-read scale übersichtliche Skala f ~
eat, to ~ essen, genießen, speisen to ~ away abfressen, anfressen, wegfressen to ~ into ausnagen, hineinfressen to ~ through durchfressen
eave Dachvorsprung m
eaves Dach-rinne f, -traufe f, -vorsprung m,
Traufe f ~ of a roof Abtraufen pl, Dachflüsse
pl ~ course Fußschicht f ~ lead bleierne Trauf-
platte f
eavesdrop, to ~ abhorchen, behorchen
ebb, to ~ zurückfließen
ebb Abfluß m, Ebbestrom m, Fallen n des Wassers ~ and flow Ebbe und Flut f ~-and-flow
gate Ebbe- und Fluttor n ~ and flow of tide
Gezeitenströmung f ~ beginning Vorebbe f ~
channel Ebberinne f ~ side of a dam Ebbeseite f
~ stream Ebbstrom m ~ tide Ebbe f, fallendes
Wasser n ~-tide channel Ebberinne f ~-tide
current Ebbestrom m ~-tide gates Ebbetor n,
Binnentor n
E-bend Krümmer m in E-Ebene (Hohlleiter)
ebigite Uranthallit m
ebonite Ebonit n, Hartgummi m ~ box Ebonit-
kapsel f ~ case Ebonit-gehäuse n, -kapsel f ~
guard insulator Ebonitschutzglocke f ~ jaw
Hartgummibacke f ~ lining Hartgummibelag
m ~ plate Hartgummiplatte f ~ press mold
Hartgummipreßform f ~ separator Hartgummi-
separator m ~ stage Harttisch m (opt.) ~ stud
Hartgummipimpel m
ebony Ebenholz n
ebulliometer Ebulliometer n
ebullioscope Ebullioskop n, Kochpunktmesser m
ebullition Aufwallung f, Kochen n, Wallung f ~
of lime Aufgehen n des Kalkes
eccenter link exzentrisches Glied n
eccentric Exzenter m, Scheibenkurbel f; außen-
mittig, außerachsig, ausmittig, exzentrisch ~
of shearing machine Exzenter m für Rundwirkmaschinen
eccentric, ~ bearing Exzenterlager m ~ bit Erweiterungsbohrer m, Exzentermeißel m ~ box
(or case) Exzenterbüchse f ~ bush Exzenterbüchse f ~ bushing Exzenterbuchse f ~ circle
Exzenterrille f (phono) ~ clamp Drahtklemme
f ~ collar Exzenter m für Druckzylinder ~
collar strap Exzenterbügel m ~ disc Exzenter-,

Kröpf-scheibe *f* ~ **drive** Exzenterantrieb *m* ~
**error** Schlagfehler *m* ~ **error measuring device**
Schlagerfehler-Prüfeinrichtung *f* ~ **gab** Ex-
zenterstangenzapfen *m* ~ **gear wheel** Exzenter-
zahnrad *n* ~ **gitting press** Exzenterabkneifpresse
*f* ~ **governor** Exzentrikregulator *m* ~ **grip**
Froschklemme *f* ~ **groove** Exzenterrille *f* ~
**guide block** Exzenterstein (Gasmaschine) *m* ~
**hoop** Exzenterring *m* ~ **housing** Exzenterhülse *f*
~ **lever** Exzenterhebel *m* ~ **load** exzentrische
Belastung *f* ~ **pin** Exzenterbolzen *m*, exzentrischer Stift *m* ~ **press** Exzenterpresse *f* ~
**pressure** Exzenterdruck *m* ~ **pump** Exzenter-
pumpe *f* ~ **relief** Hinterschliff *m* ~ **relief cutter**
hinterschliffener Fräser *m* ~ **rod** Schwingen-
stange *f* ~ **shaft** Exzenterwelle *f* ~ **shaft for
impression** Druckstellwelle *f* ~ **sheave** Exzen-
ter-, Hub-, Kröpf-scheibe *f* ~ **sleeve** Exzenterhülse *f* ~ **strap** Exzenterlagerauge *n*, Hub-
scheibenring *m* ~ **stud** Exzenter-bolzen *m*,
-stiftschraube *f* ~ **trimming press** Exzenterab-
gratpresse *f* ~ **weight** Unwuchtscheibe *f* ~
**wheel** Exzentrikrad *n*
**eccentrical** außermittig, exzentrisch ~ **(ly)**
**relieved tooth** hinterschliffener Zahn *m*
**eccentricity** Außermittigkeit *f*, außermittige
Lage *f*, Dezentrierung *f*, Exzentrizität *f* ~ **of
the center of gravity** Schwerpunktverlagerung *f*
~ **error** Exzentrizitätsfehler *m*
**eccentrics** Exzenter *pl* für Rundwirkmaschinen
**Eccles-Jodan trigger** Flip-Flop-Auslöser *m*
**echelette** Echelettegitter *n* (phys.)
**echelle grating** Stufengitter *n*
**echelon, to** ~ staffeln
**echelon** Echelon *m*, Staffel *f* **in** ~ treppenweise ~
**of radio units** Funkstaffel *f*
**echelon,** ~ **circuit** Staffelleitung *f* ~ **duplex**
Gegensprechen *n* in Staffelschaltung ~ **duplexing** Staffelgegensprechen *n* ~ **formation of
multiple echoes** stufenförmiges Echo *n* ~ **grating** Stufengitter *n* ~ **lens** Stufenlinse *f* ~ **lens
antenna** gestaffelte Linsenantenne *f* ~ **multiplex**
**telegraph** gestaffelter Mehrfachtelegraf *m* ~
**prism** Stufenprisma *n* ~ **spectroscope** Glas-
plattenstaffeln *pl* ~ **strapping** gestaffelte Koppelleitung *f* ~ **table** Staffelungstafel *f* ~ **working** Staffelbetrieb *m*
**echeloned** staffelförmig, stufenartig, treppenförmig ~ **in depth** tiefgestaffelt
**echelonment** (of guns) Staffelung *f* ~ **in depth**
Tiefenstaffelung *f*
**echo, to** ~ nachschallen, widerhallen **to** ~ **back**
zurückwerfen
**echo** Echo *n*, Nachhall *m*, Rückstrahl *m*, Schat-
tenbild *n*, Widerhall *m* ~ **of long duration** Langzeitecho *n*
**echo,** ~ **area** Echoweite *f* ~ **attenuation** Dämpfung *f* der Echoströme ~ **attenuation measuring
set** Echometer *m* ~ **box** Impulstastgerät *n* ~
**box antenna** Echoboxantenne *f* ~ **cancellation**
Echokompensation *f* ~ **chamber** Hallraum *m* ~
**checking** Echo-kontrolle *f*, -probe *f*; Schleifen-
probe *f* (comput.) ~ **current** Echostrom *m* ~
**current attenuation** Echostromdämpfung *f* ~
**depth sounder** Behm Lot *n* ~ **effect** Echowir-
kung *f* ~ **image** Strahlenfigur *f* ~ **images**
Doppelbild *n* ~ **killer** Echosperrer *m* ~ **killing**
Echounterdrückung *f* ~ **kinesis** Bewegungs-

nachahmung *f* ~ **matching** Echozeichenab-
gleich *m* ~ **meter** Echometer *n* ~ **microphone**
Echomikrofon *n* ~ **organ** Echowerk *n* ~ **path**
Echoweg *m* ~ **ranging** Echo-Ortung *f* ~ **receiver** Echoempfänger *m* ~ **signal** Doppelzeichen
*n*, Mehrfachzeichen *n* ~ **sounding** Echolot *n*
mit Ultraschall ~ **sounding device** Echolot *n* ~
**sounding oscillator** Tonlotsender *m* ~ **splitting**
Echoteilung *f* ~ **suppression** Echounterdrük-
kung *f* ~ **suppressor** Echo-absperrer *m*, -sperre
*f*, unstetig arbeitende Echosperre *f* ~ **term**
Laufzeitkorrektur *f* ~ **transmission time** Echo-
laufzeit *f* ~ **trouble** Echostörung *f* ~ **wave**
Echowelle *f* ~ **weighting term** Frasenentzerrung
*f*
**echoed sound** Doppelklang *m*
**echoing area** Rückstrahlfläche *f*
**eclimeter** Eklimeter *n*
**eclipse, to** ~ abdecken, abdunkeln, maskieren,
überdecken, verdecken
**eclipse** Verfinsterung *f*, Verdunk(e)lung *f* ~ **rays**
Strahlabblender *m* ~ **rings** Eklipseringe *pl* ~
**system** Eklipsebauart *f*
**ecliptic** Sonnenbahn *f*
**economic,** ~ **(al)** sparsam, wirtschaftlich ~ **con-
siderations** wirtschaftliche Gesichtspunkte *pl*
~ **cruising** Sparreiseflug *m* ~ **domain** Wirtschaftsraum *m* ~ **geology** Lagerstättenlehre *f*
~ **measure** Rationalisierungsmaßnahme *f* ~
**mixture for cruising** Spargemisch *n* für Reiseflug ~ **policy** Wirtschaftspolitik *f* ~ **power**
Wirtschaftskraft *f* ~ **production** wirtschaftliche
Fertigung *f* ~ **speed** Sparfluggeschwindigkeit *f* ~
**economical** ökonomisch, pfleglich, rationell,
sparsam ~ **cruising** Sparflug *m* ~ **cruising boost**
Ladedruck *m* für Sparflug ~ **cruising mixture**
Sparfluggemisch *n* ~ **equalization of energy**
energiewirtschaftlicher Ausgleich *m* ~ **filament** Sparfaden *m* ~ **fluctuation** wirtschafts-
politische Schwankung *f* ~ **load** Bestlast *f* ~
**manufacture** wirtschaftliche Fertigung *f* oder
Herstellung *f* ~ **oiling apparatus** Ölsparapparat
*m* ~ **operations** sparsames Wirtschaften *n* ~
**pressure-water valve** Druckwassersparventil *n*
~ **rating** Bestlast *f* ~ **tanning extract (or mordant)** Spar-beize *f*, -beizstoff *m* ~ **working
methods** rationeller Arbeitsablauf *m*
**economics** Wirtschaftspolitik *f* ~ **of mine** Gru-
benhaushalt *m*
**economist** Nationalökonom *m*
**economization** Rationalisierung *f*
**economize, to** ~ ersparen, nutzbar machen, sparsam anwenden
**economizer** Abgasvorwärmer (mach.), Ekono-
miser *m*, Rauchgas-speisewasservorwärmer *m*,
-vorwärmer *m*, Sauerstoffsparregler *m* (Flugzeug), Schaltuhr *f*, Spar-düse *f*, -einrichtung *f*,
vorrichtung *f*, Vorwärmer *m*, Zünd- und
Löschuhr *f* ~ **with ribbed pipes** Rippenrohr-
rauchgasvorwärmer *m* ~ **control valve** Spar-
hahn *m*
**economizing** Kostensenkung *f* ~ **furnace** Spar-
feuerung *f*
**economy** Ersparnis *f*, Sparsamkeit *f*, Wirtschaft *f*,
Wirtschaftlichkeit *f* ~ **of manpower** Kräfteer-
sparnis *f* ~ **of operation** Wirtschaftlichkeit *f* im
Betrieb ~ **of work's operation** Betriebswirt-
schaftlichkeit *f*

**economy, ~ casting** Sparguß *m* **~ circuit** Sparschaltung *f* **~ compressor** Sparkompressor *m* **~ curve** Verarmungskurve *f* **~ loop** Verarmungsschleife *f* **~ punchings** Sparschnitt *m* **~ switch** Sparschalter *m* **~ tub** Sparwanne *f* **E-core** E-Kern *m* **~ detector (unit)** E-Kern-Geber *m* **~ pickoff** E-Kern-Geber *m*
**ecrasite** Ekrasit *n*
**ectype of a coin** Abdruck *m* einer Münze
**eddy, to ~** schnell kreisen, wirbeln
**eddy** Gegenströmung *f*, Nehrstrom *m*, Strudel *m*, Turbulenz *f*, Walze *f*, Wirbel *m* **~ chamber nozzle** Wälzkammerdüse *f* **~-conductivity** Scheinleitfähigkeit *f*
**eddy-current** Nehr-, Streu-, Wirbel-strom *m* **~ brake** Wirbelstrombremse *f* **~ coefficient** Wirbelstrombeiwert *m* **~ disc** Wirbelstromscheibe *f* **~ loss** Winkelstromverlust *m* **~ plate** Wirbelstromscheibe *f* **~ revolution counter** Wirbelstromdrehzahlmesser *m*
**eddy, ~ currents** Foucaltströme *pl* **~ diffusivity** Scheindiffusionskoeffizient *m* **~ energy** Turbulenzenergie *f* **~ flux** Turbulenzfluß *m* **~ formation** Wirbelung *f* **~ freedom** Wirbelfreiheit *f* **~ making** (hydrodynamics) Ablösung *f* **~ motion** turbulente Strömung *f*, Wirbelbewegung *f* **~ sink (or sump)** Wirbelsenke *f* **~ viscosity** Scheinreibung *f* **~ viscosity coefficient** Turbulenzkoeffizient *m* **~ zone** Wirbelzone *f*
**eddying** Wirbelbildung *f*, Wirbeligkeit *f*, Wirbelung *f*; kreisend **~ at the banks caused by bank friction** Umlaufbewegung *f* **~ whirl** Wirbelbewegung *f*
**edema** Ödem *n*
**edge, to ~** (timber) abschwarten, besäumen, beschroten, bördeln, einfassen, kanten, picken, pillen, rändeln, rändern, säumen, schärfen **to ~ into line** einscheren
**edge** (leading or entering) Anblasrand *m*, Bord *m*, (trailing or rear edge) Durchflußkante *f*, Kante *f*, Kimme *f*, Naht *f*, Rand *m*, Saum *m*, (cutting) Schneide *f*, Umrandung *f*, Zarge *f* **on ~** hochkant, hochkantig, kantenweise
**edge, ~ of the armature pole shoe** Ankerpolschuhkante *f* **~ on back of stock** Schaftnase *f* **~ of bridge** Stegkante *f* **~ of cloth** Webekante *f* **~ of the compass rose** Rosenrand *m* **~ of forest** Waldrand *m* **~ of hole** Lochkante *f* **~ of an ingot** Blockrand *m* **~ of load** Lastrand *m* **~ of the magnet pole shoe** Magnetpolschuhkante *f* **~ of nail** Nagelschneide *f* **~ of the nozzle exit** Düsenaustrittskante *f* **~ of pattern** Musterkante *f* **~ of the pole pieces** Pol(schuh)kante *f* **~ of propeller blade** Schraubenblattkante *f* **~ of pulse** Impulsflanke *f* **~ of regression** Gratlinie *f*, Rückkehrlinie *f* (geom.) **~ of riveted joint** Nietrand *m* **~ about which rotation takes place** Drehkante *f* **~ and shank trimming machine** Schnitt- und Gelenkfräsmaschine *f* **~ of the tooth crest** Zahnkopfkante *f* **~ of woods** Waldrand *m*
**edge, ~ angle** Kantenwinkel *m* **~ beam** Randträger *m* **~ bevelling machine** Kantenschrägmaschine *f* **~ board** Ortsbrett *m* **~ bond** Eckverband *m* **~ breakage** Kantenriß *m* **~ cam** Manteldaumen *m* **~ condition** Randbedingung *f* **~-contact at beginning of mesh** Flankeneintrittsspiel *n* **~ cracking** Kantenriß *m* **~ cracks**

**Randrisse** *pl* **~ cutter** Ränderschneideapparat *m* **~ damping** Randdämpfung *f* (acoust.) **~ dislocations** Stufenversetzungen *pl* **~ domain** Kantenbezirk *m* **~ drop corn planter** Randtropfpflanzmaschine *f* **~ effect** Kantenwirkung *f*, Rand-, Spitzen-effekt *m* **~ fairing** Nasenkasten *m* **~ fiber** Randfaser *f* **~ filter** Spaltebenekristallfilter *n*, Stabfilter *m* **~ flare** Randaufbruch *m* (film, TV) **~ fog** Randschleier *m* (film) **~ fracture** Kantenriß *m* **~ gluing machine** Kantenanleimmaschine *f* **~ hammered down** Falz *m*, Fugenfalz *m* **~ hit by a jet** Kante *f* die von einem Strahl angeblasen wird **~ interference** Kanteneingriff *m* **~ joint** Eck-verband *m*, -verzapfung *f* **~ joint file** Scharnierfeile *f* **~ life** (tool) Standzeit *f* **~ lighted** indirekt beleuchtet **~ mill** Koller-gang *m*, -mühle *f*; Läufermühle *f* **~ mixer** Misch-koller *m*, -kollergang *m* **~-on conveyer** Hochkantförderer *m* **~ planing machine** Kantenhobelmaschine *f* **~ protection** Kantenschutz *m* **~-punched card** Randlochkarte *f* **~ rail** Kantenschiene *f* **~ runner** Kollergangsläufer *m*, Kollerläufer *m*, Rollgang *m* **~ runner mill** Koller-, Läufer-, Pfannen-, Schlagnasen-mühle *f* **~ runner mixture** Kollergang *m* **~-runner pan** Kollergangsschale *f* **~-runner plate** Kollerplatte *f* **~ runner for wet grinding** Naßkollergang *m* **~ seam** rechtes oder stehendes Flöz *n* (geol.) **~ sharpness** Berandungsschärfe *f* **~ spectrograph** Schneidenspektograf *m* **~ stiffening wire** Abschlußdraht *m* **~ stress** Grenz-, Rand-spannung *f* **~ system** Kantensystem *n* **~ tone** Schneideton *m* **~ toolmaker** Grobschmied *m* **~ tools** Schneidzeug *n* **~ trimming machine** Kantenbestoßmaschine *f* **~ turned over inside** Abkantung *f* mit innerem Falz **~ water** Randwasser *m* **~ weld** Stirnstoß *m*
**edged** gekröpft, gerändelt, gerändert **~ adjusting pin** Paßkerbstift *m* **~ beading** Randwulst *f*
**edger** Beschneidehobel *m*, Kanter *m*, Kantvorrichtung *f*
**edges, ~ of gripping tool** Greiferschneiden *pl* **~ of the pole pieces** Polschuhkanten *pl*
**edgewise** hochkant, hochkantig, kantenweise **~ bend** H-Bogen *m* **~ instrument** Profilinstrument *n* **~ winding** Hochkantentwicklung *f* **~ wound ribbon coil** hochkant gewickelte Bandspule *f*
**edging** Kanten *n*, Umrandung *f*, Vorstoß *m*, Zacke *f*, Zacken *m* **~ and glazing** (fittings) Einschleifen *n* **~ device** Kanter *m* **~ machine** Kantmaschine *f* **~ pass** End-, Stauch-stich *m*, (in rolling) Stauchkaliber *n* **~ planer** Beschneidehobel *m* **~ saw** Besäumsäge *f* **~ stand** Stauchgerüst *n* **~ tool** Beschneidemesser *n*
**edifice** Bauwerk *n*
**Edison, ~ accumulator** Eisennickelakkumulator *m* **~ effect** Edison-Effekt *m* **~ lamp holder with suspension hook** Edisonfassung *f* mit Aufhängehaken **~ sound grove** Tiefenschrift *f* **~ storage battery** Edisonbatterie *f* **~ storage cell** alkalischer Sammler *m* **~ storage cell (or battery)** Edisonsammler *m* **~ track** Tiefenschrift *f*
**edit, to ~** herausgeben, redigieren (film)
**editing** Bearbeitung *f*, Redigierung *f*, Schriftleitung *f*, Sichtung *f*, Überprüfung *f* **~ of the tape** Bandschnitt *m* **~ table** Schneidetisch *m* (tape rec.)

**edition** Auflage *f*, Ausgabe *f* ~ **de luxe** Luxus-, Pracht-ausgabe *f*

**editor** Herausgeber *m*, Schriftleiter *m*

**editorial,** ~ **office** Redaktion *f* ~ **staff** Schriftleitung *f*

**editorialize, to** ~ Leitartikel *m* schreiben

**editorially,** ~ **cut print** Schnittkopie *f* ~ **edited print** Schnittkopie *f*

**educate, to** ~ ausbilden, erziehen

**educated** gebildet

**education** Ausbildung *f* ~ **allowance** Erziehungsbeihilfe *f*

**educational** erzieherisch ~ **age** Bildungsstufe *f* ~ **establishment** Lehranstalt *f* ~ **exhibition** Lehrschau ~ **film** Kulturfilm *m* ~ **institution** Bildungsanstalt *f* ~ **motion picture** Lehrfilm *m*

**eduction** Entweichung *f*

**edulcorate, to** ~ aussüßen

**efface, to** ~ abwischen, austilgen, tilgen, verwischen

**effect, to** ~ ausführen, bewerkstelligen, bewirken, leisten, tätigen, wirken

**effect** Auswirkung *f*, Effekt *m*, Einfluß *m*, Einwirkung *f*, Leistung *f*, Wirkung *f*, Wirkungsgrad *m*

**effect,** ~ **of air pressure** Luftdruckwirkung *f* ~ **on alternating-current resistance** Einfluß *m* auf den Wechselstromwiderstand ~ **of attraction** Anziehungseffekt *m* ~ **of bomb explosion** Bombenwirkung *f* ~ **of boundary** Grenzschichtwirkung *f* ~ **of collisions** Stoßeinflüsse *pl* ~ **of compressibility** Kompressibilitätseinfluß *m* ~ **of concussion** Luftdruckwirkung *f* ~ **of contrast** Kontrastwirkung *f* ~ **of end restraint** Einspannwirkung *f* ~ **of errors** Fehlereinfluß *m* ~ **on fading** Einfluß *m* auf das Fading ~ **of firing** Schießergebnis *n* ~ **of force** Kraftäußerung *f* ~ **of forces** Kräftespiel *n* ~ **of frost** Frostwirkung *f* ~ **of gravity** Schwerkraftwirkung *f* ~ **of heat** Wärmewirkung *f* ~ **of hyperfine structure** Effekt *m* der Hyperfeinstruktur ~ **of the impulse** Stoßablauf *m* ~ **of inertia** Trägheitseffekt *m* ~ **of light** Lichteinwirkung *f* ~ **of perspective** räumliche Wirkung *f* ~ **of propulsive jet** Strahleinfluß *m* ~ **of radiation** Strahlenwirkung *f* ~ **of radium** Radiumeinwirkung *f* ~ **of repulsion** *m* Abstoßungseffekt *m* ~ **of rotation of the earth** Anstau *m* ~ **of shell burst** Granatwirkung *f* ~ **of side spray** (artil.) Breitenwirkung *f* ~ **of temperature** Temperaturgang *m* ~ **of three-phonon interaction** Dreiphononenwechselwirkung *f*

**effect,** ~ **color** Einstichfarbe *f* ~ **energies of the gyrating asymetric molecules** Effektenenergien *pl* asymmetrischer Kreiselmoleküle ~ **film** Geräuschfilm *m* ~ **lighting** Effektbeleuchtung *f* ~ **projector** Leuchteffektprojektor *m* ~ **threads** (textiles) Effektfaden *m*

**effective** echt, effektiv, nutzbar, tatsächlich, wirklich, wirksam, wirkungsvoll, zweckmäßig ~ **from** mit Wirkung vom

**effective,** ~ **address** wirkliche Adresse ~ **angular field of image** nutzbarer Bildwinkel *m* ~ **antenna area** Antennenwirkfläche *f* ~ **antenna height** effektive Höhe *f* einer Antenne ~ **aperture of a lens** nutzbarer Linsendurchmesser *m* ~ **area** Nutzfläche *f* ~ **area of grate** Rostbrennfläche *f* ~ **attenuation** Betriebsdämpfung *f* ~

**bandwidth** Nutzbandbreite *f* (acoust.) ~ **battery** fühlbare Batterie *f* ~ **beam width** wirksame Strahlbreite *f* ~ **beaten zone** bestrichener Raum *m* der Kerngabe ~ **bunching** effektiver Laufwinkel *m* ~ **call** ausgeführte Verbindung *f* ~ **capacity** wirksame Kapazität *f*, effektive Leistung *f*, Nutzleistung *f* ~ **center of gravity** Schwerpunkt *m* der wirksamen Masse ~ **charge** effektive Ladung *f* ~ **collision cross-section** Wirkungsquerschnitt *m* ~ **component** Wirkwert *m* ~ **conductivity** effektive Reichweite *f* ~ **confusion area** zweiäquivalente Düppelfläche *f* ~ **content** Wirkungsgehalt *m* ~ **coverage** wirksamer Ausbreitungsbereich *m*, Deckvermögen *n*, wirksame Überdeckung *f* ~ **cross-section(al) area** Wirkungsquerschnitt *m* ~ **current** Arbeiter-, Effektiv-strom *m* ~ **curve** Wirkungslinie *f* ~ **cut-off** effektive Grenzfrequenz *f* ~ **cut-off frequency** Nutzgrenzfrequenz *f* (acoust.) ~ **damping decrement** scheinbares logarithmisches Dämpfungsdekrement *n* ~ **date** Tag *m* des Inkrafttretens ~ **deionization** effektive Entionisationszeit *f* ~ **depth** Nutzhöhe *f* ~ **diameter** Flankenmaß *n*, lichte Weite *f* ~ **diameter of female thread** Flankendurchmesser *m* des Muttergewindes ~ **efficiency** Wirkleistung *f* ~ **field** anklingendes Feld *f* ~ **fire** wirksames Feuer *n*, Wirkungsschießen *n* ~ **floor space** bebaute Fläche *f* ~ **gap capacitance** effektiver Spaltleitwert *m* ~ **half life** reelle Halbwertzeit *f* ~ **head** Wirkdruck *m* ~ **heat drop** Nutzgefälle *n* ~ **height** scheinbare oder wirksame Höhe *f*, Effektiv-, Nutz-, Reflexions-höhe *f* ~ **height of an antenna** Strahlhöhe *f* oder Strahlungshöhe *f* einer Antenne ~ **input admittance** wirksame Eingangsadmittanz *f* ~ **input capacitance** effektive Eingangskapazität *f* ~ **input impedance** effektive Eingangsimpedanz *f* ~ **input noise** Eingangsrauschtemperatur *f* ~ **landing area** nutzbares Landungsgebiet *n* ~ **length** Knicklänge *f*, Nutzlänge *f* ~ **length of a bolt** Klemmlänge *f* eines Schraubenbolzens ~ **logarithmic decrement** scheinbares logarithmisches Dämpfungsdekrement *n* ~ **noise bandwidth** wirkliche Rauschbandbreite *f* ~ **number of neighbors** effektive Nachbarzahl *f* ~ **output** abgegebene Leistung *f* ~ **output admittance** effektive Ausgangsadmittanz *f* ~ **output impedance** wirksame Ausgangsimpedanz *f* ~ **part of cone ballistics** Kerngarbe *f* ~ **part of scale** Meßbereich *m* ~ **percentage modulation** wirklicher Modulationsgrad *m* ~ **pitch progress** Fortschritt *m* ~ **pitch of a propeller** Fortschrittssteigerung *f* ~ **pore area** Flächenporosität *f* ~ **portion of cone of fire** (ballistics) Kerngarbe *f* ~ **potential** Effektivpotential *f* ~ **power** Nutzeffekt *m*, -kraft *f*, -leistung *f*; Wirtschaftlichkeit *f* im Betrieb ~ **pressure** Arbeitsdruck *m* ~ **pressure of piston** effektiver Kolbendruck *m* ~ **pressure as shown by the manometer** Dampfüberdruck *m* ~ **pull** Belastungsspannung *f* ~ **radiated power** äquivalente Strahlungsleistung *f* ~ **radium content** effektive Radiumquantität *f* ~ **radius** Wirkungs-halbmesser *m*, -kreis *m* ~ **range** Schießweite *f*, Wirkungsbereich *m* ~ **range of fire** Feuerwirkungsbereich *n* ~ **range of gun** Schußbereich *m* ~ **reduction** einwandfreie Verarbeitung *f* ~ **reflection point** Reflexionsschwerpunkt *m* ~ **resistance** Dämpfungs-, Verlust-, Wirk-widerstand *m*, effektiver, induktionsfreier, oder nützlicher Widerstand *m* ~ **rotation** effektive Weglänge *f* ~ **separating density** Ausgleichwichte *pl* (Trennwichte) ~ **separating gravity** Teilungswichte *f* ~ **simple process factor** effektiver Trennfaktor *m* einer Stufe ~ **slit width** optimale Spaltbreite *f* (acoust.) ~ **solid angle** effektiver Raumwinkel *m* ~ **sound pressure** Schalldruck *m* ~ **strength** Effektiv-, Präsenz-stärke *f* ~ **substance** Wirkstoff *m* ~ **surface of parachute** Fallschirmwiderstandsfläche *f* ~ **target area** wirksame Fläche *f* der Treffplatte ~ **temperature** Strahlungstemperatur *f* ~ **thread diameter** Gewindeflankendurchmesser *m* ~ **transmission equivalent** Nutzdämpfung *f* ~ **transmission equivalent of a toll circuit** Nutzdämpfung *f* der Fernleitung ~ **traverse of a beam** wirksamer Winkelbereich *m* ~ **usefulness of the shutter** Nutzeffekt *m* des Verschlusses ~ **value** Effektivwert *m*, quadratischer Mittelwert *m*, Wirkungswert *m* ~ **value of a periodic quantity** Effektivwert *m* einer periodischen Größe ~ **voltage** Effektivspannung *f*, effektive Spannung *f* ~ **wavelength** effektive Wellenlänge *f* ~ **weight** Nutzlast *f* ~ **width of reinforcement** mittragende Breite *f* ~ **work** Nutzarbeit *f*

**effectiveness** Wirksamkeit *f*, Wirkung *f*, Wirkungsgrad *m* ~ **of current** Stromenergie *f* ~ **of a weapon** Waffenwirkung *f*

**effectives** Effektiv-, Präsenz-stärke *f*

**effects** Einflüsse *pl*

**effervesce, to** ~ aufbrausen, aufschäumen, efferveszieren, gären, mussieren, schäumen

**effervescence** Aufbrausen *n*, Aufschäumen *n*, Aufwallung *f*, Schäumen *n*

**effervescent** moussierend ~ **powder** Brausepulver *n* ~ **salt** Brausesalz *n* ~ **steel** unruhiger Stahl *m*

**efficacious** wirksam, wirkungsvoll

**efficacy** Wirksamkeit *f*, Wirkung *f*

**efficiency** einwandfreies Arbeiten *n*, Aufnahmefähigkeit *f*, Ausbeute *f*, Ausnutzung *f*, Empfindlichkeit *f*, Güte *f*, Güteverhältnis *n*, Kapazität *f*, Leistung *f*, (operational) Leistungsfähigkeit *f*, Nutzeffekt *m*, Wirksamkeit *f*, Wirkungsgrad *m*

**efficiency,** ~ **of airscrew** Schraubenleistung *f* ~ **of countertube** Ansprechwahrscheinlichkeit *f* (eines Zählrohres ) ~ **of economy** Ökonomiekoeffizient *m* ~ **of engine** Maschinenstärke *f* ~ **of a luminous source** Lichtausbeute *f* ~ **of pump** Antriebsleistung *f* ~ **of radiation** Strahlungsökonomie *f* ~ **of rectification** Gleichrichterwirkungsgrad *m* ~ **of separation** Trennungsgrad *m* ~ **in service** Dienstleistung *f* ~ **of the shutter** Nutzeffekt *m* des Verschlusses ~ **of transformation** Umformungswirkungsgrad *m* ~ **of transmission** (of a shutter) Durchlässigkeitsgrad *m* ~ **per unit** Einzelwirkungsgrad *m* ~ **of utilization** Nutz-effekt *m*, -wirkung *f* ~ **of working** Arbeitsgrad *m*

**efficiency,** ~ **circuit** Sparschaltung *f* (TV) ~ **dates** Gütedaten *pl* ~ **department** Leistungskontrollabteilung *f* ~ **diode** Schalter-, Spardiode *f* ~ **expert** Sparingenieur *m* ~ **factor** Güte-grad *m*, -zahl *f*, -ziffer *f* ~ **matrix** Aus-beutematrix *f* ~ **rating** Wirkungsgrad *m* ~ **report** Qualifikationsbericht *m* ~ **(or production) stoppage** Leistungsausfall *m* ~ **tensor** Ausbeutetensor *m* ~ **test** Leistungsprüfung *f*, Wirkungsgradbestimmung *f* ~ **-testing machine** Leistungsprüfmaschine *f*

**efficient** betriebswissenschaftlich, leistungsfähig, tüchtig, wirksam, zweckdienlich, zweckentsprechend ~ **utilization** Ausnutzung *f*

**effloresce, to** ~ (paint, films) abblättern, ansetzen, beschlagen, effloreszieren, verwittern

**efflorescence** Anflug *m*, Ausblähung *f*, Ausblühung *f*, Ausscheidung *f*, Ausschlag *m*, Auswittern *n* (von Salzen), Auswitterung *f*, Beschlag *m*, Verwitterung *f* **to be in** ~ ausblühen

**effluent** Ablaufwasser *n*, Ausfluß *m*, ausströmendes Mittel *n*; ausfließend, ausströmend **(outflowing)** ~ Abfluß *m* ~ **by-pass flap** (jet) Nebenauslaßklappe *f* ~ **(sewage)** Sielwasser *n*

**efflux** glatter Abfluß *m*, Ausfluß *m*, Ausflußmethode *f* ~ **of energy** Energieabfluß *m* ~ **of gas** Gasausströmung *f* ~ **condenser** Abflußkühler *m* ~ **nozzle** Ausströmdüse *f*

**effort** Anstrengung *f*, Bestreben *n*, Bestrebung *f*, Kraft *f*, Mühe *f*, Versuch *m* ~ **arm** langer Hebelarm *m*

**effusiometer** Ausströmungsmesser *m*

**effusion** Ausguß *m*, Effusion *f*, Erguß *m* ~ **phenomena** Ausströmerscheinungen *pl*

**egg** Ei *n* ~ **calipers with slider** Ellipsenzirkel *m* mit Gleitführung ~ **coal** Eierkohle *f* ~ **coke** Brechkoks *m* ~ **crate decking** Lichtgitterrost *m* ~ **crate fixture** Rasterleuchte *f* ~ **insulator** Eierisolator *m* ~ **laying (or deposit)** Eiablage *f* ~ **-shaped** eiförmig, eirund ~ **-shaped section** Eiquerschnitt *m* ~ **shell finish** Eierschalmattierung *f* ~ **-shell luster (or polish)** Eierschalenglanz *m* ~ **whisk** Schneeschläger *m*

**egress** Ausgang *m*, Austritt *m*

**Egyptian type** Blockschrift *f*

**eigen,** ~ **frequency** Eigenfrequenz *f* ~ **function** Eigen-funktion *f*, -zustand *m* ~ **-temperature** Eigentemperatur *f* ~ **value** Eigenwert *m*

**eight** Acht *f*; acht ~ **-barreled** achtläufig ~ **bells** acht Glas *n* ~ **-cylinder in-line engine** Achtzylinderreihenmotor *m* ~ **-cylinder inverted V engine** hängender Achtzylinder-V-Motor *m* ~ **-cylinder radial engine** Achtzylindersternmotor *m* ~ **-day clock** Achttageuhr *f* ~ **-electrode type of tube (or valve)** Achtpolröhre *f* ~ **-film reel (or spool)** Achtspule *f* ~ **-hour working day** Achtstundentag *m* ~ **-hourly wave** dritteltägige Welle *f* ~ **lines pica** Real *f* ~ **-lock knitting machine** Achtschloßstrickmaschine *f* ~ **-point recorder** Achtfachschreiber *m* ~ **-reel press** Achtrollenmaschine *f* ~ **-shaped pattern** Ziffernumrahmung *f* ~ **-speed transmission** Achtganggetriebe *n* ~ **-wire core** Achter *m*, Achterbündel *n*

**eightfold** achtfach ~ **twisting** (cable) Achterverseilung *f*

**eighth** Achtel *n*

**eikonal,** ~ **equation** Eikonalgleichung *f* ~ **function** Eikonalfunktion *f*

**eikonogen** Eikonogen *n*

**eikonometer** Ikonometer *n*

**einsteinium isotopes** Einsteinium-Isotopen *pl*

**Einthoven galvanometer** Saitengalvanometer *m*

**either, in** ~ **direction** vor- und rückwärts
**either-or-** exklusives oder- (data proc.)
**eject, to** ~ ausstoßen, austreiben, auswerfen, herausschleudern, vorschieben
**eject-condenser** Mischdüse *f*
**ejected,** ~ **block** (ejectum) Auswürfling *f* ~ **particle** emittiertes Teilchen *n*
**ejecter** see ejector
**ejecting,** ~ **machine** Ausstoßmaschine *f* ~ **piston** Ausstoßkolben *m* ~ **pressure** Abspritzdruck *m* ~ **seat** Schleudersitz *m*
**ejection** Auswurf *m* ~ **of cartridge case** Hülsenauswurf *m* ~ **and inflation test** Auswurf- und Aufblasversuch *m*
**ejection,** ~ **control hub** Vorschubkontrollbuchse *f* ~ **mandrel** Ausstoßdorn *m* ~ **opening** Auswurföffnung *f* ~ **seat** Schleudersitz *m* ~ **well** Auspreßsonde *f*
**ejector** Ausstoßer *m*, Auswerfer *m*, Auswerferstift *m*, Düse *f*, Ejektor *m*, Patronenauswerfer *m*, Saugstrahlpumpe *f*, Strahlabsaugepumpe *f* (aviat.), Strahlpumpe *f*, Strahlsauger *m* ~ **booster pump** Dampfstrahlpumpe *f* ~ **condenser** Dampfstrahlverdichter *m* ~ **diffusion pump** Dampfstrahlpumpe *f* (Quecksilber) ~ **exhaust** Auspuff *m* mit Abgasschubausnutzung ~ **mechanism** Auswerfervorrichtung *f* ~ **nose** Anschlagnase *f* ~ **nozzle** Strahldüse *f* ~ **nut of breech mechanism** Spannfalle *f* ~ **opening** Auswurföffnung *f* ~ **pin** Aushebestift *m*, Auswerferbolzen *m* ~ **pipe** Abgasschubrohr *n* ~ **securing bolt** Auswerfervorstecker *m* ~ **sleeve** Auswerfermuffe *f* ~ **(or pushing) slide** Ausstoßschlitten *m* ~ **spring** Ausstoßfeder *f* ~ **tube** Ausstoßrohr *n*
**elaborate, to** ~ ausarbeiten, sorgfältig ausgearbeitet
**elaborate** ausgeklügelt, kompliziert (Arbeit)
**elaboration** Ausarbeitung *f*, Erweiterung *f*
**elaidic acid** Elaidinsäure *f*
**elalometer** Eläometer *n*
**elapse, to** ~ verfließen, verstreichen
**elapsed,** ~ **time** Zeitfolge *f* ~-**time meter** Betriebsstundenzähler *m*
**elastance** Elastenz *f*, dielektrische Festigkeit *f* oder Widerstandsfähigkeit *f*
**elastic** Gummiband *n*; dehnbar, elastisch, federhart, federkräftig, fedrig, gefedert, nach-federnd, -giebig; spannkräftig **to be** ~ federn
**elastic,** ~ **abrasive head** elastischer Schmirgelkörper *m* ~ **after-effect** Nachfließen *n*, elastische oder plastische Nachwirkung *f* ~ **afterflow (or afterworking)** elastische Nachwirkung *f* ~ **axle box** federndes Achslager *n* ~-**band weaver** Gummizeugweber *m* ~ **bands and ribbons** Gummibänder *pl* ~ **belt** Kautschuktreibriemen *m* ~ **bonding** Schellackbindung *f* ~ **breakdown** Dauerbruch *m* ~ **cable** Gummiseil *n* ~ **chuck** Klemmfutter *n* ~ **collision** elastischer Stoß *m* ~ **cord** Abfederungsseil *n*, Gummischnur *f* ~ **coupling** elastische oder nachgiebige Kupplung *f* ~ **curve** Bieglinie *f*, elastische Linie *f* (mech.) ~ **deformation** federnde oder elastische Formänderung *f* ~ **diaphragm** Federungsblock *m* ~ **driving gear** elastisches Vorgelege *n* ~ **elongation** elastische Dehnung *f* ~ **fabric** Kautschukgewebe *n* ~ **flow** elastischer Strom *m* ~ **force** Feder-, Rück-

stell-, Spann-kraft *f* ~ **grommet** elastische Schutzmuffe *f* ~ **gum** Gummielastikum *n* ~ **head** Elastikkopf *m* ~-**isotropic semi-infinite body** elastisch-isotropischer Halbraum *m* ~ **limit** Dehnungsgrenze *f*, Elastizitätsgrenze *f*, elastische Grenze *f*, Streckgrenze *f* ~-**loop dynamometer** Meßschlange *f* ~ **non-rubber fabrics** (gummilos) elastisches Gewebe *n* ~ **pipe suspension** Rohraufhängung *f* (federnd) ~ **quality** Federeigenschaft *f* ~ **range** Elastizitätsgebiet *n* ~ **ratio** Elastizitätsgrad *m* ~ **recovery (or refreshment)** elastische Nachwirkung *f* ~ **resistance** elastischer Widerstand *m* ~ **restoration** elastische Rückkehr *f* zur Ursprungsform ~ **restraining force** Federfesselungskraft *f* ~ **scattering** elastische Streuung *f* ~ **shoulder** (lug) federnde Nase *f* ~ **spring** Spring-, Sprungfeder *f* ~ **strain** elastische Beanspruchung *f* oder Belastung *f*; elastische Formänderung *f* ~ **stress** elastischer Spannungszustand *m* ~ **stop nut** Mutter *f* mit Fibereinlage oder mit Sicherung durch elastische Verformung eines Fiberringes ~ **suspension** Federfesselung *f* ~ **suspension system** elastische Aufhängung *f* ~ **temper** Federhärte *f* ~ **test** elastischer Belastungsversuch *m* ~ **thread** Gummifaden *m* ~ **tie rod** elastisches Zugband *n* ~ **traction rod passing right through** durchgehende federnde Zugstange *f* (r. r.) ~ **waves** elastische Wellen *pl* ~ **web** elastisches Band *n*
**elastical** federnd
**elastically and inelastically scattered** elastisch und unelastisch gestreut
**elasticator** Elastizitätsverstärker *m*
**elasticity** Dehnbarkeit *f*, Dehnung *f*, Dehnungsfähigkeit *f*, Durchfederung *f*, Elastizität *f*, Federkraft *f*, (of flexure) Federung *f*, Fedrigkeit *f*, Prallkraft *f*, Schnellkraft *f*, Spannkraft *f*, Springkraft *f*, Verdichtsteife *f*, Weichheit *f*, Zügigkeit *f* ~ **of compression** Druckelastizität *f* ~ **of extension** Zugelastizität *f* ~ **of frame** Rahmendurchfederung *f* ~ **in shear** Gleitmaß *n*
**elasticity,** ~ **constant** Federkonstante *f* ~ **correction on gun** Ausschalten *n* der Temperatureinflüsse
**elastomeric** gummiartig ~ **compounds** weichgestellte Massen *n*
**elastometer** Elastizitätsmesser *m*
**elatation cross section** Elatationswirkungsquerschnitt *m*
**elaterometer** Dampfmesser *m*
**elbow** Ellbogen *m*, Krümmer *m*, Kröpfung *f*, Schenkelrohr *n* (mech.), Winkel *m*, Winkelstecker *m*, Winkelstück *n*, (square ~) Knierohr *n* ~ **of steam exhaust system** Abdampfkrümmer *m*
**elbow,** ~ **board** Lattenbrett *n* ~ **connection** gewinkeltes Verbindungsstück *n* ~ **connector** Anschlußkrümmer *m* ~ **cushion** Armlehne *f* ~ **duct** Leitungsknie *n*, Leitungskrümmer *m*, Winkelstück *n* ~ **grease gun nipple** Winkelnippel *m* ~ **joint** Kniegelenk *n*, Knieverbindung *f*, Krümmerverbindung *f* ~ **lever** Kniehebel *m* ~ **nipple** Winkelstutzen *m* ~ **piece** Kniestück *n*, Schenkelrohr *n* ~ **pliers** Eckrohrzangen *pl* ~ **rail** Armlehne *f* ~ **rest** Armauflage *f*, Fensterbrüstung *f* ~ **shaper** Kröpfform *f* ~ **sight** Win-

keloptik *f* ~ **telescope** Winkelfernrohr *n* ~ **tube** Schenkelrohr *n* ~ **tube pliers** Eckrohrzange *f*
**elbowed** gegliedert
**elect, to** ~ auswählen, wählen
**elector** Abzweigdose *f*
**electret** Elektret *n*
**electric,** ~ **(al)** elektrisch ~ **absorption** dielektrische Absorption *f* ~ **accounting machine** Tabelliermaschine *f* ~ **actuator** elektrischer Versteller *m* ~ **aftereffect** elektrische Nachwirkung *f* ~ **anode dissipation** Anodenverlustleistung *f* ~ **arc** Flammen-, Licht-bogen *m* ~**-arc furnace** Lichtbogenelektro-ofen *m* ~**-arc furnace for making steel** Lichtbogenelektrostahlofen *m* ~**-arc steel melting furnace** Lichtbogenstahlofen *m* ~**-arc welding** Elektroschmelzschweißung *f* ~ **beam transmitter** Stromstrahler *m* ~ **bell** Wecker *m* ~ **bill** Elektrizitätsrechnung *f* ~ **bistoury** Hochfrequenzkauter *m* ~ **braking** elektrische Bremsung *f* ~ **bulb** Birne *f* ~ **bus** Elektrobus *m* ~ **cable** Leitungskabel *n* ~ **calamine** Zink-glas *n*, -glaserz *n* ~ **capacitance of a conductor** Kapazität *f* eines Leiters ~ **capacity field** Kapazitätsfeld *n* ~ **cashregister** elektrische Registrierkasse *f* ~ **cast steel** Elektrostahlguß *m* ~ **cell** Stromzelle *f* ~ **central station** Elektrizitätswerk *n* ~ **chain hoist** Flaschenzug *m* mit elektrischem Antrieb ~ **charge** elektrische Ladung *f* ~ **chart drive** elektrischer Blattantrieb *m* ~ **circuit** Strom-kreis *m*, -netz *n* ~ **coal-face lighting in a coal mine** elektrische Abbaubeleuchtung *f* ~ **compressor** E-Verdichter *m* ~ **conduction mechanism** elektrischer Leitungsmechanismus *m* ~ **conductor** Schleifleitung *f* ~ **conduit** Isolierrohr *n* ~ **connection box** Stromanschlußkasten *m* ~ **contact** elektrische Zündung *f* ~ **control gear** elektrische Ausrüstung *f* mit Kontrollern ~ **counting tubes** elektrische Zählrohre *pl* ~ **coupling** elektrische oder kapazitive Kopplung *f* ~**-crane truck** Elektrokrankarren *m* ~ **current** (slang) Saft *m* ~ **deflection** elektrische Ablenkung *f* ~ **delay line** elektrische Verzögerungsleitung *f* ~ **demand** Energieverbrauch *m* (electr.) ~ **density** elektrische Dichte *f* ~ **detonator** Glüh-zünder *m*, -zündstück *n*, Zündmaschine *f* ~ **dipole** elektrischer Dipol *m* ~ **disc-type polisher** elektrischer Handpolierer *m* ~ **displacement** elektrische Verschiebung *f* ~ **double layer** Dipolschicht *f* ~ **drum** Elektrotrommel *f* ~ **dump truck** Elektrokarren *m* mit Kippmulde ~ **dynamometer** Bremsstand *m* mit elektrischer Leistungsaufnahme ~ **elevating-platform truck** Elektrohubwagen *m* ~ **endosmosis** karaphorische Wirkung *f* ~ **energy** Stromarbeit *f* ~ **eye** Abstimm-anzeigerröhre *f*, -zeiger *m*, optische Abstimmung *f*, Fotozelle *f*, fotoelektrische Zelle *f* ~ **fan** elektrischer Fächer *m* ~ **field** elektrisches Feld *n* ~ **field at the interface** Randfeldstärke *f* ~**-field strength** elektrische Feldstärke *f* ~ **firing** Glühdrahtzündung *f* ~ **floor polisher** Blocher *m* ~ **flux** Verschiebungsstrom *m* ~ **force** elektrische Feldstärke *f* ~ **freight truck** Elektrokarren *m* ~ **furnace** Elektro-ofen *m* ~**-furnace melting** Elektrostahlschmelzen *n* ~**-furnace steel** Elektrostahl *m* ~**-furnace voltage** Elektro-ofen-

spannung *f* ~ **generator** Generatorsatz *m*, Stromerzeugungsanlage *f* ~ **generator sets** Elektroaggregate *pl* ~**-hearth furnace** Elektroherdofen *m* ~ **heater bar** elektrischer Heizstab *m* ~ **heating ventilator** elektrischer Heizlüfter *m* ~ **hoisting gear** Elektrohubwerk *n* ~ **hygrometer** elektrisches Hygrometer *n* ~ **igniter** Glüh-zünder *m*, -zündstück *n* ~ **ignition** Kerzenzündung *f* ~**-ignition engine** Motor *m* mit Funkenzündung ~ **immersion pump** Elektro-Tauchpumpe *f* ~ **inertia starter** Schwungkraftanlasser *m* für elektrischen Betrieb ~ **influence** elektrostatische Beeinflussung *f* ~ **lamp for lighting** elektrische Ausleuchtung *f* ~ **lamp for lighting off** elektrische Ableuchtlampe *f* ~**-lamp signal** Lichtschauzeichen *n* ~ **light supply** Lichtleitungsanschluß *m* ~ **lighting** elektrische Beleuchtung *f* ~**-lighting carbon** Lichtkohle *f* ~ **line** Leitung *f* ~ **line of force** elektrische Kraftlinie *f* ~ **load** elektrische Ladung *f* ~ **lock** Glühzünderschloß *n* ~ **mains** Netz *n* ~ **master clock** elektrische Normaluhr *f* ~ **melting furnace** Elektro-schmelzofen *m*, -stahlofen *m* ~ **meter** Elektrizitätszähler *m* ~ **microtime measuring device** elektrisches Kurzzeitmeßgerät *n* ~ **motor** Dynamo *m*, elektrischer Motor *m*, Elektromotor *m*, E-Maschine *f*
**electric-motor,** ~ **drive** Elektromotorantrieb *m* ~ **driven** elektrischer Antrieb *m* ~ **oil** Elektromotorenöl *n* ~ **operated** elektrischer Antrieb *m* ~ **pulley** Elektrotrommel *f* ~ **room** E-Maschinenraum *m*
**electric,** ~ **network** elektrische Schaltung *f* ~ **osmosis** kataphorische Wirkung *f* ~ **output** elektrische Leistung *f* (in Watt) ~ **pad** Heizkissen *n* ~**-phonograph sound box** Abspieldose *f* ~ **pig** Elektroroheisen *n* ~**-potential fall (or gradient)** Potential-, Spannungs-gefälle *n* ~ **power** Elektroantrieb *m*, Stromleistung *f*
**electric-power,** ~ **company** Elektrizitätswerk *n* ~ **house** elektrische Zentrale *f* ~ **station** Elektrizitätswerk *n* ~ **supply** Stromlieferung *f* ~**-supply network** Netzspeisung *f* ~ **tool** Kraftwerkzeug *n* ~ **travelling crane** Laufkran *m* mit elektrischem Antrieb
**electric,** ~ **precipitator** Elektroabscheider *m* ~ **pressure** elektrische Spannung *f* ~ **printing punch** Magnetschreiblocher *m* ~ **projector** elektrischer Scheinwerfer *m* ~ **propulsion** Elektroantrieb *m* ~ **pulley** Elektroflaschenzug *m* ~ **punch** Magnetlocher *m* ~ **quick blender** elektrisch angetriebener Schnellmischer *m* ~ **radio receiver** Lichtnetzempfänger *m*, Netz(anschluß)empfänger *m* ~ **radiator** freistrahlender Heizkörper *m* ~**-railway line** Strecke *f* für elektrischen Betrieb ~ **recorder of a round** elektrische Wächterkontrollanlage *f* ~ **(ore) reduction furnace** Elektroreduktionsofen *m* ~ **resistance meter** Widerstandsmesser *m* ~ **rigidity** Durchschlagsfestigkeit *f* ~ **round recorder** Wächterkontrollanlage *f* ~ **screen** elektrostatischer Schirm *m* ~ **screening** elektrostatische Abschirmung *f* ~ **shaft furnace** Elektro-hochofen *m*, -schachtofen *m* ~ **shock** elektrischer Ausfall *m* (electron.), elektrischer Schlag *m* ~ **shunt** elektrischer Nebenschluß *f* ~ **sign** Leuchtschild *n* (astron.) ~ **smelting furnace** Elektrohochofen *m*, -schmelzofen *m*, elektrischer Hoch-

ofen *m* ~ **smelting plant** Elektroschmelzanlage *f* ~ **sound box** Abtastdose *f* ~ **spark** elektrischer Funken *m* ~ **speed indicator** Tachodynamo *n* ~ **steam boiler** elektrischer Dampfkessel *m* ~ **steel** Elektro-eisen *n*, -flußstahl *m*, -stahl *m* ~ **steel-making process** Elektrostahlverfahren *n* ~ **steel plant** Elektrostahl-anlage *f*, -werk *n* ~ **strength** Durchschlagsfestigkeit *f*, dielektrische Festigkeit *f* ~ **stress** Beaufschlagung *f* ~ **sump** elektrischer Sumpf *m* ~ **supply** Elektrizitätsversorgung *f* ~**-supply line** Stromnetz *n* ~ **terminal board** Kontaktleiste *f* ~ **test** elektrische Untersuchung *f* ~ **time switch** elektrischer Zeitschalter *m* ~ **torch** elektrische Handlampe *f* ~ **torque** Anlaßstrom *m* ~ **traction** elektrischer Fahrbetrieb *m* ~ **transducer** elektrischer Übertrager *m* ~ **transformer** Stromtransformator *m* ~ **traversing gear** Elektrofahrwerk *n* ~ **trolley** Elektrokarren *m* ~ **trolley system** Elektrohängebahn *f* ~**-truck trailer** Elektrokarrenanhänger *m* ~**-type telegauge** Fernmeßinstrument *n* mit elektrischer Übertragung ~ **upsetting machine** Elektrostauchmaschine *f* ~ **water heater** Warmwasserbereiter *m* ~**-wave filter** Filter *m*, Frequenzsieb *n* ~ **waves** Hertzsche Wellen *pl* ~ **wax atomizers** elektrische Wachszerstäuber *pl* ~ **welding** Elektroschweißen *n*, elektrische Schweißung *f* ~ **wind** Effluvien *pl* ~ **wire** elektrischer Leitungsdraht *m*, Leitungsmaterial *n* ~ **wiring accessories** Leitungszubehör *n* ~ **work** Stromarbeit *f*

**electrical,** ~ **accessories** elektrisches Zubehör *n* ~ **accounting machine** Tabellier-, Lochkartenmaschine *f* ~ **actuator** elektrischer Steller *m* ~ **beam swinging** elektrische Schwenkung *f* der Richtcharakteristik ~ **biasing** (in film recording) Nullinienverlagerung *f* ~ **centering** elektrische Zentrierung *f* ~ **conduction** Elektrizitätsleitung *f* ~ **conductivity** elektrische Leitfähigkeit *f* ~ **connection** Netzzuführung *f* ~ **connection for retracting mechanism** Leiterantriebsanschluß *m* ~ **contact** Kontakteinrichtung *f* ~**-contact stop** Kontaktanschlag *m* ~ **control gear** elektrische Ausrüstung *f* ~ **control panel** Schaltbrett *n* ~ **corrosion protection** elektrischer Korrosionsschutz *m* ~ **coupling rigidly in phase** phasenstarre elektrische Verkopplung *f* ~ **crimp-style terminal** Quetschkabelschuh *m* ~ **depth sounder** Elektrolot *n* ~ **director** elektrisches Feuerleitrechengerät *n* ~ **discharge wire** Sprühdraht *n* ~ **discharger gear** Elektrizitätsableitungsvorrichtung *f* ~ **dust collector** Elektroabscheider *m* ~ **echo sounding** elektrische Lotung *f* ~ **energy** elektrische Arbeit *f* (Joule, Wattstunde) ~ **engineer** Elektro-ingenieur *m*, -techniker *m* ~ **engineering** Elektrotechnik *f* ~ **equipment** elektrische Ausrüstung *f* ~ **firing mechanism** elektrische Zündung *f* ~ **flux** elektrischer Induktionsfluß *m* ~ **follow-up mechanism** elektrischer Nachlaufgeber *m* oder Rückführgeber *m* ~ **forming** elektrische Formierung *f* ~ **friction machine** Reibungselektrisiermaschine *f* ~ **fuse** Abschmelzsicherung *f*, elektrischer Zünder *m* ~ **general-utility truck** Elektrokarren *m* ~ **harness assembly** Kabelgeschirr *n* ~ **heating bath (plumbum) for Ramsbottom Test** elektrisches Bleibad *n* für

Ramsbottom ~ **heating de-icer system** elektrisch beheizter Enteiser *m* ~ **ignition (or firing)** elektrische Zündung *m* ~ **image** Ladungs-, Potential-bild *n* ~ **image of a conductor as to earth** elektrostatisches Spiegelbild *n* ~ **impulse resulting from elementary area** Bildpunktsignal *n* ~ **industrial utility truck** Elektrokarren *m* ~ **industry** Elektroindustrie *f* ~ **inertia** elektrische Trägheit *f* ~**-insulation value** Durchschlagsfestigkeit *f* ~ **interconnection** Zusammenschalten *n* ~ **interlock board** Schalterwerk *n* ~ **interlocking post** elektrische Stellwerkanlage *f* ~ **length** elektrische Länge *f* ~ **load** Strombelastung *f* ~ **matching** elektrische Anpassung *f* ~ **measuring instrument** Meßgerät *n* für elektrische Größen ~ **noise** elektrisches Rauschen *n* ~ **operating of signals and switches** elektrische Signal- und Weichenstellung *f* ~**-optical location** elektrisch-optische Ortung *f* ~ **precipitator** (for gas cleaning) Elektrofilter *n* ~**-pressure transmitter** Rechenglied *n* ~ **radar location** elektrische Ortung *f* ~ **remote control** elektrische Fernbedienung *f* (rdr) ~ **replacement drop indicator** Fallklappe *f* mit elektrischer Rückstellung ~ **salinometer** elektrischer Salzgehaltmesser *m* ~ **sensing** elektrische Abfühlung *f* ~ **servo loop** elektrischer Regelkreis *n* ~ **sheet-steel lamination** Blechschnitt *m* ~ **shielding** elektrische Abschirmung *f* ~ **(engineering) shop** Elektrowerkstatt *f* ~ **storm** Gewitter *n* ~ **strain-gauge scale** elektrotensometrische Waage *f* ~ **strength** elektrische Festigkeit *f* ~ **subway** Kabelkanal *m* ~ **supply** elektrische Zuleitung *f* ~ **synchronizer** elektrischer Synchronisator *m* ~ **tape** Isolierband *n* ~ **tension** Spannung *f* ~ **test set** elektrisches Prüfgerät *n* ~ **time distribution system** Zentraluhrenanlage *f* ~ **trajectory tracing** Funkvermessung *f* ~ **transient** Einschaltvorgang *m* ~ **vehicle** Akkumulatorenfahrzeug *n* ~ **wiring plan** Stromlaufplan *m* ~ **zero** elektrischer Nullpunkt *m*

**electrically,** ~ **actuated** elektrisch betätigt ~ **biased** elektrisch vorgespannt ~ **bonded rubber hoses** elektrisch leitfähige Gummischläuche *pl* ~ **charged entanglement** elektrisches Hindernis *n* ~ **conductive** elektrisch leitend ~ **connected** elektrisch gekuppelt ~ **controlled variable-pitch airscrew** elektrisch verstellbare Schraube *f* ~ **displacement** elektrische Verschiebung *f* ~ **driven** elektrisch angetrieben ~ **driven blower** Elektrogebläse *n* ~ **driven pendulum hydroextractor** Elektropendelzentrifuge *f* ~ **heated boots** Heizstiefel *pl* ~ **heated pitot tube** elektrisch heizbares Staurohr *n* ~ **heated suit for high-altitude flying** Heizanzug *m* für Höhenflug ~ **heated thermocouple** Thermoumformer *m* ~ **interlocked** elektrisch gekoppelt oder verblockt ~ **long** elektrisch lang ~ **operated** elektrisch angetrieben, betätigt oder betrieben ~ **operated flaps** elektrisch betätigte Klappen *pl* ~ **operated retractable undercarriage** elektrisch einziehbares Fahrgestell *n* ~ **operated rotating top turret** Drehringlafette *f* ~ **operated shutter** elektrisch betriebener Verschluß *m* ~ **powered telephone** batteriegespeister Fernsprechapparat *m* ~ **suspended gyro** elektrostatischer Kreisel *m* ~ **welded** elektrisch geschweißt

**electrician** Elektriker *m*, Elektro-mechaniker *m*,

-techniker *m*, Installateur *m* ~ **crew** Elektrotrupp *m* ~ **fitter** Elektroinstallateur *m* ~ **squad** Elektrikertrupp *m*
**electrician's,** ~ **knife** Kabelmesser *n* ~ **pliers** Blitzrohrzange *f* ~ **side-cutter pliers** Beißzange *f*
**electricity** Elektrizität *f* ~ **of opposite sign** ungleichnamig Elektrizität *f* ~ **of same sign** gleichnamige Elektrizität *f*
**electricity,** ~ **container** Elektrospeicher *m* ~ **produced by dropping water on a hot plate** Wasserfallelektrizität *f* ~**-supply meter** Elektrizitätszähler *m*
**electrifiable** elektrisierbar
**electrification** Elektrifizierung *f*, Elektrisierung *f*
**electrified** elektrisch, elektrisiert ~ **railway** elektrisierte Eisenbahn *f*
**electrify, to** ~ elektrifizieren, elektrisieren
**electro,** to ~**-deposit** galvanisch niederschlagen **to** ~**-type** klischieren **to** ~**-weld** elektrisch schweißen
**electro-acoustic,** ~ **index** elektroakustisches Verhältnis *n* eines Empfangssystems ~ **transducer** elektroakustische Kette *f*, Schallumwandlungseinrichtung *f*, elektroakustischer Wandler *m*
**electro,** ~**-acoustics** Elektroakustik *f* ~**-analogue** Elektro-Analogon *n* ~ **analysis** Elektroanalyse *f* ~ **analytical** elektroanalytisch ~ **arteriograph** Blutfarbemesser *m* ~**-automatic doubling device** Selbstdoublierer *f* ~**-caloric** elektrokalorisch ~**-capillary** elektrokapillar ~**-cardiogram** Elektrokardiogramm *n*, Herzspannungskurve *f* ~**-cardiograph** Triograf *m* ~**-cautery** Glühkauter *m*
**electro-chemical** elektrochemisch ~ **accumulator** elektrochemischer Speicher *m* ~ **pickling** elektrolytisches Dekapieren *n* oder Reinigen *n* ~ **series** elektrochemische Spannungsreihe *f*
**electro,** ~ **chemist** Elektrochemiker *m* ~ **chemistry** Elektrochemie *f* ~ **chilled-iron foundry** Elektrohartgußgießerei *f* ~ **chilling** Elektrohartguß *m* (metall.) ~**-corundum** Edelkorung *n* ~ **culture** Elektrokultur *f* ~**-deposit** galvanischer Niederschlag oder Überzug *m* ~ **deposited** galvanisch gefällt ~ **deposition** Elektronenniederschlag *m*, elektrolytische Fällung *f*, Galvanisierung *f*, Galvanostegie *f*, galvanischer Niederschlag *m* ~ **deposition of lead** galvanische Verbleiung *f* ~**-dialysis** Elektrodialyse *f* ~**-disintegration** Elektronenzertrümmerung *f*
**electrodynamic** elektrodynamisch ~ **balance** Stromwaage *f* ~ **loud-speaker** fremderregter dynamischer Lautsprecher *m* ~ **meter** elektrodynamisches Meßgerät *n* ~ **microphone** (elektro)dynamisches Mikrofon, Tauchspulenmikrofon *n*
**electrodynamics** Elektrodynamik *f* ~ **in vacuo** Vakuumelektrodynamik *f*
**electro,** ~**-encephalogram** Elektroenzephalogramm *n* (E.E.G.) ~**-graphic ink** leitfähige Tinte *f* ~ **heating** Elektrowärme *f* ~**-hydraulic brake lifter** Eldrogerät *n* ~**-hydraulic control valve** elektro-hydraulisches Umsteuerventil *n* ~**-inductive** elektroinduktiv ~ **kineties** Bewegungselektrizitätslehre *f*, Lehre *f* von der elektrischen Strömung ~**-kymograph** Elektrokymograf *m* ~**-luminescence** Elektrolumineszenz *f*
**electrolyser** Elektrolyseur *m*

**electrolysis** Elektrolyse *f* ~ **tanks** Elektrolysenbehälter *m*
**electrolyte** Akkumulatorsäure *f*, Elektrolyt *m*, Füllsäure *f*, Lauge *f* ~ **circulation** Laugenumlauf *m* ~ **purification** Laugenreinigung *f* ~ **recovery** Aufarbeitung *f*
**electrolytic,** ~ **(al)** elektrolytisch ~ **area** Elektrodenfläche *f* ~ **arm** Elektroden-arm *m*, -ausleger *m* ~ **arrangement** Elektrodenanordnung *f* ~ **bath** Elektrolysenbad *n* ~ **capacitor** Elektrolytkondensator *m* ~ **carrying superstructure** Elektrodentragvorrichtung *f* ~ **cell** Elektrolysenbad *n*, Elektrolysiszelle *f* ~ **clamp** Elektroden-anschlußstück *n*, -fassung *f*, -durchführung *f* (metal) ~ **condenser** elektrolytischer Kondensator *m* ~ **copper** Elektrolytkupfer *n* ~ **corrosion** elektrolytische Zerstörung *f* ~ **deposit** elektrolytischer Niederschlag *m* ~ **detector** Elektrolytdetektor *m*, elektrolytischer Detektor *m*, elektrolytischer Wellenanzeiger *m*, elektrolytische Zelle *f* ~ **diaphragm** Elektrodenstellungsregler *m* ~ **lead refining** elektrolytische Bleiraffination *f* ~ **meter** Elektrolytzähler *m* ~ **parting** elektrolytische Raffination *f* ~ **photocell** Elektrolytzelle *f* ~ **polishing** Blankätzen *n* ~ **rectifier** Elektrolytgleichrichter *m*, elektrolytisches Ventil *n* ~ **refining** elektrolytische (Metall-) Raffination *f* ~ **refining of lead** Bleielektrolyse *f* ~ **refining practice** Elektrolysenbetrieb *m* ~ **refining process** Elektrolyseprozeß *m* ~ **resistance** Zersetzungswiderstand *m* ~ **responder** elektrolytischer Detektor *m* ~ **silver refining** Silberelektrolyse *f* ~ **slime** Elektrolysenschlamm *m* ~ **solution pressure** elektrolytischer Lösungsdruck *m*, Lösungstension *f* ~ **surface oxidation of metals in an oxalic bath** Eloxalverfahren *n* ~ **tank** elektrolytischer Trog *m* ~ **tin** Elektrolytzinn *n* ~ **tin-refining process** Zinnelektrolyseverfahren *n* ~ **valve** Polarisationszelle *f*, elektrolytisches Ventil *n*, elektrolytische Ventilzelle *f* ~ **vessel** Elektrolysiergefäß *n* ~ **zinc plant** Zinkelektrolysenanlage *f* ~ **zinc process** Zinkelektrolyse *f*
**electrolytically deposited** elektrolytisch aufgebracht (film)
**electrolyzation** Elektrolysierung *f*
**electrolyze, to** ~ elektrolysieren, durch elektrischen Strom zersetzen
**electrolyzer** Elektrolyser *m*, Elektrolyseur *m*
**electro-magnet** Elektromagnet *m*, fremderregter Magnet *m* ~ **armature** Elektromagnetanker *m* ~ **coil** Elektromagnetspule *f* ~ **core** Elektromagnetkern *m* ~ **winding** Elektromagnetwicklung *f*
**electro-magnetic** elektromagnetisch ~ **accelerometer** Beschleunigungsmeßkopf *m* ~ **braking** elektromagnetische Bremsung *f* ~ **clutch** Elektromagnetkupplung *f*, elektromagnetische Kupplung *f* ~ **contactor** elektromagnetischer Schaltschütz *m* ~ **cutout** strommagnetischer Selbstausschalter *m* ~ **disk** Wirbelstromscheibe *f* ~ **emission** elektromagnetische Ausstrahlung *f* ~ **flaw detector** Fehlerscheibchensucher *m* ~ **holding device** elektromagnetische Ausspannvorrichtung *f* ~ **horn** Hornstrahler *m* ~ **induction** elektromagnetische Beeinflussung *f* (Induktion) ~ **leakage** elektromagnetische Streuung *f* ~ **linkage** elektromagnetische Verkettung

*f* ~ **loud-speaker** elektromagnetischer Lautsprecher *m* ~ **mirror vibrator** Lichthahn *m* ~ **multidisc test rig** Elektrolamellenkupplungsprüfstand *m* ~ **oscillation** elektromagnetische Schwingung *f* ~ **oscillator** Blattfeder-, Magnetsummer *m* ~ **percussion welding** elektromagnetische Stoßschweißung *f* ~ **pickup** elektromagnetischer Tonabnehmer *m* ~ **plate** Wirbelstromscheibe *f* ~ **plunger** elektromagnetische Weiche *f* ~ **receiver** elektromagnetischer Fernhörer *m* ~ **screening** elektromagnetische Abschirmung *f* ~ **seismometer** elektromagnetischer Seismograph *m* ~ **separator** elektromagnetischer Abscheider *m* ~ **shunt** Gegenstromrolle *f* ~ **slipper brake** elektromagnetische Schienenbremse *f* ~ **spray valve** elektromagnetisch gesteuerte Einspritzdüse *f* ~ **system of measurement** elektromagnetisches Meßsystem *n* ~ **transmitter** Bandmikrofon *n* ~ **unit** elektromagnetische Einheit *f* ~ **wave** elektromagnetische Welle *f* ~ **wave velocity** Lichtgeschwindigkeit *f*
**electromagnetism** Elektromagnetismus *m*
**electromechanic** elektromechanisch
**electromechanical** elektromechanisch ~ **brush** elektrischer oder elektromechanischer Bohrer *m* ~ **drive** elektromechanische Steuerung *f* ~ **recording** elektromechanische Aufzeichnung *f* ~ **transducer** elektromechanischer Wandler *m*
**electro,** ~ **mechanics** Elektromechanik *f* ~ **metallurgist** Elektrometallurge *m* ~ **metallurgy of iron and steel** Elektrostahlverfahren *n*
**electrometer** Elektrizitätsmesser *m* ~ **tube** Elektrometerröhre *f* ~ **valve** Elektrometerröhre *f*
**electromotive** elektromotorisch ~ **force** elektromotorische Kraft *f* ~ **force measurements** Messung *f* elektromotorischer Kräfte ~ **force at liquid junctions** Potentialdifferenz *f* an der Berührungsstelle zweier verschiedener Lösungen ~ **force of polarization** Polarisationsspannung *f* ~ **force of self-induction** elektromotorische Kraft *f* der Selbstinduktion ~ **series** Spannungsreihe *f*
**electromotor** Elektromotor *m*
**electro,** ~-**negative** negativ elektrisch, unedel ~-**negativity** Elektronegativität *f*
**electro-optic** Elektro-Optik *f* ~ **(al) cell** elektrooptisches Lichtrelais *n* ~ **shutter** elektrooptischer Verschluß *m*
**electro-optics** Elektro-Optik *f*
**electro-osmosis** Elektroosmose *f*
**electropathy** Elektrotherapie *f*
**electro permalloy film** galvanischer Permalloyfilm *m*
**electrophone,** ~ **call** Verbindung *f* für Theaterübertragung ~ **system** Drahtrundspruchanlage *f*
**electrophonic effect** elektrofonischer Effekt *m*
**electrophoresis** Elektrophorese *f*, Ionophorese *f* ~ **cell** Elektrophoresezelle *f* ~ **ionic medication** Ionentherapie *f*, Iontophorese *f* ~ **paper strips** Papierelektrophoresestreifen *pl*
**electrophorus** Elektropher *m*
**electro,** ~-**photometer** lichtelektrisches Photometer *n* ~-**physical** elektrophysikalisch ~-**physics** Elektrophysik *f*
**electroplate, to** ~ galvanisieren, plattieren

**electroplate** Galvano *n*
**electroplated** galvanisch hergestellt, verzinkt
**electroplating** Elektroplattierung *f*, Galvanisieren *n*, Galvanostegie *f*, Galvano-technik *f*, -plastik *f*, galvanische Metallüberziehung *f* oder Plattierung *f*, galvanischer Überzug *m* ~ **with chromium** Hartverchromung *f* ~ **with steel** galvanische Verstählung *f*
**electroplating,** ~ **bath** galvanoplastisches Bad *n* ~ **plant** Galvanisier-anlage *f*, -anstalt *f* ~ **process** Elektroplattierverfahren *n* ~ **shop** Galvanisierwerkstatt *f*
**electropneumatic (al)** elektropneumatisch ~ **controller** elektropneumatischer Fahrschalter *m* ~ **regulator** Pumpenselbstschalter *m*
**electro,** ~-**positive** edel, positiv elektrisch ~-**refining** Elektroraffination *f* ~ **scope** Elektrizitätszeiger *m* ~-**sheet** Elektroblech *n* ~-**slag welding** Elektroschlackeschweißen *n*
**electrostatic,** ~ **(al)** elektrostatisch ~ **actuator** Eichel-, Hilfs-elektrode *f* (acoust.) ~ **adhesion** elektrische Klebkraft *f* ~ **bond** Ionenbindung *f* ~ **calibration** elektrostatische Eichung *f* ~ **capacity** elektrostatische Kapazität *f*, elektrisches Ladungsvermögen *n* ~ **centimeter-gram-second system** elektrostatisches C.G.S System *n* ~ **coupling** kapazitive oder elektrische Kopplung *f* ~ **deflection** elektrostatische Ablenkung *f* ~ **displacement** elektrische Verschiebung *f* ~ **feedback** kapazitive Rückkopplung *f* ~ **field** elektrostatisches Feld *n*, Influenzfeld *n* ~ **generator** Influenzmaschine *f* ~ **induction** elektrische Verschiebung *f* oder Induktion *f*, Elektrisieren *n* durch Influenz ~ **induction voltmeter** Feldstärkenmühle *f* ~ **lens** elektrische Linse *f* ~ **loudspeaker** elektrostatischer Lautsprecher *m* ~ **microphone** Kondensatormikrofon *n* ~ **percussion welding** Entladestoßschweißung *f* ~ **phenomena** elektrostatische Erscheinungen *pl* ~ **precipitation** elektrostatisches Arbeitsverfahren *n* ~ **precipitator** (for glas cleaning) Elektrofilter *n* ~ **receiver** Kondensatorfernhörer *m* ~ **retentive force** elektrische Klebkraft *f* ~ **screening** elektrostatische Abschirmung *f* ~ **shield** kapazitiver Schirm *m* ~ **shielding** kapazitive Schirmung *f* ~ **storage** elektrostatische Speicherung *f* ~ **system of measurement** elektrostatisches Meßsystem *n* ~ **unit** elektrostatische Einheit *f* ~ **valence** heteropolare Valenz *f*
**electrostatically screened (or shielded)** elektrostatisch geschirmt
**electro,** ~ **statics** Elektrostatik *f*, Ruheelektrizitätslehre *f* ~ **striction** Elektrostriktion *f* ~ **strictive force** elektrostriktive Kraft *f* ~ **technical** elektrotechnisch ~ **therapeutic light apparatus** elektrischer Lichtheilapparat *m*
**electrothermal** elektrothermisch ~ **recording** elektrothermische Registrierung *f* ~ **relay** Thermorelais *n*
**electrothermics** Elektrowärmelehre *f*
**electro,** ~ **tonus** Elektrotonus *n* ~-**type** galvanographische Abbildung *f*, Elektrotyp *m*, Galvano *n* ~-**typer** Galvanoplastiker *m* ~-**typing** Galvanotypie *f* ~-**valence** Elektrovalenz *f* ~-**valent compound** elektronenvalente Bindung *f* ~-**welding** elektrisches Aufschweißen *n*, elektrische Schweißung *f*

**electrode** Drahtnetzelektrode *f*, Elektrodenkitt *m*, (wrench) Elektrodenspanner *m*, Poldraht *m* ~ **with a gasproducing coating** gaserzeugende Mantelelektrode *f*

**electrode,** ~ **adjusting gear** Elektrodenstellungsregler *m* ~ **admittance** Elektroden-Scheinleitwert *m* ~ **area** Elektrodenfläche *f* ~ **arm** Elektroden-arm *m*, -ausleger *m* ~ **arrangement** Elektrodenanordnung *f* ~ **bias** Elektrodenvorspannung *f* ~ **carrying superstructure** Elektrodentragvorrichtung *f* ~ **characteristic** Elektrodenkennlinie *f* ~ **clamp** Elektrodenanschlußstück *n*, Elektrodenfassung *f* ~ **clip** Elektrodenklammer *f* ~ **collar** Elektroden-fassung *f*, -durchführung *f* (metall.) ~ **compound** Elektrodenkitt *m* ~ **conductance** Elektrodenwirkleitwert *m* ~ **connection** Elektroden-anschluß *m*, -schaltung *f* ~ **contact** Elektrodenanschluß *m* ~ **control** Elektroden-führung *f*, -regulierung *f*, -stellungsregelung *f* ~ **control device** Elektroden-regler *m*, -regelvorrichtung *f* ~ **dissipation** Elektrodenverlustleistung *f* ~ **economizer** Elektrodenabdichtung *f*, wassergekühltes Elektrodenanschlußstück *n* ~ **end** Elektrodenende *n* ~ **force** Elektrodendruck *m* ~ **gap** Elektrodenabstand *m* ~ **gap gauge** Elektrodenabstandslehre *f* (Zündkerze) ~ **gland** Elektrodenabdichtung *f* ~ **head** Elektrodenkopf *m* ~ **hearth furnace** Elektrodenherdofen *m* ~ **holder** Elektroden-fassung *f*, -halter *m*, Schweißzange *f* ~ **impedance** Elektroden-Scheinwiderstand *m*

**electrodeless** elektrodenlos ~ **tube** Nullode *f*, elektrodenlose Röhre *f*

**electrode,** ~ **mast** (electric furnaces) Elektrodenfahrsäule *f*, Elektrodenständer *m* ~ **plate** Elektrode *f* ~ **potential** Elektrodenspannung *f* ~ **radiator** Kühlflügel *m* ~ **reactance** Elektroden-Blindwiderstand *m*, Elektrodenreaktanz *f* ~ **regulation** Elektroden-regulierung *f*, -verstellung *f* ~ **regulator** Elektrodenregeler *m* ~ **resistance** Elektroden-Wirkwiderstand *m* ~ **rod** Elektrodenstift *m* ~ **shell** Elektrodenmantel *m* ~ **short circuit** Elektrodenschluß *m* ~ **spacing** Elektrodenabstand *m* ~ **support** (electric furnace) Elektroden-fahrsäule *f*, -fläche *f*, -halter *m*, -ständer *m*, -tragkonstruktion *f*, -träger *m* ~ **surface** Elektrodenoberfläche *f* ~ **susceptance** Elektrodenblindleitwert *m* ~ **terminal** Elektrodenanschlußstücke *n* ~ **tip** Elektrodenkopf *m* ~ **tips** (spot welder) Elektrodenspitze *f* ~ **tube** Elektrodenröhre *f* ~ **voltage** Elektrodenspannung *f* ~ **waste** Elektrodenreste *pl* ~ **wear** Elektrodenverschleiß *m*

**electron** (a metal) Elektrometall *n*, (an alloy) Elektron *n*, Negaton *n* ~ **accumulation** Elektronenstauung *f* ~ **affinity** Austrittsarbeit *f*, Elektronenenergie *f* ~ **affinity spectra** Elektronenaffinitätsspektren *pl* ~ **amplification** Elektronenvervielfachung *f* ~ **attachment** Elektronen-anlagerung *f*, -einfang *m* ~ **avalanche** Elektronenlawine *f* ~ **beam** Brennpunkt *m*, Elektronen-abstaststrahl *m*, -bündel *n*, -strahl *m*, -strahlabtaster *m*, -strahlbündel *n* ~**-beam intensity** Elektronenintensität *f* ~**-beam magnetometer** Elektronenstrahlmagnetometer *n* ~**-beam scanner** Kathodenstrahlabtaster *m* ~**-beam tube** Braunsche Röhre *f*, Elektronen-, Kathoden-strahlröhre *f* ~**-beam valve** Elektronen-

strahlröhre *f* ~ **bombardment** Auftreffen *n* der Elektronen, Elektronenbeschießung *f* ~ **"brain"** elektronisches Rechengerät *n* ~ **brush** Elektronenstrahlabtaster *m* ~ **burst** Elektronenblitz *m* ~ **camera truck** Aufnahmewagen *m*, Fernsehaufnahmewagen *m* ~ **capture** Elektronen-auffang *m*, -einfang *m* ~ **casting** Elektrongußstück *n* ~ **catch-up region** Elektronenüberholungsgebiet *n* ~**-charge mass ratio** spezifische Ladung *f* des Elektrons ~ **cloud** Elektronen-hülle *f*, -stauung *f* ~ **collection** Elektronenabsaugung *f* ~ **collector** Auffangelektrode *f* ~ **computer** elektronisches Rechengerät *n* ~ **coupled oscillator** elektronengekoppelter Oszillator *m* ~ **coupling** Elektronenkuppelung *f* ~ **crankcase** Elektronkurbelgehäuse *n* ~**-current image** Elektronenbild *n* ~ **density** Elektronen-dichte *f*, -konzentration *f* ~ **detachment** Ablösung *f* eines Elektrons ~ **diffraction** Elektronenbeugung *f* ~ **diffraction camera** Elektronenbeugungskammer *f* ~**-diffraction lattice** Elektronenbeugungsgitter *n* ~**-diffraction pattern** Elektronenbeugungsdiagramm *n* ~ **discharge** Elektronenentladung *f* ~ **discharge relay** Elektronenrelais *n* ~ **drift** Elektronenfluß *m* ~ **efficiency** Elektronenausbeute *f* ~**-electron scattering** Elektron-Elektron-Streuung *f* ~ **emission** Elektronen-ausbeute *f*, -auslösung *f*, -emission *f* ~**-emitting** elektronenaussendend ~ **emitting (or emission) area** Elektronenabgabefläche *f* ~ **emitter** Elektronenquelle *f* ~ **energy** Elektronenenergie *f* ~**-etching plate** Elektronätzplatte *f* ~ **evaporation** glühelektrische Elektronenemission *f*, Elektronenverdampfung *f* ~ **flow** Elektronen-fänger *m*, -fluß *m* ~ **focusing** Elektronenkonzentration *f* ~ **gas** Elektronengas *n* ~ **gun** Elektronen-rohr *n*, -schleuder *m*, -spritze *f*, Kathodenstrahlerzeuger *m*, Sauganode *f*, Strahlerzeugung *f*, Strahlerzeugungssystem *n* ~ **halftone** Elektronautotypie *f* ~ **image** Elektronenabbildung *f* ~ **image dissector** Elektronenbildzerleger *m* ~ **image pickup** Elektronenbildfang *m* ~ **image tube** Elektronenbildröhre *f* ~ **impact** Elektronenstoß *m* ~ **incendiary bomb** Elektronbrandbombe *f* ~ **indicator tube** magisches Auge *n* ~ **inertia** Elektronenträgheit *f* ~ **injector** Elektroneninjektor *m* ~ **input** Elektronenzufuhr *f* ~ **interaction short range** kurzreichweitige Elektronenwechselwirkung *f* ~ **interference** Elektroneninterferenz *f* ~**-ion recombination** Elektron-Ion-Rekombination *f* ~**-ion wall recombination** Elektron-Ion-Wandrekombination *f* ~**-lattice model** Elektronengittermodell *n* ~ **level** Elektronenniveau *n* ~ **lever** Elektronenstrahlabtaster *m* ~ **line-plate** Elektrostrichätzung *f* ~ **locus** Elektronenort *m* ~ **machine** Elektronenbohrmaschine *f* ~ **micrograph** elektronenmikroskopische Aufnahme *f* ~ **microscopy** Übermikroskopie *f* ~ **mirror** Parallelelektrode *f*, Sekundäremissionskathode *f* ~ **mobility** Elektronenbeweglichkeit *f* ~ **multiplication** Sekundäremissionsverstärkung *f* ~ **multiplier** Sekundäremissionvervielfacher *m* ~ **multiplier phototube** Fotozellenverstärker *m*, Sekundäremissionsvervielfacher *m*, Vervielfacherzelle *f* ~ **multiplier tube** Elektronenvervielfacher *m* (Fotozelle), Sekundäremissionsvervielfacher *m* ~**-neutron interaction**

Elektron-Neutron Wechselwirkung *f* ~-optical elektronenoptisch ~-optically corrected elektronenoptisch korrigiert ~ optics Elektronenoptik *f* ~ orbit Elektronenbahn *f* ~-oscillation detector Gegentaktbremsfeldaudion *n* ~-ocsillation frequency Pendelfrequenz *f* ~-oscillation rectifier Bremsfeld-, Gegenbremsfeld-, Gegentaktbremsfeld-audion *n* ~-oscillation scheme with reflecting electrode Bremsfeldschaltung *f* ~ oscillations Elektronen-, Laufzeit-schwingungen *pl* ~ over-take region Elektronenüberholungsgebiet *n* ~ path Elektronen-bahn *f*, -strecke *f* ~-path oscillation Leitbahnwelle *f*, Magnetronleitbahnwelle *f* ~ pencil Elektronenabtaststrahl *m* ~-phonon interaction parameter Elektron - Phonon -Wechselwirkungsparameter *n* ~ plasma Elektronenplasma *n* ~ plate Elektronklischee *n* ~ position Elektronenort *m* ~ probe Elektronensonde *f* ~ process Elektronenbohrverfahren *n* ~ promotion Elektronenübergang *m*, Quantenzahlenvergrößerung *f* bei Elektronen ~ quantity modulation Elektronenstrommodulation *f* ~ ray Elektroden-, Kathoden-strahl *m* ~-ray indicator tube Abstimmanzeigeröhre *f* ~-ray path Elektronenstrahlengang *m* ~ relay Elektronenrelais *n* ~ resonant wave length Eigenwelle *f*, Eigenwellenlänge *f* ~-scan microscope Elektronenrastermikroskop *n* ~ scan microscope Rastermikroskop *n* ~ scanning brush Elektronenabtaststrahl *m* ~-scanning pencil Elektronenstrahlabtaster *m* ~ scanning spot Elektronenabtaststrahl *m* ~ screening Kernabschirmung *f* ~ sheath Elektronenschicht *f* ~ shell Elektronenhülle *f* ~ shells Elektronenschalen *pl* ~ source Elektronenquelle *f* ~ specific heat spezifische Elektronenwärme *f* ~ spectroscopy Elektronoskopie *f* ~ spin Elektronendrall *m* ~ spin resonance Elektronenspinresonanz *f* ~ spot Abtaststrahl *m*, Elektronenfleck *m*, Elektronenstrahlabtaster *m* ~ state Elektronenzustand *m* ~ stream Elektronenstrahl *m*, -strömung *f* ~-stream potential Bündelpotential *n* ~ switch Elektronenschalter *m* (rdo) ~ term Elektronenterm *m* ~ thermionic relay Elektronenrelais *n* ~ thermite bomb Elektronthermitbombe *f* ~ trajectory Elektronenbahn *f*, Bahn *f* eines Elektrons ~ transfer Elektronenübergang *m* ~ transit time Elektronenlaufzeit *f* ~ transition Elektronenübergang *m* ~ transition probability Elektronenübergangswahrscheinlichkeit *f* ~ transition (or jump) spectrum Elektronensprungspektrum *n* ~-transmitting (or permeable) elektronendurchlässig ~ trap Elektronenhaftstelle *f* ~ trapping Elektroneneinfang *m* ~ treatment Elektronenmodell *n* ~ tube Elektronenröhre *f*, Vakuumröhre *f*

electron-tube, ~ base gauge Röhrensockellehre *f* ~ generator Röhren-generator *m*, -summer *m* ~ rack Röhrengestell *n* ~ rectifier Elektronenröhrengleichrichter *m* ~ terminal Elektronenröhrenklemme *f* ~ tester Röhrenprüfgerät *n*

electron, ~-type rectifier Hochvakuumgleichrichter *m* ~ vacancy Defektelektron *n* ~ valve relay Elektronenrelais *n* ~ voltage Elektronenspannung *f* ~ warm-pressed parts Elektronwarmpreßteile *pl* ~ wave Elektronen-, Materie-,

Phasen-welle *f* ~ wave tube Elektronenwellenröhre *f*

electronic elektronisch ~ and lattice components Elektronen- und Gitterkomponente *f* ~ altimeter set elektronischer Höhenmesser *m* ~ amplification Röhrenverstärkung *f* ~ amplifier Elektronenverstärker *m* ~ autopilot elektronisch gesteuerter Kursgeber *m* ~ band spectra Elektronenbandenspektren *pl* ~ beam Elektronenbündel *n*, -strahl *m* ~ Bohr magneton Bohrsches Magneton *n* ~ bomb Elektronenbombe *f* ~ brain Elektronengehirn *n* ~ calculating punch elektronischer Rechenlocher *m* ~ calculator elektronisches Rechengerät *n* ~ charge Elementarladung *f* ~ circuit Emissionsstrom *m*, Röhrenschaltung *f* ~ commutator elektronischer Schalter *m* ~ computer elektronische Rechenmaschine *f* ~ conduction caused by intrinsic impurities Eigenstörstellenleitung *f* ~ contactor Schaltschütz *n* ~ control Elektronensteuerung *f*, elektronische Regelung *f* ~ counting register elektronisches Zählwerk *n* ~ crack detector Fehlerscheibchenspür *m* ~ cross-section Elektronenquerschnitt *m* ~ current Elektronen-, Emissions-strom *m* ~ dewpoint recorder elektronischer Taupunktschreiber *m* ~ diffraction diagram Elektronenbeugungsdiagramm *n* ~ diffraction grating Elektronenzeugungsgitter *n* ~ discharge vessel elektrisches Entladungsgefäß *n* ~ editor elektronische Dokumentationsmaschine *f* ~ efficiency elektronischer Wirkungsgrad *m* ~ energy states in crystals Energiezustände *pl* des Elektrons im Kristall ~ excitation Elektronenanregung *f* ~ fire detection device elektronischer Feuermelder *m* ~ gap admittance elektronischer Spaltleitwert *m* ~ gate Schalter *m* ~ ignition system Röhrenzündung *f* ~ insertion elektronische Einblendung *f* ~ integrators elektronische Integratoren *pl* ~ keying elektronisches Tasten *n* ~ leak detector Lecksuchgerät *n* ~ lens Bildwandler *m*, Elektronenlinse *f* ~ modes Freiheitsgrade *pl* des freien Elektrons ~ multicoupler elektronischer Antennenverteiler *m* ~ multimeter Mehrzweck-Mehrbereich-Instrument *n* ~ package Elektronikteil *n* ~ partition function Elektronen-Zustandssumme *f* ~ peak-reading voltmeter elektronisches Voltmeter *n* ~ pH-recorder elektronischer pH-Schreiber *m* ~ picture-reproducing tube Wiedergaberöhre *f* ~ potentiometer elektronische Meßbrücke *f* ~ print reader elektronischer Druckschriftleser *m* ~ probe materienlose Sonde *f* ~ profilometer elektronischer Profil- oder Rauhigkeits-prüfer *m* ~ punch-card machine elektronischer Locher *m* ~ ray indicator magisches Auge *n*, Elektronenstrahlindikator *m* ~ rectifier elektronischer Gleichrichter *m*, Sperrgleichrichter *m* ~ reverberation elektronischer Nachhall *m* ~ scaler Strahlungsmeßgerät *n* ~ scanner elektronische Abtastung *f*, Bildzerlegung *f* mit Kathodenstrahlröhre ~ shot effect Elektronenflußdiskontinuität *f* ~ supermicroscope Elektronenübermikroskop *n* ~ target-detecting control Zielkopfsteuerung *f* ~ timer Röhrenschweißtakter *m*, elektronischer Zeitschalter *m* ~ tube Absperr-glied *n*, -organ *n*, -teil *m*, -vorrichtung *f* ~ tube shield Abschirmkappe *f* für

Elektronenröhren ~ **tuning** elektronische Abstimmung f ~ **tuning range** elektronischer Abstimmbereich m ~ **ultramicroscope** Elektronenübermikroskop n ~ **vacuum tube** Hochvakuumröhre f ~ **valve** Elektronen-röhre f, -ventil n; Ionenventil n ~ **valve tube** Hochvakuumröhre f ~ **view finder** elektronischer Sucher m ~ **voltmeter** elektronisches Voltmeter n ~ **wattmeter** elektronischer Leistungsmesser m ~ **weighing** elektronische Wägung f

**electronics** Elektronen-lehre f, -physik f, -technik f, Elektronik f ~ **engineer** Elektroniker m

**electronogen** Elektronogen n

**electronogenic** lichtelektrisch

**electronography** Elektronografie f

**electrons** Elektronen pl ~ **of favorable phase** richtigphasige Elektronen pl ~ **of unfavorable phase** falschphasige Elektronen pl

**electrons,** ~ **adjacent nucleus** kernnahe Elektronen pl ~ **sticking to electronegative gases or surfaces** klebende Elektronen pl

**element** wesentlicher Bestandteil m, Element n, Glied n, Grundstoff m, Grundteil m, Kette f, Kettenglied n, Plattenblock m, Teil m, Teilchen n (math.), Urstoff m, Zelle f

**element,** ~ **of arc** Bogenelement n ~ **of area** Flächenelement n ~ **of a closed loop** Übertragungsglied n ~ **and compound-semiconductor** Element- und Verbindungshalbleiter m ~ **of construction** Konstruktionsglied n ~ **of distance** Wegabschnitt m ~ **with large number of spectrum lines** linienreiches Element n (Spektroskopie) ~ **of length** Strecke f, Wegabschnitt m ~ **of a microphone** Kapsel f ~ **of path** Wegabschnitt m ~ **of volume** Volumenelement n ~ **of a winding** Ankerspule f

**element,** ~ **glass jar** Elementglas n ~ **having capacity** Speicherelement n ~ **leads** Heizkörperanschlüsse pl ~ **(or component) producing disturbance** Störglied n ~ **wires** Glühspirale f

**elemental** elementar ~ **charge** Elementarladung f

**elementary** einfach, elementar, urstofflich ~ **in facts** grundsätzlich

**elementary,** ~ **area** Bildpunkt m, Flächenstück n, Rasterelement n, Rasterpunkt m ~ **areas** Flächenelemente pl ~ **cell** Einheits-, Elementarzelle f, Gittereinheit f ~ **charge** elektrisches Elementarquantum n ~ **coil of armature** Ankerglied n ~ **condenser** Elementarkondensator m ~ **density value** Lichttonpunkt m ~ **diffusion equation** elementare Diffusionsgleichung f ~ **material** Grundmaterial n ~ **shading value** Lichttonpunkt m ~ **system** Blockschaltbild n ~ **vortex** Elementarwirbel m ~ **wavelet** Elementarwelle f

**elements** Atmosphärilien pl ~ **of accompanying iron** Eisenbegleiter m ~ **of constancy** Konstanzelemente pl ~ **integral** ganzzahlige Elemente pl ~ **of orientation** Orientierungselemente pl

**elemi (gum)** Elemiharz n

**eleolite** Davyn m, Elaolith m, Fettstein m, Nephelin m

**eleometer** Eläometer n

**elephant-trunk chute** Elefantenrüsselrutsche f

**elevate, to** ~ erhöhen, hochfördern, kippen (Winkel über der Horizontale), veredeln

**elevated** erhaben, erhöht, hochliegend ~ **antenna** Höhenantenne f (rdr) ~ **approach** Ansteigung f

~-**cableway crane** Kabelhochbahnkran m ~ **duct** Höhenkanal m ~ **light** Überflurfeuer n ~ **pile** Stelzenunterbau m ~ **plain** Hochebene f ~ **plane** Plateau n ~ **position on building (or scaffold)** Hochstand m ~ **railway** Hochbahn f ~ **storage basin** Hochspeicherbecken n ~ **tank** Hochreservoir

**elevating,** ~ **and traversing mechanism** Richtgetriebe n ~ **arc** Höhenrichtbogen m ~-**arm shaft** Richthebelwelle f ~ **capacity** Förderleistung f ~ **chain** Lasthebekette f ~ **(or elevation) drive** Höhenwinkelantrieb m ~ **drum** Aufsatztrommel f ~ **gear** Höhenricht-maschine f, -vorrichtung f; Höhentrieb m, Richtmaschine f ~-**gear column** Richtmaschinensäule f ~ **gear of a telescope tripod** Hochstellvorrichtung f eines Fernrohrstativs ~ **handle** Höhenrichtkurbel f ~ **handwheel** Handrad zur Höhenrichtung, Höhenrichtrad n ~ **height setting** Höhenrichtung f ~ **lever** Richt-, Steuer-hebel m für die Höhenstellung ~ **link** Richtgelenk n ~ **mechanism** Höhenricht-maschine f, -werk n, Höhentrieb m, Hubmechanismus m, Hubvorrichtung f ~ **motor** Höhenrichtmotor m ~ **plant** Hebeanlage f ~-**platform truck** Hubkarren m ~ **rack** Richtbogen m, Richtzahnbogen m ~-**screw** Höhenstellspindel f, Hubspindel f, Richtschraube f ~ **screw bearing** Auftriebspindellager n ~ **screw gear** Zahnrad n zur Höhenstellung des Armes ~ **screw for table** Schraubenspindel f für die Höhenstellung des Tisches ~ **sector** Höhengradbogen m ~ **shaft** Welle f für die Auf- und Niederbewegung ~ **speed** Hubgeschwindigkeit f ~ **spindle** Hubspindel f ~ **tumble-plate segment** Stellhebel m für die Höhenstellung ~ **worm** Kippschnecke f

**elevation** Anhöhe f, Ansicht f, Aufriß m, Aufsatzhöhe f, Aufsteigen n, Aufzug m, Böschung f, Emporheben n, Erhebung f, (quadrant) Erhöhung f, (above a certain point, surveying, etc.) Festpunkt m, geographische Höhe f (nav.), (surveying) Hebung f, Heraushebung f, Hochförderung f, Höhe f, Hoheit f, Höhenrichtung f (Vermessungsgerät), Höhenwinkel m, Richthöhe f, (front or side) Riß m, (in capillary tube) Steighöhe f, Vertikalschnitt m, Wulst m ~ **of a building** Aufriß m ~ **of lobe** Höhenschwenkwinkel m (rdr) ~ **of a source of sound** Erhebung f einer Schallquelle

**elevation,** ~ **adjustment** Höhenverstellung f ~ **angle** (wave propagation) Erhebungswinkel m, Höhenwinkel m (rdr) ~ **axis** Querachse f (Flugzeug) ~ **base line** Höhenbasis f ~ **chart** Höhenplan m ~ **clamping handle** Höhenhebel m (machine gun) ~ **control wheel** Steuerradhöhe f ~ **deviation** Höhenabweichung f ~ **diagram** Höhenplan m ~ **dial** Höhenrichtschraube f ~ **error** Erhöhungsfehler m, Höhenabweichung f, (vertical) Höhenfehler m ~ **field pattern** Vertikaldiagramm n (antenna) ~ **graduations** Aufsatzeinteilung f ~ **gyro** Querachsenkreisel m ~ **hand-wheel** Handrad n, Richtmaschinenantrieb m ~ **high mode** Hochsuchen n ~ **indicator** Höhenrichtzeiger m, (gunner) Höhenweiser m, Höhenzeiger m ~ **lead** Höhenwinkelvorhalt m ~ **low mode** Tiefsuchen n ~ **measurement** Höhegebung f (rdr) ~ **meachanism on dial sight** Aufsatztrieb m ~ **movement** Höhenwinkeländerung *f* ~ **pattern** Vertikaldiagramm *n* ~ **position indicator** Höhen- und Ortsanzeiger *m* (rdr) ~ **quadrant** Höhengradbogen *m* ~ **receiver** Höhenempfänger *m* ~ **release catch** Umschaltklinke *f* ~ **scale** Erhöhungsskala *f*, Höhenteilung *f*, Höhenwinkelteilung *f* ~ **scale drum** (dial sight) Teiltrommel *f* ~-**scale drum on dial sight** Aufsatzgehäuse *n* ~ **selsyn** Geberanlagehöhe *f* ~ **setter** Höhenrichtkanonier *m* ~ **setting** Höheneinstellung (d. Befeuerung), Visierwinkel *m* ~ **setting screw** Höhenstellschraube *f* ~ **stop** Höhenbegrenzer *m* ~-**tracking telescope** Höhenrichtfernrohr *n* ~ **view of antenna field pattern** Kreisebene *f* ~-**worm knob** Triebscheibe *f* zur Höhenrichtschraube

**elevator** in der Höhe verstellbare Arbeitsbühne *f*, Aufzug *m*, Elevator *m*, Fahrgestellstuhl *m*, (cabin) Fahrstuhl *m*, Förderwerk *n*, Garbenträger *m*, Hebewerk *n*, Hebezeug *n*, Höhenförderer *m*, Höhenruder *n* (Flugzeug), Höhensteuer *n*, Speicher *m*, Tiefenruder *n* ~ **for push binder** Elevator *m* für Schubbinder oder Stoßbinder

**elevator,** ~-**action** Höhenrudervorderkante *f* ~ **apron shaft** Nadeltuchwelle *f* ~ **area** Höhenruderflächeninhalt *m* ~ **bay** Höhenruderzelle *f* ~ **belt** Elevatorgurt *m*, Gurt-förderband *n*, -förderer *m*, -transporteuer *m* ~ **boost ratio** Verstärkungsverhältnis *n* (d. Höhensteuerung) ~ **booster tab** Höhenruderausgleichklappe *f* ~ **boot** Elevatorbehälter *m* ~ **bucket** Elevatorbecher *m*, Förderkorbkübel *m* ~ **bushing** Hebebündel *n* ~ **cable** Ruderkabel *n* ~ **cage** Aufzugkabine *f* ~ **canvas** Elevatortuch *n* ~ **car** Fahrbühne *f* ~ **casing** Elevatorgehäuse *n* ~ **centering knob** (autopilot) Neigungsgeber *m* ~ **chain** Elevatorenkette *f* ~ **chord** Höhenrudertiefe *f* ~ **control(s)** Höhensteuerung *f*, Tiefenruder *n*

**elevator-control,** ~ **cable** Höhenrudersteuerseil *n* ~ **gear** Aufzugsteuerung *f* ~ **lever** Höhenrudersteuerhebel *m* ~ **rod** Höhenrudersteuerstange *f* ~ **stand** Höhenrudersteuerrad *n* ~ **wire** Höhenrudersteuerseil *n*

**elevator,** ~ **edge** Höhenruder-endbogen *m*, -randbogen *m* ~ **false spar** Höhenruderholm *m* ~ **frame** Aufzugsgerüst *n* ~ **gear** Förderhaspel *m* ~ **hinge** Höhenrudergelenk *n* ~ **hoisting cable** (or rope) Aufzugseil *n* ~ **jaw** Elevatorkopf *m* ~ **lever** Elevatorhebel *m* ~ **operator** Höhensteuermann *m* ~ **push-and-pull rod** Höhenruderstoßstange *f* ~ **rocking-shaft** Höhenruderwelle *f* ~ **scoop** Elevatorbecher *m* ~ **servomotor** Höhenrudermaschine *f* ~ **shaft** Aufzugschacht *m* ~ **slide** Elevatorschlitten *m*, Heberschieber *m* ~ **station** Höhensteuerrad *n* ~ **stringer** Höhenruder-endbogen *m*, -randbogen *m* ~ **strip** Höhenruderrandbogen *m* ~ **surface** Höhenruderflächeninhalt *m* ~ **tip** Höhenruderrandbogen *m* ~ **tip strip** Höhenruderendbogen *m* ~ **trailing edge** Höhenruderhinterkante *f* ~ **trim handwheel** Höhenrudertrimmrad *n* ~ **trim tab** Höhenrudertrimmklappe *f* ~ **trimming** Höhenrudertrimmung *f* ~ **trunking** Elevatorschlotte *f* ~ **tube** Höhenruderausgleichsklappe *f*, Ruderachsrohr *n*

**eleven-punch** X-Lochung *f*

**eleventh step contacts** Durchdrehkontakt *m*

**elever** Höhenseitenruder *m*

**elevon** Höhenquerruder *n*

**eliasite** Eliasit *m*

**eligible** berechtigt, in Frage kommend

**eliminate, to** ~ (heat) abführen, absondern, ausmerzen, ausschalten, ausscheiden, aussondern, ausstoßen, ausziehen, beheben, beseitigen, eliminieren, extrahieren, (a factor) sich fortheben, (chem.) herauslösen, vermeiden **to** ~ **bitter substances** entbittern **to** ~ **distortion** entzerren **to** ~ **disturbances in the smooth working of a plant** Betriebsstockungen *pl* ausschalten **to** ~ **heat** Wärme *f* abführen **to** ~ **iron from** enteisen **to** ~ **sources of waste** Unwirtschaftlichkeiten *pl* beseitigen

**eliminated** ausgelassen, ausgemerzt, ausgeschaltet, ausgeschieden, ausgestrichen, beseitigt, eliminiert, entfernt ~ **of induction** induktionsgeschütz ~ **carrier wave** unterdrückte Trägerwelle *f* (electr.)

**eliminating test** Ausscheidungs-, Geschwindigkeits-prüfung *f*

**elimination** Absonderung *f*, Ausschaltung *f*, Ausscheidung *f*, Aussonderung *f*, Beseitigung *f*, Elimination *f*, Wegfall *m*

**elimination,** ~ **of curvature** Bildfeldebnung *f* (opt.) ~ **of defects** Fehlerbeseitigung *f* ~ **of disturbing screen charges** Beseitigung *f* störender Schirmaufladungen ~ **of errors** Fehlerbeseitigung *f* ~ **of finishing process** Entfeinerung *f* ~ **of flicker** Beseitigung *f* des Flimmerns *n* ~ **of interference (or jamming)** Störbefreiung *f* ~ **of jamming effects** Entdüppelung *f* (rdr) ~ **of radio interference** Funkentstörung *f* ~ **of valueless products** Ausschaltung *f* wertloser Erzeugnisse

**elimination,** ~ **method** Eliminationsverfahren *n* ~ **run** Ausscheidungsrennen *n*

**eliminator** Funkentstörer *m*, Schieb *m* (rdo), Sieb *n* ~ **and stoner** Entsteiner *m* ~ **of static electricity** Entladeeinrichtung *f* ~ **section** Abscheidekammer *f*

**eliminators** Ersatzgerät *n* (battery)

**elite** Elite *f*

**ell** Elle *f*

**ellipse** Ellipse *f*, Kegelschnittlinie *f* ~ **of essential information** Informationsmarke *f* ~ **of inertia** Trägheitsellipse *f* ~-**shaped** ellipsenförmig

**ellipsoid** Ellipsoid *n* ~ **of error** Fehlerellipsoid *n* ~ **of inertia** Trägheitsellipsoid *n* ~ **of revolution** Rotations-, Umdrehungs-ellipsoid *n* ~ **oblate and oblong** Drehellipsoid *n*

**ellipsoidal** elliptisch ~ **core antenna** Ellipsoidkernantenne *f* ~ **model** ellipsoidales Modell *n*

**ellipsoids of revolution** Rationsellipsoide *pl*

**elliptic,** ~ **(al)** elliptisch ~ **curvature** elliptische Krümmung *f* ~-**hyperbolic type** elliptisch-hyperbolischer Typ *m* ~ **integrals** elliptische Integrale *pl* ~ **mapping** elliptische Abbildung *f* ~ **paraboloid** elliptisches Paraboloid *n* ~ **plane** elliptische Ebene *f* ~ **space** elliptischer Raum *m* ~ **spring** Elliptikfeder *f*

**elliptical** eiförmig, länglichrund, langrund ~ **bogie spring for rotating platforms** Doppelfeder *f* für Drehgestelle (locomotive) ~ **cross section** Ellipsenquerschnitt *m* ~ **fuselage** elliptischer Rumpf *m* ~ **orbit** Ellipsenbahn *f* ~ **speaker** Ovallautsprecher *m* ~ **wing** elliptischer Flügel *m*

**elliptically, ~ polarized light** elliptisch polarisiertes Licht *n* **~ polarized wave** elliptisch polarisierte Welle *f*

**ellipticity** Abplattung *f* **~ coefficient** Halbachsenverhältnis *n* **~ factor** Abplattungsfaktor *m*, Halbachsenverhältnis *n*

**ellsworthite** Ellsworthit *m*

**elm wood** Ulmenholz *n*

**elocution velocity** Vortragsschnelligkeit *f*

**elongate, to** ~ ausdehnen, ausstrecken, dehnen, längen, recken, strecken, verlängern

**elongated** gestreckt, langgestreckt, länglich, verlängert **~ charge** gestreckte Ladung *f* **~ crystal** langstrahliges Kristall *n* **~ hole** Langloch *n*, ausgeschlagenes Loch *n* **~ hole drilling machine** Langlochbohrmaschine *f* **~ loading coil** längliche Pupinspule *f* **~ slot** Längsschlitz *m* **~ twill** Stufenköper *m*

**elongating** Recken *n*, Strecken *n*

**elongation** Ausdehnung *f*, Bruchdehnung *f*, Dehnung *f*, Dilation *f*, Höchstausschlag *m*, Längenänderung *f*, Längsstreckung *f*, Längung *f*, Verlängerung *f* **~ at break** Zerreißdehnung *f* **~ on break** Bruchdehnung *f* **~ of cable** Seildehnung *f* **~ to failure** Reißlast *f* **~ of wires** Dehnung *f* von Drähten

**elongation, ~ curve** Längungskurve *f* **~ dial inducator** Dehnungsmeßuhr *f* **~ exterior cable** Verlängerungskabel *n* **~ indicator** Dehnungszeigerapparat *m* **~ meter** Dehnungsmesser *m* **~ portion** Dehnbereich *m* **~ speed** Dehnungsgeschwindigkeit *f* **~ ruler** Dehnungsmeßlineal *n* **~ tensor** Elongationstensor *m* **~ test** Dehnungsprüfung *f* **~ value** Dehnungswert *m*

**eloxadize, to** ~ eloxieren

**eloxation** Eloxalverfahren *n* **~ layer** Eloxalschicht *f*

**eluate grade** Eluierungsgrad *m*

**elucidate, to** ~ aufhellen, verdeutlichen

**elucidation** Erklärung *f*, Erläuterung *f*

**elude, to** ~ ausweichen, entgehen, sich entziehen, vermeiden

**elution, ~ chromatography** Spülungschromatografie *f* **~ process** Elutionsverfahren *n*

**elutriate, to** ~ abschlämmen, abschwemmen, elutrieren, schlämmen

**elutriating** Schlämmung *f* **~ apparatus** Schlämmapparat *m*, -verrichtung *f* **~ device** Schlämmvorrichtung *f* **~ process** Schlämmverfahren *n*

**elutriation** Auswaschung *f*, Schlämmen *n* **~ analysis** Schlämmanalyse *f* **~ method** Schlämm-, Spül-verfahren *n* **~ process** Schlämmvorgang *m*

**elutriator** Schlämmgerät *n*

**eluvial soil** Ausschwemmungsboden *m*

**emaciate, to** ~ abzehren

**emaciation** Abzehrung *f*

**emanate, to** ~ ausfließen, ausgehen, ausströmen, herrühren

**emanating power** Emanationskoeffizient *m*

**emanation** Ausfluß *m*, Ausstrahlung *f*, Ausströmen *n*, Ausströmung *f*, Emanation *f* **~ with diminished (or decaying) radioactivity** abgeklungene Emanation *f*

**emanometer** Emanationsmesser *m*

**embank, to** ~ eindämmen

**embankment** Abdämmung *f*, Auffüllung *f*, Auftrag *m*, Böschung *f*, Damm *m* (r.r.), Deich *m*, Eindämmung *f*, Erschütterung *f*, Erdaufwurf *m*,

Erddamm *m*, Erdwall *m*, Hafendamm *m*, Schüttdamm *m*, Schüttung *f*, Ufereinfassung *f* (Kai), Wall *m*

**embargo** Ausfuhrverbot *n*, Beschlagnahme *f*, Embargo *n*, Hafensperre *f*, Sperre *f* **~ on commerce** Handelssperre *f*

**embark, to** ~ an Bord gehen, (passengers) einschiffen, einsteigen, verladen **to ~ on** sich aufmachen

**embarkation** Einschiffung *f*, Verladung *f* **~ point** Fährstelle *f*

**embattled wall** gezinnelte Mauer *f*

**embed, to** ~ betten, einbetten, einlagern, einlassen, einlegen, einmauern, einschichten, einschließen, lagern **to ~ in concrete** einbetonieren

**embedded** eingelagert, eingelassen, (mooring sinker) eingesandet **~ rectificier** gekapselter Trockengleichrichter *m* **~ temperature detector** gekapseltes Thermoelement *m*

**embedding** Einbetten *n*, Einbettung *f*, Einmauerung *f* **~ material** Einbettungsmaterial *n* **~ medium** Einbettmittel *n* **~ resin** Einbettharz *n*

**embedment** Einlagerung *f*

**ember** glimmende Kohle *f*

**embers** Asche *f*, Glühasche *f*

**embodiment** Anwendungsform *f*, Ausführungsart *f*, Verkörperung *f* **~ of a measure** Maßverkörperung *f*

**embody, to** ~ verkörpern

**embolite** Chlorbromsilber *n*, Embolit *m*

**emboss, to** ~ abbilden, ausbuckeln, bosseln, bossieren, (sound track on film) einprägen, erhaben herausarbeiten, prägen, treiben **to ~ in relief** hochprägen

**embossed** erhaben **~ character** Prägezeichen *n* **~ film** Prägefolie *f* **~ foil** geprägte Folie *f* **~ grain** aufgepreßte Narben *pl* **~ groove recording** Gravürfilm *m* **~ hangings** gepreßte Tapete *f* **~ iron plate** getriebenes Eisenblech *n* **~ leather** gepreßtes Leder *n* **~ map** Reliefplan *m* **~ mark** Treibarbeit *f* **~ printing for blind** Blindendruck *m* **~ seal** Siegelmarke *f* **~ work** erhabene oder getriebene Arbeit *f* **~ writing picture telegraphy** Reliefverfahren *n*

**embosser** Relief-, Stift-schreiber *m*

**embossing** Aufprägung *f*, Gaufrieren *n*, Gravierung *f*, Pressen *n*, Rasterung *f* **~ and drawing machine** Präge- und Ziehautomat *m* **~ batten** Wirkladen *m* **~ calender** Gaufrier-, Profilkalender *m* **~ die** Prägestanze *f* **~ electros** Prägegalvanos *pl* **~ iron** Bossiereisen *n* **~ machine** Gaufrier-, Präge-maschine *f*, Prägepresse *f* **~ press** Gaufrierpresse *f* (metal) **~ press for hot process** Heißprägepresse *f* **~ punch** Bossierdurchschlag *m* (metal) **~ stamp hole** Prägestempel *m* **~ stylus** Stichel *m* (phono) **~ tool** Treibform *f*

**embossment** Relief *n* **~ machine** Wulstmaschine *f*

**embrace, to** ~ fassen, umfassen, umschlingen

**embracing** umfassend

**embrasure** Einschnitt *m*, Schießscharte *f*, Schießschlitz *m* **~ of a window** Fensterschmiege *f*

**embrittle, to** ~ spröde machen, verspröden

**embrittlement** Brüchigwerden *n*, Sprödewerden *n*, Versprödung *f* **~ in steam boilers** Längensprödigkeit *f* von Kesselblechen

**embrittling** Sprödewerden *n*

**embroidered** durchwirkt ~ **tapestry goods** Tapisseriewaren *pl*
**embroidering** Sticken *n* ~ **apparatus** Stickapparat *m* ~ **needle** Sticknadel *f*
**embroidery** Stickerei *f* ~ **finishing** Stickereiappretur
**embussing** Verladung *f*
**emerald** Smaragd *m* ~ **(colored)** smaragdgrün ~ **copper** Kupfersmaragd *m* ~ **green** Mitisgrün *n*, smaragdgrün ~ **nickel** Zaratit *m*
**emerge, to** ~ auftauchen, austreten, emporkommen, heraustreten, zum Vorschein kommen
**emergence of beam** Strahlausfall *m*
**emergency** kritischer Augenblick *m*, Behelf *m*, Ernstfall *m*, Not *f*, Notfall *m*, Notstand *m*, Unglück *n* **(case of)** ~ Notfall *m* **in** ~ Störungsfall *m*
**emergency,** ~ **A. C. bus** Not-Wechselstromschiere *f* ~ **A. C. generator** Not-Drehstromgenerator *m* ~ **airport** Ausweichhafen *m* ~ **airpressure control pipe** Notsteuerdruckleitung *f* ~ **alighting on water** Notlandung *f* (auf dem Wasser) ~ **anchor** Notanker *m* ~ **antenna** Behelfsantenne *f* ~ **apparatus** Not-apparat *m*, -einrichtung *f* ~ **ballast** Notballast *m* ~ **battery** Notbatterie *f* ~ **bilge pump** Bilgewasserhilfspumpe *f* ~ **brake** Notbremse *f* ~ **bridge** Behelfsbrücke *f* ~ **bridge construction** Behelfsbrückenbau *m* ~ **broadcast** Not-Rundsendung *f* ~ **button** Notzugknopf *m* ~ **cable** Hilfskabel *n* ~ **call** Feuermeldung *f* (to fire department), Notgespräch *n*, Notruf *m*, Überfallruf *m*, Unfallmeldung *f* ~ **car** Hilfswagen *m* ~ **ceiling** Gipfelhöhe *f* mit einem stehenden Motor *m* ~ **cell switch** Doppelzellenschalter *m* ~ **circuit** Notschaltung *f* ~ **cock** Nothahn *m* ~ **communication** Notverkehr *m* (rdo) ~ **compass** Notkompaß *m* ~ **compressor installation** Notkompressoranlage *f* ~ **construction** Behelfskonstruktion *f* ~ **control unit** Notsteuergeber *m* ~ **corps** Nothilfe *f* ~ **crew** Mannschaft *f* für Noteinsatz *m* ~ **cruising output** erhöhte Dauerleistung *f* (aviat.) ~ **cutout** Notausschalter *m* ~ **D. C. bus** Notgleichstromsammelschiere *f* ~ **decree** Notverordnung *f* ~**-descent** Notsinkflug *m* ~ **device** Notbehelf *m* ~ **Diesel set** Notstromdiesel *m* ~ **disconnecting switch** Notausschalter *m* ~ **dressing** Notverband *m* ~ **equipment** Ersatzhilfsanlage *f*, Notausrüstung *f* ~ **exchange** Notamt *n* (teleph.) ~ **exhaust** Notauspuff *m* ~ **exit** Not-ausgang *m*, -aussteig *m* ~ **ferry** Behelfsfähre *f* ~ **fill line** Notzapfleitung *f* ~ **filler neck** Notfüllstutzen *m* ~ **flotation gear** Notschwimmfläche *f* ~ **fuel** Fallbenzin *n* ~ **fuel control** Notkraftstoffregler *m* ~ **fuel tank** Reservebrennstoffbehälter *m* ~ **gate** Notverschluß *m*, Sicherheitstor *n* ~ **generating plant** Notstromanlage *f* ~ **governor** Schnellschlußeinrichtung *f* ~ **handle** Notzug *m* ~ **hatch** Notausstieg *m* ~ **horsepower** Notpferdekraft *f* ~ **idle needle** Notleerlaufnadel *f* ~ **jettison of fuel** Schnellentleerung *f* des Brennstoffs (aviat.) ~ **jettisoning lever** Notabwurfhebel *m* ~ **lamp fitting** Notbeleuchtungsgerät *n* ~ **landing field** Hilfslande-, Notlandeplatz *m* ~ **landing gear** Notfahrgestell *n* ~ **landing place** Hilfslandeplatz *m* ~ **landing strip** Notlandebahn *f* ~ **lantern** Notlaterne *f* ~ **lighting** Not-beleuchtung *f*, -befeuerung *f* ~

**lights** Notfeuer *pl* ~ **line** Notleitung *f* (electr.) ~ **machine** Hilfs-, Zusatz-maschine *f* ~ **means** Notmittel *n* ~ **needs** Stoßbedarf *m* ~ **opening** Notauslaß *m* ~ **outfit** Notausrüstung *f* ~ **outlet** Notauslaß *m* ~ **outlet of a dam** Notablaß *m* einer Sperrmauer ~ **parachute** Rettungsfallschirm *m* ~ **path** Notpiste *f* ~ **plant** Ersatzhilfs-, Not-anlage *f* ~ **position of automatic mixture control** Anreicherungsstellung *f* des Gemischreglers für Nothöchstleistung ~ **power** Notleistung *f* ~ **power plant (or set)** Notstromversorgung *f* ~ **power station** Aushilfskraftwerk *n* ~ **power supply** Notstrom *m* ~ **power unit** Notstromaggregat *m* ~ **pressure** Notdruck *m* ~ **procedure** Notverfahren *n* ~ **prop** Hilfsstempel *m* ~ **provision** Notversorgung *f* ~ **pump** Reservepumpe *f* ~ **radiator tank** Reservekühlwasserbehälter *m* ~ **radio channel** Notfrequenz *f* ~ **ratings** für Notfälle festgelegte Leistung *f* ~ **rations** Notproviant *m* ~ **receiver** Notempfänger *m* ~ **reception** Notempfang *m* ~ **relay valve** Notrelaisventil *n* ~ **release** Blindabwurf *m*, Schnellablaß *m* (aviat.) ~ **relief valve** Sicherheitsventil *n* (Kraftstoffanlage) ~ **salvo release** Notabwurf *m* ~ **set** Ersatzanlage *f*, Ersatzhilfsanlage *f*, Hilfsanlage *f*, Noteinrichtung *f*, Zusatz *m* ~ **sight** Notvisier *n* ~ **sluice** Notverschluß *m* ~ **starter** Aushilfsanlasser *m* ~ **station** Aushilfsstelle *f* ~ **stop button** Gefahrentaster *m* ~ **stop cock (or valve)** Notabsperrhahn *m* ~ **stretcher** Nottrage *f* ~ **suction line** Notaussaugleitung *f* ~ **supply system** Notspeiseanlage *f* ~ **switch** Notschalter *m*, Reserveschaltung *f* ~ **system** Reserveschaltung *f* ~ **take-off** Notstart *m* ~ **take-off power** erhöhte Startleistung *f* für Notfall ~ **trail** Notpiste *f* ~ **transmitter** Notsender *m* ~ **transmitter beacon** Notpeilsender *m* ~ **transportation** Notversorgung *f* ~ **valve** Notabsperrventil *n*, Notverschluß *m*
**emergent,** ~ **light** ausfallendes oder austretendes Licht *n* ~ **light ray** ausfallender Lichtstrahl *m* ~ **ray** ausfallender Strahl *m* ~ **ray point** Ausgangspunkt *m* des Zielstrahls ~ **spectrum** Endspektrum *n*
**emerging light beam** austretende Lichtbüschel *pl*
**emersion** Austauchung *f*, Auswässerung *f*, Emporsteigen *n*
**emery** körniger Korund *m*, Schmirgel *m* ~ **cloth** Schmirgel-leinen *n*, -leinwand *f*, -tuch *n* ~ **dust** Schmirgel-pulver *n*, -staub *m* ~ **grinding machine** Schmirgelschleifmaschine *f* ~ **impregnated** schmirgelimpregniert ~ **paper** Schmirgelpapier *n* ~ **papers for cooking range** Herdputzblättchen *pl* ~ **paste** Schleifpaste *f* ~ **powder** Schmirgelpulver *n* ~ **roller** Schmirgelwalze *f* ~ **stick** Mineralfeile *f*, Putzholz *n* ~ **stone** Schmirgelstein *m* ~ **wheel** Schmirgelscheibe *f*
**E. M. F. (electro-motoric force)** E. M. K. (elektromotorische Kraft *f*)
**emilite** Emilit *n*
**emission** (of radiations, electrons, etc.) Abgabe *f*, (of a sound) Aussendung *f*, Ausstrahl *m*, Ausströmung *f*, Emission *f*, Sendung *f*, Strahlung *f*, Werfen *n* ~ **of heat** Wärmeabführung *f* ~ **of oscillation (or rays)** Ausstrahlung *f* ~ **of sparks** Funkenauswurf *m* ~ **of vibration (or waves)** Ausstrahlung *f*

**emission, ~ band** Emissionsband *n* **~ capability**
Emissionsfähigkeit *f* **~ cell** Fotozelle *f* mit
äußerem lichtelektrischem Effekt **~ characteristic** Emissionskennlinien *pl* **~ coefficient** Emis-
sionskoeffizient *m* **~ contact** Emissionskontakt
*m* **~ control** Emissionssteuerung *f* **~ current**
Emissionsstrom *m* **~ curve** Emissionskennlinie
*f* **~ efficiency** Elektronen-, Emissions-wirkungs-
grad *m* **~ electron** Streuelektron *n* **~ law**
. Emissions-, Entladungs-gesetz *n*, Lambert'sches
Gesetz *n*, Raumladungsgesetz *n* **~ measuring
device** Emissionsmeßgerät *n* **~ spectra of
crystals** Emissionskristallspektren *pl*
**emissive** aussendend, ausstrahlend **~ mechanism**
Emissionsmechanismus *m* **~ power** Ausstrah-
lungs-, Emissions-vermögen *n*
**emissivity** Emissions-stärke *f*, -vermögen *n*
**emit, to ~** auslösen, aussenden, ausstoßen, aus-
strahlen, ausströmen lassen, schießen, (rays)
strahlen, werfen **to ~ flashes** aufleuchten, auflodern, blitzen **to ~ light** leuchten **to ~ thick
smoke** qualmen
**emitted** emittiert **~ particle** emittiertes Teilchen *n*
**emitter** Sender *m*, Strahler *m*, Strahlergebilde *n*
**~ branch** Emitterzweig *m* **~ efficiency** Emitter-
wirkungsgrad *m* **~ follower** Emitterverstärker
*m* **~ junction** Emitterübergang *m* **~ lead** Emit-
terklemme *f*
**emitting, ~ antenna** Sendeantenne *f* **~ area** Emis-
sions-, Strahlungs-fläche *f* **~ electrode** emittierende Elektrode **~ electron** Leuchtelektron *n* **~
layer** Emissionsschicht *f* **~ light** lichtentwik-
kelnd **~ power** Emissionsvermögen *n* **~ surface**
Emissions-, Strahlungs-fläche *f*, emittierende
Oberfläche *f*
**emmensite** Emmensit *n*
**emmentropic** normalsichtig, rechtsichtig **~ eye**
rechtsichtiges Auge *n*
**Emn mode** Emn-Typ *m*
**emollient** aufweichend **~ softener** Erweichungs-
mittel *n*
**emolument** Sondervergütung *f*
**empennage** Leitwerk *n*, Schwanz *m*, Schwanzfläche *f* **~ force** Leitwerkskraft *f*
**emphasis** Betonung *f* **~ on bass tones** Tiefenhervorhebung *f*
**emphasize, to ~** anheben, bekräftigen, betonen,
hervorheben, Nachdruck legen auf
**emphasizer** Anhebungsfilter *n*
**emphasizing** Bevorzugung *f*
**empire** Großraum *m*, Reich *n* **~ cloth** Ölstoff *m*
**empiric(al)** empirisch
**empirical** erfahrungsgemäß **~ data** Erfahrungs-
daten *pl*, -tatsachen *pl* **~ entropy** empirische
Entropie *f* **~ facts** Erfahrungstatsachen *pl* **~
formula** empirische Formel *f*, Rohformel *f* **~
mass formula** empirische Massenformel *f* **~
result** Erfahrungsergebnis *n* **~ value** Erfahrungs-
wert *m*
**empirically determined** (work function) empirisch
bestimmt
**emplace, to ~** aufstellen
**emplacement** Anlage *f*, Feuerstellung *f*, Instellungbringen *n*, Nest *n*, (fort.) Schießgerüst *n*,
Stand *m*, Stellung *f*
**emplacing** Instellungbringen *n*
**emplane, to ~** besteigen, ein Flugzeug *n* besteigen

**emplectite** Emplektit *m*, Kupferwismutglanz *m*,
Wismutkupfererz *n*
**employ, to ~** anlegen, anstellen, anwenden, aufwenden, bedienen, (workman in a mine) belegen, benutzen, beschäftigen, einstellen, gebrauchen, nutzbar machen, verbrauchen, verwenden
**employability** Verwendungsfähigkeit *f*
**employed** angestellt, beschäftigt
**employee** Angestellter *m*, Arbeitnehmer *m* **~ of
the firm** Werksangehöriger *m* **~ time recorder**
Arbeitszeitregistrieruhr *f*
**employees** Belegschaft *f*, Gefolgschaftsmitglie-
der *pl*
**employer** Arbeitgeber *m*, Auftraggeber *m*,
Dienstherr *m*, Lehrherr *m*, Lehrmeister *m*,
Unternehmer *m*
**employers', ~ liability** Berufsgenossenschaft *f* **~
liability policy** Gewerbeunfallversicherungs-
schein *m*
**employment** Anwendung *f*, Benutzung *f*, Gebrauch *n* **~ of radio** Funkeinsatz *m* **~ exchange**
Stellennachweis *m* **~ office** Arbeiterannahme-
stelle *f*
**empower, to ~** bevollmächtigen, ermächtigen
**emptied space** verhauener Raum *m* (min.)
**empties** Leer-gut *n*, -material *n*
**emptiness** Hohlheit *f*, Leere *f*
**empty, to ~** abfüllen, ablassen, ausleeren, ausschöpfen, ausschütten, ausstürzen, einmünden,
(a shell) entladen, entleeren, leeren, luftleer
machen, münden **to ~ a boiler** Kessel *m* ablassen
**empty** hohl, leer, luftleer **~ band** unbesetztes
Band *n* **~ belt** Untertrum *n* **~ car yard** Abstell-
gelände *n* **~ cartridge cases** abgeschossene
Hülsen *pl* **~ center** Leerstelle *f* **~ deposit** Leer-
lager *n* **~ ears** Abschöpfgerste *f* **~ set** leere
Menge *f* **~ truss** Leergebinde *n*, Zwischengesperre *n* **~ tube conveyor** Hülsentransport *m* **~
weight** Leergewicht *n* **~ weight plus fixed
weight** Rüstgewicht *n* (aviat.)
**emptying** Abräumung *f*, Ausleerung *f*, Entleeren
*n*, Entleerung *f*, Leerung *f* **~ contrivance** Ab-
füllvorrichtung *f* **~ grid** Auspackrost **~ hole**
Abflußloch *n* **~ jigger** Ausleerrüttler *m*
**empyreumatic** brenzlich(er Geruch *m*)
**emquad** Ganzgeviertstück *n*
**em quadrat** Geviert *n* (print.)
**emulsibility** Emulgierbarkeit *f*
**emulsification** Emulgierung *f*, Emulsionierung *f*,
Emulsionsbildung *f*
**emulsified alumina** geschlämmte Tonerde *f*
**emulsifier** Aufrahmungsmittel *n*, Emulgator *m*,
Emulgiermaschine *f*, Emulsionsbildner *m*
**emulsify, to ~** beschichten, emulgieren, emul-
sionieren, milchen, vermilchen
**emulsifying** Emulgieren *n* **~ agent** Emulgator *m*,
Emulgierungskörper *m*, Emulsionsbildner *m* **~
oil** Emulgierungsöl *n*
**emulsion** Emulsion *f*, milchige Flüssigkeit *f*,
Milch *f*, Milchschicht *f*, Mischbinder *m*, Negativmaterial *n*, Plattenschicht *f*, Schicht *f* **~ to
~ Schicht *f* auf Schicht *f*
**emulsion, ~ binder** Bindemittelemulsion *f* **~
carrier** Schichtträger *m* **~ coater** Emulsion-
aufstreicher *m* **~ intaglio printing** Emulsions-
tiefdruck *m* **~ layer** Emulsionsschicht *f* **~

**matrix** Trägersubstanz $f$ ( für Emulsionen) ~
**mixture** Schaumgemisch $n$ ~ **outlet** Austritts-
düse $f$ des Schaumgemisches ~ **paint** Emulsions-
farbe $f$ ~ **side** Schicht $f$ ~ **side of film** Film-
schichtseite $f$ ~ **stack** Emulsionspaket $n$ ~ **test**
Emulsionsversuch $m$ ~ **test flask** Schüttel-
flasche $f$ ~ **tube** (in carburetor) Düsenhütchen $n$
**enable** befähigen, ermöglichen, instandsetzen ~
**time** Zeit $f$ bis zur Erreichung der Steuerungsfähigkeit (g/m) ~ **transient** Einschaltspitze $f$
**enabled** (to steer) steuerungsfähig (g/m) ~ **balance** Symmetrie $f$ bei Steuerungsfähigkeit
**enabling,** ~ **pulse** Steuerimpuls $m$ für Torschal-
tungen ~ **transient** Einschwingvorgang $m$ (g/m)
**enact, to** ~ verordnen
**enacted, to be** ~ sich abspielen
**enalite** Enalith $m$
**enamel, to** ~ emaillieren, mit Email $n$ überziehen,
schmalten
**enamel** Amause $f$, Beglasung $f$, Email $n$, Emaille
$f$, Glasschmelz $m$, Lack $m$, Lackfarbe $f$, Schmalt
$m$, Schmelzglasur $f$ ~ **for signals** Signalfarbe $f$
~ **for tubes** Tubenemaille $f$
**enamel,** ~ **bead** Schmelzperle $f$ ~ **cleaning compound** Lackpflegemittel $n$ ~ **coat** Emaillelackierung $f$, -überzug $m$ ~ **coated** emailliert ~
**copper wire** Kupferlackdraht $m$ ~ **covered wire**
Emaildraht $m$ ~ **gauge** Emailleskala $f$ ~ **glass**
Emailglas $m$ ~ **glaze** Schmelz $m$ ~ **insulated
wire** Emaildraht $m$ ~ **laid on the back of a plate**
Gegenemail $n$ ~-**like** schmelzartig ~ **paint**
Emailfarbe $f$ ~ **painting** Schmelzmalerei $f$ ~
**paper** gestrichenes Papier $n$ ~ **polisher** Lack-
polierer $m$ ~ **thinning agent** Lackverdünner $m$
~ **varnish** Emaillack $m$ ~ **white** Barytflußspat-
weiß $n$
**enameled** emailliert ~ **brick** Glasurziegel $m$ ~
**cable** Lackkabel $n$ ~ **cardboard** Emailkarton $m$
~ **copper wire** Kupferlackdraht $m$ ~ **electric
fire** Emailleuchtofen $m$ ~ **paper** Kunstdruckpapier $n$ ~ **plate** Emailschild $n$ ~ **sheet** Email-
blech $n$ ~ **split** Lackspalt $m$ ~ **strand** Emaillitze
$f$ ~ **wire** Email-, Lack-draht $m$
**enameler** Emailleur $m$, Schmelzarbeiter $m$
**enameling** Überschiebungsschmelzung $f$ ~ **furnace** Farbenschmelzofen $m$ ~ **operation**
Schmelzarbeit $f$ ~ **oven** Emaillierofen $m$ ~
**process** Emaillierverfahren $n$, Schmelzarbeit $f$
**enantiomorphic (or enantiomorphous)** enantio-
morph
**enantiotropic** enantiotrop
**encapsulated fuel** eingehülstes Spaltstoffelement
$n$
**encase, to** ~ einhüllen, einkapseln, einlagern,
einschalen, mit einem Gehäuse versehen, ineinanderschieben, kapseln, umhüllen, ummanteln, verschalen
**encased** eingekapselt, gekapselt ~ **fixed tubular
paper (or mica-wound or wrapped) condenser**
Becherblock $m$ ~ **gold** vererztes Gold $n$
**encasement** Abdeckteil $m$ ~ **pump** Einsatzpumpe
$f$
**encasing** Verschalung $f$ ~ **tube** Hüllenrohr $n$,
Hüllröhre $f$
**encatchment area** Niederschlagsgebiet $n$
**encaustic** Wichse $f$ ~ **tile** farbig glasierter Ziegel
$m$
**enceinte** Enceinte $f$, Kernumwallung $f$

**encephalography** Encephalografie $f$
**enchase, to** ~ einlegen
**encipher, to** ~ chiffrieren, (ver)schlüsseln
**enciphered,** ~ **call sign (or name)** Namenszeichen
$n$ ~ **facsimile communications** verschlüsselter
Fäksimileverkehr $m$ ~ **telephony** verschlüsselter
Sprechverkehr $m$
**enciphering** Schlüssel $m$, Schlüsselung $f$ ~ **sheet**
Schlüsselblatt $n$
**encipherment** Chiffrierung $f$, Verschlüsseln $n$
**encircle, to** ~ einkesseln, einkreisen, umgeben,
umkreisen, umschließen, umschnüren, umzingeln, zernieren
**encirclement** Einkesselung $f$, Einkreisung $f$, Um-
klammerung $f$, Zernierung $f$
**encircling,** ~ **road** Ringstraße $f$ ~ **wall** Umfassungswand $f$
**enclave** Enklave $f$
**enclose, to** ~ anfügen, anschließen, einfassen,
einfriedigen, einhängen, einkapseln, einschleifen (in ein Gehäuse), einschließen, einwickeln,
einzäunen, kapseln, umhüllen, umschließen
**enclosed** (sea basin) abgeschnürt, beifolgend,
beiliegend, eingeschlossen, gekapseln, geschlossen, inliegend, innenliegend, umschlossen
~ **in chute** eingeschnurrt
**enclosed,** ~ **angle** Öffnungswinkel $m$ ~ **anode** gekapselte Anode $f$ ~ **arc lamp** Dauerbrandlicht-
bogenlampe $f$ ~ **blower** Kapselgebläse $n$ ~
**cabin** geschlossene Kabine $f$ ~-**cabin airplane**
geschlossenes Flugzeug $n$ ~ **compressor** Kapsel-
kompressor $m$ ~ **detonation** Detonation $f$ unter
Einschluß (semi-) ~ **dynamo** (halb)gekapselter
Dynamo $m$ ~ **gearbox** gekapseltes Getriebe $n$
~ **horizon** Dosenhorizont $m$ ~ **lamp** Dauer-
brandbogenlampe $f$ ~ **motor** gekapselter Motor
$m$ ~ **motor (or engine)** Kapselmotor $m$ ~
**pilot's cockpit** geschlossener Führersitz $m$ ~
**printing unit** Druckwerkeinkapselung $f$ ~ (**or
shrouded**) **propeller** Mantelschraube $f$ ~ **sheeting** Spundwandeinfassung $f$ ~ **space** abgeschlossener Innenraum $m$ ~ **starter** eingekapselter Anlasser $m$ ~ **switchgear** eisengekapselte
Schaltanlage $f$ ~-**type motor** geschlossener
Motor $m$ ~ **ventilated motor** ventiliert gekapselter Motor $m$
**enclosing,** ~ **sheet piling** hintere Spundwand $f$ ~
**wall** Einfassungsmauer $f$, Umfassungswand $f$
**enclosure** Abschlußkessel $m$, Anlage $f$, Anschluß
$m$, Beilage $f$, Einfassung $f$, Einfriedigung $f$, Einlage $f$, Einschließung $f$, Einschluß $m$, Einzäunung $f$, Gehege $n$, Koppel $f$, Schranke $f$, Um-
friedigung $f$, Umgitterung $f$, Umzäunung $f$,
Verkleidung $f$, Wall $m$ ~ **of cockpit** Führersitzverkleidung $f$
**encode, to** ~ kodieren
**encode** Kodierteil $m$
**encoded** geschlüsselt
**encoder** Kodierer $m$, Kodiergerät $n$, Verschlüss-
ler $m$ ~ **matrix** Verschlüsselungsmatrix $f$
**encoding** Schlüsseln $n$
**encompass, to** ~ umfassen
**encounter, to** ~ anfahren, antreffen, auftreffen,
begegnen
**encroach, to** ~ übergreifen **to** ~ **upon** beeinträchtigen
**encroachment** Spielraum $m$ ~ **by sand** Sandan-
schwemmung $f$

encrustations Beschläge *pl*
encrypt, to ~ verschlüsseln
encrypted message part verschlüsselter Teil *m* eines Spruchs
encumber, to ~ belasten, belemmern, verschütten
encumbrance Beschwerung *f*
encyclopedia Sachwörterbuch *n*
encym, ~ desizing Encymentschlichtung *f* ~ preparation Encympräparat *n*
end, to ~ aufhören, (Kabel) auslaufen, (the page) ausgehen, ausmünden, beendigen, endigen to ~ working Schicht *f* machen
end Ausgang *m*, Ende *n*, Grenze *f*, Kopfseite *f*, Ort *m*, Schluß *m*, Schwanz *m*, Trumm *n*, Ziel *n*, Zweck *m* at the ~ hinten by the ~, on ~ endweise on ~ im Lot *n*
end, ~ of adit Stollenort *m* ~ of billing period Tag *m* der Zählerstandsaufnahme ~ of the casing Stutzen *m* (Gehäusestutzen) ~ of the crystal Stirnfläche *f* des Kristalls ~ of delivery Förderende *f* ~ of dial-pulse-train relay Wählimpulsreihe-End-Relais *n* ~ of digit-selection relay Wählimpulsereihe-End-Relais *n* ~ of ebb Hinterebbe *f* ~ of the ebb Achterebbe *f*, letzte Ebbe *f* ~ of hub Nabenstirnfläche *f* ~ of a level Feldort *m* ~ of the line Zeilenschluß *m* ~ of message indication lamp Schlußlampe *f* ~ of nozzle Düsenmündung *f* ~ of page signal Bogenendsignal *n* ~ of pole Stangenfuß *m* ~ of record Satzende *n* (data proc.) ~ of scale-sensitivity Endempfindlichkeit *f* ~ and side-cutting pliers Vor- und Seitenschneider *m* ~ of the spring Federende *n* ~ of tunnel Stollenmundloch *n* ~ of word character Wortendezeichen *n*
end, ~ alkalinity Endalkalität *f* ~ amplitude Endamplitude *f* ~ apparatus Endapparat *m* ~-around carry End-, Umlauf-übertrag *m* ~ arrangement Endverschluß *m* ~ bearing Endlager *m* ~-bearing pile Spitzenbelastungspfahl *m* ~ bench Endstrosse *f* ~ block Endmaß *n*, Parallelendmaß *n* ~ bulk-head Einleitungsspant *n* ~ burner Kopfbrenner *m* ~ cap Abschlußblech *n*, Endkappe *f*, Endstößel *m*, (jacket) Mantelkopf *m* ~ cap for contact breaker Verschlußdeckel *m* mit Dichtplatte für Unterbrecher ~ capacity Endkapazität ~ caps Gehäuseschilder *pl* ~ carriage Laufradträger *m* (Laufkran) ~ cell Schalt-, Zusatz-zelle *f* ~ centered basisflächenzentriert ~ chap spindle Hinterzangenspindel *f* ~ clearance Axial-, Längs-spiel *n* ~ closing press Bombierpresse *f* ~ collar Anschlag-, Bund-ring *m* ~ collar for front guard with lever Anschlagring mit Ausrückhebel ~ cooler Endkühler *m* ~ cover Abschlußblech *n*, (of el. motor) Schild *m* ~ crater Endkrater *m* ~ cross member Schlußquerträger *m* ~ curve Schlußkurve *f* ~ cut Hirnschnitt *m* ~ cutting reamers Reibahlen *pl* mit Stirnzähnen ~ device Schlußorgan *n* ~ disc Schlußscheibe *f* ~ distortion Rückflankenverschiebung *f* (transmitter) ~ drive Endantrieb *m* ~ dump car Vorder-kipper *m*, -kippwagen *m* ~-dump truck Holländerwagen *m* ~ dumping Hinterkippung *f* ~ electrode Seitenelektrode *f* ~ elevational view Stirnansicht *f* ~ exchange Endamt *n* ~ face Endfläche *f*, Stirnseite *f* ~ face of a crystal Kristallendfläche *f* ~-face mill Walzenstirn-

fräser *m* ~-fire array Längsstrahler *m* ~ flange Abschlußflansch *m* ~ float Axialspiel *n*, Spiel *n* in Längsrichtung ~ frame Abschlußrahmen *m* ~ gas Restladungsanteil *m* ~ girder Kopfträger *m* ~ grain Hirn-holz *n*, -seite *f* ~-grained cut Hirnschnitt *m* ~-grained wood Hirn-, Stirnholz *n* ~ head Endkopf *m* (bei E-Messern) ~ hoop Achsring *m* ~ impedance Endimpedanz *f* ~ inhibitor Endring *m* ~ joint Lötstelle *f* ~ journal Stirnzapfen *m* ~ journal bearing Stirnlager *n* ~-lap weld Kopfschweiße *f* ~-loading ramp Stirnrampe *f* ~ loss Endverlust *m* ~ measure Endmaß *m* ~-measure gauge Endmaßlehre *f* ~ measuring block Parallelendmaß *n* ~-measuring rod Endmaß *n*, Endmaßeinsatz *m*, Parallelendmaß *n* ~-measuring rod with spherical end sphärischer Endmaßeinsatz *m* ~ member Endstück *n* ~ mill Dingerfräser *m*, Fräser *m* mit Stirnzähnen, (shank) Schaftfräser *m*, Stirnfräser *m*, Stirnfräsmaschine *f* ~ mill cutter Langlochfräser *m* ~ milling Fräsen *n* paralleler Stirnflächen, Langlochfräsen *n* ~-milling attachment Fingerfräseinrichtung *f* ~-milling cutter Fingerfräser *m*, Fräser *m* mit Stirnzähnen, Messerkopf *m*, Schaftfräser *m*, Stirnfräser *m*, Stirnfräsmaschine *f* ~ milling machine Endfräsmaschine *f* (fixed) ~ moment Einspannungsmoment *n* ~-on directional antenna Antennenanordnung *f* ~-on view Ansicht *f* von einem Ende aus ~-over-end type drum mixer aufrechtstehender Trommelmischer *m* ~ panel Schlußpaneel *n* ~ paper (bookbinding) Vorsatzblatt *n* ~ path Weg *m* des vom Leitungsende herrührenden Echos ~ piece Federkolben *m*, Hebelauge *n*, Kopfträger *m*, Stangenkopf *m*, Trumm *n*, Wange *f* ~ piece fork Gabelstück *n* ~ plane Stirnfläche *f* ~ plank Stirnwand *f* ~ plate Abdeckscheibe *f* (Scheibenkupplung), Abschlußdeckel *m*, Abschlußwand *f*, Flügelendscheibe *f*, Kesselboden *m*, Lagergehäusedeckel *m*, Motorstirnblech *n* ~ plate for ink duct Farbkastenseitenteil *m* ~ platform (across two or more roads) Querbahnsteig *m* ~ play Achsial-, Axial-, End-spiel *n*, leerer Gang *m*, Spiel *n* in Längsrichtung ~ point Abtitrierungs-, Äquivalenz-, End-, (distillation) Endsiede-punkt *m* ~-point color Umschlagsfarbe *f* ~-point control Güteregelung *f* ~ pole End-mast *m*, -gestänge *n* ~ pole dynamo Außenpoldynamo *f* ~ position End-lage *f*, stellung *f* (aviat.) ~ pressure Kantenpressung *f* ~ ramp Kopframpe *f* ~ recess Endnische *f* ~ relieving axiales Hinterdrehen *n* ~ result Endresultat *n*, Schlußergebnis *n* ~ rib Einleitungsrippe *f*, Endversteifung *f* (aviat.) ~ ring Abschluß-, Bund-ring *m* ~ rope Endtau *n* ~ scale value Maximalskalenwert *m* ~ screw Hinterzange *f*, Schneckentrieb *m* ~ seam Bodenrundnaht *f* (Kessel) ~ section Anlaßstrecke *f* eines Pupinkabels, Endstück *n* einer Mole ~ section with one-third of the multiple Ansatzschrank *m* für ein Drittel des Vielfachfeldes ~ shackle Endschäkel *m* ~ shield Abschlußschirm *m*, Lagerschild *n* ~ shield securing screw Schraube *f* zum Befestigen des Lagerdeckels ~ shields of ganged condenser Kondensatorabschlußwände *pl* ~-shorted circuit am Ende kurzgeschlossener Stromkreis *m* ~ sill Endschwelle *f* des Tosbeckens, Kopfstück *n* ~ **sleeve for buffing cylinder** Stirnflansch *m* ~ **span** Endfeld *n* ~ **stage** nachgeschaltete Stufe *f* ~ **stanchion** Eck-, Kopf-runge *f* ~ **station** Endstation ~ **stop** Endanschlag *m*, Feststeller *m* ~ **strake** Endplatte *f* ~ **stray** Flankenstreuung *f* ~ **support(s)** Endauflager *n*, Erdauflager *n*, Gegenlagerständer *m*, Kopfquerträger *m* ~ **support column** Stützlagerständer *m* ~ **terminal** Endpolklemme *f* ~ **thrust** Achsial-, Axial-druck *m*; Axialschub *m*, Enddruck *m*, Längsdruck *m* ~-**thrust bearing** Spurzapfenlager *n* ~-**tip wagon** Stirn-, Vorkipper *m* ~ **tippler** Kopfwippler *m* ~-**to-end measurement** Streckenmessung *f* ~-**to-end scavenging** Gleichstromspülung *f* ~-**to-end test** Streckenmessung *f* ~ **truss** Endträger *m* ~ **turns** Endwicklung *f* (electr.) ~ **value of the range** Bereichsendwert *m* ~ **view** End-, Seitenansicht *f* ~ **wall** Kopf-, Stirn-wand *f* ~-**wall tube** Wandröhre *f* ~ **way of the grain** Hirnseite *f* ~ **winding** Endwicklung *f*, Stirnverbindung *f* ~-**window counter** Fensterzählrohr *n* ~-**window Geiger counter (or tube)** Glockenzählrohr *n* ~ **window type** Endfensterausführung *f*

**endanger, to** ~ gefährden **to** ~ **human life** Menschenleben *pl* gefährden

**endangered** gefährdet

**endangering** Gefährdung *f* ~ **life** lebensgefährlich ~ **the security** die Sicherheit *f* gefährden

**endeavor, to** ~ sich bemühen, sich bestreben, trachten

**endeavor** Anstrengung *f*, Arbeitsfeld *n*, Bestrebung *f*, Streben *n*

**ending** Endung *f*, Spruchende *n* (rdo); ausgehend, auslaufend ~ **in a nozzle** in eine Düse *f* auslaufend

**endless** endlos ~ **abrasive belt** endloses Schleifband *n* (Metall) ~ **belt** Band *n* ohne Ende, endloser Riemen *m* ~ **belt system** laufendes Band *n* ~ **blanket** Fortführungstuch *n* ohne Ende ~ **bucket trencher** Grabenbagger *m* (mit Eimern) ~ **cable** Kette *f* ohne Ende, endlose oder geschlossene Kette *f*, endloses oder geschlossenes Seil *n* ~ **cloth** Rundtuch *n* ~ **conveyer** Kreistransporteur *m* ~ **film strip** endloser Filmstreifen *m* ~ **length** endlose Länge *f* ~ **paper** Papierband *n*, Rollenpapier *n* ~-**platform (or endless-roller) carrousel** Kreistransporteur *m* ~ **rope** endloses oder geschlossenes Zugseil *n* ~-**rope haulage** Körperfördermaschine *f* (min.) ~ **sanding belt** endloses Schleifband *n* (Holz) ~ **screw** Schraube *f* ohne Ende ~ **tangent screw** Seitenfeinbewegung *f* ~ **wet machine** Rundsieb-Entwässerungsmaschine *f* ~ **wire** Langsieb *n* ~ **(or continuous) worm gear** endlose Schnecken *pl*

**endodyne** Autodyn *n*, Selbstüberlagerung *f*

**endoergic** endotherm ~ **reaction** endotherme Reaktion *f*

**endogenous** endogen

**endomorphism** Endomorphismus *m*

**endorse, to** ~ (a bill) girieren, indossieren, Rückhalt *m* geben, auf die Rückseite *f* schreiben, stützen

**endorsee** Indossat(ar) *m*

**endorsement** Eintragung *f* (in einem Ausweis), Giro *n*, Indossament *n*, Nachtrag *m*

**endorser** Girant *m*, Indossant *m*, Indossent *m*

**endorsing unit** Indosserieeinrichtung *f*

**endosity** Endosität *f*

**endosmosic** endosmotisch

**endosmosis** Endosmose *f*

**endosteum** Endost *n*

**endothelium** Endothelium *n*

**endothermal** wärmeverzehrend, Wärme *f* verzehrend, wärmeverbrauchend

**endothermic** endotherm, Wärme *f* verzehrend oder aufnehmend

**endow, to** ~ ausstattung (mit etwas)

**endowment** Stiftung *f* ~ **fund** Unterstützungskasse *f*

**ends** Böden *pl* (Tankwagen) ~ **up** Duplierung *f*

**endurance** Ausdauer *f*, Dauer *f*, Dauerhaftigkeit *f*, Dauerleistung *f*, Haltbarkeit *f*, Höchstflugdauer *f*, Lebensdauer *f*, Standzeit *f* ~-**bending-test specimen** Dauerversuchsbiegeprobe *f* ~ **crack** Ermüdungsriß *m* ~ **failure** Dauerbruch *m* ~ **flight** Dauerflug *m* ~ **impact test** Schlagdauerversuch *m* ~ **limit** Dauerbruch-, Dauerbefestigkeits-grenze *f*, Dauerstandfestigkeit *f*, Ermüdungsfestigkeit *f*, Ermüdungsgrenze *f*, (strength of material) Wechselbruchspannung *f* ~ **limit of stress** Arbeits-, Dauerschwingungsfestigkeit *f* ~ **ratio** Ermüdungsverhältnis *n* ~ **record** Dauerrekord *m* ~ **(of) rigidity** Biegedauerhaltbarkeit *f* ~ **strength** Dauerhaltbarkeit *f*, Dauerstand-, Ursprungs-, Zeit-festigkeit *f* ~ **stress** Dauerstandfestigkeit *f* ~ **tensile strength (of overhead lines)** Dauerzugfestigkeit *f* ~-**tension test** Dauerzugversuch *m* ~ **tension test** Zugversuch *m* mit Dauerbeanspruchung *f* ~ **test** Dauer-abbremsung *f*, -lauf *m*, -probe *f*, -prüfung *f*, -versuch *m* (mit ruhender Beanspruchung), Ermüdungsversuch *m* ~ **testing machine** Dauerfestigkeitsprüfmaschine *f* ~-**testing machine for alternating stresses** Dauerversuchsmaschine *f* für wechselnde Beanspruchung ~-**testing mashine for repeated torsion** Drehschwingungsmaschine *f* ~-**torsion-test specimen** Dauerverdrehungs-, Dauerwinde-probe *f* ~ **transverse-stress test** Dauerbiegeversuch *m* ~ **trial** Dauerfahrt *f* ~ **under torsion stress** Verdrehungsdauerhaltbarkeit *f*

**endure, to** ~ anhalten, ausdauern, aushalten, ausstehen, bestehen, dauern, dulden, erdulden, ertragen, herhalten, währen

**enduring** dauerhaft

**endways** der Länge *f* nach

**endwise** aufrecht, mit dem Ende *n* nach vorn, in Längsrichtung *f* ~ **(or axial) movement** Längsbewegung *f*

**energetic** energiereich, tatkräftig

**energization** Erregung *f*

**energize, to** ~ anreizen, betätigen, betreiben, erregen, erregt werden, in Schwung *m* bringen, speisen **to** ~ **a relay** ein Relais *n* erregen

**energized** stromführend ~ **dipole** Erregerdipol *m*

**energizer** Verstärker *m*

**energizing** kraftgebend ~ **of a relay** Erregung *f* eines Relais ~ **circuit** Erreger-, Erregungs-kreis *m* ~ **current** Erregerstrom *m* ~ **power** Erregungsleistung *f* ~ **voltage** Erregerspannung *f*

**energy** (physikalische) Arbeit *f*, Arbeitsfähigkeit *f*, Arbeitsvermögen *n*, Energie *f*, Kraft *f*, Leistung *f*, Tatkraft *f*

**energy, ~ of blow** Schlagleistung *f*, Stoßkraft *f*

~ **braking** Bremsenergie *f* ~ **of deformation** Arbeitsvermögen *n* ~ **of dislocation** Versetzungsenergie *f* ~ **in the exhaust steam** Abdampfenergie *f* ~ **of flow** Strömungsenergie *f* ~ **of formation** Bildungsenergie *f* ~ **of fracture** Brucharbeit *f* ~ **to fracture** Schlagarbeit *f* ~ **of generation** Bildungsenergie *f* ~ **of grain boundary** Korngrenzenergie *f* ~ **of mixing** Mischungsenergie *f* ~ **of oscillation** Oszillationsenergie *f* ~ **of rotation** Rotationsenergie *f* ~ **of transfer** Überführungsenergie *f* ~ **of turbulence** Turbulenzenergie *f* ~ **of vibration** Schwingungsenergie *f*

**energy,** ~ **absorption** Arbeitsaufnahme *f* ~ **amplification** Leistungsverstärkung *f* ~ **balance** Energiebilanz *f* ~ **band broadening** Energiebandverbreiterung *f* ~ **band structure** Energiebandstruktur *f* ~ **barrier** Energieberg *m* ~ **bearing** energiereich ~ **chamber** (of klystron) Leistungskammer *f* ~ **change** Energieveränderung *f* ~ **component** Sprach-, Watt-, Wirk-, Wirkspannungs-, Wirkstrom-komponente *f* ~ **conservation** Energieerhaltung *f* ~ **conservation law** Energiesatz *m* ~ **consumption** Energie-, Kraft-, Sprach-verbrauch *m* ~-**containing** energiehaltig ~ **content** Energieinhalt *m* ~ **conversion factor** Energieumwandlungsfaktor *m* ~ **converter** Energieumformer *m* ~ **current** Verlust-, Wirk-strom *m* ~ **decrement** Energieabnahme *f* ~-**dissipating device** Energievernichtungseinrichtung *f* ~ **dissipation** Leistungsaufwand *m*, Sprachverzehrung *f* ~ **dissipator** Energievernichter *m* ~ **distribution** Energie-, Sprachverteilung *f* ~ **efficiency** energetischer Wirkungsgrad *m*, Wirkungsgrad *m* nach Menge und nach Energie ~ **elements in atoms** Quanten *pl* ~ **ellipsoid** Energieellipsoid *n* ~ **equation** Arbeitsgleichung *f*, Energiegleichrichtung *f* ~ **exchange** Energieaustausch *m*, Kraftwechsel *m*, Leistungsumsatz *m* ~ **expended** Arbeitsaufwand *m* ~ **flow** Energie-fluß *m*, -strom *m*; Sprachfluß *m* ~ **flow vector** Energiestromdichte *f* ~ **flux density** Energiestromdichte *f* ~ **gain** Energiezuwachs *m* ~ **gap** Energiesprung *f* ~-**gap model** Energielückenmodell *n* ~ **gradient** Energielinie *f* ~ **hill** Energieberg *m* ~ **impulse** Energiestoß *m* ~ **input** eingeleitete oder aufgenommene Leistung *f* ~ **level** Energie-niveau *n*, -stufe *f*; (of electrons) Niveaustufe *f*, Quantumzustand *m* ~ **level diagram** Energieniveaudiagramm *n* ~ **level system** Energieniveausystem *n* (grafische Darstellung) ~-**level width** Breite *f* des Energieniveaus ~ **loss** Verlustarbeit *f* ~ **measurements** Energiemessungen *pl* ~ **meter** Energiezähler *m*, Wirkverbrauchszähler *m* ~-**momentum density** Energie-Impuls-Dichte *f* ~ **momentum tensor** Energie-Impulstensor *m* ~ **momentum theorems** Energie-Impulssätze *pl* ~-**momentum vector** Energie-Impuls-Vektor *m* ~ **output** Energieleistung *f*, Leistungsentnahme *f* ~ **preference** energetische Bevorzugung *f* (rdo) ~ **preservation** Energieerhaltung *f* ~ **quantum** Energiequantum *n* ~ **range** Energie-band *n*, -bereich *m* ~ **recovery** Energierückgewinnung *f* ~ **recuperation** Energierückgewinnung *f* ~ **region** Reichweite *f* ~ **relation** Energie-Frequenz-Beziehung *f* ~ **release** Energiefreigabe *f* ~ **required to produce disintegration of a molecule of an explosive** Aktivierungsenergie *f* ~ **requirement** Arbeits-, Sprach-bedarf *m* ~ **resolution** Energieauflösung *f* ~-**rich** energiereich ~-**saving** Kraftersparnis *f*, Kraftsparend ~ **sensitivity** Nachweisempfindlichkeit *f* ~ **shell** Energieschale *f* ~ **shift** Energietermverschiebung *f* ~ **spacing** Energieabstand *m* ~ **states** Energiezustände *pl* ~ **storage** Arbeitsinhalt *m*, Energieaufspeicherung *f* ~-**storage braking** Speicherbremsung *f* ~-**storage electrode** Speicherelektrode *f* (kinetic) ~ **stored in flywheel** Schwungenergie *f* ~-**stress complex** Energiespannungskomplex *m* ~ **supply equipment** Energieversorgungseinrichtung *f* ~ **term** Energiestufe *f* ~ **thickness** Energiedichte *f* ~ **threshold** Schwellenenergie *f* ~ **transfer equation** Energietransportgleichung *f* ~ **transformation** Energieumwandlung *f* ~ **unit** Energie-Einheit *f* ~ **wall** Energiewall *m* ~ **zero** Energie-Nullpunkt *m*

**enervate, to** ~ entnerven
**enfilade, to** ~ der Länge *f* nach beschießen, (mit Feuer) bestreichen, enfilieren
**enfilade** Schrägfeuer *n* ~ **fire** Flanken-, Flügelfeuer *n*, Längsbestreichung *f*, Längsfeuer *n* (artil.)
**enfilading** Enfilierung *f* ~ **fire** bestreichendes oder flankierendes Feuer, Querfeuer *n* (artil.)
**enforce, to** ~ Geltung *f* verschaffen
**enforceable** vollstreckbar
**enforced** zwangsläufig ~ **surrender** Zwangsrücklauf *m*
**engage, to** ~ angreifen, ankuppeln, in Anspruch oder in Dienst nehmen, anstellen, binden, (in, with) eingreifen, in Eingriff stehen, (in a catch) einklinken, (in) sich einlassen, einrasten, einrücken, einschalten, einstellen, erfassen, in Gang setzen, hineinfassen **to** ~ **and disengage** ein- und ausrücken **to** ~ **over** übergreifen **to** ~ **over one another** übereinandergreifen
**engage switch** Kuppelschalter *m*
**engageable at-will blower** nach Belieben einschaltbarer Lader *m*
**engaged** belegt, besetzt, eingeklinkt, eingerückt, eingreifend (mach.), unfrei; (condition) Besetztsein *n* ~ **on local call** ortsbesetzt ~ **on trunk call** fernbesetzt
**engaged,** ~ **battery** Besetztspannung *f* ~ **click** Knackgeräusch *n* oder Knacken bei der Besetztprüfung ~ **jack** besetzte Klinke *f* ~ **lamp** Besetztlampe *f* ~ **line** besetzte Leitung *f* ~ **signal** Besetztsignal *n* ~ **test** Besetztprüfung *f*
**engagement** enge Anschmiegung *f*, Besetztfall *m* (teleph.), Besetztsein *n*, Eingriff *m*, Einrückung *f*, Einstellen *n*, Gefecht *n*, Gefechtshandlung *f*, Ineinandergreifen *n*, Kampf *m*, Kampfhandlung *f*, Zusammenstoß *m* **without** ~ freibleibend ~ **of claws** Eingreifen *n* der Greifer ~ **on local call** Ortsbesetztsein *n* ~ **of teeth** Zahneingriff *m* ~ **of the trunk switchboard** Belegung *f* des Fernschranks
**engagement,** ~ **(or desk) diary** Terminkalender *m* ~ **spring** Feder *f* unter dem Kupplungsbelag
**engager lever** Einrücker *m*
**engaging** Einrücken *n*, (einer Kupplung) Einschalten *n*; ineinandergreifend ~ **and disengaging** ein- und ausschaltbar
**engaging,** ~ **(gear)action** Eingriffswirkung *f* ~

**bail** Rastbügel *m* ~ **claw** Einrückklaue *f* ~ **clutch** Einfallvorrichtung *f* ~ **disc** Rastscheibe *f* ~ **dog** Mitnehmerklaue *f* ~ **gear** Einrückvorrichtung *f*, Wälzzahnrad *n* ~ **gear clutch** Kupplung *f* ~ **lever** Einrückhebel *m* ~ **period** Schlupfzeit *f* ~ **piece** Mitnehmer *m* ~ **piston** (jet) Einrückkolben *m* ~ **position** Raststellung *f* ~ **wheel** Rastscheibenrad *n*
**engine** Lokomotive *f*, Maschine *f*, Maschinenkörper *m*, Motor *m*, Triebwerk *n* ~ **accessible in flight** im Fluge zugänglicher Motor *m* ~ **for air injection of fuel** Einblasemaschine *f* (Diesel) ~ **with four cylinders horizontally opposed** liegender Vierzylinder *m* mit paarweise gegenüberliegenden Zylindern ~ **at full throttle** Vollmotor *m* ~ **with independent sets of high- and low-pressure cylinders** Verbundmotor *m* ~ **with one-piece cylinder** Blockmotor *m* ~ **with short term action** Kurzzeittriebwerk *n*
**engine, ~ accessories** Motorrüstung *f* ~ **adapter** Triebwerk-Montagebügel *m* ~ **air by-pass flap** Triebwerk-Sekundär-Luftklappe *f* ~ **air inlet duct** Triebwerkslufteintrittskanal *m* ~ **analyzer** Motorkontrolloszillograf *m* ~ **anti-icing circuit** Stromkreis *m* für Triebwerk-Vereisungsschutz ~ **attendant** Maschinist *m* ~ **base** Maschinensockel *m* ~ **bay** Triebwerkraum *m* ~ **bearer** Motor-bock *m*, -pratze *f*, -träger *m* ~ **bearer feet** Motorauflage *f* ~ **bed** Motor-bock *m*, -fundament *n* ~ **bedplate** Motortragplatte *f* ~ **bleed pressure** Triebwerkabzapfdruck *m* ~ **block** Motorblock *m* ~ **boiler** Lokomotivkessel *m* ~ **bolt** Motoranker *m* ~ **booster pump** Ladedruckpumpe *f* ~ **bracket** Motorgerüst *n* ~ **brake-H. P. test bed** Motorenbremsprüfstand *m* ~ **breakdown** Motorpanne *f* ~ **build-up truck** Triebwerkmontagewagen *m* ~ **builder** Maschinenbauer *m* ~ **building** Maschinenbau *m* ~ **car** Maschinengondel *f* ~ **cart** Maschinenkarren *m* ~ **casting** Motorenguß *m* ~ **circulation lubrication system** Motorumlaufschmierung *f* ~ **compartment** Maschinenkammer *f*, Triebwerkraum *m* ~ **compressor set** Kompressoraggregat *n* ~ **container** Triebwerkbehälter *m* ~ **control lever** Gashebel *m*, Reglerhebel *m* **control linkage** Triebwerkbedienungsgestänge *n* ~ **controls** Maschinensteuerwerk *n*, Triebwerksgestänge *n* ~ **coolant** Motorkühlstoff *m* ~ **cooling** Motorkühlung *f* ~ **cover** Motorüberzug *m* ~ **cowling** Haubenverkleidung *f*; Motorschutz *m*, Motor-verkleidung *f*, -verschalung *f*; Triebwerksverkleidung *f* ~ **cradle** Motorgestell *n* ~ **currently in production** Serienmotor *m* ~ **cycle** Motorperiode *f* ~ **data** Motorkenndaten *pl* ~ **detonation limit** Klopfgrenze *f* ~ **distillate** Destillat *n* für Maschinenreinigung ~ **drive** Motorgetriebe *n* ~ **drive shaft** Hauptantriebswelle *f* ~ **driven** motorgetrieben ~ **driven supercharger** Getriebelader *m* ~ **driver** Maschinenführer *m* ~ **duct** Triebwerkslufteintrittskanal *m* ~ **earth (or ground)** Motormasse *f* ~ **efficiency** Motorwirkungsgrad *m* ~ **equipment** Motorrüstung *f* ~ **exhaust cone** Austrittskegel *m* ~ **factory** Motorenwerk *m* ~ **failure** Ausfallen *n* eines Motors, Motorstörung *f* ~ **fairing** Triebwerksverkleidung *f* ~ **fault release** Motorstörungsentriegelung *f* ~ **fire wall** Triebwerkbrandschott *n* ~ **fitter** Maschinenschlosser *m*

~ **flange** Motoranschlußflansch *m* ~ **flywheel** Motorschwungrad *n* ~ **frame** Motorrahmen *m* ~ **fuel** Treibstoff *m* ~ **fuel flow control** Kraftstoffregler *m* ~ **gap lathe** Drehbank *f* mit Kröpfung ~ **gear wheel** Motorgetriebe *n* ~ **governor shaft** Motorregulierungswelle *f* ~ **hatch door** Triebwerksklappe *f* ~ **headlight** Lokomotivlaterne *f* ~ **hours** Motorbetriebsstunden *pl* ~**-house** Fördermaschinengebäude *n*, Maschinen-haus *n*, -raum *m* ~**-house crane** Maschinenhauskran *m* ~**-in** mitlaufendem Motor *m* ~ **inoperative** Motor ausgefallen, stillgelegter Motor *m* ~ **installation** Motoreinsetzung *f* ~ **knife** Walzenmesser (Holländer) *m* ~ **lathe** Leitspindel *f*, Leitspindeldrehbank *f*, Spitzendrehbank *f*, Spitzendrehbank *f* mit Leitspindel ~ **log** Motorprotokoll *n* ~ **logbook** Motorflugbuch *n* ~ **lubricator** Maschinenschmierer *m* ~ **magazine** Motoraufbewahrungsraum *m* ~ **man** Maschinist *m* ~ **manifold pressure** Ladedruck *m* ~ **manufacturer** Motorenbauer *m* ~ **mechanic** Motorenwart *m* ~ **mount** Motor-aufhängegerüst *n*, -bock *m*, -fuß *m*, -lafette *f*, -lager *n*, -vorbau *m*; Triebwerkaufhängung *f* ~**-mount ring** Motortragring *m* ~ **mounting** Motor-aufhängung *f*, -bockbeschlag *m*; Motoreneinbau *m*, Motorträger *m*, Triebwerkgerüst *n* ~ **mounting bracket** Motortragblock *m* ~**-mounting lug** Motorbefestigungsauge *n* ~ **nacelle** Motor(en)gondel *f*, Triebwerksverkleidung *f* ~**-nacelle bracing** Motorverstrebung *f* ~ **noise** Motorgeräusch *n* ~ **oil** Maschinenöl *n* ~ **operator** Maschinist *m* ~ **order telegraph** Maschinentelegraf *m* (shipboard) ~ **output** Motorleistung *f* ~ **overhaul(ing)** Motorüberholung *f* ~ **overspeed** Motorüberdrehzahl *f* ~ **pad** Anbauflansch *m* ~ **part** Motorteil *m* ~ **performance** Motorleistung *f* ~ **pipeline** Triebwerksleistung *f* ~ **piston** Motorkolben *m* ~ **plane** Förderweg *m* ~ **plate** flacher Motoraufhängering *m*, Motorschild *n* ~**-plate ring** Motortragring *m* ~ **pod** Triebwerksgondel *f* ~ **power** Motorleistung *f* ~**-power absorption by the airscrew** Leistungsaufnahme *f* durch die Luftschraube ~ **power control lever** Leistungs-, Gas-hebel *m* ~**-preheating device** Motoranwärmegerät *n* ~ **primer** Einspritzanlasser *m*, Triebwerksanlassereinspritzvorrichtung *f* ~ **priming** Einspritzen *n* von Anlaßkraftstoff in den Motor ~**-propeller unit** Triebwerk *n* ~ **pump** Triebwerkpumpe *f* ~ **rating** Motoreinschätzung *f*, Motorkrafteinschätzung *f* ~ **rear compartment** hinterer Geräteraum *m* (zwischen Brandschott und Motor) ~ **removal** Motorausbau *m* ~ **revolutions** Motordrehzahl *f* ~ **ring cowling** Ringverkleidung *f* ~ **road** Förderstrecke *f* mit maschineller Förderung ~ **room** Maschinen-halle *f*, -kammer *f*, -raum *m* ~**-room log** Maschinentagebuch *n* ~ **with** running mit laufendem Motor ~ **run-up** Prüf-, Warm-lauf *m* ~ **set** Maschinensatz *m* ~ **shaft** Maschinenschaft *m* ~ **shed** Remise *f* ~ **specification** vorgeschriebener Motorkennwert *m* ~ **speed** Motordrehzahl *f* ~**-speed counter** Motordrehzahlmesser *m* ~ **start-switch** Triebwerkanlaßschalter *m* ~ **starting** Anlassen *n* des Motors ~ **strut** Motorstrebe *f* ~ **strutting** Motorverstrebung *f* ~ **support** Motor-ab-

stützung f, -stütze f ~ support bracket Triebwerkaufhängebeschlag m ~ suspension Motorlagerung f (Auto), Triebwerksaufhängung f ~ tachometer circuit Drehzahlmesserstromkreis m ~-test certificate Motorprüfschein m ~-test log Motorabnahmeprotokoll n ~-test report Motorprüfbericht m ~-test stand Motorprüfstand m ~ torque Drehmoment n des Motors, Motordrehmoment ~ trouble Motorpanne f, Motorstörung f ~ trunnion Triebwerkaufhängebeschlag m ~ unit Maschinensatz m ~ valve Motorventil n ~ vibration Triebwerksschwingung f ~ warm-up Warmlaufen n ~ waste Putzbaumwolle f ~ weight per horsepower Motorgewicht n gemäß Pferdekraft ~s working on the same crankshaft Gegenzwilling m ~ works Maschinenfabrik f
engined flight Motorflug m
engineer, to ~ ausklügeln
engineer Ingenieur m, Lokomotivführer m (r.r.), Maschinenbetriebsführer m, Techniker m ~ for drainage Kulturtechniker m ~ on duty wachhabender Ingenieur m
engineer, ~ corps Ingenieurkorps n ~ platoon Fachtrupp m ~ rating Maschinist m
engineered ausgeklügelt
engineering Technik f, Verkehrswasserbau m (nav.); technisch ~ assistant technischer Gehilfe m ~ change technische Änderung f ~ circuit Dienst-kanal m, -leitung f ~ construction Maschinenbau m ~ department Konstruktionsabteilung f ~ design Konstruktionsausführung f ~ diagram technische Zeichnung f ~ division Materialverwaltung f ~ flight Erprobungsflug m ~ grades of mechanite Ausführungen pl in Mechanite ~ manager technischer Direktor m ~ material Baustoff m ~ officer technischer Beamter m ~ personnel Maschinenpersonal n ~ reactor industrieller Reaktor m ~ reactor research Ingenieurforschungsreaktor m ~ standards committee Fachnormenausschuß m ~ structural steel Maschinenbaustahl m ~ superintendent technischer Betriebsmeister m ~ survey technische Aufnahme f ~ table Konstruktionsmerkblatt n ~ time Wartungszeit f
engineer's, ~ scale Reduktionsmaßstab m ~ wrench (or spanner) Gabelschlüssel m
English englisch ~-pich lead screw Zolleitspindel f ~ red Zementrot n
engrobe Begußmasse f, Engobe f (Keramik)
engobing machine Engobiermaschine f
engram Engramm n
engrave, to ~ einritzen, einschneiden, gravieren, stechen
engraved eingeätzt ~ brass cylinder Gaufrier-, Muster-walze f ~ printing roller Gravürdruckwalze f ~ screen Gravurraster m ~ sound film Gravür-, Präge-film m
engraver Bildstecher m, Graveur m, Schneidestichel m, Zeiger m ~'s burnisher Polierstahl m
engraving Gravierung f, Gravüre f, Stich m ~ on glass Glasgravur f ~ on metals Metallschneidekunst f ~ of the roller Walzengravur f
engraving, ~ cylinder Bildzylinder m ~ establishment Gravieranstalt f ~ like print stichähnlicher Druck m ~ needle Gravierstichel m ~ plate Druckplatte f ~ roller Dessinwalze f ~ tool Gravierwerkzeug n, Schreibstift m, Zise-

lierwerkzeug n ~ wax Gravierwachs n
engross, to ~ in Anspruch nehmen
enhance, to ~ in günstigeres Licht stellen
enhanced, ~ line besonders starke Linie f ~ permeability gesteigerte Durchlässigkeit f
enharmonic enharmonisch
enigma Rätsel n ~ machine Chiffermaschine f
enlarge, to ~ aufweiten, ausdehnen, ausweiten, erweitern, verbreitern, vergrößern, weiten to ~ holes with a drift Löcher pl aufdornen
enlarged vergrößert ~ image vergrößertes Bild ~ macroscopic photograph makroskopische Vergrößerung f ~ scale (or rule) vergrößerter Maßstab m ~ view vergrößerte Ansicht f
enlargement Ausbau m, Ausbreitung f, (of a vent or orifice) Ausweitung f, Eingreifen von Zahnrädern, Erweiterung f, Umgrößerung f, Verbreiterung f, Vergrößerung f, Vermehrung f ~ factor Vergrößerungszahl f
enlarger Vergrößerungs-apparat m (photo), -gerät n
enlarging Vergrößern n (photo) ~ bailer Erweiterungsbüchse f ~ bit Flügelbohrer m ~ old shops Erweiterung bestehender Betriebe pl ~ shaft Erweiterungsschacht m
enlightening lehrreich
enlist, to ~ in Listen eintragen, eintreten, sich freiwillig melden
enliven, to ~ beleben
ennoble, to ~ veredeln
enormous übermäßig ~ cube Riesenwürfel m ~ quantity Unmasse f
enough genug, genügend to be ~ genügen
enquiry Nachfrage f
enrich, to ~ anreichern, veredeln
enriched angereichert ~ metal Spurstein m ~ pile (or reactor) angereicherter Reaktor m ~ produce angereichertes Haufwerk n
enriching Anreicherung f, (of copper) Reichfrischen n, Veredeln n ~ with oxygen Sauerstoffanreicherung f
enrichment Anreicherung f, Veredelung f ~ cam Anreicherungsnockenscheibe f ~ jet (or nozzle) Anreicherungsdüse f ~ valve Anreicherungsventil n ~ weight Anreicherungsfliehgewicht n
enroll, to ~ ausheben, einreihen to ~ the crew die Mannschaft anmustern
enrollment Aushebung, Einreihung f
en-route auf Strecke, unterwegs
ensemble Gesamtheit f ~ average Phasenmittelwert m
enshrouding cylinder Fang-, Ummantelungs--zylinder m
ensiform schwertförmig
ensign Abzeichen n, (flag) Fahne f
ensilage, to ~ einsäuern
ensilage cutter Futterschneider m, Grünfutter-Häcksler m
ensile, to ~ ensilieren
ensure, to ~ ergeben, sicherstellen
entail, to ~ zur Folge haben
entangled verwachsen
entanglement Flächendrahthindernis n, Hindernis n, Sperre f, Verflechtung f ~ wire Hindernisdraht m
enter, to ~ anfahren, buchen, eindringen, (obligations) eingehen, einlaufen, einmarschie-

ren, einsteigen, eintragen, eintreten, kartieren **to ~ into an agreement** ein Übereinkommen treffen **to ~ by air** einfliegen **to ~ the circuit** mithören **to ~ into combination** eine Verbindung eingehen **to ~ into competition with** in Konkurrenz treten mit **to ~ a connection** in eine Verbindung eintreten **to ~ in a conversation** in ein Gespräch eingreifen **to ~ into an equation** in eine Gleichung eingehen **to ~ into fermentation** in Gärung übergehen **to ~ into a formula** in eine Formel eingehen **to ~ into** einzeichnen, (in einen Plan) eintragen **to ~ on a list** vormerken **to ~ into negotiation** in Unterhandlung treten **to ~ into negotiations** in Verbindung treten **to ~ port** in einen Hafen *m* einlaufen **to ~ service** den Dienst antreten **to ~ upon** betreten

**entered** eingegangen

**entering** Einführung *f*; eintretend **~ of flood** das Einsetzen *n* der Flut **~ of the water** Wasserandrang *m*

**entering, ~ angle** Eintrittswinkel *m* **~ arrangement for fabric** Gewebeeinführapparat *m* **~ channel** Einführungsrinne *f* **~ edge** Flügeleintrittskante *f* (aviat.) **~ edge balance** Innenausgleich *m* **~ edge of wing** Eintrittskante *f* (aviat.) **~ file** Lochfeile *f* **~ gear tooth** Eingriffszahn *m* **~ guide** Einführung *f* **~ reel (or feeding) roller** Einführungswalze *f* **~ side** (in rolling) Einstich-, Eintritts-seite *f* **~ side of rools** Walzeneintritt *m* **~ thickness** Eintrittsdicke *f* **~ trough** Einführungsrinne *f* **~ wedge** Vorderkante *f*

**enterprise** Betriebsunternehmen *n*, Unternehmen *n*, Unternehmung *f* **~ of hydraulic structure** Wasserbauunternehmung *f*

**entertain, to ~** darbieten (rdo), unterhalten

**entertainment** Bewirtung *f*, Darbietung *f* (rdo, acoustic or visual), Unterhaltung *f* **~ value** Unterhaltungswert *m*

**enthalpy** Bildungswärme *f*, Enthalpie *f*, Wärmeinhalt *m*

**entire** ganz, gänzlich, gesamt, voll, völlig, vollständig **~ face plate area** Bildfläche *f* (TV) **~ pipe system** Durchhängenetz *n* **~ proceeds** Gesamtertrag *m* **~ reduction ratio** Gesamtübersetzung *f*

**entirely** lediglich

**entirety** Ganze *n*, Gesamtheit *f*, Vollständigkeit *f*

**entitle, to ~** berechtigen, überschreiben

**entitled** befugt, berechtigt, betitelt **~ to** berufen zu **to be ~ to** Anspruch *m* haben auf **to be ~ to salary** auf Gehalt *n* Anspruch haben **~ to vote** stimmberechtigt **~ to win** gewinnungsberechtigt sein

**entitlement** Bezugsberechtigung *f*

**entitling** Überschriftung *f*

**entity** Größe *f*

**entrain, to ~** einkuppeln, einladen, mitreißen, verladen

**entrainer** Entzieher *m* **~ cam** Schlagdaumen *m* **~ disc** Mitnehmerscheibe *f*

**entraining** Verladung *f* **~ point** Ladestelle *f* **~ table** Fahrtliste *f* **~ time** Lade-, Verlade-zeit *f*

**entrainment** Mitführung *f*, Überreißen *n* **~ means** Mitnehmer *m*

**entrance** Amtsantritt *m*, Anzug *m*, (vehicle) Einfahrt *f*, Einführen *n*, Einführungsstelle *f*, Eingang *m*, Eingangsöffnung *f*, Einlaß *m*, Ein-

lauf *m*, Eintritt *m*, Leitungseinführung *f* (teleph.), (upper or lower) Vorschleuse *f*, Zugang, Zulauf *m*, Zulauföffnung *f*, Zutritt *m* **~ into a gallery** Minenauge *n* (min.) **~ of a pass** Walzeneintritt *m*

**entrance, ~ bar** Einführ-, Eingangs-wange *f* **~ box for lines** Einführungskasten *m* für Leitungen **~ bucket** (gas turbines) Leitschaufel *f* **~ cone** Eingangskegel *m* **~ door** Einsteigtür *f* **~ duty** Einfuhrzoll *m* **~ examination** Aufnahmeprüfung *f* **~ exit sign** Eingangs- Ausgangszeichen *n* **~ fee** Eintrittspreis *m* **~ gallery** Eingangsstollen *m* **~ gate** (of dry dock) Dockhaupt *n* **~ hall** Vorhalle *f* **~ hatch** Einsteigluke *f* **~ ladder** Einsteigleiter *f* **~ level** Anfahrsohle *f* **~ lock** Einfahrtschleuse *f* **~ loss** Eintrittsverlust *m* **~ phase** Eintrittsphase *f* **~ pupil** Eingangs-, Eintritts-pupille *f* **~ slit** Eingangsschlitz *m* **~ tunnel** Eingangsstollen *m* (min.) **~ zone** Einflußzone *f*

**entrap, to ~** einschließen (water, oil, etc.), fangen

**entrapped air** Lufteinschluß *m*

**entreat, to ~** dringend bitten ·

**entrefer** Luftspalt *m*

**entrench, to ~** eingraben, einschanzen, schanzen

**entrenching** Schanz-arbeit *f*, -tätigkeit *f* **~ tools** Schanzzeug *n*

**entrenchment** Schanze *f*, Verschanzung *f*

**entresol** Zwischengeschoß *n*

**entries** Prüfpunkte *pl*

**entropy** Entropie *f*, Wärmegewicht *n*, Zustandsgröße *f* **~ of activation** Aktivierungsentropie *f* **~ of fusion** Schmelzentropie *f* **~ of transfer** Überführungsentropie *f*

**entropy, ~ analog** Entropieanalogon *n* **~ concept** Entropiebegriff *m* **~ flow** Entropiestromdichte *f* **~ flux** Entropiefluß *m* **~ production** Entropieerzeugung *f*

**entruck, to ~** einladen

**entrucking, ~ point** Lade-, Verlade-stelle *f* **~ table (or survey)** Einladeübersicht *f* **~ time** Ladezeit *f*

**entry** Einklarierung *f*, Einschreiben *n*, Einstieg *m*, Eintrag *m*, Eintragung *f*, Eintritt *m*, Einzug *m*, Fahrstrecke *f*, Posten *m*, Vermerk *m*, Zugang *m* **no ~** abgesperrt **~ by air** Einflug *m* **~ in a contest** Nennung *f* **~ and exit** Einstieg und Ausstieg *m* **~ on map** Karteneintragung *f*, Karteneinzeichnung *f* **~ in the minutes** Protokolleintragung *f*

**entry, ~ duct** Stirngehäuse *n* **~ end** Einlaufzeit *f* **~ fan** Eintrittsschaufel *f* **~ guide casing** Einführungsgehäuse *n* **~ guide jaw** Einführungsbacke *f* **~ guide press** Einführungspresse *f* **~ hub** Eingangsbuchse *f* **~ leg** Einflugteil *m* (aviat.) **~ pauch** Eingabelocher *m* **~ portal** Bestrahlungs-, Einfalls-feld *n* **~ side** Einlaufseite *f* **~ tables** Einlauftische *pl* **~ vane** Eintrittsblatt *n*

**entwine, to ~** verflechten

**enucleate, to ~** herausschälen

**enumerable** abzählbar (math.)

**enumerably infinite dimensional** abzählbar, unendlichdimensional

**enumerate, to ~** aufführen, aufzählen, nummern

**enumeration** Aufzählung *f*

**enumerative geometry** abzählende Geometrie *f*

**enunciation** Aussprache *f* **~ of numbers** Zahlenaussprache *f* envelop, to ~ bewickeln, einhüllen, einkreisen, umfassen, umhüllen, umschlagen, umschleiern, umwickeln, umzingeln

envelope Beplankung f, Decke f, Einhüllende f, Glasgefäß n, Hülle f, Hülle f eines Luftfahrzeuges leichter als Luft, Kolben m, Kuvert n, Mantel m, reaktionsfähiger Mantel m, Meßkolben m für Fotozelle, Überzug m, Umgrenzung f einer Kurve, Umgrenzungslinie f, Umhüllende f (rdr), Umhüllung f, Umhüllungslinie f, Ummantelung f, Umrandungskurve f, Umschlag m, Umwicklung f ~ of carbon Kohlehülle f ~ of a family of curves Berandungslinie f einer Kurvenschar

envelope, ~ curve Hüllen-, Hüll-kurve f ~ delay Gruppenlaufzeit f ~ delay/frequency distortion Laufzeitverzerrung f ~-delay meter Gruppenlaufzeitmeter n ~ detection Hüllenkurvengleichrichtung f ~ detector Hüllenkurven-Demodulator m ~ gas bag Ballonhülle f ~ magazine Hüllenmagazin n ~-making machine Umschlagfertigungsmaschine f ~ sealer Petschaft n ~ velocity Gruppengeschwindigkeit f

enveloping, ~ curve Umhüllungskurve f ~ fluid Umgebungsflüssigkeit f

envenomation (corrosion) Schädigung f

environment Milieu n, Umgebung f ~ contamination Umgebungsversuchung f

environmental, ~ chamber Klimakammer f ~ influence Umwelteinfluß m ~ sensitivity Umgebungsempfindlichkeit f

environs Weichbild n

enwrap, to ~ einhüllen

enzymatic enzymatisch ~ splitting enzymatische Spaltung f

enzyme Enzym n

eocene Eozän n

eolation Windschliff m

eolic gradation Windschliff m

eosine Eosin n

eosinophil Eosinophil

epaulement Schulterwehr f

ephantinite Ephantinit m

ephemeral locker

ephemerides Ephemeriden pl

epicadmium Epikadmium n

epicenter Epizentrum n, Oberflächenmittelpunkt m

epicycle Bei-, Neben-kreis m

epicyclic epizyklisch ~ gear Planetenrad n ~ gear(ing) Umlaufgetriebe n ~ (train of) gear Planetengetriebe n ~ gearing Planetengetriebe n ~ type Planetenbaumuster n ~ unit Planetenträger m

epicycloid Epizykloid f, Radlinie f

epidiascope Aufbildwerfer m, Epidiaskop n

epidote Eisenepidot m

epi-illumination Auflichtbeleuchtung f

epilation Enthaarung f

epiphysis Epiphyse f

epipolar, ~ axis Kernachse f ~ pencil of rays Kernstrahlenbüschel n ~ plane Kernebene f ~ ray Kernstrahl m

epipole Kernpunkt m

episcope Episkop n

epitasimeter Epitasimeter n

epithelial cells Epithelzellen pl

epithelioma Epitheliom n

epithermal epitherm

epitome Inbegriff m

epitrochoid Epitrochoide f

epizogenics Epizogenik n

E plane bend E-Bogen m

epoch Phase f, Zeit f

E.P. (extreme pressure) properties Hochdruckeigenschaften pl

Epsom salt Bittersalz n, schwefelsaures Magnesium n, englisches Salz n, Epsom Salz n

equal, to ~ gleichkommen; angemessen, gleich, gleichgroß, gleichmäßig, gleichwertig; to be ~ gleichen of ~ amplitude amplitudengetreu (of) ~ angle winkeltreu ~ area (of a map or chart) flächengetreu on an ~ footing paritatisch of ~ legs gleichschenklig ~ and apposite voltages symmetrische Ablenkspannungen pl of ~ quantity ebenbürtig to be ~ to the task der Aufgabe gerecht werden

equal, ~-angle point winkeltreuer Punkt m (Radialtriangulation) ~-area projection (mapping) flächentreue Projektion f ~-armed gleicharmig ~-energy white Weiß n gleicher Energien (der Farbkomponenten), Idealweiß n (TV) ~-falling gleichfällig ~ falling Gleichfälligkeit f ~ length multi-unit code Einheitskode m; (un) ~ letter telegraph Telegraf m mit (un)gleich langen Zeichen ~ mark Gleichheitszeichen n ~-phase gleichphasig ~ power distribution between cylinders Leistungsgleichheit f der einzelnen Zylinder ~-pressure altitude Gleichdruckhöhe f ~-ratio bevel gear Kegelradübertragung f ~-settling gleichfällig ~ settling Gleichfälligkeit f ~-sided gleichschenklig ~-sided angle iron gleichschenkliges Winkeleisen n ~-signal white Mittelweiß n (TV) ~-sized gleichgroß ~-span biplane Doppeldecker m mit gleichgroßen Flächen

equality Gleichförmigkeit f, Gleichheit f, Parität f ~ of brightness photometer Gleichheitsfotometer n ~ of impression Eindrucksgleichheit f ~ of sensation Eindrucksgleichheit f ~ sign Gleichheitszeichen n

equalization Abgleichung f, Ausgleich m, Ausgleichen n, Ausgleichung f, Egalisierung f, Einebnung f, Entzerrung f, Gleichmachung f, Gleichsetzung f, Gleichung f ~ of charges Ausgleich m der Ladungen ~ of coupling Kopplungsausgleich m ~ of energy energiewirtschaftlicher Ausgleich m ~ of levels Pegelausgleich m ~ of pressure Druckausgleich m ~ in series Längsentzerrung f ~ of temperature Temperaturausgleich m

equalization network Ausgleichsnetzwerk n

equalize, to ~ abgleichen, ausgleichen, egalisieren, entzerren, glätten, gleichmachen, gleichmäßig machen, gleichstellen, nivellieren to ~ current output den Ausgangsstrom m abgleichen; to ~ loads auslasten

equalized abgeglichen, ausgeglichen

equalizer Ausgleichapparat m, Ausgleicher m, Ausgleichhebel m, Ausgleitzeiger m, Entzerrer m (für Telefonleitungen), Gleichmacher m, Korrektionsfilter n ~ beam Ausgleichsstange f ~ disc Ausgleichsscheibe f ~ horn Ausgleichshorn n ~ layer Ausgleichsschicht f ~ spring Ausgleichs-, Einstell-feder f

equalizing, ~ amplifier entzerrender Verstärker

*m* ~ **basin** (pondage) Ausgleichsbecken *n* ~ **capacity** Glättungskapazität *f* ~ **chamber** Ausgleichskammer *f* ~ **charge** Ausgleichsladung *f* ~ **circuit** Ausgleichleitung *f* ~ **coil** Drossel *f* für Dreileitermaschinen ~ **current** Ausgleichs-, Energie-strom *m*, Energieströmung *f*, Strom *m*, Strömung *f* ~ **current impact (or current surge)** Ausgleichsstromstoß *m* (electr.) ~ **device** Aufteileinrichtung *f* ~ **evaporator** Ausgleichsverdampfer *m* ~ **fixture** selbstfixierende Vorrichtung *f* ~ **flow** Ausgleichströmung *f* ~ **layer** Ausgleichsschicht *f* ~ **levels** Pegelausgleich *m* ~ **line** Luftdruckausgleichsleitung *f* ~ **network** Ausgleichsschaltung *f*, Dämpfungsausgleicher *m*, Dämpfungsentzerrer *m*, Entzerrungs-filter *n*, -kette *f*; Kunstleitung *f* ~ **path** Ausgleichung *m* ~ **pipe** Ausgleichrohr *n* ~ **port** Ausgleichsauge *n* ~ **pressure** Ausgleichdruck *m* ~ **pulley** Ausgleichsrolle *f* ~ **pulse** Ausgleichsimpuls *m*, Trabant *m* (TV) ~ **repeater** entzerrender Verstärker *m* ~ **rolling mill** Egalisierwalzmaschine *f* ~ **screw stocks** Egalisierkluppen *pl* ~ **self inductance** Glättungsselbstinduktion *f* ~ **tank** Ausgleichbehälter *m* ~ **time** Ausgleichzeit *f* ~ **truck** Fahrschemel *m* ~ **tube** Ausgleichsrohr *n* ~ **valve** Ausgleichventil *n* ~ **wire** Ausgleichdraht *m*

**equally,** ~ **efficient** ebenbürtig ~ **loaded** gleichbelastet ~ **spaced** gleichentfernt ~ **spaced points** äquidistante Abszissen *pl* ~ **tempered scale** wohltemperierte Tonleiter *f*

**equate, to** ~ angleichen, auf ein Durchschnittsmaß bringen, auf einen Mittelwert bringen, gleichsetzen, nullsetzen

**equated,** ~ **busy hour call (E.B.H.C.)** Zweiminutenverbindung *f* ~ **value** Durchschnittswert *m*

**equating** Gleichsetzung *f*

**equation** Berechnungsformel *f*, Beziehung *f*, Gleichsetzung *f*, Gleichung *f*, (of a reaction) Reaktionsgleichung *f* **(Boltzmann)** ~ Stationäritätsbedingung *f* (Boltzmannsche) *f* **quadrat** ~ Gleichung *f* zweiten Grades

**equation,** ~ **of adiabatic change of state** Adiabatengleichung *f* ~ **of the center** Mittelpunktgleichung *f* ~ **of combination** Verbindungsgleichung *f* ~ **of condition** Bedingungs-, Bestimmungs-, Zustands-gleichung *f* ~ **of continuity** Kontinuitätsgleichung *f*, Stetigkeitsbedingung *f* ~ **of dose** Dosisgleichung *f* ~ **of equilibrium** Gleichgewichtsbedingung *f* ~ **of formation** Bildungsgleichung *f* ~ **of lift** Auftriebsformel *f* ~ **of moments** Hebelgesetz *n*, Momentengleichung *f* ~ **of motion** Beschleunigungsgleichung *f*, Bewegungsgleichung *f* ~ **of the program branch** Umlenkgleichung *f* ~ **for slowing down** Auslaufgleichung *f* ~ **of state** Zustandgleichung *f* ~ **in straight form** Gerade *f* ~ **of thermal conduction** Wärmeleitungsgleichung *f* ~ **of time** Zeitgleichung *f*

**equation,** ~ **connecting the components** Seitenkraftgleichung *f* ~ **solver** Lineargleichungslöser *m*, Polynomrechner *m* ~ **state** Zustandsgleichung *f*

**equations,** ~ **of field** Feldgleichungen *pl* ~ **grouped in families** Gleichungsfamilie *f* ~ **of motion** Bewegungsgleichungen *pl* ~ **(or solutions) in spherical coordinates** Gleichungen oder

Lösungen *pl* in Kugelkoordinaten

**equator** Äquator *m*, Gleicher *m* (aviat.) ~ **-ecliptic body** Ekliptikkugel *f*

**equatorial,** ~**(ly)** äquatorial ~ **axis** parallaktische Achse *f* ~ **belt** äquatoriale Gebiete *pl* ~ **current** Äquatorialstrom *m* ~ **grid reference system** äquatoriales Gradnetz *n* ~ **head with hour circle and declination circle** parallaktisches Achsensystem *n* mit Stundenkreis und Deklinationskreis ~ **mounting** äquatoriale Aufstellung *f* ~ **seam** Äquatornaht *f* ~ **section** Äquatorschnitt *m* ~ **semi-diameter** Erdhalbmesser *m* ~ **system of carriers of the telescope mounting** parallaktisches Tragsystem *n* der Fernrohrmontierung ~ **telescope** parallaktisches Fernrohr *n*

**equatorially mounted telescope** parallaktisch montiertes Fernrohr *n*

**equi,** ~ **angular** gleichwinklig ~ **angular figure** Gleicheck *n* ~ **angular-spiral** logarithmische Spirale *f* ~ **anharmonic** äquianharmonisch ~ **areal (or area-preserving)** Flächentreue *f* ~ **axial** gleich-achsig, -gerichtet ~ **directional** gleichgerichtet ~ **distance** gleicher Abstand, Schichtenabstand *m*; abstandsgleich, äquidistant, gleichentfernt, gleichweit, gleichweit abstehend, parallel ~ **distant curves** Kurven *pl* gleichen Abstandes ~ **frequent** von gleicher Frequenz ~ **frequent conductor** mitschwingender Leiter *m* ~ **lateral** gleichseitig ~ **lateral triangle** gleichseitiges Dreieck *n* ~ **lateral-triangular waveguide** gleichseitiger Dreiecksquerschnitt *m*

**equilibrant** ausgleichend

**equilibrate, to** ~ abgleichen, auswuchten, ins Gleichgewicht bringen, im Gleichgewicht erhalten oder sein

**equilibrating process** Ein- und Ausschwingen *n*, flüchtiger Vorgang *m*

**equilibration** Ausgleichen *n*, Ausgleichung *f*

**equilibrator,** ~ **anchor** Ausgleicherarm *m* ~ **arm** Ausgleichsarm *m* ~ **seat** Ausgleicherlager *n* ~ **spring** Ausgleichfeder *f*

**equilibrium** Ausgleich *m*, Gleichgewicht *n* ~ **of forces** Kräftegleichgewicht *n* ~ **of moments** Momentausgleich *m*

**equilibrium,** ~ **angle** Gleichgewichtswinkel *m* ~ **approach** Gleichgewichtsannäherung *f* ~ **ball float valve** Gleichgewichts-Kugelschwimmerventil *n* ~ **condition** Beharrungszustand *m*, Gleichgewichtsfall *m* ~ **constant** Gleichgewichtskonstante *f* ~ **(or transformation) curve** Gleichgewichtskurve *f* ~ **diagram** Hebelbeziehung *f*, Zustandsdiagramm *n* ~ **enrichment factor** Gleichgewichtsanreicherungsfaktor *m* ~ **field** Gleichgewichtsfeldstärke *f* ~ **luminance** Gleichgewichtsleuchtdichte *f* ~ **orbit** Gleichgewichtsbahn *f*, Sollkreis *m* ~ **position** Ruhelage *f* ~ **potential** Gleichgewichtspotential *f*, Ruhespannung *f* ~ **properties** Gleichgewichtseigenschaften *pl* ~ **ratio** Gleichgewichtsverhältnis *n* ~ **region** Zustandsfeld *n* ~ **rest potential** Ruhepotential *n* ~ **restoration** Gleichgewichtseinstellung *f* ~ **temperature** Umwandlungstemperatur *f* ~ **value** Ruhewert *m* ~ **velocity** Grenzgeschwindigkeit *f* ~ **water** normaler Wassergehalt *m*

**equimolecular** äquimolekular

**equimomental** vom gleichen Moment ~ **mass distribution** Massenverteilungen *pl* gleicher Trägheitsmomente

**equimultiple** gleichvielfach

**equinoctial** äquinoktial ~ **gale (or storm)** Äquinoktialsturm *m*

**equinox** Äquinoktium *n*, Tag- und Nachtgleiche *f*

**equip, to** ~ anschließen, armieren, aufrüsten, ausbauen, auskleiden, ausrüsten, ausstatten, einrichten, installieren, rüsten, verlegen **to** ~ **completely** vollbereiten **to** ~ **with a handle** bestielen **to** ~ **with ropes** seilen **to** ~ **with springs** abfedern

**equipartition** Gleich-teilung *f*, -verteilung *f* ~ **theorem** Äquipartitionstheorem *n*

**equiphase zone** Überschneidungsgebiet *n* (rdo, nav.)

**equipment** Anlage *f*, Anrüstung *f*, Anstellung *f*, Apparat *m*, Apparatur *f*, Armierung *f*, Ausrüstung *f*, Ausstattung *f*, Betriebsanlage *f*, (technische) Einrichtung *f*, Gerät *n*, Geschirr *n*, Material *n*, Montierung *f*, Rüstung *f*, Vorrichtung *f*, Werkgerät *n*, Zurüstung *f* ~ **for producing aluminum** Aluminiumgewinnungseinrichtung *f* ~ **in short supply** Engpaßgerät *n* ~ **with a socket** Sockelung *f* ~ **for transportation** Beförderungsanlage *f* ~ **for xylolith factory** Steinholzfabrikeinrichtung *f*

**equipment,** ~ **capacity** Aufnahmefähigkeit *f* ~ **checking book** Gerätüberwachungsbuch *n* ~**-collecting point** Gerätsammelstelle *f* ~ **condition** Rüstzustand *m* ~**-control log** Gerätüberwachungsbuch *n* ~ **costs** Einrichtungskosten *pl* ~ **dump** Feldzeuglager *n* ~ **inspector** Gerätinspizient *m* ~ **inventory** Ausrüstungsverzeichnis *n* ~ **issuing point** Geräteanschlußkabelausgabestelle *f* ~**-issuing port** Ausrüstungshafen *m* (mil.) ~ **object** Einrichtungsgegenstand *m* ~ **park of dump** Gerätepark *m* ~ **pool** Bauhof *m* ~ **schedule** Gerätenachweis *m* ~ **specification** Ausrüstungsvorschrift *f* ~ **status report** Ausrüstungsnachweis *m* ~ **truck** Gerätkraftwagen *m* ~ **wagon** Gerätewagen *m* ~ **wiring** Geräteverdrahtung *f*

**equipotential** äquipotentiell, isoelektrisch ~ **cathode** Äquipotentialkathode *f*, indirekt geheizte Glühkathode *f* ~ **connection** Ausgleichsverbindung *f* ~ **contour** Äquipotentialfläche *f* ~ **line** Äquipotentiallinie *f*, Linie *f* gleichen Potentials, Niveaulinie *f*, Potentiallinie *f* ~ **line method** Methode *f* der äquipotentialen Linien ~ **plane** Äquipotentialfläche *f* ~ **space** Raum *m* gleichen Potentials ~ **surface** Äquipotentialfläche *f*, Fläche *f* gleichen oder konstanten Potentials, Niveaufläche *f*, konstante Potentialfläche *f* ~ **surfaces** äquipotentiale Oberflächen *pl* ~ **tube** indirekt geheizte Röhre *f*

**equipped,** ~ **with** ausgerüstet mit ~ **with telescopic sight** fernrohrbesetzt ~ **capacity** Fassungsvermögen *n*

**equipresence** Äquipräsenz *f*

**equiquadrature lines** Quadraturlinien *pl*

**equiradial** ungerichtet ~ **antenna** ungerichtete Antenne *f*

**equisignal,** ~ **beacon** Leitstrahlsender *m* ~ **course (or line)** Leitlinie *f*, Leitstrahllinie *f* (rdo nav.) ~ **radio-range beacon** Leitstrahlsender *m* ~ **sector** Leitstrahl *m* ~ **track** Funk-

schneise *f* ~ **zone** Dauerstrichzone *f* (nav.), Leitlinie *f*, Leitungszone *f*, gleichsignalige Zone *f*

**equity** Gerechtigkeit *f* ~ **suit** Billigkeitsverfahren *n*

**equivalence** Äquivalenz *f*, Gegenleistung *f*, Gleichwertigkeit *f* ~ **principle** Ausschließungs-, Eindeutigkeits-prinzip *n*

**equivalency** Ersatzfähigkeit *f*

**equivalent** Äquivalent *n*, Entsprechung *f*, Gegenwert *m*, (value) Gleichwert *m*, Maß *n*; äquivalent, gleichbedeutend, gleichwertig, sinngemäß, ~ **of heat** Hitzegegenwert *m* •

**equivalent,** ~ **absorption** Äquivalenzverschluckung *f* (acoust.) ~ **acoustics** Ersatz-Schalleffekte *pl*, Geräuschkasten *m* ~ **airspeed (EAS)** äquivalente Eigengeschwindigkeit *f* ~ **amount** Äquivalenzbetrag *m* ~ **articulation loss** Ersatzdämpfung *f* ~ **binary digits** äquivalente Binärstellenzahl *f* ~ **build-up time** äquivalente Anstiegzeit *f* ~ **capacity** Äquivalent-, Ersatzschema *n*, -schaltung *f*, -stromkreis *m*; Kunstschaltung *f*, äquivalente Leitung *f*, äquivalenter Stromkreis *m* ~ **circuit diagram** Ersatzbild *n*, Ersatzschaltbild *n* ~**-circuit scheme** Ersatzschema *n* ~ **concentration** Molarität *f*, Normalität *f* ~ **conductance** Äquivalentleitvermögen *n* ~ **damping decrement** scheinbares logarithmisches Dämpfungsdekrement *n* ~ **density altitude** die einer gegebenen Luftdichte entsprechende Höhe *f* ~ **diode of a triode** äquiva-kapazität *f* ~ **circuit** Ersatz-kreis *m*, -schaltente Diode *f* ~ **diode voltage** äquivalente Diodenspannung *f* ~ **disturbing current** äquivalenter Störstrom *m* ~ **disturbing voltage** äquivalente Störung *f* ~ **earth plane** äquivalente Erdoberfläche *f* ~ **electron** Ersatzelektron *n* ~ **electron volts** Voltelementarladung *f* ~ **focal length** Äquivalenzbrennweite *f* (film) ~ **focus** Äquivalenzbrennpunkt *m* ~ **four-wire system** Zweidraht-Getrenntlagesystem *n* (film) ~ **ground plane** äquivalente Erdoberfläche *f* ~ **inductance** Äquivalentinduktivität *f* ~ **logarithmic decrement** scheinbares logarithmisches Dämpfungsdekrement *n* ~ **logarithmic increment** äquivalentes logarithmisches Inkrement *n* ~ **loudness** äquivalenter Lautheitspegel *m*, subjektive Lautstärke *f* ~ **network** Ersatzschaltung *f* ~ **piston** Äquivalenzkolben *m* (acoust.) ~ **points with respect to discontinuous groups of mappings** äquivalente Punkte *pl* bei diskontinuierlichen Abbildungsgruppen ~ **resistance** Äquivalent-, Ersatz-widerstand *m*, äquivalenter Widerstand, Widerstandsäquivalent *n* ~ **resistor** Ersatzwiderstand *m* ~ **response pulse growth-time** Anstiegzeit *f* des äquivalenten Impulses ~ **response pulse total duration** Gesamtzeit *f* des äquivalenten Impulses ~ **series resistance** Reihenverlustwiderstand *m* ~ **sine wave** äquivalente Sinuswelle *f* ~ **spark** Funkenschlagweite *f* ~ **standard sample** Eichdeckel *m* ~ **stopping power** Bremsäquivalenz *f* ~ **system** Ersatzschaltschema *n* ~ **weight** Äquivalentgewicht *n* ~ **widths** äquivalente Breiten *pl*

**equivalve** gleichschalig

**equivocal** doppelsinnig, zweideutig

**equivocation** Mehrdeutigkeit *f*

**era** Ära *f*, Phase *f*
**eradicate, to** ~ ausmerzen, tilgen
**erasable,** ~ **storage** Löschspeicherung *f*, löschbarer Speicher *m* ~ **store** löschbarer (Puffer-) speicher *m* (info proc.)
**erase, to** ~ auslöschen, ausradieren, ausreiben, austreichen, löschen, radieren, tilgen, überlochen (info proc.)
**erase,** ~ **character** Auslaßzeichen *m* ~ **cut-out key** Löschsperre *f* (tape rec.) ~ **frequency** Löschfrequenz *f* ~ **generator** Löschimpulsgenerator *m* ~ **head** Löschkopf *m* ~ **key** Irrungstaste *f* ~ **signal** Irrung *f*, Irrungszeichen *n*
**erased** gelöscht (magn. tape)
**eraser** Reibgummi *m*
**erasing** Löschung *f* ~ **field** magnetisches Löschfeld *n* ~ **head** Löschkopf *m* ~ **knife** Radiermesser *n* ~ **liquid** Radierwasser *n* ~**-proof (or grindig-proof)** radierfest ~ **shield** Radierschablone *f* ~ **speed** Löschgeschwindigkeit *f* ~ **voltage** Löschspannung *f*
**erasure** Irrung *f*, radierte Stelle *f* ~ **of errors** Löschung *f* ~ **signal** Irrungszeichen *n*
**erbium** Erbium *n* ~ **oxide** Erbinerde *f*
**erect, to** ~ aufführen, aufmontieren, aufpflanzen, aufrichten, aufschlagen, aufsetzen, aufstellen, bauen, erbauen, errichten, (Anlagen) herstellen, hochrichten, montieren, rüsten, setzen, zusammenbauen **to** ~ **a perpendicular** Senkkörper *pl* errichten, eine Senkrechte *f* errichten **to** ~ **a pole** eine Stange *f* setzen
**erect** aufrecht, gerade
**erected,** ~ **image** aufgerichtetes oder aufrechtes Bild *n* ~ **position** direkte Lage *f*
**erecting** Montierung *f* ~ **bay** Montagehalle *f* ~ **crane** Montagekran *m* ~ **frame** Kipprahmen *m* (g/m) ~ **prism** Aufrichte-, Aufrichtungs-prisma *n*, bildumkehrendes Prisma *n* ~ **scaffolding** Montagegerüst *n* ~ **shop** Montage-werkstatt *f*, -halle *f*, Rüsthalle *f* ~ **tackle** Rüstzeug *n* ~ **tool** Montagewerkzeug *n*
**erection** Aufbau *m*, Aufrichtung *f*, Aufstellung *f*, (of buildings) Ausführung *f*, Errichtung *f*, Montage *f*, Montierung *f*, Rüstung *f*, Zusammenbau *m* ~ **in the field** Aufstellung *f* am Betrieb(s)ort
**erection,** ~ **blueprint** Montagezeichnung *f* ~ **cross-talk** Aufrichtübersprechen *n* ~ **drift** Richtabdrift *f* (g/m) ~ **mast** Montagemast *m* ~ **mechanism** Aufrichtmechanismus *m* ~ **motor** Stutzmotor *m*
**erector** Monteuer *m*, Rüster *m* ~ **stop** Aufrichtungsblende *f* ~ **tube** Umkehrsystem *n*
**erector's instrument** Montageinstrument *n*
**erg** Erg *n* (cgs-Einheit der Arbeit)
**ergmeter** Ergmeter *n*
**ergodic,** ~ **hypothesis** Ergodenhypothese *f* ~ **problem** Ergodenproblem *n* ~ **system** ergodisches System *n*
**ergometer** Energie-, Leistungs-messer *m*, Ergometer *n*
**ericite** Heidenstein *m*
**Ericsson,** ~ **cupping** Ericssontiefung *f* ~ **cupping machine** Ericssonblechprüfapparat *m* ~ **cupping test** Ericssontiefziehprobe *f* ~ **ductility machine** Ericssonblechprüfapparat *m* ~ **ductility test** Ericssontiefziehprobe *f* ~ **selector**

Ericssonwähler *m*, Kulissenwähler *m* ~ **sheet-metal testing apparatus** Ericssonblechprüfapparat *m* ~ **test** Ericssonversuch *m*, Tiefziehversuch *m*
**eriometer** Eriometer *n* (opt.), Wollstärkemesser *m*
**Erlenmeyer flask** Becherglaskolben *m*, Erlenmeyerkolben *m*
**erode, to** ~ abfressen, (cavitation) anfressen, ausfressen, auskolken, auswaschen, durchfressen, einfressen, erodieren, fressen, hineinfressen, kolken, zerfressen
**eroded soil** fortgespülter Boden *m*
**eroding (or washing) bank** Abbruchufer *n*
**erosion** Abnutzung *f*, Abrasion *f*, Abtragen *n*, Abtragung *f*, Anfraß *m*, Anfressung *f*, Ausbrennung *f*, Ausfressung *f*, Auskolkung *f*, Auswaschung *f*, Durchfressen *n*, Einfressung *f*, Eintiefung *f*, Kolkbildung *f*, Verschleiß *m*, Zerfressung *f* ~ **of the bore** Rohrabnutzung *f* ~ **by slag** Schlackenangriff *m* ~ **resistor** Abbrandwiderstand *m*
**err, to** ~ abirren, fehlen, fehlgehen, irren
**errand boy** Laufbursche *m*
**errata** Druckfehlerverzeichnis *n*
**erratic** erratisch, launisch, unregelmäßig, ziellos ~ **block** Findling *m* ~ **blocks** Geschiebeblöcke *pl* ~ **value** Streuwert *m*
**erratum** Druckfehler *m*
**erroneous** fehlerhaft, irrig ~ **conclusion** Trugschluß *m* ~ **construction (or interpretation)** falsche Auslegung *f*
**erroneously** irrtümlich
**error** Druckfehler *m*, Fehler *m*, Irre *f*, Irrtum *m*, Irrung *f*, Mißgriff *m*, Unrichtigkeit *f*, Versehen *n* (pilot's) ~ **Bedienungsfehler** *m* ~ **of approximation** Verfahrensfehler *m* (info proc.) ~ **of avertence** Verschwenkungsfehler *m* ~ **of axis of tilt** Kippachsenfehler *m* ~ **in calculation** Schätzungstäuschung *f* ~**s of the calibrated circles** Nullpunktfehler *pl* der Teilkreise ~ **in card punching** Falschlochung *f* ~ **of convergence** Konvergenzfehler *m* ~ **of curvature** Krümmungsfehler *m* ~ **of demonstration** Beweisfehler *m* ~ **in depth** Entfernungsfehler *m* ~ **of direction** Richtungsfehler *m* ~ **of distortion** Verzeichnungsfehler *m* ~ **in eccentricity** (of gear) Teilkreisschlag *m* ~ **in estimating** Schätzungsfehler *m* ~ **from external sources** Fremdfehler *m* ~ **in figures** Rechnungsfehler *m* ~ **in fix (or reckoning)** Besteckfehler *m* ~ **in focusing** Einstellfehler *m* ~ **of graduation** Teilungsfehler *m* ~ **of height** Höhenfehler *m* ~ **in indication** Falschweisung *f* ~ **in judging** Schätzungsfehler *m* ~ **in judgment** Schätzungsirrtum *m* ~ **of judgment** Fehlgriff *m* ~ **of measurement** Meßfehler *m*, Versuchsfehler *m* ~ **in observation** Beobachtungsfehler *m* ~ **of phase** Phasenfehler *m* ~ **of position** Lagefehler *m* ~ **in range** Aufsatzfehler *m*, Längenabweichung *f* ~ **in range due to rain** Längenstreuung *f* ~ **in (or of) reading** Ablesefehler *m* ~ **of sighting** Visierfehler *m* ~ **due to squint** Schielfehler *m* ~ **in starting** Abgangsfehler *m* ~ **of swing** Verkantungsfehler *m* ~**s in taking bearings** Fehlweisung *f* ~ **of test** Versuchsfehler *m* ~ **of tilt** Neigungsfehler *m* ~ **in weighing** Wiegefehler *m*
**error,** ~ **card** Fehleranzeigekarte *f* ~ **character**

Auslaßzeichen $n$ ~ chart Fehlertafel $f$ ~-com-
pensation value Peilungsbeiwert $m$ ~-correcting
code fehlerkorrigierender Kode $m$ ~ detecting
code selbstprüfender Kode $m$ ~ detection
Fehlersuche $f$ ~ detector Fehlersignal-Detektor
$m$, Meßglied $n$ ~ determination Fehlerbestim-
mung $f$ ~-distribution curve Fehlerscheibchen-
verteilungskurve $f$ ~ due to jump Abgangsfehler
$m$ ~ equation Fehler-funktion $f$, -gleichung $f$
~-finding calculation Fehlerrechnung $f$ ~
function Fehler-funktion $f$, -integral $n$ ~ integral Fehlerintegral $n$ ~ limits Meßfehler-
grenzen $pl$ ~ matrix Fehlermatrix $f$ ~ propagation Fehlerfortpflanzung $f$ ~ reading fehlerhafte Ablesung $f$ (eines Meßgerätes) ~ signal
Fehler-, Korrektur-signal $n$
ersatz Austauschstoff $m$, Austauschware $f$
eruption Ausbruch $m$, Ausschlag $m$, Durchbruch
$m$, Vukanausbruch $m$
eruptive rock Eruptivgestein $n$
erythema dose Erythem-, Hauteinheits-dosis $f$
erythosin Erythosin $n$
erythrite Kobalt-blume $f$, -blüte $f$, -schlag $m$
escalator Rolltreppe $f$, Stufenbahn $f$
escape, to ~ abströmen, abziehen, auspuffen,
ausströmen, durchschlüpfen, entgehen, entkommen, entlaufen, entrinnen, entweichen,
fliehen to ~ from entströmen
escape Abgang $m$, Abziehen $n$, Auspuff $m$, Ausströmung $f$, Austritt $m$, Durchbruch $m$, Effusion $f$, Entkommen $n$, (of a gas) Entweichen $n$,
Entweichung $f$, Flucht $f$, Hilfsschacht $m$ (min.),
Rettung $f$, Verlust $m$ ~ of gas Gasausbruch $m$
~ by parachute Rettung $f$ mit dem Fallschirm
escape, ~ capsule Rettungskapsel $f$ ~ channel
Abzugskanal $m$ ~ character Umschaltzeichen
$n$ ~ groove Antriebsrille $f$ ~ hatch Rettungs-
luke $f$ ~ maneuver Auslaufmanöver $n$ ~ pipe
Abblaserohr $n$ ~ probability Überlebenswahr-
scheinlichkeit $f$ ~ shaft Not-, Sicherheits-
-schacht $m$ ~ speed Fluchtgeschwindigkeits-
grenze $f$ ~ steam Auspuffdampf $m$ ~ surface
Ausblasfläche $f$ ~ valve Abfluß-, Auslaß-,
Entweichungs-ventil $n$ ~ velocity Fliehge-
schwindigkeit $f$ ~ wheel Hemmrad $n$
escapement Anker $m$, (watch) Auslösung $f$,
Gesperr $n$, Hemmung $f$ ~ button Auslösungs-,
Auslöser-knopf $m$ ~ crank Sperrkurbel $f$ ~
knob Auslösungsknopf $m$ ~ operating bale
Schaltbügel $m$ ~ spring Auslösungsfeder $f$ ~
teletachometer zwangsläufiger Ferndrehzahlmesser $m$ ~ wheel Hemm-, Hemmungs-, Steig-
rad $n$
escaping gas Abzugsgas $n$
escarp (or counterscarp) Grabenböschung $f$
escarpment Böschung $f$, Landstufe $f$, Steilwand $f$
eschynite Eschynite $m$
E scope E-Schirm $m$ (rdr)
escort, to ~ begleiten, geleiten
escort Bedeckung $f$, Begleiter $m$, Begleitschutz
$m$, Begleitung $f$, Geleit $n$ ~ party Begleitkommando $n$ ~ personnel Bedeckungs-, Geleit-
-mannschaft $f$ ~ vessel Begleitschiff $n$, Flotten-
begleiter $m$, Geleitschiff $n$
escutcheon Verzierungsleisten $pl$ an Geräten ~
die Wappenstempel $m$
eskers Öser $pl$
esparto Faengras $n$ ~ paper Alfapapier $n$

especial besonder, erheblich
especially eigens, im besondern, vorab, vorwiegend
espionage Ausspähung $f$, Spionage $f$
esplanade Hochplan $m$
essay Aufsatz $m$
essence Essenz $f$, Natur $f$, Substanz $f$, Wesen $n$
essential hauptsächlich, wesentlich ~ feature
Wesen $n$ ~ load Grundbelastung $f$ ~ oil ätherisches oder flüchtiges Öl $n$ ~ part Kernstück
$n$ ~ point Anhaltspunkt $m$ ~ reaction Grund-
reaktion $f$
essonite Hessonit $m$
establish, to ~ anlegen, aufnehmen, aufstellen,
begründen, einrichten, feststellen, festigen,
festsetzen, gründen, Anlagen herstellen, (an
equilibrium or concentration) verlegen, vorlagern to ~ the busy condition Besetztspannung
$f$ anlegen to ~ communication (or contact) eine
Verbindung $f$ aufnehmen to ~ a connection
eine Verbindung $f$ herstellen to ~ a curve eine
Schaulinie $f$ aufzeichnen to ~ a fix by dead
reckoning koppeln to ~ one's range (gunnery)
festlegen to ~ a record einen Rekord $m$ aufstellen to ~ stereoscopic contact in räumliche
Übereinstimmung $f$ bringen
established fest, feststehend
establishing, ~ contact with Aufnahme $f$ der
Fühlung ~ shot Gesamtaufnahme $f$
establishment Anlage $f$, Anstalt $f$, Aufstellung $f$,
Ausbau $m$, Einrichtung $f$, Feststellung $f$, Gründung $f$, Niederlassung $f$ ~ of accounts Aufstellung der Rechnungen ~ of communication
Verbindungsaufnahme $f$ ~ of a connection Aufbau $m$ (Herstellung) einer Verbindung ~ of
the connections Verbindungsaufbau $m$ ~ of
contact Fühlungnahme $f$ ~ of liaison (or contact) Verbindungsaufnahme $f$ ~ of a port
Hafenzeit $f$ ~ of scholarships Stipendienstiftung
$f$ ~ of titer Titerstellung $f$
establishment, ~ charge Anlagekosten $pl$ ~
charges Anlage-, Einrichtungs-kosten $pl$
estate Anwesen $n$, Grund $m$, Landgut $n$
ester Ester $m$ ~ gum Estergummi $n$, Harzester $f$
~ interchange Umesterung $f$ ~ varnish Ester-
lack $m$
esterification Veresterung $f$
esterify, to ~ verestern
esthesiometer Ästhesiometer $n$
estimable bestimmbar
estimate, to ~ abschätzen, anschlagen, auswerten, beurteilen, einschätzen, gissen, -rechnen,
schätzen, taxieren, überschlagen, veranschlagen
estimate Angabe $f$, Ansatz $m$, Anschlag $m$, Gutdünken $n$, Kalkulation $f$, Kostenanschlag $m$,
Kostenrechnung $f$, Schätzung $f$, (rough) Überschlag $m$, (preliminary) Voranschlag $m$, Veranschlagung $f$, Wertbestimmung $f$, Wertung $f$
estimated abgeschätzt, gegißt, veranschlagt,
überschläglich ~ elapsed time voraussichtliche
Flugdauer $f$ ~ error geschätzte Abweichung $f$
~ expenditure Sollausgabe $f$ ~ life geschätzte
Lebensdauer $f$ (eines Gegenstandes) ~ value
Schätzungswert $m$
estimating Veranschlagung $f$
estimation Abschätzung $f$, Beurteilung $f$, Veranschlagung $f$ ~ of errors Fehlerabschätzung $f$
estimator Berechner $m$

estopped (law) Hindernis n, Inhibierung f
estuary Astuar n, Einmündung f, Flußhafen m,
Meeresarm m ~ part Binnenhafen m
etamine Kongreßstoff m
etch, to ~ abätzen, anätzen, ätzen, beizen,
radieren to ~ in relief hochätzen to ~ upon
aufätzen
etch Ätze f ~ pattern Ätzfigur f ~ specimen (or
test) Ätzprobe f
etched, ~ copper plate Radierung f ~ identification eingeätzte Werkstückbezeichnung f ~
lines eingeätzte Linien pl ~ plate Ätzplatte f
~ wire technique Ätzleiteverfahren n
etcher Ätzer m, Kupferstecher m, Radierer m
etching Anätzen n, Ätzdruck m, Ätze f, Ätzen n,
Ätzprobe f, Ätzung f, Beizen n, Korrosion f,
Radierung f, Schliff m (art of) ~ Radierkunst f
~ of the barrel metallografisches Ätzen n der
Laufbuchse, Korrosionsangriff m bei der Laufbuchse
etching, ~ acid Ätzflüssigkeit f, Beize f ~ action
Ätzwirkung f ~ bands Ätzstreifung f ~ bath
Ätzbad n ~ device Ätzvorrichtung f ~ equipment Ätzeinrichtung f ~ figure (glass) Ätz-
-figur f, -farbe f ~ ground Radiergrund m ~ ink
Ätztinte f ~ liquid Ätzflüssigkeit f ~ needle
Radiernadel f ~ operation Ätzen n, Ätzung f
~ paper Kupferdruckpapier n ~ paste for
incandescent lamps Ätzpaste f für Glühlampen,
Glühlampenätzpaste f ~ pit Ätz-grübchen n,
-vertiefung f ~ (medium) reagent Ätzmittel n
~ sheet Ätzblech n ~ solution Ätzlösung f,
Beize f ~ tool Metallbeschrifter m ~ treatment
Ätzen n, Ätzung f ~ trough (or box) Ätz-
kasten m ~ varnish Radierfirnis m
ethane Äthan n
ethanoic Äthansäure f
ether Äther m, Äthyläther n ~ acid Äthersäure f
~ drift Ätherverschiebung f ~ mask Narkose-
maske f ~ resistant ätherecht ~ soluble äther-
löslich ~ wave Ätherwelle f ~ waves ätherische
Wellen pl
ethereous ätherisch
etherial ätherartig, ätherisch
etherification Ätherifizierung f
etherified wood Holzäther m
etherify, to ~ ätherifizieren, veräthern
etherization Ätherifizierung f, Ätherisierung f
etherize, to ~ ätherifizieren, ätherisieren
etherous ätherartig
ethine Äthin n
ethrioscope Äthrioskop n
ethyl Äthyl n ~ abietate Abietinsäureäthylester
m ~ acetate Äthylazetat n, Essigäther m, Essig-
säureäthylester n ~ alcohol Äthylalkohol m ~
alcohol stills Brennkolonnen pl ~ aldehyde
Äthylaldehyd n ~ amine Äthylamin n ~
benzoate Benzoäther m, Benzoeäthylester n ~
bromide Brom-äther m, -äthyl n ~ butyrate
Butteräther m, Buttersäureäther m, buttersaures Äthyl n ~ chloride Äthylenchlorid n, Chloräthyl n, Chlorwasserstoffäther m, Salzäther m
~ cynide Zyanäthyl n ~ dichlorarsine Äthyl-
arsindichlorid n, Äthyldichlorarsin n ~ ether
Äthyläther n ~ formate Ameisenäther m,
ameisensaures Äthyl n, Äthylformiat n ~
hydrosulfide (mercaptan) Merkaptan n ~
iodide Äthyljodid n, Jodäthyl n ~ isobutyrate

Äthyl-i-butyrat n ~ nitrate Salpeteräther m ~
nitrite Salpetrigäther m ~ oxalate Oxaläther m
ethylated wood Äthylholz n
ethylene Äthylen n, schweres Kohlenwasserstoffgas n ~ bromide Äthylenbromid n, Bromäthy-
len n, Dibromäthan n ~ chloride Äthylenchlorid n ~ compound Äthylenverbindung f
~ diamine Äthylendiamin n ~ dichloride Azetylendichlorid n, Dichloräthylen n ~ glycol
Äthylenglykol n ~ glycol cooling Heißkühlung
f ~ maximum pressure compressor Äthylen-
höchstdruckkompressor m
ethylenic hydrocarbon Ethylenkohlenwasser-
stoff m
ethylic aldehyde Äthylaldehyd n
ethylidene Äthyliden n ~ chloride Äthyliden-
chlorid n, Chloräthyliden n
etiology Ätiologie f
eucalyptus oil Eukalyptusöl n
Euclidean euklidisch ~ affine euklidisch-affin ~
frame euklidisches Bezugssystem n ~ geometric
euklidisch-geometrisch ~ plane euklidische
Ebene f ~ space euklidischer Raum m ~ space-
time euklidisches Raum-Zeit-Kontinuum n
eudiometer Eudiometer n, Gasprüfer m, Luft-
gütemesser m
Euler's theorem on polyhedra Eulerscher Poly-
edersatz m
eulytite Eulytin m, Wismut(h)blende f
euosmite Kampferharz n
euphonious wohlklingend
euphony Wohlklang m
European base Europasockel m
europium Europium n
eustatic eustatisch
eutectic Eutektikum n; eutektisch ~ mixture
Eutektikum n, eutektisches Gemisch n ~
solder (or alloy) eutektisches Lot n
eutectoid Eutektoid n, eutektoidisch
euxenite Euxenit m
evacuate, to ~ abbefördern, abschieben, abtransportieren, ausleeren, auspumpen, ausrücken, entlüften, evakuieren, leeren, luftleer
machen, räumen
evacuated abgesaugt, ausgepumpt, evakuiert,
luftleer ~ bellows Unterdruckdose f ~ capsule
Aneorid n, Vakuumdose f ~ contacts Vakuum-
schalter m ~ glass luftleeres Glas n ~ housing
evakuiertes Gehäuse n ~ space luftleerer Raum
~ vessel evakuiertes Gehäuse n
evacuating system Evakuieranlage f
evacuation Abschub m, Abtransport m, Auspumpen n, Erzeugung f von Luftleere, Eva-
kuierung f, Räumung f ~ of floodwater spill-
ways Abführen n des Hochwassers
evacuation, ~ center Abschubstelle f ~ floodwater Entlastungswerk n ~ port Austrittkanal
m ~ trench Räumungsgraben m
evade, to ~ ausweichen, entgehen, sich entziehen, umgehen, vermeiden ~ a patent Umgehung f eines Patentes
evading movement Ausweichbewegung f
evagination Ausstülpung f
evaluate, to ~ abmustern, abschätzen, zahlenmäßig ausdrücken, auswerten, schätzen, werten
evaluated abgeschätzt
evaluating grating spectroscope Gittermeßspek-
troskop n

**evaluation** Abschätzung *f*, Ansatz *m*, Ausdeutung *f*, Auswertung *f*, Bewertung *f*, Deutung *f*, Schätzung *f*, Wertbestimmung *f*, Wertung *f* ~ **and computer stage** Rechenstufe *f* ~ **of information** Nachrichtenbeurteilung *f* ~ **of probability** Häufigkeitsauswertung *f*

**evaluation,** ~ **angle** Anschlagwinkel *m* ~ **curve** Bewertungskurve *f* ~ **equipment** Auswertgerät *n* ~ **form** Auswertevordruck *m* ~ **method** Auswerteverfahren *n* ~ **period** Verwertungsperiode *f* ~ **receiver** Vermessungsstelle *f* (g/m)

**evanesce, to** ~ verschwinden

**evanescence** Ausleuchtung *f* und Tilgung *f*, Flüchtigkeit *f*, Fluoreszenzauslöschung *f*, Tilgung *f* ~ **process** Momentprozeß *m*

**evanescent wave** abklingende Welle *f*

**evaporable** verdampfbar, verdampfungsfähig, verdunstbar

**evaporate, to** ~ abdampfen, abdampfen lassen, abdämpfen, abdünsten, abrauchen, aufgehen, ausdämpfen, ausdünsten, dämpfen, dunsten, eindampfen, einkochen, verdampfen, verdunsten, verflüchtigen **to** ~ **on** aufdampfen

**evaporate speed** Verdampfungsgeschwindigkeit *f*

**evaporated,** ~ **film** Aufdampffilm *m* ~ **gold film** Golddampfschicht *f* ~ **line** Aufdampfleiter *m*

**evaporating,** ~ **apparatus** Eindampf-, Verdampf-apparat *m* ~ **boiler** Abdampf-, Siede-pfanne *f* ~ **capacity** Dampferzeugung *f* (Kessel) ~ **compartment** Verdampferfach *n* ~ **dish** Abdampfschale *f*, Verdampfbecken *n* ~ **dish (or basin)** Dampfschale *f* ~ **flask** Abdampfkolben *m* ~ **funnel** Abdampftrichter *m* ~ **pan** Abdampf-gefäß *n*, -pfanne *f* ~ **pan with helical heating coil** Abdampfpfanne *f* mit rotierender Heizspirale *f* ~ **pan with rotating squirrel-cage heating pipes** Abdampfpfanne *f* mit rotierenden Heizröhren in Käfigform ~ **screen** Verdampferblende *f* ~ **steam** Abdampf *m* ~ **surface** Verdampfoberfläche *f* ~ **tray** Verdampfertrog *m*

**evaporation** Abdampfen *n*, Abdampfung *f*, Aufdämpfung *f*, Ausdampfung *f*, Ausdünstung *f*, Eindampfen *n*, Eindämpfung *f*, Verdunsten *n* ~ **of droplets** Tröpfchenverdampfung *f* ~ **by ebullition** Verdampfen *n*, Verdampfung *f* ~ **of water** Wasserverdampfung *f*

**evaporation,** ~ **carburetor** Verdampfungsvergaser *m* ~ **cathode** Aufdampfkathode *f* ~ **cooling** Heißkühlung *f* ~ **loss** Verdampfungsverlust *m* ~ **plant** Verdampfanlage *f* ~-**proof** gasdicht ~ **psychrometer** Aspirationspsychrometer *n*

**evaporative,** ~ **capacity** Verdampfungsfähigkeit *f* ~ **cooling** Verdunstungskühlung *f* ~ **efficiency** Verdampfungskraft *f* ~ **power** Verdampfungs-leistung *f*, -vermögen *n*

**evaporator** Eindampfgerät *n*, Verdampfapparat *m*, Verdampfer *m* ~ **for display cabinet** Vitrinenverdampfer *m*

**evaporator,** ~ **feed pump** Verdampferspeisepumpe *f* ~ **fin** Verdampferrippe *f* ~ **furnace** Verdampfungsofen *m* ~ **tower** Verdampfturm *m* ~ **unit** Meßwasserdurchfluß *m*

**evaporimeter** Ausdünstungs-, Verdampfungs--messer *m*

**evasion** Ausflucht *f*, Ausrede *f*, Ausweichen *n*, Entweichen *n* (rdr)

**evasive,** ~ **action of a plane** Abwehrbewegung *f* ~ **reply** ausweichende Antwort *f*

**evection** Evektion *f*

**evectional tides** Evektionstiden *pl*

**even, to** ~ abgleichen, ebnen, einebnen, nivellieren, planieren

**even** ausgeglichen, eben, ebenmäßig, flach, (number) gerade, glatt, gleich, gleichförmig, gleichmäßig, rund, schlicht, selbst ~ **of valence** geradwertig

**even,** ~ **charging** gleichmäßiger Satz *m* ~-~ **nucleus** gerade-gerade Kern *m*, g-g-Kern *m* ~ **fracture** feinkörnige Bruchfläche *f* ~ **function** gerade Funktion *f* ~ **harmonic** gerade (geradzahlige) Harmonische *f*, gerade Oberharmonische *f* ~ **illumination** gleichmäßige Ausleuchtung *f* ~ **integer** Gerade *f* ~ **keel** gleichbeladener Kiel *m* **on** ~ **keel** gleichlastig ~ **line interlace** geradzahliger Zellensprung *m* ~ **multiple** gerades Vielfaches *n* ~ **number** gerade Zahl *f* ~-**numbered** geradzahlig ~-**odd nucleus** gerader-ungerader Kern *m* ~ **operators** gerade Operatoren *pl* ~ **page** Kehrseite *f* ~ **position (or altitude)** Waagerechtlage *f* ~-**spangled** Frostmuster *n* (Blech) ~ **term of atom** Atomterm *m* mit geradzahligem Spin ~-**turning cycle** Gleichdruckprozeß *m* ~-**valent groups** geradwertige Gruppen *pl*

**evener** Waage *f* für Mehrgespann ~ **comb above the scale** Hacker *m* über Waage ~ **lattice** Rückstreichtisch *m* ~ **roller** Rückstreichwalze *f*

**evenly,** ~ **distributed** gleichmäßig oder stetig verteilt ~ **distributed inductance** gleichmäßig oder stetig verteilte Induktivität *f* ~ **distributed load** gleichförmige Belastung *f* ~ **heat treated throughout** gleichmäßig durchvergütet (un)~ **spaced** in (un)gleichen Abständen *pl*

**evenness** Ebenheit *f*, Gleichförmigkeit *f*, Gleichheit *f*, Gleichmäßigkeit *f* ~ **tester** Gleichmäßigkeitsprüfer *m*

**event** Begebenheit *f*, Ereignis *n*, Erscheinung *f*, Fall *m*, Geschehnis *n*, Beranstaltung *f*, Verlauf *m*, Vorgang *m* ~ **recorder** Mehrwertschreiber *m*

**eventful** ereignisreich

**ever,** ~ **increasing** dauernd zunehmen (power) ~-**ready case** Bereitschaftstasche *f* (photo)

**eversion** Auswärtşkehrung *f*

**eviction** Austreibung *f*

**evidence** Augenschein *m*, Aussage *f*, Beweis *m*, (documentary) Beweismaterial *n*, Nachweis *m*, für einen Nachweis dienende Auskünfte, Zeugnis *n* ~ **on file** vorliegende Belege *pl* ~ **impeaching a patent (or prejudicial to validity of a patent)** Beweis *m*, der die Gültigkeit eines Patentes Abbruch tut

**evident** augenfällig, klar, offenbar, offenkundig **to be** ~ auf der Hand liegen, naheliegen, zutage liegen

**evidential,** ~ **material (proof, or testimony)** Beweismaterial *n* ~ **value** Beweis-kraft *f*, -wert *m*

**evince, to** ~ bekunden

**evolute** Evolute *f*, abgewickelte Linie

**evolution** Entwicklung *f*, Radizierung *f*, Wurzelziehen *n* ~ **of heat** Wärme-entwicklung *f*, -tönung *f* ~ **of hydrogen** Wasserstoffentwicklung *f* ~ **of oxygen** Sauerstoffentwicklung *f* ~ **in tactics** Formveränderung *f*

evolutionary development (or progress) Weiterbildung *f*
evolve, to ~ entfalten, entwickeln, radizieren
to ~ gas gasen, Gas entwickeln
evolent Evolente *f*
Ewart chain Ewartskette *f*
Ewing method Isthmusverfahren *n*
exact, to ~ beitreiben
exact genau ~ date and hour Zeitangabe *f*
exacting anspruchsvoll ~ requirements hohe Anforderungen *pl*
exactly, in ~ the same shade nuancengleich
exactness of shade Musterkonformität *f*
exaggerate, to ~ übertreiben
exaggerated übertrieben ~ relief Überplastik *f*
exaggeration Aufbauschung *f*, Übertreibung *f*
exalt, to ~ erheben
exalted erhaben
examination Beobachtung *f*, Besichtigung *f*, Prüfung *f*, Überprüfung *f*, Untersuchung *f*, Verhörung *f*, Vernehmung *f* ~ of books and accounts Bilanzprüfung *f* ~ by experts Expertise *f*, Gutachten *n* eines Sachverständigen, Sachverständigenbegutachtung *f* ~ of iron Eisenuntersuchung *f* ~ of the line Durchprüfen *n* der Linie (teleph.), Linienrevision *f* ~ of surface Oberflächenbeobachtung *f*
examination, ~ bureau (or station) Untersuchungsstelle *f* ~ paper Prüfungsaufgabe *f*
examine, to ~ (the ground) auspeilen, besichtigen, betrachten, durchprüfen, durchsehen, erforschen, inspizieren, nachprüfen, prüfen, revidieren, überprüfen, untersuchen, verhören to ~ (papers) carelessly (or superficially) flüchtig durchsuchen to ~ closely mustern to ~ the formation of deposits Kontrolle *f* der Rückstandsbildung to ~ the ground den Grund *m* abloten to ~ (or measure) by photometry fotometrieren to ~ by touch abfühlen to ~ a vessel ein Schiff *n* durchsuchen
examiner Prüfer *m*, (in patent office) Vorprüfer *m*
examining, ~ board Prüfungsausschuß *m* ~ post Durchlaßposten *m*
example Aufgabe *f* (math.), Beispiel *n*, Muster *n* ~ of application Anwendungsbeispiel *n* ~ of operation Ausführungsbeispiel *n*
exanthema Exanthem
excavate, to ~ ausbaggern, ausgraben, aushöhlen, ausschachten, baggern, graben, schürfen to ~ under water ausheben
excavated gewonnen ~ material Abtrag *m*, Aushub *m* ~ peat Maschinentorf *m*
excavating, ~ a canal Aushub *m* eines Kanals ~ machine Bagger *m*
excavation Abtrag *m*, Aufgrabung *f*, Ausbruch *m*, Ausgrabung *f*, Aushub *m*, Ausschachtung *f*, Ausschaltung *f*, (for structure) Baugrube *f*, Einschnitt *m*, Durchstich *m*, (of foundation) Fundamentsaushub *m*, (of the optic disk) Gefäßtrichter *m*, Grube *f*, Hohlraum *m*, Höhlung *f*, Kammerbau *m*, Kessel *m* ~ work Erdarbeit *f*
excavator Bagger *m*, Becherwerk *n*, Exkavator *m*, Trockenbagger *m* ~ body Grundbagger *m*
exceed, to ~ überschreiten, übersteigen, übertreffen to ~ an issue überheben to ~ the limits of authority die Grenzen *pl* der Dienstgewalt *m*

überschreiten
exceeded überschritten
exceeding (a certain limit, mark, level, or value) Überschreitung *f* ~ authorized basic allowance (or strength) außerplanmäßig
exceedingly, ~ heavy überschwer ~ live room (or space) überakustischer Raum *m*
excel, to ~ übertreffen
excellent ausgeprägt, ausgezeichnet, famos, trefflich ~ point ausgezeichneter Punkt *m*
excelsior kleine weiche Holzspäne *pl*, Holz-faser *f*, -wolle *f*
excentric exzentrisch
except, to ~ ausnehmen
except außer, und nicht-, jedoch nicht- ~ gate Jedoch nicht-Schaltung *f* (data proc)
except(ing) ausgenommen, mit Ausnahme von
exception Ausnahme *f*, Ausnahme-, Sonder-fall *m* by way of ~ ausnahmsweise with ~ of ausschließlich ~ card Sonderfallkarte *f*
exceptional case Ausnahme-, Sonder-fall *m*
excerpt Extrakt *m*
excess Überfluß *m*, Übermaß *n*, Überschuß *m*, Übertreibung *f*; überreichlich, überschüssig in ~ of authorized allowance strength überplanmäßig ~ of charge Ladungsüberschuß *m* ~ of gas Gas-, Glasflaschen-überschuß *m* ~ of lime Kalküberschuß *m* ~ of material Stoffüberschuß *m* ~ of power Kraftüberschuß ~ and total meter Gesamt- und Überverbrauchszähler *m*
excess, ~ air Luftüberschuß *m* ~-air coefficient Luftüberschußzahl *f* ~ angle Winkelübertreibung *f* ~ conductor (in-type conductor) Überschuß-(halb)leiter *m* ~ constituent Übergemengteil *m* ~ current Überstrom *m* ~-current meter Spitzen-, Überverbrauchs-zähler *m* ~-current switch Überstromschalter *m* ~-current zero-voltage cutout switch Überstromnullspannungsausschalter *m* ~-defect contact Störstelleninversionszone *f* ~ delivery Mehrlieferung *f* ~ end of shank Überstand *m* ~ entropy Überschußentropie *f* ~ frequency rise Frequenzüberhöhung *f* ~ frequency Überschußfunktion *f* ~ heat Wärmeüberschuß *m* ~ horsepower Überschußpferdekraft *f* ~ hydrostatic pressure hydrostatischer Überdruck *m* ~ load Überlast *f* ~ metal at root of seam weld Schweißbart *m* ~ meter Spitzenzähler *m* ~ multiplication constant Überschußreaktivität *f* ~ network Leitungsergänzung *f*, Verlängerungsleitung *f* ~ parts überzählige Teile *pl* ~-power meter Spitzen-, Überverbrauchs-zähler *m* ~-pressure Überdruck *m* ~-pressure valve Überdruckventil *n* ~ property Mehrbestände *pl* ~ reactivity übermäßiger Neutronenfluß *m* ~ revolutions per minute Überdrehzahl *f* ~-three code Dreiüberschußkode *m* ~ voltage elektrolytische Überspannung *f* ~-voltage cutout Überspannungssicherung *f* ~-voltage wave Überspannungswelle *f* ~ weight Mehrgewicht *n*
excessive übermäßig, überreichlich, übertrieben, unmäßig ~ boiling Aufschäumen *n* ~ brightness Aufhellung *f* schwarzer Stellen ~ compensation Überregulierung *f* ~ current drain übergroßer Stromfluß *m* ~ curvature Überkrümmung *f* ~ data Überlaufdaten *pl* ~ engine speed Motorüberdrehzahl *f* ~ load Überbe-

lastung *f* ~ **load characteristic** Stromverlauf *m* bei Überbelastung ~ **lubrication** Überölung *f* ~ **neutron flux** übermäßiger Neutronenfluß *m* ~ **pressure on the valve** Ventilüberdruck *m* ~ **prolongation of decay of wave tail** Ausschwingungsverzug *m* ~ **regulation** Überregulierung *f* ~ **reverberation period** übertriebene Nachhalldauer *f* ~ **strain** Überbeanspruchung *f* ~ **stuffing** Überfettungserscheinung *f* ~ **voltage** Überspannung *f*
**exchange, to** ~ austauschen, einwechseln, tauschen, umwechseln, vertauschen, verwechseln, wechseln
**exchange** (as of coins) Austausch *m*, Auswechs(e)lung *f*, Börse *f*, Eintausch *m*, Fernsprechamt *n*, Fernsprechanlage *f*, Geldhandel *m*, Kurs *m*, Tausch *m*, Umsatz *m*, Umspeicherung *f*, Umtausch *m*, Vermitteilung *f*, Vermittlung *f*, Vermittlungsamt *n*, Vertauschung *f*, Wechsel *m*, Wechseln *n*, Zweiganlage *f* **(telephone)** ~ Zentrale *f*
**exchange, ~ of bases** Basenaustausch *m* (chem.) ~ **of costumers** Abnehmertausch *m* ~ **of experience** Erfahrungsaustausch *m* ~ **of gases** Gasaustausch *m* ~ **of ideas** Gedankenaustausch *m* ~ **of notes** Notenaustausch *m* ~ **of places** Platzwechsel *m* (electrons) ~ **of power** Energieaustausch *m* ~ **of stocks** Effektenhandel *m*
**exchange, ~ apparatus** Amtseinrichtung *f* ~ **area** Amtsbezirk *m*, Anschlußbereich *m* einer Vermittlung, (multi-office) Anschlußbereich *m*, Gebührenzone *f*, (single- or multi-office) Ortsfernsprechnetz *n*, Taxzone *f* ~ **area layout** Netzplan *m* ~ **battery** Amtsbatterie *f* ~ **cable** Anschluß-, Bezirks-, Fernleitungs-kabel *n* ~ **call** Amts-anruf *m*, -verbindung *f* ~ **charge** Austauschladung *f* ~ **collector** Umfüllsammler *m* ~ **collision** Kollision *f* mit Energieauswechslung ~ **correction** Austauschkorrektur *f* ~ **designation** Amtskennziffer *f* (teleph.) ~ **energies** Austauschkräfte *pl* ~ **equalization fund** Währungsausgleichfond *m* ~ **extension set** Amts-, Post-nebenstelle *f* ~ **fault** Fehler *m* in einem Amt ~**-fault staff** Amtsstörungspersonal *n*, Störungspersonal *n* für Amtsstörungen ~ **fluctuation** Kursschwankung *f* ~ **forces** Austauschkräfte *pl* ~ **fraction** Austauschfaktor *m* ~ **integral** Austauschintegral *m* ~ **interaction** Austauschwechselwirkung *f* ~ **jack** Amtsklinke *f* **(direct)** ~ **line** Amts-, Anschluß-leitung *f* ~ **loss** Amtsdämpfung *f* ~ **maintenance work** Amtspflege *f* ~ **mass correction** Massenkorrektur *f* für Austausch ~ **narrowing** Austauschbeschränkung *f* ~ **number** Amtsnummer *f* ~ **plant** Vermittlungseinrichtung *f* ~ **principle** Wechselsatz *m* ~ **prohibitory circuit** Verhinderungsschaltung *f* ~ **reaction** Austauschreaktion *f* ~ **ship** Austauschschiff *n* ~ **side** Amtsseite *f* (des Hauptverteilers) ~ **side** (des Hauptverteilers) (distributing frame) Innenseite *f* des Hauptverteilers ~ **stamp tax** Wechselstempelsteuer *f* ~ **switch** Amtsschalter *m* ~ **term** Austauschterm *m* ~ **testing position** Störungsüberwachungsplatz *m* ~ **troubles** Fehler *pl* in einem Amt ~ **voucher** Devisenbescheinigung *f*
**exchangeability** Auswechselbarkeit *f*
**exchangeable** austauschbar, (positive ions) aus-

tauschfähig, auswechselbar, vertauschbar ~ **cations** Schwarmionen *pl*
**exchanger** Austauscher *m* ~ **rating sheets** Austauscherwertetabellen *pl*
**exchanging** Umspeicherung *f* ~ **of money** Börsengeschäft *n*
**exchequer** Fiskus *m*
**excise** Akzise *f*, Aufschlag *m*, Steuer *f* ~ **tax** Verbrauchssteuer *f*
**excitability** Erregbarkeit *f*
**excitable** erregbar
**excitant** Erreger *m*, Erregermasse *f*
**excitation** Anregung *f*, Antennenspeisung *f* (electron.), Erregung *f*, Erregungsursache *f*, Reiz *m*, Schallerregung *f*, Speisung *f* ~ **anode** Erregeranode *f* ~ **band** Anregungsband *n* ~ **cross sections** Anregungsquerschnitte *pl* ~ **drive** Erregerspannung *f* ~ **energy** Ablösungs-arbeit, -energie *f* des Elektrons, Anregungsenergie *f* ~ **factor** Anregungsfaktor *m* ~ **output** Erregerleistung *f* ~ **set** Erreger(maschinen)-satz *m* ~ **state** Anregungsarbeit *f* ~ **system** Erregermaschine *f* ~ **winding** Erregerwickelung *f*
**excite, to** ~ (a wave, vibration, an oscillation) anfachen, anheulen, anreizen, aufregen, erregen, reizen, unter Strom setzen
**excited** angeregt ~**-atom density** Dichte *pl* der angeregten Atome ~ **intermittently** absatzweise bewegt ~ **state** angeregter Zustand *m*
**exciter** Erreger *m*, Erregermaschine *f*, Treiberstufe *f* ~ **circuit** Erregerstromkreis *m* ~ **coil** Erreger-spule *f*, -wickelung *f* ~ **lamp** Belichtungs-, Erreger- (film), Erregungs-, Ton-, Tonlicht-lampe *f*, ~ **panel** Erregerfeld *n* ~ **point** Tonsteuerstelle *f* ~ **set** Erregergruppe *f* ~ **socket** Erregermaschinensteckdose *f* ~ **stage** Erreger-, Steuer-stufe *f* ~ **tube** Erreger-, Steuer-röhre *f*
**exciting** aufregend ~ **anode** Erreger-, Hilfs-, Zünd-anode *f* ~ **circuit** Anreiz-, Erreger-kreis *m* ~ **coil** Erregerspule *f* ~ **current** Erregerstrom *m* ~ **field** Erregerfeld *n* ~ **fluid** Erregerflüssigkeit *f* ~ **flux** Erregerfluß *m* ~ **lamp** Erreger- (film), Erregungs-lampe *f* ~ **light** eingestrahltes Licht *n* ~ **mechanism** Anfachungsmechanismus *m* (electron.) ~ **oscillation** erregende Schwingung *f* ~ **paste** Erregerpaste *f* ~ **potential** Anregungsspannung *f* ~ **probe** Einkoppelsonde *f* (meas.) ~ **spark gap** Erregerfunkenstrecke *f* ~ **voltage** Anregungsspannung *f*, Erregerspannung *f* ~ **winding** Erregerwickelung *f*, Feldspule *f*, Feldwicklung *f*, Magnetwicklung *f*
**excition** Anregungswelle *f*, (in dielectric breakdown) angeregte Gruppe *f*
**exclamation point** Ausrufungszeichen *n*
**exclave** Ausschlußgebiet *n*
**exclude, to** ~ ausfiltern, ausschalten, ausscheiden, ausschließen, beanstanden, sperren
**excluder** Sperrer *m*
**excluding** ausschließlich ~ **action (or effect)** Ausscheidungseffekt *m*
**exclusion** Ausschließung *f*, Ausschluß *m*, Aussperrung *f*, Ausschaltung *f* ~ **of air** Luftabschluß *m* ~ **area** verbotenes Gebiet *n* ~ **band** Sperrbereich *m* ~ **principle** Ausschließungs-, Eindeutigkeits-prinzip *n*
**exclusive** ausschließlich, konkurrenzlos ~ **in-**

**formation** Privatauskunft f ~ **property** Allein-
besitz m ~ **right** Alleinberechtigung f, Vorrecht
n ~ **sale** Alleinverkauf m
**exclusivity rights** Ausschließlichkeitsrechte pl
**excorticate, to** ~ abrinden
**excorticated** abgerindet
**excrement** Kot m
**excrescence** (unregelmäßiger) Auswuchs m
**excretion** Exkretion f
**excursion** Abschweifung f, Ausflug m, Auslenkung f, Ausschlag m ~ **of a chord** Schwingungsausschlag m einer Saite ~ **amplitude of dia-**
**phragm** Membranschwingamplitude f ~ **power**
Leistungsausbruch m
**exducer** Austrittsschaufelrad n (Kühlturbine)
**executable** vollstreckbar
**execute, to** ~ ausführen, ausüben, durchführen,
erledigen, hinrichten, verrichten, vollstrecken
**to** ~ **a document** Urkunde f ausstellen **to** ~ **a**
**maneuver (or a movement)** eine Bewegung f
ausführen **to** ~ **a power of attorney** Vollmacht
f ausstellen
**execution** Ausführung f, Ausübung f, Durchführung f, Erfüllung f, Pfändung f, (orders)
Vollzug m ~ **of the agreement** Unterzeichnung
f des Abkommens ~ **of a repair** Ausführung f
einer Reparatur ~ **of sentence** Strafvollstrek-
kung f ~ **of a sentence** Urteilsvollstreckung f
~ **of sentences** Strafvollzug m
**executive** leitender Angestellter, Beamter m ~
**committee** Vollzugsausschuß m ~ **instruction**
Ausführungsbefehl m (data proc.) ~ **method**
Ausführungsverfahren n ~ **program** organisa-
torisches Programm n ~ **routine** Superpro-
gramm n ~ **signal** Ausführungssignal n
**executory** vollstreckbar
**exemplar** Vorbild n
**exemplarly** mustergültig, musterhaft, vorbildlich
**exemplified** beispielsweise ~ **embodiment** Ausführungsbeispiel n
**exempt, to** ~ ausnehmen, freimachen
**exempt** frei ~ **from** ausgenommen von ~ **from**
**registration** zulassungsfrei
**exempted addressee** nicht betroffener Adressat m
**exemption** Ausschließung f, Befreiung f ~ **from**
**charge (or fee)** Gebührenfreiheit f
**exercise, to** ~ üben **to** ~ **power of attorney** eine
Vollmacht f ausüben
**exercise** Akzise f, Anwendung f, Ausführung f,
Ausübung f, Übung f ~ **head** Übungskopf m
(g/m)
**exercised** ausgeübt
**exert, to** ~ anstrengen, ausüben **to** ~ **oneself**
sich anstrengen
**exertion** Anspannung f, Anstrengung f
**exfoliate, to** ~ abblättern, abplatzen, abschie-
fern, aufblättern, losblättern, sich schiefern
**exfoliation** Abblätterung f, Abschieferung f ~
**of rail** Abblätterung f der Schiene
**exhalation** Ausdünstung f, Brodem m, Dampf m
**exhale, to** ~ ausdünsten, aushauchen
**exhaling valve** Ausatemventil n
**exhaust, to** ~ abbauen, (gas) absaugen, aufbrauchen, ausblasen, auslassen, ausnutzen,
auspuffen, auspumpen, aussaugen, ausschie-
ben, entkräftigen, entleeren, ermüden, erschöpfen, luftleer machen **to** ~ **completely** klar
ausziehen **to** ~ **oneself** sich erschöpfen

**exhaust** Abdampfrohr n, (steam, air, etc) Abführung f, Absauger m, Absaugung f, Auslaß
m, Auspuff m, Ausschub m, (steam) Ausströmung f, Ausströmgas n (rocket), Austritt
m, Entgaser m ~ **accenter** Auslaßexzenter m
~ **air** Abgas n, Abluft f ~ **alarm** Auspuffpfeife
f ~ **and air-intake valve** Aus- und Einlaßventil
n ~ **aperture** Austrittsquerschnitt m ~ **arrange-**
**ment** Auspuffsystem n ~ **back pressure** Aus-
puffgegendruck m ~ **balance line** Abgasaus-
gleichsleitung f ~ **bench for cleaning castings**
Gußputztisch m mit Staubabsaugung ~ **blast**
Abgasstrom m ~ **blast momentum** Abgasstrom-
impuls m ~ **bore** Ausblasebohrung f ~ **cam**
Auslaßnocken m ~ **cam axle** Auspuffsteuer-
achse f ~ **cam shaft** Auspuffsteuerwelle f ~
**chamber** Auspuffkessel m, Auspufftopf m ~
**collector** Auspuffsammler m ~ **collector ring**
Abgassammler m, Auspuffsammelring m,
(Sternmotor) ringförmiger Auspuffsammler m
~ **conduction device** (jet) Austrittsleitapparat m
~ **cone** (jet) Austrittskegel m ~ **connection**
Saugstutzen m ~ **cycle** Auspufftakt m ~ **de-**
**pression** Heckmulde f (g/m) ~ **detonation** Auspuffknallen n ~ **diffuser** Austrittsleitapparat m
~ **dislocation** Abgastrübung f ~**-driven super-**
**charger** Abgasturbolader m ~**-driven turbine**
Abgasturbine f ~ **duct** Auspuff-kanal m,
-leitung f ~ **fan** Absauger m, Bläser m, Entlüfter m, Exhaustor m, (for air) Luftsauger m,
Sauger m, Saugventil m, Saugzugventilator m
~**-flame damper** Auspuffflammendämpfer m ~
**flange** Auspufflansch m ~ **flap** Auslaßklappe f
~ **flue** Abgaskanal m ~ **gas** Abgas n, Auspuffgas n ~**-gas analyzer** Abgasprüfgerät n, Aus-
puffgaszerleger m ~ **gas cooling coil** Abgas-
rohrschlange f ~ **gas driven compressors** Ab-
gasgebläse pl ~**-gas (or exhaust-) driven tur-**
**bine** Abgasturbine f ~**-gas jet thrust** Abgas-
strahlschub m ~**-gas outlet** Abgasabzug m
~**-gas pyrometer** Abgaspyrometer m ~**-gas**
**trunk** (airship) Gasentlüftungsschacht m ~ **gas**
**turbine blade** Abgasturbinenschaufel f ~**-gas**
**velocity** Auspuffgeschwindigkeit f ~ **(or waste)**
**gate** Abgasauslaß m ~ **gear** Auslaßgestänge n
~**-gear mechanism** Auslaßsteuerung f ~ **head**
Auspufftopf m ~**-heat boiler** Abwärmekessel m
~ **heat exchanger** Abgaswärmeaustauscher m
~ **installation** Absaugungsanlage f ~**-jacketed**
**carburetor** Vergaser m mit Auspuffgasheiz-
mantel ~ **line** Auspuffleitung f ~ **main** Saug-
leitung f ~ **manifold** Abgassammelleitung f,
Abgassammler m, Auspuff-krümmer m, -lei-
tung f, -rohr n, -sammler m, -stutzen m
~**-manifold clamp** Auspuffrohrschelle f ~
**manifold gasket** Auspuffkrümmerdichtung f
~ **muffler** Auspuff-schalldämpfer m, -topf m
~ **nipple** Ausblasemundstück n ~ **noise** Aus-
puffgeräusch n ~ **nozzle** Absaug-, Schub-düse f
~ **nozzle ring** Austrittsleitkranz m ~ **opening**
Auspufföffnung f ~**-operated air preheater**
abgasbeheizter Luftvorwärmer m ~ **passage**
Auslaßkanal m ~ **pipe** Ausfahrungsrohr n,
Auspuff-krümmer m, -leitung f, -rohr n, Ausströmungsrohr n, Exhaustrohr n, Saugleitung
f, Saugrohr n ~ **pipe line** Abdampfleitungen f
~ **plug** Ablaßschraube f, auslaßseitige Zündkerze f ~ **port** Auslaßventilkammer f, Auspuff-

kanal *m*, Auspuffschlitz *m*, Austrittkanal *m*, Austrittsöffnung *f* ~ **ports** Auspuffblitze *pl* ~ **pressure** (Dampf)austrittsdruck *m* ~ **pressure wave** Auslaßdruckwelle *f* ~ **process** Auszieh-verfahren *n* ~ **resistance** Auspuffwiderstand *m* ~ **ring** ringförmiger Auspuffsammler *m* ~ **rods** Auslaß-, Auspuff-gestänge *n* ~ **silencer** Abgas-schalldämpfer *m*, Auspuff-schalldämpfer *m*, -topf *m* ~ **sleeve** Auspuffventilkanone *f* ~ **smoke** Auspuffwolke *f* ~ **space** luftleerer Raum *m* ~ **spark catcher** Auspuff-Funkenkorb *m* ~ **speed** Austrittsgeschwindigkeit *f* ~ **stack** Abgasstutzen *m*, Auspuff-krümmer *m*, -röhre *f*, -stützen *m* ~ **stator plate** Abgasleit-, Austritts--schaufel *f* ~ **steam** Abdampf *m*, Ausblase-dampf *m*, Maschinenabdampf *m*

**exhaust-steam**, ~ **accumulator** Abdampfspeicher *m* ~ **condensing-water separator** Abdampfkondenswasserabscheider *m* ~ **driven tender** (for locomotives) Abdampftriebtender *m* ~ **oil separator** Abdampfentöler *m* ~ **pipe** Abdampf-stutzen *m* ~ **preheater** Abdampfvorwärmer *m* ~-**pressure regulator** Abdampfdruckregler *m* ~ **utilizing plant** Abdampfverwertungsanlage *f*

**exhaust**, ~ **stroke** Auslaßhub *m*, Auspuffen *n*, Auspuff-hub *m*, -periode *f*, -takt *m*; Aus-schub-hub *m*, -periode *f*, Ausstoßen *n*, Aus-strömungshub *m*, Entladehub *m* ~ (**scavenging**) **stroke** Ausströmhub *m* ~ **supercharger** Abgaslader *m*, Abgasturbolader *m* ~ **system** Abgas-führung *f*, -system *n*, Auspuffsystem *n*, Saugsystem *n* ~ **tail pipe** Abgas-abführungs-rohr *n*, -leitung *f*, Strahlrohr *n* ~ **temperature indicator** Austrittsgastemperaturanzeiger *m* ~ **test** Nachzugsmuster *n* ~-**throttle disk** Löse-düse *f* ~ **trail** Rauchfahne *f* ~ **tube** Entlüf-tungsrohr *n*, Pumpstengel *m* ~ **tumbling mill** Putztrommel *f* mit Saugleitung ~ **turbosuper-charger** abgasgetriebener Turbokompressor *m* ~-**type supercharger** Abgasturbo-, Abgastur-binen-vorverdichter *m* ~ **valve** Abblaseventil *n*, Auslaß *m*, Auslaßseite *f*, Auslaßsteuerung *f*, Auslaßventil *n*, Auspuff-klappe *f*, -ventil *n*, -ventilkanone *f*; Druckventil *n*

**exhaust-valve**, ~ **box** Abblaseventilkammer *f* ~ **chamber** (**or chest**) (internal-combustion) Aus-puffwulst *m* ~ **guide** Auspuffventilführung *f* ~ **lift** Auspuffventildrücker *m* ~ **lifter** Auspuff-ventilnehmer *m* ~ **pin** Auspuffventilbolzen *m* ~ **spring** Auspuffventilfeder *f* ~-**stem** Aus-puffventil-spindel *f*, -stange *f*

**exhaust**, ~ **velocity** (of turbines) Austritts-geschwindigkeit *f* ~ **vent** Pumpstengel *m* ~ **wheel** Niederdruckrad *m*

**exhausted** abgebaut, abgejagt, (of air) luftleer, matt, schlaff, verbraucht, vergriffen ~ **lamp bulb** entleerte Glühbirne *f*, evakuierte Lampe *f* ~ **space** luftleerer Raum *m* ~ **vain** abgebauter Gang *m*

**exhauster** Entlüfter *m*, Exhaustor *m*, Luftsauger *m*, Sauger *m*, Saug-gebläse *n*, -ventilator *m*, saugender Ventilator *m* ~ **jet** Absauger *m*

**exhausting** anstrengend ~ **to atmosphere** freies Auspuffen *n* ~ **column** Abtreibekolonne *f* ~ **plant for smoke** Rauchabsaugeanlage *f*

**exhaustion** vollkommene Abtrennung (phys.), (potential energy) Aufspaltung *f*, Auspumpen *n*, Entleerung *f*, Erschöpfung *f*, Vertaubung *f*

~ (**in the boundary layer of a reserve semi-conductor**) Störstellenerschöpfung *f* ~ **of a mine** Abbauen *n* (Erschöpfung *f*) eines Feldes **exhaustion**, ~ **machine** Evakuierungsmaschine *f* ~ **semiconductor** Dissoziationshalbleiter *m*

**exhaustive** eingehend, erschöpfend

**exhibit, to** ~ aufweisen, ausstellen, beschicken, zur Schau stellen, vorzeigen

**exhibit** ausgestellter Gegenstand *m*, (law) Be-weis *m*, Beweisstück *n*

**exhibiting for sale** Feilhalten *n*

**exhibition** Ausstellung *f*, Darstellung *f*, Schau *f* ~ **on** ausgestellt *o* ~ **hall** Ausstellungshalle *f* ~ **model** Ausstellungsmodell *n*

**exhibitor** Aussteller *m*

**exhort, to** ~ mahnen

**exigency** Anforderung *f*, Bedarf *m*, Erfordernis *n*

**exist, to** ~ bestehen

**existence** Bestand *m*, Bestehen *n*, Dasein *n*, Existenz *f*, Leben *n*, Vorhandensein *n* ~ **proof** Existenzbeweis *m* ~ **theorem** Existenzsatz *m*

**existent, to be** ~ vorliegen

**existing** vorhanden

**exit, to** ~ hinausgehen

**exit** Ausgang *m*, Auslaß *m*, Ausmündung *f*, Auspuff *m*, Austritt *m* ~ **of light** Lichtaus-tritt *m*

**exit**, ~ **angle** (gas turbine) Abströmwinkel *m* ~ **cone** Ausgangskegel *m* ~ **cone angle** Düsen-winkel *m* ~ **displacement** Austrittsverschie-bung *f* ~ **door** Aussteigklappe *f* ~ **dose** Aus-trittsdosis *f* ~ (**or delivery**) **end** Endauslauf *m* ~ **gas** Abzugsgas *n* ~ **gradient** Austrittsge-fälle *f* ~ **hub** Ausgangsbuchse *f* ~ **instruction** Ausgangsbefehl *m* ~ **loss** Austrittsverlust *m* ~ **pipe** Brüdenrohr *n* ~ **port** Ausgangsöffnung *f* ~ **portal** Austrittsfeld *n* ~ **pupil** Augenkreis *m*, Augenpunkt *m*, Austritts-, Ausgangs-pupille *f* (phys.) ~ **section** Düsenende *n* ~ **shock coeffi-cient** Ausstoßbeiwert *m* ~ **slit** (spectograph) Ausgangsschlitz *m*, Austrittsblende *f* ~ **slot** Auslaßschlitz *m* (Kühlluft), Austrittsspalte *f* ~ **tail ring** Heckring *m* ~ **velocity** Austrittsge-schwindigkeit *f*

**exitation anode** Anode *f* der Exitronröhre

**exlibris** Bucheignerzeichen *n*

**exoergic** exotherm

**exogenous** nach außen hin wachsend

**exograph** Röntgenbild *n*, Röntgenstrahlenauf-nahme *f*

**exorbitant** übermäßig, übertrieben

**exosmosis** Aussickerung *f*, Exosmose *f*

**exosporal** exospor

**exothermal** Wärme *f* abgebend, wärmegebend, wärmeliefernd

**exothermic** exotherm, exothermisch, wärme-gebend, Wärme *f* abgebend oder erzeugend ~ **nuclear disintegration** exothermer Kernzer-fall *m* ~ **nuclear reaction** exotherme Kernre-aktion *f*

**expand, to** ~ aufblähen, aufweiten, aufweitern, ausbreiten, sich ausbreiten, ausdehnen, aus-lassen, sich auslassen, ausweiten, blähen, dehnen, sich dehnen, erweitern, entwickeln, expandieren, nachlassen, spreizen (Bildbasis), (cakes of coke) treiben, verausgaben, ver-größern (Bildausschnitt), wachsen **ability to** ~ Ausdehnungsvermögen *n* **to** ~ **a boiler tube**

ein Kesselrohr *n* aufwalzen **to ~ a hole** Loch *n* auftreiben **to ~ in** einwalzen **to ~ by mandrel** dornen **to ~ in a series** entwickeln in einer Reihe *f*
**expandable rubber arbor** Gummispreizdorn *m*
**expanded** erweitert **~ clay** Blähton *m* **~ contrast** Kontrastverbesserung *f* **~ flange** Aufwalzflantsch *m* **~ material with ribs** Rippenstreckmaterial *m* **~ metal** Streckmetall *n* **~ metal with ribs** Rippenstreckmetall *n* **~ plastics materials** Kunststoffschaumstoffartikel *m* **~ rubber** Schaumgummi *n* **~ scope** vergrößerter (Schirm)Bildausschnitt *m* (rdr) **~ sweep** kompensierte Abtastung *f*, Basisspreizung *f* (rdr)
**expander** Ausdehnungsring *m*, Dehner *m*, Eintreibedorn *m*, Expander *m*, Expanderverstärker *m*, Rohrdichter *m*, Spreizstoff *m* **~ for drill pipe protector** Gestängeschutzaufziehvorrichtung *f* **~ shoe brake** Innenbackenbremse *f* **~ tube** Expanderschlauch *m*
**expanding** auftreibbar, Auftreiben *n*, ausweitbar **~ agent** Treibmittel *n* **~ arm** Spreizhebel *m* **~ band** (internal brake band) Spreizband *n* **~-band brake** Expansionsbandbremse *f* **~ band clutch** Spreizringkupplung *f* **~ block** Streckform *f* **~ bracket plate** Spreizbügelplatte *f* **~ (internal) brake** Innenbackenbremse *f* **~ bullet** Dumdumgeschoß *n* **~ bushing** Spreizdorn *m* **~ center bit** Universalzentrumbohrer *m* **~ chuck** Expansionsdorn *m*, Klemmfutter *n* zum Innenklemmen **~ combs** Expansionskämme *pl* **~ cone** Spreizkonus *m* **~ cylinder** Spreizzylinder *m* **~ device** Aufdornvorrichtung *f* **~ frill cowling** Spreizklappenring *m* **~ (spreader) lever** Spreizhebel *m* **~ lever shoe brake** Spreizhebelbackenbremse *f* **~ lock** Spreizschloß *n* **~ machine** Aufspannmaschine *f*, Aufwalzmaschine *f* **~ mandrel** verstellbarer Aufspanndorn *m*, Federdorn *m*, Spreizdorn *m* **~ mill** Aufweitewalzwerk *n* **~ nozzle** Expansionsdüse *f* **~ property** Blähungsgrad *m*, Schaumfähigkeit *f* **~ pulley** Ausdehnungsriemenscheibe *f* **~ reamer** nachstellbare Reibahle *f* **~ ring** Spreizring *m* **~ roller** Spreizrolle *f* **~ spring** Spreizfeder *f* **~ strut** Spreize *f* **~ test** (tube) Ausschwell-, Expandier-probe *f* **~-type tubing packer** Expansionsröhrengreifer *m* **~ washer** Abdichtungsring *m*
**expanser** Ausdehner *m*
**expansibility** Dehnbarkeit *f*, Dehnungsfähigkeit *f*
**expansible** ausdehnbar, dehnbar
**expansion** Aufarbeitung *f*, Aufblähen *n*, Aufdornen *n*, Auflösung *f* (math.), Aufschwung *f*, Aufweitung *f*, Ausbauchung *f*, Ausbreitung *f*, Ausdehnung *f*, Dehnung *f*, Dilatation *f*, (of steam) Entspannung *f*, Entwicklung *f*, Entzerrung *f* (Dynamik), Erweiterung *f*, Expansion *f*, Quellung *f*, Treiben *n*, Umsichgreifen *n*, Verbreiterung *f*, Zerlegung *f*, Zuwachs *m* **~ of clouds of gas (or smoke)** Schwadenausbreitung *f* **~ of the jet stream** Strahlexpansion *f* **~ in lattice spacing** Gitteraufweiterung *f* **~ in a power series** Reihenentwicklung *f* **~ of special functions** Entwicklung *f* von speziellen Funktionen **~ of waves** Abschwächung *f* des Seegangs
**expansion, ~ anchor** Fuß *m* des Bügels für Be-

festigung in Mauerwerk **~ base** Wachstumsspitze *f* **~ bearing** bewegliches Auflager *n*, Ausdehnungslager *n* **~ bellows** Druckausgleichdose *f* **~ bend** Schleifenrohr *n* **~ bolt** Bolzenschraube *f* **~ chamber** Ausdehnungs-, Entspannungs-raum *m*; Expansions-, Nebel-, Wilson-kammer *f*, Windkessel *m* **~ circuit breaker** Expansionsschalter *m* **~ circuit switch** Expansionsschalter *m* **~ clutch** Ausdehnungskupplung *f* **~ coefficient** Ausdehnungskoeffizient *m* **~ coil** Ausgleichspirale *f* **~-contraction diagram** Spannungsdehnungsschaubild *n* **~ cooler** Entspannungskühler *m* **~ coupling** Ausdehnungskupplung *f*, längsbewegliche Kupplung *f* **~ crack** Dehnungs-, Treib-riß *m* **~ end tension** Expansionsendspannung *f* **~ (or restraining) force** Verspannkraft *f* **~ formula** Divergenzformel *f*, Entwicklungssatz *m* **~ gear** Expansionssteuerung *f*, Expansionsvorrichtung *f* **~ hand reamer** nachstellbare Handreibahle *f* **~ head** Universalspannkopf *m* **~ instrument** (hot wire) Hitzdrahtinstrument *n* **~ joint** Ausdehnungs-dichtung *f*, -fuge *f*, -stoß *m*, -verbindung *f*; Dehnfuge *f*, Dehnungsfuge *f*, Dehnungskörper *m*, Dilatationsvorrichtung *f*, Expansionskuppelung *f*, Expansionsmuffe *f*, Expansionsverbindung *f*, Stoßfuge *f*, Trennungsfuge *f* **~ (construction) joint** Arbeitsfuge *f* **~ link** Expansionskulisse *f* **~ liquefier** Ausdehnungsverflüssiger *m* **~ loop** Ausdehnungsrohrbogen *m*, (steam pipes) Ausgleichschleife *f* **~ machine** Drillingsmaschine *f* **~ mandrel** Expansionsdorn *m* **~ meter** Dehnungsmesser *m* **~ orbit** Auslaufbahn *f* **~ piece** Dehnungsstück *n* **~ pipe** Ausgleichsrohr *n*, (hydraulischer Druck) Expansionsrohr *n*, Federrohr *n* **~ pipe joint** Ausdehnungsrohrverbindung *f* **~ plug** Ausdehnungszapfen *m*, federnde Verschlußscheibe *f* **~ port** Schnüffelloch *n* **~ program** Erweiterungsprogramm *n* **~ ratio** Entspannungsverhältnis *n* **~ ring** Ausdehnungsringstück *n*, Spannring *m* **~ roller** Ausbreitwalze *f* **~ screw** Spreizschraube *f* **~ seat ring** Dehnungsring *m* **~ slide valve** Expansionsschieber *m* **~ sliding block** Expansionsgleitbacke *f* **~ spring** Feder *f* unter dem Kupplungsbelag **~ stage** Expansionsstufe *f* **~ strip** Dehnungsschiene *f* **~ stroke** Arbeits-, Ausdehnungs-, Expansions-hub *m*; Expansionsschlag *m*, Verbrennungstakt *m* **~ tank** Ausdehnungsgefäß *n*, Ausgleichsbehälter *m*, Dampfdom *m* **~ tap** Expansionsschraubenbohrer *m* **~ theorem** Entwicklungssatz *m* **~ turbine** Expansionsturbine *f* **~ U-bend** (steam pipes) Federrohrbogen *m* **~ valve** Entspannungsventil *n* **~ V-belt pulley** Expansionskeilriemenscheibe *f* **~ wave** Expansionswelle *f*
**expansive** ausdehnbar, expansiv **~ capacity** Quellungsvermögen *n* **~ force** Ausdehnungs-, Expansiv-kraft *f*
**expatiate, to ~** sich auslassen
**expect, to ~** abwarten, entgegensehen, erwarten **expectancy of hitting** Treffwahrscheinlichkeit *f* **expectation** Erwartung *f* **in ~ of** in Aussicht stehend **~ value** voraussichtlicher Ertragswert *m*
**expected** voraussichtlich **as is ~** voraussichtlich **~ approach time (EAT)** voraussichtlicher An-

flugzeitpunkt *m* (aviat.) ~ **level** Meßspiegel *m*
~ **value** Erwartungs-, Nenn-wert *m*
**expediency** Zweckmäßigkeit *f*
**expedient** Aushilfe *f*, Ausweg *m*, Behelf *m*, Behelfslösung *f*, Behelfsmittel *n*, Hilfsmittel *n*, Kunstgriff *m*, Maßregel *f*, Mittel *n*, Notbehelf *m*, Notmittel *n*; angemessen, rationell, zweckmäßig
**expedite, to** ~ beschleunigen
**expediting** Versand *m*
**expedition** Streifzug *m*
**expel** (fumes) abtreiben, ausstoßen, austreiben, ausweisen, entbinden, herausspülen, verdrängen **to** ~ **humidity** Feuchtigkeit *f* austreiben
**expellant** Rückstoßmasse *f*
**expeller** Austreiber *m*
**expelling** Ausstoßen *n*
**expend, to** ~ aufwenden **to** ~ **ammunition** verschießen
**expendable,** ~ **load** Verbrauchslast *f* ~ **jammer** Verluststörsender *m* ~ **material** Verbrauchsmaterial *n* ~ **non-recoverable item** Verschleißteil *n* ~ **stocks record** Verbrauchsmittelbestandbuch *n* ~ **supply** Verbrauchs-stoff *m*, -güter *pl*
**expenditure** Aufwand *m*, Aufwendung *f*, Ausgabe *f*, Geldausgabe *f*, Verbrauch *m* ~ **of energy** Arbeits-, Energie-aufwand *m* ~ **of material** Stoffaufwand *m* ~ **of mechanical work** Arbeitsverbrauch *m* ~ **of power** Kraftaufwand *m*
**expenditure control** Ausgabenkontrolle *f*
**expense** Abgabe *f*, Ausgabe *f*, Geldausgabe *f*, Unkosten *pl* ~ **of upkeep** Unterhaltungskosten *pl* ~ **distribution** Unkostenverteilung *f*
**expenses** Aufwand *m*, Auslagen *pl*, Kosten *pl*, Spesen *pl* ~ **actually incurred** effektiv angefallene Kosten *pl* ~ **on discounted bills** Diskontspesen *pl* **all** ~ **paid** spesenfrei
**expensive** kostbar, kostspielig, teuer
**experience, to** ~ empfinden
**experience** Erfahrung *f*
**experienced** erfahren, geschäftskundig, routiniert, sachkundig ~ **man** Praktiker *m*
**experiment, to** ~ ausprobieren, experimentieren, versuchen, Versuch anstellen
**experiment** Erprobung *f*, Experiment *n*, Probe *f*, Untersuchung *f*, (wissenschaftlicher) Versuch *m*, Vorführung *f* ~ **in cultivation** Anbauversuch *m* ~ **on a model** Modellversuch *m*
**experiment,** ~ **chamber** Meß-kammer *f*, -raum *m* ~ **station** Prüfungs-, Versuchs-anstalt *f*
**experimental** experimentell, versuchsmäßig ~ **(ly)** versuchsweise ~ **arrangement** Versuchsanordnung *f* ~ **boring** Bohrversuch *m* ~ **data** Erfahrungstatsache *f* ~ **datum** Versuchswert *m* ~ **department** Versuchsabteilung *f* ~ **device** Versuchsvorrichtung *f* ~ **engine** Versuchs-maschine *f*, -motor *m* ~ **error** Versuchsfehler *m* ~ **facts** Erfahrungstatsache *f* ~ **flight** Versuchsflug *m* ~ **laboratory** Versuchslaboratorium *n* ~ **material** Versuchsmaterial *n* ~ **method** Versuchsmethode *f* ~ **object** Versuchsgegenstand *m* ~ **plane** Versuchsflugzeug *n* ~ **plant** Versuchs-anlage *f*, -anstalt *f* ~ **radio station** Versuchsfunkstelle *f* ~ **release** Versuchsaufgabenerteilung *f* ~ **sequence** Versuchsfolge *f* ~ **setup** Brettschaltung *f* ~ **stage** Versuchsstadium *n* ~

**station** Erprobungs-, Versuchs-stelle *f* ~ **test of rockets on ground** Bodenversuch *m* ~ **theorem** Erfahrungssatz *m* ~ **type** Versuchsmuster *n* ~ **unit** Erprobungskommando *n* (mil.) ~ **work** Prüfarbeit *f* ~ **works level** Versuchsbetriebsstufe *f*
**experimentation** Experimentierung *f*
**experimenter** Forscher *m* ~**'s license** Experimentierlizenz *f*, Versuchserlaubnis *f*, Versuchslizenz *f*
**expert** Fachmann *m*, Sach-bearbeiter *m*, -kenner *m*, -kundiger *m*, -verständiger *m*; fachkundig, fachmännisch, sachgemäß, sachkundig, sachverständig **among** ~**s** in Fachkreisen *pl* ~ **engineer** Fachingenieur *m* ~ **fitter** Spezialist *m* ~ **furnaceman** Ofenfachmann *m* ~ **knowledge** Sach-kenntnis *f*, -kunde *f* ~ **mechanic** Spezialmonteur *m* ~**'s opinion (or report)** Gutachten *n* eines Sachverständigen
**expertness** Geschicklichkeit *f*
**expiration** Ablauf *m*, (patent) Ablaufen *n*, Ablaufzeitpunkt *m*, Erlöschen *n*, Verfall *m* ~ **of delivery** Ablauf *m* der Lieferzeit ~ **(on maturity or notes)** Fälligkeit *f* ~ **of a patent** Erlöschen *n* eines Patentes, Patenterlöschung *f*
**expiration date** Fristablauf *m* (Lagerhaltung)
**expire, to** ~ ablaufen, ausatmen, ein- oder ausatmen, erblassen, (patents) erlöschen, verstreichen
**expired** (patents) abgelaufen, erloschen ~ **patent** abgelaufenes Patent *n*
**expiry of a patent** Erlöschen *n* eines Patentes
**explain, to** ~ aufklären, aufschließen, auseinandersetzen, auslegen, erklären, deuten, klar machen
**explainable** erklärlich
**explanation** Aufschluß *m*, Ausdeutung *f*, Auseinandersetzung *f*, Ausführung *f*, Auslegen *n*, Auslegung *f*, Deutung *f*, Erklärung *f*, Erläuterung *f*, Klarstellung *f* ~ **code** Begründungskode *m*
**explanatory remark** Begleitwort *n*
**explicit** ausdrücklich, ausführlich, explizit, unzweideutig
**explodable** explodierbar, explosibel
**explode, to** ~ bersten, detonieren, explodieren, krepieren, platzen, plotzen, sprengen, verpuffen, zerknallen, zerplatzen, zerspringen
**exploded,** ~ **view** Darstellung *f* in auseinandergezogener Anordnung ~ **wire** Metalldrahtentladung *f*
**exploding charge** Kanonenschlag *m*
**exploit, to** ~ ausbeuten, ausnutzen, auswerten, betreiben, gewinnen, in Betrieb setzen **to** ~ **something for private economic purposes** etwas privatwirtschaftlich ausbeuten
**exploitable** abbauwürdig, aufschließbar
**exploitation** Abbau *m*, Ausbeutung *f*, Ausnutzung *f*, Betrieb *m*, Gewinnung *f*, Nutzung *f*, Wirtschaftsführung *f* ~ **factor** Ausnutzungsfaktor *m* ~ **section** Auswertestelle *f*
**exploiter** Ausbeuter *m*
**exploiting right** Abbaugerechtigkeit *f* (mine, land)
**exploration** Abtastung *f*, Aufsuchung *f*, Erforschung *f*, Schürfen *n*, Schürfung *f*, Untersuchung *f* ~ **coil** Erforschungswindung *f*, Prüfspule *f* ~ **shaft** Schurfschacht *m* ~ **tunnel**

Sondierstollen *m* ~ unit Abtastvorrichtung *f*
~ well Versuchsbohrung *f* ~ work Ausrichtungsarbeit *f*
explorator Fühler *m* (einer Fühlersteuerung)
exploratory, ~ borehole Explorationsbohrloch *n* ~ drift (or gallery) Forschungsstollen *m*
explore, to ~ abtasten, aufsuchen, auskundschaften, ausrichten (min.), (picture area or screen with beam, pencil or spot) bestreichen, erforschen, erkundigen, forschen, prüfen, schürfen, untersuchen to ~ the ground by bore holes ein Terrain abbohren
explorer Entdecker *m*, Erforscher *m*, Forscher *m*, Taster *m* (TV)
exploring Abtasten *n*, Abtastung *f*, Ausrichtung *n* (min.), Bild-abtastung *f*, -punktverteilung *f*, -synthese *f*, -wiedergabe *f*, -zusammensetzung *f*, Tasten *n* ~ of image Bilderzeugung *f* ~ of picture Bildzerlegung *f*
exploring, ~ binoculars Doppelfernrohr *n* ~ coin Erforschungswindung *f*, Kopplungsschleife *f*, Suchspule *f* ~ disk Abtastscheibe *f* ~ drift Untersuchungsstollen *m*, Versuchsstrecke *f* ~ electrode Suchelektrode *f* ~ means Zerlegungsvorrichtung *f* ~ point Bildpunkt *m* ~ spot abtastender Lichtfleck *m*
explosion Detonation *f*, Explosion *f*, Knall *m*, Losgehen *n*, Plotz *m*, Sprengung *f*, Verpuffung *f*, Zerknall *m* ~ in the carburetor Knallen *n* im Vergaser
explosion, ~ chamber Explosions-kammer *f*, -raum *m*, Löschkammer *f* ~ charge Ausstoßladung *f* ~ door Explosions-klappe *f*, -tür *f* ~ engine gemischverdichtende Maschine *f*, Verpuffungsmaschine *f* ~ gas turbine Expansionsgasturbine *f* ~ hazard Explosionsgefahr *f* ~ limit Berstdruck *m* ~ limits Explosionsgrenzen *pl* ~ motor Verpuffungsmotor *m* ~-pipe eruption Punkteruption *f* ~ pressure in engines Explosionsdruck *m* ~ prevention Explosionsverhütung *f* ~-proof (apparatus) explosionsgeschützt, explosionssicher, luftdicht gekapselt ~-proof heating member explosionssicherer Heizkörper *m* ~-proof plant explosionssichere Anlage *f* ~-proof tank explosionssicheres Gefäß *n* ~-proof vessel explosionssicheres Gefäß *n* ~ pyrometer Knallpyrometer *n* ~ relief valve Explosionskappe *f* ~ rivet Sprengniet *n* ~ stroke Explosions-periode *f*, -takt *m*
explosive Sprengstoff *m*; explodierbar, explosiv ~ action Sprengwirkung *f* ~ ammunition Sprengmunition *f* ~ atmosphere Schlagwetter *n* ~ bolt Sprengbolzen *m* ~ bomb Sprengbombe *f* ~-bomb effect Sprengbombenwirkung *f* ~ bonding Explosionsverbinden *n* ~ box Ladungskasten *m* ~ bullet Explosiv-, Spreng--geschoß *n*, Sprengstoffkugel *f* ~ canister Sprengbüchse *f* ~ cap Streichkappe *f* ~ cartridge Bohrpatrone *f* ~ charge Ladung *f*, Spreng-füllung *f*, -körper *m*, -ladung *f*, -masse *f*, -mittel *n*, -stoffüllung *f*, Zündladung *f* ~ charge set off by detonation of another charge Folgeladung *f* ~ consonant explosiver Konsonant *m* ~ constant Schießkonstante *f* ~ container Ladungsgefäß *n* ~ crater Sprengtrichter *m* ~ dump Sprengstofflager *n* ~ effect Brisanz-, Spreng-wirkung *f* ~ flame Stichflamme *f* ~

field Entladefeld *n* (electr.) ~ fission explosive Spaltung *f* ~ force Sprengkraft *f* ~ forming Explosivformgebung *f* ~ gas zerknallbares Gas *n* ~ missile Sprenggeschoß *n* ~ mixture Entzündungs-, Explosions-gemisch *n* ~ powder Füllpulver *n* ~ projectile Sprengstoffkugel *f* ~ pyroxyline fiber Pyroxylinfaden *m* ~ rivet Sprengniet *n* ~ smoke-producing observation cartridge Beobachtungspatrone *f* ~ sound Explosivkonsonant *m*, explosiver Laut *m* ~ thrust Verbrennungsdruck *m* ~ train (in measuring detonation velocity) Meßstrecke *f*, Zündfolge *f*
explosiveness Explosionsfähigkeit *f*
explosives Sprengbedarf *m* ~ used in mining Wetterspreng-mittel *n*, -stoff *m* ~ with toluol bases Dinitrotoluolbase *f*
exponent Anzeiger *m* (math.), Exponent *m*, Hochzahl *f*, Potenz *f*, Verhältnisanzeiger *m* (math.) ~ of power Exponent *m*, Potenzexponent *m*
exponential exponential, exponentiell ~ curve Exponential-kurve *f*, -linie *f* ~ decay exponentieller Abfall *m* ~ equation Potenzansatz *m* ~ expression Exponentialausdruck *m* ~ function Potenzgesetz *n* ~ horn Exponential-trichter *m*, -horn(strahler) *m* ~ law Exponential-, Potenz--gesetz *n* ~ pentode Bremsgitterregelröhre *f* ~ resistor spannungsabhängiger Widerstand *m* ~ series Potenzreihe *f* ~ sweep exponentielle Ablenkung *f* ~ tail exponentieller Flankenabfall *m* ~ tube Exponential-, Regel-röhre *f*, Röhre *f* mit variablem Verstärkungsfaktor, Röhre *f* mit veränderlichem Durchgriff ~ well Exponentialkasten *f*
exponentially nach einem Exponentialgesetz *n*
export, to ~ ausführen, herausbringen
export, ~(ation) Ausfuhr *f* ~ of foreign exchange Devisenexport *m*
export, ~ activity Ausfuhrtätigkeit *f* ~ business Ausfuhrgeschäft *n* ~ certificate Ausfuhrschein *m* ~ country Ausfuhrland *n* ~ duty Ausgangszoll *m* ~ paid through clearing Kompensationsgeschäft *n* ~ permit Ausfuhrschein *m* ~ trade Ausfuhr *f*, Ausfuhrhandel *m*, Export *m*, Exportgeschäft *n* ~ version Exportausführung *f*
exportable ausführbar
exportation Ausführung *f*
exporter Exporteur *m*
exports Außenhandel *m*, Exportartikel *pl*
expose, to ~ aufdecken, auslegen, aussetzen, beschicken, bestrahlen, bloßlegen, bloßstellen, entblößen, exponieren, freilegen to ~ to the air der Luft *f* aussetzen to ~ the flank die Flanke *f* aufdecken to ~ goods for sale Waren *pl* zum Verkauf aufstellen to ~ to light belichten to ~ to weather bewettern
exposed belichtet (photo), blank, bloß, entblößt, freistehend, offen, ungeschützt to be ~ ausgesetzt sein, freiliegen, freistehen to be ~ to einwirken lassen ~ to the elements dem Wetter ~ ausgesetzt ~ to the fire feuerberührt ~ to stress stark beansprucht
exposed, ~ filament chamber Meßzelle *f* mit Drähten im Verbrennungsraum ~ film belichteter Film *m* ~ flank unangelehnte Flanke *f*, offener Flügel *m* ~ position offene Stellung *f* ~ wing offener Flügel *m* ~ wiring Verlegung *f* über Putz

exposing Freilegung *f* ~ to weather Bewetterung *f*
exposition Schaustellung *f*
exposure Aufnahme *f* (photo), Aufschluß *m*,
Aufstellung *f*, Auslieferung *f*, Aussetzen *n*, Aussetzung *f*, Belichtung *f* (photo), Bestrahlung *f*,
Exposition *f*, Interferenzaufnahme *f*, Witterungseinfluß *m* ~ of electron diffraction Elek-
tronenbeugungsaufnahme *f* ~ to light Belichtung *f* ~ to radiation Bestrahlungsdosis *f* ~ to
solar radiation Bestrahlung *f* ~ against the sun
Gegenlichtaufnahme *f* (photo) ~ to sunlight
Sonnenbestrahlung *f*
exposure, ~ box Belichtungskasten *m* ~ clock
Belichtungsschaltuhr *f* ~ control unit Licht-
dosiergerät *n* ~ counter Aufnahmezähler *m*
~-density relationship Schwärzungskurve *f* ~
difference Belichtungsdifferenz *f* (photo) ~
drum Belichtungstrommel *f* ~ factor Schwellen-
wert *m* (photo) ~ indicator Aufnahmeanzeige-
,lampe *f* ~ latitude Belichtungsspielraum *m* ~
level Belichtungsmaß *n* ~ lid Belichtungsschieber *m* ~ meter Belichtungsmesser *m* ~
-proof material beständig gegen physikalische
Einflüsse *pl* ~ recording slide Belichtungs-
reihenschieber *m* ~ scale Expositionszeit *f* ~
shutter Belichtungsschieber *m* ~ station Aufnahmeort *m* ~ test Beregnung *f*, Lichtecht-
heitsprobe *f* ~-time table Belichtungstabelle *f*
~ timer Belichtungsmesser *m*
express, to ~ ausdrücken, aussprechen, äußern
to ~ in round figures (or numbers) eine Zahl *f*
abrunden to ~ by symbols versinnbildlichen
express per Eilboten *m*; dringend ~ air liner
Schnellverkehrsflugzeug *n* ~ call dringender
Anruf *m*, dringendes Gespräch *n* ~ cargo boat
Eilgüterboot *n* ~ delivery Eilbeförderung *f*,
Eilbotenbestellung *f* ~ freight Eilfracht *f*
~-freight tariff Eilgebührensatz *m* ~-freight
traffic Eilgutverkehr *m* ~ freight train Eilgüter-
zug *m* ~ goods Eilgut *m* ~ highway Autobahn *f*
~ messenger Kurier *m* ~ pictorial report
Schnellbildsendung *f* ~ road Schnellstrasse *f*
~ telephone service Fernsprechschnellverkehr
*m* ~ train D-Zug *m*, Eilzug *m*, Schnellzug *m* ~
way Brückenstraße *f*, Hochstraße *f*
expressed ausgedrückt ~ liquor Preßlauge *f*
expressible oil and moisture Öl- und Feuchtigkeitsgehalt *m*
expressing Spedition *f*, Verfrachterei *f*, Verfrachtung *f*
expression Ansatz *m*, Ausdruck *m*, Ausdrucksweise *f*, Bezeichnung *f*, Funktion *f*, Strangpreßprofil *n* ~ representing the work Arbeits-
ausdruck *m*
expressions Formelausdrücke *pl*
expressive ausdrucksvoll
expressly ausdrücklich
expropriate, to ~ enteignen
expropriation Beschlagnahme *f*, Enteignung *f*,
Grundablösung *f*
expulsion Ausstoß *m*, Ausstoßen *n*, Ausstoßung
*f*, Ausweisung *f*
expunged ausgestrichen
exquisite auserlesen
exsiccate, to ~ austrocknen
exsiccation Austrocknung *f*
exsiccator Austrockner *m*, Exsikkator *m*
extemporaneous (speech) aus dem Stegreif *m*

extemporization Extemporierung *f*
extemporize, to ~ extemporieren
extend, to ~ ausbauen, ausbreiten, ausdehnen,
auseinanderbiegen, auseinanderziehen, ausfahren, auslegen, ausrecken, (flaps) ausschlagen,
ausschmieden, ausspannen, ausstrecken, ausweiten, breiten, dehnen, (credit) erhöhen, erstrecken, erweitern, herragen, längen, laufen,
recken, spannen, strecken, verlängern ability
to ~ Ausdehnungsvermögen *n* to ~ a call
eine Verbindung *f* weiterleiten to ~ a call
beyond three minutes Gespräch *n* über drei
Minuten verlängern to ~ a circuit eine Leitung
*f* verlängern to ~ the flaps die Spreizklappen *pl*
öffnen to ~ to infinity sich ins Unendliche *n*
erstrecken to ~ a line eine Leitung *f* weiter-
führen to ~ a spring (eine) Feder *f* spannen
to ~ a subscriber's circuit to the first group
selector eine Teilnehmerleitung *f* zum ersten
Gruppenwähler durchschalten to ~ through
(a call) durchverbinden to ~ to durchschalten
to ~ the undercarriage das Fahrwerk *n* ausfahren
extend line Ausfahrleitung *f*
extendable mount Klapplafette *f*
extended ausgebreitet, geöffnet, länger, langgestreckt, verlängert ~ concession Gebietfeld *n*
~ cross slide durchgehender Unterschieber *m*
~ derrick-type belt conveyer Förderbandaus-
leger *m* ~ guard verlängertes Besetzthalten *n*,
verzögertes Trennen *n* ~ ion source ausgedehnte Ionenquelle *f* ~ leave Nachurlaub *m*
~-order construction getrennter Bau *m* of ~
oval shape langeiförmig ~ position of the undercarriage Landestellung *f* ~ quadruplex Doppelgegensprechen *n* für Staffelverkehr ~ subscription call Zusatzgespräch *n* ~ surface pipe
Lamellenrohr *n* ~ time test Langzeitversuch *m*
~ T-shaped antenna verlängerte T-Antenna *f*
~ undercarriage ausgefahrenes Fahrgestell *n*
~ value Gesamtwert *m* einer Position (logistics)
extender Streckmittel *f*, Zug *m*
extending Strecken *n*, Streckung *f*
extensibility Dehnbarkeit *f*, Dehnungsfähigkeit
*f*, Spannbarkeit *f*, Spannkraft *f*, Streckbarkeit *f*
extensible ausbaubar, ausdehnbar, auslaßbar
(landing gear), ausweitbar, ausziehbar, dehnbar, spannbar, streckbar, (tripod leg) zusammenschiebbar not ~ unstreckbar ~ applications
Einsatz- und Erweiterungsmöglichkeiten *pl* ~
power cables Starkstromdehnungskabel *pl*
extension Ansatz *m*, Ausbau *m*, Ausbreitung *f*,
Ausdehnung *f*, Ausläufer *m*, (Außen)neben-
stelle *f*, Ausziehung *f*, Auszug *m*, Dehnung *f*,
(pair of compasses) Einsatz *m*, Erstreckung *f*,
Erweiterung *f*, zweiter Fernsprechapparat *m*,
verlorener Kopf *m*, (linear) Längenausdeh-
nung *f*, Längsstreckung *f*, Längung *f*, (circuit)
Nebenanschlußleitung *f*, Verlängerung *f*, Vergrößerung *f*, Zuwachs *m*
extension, ~ in depth Tiefenausdehnung *f* ~ of
dry dock Dockverlängerung *f* ~ of leave Nachurlaub *m*, Urlaubsverlängerung *f* ~ of measuring ranges Meßbereichsumfang *m* ~ of piston
Kolbenansatz *m* ~ in phase Phasenraum *m*
~ of term (or time-limit) Fristverlängerung *f*
~ of time Fristgewährung *f* ~ under own
weight Ausfahren *n* durch Eigengewicht ~ per

unit length Schubzahl *f* ~ **in velocity** Geschwindigkeitsausdehnung *f* ~ **for vent** Entgasungsrohr *n* ~ **of width** Breitenausdehnung *f* ~ **of works** Betriebserweiterung *f*
**extension,** ~ **angle lugs** überstehende Winkelsporen *pl* ~ **arm** einseitiger Querträger *m*, Klapparm *m* ~ **bell** Außenwecker *m*, zweiter Wecker *m* ~ **board** Anbaugruppe *f* ~ **bolt** Nachlaßschraube *f* ~ **bracket (or arm)** Verlängerungsschiene *f* ~ **camera** Spreizenkamera *f* ~ **chain** Nachlaßkette *f* ~ **circuit** Leitungs-ergänzung *f*, -verlängerung *f*, Verlängerungsleitung *f* ~ **crane** Ausfahr-katze *f*, -schiene *f* ~ **flap** Aufsatzklappe *f* ~ **funnel** Verlängerungstrichter *m* ~ **gate** Anguß *m* ~ **handwheel block** Flaschenzug *m* mit verlängerter Antriebswelle ~ **indicator** Nebenstellenklappe *f* ~ **ladder** Bauleiter *f* ~ **lead** Ausgleichsleitung *f* ~ **line** (artificial) Leitungsergänzung *f*, Nebenanschluß *m*, Nebenanschlußleitung *f*, Nebenstellenleitung *f* ~**-line jack** Nebenstellenklinke *f* ~ **link** Gelenkzwischenstück *n* (helicopter) ~ **mast** Kurbelmast *m* ~ **means** Posaunenzug *m* ~ **multiple section** Ansatzplatz *m* (teleph.) ~ **pad** (artificial) Verlängerungsleitung *f* ~ **piece** Ansatzstück *n* ~ **prolongation** Fortsatz *m* ~ **rim** Radverbreiterung *f*, Verbreiterungsreifen *m* ~ **rod** Aufsteckrohr *n* ~ **rod for sighting** Richtlatte *f* ~ **ruler (or straightedge)** Verlängerungslineal *n* ~ **shaft** Aufsteckachse *f*, Fernwelle *f*, Nachlaßkettenwinde *f*, Verlängerungswelle *f*, Wellenleitung *f* ~ **socket (or part)** Verlängerer *m* ~ **spring** Torsions-, Zug-feder *f* ~ **station** (off-premises) Außennebenstelle *f*, Nebenstelle *f* ~ **station without exchange facilities** Hausstelle *f* ~**-station service** Nebenstellenbetrieb *m* ~ **stem** verlängerte Spindel *f* ~ **swivel** Nachlaßschraubenwirbel *m* ~ **telephone** Neben-anschluß *m*, -stelle *f* ~ **tube** Ansatztubus *m*, Aufsteck-rohr *n*, -hülse *f* ~ **user** Nebenstellenteilnehmer *m* ~ **winding gear** Nachlaßwinde *f* ~ **wire** Nebenanschlußleitung *f* ~ **(lead) wire** Kompensationsdraht *m*
**extensive** ausgedehnt, ausgreifend, umfangreich, umfassend, vielseitig, weitgehend, weitläufig ~ **mine explosion (or fire)** ausgedehnte Grubenexplosion *f* ~ **natural obstacle** Barriere *f* ~ **shower** explosiver Schauer *m* ~ **smoke screen** Großvernebelung *f* ~ **target** Flächenziel *n*
**extensometer** Ausdehnungs-, Deformations-, Dehnbarkeits-, Dehnungs-messer *m*, Extensometer *m*, (dial) Meßuhr *f*
**extent** Ausdehnung *f*, Aushalten *n* (geol.), Ausmaß *n*, Bereich *m*, Erstreckung *f*, Größe *f*, Maß *n*, Strecke *f*, Umfang *m* ~ **of cloudiness** Bewölkungs-grad *m*, -größe *f* ~ **of knowledge** Wissensbereich *n* ~ **of losses** Verlustmaß *n* ~ **of object** objektives Sehfeld *n* ~ **of the open sea** Fetch *f* ~ **of a port** Ausdehnung *f* eines Hafens ~ **of torsion** Verdrehungsspanne *f* ~ **of the variations** Schwankungsbreite *f* ~ **of view** Objektausschnitt *m* ~ **of vortex sheet (or discontinuity)** Unstetigkeits-, Wirbel-strecke *f* ~ **to which level is populated by electrons** Besetzungszahl *f* der Elektronen in einem Niveau
**extenuate, to** ~ sich auslassen
**extenuating circumstances** Milderungsgründe *pl*,

Strafmilderungsgrund *m*, mildernde Umstände *pl*
**exterior** äußerlich ~ **angle of a triangle** Außenwinkel *m* eines Dreiecks ~ **antenna** Außen-, Frei-antenne *f* ~ **ballistics** Außenballistik *f*, Witterungseinfluß *m* ~ **ballistics and interior ballistics** besondere und Witterungseinflüsse *pl* ~ **ballistics institute** flugmechanisches Institut *n* ~ **bracing** Außenverspannung *f* ~ **flash** Außenüberfang *m* ~ **grid** Außengitter *n* ~ **guard** Außenwache *f* ~ **illumination** Außenleuchten *n* ~ **line** Außenlinie *f* ~ **load** äußere Belastung *f* ~ **part of an ingot** Blockrand *m* ~ **picture** Außenaufnahme *f* ~ **plywood layers** Außenschichten *pl* aus Sperrholz ~ **shooting** Freilichtaufnahme *f* ~ **shot** Außen-, Freilicht-aufnahme *f* ~ **wood construction** Holzummantelung *f*
**exterminate, to** ~ ausrotten
**extermination** Vernichtung *f*
**external** außenliegend, außenseitig, außer, äußer, äußerlich, (radiators) freifahrend ~ **aileron** äußerer Hilfsflügel *m* ~ **angle** ausspringende Ecke *f* ~**-armature generator** Innenpoldynamo *m* ~ **audit** betriebsfremde Revision *f* ~ **auditor** betriebsfremder Prüfer *m* ~ **auto drop switch** Selbstabwurfschalter *m* (Flugz.) ~ **autoignition** Fremdentzündung *f* ~ **bearing** Außenlager *n* ~ **bilge skin** (of ship) Außenhautboden *m* ~ **bracing** Außenverstrebung *f* ~ **brickwork** Außenmauerwerk *n* ~ **caliper gauge** Gabellehre *f* ~ **causes of vibration** Schwingungsursache *f* ~ **cone angle** Kopfkegelwinkel *m* ~ **contact base** Außenkontaktsockel *m* ~ **contracting band brake** Außenbandbremse *f* ~ **contracting brake** Außenbackenbremse *f* ~ **control devices** sekundäre Regeleinrichtung *f* ~ **control tube** Außensteuerröhre *f* ~ **control vane** Luftsegel *n*, Trimmruder *n* (g/m) ~ **controls** Außensteuerung *f* ~ **copying** Außenkopieren *n* ~ **diameter** Außendurchmesser *m* ~**-dimension measurement** Außenabmessung *f* ~ **disc** Außenlamelle *f* (Kupplung) ~ **drag body** Widerstandskörper *m* ~ **drag wire** Schrägkabel *n* ~ **effect** Außenwirkung *f* ~ **efficiency** äußerer Wirkungsgrad *m* ~ **exhaust collector** Außensammler *m* ~ **extension** Außennebenstelle *f* ~ **faultsman** Störungssucher *m* im Außendienst ~ **fault staff** Störungspersonal *n* für Außenstörungen ~ **field** Außenfeld *n* ~ **field influence** äußerer Widerstand *m* ~ **fields** fremde Felder *pl* ~ **force** Beanspruchung *f*, äußere Kraft ~ **frame** Außenarmierung *f* ~ **fuel transfer** Außentankumfüllung *f* (Flugz.) ~ **gauge** Lehrring *m*, Rachen-, Wellen-lehre *f* ~ **gauging** Außenmessung *f* ~**-gauging contact tip** Meßhütchen *n* für Außenmessungen ~**-gauging optimeter** Optimeter *m* für Außenmessungen ~ **gear** außen verzahntes Zahnrad *n* ~ **grinding machine** Außenschleifmaschine *f* ~ **gripping device** Umspannzeug *f* ~ **idle time** externe Leerlaufzeit *f* ~ **lap** Läppring *m* ~ **leads** Außenleitung *f* ~**-limit (snap) gauge** Grenzrachenlehre *f* ~ **line-balancing resistance** außenliegender Widerstand *m* ~ **load circuit** Belastungskreis *m* ~ **loop** Außenrahmen *f* ~ **matter** Fremdbestandteil *m* ~ **measurement** Außenmaß *n* ~ **memory** Ausgabespeicher *m* ~

**micrometer** Bügelmeßschraube *f* (Mikrometer) **~ moment** äußerer Moment *m* **~ notch** Außenkerbe *f* **~ oil system** äußere Ölleitungen *pl* **~ operating ratio** externe Maschinenausbeute *f* **~ perspective center** (of the lens) dingseitiger (vorderer) Hauptpunkt *m* **~ photoelectric effect** lichtelektrische Ausbeute *f*, äußerer lichtelektrischer Effekt *m*, lichtelektrische Elektronenemission *f*, Hallwachseffekt *m*, äußerer Fotoeffekt *m* **~ plant** Außenanlage *f* **~ pole dynamo** Außenpoldynamo *f* **~ pole generator** Außenpolgenerator *m* **~ power cart** Generatorwagen *m* **~ power plug** Außenbordanschlußstecker *m* **~ power receptable** Außenbordsteckdose *f* **~ power supply** Fremdstromversorgung *f* **~ quenching** externe Löschung *f* **~ relationship** äußerer Zusammenhang *m* **~ resistance** Leistungswiderstand *m* **~ safety ring** Außensicherungsring *m* **~ screw part** Einschraubstück *n* **~ shakes** Spiegelkluft *f* **~ shape of pattern** Modellform *f* **~ shell** Rumpfaußenwand *f* **~ shoe brake** Außenbackenbremse *f* **~ storage** Außenspeicher *m* **~ store** äußerer Pufferspeicher *m* (info proc.) **~ stowage** Außenaufhängung *f* **~ strutting** Außenverstrebung *f* **~ suspension** Außenaufhängung *f* **~ tamping bag** Außenbesatzbeutel *m* **~ tank** Außentank *m* **~ thread** Außen-, Bolzen-gewinde *n* (auf Drehbänken) **~ thread and form grinding machine** Außengewinde- und Profilschleifmaschine *f* **~ thread cutting machine** Außengewindeschneidmaschine *f* **~ toothing of gears** Außenverzahnung *f* **~ upset** außen verstärkt **~ voltage** Fremdspannung *f* **~ world** Außenwelt *f* **~ zone** Außenzone *f*, Rand-schicht *f*, -zone *f*

**externally, ~ finned pipe** Rippenrohr *n*, außen geripptes Rohr *n* **~ heated arc** Lichtbogen *m* mit Fremdheizkathode **~ heated rotary drier** Trommeltrockner *m* mit Außenbeheizung **~ mounted radiator** freifahrender Kühler *m* **~ ribbed pipe** außen geripptes Rohr *n*, Rippenrohr *n* **~ ventilated** fremdgelüftet

**exterpolate, to ~** extrapolieren

**extinct** ausgestorben, erloschen **~ natural radionuclides** zerfallene natürliche Radionuklide *pl*

**extinction** Ablöschung *f*, (of fluorescence) Ausleuchtung *f* und Tilgung *f*, Auslöschung *f*, Ausrottung *f*, Erlöschen *n*, Erlöschung *f*, Löschung *f* **~ of oscillation procedure** Ausschwingverfahren *n*

**extinction, ~ angle** Auslöschungsschiefe *f* **~ coefficient** Extinktionskoeffizient *m* **~ law** Schwächungsgesetz *n* **~ position** Auslöschungslage *f* **~ potential** Löschspannung *f* **~ pulse** Löschstoß *m* **~-striking potential difference** Löschzündspannungsdifferenz *f* **~ theorem** Auslöschungssatz *m* **~ voltage** Löschspannung *f*

**extinguish, to ~** auslöschen, dämpfen, zum Erlöschen *n* bringen

**extinguish fluid** Löschflüssigkeit *f*

**extinguished** gelöscht **to be ~** erlöschen, verlöschen **to become ~** ausgehen

**extinguishing** Ablöschung *f*, Erlöschen *n*, Löschung *f* **~ of the engaged lamps** Erlöschen *n* der Besetztlampen

**extinguishing, ~ agents** Löschmittel *pl* **~ plant for coking coal** Kokslöschanlage *f* **~ tube for**

**gas-blast switch** Gasschalterlöschrohr *n* **~ voltage** Löschspannung *f*

**extirpate, to ~** extirpieren

**extirpator** Scharegge *f*

**extort, to ~** herauspressen

**extra**Nebenlieferung *f*, Zugabe *f*, Zulage *f*, Zuschlag *m*; extra **~ air duct** Zusatzluftkanal *m* **~-axial** außer Achse *f*, außerachsial **~-axial aberration** Aberration *f* außerhalb der Achse **~-axial mirror zone** außerachsialer Spiegelabschnitt *m* **~-axial ray** außerachsialer Strahl *m* **~-axial rays** außerachsiale Strahlen *pl* **~ burst** Aufschlag *m* **~ charge** Aufschlag *m*, gesonderte Kostenberechnung *f*, Mehrpreis *m* **~ charge for overdue payment** Säumniszuschlag *m* **~ cost** Extrakosten *pl* **~ equipment** Sonderausstattung *f* **~-European range of** (international) **rules** außereuropäischer Vorschriftenbereich *m* **~ expenditure** Mehraufwand *m* **~ fee** Zuschlaggebühr *f* **~ freight** Frachtzuschlag *m* **~ fuel** zusätzliche Kraftstoffmenge *f* **~ fuel tank** Kraftstoffzusatzbehälter *m* **~-glossy paste** Hochglanzpaste *f* **~ heating** Rückheizung *f*

**extra-high, ~ tension** Extrahoch-, Hoch-, Höchst-spannung *f* **~ tension cables** Höchstspannungskabel *n* **~-tension plant** Höchstspannungsanlage *f* **~ tension unit** Hochspannungsgerät *n* **~-voltage system** Höchstspannungsnetz *n*

**extra, ~ keyboard** Zusatztastatur *f* **~-light-loaded** besonders leicht pupinisiert **~-light loading** besonders leichte Pupinisierung *f*, leichte Bespulung *f* **~ magnification** Übervergrößerung *f* **~ mileage rate** Leitungszuschlag *m* (teleph.) **~ pay** Zuschuß *m* **~ play tape** Langspielband *n* **~ price** Aufpreis *m* **~ radiator** Zusatzkühler *m* **~ recording image** Zwischenregistrierung *f* **~ rim** Hilfs-, Reserve-felge *f* **~-sensitive orthotest** Orthotest *m* mit erweitertem Teilstrich-Abstand **~ solution** Extralösung *f* **~ task** Überschicht *f* (min.) **~ tax** Zuschlaggebühr *f* **~ value** Mehrwert *m* **~ weight** Übergewicht *n* **~ width** Verbreiterung *f*

**extract, to ~** abreißen, abscheiden, ausfeilen, ausfördern, (gaseous or liquid) ausgaren, auskochen, auslaugen, ausscheiden, ausziehen, (energy from an electron beam) entnehmen, entziehen, extrahieren (data proc.), gewinnen, herausgreifen, herausziehen (nails, spikes), (malleable iron directly from ore) rennen, substituieren **to ~ alcohol** entalkoholisieren **to ~ bungs** entspunden **to ~ copper** entkupfern **to ~ the cube root** Kubikwurzelziehen **to ~ dust** abstäuben, ausstäuben **to ~ gas** entgasen **to ~ moisture** entfeuchten **to ~ ore** Erz *n* gewinnen **to ~ resin** entharzen **to ~ the root of** radizieren, die Wurzel *f* ziehen **to ~ square root** Quadratwurzel *f* ziehen

**extract** Auszug *m*, Extrakt *m*, (substance or liquid) Geist *m*, Übersicht *f* **in ~s** auszugsweise **~ contents** Extraktgehalt *m* **~ layer** Extraktionsschicht *f* **~ measuring instrument** Extraktbemesser *m* **~ order** Einzelbefehl *m* **~ yield** Extraktausbeute *f*

**extractable radiator** ausziehbarer Kühler *m*

**extractant** Extraktionsmittel *n*

**extractible** extrahierbar

**extracting, ~ collet** Abziehzange *f* **~ mandrel**

Ausdruckform *f* ~ **plant** Extraktionsanlage *f* ~ **rim** Hülsenrand *m* ~ **a sheet pile** Ausziehen *n* einer Spundbohle ~ **(or bleeder type) steam engine** Anzapfdampfmaschine *f* ~ **tool** Ausziehwerkzeug *n*

**extraction** Abscheidung *f*, Absonderung *f*, (wet) Auslaugung *f*, Ausziehen *n*, Ausziehung *f*, Auszug *m*, Bluten *n*, Entnahme *f*, Entziehung *f*, Erzielung *f*, Extraktion *f* (data proc.), Gewinnung *f*, Scheidung *f*

**extraction,** ~ **of copper** Kupfergewinnung *f* ~ **of cream** Rahmgewinnung *f* ~ **of cube root** Kubikwurzelziehen *n* ~ **by fusion** Röstreduktionsarbeit *f* ~ **of gold** Goldgewinnung *f* ~ **of helium** Heliumgewinnung *f* ~ **of iron from water** Wasserenteisenung *f* ~ **of resin** Entharzung *f*, Harzentziehung *f* ~ **of roots** Wurzelziehen *n* ~ **of sugar** Entzuckerung *f* ~ **of sugar from molasses** Melasseentzuckerung *f*

**extraction,** ~ **apparatus** Auslaugeapparat *m* ~ **cartridge** Extraktionshülse *f* ~ **energy** Ablösungs-arbeit *f*, -energie *f* des Elektrons ~ **flask** Extraktionskolben *m* ~ **parachute** Auszieh-schirm *m* ~ **plant** Extraktionsanlage *f* ~ **process** Gewinnungsverfahren *n* ~ **shell** Extraktionshülse *f* ~ **solvent** Extraktionsmittel *n* ~ **system** Abluftanlage *f* ~ **test** Extraktionsprobe *f* ~ **thimble** Extraktionshülse *f* ~ **turbine** Anzapfturbine *f* ~ **(or bleeder) valve** Entnahmeventil *n*

**extractive,** ~ **distillation** Extraktionsdestillation *f* ~ **matter** Extraktivstoff *m*

**extractor** Abziehvorrichtung *f* (für Schleifscheibe), Auskocher *m*, Auswerfer *m*, Auswerferbolzen *m*, Ausziehvorrichtung *f*, Entlader *m*, Extraktionsapparat *m*, Fanghaken *m* (Bohrgerät), Herausheber *m*, Hülsenzieher *m*, Patronenauszieher *m*, Patronenhalter *m*, Schieber *m* ~ **barb** Ausziehkralle *f* ~ **basket** Zentrifugenkorb *m* ~ **cowl** Absaughaube *f* ~ **equipment** Extraktionsanlage *f* ~ **fan** Absauggebläse *n* ~ **groove in a cartridge case** Eindrehung *f* ~ **hook** Auszieh-kralle *f*, -zahn *m* ~ **socket** Ausblasstutzen *m* ~ **spring** Ausziehfeder *f* ~ **tooth** Ausziehzahn *m*

**extradite, to** ~ ausliefern

**extradition** Auslieferung *f* ~ **treaty** Auslieferungsvertrag *m*

**extrados** äußere Bogenfläche *f*, äußerer Rand *m*

**extrajudicial** außergerichtlich

**extraneous,** ~ **donation** außerordentliche Zuwendungen *pl* ~ **field** äußeres Feld *n*, Streufeld *n* ~ **items** außerordentliche Aufwendungen *pl*, externe Posten *pl* ~ **light** störendes Licht *n* ~ **matter** Fremdbestandteil *m* ~ **noise** Fremd-, Neben-geräusch *n* ~ **rust** Fremdrost *m*

**extranuclear,** ~ **electron** Hüllenelektron *n* ~ **fields** kernfremde Felder *pl* ~ **process** Prozeß *m* außerhalb des Kerns

**extraordinary** außerordentlich, außerplanmäßig ungewöhnlich ~ **charge** Nebenausgabe *f* ~ **component (of rays)** außerordentliche Komponente *pl* ~ **expense** Nebenausgabe *f* ~ **ray** außerordentlicher Strahl *m* ~ **wave** außerordentliche Welle *f*

**extrapolate, to** ~ extrapolieren

**extrapolated,** ~ **boundary** extrapolierte Reaktorbegrenzung *f* ~ **tangent** extrapolierte Tangente *f*

**extrapolation** Extrapolation *f*

**extraterrestrial,** ~ **disturbance** außerirdische Störung *f* ~-**space disturbances (or -space noise)** außerordentliche Störungen *pl*

**extraterritorial waters** Außengewässer *pl*

**extremal** Extremale *f* ~ **property** Extremaleigenschaft *f*

**extreme** außerordentlich, übermäßig, übertrieben ~ **emergency** äußerste Not *f* ~ **fiber stress** Randspannung *f* ~ **low water** Niedrigwasser *n* ~ **maximum diameter** Grenzmaß *n* ~ **position** Endlage *f*, Endstellung *f* (aviat.) ~-**pressure lubricant** Hochdruckschmiermittel *n* ~ **problem** Extremalproblem *n* ~ **range** Höchstschußweite *f* ~ **record performance** Höchstleistung *f* ~ **terms** äußere Glieder *pl* (Proportion) ~ **tip of the pole shoe** Polhorn *n* ~ **tolerance conditions** besonders ungünstige Abmaßsummierungsverhältnisse *pl* ~ **value** Extremwert *m*

**extremely,** ~ **high frequency (EHF)** Millimeterwellenfrequenz *f* ~ **light** ultraleicht ~ **sensitive** höchstempfindlich (chem.) ~ **small** winzig

**extremes** äußere Glieder *pl* (Proportion), Extreme *pl*

**extremity** Abschram *m* (min.), äußerstes Ende *n*, Ende *n*, Endpunkt *m*

**extricate, to** ~ freimachen, (from an entanglement) herauswickeln, losmachen **to** ~ **from a spin** herausfangen (aviat.)

**extrinsic,** ~ **conduction** Störleitung *f* ~ **load** äußere Belastung *f* ~ **semi-conductor** Störhalbleiter *m*, Störstellenhalbleiter *m*

**extrovert, to** ~ herausstülpen

**extrudable material** verpreßbares Material *n*

**extrude, to** ~ ausstoßen, herauspressen, pressen, spritzen, strangpressen

**extruded** gespritzt, stranggepreßt ~ **shapes** Preßprofil *n* ~ **sheathing** direkt umpreßter Mantel *m* ~ **spar** gezogener Holm *m* ~ **type of electrode** Preßmantelelektrode *f*

**extruder** Spritzmaschine *f*, Strangpresse *f* ~ **screws** Extruderschnecken *pl*

**extruding,** ~ **of cannister grain** Pressen *n* des Raketentreibsatzes ~ **die** Preßmatrize *f* (Strangpresse) ~ **press** Lochblechpresse *f* ~ **speed** Ausströmungsgeschwindigkeit *f*

**extrusion** Aufdornen *n*, Ausstoßung *f*, Pressen *n*, Spritzguß *m*, Stahlspritzen *n*, Strang- und Fließpressen *n*, Strangpreßprofil *n* ~ **compound** Strangpreßmischung *f* ~ **die** Preßmatrize *f*, Preßstempel *m* ~ **die for hot work** Warmziehring *m* ~ **machine** Schneckenpresse *f* ~ **molding** Fließpressen *n* ~ **press** Metallstangen-, Rohrzieh-, Stangen-, Zieh-presse *f* ~ **press for metal bars** Metallstangenpresse *f* ~ **press for tubes** Rohrpresse *f* ~ **press tools** Strangpreßwerkzeuge *pl* ~ **process** Stangenpressen *n* ~ **screw** Extruderschnecke *f*

**extrusive (or effusive) rocks** Ergußgesteine *pl*

**exudation** Ausscheidung *f*, Ausschlag *m*, Ausschwitzung *f*

**exude, to** ~ ausgeschieden werden, ausschlagen, ausschwitzen, sickern, spritzen

**ex works supplies** Werksablieferungen *pl*

**eye** Auge *n*, Bunzen *m*, Loch *n*, (of needle) Öhr *n*, Öse *f* ~ **(or inlet)** (of a centrifuge or compressor) Einlauf *m* einer Schleuderpumpe, Einlaufdüse *f* ~ **of connecting rod** Gabelöse *f*

~ of an eyebolt (or fork) Gabelauge *n* ~ of laminated springs Federbüchse *f* ~ of the millstone Loch *n* eines Mühlsteins ~ of a shaft Hängebank *f* ~ of side bar Kettenschenkelauge *n* (elektr.), Schenkelauge *n* ~ of the storm die Mitte *f* oder Auge *n* des Sturms ~ of the supercharger Ladereinlaß *m*
eye, ~ accommodation (or adaption) Augenanpassung *f* ~ bar Augenstab *m* ~ base Augenbasis *f*
eyebolt Augbolzen *m*, Augenschraube *f*, Gewindeöse *f*, Händelschraube *f*, Hebeauge *n*, Ösenbolzen *n*, Ösenschraube *f*, Ringbolzen *m*, Ringschraube *f*, Spannschloßöse *f*, Splintbolzen *m* ~ and key Schließbolzen *m* ~ with a forelock Schotbolzen *m* ~ pad Ösenbolzenflansch *m* ~ sleeve Augenbolzenmuffe *f*
eye, ~ catcher Blende *f*, Blickfang *m* ~-cutting pliers Kneifzange *f* ~-distance scale (binoculars) Teilscheibe *f* ~ dropper Tropfenzähler *m* ~ guard Augenmuschel *f* ~ hole Ösenloch *n* ~ hook Ösenhaken *m* ~ lens Augenlinse *f* ~-lens slide Okularauszug *m* ~ level Augenhöhe *f* ~-level finder Durchsichtssucher *m* ~-lid . . . kugelschalenförmig ~-lid elevator Sperrelevator *m* ~ nut Ring-, Ösen-mutter *f* ~'s own light Eigenlicht *n* des Auges, Augeneigenlicht *n*
eyepiece (gas mask) Augenfenster *n*, Einblick *m* (opt.), Lupe *f*, Okular *n*, Okularende *n*, Okularöffnung *f*, Schauglas *n* ~ with aspherical surface deformiertes Okular *n* ~ of long focal length langbrennweitiges Okular *n*, Okular *n* mit langer Brennweite

eyepiece, ~ attachment Okularansatz *m* ~ correcting lens Okularaufsteckglas *n* ~ cup Okularmuschel *f* ~ crossline micrometer Okularnetzmikrometer *n* ~ diameter Okularblende *f* ~ draw-tube extension Okularauszugsverstellbarkeit *f* ~ end Einblickkopf *m* (des Sehrohrs) ~ flange Augenring *m* ~ gasket Fensterring *m* ~ lens Einblick-, Okular-linse *f* ~ mark Okularmarke *f* ~ micrometer Okularmikrometer *n* ~ mount Einblickstutzen *m* ~ ocular disk (or body) Okularrevolverscheibe *f* ~ optical system Okularoptik *f* ~ registering surface Okularauflagefläche *f* ~ screw Okularschraube *f* ~ sleeve Okularschiebhülse *f*, Okularsteckhülse *f* ~ slit Okularspalt *m* ~ socket Einblickstutzen *m* ~ square pattern micrometer Okularnetzmikrometer *n* ~ stop Okularblende *f* ~ suppression filter Okularsperrfilter *n* ~ tube Einblickrohr *n*, Okularstutzen *m*
eye, ~ pit Augenpunkt *m* ~ point Augenpunkt *m* ~-pointed needle Lochnadel *f* ~-radiation apparatus Augenbestrahlungsgerät *m* ~ ring Ringöse *f* ~ rod Ösenstange *f* ~ screw Schraubenring *m* ~ shield Augenmuschel *f* ~ splice Augspleiß *m* ~ strain Augenanstrengung *f*
eyelet Anschlußöse *f*, Auge *n*, Gat(t) *n*, Gatjen *n*, Öhr *n*, Öse *f*, Ösenloch *n*, Schnürloch *n* ~ attachment Befestigungsöse *f* ~ hole Dachluke *f* ~ pliers Ösenzange *f* ~ spring Ösenhalter *m* ~ strip Ösenstreifen *m* ~ thimble Rundkausche *f*
eyeletting machine Öseneinsetzmaschine *f*
eyes Augen *pl* ~ of the drums (paper) Auslauf *m* der Waschtrommel

# F

**fabric** Bespannstoff *m*, Erzeugnis *n*, Gespinst *n*, Tuch *n*, Bespannung *f*, (airplane) Bespannungsstoff *m* ~ **for covering plane parts** Stoff *m* für die Bespannung ~ **of (gas) mask** Maskenstoff *m*
**fabric,** ~ **binding** Leinenumwicklung *f* ~ **cap** Leinwandkappe *f* ~ **cover** Stoffdeckel *m* ~**-covered** stoffbespannt ~ **covering** Stoffbespannung *f* ~ **dope** Cellon *n* ~ **porosity** Luftdurchlässigkeit *f* des Gewebes ~ **reinforcement** Bewehrungsnetz *n* (im voraus hergestelltes) ~ **ring** Gewebereifen *m* ~ **serving** Leinenumwicklung *f* ~ **shop** Bespannerei *f* ~ **tension (or tautness)** Stoffspannung *f* ~ **test** Stoffprüfung *f* ~ **thread tire** Kreuzgewebereifen *m* ~**-type filter** Trockenluftfilter *n*
**fabricate, to** ~ durch maschinelle Bearbeitung herstellen, fabrizieren, verarbeiten, verfertigen, zimmern, aus Teilen zusammensetzen
**fabricated piping** fertige Rohre *pl*
**fabricating operation** Fabrikationsvorgang *m*
**fabrication** Ausführung *f*, Erdichtung *f*, Erzeugung *f*, Fabrikation *f*, Fertigung *f*, Verfertigung *f* ~ **requirements** Fabrikationsanforderung *f*
**Fabry-Pérot interferometer** Luftplattenspektroskop *n*
**facade** Brust *f* ~ **wall** Fassade *f*
**face, to** ~ abfasen, abflachen, (in lathe) abflächen, auftragen, bekleiden, drehen, Enden schleifen (mech.), entgegentreten, (in lathe) flachdrehen, fräsen, (in lathe) freidrehen, plandrehen, planen, planfräsen, schlichten, Stirnfläche bearbeiten, (wall) verblenden, verkleiden, zukehren **to** ~ **about** kehrtmachen **to** ~ **grind** abstirnen
**face** Abbaufront *f*, Antlitz *n*, Aufsatzbacke *f*, Bahn *f*, Fläche *f*, Front *f*, Gesicht *n*, Jochleiste *f*, Kohlenstoß *m*, Kranz *m*, Ort *m*, Ortsstoß *m*, Rücken *m* (Fräserzahn), Scheibe *f*, Schirm *m* des Braun'schen Rohres, (value) Sitzfläche *f*, Stirnfläche *f*, Stirnseite *f*, Strosse *f*, Verblendung *f* **in the** ~ **of** angesichts on the ~ rechtsseitig ~ **to** ~ (petroleum) Baulänge *f*
**face,** ~ **of an arch** Bogenfläche *f* ~ **of the ascending step** Firstenstirn *f* ~ **of a building** Facade *f* eines Gebäudes ~ **and circular grinding machine** Plan- und Rundschleifmaschine *f* ~ **of column** Gleitfläche *f* der Säule ~ **with curved profile** Luftseite *f* mit Kurvenprofil ~ **of the face plate** Werkstückanlagefläche *f* der Planscheibe ~ **of flute** Spanfläche *f* ~ **of gear** Radbreite *f* ~ **of a housing slide** (tool equipment) Gleitbahn *f* des Ständers ~ **of joint** Dichtungsleiste *f*, Fläche *f* der Fuge, Liderungsfläche *f* ~ **of muzzle** Mündungsfläche *f* ~ **of the objective** Objektivoberfläche *f* ~ **of a rim** Kranzfläche *f* ~ **of a shaft** Schachtstoß *m* ~ **and side cutter** Schneidscheibe *f* ~ **of a slope** (railway cutting) Durchstichböschung *f*, Einschnittsböschung *f* ~ **of standard** (tool equipment) Gleitbahn *f* des Ständers ~ **of a tool** Schneid-

brust *f* ~ **of a tooth** Kopfflanke *f* eines Zahnes, Zahnkopf *m* ~ **of uprights** Gleitfläche *f* der Ständer ~ **of workings** Abbaustoß *m*
**face,** ~ **angle** Spannwinkel *f* (b. Räumnadel) ~ **bank** Stirnböschung *f* ~ **bearing emulsion** Emulsionsseite *f* (film) ~ **brick** Klinker *m* ~ **cam** achsial wirkender Daumen *m*, Plankurve *f* ~**-centered** flächenzentriert (cryst.) ~**-centered cubic lattice** kubisch flächenzentriertes Gitter *n* ~**-centered cubic structure** flächenzentrierte Struktur *f* ~**-centered lattice space** flächenzentriertes Raumgitter *n* ~ **connector** Stirnverbindung *f* ~ **contact** Kopfflankeneingriff *m* (mech.) ~ **conveyor** Abraum-Strossenband *n*, Kohlenrutsche *f* ~ **copying** Plankopieren *n* ~ **form** Maskenspanner *m* ~ **gear** Kron-, Kronen-, Plan-rad *n*; Zahnscheibe *f* ~**-gear guard** Stirnradschutzkappe *f* ~ **grinding chuck** Spannfutter *n* zum Planschleifen ~ **hammer** Bahnhammer *m* ~**-hardened** nitriert gehärtet ~ **height** Abtragshöhe *f* ~ **mill** Planfräser *m* ~ **milling** Planfräsen *n* ~**-milling cutter** Flächen-, Walzenstirn-fräser *m* ~ **page** Vorderseite *f* ~ **piece** Stoffteil *m* (d. Gasmaske) ~ **piece of respirator** Maskenkörper *m*
**face-plate** Auflagerblech *n*, Frontplatte *f* (TV), Planscheibe *f* (einer Drehbank), Richtscheibe *f*, Schirmträger *m* (CRT), Schutzplatte *f*, Spannfutter *n* ~**-controller** Rundzellenkontroller *m* ~ **jaw** Planscheiben-klaue *f*, -kloben *m*, Spannkloben *m* ~ **starter** Flachbahnanlasser *m* ~**-type tailstock** Reitstockplanscheibe *f* ~ **works** auf Planscheibe eingespannte Werkstücke *pl*
**face,** ~ **pressure** Flächenpressung *f* ~ **profile** Plankontur *f* ~ **relieving** stirnseitiges Hinterschleifen *n* ~ **screen** Gesichtsschirm *m* ~ **shield** Gesichts-maske *f*, -schutz *m* ~ **shield mat** (in film recording) Maske *f* ~ **silver coating** Oberflächenversilberung *f* ~ **silvering** Oberflächenversilberung *f* ~ **support** Strebausbau *m* ~**-up printing frame** Horizontalkopierrahmen *m* ~ **value** Gesichts-, Nenn-, Soll-wert *m* ~ **velocity** Anstrom-Geschwindigkeit *f* ~ **wheel** Kron-, Kronen-rad *n* ~ **width** Spurweite *f*, Zahnbreite *f* ~ **workings** Örterbau *m*
**faced** (on lathe) abgedreht ~**with clinkers** mit Klinkerverkleidung *f* ~ **end journal** Stirnzapfen *m*
**facet, to** ~ fazettieren
**facet** Abschrägung *f*, Anschrägung *f*, Fläche *f*, Schleifseite *f*, Schrägung *f*, Seitenfläche *f*
**facial,** ~ **brightness** Gesichtswert *m* ~ **suture** Gesichtsnaht *f*
**facilitate, to** ~ erleichtern
**facilitating filtration** Filtrationserleichterung *f*
**facilities** Einrichtungen *pl* ~ **for trade** Verkehrserleichterung *f*
**facility** Einrichtung *f*, Erleichterung *f*, Leichtigkeit *f* ~ **of inspection** Übersichtlichkeit *f* ~ **of operation** leichte Bedienung *f* ~ **of replacement** Auswechslungsfähigkeit *f*, Ersatzbequemlichkeit *f*

**facility,** ~ **program** Programmprüfanordnung *f*
(data proc.) ~ **requirements** Potentiale *pl*
**facing** Aufschlag *m*, Belagring *m*, Fassadenver-
kleidung *f*, Fräsen *n*, (foundry) Gießerschwärze
*f*, (of steel) Hobelspan *m*, Kupplungsscheibe *f*,
Plandrehen *n*, Planfräsen *n*, Schlichte *f*, Steintafel *f*, Verblendung *f*, Verkleidung *f*, Wendung
*f* ~ **of dam** Böschungsneigung *f* ~ **with fascines** Berauhwehrung *f* ~ **of natural stone**
Quaderverblendung *f*, Steinverblendung *f* ~
**the object** objektseitig ~ **of pitching** Außenfläche *f* ~ **of varying thickness** Verkleidung *f*
von abwechselnder Dicke ~ **of weir** Böschungsneigung *f*
**facing,** ~ **attachment** Plandrehvorrichtung *f* ~
**board** Blendholz *n* ~ **brick** Verblender *m*, Verblendstein *m* ~ **(or brick) clinker** Fassaden-
klinker *m* ~ **cut** Planschnitt *m* ~ **end of a rail**
Anlaufende *f* einer Schiene ~ **forward** vom
Fahrtwind beaufschlagt ~ **head** Plansupport *m*
~ **lathe** Kopf-, Plan-drehbank *f* ~ **page** Umschlagseite *f* (Buch) ~ **plastering** Edelputz *m*
~ **sand** Anlagesand *m*, feingesiebter Formsand
*m*, Modellsand *m* ~ **slab** Frontplatte *f* ~ **stone**
Verblendstein *m* ~ **tool** Plandrehwerkzeug *n*
**facings** Plandrehspäne *pl*
**facsimile** Bildtelegrafie *f*, Fernbildschrift *f*, Fernfotografie *f*, Nachbild *n* ~ **apparatus** Bildfunkanlage *f* ~ **apparatus radio** Bildanlage *f* ~
**equipment** Bildübertragungsgerät *n* ~ **posting
machine** Postenumdrucker *m* ~ **receiving office
(or station)** Bildempfangsstelle *f* (telegr.) ~-**record member (drum)** Aufzeichnungsträger *m*
(Bildtelegrafie) ~ **recorder** Faksimileschreiber
*m* ~ **set** Bildübertragungsgerät *n* ~ **signal level**
Faksimilesignalpegel *m* ~ **transceiver set** Bildsendeempfangsanlage *f* ~ **transient** Faksimile-
Einschwingvorgang *f* ~ **transient distortion**
Ausschwingungsverzerrung *f* ~ **transmission**
Bildfunk *m*, Bildübertragung *f* ~ **transmission
by electro-mechanical (or electrochemical)
means** elektromechanische oder elektrochemische Bildübertragung *f* ~ **transmitter** Bildsender *m*
**fact** Tatsache *f* ~-**finder** Prüfer *m*
**factice** Faktis *m*, Ölkautschuk *m*
**factor** Beiwert *m*, Beizahl *f*, Bestimmungsstück
*n*, Faktor *m*, Mehrer *m*, Moment *n*, Wertezahl
*f*, Wertziffer *f*, Zahl *f*, Ziffer *f*
**factor,** ~ **of correction** Berichtigungsfaktor *m*
~ **of demagnetization** Entmagnetisierungsfak-
tor *m* ~ **of effiency** Abminderungsfaktor *m* ~
**of expansion** Dehnungsfaktor *m* ~ **of maintenance** Unterhaltungsfrage *f* ~ **of merit** Güte-
faktor *m*, Kreisgüte *f* ~ **of probability** Wahr-
scheinlichkeitsfaktor *m* ~ **of quality** Kreisgüte
*f* ~ **of safety** Bausicherheit *f*, Sicherheits-grad
*m*, -koeffizient *m*, -zahl *f* ~ **of visibility** Sichtig-
keitsbeiwert *m*
**factor,** ~ **defining variation** Änderungsfaktor *m*
~ **group** Faktorgruppe *f*
**factorial** Faktorielle *f* ~ **function** Fakultät *f* ~
**magnification** Lupenvergrößerung *f*
**factorization** Faktorenzerlegung *f*
**factors,** ~ **inhibiting corrosion** korrosionshem-
mende Faktoren *pl* ~ **stimulating corrosion**
korrosionsfördernde Umstände *pl*
**factory** Betrieb *m*, Fabrik *f*, Fabrikgebäude *n*,

Fertigungshalle *f*, Handelsniederlassung *f*,
Werk *n*, Werkstatt *f* ~ **for constructing ma-
chines** Maschinenfabrik *f* ~ **and manufacturing
costs** Betriebskosten *pl* ~ **with scalding process**
Brühfabrik *f* **on a** ~ **scale** fabrikmäßig
**factory,** ~-**acceptance gauge** Revisionslehre *f*,
Werkstattabnahmelehre *f* ~ **airfield** Industrie-,
Werk-flugplatz *m* ~ **cost** Gestehungs-, Selbst-
-kosten *pl* ~ **costs** Fabrikkosten *pl* ~-**finished
fabrikfertig** ~ **guard** Werkschutz *m* ~ **hand**
Fabrikarbeiter *m* ~ **laws** (social legislation)
Arbeiterschutzgesetzgebung *f* ~ **length** Her-
stellungslänge *f* ~ **manager** Fabriksdirektor *m*,
Fabrikleiter *m* ~ **molasses** Rohzuckermelasse *f*
~ **order blank** Betriebsbestellschein *m* ~
**passive air-defense system** Werkluftschutz *m*
~ **price** Fabrikpreis *m* ~ **process** Herstellungsverfahren *n* ~ **railway** Industriebahn *f*, Werk-
bahn *f* ~ **regulation** Arbeitsordnung *f* ~ **selling
expenses** Werkvertriebskosten *pl* ~ **serial
number** Werknummer *f* ~ **standard** Werknorm
*f* ~ **stores** Fabriklager *n* ~ **test** Abnahme-,
Fabrik-messung *f* ~ **testing** Neueichung *f* ~
**yard** Fabrikhof *m*
**facts** Daten *pl* ~ **of a case** Tatbestand *m* ~ **of
the case** Sachverhalt *m*
**factual** tatsächlich ~ **findings** Tatbestand *m* ~
**report** Tatbestandaufnahme *f* ~ **situation**
Sach-lage *f*, -verhalt *m* ~ **statement** Tatbe-
standsaufnahme *f*
**faculty** Fakultät *f*
**fade, to** ~ abblenden, (ab)bleichen, (print) ab-
schwimmen, dahinschwinden, erblassen, verblassen, (grow pale) verbleichen, verschwinden
**to** ~ **away** schwinden, (sound) verhallen, verklingen, zerrinnen **to** ~ **in** aufblenden (film),
aufkommen lassen, einblenden **to** ~ **out** abklingen, ausblenden (film), ausschwingen, verklingen **to** ~-**over** überblenden **to** ~ **up and
down** überblenden
**fade,** ~ **down** Pegel *m* langsam absenken und
halten ~-**in** Aufblendung *f* ~-**in and mixing of
microphones** Einblenden *n* der Mikrofone,
Mikrofoneinblendung *f* ~-**in**-~-**out device** Ton-
überblendungseinrichtung *f* ~-**out** Abblendung
*f*, Ausblenden *n* (electr.), (of radio signals)
Funkschwund *m* ~-**out control** Tonumschalter
*m* ~-**out time** Abklingzeit *f* ~-**over (in)** weiche
Überblendung *f* ~-**reducing antenna** schwundmindernde Antenne *f* ~-**up** Pegel *m* langsam anheben und halten
**faded** fahl ~ **parts** verschossene Stellen *pl*
**fadeometer** Farbenlichtechtheitsprüfer *m*
**fader** Attenuator *m*, Lautstärkenregler *m*,
Mischer *m*, Regler *m*, Tonblende *f*, Überblender *m* (film) ~ **setting** Reglerstellung *f*
**fading** Abklingen (photo), Abklingen einer
Schwingung (rdo), Fading *n*, Funkschwund *m*,
Intensitätsausfall *m*, allmähliche Schichtzer-
setzung *f* (photo), Schwund *m*, Schwunderscheinung *f*, (effect) Schwundwirkung *f*, Störbefreiung *f*, Überblendung *f*, Verschwinden *n*
~ **of radio signals** Schwächerwerden *n* der
Radiosignale
**fading,** ~ **amplifier** Überblendverstärker *m* ~
**control** automatische Schwundregelung *f*,
Überblendungseinrichtung *f* ~-**control unit**
Übergangsschaltung *f* (film) ~ **device** Tonum-

schaltungseinrichtung *f* ~ **effect** Fadingeffekt *m*, Schwinderscheinung *f* ~ **hexode** Fadinghexode *f*, Regelröhre *f*, Sechspolregelröhre *f*, Viergitterregelröhre *f* ~**-mix hexode** Regelmischröhre *f*, Viergitterregelmischröhre *f* ~ **period** Schwächungs-, Schwund-periode *f* ~**-reducing antenna** schwundmindernde Antenne *f*

**fadings** Lautstärkeschwankungen *pl*

**fag end** (of rope) Kuhschwanz *m*

**faggot, to** ~ paketieren

**faggot** paketierte Luppeneisenstäbe *pl*, Schweißeisen-, Schweiß-paket *n* ~**-type cooler** Reisiggradierwerk *n*

**faggoting** Paketierung *f*, Paketierverfahren *n*

**faggots** Reisig *n*

**fagot** Faschine *f*, Garbe *f*, (for smelting) Gespann *n*, Holzbündel *n*, Rohschienenpaket *n*, Schienenpaket *n* ~ **of steel bars** Bundstahl *m*

**fagot,** ~ **fitting** Faschinenausfüllung *f* ~ **magnet** Magnetbündel *n* ~ **steel** Bundstahl *m* ~ **furnace** Alteisenofen *m*

**fagoting** Schienenpaketierung *f*

**Fahrenheit scale** Fahrenheitsskala *f*

**fail, to** ~ aussetzen, brechen, fallieren, fehlen, fehlgehen, fehlschlagen, mißlingen, scheitern, unterlassen, versagen **to** ~ **altogether** gänzlich ausfallen **to** ~ **to come** ausbleiben **to** ~ **to** verabsäumen

**fail-safe** ausfallsicher; Selbstschutz *m* ~ **circuitry** (or **monitor**) Ausfallüberwachungsschaltung *f* ~ **servo disabling** Sicherheitskreis *m* für Servoblockierung (g/m)

**failed piece** mißratenes Stück *n*

**failing** mangels, in Ermangelung von ~ **of an embankment** (or **dike**) Deichbruch *m*

**failure** Ausfall *m*, Aussetzen *n*, Aussetzer *m*, Bruch *m*, Falliment *n*, Fallissement *n*, Fehlen *n*, Fehler *m*, Fehlstellen *n*, Konkurs *m*, Mißerfolg *m*, Unzuträglichkeit *f*, Untergehen *n*, Verfall *m*, Versagen *n*, Versager *m*, Zerstörung *f*, Zurückgehen *n*, Zusammenbruch *m*

**failure,** ~ **in aircraft wiring** Bordnetzausfall *m* (electr.) ~ **of audible ringing signal** kein Ruf *m* ~ **of communication** Ausfall *m* der Fernmeldeverbindung ~ **to deliver** Nichtlieferung *f* ~ **to drill oil** Fehlbohrung *f* ~ **to fire** (rectifier, ignitron) Fehlzündung *f* ~ **to ignite** (rectifier, ignitron) Fehlzündung *f* ~ **of oscillation** Aussetzen *n* der Schwingungen ~ **of pressure** Aussetzen *n* des Druckes ~ **to release** Auslösefehler *m* ~ **to start** Anlaßversager *m* ~ **of a wall** Mauereinsturz *m*

**failure,** ~ **condition** Bruchbedingung *f* ~ **load** Bruchdehnung *f*, Reißlast *f* ~ **rate** Ausfallquote *f* ~ **signal** Ausfallsignal *n* ~ **warning light** Ausfallwarnleuchte *f*

**faint, to** ~ abbauen, das Bewußtsein verlieren, ohnmächtig werden

**faint** ohnmächtig, schwach, undeutlich ~ **luminosity** lichtschwache Erscheinung *f* ~ **negative** flaues Negativ *n*

**faintness** Schwäche *f*

**faints** Nachlauf *m*

**fair, to** ~ abschärfen, verkleiden (aviat.) **to** ~ **into the fuselage** glatt in den Rumpf *m* übergehen lassen

**fair** Ausstellung *f*, Mustermesse *f*; angemessen,

anständig ~ **faced plaster** Glattputz *m* ~**-lead** Führungsrolle *f*, (naut.) Rollklampe *f*, (naut.) Verholklampe *f* ~ **price** annehmbarer Preis ~ **wear and tear** normaler Verschleiß *m*

**faired** aerodynamisch günstig verkleidet, windschnittig ~ **curves** durch streuende Versuchspunkte gelegte berichtigte Kurven *pl* ~ **undercarriage** verkleidetes Fahrgestell *n*

**fairing** Füllstück *n*, Profilierung *f*, Profillagerholz *n*, Umkleidung *f*, (aviat.) Verkleidung *f*, Verkleidungsübergang *m* ~ **of telescopic leg** Federstrebenverkleidung *f*

**fairing,** ~ **former** Verkleidungsaussteifung *f* ~ **joggle** Absetzung *f* in der Vertikalverkleidung (Flugz.) ~ **plate** Verkleidungsblech *n* ~ **wire** Formhaltungsdraht *m*

**fairlead** Halterung *f*, Leitungshalterung *f* ~ **support** Mehrfachschelle *f*

**fairway,** ~ **channel** Fahrwasser *n* ~ **buoy** Ansegelungstonne *f*, Warnboje *f*

**faithful** (in amplitude or phase of signals) form-gerecht, -getreu, (reproduction) genau, getreu, gläubig, sinngetreu, treu ~ **reproduction** genaue oder getreue Wiedergabe *f*

**faithfulness** Genauigkeit *f*, Naturtreue *f*, Richtigkeit *f* ~ **of reproduction** Natürlichkeit *f* der Wiedergabe, Wiedergabequalität *f*

**fake echo** Täuschecho *n*

**fall, to** ~ entfallen, fallen, reißen, stürzen, verfallen, (wind) sich legen **to** ~ **apart** auseinandergehen **to** ~ **away** in Fortfall oder in Wegfall kommen **to** ~ **back** ausweichen, weichen, zurückgehen, zurückweichen, sich zurückziehen **to** ~ **back upon a prepared position** auf eine rückwärtige Stellung *f* zurückgehen **to** ~ **behind** in Rückstand kommen **to** ~ **below** unterschreiten **to** ~ **calm** sich legen **to** ~ **down** einbrechen, herunterstürzen, niederbrechen, niederfallen, zusammenstürzen, in Verzug sein **to** ~ **due** fällig werden **to** ~ **in** einbrechen, einfallen, einklinken, einschnappen, einstürzen, (min.) zu Bruche gehen **to** ~ **into** einfallen **to** ~ **to leeward** seewärts abtreiben **to** ~ **into a notch** einklinken **to** ~ **off** abfallen, abnehmen, seitlich abrutschen, in Fortfall *m* kommen, (in value) an Größe *f* oder Wert *m* verlieren, in Wegfall *m* kommen, sinken, vermindern **to** ~ **out** ausfallen, ausscheren **to** ~ **overboard** über Bord fallen **to** ~ **to pieces** auseinanderfallen; verstürzen **to** ~ **into the sea** ins Meer *n* stürzen **to** ~ **short** unterschreiten, zurückbleiben **to** ~ **shut** zuklappen **to** ~ **into a spin** ins Trudeln *n* kommen **to** ~ **in step** in Tritt kommen **to** ~ **out of a step** außer Tritt fallen **to** ~ **through** durchfallen **to** ~ **upon** befallen, einfallen

**fall** Abdachung *f* (Gelände), Abfall *m*, Abfallen *n*, Abnahme *f*, Abschwächung *f*, Absturz *m*, Abwärtsgang *m* (aviat.), Abwärtsgehen *n*, Einfall *m*, Fall *m*, Gefälle *n*, Herbst *m*, Neigung *f*, Niedergang *m*, Sinken *n*, Sturz *m*, Verminderung *f*

**fall,** ~ **of country** Gebirgsschlag *m* ~ **of current** Stromabnahme *f* ~ **of potential** Potential-abfall *m*, -fall *m* ~ **of a river** Stromgefälle *n* ~ **of rope** Seilstrang *m* ~ **of a ship's deck** Einsenkung *f* ~ **in tension** Abnahme *f* der Spannung (electr.) ~ **of a wall** Mauereinsturz *m* ~ **of water** Fall *m*,

(on weir) Stauhöhe *f* ~ **of the water** Fallen *n* des Wassers ~ **of weir** Fall *m*
**fall,** ~ **back circuit** Hilfskreis *m* ~**-back possibility** Betriebsfortsetzungsmöglichkeit *f* ~ **breaking** fallbremsbar ~ **fit** Gewindepassung *f* ~ (of water) **head** Fallhöhe *f* ~**(ing)-off** Sinken *n* ~**-out** radioaktiver Niederschlag *m* ~**-out disposal** Abfallbeseitigung *f* ~ **streaks (or streamer)** Fallstreifen *pl* (clouds) ~ **time** Abfallzeit *f* ~ **tubes** Fallröhren *pl* ~ **wind** herunterwehender Wind *m* (aviat.)
**fallacious argument** Trugschluß *m*
**fallacy** Trugschluß *m*
**faller,** ~ **motion** Aufwinderbewegung *f* ~ **shaft** Aufwindewelle *f* ~ **washing machine** Stampfwaschmaschine *f* ~ **wire** Aufwindedraht *m*
**falling** Niedergehen *n*; fallend ~**-ball method** (viscometry) Kugelfallmethode *f* ~ **ball viscosimeter** Kugelfallviskosimeter *n* ~ **body** freier Fall *m* ~ **characteristic** fallende Charakteristik *f* ~ **disk** Fallscheibe *f* ~ **drop indicator** Falltropfenanzeiger *m* ~ **head** mit abnehmendem Wasserdruck *m* ~**-in** Einfall *m*, Verbruch *m*, Zubruchgehen *n* ~**-off** (current) Abfall *m*; abfallend ~**-off burner** Abschmelzbrenner *m* ~**-off production** abnehmende Erzeugung *f* ~ **out** Außertrittfallen *n* ~ **out-of-step** Außertrittfallen *n* ~ **latch** Fallriegel *m* ~ **leaf** welkes oder fallendes Blatt *n* (aviat.) ~ **roller gate** versenkbare Walze *f* ~ **shoulder** abfallender Absatz *m* ~ **slope** abfallendes Gelände *n* ~ **sluice** Regulierfallenzug *m* ~ **speed** Fallgeschwindigkeit *f* ~ **sphere viscosimeter** Kugelfallviskosimeter *n* ~ **time** Fallzeit *f* ~ **time of bomb** Bombenfallzeit *f* ~ **tup** Fallbär *m* ~ **velocity** Fall- oder Sinkgeschwindigkeit *f* ~ **water** Fallwasser *n* ~ **weight** Fallgewicht *n*, (explosives) Fallhammer *m* ~**-weight test** Fallgewichtsversuch *m*, (explosives) Fallhammerprobe *f*, Fall-, Schlag-probe *f*, Schlagversuch *m*
**fallow** fahl, falb
**falls** Fälle *pl*
**false** falsch, irrig, nachgemacht, unwahr ~ **alarm** blinder Alarm *m* ~ **anchorage** Blockierung *f*, falsche Radstellung *f* ~ **bedding** Diagonal-, Kreuz-schichtung *f* ~ **bottom** doppelter Boden *m*, Formboden *m*, Zwischenboden *m* ~ **ceiling** Einschub *m*, Fehlboden *m*, Doppeldecke *f* ~ **channel** blinder Arm *m*, Außendeichsland *n*, Groden *m* ~ **cirrus** falscher Zirrus *m* ~ **collar** Halsbinde *f* ~ **conclusion** Trugschluß *m* ~ **cone of silence** falscher Lautlosigkeitskegel *m* ~ **echoes** Falschechos *pl* (rdr) ~ **flames** bengalisches Feuer *n* ~ **flash** Nebenblitz *m*, Reflex *m* ~ **floor** Fehlboden *m*, Zwischen-boden *m*, -decke *f* ~ **frame** Hilfs-joch *n*, -kranz *m* (min.), (building construction) Notjoch *n* ~ **front** Hintersetzer *m* ~ **grain** Feinkorn *n* ~ **grain formation** Feinbildung *f* ~ **image** Scheinbild *n* ~ **jaws** Einsatzbacken *pl* ~ **keel** Loskiel *m* ~ **key** Dietrich *m*, Nachschlüssel *m* ~ **leader** Einsetzführung *f* ~ **light** (opt.) Irrstrahl *m*, Nebenlicht *n* ~ **lock** blindes Schloß *n* ~ **mildew** falscher Mehltau *m* ~ **mold** falsches Teil *n* ~ **ogive** Zünderdeckel *m* ~ **ogive type of shell** Haubengeschoß *n* ~ **panel** blinde Füllung *f* ~ **pass** blindes Kaliber *n* ~ **pattern** Störmuster *n* ~ **report** Ente *f*, Tatarennachricht *f*

~ **ring** Falschanruf *m* ~ **rib** Hilfsrippe *f*, Nebenrippe *f*, falsche Rippe *f* ~ **roof work** Doppeldach *n* ~ **sap** Wasserstreifen *m* ~ **sap ring** Mondring *m* ~ **seizure** Fehlbelegung *f* (teleph.) ~ **spar** falscher Holm, Zwischenholm *m* ~ **start** Fehlstart *m* ~**-start call** Fehlanruf *m* ~ **start release** Fehlstartentriegelung *f* ~ **wing rib** Hilfsflügelrippe *f* ~ **work** Lehrgerüst *n*, Stützwerk *n*
**falsificate** Falsifikat *n*
**falsification** Fälschung *f*, Urkundenfälschung *f*
**falsify, to** ~ fälschen
**faltboat** Faltboot *n*
**falter, to** ~ stocken
**familiar** vertraulich, vertraut ~ **with an area (or with a locality)** ortskundig **to get** ~ **with** einarbeiten
**familiarization** Einweisung *f* ~ **flight** Einweisungsflug *m*
**familiarize, to** ~ vertraut machen
**family** Familie *f* ~ **of algebraic loci** Schar *f* algebraischer Gebilde ~ **of characteristics** Kennlinien-bündel *n*, -feld *n*, -schar *f* ~ **of curves** Kennlinien-bündel *n*, -feld *n*, Kurvenschar *f* ~ **of graphs** Kurvenschar *f* ~ **of plane surfaces** Ebenenschar *f*
**fan, to** ~ fächeln **to** ~ **out** auffächern, (a cable) aufspleißen, (storage cell plates) Auswüchse bilden, fächerförmig spleißen, seitlich einschwenken, wachsen
**fan** Bastler *m*, Bläser *m*, Fächer *m*, Flügelgebläse *n*, Gebläse *n*, Lüfter *m*, Ventilator *m*, Wedel *m*, Wetterrad *n*, Windzeuger *m*, Windfächer *m*, Windmaschine *f*, Zielspinne *f* ~ **with four whiskers** Flügelmutter *f* mit Stoßarmen
**fan,** ~ **air** Gebläsestrahl *m* ~ **antenna** Harfen-, Fächer-antenne *f* ~ **baffle** Ventilatorflügel *m* ~ **base with tilting device** Ventilatorfuß *m* mit Gelenk *n* ~ **bearing** Lüfterlagerung *f* ~ **belt** Lüfter-, Ventilator-riemen *m* ~ **blade** Gebläseflügel *m*, Windflügel *m*, Windflügelblech *n* ~ **blast** Gebläse-, Ventilator-wind *m* ~ **blower** Flügelgebläse *n*, Schleuderrad *n*, Ventilatorgebläse *n* ~**-blower mixer** Schleuderradmischer *m* ~ **boss** Windflügelnabe *f* ~ **bracket (or carrier)** Lüfterträger *m* ~ **brake** Bremsluftschraube *f*, Flügelbremse *f*, Klatsche *f*, Luftbremse *f* ~**-brake blade** Luftbremsflügel *m* ~ **casing** Ventilatorgehäuse *n* ~**-cooled** außenbelüftet (Motor) ~ **cooler** Ventilatorkühler *m* ~ **cooling** Luftkühlung *f* ~ **coupling** Lüfterkupplung *f* ~ **drive** Ventilatorantrieb *m* ~ **drive pulley** Lüfterantriebsriemenscheibe *f* ~**-driving pulley** Windflügelriemenscheibe *f* ~ **engine** Fächermotor *m* ~ **fade-in** Fächeraufblendung *f* ~**-fading shutter** Fächerblende *f* ~**-felt fastener** Ventilatorriemenschloß *n* ~ **fitter** Fächerzurichter *m* ~**-forced heater** Heizlüfter *m* ~ **governor** Windfangregler *m* ~ **guard** Ventilatorschutzring *m* ~ **housing** Ventilatorgehäuse *n* ~ **hub** Nabe *f* des Windflügels, Windflügelnabe *f* ~ **inlet** Ventilatoreinströmöffnung *f* ~**-light shutter** Schalterladen *m* ~ **marker** Fächerfunkfeuer *n* (rdo nav.) ~ **marker beacon** Fächerfunkfeuer *n* ~ **outlet** Gebläseöffnung *f* ~ **pressure** Ventilatorpressung *f* ~ **pulley** Ventilatorriemenscheibe *f* ~ **ring** Wind-

flügelring *m* ~ **running noise** Lüftergeräusch *n* ~ **screw for defroster** Luftschraube *f* zum Entfroster ~ **shaft** Lüfterwelle *f* ~**-shaped antenna** Fächerantenne *f* ~**-shaped fold** Fächerfalte *f* ~**-shaped gate** Fächertor *n* ~ **shell** Lüfterhaube *f*, Ventilatortrichter *m* ~ **shroud** Lüfterverkleidung *f* ~ **support** Windflügelstütze *f* ~**-tail burner** Fächerbrenner *m* ~ **test rig** Ventilatorprüfstand *m* ~ **turn** Fächerturn *m* (aviat.) ~**-type combustion chamber** Fächerkammer *f* ~ **wheel** Flügelrad *n* ~**-wheel anemograph** Windrad-Anemograf *m* ~**-wheel anemometer** Flügelradanemometer *n* ~**-wheel blower** Gebläserad *n* ~ **wind** Lüfterwind *m*

**fancy,** ~ **articles in cardboard** Pappgalanteriewaren *pl* ~ **cloth** dessiniertes Gewebe *n* ~ **frame** Fantasieeinfassung *f* ~**-goods turner** Kunstdrechsler *m* ~ **leather** Taschnerleder *n* ~ **lighting** Effektbeleuchtung *f* ~ **line** Zierlinie *f* (print.) ~ **paper** Ausstattungspapier *n* ~ **roller** Aushebewalze *f* ~ **thread** Zierfaden *m* ~ **threader** Drahtumspinner *m*

**fanfare set** Fanfarenanlage *f*

**fang head** Kerbkopf *m* (mech.)

**fanned beam antenna** Fächerstrahlantenne *f*

**fanning** Anfachung *f*, Aufdrehen *n*, Auffächern *n* ~**-out** Ausbreitung *f*, Wachsen *n* ~ **strip** Kamm *m* (am Lötösenstreifen)

**fans connected in series** hintereinander geschaltete Ventilatoren *pl*

**far** fern, weit **as** ~ **as** bis **by** ~ bei weitem **so** ~ bisher **so** ~ **as** sofern

**far,** ~**-end cross talk** Fernnebensprechen *n*, Gegennebensprechen *n* (teleph.) ~**-end crosstalk attenuation** Gegennebensprechdämpfung *f* ~ **field** Fernfeld *n* ~ **point** Fernpunkt *m* (Stereoaufnahmen) ~**-reaching** umfassend, weit ausgreifend, weitgehend, weitreichend ~ **sighted** übersichtig, weitblickend, weitsichtig ~ **sighted eye** fernsichtiges oder weitsichtiges Auge *n* ~ **sightedness** Weitsichtigkeit *f*, Übersichtigkeit *f*

**farad** Farad *n* ~ **bridge** Kapazitätsbrücke *f* ~ **meter** Kapazitätsmesser *m*

**Faraday,** ~ **cell** elektrooptisches Lichtrelais *n* ~ **collector** Faradayscher Käfig *m* ~ **cylinder** Faradaysches Gefäß *n* ~ **dark space** Faradayscher Dunkelraum *m* ~ **effect** magneto-optische Drehung *f* ~ **ice pail** Faradaysches Gefäß *n* oder Käfig *m* ~**'s law** Induktionsgesetz *n* ~**'s law of electrolysis** Faradaysche Gesetze *pl* ~ **screen (or shield)** Faradayscher Käfig *m*

**faradic** faradisch ~ **current** faradischer Strom *m*

**faradization** Faradisation *f*

**fare, to** ~ abschneiden **car** ~ **for workers living far from place of work** Weggeld *n*

**Farewell rock (or sand)** Farewell Sand *m*

**farina** Kartoffelstärke *f*

**farinaceous** mehlartig

**farm, to** ~ pachten

**farm** Bauerngehöft *n*, Gehöft *n*, Meierei *f* (dairy) Meierhof *m*, Pachthof *m* ~ **building** Wirtschaftsgebäude *n* ~ **hand** Knecht *m* ~ **implements** Ackergeräte *pl* ~ **laborer** Feldarbeiter *m* ~ **land** Ackerland *n* ~ **wagon** Acker-, Bauern-wagen *m*

**farmer** Bauer *m*, Landwirt *m*, Pächter *m* ~ **line** Farmerleitung *f*

**farming** Acker-, Land-bau *m*

**farmyard** Hof *m*, Meierhof *m*, Wirtschaftshof *m*

**Farnsworth,** ~ **centers** Farbzentren *pl* ~ **dissector of electron camera** Bildabtasterröhre *f* mit mechanischer Blende ~ **image dissector tube** Farnsworthsche Bildfangröhre *f* ~ **multiplier tube** Farnsworthscher Vervielfacher *m*

**farrier** Hufschmied *m*

**fascia** Stirnbrett *n*

**fasciculation** Konzentrieren *n*

**fascine** Faschine *f*, Strauchbündel *n* ~ **dam** Faschinendamm *m*, Senkstück *n* ~ **dam sunk by means of stones** Senkfaschine *f* ~ **filling** Faschinenausfüllung *f*, Packwerk *n* ~ **footpath** Faschinenbahn *f* ~ **mattress** Sinkstück *n* ~ **poles** Wurst *f* ~**-reveted dam** Senkfaschinendamm *m* ~ **revetment** Faschinen-bekleidung *f*, -verblendung *f* ~ **road** Faschinenbahn *f* ~ **trestle** Faschinenbock *m* ~ **work** Faschinen-lage *f*, -packwerk *n*, -werk *n* (hydraul.)

**fascines** Busch-, Faschinen-packung *f*

**fash** Bart *m*, Gußnaht *f*

**fashion, to** ~ bilden, (or shape in outline) façonnieren

**fashion** Art *f* und Weise *f*, äußere Form *f* **out of** ~ unmodern ~ **parts** Façonstücke *pl*

**fashioned, capability of being** ~ (by pressure and/or heat application) Bildsamkeit *f* ~ **bar iron** Fassoneisen *n* ~ **wire** Fassondraht *m*

**fashioning** Formgebung *f*, Gestaltung *f* ~ **by machining with removal of material by cutting tools** spanabhebende Formgebung *f*

**fast** befestigt, (colors) beständig, dauerhaft, (colors) echt, fest, feststehend, flott, geschwind, haltbar, (color) lichtecht, schnell, waschecht **not** ~ **color** unechte Farbe *f* **of** ~ **color** farbecht

**fast,** ~ **to acid milling** säurewalkecht ~ **to boiling** kochecht ~ **to crabbing** crabbecht ~ **to fulling** walkecht ~ **to light** lichtecht ~ **and loose escapement** Gesperr *n* mit fester und lose Klinke ~ **and loose pulley** Fest- und Losscheibe *f*, Fest- und Leerlaufscheibe *f* ~ **to scraping** scheuerfest

**fast,** ~ **access storage** Schnellspeicher *m* ~ **breeder** schneller Brutreaktor *m* ~ **charger** Schnellader *m* ~ **charging current** Schnellladestrom *m* ~ **chopper** Schnellunterbrecher *m*, Schnellzerhacker *m*, mit hoher Geschwindigkeit arbeitender periodischer Unterbrecher *m* (nucl.) ~ **color base** Echtbase *f* ~**-curing molding compound** Schnellpreßmasse *f* ~ **cycling injection machine** Schnelläufer-Spritzgießmaschine *f* ~**-dyed** farbecht ~ **dyeing** Fixfärberei *f* ~ **effect** Schnellspaltungseffekt *m* ~ **fission** Schnellspaltung *f* ~ **fission effect** Schnellspaltungsfaktor *m* ~ **fission reactor** Schnellneutronenreaktor *m* ~**-flowing** schnellfließend ~ **forward** Schnellvorlauf *m* (tape rec.), rascher Vorlauf *m* ~ **fragment** schnelles Bruchstück *n* ~ **landing airplane** schnell landendes Flugzeug *n* ~ **lens** lichtstarkes Objektiv *n* ~ **mordant dyestuff** Echtbeizenfarbstoff *m* ~ **motion effect** Zeitraffung *f* (film) ~ **neutron** schnelles Neutron *n* ~ **neutron fission** Spaltung *f* durch schnelle Neutronen ~ **neutron range** Bereich *m* der schnellen Neutronen ~ **neutron reactor** schneller Reaktor *m* ~ **objective** lichtstarkes Objektiv *n* ~ **printing color** Echtdruckfarbe *f*

**~ profile** Schnellprobe *f* **~ pulley** Fest- oder Vollscheibe *f* **~ register** Register *n* (info proc.) **~ reverse** rascher Rücklauf *m* **~ rewind** Schnellrücklauf *m* (tape rec.) **~-running** schnellgehend **~ stonehead** festes Gestein *n* unter dem Abraumgebirge **~ store** Schnellspeicher *m* **~ time constant (FTC)** Enttrübungsschaltung *f* (rdr) **~ tractor** Schnellzugmaschine *f* **~ train** Eilzug *m* **~ vehicle** Flitzwagen *m* **~ welding** Kurzzeitschweißung *f* **~ white** Permanentweiß *n* **~ wind** Umspulen *n* des Bandes (tape rec.) **~-working furnace** Ofengang *m*

**fasten, to ~** anheften, (screw) anschrauben, anstechen, anstecken, befestigen, festbinden, festlegen, festmachen, (draw fast) festziehen, heften, sichern, verfestigen, verkeilen, vernieten **to ~ with adhesive** ankleben **to ~ with blows** festschlagen **to ~ with cords** beschnüren **to ~ with hooks** anhaken **to ~ with iron bands** anhaspen **to ~ a lock** (to a door) ein Schloß *n* anschlagen **to ~ with pales** anpfahlen, verpfählen **to ~ with piles** pfählen **to ~ with screws** festschrauben **to ~ with stakes** anpfahlen, pfählen, verpfählen **to ~ with a wedge** ankeilen **to ~ by wedges** festkeilen

**fastened** befestigt, eingeklemmt

**fastener** Befestigungselement *n*, Befestigungsmittel *n*, Halter *m*, (slide) Reißverschluß *m*, Wirbel *m* **~ setter** Ösenschlager *m*

**fastening** Anknüpfung *f*, Befestigung *f*, Einklemmen *n*, Haken *m*, Riegel *m*, Schloß *n*, Verband *m*, Verfestigung *f*, Verschluß *m*

**fastening, ~ angle** Befestigungswinkel *m* **~ the chairs to the ties** Aufsetzen *n* der Stühle auf Querschwellen **~ cord** Befestigungsschnur *f* **~ fishplates to the ends of the rail** Befestigung *f* der Laschen an den Schienenenden **~ hole** Befestigungsloch *n* **~ iron** Moniereisen *n* **~ (or fixing) lug** Befestigungslappen *m* **~ means** Befestigungsvorrichtung *f* **~ part (or fitting)** Befestigungsteil *m* **~ pin** Befestigungsnadel *f* **~ place** Anknotstelle *f* **~ plate** Befestigungs-, Druck-platte *f* **~ point** Befestigungspunkt *m* **~ possibility** Befestigungsmöglichkeit *f* **~ rail** Anschraubleiste *f* **~ the rails to the sleepers with bolts and spikes** Nagelung *f* der Schienen **~ ring** Halte-, Schließ-ring *m* **~ screw** Befestigungsschraube *f* **~ spring for head lamp** Bügelfeder *f* für Scheinwerfer **~ strip** Befestigungsstreifen *m*

**fastenings** Schließbeschlag *m*

**faster, ~ lens** Objektiv *n* mit höherer Anfangsöffnung **~ moving** schneller bewegt

**fastidious** anspruchsvoll, wählerisch, schwer zufriedenzustellen

**fastness** (dye) Echtheit *f* **~ to buffing** Schleifechtheit *f* **~ of color** Farbbeständigkeit *f* **~ to friction glazing** Stoßechtheit *f* **~ to hot pressing** Bügelechtheit *f* **~ to light** Lichtbeständigkeit *f* **~ to lime** Kalkechtheit *f* **~ to processing** Fabrikationsechtheit *f* **~ to storing** Lagerbeständigkeit *f* **~ to vulcanizing** Vulkanisierechtheit *f*

**fastness, ~ demanded (or required)** Echtheitsanspruch *m* **~ grade** Echtheitsgrad *m*

**fat, to ~ liquor** lickern

**fat** (of clay) Fett *n*, Schmiere *f*, Talg *m*; fett,

fettig, ölhaltig, ergiebig **~ of coconut-oil and palm-kernel-oil group** Leimfett *n*

**fat, ~ boiler** Fettkessel *m* **~ clay** Letten *m* **~ cleaving** fettspaltend **~ content** Fettgehalt *m* **~ coal** Back-, Fett-kohle *f*, fette Kohle *f* **~ diagram** völliges Diagramm *n* **~ dipole** dicker Dipol *m* **~ dissolving soap** Fettseife *f* **~ extraction apparatus** Entfettungsapparat *m* **~ hardening** Fetthärtung *f* **~ knife** Fettstecher *m* **~ lime** Fettkalk *m* **~ liquor** Fettschmiere *f* **~-liquoring process** Lickerverfahren *n* **~ mixture** fette Mischung *f* **~ resin resist** Fettharzreserve *f* **~ soluble black** Fettschwarz *n* **~ solvent** Fettlöser *m* **~ tight** fettdicht **~-wrinkle crease** Mastfalte *f*

**fata morgana** Luftspiegelung *f*

**fatal** tödlich **~ accident** tödlicher Unfall *m* **~ crash** Todessturz *m*

**fathom, to ~** eine Tiefe *f* abmessen, ergründen, klaftern, sondieren

**fathom** Faden *m*, Fadon *n*, Klafter *n*, Lachter *f* (min.), Längen- und Tiefenmaß *n* **~ work** Längengedinge *n*

**fathometer** Behm-Lot *n*, Echolot *n*, Fadenmeter *n*

**fatigue, to ~** altern, ermüden

**fatigue** Abschwächung *f*, Erlahmung *f*, Ermüdung (metal), Strapaze *f*, Zermürberscheinung *f* **~ of material** Material-, Werkstoff-ermüdung *f*

**fatigue, ~ bend test** Dauerschlagbiegeversuch *m* **~ bending** Dauerbiegung *f* **~ bending machine** Dauerbiegemaschine *f* **~ bending test** Dauerbiegeversuch *m* **~ bendingtest specimen** Dauerbiegeprobe *f* **~ breakdown** Ermüdungsbruch *m* **~ crack** Dauerbruch *m*, Ermüdungsriß *m* **~ crack(ing)** Dauerriß *m* **~ detail** Arbeitsabteilung *f* **~ failure** Dauerbruch *m*, Ermüdungsbruch *m* **~ fracture** Dauerbruch *m* **~ impact strength** Dauerschlag-festigkeit *f*, -versuch *m* **~-impact-test machine** Dauerschlagwerk *n* **~ inhibitor** Ermüdungsschutzmittel *n* **~ life** Dauerfestigkeit *f* **~ limit** Arbeits-, Dauer--festigkeit *f*; Dauerfestigkeitsgrenze *f*, Dauerhaltbarkeit *f*, Ermüdungsgrenze *f*, Schwellfestigkeit *f*, Ursprungsfestigkeit *f* **~ loading** Dauer-beanspruchung *f*, -belastung *f* (electr.) **~ machine for alternating tension and compression stresses** Zugdruckschwingungsprüfmaschine *f* **~ party** Arbeits-kommando *n*, -trupp *f* **~ phenomenon** Ermüdungserscheinung *f* **~-proof** ermüdungsfrei **~ range** Ermüdungsgrenze *f* **~ ratio** Dauerfestigkeits-Verhältnis *n* **~ resistance** Dauerfestigkeit *f*, Ermüdungswiderstand *m* **~ shock test** Dauerschlagprobe *f*

**fatigue-strength** Arbeitsfestigkeit *f*, Biegefestigkeit *f*, Dauer-festigkeit *f*, -festigkeitsgrenze *f*, -haltbarkeit *f*, -schwingungsfestigkeit *f*, -standfestigkeit *f*, Ermüdungsfestigkeit *f*, Schwingungsfestigkeit *f*, Wechselfestigkeit *f* **~ of alternating tensile stresses** Dauerfestigkeit *f* im Zug-Wechselbereich **~ under corrosion for finite life** Korrosionszeitschwingfestigkeit *f* **~ for finite life** Zeitschwingfestigkeit *f* **~ under pulsating (oscillating or fluctuating) compressive stress** Dauerfestigkeit *f* im Druck-Schwellbereich **~ under repeated (or fluctuating) bending stresses** Biegewechselfestigkeit *f*

**fatigue, ~-strength diagram** Dauerfestigkeits-schaubild *n* ~ **stress** Dauerstandfestigkeit *f* **~-stress concentration factor** Kerbwirkungs-zahl *f* ~ **striations** Ermüdungsstriemen *pl* ~ **suit** Drill-, Trainings-anzug *m* ~ **tension test** Zugversuch *m* mit Dauerbeanspruchung
**fatigue-test** Dauer-prüfung *f*, -schwingversuch *m*, -standversuch *m*, -versuch *m*, Ermüdungs-versuch *m*, Wechselfestigkeitsprüfung *f*, Wöh-lerversuch *m*, Zermürbversuch *m* ~ **under actual service conditions** Betriebsschwingver-such *m* ~ **under corrosion** Dauerschwingver-such *m* unter Korrosion ~ **at elevated tem-perature** Dauerschwingversuch *m* in der Wärme ~ **in one load stage (or step)** Einstufen-Dauer-schwingversuch *m* ~ **at low temperature** Dauer-schwingversuch *m* in der Kälte ~ **under rotary bending loads** Umlaufbiegeversuch *m*
**fatigue, ~ tested (or fatigued) specimen without rupture** Durchläufer *m* **~-testing machine** Dau-erprüf-, Dauerversuchs-maschine *f* **~-testing machine for alternating-impact stresses** Dauer-versuchsmaschine *f* für wechselnde Schlagbe-anspruchung ~ **uniform** Arbeits-, Drillich--anzug *m*
**fatigued** ermüdet
**fatness** Ausgiebigkeit *f*, Fettheit *f*, Fettigkeit *f*, Öligkeit *f*
**fatten, to** ~ mästen
**fattening (pasture)** Mast *f*, Mästung *f* ~ **stable** Mastkäfig *m*
**fattiness** Fettigkeit *f*
**fatty** fett, fetthaltig, fettig, ölig ~ **acid** Fettsäure *f* **~-acid still** Fettsäuredestillieranlage *f* ~ **aspect** Fettglanz *m* ~ **basis** Fettbasis *f* ~ **matter** Fettstoff *m* ~ **oil** Fettöl *n*, Öl *n* hohen Schlüpf-rigkeitsgrades ~ **substance** Fettbestandteil *m*
**faucet** Absperr-glied *n*, -organ *n*, -teil *m*, -vor-richtung *f*; Faßspund *m*, Flasche *f*, Hahn *m*, Kran *m*, Spund *m*, Zapfen *m* ~ **joint** Randver-bindung *f* ~ **plug** Ansteckspund *m*
**fault** Abgrund *m*, Fehler *m*, Fehlstelle *f*, Schuld *f*, Sprung *m*, Störstelle *f*, Störung *f*, Verschul-den *n*, Versehen *n*, Verwerfung *f*, Verwerfungs-kluft *f* ~ **of aim** Zielfehler *m* ~ **of balancing** Nachbildungsfehler *m* (teleph.) ~ **of (or in) construction** Baufehler *m*, Konstruktionsfehler *m* ~ **of current** Stromstörung *f* ~ **in the glass** Glasblase *f* ~ **on a line** Fehler *m* oder Störung *f* auf einer Leitung ~ **in the pipe** Leitungsstö-rung *f* ~ **due to pressing** Preßlunker *m* ~ **due to torsion** Drehverwerfung *f* ~ **in weaving** Webfehler *m*
**fault, ~ block** verworfenes Gestein *n* ~ **breccia** Reibungsbreccie *f* **~-clearing device** Fehler-scheibchenschutzeinrichtung *f* ~ **clerk** Stö-rungsaufsicht *f* ~ **complaint service** Störungs-dienst *m* ~ **conglomerate** Reibungskonglomerat *n* ~ **current** Erdgeschoß-, Fehler-strom *m* ~ **dip** Auerverwerfungswinkel *m* (geol.) ~ **docket** Störungsmeldung *f* ~ **electrode current** Über-laststrom *m* bei Kurzschluß ~ **indication** Alarmanzeige *f* ~ **indicator** Erdschluß *m*, Störungsanzeiger *m* ~ **lamp** Warnleuchte *f* ~ **line** Bruch-, Verwerfungs-linie *f* ~ **localization** Fehlereingrenzung *f* **~-localization bridge** Fehlerortmeßbrücke *f* ~ **locating** Fehlerortung *f* ~ **location** Fehlereingrenzung *f* **~-location**

**test** Fehlerortsmessung *f* ~ **log** Störungsbuch *n* ~ **normal (petroleum)** Abschiebung *f* ~ **over-thrust** Überschiebung *f* ~ **plane** Kluftfläche *f* ~ **relay** Stör(ungs)relais *n* ~ **release** Störungs-entriegelung *f* ~ **resistance** Fehlerwiderstand *m* ~ **section** Störungsstelle *f* ~ **sectionalizing** Fehlereingrenzung *f* ~ **staff** Störungspersonal *n* ~ **surface (or plane)** Verbergungsfläche *f* ~ **valley** Verwerfungstal *n* (geol.)
**faultiness** Mangelhaftigkeit *f*
**faulting** Fehlersuchen *n*, Verwerfung *f*
**faultless** fehlerfrei, tadellos, untadelhaft, ver-windungsfrei ~ **grounding** einwandfreie Erdung *f* (electr.)
**faulty** fehlerhaft, gestört, mangelhaft, rissig, schadhaft ~ **alignment** Fehlabgleichung *f* ~ **annealing** Verglühen *n* ~ **casting** Fehlgußstück *n* ~ **circuit** gestörte Leitung *f* ~ **concentric running** Unrundlauf *m* ~ **condition** Mißstand *m* ~ **control** Fehlschaltung *f* ~ **cut** Fehlschnitt *m* ~ **delineation of a picture** Verzeichnung *f* ~ **design** Konstruktionsfehler *m* ~ **dimension** Maßfehler *m* ~ **dying** mißlungene Färbung *f* ~ **focusing control** Einstellfehler *m* ~ **forging** Zerschmiedung *f* ~ **line** gestörte Leitung *f* ~ **machining** Bearbeitungsfehler *m* ~ **material** Materialfehler *m* ~ **pitch** Teilungsfehler *m* (TV) ~ **posture** Haltungsschwäche *f* ~ **re-creation of a picture** Verzeichnung *f* ~ **section** Fehler-strecke *f* ~ **selection** Falschwahl *f*, verzerrte Nummernwahl *f* ~ **sheet** Defektbogen *m* ~ **soldered joint** fehlerhafte oder kalte Lötstelle *f* ~ **spot control** Einstellfehler *m* ~ **stamping** Fehlprägung *f* ~ **switching** Fehlschaltung *f* ~ **wire** Fehlerader *f* ~ **wiring** Verdrahtungsfehler *m*, Verschalten *n*
**Faure plate (pasted plate)** geschmierte Platte *f*
**favor, to** ~ begünstigen, bevorzugen, fördern, unterstützen
**favor** Gefälligkeit *f*, Vergünstigung *f*
**favorable balance of trade** aktive Handelsbilanz *f*
**favored** beliebt
**favoring high lights over shadows** lichtfreundlich
**fawn** fahl
**fay, to** ~ zum Fluchten bringen
**fayalite** Eisenchrysolith *m*, Eisenglas *n*, Eisen-peridot *n*
**faying surface** Dichtungsfläche *f*, anpassende Oberfläche *f*
**F-band** F-Bande *f*
**F-center** F-Farbzentrum *n*
**fear, to** ~ befürchten, scheuen
**fear** Angst *f* ~ **of personal danger** Besorgnis *f* vor persönlicher Gefahr
**feasability** Ausführbarkeit *f*
**feasible** ausführbar, durchführbar
**feat of endurance** Dauerleistung *f*
**feather, to** ~ den Einfallswinkel *m* ändern, (oars, paddles, fan blades) federn, Blattwinkel *m* periodisch verstellen ~ **a propeller** Ver-stelluftschraube *f* in Segelstellung bringen
**feather** Feder *f*, Feder *f* für Keilnut, Federkeil *m*, Federn *n* (den Einfallswinkel *m* ändern), Gußnaht *f*, Keil *m*, künstliches Echo *n*, Naht *f*, Rippe *f*, Verstärkungsrippe *f* ~ **of a flame** Flammensaum *m*
**feather, ~ beard** Federfahne *f* ~ **broom** Feder-wisch *m* ~ **duster** Staubwedel *m* ~ **edge** zuge-

schärfte Kante f ~-edged file Einstreichfeile f
~ key (fixed on shaft) Federkeil m, Längs-,
Paß-feder f ~ oil Lickeröl n ~ ore Federerz n
**Feather rule** Feathersches Gesetz n
**feather,** ~ **weight** federleicht; Leichtgewicht n
~ **zeolite** Radiolit m
**featherable** verstellbar
**feathered** befiedert, gefiedert ~ **bolt** Rippenschraube f ~ **pitch** Fahnenstellung f (des Propellers), Segelstellung f ~ **position of propeller**
Segelstellung f der Luftschraube ~ **(variable-
pitch) propeller** Verstellpropeller m ~-propeller
**position** Luftschraubensegelstellung f
**feathering** schlingern (b. Turbinenschaufeln);
Luftschraube f in Segelstellung ~ **axis** Achse f
für periodische Blattverstellung ~ **bearings**
Lager pl für periodische Blattwinkelverstellung
~ **float** bewegliche Schaufel f ~ **operation** Verstellung f in Segelstellung ~ **paddle** bewegliche
Schaufel f ~ **position** Leerstellung f ~ **propeller** verstellbare Luftschraube f
**feathery** verästelt (metall.)
**feature** Besonderheit f, Eigenschaft f, äußere
Erscheinung f, Gesichtspunkt m, Grundzug m,
Merkmal n, Zug m ~ **of construction** Konstruk-
tionsmerkmal n ~ **article** Leitartikel m ~-edge
**file** Schwertfeile f
**feaze, to** ~ ausfasern
**feces** Abgang m
**fed** gespeist ~ **reflector** direkt erregter oder ge-
speister Reflektor m
**Federal German Patent** Deutsches Bundes-
Patent n (DBP)
**Federal German Patent pending** Deutsches
Bundes-Patent n angemeldet (DBPa)
**Federal German Registered Design** Deutsches
Bundes-Gebrauchs-Muster n (DBGM)
**federation** Bund m, Verband m
**fee** Abgabe f, Gebühr f, Honorar n ~ **for a visa**
Überschreitungsgebühr f
**feeble** dünn, matt, schwach, schwächlich ~
**current** schwacher Strom m
**feebleness** Kraftlosigkeit f, Mattheit f
**feed, to** ~ abfüttern, (machine) anlegen,
(furnace) aufschütten, beliefern, beschicken,
chargieren, einfüllen, eingeben, einschütten,
(into the bowl) einstreuen (Material), eintragen,
ernähren, füttern, nähren, schalten, speisen,
verfüttern, versorgen, vorlegen, (forward) vorrücken, (engin.) vorschieben, Vorschub m
geben, weiterbewegen, zuführen, zugeben,
zuteilen to ~ **back** rückkoppeln, zurückleiten,
entgegengesetzt zur Vorschubrichtung f zurückziehen to ~ **current** Strom m senden to ~
**from** speisen aus to ~ **into** einführen, einleiten
**feed** Aufgabegut n, Beschickung f, Einspeisung
f (current), Eintrag m, Förderung f, Fortrückung f, Futter n, Haufwerk n, Leitvorschub m, Nachschub m, Nahrung f, Speisung
f, (rod) Stangenvorschub m, Vorschub m, Vor-
schubarbeit f, Vorschubgeschwindigkeit f,
Vortrieb m, Werkstoffzuführ f, Zufluß m, Zufuhr f, Zuführung f, Zulauf m, Zuteilung f
**feed,** ~ **of carbons** Kohlenvorschub m ~ **and
delivery** (sugar mill) Abführtisch m ~ **and
delivery sides** Ein- und Ausgang m ~ **of frame**
(motion pictures) Bildschaltung f ~ **of material**
Werkstoffzuführung f ~ **and take-up sprocket**

**mechanism** Schleifenfänger m ~ **for the turntable** Revolvertellerzuführung f
**feed,** ~ **adjuster** Vorschubregelung f ~ **adjustment** Beistellung f ~ **apparatus** Speiseanschluß
m, Vorschubschlitten m
**feedback** Rückführung f (aut. contr.), Rückkopplung f, Rückwirkung f ~ **amplifier** gegen-
gekuppelter oder rückgekoppelter Verstärker
m ~ **circuit** Rückkopplungs-kreis m, -netzwerk
n ~ **connection** Rückkuppelungsschaltung f
~ **coil** Rückkuppelungsspule f ~ **control** Regelung f, Rückkopplungsreg(e)lung f ~ **control
system** Regelung f mit Rückführung ~ **controller** Rückführungsregler m ~ **coupling** Rückkopplung f ~ **current** Rückstrom m ~ **formula**
Rückkuppelungsformel f ~ **method** Rück-
kuppelungsmethode f ~ **network** Rückkopp-
lungs-glied n, -schaltung f ~ **oscillation** Rückkopplungsschwingung f ~ **oscillator** Rück-
kopplungs-generator m, -magnetron n, -sum-
mer m ~ **power amplifier** Verstärker m mit
Rückkopplung ~ **ratio** Rückkopplungsgrad m
~ **receiving circuit** Rückkuppelungsempfangs-
schaltung f ~ **resistance** Gegenkopplungs-
widerstand m ~ **suppressor** Rückkuppelungs-
sperre f ~ **transformer** Rückkuppelungstrans-
formator m ~ **transmitter** rückgekoppelter
Sender m
**feed,** ~ **bar** Vorschubstange f ~-**bar lever** Ge-
fühlshebel m ~ **bin** Schüttrumpf m ~ **board**
Einlegetisch m ~ **bolt** Vorschubstift m ~ **box**
Schaltgehäuse n, Schalträderkasten m, Vor-
schubkasten m ~ **box control levers** Schalthebel m für Vorschubgetriebe ~ **bracket** Konsollager n für den Selbstgang ~ **bush** Anguß-
buchse f ~ **cable** Zuleitungskabel n ~ **cam**
Vorschub-kurve f, -nocken f ~ **cam drum** Vor-
schubkurventrommel f ~ **change** Vorschubwechsel m ~ **change box** Vorschubwechsel-
räderkasten m ~ **change gear** Vorschubwech-
selräderkasten m ~-**change lever** Hebel m für
den Vorschubwechsel **(mud)** ~ **channel** Einführungsrinne f, Schlammkanal m ~-**check
valve** Speiseabsperrventil n ~ **chuck** Vorschub-
patrone f ~ **clutch** Selbstgang-, Vorschub-
-kupplung f ~-**clutch lever** Selbstgang-, Vor-
schub-kuppelungshebel m ~ **cock** Füll-,
Speise-hahn m ~ **coil** Speisedrossel f ~ **collect**
Vorschubzange f ~ **control** Schaltung f, Vorschubschaltung f ~ **control bar** Tasterstange f
~ **control lever** Tasthebel m ~ **control unit**
Auswahleinheit f ~ **current** Anodenruhe-,
Ruhe-strom m ~ **cycle** Förderkorbkreis m ~
**cylinder** Vorschubzylinder m ~ **direction** Vorschubrichtung f ~ **drive** Vorschubantrieb m
~ **driving cone** Selbstgangstufenscheibe f ~
**drum** Verteilerband m (Versatzvorrichtung) ~
**eccentric** (textiles) Fortrückexzenter m ~ **elevator** Zuführungsglied n (film) ~ **end** Einlaufseite f (film), Eintrag(s)ende n ~ **engaging
button** Kupp(e)lungsknopf m (teletype) ~
**entrance** Eintragsöffnung f ~ **fittings** Einlaufarmatur f ~ **forward** Rückkopplung f ~
**friction nut** Friktionseinrückmutter f ~ **gauge**
Druckanlage f ~ **gear** Übertragungsrad n,
Vorschub-getriebe n, -zahnrad n ~ **gear box**
Vorschubräderkasten m ~ **gearing** Vorschub-
räderwerk n ~ **grinder** Schrotmühle f ~ **guide**

(of gun) Führungsstück *n* ~ **head** Anguß *m*, Haube *f*, verlorener Kopf *m*, Massel *f*, Saugmassel *f* ~ **heater** Anwärmapparat *m* ~ **heater with ribbed pipes** Rippenrohrvorwärmer *m* ~ **hole** Führungsloch *n*, Transportloch *n* (teletype) ~**-hole space** Führungslochabstand *m* (teleph.) ~ **hopper** Aufgabe-, Aufgabe-trichter *m*, Aufschüttrumpf *m*, Einwurftrichter *m*, Füllrumpf *m* ~ **(or mill) hopper** Einlauftrichter *m* ~ **horn** Einspeisungshohlleiter *m*, Speisehorn *n* ~ **inlet** Einlaßöffnung *f* ~ **lattice-table** Zuführtisch *m* ~ **length** freie Abbrandlänge *f* ~ **lever for cross-slide** Vorschubhebel *m* für Kreuzschlitten ~ **line** Speise-, Zubringer-, Zulauf-leitung *f*, Zulaufrohr *n*, Zuleitung *f* ~ **liquor** Speisewasser *n* ~ **locomotive** Zubringerlokomotive *f* ~ **lye** Speisewasser *n* ~ **make-up water** Speisezusatzwasser *n* ~ **marking** Vorschubmarkierung *f* ~ **mechanism** Schaltwerk *n*, Transportmechanismus *m*, Vorschubapparat *m*, Zuführer *m* ~**-mechanism housing** Zuführungsgehäuse *n* ~ **mill** Vorwärmewalze *f* ~ **motion (or path)** Schaltbewegung *f*, Schaltung *f*, Schaltweg *m*, Zuführungsbewegung *f* ~ **motor** Wälzmotor *m* ~ **pawl** Vorschub-klinke *f*, -stift *m* ~ **performance** Förderleistung *f* ~ **pile board** Stapeltisch *m* ~ **pilot valve** Vorschubschaltventil *n* ~ **pin** Vorschubstift *m* ~ **pipe** Förderleitung *f*, (in charging converters) Füllrohr *n*, Speiserohr *n*, Steig-leitung *f*, -rohr *n*, Zuflußrohr *n*, Zuführungsrohr *n*, Zuleitung *f*, Zuleitungsrohr *n* ~ **point** Speisepunkt *m* ~ **position** Vorschubstellung *f* ~ **pressure** Förder-, Vorschub-druck *m* ~ **process** Fördervorgang *m* ~ **pump** Arbeits-, Förder-, Speise-pumpe *f* ~ **rack** Vorschubzahnstange *f* ~**-rack worm shaft** Tiefschaltungsschneckenwelle *f* ~ **range** Vorschub-bereich *m*, -reihe *f* ~ **ratchet** Spannkreuz *n* ~ **rate** Einlauf *m* ~ **reduction control** Vorschubreduzierung *f* ~ **reel** Abwickelspule *f* (film) ~ **regulating valve** Speiseregelventil *n* ~ **regulation** Füllungsregelung *f* ~ **regulator** Speise-, Vorschub-regler *m*, Tropfventil *n* ~**-retardation coil** Speisebrücke *f* ~ **reverse** Selbstgangumsteuerhebel *m*, Vorschubumsteuerung *f* ~ **reverse lever** Vorschubumkehrhebel *m* ~ **rod** Antriebsstange *f*, (textiles) Schaltstange *f*, Vorschubrohr *n*, Vorschubwelle *f*, Zugspindel *f* ~ **rod lathe** Zugspindeldrehbank *f* ~ **roll** Andrückwalze *f*, Aufgebe-, Speise-walze *f*; Vorschubrolle *f*, Zugwalze *f* ~ **roller** Zuführungswalze *f* ~ **rollers** (textiles) Einführungswalzen *pl*, Einziehwalzen *pl* ~ **screw** Aufgabe-, Eindreh-, Eintrags-, Speise-, Transport-schnecke *f*, Vorschubspindel *f*, Zugwelle *f*, Zuteilschnecke *f* ~ **selection dial** Vorschubwählskala *f* ~ **setting** Vorschubeinstellung *f* ~ **shaft** Schaltwelle *f*, Zug-welle *f*, -spindel *f* ~ **shaft drive** Zugspindelantrieb *m* ~ **side** Einlauf *m*, Zwischenständer *m* für Anlage ~ **sleeve** Vorschubmuffe *f* ~ **slide** Vorschubschlitten *m* ~ **spindle** Abwickelachse *f* ~ **spool** Abwickelspule *f* ~ **sprocket** Vorwickel--rolle *f*, -trommel *f* (film) ~ **star** Einführungsstern *m* ~ **stock** Ausgangsmaterialien *pl* ~ **stop** Vorschubanlage *f* ~**-stop valve** Speiseabsperrventil *n* ~ **stroke** Schaltperiode *f* (film) ~ **system** Förder- oder Speiseanlage *f* ~ **table**

Transporttisch *m* ~ **tank** Betriebs-, Speise--behälter *m* ~**-through capacitor** Durchführungskondensator *m* ~**-through insulator** Durchführungs-, Einführungs-isolator *m* ~ **travel** Förderweg *m* ~**-trip lever** Hebel *m* für die Auslösung des Selbstganges, Hebel *m* zum Auslösen des Vorschubes ~**-trip plate** Drehplatte *f* für die selbsttätige Auslösung des Tischvorschubes ~ **trough** Einführungsrinne *f* ~ **tube** Vorschubrohr *n* ~ **unit** Beschickungsvorrichtung *f* ~ **valve** Speiseventil *n* ~ **velocity** Liefergeschwindigkeit *f* ~ **voltage** Vorschubspannung *f* ~ **water** Speisewasser *n*

**feedwater,** ~ **conditioning** Speisewasseraufbereitung *f*, (boiler) Kesselspeisewasservorwärmung *f* ~ **de-aeration** Speisewasserentgasung *f* ~ **heater** Vorwärmer *m* für Speisewasser ~ **heating stages** Vorwärmestufen *pl* ~ **preheater** Speisewasservorwärmer *m* ~ **pump** Speisewasserpumpe *f* ~ **regulator** Speisewasserregler *m*, Wasserspeiser *m* ~ **system** Speisewasserkreislauf *m* ~ **treater** Speisewasseraufbereiter *m* ~ **treatment** Speisewasseraufbereitung *f*

**feed,** ~ **wheel** Vorschubrad *n* ~ **winch** Zubringerhaspel *m* ~ **worm** Eintrags-, Langzug-, Vorschub-schnecke *f* ~ **worm gear shaft** Vorschubschneckenradwelle *f* ~ **worm shaft** Vorschubschneckenwelle *f* ~**-worm wheel** Langzugschneckenrad *n*

**feeder** Abstichgraben *m*, Abzweig *m* (power system), Anleger *m* (print.), Aufgabevorrichtung *f*, Aufgeber *m*, (foundry) Aufläufer *m*, Einguß *m*, Einleger *m* (print.), Feeder *m*, Füllvorrichtung *f*, wasser- oder erzführende Kluft *f*, Massel *f*, Saugmassel *f*, Speiseleitung *f*, Speiser *m*, Stromader *f*, Verbindungsleitung *f*, Vorholer *m* (Pilgerwalzwerke), Vorschubeinrichtung *f*, Zapfer *m*, Zubringer *m*, Zuleitungskabel *n*, Zuteiler *m* ~ **and delivery table** An- und Ablegetisch *m* (print.) ~ **of gas** Bläser *m* ~ **from headrace** Oberwassergraben *m*

**feeder,** ~ **aircraft** Zubringerflugzeug *n* ~ **air service** Zubringerluftverkehr *m* ~ **arm** Nüßchenschalter *m* ~ **bar** Speiseschiene *f* ~ **cable** Energieleitung *f*, Speisekabel *n*, Stromabnahmekabel *n* ~ **canal** Speisekanal *m* ~ **circuit** Verbraucherstromkreis *m* ~ **ditch** Speisegraben *m* ~ **down-lead** Energieleitung *f* ~ **drum** Speisetrommel *f* ~ **frame** (for logs) Dockenwagen *m* ~ **head** Saugmassel *f*, (riser) Steiger *m* ~ **hopper** Fülltrichter *m* ~ **hoses** Bogenanlegerschläuche *pl* ~ **junction point** Speisepunkt *m* (energy) ~ **lead** Energieleistung *f* ~ **line** Anschlußstrecke *f*, Antennenzuleitung *f*, Energieleitung *f*; (air traffic) Zubringerlinie *f* ~ **line airline** Zubringerluftlinie *f* ~**-line system** Speiseleitungssystem *n* ~ **pillar** Schaltgerüst *n* ~ **plug box on the ground** Erdanschlußkasten *m* ~ **pot** Einschüttkasten *m* ~ **rail** Speiseschiene *f* ~ **reactors** Anzweigreaktanzspulen *pl* ~ **reservoir for wells** Behälter *m* zum Speisen der Brunnen ~ **road** Zubringerstraße *f* ~ **rod** Tragstange *f* ~ **roller** Speisewalze *f* (Kohlenförderer) ~ **service** Zubringerdienst *m* ~**-service airplane** Zubringerflugzeug *n* ~ **skid** Aufgabeschlitten *m* ~ **spring** Zubringerfeder *f*

**feeders** (textiles) Einführungswalzen *pl*, Einziehwalzen *pl*

**feeding** Beschickung *f*, Einlegen *n*, Einschüttung *f*, Eintragsöffnung *f*, Nachschub *m*, Vorschiebung *f* (engin.) **~ by bridges** Brückenspeisung *f* (teleph.) **~ of carbons** Kohlenvorschub *m* **~ and delivery end** Warenein- und Auslauf *m* **~ and delivery rollers** Ein- und Ausführwalzen *pl*

**feeding, ~ action (or filling effect)** Füllwirkung *f* **~ addition** Zubesserung *f* **~ apparatus** Anlegeapparat *m*, Bogenanleger *m*, Speisegerät *n* **~ attachment (or arrangement)** Zuführungseinrichtung *f* **~ back** Rückspeisung *f* **~ belt** Aufgabeband *n* **~ board** Anlegebrett *n* **~ bottle** Saugflasche *f* **~ box** Einschüttkasten *m* **~ chamber** Zulaufraum *m* **~ check** Förderkontrolle *f* **~ circuit** Speisebrücke *f* **~ cloth** Einlaßtuch *n* **~ collet** Vorschub-patrone *f*, -zange *f* **~ current** Speisestrom *m* **~ device** Aufgabe-, Einfüll-, Speise-vorrichtung *f*; Vorschub--antrieb *m*, -einrichtung *f*, -mechanismus *m*; Zufuhrvorrichtung *f*, Zuführungsvorrichtung *f* **~ disc** Aufgabetisch *m* **~ element** Vorschuborgan *n* **~ elevator** Aufgabebecherwerk *n* **~ end** Einlaßende *n* **~ equipment** Speisevorrichtung *f* **~ film from magazine (or feed spool)** Abspulen *n* **~ funnel** Aufgabetrichter *m* **~ furnishing by roller** Walzenauftrag *m* **~ gate** Massel *f*, Saugmassel *f* **~ gate of a furnace** Einsatzöffnung *f* **~ gear** Nachschubvorrichtung *f* **~ hand lever** Handspannhebel *m* **~ head** Gußkopf *m*, verlorener Kopf *m*, Saugmassel *f* **~ height** Zulaufgefälle *n* **~ hole** Einschüttöffnung *f* **~ hopper** Beschickungstrichter *m*, Einschüttkasten *m*, Fülltrichter *m* **~ housing** Zulaufgehäuse *n* **~-in arrangement** Wareneingang *m* **~-in winch** Einziehhaspel *f* **~ installation** Beschickungsanlage *f* **~ liquor (or solution)** Nachsatzlösung *f*, Zulaufflotte *f* **~ magnet** Fortschalt-, Vorschub-magnet *m* **~ mechanism** Vorschubschaltung *f* **~ movement (or motion)** Vorschubbewegung *f* **~ plunger** Zubringerkolben *m* **~ point** Speise-punkt *m*, -stelle *f* **~ power** Vorschubkraft *f* **~ pressure** Schleifdruck *m* **~ pulley** Zuführungsrolle *f* **~ reel** Filmabwickler *m* **~ roller** Aufgabewalze *f* **~ roller table** Aufgabeteppich *m* **~ screws** Förderschnecke *f*, Nachschubspindeln *pl* **~ shoe** Beimischer *m* **~ skid** Aufgabeschlitten *m* **~ spring** Drahtführungsfeder *f* **~ station** Aufgabezone *f* **~ tapes** Bogenzuleitungsbänder *pl* (print.) **~ time** Bildwechselzeit *f* (film)

**feel, to ~** anfühlen, betasten, empfinden, fühlen **to ~ about (or around)** herumfühlen **to ~ greasy** fettig anfühlen **to ~ one's way forward** vorfühlen **to ~ the roof of a vein for its firmness** das Hängende eines Flözes auf seine Festigkeit hin abklopfen (min.)

**feel** Anstrich *m*, Fühlung *f*, (of paper) Griff *m* **~ force** künstlicher Steuerdruck *m* **~ (or probe) surface** Fühlfläche *f*

**feeler** Fühllehre *f*, Taster *m*, Taststift *m* **~ arm** Tastarm *m* **~ cap** Meßhütchen *n* **~ device** Tastvorrichtung *f* **~ gauge** Abstandsmesser *m*, Blech-, Einstell-, Fühl-, Spalt-lehre *f*, (slang) Spion *m* **~ lever** Fühlhebel *m* **~ pin** Tasterstift *m* **~ roll** Tastrolle *f* **~ scale** Tasterskala *f*

**feelers** (fitter's gauge strips) Fühler *m*

**feeling** Tasten *n* **~ apparatus** Tastapparat *m* **~**

**body (or member)** Fühlkörper *m* **~ index** Fühlindex *m* (Schaltschutz) **~ tube** Fühlrohr *n*

**feet, ~ of a plate** Füßchen *pl* **~ lugs** Ankernocken *pl*

**feign, to ~** fingieren

**feigned** blind, fingiert, nachgeahmt

**feint** Ablenkungsmanöver *n*, Demonstration *f*, Scheinbewegung *f* **~ plug** Leerstecker *m*

**feldspar** Euritporphyr *m*, Feldspat *m*

**feldspathic** feldspatartig, feldspathaltig **~ rock** Feldspatgestein *n* **~ stoneware** Halbporzellan *n*

**felite** Felit *m*

**fell, to ~** fällen, (of trees) schlagen, umschlagen

**fellable** haubar

**felling, ~ brush** Walkbürste *f* **~ saw** Waldsäge *f* **~ season** Wadelzeit *f*

**fellmongered wool** Blutwolle *f*

**felloe joint** Felgenstoß *m*

**fellow** Gespann *n* (print.)

**felly** Felge *f*

**felsite** dichter Feldspat *m*

**felt, to ~** greifen, verfilzen

**felt** Filz *m*, Pappe *f* **~ board** Filzpappe *f* **~ board for printing shops** Bunzenauslegekarton *m*, Filz *m* **~ boot** Filz-schuh *m*, -stiefel *m* **~ covering (boilers, cables and pipes)** Filzumkleidung *f* **~ cushion** Filzpolster *n* **~ disk** Filzscheibe *f*, Filzunterlegscheibe *f* **~ dressing** Filzaufzug *m* **~ element** Filzeinsatz *m* **~ fuller** Filzwalker *m* **~ grinding disk** Filzschleifteller *m* **~ guide roll (or roller)** Filzleitwalze *f* **~ jacket** Filzmantel *m* **~ (under)layer** Filzunterlage *f* **~ pad** Filz-dichtung *f*, -kufe *f*, -platte *f*, -röllchen *n* (tape rec.), -unterlage *f* **~ pad element** Filzplatteneinsatz *m* **~ pad filter** Filzplattenfilter *m* **~ pad insert** Filzplatteneinsatz *m* **~ plate** Filzplatte *f* **~ plate filter** Filzplattenfilter *m* **~ polishing wheel** Filzpolierscheibe *f* **~ ring** Filzring *m* **~ seal** Filzring *m* **~ seal retainer** Filzdichtungsring *m* **~ side** Papierüberseite *f* (paper mfg.) **~ sleeve** Filzärmel *m* **~ spigot** Filzstopfen *m* **~ strip** Filz-, Tuch--streifen *m* **~ suction** Filzsauger *m* (paper mfg.) **~-tube insert** Filzrohreinsatz *m* **~ washer** Filz-platte *f*, -ring *m*, -scheibe *f*, -unterlegscheibe *f*

**felted** mit Filz *m* ausgelegt oder unterlegt, verfilzt

**felting** Filzen *n* **~ property** Filzbarkeit *f*, Verfilzungsfähigkeit *f*

**felucca** Feluke *f*

**female** weiblich; Mutterkaliber *n* (mach.) **~ collars** Matrizen *pl* **~ coupling "swaged on type"** Kupplungsteil *m* mit Innengewinde **~ dies** Unterschnitte *pl* **~ groove** Mutterfurche *f* **~ ground joint** Hülsenschliff *m* **~ half** Gegenstückhälfte *f* **~ hexagon screw** Sechskanthohlschraube *f* **~ insulator plate** Steckdosenplatte *f* **~ mold** Matrize *f* **~ nozzle** Fangdüse *f* **~ part of ground joint** Hülsenschliff *m* **~ part of magnetic plug** Abreißsteckergegenstück *n* **~ piece** aufnehmender Teil *m*, Matrize *f* **~ roll** Mutterfurche *f* **~ screw** Hohl-, Mutter--schraube *f* **~ screw thread** Muttergewinde *n* **~ thread** (as of pipes) Hohlgewinde *n*, Innengewinde *n* **~ trunnion** Zapfenlager *n*

**femic** femisch

**fen** Grat *m*

**fence, to ~** einhegen, fechten **to ~ in** befrieden,

einfriedigen, einhägen, einzäunen, hegen **to ~ with pales (or stakes)** einpfählen

**fence** Einfriedigung *f*, Gehege *n*, Gitter *n*, Hag *m*, Hecke *f*, Umzäunung *f*, Zaun *m ~* **of a plane** verstellbarer Hobelanschlag *m*

**fence, ~ netting** Zaungitter *n ~* **post** Zaunpfosten *m ~* **row** Feldrain *m ~* **wire** Gitter-, Zaun-draht *m*

**fencing** Umpfählung *f ~* **equipment** Elektrozaunanlagen *pl* (electr.) **~ plate for temperature control** Temperaturmeßplatte *f*

**fender** Abweiser *m*, Bergholz *n*, (of boat) Fender *m*, Freihalter *m*, Grießholm *m*, (auto) Kotflügel *m*, Kotschutz *m*, (pile) Prellbock *m*, Prellpfahl *m ~* **beam** Eisbalken *m ~* **iron** eiserne Scheuerleiste *f ~* **pile** Reib-, Schutz-, Streich-pfahl *m*

**fenders** Leitwerk *n*

**fenestration operation** Fensteroperation *f*

**fennel oil** Fenchelöl *n*

**ferberite** Ferberit *m*

**ferghanite** Ferghanit *m*

**fergusonite** Fergusonit *m*

**ferment, to ~** aufgehen, fermentieren, gären, treiben, vergären

**ferment** Ferment *n*, Gärstoff *m*, Gärungs-pilz *m*, -stoff *m*, Regstoff *m ~* **action** Fermentwirkung *f*

**fermentability** Gärungsfähigkeit *f*

**fermentable** gärbar, gärfähig, vergärbar

**fermentation** Angärung *f*, Fermentwirkung *f*, Gärung *f*, Vergärung *f ~* **in open tubs** offene Bottichgärung *f*

**fermentation, ~ cellar** Gärkeller *m ~*-**inhibiting** gärungshemmend **~ pond** Gärteich *m ~* **product** Gärerzeugnis *n ~*-**putrefaction process** Gärfaulverfahren *n ~* **vat** Gärbottich *m*

**fermentative** gärfähig **~ activity** Gärtätigkeit *f* **~ energy (or power)** Gärkraft *f ~* **test** Gärkraftbestimmung *f*

**fermented** vergoren

**fermenting, ~ plant** Göranlage *f ~* **room** Gärkeller *m ~* **tank** Gärtank *m*

**Fermi, ~ age** Fermi-Alter *n ~* **characteristic-energy-level** Fermi-Kante *f ~* **level** Fermikante *f*, (in semiconductors) elektrisches Potential *n*

**fermium isotope** Fermium-Isotop *n*

**fern-leaf crystal** Dendrit *m*

**Ferranti furnace** Ferranti-Ofen *m*

**ferrate of potash** Kaliumferrat *n*

**ferrel** Hirnring *m*

**ferret** (in glass) Randkolben *m*

**ferri, ~ chloric acid** Eisenchlorwasserstoff *m*, Ferrichlorwasserstoffsäure *f ~* **cyanic acid** Ferrizyanwasserstoffsäure *f ~* **cyanic compound** Ferrizyanverbindung *f ~* **cyanide** Ferrizyanverbindung *f ~* **cyanogen compound** Eisenzyanverbindung *f ~*-**electric** ferrielektrisch

**ferric** eisengeschlossen, eisenhaltig **~ ammonium citrate** zitronensaures Eisenoxydammon *n ~* **bromide** Ferribromid *n ~* **chloride** Chloreisen *n*, Chloreisenoxyd *n*, Eisenblumen *pl*, salzsaures Eisenoxyd *n*, Eisensublimat *n ~* **cyanogen** Ferrizyan *n ~* **disulfide** Pyrit *m ~* **ferrocyanide** Ferriferrozyanid *n ~* **inductance coil** Induktanzspule *f* mit Eisenkern **~ iodate** Eisenjodat *n ~* **magnetic circuit** Eisenkreis *m ~* **nitrate** salpetersaures Eisen(oxyd) *n ~* **oxalate** oxalsaures Eisenoxyd *n ~* **oxide** Eisenoxyd *n*,

**Ferrioxyd** *n ~* **phosphate** Blau-eisenerde *f*, -eisenerz *n*, -eisenspat *m*, phosphorsaures Eisenoxyd **~ potassium ferrocyanide** Ferrikaliumferrozyanid *n ~* **potassium sulfate** Kaliumferrisulfat *n ~* **sulfate** schwefelsaures Eisenoxyd *n ~* **sulfide** Eisensulfid *n ~* **sulphocyanate** Eisensulfozyanid *n*, Ferrirhodanid *n ~* **thiocynate** Eisensulfozyanid *n*, Ferrirhodanid *n*, Sulfozyaneisen *n ~* **valerate** Eisenvalerianat *n*

**ferriferous** eisenführend, eisenhaltig, eisenschüssig **~ cassiterite** Eisenzinnerz *n ~* **clay** eisenschüssiger Letten **~ gold sand** Eisen-hardt *m*, -hart *m ~* **quartz** Eisenquarz *m ~* **smithsonite** Zinkeisenspat *m*

**ferrimagnetism** Ferrimagnetismus *m*

**ferrite** Ferrit *n ~* **antenna** Ferritantenne *f ~* **base structure** ferristisches Grundgefüge *n ~* **core** (erase head) Ferritbügel *m ~*-**core switch matrix** Ferritkernschaltmatrix *f ~* **filled rod radiator** ferritgefüllter dielektrischer Strahler *m ~* **rod** Ferritstab *m ~* **sphere** Ferritkugel *f*

**ferritic** ferritisch

**ferro, ~ alloy** Eisen-, Ferro-legierung *f ~* **aluminum** Aluminiumeisen *n*, Ferroaluminium *f* **~ bromide** Eisendibromid *n ~* **carbon-titanium** Ferrokohlentitan *n*

**Ferrocart core** Siruferkern *m*

**ferro, ~ chrome** Ferrochrom *n ~* **chromium** Chromeisen *n*, Eisen-, Ferro-chrom *n ~* **columbium** Ferrocolumbium *n ~* **concrete** Eisenzement *m ~* **cyanic acid** Eisenblausäure *f*, Ferrozyanwasserstoffsäure *f ~* **cyanide** Blausäure *f*, Ferro-zyanid *n*, -zyanverbindung *f ~* **cyanogen** Ferrozyan *n ~* **cyanogen compound** Eisenzyanverbindung *f ~* **dynamic instrument** eingeschlossenes Dynamometer *n ~*-**electricity** Ferroelektrizität *f ~*-**electricity hafnate** Ferroelektrizität *f* der Hafnate **~-flux tester** Ferro-Flux-Gerät *n ~* **magnesium** Eisenbitterkalk *m*

**ferromagnetic** eisen-, ferro-magnetisch **~ circuit** Eisenweg *m*

**ferromagnetics** Ferromagnetik *f*

**ferro, ~ magnetism** Ferromagnetismus *m ~* **manganese** Eisenmangan *n*, Manganeisen *n*, Manganoferrum *n ~* **manganese tungstate** wolframsaures Eisenmangan *n ~* **meter** Ferrometer *n ~* **molybdenum** Ferromolybdän *n*, Molybdäneisen *n ~* **nickel** Nickeleisen *n ~* **pentacarbonyl** Eisenpentacarbonyl *n ~* **phosphorus** Phosphoreisen *n ~* **prussiate negative paper** Eisenblaunegativpapier *n ~* **resonance** Ferroresonanz *f ~* **silicon** Ferrosilizium *n*, Siliziumeisen *n*

**ferrosoferric, ~ bromide** Eisenbromürbromid *n* **~ chloride** Eisenchlorürchlorid *n ~* **iodide** Eisenjodürjodid *n ~* **oxide** Ferriferrooxyd *n*, Ferroferrioxyd *n*

**ferrospinel** Ferrospinell *m*

**ferro, ~ titanium** Eisentitan *n ~* **tungsten** Eisenwolfram *n*, Ferrowolfram *n ~* **type** Weicheisenblech *n ~* **type diaphragm** Eisenblechmembran *f*, Weichbleimembran *f ~* **type method** Hochglanztrocknung *f* (photo)

**ferrous** eisenhaltig, eisern **~ bromide** Eisendibromid *n*, Ferrobromid *n ~* **carbonate** Eisencarbonat *n ~* **chloride** Chloreisenoxydul *n*, Eisendichlorid *n*, salzsaures Eisenoxydul **~ coated tape** Schichtband *n* (tape rec.) **~ com-**

**pound** Eisenoxydulverbindung *f* ~ **cyanide** Eisenzyanür *n*, Ferrozyanid *n* ~ **iodide** Eisendijodid *n* ~ **metal** Eisenmetall *n* ~ **oxide** Eisenoxydul *n*, Ferrooxyd *n* ~ **phosphate** phosphorsaures Eisenoxydul *n* ~ **potassium sulfate** Ferrokaliumsulfat *n* ~ **selenide** Seleneisen *n* ~ **sulfate** schwefelsaures Eisenoxydul *n*, Eisensulfat *n* ~ **sulfate-lime vat** Eisenvitriolkalkkupe *f* ~ **sulfate-tin crystals vat** Eisenvitriolzinnsalzkupe *f* ~ **sulfide** Einfachschwefeleisen *n*, Eisensulfür *n*
**ferrovanadium** Eisenvanadin *n*, Vanadineisen *n*
**ferrozirconium** Ferrozirkon *n*
**ferruginous** eisenähnlich, eisenartig, eisenhaltig, eisenschüssig ~ **antimony** Eisenantimon *n* ~ **calamine** Eisenzinkspat *m* ~ **dolomite** Eisenbitterkalk *m* ~ **outcrop of a lode** eiserner Hut *m* ~ **quartz** Eisenkiesel *m* ~ **red oxide of copper** Kupferbraun *n* ~ **sand** Eisensand *m*
**ferrule, to** ~ fretten
**ferrule** Eisenband *n*, Eisenbeschlag *m*, Eisenring *m*, Endring *m*, Frette *f*, Hirnring *m*, Kapsel *f*, Klinkring *m*, Sperring *m*, Stockzwinge *f*, Zwinge *f*
**ferruled pile** beschuhter Pfahl *m*
**ferry, to** ~ überführen
**ferry** Fähre *f*, Übersetzfähre *f* ~ **for vehicles** Wagenfähre *f* ~ **boat** Fähre *f*, Trajektschiff *n* ~ **bridge** Fährbrücke *f*, Trajekt *n* ~ **cable** Fahrseil *n* ~ **cable gear** Fahrseilgerät *n* ~ **rope** Fahrseil *n* ~ **site** Fährstelle *f*
**ferrying,** ~ **operation(s)** Fährbetrieb *m* ~ **point** Übersetzstelle *f* ~ **station** Überführungsstelle *f*
**fertile** ergiebig, fett, fruchtbar ~ **material** Spaltrohstoff *m* ~ **materials** Brutstoffe *pl*
**fertility** Ausgiebigkeit *f*, Ergiebigkeit *f*
**fertilize, to** ~ düngen
**fertilizer** Düngemittel *n*, Dünger *m* ~**dispersing attachment (or device)** Düngerstreu-apparat *m*, -einrichtung *f* ~**dispersing and grain drill machine** Düngerstreuer- und Getreidedrillmaschine *f* ~ **drill** Getreidedrillmaschine *f* mit Düngerstreuvorrichtung ~ **mixer** Düngermischmühle *f* ~**producing plant** Düngemittelfertigungsanlage *f* ~ **sower (or spreader)** Düngerstreuer *m*
**ferulic acid** Ferulasäure *f*
**fescolizing** Aufbringung *f* eines galvanischen Niederschlages als Korrosionsschutz (aviat.), Fescolisieren *n*
**festoon** Gehänge *n* ~ **cloud** Festonwolke *f* ~ **dryer** Hängebandtrockner *m* ~ **frame** Auszack-, Ausbiege-maschine *f* ~ **lamp** Röhrenformlampe *f* ~ **lighting** Sofittenbeleuchtung *f*
**festooning** Runzelbildung *f* (paint)
**fetch, to** ~ abholen, anheben, herbeiholen, holen, suchen gehen **to** ~ **up** aufholen
**fetch** Fetch *f*
**fetid,** ~ **air** Mief *m* (slang) ~ **shale** Stinkschiefer *m*
**fetlock** Köte *f* ~ **joint** Fesselgelenk *n*
**fetter, to** ~ fesseln
**fetters** Fesseln *pl*
**fettle, to** ~ (castings) abrauhen, ausbessern, ausfüttern, beschneiden, mischen (metall.)
**fettling** Abschroten *n*, Ausfütterung *f*, Putzen *n*
**fiber (or fibre)** Faden *m*, Faser *f*, Faserstoff *m*, Fiber *f*, Sehne *f*, Zaser *f* **having lost** ~**s (or filaments)** abgefasert ~ **of agave** Agavenfaser *f*

~ **of latitudes on the surface of a shell** (sphere cylinder hemisphere, dished boiler, tank head, etc.) Breitenkreisfaser *f*
**fiber,** ~ **artificial leather** Faserkunstleder *n* ~ **board** Faser-, Farb-platte *f*, Hartpappe *f* ~ **braid** Faserstoffbeflechtung *f* ~ **button** Fiberklötzchen *n* ~**catching capsule** Faserfänger-Kapsel *f* ~**catching cartridge** Faserfänger-Patrone *f* ~**catching device** Faserfänger-Vorrichtung *f* ~ **conduit** Fiberrohrstrang *m* ~ **core** Faserseele *f* ~**covered cable** Faserstoffkabel *n* ~ **disc for ignition distributor** Anlaufring *m* aus Fiber für Zündverteiler ~ **duct** Fiberrohrstrang *m* ~ **dust** Faserstaub *m* ~ **gasket** Fiberdichtung *f* ~ **gauge** Fadenmanometer *n* ~ **glass** Glas-faser *f*, -seide *f*, gesponnenes Glas *n* ~ **glass radome face plate** Kunststoff-Frontplatte *f* (Flugzg.) ~ **grease** fadenziehendes (Stauffer)-fett *n* ~ **hardening** Faserverfestigung *f* ~ **lap** Faserwickel *n* ~ **length** Stapel *m* ~ **packing** Fiberpäckung *f* ~ **paper** Vulkanpapier *n* ~ **pin** Fiberbolzen *m* ~ **pinion** Fiber-rad *n*, -ritzel *n*, -zahnrad *n* ~ **plate** Fiberscheibe *f* ~ **reinforcing tube** Versteifungsrohr *n* aus Fiber ~ **ring** Fiberring *m* ~ **saturation** Fasersättigung *f* ~ **sleeve** Fiberhülse *f* ~ **slurry** Faserbrei *m* ~ **strength** Faserstärke *f* ~ **structure** Faserstruktur *f* ~ **suspension** Fadenaufhängung *f* ~ **test** Faserprüfung *f* ~ **tuft** Faserbüschel *m* ~ **washer** Fiberdichtung *f* ~ **wedge** Fiberkeil *m* ~ **yield** Fasergehalt *m*
**fibering** Faserstruktur *f*, Faserung *f*
**fiberising** Abspülen *n* von Fasern
**fiberlike** faserähnlich
**fibers of sinews** Sehnenfasern *pl*
**fibery** faserig
**fibrella** Fäserchen *n*
**fibriform,** ~ **hairlike crystal** Trichit *m* ~ **structure** faserige Struktur *f*
**fibrillation** Faserung *f* ~ **of lens** Augenlinsenfaserung *f*
**fibrine** Faserstoff *m*
**fibroid** faserähnlich
**fibroin** Fibroin *n*
**fibrolite** Faserkiesel *m*
**fibrous** fadenförmig, faserförmig, faserig, fibrös, krätzig, sehnig, (geol.) verworrenfaserig ~ **asbestos** Faserasbest *m* ~ **blende** Strahlenblende *f* (min.) ~ **cassiterite** Holzzinnerz *n* ~ **coal** (soft, humified, bituminous coal) Faserkohle *f*, Faser-Steinkohle *f* ~ **crystal** Trichit *m* ~ **fracture** faserige oder splitterige Bruchfläche *f* ~ **gypsum** Federgips *m* ~ **insulating materials** Faserdämmstoffe *pl* ~ **iron** Sehneisen *n* ~ **lignite** Faserbraunkohle *f* ~ **limestone** Faserkalk *m* ~ **linen disc** Fiberleinenscheibe *f* ~ **manganese oxide** schwarzer Glaskopf (Ziegel) *m* ~ **olivenite** Holzkupfererz *n* ~ **peat** Faser-, Wurzel-topf *m* ~ **red iron ore** Adlerstein *m*, roter Glaskopf *m* ~ **silica** Eisen-amiant *m*, -asbest *m* ~ **sphalerite** Schalenblende *f* ~ **structure** Faserstruktur *f*, faseriges Gefüge *n*, faserige Struktur *f*
**fictile** formbar ~ **insulating substance** Isoliermasse *f* ~ **material** Preßstoff *m* ~ **substance** Preßmasse *f*
**fictility** Bildsamkeit *f*
**fictitious** fiktiv, gedacht, unwahr ~ **bill** Reit-

wechsel *m* ~ **boundary** scheinbare Grenze *f* ~
**forces** Scheinkräfte *pl* ~ **load** fiktive Belastung *f*
**fictive** fiktiv
**fiddle bow (or drill)** Fiedelbogen *m*
**fidelity** Genauigkeit *f* (Treue der Wiedergabe),
Naturtreue *f*, Pflichttreue *f*, (of reproduction)
Richtigkeit *f*, Treue *f*, Wiedergabetreue *f* ~ **in
reproduction** getreue Wiedergabe *f* ~ **of
reproduction** Natürlichkeit *f* (Reinheit *f* oder
Treue *f*) der Wiedergabe, Wiedergabennatür-
lichkeit *f*
**field** Acker *m*, Baustrecke *f*, (of study) Fach *n*,
Feld *n*, Gebiet *n*, Körper *m* (math.), (of
forces) Kraftfeld *n*, Lagerstätte *f*, Strecke *f*,
Rasterbild *n* (TV), Teilbild *n* (TV), Stromfeld *n*
(electr.), Teilraster *m*, Wesen *n*, Wicklung *f*
(inductor)
**field**, ~ **of action** Arbeitsfeld *n*, Gefechtsfeld *n* ~
**of activity** Arbeitsgebiet *n* ~ **of application** Anwendungsgebiet *n* ~ **of attraction** anziehender
Bereich ~ **of combat** Kampffeld *n* ~ **of competency** Fach *n* ~ **of condition** Zustandsfeld *n*
~ **that draws away** Saugfeld *n* ~ **of fire** Feuerbereich *m*, Richtfeld *n*, Schußfeld *n*, Seitenrichtfeld *n* ~ **of flow** Strömungsfeld *n* ~ **of
force** Kraftlinienfeld *n* ~ **of force on a model**
Stromlinienaufzeichner *m* ~ **for glider training**
Fluggelände *n* ~ **of gravity** Schwerefeld *n* ~ **of
headwater** Oberwassergebiet *n* ~ **of illumination**
Leuchtfeld *n* ~ **of knowledge** Wissensbereich *n*
~ **of lines of force** Kraftlinienfeld *n* ~ **for
making tests and experiments** Versuchsfeld *n*
~ **of measurement** Meßfeld *n* ~ **of operations**
Betriebsfeld *n* ~ **of research** Prüffeld *n* ~ **of
selection** Wählerkontaktfeld *n* ~ **and space
charge ratio** Feld- und Raumladeverhältnis *n*
(teleph.) ~ **of state** Zustandsfeld *n* ~ **of tail
water** Unterwassergebiet *n* ~ **of traverse**
Schwenkbereich *m*, Seitenrichtfeld *n* ~ **of
view** Blickfeld *n* (opt.), Gesichts-feld *n*, -kreis
*m* ~ **of view index** Sehfeldzahl *f* ~ **of view
limitation** Gesichtsfeldbeschränkung *f* ~ **of
view of oscillogram** Blickfeld *n* des Oszillo-
grammes ~ **of vision** Gesichts-, Seh-, Sicht-feld *n*
**field**, ~ **adjusting device** Feldjustiervorrichtung *f*
(photo) ~ **airdrome** Feldflughafen *m* ~ **altimeter
setting** Platzhöheneinstellwert *m* ~ **ammeter**
Feldampèremeter *n* ~ **amplification** Feldstrom-
verstärkung *f*, Feldverstärkung *f* ~ **apparatus
for ground photogrammetry** Erdbildaufnahme-
gerät *n* ~ **aspect of matter** Feldbild *n* der Materie ~ **barometric pressure** Luftdruck *m* am
(Flug-)Platz ~ **belt** Feldbinde *f* ~ **bend** Teil-
bildkorrektion *f* ~ **blanking impulse** Zeilenaustastimpuls *m* ~ **book** (of surveyor) Handregister *n*, Manual *n*, Meßbuch *n* ~ **book for
calculations** Winkelbuch *n* ~ **break switch**
Magnetausschalter *m* ~ **break switch with
damping rheostat** Feldunterbrechungsschalter
*m* mit Dämpfungswiderstand ~ **broadcasting**
Außenübertragung *f* ~ **cable** Feldkabel *n* ~
-**cable line** Feldkabelleitung *f* ~ **calibration**
Nacheichung *f* ~ **camera** Geländekamera *f* ~
**cart** Ackerwagen *m* ~ **(instrument) case** Feld-
besteck *n* ~ **cell** Feldelement *n* ~ **chain** Geländekette *f* ~ **(frame) change** Teilbildwechsel *m* ~
**changes** Feldsprünge *pl* ~ **changing** Feldverlauf
*m* ~ **circuit** Erregernetz *n* ~ **coil** Erregerwicke-

lung *f*, Feldspule *f*, Feldwicklung *f*, Magnetspule *f*, Magnetwicklung *f* ~ **(exciting) coil**
Felderregerspule *f* ~ **coil of a moving-coil
speaker** Feldspule *f* eines elektrodynamischen
Lautsprechers ~ **compensating winding** Feld-
ausgleichwicklung *f* ~ **configuration** Feldverlauf *m* ~ **covered** Meßfeld *n* ~ **cultivator** Feld-
kultivator *m* ~ **current** Erregerstrom *m*, Feldstrom *m*, Feldstromkreis *m* ~ **cylinder roller**
Ackerwalze *f* ~ **de-excitation switch** Feld-
entregungsschalter *m* ~ **deflection** Teilbild-
ablenkung *f* ~ **deformation** Feldverzerrung *f*
~ **density** Feld-dichte *f*, -stärke *f* ~ **dependence**
Feldabhängigkeit *f* ~ **depot** Feld-magazin *n*,
-speicher *m* ~ **designation** Lochfeldbezeich-
nung *f* ~ **designed to concentrate (draw away or
drive) electrons** Absaugfeld *n* ~ **desorption** Feld-
desorption *f* ~ **diaphragm** Sehfeldblende *f* ~
**direction** Feldrichtung *f* ~ **discharge switch**
Nebenschlußunterbrecher *m* ~ **distortion**
Bild-, Feld-verzerrung *f* ~ **(strength) distribution** Feldverteilung *f* ~ **divider** Teilbildfrequenz-
teiler *m* ~ **dose** Oberflächendosis *f* ~ **dressing**
Sanitätsverbandzeug *n* ~**-dressing station** Ver-
bandplatz *m* ~ **effect transistor** Feldeffekt-
transistor *m* ~ **electrons** Feldelektronen *pl* ~
**elevation** Platzhöhe *f* über NN ~ **emission**
autoelektrische Entladung *f*, Feldelektronenemission *f*, Feldstrom *m* ~ **emission of electrons**
Autoelektronenemission *f* ~ **emission microscope** Feldemissionsmikroskop *n* ~ **emitter
sources** Feldemissionsquellen *pl* ~ **engineer**
Baustelleningenieur *m* ~ **equations** Feldgleichungen *pl* ~ **equipment** Feld-ausrüstung *f*,
-ausstattung *f*, -gerät *n* ~ **excitation** Felderre-
gung *f* ~ **exciter rheostat** Sollwerteinsteller *m*
~-**flattening** krümmungsreduzierend; Feld-
krümmungskorrektur *f*, Krümmungskorrektur
*f* ~ **flattener** Feldebenungslinse *f* (opt.) ~
**flood lighting** Landefeldbeleuchtung *f* ~ **fortification** Feld-befestigung *f*, -schanze *f* ~**-free**
feldfrei ~**-free emission current** thermischer
oder feldfreier Emissionsstrom *m* ~ **frequency**
Raster-, (in interlaced scanning) Rasterwech-
sel-, Teilraster-frequenz *f*, Zeilenzugwechsel-
periode *f* ~ **frequency control** Bildwechsel-
frequenzregelung *f* ~**-frequency synchronizing
impulse** Teilrastergleichlaufimpuls *m* ~ **funnel
theory** Feldrichtertheorie *f* ~ **fuse** Feldsiche-
rung *f* ~ **fuse box** Feldsicherungskästchen *n* ~
**generator** Maschinensatz *m* ~ **glass** Doppel-
fern-glas *n*, -rohr *n*, Feldstecher *m*, Fernglas *n*,
Glas *n*, Krimstecher *m* ~-**glass magnifier** Feldstecher *m* ~ **gun** Feld-geschütz *m*, -kanone *f* ~
**handling frame** Haltegestell *n* ~ **handset** Feld-
handapparat *m* ~ **howitzer** Feldhaubitze *f* ~
**imperfection** Feldunvollkommenheit *f* ~ **implements** Feldgerät *n* ~ **index** Feldexponent *m* ~
**intensity** Feld-intensität *f*, -stärke *f*; Nutzfeldstärke *f* ~ **intensity of the television transmitter**
Feldstärke *f* des Fernsehsenders ~ **intensity
measurement** Feldstärkemessung *f* ~ **intensity
meter** Feldstärke-messer *m*, -meßgerät *n*, Gitterableitwiderstand *m* ~ **ion emission** Feld-
ionenemission *f* ~ **ion mass spectroscopy** Feld-
ionenmassenspektroskopie *f* ~ **keystone waveform** Aufhebung *f* der geometrischen Teilbild-
verzeichnung ~ **kiln** Feldbrand *m* ~ **lens** Feld-,

Vorder-linse *f*, Kollektiv *n* (opt.) ~ **lens of eyepiece** Kollektiv *n* des Okulars, Okularkollektiv *n* ~ **lightning arrester** Feldblitzableiter *m* ~ **linearity control** Linearitätsregelung *f* ~ **litter** Feldtrage *f* ~ **magnet** Feldmagnet *m* (electr.) ~ **maintenance** laufende Wartung *f* (aviat.) ~ **map** Feldverlauf *m*, Flurkarte *f* ~ **mapping section** Kartenfelddruckerei *f* ~ **maps** Feldbilder *pl* ~ **mattress** Feldmatratze *f* ~ **measuring** Flurvermessung *f* ~ **monitoring tube** Teilbildkontrollröhre *f* ~-**neutralizing coil** Abschirmspule *f* (TV) ~ **operations** Geländearbeit *f* ~ **parallel to axis** Längsfeld *n* ~ **pattern** Feldprofil *n*, Feldstärkendiagramm *n*, Feldstärkenprofil *n*, Feldverlauf *m*, Strahlenkennlinie *f*, Strahlungscharakteristik *f*, Strahlungsdiagramm *n* ~ **periscope** Feldperiskop *n* ~ **photogrammeter** Feldfotogrammeter *m* ~ **phototheodolite** Feldfototheodolit *m* ~ **pick-up** Außenaufnahme *f* ~ **piece** Feldgeschütz *n* ~ **pin** Dorn *m* (Brückenbau) ~ **pitch** Feldschritt *m* (electr.) ~ **point** Aufpunkt *m* ~ **pole** Spulenkasten *m* ~ **projection** Magnetzahn *m* ~ **protection resistor** Feldschutzwiderstand *m* ~ **quantization** Feldquantisierung *f* ~ **quantum** Feldquant *m* ~ **radio set (or telephone)** Feldfunksprecher *m* ~-**radio station** Kleinfunkstelle *f* ~ **railroad** Feldeisenbahn *f* ~ **regulating** Feldregulierung *f* ~ **regulating switch** Feldregulierschalter *m* ~ **regulator** Feld-regler *m*, -regulator *m*, -regulierwiderstand *m*, -widerstand *m* ~ **relay** Feldschütz *m* ~ **repair shop** Feldwerkstätte *f* ~ **repeating coil** Feldringübertrager *m* (electr.) ~ **rheostat** Feld-regler *m*, -widerstand *m* ~ **rivet** Montageniet *n* ~ **rod** Fluchtstab *m* ~ **roller** Ackerwalze *f* ~ **selection** Lochfeldsteuerung *f* ~ **sensitivity** Feldempfindlichkeit *f* (microph.) ~ **sequential system** Feldfolgesystem *n* ~ **spider with pole ring** Erregerteil *m* (electr.) ~ **spider separator** Polradscheider *m* ~ **starter** Feldanlasser *m* ~ **stop** Blickfeldblende *f*, Feldblende *f*, Feldpunkt *m*, Gesichtsfeld-, (diaphram) Leuchtfeld-, Sehfeld-, Vorder-blende *f* ~ **strength** Feld-dichte *f*, -intensität *f*, -stärke *f* ~-**strength distribution** Feldverlauf *m* ~-**strength measurement** Feldstärkemessung *f* ~-**strength measuring set** Feldstärkemeßgerät *n* ~-**strength meter** Feldstärkemesser *m* ~ **stress** Streckgrenzenspannung *f* ~ **stretcher** Feldsystem *n* ~ **subfluvial cable** Feldflußkabel *n* ~ **sweep** Teilbildabtastung *f* ~ **synchronizing signal** Teilbildsynchronsignal *n* ~ **system** Feldsystem *n* ~ **telephone** Feldfernsprecher *m* ~-**telephone exchange** Feldklappenschrank *m* ~ **terminal** Feldklemme *f* ~ **test** Betriebsversuch *m*, Streckenversuch *m*, Versuch *m* auf der Baustelle ~ **test panel** Feldprüfschrank *m* ~ **testing set** Feldmeßkästchen *n*, Feldprüfschrank *m* ~ **tilt** Rasterverformungsentzerrung *f* ~ **time base** Teilbildzeitbasis *f* ~ **toroidal repeater coil** Feldringübertrager *m* ~ **train** Tross *m* ~ **trench** Feldschanze *f* ~ **trial** Betriebsversuch *m* ~ **tripod** Feldstativ *n* ~ **tripod with adjustable stays** Feldstativ *n* mit verstellbaren Spreizen ~ **trunk cable** Feldfernkabel *n* ~ **trunk wire** Feldfernkabel *n* ~ **umbrella** Schirm *m* (Geometer) ~ **value** Feldwert *m* ~ **variation** Feldregelung *f* ~ **vehicle** Feldfahrzeug

*n* ~ **wagon** Feldwagen *m* ~ **wave** Feldwelle *f* ~ **well** produktives Bohrfeld *n*, Feldbrunnen *m* ~ **winding** Erregerwickelung *f*, Feldspule *f*, Feld-, Magnet-wicklung *f* ~ **wire** Feldkabel *n* ~ **wire laying** Feldkabelbau *m* ~-**wire line** Feldkabelleitung *f* ~-**wire strand** Feldkabelader *f* ~ **work** Feld-, Gelände-arbeit *f*, Feldwerk *n*, Schanze *f* ~ **works** Geländeeinrichtungen *pl* ~ **workshop** Feldwerkstatt *f*

**fiery** feurig ~ **cyclone** Glutwirbel *m* ~ **red** feuerfarben

**fife** Querpfeife *f*, Trommelflöte *f*

**fifteen** Mandel *f* by ~**s** mandelweise

**fifth**, ~ **order theory** Theorie *f* fünfter Ordnung ~**wheel** Drehkranz *m* (bei Sattelschleppern) ~-**wheel device** Aufsattelvorrichtung *f* ~-**wheel load** Aufsatteldruck *m* (beim Sattelschlepper ~-**wheel steering** Drehschemellenkung *f*

**fifties-selector** Fünfziger-Gruppen-Wähler *m*

**fight, to** ~ bekämpfen, fechten **to** ~ **to a decision** ausfechten

**figurative** bildlich

**figure, to** ~ errechnen **to** ~ **out** berechnen **to** ~ **a graduation** eine Teilung beziffern

**figure** Abbildung *f*, Bild *n*, Gebilde *n*, Gestaltung *f*, Figur *f*, Form *f*, Illustration *f*, Schaubild *n*, Schnittzeichnung *f*, Zahl *f*, Ziffer *f* ~ **of eight** Achterfigur *f*, Lemniskatenkennlinie *f*, Loopingacht *f* (aviat.) ~ **of merit** (relay) Ansprechstromstärke *f*, Flugzahl *f*, Güte *f* (der Röhre), Güte-faktor *m*, -wert *m*, -zahl *f*, -ziffer *f*, Stromstärke *f* für den Ausschlag, Wertungsziffer *f*

**figure**, ~ **bar iron** Façoneisen *n* ~ **blank** Zahlenblank *n*, Zifferweiß *n* (teleph.) ~ **dial** Zifferscheibe *f* ~ **mask** Ziffernmaske *f* ~-**of-eight characteristic** achtförmige Richtcharakteristik *f* ~-**of-eight diagram** Doppelkreisdiagramm *n* ~-**of-eight pattern** achtförmige Richtcharakteristik *f* ~-**of-eight polar diagram** (direction finder) Doppelkreischarakteristik *f* ~-**of-merit curve** Gütekurve *f* ~ **pattern** Ziffernumrahmung *f* ~ **printing** Figurendruck *m* ~ **punch** Zahlenpunze *f* ~ **reading electronic device** elektronisches Ziffernlesegerät *n* ~ **shift** Zahlen-umschaltung *f*, -wechsel *m* ~-**shift key** Zi-Taste *f* (teletype) ~**space** Zahlen-blank *n*, Zeichenabstand *m* ~ **stamps** Schlagzahlen *pl* ~ **stone** Agalmatolith *m*, Bildstein *m* ~ **values** Zahlenwerte *pl*

**figured** Fassondraht *m*; gemustert ~ **bar iron** Fassoneisen *n* ~ **dimension** eingeschriebenes Maß *n* ~ **gauze** broschierte Gaze *f* ~ **net** Mustertüll *m* ~ **surface** deformierte Fläche *f*

**figures** Bezifferung *f*, Werte *pl* ~ **case** Zahlengruppe *f* ~ **position** Ziffernstellung *f* ~ **shift** Zahlenumschaltung *f* ~-**shift signal** Zahlenumschaltungszeichen *n*

**figuring**, ~ **of graduation** Teilungsbezifferung *f* ~ **machine** Dessinmaschine *f* ~ **weft** Figurschuß *m*

**filament** Draht *m*, Drathstück *n*, Durchschlagskanal *m*, Einzeltiter *m*, Faden *m*, Faser *f*, Faserbrücke *f*, (heated) Glühdraht *m*, (heated) Glühfaden *m*, Heiz-faden *m*, (of valve) -kathode *f*, (coiled) -wendel *m*, -regler *m*, Leuchtdraht *m*, Schnur *f*, Zaser *f* **having lost fibers or** ~**s** abgefasert ~ **of water** Wasserfaden *m* filament, ~ accumulator Heiz-akkumulator m, -sammler m ~-activity test Röhrenprüfung f (im Betriebe) ~ battery A-Batterie f, A-Verstärker m, Heiz-batterie f, -sammler m ~ break Heizfadenbruch m ~ circuit Heizkreis m, (heater) Heizstromkreis m ~ current Faden-, Heiz-strom m ~ current power Heizleistung f ~-current regulation Heizstromregler m ~ data Heizdaten pl ~ electrometer Fadenelektrometer n ~ emitter (of electrons) Kathode f ~ energy Heizenergie f ~ energy consumption Heizleistung f ~ fiber Haarfaser f ~ generator Heizmaschine f ~ heater circuit Heizkreis m ~ holder Glühdrahthalter m (rdo) ~ power Heizleistung f ~ resistance Faden-, Heizwiderstand m ~ rheostat Heizwiderstand m ~ screen Abdeckschirm m (Bilux-Lampe) ~-screening grid Raumladegitter n ~ seal Einschmelzstelle f des Glühfadens ~ seals Einschmelzstellen pl des Glühfadens ~ supply Heizfadenspeisung f, Heizsammler m, Röhrenheizung f ~ supply transformer Heiztransformator m ~ swing (magnetron) schwingender Draht m ~ temperature Glühfadentemperatur f ~ tension (or stretching) device Fadenspannvorrichtung f ~ transformer Heiztransformator m ~ voltage Fadenspannung f, Heizseite f, (heater) Heizspannung f ~ wattage Heizleistung f ~ winding Heizwickelung f

filamentary drahtförmig, fadenförmig ~ cathode Fadenkathode f ~ molecule Fadenmolekül m ~ rectifying tube direktgeheizte Gleichrichterröhre f ~ transistor Fadentransistor m

filamentous haarfaserig

filar, ~ illumination Fadenbeleuchtung f ~ micrometer Fadenmikrometer n

file, to ~ ablegen, einordnen, einreichen, feilen, hinterlegen, (a petition) nachsuchen, registrieren to ~ across querfeilen to ~ an action erheben to ~ an application ein Patent n anmelden to ~ off (or away) abfeilen to ~ out ausfeilen to ~ a patent application eine Patentanmeldung einreichen to ~ petition (or application) to be heard um Gehör n ersuchen to ~ petition in bankruptcy Konkurs anmelden to ~ smooth schlicht feilen to ~ a suit erheben, klagen to ~ a toll call ein Gespräch anmelden to ~ into wrong boxes or places falsch ablegen

file Ablage f (documents), Ablegemappe f, Aktenbündel n, Aktenstück n, Blattordner m (print.), Bündel n (paper), Feile f, Ordner m, Reihe f, Rolle f, Rotte f, Stapel m, Stoß m, Zettelspieß m ~ of documents Aktenstoß m ~ for filing machines Feilmaschinenfeile f ~ for removing burred edges of rifling Felderfeile f

file, ~ brush (or card) Feilenbürste f ~ cards Feilkarden pl ~ cleaners Feilenkratzbürste f ~ closer Schließender m ~ code notation Rubrum n ~ consolidation Datenverschmelzung f ~ cutter Feilenhauer m ~ cutting anew Feilenaufhauen n ~-cutting chisel Feilenhauermeißel m ~ cutting machine Feilenhaumaschine f ~-cutting tool Feilenhaue f ~ dust Feilstaub m ~ handle Feilenheft n ~ hard feilenhart ~ hardness Feilenhärte f ~ index Aktenverzeichnis n ~ leader Vordermann m ~ made of cardboard Ablegekarton m ~ number (of a patent) Aktenzeichen n, Systemzahl f ~ steel Feilenstahl m

~ stroke Feil-strich m, -stück n ~ tang Feilenangel f ~ test Feilenprobe f ~-testing machine Feilenprüfmaschine f ~ transporting elevators Aktenumlaufaufzug m ~ wrapper Anmeldungsakten pl, Erteilungsakten pl, Rolle f

filed angemeldet, (patent) eingereicht, gefeilt ~ patent angemeldetes Patent n

files, in ~ gliederweise

fileting Ebnen n

filiform faserförmig ~ corrosion Fadenkorrosion f

filigree Drahtarbeit f, Filigran n ~ corrosion Filigrankorrosion f ~ glass Fadenglas n

filing Anmeldung f, (of a message) Aushändigung f, (of message) Zustellung f einer Botschaft ~ of a message Aushändigung f einer Nachricht f

filing, ~ block Feil-block m, -holz n ~ cabinet Aktenschrank m, Fach n, Kartei f, Kartenkasten m ~ card Karteikarte f ~ dust Feilicht n, Feilsel n ~ fee Anmeldegebühr f ~ system Registratursystem n ~ table Karteitisch m ~ time Eingangszeit f (bei der Durchgangsanstalt) ~(booking) time Anmeldezeit f ~ time of a call Anmeldezeit f eines Gespräches ~ vise Hand-, Feil-kloben m

filings Feilicht n, Feilsel n, Feilspäne pl, Feilstaub m

fill, to ~ aufschütten, auskitten, beladen, beschicken, füllen, strecken, vollfüllen ~ and dip caps Füll- und Peilverschlüsse pl to ~ in ausfüllen, einfüllen, einschaufeln, vergießen, zuschmieren to ~ up irregularities Auftrag schweißen to ~ in the joints die Fugen pl mit Mörtel vergießen to ~ up the joints with blocks (or wedges) verzwicken to ~ up with mud verschlämmen to ~ up by pouring zugießen to ~ the sails abbrassen to ~ up storage cells Sammler nachfüllen to ~ up abfüllen, anfüllen, anschütten, auffüllen, ausfüllen, einfüllen, füllen, nachfüllen, vergießen, verstreichen

fill kegelförmige Anschüttung f, Aufschüttung f, Damm m, Erschüttung f, Füllmaterial n, Schüttung f, (earth) Überschüttung f ~ box Einfüllkasten m, Hydrant m ~ cap Füllklappe f ~ hole Füll-einlaß m, -loch n, -öffnung f ~ hole cover Füllochabdeckung f ~-in light Aufheller m ~-in sheets Wechselbogen m ~ mass Füllmasse f ~ nozzle Zapfventil n ~ pan Füllpfanne f, -wanne f ~ pipe (in charging converters) Füllrohr n ~ plate Futterblech n ~ plug Füllstopfen m ~-up Beifilm m ~-up valve Füllventil n

filled abgefüllt, gefüllt, (said of shell) schußfertig ~ with air lufterfüllt ~ up with liquid hot material mit flüssiger heißer Masse vergossen

filled, ~ band vollbesetztes Energieband n ~ equilibrium volles Gleichgewicht n ~-in ground aufgeschütteter Boden m ~ joint ausgegossene Kabelmuffe f ~ lattice energy bands besetzte Gitterenergiebänder pl ~ nigre Leimseife f ~-section rail Block-, Voll-, Zungen-schiene f ~ soap Leimseife f

filler Einlage f, Fördermann m, Formstück n, (in smoothing surfaces for paint) Füller m, Füll-einlage f, -klotz m (aviat.), -masse f (sugar mfg.), -material m, -stoff m, -stück n, Grundierlack m, Grundlack m (lacquer),

Grundmasse *f* (for lacquers), Polsterholz *n*, Porenfüller *m*, Spachtelmasse *f*, Stoffeinlage *f*, Stopfholz *n*, Trense *f*, (for lacquers, etc.), Verschnittmittel *n*, (cable) Zwickel *m*
**filler**, ~ **cap** Einfüllverschluß *m*, Eingußkappe *f*, Füll-schraube *f*, -stutzen *m* ~ **cap bottom part** Einfüllverschlußunterteil *n* ~ **cap filter** Einfüllverschluß-Kraftstoffsieb *n* ~ **cap top part** Einfüllverschlußoberteil *n* ~ **clay** Füllton *m* ~ **hose** Betankungsschlauch *m* ~ **material** Füllmittel *n*, Zusatzmaterial n ~ **metal** Elektrodenmetall *n*, Zusatz-material *n*, -metall *n* ~ **neck** Betankungs-, Einfüll-, Eintritts-stutzen *m* ~ **neck coupling rockets** Betankungskuppelung *f* ~ **piece** Abstandsflansch *m* (petroleum) ~ **plate** Füllplatte *f* ~ **plug** Einfüllschraube *f* ~ **rod** Schweiß-draht *m*, -stab *m*, Zusatzstab *m* ~ **screw** Einfüllschraube *f* ~ **socket (or connection)** Füllstutzen *m* ~ **speck** Materialfehler *m* ~ **stripper** (tobacco) Einlageripper *m* ~ **top** Fülleroberteil *n* ~ **tube** Eingußstutzen *m* ~ **vent** Einfüllung *f* ~-**well adaptor assembly** Auffüllarmatur *f* ~ **on wheels** Gießkarren *m* ~ **wire** Zusatzdraht *m*
**fillers** Mischungszusätze *pl*, Quirlsteine *pl*
**fillet** (band) Astragal *n*, (beading) Astragal *n*, (collar) Astragal *n*, (molding) Astragal *n*, (outside) äußere Ausbuchtung *f*, (inside) innere Ausbuchtung *f*, Auskehlung *f*, Ausrundung *f*, Aussparung *f*, Eckverstärkung *f*, Hohlkehle *f*, Holzkeil *m*, Kehle *f*, Leistchen *n*, Leiste *f*, Plättchen *n*, Randleiste *f* um Tischplatte, Schraubengewinde *n*, Steg *m*, Streifen *m*, Übergangsstück *n*, Verkleidungsübergang *m* ~ **of cement** Zementputzband *n* ~ **in parallel shear** Flankenkehlnaht *f* ~ **between wing and fuselage** Auskehlung *f* zwischen Flügel und Rumpf
**fillet,** ~ **airplane** Kehlflugzeug *n* ~ **band** vollbesetztes Energieband *n* ~ **joint** Kehlnaht *f* ~ **radius** Hohlkehlenhalbmesser *m* ~ **weld** Kehlnaht *f*, -nahtschweißung *f*, -schweißung *f* ~-**weld seam** Kehlnaht *f* ~ **welding** Kehlschweißung *f* ~ **welds** Kehlnahtschweißung *f*
**filleting** Glätten *n*
**filling** Anfüllung *f*, Appretur *f*, Auffüllen *n*, Auffüllung *f*, (in earthwork) Auftrag *m*, (for wood surfaces) Ausfüllung *f*, Begichtigung *f*, (mine) Berg(e)versatz *m*, (in fabric) Eintraggarn *n*, Fülle *f*, Füllung *f*, (brickwork) Gußmauerwerk *n*, (pressure) Luftdruckauffüllung *f*, (fabric) Schuß, Schußgarn *n*, Spülbetrieb *m*, ~ **a caisson** (concrete) Füllung *f* des Senkkastens ~ **a swamp** Sümpfen *n*
**filling,** ~ **agent** (loading material) Füllmittel *n* ~ **apparatus** Abfüllapparat *m* ~ **bays** Abfüllstellen *pl* ~ **block (or board)** Feilfutter *n* ~ **body** Füllkörper *m* ~ **bowl** Füllbehälter *m* ~ **capacity** Füllmenge *f* ~ **(or cleansing) cellar** Abfüll-keller *m*, -raum *m* ~ **chamber** Zulaufkammer *f* ~-**chest process** Füllkastenverfahren *n* (r. r.) ~ **chute** Füllkanal *m* ~ **compound** Ausguß, Füll-masse *f* ~ **connection (or coupling)** Füllanschluß *m* ~ **course** Schüttlage *f* ~ **cylinder** Füllflasche *f* ~ **device** Einfüll-, Füll-vorrichtung *f* ~ **earth** Füllerde *f* ~ **effect** Fülleffekt *m* ~ **efficiency** Füllleistung *f* ~ **extract** Füllextrakt *m* ~ **factor** Füllfaktor *m* ~ **frame** Aufsetzrahmen

*m*, Füllspant *n* ~ **funnel** Fülltrichter *m* ~ **heads** Füllstellen *pl* ~ **height** Füllhöhe *f* ~ **height accuracy** Füllhöhengenauigkeit *f* ~ **hole** Einfüllloch *n*, -öffnung *f*, Füllöffnung *f* ~-**hole plug (or screw)** Füllochschraube *f* ~ **hose** Betankungsleitung *f*, Zapfschlauch *m* ~-**in cam** Führungsschloßteil *n* ~-**in paste** Füllmasse *f* ~-**in work** Füllmauerwerk *n* ~ **lattice** Füllungslattentuch *n* (textiles) ~ **level accuracy** Füllhöhengenauigkeit *f* ~ **limit** Füllgrenze *f* ~ **machine** Füllmaschine *f* ~ **mark** Füllstrich *m* ~ **mass** Ausfüllmasse *f* ~ **material** Füll-körper *m*, -material *n*, -mittel *n*; Versatz-berg *m*, -stoff *m* ~ **neck** Einfüllstutzen *m* ~ **neck plug** Einfüllschraube *f* ~-**out** Ausfüllung *f* ~ **paste** Füllmasse *f* ~ **pencil** Füllbleistift *m* ~ **piece** Füllstück *n*, Futter *n* ~ **pile of a cofferdam** Füllpfahl *m* eines Fangdammes ~ **pipe** (petroleum) Füllungsleitung *f* ~ **pipe line** Fülleitung *f* ~ **pipette** Einfüllpipette *f* ~ **plant** Füllanlage *f* ~ **plug** Füllzapfen *m* ~ **pole** Zapfmast *m* ~ **position** Füllstellung *f* ~ **power** Füllvermögen *n* (Farbe) ~ **pressure** Abfülldrücken *n*, Füllungsdruck *m* ~ **properties** Füllkraft *f* ~ **ratio** Füllungsverhältnis *n* ~ **regulation** Füllungsregelung *f* ~ **room** Füllstube *f* ~ **screw** Füll-schraube *f*, -verschraubung *f* ~ **sleeve** Füllansatz *m* ~ **sleeve cord** Füllansatzleine *f* ~ **sleeve valve** Füllansatzventil *n* ~ **space** Aufgußraum *m*, Trübraum *m* ~ **spout** Füllstutzen *m* ~ **station** Abfüllstation *f*, Füll-station *f*, -stelle *f*, Tank-station *f*, -stelle *f*, Zapfstelle *f* ~ **tap** Abfüllhahn *m* ~ **technique** Abfülltechnik *f* ~ **timber** Füllspant *n* ~ **timbers** Binnenpfähle *pl* ~ **tube** Einfüllstutzen *m* ~ **unit** Füllerteil *m*, Nachfülleinrichtung *f* ~-**up** Nachfüllen *n*, Versetzen *n* ~-**up cables** Ausgießen *n* der Kabel (electr.) ~-**up pipe** Abfüllschlauch *m* ~ **valve** Füllventil *n* ~ **valve control** Füllhahnsteuerung *f* ~ **vent** Einfüllstutzen *m* ~ **weight** Füllgewicht *n*
**fillister** (plane) Falzhobel *m* ~ **head** Linsenkopfschraube *f*, runder Kopf *m*, Rundkopf *m* ~-**head screw** Linsen-, Zylinder-schraube *f* ~-**headed screw with undercut neck** Linsenhalsschraube *f* ~ **socket head screw** Zylinderschraube *f* mit Innensechskant
**film, to** ~ beschlagen, filmen, (Häutchen) überziehen
**film** Anstrichfilm *m*, Belegung *f*, Film *m*, Folio *n*, laminare Grenzschicht *f*, Haut *f*, Negativmaterial *n*, Schicht *f*, Überzug *m* ~ **of foreign material** (at rectifying junction) Fremdschicht *f* ~ **of getter metal** Gettermetallspiegel *m* ~ **of oil** Ölhauch *m* ~ **and outdoor direct pickups** Film- und Freilichtaufnahmen *pl* ~ **with photographed sound** Lichttonfilm *m* ~ **of radioactive matter** radioaktiver Niederschlag *m* ~ **and sound cutting** Film- und Tonbandschnitt *m*
**film,** ~ **adhesive** Filmklebemittel *n* ~ **arrays** Dünnfilm-Matrix *f* ~ **badge** fotografische Platte *f* ~ **band** Filmband *n* ~ **base** Blankfilm *m*, Filmträger *m*; Schichtträger *m* ~ **base face (or side)** Trägerseite *f* ~ **buckling** Filmwölbung *f* ~ **calender** Folienkalander *m* ~ **camera** Filmkammer *f* ~ **carrier** (TV) Film-träger *m* ~ **cartridge** Filmpatrone *f* ~ **cassette** Filmkassette *f* ~ **casting solution** Filmgießlösung *f* ~ **channel** (camera) Belichtungskanal *m*, Film-bahn *f*,

-kanal *m*; Führungskanal *m* ~ **coat** Filmbildschicht *f* ~ **coefficient** Wärmeübergangszahl *f* ~ **creep** Kriechen *n* dünner Flüssigkeitsschichten ~ **cup** Filmschale *f* ~ **cutter lamp** Filmprüflampe *f* ~ **demonstration** Filmvorführung *f* ~ **density** Filmschwärzung *f* ~ **dosimeter** Filmdosimeter *n* ~ **drive** Filmlaufwerk *n* ~ **drive coupling pin** Kassettenschlüssel *m* ~ **drive sprocket** Filmtransportrolle *f* ~ **drum** Filmtrommel *f* ~ **duplicating** Filmdoppeln *n* ~ **editing** Filmschneiden *n* ~ **element** Folioelement *n* ~ **emulsion** Filmbildschicht *f* ~ **exposure** (in gate) Filmvorführung *f* ~ **feed** (from magazine spool) Ablauf *m*, Film-fortschaltung *f*, -führung *f*, -schaltung *f*, -transport *m*, -vorschub *m*, Schaltwerk *n* ~ **feed mechanism** Laufwerk *n* ~-**feeder pin** Greiferstift *m* ~-**footage counter** Filmzählwerk *n*, Meterzähler *m* ~ **formation** Hautbildung *f*, Oberflächenfilmbildung *f* ~ **former** filmbildendes Material *n* ~**free** schleierfrei ~ **gate** Abtaststelle *f*, Bildabtaststelle *f*, Bildfenster *n*, Film-belichtungsstelle *f*, -ebene *f*, -fenster *n*, -führung *f*, -tür *f* ~ **gate with two frames** Bildfenster *n* mit zwei Ausschnitten ~ **glue** Klebefilm *m* ~ **grain** Filmkorn *n* ~-**grain noise** Kornrauschen *n* ~ **guide roller** Filmführungsrolle *f* ~ **guiding** Filmführung *f* ~ **holder** Kassette *f* ~ **illuminator** Negativschaukasten *m* ~ **indicator** Filmmerkscheibe *f* ~ **layer** Halbleiterschicht *f* ~ **lending (institute)** Filmverleih *m* ~ **library** Filmarchiv *n* ~ **loop** Filmschleife *f* ~ **magazine** (roll film) Filmkassette *f*, Filmmagazin *n* ~-**magazine roll** Filmvorratsspule *f* ~ **marker** Filmbezeichnungsgerät *n* ~ **medium** Schallträger *m* ~ **memory** Filmspeicher *m* ~-**memory driver** Filmspeicherschreibstufe *f* ~ **motion** Filmfortschaltung *f* ~ **movement** Filmschaltung *f* ~-**movement mechanism** Filmmechanismus *m* ~ **negative** Kehrbild *n* ~ **output** Filmwiedergabeleistung *f* ~ **pack** Filmpack *n* ~-**pack magazine** Filmpackkassette *f* ~ **pickup** Filmgeber *m* ~ **plane** Filmebene *f* ~-**polish measurement** Mattierungsmessung *f* ~ **processing** Filmverarbeitung *f* ~-**processing equipment** Filmentwicklungsvorrichtung *f* ~ **processor** Filmentwicklungsvorrichtung *f* ~ **projection** Filmvorführung *f* ~ **projector** Filmprojektor *m* ~ **projector lamp** Filmprojektionslampe *f* ~ **properties** Filmeigenschaften *pl* ~ **pull** Filmzug *m* ~ **pulldown** Fortschaltung *f* des Filmes ~ **reel** Filmrolle *f* ~ **resulting from developing by reversal** Umkehrfilm *m* ~ **ring** Filmring *m*, Fingerring *m* mit Filmplatte ~-**roll cutting machine** Filmrollenschneidemaschine *f* ~-**roll dark slide** Rollkassette *f* ~-**roll holder** Filmspulenhalter *m* ~ **run** Filmlauf *m* ~-**run control** Filmlaufkontrolle *f* ~ **rupture** Reißen *n* des Filmes ~ **rust** Flugrost *m* ~ **scan disk** Ringlochscheibe *f* ~ **scanner** Film-abtaster *m*, -geber *m* ~ **scanning** Filmabtastung *f*, -übertragung *f* ~-**scanning disk with apertures circularly rather than spirally arranged** Kreislochscheibe *f* ~ **screen** Filmschablone *f* ~ **setting** Licht-satz *m*, -setzen *n* ~ **size** Filmformat *n* ~ **slide** Filmkufen *pl* ~ **specification** Dünnfilmvorschrift *f* ~ **speed** Filmgeschwindigkeit *f* ~ **speed control** Filmgeschwindigkeitsregler *m* ~ **speed indicator**

Tachometer *n* ~ **splicer** Klebeapparat *m* ~-**splicing gauge** Filmkittlehre *f* ~ **spool** Filmspule *f* ~ **(-feed) sprocket** Filmtransportrolle *f* ~ **station** Filmstelle *f* ~ **strip** Film-band *n*, -streifen *m*, Tonspur *f*, Trum *n* ~ **support** Filmträger *m* ~ **-take-up reel** Filmauflaufspule *f*, Filmaufwickelspule *f*, Filmrolle *f* ~-**take-up spool** Film-auflaufspule *f*, -aufwickelspule *f*, -führungsrolle *f*, -rolle *f* ~ **tape** Filmband *n* ~ **television** Fernkinematografie *f*, Fernkino *n* ~ **tension** Filmspannung *f* ~ **theory** (of passivation of metals) Bedeckungstheorie *f* ~ **thickness** Schichtdicke *f* ~ **threading** Durchsetzen *n* des Films, Filmeinfädelung *f* ~ **track** Belichtungskanal *m*; Film-bahn *f*, -kanal *m*, Führungskanal *m* ~ **traction** Filmspannung *f* ~-**traction regulator (stabilizer)** Filmregler *m* ~ **transmitter** Fernkinosender *m*, Filmgeber *m* ~ **transport mechanism** Kassettenlaufwerk *n* ~ **trap** Bildbühne *f*, Bildfenster *n*, (projection work) Druckrahmen *m*, Filmfenster *n* ~ **travel** Filmfortschaltung *f* ~ **uniformity** Filmgleichmäßigkeit *f* ~ **viewer** Durchleuchtungskasten *m* (photo) ~ **viewer lamp** Filmprüflampe *f* ~-**viewing machine** Betrachtungstisch *m* ~-**winding mechanism** Filmaufwickelvorrichtung *f* ~ **wise** filmartig

**filter, to** ~ abfiltern, abfiltrieren, abflachen, absperren, durchfiltern, durchgeben, durchgießen, durchlassen, durchlaufen, durchseihen, filtern, filtrieren, kolieren, (a liquid) läutern, sieben to ~ **away** wegfiltrieren **to** ~ **by means of suction** abnutschen **to** ~ **off** abfiltern, abfiltrieren, abseihen, wegfiltrieren **to** ~ **out** abfiltern, ausfiltern **to** ~ **by suction** nutschen

**filter** Abscheider *m*, Durchguß *m*, Durchlaß *m*, Durchschlag *m*, Filter *m*, Filtervorsatz *m*, Kettenleiter *m*, Seiher *m*, Sieb *n*, (circuit or device) Siebgebilde *n*, Siebkettenleiter *m*, (circuit) Siebschaltung *f*, Störschutz *m* (electr.), Weiche *f*, (wave) Wellenfilter *n*, Wellensieb *n*, ~ **for band elimination** Bandsperrfilter *n* ~ **in gas mask** Atemfilter *m* ~ **for micrographic work** Mikrofilter *n* ~ **for monochromator entrance slit** Vorzerleger *m* ~ **for secondary oil circulation** Nebenstromölfilter *m* ~ **for selective contrasts** Kontrastfilter *n*

**filter,** ~ **acorn** Filterziermutter *f* ~ **aid** Filtermittel *n*, Hilfsmittel *n* für die Filtration ~ **area** Filterfläche *f* ~ **arrangement** Filterabgriff *m* ~ **attenuation band** Filtersperrbereich *m* ~ **bag** Filter-beutel *m*, -sack *m* ~ **band width** Frequenzsieblochbreite *f* ~ **basin** Sickerbecken *n* ~ **bed** Filter-bett *n*, -schicht *f*, Filtrationsschicht *f* ~ **block** Filtermasse *f* ~ **box** Filterkasten *m* ~ **bracket** Filterbrücke *f* ~ **cake** Filterkuchen *m* ~-**cake removing device** Kuchenabnahmevorrichtung *f* ~ **canister** Büchse *f* ~ **capacitor** Siebkondensator *m* ~ **capacity** Siebkapazität *f* ~ **cartridge** Filterkerze *f*, Stabfilter *n* ~ **case** Filtergehäuse *n* ~ **cavity** Filterschacht *m* ~ **cell** Filtertasche *f* ~ **cell unit** Zellenfiltereinsatz *m* ~ **chamber** Filterkammer *f* ~ **chain** Siebkette *f* ~ **choke** Siebdrossel *f*, Verdrosselung *f* ~ **circuit** Filter-, Reinigungs-, Saug-, Sieb-kreis *m* ~ **cleaning liquid** Filterreinigungsflüssigkeit *f* ~ **cloth** Filter-tuch *n*, -stein *m*, Koliertuch *n* ~ **cloth consumption** Filtertuchverbrauch *m* ~ **cloth support** Filtertuchunterlage *f* ~ **coke**

Preßkuchen *m* ~ **condenser** Glättungs-, Siebkondensator *m* ~ **cone** Filter-einlage *f*, -kegel *m*, -konus *m*, -schoner *m*, Trichtereinlage *f* ~ **container** Filtergehäuse *n* ~ **core** Siebkern *m* ~ **crossbar** Filterkreuzschiene *f* ~ **cup** Filterdeckel *m* ~ **cylinder** Filterzylinder *m* ~ **discrimination** Selektivität *f* oder Trennschärfe *f* eines Filters ~ **draw-off** Filterablaß *m* ~ **drum** Filtertrommel *f* ~ **drum cover** Bespannung *f* der Filtertrommel ~ **effect** Filterwirkung *f* ~ **element** Filter-einsatz *m*, -satz *m*, Siebmittel *n* ~**-element container** Einsatztopf *m* ~ **element in gas mask** Atemeinsatz *m* ~ **end cap** Filterpatrone *f* ~ **factor** Filterfaktor *m* ~ **flask** Absaugekolben *m* ~ **floor** Filterbelag *m* ~ **foil** Filterfolie *f* ~ **frame** Filterrahmen *m* ~ **funnel** Filter-, Filtriertrichter *m* ~ **gauze** Filtergewebe *n* ~ **gravel** Filter-kegel *m*, -kies *m* ~ **handle** Filterbetätigung *f* ~ **holder ring** Filterring *m* ~ **hood** Filterglocke *f* ~ **hose** Filterschlauch *m* ~ **housing** Filtergehäuse *n* ~ **insert** Filtereinsatz *m* ~ **jumper ring** Filtertragring *m* ~ **law** Filtergesetz *n* ~ **layer** Filterlage *f*, Filtrationsschicht *f* ~ **lubricating oil** Filterschmieröl *n* ~ **material** Siebmittel *n* ~ **medium** Auflagefilter *f* ~ **mesh** Filtergewebe *n*, -glied *n* ~ **mud** Filterschlamm *m* ~ **output** Filterleistung *f* ~ **pad** Filterkörper *m* ~ **paper** Filter-blatt *n*, -papier *n* ~ **pass band** Filterdurchlaßbereich *m* ~ **pipe** Filterrohr *n* ~ **plate** Filter-belag *m*, -scheibe *f* ~ **plate with raised edges** Filterschale *f* ~ **plates** Filterplattenpaket *n* ~ **pocket** Filtertasche *f* ~ **pouch** Filtertasche *f* ~ **press** Filterpresse *f*, Preßfilter *m* ~ **press cake-removing device** Filterkuchenabnahmevorrichtung *f* ~ **press with central feed** Filterpresse *f* mit zentraler Schlammzuführung ~ **press with hydraulic closing device** Filterpresse *f* mit hydraulischem Verschluß ~ **press with side-feed channel** Filterpresse *f* mit seitlichem Schlammkanal ~ **pulp** Filterschleim *m* ~ **range** Durchlässigkeitsbereich *m* eines Filters ~ **regulator (for pneumatic control system)** Reduzierstation *f* ~ **replacement** Filterwechsel *m* ~ **resistance (or resistor)** Siebwiderstand *m* ~ **roller arms** Pendelrollen *pl* des Tonlaufwerks ~ **run-around check valve** Filterumgehungsventil *n* ~ **safety-cock fitting (armature)** Filterbrandhahn *m* ~ **safety device** Filtersicherung *f* ~ **screen** Filtersieb *n* ~ **section** Filterglied *n*, Glied *n* eines Filters, Kettenglied *n*, Kettenleiterglied *n*, Siebpartie *f* ~ **slot** Sickerschlitz *m* ~ **spiral** Filterspirale *f* ~ **spout** Filterschlauch *m* ~ **stand** Filter-gestell *n*, -stativ *n* ~ **strainer** Filterkorb *m* ~ **support** Filterstütze *f* ~ **tank** Reinigungsbehälter *m* ~ **termination** Filterabschluß *m* ~ **trough** Filtertrog *m* ~ **tube** Filterröhre *f*, -stab *m* ~ **type equalizer** Entzerrungsfilter *n*, -kette *f* ~ **unit** Filter-einheit *f*, -einsatz *m*, -glied *n* ~ **washing plant** Waschfilterstraße *f*

**filterable** filtrierbar

**filtered** gesiebt ~ **display** gefilterte Anzeige *f* (rdr) ~ **syrup** Kochklare *f*

**filterer** Filtrierer *m*

**filtering** Einsickern *n*, Filterung *f*, Filtrierung *f*, Siebung *f*, Trennung *f* ~ **by means of a funnel** Abnutschen *n*

**filtering,** ~ **action** Ausscheidungseffekt *m*, Filterwirkung *f* ~ **apparatus** Filtereinsatz *m*, Fil-

trierapparat *m* ~ **bag** Filtrierbeutel *m*, Seihesack *m* ~ **basin** Filterbecken *n* ~ **basket** Filterkorb *m* ~ **candle** Spinnkerze *f* (rayon) ~ **charcoal** Filterkohle *f* ~ **cloth** Filtertuch *n*, Filtriertuch *n* ~ **cone** Filtrierkonus *m* ~ **disc** Filtrierscheibe *f* ~ **dish** Filtrierschale *f* ~ **driving shoe** Filterrammspitze *f* ~ **effect** Ausscheidungseffekt *m* ~ **efficiency** Filterleistung *f* ~ **flask** Nutsche *f* ~ **fountain** Filterrohrbrunnen *m* ~ **funnel** Filter-, Filtrier-, Seihe-trichter *m* ~ **grade** Durchlässigkeitsgrad *m* ~ **jug** Filtertopf *m* ~ **layer** Filterschicht *f* ~ **material** Bodenschicht *f* ~ **medium** Filterstoff *m* ~ **number** Durchlässigkeitsnummer *f* ~ **pellicle** Filterhäutchen *n* ~ **property** Filtrierbarkeit *f* ~ **sieve** Filtriersieb *n* ~ **stone** Filterstein *m* ~ **tissue** Filtergewebe *n* ~ **tube** Filterrohr *n*

**filth** Dreck *m*, Kot *m*, Schmutz *m*

**filthy** schmutzig

**filtrability** Filtrierbarkeit *f*

**filtrate, to** ~ durchfiltern

**filtrate** Filter-gut *n*, -wasser *n*; Filtrat *n*, Filtricht *n*, Kolatur *f* ~ **centrifugal pump** Filtratkreiselpumpe *f* ~ **output** Filtratleistung *f* ~ **suction pipe** Filtratsaugleitung *f*

**filtration** Durchguß *m*, Filtration *f*, Filtrierung *f*, Siebung *f* ~ **plant** Massenfilter *m* ~ **velocity** Filtergeschwindigkeit *f*

**fin** Bart *m*, Finne *f*, Flosse *f*, Floßfeder *f*, (motar shell) Flügelschaft *m*, Grat *m*, Gußfuge *f*, Gußnaht *f*, Keilfuge *f*, Lappen *m* (mech.), Leitblech *n*, (aviat) Leitfläche *f*, (tail) Leitwerksflosse *f*, Naht *f*, Rippe *f*, Ruder *n*, Sieke *f*, Stanzfuge *f*, Walzgrat *m* ~ **adjusting instrument** Flossenverstellgerät *n* ~ **antenna** Flossenantenne *f* ~ **area** Feld *n* der Flossen (aviat.) ~ **assembly** (rocket, bomb) Leitwerk *n*, (aircraft) Seitenflosse *f* ~ **carrier** Flossenträger *m* ~ **contact wire** Leitung *f* zum Ruderkontakt (g/m) ~ **fuse crystal** Ruder-Zündkristall *n* (g/m) ~ **jitter** Ruderflattern *n* ~ **keel** Flossenkiel *m* ~ **less** gratlos ~ **pitch** Rippen-abstand *m*, -teilung *f* ~ **removal by heat** Warmentgratung *f* ~ **restraint** Ruderfesselung *f* (g/m) ~ **root** Rippenwurzel *f* ~ **shaft for mortar shell** Minenschaft *m* ~ **spacing** Rippenteilung *f* ~ **stabilized** pfeilstabil ~ **stabilized missile** Flossengeschoß *n* ~ **stabilized rocket non-rotating in flight** flossengedämpfte Rakete *f* ~ **tip** Rippenspitze *f* ~ **width** Rippen-breite *f*, -höhe *f*

**finagle, to** ~ ausklügeln

**final, to give a** ~ **polish** nachpolieren

**final** Endamplitude *f*; abschließend, endgültig ~ **acceptance** endgültige Abnahme *f* ~ **account** Abschlußrechnung *f* ~ **amplifier** Endverstärker *m*, Leistungsstufe *f* ~ **amplifier stage** Endverstärkerstufe *f* ~ **analysis of intelligence** Endauswertung *f* ~ **anode** Endanode *f* ~ **approach** Endanflug *m* ~ **approach level** Endanflughöhe *f* ~ **assembly** Endaufbau *m*, Endmontage *f*, Fertigmontage *f*, Rückmontage *f* eines Motors (nach der Zerlegung) ~ **attenuation** Endvergärung *f* ~ **balance** Endgleichgewicht *n* ~ **balance sheet** Hauptabschlußbogen *m* ~ **boiling point** Endsiedepunkt *m* ~ **cleaning** End-, Nachreinigung *f* ~ **clearance** Schlußabfertigung *f* ~ **coat** Schlußanstrich *m* ~ **combustion** Ausbrand *m* ~ **condition** Endzustand *m* ~ **connector**

Leitungswähler *m* (L. W.) ~ **control desk** Endkontrollpunkt *m* ~ **control element** Endregelungsgerät *n*, Regel-, Stell-glied *n* ~ **controlled condition** Sollwert *m* der Regelgröße ~ **controller** Landeradarlotse *m* ~ **copy** Reinschrift *f* ~ **costs** Endkosten *pl* ~ **date of packing parachute** Packtermin *m* ~ **desulfurization** Endschwefelung *f* ~ **diaphragm** Schlußblende *f* ~ **drive** Endtrieb *m*, Gelenkwelle *f* (Auto), Seitengetriebe *n* ~ **drive housing** Achsantriebsgehäuse *n*, Achstrichter *m*, Seitengehäuse *n* ~ **drive shaft** Achsabtriebvorgelegewelle *f*, Endtriebwelle *f* ~ **etching** Nachätzen *n* ~ **expansion pressure** Expansionsenddruck *m* ~ **felling** gänzlicher Abhieb *m* ~ **finishing** Fertig-polieren *n*, -schlichten *n* ~ **forging** Weiterverschmieden *n* ~ **gettering** Nachgetterung *f* ~ **hardening** Schlußhärten *n* ~ **image luminescent screen** Endbildleuchtschirm *m* ~ **inspection** Endkontrolle *f*, Nachrevision *f* ~ **judgment** Endurteil *n* ~ **kilning temperature** Abdarrtemperatur *f* ~ **liquor** Endauflage *f* ~ **machining** Fertigbearbeitung *f* ~ **magnification** Endvergrößerung *f* ~ **mass (of rocket at end of flight)** Endmasse *f* ~ **modulation** Direktmodulation *f* (Richtfunk) ~ **number of revolutions** Enddrehzahl *f* ~ **objective** Endziel *n* ~ **operation** Endarbeitsgang *m* ~ **ordinate** Endordinate *f* ~ **oscillating** Auspendeln *n* des Motors ~ **output transmission angle** abtriebsseitig wirksamer Gesamtübertragungswinkel *m* ~ **pass** End-, Polier-, Schlicht-stich *m*, Schlußkaliber *n* ~ **picture (or end image)** Endbild *n* ~ **plot plans** endgültige Lagerpläne *pl* ~ **point** Endpunkt *m* (einer Kurve) ~ **polishing pass** Fertigpolier-, Fertigschlicht-kaliber *n* ~ **polishing pass** Fertigschlichtstich *m* ~ **position** Endlage *f* **in ~ position** in die endgültige Lage *f* ~ **pressure** Enddruck *m* ~ **processing object** Verarbeitungsziel *n* ~ **product** End-erzeugnis *n*, -produkt *n* ~ **protective fire** Feuersperre *f* ~ **provision** Schlußbestimmung *f* ~ **purification** Fertigfrischen *n*, Nachreinigung *f* ~ **purifier** Nachreiniger *m* ~ **purpose** Endzweck *m* ~ **range value** Bereichsendwert *m* ~ **reaction** Endreaktion *f* ~ **regulator** Summenregler *m* (Mischpult) ~ **rejection** endgültiger Versagungsschluß *m* ~ **remark (or sentence)** Schlußsatz *m* ~ **report** Abschlußmeldung *f*, Schlußbericht *m* ~ **result** Endergebnis *n* ~ **safety trip** Endausklinkung *f* ~ **saturation** Nachsaturation *f* ~ **screen** Endschirm *m* ~ **selector** Leitungswähler *m* (L. W.), Linienwähler *m* ~ **selector (bank) multiple** Leitungswählervielfach(feld) *n* ~-**sequence switch and relay set** Relaissatz *m* und Folgeschalter *m* für Leitungswähler ~-**sequence switch-and-relay set** Folgeschalter *m* und Relaissatz *m* für Leitungswähler ~ **set** Ende *n* des Abbindens, Endsatz *m* ~ **settlement** Schlußabrechnung *f* ~ **shade** Endnuance *f* ~ **shipment** Restlieferung *f* ~ **shot** heiße Probe *f* ~ **sintering operation** Fertig-, Hoch-sinterung *f* ~ **size desired** Endfeinheit *f* ~ **slag** End-, Feinungs-, Reduktions-schlacke *f* ~ **stage** Abtriebsschlag *m* (metal), Kraftstufe *f* ~ **stage valve** Endröhre *f* ~ **state** Endzustand *m* ~ **steam temperature** Überhitzungstemperatur *f* ~ **sulfitation of thin juice** Dünnsaft-endschwefelung *f*, -schlußschwefelung *f* ~ **surfacing** endgültige Ober-

flächenausführung *f* ~ **switch** Leitungswähler *m* ~ **switch rack** Leitungswählergestell *n* ~ **tanning** Ausgerbung *f* ~ **temperature** Endtemperatur *f* ~ **test** Endabnahme *f*, (smelting) Fertigprobe *f*, Generaldurchschaltung *f* ~ **test flight** Nachfliegen *n* ~ **treatment** Nachbehandlung *f* ~ **turn** Endanflugkurve *f* ~ **value** Endwert *m* ~ **value theorem** Ähnlichkeitssatz *m* ~ **velocity** Auftreffgeschwindigkeit *f* ~ **voltage** Endspannung *f*, Spannung *f* am Ende ~ **welding** Fertigschweißen *n* ~-**welding jig** Endzusammenschweißvorrichtung *f*

**finance,** ~ **accounts** Kassenwesen *n* ~ **regulation** Kassenordnung *f* ~ **section** Kassenabteilung *f* ~ **service** Kassendienst *m*

**financial** finanziell ~ **administration** Kassenverwaltung *f* ~ **backer** Geldgeber *m* ~ **business** Geldwesen *n* ~ **embarrassment** Geldverlegenheit *f* ~ **index (or numbers)** betriebswirtschaftliche Kennzahlen *pl* ~ **year** Geschäftsjahr *n*

**financially strong** kapitalkräftig

**find, to** ~ abtasten, aufsuchen, befinden, entdecken, (law) erkennen, frei suchen to ~ **acceptance** Anklang *m* finden **to ~ by boring** erbohren **to ~ direction** peilen **to ~ by feel (or touch)** ertasten **to ~ fault with** bemängeln, monieren **to ~ a layer by boring** eine Schicht *f* anbohren **to ~ out** auffinden, ausfindigmachen, ergründen, ermitteln, eruieren, konstatieren, suchen **to ~ a position** sich orientieren **to ~ a way into** Eingang *m* finden **to ~ one's way about** zurechtfinden

**find** Fund *m*

**finder** Finder *m*, Sucher *m* ~ **circle** Finderteilkreis *m* ~ **frame** Bildsucher *m*, Rahmensucher *m* ~ **lens (or objective)** Sucherobjektiv *n* ~ **level (or bubble)** Sucherlibelle *f* **(line)** ~ Anrufsucher *m* ~ **parallax** Sucherparallaxe *f* ~ **rack** Anrufsuchergestell *n* ~ **switch** Anrufsuchen *n*, -sucher *m* ~ **telescope** Suchfernrohr *n*

**finding** Aufsuchen *n*, Bestimmung *f*, (law) Erkenntnis *f*, Suchen *n* ~ **or course** Fahrtbestimmung *f* ~ **action** Anrufsuchen *n*, Vorwahl *f* ~ **extreme values** Extremierung *f* ~ **process** Frischarbeit *f*

**findings** Befunde *pl*, Ermittlungen *pl* ~ **of studies** Ergebnisse *pl* von Untersuchungsarbeiten

**fine, to** ~ feinen, frischen, gerben, verurteilen **to ~-grind** savonieren

**fine** engmaschig, dünn, fein, (gravel) klein, rein, schlicht **as ~ as dust** staubfein

**fine** Geldstrafe *f*, Strafe *f* (by law) ~ **for nonperformance of contract** Konventionalstrafe *f*

**fine,** ~ **adjusting device** Feinregulierung *f* ~ **adjusting screw** Feineinstellschraube *f* ~ **adjustment** Fein-einstellung *f*, -richtung *f* (of a tool or cutter), -stellen *n* (of a tool or cutter), -stellung *f* (of microscope), Triebbewegung *f* ~ **adjustment device** Feinstellvorrichtung *f* ~ **adjustment screw** Fein-schraube *f*, -stellspindel *f* ~ **annual** feinjährig ~ **beam** Feinkennung *f* (rdo nav.) ~ **boring block** Feinbohrblock *m* ~ **boring tool** Feinbohrmesser *n* ~**brass wire** Messingfeindrähte *pl* ~ **brushes** Feinbürsten *pl* ~ **cardboard box** Feinkartonage *f* ~ **casting** Edelguß *m* ~-**celled** feinzellig ~ **coal** Fein-, Staub-kohle *f* ~ **coal jig** Feinkornmaschine *f* ~ **coal sump** Feinkohlensumpf *m* ~ **coke** Koksgrus *m*

~ **control** Feineinstellung *f* ~ **control rod** Fein-regelstab *m* ~ **crusher** Feinbrecher *m* ~ **crushing** Fein-mahlen *n*, -walzen *n*, -zerkleinern *f*, -zerkleinerung *f* ~ **details** Feinheiten *pl* ~ **dispersion** Vernebelung *f* ~ **displacement** Fein-verschiebung *f* ~ **division** Feinverteilung *f* des Eisens ~ **drawer** Feinstrecker *m* ~ **dust** Flug-staub *m* ~ **error voltage bridge** Einer-Einstell-brücke *f* (Tacan) ~ **etching** Feinätzung *f* ~ **feed** geringer Vorschub *m* ~ **fibered (or fibrous)** fein-faserig ~ **file** Schlichtfeile *f* ~ **filter sheets** Fein-filterschichten *pl* ~-**filtering component** fein-filtrierender Anteil *m* ~ **finish** Glättung *f*, hohe Oberflächengüte *f* ~ **finishing** Feinschlichten *n* ~ **focus diaphanoscopes** Feinfokus-Durchleuch-tungsgeräte *pl* ~-**gauge (wire)** schwach ~ **gold** Fein-, Kapellen-gold *n* ~ **graduated circle** Fein-teilkreis *m* ~ **grain** Feinkorn *n*, feines Korn *n* ~ **grain data** Feindaten *pl*

**fine-grain,** ~ **developer** Feinkornentwickler *m* ~ **film** Feinkornfilm *m* ~ **quality** Feinkörnigkeit *f* ~ **sand** feinkörniger Sand *m* ~ **washing machine** Feinkornsetzmaschine *f*

**fine-grained** fein-faserig, -körnig, -kristallinisch, -porig ~ **density** feine Dichte *f* ~ **fracture** fein-körnige Bruchfläche *f* ~ **iron** Feineisen *n* ~ **river sand** Silbersand *m* ~ **sand** feinkörniger Sand *m*

**fine,** ~-**granular** feinkörnig ~ **granulation** Fein-bruch *m* ~ **gravel (coal slack)** Grus *m*, Splitt *m* ~ **grid** engmaschiges Gitter *n* ~ **grinding** Fein-mahlen *n*, -mahlung *f*, -schleifen *n*, -zerklei-nern *n*, -zerkleinerung *f* ~ **grist** Feinschrot *n* ~ **hair brush** Haarpinsel *m* ~ **indentations** feine Zahnung *f* ~ **levelling** Feinhorizontierung *f* ~ **linen** Feinleinen *n* ~ **liquor (sugar)** Klärsel *n* ~ **mechanics** Feinmechanik *f* ~ **(coarse) mesh** feines (grobes) Geflecht *n* ~-**mesh filter** Fein-filter *n* ~-**mesh screen** engmaschiges Sieb *n* ~-**meshed** englochig, feinmaschig ~ **motion** Fein-bewegung *f*, -trieb *m* ~ **muslin** Mull *m*, Nessel-tuch *n* ~ **ore** Feinerz *n* ~-**ore furnace** Feinerz-ofen *m* ~ **pitch** Flacheinstellung *f* (propeller), kleine Steigung *f* ~ **pitch screen** Feingewinde-spindel *f* ~-**pointed flame** Stichflamme *f* ~ **polishing** Genauigkeitspolieren *n* ~-**pored** fein-porig **of** ~ **porosity** kleinluckig ~ **progression** feingestuft ~ **quantum number** Feinstruktur-quantenzahl *f* ~-**range oscilloscope** Entfer-nungsfeinmeßröhre *f* ~-**range presentation unit of radar equipment** Beobachtungsteil *n* ~-**ranging unit** Emilentfernungsanzeige *f* (rdr.), Entfernungsanzeigegerät *n* ~ **rasp** Feinraspel *f* ~ **reduction jaw crusher** Feinbrecher *m* ~-**regulating rod** Feinregelstab *m* ~ **regulation** Feinregulierung *f* ~ **sand** Flußkies *m* ~ **sand slope** Böschung *f* aus Feinsand ~-**scaled magnetic structure** feinschuppige magnetische Struktur *f* ~ **screen** Feinsieb *n* ~-**screened** fein-gesiebt ~ **sea gravel** Fluß-kies *m*, -schotter *m* ~ **sea sand** Flußschotter *m* ~ **section** Feinprofi-lierung *f* ~ **setting** Feineinstellung *f* ~ **sight** Feinkorn *n* ~ **silver** Kapellensilber *n* ~ **small coal** Feinförderkohle *f* ~ **small duff** gewaschene Feinkohle *f* ~ **smudge** Feinkohle *f* ~ **steel** Edel-, Herd-stahl *m* ~ **splitting machine** Feinspalt-maschine *f* ~ **stoneware** Feinsteinzeug *n* ~-**stranded** dünndrähtig ~ **structure** Fein-gefüge

*f*, -**struktur** *f*; Kleingefüge *n* ~ **structure doublet** Feinstrukturdublett *n* ~ **thread** Feingewinde *n*, scharfgängiges Gewinde *n* ~ **tin** Fein-, Klang-zinn *n* ~-**toothed saw** feingezähnte Säge *f* ~ **T. T. buying rate** Feingoldkaufsatz *m* für tele-grafische Überweisungen ~ **tuning** feine Ab-stimmung, scharfe Abstimmung *f*, Antrieb *m*, Feinabstimmung *f*, Feineinstellung *f*, Nach-stimmung *f* ~-**tuning antenna** feine Antenne *f* ~ **tuning control** Feinregelung *f* ~-**tuning lever** Antriebshebel *m* ~-**tuning shaft** hohle Antriebs-welle *f* (rdo) ~ **turning** Feindrehen *n* ~ **voltage bridge** Einer-Einstellbrücke *f* (Tacan) ~ **wire coil winding machine** Feindrahtwickelmaschine *f* ~-**wire fuse** Feinsicherung *f* ~ **wire guide** Feindrahtführer *m* ~ **wood** Edelholz *n* ~ **work** Feinarbeit *f*

**fined iron** Frischfeuereisen *n*

**finely,** ~ **adjusting screw** Feinstellschraube *f* ~ **crushed rock** Kleinschlag *m* ~ **crystalline** fein-kristallin(isch) ~ **dispersed (or distributed)** fein verteilt ~ **divided** feinverteilt, feinzerteilt, staub-förmig ~ **divided carbon** Kohlengrieß *m*, Koh-lenpulver *n* ~ **divided coal** staubförmige Kohle *f* ~ **dividing** Feinzerkleinern *n* ~ **engraved roller** Tausendpunktwalze *f* ~ **fluted** feingeriffelt ~ **grained** feinkörnig (film) ~ **granular** feinkörnig (film) ~ **ground charcoal** Holzkohlenmehl *n* ~ **ground coke** Koksmehl *n* ~ **ground concentrate** feingemahlene Erzkonzentrate *pl* ~ **ground fire clay** Tonmehl *n* ~ **ground fluorite** Flußspatmehl *n* ~ **ground granulated sugar** Kristallpuder-zucker *m* ~-**ground gypsum** Feingips *m* ~ **po-rous** feinporig ~ **powdered** feingepulvert ~ **sub-divided** fein unterteilt ~ **threaded screw** Feinge-windeschraube *f*

**fineness** Feine *n*, Feingehalt *m*, Feinheit *f* (des Kornes), Schlankheit *f* ~ **of comminution** Mahl-feinheit *f* ~ **of grain** Kornfeinheit *f*, Mahlfein-heit *f* ~ **of grain in film** Auflösungsvermögen *n* ~ **of grinding** Mahlfeinheit *f* ~ **of scanning** Rasterfeinheit *f*

**fineness,** ~ **coefficient** Feinheitskoeffizient *m* ~ **grade** Feinheitsstufe *f* ~ **number (swedish)** Fin-lekstal ~ **ratio** Gleitzahl *f*, Schlankheits-grad *m*, -verhältnis *n*, Streckungsverhältnis *n*, Zu-schärfungsverhältnis *n* ~ **test** Feinheitsprobe *f*

**finer** Frischer *m*, Frischmeister *m*

**finery** Flitter *m* ~ **fire** Feinfeuer *n* ~ **hearth** Fein-feuer *n*, -ofen *m* ~ **process** Frischprozeß *m*

**fines** Abrieb *m*, Feine *pl*, Feinkohle *f*, Fein-schlag *m*, Grus *m*, fein gesiebtes Material *n*, Rückfälle *pl* (metal) ~ **jig built** Feinkohlensetz-maschine *f*

**finesse** Finesse *f*

**finest adjustment** Feinsteinstellen *n*

**finger, to** ~ befühlen, befingern, berühren, be-tasten, spielen mit . . . **to** ~ **print** daktylosko-pieren

**finger** Finger *m* ~ **of light** Lichtfinger *m*

**finger,** ~ **bit** Fingerkrone *f* ~ **board** Gestänge-rechen *m*, Griffbrett *n*, Klaviatur *f*, (of string instrument) Tastenbrett *n* ~ **contact** (control-ler's) Fingerkontakt *m* ~ **cot** Fingerling *m* ~ **disk** Fingerscheibe *f* ~ **gauges** (for thickness) Fühler *m* ~ **grip** Fingergriff *m*, Glückshaken *m* ~ **guard** Handschutzbügel *m* ~ **hole** Finger-öffnung *f*, Greifloch *n* ~ **marks** Berührungs-

stellen pl (mit den Fingern) ~ **nut** Korbmutter f
~ **pad** Fingerhut m ~ **paddle agitator** Finger-
rührer m ~ **patch** fingerförmiges Befestigungspflaster n ~ **print** Fingerabdruck m ~ **protector
for feeders** Fingerschutzvorrichtung f (print.)
~ **rim** Lupendeckel m ~**-shaped** fingerförmig
~**-shaped gate** (in casting) Fingertrichter m ~
**stall** Fingerfutter n, Fingerling m ~ **stone** Fingerstein m ~ **stop** (of dial switch) Fingeran-
schlag m ~**-tight nut** mit den Fingern angezo-
gene Schraubenmutter f ~**-type (combustion)
chamber** Fingerkammer f ~ **wheel** Fingerscheibe f
**fingering** Fingersatz m
**fingers** Finger pl
**finial** Aufsatz m der Stange, Verschlußkappe f
des Rohrständers
**fining** Frischen n, Schönen n **(re)** ~ Feinarbeit f,
Scheiden n ~ **in a charcoal bed** Löscharbeit f
**fining,** ~ **process** Frischarbeit f (metal), Frischungs-prozeß m, -verfahren n ~ **sirup man**
Deckarbeiter m ~ **substance** Klär-, Klärungsmittel n
**finish, to** ~ (a task) abarbeiten, abschleifen, abtun, aufarbeiten, aufhören, avivieren, bearbeiten, beendigen, dressieren, endbearbeiten,
enden, endigen, erledigen, fertigbearbeiten,
fertigen, fertigmachen, fertigschlichten, fertigstellen, fertigwalzen, glätten, nachbearbeiten,
nachwalzen, schlichten, schließen, veredeln,
vollenden, zubereiten, zurichten **to** ~ **bore** fer-
tigbohren **to** ~ **extrabright** hochglanzpolieren
**to** ~ **face** planschlichten **to** ~ **to gauge** auf genaues Maß n bringen **to** ~ **grinding** glattschlei-
fen **to** ~ **to hole** die Bohrung f schlichten **to** ~
**mill** nachfräsen **to** ~ **off a planed surface of
wood** das Holz abschlichten **to** ~**-ream** fertig-
reiben, nachreiben **to** ~**-rolling** fertigwalzen **to**
~ **to size** auf Fertigmaß n bearbeiten, fertigschlichten **to** ~ **to template** auf genaues Maß n
bringen **to** ~ **turn** fertigdrehen **to** ~ **turning**
schlichten
**finish** Anstrich m, Ausbringung f, Ausführung f,
äußerliche Ausführung f, Aussehen n, Bearbeitung f, Bearbeitungsgüte f, Deckanstrich m,
Ende n, Fertigstellen f, Fertigung f, Lackierung
f, Oberflächengüte f, Oberflächenzustand m,
Politur f, (glaze) Satinage f, Schluß m, Verputz
m, Vollendung f **(floor)** ~ **on face
and back** zweiseitige Appretur f ~ **and finishing
of machine parts** Ausarbeitung f ~ **(ing) of paper** Appretur f ~ **of surface in plastics** Flächen-
anziehungsbearbeitung f
**finish,** ~ **appearance** Außenbeschaffenheit f ~
**coat** Fertiganstrich m ~**-cutting** Fertigfräser m
~ **draw punch** Fertigzugstempel m ~ **grinding**
Fertig-, Nach-schleifen n ~ **line** Ziellinie f ~
**milling** Fertigfräsen n, Schlichtfräsen n ~
**planing** Fertighobeln n ~ **product** Enderzeugnis
n ~ **seasoning** Appretur f ~ **strip** verlorener
Kopf m ~ **tester** Oberflächen-prüfer m (-prüf-
gerät) ~ **(or processing) traces** Bearbeitungs-
spuren pl ~ **turning** Fertigdrehen n
**finished** (task) aufgearbeitet, fertig, fertig ge-
macht, fertiggestellt ~ **and semifinished products** ganz und halb fertige Waren pl ~ **article**
Fertigerzeugnis n, Ganzfabrikat n ~ **blank** fertiger Quarzkristall m ~ **bolt** bearbeiter Bol-

zen m ~ **concentrate** Fertigkonzentrat n ~ **flat
(plate)** geschliffen ~ **nut** bearbeitete Mutter f
~ **part** Fertigteil m ~ **product** Enderzeugnis n,
Endprodukt n, Fertig-erzeugnis n, -fabrikat n,
-produkt n, -ware f, Ganzzeug n ~ **rolled** fertig-
gewalzt ~ **section** Fertigprofil n ~ **size** Fertigmaß n ~ **state** Gare f ~ **stock** Fertigwarenbe-
stand m ~ **stores** Fertigteillager n ~ **wrought**
durchgeschmiedet
**finisher** Appretierer m, Ausrüster m, Feinzeugholländer m (paper mfg.), Fertiger m (maker),
(die) Fertiggesenk n, Nachbearbeitung f,
Ganzzeugholländer m, Polierwalze f
**finishes** Ausführungsarbeiten pl
**finishing** Appretieren n, (paper) Ausrüstung f,
Avivage f, Bearbeitung f, Feinschliff m, (of
threads) Fertigschneiden n, Nachbearbeitung f,
(photographic emulsion) Nachreifung f,
Schlichtarbeit f, Tuschieren n, Veredeln n, Zurichtung f ~ **by handlapping** Handläppschliff
m ~ **by machine lapping** Maschinenläppschliff
m ~ **in the padding machine** Klotzappretur f ~
**to size** Fertigpolieren n, Fertigschlichten n ~ **of
tissues** Appretieren n von Geweben
**finishing,** ~ **agent** Veredelungsprodukt n ~-
**allowance equalizing gauge** Materialaufteillehre
f ~ **bit** Kaliber-, Schlicht-bohrer m ~ **block**
Feinzug m ~ **blow** Schlichtschlag m ~ **carder**
(textiles) Feinkrempel m ~ **coat** Deckanstrich
m ~ **cut** Feinhieb m, Fertigschnitt m, Schlicht-
schnitt m ~ **cutter** Fertig-, Nach-, Schlichtfräser m ~ **diamond pass** Schlichtraute f ~ **die**
Endstein m, Feinzug m ~ **edger** Fertigstaucher
m ~ **flier frame** (textiles) Feinspindelbank f ~
**fly frame** Feinfleier m ~ **frame** Nachwalzgerüst
n ~ **furnace** Fertigfrischofen m ~ **groove** Polierkaliber n ~ **heat** Abdarrtemperatur f ~ **hob**
Fertigwälzfräser m ~ **industry** Veredelungs-
industrie f ~ **kiln** Glattofen m ~ **lathe** Fertig-
drehbank f ~ **machine** Appreturmaschine f,
Deckenfertiger m, Feinspinn-, Fertigungs-,
Schlicht-maschine f ~ **machine for wax cloth**
Wachstuchfertigungsmaschine f ~ **metals** Zusätze pl ~ **mill** Nachwerk n, Schlichtfräser m
~**-mill train** Fertig-straße f, -strecke f ~ **mold**
(glass-bottle making) Fertigform f ~ **nail** Ab-
arbeitungsnagel m ~ **operation** Zurichtungs-
prozedur f ~ **oval pass** Schlichtoval n ~ **pass**
End-kaliber n, -stich m, (drawing) Feinzug m,
Fertigstich m, (drawing) Fertigzug m, Polier-
stich m, Schlichtkaliber n, Schlichtstich m ~
**plant** Weiterverarbeitungsbetrieb m ~ **(sizing)
preparation** Appreturmittel n ~ **pressing tool
steel** Werkzeugstahl m für spanabhebende Formung ~ **process** Ausrüstungsverfahren n, End-
bearbeitung f ~ **reamer** Fertig-, Nach-,
Schlicht-reibahle f ~ **roasting** Garrösten n ~
**roll** Fertig-, Schlicht-walze f ~**-roll line** Fertig-
strang m ~ **roll (or mill) train** Fertigwalzstraße
f ~ **rollers** Reckwalzwerk n ~ **rolling mills** Fein-
walze f ~ **rolls** Feinwalzwerk n, Fertigstrecke f,
Reckwalzwerk n ~ **screen** Feinsieb n ~ **shop**
Fertigungshalle f ~ **size** volles Maß n, Vollmaß
n ~ **stand** Endgerüst n, Polierständer m ~ **stand
of rolls** Fertiggerüst n ~ **stands in train** Fertigstrecke f ~ **strip** Abdeckleiste f ~ **surface**
Schlichtoberfläche f ~ **the surface** Oberflächen-
bearbeitung f, -behandlung f ~ **system** Fertig-

verarbeitung *f* ~ **technique** Fertigungstechnik *f*
~ **temperature** Abdarrtemperatur *f* ~ **tool**
Stechstahl *m* mit Rundschneide ~ **train** Fertig-
straße *f*, -strecke *f* ~ **trowel** Spachtel *f* ~ **varnish**
Überzugslack *m* ~ **wool card** Wollkrempel *m*
**finite** begrenzt, endlich ~ **amplitude** endliche
Amplitude *f* ~ **dimensional** endlichdimensional
~ **line** endliche Leitung *f*, Leitung *f* von endlicher Länge **of** ~ **number** zählbar ~ **slab**
Schicht *f* endlicher Dicke ~ **square well** endlicher Potentialtopf *m* ~ **steady-state output**
endliche Statik *f* ~ **stretch** endliche Dehnung *f*
~ **value** endlicher Wert *m*
**finitely elastic** nichtlinear elastisch
**finiteness** Endlichkeit *f*
**finned** gerippt ~ **cylinder** Rippenzylinder *m* ~
**heat exchanger** Rippenheizkörper *m* ~ **pipe**
Lamellenrohr *n* ~ **radiator** Lamellen-, Streifen-
-kühler *m* ~ **spark plug** mit Kühlrippen versehene Zündkerze *f* ~ **surface** Rippenober-
fläche *f* ~ **tube** Rippenrohr *n*
**finning** (of cylinders) Verrippung *f*
**fins cast integral** angegossene Rippen *pl*
**fiord** Fjord *m*
**fir** Föhre *f*, Kiefer *f*, Tanne *f*, Tannenholz *n*
**(wood of)** ~ Föhrenholz *n* ~ **crystal** Dendrit *m*
~ **wood** Kiefer(n)holz *n*, Kienholz *n*
**fire, to** ~ abdrücken, abkrümmen, abschießen,
abtun (von Schüssen), brennen, feuern, (torpedoes) lancieren, schießen, verfeuern, verpuffen, zünden **to** ~ **across** überschießen **to** ~
**at** berasen **to** ~ **away** verschießen **to** ~ **beyond**
**target** überschießen **to** ~ **boilers** anheizen **to** ~
**a camera** Kamera *f* auslösen **to** ~ **a torpedo** ein
Torpedo *m* feuern **to** ~ **with combination fuse**
Schießen *n* mit Doppelzünder **to** ~ **at high**
**angle** steilschießen **to** ~ **with high bursting points**
Schießen *n* mit hohen Sprengpunkten **to** ~ **on**
beschießen **to** ~ **outside** vorbeifeuern **to** ~ **over**
überschießen **to** ~ **past** vorbeifeuern **to** ~ **up**
entflammen **to** ~ **upon** unter Feuer nehmen
**to be under** ~ Feuer erhalten
**fire** Beschuß *m*, Brand *m*, Feuer *n*, Feuerung *f* ~
**in the carburetor** Vergaserbrand *m* ~ **on crash**
Aufschlagbrand *m* ~ **by indirect laying** indirekter Beschuß *m* (indirektes Beschießen) **on** ~
brennend
**fire,** ~ **alarm** Feuerlärm *m* (instrument), Feuer-
meldeapparat *m*, (signal box) Feuermelder *m*
~ **alarm box** Feuermelder *m* ~ **alarm circuit**
Feuermelderleitung *f* ~ **alarm equipment**
Schiffsfeuermeldeanlage *f* ~ **alarm signal** Feuersignal *n* ~ **alarm siren** Feuersirene *f* ~-**alarm**
**telegraph** Feuerwehrtelegraf *m* ~ **arch** Frittofen
*m* ~ **assay** Brandprobe *f* ~ **ball** Feuer-, Glüh-
kugel *f* ~ **bar** Roststabeisen *m* ~-**bar bearer** Rostbalken *m* ~-**bar frame** Rostlager *n* ~ **bar spac-**
**ing** Rostspaltweite *f* ~ **bearer** Rostbalken *m* ~
**blende** Feuerblende *f*
**firebox** Brennkammer *f*, (of boiler) Feuerbüchse
*f*, Feuerkiste *f*, Feuerraum *m*, Feuerstelle *f*,
Feuerung *f*, Flammkammer *f*, Heizkammer *f*,
Heizraum *m*, Herd *m*, Verbrennungsraum *m* ~
**of boiler** Feuerkammer *f* ~ **outside sheet** Steh-
kesselmantel *m*
**firebox,** ~ **bottom flange** Feuerbuchsbodenring
*m* ~ **crown** Feuerbuchsdecke *f* ~ **drilling ma-**
**chine** Feuerbuchsbohrmaschine *f* ~ **shell**

Brennkammer *f*, Feuerbuchsmantel *m* ~ **stay**
Feuerbüchsenstrebe *f*
**fire break** Waldschneise *f*
**fire-brick** Brandziegel *m*, Chamottestein *m*,
Feuerstein *m*, Ofenziegel *m*, Schamotte-stein
*m*, -ziegel *m*, feuerfester Stein *m*, feuerfester
Ton *m* oder Ziegel *m*, feuerbeständiger Ziegel *m*
~ **lining** Schamottegemäuer *n* ~ **sleeve** Gießlochstein *m*, Lochstein *m* ~ **wall** feuerfeste
Wand *f*, Ziegelmauerwerk *n*
**fire,** ~ **bridge** Feuer-bock *m*, -brücke *f*, Herd-
brücke *f* ~ **bridge with swingtype ash bars** Stau-
pendelfeuerbrücke *f* ~ **bridge water boxes** Feuer-
brückenkühlrohre *pl* ~ **brigade** Feuerwehr *f* ~
**bucket** Löscheimer *m* ~ **call** Feuersignal *n* ~ **cham-**
**ber** Verbrennungsraum *m* ~ **checking** Brandrißbildung *f* ~ **classification** Feuerbegriffe *pl*
**fire-clay** Feuerton *m*, Schamotte *f*, Schamotte-
stein *m*, Schamotteton *m*, feuerfester Ton *m*,
gebrannter Ton *m* ~ **brickbat** Schamotte-
brocken *m* ~ **crucible** Schamottetiegel *m* ~
**lining** Schamotteausfütterung *f* ~ **mortar** Scha-
mottemörtel *m* ~ **nozzle** Gießlochstein *m*,
Schamottehaube *f* ~ **plate** Schamotteplatte *f*
~ **retort** Schamotteretorte *f* ~ **sleeve** Schamottehaube *f*
**fire,** ~ **cock** Brand-, Feuer-hahn *m* ~-**command**
**post** Feuerleit-stand *m*, -stelle *f*
**fire-control** Feuerleitung *f*, Feuerleit . . . ~ **aid**
Schieß-behelf *m*, -hilfsmittel *n* ~ **equipment**
Feuerleitgerät *n* ~ **exercise** Feuerleitungsübung
*f* ~ **grid** Gitterschießplan *m* ~ **indicator** Kommandotafel *f* ~ **instrument** Feuerleitgerät *n* ~
**instruments** Feuerleitungsinstrumente *pl* ~ **map**
Feuerleitungsplan *m*, Schießplan *m* ~ **method**
Feuerleitverfahren *n* ~ **net** Feuerleitungsnetz *n*
~ **radar** Feuerleitradar *n*
**fire,** ~ **controller** Feuerleiter *m* ~ **controller's**
**tables** Feuerleitertafeln *pl* ~ **cracker** Kanonenschlag *m* mit Knallerscheinung, Knallkörper *m*
~ **cracking** Brandrißbildung *f* ~ **crest** Brust-
wehrkrone *f*, Gewehrauflage *f*, Krete *f* ~ **dam**
Branddamm *m* ~ **damage** Brandschaden *m*
**fire-damp** Feuerschwaden *m*, Grubengas *n*,
Schlagwetter *pl*, schlagende Wetter *n*, schlagen-
des Wetter *n* ~-**consuming lamp** Schlagwetter-
zehrapparat *m* ~ **safety device** Schlagwetter-
sicherung *f*
**fire,** ~ **department** Feuerwehr *f* ~ **detection**
**device** elektronisches Feuermeldegerät *n* ~
**detector** Feuer-anzeiger *m*, -fühler *m*, -melder
*m* ~ **devil** Lötofen *m* ~ **directed on aircraft**
Fliegerbeschuß *m* ~ **directed on a single point**
Punktschießen *n* ~-**direction center** Leitstand
*m* ~-**direction chart** Schießplan *m* ~-**direction**
**net** Feuerleitungsnetz *n* ~-**director data** Kom-
mandowerte *pl* ~ **distribution** Feuerverteilung *f*
~ **door** Brand-, Feuer-, Feuerungs-tür *f*, Heizloch *n* ~-**door hole ring** Türlochring *m* ~-**dried**
feuergetrocknet ~-**dried beet pulp** feuergetrock-
nete Schnitzel *pl* ~ **effect** Feuer-, Geschoßwirkung *f* ~ **end** Eintauchende *n* ~-**engine**
Feuerspritze *f* ~ **engine** Kraftspritze *f* ~ **engine**
**mounted on a truck** Abprotzspritze *f* ~ **equip-**
**ment** Feuerungsanlage *f* ~ **escape** Feuerleiter *m*
~-**escape ladder** Anstelleiter *f* ~ **extinguisher**
Feuer-löschanlage *f*, -löscher *m*, -spritze *f*,
Löschgerät *n*

**fire-extinguishing, ~ appliances** Feuerlöschgerät n, Löschgerät n ~ **cover** Feuerlöschdecke f, Löschdecke f ~ **equipment** Feuerlöschanlage f ~ **fluid** Feuerlöschflüssigkeit f ~ **medium** Feuerlöschmittel n ~ **pump** Feuerlöschpumpe f

**fire-fighting** Brandbekämpfung f ~ **battalion** Feuerschutzabteilung f ~ **party (or squad)** Löschtrupp m ~ **water truck** Tankspritze f

**fire, ~ float** Spritzenboot n ~-**float (boat)** Feuerlöschboot n ~ **gases** Feuergase pl ~-**gilt** feuervergoldet ~ **grate** Feuer-, Feuerungs-rost m ~ **guard** Feuerwache f ~ **hazard** Feuergefährlichkeit f, Feuergefahr f ~ **hook** Schürhaken m ~ **hose** Feuer-schlauch m, -spritze f, -spritzenschlauch m, Spritzenschlauch m ~ **hydrant** Feuer-hahn m, -löschhydrant m, -pfosten m ~ **insurance** Feuerversicherung f ~ **iron** Schüreisen n ~ **lighter** Zündbüchse f ~ **loss** Abbrand m ~ **main** Feuerlöschhauptleitung f ~ **man** Heizer n, Stocher m ~ **marshal** Brandmeister m ~ **mission** Feuer-auftrag m, -befehl m ~ **place** Feuerraum m, Feuerung f, Feuerungsanlage f, Herd m, Kamin m ~ **plate** Aufschüttplatte f, Brandloch n ~ **plug** Feuer-hahn m, -pfosten m ~ **point** Brenn-, Flamm-punkt m ~-**polish (of glassware)** Feuerpolitur f ~-**polished** abgerundet ~-**polished glass** feuerpoliertes Glas n ~ **power** Feuerkraft f ~-**prevention mechanism for crashes** Aufschlagschutz m (aviat.) ~-**prevention plan** Feuerlöschordnung f

**fireproof** brandfest, explosionsgeschützt, feuerbeständig, feuerfest, feuersicher, unverbrennbar ~ **building material** feuerfester Baustoff m ~ **bulkhead** Brand-schott n, -spat m, Wärmeschott m ~ **casing (of a blast furnace)** feuerfeste Auskleidung f ~ **cement** feuerfester Kitt m ~ **clay** Schamotte f ~ **clothing** feuerfester Anzug m ~ **coating (or dope)** feuerfester Anstrich m ~ **material** feuerfester Baustoff m ~ **mortar (for setting firebricks)** Feuerkitt m ~ **suit** feuerfester Anzug ~ **wall** Brandmauer f ~ **wire** feuersicherer Draht m

**fireproofing** Umkleidung f ~ **flame dampener** Flammenabdeckung f

**fireproofness** Feuerbeständigkeit f

**fire, ~ protection** Brandschutz m, Feuerverhütung f ~-**protection police** Feuerschutzpolizei f ~ **rake** Feuer-haken m, -harke f, -krücke f ~ **recording equipment** Geschützaufnahmegerät n ~-**resistant** feuer-hemmend, -sicher ~-**resisting** feuer-beständig, -sicher ~-**resisting paint** feuersicherer Anstrich m ~ **ring** Feuerring m ~ **risk** Feuergefahr f, Feuerrisiko n ~ **screen** Brandschutz-, Ofen-schirm m ~ **setting** Feuersetzen n ~ **shovel** Feuer-, Kohlen-schaufel f ~ **shrinkage** Feuerschwindung f ~-**shutter control** Fallklappenauslösung m (film) ~ **station** Feuerwache f, -wehrposten m ~ **stone** Feuerstein m ~ **stroke** Arbeitshub m ~ **swatter used in fire fighting** Feuerpatsche f ~ **test** Brand-, Feuerprobe f, Probebrand m ~ **tile block** Ziegelblock m ~-**tinned** feuerverzinnt ~ **tongs** Feuer-, Glutzange f ~ **trench** Kampfgraben m ~ **tube** Feuer-, Flamm-, Heiz-, Rauch-rohr n ~-**tube boiler** Feuer-, Flamm-rohrkessel m, Heizröhrenkessel m, Rauchrohrkessel m ~-**tube lining** Flammrohrfutter n ~ **unit** Feuereinheit f ~ **wall** Brandschott n, Schottwand f ~-**wall bulkhead**

**plug** Brandschottstecker m ~-**wall seal** Brandschottring m ~ **water** Feuerlöschwasser n ~ **weld** Schweißung f im Feuer ~ **wood** Brennholz n ~ **works** Feuerwerk n ~ **zone** Brandfeld n, Feuerbereich m

**fired** beheizt ~ **with powdered fuel** staubbeheizt

**fires** Brandfälle pl

**firing** Abfeuerung f, Aktivierung f, Befeuerung f, Beheizung f, Feuerführung f, Feuerung f, Heizung f, Schießen n, Verfeuerung f, Zündung f ~ **and cocking switch box** Abfeuerdurchladeschaltkasten m ~ **with powdered fuel** Staubfeuerung f ~ **on stepped grate bars** Pult-(be)feuerung f

**firing, ~ accessories** Schießzubehör n ~ **accessory** Schießhilfsmittel n ~ **activity** Feuertätigkeit f ~ **angle** Sättigungswinkel m ~ **angles** Winkelgruppe f ~ **arc** Bestreichungswinkel m ~-**arm trigger piece** Abzugsstück n ~ **azimuth** Schußseitenwinkel m ~ **base** Lafettenstütze f ~ **bell** Feuerglocke f ~ **brake** Schießbremse f ~ **chart** Batterie-, Feuer-plan m ~ **circuit** Zündleitung f ~ **command** Feuerkommando n ~ **connector (explosives)** Abzugsschalter m ~-**control-station transmitter** Befehlsübertragungsanlage f ~ **cord loop** Abreißschlaufe f ~ **data** Schießgrundlage f, Schießunterlagen pl, Schußelement n, Schußwerte pl ~-**data computer** Schußwerteberechner m ~ **electrode** Zündelektrode f ~ **equipment** Feuerung f (Gas- und Ölfeuerung) ~ **floor** Heizerstand m ~ **hole bush** Zündstollen m ~ **interval** Schußfolge f ~ **jack** Geschützwinde f ~ **key** Abfeuerungstaste f ~ **lanyard** Abzugsschnur f ~ **lever** Abfeuerungs-, Abzugs-, Lade-hebel m ~ **limitation** Feuer-begrenzung f, -beschränkung f ~ **line (explosives)** Abzugsleine f, Feuer-, Schützen-linie f ~ **machine** Bekohlungsanlage f ~ **mechanism** Abdrückvorrichtung f, Abfeuerungseinrichtung f, Abfeuerungs-, Abzugs-vorrichtung f ~ **notch** Sperrnase f ~ **order** Zündfolge f, Zündzeitfolge f

**firing-pin** Bolzen m, Federkolben m, Nadel f, Schlagbolzen m, (of rifle) Schlagstift m, Zündstift m ~ **for guns** Schlagbolzen m für Gewehre ~ **guide (fuse)** Nadelstück n ~ **spring** Schlag-(bolzen)feder f ~ **spring washer** Bolzenfederscheibe f ~ **support** Nadelstück n

**firing, ~ plane** Schußebene f ~ **platform** Pritsche f, Schieß-gerüst n, -gestell n ~ **point** Auslösepunkt m, Schützenstand m, Zünd(zeit)punkt m ~ **port** Schießschlitz m ~-**port shield** Schartenschild n ~-**port shutter** Schartenblende f ~ **position** Abschußpunkt m, Anschlag m, Feuerstellung f ~ **in ~ position** feuerbereit, in Feuerstellung f ~ **post** Gefechtsstand m ~ **potential** Ionisierungsspannung f ~ **practice** Feuerungstechnik f, Schießübung f ~ **procedure according to map** Kartenrichtpunktverfahren n ~ **range** Schießplatz m ~-**readiness time** Feuerbereitungszeit f ~ **restriction** Feuerverbot n ~ **rod** Abzugswelle f ~ **schedule** Feuerführung f ~ **scoop** Heizerschaufel f ~ **sequence** Feuerfolge f ~ **spring (explosives)** Abzugsbogen f ~ **stand** Schießgerüst n ~ **step** Grabenstufe f ~ **stroke** Verbrennungshub m ~-**table** schußtafelmäßig ~ **table** Schußtafel f ~-**table elevation** schußtafelmäßige Erhöhung f ~ **tape**

Zündschnur *f* ~**-test result** Beschußergebnis *n* ~ **top center** oberer Totpunkt *m* ~ **triangle** Schußdreieck *n* ~ **voltage** Zündspannung *f* ~ **wire** Zünderdraht *m*

**firm** Firma *f*, Geschäft *n*, Haus *n* ~ **of publishers** Verlags-buchhandlung *f*, -haus *n*

**firm** derb, dicht, entschlossen, fest, feststehend, fix, hart, kräftig, stabil, standfest, standhaft, steif, stetig ~**-joint caliper** Scharnierkaliber *n* ~**-joint calipers** Außentaster *m*, Taster *m* mit Schraubenscharnier

**firmer,** ~ **chisel** Stechbeitel *m* ~ **chisel with socket** Düllstechbeitel *m*

**firmly,** ~ **adherent** fest anhängend ~ **consolidated** festgelagert

**firmness** Dichte *f*, Dichtheit *f*, Dichtigkeit *f*, Festigkeit *f*

**first, to** ~**-finish** verschlichten **to** ~**-pickle** vorbeizen **to** ~**-polish** vorschlichten

**first** ursprünglich **at** ~ vorderhand **from the** ~ von vornherein ~ **of all** vorab ~ **and second sprocket wheel** Vor- und Nachwickelrolle *f*

**first accelerator** Sauganode *f*

**first-aid** Erste Hilfe *f*, Hilfeleistung *f* ~ **bag** Verbandtasche *f* ~ **box** Verbandkasten *m* ~ **call installation** Unfallmeldeanlage *f* ~ **dressing** Notverbandpäckchen *n* ~ **kit** Sanitätspack *m*, Verbands-kasten *m*, -päckchen *n* ~ **outfit** Verbandzeug *n* ~ **packet** Notverbandpäckchen *n* ~ **pouch** Notverbandtasche *f* ~ **station** Rettungsstelle *f*

**first,** ~ **air supply** Erstluftzufuhr *f* ~ **annealing** erste Glühung *f* ~ **anode** Anodenblende *f*, Saug-, Vor-anode *f* ~ **approximation** erste Näherung *f* ~ **ball cup** erste Kugelschale *f* ~ **blow** Anhieb *m* ~ **boiling** Vorsud *n* ~ **bowl** Eingangsbehälter *m* ~ **breaking-up** Rohraufbrechen *n* ~ **cabinet** Vorgestell *n* (g/m) ~ **charge of secondary (or storage) cell** erste Ladung *f* eines Sammlers ~ **choice trunkgroup** Querleitungsbündel *n* (teleph.) ~ **circuit** Trennschnitt *m* ~ **class acceptance** glatte Annahme *f* ~ **coat** Bewurf *m*, Grundfirnis *m*, Grundierung *f*, Rohputz *m* ~ **coat of paint** Grundierfarbe *f* ~ **coil** Ausgangswicklung *f* ~ **cost** Anlage-, Anschaffungs-, Herstellungskosten *pl* ~ **cost of installation** Gesamtanlagekosten *pl* ~ **course of a file** Unterhieb *m* einer Feile ~ **crusher** Vormühle *f* ~ **crushing rollers** Aufbereitungswalzwerk *n* ~ **cut** Anhieb *m*, (files) Grundhieb *m* ~ **detector** erster Detektor *m*, Mischstufe *f*, Überlagerungsstufe *f* ~ **diamond pass** (in rolling) Vorraute *f* ~ **draft** Grundentwurf *m*, Urschrift *f* ~ **drawing** Grobzug *m* ~ **dressing mill** Aufbereitungswalzwerk *n* ~ **edition** Originalausgabe *f* ~ **equipment** Erstbestückung *f* **on** ~ **establishment** im ersten Ausbau *m* ~ **etching** Anätzung *f* ~ **felling** Anhieb *m* ~ **fill mass** Erstproduktfüllmasse *f* ~**-finished section** Vorschlichtprofil *n* ~**-finishing pass** Vorpolier-kaliber *n*, -stich *m*, Vorschlichtkaliber *n* ~ **focal point** vorderer Brennpunkt *m* ~**-focus action** Vorkonzentration *f* ~**-focusing lens** Vorsammellinse *f* ~ **form** Schöndruck *m* ~ **frame of a mine shaft** Füllbaum *m* ~**-grade** erstrangig ~ **group selector** erster Gruppenwähler *m* ~ **gun** Sauganode *f* ~ **harmonic** Grund-frequenz *f*, -schwingung *f*, erste Harmo-

nische *f* ~ **image** Zwischenbild *n* **of** ~ **importance** erstrangig ~ **impression** Vordruck *m* ~ **installment** Angeld *n*, Anzahlung *f* ~ **law of magnetism** erstes Gesetz *n* der Anziehungskraft ~ **lead removed in the Carinthian process** Rührblei *n* ~ **letter** Anfangsbuchstabe *m* ~ **line** Vordertreffen *n* ~ **line finder** erster Anrufsucher *m* ~ **loading coil section** Anlauflänge *f* ~ **lump** Urdeul *m* (metal.) ~ **marking signal** Vorsignal *n* ~ **matte** Rohstein *m* ~ **member of an equation** linke Seite *f* einer Gleichung ~ **minerals to separate out** Erstausscheidungen *pl* ~ **mold of vellum** Dickquetsche *f*

**first-order** erster Ordnung **to a** ~ auf die erste Dezimale *f* ~ **equation** Gleichung *f* ersten Grades ~ **significance** ausschlaggebender Einfluß *m*

**first,** ~**-party release** Auslösung *f* einer Verbindung, sobald einer von beiden Teilnehmern einhängt; Vorwärts- und Rückwärtsauslösung *f* ~ **pass** Anstich *m* (Walzwerk) ~ **phase** Startvorriegelung *f* (film) ~ **point of contact** Auflaufpunkt *m* ~ **printed (or preprinted) resist** Vordruckreserve *f* ~ **product** Ersterzeugnis *n* ~**-product crystallizer** Erstproduktmaische *f* ~**-product sugar** Erstzucker *m* ~**-product vacuum** Erstproduktvakuum *n* ~ **proof** Hauskorrektur *f*, erste Korrektur *f*, Korrekturbogen *m* ~ **quantum number** Hauptquantenzahl *f* ~ **quarter of the ebb** Vorebbe *f* ~ **rank** Vorrang *m* ~**-rate** erstklassig, hochfein ~ **raw sugar** Rohzuckererstprodukt *n* ~**-refined copper** rohgares Kupfer *n* ~ **refining** Rohfrischen *n* ~ **reflection** erste Reflektion *f* ~ **report** Vorbericht *m* ~ **resonating frequency** Resonanzgrundfrequenz *f* ~ **ripping** Grobzug *m* ~ **route** Regelweg *m* ~ **rumpling** Grobzug *m* ~ **run** Versuchslauf *m* ~**-run slag** Topfschlacke *f* ~ **runnings** Vorlauf *m*, Vorprodukt *n* ~**-runnings product** Anfangsergebnis *n*, -erzeugnis *n* ~ **section** Anlaßstrecke *f* eines Pupinkabels ~ **section to be constructed** erster Ausbau *m* ~ **slag** Einschmelzschlacke *f* ~ **sound in liquid helium** Schall- und Wärmewellen *pl* erster Art in flüssigem Helium ~**-speed sliding gear** Schubgetriebe *n* des ersten Ganges ~ **spray** Vorspritzung *f* ~ **stage** (of amplifier) Vorstufe *f* ~ **stage climb performance** Steigleitung *f* in der ersten Steigflugphase ~ **stage compressor** Vorverdichter *m* ~ **step** Antritt *m*, Vorstufe *f* ~ **stuff** Halbzeug *n* (paper mfg.) ~ **subscriber release** Auslösung *f* durch den Teilnehmer, der zuerst anhängt ~ **test** Einschmelzprobe *f* ~ **upsetter** Vorstauchstempel *m* ~ **visible red** beginnende Glut *f* ~ **wave of harmonic** Fundamentalschwingung *f* ~ **wet-felt** Preßfilz *m* ~ **winding** Ausgangswicklung *f* ~ **zero voltage** erste Nullspannung *f*

**fiscal year** Berichts-, Geschäfts-, Haushalts-, Rechnungs-jahr *n*

**fise** Fiskus *m*

**fish, to** ~ fischen, verlaschen (r. r.)

**fish,** ~ **of a trailing antenna** Antennenendgewicht *n* ~ **back spring leaf** Federblatt *n* mit Längsnut und Rippe ~**-bellied** Fischbauchträgerbrücke *f* ~**-bellied antenna** Fischbauchprofil *n* ~**-bellied girder bridge** Fischbauchantenne *f* ~ **belly flap** Fischbauchklappe '

**fishbone** Fischgräte *f*, Gräte *f* ~ **antenna** Tannen-

baum-, Fischgräten-antenne *f* ~ **obstacle pattern** Fischgrätenhindernis
**fish,** ~ **chair** Stuhllasche *f* (r.r.) ~ **curing house** Fischkonservenfabrik *f* ~ **dock** Fischereibecken *n* ~ **eye** Materialfehler *m* ~ **glue** Fischleim *m* ~ **hook** Fischhaken *m* ~ **joint** Fischband *n* (mech.), Stoßverbindung *f* ~ **ladder** Fischleiter *f*, -paß *m* ~ **line problem** Angelschnurproblem *n* ~ **market** Fischhalle *f* ~-**mouth shaped** fischmaulförmig ~-**mouth splice** fischmaulförmige Spleißung *f* ~ **oil** Fisch-öl *n*, -tran *m*, Tran *m* ~-**oil running plant** Tranraffinieranlage *f* ~ **oil tannage** Trangerbung *f* ~ **paper** Isolationspappe *f*
**fishplate, to** ~ verlaschen
**fishplate** Fischplatte *f*, Knotenblech *n*, Lasche *f*, Laschenblech *n*, Schienenlasche *f*, Stecklasche *f*, Stoßblech *n*, Stoßlasche *f* ~ **drilling machine** Laschenbohrmaschine *f* ~ **pass** Laschenkaliber *n* ~ **punching machine** Laschenlochmaschine *f* ~ **seating** Laschengehäuse *n* ~ **section** Laschenprofil *n*
**fishtail, to** ~ abbremsen
**fishtail,** ~ **bit** Fischschwanz-bohrer *m*, -meißel *m* ~ **burner** (gas) Fischschwanzbrenner *m*, Flachbrenner *m* ~ **die** Breitschlitzdüse *f* ~ **type** Fischschwanz-Type *f* ~ **type kneader** Fischschwanzkneter *m*
**fishtailing** Fischschwanzflugmanöver *n*, den Schwanz *m* von einer Seite zur anderen schwingen
**fish,** ~ **tracks** kurze Nebelspuren *pl* ~ **welding** Abbrennschweißung *f*
**fisherman** Fischer *m*
**fishery** Fischerei *f* ~ **patrol boat** Fischereischutzboot *n*
**fishing** Fangarbeit *f*, Fanggeräte *pl*, Fischerei *f*, Fischfang *m*; Laschen *n*, Laschenverbindung *f*, Verlaschen *n* ~ **appartus** Fangapparat *m* ~ **bell** Fangglocke *f* ~ **chain** Kettenlasche *f* ~ **cone** Fangtrichter *m* ~ **grab** Bohrfänger *m* ~ **(or fishery) harbor** Fischereihafen *m* ~ **hook** Angel *f*, Fang-, Glücks-haken *m* ~ **hook for bits** Bohrmeißelhaken *m*, Hakenfänger *m*, Meißelfanghaken *m* ~ **hook for collars** Hülsenhaken *m* ~ **hook for poles** Gestängefanggerätsförderstuhl *m* ~ **hook for rope** Seilfanghaken *m* ~ **jars** Fallfangschere *f*, Fangrutschschere *f* ~ **jars for ropes** Fangschere *f* ~ **job** Fangarbeit *f* ~ **product** Fischereierzeugnis *n* ~ **ring** Fangring *m* ~ **rod** Angel *f* ~ **rods** Fanggestänge *n* ~ **socket** Fang-hülse *f*, -krone *f*, Keilfänger *m* ~ **socket for hollow rods** Gestängerohrfangkrone *f* ~ **socket for poles** Gestängefangkrone *f* ~ **stirrup** Fangbügel *m* ~ **tackle** Fischereigerät *n* ~ **tap** Fangdorn *m* ~ **(screw) tap** Spitzfänger *m* ~ **tongs** Fangzange *f* ~ **tool** Fang-gerät *n*, -werkzeug *n* ~ **tool for rivets** Nietenfänger *m* ~-**tool socket** Dornfänger *m* ~-**valve socket** Fangbüchse *f* ~ **wire** Durchziehband *n*
**fissile material** Spaltmaterial *n*
**fission** Spaltung *f*, Zertrümmerung *f* (atom) ~ **algae** Spaltalgen *pl* ~ **background** Spaltungsuntergrund *m* ~ **chain** Spaltungskette *f* ~ **counter** Spaltzähler *m* ~ **decay chain** Spaltungskette *f* ~ **detection** Spaltungsnachweis *m* ~ **fragments** Spaltbruchstücke *pl*, Trümmer *pl* ~ **process** Spaltprozeß *m* ~ **product** Spaltungs-

produkt *n* ~ **products** Bruchstücke *pl*, Trümmer *pl* ~ **recoil particles** Rückstoßteilchen *n* bei der Spaltung ~ **threshold** Spaltschwelle *f* ~-**yield curve** Spaltungsausbeutekurve *f*
**fissionable** spaltbar, zertrümmerbar ~ **material** Spaltmaterial *n* ~ **nuclides** spaltbare Kerne *pl* ~ **substance** zertrümmerbares Material *n*
**fissuration** Rißbildung *f* ~ **due to welding** Schweißrissigkeit *f*
**fissure** Bruch *m*, Bruchstelle *f*, Einriß *m*, Kluft *f*, Naht *f*, Riß *m*, Ritze *f*, Spalt *m*, Spalte *f*, Sprung *m*, Ziehriefe *f* ~ **caused by frost** Frostspalte *f* ~ **of overthrust** Überschiebungsspalte *f*
**fissure,** ~ **craze** Haarriß *m* ~ **eruption** Spalteneruption *f* ~ **filling** Spaltenfüllung *f* ~ **vein** echter Gang *m*
**fissured** rissig
**fissuring** Zerklüftung *f*
**fist** Faust *f*, Handzeichen *n*
**fistula** Fistel *f*
**fit, to** ~ anbringen, anliegen, anmessen, anpassen, anschließen, aufmontieren, aufpassen, ausrüsten, beschlagen, einlegen, einpassen, einrichten, gängig machen, hineinpassen, passen, stehen, zurichten, zusammengehören, aus Teilen *pl* zusammensetzen **to** ~ **an equation** einer Gleichung *f* entsprechen **to** ~ **in** einbauen, einfügen **to** ~ **into** hineinpassen **to** ~ **into each other** ineinanderpassen **to** ~ **into one another** ineinander stecken **to** ~ **a key** aufkeilen **to** ~ **loosely** schlottern **to** ~ **on** anproben, (tire) aufziehen **to** ~ **out** ausrüsten **to** ~ **a relation** eine Beziehung *f* erfüllen **to** ~ **snugly** eng einpassen **to** ~ **tightly** dichten, (clothing) eng anschließen **to** ~ **together** zusammenpassen **to** ~ **together by grinding one on the other** aufeinander schleifen **to** ~ **together without cement** zusammensprengen **to** ~ **up** aufstellen
**fit** enge Anschmiegung *f*, Passung *f*, Sitz *m*; anstellig, brauchbar, dienlich, geeignet, gebührend, gehörig, ratsam, tauglich, zweckmäßig **by** ~**s and jerks** schubweise **not** ~ **to use** gebrauchsunfähig
**fit,** ~ **for bottling** flaschenreif ~ **for casting** gießbar ~ **for cutting** haubar ~ **for duty** dienstfähig ~ **for elimination** aussonderungsreif ~ **for founding** gießbar
**fit,** ~ **pin** Spannstift *m* ~ **size** Paßmaß *n* ~ **tolerance zone** Paßtoleranzfeld *n*
**fitment** Ausstellungsstück *n* ~ **(slide)** Schlittenstück *n* (Plankton-Mikroskop)
**fitness** Brauchbarkeit *f*, Eignung *f*, Geeignetheit *f*, Schick *m*, Tauglichkeit *f*
**fitness,** ~ **for flying** Fliegertauglichkeit *f* ~ **for service in tropical climate** Tropenverwendungsfähigkeit *f* ~ **for his work** Arbeitstüchtigkeit *f*
**fits** Einpaßgrößen *pl*
**fitted** eingebaut, eingepaßt, eingerichtet ~ **on an apparatus** eingebaut in einen Apparat ~ **with lock and key** verschließbar ~ **by shrinking** aufgeschrumpft *m* ~ **with** ausgerüstet mit
**fitted,** ~ **bolt** Paß-bolzen *m*, -schraube *f*, eingepaßte Schraube *f* ~ **pattern** Paßmuster *n* ~ **piece** Einlegling *m* ~ **shank screen** Paßbolzen *m* mit Gewindezapfen *m*
**fitter** Arbeiter *m*, Maschinenschlosser *m*, Monteur *m* (engine), Rüster *m*, Rüstmann *m*, Schlosser *m* ~ **of dust pipes** (watchmaking) Hutaufsetzer *m*

**fitter's, ~ hammer** Schlosserhammer *m* **~ ordinary flat cold chisel** Flachmeißel *m*
**fitting** geeignet, passend; Anpassung *f*, Anprobe *f*, Armatur *f*, Aufstellung *f*, Ausrichten *n*, Ausrüstung *f*, Einbau *m*, Einlegling *m*, Fassung *f*, Formstück *n*, Herstellung *f*, Leuchte *f*, Montage *f*, Montierung *f*, Rohverbindungsstück *n*, Trageweise *f* (how to wear), Verbundstück *n*, Zubehörteil *m*, Zusammenbau *m*, Zusammensetzen *n* **~ for attachment to fuselage** Rumpfanschlußstück *n* **~ the blading** Beschaufeln *n* **~ of crossing point of wires** Seilkreuzungshülse *f* **~ of delivery wagon** Förderwagenbeschlagteil *m* **~ for lye pipe line** Laugenleitungsarmatur *f*
**fitting, ~ allowance** Zugabe *f* zum Einpassen **~ area** Paßfläche *f* **~ bolt** Montageschraube *f* **~ curve** Ausgleichkurve *f* (math.) **~ device** Einbaulage *f* **~ dimension** Paßmaß *n* **~ factor** Zuschlagfaktor *m* für Beschläge **~ edge** Anlagekante *f*, Anschlag *m* **~ fixture** Anschlußvorrichtung *f* **~-in** Einfügen *n*, Einfügung *f* (eines Getriebes) **~-in place** Einpaßstelle *f* **~ instruction** Einbauanleitung *f* **~ jig** Anschlußgroßvorrichtung *f*, Anschlußvorrichtung *f* **~ key** Paßfeder *f* **~ limit** Paßgrenze *f* **~ members** Paßteile *pl* **~-out wharf** Ausrüstungshafen *m* **~ piece** Paßstück *n* **~ pin** Paßstift *m* **~ plate** Befestigungslasche *f* **~ point** Anhaltspunkt *m* **~ practice for antifriction bearings** Passungsrichtlinien *pl* für Wälzlager **~ proposal (or suggestion)** Einbauvorschlag *m* **~ rig** Anschlußgroßvorrichtung *f* **~ surface** Paß-fläche *f*, -stelle *f* **~ wedge pieces** Ausrichtkeile *pl* **~ work** Paßarbeit *f*
**fittings** Apparatur *f*, Beschlag *m*, Beschlagteile *pl*, (of an insulator) Bewehrung *f*, Einrüstung *f*, Ersatzteil *m*, Fassonstücke *pl*, Feinausrüstung *f* (Kessel), Fittings *pl*, Garnitur *f*, Garniturteile *pl*, Mastarmatur *f*, Zubehör *n* **~ for railroad cars** Eisenbahnwagenarmatur *f* **~ for rolling** Walzarmaturen *pl* **~ for safes** Geldschrankbeschläge *pl*
**five, ~-bearing crankshaft** fünfmal gelagerte Kurbelwelle *f* **~-channel mixer** fünfgliedriger Mischer *m* **~-cylinder radial engine** Fünfzylindersternmotor *m* **~-electrode tube** Fünfpolröhre *f*, Schutzgitterfanggitterröhre *f* **~-electrode valve** Fünfelektrodenröhre *f* **~-engined** fünfmotorig **~-figure accuracy** fünfstellige Genauigkeit *f* **~-fold** fünf-fach, -fältig **~-key transmitter** Fünftastengeber *m* **~ lens combination** Fünflinsenobjektiv *n* **~-pin base** Fünfsteckersockel *m*, fünfpoliger Sockel *m* **~ point draft gauge** Fünffachzug-Anzeiger *m* **~-point switch** Schalter *m* mit fünf Ausgängen **~-pole** fünfpolig (bei Steckkontakt) **~-prong base** Fünfsteckersockel *m*, fünfpoliger Sockel *m* **~-roller mill** Fünfwalzwerk *n* **~-step weakener** Fünfstufenschwächer *m* **~-roller sugar-cane mill** Fünfwalzanmühle *f* **~-tuck splice** fünfgängige Spleißung *f* **~-unit alphabet (or code)** Fünferalphabet *n*, Fünfströme-Alphabet *n* **~-wire network** Fünfleiternetz *n* **~-wire system with direct current** Fünfleitersystem *n*
**fix, to ~** (limits) abgrenzen, anberaumen, anhalten, anstecken (a rope), arretieren, aufspannen, aufstechen (print), befestigen, bestimmen, bremsen, einsetzen, einspannen,

fesseln, festhalten, festigen, festlegen, festmachen, (concrete) festsetzen, feststellen, fixieren, lagern, normieren
**fix, to ~ damage** (indemnity) Entschädigung *f* festsetzen **to ~ date for hearing a case** Termin *m* zur Verhandlung einer Sache anberaumen **to ~ by firing** anbacken **to ~ with a grapnel** aufklauen **to ~ with hemp** (insulators) aufhanfen (Isolatoren) **to ~ a nut with a split pin** eine Mutter *f* versplinten **to ~ a position** sich orientieren **to ~ as a support** zugrunde legen **to ~ with a toggle** knebeln
**fix** Beizen *n*, Festpunkt *m* (ATC), Koppel-, Koppelnavigations-ort *m*, Kreuzpeilung *f* (nav.), Ortsfeststellung *f*, Standortbestimmung *f*
**fixable** feststellbar
**fixation, ~-clamping section** Einspannquerschnitt *m* **~ forceps** Feststellpinzette *f* **~ lamp** Fixierleuchte *f*
**fixative** Fixiermittel *n*
**fixed** bestimmt, fest eingebaut, eingespannt, fest, festgelagert, festgesetzt (rdr), feststehend, feuerbeständig, fix, nicht flüchtig, gebunden (binden), konstant, ortsfest, ständig, starr, stationär, stehend, unbeweglich, unverstellbar **~ in aircraft (or in body)** flugzeugfest **~ and flashing light** Festfeuer *n* mit Blitzen **~ and group occultating light** Festfeuer *n* mit Gruppenblinken **~ with respect to a body** körperfest **~ at right angle to** rechtwinklig angesetzt an **~ in space** raumfest
**fixer, ~ air inlet** Hauptlufteinlaß *m* **~ air-screw (or blade)** unverstellbare Luftschraube *f* **~ ammunition** Einheits-, Patronen-munition *f* **~ anchorage** fester Anker *m* **~ antenna** Festantenne *f* **~ armament** feste Waffe *f* **~ armature (rotation)** feststehender (umlaufender) Anker *m* **~ assets and movables** feste und bewegliche Anlagen *pl* **~ auxiliary wing slot** fester (unbeweglicher) Hilfsflügelschlitz *m* **~ axle** feststehende Achse *f* **~ ballast** fester Ballast *m* (aircraft) **~ bank** Stator-paket *m*, -platten *pl* (eines Drehkondensators) **~-base rotation** Zahlendarstellung *f* mit fester Basis **~ bayonet** aufgepflanztes Bayonett *n* **~ beam** eingespannter Balken *m* **~ bed adsorption unit** Festbettadsorber *m* **~ bell** Ventilhaube *f* **~ bench mark** Festpunkt *m* **~-blade propeller** Starrschraube *f* **~ block** fester Block *m* **~ bridge** Brücke *f* mit festen Stützen **~-caliper gauge** Loch- und Tasterlehre *f* **~ (aircraft) camera mounting** Kammergestell *n* für den festen Einbau (in der Flugzeug) **~ cannon** fest eingebaute Kanone *f* **~ capacitor** Block-, Fest-kondensator *m*, fester Kondensator *m* **~ carbon** fester Kohlenstoff *m* **~ cathode spot** fixierter Kathodenfleck *m* **~ chute** feste Rinne *f* **~ coil** feste Spule *f*, Statorspule *f* **~ condenser** Blockkondensator *m*, fester Kondensator *m*, Festkondensator *m* **~ contact between rails** fester Kontakt *m* in der Gleismitte **~ controls** festgehaltenes Ruder *n* **~ converter** feststehender Konverter *m* **~ conveying plant** ortsfeste Förderanlage *f* **~ coupling** starre Kupplung *f* **~-coupling capacitor** Kopplungsblock *m* **~ cowl** unverstellbare feste Motorhaube *f* **~-cycle control** Taktsteuerung *f* **~ cycle operation** Takt-, Zeitgeber-betrieb *m*

**dam** festes Wehr n ~ **datum mark** Festpunkt m ~ **distance** Festabstand m ~**-echo elimination** Festzeichenbefreiung f ~ **electrons** kernnahe Elektronen pl ~ **emitting station** feste Funkstelle f, fester Sender m ~ **end** Aufhängepunkt m, (of a cable) Fixpunkt m, (of cable) Gegenzapfen m ~**-end moment** Einspannmoment m **with** ~ **ends** eingespannt ~ **engine** ortsfester Motor m ~ **error** fester Fehler m ~ **eye** akkomodationsloses Auge n ~ **field** feststehendes oder ruhendes Feld n ~ **fire** fester Feuer n ~ **flagpole** Fahnenmast m ~ **flange** fester Flansch m ~ **focus** konstante Brennweite f ~**-focus pyrometer** Pyrometer m mit konstanter Brennweite ~ **force** feststehende Kraft f ~ **frequency** feste Frequenz f (rdr) ~ **frequency generator** Generator m mit konstanter Frequenz ~ **fulcrum** fester Drehpunkt m ~ **fuse** Fertigzünder m ~ **gas** verflüssigtes Gas n, Treibgas n ~ **gauge** Grenzlehre f ~ **grate-type gas producer** Gaserzeuger m mit fixem Rost ~**-grate-type producer** Generator m mit fixem Rost ~ **gripper lever** Greiferhebel m ~ **guide** feste Führung f ~ **gun** starre Kanone f ~ **gun mounting** starre Lafette f ~ **horizontal fin** Höhenflosse f ~ **ignition** Festzündung f ~ **intake screen** festeingebautes Lufteintrittsgitter n ~ **jib** starrer Ausleger m ~ **kettle** Standkochkessel m ~ **landing gear** festes oder nicht einziehbares Fahrwerk n ~ **launching tub** (torpedo) fest eingebauter Lancierapparat m ~ **layer resistances** Schichtfestwiderstände pl ~ **length record system** System n mit festen Wortlängen (data proc.) ~ **light** (lighthouse) Dauerfeuer n, Festfeuer n, festes Lichtsignal n ~ **line** feste Leitung f ~**-line posting** feste Zeileneinstellung f ~ **loop antenna** Festrahmenantenne f ~ **machine gun** starres Maschinengewehr n ~ **magnification value** Vergrößerungsstufe f ~ **mandrel** Dockenstock f ~ **mooring** Ankerklotz m ~ **mount** starre Aufhängung f (Triebwerk) ~ **mounting** fester Einbau m ~ **parts** feststehende Teile pl ~ **pawl** feste Einfallklinke f ~ **percussion fuse** Fertigaufschlagzünder m ~ **phase winding** Bezugswicklung f ~ **pipe suspension** feste Rohraufhängung f ~ **pitch** feste oder unverstellbare Steigung f ~ **pitch metal propeller** Metallfestpropeller m ~**-pitch propeller** feste Luftschraube f, Luftschraube f mit unbeweglicher Gierung, Starrschraube f ~ **plane** Steuerorgan n ~ **planes** Steuer n ~ **plate** feste Scheibe f

**fixed-point** Festpunkt m, Fixierloch n, festes Komma n (info proc.) ~ **for line crossings** Festpunkt m für Leitungskreuzungen ~ **for loading coil spacing** Festpunkt m für Spulenabstände ~ **of a mapping** Fixpunkt m einer Abbildung

**fixed-point,** ~ **computation** Festkomma-Rechnung f ~ **control** Festwertregelung f (aut. contr.) ~ **representation** Festpunktdarstellung f ~ **theorem** Fixpunktsatz m

**fixed,** ~ **pulley** Fest-rolle f, -scheibe f ~ **pump** feststehende Pumpe f ~ **radial engine** Sternstandmotor m ~ **radio station** feste Funkstelle f ~ **radix rotation** Zahlendarstellung f mit fester Basis ~ **rail** feste Auflager n (Schine) ~ **recticle circle image** feststehendes Leuchtkreisbild n ~ **resistance (or resistor)** Festwiderstand m ~ **rule** Satzung f ~ **salt** fixes Salz n ~

**section** fester Teil m ~ **sight** Standvisier n, starres Visier n ~ **spanner** Maulschlüssel m ~**-star body** (fixed-star ball) Fixsternkugel f **the** ~ **stars** Fixsternhimmel m ~ **station** ortsfeste Funkstelle f ~ **store** Festspeicher m ~ **support** Fußeinspannung f ~ **surface** Flosse f, unbewegliche Oberfläche f ~**-table feed trips** Einstellknaggen pl zum Tischvorschub ~ **target** Festpunkt m, Standziel n, feststehendes oder festes Ziel n ~ **throttle** unveränderte Drosseleinstellung f ~ **thrust** fester Schub m ~**-time call** Ferngespräch n zu bestimmter Zeit ~ **tommy head** Ohrenkopf m ~ **track** festes Gleis n ~ **tool-post** Herzklaue f ~ **transmitter** ortsfester Sender m ~ **trim** Trimmblech n ~ **trim tab** Trimmkante f ~ **trimming tab** Trimm-blech n, -kante f ~**-type landing gear** unbewegliches Landungsgestell n ~ **undercarriage** Festfahrwerk n, festes oder nicht einziehbares Fahrgestell n ~ **unit** bodenständige Formation f ~ **value** Fixwert m ~ **voltage** Festspannung f ~ **wages** Festlohn m ~ **wave** Fest-, Insel-welle f (rdr) ~ **weight** festgesetztes Gewicht n ~ **wheel leaf gate** Rollschütze m ~**-wing aircraft** Starr-flügelflugzeug n, -flügler m ~**-wing airplane** Flugzeug n mit nicht abnehmbaren Tragflächen ~ **wire resistance** Drahtfestwiderstand m ~ **wireless-telegraphy antenna** Festantenne f ~ **word length** feste Wortlänge f

**fixedly (or permanently) connected** dauernd gekuppelt

**fixer** Fixier-bad n, -mittel n

**fixing** Anschlag m, Aufspannschlitz m, Befestigung f, Einspannung f, Festlegung f, Festsetzung f, Verbleien n, Ortung f (rdo nav.), Standortbestimmung f (rdo nav.) ~ **a position** Besteckrechnung f, Lagebestimmung f ~ **the threshold** Schwellenfestlegung f

**fixing,** ~ **agent** Fixiermittel n ~ **aircraft position by soundlocater devices** Abhorchen n von Flugzeugen ~ **bath** Fixierbad n, Tonfixierbad n ~ **bolt** Befestigungsschraube f, Feststellriegel m ~ **bracket** Halteklammer f ~ **catch** Befestigungsriegel m ~ **clamp** Montagehaken m ~ **clay** Fixierton m ~ **clip** Befestigungsschelle f, Feststellklammer f ~ **flange** Aufspannflantsche f ~ **flap** Befestigungslappen m ~ **hole** Befestigungsloch n ~ **key** Verbindungskeil m ~ **liquid (or liquor)** Fixierflüssigkeit f ~ **lug** Befestigungsknagge f ~ **needle** Aufstechnadel f ~ **part (or fitting)** Befestigungsteil m ~ **piece** Anschraubstück n ~ **plate** Aufspann-, Feststell-platte f ~ **point** Einspannstelle f ~ **salt** Fixiersalz n ~ **screw** Befestigungs-, Fixier-schraube f ~ **slot** Aufspannschlitz m ~ **solution** Fixierlösung f ~ **strap** Befestigungsbügel m, (lug) Befestigungseisen n ~ **strip** Befestigungsstreifen m ~ **tape** Befestigungsband n ~**-up** Herstellung f

**fixture** Abmachung f, Aufsatz m, Aufspanntisch m, Bearbeitungsvorrichtung f, Futter m, Halter m, Haltevorrichtung f, Unterbauvorrichtung f, Vorrichtung f, Zusammenbaulehre f

**fixtures** Beschläge pl

**fizz** Zischen n

**fizzing** Aufzischen n

**fizzler** Trotler n

**flabby** lappig, schlaff ~ **ballonet** unpralles Ballonet n

**flaccid** lappig, schlaff
**flag** Fahne *f*, Flagge *f*, Fliese *f*, Linsenschirm *m*, (instrument) Marke *f*, Plattenbelag *m* ~ **of distress** Notflagge *f* ~ **of truce** Friedensflagge *f* **flag,** ~ **guidon** Fahne *f* ~ **locker** Flaggenspind *m* ~ **signal** Flaggensignal *n*, Winkerzeichen *n* ~ **signalman** Winker *m*
**flagged floor** Fliesboden *m*, Fliesenpflaster *n*
**flagging** Fliesboden *m*, Fliesenpflaster *n*
**flagpole** Balken *m*, Fahnenstange *f* ~ **antenna** Stabantenne *f*
**flagstaff** Flaggen-stange *f*, -stock *m*
**flagstone** Fliese *f*, Fußbodenplatte *f* ~ **pavement** Plattenpflaster *n*
**flake, to** ~ abblättern, sich schichtweise ablösen (aviat.), blättern, flocken **to** ~ **off** abbröckeln, abschuppen
**flake** Blättchen *n*, Flocke *f*, Flockenriß *m*, Schuppe *f* ~ **discharger** Flockenabscheider *m* ~-**ice maker** Flockeneisenerzeuger *m* ~ **litharge** Schuppenglätte *f* ~ **powder** Platten-, Blättchen-, Platten-pulver *n* ~-**shaped** blättchenförmig ~ **shellac** Blattschellack *m* ~ **soap** schuppige Seife *f* ~ **twist with long or short slubs** Flammenzwirn *m*
**flaked** schuppig, zerflockt ~ **asbestos** Flockenasbest *m* ~ **graphite** flockiger Graphit *m* ~ **gunpowder** Blättchenpulver *n*
**flakes** (fisheyes) Flocken *pl*
**flakily** flockenrissig
**flaking** (protective coatings) Abkreiden *n*, blätten, Schuppelbildung *f* ~ **mill** Flockenwalze *f* ~ **speed** Ausflockungsgeschwindigkeit *f*
**flaky** flockenartig, flockenrissig, flockig, geschichtet, schuppig ~ **arsenic** Fliegenstein *m* ~ **graphite** Flockengraphit *m*, Schuppengraphit *m*
**flame, to** ~ flammen, sengen
**flame** Flamme *f* **to go down in** ~**s** brennend abstürzen ~ **playing at the mouth** Mündungsflamme *f*
**flame,** ~ **adjustment** Flammeneinstellung *f* ~ **arc** Flammenbogen *m* ~ **arrester** Flammsperre *f* ~ **attachment for spectrophotometer** Flammenzusatzgerät *n* zum Spektralphotometer ~ **baffle** Flammen-dämpfer *m*, -vernichter *m*, -verzehrer *m* ~ **baffling** flammendämpfend, Flammen-einstellung *f*, -führung *f* ~ **bridge** Feuerbrücke *f* ~ **carbon** Effektkohle *f* ~-**colored** flammenfarbig ~ **cone** (of a Bunsen burner) Brennkegel *m* ~-**control magnet** Blasmagnet *m* ~-**cutting nozzle** Brennerdüse *f* ~ **damper** Blendblech *n*, Flammendämpfer *m* ~ **effect** Flammeneffekt *m* (Zebra) ~ **emission continuum** Flammenemissionskontinuum *n* ~ **flue** Flammenzug *m* ~ **formation** Flammenentwicklung *f* ~ **front** Flammenfront *f* ~ **gases** Feuergase *pl* ~-**gouging** Fugenhobeln *n* ~ **hardening** (by oxyacetylene flame) autogene Oberflächenhärtung *f* ~ **height** Flammenform *f* ~ **holder** Flammenhalter *m* (Triebwerk) ~ **ignition** Flammen-, Heizlampen-zündung *f* ~ **jet** Flammenstrahl *m* ~ **lamp** Effektbogenlampe *f* ~ **lighter** Anzünder *m* ~-**like** flammenfarbig ~ **microphone** Flammenmikrofon *n*, Lichtbogenmikrofon *n* ~ **movements** Flammenbewegungen *pl* ~ **obstacle** Flammenhindernis *n* ~-**out** Brennschluß *m* (rocket), Flammabriß *m*

~ **photometry attachment** flammenfotometrischer Zusatz *m* ~ **pictures** Flammenaufnahmen *pl* ~ **plating process** Flammplattierverfahren *n*
**flameproof, to** ~ feuersichermachen
**flameproof** feuersicher, flammenbeständig, flammsicher, schlagwettersicher, unentflammbar ~ **appliance** schlagwettersicherer Apparat *m* ~ **motor** druckfester Motor *m* ~ **wire** feuersicher Draht *m*
**flame,** ~ **propagation** Flammenfortpflanzung *f* ~-**propagation velocity** Zündgeschwindigkeit *f* ~ **resistance** schwere Brennbarkeit *f* ~ **resistant** schwer entflammbar, flammenbeständig ~ **scarfing** Abflämmen *n* ~ **softening** Flammenglühen *n* ~-**splitting tube** Flammenspaltrohr *n* ~ **spraying equipment** Flammspritzgerät *n* ~ **spreader** (in heating devices) Prallstift *m* ~ **temperature** Flammentemperatur *f* ~ **thrower** Flammenwerfer *m* ~-**thrower fuel** Flammenöl *n* ~-**tight motor** schlagwettergeschützter Motor *m* ~ **transmitter** Flammenmikrofon *n* ~ **trap** Flammenrückschlagsicherung *f* (in der Ladeleitung), Flammenschutz *m* ~ **traveling in a horseshoe direction** Hufeisenflamme *f* ~ **tube** Flammrohr *n*, Muffe *f*, Muffel *f* ~ **turbulence** Verbrennungswirbel *m* ~ **vortex** Flammenring *m* ~ **washing** Abschweißen *n*
**flaming** flammig ~ **arc** Flamm-, Flammenbogen *m* ~ **coal** Flammkohle *f*
**flammability, (in)** ~ Feuergefährlichkeit *f*
**flammable** entflammbar, entzündlich
**flange, to** ~ anflanschen, bördeln, börteln, flanschen, krämpen, kümpeln, (a tube) umbördeln, umkrempeln **to** ~ **the edge over** flanschen **to** ~ **the tube (or pipe)** das Rohr umbördeln
**flange** Backen *m*, umgezogener Bord *m*, Bördel *n*, Bordflansch *m*, Bordrand *m*, Bund *m*, Endflügel *m* (outer wing), Facettenhalter *m*, Flansch *m*, Gleitschuh *m*, Gurt *m*, Gurtung *f*, Kranzstück *n*, Lappen *m*, Ösenblatt *n*, Rand *m*, vorstehender Rand, Scheibe *f*, Spurkranz *m*, Steg *m*, Stutzen *m*
**flange,** ~ **of a beam** Trägerflansch *m* ~ **of boiler end** Krempe *f* des Kesselbodens ~ **of brass bearings** Lagerschalenbund *m* ~ **of cartridge case** Bodenrand *m* ~ **of drum** Seitenwand *f* der Kabeltrommel ~ **for fastening the brake drum** Bremsbefestigungsflansch *m* ~ **of the feed bar** Abzugsleiste *f* ~ **with feet** Fußflansche *m* ~ **of the fishplate** Laschenschenkel *m* ~ **with plain face** glatter Rohrflansch *m* ~ **of a rail** Schienenfuß *m* ~ **of the spar** Holmgurtung *f* ~ **of a truss** Gurtung *f* eines Brückenträgers
**flange,** ~ **angle** Gurtungswinkel *m* ~ **axle collar** Flanschstoßscheibe *f* ~ **bearing with stop** Flanschlager *n* mit Anschlag ~ **bolt** Bundbolzen *m*, Flanschen-bolzen *m*, -schraube *f* ~ **box** Flanschkappe *f* ~ **bushing** Bundbuchse *f* ~ **butt-welding joint** Bördelschweißnaht *f* ~ **clinch** Felgenohr *n* ~ **connected** angeflanscht ~ **coupling** Flansch-kupplung *f*, -verbindung *f* ~ **(-face) coupling** Scheibenkupplung *f* ~ **dog (or yoke)** Flanschmitnehmer *f* ~ **edge joint** Stirnstoß *m* ~ **facing** Flanschfläche *f* ~ **fin** Wulstflosse *f* ~ **gate valve** Flanschabsperrschieber *m* ~ **groove** Spurrille *f* ~ **housing** Anschlußgehäuse *n* ~ **hub** Flanschennabe *f* ~

iron Gurtungseisen *n* ~ joint Flanschenverschraubung *f* ~-joint welding Bördelnahtschweißung *f* ~ lug Flanschlappen *m* ~ member Gurtstab *m* ~-mounted angeflanscht ~-mounted three-phase motor Drehstromflanschmotor *m* ~ nut Bundmutter *f* ~ packing Flanschdichtung *f* ~ plate Gurtplatte *f*, Gurtungslamelle *f* ~ riveting Bördelnietung *f* ~ rolling-on machine Flanschenwalzmaschine *f* ~ sheet Gurtplatte *f*, Gurtungslamelle *f* ~ socket Flansch-dichtung *f*, -muffe *f* ~ spigot Flanscheinpaß *m* ~ stifening (or strengthening) Gurtversteifung *f* ~ surface Flanschauflagefläche *f* ~ tool Flanschwerkzeug *n* ~ tube (or connection) Flanschstutzen *m* ~-turning jig Flanschendrehsupport *m* ~-type Anflansch *m* (bei Geräten) ~-type cock Flanschenhan *m* ~-type cock with bent outlet Flanschenhan *m* mit gebogenem Auslauf ~-type cock with cap screw Flanschenhan *m* mit Überwurfmutter ~-type weld Bördelnahtschweißung *f* ~ way clearance Zungenrille *f* ~ wrench Flanschenschraubenschlüssel *m* ~ yoke Flanschmitnehmer *m*

flanged geflanscht, gerippt, umgebördelt, verflanscht ~ ball-bearings Schulterkugellager *n* ~ bearing Flanschlager *n* ~ bend Krümmer *m* mit Flanschen ~ bobbin Scheibenspule *f* ~ bush Flanschenbuchse *f* ~ connector Flanschenkopplung *f* ~ coupling Flanschenverbindung *f*, Flanschett *n* ~ coupling joint Flanschenverbindung *f* ~-coupling pulley Bordscheibe *f* ~ edge Bördelrand *m* ~ end gebördelter oder umgebördelter Boden *m* ~ ends mit Flanschen *pl* ~ fire box umgebördelte Feuerbüchse *f* ~ gear Zahnrad *n* mit Flansch ~ hub Flanschnabe *f* ~ idler roller Umlenkrolle *f* ~ joint Flanschenverbindung *f* ~ mild-steel plate Bördelblech *n* aus Flußeisen ~-on pump Flanschpumpe *f* ~ pipe Rohr *n* mit Flanschen ~ pipe (welded or cast on) Flanschen-rohr *n*, -röhre *f* ~ pipe with loose back flange Bördelflanschrohr *n* ~-planing and -trimming machine Flanschendrehbank *f* ~ plate Kümpelblech *n* ~ pressure pipe Flanschendruckrohr *n* ~ pulley Flanschenscheibe *f* ~ radiator Lamellen-Streifen-kühler *m* ~ ring Lappenring *m* ~ roller Bundrolle *f* ~-roller gravity conveyer Rollenbahn *f* mit Bundrollen ~ seam (on a boiler) Flanschnaht *f* ~ shaft Flanschwelle *f* ~ sheet Bördelblech *n* ~ sheet-work Bördelarbeit *f* ~ specials Flanschenformstücke *pl* ~ spigot Einflanschstück *n* ~ spool Scheibenspule *f* ~ starter Flanschanlasser *m* ~ threephase current motor with change of pole connections polumschaltbarer Drehstromflanschmotor *m* ~ toolholder Schafthalter *m* mit Flansch ~ T's Flansch T-Stücke *pl* ~ tube Flanschrohr *n* ~-tube radiator Rippenrohrkühler *m* ~ tubing Flanschrohr *n* ~ union Flanschanschluß *m*

flangeless bundlos, flanschenlos

flanging Umbördelung *f* ~ attachment Bördeleinrichtung *f* ~ forces Bördelkräfte *pl* ~ machine Bördel-, Flansch-maschine *f* ~ press Bördel- und Flanschpresse *f*, Börtel-, Kümpelpresse *f* ~ radius Eckradius *m* ~ roll Bördelwalze *f* ~ roller Falzrolle *f* ~ test Bördelprobe *f*

flank, to ~ begrenzen, flankieren, umgehen

flank Ende *n*, Flanke *f*, Flügel *m*, Schenkel *m*, vordere Schneidfläche *f*, Seite *f*, Weiche *f* ~ of a fold (geol.) Faltenschenkel *m* ~ of thread Gewindeflanke *f* ~ of a tooth Zahnflanke *f* ~ of a worm Schneckenflanke *f*

flank, ~ clearance Flankenspiel *n* ~ column Flügelkolonne *f* ~ escape Flutauslaß *m*, Freiarche *f* ~-group position Riegel *m* ~ guard Seiten-deckung *f*, -hut *f* ~ protection Flankendeckung *f*, -schutz *m*; Seiten-schutz *m*, -sicherung *f* ~ security Flankensicherung *f*, Seiten--deckung *f*, -schutz *m*

flannel Flanell *m* ~ polishing wheel Flanellpolierscheibe *f*

flap, to ~ klappen

flap Briefverschluß *m*, (inserting) Einsteckklappe *f*, Flügel *m*, Klappe *f*, Lappen *m*, lose herabhängender Lappen *m*, Lasche *f*, (split) Spreizklappe *f*, Zwischenflügel *m* (aviat.) ~ on engine cowling Abgasregelklappe *f*

flap, ~ actuator Klappenbetätigung *f* ~ angle Spreizklappenöffnungswinkel *m* ~ attenuator Streifenabschwächer *m* ~ butt angle Stoßwinkel *m* ~ chord Klappentiefe *f* ~ control Klappenbetätigung *f* ~-covered lubricator Klappöler *m* ~ displacement Klappenausschlag *m* ~ gate Drehschützklappe *f*, Klapptor *n* ~ guide Klappenführung *f* ~ hinge Klappenscharnier *n* ~ position indicator Anzeiger *m* für Landeklappenausschlag ~ shutter Klappenverschluß *m* ~-slot Einsteckloch *n* ~ track Klappenführung *f* ~-type shutter Dreh-klappenabdeckung *f*, -klappenjalousie *f* ~ vacuum valve Unterdruckklappe *f* ~ valve Klappenventil *n* ~ valve with levers Hebelklappenventil *n* ~ valves hinged at top hängende Ventile *pl*

flapper Prallplatte *f* ~ check valve Klappenrückschlagventil *n*

flapping Schlagen *n* (helicopter) ~ of the belt Schlagen *n* des Riemens

flapping, ~ bearings Schlaggelenklager *pl* ~ flight Schwingenflug *m* ~ hinge Schlaggelenk *n* (helicopter) ~ link Schlaggelenkstück *n* ~ object (or part) Flatterer *m* ~ stop pin Schlagbegrenzungsstift *m* ~-wing machine Schwingenflugzeug *n* ~-wing model Schwingenflugmodell *n*

flare, to ~ abfangen (Flugz.), nach außen erweitern, flackern, kelchen to ~ back zurückschlagen to ~ out das Flugzeug bei der Landung abfangen to ~ up aufflackern, aufflammen

flare Abwerfleuchte *f*, Blendfeuer *n*, Fackel *f*, Flackerfeuer *n*, Hornstrahler *m* (rdr), Leuchtapparat *m*, -bombe *f*, -kugel *f*, -mittel *n*, Lichtrakete *f*, Markierungsleuchte *f*, Öffnungsweite *f*, Reflexionsfleck *m* (photo) ~ on lamp Landefeuer *n* (aviat) ~ of ship's frames (ribs, stem posts, or stern posts) Ausfallen *n* ~ and signal material Leucht- und Signalmittel *n*

flare, ~ angle Kelch-, Öffnungs-winkel *m* (antenna) ~ back Flammenrückschlag *m*, Stichflamme *f*, (of fire arms) Vorbeischlagen *n* ~ bouy Luxboje *f* ~ cartridge Leuchtpatrone *f* ~ composition Leuchtsatz *m* ~ ghost Reflexions- ~-out Abfangen *n* (Aushungern) ~ path Leuchtpfad *m* ~-path lighting Befeuerung *f* ~ point (of oil) Entzündungspunkt *m* ~ projectile Signalgeschoß *n* ~ shell Leucht-geschoß *n*, -granate *f* ~ signal Leucht-patronensignal *n*, -zeichen *n*, Raketenzeichen *n* ~ work Strebewerk *n*

flared, ~ entrance of baffle sich nach außen erweiternder Lufteintritt *m* eines Leitbleches ~ nozzle nach außen erweiterte Düse *f* ~-out konisch verlaufen ~ radiating guide Hohlleiter *m* mit Leitblechen ~ throat erweiterter Trichterhals *m* (acoust) ~ tube ausgeweitetes oder gekelchtes Rohr *n* ~ tube of rear axle Hinterachstrichter *m*

flaring Aufleuchten *n*, (petroleum) Auswalzen *n*, Erweiterung *f* ~ of excess gas Vernichtung *f* (oder Abfackeln) überschüssigen Gases ~ die Kelchmatrize *f* ~ expander Aufweitedorn *m* ~ machine Flanschmaschine *f* ~ punch Kelchstempel *m* ~ tool Aufweitewerkzeug *n*

flaser Flaser *f*

flash, to ~ abblitzen, auflodern, blinken, Blinkzeichen geben, blitzen, feuern, flackern, schießen, (comput) to ~ back zurückschlagen (Flammen to ~ in the exchange durch Blinken *n* das Amt zum Eintreten veranlassen to ~ off abschmelzen to ~ operator Flackerzeichen *pl* geben to ~ over überschlagen to ~ up aufblitzen, aufflackern, aufflammen, aufleuchten

flash Aufblitzen *n*, Austrieb *m* (molding), Blinkgarbe *f*, Blitz *m* (film), Druckverminderung *f*, Feuerstrahl *m*, Flackern *n*, Funken *pl*, Gesenkfuge *f*, Grat *m*, Schein *m*, Strahl *m* (of lightning), Strahlung *f*, Überlauf *m* ~ of fire Feuergabe *f* ~ of light Lichtblitz *m*

flash, ~ arc Rocky-Point-Effekt *m* ~ back Flammen-durchschlag *m*, -rückschlag *m*, Rückschlaglöschung *f* ~ barrier Schutzwand *f* ~ beacon Blinkbake *f* ~ boiler Blitzkessel *m*, Kessel *m* für Augenblicksverdampfung, Schnellverdampfer *m* ~ bomb Leuchtbombe *f* ~ bulb Blitz-lampe *f*, -licht *n*, -lichtlampe *f* ~ burn Strahlungsbrandwunde *f* ~ butt welding Abbrennschweißung *f* ~ combustion unit Ausbrenner *m* ~ composition Momentzündpille *f* ~ contact Blitzlichtanschluß *m* ~ cooler Vakuumkühler *m* ~ drum Entspannungs-kammer *f*, -trommel *f* ~-drying Entspannungstrocknung *f* ~ edge (leak) Abquetschfläche *f* ~ equipment Blitzlichtausrüstung *f* ~ evaporation Verdampfung *f* durch Entspannung ~ figure Glühspuk *m* (metall.) ~ gas Spülgas *n* ~ groove (molding) Austriebsbrille *f*, Stoffabflußnute *f* ~ hider Feuerdämpfer *m*, Mündungsfeuerdämpfer *m*, Vorlage *f* ~ hole Zündkanal *m* ~ indication (direction finding) Zuckanzeige *f* ~ lamp Blink-, Morse-lampe *f*

flashlight Batterielampe *f*, Blitzlampe *f*, Blitzlicht *n* (photo), unterbrochenes Feuer *n*, ununterbrochenes Feuer *n*, Leuchtstab *m*, Taschenlampe *f* ~ battery Taschenlampenbatterie *f* ~ bomb Blitzlichtbombe *f* ~ bulb Taschenlampenbirne *f* ~ capsule Kapselblitz *m* ~ lamp (bulb) Blitzlicht *n* ~ powder Blitzlichtpulver *n*

flash, ~ method of welding Abschmelzschweiß-

flashover Funkenüberschlag *m*, Kriechfunken *m*, Rundfeuer *n*, Selfbogen *m*, Überschlag *m* elektrischer Funken ~ distance (of a gap)

Schlagweite *f* ~ potential Durchschlags-, Einsatz-spannung *f* ~ test Überschlagsversuch *m* (electr.) ~ voltage Überschlagspannung *f*, Überspannung *f*

flash, ~ pane (of a hammer) Finne *f* ~ period Blinkfolge *f* ~ point Entflammungs-punkt *m*, -temperatur *f*, (of oil) Entzündungs-, Flamm-, Leucht-, Zünd-punkt *m* ~-point apparatus Flammpunktapparat *m* ~-point tester Flammpunktprüfer *m* ~-point tester (or testing apparatus) Ölflammpunktprüfer *m* ~ radiography Blitzaufnahme *f* ~ ranging Lichtmessen *n*, -messung *f*, meßverfahren *n* ~ ranging theodolite Lichtmeßtheodolit *m* ~ reading (direction finding) Zuckanzeige *f* ~ reducer Mündungsfeuerdämpfer *m* ~-reducing wad Kartusch-, Salz-vorlage *f* ~ removing lathe Abgratbank *f* ~ resistant brandträge ~ ridge Abquetschrand *m* ~ signal Blinkzeichen *n* ~ (ing) signal lamp transmission Blinken *n* ~ spotting Lichtmeßverfahren *n* ~ spotting by leading ray Hauptschnittverfahren *n* ~ test Bestimmung *f* des Flammpunktes, Flammpunktbestimmung *f*, Impulsprüfung *f* ~ tin-plating kurzzeitiges oberflächliches galvanisches Verzinnen *n* ~ tower Entspannungsturm *m* ~ trap Entspannungstopf *m* ~ trimmer Abgratmaschine *f* ~ tube (Elektronen)blitzröhre *f*, Kammerhülse *f*, Zündrohr *n* ~-tube charge Kammerhülsenladung *f* ~-type evaporator Unterdruckverdampfer *m* ~-type intercoller Mitteldruckflasche *f* ~-up (power) Leistungsspitze *f* ~ vent Zündloch *n* ~ V-ray tube Röntgen-Blitzröhre *f* ~ welding Abbrennstumpf-, Abschmelz-schweißung *f*, Brennschweißen *n*, Widerstandsabbremsschweißung *f* ~-welding method Abschmelzverfahren *n*

flashed glass Überfangglas *n*, überfangenes Glas *n*

flasher, ~ device Blink-anlage *f*, -lampe *f*, Würfelreflektor *m* ~ relay Blinkerrelais *n*

flashing Aufleuchten *n*, Blinkzeichen *n*, (lamp), Feuern *n*, Flackern *n*, (tungsten) Glühdrahtverdampfung *f*, thorierte Kathode *f*, Metalleinfassung *f*, Selbstverdampfung *f*, Überzug *m*, Verkleidungsdeckel *m*, Wandkehle *f*, Wiederbelebung *f* der thorierten Kathode ~ of flashlight signal Blitzlichtzeichen *n* ~ of image Bildausleuchtung *f* ~ of a tube Zünden *n* einer Röhre

flashing, ~ approach light Blitzanflugleuchten *pl* ~ back of flame zurückschlagende Flamme *f* ~ beacon Blitzfeuer *n*, Blitz(leucht)feuer *n* ~ chamber Entspannungsverdampfer *m* ~ code beacon Blitz-, Blink-kennfeuer *n* ~ cut-in and error button Flacker-, Aufschalte- und Irrungstaste *f* ~ gas beacon Gasblickfeuer *n* ~ getter abschießbarer Getter *m* ~ lamp (in sound recording) Glimmlampe *f* ~ light Blink-, Blitz-, Drehblick-feuer *n*, unterbrochenes Feuer *n* (aviat.), regelmäßig aufblitzendes Licht *n* ~-light signal Blinksignal *n* ~ light system Flakkerlichtanlage *f* ~-off Abschmelzen *n* ~-over Überspringen *n* ~ period Blindauer *f* ~ point Flammpunkt *m* ~ relays Blink-, Flackerrelais *n* ~ signal Besetztprüfung *f* mit Flackerzeichen ~ time Anbrennzeit *f* ~ unit Blinkautomat *m*, -geber *m*

**flashings** Abweis(e)bleche *pl*
**flashless,** ~ **injection molding** gratlose Spritz-preßformung *f* ~ **powder** mündungsfeuerfreies Pulver *n*
**flashograph** Abstimmglimmröhre *f*, Amplituden-anzeiger *m*, -glimmröhre *f*, Resonanzröhre *f*
**flashometer** Lichtblitzanalysator *m*
**flask** Fläschchen *n*, Flasche *f* (Labor), Flaske *f*, (founding) Gußeisenformkasten *m*, Kasten-gußform *f*, (for boiling) Kochflasche *f*, Kolben *m* ~ **for casting** Formkasten *m* ~ **with con-stricted neck** Schnürkolben *m* ~ **with flange** Formkasten *m* mit Sandrippe ~ **with a ground-in stopper** Schliffkolben *m* ~ **with pinhold** Form-kasten *m* mit Stiftenführung ~ **with ring neck** Kochflasche *f* mit umgelegtem Rand ~ **with sand rib** Formkasten *m* mit Sandrippe
**flask,** ~ **band** Formkastenband *n* ~ **bar** Form-kasten-schore *f*, -traverse *f* ~ **board** Form-kastenbrett *n* ~ **-bottom board (or bottom plate)** Formkastenboden *m* ~ **clamp** Formkastenfest-halter *m* ~ **holder (axle)** Flaschenhalter *m*
**flask-lift,** ~ **machine** Formmaschine *f* mit Stif-tenabhebung ~ **machine with air pressure (or air squeezing)** Abhebeformmaschine *f* mit Druck-luftpressung ~ **mechanism** Abhebevorrich-tung *f* ~ **stripping machine** Abhebeformma-schine *f* mit Abstreifkamm ~ **stripping-plate machine** Abhebe- und Durchzugformmaschine *f* ~**-type molding machine** Abhebeformma-schine *f*
**flask,** ~ **lifting machine with hand-lever pressure** Abhebeformmaschine *f* mit Handhebelpres-sung ~**-lifting post machine** Stiftenabhebeform-maschine *f* ~**-lowering device** (foundry) Kasten-absenkvorrichtung *f* ~ **lug** Formkastenlappen *m* ~ **nitrogen** Bombenstickstoff *m* ~ **pin** Form-kasten-, Führungs-stift *m* ~ **section** Formka-stenhälfte *f*
**flat** Altan *m*, Anflächung *f*, Deckelgleitfläche *f*, (music) Erniedrigungszeichen *n*, Etage *f*, Reifenpanne *f*, Schwimmer *m*, Schlüsselfläche *f* (Mutter), (naut) Untiefe *f*, Wohnung *f*
**flat** eben, flach, flau (photo), gestreckt, leblos, platt, (trajectory) rasant, reflexionsfrei (transm. line), tafelförmig, tellerförmig, unscharf, wöl-bungsfrei **on the** ~ flachnebeneinander ~ **with side rollers** Plattenband *n* mit Rollenführung
**flat,** ~ **air valve spring** Blattfeder *f* (print.) ~**-and--edge box-piled faggot** auf Sturz paketierte Luppenstäbe *pl*, auf Sturz paketiertes Schweiß-eisenpaket *n* ~**-and-edge-box piling** Paketierung *f* auf Sturz ~ **anvil** Flachamboß *m* ~ **arch** Flachbogen *m*, Streifen *m* ~ **armature** Flach-anker *m* ~ **back stoping** Firstenbau mit waagerechter Firste ~ **band arch** Strei-fen *m* ~ **bank** flache Kurve *f* oder Kehre *f* (aviat.) ~ **bar** Flacheisen *n* ~**-bar** Flachstab *m* ~ **bars** Flacheisen *n* ~**-bar steel** Flachstahl *m* ~**-base annulus around jet** Heckring *m* ~**-base area** Heckring *m* ~**-base rim** Flachbettfelge *f* ~**-base-type instrument** Flachsockelbauart *f* (electr.)
**flatbed,** ~ **machine** Flachformmaschine *f* ~ **offset machine** Flach-Offsetpresse *f* ~ **offset proofing** Offset-An- und Flachdruckpresse *f* ~ **press** Flachdruckpresse *f* ~ **rotary machine** Flachdruckrotationsmaschine *f* ~ **trailer** Tief-

ladeanhänger *m* ~ **web machine** Flachdruck-rotationsmaschine *f* (conveyor)
**flat,** ~ **bedded rubble** Bruchsteinmauerwerk *n* mit dünnen Steinlagen ~ **bedway** Flachbahn *f* ~ **belt** Flach-band *n*, -riemen *m* ~ **belt drive** Flachriemenantrieb *m* ~**-bending test** Flach-biegeprobe *f* ~ **bending torsion testing machine** Flachbiegetorsionsmaschine *f* ~ **bevel** Flach-facette *f* ~ **bit** flacher Bohrmeißel *m* ~ **blade holder** Flachstahlhalter *m* ~ **blade paddle agitator** Blattrührer *m* **with** ~ **bottom** flach-rund ~**-bottom bag** Bodenbeutel *m* ~**-bottom buoy** Flachwassertonne *f* ~ **bottom plane** Boden-fläche *f*, Flachwassertonne *f* ~**-bottomed flask** Stehkolben *m* ~ **box pile** flachliegende Paketie-rung *f* ~**-box-piled faggot** auf Sturz paketierte Luppenstäbe *pl* ~**-box piling** flachliegende Pake-tierung *f* ~ **bridge** Flachsteg *m* ~ **brush** Flachpin-sel *m* ~ **bulbs** Flachwulsteisen *pl* ~ **burner** Flach-brenner *m* ~ **cable** Bandseil *n*, Flach-kabel *n*, -seil *n* ~**-cable drum** Bandseiltrommel *f* ~ **cable plug** Flachkabelstecker *m* ~ **camera** Flachkamera *f* ~ **car** Lore *f*, Plattformwagen *m* ~ **card resolver** sin-cos-Flächenpotentiometer *m* ~ **charac-teristic** flache oder flaue Kurve *f* ~ **chisel** Flach-meißel *m*, Meißelbohrer *m* ~ **coat** Grund-, Grundier-anstrich *m* ~ **coil** Flachspule *f* ~**-coil antenna** Spiralantenne *f* ~ **coil measuring instrument** Flachspulinstrument *n* ~ **coil spring** Uhrfeder *f* ~ **coils** Flachteilspulen *pl* ~ **collar nut** Flachbindmutter *f* ~ **color** Grund-, Grun-dier-farbe *f* ~ **condition** Flachheit *f* (photo) ~ **copper bar (or strip)** Flachkupfer *n* ~ **copy** flache Vorlage *f* ~ **core** Flachkern *m* ~ **cost** Gestehungskosten *pl* ~ **cotter** Blattkiel *m* ~ **country** Ebene *f*, Flachland *n* ~ **cross section** Flachquerschnitt *m* ~ **curve** flache oder niedri-ge Kurve *f* ~ **die for thread rolling** Gewinderoll-backe *f* ~ **drill** Flach-, Spitz-bohrer *m* ~ **drilling bit** flacher Spatenmeißel ~ **drive** Deckelantrieb *m* ~**-drop attachment** Flachdibbelapparat *m* ~**-drop corn planter** Flachdibbelmaisdrill *m*, Flachtropfpflanzmaschine *f* ~ **edgewise pattern meter** Flachprofilinstrument *n* ~ **face** Schlüssel-fläche *f* (der Mutter) ~ **face-plate** Planplatte *f* ~**-face pulley** Flachscheibe *f* ~**-field achromat** planachromatisches Objektiv *n* ~ **file** Ansatz-, Flach-feile *f* ~ **finish** Mattlack *m* ~ **fit** Flach-passung *f* ~**-floored (flat-top)** flachbodig ~ **flute** Einziehung *f*, Halskehle *f* ~ **foot** Flach-, Platt-fuß *m* ~ **foundation** Flachgründung *f* ~**-furnace mixer** Flachherdmischer *m* ~ **gasket ring** Flachdichtung *f* ~ **gib key** Nasenflachkeil *m* ~**-glass industry** Flachglaserzeugung *f* ~ **grain** Flader *m* ~**-grid type** flachgewickelter Typ *m* ~ **grinding arrangement** Deckelschleif-vorrichtung *f* ~ **groove** Flachkaliber *n* ~ **ground** ebenes Gelände *n* **to** ~ **hammer** glatt hämmern ~ **hammerer** Plattenschleifer *m*
**flathead** Flachkopf *m* ~ **cap screw** Senkschraube *f* ~ **screw** Flachkopf-, Kegelsenk-schraube *f* ~ **wood screw** Senkkopfholzschraube *f*
**flat,** ~**-headed bolt** flachköpfiger Bolzen *m*, Bolzen *m* mit flachem Kopf ~**-headed rail** Flachkopfschiene *f* ~ **headed rivet** flachköpfiges Niet *n* ~**-headed screw** Flachkopfschraube *f* ~ **hollow** Einziehung *f*, Halskehle *f* ~ **idler** Flachrolle *f*

**flat-iron** Flach-, Bügel-, Plätt-eisen *n* ~ **holder** Flacheisenhalter *m* ~ **shearing machine** Flacheisenschere *f* ~ **train** Bandeisenstraße *f*
**flat,** ~**-jawed pliers** Flachzange *f* ~ **joint** (flush joint) Flachlasche *f* ~ **joint cover** Flachlaschendeckel *m* ~ **keep spring** Federband *n* ~ **key** (shaft) Flachkeil *m* ~ **knurled nur** flache Rändelmutter *f* ~**-link articulated chain** Gelenkkette *f* ~ **lock floor** ebene Schleusensohle *f* ~ **loudspeaker** Flachlautsprecher *m* ~ **lug** abgeflachter Teil *m* von Rundeisen ~ **lugs** Flachgreifer *m*, flache Sporen *pl* ~ **map** Plattkarte *f* ~ **mattock** Flachhacke *f* ~ **mold attachment** Flachstabstreichblech *n* ~ **nose pliers** Flach-, Platt-zange *f* ~**-nosed pliers** Drahtzange *f* mit flachem Maul ~ **paint** Grundier-, Matt-farbe *f* ~ **parachute** Plankappenfaltschirm *m* ~ **part** flächiges Werkstück *n* ~ **pass** Flachstich *m* ~ **pick** Kreuzhacke *f* ~ **picture** flaches Bild *n* ~ **pin** Flachstiftstecker *m* ~ **plain key** Flachkeil *m* ~ **planting** Flachbau *m* ~ **plate** Drehplatte *f* ~ **plate at zero incidence** längsangeströmte Platte *f* ~ **plate caster** Flachgießinstrument *n* ~**-plate glow lamp** Flächenglimmlampe *f* ~ **plate stereotyping apparatus** Flachstereotypieapparat *m* ~**-point screw** Schraube *f* ohne Kuppe ~ **pointing** volle Fuge *f* ~ **quality** Flachheit *f* (photo) ~ **quality of a picture** Bild-flachheit *f*, -flauheit *f* ~ **rail** Flachschiene *f* ~ **rate** (tariff) Einfachtarif *m*, Pauschalgebühr *f* ~**-rate subscriber** Teilnehmer *m* mit Pauschgebührenanschluß ~**-rate subscription** Pauschgebührenanschluß ~**-rate tariff** Pauschalgebühr *f* ~ **reproduction** flache Wiedergabe *f* ~ **response** flache Wiedergabe *f* (acoust) ~**-ring side valve** Flachringschieber *m* ~ **roll** glatte Walze *f* ~ **roof** Altan *m* ~ **rope** Band-, Flach-seil *n* ~**-rope drum** Flachseiltrommel *f* ~ **router** Flachfräser *m* ~**-routing machine** Planfräsmaschine *f* ~ **rubber gasket ring** Flachgummidichtring *m* ~ **scraper** Flachschaber *m*, -stichel *m* ~ **(or plane) screen** ebener Schirm *m* ~ **seat nozzle** Flachsitzdüse *f* ~**-section instrument** Flachprofilinstrument *n* (electo) ~**-section jet burner** Flachstrahlbrenner *m* ~**-shaped ingot** Flachblock *m* ~**-sheet covering** Glattblechbeplankung *f* ~ **side** Seitenfläche *f* ~ **slab** trägerlose Decke *f*, Plizdecke *f* ~ **slide valve** Flachschieber *m* ~ **sliding bolt** Vierkantschubriegel *m* ~ **smooth file** Flachschlichtfeile *f* ~ **spaces** ebene Räume *pl* ~ **spin** Flachtrudeln *n* (aviat.), Tellertrudeln *n* ~ **spiral** Flachspirale *f* ~ **spiral antenna** Spiralantenne *f* ~ **spiral coil** Spiralantenne *f*, flache Spiralrohrschlange ~ **spiral spring** Flachfederspirale *f* ~ **spring** Blatt-, Flach-feder *f* ~ **spring washer** Blattfederscheibe *f* ~**-square coil** quadratische Flachspule *f* ~ **squeegee** Flachstreicher *m* ~ **steel** Flacheisen *n* ~ **stitching** Flachheftung *f* ~ **stock** flaches Erzeugnis *n* ~ **stone chisel** flacher Steinmeißel *m* ~ **stone mill** Trichtermühle *f* ~ **strand rope** Flachlitzenseil *n* ~**-strip windings** Flachbandwindungen *pl* ~ **strut** Flachstrebe *f* ~ **support** Flachträger *m* ~ **surface** ebene Platte *f* ~ **switchboard** Flachschalttafel *f* ~ **terminal** Flachkern *m* (electr.) ~ **thread** Flachgewinde *n* (mach) ~**-threaded screw** Flachgang *m* ~ **tile** Flachziegel *m*, Ochsenzunge *f*, Zungenstein *m* ~ **tilter** Flachkipper *m* ~ **tire** flacher Reifen *m*

~ **tone** leerer Ton *m* ~**-top** (piston) Flachboden *m*, flacher Boden *m* ~**-top antenna** Flachantenne *f*, horizontale Langdrahtantenne *f*
**flat-topped** oben abgeflacht, (curve) flach ~ **combustion chamber** Verbrennungsraum *m* mit flachem Boden ~ **curve** flache Kurve *f* (rdo) ~ **electrode** Elektrode *f* mit flachem Ende (Zündkerze) ~ **electromotive-force curve** rechteckige Kurvenform *f* der elektromotorischen Kraft ~ **piston** Flachbodenkolben *m* ~ **potential** (wave) abgeflachte Spannung *f* ~ **resonance crest** abgeflachte Resonanzspitze *f*
**flat,** ~ **trailer** Troganhänger *m* ~ **trajectory** Flachbahn *f*, gestreckte oder rasante Flugbahn ~ **trajectory fire** Flachfeuer *n* ~ **trajectory gun** Flachbahngeschütz *n* ~ **triblet** flacher Dorn *m* ~**-tube radiator** Flachrohrkühler *m* ~ **tubular braided cordage** geklöppeltes Schlauchband *n* ~**-tuned** unscharf abgestimmt ~ **tuning** rohe oder unscharfe Abstimmung *f* ~ **tuning condition** Unschärfe *f* ~ **turn** Hängen *n*, flache Kurve *f*, Schieben *n* in der Kurve ~ **TV tube** Flachröhre *f* ~**-type engine** flacher Motor *m* ~**-type relay** Flachrelais *n* ~**-twin engine** Boxermotor *m*, Zweizylinderboxermotor *m* ~ **valuation** Anschlag *m* ~ **valve-type nozzle** Plansitzlochdüse *f* ~ **V-belt** Flachkeilriemen *m* ~ **vertical** eben, senkrecht ~ **vortex hydro-sifter** Hydro-Spiralflachsichter *m* ~ **wall paint** Wandfarbe *f* ~ **washer** Unterlegscheibe *f* ~ **way** Flachbahn *f* ~ **weight** Plattengewicht *n* ~ **weld** Schweißung *f* von oben ~**-winding coil (or winding drum)** Flachseiltrommel *f* ~ **wire** Flachdraht *m* ~ **wire pincers** Drahtzange *f* ~ **wire rope** Flachdrahtkabel *n* ~**-wire sheathing** Flachdrahtbewehrung *f* ~ **wise bend** E-Bogen *m*, E-Krümmer *m* ~ **work** lagenartige Erzausscheidung *f*
**flatly,** ~ **conchoidal** flachmuschelig ~ **tuned** unscharf abgestimmt
**flatness** Ebenheit *f*, Fläche *f*, Flachung *f*, (ballistics) Rasanz *f*, Stille *f* ~ **of field** Bildfeldebnung *f* (opt.) ~ **tolerance** Unebenheits-Toleranz *f*
**flats** Flacheisen *n*; Flachland *n*; flacher Gang *m* (geol.) ~ **and grooves in rifling of barrel** Felder *pl* und Züge *pl*
**flatted thread** angeflanschtes Gewinde *n*
**flatten, to** ~ abflachen, abplatten, ausbeulen, ausbreiten, ausglatten (graphs), ausschlagen, breiten, ebnen, egalisieren, eindrücken, einebnen, flachschlagen (by impact), glatthämmern, glattstreichen (glass mfg), lamellieren, nachhämmern, plattdrücken, recken, strecken, verflachen to ~ out after a dive Kurve *f* reißen (aviat.) to ~ out of a dive abfangen (aviat.) to ~ in focal plane planliegen in Fokusebene to ~ iron das Eisen abbreiten to ~ out aus dem Sturzflug abfangen; ausplatten, ausschweben (aviat.), ausschweben lassen, flau werden, im Gleitflug heruntergehen
**flattened** geebnet, wölbungsfrei ~ **curve** flaue Kurve *f* ~ **ellipsoid of rotation** abgeplattetes Rotationsellipsoid *n* ~ **field** (by flattening lens) krümmungsfreies Feld *n*, wölbungsfreies Feld *n* ~ **image field** geebnetes Bildfeld *n* ~ **plank** vollkantiges Brett *n*
**flattener** Breitschläger *m*
**flattening** Abflachung *f*, Abplattung *f*, Aus-

platten *n*, Breiten *n*, Dämpfung *f* (eines Schwingungskreises), Planlegung *f* (film) ~of a curve Kurveneinebnung *f* ~ of the image field Bildfeldebnung *f* ~ of a minimum (direction finder) Trübung *f* eines Minimums ~ of threads Abflachen *n* von Gewinden

**flattening, ~ arrangement** Abflachschaltung *f* (teleph.) ~ **hammer** Plätthammer *m* ~-**out** Abfangen *n* (aviat.) ~ **pressure** Flachdruck *m* ~ **test** Ausbreiteprobe *f* ~ **tool** (for window glass) Krappen *n*, Streckeisen *n*

**flatting, ~ agent** Mattierungsmittel *n* ~ **mill** Plättmaschine *f* ~ **paste** Schleifwachs *n* ~ **varnish** Politurlack *m*

**flavor, to** ~ abschmecken

**flavor** Geschmack *m*

**flavorless** geschmacklos

**flaw** Anriß *m*, Blase *f*, (defect) Einriß *m*, Fleck *m*, Flecken *m*, Flinse *f*, (defect) Gußnarbe *f*, Narbe *f*, Riß *m*, Sprung *m*, Windriß *m* ~ **in casting** Gußfehler *m* ~ **in a casting** Gußblase *f* ~ **of material** Stoffehler *m* ~ **in weaving** Webfehler *m*

**flaw, ~ defect** Gußschaden *m* ~ **detector** Gerät *n* zur Feststellung von Rissen in metallischen Werkstoffen ~ **flaw sensitivity** Empfindlichkeitsgrad *m*

**flawed** brüchig, fehlerhaft, gerissen

**flawless** fehlerlos, rißfrei, tadellos

**flaws, having** ~ fehlerhaft

**flawy coin plate** ungange Münzplatte *f*

**flax** Flachs *m* ~ **brake** Flachsbreche *f* ~ **buncher** Flachs-bindeeinrichtung *f*, -häufler *m* (Flachsbindeeinrichtung) ~ **drawer** Flachsstrecker *m* ~ **rippling** Flachsriffeln *n* ~ **rougher** Flachsvorhechler *m* ~ **scutching** Schwingen *n* des Flachses ~ **seed ore** Clinton-Erz *n* ~ **spreader** Flachsaufleger *m* ~-**swathing attachment** Flachsschwadenablage *f* ~ **yarn** Leinengarn *n* ~

**flaxen tow yarn** Leinenwerggespinst *n*

**flay, to** ~ **cattle** Vieh abdecken

**flayer** Abdecker *m*

**F layer** F-Schicht *f*

**flaying house** Abdeckerei *f*

**fleck** Spinnknötchen *n* ~ **shiefer** (slate with minute spots) Fleckschiefer *m*

**flection** Biegung *f*

**fleece roller** Aufroller *m*

**fleeced goods** gerauhte Ware *f*

**fleecy** wollig ~ **clouds** Schäfchenwolken *pl* ~ **paper** Wolkenpapier *n*

**fleet, to** ~ (marine engin) abschricken

**fleet** Aviso *m*, Flotte *f*, Wagenpark *m* ~ **angle** Anzugswinkel *m*

**fleeting** flüchtig ~ **knife** Abweismesser *n* ~ **ring** Abdrängring *m* ~ **target** Augenblicks-, Gelegenheits-ziel *n*

**Fleming, ~ 's rule** Dreifingerregel *f* ~ **valve** Diode *f*, Zweielektrodenröhre *f*

**flemish, to** ~ **down** aufschießen

**Flemish window** Halbgeschoßfenster *n*

**flesh** Fleisch *n* ~ **charcoal** Fleischkohle *f* ~ **side of the belt** Fleischseite *f* des Riemens ~ **wound** Fleischwunde *f*

**fleshing, ~ cylinder** Entfleischzylinder *m* ~ **knife** Schabeisen *n*, Schab(e)messer *n*, Schereisen *n*

**fleshings** Schnitzel *m*

**Flettner, ~ balance** Flettnerhilfsruder *n* ~ **control**

**surface** Hilfsruder *n*

**flex, to** ~ biegen **to** ~ **a spring** Feder *f* betätigen

**flexibility** Anpassungsfähigkeit *f*, Beweglichkeit *f*, Biegbarkeit *f*, Biegsamkeit *f*, Biegungsvermögen *n*, Dehnbarkeit *f*, Dehnungsfähigkeit *f*, Elastizität *f*, Fedrigkeit *f*, Geschmeidigkeit *f*, Knickfestigkeit *f* (of a film), Unstarrheit *f*

**flexible** anpassungsfähig, atmende Membran(e) *f*, biegbar, biegsam, biegungsfähig, dehnbar, durchbiegungsfähig, elastisch, federnd, fedrig, flexibel, fügsam, gefedert, gelenkig, geschmeidig, nachgiebig, pliant, schmiegsam, unstarr

**flexible, ~ axle** Vereinslenkachse *f* ~ **axle frame** Lenkachsuntergestell *n* ~ **bend** Flexibelbogen *m* ~ **braided metal sleeve** Metallgeflechtsschlauch *m* ~ **cable** Geräteanschlußschnur *f*, hoch biegsames Seil *n* ~ **cable with cab-tire** Gummischlauchleitung *f* ~ **cable conduit** Seilzugrohr *n* ~ **characteristic curve** nachgiebige Kennlinie *f* ~ **connection** Federungskörper *m*, Verbindungsschlauch *m* ~ **contact** federnder oder weicher Kontakt *m* ~ **control** Duzzing *n* ~ **control cables** Duzgestänge *n* ~ **cord** Leitungsschnur *f*, biegsame Schnur *f* ~ **coupling** bewegliche oder elastische Kupplung *f*, Elastkupplung *f*, Gelenk *n*, Kreuzgelenkkupplung *f*, nachgiebige Kupplung *f*, biegsame Verbindung *f* ~ **diaphragm** atmende Membran *f* ~ **disk** (motor transport) Hardyscheibe *f* ~ **disk coupling** elastische oder nachgiebige Scheibenkupplung ~-**disk joint** Trockengelenk *n* ~ **fastening** gelenkige Befestigung *f* ~ **fixing** gelenkige Befestigung *f* ~ **frontdrive axle** Schwingachse *f* ~ **fuel cell** Falttank *m* ~ **grinding** Polierschleifen *n* ~ **gun mount** Schwenklafette *f* ~ **high-pressure tubing** Hochdruckschlauch *m* ~ **hose** biegsames Rohr *n* ~ **hub** Federnabe *f* ~ **hull** nachgiebiger Bootskörper *m* ~ **joint** Universalgelenk *n*, biegsame Verbindung *f* ~ **knife** Abschabspachtel *f* ~ **lead** Anschlußschnur *f*, biegsamer Leiter *n* ~ **low-pressure tubing** Niederdruckschlauch *m* ~ **machine gun** bewegliches Maschinengewehr *n* ~ **metal hose** Stahlschlauch *m* ~ **metallic tube** Metallschlauch *m* ~ **pavement** nachgiebiger Belag, flexible Decke *f* ~ **pin coupling** elastische Bolzenkupplung *f* ~ **pipe** elastischer Schlauch *m* ~ **plunger molding** Kolbenpreßspritzen *n* ~ **quartz** (or sandstone) Itakolumit *m* ~ **roller bearing** elastisches Rollenlager *n* ~ **sandstone** Gelenkquarz *m* ~ **shaft(ing)** Gelenkwelle *f*, biegsame Welle *f* ~ **shaft attachment** Biegewellenvorsatz *m* ~ **spar** Federstrebe *f* ~ **speaking tube** biegsames Sprachrohr *n* ~ **spindle** Federwelle *f* ~ **stay bolt** Gelenkbolzen *m*, biegsame Strebe *f* ~ **steam pipe** (articulated) Gelenkschlauch *m* ~-**steel cable** Stahldrahtseil *n* ~ **steel tape** Stahlbandmaß *n* ~ **steel wire rope** biegsames Stahldrahtseil *n* ~ **support** Rippenträger *m* ~ **suspension** (of trolley wire) freibewegliche Aufhängung *f* ~ **telephone cord** Telefonschnur *f* ~ **tubing** biegsame Leitung *f* ~ **waveguide** flexibler Hohlleiter *m* ~ **wire** Litze *f*, Litzendraht *m*

**flexibly, ~ jointed** nachgiebig angelenkt ~ **mounted engine** elastisch aufgehängter Motor *m*

**flexing, ~ die** Biegeform *f* (Bif) ~ **stress** Scherbeanspruchung *f* ~ **test** Biegungsversuch *m*

**flexographic printing** Anilindruck *m*
**flexural, ~ fatigue strength** Biegungsdauer-schwingfestigkeit *f*, Dauerbiegefestigkeit *f* ~ **gliding** Biegegleitung *f* ~**-impact fatigue test** Dauerschlagbiegeversuch *m* ~ **mode of vibration** Biegungsschwingungsart *f* ~ **properties** Biegeeigenschaften *pl* ~ **stiffness** Biegesteifigkeit *f* ~ **strength** Druckdehnung *f* ~ **stress** Biegungsbeanspruchung *f* ~ **tensile stress** Biegezugspannung *f* ~ **test** Biegeprüfung *f* ~ **torque** Biegemoment *m* ~ **vibration** Biegungsschwingung *f* ~ **waves** Biegewellen *pl*
**flexure** Abbiegung *f*, Beugen *n*, Beugung *f*, Biegebeanspruchung *f*, Biegefestigkeit *f*, Biegung *f*, Durchbiegung *f*, Einbiegung *f*, Falte *f* ~ **level gauge** Biegepfeilmesser *n* ~ **modes** Biegungsschwingungen *pl* ~ **test** Knickversuch *m* ~ **yield point** Biegefließgrenze *f*
**flick roll** schnelle Rolle *f* (aviat)
**flicked off** zerhackt
**flicker, to** ~ (instrument) kurz ausschlagen, blinken, flackern (teleg), flattern (of light), flimmern
**flicker** Flimmern *n*, Spratzen *n* ~ **of cathode** Kathodenspratzen *m* ~ **control** Zweipunktregelung *f*, -steuerung *f* ~ **direction-finding method** Flicker-, Flimmer-peilung *f* ~ **effect** (of a cathode) Fackelerscheinung *f*, Flacker-, Funkel-effekt *m* (electr. tube), Funkelrauschen *n*, Funkenrauschen *n*, Rauscheffekt *m* ~**-free reproduction** flimmerfreie oder kontinuierliche Wiedergabe *f* ~ **microscope** Blinkmikroskop *n* ~ **noise** Flackerrausch *m* ~ **photometer** Flacker-, Flicker-, Flimmer-, Schwankungs-photometer *n* ~ **shutter** flackerfreie Blende *f* (film) ~**-signal calling** Flackertaste *f* ~ **switch** Flimmerschalter *m*
**flickering** Flackern *n* ~ **of the flame** Flackern *n* oder Zucken *n* der Flamme
**flickering, ~ device** Blinkeinrichtung *f* ~ **flame** flackernde Flamme *f* ~ **lamp** Flackerlampe *f* ~ **signal** Blinkzeichen *n*, Flackerzeichen *n* ~ **(or flashing) signal** Flackersignal *n*
**flickerless** flimmerfrei
**flier** Flieger *m*; Freitreppe *f* ~ **lathe** Federlade *f*
**fliers** freitragende Treppe *f*
**flight** Flucht *f*, Flug *m*, Luftreise *f*, Kette *f*, Plattenband *n*, Staffel *f* ~ **along angle of sideslip** Messerflug *m* ~ **of electrons** Elektronenbahn *f* ~ **with full throttle** Vollgasflug *m* ~ **of locks** Schleusentreppe *f* ~ **away from object** Zielabflug *m* ~ **of planes** Flugzeugkette *f* ~ **of stairs** gerade Treppe *f* ~ **of steps** Treppenflucht *f* ~ **with wide-open throttle** Flug *m* mit Vollgas
**flight, ~ altitude** Flughöhe *f* ~ **analyzer** Flugschreiber *m* ~ **attachment** (of a chain) Mitnehmer *m* ~**-attachment chain** Mitnehmerkette *f* ~ **attitude** Fluglage *f* ~ **axis** Bahnachse *f* ~ **base line** Anfluggrundlinie *f* ~ **characteristic** Flugverhalten *n* ~ **characteristics** Flugeigenschaften *pl* ~ **check** Überprüfung *f* im Flug ~ **conditions** Flugeigenschaften *pl* ~ **control** Bewegungskontrolle *f*, Flugzeugsteuerung *f* ~**-control surface used for trimming** Trimmvorrichtung *f* ~ **controls** Steuerorgane *pl* ~ **converging on one point** Sternflug *m* ~ **crew** fliegende Besatzung *f*, Flugbesatzung *f* ~ **deck** Abflug-, Ablauf-, Flug-, Lande-deck *n* ~

**diagram** Auftragskarte *f* ~ **diving speed** Stechfluggeschwindigkeit *f* ~ **engineer** Bord-mechaniker *m*, -monteur *m* ~ **equilibrium** Gleichgewichtsflugzustand *m* ~ **equipment** Bordgerät *n* ~ **height** Flughöhe *f* ~ **identification number** Flug(kenn)nummer *f* ~ **indicator** Fluganzeiger *m* ~ **information center** Fluginformationszentrale *f* ~ **instructor** Fluglehrer *m* ~ **instrument** Bordgerät *n*, Flugüberwachungsinstrument *n* ~**-instrument layout** Bordgeräteanordnung *f* ~ **instruments** Fluginstrumente *pl*, Flugüberwachungsinstrumente *pl* ~ **interruption** Flugstörung *f* ~ **kit** Bordwerkzeugtasche *f* ~ **lead** Kettenführung *f* ~ **level** Flugfläche *f* ~ **like** fluchtartig ~ **line** Flugweg *m* ~ **load(ing) conditions** Flugbelastungszustände *pl* ~ **loads** Luftlasten *pl* ~**-log** Flugwegschreiber *m* ~ **map** Flugwegkarte *f* ~ **mechanic** Bord-, Flugzeugmechaniker *m* ~ **mission** Flugaufgabe *f* ~ **navigator** Flugnavigator *m* ~ **notification** Fluganmeldung *f* ~ **operational control** Flugleitung *f* ~ **operations manual** Flugbetriebshandbuch *n*
**flight-path** Bahnrichtung *f*, Flugbahnrichtung *f*, Flugroute *f*, Wurfbahn *f* ~ **of the bomb** Wurfbahn *f* der Fallkurve ~ **angle** Flugbahnwinkel *m*, Steigwinkel *m* des Flugzeuges ~ **computer** Flugwegrechner *m* (rdo nav) ~ **correction** Flugbahnkorrektur *f* ~ **data** Flugwegdaten *pl* ~ **speed** Flugbahngeschwindigkeit *f*
**flight, ~ performance** (aircraft) Flugleistung *f* ~ **plan** Flugplan *m* ~ **position** Flugzustand *m* ~ **progress board** (Flug)Bewegungskontrollpunkt *n* (ATC) ~ **properties** Flugeigenschaften *pl* ~ **radio operator** Bordfunker *m* ~**-range route of flight airway** Flugstrecke *f* ~ **ration** Bordverpflegung *f* ~ **recorder** Flugleistungsschreiber *m* ~ **regularity message** Flugbetriebsmeldung *f* ~**-reporting station** Flugmeldestelle *f* ~ **rules** Flugregeln *pl* ~ **safety** Flugsicherheit *f* ~ **security service** Flugsicherung *f* ~ **simulator** Flugnachahmungsgerät *n* ~ **speed** Fahrtgeschwindigkeit *f* ~ **station** Fliegerkabinenplatz *m* ~ **strip** Flugstreifen *m* ~**-supervision** Flugüberwachung *f* ~ **test** Flugerprobung *f*, Probe-, Prüf-flug *m* ~ **testing** Flugerprobung *f* ~ **time** Flugzeit *f* ~ **track** zurückgelegter Flugweg *m* ~**-tracking receiver** Vermessungsempfänger *m* ~ **training** Flugausbildung *f* ~ **vector** Flugvektor *m* ~ **visibility** Flugsicht *f*
**flighted drier** (drier with inside flights) Rieseltrockner *m*
**flimsy** Laufkarte *f*, einer stärkeren Beanspruchung nicht gewachsen
**fling, to** ~ schlenkern, schleudern, werfen
**fling** Schleudern *n*
**flint** Flint *n*, Flintstein *m*, Feuerstein *m*, Glanzdeckel *m*, Kiesel *m*, Kieselstein *m*, Zündstein *m* ~ **clay** Schieferton *m* ~**-dry** scharf getrocknet ~ **glass** Flint-, Kiesel-, Weiß-glas *n* ~**-glass prism** Flintglasprisma *n* ~ **glazing** Steinglättung *f* ~**-hard** kieselhart ~ **mill** Stahlmühle *f* ~ **stone** Kiesel *m* ~**-stone glazing machine** Steinglättmaschine *f* ~ **tube mill** Flintmühle *f* ~ **work** Geröllmauerwerk *n*
**flintiness** Glasigkeit *f*
**flinty** kieselartig, kieselig, muschelig ~ **ground** Kegel-, Kei-grund *m* ~ **malt** Glasmalz *n*

**flip coil** Feldinduktionsspule *f*, (cathode-ray oscillator) Kippspule *f*, Suchspule *f*
**flipflap telegraph** Klippklapptelegraf *m*
**flip-flop**, ~ **circuit** (staticisor) Flip-Flop-Schaltung *f*, binäre Zählstufe *f*, Zweifachuntersetzer *m* ~ **control** Zweipunktregelung *f* ~ **multivibrator** bistabiler Multivibrator *m* ~ **oscillator** bistabiler Oszillator *m* ~**-relay** Flip-Flop-Relais *n* ~ **storage** (or store) Flip-Flop-Speicher *m*
**flip-over process** Umklappprozeß *m*
**flipper** Höhensteuer *n*, Tiefenruder *n*
**flitch plate** Verstärkungslasche *f*
**flitch(ed) beam** Verbundbalken *m*
**flitting** Umsetzen *n*
**float, to** ~ aufschwemmen, aufspachteln, flößen, flotieren, flott machen, puffern (a battery), schweben, schwemmen, schwimmen, schwingen, treiben, auf dem Wasser treiben, triften, umherschwimmen, verstreichen **to ~ back** zurückschwimmen **to ~ before** vorschweben **to ~ in cement** in Zement einschwemmen **to ~ freely** freifliegen **to ~ in** anschweben **to ~ prepared bridge section into place** einfahren **to ~ a stranded ship** ein gestrandetes Schiff abbringen
**float** Ausfließen von Pigmenten, (in a glass tank) Brückenstein *m*, Floß *n*, Flosse *f*, Fußteller *m*, Schlauchboot *n*, Schlepper *f*, Schwebekörper *m* (gyro), Schwimmer *m*, Schwimmkörper *m*, Spachtel *m*, Tauchschwimmer *m* ~ **and sink analysis** Wichteanalyse *f*, Wichte-Schwimm- und Sink-Analyse *f* ~ **of seaplane** Pallung *f*
**float,**~**-and-sink analysis by sizes** Wichte-Sieb-Analyse *f* ~**-and-sink analysis by sizes with ash contents** Wichte-Sieb-Asche-Analyse *f* ~ **arrangement** Schwimmereinrichtung *f* ~ **axis** Schwimmerachse *f* ~ **bottom** Schwimmerboden *m* ~ **bowl** Schwimmergehäuse *n* ~ **chamber** Schwimmer-gehäuse *n*, -kammer *f*, -raum *m* ~ **check valve** Ventil *n* mit Schwimmer ~**-circle sight** Schwebekreisvisier *n* ~ **column** (gun outrigger) Tellersäule *f* ~ **construction** Schwimmeraufbau *m* ~ **control** Schwimmerregulierung *f* ~ **deck** Verdeck *n* des Schwimmers ~ **device** Schwimmvorrichtung *f* ~ **fraction** Abschwimmschicht *f* ~ **fuel gauge** Fadenschwimmuhr *f*, Schwimmer *m* ~ **(landing) gear** Schwimmergestell *n* ~ **gold** Flutgold *n* ~ **guard** Schwimmerpolster *n* ~ **guide** Schwimmerführung *f* ~ **guide bar** (or rail) Schwimmerführungsschiene *f* ~ **holder** Pendeleinsatz *m* ~ **hull** (or outer skin) Schwimmerhülle *f* ~ **landing gear** Schwimmwerk *n* ~ **level indicator** Schwimmerfüllungsanzeiger *m* ~ **light** Rauch-ball *m*, -bombe *f* ~ **method** (hydrometer) Auftriebmethode *f* ~ **(valve) needle** Schwimmernadel *f* ~**-operated switch** Schwimmerschalter *m* ~ **ore** Erzstück *n* (lose, an der Oberfläche) ~ **overflow** Schwimmerüberlauf *n*
**floatplane** Schwimmerflugzeug *n* ~ **amphibian** Amphibium *n* mit Schwimmern ~ **model** Schwimmermodell *n*
**float,** ~ **planking** Schwimmerhaut *f* ~ **plate** Ansteckblech *n* ~ **position** Unterlegestellung *f* der Nadel ~ **seaplane** Schwimmerflugzeug *n* ~ **skimming device** Schaumabstreifer *m* ~ **skin** Schwimmerhaut *f* ~ **spindle** Schwimmerstange *f* ~ **step** Schwimmerstufe *f* ~ **strut** Schwimmer-

strebe *f* ~ **support** Schwimmerlager *n* ~ **switch** Schwimmerschalter *m* ~ **switch controlled** schwimmergesteuert ~ **test** Schwimmerprobe *f* ~ **trap** Schwimmerkondenstopf *m* ~ **tube** Regelhülse *f* ~**-type landing gear** Schwimmer *m* ~ **undercarriage** Schwimmerwerk *n* ~ **valve** Schwimmerventil *n* ~**-valve needle** Schwimmerventilnadel *f* ~ **vent** Schwimmerentlüftung *f* ~ **well** Schwimmerraum *m* ~ **wood** Flößholz *n*
**floatability** Schwimmfähigkeit *f*
**floatable** flößbar, flotationsfähig, flotierbar
**floatation stability** Schwimmstabilität *f*
**floated, (to be)** ~ gepuffert (werden) ~ **battery** Pufferbatterie *f* ~ **element** Schwebekörper *m* (im Kreisel) ~ **gyro** schwimmender Kreisel *m*
**floater** Pegel *m*, Peil *m*, Schwimmer *m*
**floating** Ausschwitzen *n* (von Farbe), Flotation *f*, Pufferbetrieb *m*, Schwimmen *n*, Trift *f*; schwebend, schwebend fliegen (aviat), schwimmend ~ **of piston** Freigang *m* des Kolbens
**floating,** ~ **action** I-Regelwirkung *f*, integrierende Regelung *f* ~ **action controller** I-Regler *m* ~ **action governor** integral wirkender Regler *m* ~ **address** symbolische Adresse *f*, Pseudo-Adresse *f* (info proc) ~ **aileron** Zwischen-klappe *f*, -ruder *n* ~ **air base** schwimmender Flugstützpunkt *m* ~ **apparatus** Schwimmapparat *m* ~ **axle** drehende Achse *f* ~ **base** Stützpunktschiff *n* (aviat.) ~ **battery** Batterie *f* in Dauerladeschaltung, schwimmende Batterie *f*, Puffbatterie *f* ~**-battery operation** Pufferbetrieb *m* ~ **beacon** Schwimm-, Treib-bake *f* ~ **bell** schwimmende Glocke *f* ~ **body** schwimmender Körper *m*, Schwimmkörper *m* ~ **boom** schwimmende Balkensperre *f*, Treibbalken *m* ~ **bridge** Brücke *f* mit schwimmenden Stützen, Floßbrücke *f*, Zugfähre *f* ~ **bush(ing)** Gleitbüchse *f* ~ **caisson** Schwimmkasten *m* ~ **calandria** eingehängter Heizkörper *m* ~ **capacity** Schwebefähigkeit *f* ~ **capital** umlaufendes Kapital *n* ~**-carrier control** Hapug-Modulation *f* ~ **carrier control** Modulation *f* mit veränderlichem Trägerwert, Trägersteuerung *f* ~**-carrier modulation** HAPUG-Modulation *f* ~ **cathode** fliegende Kathode *f* ~**-chain conveying installation** Oberkettenförderung *f* ~ **chamber** Schwimmkasten *m* ~ **chase** beweglichere Formrahmen *m* ~ **compass** Flüssigkeitskompaß *m* ~ **controller action** I-Regelung *f* ~ **crane** Ponton-, Schwimmkran *m* ~ **dock** Dockschiff *n*, Schwimmdock *n* ~ **electromotive force** Pufferspannung *f* ~ **end cover** Schwimmdeckel *m* ~ **fender** Treibbalken *m* ~ **grid** freies Gitter *n* ~ **gudgeon pin** schwimmender Kolbenbolzen *m* ~ **head exchanger** Schwimmkopf *m* ~ **holder** pendelnder Halter *m* ~ **ice** Eisscholle *f*, Treibeis *n* ~ **ice barricade** (or boom) Eisbaum *m* ~ **island** schwimmende Insel *f* ~**-kidney vibration absorber** Pendelschwingungsdämpfer *m* ~ **laundry** Waschboot *n* ~ **lever** Füllungshebel *m* (Diesel) ~ **light** Feuerschiff *n*, Leuchtfeuer *n* ~ **line** Treibleine *f*, Wasserlinie *f* ~ **mandrel** fliegender Dorn *m* ~ **mark** schwebende oder wandernde Marke *f* ~ **mine** Streu-, Treib-mine *f* ~ **mounting** schwimmende Lagerung *f* ~ **nut** Schwimmutter *f* ~ **pile** schwebender Pfahl *m* ~ **pile foundation** schwebende oder schwimmende Pfahlgründung *f*
**floating-point** bewegliches oder gleitendes Kom-

ma *n* (info proc.) ~ **calculation** Gleitkomma-rechnung *f* ~ **computation** Gleitkommarech-nung *f* ~ **notation** halblogarithmische Schreib-weise *f* ~ **representation** Gleitpunktdarstellung *f*, halblogarithmische Zahlendarstellung *f* ~ **routine** Gleitkommaprogramm *n*

**floating,** ~ **position** Freigangstellung *f* ~ **potential** Schwimmspannung *f* ~ **power** elastische weiche Lagerung *f* des Motors in Gummi, Schwebe-fähigkeit *f* ~ **radiometer** Schwimmradiometer *n* ~ **rate** Arbeits-, Schließ-geschwindigkeit *f* ~ **reamer** Pendelreibahle *f* ~ **ring** Laufring *m* (Kompressor) ~ **seat** schwimmender Sitz *m* ~ **shears** schwimmender Mastkran *m* ~ **sheave** wandernde Seilrolle *f* ~ **slewing crane** Schwimm-drehkran *m* ~ **sluice** Schwimmerschleuse *f* ~ **smoke pot** Rauchschwimmer *m* ~ **speed** Stell-geschwindigkeit *f* ~ **stage** Kalfaterfloß *n* ~ **(temporary) staging** schwimmendes Gerüst *n* ~ **suction** Schwimmerabsaugung *f* ~ **support** Brückenfloß *n* ~ **switch** Schwimmerschalter *m* ~ **test** Schwimmprobe *f* ~ **tool** Pendelwerkzeug *n* ~ **tool holders** Pendelfutter *pl* ~ **tube sheet cover** Schwimmrohrplattendeckel *m* ~ **vibrator** schwimmender Rüttelapparat *m* ~ **voltage** erd-freie Spannung *f* ~ **weight** Fallgewicht *n*

**floats** Rampenbeleuchtung *f* (film), Schwimm-gut *n*

**flocculate, to** ~ flocken, flockig machen (de) ~ ausflocken

**flocculated sol** ausgeflockter Sol *n*

**flocculation, (de)** ~ Ausflocken *n*, Ausflockung *f*, Entschuppen *n*, Flockenbildung *f*, Flockung *f*

**floccule** Flocke *f*

**flocculence** wollige Beschaffenheit *f*, Flockigkeit *f*

**flocculent** flockenartig, flockig ~ **structure** Flockenstruktur *f*

**flock, to** ~ strömen, zulaufen

**flock** Ausschußwolle *f*, Flocke *f*, Herde *f*; Reiterrudel *n* ~ **silk** Abseide *f* ~ **test** Flocken-versuch *m*

**flocking,** ~ **apparatus** Beflockungsgerät *n* ~ **speed** Ausflockungsgeschwindigkeit *f*

**flocky** flockig

**floe** Treibeis *n*

**flong** Fladen *m*; Matrizenkarton *m*, Stereotypie-papier *n* ~ **drying apparatus** Materntrocken-apparat *m*

**flood, to** ~ anstauen, überfluten, überschwem-men **to** ~ **a mine** eine Grube *f* unter Wasser setzen, ersäufen **to** ~ **tanks** (of submarines) fluten

**flood** Ausfließen von Pigmenten, Flut *f*, Flut-strom *m*, Gewässer *n*, Hochwasser *n*, Schwall *m* (surge), Steigen *n* des Wassers, Überflutung *f*, Überschwemmung *f* ~ **arch** Flutbrücke *f* ~ **bank** Hochwasserschutzdamm *m* ~ **basin** Ein-gußtümpel *m*, Sumpf *m* ~ **beginning** Vorflut *f* ~ **channel** Flutrinne *f* ~ **cock** Fluthahn *m* ~ **gate** Flut-, Ober-tor *n*, Schleuse *f*; Schleusen-schütze *n*, -schütze *f*

**floodlight, to** ~ anstrahlen

**floodlight** Flutleuchte *f*, Flutlicht *n*, Lichtstrah-ler *m*, Reflektor *m*, Scheinwerfer *m*, Streulicht *n*, Tiefstrahler *m* ~ **s for incandescent lamps** Glühlampenscheinwerfer *m* ~ **head** Reflektor-kopf *m* ~ **scanning** Rampenabtastung *f* ~ **unit** Glühlampenstrahler *m*

**flood,** ~ **lighting** Fassadenbeleuchtung *f* ~ **lubri-cation** Eintauchschmierung *f* ~ **oil system** Öl-spülung *f* ~ **-oiling** Ölberieselung *f* ~ **plain deposit** Hochwasserablagerung *f* ~ **pool** Hoch-wasserbecken *n* ~ **proof** überflutungssicher ~ **span** Flutöffnung *f* ~ **tide** Flut *f*, steigendes Wasser *n* ~ **-tide channel** Flutrinne *f* ~ **-tide current** Flutstrom *m* ~ **-tide gate** Außentor *n*

**flooded** überflutet ~ **area** Anstauung *f*, Über-schwemmungsgebiet *n*

**flooding** Anschwöden *n* (mit Kalk und Arsenik), Anstauung *f*, Überfluten *n*, Überflutung *f*, Überlaufen *n*, Überschwemmung *f*, Versump-fung *f*, (petroleum) Verwässerung *f*, Wasser-treibverfahren *n* ~ **device** Wasserfülleinrich-tung *f* ~ **flap** Flutklappe *f* ~ **point** Verflüssi-gungspunkt *m* ~ **valve** Flut-hahn *m*, -ventil *n*

**floodometer** Flutometer *n*

**floor** Arbeitsbühne *f*, Boden *m*, Bodenwrange *f*, Bohlenbelag *m*, Deck *n*, flaches Erztrumm, Flur *m*, Fußboden *m*, Geschoß *n*, Laufboden *m*, Schicht *f* (min.), Stockwerk *n*, Trumboden *m* **on the first** ~ im ersten Stockwerk

**floor,** ~ **with boards** den Fußboden *m* täfeln ~ **to (lock) chamber** Sohle *f* der Schleusen-kammer ~ **of cut** Schnittsohle *f* ~ **of dry dock** Docksohle *f* ~ **of fuselage** Boden *m* des Rump-fes, Rumpfboden *m* ~ **of lime grains** Kalk-krumpenestrich *m* ~ **of the lock entrance** (upstream or downstream) Vorkammerboden *m* ~ **of manhole** Schachtsohle *f* ~ **of a seam** Liegendes *n* eines Flözes ~ **of slabs** Platten-belag *m* ~ **of trench** Grabensohle *f*

**floor,** ~ **beam** Fahrbahnträger *m*, Fußboden-balken *m*, -lager *n*, -schwelle *f* ~ **board** Boden-brett *n*, Fußboden *m*, Trittleiste *f* ~ **boards** Fußbodenbrett *n* ~ **clearer** (of kiln) Abräumer *m* ~ **cloths** Scheuertücher *pl* ~ **column** Spindel-ständer *m* ~ **contact** Fußumschalter *m* ~ **control panel** Flurbedienungsstand *m* ~ **covering** Bodenbelag *m*, Fußbodenbelag *m* ~ **drain** Ab-fluß *m* ~ **dressing** Parkettabhobeln *n* (inlaid-) ~ **dressing** Abhobeln *n* des Parketts ~ **enamel** Fußbodenlackfarbe *f* ~ **flange** runde Boden-platte *f* ~ **frame** Lager-bock *m*, -stuhl *m*, Steh-bock *m* ~ **grating** Bodenrost *m* ~ **grinder** Plan-schleifmaschine *f* ~ **gun mounting** Bodenlagette *f* ~ **heads** Bilge *f*, Kimmung *f* ~ **joist** Decken-balken *m* ~ **lamps** Ständerleuchten *pl* ~ **leads** Bodenleitung *f* ~ **level** Bodenhöhe *f*, Flurebene *f*, Sohle *f* ~ **line** Flurebene *f* ~ **monitor** Fuß-bodenkontrollgerät *n* ~ **-mounted shears** Über-flurschere *f* ~ **oil** Boden-, Staub-öl *n* ~ **outlet** Bodensteckdose *f* ~ **painting** Bodenanstrich *m* ~ **pan** Bodenwanne *f* ~ **paving** Fußbodenbelag *m* (aus Ziegeln oder Platten) ~ **plan** Grundriß *m*, Raumaufteilung *f* ~ **plate** Abdeck-, Flur-, Fußboden-platte *f*, (of a ship) Bodenwrange *f* ~ **-plate transoms** Bodenwrangenplatten *pl* ~ **polisher** Bohnermaschine *f* ~ **-polishing paste** Bohnermasse *f* ~ **saturant** Fußbodenöl *n* ~ **shaft** Schachtsohle *f* ~ **slab form** Deckenscha-lung *f* ~ **space** Boden-, Grund-fläche *f* ~ **space required** Platz-, Raum-bedarf *m* ~ **stain** Boden-beize *f* ~ **stand** Bodengestell *n*, Lagerbock *m*, Lagerstuhl *m*, Stehbock *m*, Stehlagerbock *m* ~ **-standard** Stehlampe *f* ~ **stands** Bedienungs-säule *f* ~ **strop** Fußbodenleiste *f* ~ **surfacing**

Fußbodenbelag *m* ~ **tile** Flur *m*, Fußboden-platte *f* ~ **tiler** Plattenleger *m* ~-to-~ **time** Gesamtarbeitszeit *f* ~ **topping** Fußbodenbelag *m* ~ **tube** Bodenrohre *n*, Sohlenrohr *n* ~ **type** Ständerbauart *f* ~**-type boring machine** Bodenbohrwerk *n* ~ **varnish** Fußbodendecklack *m* ~ **wax** Bohnerwachs *n*

**flooring** Belag *m*, Fußboden *m*, Fußbodenbelag *m*, Fußbodenblech *n* ~ **of a bridge** Bedielung *f* ~ **board** Diele *f* ~ **clamps** Fußbodenpressen *pl* ~ **plank** Fußbodenbrett *n* ~ **plaster** Estrichgips *m* ~ **slab (or tile)** Fußbodenplatte *f*

**floret** Florett *n*

**Florian net** Floriannetz *n*

**florist's wire** Blumendraht *m*

**flos ferri** Eisenblüte *f*, Faseraragonit *m*

**floss** Florettseide *f* ~ **hole** Fuchsöffnung *f*

**flotable** schwimmfähig

**flotation** Flotation *f*, Schaumschwimm-, Schwemm-aufbereitung *f*, Schwimmverfahren *n* ~ **agent** Flotierungsmittel *n* ~ **air bag** Schwimmsack *m* ~ **blende** Flotationsblende *f* ~ **concentrate** Flotationskonzentrat *n* ~ **gear** Behelfsschwimmwerk *n* für Landflugzeuge, schwimmfähiges Gestell *n*, Schwimmerwerk *n*, Schwimmvorrichtung *f*, Schwimmwerk *n* ~ **machine** Flotationsmaschine *f*, Flotator *m* ~ **method** Aufschlemm-, Schwemm-verfahren *n*, Schwimmethode *f* ~ **oil** Flotationsöl *n* ~ **plane** Schwimmfläche *f* ~ **plant** Flotationsanlage *f*, Schwimmaufbereitungsanlage *f* ~ **process** Flotationsverfahren *n* ~ **rougher** Vorschaumer *m* ~ **tube** Schlauchkörper *m*

**flotilla** Flotille *f*

**flour** Mehl *n* ~ **dust** Mehlstaub *m* ~ **gold** Goldstaub *m* ~ **mill** Getreidemühle *f* ~ **train** Mehlzug *m* ~ **worm conveyer** Mehlförderschnecke *f*

**flourish** Randverzierung *f* ~ **of pen** Federzug *m*

**flourishing** schwunghaft

**floury** mehlig

**flow, to** ~ einmünden, entströmen, fließen, laufen, quellen, rinnen, schießen, strömen **to** ~ **against** anströmen (of gases) **to** ~ **along** entlangstreichen **to** ~ **around** umspülen, umströmen **to** ~ **around with** umfließen **to** ~ **away** wegfließen **to** ~ **away (or off)** abströmen **to** ~ **back** zurück-fließen, -strömen **to** ~ **in contact with** umfließen, umströmen **to** ~ **down** einfallen **to** ~ **down a river with the tide** mit der Flußströmung absacken **to** ~ **downward** abwärtsstreichen **to** ~ **by heads** unregelmäßig hervorquellen **to** ~ **in** einlaufen, einströmen, zuströmen **to** ~ **off** ablaufen, ab-, weg-fließen **to** ~ **out** abfließen, ausfließen, auslaufen, ausströmen **to** ~ **out laterally** ausquetschen **to** ~ **over** überströmen **to** ~ **together** zusammenfließen **to** ~ **through** durchfließen, durchstreichen, durchströmen **to** ~ **toward** zufließen **to** ~ **upward** aufwärtsströmen

**flow** Ablauf *m*, Auftrieb *m*, Ausbruch *m*, Ausfluß *m*, Ausströmung *f*, (of radio frequency) Durchflutung *f*, Durchströmung *f*, Fehler *m*, Fehlstelle *f*, Fließen *n*, Fluß *m*, (Gegensatz zu stocking) Fluß *m*, Strom *m*, Strömung *f*, Strömungsverlauf *m*, Verkehrszufluß *m* (of traffic), Zufluß *m*, Zufluß *m* an Gesprächsanmeldungen (of telephone calls) **ability to** ~ Fließfähigkeit *f* ~ **around edge** Kantenumströ-

mung *f* ~ **around a parabola** Parabelumströmung *f*

**flow,** ~ **of chips** Späneabfall *m* ~ **of control agent** Stellstrom *m* ~ **of current** Strom-fluß *m*, -lauf *m*, -durchgang *m* ~ **of electrons** Elektronenfluß *m* ~ **of the film** Verlauf *m* des Films (Lack) ~ **of gas** Gasstrom *m* ~ **of gypsum** Stuckgips *m* **having parallel** ~ wirbelfrei ~ **of liquid** Flüssigkeitsverlauf *m* ~ **of material** Materialwanderung *f*, Stoff-fluß *m*, -wanderung *f* ~ **of material fibres** Verlauf *m* der Materialfasern ~ **in multiply bodies** Fluß *m* in mehrfach verbundenen Körpern ~ **of production** Produktionsfluß *m* ~ **of a tensor field** Linienintegral *n* eines Tensorfeldes ~ **of the torque** Drehmomentverlauf *m* ~ **of traffic** Verkehrsablauf *m* ~ **with velocity less than the critical value** Fließen *n* ~ **of work** Arbeitsfluß *m*

**flow,** ~ **angle** Strömungswinkel *m* ~ **angle indicator** Winkelsonde *f* ~ **area** Durchflußquerschnitt *m* ~ **assembly** Durchströmgefäß *n* ~ **bean** Ausbruchsdrosselhahn *m*, Förderdüse *f* (Erdöl) ~ **becoming nonlaminar** Ablösung *f* ~ **cell** Durchflußküvette *f* ~ **chamber** Durchlaufgefäß *n* ~ **chart** Ablaufplan *m*, Arbeitsplan *m*, Flußdiagramm *n* (info proc), Flußplan *m* ~ **colorimeter** Durchflußkolorimeter *m* ~ **condition** Strömungsverhältnis *n* ~ **constriction** Klemmung *f* ~ **control** Mengenregelung *f* ~ **-control valve** Druckflußregler *m*, Regulierung *f* zur Regelung des Zuflusses ~ **controller** Durchflußregler *m* ~ **cooling** Durchflußkühlung *f* ~ **correction** Strahlberichtigung *f* ~ **counter** Durchflußzählrohr *n* ~ **cross-section** Durchflußquerschnitt *m* ~ **deformation** Fließverzug *m* ~ **diagramm** Ablaufplan *m*, Fließbild *n*, Flußplan *m* ~ **direction** Strömungseinrichtung *f* ~ **direction change** Flüssigkeitsstrom-Umlenkung *f* ~**-direction indicator** Durchflußrichtungsanzeiger *m* ~ **distortion** Fließverzug *m* ~ **distributor** Kraftstoffverteiler *m* (Triebwerk) ~ **duration** Strömungsdauer *f* ~ **electric wire lead** Ader *f* ~ **equalizer valve** Durchfluß-Ausgleichventil *n* ~ **equations** Strömungsgleichungen *pl* ~ **filament** Wasserfaden *m* ~ **formation** Strömungsgebilde *n* ~ **gate** Steigetrichter *m* ~ **glaze** Laufglasur *f* ~**-guide baffles** Leitwerk *n* ~**-guide blades** Schaufelblech *n* ~ **head** Ausbruchs-, Ausfluß-kopf *m*, Flüssigkeitsfänger *m* (für Gasbrunnen) ~ **heater** Durchflußerhitzer *m* ~ **indicator** Durchflußanzeiger *m*, Gefälleanzeiger *m* ~**-in pressure** Einströmdruck *m* ~ **interference** Strömungsbeeinflussung *f* (aerodyn.) ~ **limit** Fließgrenze *f* ~ **limiter** Durchflußbegrenzer *m* ~ **line** Streckfigur *f*, Stromfläche *f*, Stromlinie *f*, Strömungslinie *f* ~ **line (or nipple)** Abflußleitung *f* ~**-line pattern** strömungstechnische Erkenntnisse *pl* ~ **lines** Faserverlauf *m*, Fließ-figuren *pl*, -linien *pl* ~ **mark** Fließmarkierung *f* ~ **medium** Fördermedium *n*

**flowmeter** Durchfluß-, Durchlauf-, Flüssigkeits-, Mengen-messer *m*, Staurohr *n*, Strömungsmesser *m*, Strömungsuhr *f*

**flow,** ~ **mixer** (pipeline mixer) Durchfluß-, Flüssigkeits-mischer *m* ~ **molding** Preßspritzen *n* ~ **net** Strömungsbild *n* ~ **nipple** Ausbruchdrosselhahn *m*, Förderdüse *f* (Erdöl) ~ **number** Fließzahl *f* ~**-off** Abfluß *m* ~**-off pressure** Ab-

strömdruck *m* ~ **opening** dichter Durchfluß *m*
~ **passage** Durchflußöffnung *f* ~ **path** Strombahn *f* ~ **pattern** Stromlinien-, Strömungs-bild
*n* ~ **plane** Stromfläche *f* ~ **point** Fließpunkt *m*
~ **potential** Strömungspotential *n* ~ **pressure**
Strömungsdruck *m* ~ **process** Fließvorgang *m*
~ **property** Formänderungsfähigkeit *f* ~ **property rise in U-tube** Fließfähigkeit *f* im U-Rohr
~ **quantity** Luftdurchlaß *m* ~ **rate** Durchfluß-
leistung *f* ~ **regulation** Durchflußregulierung *f*
~ **regulator** Strömungsregler *m*, Zuflußregler
*m* ~ **research** Strömungsforschung *f* ~ **resistance** Flüssigkeits-, Strömungs-, Wasserdurch-
fluß-widerstand *m* ~ **restrictor** Durchflußbegrenzer *m* ~ **route** Strömungsweg *m* ~
**sample** Fließprobe *f* ~ **sensing line** Durchfluß-
steuerleitung *f* ~ **sensitivity** Dickenempfind-
lichkeit *f* ~ **separation** Strömungsablösung *f* ~
**sheet** schematischer Arbeitsplan *m*, Mühlen-
diagramm *n*, Stammbaum *n*, Stammtafel *f*,
Strombild *n* ~ **straightener (or honeycomb grid)**
Gleichrichtergitter *n* ~ **straightener vane** radiale
Leitschaufel *f* ~ **stream function** Stromfunktion
*f* ~ **stress** Fließspannung *m* ~ **stress ratio**
Fließspannungsverhältnis *n* ~ **structure** Fluidal-
textur *f* ~ **switch** durch Strömung (Luft, Wasser, etc.) bestätigter Schalter *m* ~ **test** Aus-
breite-, Fließ-probe *f* ~ **test with air** Luftströ-
mungsversuch *m* ~**-through cell** Durchflußküvette *f* ~**-through centrifuge** Durchfluß-
zentrifuge *f* ~**-through principle** Durchfluß-
betrieb *m* ~**-turning** Fließdrehen *n* ~**-type
electrode** Durchflußelektrode *f* ~ **velocity** Anströmungsgeschwindigkeit *f* ~**-volume regulator** Durchflußmengenregler *m*
**flowability** Fließfähigkeit *f*, Gußbarkeit *f*
**flowable** fließfähig
**flowed wax** Aufschmelzwachsplatte *f*
**flower** Blume *f*
**flowered** gemustert
**flowers,** ~ **of sulphur** Schwefelblüte *f* ~ **of zinc**
Zinkblumen *pl*
**flowing** Fließen *n*, (of lacquer) Verlauf *m*, Wirbelströmung *f*; durchlaufend, fließend ~ **in a
circular path** Kreisströmung *f* ~ **by heads** stoß-
weiser Ausbruch *m* ~ **out** Ausströmen *n* ~
**through** hindurchströmend, Durchfluß *m*
**flowing,** ~ **agent (or medium)** Strommittel *n* ~
**furnace** Blauofen *m* ~**-in of air** Lufteintritt *m*
~**-in vapor** einströmender Dampf *m* ~ **life**
Dauer *f* des natürlichen Ausflusses ~**-off** Abfluß *m* ~ **pressure** Abflußdruck *m* ~ **well** natürlicher Ausfluß *m*
**flucan** Salband *n*
**fluctuate, to** ~ fluktuieren, schwanken, streuen
**fluctuating** veränderlich ~ **current** unregelmäßiger Strom *m* ~ **density** Schwankungsdichtefeld
*n* ~ **power** Schwingleistung *f* ~ **speed** schwankende Drehzahl *f* ~ **voltage** Spannungsschwankung *f* ~ **zero** wandernder Nullpunkt *m*
**fluctuation** Abweichung *f*, Fluktuation *f*, Fluktuieren *n*, Schwanken *n*, Schwankung *f*, Unruhe *f*, Veränderlichkeit *f* ~ **of anode feed
current** Anodenunruhe *f* ~ **of the course** Kursschwankung *f* ~ **in heating current** Heizunruhe
*f* ~ **in level (or height)** Niveauschwankung *f* ~
**of list** Auftriebsschwankung *f* ~ **of noise** Stör-
geräuschatmen *n* ~ **of plate** Anodenunruhe *f*

~ **of pressure** Druckschwankung *f* ~ **of sensitivity** Empfindlichkeitsschwankung *f* (rdr) ~ **in
temperature** Temperaturschwankung *f* ~ **of
turnover** Umsatzschwankung *f*
**fluctuation,** ~ **dissipation theorem** Schwankungs-
Dissipationstheorem *n* ~ **noise** Schroteffekt *m*,
Stromverteilungsrauschen *n* ~ **theorem** Schwan-
kungstheorem *n* ~ **time** Schwankungszeit *f* ~
**voltage** unregelmäßige Spannung *f*
**fluctuations,** ~ **of density** Dichteschwankungen
*pl* ~ **in the main voltage** Netzspannungsschwankungen *pl*
**flue** Feuerkanal *m*, (metal tube in boiler) Feuerrohr *n*, Feuerzug *m*, (of boiler, metal tube in
boiler) Flammrohr *n*, Fuchs *m*, (heating) Heizkanal *m*, Kamin *m*, Leitung *f*, Ofenzug *m*,
Rauchabzugskanal *m*, Rauchfang *m*, Rauch-
gasfuchs *m*, Rohr *n*, Schornstein *m*, Zug *m*,
(chimney) Zugkanal *m*
**flue,** ~ **boiler** Flammrohrkessel *m* ~ **bridge**
Fuchsbrücke *f* ~ **brush** Feuerrohrbürste *f* ~
**chimney** Abzugskamin *m* ~ **draft** Schornstein-
zug *m* ~ **dust** Flug-asche *f*, -staub *m*, Gichtgas-
staub *m*, (blast-furnace) Gichtstaub *m* ~**-dust
briquetting plant** Gichtstaubbrikettierungs-
anlage *f* ~**-dust catcher** Flugaschefänger *m*
~**-dust condensation** Flugstaubkondensation *f*
~**-dust separation** Gichtstaubabscheidung *f*
~**-dust separator** Flugaschenabscheider *m* ~
**end** Fuchsende *n* ~ **expander** Rohrweiter *f* ~
**gas** Ab-, Abzugs-, Essen-, Fuchs-, Rauch-gas *n*
~ **gas extraction** Rauchgasentnahme *f* ~ **gas
flap** Rauchgaskappe *f* ~ **gas plumes** Rauchgas-
fahne *f* ~**-gas register** Rauchgasschieber *m*,
Rauchschieber *m* ~**-gas utilization** Abhitzeverwertung *f* ~ **gases** Feuergase *pl* ~ **losses** (brick
flue) Fuchsverluste *pl* ~ **opening** Fuchsöffnung
*f* ~ **pipe** (of organ) Labialpfeife *f*, Lippenpfeife
*f* ~**-sheet clinker** dünnflüssige Schlacke *f* ~ **stop**
(of flue pipe) Schneideton *m* ~ **system** Heiz-
röhrensystem *n*
**fluency** Fertigkeit *f*
**fluff** Flaumflocke *f*, flockige Schicht *f*, Staub-
flocke *f*
**fluffing (or grinding) machine** Schleifmaschine *f*
**fluid** Fluidum *n*, Flüssigkeit *f*; fließend, fluid,
flüssig, rieselfähig ~ **for filling recoil buffer**
Rohrbremsflüssigkeit *f*
**fluid,** ~ **bath** Flüssigkeitsbad *n* ~ **body** Fließkörper *m* ~ **clinker** flüssige Schlacke *f* ~
**compass** Flüssigkeitskompaß *m*, Schwimm-
kompaß *m* ~**-compression process** Blockpreß-
verfahren *n* ~ **cooling** Flüssigkeitskühler *m* ~
**coupling** Flüssigkeitskupplung *f* ~ **damping
(or dashpot) action** Flüssigkeitsdämpfung *f* ~
**drive** Flüssigkeitsgetriebe *n* (theoretical) ~
**dynamics** Strömungslehre *f* ~**-energy mill**
Strahlmühle *f* ~ **equalizer** Flüssigkeitsausgleicher *m* ~ **filament** Wasserfaden *m* ~ **flow**
Flüssigkeitsströmung *f* ~ **flow without compressibility** Potentialströmung *f* ~ **flow engine**
Strömungsmaschine *f* ~ **flywheel** Flüssigkeitskupplung *f*, -getriebe *n* ~ **friction** Flüssigkeitsreibung *f* ~ **friction damping** Flüssigkeitsdämpfung *f* ~ **lens** Flüssigkeitslinse *f* ~ **level
gauge** Füllungsanzeiger *m* ~ **level indicator**
Hydrauliköl-Vorratsanzeiger *m* ~ **like** fließ-
gerecht ~ **manometer** Druckmesser *m* ~ **metal**

Flußmetall *n* ~-metal charge flüssiger Einsatz *m* ~ mixing tank Mischbehälter *m* ~ motion Flüssigkeitsbewegung *f* ~ mud Dickspülung *f* ~ packing Dichtung *f* durch Sperrflüssigkeit ~ pipe Flüssigkeitsleitung *f* ~ pressure Flüssigkeits-druck *m*, -pressung *f* ~ reservoir Ausgleichsbehälter *m* ~ slag dünnflüssige Schlacke *f* ~ solution Schmelze *f* (metall) ~ sphere gyro Kreisel *m* mit Flüssigkeitsrotor ~ tar Dünnteer *m* ~ torque converter Flüssigkeitsdrehmomentwandler *m* ~ transmission Flüssigkeitsübertragung *f* ~-type equilibrator Flüssigkeitsausgleicher *m* (artil) ~ welding Abschmelz-, Schmelz-schweißung *f*

**fluidal structure** Fluidalstruktur *f*
**fluidify, to** ~ flüssig machen
**fluidimeter** Fluidometer *n*
**fluidity** (being the reciprocal of dynamic) Beweglichkeit *f*, Dünnflüssigkeit *f*, (Gegensatz zu stocking) Fluß *m*, Flüssigkeit *f*, (degree of) Flüssigkeitsgrad *m*, Zähigkeitswert *m* ~ coefficient reziproker Viskositätskoeffizient *m*
**fluidized,** ~ bed Wirbelbett *n* ~ bed drier Fließbettrockner *m* ~ dust Flugstaub *m* ~ reactor Flüssigspaltstoffreaktor *m*
**fluke** Ausräumer *m*, (of an anchor) Flügel *m*, Fuchs *m*, (navy) Klaue *f*, Schaufel *f*
**flukes of an anchor** Ankerhand *f*
**flume** Flutgerinne *n*, Gefluder *n*, Gerinne *n*, Schwemme *f*
**fluo-aluminic acid** Aluminiumfluorwasserstoffsäure *f*
**fluoboric** borflußsauer ~ acid Borflußsäure *f*
**fluor,** ~ earth Flußerde *f* ~-spar Fluorit *n*, Flußspat *m* ~ spar quarry Flußspatgrube *f*
**fluoresce, to** ~ fluoreszieren
**fluorescein** Fluoreszein *n* ~ bag Farb-, Meldebeutel *m*
**fluorescence** Fluoreszenz *f* ~ brightness Fluoreszenzbildhelligkeit *f* ~ extinction (or quenching) Fluoreszenzauslöschung *f* ~ yield Fluoreszenzausbeute *f*
**fluorescent,** ~ bulb (or lamp) Leuchtstoff-Röhre *f* ~ fitting Beleuchtungskörper *m* für Leuchtstoffröhre ~ glow Fluoreszenzleuchten *n* ~ illumination Leuchtstofflampenbeleuchtung *f* ~ lamp Glimm-, Leuchtstoff-lampe *f* ~ (screen or lamp) materials Phosphore *pl* ~ poster Leuchtplakat *n* ~ radiation Eigenstrahlung *f* ~ response Fluoreszenz-Ansprechvermögen *n* ~ screen Fluoreszenz-, Leucht-, Glüh-schirm *m* (electr.), Röntgenschirm *m* ~-screen scanning Leuchtschirmabtastung *f* ~-screen scanning of persons Leuchtschirmpersonenabtaster *m* ~ screen line Fluoreszenzstrich *m* ~ screen tracing Fluoreszenzstrich *m* ~ spot Leuchtfleck *m* ~ strip lamp Leuchtstoffröhre *f* ~ substance Leuchtschirmsubstanz *f*, Leuchtstoff *m*, Selbstleuchter *m* ~ substances for screens Schirmluminophore *pl* ~ target current Leuchtschirmstrom *m* ~ tube Leuchtröhre *f*
**fluoride** Fluorid *n*, Fluorsalz *n*
**fluorimeter** Fluoreszenzmesser *m*
**fluorimetry** Fluorimetrie *f*
**fluorine** Fluor *m*
**fluorite** Fluorkalzium *n*, Flußspat *m* ~ lens Flußspatlinse *f* ~ phosphors Flußspatleuchtstoffe *pl* ~ plate Flußspatplatte *f*

**fluoroborate** Boriumfluorid *n*
**fluorogen** fluoreszenzerregender Stoff *m*
**fluorogenic** fluoreszenzerzeugend, Leuchtenergie *f* erzeugend
**fluoro,** ~ meter Fluoreszenzmesser *m* ~ photometer Fluoreszenzmesser *m*
**fluoroscope** Durchleuchtungsapparat *m*, Fluoreszenzschirm *m*
**fluoroscopic picture** Durchleuchtungsentnahme *f*
**fluoroscopy** Durchleuchtung *f*
**fluosilicate** Fluorsilikat *n*, Kieselsalz *n* ~ elextrolyte kieselflußsaurer Elektrolyt *m*
**fluosilicic,** ~ acid Kieselflußsäure *f*, Siliziumfluorwasserstoffsäure *f* ~ electrolyte kieselflußsaurer Elektrolyt *m*
**fluotitanous acid** Titanfluorwasserstoffsäure *f*
**flush, to** ~ Bündel abschneiden; spülen to ~ again nachspülen to ~ away wegspülen to ~ off abschlacken to ~ out (drains) ausaalen, ausschwemmen, ausspritzen, ausspülen to ~ -rivet versenknieten
**flush** Abschlackung *f*, Baufluchtlinie *f*, Spülen *n*; abgeglichen, bündig abschneidend mit, bündig, glatt eingelassen, (with) fluchtgerecht, fluchtrecht, gerade, glatt, in gleicher Ebene liegend, stumpf ~ with in einer Ebene mit ~ with the ground ebenerdig
**flush,** ~ back cones Rückenbündigkeit *f* ~ bolt of door Kantenriegel *m* ~ boring tools Spülbohrgerät *n* ~ box kleiner Kabelbrunnen *m* ~ cleaning Spülung *f* ~ deck Glattdeck *n* ~ encased lock eingelassenes Schloß *n* ~-fitting head lamp Einbauscheinwerfer *m* ~ head Versenkkopf *m* ~ joint bündiger Anstoß *m*, bündige Fuge *f*, Spülkopf *m* ~-joint drill pipe Gestänge *n* mit glatten Verbindungen ~ light fitting Unterflurleuchte *f* ~ lights Unterflurfeuer *pl* ~-mounted eingelassen (in die Wand) ~-mounted instrument versenktes Einbauinstrument *n* ~ mounted switch eingebauter Schalter *m* mit Frontplatte ~ mounting eingebaut (Maschinenbau), versenkter Einbau *m* ~ mounting instrument Einbaugerät *n* ~ panel bündige Füllung *f* ~-pin gauge Tiefenlehre *f* ~ pointing volle Fuge *f* ~-pressure switch Anstoßdruckschalter *m* ~ production Ausbruchsproduktion *f* ~ ring Einbauring *m* ~ rivet versenktes Niet *m*, Niete *f* mit versenktem Kopf, Senkniet *n*, Versenkniet *n* ~ rivet head Versenkkopf *m* ~ riveted versenkt genietet ~ riveting Glatthaut-, Senk-, Versenk-nietung *f* ~ socket Unterputzsteckdose *f* ~ sprinkler versenkter Sprinkler *m* ~-type switch eingelassener Schalter *m* ~ weld Flachnaht *f* ~ wood screw Senkholzschraube *f*
**flushed** durchgespült ~ colors Direktfarben *pl*
**flusher** Dipper *m*
**flushing** Ausschwemmen *n*, Ausspülen *n*, Baufluchtlinie *f*, Wasserspülung *f* ~ and scouring sluices Ausspülschleusen *pl*
**flushing,** ~ arrangement Spülanlage *f* ~ auger Spülschappe *f* ~ back Rückspülung *f* ~ basin Spülbecken *n* ~ gate Spül-becken *n*, -ventil *n* ~ head for hollow rods Gestängerohrspülkopf *m* ~ hole Schlacken-form *f*, -loch *n*, -öffnung *f*, -stich *m*, -stichloch *n* ~ line Waschleitung *f* ~ oil Spülöl *n* ~-out Ausspülung *f* ~-out apparatus Ausspüler *m* ~ plate Abdichtungsblech *n* ~

**sluice** Spülschleuse *f* ~ **surface** Spülfläche *f* ~ **tank** Schwimmer-, Überlauf-behälter *m* ~ **wimble** Spülschappe *f*
**flute, to** ~ auskehlen, einkehlen, kandeln, kannelieren, kehlen, nuten, riefen, riffeln
**flute** Auskehlung *f*, Falz *m*, Flöte *f*, Gewindenute *f*, Kannelierung *f*, Kehle *f*, Nut *f*, Nute *f*, Pfeife *f*, Riefe *f*, Riffel *m*, Rinne *f*, Rippe *f*, Schratte *f*, Spannut *m* ~ **of drill** Bohrernute *f* ~ **with tuning slide** Flöte *f* mit Stimmzug
**flute,** ~ **cutter** Gewindenutenfräser *m* ~-**type** rillenförmig
**fluted** ausgekehlt, genutet, gereifelt, gerieft, geriffelt, gerillt, kanneliert, kehlig ~ **bar iron** Hohlkanteisen *n* ~ **delivery roller** Lieferzylinder *m* ~ **expander** geriffelter Ausbreitstab *m* ~ **filter** Faltenfilter *n* ~ **handle** gerillter Griff *m* ~ **head cap screw** Zylinderschraube *f* mit Innensternnut ~ **hub** Riffelnabe *f* ~ **roll** geriffelte Walze *f* ~ **roller** Riffelwalze *f* ~ **spike** eingekerbter Nagel *m* ~ **surface** geriffelte Oberfläche *f*
**fluting,** ~ **mortise** Ausfalzung *f* ~ **steel** Riffelstahl *m*
**flutter, to** ~ flattern, kochen, schwanken, schwänzeln, vibrieren **to** ~ **freely** frei bewegen (tape rec.)
**flutter** Flackereffekt *m*, Flattern *n* (aerodyn), Flatterschwingung *f* (acoust), (speed) Geschwindigkeitsschwankungen *pl*, Gleichlauf *m* (performance), Instabilität *f*, Rüttelerscheinung *f*, (in wings) Schwingung *f*, Übertragungsverlustschwankung *f*, (film pull) Unstetigkeit *f*, Vibrieren *n*, Zittern *n*
**flutter,** ~ **echo** Flatterwiderhall *m*, Mehrfach-, Vielfach-echo *n* ~ **effect** Flatter-, Flutter-effekt *m* ~ **equipment** Flattereinrichtung *f* (rdr) ~ **measuring instrument** Tonschwankungsmesser *m* (film) ~ **meter** Tonschwankungsmesser *m* ~ **testing** Gleichlaufprüfung *f* ~ **valve** Ausatem- (gas mask), Teller-ventil *n*
**fluttering** schnelle unregelmäßige Bewegung *f*, flatterig, Flattern *n*, Schwanken *n* ~ **effect** Flatterwirkung *f*
**fluvatic hydraulics** Flußwasserbau *m*
**fluviograph** selbstregistrierender Pegel *m*
**flux, to** ~ in Fluß bringen, flüssig machen, schmelzen, verschlacken
**flux,** Fluß *m*, Flußmittel *n*, Flux *m*, Kraftfluß *m*, Kraftströmung *f* (electr.), Magnet-, Schmelzfluß *m*, Schmelzflux *m*, Schmelzmittel *n*, Schmelzpulver *n*, Schweißpaste *f*, Strömung *f*, Strömungsvorgang *m*, Vorschlag *m*, Zufluß *m*, Zuschlag *m* ~ **of force** Kraftfluß *m* **(magnetic)** ~ **in iron** Eisenkraftfluß *m* ~ **for making slag** Schlackenbildner *m* ~ **of neutrons** Stromdichte *f* der Neutronen ~ **of a vector** Fluß *m* eines Vektors
**flux,** ~ **bin** Bunker *m* für Zuschläge ~-**core-type electrode** Seelendraht *m* ~-**current loop** dynamische Hystereseschleife *f* ~ **density** Feldstärke *f*, Flußdichte *f*, magnetische Induktion *f*, Kraftfluß-, Kraftlinien-dichte *f* ~ **density in airgap** Luftspaltinduktion *f* ~ **gate** Gaberelement *n* ~-**gate compass** Drosselkompaß *m* ~-**gate magnetometer** Luftspaltmagnetometer *n* ~ **level** Kraftflußhöhe *f* ~ **line** Flußlinie *f* ~ **linkage** Flußverkettung *f*, induktive Kopp-

lung *f* (rdo) ~ **linking a coil** Spulenfluß *m* ~ **material** Zuschlagmaterial *n* ~ **meter** Flußmesser *m* (magnetic), Kriechgalvanometer *n*, Strommesser *m* ~ **oil** Durchlauföl *n* ~ **path** Kraftflußweg *m* ~ **plate** Flußbügel *m* ~ **plot** Flußdiagramm *n* ~ **powder** Flußpulver *n* ~ **stone** Flußkalkstein *m* **the** ~ **threads with the turns** der Fluß durchsetzt die Windungen ~ **trapping** Stromfalle *f* ~ **turn** Flußverbindung *f* ~ **turns** Kupplung *f*, Windungsfluß *m* ~ **valve** Gaberelement *n*, Magnetfeldsonde *f*
**fluxed,** ~ **electrode** Schmelzmantelelektrode *f* ~ **pitch felt** Teerpappe *f*
**fluxing** Flußmittel *n*, Plastifizieren *n*, Weichmachen *n* ~ **action of slag** Schlackenangriff *m* ~ **agent** Flußmittel *n* ~ **material** Flußmittel *n* ~ **ore** Zuschlagerz *n* ~ **power** Verschlackungsfähigkeit *f*
**fly, to** ~ fliegen, schießen lassen **able to** ~ flugfähig **to** ~ **automatically** vollkurssteuern **to** ~ **blind** blind fliegen **to** ~ **the great circle** innerhalb des großen Kreises *m* fliegen **to** ~ **(with) hands off** freihändig oder mit losgelassenem Steuer fliegen **to** ~ **into a hill** gegen einen Hügel *m* fliegen **to** ~ **low** niedrig fliegen **to** ~ **nose high** das Flugzeug überziehen **to** ~ **off** abspringen, ausspringen **to** ~ **over** überfliegen **to** ~ **solo** allein fliegen **to** ~ **straight** geradeaus fliegen **to** ~ **on a straight course** Strich fliegen **to** ~ **with tail wind** mit dem Wind *m* fliegen **to** ~ **at the target** das Ziel anfliegen **to** ~ **through** (a cloud bank) durchstoßen **to** ~ **toward** anfliegen **to** ~ **upwind** gegen den Wind *m* fliegen **to** ~ **without use of controls** freihändig fliegen
**fly** Fliege *f* ~ **ash** (sugar refining) Flugasche *f* ~-**away factory (f.a.f.)** flugfertig ab Herstellerwerk n
**fly-back** Rückbewegung *f* (TV), Rückführung *f*, Rücklauf *m* (TV, CRT), Heimlauf *m*, Strahlrücklauf *m* (rdr) ~ **of electric high tension** (aus) Zeilenrücklauf gewonnene Hochspannung *f* ~ **of sweep potential** Kippspannungsrücklauf *m*
**fly-back,** ~ **elimination** Rücklaufverdunkelung *f* ~ **period** Rücklaufzeit *f* ~ **speed** Rückkippgeschwindigkeit *f* ~ **suppression** Rücklaufunterdrückung *f*, -verdunkelung *f* ~ **suppressor** Rücklaufsperre *f* ~ **voltage** Zeilen-Rücklaufspannung *f*
**fly-ball** Fliehgewicht *n*, Reglerkugel *f*, Schwungkugel *f* ~ **governor** Fliehkraft-, Schwungkugelregler *m*
**fly,** ~ **bar** Grundwerkschiene *f*, Holländer-messer *n*, -schiene *f* ~ **bars** Blätter *pl* ~ **cutter** Schlagformmesser *n* ~ **frame** Vorspinnmaschine *f* ~-**frame helper** Fleiergehilfe *m* ~-**frame spinner** Flügelspinner *m* ~ **frame tenter** Fleier-Arbeiter *m*, Fleier-Bediener *m* ~ **hammer** Aufwerfhammer *m* ~ **leaf** Allonge *f*, Deckblatt *n*, Schutzblatt *n* ~ **lever** Schwunghebel *m* ~ **nut** Flügelmutter *f* ~ **piston** Flügelkolben *m* ~ **press** Stoßwerk *n* ~ **pump** Flügelpumpe *f* ~ **sheet** Flugblatt *n* ~ **switch** Schnepperweiche *f* ~ **trunnion (or weight)** Fliehgewichtsbolzen *m* ~ **waste** Streufaden *m* ~ **weight** Flieh-gewicht *n*, -pendel *n* (jet)
**fly-wheel** (inertial system) Kreiselrad *n*, Schwungkranz *m*, Schwungrad *n*, Treibrad *n* ~ **between two bearings** beiderseitig gestütztes Schwungrad *n*

**fly-wheel, ~ action** Schwungmoment n **~ blower** Radialgebläse f **~ box** Endlagerbüchse f **~ brake** Getriebebremse f **~ circuit** Schwungradkreis m, -schaltung f **~ connection** Schwungradschaltung f **~ converter** Schwungradumformer m **~ cover** Schwungradschutz m **~ drive** Schwungradantrieb m **~ dynamo starter and battery ignition unit** Schwunglichtanlaß(batterie)zünder m **~ engine** Schwungradmaschine f **~ force** Fliehkraft f **~ gear ring** Schwungradzahnkranz m **~ generator magneto** Schwungradlichtmagnetzünder m **~ housing** Schwungradgehäuse n **~ hub** Schwungradnabe f **~ magneto** Schwungradmagnet m **~ marking** Schwungradmarke f **~ mass** Schwungmasse f **~ moment** Schwungmoment n **~ pit** Schwungradgrube f **~ pulles** Riemenscheibenschwungrad n **~ pump** Schwungradpumpe f **~ resistance** Schwungradwiderstand m **~ rim** Schwungradkranz m **~ ring gear** Schwungradzahnkranz m **~ side** Schwungradseite f **~ speed** Schwungradumdrehung f **~ starter** Schwungmoment-, Schwungrad-anlasser m, Schwungradstarter m **~ starter gear** Schwungverzahnung f **~ toothing (or gearing)** Schwungradverzahnung f **~-type magneto** Schwungradmagnetapparat m

**flyer, ~ sheet deliverer** Bogenausleger m **~ spindle** Flügelspindel m

**flying, ~** Fliegerei f, Flug m (aviat.), (of pencil) Rückgang m; fliegend **~ on the beam** Leitstrahlverfahren n **~ when controls go sluggish** unbeständiges (pilziges) Fliegen n

**flying, ~ ability** fliegerische Fähigkeit f **~ accident** Flugunfall m **~ allowance** Fliegerzulage f **~ boat** Bordflugzeug n, Flugboot n, Flugschiff n **~ boom** Tankerrohr n (Luftbetankung) **~ boots** Fliegerstiefel pl **~ boxcar** Frachtflugzeug n **~ bridge** fliegende Brücke f, Gierbrücke f **~ buttress** Bogen-, Gewölbe-pfeiler m **~ cap** Kopfkappe f **~ capacity** fliegerische Fähigkeit f **~ characteristics** Flug-eigenschaften pl, -verhalten n **~ clothing** Flugkleidung f **~ column** fliegende Kolonne f **~ competition** Flugwettbewerb m **~ condition** Flugzustand m **~ conditions** Flugbedingungen pl **~ control instruments** Flugüberwachungsgerät n **~ course** Flugziel n **~ cradle** fliegende Bühne f **~ day** Flugtag m **~ demonstration** Vorführung f im Fluge **~ drum** Bandrolle f **~ dust** Flugstaub m **~-endurance contest** Dauerflugwettbewerb m **~ experience** Flugerfahrung f **~ ferry** Gierfähre f **~ field** Flugplatz m **~ formation** Flug-form f, -formation f **~ fortress** fliegende Festung f **~ gloves** Fliegerhandschuhe pl **~ goggles** Fliegerbrille f **~ height** Aufnahmehöhe f **~ helmet** Fliegerhaube f **~ helmet with earphones** Hörkappe f **~ hour** Flugstunde f **~ instruction** Flugunterricht m **~ instructor** Fluglehrer m **~ instrument** Bordgerät n zur Flugüberwachung **~ jib** Außenklüver m **~ kit** Fliegerschutzanzug m **~ lane** Einflugschneise f **~ machine** Flieger m, Flugapparat m, Flugmaschine f, Flugzeug n **~ map** Fliegerkarte f **~ model** Flugmodell n **~-off turntable** Startdrehscheibe f **~ operations** Flugbetrieb m **~ party** fliegende Kolonne f **~ performance** (aircraft) Flugleistung f **~-performance testing apparatus** Flugleistungsmesser m **~ (or flight) personnel** Flugpersonal n **~ pin**

fliegender Dorn m **~ position** Flug-haltung f, -lage f, -stellung f **~ print** Abdruck m im Fluge **~ pulley** bewegliche Rolle f **~ qualities** Flugeigenschaften pl **~ range** Aktionsradius m, Flug-bereich m, -radius m, -reichweite f, -weite f **~ range at economic speed** Flugbereich m bei Sparfluggeschwindigkeit **~-range light** Flugbereichlicht n **~ range at maximum (or full) speed** Flugbereich m bei größter (voller) Geschwindigkeit **~ reef** unterbrochener oder unregelmäßiger Gang m **~ regulations** Flugordnung f **~ restriction** Startverbot n **~ risk** Flugrisiko n **~ safety** Flugsicherheit f **~ saucer** fliegende Untertasse f **~ scale model** fliegendes maßstäbliches Modell n **~ schedule** Flugplan m **~ school** Flieger-, Flug-schule f **~ senses** Flugsinne pl **~ service** Flugdienst m **~ shears** fliegende Schere f **~ ship** Flugschiff n **~ shuttle** Schnellschütze m **~ sollar** fliegende Bühne f **~ speed** Fluggeschwindigkeit f **~ spot** wandernder Lichtfleck m, Lichtpunkt m oder Lichtstrahl m **~ spot film scanners** Filmübertragungsanlage f **~-spot method of scanning** Lichtpunktabtastung f **~ spot scanner** Lichtpunktabtaster m **~ suit** Fliegerschutzanzug m, Kombination f **~ target** Flug-scheibe f, -ziel n **~ technique** Flugtechnik f **~ terrain** Fluggelände n **~ time** Flug-stunde f, -zeit f **~ training** fliegerische Ausbildung f **~ unit** fliegender Verband m **~ view** Flugansicht f **~ weather** Flieger-, Flug-wetter n **~ wing** fliegender Flügel m, Nurflügel m **~-wing missile** Nurflügelaggregat n **~ wing plane** Nurflügelflugzeug n **~ wire** Abfangkabel f, Fliegedraht m, Tragdraht m, Tragkabel n

**FM (frequency modulation or modulated)** Frequenzmodulation f, frequenzmoduliert **~ cyclotron** Synchrozyklotron n **~ detector** Frequenzmodulationsdetektor m **~ jamming** frequenzmodulierte Störung f

**F number** (of a lens) Lichtstärke f

**foam, to ~** gischen, schäumen **to ~ over** überschäumen **to ~ up** aufschäumen

**foam** Schaum m **~ air** Sumpf m **~ catcher** Schaumfänger m **~ collapse** ungenügendes Schäumen n **~ concrete** Zellbeton m **~ core** Schaumkern m **~ (fire) extinguisher** Schaumfeuerlöscher m **~ formation** Schaumbildung f **~ generator** Schaumgenerator m **~ lance** Schaummundstück n **~ mixing chamber** Schaummischkammer f **~ mixture** Schaumgemisch n **~ removal** Schaumentnahme f **~-rubber articles** Schaumgummiartikel pl **~ rubber gasket** Moosgummidichtung f **~ rubber sleeve** Schwammgummihülse f **~(-mixing) station** Schaumzentrale f **~ test** Schaumprüfung f

**foamed, ~ ceramics** Schaumkeramik f **~ latex** Schwammgummi m **~ structure** Schaumstoffbauteil n

**foamer** Schaumbildner m

**foaming** Aufschäumen n, Schaumbildung f, Schäumen n **~ over** Überschäumen n

**foamlike** schaum-ähnlich, -artig

**foamy** schaum-ähnlich, -artig, schaumig

**fob** Täschchen n

**F.O.B. factory** ab Werk

**focal, ~ adjustment** Okulareinstellung f **~ adjustment clamp** Klemmung f der Fokussierung **~ aperture** Blendenöffnung f (film) **~ axis**

Brennachse f ~ curves of a surface of second order Fokalkurven pl einer Fläche zweiter Ordnung ~ depth Herdtiefe f ~ distance Brennabstandspunkt m, Brennweite f, Fokal-distanz f, -entfernung f ~ field Schärfenfeld n ~ film register Filmrahmen m ~ length Brenn-, Treffweite f (of electrons) ~ length of the photograph Aufnahmebrennweite f ~ length of plotting lens Auswertebrennweite f ~ length value Nennwert m ~ line Bild-, Brenn-linie f ~ plane Brennebene f, -fläche f, -punktebene f, Fokalebene f, Mattscheibenebene f ~ plane cover Rahmenschutzdeckel m ~ plane frame Meßrahmen m ~-plane guard Kameraschutzdeckel m (Luftbild) ~-plane shutter Fokalschlitzverschluß m, Schlitzverschluß m ~ point Brennpunkt m, Fokalpunkt m, Metapol m ~ point of the reflector Spiegelbrennpunkt m (Antenne) ~ range Einstellweite f ~ region Herdgebiet n ~ spot Brennfleck m, Fokus m ~ surface Fokalfläche f ~ surface of a surface Brennfläche f einer Fläche

foci Brennpunkte pl ~ of a conic Brennpunkte pl eines Kegelschnittes ~ of a surface normal Brennpunkte pl einer Flächennormalen

focimeter (focometer) Brennweitenmesser m, Fokometer n, Fokusmesser m

focus, to ~ bündeln, einstellen, in den Brennpunkt einstellen (lamps), auf richtige Entfernung einstellen, scharf einstellen, auf Sehschärfe einstellen, entwerfen, fokussieren, justieren (die Optik), sammeln, im Brennpunkt vereinigen

focus (of X rays) Brennfleck m, Brennpunkt m, Feuerpunkt m (opt.), Fokus m, Lichtpunkt m, Schärfepunkt m, Strahlenbündelung f in ~ scharf eingestellt ~ on the image side bildseitiger Brennpunkt m ~ of a mine Minenherd m ~ on the object side dingseitiger Brennpunkt m out of ~ unscharf

focus, ~ anode Fokussieranode f (TV) ~ contrast Bildschärfe f ~ control Scharfeinstellung f ~ correction (or control) Nachfokussierung f ~-defocus mode Fokus-Defokus-Verfahren n in-and-out-of-~ effects Filmatmen n ~ end fokusnahe Seite f ~ lamp Punktglimmlampe f, punktförmige Quelle f ~ plane Fokusebene f ~ voltage Fokussierspannung f

focused, ~ for infinity auf unendlich eingestellt ~ beam gebündelter und gerichteter Strahl m ~ cathode ray gerichteter Kathodenstrahl m

focusing Ausrichten n, Bildeinstellung f, Bündelung f, Einblendung f, Einstellung f, Konzentrieren n, Linsenkonzentration f, Optierung f, Scharfeinstellung f, Strahlkonzentrierung f, (of rays, electrons) Wiedervereinigung f ~ of an electron beam Konzentration f eines Elektronenbündels ~ of the light spot Schärfe f des Lichtstrahls ~ of prismatic elements of an optic Einsetzen n und Verkeilen n der Glasringe

focusing, ~ adjustment Fokussierungseinstellung f ~ anode Bündelungs-, Fokussier-anode f ~ aperture lens (or apertured disk) konzentrierende Lochscheibe f ~ arc lamp Fixpunktbogenlampe f ~ board Einstelltafel f ~ coil Fokussier-, Konzentrations-, Konzentrier-, Striktions-spule f ~ control Fokussierknopf m ~ collar Sehschärfeneinstellring m ~ condition

Fokussierbedingung f ~ coupling Fokussierungskopplung f ~ cup Kathodenbecher m ~ cylinder Sammel-, Steuer-zylinder m ~ defect Fokussierfehler m ~ dial Sehschärfeneinteilung f ~ electrode Bündelungs-, Fokussierungselektrode f ~ eyepiece Einstellokular n ~ field Fokussierungsfeld n ~ hood unit Lichtschachtaufsatz m ~ knob Einstellknopf m (of camera) ~ lamp Fokuslampe f ~ lens Fokussierungslinse f, Sucherokular n ~ magnet Abbildungs-, Fokussier-magnet m ~ magnifier Einstellupe f ~ mark Einstellmarke f ~ mechanism Einstellvorrichtung f ~ microscope Einstellmikroskop n ~ mount (of lens) Einstellfassung f ~ mount ring Schneckengangring m ~ optics Abbildungsoptik f ~ plane Einstell-, Einstellungs-ebene f ~ point Brennpunkt m ~ ring Fokussierhülse f, abbildende Ringelektrode f ~ rings abbildende Ringelektroden pl ~ scale Einstellskala f, Sehschärfeneinteilung f ~ screen Visiermattscheibe f, Visierscheibe f ~ screw Okularschraube f ~ sleeve Einstellhülse f ~ telescope Einstellfernrohr n ~ voltage Fokussierungs-, Scharfstellspannung f

foehn Föhn m ~ air Föhnluft f

Foettinger, ~ fluid torque converter Föttinger-Getriebe n ~ transformator Foettinger-Getriebe n

fog, to ~ anlaufen, verschleiern, sich verschmieren

fog Daak m, Grauschleier m, Nebel m, Schleier m (TV, photo), Verschleierung f, Wasserdampf m

fog, ~ area Nebelgebiet n ~ bow Nebelbogen m, Nebelregenbogen m ~ chamber unit Nebelkammeransatz m ~ cloud(iness) Nebel m ~ density Schleierschwärzung f ~ drop Nebeltropfen m ~ formation (on window) Beschlagen n ~ grain Schleierkorn n ~ horn Nebelhorn n, Schalltrichter m ~ image interaction Nebel-Bild-Wechselwirkung f ~ lamp Breitstrahler m, Nebelscheinwerfer m ~ landing Nebellandung f ~ mist Dunstnebel m ~ particle Nebeltropfen m ~ precipitation Wolkenzerstreuung f ~ repeater Nebellichtsignal n ~ signal Nebel-signal n, -zeichen n ~ signalling equipment Nebelsignalgerät n ~ tracks Nebelspuren pl ~ value Schleierwert m ~ whistle Nebelpfeife f

fogged schleierig ~ condition Verschleierung f (film)

fogger Registrierlampe f (film)

foggy nebelig, nebelreich, neblig, schleierig, verschwommen ~ appearance verschwommenes Aussehen n (of image)

foil Blatt n, Blättchen n, Blattmetall n, Blechfolie f, Florett n, Folie f, Metallfolie f, Spiegelbelag m, Unterlage f

foil, ~ brush Blätterbürste f ~ calenders Folienkalander pl ~ condenser (or capacitor) Folienkondensator m ~ detector Aktivierungsfolie f ~-embossing machine Folienprägemaschine f ~-jamming Verdüpplung f ~-laminating machine Folienrollenkaschiermaschine f ~ metal Blattmetall n ~ mills Folienwalzwerk n ~-paper capacitor Folienpapierkondensator m ~ strip Bandfolie f, Folienband n ~ tab Anschlußstreifen m ~ thickness Folienstärke f

foiled arch Kleebogen m

**fold, to** ~ bördeln, einknicken, falten, falzen, klappen, runzeln, übereinanderlegen, umbiegen, zusammenklappen **to** ~ **back** zurückschlagen **to** ~ **in** einfalzen **to** ~ **out (down, or up)** ausklappen **to** ~ **together** zusammen-falten, -legen

**fold** (bending crease) Biegewulst *m*, Falte *f*, Faltung *f*, (of strata) Faltung *f* der Schichten, Falz *m*, Gebirgsfalte *f*, Geländefalte *f*, Hürde *f* **having** ~s faltig

**fold**, ~ **break** Falzbuch *m* ~ **here** Faltstelle *f* ~ **mountains** Faltengebirge *n* ~-**out tread** aufklappbare Stufe *f* ~-**over multipath effect** Echo *n* ~ **test** Knittezahl *f*

**foldability** Faltbarkeit *f*

**foldable** faltbar, zusammenklappbar

**foldableness** Falzfähigkeit *f*

**folded** beigeklappt, faltenreich, gefalzt, zusammengefaltet ~ **blade** Faltschaufel *f* ~ **cavity** Faltenhohlraum *m* ~ **dipole antenna** Faltdipol *m*, Schleifendipolantenne *f* ~ **down on hinges** vorschlagbar ~ **edge protection** Kantenschutz *m* ~ **exponential horn** (of loud-speaker) gefalteter Exponentialtrichter *m* ~ **filter** Faltenfilter *n* ~ **horn** (of a loud-speaker) Faltenhorn *n*, gefalteter Trichter *m* (acoust) ~-**horn loudspeaker** Faltenlautsprecher *m* ~ **loud-speaker horn** gefalteter Lautsprechertrichter *m* ~-**over multipath effect** Geisterbild *n* ~ **packing ring** Faltdichtring *m* (für Kerzen) ~ **plate structure** Faltwerk *n* ~ **profile** Abkantprofil *n* ~ **seam** Falznaht *f* ~ **sheet** gekantetes Blech *n* ~ **wing** Schwenkflügel *m*

**folder** Bördelmaschine *f*, Broschüre *f*, Falter *m*, Faltprospekt *m*, Falzbein *n*, Falzer *m*, Falzmaschine *f*, Glätter *m*, Klappkneifer *m*, Mappe *f* ~ **with few tapes** bänderarmer Falzapparat *m* ~ **stock** Aktendeckelpapier *n* ~ **stop** Klappblende *f*

**folding** Faltung *f*; beiklappbar, einklappbar, zusammen-klappbar, -legbar ~ **down** herunterklappbar ~ (**convolutionä of functions** Faltung *f* von Funktionen ~ **and grooving press** Langfalzzudrückpresse *f* ~ **by hand** Handfalzung *f* ~ **in quires** Langenfalzung *f* ~ **of spectral lines** Spektrallinienfaltung *f*

**folding** ~ **blade** Falzklappe *f* ~ **boat** Faltboot *n* ~ **border** Falz *m*, Fugenfalz *m* ~ **box** Faltschachtel *f* ~ **cage** Klappkäfig *m* ~ **chair** Klappstuhl *m* ~ **cot** Klappbett *n* ~ **crank** umlegbare Kurbel *f* ~ **die** Umschlagform *f* ~ **direction** Falzlinie *f* ~ **door** Flügeltür *f* ~ **dormer window** Dachklappe *f* ~ **endurance** Falz-beständigkeit *f* (of film), -festigkeit *f*, Knickwiderstandsfähigkeit *f* (of film) ~ **finder** umlegbarer Klappsucher *m* ~ **job** Abkantarbeit *f* ~ **ladder staircase** einschiebbare Bodentreppe *f* ~ **lay** Falzanlage *f* ~ **leg** Klappsporn *m* ~ **machine** Faltemaschine *f*, Falzeinbrennmaschine *f*, Landfalz-, Biege- und Zudrückmaschine *f* ~ **magnifier** aplanatische Einschlaglupe *f* ~ **magnifying glass** Einschlaglupe *f* ~ **mirror stereoscope** Klappspiegelstereoskop *n* ~ **mold** Klappform *f* ~ **mount** Einschlagfassung *f* ~ **peep sight** Umlegediopter *m* ~ **press** Abkantmaschine *f*, -presse *f* (bei Blechen) ~ **reflex camery** Spiegelklappkamera *f* ~ **resistance** Falzwiderstand *m* ~ **roof** zusammenklapp-

bares Verdeck *n* ~ **rule** Gliedermaßstab *m*, Kluft *f* ~ **ruler** Gelenkmaßstab *m* ~ **safety device** Klappensicherung *f* ~ **screen** Wandschirm *m* ~ **seat** Klappsitz *m* ~ **sheets** Faltschichten *pl* ~ **spring** Federkorn *n* ~ **strength** Falzfestigkeit *f* ~ **stress** Faltbeanspruchung *f* ~ **strip** Faltleiste *f* ~ **table** Falzbrett *n*, Klapptisch *m* ~-**tail spade** Federsporn *m* ~ **test** Faltversuch *m* ~ **tongs** Falzzange *f* ~ **tool** Deckschaufel *f*, Schall-, Scholl-eisen *n* ~ **trail spade** Klappsporn *m* ~ **wedge** Gegenkeil *m* ~ **wing** Klappflügel *m*, zusammenlegbarer Flügel *m* ~ **wing aeroplane** Flugzeug *n* mit Klappflügeln ~ **wings** beiklappbare oder zusammenlegbare Flügel *pl*

**foliage** Baumbewachsung *f*

**foliated** blätt(e)rig, geblättert, geschiefert, schalig, schieferig, schiefrig ~ **coal** Blatt-, Blätter-, Büschel-, Schichten-kohle *f* ~ **granular gypsum** schuppigkörniger Gips *m* ~ **gypsum** Schneegips *m* ~ **joint** überfalzte Fuge *f* ~ **silver** Silberschaum *m* ~ **tellurium** Blätter-erz *n*, -tellur *m* ~ **zeolite** Blätterzeolith *m*

**foliation** schichtenförmige Lagerung *f*

**folio** Blatt *n*, Kolumnenziffer *f* in ~ Folioformat *n* ~ **galley** Folioschiff *n* ~ **size** Folioformat *n* ~ **volume** Foliant *m*

**folium** Blattkurve *f*

**follow, to** ~ sich anhängen, folgen, nacheilen, nachfolgen, nachkommen **to** ~ **closely** kleben **to** ~ **run on the same track** spuren

**follow**, ~ **dies** Folgeschnitt *m* ~-**focus device** mitlaufender Sucher *m* (film) ~-**on copying** Folgekopieren *n* ~-**on current** nachfolgender (Netz-) Strom *m* ~-**on tool** Folgewerkzeug *n* ~ **shot** Fahraufnahme *f* ~-**the-pointer drive** Folgezeigerantrieb *m* ~-**the-pointer mechanism** Folgezeigereinrichtung *f* ~-**the-pointer sight mechanism** Zeigerzieleinrichtung *f* ~-**the-pointer-type fuse setter** Folgezeigerempfänger *m*

**follow-up** Rückführung *f* ~ **circuit** Nachlaufschaltung *f* ~ **control** Folgesteuerung *f* ~ **device** Nachführung *f* oder Rückführung *f* (des Regler) ~ **letter** Nachfaßbrief *m* ~ **lever** Rückführhebel *m* (jet) ~ **link** Rückführhebel *m* ~ **notice** Mahnung *f* ~ **piston** Folgekolben *m* (Regeltechnik) ~ **servo** Nachlaufregelkreis *m* ~ **shaft** Nachlaufwelle *f* ~ **study** Erfolgskontrolle *f* ~ **system** Folgezeigersystem *n* ~ **transmission linkage** Nachsteuergestänge *n* ~ **voltage** Nachlauf- oder Rückführspannung *f*

**follower** Anhänger *m*, Getriebe *n*, Gewindebacke *f* (des Leitapparates), Gewindeleitbacke *f*, Kettenspanner *m*, Kolbendeckel *m*, Leitbacke *f*, Manschette *f*, mitlaufende Brille (aviat.), Mitnehmer *m*, Nachfolger *m*, Nebenrad *n*, Nockenstößel *m*, Stern *m* (bei Drehbänken), Stopfbüchsdeckel *m*, Zubringer *m* (magazine) ~ **of eccentric disc** Exzenterrolle *f*

**follower**, ~ **control** Folgeregelung *f* (aut, contr.) ~ **control device** Nachlaufsteuerung *f* ~ **pin** Mitnehmerstift *m* ~ **potential** mitlaufende Spannung *f* ~ **pulley** getriebene Scheibe *f* ~ **spring** Zubringerfeder *f* ~-**spring stud** Federkopf *m* ~ **stage** Nachlaufstufe *f* ~ **wheel** getriebene Scheibe *f*

**following** Befolgung *f*, Kundschaft *f*, Nachfall *m*; mitlaufend, nachfolgend, nachstehend ~ **with**

the casing Nachführung *f* (der Rohre) not ~ a
rule regellos ~ of the telescope Nachführung *f*
des Fernrohres
following, ~ lever Steuerknagge *f* ~ microphone
bewegliches oder entfesseltes Mikrofon *n* ~
wind Schiebewind *m*
foment, to ~ blähen
foment action Blähung *f*
font of letter Schriftzettel *m*
fontactoscope Fontaktoskop *n*
food Beköstigung *f*, Essen *n*, Futter *n*, Kost *f*,
Nahrung *f*, Nahrungsmittel *n*, Speise *f* ~ chemist Nahrungsmittelchemiker *m* ~ defrosting
Auftauen *n* tiefgekühlter Lebensmittel ~ freezer
Kühltruhe *f* ~ package Nahrungsmittelver-
packung *f* ~ products industry Nahrungsmittel-
industrie *f* ~ research Lebensmitteluntersu-
chung *f* ~ stuff preparation Nährpräparat *n* ~
stuffs Lebensmittel *pl*, Nährmittel *pl* ~ supplies
in kind Naturalverpflegung *f* ~ supply Nahrungsmittel *n*, Verpflegung *f*
foolproof betriebsicher, narrensicher, verläßlich
~ apparatus gegen falsche Bedienung gesicher-
ter Apparat *m*
foolscap Kanzlei-format *n*, -papier *n*
foot, to ~ anstricken, (stockings) anwirken
(Strumpf), (bill) (be)zahlen, (column of figures)
summieren
foot Anlage *f* (arch.), Fuß *m*, Fußpunkt *m*,
Schenkel *m*, Tragfuß *m*, Unterteil *n* ~ of jib
Auslegerfuß *m* ~ of magnet Magnetfuß *m* ~ of
the optical axis upon the plate Achsendurchstoßpunkt *m* mit der Platte ~ of a perpendicular
Fußpunkt *m* des Lotes ~ of rail Fuß *m* der
Schiene ~ of the rear sight Visiersattel *m* ~ of
the rest Fuß des Supports ~ of slope Böschungsfuß *m*
foot, ~ bar Fußschaltleiste *f* ~ bellows Tretgebläse *n* ~ bend Fußkrümmer *m* ~ block Sohlen-
quader *m* ~ blower Blasebalg *m* mit Tretvorrichtung, Tretgebläse *n* ~ board Fuß-, Trittbrett *n* ~ board bracket Trittbrettträger *m* ~
brake Fußbremse *f* ~ bridge Brückensteg *m*,
Fußgängerbrücke *f*, Laufbrücke *f*, Laufsteg *m*,
Schnellsteg *m*, Steg *m* ~ bubble Fußlibelle *f*
~-candle Fußkerze *f* ~ capping Fußkappe *f* ~
caster of a stand Fußrolle *f* eines Stativs ~
change gearbox Getriebekasten *m* mit Fußschaltung ~ control Fußsteuerung *f* ~ controls
Fußhebelwerk *n* ~ drive Fußantrieb *m* ~ eye
Fußöse *f* ~ fall sound attenuation Trittschall-
dämpfung *f* ~ feed Beschleunigungspedal *n*
~-feed spring Gasfußhebelfeder *f* ~-fishing
joint Fußlaschenverbindung *f* ~ fishplate Fußlasche *f* ~ generator Tretsatz *m* (electr.) ~ hill
Vorgebirge *n* ~ hold Anhalt *m*, Auftritt *m*, fester
Standpunkt *m*, Stelle *f* an der man Fuß fassen
kann ~ interrupter Fußunterbrecher *m* ~ lathe
Fußdrehbank *f* ~ lever Fuß-hebel *m*, -tritt *m*,
Tritthebel *m* ~-lever press Fußpendelpresse *f* ~
lights (theater) Bühnenbeleuchtung *f*, Fuß-
lichter *pl*, Rampenbeleuchtung *f*, Rampenlicht
*n* ~ line Unterschlagszeile *f* ~ lugs Ankernocken
*pl* ~ measure (or measurement) Fußmaß *m*
~-mounted (or standard) motor Fußmotor *m*
~ muff Fußsack *m* ~ note Anmerkung *f*, Fuß-
note *f*
foot-operated fußgeschaltet ~ antidazzle switch

Fußabblendumschalter *m* ~ brake Fußtritt-
bremse *f* ~ dimming switch Fußabblendschalter
*m* ~ starting switch Fußanlaßschalter *m* ~
switch Fußumschalter *m*
foot, ~ operation Fuß-betätigung *f*, -steuerung *f*
~ operation lever Fußbedienungshebel *m* ~
path Fuß-steig *m*, -weg *m*, Laufsteg *m*, Steg *m*,
Steig *m* ~ pedal Fahrfußhebel *m*, Fußhebel *m*,
-pedal *n*, -tritt *m*, Tretkurbel *f* ~-pedal pad
Pedalauflage *f* ~ plate Fußscheibe *f*, Latten-
untersatz *m*, Plattform *f* ~-pound Fußpfund *n*
~-power attachment Ausrüstung *f* für Fußantrieb ~-power knife grinder Schleifapparat *m*
für Fußbetrieb ~ press Fußspindelpresse *f*
~ pressure Fußkraft *f* ~ release Trittklinke *f*
~ rest Fuß-brett *n*, -leiste *f*, -raste *f*, -rest *m*,
-stütze *f* ~ rule Zollstock *m* ~ screw of a stand
Fußschraube *f* eines Stativs ~ shifter Fußschaltung *f* ~ snare mine Fußschlingenmine *f* ~
starter Fußanlasser *m*, -schalter *m* (electr.) ~
step Aufstieg *m* (aviat.), Fuß-bank *f*, -tritt *m*;
Spur-platte *f*, -zapfen *m*, Stützzapfen *m* ~ step
bearing Stehlager *n* ~ step plate Spurlatte *f* ~
stock Reitstockspitze *f* ~ stock spindle Reit-
stockspindel *f* ~ stool Fuß-bank *f*, -schemel *m*
~ switch Tretschalter *m* ~ (-operated) switch
Fußschalter *m* ~ throttle Fußgashebel *m* ~
trail Fußschiene *f* ~ treadle Fußtritt *m* ~ valve
Bodenventil *n*, Fußrückschlagventil *n*, (of air
pump, feed pump) Fußventil *n*, Grundventil *n*
~-valve screen Sieb *m* am Fußventil ~ walk
Laufgang *n* ~ wall (of embankment) Fuß-
mauer *f* ~ way Bürgersteig *m* ~ way jointing
chamber Kabelschacht *m* für Gehbahn ~ wear
Fußbekleidung *f*, Schuhzeug *n* ~ white Fuß-
setzlinie *f* (print.)
footage Metergedinge *n* ~ counter Film-messer
*m*, -merterzähler *m*, -uhr *f*, -zähler *m*
footing Einzelfundament *n*, Fuß *m*, Gründung *f*
~ of buttress Fundament *n* des Strebepfeilers ~
of a wall Latsche *f* ~ of walls Grundschicht *f* ~
trench Fundamentgraben *m*
footings Bankett *n*
foots oil Ablauföl *n* (Entparaffinierung),
Schwitzkammerablauf *m*, Schwitzöl *n*
forage, to ~ furagieren
forage Furage *f*, Futter *n* ~ ladder Schoßkelle *f*
forbid, to ~ untersagen, verbieten
forbidden unerlaubt, verboten ~ band verbotenes
Energieband *n* ~ line verbotene Linie *f*
nuclear transistion verbotener Übergang *m* ~
transition unerlaubter oder verbotener Übergang *m*
forbidding Untersagung *f*
force, to ~ beschleunigen, erzwingen, forcieren,
nötigen, quälen (photo), treiben to ~ against
andrücken to ~ ahead aufkommen to ~ apart
spreizen to ~ away abdrängen to be in ~ es
gelten, gültig sein to come into ~ in Kraft treten
to ~ the crank by pressure die Kurbel aufpressen to ~ down herunterdrücken, niederdrücken
to ~ in einarbeiten, einpressen, einzwängen,
einzwingen to ~ into eindrücken, hineintreiben,
hineinzwingen to ~ into engagement in Eingriff
bringen to ~ off abtreiben to ~ on aufpressen,
aufziehen to ~ on (or up) auftreiben to ~
oneself sich aufdrängen to ~ one's way through
sich durcharbeiten, durchdringen to ~ one's

way into hineindringen **to ~ open** aufbrechen **to ~ out** ausdrücken, auspressen, herausdrücken **to ~ through** durchdrücken, durchpressen, durchtreiben, treiben **to ~ together under pressure** gegeneinander stauchen **to ~ up** steigern **to ~ upon** aufzwingen

**force** Auftrieb *m*, Beanspruchung *f*, Energie *f*, Gewalt *f*, Kraft *f*, Macht *f*, Mannschaft *f*, Stärke *f*, Truppenabteilung *f*, Umdrehungskraft *f* (propelling), Winddruck *m*, Wirkungsgrad *m*, Zwang *m* **~ acting on tail unit** Leitwerkskraft *f* **~ acts on a point** die Kraft greift in einem Punkte an **~ of attraction** Anziehungskraft *f* **~ of the blow** (typewriter) Anschlagstärke *f*, Schlagstärke *f* **~ on control surface** Ruderkraft *f* **~ at edge** Randkraft *f* **~ of expansion** Dehnungskraft *f* **~ of the fall** Fallwucht *f* **~ due to gravity (or due to inertia)** Massenkraft *f* **~ of gravity** Schwerkraft *f* **~ of impact** Auftreffwucht *f* **~ of induction** Induktionskraft *f* **~ of inertia** Trägheitskraft *f* **~ of nature** Naturkraft *f* **~ of propeller thrust** Propellerschubkraft *f* **~ of reaction** Reaktionskraft *f* **~ of spring** Federungskraft *f*

**force, ~-account work** Regiearbeit(en) *f* **~ acting on a section** Schnittkraft *f* **~ balance** (in measurement) Kraft-auswägung *f*, -vergleich *m* **~-balance accelerometer** Kraftausgleich-Beschleunigungsmesser *m* **~ calculation** Kräfteberechnung *f* **~ circulation** Druckumlauf *m* **~-circulation cooling system** Wasserpumpenkühlung *f* **~ component** Kraft-anteil *m*, -komponente *f* **~ constant** Kraftkonstante *f* **~ constants of linkages** Kernbindungskräfte *pl* **~ couple** Kräftepaar *n* **~ density** Kraftdichte *f* **~ diagram** Kraft-eck *n*, -plan *m* **~ distribution** Kraftverlauf *m* **~ equation** Kraftgleichung *f* **~ equilibrium** Kräfteausgleich *m* **~ factor** Kraftfaktor *m*, -linie *f* **~-feed lubrication** Preßschmierung *f* **~-feed oiler** Ölspritzkanne *f* **~ feed oiler** Schmierpresse *f* **~ field** Kraftfeld *n* **~ fit** Klemmenspannung *f*, Klemmgesperre *n*, Klemmsitz *m*, Preßsitz *m* **~ function** Kräftefunktion *f* **~-limiting device** Lastbegrenzer *m* **~ link** Kraftübertragungsgestänge *n* **~-locking** kraftschlüssig **~ majeur** höhere Gewalt *f* **~ model** Tensorkraftmodell *n* **~ normal to chord** Pfeilkraft *f* **~ parallel along (or to) chord** Sehnenkraft *f* **~ polygon** Krafteck *n*, Kraftvieleck *n*, Zusammensetzung *f* der Kräfte **~ rebalancing accelerometer** Kraftausgleich-Beschleunigungsmesser *m* **~ system** Kraftsystem *n* **~ vector** Kraftvektor *m*

**forced** aufgedrückt, erzwungen, forciert, notgedrungen, zwangsläufig (en) **~** gezwungen **~ to land** zur Landung *f* gezwungen

**forced, ~-air cooling** Gebläsekühlung *f* **~ alighting** Notwässerung *f* **~ alternating current** erzwungener Wechselstrom *m* **~ circulating mixer** Zwangsmischer *m* **~-circulation boiler** Zwangsumlaufkessel *m* **~ circulation cooling** Zwangsumlaufkühlung *f* **~-circulation evaporator** Zwangsumlaufverdampfer *m* **~ circulation steam generator** Zwangsumlauf-Dampferzeuger *m* **~ coding** Bestzeit-, Schnell-programmierung *f* **~ convection** Druckkonvektion *f* **~ convection heating** Zwangsumlauferhitzung *f* **~ draft** Druckluftstrom *m*, Unterwind *m*, künstlicher

Zug *m*, verstärkter Zug *m* **~-draft cooling** Druckluftkühlung *f* **~ draft duct** Druckluftkanal *m* **~ draft fan** Unterwindventilator *m* **~-draft furnace** Feuerung *f* mit Druckluft, Unterwindfeuerung *f* **~-draft supply** Unterwindzufuhr *f* **~ ejection** Hinausschleudern *n* **~ entry** gewaltsames Eindringen *n* **~-fan unit heaters** Luftheizkörper *m* (electr.) **~ feed** zwangsläufige Zuführung *f* **~-feed grates** Vorschubroste *pl* **~-feed lubrication** Druck-ölung *f*, -umlaufschmierung *f*, Hubaktschmierung *f*, Zwangsschmierung *f* **~-feed lubricator** Ölpresse *f* **~ fit** Kraftsitz *m* **~ flapping** Schlagantrieb *m* **~-flow boiler** Zwangslaufkessel *m* **~ hot water heating** Heizwasserheizung *f* **~ induction** zwangsläufige Gaszuführung *f* **~-induction engine** Gebläsemotor *m* **~ labor** Zwangsarbeit *f* **~ landing** Notlandung *f* **~ loan** Zwangsanleihe *f* **~ (-feed) lubrication** Druckschmierung *f* **~-lubrication bearing** Drucköllager *n* **~ march** Dauer-, Eil-, Gewalt-marsch *m* **~-on gear ring** aufgesetzter Zahnkranz *m* **~ oscillation** aufgedrückte oder erzwungene Schwingung *f* (acoust.), fremderregte Schwingung *f* **~ oscillations** aufgedrückte oder erzwungene Schwingungen *pl* **~ rate of exchange** Zwangskurs *m* **~ release** Zwangsauslösung *f* **~ rotation** Zwangsumlauf *m* **~ rupture** Gewaltbruch *m* **~ sale** Zwangsverkauf *m* **~-through-flow boiler** Zwangsdurchlaufkessel *m* **~ transition** erzwungener Übergang *m* **~ upcurrent** Hangwind *m* **~ ventilation** Bewetterung *f*, Druckzugventilation *f* **~ vibration** aufgedrückte oder erzwungene Schwingung *f* **~ water cooling** Pumpenkühlung *f* **~ work** forcierter Betrieb *m*

**forceful** gewaltsam

**forceps** Kluppzange *f*, Pinzette *f*, Zange *f* **~ with teeth** gezahnte Pinzette *f* **~ with three teeth** dreizähnige Pinzette *f*

**forces** Kräfte *pl*, Truppen *pl* **~ acting on a wall** Beanspruchung *f* (einer Mauer) **~ of changing magnitude and direction** Kräfte *pl* von veränderlicher Größe und Richtung **~ due to mass** Massenkraft *f* **~ of repulsion** Abstoßungskräfte *pl*

**forcible** heftig, kräftig

**forcibly actuated** zwangsläufig betätigt

**forcing, ~ of projectile** Geschoßführung *f* **~ cone** Übergangskegel *m* **~ disc** Abdruckscheibe *f* **~ lever** Druckbaum *m* **~ machine** Spritzmaschine *f* **~ pump** Druckpumpe *f* **~ resistance** Beschleunigungswiderstand *m*

**ford, to ~** durchwaten, überschreiten

**ford** Flußübergang *m*, Furt *f*, Schwelle *f*, Stromübergang *m*

**fording, ~ ability** Watvermögen *n* **~ power of a motor vehicle without wetting the motor** Watfähigkeit *f*

**fore, to ~ shorten** perspektivisch zeichnen

**fore** vorn, Vorderfront *f* **~-aft line** Längsachse *f* **~-aft line axis of airplane** Flugzeuglängsachse *f*

**fore-and-aft** längsschiffs **~ control** Höhensteuerung *f*, Längssteuerung *f* (helicopter), Nicksteuerung *f* **~ corrugation** Längsriffelung *f* **~ force** Längsschiffskraft *f* **~ iron mass** Längsschiffseisenmasse *f* **~ level indicator** Längsneigungsmesser *m* **~ oscillation** Längsschwingung *f* **~ pole** Längsschiffspol *m* **~ ship's**

magnet Längsschiffsmagnet *m* ~ **stability**
Längsstabilität *f* ~ **tilt** Nadirdistanz *f* in Flugrichtung
**fore,** ~ **arm** Vorarm *m* ~ **bay** Abfallmauer *f*,
oberer Vorhafen *m*, Vorhof *m* ~ **boiler** Vor-
kocher *m* ~**-bolster** (on front-axle drive of car)
Vorderachsschale *f* ~ **bowling** Fockbulin *f* ~
**brace** Fockbraß *f* ~ **bunt whip** Fockbuktalje *f*
~ **buntline** Fockbukgording *f* ~ **carriage** Vor-
dergestell *n*
**forecast, to** ~ vorhersagen
**forecast** Prognose *f*, Voraussage *f*, Voraussa-
gung *f*, Vorhersage *f*
**fore,** ~ **castle** Back *f*, Vordeck *n*, Vorschiff *n* ~
**cellar** Vorkeller *m* ~ **channels** Fockrüst *f*, Vor-
rüst *f* ~ **chill room** Vorkühlraum *m* ~ **clew**
**garnet** Fockgeitau *n* ~ **closure** Beschlagnahme
*f*, Zwangsversteigerung *f* ~ **cooler** Vorkühler *m*
~ **deck** Vordeck *n* ~ **edge of system** Vorder-
kante *f* des Vorstevens ~ **finger rest** Zeige-
fingerauflage *f* ~ **foot** Unterlauf *m* ~ **gaff** Vor-
gaffel *f*
**foreground** Vordergrund *m*, Vorgelände *n* ~
**collimating mark** Rahmenvordergrundmarke *f*
~ **collimating point** Bildvordergrundmarke *f*
**fore,** ~ **hammer** Vorschlaghammer *m* ~ **hand**
Vorhand *f* ~ **hand welding** nach links schweißen
**forehead** Abbaustoß *m*, Ort *m*, Stirn *f*, Strecken-
ort *m* (min.) ~ **pad** Stirnwulst *m* ~ **rest** Stirnstütze *f*
**fore,** ~ **hearth** Spurofen *m*, Vorderherd *m*, Vorherd *m*, Vorkammer *f* (oven), Vortiegel *m*, Vorwärmeofen *m* ~ **land** Huk *f* ~ **lock** Achsnagel
*m* ~ **lock bolt** Schießbolzen *m*
**foreman** Aufseher *m*, Bauführer *m*, Betriebsmeister *m*, Faktor *m*, Krückelführer *m*, Meister
*m*, Obersteiger *m*, Polier *m*, Vorarbeiter *m*,
Vormann *m*, Vorsitzender *m*, Werkführer *m*,
Werkmeister *m* ~ **in charge of blower** Blasemeister *m* ~ **of gang** Rottenführer *m* ~ **of plate**
**layers** Kolonnenführer *m* ~ **driller** Bohrmeister
*m*
**fore,** ~ **marker** Voreinflugzeichen *n* ~ **mast**
Fock *f*, Fockmast *m* ~ **noon altitude** Vormit-
tagshöhe *f* ~ **noon fix** Vormittagsbesteck *n* ~
**noon watch** Vormittagswache *f* ~ **part** Front *f*
~ **plate** Brustzacken *m* ~ **pressure tolerance**
Vorvakuumbeständigkeit *f* ~ **prism** Vorprisma
*n* ~ **pump** Vorpumpe *f*, Vorvakuumpumpe *f*
~**-pumping** Vorpumpen *n* ~ **runner** Vorläufer
*m* ~ **sail** Fock *f* ~ **sail-sheet** Fockschot *f*
~**-scattering** Vorwärtsstreuung *f* ~ **screen** Vor-
sieb *n* ~ **shaft** Vorschacht ~ **side** Vorderseite
*f* ~ **side front** Ostseite *f* ~ **stalling** Unterbindung
*f* ~ **strut** Vorderstrebe *f* ~ **studding sail** Focklee-
segel *n* ~**-studding-sail halyard** Fockleesegelfall
*m* ~ **tack** Fockhals *m* ~ **top** Fock-, Vor-mars *m*
~**-vacuum connection** Vorvakuumanschluß *m*
~**-vacuum sides** Vorpumpenseite *f* ~**-vacuum**
**valve** Vorvakuumventil *n* ~ **warming** Vorwärmung *f*
**foreign** ausländisch, auswärtig, fremd ~ **admix-**
**ture** Begleitkörper *m* ~ **bill** Devise *f* ~ **body**
Fremd-körper *m*, -stoff *m* ~ **business** Auslands-
geschäft *n* ~ **currency** Valuta *f* ~ **current**
Außen-, Fremd-strom *m* (stray) ~ **delivery**
Auslandslieferung *f* ~ **exchange** Devisen *pl*
~**-exchange broker** Devisenhändler *m* ~

**exchange certificate** Devisenbescheinigung *f* ~
**exchange operations fund** Devisenbetriebs-
fonds *pl* ~ **excitation** Fremderregung *f* ~ **gas**
Fremdgas *n* ~**-gas additions** Fremdgaszusatz *m*
~ **language** Fremdsprache *f* ~ **language type-**
**setting and printing** Fremdsprachensatz und
-druck *m* ~ **literature** Auslandsschrifttum *n* ~
**magnetic field** magnetische Fremdfelder *pl* ~
**manufacture** Fremdfertigung *f* **of** ~ **manufac-**
**ture** fremdländisch ~ **matter** fremder Bestandteil *m*, Fremdbestandteil *m*, Fremdkörper *m*,
Fremdstoff *m* ~ **message** Auslandstelegramm *n*
~ **metal** Begleitmetall *n* ~ **news** Auslandsnachricht *f* ~ **service** Auslandsdienst *m* ~ **ship** Aus-
landsschiff *n* ~ **shipping** Auslandsversand *m* ~
**substance** (admicture) Begleitkörper *m*, Begleitstoff *m*, fremder Bestandteil *m*, Fremdbestandteil *m*, Fremdstoff *m* ~ **trade** Außen-
wirtschaft *f*, Überseehandel *m*
**forest** Forst *m*, Wald *m* ~ **clearing** Wald-blöße *f*,
-lichtung *f* ~ **compartment (or division)** Wald-
abteilung *f* ~**-patrol plane** Forstüberwachungs-
flugzeug *n* ~ **ranger** Forstaufseher *m*, Förster *m*,
Waldhüter *m* ~**-ranger's house** Forsthaus *n* ~
**road** Waldweg *m* ~ **stand** Waldbestand *m* ~
**surveying** Forstvermessung *f* ~ **wool** Waldwolle
*f*
**forester** Förster *m* ~**'s district (or house)** Förste-
rei *f*
**forestry** Waldwirtschaft *f* ~ **association** Wald-
genossenschaft *f*
**forfeit, to** ~ aufgeben, einbüßen
**forfeit** Abstandsgeld *n*
**forfeiture** Aberkennung *f*
**forge, to** ~ ausschmieden, fälschen, schmelzen,
schmieden, stauchen, verschmieden **to** ~
**ahead** sich vorarbeiten, vordringen **to** ~ **badly**
zerschmieden **to** ~ **on** **to** anschmieden **to** ~ **out**
ausschmieden **to** ~ **in the rough** vorschmieden
**to** ~ **in the solid** massiv schmieden, vollschmie-
den
**forge** Herd *m*, Schmiede *f*, Schmiede-esse *f*,
-werkstatt *f* ~ **coal** Eß-, Essen-, Schmiede-kohle
*f* ~ **fire** Schmiedefeuer *n* ~ **hammer** Aufwerf-,
Schmiede-hammer *m* ~ **lathe** Schmiededreh-
bank *f* ~ **pig** Puddelroheisen *n* ~ **pig iron**
Frischerei-, Uddel-roheisen *n* ~ **scrap** Schmie-
deabfall *m* ~ **shop** Hammerwerk *n* ~ **welding**
Feuer-, Hammer-schweißung *f*
**forgeability** Hämmerbarkeit *f*, Schmiedbarkeit
*f*, Verschmiedbarkeit *f*, Warmbildsamkeit *f*
**forgeable** schmiedbar ~ **alloy** Knetlegierung *f* ~
**casting** schmiedbarer Guß *m*
**forged** geschmiedet ~ **iron** Hammereisen *n*
~**-metal propeller** geschmiedete Metallschraube
*f* ~**-on shaft collar** angeschmiedete Wellen-
flantsch *m* ~ **steel** Schmiedestahl *m* ~**-steel**
**chisel** Meißel *m* ~**-steel flanges** geschmiedete
Flantsche *pl* ~ **through** durchgeschmiedet
**forgeman** Grob-, Hammer-schmied *m*
**forger** Fälscher *m*
**forgery** Fälschung *f* ~ **of documents** Urkundenfälschung *f*
**forging** Freiformschmiede *f*, Schmieden *n*,
Schmiede-stück *n*, -teil *m*, Schmiedling *m*,
Schmiedung *f*, spanlose Formgebung *f*, Warmverformung *f* ~ **against back pressure** Gegen-
druckschmieden *n* ~ **bursts** Kernzerschmiedung

*f* ~ **coal** fette Kohle *f* mit kurzer Flamme ~ **crack** Schmiedriß *m* ~ **die** Gesenk *n* ~ **furnace** Schmiedeofen *m* ~**-grade steel** schmiedbare Eisensorte *f* ~ **hammer** Schmiedehammer *m* ~ **machine** Schmiedemaschine *f* ~ **metal** Schmiede-metall *n* ~ **operation** Schmiede-arbeit *f*, -vor-gang *m* ~ **practice** Schmiedetechnik *f* ~ **press** Gesenk-, Schmiede-, Warm-presse *f* ~**-press pump** Schmiedepreßpumpe *f* ~ **property (or** ~ **quality)** Schmiedbarkeit *f* ~ **scale** Schmiede-sinter *m* ~ **steel** schmiedbares Eisen *n*, Schmie-deeisen *n* ~ **strain** Schmiedespannung *f* ~ **temperature** Schmiede-hitze *f*, -temperatur *f* ~ **test** Schmiedeprobe *f* ~ **tool** Schmiedewerkzeug *n*

**fork, to** ~ gabeln, sich gabeln, gabelschaufeln

**fork** Abfanggabel *f*, Bolzenzieher *m*, Dreistachel *m* (agr.), Gabel *f*, Gabelschlüssel *m*, Gabelung *f*, Krückchen *n* (Weberei), Maulschlüssel *m*, Verzweigung *f* ~ **of mixing broken stones** Stein-schlaggabel *f* ~ **of a river** Flußgabelung *f* ~ **for running boards** Gabel *f* für Laufbretter ~ **of saddle** Vorderzwiesel *m* ~ **for separating malt** Greifhaufenschüttelgabel *f*

**fork,** ~ **arm** Gabelarm *m* ~ **assembly** Radgabel *f* ~ **bolt** Gabelbolzen *m* ~ **connecting rod** Gabel-verbindungsstange *f* ~ **connection** Sechsphasen-Gabelschaltung *f* ~ **girder (or blade)** Gabel-scheide *f* ~ **guide** Gabelführung *f* ~ **head (of connecting rod)** Gabelkopf *m* ~ **joining** Gabel-gelenk *n* ~ **lift** Gabelstapler *m* ~ **lift truck** Gabel-, Hub-stapler *m* ~ **mechanism** Gabel-werk *n* ~ **member** Gabelfalle *f* ~ **mounting** Gabelmontierung *f* ~ **oscillator** Stimmgabel-oszillator *m* ~ **rod** Gabestange *f*, Hauptpleuel *n* ~**-shaped agitator** Gabelrührer *m* ~ **shifter** Hebelschalter *m* ~ **shovel** Gabelschaufel *f* ~ **spanner** Gabelschlüssel *m* ~ **steel** Gabelstahl *m* ~ **support** Gabelträger *m* ~ **tines** Gabelzinken *pl* ~ **truck** Gabelstapler *m* ~**-type joint** Gelenk-gabel *f* ~**-type tedder** Gabelheuwender *m* ~ **wire guide** Gabeldrahtführer *m* ~ **wrench** Gabelschlüssel *m*

**forked** doppelgängig, gabelartig, gabelförmig, gegabelt, (lightning) gezackt ~ **axle** Gabel-achse *f* ~ **bed** Gabelrahmen *m* ~ **cable lug** Krallenkabelschuh *m* ~ **circuit** gebelte Lei-tung *f* ~ **clamp** Gabelkammer *f* ~ **connecting rod** Gabelpleuel *n*, Gabelpleuelstange *f*, Gabel-stange *f* ~ **end** Gabelstück *n* ~ **ends (or horns)** Zahnstangengabel *f* ~ **end-turnbuckle** Gabel-spannschloß *n* ~ **fitting** Gabelschuh *m* ~ **head-light bracket** Gabelhalter *m* für Scheinwerfer ~ **hoe** Gabelhaue *f* ~ **hoes** Doppelhaue *f* ~ **lever** Gabelhebel *m*, Klauenkurbel *f*, Zangenhebel *m* ~ **lightning** Gabel-, Zickzack-blitz *m* ~ **link** Gabelgelenkstück *n* ~**-multiples (or multiplex) telegraph** Mehrfachtelegraf in Gabelschaltung ~ **peep sight** Gabeldiopter *m* ~ **pipe** Gabelrohr *n* ~ **projection** gegabelte Nase *f* ~ **prop** Gabel-stütze *f* ~ **repeater** Gabelübertragung *f* ~ **(connecting) rod** Gabelschubstange *f* ~ **sight** Gabelvisier *n* ~ **stand** Gabelbock *m* ~ **steering-arm shaft** Gabelhebelwelle *f* ~ **strut** Gabelstiel *m* ~ **tie** Gabelanker *m* ~ **tube** Gabelrohr *n*

**forking** Gabelung *f*

**form, to** ~ abformen (aviat.), aufziehen, aus-bilden, bilden, einschließen (an angle), ent-stehen, (an image) entwerfen, fassonieren, formen, formieren, gestalten, profilieren, ver-formen to ~ **again** zurückbilden to ~ **anew** neubilden to ~ **the average** ausmitteln to ~ **into a ball** kugeln to ~ **balls** ballen to ~ **bays** sich ausbuchten to ~ **beads** perlen to ~ **branches** sich verzweigen to ~ **bubbles** perlen to ~ **out a cable** ein Kabel ausformen to ~ **the chime** (of a cask) abkimmen to ~ **column of twos** in Doppel-reihe *f* antreten to ~ **compartments** abfachen to ~ **a compound** eine Verbindung *f* eingehen to ~ **a crust** verharschen to ~ **crystals** impfen (kristallisieren) to ~ **drops** perlen to ~ **an equation** eine Gleichung *f* ansetzen to ~ **a film** beschichten to ~ **flakes** flocken to ~ **flocks** flocken to ~ **fluosilicate** fluotieren to ~ **heads** (of screws, bolts) anstauchen to ~ **a hydrogen skin** eine Oberfläche *f* mit Wasserstoff belanden to ~ **an integral part** organisch einbauen to ~ **a knee** kröpfen to ~ **a layer** beschichten to ~ **line abreast** Dwarslinie *f* bilden to ~ **out** (a cable) ausformen **to put the** ~ (in the Press) die Form einheben to ~ **ranks in depth** tiefgliedern to ~ **slag** Schlacke *f* bilden, schlacken to ~ **steps** ab-stufen to ~ **up** aufmarschieren

**form** Fasson *f*, Form *f*, Format *n*, Formular *n*, Gebilde *n*, Gestalt *f*, Linienführung *f*, Profil *n*, Schablone *f*, (concrete) Schalung *f* Schema *n*, Verschalung *f*, Vordruck *m*

**form,** ~ **having a neat pure aerodynamic** ~ aerodynamisch günstig ~ **of application** An-tragsformular *n* ~ **for blockmaking** Gießform *f* ~ **of construction** (cited by way of example in patent specifications) Ausführungsbeispiel, Verwirklichungsform *f* ~ **of cross section** Quer-schnittausbildung *f* ~ **of crystal** Kristallform *f* ~ **and distributor roller journals** Verreib- und Auftragwalzen *pl* **in the** ~ **of dust** staubförmig ~ **of energy** Energieform *f* ~ **of fire** Feuerform *f* **in the** ~ **of flakes (or leaves)** in Blättchenform *f* ~ **of groove** Kaliberform *f* ~ **of insurance** Ver-sicherungsart *f* ~ **of origin** Ursprungsform *f* **for** ~**'s sake** der Ordnung halber ~ **of tooth profile** Zahnflankenform *f* ~ **trajectory be-havior of ballistics variables** Bahnverlauf *m* ~ **of use** Verwendungsform *f* ~ **of vibration** Schwin-gungsform *f*

**form,** ~ **boards** Schalenbretter *pl* ~**-cast piston ring** Formgußkolbenring *m* ~ **change** Form-veränderung *f* ~**-closed** formschlüssig (elec-tron.) ~ **cutter** Fasson-, Form-fräser *m*, Form-schneider *m*, Fräser *m* mit hinterdrehten Schneidezähnen, Profilfräser *m* ~ **drilling and boring tool** Formbohrer *m* ~ **factor** (of a symmetrical alternating quantity) Formfaktor *m*, Formzahl *f* ~ **feeding devices** Formular-führungseinrichtungen *pl* ~ **grinder** Profil-schleifmaschine *f* ~ **guard** Schutzbügel *m* ~ **lines** Formlinien *pl* ~ **lock** Formhalter *m* ~**-locking** formschlüssig (electron.) ~**-locking of bolts** formschlüssige Sicherung *f* ~ **plate** Fassonschiene *f*, Kopierlineal *n* ~ **plateholder** Kopierschienenhalter *m* ~ **positioning device** Formeinpaßvorrichtung *f* ~ **pressure drag** Formanteil *m* des Gesamtwiderstandes ~ **quotient** Formzahl *f* ~ **rack** Formenregal *n* ~ **rib** Nebenrippe *f* ~ **setter** Formenausschießer *m* (print.) ~ **stability** Gestaltfestigkeit *f* ~

**wheel** Profilschleifscheibe *f* ~ **work** Fasson-arbeit *f*, Schalung *f*

**formability** Bildsamkeit *f*, Formveränderungs-vermögen *n* (**de**) ~ Formbarkeit *f*

**formal** formal ~ **defect** Formfehler *m* ~ **logic** formale Logik *f*

**formaldehyde** Ameisen-, Form-aldehyd *n*

**formalism** Schematismus *m*

**formanite** Formanit *m*

**formant** Formant *m*

**formants** Formanten *pl*

**format** Format *n*, Größe *f* ~ **of picture** Bild-größe *f* ~ **control** Formatsteuerung *f*

**formate** Formiat *n* ~ **of** ameisensau(e)r

**formation** Aufstellung *f*, Ausbildung *f*, Bildung *f*, Entstehung *f*, Entwicklung *f*, Formation *f*, Formieren *f*, Formung *f*, Gebilde *n*, Gebirgs-schicht *f*, Gestaltung *f*, Gliederung *f*, Lauf-richtung *f*, Planie *f*, Schichtenbildung *f* (geol.), Schichtengruppe *f*, Verband *m*, Zustande-kommen

**formation,** ~ **of air pockets** Luftsackbildung *f* ~ **of air shadows** Luftschattenbildung *f* ~ **of arcs** Bogenbildung *f* ~ **of beardlike clusters of filings** Bartbildung *f* (Magnetpulver) ~ **of blowholes** Blasenbildung *f* ~ **of bubbles** Dampf-blasenbildung *f* ~ **of burr** Gratbildung *f* ~ **of cakes** Kuchenbildung *f* ~ **of chlorine** Chlor-entwicklung *f* ~ **of clay** Vertonung *f* ~ **of clinker** Schlackenbildung *f* ~ **of cracks** Riß-bildung *f* ~ **of a crystalline grain** Kristallkorn-bildung *f* ~ **of crystals** Kristallausbildung *f* ~ **of a cutting** Spanbildung *f* ~ **of dew** Taubildung *f* ~ **of drops** Tropfenbildung *f* ~ **of edge** Grat-bildung *f* ~ **of fibers** Lagerung *f* der Fasern ~ **of fins** Gratbildung *f* ~ **of fog** Nebelbildung *f* ~ **of froth** Schaumbildung *f* ~ **of gas** Gasentwick-lung *f* ~ **of gritty particles** Grießbildung *f* ~ **of ground** Bodengestaltung *f* ~ **of haze** Schleier-bildung *f* (Lack) ~ **of holes** thermische Fehl-ordnungserscheinung *f* (cryst.) ~ **of an image** Bildentwerfung *f*, Bilderzeugung *f* ~ **of knots** Knäuelbildung *f* ~ **of lunar seas** Maarbildung *f* ~ **of masses** Massenbildung *f* ~ **of oxygen** Sauerstoffentwicklung *f* ~ **of pairs** Paarbildung *f*, Paarigkeit *f* ~ **of peat** Vertorfung *f* ~ **of pits** Grübchenbildung *f* ~ **of ridge** Gratbildung *f* ~ **of rust** Korrosions-, Rost-bildung *f* ~ **of scale** Zunderbildung *f* ~ **of a shoulder** Gratbildung *f* ~ **of slag** Schlackenbildung *f* ~ **of snarls** Schlingenbildung *f* ~ **of soap** Seifenbildung *f* ~ **of the soil** Bodenbildung *f* ~ **of sounds** Laut-bildung *f* ~ **of sparks** Funkenbildung *f* ~ **of steam** Dampfblasenbildung *f* ~ **of strong point** Schwerpunktbildung *f* ~ **of a substance** Stoff-bildung *f* ~ **of sums** Summenbildung *f* ~ **of support aircraft** Frontverband *m* ~ **of temper carbon** Temperkohlebildung *f* ~ **of a thunder-storm** Gewitterbildung *f* ~ **of a waist** (in test piece under tension or tensile stress) Quer-schnittszusammenziehung *f* ~ **of whiskerlike clusters of filings** Bartbildung *f* (Magnetpulver) ~ **of whiskerlike clusters of magnaflux powders** Bartbildung *f* ~ **in width** Breitengliederung *f*

**formation,** ~ **condition** Entstehungsbedingung *f* ~ **flying (or flight)** Fliegen *n* im Verband, For-mationsfliegen *n*, Verbandsflug *f* ~ **heat** Bil-dungswärme *f* ~ **level** Kronlinie *f* der Erd-

arbeiten, Planie *f*, Planumsohle *f* ~ **lights** For-mationslampe *f* ~ **line** Planumsohle *f* (r.r.) ~ **plotting** Verbandsmessung *f* (electron.) ~ **release** Verbandswurf *m* ~ **shutoff** Gebirgs-sperre *f* ~ **tester** Flüssigkeitsprobeentnehmer *m* ~ **time lag** Aufbauzeit *f* ~ **time of spark discharge** Funkentladungsaufbauzeit *f*

**formative,** ~ **element** Bildungselement *n* ~ **time** Bildungsdauer *f* ~ **time lag** Aufbauzeit *f*

**formed** gestaltet **to be** ~ sich bilden ~ **aileron tip strip** Querruderrandbogen *m* **capability of being** ~ (by pressure and/or heat application) Bildsamkeit *f* ~ **countersink** Formstirnsenker *m* ~ **cutter** hinterdrehter Fräser *m*, Profilmesser *n* ~ **head** Schließkopf *m* ~ **milling cutter** Form-fräser *m* ~ **plate** formierte Platte *f* ~ **wood** Profilholz *n*

**former** Bildner *m*, falsche Flügelrippe (a shaping strip of a plane), Formblock *m* (Schärfma-schine), Former *m*, Formgesenk *n*, Gestalter *m*, Hilfsflügelrippe *f* (rip), Kopierlineal *n*, (Kopier) Schablone *f*, Preßklotz *m*, Profilstahl *m*, Spant *n*, Spantring *m*, Spulenkörper *m* (coil), Stech-beitel *m*, Wicklungshalter *m*; bisher, früher ~ **of coil** Körper *m* der Spule, Spulenkasten *m*

**former,** ~ **pin** Kopierfinger *m* ~ **plate** Schablone *f* (f. Kopiervorrichtg.) ~ **plate holder** Form-plattenhalter *m* ~ **rib** Neben-, Zwischen-rippe *f* ~ **roll** Patrizenwalze *f* ~ **tongue** Patrizenwalze *f* ~ **winding** Schablonenwickelung *f*

**former's tools** Formerwerkzeug *n*

**formers** Formanten *pl*

**formic,** ~ **acid** Ameisen-, Formyl-säure *f* ~ **ether** Ameisenäther *m*

**formica** Resopal *n*

**formin** Hexamethylenamin *n*

**forming** Fassonieren *n*, Formen *n*, (of a valve or cell) Formierung *f*, Formung *f*, Profilgestal-tung *f*, Verformung *f* ~ **with clay sheets** Schwartenformerei *f* ~ **of an image** Abbildung *n* (opt.) ~ **in** Einformung *f* ~ **of piping** Lunker-bildung *f* ~ **a quorum** beschlußfähig ~ **of section** profilieren

**forming,** ~ **attachment** Fassoniereinrichtung *f* ~ **battery** Formierbatterie *f* ~ **board** Kabelbrett *n* ~ **cam** Formkurve *f* ~ **condition** Verformungs-verhältnis *n* ~ **cut** Fassonschnitt *m* ~ **die** An-paß-, Aufsatz-form *f*, Formstanze *f* ~ **dies** Biegegesenke *pl* ~ **efficiency** Formänderungs-wirkungsgrad *m* ~ **end-pressure** Fassonend-druck *m* ~ **lathe** Fassondrehbank *f* ~ **operation** Fassonierarbeit *f* ~-**out** Ausformen *n* ~ **pliers** Drückzange *f* ~ **potential** Formierspannung *f* ~ **process** Verformungsvorgang *m* ~ **property** Verformbarkeit *f* ~ **rack** Formierrahmen *m* ~ **ratio** Verformungswert *n* ~ **rest** Fasson-, Formdreh-support *m* ~ **roll** Patrizenwalze *f* ~ **sediments** satzbildend ~ **tool** Fasson-, Form-stahl *m*, Schellhammer *m* ~ **tough clinker (or slag)** zähschlackig ~ **voltage** Formierspannung *f*

**formless** formlos, gestaltlos

**forms of joints** Stoßverbindungsart *f*

**formula** Ansatz *m*, Aufstellung *f*, Ausführungs-form *f*, Bauart *f*, Formel *f* ~ **for attenuation of light in air** Luftlichtformel *f* ~ **for calculation of yield** Ausbeuteformel *f* ~ **of designation (or inscription)** Beschriftungsformel *f* ~ **for gui-**

dance into line of sight (of guided missiles) Einlenkprogramm *n* ~ of parts Teileformel *f* ~ for a rough approximation Faustformel *f*
formula, ~ relating to rigid system Rahmenformel *f* ~ translation (FORTAN) Formelübersetzung *f* (data proc.) ~ translator Formelübersetzer *m* ~ weight Formelgewicht *n*
formulas Formelausdrücke *pl*
formulate, to ~ ansetzen, festlegen
formulating the mixing Einstellung *f* der Mischung
formulation Abfassung *f*, (of solvent or diluent mixtures) Abstimmung *f*, Ansatz *m*, Formulierung *f* ~ of decision Entschlußfassung *f* ~ of standard norms (or rules) Normenaufstellung *f*
forshorten, to ~ mit perspektivischer Verkürzung *f* zeichnen
fort Feste *f*, Festung *f*
Forter (reversing) valve Forter-Ventil *n*
fortification Befestigung *f*, Befestigungs-anlage *f*, -werk *n*, Festung *f*, Festungsanlage *f*, Verschanzung *f* ~ of dunes Dünenschutzwerk *n* ~ chambers Hohlbauten *pl* ~ work Befestigungsarbeit *f*
fortified befestigt ~ area Lagerfestung *f* ~ camp verschanztes Lager *n* ~ place fester Platz *m* ~ position befestigte Stellung *f* ~ zone Befestigungszone *f*
fortify, to ~ befestigen, verstärken
fortnightly halbmonatlich
fortress Festung *f* ~ artillery Festungsartillerie *f* ~ covered by outer belt of fortification Gürtelfestung *f* ~ engineer Festungspionier *m*
fortuitious zufällig ~ distortion unregelmäßige Verzerrung *f* ~ finding (or result) Zufallsergebnis *n*
fortuity Zufälligkeit *f*
forward, to ~ abgehen lassen, absenden, befördern, beschleunigen, einsenden, nachschikken, transportieren, verladen, weiterbefördern, weitergeben, jemandem etwas zustellen
forward voran, voraus, vordringlich, vorn, vorwärts ~ acting regulation Vorwärtsregelung *f* ~-and-backward-bending tester Hin- und Herbiegeprüfer *m* ~-area circular sight for light and medium guns Schwebekreisvisier *n* ~ area warning Vorausraumabsuchung *f* ~ bridge bulkhead Brückenfrontschott *n* ~ conductance Flußleitwert *m* ~ control Drücken *n* (Flugzeug), Stirnsatz *m* ~ control drive Stirnsitzantrieb *m* ~ cross traverse Plan-vor-Bewegung *f* ~ currency purchase Devisenterminkauf *m* ~ current Durchlaßstrom *m* ~ current density Stromdichte *f* in Flußrichtung ~ difference aufsteigende Differenz *f* ~ differential equation Vorwärts-Differentialgleichung *f* ~ direction (rectifier) Durchgangsrichtung *f*, Durchlaßrichtung *f*, (of a dry or oxide rectifier) Flußrichtung *f* ~ displacement center of gravity Schwerpunktsvorlage *f* ~ drive omnibus Frontlenkomnibus *m* ~-driving pulley Vorlaufscheibe *f* ~ end Auflauf *m* (paper mfg.) ~ end of a telephone line under construction (signal) Bauspitze *f* ~ flow angle Stromflußwinkel *m* ~ frame vorderer Gehäuseflansch *m* (Triebwerk) ~ fuselage tank vorderer Rumpftank *m* ~ gear Vorwärtssteuerung *f* ~ lead of brushes Bürstenvorschub *m* ~ light field Leuchtvorfeld *n* ~

machine-gun station Buggefechtsstand *m* ~ motion Vorwärtsbewegung *f* ~ move Hinlauf *m* ~ movement Weiterbewegung *f* ~ movement of the film Filmtransport *m* (in rolling) ~ movement (or motion) traverse (or run) Vorwärtslauf *m* ~ pass Hingang *m* ~ piston stroke Kolbenvorlauf *m* ~ portion of fuselage Rumpfvorderteil *n* ~ pull Haspelzug *m* ~ recall signal Vorwärtsnachruf *m* ~ region Durchlaßbereich *m* (bei Richtleiter) ~ resistance Durchlaß-, Flußwiderstand *m* ~ scatter(ing) Vorwärtsstreuausbreitung *f*, Vorwärtsstreuung *f* ~ signal Anregungsgröße *f* ~ slip Voreilen *n* ~ slope Vorderhang *m* ~ slope of trench Eskarpe *f* ~-slope position Vorderhangstellung *f* ~ slot Vorderschlitz *m* ~ speed Vortriebsgeschwindigkeit *f*, Vorwärtsgang *m* ~ stagger Vorwärtsstaffelung *f* ~ step position in flight formation Vorderstufenabstand *m* ~ stroke Vorwärtshub *m* ~ tail group Kopfleitwerk *n* ~ tension Haspelzug *m* ~ thrust Vortrieb *m* ~-to-back ratio Vor-Rück-Verhältnis *n* (TV) ~ trace sweep Hinlauf *m* (TV) ~ transadmittance Übertragungsleitwerk *m* ~ travel Hingang *m*, Wurfweite *f* ~ view (or vision) Sicht *f* nach vorn ~ voltage Durchlaßspannung *f* ~ wave Vorwärtswelle *f* ~ wing Vorderflügel *m* ~-working stroke Hinlauf *m*
forwarded versandt
forwarder Transporteur *m*
forwarding Beförderung *f*, Transport *m*, Übermittlung *f*, Wanderung *f* ~ of freight (or of goods) Güterbeförderung *f*, Spedition *f*, Verfrachterei *f*, Verfrachtung *f* ~ of telegrams Nachsendung *f* von Telegramme *n*
forwarding, ~ agent Transportmakler *m* ~ capability Transportfähigkeit *f* ~ instruction Versandvorschriften *pl* ~ point (or station) Weiterleitungsstelle *f* (r.r.)
fosse of a fort Festungsgraben *m*
fossil Fossil *n*, Petrefakt *n*, Versteinerung *f* ~ content Fossilinhalt *m* ~ meal Bergmehl *n* Kieselgur *f* ~ remains Abdruck *m* einer Versteinerung
fossiliferous rocks fossilführende Gesteine *pl*
fossula Septalgrube *f*
Foucault currents Foucaultströme *pl*, Wirbelströme *pl*
fougasse Fugasse *f*, Steinmetzmine *f*
foul, to ~ (firearms) verschleimen, verschmutzen
foul Zusammenstoß *m*; (of pipes, ships) faul, garstig, schlecht, verschmutzt ~ air verdorbene Luft *f*, schlechtes Wetter *n* ~ anchor unklarer Anker *m* ~ bottom bewachsener, schmutziger Boden *m* ~ electrolyte Endlauge *f* ~ pit Schlagwettergrube *f* ~ pump unklare Pumpe *f* ~ water Schwelwasser *n*
fouling beschlagen (photo)
fouling (of plugs) Verölung *f*, Verschmutzen *n*, (of gun barrels) Verschmutzung *f* ~ of nozzle Düsenverschmutzung *f*
found, to ~ abgießen, begründen, gießen, gründen
foundation Bettung *f*, Fundament *n*, Fundamentierung *f*, Fundierung *f*, Grund *m*, Grund-bau *m*, -lage *f*, -sohle *f*, -stock *m*, Gründung *f*, Sockel *m*, Sohle *f*, Träger *m*, Unterbettung *f*, Unter-bau *m*, -lage *f*, -lager *n*, Wehrsohle *f* the

~ **is blown up by water (or underminded by water)** der Grund ist unterspült ~ **on caissons** Senkkastengründung *f* ~ **between crib cofferdams** Gründung *f* mit Fangdämmen ~ **on a grillage** Pfahlrostgründung *f* ~ **of masonry** Grundmauer *f* ~ **for screen** Schirmträger *m* ~ **on timber platform** Rostgründung *f* ~ **on wells** Brunnengründung *f*

**foundation,** ~ **base** Fundament-sockel *m*, -sohle *f* ~ **block** Fundamentklotz *m* ~ **(or toe) block** Fundamentblock *m* ~ **bolt** Ankerschraube *f*, Fundament-anker *m*, -bolzen *m*, -schraube *f*, Grundbolzen *m* ~ **brickwork** Fundament-, Grund-mauerwerk *n* ~ **cap** Fundamentklotz *m* ~ **course** Unterbau *m* ~ **(or toe) course** Fundamentblock *m* ~ **depth** Gründungstiefe *f* ~ **ditch** Fundament-, Grund-graben *m* ~ **engineering** Grundbau *m* ~ **equipment** Gründungsausrüstung *f* ~ **excavation** Fundamentaushub *m* ~ **failure** Grundbruch *m* ~ **footing** Bauwerksohle *f* ~ **frame** Grundbaurahmen *m* ~ **hole** Spur *f* ~ **level** Bauwerk-, Gründungs-sohle *f* ~ **loading diagram** Fundamentbelastungsschema *n* ~ **operation** Gründungsverfahren *n* ~ **pile** Grundpfahl *m* ~ **piles** Gründungspfähle *pl* ~ **piling** Grundpfählung *f* ~ **plan** Aufstellungsplan *m*, Fundamentzeichnung *f* ~ **plate** Fundamentplatte *f*, Grund-fläche *f*, -pfeiler *m*; Sohlenplatte *f* ~ **plinth** Fundamentsockel *m* ~ **practice** Gründungstechnik *f* ~ **sketch** Fundamentzeichnung *f* ~ **slab** Fundamentsohle *f*, Sohlenplatte *f* ~ **slide rail** Fundamentschiene *f* ~ **soil** Bau-, Unter-grund *m* ~ **stone** Auflagerstein *m*, Fundament-sockel *m*, -stein *m*; Grundstein *m* ~ **stops (organ)** Grundstimmen *pl* ~ **structure** Grundbau *m* ~ **substance** Grundkörper *m* ~ **substructure of bridge** Brückenunterbau *m* ~ **timber** Pochklotz *m* ~ **trench** Fundamentgraben *m* ~ **trench sheeting** Baugrubenverkleidung *f* ~ **wall** Fundament-, Grund-mauer *f*, Grundmauerwerk *n* ~ **washer** Ankerplatte *f* ~ **work under compressed air** Druckluftgründung *f*

**founder, to** ~ scheitern

**founder** Erbauer *m*, Gießer *m*, Gründer *m*, Schmelzarbeiter *m*, Schmelzer *m*, Urheber *m* ~ **'s art** Gießkunst *f* ~ **'s black** Gießerschwärze *f* ~ **shaft** Fundgrube *f* ~ **'s truck** Schleppe *f* ~ **'s type** Handersatztype *f*

**founding** Abguß *m*, Gießereiwesen *n*, Gießkunst *f*, Guß *m* ~ **of rollers** Walzenguß *m* ~ **furnace** Gießofen *m* ~ **metal** Gußmetall *n* ~ **work** Hüttenarbeit *f*

**foundry** Gießerei *f*, Gieß-haus *n*, -hütte *f*, Hüttenwerk *n*, Schmelzanlage *f*, Schmelzerei *f*, Schmelzwerk *n* ~ **barrow** Gießereikarren *m* ~ **bronze** Gußbronze *f* ~ **cart** Gießereikarren *m* ~ **charcoal pig** Gießereiholzkohlenroheisen *m* ~ **coke** Gießereikoks *m*, Gießkoks *m*, Schmelzkoks *m* ~ **course** Gießereikursus *m* ~ **crane** Gießkran *m* **crucible-steel** ~ Schmelzbau *m* für Tiegelguß ~ **cupola** Gießerei-kupolofen *m*, -schachtofen *m* ~ **device** Gießereieinrichtung *f* ~ **engineer** Gießereiingenieur *m* ~ **equipment** Gießereieinrichtung *f* ~ **fettling shop** Gußputzerei *f* ~ **field** Gießereiwesen *n* ~ **flask** Form-flasche *f*, -kasten *m* ~ **floor (or floor level)** Gießereisohle *f* ~ **furnace** Gießereiflammofen *m*, -ofen *m* ~ **ingot** C-Barren *m*,

Gußblöckchen *n* ~ **ingots** Rohmasseln *pl* ~ **iron** Gießereieisen *n* ~ **jib crane** Gießereidrehkran *m* ~ **ladle** Füllöffel *m*, (large) Gabelpfanne *f*, Gießpfanne *f* ~ **machine** Gießereimaschine *f* ~ **man** Eisengießer *m*, Gießerei-ingenieur *m*, -mann *m* ~ **material** Gießereibedarf *m* ~ **matters** Gießereiwesen *n* ~ **molding machine** Gießereiformmaschine *f* ~ **nail** Formnagel *m*, Stift *m* ~ **patternmaker** Gießereimodelltischler *m* ~ **pig** Gießereieisen *n* ~ **pig iron** Gießereiroheisen *n* ~ **pin** Führungsstift *m* ~ **pit** Damm-, Gieß-grube *f*, Gießplatz *m* ~ **plant** Gießereianlage *f* ~ **practice** Gießerei-betrieb *m*, -technik *f*, -wesen *n* ~ **requisite** Gießereibedarfsartikel *m* ~ **sand** Gießereisand *m* ~ **superintendent** Gießereileiter *m* ~ **supervisor** Gießereiinspektor *m* ~ **traveling crane** Gießereilaufkran *m* ~ **work** Gießerei *f*, Gießereiwesen *n*, Hüttenarbeit *f* ~ **worker** Hüttenarbeiter *m*

**fount** Guß *m* ~ **-case** Defektenkasten *m*

**fountain** Brunnen *m*, Fontäne *f* ~ **head** Quelle *f* ~ **pen** Fließ-feder *f*, Füllfeder *f*, Füllfederhalter *m* ~ **roller complete** Duktrowalze *f*

**four,** ~ **-address instruction** Vieradressenbefehl *m* ~ **-axle high-speed tractor** Vierachseilschlepper *m* ~ **-band light valve** Vierbandlichtschleuse *f* ~ **-bar-chain** Viergelenkkette *f* ~ **-bar link** Gelenkviereck *n* ~ **-barreled antiaircraft gun** Flakvierling *m* ~ **blade** Vierblatt *n* ~ **-blade propeller** Kreuzpropeller *m*, Vierblattschraube *f* ~ **-bladed fan brake** vierflügelige Bremsluftschraube *f* ~ **-bladed propeller** vierflügelige Luftschraube *f* **with** ~ **blades** vierschneidig ~ **-block brake** Vierklotzbremse *f* ~ **-brush commutator bearing** Vierbürstenlager *n* ~ **-chamber engine** Vierkammermotor *m* ~ **-collar piston** Vierbundkolben *m* ~ **-color equation** viergliedrige Farbgleichung *f* ~ **-color printing** Vierfarbendruck *m* ~ **-color problem** Vierfarbenproblem *n* ~ **-color process plates** Vierfarbsatz *m* (print.) ~ **-column hydraulic press** Viersäulenpresse *f* **in** ~ **columns** vierspaltig (print.) ~ **-component alloy** quaternäre Legierung *f* ~ **-component theory** Vierkomponententheorie *f* ~ **-component vector** Vierervektor *m* ~ **-conductor cable terminal** Vierfachkabelklemme *f* ~ **-connection small-size lubricator** vierstellige Kleinschmierpumpe *f* ~ **-cornered** viereckig ~ **-cornered shaft** Vierkant *m* ~ **-cusped** vierlappig ~ **-cycle piston** Viertaktkolben *m*

**four-cylinder,** ~ **distributor** Vierzylinderverteiler *m* ~ **engine** Viertaktmotor *m* ~ **ignition (or timing) distributor** Vierzylinderzündverteiler *m* ~ **injection pump** Vierzylindereinspritzpumpe *f* ~ **motor** Vierzylindermotor *m* ~ **opposed arrangement** Boxeranordnung (4-Zylinder...)

**four,** ~ **digit** vierstellig ~ **-dimensional space** vierdimensionaler Raum *m* ~ **-dimensional vector** vierdimensionaler Vektor *m* ~ **-disk shutter** Vierscheibenverschluß *m* ~ **-electrode tube** Vierpolröhre *f* ~ **-electrode vacuum tube** Doppelgitterröhre *f* ~ **-electrode valve** Vierelektroden-, Vierpol-röhre *f* ~ **-engined** viermotorig ~ **-faced** vierflächig ~ **-figure number** vierstellige Zahl *f* ~ **-flanged shutter** Vierscheibenverschluß *m* ~ **fold** vierfältig ~ **fold spiral** vierfache Spirale *f* ~ **-force** Vierer-kraft *f*, -vektoren *pl* ~ **-frequency dialing** Fernwahl *f* mit vier Frequenzen,

Vierfrequenzfernwahl *f* ~-**frequency key sending** Fernwahl *f* mit vier Frequenzen ~-**frequency signaling** Vierfrequenzfernwahl *f* ~-**gimbal platform** Vierrahmen-Plattform *f* ~-**girder design** Vierträgerbauart *f* ~-**high mill** Quartowalzwerk *n* ~-**high (mill) stand** Vierwalzengerüst *n* ~-**hole cooking range** Vierstellenkochherd *m* ~-**in-line engine** X-Motor *m* ~-**jaw chuck** Vierbackenfutter *n* ~-**lipped twist drill** Vierlippenbohrer *m* ~-**lobed revolving cam** vierfacher, umlaufender Nocken *m* ~-**membered** viergliedrig ~-**momentum** Viererimpuls *m* ~-**page folder** Doppelblatt *n* ~-**paneled door** Vierfüllungstür *f* ~-**part operation table** vierteiliger Apparattisch *m* (teleph.) **in** ~ **parts** vierteilig ~-**party line** Gesellschaftsleitung *f* für 4 Anschlüsse ~ **phase half wave** Vierphasen-Halbwellenbetrieb *m*
**four-point** vierpolig, vierteilig ~ **attachment** Motoraufhängung *f* in vier Punkten, Vierpunktaufhängung *f* ~ **bearing** Vierstrichpeilung *f* ~ **fixing (or fastening)** Vierpunktbefestigung *f* ~ **method** Vierpunktverfahren *n* ~ **recorder** Vierfachschreiber *m* ~ **switch** vierpoliger Schalter *m*

**four,** ~-**pointed knife rest** Vierspitz *m* ~-**pole** vierpolig ~-**pole circuit (or network)** Vierpol *m* ~-**potential** Viererpotential *n* ~-**product** Viergut *n* ~-**quad cable** Kabel *n* mit vier Vierern **with** ~ **rails** viergleisig ~-**reel illustration rotary web press** Vierrollen-Illustrations-Rotationsdruckmaschine *f* ~-**ribbed nut** Kreuzrippenmutter *f* ~-**ribbon light valve** Vierbandlichtschleuse *f* ~-**roller cane mill** Vierwalzenmühle *f* ~-**roller mill** Vierwalzwerk *n* ~-**roller plate-bending machine** Vierwalzen-blechbiegemaschine *f*, -blechrundmaschine *f* ~-**seater** Viersitzer *m* ~-**seating control** Vierkantensteuerung *f* ~-**shoe brake** Vierklotzbremse *f* ~ **sided** vierseitig ~-**speed gear** Vierganggetriebe *n* ~-**speed planetary gear** Viergangplanetengetriebe *n* ~-**spindle planer-type milling machine** vierspindlige Langfräsmaschine *f* ~-**spindle tracer milling machine** Vierspindelnachform-Fräsmaschine *f* ~-**square** rechteckig ~-**square lock** Kesselschleuse *f* ~-**stage** vierstufig ~-**start worm** viergängige Schnecke *f* ~-**step V-pulley** Vierstufenkeilriemenscheibe *f* ~-**stimulus equation** viergliedrige Farbgleichung *f* ~-**stroke cycle** Viertakt *m*, Viertakt-motor *m*, -verfahren *n* ~-**stroke (cycle) engine** Viertaktmotor *m* ~-**stroke unit** Viertaktmotor *m* ~-**terminal network** Vierpol *m* ~-**terminal transmission** Vierpol *m* ~-**terminal-type equalizer** Entzerrungsfilter *n*, -kette *f* ~ **throw** vierfach gekröpft ~-**tone print** Vierfarbendruck *m* ~ **track** Vierteilspur *f* (tape rec.) ~-**track magnetic soundhead** Vierkanalmagnetkopf *m* ~ **valve apparatus** Vierröhrengerät *n* ~-**variable recorder** Vierfachschreiber *m* ~-**vector** Vierer-Vektor *m* ~-**velocity** Vierergeschwindigkeit *f*
**four-way,** ~ **cock** Kreuzhahn *m* ~ **conduit** vierfaches Fassonrohr *n* (electr.) ~ **jack** Anschaltklinke *f* ~ **platform** Vierwegplattform *f* ~ **stop cock** Vierweghahn *m* ~ **T** Doppel-T-Verteilungsstück *n* ~ **tool block (or tool-post)** Vierfachstahlhalter *m*, Vierkant-Revolversupport *m* ~ **unit** Vierwegeeinheit *f* ~ **wheel brace**

Kreuzschlüssel *m* (Radmutterschlüssel)
**four-wheel** Vierrad *n* ~ **brake** Vierradbremse *f* ~ **control** Vierradlenkung *f* ~ **drive** Vierradantrieb *m* ~ **steering** Vierradlenkung *f* ~ **vehicle** Vierradfahrzeug *n*, Zweiachser *m*
**four-wire** vieradrig ~ **arm** Querträger *m* für vier Leitungen ~ **cable** vieradriges Kabel *n*, Vierdrahtkabel *n* ~ **circuit** Vierdrahtleitung *f* ~ **connection** Gabel *f*, Vierdrahtschaltung *f* ~ **core** Viererbündel *n* ~ **double current working** Vierdrahtdoppelstrombetrieb *m* ~ **fork** Drahtgabeln *pl*, (Vier)-Einheitsgabeln *pl* ~ **intermediate repeater** Vierdrahtzwischenverstärker *m* ~ **operation** Vierdrahtbetrieb *m* ~ **repeater** Vierdrahtverstärker *m* ~ **side circuit** Vierdrahtstammleitung *f* ~ **terminating set** Gabelschaltung *f*, Vierdrahtgabel *f* ~ **termination** Vierdrahtgabelschaltung *f* ~ **three-phase alternating current** Vierleiterdrehstrom *m* ~ **unit** Vierer *m*
**fourble,** ~ **board** Plattform *f* mit vierfachem Durchgang
**Fourdrinier** Langsiebmaschine *f* ~ **drier** Fourdriniertrockner *m*
**Fourier's's** ~ **analysis** Fourieranalyse *f*, Fouriersche Zerlegung *f* ~ **series** Fouriersche Reihe *f* ~ **theorem** Fourierscher Satz *n*
**fourragère** Fangschnur *f*
**fourth** Viertel *n* ~ **part of circumference** Viertelkreisbogen *m* ~ **power** Biquadrat *n*
**foveal vision** Netzhautzentersehen *n*
**Fowler,** ~ **flap** Fowlerklappe *f* ~ **wing-flap** Fowler-Flügelklappe *f*
**fox** Fuchs *m* ~ **bolt** Bolzen *m* mit gespaltenem (Anker) Ende und Keil ~ **hole** Deckungs-, Fuchs-, Schützen-loch *n* ~ **saw** Fuchsschwanzsäge *f* ~ **wedge** Gegenkeil *m*
**foxed** moderfleckig ~ **paper** moderfleckiges Papier *n*
**foxy paper** moderfleckiges Papier *n*
**fraction** Bruch *m* (math.), Bruchteil *m*, Fraktion *f* ~ **of atomic ions** Atom-Ionenanteil *m* ~ **of a gram** Bruchgramm *n*, Grammbruchteil *m* ~ **of length** Wegabschnitt *m* ~ **of saturation** Sättigungsanteil *m* ~ **of time** Bruchteil *m* der Zeit
**fraction,** ~ **exchange** Austauschfaktor *m* ~ **mark** Bruchzeichen *n* ~ **seismics** Refraktionsseismik *f* ~ **stroke** Bruchstrich *m* ~ **term** Brechungsglied *n*
**fractional** fractioniert, gebrochen, nicht ganz ~ **boiling point** Siedepunkt *m* eines Kraftstoffanteils bei der Siedetrennung ~ **card** Mehrfachkarte *f* ~ **change** partielle Veränderung *f* ~ **coil** Teilspule *f* ~ **condensation** Fraktionierkondensation *f* ~ **cooling-power loss** Anteil *m* des Leistungsaufwandes für die Kühlung ~ **crystallization** Umkristallisation *f* ~ **distillation** absatzweise Destillation *f*, Fraktionieren *n*, Fraktionierdestillation *f* ~-**distillation flask** Fraktionskolben *m* ~ **focus** Feinfokus *m* ~-**horsepower motor** Kleinmotor *m* mit Bruchteilen einer Pferdestärke ~ **isotopic abundance** relative Isotopenhäufigkeit *f* ~ **motor** Handmotor *m* ~ **number** gebrochene Zahl *f* ~ **part** Bruchteil *m* ~ **pitch** Teilschritt *m* ~ **scan** Teilabtastung *f*, -raster *m* ~-**scan impulse** Zeilenfolgeimpuls *m* ~ **sizes** Zwischenabmessungen *pl* ~-**slot winding** Bruchlochwicklung *f* ~ **sweep**

(in interlaced scanning) Teilraster *m* ~ **washing**
Stufenwäsche *f* (paper mfg.) ~ **weight** Bruch-
gewicht *n* ~ **yield** Ausbeute *f* pro Stufe
**fractionate, to** ~ fraktionieren
**fractionated treatment** Fraktionierung *f*
**fractionating** Fraktionierung *f* ~ **column**
Dephlegmator *m*, Destillieraufsatz *m*, Frak-
tionierkolonne *f* ~ **factor** Trennungsfaktor *m*
**fractionation** Fraktionierung *f*, Siedetrennung *f*
**fractionize, to** ~ brechen
**fractions** Bruchziffern *pl*
**fractostratus** Fraktostratus *m*
**fracture, to** ~ brechen, reißen, springen, zerbrechen
**fracture** Anbruch *m* (metall), Brechen *n*, Brechung *f*, Bruch *m*, Bruchbildung *f*, Bruchfläche
*f* (pig iron), Fraktur *f*, Spalte *f*, Unterbrechung
*f*, Zerbrechen *n* ~ **of black-heart malleable iron**
Bilderrahmenbruch *m* ~ **by crushing off** Ab-
quetschfraktur *f* ~ **of mineral** Querriß *m*
**fracture,** ~ **origin** Bruchentstehung *f* ~ **pieces**
Bruchstücke *pl* ~ **plane** Reißebene *f* ~ **residual**
Restbruch *m* ~ **test** Bruchversuch *m*
**fractured** brüchig ~ **deflection** Bruchablenkung *f*
~ **plane (or surface)** Bruchfläche *f*
**fracturing** Zerstückelung *f* ~ **load** Knickkraft *f*
**fragile** brechbar, bröck(e)lig, brüchig, mor-
sch(ig), zerbrechlich
**fragility** Brechbarkeit *f*, Brüchigkeit *f*, Zerbrechlichkeit *f*
**fragment** Abbruch *m*, Brocken *m*, Bruch *m*,
Bruchteil *m*, Fragment *n*, Scherbe *f*, Splitter *m*,
Stück *n*, Trumm *n* ~ **density** Splitterdichte *f*
**fragmental** aus Bruchstücken bestehend, fragmentarisch, klastisch ~ **chip** Abreißspan *m* ~
**order** Einzelbefehl *m*
**fragmentary** fragmentarisch
**fragmentation** Spaltung *f*, Zerteilung *f*, Zertrümmerung *f* ~ **of nucleus** Kernexplosion *f*
**fragmentation,** ~ **effect** Splitterwirkung *f* ~
**effect of highexplosive shell** Splitterwirkung *f*
der Granate ~ **effect of ricochet burst** Splitterwirkung *f* des Abprallers ~ **power** Zerteilungs-
kraft *f* ~ **shell** Splittergranate *f* ~ **test circle**
Sprengzirkel *m*
**fragments** Bruchstücke *n*, Trümmer *pl*
**fragrance** Duft *m*
**frail** hinfällig, morsch(ig)
**frame, to** ~ abfassen, aneinanderfügen, einfassen, fassen, umrahmen, verblatten, zimmern
**to** ~ **poles** Telegrafenstangen verkuppeln
**frame** Abtastfeld *n*, Bandsprosse *f* (tape rec.),
Bild *n* (film, TV), Bildausschnitt *m* (of camera),
Bilderrahmen *m*, Bildfeld *n*, Binder *m*, Bock *m*
(beim Montagegerüst), Bügel *m*, Einzelbild *n*,
Fassung *f*, Feld *n* (film pictures), Fundamentrahmen *m*, Gatter *n*, Gestell *n*, Geviert *n*, In-
holz *f*, Klaubherd *m*, Maschinenplatte *f*, Rahmen *m* (rigid), Spant *n*, Spantring *m*, Ständer *m*,
Teilraster *m* (TV), Verteiler *m*, Vollraster *m*
(TV), Wange *f*, Wehrbock *m*, Zarge *f* ~ **(work)**
Gebälk *n*
**frame,** ~ **of airplane** Flugzeugrumpf ~ **for**
**arcing plates** Funkenrechen *m* ~ **for belt-**
**loading machine** Einheitsrahmen *m* für die
Gurtfüllmaschine ~ **of boring rig** Bohrturm *m*
~ **of convenient working height** gebrauchshoher
Ständer *m* ~ **for incendiary bombs** Schütte *f* ~ **of**

**the lens attachment** Vorhängergestell *n* ~ **of**
**machine** Körper der Maschine ~ **of the machine**
Maschinengestell *n* ~ **for manhole cover** Deckel-
rahmen *m* ~ **of a mold** Gießkasten *m* ~ **of**
**the outer hull of a submarine** Außenspant *n*
~ **of plates** Plattenrahmen *m* ~ **of reference**
Bezugssystem *n* ~ **with rigid joints** Stabwerk *n*
~ **of a roof** Dachgerüst *n* ~ **for scoop wheel**
Heberadgerüst *n* ~ **of trapezoidal shape**
Trapezrahmen *m*
**frame,** ~ **adjusting device** Rahmenrichtgerät *n* ~
**aligning device** Rahmenrichtgerät *n* ~ **antenna**
Rahmen-, Schleif-antenne *f* ~ **arrangement**
Gerüstanordnung *f* ~ **assembly** Rahmen *n*,
Rumpf-gerüst *n* (Flugzeug), -gehäuse *n*
~-**blanking value** Rasteraustastung *f* ~ **bridge**
Lattensteg *m* ~ **bulkhead** Rahmenquerspant *n*
~ **charge impulse** Rasterwechselimpuls *m* ~
**circuit** Rahmenkreis *m* ~ **coil** Bildeinstellungs-
spule *f* ~ **connection** Rahmenverbindung *f* ~
**construction** Holzbauweise *f* ~ **counter** Bild-
zähler *m* ~ **deflection** Ablenkungswechsel *m* ~
**design** Körperkonstruktion *f* ~ **designed to**
**make minimum (or zero point) sharp** Enttrübungsrahmen *m* ~ **distortion** Teilbildverzerrung
*f* ~ **drive** Peilantrieb *m* ~ **fillet** Rahmenstab *m*
~ **filter** Rahmenfilter *m* ~ **flyback** Bildrücklauf
*m* ~ **format** Bildformat *n* ~ **frequency** Bild-,
Bildwechsel-frequenz *f*, Bildwechselzahl *f*, Bildzahl *f*, Helligkeitswechsel *m* ~ **gauge** Bildschritt
*m* ~ **girder** Fachwerk-, Gitter-träger *m* ~-**grid**
Spanngitter *n* ~-**grid tube** Spanngitterröhre *f* ~
**handle** Daumendrücker *m* ~ **impulse** Bild-,
Bildwechsel-impuls *m* ~-**impulse lamp** Bildim-
pulslampe *f* ~ **impulses** Bildstöße *pl* ~ **incidence**
**panel** Rahmentiefenkreuz *n* ~ **indicator** Bild-
zähler *m* ~ **joint** Scharnier *n* ~-**leader** Läuferrute *f* ~ **ledge** Schlagleiste *f* ~ **line** Bildbühneneinstellung *f*, Bildstrich *m* (film) ~ **line noise**
Bildstrichgeräusch *n* (film) ~ **loop** Rahmenantenne *f* ~ **member** Rachwerkstab *m*, Spant-
profil *n* ~ **noise** Bildgeräusch *n* (film) ~ **opening**
Rahmenausschnitt *m* ~ **period** Rasterperiode *f*
~ **picture** vollständiges Bild *n* ~ **piece** Rahmen-
schenkel *m*, Seite *f* eines Rahmens ~ **post** Ge-
rüstpfosten *m* ~ **pulse** Rasterimpuls *m* ~
**racheting** Bildschaltung *f* ~-**repetition frequency**
Bildwechselfrequenz *f*, Bildzahl *f* ~ **repetition**
**rate** Vollbildfrequenz *f* ~ **ring** Rahmenring *m*
~ **saw** Furnier-, Gatter-, Spann-, Zuschneide-
säge *f* ~ **saw with two blades** Doppelgatter *n* ~
**scan** Bild-ablenkung *f*, -abtastung *f* (rdr) ~
**scanning array** Rasterzaun *m* (für Selbstrücklauf), (TV) ~ **section** Rahmenstück *n* ~ **seg-**
**ment** Spantsegment *n* ~ **shaft** Rahmenschaft *m*
~ **shim** Futterholz *n* ~ **size** Bildformat *n* ~
**slotting machine** Rahmenstoßmaschine *f* ~
**splice** Rahmensplissung *f* ~ **spring** Trag-feder *f*,
-leitfeder *f* ~ **stacking of beat tops** Aufreutern *n*
von Rübenblättern ~ **stand** Gerüstständer *m* ~
**stay** Rahmenverstrebung *f* ~ **stencil mark**
Rahmen(meß)marke *f* ~ **straightening device**
Rahmenrichtgerät *n* ~ **structure** Rahmenwerk *n* ~
**support** Rahmenträger *m* ~ **suppression** Teilbildaustastung *f* ~ **sweep apparatus** Bildkippgerät *n*
~-**sweep scansion** Bildvorschubbewegung *f* ~
**sweep voltage** Bildkippspannung *f* ~ **synchroni-**
**zation signal** Rastersynchronisiersignal *n*

**frame-synchronizing,** ~ **cycle** Vertikalwechsel *m*
~ **impulse** Bildsynchronisierimpuls *m*, Bild-
synchronisierungsimpulsverstärker *m* ~ **pulse**
Bildsynchronisierimpuls *m*, Vertikalwechsel *m*
~-**pulse amplifier** Bildsynchronisierungsimpuls-
verstärker *m*
**frame,** ~ **timber** Holzgerippe *n* ~ **time base** Bild-
kipper *m*, -schaltung *f* (TV), Rasterkippspan-
nung *f* ~ **time-base impulse (or oscillation)**
Bildkippschwingung *f* ~ **trussing** Rahmenunterzug *m* ~-**type rocket projector** Werferrahmen *m*
~ **view finder** Rahmensucher *m* ~ **winding**
Gestellförderung *f* ~ **window** Zargenfenster *n*
**frame-work** Aussteifung *f*, Diagonalverband *m*,
Eisenbeschlag *m*, Fachwerk *n*, Gebälk *n*, Gerippe *n*, Gezimmer *n*, Rahmen *m*, Rahmenwerk
*n*, Riegelwand *f*, Riegelwerk *n*, Rohbaugerüst *n*,
Werksatz *m* ~ **(or timberwork)** Holzwerk *n*
**(supporting)** ~ Gerüst *n* **(wooden)** ~ hölzerne
Unterzüge *pl*
**frame-work,** ~ **of car** Gondelgerippe *n* ~ **for**
**covering** Bespannungsgerüst *n* ~ **of a crane**
Krangerüst *n* ~ **of fixed points** Festpunktnetz *n*
~ **of shutter** Rahmen *m* der Tafel ~ **of tail** Leit-
werkgerippe *n* ~ **of tubs** rohrgeflochtenes Fachwerk *n* ~ **of a window** Fenstergerähme *n*
**frame-work,** ~ **arrangement** Gerüstanordnung *f*
~ **body (or fuselage)** Fachwerkrumpf *m*
(Flugz.) ~ **cutout** Ausschaltung *f* ~ **disk** Trag-
scheibe *f* ~ **inside a pit (or well)** Getriebezim-
merung *f* ~ **stay** Rahmenstab *m* ~-**type projector for high explosive (or incendiary) rockets**
Wurfrahmen *m*
**framed,** ~ **building** Fachwerkbau *m* ~ **crosscut**
**saws** Bügelsägen *pl* ~ **floor** Friesfußboden *m* ~
**whipsaw** Örtersäge *f*
**frameless** rahmenlos, ungefaßt
**framer** Bildeinstellschaltung *f*
**frames** Bildreihe *f* (film)
**framing** Bildbühneneinstellung *f* (film), Bildeinstellung *f* (TV), Bildnachstellung *f*, Fachwerk *n*,
Pfahlrost *m*, Umrahmung *f*, Zulage *f* ~ **of**
**image** Bildverstellung *f* (film) ~ **of joists** Balken-
lage *f* ~ **of a mill** Mühlgerüst *n* ~ **of picture**
Bildstricheinstellung *f*
**framing,** ~ **adjustment** Bildverstellung *f* ~ **device**
Bildverstellvorrichtung *f* (film) ~ **lamp** Bild-
einstellampe *f* ~ **lever** Bildeinstellhebel *m* (film)
~ **mask** Bildmaske *f* ~ **pulses** Rahmenimpulse
*pl* (sec. rdr.) ~ **timber** Verbandholz *n*
**Francis turbine** Francisturbine *f*
**francium** Francium *n*
**franco** franko (frei)
**frangibility** Brechbarkeit *f*
**frangible** brechbar ~ **disk** Zerreißscheibe *f* ~
**glass smoke grenade** Blendkörper *m* ~ **grenade**
Brandflasche *f*
**frank, to** ~ frankieren, frei machen (telegrams)
**frank** aufrichtig, offen, portofrei
**Franke machine** Frankesche Maschine *f*
**franking machine** Frankiermaschine *f*
**franklinite** Franklinit *m*, Zink-eisenerz *n*, -eisen-
stein *m*
**franklinization** Franklinisation *f* (electr.)
**fraud** Betrug *m*, Betrügerei *f*, Fälschung *f*,
Prellerei *f*, Schwindel *m*, Täuschung *f*
**fraudulent** arglistig, betrügerisch, schwindelhaft
**Fraunhofer,** ~ **(absorption) lines** Fraunhoferlinie

*f* ~ **region** Fraunhofersches Gebiet *n*
**fray, to** ~ abfasern, sich abnutzen, abreiben,
abscheuern, ausfransen
**frazil** Siggeis *n*
**freak value** Zufallswert *m*
**freckle, to** ~ tüpfeln
**freckled** gesprenkelt
**free, to** ~ befreien, freigeben, frei machen,
gängig machen **to be** ~ freistehen **to** ~ **from**
**branches** entästen **to** ~ **from dross** entschlacken
**to** ~ **an engine** Motor *m* durch Durchdrehen
losbrechen **to** ~ **from ice** abeisen, enteisen **to** ~
**the line** entblocken **to** ~ **from mud** entschlammen **to** ~ **a nut** eine Schraubenmutter *f* etwas
lösen **to** ~ **from slag** entschlacken **to** ~ **of**
**sulphur (ores)** abschwefeln **to** ~ **spirits from**
**amilic alcohol** den Branntwein *m* entfuseln **to** ~
**from tannic acid** entgerben
**free** frei, losgelassen, unbesetzt, (of charge) unentgeltlich, unverbindlich
**free,** ~ **alongside ship** Ausladungsplatz *m*, frei
auf dem Kai ~ **on board** bahnfrei, frei, franko
Bord, Lieferung frei Schiff ~ **from carbon**
kohlenstofffrei ~ **of charge** spesenfrei, ohne
Berechnung ~ **to choose** wählerfrei ~ **from**
**click** prellfrei ~ **of clinker** schlacken-frei, -rein
~ **of coating** belagfrei ~ **from contagion** keimfrei
~ **from contamination** beimischungsfrei ~ **of**
**cracks** rißfrei ~ **from defects** fehlerfrei ~ **from**
**delay** verzögerungsfrei ~ **of delivery charge**
frachtfrei ~ **of distortion** verzerrungsfrei ~ **from**
**duty** abgabefrei ~ **from eddies** wirbellos ~ **from**
**end play** frei von totem Gang ~ **from faults**
fehlerfrei ~ **from feed-back action** rückkuppelungsfrei ~ **from flaws** fehlerfrei ~ **of flaws**
rißfrei ~ **from frequency effect** frequenzunabhängig ~ **of gases** gasfrei ~ **of glare** blendfrei
~ **from grinding marks** reifenfrei ~ **from hair**
**lines** reifenfrei ~ **of hydrocarbons** kohlenwasser-
stofffrei ~ **from ice** eisfrei ~ **from impurities**
beimischungsfrei ~ **from inertia** trägheitslos
~ **from lag** verzögerungsfrei ~ **from lead** blei-
frei ~ **from lime** kalkfrei ~ **of losses** verlustfrei,
verlustlos ~ **from mist (or fog)** schleierfrei ~
**from night effect** nachteffektfrei ~ **of oil** entölt ~
**from oxides** oxydfrei ~ **of phosphorus** phosphor-
frei ~ **from play** spielfrei ~ **from polarization**
**error** (direction finder) polarisationsfehlerfrei ~
**from polarization errors** nachteffektfrei ~ **of**
**radiation** strahlungsfrei ~ **of ripples** welligkeits-
frei ~ **of scale** zunderfrei ~ **of shrink holes** lunkerfrei, lunkerlos ~ **of slag** schlackenrein ~ **of**
**slip** ruchtlos ~ **from slipping** schlupffrei ~ **from**
**sluggishness** verzögerungsfrei ~ **of smoke**
rauchfrei ~ **from stamp duty** stempelfrei ~
**from strain** spannungsfrei ~ **of stress** spannungsfrei ~ **from tar** teerfrei ~ **of tears** rißfrei ~ **of**
**tension** spannungsfrei ~ **from thumping** prellfrei ~ **of turbidity** trübstoff-frei ~ **from under-**
**cut** unterschnittfrei ~ **from vibration** erschütt-
terungsfrei ~ **from zones** zonenfrei
**free,** ~ **air** Saugzustand *m* ~ **air capacity** Ansaug-
menge *f*, Saugleistung *f* ~ **air dose** Luftdosis *f*
~-**air thermometer** Außenluftdruckthermo-
meter *n* ~ **area** Durchgangsquerschnitt *m*
(Klimaanlage) ~ **baggage** Freigebäck *n* ~
**balloon** Freiballon *m* ~-**beaten** rösch gemahlen
~ **board** Abstand *m* zwischen Wasser und

Krone, (ship) Bordhöhe *f*, Freibord *n*, Höhe *f* über dem Wasserspiegel, Tiefladelinie *f* ~-**body diagram** Kraftplan *m* ~-**bond transition spectrum** Elektronenkontinuum *n* ~-**burning** leichtverbrennlich ~-**burning coal** Sandkohle *f* ~ **call** gebührenfreier Anruf *m*, nicht berechnetes Gespräch *n* ~ **carbon** ungebundener Kohlenstoff *m* ~ **charge** freie oder ungebundene Ladung *f* ~ **controls** losgelassenes Ruder *n* ~ **copy** Freiexemplar *n* ~ **copy wheel** Freilaufrad *n* ~ **crystal** freigewachsener Kristall *m* ~ **current operation** Betrieb *m* mit natürlicher Erregung ~ **cutting** gut spanabhebend ~-**cutting bit** Freischneidebohrer *m* ~ **cutting brass** Automatenmessing *n* ~-**cutting machinability** spanabhebende Bearbeitbarkeit *f* ~ **cutting and machining property** Zerspannbarkeit *f* ~-**cutting steel** Automatenstahl *m* (mit kurzbrechendem Span) ~ **cylinder** Schubzylinder *m* ~ **delivery zone for telegrams** Bezirk *m* der gebührenfreien Telegrammzustellung ~-**deviation-action turbine** Freistrahlturbine *f* ~-**digging capacity** Spitzenleistung *f* (Kran) ~-**discharge** freie Abströmung *f* (GEM), unvollkommener Überfall *m* ~ **discharge over a weir** vollkommener Überfall *m*

**freedom** Freiheit *f* **having one degree of** ~ einfachfrei ~ **of action** Aktionsfreiheit *f*, Bewegungsfreiheit *f* ~ **from angular differences** Phasenreinheit *f* ~ **from lens distortion** Verzeichnungsfreiheit *f* ~ **of maneuver** Aktions-, Bewegungs-, Manövrier-freiheit *f* ~ **to move from place to place** Freizügigkeit *f* ~ **of movement** Bewegungsfreiheit *f* ~ **of trade** Handelsfreiheit *f* ~ **to organize** Koalitionsfreiheit *f* ~ **from phase shift** Phasenreinheit *f* ~ **for play** Spielfreiheit *f* ~ **from poles** (uniformity of normal component of magnetic induction) Quellenfreiheit *f* ~ **from transients** Beharrungszustand *m*

**free,** ~ **draining** frei strömend ~ **drop** freier Fall *m*, Freifall *m* ~ **drum** abkuppelbare Trommel *f* ~-**electron approximation** freie Elektronen-Näherung *f* ~ **electron gas** freies Elektronengas *n* ~ **end** freies oder kurzes Ende *f* (eines Seils) ~-**energy change** Austausch *m* freier Leistung ~-**energy function** Funktion *f* der freien Leistung ~ **fall** Freifall *m* ~-**fall apparatus** Freifallapparat *m* ~-**falling cutter** Freifallbohrer *m* ~-**falling hammer** frei fallender Rammbär *m* ~-**falling mixer** Freifallmischer *m* ~-**field pressure** freies Druckfeld *n* ~ **fit** Gewindepassung *f* (mittel) ~ **flight** Freiflug *m* ~-**flight angle** Freiflugwinkel *m* (of a rock) ~ **flight trajectory** antriebsfreie Bahn *f* ~ **flight weight** Freifluggewicht *n* ~ **flow** freie Strömung *f* ~ **flow position** Freigangstellung *f* ~-**flow-type interconnect** ventillose Verbindungsleitung *f* ~-**free electron transition** Frei-Frei-Elektronenübergänge *pl* ~-**free-radiation** Frei-Frei-Strahlung *f* ~ **gold** amalgamierbares Gold *n*, Berg-, Frei-gold *n* ~ **grid** freies Gitter *n* ~ **guns** (as opposed to fixed) gesteuerte Waffen *pl* ~ **gyro** freier Kreisel *m*, Lagekreisel *m*

**freehand** freihändig ~ **drawing** Freihandzeichnung *f* ~ **grinding** Handschleifarbeit *f*

**free,** ~ **height** ungespannte Länge *f* (Feder) ~ **impedance** Eingangsimpedanz *f* ~ **ionization**

**chamber** offene Ionisationskammer *f* ~ **jet** Freistrahl *m*, (aerated) freistrahlgelüfteter Überfall *m* ~-**jet air blast breaker** Freistrahldruckgasschalter *m* ~-**jet tunnel** Freistrahlkanal *m* ~-**jet turbine** Freistrahlturbine *f* ~ **length of rope** Seilpendellänge *f* ~-**line signal** Freizeichen *n* ~ **moisture** freier Wassergehalt *m* ~ **molecule diffusion** Diffusion *f* freier Moleküle ~ **motion** Spiel *n*, Spielraum *m* ~ **motional impedance** freie Bewegungsimpedanz *f* ~-**of-charge** kostenfrei, -los ~ **one-dimensional model** eindimensionales Modell *n* ~ **opening** freier Durchflußquerschnitt *m* ~ **oscillation** freie oder selbststeuernde Schwingung *f* ~-**oscillation test** Ausschwingungsversuch *m* ~-**overfall** weir vollkommener Überfall *m*, vollkommenes Überfallwehr *n* ~ **pass** freie Fahrt *f* ~ **path** freie Weglänge *f* ~ **pedal travel** Bremsleerweg *m* ~ **period** Eigenperiode *f* (acoust.) ~-**piston engine** Flugkolbenmaschine *f*, Freikolbenmotor *m* ~-**piston engine compressor** Freiflugkolbenverdichter *m* ~ **piston gauge** Freikolbenanometer *n* ~-**piston-type engine** Flugkolbenmotor *m* (Freikolben) ~-**point tester** Röhrenprüfgerät *n* ~ **port** Freihafen *m* ~ **progressive wave** fortschreitende Welle *f* ~ **prospecting** Freischürfen *n* ~-**prospecting territory** Freischürfgebiet *n* ~ **radical** freies Radikal *n* ~ **rolling test** (for ships) Ausschlinger *m* ~-**rotor gyro** lagerfreier Kreisel *m* ~-**run driers** Freilauftrockner *pl* ~-**run test** Auslaufverfahren *n* ~-**run test inertia disk** Auslaufscheibe *f*

**free-running** freilaufend; rösch gemahlen ~-**drive** Freilaufgetriebe *n* ~ **frequency** Eigenfrequenz *f* ~ (**or astable**) **multivibrator** astabiler Multivibrator *m* ~ **speed** Endgeschwindigkeit *f* ~ **sweep** nichtsynchronisierte Ablenkung *f* ~ **switching operation** Freilauf *m*

**free,** ~ **share** Freikux *m* ~ **sketch** Augenmaßaufnahme *f* ~ **space** freier Raum, Schallfeld *n*, Spiel *n*, Spielraum *m* ~ **space charge wave** freie Raumladungswelle *f* ~ **space field intensity** Freiraumfeldstärke *f* ~ **space pattern** Freiraum-Ausbreitungsdiagramm *n* ~-**space sectional area** freier Querschnitt *m* ~ **spinning tunnel** senkrechter Windkanal *m* ~-**standing** freistehend ~ **state** ungebundener Zustand *m* ~ **stone** Haustein *m*, Quadersandstein *m*, Werkstein *m* ~-**stream velocity** Anströmgeschwindigkeit *f* ~ **supplement** Gratisbeilage *f* ~-**swinging drive** freischwingender Antrieb *m* ~-**swinging loud-speaker** freischwingender Lautsprecher *m* ~ **trade** Freihandel *m*, Handelsfreiheit *f* ~-**transmission range** Durchlässigkeitsbereich *m* eines Filters ~ **travel** Totgang *m* ~ **trip** (**or flight**) Freifahrt *f* ~ **trunnion** schwimmender Zapfen *m* ~ **turbine** Freifahrtturbine *f* ~ **valence** freie Wertigkeit *f* ~ **venting** (in molding sand) luftdurchlässig ~-**vibration test** Ausschwingungsversuch *m* ~ **vibrations** Ausschwingen *n* ~ **water** nicht-gebundenes Wasser *n*

**free-wheel** ungebremst; Walzenfreilauf *m* (Walzenlöser *m*) ~ **with ball bearing** Kugellagerfreilaufrad *n* ~ **bicycle** Freilaufrad *n* ~ **clutch** Freilaufkupplung *f* ~ **crank** Freilaufkurbel *f* ~ **disc** Freilaufscheibe *f* ~ **gear** Freilaufeinrichtung *f* ~ **hub** Freinabe *f* ~ **overrunning clutch** Freilaufüberholkupplung *f*

**free-wheeling** Freilauf *m*; freilaufend ~ **axle** Freilaufachse *f* ~ **brake** Freilaufbremse *f* ~ **clutch** Freilauf-, Leerlauf-kupplung *f* ~ **gear** Freilaufzahnrad *n* ~ **propeller** Luftschraube *f* mit Freilauf ~ **sleeve** Freilaufmuffe *f*

**freed** befreit, losgelassen

**freeing** (from legal charges or arrest) Freierklärung *f*, Freifahrung *f* ~ **from interference** Störungsbehebung *f*

**freeness** (pulp) Mahlungsgrad *m*, (in papermaking) Röschheit *f* ~ **from flicker** Flimmerfreiheit *f* ~ **degree of grinding** (wood pulp) Schmierigkeitsgrad *m* ~ **tester** Mahlungsgradprüfer *m* (paper mfg.)

**freezable** gefrierbar

**freeze, to** ~ erfrieren, (liquids) erstarren, festfressen, frieren, gefrieren, (Anker) kleben, ratinieren, vereisen **to** ~ **to death** erfrieren **to** ~ **fast** festfrieren **to** ~ **in** einfrieren **to** ~ **out** (in liquidair trap) ausfrieren **to** ~ **together** (of contacts) verschmoren, zusammenfrieren **to** ~ **up** einfrieren, zufrieren

**freeze** Frieren *n* ~**-drying** Gefriertrocknung *f* ~**-out** Anschlußsperrung *f* ~ **point** Frostpunkt *m* ~ **proofness** Frostbeständigkeit *f*

**freezer** Eismaschine *f*, Gefrierapparat *m* ~ **for ice goblets** Gefrierer *m* für Eisbecher ~ **compartment** Gefrierfach *n*

**freezing** Einfrieren *n*, Erstarrung *f*, Erstarrungspunkt *m*, Frieren *n*, Frost *m*, Gefrieren *n*, Kleben *n*, (in boring) Klemmen *n*, (of engines) Verschweißerscheinung *f* ~ **apparatus** Gefrierapparat *m*, -maschine *f* ~ **container** Ausfriergefäß *n* ~**-in temperature** Einfriertemperatur *f* ~ **method** Erstarrungsverfahren *n* ~ **mist** gefrierender Nebeldunst *m* ~ **mixture** Frost-, Gefrier-mischung *f*, Kälte-gemisch *n*, -mischung *f*, -mittel *n*, Kühlmittel *n* ~ **nuclei** Gefrierkerne *pl* ~ **pipe** Gefrier-rohr *n*, -röhre *f* ~ **plant** Gefrieranlage *f* ~ **point** Eis-, (of the thermometer) Frier-, Frost-, Gefrier-punkt *m*, Gefriertemperatur *f*, Kälte-, Null-, Stockpunkt *m* ~**-point curve** Erstarrungskurve *f* ~ **point depression** Gefrierpunkterniedrigung *f* ~ **process** Ausfrier-, Gefrier-verfahren *n* ~**-shaft** Gefrierschacht *m* ~ **solid in boring** Festwerden *n* ~ **test** Frostversuch *m* ~ **together of contacts** Kontaktzusammenschmoren *n* ~ **tool** Körnchenpunze *f* ~**-up** Vereisen *n*

**freieslebenite** Schilfglaserz *n*

**freight, to** ~ beladen, frachten

**freight** Fracht *f*, Fuhrlohn *m*, Güter *pl* ~ **by measure** Verfrachtung *f* nach Maß

**freight,** ~ **airplane** Frachtflugzeug *n*, Güterflugzeug *n* ~ **capacity** Frachtraum *m* ~ **car** Güter-, Last-wagen *m* ~**-car scales** Waggonwage *f* ~ **car tipper** Waggonkipper *m* **big** ~ **carrier** Großfrachtflugzeug *n* ~ **carrying aircraft** Fracht-, Last-flugzeug *n* ~ **category** Verladeklasse *f* ~**-charging** Ladung *f* ~ **compartment** Frachtabteil *n* ~ **contract** Befrachtungsvertrag *m* ~ **depot** Güterabfertigungsstelle *f*, Güterbahnhof *m* ~ **elevator** Güter-, Lasten-, Waren-aufzug *m* ~ **glider** Lasten-segelflugzeug *n*, -segler *m*, Lastsegelflugzeug *n* ~**-glider towing** Lastenseglerschleppen *n* ~ **handling** Frachtabfertigung *f* ~ **locomotive** Güterzuglokomotive *f* ~ **office** Güterabfertigung *f* ~ **platform** Güterladeplatz

*m* ~ **rate** Frachtsatz *m* ~ **rates** Bahnfrachten *pl* ~ **shed** Güterschuppen *m* ~ **space** (capacity) Laderaum *m* ~ **storage** Frachtlagerung *f* ~ **tariff** Frachtsatz *m* ~ **traffic** Frachtgutverkehr *m*, Güterverkehr *m* ~ **train** Güter-, Last-zug *m* ~ **transport** Frachtbeförderung *f* ~ **yard** Auslade-, Güter-bahnhof *m*

**freightage** Frachtgeld *n*

**freighter** Frachtdampfer *m*, Frachter *m*, Frachtschiff *n*, Verfrachter *m*

**French,** ~ **brandy** Franzbranntwein *m* ~ **calorie** Kikokalorie *f* ~ **casement** Flügelfenster *n* ~ **chalk** Federweiß *n*, Talkum *n*, Wiener Kalk *m* ~ **curve** Bogenlineal *n* ~ **horn** Waldhorn *n* ~ **old style** Elzevir *f*

**Frenkel disorder (or defect)** Frenkel-Fehlordnung *f*

**freon,** Freon *n*, Frigen *m* (electr.) ~ **filling** Frigenfüllung *f* ~ **gas** Freongas *n* ~ **operation (or service)** Frigenbetrieb *m* ~ **proof varnished wire** frigenbeständiger Lackdraht *m* ~ **refrigerator** Frigenkühlmaschine *f*

**frequencies of image current** Bildstromfrequenzen *pl*

**frequency** Bereich *m*, Frequenz *f* (rdo), Häufigkeit *f*, Periodenzahl *f*, Periodizität *f*, Polwechselzahl *f*, (in radians) Wechselgeschwindigkeit *f*, Wechselzahl *f*

**frequency, to be of the same** ~ in der Frequenz *f* übereinstimmen ~ **of accident** Unfallhäufigkeit *f* ~ **of beats** Interferenzfrequenz *f* ~ **for comparison** Vergleichsfrequenz *f* ~ **of conversation** Gesprächsdichte *f* ~ **of conversations** Gesprächsfrequenz *f* ~ **of deflecting voltages** Frequenz *f* der Ablenkspannungen ~ **of excitation** Erregerfrequenz *f* ~ **of flutter** Schwankungsfrequenz *f* ~ **of load cycles** Lastspielfrequenz *f* ~ **of natural torsional** Verdrehungsfrequenz *f* ~ **of the oscillation** Schwingungsfrequenz *f* ~ **of period of light** Sichtbarkeitshäufigkeit *f* ~ **at which phase shift is zero** Nullfrequenz *f* ~ **in radians** Kreisfrequenz *f* ~ **of rain** Regenhäufigkeit *f* ~ **of recurrence** Wiederholungsfrequenz *f* ~ **of sampling** Entnahmehäufigkeit *f* (Proben) ~ **of vibration** Schüttelfrequenz *f* ~ **and waveband designation** Wellenabgrenzung *f*

**frequency,** ~ **adjustment** Frequenz-ausgleich *m*, -einstellung *f* ~ **allocation** Frequenzzuweisung *f*, Zuweisung *f* der Wellenlängen ~ **allocation scheme** Frequenzraster *m* ~ **allotment plan** Frequenzverteilungsplan *m* ~ **alteration** Frequenzpendelung *f* ~ **analysis** Frequenzanalyse *f* ~ **assignment** Zuweisung *f* der Wellenlängen ~ **band** Frequenz-, Wellen-band *n*; Wellenbereich *m* ~ **band covered by a capacitor (or condenser)** Frequenzbereichüberstreichung *f* durch Kondensator ~ **band of emission** Übertragungsband *n* ~ **band width** Frequenzbandbreite *f* ~ **branching filter** Frequenzweiche *f* ~ **bridge** Frequenzbrücke *f* ~ **calibration** Frequenzprüfung *f* ~ **calibration meter** Frequenzeichgerät *n* ~ **change** Frequenz-änderung *f*, -gleiten *n*, -wanderung *f*, Wellenwechsel *m* ~**-change oscillator** Oszillatorstufe *f* ~ **changeover relay** Frequenzumschaltrelais *n* ~ **changer** Bandumsetzer *m*, Frequenz-transformator *m*, -umformer *m*, -wandler *m*, Kaskadenumformer *m*, Mischerröhre *f*, Mischerstufenröhre *f*,

Wechselstromfrequenzwandler *m*, Wechselstromrichter *m*, Wechselumrichter *m* ~**-changer crystal** Kristallmodulator *m* ~**-changer valve** Mischröhre *f* ~ **channel** Frequenzkanal *m* ~ **characteristic** Frequenzkennlinie *f* ~ **characteristics** Frequenzverlauf *m* ~ **charger** Frequenzübersetzer *m* ~ **check** Frequenzkontrolle *f* ~ **control** Bandbremse *f*, Frequenz-angleich *m* (rdr), -einstellung *f*, -kontrolle *f*, -stabilisierung *f* ~ **control coupling** Bandbremskupplung *f* ~ **converter** Frequenz-transformator *m*, -umformer *m*, -wandler *m*, Wechselumrichter *m* ~ **correction** Scharfabstimmung *f* ~ **counter** Signalfrequenzmesser *m* ~ **coverage** Frequenzumfang *m* ~ **critical** frequenzempfindlich ~ **cut-off** Grenzfrequenz *f* ~ **data** Frequenzangaben *pl* ~ **decrease** Frequenzabsenkung *f* ~ **demodulator** Frequenzmodulator *m* ~ **departure** Frequenz-abweichung *f*, -schwankung *f* ~ **dependence** Frequenz-abgrenzung *f*, -abhängigkeit *f* ~ **designation** Frequenzangabe *f* ~ **deviation** Frequenz-abweichung *f*, -hub *m* ~ **deviation meter** Hubmeßgerät *n* ~ **discrimination** Frequenzunterscheidung *f* ~ **distortion** Frequenz-verschiebung *f*, -verzerrung *f* ~ **distribution** Frequenzverteilung *f* ~ **diversity** Frequenzmehrfachempfang *m* ~ **divider** Frequenz-einteiler *m*, -teiler *m* ~ **divider stages** Frequenzteilerstufen *pl* ~ **division** Frequenz-abbau *m*, -erniedrigung *f*, -teilung *f* ~ **division multiplex** Frequenzvielfach *n* ~ **doubler** Frequenzverdoppeler *m* ~ **doubler tube** Verdoppelungsröhre *f* ~ **drift** Frequenz-abweichung *f*, -auswanderung *f* ~ **effect** Frequenzgang *m* **having a** ~ **effect** Frequenzabhängigkeit *f* ~ **equation** Frequenzgleichung *f* ~ **error** Frequenzunsicherheit *f* ~**-exchange keying** Tastung *f* durch Frequenzwechsel ~ **factor** Häufigkeitsfaktor *m* ~ **film** Frequenzfilm *m* ~ **fluctuation** Frequenzhub *m* ~ **flutter** Frequenzflattern *n*, -schwankung *f* ~ **fraction** Teilfrequenzen *pl* ~ **frogging** Frequenzaustausch *m* ~ **gain curve** Frequenzdämpfungskurve *f* ~ **group-selector plug** Gruppenwähler *m* ~ **hysteresis** Effekt *m* einer langen Leitung ~ **indicator** Frequenzanzeiger *m* ~ **interlacing** Frequenzverschränkung *f* (TV) ~ **inversion** Frequenzumkehrung *f* ~ **inverter** Frequenzwende *f* ~ **jumping** Frequenzsprung *m* ~ **level** Frequenzniveau *n* ~ **limit of equalization** Entzerrungsgrenze *f* ~ **measuring bridge** Frequenzmeßbrücke *f* ~ **memory** Frequenzspeicher *m* ~ **meter** Frequenz-messer *m*, -zeiger *m*, Wellenmesser *m* ~ **meter change-over switch** Frequenzmesserumschalter *m* ~ **modulated exciter** frequenzmodulierte Steuerstufe *f* (eines Senders) ~ **modulated exciter** FM-Steuerstufe *f* ~ **modulated transmitter** FM-Sender *m* ~ **modulation** Frequenz-modelung *f*, -modulation *f* ~ **modulation tube** Laufzeitröhre *f* ~ **modulator** Frequenzmodler *m* ~ **multiplication** Frequenzmultiplikation, -steigerung *f*, -vervielfachung *f* ~ **multiplier** Frequenz-vervielfacher *m*, -wandler *m* ~ **passing characteristic** Frequenzdurchlässigkeit *f* ~ **probability (or curve)** Häufigkeitskurve *f* ~ **pull-in range** Einfanggebiet *n* ~ **pulling** Mitziehen *n* der Frequenz *f* **frequency-range** Bandbreite *f* (rdo), Frequenz-

bereich *m* ~ **of equalization** Entzerrungsbereich *m* ~ **of simulation** Frequenzbereich *m* (Nachbildungsbereich) ~ **of voice** Tonbereich *m*
**frequency-range,** ~ **control** Bandbreiteeinstellung *f* ~ **covered by a capacitor (or a condenser)** Frequenzbereichüberstreichung *f* durch Kondensator ~ **indicator** Frequenzbereichanzeiger *m* ~ **switch** Frequenzbereichumschalter *m*
**frequency,** ~ **record** Frequenztafel *f*, Meßschallplatte *f* ~ **recorder** Frequenzschreiber *m* ~ **regulator** Frequenzregler *m* ~ **relay** Frequenzrelais *f* ~ **research statistics** Großzahlforschung *f*
**frequency-response** Frequenzgang *m* ~ **characteristic** Frequenz-gang *m*, -kennlinie *f*, -kurve *f*, -wiedergabe *f* ~ **characteristic of attenuation** Dämpfungsgang *m* ~ **curve** Frequenz-kennlinie *f*, -kurve *f* ~ **investigation** Frequenzganguntersuchung *f* (aut. contr.)
**frequency,** ~ **revolution speed counter** Frequenzdrehzahlmesser *m* ~ **rise** Frequenzerhöhung *f* ~ **scale** Frequenzskala *f* ~ **scanning** elektronische Abtastung *f* (rdr), Wobbeln *n* ~ **shift** Frequenz-umtastung *f*, -verschiebung *f*, -verwerfung *f* ~ **shift diplex operation** Twinplex *m* ~ **shift keying** Frequenzumtastung *f* **F-1** ~ **shift telegraphy** F-1 Morse-Telegrafie *f* durch Frequenztastung *f* ~ **shifter** Frequenzsieb *n*, Wellenschlucker *m* ~ **sliding** Frequenzkonstanz *f* ~ **spacing** Frequenzabstand *m* ~ **spectrum** Frequenz-gemisch *n*, -skala *f*, -spektrum *n*, Wellenspektrum *n* ~ **splitting** Frequenzaufspaltung *f*, Parasitärfrequenz *f* (rdr) ~ **stability** Beständigkeit *f* der Frequenz ~ **stabilization** Frequenz-konstanthaltung *f*, -steuerung *f* ~ **stabilized transmitter tube** frequenzgesteuerte Senderöhre *f* ~ **standard** Frequenznormal *n* ~ **stress number** Lastwechselzahl *f* ~ **submultiplication** Frequenzabbau *m* ~ **submultiplier** Frequenzteiler *m* ~ **sweep** Frequenzhub *m* ~ **sweeping** Frequenzwobbelung *f* ~ **swept** frequenzgewobbelt ~ **swing** Frequenzhub *m* ~ **table** Klassenhäufigkeitstabelle *f* ~ **tester** Frequenzprüfgerät *n* ~ **tolerance** Frequenztoleranz *f*, zulässige Frequenz *f* ~ **transducer** Frequenzumsetzer *m* ~ **transfer** Frequenzmeßwandler *m* ~ **transformation** Frequenz-transformierung *f*, -übersetzung *f*, -umformung *f*, -wandlung *f* ~ **transformer** Frequenzwandler *m*, Wechselstromrichter *m*, Wechselumrichter *m* (static) ~ **transformer** Frequenz-transformator *m*, -umformer *m* ~ **translation** Frequenztransponierung *f* ~ **trap** Wellenschlucker *m* ~ **trebler** Dreifachfrequenztransformator *m* ~ **tripler** Frequenzverdreifacher *m* ~ **triplication** Frequenzverdreifachung *f* ~ **tuning** Frequenzabstimmung *f* ~ **variation** Frequenzhub *m* ~ **wow** Frequenzschwankung *f* (acoust.)
**frequent, to** ~ frequentieren
**frequent** häufig, vielfach ~ **tapping** häufiger Abstrich *m*
**frequently** ~ vielfach
**fresco** Fresko *n*
**fresh** frisch, neu, verschieden
**fresh-air** Frischluft *f* ~ **charge** Frischluftfüllung *f* ~ **duct** Frischluftkanal *m* ~ **entrance** Frischlufteintritt *m* ~ **intake tube** Frischlufteintrittrohr *n* ~ **scoop** Frischlufteintritt *m*

fresh, ~ cell preparations Frischzellenpräparat *n*
~ fallen snow Neuschnee *m* ~ feed (petroleum)
frische Beschickung *f* ~ grouping of the supply
area Neugliederung *f* des Stromversorgungs-
gebietes ~ lining Neuzustellung *f* ~ oil Frischöl
*n* ~ oil suction pipe Frischölsaugleitung *f* ~ oil
suction pump Frischölpumpe *f* ~ steam
pressure Frischdampfdruck *m*
fresh-water Fluß-, Frisch-, Süß-wasser *n* ~
cooler with brine cooling Süßwasserkühler *m*
mit Solekühlung *f* ~ cooling Frischwasser-
kühlung *f* ~ deposit Süßwasserablagerung *f* ~
inlet Frischwassereintritt *m* ~ lake Haff *n* ~
storage tank Frisch-, Süß-wasserbehälter *m* ~
supply Frischwasserzuleitung *f*
freshen, to ~ (a rope) auffieren to ~ up auffrischen
freshet Überschwemmung *f*
freshly neuerdings ~ lined furnace neuzugestell-
ter Ofen *m* ~ mined grubenfeucht ~ mined coal
grubenfeuchte Kohle *f* ~ precipitated frisch gefällt ~ quarried sand grubenfeuchter Sand *m*
freshness Frische *f*
Fresnel Fresnel-Einheit *f* ~ lens Fresnelsche
Linse *f*, Stufenlinse *f* ~ region Fresnelsches
Gebiet *n* ~ zone Fresnelsche Wellenfläche *f*,
Ringfigur *f*
fret, to ~ durch Reibung *f* abnutzen, sich abscheuern
fret Abnutzung *f* durch Reibung, Fundpunkt *m*
weggeschwemmter Erze ~ saw Frett-, Laub-,
Loch-, Schweif-säge *f* ~ work Laubsägearbeit *f*,
Millefiori *pl*, ausgeschnittenes Motiv *n*
fretting corrosion Reibkorrosion *f* (metal),
Passungsrost *m*
freyalite Freyalith *m*
friability Auslösung *f*, Brüchigkeit *f*, Sprödigkeit *f*, Zerreiblichkeit *f*
friable bröck(e)lig, brüchig, morsch(ig), mürbe,
zerbrechend, zerbrechlich, zerbröckelnd, zerreibbar, zerreiblich ~ condition (of materials)
Rieseln *n* ~ soil krümeliger Boden *m*
friar Mönch *m*, Mönchsbogen *m*
friccative Zischlaut *m*
Frick (induction) furnace Frickofen *m*
friction Friktion *f*, Reibung *f* ~ at the pin Pinnen-
reibung *f* ~ of rest Haftreibung *f*, Reibung *f* der
Ruhe *f* ~ of slides Fressen *n* der Gleitsteine
friction, ~ action Bremswirkung *f* ~ band Bremsband *n*, Reibschiene *f*, Reibungsband *n* ~
bearing Gleitlager *n* ~ board gewalzte Graupappe *f* ~ brake Backen-, Reibungs-bremse *f*
~ burns Reibungs-brandstelle *f*, -schmelzstelle *f*
(Kunstfaser) ~ calender Glätt-, Glanz-kalander
*m* ~ cap Streichkappe *f* ~ change gear Rei-
bungswendegetriebe *n* ~ characteristics Rei-
bungsverhalten *n* ~ circle Gleitkreis *m* ~ clip
Hülsenkupplung *f*
friction-clutch Friktionskegel *m*, Reib-, Rei-
bungs-kupplung *f* ~ coupling Reibungskupplung *f* ~ cutoff coupling lösbare Reibungskupplung *f* ~ handlever Eindrückhebel *m* zur
Reibungskupplung für das Vorgelege ~ lever
Friktionseinrückhebel *m* ~ sleeve Reibungs-
kupplungsmuffe *f*
friction, ~ constant Reibungskoeffizient *m* ~
cone Reib-, Reibungs-kegel *m* ~ contact Reibfläche *f* ~ countershaft Friktionsvorgelege *n* ~

coupling Friktionskupplung *f*, Reibungskopp-
lung *f* (electr.) ~ covering (or lining) Reibungs-
belag *m* ~ cylinder Bremszylinder *m* (telet.) ~
damper Reibungsdämpfer *m* ~ disk Brems-,
Friktions-scheibe *f*, Mitnehmerring *m*, Reibrolle *f*, -scheibe *f*, -teller *m* ~ disk clutch Rei-
bungslamellenkupplung *f* ~ disk shaft Rei-
bungsscheibenwelle *f* ~ drag Reibungswiderstand *m* ~ drive Antrieb *m* durch Reibräder,
Friktions-, Gleit-, Reib-antrieb *m*, Reib-ge-
triebe *n*, -räderantrieb *m*, Reibungsantrieb *m*,
Reibungsgetriebe *n* ~-driven press Reibtrieb-
presse *f* ~-driven roll Schleppwalze *f* ~ drum
Reibtrommel *f*, Reibungswalze *f* ~ effect Rei-
bungseinfluß *m* ~ facing Reibbelag *m* ~ factor
Reibungsfaktor *m* ~ feed action Friktionsvor-
schub *m* ~ finish Stoßappretur *f* ~ forging
hammer Friktionsschmiedehammer *m* ~ gear
Reibgetriebe *n*, Reibradgetriebe *n* ~-geared
winch Friktionswinde *f* ~ gearing Friktions-
getriebe *n*, Friktionsräder *pl*, Friktionsscheiben-
getriebe *n* ~ glaze Stoßglanz *m* ~ governor Frik-
tions-regler *m*, -regulator *m* ~ hanger bearing
Gleithängelager *n* ~ harness strap Karabinerhaken *m* mit Klemmschnalle ~ head Widerstandhöhe *f* ~ held cap Kappe *f* zum Festklemmen ~ held sleeve Steckhülse *f* (Einsteckhülse)
~ hoist Friktionsaufzug *m* ~ horsepower Rei-
bungs-leistung *f*, -pferdekraft *f* ~ igniter (pin of
hand grenade) Abreißzünder *m*, Abzieh-,
Friktions-zünder *m*, Schlagröhre *f* ~ layer
Reibungsschicht *f* ~ lever Reibungshebel *m* ~
lining Reibbelag *m* ~ load Reibungs-last *f*, -ver-
lust *m* ~ loss Reibungsverlust *m* ~ marks Reib-
stellen *pl* ~ match Reibzündhölzchen *n* ~
mean effective pressure dem Reibungsleistungs-
anteil *m* entsprechender Mitteldruck ~ meter
Friktionsmesser *m* ~ nut Reibungsmutter *f* ~
pads Backensetzstock *m* ~ pawl Reibungsklinke
*f* ~ peak Reibungsspitze *f* ~ pendulum (explosives) Fallpendel *n* ~ pile foundation Mantel-
pfahlgründung *f* ~ pillar (screw) press Frik-
tionssäulenpresse *f* ~ pillow block Gleitsteh-
lager *n* ~ plate Friktionsscheibe *f* ~ pointer
Schleppzeiger *m* ~ primer Friktionszünd-
schraube *f*, Reibzunder *m* ~ priming Frik-
tionszündung *f* ~ pulley Friktions-rad *n*, -rolle *f*,
(for friction hammer) Führungsrolle *f*, Reibungsrad *n* ~ ratchet gear Klemmsperre *f* ~
reserving gear Friktionswendegetriebe *n*, Frik-
tionswendekupplung *f* ~ revolution counter
Reibungsdrehzahlenmesser *m* ~ roll (for friction hammer) Führungsrolle *f*, Schleppwalze *f*
~ roller Antriebsrolle *f*, Reibrolle *f*, Reibtrommel *f*, Reibungswalze *f* ~ run Motorlauf *m*
zur Ermittlung der Reibungsleistung ~ saw
Reibsäge *f* ~ screw-drawing press Reibtrieb-
spindelziehpresse *f* ~ screw-driven press Reib-
spindelpresse *f* ~ screw press Friktionsspindel-
presse *f* ~ socket Friktionshülse *f*, Rohrreibungsglocke *f* ~ spindle press Friktionsspindel-
presse *f* ~ starting clutch Antriebsreibungs-
kupplung *f* ~ stress Reibungsspannung *f* ~
strip Isolierstreifen *m* ~ surface Lauf-, (of a
brake) Gleit-, Reibungs-fläche *f* ~ tape Isolier-,
Kleb-, Teer-band *n* ~ tapping device Reibungs-
gewindeschneidvorrichtung *f* ~ tensor Rei-
bungstensor *m* ~ test Reibungsprobe *f* ~

**thimble** Fühlschraube *f* ~ **tight** im Reibsitz *m* sein ~ **torque** Reibungsmoment *n* ~ **track** Friktionsschiene *f* ~ **tube** Friktionsschlagrohr *n*, Schlagröhre *f* ~**-type shock absorber** Reibungsschwingungsdämpfer *m* ~ **value** Reibungswert *m* ~ **velocity** Schubspannungsgeschwindigkeit *f* ~ **vibration damper** Reibungsschwingungsdämpfer *m* ~ **vortex** Reibungswirbel *m* ~ **welding** Reibungsschweißen *n* ~ **wheel** Friktions-förderung *f*, -rad *n*, -rolle *f*, Reibrad *n*, Reibungsrad *n* ~**-wheel reversing gear** Reibungsräderwendegetriebe *n* ~ **winch** Friktionswinde *f*

**frictional,** ~ **connection** Kraftschluß *m* ~ **contact** Reibkontakt *m* ~ **countershaft** Reibungsvorgelege *n* ~ **couple** Reibungsdrehmoment *n* ~ **drag** durch Luftreibung *f* hervorgerufener Widerstand ~ **electricity** Reibungselektrizität *f* ~ **error** Reibungsfehler *m* (aut-contr.) ~ **force** Reibungskraft *f* ~ **grooved gearing** Keilrädergetriebe . *n* ~ **heat** Reibungswärme *f* ~ **load** Reibungslast *f* ~ **loss** Reibungsverlust *m* ~ **oxydation** Reibungsoxydation *f* ~ **portion** Reibungsanteil *m* ~ **resistance** Reibungswiderstand *m* ~ **torque** Reibungsdrehmoment *n* ~ **winding-on machine** Friktionsaufwickelapparat *m* ~ **work** Reibungsarbeit *f*

**frictionally connected** kraftschlüssig verbunden

**frictionless** reibungslos ~ **transmission of operating force** reibungsfreie Übertragung *f* der Bestätigungskraft (Bremsen)

**frictions windlass** Friktionswinde *f*

**friemel tilter** Friemelkanter *m*

**frieze** Fries *m*

**friezing** Kräuseln *n*

**frigate** Fregatte *f*

**frigid** kalt

**frill, to** ~ kräuseln

**frill** Faltenkrause *f*

**frilling** Ablöschen *n* (photo) ~ **machine** Faltermaschine *f*

**fringe** Franse *f*, Saum *m*, Streifen *m* ~ **of forest (or wood)** Waldrand *m* ~ **area** Randgebiet *n* (electr.) ~ **effect** Kantenwirkung *f*, Randeffekt *m* ~ **patterns** Streifenbilder *pl*

**fringing** Farbsaum *m* (film) ~ **reef** Saumriff *n*

**Frise aileron** Frisehilfsflügel *m*

**frisket** Rahmen *m*

**frit, to** ~ einfritten, fritten, schmelzen, sintern, **to** ~ **together** festbacken, zusammenbacken, zusammenfritten

**frit** Fritte *f*, Glasmasse *f*, Weichporzellanmasse *f* ~ **kiln** Äscherofen *m* ~ **porcelain** Frittporzellan *n*

**fritted,** ~ **glass plate provided for catching fibers** Faserfänger-Fritte *f* ~ **hearth bottom** Sinterherd *m* ~ **Jena glass** Glasfritte *f* aus gesintertem Jenaer Glas

**fritting** (together) Festbacken *n*, Fritten *n*, Frittung *f*, Sintern *n*, Sinterung *f*

**fritscheite** Fritzscheit *m*

**frog** Aushöhlung *f*, Austiefung *f*, (of voilin bow) Frosch *m*, Herzstück *n*, Strahl *m* ~ **with forged-steel point** Herzstück *n* mit geschmiedeter Flußstahlspitze (r.r.) ~**'s leg experiment** Froschschenkelexperiment *n* ~**-leg winding** Froschbeinwicklung *f* ~ **point** Herzspitze *f* (r.r.) ~ **spawn** Froschlauchpilz *m* ~ **suit** Schutzanzug *m*

**front** Front *f*, Stirn *f* **from** ~ **to back** von vorn nach hinten ~ **and back elevator** doppelseitiger Hebetisch *m* ~ **of a building** Fassade *f* eines Gebäudes *n* ~ **along which combustion takes place** Umsetzungsfront *f* ~ **of the foot** Fußdecke *f* ~ **of glasses** Brillenmittelteil *m* **in** ~ vorn **in the** ~ vorder **in** ~ **of** vorgebaut ~ **of thunderstorms** Gewitterfront *f*

**front,** ~ **arc** Frontbogen ~ **arch of sadle** Vorderzwiesel *m* ~ **armament** Bugbewaffnung *f* ~ **armor** Frontpanzer *m* ~ **assembly** Vorbau *m* ~ **attachment prism** Doppelbildvorsatz *m* ~ **axial rotor** Voraxialrad *n*

**front-axle** Vorderachse ~ **beam** Vorderachskörper *m* ~ **center** Vorderachsenmittelstück *n* ~ **differential casing** Vorderachsbrücke *f* ~ **gear** Vorderachstriebwerk *m* ~ **load** Vorderachsdruck *m* ~ **part** Vorderachsteil *m* ~ **relief** Vorderachsenentlastung *f* ~ **suspension** Vorderachsaufhängung *f*

**front,** ~ **bars** Vorstellrost *f* ~ **bearing** vorderes Lager *n* ~ **board stake** Kopfwandrunge *f* ~ **body pillar** Vorderpfeilerfuß *m* (chassis) ~ **bottom roller** Vorderzylinder *m* ~ **building** Vordergebäude *n* ~ **bumper** vorderer Abweiser *m* ~ **bush (or supplementary bush)** Vorsatzbüchse *f* ~ **chop spindles** Vorderzangenspindeln *pl* ~ **chord** Verbindungsschnur *f* ~ **clearance angle** Freiwinkel *m* (Drehstahl) ~ **column** Vordersäule *f* ~ **connection** vorderseitiger Anschluß *m* ~ **contact** Arbeitskontakt *m*, Schließstelle *f* ~ **contact of the key** vordere Kontaktschiene *f* der Taste ~ **control** Kopfsteuerung *f* ~ **control surface** Kopfruder *n* ~ **controls** Kopfleitwerk *n* ~ **cover** Gehäusenase *f* ~ **cover of cylinder** Zylindervorderdeckel *m* ~ **crank** Stirnkurbel *f* ~ **cross member of frame** vorderer Kopfträger *m* (g/m) ~ **delivery** Frontbogenausleger *m* (print.) ~ **diaphragm** Abschlußblende *f* ~ **down** Vordüne *f* ~ **driver** Vorwärtsfahrer *m* ~ **edge of sheet** Bogenanfang *m* ~ **edge of spindle** Spindelvorderkante *f* ~ **effect cell** Stirnwandzelle *f* ~ **electrode** Stirnelektrode *f* ~ **elevation** Stirn-, Vorder-ansicht *f* ~ **elevation drawing** Aufriß *m* ~ **elevator** Kopfhöhenruder *n*

**front-end** Stirnseite *f* ~ **of assembly line** Bandanfang *m* ~ **plate** Vorderplatte *f* ~ **plate of a boiler** Kesselstirnwand *f* ~ **volatility of a fuel** Anfangswert *m* der Flüchtigkeit an der Verdampfungskurve eines Brennstoffs

**front,** ~ **face** Vorderfläche *f* ~ **face of wedge** vordere Keilfläche *f* ~ **flange** Stirnring *m* ~ **focal plane** vordere Brennebene *f* ~ **focus** dingseitiger oder vorderer Dingbrennpunkt *m*, Objektbrennpunkt *m* ~ **fork** Vordergabel *f* ~ **fork spring** Vordergabelfeder *f* ~ **frame** Vorderrahmen *m* ~ **gate** Portal *n* ~ **guard rail** Greiferschutzstange *f* ~ **guide lay gauge** Anlegemarke *f* ~ **guide plate** Anschlagplättchen *n* ~ **gun** Bugkanone *f* ~ **hammer** Stirnhammer *m* ~ **horizontal stabilizing surface** Kopfhöhenflosse *f* ~ **hub** Vorderradnabe *f* ~ **joint** Stirnstoß *m* ~ **landing gear strut** Fahrgestellvorderstrebe *f* ~ **lay gauge** Anschlagwinkel *m* ~ **lens** Vorderlinse *f* ~ **lens attachment** Vorsatzlinse *f* ~ **lens cap** Objektivschutzdeckel *m* ~ **lens mount** Vorderlinsenfassung *f* ~ **light** Bug-,

Vorder-licht *n*, (of lights in line) Unterfeuer *n* ~ **loader** Frontlader *m* ~ **meniscus** Frontlinsenmeniskus *m* ~ **milling tool** Stirnfräser *m* ~ **nodal point** (vorderer) Knotenpunkt *m* ~ **outrigger** Vorderholm *m* ~ **page** Titelblatt *n*, Vorderseite *f* ~ **panel** Frontverkleidung *f* ~ **panel of gas bag (or cell)** Zellenstirnwand *f* ~ **piece of toggle joint** Vordergelenk *n* (Pistole) ~ **plate** vordere Scheibe *f*, Stirn-blech *n*, -platte *f*, -wand *f* ~ **plate of a firebox** Stehkesselvorderwand *f* ~ **porch** vordere Schwarzschulter *f* ~ **position control valve** vorgeschaltetes Steuergerät *n* ~ **post** Vorderpfeilerfuß *m* (Chassis) ~ **projection** Aufprojektion *f* ~ **propeller** Vorderschraube *f* ~ **radial cooler** Stirnringkühler *m* ~ **radiator** Stirnkühler *m* ~ **rake** Brust-, Keilwinkel *m*, Projektion *f* des Spannwinkels auf eine senkrechte Längsebene des Drehstahles ~ **roll** Vorderwalze *f* ~ **roller** Abfang-, Kletterwalze *f* ~ **row** Stirnreihe *f* ~ **rubber hook** Haken der Luftschraubenwelle ~ **rudder** Kopfseitenruder *n* ~ **scanning** Vorderabtastung *f* (film) ~ **seam** Stirnnaht *f* ~ **seat** Vordersitz *m* ~ **-seat canopy (or hood)** Vordersitzverdeck *n* ~ **sector** Frontabschnitt *m* ~ **shed** Vorderfach *n* ~ **shield** Windscheibe *f* ~ **shutter** Vorderblende *f* (film) ~ **side** Stirnseite *f* ~ **sight** Diopterkorn *n*, Korn *n*, Richtstrich *m* ~ **-sight base** Kornfuß *m* ~ **-sight base ramp** Kornsattel *m* ~ **-sight holder** Kornhalter *m* ~ **-sight piece** Fadenstück *n* ~ **-sight stud** Kornwarze *f* ~ **silhouette** Stirnschattenriß *m* ~ **spar** Vorderholm *m* ~ **spring** Vorderfeder *f* ~ **-spring (hanger) bracket** Vorderfederbock *m* ~ **-spring clamp** Vorderfederklampe *f* ~ **spring suspension** Vorderfederaufhängung *f* ~ **stabilizing surface** Kopfflosse *f* ~ **stage** Proszenium *n* ~ **standard** (textiles) Polflügel *m* ~ **stops** Vordermarken *pl* ~ **strut** Vorderstrebe *f* ~ **surface** Stirnfläche *f* ~ **thunderstorm** Frontgewitter *n* ~ **tipper** Vorder-kipper *m*, -kippwagen *m* ~ **-to-back ratio** (antenna) Rückdämpfung *f* ~ **-to-rear ratio** Vorwärts-zu-Rückwärtsverhältnis *n* ~ **tripod leg** Vorderstütze *f* ~ **view** Stirn-, Vorder-ansicht *f* ~ **(wall)** Brust *f* ~ **wall** Fassade *f*, Stirn-, Vorder-wand *f* ~ **wall of firebox** Feuerbuchsrohrwand *f* ~ **wall rectifier** Vorderwandgleichrichter *m* ~ **warp** Beiende *n* ~ **wave** Kopf-, Stirn-welle *f*

**front-wheel** Bug-, Vorder-, Well-rad *n* ~ **assembly** Vorderrad-bremszylinder *m*, -lagerung *f* ~ **brake** Vorderradbremse *f* ~ **drive** Front-, Vorderrad-, Vorderachsen-antrieb *m* ~ **fork** Vorderradgabel *f* ~ **pair** Vorderradsatz *m*

**front yard** Vorhof *m*

**frontage** Ausdehnung *f*, Mittelbau *m* ~ **alignment** Fluchtlinienplan *m*

**frontal** frontal, Stirn ..... ~ **area** Stirnfläche *f* ~ **area of motor** Stirnfläche *f* des Motors *m* ~ **bank** Stirnböschung *f* ~ **conveyer** Stirnband *n* ~ **cyclone** Frontalzyklon *m* ~ **fields** Stirnfelder *pl* ~ **formation** Frontbildung *f* ~ **joint** Stirnfuge *f* ~ **resistance** Stirnwiderstand *m* ~ **screen** Vordersieb *n* ~ **soaring** Fronten-flug *m*, -segeln *n*, Gewitterflug *m* ~ **viewing** Aufsichtbetrachtung *f* (TV) ~ **wave** Frontalwelle *f* ~ **weather** Stirnwetter *n* ~ **zone** Frontalzone *f*

**frontier** Grenze *f* ~ **area** Grenzgebiet *n* ~ **district** Grenzbezirk *m* ~ **guard** Grenz-schutz *m*

-wache *f*, -wacht *f* ~ **point** Randpunkt *m* ~ **protective force** Grenzschutz *m* ~ **rectification** Grenzbereinigung *f* ~ **sentry** Grenz-wache *f*, -wächter *m* ~ **traffic** Grenzverkehr *m*

**frontogenesis** Frontogenese *f*

**frontolysis** Frontolysis *f*

**frost, to** ~ gefrieren, mattieren, (Eisblumeneffekt) vereisen, **to ~ with sand and water** mattschleifen

**frost** Frost *m*, Kälte *f* ~ **in the cracks of rocks** Spaltenfrost *m*

**frost,** ~ **cleft** eiskluftig ~ **cleft in wood** Eiskluft *f*, Frostriß *m* ~ **-cracked** frostrissig ~ **goggle** Kälteschutzbrille *f* ~ **heave** Frosthebung *f* ~ **period** Frostzeit *f* ~ **-proof** frostbeständig ~ **-protection agent** Frostschutzmittel ~ **protective** Frostschutzmittel *n* ~ **resistant** frostsicher ~ **shake in wood** Eiskluft *m*, Frostriß *m* ~ **shield** Frostschutzscheibe *f* ~ **smoke** Frostdunst *m*, -nebel *m*

**frosted** matt, mattgeschliffen, mattiert, rauh ~ **bulb** mattierte Birne *f* ~ **finish** Eisblumenlack *m* ~ **glass** Eis-, Eisblumen-, Matt-glas *n*, Mattscheibe *f*, Opal-, Riffel-, Trüb-glas *n* ~ **-glass hood (or screen)** Mattglaskalotte *f* ~ **globe** mattierte Glocke *f* ~ **inside** innenmattiert (Birne) ~ **lamp** mattierte Lampe *f*

**frosting** (of glass) Anlaufen *n*, Beglasung *f*, Mattierung *f*

**frosts** Mattglasschirme *pl* (film)

**frosty weather** Frostwetter *n*

**froth, to** ~ aufschäumen, brausen, gären, gischen, schäumen **to ~ over** überschäumen

**froth** Schaum *m* ~ **breaker** Schaumbrecher *m* ~ **formation** Schaumbildung *f* ~ **killer** Schaumbrecher *m* ~ **mixture** Schaumgemisch *n* ~ **skimmer** Schaumabheber *m*, (flotation) Schaumabstreifer *m* ~ **skimming** Schaumabheben *n* ~ **stains** Schaumflecken *pl*

**frother** Schäumer *m*

**frothing** Aufschäumen *n*, Schaumbildung *f*, Schäumen *n* ~ **agent (or oil)** Schaumbildner *m* ~ **quality** Schaumbildungsvermögen *n*

**frothy** schaumig

**Froude principle of wind-tunnel tests** Froudescher Satz *m*

**frozen, to be** ~ (vessel) einfrieren ~ **equilibrium** gehemmtes Gleichgewicht *n* ~ **fog** Reif *m* ~ **fog formation** Reifbildung *f*, Rauhreifbildung *f* ~ **food** tiefgekühlte Lebensmittel *pl* ~ **food chest** Tiefkühltruhe *f* ~ **-in** eingefroren ~ **meat** Gefrierfleisch *n* ~ **section** Frost-, Gefrierschnitt *m* ~ **shaft** Gefrierschacht *m*

**fruchtschiefer** Fruchtschiefer *m* (geol.)

**frugality** Mäßigkeit *f*

**fruit** Frucht *f*, Obst *n* ~ **acid** Fruchtsäure *f* ~ **-acid-proof** fruchtsäurebeständig ~ **drying plant** Dörranstalt *f* für Obst *n* ~ **ether** Fruchtäther *m* ~ **preserved in sugar** verzuckerte Frucht *f* ~ **presser** Fruchtkelterer *m*

**fruitful** fruchtbar

**frustrate, to** ~ (plans) Pläne *pl* durchkreuzen, vereiteln

**frustum** Kegelstumpf *m*, stumpfe Pyramide *f*, Pyramidenstumpf *m*, Stumpf *m*, Stumpfkegel *m* ~ **of cone** abgestumpfter Kegel *m*, stumpfer Kegel *m* **with** ~ abgestumpft

**frying** (noise) Knallgeräusche *pl*, Kratzgeräusch

n, Nebengeräusch n, Prasseln n (rdo) ~ **noise**
Bratpfannengeräusch n ~ **pan** Brat-, Röst-
pfanne f
**F scope** F-Schirm m (rdr)
**F-sharp key** Fisklappe f
**f-sum rule** Summensatz m
**F-1-Telegraphy without modulation** F-1-tonlose
Telegrafie f
**fuchsin** Fuchsin n
**fuel, to** ~ (ship) Brennstoff m einnehmen,
tanken, verbleien
**fuel** Benzin n, Betriebsstoff m, Brennmaterial n,
Feuerungsmaterial n, Heiz-material n, -mittel
n, -stoff m, Kraftstoff m, Spaltstoff m, Treibmittel n, Treibstoff m ~ **and oil** Betriebsstoff m
~ **and oil records** Kraft- und Schmierstoffauf-
zeichnungen pl ~ **of recent geological formation**
junger Brennstoff m
**fuel,** ~ **accumulator** Kraftstoffanhäufer m ~
**additive** Kraftstoffzusatzmittel n ~ **adjusting
screw** Kraftstoffeinstellschraube f ~ **adjustment**
Kraftstoffeinstellung f ~ **air mixture** Kraftstoff-
Luft-Gemisch n, Luftbrennstoffgemisch n
~**-air mixture analyzer** Brennstoff- und Luft-
mischungszerleger m ~**-air ratio** Brennerkenngröße f, Kraftstoff-Luft-Verhältnis n, Verhältnis n von Luft zu Brennstoff ~**-air ratio indicator** Meßgerät m zur Bestimmung des Kraft-
stoff-Luft-Verhältnisses ~ **alcohol** Brenn-
spiritus m ~ **ascending line** Brennstoffsteig-
leitung f ~**-atomizer system** Brennstoffzer-
stäubersystem n ~ **bed** Brennstoffschichte f ~
**bed regulator** Einstellschieber m ~ **bin** Brennstoffbehälter m ~ **boost pump** Kraftstoff-
förderpumpe f, (jet) -überladungspumpe f ~
**breeding cycle** Spaltstoffkreislauf m ~**-burning
engine (or machine)** Brennkraftmaschine f ~
**burnt per kilo of water evaporated** Brennstoffverbrauchsprobe f ~ **by-pass line** Brennstoff-
umwälzleitung f ~ **by-pass regulator** Brennstoff-
nebenwegregler m ~ **cam** Brennstoffnocken m
~ **cell** Kraftstoff-tank m, -zelle f ~ **cell manhole
cover** Tankmannlochdeckel m ~ **charging**
Brennstoffbeschickung f ~ **chemistry** Brennstoffchemie f ~ **cock** Benzinhahn m ~**-collector
box** Kraftstoffsammelbehälter m ~ **consumption**
Betriebsstoffverbrauch m, Brennstoff-aufwand
m, -verbrauch m, Kraftstoffverbrauch m
~**-consumption gauge** Brennstoff-, Kraftstoffverbrauchsmesser m ~**-consumption loop** schlei-
fenförmige Kraftstoffverbrauchskurve f ~**-con-
sumption meter** Kraftstoffverbrauchsmesser m
~ **consumption per unit horsepower-hour** Ein-
heitsbetriebsstoffverbrauch m ~**-consumption
test** Brennstoffverbrauchsprobe f ~ **control**
Kraftstoff-regelung f, -regler m, -zumessung f
~ **creepage** Durchtritt m von Leckkraftstoff f ~
**cutoff valve** Kraftstoffabsperrschieber m ~
**cycle** Spaltstoffkreislauf m ~ **de-aerator** Kraftstoffentlüfter m ~ **delivery pipe** Kraftstoffab-
laßleitung f ~ **depot** Tanklager n ~ **distributing
ring** Brennstoffeinlaufsicke f ~ **drain valve**
Brennstoffentleerungsventil n ~ **draining** Brenn-
stoffenttanken n ~ **drum** Kraftstoffbehälter m
~ **dump** Brennstoff-, Kraftstoff-, Treibstoff-
lager n ~ **dump valve** Kraftstoffablaßventil n ~
**dumping** Kraftstoffschnellablaß m ~ **dust**
Brennstaub m ~ **economizer** Brennstoffsparer

m ~ **economy** Kraftstoffausnutzung f ~ **engineering** Brennstofftechnik f, Feuerungs-kunde
f, -technik f ~ **feed** Kraftstoff-förderung f,
-zufuhr f ~ **feed line** Brennstoffdruckleitung f ~
**feed pipe** Kraftstoffzulaufleitung f ~ **feed pump**
Brennstoffförderpumpe f ~ **feeder (or feeding
apparatus)** Brennstoffspeisevorrichtung f ~
**filler cap** Brennstofffüllstutzen m ~ **filler hose**
Brennstoffbetankungsschlauch m ~ **filler neck**
Kraftstoffeinfüllstutzen m ~ **filling** Brennstoff-
tanken n ~ **filter** Benzinreiniger m, Kraftstoff-
filter m, Kraftstoffreiniger m ~**-fired** feuerbe-
heizt ~ **flask** Kraftstoffeinspritzzylinder m ~
**flow-meter** Brennstoffverbrauchsmesser m,
Kraftstoff-, Treibstoff-durchflußmesser m ~
**flowrate** Treibstoffdurchsatz m ~ **gas** Betriebs-,
Heiz-, Kraft-, Treib-gas n ~ **gas burner** Heiz-
gasbrenner m ~ **gas cell** Treibgaszelle f ~ **gas
current** Heizgasstrom m ~ **gases** Brenngase pl
~ **gauge** Benzindruckmesser m, Benzinuhr f,
Betriebsstoffmesser m, Inhaltanzeiger m ~
**hose** Kraftstoffschlauch m ~ **inhibitor** Klopfbremse f ~ **injection** Kraftstoffeinspritzung f ~
**injection engine** Einspritzmotor m ~ **injection
nozzle** Einspritzdüse f ~ **injection pump** Brenn-
stoffeinspritzpumpe f ~ **injector** Benzininjektor
m, Brennstoffeinspritzer m, Kraftstoffeinspritz-
vorrichtung f ~ **inlet** Betankungsöffnung f ~
**inlet control** Füllungsregulierung f ~ **inlet valve
needle** Benzineinführungsnadel f ~**-intake
connection** Brennstoffeinlaufstutzen m ~**-intake
fold** Brennstoffeinlaufsicke f ~**-intake port**
Brennstoffsaugstutzen m ~ **jet** Brennstoff-
strahl m, Kraftstoffdüse f ~**-jet adjuster** Düsenkorrektor m (aviat.) ~ **jettisonning** Brennstoff-,
Kraftstoff-schnellentleerung f, Treibstoff-
schnellablaß m ~ **lead** Brennstoffvoreilung f ~
**leak** Benzinleck n ~ **leak line** Brennstoffleck-
leitung f ~ **level** Kraftstoffspiegel m ~ **level of
tank** Inhalt m des Kraftstoffbehälters m ~**-level
gauge** Betriebsstoffvorratsmesser m, Brenn-
stoffuhr m ~**-level indicator** Inhaltsanzeiger m
~ **level sensor** Kraftstoffspielabtaster m ~ **line**
Kraftstoffleitung f ~ **load** Brennstoffgewicht n
~ **manifold** Kraftstoffleitung f, Kraftstoffverteiler m ~ **measuring** Kraftstoffverbrauchs-
messung f ~ **meter** Kraftstoffmeßuhr f
~**-metering jet** Kraftstoffzumessung f ~**-meter-
ing nozzle** Kraftstoffdüse f ~ **mixed with air**
Brennstoffluftgemisch n, Gasgemisch n ~
**mixture** Brennstoffgemisch m ~ **nozzle** Brenn-
stoffeinspritzdüse f, Kraftstoffdüse f
**fuel-oil** Brennöl n, Brennstofföl n, Heizöl n,
Treiböl n ~ **leakage** Leckbrennstoff m ~ **meter**
Heizölmesser m ~ **preheater** Ölvorwärmer m ~
**supply** Heizölzufuhr f ~ **supply tank** Brennstoff-
vorratsbehälter m ~ **valve** Brennstoffventil n
**fuel,** ~ **permit** Tankausweis m ~ **pipe** Benzinrohr
n, Kraftstoffleitung f ~ **pipe line** Benzinleitung f
~ **plate** Brennstoffplatte f ~ **pressure** Kraft-
stoffdruck m ~**-pressure gauge** Kraftstoff-
druckmesser m, -verbrauchsmesser m ~
**pressure system** Brennstoffdrucksystem n ~
**priming-system** doper Anlaßeinspritzvorrich-
tung f ~**-pulverizing mill** Brennstaubanlage f ~
**pulverizing scroll** Benzinvernebler m
**fuel-pump** Benzinpumpe f, Brennstoffpumpe f,
Förderpumpe f, Kraftstoff-förderpumpe f,

-pumpe *f* ~ **assembly** Brennstoffpumpen-maschinensatz *m* ~ **gasket** Kraftstoffpumpen-dichtung *f* ~ **liner** Brennstoffpumpenlauf-büchse *f* ~ **tappet** Brennstoffpumpenstößel *m*
**fuel,** ~ **quantity indicator** Treibstoffvorratsan-zeiger *m* ~ **rack** Brennstoffgestell *n* ~ **rating** Klopfwertbestimmung *f* ~ **ratio** Heizwertzahl *f* ~ **reclaiming tank** Kraftstoffrücklaufbehälter *m* ~ **reconditioning** Spaltstoffaufarbeitung *f* ~ **regeneration** Spaltstoffaufarbeitung *f* ~**-release valve** Enttankungsventil *n* ~ **reserve** Brennstoff-reserve *f* ~**-reserve tank** Hilfsbrennstoffbehälter *m* ~ **reservoir** Brennstoffbehälter *m* ~ **residue** Kraftstoffrückstand *m* ~ **return line** Brenn-stoffumwälzleitung *f* ~ **return needle** Kraft-stoffrücklaufnadel *f* ~ **rod** Spaltmaterialstab *m* ~ **saving** Brennstoffersparnis *f* ~ **sediment** Kraftstoffablagerung *f* ~ **selector switch** Kraft-stoffwahlschalter *m* ~ **sensivity of an engine** Kraftstoffempfindlichkeit *f* eines Motors ~ **ship** Heizölschiff *n*, Tanker *m* ~ **shortage** Brenn-stoffmangel *m* ~ **shut-off valve** Kraftstoffab-stellhahn *m* ~ **solution** Spaltstofflösung *f* ~ **source** Brennstoffquelle *f* ~ **spillage** Überlaufen *n* von Kraftstoff ~ **spray entering cylinder of Diesel engine** Brennstoffschleier *m* ~**-spray-nozzle multiple** Brennstoffdüse *f* ~ **station** Tankanlage *f* ~ **stop cock** Brennstoffab-sperrhahn *m* ~ **storage** Brennstofflagerung *f* ~ **strainer** Brennstoff-, Kraftstoff-filter *m*, Kraftstoffreiniger *m*, Kraftstoffsieb *n* ~ **structure ratio** Kraftstoff-Gewicht-Verhältnis *n* ~ **supply** Benzinförderung *f*, Betriebsstoffver-sorgung *f*, Kraftstoffvorrat *m* ~**-supply gauge** Kraftstoffvorrats-anzeiger *m*, -messer *m* ~**-sup-ply indicator** Kraftstoffvorratsmesser *m* ~**-sup-ply line** Brennstoffleitung *f* ~ **system** Brenn-stoffleitung *f*, Kraftstoffanlage *f*, (die gesamte) Kraftstoffzuführungseinrichtung *f*, Verbren-nungssystem *n* ~**-system diagram** Brennstoff-leitungsschema *n*
**fuel-tank** Behälter *m* für Brennstoff *m*, Benzin-tank *m* ~ **baffle** Brennstofftankschott *m* ~ **cleaning squad** Tankreinigungstrupp *m* ~ **cover** Kraftstoffbehälterdeckel *m* ~ **vent** Brennstoff-behälterauslaßröhre *n*
**fuel,** ~ **tankage** Treibstoffassungsvermögen *m* ~ **tape** Brennstoffband *n* ~ **train** Kraftstoffzug *m* ~ **transfer pump** Kraftstoffausfüllpumpe *f* ~ **transfer tap** Kraftstoffumschalthahn *m* ~ **trap** Kraftstoffabscheider *m* ~ **trouble** Benzinstö-rung *f* ~ **truck** Kesselwagen *m*, Kraftstoff-kesselkraftwagen *m* ~ **value** Brennwert *m* ~ **vapor lock** Dampfblasenbildung *f* (in der Kraft-stoffleitung) ~ **volume switch** Kraftstoff-volumenschalter *m* ~**-weight ratio** Kraftstoff-Gewicht-Verhältnis *n*
**fueling** Tanken *n* ~ **assistant (or orderly)** Tank-wart *m* ~ **bridge** Kraftstoffbrücke *f* ~ **facilities** Tankanlage *f* ~ **pit** Zapfgrube *f* ~ **stop** Tank-rast *f*
**fugacity** Flüchtigkeit *f*, Fugazität *f*
**fugative** flüchtig, unecht ~ **color** unechte Farbe *f* ~ **elasticity** Flußelastizität *f*
**fugue** Fuge *f*
**fulcrum, to** ~ anlenken, schwenken
**fulcrum** Anhältspunkt *m*, Anlenkungspunkt *m*, Drehachse *f*, Dreh-punkt *m*, Gelenkpunkt *m*,

Hebelunterlage *f*, Hebe-, Ruhe-, Stütz-punkt *m* ~ **of lever** Hebeldrehpunkt *m* ~ **for pumps** An-triebsbock *m* (für Pumpen) ~ **of suspension** Aufhängepunkt *m*
**fulcrum,** ~ **bar** Lagerschiene *f* ~ **bracket** Hebel-träger *m* ~ **pin** Drehbolzen *m* ~ **pivotal point** Hebeldrehpunkt *m*
**fulcrumed,** ~ **(at)** drehbar gelagert ~ **lever** dreh-bar gelagerter oder zweiarmiger Hebel *m*
**fulfill, to** ~ erfüllen, vollfüllen
**fulguration** Aufblick *m*
**fuligious** rußig
**full, to** ~ walken **to** ~ **up** (tire) pumpen
**full** eingehend, voll, völlig **of** ~ **age** volljährig ~ **of bumps** beulig ~ **of (saturned) color** vollfarbig ~ **of crevices** rissig ~ **of energy** energiereich ~ **of fissures** kluftreich, zerklüftet ~ **at interval** geöffnet **at** ~ **load** vollbelastet **on (or with)** ~ **load** mit voller Last *f* ~**-of-shot indicator clock** Aufschlagmeldeuhr *f*
**full,** ~**-admission turbine** Turbine *f* mit voller Beaufschlagung ~ **analysis** Vollanalyse *f* ~ **annealing** Hochglühen *n*, Tempern *n* ~ **auto-matic bar machine** Stangenvollautomat *m* ~**-availability group** vollkommenes Bündel *n* ~ **band** besetztes Band *n* ~ **boost** Volldruck *m* ~ **bore (blasting)** Vollausbruch *m* ~**-bore valve** Freiflußventil *n* ~ **braking** Vollbremsung *f* ~ **cantilever** vollfreitragend ~ **cargo** volle Fracht *f*, ganze Ladung *f* ~ **clear reproduction** satte Wiedergabe *f* ~ **cock** Hahn *m* in Spannrast ~ **compartment method** Meßkammersystem *n* ~ **connections** Gesamtschaltbild *n* ~ **correction** Vollberichtigung *f* ~ **cut** Vollschnitt *m* ~ **deflection** Endausschlag *m* ~ **delivery** Voll-förderung *f* ~ **development** Ausentwicklung *f* ~ **dial service** Wählbetrieb *m* zwischen Teilneh-mern ~**-duplex operation** Gegenschreiben *n* ~ **earth** glatte Erde *f* (teleph.) ~**-edged** vollkantig ~ **excursion** Endausschlag *m* ~ **exposure** Um-strahlung *f* ~**-feathering propeller** 180-gradige Einstelluftschraube *f*; Luftschraube *f*, die sich in volle Segelstellung bringen läßt ~ **(or rough) file** Schruppfeile *f* ~ **fire** Vollfeuer *n* ~**-floating axle** vollfliegende Achse *f* ~**-floating back axle** vollschwingende Hinterachse *f* ~**-floating piston pin** schwimmender Kolbenbolzen *m* ~**-flow condition** Sollförderung *f* (hydraul. pump) ~**-flow filter** Filter *m* ohne Verminde-rung des Durchtrittsquerschnittes ~ **gantry crane** Vollportalkran *m* ~**-grown** volljährig ~ **heat** Vollfeuer *n* ~ **hole** Außenmuffe *f* ~ **jacket drum** Vollmanteltrommel *f* ~**-length top** durch-gehendes Verdeck *n* ~ **line** (voll) ausgezogene Linie *f* ~ **load** volle Belastung *f*, Dauerbetrieb *m*, volles Gewicht *n* mit voller Belastung *f*, Nennlast *f*, Vollbelastung *f*, Vollast *f* ~ **load self-regulating (or selftiming) curve** Vollast-selbstverstellinie *f* ~ **load speed** Vollastdreh-zahl *f* ~ **nose-down trim** Volltrimmausschlag *m* (kopflastig) ~ **nose-up trim** Volltrimmaus-schlag *m* (schwanzlastig) ~**-open throttle** voll geöffnete Drossel *f* ~ **page** ganzseitig ~ **panel flight** Vollinstrumentanflug *m* ~**-parabolic** rotationsparabolisch ~ **particulars** alle Einzel-heiten *pl* ~**-pitch winding** Durchmesserwick-lung *f*, ungesehnte Wicklung *f* ~ **power** Vollgas *n* ~ **power of attorney** Generalvollmacht *f*

~-power dive Sturzflug *m* mit Vollkraft ~-power output Volleistung *f* ~-power trial Probe *f* mit äußerster Kraft ~ pressure Volldruck *m* ~-pressure altitude Volldruckhöhe *f* ~-pressure ratio Volldruckverhältnis *n* ~ (or heavy) print satter Druck *m* ~ (or full strength) print Volldruck *m* ~ radiator (Planckian) schwarzer Körper *m*, Planckscher Strahler *m* ~-range amplifier Vollverstärker *m* ~ rate volle Gebühr *f* ~ resolution gute Auslösung *f* (TV) ~ rotation of a selector Durchdrehung *f* eines Wählers ~ rudder voll angewandtes Ruder *n* ~ satellite exchange Teilamt *n*
full-scale maßstäblich, natürliche oder wahre Größe *f* ~ meter Weitwinkelinstrument *n* ~ model Vollmaschine *f* ~ production Serienherstellung *f* ~ reading Skalenendwert *m*, Vollausschlag *m* ~ test naturgroßer Versuch *m* ~ value maximal zulässiger Ablesewert *m* ~ wind tunnel Windkanal *m* für Modelle in natürlicher Größe
full, ~ shade (Lack) Vollton *m* ~ shot Totale *f* (film) ~ shot current Schrotstrom *m* ~ sight Vollkorn *n*
full-size maßstäblich, natürliche Größe *f*, natürlicher Maßstab *m*, naturgetreu ~ (or scale) drawing Bau *m* in natürlicher Größe ~ test Prüfung *f* am ganzen Versuchsstück ~ test area Vollprüffeld *n* ~ test stand Vollprüffeld *n* ~ view Ansicht *f* in voller (natürlicher) Größe
full, ~-sized fang Vollmaßzapfen *m* ~-skirted-type piston Topfkolben *m* ohne Aussparungen ~ solution Durchrechnung *f* ~ speed volle Fahrtgeschwindigkeit *f*, in vollem Gang *m* ~ speed ahead die Pulle reinschieben ~-squared sleeper (or tie) vollkantige Schwelle *f* ~-stall landing Landung *f* im überzogenem Flugzustand ~ steam Volldampf ~ stop Punkt *m* (Schriftzeichen), Vollanschlag *m* ~ strength vollzählig ~ string of casing volle Verrohrung *f* ~ stroke Vollhub *m* ~-supply turbine Turbine *f* mit voller Beaufschlagung ~ swing rest drehbarer Support *m* ~ swivel Vollkreis *m* ~ thread Gewindedurchmesser *m* ~-throttle altitude Vollgas-, Volleistungs-höhe *f* ~-throttle bench revolutions per minute Vollgasstanddrehzahl *f* throttle calibration Vollgasmeßwerte *pl* ~-track armored vehicle Gleiskettenpanzerfahrzeug *m* ~-track head Vollspurkopf *m* ~-track vehicle Raupen-, Vollketten-fahrzeug *n* ~-travel position maximaler Ausschlag *m* (d. Steuerorgane) ~ trim Volltrimmausschlag *m* ~ utilization of screen Schirmaussteuerung *f* ~ value Versicherungswert *m*, Vollgehalt *m* ~-view windscreen Panoramascheibe *f* ~ vision cab(in) Vollsichtkanzel *f* ~ vision turret Vollsichtkuppel *f* ~ voltage volle Spannung *f*
full-wave Doppelweg ...., Vollweg .... ~ rectification Doppelweggleichrichtung *f*, Gleichrichtung *f* beider Halbwellen, vollständige Gleichrichtung *f*, Vollweggleichrichtung *f* ~ rectifier Doppelgleichrichter *m* (der Wellen), Einweggleichrichter *m* (der Wellen), Glühkathodenquecksilberdampf (wellen) gleichrichter *m*, Quecksilberdampf-, Vollweg-, Zweiweggleichrichter *m* ~ rectifier circuit Vollweggleichrichterschaltung *f* ~ rectifier circuit organization Doppelwegschaltung *f* ~ rectifying valve Doppelweg-, Vollweg-gleichrichterröhre *f*

full, ~-web plate construction geschlossene Vollwandkonstruktion *f* ~-width acidulating device Breitsäureanlage *f* ~-width freezer evaporator with compartment Gefrierfachverdampfer *m* ~-width hydroextractor Breitschleuder *f* ~ wing Knickflügel *m*
fuller Füllungseisen *n* (naut.), Streckgesenk *n*, Walker *m*, to a ~ extent in höherem Maße *f* ~'s earth Bleicherde *f*, Fetton *m*, Fullerde *f*, Schaumton *m*, Seifen-erde *f*, -ton *m*, Walkerde *f*
Fuller, ~ board Preßspan *m* ~ (ring-roll) mill Fullermühle *f* ~ phone Utelapparat *m*
fullering Stemmen *n*, Verstemmen *n*, Verstemmung *f*
fulling Abrecken *n* ~ fat Walkfett *n* ~ machine Walke *f* ~ mill Hammer-, Loch-, Stock-walke *f*, Walke *f*, Walkerei *f*, Walkmühle *f* ~ mills Filzmühle *f* ~ trough Walktrog *m*
fullness Fülle *f*, Füllkraft *f*, Völligkeit *f* ~ of a design Schwere *f* des Musters ~ factor Ausbauchungsfaktor *m*, (curve ferromagnetism) Kurvenfüllfaktor *m* ~ indicator Prallanzeiger *m*
fully, to ~ expose durchbelichten to ~ irrediate durchbelichten
fully, ausführlich, völlig ~ advanced ignition äußerste Frühzündung *f* ~ articulated rotor vollgelenkiger Rotor *m* ~ automatic ganz-, voll-automatisch ~ automatic(al) vollselbstständig ~ automatic asynchronous power station Asynchronkraftwerk *n* ~ automatic machine ganzselbsttätige Maschine *f* ~ bored voll ausgebohrt ~-braided cordage vollgeklöppelte Leine *f* ~ controlled airscrew (or propeller) Vollverstellschraube *f* ~ dimensioned volldimensioniert ~ exposed ausexponiert ~ exposed radiator Kühler *m* ganz im freien Fahrtwind ~ feathering type airscrew Verstellluftschraube *f* ~ grown hochstämmig ~ hardenable aushärtbar ~ intermeshed network Maschennetz *n* ~ mechanical (or mechanized) vollmechanisch ~ perforated tape durchgelochter Streifen *m* ~ pot-galvanized vollbadfeuerverzinkt ~ protected section Schutzstrecke *f* ~ restrained eingespannt ~ supercharged engine Höchstdrucklademotor *m* ~ transistorized volltransistorisiert ~ wired receiver set verdrahteter Empfänger *m* ~ wound vollgespult
fulminate, to ~ blitzen und donnern, unter Knall *m* und Feuererscheinung explodieren, (with a flame) verpuffen
fulminate, ~ Reibsatz *m* ~ compound Anfeuchtfeuerung *f*
fulminating (or fulminant) sich entladend ~ powder Knallpulver *n* ~ silver Knallsilber *n*, Silberoxydammoniak *n*
fulminic acid Knallsäure *f*
fumble, to ~ Arbeiten ungeschickt und ohne Sachkenntnis ausführen, fummeln, herumfühlen, herumkramen
fume, to ~ dämpfen, rauchen to ~ off abrauchen
fume Abgas, Dampf *m*, Dunst *m*, Rauch *n* ~ chambers Abzüge *pl* ~ condenser Abgasverdichter *m* ~ dispersion equipment Entnebelungsanlage *f* ~-dispersion installation Entnebelungsanlage *f* ~ extractor Entgaser *m* ~ nave Rauchschiff *n* ~ removal Entnebelung *f* ~ seal Gasdichtung *f*
fumeless dunstlos

**fumers** Brodem *m*, Dämpfe *pl*, Rauchgas *n*
**fumigate, to** ~ ausräuchern
**fumigation** Anräucherung *f*, Räucherung *f*
**fumigator** Räucheranlage *f*
**fuming** Verqualmung *f*; kochend
**function, to** ~ arbeiten, bewähren, funktionieren, laufen, wirken
**function** (motor) Ansprechvermögen *n*, Funktion *f*, Gang *m*, Veranstaltung *f* **ability to** ~ (motor) Ansprechvermögen *n* **as a** ~ **of** in Abhängigkeit von **to be a** ~ **of** empfindlich **being a** ~ **of pressure** druckabhängig ~ **of representation** Abbildungsfunktion *f* ~ **of scattering angle** Funktion *f* des Streuwinkels ~ **of strain** Verformungsfunktion *f* ~ **of temperature** Temperaturgang *m*
**function,** ~ **circuit** Dekodierschaltung *f* ~ **digit** funktionelle Ziffer *f* ~ **generator** Drehmelder *m* ~ **keys** Funktionstasten *pl* (telet.) ~ **lever** Funktionsgabel *f* ~ **lever-spacing suppression** Vorschubausschaltkamm *m* (telet.) ~ **multiplier** Funktionsmultiplizierer *m* ~ **selector switch** Betriebsartenschalter *m*, Flugregler-Bediengerät *n* ~ **showing tendency of variation** Änderungstendenzfunktion *f* ~ **switch** Betriebsarten-, Funktions-schalter *m* ~ **table** Funktionstabelle *f* (math.), Funktionstafel *f* (info proc.) ~ **unit** Funktionsbaugruppe *f* ~ **value** Gebrauchswert *m*
**functional** betrieblich, funktionell, zweckbedingt, zweckmäßig ~ **character** Steuerzeichen *n* ~ **check** Funktionsprüfung *f* ~ **data flow** Signalwege *pl* (in Schaltbildern) ~ **derivative** Funktionalableitung *f* ~ **diagram** Funktionsschema *n*, Pfeilzeichnung *f*, Stromlaufplan *m* ~**-differential equation** Funktional-differential *n*, -gleichung *f* ~ **relation** Abhängigkeitsbeziehung *f* ~**-relationship of temperature** Temperaturabhängigkeit *f* ~ **symbol** Funktionalsystem *n*, logisches Symbol ~ **test** Funktionsprüfung *f* ~ **unit** Funktionseinheit *f*
**functioning** Arbeiten *n*, Funktionierung *f*, Wirken *n* ~ **action** Arbeitsspiel *n*
**functions of formal** Wahrheitsfunktionen *pl*
**fund** Grundstock *m*, Kapital *n* ~ **raising** Umlage *f*
**fundament** Fundament *n*
**fundamental** Grund *m*, Grundlage *f*; fundamental, grundlegend ~ **in facts** grundsätzlich
**fundamental,** ~ **absorption** Fundamentalabsorption *f* ~ **band** Grundband *n*, Grundschwingungsband *n* ~ **bridge equation** allgemeine Gleichung *f* der Meßbrücke ~ **building block** Fundamentalbausteine *pl* ~ **circuit** Grundstromkreis *m* ~ **component** Grundcomponente *f* ~ **component distortion** Amplitudenverzerrung *f* ~ **concept** Grundgedanke *m* ~ **constituent** Urbestandteil *m* ~ **domain of a discontinuous group of transformations** Fundamentalbereich *n* einer diskontinuierlichen Abbildungsgruppe ~ **entity** Fundamentalgröße *f* ~ **equation** Grund-, Haupt-gleichung *f* ~ **field particle** Feldteilchen *n* ~ **form** Grundform *f* ~ **formula** Grundformel *f* ~ **frequency** Grund-frequenz *f*, -harmonische *f*, -periodenzahl *f* ~ **(or natural) frequency** Eigenfrequenz *f* ~ **group of surface** Fundamentalgruppe *f* einer Fläche ~ **idea** Grundgedanke *m* ~ **interval** Fundamentalabstand *m* ~ **lattice**

**absorption** Grundgitterabsorption *f* ~ **law** Grundgesetz *n*, Hauptsatz *m* ~ **layer** Grundschicht *f* ~ **(basic) method** prinzipielle Methode *f* ~ **mode** Hauptschwingbereich *m*, unterster Schwingbereich *m* ~ **note** Grundton *m* ~ **oscillation** Grund-schwingung *f*, -welle *f*, erste Harmonische *f* ~ **particle** Grundteilchen *n* ~ **period** Grundschwingung *f* ~ **periodic parallelogram** Periodenparallelogramm *n* ~ **principle** Fundamentalsatz *m*, Grundbegriff *m*, Hauptsatz *m* ~ **quantities** Grundgrößen *pl* ~ **research** Grundlagenforschung *f* ~ **resonance** Grundresonanz *f* ~ **shape of form** Ausgangsform *f* ~ **standard** Grundnorm *f* ~ **strain theorems** Fundamentalsätze *pl* der Verformung ~ **substance** Grundkörper *m* ~ **tensor** Fundamentaltensor *m* ~ **tolerance group** Grundtoleranzreihe *f* ~ **tone** Grundton ~ **unit** Fundamental-, Grundeinheit *f* ~ **wave** (length) Grundwelle *f* ~ **wave of harmonic** Fundamentalschwingung *f* ~ **wave of voltage** Grundwelle *f* der Spannung ~ **wave length** Grundwellenlänge *f*
**fundamentally speaking** an sich, im Grunde genommen
**funds** Fond *m*
**fungicidal** schimmeltötend
**fungicide** Konservierungsmittel *n*
**fungistatic** pilzwuchshemmend
**fungus** Schwamm *m*
**funicular** Berg- und Talbahn *f*; faserig ~ **force** Seilkraft *f* ~ **line** Seilstrahl *m* ~ **polygon** Seil-eck *n*, -polygon *n*
**funnel** Fülltulle *f*, Schiffsschornstein *m*, Schornstein *m*, Spitztrichter *m*, Trichter *m*, Trombe *f*, Tülle *f* ~ **for adding dyestuff** Farbstoffzusatztrichter *m* ~ **of a charcoal pile** Zug *m* eines Kohlenmeilers
**funnel,** ~ **air case** Rauchfangmantel *m* ~ **apparatus** Spitztrichterapparat *m* ~ **cloud** Rüsselwolke *f*, Schlauch *m*, Trichter *m* (meteor.) ~ **door** (or valve) Trichterverschluß *m* ~ **flange** Trichterflansch *m* ~ **flask** Trichterrohr *n* ~**-shaped** trichterförmig ~**-shaped antenna** Trichterantenne *f* ~**-shaped auger** Trichterbohrer *m* ~**-shaped lubricator** Trichterformöler *m* ~**-shaped pit** Binge *f* ~ **tube** Einfülltrichter *m*, Trichterrohr *n* ~**-type antenna** Trichterantenne *f*
**funnels** Spritztrichterapparat *m*
**fur, to** ~ **the boiler** den Kessel *m* ausklopfen
**fur** Kesselstein *m*, Pelz *m* ~**-lined** pelzgefüttert ~**-red cylinder jacket** Kühlwassermantel *m* mit Kesselsteinablagerung *f*
**furan** Furan *n*
**furbish, to** ~ abschleifen, polieren
**furbisher** Drahtscheurer *m*
**furca** Forkel *f*
**furcated** gabelartig
**furcation** Gabelteilung *f*
**furlough** Urlaub *m*
**furnace** Brandherd *m*, Brennofen *m*, Feuerbüchse *f*, -kiste *f*, -raum *m*, Feuerung *f*, Feuerungsanlage *f*, Herd *m*, Lötofen *m*, Ofen *m* **one-story** ~ Einetagenofen *m* ~ **for the distillation of sulfur** Galeerenofen *m* zur Läuterung *f* des Schwefels *m* ~ **for first refining** Stückofen *m* ~ **below floor level** Unterflurofen *m* ~ **for heating plates** Blechglühofen *m* ~ **with recirculating air** Luftumwälzofen *m* ~ **with stepped**

grate Treppenrostfeuerung *f* ~ **with trough grate**
Muldenrostfeuerung *f* ~ **with two pits** Brillenofen *m* ~ **for ventilating a mine** Wetterofen *m*
**furnace,** ~ **addition** Schmelzzuschlag *m* ~ **air**
Verbrennungsluft *f* ~ **attendant** Ofenwärter *m*
~ **attendants** Ofenkolonne *f* ~ **blast** Hochofen-
wind *m* ~ **body** Ofenkörper *m* ~ **bosh** Ofenrost
*m* ~ **bottom** Herd-, Ofen-sohle *f* ~ **brick lining**
Ofen-ausmauerung *f*, -mauerung *f*, -mauerwerk
*n* ~ **brickwork** Ofenmauerwerk *n* ~ **builder**
Ofenbauer *m* ~ **cadmia (or calamine)** Gichtschwamm *m*, Tutia *f* ~ **capacity** Ofenfassung *f*
~ **carriage** Gichtwagen *m* ~ **casting** Herdguß
*m* ~ **chamber** Ofenraum *m* ~ **charge** Ofenbe-
schickung *f* ~-**charging carriage** Gicht-fahrzeug
*n*, -wagen *m* ~ **charging crane** Gichtbühnen-
kran *m* ~-**charging gear** Gichtaufzug *m* ~
**coiler** Haspelofen *m* ~ **coke** Groß-, Schmiede-,
Zechen-koks *m* ~ **column** Tragkranzsäule *f* ~
**construction** Ofenkonstruktion *f* ~-**control test**
Einschmelz-, Vor-probe *f* ~ **cooling tube** Seiten-
wandkühlrohr *n* (Kessel) ~ **crew** Ofenkolonne *f*
~ **crucible** Ofengefäß *n* ~ **design** Ofenkonstruktion *f* ~ **door** Feuertür *f* ~ **drying** Ofentrocknung *f* ~ **efficiency** Ofen-ausnutzung *f*, -leistung
*f*, -wirkungsgrad *m* ~ **end of tuyère** Formrüssel
*m* ~ **expert** Ofenfachmann *m* ~-**filling counter**
Gichtenzähler *m* ~ **flue** Ofenkanal *m* ~ **front**
Feuergeschränk *n* ~ **gases** Feuergase *pl* ~ **grate**
Feuerungs-, Ofen-rost *m* ~ **hearth** Ofenherd *m*
~-**heat capacity** Feuerraumbelastung *f* ~-**hoist**
Gichtaufzug *m* ~-**hoist carriage** Gichtwagen *m*
~-**hoisting machine** Gichtaufzugwinde *f* ~
~-**installation** Ofenanlage *f* ~ **lead** Herdblei *n* ~
**life** Ofenhaltbarkeit *f* ~ **lines** Ofenprofil *n* ~
**lining** Ofen-auskleidung *f*, -futter *n*, -zustellung
*f* ~ **man** Ofenmann *m* ~ **maneuver** Ofenarbeit *f*
~ **operation** Ofen-arbeit *f*, -betrieb *m* ~ **operator**
Ofenwärter *m*, Öfner *m* ~ **output** Ofen-erzeu-
gung *f*, -leistung *f* ~ **plant** Ofenbetrieb *m* ~
**plates** Feuerblech *n* ~ **practice** Ofenbetrieb *m* ~
**pressure** Ofenzug *m* ~ **proper** Herdkörper *m* ~
**rake** Feuerbrücke *f* ~ **refining** Raffination *f* im
Schmelzfluß ~ **residue** Feuerungsrückstände *pl*
~ **roof** Feuerraumdecke *f*, Ofenabdeckung *f* ~
**shaft** Ofenschacht *m* ~ **shell** Ofenmantel *m* ~
**soot** Ofenruß *m* ~ **sow** Bodensau *f*, Ofenwolf *m*
~ **stack** Ofenschacht *m* ~ **structual steel** Verankerung *f* des Ofens ~ **tapped** Hütte *f* gezogen ~ **throat** Gichtöffnung *f*, Ofengicht *f*
~-**throat stopper** Gichtpfropfen *m* ~ **tilting**
**control** Ofenkippsteuerung *f* ~ **tool** Schürgerät
*n* ~ **tools** Schürzeug *n* ~ **top** Gichtöffnung *f*
~-**top bell** Gichtglocke *f* ~-**top distributing**
**gear (or top distributor)** Gichtverschluß *m*
~-**top hopper** Gichttrichter *m* ~ **water walls**
Feuerraumkühlsystem *n* ~ **well** Ofensack *m* ~
**work** Ofenarbeit *f* ~ **working** Ofengang *m*
**furnish, to** ~ ausfüllen, einräumen, einrichten,
erteilen, gewähren, versehen to ~ **bail**
**(or bond)** Kaution *f* stellen to ~ **the demand**
Bedarf *m* decken to ~ **evidence** tatsächliche
Unterlagen *pl* liefern to ~ **with glass** die Scheiben *pl* in Kitt verglasen to ~ **with a handle**
schäften to ~ **with points** zacken to ~ **security**
Kaution *f* stellen to ~ **with a stock** schäften
**furnish** Eintrag *m* (paper mfg.)
**furnishing** Einrichtung *f* ~ **with boards** Beboh-

lung *f* **the** ~ **of provisions** Proviantlieferung *f* ~
**with a stock** Schäftung *f*
**furnishing,** ~ **fabrics** Dekorationsstoffe *pl* ~ **pan**
Leerbecher *m*
**furniture** Ausrüstung *f*, Möbel-, Zimmer-aus-
stattung *f* ~ **set** Garnitur *f*
**furring** Futterholz *n*
**furrow, to** ~ durchfurchen, riffeln
**furrow** Ackerfurche *f*, Bodenfalte *f*, Furche *f*,
Nut *f*, Nute *f*, Riefe *f*, Rille *f*, Rinne *f*, Schleifrille *f*, Schram *m*
**furrowed** furchig
**furrowing** Furchung *f*
**further, to** ~ fördern
**further** noch, weiter ~ **development** Weiterbildung *f* ~ **end** hinteres Ende *n* ~ **particulars**
weitere Aufgaben *pl*
**furtherance** Vorschub *m*
**furthering** Förderung *f*
**fusain** Faser *f*, Steinkohle *f*
**fuse, to** ~ anschmelzen, aufschmelzen, aufschweißen, auslaugen, durchbrennen, (Sicherung) durchschmelzen, erweichen, schmelzen,
schmieden, schweißen, sichern, sintern, rake-
tenartig verbrennen, verschmelzen, zünden,
zusammenschmelzen to ~ **off** abschmelzen,
ausschmelzen, niederschmelzen to ~ **a shell**
eine Granate *f* scharf machen to ~ **together**
zusammensintern
**fuse** Abschmelzsicherung *f*, Brander *m* (electr.),
Brennzünder *m*, Folgezeigerempfänger *m*,
Halm *m*, Lunte *f*, Schmelzeinsatz *m*, Schmelzsicherung *f*, Sprengzünder *m*, Überspannungssicherung *f*, Verschlußschraube *f*, Zünder *m*,
Zündrohr *n*, Zündschnur *f*, Zündung *f*
**fuse,** ~ **for commutating device** Sicherung *f* für
Umschalteeinrichtung *f* ~ **for control leads**
Sicherung *f* für Steuerleitung ~ **for distance**
Fernzünder *m* ~ **for pilot relay** Sicherung *f* für
Überwachungsrelais ~ **for power supply**
**connection** Sicherung *f* für Kraftanschluß ~ **for**
**stopping and starting device** Sicherung *f* für Ab-
stell- und Anlaßgerät ~ **for sychronization**
Synchronisierungssicherung *f* ~ **for three-phase**
**A. C. excitation** Sicherung *f* für Drehstromer-
regung ~ **for use at sea** Seezünder *m*
**fuse,** ~ **alarm** Sicherungsüberwachungseinrich-
tung *f* ~-**alarming gear** Blindscharfgestänge *n*
~ **bases** Sicherungsträger *pl* ~ **blow lamp**
Sicherungssignallampe *f* ~ **board** Sicherungsbrett, -gestell *n*, -tafel *f* ~ **body** Zünder-gehäuse
*n*, -hülse *f* ~ **box** Sicherungs-dose *f*, -kasten *m*
~ **cap** Deplattung *f*, Verschluß-, Vorschuß-,
Zünder-kappe *f* ~-**carrier** Sicherungshalter *m*
~ **casing** Zündergehäuse *n* ~ **check** Zünder-
kontrolle *f* ~ **circuit** Zündernetz *n* ~ **composi-**
**tion** Zünderfüllmasse *f* ~ **cord** (hand grenade)
Abreißkopf *m*, Leitfeuer *n*, Zündschnur *f* ~
**cover** Zünder-gehäuse *n*, -hülse *f* ~ **cutout**
Stromsicherung *f* ~ **date** Zünderanzeige *f* ~
**deviation** Seitenablage *f* ~ **element** Sicherungselement *n* ~ **equipment** Sicherungsanlage *f* ~
**fitting** Sicherungsleiste *f* ~ **holder** Einsatzschuh
*m* (electr.) ~ **hole** Mundloch *n* ~ **hole bush**
Mundlochbuchse *f* ~ **indicator** Brennlängen-
scheibe *f* ~ **insulator** Isolator *m* mit eingebauter
Sicherung, Sicherungstrenner *m* ~ **key** Zünderschlüssel *m* ~ **length** Brennzeit *f* des Zünders

~ **lighter** Zündschnuranzünder *m* ~ **link** Schmelzeinsatz *m*, Sicherungsdraht *m* ~ **mounting** Sicherungssockel *m* ~ **panel** Sicherungsbrett *n*, -tafel *f* ~ **plug** Sicherungsstöpsel *m* ~ **rack** Sicherungsgestell *n* ~ **range** Wirkungsbereich *m* ~ **receptable** Sicherungselement *n* ~ **safety cap** Sperrhülse *f* ~ **safety pin** Vorstrecker *m* ~ **scale** Zündereinteilung *f* ~-**setter support** Zünderstellsitz *m* ~ **setting** Brennlänge *f*, Schußzünderstellung *f*, Zünder-einstellung *f*, -laufzeit *f*, -stellung *f* ~-**setting ring** Zünderring *m* ~ **shelf** Sicherungsschiene *f* ~ **steam** Zünderstiel *m* ~ **striker** Nadel *f* ~ **strip** Abschmelzstreifen *m*, Sicherungs-leiste *f*, -streifen *m* ~ **switchbox** Zünderschaltkasten *m* ~-**time measuring device** Zeitdrucker *m* ~ **tongs** Sicherungszange *f* ~ **train** Brennzünder *m* ~ **welding** Schmelzschweißung *f* ~ **wire** Abschmelz-, Schmelz-, Schmelzsicherungs-, Sicherungsdraht *m* ~ **wire insert** Sicherungseinsatz *m*

**fused** (coil) durchgebrannt, feuerflüssig, geschmolzen, geschweißt, zündfertig ~ **alumina** geschmolzenes Aluminiumoxyd ~ **area** Schmelzstelle *f* ~ **hearth bottom** Sinterherd *m* ~-**in wire** Einschmelzdraht *m* ~ **junction** Rekristallisationsschicht *f* ~ **mass** Schmelze *f*, Schmelzfluß *m* ~ **oil** Fuselöl *n* ~ **quarts** geschmolzener Quarz *m* ~ **salt bath** Salzschmelze *f* ~ **silica** Hartfeuerporzellan *n* ~ **zone** Schweißlinse *f*

**fuselage** Flugzeugrumpf *m*, Rumpf, Rumpfwerk *n* ~ **of circular section** kreisrunder Rumpf *m* ~ **of oval section** ovaler Rumpf *m* ~ **of rectangular section** rechteckiger Rumpf *m* ~ **of steel-tubing framework** Stahlrohrrumpf *m*

**fuselage, ~ accessories** zellenseitiges Gerät *n* ~ **aft section** Rumpf-heck *n*, -hinterteil *n* ~ **bay** Rumpffeld *n* ~ **bracing strut** Rumpfverstärkung *f* ~ **break** Rumpftrennstelle *f* ~ **compass** Rumpfkompaß *m* ~ **covering** Rumpf-beplankung *f*, -verkleidung *f* ~-**covering fabric** Rumpfbespannung *f* ~ **cutout section** Rumpfaussparung *f* ~ **fitting** Rumpfbeschlag *m* ~ **forward section** Rumpfvorderteil *m* ~ **frame** Rumpfgerüst *n* ~ **framework** Rumpf-fachwerk *n*, -gerippe *n* ~ **heater** Rumpfbeheizungskörper *m* ~ **installation** Rumpfeinbau *m* ~ **joint** Knotenpunkt *m* ~ **mid section** Rumpfmittelteil *n* ~ **model** Rumpfmodell *n* ~ **nose (section)** Rumpfbug *m*, -vorderteil *m* ~ **radiator** Rumpfkühler *m* ~ **recess** Rumpfaussparung *f* ~ **shower** Zellendusche *f* ~ **side** Rumpf-seite *f*, -seitenwand *f* ~ **skin** Rumpfbeplankung *f* ~ **splice** Rumpftrennstelle *f* ~ **tank** Rumpfbehälter *m*

**fuses** Sicherungsmaterial *n*

**fusetron** Elektronensicherung *f*

**fusibility** Fließfähigkeit *f*, Schmelzbarkeit *f* ~ **curve** Zustandsdiagramm *n*

**fusible** Schmelzsicherung *f*; schmelzbar, schmelzflüssig, schmelzig ~ **alloy** schmelzbare Legierung *f*, Schmelzlegierung *f* ~ **cone** Schmelzkegel *m* ~ **cutout** Abschmelzsicherung *f* ~ **earth** Schmelzerde *f* ~ **glass** Emailglas *n* ~ **metal** Schnellot *n* ~ **plug** Bleisicherung *f*, Schmelz-stift *m*, -pfropfen *m* ~ **strip** Schmelzstreifen *m*

**fusiform** spindelförmig

**fusing** Abschmelzen *n*, Ansprechen *n*, Aufschmelzen *n*, Durchbrennen *n*, Durchschmelzen *n*, Erweichung *f*, Flüssigwerden *n*, Schmelzen *n*, Sintern *n*, Sinterung *f* ~ **of a fuse** Abschmelzen *n* einer Sicherung

**fusing, ~ battery** Zündmagnet *m* (Triebwerk) ~ **circuit switch** Zündstromklappenschalter *m* ~ **current** Abschmelz-strom, -stromstärke *f*, Schmelzstrom *m* ~ **disk** Reibsäge *f* ~ **point** Schmelzpunkt *m* ~-**point pyrometry** Schmelzpyrometrie *f* ~ **temperature** Erweichungstemperatur *f* ~ **time** Abschmelzdauer *f* ~ **welding** Abschmelzschweißung *f*

**fusion,** Anschluß *m*, Autogene *f*, Bindung *f*, Durchschmelzung *f*, Fluß *m*, Flüssigwerden *n*, Fusion *f*, Schmelz *m*, Schmelze *f*, Schmelzen *n*, Schmelzfluß *m*, Schmelzung *f*, Vereinigung *f*, Verschmelzung *f*, Zusammenfassung *f*, Zusammenschluß *f* ~ **of rays** Strahlenvereinigung *f*

**fusion, ~ adhesive** Schmelzkleber *m* ~ **cones** Brennkegel *m* ~ **curve** Schmelz-kurve *f*, -linie *f*, Schmelzungslinie *f* ~ **electrolysis** Schmelzflußelektrolyse ~ **frequency** Beruhigungs-, Flimmer-frequenz *f* ~ **kettle** Schmelzkessel *m* ~ **point** Erweichungs-, Fusions-, Schmelz-punkt *m*, Schmelzwärmegrad *m* ~ **rate** Einbindungsgrad *m* ~ **temperature** Schmelz-temperatur *f*, -wärmegrad *m* ~ **welding** Abschmelzverfahren *n*, Schmelz-schweißen *n*, -schweißung *f* ~ **zone** Schmelzzone *f*

**fusionable material** schmelzbares Material *n*

**fustic** (wood) Gelbholz *n*

**futtock** Auflanger *m*, Gabelholz *n*

**future** Folge *f*, kommende Zeit *f*, Zukunft *f* ~ **in the** ~ künftig ~ **bearing** Treffseitenwinkel *m* ~ **position** Treff-, Vorhalte-punkt *m* ~-**position triangle** Treff-, Treffer-dreieck *n* ~ **range** Treffentfernung *f* ~ **relative (or target) position** errechneter Treffpunkt *m*

**futures** Termingeschäft *n* ~ **trading** Terminhandel *m*

**fuzz, to** ~ abfasern

**fuzz** Flaum *m*, Franse *f*

**fuzziness** (picture) Bild-flachheit *f*, -flauheit *f*; Faserigkeit *f*; faserig

**fuzzy** Faserigkeit *f*; faserig, flau, unscharf (photo) ~ **heads** Braunkräusen *pl*

# G

**gabardine** Gabardine *f*
**gabbro** Gabbro *m*
**gabion** Korbbodenspule *f*, Schanzkorb *m*
**gable** Giebel *m* ~ **bottom** Sattelboden *m* ~ **roof**
Giebeldach *n* ~ **wall** Giebel-mauer *f*, -wand *f*
**gabled window** Giebelfenster *n*
**gad** (tool, textiles) Bohrnadel *f*, Fimmel *m*
**gadolinite** Gadolinerde *f*, Ytterbit *m*
**gadolinium,** ~ **nitrate** Gadoliniumnitrat *n* ~
**oxide** Gadolinerde *f*
**gaffel** Gaffel *f* ~ **of climbers** Klettersporn *m* ~
**end** Gaffelnock *f* ~**-head rope** Gaffelliek *n*
~**-jaw** Gaffelklaue *f* ~**-jaw rope** Gaffelrack *n* ~
**peak** Gaffelnock *f*
**gag** Knebel *m* ~ **press** Richtpresse *f*, Stempel-
richtpresse *f*
**gagger** Einkerbung *f*, Sandhaken *m*
**gahnite** Automolit *m*
**gain, to** ~ einbringen, erzielen, gewinnen **to** ~ **a**
**basis** Anhalt *m* gewinnen **to** ~ **a foothold** festsetzen **to** ~ **by force** erkämpfen, erzwingen **to**
~ **a foundation** Anhalt *m* gewinnen **to** ~ **ground**
Boden *m* gewinnen **to** ~ **height** Höhe *f* gewinnen (aviat.) **to** ~ **information** erkunden **to** ~ **on**
aufholen **to** ~ **speed** Fahrt *f* aufnehmen **to** ~
**strength** erstarken
**gain** Ausbeute *f*, Brust *f*, Dämpfungsverminderung *f*, Energiegewinn *m*, Gewinn *m* (antenna), (of fuse) Mundlochfutter *n*, Pegel *m*,
Verstärkung *f*, Verstärkungsgrad *m*, (repeater)
Verstärkungsüberschuß *m*, Vorteil *m* ~ **in**
**altitude** Höhengewinn *m* ~ **of energy** Energiezuwachs *m* ~ **of input stage** Verstärkungsgrad
*m* der Eingangsstufe ~ **and modulation width**
Verstärkungsgrad *m* und Modulationsbreite *f*
~ **in sensitivity** Empfindlichkeitsgewinn *m* ~ **in**
**weight** Gewichtszunahme *f*
**gain,** ~ **adjustment** Verstärkungsregelung *f* ~
**amplification** Vorverstärkung *f* ~**-band merit**
Bandbreite *f* bei Verstärkung ~ **checking** Verstärkungsprüfung *f* ~ **control** Verstärkungs-
reglung *f* ~ **controller** Schwächungswiderstand
*m*, Verstärkungsregler *m* ~ **equation** Ausbeute-
gleichung *f* ~ **factor** Gewinn-, Verstärkungsfaktor *m* ~**-frequency curve** Entdämpfungs-
frequenzkurve *f* ~**-measuring device** ~**-measuring set** Ver-
särkungsmeßgerät *n* ~ **monitor** Mithorver-
stärker *m* ~ **regulator** Schwächungswiderstand
*m* ~ **sensitivity control** selektiver Abschwächer
*m* ~ **set** Verstärkungs-meßeinrichtung *f*,
-messer *m*, -zeiger *m* ~ **spoiler** Verstärkungs-
begrenzer *m*, -unterdrücker *m* ~ **time control**
(G.T.C.) Nahantwortdämpfer *m* (sec. rdr.) ~
**tracing** Verstärkungsprüfung *f* (stufenweise)
**gainfully employed** erwerbstätig
**gaining,** ~ **of an austenitic structure** Austempe-
rierung *f* ~ **of ground** Raumgewinn *m*
**gait** Gang *m*, Gangart *f*
**gaiter** Webstuhlvorrichter *m*
**galactic** Milchstraßen ...
**galactometer** Galaktometer *n*

**galactose** Laktoglukose *f*
**galacturonic acid** Galakturonsäure *f*
**galalith** Galalith *n*
**galaxy** Milchstraße *f*
**gale** Kühle *f*, Mutung *f*, starker Sturm *m*,
Sturmwind *m* ~ **warning** Sturmwarnung *f*
**galena** Bleiglanz *m*, Galenit *n*, Glanz-, Grau-erz
*n* ~ **detector** Bleiglanz-, Röhren-detektor *m* ~
**crystal** Bleiglanzkristall *m* ~ **receiver** Detektorenempfänger *m*, Röhrenempfänger *m*
**galenical preparation** galenisches Präparat *n*
**galenobismutite** Selenbleiwismutglanz *m*
**galipot** Galipot *n*
**gall, to** ~ durch Reibung *f* abnutzen, beschädigen, fräsen, (wires, ropes etc.) frottieren
**Gall,** ~ **chain** Gallsche Kette *f* ~**-chain-link**
**belting (or steel bushed chain)** Gallsche Gelenkkette *f*
**gall** Abnutzung *f* durch Reibung (aviat.), Galle *f*
~**extract** Gallapfelauszug *m* ~ **nut** Gallapfel *m*,
Knopper *m* ~ **soap** Gallenseife *f* ~ **tannin**
Gallengerbstoff *m*
**galleries** Förderstollen *m pl*
**gallery** Bühne *f*, Empore *f*, Galerie *f*, (of a mine)
Gezeugstrecke *f*, Grund- und Sohlenstrecke *f*,
Kabeltunnel *m*, Minengang *m*, Sohlenstrecke *f*,
Stollen *m* (min.), Strecke *f*, Tunnel *m* ~ **con-**
**struction** Stollenbau *m* ~ **dugout** Korridor-
unterstand *m* ~ **frame** Türgerüst *n* ~ **ports**
Gallerieanordnung *f* ~ **rifle** Zimmerstutzen *m*
**galley** Bürsten- oder Fahnenabzug *m* (print.),
(on ship) Kombüse *f*, Küche *f*, Setzschiff
*n* ~ **press** Abziehpresse *f* ~**-slug** Nummern-
reglette *f*
**gallic acid** Gallapfelsäure *f*, Gallussäure *f*
**galling** Fressen *n* der Oberfläche (durch Druck),
Grübchen *n* (aviat.), Kerbe *f*, Passungsrost *m*,
(of engines) Verschweißerscheinung *f*, (Narben)
im Unterbau *m* ~ **engines** Freßerscheinung *f*
**gallium** Gallium *n* ~ **chloride** Galliumchlorid *n*
**gallnut** Gallapfel *m*
**gallon** Gallone *f*
**galloon** Borte *f* ~ **maker** Bortenwirker *m*
**gallop** Galopp *m*
**galloping** Galoppieren *n*
**gallotannic acid** Gallapfelgerbsäure *f*
**gallows** Deckelstuhl *m* (print.), Galgen *m*; Sattel
*m*
**Galton pipe** Galtonpfeife *f*
**galvanic,** ~ **(al)** galvanisch ~ **battery (or cell)**
galvanisches Element *n*, Primärelement *n* ~
**coloring** Galvanochromie *f* ~ **coupling** galvanische Kopplung *f* ~ **etching** Galvanokaustik
*f* ~ **metallization** galvanische Metallisierung *f*
~ **plant** galvanische Anlage *f*
**galvanism** Galvanismus *m*
**galvanization** Galvanisation *f*, Galvanisierung *f*,
Verzinkung *f*
**galvanize, to** ~ (hot) feuerverzinken, galvanisieren, verzinken
**galvanized** galvanisiert, verzinkt ~ **by the spray-**
**ing method** spritzverzinkt

galvanized, ~ iron galvanisiertes Eisen n ~-iron wire galvanisierter oder verzinkter Eisendraht m ~ pipe galvanisiertes Rohr n ~ sheet metal verzinktes Blech n ~ steel plate verzinktes Eisenblech n ~ wire verzinkter Draht m
galvanizer Galvaniseur m, Verzinker m
galvanizing Galvanisieren n, Verzinken n, Verzinkung f ~ bath Verzinkungs-bad n, -wanne f ~ furnace Verzinkungsofen m ~ kettle Zinnkessel m ~ plant Galvanisier-anlage f, -anstalt f, Verzinkerei f, Verzinkungsanlage f ~ pot Zinnkessel m ~ process Galvanisierprozeß m ~ sheet Zinkblech n
galvanomagnetic galvanomagnetisch
galvanometer Galvanometer m, Strom-anzeiger m, -messer m, -zeiger m ~ with moving magnet Nadelgalvanometer m
galvanometer, ~ circuit Galvanometerkreis m ~ coil Galvanometerspule f ~ constant Eichzahl f, Galvanometerkonstante f ~ indicator Galvanometerablesegerät n ~ mirror Galvanometerspiegel m ~ recorder Lichtbahn m, Registriergalvanometer n ~ shunt Galvanometernebenschluß m
galvanoplastic galvanoplastisch ~ (art) Galvanoplastik f ~ copper bath Kupferplastikbad n
galvanoplastics Galvanisieren n, Galvanostegie f, Galvanotechnik f
galvanoscope Galvanoskop n
gamete Gamet m, Keimzelle f
gamma Gamma n ~ background natürliche Gammastrahlung f ~ correction Dynamikentzerrung f, Gammaregelung f ~ counter Gammazählrohr n ~ function Gammafunktion f
gammagraphy Gamma-Autoradiografie f
gamma, ~ iron Austenit n, Gammaeisen n too low a ~ zu flach ~ quantum Gamma-Foton n ~ radiation Gammaaktivität f ~-ray capsule Gammakapsel f ~-ray photon Gammafoton n ~-ray source strength Intensität f einer Gammastrahlenquelle f ~ rays Gammastrahlen pl ~ samples Gammapräparate pl ~ spectrometry Gammaspektrometrie f ~ spectroscopy Gammaspektroskopie f ~-type antenna L-Antenne f
gang, to ~ kuppeln, (condensers) verriegeln
gang Arbeiterkolonne f, Belegschaft f, Bergart f, Drittel n, Horde f, (of men) Kolonne f, Mannschaft f, Rotte f, Satz m, Schicht f, Trupp m, (aviat. tools) Werkzeugsatz m ~ punches Lochstempelsatz m, Stempelsatz m ~ of workers Arbeitsdrittel n ~ of workmen Arbeiterrotte f
gang, ~ blanking tool Mehrfachausbauwerkzeug n ~ board Laufpaß m ~ boss Vorarbeiter m ~ condenser (or capacitor) Gangkondensator m, gekuppelter Kondensator m, Gleichgang-, Mehrfach-, Mehrfachdreh-kondensator m ~ cutter Salzfräser m ~ cutters zusammengesetzter Fräser m, Fräsersatz m, Gruppenfräser m ~ drill Gruppenbohrmaschine f ~ plank Landplanke f, Lauf-brücke f, -planke f ~ plow mehrschariger Pflug m ~ press Stufenpresse f ~ press wheel Druckrolle f (agr.) ~ punch Schnellstanzer m ~ saw Gattersäge f ~ summary punch Summenstanzer m ~ switch Gruppenschalter m, Schalter m mit mehreren Schaltebenen ~ tuning capacitor Mehrfachabstimmkondensator m ~ type milling cutters Gruppenfräser m

gangway Durchgang m, Fahrgaststeg m, (ladder) Fallreep n, Gang m, Laufbrücke f, Laufgang m, Laufsteg m, Mittelgang m ~ bellows Faltenbalg m
ganged gemeinschaftlich betrieben, gleichlaufend, mechanisch gekoppelt it is ~ es befindet sich im Gleichlauf m ~ capacitor Mehrgangkondensator m ~ condenser gleichlaufender Kondensator m, mehrfacher Kondensator m, Kondensatorwanne f, Mehrgangkondensator m ~ condensers gleichlaufende Kondensatoren pl, Kondensatoren pl in Reihe ~ control mechanische blockierte Steuerung f ~ cutters zusammengesetzter Fräser m ~ potentiometer Tandempotentiometer m
ganging mechanische Kupplung f oder Kopplung f, Gleichlauf m
gangrene Holzbrand m, Wundbrand m
gangue Bergart f, Gangart f, Ganggestein n the ~ changes das Gestein setzt ab
ganister Flickmasse f, Ganister m, Kalkdinas m ~ brick Kalkdinasstein m
gantries Beleuchtungsbrücke f (film)
gantry Bockkran m, Krangerüst n, (of a crane) Portal n ~ crane Bock-, Gerüst-, Portal-, Überlade-kran m ~ girder Torgerüst n (Kran) ~ span Stützentfernung f ~ travel gear Portalfahrwerk n ~ travel wheels Portallaufräder pl
gap, to ~ verhacken, verhauen, versetzen
gap Ausladung f, Fehlstelle f, Fuge f, (in aerophotographic plans) Klaffung f, Kröpfung f, Leerstelle f, Loch n, (in wood) Lücke f, Lückensynchronisierungsverfahren n, (in a cylinder) Ringstoß m, Riß m, Ritze f, Spalt m, Stoß m des Kolbenringes, toter Raum m (photo), Zwischenraum m ~ in a cylinder Stoßspielkolben m ~ in the map Lücke f in der Karte oder Aufnahme ~ between sheets Blechabstand m ~ in the survey Lücke f in der Karte oder Aufnahme ~ of tooth Zahnlücke f
gap, ~ bed ausgespartes oder gekröpftes Bett n (Drehbank) ~ bridging Brückenbildung f ~ coding Pausenkodierung f ~ cut Kröpfschnitt m ~ filler Diagrammausfüllung f ~ filling adhesive Fugenkitt m ~ gauge Rachenlehre f ~ length Spaltbreite f (acoust.) ~ piece drilling fixture Einsatzbrückenbohrvorrichtung f ~ piece filler block Einsatzbrücke f (Drehkran) ~ range Unterbrechungsbereich m ~ scatter Spaltlagenstreuung f (tape rec.) ~ separation Elektrodenabstand m, Spaltbreite f ~ test Überschlagsprobe f ~-type distributor Verteiler m mit springenden Funken ~ undershaft gekröpfte Unterwelle f ~ width Spaltbreite f
gape, to ~ klaffen
gaping Klaffen n
gapless traverse chain fortlaufender Polygonzug m
garage, to ~ Kraftfahrzeuge pl unterstellen
garage Einstand m für Kraftwagen, Einstellraum m, Garage f, Halle f, Kraftwagenwerkstatt f, Schuppen m, Wageneinstellhalle f, Wagenhalle f, Wagenraum m
garancine Garanzin n
garbage Müll n, Schund m ~-and refuse-burning furnace Müll- und Abfallverbrennungsofen m ~-collecting cart Abfuhrwagen m (Kehricht) ~ truck Müllast(kraft)wagen m

**garbenschiefer** (spotted rock in which the spots resemble caraway seeds) Garbenschiefer *m*
**garble** Fremdstörung *f* (rdo), Verstümmelung *f* ~ **table** Entstümmelungstabelle *f*
**garbled,** ~ **message** verstümmelte Fernmeldung *f* ~ **telephony** Geheimtelefonie *f*
**garbling** sinnentstellend; Verwirrung *f* (sec. rdr.)
**garden wall bond** gotischer Verband *m*
**gardener** Gärtner *m*
**gardening** Gartenbau *m*, Gärtnerei *f* ~ **knife** Hippe *f*, Saß *m* ~ **tools** Gartengeräte *pl*
**gargle** Geschwindigkeitsschwankungen *pl*
**gargoyle** Abtraufe *f* der Dachrinne, Wasserspeier *m*
**garland** Girlande *f*, bunter Lappen *m*
**garnet** Granat *m*, Granatstein *m*, Schleifmittel *n* aus komplexen Eisen- und Aluminiumsilikaten ~ **carbuncle** Karfunkelstein *m* ~ **cutter** Granatschleifer *m* ~ **shade** Granatton *m* ~-**studded articles (or work)** Granatwaren *pl*
**garnierite** Garnierit *m*
**garnish, to** ~ ausstaffieren
**garnish** bunter Lappen *m*, Umlage *f* ~ **plates** Zierbilder *pl*
**garniture** Beschlag *m*, Garnitur *f*
**garret,** ~ **door** Bodentür *f* ~ (**looping-rod**) **mill** Garretstraße *f* ~ **staircase** Bodentreppe *f* ~ **story** Boden-, Dach-geschoß *n* ~ **window** Luke *f*
**garrison, to** ~ besetzen, rechnen
**garrison** Besatzung *f*, Garnison *f*, Standort *m*
**garter spring** Ringbandfeder *f*
**gas, to** ~ (a balloon) füllen, gasen, gasieren **to** ~ **the thread** den Faden *m* brennen
**gas** Gas *n* ~ **of high calorific value** heizkräftiges Gas *n* ~ **from low-temperature distillation** Schwelgas *n* ~ **from roasting** Röstgas *n*
**gas,** ~ **alarm** Gasalarm *m*, Grubengasanzeiger *m* (min.) ~-**alarm siren** Gasalarmsirene *f* ~ **alarm signal** Gasalarmzeichen *n* ~ **alert** Gasbereitschaft *f* ~ **amplification factor** Gasverstärkungsfaktor *m* ~ **analysis** Gas-analyse *f*, -untersuchung *f* ~ **analyzer** Gasprüfer *m* ~ **anchor** Gasanker *m* ~ **appliances** Gasgeräte *pl* ~-**atomizing oil burner** Ölbrenner *m* für Dampfzerstäubung
**gas-bag** (balloon) Gashülle *f*, Gas-sack *m*, -zelle *f*, Traggaszelle *f* ~ **envelope** Zellenmantel *m* ~ **fabric** Zellenstoff *m* ~ **suspension** Zellenaufhängung *f*
**gas,** ~ **balance** Gaswaage *f* ~ **ballast regulator** Gasballaststeuerung *f* ~ **barrier** Gassperre *f* ~ **bearing gyro** gasgelagerter Kreisel *m* ~ **bell** Gasflaschenglocke *f*, (small holder) Gasglocke *f* ~ **blanket** Gasschutzvorhang *m* ~-**blast burner** Gasgebläselampe *f* ~-**blast switch** Druckgasschalter *m* ~ **blow** Durchblasen *n* des Kolbens ~ **blower** Gasgebläse *n* ~-**blowing engine** Gasmaschinengebläse *n* ~ **blowpipe** Löt(blei)brenner *m* ~ **boiler** Gaskessel *m* ~ **bomb** Druckgasflasche *f*, Gasbombe *f* ~ **booster** Gaszusatzmaschine *f* ~ **bottle** Gasentbindungsflasche *f*, Gasflasche *f* ~ **bracket** Gasarm *m* ~ **breather** Gasmaske *f* ~ **bubble** Gasblase *f* ~ **bubbles** Gasbläschen *pl* ~ **buoy** Gastonne *f* ~ **burette** Gasbürette *f* ~ **burner** Gasbrenner *m* ~ **can** Kanister *m* ~ **candle** Giftnebelkerze *f*, -rauchkerze *f* ~ **capacity** Gasinhalt *m* ~ **capsule** Riechwürfel *m* ~-**carburetion**

**oil** Vergasungsöl *n* ~-**carburized** oberflächengehärtet ~ **casehardening process** Druckgaseinsatzverfahren *n*, Gaseinsatzhärtung *f* ~ **cavity** Gasblase *f* ~-**cell alarm** (airship) Prallanzeiger *m* ~-**cell fabric** Zellenstoff *m* ~ **chamber** Gas-kammer *f*, -raum *m* ~-**chamber kiln** Gaskammerofen *m* ~-**chamber test** Gasraum-probe *f*, -prüfung *f* ~-**changing process** Gaswechselarbeit *f* ~ **check** Gasabdichtung *f*, Liderung *f* ~-**check pad** Kissen *n* ~-**checker chamber** Gaskammer *f* ~ **cleaner** Gasreiniger *m* ~ **cleaning** Glasreinigung *f* ~-**cleaning installation (or plant)** Gasreinigungsanlage *f* ~ **cleanup** Gasaustreibung *f* ~ **clothing** Gasschutzanzug *m* ~ **cloud** Giftnebelwolke *f* ~ **coal** Gasförderkohle *f*, (highgrade bituminous coal) Gaskohle *f*, gasreiche Kohle *f* ~ **cock** Gashahn *m* ~ **coke** Gaskoks *m* ~ **collecting main** Gassammelleitung *f* ~-**collecting tube** Gassammelrohr *n* ~ **collector** Gasfang *m* ~ **compound** Gasgemisch *n* ~-**compression pump** Gasladepumpe *f* ~ **compressor** Gasverdichter *m* ~ **concentration** Gas-ballung *f*, -konzentration *f*, -konzentrierung *f*, Kampfstoffgehalt *m* ~ **conditioning** Gasreinigung *f* ~ **conduit** Gaskanal *m*, -leitung *f* ~ **constant** Gaskonstante *f* ~ **containing** gasführend ~-**contaminated surface** gasbehaftete Oberfläche *f* ~ **contamination** Gasvergiftung *f* ~ **content** Gas-füllung *f*, -gehalt *m*, -haushalt *m*, -inhalt *m* ~-**content tube** gasgefüllte Röhre *f* ~ **control** Gassteuerantrieb *m* ~ **cure** Gasvulkanisation *f* ~ **current** Gas-strom *m*, -strömung *f*; Ionenstrom *m* ~ **curtain** Gasvorhang *m* ~ **cushion** Gaskissen *n* ~ **cut** Gasbrennschnitt *m* ~ **cutting** Brennschneiden *n* ~ **cylinder** Blasflasche *f*, Druckluftflasche *f*, Gas-bombe *f*, -flasche *f*, (recoil) Luftbehälter *m* ~-**cylinder manifold** Füllstutzen *m* ~-**cylinder nozzle** Gasdüse *f* ~ **delivery** Gasabgang *m* ~-**density gauge (or meter)** Gasdichtemesser *m* ~ **desulfurisation plant** Gasentschwefler *m* ~-**detecting service** Gaserkennungsdienst *m* ~ **detection** Abspüren *n* ~-**detection apparatus** Gas-spürer *m*, -spürgerät *n* ~ **detection compound** Gasspürmittel *n* ~ **detection by smelling** Gasspüren *n* ~ **detection squad** Gasspürtrupp *m* ~ **detector** Gaserkennungsmittel *n* ~-**detector kit (or set)** Gasanzeiger *m* ~-**discharge** Gasentladung *f* ~-**discharge gauge** Glimmentladungsmanometer *m* ~-**discharge path** Gasentladungsstrecke *f* ~-**discharge relay** Gasentladungsrelais *n* ~-**discharge tube** Entladungsgefäß *n*, Gasentladungs-gefäß *n*, -rohr *n*, -röhre *f*, Ionenröhre *f* ~-**discharge valve** Gasentladungsrohr *n* ~ **disintegrator** Gas-reiniger *m*, -waschapparat *m*, -wäscher *m* ~-**distributing channel** Gasverteilungskanal *m* ~-**distributing pipe** Gasteilungsröhre *f* ~ **distribution** Gasführung *f* ~ **distributor** Gasteilungsröhre *f* ~ **door** Gasschutzvorhang *m* ~ **downcomer tube (or downtake tube)** Gichtgasabzugsrohr *n* ~ **drain** Gasabzug *m* ~-**draining vent** Gasabzugsöffnung *f* ~ **drilling derrick** Gasbohrturm *m* ~-**driven blowing engine** Gichtgasgebläse *n* ~-**driven engine blower** Gasgebläse *n* ~-**drying apparatus** Gastrockner *m* ~ **duct** Gaskanal *m* ~ **emitted in discharging** Abgasbrüden *n* ~ **engine** Gas-, Gaskraft-maschine *f*, Gasmotor *m*, gemischverdichtende Maschine *f* ~-engine blower Gichtgasgebläse *n* ~ engine driven blower Gasgebläsemaschine *f* ~ engine generator Gasdynamo *m* ~ engineer Gas-fachmann *m*, -techniker *m* ~ engineering Gas-fach *n*, -technik *f* ~ envelope Dampfhülle *f*, Gasmantel *m* ~ eruption Gasausbruch *m* ~ evolution Gasentwicklung *f* ~ exhaust Gasabführung *f* ~ exhauster Gassauger *m* ~ expeller Entgaser *m* ~ expulsion Gasaustreibung *f* ~ factor Gasgehalt *m* ~ feed blower Gasfördergebläse *n* ~ field Erdgasfeld *n*

gas-filled (of a lamp bulb) gasgefüllt ~ bulb gasgefüllte Birne *f* ~ housing Gas-gehäuse *n*, -haus *n* ~ lamp gasgefüllte Lampe *f* ~ photocell Gaszelle *f*, gasgefüllte Fotozelle *f* ~ photo-emission cell Gaszelle *f* ~ rectifier gasgefüllter Gleichrichter *m* ~ relay Gasentladungs-, Ionen-relais *n*, Ionenschalter *m* ~ tube gasgefüllte oder weiche Röhre *f* ~ valve Gasentladungsröhre *f*

gas, ~ filling Gasfüllung *f* ~ filter Gasfilter *m*, Gassieb *n*, (mask) Nebelfilter *m*

gas-fired, gasgefeuert ~ (air) furnace Gasflammofen *m* ~ calcining kiln Gasröstofen *m* ~ drying furnace (or oven) Gastrockenofen *m* ~ furnace gasgefeuerter Ofen *m* ~ furnace for hardening Gashärteofen *m* ~ heat treating furnace (or oven) Gaseinsatzofen *m* ~ muffle furnace gasgeheizter Muffelofen *m* ~ reverberatory furnace Gasflammofen *m*

gas, ~ firing Gasfeuerung *f* ~ fitter Gasinstallateur *m* ~ fitters Gasrohrzange *f* ~ fitting Gasarmatur *f*, -einrichtung *f* ~-fitting work Gasinstallationswerk *n* ~-flash test Gasflammenprobe *f* ~-flask container Gasflaschenbehälter *m* ~ flat-iron Gasbügeleisen *n* ~ flow Gas-fluß *m*, -richtung *f*, -strömung *f* ~-flow control Gasströmungsregelung *f* ~-flow counter Gasdurchflußzählrohr *n* ~ flue Gas-kanal *m*, -leitung *f*, -zuführungszug *m* ~-flue leading Gasleitung *f* ~ focusing Gas-konzentration *f*, -konzentrierung *f*, Ionenfokussierung *f* ~ fumes Gasschwaden *pl* ~ furnace Gasfeuerung *f*, (coke oven) Gasofen *m*, gasgefeuerter Ofen *m* ~ gangrene Gasbrand *m* ~ generation Gas-erzeugung *f*, -gewinnung *f* ~ generator Gas-entwickler *m*, -entwicklungsapparat *m*, -erzeuger *m*, -generator *m* ~ generator with suction and with pressure Druckgasgenerator *m* ~ gettering Gasaustreibung *f* ~ geyser Gasbadeofen *m* ~ given off during decomposition Faulgas *n* ~ globe produced by underwater explosion Schwadenblase *f* ~ goggles Gasbrille *f* ~ governor Gas-regler *m* -regulator *m* ~ grid system Gasversorgungssystem *n* ~ hammer Gashammer *m* ~ heater gasbetriebenes Heizgerät *n* ~ heating Gasheizung *f* ~-heating jacket Gasanwärmegehäuse *n* ~-heating operation Gasführung *f* ~ high-holder Gas-behälter *m*, -glocke *f*, Gasometer *m*, Gassammler *m* ~ hole Gasblase *f* ~ house Gaswerk *n* ~ house coke Gasanstaltskoks *m* ~ houseliquor leaching Ammoniaklaugung *f* ~ house retort Gasanstaltsretorte *f* ~ igniter Gasanzünder *m* ~ inlet valve Gaseinlaßventil *n* ~ intake Gaszuführungsrohr *n* ~-intake pipe Gasentnahmestutzen *m* ~ issue Gasabgang *m* ~ jet Gasdüse *f* ~ leakage Gasaustritt *m* ~ leases Gasschürfrecht

*n* ~ lift Gas-heberanlage *f*, -hub *m* ~ lighter Gasanzünder *m* ~ lighting Gas-beleuchtung *f*, -befeuerung *f* ~ line Gasleitung *f* ~-liquefying plant Gasverflüssigungsanlage *f* ~ liquor Gasreinigungswasser *n* ~ machine Gasmotor *m* ~ magnification Gasverstärkung *f* ~ main Gasleitung *f*, Gasleitungsrohr *n*, Hauptgasleitung *f* ~-main fiting Gasleitungsarmatur *f* ~ mains Gashauptleitungen *pl* ~-making plant Gasanstalt *f* ~ mantle Gasstrumpf *m* ~ mask Gasmaske *f* ~-mask canister Filterbüchse *f* ~-mask carrier Büchse *f* ~ mask-testing apparatus Gasmaskenprüfgerät *n* ~-measuring tube Gasmeßrohr *n* ~ meter Gas-meßapparat *m*, -messer *m*, -uhr *f*, -zähler *m* ~ mileage Kraftstoffverbrauch *m* (Kraftwagen) ~ mixture Gas-gemisch *n*, -mischung *f* ~ off-take Gasabgang *m* ~ off-take main Gasvorlage *f* ~ off-take tube Gichtgasabzugsrohr *n* ~ oil Gasöl *n* ~-oil ratio Gas-gehalt *m*, -inhalt *m*, -ölverhältnis *n* ~-operated gun Gasdrucklader *m* ~ outburst Gasausbruch *m* ~ outlet Gas-abführung *f*, -abzug *m*, -ausgang *m*, -austrittsöffnung *f* ~ output Gasausbeute *f* ~-oven construction Gasofenbau *m* ~ passage Gasdurchgang *m* ~ phototube Gaszelle *f*, gasgefüllte Fotozelle *f* ~ pipe Gas-leitung *f*, -leitungsrohr *n*, -rohr *n*, -röhre *f* ~-pipe handle Gasrohrstiel *m* ~-pipe stock Gaskluppe *f* ~-pipe tap Gasgewinde(schneide)bohrer *m* ~-pipe thread Gasgewinde *n* ~-pipe vice Gasrohrschraubstock *m* ~ pipette Gaspipette *f* ~ piping Gasrohrleitung *f* ~ plant Gas-anstalt *f*, -werk *n* ~ plant for longdistance supply Gasfernversorgungs-anlage *f*, -netz *n* ~ pliers Gas-rohrzange *f*, -zange *f*; Kugelzange *f* ~ pocket Gasblase *f*, -sumpf *m* ~-poisoned gaskrank ~ poisoning Gasvergiftung *f* ~ polarization Gaspolarisation *f* ~ port Gas-anschluß *m*, -öffnung *f*

gas-pressure, ~ gauge Gasdruckmeßgerät *n* ~ igniter at distance Gasdruckfernzünder *m* ~ indicator Prallanzeiger *m* ~ meter Gasdruckmesser *m* ~ reducing valve Gasdruckminderer *m* ~ regulator Gasdruckregler *m* ~ relief Gasentlastung *f* ~ tank Druckgastank *m* ~ test Gasdruckprüfung *f*

gas, ~ producer Gas-entwickler *m*, -erzeuger *m*, -generator *m*; Generatorgaserzeuger *m*, Kraftgaserzeuger *m* ~-producer efficiency Gaserzeugerwirkungsgrad *m* ~ producer pit Gaserzeugergrube *f* ~-producer plant Gaserzeugeranlage *f* ~-producer shell Gaserzeugermantel *m* ~-producer unit Gaserzeugeranlage *f* ~-producing plant Gaserzeugungsanlage *f* ~ production Gas-erzeugung *f*, -gewinnung *f* ~-proof gassicher, schlagwettersicher ~-proofed gasgeschützt ~-proofing Abdichtung *f* gegen Kampfgas ~ proofness Gasfestigkeit *f* ~-protected rescue gang Gasschutztruppe *f* (anti)~protection Gasschutz *m* ~-protection suit Gasanzug *m* ~ puddling Gasflammofenfrischen *n*, Gasfrischen *n* ~ pump Gaspumpe *f* ~ purifier Gas-reiniger *m*, -waschapparat *m*, -wäscher *m* ~ pyrometer Gaspyrometer *m* ~ quality Gasbeschaffenheit *f* ~-rate collector Gasgebührenerheber *m* ~ ratio Vakuumfaktor *m* ~ receiver Gasometer *m* ~-regenerating chamber Gaskammer *f* ~ **regenerator** Gaswärmespeicher *m* ~ **regulation** Gasführung ~ **regulator** Gas-einteiler *m*, -regler *m*, -regulator *m* ~ **regulator tube** Glimmstabilisator *m* ~ **relay** Gastriode *f* ~**-release method** Gasabblaseverfahren *n* ~ **residue** Gasreste *pl* ~ **retaining rubber film** gasundurchlässige Gummischicht *f* ~ **retort** Gasretorte *f* ~**-retort operation (or practice)** Gasretortenbetrieb *m* ~**-reversing valve** Gaswechselklappe *f* ~ **ring** (heater) Gaskranz *m* ~**-rolling mill engine** Gaswalzenzugmaschine *f* ~**-safety device** Gassicherheitsapparat *m* ~**-sample collector** Gassammelrohr *n* ~ **sampling tube** Gaspipette *f* ~ **scale** Gaswaage *f* ~ **scrubber** Gas-reiniger *m*, -waschapparat *m*, -wäscher *m*, -waschvorrichtung *f* ~ **scrubbing** Gaswäsche *f* ~ **seal** Gasverschluß *m* ~ **section** Gastrupp *m* ~ **sentry** Gasspürer *m* ~ **separation** Gasreinigung *f* ~ **separator** Gasabscheider *m* ~**-shaft** Entlüftungshutze *f* ~**-shell filler** Füllstoff *m* ~ **singeing machine** Gassenge *f* (Gassengmaschine) ~ **slack** Gas-grus *m*, -gruskohle *f* ~ **slide valve (or sluice valve)** Gasschieber *m* ~ **soldering copper** Gaslötkolben *m* ~ **source** Gasquelle *f* ~ **spray** Gasnebel *m*, Spritzgas *n* ~ **starter** Gasanlasser *m* ~ **stove** Gasheizofen *m* ~ **stove for baths** Gasbadeofen *m* ~ **stream** Gasstrom *m* ~**-suction plant** Gasabsaugungsanlage *f* ~ **supply** Gaszufuhr *f* ~**-supply pipe** Gaszuführungsrohr *n* ~ **take** Gasfang *m* ~ **take of blast furnace** Gichtgasfang *m* ~ **take-off (in pipe)** downcomer Gasableitung *f* ~ **tank** Gasbehälter *m* ~ **tap** Gashahn *m* ~ **tension** Gasspannung *f* ~**-testing compartment** Gasraum *m* ~ **thermometer** Gasthermometer *n* ~**-thread pipe stock** Gasgewindeschneidekluppe *f* ~ **throttle** Gas-regler *m*, -regulator *m* ~**-throttle controls** Gasdrosselgestänge *n* ~ **throttle screw** Gasdrosselschraube *f* ~**-throttle valve** Gasdrosselklappe *f* **gas-tight** gas-dicht, -undurchlässig ~ **membrane (or partition)** gasfeste Grenzfläche *f* ~ **seal** gasdichter Abschluß *m* ~ **wall** gasfeste Grenzfläche *f* **gas,** ~ **tongs** Gaszange *f* ~**-torch welding** Gasschmelzschweißung *f* ~ **trap** Gas-abscheider *m*, -sack *m*, -schleuse *f* ~ **tube** Entladungsgefäß *n*, Gas-entladungsgefäß *n*, -entladungsröhre *f*, -rohr *n*, -röhre *f*, -schlauch *m*; Ionenröhre *f* ~**-tube vice** Gasrohrschraubstock *m* ~ **turbine** Axial-kompressor *m*, -turbine *f*; Gasturbine *f* ~**-turbine power-plant** tulizing regeneration Regeneratorengasturbotriebwerk *n* ~**-turbine rotor** Kreiselrad *n* ~**-type arrester** Luftleerblitzableiter *m* ~ **up-and-downtake flue** Gaszuführungszug *m* ~ **uptake tube** Gichtgasabzugsrohr *n* ~ **valve** Gasventil *n*, (balloon) Manövrierventil *n* ~ **vibrations** Gasschwingungen *pl* ~ **volume** Gasvolum(en) *n* ~ **volumeter** Gasvolumeter *n* ~ **warning** Gaswarnung *f* ~ **washer** Gas-reiniger *m*, -waschapparat *m*, -wäscher *m* ~**-washing bottle** Gaswaschflasche *f* ~ **welder** Autogenschweißapparat *m* ~ **welding** autogene Schweißung *f*, Gas-schmelzschweißung *f*, -schweißung *f*, -schweißverfahren *n* ~**-welding apparatus** Autogengerät *n* ~**-welding method** Abbrennschweißverfahren *n* ~ **well** Gasbohrung *f*, (natural) Gasquelle *f* ~ **well tubing**

Rohre *pl* für Erdgasquelle ~ **withdrawal** Gasabgang *m* ~ **works** Gas-fabrik *f*, -werk *n* ~ **yield** Gas-abfall *m*, -ausbeute *f*, -ergiebigkeit *f* **gaseous** gasartig, gasförmig, gashaltig, gasig ~ **atmosphere** Gasatmosphäre *f* ~ **compound** gasförmige Verbindung *f* ~ **decomposition products** Zersetzungsgase *pl* ~ **diffusion** Gasdiffusion *f* ~ **discharge lamp** Gasentladungslampe *f* ~ **envelope** Gashülle *f* ~ **film** molekulare Gasschicht *f* ~ **fuel** gasartiger Brennstoff *m* ~ **gap** (in a tube) Gasstrecke *f* ~ **hydrochloric acid** Salzsäuregas *n* ~ **inclusion** (in glass) Blase *f*, Gipsen *m* ~ **mixture** Gasgemisch *n* ~ **particle** Schwebstoffteilchen *n* ~ **path** Gasstrecke *f*, -strom *m* ~ **rectifier tube** gasgefüllte Gleichrichterröhre *f* ~ **tube** gasgefüllte Röhre *f*

**gaseousness** Gas-förmigkeit *f*, -zustand *m* **gash, to** ~ mit Einschnitten *pl* versehen **gash** Spanlücke *f*, klaffende Wunde *f* ~ **vein** nach der Tiefe sich auskeilender Gang *m* **gashing,** ~ **angle** Zahnlückenwinkel *m* ~ **method** Einstechverfahren *n* **gasifiable** vergasbar **gasification** Gas-bildung *f*, -erzeugung *f*, Vergasung *f* ~ **of coal** Kohlenvergasung *f*, Steinkohlengaserzeugung *f* ~ **industry** Umwandlungstechnik *f* **gasifier** Vergaser *m* **gasiform** gasartig **gasify, to** ~ vergasen **gasket** Abdichtung *f*, Beschlagzeising *f*, Dichtring *m*, Dichtungsmanschette *f*, Dichtungsscheibe *f*; Einsatz-, Flansch-, Flanschendichtung *f*; Packung *f*, Scheibe *f*, Unterlagscheibe *f* ~ **for housing half-member** Dichtung *f* für Gehäusehälfte (Magnetseite) **gasket,** ~ **board** Dichtungspappe *f* ~ **cap** Dichtungsschutzkappe *f* ~ **coating** Düsenlackierung *f* (durch Filz für Drahtlacke) ~ **paste** pastenförmiges Dichtungsmittel *n* ~ **ring** Dichtungsring *m* ~**-sealing washer** Dichtung *f* **gasoline** Benzin *m*, Gasolin *n*, Krystallöl. *n*, Leichtbenzin *n* ~ **and lubricant bulk-storage plant** Betriebsstofflager *n* ~ **and lubricant railhead** Eisenbahntankstelle *f* ~ **and lubricants supply column** Betriebsstoffkolonne *f* **gasoline,** ~ **balance** Benzinwaage *f* ~ **can** Benzinkanne *m* ~ **carried in the trim tank** Trimmbenzin *n* ~ **carried in the wing** Flügelbenzin *n* ~ **chamber** Benzinsack *m* ~ **consumption** Benzinverbrauch *m* ~ **container** Benzinkanister *m* ~ **dump** Brennstofflager *n* ~ **electric generator (or power unit)** Benzinaggregat *n* ~ **engine** Benzinmotor *m*, Brennkraftmaschine *f*, Treibgasmotor *m* ~ **feed** Benzinzufuhr *f* ~ **filter** Benzinfilter *m* ~ **flowmeter** Benzinströmungsmesser *m* ~ **gauge** Benzinstand-anzeiger *m*, -messer *m* ~ **generator** Benzinaggregat *n* ~ **hydro-treater** Benzinhydrieranlage *f* ~ **injection system** Benzineinspritzverfahren *n* ~ **level gauge** Brennstoffvorratsmesser *m* ~ **motor** Benzinmotor *m* ~**-oil mixture** Kraftstofföl mischung *f* ~**-pressure gauge** Benzin-druckmesser *m*, -manometer *n* ~ **propulsion for engines** Benzinbetrieb *m* ~ **pump** Benzin-, Brennstoff-pumpe *f* ~ **pump with liquid level controller** Benzinpumpe *f* mit sichtbarem Meß-glas *n* ~ **scale** Benzinwaage *f* ~ **separator** Benzinreiniger *m* ~ **service truck** Abfüllwagen *m* ~ **station** Benzintankstelle *f* ~ **supply** Benzinvorrat *m* ~ **tank** Benzin-behälter *m*, -tank *m* ~ **tank hose** Benzingießer *m* ~ **tank-meter** Benzinuhr *f* ~ **tank truck** Betriebsstoffkesselkraftwagen *m*, Brennstofftankwagen *m* ~ **tractor** Benzinschlepper *m* ~ **used for cleaning purposes** Waschbenzin *n* ~ **wheel tank** fahrbare Benzintank *m*

**gasometer** Ausgleichbehälter *m*, Gas-behälter *m*, -meßapparat *m*, -messer *m*, Gasometer *m*, Gassammler *m*

**gasometric** gasvolumetrisch

**gassing** gasend, Gasen *n* (im Ruhezustand), Gasentwicklung *f* ~ **a balloon** Ballon *m* mit frischem Gas füllen ~ **after the charge** Nachkochen *n*

**gassing,** ~ **accumulator** kochender Akkumulator *m* ~ **factor** Gas-faktor *m*, -koeffizient *m* ~-**up** Nachfüllen *n*

**gassy** gasführend ~ **magnetron** Magnetron *n* mit ungenügendem Vakuum ~ **tube** gasgefüllte oder weiche Röhre *f*

**gastropod shell** Gastropodenschale *f*

**gastunite** Gastunit *m*

**gat** Seegatt *n*

**gate, to** ~ abblenden, (pencil or beam on flyback) austasten, einblenden (electron.), mit Torimpuls *m* steuern oder tasten **to** ~ **spots** abdunkeln (TV)

**gate** Ablaufpunkt *m* (aviat.), Abschirmdeckel *m*, (founding) Anschnitt *m*, Auftastimpulskreis *m*, Auslauf *m*; (hopper) Einguß *m*, Einguß-stelle *f*, -trichter *m*; (for pouring metal) Einlaufrinne *f*, Fenster *n* (film), Filmprojektionsfenster *n*, Führungsschaltung *f* (LKW-Getriebe), Führungsschlitz *m*, Gattertor *n*, Gießtrichter *m*, Kulissenplatte *f* (Auto), Pforte *f*, schwenkbarer Rahmen *m*, Reintonblende *f*, schwenkbares Gestell *n*, Sektor *m*, Spalt *m* des Stoffauflaufs (paper mfg.), Tor *n*, Tor-impuls *m*, -portal *n*; Trichter *m*, Trichter-einlauf *m*, -mündung *f*, (founding) Trichterkopf *f*

**gate,** ~ **of entry** Einfallsfeld *n* ~ **in headrace** Oberklappe *f* ~ **of mill dam** Freigerinne *n*, Freischütze *n* *f* ~ **of railroad** Eisenbahnwegschranke *f* ~ **for retaining slag and dirt** Schaumfang *m* ~**s and risers** Gießlingkopf *m* ~**s for waterwork** Abschlußwerk *m*

**gate,** ~ **angle** Sättigungswinkel *m* ~ **bearing** Klappenachse *f* ~ **chain gallery** Kabelkanal *m* ~ **chamber** Torkammer *f* ~ **change** Kulissenschaltung *f* ~ **circuit** Eingangskreis *m*, Durchlaß-, Tor-schaltung *f* ~ **connecting rod (or driving rod)** Gießtrichter *m* ~ **end-plate** Drehscheibe *f* ~ **friction** Fensterdruck *m* ~ **generator** Auftast-, Torimpuls-generator *m* ~ **mechanism** Leitapparat *m* (Kaplan-turbine) ~ **opening** Trichter-kopf *m*, -loch *n*, -mündung *f* ~ **paddle agitator** Gitterrührer *m* ~ **pipe** Trichterrohr *m* ~ **pivoting about horizontal axis along base along top** Klapptor *n* ~ **pivoting about vertical axis along one edge** Stemmtor *n* ~ **pressure** Fensterdruck *m* ~ **pulse** Torimpuls *m* ~ **rays** Strahlabblender *m* ~ **recess** (of lock) Tornische *f*, (balancing chamber) Gleichgewichtskammer *f* ~ **resistance** Einflußwider-

stand *m* ~ **road** Förderstrecke *f* ~ **selector** Gangführungsplatte *f* ~ **shaft** Schacht *f* des Verschlusses ~ **shears** Blechschere *f* mit geschlossenem Gestell ~ **stick** Eingußstock *m*, (founding) Holztrichter *m* ~ **strut** Gegentor *n* **the** ~ **tilts forward** das Tor *n* versackt ~ **track** Laufschiene *f* ~ **trigger diode** Auftastimpulsauslösediode *f* ~ **tube** Torröhre *f*

**gate-type,** ~ **gas-regulating valve** Gasschieber *m* ~ **shut-off valve** Absperrschieber *m* ~ **slide valve** Flachschieber *m* ~ **steam-shutoff valve** Dampfabsperrschieber *m* ~ **valve joint** Schieberverschluß *m*

**gate,** ~ **valve** Absperr-schieber *m*, -ventil *n*, Durchgangsschieber *m*, Durchlaßventil *n*, Schieber *m*, (clamp) Schieberventil *n*, Schleuse *f* ~ **valve with bell ends** Muffenabsperrschieber *m* ~ **video** übergelagerter Meßimpuls *m* ~ **voltage** Tor-Steuerspannung *f* ~ **way** Blindort *m* (min.), Seegatt *n*, Torweg *m* ~ **wing** Torflügel *m*

**gated,** ~-**beam tube** Pentode *f* mit konstanter Steilheit ~ **sweep** torgesteuerte Ablenkung *f*

**gather, to** ~ ansammeln, auflesen, (speed) entwickeln, Lagen machen, sammeln **to** ~ **speed** Fahrt *f* aufnehmen **to** ~ **together** zusammenhäufer

**gathering** Versammlung *f*, Zubrand *m* ~ **anode** Auffang-, Fang-, Sammel-anode *f* ~ **drum** Auflaufhaspel *m* ~ **electrode** Sammelelektrode *f* ~ **iron** Angangeisen *n* ~ **line** Zuleitung *f* ~ **mould** (glass mfg.) Vorform *f* ~ **ring** (glass blowing) Kranz *m* ~ **system** Sammelnetzsystem *n*

**gating** Anschnitt *m*, (founding) Art *f* des Anschneidens der Eingüsse, Signalauswertung *f*, Tastung *f*, Torsteuerung *f* ~ **of the beam** Strahlsperrung *f* ~ **at the joint** Gießen *n* an der Trennlinie ~ **of pulses** Austasten *n* von Impulsen

**gating,** ~ **network** Torschaltung *f* ~ **practice** Ausschnittechnik *f* ~ **pulse** Auswerte-, Toröffnungs-impuls *m*

**gauge, to** ~ ablehren, abmessen, beurteilen, eichen, justieren, kalibrieren, lehren, messen, normieren, peilen, vermessen, visieren, zurichten **to** ~ **a form** Format *n* machen

**gauge** Breite *f*, Dicke *f*, Kaliber *n*, Kernmaß *n* (Schriftgießerei), Kreuzmaß *n*, Lehre *f*, Maschenzahl *f* (Strumpffabrikation), Maß *n*, Maßlatte *f*, Maßstab *m*, Maßstock *m*, Messer *m*, Meß-, Paß-lehre *f*, Pegel *m*, Pegellatte *f*, Röhre *f*, Spur *f*, (track) Spurweite *f*, Stärke *f*, Stichmaß *n*, Streichmodel *m*, Vermessung *f*,

**gauge,** ~ **for base** Kaliberring *m* ~ **of excellence** Gütemaß *n* ~ **for glasses** Spiegelmesser *m* ~ **of laths** Lattenprofil *n* ~ **for measurement of thickness** Dickenmeßanlage *f* ~ **of a plate** Blech-dicke *f*, -stärke *f* ~ **of railway** Spurweite *f* des Geleises ~ **of a sheet** Blech-dicke *f*, -stärke *f* ~ **of wire** Draht-dicke *f*, -durchmesser *m*, -maß *n*, -nummer *f*

**gauge,** ~ **allowance** Lehrenabmaß *n* ~ **block** Meß-klotz *m*, -scheibe *f*; Parallelendmaß *n* ~ **block interferometer** Endmaßkomparator *m* ~ **cock** Probierhahn *m*, Standmesserhahn *m* ~ **condition** Eichbedingung *f* ~ **distance** Wurzelmaß *n* ~ **door** Drosseltür *f* ~ **engineer** Lehren-

bauer *m* ~ **equipment** Flüssigkeitsanzeiger *m*, Niveaumeßapparat *m* ~ **film spool** Schmalfilmspule *f* ~ **finger** Tastbolzen *m* ~ **flask** Meßkolben *m* ~ **glass** Meß-, Peil-, Schau-, Stand-glas *n*; Wasserstands-rohr *n*, -zeiger *m* ~- -**glass reflector** Reflektor *m* für Standrohr ~ **grinding and polishing machine** Lehrenschleif- und Poliermaschine *f* ~ **head** Standglaskappe *f* ~ **holder (or support)** Meßdosenträger *m* ~ **intervals** Endmaßstufen *pl* ~ **invariance** Eichinvarianz *f* ~ **knob** Begrenzungskugel *f* ~ **lamp** Begrenzungs-lampe *f*, -leuchte *f* ~ **length** Meß-länge *f*, -lineal *n*; Reiß-, Streck-länge *f* ~ **line** Füllstrich *m* ~ **manufacturer** Lehrenbauer *m* ~ **mark** Einstellmarke *f*, Füll-, Index-strich *m* ~ **number** Stärkenabmessung *f* ~ **pin** Kaliberbolzen *m* ~ **pipe** Inhaltsanzeigerohr *n*, Manometerleitung *f*, Peil-, Stand-rohr *n* ~ **plate** Ausmeßplatte *f* ~ **point** Endmarke *f*, Körner *m* ~ **pressure** Manometer-, Meß-, Über-druck *m* ~ **pressure in atmospheres** Atü (Atmosphärenüberdruck) ~ **propeller** Eichschraube *f* ~ **ring** Paßring *m* (electr) ~ **rod** Peilstab *m*, Spurstange *f*, Visierstab *m* ~-**rod cap** Peilstabkappe *f* ~ **rod of a pump** Pumpenpeilstock *m* ~-**setting device** Spurrichter *m* ~ **shoes** Spurschuhe *pl* ~ **slide** Meßeinsatz *m* ~ **stick** Meß-, Peil-stab *m* ~ **tolerance** Lehrentoleranz *f* ~ **tool** Lehrwerkzeug *n* ~ **(group) transformer** Maßstabtransformer *m* ~ **tube** Meßrohr *n* ~ **valve** Probierventil *n* ~ **wear** Lehrenverschliß *m* ~ **wheel** Spurrad *n*
**gauged** geeicht ~ **pile of a cofferdam** Bordpfahl *m*
**gauger** Eichmeister *m* ~'s **fee** Eichgebühr *f*
**gauging** Abtasten *n*, Eichung *f*, Messung *f*, Normierung *f* ~ **of capacitance** Kapazitätsmessung *f*
**gauging**, ~ **apparatus** Kubizierapparat *m* ~ **block** Kalibrierungsblock *m* ~ **calipers** Taster-, Visier-zirkel *m* ~ **hatches** Meßrohrleitung *f* ~ **hole** Pegelloch *n* ~ **instrument** Breitenmaß *n* ~ **lines** geeichte Strichmarken *pl* ~ **method** Peilmethode *f* ~ **office** Eichamt *n* ~ **(or mixing) platform** Mischbühne *f* ~ **plug** Eichpfropfen *m* ~ **(contact) pressure** Meßdruck *m* ~ **range** Meßbereich *m* ~ **rod** Peilstab *m* ~ **sheet** Meßzettel *m* ~ **station** Lage *f* eines Pegels, Messungsstelle *f* (hydr.)
**gauntlet** Überhandschuh *m*
**gauss** Gauß *n* (Einheit der magnetischen Feldstärke) ~ **error-distribution law** Gaußverteilung *f* ~ **meter** Gaußmeter *n*, Induktionsmeßgerät *n* ~ **ocular** Gaußokular *n*
**Gaussian** Gaußsch . . . (e, er, es) ~ **number field** Gaußscher Zahlenkörper *m* ~ **plane** Zahlenebene *f*
**Gautier grid** Gautiergitter *n*
**gauze** Flor *m*, Gaze *f*, Gewebe *n*, Netz *n* ~ **with damask figures** Damastgaze *f*
**gauze**, ~ **bandage** Gazebinde *f* ~ **bottom** Gazeboden *m* ~ **brush** Gazebürste *f* ~ **filter** Filtersieb *n*, Maschenfilter *n* ~-**like** siebartig ~ **loom** Gazegeschirr *m* ~ **ribbon** Dünntuch-, Gazeband *n* ~ **sieve** Gazesieb *n* ~ **stretching device** Gazeaufspannvorrichtung *f* ~ **top** Drahtnetzkappe *f*, Siebaufsatz *m* ~ **washer** Drahtgewebescheibe *f* ~ **wire** Drahtgeflecht *n*

**gay** lebhaft (color)
**gaylussite** Gaylüssit *m*, Natrokalzit *m*
**gear, to** ~ ineinandergreifen, in Eingriff *m* nehmen, mit Getriebe *n* versehen, übersetzen,
**gear, to** ~ eingreifen, ineinandergreifen, in Eingriff *m* nehmen, mit Getriebe *n* versehen, übersetzen, verzahnen
**gear** Gang *m*, Gerät *n*, Geschirr *n*, Gestänge *n*, (meshing) Getriebe *n*, Gezähne *n*, Mechanismus *m*, Rad *n*, Rädergetriebe *n*, Verzahnung *f*, Werk *n*, Zahnrad *n*, Zahnrädergetriebe *n*, Zahnräderwerk *n* **in** ~ im Betrieb *m*, eingerückt, eingreifend (mach.) **to be in** ~ in Eingriff *m* stehen **out of** ~ in Unordnung *f*, ausgerückt, ausgeschaltet **to be out of** ~ außer Gang *m* sein ~ **with cogging** Stirnrad *m* mit Schrägverzahnung ~ **for launching small boats** (booms, derricks or davits) Bootheißvorrichtung *f* ~ **and mating gear** Rad *n* und Gegenrad *n* ~ **in a shell** Hülsengetriebe *n* ~ **for sinking shaft** Senkvorrichtung *f* ~ **with tooth correction** V-Rad *n* ~ **in train** Langzugübersetzungsrad *n*
**gear**, ~ **assembly (or set)** Sammelgetriebe *n* ~ **blank** Zahnradkörper *m*
**gear-box** Getriebe-gehäuse *n*, -kasten *m*, Kegelgehäuse *n*, Räderkasten *m*, Schalträderkasten *m*, Zahnradgehäuse *n*, Zahnradkasten *m* ~ **brake** Triebwerkbremse *f* ~ **cover** Getriebegehäuse-, Räderkasten-deckel *m* ~ **drive** Räderkastenantrieb *m*
**gear**, ~ **case** Getriebe-gehäuse *n*, -kasten *m*; Trieb-, Zahnrad-, Zahnrädergetriebe-gehäuse *n*; Zahnradverkleidung *f* ~ **casing** Rädergehäuse *n*, -kasten *m*, Schaltgehäuse *n* ~ **chain** Zahnkette *f* ~ **chamfering cutter** Abrundfräser *m* ~ **change** Wechselgetriebe *n* ~-**change diagram** Ganganordnung *f* ~ **change gate** Kulissenplatte *f* ~ **changing** Geschwindigkeitsänderung *f*, Umschalten *n* ~-**control hand lever** Gangwählhebel *m* ~ **coupling** Zahnkupplung *f* ~ **cover** Getriebe-deckel *m*, -gehäuse *n* ~ **cutter** Fräser *m* für Zahnräder, Zahnform-, Zahnrad-fräser *m* ~-**cutting division** Zahnradabteilung *f* ~-**cutting machine** Räderfräs-, Verzahnungs-maschine *f* ~ **deburring machine** Zahnradentgratmaschine *f* ~ **dividing head** Zahnräderteilkopf *m* ~ **downline** Fahrwerk-Ausfahrleitung *f* ~ **drive** Rad-, Räder-antrieb *m*, Zahnkupplung *f*, Zahnradgetriebe *n* ~- **driven faceplate** Zahnkranzplanscheibe *f* ~- **driven impulse starter** Zahnradschnapper *m* ~-**driven supercharger** Getriebelader *m*, direkt angetriebener Vorverdichter *m* ~ **dynamometer** Zahndruckdynamometer *n* ~ **extension hose** Fahrwerk-Ausfahrleitung *f* ~ **fluid** (automatic transmission fluid) Getriebeflüssigkeit *f* ~ **forging** Zahnradschmiederohteil *m* ~ **frame** Getriebegestell *n* ~ **generator** Zahnradwalzmaschine *f* ~ **grease** Zahnradfett *n* ~ **guard** Zahnradschutz *m* ~ **hob** Zahnradabwälzfräser *m* ~ **hobber** Abwälzfräser *m* (aviat.), Zahnradabwälzfräser *m*, -abwälzfräsmaschine *f* ~ **hobbing** Zahnradfräsmaschine *f* ~ **housing** Getriebegehäuse *n*, Kettenschutz *m*, Triebgehäuse *n*, Zahnrädergetriebegehäuse *n* ~ **hub** Zahnradnabe *f* ~ **idling** Getriebeleerlauf *m* ~ **involute** Evolventenprüfgerät *n* ~ **leg** Fahrwerkbein *n* ~ **lever** Getriebearm *m* ~ **lock** Getriebesperre *f*

*f* ~ **(type) oil pump** Zahnradölpumpe *f* ~
**pattern** Modellzahnrad *n* ~**-preserving** getriebe-
schonend ~ **pump** Getriebepumpe *f* ~ **quadrant**
Räderschere *f* ~ **rack** Zahnstange *f* ~ **rating**
(segment gear) Zahnsegment *n* ~ **ratio** Ge-
triebeübersetzungsverhältnis *n*, Übersetzung *f*,
Übersetzungsverhältnis *n* ~ **reduction** Räderübersetzung *f*, -untersetzung *f* ~ **retraction**
**hose (or up-line)** Fahrwerkeinfahrleitung *f* ~
**rig** Verspannungsgetriebe *n* ~ **rim** Zahnkranz *m*
**gear-ring,** ~ **clearance** Zahnkranzabstand *m* ~
**cranking starter** Zahnkranzdurchdrehanlasser
*m* ~ **hub** Zahnkranznabe *f* ~ **inertia starter**
Zahnkranzschwungkraftanlasser *m* ~ **profile**
**(or face)** Zahnkranzflanke *f* ~ **rocker starter**
Zahnkranzpendelanlasser *m*

**gear,** ~ **seat** Getriebekastenunterlage *f* ~
**selector** Gangwähler *m* ~ **shaft** Getriebe-,
Rad-, Transmissions-welle *f* ~ **shaper** Zahn-
radstoßmaschine *f* ~ **shaper cutter** Schneidrad-
schärfmaschine *f*

**gear-shift** Gangwechsel *m*, Gangwechselgetriebe
*n*, Getriebeumschaltung *f*, Schaltung *f* ~ **bar**
Schaltstange *f* ~ **cap** Kugelhaube *f* ~ **lever**
Schalthebel *m* ~ **lever bracket** Schaltbock *m* ~
**lever housing** Schaltblock *m* ~ **lever shaft**
Schaltwelle *f* ~ **mechanism** Schaltgetriebe *n* ~
**sleeve** Schaltmuffe *f*

**gear-shifting** Getriebeschaltung *f* ~ **fork** Schalt-
stangenhebel *f* ~ **gate** Schaltführung *f* (Auto)
~ **quadrant** Schaltsegment *n*

**gear,** ~ **sleeve** Zahnradbüchse *f* ~ **stud** Zwi-
schenradbolzen *m* ~ **switch mechanism** Gang-
schaltgetriebe *n* ~ **teeth** Zahnradzähne *pl* ~-
**testing apparatus** Zahnräderprüfapparat *m*
~**-tipping appliance (or tilting device)** Getriebekippvorrichtung *f* ~**-tooth cutter** Verzahnungsfräser *m* ~**-tooth system** Verzahnung *f* ~-
**toothed calipers** Zahnmeßschraublehre *f* ~-
**toothed chamfering machine** Zahnräderabrund-
maschine *f* ~**-toothed vernier** Zahnmeß-schieb-
lehre *f*, -schublehre *f* ~ **train** Getriebe *n*, Zahn-
radverbindung *f*, Zahnradvorgelege *n* ~ **transmission** Zahnradvorgelege *n* ~ **transmission**
**with V-belt** Keilriemenübersetzung *f* ~**-type**
**clutch** Zahnradkupplung *f* ~ **unit** Getriebe *n*
~ **wheel** Getriebezahnrad *n*, Zahnrad *n* ~
**wheel air turbines** Druckluftzahnradmotoren
*pl* ~**-wheel engagement** Radeingriff *m* ~-
**wheel molding machine** Zahnradformmaschine *f*
~**-wheel pump** Zahnradpumpe *f* ~**-wheel rim**
**chuck** Zahnkranzfutter *n* ~**-wheel transmission**
Zahnradübertragung *f* ~ **wheels** Getrieberäder
*pl*

**geared** verzahnt, übersetzt ~ **down** untersetzt
~ **and locked together** gemeinschaftlich betrieben ~ **together** gekuppelt durch Rädergetriebe

**geared,** ~ **chain block (or hoist)** Zahnradfla-
schenzug *m* ~ **crane ladle** Krangießpfanne *f*
mit Getriebekippvorrichtung ~ **drive** schlupffreier Antrieb *m* ~ **engine** Getriebemotor *m* ~
**fan** Lader *m* mit übersetztem Antrieb *m* ~
**headstock** Räderkastenspindelstock *m* ~ **hind**
**axle** Zahnradbrücke *f* ~ **ladle** Gießpfanne *f*
oder Pfanne *f* mit Getriebekippvorrichtung ~
**lubricating pump** Zahnradschmierölpumpe *f* ~
**motor** untersetzter Motor *m* ~ **power pump**

Zahnradpumpstation *f* ~ **(down) propeller**
untersetzte Luftschraube *f* ~ **pump** Zahnradpumpe *f* ~ **reducing press** Räderziehpresse *f* ~
**rubber motor** Gummimotor *m* mit Getriebe ~
**scroll lathe chucks** Drehbankspannfutter *n* ~
**servomotor** Stellmotor *m* mit Getriebe ~
**spindle drive** Spindelgetriebe *n* ~ **transmission**
Räderübersetzung *f* ~ **trolley** Katze *f* mit
Rädervorgelege, Laufkatze *f* mit Rädervorge-
legevorschub ~ **turbine** Getriebeturbine *f* ~
**wheels** verzahnte Räder *pl*

**gearing** Antrieb *m*, Eingreifen *n*, Eingriff *m*,
Getriebe *n*, Ineinandergreifen *n*, Räder-getriebe
*n*, -übersetzung *f*, -übertragung *f*, -werk *n*,
Schalten *n*, Transmission *f*, Übersetzung *f*,
Verzahnung *f*, Vorgelege *n*, Zahnradgetriebe *n*,
Zahnradübersetzung *f*, Zahnwerk *n*

**gearing,** ~ **and controlling mechanism** Antriebs-
und Steuerungsmechanismus *m* ~ **of draw**
**frame** (textiles) Streckengetriebe *n* ~ **of drawing**
**frame** (textiles) Streckentriebwerk *n* ~ **with**
**jaw-type clutch** Getriebe *n* mit Klauenschal-
tung ~ **of teeth** Zahneingriff *m*

**gearing,** ~ **end frame** Antriebsbock *m* ~**-in** Eingreifen *n* von Zahnrädern ~ **layout work diagram** Getriebeplan *m* ~ **loads** Getriebekräfte *pl*
~ **tooth crest track** Kopfbahn *f* ~ **top relief**
Kopfabrundung *f* ~ **unit** Getriebeaggregat *n*

**gearless** getriebelos ~ **hay loader** getriebeloser
Heuauflader *m*

**gears** Getrieberäder *pl*, Räderwerk *n*

**gedrite** Gedrit *m*

**Geiger,** ~ **counter** Geiger-Müller Zählrohr *n*,
Geigerzähler *m* ~ **dip tube** Flüssigkeitszählrohr
*n* ~ **region** Geigersches Gebiet *n*

**gel,** ~ **formation** Gelbildung *f* ~**-like** gelartig

**gelatinate** gelieren

**gelatin(e)** Gallerte *f*, Gelatine *f*, reiner Knochenleim *m* ~ **board** Lichtdruckkarton *m* ~ **capsule**
Arzneikapsel *f* ~ **chloride paper** Deltapapier *n*
~**-coated film base** gelatinebeschichteter Blankfilm *m* ~ **coating** Gelatineüberzug *m* ~ **film**
**for tracing** Gelatinepaushaut *f* ~ **hardening**
Gelatineschichtgerbung *f* ~ **picture** Hauchbild
*n* aus Gelatine ~ **spangle** Gelatineflitter *m* ~
**tracing paper** Gelatinepauspapier *n*

**gelatinization** Behandlung *f* mit Gelatine, Gelie-
rung *f*

**gelatinize, to** ~ gelatinieren

**gelatinizer** Gelatinier(ungs)mittel *n*

**gelatinous** gallertartig, gelartig, gelatinös, leimartig

**gelation** Gelatinierung *f*

**gell of antimony trisulphide** Antimontrisulfidsol
*n*

**gelling point** Gel-Bildungstemperatur *f*

**gem** Edelstein *m*

**Gemini** die Zwillinge *pl*

**gemmology** Edelsteinkunde *f*

**gemstone** Edel (Halbedel)-stein *m*

**gender** Gattung *f*, Geschlecht *n*

**gene** Gen *n*, Erbeinheit *f*

**genealogical tree** Stammbaum *m*

**general** allgemein, gesamt, gewöhnlich, üblich
**in** ~ allgemein

**general,** ~ **apparatus and equipment** Geräte *pl*
(allgemeine Ausrüstung) ~ **assembly** Haupt-
baugruppe *f* ~ **behaviour** Gesamtverhalten *n* ~

**call "to all stations" (CQ)** allgemeiner Anruf „an Alle" (rdo) ~ **call pilot lamp** Anrufkontrollampe f ~ **cargo** Stückgutladung f ~ **concept (or designation)** Dachbegriff m ~**-concession map** Mutungsübersichtskarte f ~ **contractor** Gesamtunternehmer m ~ **damage** gewöhnliche Havarie f c/o ~ **delivery** postlagernd ~ **design (or construction) data** allgemeine Konstruktionsdaten pl ~ **design feature** allgemeines bauliches Kennzeichen n ~ **direction of strike** Hauptstreichrichtung f (min.) ~ **drawing** Zusammenstellungszeichnung f ~ **ellipsoid** dreiachsiges Ellipsoid n ~ **equation** allgemeine Gleichung f ~ **expenditures** Allgemeinkosten pl ~ **expenses** Allgemeinkosten pl, Generalunkosten pl ~ **inference** allgemeine Wetterlage f ~ **interrogation** Generalabfrage f ~ **layout of models** Bedienungsschema n ~ **maintenance** großer Unterhalt m ~ **manager** Generaldirektor m ~ **map** General-, Übersichtskarte f ~ **meeting** Plenarsitzung f ~ **message** Spruch m mit Sammeladresse (rdo) ~ **meteorological situation** Großwetterlage f ~ **offices** Hauptverwaltung f ~ **orders** allgemeingültige Anordnungen pl ~ **overhauling** Grundüberholung f ~ **plan** Gesamt-anlageplan m, -anordnung f, Übersichtsplan m ~ **plan of arrangement** Grundrißplan m ~ **principle for action (or thought)** Anhalt m

**general-purpose** Mehrzweck . . . ~ **airplane** Gebrauchsflugzeug n ~ **computer** Universalrechenmaschine f ~ **diode** Allzweckdiode f ~ **instrument** Universalinstrument n ~ **microscope** Gebrauchsmikroskop n ~ **plow** Pflug m für allgemeine Zwecke

**general, ~ rain** Landregen m ~ **routine** allgemeines Programm n ~ **rules** allgemeine Bestimmungen pl ~ **sketch** Übersichtsskizze f ~ **solution** allgemeine Lösung f (math.) ~ **spectrum** Übersichtsspektrum n ~**-staff map** (scale 1 : 100 000) Generalstabskarte f ~ **validity** Allgemeingültigkeit f ~ **view** allgemeine Ansicht f, Gesamt-anlage f, -ansicht f, -darstellung f; Überblick m, Übersichtsbild n ~**-view oscilloscope** Übersichtsrohr n ~**-view tube** Übersichtsrohr n ~ **viewing tube** Übersichtsröhre f

**generality** Allgemeingültigkeit f, Allgemeinheit f, Gesamtheit f

**generalization** Verallgemeinerung f

**generalize, to** ~ verallgemeinern

**generally** die Allgemeinheit f betreffend, durchgängig ~ **speaking** allgemein gesprochen ~ **valid** allgemeingültig

**generant** erzeugend

**generate, to** ~ abgeben, entwickeln, (gases) erzeugen, hervorbringen, hervorrufen, zeugen **to ~ steam** Dampf abgeben

**generated, ~ address** synthetische Adresse f ~ **power** Antriebsleistung f ~ **surface** Mantelfläche f

**generating** erzeugend ~ **angle** Öffnungswinkel m ~ **angle of gear-tooth profile** Eingriffswinkel m der Zahnflanke ~ **change gears** Wälzwechselräder pl ~ **efficiency** Erzeugungswirkungsgrad m ~ **feed** Wälzvorschub m ~ **flask** Entbindungsflasche f ~ **function** erzeugende Funktion f ~ **gear planer** Walzhobelmaschine f ~ **hob-**

**bing method** Abwälzverfahren n ~ **line** Zeugelinie f ~ **line of the back cone** Erzeugende f des Ergänzungskegels ~ **link** Wälz-hebel m, -bewegung f, -verhältnis n ~ **method** Wälzhobelverfahren n ~ **milling cutter** Abwälzfräser m ~ **motion** Wälzung f ~ **motion cradle** Wälzkörper m ~ **planing method** Abwälzhobeln n ~ **plant** Kraftwerk n, Strom- (electr.), Stromerzeugungs-, Stromversorgungs-anlage f ~ **program** erzeugendes Programm n (info proc.) ~ **roll feed change gears** Wälzvorschubwechselräder pl ~ **roll feed motor** Motor m für den Wälzvorschub ~ **routine** erzeugende Programm n (info proc.) ~ **set** Generatorsatz m, Stromerzeugungsanlage f ~ **space** Arbeitsraum m ~ **station** Kraftwerk n ~ **unit** Erzeugeranlage f, Kraftquelle f, Maschinensatz m ~ **valve** Generatorröhre f ~ **weight** Fliehgewicht n

**generation** Ausscheidung f, (of gases) Entwicklung f, Erzeugung f, Zeugung f ~ **of beams** Strahlerzeugung f ~ **of chlorine** Chlorentwicklung f ~ **of current** Stromerzeugung f ~ **of draught** Zugerzeugung f ~ **of electricity** Elektrizitätserzeugung f ~ **of gas** Gasentwicklung f ~ **of heat** Hitze-, Wärme-erzeugung f ~ **of hydrogen** Wasserstoff-gaserzeugung f, -herstellung f ~ **of oscillations** Schwingungserzeugung f ~ **of smoke** Nebelentwicklung f ~ **of water gas** Wassergas-erzeugung f, -herstellung f

**generation, ~ rate** Paarbildungsgrad m ~ **time** Generationsdauer f

**generator** Dynamo m, Dynamomaschine f, Entwickler m, Entwicklerbehälter m, Erzeuger m, Erzeugungsapparat m, Generator m, Lichtmaschine f, Oszillator m, (ringing) Rufsatz m, (current or electric) Stromerzeuger m ~ **for calibrating** Eichimpulsgenerator m ~ **for the filament supply** Heizmaschine f ~ **of rectangular (or square) waves** Quadratwellengenerator m

**generator, ~ braking** Senkbremsschaltung f ~ **cable** Generatorkabel n (Triebwerk) ~ **call** Induktoranruf m ~ **contactor** Generatorschütz m ~ **control relay** Generatorsteuerrelais n ~ **cooling tube** Generator-Kühlluftleitung f ~ **(magneto) crank (or handle)** Induktorkurbel f ~ **drive** Stromerzeugerantrieb m ~ **end** Erzeuger-, Speise-, (of cable) Sende-seite f ~ **fittings** Generatorarmatur f ~ **gas** Apparate-, Generator-, Saug-gas n ~ **(dynamo) gear wheel** Lichtmaschinenrad n ~ **hum** Maschinengeräusch n ~ **machine** Kesseldampfmaschine f ~ **plant** Generatoranlage f **(hand-)**~ **ringing** Induktoranruf m ~ **rotor** Generatorläufer m ~ **set** Stromerzeugungsaggregat n ~ **signaling working with individual line group** Anrufbetrieb m auf Leitungen, die von einer Beamtin bedient werden ~ **socket** Generatorsteckdose f ~ **space** (beam tube) Anfachraum m ~ **switch** Generatorschutzschalter m ~ **trailer** Generatoranhänger m ~ **triode** Senderöhre f ~ **unit** Generatoranlage f ~ **valve** Schwingröhre f ~ **voltmeter tachometer** Tachodynamo n

**generatrix** Erzeugende f, Erzeugungslinie f, (of cylinder) Mantellinie f, Zeugelinie f

**generic name** Gattungsname m

**genesis** Entstehung f, Entstehungsgeschichte f

**Geneva, ~ convention** Genfer Abkommen n ~

cross Genferkreuz n ~ lock mechanism Malte-
serkreuzsperre f ~ Red Cross Genfer Rotes
Kreuz n ~ stop Malteserkreuz n (Werkzeug-
maschinenteil) ~-stop mechanism Malteser-
kreuzgesperre n ~ wheel Malteserkreuz n
gentle mild, sacht, sanft, weich ~ bank große
Kurve f (aviat.) ~ breeze leichte Brise f ~ return nachgiebige Rückführung f ~ slope flache
Böschung f ~ start of oscillations weicher Einsatz m der Schwingungen, weicher Schwingungseinsatz m ~ turn schwache Kurve f
gently, ~ dipping seam flach einfallendes Flöz m
~ heated schwacherhitzt
genuine echt, gediegen, lauter, wahr, wirklich
~ absolute pitch absolutes, echtes Gehör n
genuineness Echtheit f
genus Gattung f, Geschlecht n
geoanticline geoantiklinal
geocentric geozentrisch ~ parallax tägliche
Parallaxe f
geochemical geochemisch
geochemistry Geochemie f
geochronology Geochronologie f
geodesic, ~ curvature geodätische Krümmung f
~ lines geodätische Linien pl ~ mapping
geodätische Abbildung f ~ parallel geodätisch
parallel
geodesics geodätische Linien pl
geodesist surveyor Erdmesser m
geodesy Erdmeßkunst f, mathematische Erdkunde f, Geodäsie f, Vermessungskunde f
geodetic, ~ construction geodätische Bauweise f,
Netzwerkbauweise f ~ control geodätische Vermessung f ~ engineer Vermessungsingenieur
m ~ head geodätisches Gefälle n ~ latitude
geodätische Breite f ~ survey Landesaufname
f
geodetical geodätisch
geodynamics Dynamik f fester Körper
geognosy geognostische Wissenschaften pl,
Gesteinskunde f
geographic, ~ index number Kartenblattnum-
mer f ~ north geografisch-Nord ~ position
geografischer Ort m
geographical geografisch ~ circuiting principle
Spurplantechnik f ~ engraver Kartenstecher m
~ map Erdkarte f ~ meridian geografischer
Meridian m ~ north geografischer Norden m
~ range Sichtweite f
geography Erdkunde f, Geografie f
geoid warping Geoid-Ondulation f
geologic, ~ age determination Altersbestimmung
f von Gesteinen ~ shelf Abschnitt m, Fest-
landssockel m, Schelf n
geological geologisch ~ methods employing
electronic apparatus Funkgeologie f ~ section
geologisches Profil ~ specimen Handstück n
geologist Geolog(e) m
geology Geologie f
geomagnetic field erdmagnetisches Feld n
geometric geometrisch ~ buckling geometrische
Wölbung f ~ center (airport) geometrischer
Mittelpunkt m ~ configuration of rays Strahlengang m ~ derivation of wave equation Geo-
metrisierung f der Wellenmechanik ~ distortion Geometriefehler m ~ distribution räumliche Verteilung f, Raumverteilung f ~ interference theory wellenkinematische Theorie f ~

location error Ortsfehler m ~ mean geometri-
sches Mittel n ~ optics Strahlenoptik f ~ pitch
(propeller) geometrische Steigung f
geometrical geometrisch ~ attenuation geometrische Dämpfung f ~ fundamental geometrische Grundform f ~ intersection geometrischer Strahlenschnitt m ~ iterated geometrisch
iteriert ~ luminosity relative brightness geometrische Fernrohrlichtstärke f ~ mean mittlere
Proportionale f ~ offset geometrische Schränkung f ~ optics of light rays Lichtoptik f ~
position geometrischer Ort m ~ progression
geometrische Reihe f
geometrically, ~ correct lagenrichtig ~ unstable (structure) statisch unterbestimmt
geometrician Geometer m
geometry Feldmesser n, Geometrie f, Größenlehre f, Raumlehre f ~ of coincidence Inzidenz-
geometrie f ~ hum Geometriebrumm m
geomorphology Erdoberflächenlehre f
geophone Erdhörer m, Geofon n, Seismograf m
geophysical geophysikalisch ~ exploration geophysische Untersuchung f ~ method of prospecting geophysikalisches Schürfverfahren n
~ prospecting geophysikalische Forschung f
geophysics mathematische Erdkunde f
geopolitics Geopolitik f
geostrophic geostrofisch ~ wind Passatwind m
geosyncline Geosynklinal n
geothermal, ~ gradient Erdwärmetiefenstufe f,
geothermischer Gradient m ~ method Methode
f des geothermischen Grades
geothermic geothermisch
geothermometer Erdwärmemesser m
Gerber, ~ design (a type of cantilever construction) Gerbersche Anordnung f ~'s diagram
(parabola) Dauerfestigkeitsschaubild n nach
Gerber ~ hinge pin-pointer spar Gerbergelenk-
holm m
germ Keim m, Keimling m, Mikrobe f, Mikroorganismus m, Sprosse f ~ process Germver-
fahren n
German deutsch of ~ make deutsches Erzeugnis
n ~ Contract Committee Deutscher Verdin-
gungsausschuß m ~ contract procedure in the
building industry Verdingungsordnung f für
Bauleistungen (VOB) ~ Industrial Standard
fitting DIN-Passung f ~ Industrial Standards
DIN (Deutsche Industrie-Norm) ~ Patent
Deutsches Bundes(republik)patent n (D.B.P.)
~ silver Argentan n, Neusilber n, Weißkupfer
n ~-silver fittings Weißgußarmatur f ~ silver
lining Neusilberspan m ~ spring lock Halbtourschloß n ~ steel Schmelzstahl m ~ text
Fraktur f ~ type Frakturschrift f ~ utility
model deutsches Gebrauchsmuster n ~ writing
Frakturschrift f
germanium Germanium n
germanous, ~ chloride Germaniumchlorür m ~
oxide Germaniumoxydul m ~ sulfide Germani-
umsulfür m
germicidal keimtötend
germinate, to ~ (seed) aufgehen, auskeimen,
(seed) auslaufen, auswaschen
germinating power Keimkraft f
germination (seed) Aufgang m, Kernbildung f,
Kristallisationswachstum n, Kristallkernbildung, Sprossung f

**germinator** Keimschale *f*
**gersdorffite** Arsennickel-glanz *m*, -kies *m*, Gersdorffit *m*, Nickel-arsenkies *m*, -glanz *m*
**get, to** ~ bekommen, beschaffen, besorgen, sich einklemmen, hereingewinnen, holen, verschaffen, werden, zulegen **to** ~ **(a ship) afloat** abarbeiten **to** ~ **away** entkommen **to** ~ **brittle** rissig werden **to** ~ **dark** dunkeln **to** ~ **done** bewerkstelligen **to** ~ **a footing** Anhalt *m* gewinnen **to** ~ **free of the aircraft** freibekommen, vom Flugzeug *n* freikommen **to** ~ **hold** fassen **to** ~ **hold of a buoy** sich an einer Boje *f* vertauen **to** ~ **hot** sich erhitzen **to** ~ **in(to)** einsteigen **to** ~ **into** geraten **to** ~ **into line** sich einfädeln **to** ~ **loose** lockerwerden, los werden, Spiel *n* haben **to** ~ **lost** verlorengehen **to** ~ **moldy** verschimmeln **to** ~ **off** aussteigen **to** ~ **out** aussetzen **to** ~ **out of alignment (or out of line)** ecken **to** ~ **out of range** auswandern **to** ~ **out of true** ungenau werden **to** ~ **the range** sich einschießen **to** ~ **ready** zurechtmachen **to** ~ **rotten** faulen **to** ~ **rough** aufrauhen **to** ~ **a set** krumm werden (beim Härten) **to** ~ **the tail up** Schwanz *m* hochnehmen **to** ~ **under way** abfahren **to** ~ **up** aufstehen, **(the tail)** hochnehmen **to** ~ **warped** sich ziehen
**getaway** Studiokulisse *f* (film) ~ **speed** Abfluggeschwindigkeit *f*
**getter** Fangstoff *m*, Füllung *f*, Getter *n*, Hauer *m* ~ **(ing)** Gasbindung *f* ~ **(ing substance)** Fangmittel *n* ~ **(ing) of gases** Gasaufzehrung *f*
**getter,** ~ **effect** Getterwirkung *f* ~ **film** Getterspiegel *m* ~ **metal vaporizer** Gettermetallverdampfer *m* ~ **patch** Getter-pille *f*, -spiegel *m* ~ **pill (or tab)** Getterpille *f* ~ **plate** Getterplatte *f*
**gettered** gegettert
**getting,** ~ **of coal** Kohlengewinnung *f* ~ **face** Betriebspunkt *m* (min.), Gewinnungspunkt *m* (min.)
**geyser** Geiser *m*, Geyser *m*, Springquelle *f*
**geyserite** Kieselinter *m*
**G factor** G-Faktor *m*
**ghost** Achter *m*, Achterleitung *f*, Doppelbild *n*, falsche Linie *f* (opt.), Faserstreifen *m*, Geist *m*, Geisterbild *n*, unscharfes Bild *n* oder Echo *n* (rdr) ~ **current** Gespensterstrom *m* ~ **echo** Geisterecho *n* ~ **image** Geister-, Neben- *n*, Relief-bild *n* ~ **image causing veiling** (in images or pictures) Schleier *m* ~ **line** Faserstreifen *m*, Längszeile *f*, Schattenstreifen *m*, Schleifriß *m*, Seigerungsstreifen *m* ~ **picture** Bildplastik *f* ~ **signal** Geistersignal *n*, Irr-zacke *f*, -zeichen *n*
**giant,** ~ **air shower** explosiver Schauer *m*, Riesenschauer *m* ~ **antenna** Großantenne *f* ~ **boring mill** Großbohrwerk *n* ~ **bucket wheel excavator** Großschaufelradbagger *m* ~ **crane** Schwerlastkran *m* ~ **plane** Riesenflugzeug *n* ~ **powder** Dynamit *n* ~ **redwood** Mammutbaum *m* ~ **resonance** Riesenresonanz *f* ~ **shovel** Riesenbagger *m* ~ **tire** Riesenluftreifen *m* ~ **vertical boring mill** Riesenkarusseldrehbank *f*
**gib** Beilagekeil *m*, Führungslineal *n*, Gegenkeil *m*, Hakenkeil *m* ~ **and cotter** Doppelkeil *m* ~-**head key** Nasenhaubenkeil *m* ~-**headed saddle key** Nasenhohlkeil *m* ~ **key** Nasenkeil *m* ~-(**head) key** Nasenkeil *m*
**Gibbs energy** Enthalpie *f*
**gibbsite** Hydrargillit *m*

**gibe, to** ~ piepen
**gibs** Leisten *pl* zum Festhalten von einem Bettschlitten auf der Flachführung ~ **for holding the table** Leisten *pl* zum Festhalten des Tisches auf den Führungen ~ **for slide guides** Stößelführungsleisten *pl*
**giddy** schwindelig
**gift** Gabe *f*, Geschenk *n*, Schenkung *f*, Zugabeartikel *m*
**gifted** begabt
**gigantic** riesig
**gigantolite** Gigantolith *m*
**gilbert** (unit of magnetomotive force) Gilbert *n*
**gild, to** ~ übergolden, vergolden
**gilded** vergoldet
**gilder** Vergolder *m*
**gilding** Goldanstrich *m*, Übergoldung *f*, Vergoldung *f* ~ **and embossing press** Vergolde- und Prägepresse *f* ~ **with gold leaf** Blattvergoldung *f* ~ **size** Vergoldergrund *m*
**gill** Kieme *f*, Luftregelklappe *f*, Spreizklappenhaube *f*, Viertelpinte *f* ~-**box minder** Rohstrecker *m* ~ **exit** Austrittsklappe *f* der Kühlluftführung, Luftaustritt *m* der Spreizklappenhaube
**Gill-Morell oscillations** Gill-Morell-Schwingungen *pl*
**gill rubber drawing frame** Heckelstrecke *f*
**gilled** gerippt ~ **cowling** Haube *f* mit verstellbaren Luftabflußklappen ~ **pipe heating surface** Rippenheizfläche *f* ~ **radiator** Lamellenkühler *m* ~ **tube** Lamellenrohr *n* ~-**tube radiator** Rippenrohrkühler *m*
**gilpinite** Gilpinit *m*, Uranvitriol *n*
**gilsonite** Gilsonit *m*
**gilt** vergoldet ~ **bronze** Gondbronze *f* ~ **edge** Goldschnitt *m* ~ **paper** Goldpapier *n* ~ **silver** Vermeil *n*
**gimbal** Kardanrahmen *m*; Kompaß-bügel, -gabel *f*; Tragbügel *m*; kardanisch ~ **bearings** Kreuzgelenkringlager *n* ~ **deflection** Rahmenausschlag *m* (gyro) ~ **frame** Balanzier-bügel *m*, -ring *m* ~ **gain** Übertragungseigenschaften *pl* des Rahmens ~ **joint** Knochengelenk *n* ~ **lock** (Kardan-)Rahmensperre *f* ~ **moment of inertia** Rahmenträgheitsmoment *n* ~ **mounting** Kardanaufhängung *f* ~ **ring** Kardanring *m*, Kreuzgelenkring *m* ~ **servo motor** Kardanrahmenstellmotor *m* ~ **suspension** Kardanaufhängung *f* ~ **system** Kardansystem *n* ~ **torquer** Kardanrahmenstellmotor *m*
**gimbals** Aufhängebügel *m*
**gimlet** Bohrer *m*, Fritt-, Hand-, Holz-, Nagel-, Schnecken-, Vor-bohrer *m* ~ **of auger type** Frettbohrer *m*
**gin, to** ~ entkörnen **to** ~ **cotton** die Baumwolle *f* entkörnen **to** ~ **(or clean) the cotton** die Baumwolle egrenieren
**gin** Egreniermaschine *f*, Förderwerk *n*, Göpel *m* ~ **block** Baurolle *f* ~ **pole** Bock *m*, Galgen *m* (Tiefbohrer) ~ **saw for cotton cleaning** Egrenierkreissäge *f*
**Gin resistance furnace** Gin-Widerstandsofen *m*
**ginning** Egrainieren *n*
**giratory breaker** Walzenbrecher *m*
**gird, to** ~ gürten
**girder** Balken *m*, Binder *m*, Deckenträger *m*, Durchzug *m*, Galgen *m*, Holm *m* (aviat.),

Querriegel *m*, Riegel *m*, Stab *m*, Tragbalken *m*, Tragebaum *m*, Träger *m*, Traverse *f*, Unterzug *m*, Zug *m* ~ **with Gerber joints (or Gerber beam)** Gerbergelenkträger *m* ~ **across landing station** Hängebankunterzug *m* (min.) ~ **and section mill** Formwalzwerk *n*

**girder,** ~ **axis** Träger-achse *f*, -mittellinie *f* ~ **beam** Trägerbalken *m* ~ **bridge** Balkenbrücke *f* ~ **casing** Balkenschalung *f* ~ **fishplate** Gurtlasche *f* ~ **grillage** Trägerrost *m* ~ **iron** Trägereisen *n* ~ **mat** Gurtmatte *f* ~ **mill** Trägerwalzwerk *n* ~ **mill train** Trägerstraße *f* ~ **mold** Trägerschalung *f* ~ **pass** Trägerkaliber *n* ~ **pole** Gittermast *m* ~ **rolling mill** Trägerwalzwerk *n* ~ **section** Träger-profil *n*, -querschnitt *m* ~ **shaped like a gable-end roof** zerbrochener Riegel *m* ~ **structure** Gitter-konstruktion *f*, -rumpf *m*

**girdle, to** ~ gürten

**girdle** Gesteinsschicht *f*, Gurt *m*, Gürtel *m*

**girt** Untergurt *m*

**girth** Gurt *m*, Gurtbrett *n*, Riegel *m*, Rippe *f*, Sattelgurt *m*, Umfang *m*, Untergurt *m* ~ **pulley** Gurtscheibe *f* ~ **ring** Laufring *m* ~ **scab** Gürtelschorf *m*

**git** Eingußtrichter *m* ~ **cutter** Eingußabschneider *m*

**give, to** ~ abgeben, ausliefern, ergeben, erteilen, geben, gewähren **to** ~ **back** herausgeben **to** ~ **backing** Rückhalt *m* geben **to** ~ **batter** abböschen **to** ~ **change** herausgeben **to** ~ **a coating of glue** leimtränken **to** ~ **consideration** to gerecht werden **to** ~ **on credit** kreditieren **to** ~ **the engine 1,000 revolutions per minute** die Drehzahl *f* des Motors auf 1,000 Umdrehungen per . Minute einregeln **to** ~ **evidence** aussagen **to** ~ **in exchange** einwechseln **to** ~ **an expert opinion** begutachten **to** ~ **the first grinding** rauhschleifen **to** ~ **ground** weichen, zurückweichen **to** ~ **a half turn** kanten **to** ~ **an incrustation** beschlagen **to** ~ **information** hinterbringen **to** ~ **instant response** sofort ansprechen **to** ~ **the last dye** ausfärben **to** ~ **milk** milchen **to** ~ **notice** aufkündigen, kündigen **to** ~ **off** ausliefern, (electrons) auslösen, aussenden **to** ~ **off the dash signal** aufstreichen **to** ~ **off steam** Dampf *m* abgeben **to** ~ **oneself up** (to police) sich stellen **to** ~ **out** abgeben, aussetzen, herausgeben, vergeben **to** ~ **out heat** Wärme abgeben **to** ~ **protection** honorieren, Schutz *m* gewähren **to** ~ **a quarter turn** kanten **to** ~ **a receipt** quittieren **to** ~ **rise to** Anlaß *m* geben zu **to** ~ **satisfactory results** bewähren **to** ~ **security** sicherstellen **to** ~ **a task** beauftragen **to** ~ **testimony** zeugen **to** ~ **up** aufgeben, herausgeben, preisgeben, überlassen **to** ~ **warning** aufsagen, kündigen **to** ~ **way** ausweichen, einbrechen, nachgeben, sich geben, zurückweichen, weichen **to** ~ **welding heat** schweißwarm machen

**given** angegeben, gegeben, vorgelegt ~ **reasonable care** bei richtiger Behandlung *f*

**giving,** ~ **off of gas** Gasentwicklung *f* ~ **up** Übergabe *f*

**Gjers soaking pit** Gjerssche-Ausgleichgrube *f*

**glacé** schillernd ~ **cotton** Glanzgarn *n* ~ **thread** Glanzzwirn *m*

**glacial** Eis . . ., eisig, eiskalt, kristallisiert,

Glazial . . ., Gletscher . . . ~ **acetic acid** Eisessig *m* ~ **drift** Anschwemmung *f* der Eiszeit ~ **loam** Geschiebelehm *m* ~ **marl** Geschiebemergel *m* ~ **polish** Spiegelpolitur *f*

**glacier** Gletscher *m* ~ **blue** Eisblau ~ **breeze** Gletscherwind *m* ~ **ice** Firneis *n* ~ **snowfield** Firn *m*

**glacis** flache Abdachung *f*, Böschung *f*

**glance, to** ~ blicken, streifen, vorbei-streichen, -streifen

**glance** Blende *f*, Blick *m*, flüchtiger Blick *m*, Glanz *m* (Mineral), plötzlicher Lichtstrahl *m* ~ **coal** Glanzkohle *f* ~ **cobalt** Glanzkobalt *m*

**glancing,** ~ **angle** Glanzwinkel *m* ~ **incidence** schiefer oder streifender Einfall *m* ~ **incidence of rays** streifender Strahleneinfall *m* ~ **incidence of sound wave** streifender Schalleinfall *m* ~ **ray** streifender Strahl *m*

**gland** Kappe *f*, Stopfbüchse *f*, Stopfbüchsenbrille *f* (mech.), eichelförmig gestalteter Teil *m* ~ **of stuffing box** Brille *f* der Stopfbüchse (aviat.)

**gland,** ~ **clearance** Radialspiel *n* (Turbokompressor) ~ **cock** Packhahn *m* ~ **leak-off condenser** Leckdampf-, Stopfbüchsen-kondensator *m* ~ **nut** Stopfbuchsen-, Überwurf-mutter *f* ~ **oil** Kühlungsöl *n* für Stopfbüchse, Stopfbüchsenkühlungsöl *n* ~ **ring** Packring *m* ~ **screw** Gasverschlußschraube *f* ~ **seal** Spritzdichtung *f*, ~ **sealing ring** Stopfbüchsendichtungsring *m* ~ **stud** Stopfbüchsenschraube *f* ~ **washer** Laufscheibe *f* (Kompressor)

**glandless pump** Pumpe *f* ohne Packungsstoff

**glare, to** ~ blenden, spiegeln

**glare** Blendung *f*, blendender Glanz *m*, (of projected image) Verschleierung *f* ~ **of fire** Feuerschein *m* ~ **of glass surfaces** Glasflächenspiegelung *f*

**glare,** ~**-free** blend-, reflektions-frei ~ **prevention** Blendungsschutz *m*

**glarimeter** Glanzmesser *m*

**glaring** blendend, grell ~ **light of arc** Blendwirkung *f*

**glary** blitzend, brennend

**glass, to** ~ reflektieren, spiegeln, verglasen

**glass** Beglasung *f*, Glas *n*, Glocke *f*, (lamp) Kugelglocke *f*, Linse *f*, Lupe *f*, Politur *f* ~ **for electric lighting purposes** Elektrobeleuchtungsglas *n* ~ **which has flown into the hearth** (glass mfg.) Herdglas *n* ~ **for industrial purposes** technisches Glas *n* ~ **with screw necks** Schraubenführungsglas *n*

**glass,** ~ **accumulator box** Akkumulatorglas *n*, Glasakkumulatorgefäß *n* ~ **accumulator jar** Glasakkumulatorgefäß *n* ~ **advertising goods** Glasreklameartikel *pl* ~ **air surfaces** Glasluftflächen *pl* ~ **ball (or bulb)** Glaskugel *f* ~ **balloon for the transport of sulfuric acid** Schwefelsäureballon *m* ~ **base** Glasfuß *m* ~ **batch house** Gemenghaus *n* (Glasindustrie) ~ **head** Glaskoralle *f*, -perle *f* ~ **bell** Glas-glocke *f*, -sturz *m* ~**-bending factory (or workshop)** Glasbiegerei *f* ~ **block** Glasbaustein *m* ~ **blow pipe** Glasgebläse *n* ~ **blower** Glasbläser *m* ~ **blowing** Glasblasen *n* ~**-blowing plant** Glasbläserei *f* ~ **brick** Glasbau-, Glas-stein *m* ~ **bubble** Glasblase *f* ~ **bubbles** Glasfehler *pl* ~ **bulb** Glaskolben *m* ~**-bulb rectifier** Glaskolbenventil *n* ~ **bulb**

**sprinkler** Glasfaßsprinkler *m* ~ **burnisher** Polierreiber *m* ~**-button base** Preßglassockel *m* ~ **cap** Glashaube *f* ~ **capacitor** Glaskondensator *m* ~ **capillary pen** Glaskapillarfeder *f* ~ **cartridge** Glaspatrone *f* (Schmelzsicherung) ~ **case** (showcase) Glasgehäuse *n*, Vitrine *f* ~ **cell** Glasküvette *f* ~ **chimney** Lampenzylinder *m* ~ **chord** Glasfehler *m* ~ **cloth** Glas-kattun *m*, -leinen *n* ~ **cloudiness** Glas-fehler *m*, -trübung *f* ~**-coating factory** (or **shop**) Glasbelegerei *f* ~ **cock** Glashahn *m* ~ **colored in the process** in der Masse gefärbtes Glas *n* ~ **container** Glasbehälter *m* ~ **cooling furnace** Glaskühlofen *m* ~ **cover** Deckglas *n*, Glas-decke *f*, -deckel *m*, -scheibe *f* ~ **crack** Glasfehler *m* ~ **crucible** Glashafen *m* ~ **crystallization bodies** Glasfehler *m* ~ **cutout** (or **fuse**) Glassicherung *f* ~ **cutter** Glaserdiamant *m*, Glasschneider *m* ~**-cutting knife** Glasschneidemesser *m* ~**-cutting wheel** Rollglasschneider *m* ~ **cylinder** Dute, Düte *f*, Glaszylinder *m*, Tute *f*, Tüte *f* ~ **decorating powder** Glasschnee *m* ~ **defect** Glasfehler *m* ~ **diffusing light** Trübglas *n* ~ **disk** Glasscheibe *f* ~ **embaling** Glaspackung *f* ~**-enclosed** verglast ~ **envelope** Glaswand *f* ~**-etching workshop** Glasätzerei *f* ~ **fiber** Glasfaser *f*, -fiber *n*, -gespinst *n* ~ **fiber brush** Glaspinsel *m* ~ **fiber quilt** Glasvlies *n* ~ **fiber winding machine** Glasfaserwickelmaschine *f* ~ **filter** Glasfilter *m* ~ **flask** Glaskolben *m* ~ **flaw** Glasfehler *m* ~**-fuse-link** Glasschmelzeinsatz *m* ~ **gall** Glas-galle *f*, -schaum *m*, -schmutz *m*, -schweiß *m* ~ **gauge** Wasserstandsglas *n* ~ **globe** Beleuchtungskuppel *f*, Glaszylinder *m* ~ **graduated circle** Glasteilkreis *m* ~ **grain** Glaskorn *n* ~ **granulations** Glaskörnung *f* ~ **grease cup** Nadelschmierglas *n* ~ **grinder** (tool) Glaserdiamant *m* ~**-grinding workshop** Glasschleiferei *f* ~ **guard of glow lamp** Glühlampenschutzglas *n* ~ **hard** glashart ~ **hardness** Glashärte *f* ~ **horizon** Glashorizont *m* ~ **insulator** Glasisolator *m* ~ **jewelry** Glaskurzwaren *pl* ~ **light fitting** Glasbeleuchtungsgegenstand *m* ~ **like** glasähnlich, glasartig ~ **liner** Überfangsschicht *f* ~ **lining** Glasüberfang-, Überfang-schicht *f* ~ **lozenge** Rautenglas *n* ~ **making** Glasfertigung *f* ~ **mastic layer** Glasverkitter *m* ~ **metal** Glasmasse *f* ~ **melter** Glasschmelzer *m* ~ **melting pot** Glastiegel *m* ~ **mirror** Glasspiegel *m* ~ **mold** Glasgießform *f* ~ **mosaics** Millifiori *pl* ~ **muddiness** Glastrübung *f* ~ **pane** Glasscheibe *f* ~ **panel** Glasfüllung *f* (Tür) ~ **paper** Glaspapier *n* ~ **partition** Glasverschlag *m* ~ **paving** Glaspflaster *n* ~ **pencil** Glaspinsel *m* ~ **pinch** Quetschfuß *m* (Elektronenröhre) ~ **plate** Glasdecke *f*, -plakat *n* ~ **plate-brake** Glasplattenbremse *f* ~ **plug** Abschlußglas *n* ~ **poster** Glasplakat *n* ~ **printing** Glasdruck *m* ~ **prism** Glaskeil *m*, -prisma *n* ~ **probe** Glasfühler *m* ~ **refining material** Glasläuterungsmittel *n* ~ **rod** Glasstab *m* ~**-rod separator** Glasrohrscheider *m* ~**-roll dampener** Glaswalzenanfeuchter *m* ~ **roofing** Glasbedachung *f* ~ **scale** Glasmaßstab *m* ~ **shade** Glassturz *m* ~ **shield** Schutz-glas *n*, -scheibe *f* ~ **slide** Glasschieber *m* ~ **slip** Deckglas *n* ~ **spark gap** Glasfunkenstrecke *f* ~ **spectrograph** Glasspektrograf *m* ~ **spere** Glaskugel *f* ~ **spiral** Glasschlange *f* ~ **stopper** Glas-

stopfen *m*, -stöpsel *m* ~ **strain** (or **stria**) Glasfehler *m pl* ~ **suitable for laboratory ware** Geräteglas *n* ~ **syringe** Glasspritze *f* ~ **system** Glasapparatur *f* ~ **thread** Glas-faden *m*, -gespinst *n* ~ **tile** Glasbaustein *m* ~**-to-metal seal** Glas-Metall-Verschmelzung *f* ~ **top** Glasdeckel *m* ~**-topped** mit Glasdeckel *m* versehen ~ **transition** Glasumwandlung *f* ~ **transparency** (picture thrown from rear on frosted-glass pane) Durchsichtsbild *n* ~ **trough** Glastrog *m* ~ **tube** Glaskolbenröhre *f*, Glasrohr *n* ~**-tube fuse** Glasrohrsicherung *f*, Grobsicherung *f* ~ **units** Glasbauteile *pl* ~**(melting) vase** Glashafen *m* ~ **wall(s)** Glaswandung *f* ~ **ware** Glasartikel *pl* ~ **ware etcher** Glasätzer *m* ~ **window** Glasfenster *n* ~ **wool** Glas-wolle *f*, -vlies *n* ~ **work** Hintergrunddiapositiv *n* ~ **works** Glasfabrik *f*, -hütte *f*

**glasses** Brille *f*

**glassine** Dünnpergamin *n*, Pergamin *n*, Pergamin-papier *n*, -seide *f*

**glassiness** Glasigkeit *f*

**glassing** Pressen *n*

**glassy** glasähnlich, glasartig, gläsern, glasig ~ **feldspar** glasiger Feldspat ~ **malt** Glasmalz *n*

**Glauber's salt** Glaubersalz *n*

**glauberite** Glauberit *n*

**glaucodot** Glaukodot *m*

**glauconite** Glaukonit *m*, Grünerde *f*

**glaucophane** Glaukophan *n*

**glaze, to** ~ beglasen, blank machen, Glasscheiben *pl* einsetzen, glanzschleifen, glasartiges Vereisen, (earthenware etc.) glasuren, lasieren, polieren, satinieren, überglasen, verglasen, die Scheiben *pl* in Kitt verglasen

**glaze** Anrann *m*, Glasierung *f*, glasige Oberfläche *f*, Glasur *f*, Lasur *f*, Überglasung *f*, Walzen *n* ~ **for mo(u)lds** Modellglasur *f*

**glaze, ~ baking** Glasurbrand *f* ~ **board** Preßspan *m* ~ **mill** Glasurbrecher *m* ~ **producing** glasurbildend ~ **wage** Glanzwelle *f*

**glazed** glasiert, lasiert, satiniert, verglast ~ **board** Glanzpappe *f* ~ **brick** Glasurstein *m* ~ **casings** satiniertes Packpapier *n* ~ **cotton** Glanzgarn *n* ~**-cotton braiding** Glanzgarnumklöppelung *f* ~ **fiber** Glanzfiber *n* ~ **finish** Satinappretur *f* ~ **frost** Glatteis *n* ~ **linen** Glanztuch *n* ~ **paper** Atlas-, Glacé-, Glanz-, Glas-papier *n* ~ **yarn** Glanzgarn *n*

**glazier** Glaser *m* ~**'s diamond** Glaserdiamant *m* ~**'s putty** Fensterkitt *m* ~**'s scraper** Ofenkrücke *f*

**glazing** (putting in windowpanes) Beglasung *f*, Glanz *m*, Glaserei *f*, Glasierung *f*, Glasur *f*, Glätten *n*, Schmirgeln *n*, Verglasen *n*, Verglasung *f*, Walzen *n*, Warmkalandrieren *n*; glasurbildend ~ **of paper** Appretur *f* ~ **of photographs** fotografiesatinieren

**glazing, ~ barrel** (for grain etc.) Glättfaß *n* ~ **calender** Glanz-, Glätt-kalander *m* ~ **cylinder** Blankstoßzylinder *m* ~ **filler** Ziehspachtel *f* ~ **jack** (or **machine**) Abglasmaschine *f* ~ **knife** Spachtelmesser *n* ~ **machine** (for rice or grain) Glanz-, Glanzstoß-, Satinier-maschine *f* ~ **machine for rice and grain** Glättwerk *n* ~ **mill** Glasurmühle *f*, Rundschmelzmaschine *f* ~ **penetrated into the body of pottery** in das Geschirr eingedrungene Glasur *f* ~ **process**

Glättprozeß *m* ~ **roller** Glanzstoßpendel *m*, Glattwalze *f* ~ **rollers** Kalander *m*, Satiniermaschine *f* ~ **rolls** Glasierwalzen *pl*

**gleam, to** ~ leuchten

**gleam** Schimmer *m*

**glean, to** ~ nachlesen, Ähren *pl* nachlesen, mühsam zusammentragen

**gleaning cylinder** Sammelzähler *m*

**glen** Klamm *f*, Tobel *m*

**glide, to** ~ mit einem Segelflugzeug *n* fliegen, gleiten, rutschen **to** ~ **in** anschweben **to** ~ **over** übergleiten

**glide** Gleitflug *m*, Traversengleitstück *n* ~ **band formation** Gleitbandbildung *f* ~ **element** Gleitelement *n* ~ **landing** Gleitlandung *f* ~ **path** Gleitpfad *m* (ILS), Gleitweg *m* (des Flugzeuges) ~**-path beacon** Gleitwegbake *f* ~**-path beam** Funkleitstrahl *m* (Blindlandung), Funkleitstrahl *m* zur Festlegung des Landewinkels bei der Blindlandung ~ **path transmitter** Gleitwegsender *m* (ILS) ~ **plane** Gleitebene *f* ~ **ratio** Gleitzahl *f* ~ **reflection** Gleitspiegelung *f* ~ **slope** Gleitweg *m* (ILS) ~ **zone** Leitungszone *f*

**glider** Gleitflugzeug *n*, Luftsegler *m*, (plane) Schleppflugzeug *n*, Segelflugzeug *n*, Segler *m* ~ **with seat** Sitzgleiter *m*

**glider,** ~ **bomb** Gleitbombe *f* ~ **factory** Segelflugzeugfabrik *f* ~ **kite** Gleitflugdrachen *m* ~ **landing parachute** Lastensegler-Bremsschirm *m* ~ **pilot** Gleit-, Segel-flieger *m*

**gliding** Ableitung *f* (cryst.), Abwärtsgleiten *n* (aviat.); gleitend ~ **(or sliding) ability** Gleitfähigkeit *f* ~ **altitude record** Segelflughöhenrekord *m* ~ **angle** Gleitflugwinkel *m* ~ **approach** Anschweben *n* (aviat.) ~ **boat** Wassergleitboot *n* ~**-distance** Gleitflugstrecke *f* ~**-duration record** Segelflugdauerrekord *m* ~ **field** Segelfluggelände *n* ~ **flight** Gleit-, Segelflug *m* ~ **mark scale** Segelflugbewegung *f* ~ **path** Gleiterbahn *f* ~ **plane** Gleit-bahn *f*, -fläche *f*; Gleitungsfläche *f* ~ **plane of crystals** Kristallwachstumsfläche *f* ~ **quality** Gleitvermögen *n* ~ **range** Gleitflug-, Gleitreich-weite *f* ~ **ratio** Gleitverhältnis *n* ~**-scale duty system** Gleitzollsystem *n* ~ **site** Segelfluggelände *n* ~ **speed** Gleitfluggeschwindigkeit *f* ~ **surface** Gleitfläche *f* ~ **test** Gleitflugversuch *m* ~ **turn** gleitender Kurvenflug *m*, Gleitflugkurve, Gleitwendeflug *m*

**glimmer, to** ~ flimmern

**glimpse** Schimmer *m*

**glisten, to** ~ gleißen, glitzern, schimmern

**glitter, to** ~ funkeln, spiegeln

**glitter** Funkeleffekt *m*, Glänze *f*

**global,** ~ **communication network** Weltverkehrsnetz *n* ~ **radiation** Globalstrahlung *f*

**globe** Ball *m*, Globus *m*, Glocke *f*, Erdkugel *f*, Kuppe *f*, Kugel *f*, Kugelglocke *f* **(lamp** ~ Lampenglocke *f* ~ **caliper** Kugeltaster *m* ~ **mounting** Kuppelfassung *f* ~ **photometer** Kugelfotometer *n* ~**-shaped condenser** Kugelkühler *m* ~ **(flanged)** T Kugel-T-Stück *n* ~ **valve** Durchgangsventil *n*; Kugel-ventil *n*, -schieber *m*; Niederschrauben-, Teller-ventil *n*

**globigerina ooze** Globigerinenschlamm *m*

**globoidal worm gear** Globoidschneckentrieb *m*

**globose** kugelförmig, kugelig

**globular** globulitisch, körnig, kugelartig, kugelförmig, kugelig, rundkörnig ~ **chart** Weltkarte *f* ~ **form** Glaskopf *m* (Ziegel) ~ **lens** Muschelglas *n* ~ **structure** Globulitengefüge *n*

**globule** Globulit *m*, Kügelchen *n*, Metallkorn *n*, Schweißperle *f*, Tröpfchen *n* ~ **of fog** Nebelkugel *f* ~**s of mosaic screen** Mikrofotozellen *pl*

**globulite** Kügelchen *n*, Kugelkörperchen *n*, rundes Kristallkörperchen *n*

**globulitic** globulitisch

**glonoin** Glonoin *n*

**gloom** Dunkel *n*

**glory** Aureole *f*, Brockengespenst *n*, Glanz *m*, Glorie *f*, Ruhm *m* ~ **hole** (an auxiliary furnace for finishing glassware) Auftriebofen *m*, Einbrennofen *m* (glass mfg.), Hauptbestrahlungskanal *m*

**gloss, to** ~ over vertuschen

**gloss** Glanz *m*, Glänze *f*, Glasierung *f*, Glasur *f*, Politur *f*, Überglasung *f* ~ **of cloth** Preßglanz *m* eines Tuches

**gloss,** ~ **coating lacquer** Glanzdecklack *m* ~ **ink** Glanzfarbe *f* ~ **meter** Glanzmesser *m*, Glattheitsprüfer *m*, Politurmesser *m*, Rauheitsprüfer *m* ~ **oil** Glanzöl *n*, Harttrockenglanzöl *n*, Lacköl *n* ~ **starch** Glanzstärke *f* ~ **value** Glanzzahl *f*

**glossary** Nomenklatur *f*, Spezialwörterbuch *n*

**glossimeter** Glanzmesser *m*

**glossimetric measurement** Mattierungsmessung *f*

**glossy** glänzend, glattbeschoren, geschoren ~**-black bituminous coal** Glanzkohle *f* ~ **coal** Moorkohle *f* ~ **effect** Glanzwirkung *f* ~ **finish** Glacéappretur *f* ~ **print** Hochglanzabzug *m* (photo)

**glottic catch (or cleft)** Stimmritze *f*

**glottis** Stimmritze *f*

**glove** Handschuh *m* ~ **box** Handschukasten *m* ~**-finger knitting machine** Fingerstrickmaschine *f*

**glow, to** ~ aufleuchten, flammen, glimmen, glühen, leuchten, verglühen **to** ~ **up** (sun) aufglühen

**glow** Beleuchtung *f*, Glimmen *n*, Glimmerscheinung, (light) Glimmlicht *n*, (phenomenon) Glühen *n*, Glut *f*, Schein *m* ~ **column** Glimmsäule *f*, Leuchtfaden *m* ~ **crystal** Leuchtquarz *m* ~ **current** Glimmstrom *m* ~ **discharge** Glimmentladung *f* ~**-discharge lamp** Gasentladungslampe *f*, Glimmstabilisator *m* ~**-discharge microphone** Glimmlichtmikrofon *m*, Kathodofon *m*, membranloses Mikrofon *n* ~**-discharge rectifier** Glimmentladungsventil *n* ~**-discharge tube** Gasentladungslampe *f*, Glimmlampe *f*, Glimmröhre *f* ~**-gap divider** Glimm-, Glimmlichtspannungs-, Glimmstreckenspannungs-teiler *m*, Glimmstrecke *f* (aut. contr.) ~ **ignition contact** Glimmzündkontakt *m*

**glow-lamp** Gasentladungs-, Glimm-, Glühlampe *f*, Glimmröhre *f* ~ **with plate-shaped cathode** Flächenglimmlampe *f* ~ **for switch** Schalterglimmlampe *f*

**glow-lamp,** ~ **base** Glühlampensockel *m* ~ **fittings** Glühlampenarmatur *f* ~ **reflector** Glühlampenlichtspiegler *m* ~ **sound recording** Glimmlampenschallaufnahme *f*

**glow,** ~ **plug** Glüh-kerze *f*, -körper *m* ~ **point (or potential)** Glimmspannung *f* ~**(-discharge)**

rectifier Glimmlichtgleichrichter *m* ~-tetrode
relay Vierelektrodenglimmrelais *n* ~ transfer
tube Übertragsglimmröhre *f* ~ transmitter
Glimmlichtsender *m* ~-triode relay Dreielek-
trodenglimmrelais *n*
glow-tube, ~ amplitude Amplituden-anzeiger *m*,
-glimmröhre *f* ~ potentiometer Glimmstrek-
kenspannungsteiler *m* ~ resonance indicator
Leuchtresonator *m* ~ stabilizer Glättungsröhre
*f* ~-type buzzer oscillator Glimmsummer *m* ~
voltage regulator Glimmstreckenspannungs-
teiler *m*
glowing Abatmen *n*, Glimmen *n*, Glühen *n*;
glühend ~ ashes Glühasche *f* ~ cathode Glühkathode *f* ~ coil Glühspule *f* ~ fire (annealing
furnace) Glühfeuer *n* ~ furnace Glühofen *m* ~
heat Glühfrischhitze *f* ~ iron Glüheisen *n* ~
red glutrot ~ transmitter Glühsender *m* ~ wire
Glühdraht *m*
glucina Beryllerde *f*
glucinum Beryllium *n*
glucometer Glukometer *n*
gluconic acid Glukonsäure *f*
glucosamine Glucosamin *n*
glucose Kartoffel-, Krümmel-, Stärke-, Trau-
ben-zucker *m*
glue, to ~ kitten, kleben, leimen, verkleben, verleimen to ~ in einkleben, einleimen to ~ on
ankleben, anleimen, aufleimen
glue Kleister *m*, Leim *m*, Leimung *f*, animalische
Leimung *f* ~ applicator Kleisterauftragmaschi-
ne *f* ~ boiler Leimsieder *m* ~ brush Kleisterstreichpinsel *m*, Schlichtbürste *f* ~ clamp
spindles Furnierbockspindeln *pl* ~ drying plant
Leimtrockenanlage *f* ~ enamel Leimemail *n* ~
factory Leimfabrik *f* ~ filter Kleistersieb *n* ~
foil Leimfolie *f* ~ penetration Ausschwitzen *n*
(Leim) ~-pot Leimbecken *n* ~ powder Leim-
pulver *n* ~ press Leimknecht *m* ~ spreader
Leimausbreiter *m* ~ stock Leimrohstoff *m*
glued, ~ with caseine kaseinverleimt ~ joint
Leimstelle *f*
gluey klebrig, leimig
gluing, to ~ leimen, kleben
gluing, ~ ability to lend itself to ~ Leimfähigkeit
*f* ~ and labeling machine Gummier- und Bekle-
bemaschine *f* ~ apparatus Klebeapparat *m* ~
component Klebemasse *f* ~ cramp Leimknecht
*m* ~ device Anleimvorrichtung *f* ~ edge Lappen
*m* ~ fixture Leimvorrichtung *f* ~ plant Ka-
schieranlage *f* ~ seam Klebenaht *f* ~ strip
Leimleiste *f*
glut Sättigung *f*, Stillung *f*, Überangebot *n*,
Schwemme *f*
glutamic acid Aminoglutarsäure *f*
gluten Gluten *n*, Kleber *m*
glutin-turbid glutintrüb
glutinization Verkleisterung *f*
glutinous klebrig, leimartig
glutted überhäuft
glyceric acid Glyzerinsäure *f*
glycerin Glyzerin *n*, Ölsuß *n* ~ brake Glyzerin-
bremse *f* ~ coating Glyzerinüberzug *m*
glycerol Glycerin *n*
glycine Glycine *n*
glycol Glycol *n* ~ installation Heißkühlungsvor-
richtung *f*, Motor *m* mit Heißkühlung *f*
glycosuria Glykosurie *f*

glyoxal acid Glyoxalsäure *f*
glyoxime Diacetyldioxim *n*
glyoxylin Glyoxylin *n*
G. M.-counter Geiger-Müller-Zählrohr *n*
gnarled knotig
gnat Schnake *f*
gnaw, to ~ ausnagen, knabbern, zerfressen to ~
through zernagen
gnawing from inside out Ausnagung *f*
gneiss Gneis *m*
gneissic formation Gneisformation *f*
gnomon Gnomon *m*, Sonnenzeiger *m*
gnomonic gnomonisch ~ chart gnomonische
Karte *f* ~ projection gnomonische Projektion
*f*, Mittelpunktprojektion *f* ~ reciprocal projection gnomonische Reziprokalprojektion *f*
go, to ~ arbeiten, gehen, laufen, ziehen to ~
ahead in Angriff *m* nehmen to ~ around durchstarten (Flugzeug), umgehen to ~ back zurückgehen to ~ behind nacheilen, nachgeben, nachlaufen to ~ down fallen, niedergehen, sinken,
vermindern to ~ into hineinwandern to ~ into
a nose dive abdrehen (aviat.) to ~ off abgeben
to ~ out ausgehen, ausströmen, austreten to ~
over durchlesen, durchsehen to ~ through
durchgehen to ~ up ansteigen, anziehen, aufgehen, aufsteigen to ~ without saying auf der
Hand *f* liegen
go, "~" and "not go" dimension Gut- und Aus-
schußmaß *m* ~-and return line Hin- und Rückleitung *f* ~-and-return measurement Schleifenmessung *f* ~-and-return test Schleifenmessung
*f* ~-around Durchstarten *n* (aviat.) ~-devil
Laufteufel *m* ~-no-go-circuit Gut-Schlecht-
Prüfkreis *m* ~-no-go-distance Startabbruch-
entfernung *f* ~-no-go-gauge Vorrichtungs-
anschlaglehre *f* ~-no-go-point Startabbruch-
punkt *m* ~-no-go-speed Startkontrollgeschwin-
digkeit *f* ~-no-go-test Funktionsprüfung *f*
"~"-plug screw gauge Gutlehrdorn *m* "~"-
ring gauge Gutlehrring *m* "~"-screw plug
member Gut-Gewindelehrzapfen *m* "~"-
setting gauge Guteinstellehre *f* "~"-thread
ring gauge Gutgewinderinglehre *f*
goal Ziel *n*, Zielpunkt *m* ~ flight Zielflug *m*
gobbing Berg *m*, Berg(e)versatz *m*, Versatzung *f*
goblet Becher *m*
gobo Blendenschirm *m*, Blendschirm *m* (acoust.,
film), Schalltilgungsmittel *n* (film), Stoffbedeckung *f*
Goerz-effect shutter Kombinationsblende *f*
goffering Gaufrieranstalt *f*, Gaufrieren *n*
goggles Autobrille *f*, Brille *f* ~ with gauze frame
Siebbrille *f*
going, ~ ahead vorwärtsstrebend ~ down untergehen ~ forward and backward Vor- und Rückwärtslauf *m* ~ into position Instellunggehen *n*
~ to press Drucklegung *f*
goiter Kropf *m*
gold Gold *n* ~ alloy Goldlegierung *f* ~ amalgam
Goldamalgam *n*, Quickgold *n* ~-and alumi-
nium-leaf electroscope Goldblatt- und Alu-
miniumblattelektroskop *n* ~ assay Goldprobe
*f* ~ basis Goldbasis *f* ~ bath Goldbad *n* ~-
bearing gold-führend, -haltig ~ beater Gold-
schläger *m* ~ beater's skin Goldschlägerhaut *f*
~ beater's waste Schabin *n* ~ beating Blatt-
goldschlägerei *f* ~ blocking Blattvergoldung *f*

~ blocking press Goldprägepresse f ~ bromide Goldbromür n ~ bronze Goldbronze f ~ buillion Rohgold n ~ carat Goldfeingewicht n ~ chips Goldabfall m ~ chloride Gold-chlorid n, -chlorür n ~ coating Gold-anstrich m -belag m, -beleg m ~ coin Goldstück n ~ color Goldfarbe f ~-colored goldfarbig ~ content Goldgehalt m ~ cup Goldschale f ~ cupel Goldschale f ~ cyanide Goldzyanid n, Zyangold n ~ digger Goldgräber m ~ dish Goldschale f ~ doping Goldzusatz m ~ dredging Goldbaggern n ~ dross Goldkrätze pl ~ dust Goldstaub m, Streugold n ~ embossing Goldprägung f ~ figure Goldzahl f ~ filigree threads Goldgespinst n ~ fixative bath Goldfixierbad n ~ foil Blatt-, Blüten-gold n; Gold-belag m, -beleg m, -blatt n, -folie f ~ fountain pen Goldfüllfederhalter m ~ grains Goldkörner pl ~-lace maker Goldbortenwirker m ~ leaching plant Goldlaugerei f ~ leaf Blatt-, Blüten-gold n; Gold-blatt n, -schaum n, -schlag m ~-leaf electrometer (or electroscope) Goldblattelektrometer n ~ like goldartig ~ litharge Goldglätte f ~ metal Goldmetall n ~ mine Gold-bergwerk n, -grube f ~-mining machinery Goldaufbereitungsanlage f ~ nugget Goldklumpen m ~ number Goldzahl f ~ ore Golderz n ~ oxide Gold-kalk m, -oxyd n ~ paper Goldpapier n ~ particle Goldteilchen n ~ parting Gold-scheiden n, -scheidung f ~ piece Goldstück n to ~-plate vergolden ~ plate Goldplattierung f ~-plated goldplattiert, vergoldet ~-plated contacts vergoldete Kontakte pl ~-plated metal Golddoublé n ~ plating Gold-belag m, -beleg m, -plattierung f, Vergoldung f ~ powder Goldpulver n ~ precipitate Goldniederschlag n ~ printer Goldschreiber m ~ quartz Goldkies m ~ refiner Goldscheider m ~ refinery Goldscheideanstalt f ~ refining Gold-raffination f, -scheiden n, -scheidung f ~ rolling Goldwalze f ~ salt Goldsalz n ~ size Gold-grund n, -leim m ~ sizer Vergoldungsgrundierer m ~ slime Goldschlich m ~ smalts Goldamalten pl ~ smith Goldschmied m ~ sodium chloride Goldnatriumchlorid n ~ solution Gold-auflösung f, -lösung f ~-stamping works Goldprägeanstalt f ~ standard (of coinage) Gold-basis f, -währung f ~ sulfate Goldvitriol n ~ sulfide Goldsulfid n ~ test Gold-probe f, -streichen n ~ trimming Goldtresse f ~ varnish Gold-firnis m, -lack m ~ vein Goldader f ~ weight Goldgewicht n ~ wire (or wire relay) Golddraht m ~ yield Goldausbringen n

**Goldberg wedge** Goldbergkeil m, Graukeil m

**golden** goldartig, golden ~ bright goldhell ~ brown goldgelbbraun n ~ luster Goldglanz m ~ yellow goldgelb

**Goldschmidt, ~ alternator** Goldschmidthochfrequenzmaschine f ~ tone wheel Goldschmidttonrad n

**goliath crane** Schwerlastkran m

**gonad** Gonade f

**gondola** Frachtwagen m, Gondel f, Luftschiffsgondel f, offener Frachtwagen m ~ car Flachwagen m ~ guide trolley Gondelwagen m

**gone** zurückgelegt (Strecke)

**gong** große flache Glockenschale f ~ bell Schalmeiwecker m ~ signal Glocken-halter m,

-zeichen n ~ support Schalenhalter m

**goniometer** Gesichtsfeldmesser m, Goniometer n, Gonios n, Peiler m, Winkel-messer m, -meßinstrument n

**goniometric** goniometrisch, winkelmessend ~ sight Winkelvisier n

**goniometry** Goniometrie f, Winkelmessung f

**good** tauglich on ~ authority quellenmäßig in ~ bond (masonry) verbandsmäßig

**good** Nutzen m, Wert m ~ appearance repräsentable Erscheinung f ~ articulation deutliche Aussprache f ~ balance (uniformity) Ausgeglichenheit f ~ definition (of image) getreue Wiedergabe f (TV) ~ finish gutes Aussehen n ~ ignition quality Zündwilligkeit f ~ keeping firmness Haltbarkeit f ~ keeping qualities Haltbarkeit f ~ keeping strength Haltbarkeit f ~ point Gutpunkt m ~ prize gesetzmäßige Prise f ~-quality trough Melierte f ~ sense Klugheit f ~ traction Haften an der Straße f ~ view (or visibility) gute Sicht f ~ working garer Gang m

**goodman's diagram** (for effect of range of stress) Dauerfestigkeitsschaubild n nach Goodman

**goodness** Arbeitssteilheit f (Röhre) ~ of a thermionic valve Güte f einer Röhre f

**goods** Artikel m, Gut n, Güter pl, Werkstoff m, Zeug n ~ commanding a ready sale gangbare Ware f ~ to be conveyed Transportgut n ~ to be dried Trocknungsgut n ~ of high nutritive value hochwertige Lebensmittel pl ~ of a ready sale gangbarer Artikel m, gangbare Ware f ~ shipped by express Expressgut n ~ sold for cash gegen bar verkaufte Ware f the ~ sweats die Ware f beschlägt ~ to be transported Fördergut n

**goods traffic** Güterverkehr m

**goodwill** Geschäftswert m

**gooseneck** Anschlußstück n, Schwanenhals m, Spinnstutzen m, Spülkopfknie f ~ tiller gebogene Ruderpinne f

**Gordon** (gashouse-liquor) **process** Gordonprozeß m

**gore, to** ~ aufspießen, durchbohren, zuwickeln

**gore** Bahn f (parachute), Gehre f, Keil m, Rockbahn f, Zwickel m ~ folding (parachute) Legen n der Bahnen

**gorge, to** ~ kehlen

**gorge** Hohlweg m, (fort.) Kehle f, Rinne f, Schlucht f ~ wheel Kehlrad n

**goslarite** Galitzenstein m

**gosport** biegsames Sprachrohr n

**gossan** Eisenhut m, eiserner Hut m

**gossaniferous clay** ockerhaltiger Letten m

**Gothic** gotisch ~ arch Spitzbogen m ~ character Frakturschrift f ~ letter Fraktur f ~ minuscule gotische Minuskel f ~ type (or writing) Frakturschrift f

**göthite** Eisenglimmer m, Göthit m, Nadeleisenerz n

**gouge** Boussiereisen n, Güdse f, Gutsche f; Hohl-beitel m, -eisen n, -meißel m; Letten m (min.), Rundmeißel m, Salband n (geol.), Verwerfungslette f ~ water seepage into underground workings Flukkan n (min.)

**gouging chisels (or planes)** Kehlzeug n

**Goulard water** Bleiwasser n

**govern, to** ~ herrschen, leiten, regeln, regulieren

**governing** Kontrolle *f*; maßgebend ~ **by opening series of nozzles in rotation** Düsengruppensteuerung *f*
**governing,** ~ **control** Steuerung *f* ~ **impulse** Gleichlaufstromstoß *m* ~ **lever** Regelhebel *m* ~ **rack travel** Regelstangenweg *m* ~ **shaft** Regler-, Steuer-welle *f* ~ **spring** Regelfeder *f*
**government** Regierung *f* ~ **agency** Behörde *f* ~ **authority** Regierungsbehörde *f* ~ **call** Staatsgespräch *n* ~ **message** Staats-gespräch *n*, -**telegramm** *n* ~ **priority call** dringendes Staatsgespräch *n* ~ **property** Staatseigentum *n* ~ **tax** Staatsabgabe *f*
**governor** Drehzahlregler *m*, (speed) Geschwindigkeitsregler *m*, (engine) Maschinenregulator *m*, Motorregler *m*, Regelgerät *n*, Regler *m*, Regulator *m*, Sicherheitsregler *m*, Ventil *n* ~ **and racker** Spar- und Abfüllapparat *m* ~ **arms** Reglergabel *f* ~ **ball** Schwungkugel *f* des Regulators ~ **base** Reglersockel *m* ~ **bracket** Reglerwinkel *m* ~ **centrifugal pendulum** Fliehkraftpendel *n* ~ **control** Reglerschaltung *f* ~ **control handle** Reglerkontrollhebel *m* ~ **droop characteristics** dauernde Ungleichförmigkeit *f* des Reglers ~ **eccentric control** Regulierexzenter *m* ~ **fitted with adapting device** Regler *m* mit Angleichung ~ **force** Reglerkraft *f* ~ **gearing** Reglergetriebe *n* ~ **housing** Reglergehäuse *n* ~ **hub** Reglernabe *f* ~ **level** Verstellhebel *m* ~ **regulation** Ungleichförmigkeitsgrad *m* ~ **relief valve** Überdruckventil *m* ~ **rod** Regelgestänge *n* ~ **rod sleeve** Regelstangenhülse *f* ~ **rod stop** Regelstangenanschlag *m* ~ **shaft** Reglerachse *f* ~ **sleeve** Reglermuffe *f* ~ **slide valve** Reglerschieber *m* ~ **spindle** Regulatorachse *f* ~ **spring** Reglerfeder *f* ~ **tap pen** Regulierstößel *m* ~ **valve** Reglerventil *n* ~ **weight** Reguliergewicht *n*
**grab** Exkavator *m*, Fänger *m*, Geschoßheber *m*, (of a crane) Greifer *m*, Greiferkübel *m*, Greifkorb *m*
**grab,** ~ **and bucket service** Greifer- und Kübelbetrieb *m* ~ **for valve clacks** Ventilkapselfänger *m*
**grab,** ~ **blade of a dredger** Greiferschale *f* ~ **bucket** Exkavator *m*, Greifer, Greifer-gefäß *n*, -korb *m*, -kübel *m*; Greifschaufel *f*, Krangreifer *m* ~-**bucket conveyer** Greiferbagger *m* ~ **crane** Greiferkran *m* ~ **dredge** Greifbagger *m* ~-**equipped trolley** Greiferkatze *f* ~ **excavator** Greifbagger *m* ~ **filling** Greiferfüllung *f* (Inhalt) ~ **hoist** Greiferwindwerk *n* ~ **hook** Greifhaken *m* ~ **jaw** Greifschaufel *f* ~ **shell of a dredger** Greiferschale *f* ~ **spade** Greiferschaufel *f* ~ **transporter** Greif-, Greifschaufelförderer *m*
**grabbing** Greiferwinde *f* ~ **excavator with luffing jib** Greifbagger *m* mit Wippausleger *m* ~ **gear** Greifereinrichtung *f* ~ **service** Greiferbetrieb *m*
**grace** Frist *f*, Fristverlängerung *f* ~ **period** (patents) Respiro *m*
**gradal lens** Gradalglas *n*
**gradation** Abstimmbarkeit *f* (photo), Abstufung *f*, Grad *m*, Gradation *f*, Stufe *f*, Stufenfolge *f* **by** ~ **of** in Stufen von ~ **of colors** Farbenabstufung *f* ~ **of grain sizes** Kornzusammensetzung *f* ~ **of light** Lichtabstufung *f* ~ **of shade** Nuancenabstufung *f* ~ **of tones** Abstufung *f* der Helligkeitsunterschiede

**gradation correction** Gradationsentzerrung *f*
**grade, to** ~ abstufen, auslesen, (by quality) einstufen; einteilen, klassieren, klassifizieren, nivellieren, planieren, separieren, sichten, sondern, sorten, sortieren, staffeln
**grade** Dienstgrad *m*, Grad *m*, Güte *f*, (slope, quality) Gütestufe *f*, Klasse *f*, Marke *f*, Qualität *f*, Rang *m*, Stufe *f*
**grade,** ~ **of accuracy** Genauigkeitsstufe *f* ~ **of brick** Ziegelsorte *f* ~ **of coal** Kohlensorte *f* ~ **of coke** Koksgüte *f* ~ **of concentration** Konzentrationsgrad *m* ~ **of copper** Kupfersorte *f* ~ **filtration** Filterfeinheit *f* ~ **of fit** Art *f* oder Gütegrad *m* einer Passung ~ **of ingot steel** Flußstahlart *f* ~ **of iron** Eisensorte *f* ~ **of porosity** Durchlässigkeitsgrad *m* ~ **of service** Betriebsgüte *f*, Verlustziffer *f* ~ **of steel** Eisensorte *f* **at the same** ~ in gleicher Höhenlage *f*
**grade,** ~ **chevron** Gradabzeichen *n* ~ **crossing** Bahn-, Niveau-übergang *m*, Überführung *f* in gleicher Weghöhe ~ **line** Gradientlinie *f*
**graded** abgestuft, sortiert, stufenartig ~ **in echelon formation** gestaffelt
**graded,** ~ **circle** Gradkreisteilung *f* ~ **logarithmically** logarithmisch abgestuft ~ **multiple** Teilvielfachfeld *n* ~ **post fixing maximum permissible elevation of reservoir-water surface** Heimpfahl *m* ~ **potentiometer** nichtlinearer Spannungsteiler *m* ~ **sand** kalibrierter oder klassifizierter Sand *m* ~ **stops** stufenförmig angeordnete Blenden *pl*
**grader** Planierer *m*, Scheider *m*, Schleuder *f*, Separator *m*, Sichter *m*, Sortierer *m*, Trenner *m*
**gradiated rate** Staffelgebühr *f*
**gradient** Anstieg *m* (math.), Bodenneigung *f*, Böschungswinkel *m*, Gefälle *n*, Geländeneigung *f*, geneigte Fläche *f*, Gradient *m*, Neigungslinie *f*, Neigungsverhältnis *n*, schiefe Ebene *f*, Steigung *f*, Steilheit *f*, Tiefenstufe *f*; steigend ~ **of climb** Steigwinkel *m* ~ **of slope** Hangwinkel *m*
**gradient,** ~ **board** Neigungsanzeiger *m* ~ **curve** Gefällkurve *f* ~ **indicator (or meter)** Steigungsmesser *m* ~ **method** Stufenmethode *f* ~ **microphone** Schalldruckgradientmikrofon *n* ~ **profile** Streckenprofil *n* ~ **recorder** Gefällemesser *m*, Gradometer *n* ~ **resistance** Steigungswiderstand *m* ~ **strip** Gradientstreifen *m* ~ **variation** Gefällsgang *m*
**grading** Anreicherung *f*, Ansteigung *f*, Aufbereitung *f*, Einteilung *f*, Güteeinteilung *f*, Klasierung *f*, Klassifizierung *f*, Körnung *f* (Sand), Planierarbeiten *pl*, Separation *f*, Sortierung *f*, Staffeln *n*, Staffelung *f* ~ **of grain** Korngrößentrennung *f* ~ **of tones** Tonabstufung *f*
**grading,** ~ **chart** Einteilungsplan *m* ~ **(or sizing) coal** Kohleklassierung *f* ~ **plant** Klassieranlage *f* ~ **plow** Pflug *m* für Landstraßenbauzwecke ~ **room** Sortierraum *m* ~ **scheme** Staffelungsplan *m* ~ **screen** Klassiersieb *n*
**gradiometer (or gradometer)** Gradio-, Gradometer *n*, Neigungsmesser *m*
**gradual** fortschreitend, stufenartig ~ **application of stress** allmähliche Belastung *f* (Werkstoffprüfung) ~ **blending** Stufenverschmelzung *f* ~ **breaking** abgestufte Bremswirkung *f* ~ **change of direction** allmähliche Richtungsänderung *f* ~ **drying** Stufentrocknung *f* ~ **measuring** punktweise Ausmessung *f* ~ **rate of delivery**

Stufenförderhöhe *f* ~ **rise** Anklingen *n* ~ **rise in resonance** Hinaufpendeln *n* ~ **waxing** Anklingen *n*
**gradually** absatzweise, allmählich, stufenweise
**graduate, to** ~ abstufen, einteilen, in Grade *pl* einteilen, (a scale) gradieren, graduieren, mit Meßeinteilung *f* versehen, teilen
**graduate** Absolvent *m*, Mensur *f*
**graduated** abgestuft, stufenartig, stufig ~ **arc** Grad-, Skalen-bogen *m* ~ **bank** Stufenbank *f* ~ **burette** Meßbürette *f* ~ **card** Gradrose *f* ~ **circle** Kreis-einteilung *f*, -teilung *f*; Meß-, Teil-kreis *m* ~ **circle of the compass** Stundenring *m* ~ **cylinder** Maßglas *n*, Meßzylinder *m* ~ **dial** Strichplatte *f* ~ **disk** Skalascheibe *f* ~ **drawtube** geteilter Tubusauszug *m* ~ **drum** Meßtrommel *f*, -walze *f* ~ **engineer** Diplomingenieur *m* ~ **flask** Meß-flasche *f*, -kolben *m* ~ **glass** Paßglas *n* ~ **glass jar** graduierte Glocke *f* ~ **hydrometer** Skalenaräometer *n* ~ **jar** Paßglas *n* ~ **measure** Meßgefäß *n* ~ **measuring glass** Meßglas *n* ~ **metal gauge rod** kalibrierter Metallpeilstab *m* ~ **oil dispenser** Ölmeßbecher *m* ~ **pipette** Meßpipette *f* ~ **plate** Strichplatte *f*, Teilscheibe *f* ~ **ring** Marken-, Seitenteil-, Skalen-, Teil-ring *m* ~ **rule** Skalalineal *n* ~ **scale** Gradabteilung *f*, kalibrierte Skala *f*, Teilstrichwaage *f* ~ **scale gear** Seitenrichtteilung *f* ~ **slide-wire** kalibrierter Schleifdraht *m* ~ **straight-edge** Meßschiene *f* ~ **tariff** Staffeltarif *m* ~ **traversing gear** Seitenrichtteilung *f* ~ **tube** Maßnahme *f*, Maßregel *f*, Maßröhre *f*, Meßrohr *n* ~ **vessel** Maßgefäß *n*, Mensur *f*, Meßgefäß *n*
**graduating,** ~ **machine** Anreißteil-, Masseneinteilungs-maschine *f* ~ **stem guide** Graduierstangenführung *f*
**graduation** Abstufung *f*, Ausgleichung *f* (math.), Eichstrich *m*, Einteilung *f*, Gradeinteilung *f*, Gradierung *f* (chem.), Gradmessung *f*, Gradteilung *f*, Graduierung *f*, Kreisteilung *f* (opt), Maßteilung *f*, Skala *f*, Staffelung *f*, Stricheinteilung *f* (scale), Teilstrich *m* (on cale), Teilstrichteilung *f*, Teilung *f*, Verleihung *f* eines akademischen Grades
**graduation,** ~ **for angular setting of wheel head** Teilung *f* für die Winkeleinstellung des Schleifscheibenkopfes ~ **into degrees** Graduierung *f* ~ **of grain sizes** Kornzusammensetzung *f* ~ **on head** (of panoramic telescope) Kopfteilung *f* ~ **of scale** Skaleneinteilung *f* ~ **by ten-degree divisions** Zehngradteilung *f* ~ **of tonal intensities** Tonabstufung *f*
**graduation,** ~ **certificate** Lehrzeugnis *n* ~ **house** Gradierwerk *n* ~ **mark** Ablesestrich *m*, Gradbogenteilstrich *m*, Tiefenzahl *f* ~ **photometer** Stufenfotometer *n*
**graduator** Graduator *m*, Induktanzspule *f*
**graft, to** ~ aufpropfen, pfropfen
**grafting** Aufpfropfung *f* ~ **knife** Okulier-, Pfropf-messer *n* ~ **saw** Pfropfsäge *f*
**Graham escapement** Grahamgang *m*
**grain, to** ~ ädern, aussalzen, (Leder) enthaaren, granulieren, körnen, masern, (Papier) narben
**grain** Faden *m*, Faser *f* (Holz), Faserung *f*, Gefüge *n*, Getreide *n*, Graupe *f*, Kern *m*, Korn *n* (eines Negativs), Körnchen *n*, Körner *pl*, Körnung *f*, Krebs *m* (im Ton), Maserung *f*,

Narbe *f*, (leather) Narbung *f*, Strich *m*, Struktur *f*
**grain,** ~ **in the mortar** Kalkkrumpe *f* ~ **of emulsion** Plattenkorn *n* (phot.) ~ **of the film** Körnigkeit *f* (des Films) ~ **of grinding band** Bandkörnung *f* ~ **in the mortar** Kalkkrumpe *f* ~ **of ore** Erzkorn *m* ~ **of a photographic layer** Korn *n* einer fotografischen Schicht ~ **of sleet** Graupelkorn *n* ~ **of wood** Holzmaserung *f*, Längenfaser *f* im Holz
**grain,** ~ **aggregation** Kornanhäufung *f* ~ **alcohol** Äthylalkohol *m* ~ **binder** Bindemähmaschine *f*, Garbenbinder *m* ~ **boundary** Korn-begrenzung *f*, -grenze *f*, -grenzlinie *f*, Kristallgrenze *f* ~ **box with register** Getreidekasten *m* mit Registrierungsapparat ~ **center** Keim-zenter *m*, -zentrum *n* ~**-cleaning machine** Getreidereinigungsmaschine *f*, Trior *m* ~ **colony** Kristallgebilde *n* ~ **counter** Granometer *m* ~ **deformation** Kornverzerrung *f* ~ **density** Korndichte *f* ~ **diameter** Korndurchmesser *m* ~ **direction** Laufrichtung *f* (paper mfg.) ~ **disintegration** Kornzerfall *m* ~ **drier** Getreidetrockner *m* ~ **drier firebox shell** Getreidetrocknerbrennkammer *f* ~ **drill** Getreidedrillmaschine *f* ~ **elevator** Getreide-elevator *m*, -heber *m*, -speicher *m* ~ **field** Getreidefeld *n* ~ **filler** Porenfüller *m* ~ **flow** Faserverlauf *m* ~ **formation** Kornbildung *f* ~ **gauge** Schrotstärke *f* ~ **growth** Kornwachstum *n* ~ **junction line** Korngrenzlinie *f* ~ **lead** Kornblei *n* ~ **lifter** Ährenheber *m*, Garbenträger *m* ~ **mixer** Getreidemischvorrichtung *f* ~**-oriented** (laminations) kornorientiert ~ **refinement** Kornverfeinerung *f* ~ **scales** Gedreidewaage *f* ~ **shocker** Garbenhocker *m* ~ **side** Haarseite *f* ~ **size** Korngröße *f* ~**-size classification** Korngrößeneinteilung *f* ~**-size curve** Siebkurve *f* ~**-size distribution** Korn-anteil *m*, -größeneinteilung *f*, -verteilung *f* ~**-size distribution curve** Kornverteilungskurve *f* ~**-size fraction** Kornanteil *m* ~ **spelter** Stechlot *n* ~ **storage** Getreidelagerung *f* ~ **structure** Härtebild *n*, Korn-gefüge *n*, -gestalt *f*, -struktur *f* ~ **tester** Farinatom *n* ~ **tin** Kornzinn *n*, Zinnpulver *n*
**grained** echtgefärbt (Wolle), faserig, gefasert, gekörnt, (of wood) geflammt, gemasert, genarbt (Leder), körnig, maserig ~ **enamel paper** Kreidekornpapier *n* ~ **leather** genarbtes Leder *n* ~ **paper** gemasertes Papier *n* ~ **wood** Maserholz *n*
**graininess** Körnigkeit *f*
**graining** Kornbildung *f*, Maserung *f*, Rauhen *n* ~ **in a sprayed film** Spritznarben *n*
**graining,** ~ **board** Krispelholz *n* ~ **color** Einlaßfarbe *f* (Antikleder) ~ **machine** Narbenpreßmaschine *f* ~ **machine for offset zinc-plates** Zinkplatten-Schleif- und Kernmaschine *f* ~ **marbles** Märbeln *pl* ~ **sand** Kornsand *m* ~ **sieve** Granuliersieb *n* ~ **steel** Granierstahl *m*
**grains** Körnerpräparat *n*
**gram** (metric unit of weight) Gramm *n* ~ **atom** Atomgramm *m*, Grammatom *n* ~ **calorie** Grammkalorie *f*, kleine Kalorie *f* ~ **degree** Grammgrad *m* ~**-element specific** spezifische Aktivität *f* pro Grammatom ~ **equivalent** Grammäquivalent *n* ~**-gage** Prüfgewicht *n* ~**-molecular volume** Grammolekül-, Mol-

volumen *n* ~-molecular weight Mol-, Mole-
kular-gewicht *n* ~ molecule Grammol *n*,
Grammolekül *n*, Mol *n*
gramophone Grammofon *n* ~ disk Grammofon-
platte *f* ~ needle Grammofonnadel *f* ~ socket
Plattenspieleranschluß *m* ~ sound box Gram-
mofonschalldose *f*
granary Frucht-haus *n*, -kammer *f*, Speicher *m*
grand großartig ~ piano Flügel *m* ~ stand
Tribüne *f* ~ total Gesamtsumme *f*
granite Granit *m* ~ ax Granit-, Hart-haue *f*
(min.) ~ drill Gesteinbohrer *m* für hartes
Gestein ~ paper Granitpapier *n* ~ pegmatite
Granit-Pegmatit *m* ~ roller Granitwalze *f* ~
sets Granitpflastersteine *pl* ~ weave Granit-
bindung *f*
granitic granit-ähnlich, -artig ~ mountain (or
ridge) granitischer Gipfel *m* ~ rock Granit-
gestein *n*
grant, to ~ bewilligen, erteilen, gestatten, gewähren, übertragen, zugeben, zusprechen
to ~ abatement Rabatt *m* bewilligen oder
geben to ~ asylum Schutz *m* gewähren to ~
leave of absence beurlauben to ~ a patent gewähren to ~ respite for payment stunden to ~
security Bürgschaft *f* gewähren
grant Beihilfe *f*, Beisteuer *f*, Bewilligung *f*, Verleihung *f* ~ of letters patent Erteilung *f* eines
Patentes ~ of license Verleihung *f* einer Genehmigung ~ of the minerals Bergeverleihung *f*
~ of respite Fristgewährung *f*
grant ~-in-aid Hilfeleistung *f*, Subvention *f*,
Zuschuß *m*
grantable gewährbar
granted (patents) erteilt
grantee Patentinhaber *m*
granting Gewährung *f* ~ a license Erteilung *f*
einer Konzession
grantor Lizenzerteiler *m*, Übertragender *m*
granular gekörnt, graupig, körnig ~ bog-iron
ore Eisengraupen *pl* ~ cementite körniger
Zementit *m* ~ coherer Körnerfritter *n* ~-
crystalline grobkristallin, körnig, kristallinisch
~ crystalline fracture grobkörnige Bruchfläche
*f* ~ form Körnerform *f* ~ noise Eigenrauschen
*n* (electron.) ~ ore Graupenbett *n* ~ oxidation
Granulier-, Korn-rost *m* ~ snow körniger
Schnee *m* ~ structure Kornstruktur *f* ~ tin
Kornzinn *n* ~ transmitter Körnermikrofon *n*
granularity Körnigkeit *f*
granulate, to ~ aufrauhen, granulieren, körneln,
körnen, schroten, verkörnen, zerkörnen
granulated gekörnt ~ character griesiges Aussehen *n* ~ hammer Stockhammer *m* ~ material
Granalien *pl* ~ metal Granalien *pl*, Körner *m* ~
powder Kornpulver *n* ~ screen Kornraster *m* ~
slag Schaumschlacke *f* ~ structure Konstruktur
*f* ~ sugar Kornzucker *f* ~ tin Tropfzinn *n*, Zinn-
granalien *pl* ~ zinc Zinkgranalien *pl*
granulating Rauhen *n* ~ ferric hydroxide Schlak-
kenrost *m* ~ hammer Stockhammer *m* ~
plant Granulationsanlage *f* ~ roller machine
Walzenkörnmaschine *f*
granulation Granulation *f*, Granulierung *f*,
Kornbildung *f*, Körnen *n*, Körnigkeit *f*,
Körnung *f*, (of sugar) Kristallbildung *f*,
Schroten *n*, Umkristallisation *f* ~ plant Granu-
lierungsanlage *f*

granulator (for chemicals, etc.) Granulier-
apparat *m*
granule Granüle *f*, Korn *n*, Körnchen *n* ~
microphone Schüttelmikrofon *n*
granules Körner *pl*
granulite Granulit *m*, Weißstein *m*
granulometer Körnigkeitsmesser *m*
granulometric, ~ analysis Korngrößenbestim-
mung *f* ~ gradation Kornabstufung *f*
granulometry Kornverteilung *f*
granulopenia Granulopenie *f*
granulose Granulose *f*
grape, ~ oil Traubenöl *n* ~ trough (or press)
Garküfe *f*
graph, to ~ grafisch darstellen
graph Aufstellung *f*, Band *n*, Darstellung *f*,
Diagramm *n*, grafische Darstellung *f* oder
Zeichnung *f*, Kurve *f*, Kurven-bild *n*, -blatt *n*,
-darstellung *f*, -schaubild *n*, -tafel *f*; Schaubild *n*, -linie *f*, -tafel *f*, Tafel *f* ~ of arithmetrical
probabilities Gaußsches Verteilungsnetz *n* ~ of
errors grafische Fehlertafel *f*
graph paper Koordinaten-, Millimeter-papier *n*
graphecon Bildspeicherröhre *f*
graphic bildlich, grafisch, schaubildlich, (granite)
schriftgranitisch, zeichnerisch
graphic, ~ arts Grafik *f* ~ course finder Kursschieber *m* ~ determination grafische Bestimmung *f* ~
formula Strukturformel *f* ~ granite Pegmatit *m* ~
institute grafische Anstalt *f* ~ instrument Schreib-
instrument *n* ~ kinematics geometrische Bewegungslehre *f* ~ panel Betriebsschaubild *n* ~
point grafischer Punkt *m* ~ range table grafische Schußtafel *f* ~ recorder Bandchrono-
graf *m* ~ recording Registrierung *f* ~ representation grafische Darstellung *f*, Kurvenbild *n*
~(al) representation Kurvendarstellung *f* ~
scale grafischer Maßstab *m*, Maßstableiste *f* ~
solution zeichnerische Auswertung *f* ~ stone
Pegmatit *m* ~ tellurium Weißtellur *n*
graphical bildlich, grafisch ~ calculation Kur-
venberechnung *f* ~ composition of forces grafische Kräftekomposition *f* ~(ly) determined
zeichnerisch ermittelt ~ firing table Artillerierechenschieber *m* ~ interpolation grafische
Interpolation *f* ~ method (Müller-Breslau) for
determining joint deflections in a truss Stabzug-
verfahren *n* ~ representation grafische Darstellung *f* ~ scale grafischer Maßstab *m* ~
solution zeichnerische Lösung *f* oder Auswertung *f* ~ static Graphostatik *f*
graphite, to ~ graphitieren, mit Graphit *m* behandeln
graphite Eisen-schwarz *n*, -schwärze *f*, Graphit
*m*, graphitischer Kohlenstoff *m*, Pottlot *n*,
Reiß-, Schwarz-blei *n* ~ in flocks flockiger
Graphit *m*
graphite, ~ blacking Graphitschwärze *f* ~
brick Graphitstein *m* ~ carbon Graphitkohle *f*,
graphitischer Kohlenstoff *m* ~-coated bronze
graphitierte Bronzebüchse *f* ~ crucible Graphittiegel *m* ~ deposit Graphit-ablagerung *f*,
-lager *n* ~ electrode graphierte Elektrode *f*,
Graphitelektrode *f* ~ eutectic Graphiteutekti-
kum *n* ~ flake Graphit-flocke *f*, -schuppe *f* ~
formation Graphitbildung *f* ~ grease graphitiertes Fett *n* ~ lattice Graphitgitter *n* ~ layer
structure Graphitschichtstruktur *f* ~ lubricant

Graphitschmiermittel *n* ~ **lubrication** Graphit-schmierung *f* ~ **mine** Graphitgewinnung *f* ~**-moderated pile** Graphit-Reaktor *m* ~**(-treated) oil** graphitiertes Schmiermittel *n* ~ **pepples** Graphitkugeln *pl* ~ **pot** Graphittiegel *m* ~ **resistance** Graphitwiderstand *m* ~ **retouching** Graphitretusche *f* ~ **slip ring** Graphitschleifring *m* ~**-treated** graphitiert
**graphited rubber** Graphitkautschuk *m*
**graphitic** graphitisch ~ **carbon** Graphit *m*, graphitischer Kohlenstoff *m*, Temperkohle *f* ~ **corrosion** faulende Rostanfressung *f*, Spongiose *f* ~ **iron** graues Roheisen *n*
**graphitiferous** graphithaltig
**graphitization** faulende Rostanfressung *f*, Graphitausbildung *f*, Graphitbildung *f*, Graphitierung *f*, Temperkohleabscheidung *f*
**graphitize, to** ~ graphitisieren
**graphitized fats** Graphitfett *m*
**graphitizing** Graphitglühen *n*
**graphitoid** graphitisch
**graphometer** Graphometer *m*, Winkelmesser *m*
**graphometry** Graphometrie *f*
**grapnel** Anker *m*, Ankersucher *m*, Bootsanker *m*, Dragganker *m*, Dragge *f*, Dregg-anker *m*, -haken *m*, Fanghaken *m*, Kletter-, Kropf-, Steig-eisen *n*, Suchanker *m* ~ **rope (of boat)** Dreggtau *n*
**grapple, to** ~ **(a submarine cable)** anklammern, dreggen, Greifklaue *f*
**grapple** ~ **claw (of in-and-out movement or feeder mechanism)** Greifer *m*
**grappling** Dreggen *n*, Verankerung *f* ~ **of arch** Bogenverankerung *f* ~ **hook** Eisenfänger *m*, Staken *m* ~ **irons** Finger *pl* ~ **rope** Fangleine *f*
**grasp, to** ~ anfassen, begreifen, erfassen, fassen, greifen
**grasp** Griff *m*, Klaue *f*
**grasped** gefaßt
**grass** Gras *n*, Grieß *m* (TV), Rasen *m*, Schnee *m* (CRT), Sprühen *n* (rdr) ~ **bleacher** Naturbleicher *m* ~ **bleaching** Grasbleiche *f* ~ **cutter** Grasschneider *m* ~ **duster** Hadernstäuber *m*, Strohsichter *m*
**grasshopper** Heuschrecke *f* ~ **conveyer for sugar** Zuckertransportrinne *f* ~ **joints** Gelenkrohre *pl* ~ **spring** Flachfeder *f*
**grass,** ~ **land** Wiese *f* und Weide *f* ~**-seed attachment** Grassäeapparat *m* ~ **temperature** Erdbodentemperatur *f*
**grassy** gras-artig, -bedeckt, -grün ~ **median strip** Grünstreifen *m* (Straße)
**grate, to** ~ gittern, kratzen, reiben, schaben, zermahlen
**grate** Fangrechen *m*, Feuer-, Gitter-rost *m*, Netz *n*, Rost *m* ~ **for vertical mill** Rostplatte *f*
**grate,** ~ **area** Rostfläche *f* ~ **band** Rostband *n* ~ **bar** Roststab *m*, Schienenrost *m* ~ **bar shaker** Roststabklopfvorrichtung *f* ~ **bearer** Rostträger *m* ~ **cooler** Rostkühler *m* ~ **core** Siebkern *m* ~ **cover** Rosteinkapselung *f* ~ **duty** Rostbelastung *f* ~ **fire** Rost-feuer *n*, -feuerung *f* ~ **firing** Rostfeuerung *f* ~ **frame** Tragrost *m* ~ **inset** Rosteinsatz *m* ~**like** netzartig ~ **opening** Rostspalt *m* ~**-shaped** rostartig ~ **stoker** Rostbeschickungsanlage *f* ~ **surface** Rostfläche *f* ~ **water boxes** Rostkühlrohre *pl*

**grater** Raspel *f*, Reibe *f*, Reibeisen *n* ~ **file** Feinraspel *f*
**graticule** (beleuchtetes oder geritztes) Fadenkreuz *n*, Fadenkreuzplatte *f*, Fadenplatte *f*, Gradabteilung *f*, Gradnetz *n*, Kartennetz *n*, Netz *n*, Strichkreuzplatte *f*, Strichplatte *f* ~ **value** Strichwert *m*
**gratinate, to** ~ bekrusten
**grating** Bildraster *m*, (for fish, or in a well) Fischpaß *m*, (for fish, or in a well) Fischweg *m*; Gatter *n*, Gitter *n*, Gitterwerk *n*, Gräting *f*, Punktraster *m*, Raster *m*, Rostwerk *n*, Vergitterung *f*, klimpernd ~ **for floating wood** Floßrechen *m* ~ **of the foundation** Schwell-rost *m*, -werk *n* ~ **of timbers** Balkenrost *m*
**grating,** ~ **beam** Rostschwelle *f* ~ **constant** Gitter-abstand *m*, -konstante *f*, -mittenabstand *m*, Netzebenenabstand *m* ~ **converter** Gitterumformer *m* ~ **diaphragm** Gitterblende *f* ~ **evaluation** Gittermessung *f* (opt.) ~ **floor** Lichtgitterbelag *m* ~ **formula** Gitterformel *f* ~ **ghost** falsche Linie *f* ~ **groove** Gitterfurche *f* ~ **line** Gitterstrich *m* ~ **nippers** Käfigbauerzange *f* ~ **reflector** Rasterreflektor *m* ~ **reflexes** Gitterreflexe *pl* ~ **replica** Gitterkopie *f* ~ **ruling** Gitterteilung *f* ~ **space** Gitter-abstand *m*, -konstante *f* ~ **spectrometer** Gitterspektrometer *m* ~ **spectrum** Gitterspektrum *n* ~ **walk** Lattensteg *m*
**gratings** Reibsel *n* ~ **in x-ray spectroscopy** Gitter *n* in der Röntgenspektroskopie
**gratis** umsonst, unentgeltlich
**gratuities** Geldzuwendungen *pl*
**gratuitous** unentgeltlich
**gratuitously** umsonst
**graupack** Schranzpapier *n*
**graupel shower** Graupelschauer *m*
**grave** Grab *n*; bedenklich, ernst, ernsthaft, schwerwiegend ~ **bodily injury** schwere Körperverletzung *f* ~ **marker** Grab-kreuz *n*, -tafel *f* ~ **stone** Grabstein *m*
**gravel, to** ~ aufschottern **to** ~ **the pavement** das Pflaster *n* bekiesen **to** ~ **a street** eine Straße *f* beschottern
**gravel** Fluß-kies *m*, Fluß-sand *m*, -schotter *m*, Grand *m*, Grieß *m*, Grobsand *m*, (fine) Kies *m*, (sand) Kiessand *m*, Schotter *m*, Steinschlag *m* ~ **for blasting engines** Gebläsekies *m* ~ **for concrete mixing** Betonschotter *m* ~ **and small boulders** Geschiebe *n*
**gravel,** ~ **ballast** Kiesbettung *f* ~ **bank** Kiesbank *f* ~ **concrete** Kiesbeton *m* ~ **dredging** Sandbaggerei *f* ~ **fraction** Kieskörnung *f* ~ **pit** Kiesgrube *f*, Pinge *f* ~ **rake** Kies-harke *f*, -rechen *m* ~ **road** Schotterstraße *f*, unbefestigter Weg *m* ~**-scattering equipment** Splittstreuer *m* ~ **soil** Kiesboden *m* ~ **trap** Kiesfang *m*
**gravimeter** Dichtemesser *m*, Gravimeter *n*, Schweremesser *m*
**gravimetric** gewichtsanalytisch, gravimetrisch ~ **analysis** Gewichtsanalyse *f*, gravimetrische Untersuchung *f* ~ **effect** Schwerewirkung *f*
**gravimetry** Gravimetrie *f*
**graving dock** Trockendock *m*
**gravitate, to** ~ sich durch Schwerkraft *f* fortbewegen, sinken
**gravitation** Anziehungskraft *f*, Erdanziehung *f*, Gravitation *f*, Gravitieren *n*, Massenanziehung

*f*, Schwere *f* ~ **constant** Gravitationskonstante *f* ~ **conveyor** Schwerkraftförderer *m* ~ **thickening** Klassierung *f*
**gravitational,** ~ **compressibility waves** gravoidelastiode Wellen *pl* ~ **constant** Fallbeschleunigung *f* ~ **field** Gravitations-, Schwerkraft-, Schwere-feld *n* ~ **force** Erdanziehungskraft *f* ~ **momentum** Fallmoment *n* ~ **potential** Gravitationspotential *n* ~ **settler** Gleichfälligkeitsapparat *m*
**graviton** Graviton *n*
**gravity** (force of) Erdschwere *f*, Extraktgehalt *m*, Schwere *f*, Schwerkraft *f*; nach dem Gesetz der Schwerkraft arbeitend **by** ~ durch Gefälle *n* **center of** ~ Straßenlage *f* (eines Fahrzeuges)
**gravity,** ~ **abutment** Schwerewiderlager *n* ~ **acceleration** Schwerkraftbeschleunigung *f* ~ **ammeter** Gewichtsspannungs-, Gewichtsstrommesser *m* ~ **anomalies** gravimetrische Unregelmäßigkeiten *pl* ~ **balance** gravimetrische Waage *f*, Gravitationsdrehwaage *f* ~ **battery** Batterie *f* aus Daniellschen Elementen ~ **cable** Riese *f*, Seilriese *f* ~ **cell** Zelle *f* verschiedener Elektrolyten, Zweischichtenelement *n* ~ **change** Schwereschwankung *f* ~ **circulation** Thermosiphonumlauf *m* ~ **concentration** Gravitationsaufbereitung *f* ~**-controlled armature** Anker *m* mit Gegengewicht ~ **conveyer with staggered rollers** Rollenbahn *f* (oder Rollenband *n*) mit versetzten Rollen ~ **dam** Gewichtsstau-, Schwergewichts-mauer *f* ~ **dam of triangular cross section** Schwergewichtsmauer *f* von dreieckigem Durchschnitt ~ **determination** Dichtebestimmung *f* ~**-discharge furnace** Durchstoßofen *m*, Stoßofen *m* ~ **dressing** Gravitationsaufbereitung *f* ~ **drive** Antrieb *m* durch Schwerkraft, Schwerkraftantrieb *m* ~ **drop** Schwerkraftkomponente *f* ~ **emptying device** Entleerungsvorrichtung *f* für Gefälle ~ **fault** Abgleitung *f* (geol.), normale Verwerfung *f* ~ **feed** Abfließen *n* durch natürliches Gefälle, Fallspeisung *f*, Förderung *f* durch Schwere, Zuführung *f* durch Gefälle ~**-feed arc lamp** Freifallbogenlampe *f* ~**-feed lubricator** Tropfölapparat *m* ~**-feed pipe** Falleitung *f* ~**-feed system** Fallförderanlage *f*, Schwerkraftförderanlage *f* ~**-feed-type sandblast unit** Sandstrahlgebläse *n* nach dem Schwerkraftsystem ~ **field** Schwerefeld *n* ~ **filler unit** Fall-Betankungsstutzen *m* ~ **force** Schweredruck *m* ~ **fuel feed** Brennstoffzuführung *f* mit Gefälle ~ **fuel system** Fallkraftstoffanlage *f* ~ **gas feed** Fallbenzinförderung *f* ~ **gravity die casting** Dauerguß *m* ~ **hose** Pendelschlauch *m* ~**-network** Schwerenetz *n* ~ **pendulum** Schwerependel *n* ~ **plane** Gegengewicht *n* im Bremsperg ~ **ram** Zugramme *f* ~ **retaining wall** Schwergewichtsmauer *f* ~ **roll carrier (or roller conveyer)** Rollenbahn *f* ~**-roller conveyer with herringbone arrangement of rollers** Rollenbahn *f* mit versetzten Rollen ~**-roller conveyer with inclined-slat-platform conveyer** Rollenbahn *f* mit schrägliegendem Plattenbandförderer ~ **segregation** Gravitationseigerung *f* ~ **separation** Gravitationstrennung *f* ~**-settling process** Absitzverfahren *n* ~ **spiral conveyer** Wenderutsche *f* ~ **surveys** Schwerevermessungen *pl* ~ **system** Schwerkraftsystem *n* ~ **tank** Fall-

behälter *m*, -benzintank *m*, -tank *m*; Hochbehälter *m* ~ **tube** Fallrohr *n* ~**-type** gewichtsbelastet ~**-type fuel tank** Fallkraftstofftank *m* ~**-type scribe** Fallbügelschreiber *m* ~ **waves** Schwerewellen *pl*
**gravure** Gravierung *f*, Gravüre *f*, Klischee *n* ~ **etchant** Tiefdruckätze *f* ~ **print** Tiefdruck *m*
**gray** farblos, grau, mattiert (photo), neutral, trübe ~ **antimony ore** Grauspiegelglanzerz *n* ~ **bands** Grausandstein *m* ~ **body** Graustrahler *m* ~ **cast iron** graues Gußeisen *n*, Grauguß *m* ~ **chalk** Graukalk *m* ~ **color graduation (or scale)** Grautonskala *f* (Rauchgasprüfer) ~ **copper ore** Fahlerz *n*, Kupferfahlerz *n* ~ **filterglass** Grauglas *n* ~ **iron** Gußeisen *n* ~ **iron castings** Graugußteile *pl* ~ **-iron foundry** Graugießerei *f* ~**-iron scrap** Gußeisenschrott *m* ~**-mottled** graumeliert ~ **pig iron** graues Roheisen *n* ~ **radiation** Graustrahlung *f* ~ **sandstone** Grausandstein *m* ~ **scale** Grauleiter *f* (Ostwalds Farbentheorie) ~ **signal** Treppensignal *n* (TV) ~ **silver** Grausilber *n* ~ **spiegel iron** Grauspiegel *m* ~**-tin modification** graues Zinn *n* ~ **tints** Grautöne *pl* ~ **wake** Graugestein *n* ~ **wedge** Grautreppe *f*
**grayed** mattiert ~ **finish** Mattierung *f* (photo)
**grayish (grizzly)** graulich, Grau . . .
**graze, to** ~ bestreichen, entlangstreifen, streichen, streifen, vorbei-streichen, -streifen ~ **of a projectile** Aufschlag *m* eines Geschosses ~ **fuse** empfindlicher Zünder *m*
**grazing** Streifen *n*, streifender Einfall *m*; rasant ~ **bullet** Streifkugel *f* ~ **fire** bestreichendes oder rasantes Feuer *n* ~ **incidence** Glanz-, Schief-einfall *m*, schiefer oder streifender Einfall *m*
**grease, to** ~ bestreichen mit, einfetten, einschmieren, fetten, Fett einspritzen, ölen, schmieren, das Lager *n* schmieren **to** ~ **leather over a charcoal fire** Leder *n* abflammen **to** ~ **thoroughly** abschmieren
**grease** Fett *n*, konsistentes Fett *n*, Starrschmiere *f*, Talg *m* **(lubricating)** ~ Schmierfett *n* ~ **battle plate** Schleuderblech *n* ~ **bearing** Schmierlager *n* ~ **boiler** Fettschmelzer *m* ~ **box** Fettkasten *m*, Schmierbehälter *m*, Schmierbüchse *f*, Teerbuchse *f* ~ **brick** Glättstein *m* ~ **chamber** Fettkammer *f* ~ **coating** Fettüberzug *m* ~ **cock** Schmierhahn *m* ~ **compound** Schleifpaste *f* ~ **cup** Büchse *f* für Fettschmierung *f*, Fett-, Schmier-büchse *f*, Schmiere *m*; Schmierglas *n*, -vase *f*; Staufferbüchse *f* ~ **extractor** Fettabscheider *m* ~ **film** Fettanstrich *m* ~ **filter** (for greasy water) Fettfänger *m* ~ **fitting** Schmiervorrichtung *f* ~**-free** fettfrei ~ **gun** Druck-fettpresse *f*, -schmierpresse *f*; Fettnippel *n*, -presse *f*, -schmierpresse *f*, -spritze *f*, Presser *m*, Schmier-pistole *f*, -pumpe *f* ~ **jack** Abstoßmaschine *f* ~ **lubrication** Fettschmierung *f* ~ **milling** Schmutzwalke *f* ~ **monkey** Abschmierer *m* ~ **nipple** Druckschmierkopf *m*, Schmiernippel *n* ~ **(inspection) pit** Grube *f* zur Reparatur ~ **pocket** Fett-kammer *f*, -tasche *f* ~ **proof** fettdicht ~ **proof paper** fettdichtes Papier *n*, Pergamentpapier *n* ~ **proportioning valve** Fettzumeßventil *n* ~ **pump** Schmierpumpe *f* ~ **resistance** Fett- und Ölbeständigkeit *f* ~ **retainer ring** Fettdichtungsstützring *m*

~ **return ring** Fettrückförderring *m* ~ **seal**
Fettdichtung *f* ~ **separator** Fett-abscheider *m*,
-fang *m* ~ **solvent** Fettlösungsmittel *n* ~**-spot
photometer** Bunsen-Fotometer *m*, Fettfleckfotometer *n* ~ **sump** Schmierstoffbehälter *m* ~
**tight** öldicht ~ **trap** Fettfang *m* ~ **wrapper**
Fetthülse *f*
**greased** geschmiert ~ **(ground-in) joint** Fett-
schliff *m*
**greaser** Schmiervorrichtung *f*, Wagenschmierer
*m*
**greasing** Einfettung *f*, Fettung *f*; bestreichbar ~
**apparatus** Anfettapparat *m* ~ **circle** Schmier-
kreis *m* ~ **set** Abschmiergerät *n*
**greasy** fett, fettig, ölig, speckig ~ **luster** Fettglanz *m* ~ **mist** Fettdunst *m* ~ **paste** schmierige
Masse *f* ~ **wool** Schweißwolle *f*
**great** groß, stark ~ **auger** Stangenbohrer *m* ~
**calorie** große Kalorie *f* ~ **circle** großer Kreis
*m*, großer Zirkel *m*, Großkreis *m*, Orthodrome
*f*
**great-circle,** ~ **bearing** Standlinienpeilung *f*,
Standortbestimmung *f* ~ **chart** Karte *f* des
Großkreises ~ **course (or direction)** Groß-
kreiseinrichtung *f* ~ **distance** Entfernungen *pl*
(Abstände) im großen Zirkel, Großkreisent-
fernung *m* ~ **route** Kurs *m* im größten Kreise
~ **sailing** Segeln *n* im größten Kreise
**great,** ~ **directivity** scharfe Bündelung *f* ~ **fire**
Feuerbrunst *f* ~ **rubble stone** Feldstein *m* ~
**span saw** Örtersäge *f* ~ **success** Treffer *m* ~
**wheel** Schloßscheibe *f*
**greater,** ~ **performance** Mehrleistung *f* ~ **wheel**
Großrad *n* (Getriebe)
**greatest,** ~ **altitude** größte Höhe *f* ~ **gradient**
Maximalgefälle *n* ~ **head flow to the barrage**
höchster Zufluß *m* zur Talsperre ~ **length
between points** größte Länge *f* zwischen den
Spitzen ~ **load carried at an altitude of n meters**
höchste Nutzlast *f* auf n Meter Höhe ~ **measure**
Höchstmaß *n* ~ **speed** größte Geschwindigkeit
*f* ~ **(or maximum) value** Größtmaß *n*
**greatness** Größe *f*
**Grecian type antenna** Bruce-Antenne *f*
**greedy** begierig, gefräßig, gierig
**green (coal)** grubenfeucht, grün ~ **and chemical
cleansing of stems (stalks)** Entbasten *n* der
Stengel ~ **assembly** Erstmontage *f* ~ **auroral
line** Nordlichtlinie *f* ~ **coal** grubenfeuchte
Kohle *f* ~ **core** grüner Kern *m*, Kern *m* aus
grünem Sande, ungebackener Kern *m* ~**-core-
making method** Grünkernformverfahren *n* ~
**earth** Grünerde *f* ~ **ebony** Grünebenholz *n* ~
**electron gun** Grünstrahlsystem *n* ~ **feldspar**
Smaragdspat *m* ~ **film** neuentwickelter Film *m*
~ **flash** grüner Strahl *m*
**greenheart** Grün- oder Gelbharzholz *n*
**green,** ~ **house** Gewächs-, Treib-haus *n* ~ **iron
ore** Dufrenit *m* ~**-lead ore** Grünbleierz *n* ~**-
mold casting** Naßguß *m* ~ **mud** Grünschlick *m*
~ **oak** Jungeiche(nholz) . . . ~ **ore** Roherz *n* ~
**pyromorphite** Grünbleierz *n* ~ **record** Grünaus-
zug *n* ~ **rot** Grünfäule *f* ~ **run** Ein-, Jungfern-
(aviat.), Probe-lauf *m*
**greensand** Grünsand *m*, Haufensand *m*, magerer
oder nasser Formsand *m*, Naßsand *m*, Naß-
gußsand *m*, magerer oder nasser Sand *m* ~
**core** grüner Kern *m*, Kern *m* aus grünem Sand

~ **mold** Form *f* aus grünem Sand, Naßguß-
form *f* ~ **molding** Naßgußformerei *f*
**green,** ~ **sensitive** grünempfindlich ~ **sirup**
Grünablauf *m* ~ **slate** Diabasschiefer *m* ~
**soap** Kaliseife *f* ~ **stone** Diabas *m*, Diorit *m*,
Grün-, Schal-stein *m* ~ **timber** saftfrisches
Holz *n* ~ **traffic light** Einfahrtszeichen *n* ~
**verditer** Berggrün *n* ~ **vitriol** Eisensulfat *n*,
Eisenvitriol *n* ~ **weight** Grüngewicht *n* ~ **wood**
grünes Holz *n* ~ **zone of intersection** grüne
Schnittlinienzone *f*
**greenish** grünlich ~ **gold** Grüngold *n* ~ **tint**
Grünstich *m*
**greenockite** Greenockit *m*
**greenovite** Greenovit *m*
**Greenwich,** ~ **hour angle** Stundenwinkel *m*
i. Gr. ~ **mean time (GMT)** mittlere Greenwich-
Zeit *f* (MGZ)
**grége** Rohseide *f*, Rohton *m* ~ **width** Rohbreite *f*
**greisen** Greisen *m*
**grenade** Granate *f*, Granatgeschoß *n* ~ **charge**
Granatenfüllung *f*
**grenadine** Gitterstoff *m*
**grenz,** ~ **rays** Grenzstrahlen *pl* ~ **tube** Grenz-
strahlenröhre *f*
**Grey,** ~ **beam** Greyträger *m* ~ **girder** Greyträger
*m*
**grey,** ~ **film** Grauschleier *m* ~**-iron foundry**
Graugießerei *f* ~**-iron piston** Graugußkolben *m*
~ **scale** Grauskala *f*
**grid** Akkumulatorplatte *f*, Fadenplatte *f*, Gitter
*n*, (map) Gradnetz *n*, Liniennetz *n*, Netz *n*,
(as used in plotting) Plantrapez *n*, Raster *m*,
Reuse *f*, Rost *m*, Sperrgitter *n*, Stahldrahtgeflecht *n*, Vergitterung *f*; (measuring) ~
Gitternetz *n* ~ **of reference** Bezugsnetz *n*
**grid,** ~ **alternating-current voltage** Gitterwechselspannung *f* ~ **anode** Netzanode *f* ~ **aperture**
Gitteröffnung *f* ~ **axis** Gitterachse *f* ~ **backing**
Gitterzurückspannung *f* ~ **base** Aussteuerungsbereich *m*, Gittersperrspannung *f* ~ **battery**
Gitterbatterie *f* ~ **(base-line) bearing** Gitterrichtung *f* ~ **bias** Gitterverschiebungsspannung *f*, Gittervorspannung *f*, Gitterwiderstand
*m*, Steuergitterspannung *f*
**grid-bias,** ~ **battery** Gitterbatterie *f* ~ **condenser** Gitter-, Gitterblock-kondensator *m* ~
**modulation** Gittergleichstrom-, Gitterstrom-
modulation *f* ~ **rectifier** Gitterstromgleich-
richter *m* ~ **surface** Vorgitterfläche *f* ~ **tube**
Hilfsfrequenzröhre *f*
**grid,** ~ **blocking** Gitterblockierung *f* ~ **blocking
capacitor (or condenser)** Gitterblockkondensator *m*, Gitterkondensator *m* ~ **bridge** Gittersteg *m* ~ **cap** Gitterkappe *f* ~ **capacitance**
Gitterkapazität *f* ~ **capacitor** Gitterkondensator *m* ~**-cathode path** Gitterkathodenstrecke
*f* ~**-cathode space** Gitterkathodenraum *m* ~
**characteristic** Gitterkennlinie *f*, (curve) Gitter-
spannungskennlinie *f*, Gitterstromkennlinie *f*
~ **chart** Gitternetzkarte *f* ~ **circuit** Gitterleitung
*f* ~**(-filament) circuit** Gitterkreis *m* ~**-circuit
detector** Audion *n* ~**-circuit modulation** Gitter-
besprechung *f*, -modulation *f* ~**-circuit rectification** Audiongleichrichtung *f* ~ **clip** Gitter-
clip *m* ~ **coil** Gitterspule *f* ~ **coil wires** Gitter-
wickeldrähte *pl* ~ **condenser** Gitter-block *m*,
-blockkondensator *m*, -kondensator *m*, (in

gridcurrent detector) Übertragerkondensator *m* ~-**condenser detection** Audiongleichrichtung *f* ~ **condenser and grid leak connected in parallel** Kondensatorwiderstandkombination *f* ~ **conductance** Gitterleitwerte *pl* ~ **connecting ring** Gitteranschlußring *m* ~ **connection** Gitterleitung *f*, Querverbindung *f* ~ **construction** Gitterfachwerk *f* ~ **control** Gitter-beeinflussung *f*, -besprechung *f*, -steuerung *f* (rdo), -tastung *f* ~ **control voltage** Gitterwechselspannung *f* ~ **controllance** Gitterdurchgriff *m* ~-**controlled current** gittergesteuerter Strom *m* ~ **coordinates** Gitterkoordinaten *pl* ~ **corner point** Netzpunkt *m* ~ **coupling** Gitterkreiskupplung *f* ~-**coupling control** Rückkupplung *f* ~ **course** Gitterkurs *m* ~ **current** Elektronengitterstrom *m*, Gitterstrom *m*, positiver Gitterstrom *m*

**grid-current,** ~ **characteristic** Gitterstromkennlinie *f* ~ **detection** (in audion value) Gittergleichrichtung *f*, Gitterstromgleichrichtung *f* ~ **detector** Audiongleichrichtung *f* ~-**grid voltage characteristic** Gitterkennlinie *f*, Gitterstromkennlinie *f* ~ **point** Gittereinsatzpunkt *m*, Gitterstrompunkt *m* ~ **rectification** (in audion value) Gittergleichrichtung *f* ~ **sweep (or swing)** Gitterstromaussteuerung *f*

**grid,** ~ **cut-off** Gitterblockierung *f* ~ **cut-off voltage** Steuergittereinsatzspannung *f* ~ **data** Gitterwerte *pl* ~ **declination** Meridiankonvergenz *f* ~ **detection** Audiongleichrichtung *f* ~ **detector** Audion *n*, Gittergleichrichter *m* ~-**dip meter (or oscillator)** Absorptionswellenmesser *m*, Resonanzmeßgerät *n* ~ **direction** Gitterrichtung *f* ~-**driving power** Gittersteuerleistung *f* ~ **electrode** Gitterelektrode *f* ~ **emission** Gitteremission *f* ~ **excitation (or excursion)** Gittersteuerung *f* ~ **extension** Netzzusammenschluß *m* ~ **faults** Gitterstörungen *pl* ~ **(to) filament circuit** Eingangskreis *m* ~ **filling** Gitterfüllung *f* (electr.) ~ **flow** Gitterströmung *f* ~ **gas supply** Ferngaswirtschaft *f* ~ **glow tube** Gitterglimmröhre *f* ~ **heading** Gittersteuerkurs *m* ~ **hum** Gitterbrumm *m* ~ **indicator** Gitterschauzeichen *n* ~-**influence coefficient (of the airfoil)** Gittereinflußbeiwert *m* (des Tragflügels) ~ **interferometer** Gitterinterferometer *n* ~-**iron** Bratrost *m*, Kielbank *f*, Kielgording . . ., Netz(werk) *n*, Rost *m* ~-**iron sidings** Gleisharfe *f* ~-**iron valve** (slide or ordinary) Gitterventil *n* ~-**iron valves** Gitterschieber *m* ~ **keying** Gittertastung *f* ~ **lead** (resistance) Gitterableitung *f*, Gitter-, Gitterableit-, Gitterableiter-, Gitterableitungs-widerstand *m* (rdo), (resistance) Gitternebenschluß *m* ~- **leak detection** Audiongleichrichtung *f* ~ **leak resistor** Gitterableitwiderstand *m* ~-**like** rostartig (roststabartig) ~ **line** Gitterlinie *f*, Gitternetzlinie *f* ~ **load** Gitterlast *f* ~ **locking** Effekt *m* übermäßiger Gitteremissionen ~ **map** Quadratkarte *f* ~ **mesh** Gittermasche *f* ~ **method** Netzverfahren *n* ~ **modulation** Gitter-beeinflussung *f*, -besprechung *f*, -modulation *f*, -steuerung *f*, Modulation *f* durch Änderung der Gitterspannung ~ **multiplier** Gittervervielfacher *m* ~ **net** Quadratnetz *n* ~ **noise resistance** Gitterrauschwiderstand *m* ~ **north** Gitternord *m* ~ **numbers** Gitterwerte *pl* ~ **penetration**

**factor** Durchgriff *m*, Gitterdurchgriff *m* ~ **pin** Gitterstift *m* ~ **pitch** Steigung *f* des Gitters ~ **plate** Gitterplatte *f*

**grid-plate,** ~ **capacitance** Anodengitter-, Gitteranoden-kapazität *f* ~ **characteristic** Anodenstrom-Gitterspannungskennlinie *f*, normale Kennlinie *f* ~ **grid voltage characteristic** Charakteristik *f* der Röhre ~ **transconductance** gegenseitige Leitfähigkeit *f*, Steilheit *f*

**grid,** ~ **polarization voltage** Gittervorspannung *f* ~ **potential** Gitter-, Vor-spannung *f* ~-**potential-anodecurrent characteristic** Anodenstrom-Gitterspannungskennlinie *f*, Charakteristik *f* der Röhre, normale Kennlinie *f* ~-**potential curve** Gitterpotentialkurve *f* ~ **priming** Gitterverschiebungsspannung *f* ~ **priming voltage** Gittervorspannung *f* ~ **pulse modulation** Gittermodulation *f* durch Impulse ~ **ratio** Rasterverhältnis *n* ~ **rectification** Audiongleichrichtung *f* ~ **rectifier** Gittergleichrichter *m* ~ **reference** Gradnetzanmeldepunkt *m* (mil.) ~ **resistance** Gitterableitwiderstand *m*, Gitterdrossel *f* ~ **(leak) resistance** Gitterwiderstand *m* ~ **resistor** Gitter-ableitwiderstand *m*, -drossel *f* ~ **resonant circuit** Gitterkreis *m* ~ **retardation** Zündverzögerung *f* ~ **return** Gitterrückleitung *f* ~ **return resistor** Gitterableitwiderstand *m* ~ **ring** Kursring *m* ~ **scale** Gittermaßstab *m* ~-**seal** Gitterdurchführung *f* ~ **sector** Plantrapez *n* ~ **separator** Gitterwerkseparator *m* ~-**shaped** siebförmig ~ **sheet** Kartenblatt *n* mit Gitter ~ **shelf** Abstellrost *m* ~ **shielding can** Gitterabschirmhaube *f* ~ **shovel** Gitterschaufel *f* ~**skirt** Gittersaum *m* (TV) ~-**socket connection** Gittersockelfeld *n* ~ **space** Aussteuerungsbereich *m* (el. tube) ~ **square** Meldequadrat *n* ~ **stay** Gitter-steg *m*, -strebe *f* ~ **stay member** Gitterholm *m* ~ **stay wire** Gitterhaltedraht *m* ~ **stopper** Gitterschutzwiderstand *m* ~-**stopping capacitor** Gitterblockkondensator *m* ~ **stretcher** Reckdorn *m* ~ **strip used in lieu of wire** Gittersteg *m* ~ **substation** Überspannstation *f* ~ **supply** Überlandanlage *f* ~ **support wires** Gittersteg *m* ~-**supporting means** Gitterstrebe *f* ~ **supporting member** Gitterholm *n* ~ **supporting wire** Gitterhaltedraht *m* ~ **suppressor** Gitterwiderstand *m* ~ **sweep (or voltage swing)** Gitteraussteuerung *f* ~ **(voltage) sweep** Gittersteuerung *f* ~ **swing** Aussteuerungsbereich *m*, Gittersteuerung *f* ~ **system** Gitter-, Leitungs-netz *n* ~ **track required** beabsichtigter Gitterkurs *m* über Grund ~ **transition time** Gitterlaufzeit *f* ~ **transparency** Durchgriff *m*, Gitterdurchgriff *m* ~ **tube** Röntgenröhre *f* mit Gitter ~ **tuning** Gitterkreisabstimmung *f* ~-**type shuttering** Rostschalung *f* ~-**type slide valve** Gitterschieber *m* ~ **valve** Gitterschieber *m* ~ **voltage** Gitterspannung *f* ~-**voltage modulation** Gitterspannungsmodulation *f* ~ **winding machine** Gitterwickelmaschine *f* ~ **wire** Steuerdraht *m* (am Leuchtstoffschirm)

**gridded map** Gitternetzkarte *f*
**grief stem** Mitnehmerstange *f*
**grievance** Beschwerde *f*, Klage *f*, Klagegrund *m*
**grill** Bratrost *m*, Gitter *n* ~-**type microphone** Gittermikrofon
**grillage** Gatter *n*, Rost *m*

**grille** Gitter *n* (Klimaanlage) ~ **bar** Zierschiene *f*
**grilled** mit einem Gitter *n* versehen ~ **cooler**
Rippenkühler *m* ~ **radiator** Streifenkühler *m* ~
**tube** Rippenrohr *n* ~ **tube air preheater** Rippen-
rohrerhitzer *m*
**grim** grimmig
**grind, to** ~ (colors) abreiben, anreiben, anschleifen, auftreiben, einschleifen, knirschen, körnen,
mahlen, (glass) mattieren, pulverisieren, reiben,
schärfen, schleifen, vermahlen, walzen, zerdrücken, zerkleinern, zermahlen,
**grind, to** ~ **across the grain** querschleifen **to** ~
**beforehand** vorreiben **to** ~ **between centers**
Spitzenschleifen *n* **to** ~ **centerless** spitzenlos
schleifen **to** ~ **coarsely** schroten **to** ~ **down** abschleifen **to** ~ **with emery** abschmirgeln **to** ~ **fine**
feinmahlen, verreiben **to** ~ **glass** Glas *n* schleifen **to** ~ **in a mortar** mörsern **to** ~ **and mix**
kollern **to** ~ **off** abschaben **to** ~ **out** ausschleifen **to** ~ **partially** anschleifen **to** ~ **on a piece**
ein Stück *n* aufschleifen **to** ~ **with sand and**
**water** matt schleifen **to** ~ **out into a spherical**
**curvature** kugelförmig ausschleifen **to** ~ **super-**
**ficially** aufrauhen **to** ~ **together** zusammen-
schleifen **to** ~ **the valve** Ventil *n* einschleifen
**to** ~ **the welding seams** die Schweißnaht
schleifen
**grind** Knirschen *n*, Mahlen *n*, Schinderei *f*,
Schleifen *n*, Wetzen *n*, Zerreiben *n* ~-**down test**
Schleifprobe *f* ~ **stone** Wetzstein *m*
**grindability** Mahlbarkeit *f*, Vermahlungsfähig-
keit *f*
**grinder** Mahlstein *m*, Mahlwerk *n*, Mühle *f*,
Planmühle *f*, Reibkasten *m*, Schleifer *m*, Schleifmaschine *f*, -trommel *f*, L-walze *f* ~ **for paint**
Farbenreiber *m*, Farbmühle *f* ~ **spindle** Schleif-
welle *f*
**grinders** Knirschen *n*
**grindery** Schleiferei *f*
**grinding** Ausmahlung *f*, Einschleifen *n*, Feinzerkleinerung *f*, Knirschen *n*, Mahlung *f*, Polieren
*n*, Schärfung *f*, Scheuern *n*, Schleifen *n*, Schliff
*m*, Vermahlung *f*, Walzen *n*, Zermahlen *n*, Zermalmung *f*, Zermürbung *f*, Zerschlagen *n* ~
**and fluting machine with a column saddle**
Pilasterschleif- und Riffelmaschine *f* ~ **the**
**interior circulary** Innenrundschleifen *n* ~ **and**
**mixing machines** Verreibungs- und Misch-
maschinen *pl* ~ **and polishing machine** Polier-
schleifmaschine *f*, Schleif- und Poliermaschine
*f* ~ **and sifting plant** Mahl- und Siebanlage *f* ~
**through** Durchschlagen *n* (Farbe) ~ **the tooth**
**lands** Rückenschliff *m*
**grinding,** ~ **addition (or allowance)** Schleif-
zugabe *f* ~ **apparatus** Mahlvorrichtung *f* ~
**attachment** Schleifzusatzvorrichtung *f* ~ **at-**
**tachment for threading dies** Gewindeschneide-
backenschleifvorrichtung *f* ~ **ball** Schleifkugel
*f* ~ **blade** Abschleifmesser *n* ~ **bomb** Mahlbom-
be *f* ~ **box** Mahlkranz *m* ~ **capacity** Mahlleistung *f* ~ **center** Schleifzentrierung *f* ~ **chamber**
Mahlkammer *f*, Schneideholländer *m* ~ **chatter**
**mark** Schleifstelle *f* ~ **check** Schleifriß *m* ~
**chuck** Schleiffutter *n* ~ **clearance** Schleifwinkel
*m* ~ **cloth** Schleifleinwand *f* ~ **compound** Ein-
schleifpaste *f* ~ **cone** Mahlkegel *m* ~ **crack**
Schleifriß *m* ~ **cycle** Mahldauer *f* ~ **device**
Schleif-, Zerkleinerungs-vorrichtung *f* ~ **disk**

Läufer *m* zum Erzmahlen, Mahl-, Schleifscheibe *f* ~ **drum** Mahltrommel *f* ~ **dust** Schleifstaub *m* ~ **efficiency** Schleifleistung *f* ~ **ele-**
**ment** Mahl-körper *m*, -organ *n* ~ **face** Mahl-
bahn *f*, -fläche *f* ~ **gauge** Schleiflehre *f* ~ **head**
**stock** Schleifspindelkasten *m* ~ **knife** Abschleifmesser *n* ~ **machine** Schleifmaschine *f* ~
**machine for cleaning castings** Gußputzschleif-
maschine *f* ~ **machine for cycle lighting mirrors**
Radlichtspiegelschleifmaschine *f* ~ **machine**
**for gold leaf** Blattgoldanreibemaschine *f* ~
**machine for patent leather** Lacklederschleif-
maschine *f* ~ **mandrel** Einschleifdorn *m* ~
**marbles** Glasmärbeln *pl* ~ **material** Schleifmaterial *n*, -mittel *n* ~ **mill** Feinzerkleinerungs-
mühle *f*, Mahlwerk *n*, Planmühle *f*, Reibkasten
*m* ~ **motor** Schleifmotor *m* ~ **needle** Schleif-
nadel *f* ~ **operation** Mahl-gang *m*, -vorgang *m*;
Schleif-arbeit *f*, -vorgang *m* ~ **pan** Kollergang-
schale *f*, Pfannenmühle *f* ~ **paste** Einschleif-
masse *f* ~ **plant** Feinmahlanlage *f*, Mahlanlage
*f* ~ **plate** Mahlplatte *f* ~ **point** Schleifstift *m* ~
**powder** Schleifpulver *n* ~ **practice** Mahltechnik
*f* ~ **process** Mahl-, Schleif-verfahren *n* ~ **pro-**
**perty** Mahlfähigkeit *f* ~ **ring** Mahl-kranz *m*,
-ring *m*, -teller *m* ~ **rod** Schleifstab *m* ~ **roll**
Mahlwalze *f*, Schleif-trommel *f*, -walze *f* ~
**roller** Schleifzeug *n* ~ **shaft** (bridge, or yoke)
Schleifspindelaufsatz *m* ~ **shop** Schleiferei *f*
~ **size** Schleifmaß *n* ~ **stone** Schleifstein *m* ~
**surface** (upper) Läufer *m* einer Erzmühle,
Mahlfläche *f* ~ **technique** Mahl-, Reib-technik
*f* ~ **test** Mahlversuch *m*, Schleifprobe *f* ~ **tool**
Schleifwerkzeug *n* ~ **track** Polierbahn *f* ~
**wheel** Formscheibe *f*, Schleif-kopf *m*, -rad *n*,
-scheibe *f* ~ **wheel dressing device** Schleifscheibenabrichtvorrichtung *f* ~ **wheel head** Schleif-
spindelstock *m* ~ **work** Schleifarbeit *f* ~ **worm**
Brechschnecke *f*
**grindings** Schleifsel *n*
**grinds** Schleifschlamm *m*
**grip, to** ~ anwürgen, erfassen, fassen, (of
bearings) sich festlaufen, klemmen, packen,
spannen **to** ~ **in** festklammern, greifen
**grip** Einspannklemme *f*, Einspannkopf *m*, Fassen *n*, Greifer *m*, Greifklaue *f*, Griff *m*, Griffigkeit *f*, Griffstück *n*, Heft *n*, Klemme *f*, Klemmhülse *f*, (of rivet) Klemmlänge *f*, Kupplung *f*,
Spannkopf *m*, Verbindungsstück *n*, Widerhalt
*m*, Ziehstrumpf *m*, Zugriff *m* ~ **for clamp**
**screw** Sterngriff *m* ~ **of control column** Knüp-
pelgriff *m* ~ **for tools** Eisenfänger *m*
**grip,** ~ **adjuster** Griffbefestiger *m* ~ **block** Fang-
klotz *m* ~ **current tester** Zangenstromwandler
*m* ~ **device** Fang-haken *m*, -schere *f*, -vorrich-
tung *f* ~ **gear of lift** Froschklemme *f* ~ **handle**
**for emergency brake** Notbremshandgriff *m* ~
**handle with (momentary) trigger switch** Handgriff *m* mit Drückerschalter ~ **holder** Klemmbacke *f* ~ **length** Klemmlänge *f* ~ **pliers** Greifzange *f* ~ **roller and expanding friction clutch**
Klemmrollenkupplung *f* ~ **screw** Faßschraube
*f*, Greif-ring *m*, -schraube *f* ~ **sleeve** Griffhülse
*f* ~-**spring tensioning element** Spannelement *n*
(Ringfeder) ~-(**ping) tongs** Greifzange *f* ~
**traction** Griffigkeit *f* ~-**type cock with bent**
**outlet** Griffhahn *m* mit gebogenem Auslauf
~-**type fuse** Griffsicherung *f*

**gripes** (naut.) Bootsklampen *pl*, Bootskrabben *pl* (naut.), Kolik *f*

**gripper** Greifer *m*, Halter *m*, Mitnehmer *m*, Zangenapparat *m* ~ **bar** Greiferstange *f* ~ **block** Greiferanschlag *m* ~ **bolt** Greiferschraube *f* ~ **carrier** Greiferträger *m* ~ **feed** Greifer-, Zangen-vorschub *m* ~ **guide spindle** Greiferführungsspindel *f* ~ **housing** Greifergehäuse *n* ~ **locking disk** Teilscheibe *f* ~ **shaft** Vorgreiferwelle *f* ~ **system** Greifvorrichtung *f*

**gripping** Einspannen *n*, Einspannung *f*, Festspannen *n*, Greifen *n*, Spannen *n*; greifend, spannend ~ **action** Spannvorgang *m* ~ **angle** Greifwinkel *m* ~ **appliance** Greifvorrichtung *f*, Spannwerkzeug *n* ~ **arrangement** Spanneinrichtung *f* ~ **attachment** zusätzliches Spannwerkzeug *n* ~ **capacity** Greifvermögen *n*, Griffigkeit *f* ~ **chuck** Einspannkopf *m*

**gripping-device** Einspannvorrichtung *f*, Fanghaken *m*, Fangschere *f*, Fangvorrichtung *f*, Greifer *m*, Greifwerkzeug *n*, Klemmbacke *f*, Klemmvorrichtung *f* ~ **for bending** Einspannkopf *m* für Biegeversuche ~ **for compression springs** Einspannvorrichtung *f* für Druckfedern ~ **for testing chains in tension** Einspannvorrichtung *f* für Kettenzugversuche

**gripping,** ~ **diameter** Spanndurchmesser *m* ~ **effect** Spannwirkung *f* ~ **form** Halteform *f* ~ **head** Einspannkopf *m* ~ **instrument** Fanginstrument *n* ~ **jaw** Spannbacke *f* ~ **lever** Spannhebel *m* ~ **part** Greiferteil *m* ~ **plate** Spannplatte *f* ~ **pliers** Blitzzange *f* ~ **position** Spannstelle *f* ~ **power** Greiffähigkeit *f*, Greifvermögen *n* ~ **power of brake** Ansprechkraft *f* der Bremse ~ **surface** Greif-, Spann-fläche *f* ~ **wedge** Greifbacke *f* ~ **yoke** Spannbrücke *f*

**grist** Feinheit *f* (des Kornes)

**gristle** Knorpel *m*

**grit** Grieß *m*, Kies *m*, Kiessand *m*, Schleifabtrieb *m*, Schrot *m*, Schrott *m* ~ **layer** Steinbewurf *m* ~ **mill** (or **tube**) Grießmühle *f* ~-**stone** feinkörniger Sandstein *m*

**grits** Schleif-schlamm *m*, -staub *m* ~ **bolter** Grießsieb *n* ~ **mill** Grützmühle *f*

**gritty** grieselig, griesig, grießig, kiesig, körnig

**grivation** Gittermißweisung *f*, Grivation *f* ~ **index** Grivationsmarke *f*

**grizzly** Gittersieb *n*, Grubenrost *m*, Rost *m* (min.), Sturzsieb *n*

**groats** Grütze *f* ~ **mill** Grützmühle *f*

**grog** gebrannter Ton *m*, gemahlene Schamotte *f* ~ **particle** (in a firebrick) Magerkorn *n*

**groin** Buhne *f*, Flutbrecher *m*, Grat *m*, Leistengegend *f*, Schlange *f*, Stake *f*, Strandbühne *f* ~ **structures** Bühneneinbauten *pl*

**groined,** ~ **arch** Bogenrippe *f*, Gradbogen *m* ~ **ceiling** gerippte Decke *f* ~ **vaulting** Kreuzgewölbe *n*

**grommet** Auge *n*, Augenring *m*, Durchführungshülse *f*, Grummetstropp *m*, Öse *f*, Puffer *m*, Schlauchtülle *f*, Seilring *m*, Taukranz *m* ~ **for electrical system** Kabeldurchführung *f* ~ **thimble** Kausche *f*

**groom, to** ~ putzen

**groove, to** ~ aushöhlen, auskehlen, einkehlen, einkeilen, mit Einschnitten *pl* versehen, einstechen, falzen, (of rollers) kalibrieren, kandeln, kehlen, kerben, riefeln, riefen, riffeln, nuten

**to** ~ **a cask** ein Faß *n* gargeln oder gergeln **to** ~ **and tongue** spunden **to** ~ **for tongue-and-groove joint** (aus)spunden

**groove** Ausbauchung *f*, Ausdrehung *f*, Aushöhlung *f*, Auskehlung *f*, Aussparung *f*, (of a hammer) Bahn *f*, Drallzug *m*, Einkerben *n*, Einschnitt *m*, Falz *m*, Federnute *f*, (including grooving, fluting, or channeling generally) Fitze *f*, Fuge *f*, Führungsrille *f* (phono), Furche *f*, Hohlkehle *f*, Höhlung *f*, Kaliber *n*, Kaliberröhre *f*, Kannelierung *f*, Keilnut *f*, Keilrille *f*, Kerb *m*, Kerbe *f*, Kordel *f*, Leitschiene *f*, Niederlegung *f*, Nut *f*, Nute *f*, Riefe *f*, Riffelung *f*, Rille *f*, Rinne *f*, Schallrille *f* (phono), Schere *f*, Schleifrille *f*, Schlitz *m*, Seitenfalz *m*, Spannute *f*, Spur *f*, Vertiefung *f*, Walzenfurche *f*, Walzenkaliber *n*,

**groove,** ~ **of bearing** Laufrille *f* ~ **at the end of the barrel** Randkaliber *n* ~ **in the floor** Sohlenfalz *m* ~ **of insulator** Drahtlager *n* der Doppelglocke ~ **in which material is rolled on edge** Stauchkaliber *n* ~ **of pulley** Rollennute *f* ~ **for shrunk-on ring** Schrumpfnut *f* (Rennkerze)

**groove,** ~-**and-tongue joint** Nut und Feder *f*, Verspundung *f* ~ **angle** Rillenwinkel *m* (phono) ~ **chisel** Nutenmeißel *m* ~-**cutting chisel** Nuteisen *n* ~ **designing** Kalibrierung *f* ~ **disc** Mittelscheibe *f* ~ **flute** Schnittnute *f* ~ **holder** Schriftschablonenleiste *f* ~ **hole** (in tube sheet) Nut *m* (Nute *f*) in der Rohrplatte ~ **grinding machine** Nutenschleifmaschine *f* ~ **jumping** Überspringen *n* (phono) ~ **pin** Kerbbolzen *m* ~ **position** Nutenstellung *f* ~ **shape** Rillenquerschnitt *m* (phono) ~ **spacing** Füllgrad *m* einer Platte (phono) ~ **speed** Rillengeschwindigkeit *f* (phono) ~ **stem** Mitnehmerstange *f* ~ **wall stiffness** Rillenwandsteifigkeit *f* ~ **wheel** Kehlrad *n*, Schnurlauf *m* ~ **width** Rillenbreite *f*

**grooved** furchig, gefurcht, genutzt, gereifelt, gerieft, geriffelt, gerillt, gerippt, (for slides) gespundet, mit Kerben *pl* versehen, mit Nuten *pl* versehen ~ **and tongued** mit Nut *m* und Feder *f*

**grooved,** ~ **anvil** Gesenkamboß *m*, Senkstock *m* ~ **armature** Nutanker *m* ~ **ball bearing** Rillenkugellager *n* ~ **beam** ausgefalzter Balken *m* ~ **cam** Nutenscheibe *f* ~ **chase** Falzrahmen *m* ~ **collars** Matrizen *pl* ~ **cover** Falzdeckel *m* ~ **disc** Kanalscheibe *f* ~ **drum** Rillentrommel *f*, profilierte Trommel *f* ~ **drum dryer** Rillenwalzentrockner *m* ~ **finishing roll** Kaliberfertigwalze *f* ~ **friction wheel** Keilrad *n*, Rillenreibrad *n* ~ **glass** Riffelglas *n* ~ **insulator** Rillenisolator *m* ~ **mandrel** Riffelkloben *m* ~ **muller** (of an edge mill) Rillenläufer *m* ~ **pile** Sprundpfahl *m* ~ **pin** (type E) Knebelkerbstift *m*, (type B) Paßkerbstift *m*, Rillenbolzen *m*, gekehltes Rohr *n*, (type D) Steckkerbstift *m*, (type C) Zylinderkerbstift *m* ~ **piston ring** Ölschlitzring *m* ~ **pulley** gekerbte Rolle *f*, Keilnutscheibe *f*, Nutrolle *f* ~ **rail** Rillenschiene *f* ~ **roll** Form-, Kaliber-walze *f*, kalibrierte oder profilierte Walze *f* ~ **roller** Molette *f*, Nutenwalze *f*, Rillen-, Schnur-rolle *f*; Seil-scheibe *f*, -rille *f* ~ **roller gear** Kammwalzengetriebe *n* ~ **rollers** Kammrollen *pl* ~ **roof tile** Dachfalzziegel *m* ~ **slices** Rinnenschnitzel *n* ~ **wedge** Keilnut *f*

**groover** Einfeiler *m*, Falzmaschine *f*

**grooving** Nutung *f*, Riefung *f*, Rille *f*, Spundung *f* ~ **at back of thread** Gewindehinterstechen *n* ~ **of rollers** Walzenriffelung *f* ~ **of rolls** Walzkalibrierung *f*

**grooving, ~ block** Gegenstanze *f* ~ **chisel** Nuteneisen *n* ~ **comb** Gargelkamm *m* ~ **cutter** (tongue-and-groove woodwork) Auskehlfräser *m* ~ **iron** Kröseleisen *n* ~ **machine** Langfalz-, Biege- und Zudrück-maschine *f*, Nutenstoßmaschine *f* ~ **plane** Falt-, Nut-, Spund-hobel *m* ~ **tool** Riffelstahl *m*

**gross** (weight) brutto, gesamt, grob, roh; (twelve dozen) Gros *n*

**gross, ~ amount** Bruttobetrag *m* ~ **buoyancy** Gesamttragkraft *f* ~ **calorific value** oberer Heizwert *m* ~ **density** Rohdichte *f* ~ **earnings** Bruttoverdienst *m* ~ **efficiency** gesamter Wirkungsgrad *m* ~ **formula** Bruttoformel *f* ~ **heating value** oberer Heizwert *m* ~ **information content** Gesamtinformationsinhalt *m* ~ **lift** Gesamtauftrieb *m* ~ **load** Bruttobelastung *f*, Rohlast *f* ~ **loading** Fluggewicht *n* ~ **negligence** grobfahrlässiges Verhalten *n* ~ **polar** Gesamtpolare *f* ~ **power** Rohleistung *f* ~ **proceeds** Rohertrag *m* ~ **profit** Bruttogewinn *m* ~ **reactions** Bruttoreaktionen *pl* ~ **receipts** Betriebseinnahmen *pl*, Bruttoertrag *m* ~ **registered tons** Brutto-registertonnen *pl*, -registriertonnen *pl* ~ **sample** Rohprobe *f* (Sammelprobe) ~ **structure** Grobgefüge *n*, Großstruktur *f*, (band spectrum) allgemeine Struktur *f* ~ **structure analysis** Grobstrukturanalyse *f* ~ **tonnage** Bruttotonnengehalt *m* ~ **vehicle weight (GVW)** Gesamtgewicht *n* ~ **weight** Brutto-, Dienst-, Flug-, Gesamt-gewicht *n* (aviat.), höchstzulässiges Fluggewicht *n*, Lade-, Roh-gewicht *n* ~ **weight to be braked** Bremsbrutto *n*

**grossly** gröblich

**ground, to** ~ erden, an Erde *f* legen, gründen, grundieren, kurzschließen **to** ~ **an aircraft** Flugzeug *n* erden, Startverbot *n* erlassen **to** ~ **the current** den Strom *m* in die Erde ableiten **to** ~ **in** eindrucken **to** ~ **a line** eine Leitung *f* erden, an Erde *f* legen **to** ~**-wire** mit einem Erddraht *m* versehen

**ground** Boden *m*, Deckmittel *n*, Erdanschluß *m*, Erdboden *m*, Erde *f*, Erder *m* (electr.), Erdreich *n*, Erdung *f* (electr.), Fußboden *m*, Gelände *n*, Gestein *n*, (bottom) Grund *m*, Terrain *n*; eingeschliffen, geschliffen, mattgeschliffen, zerrieben, zugeschliffen **near the** ~ bodenseitig, am Boden *m*, in Bodennähe *f* **on the** ~ am Boden, bodenseitig ~ **in ball and socket joint** Kugelschliff *m* ~ **between delaying positions** Zwischenfeld *n* **the** ~ **dries out gradually** der Boden *m* wird allmählich entwässert ~ **of a mine** Grubenfeld *n* ~ **out** ausgeschliffen ~ **and polished surface** (under microscopic examination) Anschliff *m*

**ground, ~ absorption** Erdabsorption *f* ~ **acquisition** Grunderwerb *m* ~ **adhesion** Bodenhaftung *f* ~ **aids** Bodenhilfen *pl* (aviat.) ~**-air density** Bodenluftdichte *f* ~**-air traffic** Bodenbordverkehr *m* ~ **alert** Startbereitschaft *f* (aviat.) ~ **anchor** Erd- (aviat.), Grund-anker *m* ~ **antenna** Boden-, Erd-antenne *f* ~ **area** be-

baute Fläche *f* ~ **arrester** Erdleitungsunterbrecher *m* ~ **bar** Erdschiene *f* ~ **basalt** Basaltmehl *n* ~**-based duct** bodennaher Kanal *m* ~ **basic slag** Thomasmehl *n* ~ **beam** Bodenschwelle *f*, Grundbalken *m*, Schwelle *f* ~ **binding post** Erdanschluß *m* ~ **brake** Erdlandebremse *f* ~ **breaking ceremony** erster Spatenstich *m* ~ **burst** Aufschlagzündung *f* ~ **bushing** Erdungsbuchse *f* ~ **cable** Erdkabel *n*, Fesseltau *n*, Körperschluß-, Rückschluß-kabel *n* ~ **cable bond** Kabelerdungsschelle *f* ~ **camera** terrestrische Kamera *f* ~ **canvas** Unterlegeplane *f* ~ **carbon** Massenanschluß-, Rückschluß-kohle *f* ~ **circuit** Erdschleife *f*, geerdeter Kreis *m* ~ **clamp** Erd-klemme *f*, -schelle *f* ~ **clearance** (of propeller) Bodenabstand *m*, (vehicle) Bauch-, Boden-freiheit *f* ~**-clearance indicator** Funkecholot *n* ~ **cloth** Unterlegeplane *f* ~ **clutter** (unerwünschte) Bodenechos *pl* (rdr) ~ **color** Fond-, Grund-farbe *f* ~ **conditions** Geländeverhältnisse *pl* ~ **conductor** Erdungsader *f* ~ **connecting angle** Massekontaktwinkel *m* ~ **connection** Erdanschluß *m*, Erdenschluß *m*, Erd-leitung *f*, -schluß *m*, -verbindung *f*, Masseanschluß *m*, Massenanschluß *m* ~ **constant** Bettungsziffer *f* ~**-contact point** Aufsetzpunkt *m*, Flugzeugaufsetzpunkt *m* ~ **contaminating agent** Geländekampfstoff *m* ~ **control** terrestrische Vermessung *f* ~ **control box** Ventilkasten *m*, Steuerkasten *m* (g/m) ~**-control operator** Bodenflugleiter *m* ~**-controlled approach** GCA-Anflug *m* (aviat.) ~ **cork** Korkstaub *m* ~ **counterpoise** ausgleichender Erdschluß *m* ~ **coverage** Geländeausschnitt *m* ~ **crew** Bodenpersonal *n*, Lande-, Landungs-mannschaft *f* ~ **current** Erdplattenstrom *m* ~ **cushion** Grundwirkung *f* ~ **decontamination** Geländeentgiftung *f* ~ **deposit** Bodensatz *m* ~ **design** Fondmuster *n* ~ **detector** Erdschluß-anzeiger *m*, -prüfer *m* ~ **direction-finder station** Bodenpeilstelle *f*, Leitstelle *f* ~ **direction finding** Fremdpeilung *f* ~ **direction-finding station** Bodenpeilstelle *f* ~**-dismissal** Kündigungsgrund *m* ~ **disposal** Vergrabungsplatz *m* ~ **distance** Horizontalkomponente *f* (rdr) ~ **distance measurement** Erdstreckenmessung *f* ~ **distributor** Grundverteiler *m* ~ **drain connection** Bodenablaß *m* ~ **dyeing** Vorfärbung *f* ~ **echo** Bodenecho *n* (rdr) ~ **effect** Bodeneinfluß *m* ~ **effect machine (GEM)** Bodeneffektfahrzeug *n* (BEF) ~ **electrode** Masseelektrode *f* ~ **elevation** geografische Höhe *f* ~ **engineer** Abnahme- und Bauaufsichtsingenieur *m* ~ **equalizer conductor** Abgleicheinrichtung *f* ~ **equipment** Landausrüstung *f* ~ **fault** Erdschlußfehler *m*, Störung *f* durch Erdschluß ~**-fault neutralizer** Erdschlußlöschvorrichtung *f* ~**-fault supervising equipment** Erdüberwachungseinrichtung *f* ~ **feature** Geländepunkt *m* ~ **finish** Glättung *f* ~ **finishing** Grundschleifen *n* ~ **flask** Schliffkolben *m* ~ **floor** Erdgeschoß *n*, Straßenhöhe *f*, Untergeschoß *n* ~ **fog** Boden-, Grund-nebel *m*, Grundschleier *m* ~ **frost** Bodenfrost *m* ~ **giving way** Bodensenkung *f* ~ **glass** Matt-glas *n*, -scheibe *f* **ground-glass** ~ **focusing screen** Einstellmattscheibe *f* ~ **frame** Mattscheibenrahmen *m* ~ **image** Mattscheibenbild *n* ~ **joint** Glasschliff-

verbindung f ~ **overlap regulator** Mattscheiben-
überdeckungsregler m ~ **plate** Mattscheibe f
~ **scale** Milchglasskala f ~ **screen** Projektions-
mattscheibe f ~ **view finder** Mattscheiben-
sucher m ~ **(or frosted glass) window** Matt-
scheibenfenster n
**ground,** ~-**grip tire** Geländereifen m ~ **guide bar**
Grundschiene f ~ **gust** Boden-, Gelände-bö f
~ **handling** Abfertigung f, Bodenbetrieb m
(aviat.) ~ **handling characteristics** Bodenbe-
triebseigenschaften pl ~ **haze** Bodendunst m
~ **humidity** Bodenfeuchtigkeit f ~ **ice** Grund-,
Schlamm-eis n ~ **idling conditions** Bodenleer-
laufzustand m (eines Flugzeugmotors) ~
**illumination** Bodenleuchten n ~ **impact** Aufprall m, Bodenstoß m ~-**in joint** Schliff m
~-**in stopcock** eingeschliffener Hahn m ~ **indicator** Erdschlußanzeiger m ~ **installation** Boden-
anlage f ~ **insulation** Bodenisolation f ~
**interference** Störung f durch Erdschluß ~
**interrogator** Bodenabfragestelle f (sec. rdr) ~
**lag** räumliche Rücktrift f ~ **layout** Grundaus-
richtung f ~ **lead** Erd-, Erdungs-leitung f, Erd-
zuleitung f (electr.) ~ **leak** Störung f durch
Erdschluß m ~ **leakage on two phases** Doppel-
erdschluß m (electr.) ~ **level** Boden-höhe f,
-nähe f; Erdgleiche f, Flurebene f ~-**level
power** Bodenleistung f ~ **leveling** Erdbewegungen pl ~ **line** Grundlinie f ~-**line section** Erd-
zone f ~ **load** Bodenkraft f ~ **loads** Belastung f
durch Bodenkräfte ~ **location** Erdschlußbe-
stimmung f ~ **loop** Grundschleifenzug m,
Ringelpietz m (aviat.), Überschlag m, Überschlag m auf dem Boden ~ **lug** Erdungsöse f
~ **magnesite** Magnesitmehl n ~ **mapping**
Bodenbilddarstellung f ~ **mass** Grundmasse f,
Magma n ~ **mat** Erddrahtnetz n ~ **material**
Grundmaterial n ~-**mechanic maintenance man**
Wart m ~-**metal surface** Metallschliff m ~ **mill**
Bodenfräse f ~ **mine** Erd-, Grund-mine f ~
**mist** Bodendunst m ~-**mixing plant** Mischanlage f für Bodenarten ~ **mooring chain** Schlaf-
kette f ~ **movement control** Kontrolle f der
Bewegungen am Boden (aviat.) ~ **network**
Erddrahtnetz n, Erdnetz n ~ **noise** Eigen-
rauschen n (acoust.) ~ **noise-reduction circuit**
Grundstromkreis m ~ **oak-bark tanned leather**
lohgrubengegerbtes Leder n ~ **objective** Boden-
ziel n ~ **observation** Erdbeobachtung f ~-**observed fire** Schießen n mit Erdbeobachtung ~
**panel** Auslegezeichen n, Erkennungstuch n,
Fliegersichtstreifen m (aviat.), Fliegertuch n,
Grundtuch n, Landekreuz n, Tuchzeichen n
(aviat.) ~ **peg (or stake)** Erdpfahl m ~ **personnel** Bodenpersonal n ~ **photogrammetry**
Erdbildmessung f, Geländefotogrammetrie f,
Geofotogrammetrie f, terrestrische Fotogrammetrie f ~ **photograph** Erdbild-, Gelände-
-aufnahme f, terrestrische Aufnahme f ~ **pipe**
Erderplatte f, Erdleitungsrohr n (rdo) ~ **plan**
Grundriß m, Lageplan m, Umriß m ~ **plane**
Erdoberfläche f, Geländeebene f, Karten-
ebene f ~ **planing knife** Schlenmesser n ~ **plate**
Erdelektrode f, Erder m, Erderplatte f, Unterlagsplatte f ~ **plate of a framework** Bundschwelle f ~ **platform** Geschützbettung f ~
**plumb bearing** (geod.) Azimut m bezogen auf
die Hauptvertikalebene der Aufnahme ~

**plumb point** Geländenadir m ~ **point of control**
Fest-, Fix-punkt m ~ **position** Standort m über
Grund (aviat.) ~ **potential** Erdpotential n ~
**power supply** (to guided missile) Bodenspeisung
f ~ **pressure** Boden-beanspruchung f, -druck m
~ **pressure per unit area** spezifischer Bodendruck m ~ **pressure pick up** Bodendruckgeber
m ~ **projector** Bodenscheinwerfer m ~ **protection** Schutz m gegen die Erdverluste ~
**radiation** Bodenstrahlung f ~ **radio beacon**
Bodenfunkfeuer n ~ **radio bearing station**
Bodenpeilstelle f ~ **radio equipment** Boden-
funkgerät n ~ **radio service** Bodenfunkdienst
m ~ **radio station** Bodenfunkstelle f ~ **ray**
Bodenstrahl m, Nullzacke f ~ **reactions** Bodenkräfte pl ~ **reception** Grundempfang m ~ **reconnaissance** Geländeerkundung f ~ **reflection**
Bodenreflexion f ~ **rent** Erbzinsvertrag m,
Grundpacht f ~ **resistance** Erdbodenerschütterung f, Erd-, Erdungs-widerstand m ~-**resistance unrest** Erdbodenwiderstand m ~
**return** Bodenecho n, Erdleitung f
**ground-return,** ~ **circuit** Einzelleitung f, mit
Erde arbeitende Leitung f, Stromkreis m mit
Erdrückleitung ~ **signaling system** Erdsystem n
~ **telegraph circuit** Erdtelegrafieverbindung f
~ **telephone circuit** Erdfernsprechverbindung f
**ground,** ~ **returns** Bodenechos pl, Festzeichen pl
(rdr) ~ **rod** Erderplatte f, Erdschlußstange f,
Erdungsstab m ~(**ing**) **rod** Erdstecker m
(electr.) ~ **rod for telephone interception** Such-
erde f ~ **rolls** Bodenrollen pl ~-**rope traction**
Unterseilführung f ~ **run** Startlauf m ~ **run-up**
Prüflauf m (Motor) ~ **screen** Erdplatte f,
Erdungsnetz n ~ **sea** Grundsee f ~ **section**
Schliff m ~ **section prepared for etching of
metals** Ätzschliff m ~ **sector** Geländeausschnitt m ~-**set circuit** grundsachliches Schild-
bild n ~ **shape** Geländegestalt f ~ **shot** Erdschluß m ~ **signal** Sichtzeichen n, Weichen-,
Boden-signal n ~-**signal flare** Handleuchtzeichen n ~ **signal station** Erdfunkstelle f ~
**sill** Grundschwelle f ~ **sketch** Geländeskizze f
~ **slag** Schlackenmehl n ~ **slide** Erdriese f,
Schliff m ~ **sluice** Grundgerinne n ~ **spear**
Senkbaum m ~ **speed** Absolutgeschwindigkeit
f (aviat.), (of an airplane) Bodengeschwindigkeit f, (Motor) Drehzahl f im Leerlauf am
Boden, Fahrt f über Grund, Fluggeschwindigkeit f gegenüber dem Erdboden, Fluggeschwindigkeit f über Grund, Geschwindigkeit
f in Bodennähe, Geschwindigkeit f über Grund,
Grundgeschwindigkeit f, relative Geschwindigkeit f (aviat.) ~-**speed indicator** Übergrund-
geschwindigkeitsmesser m ~-**speed meter** relativer Geschwindigkeitsmesser m ~ **spike** Erd-
spieß m ~ **squad** Haltemannschaft f ~ **squall**
Geländeebbe f ~ **staff** Bodenpersonal n ~
**starter** Bodenanlasser m ~ **state** (of an atom)
Grundzustand m ~ **station** Bodenstelle f, Erd-
station f ~ (**radio**)**station** Landfunkstelle f
~-**station direction finder** Bodenpeilapparat m
~-**station transmitter** Bodensender m ~ **stop-
lock** Lahnstopfen m ~(-**in**) **stopper** Schliff-
-kolben m, -stopfen m ~ **subsidence (or submergence)** Bodensenkung f ~ **supercharged
engine** Bodenlademotor m ~ **surface** Boden-
fläche f, Tagesoberfläche f ~ **surveillance radar**

Bodensuchradar *n* ~ **survey** Gelände-, Terrain-
-aufnahme *f* ~ **swell** Dünung *f*, Grunddünung
*f*, Grundsee *f* ~ **switch** Erdungsschalter *m* ~
**system of an antenna** Erdleitung *f* ~ **take-off**
Landstart *m* ~ **target** Boden-, Erd-ziel *n* ~
**temperature** Boden-temperatur *f*, -wärme *f* ~
**terminal** Erdklemme *f*, Körperschluß-, Kurz-
schluß-, Rückschluß-klemme *f* ~ **timber** Bodenschwelle *f*, Grundbalken *m*, Schwelle *f* ~
**-to-air communication** Boden-Bordverkehr *m* ~
**triangulation** terrestrische Triangulation *f* ~
**unrest** Erdbodenerschütterung *f* ~ **varnish**
Grundierfirnis *m* ~ **visibility** Bodensicht *f* ~
**wall of a furnace** Ofenstock *m* ~ **warp** Unterkette *f* ~ **water** Grund-, Schicht-, Tiefen-
-wasser *n*
**ground-water**, ~ **level** Grundwasser-spiegel *m*,
-stand *m* ~ **movement** Grundwasserbewegung
*f* ~ **table** Grundwasserspiegel *m*
**ground-wave** Bodenwelle *f*, direkte Welle *f*, Erd-,
Grund-, Oberflächen-welle *f* ~ **direction finder**
Bodenwellen-, Nahefeld-peiler *m* ~ **reception**
**zone** Bodenwellenempfangszone *f* ~ **resulting**
**in undistorted bearing** peiltreue Bodenwelle *f*
**ground**, ~ **weather chart (or map)** Bodenwetter-
karte *f* ~ **wind** Bodenwind *m* ~ **wire** Erd-ader *f*,
-draht *m*, -leiter *m*, -leitung *f* ~ -leitungsdraht *m*,
Erdungsleitung *f* ~ **wire for a plug** drahtförmige
Masseelektrode *f* einer Zündkerze ~ **work**
Feldarbeit *f*, Gründung *f* ~ **zero** Hypozentrum *n* ~ **zone** Geländeausschnitt *m*, Schliff-
zone *f*
**grounded** an Erde *f* liegend, erdgebunden, geerdet, gemahlen ~ **anode amplifier** Anodenbasisverstärker *m* ~ **-anode circuit** Anoden-
basisschaltung *f* ~ **base arrangement** Block-
basisgrundschaltung *f* (Transistor) ~ **base**
**connection** Basisschaltung *f* ~ **battery** geerdete
Batterie *f* ~ **cathode amplifier** Kathodenbasisverstärker *m* ~ **circuit** Stromkreis *m* mit Erdrückleitung, geerdeter Stromkreis *m* ~ **collector**
**arrangement** Kollektorbasisschaltung *f* (Transistor) ~ **collector connection** Kollektorgrundschaltung *f* ~ **emitter arrangement** Emitter-
Basis-Schaltung *f* (Transistor) ~ **emitter con-**
**nection** Emittergrundschaltung *f* ~ **grid** Brems-,
Fang-gitter *n*, Gitterbasisschaltung *f* ~ **-grid**
**triode** Gitterbasistriode *f*, Friode *f* für Katho-
densteuerung ~ **-grid type circuit** Gitterbasisschaltung *f* ~ **guard ring** Schutzerdungsring *m*
~ **guard strip** Schutzerdungsleiste *f* ~ **guard**
**wire** geerdeter Schutzdraht *m* ~ **line** Einzelleitung *f* ~ **neutral** geerdeter Mittelleiter *m* ~
**neutral point** geerdeter Nullpunkt *m* ~ **-plate**
**amplifier** Kathodenverstärker *m*
**grounding** Ableitung *f*, Erden- *n*, Erdschluß *m*,
Erdung *f* (aviat.) ~ **of working batteries** Erdung
*f* von Betriebsbatterien
**grounding**, ~ **anchor antenna** Erdungsanker *m*
~ **block** Erdungsschiene *f* ~ **brush** Massebürste
*f* ~ **circuit** Erdungsleitung *f* ~ **clamp** Erdungs-
schelle *f* ~ **contact** Erdkontakt *m* ~ **device**
Erdung *f* ~ **jack** Erdungsanschluß *m* ~ **key**
Erdtaste *f* ~ **machine** Fonciermaschine *f* ~
**sleeve** Erdbuchse *f* ~ **strand** Erdungslitze *f* ~
**strap** Masseschiene *f* ~ **switch** Erdschalter *m*
~ **terminal** Erdungsklemme *f*
**groundless** gegenstandslos

**group, to** ~ anordnen, (motors, cells, pipes)
gruppieren
**group** Bündel *n*, Gruppe *f*, Konsortium *n*,
Plattensatz *m*, Rotte *f* ~ **of apparatus** Teilgruppe *f* ~ **of battery of siphons** (large dams)
Hebergruppe *f* ~ **of bearing recordings** Serienpeilungen *pl* ~ **of cams** Nockenbündel *n* ~ **of**
**conductors** Adernbündel *n* ~ **of contacts**
Kontaktkranz *m* ~ **of curves** Kurvenschar *f*
~ **of display units** Sichtgerätegruppe *f* ~ **of**
**four wires with twist system** Viererlage *f* ~ **of**
**graphs** Kurvenschar *f* ~ **of lines** Leitungsbündel *n* ~ **of mappings** Abbildungsgruppe *f* ~ **of**
**plates** Plattensatz *m* ~ **of power stations** Kraft-
werkgruppe *f* ~ **of translations** Bewegungs-,
Translations-gruppe *f*
**group**, ~ **accounts** Gruppenabschlüsse *pl* ~
**allocation** Gruppenverteilung *f* ~ **alternating**
**light** (navy) Gruppenwechselfeuer *n* ~ **ampli-**
**fier** Gruppenverstärker *m* ~ **assembly** Grup-
penmontage *f* ~ **assigned to duty** Dienstgruppe
*f* ~ **busy lamp** Drängelampe *f* im Fernverkehr
~ **busy signal** Anhäufungszeichen *n*, Drängelampe *f* im Fernverkehr ~ **casting** Gießen *n* im
Gespann, steigender Guß *m* ~ **center** Verteil-
zentrum *n* ~ **-combining drawer** Gruppenkopp-
lungslade *f* ~ **consolidations** konsolidierte
Gruppenabrechnung *f* ~ **count** Gruppenzahl *f*
~ **delay** Gruppenlaufzeit *f* ~ **delay/frequency**
**characteristic** Phasen/Frequenz-Charakteristik
*f* ~ **delay variation** Gruppenlaufzeitverlauf *m* ~
**detector** Gruppensucher *m* ~ **-diffusion method**
Gruppendiffusionsverfahren *n* ~ **distribution**
**frame** Gruppenverteiler *m* ~ **drive** Gruppenantrieb *m*, gruppenweiser Antrieb *m* ~ **flashing**
**light** Gruppen-blitzfeuer *n*, -drehblinkfeuer *n*,
regelmäßig aufblitzendes Gruppenlicht *n* ~
**frequency** Gruppen-, Wellenzug-frequenz *f* ~
**grading** Bündelstaffelung *f* ~ **indication** Grup-
penanzeige *f* ~ **indication cycle** Gruppenan-
zeigegang *m* ~ **key** Gruppenschlüssel *m* ~ **line**
(or conductor) Gruppenleitung *f* ~ **link** Gruppenverbindung *f* ~ **locking** Gruppensperrung *f*
~ **mark(er)** Gruppenmarke *f* ~ **modulator**
Gruppenumsetzer *m* ~ **occulting light** Grup-
penblinkfeuer *n* ~ **oiling system** Gruppen-
schmierung *f* ~ **phenomena** Gruppenerschei-
nungen *pl* (cryst.) ~ **piece work** Gruppenakkord
*m* ~ **posted piece** Gruppennotierungen *pl* ~
**relay** Dekadenrelais *n* ~ **relay for the numbers**
**with even (odd) units digits** Gruppenrelais *n* für
gerade (ungerade) Einer ~ **repeater drawer**
Gruppenverstärkerlade *f* ~ **retardation** Gruppengeschwindigkeit *f* ~ **selection** Gruppenwahl
*f* ~ **selector** Gruppenwähler *m* ~ **sorting device**
Leitkartensortiereinrichtung *f* ~ **switch** Gruppenschalter *m* ~ **theoretical treatment** gruppen-
theoretische Behandlung *f* der Kristallspektren
~ **theory** Gruppentheorie *f* ~ **transfer point**
Gruppenübergangspunkt *m* ~ **translating**
**equipment** Gruppenmodulationsgeräte *pl* ~
**translation** Gruppenumsetzung *f* ~ **velocity**
Gruppengeschwindigkeit *f*
**grouped**, ~ **flashes** Gruppenblitz *m* ~ **positions**
zusammengelegte Plätze *pl* ~ **records** grup-
pierte Aufzeichnungen *pl* ~ **style** Blocksatz *m*
**grouping** Anordnung *f*, Füllschrift *f* (phono),
Gruppierung *f*, Zusammenstellung *f* ~ **of**

**positions** Platzzusammenschaltung *f* (mit Platzschalter) **~ key** Konzentrationsschlüssel *m*, Platzschalter *m*
**groups, ·by ~** gruppenweise, nestweise **~ of motions** Bewegungsgruppen *pl* **~ of transformations** Abbildungsgruppen *pl*
**grouser** (track shoe) Gleiskettenschuh *m*
**grout, to ~** einspritzen, (joints) untergießen **to ~ with cement** mit Zement *m* ausgießen **to ~ the floor** den Boden *m* tränken
**grout** Mörtelschlamm *m*, Vergußmaterial *n*, Zementmilch *f*
**grouting** flüssige Mörtelarbeit *f*, Verguß *m* **~ hole** Einspritzloch *n* **~ tube** Einspritzrohr *n* **~ valve** Einspritzventil *n*
**Grove cell** Grove Element *n*
**grow, to ~** anbauen, anwachsen, schwellen, wachsen, werden, züchten, zunehmen **to ~ cold** erkalten **to ~ crooked** auswaschen **to ~ dull** abblicken, erlöschen **to ~ fat** (soap) zäh werden **to ~ hot** erhitzen **to ~ rich** (in) anreichern **to ~ through** durchwachsen **to ~ together** verwachsen, zusammenwachsen **to ~ viscid** gerinnen
**grower** Anbauer *m*, Federring *m*, (washer) federnde Unterlegscheibe *f*
**growing** Kultur *f*, Volumenzunahme *f*, Züchtung *f* (cryst.); entstehend, wachsend **~ crops** Saat *f* **~ vessel** Züchtungsgefäß *n*
**growl, to ~** knurren
**grown, ~ junction** gewachsene Schicht *f* **~ junction transistors** gezogene Transistoren *pl* **~-out** ausgewachsen
**growth** Anwachsen *n*, Anwuchs *m*, Gedeihen *n*, Gewächs *n*, Wachsen *n*, Wachstum *n*, Wuchs *m*, Züchtung *f* (cryst.), Zunahme *f*, Zuwachs *m* **~ of crystal** Kornwachstum *n* **~ in depth** Tiefenwachstum *n* **~ of films** Schichtwachstum *n* **~ in length** Wachstum *n* in der Länge **~ in production** Erzeugungssteigerung *f*
**growth, ~ curve** Wachstumskurve *f* **~ step** Wachstumstreppe *f* **~ time** Wachstumszeit *f*
**grub, ~ hoe** Lettenhaue *f* **~ screw** Gewindestift *m*, Made *f*, Maden-, Schnitt-schraube *f*
**grubbers and weeders** Grubber *pl*, Rodewerkzeuge *pl* (agr.)
**grummet** Dichtkegel *m*, Durchgangsdichtung *f*, Endhülle *f*, Kardeelstropp *m*, Metallöse *f*
**G scope** G-Schirm *m*
**G-sharp key** Gisklappe *f*
**GS marker** Gruppenwähler-Markierer *m*
**guaiac** Guajak *n*
**guaiacol** Guajakol *n*
**guaiacum resin** Guajakharz *n*
**guanajuatite** Selenwismutglanz *m*
**guarantee, to ~** avalieren, bürgen, garantieren, gewähren, gewährleisten, gutsagen, haften, sichern, sicherstellen, verbürgen
**guarantee** Bürgschaft *f*, Garantie *f*, Gewähr *f*, Sicherung *f* **~ of payment (or security)** Delkredere *n* **~ bill** Kautionswechsel *m* **~ certificate** Gütezeugnis *n* **~ claim** Gewährleistungsanspruch *m*
**guaranteed, ~ full load fuel consumption** garantierter Brennstoffverbrauch *m* **~ mirrors** Garantiespiegel *m* **~ quality** gewährleistete Güte *f* **~ value** Garantiewert *m*
**guarantor** Gewährsmann *m*

**guaranty** Delkredere *n*
**guard, to ~** (a circuit) besetzt halten, (a circuit) besetztmachen, bewachsen, schützen, sichern, sperren, überwachen **to ~ against** verbauen **to ~ a position** einen Platz *m* sperren
**guard** Abdeckung *f*, Abstreifmeißel *m*, Bedekkung *f*, Bewacher *m*, Bewachung *f*, (fencing) Deckung *f*, Funkbereitschaft *f*, (in rolling) Hund *m*, Posten *m*, Schaffner *m*, Schutz *m*, Schutzblech *n*, Schutzdeckel *m*, Schutzvorrichtung *f*, Wache *f*, Wacht *f*, Wärter *m* **to be on ~** auf der Hut *f* sein **to be on one's ~** sich in Acht *f* nehmen **~ of gauze** Drahtnetzhülse *f* **~ for rack** Zahnstangenschutz *m* **~ of wire netting** (for gauge glass) Drahtnetzhülse *f*
**guard, ~ balanced by a counterweight** Abstreifmeißel *m* oder Hängemeißel *m* mit Gewichtsausgleichung *f* **~ balanced by a spring** Abstreifmeißel *m* oder Hängemeißel *m* mit Federausgleichung **~ band** Schutzfrequenzband *n* **~ bar** Schutzstange *f* **~ board** Filzschaber *m* **~ cam** Sicherungsschloßteil *m* **~ circle** End-, Schutz-rille *f* (phono) **~ circuit** Überwachungskreis *m* **~ ear** Schutzöse *f* **~ fence** Führungsplanke *f*, Sicherheits-Leitplanke *f* (Straße) **~ frequency** Wachfrequenz *f* **~ gate** Schutzschleuse *f* **~ gates** Sperrschleuse *f* **~ grille** Schutzgitter *n* **~ house** Wachgebäude *n* **~ interval** Schutzzeit *f* (Impulstechnik) **~ net** Schutzerdungsnetz *n* **~ plate** Anlaufscheibe *f*, (over the cinder hole of a blast furnace) Schlackenschutzkasten *m*, Schutzblech *n*, Schutzerdungsplatte *f*, Schutzplatte *f* **~ position** Sparstelle *f* **~ rail** Brüstung *f*, Flügelschiene *f*, Geländer *n*, Leitzwangschiene *f*, Reling *f*, Schutzgeländer *n*, Schutzschiene *f*, Sicherheits-Leitplanke *f* (Straße); Sicherheitsschiene *f* **~ relay** Halte-, Sperr-relais *n* **~ relay against intrusion of other calls** Belegungs-Sperr-Relais *n* **~ ring** Abstandlehre *f*, Abwehrring *m* (electr.), Haltering *m*, Schutzring *m*, Sicherungsring *m*, Sprühschutzwulst *f* **~ room** Wach-lokal *n*, -stube *f* **~ rope** Geländerleine *f* **~ service** Sicherungsdienst *m* **~ sleeper** Schutzschwelle *f* **~ sleeve** Lichtschutzkappe *f* **~ socket** Schutztülle *f* **~ tooth** Schutzzinke *f* **~'s van** Bremswagen *m* **~ wall** Abschlußschwelle *f*, Leitwerk *n* **~ wire** Fang-, (under high-voltage cables) Prell-, Schutz-draht *m* **~ wires** Schutznetz *n*
**guarded, ~ frequency** überwachte Frequenz *f* **~ switch** Schalter *m* mit Schutzkappe
**guarding** Schutzerdungsvorrichtung *f* **~ effect** Abschirmwirkung *f*
**gudermannian amplitude** Hyperbelamplitude *f*
**gudgeon** Bolzen *m*, Drehzapfen *m*, eingelassener Zapfen *m*, Schildzapfen *m*, Stirnzahn *m*, Stirnzapfen *m*, Zapfen *m* **~ of the rudder** Ruderöse *f*
**gudgeon, ~ pin** Halszapfen *m*, Kolbenbolzen *m* **~-pin boss** Kolbenbolzenauge *n*, Kolbenbolzennabe *f* **~ pin lock** Kolbensicherung *f*
**guhr** Gur *f*
**guidance** Anführung *f*, Anleitung *f*, Führung *f*, Richtschnur *f* **~ system** Lenksystem *n* **~ unit program clock-work** Umlenkgerät *n* (g/m)
**guide, to ~** anleiten, führen, leiten, lenken **to ~ into line of sight (or into path)** einlenken
**guide** Anhalt *m*, Anhaltspunkt *m*, Anleitung *f*,

Anschlag *m*, Aufleiter *m*, Einweiser *m*, Fadenführer *m*, Führer *m*, Führerboot *n*, Führungskörper *m*, (crosshead) Geradführung *f*, (crosshead) Gleitbahn *f*, Gleitbügel *m*, Gleitstein *m*, Leiste *f*, Leitartikelauge *n*, (book) Leitfaden *m*, (gun) Lenker *m*, Nadelschwimmer *m*, Verschlußschiene *f*, Weiser *m*

**guide,** ~ **for band saw** Blattführung *f* ~ **for blower bar** Bläserführung *f* ~ **for hollow rods** Gestängerohrführung *f* ~ **with projecting teeth** gezahnte Gleitbahn *f* ~ **of reversing lever** (locomotive) Führungsbogen *m* des Steuerungshebels ~ **of the tabulator-key levers** Tastenhebelführungsblech *n* ~ **for valve cone** Ventilkegel *m* mit Schaft

**guide,** ~ **analysis** Richtanalyse *f* ~ **angle iron** Führungswinkeleisen *n* ~ **arm** Führungslineal *n* ~ **arrangement** Leitvorrichtung *f* ~ **baffles** Leitgitter *n* ~ **bar** Führungsschiene *f*, (round or square) Führungsstange *f*, Lineal *m* (b. Manipulator), Stütze *f* (print.), Verschiebelineal *n* ~ **bar bracket** Kulissenführung *f* ~ **bar swing shaft** Legschienenschwungexzenter *m* ~ **bars** (general) Gleitschiene *f* ~ **basin** Führungsschale *f* ~ **beam** Fahrstrahl *m*, Laufrute *f*, Leitstrahl *m*, Seitenführungsstrahl *m*

**guide-beam,** ~ **fan** Leitstrahlfächer *m* (rdr) ~ **focus** Leitstrahlschärfe *f* ~ **installation** Leitstrahlanlage *f* ~ **plane** E-ebene *f* ~ **receiver** Leitstrahlbordgerät *n* (g/m) ~ **receiver time switch** Leitstrahlzeitschalter *m* ~ **rotation** Leitstrahlschwenkung *f* ~ **superposition** Leitstrahlschaltung *f* ~ **unit** Leitgerät *n*

**guide,** ~ **bearing** Führungs-hülse *f*, -lager *n* ~ **bit** Zuführungsbohrer *m* ~ **blade** Führungsschaufel *f* ~ **block** Führungs-böckchen *n*, -klötzchen *n*, -schlitten *m*, Geradeführungsbacke *f*, Gleitbacken *m*, Gleitklotz *m*, Leitklotz *m*, Seilumlenkrolle *f* ~ **block for sucker bar** Saugstangenführung *f* ~ **bolt** Abhebestift, Führungs-bolzen *m*, -schraube *f* ~ **book** Führungsbuch *n* ~ **box** Führungsbuchse *f* ~ **braces** Führungsstützen *pl* ~ **bracket** (of a forming attachment) Kopierschienenhalter *m* ~ **brush** Futterrohr *n* ~ **bucket** Leit-korb *m*, -zelle *f* ~ **bush** Führungs-buchse *f*, -hülse *f* ~ **bushing** Rastbuchse *f* ~ **cage** Führungskasten *m* ~ **cam** Führungsnocken *m* ~ **card** Indexkarte *f* ~ **casing** Führungsbock *m* (g/m) ~ **check** Tragbacke *f* ~ **claw** Führungs-, Luft-klaue *f* ~ **components** Führungsteile *pl* ~ **cone** Führungspilz *m* ~ **curve** Führungskurve *f* ~ **cylinder** Führungszylinder *m* ~ **edge** Spurleiste *f* ~ **eye** Führungsauge *n* ~ **facings** Arbeitsleisten *pl* ~ **fence** Leitzaun *m* ~ **field** Führungsfeld *n* ~ **finger** Führungs-, Kopier-finger *m* ~ **flange** Führungsflansch *m* ~ **flap** Lenkklappe *f* ~ **fork** Führungsbügel *m* ~ **frame** Führungsrahmen *m*, Waagebalken *m* ~ **groove** Führungsnute *f* ~ **hole** Führungsloch *n*, Ventilführungsloch *n* ~ **idler** Führungsrolle *f* ~ **iron** Führungshülse *f* ~ **jaw** Führungsbacke *f* ~ **line** Richtlinie *f* ~ **liner** Gleitfutter *n* ~ **lining** Klauenfutter *n* ~ **links** Gelenkstangenführung *f* ~ **lug** Führungsarm *m* (Fräsmaschine), Führungs-auge *n*, -warze *f* ~ **mark** Ausrichte-, Leit-marke *f* ~ **mechanism** Leitvorrichtung *f* ~ **member** Umlenkkranz *m* ~ **mill** Drahtwalzwerk *n* mit

mechanischen Ein- und Ausführungen, Handels-, Universal-eisenwalzwerk *n* ~ **motion** Bewegungsrichtung *f* ~ **nozzle** Leitdüse *f* ~ **order** Führungsleiste *f* ~ **perforation** (for engagement of sprocket) Führungsloch *n* ~ **piece** Führungsstück *n* ~ **pieces** Führungssteine *pl* ~ **pile** Führungspfahl *m* ~ **pilot** Führungszapfen *m* ~ **pin** Abbiege-, Abhebe-, Führungs-, Steck-stift *m* ~ **pin bushing** Führungsbolzenbuchse *f* ~ **pin-spring** Rastfeder *f* ~ **pivot** Spitzenführung *f* ~ **plate** Führungs-blech *n*, -platte *f*, Kopierschiene *f*, Lasche *f*, Spurplatte *f* ~ **plate of caterpillar track** Führungsplatte *f* der Raupe ~ **plate templet** Kopierlineal *n* ~ **pole of a pack animal** Führstange *f* ~ **post** Läuferrute *f* ~ **pulley** Ablenk-, Führungs-rolle *f*, Leitblock *m*, Packrolle *f*, Riemenleiter *m*, Umlenkrolle *f* ~ **-pulley drive** Winkeltrieb *m* ~ **-pulley sheave** Gleitrolle *f* (electr.) ~ **rail** Anlege- (print.), Führungs-, Lauf-, Leit--schiene *f*, Radlenker *m*, Spurlatte *f*, Spurplatte *f* ~ **rib** Verschlußschiene *f* ~ **ring** Führungs-, Klauen-, Nuten-, Stell-ring *m* ~ **rings on a projectile** Zwangführung *f* ~ **rod** Führungsbacke *f*, Gleitstab *m*, Leitbaum *m*, Leitstange *f*, Lenker *m*, Lenkergestänge *n* ~ **roll** Führungswalze *f*, Siebwalze *f* ~ **roller** Führungs-rolle *f*, -walze *f*; Kurven-, Leit-rolle *f*; Leit-, Siebleit-walze *f*, Umlenkrolle *f* ~ **rope** Führungs-kabel *f*, -seil *n* ~ **screw** Leitspindel *f*, Schraubenpatrone *f* ~ **section** Führungsteil *m* ~ **shaft** Führungsschaft *m* ~ **shoe** Gleitklotz *m* ~ **sleeve** Führungs-buchse *f*, -muffe *f*, Zentrierhülse *f* ~ **sleeve of the spindle** Gleitauge *n* ~ **socket** Führungsschuh *m* ~ **spring** Scherenfeder *f* ~ **sprocket wheel** Umlenkkettenrad *n* ~ **stem** Führungsstift *m* ~ **surface** Führungsfläche *f* ~ **traverse** Führungstraverse *f* ~ **tube** Führungsrohr *n* ~ **tube for antenna** Antennenführungsrohr *n* ~ **value** Richtwert *m* ~ **vane** (jet) Ablenkfläche *f*, Lauf-, Leit-, Umlenk-schaufel *f* ~ **-vane arrangement** Turbinengitteranordnung *f* ~ **-vane ring** (gas turbines) Leitkranz *m* ~ **vanes** (turbine) Direktionsschaufeln *pl* ~ **washer** Führungsscheibe *f* ~ **wave impedance** Feldwellenwiderstand *m* im Hohlleiter ~ **wavelength** Wellenlänge *f* im Hohlleiter (od. Wellenleiter) ~ **way** Bettführung(en) *f* (Drehbank), Führungs-bahn *f*, -schiene *f* ~ **wheel** Führungs-rad *n*, -rolle *f* (film), Leitrad *n*, Leitring *m* ~ **-wheel carrier** Leitradträger *m* ~ **-wheel spindle** Leitradwelle *f* ~ **winch** Leithaspel *f* ~ **wing** Richtungsflügel *m* ~ **wire** Eindrahtwellenleiter *m* ~ **yoke** (slide bars) Gleitschienenjoch *n*

**guided** geführt, gelenkt, gesteuert, zwangläufig, zwangsläufig **to be** ~ **by** sich richten nach ~ **missile** Fernlenkwaffe *f* ~ **motion** Zwangsläufigkeit *f* ~ **wave** geführte oder leitungsgebundene Welle *f*

**guides** (crosshead) Geradführung *f*, Leitepunkte *pl* (print.)

**guiding** Lenkung *f*, Weisung *f*; führend, leitend ~ **to and from** Hin- und Herführen *n* ~ **of the sliver** Bandführung *f*

**guiding,** ~ **arm** Führungsarm *m* ~ **bar grip** Lenkergriff *m* (Motorrad) ~ **belt** Haltegurt *m* ~ **cable** Schwenk-, Schwung-seil *n* ~ **calculation**

Richtkalkulation *f* ~ **cam** Kurvenschleife *f* ~ **channel** Leitkanal *m* ~ **control** Lenkvorrichtung *f* ~ **device** Leit-, Lenk-vorrichtung *f* ~ **edge** Führungs-, Steuer-kante *f* ~ **eyepiece head** Pointierungsokularkopf *m* ~ **groove** Führungs-nut *f* ~ **lights** Leitfeuerkette *f* ~ **method** Lenkungsverfahren *n* ~ **microscope** Pointierungsmikroskop *n* ~ **plate** Stützblech *n* ~ **principle** Leitsatz *m* ~ **projection** Führungsknaggen *m* ~ **pulley** Führungsrolle *f* ~ **rods** Gestänge *n* ~ **rule** Führungslineal *n* ~ **screw shaft** Leitwelle *f* ~ **shield** Führungsrohr *n* und Schutzschild *n* ~ **shoulder** Führungshals *m* ~ **spindles of the carbon holders** Spindeln *pl* für die Kohlenhalter ~ **stud** Führungsbeschlag *m* (g/m) ~ **surface** Führungsfläche *f* ~ **tapes** Einführbänder *pl* ~ **telescope** Leitfernrohr *n* ~ **theme** Leitmotiv *n* (film) ~ **tongue** (on a frog) Führungszunge *f* ~ **wall** Leitwand *f*
**guild** Gilde *f*, Innung *f*, Zunft *f*
**guilloche** Guilloche *f*, Schlangenverzierung *f* (arch.)
**guilloshing** Guillochieren *n*
**guillotine,** ~ **cutting machine** Formatschneider *m*, Riesbeschneidemaschine *f* ~ **plate shears** Rahmenblechschere *f* ~ **shear** Kurbelblechtafelschere *f* ~ **shears** Blechschere *f* mit geschlossenem Gestell, Parallel-, Rahmen-schere *f*
**gulch** Tobel *m*
**Guldin's rule** Guldinsche Regel *f* (math.)
**gulf** Bucht *f*, Golf *m*, Meerbusen *m*
**gull wing** Knick-, Möwen-flügel *m*
**gullet** (for pouring metal) Einlaufrinne *f*, Kluft *f*, Nische *f* ~ **of a bridge** Brückenbogen *m* ~ **of a saw tooth** Einschweifung *f* des Sägezahnes ~ **tooth** Hakenzahn *m*, (saw) Wolfszahn *m*
**gully** Absturzschacht *m*, Bodenfalte *f*, Regeneinlaß *m*, Runse *f*, Wassertopf *m*
**Gulstad relay** Gulstadrelais *n*
**gum, to** ~ aufkleben **to** ~ **to a form** auf ein Formular kleben **to** ~ **silk** Seide *f* gummieren
**gum** Dammarharz *m*, Klebstoff *m*, Kleister *m*, Gummi *m* und harziger Rückstand *m* im Kraftstoff ~ **accaroid** Accaroidharz *n* (Lack) ~ **arabic** Acacin *n*, Gummiarabikum *n* ~-**arabic process** Chromgummiverfahren *n* ~ **copy** Gummikopie *f* ~ **dragon** Gummitragant *m* ~ **guaiak** Guajakharz *n* ~ **printing varnish** Gummidrucklack *m* ~ **process** Leimkopierverfahren *n* ~ **resin** Gummiharz *n* ~ **reversal** Positivkopie *f* ~ **rosin** Zapfkolophonium *n* ~ **solvent** Gummilösung *f* ~ **test** Untersuchung *f* auf Gehalt an harzigen Rückständen ~ **thickening** Gummiverdickung *f* ~ **wood** Fieberbaum(-holz) *m*
**gumbo sticky shale** haftender (schmieriger) Schieferton *m*
**gumlac** Gummilack *m*
**gummed** gummiert ~ **edge** Klebrand *m*, Leimkante *f* ~ **label** Klebezettel *m* ~ **marble paper** Kleistermarmorpapier *n* ~ **paper base** Gummierrohpapier *n* ~ **piston ring** verklebter Kolbenring *m* ~ **scaling reel** Klebeverschlußrolle *f* ~ **tapes and films** Klebebänder und -folien *pl* ~ **veneer tape** Furnierkleberolle *f*
**gummer** Gummisammler *m*, Kleber *m*, Sägefeile *f*
**gumminess** Neigung *f* zum Verharzen
**gumming** Gumbildung *f*, Gummierung *f*, Harz-

bildung *f*, Kleben *n*, (valve) Verharzung *f* ~ **machine for paper in rolls** Rollengummiermaschine *f*
**gummite** Urangummi *n*
**gummous** gummiartig
**gummy** gummiartig ~ **disease** Bakteriose *f* ~ **resin** gummiartiges Harz *n*
**gums** Harz *n*
**gun** Anodenloch *n* (CRT), Geschütz *n*, Kanone *f*, (grease) Presse *f* ~ **with chase rings** Ringkanone *f* ~ **(or cannon) for point-blank firing** Flachbahngeschütz *n*
**gun,** ~ **angle of elevation** Höhenrichtfeld *n* ~ **anode** Beschleunigungsanode *f* ~ **barrel** Flintenlauf *m*, Geschützrohr *n*, Gewehrlauf *m* ~-**barrel drill** Gewehrlaufbohrer *m* ~-**barrel oxydizing** Brünieren *n* ~ **bay** Waffenraum *m* ~ **bearing** Schußseitenwinkel *m* ~ **boat** Kanonenboot *n* ~ **bronze** Kanonenbronze *f* ~ **car** Geschützwagen *m* ~ **carriage** Kastenlafette *f*, Lafette *f*, Lafettenfahrzeug *n* ~-**carriage axle** Lafettenachse *f* ~-**carriage bed** Lafettentisch *m* ~-**carriage construction** Lafettenaufbau *m* ~-**carriage side plate** Lafettenwand *f* ~ **carriage with shafts** Gabellafette *f* ~-**carriage wheel chock** Schießklotz *m* ~ **chamber** Geschoßraum *m* ~ **clip** Patronenrahmen *m* ~-**core drill** Kanonenbohrer *m* ~ **cotton** Kollodiumwolle *f*, Nitrozellulosepulver *n*, Schießbaumwolle *f*, Schießwolle *f* ~-**directing radar** Feuerleitradar *n* ~ **dispersion** Streuung *f* des Geschützes, Waffenstreuung *f* ~ **emplacement** Geschützaufstellung *f*, -bettung *f*, -stand *m*, -stellung *f* ~ **flash** Aufleuchten *n*, Feuerschein *m*, Feuerstrahl *m*, Mündungsfeuer *n* ~ **foundry** Geschütz-, Kanonen-gießerei *f* ~ **grease** Waffen-fett *n*, -schmiere *f* ~ **hoist** Batteriekran *m*, Geschoß-heber *m*, -kran *m* ~-**lifting gear** Geschützhebezeug *n* ~ **line** Seelenachse *f* ~ **lock** Flintenschloß *n* ~ **mantlet** Waffenblende *f* ~ **metal** Bronze *f*, Geschütz-bronze *f*, -bronzeguß *m*, -metall *n*, Kanonen-bronze *f*, -metall *n*, Rotguß *m*, Speise *f*, Stückmetall *n* ~ **mount** Lafette *f*, Waffenaufnahme *f* ~ **mount on airplane (or ship)** Bordlafette *f* ~-**mount outrigger** Lafettenholm *m* ~ **mounting** Geschütz-, Kanonen-lafette *f*, Lafette *f* ~ **oil** Waffenöl *n* ~ **pit** Geschützeinschnitt *m* ~ **platform** Feuerplattform *f*, Geschützbettung *f* ~-**platform excavation** Bettungsloch *n* ~ **position** Gefechtsstand *m*, Geschütz-stand *m*, -stellung *f* ~ **potential** Beschleunigungsspannung *f* ~ **powder** Pulver *n*, Schießpulver *n* ~ **powder ore** Pulvererz *n* ~ **recoil** Rohrrücklauf *m* ~ **rest** Astgabel *f* ~ **shaft** Aufstiegschacht *m* ~ **shield** Geschütz-, Schutz-schild *n*, Waffenblende *f* ~ **shot** Flinten-schrot *n*, -schuß *m*; Kanonenschuß *m* ~ **sight** Aufsatz *m*, Korn *n*, Richtaufsatz *m*, Visier *n* ~-**sighting telescope** Gewehrzielfernrohr *n* ~ **slide** Gleitschuh *m* ~ **smith** Büchsenmacher *m* ~ **stock** Gewehrschaft *m* ~ **system** Strahlerzeugungssystem *n* ~-**trail crosspiece** Lafettenholm *m* ~ **tube** Kanonen-, Mutter-rohr *n* ~ **turret** Maschinengewehrdrehturm *m*, Waffenkuppel *f* ~ **wale** Dollbord *m*, Dollebord *n*, Schandeck *n* ~ **welder** Punktschweißzange *f* (-bügel), Spreizelektrode *f*

**gunite, (coat of)** ~ Torkretputz m ~ **pneumatic stowing machine** Torkretversatzmaschine f
**gunner,** ~'s **bay** Krähennest n ~'s **blister** Bodenlafette f ~'s **seat** Lade-, Lafetten-, Richt-sitz m ~'s **store** Feuerwerkhellegat n ~'s **turret** Gefechtsstand m, Kanzel f (aviat.)
**gurgle, to** ~ rauschen
**gush, to** ~ quellen **to** ~ **forth** hervorquellen
**gusher** Ausbruch m, Sprudel-bohrung f, -brunnen m ~ **oil well** Springer m (Erdöl)
**gushing** Rieseln n
**gusset, to** ~ mit einem Keil m oder Zwickel m versehen, zwickeln
**gusset** Anschluß m, Anschlußblech n, Eckblech n, Ecksitzplatte f, Gehren m, Knoten m, (plate) Zwickel m ~ **bag** Faltenbeutel m ~ **plate** Eckblech n, Eckplatte f am Rahmen, (ship) Fächerplatte f, Knotenblech n ~ **plate of frame** Rahmeneckbeschlag m ~ **stay** Eckverstrebung f ~ **stays of boiler** Blechanker m
**gust** Bö f, Luftbö f, Stoßwind m, Windbö f, Winddalle f, Windstoß m ~ **due to eddy** Wirbelbö f ~ **of wind** Sturzwind m
**gust,** ~ **envelope** Böen-V-n-Diagramm n ~ **lock** Böen-Verriegelung f ~ **meter** Böenmesser m ~ **recorder** Böenschreiber m, Registrieranemometer n ~ **tunnel** Böenwindkanal m
**gusty** böig, stürmisch, windig ~ **air** böige Luft f
**gut string** Darmsaite f
**guttapercha** Guttapercha f ~ **coat(ing)** Guttaperchaschicht f ~ **fuse** Guttaperchazündschnur ~ **paper** Guttaperchapapier n
**gutter, to** ~ kandeln, rinnen, strömen
**gutter** Abfluß-graben m, -rinne f, Abstichrinne f, Ausflußrohr n, Dachrinne f, Durchguß m, Einflußrinne f, Flutgerinne n, Gassenrinne f, Gerinne n, Gosse f, Gußrinne f, (print.) Hohlsteg m, Nonne f, Rinne f, Rinnsal n, Spur f, Straßenrinne f, Traufrinne f, Wasserrinne f ~ **pipe** Dachröhre f ~ **ring** Kausche f ~-**shaped rolled member** Rinnensprosse f ~ **sticks** Formatquadrate pl (print.) ~ **stone** Gossenbordstein m, Gossenstein m, Rinnstein m, Traufstein m ~ **tile** Deck-, Falz-, Kappen-ziegel m, Kehlrinne f
**guy, to** ~ abspannen, (a pole) verankern, verspannen, siehern
**guy** Abspannung f, Anker m, Ankerseil n, Drahtseilanker m, Halter m, Leitstrick m, Spanndraht m, Verspannung f, Widerhalt m
**guy,** ~ **attachment** Ankerstütze f ~ **backstay** Achterholer m ~ **cable** Abspannseil n ~ **clamp** Tragseilklemme f ~ **(wire) hook** Ankerhaken m ~ **line** Halteseil n ~-**line anchorage** Landverankerung f ~ **rope** Abspannseil n, Haltetau n, Spannseil n ~ **thimble** Ankerkausche f, Kausche f (für Ankerseile) ~ **wire** Abspanndraht m, Drahtanker m, Gei f, Pardun n, Pardune f, Verspannungsseil n
**guyed,** ~ **derrick** Mastenkran m ~ **mast** freitragender Mast m ~ **pole** Abspannmast m, Stange f mit Anker ~ **stack** verspannter Schornstein m
**guying** Verspannung f
**guys** verspannte Bohrtürme pl, Verspannung f
**G value** G-Faktor m
**gyle** Anstellbottich m
**gymnastic apparatus** Turngerät n

**gypseous,** ~ **marl** Gipsmergel m ~ **spar** Marienglas n
**gypsometer** Gypsometer n
**gypsoplast** Gipsabguß m
**gypsum** Gips m, Gyps m ~ **burning** Gipsbrennerei f ~ **crusher** Gipsbrecher m ~ **deal** Gipsdiele f ~ **plaster** Stuckgips m ~ **plate** Gipsblatt n ~ **quarry** Gipsgrube f ~ **wedge** Gipskeil m
**gyrate, to** ~ sich drehen, schnell kreisen, umkreisen, (spirally) umlaufen
**gyrating** umlaufend ~ **mass** Schwungmasse f
**gyration** Drehung f, Elektronenhülle f, schnelle Kreisbewegung f, Wirbel m ~ **with apparently inverted control** Steuerwechselkreis m ~ **radius** Elektronenhüllenradius m
**gyratory** kreisend ~ **crusher** Drehmühle f, Kreiselbrecher m, Rundbrecher m ~ **cycle motion** Kreiselung f ~ **motion of the earth** (precession) Präzisionsgang m
**gyro** Drehung f, Kreisel m, Ring m, Spirale f ~ **axis** Kreiselachse f ~ **bearing** korrigierte Peilung f ~ **cap (or cone)** Kreiselkappe f ~ **case (or compartment)** Abschlußkessel m ~ **centering device** Kreiselzentriervorrichtung f ~ **compass** Fernkurskreisel m, Kreiselkompaß m, Kurskreisel m ~ **compass autopilot** Kreiselüberwachungsschalter m ~ **contact for restoring** Fesselungskontakt m ~-**controlled** kreiselgesteuert ~ **drift** Kreiselauswanderung f ~ **dynamics** Kreiseldynamik f ~ **engine** Kreiselmaschine f ~ **flux gate compass** Gyroflux-Fernkompaß m ~-**frequency** Gyrofrequenz f ~-**gain** Kreiselsignal-Verstärkung f ~ **horizon** gyroskopischer Horizont f, Kreiselhorizont m
**gyrolite** Gyrolith m
**gyromagnetic** gyromagnetisch ~ **anomaly** kreismagnetischer Effekt m ~ **compass** Kreiselmagnetkompaß m ~ **resonance** gyromagnetische Resonanz f ~ **resonance uniline** Resonanzeinwegleitweg m ~ **ratio** gyromagnetisches Verhältnis n
**gyro,** ~ **meter** Gyrometer n ~ **mount** Kreiselflansch m ~ **pendulum recording gauge** Anschütz-Punkter m ~ **pilot** Kreiselpilot m, Steuergerät n ~ **plane** Autogiro m, Drehflügelflugzeug n, Hubschrauber m, Tragschrauber m, Windmühlenflugzeug n ~ **rector** Kreiselgradflugweise m
**gyroscope** (Obry's gear) Geradlaufapparat m, Gyroskop n, Kreisel m, Kreiselachse f ~ **with angle pickoff** Meßkurskreisel m ~ **bank indicator** Kreiselneigungsmesser m ~ **control** Kreiselsteuerung f ~ **disc** Kreiselscheibe f ~ **failure** Geradlaufversager m ~ **pendulum** Kreiselpendel m
**gyroscopic** gyroskopisch, kreiselgestützt ~ **couple** Kreiselmoment n ~ **course indicator** Kurskreisel m ~ **device** Kreiselgerät n ~ **flight direction indicator** Kreiselgradflugweiser m ~ **flight indicator** Gyrorektor m ~ **force of the propeller** Kreiselwirkung f ~ **horizon** Horizontkreisel m ~ **oscillation** Kreiselschwingung f ~ **resistance** Kreiselwiderstand m ~ **roll** (acrobatics) Rollenkreis m ~ **sextant** Kreiselkompaß m ~ **stabilizer** Gyroskopstabilisator m ~ **theory** Kreiseltheorie f ~ **torque** Kreiselbelastung f ~ **turn indicator** Kreiselkurvenzeiger m

gyro, ~-setting contact Stützkontakt *m* ~ slow-
down Kreiselauslauf *m* ~ spin Kreisel-drall *m*,
-drehung *f* ~ spin motor run Kreiselbetriebs-
spannung *f* ~ stabilizer Kreisel *m*
gyrostat Kreisel *m*
gyrostatic, ~ compass Kreiselkompaß *m* ~
effect Kreiselwirkung *f* ~ stabilizer Schiffs-

kreisel *m* ~ thermometer Schleuderthermometer *n*
gyro-supporting platform Richtgeberplatte *f*
gyrosyn compass Erdinduktionskompaß   *m*,
Kompaß-Kurskreisel *m*
gyro, ~ system Kreiselverfahren *n* ~ vector axis
Drallachse *f*

# H

**Habann, ~ oscillation** Habannschwingung *f* ~
**tube** Habannröhre *f*
**habit, ~ effect** Habituswirkung *f* ~ **plane** Habitusebene *f*
**hachure, to ~** schraffieren
**hachure** (on a printing roll) Haschur *f*, Schattenstrich *m*, Schraffe *f*, Schraffierung *f*
**hachures** Bergstriche *pl* (auf Landkarten)
**hack, to ~** hacken **to ~ the ground** den Erdboden *m* loshauen
**hack, ~ file** Schneidefeile *f* ~ **knife** Kabelmesser *n* ~ **saw** Bügel-, Drill-, Metall-säge *f*
**hackle** Hechel *f* ~ **tow** Hechelwerg *n*
**hackler** Hechler *m*
**hackling, ~ block** Hechelbrett *n* (Flachs *m*) ~
**room** Hechelei *f*
**hackly** hakig
**hade, to ~** von der Senkrechten *f* abweichen, gegen die Senkrechte *f* einfallen
**hade** Neigungswinkel *m* ~ **of fault** Einfallen *n* eines Ganges ~ **of the fault** Fallwinkel *m*
**Hadley's, ~ quadrant** Spiegelquadrant *m* ~ **sextant** Spiegelsextant *m*
**haematachometer** Hämatachometer *m*
**haematometer** Hämatometer *n*
**haemoglobinometer** Hämoglobinometer *n*
**hafnium** Hafnium *n*
**haft** Heft *n*, Stiel *m*
**hail, to ~** anrufen, begrüßen, hageln **to ~ a boat** ein Boot *n* anrufen
**hail** Hagel *m*, Ruf *m* ~ **clouds** Hagelwolken *pl* ~ **formation** Hagelbildung *f* ~ **shower** Graupelschauer *m* ~ **squall** Hagelbö *f* ~ **stone** Hagelkorn *n*, Schlosse *f* ~ **storm** Hagel-bö *f*, -schauer *m*, -schlag *m*, -sturm *m*, -wetter *n*
**hair** Haar *n* **to a ~** haargenau
**hair, ~ carder** Haarkrempler *m* ~ **cloth** Haargewebe *n*, Haartuch *n* ~ **copper wires** Kupferhaarlitzen *pl* ~ **crack** Haarsprung *m* ~ **grout** Haarkalk *m* ~ **hygrograph** Haarfeuchtigkeitsschreiber *m* ~ **hygrometer** Haar-feuchtigkeitsmesser *m*, -hygrometer *n* ~ **like** haarartig ~ **line** Faden *m*, Haarstrich *m* ~ **line crack** Haarriß *m* ~ **line indicator** sehr feine Maßbegrenzung *f* ~ **lines** Fadenkreuz *n*
**hairpin** Haarnadel *f* ~ **bend** (curve, or turn) Haarnadelkurve *f*, S-Kurve *f* ~ **spring** Haarnadelfeder *f*
**hair, ~ pointer** Fadenzeiger *m* ~ **pyrites** Millerit *m* ~**-shaped** haarförmig ~ **side** (horse) Haarseite *f* ~ **sieve** Haarsieb *n* ~**-slippery** haarlässig ~ **space** Haarspatium *n* ~ **spring** feine Feder *f* ~ **(or thin) stroke** Haarstrich *m* ~ **trigger lock** Stechschloß *n*
**hairless** unbehaart
**hairy** behaart, haarig
**halation** Halobildung *f* (TV), Lichthof *m*, Lichthofbildung *f*, Überstrahlung *f* ~ **by reflection** Reflexionslichthof *m*
**half** Halb . . ., Halbscheid *n*, Hälfte *f*; halb, unvollkommen ~ **of axle** Achsschenkel *m* ~ **of a beat** (or of a pontoon) Kaffe *f* ~ **of casing**

Gehäusehälfte *f* **at** ~ **cock** in Ruhrast *f* **at** ~ **the depth** halbtief ~ **of the girder** Riegelhälfte *f* ~ **the girder** Halbriegel *m* **at** ~**-mast** halbstocks **of** ~ **an ounce** lötig ~ **of pattern** Modellhälfte *f* **on** ~ **pay** zur Disposition *f* (z.D.) ~ **a section** Halbzug *m* ~ **of the system** Netzhälfte *f* (electr.)
**half, ~ adder** Halbaddierwerk *n*, symbolischer Addierer *m* (data proc.) ~ **angle** Halbwinkel *m* ~**-back** Seitenbeleuchtung *f* (film) ~**-back illumination** Hinterbeleuchtung *f* ~**-beam** Halbbalken *m* ~ **binding** Halbfranzband *n* ~ **bleach** halbweiß ~ **brace head** Halbkrückel *m* ~**-bright** halbblank ~**-calf** Halbfranzband *n* ~ **cell** Halbzelle *f* ~ **chess** (in bridge construction) Halbbrett *n*, Halbpfosten *m* ~**-close** Halbschluß *m* (unvollkommene Kadenz) ~ **cock** Hahn *m* in Ruhrast ~**-compression cam** Dekompressionsnocken *n* ~**-compression device** Dekompressionseinrichtung *f* ~**-convergence error** Winkelkonvergenz *f* ~**-countersunk rivet** Halbversenkniet *m* ~ **cover** Halbverdeck *n* ~**-crest-value time** Halbwertzeit *f* ~**-curd soap** Halbkern *m* ~ **cycle** Halb-periode *f*, -welle *f* ~**-dark** halbfett ~**-decay period** Halbwertsperiode *f* ~ **delivery** Halbförderung *f* ~**-diamond-point tool** Seitenschruppstahl *m* mit Vierkantspitze ~**-discharge** Halbätze *f* ~**-dislocation** Halbversetzung *f* ~**-double plate wheel** Halbdoppelscheibenrad *n* ~ **dry** halbtrocken ~ **duplex operation** einseitiger Duplexbetrieb *m*, einseitiger Betrieb *m* in Gegensprechschaltung *f*, einseitiger Gegensprechbetrieb *m*, einseitiges Gegensprechen *n* ~**-elliptic** halbelliptisch ~**-finished steel product** A-Produkt *n* ~**-fused** halbgesintert ~**-gas firing** Halbgasfeuerung *f* ~ **gate** Halbschranke *f* (r.r.) ~**-hard tempered** halbhart entspannt ~ **hitch** Halbsteg *m*, Halbstich *m* (naut.) ~**-hitch knot** Halbstichschurz *m*, Schlüsselknoten *m* ~ **image** (of stereophotographs) Halbbild *n* ~**-inch plank** Dünnbrett *n*, Durchschnittbrett *n*, Gemeinlade *f* ~**-inch shelf** Durchschnittsbrett *n* ~ **integer** Halbzähligkeit *f* ~**-integral spin** halbzähliger Spin *m* ~**-intensity period** halbe Ausschwingzeit *f* ~**-lap joint** Anblattung *f* ~**-lense** Halblinse *f* ~**-life period** Halbwertzeit *f* ~ **lives** Halbwertzeiten *pl* ~ **load** Halblast *f* ~ **location** Teilversetzung *f* (unvollständige Versetzung) ~ **loop** halber Looping *m* ~ **mast** halbmast ~**-matt glaze** halbmatte Glasur *f* ~**-matt glazing** Halbmattglasur *f* ~**-maximum point** Halbwertpunkt *m* ~ **measure** Halbheit *f* ~**-moon dike** Kesseldeich *m* ~**-moonshaped** mondsichelförmig ~ **nut** Mutterschloßhebel *m* ~**-nut cam** Mutterschloßexzenter *m*, Schloßhebel *m* ~**-odd spin** halbzähliger Spin *m* ~**-oval steel** Fluchhalbrundstahl *m* ~ **parameter** Halbparameter *m* ~**-part molding box** Formkastenhälfte *f* ~ **pay** halbes Gehalt *n* ~ **period** Halbperiode *f*, Halbwelle *f* ~ **pint** Schoppen *m* ~ **plane** Halbebene *f* ~**-plank** Gemeinlade *f* ~

plank Dünn-, Durchschnitts-brett *n* ~-power angle Halbwertwinkel *m* ~-power point Halbwertspunkt *m* (der Leistung) ~-power width Halbwertbreite *f* (antenna) ~-quantum number Halbquantenzahl *f* ~ resist Halbreserve *f* ~ roll halbe Rolle *f* ~ round halbrund ~-round bar Halbrundstange *f* ~ round file Halbrundfeile *f* ~-round iron (bar) Halbrundeisen *n* ~-round reamer bit halbrunder Bohrer *m* ~-round wood rasp Halbrundholzraspel *f* ~-rounds Halbrundeisen *pl* ~ saturated halbgesättigt ~ shade Halbschatten *m* ~-shade apparatus (or -shadow analyzer) Halbschattenapparat *m* ~-shadow polarimeter Halbschattenpolarisationsapparat *m* ~ shell Halbschale *f* ~-shell interior Behälterinnenseite *f* ~-shrouded halbummantelt ~ shut halbgeschlossen ~-side mill einseitiger Scheibenfräser *m* ~-silvered (mirror) halbdurchlässig (Spiegel *m*) ~-sized halbgeleimt ~ snap roll halbe schnelle Rolle *f* ~ space Halbraum *m* ~ span model Halbflügelmodell *n* ~ speed halbe Fahrt *f* ~-speed shaft Halblaufwelle *f* ~-steel Mockstahl *m* ~-stitch (of rope) Zustich *m* ~ stop Halbblende *f* ~-strut halbstielig ~ stuff Halbzellstoff *m* (paper mfg.), Halbzellulose *f* (in paper mfg.) ~ thickness Halbwertschicht *f* ~ tide Halbgezeit *f*, Mittelhöhe *f* der Ebbe *f*, Tide *f* ~-tide bassin Halbtidebecken *n* ~-timber construction Riegelbau *m* ~-time of exchange Halbwertzeit *f* ~ tin Halbzinn *m*

**halftone** Halbschatten *m*, (graphic) Halbton *m*, Halbtonverfahren *n*, Mittellicht *n* ~ engraving Halbtonätzung *f* ~ engraving (on brass) Messing-Autotypie-Ätzung *f* ~ engraving on copper Kupfer-Autotypie-Ätzung *f* ~ etching Autotypie *f* ~ form Bilderform *f* ~ image Halbtonbild *n* ~ phototypography Halbtonätzung *f* ~ picture Halbtonbild *n* ~ plate Halbtonklischee *n* ~ printing Illustrationsdruck *m* ~ process Halbtonverfahren *n* ~ process image gerastertes Bild *n* ~ screen Autotypie-Raster *m*

**halftrack** Doppelspur *f* (tape rec.), Halbkettenantrieb *m* ~ drive Halbkettenantrieb *m* ~ motorcycle Kettenkraftrad *n* ~ motor vehicle Halbketten-lastkraftwagen *m*, -kraftfahrzeug *n* ~ prime mover Zugkraftwagen *m* ~ recorder Doppelspurtonbandgerät *n* ~ vehicle Halbkettenfahrzeug *n*, Räderkettenfahrzeug *n*, Zwitterfahrzeug *n*

**half,** ~-truss (or principal) Halbbinder *m* ~-turn socket Halbschappe *f* ~-turning lock Halbtourschloß *n*

**half-value,** ~ breadth Halbwertbreite *f* (opt.) ~ layer Halbwert-dicke *f*, -schicht *f* ~ life Halbwertzeit *f* ~ period Halbwertsperiode *f* ~ pressure Halbwertdruck *m* ~ thickness Halbwertdicke *f*

**half-watt,** ~ bulb Halbwattbirne *f* ~ filament lamp Halbwattfadenlampe *f* ~ lamp Halbwattlampe *f*

**half-wave** Halbwelle *f* ~ antenna Halbwellenantenne *f* ~ length Halbwellenlänge *f* ~ plate Halbwellenlängenplättchen *n*, Lambda/zwei Blättchen *n* ~ potential Halbwellenpotential *n* ~ rectification Einweggleichrichtung *f*, Gleichrichtung *f* einer Halbwelle *f*, Halbweggleichrichtung *f* ~ rectifier Einweg-gleichrichter *m*,

-schalter *m*, Halbweggleichrichter *m*, Halbwertwiderstand *m* ~ rectifying valve Einweggleichrichterröhre *f*

**halfway** halbwegs ~ step Halbheit *f* ~ throttle Halbgas *n*

**half,** ~ white halbweiß ~ width Halbwertbreite *f* ~ width of interference maxima Halbwertsbreite *f* der Interferenzen ~ wing Tragdeckhälfte *f* ~-wing test Halbflügelmodell *n* ~ wool Halbwolle *f* ~ write pulse Rückschreibimpuls *m* ~ yearly halbjährlich

**halide** Halogenid *n*, Haloidsalz *n* ~ of silver Silberhalogen *n* ~ torch Freonprüflampe *f*

**hall** Flur *m*, Halle *f*, Saal *m*

**Hall-effect** Halleffekt *m*

**hallmark** Feingehaltsstempel *m*

**Hall mobility** Trägerbeweglichkeit *f*

**hall noise** Saalgeräusch *n* (film)

**Hall wachs effect** lichtelektrische Ausbeute *f*, äußerer lichtelektrischer Effekt *m*, lichtelektrische Elektronenmission *f*, Hallwachseffekt *m*, äußerer Fotoeffekt *m*

**hall way** Hausflur *m*

**halo** Aura *f*, Glorienschein *m*, Halo *m*, Halo-erscheinung *f*, -phänomen *n*, Heiligenschein *m*, Hof *m*, Kranzerscheinung *f*, Lichthof *m* (film), Schwärzungshof *m* ~ chemistry Halochemie *f* ~-coupled loop Endkoppelschleife *f* ~ disturbance Lichthofstörung *f* ~ effect Punktscheinwerfer *m*

**halogen** Salzbildner *m* ~ leak detector Halogen-Lecksucher *m*

**halogenate, to** ~ halogenieren

**halogenized** halogensubstituiert

**halogenous** salzbildend

**haloid salt** Haloidsalz *n*

**halometer** Salinometer *n*

**halotrichite** Bergbutter *f*

**halt, to** ~ parieren, rasten

**halt** Aufenthalt *m*, Halt *m*, Haltezeit *f* ~ area Halteplatz *m*

**halter** Halfter *m*, Reithalfter *m*, Stallhalfter *m* ~ rope Halfterriemen *m*

**halting point** Haltepunkt *m*

**halve, to** ~ halben, halbieren, hälften, überblatten, verblatten

**halved joint** Anblattung *f*

**halving** Anblattung *f* ~ adjustment head (or knob) Höhenberichtigungsknopf *m* ~ adjustment on range finder Ausschaltung *f* der Schnittbildversetzung *f* ~ adjustment roll Höhenberichtigungswalze *f* ~ adjustment scale Höhenberichtigungsteilung *f* ~ together Überblattung *f*

**halyard** Ausholer *m*, Flaggleine *f*, Hißtau *n*

**ham** Funkamateur *m*

**hame** Kummetholz *n*

**hammer, to** ~ ausbreiten, ausschmieden, hämmern, recken, schmieden, treiben (metal) to ~ dress hämmern to ~ even (or flat) aushämmern to ~-forge hämmern to ~-harden hartschlagen to ~ quickly kurz zuschlagen to ~ roughly grob abhämmern

**hammer** Hammer *m*, Klöppel *m*, Schlagstück *n* ~ of a pile driver Bär *m* ~ for removing bumps Beulenklopfer *m*

**hammer,** ~ band Klappeisen *n* ~ bolt Hammerkopfschraube *f* ~ break Hammerunterbrecher *m* ~-break spark coil Funkeninduktor *m* mit Hammerunterbrecher *m*, Hammerinduktor *m* ~ bursts Hammerrisse *pl* ~ claw Hammerklaue *f* ~ clip Hammerbeschlag *m* ~ crusher Hammerbrecher *m* ~ cylinder Bärzylinder *m* ~ dressed mit dem Hammer *m* bearbeitet ~ dressing Hämmern *n* ~ drill Bohrhammer *m* ~ driver Hammerführer *m* ~ drop Fallhammer *m* ~ end of the pick Nacken *m* der Keilhaue *f* ~ face Hammer-bahn *f*, -fläche *f* ~ faces Hammerkerne *pl* ~ finish Hammerschlag *m* ~-forged freiformgeschmiedet ~ forging Hämmern *n* ~ forgings Freiformschmiedestücke *pl* ~ fulling mill Stampfwalke *f* ~ gun Hahngewehr *n* ~ handle Hammer-griff *m*, -stiel *m* ~ hardened iron hargeschlagenes Eisen *n* ~-hardening Kaltschmieden *n* ~ hatchet Hammerbeil *n* ~ head Hammer-bär *m*, -kopf *m* ~-head crane Hammerkran *m* ~-head felt (piano) Hammerkopffilzer *m* ~-head inserts Hammerkopfeinsätze *pl* ~ head soldering iron Hammerlötkolben *m* ~-head stop Bäraufhaltevorrichtung *f* ~ interruptor contact point Unterbrecherhammerkontakt *m* ~ lock Hammersperre *f* ~ lock control switch Hammersperrschalter *m* ~ mechanism Schlagwerk *n* ~ mill Hammer-mühle *f*, -werk *n*, Schlagkreuzmühle *f*, Schlagmühle *f* ~ operated on the principle of a lever Hebelhammer *m* ~ peen Finne *f* des Hammers *m*, Hammerfinne *f* ~ pin Hahnbolzen *m* ~ piston Bärkolben *m* ~-press process (plastics) Schlagpreßverfahren *n* ~ release lever Abspannhebel *m* ~ riveting Hammer-nieten *n*, -nietung *f* ~ scale Glühspann *m*, Schmiedesinter *m*, Zunder *m* ~ shaft Hammer-helm *m*, -stiel *m* ~-shaped hammerförmig ~-shaped iron plug Anfangsbohrer *m* (Steinbrechen *n*) ~ shears Schlagschere *f* ~ slag Hammerschlacke *f* ~ sledge Fäustel *n*, (hammer) Kanthammer *m* ~ smith Grob-, Hammer-schmied *m* ~ strut Spannklinke *f* ~ test Schlagversuch *m* ~ (ing) test Schlagprobe *f* ~ tracks Spuren *pl* des Lithiumkerns *m* ~ tup Hammer-bär *m*, -klotz *m* ~ tup guide Bärführung *f* ~ welding Hammerschweißung *f*

**hammered** gehämmert ~ (or beaten) aluminum Schlagaluminium *n* ~ iron Hammereisen *n* ~ plate geschlagenes Blech *n*

**hammering** Hämmern *n* ~ motion Schlagbewegung *f* ~ press Schlagpresse *f* ~ spanner Schlagschlüssel *m* ~ test Ausbreite-, Polter-, Schlag-, Stauch-probe *f* ~ wheel for boiler cleaning Schlagrädchen *n* für Kesselreinigung *f*

**hammerless** hahnlos

**hammock** Hängematte *f* ~ netting Finknetzkasten *m* ~-seat cultivator Hackmaschine *f* mit Hängesitz *m* (zwischen Rahmen)

**hamper** Korb *m*, Packkorb *m*

**hand, to** ~ überreichen, zureichen **to** ~ **in** einreichen **to** ~ **lap** handläppen **to** ~ **over** aushänden, aushändigen, preisgeben, übergeben **to** ~ **in a telegram** ein Telegramm aufgeben

**hand** Arbeiter *m*, Hand *f*, Nadel *f*, Weiser *m*, (of watch) Zeiger *m* **at** ~ vorhanden **on** ~ vorrätig **to be on** ~ vorliegen, lagern **to go into** ~ den Handbetrieb *m* einschalten **on the other** ~ andererseits, in entgegengesetztem Sinne *m* **of a** ~'s breadth handbreit ~ of a clock Uhr-

zeiger *m* ~ **and foot monitor** Hand-Fuß-Monitor *m* ~ **of spiral gear** Zahnkrümmungssinn *m* eines Spiralkegelrades *n*

**hand,** ~ **adjustment** Handeinstellung *f* ~ **agitator** handgetriebenes Rührwerk *n* ~ **air pump** Handluftpumpe *f* ~ **arm** Handwaffe *f* ~ **auger** Handbohrer *m* ~ **bag** Handtasche *f* ~-**ball thrust apparatus** Handkugeldruckprüfgerät *n* ~ **barrow** Handkarren *m*, Schiebkarre *f*, Tragbahre *f* ~ **base pole (or staff)** Handlatte *f* ~ **basin** Waschbecken *n* ~-(operated) **battery switch** Handzellenschalter *m* ~ **beader** Handsickmaschine *f* ~ **belt sander** Handbandschleifmaschine *f* ~ **bill** Flugblatt *n* ~ **blower** Handventilator *m* ~ **blowpipe** Handgebläse *n* ~ **book** Handbuch *n* ~ **booster magneto for starting** Handanlaßmagnetzünder *m* ~ **brace** Bohrkurbel *f*, Boorwinder *m*, Brustbohrer *m*, Faustleier *f* ~ **bracket** Bügel *m* ~ **brake** Bremshebel *m*, Handbreche *f*, Handbremse *f* ~-**brake lever** Handbremshebel *m* ~ **brake (lever) shaft** Handbremshebelwelle *f* ~ **bucket** Handeimer *m* ~-**burned face brick** Hartbrandziegel *m* ~ **cable winch** Handkabelwinde *f* ~ **capacity** Handkapazität *f* ~-**capacity effect** Handkapazität *f* ~ **(or body) capacity effect** Handempfindlichkeit *f* (electron.) ~ **cart** Flurfördermittel *n*, Handwagen *m* ~-**cast** Handguß *m* ~-**casting machine** Handgießmaschine *f* ~ **chain** Handkette *f* ~-**chain guide** Handkettenführung *f* ~ **charging** Handbeschickung *f* ~ **chisel** Handmeißel *m* ~ **clamping** Handbügelbefestigung *f* ~ **composition** Handsatz *m* ~ **compositor** Handsetzer *m* ~ **control** Handbedienung *f* ~ **controlled ribbon reverse** handbetätigte Bandumschaltung *f* ~ **cork board** Handkrispelholz *n* ~ **corn sheller** Maiskolbenschäler *m* für Handbetrieb *m* ~ **crane** Arm-, Hand-kran *m* ~ **crank starter** Kurbelwellenhandanlasser *m* ~-**cranking** (oilpump) Nachkurbelung *f* ~ **cuff** Hand-fessel *f*, -schelle *f* ~-**cut** hand-geschliffen, -geschnitten ~-**cut glue stock** handgeschorenes Leimleder *n* ~-**cut overlay** Handausschnitt *m* ~ **disc sander** Handtellerschleifmaschine *f* ~-**drawn scale** handgezeichnete Skala *f* ~ **dredge** (spoon bag) Handbagger *m*, Sackbagger *m* ~ **dressing** Hand-scheiden *n*, -scheidung *f* ~ **drill** Hand-bohrer *m*, -leier *f*, Sondierhandbohrer *m* ~ **(operated) drilling machine** Handbohrmaschine *f* ~ **drive** Handbetrieb *m* ~-**driven ventilator** Entlüfter *m* mit Handantrieb *m* ~-**fed** mit Handnachschub *m*

**hand-feed** Hand-aufgabe *f*, -vorschub *m* ~ **arc lamp** Handregulierbogenlampe *f* ~ **attachment** Handeinleger *m* ~ **pump** Handspeisepumpe *f* ~ **punch** Handlocher *m* ~ **wheel** Vorschubhandrad *n*

**hand,** ~ **feeding** Handaufgabe *f* (metal) ~ **file round edges** Flachstumpffeile *f* mit runden Kanten ~ **finishing** Nacharbeit *f* von Hand ~ **fire extinguisher** Handfeuerlöscher *m* ~ **firearm** Faustschußwaffe *f*, Hand-feuerwaffe *f*, -schußwaffe *f* ~ **firing** Handfeuerung *f* ~ **flare** Handleuchtzeichen *n* ~ **forging** Freiformschmiede *f*, Freihandschmiedestück *n* ~ **formed brick** Hand(form)ziegel *m* ~ **fuse setter** Stellschlüssel *m*, Zünderschlüssel *m* ~ **gag** Handknebel *m* ~ **gear** Handsteuerung *f* ~ **generator** Kurbel-dynamo *m*, -induktor *m*, Magnetinduktor *m* ~-**generator powered field telephone** Erdsprechgerät *n* ~-**grating spectroscope** Gitterhandspektroskop *n* ~ **grenade** Handgranate *f* ~-**grenade casing** Topf *m* ~ **grenade without detonator** entschärfte Handgranate *f* ~ **grip** Griff *m*, Handgriff *m*, Kurbelheft *n* ~ **ground** handgeschliffen ~ **guard** Handschutz *m* ~ **hammer** Fäustel *n*, Handhammer *m*, Schlägel *m* ~ **hold** Handgriff *m* ~ **hold handle** Haltegriff *m* ~ **hole** Hand-, Mann-loch *n* ~ **hole cover** Hand-, Schau-lochdeckel *m* ~ **hole stoppage** Handlochverschluß *m* ~-**indexing device** (tool machine) Handteilgerät *n* ~ **inertia starter** Handschwungkraft-, Schwungkraft-anlasser *m* für Handbetrieb *m* ~ **inking roller** Handauftragwalze *f* ~ **interrupter** Handunterbrecher *m* ~ **jack** Fußwinde *f* ~-**jarring pattern draw machine** Handrüttler *m* ~-**jarring turnover pattern draw machine** Handrüttler *m* mit Wende- und Abhebeeinrichtung *f* ~ **jig** Handsetzmaschine *f* ~ **jigging sieve** Stauchsieb *n* ~-**labor Pattinson process** Hand-Pattinson-Verfahren *n* ~ **ladle** Hand-gießlöffel *m*, -pfanne *f* ~ **ladling** Gießen *n* mittels Gießlöffel ~-**laid rock in courses** gemauerter Stein *m* ~ **lamp** Gruben-, Hand-lampe *f*, Hand--laterne *f*, -leuchte *f* ~ **lamp cable** Handlaternenleitung *f* ~-**lapped finish** Handläppschliff *m* ~ **lapping** Handläppen *m* ~ **lathe** Handdrehbank *f* ~ **launching** Handstart *m* ~ **lead** Handlot *n* ~ **lead seal press** Plombenzange *f* ~ **lens** (magnifier) Handlupe *f* ~ **level** Freihandnivellierinstrument *n*

**hand-lever** Bedienungs-, Griff-, Hand-hebel *m*, Handschwengel *m* ~ **bearing** Lagerblock *m* ~ **control** Handhebelschaltung *f* ~ **feed** Handhebelvorschub *m* ~ **milling machine** Handhebelfräsmaschine *f* ~-**operated tailstock** Handhebelreitstock *m* ~ **pattern lift (or draw)** Handhebelmodellaushebung *f* ~ **shifter** Handhebelschalter *m* ~ **squeezing** Handhebelpressung *f*

**hand,** ~ **lift truck** Handschubkarren *m* ~-**lifting process** Aushebeverfahren *n* ~ **line** Handlotleine *f* (naut.)

**handmade,** ~ **board** Büttenkarton *m* ~ **lace** handgeklöppelte Spitze *f* ~ **paper** Büttenpapier *n*, geschöpftes Papier *n* ~ **wire cloth** Handdrahtgeflecht *n* ~ **wood pulp board** Handholzpappe *f*

**hand,** ~ **mapping camera** Handmeßkammer *f* ~ **metal punch** Plombenzange *f* ~ **method of working** Handbearbeitung *f* ~ **mold** Handgußinstrument *n* ~ **molded** handgeformt ~ **molder** Handformer *m* ~ **molding** Handstrich *m* ~-**molding machine** Handformmaschine *f* ~-**molding shop** Handformerei *f* ~-**molding work** Handformerei *f* ~ **net** Fanggarn *n*

**hand-operated** hand-angetrieben, -bedient, -betätigt ~ **antidazzle switch** Handabblendschalter *m* ~ **block and tackle** Handflaschenzug *m* ~ **blower** Handlüfter *m* ~ **camera** handbediente Kamera *f*, Handkamera *f* ~ **chuck** Handspannfutter *n* ~ **firing mechanism** Handabfeuerung *f* ~ **flasklifting molding machine** Handformmaschine *f* mit Stiftenabhebung, Stiftenabhebehandformmaschine *f* ~ **oil pres-**

sure **pump** Handöldruckpumpe *f* ~ **plate shears** Handblechschere *f* ~ **pressmolding machine** Handpreßformmaschine *f* ~ **pump** Pumpe *f* für Handbetrieb *m* ~ **screw press** Handspindelpresse *f* ~ **squeezing machine** Handpreßformmaschine *f* ~ **trolley conveyer** Hängebahn *f* für Handbetrieb ~ **turnover molding machine** Wendeplattenformmaschine *f* für Handbetrieb ~ **valve** Handabsperrventil *n* ~ **ventilator** Handlüfter *m* ~ **wheeler** Handschleuderapparat *m*

**hand,** ~ **perforator** Handlocher *m* ~-**picked coal** von Hand geschiedene Kohle *f* ~ **picking** Klaubarkeit *f*, Klaubung *f* ~ **pile driver** Hand-, Zug-ramme *f* ~ **pinion** Ritzel *m* für Handlangzug ~ **pit** Handschacht *m* ~ **plane** Handhobel *m* ~ **power** Handbetrieb *m* ~-**power crane** Handkran *m* ~-**power knife grinder** Schleifapparat *m* für Handbetrieb ~-**power squeezing molding machine** Formmaschine *f* mit Handpressung ~-**power trolley** Handkatze *f* ~ **press** Hand-, Stempel-presse *f* ~ **press printing** Handpressendruck *m* ~ **priming** (oil pump) Nachkurbelung *f* ~ **priming unit** Handpumpenvorrichtung *f* ~ **printing** Handdruck *m* ~ **pull** Handabzug *m* ~ **pump** Handpumpe *f* ~ **punch** Handlochmaschine *f* ~ **puncher** Handlocher *m* ~-**rabbled furnace** Handfortschaufelungsofen *m* ~ **rail** Brüstung *f*, Geländer *n*, Geländerstange *f*, Handlauf *m*, Handleiste *f* (r.r.), Haltestange *f*, Laufstange *f* ~ **rail post** Geländerstütze *f*, Holzschaft *m* ~ **rail standard** Geländerpfosten *m* ~ **rake** Baubrücke *f* ~ **rammer** Handramme *f*

**hand-ramming** Handstampfung *f* ~ **flask-lift machine** Abhebeformmaschine *f* mit Handstampfung ~ **flask-lifting post machine** Stiftenabhebeformmaschine *f* mit Handstampfung ~ **machine** Handformmaschine *f* ~ **turnover foot-draw machine** Wendeplattenformmaschine *f* mit Handstampfung und Fußhebelmodellabhebung

**hand,** ~ **reamer** Handreibahle *f* ~ **receiver** Fernhörer *m* ~ **reel** Handhaspel *f* ~ **refractionometer** Skiaskop *n* (Handrefraktometer *m*) ~-**regulated** mit Handnachschub *m* ~ **regulation** Handregulierung *f* ~ **rejector** Handabweiser *m* ~ **release** (gun) Handabzug *m* ~ **reverberatory calciner** Handfortschaufelungsofen *m* ~ **reverberatory furnace** Handflammofen *m* ~ **reverberatory roaster** Handfortschaufelungsofen *m* ~ **reversing mechanism** Handumsteuerung *f* ~ **revolution counter** Handdrehzähler *m* ~ **riveting** Handnieten *n* ~ **roller** Handabroller *m* ~ **rubble** Schotter *m* von Faustgröße *f* ~ **safety guard** Handschutzvorrichtung *f* ~ **saw** Fuchsschwanz *m*, Handsäge *f* ~ **saw jointer** Handsägeabrichter *m* ~ **scoop** Wurfschaufel *f* ~ **screw press** Handspindelpresse *f* ~-**screw pump** Handschraubenpumpe *f*

**handset** Handapparat *m*, Hörer *m* (teleph.), Mikrotelefon *n* ~-**off-hook relay** Hakenumschalterrelais *n* ~ **switch** Lauthörknopf *m*

**hand,** ~ **setting** Zeigerstellung *f* ~ **shears** Handschere *f* ~ **shield** Schweißschirm *m* ~ **shift** Handsteuerung *f* ~ **shovel** Handschaufel *f* ~ **sizing** Handleimen *n* ~ **smoke signal** Handrauchzeichen *n* ~ **sorting** Hand-scheiden *n*,

-scheidung *f* ~ **sounding machine** Handlot-
maschine *f* ~ **spike** Hebebaum *m*, Spake *f* ~
**splitter** Handaufwinder *m* ~ **squeezer** Handpreßformmaschine *f* ~**-squeezing core machine**
Kernformmaschine *f* mit Handhebelpressung
~**-squeezing machine** Formmaschine *f* mit
Handpressung ~ **starter** Durchdrehungsvorrichtung *f*, Handanlasser *m* ~ **starting** Hand-
-anlaß *m*, -start *m* ~**-starting magneto** Hand-
anlaßmagnet *m* ~**-steering tongue** Handlenk-
deichsel *f* ~ **stoking** Handbeschickung *f* ~
**stuffing** Handschmierverfahren *n* ~ **switch**
Hand(um)schalter *m* ~ **tap** Handgewinde-
bohrer *m* ~ **throttle** Handgaszug *m* ~ **throttle
adjustment** Handgaseinstellung *f* ~ **tiller** Hand-
pinne *f* ~**-tilting (or tipping) device** Handkippvorrichtung *f* ~**-tipping-type ladle** Gießpfanne
*f* mit Handkippvorrichtung ~ **trailer** Transportwagen *m* ~ **traversing gear** Handfahrwerk
*n* ~ **truck** Handtransportkarren *m*, Handwagen *m* ~**-turning gear** von Hand betätigte
Durchdrehungsvorrichtung *f* ~ **turnover molding machine** Handwendeplattenformmaschine *f*
~ **valve** Hand-rad *n*, -ventil *n* ~ **valve lever
(balloon)** Ventilleine *f* ~ **ventilator** Entlüfter *m*
mit Handbetrieb ~ **vise** Feil-, Hand-, Handfeil-
-kloben *m*, Handfeilenklobe *f*, Handschraub-
stock *m*, Schraubenzange *f*, Spannkloben *m*,
Stielfeilkloben *m* ~ **walking beam** Handschwengel *m*
**handwheel** Handrad *n*, Handradwelle *f*, Kurbel-,
Stell-rad *n* ~ **for crossfeed** Handrad *n* für den
Planzug ~ **on dial sight** Aufsatztrieb *m* ~ **for
dressing** Richthandrad *n* ~ **to move table** (of a
machine) Handrad *n* für die Tischbewegung
~ **for traverse feed** Handrad *n* für den Langzug
**handwheel,** ~ **control** Handradsteuerung *f* ~
**shaft** Handradwelle *f*
**hand,** ~ **winch** Handwinde *f* ~**-worked block
system** handbedienter Block *m* ~ **wrench adjusting gear** Schlagschlüsselanstellung *f* ~
**writing** Handschrift *f*, Manuskript *n*, Schrift *f*
~ **written copy** Abschrift *f* mit der Hand
**handed,** ~ **in** eingereicht ~ **propellers** gegenläufig drehende Propeller *pl*
**handicap** Hemmnis *n*
**handicapped** beeinträchtigt, behindert
**handicraft** Gewerbe *n*, Gewerbe-art *f*, -zweig *m*,
Handfertigkeit *f*, Handwerk *n*
**handie-talkie** Sprechfunkgerät *n*
**handiness** Handlichkeit *f*
**handing,** ~ **in** (of message) Aushändigung *f*,
Zustellung *f* einer Botschaft *f* ~**-in of telegrams**
Telegrammaufgabe *f* ~ **over** Aushändigung *f*,
Übergabe *f*
**handle, to** ~ (affairs) abwickeln, anrühren, bedienen, behandeln, betasten, bewältigen, erledigen, gebrauchen, handhaben, hantieren,
manipulieren, verarbeiten, verladen, werken
**to** ~ **the traffic** den Verkehr *m* abwickeln
**handle** Griff *m*, Griffbügel *m*, Griffstück *n*,
Halter *m*, Heft *n*, Henkel *m*, Klinke *f*, (crank)
Kurbel *f*, Löffelstange *f*, Schaft *m*, Schaltgriff
*m*, Stiel *m* ~ **of a bucket** Bügel *m* ~ **of key**
Tastenschlüssel *m* ~ **of a pail** Bügel *m* ~ **for
removing shells** Geschoßheber *m* ~ **of tongs**
Zangengriff *m* ~ **of tools** Werkzeugheft *n* ~ **at
top** Obensteuerung *f* (elektr. Kontroller)

**handle,** ~ **attachment** Hebelausrüstung *f* ~ **bar**
Anhebe-, Gelenk-, Griff-, Leit-, Lenk-stange *f*
~**-bar stem** Lenkstangenrohr *n* ~ **controlling
the gears for raising crossrail** Hebel *m* zur
Hebevorrichtung ~ **grip** Kurbelgriff *m* ~ **hole
of a hammer** Hammerauge *n* ~ **wrench** Griff-
schlüssel *m*
**handles** Holme *pl*
**handling** Abwicklung *f*, Bearbeitung *f*, Bedienung *f*, Behandlung *f*, Betriebsabwicklung *f*,
Erledigung *f*, Handhabung *f*, Hantierung *f*,
Steuerbarkeit *f* ~ **of calls** Gesprächs-, Ver-
kehrs-abwicklung *f* ~ **of camera** Kamerabe-
dienung *f* ~ **of traffic** Gesprächs-, Verkehrs-
-abwicklung *f*
**handling,** ~ **bridge** Entlade-, Lade-brücke *f* ~
**characteristic** Steuerungseigenschaft *f* ~ **crane**
Kübelstange *f* ~ **crew** Haltemannschaft *f* ~
**equipment** Verladeanlage *f* ~ **guy** Hochlaßtau
*n* ~ **instructions** Behandlungsanweisungen *pl* ~
**iron** Handeisen *n* ~ **line** Halteleine *f* ~ **machinery** Verladeanlage *f* ~ **operation** Fördervorgang *m* ~ **place** Abfertigungsstelle *f* ~
**platform** Ausladebrücke *f*, Verladegerüst *n* ~
**pole** Wendestock *m* (Färberei) ~ **problem**
Transportproblem *n* ~ **qualities** Steuereigen-
schaften *pl* ~ **rope** Transporttau *n* ~ **stick**
Wendestock *m* (Färberei) ~ **time** Abferti-
gungs-, Abwicklungs-, Bedienungs-zeit *f* ~
**wheels** Bodenrollräder *pl*
**hands off** freihändig
**handy** handlich ~ **device** Kniff *m* ~ **man** Handlanger *m* ~ **reference** handliche Unterlage *f*
~**-talkie** Sprechfunkgerät *n*
**hang, to** ~ aufhängen, einhängen, hängen, hangen **to** ~ **back** zaudern **to** ~ **behind** abhängen
**to** ~ **by (from) gymbals** kardanisch aufhängen
**to** ~ **on** anhangen, anhängen **to** ~ **over** überhängen **to** ~ **with tapestry** tapezieren **to** ~ **up**
abhängen, anhängen, aufhängen, den Hörer
*m* auflegen **to** ~ **up the receiver** den (Fern)hörer
*m* anhängen
**hangar** Flughalle *f*, Flugzeughalle *f*, Flugzeug-
schuppen *m*, Halle *f*, Hangar *m*, Schuppen *m*
~ **fee** Unterstellgebühr *f* ~ **mechanic** Hallen-
monteur *m* ~ **shed** (for gliders) Segelflugzeug-
halle *f* ~ **skylight** Hallendachlicht *n*
**hangarage** (of airplane) Unterbringung *f*
**hanger** Aufhänger *m*, Aufhängering *m*, (on
mast) Ausleger *m*, Hängeeisen *n*, Hirschfänger
*m*, (hydraul. pump), Hubscheibe *f*, Kabelhalter *m*, Unterlitze *f* ~ **bar** Auflager *n*,
Trachte *f* ~ **bearing** Hänge-, Hubscheiben-
-lager *n* ~ **bearing box** Hängelagerschale *f* ~
**bracket** Hängebock *m* ~ **frame** Hängelager-
bock *m* ~ **link** Pendellasche *f* ~**-on** Anhänger
*m*, Anschläger *m*, Aufschieber *m* ~ **plate** Auf-
hängerblech *n* ~ **point** Kegelzapfen *m* ~**-point
set-screw** Druckschraube *f* mit Kegelzapfen ~
**rod** Aufhängeeisen *n* ~ **shaft** Gehängewelle *f*
~ **shoe** Hängelagerfuß *m*
**hangers** Gehänge *n*
**hangfire** verspätetes Losgehen *n* ~ **cartridge
(or shell)** Nachbrenner *m*
**hanging** Aufhängung *f*, Hängendes *n* (geol.);
hängend ~ **up telephone receiver** Einhängen *n*
des Hörers
**hanging,** ~ **angle** Hängewinkel *m* ~ **beam**

Hängebalken *m* ~ **compass** Hängekompaß *m*
~ **glider** Hängegleiter *m* ~ **guard** Hängemeißel
*m* ~ **mine** Hängemine *f* ~ **room** Aufhänge-
-boden *m*, -saal *m* (print.), Trockenboden *m* ~
**stage** Hängegerüst *n* ~ **test** Reißversuch *m*
~ **tie** Hänge-band *n*, -schiene *f* ~ **valley** Hänge-
tal *n* ~ **wall** hängende Schicht *f*
**hangings** Tapeten(roh)papier *n*
**hang-on** Bodenhaftung *f*
**hangover** Ausschwingungsverzug *m* (electr.) ~
**effect** Nachwirkung *f* eines Effekts oder eines
Einflusses ~ **time** Nachwirkzeit *f*
**hank** Gebund *n*, Strähne *f*, Strang *m*, Wickel *m*
~ **knotter** Dockenknüpferin *f* ~ **mercerizing**
**machine** Garnmerzerisiermaschine *f* ~ **polisher**
Garnglätter *m* ~ **winder** Dockerin *f*
**Hankel function** Hankelsche Funktion *f*
**haphazard** aufs Geratewohl, beliebig, willkürlich, zufällig ~ **arrangement** regellose Anordnung *f* ~ **problem** Regellosigkeitsproblem *n*
**haphazardly** wahllos
**harass, to** ~ bedrängen, beunruhigen, stören
**harbor, to** ~ beherbergen, Schutz gewähren
**harbor** Hafen *m*, Hafenbecken *n* ~ **of refuge**
Nothafen *m*, Notreede *f*, Schutz-, Zufluchts-
-hafen *m* ~ **of shelter** Schutzhafen *m*
**harbor,** ~ **area** Hafengebiet *n* ~ **attendant** Hafen-
wächter *m* ~ **bar shoal** Barre *f* ~ **barrier** Hafensperre *f* ~ **boom** Hafenschwengel *m* ~ **channel**
Hafen-einfahrt *f*, -mündung *f* ~ **construction**
Hafenbau *m* ~ **defense** Hafenschutz *m* ~
**defense boat** Hafenschutzboot *n* ~ **district**
Hafenbezirk *m* ~ **dues** Hafen-abgabe *f*, -ge-
bühren *pl* ~ **engineering** Seehafentechnik *f* ~
**entrance** Hafen-einfahrt *f*, -mündung *f* ~ **lock**
Hafen-, Meer-schleuse *f* ~ **master** Hafen-
-kapitän *m*, -meister *m* ~ **net** Fangnetz *n* ~
**operation** Hafenbetrieb *m* ~ **patrol** Hafen-
polizeidienst *m* ~ **police** Hafenwache *f* ~ **rail-**
**way** Hafenbahn *f* ~ **ship** Depotschiff *n* ~
**structure** Hafenbau *m* ~ **tugboat** Hafen-
schlepper *m* ~ **warehouse** Hafenlagerhaus *n*
**hard** (Zement) abgebunden, hart, schwer,
schwierig, stark, stramm ~ **to follow up (or**
**inspect)** unübersehbar ~ **as iron** eisenhart ~ **to**
**sell** schwer verkäuflich **to** ~ **solder** hartlöten
**as** ~ **as steel** stahlhart ~ **as stone** steinhart
**hard,** ~-**alloy grinder** Horizontalschleifständer
*m* ~-**alloy inserted tooth** Hartmetallmesserkopf
*m* ~ **axle** Hartachse *f* ~ **board** Hartfaserplatte
*f* ~ **brass** Hartmessing *n* ~-**burned brick** Klinker *m* ~ **caoutchouc** Hartkautschuk *m* ~ **cast**
**iron blanks** Hartgußkörper *m* ~-**cemented**
**carbide grade** gesinterte Hartmetallsorte *f*
~-**chine hall** Doppelkeilspanten *pl* ~-**chromium**
**plating** Hartverchromung *f* ~ **coal** magere
Kohle *f*, Mager-, Sand-, Sinter-kohle *f* ~-**coal-**
-**bearing strata** flözführendes Steinkohlengebirge *n* ~ **coal briquetting** Steinkohlenbrikettie-
rung *f* ~ **component** harte Komponente *f* ~
**component of cosmic rays** harte Komponente *f*
der Höhenstrahlung ~-**cutting alloy** Hart-
schneidemetall *n* ~-**drawn** hart gezogen, hart-
gezogen ~-**drawn copper** Hartkupfer *n* ~-**drawn**
**copper wire** Hartkupferdraht *m* ~-**drawn wire**
hartgezogener Draht *m* ~ **driven** heiß ausgezogen ~ **drying** (Farbe) Durchhärten *n* ~-**dry-**
**ing brillant oil** Harttrockenglanzöl *n* ~ **etching**

**ink** Nachätzfarbe *f* ~ **facing** Auftragen *n* von
Hartmetall, Besetzung *f*, Hartauftragsschweis-
sen *n*, Hartmetallauflage *f*, Verstählen *n*
~-**facing alloy** Aufschweißlegierung *f* ~ **fiber**
**cover** Vulkanfiberauflage *f* ~ **flow** schlechter
Fluß *m* ~-**formation cutting head** Bohrer *m* für
hartes Gestein ~-**glass insulator bushing** Hart-
glasdurchführungsisolator *m* ~-**glazed porcel-**
**laine** Hartporzellan *n* ~ **gloss** Hartglanz *m* ~
**grain** harte Frucht *f*, Krebs *m* im Marmor
~-**ground miner** Gesteinshauer *m* ~ **head** Härtling *m* ~ **image** hartes Bild *n* ~ **iron** Harteisen
*n* ~ **lead** Hartblei *n* ~-**lead ball ammunition**
Hartbleimunition *f* ~ **light** Kurzflammlampe *f*
~ **metal** Hartmetall *n* ~-**metal alloy** Hartlegierung *f* ~-**metal crushing tool** Hartmetallknäp-
perbohrkrone *f* ~ **negative** hartes Negativ *n*
~-**nickel plating** Hartvernickelung *f* ~-**over**
**check** Prüfung *f* des Dämpfungsendausschlages
*m* ~ **packed material** festes Gut *n* ~ **pan** fester
Untergrund *m* ~ **paper** (bakelized paper) Hartpapier *n* ~-**paper insulation** Hartpapierisolation
*f* ~ **paper pressing** Hartpapierumpressung *f* ~
**picture** hartes Bild *n* ~ **piece** Hartstück *n* ~
**pitch** Hartpech *n* ~ **plastic** Hartgewebe *n* ~
**porcelain** Hart-, Steinmatz-porzellan *n* ~
**pressboard** Hartpreßspan *m* ~ **pulling** Durch-
zugsvermögen *n* ~ **rays** harte Strahlen *pl* ~
**rebel rock** Hartklamm *n*, Knauer *m* ~ **resin**
**lacquer** Hartharzlack *m* ~ **rock** Hartklamm *m*
(min.), Knauer *m* ~-**rolled** walzhart ~ **rubber**
Hartgummi *m*, wiederbearbeiteter Gummi *m*
~-**rubber coating** Hartgummiüberzug *m* ~-**rub-**
**ber disk** Hartgummischeibe *f* ~ **salt** Hartsalz *n*
~ **semolina mill** Hartgrießmühle *f* ~-**ship case**
Härtefall *m* ~-**sized** stark geleimt (Papier) ~
**slag** Härtling *m* ~ **solder** Hart-, Schlag-,
Streng-lot *n* ~-**solder wire** Hartlotdraht *m*
~-**soldered** hartgelötet ~-**soldering** Hartlötung
*f* ~ **standing** fester Untergrund *m* ~ **start of**
**oscillations** harter Einsatz *m* der Schwingungen,
harter Schwingungseinsatz *m* ~ **starting**
schweres Anlassen *n* ~ **starting of the engine**
schlechtes Anspringen *n* des Motors ~ **steel**
Hartstahl *m* ~-**steel roller** gehärtete Stahlrolle
*f* ~ **stone** Hartklamm *m*, Knauer *m* (min.) ~
**stony ground** fester, steiniger Boden *m* ~
**stroke** starker Anschlag *m* ~ **substances** Feststoffe *pl* ~ **surfacing** Aufbringen *n* eines harten
Überzuges, Auftragsschweißung *f*, Hartauf-
tragstechnik *f* ~ **tissue** Hartgewebe *n* ~ **tube**
harte Röhre *f*, Hochvakuumröhre *f* ~ **vacuum**
(electronic tubes) Hochvakuum *n* ~ **valve**
harte Röhre *f* ~ **vein** Faser *f*
**hardware** Beschlag *m*, Beschläge *pl*, Eisenwaren
*pl* Gerät *n*, Kleinteile *pl*, Kurzwaren *pl*, Metallwaren *pl* ~ **check** Selbstprüfung *f* (data proc.)
~ **representation** konkrete Darstellung *f* (info
proc.), Maschinensprache *f*
**hardwood** Hartholz *n*, hartes Holz *n* ~ **block**
**paving** Hartholzpflaster *n* ~ **handle** Hartholz-
griff *m* ~ **slat** Holzlatte *f*
**harden, to** ~ abbinden, abhärten, aushärten,
binden, erhärten, erstarren, härten, hart werden, verfestigen, vergüten, verhärten **to** ~
**metal** (by sudden cooling) abschrecken
**hardenability** Härtbarkeit *f*
**hardenable steel wire** härtbarer Stahldraht *m*

**hardened** abgehärtet, gehärtet ⁓ **bed V(ee)** Führungsprisma *n* ⁓ **case** Härteschicht *f* ⁓ **coin** stempelharte Münze *f* ⁓ **dowel pin** gehärteter Zylinderstift *m* ⁓ **glass** Hartglas *n* ⁓ **resin** Hartharz *n* ⁓ **steel** Compoundstahl *m*, gehärteter Stahl *m* ⁓ **steel ball** Kugel *f*

**hardener** Härter *m*, Vor-, Zusatz-legierung *f*, Zwischen-legierung *f*

**hardening** Abschreckung *f*, Erhärtung *f*, Erstarrung *f*, (martensitic hardening) Härten *n*, Härterwerden *n*, (temper) Härtung *f*, Stählung *f*, Steifwerden *n*, Verhärtung *f* ⁓ **of filter cloth** Hartwerden *n* der Filtertücher ⁓ **by high-frequency current** Hochfrequenzhärtung *f* ⁓ **by nitridation** Nitrierhärtung *f* ⁓ **of steel** Stahlhärtung *f* ⁓ **of an X-ray tube** Hartwerden *n* einer Röntgenröhre

**hardening,** ⁓ **agent** Härte-, Härtungs-mittel *n* ⁓ **capacity** Härtbarkeit *f*, Härtungs-fähigkeit *f*, -vermögen *n* ⁓ **carbide (or carbon)** Härtungskohle *f* ⁓ **compound** Härtmittel *n* ⁓ **constituent** Härtebildner *m* ⁓ **depth** Einhärtetiefe *f* ⁓ **installation** Härteanlage *f* ⁓ **layer** Einsatzschicht *f* (metal) ⁓**-on** Starkbrennen *n* ⁓ **operation** Härtevorgang *m* ⁓ **plant** Härte-anlage *f*, -vorrichtung *f* ⁓ **plaster** Gipserhärtung *f* ⁓ **process** Härte-verfahren *n*, -vorgang *m*, Härtungsverfahren *n* ⁓ **room** Erstarrungs-kammer *f*, -raum *m*, Härterei *f* ⁓ **shop** Härterei *f* ⁓ **strain** Härte-, Härtungs-spannung *f* ⁓ **test (or sample)** Erhärtungsprobe *f* ⁓ **trough** Härtetrog *m*

**hardenite** Hardenit *m*

**hardest material** dauerhaftester Werkstoff *m*

**hardie** Schröter *m* (Werkzeug)

**hardly fusible glass** schwer schmelzbares Glas *n*

**hardness** Festigkeit *f*, Härte *f*, Widerstandsfähigkeit *f*, Zähigkeit *f* ⁓ **due to carbonates** Karbonhärte *f* (Wasser) ⁓ **of great cutting effect** schneidhaltende Härte *f* ⁓ **due to lime** Kalkhärte *f* ⁓ **of radiation** Strahlenhärte *f* ⁓ **of steel** Stahlhärte *f* ⁓ **of water** Härte *f* des Wassers ⁓ **of an X-ray tube** Härte *f* einer Röntgenröhre

**hardness,** ⁓ **capacity** Härtefähigkeit *f* ⁓ **cooling-rate curve** Härtbarkeitskurve *f* ⁓ **curve** Härteverlauf *m* ⁓ **determination** Härtebestimmung *f* ⁓**-drop tester** Fallhärteprüfer *m* ⁓ **gauge** Härtemesser *m* ⁓ **gradient** Härteverlauf *m* ⁓ **increment** Aufhärtung *f* (Schweißen) ⁓**-measuring apparatus** Härtemaß *n* ⁓**-measuring magnifier** Härtemeßlupe *f* ⁓ **number** Härte-wert *m*, -zahl *f* ⁓ **number in terms of Mohs' scale** Härtestufe *f* ⁓ **range** Härtebereich *m* ⁓ **test** Härte-probe *f*, -prüfung *f*, -versuch *m* ⁓**-test procedure** Härteprüfverfahren *n* ⁓ **tester** Härte-prüfer *m*, -prüfgerät *n*, Schleifmaschine *f* ⁓ **testing (sclerometry)** Härte-messung *f*, -prüfung *f* ⁓**-testing of material under drill** Bohrleistungsversuch *m* ⁓**-testing instrument** Härteprüfer *m* ⁓ **testing machine** Härteprüfmaschine *f*

**hards** Hartkohle *f*

**hardy** Abschröter *m* (metal), Blockmeißel *m*, Meißel *m*

**harm, to** ⁓ schaden

**harm** Schaden *m*

**H armature** Doppel-T-Anker *m*, I-Anker *m*

**Harmet,** ⁓ **electric-smelting process** Harmet-Verfahren *n* ⁓ **fluid-compression method** Harmet-Blockpreßverfahren *n* ⁓ **repouring (or transfer) process** Harmet-Prozeß *m*

**harmful** nachteilig

**harmless** hygienisch unbedenklich, unschädlich

**harmonic** harmonische Oberwelle *f* oder Oberschwingung *f*; harmonisch ⁓ **analyzer** harmonischer Analysator *m* ⁓ **antenna** in Oberwellen erregte Antenne *f*, Oberwellenantenne *f* ⁓ **approximation** harmonische Näherung *f* ⁓ **balancer** harmonischer Stabilisator *m* ⁓ **band** harmonisches Band *n* ⁓ **bar** (piano) Drucksteg *m* ⁓ **components** harmonische Teilschwingungen *pl* ⁓ **content** Oberwellengehalt *m* ⁓ **conversion transducer** Frequenz-teiler *m*, -vervielfacher *m*, harmonischer Frequenzwandler *m* ⁓ **current** sinusförmiger Strom *m* ⁓ **curves** harmonische Sinusschwingungen *pl* ⁓ **determining timbre color (or tone color)** Formant *m* ⁓ **dial** Periodenuhr *f* ⁓ **distortion** nichtlineare oder ungradlinige Verzerrung *f*; Form-, Klirr-verzerrung *f* ⁓ **distortion expressed as an attenuation** Klirrdämpfung *f* ⁓ **eleminator (or excluder)** Stromreiniger *m* ⁓ **filter** harmonischer Löser *m* ⁓ **frequency** Oberschwingungsfrequenz *f* ⁓ **generator** Oberschwingungsgenerator *m*, Oberschwingungserzeuger *m*, Röhrengenerator *m*, Röhrengenerator *m* zur Erzeugung von Oberschwingungen, Sinuswellenerzeuger *m* ⁓ **interference** Oberwellenstörungen *pl* ⁓ **motion** sinusförmige (Wellen)bewegung *f*, Wellenbewegung *f* ⁓ **movement** sinusförmige Bewegung *f* ⁓**-multiple telegraph** Mehrfachtelegraf *m* mit abgestimmten Wechselströmen ⁓ **oscillation** harmonische Schwingung *f* ⁓ **points** harmonische Punkte *pl* ⁓ **reducing circuit** Zwischenkreis *m* ⁓ **response** Frequenzgang *m* ⁓ **response characteristic** harmonische Frequenzcharakteristik *f* ⁓ **ringing** abgestimmter Anruf *m* ⁓ **selective ringing** Wahlanruf *m*, Einzelanruf *m* mit bestimmten Frequenzen, Wahlanruf *m* mit abgestimmten Einrichtungen, wahlweiser Ruf *m* mit abgestimmten Einrichtungen ⁓ **selective signaling** Signalisierung *f* mit abgestimmten Einrichtungen ⁓ **suppressor** Zwischenkreis *m* ⁓ **telegraphy** Tonfrequenztelegrafie *f* ⁓ **tone creation** Obertonbildung *f* ⁓ **vibration** Oberschwingung *f* ⁓ **voltage** sinusförmige Spannung *f* ⁓ **wave attenuation** Oberwellenleistung *f*

**harmonics** Flageolettöne *pl*, Harmonische *f*, harmonische Teilschwingungen *pl* (Komponenten)

**harmonious** zusammenstimmend

**harmonization** Gesamtjustierung *f*, Harmonisierung *f*, Ineinklangbringen *n*

**harmonize, to** ⁓ abstimmen, übereinstimmen

**harmonizing** Gesamtjustierung *f*

**harmony** Einklang *m* ⁓ **whistle** Zweiklangpfeife *f*

**harmotone** Barytkreuzstein *m*

**harness, to** ⁓ anschirren, aufschirren, schirren **to** ⁓ **together** koppeln

**harness** (of headphones) Bügel *m*, Fallschirmgurt *m*, (textiles) Getriebe *n*, (horses) Geschirr *n*, Gurtzeug *n*, (weaving) Harnisch *m*, Kabelbaum *m*, Kabelgeschirr *n*, Kopfhörerbügel *m*, Traggurt *m*, Webgeschirr *n*, Webschütze *m*

**harness,** ⁓ **fixer** Blatteinrichter *m* ⁓ **maker** Ge-

schirrmacher *m* ~ **ring** Gürtelring *m* ~ **snap** Karabinerhaken *m*
**harnesser of atomic energy** Atomkraftbändiger *m*
**harp antenna** Fächer-, Harden-antenne *f*
**harrow, to** ~ eggen
**harrow** Egge *f* ~ **equipment** Einrichtung *f* zum Eggen
**harrowing** Eggen *n*
**harsh** gerb, mißtönend, streng ~ **feel** harter Griff *m* ~ **grained** rauhnarbig ~ **hand** harter Griff *m* ~ **image (or picture)** hartes Bild *n*
**harshness** Härte *f*
**Hartley,** ~ **circuit** Dreipunkt-, Röhren-schaltung *f* mit induktiver Rückkopplung *f* ~ **oscillator** Dreipunktkreisschwingungserzeuger *m* ~ **oscillator circuit** Dreipunkt-, Hartley-Schaltung *f*
**harvest, to** ~ ernten
**harvest** Ernte *f* ~ **time** Erntezeit *f*
**harvester** Erntemaschine *f*, Mäher *m*, Selbstbinder *m* ~ **and binding machine** Bindemähmaschine *f* ~**-binder** Bindemäher *m* ~ **combine** Mähdrescher *m* ~ **ganger** Vorschnitter *m* ~ **thresher** Mähdrescher *m*, kombinierte Mäh- und Dreschmaschine *f*
**harvesting,** ~ **machine for grass** Graserntemaschine *f* ~ **machine for hay** Heuerntemaschine *f*
**hash, to** ~ haschieren, zerstückeln
**hash** Störungszeichen *n* ~ **filter** Störschutzfilter *m*
**hasp** (French window) Drehriegel *m*, Haspe *f*, Kettel *f*, Überwurf *m* ~ **width** Bügelweite *f*
**hasty** fluchtartig, flüchtig, hastig, übereilt ~ **blasting charge** Schnelladung *f* ~ **landing bridge** Uferschnellsteg *m* ~ **obstacle** Schnellhindernis *n* ~ **sap** flüchtige Erdsappe *f* ~ **single-span footbridge** Uferschnellsteg *m*
**hat** Hut *m*, (tanning) Lohschicht *f* ~ **nut** Kappenmutter *f*
**hatch, to** ~ hachieren, schattieren, schraffen, schraffieren, stricheln
**hatch** Durchlaßöffnung *f*, Gatter *n*, (relief and gauging) Klappe *f*, (seaplane) Luk *n*, Luke *f*, Sicherheits- oder Meßklappe *f*, Stauschütz *n*, Ziehschütz *n* ~ **cover** Einstiegklappe *f*, Lücken-, Luken-deckel *m* ~ **way** Niedergang *m*, Schiffsluke *f*
**hatched, (cross)** ~ schraffiert ~ **drawing** schraffierte Zeichnung *f* ~ **stone** gezähnelter Stein *m*
**hatchel** Hechel *f*
**hatchet** Axt *f*, Beil *n*, Hacke *f*, Handbeil *n*
**hatchettine** Berg-, Mineral-talg *m*
**hatchettolite** Hatchettolit *m*
**hatching** (on a printing roller) Hachure *f*, Schattierung *f*, (in drawings) Schraffierung *f*, (of maps) Schummerung *f*
**haul, to** ~ befördern, einziehen, fördern, schleppen, transportieren, trecken **to** ~ **down** herunterholen, (sails or flags) streciehn **to** ~ **hand over first (or hand over hand)** palmen **to** ~ **in** (Antenne) einziehen **to** ~ **in (or down)** einholen **to** ~ **out** ausholen
**haulage** Förderung *f*, Transport *m*, Treideln *n*, Wanderung *f*, Ziehen *n* ~ **in exhaustions** Förderung *f* im Abbauen (min.) ~ **equipment** Förder-, Transport-anlage *f* ~ **rope** Förder-, Schacht-seil *n* ~**-way** Förderstrecke *f* ~ **winch** Schlepperhaspel *f*
**hauler** Frachtführer *m*, Schlepper *m* (min.)

**hauling** Fuhrgewerbe *n*, Schleppen *n*, Ziehen *n* ~ **away** Abfuhr *f* ~ **cable** Förderkabel *n*, Zugseil *n* ~ **chain** Zugkette *f* ~ **contractor** Transportunternehmer *m* ~**-down cable** Abzugskabel *n*, Haltetau *n* ~ **engine** Haspelstreckenförderung *f* ~ **equipment** Transporteinrichtung *f* ~ **gang** Schlepptrupp *m* ~ **(or conveying) installation** Fördereinrichtung *f* ~ **mechanism** Haspel *m* einer Schremmaschine ~**-off** Abholung *f* ~ **output** Förderleistung *f* ~ **practice** Förderbetrieb *m* ~ **rope** Schlepp-, Zug-seil *n* ~ **shaft** Förderschacht *m* ~ **steam engine** Dampffördermaschine *f* ~**-up slip** Aufschlepphelling *f* ~ **winch** Förderwinde *f* ~ **windlass** Förderhaspel *m*
**haunch** Hüfte *f*, (of beam) Kehlung *f*, Schenkel *m*, (on beam) Schräge *f*, (of beam) Voute *f*
**haunched slab** Voutenplatte *f*
**hausmannite** Glanzbraunstein *m*, Hausmannit *m*, Scharfmanganerz *n*, schwarzer Braustein *m*, schwarzes Manganerz *n*, Schwarzmanganerz *n*
**have, to** ~ aufweisen, haben ~ **appellate jurisdiction** zuständig sein als Berufungsinstanz *f* **to** ~ **a blowout (or breakdown)** eine Panne *f* haben **to** ~ **on hand** auf Lager *n* halten **to** ~ **an interest** teilhaben **to** ~ **steam up** Dampf aufhaben **to** ~ **in stock** auf Lager halten **to** ~ **to** müssen
**hawse,** ~ **hole** Klüse *f* ~ **pipe** Ankerklüse *f* (naut.)
**hawser** Festmachetau *n*, Halteleine *f*, Hanftau *n*, Kabeltau *n*, Tau *n*, Tresse *f*, Trosse *f*, Zugseil *n* ~ **barrier** Trossensperre *f* ~**-laid rope** Kardeel *n* ~ **roller** Trossenrolle *f*
**hay** Heu *m* ~ **(press) baler** Heupresse *f* ~ **barn** Heuscheuer *f* ~ **crop** Mahd *f* ~ **fork** Heugabel *f* ~ **harvest** Mahd *f*
**haying machine** Heumaschine *f*
**hay,** ~ **loft** Scheune *f* ~ **maker** Mäher *m* ~**-making machine** Heuerntemaschine *f* ~ **rake** Heurechen *m* ~ **stacker** Heustapler *m* ~ **tedder** Wendemaschine *f* für Heu
**hazard, to** ~ aufs Spiel *n* setzen, riskieren
**hazard** Gefahr *f*, Risiko *n* ~ **of shock** Berührungsgefahr *f* (electr.) ~ **beacon** Gefahrenfeuer *n* (aviat.)
**hazardous** gefährlich, gewagt, riskant ~ **flight conditions** gefährliche Flugbedingungen *pl* ~ **item** hochgefährlicher Artikel *m*
**hazardousness** Gefährlichkeit *f*
**haze, to** ~ anlaufen
**haze** diesiges Wetter *n*, Dunst *m*, (on surfaces) Hauch *m*, Höhenrauch *m*, Nebel *m*, Schleier *m*, Wolkenschleier *m* ~ **atmosphere** Dunstatmosphäre *f* ~ **formation** Dunstbildung *f* ~ **scattering** Dunststreuung *f*
**hazelnut oil** Haselnußöl *n*
**hazel-rod fender** Rohrfender *m*
**haziness** Diesigkeit *f*, Dunstigkeit *f*, Unschärfe *f*
**hazy** dunstig, schleierig, unscharf, unsichtig
**H bar control** Horizontalbalkenregler *m*
**H beam** H-förmiger Balken *m*, H-Formszahl *m*
**H bend** H-Bogen *m*
**H circuit** H-Leitung *f*
**H corner** H-Winkel *m*
**head, to** ~ anköpfen, anstauchen, stauchen **to (be)** ~ köpfen **to** ~ **a cask** ein Faß *n* ausböden **to** ~ **for** abfliegen nach, ansteuern **to** ~ **into the wind** dem Wind *m* entgegendrehen

**head** Abschluß *m*, Anschlag *m*, Aufgabegut *n*, Boden *m*, Dach *m*, Druck *m*, Fallhöhe *f*, (cask) Faßboden *m*, First *m*, Gefälle *n*, Geißkopf *m*, Gußkopf *m*, Haupt *n*, (of still) Helm *m*, Höhe *f*, (of piston rod) Kolbenstangenkopf *m*, Kopf *m*, Kopfstück *n*, (connecting rod) Pleuelstangenkopf *m*, Rubrik *f*, Saugmassel *f*, Saugwirkung *f*, (of water) Säule *f*, (draft) Saugzug *m*, Spitze *f*, Stirn *f*, Support *m*, Topp *m*, Verwalter *m*, Vorderteil *n*, Vorsatzgerät *n* (film), Zunahme *f*

**head, at the ~** vorn **to be at the ~** of an der Spitze *f* stehen **~ of angle-measuring instrument** Winkelkopf *m* **~ around the well** Brunnenkranz *m* **~ away from port** Gegenkurs *m* **~ of bell** Glocke *f* (mach) **~ and chest set** Brustfernsprecher *m* **~ of column** Kolonnenspitze *f* **~ of combustion unit** Heizbehälterkopf *m* **~ of connecting rod** Pleuelfuß *m* **~ of control column** Steuersäulenkopf *m* **~ on crest (or on weir)** Überfallhöhe *f* **~ of department** Abteilungsleiter *m* **~ of a dike** Buhnenkopf *m* **~ of the discharge** Schopfhöhe *f* **~ of the grab** Greiferkopf *m* **~ of a groin** Buhnenkopf *m* **~ of liquid** Flüssigkeitssäule *f* **~ made during process of riveting** Schließkopf *m* **~ of needles** Nadelkopf *m* **~ of pile** Pfahlkopf *m* **~ of a plane** Hobelkopf *m* **~ of projectile** Geschoßkopf *m* **~ of the ram** Stößelkopf *m* **~ of a river** Flußquelle *f* **~ of sponge** Wischerkolben *m* **~ of stand** Stativkopf *m* **~ of state** Staatsoberhaupt *n* **~ of static rod** Winkelkopf *m* **~ of a surge** Lawinenkopf *m* **~ with tommy bar** Knebelkopf *m* **~ of tripod** Stativkopf *m* **~ of water** Fall *m*, (on weir) Stauhöhe *f* **~ of water vaporized** Verdunstungshöhe *f* **~ of weir** Fall *m* **~ of a window** Fensterschluß *m*

**head, ~ amplifier** Bild-, Lichtton-verstärker *m* **~-and-chest set** Brustsprechsatz *m* **~-and-stave jointing machine** Boden- und Daubenfügemaschine *f* **~-and tail effect** Richtungseffekt *m* **~ arch** Obergurt *m* **~ assembly** (books) Kapitel *n*, Kopfhörerbügel *m*, Kopfträger *m*, Stirn--band *n*, -reifen *m* **~ bay** Oberhaupt *n* **~ beam** Deckschwelle *f*, Kron-holz *n*, -schwelle *f*, Oberholm *m* **~ bearing** Kursrichtung *f*, Lagerkopf *m* **~ bolt for capitals** Kapitalkeil *m* **~ bolting** Deckelverschraubung *f* **~ box** (refinery) Umkehrkammer *f* des Rücklaufstückes *n* **~ brace** Brustleier *f* **~ capacity curves for pumps** Pumpenkennlinien *pl* (Förderhöhe) **~ center** Spindelkopfspitze *f* **~-clamp bolt** Spindelkopfklemmschraube *f* **~ coil** Kopfspule *f* (g/m) **~ core** Topfkern *m* **~ counterbores** Kopfsenker *pl* **~ countersink** Kopfsenker *m* **~-dress** Kopfbedeckung *f* **~ drive** Kopfantrieb *m* **~ element** Brennkopf *m* (g/m) **~ end** Aufnahme-, Belade--station *f*, Kopf-ende *n*, -seite *f* **~ (receiving) end** Aufgabestation *f* **~-feed gear** Vertikalgetriebe *n* **~ foreman** Obermeister *m* **~ frame** Förder- (min.), Kopf-gerüst *n* **~ framework** Förderschacht *m* **~ gate** Obertor *n*, Saugmassel *f*, Vordertor *n* **~ gauge** Kopfteilung *f* **~ gear pulley** Turmscheibe *f* **~ gear receiver** Kopfhörer *m* **~ guy** Linienanker *m* **~ harness** Kopfbänderung *f* **~ jib** Spitzenausleger *m* **~-jointing machine** Boden-, Bodenbretter--fügemaschine *f*

**headlamp** Scheinwerfer *m* **~ with tachometer** Tachometerscheinwerfer *m* **~ adjusting device** Scheinwerfereinstellgerät *n* **~ black-out hood (or dimming hood)** Scheinwerferblendkappe *f* **~ carrier** Scheinwerferstütze *f* **~ cone** Scheinwerferkegel *m* **~ reflector** Scheinwerferspiegel *m* **~ support tie rod** Scheinwerferverbindungsstange *f* **~ unit** (lamp insert) Scheinwerfereinsatz *m*

**headland** Kap *n*, Landzunge *f*, Vorgebirge *n*

**headless screw** Gewindestift *m*, Made *f*, Maden-, Schnitt-schraube *f*

**headlight** Bug-, Fern-licht *n*, Scheinwerfer *m*, Stirnlampe *f*, Topplicht *n* **~ for lighting identification** Kennlichtscheinwerfer *m* **~ dimmer** Blendschutzvorrichtung *f* bei Scheinwerfern **~ mask** Abblendkappe *f* **~ shield** Blendkappe *f* **~-support tie rod** Verbindungsstange *f* der Scheinwerfer

**head, ~ line** Kolumnentitel *m*, Rahseil *n*, Schlagzeile *f*, Überschrift *f* **~ long movement** Drauflosfahrt *f* **~ loss** Verlusthöhe *f* **~ meter** Drosselgerät *n* **~ molding** Türverdachung *f* **~-office building** Hauptverwaltungsgebäude *n* **~-on** gegeneinander **~-on collision** Frontalzusammenstoß *m*, gerader zentrischer Stoß *m* **~-on course** recht-voraus Kurs *m*, Grundkurs *m* **~ ovaling machine** Bodenovalschneidemaschine *f* **~ phone** Kopf-fernhörer *m*, -hörer *m*, -telefon *n* **~ phones** Doppelkopf(fern)hörer *m* **~ piece** Aufsatz *m*, Aufsatz *m* der Stange *f*, Kopfteil *n* **~ piercer** Hilfsspinner *m* **~ pin** Kopfpinne *f* **~ pitman** Hängebankarbeiter *m* **~ planing, jointing, and doweling machine** Hobel- und Fügemaschine *f* mit Dübellochbohrapparat (für Bodenbretter) **~-planing machine** Bodenhobelmaschine *f* **~ pressure** Kopfdruck *m*

**headrace** Fallwasser *n*, Obergraben *m*, Speisekanal *m* **~ channel** Oberwasserkanal *m* **~ feeder** Oberkanal *m* **~ reference mark** Oberspiegelhöhe *f*

**head, ~ rail** (ships) Gallionsleiste *f*, Türriegel *m* **~ receiver** Kopfhörer *m* **~ resistance** Stirnwiderstand *m* **~ rest** Kopf-halter *m*, -raste *f*, -stütze *f*, Stirnbügel *m*, Stirnstütze *f* **~ rivet** Setzkopf *m* **~ room** Bau-, Raum-höhe *f* **~ room under a bridge** lichte Höhe *f* **~ rope** (of sail) Anschlagliek *n* **~-rounding machine** Bodenrundschneidemaschine *f* **~ screen** Schweißerkappe *f* **~ sea** Gegensee *f* **~ separation** Prüflänge *f* **~ set** Kopfhörer *m*, kopfbefestigter Fernsprecher *m* **~ setter** Titelsetzer *m* **~ shaft** Hauptantriebswelle *f* **~-shaft line** Hauptwellenstrang *m* **~ side** Anlegesteg *m* (print.) **~ space** Verschlußabstand *m* **~ spar** Stirnholm *m* **~ stack** Mehrspurknopf *m* (tape rec.) **~ stall** Hauptgestell *n*, (harness) Kopfgestell *n*, Stallhalfter *m* **~ start** Vorsprung *m* **~ stick for capitals** Kapitalsteg *m*

**headstock** (of lathe) Reitstock *m*, Spindel--backen *m*, -kasten *m*, -stock *m*, Triebgestell *n* **~ base** Spindelstockuntersatz *m* **~ bearing** Triebwerkslager *n* **~ center** Spindelspitze *f* **~ collet** Spindelstockspannpatrone *f* **~ compartment** Spindelstockgehäuse *n* **~ index finger** Spindelkopfwinkelstellungsanzeiger *m* **~ main gearing** Hauptspindelgetriebe *n*

**head,** **~-straightening and -jointing machine**
Bodenabricht- und Fügemaschine *f*, Boden-
brettabrichthobel- und Fügemaschine *f* **~-**
**straightening machine** Bodenabrichtmaschine *f*
**~ strap** (mask) Kopf-band *n*, -latte *f*, Stirnband
*n* **~-strap cushion** Kopfplatte *f* **~-swiveling**
**worm** Schnecke *f* zum Drehen des Spindel-
kopfes **~ tank** Vorlaufbehälter *m*, Wasserturm
*m* **~ timbers** Gallionsstütze *f* **~ water** Oberwasser *n*, Quelle *f* **~ water elevation** gestauter
Wasserspiegel *n*, Oberwasserspiegel *m* **~ wave**
Kopfwelle *f*
**headway** Fahrt *f*, Geschwindigkeit *f*, lichte Höhe
*f* (arch.), Vorausgang *m*, Vortriebsstollen *m*
(min.), Zeitabstand *m* **~ conveyors** Abbau-
streckenbänder *pl*
**head,** **~ welding** Kopfschweißung *f* **~ wheel**
Seilscheibe *f*, Sperre *f* **~ wind** Gegen-, Vor-
-wind *m* **~ windlass** Kopfwinde *f*
**headed** betitelt **~ fascine** Kopffaschine *f* **~ test**
**bar** Probestab *m* mit Stabköpfen **~ test piece**
Stabkopfprobe *f*
**header** Ährenmähmaschine *f*, (brick) Binder *m*,
Bindestein *m*, Druckkessel *m*, Fertigstauch-
stempel *m*, Kopfmacher *m*, Kopfstaucher *m*,
Kopfstein *m*, Kopfstück *n*, Massel *f*, oben
liegender Flüssigkeitsbehälter *m*, Rücklaufbüchse *f*, Sammelrohr *n*, (of boiler) Sammler
*m*, Sammlerstück *n*, Saugmassel *f*, Scheinbin-
der *m*, (in bolt manufacturing) Stauchstempel
*m*, Vollbinder *m*
**header,** **~ attachment for binder** Ährenmähvor-
richtung *f* für Binder (agr.) **~ bond** Kopfver-
band *m* (Mauerwerk) **~ box** (petroleum) Umkehrkammer *f* des Rücklaufstückes, Umkehr-
stück *n* **~ course** Kopfsteinschicht *f* **~ tank**
Ausgleichsbehälter *m*, Fall-behälter *m*, -tank *m*
**heading** Abteilung *f*, Anbringung *f* eines Kopfes,
Anköpfen *n*, Anstauchen *n* eines Kopfes, Betriebspunkt *m*, Bodenholz *n*, Decklohe *f*, Kolumnentitel *m*, Kurs *m*, Ort *m*, (tunnel) Ortsstoß *m*, Richtstrecke *f*, Rubrik *f*, Steuerkurs *m*
(nav.), Stollen *m*, Stollenvortrieb *m*, Titel *m*,
Überschrift *f* **~ of a message** Übermittlungs-
vorsatz *m* **~ on radio message** Funkspruchkopf *m*
**heading,** **~ advances** fortschreitender Vorhieb *m*
**~ bond** Kopfsteinverband *m* **~ course** Binder-,
Kopf-schicht *f* **~ die** Preßbacke *f*, Stauchmatrize *f* **~ dies** Stauchbacken *pl*, Stoßköpfe *pl*
**~ indicator** (Steuer-) Kursanzeiger *m* **~ stone**
Bogenschluß *m* **~ tool** Anköpfer *m*, Nageleisen
*n*, Nagelkopfeisen *n* **~ weaving** Zeugenden-
weberei *f*
**headings of costs** Kostenbezeichnungen *pl*
**heads** (roof) Fußschicht *f*
**heal, to** **~ up** ausheilen, auswaschen
**heald** Häfel *n*, Helfe *f*, Schaftlitze *f* **~ machine**
Schaftmaschine *f*
**health** Gesundheit *f* **~ commission** Sanitätsbe-
hörde *f* **(national)** **~ insurance** Krankenkasse *f*
**heap, to** **~** häufeln **to** **~ in a dry state** trocken
aufschütten **to** **~ up** anhäufen, aufhäufen, aufstapeln, conglobieren, häufen, schütten, zusammenhäufen **to** **~ up against** anschütten
**heap** Haufen *m*, Stapel *m*, Stoß *m* **~ of debris**
Haufwerk *n* **~ of earth** Erdaufwurf *m* **~ of**
**gravel** Schottermasse *f*

**heap,** **~ chlorination** Haufenchlorierung *f* **~**
**leaching** Haufenlaugen *f* **~ roasting** Haufen-,
Hauf-röstung *f*, Röstung *f* im Haufen **~ sand**
Haufensand *m* **~ stead** Füllort *m*
**heaped,** **~ capacity** gehäufte Leistung *f* **~ con-**
**crete** Schüttbeton *m*
**hear, to** **~** hören
**hearing** Audienz *f*, Gehör *n*, Gerichtsverhandlung *f*, Hören *n*, Verhandlung *f*, Verhör *n*,
Vernehmung *f* **~ ability** Hörempfindlichkeit *f*
**~ aid** Hörgerät *n* **~ distance** Hörweite *f* **~**
**impairment** Hörschädigung *f* **~ interval** Hör-
lücke *f* **~ sound twice** Doppelklang *m*
**hearings** mündliches Verfahren *n*
**heart** Herz *n*, Kernholz *n*, Mark *n* **~ of a thun-**
**derstorm** Gewittersack *m* **~ cam** Herzexzenter
*m* **~ cut** (petroleum) Herzfraktion *f* **~ shake**
(in wood) Kernriß *m*
**heart-shaped** herzförmig **~ curve** Kardioidenkennlinie *f* (rdo) **~ dead-eye** Herzkausche *f* **~**
**diagram** Kardioidenkennlinie *f* (rdo) **~ thimble**
Herzkausche *f*
**heart wheel** Herzscheibe *f*
**heartwood** Stammholz *n*, Kernholz *n* **~ tree**
Kernbaum *m*
**hearth** Erhitzungskammer *f*, Feuerpunkt *m*
(min.), Feuerraum *m*, Feuerung *f*, Feuerungsanlage *f*, (blast-furnace) Gestell *n*, Herd *m*,
Schmelzraum *m* **~ of a mine** Minenherd *m*
**hearth,** **~ accretion** Herdansatz *m* **~ area** Herd-
fläche *f* **~-area efficiency** Herdflächenleistung *f*
**~ ashes** Herdasche *f* **~ bottom** Bodenstein *m*,
(blast-furnace) Gestellboden *m*, Herd-boden
*m*, -sohle *f* **~ casing** Herdeinsatz *m* **~ casting**
(blast-furnace) Gestellpanzer *m* **~ contact**
Bodenkontakt *m* **~ dimensions** Herabmessung
*f* **~ electrode** Bodenelektrode *f* **~ fettling**
Herdfutter *n* **~ flue** Herdzug *m* **~ furnace**
Gefäßofen *m* **~ jacket** Gestell-mantel *m*
(Hochofen), -panzer *m* **~ lining** Herdaus-
-kleidung *f*, -futter *n* **~ material** Herd-
material *n* **~ mold** Herdform *f* **~ opening**
(blast-furnace) Brust *f* **~ patching** Herdflicken
*n* **~ plate** Herd-platte *f*, -zacken *m*, Legeeisen *n*
**~ refining** Herdfrischen *n* **~ roaster** Röstherd
*m* **~ smelting** Herdarbeit *f* **~ sole** Herdsole *f*
**~-type furnace** Herdofen *m* **~-type reverbera-**
**tory furnaces** Herdflammöfen *pl* **~-type smelt-**
**ing furnace** Herdschmelzofen *m* **~ wall** Herd-
wandung *f* **~ zone** Schmelzzone *f*
**heat, to** **~** anwärmen, ausheizen, beheizen, erhitzen, (up) sich erhitzen, erwärmen, feuern,
heizen, vergüten, wärmen **to** **~ body** durch
Hitze *f* eindicken **to** **~ up boilers** anheizen
**to** **~ to red heat** auf Rotglut *f* erhitzen **to** **~**
**red-hot** ausglühen **to** **~ to redness** rotglühen
**to** **~ seal** (plastics) verschweißen **to** **~ slightly**
aufwärmen **to** **~-treat** vergüten (Stahl) **to** **~ up**
aufheizen, warm werden
**heat** Chargengang *m*, Charge *f*, (of a melt)
Gang *m*, Glut *f*, Hitze *f*, Ofengang *m*, Schmelzgang *m*, (run) Schmelzung *f*, Wärme *f*
**heat** **~ of absorption** Bindungswärme *f* **~ of**
**combination** Bindungs-, Verbindungs-wärme *f*
**~ of combustion** Verbrennungs-, Verdichtungs-
wärme *f* **~ of condensation** Kondensations-
wärme *f* **~ of conduction** Leitungswärme *f*
**~ in cooling water** Kühlwasserwärme *f* **~ of**

decomposition Trennungs-, Zersetzungs-wärme *f* ~ of degasification Entgasungswärme *f* ~ of dilution Verdünnungswärme *f* ~ of dissociation Dissoziations-, Ionisations-, Zerlegungs-wärme *f* ~ of elastic extension Dehnungswärme *f* ~ of an electric arc Lichtbogenwärme *f* ~ of emission Emissionswärme *f* (latent) ~ of evaporation Verdampfungswärme *f* ~ of fermantation Gärwärme *f* ~ of formation Bildungswärme *f* ~ of fusion Schmelzwärme *f* ~ of fusion (or solidification) Erstarrungswärme *f* ~ of fusion of atomic weight Atomschmelzwärme *f* ~ of ionization Ionisationswärme *f* ~ of precipitation Präzipitationswärme *f* ~ of radioactivity Zerfallswärme *f* ~ of reaction Reaktionswärme *f* ~ of recalescence Umwandlungswärme *f* ~ of reduction Reduktionswärme *f* ~ of separation Trennungswärme *f* ~ of setting (of cement) Abbindewärme *f* ~ of solution Auflösungs-, Lösungs-wärme *f* ~ of sublimation Sublimationswärme *f* ~ of thermal cumulus Wärmekumulus *m* ~ of transfer Überführungswärme *f* ~ of transformation Umwandlungswärme *f* ~ of transport zugeführte Wärme *f* ~ of vaporization Verdampfungswärme *f* ~ of wetting Benetzungswärme *f*

heat, ~ abduction Wärmeabführung *f* ~ absorbed in reduction Reduktionswärme *f* ~-absorbing wärmeaufnehmend ~ absorbing filter Wärmeschutzfilter *m* ~-absorbing surface Wärmeaufnahmefläche *f* ~ absorption Wärme-aufnahme *f*, -bindung *f*, -einsaugung *f*, -einstrahlung *f* ~-absorption capacity Wärme-aufnahmefähigkeit *f*, -kapazität *f* ~-abstracting characteristic Wärmeableitfähigkeit *f* ~ accumulator Wärmespeicher *m* ~ actuated device thermopneumatischer Auslöser *m* ~-affected zone Nachbarzone *f* ~ balance Hitzeausgleich *m*, Wärmebilanz *f* ~ barrier Wärmemauer *f* ~ breakdown Durchschlag *m* durch Erhitzung ~ calculation Wärmerechnung *f* ~ capacity Wärme-inhalt *m*, -kapazität *f*, -vermögen *n* ~ carrier Wärmeträger *m* ~ change Wärmetönung *f* ~ changer Wärmewechselvorrichtung *f* ~ checking Brandrißbildung *f* ~ coil Feinsicherung *f*, (fuse) Feinsicherungs--einsatz *m*, -patrone *f*, Hitzrolle *f*, Hitzdrahtspulensicherung *f* ~ coil and fuse Sicherungskästchen *n* ~ compensator Wärmeaustauscher *m* ~ conducted away (or off) abgegebene Wärme *f* ~-conducting wärmeleitend ~ conduction Wärme-ableitung *f*, -leitung *f* ~-conduction error Wärmeableitungsfehler *m* ~ conduction path Wärmeleitung *f* ~ conductivity Wärme-leitfähigkeit *f*, -leitungsvermögen *n*, -leitvermögen *n* ~-conductivity curve Wärmeleitungskurve *f* ~ conductor Wärmeleiter *m* ~ connection Wärmeanschluß *m* ~ consuming Wärme *f* verzehrend oder verbrauchend ~ consumption Wärmeverbrauch *m* ~ content Wärme-gehalt *m*, -inhalt *m*, -tönung *f* ~ (or thermal) control Wärmesteuerung *f* ~ control knob Reglerknopf *m* ~ convection Wärmekonvektion *f* ~-convection constant Wärmeübergangszahl *f* ~ convention Wärmeströmung *f* ~ crack Vielhärtungs-, Wärme-, Wärm-riß *m* ~ cracking Brandrißbildung *f* ~ creases Hitze-

falten *pl* ~ cure Heißvulkanisation *f* ~ current Faden-, Heiz-strom *m* ~ currents Wärmeströmung *f* ~ curve Wärmekurve *f* ~ damper Hitzedämpfer *m* ~ death Wärmetod *m* ~ deficit Wärmemanko *n* ~ devaluation Wärmeabwertung *f* ~ direction finder Wärmesuchmesser *m* ~ dissipation Wärme-verlust *m*, -zerstreuung *f* ~ distortion Wärmestörung *f* ~ drop Wärme-abfall *m*, -gefälle *n* ~ dummy coil Blindsicherung *f* ~ dynamometer Wärmedynamometer *n* ~ economy Wärmewirtschaft *f* ~ eddy Sonnenbö *f*, Temperatur-, Wärme-wirbel *m* ~ effect Wärmetönung *f* ~ efficiency Wärmeausnutzung *f* ~ element Ofentopf *m* (g/m) ~ embossing press Heißprägepresse *f* ~-emitting surface Wärmeabgabefläche *f* ~ energy Wärmeenergie *f* ~-engine generating station Wärmekraftwerk *n* ~-engine set Wärmekraft-(maschinen)satz *m* ~ engineering Wärmetechnik *f* ~ equation Wärmegleichung *f* ~ equivalent Wärme-äquivalent *n*, -wert *m* ~ evacuation Wärmeabführung *f* ~ exchange Wärmeaustausch *m* ~ exchanger Temperaturwechsler *m*, Wärme-austauschapparat *m*, -austauscher *m*, -übertrager *m*, -verteiler *m*, -wechselgefäß *n* ~-flash Wärmeblitz *m* ~ flow Wärme-abfuhr *f*, -fluß *m*, -strom *m* ~ flow diagram Wärmeschaltbild *n* ~ flow vector Wärmestromdichte *f* ~ fluctuation Wärmeschwankung *f* ~ flush Wärmestoß *m* ~-flux density Wärmestromdichte *f* ~ frequency Überlagerungsfrequenz *f* ~-frequency oscillator Regelstufe *f* ~ given off abgegebene Wärme *f* ~ graph Schmelzschaubild *n* ~ input Wärmezufuhr *f* ~ installation Heizanlage *f* ~-insulating adiatherman, wärmeundurchlässig ~ insulating blanket Wärmeisoliermatte *f* (g/m) ~ insulation Wärme-isolierung *f*, -isolation *f* ~ insulator Wärme-isolator *m*, -schutzmasse *f*, -schutzmittel *n* ~ interchanger Wärme-austauschapparat *m*, -verteiler *m* ~ light Temperaturkontrolleuchte *f* ~ lightning Wärmegewitter *n*, Wetterleuchten *n* ~ load Aufheizung *f* ~ log Schmelzkarte *f* ~ loss Erwärmungs-, Heiz-, Wärme-verlust *m* ~ measurement Wärmemessung *f* ~-measuring device Wärmemeßeinrichtung *f* ~ motion Wärmebewegung *f* ~ motor Wärmekraftmaschine *f* ~ producer Wärme-bildner *m*, -erzeuger *m* ~ production Wärme-erzeugung *f*, -tönung *f* ~ proof feuerfest, hitzebeständig, wärme-beständig, -sicher ~-proof alloy hochhitzebeständige Legierung *f* ~ proof cast iron feuerbeständiger Guß *m* ~ proof quality Wärmebeständigkeit *f* ~ protective covering (heat insulation) Wärmeschutzpanzer *m* ~ quantity Wärmetönung *f* ~ radiation Dunkelstrahlung *f*, Wärmeausstrahlung *f* ~-radiation pyrometer Strahlungspyrometer *m* ~ radiation sensing device Wärmestrahlungsfühler *m* ~ range Wärmebereich *m* ~ ray Wärmestrahl *m* ~ received zugeführte Wärme *f* ~ record Schmelzbericht *m* ~ recovery Wärmerückgewinn *f* ~ reflection Wärmerückstrahlung *f* ~-regulated heizabgestimmt ~ regulation Flammenführung *f* ~ regulator Thermo-, Wärme-regler *m* ~ rejection Wärme-abfuhr *f*, -ableitung *f* ~ release per unit grate area Feuerraumbelastung *f* ~ requirement Wärmebedarf *m* ~

**reservoir** Wärmespeicher *m* ~ **resistance** Erhitzungswiderstand *m*, Feuerbeständigkeit *f*, Hitze-beständigkeit *f*, -festigkeit *f* ~-**resistant** feuerbeständig, wärmewiderstehend, warmfest ~-**resistant steel** hitzebeständiger Stahl *m*

**heat-resisting** hitze-,· wärme-beständig, wärmesicher ~ **paint** hitzefeste Farbe *f* ~ **quality** Hitzebeständigkeit *f* ~ **varnish** Heizkörperlack *m*

**heat retainer** Wärmespeicherungsmasse *f*

**heat-retaining** wärme-beständig, -speicherungsfähig ~ **capacity** Wärmebeständigkeit *f* ~ **function** Wärmespeicherwirkung *f* ~ **mass** Wärmespeicherungsmasse *f*

**heat,** ~ **retention** (by adjacent masses) Ansammlung *f* ~-**rigor** Wärmestarre *f* ~ **run** Chargengang *m*, Erwärmungslauf *m* ~-**sealing** Heiß-verkleben *n* ~-**sealing paper** heißsiegelfähiges Papier *n* ~ **sealing of plastics** Schweißen *n* von Kunststoff ~-**setting** Heißfixierung *f* ~ **shield** Hitzeschutzblech *n*, Strahlschutz *m* (of a cathode), Wärmeschild *m* ~ **shock** Wärmestoß *m* ~ **shrinking** Aufschrumpfen *n* ~ **sink** Kühlblech *n* ~ **source** Wärmequelle *f* ~ **spectrum** Wärmespektrum *n* ~ **spot** Wärmeempfindung *f* ~ **stable** hitzebeständig ~ **storage** Ansammlung *f*, Wärme-speicher *m*, -speicherung *f* ~ **storage capacity** Anodenwärmekapazität *f* ~-**strength test** Warmzerreißversuch *m* ~ **stress** Wärmespannung *f* ~ **stroke** Hitzschlag *m* ~ **test** Hitze-probe *f*, -prüfung *f* ~ **thunderstorm** Wärmegewitter *n* ~ **tone** Wärmetönung *f* ~ **transfer** Temperaturüberführung *f*, Wärme-bewegung *f*, -durchgang *m*, -übergang *m*, -übergangszahl *f*, -übertragung *f*

**heat-transfer,** ~ **coefficient** Wärme-leitzahl *f*, -übergangszahl *f* ~ **cycle** Wärmeübergangskreislauf *m* ~ **fluid** Wärmeträger *m* ~ **rate** Wärmedurchgangssatz *m*

**heat,** ~ **transferred per unit surface** Heizflächenbelastung *f* ~ **transformation** Wärmeumsatz *m* ~ **transmission** Wärme-abführung *f*, -durchgang *m*, -übergang *m* ~-**transmission resistance** Wärmeübergangswiderstand *m* ~ **transmitting** wärmedurchlassend ~-**treatable** vergütbar ~-**treatable steel** Vergütungsstahl *m*

**heat-treated** wärmebehandelt, (steel) vergütet ~ **bar** vergütetes Stangenmaterial *n* ~ **steel** gehärteter Stahl *m* ~ **stock** vergüteter Stahl *m*

**heat-treating** Vergüten *n*, Wärmebehandlung *f* ~ **department** Härterei *f* ~ **furnace** Hitzbearbeitungsofen *m* ~ **plant** Vergütungsanlage *f* ~ **practice** Wärmebehandlungsverfahren *n* ~ **quality** Vergütungsfähigkeit *f*

**heat,** ~ **treatment** thermische Behandlung *f*, Vergütung *f*, Warmbehandlung *f*, Wärme-behandlung *f*, -vergütung *f* ~-**treatment department** Vergüterei *f* ~ **trip** Wärmeauslöser *m* ~ **utilization** Wärmeausnutzung *f* ~ **value** Wärmewert *m* ~ **variation** Wärmeschwankung *f* ~ **vibration** Wärmeschwingung *f* ~ **waster** Wärmevernichter *m* ~ **wave** Hitzewelle *f* ~-**yielding** wärmegebend

**heatable electric stage** elektrischer Heiztisch *m*

**heated** beheizt, erhitzt, geheizt ~ **to glowing** rotglühend ~ (**hot**) **bearing** heißgelaufenes Lager *n* ~ **carburetor** Vergaser *m* mit Vorwärmung ~ **cathode** Glühkathode *f* ~ **filament** Brenner *m*,

Glühfaden *m*, Heizdraht *m* ~ **wire** Heiz-, Hitz-draht *m*

**heater** Anwärmer *m*, Erhitzer *m*, Heiz-apparat *m*, -draht *m*, -körper *m*, -ofen *m*, -spule *f*, -vorrichtung *f*, (of cathode) Strahler *m*, Vorwärmer *m*, Wärmer *m* (**pre**)~ Anwärmvorrichtung *f* ~ **for flat-iron** Plätteisenbolzen *m*

**heater,** ~ **adapter** Heizeinsatzstück *n* ~ **battery** A-Batterie *f*, Heizsammler *m* ~ **coil** Heizwendel *m* ~ **control** Heizdrossel *f* ~ **current** Heizfadenstrom *m*, Heizstrom *m* ~ **fan** Heizgebläse *n* ~ **motor** Heizmotor *m*

**heater-plug** Glühkerze *f* ~ **controller** Glühüberwacher *m* ~ **filament** Glühdraht *m* (Glühkerze) ~ **indicator** resistor Glühkerzenanzeigewiderstand *m* ~ **installation** Glühkerzenanlage *f* ~ **resistance** (or resistor) Glühkerzenwiderstand *m* ~ **time switch** Glühkerzenzeitschalter *m*

**heater,** ~ **power** Heiz-leistung *f*, -leistungsbedarf *m* ~ **spiral** Heizwendel *m* ~ **supply** Röhrenheizung *f* ~ **temperature** Heizfadentemperatur *f* ~-**type (vacuum) tube** indirekt geheizte Röhre *f* ~ **valve** Warmluftklappe *f* ~ **voltage** Heizspannung *f* ~ (**filament**) **voltage** Heizfadenspannung *f* ~ **wire** Heizdrahtfaden *m*

**heating** Anwärmung *f*, Befeuerung *f*, Beheizung *f*, Durchwärmung *f*, Erhitzen *n*, Erwärmen *n*, Erwärmung *f*, Feuerung *f*, Heißwerden *n*, Heizkraft *f*, Heizung *f*, Verfeuerung *f*, Wärmung *f*

**heating,** ~ **from bottom** Bodenheizung *f* ~ **by circulating air** Umluftheizung *f* ~ **by coke** Koksheizung *f* ~ **with coke** Koksheizung *f* ~ **of conductor** Leitungserwärmung *f* ~ **of cossettes** Schnitzelanwärmung *f* ~ **of image** Abbildungswärme *f* ~ **under pressure** Druckerhitzung *f* ~ **of trains** Zugheizung *f* ~ **up** Erhitzen *n*, Warmlaufen *n*, Warmwerden *n* ~ **of the upper plates** Oberplattenbeheizung *f*

**heating,** ~ **apparatus** Heiz-apparat *m*, -gerät *n*, -vorrichtung *f*, Heizungsvorrichtung *f* ~ **blower** Heizungsgebläse *n* ~ **body** Heizkörper *m* ~ **boiler plant** Heizkesselanlage *f* ~ **capacity** Heiz-fähigkeit *f*, -vermögen *n* ~ **chamber** Erhitzungskammer *f*, Glühraum *m*, Heiz-kammer *f*, -raum *m*, -schrank *m*, Herdraum *m* ~ **circuit** Heizkreis *m* (rdo) ~ **coil** Anwärmschlange *f*, Heiz-schlange *f*, -spule *f*, -spulensicherung *f*, Rippenheizrohr *n* ~ **collar** Heizring *m* ~ **conductor** Heiz-leiter *m*, -widerstand *m* ~ **connection** Heizanschluß *m* ~ **contactor** Heizungsschütz *m* ~ **cord carrier** Heizkordelträger *m* ~ **crack** Warmriß *m* ~ **current** Faden-, Heiz-strom *m* ~-**current intensity** Heizstromstärke *f* ~-**current supervisory relay** Heizstromkontrollrelais *m* ~-**current transformer** Heiztransformator *m* ~-**current variation** Heizstromänderung *f* ~ **curve** Wärmekurve *f* ~ **device** Heiz-, Heizungs-vorrichtung *f* ~ **effect** Heiz-effekt *m*, -wirkung *f* ~ **efficiency** Heizwirkungsgrad *m*, Wärmeleistung *f* ~ **element** Heiz-element *n*, -körper *m* ~ **equation** Erhitzungsgleichung *f* ~ **fabric** Heizgewebe *n* ~ **filament** (in indirectly heated tubes) Heizfaden *m* ~ **flame** Heizflamme *f* ~ **flue** Feuer-, Feuerungs-, Heiz-zug *m* ~ **furnace** Glüh-, Heiz-ofen *m*, langsam gehender Ofen *m*, Schweiß-, Wärme-ofen *m* ~-**furnace cinder**

Schweißofenschlacke *f* ~ **gas** Heizgas *n* ~ **grid** Widerstandselement *n* ~ **grill** Glührost *m*, Heizgitter *n* ~ **inset** Heizeinsatz *m* (electr.) ~ **installation** Heiz-anlage *f*, -einrichtung *f* ~ **jacket** Gasanwärmer *m*, Heizmantel *m* ~ **lamp** Heizlampe *f* ~ **liquid** Heizflüssigkeit *f* ~ **mat** Heizteppich *m* ~ **method** Heizverfahren *n* ~ **motor** Wärmemitführung *f* ~ **muff** Heizmuffe *f* ~ **oven** Heizofen *m* ~ **pad** Heizkissen *n* ~ **period** Heizzeit *f* ~ **pin** Wärmeleitstift *m* ~-**pipe coupling** Heizkupplung *f* ~-**pipe system** Kaloriferendarre *f* ~ **plant** Heizungsanlage *f* ~-**plant construction** Feuerungsbau *m* ~ **plate** Heizplatte *f* ~ **point** Heizpunkt *m* ~ **power** Erwärmungskraft *f*, Heizwert *m* ~-**power machine** Abwärmekraftmaschine *f* ~ **problem** Beheizungsfrage *f* ~ **process** Heizvorgang *m*, Versud *m* ~ **recess** Wärmenische *f* ~ **sample** Glühprobe *f* ~ **section** Heizschacht *m* ~ **sleeve (or jacket)** Heizmantel *m* (durch Wasser geheizt) ~ **station** Heizwerk *n* ~ **stove** Warmluftofen *m* ~ **surface** Heiz-fläche *f*, -oberfläche *f* ~ **switch** Heizungsschalter *m* ~ **system** Heizsystem *n* ~ **test** Glühprobe *f* ~ **tongs** Nietzange *f* ~ **transformer** Heiztrafo *m* ~ **transmitter** Heizungssender *m* ~ **tube** Heiz-, Kessel-, Siede-rohr *n* ~-**up efficiency** Anheizwirkungsgrad *m* ~-**up period** Anheiz-periode *f*, -zeit *f*, Einbrennzeit *f* (electron.) ~-**up zone** Vorwärmungszone *f* ~ **value** Heizwert *m* ~-**value determination** Heizwertbestimmung *f* ~ **value of gas** Gasheizwert *m* ~ **value per unit weight** Heizwertzahl *f* ~ **value per unit weight of gas** Gasheizwertzahl *f* ~ **winding** Heizwicklung *f* ~ **wire** Heizdraht *m* ~ **worm** Heizschlange *f* ~ **zone** Heizzone *f*

**heave, to** ~ einhieven, heben, hochwinden, hochziehen, schwellen **to** ~ **the cable** das Ankertau einwinden **to** ~ **in** hieven **to** ~ **in cable** einhieven **to** ~ **the log** das Log werfen **to** ~ **taut** steif holen **to** ~ **to** backbrassen, beidrehen, lösen, losreißen **to** ~ **up** schleppen

**heave** Aufwinden *n*, Heben *n*, Hebung *f*, (transversale) Horizontalverschiebung *f*, Hub *m*, Verwerfung *f* ~ **(gauge) line** starkdrähtige Leitung *f*

**heaved** gehoben, verworfen ~ **block** Horst *m*

**heaver** Hebebaum *m*, Heber *m*, Winde *f*

**heavier-than-air** schwerer als Luft *f* ~ **aircraft** Aerdyn *n*, Luftfahrzeug *n* schwerer als Luft

**heavily** lastend, schwer ~ **armed** stark bewaffnet, stark bewehrt ~ **coated electrode** Dickmantelelektrode *f* ~ **loaded** stark belastet ~ **stressed** hoch beansprucht (Maschinenteile) ~ **tapered** stark zugespitzt ~ **wooded** dichtbewaldet ~ **worked material** stark durchgearbeiteter Werkstoff *m*

**heaviness** Dumpfheit *f*, großes Gewicht *n*, Schwere *f*, Schwerfälligkeit *f*

**heaving** (frost) Hebung *f* ~ **of the floor** Quellen *n* (des Liegenden)

**heaving,** ~ **effect** schiebende Wirkung *f* ~ **line** Wurfleine *f* ~ **motion** Vertikalbewegung *f* ~ **rock** quellendes Gebirge *n* ~ **shale** drückender Schiefer *m* ~-**up** Losreißen *n*, Schleppen *n*

**Heaviside,** ~ **effect** Stromverdrängung *f* ~ **function** Stammgleichung *f* (teleph.) ~ **layer** Heavisideschicht *f* ~ **unit function** Heaviside'sche Einheitsfunktion *f*

**heavy** (paper) dicht, dick, drückend, dumpf, massig, massiv, schwer, schwerfällig, stabil, standfest, stark, stetig **of** ~ **precipitation** niederschlagsreich **to be of** ~ **sale** schlechten oder schwierigen Absatz *m* haben

**heavy,** ~ **aggregate concrete** Schwerbeton *m* ~ **alloy** Wolframlegierung *f* ~ **anode** Vollanode *f* ~ **armament** schwere Bewaffnung *f* ~-**armature relay** Relais *n* mit schwerem Anker ~ **armor** schwere Armierung *f* oder Bewehrung *f* ~ **atom method** Röntgenstrukturanalyse *f* ~ **axle-journal grinding machine** Schwerachsschenkelschleifmaschine *f* ~ **benzene** Schwerbenzin *n* ~ **benzol** Schwerbenzol *n* ~ **bomber** Großbombenflugzeug *n* ~ **box-type construction** kastenförmige Konstruktion *f* ~ **contact arm** großer Meßbügel *m* ~-**core casting** Hohlform *f* ~ **current** Starkstrom *m* ~-**current arc** Hochstrombogen *m* ~-**current discharge** Hochentladung *f* ~-**current engineering** Starkstromtechnik *f* ~ **drawing and wrapping paper** Tauenpapier *n* ~ **dredging equipment** Großbaggergerät *n*

**heavy-duty** dauerhaft, Hochleistungs . . . ~ **boiler** Hochleistungskessel *m* ~ **boring and drilling machine** Hochleistungsbohrmaschine *f* ~ **chuck** Kraftspannfutter *n* ~ **design** schwere Bauart *f* ~ **drive** Hochleistungsantrieb *m*, Trieb *m* für schwere Beanspruchung ~ **folder** Hochleistungsfalzapparat *m* ~ **gear** Kraftzahnrad *n* ~ **jaw crusher** Großbackenbrecher *m* ~ **lathe** Hochleistungsdrehbank *f* ~ **lorry** Schwerlastwagen *m* ~ **machine** Hochleistungsmaschine *f* ~ **machine for accurate work** Hochleistungsgenauigkeitsmaschine *f* ~ **measuring** Schwersonde *f* ~ **milling machine** Hochleistungsfräsmaschine *f* ~ **operation** Schwerbetrieb *m* ~ **potentiometer** Hochlast *f* ~ **relay made by Bosch** Bosch-Schütz *n* ~ **saw** Säge *f* für Höchstleistung *f* ~ **service** Schwerbetrieb *m* ~ **speaker** Hochleistungslautsprecher *m* ~ **surface milling machine** Hochleistungsflächenfräsmaschine *f* ~ **switch** Leistungsschalter *m* ~ **thread** Kraftgewinde *n* ~ **wire wound resistor** Hochlastdrahtwiderstand *m*

**heavy,** ~ **electron** schweres Elektron *n* ~ **(wet) fog** schwerer, nasser Nebel *m* ~ **force fit** Edelschiebesitz *m* ~ **(pump) fuel feed** Schwerölförderpumpe *f* ~ **fuels** Schwerkraftstoffe *pl* ~ **gale** orkanartiger Sturm *m* ~ **-gauge wire** dickdrähtig, starker Draht *m* ~ **-gauge wire cable** dickdrähtiges Kabel *n* **of** ~ **gauge wire** starkdrähtig ~ **gearings** Großgetriebe *n* ~ **gun** Schwergeschütz *n* ~ **-handed** behindert ~ **high-explosive bomb** Minenbombe *f* ~ **hours** verkehrsstarke Zeit *f* ~ **hydrogen** schwerer Wasserstoff *m* ~ **interference** starke Störung *f* ~ **landing** Bumslandung *f*, harte Landung *f* ~ **liquid test unit** Schwerflüssigkeitsversuchsaggregat *n* ~ **-load capacity** Hochbelastbarkeit *f* ~ **loading** starke Pupinisierung *f* ~ **machine construction** Schwermaschinenbau *m* ~ **machine tool** Schwerwerkzeugmaschine *f* ~ **medium ore** Erztrübe *f* ~ **medium separator** Sinkscheider *m* ~ **mesons** schwere Mesonen *pl* ~ **-metal activator** Schwermetallaktivator *m* ~ **oil** schwerflüssiges Öl *n*, Schweröl *n* ~-**oil caburetor** Schwerölvergaser *m* ~ **-oil engine** Schwerölmotor *m* ~ **plate** Grobblech *n* ~ **rain** heftiger Regen *m* ~ **sea** hochgehende See *f*, hoher Seegang *m* ~ **seas** schwere Ozeanseen *pl* ~ **spar** Baryt *m* ~ **squall** schwere Bö *f* ~ **storm** orkanartiger Sturm *m* ~ **supporting structure** Schwerbauteil *m* ~ **tar** Dickteer *m* ~ **traffic** starker Verkehr *m* ~ **-walled** dickwandig ~ **water solution** Lösung *f* in schwerem Wasser ~ **weave fabrics** Schwergewebe *n*

**Heberlein process** (sintering) Heberlein-Verfahren *n*

**heck of a spinning wheel** Gabel *f* eines Spinnrades

**hectare** Hektar *n*

**hectogram** Hektogramm *n*

**hectograph** Hektograf *m* ~ **mass (or pulp)** Hektografenmasse *f*

**hectographic copy** hektografischer Abzug *m*

**hectographing paper** Hektografenblatt *n*

**hectoliter** Hektoliter *m*

**hectometric waves** Hektometerwellen *pl*, MF-Bereich *m*, Mittelwellen *pl*

**hectowatt** Hektowatt *n*

**heddle** Häfel *n*, Helfe *f*, Weblitze *f* ~ **hook** Einziehhaken *m*

**hedenbergite** Hedenbergit *m*

**hedge, to** ~ **with pales (or stakes)** einpfählen

**hedge** Hecke *f*

**hedgehog transformer** Igeltransformator *m*

**hedgehop, to** ~ dicht über den Boden fliegen

**hedgehop** Buschhopsen *n* (aviat.)

**hedgehopper airplane** Buschhopserflugzeug *n*

**hedgehopping** Hecken-springen *n*, -sprung *m*; Tiefflug *m*

**hedging,** ~ **bill** Faschinenmesser *n* ~ **knife** Hippe *f*

**hedyphane** Hedyphan *m*

**heedlessness** Unachtsamkeit *f*

**heel, to** ~ **krängen to** ~ **over** kentern, überliegen

**heel** Ferse *f*, (of mast) Fuß *m*, (of shoe) Hacke *f*, (of keel) Hieling *f*, Rückstand *m*, Seitenneigung *m* (ship) **to put on a** ~ mit einem Absatz *m* versehen ~ **of switch** Zungenstuhl *m* ~ **of a well** Bohrlochöffnung *f*

**heel,** ~ **chair** Zungendrehstuhl *m* (r.r.) ~ **effect** Anodenschatten *m* ~ **nail** Absatzbaustift *m* ~ **plate** Kappe *f* ~ **strap** Hängebügel *m* ~ **tire** Wulstreifen *m*

**heeling** Krängung *f* ~ **error** Krängungsfehler *m* ~ **magnet** Krängungsmagnet *m* ~ **moment due to wind pressure** (of ship) Winddruckkrängungsmoment *n*

**Hefner candle** Hefnerkerze *f* ~ **unit** Hefnereinheit *f*

**Hegeler furnace** (semimechanical rabble) Hegeler-Ofen *m*

**height** Erhebung *f*, Größe *f*, Höhe *f* (über Bezugswert) Höhenmaß *n*, Pfeilhöhe *f* (arch.) **at a safe** ~ in sicherer Höhe *f*

**height,** ~ **of airplane** Flugzeughöhe *f* ~ **of arc** Pfeilhöhe *f* des Bogens, Bogenhöhe *f* (geom.) ~ **of an arch** Bogentisch *m*, Pfeilhöhe *f* (**overall**) ~ **of a barge** Schiffshöhe *f* ~ **of barometer** Luftdruckstand *m* ~ **of bed** Schicht-, Schütthöhe *f* ~ **of the border** Bördelbreite *f* ~ **of burst** Explosionshöhe *f* ~ **of bursting pressure** Platzhöhe *f* (tire) ~ **of camber** Pfeilhöhe *f* ~ **of center** Achshöhe *f* ~ **of centers** Spitzenhöhe *f*

~ **of the center of gravity** Höhe *f* des Schwerpunktes ~ **of cheek** Wangenhöhe *f* ~ **of crown** Pfeilhöhe *f* (des Bogens) ~ **of a curve** Pfeil *m* ~ **of the day** lichte Höhe *f* ~ **of delivery** Ausgußhöhe *f* ~ **of dip** Abdachungs-, Pfeil-höhe *f* ~ **of discharge** Austragshöhe *f* ~ **of drop** Fallhöhe *f* ~ **of equilibrium** Gleichgewichtshöhe *f* ~ **of fall** Fallhöhe *f* ~ **of feature** Geländepunkthöhe *f* ~ **of flood bank above water line** Höhe *f* des Ufers über dem Wasserspiegel ~ **of free fall** Freifallhöhe *f* ~ **of gib head** Nasenhöhe *f* (des Keils) ~ **above ground** Höhe *f* über Grund ~ **of image and line length** Bildhöhe *f* und Zeilenlänge *f* ~ **of incidence** Einfallshöhe *f* (opt.) ~ **of instrument** Instrumentenhöhe *f* ~ **of land** Wasserscheide *f* ~ **of layer** Schicht-, Schütt-höhe *f* ~ **of lens** Objektivhöhe *f* ~ **of level** Hubhöhe *f*, Wasserstand *m* ~ **of lift** Hub-höhe *f*, -weg *m* ~ **of mask** Deckungshöhe *f* ~ **of the meniscus** Pfeilhöhe *f* ~ **of mountains** Gebirgshöhe *f* ~ **of muzzle** Feuerhöhe *f* ~ **of objective** Objektivhöhe *f* ~ **of oscillation** Schwinggröße *f* ~ **of the potential barrier** Diffusionsspannung *f* ~ **of projection** Wurfhöhe *f* ~ **of quay surface above high water** Höhe *f* über dem Wasserspiegel ~ **of rail** Schienenhöhe *f* ~ **with reference to a bench mark** Höhenlage *f* gegen einen höher gelegenen Festpunkt ~ **of rib web** Rippensteghöhe *f* ~ **of rise** Steighöhe *f* ~ **of sag** Pfeilhöhe *f* ~ **above sea level** absolute Höhe *f*, Meereshöhe *f* ~ **of shank** Schafthöhe *f* ~ **of stratum** Schichthöhe *f* ~ **of the strip** Bahnhöhe *f* ~ **of suspension** Aufhängehöhe *f* ~ **of target** Zielhöhe *f* ~ **of the tide** Fluthöhe *f* (**over-all**) ~ **of a vessel** Schiffshöhe *f* ~ **of water** Druckhöhe *f* ~ **of weir** Wehrhöhe *f*

**height,** ~ **-adjusting knob** Meßwalze *f* (height finder) ~ **adjustment screw** Höhenstellschraube *f* ~ **arm** Höhenlineal *n* ~ **bridge** Höhenbrücke *f* ~ **capacity** Spitzenhöhe *f* ~ **-change-over switch** Höhenumschaltung *f* (rdr) ~ **circle** Höhenkreis *m* ~ **computation** Höhenberechnung *f* ~ **control** Bildeinstellung *f* ~ **convenient to the eye** bequeme Sichthöhe *f* ~ **dependance** höhenabhängig ~ **determination** Höhenlotung *f* ~ **difference of water near floodgate** Schleuseneinsatz *m* ~ **factor** Hochflugzahl *f* (figure) ~ **finder** E-messgerät *n*, Höhen-finder *m*, -meßradar *n* ~ **gauge** Höhenmaßstab *m* ~ **indicator** Höhenanzeiger *m* ~ **level** Höhenregion *f* ~ **mark** Höhenmarke *f* ~ **measurement** Höhenvermessung *f* ~ **-measuring device** Höhenmeßvorrichtung *f* ~ **overall** Gesamthöhe *f* ~ **slide** Höhenschlitten *m* (foto) ~ **-to-paper shootboard** Höhehobel *m* ~ **wheel** Höhenrad *n* (Bildkartiergeräte)

**Heising,** ~ **modulation** Anodenspannungsmodulation *f*, Heising-Modulation *f*, Parallelröhrenmodulation *f* ~ **modulation method** Absorptionsmodulation *f*

**Helberger furnace** Helberger Ofen *m*

**held** eingespannt, gefaßt ~ **back** gehemmt (e.g. a trigger) ~ **solid (with)** befestigt ~ **water** Haftwasser *n*

**helical** schnecken-, schrauben-, spiral-förmig, spiralig ~ **antenna** Wendelantenne *f* ~ **auger** Spiralbohrer *m* ~ **band friction clutch** Spiralbandkupplung *f* ~ **bevel gear** spiralverzahntes Kegelrad *n* ~ **blower** Propellergebläse *n* ~

**chopper** schraubenförmiger Zerhacker *m* ~ **coil air filter** Spannwickelluftfilter *m* ~ **coil insert** Schraubenfeder-Gewindeeinsatz *m* ~ **distributor** Benzinvernebler *m* ~ **elevator** Förder-rohr *n*, -röhre *f* ~ **fibering** Spiralfaserstruktur *f* ~ **gear** Schneckengetriebe *n*, Schrägverzahnungsgetriebe *n*, Schrägzahnrad *n*, Schrauben-rad *n*, -zahngetriebe *n*, -zahnrad *n* ~ **gear-wheel** Schrägverzahnung *f* ~ **gearing** Getriebe *n* mit Schrägverzahnung ~ **gears** Schraubenradverzahnung *f* ~ **grid** Wickelgitter *n* ~ **groove** Drall-, Schrauben-nut *f* ~ **inlet port** Einlaßdrallkanal *m* ~ **lens mount** Schneckengangfassung *f* ~ **line** Schnecken-, Spiral-linie *f* ~ **pinion planet** schrägverzahntes Planetenrad *n* ~ **reinforcement** Spiralbewehrung *f* ~ **scanning** Schraubenlinienabtastung *f* (rdr), Wendelabtastung *f* ~ **slide** Schrägzahnschieber *m* ~ **slot** Schraubennut *f* ~ **spring** Schrauben-, Spiral-feder *f* ~ **spring plate** Federteller *m* für Schraubenfeder ~ **spur gear** Schrägzahnstirnrad *n* ~ **tape** Spiralband *n* ~ **-toothed bevel gears** Schrägverzahnung *f* ~ **-toothed spur gear** Schrägzahnstirnrad *n* ~ **torsion springs** gewundene Biegungsfedern *pl* ~ **tube** Wendelrohr *n* ~ **-type cutter bar** Schlangenbohrstange *f* ~ **vault** Schneckengewölbe *n* ~ **wire** Drahtspirale *f*
**helically toothed wheel** Rad *m* mit Schrägverzahnung
**helicity** Helizität *f*
**helicoid** schneckenartig, schraubenförmige Fläche *f*
**helicoidal** schraubenartig ~ **spreader** Spiralwattenmaschine *f* ~ **surface** Schraubenfläche *f*
**helicometer** Helikometer *f*
**helicopter** Drehflügelflugzeug *n*, Hubschrauber *m*, Schrauben-flieger *m*, -flugzeug *n*, Steilschrauber *m* ~ **screw** Hubschraube *f*, Hubschrauber *m*
**heliochromogravure** Heliochromogravüre *f*
**heliograph** Heliograf *m*, Lichtsprechgerät *n*, Sonnenlichttelegraf *m*, Sonnenscheinautograf *m*, Spiegeltelegraf *m* ~ **signaling** Lichttelegrafie *f*
**heliographic**, ~ **print** Lichtpause *f* ~ **printing** Farbenlichtdruck *m*, Lichtpauseverfahren *n*
**heliography** Heliografie *f*, Heliogravüre *f*
**heliometer** Heliometer *n*, Sonnenmesser *m*
**helioscope** Sonnen-fernrohr *n*, -glas *n*
**heliostat** Heliostat *n*
**heliport** Hubschrauberlandeplatz *m*
**helipot** Wendelpotentiometer *m*
**helium** Helium *n* ~ **continuous spectrum of principal series** Helium-Hauptserienkontinuum *n* ~ **film** dünne Heliumschicht *f* ~ **gas** Heliumgas *n* ~ **ion collisions** Helium-Ionenstöße *pl* ~ **level** Heliumniveau *n* ~ **tube** Heliumröhre *f*
**helix** Kehrwendel *n*, Raumspirale *f*, Schnecke *f*, Schnecken-gewinde *n*, -linie *f*, Schraubenlinie *f*, Seele *f* (Kabel), Solenoid *n*, Spirale *f*, Spiral-linie *f*, Wendel *f* ~ **of constant lead** Schraubenlinie *f* gleichbleibender Steigung ~ **of tape-wire** Bandschraube *f* ~ **of thin wire** Drahtschraube *f*
**helix**, ~ **angle** Steigungswinkel *m* ~ **antenna** Wendelantenne *f* ~ **wave guide** Wendelhohlleiter *m*
**Hell printer** Hellschreiber *m*

**helm** Helm *m*, Helmstock *m*, Ruder *n*, Ruderpinne *f* ~ **port** Hennegatt *n*
**helmet** Helm *m*, Kopfbedeckung *f*, Kopfhaube *f*, Kopfschutzhaube *f*, Pfahlkopf *m*, Schutzhaube *f*, Schweißkappe *f* (welding) ~ **cover** Helmüberzug *m*
**Helmholtz circles** Helmholtzsche Kreise *pl*
**helmsman** Rudergänger *m*, Steuermann *m*
**help, to** ~ beistehen, beitragen, fördern, helfen, mitwirken, raten, unterstützen
**help** Abhilfe *f*, Beistand *m*, Erleichterung *f*, Hilfe *f*, Hilfeleistung *f*, Unterstützung *f*, Vorschub *m*
**helper** Arbeiter *m*, Bohrhauer *m*, Gehilfe *m*, Handlanger *m*, Helfer *m*, Hilfsarbeiter *m*, Lehrhauer *m* ~ **spring** Stufen-, Stütz-feder *f*
**helpful hint** Kniff *m*
**helve, to** ~ bestielen **to** ~ **an ax** ein Beil anschäften
**helve** Hammer *m* ~ **hammer** Aufwerfhammer *m* ~ **hoop (or ring)** Hammerbeschlag *m*
**helvite** Helvit *m*
**hem, to** ~ besäumen, einfassen, säumen **to** ~ **in** beschränken, einengen
**hem** Rand *m*, Saum *m*, Umschlag *m* ~ **stitch** Hohlsaum *m*
**hemaraphotometer** Tageslichtfotometer *n*
**hematine** Blauholzextrakt *m* (gereinigt)
**hematite** Blutsein *m*, Eisen-glanz *m*, -glimmer *m*, Glanzeisenerz *n*, Glanzerz *n*, Hämatit *m*, Spekularit *m* ~ **deposit** Hämatitvorkommen *n* ~ **iron ore** roter Glaskopf *m* ~ **ore** Hämatiterz *n*
**hematoxylin** Hämatoxylin *n*
**hemicellulose** Halbzellulose *f*
**hemichloride of copper** Kupferchlorür *n*
**hemicircular lens** Halbkugellinse *f*
**hemicycle** Halbkreis *m*
**hemicyclic(al)** hemizyklisch
**hemihedral** halbflächig, hemiëdrisch ~ **crystal (or form)** Hemiëder *n* ~ **form** Hemiëdrie *f*
**hemihedrism** Hemiëdrie *f* Hemimorphismus *m*
**hemihedron** Halbflächner *m*, Hälftflächner *m*, Hemiëder *n*
**hemimorphism** Hemimorphismus *m*
**hemiprism** Halbprisma *n* (cryst.)
**hemisphere** Erdhalbkugel *f*, Halbkugel *f*, Hemisphäre *f*
**hemispheric**, ~ **diaphragm** Kalottenmembran *f* ~ **front lens** frontale Halbkugellinse *f* (electron.)
**hemispherical** halbkugelig, halbrund ~ **-bearing support** Spurlager *n* ~ **cupping bowl** Stößel *m* ~ **depression in the piston crown** halbkugelförmige Aussparung *f* im Kolbenboden ~ **flux** hemisphärischer Lichtstrom *m* ~ **front lens** frontale Halbkugellinse *f* ~ **lens** Halbkugellinse *f* ~ **shape** Halbkugelgestalt *f* ~ **stage** Kugeltisch *m* ~ **vault** Kugelgewölbe *n*
**hemitrope** Zwillingskristall *m*
**hemitropic** hemitrop
**hemming machine** Saummaschine *f*
**hemocytometer** Blutkörperchen-zähler *m*, -zählkammer *f*
**hemostatic forceps** Arterienklemme *f*, Blutgefäßklemme *f*
**hemp** Hanf *m*, Hede *f* ~ **belt (or belting)** Hanfriemen *m* ~ **center** Hanfseele *f* ~ **core** Hanfeinlage *f*, -seele *f* (of a rope) ~ **fiber** Hanffaser *f* ~ **insulation** Hanfisolation *f* ~ **linen** Hanfleinwand *f* ~ **packing** Hanf-dichtung *f*, -liderung *f*,

-umwicklung *f*, Wergdichtung *f* ~ **rope** Hanf-
leine *f*, -seil *n*, -tau *n*, Spitzstrang *m* (Montage)
~ **rope holders** Hanfseilspannbacken *pl* ~ **rope
sheave** Hanfseilscheibe *f* ~ **strand** Hanflitze *f*
~ **thread** Hanfzwirnfaden *m*
**hempen, to** ~ hanfen
**hendecane** Undecan *n*
**Henry's law** Henrysches Gesetz *n*
**hepatic,** ~ **cinnabar** Quecksilberlebererz *n* ~ **iron
ore** Eisenlebererz *n* ~ **ore** Lebererz *n*
**hepatite** Hepatit *m*, Leberstein *m*
**heptagon** Siebeneck *n*
**heptagonal** siebeneckig
**heptahedron** Heptaeder *n*
**heptaldehyde** Heptaldehyd *n*
**heptane** Heptan *n*
**heptavalent** siebenwertig
**heptene** Hepten *n*
**heptode** Hepthode *f*, Pentagridröhre *f*, Sieben-
elektrodenröhre *f*, Siebenpolröhre *f*
**heptylalcohol** Heptylalkohol *m*
**heptylene** Heptylen *n*
**heptyne** Heptin *n*
**herald engraver** Wappenstecher *m*
**heraldic chaser** Wappenziselör *m*
**herapatite** Herapatit *n*
**Herbert bridge unit** Herbertgerät *n*
**herd** Herde *f*, Schar *f*
**hereditary materials** Nachwirkung *f* als Material-
eigenschaft
**hermetic** dicht, hermetisch, luftdicht, vakuumdicht, vollständig dicht ~ **plumbing** Abdichtung
*f* ~ **seal** hermetischer oder luftdichter Verschluß *m*, Luftabschluß *m*
**hermetical** hermetisch, vollständig dicht ~
**fitter** Dosenschließer *m*
**hermetically sealed** luftdicht verschlossen
**hermeticity** Hermitizität *f*
**Heroult,** ~ **electric-arc furnace** Heroult-Licht-
bogenofen *m* ~ **electrode-hearth or direct-
heating arc furnace** Heroult-Lichtbogen-Widerstandsofen *m* ~ **ore-smelting furnace** Heroult-
Schachtofen *m* ~ **resistance furnace** Heroult-
Widerstandsofen *m* ~ **series arc furnace**
Heroult-Lichtbogen-Widerstandsofen *m*
**herpolhode** Herpolhodie *f*, Rastpolkegel *m*
**herringbone** fischgrätenartige Anordnung *f*,
Fischgräten-muster *n*, -verzierung *f* ~ **bond**
Fischgrätverband *m* ~ **gear** Pfeil-rad *n*, -rad-
getriebe *n*, -zahnrad *n*, Rad *n* mit Winkelzähnen, Stirnrad *n* mit Winkelverzahnung ~
**gear(ing)** Getriebe *n* mit Pfeilverzahnung,
Pfeilverzahnung *f* ~ **teeth** Pfeilverzahnung *f* ~
**tooth** Pfeil-, Winkel-zahn *m* ~ **wheel** Pfeilrad *n*,
Rad *n* mit Winkelzähnen, Winkelzahnrad *n*
**herring,** ~ **disc wheel** Hering-Scheibenrad *n* ~
**work** Fischgrätenbau *m* (arch.)
**Hertz effect** fotoelektrische Wirkung *f*
**hertzian,** ~ **doublet** Hertzscher Dipol *m*, Hertzscher Doppelpol *m* ~ **oscillator** Hertzscher Sender *m*, Schwinger *m* ~ **radiation** Hertzsche Wellen *pl* ~ **radiation integral** Hertzsche Funktion *f*
~ **unit** Hertz-Einheit *f* ~ **waves** Hertzsche Wellen *pl*
**Hessian,** ~ **radiation** Heßsche Strahlung *f* ~
**sack cloth** Rupfen *n*
**heterocharge** Heteroladung *f*
**heterochromatie** heterchromatisch ~ **lights** ver-

schiedenfarbige Lichter *pl* ~ **radiation** Bremsstrahlung *f* ~ **X-radiation** heterogene Röntgenstrahlung *f*
**heterocyclic** heterozyklisch
**heterodyne, to** ~ mittels Schwebungen oder
Überlagerung empfangen, superponieren, überlagern
**heterodyne** Heterodyn(e) *m*, Oberlagerer *m*,
Überlagerer *m*, Überlagerungs . . .; heterodin
~ **action** Frequenzwandlung *f* ~ **amplifier**
detonierender Verstärker *m*, Schwebungsver-
stärker *m* ~ **analyzer** Suchtonanalysator *m* ~
**circuit** Überlagerungskreis *m* ~ **detector** erster
Detektor *m* ~ **frequency** Überlagerungsfrequenz *f* ~ **frequency meter** Überlagerungs-
frequenzmesser *m* ~ **mixer** Überlagerungs-
gerät *n* ~ **note** Schwebungston *m* ~ **oscillation**
Überlagerungsschwingung *f* ~ **oscillator** Schwe-
bungs-, Schwingungs-summer *m*; Überlagerungs-empfänger *m*, -summer *m* ~ **oscillator
stage** Schwingstufe *f* ~ **receiver** Heterodyn-,
Interferenz-, Schwebungs-, Überlagerungsempfänger *m*, Superhet *m* ~ **sound analyzer**
Tonanalysator *m* ~ **tub** Heterodynröhre *f* ~
**warbler** Schwebungssummer *m* ~ **wavemeter**
Interferenz-, Interfrequenz-wellenmesser *m*
**heterodyning** Überlagern *n*, Überlagerung *f*
**heterogeneity** Fremdartigkeit *f*, Heterogenität *f*,
Uneinheitlichkeit *f*, Ungleichförmigkeit *f*, Ungleichmäßigkeit *f*, Verschiedenartigkeit *f*
**heterogeneous** artfremd, fremdstoffig, heterogen,
inhomogen, ungleichartig, ungleichmäßig, verschiedenartig, von fremdem Ursprung ~ **beam
of electrons** weißer Elektronenstrahl *m* ~
**radiation** heterogene Strahlung *f* ~ **reaction in
combustion** heterogene Reaktion *f* bei der Verbrennung ~ **X-radiation** heterogene Röntgenstrahlung *f*
**heterogeneousness** Heterogenität *f*, Inhomogenität *f*
**heteroion** Ion-Molekül-Komplex *m*
**heteromorphic** heteromorph
**heteromorphite** Federerz *n*
**heteropolar** wechselpolig ~ **alternator** Wechsel-
polalternator *m* ~ **bond** heteropolare Bindung *f*
**heterostatic** heterostatisch (Schaltung)
**heterotopic** heterotopisch
**heterotropic** heterotrop
**heulandite** Blätterzeolith *m*
**heuristic** heuristisch (methodisch probieren)
**Heurtley's magnifier** Heurtleyrelais *n*
**hew, to** ~ abschroten, abschruppen, hacken,
hauen **to ~ an ashlar with the pickhammer** einen
Stein *m* bespitzen **to ~ trenches** schrämen (to
ore veins)
**hew** Abhieb *m* (stone)
**hewer** Hauer *m*
**hewing** Hau *m*, Hauerarbeit *f*, Schrämen *n* **fit for**
~ haubar ~ **putter** Fördermann *m*
**hewn timber** glattbehauenes Bauholz *n*
**hex,** ~ **nut** Sechskantmutter *f* ~ **wheel dresser**
Schleifscheibenabrichter *m*
**hexachlorethane** Hexachloräthan *n*
**hexadecimal** hexadezimal (info proc.), sedezimal
**hexadiene** Hexadien *n*
**hexagon** Hexygon *n*, Sechs-eck *n*, -flach *n*, -kant
*m* ~ **and pentagon broaches** Fünfkantreibahle *f*

~ **bar (or rod)** Sechskantstab *m* ~ **bar iron** Sechskanteisen *n* ~ **bolt** Sechskantschraube *f* ~ **box spanner (or wrench)** Sechskantsteckschlüssel *m* ~ **cap screw** Sechskantschraube *f* ~ **coach (or wood) screw** Sechskantholzschraube *f* ~ **dimension** Sechskantschlüsselweite *f* ~ **dowel bolt** Sechskantpaßschraube *f* ~ **head** Sechskantkopf *m* ~ **head with collar** Sechskant *n* mit Ansatz ~ **head screw** Sechskantschraube *f* ~ **headed bolt** Bolzen *m* mit sechseckigem Kopf ~ **nut** sechseckige Schraube *f* ~ **(al) nut** Sechskantmutter *f* ~ **pin spanner** Sechskantstiftschlüssel *m* ~ **screw** sechseckige Schraube *f* ~ **spanner** Sechskantschlüssel *m* ~ **turret** Sechskantrevolverkopf *m*

**hexagonal** hexagonal, sechs-eckig, -kant, -seitig, -winklig ~ **barrel mixer** Sechskanttrommelmischer *m* ~ **bolt** sechskantiger Bolzen *m* ~ **coil** sechseckige Spule *f* ~ **collar** Sechskantbund *m* ~ **fuselage** sechseckiger Rumpf *m* ~ **head cup screw** Sechskantschraube *f* ~ **head machine screw** Sechskantschraube *f* ~ **jam nut** flache Sechskantmutter *f* ~ **nipple** Doppelnippel *n* mit Sechskant ~ **nipple nut** flache Sechskantmutter *f* ~ **nut (finished)** flache Sechskantmutter *f* ~ **recess** Innensechskant *n* ~ **wire** Sechskantdraht *m*

**hexahedral** hexaedrisch, sechsflächig
**hexahedron** Hexaeder *n*, Kubus *m*, Sechsflach *n*
**hexahydrobenzene** Hexahydrobenzol *n*
**hexakistetrahedron** Hexakistetraeder *n*
**hexamethylenamine** Hexamethylenamin *n*
**hexane** Hexan *n*
**hexangular** sechswinklig
**hexatomic** sechs-atomig, -wertig
**hexavalent** sechswertig
**hexene** Hexen *n*
**hexode** Hexode *f*, Sechs-gitterröhre *f*, -polröhre *f*, Viergitterröhre *f* ~ **mixing valve** Mischhexode *f*, Sechspolmischröhre *f* ~ **-triode tube** Sechspoldreipolröhre *f*

**H. F. alternator** Hochfrequenzmaschine *f*
**H-fixture** H-förmiges Gestänge *n* ~ **line** Doppelstangenlinie *f*
**H formation** H-Motor *m* ~ **engine** Motor mit H-förmiger Anordnung der Zylinder
**H girder** H-Träger *m*
**hiatus clouds** Lückenwolken *pl*, Mammaturestform *f*
**hickory** Hickorynußbaum *m*
**hidden** versteckt ~ **intention** Nebenabsicht *f* ~ **margin** stille oder versteckte Reserve *f*
**hide, to** ~ eingraben, verbergen, verdecken, verstecken
**hide** Fell *n*, Haut *f*, Leder *n*, Tierhaut *f* ~ **mill** Hammergeschirr *n* ~ **parings** Leimleder *n* ~ **powder analysis** Hautpulververfahren *n* ~ **powder shaking method** Hautpulverschüttelmethode *f* ~ **scrapings (or shreds)** Leimleder *n* ~ **substance** Hautsubstanz *f*
**hiding** Verborgenheit *f* ~ **pigment** deckendes Pigment *n* ~ **power** Deckkraft *n*, (of paint) Farbdeckfähigkeit *f* ~ **(or covering) power** Deckfähigkeit *f*
**hiduminium** Y-Legierungen *pl*
**hieroglyphics** Bilderschrift *f* (print.)
**hi-fi (high-fidelity)** getreue Tonwiedergabe *f*, hohe Wiedergabegüte *f* ~ **panorama speaker**

Hi-Fi-Strahler *m* (tape rec.)
**high** hoch, hochliegend, mächtig **too** ~ zu steil **of** ~ **carbon content** hochgekohlt **with** ~ **climbing capacity** mit großem Steigvermögen **of a** ~ **degree** hochgradig **of** ~ **early strength** frühhochfest **of** ~ **fidelity** klanggetreu **for** ~ **frequency** hochfrequenzmäßig **of** ~ **frequency** hochfrequent **of a** ~ **grade** hochgradig **as** ~ **as a house** haushoch **of** ~ **humidity** wasserreich ~ **in inerts** ballastreich **of** ~ **intensity** lichtstark **of** ~ **percentage** hochprozentig **of** ~ **quality** hochwertig **in** ~ **relief** hocherhaben **on the** ~ **seas** auf Hochsee **at** ~ **speed** mit hoher Geschwindigkeit **of** ~ **speed** hochtourig **of** ~ **strength** hochwertig **at** ~ **temperature** geglüht **on the** ~**-tension side** oberspannungsseitig **of** ~ **valence** hochwertig **on the** ~**-voltage side** oberspannungsseitig ~ **out of water** mit großem Freibord

**high,** ~ **acceleration** Schnellanlauf *m* ~**-activity wastes (or waters)** hochradioaktive Abwässer *pl* ~**-air burst point** Höhensprengpunkt *m* ~ **aircraft** Höhenflugzeug *n* ~ **alloyed** hochlegiert ~ **altitude (or air)** Höhen ..., Höhenluft *f* ~ **altitude balloon barrage** Hochsperre *f* ~ **altitude detector** Kraftdetektor *m* ~ **altitude effects** Höhenwirkung *f* ~ **altitude engine** Höhenmotor *m* (für große Nennhöhe), übermessener Motor *m* ~ **altitude engine coupled with independent turbosuper-charger** Verbundhöhenmotor *m* ~ **altitude flying suit** Höhenfluganzug *m* ~ **altitude gas** Höhengas *n* ~ **altitude heat phenomenon** Höhenthermik *f* ~ **altitude isothermal chart** Höhenisothermenkarte *f* ~ **altitude mixture control** Höhenmischreglung *f* ~ **altitude nausea** Höhenkrankheit *f* ~ **altitude operations** Flugbetrieb *m* in großen Höhen ~ **altitude supercharger** Höhenlader *m* ~ **altitude thunderstorm** Hochgewitter *n* ~**-alumina brick** hochtonerdehaltiger Stein *m* ~ **alumina cement** Tonerdezement *m* ~**-amperage conductors** Hochstrombahn *f* ~**-and low-frequency amplification** Hoch- und Niederfrequenzverstärkung *f* ~**-angle fire** Bogenschuß *m*, Senk-, Steil-, Wurf-feuer *n* ~**-angle gun** Steilfeuergeschütz *n* ~**-angle shot** Bogenschuß *m* ~ **antenna** Hochantenne *f* ~**-ash coal** aschereiche Kohle *f* ~**-back thrusts** Stahlrückdrücke *pl* ~ **bar armature** Hochstabanker *m* ~ **bar rotor** Hochstabläufer *m* ~ **bed** Untiefe *f* ~**-bituminous lignite coal** Schwelkohle *f* ~**-boiled** hochbleichfähig ~**-boiling** hochsiedend ~**-boiling-point cooling system** Heißkühlsystem *n* ~ **boost** Hochfrequenzausgleich *m* ~ **bow** Hochbügel *m* ~**-breast water wheel** rückschlächtiges Wasserrad *n* ~**-burst ranging** Einschießen *n* mit hohen Sprengpunkten

**high-capacity** von hohem Aufnahmevermögen *n*, leistungsfähig ~ **circuit breaker** Hochleistungsschalter *m* ~ **gas engine** Großgasmaschine *f* ~ **nibbling machine** Hochleistungsdekupiermaschine *f* ~ **plate-straightening machine** Hochleistungsblechrichtmaschine *f* ~ **telegraph** Massentelegraf *m*

**high,** ~**-carbon steel** Flußstahl *m* mit hohem Kohlenstoffinhalt, hochgekohlter Stahl *m* ~**-(low) carbon steel** Stahl *m* mit hohem (niedrigem) Kohlengehalt *m* (Kohlenstoffgehalt *m*) ~**-carbon steel plates** Hartstahlbleche *pl* ~

**charging** Schnelladung *f* **~-class** hochwertig **~-class cement** hochwertiger Zement *m* **~ -class steel** Edelstahl *m* **~-compressing** hochverdichtend **~-compression engine** hochverdichteter Motor *m* **~-contrast emulsion film** Spezialfilm *m* hoher Steilheit **~-contrast image** hartes Bild *n* (TV) **~-contrast picture** hartes Bild *n* **~-copper** kupferreich **~ current carbon arc** Hochstromkohlebogen *m* **~ current discharge** Entladung *f* mit hoher Stromdichte **~ current test** Hochstromprüfung *f* **~ cutting mower bar** Hochschnittmähbalken *m* **~ definition** scharfes Bild *n* (film) **~-definition image (or picture)** hochzeiliges Bild *n* (TV) **~-definition scan** Feinrasterung *f* (of picture) **~-definition television** Fernsehen *n* hoher Güte **~ degree of boost** hoher Grad *m* der Aufladung **~ deposition rate** schnellfließend **~ discharge capacity** Ableitung *f* großer Wassermengen (of a river) **~-diving board** Sprungturm *m* **~-domed** hochgewölbt **~-dosage counter tube** Hochdosiszählrohr *n* **~ draft system** Höchstverzugsstreckwerk *n* **~-drafting** Hochverzug *m* **~ dredger** Hochbagger *m*

**high-duty** hoch beansprucht **~ anodes** hochbelastbare Anoden *pl* **~ gas engine** Großgasmaschine *f* **~ lathe** Hochleistungsdrehbank *f* **~ machine** Hochleistungsmaschine *f* **~ milling cutters** starkspiraliger Hochleistungsfräser *m* **~ running of an engine** angestrengter Betrieb *m* des Motors **~ service** Schwerbetrieb *m*

**high,** **~ early strength cement** Zement *m* mit hoher Anfangsfestigkeit **~ effect tentering and drying machine** Hochleistungsspannrahmenmaschine *f* **~ efficiency milling machine** Hochleistungsfräsmaschine *f* **~ efficiency speaker** Hochleistungslautsprecher *m* **~ electric conduit** Hochleitung *f* **~-end response** Frequenzgang *m* der Höhen

**high-energy,** **~ alpha particle** energiereiches Alphateilchen *n* **~ edge** (emission band) kurzwellige Kante *f* **~ fission** Spaltung *f* durch schnelle Neutronen **~ level reactor** Hochleistungsreaktor *m* **~ propellants** energiereiche Treibstoffe *pl* **~ rate forming** Hochenergieformgebung *f*

**high explosive** Brisanz *f*, Hochleistungssprengstoff *m*, hochexplosiver Sprengstoff *m*

**high-explosive** brisant, hochexplosiv **~ ammunition** Brisanzmunition *f* **~ bomb** Brisanzbombe *f* **~ powder** Brisanzpulver *n* **~ rifle grenade** Gewehrsprenggranate *f* **~ shell** Brisanz-geschoß *n*, -granate *f*, Sprenggranate *f* **~ shrapnel** Brisanzschrapnell *n*

**high,** **~ fidelity** hohe Wiedergabegüte *f*, getreue Tonwiedergabe *f* **~-field-emission arc** Feldbogen *m* **~-flux reactor** Hochflußreaktor *m*, Reaktor *m* mit hohem Neutronenfluß **~ frame** Hochspant *n* **~ (audio) frequencies** Höhen *pl* **~ frequency** Dekamenterwellenbereich *m*, Hochfrequenz *f*, hohe Frequenz *f*, Kurzwellenbereich *m*

**high-frequency** hochfrequent, Hochfrequenz ... **~ in shifting** Schalthäufigkeit *f* **~ alternator** Hochfrequenz-generator *m*, -maschine *f* **~ amplification** Hochfrequenzverstärkung *f* **~ amplification stage** Hochfrequenzverstärkungsstufe *f* **~ amplifier** Hoch-, Hochfrequenzverstärker *m* **~ anode current** Anodenhoch-

frequenzstrom *m* **~ bridge** Hochfrequenzmeßbrücke *f* **~ buzzer** Töner *m* (rdo) **~ cable** Hochfrequenz-, Schalen-kabel *n* **~ carrier cable** Breitbandkabel *n* **~ carrier-current telephone channel** Hochfrequenz-sprechkanal *m*, -sprechweg *m*, -weg *m* **~ carrier wave** Hochfrequenzübertragerwelle *f* **~ choke** Hochfrequenzdrossel *f* **~ circuit** Hochfrequenz-kreis *m*, -leitung *f* **~ coil** Hochfrequenzspule *f* **~ commutator** Hochfrequenzunterbrecher *m* **~ compensation** Anhebung *f* der hohen Frequenzen **~ (alternating) current** Hochfrequentstrom *m* **~ cut-off** Tonveredler *m* **~ cut-out** Hochfrequenzsicherung *f* **~ (electrical) engineering** Hochfrequenztechnik *f* **~ equipment** Hochfrequenzeinrichtung *f* **~ field** Hochfrequenzfeld *n* **~ filter** Tonveredler *m* **~ furnace** Hochfrequenz-, Wirbelstrom-ofen *m* **~ fuse** Hochfrequenzsicherung *f* **~ generator** Hochfrequenz-erzeuger *m*, -generator *m*, -maschine *f* **~ indicator** Hochfrequenzanzeiger *m* **~ induction furnace** Hochfrequenzinduktionsofen *m* **~ insulator** Hochfrequenzisolator *m* **~ interference** Hochfrequenzstörung *f* **~ interrupter** Hochfrequenzunterbrecher *m* **~ iron core** Hochfrequenzeisenkern *m* **~ loudspeaker** Hochtonlautsprecher *m* **~ measurement** Hochfrequenzmessung *f* **~ muffler** Tieftondurchlasser *m* **~ multiple telephony** Hochfrequenzmehrfachfernsprecher *n*, Mehrfachsprechen *n* oder Mehrfachtelefonie *f* mit hochfrequenten Trägerströmen **~ note** hoher Ton *m* **~ (harmonic) oscillation** Hochfrequenzschwingung *f*, hochfrequente harmonische Schwingung *f* **~ pentode** Fünfpolschirmröhre *f*, Hochfrequenzpenthode *f* **~ power** Hochfrequenzleistung *f* **~ transmitter** Hochfrequenzsender *m* (r.f.) **~ resistance** Hochfrequenzwiderstand *m* **~ stage** Hochfrequenzstufe *f* **~ tachometer** Tourendynamo *n* **~ telegraphy along lines** leitungsgerichtete Trägerwellentelegrafie *f* **~ telephony** Hochfrequenztelefonie *f* **~ telephony alohg lines** leitungsgerichtete Hochfrequenztelefonie *f* **~ transformer** Hochfrequenztransformator *m* **~ vacuum tester** Hochfrequenz-Vakuumprüfer *m* **~ warble tone** Heulton *m* **~ wave** kurze Welle *f* **~ wire strand** Hochfrequenzlitzendraht *m*

**high,** **~ gain** hochverstärkend(e) .... **~-gamma picture** kontrastreiches Bild *n* **~ gamma tube** Röhre *f* mit harmonischer Gradation **~ gear** großer Gang *m* **~-glaze calendering** Hochglanzsatinage *f*

**high-gloss,** **~ color** Hochglanzfarbe *f* **~ finish** Hochglanzzurichtung *f* **~ lacquer** Hochglanzlack *m* **~ lamination** Hochglanzkaschierung *f* **~ transfer varnish** Hochglanzüberdrucklack *m* **~ varnish** Brillantlack *m*

**high-grade** erstklassig, hochhaltig, hochwertig, mit hoher Festigkeit **~ alloy steel** hochlegierter Stahl *m* **~ cast iron** hochwertiger Guß *m* **~ cast steel** Edelstahlguß *m* **~ copper matte** kupferreicher Stein *m* **~ massecuite** Erstproduktfüllmasse *f* **~ matte** reicher Stein *m* **~ melting scrap** Kernschrott *m* **~ nuclear fuel** hochgradiger Spaltstoff *m* **~ ore** hochhaltiges oder reichhaltiges Erz *n* **~ paper machine** Feinpapiermaschine *f* **~ steel** Edelstahl *m*, Qualitätsstahl *m*, Stahl *m* mit hoher Festigkeit

high, ~ green sirup Fein-, Weiß-ablauf *m* ~-hat-tripod Kleinstativ *n* (film) ~-impedance charac-ter Hochohmigkeit *f* ~-impedance relay Dros-selrelais *n*, hochinduktives Relais *n* ~-inertia blades Blätter *pl* mit hoher Massenträgheit ~ -initial-oil-pressure system Hochdruckschmier-system *n* für Startzwecke ~-intensity carbon Hochintensitätskohle *f* ~-intensity lamp Hoch-leistungslampe *f* ~-intensity lighting system Hochleistungsbefeuerung *f* (airport) ~-internal-resistance tube hochohmige Röhre *pl* ~-key picture überhelles Stimmungsbild *n* ~ kilo-voltage radiography Hartstrahlaufnahmen *pl* ~ land Hochland *n* ~ lands Mittelgebirge *n*
high-level, ~ blower hohe Druckstufe *f* (Lader), Lader *m* für große Nennhöhe ~ camshaft oben-liegende Nockenwelle *f* ~ cave Kilocurie-Zelle *f* ~ flight Flug *m* in großen Höhen ~ frame Hochrahmen *m* ~ impulse Hochtastrausch-diode *f* ~ modulation Endstufenmodulation *f* ~ portion of the cam Kopfkreis *m* des Nockens, Teil *n* des Nockens auf dem der Hub erfolgt ~ radiation Strahlung *f* mit hohem Energieniveau ~-rated engine Höhenmotor *m* (für große Nennhöhe) ~ sources of neutrons hochgradige Neutronenquelle *f* ~ tank Hochbehälter *m*
high-lift, ~ device Aufstiegvorrichtung *f*, Auf-triebvorrichtung *f*, Hochauftriebseinrichtung *f*, Landehilfe *f*, Vorrichtung *f* zur Steigerung des Auftriebs ~ fork stacking truck Gabelstapler *m* ~ platform truck Hochschubkarren *m* ~ safety valve Hochhubsicherheitsventil *n* ~ valve Hoch-hubventil *n* ~ wing hochtragender Flügel *m*
highlight, to ~ hervorheben, betonen
highlight Glanzlicht *n* (photo), hellster Bildpunkt *m*, Schlaglicht *n*, Spitzenlicht *n* ~ power licht-stark
highlights hellste Bildpunkte *pl* (film), Spitzen-lichter *pl* (art)
high, ~ limit Größtmaß *n* ~-limit gauge Größt-maßlehre *f* ~-load capacity Hochbelastbarkeit *f* ~ lookout post Hochstand *m* ~ luminous effi-ciency lichtstark ~-luster polish Hochglanz-politur *f* ~-melting hochschmelzend ~-melting-point alloy hochschmelzende Legierung *f* ~ melting-point grease Heißlagerfett *n* ~ mill Stopfenwalzwerk *n* ~ mirror finish Hochglanz *m* ~ moorland Hochmoor *n* ~ mountains Hoch-gebirge *n* ~ note attenuation Benachteiligung *f* der hohen Töne ~ number of revolutions hohe Drehzahl *f* ~ oblique photograph Flachauf-nahme *f* ~-octane hochklopffest, mit hoher Oktanzahl ~-octane gasoline Benzin *m* mit hoher Oktanzahl ~-ohmic hochohmig ~ ohmic resistor Hochohmwiderstand *m* ~-output air-craft engine Hochleistungsflugmotor *m* ~ -output engine Hochleistungsmotor *m* ~-pass filter Hochfrequenzsiebkette *f*, Hoch-filter *m*, -paß *m*, Kondensator-kette *f*, -leitung *f*; Niederfrequenz-sperrkette *f*, -sperrkreis *m* (circuit); Kondensatorleiter *m* (teleph.) ~-pass selective circuit (or filter) Hochfrequenzsieb-gebilde *n* ~ pentode frequency Hochfrequenz-penthode *f*
high-performance hochleistungsfähig ~ aircraft Hochleistungsflugzeug *n* ~ microscope lamp Hochleistungsmikroskopierleuchte *f* ~ set Hochleistungssatz *m*

high, ~ pile chain delivery Hochstapelkettenaus-leger *m* ~ pile combination stream Großstapel-anleger *n* ~ pitched helltönend ~-pitched note hoher Ton *m* ~-pitched whine Heulton *m* ~ pitches Höhen *pl* ~-point piston ring Kolben-ring *m* mit erhöhtem Radialdruck an der Stoß-stelle ~ polish Hochglanz *m* ~ polish galvanized glanzverzinkt ~-polish lead crystal glass Hoch-glanzbleikristall *n* ~ position Hochlage *f* ~ potential Hochspannung *f* ~ potential capacitor Hochspannungskondensator *m* (fixed) ~ -potential condensor Hochspannungskonden-sator *m* (fixed) ~-potential transformer and vacuum tube Beschleunigungseinrichtung *f* ~ power Hochbelastung *f* (electr.), hohe Leistung *f*
high-power, ~ broadcasting station Großrund-funksender *m* ~ coil ignition Hochleistungs-zündspule *f* ~ electrolytic capacitor Hochvolt-elektrolytkondensator *m* ~ gas engine Groß-gasmaschine *f* ~ lens lichtstarke Linse *f* ~ loud-speaker Hochleistungslautsprecher *m* ~ magnifier stark vergrößernde Lupe *f* ~ modu-lation Endstufenmodulation *f* ~ objective licht-starkes oder starkes Objektiv *n* ~ process In-tensivverfahren *n* ~ transmitter Großsender *m*, Starkstrommikrofon *n* ~ (vacuum) tube Großleistungsröhre *f*, Hochleistungsröhre *f* ~ wireless (or radio) station Großfunkstelle *f*
high, ~-powered stark, starkmotorig ~-powered engine Hochleistungsmotor *m*, starker Motor *m* ~-powered incendiary bomb Intensivbrand-bombe *f* ~-precision lathe Präzisionsdrehbank *f* ~-precision leveling staff Feinnivellierlatte *f* ~ pressure Hoch-druck *m*, -spannung *f* ~ -pressure (area) Lufthochdruck *m*
high-pressure, ~ air-line blowing system Aus-drückverteiler *m* ~-air system Hochdruckluft-system *n* ~ apparatus Hochdruckapparaturen *pl* ~ area Hochdruckgebiet *n*, Hochluftdruck-gebiet *n* ~ atomizer Hochdruckzerstäuber *m* ~ ballast pump Hochdruckballastpumpe *f* ~ belt Hochdruckgürtel *m* ~ blade Hochdruckschau-fel *f* ~ blower Hochdruckgebläse *n* ~ boiler Hochdruckdampfkessel *m* ~ burner Hoch-druckbrenner *m* ~ cabin Überdruck-kabine *f*, -kammer *f* ~ cock Überdruckhahn *m* (jet) ~ compressor Hochdruckkompressor *m* ~ con-densing turbine Hochdruckkondensationsma-schine *f* ~ conveyance Hochdruckförderung *f* ~ cycle Hochdruckkreis *m* ~ cylinder Hoch-druck-flasche *f* (vessel), -zylinder *m* ~ diving suit Überdruckanzug *m* ~ drilling oil Hoch-druckbohröl *n* ~ engine Hochdruckmaschine *f* ~ exhauster Gasgebläse *n* ~ fuel strainer Kraftstoffhochdruckfilter *n* ~ glow discharge Hochdruckglimmentladung *f* ~ hose Hoch-druckschlauch *m* ~ line (piping) Hochdruck-leitung *f* ~ lubricant Hochdruckschmiermittel *n* ~ lubricating pump Hochdruckölschmier-pumpe *f* ~ lubrication Hochdruckschmierung *f* ~ lubrication set Hochdruckabschmiergerät *n* ~ mercury-discharge tube Quecksilberhoch-druckentladungsrohr *n* ~ mercury vapor lamp Quecksilberdampfhochdruckleuchte *f* ~ oil gun Hochdruckölpresse *f* ~ oil pump Hochdrucköl-schmierpumpe *f* ~ (rubberized-asbestos) pack-ing sheets It-Platten *pl* ~ pipes Druckwasser *n* ~ piston Hochdruckkolben *m* ~ pump Hoch-

druckpumpe f ~ relief valve Überdruckventil
n (jet) ~ rotary blower Hochdruckkapselgebläse
n ~ sensing line Hochdruck-Geberleitung f ~
side Druckscheibe f (Kühlmaschine) ~ slide
valve Hochdruckschieber m ~ spring Über-
druckfeder f ~ stage Hochdruckstufe f ~ stall
Hochdruckkammer f ~ steam Hochdruck-
dampf m, hochgespannter Dampf m ~ steam
boiler Hochdruckheißdampfkessel m ~ tank
Hochdruckkessel m ~ tire Hochdruckreifen m
~ transport Hochdruckförderung f ~ tube
Hochdruckrohr n ~ turbine Überdruckturbine
f ~ unit Hochdrucksatz m ~ vacuum pump Vorvakuumpumpe f ~ valve Hochdruckventil n ~
valve filler Hochdruck-Ventilfüller m ~ valves
and fittings Hochdruckarmaturen pl ~ zone
Hochdruckgürtel m
high, ~-priority material Sparstoff m ~-produc-
tion automatic turret screw machine Hochlei-
stungsrevolverautomat m ~-production equipment Hochleistungseinrichtung f
high-quality, ~ cast iron hochwertiger Guß m,
hochwertiges Gußeisen n, Qualitätsguß m ~
castings Feinguß m ~ gray iron hochwertiger
Grauguß m ~ objective Hochleistungsobjektiv
n ~ sheet steel Qualitätsblech n ~ steel Edelstahl m, hochwertiger Stahl m
high, ~ radiation flux intensiver Strahlungsfluß
m ~-radio transformer Übertrager m mit hohem
Umsetzungsverhältnis ~-rate discharge Stoßentladung f ~-rate series production Großreihenfertigung f ~ ratio pulsing hohes Impulsverhältnis n ~ reference point Hochpunkt m
~-resistance direction Rückwärtsleitung f,
Sperrichtung f ~ resistance noises Knallgeräusche pl ~-resistance winding hochohmige
Wick(e)lung f ~ resolution hohes Auflösungsvermögen n ~ resolution camera Feinstrahl-
anordnung f (electron) ~ road Dammstraße f
~ rudder Hochseitenruder n ~-safety drums
Hochsicherheitstrommel f ~ sea Hochsee f ~
seas freie See f ~ shear rivet hochscherfester
Niet n ~ shot Steilzielung f ~ side heißes Ende
n ~ side drum Wirbeltrommel f ~-silicon iron
hochsiliziertes Eisen n ~-silicon (pig) iron
Glanzeisen n ~-silicon pig iron hochsiliziertes
Roheisen n, Schwarzeisen n ~-silicon softener
Ferrosilizium n ~ size Hochform f (Reklame),
Hochformat n ~ solidity rotor Rotor m mit
großer Blattdichte
high-speed von großer Geschwindigkeit f,
Schnellgang m, schnellaufend ~ adjustable
hand reamers Schnellverstellhandreibahlen pl
~ aircraft Schnellluftfahrzeug n ~ airplane
Geschwindigkeitsflugzeug n ~ auxiliary spindle
Hilfsspindel f ~ bomber Schnellbomber m ~
brass Schnelldrehmessing n ~ bullet Ultra-
geschoß n ~ camera Hochfrequenzkamera f,
Zeitdehner m ~ camera shooting Zeitdehnaufnahme f, Zeitdehnung f ~ carry Neuenüber-
trag m ~ centrifuge Hochleistungs-, Ultrazentrifuge f ~ charged particle schnelles La-
dungsteilchen n ~ circuit breaker Zeitschnellschalter m ~ circuit switch Zeitschnellschalter
m ~ coasting vessels Bäderdampfer m ~ commercial airplane Schnellverkehrsflugzeug n ~
computer Schnellrechner n ~ cutter Schnell-
schneidemaschine f ~ cutting metals Hochlei-

stungsschneidemetalle pl ~ development Schnell-
entwicklung f ~ Diesel engine schnellaufende
Dieselmaschine ~ digital storage Ziffern-
Schnellspeicher m ~ drawing mill Schnellzieh-
walzwerk n ~ (steel) drill Schnellstahlbohrer
m ~ drilling attachment Schnellbohreinrichtung f ~ drop hammer Schnellgesenkhammer
m ~ electron schnelles Elektron n ~ engine
Schnelläufer m, Schnelläufermotor m ~
extrarapid hochempfindlich (photo) ~ feed
beschleunigte Zuführung f ~ fission neutron
sehr schnelles Spaltungsneutron n ~ flash tube
Lichtblitzröhre f ~ flight Schnellflug m ~
forging hammer Hochleistungsschmiedeham-
mer m ~ gear Schnellgang-, Schongang-
getriebe n ~ headstock Schnelläuferspindel-
kasten m ~ lathe Schnelldrehbank f, Schnell-
schnittdrehbank f ~ lens hochlichtstarkes Objektiv n, lichtstarke Linse f, Momentobjektiv n
~ lever Eilganghebel m ~ lock nut Schnell-
spannmutter f ~ method Schnellmethode f ~
milling Hochgeschwindigkeitsfräsen n ~ milling attachment Schnellfräskopf m ~ motion
picture Zeitdehneraufnahme f ~ objective
hochlichtstarkes oder lichtstarkes Objektiv n
~ paper drilling machine Papierschnellbohr-
maschine f ~ photometer Schnellfotometer m
~ photorecording gauge (for echo-sounding
devices such as Behmlot) Lichtzeigerkurzzeit-
messer m ~ picture Zeitdehnaufnahme f ~
precision lathe Genauigkeitsschnelldrehbank f
~ press Schnelläuferpresse f ~ printing press
Schnellpresse f ~ processing Schnellverfahren
n ~ pump Hochleistungspumpe f ~ reception
for Morse code Schnellschreibempfang m ~
relay Schnellrelais n ~ research plane Geschwin-
digkeitsversuchsflugzeug n ~ rolling mill
Schnellwalzwerk n ~ rotary air motion kreisende Luftströmung f ~ roughing lathe Schnell-
schruppbank f ~ selector Schnelläufer m ~
sewing machine Schnellnähmaschine f ~ shaker
screen Schnellschwingsieb n ~ shaper Schnell-
hobler m ~ shutter Momentverschluß m ~
spark camera Funkenzeitlupe f ~ spark photograph Funkenzeitlupenaufnahme f ~ spindle
Schnellaufspindel f ~ spur-geared block Schnell-
flaschenzug m ~ stall Strömungsabriß m bei
hoher Geschwindigkeit ~ start Schnellanlauf
m ~ steel Rapid-, Schnell-, Schnellschneide-
stahl m ~ (tool) steel Schnelldrehstahl m ~
steel cutter Schnellstahlfräser m ~ steel tap
Schnellstahlgewindebohrer m ~ steel tip
Schnellstahlschmiede f ~ steel tool Schnell-
arbeitsstahl m ~ stranding machine Schnell-
verseilmaschine f ~ switch schnellaufender
Wähler m, Schnelläuferwähler m ~ tape punch
Hochleistungslocher m ~ tape reader Hoch-
leistungsabfühlvorrichtung f ~ tape transmitter
Lochstreifenschnellsender m ~ telegraph
Reihen-, Schnell-telegraf m ~ telegraphy
Schnelltelegrafie f ~ test Beschleunigungs-
untersuchung f ~ tool steel Schnellschnittstahl
m, Schnellstahl m, Schnellstahlwerkzeug n ~
tractors Schnelltransporter pl ~ traffic Schnellverkehr m ~ trimmer Schnellschneider m ~
tripleblade trimmer Dreimesserschnellschnei-
der ~ turret lathe Schnellauf-Drehbank f ~
upwinds Windschutzthermik f ~ vertical spindle

**head** schnellaufender Senkrechtfräskopf *m* ~ **wind tunnel** Windkanal *m* für hohe Geschwindigkeiten ~ **zinc-printing machine** Zinkdruckschnellpresse *f*
**high spot** Ort *m* hoher Strahlungsdichte
**high-strength** Durchschlagsfestigkeit *f* ~ **cast iron** Festigkeitsguß *m*, hochwertiger Grauguß, Qualitätsguß *m* ~ **rayon** Festreyon *n*
**high surface reflex** Hochglanzoberflächenspiegel *m*
**high-temperature** warmfest (Rohr) ~ **air filter** hochwarmfestes Lichtfilter *n* ~ **approach** Hochtemperatur-Entwicklung *f* ~ **coal tar** Steinkohlenhochtemperaturteer *m* ~ **(carbonization) coking** Hochtemperaturverkokung *f* ~ **corrosion-resistant** heißkorrosionsbeständig ~ **creep strength** Warmdauerstandfestigkeit *f* ~ **distillation** Hochtemperaturdestillation *f* ~ **liquid-cooling** Heißkühlung *f* ~ **oil** Sommeröl *n* ~ **processing** Hochtemperaturbehandlung *f* ~ **stability** Warmfestigkeit *f* ~ **steel castings** warmfester Stahlguß *m* ~ **tar** Hochtemperaturteer *m* ~ **yield point** Warmstreckgrenze *f*
**high-tensile,** ~ **steel** hochwertiger Stahl *m* ~ **steel hoops** Verpackungshartbandstahl *m*
**high-tension** Anodenspannung *f*, hochgespannt, mit hoher Spannung *f* ~ **battery** Anodenbatterie *f* ~ **bridge** Hochspannungsmeßbrücke *f* ~ **cable outlet** Hochspannungsleitungsabführung *f* ~ **cap** Spaltglühzünder *m* ~ **condenser** Hochspannungskondensator *m* ~ **cross** Höchstspannungskreuz *n* ~ **current** hochgespannter Strom *m*, Hochspannungs-, Stark-strom *m* ~ **current impulses (or wave)** Hochspannungsstoß *m* ~ **cycle with terminals** Hochspannungskabel *n* mit Stromabnehmer ~ **direct current** Hochspannungsgleichstrom *m* ~ **distributor** Hochspannungsverteiler *m* ~ **generator** Hochspannungs-erzeuger *m*, -generator *m* ~ **ignition** Summerzündgerät *n* ~ **ignition cable** Hochspannungszündleitung *f* ~ **ignition unit** Hochspannungszündanlage *f* ~ **insulator** Hochspannungsisolator *m* ~ **keying** Anodenspannungstastung *f* ~ **lead** Hochspannungskabel *n* ~ **lead-in** Hochspannungszuführung *f* ~ **line** Hochspannungsleitung *f* ~ **magneto** Hochspannungsmagnet *m* ~ **network** Hochspannungsnetz *n* ~ **outlet** Hochspannungsausführung *f* ~ **(voltage) plant** Hochspannungsanlage *f* ~ **power plant** Überlandzentrale *f* ~ **probe** Hochspannungsmeßkopf *m* ~ **remote-control switch** Hochspannungsfernschalter *m* ~ **side** Hochspannungsseite *f* ~ **steel** Hochspannungsstahl *m* ~ **supply** Anodenspannungsversorgung *f* ~ **supply unit** Anodenspannungsapparat *m* ~ **switch** Hochspannungsschalter *m* ~ **tester** Hochspannungsprüfgerät *n* ~ **testing equipment** Hochspannungsprüfeinrichtung *f* ~ **test(ing) field** Hochspannungsprüffeld *n* ~ **transformer** Hochspannungstransformator *m* ~ **unit** Hochspannungsgerät *n* ~ **voltage** Oberspannung *f*
**high,** ~**-test cast iron** hochwertiges Gußeisen *n*, Qualitätsguß *m* ~**-test gray iron** hochwertiger Grauguß *m* ~ **test sprinkler** hochgrädiger Sprinkler *m* ~ **tide** Flut *f*, Hochflut *f*, Hochwasser *n* ~ **top heat** Oberfeuer *n* ~ **vacuum** Hochvakuum *n* (electron.)
**high-vacuum,** ~ **amplifier** Hochvakuumverstär-

ker *m* ~ **breakdown** Hochvakuumdurchschlag *m* ~ **coater** Hochvakuumbedämpfungsanlage *f* ~ **electron valve** Hochvakuumelektronenröhre *f* ~ **engineering** Hochvakuumtechnik *f* ~ **furnace** Hochvakuumofen *m* ~ **phototube** Hochvakuumzelle *f* ~ **rectifier** Glühkathode *f*, Glühkathodengleichrichter *m* ~ **rectifier valve** Hochvakuum-gleichrichterröhre *f*, -glühkathodengleichrichterröhre *f*, Kenotron *n* ~ **television tube** Hochvakuumfernsehröhre *f* ~ **tube** Elektronenröhre *f*, harte Röhre *f* ~ **valve** Hochvakuumröhre *f* (Elektronenröhre)
**high,** ~**-velocity airfoil** Schnellflußprofil *n* ~**-velocity clarified juice heater** Schnellstromdünnsaftvorwärmer *m* ~ **voltage** Hochspannung *f*
**high-voltage,** ~ **battery** Anodenbatterie *f*, B-Batterie *f* ~ **cable** Zählrohranschlußkabel *n* ~ **current** Hochspannungs-, Stark-strom *m* ~ **discharge tube** Hochspannungsentladungsröhre *f* ~ **duct** Hochspannungsdurchführung *f* ~ **fence** Starkstromzaun *m* ~ **grid (or system)** Hochspannungsnetz *n* ~ **insulation support** Hochspannungsstütze *f* ~ **line** Hochspannungsleitung *f* ~ **obstacle** Starkstromsperre *f* ~ **panel** Hochspannungsteil *m* (rdr) ~ **plate** hochsperrende Platte *f* ~ **plug** Starkstromstecker *m* ~ **plug-in unit** Hochspannungseinschub *m* ~ **power pack** Hochspannungsnetzgeräte *pl* ~ **power station** Hochspannungsanlage *f* ~ **protector** Grobspannungsschutz *m* ~ **rectifier valve** Hochspannungsgleichrichter *m* ~ **source** Hochspannungsquelle *f* ~ **spark plug** Hochspannungszündkerze *f* ~ **supply** Hochspannungsversorgung *f* ~ **surge-limiting diode** Spannungsstoßbegrenzungsdiode *f* für hohe Spannung ~ **switch** Überstromschalter *m* ~ **system** Hochspannungsnetz *n* ~ **tested** mit Hochspannung *f* geprüft ~ **testing equipment** Hochspannungsprüfeinrichtung *f* ~ **transformer** Hochspannungstransformator *m* ~ **warning device** Berührungswarngerät *n*
**high,** ~ **volts** Hochvolt *n* ~**-volume bell** Starktonglocke *f* ~**-(twist) warp** Hochkette *f* ~**-warp tapestra** hochschäftige Tapete *f* ~ **water** Hochwasser *n*
**high-water** Hochwasser ... ~ **of ordinary neap tide** Nipphochwasser *n* ~ **of spring tides** Springhochwasser *n*
**high-water,** ~ **arch** Flutöffnung *f* ~ **gauge** Hochseepegel *m* ~ **level** Hochwasserstand *m* (the) ~ **lunitidal interval at full change of the moon** Hafenzeit *f* ~ **mark** Flut-grenze *f*, -wassermarke *f* ~**-mark gauge** Hochwasserstandsmesser *m* ~ **turbine** Hochwasserrad *n*
**highway** Chaussee *f*, Fahr-damm *m*, -straße *f*, Heerstraße *f*, Landstraße *f*, öffentliche Straße *f*; Vielfachleitung *f* (data proc) ~ **bridge** Straßenbrücke *f* ~ **construction (or engineering)** Straßenbau *m* ~ **route marker** Landstraßenwegweiser *m* ~ **surface** Straßendecke *f*
**high,** ~**-webbed** (girder) hochstetig ~**-wheeled** hochräderig ~ **wind** starker Wind *m*
**high-wing,** ~ **airplane** Hochdecker *m* (semi) ~ **monoplane** Schulterdecker *m* ~ **strut-braced monoplane** abgestrebter Hochdecker *m*
**high,** ~ **working rate** starke Beanspruchung *f* ~**-yielding beet** Ertragsrübe *f*

**higher** höher, ober ~ **bid** Übergebot *n* ~ **even harmonics** gerade Oberharmonische *pl* ~ **harmonics** gerade (geradzahlige) Harmonische *pl*, höhere Harmonische *pl*, Oberharmonische *pl*, Oberwelle *f* ~ **isotope** schweres Isotop *n* ~ **limit** Größtmaß *n* ~ **limiting filter** Hochfrequenzsperrkette *f* ~ **speed motor** Schnellgangmotor *m* ~ **stage** nachgeschaltete Stufe *f* ~ **tensile value** höhere Festigkeit *f* ~ **timbre** hellere Klangfarbe *f*
**highest** oberst, optimal ~ **bid** Höchst-, Meistgebot *n* ~ **Mach number** maximale Mach-Zahl *f* ~ **navigable flood stage** höchster schiffbarer Wasserstand *m* ~ **number** Höchstzahl *f* ~ **offer** Höchstgebot *n* ~ **polarization color** Höhe *f* der Interferenzfarbe ~ **pressure** Maximalspannung *f* ~ **references** erstklassige Referenzen *pl* ~ **resiliency** höchster elastischer Ausgleich *m* ~ **stress** Höchstbeanspruchung *f* ~ **tender** Höchst-angebot *n*, -gebot *n* ~ **upper pool elevation** höchster gestauter Wasserspiegel *m*
**highly** hoch, höchst, sehr ~ **active** hochwirksam ~ **aluminous** hochtonerdehaltig ~ **argentiferous** silberreich ~ **colored** hochfarbig ~ **compressed** hochverdichtet ~ **concentrated** hochkonzentriert ~ **conductive** hochleitfähig ~ **damped** stark gedämpft ~ **damped instrument** aperiodisches Meßinstrument *n* ~ **developed** hochentwickelt ~-**developed controller** Großregler *m* ~ **dilute** hochverdünnt ~ **directive** stark richtfähig ~ **dispersed** hochdispers ~ **dispersive prism** stark fächerndes Prisma *n* ~ **effective** hochwirksam ~ **efficient hydro-extractor** Hochleistungspendelzentrifuge *f* ~ **evacuated tube** harte Röhre *f* ~ **excited states** hochangeregte Zustände *pl* ~ **explosive** hochbrisant ~ **fluid** dünnflüssig ~ **glazed** hochsatiniert ~ **heated** hochbeheizt ~ **inductive shunt** Shunt *m* mit hoher Selbstinduktion ~ **loaded** stark belastet ~ **machine-finished** hoch maschinenglatt ~ **magnifying telescope** stark vergrößerndes Fernrohr *n* ~ **maneuverable** sehr wendig ~ **mill-finished** hoch maschinenglatt ~ **mobile task force** Schnelltruppe *f* ~ **opaque** starkverdurchlässig ~ **polished** hochpoliert, spiegelblank ~ **polished surface** spiegelblanke Oberfläche *f* ~ **radioactive** hochradioaktiv ~ **refractive** hochbrechend ~ **refractory** hochfeuerfest ~ **resinous** harzreich ~ **resistant (or resistive)** hochohmig ~ **saturated** hochgesättigt ~ **selective** (of resonance) scharfbegrenzt ~ **sensitive** hochempfindlich ~ **stressed** hochgespannt ~ **transparent film** transparente oder glasklare Folie *f* ~ **viscous** hochviskos ~ **volatile** hoch-, leicht-flüchtig ~ **wear-resistant** hochverschleißfest
**hill** Anhöhe *f*, Berg *m*, Höhe *f*, Hügel *m* ~-**and-dale track (or recording)** Tiefenschrift *f* (phono) ~ **climb** Bergfahrt *f* ~-**drop corn planter** Häufeltropfpflanz-, Maisdibbel-maschine *f* ~-**side** hügelseitig (hängig) ~ **side** Abhang *m*, Berghang *m* ~ **side plow** Wendepflug *m* ~ **water** Hangwasser *n*
**hiller** Häufler *m* (agric. mach) •
**hills** Mittelgebirge *n*
**hilly** bergig, hügelig ~ **country** welliges Gelände *n* ~ **ground** Gebirgsgelände *n*
**hi-lo-check** Hoch-Niedrig-Prüfung *f*
**hilt** Heft *n*

**hind,** ~ **board** Hinterwand *f* ~ **bolster** Hinterachsschale *f* ~ **carriage** Hintergestell *n* ~ -**quarter** Hinterhand *f* ~ **trace** Hinterstrang *m*
**hinder, to** ~ behindern
**hindered,** ~ **falling** Fall *m* im beengten Raume ~ **rotation** (of molecules) behinderte Rotation *f* ~ **settling** verzögerte Fällung *f*
**hindering** Behinderung *f*
**Hindley's screw** Globoidschneckentrieb *m*
**hindrance** Behinderung *f*, Hemmung *f*
**hinge, to** ~ anlenken, drehbar anlenken **to** ~ **away** abklappen **to** ~ **down** abklappen, herunterklappen **to** ~ **out** abklappen
**hinge** Angel *f*, Anlenkungspunkt *f*, Drehachse *f*, Falte *f*, Gelenk *n*, Gelenkpunkt *m*, Knickbrücke *f* Scharnier *n*, (Tür-)Angel *f* ~ **with hook** Aufsetz-, Kegel-band *n* ~ **with hooks** Hakenband *n*
**hinge,** ~ **bellows unit** Ausdehnungsscharnier *n* ~ **bolt** Scharnierriegel *m*, -stift *m*, Überwurfmutter *f* ~ **butt** Scharnierplatte *f* ~ **center line** Drehachse *f* ~ **connection** Gelenkverbindung *f* ~ **coupling** Gelenkkupplung *f* ~ **cover** Scharnierverkleidung *f* ~ **fitting** Gelenkanschluß *m* ~ **joint** Drehgelenk *n*, Kippe *f* ~ **line** Schloßrand *m* ~ **moment** Ruder-, Scharnier-moment *n* ~ **piece** Klappstück *n* ~ **pin** Federbolzen *m*, Kippachse *f*, Lagerstift *m*, Scharnier-bolzen *m*, -stift *m* ~ **plate** Gehänge *n*, Schloßplatte *f*, Türband *n* ~ **rod** Gelenkstange *f* ~ **saw** Gelenksäge *f* ~ **shaft bearing** Kipplager *n* ~ **sleeve** Scharnierhülse *f* ~ **socket** Halteblech *n* (Stativeinfassung mit Scharnier) ~ **stocks** Scharnier-, Scher-kluppe *f* ~ **teeth** scharnierartige Zähne *pl*
**hinged** aufklappbar, einklappbar, herunterklappbar, klappbar, schwenkbar, mit Scharnieren versehen **with** ~ **supports** gelenkig gelagert
**hinged,** ~ **apron** aufklappbarer Ausleger *m* ~ -**arch bridge** Gelenkbogenbrücke *f* ~ **arm** Gelenkausleger *m*, Klapparm *m* ~ **back** umlegbare Rücklehne *f* ~ **back-plate** Flügelträger *m* ~ **bearing** Gelenk *n*, Kipplager *n* ~ **block** Scharnierflasche *f* ~ **body** Klappteil *n* ~ **bolt** Gelenk-, Klapp-schraube *f* ~-**bottom door** Bodenklappe *f* ~ **bracket** Gelenkarm *m* ~ **bucket** Gelenkbecher *m*, Klappkübel *m* ~ **bulkhead** Klappspant *n* ~-**cap bearing (or seat)** aufklappbares Lager *n* ~-**coil** variometer Klappvariometer *n* ~ **core box** Scharnierkernkasten *m* ~ **cover** Deckel *m* mit Scharnier, Klappdeckel *m*, Klappenverschluß *m*, Verschlußklappe *f* ~ **curtain** Rolljalousie *f* ~ **dark slide** Klappkassette *f* ~ **door** Drehtür *f*, Schwenkklappe *f* ~ **feeder** Scharniernüßchen *n* ~ **frame** (with knuckle joint) Gelenkstrebe *f*, Klapp-, Scharnier-rahmen *m* ~ **girder** Schleppträger *m* ~ **girder bridge** Gelenkträgerbrücke *f* ~ **handle** Klappgriff *m* ~ **hatch** Klappluke *f* ~ **jib** Klappausleger *m* ~ **joining member** gelenkiger Stoßriegel ~ **leading edge** Knicknase *f* ~ **lid** Klappdeckel *m* (tap rec) ~ **plateholder** aufklappbare Kassette *f* ~ **pushing arm** umlegbare Ausdrückstange *f* ~ **ring (or ring dish holer)** ringförmiger Halter *m* ~ **screen** klappbares Siebblech *n*, Lichtklappe *f* ~ **spoiler** Störkante *f* ~ **support** Bolzenlager *n*, festes Gelenk *n*, Fußgelenk *n*,

Pendel-, Trag-stütze *f* (sewing machine) ~
**switchboard** aufklappbarer Klappenschrank *m*,
Klappschalttafel *f* ~ **table guide** Schwenkfüh-
rung *f* ~**-telescope camera** Scherenfernrohr-
kamera *f* ~ **tie bar** Gelenkstange *f*, Zugstrebe *f*
~ **tracing lever** (of planimeter) Fahrarm *m*
~**-type float** Kippschwimmer *m* ~ **valve** Drehklappe *f* ~ **vice** Gelenkschraubstock *m*
**hinging bolt** Anlenkbolzen *m*
**hint, to** ~ einen Wink geben **to** ~ **at** anspielen
auf
**hint** Andeutung *f*, Anspielung *f*, Fingerzeig *m*,
Hinweis *m*, Wink *m*
**hinterland** Hinterland *n*
**Hiorth (induction) furnace** Hiorth-Ofen *m*
**hip** Eckfirst *m*, Grat *m*, Hüfte *f* ~ **bead** Gratwulst
*f* ~ **knob** Helmstangenspitze *f* ~ **lead** Gratblech *n* ~ **pad** (harness) Hinterzeug *n* ~ **rafter**
Schrifter *m*, Schrift-, Walm-sparren *m* ~ **roof**
Walmdach *n* ~ **sheet** Gratblech *n* ~ **side of a**
**roof** Gratseite *f* eines Daches ~ **tile** Gratziegel *m*
**hippuric acid** Hippursäure *f*
**hire, to** ~ anheuern, heuern **on** ~ leihweise
**hired,** ~ **help** Arbeitskraft *f* ~ **plane** Mietflugzeug *n*
**hiring** (a miner) Anlegen *n* eines Bergarbeiters ~
**out** Verdingung *f*
**H iron** H-Eisen *n*
**hirsute** behaart, haarig
**Hirth-type serrations** Hirth-Verzahnung *f*
**hisingerite** Gillingit *m*
**hiss, to** ~ fauchen, zischen
**hiss,** ~ **effect** Zischeffekt *m* (film) ~ **random**
**noise** Rauschen *n* (el tube)
**hissing** Gezisch *n*, Knallgeräusche *pl*, Nebengeräusch *n*, Zischen *n* ~ **sound** Zischlaut *m*
**histochemistry** Histochemie *f*
**histogram** Rechteckdarstellung *f*, Säulendia-
gramm *n*
**hit, to** ~ aufschlagen, schlagen, stoßen, treffen,
verfallen **to** ~ **back** zurückschlagen **to** ~ **hard**
zuschlagen **to** ~ **a mine** auf eine Mine *f* laufen
**to** ~ **the target** das Ziel *n* treffen **to** ~ **upward**
aufhauen
**hit** Aufschlag *m*, Granattreffer *m*, Hieb *m*, Stoß
*m*, (firing) Treffer *m* ~ **by indirect fire** Prall-,
Prell-schuß *m* ~**s per scan** Trefferzahl *f* pro
Abtastung (rdr.) ~**-and-miss governing** Aus-
setzerregelung *f* ~ **theory** Treffertheorie *f*
**hitch, to** ~ (couple) anhaken, (animals) anspannen, anstechen, anstecken, einhaken, kuppeln **to** ~ **to** ankuppeln
**hitch** Bühnloch *n*, Festmachen *n*, Halt *m*, Ruck
*m*, Stich *m*, Zug *m* ~ **for hooking a tackle to a**
**rope** Hakenschlag *m* ~ **bracket** Anhänger-
stütze *f* ~ **hook** Anhängerhaken *m* ~ **roll**
Spannwalze *f* (paper mfg.)
**hitherto** bisher
**Hittorf tube** Hittorfsche Röhre *f*
**hjelmite** Hjelmit *m*
**H.M. Stationery Office** amtliche Druckschrif-
tenvertriebsstelle *f* des Englischen Schatzamtes
**H-network** H-Filterglied *n*
**hoar-frost** Boden-, Rauh-frost *m*, Rauhreif *m* ~
**deposit** Rauhreif *m* (auf Leitungen)
**hoarse** heiser, rauh
**hoarseness** Heiserkeit *f*, Rauheit *f*

**hob, to** ~ (ab)wälzen, verzahnen, wälzfräsen
**hob** Abwälz-, Gewinde-fräser *m*, Holzpflock *m*,
Originalbohrer *m*, Schneckenfräser *m*, Strehlbohrer *m*, Wälzfräser *m* **master** ~ Eindrück-
pfaffe *f* (plastics), Pfaffe *f* ~ **cutter grinder** Ab-
wälzfräserschleifmaschine *f* ~ **method** Abwälzverfahren *n* ~ **nail** grober Schuhnagel *m* ~
**sharpening machine** Abwälzfräserschärfma-
schine *f* ~ **spindle** Frässpindel *f* einer Abwälz-
fräsmaschine ~**-type milling cutter** Abwälzfräser *m*
**hobbing** Fräsen *n* nach Abwälzverfahren ~
**attachment** Wälzeinrichtung *f* ~ **carriage** Wälz-
frässupport *m* ~ **head** Fräskopf *m* einer Wälzfräsmaschine ~ **machine** Wälzfräsmaschine *f*
~ **method** Wälzfräsverfahren *n* ~ **process**
Wälzvorgang *m*
**hobbyist** Bastler *m*
**hock** Hacke *f*, Sprunggelenk *n*
**hod** Tragmulde *f*, Trog *m* ~ **man** Handlanger *m*
~ **trough** Mörteltrog *m*
**hodograph** Hodograf *m*, Wegkurve *f* ~ **(diffe-**
**rential) equation** Hodografengleichung *f*
**hodographic** hodografisch
**hodometer** Meßrad *n*, Wegmesser *m*
**hodoscope** Hodoskop *n*
**hoe, to** ~ hacken
**hoe** Hacke *f*, Haue *f* ~ **broadcast seeder** Breit-
saatsä(e)- und Hackmaschine *f*, Breitsä(e)- und
Hackmaschine *f* ~ **drill** Drillmaschine *f* mit
Hackschare
**hoeing** Hacken *n*
**hoist, to** ~ anheben, aufhaspeln, aufhissen, aufstecken, aufwinden, aufziehen, fördern, heben,
hissen, hochrichten, hochwinden, in die Höhe
ziehen, winden **to** ~ **a boat** ein Boot einsetzen
**to** ~ **out** aussetzen **to** ~ **a seaplane** ein Segelflugzeug einsetzen
**hoist** Aufzug *m*, Fahrstuhl *m*, Flaschenzug *m*,
Förder-korbkran *m*, -maschine *f*, -werk *n*,
Haspelwinde *f*, Hebewerk *n*, Heh (Heißvorrichtung), Kranlaufwinde *f*, Winde *f*, Windwerk *n*, Zug *m* ~ **boom** Windenausleger *m* ~
**bridge** Schrägaufzugbahn *f* ~ **carriage** Aufzug-
wagen *m* ~ **chain** Flaschenzugkette *f* ~ **crane**
Heißvorrichtung *f* ~ **direction** im Hubsinn ~
~ **drive** Hebeantrieb *m* ~ **fitting** Heißbeschlag
*m* ~ **frame** Förderturm *m* ~ **gear for grab**
**service** Greiferwindwerk *n* ~ **gearing** Hubwindwerk *n* ~**-line** Heißleine *f* ~ **motor** Windwerk-
motor *m* ~ **rope** Flaschenzugseil *n* ~ **structure**
Aufzugs-, Förder-gerüst *n* ~ **way** Aufzugs-
schacht *m*
**hoisting** Aufziehen *n*, Hissen *n*, Hochheißen *n*
~ **apparatus** Hebe-vorrichtung *f*, -zeug *n* ~
**block** Aufzugkloben *m*, Förderhaken *m* ~
**bucket** Aufzugkasten *m*, Gichtkübel *m* ~ **cable**
Flaschenzugseil *n*, Förderseil *n*, Förderzug *m*,
Hubseil *n* ~ **capacity** Hubvermögen *n*, Lasthub
*m* ~ **carriage** Laufwagen *m* ~ **chain** Hubkette *f*
~ **clevis** Hebeklaue *f* ~ **crab** Bock-, Hebe-,
Montage-winde *f* ~ **crane** Hebekran *m* ~
**depth** Förderhöhe *f* ~ **device** Aufzugöse *f*,
Hebewerk *n*, Heißvorrichtung *f* ~ **diagram**
Kranhebediagramm *n* ~ **direction** Hubsinn *m*
~ **drum** Förder-, Seil-, Winden-, Winde-trom-
mel *f* ~ **engine** Fördermschine *f* (min.), Hebewerk *n*, Ladekran *m*, Heißmaschine *f*, Hub-

motor m ~ **equipment** Hebezeug n, Windwerk n ~ **force** Lasthebekraft f ~ **gear** Förderschacht m, Windwerk n ~-**gear train** Hubwerk n, Hubwindwerk n ~ **hook** Aufzug-, Förder-haken m ~ **jack** Förderwinde f, Hebeflasche f ~ **link** Hebekloben m ~ **loop** Förderöse f ~ **machine** Hebemaschine f ~ **magnet** Hebemagnet m ~ **mechanism** Hubmechanismus m, Windwerk m ~ **motion** Hubbewegung f ~ **motor** Aufzug-, Hub-motor m ~ **points** Heißpunkte pl (parachute) ~ **ring** Aufzugring m ~ **rope** Aufzug-, Förder-, Hebezug-, Hub-, Last-, Zug-seil n ~ -**rope sheave** Hubseilrolle f ~ **shaft** Förderschacht m ~ **sling** Aufhängeschlaufe f, Seilgehänge n ~ **speed** Hubgeschwindigkeit f ~ **tackle** Flaschenzugkloben m ~ **time** Förder-, Hub-zeit f ~ **unit** Hubwerk n ~ **wheel** Heberad n ~ **winch** Aufzugwinde f ~ **work** Förderarbeit f
**hold, to** ~ anhalten, anpacken, (communication) Zeitdauer belegen, besetztmachen, (a circuit) besetzt halten, fassen, festhalten, gelten, halten, klemmen, spannen **to** ~ **in abeyance** aussetzen **to** ~ **back** arretieren, bremsen, festhalten, feststellen, hinterhalten, zurückhalten **to** ~ **a call** eine Anmeldung zurückstellen **to** ~ **a circuit** eine Leitung (Teilnehmerleitung) besetzt halten **to** ~ **down** niederhalten **to** ~ **it** den Druck aushalten **to** ~ **the line** in (oder an) der Leitung bleiben (teleph.) **to** ~ **off** abhalten, ausschweben lassen **to** ~ **out** aushalten, widerstehen **to** ~ **over** offenlassen **to** ~ **a position** Stellung behaupten **to** ~ **a record** einen Rekord halten **to** ~ **a relay** ein Relais halten **to** ~ **security** gesichert sein **to** ~ **a subscriber's line** eine Leitung (oder eine Teilnehmerleitung) besetzt halten **to** ~ **below the target in aiming** das Ziel aufsitzen lassen **to** ~ **together** zusammenhalten **to** ~ **true** gültig sein, zutreffen **to** ~ **up** stocken
**hold** Griff m, Haft f, Halt m, (ship) Laderaum m, Ladungsraum m, Schiffsraum n, Widerhalt m ~-**back agent** Rückhaltemittel n ~-**back carrier** Rückhalteträger m ~-**back sprocket** Nachwickelrolle f (film) ~ **capture** Lochfang m (transistor) ~ **circuit** Haltestromkreis m ~ **concentration** Löcherkonzentration f (transistor) ~ **conduction** Löcherleitung f (transistor) ~ **control** Regelung f der Kippfrequenz ~ **current** Haltestrom m ~-**down spring** Haltefeder f (federnder Niederhalter) ~ **downs** Halteklammer f ~ **ejection** Lochwanderung f (transistor) ~ **electron** Mangelelektron n (transistor) ~ -**fast for boat hocks** Kranschiene f ~ -**in range** Haltebereich m ~ **lamp** Besetztlampe f ~-**over key** Haltetaste f (teleph.) ~-**over position** Haltestellung f (teleph.) ~ **range** Haltebereich m ~ **signal (or lamp)** Wartelampe f ~ **take** Reserveaufnahme f (film) ~-**up** Materialeinsatz m; Stauung f ~-**up attachment (or lock)** Dornverriegelung f ~-**up time** Aufenthaltsdauer f
**holder** Aufsteckhalter m, Besitzer m, Bügel m, Dose f, Drehsupport m, Fassung f, Halter m, Halterung f, Halterungssystem n, Klemme f, Sockel m, Ständer m, Träger m, Zwinge f ~ **for ammunition drums** Trommelhalter m ~ **of a bill** Wechselgläubiger m ~ **for (screwing) dies** Gewindeschneidekluppe f ~ **for gripping tape** Halter m für Bandfeder ~ **for gypsum and mica plates** Kompensatorhalter f für Gips- und

Glimmerplättchen ~ **of letters patent** Patentinhaber m ~ **for lever** Hebelauge n ~ **of a license** Inhaber m einer Verleihungsurkunde ~ **of a record** Rekordinhaber m
**holder,** ~ **claw** Halterklaue f ~-**on (or stationary) die** Anpreßstempel m ~ **profile** Halterprofil n (für Düse)
**holders** Halterungsteile pl
**holding** Belegung f, Bestand m, Einspannen n, Festspannen n, Halterung f, Spannen n; haltig, hältig, hinhaltend ~ **of course** (automatic pilot) Kurssteuerung f ~ **with stopper** (of a chain) Haltekette f ~ **a subscriber line** (by toll operator) Besetzthaltung f ~ **a subscriber line by toll operator** Belegung f der Teilnehmerleitung durch die Fernbeamtin
**holding,** ~ **angle** Haltewinkel m ~ **apparatus** Aufsetzvorrichtung f ~ **apparatus for elevator cages** Förderkorbaufsetzvorrichtung f ~ **attack** Fesselungsvorstoß m ~ **back** Zurückhaltung f ~ **bar magnet** Brückenmagnet m ~ **bar switching relay** Brückenanschalterrelais n ~ **bases** Halterfüße pl ~ **battery** Ergänzungsbatterie f ~ **beacon** Warteraum-Funkfeuer n ~ **beam** Haltestrahl m, Testbildstrahl m ~ **button** (trunk or service line) Haltetaste f ~ **cap** Haltekappe f ~ **capacity** Fassungsvermögen n ~ **circuit** Haltestromkreis m ~ **clamp** Spannblock m ~ **coil** Haltewicklung f (teleph.) ~ **collar** Haltemanschette f ~ **column** Spannsäule f (Bohrhammer) ~ **company** Gesamtgesellschaft f (Dachgesellschaft) ~ **contact** Selbsthaltekontakt m ~ **control** Regelung f der Kippfrequenz ~ **current** Halte-, Ruhe-strom m ~ **detent** Sperrklinke f ~ **device** Haltevorrichtung f ~ **disk** Haltescheibe f ~ **dog** Sperrklinke f ~ -**down attachment** Anpreßvorrichtung f ~ -**down bars** Bügelstangen pl ~-**down beam** Drucksäule f ~-**down bolt** Anker-, Fundamentbolzen m, Lagerfußschraube f, Verankerungsbolzen m ~-**down clamp for plates** Blechfesthaltung f ~-**down sinker** Einschließplatine f ~ **drum** Haltetrommel f ~ **fixture** Spannzeug n ~ **furnace** Warmhalteofen m ~ **head** (nozzle holder) Halteknopf m ~ **hook** Kurbelhalterhaken m ~ **jack** Halteklinke f ~ **key** Halte-, Sperr-schalter m ~ **lug** Befestigungsknagge f, Haltelappen m ~ **magnet** Haltemagnet m ~ **means (or arrangement)** Haltevorrichtung f ~ **nut** Schraubenmutter f zum Festhalten ~-**on coil** Haltespule f ~-**on point** Anklammerungspunkt m ~-**on tool** Vorhalter m ~ **pattern** Warteschleife f (aviat.) ~ **(the) pirn** Spulenlagerung f ~ **plate** Griffplatte f, Tragblech n ~ **plug** Haltezapfen m ~ **point** (airport) ~ **position marking** Rollhaltemarkierung f (airport) ~ **ratio** Halteverhältnis n ~ **relay** Besetzt-, Halte-relais n ~ **ring** Haltering m ~ **ring for ball bearing** Fassungsring m (Kugellager) ~ **rope** Halteseil n ~ **screen** Sperrechen m ~ **screw** Halte-schraube f, -stift m ~ **sequence** Wartefolge f (aviat.) ~ **shank** Halterschaft m ~ **(or retaining) spanner** Halteschlüssel m ~ **stack** Wartestapel m (aviat.) ~ **strap** Fixierbügel m ~ **time** Belegungsdauer f ~ **under** Unterhaltung f ~ **winding** Haltewicklung f (teleph.)
**hole, to** ~ aushöhlen, ausschrämen, ein Loch herstellen **to** ~ **the trenches** unterschrämen

**hole** Auskolkung *f*, Auslösung *f*, Ausschnitt *m*, Defektelektron *n*, Gat(t) *n*, Gatjen *n*, Grube *f*, Höhle *f*, Leerstelle *f*, Loch *n*, (in nuclear theory) freier Platz *m*, tote Zone *f* (rdo) ~ **for axle** Achsloch *n* ~ **of the drum axle** Trommelachsenöffnung *f* ~ **in the fishplate for the bolt** Laschenloch *n* ~ **of a hollow bit** Bohrmeißelkanal *m* ~ **for pinning** Steckloch *n* ~ **for the ram** Kropfloch *n* ~ **for the retractable undercarriage** Fahrgestelleinziehschacht *m* ~ **through spindle** Hauptspindelbohrung *f*

**hole,** ~ **absorption** Absorption *f* durch Defektelektronen ~**-and slot magnetron** Schlitz- und Loch-Magnetron *n* ~ **basis system** Einheitsbohrung *f* ~ **capture** Lochfang *m* ~ **conduction** Defekt-, Löcher-leitung *f* ~**-cored** vorgegossen ~**-cutting machine** Lochschneidemaschine *f* ~ **deviation measuring device** Bohrlochneigungsmesser *m* (Tiefbohranlage) ~ **diaphragm** Lochblende *f* ~ **free steel** lunkerfreier Stahl *m* ~ **furnace** Unterflurtiegelofen *m* ~ **injection** Lochwanderung *f* ~ **mark** Lochmarke *f* ~ **model** Löchermodell *n* ~ **-piercing apparatus for blocks** Klischeelochapparat *m* ~ **punch** Schnapplocker *m* ~ **punching and eyeletting machine** Loch- und Öseneinsetzmaschine *f* ~ **siphoning** Lochabsaugung *f* ~ **socket** Lochstutzen *m* ~ **spectra** Löcherspektren *pl* ~ **template** Lochschablone *f* ~ **theory** Löchertheorie *f* ~ **trap** Eingangszentrum *n* für Defektelektronen ~ **-type conductivity** Fehlstellenleitung *f* ~**-type nozzle** Lochdüse *f* ~ **wall** Düsenwand *f*

**holed** gelocht

**holes** Löcher *pl* **having** ~ durchlöchert

**holing** Bühnloch *n*, Pfeilerdurchhieb *m*, Schram *m*

**hollander-beating** Holländermahlung *f*

**hollandite** Hollandit *m*

**hollow, to** ~ (turn) ausdrehen, unterhöhlen, vertiefen **to** ~ **out** sich ausbauchen, aushöhlen, aussparen, austiefen **to** ~ **a precious stone** einen Edelstein ausschlägeln

**hollow** Ausbauchung *f*, Aussparung *f*, Falte *f*, Höhle *f*, Höhlung *f*, Kugel-kaliber *n*, -lehre *f*, Mulde *f*, Senkel *m*, Senkung *f*, Talmulde *f*; hohl, leer ~ **adze** Hohldechsel *f* ~ **billet** Hohlblock *m* ~ **bit** Hohlmeißel *m* ~**-bit tongs** Hohlmaulzange *f* ~ **blade** Hohlschaufel *f* ~ **blank** Hohlblock *m* (Rohrwalzwerke) ~ **block** Hohlstein *m* ~ **body** Hohlkörper *m* ~ **bolt** Durchflußbolzen *m* ~ **brace** Krückelrohr *n* ~ **braided line** hohlgeklöppelte Leine *f* ~ **brass pin** Messinghohlstift *m* ~ **brick** Hohlziegelstein *m* ~ **building block** Hohlblockstein *m* ~ **cable** Hohlseil *n* ~ **cam shape** Hohlnockenform *f* ~**-cast quads** Hohlgußquadrate *pl* ~ **casting** Hohlguß *m* ~ **cathode** Hohlkathode *f* ~ **cathode discharge** Hohlkathodenentladung *f* ~ **center** Hohlkörnerspitze *f* ~ **chamfer fillet** Hohlkehle *f* ~ **(explosive) charge** Hohlladung *f* ~**-charge ammunition** Hohlladungsmunition *f* ~**-charge projectile** Hohlkopfgeschoß *n* ~ **chisel** Hohleisen *n* ~ **chisel bit** Hohlmeißelbohrer *m* ~ **-chisel mortising machine** Hohlmeißelstemmmaschine *f* ~ **column** Hohlständer *m* ~ **concave grinding** Hohlschliff *m* ~ **concrete block** Hohlblockstein *m*, Zehnerstein *m* ~**-concrete slab** Hohldielezement *m*, Zementhohldiele *f* ~ **conductor** Hohlkabel *n* ~ **connection plug** Hohl-

schraubenstutzen *m* ~ **container** Hohlgefäß *n* ~ **core** hohler Kern *m*, Hohlkern *m* ~ **crankpin** durchbohrter Kurbelzapfen *m* ~ **cross-section area** Hohlquerschnittsfläche *f* ~ **cylinder** Hohlzylinder *m* ~**-disk reversing valve** Doppelsitzventil *n* ~**-drawn article** gezogener Hohlkörper *m* ~ **drift** Hohl-setzen *n*, -stempel *m* ~ **drill** Hohlbohrer *m* ~ **drilling rods** Hohlbohrgestänge *n* ~ **driving shaft** Antriebsstahlhohlwelle *f* ~ **edged grinding** Hohlschliff *m* ~ **filler tile** Deckenhohlstein *m* ~ **flat-bar steel** Hohlflachstahl *m* ~ **flier with spring finger** hohler Flügel *m* mit Preßfinger ~ **formed by wind** Windmulde *f* ~ **frame** Hohlrahmen *m* ~**-frame construction** Hohlträgerkonstruktion *f* ~ **girder** Hohlträger *m* ~ **glass** Hohlglas *n* ~ **-ground circular saw** Hohlkreissäge *f* ~**-ground edge** eingeschliffene Hohlkehle *f* ~**-ground tool** Hohlkehlenstahl *m* ~ **ingot** Hohlblock *m* ~ **key** Hohlkeil *m*, Rundschlüssel *m* ~ **masonry block** Hohlblockstein *m* ~ **mill** Außen-, Hohl-, Kronen-, Stift-fräser *m* ~**-mold press** Hohlformenpresse *f* ~ **mount casting** Hohlfußguß *m* ~ **(or tubular) needle** Nadelkanüle *f* ~**-nosed plane** Rundhobel *m* ~ **piece** Hohlkörper *m* ~ **pipe line** Hohlrohrleitung *f* ~ **piston** durchbrochener Pumpenkolben *m* ~ **plug** Hohlstopfen *m* ~ **point** Hohlspitze *f* ~ **prism** Hohlprisma *n* ~ **projectile** Hohlgeschoß *n* ~ **punch** Loch-eisen *n*, -pfeife *f*, -stanze *f* ~ **quoin** Eckstein *m* mit Hohlkehle, Wendenische *f* ~ **rayon** Luftseide *f* ~ **reamer** Hohlräumer *m* ~**-ring charge** Hohlringladung *f* ~ **rivet** Lochniete *f* ~**-rod clamp** Gestängerohrklammer *f* ~**-rod-collar thread** Gestängerohr- und Muffengang *m* ~**-rod cutter** Gestängerohrschneidezeug *n* ~ **rods** Gestängerohr *n* ~ **rope** Hohlseil *n* ~ **saw** Kreisnutenfräser *m* ~ **screw** Hohl-, Inbus-schraube *f*, Winkelrohrstutzen *m* ~ **section** Hohlquerschnitt *m* ~ **set-screw** Innenvierkant- oder Sechskant-schraube *f* ~ **shaft** hohle Welle *f*, Hohlwelle *f* ~ **slot** Ausschnitt *m* ~ **space** Hohlraum *m* ~ **sphere** Hohlkugel *f* ~ **spindle** Hohlspindel *m* ~ **splint** Lagerungsschiene *f* ~ **stay** Hohlsteg *m* ~ **steel shaft** (Stahl)hohlwelle *f* ~ **strut** Hohl-stiel *m*, -strebe *f* ~ **support** Hohlstütze *f* ~ **supporting column** Hohlsäule *f* ~ **tile** Falzziegel *m*, Hohlstein *m*, Nonne *f* ~ **tube** Hohlröhre *f* ~ **turbine bucket** Topfschaufel *f* ~**-type rivet** Hohl-niet *m*, -niete *f* ~ **upright** Hohlgußständer *m* ~ **wall** Hohlmauer *f* (Hohlwand) ~**-walled** doppelunterbrechungs-, hohl-wandig ~ **ware** Blechgeschirr *n*, Hohlkörper *m* ~ **wave guide** Hohlrohrleitung *f* ~ **wheel set** Hohlradsatz *m*

**hollowed** ausgehöhlt, ausgespart ~ **out** ausgeschlitzt

**hollower** Hohlkehlendrechsler *m*

**hollowing,** ~ **hammer** Klopfschlägel *m* ~ **knife** Krummeisen *n* ~ **plane** Rundhobel *m*

**hollowness** Höhlung *f*, Kavitation *f*

**Holmann projector** Regenspritze *f*

**holmium** Hohlmium *n*

**holoaxial** holoachsial

**holocrystalline** holo-, voll-kristallin

**holoheder** Vollflächner *m*

**holohedral** holoedrisch, vollflächig ~ **form of holohedrism** Holoedrie *f*

**holohedron** Holoeder *m*, Vollflächner *m*
**holohedry** Vollflächigkeit *f*
**holoisometric** holoisometrisch
**holometer** Holometer *n*
**holomorphic** holomorph
**holonomic** holonom
**holophane glass** Riffelglas *n*
**holophotal optical apparatus** Linsenleuchte *f*
**holster** Halftertasche *f*; Spurplatte *f*, Walzenständer *m*
**Holt, ~-Dern** (chloride-roasting and -leaching process) Holt-Dern-Prozeß *m* ~ **flare** Holt-Fackel *f*
**home, to ~ on a beacon** Eigenpeilen *n* auf Funkfeuer (aviat.)
**home** Heimat *f*, Wohnort *m* ~ **address (or station)** Kennort *m* ~ **base** Heimat-flughafen *m*, -horst *m* ~ **constructor** Radiobastler *m* ~ **contact** Wählerruhekontakt *m* ~ **converser and entrance telephone system** Heim- und Tor-Fernsprechanlage *f* ~ **demand** Inlandnachfrage *f*, Nachfrage *f* im Inland ~ **economy** Binnenwirtschaft *f* ~ **exchange** Eigenamt *n* ~ **field** Einsatzhafen *m* ~ **freezer** Kühltruhe *f* ~**-grown raw materials** heimische Rohstoffe *pl* ~ **indicator** Anrufzeichen *n* beim Vielfachschrank ~ **jack** Abfrageklinke *f* ~ **land** Heimatgebiet *n* ~ **mechanic** Bastler *m* ~ **office** Stammbüro *n* ~ **port** Heimathafen *m* ~ **position** Abfrageplatz *m*, (of a wiper) Ausgangsstellung *f*, Nullstellung *f*, Ruhestellung *f*, Teilnehmerplatz *m* ~ **record** Kontroll-streifen *m*, -schrift *f* ~ **signal** Ankunft-, Einfahr-, Einfahrt-signal *n* ~ **station** eigenes Amt *n*, Heimatsstation *f* ~ **telephone** Haustelefon *n* ~ **television receiver** Heimfernsehempfänger *m* ~ **television set** Heimfernseher *m* ~ **trade** Binnenhandel *m* ~ **traffic** Binnenverkehr *m* ~ **wiring** Hausinstallation *f* ~ **work** Heimarbeit *f*
**homentropic** homöoentrop
**homing** Zielflug(verfahren) *n*, Ziel-fahrt *f*, -peilung *m*, Rücklauf *m* (Wähler, Schalter) ~ **action** Rücklauf *m* eines Wählers ~ **adapter** Zielfluggerät *n* ~ **apparatus** Zielfluggerät *n* ~ **beacon** Zielflugfunkfeuer *n* ~ **device** Peilgerät *n*, Zielfluggerät *n* ~ **direction** (of an airplane) Anflugrichtung *f* ~ **flight** Zielflug *m* ~ **guidance** Zielsuchlenkung *f* ~ **head** Zielsuchknopf *m* (g/m) ~ **loop** Sucherkreis *m* ~ **method** Zielflugverfahren *n* ~ **station** Zielflugfunkstelle *f* ~ **system** Zielsuchverfahren *n*
**homocellular** gleichzeitig
**homocentric** punktzentrisch
**homocharge** Homöoladung *f*
**homochromatic** gleichfarbig, homochrom ~ **lights** gleichfarbige Lichter *pl*
**homochromous** homochrom
**homocline** Monoklinale *f*
**homodyne** Trägerfrequenzüberlagerer *m*; **homodyn ~ reception** Empfang *m* mit schwingendem Audion im Schwebungsnull, homodiner Empfang *m*, homodine Verständigung *f*, Superheterodynempfang *m* mit selbsterregter Trägerwelle
**homogeneity** Einheitlichkeit *f*, Gleichartigkeit *f*, Gleichstoffigkeit *f*, Homogenität *f*
**homogeneous** einheitlich, gleich-artig, -förmig -mäßig, -stoffig, homogen ~ **carbon rod** Ho-

mogenkohlestift *m* ~ **field** homogenes Feld *n* ~ **light** einfarbiges Licht *n* ~ **line** homogene Leitung *f* ~ **lubricant** einheitlicher Schmierstoff *m* ~ **platoon** Teileinheit *f* ~ **radiation** homogene Strahlung *f* ~ **solution-type** homogener Reaktor *m* vom Lösungstypus ~ **steel** Homogenstehl *m* ~ **strain** homogene Formänderung *f*
**homogeneously, ~ mixed** innig vermischt ~ **strained** homogen verzerrt
**homogeneousness** Gleichförmigkeit *f*, Homogenität *f*
**homogenization** Homogenisierung *f*
**homogenize, to ~** gleichstoffen, homogenisieren
**homogenizer** Homogenisier-apparat *m*, -maschine *f*
**homographies in the plane** Kreisverwandtschaften *pl*
**homography in space** Kugelverwandtschaft *f*
**homologate, to ~** amtlich anerkennen **to ~ (a record)** anerkennen
**homologation of a record** amtliche Anerkennung *f* eines Rekordes
**homologous** gleichnamig, homolog, spiegelbildlich ~ **condition (or relation)** Spiegelung *f*
**homophonous** gleichlautend
**homopolar** Gleichpol *m*; elektrisch symmetrisch, gleichpolig ~ **alternator** Gleichpolalternator *m* ~ **dynamo** Unipolardynamo *m* ~ **generator** Unipolarmaschine *f* ~ **synchronous machine** Gleichpolsynchronmaschine *f*
**hone, to ~** abziehen (mit dem Ölstein), honen, wetzen, ziehschleifen
**hone stone** Wetzschiefer *m*
**honeycomb** Bienenwabe *f*, bienenkorbartiges Gitter, Blase *f*, dünnflüssige Schlacke, Sand- oder Kiesnest *n* (im Beton), Wabe *f*, zellenartige Struktur *f* ~ **clinker** Schwalbennest *pl* ~ **coil** Honigwabenspule *f*, Wabenspule *f* ~ **coil with banked winding** Mehrlagenspule *f* ~ **corrosion** narbenartige Anfressung *f*, Rostanfressung *f* bis zu tiefen Grübchen ~ **covering** Waffeldecke *f* ~ **formation** Schwalbennestbildung *f* ~ **grill** Zellengleichrichter *m* ~**-like** zellähnlich ~ **nylon** Nylonwaben *pl* ~ **principle** Zellenbauweise *f* ~ **radiator** Bienenkorb-, Waben-, Zellen-kühler *m* ~ **straightener** (in wind tunnel) Strom-, Waben-gleichrichter *m* ~ **structure** Wabenstruktur *f* ~**-tube coil** Honigwabenspule *f* ~ **weave** Waffelbindung *f* ~ **winding** Wabenwicklung *f*
**honeycombed** blasig, höhlig, löcherig, lückig, luckig, lunkerig, wabenartig, zellig ~ **fabric** Waffelgewebe *n* ~ **structure** zellenartige Struktur *f*
**honeycombing** Aschenverflüssigung *f*, lochfräßähnliche Zerstörung *f*, Platzverschwendung *f* (im Lager), Rostanfressung *f*, Schwalbennestbildung *f*
**honeycombs** Schwalbennester *pl*
**honeystone** Mellit *m*
**honey strainer** Honigschleudergerät *n*
**honing** Abziehen *n*, Schleifen *n* ~ **device** Schleifvorrichtung *f*, Hon-, Ziehschleif-maschine ~ **tool** Honwerkzeug *n*
**honor, to ~** honorieren **to ~ a bill (or a draft)** einen Wechsel einlösen
**honorarium** Honorar *n*

honorary, ~ council(lor) Ehrenrat *m* ~ member Ehrenmitglied *n* ~ position Ehrenamt *n* ~ rank Ehrenrang *m*

hood (chem.) Abzug *m*, Abzugs-haube *f*, -schrank *m*, Decke *f*, (for fumes) Dunstabzug *m*, Haube *f*, Kappe *f*, Kapuze *f*, Verdeck *n*, (to protect valve) Schutzkorb *m* ~ of a car Wagenplane *f* ~ of cockpit Führersitzverkleidung *f* ~ for limestone spreader Haube *f* für Kalksteinstreuer

hood, ~ and magnifier release Lichtschachtverriegelung *f* ~ catch Hauben-halter *m*, -schloß *n* ~ cover Verdeckschutzdecke *f* ~ fastener Haubenhalter *m*, -schloß *n*, Motorhaubenverschluß *m* ~ handle Haubengriff *m* ~ strap Haubenriemen *m* ~-type annealing furnace Haubenglühofen *m*

hooded, ~ bead sight Hülsenperlkorn *n* ~ front sight Blenden-, Tunnel-korn *m* ~ revolving front sight Tunnelsternkorn *m*

hoof Huf *m* ~ ball Ballen *m* ~ pick (or scraper) Hufkratzer *m*

hook, to ~ abbiegen, anhaken, anheften, anspannen, (to) einhaken to ~ with iron bands anharpen to ~ on (watchmaking) quälen to ~ up abketteln, kombinieren

hook (spinning defect) Flagge *f*, (of a chain) Gliederkopf *m*, Haken *m*, Hängehaken *m*, Henkel *m*, Öse *f*, (in milling) Schlagen *n*, Stange *f* ~ of the receiver Hörgabel *f* (teleph.) ~ on the tumbler (lock) Zuhaltungshaken *m*

hook, ~-and-butt joint Hakenlaschung *f* ~ belt fastener Riemenhaken *m* ~ bolt Hakenbolzen *m* ~ bracket for insulator (with wedge lug) Keilstütze *f* ~ engagement Hakeneingriff *m* ~ fittings Hakengeschirr *n* ~ guard Sicherheitsstütze *f* ~ harp Hakenharfe *f* ~ head (of cable) Hakenkrone *f* ~ latch Hakenlatte *f* ~-like hakenartig ~-link chain Haken-kette *f*, -glied *n*, -verbindung *f* ~ lock Hakenschloß *n* ~ nail Hakennagel *m* ~ neck Hakenhals *m* ~ opening Hakenmaul *m* ~-plate track Hakenplattenoberbau *m* ~ screw Schraubhaken *m* ~-shaped hakenförmig ~-shaped bracket Hakenstütze *f* ~ sluice Hakenschütze *f* ~ spanner Hakenschlüssel *m* ~ spring ring Hakensprengring *m* ~ stick Hakenstange *f* ~ switch Gabelumschalter *m* ~ thread guide Hakenfadenführer *m* ~ type bottom block Hakenflasche *f* ~-up Leitungsschema *n*, Schaltschema *n*, Schaltung *f*

Hooke's, ~ joint Gelenkkupplung *f*, Kardan-Kreuz-gelenk *n* ~ law Elastizitätsgesetz *n*, Hookesches Gesetz *n*, Proportionalitätsgesetz *n*, Proportionalitätsgrenze *f*

hooked gebogen, hakenförmig ~ cable lug Winkelkabelschuh *m* ~ fish joint (or fishplate) Hakenlasche *f* ~ forceps Hakenpinzette *f* ~ nails Hakenstifte *pl* ~ tie plate Hakenplatte *f* ~ tie plate with tenon Hakenzapfenplatte *f*, Zapfenplatte *f* ~ wrench Hakenschlüssel *m*

hooker Paketierofenmann *m*

hooking-on-arm Greifer *m*

hooks and eyes Haken *pl* und Ösen *pl*

hoop, to ~ (cask) bereifen to ~ a pile einen Pfahl *m* beringen oder rinken

hoop (for cask) Band *n*, Bügel *m*, Busche *f*, Daube *f*, (cask) Gebinde *n*, Hirnring *m*, Mantel-

band *n*, Reif *m*, Reifen *m*, (for hood or tilt) Spriegel *m*, Zwinge *f* ~ of a gun Frette *f*

hoop, ~ actuated by compressed air for overhead contact Druckluftbügelbetätigung *f* für Stromabnehmer ~ cramp Schraubwinde *f* ~-driving machine Antreibmaschine *f* für Faßreifen, Faßreifenauftreibmaschine *f* ~-drop relay Fallbügelrelais *n* ~ iron Bandeisen *n* ~-iron binder Bandeisenbügel *m* ~-iron reel Bandeisenhaspel *f* ~-iron sheathing Bandeisenbewehrung *f* ~ mill Bandstahlwalzwerk *n* ~ milling Bandwalzerei *f* ~ net Senke *f* ~ steel Reifenstahl ~ tension Ringspannung *f*

hooped, ~ bed cradle Reifenbahre *f* ~ concrete core (of a column) umgeschnürter Betonkern *m* ~-up engine beschleunigter Motor *m*

Hooper wire Hooperscher Draht *m*

hooping Faßbinden *n*, Frettage *f*, Umschnürung *f*, Umwicklung *f*

hooter Hupe *f*

hooting signal Hupensignal *n*

hop, to ~ hüpfen to ~ off starten

hop Hops *m*, Sprung *m*, Tappe *f* (aviat.) ~ back Ausschlagbottich *m* ~-on resistance Anpringwert *m* ~-plucking apparatus Zerblätterungsapparat *m*

hopeite Zinkphyllit *m*

hopper Behälter *m*, Einfüllrumpf *m*, Einfülltrichter *m*, Einwurf *m*, Füll-gefäß *n*, -kasten *m*, -rumpf *m*, -schacht *m*, -stutzen *m*, Laderaum *m*, Speisetrichter *m*, Springer *m*, (for crop dusting) Stäuber *m*, Trichter *m*, trichterförmiger Behälter *m*, Trog *m* (Erdöl), Vorratstrichter *m* ~ with bell Glockenrumpf *m*

hopper, ~ apron Bodentisch *m* ~ barge Klappschute *f* ~ beam Rumpf-baum *m*, -leiter *f* ~ bottom Bodentrichter *m* ~ car Füll-, Trichterwagen *m* ~-charging bucket Trichterfüllgefäß *n* ~-cock Wasserschlußventil *n* ~ conveyor Förderrinne *f* ~ cooler Kühlsieb *n* ~ dredger Prahmbagger *m* ~ drier Trichtertrockner *m* ~ feed Kastenspeiser *m* ~ feeder Kastenöffner *m* ~ mill Trichtermühle *f* ~ mixer with staggered baffles Freifallmischer *m* ~ outlet Trichterauslauf *m* ~ tank Füll-rumpf *m*, -tank *m* ~ trailer Anhängewagen *m* mit Kippkasten ~ truck Füll-, Mulden-wagen *m* ~-type car Kübelwagen *m* ~-type truck Förderwagen *m* ~ wagon Trichterwagen *m*

horizon Bezugshorizont *m*, Gesichtskreis *m*; Horizont *m*, (geol. u. stratigraphical) Zone *f* ~ bar Horizontbalken *m* ~ camera Horizont-(zusatz)kamera *f* ~ glass Kimmspiegel *m* ~ gyro Horizontkreisel *m* ~-inclination meter Kimmtiefenmesser *m* ~ index Horizontzeiger *m* ~ lamp Horizontleuchte *f* ~ light Horizontfeuer *n* ~ line Horizontlinie *f* ~ pointer Horizontzeiger *m* ~ record Horizontbild *n* ~ repeater Horizonttochter *f* ~ trace Bildhorizont *m*, Haupthorizont *m* (photo)

horizontal Horizontale *f*, Waag(e)rechte *f*, Zielwaagerechte *f*; horizontal, liegend, söhlig, waag(e)recht ~ and vertical planes Horizontal- und Vertikalebenen *pl* (navig.lights)

horizontal, ~ abrasive machine Horizontalschleiftisch *m* ~ adjustment Waagerechteinstellung *f* ~ angle Horizontalwinkel *m* ~ angular acceleration Seitenwinkelbeschleunigung *f* ~

**angular velocity** Seitenwinkelgeschwindigkeit *f* ~ **anvil** Längsamboß *m* ~ **arm** Querarm *m* ~ **axis** Horizontal-, Kipp-achse *f* ~ **azimuth circle** Horizontalteilkreis *m* ~ **ball mill** Roulette *n* ~ **bar** Reck *n*, Schwebebalken *m* ~**-bar oscillator** Horizontalbalkengenerator *m* (TV) ~ **base** Langbasis *f* ~**-belt conveyer** Horizontalförderband *n* ~ **blanking** Teilbildaustastung *f* ~ **blanking impulse** Zeilenaustastimpuls *m* ~ **boring machine** Horizontalbohrmaschine *f* ~ **boring mill** Waagerechtenbohrwerk *n* ~ **broaching machine** Horizontalräumungsmaschine *f* ~ **capsule (or carrier) ejection** waagerechte Ausschleusung *f* der Rohrpostbüchsen ~ **centering** Horizontalregelung *f* ~ **centering control** Horizontalzentrierregler *m* ~ **chamber** liegende Kammer *f* ~ **circle** Horizontalkreis *m*, Limbus *m* ~ **clearance** Durchfahrtbreite *f* unter einer Brücke ~ **collimation** Seitenkollimation *f* ~ **component** Horizontalkomponente *f*, waagerechte Teilkraft *f* ~ **component of target travel during time of flight** horizontale Auswanderungsstrecke *f* ~ **control** Bildeinstellung *f*, Lagemessung *f* (Kartografie) ~ **converter** liegender Konverter *m* ~ **conveyance** Horizontaltransport *m* ~ **crystallizer** liegender Kristallisator *m* ~ **cycle** Horizontalwechsel *m* ~ **deal frame** Horizontalgatter *n* ~ **deep hole boring lathe** Tieflochbohrbank *f* ~ **deflecting electrodes** Horizontalablenkelektroden *pl* ~ **deflection** Horizontalablenkung *f*, horizontale Ablenkung *f* (sweep) Zeilenablenkung *f* ~ **deflection amplifier** Horizontalablenkverstärker *m* ~ **drilling machine (or drill press)** Waagerechtbohrmaschine *f* ~ **edge** Blendenscheibe *f* (film) ~ **elevators** Höhenleitwerk *n* (aviat.) ~ **engine** liegender Motor *m* ~ **face grinding machine** Waagerechtplanschleifmaschine *f* ~ **feed mechanism** Quergurtzugführung *f* ~ **fiber** Breitkreisfaser *f* ~ **fin** Stabilisierungsflosse *f* ~ **fine boring machine** Waagerechtfeinbohrwerk *n* ~ **flight** Geradeaus-, Horizontal-flug *m* ~ **-flight behavior** Horizontalflugverhalten *n* ~ **flight indicator** Fliegerhorizont *m* ~ **flight position** Waagerechtfluglage *f* ~**-flue coke oven** horizontaler Koksofen *m* ~ **flyback** horizontaler Rücklauf *m* (TV) ~ **flying** Geradeausflug *m* ~ **force** Horizontalintensität *f*, Waagerechtkraft *f* ~ **grate** Planrost *m* ~ **gyro** Hgrizontalkreisel *m* ~ **hinge** Horizontalgelenk *n* ~ **hot-air tentering machine** Etagenspannrahmen *m* ~ **induction** Waagerechtinduktion *f* ~ **intensity** Horizontalintensität *f* ~ **light** Unterlicht *n* ~ **line** Horizontale *f* ~ **machine** Bockfräse *f* ~ **magazine** Horizontalmagazin *n* ~ **magnet** Horizontal-, Waagerecht-magnet *m* ~ **mark** Horizontalstrich *n* ~ **measure of a slope** Fuß *m* einer Böschung ~ **member** Riegel *m* ~ **milling machine** Horizontal-, Plan-, Waagerecht-fräsmaschine *f* ~ **oscillation** Waagerechtschwingung *f* ~ **parallax** Horizontalparallaxe *f* ~ **parallax slide** Horizontalparallaxenschlitten *m* ~ **pattern** Horizontaldiagramm *n* ~ **picture** Queraufnahme *f* ~ **plane** Horizontal-, Niveauebene *f* ~ **plane antenna** horizontale Flächenantenne *f* ~ **plane directional pattern** horizontales Strahlungsdiagramm *n* ~ **plane milling machine** Langfräsmaschine *f* ~ **plane through**

**axis of bore at zero-degree elevation** Mündungswaagerechte *f* ~ **plane through target** Höhenebene *f* ~ **planing machine** Waagerechthobler *m* ~ **position** Waagerechteinstellung *f* ~ **power** Zeilenfrequenz *f* ~ **projection** Grundriß *m*, Horizontalprojektion *f* ~ **projection of the base** Grundlinie *f* ~ **rack** Horizontalträger *m* ~ **radiation pattern** horizontales Strahlungsdiagramm *n* ~ **range** Kartenentfernung *f*, waagerechte Schußweite *f* ~ **range to target** Treffkartenentfernung *f* ~ **range to target at firing point** Abschußkartenentfernung *f* ~ **range of water jet** Sprungweite *f* des freien Wasserstrahls ~ **recording** Seitenschrift *f* ~ **resolution** Horizontalauflösung *f* (rdr.) ~ **retort** horizontale Retorte *f* ~ **row** waagerechte Reihe *f* ~ **rudder** Höhensteuer *n* ~ **saw-mill** Horizontalgatter *n* ~ **saw-tooth** horizontaler Sägezahn *m* ~ **scan** Horizontalablenkung *f*, horizontale Ablenkung *f* ~ **scan generator** Horizontalablenkgerät *n* (rdr.) ~ **scanning frequency** Zeilenfrequenz *f* ~ **scanning frequency** Zeilenfrequenz *f* (TV) ~ **screen** Flachsieb *n* ~ **section** Flach-, Horizontal-schnitt *m* ~ **shaft** liegende Welle *f* ~ **shear** Längsschubkraft *f* ~ **shot** Bodentreffbild *n* ~ **size control** Zeilenbreitenregler *m* (TV) ~ **sluice** horizontale Schütze *f* ~ **speed** Horizontalgeschwindigkeit *f* ~ **spindle arm** (in molding) Schablonenträger *m* ~ **spiral** gerissene oder ungesteuerte Rolle *f* (aviat.) ~ **spotting** Beobachtung *f* der Längenabweichung *f* ~ **stabilizer** Höhenflosse *f* ~ **stabilizer winds** horizontale Verdrängungswinde *pl* ~ **stabilizers** Höhenleitwerk *n* ~ **steam engine** liegende Dampfmaschine *f* ~ **steering lever** Schiefensteuerung *f* ~ **stress** Bogen-, Horizontal-schub *m*, waagerechter Seitenschub *m* ~**-stripe pattern** Ringelmuster *n* ~ **surface** Horizontalfläche *f* (airport) ~ **surface grinding machine** Waagerechtflächenschleifmaschine *f* ~ **sweep** Horizontalablenkung *f*, horizontale Ablenkung *f* ~ **sweep of gun** Bestreichungswinkel *m* ~ **swing** Verschwenkung *f* ~ **swivel mount index drum** Teilungsring *m* für Waagerechtschwenklager ~ **synchronizing pulse** Horizontalwechsel *m* ~ **synchronizing impulse** Zeilengleichlaufstoß *m* ~ **tail area** horizontale Schwanzfläche *f* ~**-tail-surface force** Höhenleitwerkkraft *f* ~ **thrust** Bogen-, Horizontal-schub *m*, Horizontalverschiebung *f*, waagerechter Seitenschub *f* ~ **timber logging** Holzauskleidung *f* ~ **timber sheeting** Holzauskleidung *f* ~ **track** Seitenschrift *f* ~**-type** liegende Bauart *f* ~ **vacuum pan** liegender Vakuumapparat *m* ~ **valve** horizontale Schütze *f* ~ **velocity (or speed)** Waagerechtgeschwindigkeit *f* ~ **wave** liegende Welle *f* ~ **weld** liegende Naht *f* ~ **wire** Horizontal-, Waagerecht-faden *m*

**horizontally,** ~ **opposed engine** liegender Motor *m* ~ **opposed twin cylinder** Zweizylinderboxermotor *m* ~ **striped goods** geringelte Ware *f*

**horn** Arm *m*, Elektrodenarm *m* (welding), Hebel *m*, Horn *n*, Hupe *f*, (of mine) Kappe *f*, Leitflächenhebel *m* (aerodyn.), Schalltrichter *m*, Signalhorn *n*, nasenförmiger Teil, Trichter *m*

**horn,** ~ **antenna** Hornantenne *f* ~ **bar** Querriegel *m* ~ **beam** Hainbuche *f* ~ **blende** Hornblende *f* ~ **blendic schist** Hornblendenschiefer *m* ~ **block**

Bügelgleitbacke f ~-block pedestal gerade Achslagerführung f ~ bulb Hupen-ball m, -birne f ~ button Druckknopf m zur Hupe, Signalknopf m ~ chips Hornspäne pl ~ coal Hornkohle f ~ control Hornbestätigung f ~ control ring for steering wheel Signalring m für Lenkrad ~ cut-off relay Hornabschaltrelais n ~ dust shield Hupenstaubschutz m ~ feed Hornspeisung f (rdr.) ~ fuse Hörnersicherung f ~ gap Hupenzwischenraum m ~ gate Horntrichter m ~ ladle Hornlöffel m ~ lead Hornblei n ~ like hornig ~-loaded loudspeaker Lautsprecher m mit Klangverteiler ~ meal Hornmehl n ~ mouth Trichteröffnung f ~ parings Hornabfall m ~ plate Achsenschiene f, Achsgabel f ~-plate stay Achsgabelsteg m ~ press Hornpresse f ~ -press ring Horndruckring m ~ push-button Horndruckknopf m ~ push-button box Signalkasten m ~ quicksilver Merkur-, Quecksilberhornerz n ~ radiator Hornstrahler m (rdr.) ~ -shaped lightning arrester Hörnerblitzableiter m ~-shaped pole Hörnerpol m ~ shavings Hornabfall m ~ signal Hornsignal n ~ silver Hornsilber n, Kerat n ~ slate Hornschiefer m ~ socket Hornbüchse f ~ sprue Horntrichter m ~ stone Hornstein m ~-stopping switch Hornabschalter m ~ switch Hornschalter m ~ throat Trichterhals m
horn-type Hohlraumstrahler m ~ antenna Trichterantenne f ~ balancing Seitenausgleich m ~ ~ loud-speaker Trichterlautsprecher m ~ pole relay Hörnerpolrelais n ~ switch Hornausschalter m
horn welding Hammerschweißung f
horned gekrümmt, hornförmig ~ nut Kronenmutter f
horning knife Aushornmesser n
horns Arme pl ~ of a tube Anschlußkontakt m an der Spitze des Glasballons, Elektrodenanschluß m
horny hornartig, hornig
horological school Uhrmacherschule f
horse Anlegetisch m (print.), Auflagebock m (aviat.), Bock m, Bodensau f, Giermast m (hydraul.), Pferd n, Strebe f, Stütze f ~ capstand Göpel m ~ collar Kummet n ~-drawn pferdebespannt ~-drawn lifting plow Gespannrodepflug m ~-driven rig (or gear) Göpelbetrieb m ~ hair Roßhaar m, Schweifhaar n ~ hoe Pferdehacke f ~ latitudes Hitzebreiten pl, Pferdebreiten pl, Roßbreiten pl
horsepower Pferdekraft f ~ available vorhandene Pferdekraft f ~ coeficient Flugzahl f ~-hour Pferdekraftstunde f, PS-Stunde f ~ output Pferdekraftertrag m ~ per square foot of wing area Flächenleistung f ~ rating Einschätzung f gemäß der Pferdekraft
horseshoe Hufeisen n ~ with calkings Stolleneisen n ~ bearing Segmentdrucklager n ~ curve Schleife f ~ filament Hufeisenheizfaden m ~ furnace Hufeisenofen m ~ girders hufeisenförmiger Eisenausbau m ~ magnet Hufeisenmagnet m ~ mixer Ankerrührer m ~-shape hufeisenförmig ~ vortex Hufeisenwirbel m
horseshoeing Hufbeschlag m
horse track Saumpfad m
horst Horst m
horticulture Garten-bau m, -kunst f

hose Schlange f, Schlauch m (air) ~ Luftsack m ~ on breathing apparatus Atemschlauch m ~ on gas mask Atemschlauch m ~ for hydraulic brake Bremsschlauch m
hose, ~ adaptors Schlauchverbindungen pl ~ cable Schlauchkabel n ~ clamp Schlauchbinder m, Spannband n ~ clip Schlauch-klemme f, -schelle f, -verbinder m ~ cock Schlauchhahn m ~ collar Schlauchmanschette f ~ connection Schlauchklemme f ~ connection and cap Schlauchanschluß m mit Verschlußklappe ~ connector Schlauchverbinder m ~ coupling Schlauch-anschluß m, -kupplung f, -stück n, -verschraubung f ~-draining attachment Schlauchentleerungsaufsatz m ~-draining valve Schlauchentleerungsventil n ~ drum Schlauchtrommel f ~ fabric Schlauchgewebe n ~ hook Schlauchhaken m ~ joint Schlauchverbinder m ~ leather for pumps Manschettenleder n (für Pumpen) ~ leveling instrument Schlauchwaage f ~ line Schlauchleitung f ~ liner kurzes Rohrstück n auf das der Schlauch aufgezogen wird ~ nipple Schlauchverschraubung f ~ nozzle Schlauchdüse f ~ pipe Schlauch-rohr n, -röhre f ~-pipe trailer Schlauchtender m ~-proof enclosure Schallwasserschutz m ~ reel Schlauchtrommel f ~ reel unwinding from center Schlauchtrommel f für zentrale Entnahme ~ retracting device Schlauchrückhaltsvorrichtung f ~ sandblast Freistrahlgebläse n ~ slips Schlauchbänder pl ~ spigot Schlauchzapfen m ~ strap Schlauch-band n, -riemen m ~ strap fastener Schlauchbandschloß n ~ tip Schlauchmundstück n ~-type sandblast gun Freistrahlrohr n ~-type-sandblast tank machine Freistrahlgebläse n ~ valve Schlauch-ventil n, -verbinder m ~ woven around with wire Panzerschlauch m
hosiery Kulier-, Strumpf-ware f ~ finishing Strickwarenappretur f ~ fulling Wirkwarenwalke f ~ machine Wirkmaschine f ~ manufacture Wirkwarenfertigung f ~ pressing Wirkwarenpresserei f
hospital Lazarett n, Spital n ~ aircraft Sanitätsflugzeug n ~ ship Lazarettschiff n ~ train Kranken-, Lazarett-zug m
host Menge f ~ crystal Grundmaterial n
hot, to ~ cadmium-plate feuerverkadmen to ~ draw warmziehen to ~ press (cloth) dekatieren, warmpressen to ~ roll warmwalzen to ~ saw warmsägen to ~ solder heißlöten to ~ strain warmrecken to ~ work warm bearbeiten, warmrecken
hot brüchig, erhitzt, heiß, warm ~ air Heißluft f
hot-air heiße Gebläseluft f, Heißluft f ~ apparatus Heißluftapparat m ~ balloon Heißluftballon m ~ blast Heißwind m ~ blower Föhn m, Heißluftbläser m ~ chamber Heißluftkammer f ~ conduction Heißluftzuführung f ~ conduit Warmluftkanal m ~ de-icer Heißluftenteiser m ~ delivery Heißluftzuführung f ~ drying loft Heißluftmansarde f ~ drying machine Heißlufttrockenmaschine f ~ duct Warmluftschlauch m ~ engine kalorische Maschine f ~ generator Luftwärmespeicher m ~ heating Warmluftheizung f ~ intake Warmlufteintritt m ~ slot Heißluftschlitz m ~ sprayer Heißluftsprüher m ~ stove Cowper m, Winderhitzer m ~ stream Heißluftstrom m ~ suppliers Heißluftduschen

pl ~ **supply pipe** Heißluft-förderleitung f,
-leitung f, -speiseleitung f ~ **tentering machine**
Trockenspannrahmen m
**hot,** ~ **ashes** Grude f ~ **atom** hochangeregtes
Atom n ~ **avalanche** Glutwolken pl ~**-band
ammeter** Hitzbandstrommesser m ~ **bearing**
warmgelaufenes Lager n ~ **bed** Warm-bett n,
-lager n ~ **bend test** Warmbiegeversuch m ~
**-bending test** Rotbruchversuch m, Rotwarm-
biegeprobe f ~ **bending test** Warmbiegeprobe
f ~ **blast** Föhn m, Heißluftgebläse n, Heißwind
m, Warm-blasen m, -gebläse n
**hot-blast,** ~ **cast iron** unter Warmluft erzeugtes
Eisen n ~ **furnace** Heißwindofen m ~ **gas**
Warmblasegas n ~ **main** Heißwindleitung f ~
**outlet** Heißwindaustritt m ~ **period** Warmblase-
periode f ~ **pig iron** heiß erblasenes Gußeisen n
oder Roheisen ~ **side valve** Heißwindschieber
m ~ **stove** Winderhitzer m, Winderhitzungsapparat m ~**-stove casing** Winderhitzermantel
m ~**-stove checkerwork** Winderhitzfachwerk n
~ **valve** Heißwindventil n ~ **valve joint** Heiß-
windverschluß m
**hot,** ~ **blow** Warmblasen n ~ **box process** Heiß-
kastenverfahren n ~**-braking test** Rotbruchversuch m ~ **bulb** Glüh-kerze f, -kopf m ~**-bulb
and starter switch** Glüh- und Anlaßschalter m
~**-bulb engine** Glühkopf-maschine f, -motor m
~**-bulb ignition** Glühkopfzündung f ~**-bulb
monitor** (light) Glühkerzenüberwacher m ~
**burnishing press** heiße Satiniermaschine f ~
**cathode** Glühkathode f ~ **cathode discharge**
Glühkathodenentladung f ~ **cathode-gas-filled
rectifier tube** Glühkathodengleichrichterröhre
f mit Gasfüllung ~ **cathode mercury-vapor
rectifier** Glühkathodenquecksilberdampf-
wellen) Quecksilberdampfglüh-gleichrichter m
~ **cathode rectifier** Glühkathoden-, Quecksil-
berdampf-gleichrichter m ~ **cathode X-ray tube**
Heizkathodenröntgenröhre f, Röntgenröhre f
mit Heizkathode ~ **charge** flüssiger Einsatz m
(Ofen) ~ **chisel** Warmschrottmeißel m ~
**clearance** Warmspiel n ~ **conductor** (with negative temperature coefficient) Heißleiter m ~
**-cooled** heißgekühlt ~**-cooled engine** heißgekühl-
ter Motor m ~ **crack** Warmriß m ~ **crimping**
Warmstauchung f ~ **cure (or hot vulcanization)**
Warmvulkanisation f ~ **die** Warmmatrize f
~**-dip calorizing** Tauchalitieren n ~**-dip-silver-
plating** Feuerversilberung f ~**-dip tinning**
Feuerverzinnung f ~ **dipped galvanized** feuer-
verzinkt ~ **dissolving process** Heißlöseverfahren
n ~ **drain** aktiver Kanal m ~ **drawing** Warmziehen n ~ **drawn** warmgezogen ~ **drop hammer** Warmhammer m ~ **drop saw** Pendel-
warmsäge f ~**-ductile** wärmedehnbar ~ **ductility** Warmdehnbarkeit f ~ **electrode** Glühkathode f ~ **end** heiße Lötstelle f, (in pyrometry)
Warmlötstelle f ~**-etching test** Heißsetzversuch
m ~ **fire-clay sleeve** Wärmhaube f ~ **foil
stamping press** Folien-Wärmprägepresse f ~
**forging** Warmschmieden n ~**-forging dies**
Warmarbeitsgesenk n, Warmpreßstempel m
~**-formed** warmverformt ~ **forming** Warm-
formgebung f, Warmverformung f ~**-forming
property** Warmverformbarkeit f ~ **galvanized**
verzinkt ~ **galvanizing** Feuer-verzinken n,
-verzinkung f ~ **galvanization** Feuerverzinkung

f ~ **gas** Heißgas n ~**-gas valve** Heißgasschieber
m ~ **hardness** Warmhärte f ~ **heading** Warm-
tauchen n ~ **house** Gewächs-, Treib-haus n ~
**ingot shear** Heißeisenschere f ~ **inspection**
Warmprüfung f ~ **iron** heißes Eisen n ~**-iron
lever-type saw** Heißeisenhebelsäge f ~**-iron saw**
Heißeisensäge f ~ **junction** heiße Lötstelle f,
(of a thermocouple) Heizlötstelle f ~ **layer**
Heizschicht f ~**-light** Hauptbeleuchtung f (TV)
~ **line (or loop)** Direktleitung f (teleph.) ~
**-loop telephone** Ringleitung f ~ **mandrel** Warm-
dorn m ~ **melt coating** Aufschmelzüberzug m
~ **metal** heißer oder warmer Einsatz m
**hot-metal,** ~ **charging crane** Roheiseneinsetzkran
m ~ **ladle** Roheisenpfanne f ~**-ladle and car**
Roheisenpfannenwagen m ~ **ladle crane** Gießkran m ~**-ladle truck** Gießwagen m ~ **mixer**
Roheisenmischer m ~ **receiver** Roheisenmischer m
**hot,** ~ **milling and sawing machine** Warmfräs-
und Sägemaschine f ~ **milling machine** Warm-
fräse f ~ **neck grease** Heißzapfenfett n ~ **nozzle**
Wärmhaube f ~ **nut press** Warmmutternpresse
f ~ **penetration test** Wärmedruckprüfung f ~
**permeameter** Permeameter m für Messungen
bei erhöhter Temperatur ~**-piercing die** Warmziehring m ~ **plasma** energiereiches Plasma n
~ **plate** Kocher m ~**-plate straightening machine**
Warmblechrichtmaschine f ~ **plug** heißer Stöpsel m, heiße Zündkerze f ~**-press-fitted** warmgepreßt ~**-press method** Warmpreßverfahren n
~**-press punch** Warmpreßstempel m ~**-pressed**
warmgepreßt ~ **pressing** Warmpressen n ~
**-pressing dies** Warmpreßformen pl, Warmpreßstempel m ~**-pressing tool** Warmpreßstahl m ~
**pull** (of a cast block) Warmbruch m ~ **punch
die** Warmschlagmatrize f ~ **pushing** Warm-
stoßen n ~ **quenching** Stufenhärtung f (Thermalhärtung) ~**-quenching furnace** Warmbadhärte-
ofen m ~ **quenching method** Thermalhärtung f
~ **riveting** Warm-nietung f, -vernietung f ~**-roll
neck grease** Fett n für Heißwalzen ~**-rolled**
warmgewalzt, warm satiniert ~ **rolling** Warmwalzen n ~ **rolling mill** Warmwalzwerk n ~ **run**
Heißlauf m ~ **running** hitziger Gang m (metall.),
Warmlaufen n ~ **sealing** Heißsiegelverfahren n
~ **section** heißer Triebwerkteil m ~ **setting**
Warmabbinden n ~**-setting adhesive** Heiß-
kleber m ~ **shaping** Wärmeformgebung f ~
**shearing** Warmscheren n ~**-short** heißbrüchig,
warmbrüchig, warmrissig ~**-shortness** geringe
Warmfestigkeit f, Rotbruch m, Warm-brüchig-
keit f, -sprödigkeit f ~ **spa** Thermalquelle f ~
**spot** Gemischvorwärmer m, (on film) Lichtfleck m, Ort m hoher Strahlungsdichte, Seiger-
stelle f, (of combustion chamber) überhitzte
Stelle f, Vorwärmeeinrichtung f ~**-spotting
arrangement** Gemischvorwärmeeinrichtung f
~ **spraying process** Heißspritzverfahren n ~
**-stage microscope** heizbares Mikroskop n ~
**stamping phosphorus nut bars** Warmpreßmut-
tereisen n ~ **start** Anlaßüberhitzung f ~ **store**
Heißlagerplatz m **in the** ~ **state** heißbrüchig
~**-straightening** (rails) Heißrichten n ~ **straightening** Warmrichten n ~ **straining** Warmrecken
n ~ **strength** Warmfestigkeit f ~ **strip mill**
Warmbandwalzwerk n ~ **stuffing process** Ein-
brennverfahren n ~ **swaging machine** Warm-

gesenkdrückmaschine *f* ~ **temperature zone** Heißtemperaturzone *f*, Vollhitzezone *f* ~ **template** Warmschablone *f* ~ **tensile test** Warmzerreiß-probe *f*, -versuch *m* ~ **thread-rolling machine** Warmgewindewalze *f* ~ **tin plating** Feuerverzinnung *f* ~ **tinning** Feuerverzinnen *n* ~ **top** Wärmehaube *f* ~ **transfer** Abbügeletikett *n* ~ **transfer film** Aufbügelfilm *m* (print.) ~ **trimming die** Warmabgratwerkzeug *n*
**hot-tube,** ~ **detonating head** Glühköpfen *n* ~ **detonator** Glühzündapparat *m* ~ **detonator fuse** Glühzündstück *n* ~ **igniter** Glührohrzündapparat *m* ~ **ignition** Glührohr-, Heizlampenzündung *f*
**hot up-setting** Warmstauchen *n*
**hot-water,** ~ **apparatus** Heißwasserapparat *m* ~ **bag** Thermophor *n* ~ **bottle** Wärmflasche *f* ~ **drain** Heißwasserablauf *m* ~ **heating** Heißwasserheizung *f* ~ **heating plant** Warmwasserheizungsanlage *f* ~ **jacket** Heizmantel *m* (durch Wasser geheizt) ~ **plant** Warmwasserbereitungsanlage *f* ~ **pump** Warmwasserpumpe *f* ~ **radiator** Warmwasserheizkörper *m* ~ **reservoir** Heißwasserspeicher *m* ~ **supply** Heißwasserentnahme *f* ~ **system** Warmwasserraumversorgung *f* ~ **tank** Heißwasserbehälter *m*, Warmwasserzisterne *f*
**hot,** ~ **wave** heiße Welle *f* ~ **well** Ausgußraum *m* des Kondensators, Fallwasserkasten *m*, Wärmespeicher *m* ~ **well water** Fallwasser *n* ~ **wire** Glühdraht *m*, spannungsführender Draht *m*
**hot-wire,** ~ **ammeter** Hitzband-, Hitzdrahtamperemeter *n* ~ **anemometer** Hitzdrahtanemometer *n*, -sonde *f* ~ **blinker unit** Hitzdrahtblinkgeber *m* ~ **bridge head** Glühköpfchen *n* ~ **gauge** Heizdrahtmanometer *n* ~ **instrument** (blasting) Hitzdrahtinstrument *n* ~ **(measuring) instrument** Hitzdrahtmeßgerät *n* ~ **meter** Hitzdrahtmeßgerät *n* ~ **microphone** Hitzdrahtmikrofon *n* ~ **voltmeter** Hitzdrahtspannungsmesser *m*, -voltmeter *n*
**hot,** ~ **work hardening** Heißverfestigung *f* ~ **workability** Warmverarbeitungsfähigkeit *f* ~ **workable** warmverformbar ~ **worked** warmverformt ~ **-worked steel** warmgeformter Stahl *m* ~ **working** heißer Gang, Warm-bearbeitung *f*, -formgebung *f*, -recken *n*, -verformung *f* ~ **working die** Warmarbeitswerkzeug *n*
**hour** Stunde *f* **by the** ~ stundenweise **per** ~ stündlich ~ **of relief** Löse-, Wechsel-stunde *f* (min.)
**hour,** ~ **angle disk** Stundenwinkelscheibe *f* (astron.) ~ **axis** Stundenachse *f* ~ **circle** Stundenkreis *m* ~ **counter** Betriebsstundenzähler *m* ~ **division** Stundeneinteilung *f* ~ **drive** Stundenantrieb *m* ~ **glass** Sanduhr *f*, Stundenglas *n* ~ **-glass effect** Sanduhreffekt *m* ~ **glass worm gear** Globoidschneckenbetrieb *m* ~ **hand** Stundenzeiger *m* ~ **mechanism** Stundenuhrwerk *n* ~ **meter** Stundenmesser *m*
**hourly** stundenweise, stündlich ~ **output** Stundenleistung *f* ~ **rate** Stundensatz *m*
**hours,** ~ **of darkness** die Stunden *pl* der Dunkelheit ~ **of daylight** Tagesstunden *pl* ~ **of duty** Wachzeiten *pl* ~ **of lighting** Brenndauer *f* ~ **of service of a central telephone office** Dienststunden *pl* einer Fernsprechanstalt ~ **of undisturbed direction finding** ungestörte Peilzeiten *pl*
**house, to** ~ enthalten, unterbringen

**house** Haus *n* ~ **cable** Innenkabel *n* für Sprechstellen ~ **carpenter** Bauzimmerer *m* ~ **carpentry** Bautischlerei *f* ~ **connecting box** Hausanschlußkasten *m* ~ **consumption** Eigenbedarf *m* ~ **exchange equipment** Hausanlage *f*
**household** Familie *f*, Haushalt *m* ~ **appliance** Hausgerät *n* ~ **implement** Wirtschaftsgerät *n*
**house,** ~ **keeper** Haushalter *m* ~ **keeping** Haushalt *m* ~ **lead-in** Hauseinführung *f* ~ **line** Hüsing *f*, dünne Linie *f* ~ **magazine** Firmenzeitschrift *f* ~ **organ** Werkzeitschrift *f* ~ **pipe drains** Hausentwässerung *f* ~ **pole** Dachgestänge *n* ~ **telephone** Haus-fernsprecher *m*, -telefon *n* ~ **-telephone call** Hausruf *m* ~ **-telephone plant** Fernsprechreihen-, Hausfernsprech-, Privatfernsprech-anlage *f* ~ **-telephone system** Reihenanlagefernsprechreihen *pl* **(pneumatic)** ~ **tube(s)** Hausrohrpost *f* ~ **water pipe (or supply)** Hauswasserleitung *f* ~ **wiring** Hausleitung *f*
**housed** eingeschlossen ~ **wicket** niedergelegte Schütztafel *f*
**housing** Einschließung *f*, Gehäuse *n*, Gelenkrohr *n*, Gerüst *n*, Lagerkörper *m*, Rahmen *m*, Ständer *m*, Unterbringung *f*, Wohnungsbeschaffung *f* ~ **for combined switch arrangement** Schalterkombinationsgehäuse *n* ~ **of the gyro(scope)** Kreiselgehäuse *n* ~ **of injection timing device** Spritzverstellergehäuse *n* ~ **over point of discharge from hopper to sluiceway** Fangtrichter *m* (Entaschung) ~ **for two rolls** Zweiwalzengerüst *n*
**housing,** ~ **back plate** Bremsschild *n* (Scheibenbremse) ~ **base** Gehäusefuß *m* ~ **bearer** Gerüstständer *m* ~ **cap** Ständerkopf *m* ~ **case** Hülse *f* ~ **collar** Gehäusehals *m* ~ **column** Ständerpfosten *m* ~ **core** von Gießmetall umschlossener Kern *m* ~ **cover** Gehäusedeckel *m* (Kupplungsseite) ~ **crest** Gehäusescheitel *m* ~ **floor plate** Kastenboden *m* ~ **frames** Ständer-, Walzen-gerüst *n* ~ **half-member** Gehäusehälfte *f* (Kupplungsseite) ~ **pin** Druckspindel *f* ~ **post** Ständerholm *m* ~ **screw** Anstell-schraube *f*, -spindel *f*, -vorrichtung *f*, Druckspindel *f* ~ **shortage** Wohnungsnot *f* ~ **slide** Ständerführungsfläche *f* ~ **upright** Ständerpfosten *m* ~ **wall** Kastenwand *f* ~ **window** Ständerfenster *n*
**hove to** beigedreht
**hovel kiln** Töpferofen *m*
**hover, to** ~ kreisen, schweben
**hover,** ~ **ceiling** Gipfelhöhe *f* (helicepter) ~ **flight** Schwebeflug *m* ~ **height** Schwebehöhe *f* ~ **plane** Hubschrauber *m*
**hovering** Schweben *n*, Schwebeflug *m*; schwebend ~ **airplane** Flugzeug *n* im Schwebeflug ~ **propulsion** Standschub *m*
**howel, to** ~ (a cask) abputzen
**howitzer** Haubitze *f*
**howl, to** ~ heulen, pfeifen
**howler** Heuler *m* (electron.), starker Summer *m* ~ **connection** Heuleranruf *m* (teleph.)
**howling** Heulen *n*, Pfeifen *n*, Pfeifen *n* der Verstärker ~ **buoy** Heul-boje *f*, -tonne *f* ~ **condensor** Heulkondensator *m* ~ **sound** Heulton *m*
**hp (horse-power)** PS (Pferdestärke)
**H particle** H-Teilchen *n*
**H plane** Halbebene *f*

**H-pole** Doppelgestänge *n*, (two poles connected) H-Mast *m* ~ **line** Doppelstangenlinie *f*
**H ray** (consisting of H particles, positive hydrogen particles or protons) H-Strahl *m*
**H scope** H-Schirm *m*
**H section** H-Profil *n* ~ **armature** Doppelanker *m* ~ **beam** H-Träger *m* ~ **iron** Doppel-T-Eisen *n*
**H-shaped** H-förmig ~ **engine** H-Motor *m*
**HT (high tension)** Anodenspannung *f* ~ **circuit** Hochspannungsstromkreis *m* ~ **ignition** Hochspannungszündung *f* ~ **machine** Hochspannungsmaschine *f* ~ **magneto** Hochspannungszündmagnet *m*, -magnetapparat *m*
**H-type,** ~ **attenuator network** Entzerrungskette *f* mit Brücken-H-Schaltung *f* ~ **network attenuator** Entzerrungskette *f* mit Brücken-H-Schaltung
**hub** Knotenpunkt *m*, Lagerbüchse *f*, Narbe *f*, Spulenkern *m*, Winkelkern *m* (für Bandgeräte) ~ **of armature** Ankernabe *f* ~ **for buffing cylinder** Schleifwalzengehäuse *n* ~ **for cam piece** Körper *m* zur Mantelkurve ~ **with conical bore** Konusnabe *f* ~ **of drum** Trommelnabe *f* ~ **with three-speed gear** Dreifachübersetzungsnabe *f*
**hub,** ~ **attaching part** Nabenanbauteil *m* ~ **-attachment nut** Kurbelwellenmutter *f* (zur Befestigung der Luftschraubennabe) ~ **barrel** Nabenschale *f* ~ **bearing** Radnaben-, Rotorkopf-lager *n* ~ **body** Nabenmittelstück *n* ~**-bolt bore** Nabenholzbohrung *f* ~ **brake** Nabenbremse *f* ~ **cap** Rad-kappe *f*, -kapsel *f*, Spannkegel *m* ~ **casing** Nabengehäuse *n* ~ **core** Nabenkern *m* ~ **cover** Nabentopf *m* ~ **diameter** Nabendurchmesser *m* ~ **drawer** Nabenabzieher *m* ~ **dynamometer** Kraftmesser *m* in der Luftschraubennabe ~ **extractor** Nabenauszieher *m* ~ **face** Nabenfläche *f* ~ **fittings** Spulenkernbefestigung *f* ~ **flange** Nabenflansch *m* ~ **gudgeon** Hauptlagerzapfen *m* ~ **mortising machine hand** Nabenbohrer *m* ~ **motor** Achsenmotor *m* ~ **nut** Nabenwellenmutter *f* ~ **plate** Stirnblatt *n* ~ **ratio** Nabenverhältnis *n* ~ **ring** Deckring *m* ~ **spider** (helicopter) Nabenstern *m* ~ **vortex (or eddy)** Nabenwirbel *m* ~ **-widening test** Nabenaufweiteprobe *f* ~ **yoke** Rotorkopfjoch *n*
**hubbing die** Pfaffe *m*
**hubbub** Geräusch *n*, Getöse *n*, Lärm *m*
**hübnerite** Manganwolframat *n*
**huckaback weave** Huckbindung *f*
**hue** Farbe *f*, Farbton *m*, Färbung *f* ~ **color** bunte Farbe *f* ~ **control** Farbwertregler *m* (TV)
**hueless** farblos, unbunt ~ **color** unbunte Farbe *f*
**hues** Pigmentfarben *pl*
**hug, to** ~ **the land** unter Land halten
**hug** enge Anschmiegung *f*
**huge** haushoch, ungeheuer
**Hughes,** ~ **silencer** Klopfer mit trägem Rade ~ **unison lever** Einstellhebel *m*
**hulking hammer** Abbauhammer *m*
**hull, to** ~ aushülsen, enthülsen
**hull** Boot *n* (aviat.), Boots-körper *m*, -rumpf *m*, Hülle *f*, Hülse *f*, Kasko *n*, Körper *m*, Rumpf *m*, Schiffskörper *m* ~ **of center section** Mittelteilzelle *f* ~ **of flying boat (or seaplane)** Flugboot-körper *m*, -rumpf *m* ~ **of a vessel** Unterschiff *n*

**hull,** ~ **fiber of cotton** Samenhaar *n* der Baumwolle ~ **floor** Wannenboden *m* ~ **plating** Außenhaut *f* ~ **resistance** Körperwiderstand *m* ~ **shape** Rumpfform *f* ~ **step** Bootrumpfstufe *f* ~ **valve of submarines** Bordventil *n*
**hulled grain** Graupe *f*
**hulling,** ~ **mill** Enthülsungsmaschine *f* ~ **mills** Graupengang *m*
**hum, to** ~ brummen, summen, tönen
**hum** Brumm-spur *f*, -streifen *m*, -ton *m*, (of alternators) Gebrumme *n*, Gesumme *n*, Netzbrummen *n* ~ **(ming)** Brummen *n*, Summen *n*, Tönen *n* ~ **of motors (or engines)** Motorengebrumm *n* ~ **and noise** Fremdspannung *f* ~ **of selectors** Wählergeräusche *pl*
**hum,** ~ **balancing potentiometer** Entbrummer *m*, Entbrummpotentiometer *n* ~ **band** Netzbrummstreifen *n* ~ **bucking coil** Brummkompensationsspule *f* (electr.) ~ **-dinger** Entbrummer *m* ~ **effect** Brummeffekt *m* ~ **eliminator** Entbrummer *m*, Netz-, Störungs-filter *n* ~ **-eliminator choke** Tonbeseitigungsdrossel *f* ~ **-eliminator coil** Netzdrossel *f* ~ **frequency** Brummfrequenz *f* ~ **level** Brummpegel *m* ~ **potential (or voltage)** Brummspannung *f* ~ **ripple** Brummwelligkeit *f*
**humboldtine** Eisenresinit *m*, Humboldtin *m*, Oxalit *m*
**humectant** Anfeuchter *m*
**humic** humin ~ **acid** Humin-, Humus-säure *f* ~ **carbon** Humuskohle *f*
**humid** dampfhaltig, feucht ~ **air** feuchte Luft *f* ~ **room test** Feuchtraumprüfung *f*
**humidification** Befeuchtung *f*, Feuchtegraderhöhung *f*
**humidifier** Luftbefeuchtungsanlage *f* ~ **for air conditioner** Berieselungsverflüssiger *m*
**humidifying** Befeuchtung *f*, Feuchtegraderhöhung *f*
**humidity** Feuchtigkeit *f*, Wassergehalt *m* ~ **of the air** Luftfeuchtigkeit *f*
**humidity,** ~ **cell** Feuchtigkeitsgeber *m* ~ **compensation** Feuchtigkeitsausgleich *m* ~ **content** Feuchtdruck *m*, Feuchtigkeitsgehalt *m* ~ **derivation** Feuchtigkeitsnebenschluß *m* ~ **effects** Feuchtigkeitseinflüsse *pl* ~ **indicator** Feuchtigkeitsanzeiger *m* ~ **protection** Feuchtigkeitsschutz *m* ~ **recorder** Feuchteschreiber *m*
**humming** Brummen, Summen *n*, Surren *n*, Tönen *n* ~ **of gears** Heulen *n* der Zahnräder ~ **of wires** Tönen *n* der Drähte, Tönen *n* von Freileitungen
**humming,** ~ **sound** Summerzeichen *n* ~ **tone** Summer-ton *m*, -zeichen *n* ~ **top** Brummkreisel *m*
**hump** Ablaufberg *m*, Anschwellen *n*, Buckel *m*, (of resonance curve) Höcker *m*, Überhöhung *f*, Unebenheit *f* ~ **of a curve** Buckel *m* einer Kurve ~ **of a wave** Wellengipfel *m*
**hump,** ~ **back** Buckel *m* ~ **resistance** (in a seaplane) Wellenwiderstand *m* ~ **speed** Abstufgeschwindigkeit (aviat.), kritische Geschwindigkeit *f* ~ **yard** Ablaufanlage *f* (r.r.)
**humpiness** Welligkeit *f*
**humus** Humus *m* ~ **colloid** Humuskolloid *n*
**hundred** hundert ~**-call second** Zweiminutenverbindung *f* ~**-degree point** Hundertpunkt *m* ~ **weight** Zentner *m*

**hundreds, ~ digit** Hunderterstufe *f* **~ place** Hundertstelle *f*

**hung, ~ on gimbals** kardanisch aufgehängt **~ striker** (on the fuse) Schlagbolzenversager *m*

**hunt, to ~** abtasten, (for) frei suchen, hetzen, jagen, nachlaufen, pendeln, überschwingen (Regeltechnik) **to ~ over a bank** freiwählen **to ~ over a complete level** durchdrehen

**hunting** Bildverschiebung *f* (film), (in airship) Dauerschwingung *f*, freie Wahl *f* (Vermittlung), Instabilität *f* (mech.), Pendelschwingung *f*, Pendelung *f*, (in airship) Phugoidbewegung *f*, Regelschwankung *f* (electr.), Regelschaltung *f* (electr.), Suchen *n*, Tanzen *n* (Regler) **~ of image** Bildverschiebung *f*

**hunting, ~ contact** Mehrfach-, Sammel-kontakt *m* **~ device** Abtastvorrichtung *f* **~ movement** Pendelbewegung *f* **~ operation** freier Wahlvorgang *m* **~ probe** auf- und niedergehende Sonde *f* **~ step** Suchschritt *m* (teleph.) **~ switch** Mischwähler *m* für Sammelanschlüsse, Nachwähler *m*, Wähler *m* mit freier Wahl **~ valve** Nockelschieber *m*

**hurdle** Horde *f*, Hürde *f* **~ bar** Hindernis *n* **~ filter** Hordenfilter *m* **~-type scrubber (or washer)** Hordenwascher *m*

**hurdles** Lumpensortiertisch *m* (paper mfg.)

**hurl, to ~** entsenden, schleudern

**hurricane** Orkan *m*, Wirbelsturm *m*, Zyklon *m* **~ lamp** Sturmlaterne *f* **~ zone** Orkanzone *f*

**hurry, to ~** sich beeilen

**hurry** Eile *f*, Hast *f* **~ print** Schnellkopie *f* **~-up stick** Beschleunigungsstange *f*

**hurt, to ~** schaden, verletzen

**hushed, in the ~ condition** mit unterdrücktem Nullpunkt *m* (meas.)

**husk, to ~** auspellen, ausschroten

**husk** Hülse *f*, Schale *f*

**husky** dickschalig

**hut** Baracke *f* **~ at pit's mouth** Kaue *f* **~ doctor** Gegenschaber *m* (textiles)

**hutch** Schachtfördergefäß *n*, Setzfaß *n*, Setzkasten *m*

**hutching** Siebwäsche *f* (min.)

**Huygens, ~ walvelets** Elementarwelle *f* **~ zone** Ringfigur *f*

**H.V. position** Wannenlage *f*

**hyalite** Gummistein *m*

**hyalography** Glasätzung *f*

**hyalosiderite** Eisenchrysolith *m*

**hyalurgy** Glaschemie *f*

**hybrid** Gabel(schaltung) *f*, Hybridrohöl *n* (gemischtes Rohöl **~ and terminating networks** Gabel- und Endschaltungen *pl*

**hybrid, ~ balance** Differenzfaktor *m*, Gabelabgleich *m* (teleph.), -symmetrie *f* **~ binding** Hybridbindungen *pl* **~ coil** Ausgleichs-transformator *m*, -übertrager *m*, Differentialübertrager *m* (dreispuliger), Zwitterspule *f* **~ coils** Drahtgabeln *pl*, (Vier)-Einheitsgabeln **~ computer** Hybridrechner *m* **~ coupler** Hybrid-Richtkoppler *m* **~ fractures of steel** Brucharten *pl* des Stahls **~ ion** Zwitterion *n* **~ junction** Hybridverbindung *f*, Ringverzweigung *f* **~ matrix** hybridische Matrize *f* **~ repeater** Gabelverstärker *m* **~ set** Differentialglied *n*, Gabelschaltung *f* **~ T** Brückenverzweigung *f*, Hybrid-Doppel-T *n* **~ transformer** Ausgleichs-,

Dreiwicklungs-transformator *m*, Gabelübertrager *m*

**hybridization** Hybridation *f*, Kreuzung *f*

**hydracid** Wasserstoffsäure *f*

**hydrant** Hydrant *m*, Wasserstock *m* **~ casing** Hydrantengehäuse *n* **~ dispenser** (pit fueling) Zapfkarren *m* **~ pillar** Hydrantensäule *f* **~ valve** Hydrantenventil *n*

**hydrate, to ~** hydratisieren, wässern

**hydrate** Hydrat *n* **~ envelopes** Hydrathüllen *pl* **~ water** Hydratwasser *n*

**hydrated** hydratiert, wasserhaltig **~ cement** gelöschter Zement *m* **~ iron peroxide** Eisenkalk *m* **~ lime** gelöschter Kalk *m*, Kalkhydrat *n*, Luftkalk *m*

**hydration** Hydration *f*, Verbindung *f* mit Wasser

**hydratization equilibrium** Hydratationsgleichgewicht *n*

**hydraulic** (fill dam) gespült, hydraulisch; hydraulische Kraft *f* **~ accumulator** Druckwasserakkumulator *m*, -speicher *m*, Hydraulik-, Preßwasser-speicher *m* **~ actuator** hydraulischer Versteller *m* **~ architecture** Wasserbaukunst *f* **~ ash handling plant** Wasserspülaschungsanlage *f* **~ axle** Gefällinie *f* **~ balance line** Überstromleitung *f* **~ bedplates** hydraulisch zu hebendes Grundwerk *m* **~ bevel gear planer** hydraulischer Kegelradhobler *m* **~ boost control** hydraulische Kraftsteuerung *f* **~ boring** Spülbohrung *f* **~ brake** Drucköl-, Flüssigkeits-, Wasserdruck-bremse *f* **~ brake hose** Öldruckbremsschlauch *m* **~ brake test bench (or torque stand)** Wasserprüfstand *m* **~ cement** Wasserkitt *m* **~ central station** Wasserelektrizitätswerk *n* **~ circulating system** Spülbohren *n* **~ clutch** hydraulische Kupplung *f* **~ control unit** Hydraulikaggregat *n* **~ core extractor** hydraulische Kernpreßvorrichtung *f* **~ coupling** Strömungskupplung *f* **~ crane** Druck-, Preß-wasserkran *m* **~ cylinder** Druckzylinder *m* **~ cylinder frame** Zylinderrahmen *m* **~ design** strömungstechnischer Entwurf *m* **~ drive** Flüssigkeitsgetriebe *n*, hydraulischer Antrieb *m* **~ efficiency** hydraulischer Wirkungsgrad *m* **~ elevating platform** hydraulische Hebebühne *f* **~ elevator** bewegliche Schleuse *f* **~ end lift hoist** hydraulischer Einachssheber *m* **~ engine** Wassermotor *m* **~ engineering** Wasserbaukunst *f* **~ equipment** hydraulische Ausrüstung *f* **~ fill** eingewaschener Boden *m*, Spülen *n* **~ filling and bleeching stand** Hydraulikfüllwagen *m* **~-flask-lifting molding machine** Druckwasserbetrieb *m* **~ fluid** Druckflüssigkeit *f*, Preßöl *n* **~ forging die** Preßdorn *m* **~ fuse** Rohrbruchventil *n* **~-gate jack** Druckwasserzylinder *m* **~ gauge** Hydrometer *n*, Meßdose *f* **~ gearing unit** Strömungsgetriebe *n* **~ grade line** Standrohrspiegellinie *f* **~ gradient** Druck-höhengefälle *n*, -linie *f* **~ head** (of water) Gefällhöhe *f*, Wassersäulendruck *m* **~ heat exchanger** Druckkühler *m* **~ hoist** Druckwasserhebezeug *m* **~ inclined car lift** hydraulischer Waschschrägheber *m* **~ intensifier** Druckdose *f* **~ ironwork** Eisenwasserbau *m* **~ jack** Druckwasserwinde *f*, Hubwinde *f*, hydraulischer Hebebock *m* oder Heber, hydraulische Spindel *f* **~ jump** Wassersprung *m* **~ lead filter** Hydraulikleitungsfilter *m* **~ leather** Leder *n* für

Dichtungen ~ **lever** Wasserdruckheber *m* ~ **lift** hydraulischer Aufzug *m* ~ **lift unit** hydraulischer Kraftheberblock *m* ~ **lime** Wasser-, Zement-kalk *m* ~ **limestone** Zementkalkstein *m* ~ **liquid** Druckflüssigkeit *f* ~ **main** Druckwasserleitung *f*, Hydraulik *f* ~ **measuring device** Meßdose *f* ~ **mining** hydraulischer Abbau *m* ~ **motor** Flüssigkeitsmotor *m*, Wasserkraftmaschine *f*, -motor *m* ~ **oil reservoir** Druckölbehälter *m* ~ **operation** hydraulische Bewegungsvorrichtung *f* ~ **operation of gates** Torbewegung *f* mit Druckwasserantrieb ~ **packing** Spülversatz *m* ~ **packing method** Spülversatzverfahren *n* ~ **padding machine** hydraulischer Abquetschfoulard *n* ~ **pipes** Druckwasserleitung *f* ~ **piston** Druckwasserkolben *m* ~ **power** hydraulische Kraft *f* ~ **power station** Wasserkraftanlage *f* ~ **press** hydraulische Presse, Säulen-, Wasserdruck-presse *f* ~ **press-finish** Plattendekatur *f* ~ **pressure** Druckwasserantrieb *m*, Fließ-, Flüssigkeits-druck *m*, Flüssigkeitspressung *f*, Wasserdruck *m* ~ **pressure drive** Antrieb *m* durch Druckwasser ~ **pressure-gauge** Wasserdruckmanometer *n* ~ **pressure head** Wasserdruckhöhe *f* ~ **pressure return pipe** Abflußrohr *n* des verbrauchten Druckwassers
**hydraulic,** ~ **propeller** Reaktionspropeller *m* ~ **pump** Druckwasser-, Spül-pumpe *f* ~ **pump gearbox** Hydraulikpumpenantrieb *m* ~ **radius** (of cross-sectional area as opposed to mean hydraulic radius of reach) Profilradius *n* ~ **ram** Bohrwidder *m*, hydraulische Kolbenstange *f*, Wasserdruckkolben *m* ~ **ram for water lifting** hydraulischer Widder *m* ~ **ram (or brake) rod** Kolben(brems)stange *f* ~ **release cylinder** Öldrucklösezylinder *m* ~ **relief valve** Hydraulikeüberdruckventil *n* ~ **remote control** Fernhydraulik *f* ~ **riveting** Druckwassernietung *f* ~ **screw** Wasserschnecke *f* ~ **set** Wasserkraftsatz *m* ~ **shearing machine** Wasserdruckschere *f* ~ **shock absorber** hydraulischer Stoßdämpfer *m*, Öldruckstoßdämpfer *m* ~ **snubber** hydraulischer Dämpfer *m* ~ **spanpress plant** hydraulische Spannpreßanlage *f* ~ **specialist** Hydrauliker *m* ~ **steel structures** Stahlwasserbauten *pl* ~ **stowing** Spülversatz *m* ~ **stretching press** hydraulische Streckpresse *f* ~ **strut** Ölfederbein *n*, Ölfederstrebe *f* ~ **supply** Druckwasser *n* ~ **supply pipes** Zuleitung *f* ~ **tachometer** Staudrucktachometer *n* ~ **technician** Hydrauliker *m* ~ **test** Abpressung *f*, Wasserdruckprobe *f* ~ **tester** Hydraulikprüfer *m* ~ **thruster** Lüftgerät *n* ~ **transmission** Flüssigkeits-getriebe *n*, -übertragung *f* ~ **transmission of power** hydraulische Kraftübertragung *f* ~ **-tubing rods** hydraulisches Spülrohrgestänge *n* ~ **-type dynamometer** Wasserbremse ~ **valve** Hydraulikschieber *m* ~ **variable speed transmission** Flüssigkeitsregelgetriebe *n* ~ **waste pipe** Abflußrohr *n* des verbrauchten Druckwassers
**hydraulicking** hydraulische Hereingewinnung *f*
**hydraulics** Hydraulik *f*, Wasserkraftwissenschaft *f*

**hydrazine hydrate** Hydrazinhydrat *n*
**hydrazoate** Azid *n*
**hydrazobenzene** Hydrazobenzol *n*
**hydride** Hydrid *n*

**hydriodate of** . . . jodwasserstoffsauer
**hydriodic** jodwasserstoffsauer
**hydro-adjustable** hydraulisch verstellbar
**hydroblast switch** Wasserschalter *m*
**hydrobromic acid** Bromwasser-stoff *m*, -stoffsäure *f*
**hydrobromide** Bromhydrat *n*
**hydrocal(ly)** flüssigkeitsgesteuert
**hydrocarbide** Kohlenwasserstoffverbindung *f*
**hydrocarbon** Kohlenwasserstoff *m* ~ **aceous** kohlenwasserstoffhaltig ~ **combustion** Kohlenwasserstoffverbrennung *f* ~ **gas** Kohlenwasserstoffgas *n* ~ **oil** Kohlenwasserstofföl *n* ~ **pavement** Schwarzdecke *f*
**hydrocell** Hydrokette *f* (electr.)
**hydrocloric acid** Chlorwasserstoff *m*, Chlorwasserstoffsäure *f*, Salzsäure *f* ~ **salt (hydrochloride)** Chlorhydrat *n*
**hydrocloride** Chlorhydrat *n*
**hydrocontrollable** druckölgesteuert ~ **propeller** druckölgesteuerte Verstelluftschraube *f*
**hydrocontrolled** druckölgesteuert
**hydrocyanic acid** Blau-, Cyanwasserstoff-, Hydrozyan-, Zyanwasserstoff-säure *f*
**hydro-cyclone** Hydrozyklon *m*
**hydrodiffusion** Hydrodiffusion *f*
**hydrodynamic tank** Wasserkanal *m*
**hydrodynamical derivative** substantielle Ableitung *f*
**hydrodynamics** Dynamik *f* flüssiger Körper, Hydrodynamik *f*, Wasserkraftlehre *f*
**hydroeconomy** Wasserwirtschaft *f*
**hydroelectric,** ~ **(al)** hydroelektrisch ~ **cell** nasses Element *n* ~ **generating station** Wasserkraftwerk *n* ~ **machine** Dampfelektrisiermaschine *f* ~ **plant of low head** Niederdruckanlage *f* ~ **plant of moderate head** Mitteldruckanlage *f*
**hydroextract, to** ~ abschwingen, schwingen (schleudern)
**hydro-extracted block** ausgeschleuderter Materialblock *m*
**hydroextractor** Trockenmaschine *f*, Zentrifugaltrockenmaschine *f*, Zentrifuge *f* ~ **with siekra drive** Siekrazentrifuge *f*
**hydrofluoboric acid** Borfluorwasserstoffsäure *f*
**hydrofluoric acid** Fluor-wasserstoff *m*, -wasserstoffsäure *f*, Flußsäure, Flußwasserstoffsäure *f*
**hydrofluosilicic acid** Kieselfluorwasserstoffsäure *f*
**hydrofoil** Gleitfläche *f*, Stromlinienfläche *f*, Trag-, Trag-flügel *m*, Wasserventilator *m* (Tragflächenboot *n*) ~ **to facilitate take-off** unter dem Rumpfboden eines Flugbootes angeordnetes Tragflügelprofil *n* zur Erleichterung des Abhebens ~ **vessel** Tragflügelboot *n*
**hydrogen** Wasserstoff *m* ~ **arc** Wasserstofflichtbogen *m* ~ **bond** Wasserstoffbrücke *f* ~ **bottle** Wasserstoffflasche *f* ~ **bromide** Bromwasserstoff *m* ~ **bubble chamber** Wasserstoffblasenkammer *f* ~ **chimney shaft** Wasserstoffschacht *m* ~ **chloride** Chlorwasserstoff *m* ~ **compound** Wasserstoffverbindung *f* ~ **container** Wasserstoffbehälter *m* ~ **-containing** wasserstoffhaltig ~ **content** Wasserstoffgehalt *m* ~ **cyanide** Zyanwasserstoff *m* ~ **cylinder** Wasserstoffflasche *f* ~ **embrittlement** Wasserstoff-krankheit *f*, -versprödung *f* ~ **engine** Wasserstofftriebwerk *n* ~

**flame** Wasserstoffflamme f ~ **gas** Wasserstoffgas n ~-**gas plant** Wasserstoffgasanlage f ~ **generating plant** Wasserstoffanlage f ~ **generation** Wasserstoffbereitung f ~ **generator** Wasserstoff-erzeuger m, -erzeugungsapparat m ~ **halides** Halogenwasserstoff m ~ **helium gas** Wasserstoffheliumgas n ~ **iodide decomposition** Jodwasserstoffzersetzung f ~ **iodide formation** Jodwasserstoffbildung f ~ **ion** Wasserstoffion n ~ **ion concentration** Titrierexponent m, Wasserstoff-ionenkonzentration f, -zahl f ~ **mixture** Wasserstoffgemisch n ~ **nucleus** Wasserstoffkern m ~ **oven** Wasserstoffofen m ~ **particle** Wasserstoffkern m ~ **peroxide** Perhydrol n, Wasserstoffhyperoxyd n, Wasserstoffsuperoxyd n ~ **peroxide bleach** Wasserstoffsuperoxydbleiche f ~ **persulphide** Wasserstoffsupersulfid n ~-**pressure regulator** Wasserstoffventil n ~-**producing plant** Wasserstoffgewinnungsanlage f ~ **ray** Wasserstoffstrahl m ~ **salt** Wasserstoffsalz n ~ **selenide** Selenwasserstoff m ~ **silicide** Kieselwasserstoff m ~ **soldering** Wasserstofflötung f ~ **spark gap** Wasserstoffunkenstrecke f ~ **starter** Wasserstoffanlasser m ~ **sulfide** Schwefelwasserstoff m, Wasserstoffsulfid n ~ **tank** Wasserstoffbehälter m ~ **thyratron** Wasserstoffthyratron n ~ **valve** Wasserstoffventil n

**hydrogenate, to** ~ berginisieren (von Bergius), hydrieren

**hydrogenated** mit Wasserstoff m verbunden ~ **oil** gehärtetes Öl n (trade name) ~ **peanut oil** Astrafett n

**hydrogenation** Hydrierung f, Wasserstoffanlagerung f ~ **of coal** Kohlenverflüssigung f ~ **process** Hydrierungsverfahren n

**hydrogenerization** Hydrogenerisation f

**hydrogenic ions** wasserstoffähnliche Ionen pl

**hydrogenize, to** ~ hydrieren

**hydrogenous** wasserstoffhaltig

**hydroglider** Wassersegelflugzeug n

**hydrograph** Hydrograf m ~ **curve of discharges** Abflußmengenkurve f

**hydrographic** hydrografisch ~ **atlas** Stromatlas m ~ **chart** See-, Wasser-karte f ~ **map** hydrografische Karte f, Paß-, See-karte f ~ **office** nautische Abteilung f des Marineamtes

**hydrography** Gewässerkunde f, Hydrografie f, Meereskunde f

**hydrokinetics** Hydrokinetik f

**hydrolizing,** ~ **agent (for sizes)** Aufschlußmittel n ~ **process** Aufschlußverfahren n

**hydrology** Gewässerkunde f, Hydrologie f, Wasserkunde f

**hydrolysis** Hydrolyse f

**hydrolytic** hydrolytisch ~ **degradation by means of acids** azidolytischer Abbau m

**hydrolyze, to** ~ (starch) aufschließen, hydrolisieren

**hydrolyzer** Fällkessel m

**hydromalium casing** Hydromaliumgehäuse n

**hydrometallurgy** Naßmetallurgie f

**hydrometeor** Niederschlag m

**hydrometer** Aerometer n, Aräometer n, Dichtemesser m, Dichtigkeitsmesser m, Flüssigkeitsdichtemesser m, Gewichtsaräometer n, Hydrometer n, Schwermesser m, Senkspindel f, Senkwaage f, Tauchwaage f, Volumaräometer n,

Wassermesser m ~ **analysis** Schlammanalyse f ~ **syringe** Hebersäuremesser m

**hydrometric** aräometrisch, hydrometrisch ~ **gauge with cup-type water wheel** Becherradflügel m ~ **vane** Strömungsmesser m

**hydrometry** Spindelung f

**hydropathic establishment** Wasserheilanstalt f

**hydrophilic** wassersüchtig

**hydrophobe** wasserscheu

**hydrophone** (navy) Geräuschempfänger m, Horchapparat m, Unterwasser-horchgerät n, -schallapparat m ~ **buoy** Geräuschboje f ~ **contact** Geräuschortung f ~ **gear** Horchgerät n

**hydrophonics** Wasserkultur f

**hydrophylic** wasserbindend

**hydroplane** Gleitboot n, Schwimmer-, See-, Wasser-, Wassersegel-flugzeug n

**hydro-pneumatic brake** Öldrucksaugluftbremse f

**hydropyrometer** Wasserpyrometer n

**hydroquinone** Hydrochinon n

**hydroselenic acid** Kieselwasserstoffsäure f

**hydrosphere** Hydrosphäre f

**hydrospinning** hydraulische Schleuder f

**hydrostabilizer** Flossenstummel m (aviat.)

**hydrostatic** hydrostatisch ~ **compass** Mutterkompass m ~ **fuse** Wasserdruckzünder m ~ **gauge** Druckbenzinuhr f ~ **head** Flüssigkeits-, Gefälle-druck m, Wassersäule f ~-**level gauge** Aräometer n ~ **pressure** Flüssigkeits-, Gefälle--druck m, hydrostatischer Druck m, Wasserdruck m ~-**pressure excess** hydrostatischer Überdruck m

**hydrostatics** Hydrostatik f

**hydrosuction machine for drying all kinds of fabrics** Absaugmaschine f zum Entnässen von Geweben aller Art

**hydrosulphide** Sulf-hydrat n, -hydrid n, Sulfohydrat n

**hydrotechnique** Hydrotechnik f

**hydrotimeter** Hydrotimeter n

**hydroturbine** Wasserturbine f

**hydrous** wasserhaltig, wässerig ~ **ferrous sulphate** Eisen-sulfat n, -vitriol n ~ **silicate of zine** Hemimorphit m ~ **zinc sillicate** Zink-glas, -glaserz n

**hydroxanthane** Xanthanwasserstoff m

**hydroxide** Hadroxyd n, wasserhaltiges Oxyd n ~ **of potash** Kaliumoxydhydrat n

**hydroxyl** Hydroxyl n ~ **amine nitrate** Hydroxylaminnitrat n

**hydroxylation** Hydroxylierung f

**hydroxyquinoline** Oxychinolin n

**hydroxytoluence** Oxytoluol n

**hydroxytoluol** Hydroxytoluol n

**hydrozincite** Zinkblüte f

**hydrozy acids** Oxydsäuren pl

**hyetograph** Regenschreiber m

**hyetometer** Regenmesser m

**hygiene** Gesundheitspflege f, Hygiene f

**hygienic** gesundheitlich, hygienisch ~ **laboratory** hygienische Untersuchungsstelle f

**hygienist** Hygieniker m

**hygrodeik** Kondensationshygrometer n

**hygrograph** Feuchtigkeitsschreiber m, Hygrograf m, registrierender Feuchtigkeitsmesser m

**hygrometer** Feuchtemesser m, Feuchtigkeitsmesser m, Hygrometer n, Luftfeuchtigkeitsmesser m

**hygrometric condition** Luftfeuchtigkeit *f*
**hygrometry** Hygrometrie *f*
**hygroscope** Feuchtigkeits-anzeiger *m*, -messer *m*, -zeiger *m*, Luftfeuchtigkeitszeiger *m*
**hygroscopic** hygroskopisch, wasser-aufnehmend, -anziehend, -gierig ~(al) hygroskopisch ~ **element** (to prevent mist-formation on lenses) Trockenpatrone *f*
**hygroscipicity** Hygroskopizität *f*, Wasseranziehungsvermögen *n*
**hygrothermograph** Hygrothermograf *m*
**hypabyssal** hypoabyssisch
**hyperacidity** Perazidität *f*
**hyperbola** Hyberbel *f* ~ **like** hyperbelähnlich
**hyperbolic** hyperbolisch ~ **angle** hyperbolischer Winkel *m* ~ **characteristic** hyperbelförmige Kennlinie *f* ~ **chart** Tangenskarte *f* ~ **curvature** hyperbolische Krümmung *f* ~ **cylinder** hyperbolischer Zylinder *m* ~ **distance** hyperbolischer Abstand *m* ~ **function** Hyperbel-, Hyper-funktion *f*, hyperbolische Funktion *f* ~ **geometry** hyperbolische Geometrie *f* ~ **horn** Hyperbelhorn(strahler) *m* ~ **inverter** Hyperbelinvertor *m* ~ **motion** hyperbolische Bewegung *f* ~ **navigation** Hyperbelnavigation *f* ~ **orbits** Hyperbelbahnen *pl* ~ **paraboloid** (monkey saddle) Affensattel *m* (geom), hyperbolisches Paraboloid *n* ~ **plane** hyperbolische Ebene *f* ~ **sine** hyperbolischer Sinus *m* ~ **space** hyperbolischer Raum *m* ~ **tangent** hyperbolischer Tangens *m*
**hyperboloid** Hyperboloid *n* ~ **of one sheet** einschaliges Hyperboloid *n* ~ **of revolution** Rotationshyperboloid *n* ~ **of two sheets** zweischaliges Hyperboloid *n*
**hyperboloidal,** ~ **gear** Hyperbelrad *n*, Zahnradübertragung *f* durch Hyperboloidräder ~ **position of n straight lines** hyperboloidische Lage *f* von n Geraden
**hyperbromous acid** unterbromige Säure *f*
**hypercomplex** hyperkomplex
**hyper-compressor** Höchstdruckkompressor *m*
**hyperconductivity** Überkonduktivität *f*
**hyper-conical** hyperkonisch
**hyperconjugation** Hyperkonjugation *f*
**hypereutectic** hyper-, über-eutektisch
**hypereutectoid** hypereutektoidisch, Übereutektoid *n* ~ **steel** übereutektoider Stahl *m*
**hyperfine,** ~ **coupling** Hyperfeinstruktur-Kopplung *f* ~ **structure** überfeine oder hyperfeine Struktur
**hyperfrequency wave** Mikrowelle *f*
**hypergeometric** hypergeometrisch ~ **series** hypergeometrische Reihe *f*
**hypergolic** hypergolisch
**hyperon detection** Hyperonennachweis *m*
**hyperopia** Über-, Weit-sichtigkeit *f*
**hyperopic** übersichtig ~ **eye** weitsichtiges oder fernsichtiges Auge *n*
**hyperoxygenated salt** Azidul *n*
**hyperphosphorous** unterphosphorig
**hyperplane** Hyperebene *f*
**hyperquantization** zweite Quantelung *f*
**hyperscopic view** hyperskopische Projektion *f*
**hypersensible** überempfindlich
**hypersonic,** ~ **aircraft** Hyperschallflugzeug *n* ~ **flow** Hyper-, Mehrfach-schallströmung *f*

**hyperstatic** statisch unbestimmt (structure)
**hypersthene** Paulit *m*
**hypersurface** Hyperfläche *f*
**hypersynchronous** über-, unter-synchron
**hyperventilation** Hyperventilation *f*, Überlüftung *f*
**hyphen** Bindestrich *m*
**hypo** Fixiernatron *n*, Fixiersalz *n*, Natriumthiosulfat; abnorm gering, schwach ~ **bromous** unterbromig ~ **center** Hypozentrum *n* ~ **chlorid acid** Unterchlorsäure *f* ~ **chlorite of lime bleach** Fixbleiche *f* ~ **chlorous** unterchlorig ~ **chlorous acid** unterchlorige Säure *f* ~ **cycloid** Hyperzykloide *f* ~ **dermic needle** Injektionskanüle *f* ~ **dermic product** Subkutanpräparat *n* ~ **eutectic** untereutektisch ~ **eutectoid** Untereutektoid *n*
**hypoid,** ~ **bevel gear** Hypoidkegelrad *n* ~ **gear** Hypoidgetriebe *n* ~ **garing** Hypoidverzahnung *f* ~ **oil** Hypoidöl *n*
**hypophosphate** Hypophosphat *n*
**hypophosphite** Hypophosphit *n*, unterphosphoriges Salz *n*, unterphosphorigsaures Salz *n* ~ **acid** unterphosphorige Säure *f*
**hyposcopic view** hyposkopische Projektion *f*
**hyposulphate** unterschwefelsaures Salz *n*
**hyposulphuric acid** Unterschwefelsäure *f*
**hyposulphurous** unterschweflig
**hypotenuse** Hypotenuse *f*
**hypothesis** Annahme *f*, Hypothese *f*, Voraussetzung *f* ~ **of compound nucleus** Sandsackmodell *n* des Atomkerns
**hypothetical** hypothetisch ~ **central office** fiktives Vermittlungsamt *n* ~ **charge** angenommene Durchgangsgebühr *f* ~ **exchange** fiktives Vermittlungsamt *n* ~ **transit charge** angenommene Durchgangsgebühr *f* ~ **transit quota** angenommener Anteil *m* an der Durchgangsgebühr, hypothetischer Durchgangsanteil *m*
**hypotrochoid** Hypotrochoide *f*
**hyposogram** Pegellinie *f*
**hypsography** Höhenbeschreibung *f*
**hypsometer** Hypsometer *n*, Pegelmesser *m*, Siedethermometer *n*, Siedewärmegradmesser *m*, Wassersiedemesser *m*
**hyposometry** Höhenmessung *f*
**hysteresigraph** Hystereseschreiber *m*, schreibender Hysteresemesser *m*
**hysteresis** Hysterese *f*, Hysteresis *f*, magnetische Reibungssatz *f*, magnetische Rücktrift *f*, Nach-federung *f*, -hinken *n*, -wirkung *f*, -wirkungsveränderung *f* ~ **of a counter** Alterung *f* eines Zählrohrs ~ **of the selenimum cell** (photo electr.) Nachwirkung *f* des Selens
**hysteresis,** ~ **coefficient** Hysterese-verlustzahl *f*, -zahl *f* ~ **constant** Hysteresekonstante *f* ~ **curve** Hystereseschleifenschreiber *m*, jungfräuliche Kurve *f* ~ **cycle (or loop)** Hystereseschleife *f* ~ **error** Hysteresefehler *m* ~ **factor** Hysteresebeiwert *m* ~ **loop** Magnetisierungsschleife *f* ~ **loss** Hystereseverlust *m* ~ **meter** Hysteresemesser *m* ~-**type** hystereseartig
**hysteretic** hysteretisch ~ **lag** hysteretische Nacheilung *f* ~ **loss** Nachwirkungsverlust *m*
**hystero-salpingography** Hysterosalpingografie *f*

# I

I beam Deckenträger *m*, Doppel-T-Eisen *n*, Doppel-T-Träger *m*, I-Eisen(stange) *f*, I-förmiger Balken *m*, I-Formstahl *m*, I-Träger *m*
I-beam, ~ girder Stegträger *m*, vollwandiger Träger *m* ~ punch and coper Stanze *f* und Ausklinkmaschine *f* für I-Träger ~ punching and coping machine Loch- und Ausklinkmaschine *f* für I-Träger ~ section, I-Profil *n*
I bolt I-Stütze *f*
ice, to ~ mit Eis kühlen, überglasen
ice Eis *n* ~ and cooling installation Eis- und Kühlanlage *f* ~ accretion Eisansatz *m* ~ adz Eispickel *m* ~ age Eiszeit *f* ~ anchor Eisanker *m* ~ apron Eisbrecher *m* ~ ax Eispickel *m* ~ barrage Eisbarre *f* ~ berg Eisberg *m* ~ berg-observation station Eisbeobachtungsstation *f* ~ berg patrol Eisdienst *m* ~ berg service Eisnachrichtendienst *m* ~ blink Eisblink *m*, Eisreflexion *f* am Horizont ~-bound eingeeist ~ box Eis-, Gefrier-schrank *m* ~ breaker Eisbrecher *m* ~ breaking Eisbrechen *n* ~-breaking ram Eisbrechpflug *m* ~-breaking structure to protect bridges Eisbrecher *m* ~ calorimeter Eiskalorimeter *m* ~ can Eiszelle *f* ~ cellar Eiskeller *m* ~ chute Eisrutsche *f* ~ coating Eis-belag *m*, -decke *f*, -überzug *m* ~ coating (on bridges or aircraft etc) Rauhreif *m* ~-cold eiskalt ~ concrete Sandeis *m* ~-crushing machine Eiszerkleinerungsmaschine *f* ~ crystal Eiskristall *m* ~ cube tray Eisschublade *f* ~ dam Eisbarre *f* ~ deposit on wires Rauhreif *m* ~ detector Eisfühler-, Eisfühl-gerät *n* ~ detector probe Eismelder *m*, Vereisungs-fühler *m*, -sonde *f* ~-eliminating system Enteisungsanlage *f* ~ field Eisfeld *m* ~ flap on a gate of a moveable dam Eisklappe *f* ~ float Eisschwimmer *m* ~ floe Eisscholle *f* ~ flow Eistrieb *m* ~ fog Eisnebel *n* ~ formation Eisbildung *f* ~-free vereisungsfrei ~ freezing tank Eiserzeuger *m* ~ gate Eisklappe *f* ~ guard Enteiser *m* ~ guard on the leading edge Enteiser *m* in der Flügelnase ~ house Eiskeller *m* ~ impact Eisstoß *m* ~ indicator Vereisungsanzeiger *m* ~ jam Eis-barre *f*, -stoß *m* ~ load Eis-belastung *f* (electr.), -last *f* ~ machine Eis-, Eiserzeugungs-maschine *f* ~-machine oil Eismaschinenöl *n* ~-making capacity Eisleistung *f* ~ mold Eiszelle *f* ~ needles Eisnadeln *pl* ~ nuclei Eiskerne *pl* ~-observation post Eiswarte *f* ~-observation station Eisbeobachtungsstation *f* ~ pack Eisfeld *n* ~ point Gefrierpunkt *m* ~ production Eisleistung *f* ~-protection pipe Vereisungsschutzrohr *n* ~ pyrheliometer Eis-Pyrheliometer *n* ~ rain Eisregen *m* ~-removing rubber device eisentfernender Gummimechanismus *m* ~ scraper Reifkratzer *m* ~ sheet Eis-decke *f*, -schild *n* ~ skimmer on a gate of a moveable dam Eisklappe *f* ~ space (trail spade) Eissporn *m* ~ trap Eisfänger *m* ~ valve Eisklappe *f* ~-warning device Eiswarngerät *n* ~-warning indicator Eisbildungswarn-, Eiswarnungs-apparat *m* ~-warning unit Eiswarngerät *n*, Ver-

eisungsanzeiger *m* ~-water tank Eiswasserkasten *m*
Iceland spar Doppelspat *m*, isländischer Doppelspat *m*
icer Fruchteiserzeuger *m*
icicle Eiszapfen *m*, (weld) Schweißbart *m*
icing Vereisen *n*, Vereisung *f* ~ measuring device Vereisungsmeßgerät *n* ~-warning device Vereisungswarngerät *n* ~ zone Vereisungszone *f*
I circuit H-Leitung *f*
iconometer Ikonometer *n*, Rahmensucher *m*
iconoscope Bildabtastrohr *n*, Bildwandlerröhre *f*, Elektronenstrahlabtaster *m*, Fernsehaufnahmeröhre *f*, Ikonoskop *n*, Kathodenstrahlröhre *f*
icosahedron Ikosaeder *n*, Zwanzigflächner *m*
icositetrahedron Ikositetraeder *n*
icy eisartig, eisig
idea Begriff *m*, Vorstellung *f*
ideal ideal, ideell, mustergültig, vorbildlich ~ bunching ideale Ballung *f* ~ condition Idealbedingung *f* (of) ~ conductance ideal leitend ~ gas ideales oder vollkommenes Gas *n* ~ line ideale Leitung *f* (teleph.) ~ position Sollage *f* ~ process factor theoretischer Trennfaktor *m* ~ rectifier idealer Gleichrichter *m* ~ saturable reactor idealer sättigbarer Reaktor *m* ~ simple pendulum mathematisches Pendel *m* ~ simple process factor theoretischer Trennfaktor *m* einer Stufe ~ transducer idealer Wandler *m* (acoust.) ~ value Istwert *m* ~ vibration Musterschwingung *f*
idealization Vereinfachung *f*
idealized characteristic idealisierte Kennlinie *f*
ideally, ~ imperfect ideal fehlgeordnet (cryst.) ~ imperfect crystal Mosaikkristall *m*
idempotent idempotent; Einzelmatrix *f*
identical identisch, übereinstimmend ~ in nature wesensgleich ~ particle collisions Stöße *pl* identischer Teilchen
identification Ausweis *m*, Feststellung *f*, Identifizierung *f*, Kennung *f* (beacon), Nachweisung *f* ~ of boundary points Randzuordnung *f*
identification, ~ apparatus Kenngerät *n* ~ beacon Kennfeuer *n* (aviat.) ~ card Ausweiskarte *f* ~ colors (of fuses) Kennfarben *pl* ~ friend-foe (IFF) Identifizierung *f* Freund-Feind ~ group Kenngruppe *f* ~ head leader Startband *n* (film) ~ keyer Kennungsgeber *m* ~ leader Startband *n* (film) ~ letter Kennbuchstabe *m* ~ light Erkennungslampe *f*, Identifizierungs-, Kenn-licht *n*, Kennscheinwerfer *m* ~ mark Erkennungszeichen *n*, Kennmarke *f* ~ marker Filmbezeichnungsgerät *n* ~ number Kennummer *f* ~ panel Erkennungstuch *n* ~ plate Kennzeichnungsschild *n* ~ signal Erkennungssignal *n*, Kennung *f* (telegr.) ~ sleeve Bezeichnungshülse *f* ~ strip Kennstreifen *m* ~ tag Erkennungsmarke *f* ~ tail leader Endband *n* (film) ~ tape Kennband *n* ~ thread Markenfaden *m* ~ trailer Endband *n* (film)
identify, to ~ ausmitteln, fassen, feststellen, identifizieren, nachweisen to ~ oneself sich ausweisen to ~ wires Adern *pl* ausprüfen

**identity** (map) Flächentreue *f*, Identität *f*, Wesenseinheit *f* ~ **transformation** identische Abbildung *f*
**I.D.F.** (**intermediate distribution frame**) Zwischenverteiler *m*
**idioelectric** idioelektrisch
**idiomorphic** idiomorph
**idiostatic circuit** (**method, or mounting**) idiostatische Schaltung *f*
**idle, to** (**run**) ~ leerlaufen
**idle** brach, frei, leer, in Ruhe befindlich, in Stellung, in der Stellung müßig, träge, unbesetzt, untätig, wählerfrei **to be** ~ außer Gang sein
**idle,** ~ **adjustment wheel** Leerlaufeinstellrad *n* ~ **component** Blindkomponente *f* ~ **component of the current** Blindstromkomponente *f* ~ **component of voltage** Blindspannungskomponente *f* ~ **current** Blindstrom *m*, nutzloser Strom *m*, Querstrom *m*, wattloser Strom *m* ~ **current at plate** Anodenruhestrom *m* ~ **engine** Langsamlauf *m* des Motors ~ **gear** Leergang *m* ~ **indicating signal** Freilampe *f* ~ **jack** freie Klinke *f* ~ **jet** Leerlaufbrennstoffdüse *f* ~ **junction** freie Verbindungsleitung *f* ~**-metering jet** Leerlaufzumeßdüse *f* ~ **motion** leerer Gang *m*, Leerlauf *m* ~ **movement** toter Weg *m* ~ **operator** freie Beamtin *f* ~ **path** Brems-, Leer-weg *m* ~ **period** Neben-, Sperr-zeit *f* ~**-phase shifter** Blindwellenschieber *m* ~ **portion of a coil** tote Windungen *pl* ~ **position** Ruhestellung *f* ~ **power** Blindleistung *f* ~ **revolution** Ansaug- und Verdichtungshub umfassende Kurbelwellenumdrehung *f* ~ **roll** Blind-, Schlepp-walze *f* ~ **rollers** lose Rollen *pl* ~ **running device** Freilaufeinrichtung *f* (Walzwerk) ~ **signals** Gleichlauf-stöße *pl*, -zeichen *pl* ~ **spark** falscher Funke *m* ~ **stop lever** Leerlaufanschlaghebel *m* ~ **stroke** Aussetzer *m*, Leer-hub *m*, -laufhub *m*, -takt *m* ~ **time** Leerlauf-, Neben-, Tot-zeit *f* ~ **travel of ram** Stößelleerlauf *m* ~ **trunk** freie Verbindungsleitung *f* ~ **tube** kalte Röhre *f* ~ **turn(s)** tote Windung(en) *pl*
**idleness** Trägheit *f*, Untätigkeit *f*
**idler** Führungsrolle (Förderanlage) *f*, Leerlauf *m*, Leitrolle *f*, Spannrad *f* (Kettenantrieb), Spannrolle *f* (Riemenantrieb *m*), Stützhebel *m* (Steuerung), Vorlagezahnrad *n*, Zwischenrad *n* ~**-bellcrank** Vorlegehebel *m* (**front**) ~ Leitrad *n* ~ **gear** loses mitgenommenes Zahnrad *n*, Zwischenzahnrad *n* ~ **pulley** Leerlaufriemenscheibe *f*, Riemenleitrolle *f*, Spannrolle *f* ~**-pulley drive** Spannrollentrieb *m* ~ **rope sheave** Leerlaufseilscheibe *f* ~ **shaft** Leerlaufwelle *f* ~ **sprocket** Ausgleichszahnrad *n* ~ **wheel** Leerscheibe *f*, Leitrad *n*, Losscheibe *f* ~**-wheel shaft** Leitradachse *f*
**idlers** Riemenspanner *m*, Tragrollen *pl*
**idling** Leerlauf *m*; langsam, leerlaufend ~ **with car overrunning the engine** Schiebeleerlauf *m*
**idling,** ~ **adjustment** Leerlaufeinstellung *f* ~ **air** Leerlaufluft *f* (Vergaser) ~ **device** Leerlaufvorrichtung *f* ~ **dike** Leerlaufschütz *n* ~ **duct** Leerlaufkanal *m* ~ **engine** leerlaufender Motor *m* ~ **jet** Leerlaufdüse *f* ~ **loss** Leerlaufverlust *m* ~ **nozzle** Leerlaufdüse *f* ~ **setting** Leerlaufstellung *f* ~ **slot** Leerlaufaustritt *m* ~ **speed** Leerlaufdrehzahl *f* (rpm) ~ **spring** Leerlauffeder *f* ~

**stop** Leerlaufanschlag *m* ~ **stop lever** Füllungsregulierhebel *m* ~ **test** Leerlauf-, Übergangsprüfung *f* ~ **tube** Leerlaufrohr *n* (Vergaser) ~ **valve** Leerlaufventil *n*
**idrialite** Quecksilberbranderz *n*
**I. F., i. f.** (**intermediate frequency**) Zwischenfrequenz *f* (ZF) ~ **band filter** Zwischenfrequenzbandfilter *n* ~ **circuit** Zwischenfrequenzkreis *m* ~ **jamming** ZF-Störung *f* ~ **preamplifier** ZF-Vorverstärker *m* ~ **rejection ratio** ZF-Selektion *f* ~ **strip** ZF-Teil *m* ~ **tube** Zwischenfrequenzröhre *f* ~ **wave trap** ZF-Saugkreis *m*
**IFR** (**instrument flight rules**) Instrumenten--Flugregeln *pl*
**I girder** Doppel-T-Träger *m*, I-Träger *m*
**igneous** durch Feuer gebildet, vulkanisch ~ **dikes** Eruptivgänge *pl* ~ **intrusion** (petroleum) durch Feuer gebildetes Eindringen *n* ~ **rock** Eruptiv-, Ur-gestein *n* ~ **veins** Eruptivgänge *pl*
**ignis fatuus** Irrlicht *n*
**ignitability** Entzündlichkeit *f*, Zündfähigkeit *f*
**ignitable** zündbar, zündfähig
**ignite, to** ~ anzünden, entzünden, erhitzen, glühen, zünden
**igniter** Anzünder *m*, Beiladung *f*, Zünd-apparat *m*, -elektrode *f*, Zünder *m*, Zünd-mittel *n*, -vorrichtung *f* ~ **cable** Zündkabel *n* ~ **flask** Zündflasche *f* ~ **liquid** Zündstoff *m* ~ **plug** Zündkerze *f* ~ **terminal** Zündstiftanschluß *m*
**igniting,** ~ **charge** Beiladung *f* ~ **device** Zeitzünder *m* ~ **flame** Zünd-flämmchen *n*, -flamme *f* ~ **flash** Zündstrahl *m* ~ **mixture** Zündmittel *n* ~ **power** Zündkraft *f* ~ **valve** Zündventil *n*
**ignition** Entzündung *f*, Glimmen *n*, (point) Zündpunkt *m*, Zündung *f* (**self-**) ~ Selbstentzündung *f* ~ **and extinction voltage** Zünd- und Löschspannung *f* ~ **by incandescence** Glühzündung *f* **the** ~ **is intermitting** die Zündung setzt zeitweilig aus ~ **of isolated droplets** Zündung *f* isolierter Tröpfchen ~ **and starting lock** Zündanlaßschloß *n*
**ignition,** ~ **anode** Erreger-, Hilfs-, Zünd-anode *f* ~ **arch** Zündgewölbe *n* ~ **armature** Zündanker *m* ~ **band** Zündband *n* ~**-battery box** Zündbatteriekasten *m* ~ **cable** Zündkabel *n* ~ **cable plug-type suppressor** Zündleitungsentstörstecker *m* ~ **cable sleeve type suppressor** Zündleitungsentstörmuffe *f* ~ **cam** Nocken *m* für die Zündung, Unterbrecherscheibe *f* ~ **cartridge** Zündpatrone *f* ~ **characteristic** Zündkennlinie *f* ~ **circuit** Zünd-kreis *m*, -stromkreis *m* ~ **coil** Glühspirale *f*, Induktor *m*, Summerzündspule *f*, (jet) Zündgerät ~ **composition** Zündmasse *f* ~ **control** Zündkontrolle *f*, Zündzeitpunkteinstellvorrichtung *f* ~**-control lever** Zündungshebel *m* ~**-control push-and-pull rod** Zündverstellstoßstange *f* ~ **core** Zündseele *f* ~ **current** Zündstrom *m* ~**-current key** Zündtaste *f* ~ **current take-off** Zündstromabnahme *f* ~ **cutout** Zündstromausschalter *m* ~ **dead center** Zündtotpunkt *m* ~ **delay** Zündverzug *m* ~ **device** Zünd-anlage *f*, -apparat *m*, -einrichtung *f*, -vorrichtung *f* ~ **disc** Zündscheibe *f* ~ **due to impact** Aufschlagzündung *f* ~ **dynamo** Zünddynamo *m* ~ **electrode** Zündelektrode *f* ~ **exciter** Zünderreger *m* **the** ~ **fails** die Zündung *f* bleibt aus, die Zündungen *pl* bleiben aus ~ **flange for make-and-break ignition** Zündflansch

*m* für Abreißzündung ~ **gap** (or **interval**) Zündabstand *m* ~ **handle** Handgriff *m* für die Zündung ~ **harness** Zündverteileranlage *f*, Zündungsgeschirr *n* ~ **hazard** Entzündungsgefahr *f* ~ **heads** Zündköpfe *pl* ~ **heat** Zündwärme *f* ~ **interference** Zündstörung *f* ~ **key** Schalt-, Zünd-schlüssel *m* ~ **knock** Zündungsklopfen *n* ~ **lag** Zündverzug *m* ~ **lever** Zünd-(verstell)-hebel *m* ~ **limit** Zündgrenze *f* ~ **lock** Zündschluß *m* ~ **loss** Brennverlust *m* ~ **nozzle** Zünddüse *f* ~ **order** Zündzeitfolge *f* ~ **paper** Zündpapier *n* (photo) ~ **peak** Zündspitze *f* ~ **pellet** Zündkirsche *f* ~ **performance** Zündleistung *f* ~ **pin** Zündstift *m* ~ **plug** Funkkerze *f* (jet), Zünd-kerze *f*, -stöpsel *m* ~ **point** Entzündungsstelle *f*, Zündstelle *f*, Zündzeitpunkt *m* ~**-point tester** Zündpunktprüfer *m* ~ **primer** Initiator *m* ~ **process** Zündvorgang *m* ~ **quality** of fuel Zündwilligkeit *f* ~ **quality improvers** Zusätze *pl* zur Erhöhung der Zündwilligkeit ~ **range** Zündgebiet *n* ~**-ray process** Zündstrahlverfahren *n* ~ **rod** Zündstab *m* ~ **screening** Abschirmen *n* der Zündung, Funkabschirmung *f* der Zündeinrichtung ~ **shield** Entstörungsklappe *f* ~ **source** (or **focus**) Zündherd *m* ~ **spark** Entzündungsfunke *m*, Zündfunken *m* ~ **spark amplifier** Zündfunkenverstärker *m* ~**-spark control** Zünderhebel *m* ~ **spark length** Schlagweite *f* des Zündfunkens ~ **spring** Zündfeder *f* ~ **stimulant** Zündpeitsche *f* ~ **stroke** Zünd-, Zündungs-hub *m* ~ **switch** Kurzschließer *m*, Magnet-, Zünd-, Zündungs-schalter *m* ~ **switchboard** Zünd-anlage *f*, -lage *f*, -schaltbrett *n*, Zündungssystem *n*, Zündvorrichtung *f* ~ **system** Zünd(strom)anlage *f* ~ **temperature** Entzündungs-, Zünd-, Zündungs-temperatur *f* ~ **tension** Zündspannung *f* ~ **test** Entzündungs-, Glühprobe *f* ~ **timer** (piston-engine) Zündverteiler *m*, (jet engine) Zündzeitgeber *m*
**ignition-timing** Zündpunkteinstellung *f* ~ **adjustment** Zündzeitpunktverstellung *f* ~ **control** Verstellung *f* des Zündzeitpunktes ~ **curve** Zündverstellinie *f* ~ **device** (or **timer**) Zündversteller *m* ~ **range** Zündverstellbereich *m*
**ignition,** ~ **trouble** Zündungs-schwierigkeit *f*, -störung *f* ~ **tube** Glüh-, Zünd-rohr *n* ~ **unit** Zünd-gerät *n*, -spule *f*, -stromerzeuger *m* ~ **vibrator** Zündstromzerhacker *m* ~ **voltage** Zündspannung *f* ~ **winding** Zündwicklung *f* ~ **wire** (blasting) Knallzündschnur *f*, Zünddraht *m*, -kabel *n*, -kerzenkabel *n*, Zündungsdraht *m* ~**-wire manifold** Zündkabel-geschirr *n*, -rohr *n* ~ **wiring** Zündleitungssystem *n* ~**-wiring harness** Zündkabelgeschirr *n*
**ignitor** Zündstab *m*
**ignitron** Gleichrichter *m* mit Zündstift, Ignitron *n*
**ignorable** zyklisch ~ **co-ordinates** zyklische Koordinaten *pl*
**ignore** Auslaßzeichen *n*, Leerzeichen *n* (comput.) ~ **character** Auslaßzeichen *n* ~ **instrument** Negierbefehl *m*
**I-head motor** Motor *m* mit hängenden Ventilen
**I iron** I-Eisen *n*
**ill** krank ~ **effect** nachteilige Wirkung *f*
**illegal** ungesetzlich, rechtswidrig, ungesetzlich, unrechtmäßig, widerrechtlich ~ **measure** widerrechtliches Mittel *n*
**illegibility** Undeutlichkeit *f*

**illegible** undeutlich
**illicit trade** Schleichhandel *m*
**illness** Krankheit *f*
**illuminant** Beleuchtungsmittel *n*, Leucht-körper *n*, -mittel *n*, Lichtquelle *f*, Weißpunkt *m* im Farbtondiagramm
**illuminate, to** ~ anleuchten, aufklären, beleuchten, bestrahlen, erhellen, erleuchten, illuminieren, leuchten **to** ~ **a cask** ein Faß ausleuchten **to** ~ **color** kolorieren
**illuminated** belichtet, mit Nachtbeleuchtung *f*, nachtleuchtend ~ **compass** Leuchtkompaß *m* ~ **cross wires** beleuchtetes Fadenkreuz *m* ~**-field stop** Leuchtfeldblende *f* ~ **landing T** nachtleuchtendes Lande-T ~ **pannel** Durchleuchtungseinrichtung *f* ~ **period** Aufleucht-, Leuchtdauer *f* ~ **planning table** Montagetisch *m* ~ **region** ausgeleuchtete Zone *f* ~ **searching** Hellsuchen *n* ~ **sight** Leucht-korn *n*, -zielvorrichtung *f* ~ **sign** Leuchtschild *n* ~ **track diagram** Fahrschautafel *f*
**illuminating,** ~ **apparatus** Leuchtgerät *n* ~ **base** Beleuchtungsuntersatz *m* ~ **color** Begleitfarbe *f* ~ **cone** Beleuchtungskegel *m* ~ **equipment** (or **plant**) Beleuchtungseinrichtung *f* ~ **fixture** Leuchtkörper *m* ~ **gas** Leuchtgas *n* ~**-gas coal** Leuchtgaskohle *f* ~**-gas engineering** Leuchtgastechnik *f* ~**-gas poisoning** Leuchtgasvergiftung *f* ~ **lens** Beleuchtungslinse *f* ~ **magnifier** Leuchtlupe *f* ~ **mark** leuchtende Marke *f* ~ **oil** Leuchtöl *n* ~ **oil for railways** Petroleum *n* für Eisenbahnlampen ~ **outfit** Beleuchtungsausrüstung *f* ~ **paraffin** Leuchtöl *n* ~ **pencil** Lichtkegel *m* ~ **pencil of rays** Beleuchtungsbüschel *n* ~ **power** Leuchtkraft *f* ~ **purpose** Beleuchtungszweck *f* ~ **time** Aufhellungsdauer *f* ~ **torch** Leuchtfackel *f* ~ **value** Leuchtvorfeld *n*
**illumination** Aufleuchten *n*, Ausleuchtung *f*, Beleuchtung *f*, Beleuchtungsstärke *f*, Belichtung *f*, Bestrahlung *f*, Brennen *n* (lamps), Erleuchtung *f*, Licht *n*
**illumination** ~ **with contrasting colors** Kontrastfarbenbeleuchtung *f* ~ **on curves** Kurvenlicht *n* ~ **of the entrance** Eingangsbeleuchtung *f* ~ **of fiducial lines** Fadenbeleuchtung *f* ~ **by means of built-in intensive fittings** Tiefstrahler *m* ~ **of picture** Bildhelligkeit *f* ~ **at a point on a surface** Beleuchtungsstärke *f* ~ **of sight** Nachtvisierbeleuchtung *f*
**illumination, advertisements** Lichtreklame *f* ~ **attachment to field of view** Gesichtsfeldbeleuchtungseinrichtung *f* ~ **level** Beleuchtungsstärke *f* ~ **(exposure) meter** Belichtungsmesser *m* ~ **photometer** Beleuchtungsstärke-messer *m*, -fotometer *n* ~ **prohibition** Leuchtverbot *n*
**illuminator** Beleuchtungs-körper *m*, -radar(gerät) *n*, Illuminator *m* ~ **diaphragm** Illuminatorblende *f*
**illuminometer** Beleuchtungsmesser *m*, Beleuchtungsstärke-messer *m*, -fotometer *n*
**illusion** Täuschung *f*, Trugbild *n* ~ **of depth** (of picture) plastische Bildwirkung *f* ~ **of motion** Bewegungseindruck *m* ~ **of visual size** Sehgrößentäuschung *f*
**illusory** illusorisch, täuschend, trügerisch
**illustrate, to** ~ bebildern, erläutern, illustrieren, verdeutlichen **to** ~ **graphically** auftragen

**illustrate part breakdown** illustrierter Telekatalog *m*
**illustrated** erläutert
**illustrating purpose** Veranschaulichung *f*
**illustration** Abbildung *f*, Beispiel *n*, Erläuterung *f*, Illustration *f*, Schnittzeichnung *f*, Veranschaulichung *f*, Zeichnung *f* ~ **of a problem in which actual figures are given** Zahlenbeispiel *n*
**illustrative** abbildend, erläuternd ~ **material** Anschauungsmittel *n*
**ilmenite** Eisentitan *n*, Ilmenit *m*, Titaneisen-erz *n*, -stein *m*
**ilvaite** Kieselkalkeisen *n*, Yenit *m*
**image, to** ~ abbilden, bildlich darstellen
**image** Abbild *m*, Abbildung *f*, Bild *n*, Ebenbild *n*, Gebilde *n*, (on screen) Schirmbild *n* ~ **of the crater** Kraterbild *n* ~ **of interference** Interferenzbild *n* ~ **of natural size** Bild *n* von natürlicher Größe ~ **on saddle stand** Bild *n* auf Reiter **not sharply defined** ~ unscharfes Bild *n* ~ **of stop** Blendenbild *n*
**image,** ~ **aberration** Bildfehler *m* ~ **amplifier** Bildverstärker *m* ~ **antenna** Schein-, Spiegelbild-antenne *f* ~ **aperture** Bildfenster *n* ~ **area** (on scanning disk) Bildausschnitt *m*, Bildfläche *f* ~ **attenuation coefficient** Spiegeldämpfungskoeffizient *m* ~**-attenuation constant** Vierpoldämpfung *f* ~**-attenuation factor** Vierpoldämpfungsfaktor *m* ~ **band** Bildband *n* ~**-carrier sound film** Bildträger *m* ~ **changer** Bildwandler *m* (TV) ~ **changes** (frame change) Bildwechsel *n* ~ **circle** Bildkreis *m* ~ **coil** Abbildspule *f* ~ **collector** Bildfänger *m* ~ **component** Farbkomponente *f* ~ **contraction** Bildschrumpfung *f* (film) ~ **control at the transmitter** Bildkontrolle *f* am Geber ~ **converter** Bildwandler *m* (TV) ~**-converter tube** Bildwandlerröhre *f* (TV) ~ **coordinate** Bildkoordinate *f* ~ **defects** Bildfehler *m* ~ **definition** Bildschärfe *f* ~ **(field) definition** Bildfeldzerlegung *f* ~**-deflection yoke** Bildablenkjoch *n* ~ **displacement** Bildpunktverschiebung *f* ~ **dissector** Bildzerlegerröhre *f* ~ **distance** Bildweite *f* ~ **distortion** Bildverzerrung *f* ~**-drift** Bildschaukeln *n* ~ **duration** Bilddauer *f* ~ **effect** (antenna) Spiegeleffekt *m* ~ **element** Bildelement *n* ~ **(spot) element** Bildpunkt *m* ~ **erecting** bildaufrichtend ~**-erecting corneal microscope** bildaufrichtendes Hornhautmikroskop *n* ~ **erection** Bildaufrichtung *f* ~ **field** Bildfeld *n* ~ **field curvature** Bildfeldwölbung *f* ~**-field dissection** Bildfeldzerlegung *f* ~**-film** Bildträger *m* ~ **flyback** Bildrücklauf *m* ~ **force** Bildkraft *f*, Bildpunktkraft *f* ~**-force calculation** Bildkraftberechnung *f* ~ **formation** Abbildung *f* (bei Brillengläsern) ~ **formed by divergent lens** Zerstreuungsbild *n*
**image-forming** abbildend ~ **defect of the eye** Abbildungsfehler *m* des Auges ~ **factors** abbildende Faktoren *pl* ~ **objective** Abbildungsobjektiv *n* ~ **optical system** abbildendes optisches System *n* ~ **system** abbildendes System *n*
**image,** ~ **frame** Bildfeldrahmen *m* ~**-frame unsteadiness** Bildverwacklung *f* ~**-framing knob** Bildverstellungsknopf *m* ~**-framing shaft** Bildverstellungswelle *f* ~ **frequency** Bild-modulationsfrequenz *f*, -punktfrequenz *f*, -wechselfrequenz *f*, -wechselzahl *f*, -zahl *f*, Spiegelfre-

quenz *f* ~ **frequency jammer** Spiegelfrequenzstörsender *m* ~ **frequency rejection** Spiegelfrequenzunterdrückung *f* ~ **frequency rejector** Spiegelfrequenzsperre *f* ~ **frequency response** Spiegelfrequenzgang *m* ~ **frequency stopper** Spiegelfrequenzsperre *f* ~ **growth** Bildzuwachs *m*
**image-height,** ~ **adjustment** Höhenberichtigung *f* (bei Stereo-E-Messern) ~ **adjustment head (knob)** Höhenberichtigungsknopf *m* ~ **adjustment roll** Höhenberichtigungswalze *f* ~ **adjustment scale** Höhenberichtigungsteilung *f*
**image,** ~ **iconoscope** Zwischenbildikonoskop *n* ~ **impedance** Kennwiderstand *m* ~ **impulse** Bildwechselimpuls *m* ~ **intensifier** Bildverstärker *m* ~**-intensifying screen** Bildverstärker *m* ~ **intercept** Bildweite *f* ~ **interference** Spielfrequenzstörung *f* ~ **intergretation** Schirmbildauswertung *f* ~ **intermediate frequency** Bildzwischenfrequenz *f* ~ **inversion** Bildumkehrung *f* ~ **likeness** Ebenbild *n* ~ **line** Bild-, Brennlinie *f*, Spiegelleiter *m* ~ **margin of** ~ Bildrand *m* ~ **measuring** Bildmessung *f* ~**-measuring apparatus** Bildmeßgerät *n* ~ **modulation** Bildmodulation *f* ~ **motion** Bildwanderung *f* ~**-multiplier iconoscope** doppelseitige Mosaikröhre *f* ~ **orthicon** Bildaufnahmeröhre *f*, Zwischenbildorthikon *n* ~**-parameter filter** Wellenparameterfilter *n* ~ **pattern** Ladungsbild *n* ~ **period** Bildwechselzeit *f* (film) ~ **persistence** Bilddauer *f* ~**-phase constant** Vierpolwinkelmaß *n* ~ **phase factor** Vierpolphasenfaktor *m* ~ **pickup tube** Bildabtast(er)-rohr *n*, -röhre *f* ~ **picture amplification (or intensification)** Bildverstärkung *f* ~ **plane** Bild-ebene *f*, -fläche *f* ~ **point** Bildpunkt *m* ~ **potential** Bildkraft *f* ~ **power amplifier stage** Bildendstufe *f* (TV) ~**-producing tube** Bildschreibrohr *n* ~ **propagation factor** Übertragungsfaktor *m* ~ **ratio** Spiegelverhältnis *n* ~ **ray** Bildstrahl *m* ~ **reactor** Bildreaktor *m* ~ **receiver** Bildfänger *m* ~**-receiver tube** Bildfangrohr *n* ~ **reception with Braun's cathode-ray tube** Bildempfang *m* mit Braunscher Röhre ~ **rejection** Spiegelwellenabschwächung *f* ~ **rejection ratio** Spiegelfrequenzselektion *f* ~ **rejector** Spiegelfrequenzsperre *f* ~**-reproducing system** Abbildungssystem *n* ~ **response** Spiegelfrequenz-empfindlichkeit *f*, -sicherheit *f* ~ **retention** Bildkonservierung *f* (TV) ~ **reversion** Bildumkehrung *f* ~ **rotation** Bilddrehung *f* ~**-rotation correcting prism** Aufrichteprisma *n* ~ **scale** Abbildungsmaßstab *m* ~ **selector** Bildwähler *m* ~ **shell** Bildschale *f* ~ **shift** Bildverschiebung *f* ~ **space** Bildraum *m* ~**-spot size** Bildpunktgröße *f* ~ **spread** Bildverbreiterung *f* ~**-storing tube** Bildspeicherröhre *f* ~ **suppression** Spiegelselektion *f* ~ **surface** Bildebene *f* ~ **(frame)sweep frequency** Bildkippfrequenz *f* ~ **sweep point** Bildkippunkt *m* ~ **telegraphy** Bildtelegrafie *f* ~ **test** Probebild *n* ~ **time** Bildwechselzeit *f* (film) ~ **transfer** Übertragungsmaß *n* ~ **transfer coefficient** Vierpolübertragungsmaß *n* ~**-transfer constant** Übertragungsfaktor *m*, Vierpolübertragungsmaß *n* ~ **transmission** Bildfunk *m*, Bildübertragung *f* ~ **transmitter** Bildsender *m* ~ **transmitter of persons** Personenbildgeber *m* ~**-viewing tube** Bildwandlerröhre *f* ~ **voltage** Bildspannung *f*

imagery Abbildung *f*, Bildentwerfung *f*, Bilderzeugung *f*

imaginary fiktive, gedacht, imaginär ~ component of electric values Blindwert *m* elektrischer Größen ~ number imaginäre Zahl *f* ~ part Imaginärteil *m* ~ unit imaginäre Einheit *f*

imaging Bildentwerfung *f* ~ equation Abbildungsgleichung *f* ~ object Abbildungsgegenstand *m* ~ optics Abbildungsoptik *f*

imbalance Gleichgewichtsfehler *m*, Ungleichgewicht *n*, Unwucht *f*

imbedded pillar flacher Pilaster *m*

imbedding theory Einbettungstheorie *f*

imbibe, to ~ einsaugen, eintränken, saugen

imbibent Schluckstoff *m*

imbibition Einfärbung *f*, Quellung *f* ~ power Saugfähigkeit *f*

imbricated ziegeldachförmig

imbue, to ~ tränken

IMC (instrument meteorological conditions) Instrumentenwetterbedingungen *pl* (aviat.)

Imhoff tank Emscherbrunnen *m*

imitate, to ~ kopieren, nachahmen, nachbilden, nachmachen to ~ flawlessly faksimilieren

imitated imitiert, künstlich, nachgeahmt, unecht

imitation Nachahmung *f*, Nachbild *n*, Nachbildung *f*; flasch ~ art paper Natur(kunstdruck)-papier *n* ~ gold Halb-, Schein-gold *n* ~ gold leaf Goldschaum *n* ~ leather Ersatz-, Kunstleder *n* ~ leather roller Lederwalzenersatz *m* ~ parchment Pergamentersatzpapier *n*, Pergamin *n*, Pergaminpapier *n* ~ silk Kunstseide *f* ~ silver foil Rauschsilber *n* ~ wood Kunstholz *n* ~ wool fiber Luftzellwolle *f*

imitative deception Täuschung *f* durch Nachahmung

immaculate tadellos

immaterial stofflos, unkörperlich, untergeordnet, unwesentlich

immature unentwickelt, unreif, unzeitlich ~ residual soil reifer Verwitterungsboden *m*

immeasurable unermeßlich, unmeßbar

immediate alsbaldig, unverzüglich ~ action Beseitigen *n*, Beseitigung *f* von Hemmungen ~-action alarm sofortige Zeichengebung *f* ~ reply umgehende Antwort *f* ~ ringing erster Ruf *m* ~ stopping Momentstillstand *m*

immediately direkt, unmittelbar, verzögerungsfrei

immense unermeßlich, ungeheuer groß

immerged eingetaucht, versenkt liegend ~ chain Tauereikette *f* ~ groin Tauchbühne *f*

immerse, to ~ einsenken, eintauchen, tauchen, untertauchen

immersed eingesenkt, eingetaucht, versenkt ~ body eingetauchter Körper *m* ~ compass Flüssigkeits-, Schwimm-kompaß *m* ~ object eingetauchter Körper *m* ~ part of a filter surface Eintauchfläche *f*

immersion, to ~ paint tauchlackieren

immersion Eintauchen *n*, Eintauchung *f*, Untertauchung *f*, Versenken *n* ~ battery Tauchbatterie *f* ~ belt separator Tauchbandscheider *m* ~ counter Eintauchzähler *m* ~ furnace for ignition coils Tauchofen *m* für Zündspulen ~ heater Tauch-brenner *m*, -sieder *m* ~ hydroextractor for impregnating Tauchzentrifuge *f* ~ lens Immersions-linse *f*, -objektiv *n* ~ liquid

(microscope) Füllflüssigkeit *f* ~ mark Eintauchmarke *f* ~ meshed bag hydro extractor Tauchnetzzentrifuge *f* ~ method Eintauchverfahren *n* ~ objective Flüssigkeitslinse *f*, Immersions-linse *f*, -objektiv *n* ~ passage Unterwasserpassage *f* ~ pipe Eintauchrohr *n* ~ quench Tauchabschreckung *f* ~ refractometer Eintauchfraktometer *m* ~ refractometer with flow-cell attachment Eintauchrefraktometer *m* mit Durchflußküvette ~ shell Tauchhülse *f* ~ soldering bath Tauchlötbad *n* ~ tank Netzkasten *m* ~ test Tauchkorrosionsversuch *m* ~ time Eintauchzeit *f* ~ tube Tauchrohr *n*

imminent drohend, überhängend, unmittelbar bevorstehend, vorspringend

immiscibility Nichtmischbarkeit *f*, Unvermischbarkeit *f*

immiscible entmischt, nichtmischbar, nicht mischbar, unmischbar, unvermischbar, unverträglich

immobile immobil, unbeweglich

immobility Unbeweglichkeit *f*

immobilization Festlegung *f*, Unbeweglichmachung *f*

immobilize, to ~ lähmen, niederhalten, unbeweglich machen

immobilized bewegungsunfähig

immobilizing brake Rastung *f*

immoderate unmäßig

immovable unbewegbar, unbeweglich, unverrückbar ~ pin Schleifspule *f*

immune unempfindlich ~ to interference störfrei

immunity Immunität *f*, Straflosigkeit *f* ~ to corrosion Rostfreiheit *f* ~ from distortion (or disturbance) Störfestigkeit *f* ~ from interference Störfreiheit *f* ~ from noise Störfestigkeit *f*

immuration Einmauerung *f*

immutable unwandelbar

imp Gerüststange *f* ~-pole Rüstbaum *m*

impact, to ~ aufschlagen, einklemmen, zudrücken, zusammenpressen

impact Anprall *m*, Anschlag *m*, Aufprall *m*, Aufschlag *m*, Auftreffen *n*, Fallwucht *f*, (test) Kerbschlag *m*, Schlag *m*, Stoß *m*, Stoßfestigkeit *f*, Wucht *f*, Zusammenstoß *m* ~ of projectile Geschoß-aufschlag *m*, -einschlag *m* ~ of a projectile Aufschlag *m* eines Geschosses ~ and recombination Stoß *m* und Wiedervereinigung *f* ~ of the type bar Anschlag *m* des Typenträgers

impact, ~ absorption system Aufprall-Dämpfungsanlage *f* ~ anvil Schlagbär *m* ~ area Aufschlaggelände *n*, Trefffläche *f* ~ atomization Schlagzerstäubung *f* ~-ball hardness Fallhärte *f* ~ bend strength Schlagbiegefestigkeit *f* ~ bending test Schlagbiege-probe *f*, -versuch *m* ~ broadening Stoßverbreiterung *f* ~ buckling test Schlagknickversuch *m* ~ chisel Stoßmeißel *m* ~ cleaving Schlagspaltung *f* ~ compression test Schlagdruck-, Schlagtauch-versuch *m* ~ crusher Prallbrecher *m* ~ crushing test Schlag-probe *f*, ~ detonator Aufschlagzünder *m* ~ diagram (firing) Treffbild *n* ~ effect Schlagwirkung *f*, Stoßwirkung *f* ~ elasticity Schlagelastizität *f* ~-endurance test Dauerschlagfestigkeit *f* ~ energy Schlagarbeit *f* ~ extrusion Kalt-spritzen *n*, -strangpressen *n* ~-fatigue endurance Dauerschlaghaltbarkeit *f* ~ flexure strength Schlagbiegefestigkeit *f* ~ fluorescence Stoßfluoreszenz *f* ~ **force** Schlagkraft *f* ~ **fuse** Aufschlagzünder *m* ~ **hardness** Schlaghärte *f* ~**-hardness test** Einhiebverfahren *n* ~ **hardness tester** Schlaghärteprüfer *m* ~ **impact ionization rate** Stoßionisations-Wahrscheinlichkeit *f* ~ **knife-edge** Schlagschneide *f* ~ **load** Schlagbeanspruchung *f* ~ **magnet** Anschlagmagnet *m* ~ **mixer** Turbomischer *m* ~ **molding** Schlagpressen *n* ~ **multiple** Stoßvielfache *n* ~ **number** Stoßzahl *f* ~ **nut** Anschlagmutter *f* ~ **nut runner** Schlagmutteranzeiger *m* ~ **plate** Anschlagplatte *f* ~ **point** Anschlagpunkt *m* ~ **potential** Stoßspannung *f* ~ **pressure** Stau-, Stoß-druck *m* ~ **pulling test** Schlagzerreißversuch *m* ~ **pulverizer** Schlagmühle *f* ~ **radiation** Stoßstrahlung *f* ~ **rail** Laschenschiene *f* ~ **ram compression** Stauverdichtung *f* ~ **rate** Stoßwahrscheinlichkeit *f* ~ **resistance** Schlagfestigkeit *f*, spezifische Kerbschlagarbeit *f* ~ **resistant** schlagfest ~ **sound** Trittschall *m* ~ **spring** Schlagfeder *f* ~ **strength** Kerbschlagfestigkeit *f*, Schlagfestigkeit *f* ~ **stress** Schlag-, Stoß-beanspruchung *f*, Schlagkraft *f* ~ **surface** Schlagfläche *f* ~ **tearing test** Schlagzerreißversuch *m* ~ **tensile stress** Schlagzugbeanspruchung *f* ~ **tensile test** Schlagzugversuch *m* ~ **tension test** Schlagzerreißversuch *m*, Zugversuch *m* mit Schlagbeanspruchung ~ **test** Dauerschlagprobe *f*, dynamische Prüfung *f*, Eindruckprüfung *f*, Fall-, Kerb-versuch *m*; Schlag-probe *f*, -versuch *m* ~**-testing apparatus** Fallwerk *n* ~**-testing field** Stoßprüffeld *n* ~ **testing machine** Schlagwerk *n* ~ **tube** (dynamisches) Pitotrohr *n*, Staurohr *n* ~ **value** Kerbschlagmeßwert *m*, Kerbzähigkeit *f* ~ **velocity** Auftreffgeschwindigkeit *f* ~ **wave** Front-, Kopf-, Stirn-, Stoß-welle *f* ~ **wave resistance** (in supersonics) Wellenwiderstand *m*

**impacting** (by electrons) Beaufschlagung *f* ~ **rod** Stößel *m*

**impactor,** ~ **anode** Prallelektrode *f* ~ **electrode** Rückprallelektrode *f* ~ **plate** Auffänger-, Fangplatte *f*

**impair, to** ~ beeinträchtigen, beschädigen, schwächen, verschlechtern

**impaired** beeinträchtigt, fehlerhaft, vermindert

**impairing** Benachteiligung *f*

**impairment** Beeinträchtigung *f*, Verschlechterung *f* ~ **of hearing** Leitungstaubheit *f*

**impalpable** ungreifbar, durch Fühlen nicht wahrnehmbar

**impart, to** ~ ausliefern, (knowledge) etwas beibringen, erteilen, geben, (to) einen Zustand mitteilen

**impassable** nicht umlauffähig, ungangbar

**impeccable** einwandfrei

**impedance** Beschwerung *f*, Dämpfung *f*, Drosselspule *f*, Drosselung *f*, Gegeninduktivität *f*, Impedanz *f*, scheinbarer Widerstand *m*, Scheinwiderstand *m*, Wechselstromwiderstand *m*, Widerstand *m*, wirksamer Widerstand *m* ~ **of the grid circuit** Gitterkreisimpedanz *f*

**impedance,** ~ **amplifier** Impedanzverstärker *m* ~ **angle** Phasen-maß *n*, -winkel *m*, Winkelmaß *n* ~ **bond** Drosselstoß *m* ~ **bridge** Scheinwiderstandsmeßbrücke *f* ~ **coil** Drossel *f*, Drosselspule *f* ~ **(choking) coil** Bogenlampendrosselspule *f* ~ **compensator** Scheinwiderstandsangleicher *m* ~ **corrector** Anpassungstransforma-

tor *m* ~**-coupled amplifier** Drossel-spulenverstärker *m*, -verstärker *m* ~ **coupling** direkte Kopplung *f* ~ **device** Hemmvorrichtung *f* ~ **discontinuities** (in slotted line measurements) Stoßstellen *m* ~ **disk** Widerstandsscheibe *f* ~ **factor** Phasenwinkel *m* ~ **imitation** Scheinwiderstandsnachbildung *f* ~ **irregularities** Schwankungen *f* im Wellenwiderstand ~ **irregularity** Ungleichförmigkeit *f* des Wellenwiderstandes ~**-matching transformer** Anpassungstransformator *m* ~ **method** Impedanzmethode *f* ~ **mismatch** Impedanzfehlanpassung *f* ~ **network** aus Impedanzen gebildeter Kettenleiter *f* ~ **ratio** Scheinwiderstandsverhältnis *n* ~ **transformer** Impedanzwandler *m* ~ **unbalance measuring set** Nachbildungsmesser *m* ~ **voltage** Kurzschlußspannung *f* ~ **voltage of a transformer** Nennkurzschlußspannung *f* ~ **wave trap** Saugkreis *m* ~ **wheel** Ausgleichschwungscheibe *f*, Schwungmassenrolle *f*

**impede, to** ~ anhalten, arretieren, bremsen, erschweren, festhalten, feststellen, verhindern, Widerstand entgegensetzen

**impeded harmonic operation** erzwungene Magnetisierung *f*

**impediment** Behinderung *f*, Hindernis *n*, Verhinderung *f* ~ **on the line** Hindernis *n* auf der Strecke ~ **to movement** Bewegungshemmung *f*

**impedometer** Impedanzmesser *m*

**impedor** Impedanz *f*, Scheinwiderstand *m*

**impel, to** ~ antreiben, treiben, vorwärtstreiben **to** ~ **forward** vorwärtsschnellen

**impellant** Nutzgas *n*

**impeller** Antreiber *m*, Flügel *m*, Flügelrad *n*, Gebläserad *n*, Kreiselrad *n*, (of a ventilator) Laufrad *n*, Leitrad *n*, loses Rad *n*, Luftflügel *m*, Maschinenelement *n* das einen Impuls erteilt, Propeller *m*, Rad *n*, Schaufelrad *n*, Windflügel *m*, Windrad *n*, Zentrifugalladerrad *n* ~ **of pump** Pumpenflügelrad *n*

**impeller,** ~ **blade** Laufradflügel *m*, Windflügel *n* ~ **blades of a turbine** Bewegungsschaufeln *pl* ~ **breaker** (cutting mill) Pralltellermühle *f* ~ **channel** Laufradkanal *m* ~ **compressor** Verdichterrad *n* ~ **drive** Flügelrad-, Laderlaufradantrieb *m* ~ **head** Schleuderkopf *m* ~ **intake guide vane** Laufradeinleitschaufel *f* ~ **mixer** Kreiselrührer *m* ~ **passage** Laufradkanal *m* ~ **pressure** Laderdruck *m* ~ **pump** Flügelradpumpe *f* ~ **rotor** Laderlaufrad *n* ~ **set** Ventilatorrad *n* (Kühlturbine) ~ **shaft** (of spindle jet) Laufradwelle *f*, Vorverdichterwelle *f* ~ **vane** Laufradschaufel *f* (Ventilator)

**impending failure** drohender Effekt *m*

**impenetrability** Undurchdringlichkeit *f*

**impenetrable** undurchdringlich, undurchlässig

**imperfect** abweichend (paper mfg.), fehlerhaft, geringwertig, mangelhaft, unvollendet, unvollkommen, unvollständig ~ **capacity** Verlustkapazität *f* ~ **combustion** unvollkommene oder unvollständige Verbrennung *f* ~ **dielectric** unvollkommener Isolator *m* ~ **elasticity** unvollkommene Elastizität *f* ~ **ink coverage** blaßgedruckte Stelle *f* (print.) ~ **insulation** unvollkommener Isolator *m* ~ **reaction** Teilreaktion *f* ~ **sheet** Fehlbogen *m* ~ **structure** Lockerstelle *f* (Smekal) ~ **tape** Fehlband *n* ~ **tuning** unscharfe Abstimmung *f*

imperfection Halbheit *f*, Mangel *m*, Unvoll-
kommenheit *f*
imperil, to ~ gefährden
imperishable unvergänglich, unwandelbar, unzerstörbar ~ photograph (or picture) unver-
gängliches Lichtbild *n*
imperishableness Unvergänglichkeit *f*
impermeability Undurchdringlichkeit *f*, Un-
durchlässigkeit *f* ~ to gas Gasdichtigkeit *f* ~ to
water Wasserdichtigkeit *f*
impermeabilization Dämmung *f*
impermeabilizing, ~ of joints (highway construction) Fugendichtung *f* ~ bar Dichtungs-
stab *m* ~ sheet Dichtungsblatt *n*
impermeable abgedichtet, undurchdringbar, undurchdringlich, undurchlässig, wasserdicht ~
to air gasdicht, luftundurchlässig ~ to gas gas-
undurchlässig to make ~ (or waterproof)
wasserdicht ausrüsten
impermeable, ~ core wasserdichter Kern *m* ~
facing wasserdichte Verkleidung *f* ~ material
Verdichtungsmaterial *n* ~ screen wasserdichte
Verkleidung *f* ~ stratum undurchlässige Bodenschicht *f*
impervious dicht, lichtdicht, undurchdringlich,
undurchlässig, wasserundurchlässig ~ to acids
säurefest ~ to heat tropenfest, wärmeundurchlässig ~ to light lichtundurchlässig, undurchsichtig
impervious blanket Abdichtungsteppich *m*
imperviousness Undurchdringlichkeit *f*, Un-
durchlässigkeit *f* ~ due to sizing Leimfestigkeit *f*
impetuous heftig, rasch, ungestüm
impetus Antrieb *m*
impinge, to ~ anprallen, anstoßen, aufprallen,
aufschlagen, (upon) auftreffen
impingement Aufprall *m*, Auftreffen *n*, Beaufschlagung *f* ~ of drops Tropfenschlag *m* ~ on
the spot (cathode) Belegung *f* des Fleckes
impingement, ~ black aktiver Gasruß *m* ~ test
Spritzversuch *m* ~-type air filter Prallflächen-
Luftfilter *m*
implant Einlage *f*, Implantat *n*
implantation Implantation *f*, Ein-, Ver-, Über-
pflanzung *f*
implement Gerät *n*, Werkzeug *n*, Werkzeugvor-
richtung *f*, Zeug *n* ~ for drain cleansing Kanalreinigungsgerät *n*
implementation Verwirklichung *f*
implements Arbeitsgerät *n*, Gerätschaft *f*,
Gezäh(e) *n*, Handwerkszeug *n* (farming) ~
Gerät *n* ~ for dismantling a nozzle Düsenaus-
ziehvorrichtung *f*
implicate, to ~ mit einbegreifen, verflechten
implication Miteinbegriffensein *n*
implicit stillschweigend (inbegriffen) ~ function
unentwickelte Funktion *f* (math.) ~ method
Nahmethode *f*
implied einbegriffen, stillschweigend
implode, to ~ platzen, zusammenbrechen
implosion Einfallen *n* ~ of a vacuum tube Röhren-
zerplatzen *n*
imply, to ~ einbeziehen
import, to ~ einführen
import Einfuhr *f*, Einfuhr-artikel *pl*, -waren *pl* ~
duty Einfuhrzoll *m* ~ paid through clearing
Kompensationsgeschäft *n* ~ quotas Einfuhr-
quoten *pl* ~ trade Einfuhrhandel *m*

importance Bedeutung *f*, Belang *m*, Tragweite *f*,
Wichtigkeit *f* of ~ von Belang *m* of no ~ nicht
von Belang *m* to be of ~ von Bedeutung *f* sein
important bedeutend, beträchtlich, erheblich,
groß, wichtig ~ for life lebenswichtig
importation Einfuhr *f*, Zufuhr *f*
impose, to ~ auferlegen, ausschießen (print.) to
~ duty bezollen to ~ the pages anew die Spalten
*pl* umschießen (print.)
imposed oscillation (or vibration) aufgedrückte
Schwingung *f*
imposing eindrucksvoll; Einschießen *n* (print.) ~
board Ausschießbrett *n* (print.) ~ stone Setzstein *m* ~ surface Formatplatte *f*
imposition Format-bildung *f* (print.), -einrich-
tung *f*, -machen *n*, Formeinrichtung *f* ~ of a
signal Aufschaltung *f*
impossibility of delivery Unzustellbarkeit *f*
impossible unmöglich ~ to execute unausführbar
impound, to ~ aufspeichern, aufstauen, einsperren, stauen
impoundage Eindämmung *f*
impounded water level Stauspiegel *m*
impounding of water Aufspeicherung *f* des Wassers
impoverish, to ~ auszehren, verarmen
impoverished material abgereichertes Material *n*
impoverishment Aussaugung *f*, Erschöpfung *f*,
Verarmung *f*, Vertaubung *f* (ore)
impracticability Unausführbarkeit *f*, Unbrauchbarkeit *f*
impracticable unausführbar, unzweckmäßig
impreg Preßholz *n* (harzbehandelt)
impregnate, to ~ durchtränken, erfüllen, sättigen, tränken to ~ timber with soluble glass das
Holz verkieseln
impregnated durchtränkt, getränkt ~ cable
getränktes oder imprägniertes Kabel *n*, Masse-
kabel *n* ~ carbon salzgetränkte Kohle *f* ~
fabric imprägnierter Stoff *m*, Hartgewebe *n* ~
tape Einschichtband *m*
impregnating, ~ agent Imprägnier-, Imprä-
nierungs-mittel *n* ~ coat Imprägnierschicht *f* ~
compound Imprägniermittel *n* ~ device Tränk-
vorrichtung *f* ~ liquor Imprägnierflotte *f* ~
oil Tränköl *n* ~ pan Imprägnierpfanne *f* ~
preparation Tränkungsmittel *n* ~ tank Tränk-
gefäß *n*, -kessel *m* ~ varnish Tränklack *m* ~ vat
Imprägniertrog *m*
impregnation Imprägnierung *f*, Tränkung *f* ~ of
timber Holztränkung *f*
impress, to ~ abdrucken, abdrücken, eine Spannung anlegen, aufbringen (signal), aufdrücken,
aufprägen, eindrucken (print.) to ~ somebody un-
favorably auf jemanden ungünstig einwirken to
~ upon einschärfen to ~ a voltage upon a circuit
einem Stromkreis eine Spannung *f* aufdrücken
impressed, ~ pressure aufgedrückte Spannung *f* ~
voltage aufgeprägte oder eingeprägte Spannung *f*
impressibility Eindruckempfänglichkeit *f*
impression Abdruck *m*, Abklatsch *m* (print.),
Aufdrücken *n*, Druck *m*, Drucklegung *f*, Eindruck *m*, Gegenabdruck *m* (print.), Gepräge *n*
~ of colors Farbendruck *m* ~ of depth Tiefen-
eindruck *m* ~ of the edge of a coin Kräuselung *f*
einer Münze ~ of a photograph fotografische
Aufnahme *f* (Vorgang) ~ of space Raumein-
druck *m*, räumliche Vorstellung *f*

**impression, ~ according to scale** Mastabdruck *m*
**~ block** Abdruck-büchse *f*, -stempel *m* **~**
**control lever** Druckeinstellhebel *m* **~ cylinder**
**dressing** Druckzylinderbekleidung *f* **~ cylinder**
**gripper** Druckzylindergreifer *m* **~ die** Petschaft
*n* **~ method** Abdruckverfahren *n* **~ molding**
Pressen *n* ohne Druck **~ operating rack** Doppel-
zahnstange *f* **~ roll** Gegendruckzylinder *m*
(print.) **~ roller** Druckrolle *f* **~ setting** Druck-
bestimmung *f* **~ support bar** Abstützleiste *f* **~**
**time** Druckzeit *f* (print.)
**impressionable** eindrucksfähig
**impressional, ~ force regulation** Druckstärke-
regulierung *f* **~ strength** Druckkraft *f* (print.)
**impressive** eindringlich
**imprint, to ~** abdrucken, aufdrucken, eindrük-
ken, einprägen, prägen
**imprint** Abdruck *m*, Aufdruck *m*, Eindruck *m*,
Zeichen *n* **~-depth indicator** Eindrucktiefenmesser *m*
**imprinter** Druckgerät *n*
**imprinting method** Abdruckmethode *f*
**improper** unanständig, ungeeignet, unpassend,
unsachgemäß, unzweckmäßig **~ fraction** unechter Bruch *m*
**improperly** unsachgemäß **~ trued** schlecht ausgerichtet
**improve, to ~** aufbessern, ausbauen, ausbessern,
ausbilden, bessern, sich bessern, durchbilden,
entdämpfen, verbessern, veredeln, verfeinern,
(quality) vergüten, vervollkommen, (lead) zunehmen **to ~ the precision (or the selectivity) of**
**tuning** die Abstimmschärfe erhöhen
**improved** verbessert **~ aplanatic magnifier** verbesserte aplanatische Lupe *f* **~ road** unterhal-
tener Fahrweg *m* **~ wood** vergütetes Holz *n*
**improvement** Aufschwung *m*, Belebung *f*, Benutzung *f*, Besserung *f*, Dämpfungsverminderung *f*, Entdämpfung *f*, Fortschritt *m*, Verbesserung *f*, Vered(e)lung *f*, Verfeinerung *f*
**improvement, ~ of the alignment** Regelung *f* des
horizontalen Verlaufs **~ of a device to a degree**
**where its use on shipboard is justified** bordreif **~**
**of the ground** Grundverbesserung *f* **~ of land**
**for cultivation by means of drainage** Kulturtechnik *f* **~ of land for cultivation by means of**
**irrigation** Melioration *f* **~ of land for culti-**
**vation by means of irrigation, (or by means of**
**conservation)** Kulturtechnik *f* **~ of a position**
Einrichten *n* einer Stellung **~ in quality** Quali-
tätsverbesserung *f* **~ in resolution and black-**
**white contrast** Konturenversteilerung *f* (TV) **~**
**of rivers and modification of the bed** Fluß-
regelung *f* **~ of a river channel** Flußregulierung
*f* **~ of river channels** Flußkorrektion *f*
**inprovement, ~ cutting** Durchhieb *m* **~ system**
**time** Ausbesserungszeit *f* des Systems
**improving** Veredeln *n*, (lead) Vorraffination *f* **~**
**furnace** Vorraffinierofen *m*
**improvisation** Aushilfe *f*, Notbehelf *m*
**improvised** behelfend, behelfs-; **~ mount** Hilfslafette *f* **~ obstacle** Schnellsperre *f* **~ plumb bob**
behelfsmäßige Setzwaage *f* **~ sight** Notvisier *n*
**~ sledge** Behelfsschlitten *m*
**imprudence** Leichtsinn *m*, Unbedachtsamkeit *f*,
Unvorsichtigkeit *f*
**impulse, to ~** einen Stromkreis *m* anstoßen,
einen Impuls *m* erteilen

**impulse** Anregung *f*, Anstoß *m*, Antrieb *m*, Bewegungsgröße *f*, Drang *m*, Impuls *m*, Regung *f*,
Reiz *m*, Schaltstoß *m*, Stoß *m*, Stromreiz *m*,
Trieb *m*, Unruhefeder *f*, Zug *m* **~ by hand**
Handantrieb *m*
**impulse, ~ action** Nummer(n)wahl *f* **~ approxi-**
**mation** Stoßapproximation *f* **~ breakdown**
Stoßdurchschlag *m* **~ circuit** Einstellweg *m*
**~-condensing turbine** Gleichdruckkondensa-
tionsturbine *f* **~ contact of the dial** Nummernscheibenkontakt *m* **~ contactor** Impulsschütz
*m* **~ control** Impulssteuerung *f* (bei Reihen-
meßkammern) **~ counter** Impulszähler *m* **~**
**counting relay for automatic message account-**
**ing** Impulszählrelais *n* bei Gebührenerfassung
**~ coupling** Abschnappkupplung *f* **~ coupling**
**operation** Schnapperbetrieb *m* **~ coupling**
**releasing speed** Schnapperauslösedrehzahl *f* **~**
**current** Stoßkurzschlußstrom *m* **~ direction-**
**finding method** Impulspeilverfahren *n* **~**
**discharge** Stoßentladung *f* **~ distortion** Impulsverzerrung *f* **~ emission** Impulsgabe *f* **~ excita-**
**tion** Stoßerregung *f* **~ exciter** Stoßsender *m* **~**
**firing** Impulszündung *f* **~ frequency** Impulsfrequenz *f*, Stromstoßgeschwindigkeit *f*, Tastfrequenz *f* **~-frequency telemetering** Impulsfre-
quenzfernmessung *f* **~ generation** Impulsge-
winnung *f*, Taktgebung *f* **~ generator** Impuls-
formerstufe *f*, -geber *m*, -generator *m*, Kippen-
generator *m*, Stoßgenerator *m*, Taktgeber *m*,
Wanderwellengenerator *m* **~ image** Impulsbild
*n* **~ intervalometer** Impulsregler *m* **~ machine**
Unterbrechermaschine *f* **~ magnet** Schnapper-
magnet *m* **~ maker** Impulsgeber *m* **~ meter**
Impulszähler *m* **~ microphone** Abgangsmikrofon *n* **~ moment** Flächenträgheitsmoment *n* **~**
**motion** Anlaufbewegung *f* **~ noise** Impuls-
rauschen *n* **~ period** Stromstoßdauer *f* **~ polar**
**duplex telegraphy** Stromstoßunterlagerungs-
telegrafie *f* **~ ratio** Impulsverhältnis *n*, Strom-
stoßteilung *f*, -verhältnis *n* **~ reaction turbine**
Gleichdrucküberdruckturbine *f* **~ receiver**
Schallabgangs-, Stromstoß-empfänger *m* **~-re-**
**ceiving place (or point)** Steuerstelle *f* **~ recorder**
Mittelwertschreiber *m* **~ relay** Impuls-, Strom-
stoß-relais *n* **~ repeater** Impulsübertrager *m* **~**
**repeating** (in automatic teleph.) Impulsübertragung *f* **~ screw** Impulsschraube *f* **~ selection**
Impulswahl *f* **~ sender** Impulsgeber *m* **~ sender**
**method** Impulsumrechnungsverfahren *n* **~-send-**
**ing key** Tastenimpulsgeber *m* **~ sequence** Im-
pulsfolge *f* **~ series** Impulsreihe *f* **~ spark**
Schnappfunke *m* **~ spark gap** Schnapper-
schlagweite *f* **~ spring** Kontaktfeder *f* des
Stromstoßgebers, Stromstoßfeder *f* **~ stage**
Gleichdruckstufe *f* **~ starter** Abschnappkupplung *f*, (beim Magnetzünder) Antriebs-
starter *m* **~ stepping** Nummerwahl *f*, Schnapper
*m* eines Anlaßmagnetes **~-storing device** Im-
pulsspeicher *m*, Register *n*, Stromstoß-empfänger *m*, -speicher *m* **~ switch** Impulsgenerator *m* **~ test** Impulsprüfung *f* **~-testing field**
Stoßprüffeld *n* **~ train** Impuls-folge *f*, -gruppe *f*,
-reihe *f*, -zug *m* **~ transmission** Impulsgabe *f* **~**
**transmitter** Impulsgeber *m*, Stoßsender *m*,
(supervisory control) Zahlengeber *m* **~ trans-**
**mitter for revolution indicator** Drehzahlgeber
*m* **~ turbine** Aktions-, Freistrahl-, Gegendruck-,

Gleichdruck-turbine f ~ voltage Spannungs-stoß m, Stoßspannung f ~-voltage break-down strength Stoßdurchschlagfestigkeit f ~ wave Stoßwelle f ~ wheel Gleichdruckrad n ~ width (or length) Impulsbreite f
**impulser** Taktgeber m
**impulsing** Impulsgabe f, Stromstoßgabe f ~ circuit Stoßkreis m ~ device Impulszentrale f ~ means Taktgeber m ~ relay Fortschalt-, Impuls-relais n ~ switch Stromstoßschalter m ~ transmitter Kommandostelle f (remote control)
**impulsion** Antrieb m, Schwung m
**impulsive** anstoßend, antreibend ~ balance Stoß-bilanz f ~ discharge aperiodische Entladung f ~ force Stoßkraft f ~ moment Stoßmoment n ~ motion Anlaufbewegung f ~ noise unter-brochenes Rauschen n ~ reaction Impulsüber-tragungsreaktion f
**impure** unrein ~ iron alum Bergbutter f ~ ore Pochgänge pl
**impureness** Unreinheit f
**impurities** Einschlüsse pl (Gas)
**impurity** Begleit-körper m, -stoff m; Beimengung f, Unreinheit f, Unreinigkeit f, Verunreinigung f ~ additions Fremdatomzusätze pl ~ band conduction Störbandleitung f ~ center Fremd-störstelle f ~ concentration Fremdstoffkon-zentration f ~ nucleus Verunreinigungskern m ~ photo-conduction Verunreinigungsfotoeffekt m ~ scattering Streuung f an Fremdatomen ~ semiconductor Störstellenhalbleiter m ~ spot Störstelle f ~ substance Störsubstanz f (chem.)
**imputability** Zurechenbarkeit f
**impute, to** ~ beimessen, zur Last legen, unter-stellen, zumessen, zuschreiben
**imputrescible** unverfaulbar, unverweslich
**in,** ~-air dose Luftdosis f ~-and-~ (method of packing in) ineinandergeschlagen (paper mfg.) ~-and-out-of-focus effect Atmen n des Films ~-and-out movement of claws Eingreifen n der Greifer ~-call terminal call Endgespräch n ~-line in der Leitung liegend ~-line engine Reihenmotor m, Reihenstandmotor m ~-line motor Reihenmotor m ~-line multibank engine Reihenmotor m mit mehreren Zylinderreihen ~-range computer Schlußbereichrechner m ~-series reihenweise
**inability** Unfähigkeit f ~ to deform Unverform-barkeit f
**inaccessibility** Unzugänglichkeit f ~ axiom Unerreichbarkeitsaxiom n
**inaccessible** unzugänglich
**inaccuracy** Ungenauigkeit f
**inaccurate** ungenau
**inaction** Bewegungslosigkeit f, Untätigkeit f
**inactivate, to** ~ unwirksam machen
**inactivation** Inaktivierung f
**inactive** außer Dienst, leblos, (exchange) lustlos, träge, untätig, unwirksam, wirkungslos ~ material totes Material n ~ mixer Rundherd-mischer m ~ status Wartestand m
**inactivity** Beschäftigungslosigkeit f, Trägheit f (chem.), Untätigkeit f
**inadequacy gap** Nachholbedarf m
**inadequate** unzulänglich
**inadvertence** Nachlässigkeit f, Unachtsamkeit f, Versehen n
**inadvertent mistake** Flüchtigkeitsfehler m

**inadvertently** irrtümlich
**inalienable** unübertragbar
**inanimate picture** Standbild n
**inarticulate** undeutlich
**inarticulateness** undeutliche Aussprache f, Un-deutlichkeit f
**inattentive** achtlos, unachtsam, unaufmerksam
**inaudibility** Unhörbarkeit f
**inaudible** unhörbar
**inauguration** Amtseinführung f
**inboard** binnenbords, innerer ~ float Innenbord-schwimmer m ~ hinge fitting Innenbordschar-nier n ~ stabilizing float stabilisierender Innen-bordschwimmer m
**inbound** Anflug .... ~ tracking Kursanflug m (nav.)
**inbye fan** Sonderventilator m
**incalculable** unberechenbar
**incandescence** Glühen n, Glüherscheinung f, Glut f, Weiß-glühen n, -glühhitze f, -glut f ~ of filament Glühen n des Fadens
**incandescent** glühend, weißglühend, weißwarm ~ acetylene burner Azetylenglühlichtbrenner m ~ acetylene lighting Azetylenglühlicht n ~ article Glühstoff m ~ body Glühkörper m ~ bulb Glühbirne f ~ burner Glühlichtbrenner m ~ cartridge Glimmstift m ~ cathode glühende Kathode f, Glühkathode f ~ cathode tube Glüh-kathodenröhre f ~ exploder Glühzünder m ~ filament Glühkathode f ~-filament image Glühfadenbild n ~ gas burner Gasglühlicht-brenner m ~ gas light Gasglühlicht n ~ gas lighting Gasglühlichtbeleuchtung f ~ lamp Glüh-birne f, -lampe f, Vakuumglühlampe f ~ lamp base Glühlampensockel m ~-lamp bulb Glühlampenkolben m ~-lamp filament Glüh-lampenfaden m ~ light Glühlicht n ~ lighting Glühlichtbeleuchtung f ~ luminous radiator Temperaturstrahler m ~ mantle Glühkörper m ~ mass Glühmasse f ~ metallic oxide cathode glühende Metalloxydkathode f ~ oil-gas burner Ölgasglühlichtbrenner m ~ oil-gas lighting Ölgasglühlicht n ~ petroleumvapor burner Petroleumglühlichtbrenner m ~ petroleum-vapor lighting Petroleumglühlicht n ~ plug switch Glühkerzenschalter m ~ spiral Glüh-schleife f (Kerze) ~ tube Glührohr n ~ welding Widerstandsschweißung f ~ wire (for light-fuses or explosives) Glühkörper m
**incapability** Unfähigkeit f
**incapable** unfähig ~ of moving bewegungsun-fähig
**incapacity** Unfähigkeit f
**incase, to** ~ einschleifen (in ein Gehäuse)
**incendiary,** ~ agent Brand-mittel n, -stoff m, Flammenmittel n ~ ammunition Brandmuni-tion f ~ bomb Brandbombe f ~ bottle Brand-flasche f ~ bullet Brandgeschoß n ~ composi-tion Brandmasse f ~ damage Brandschäden (aviat.) pl ~ effect Brandwirkung f ~ rocket Brandrakete f ~ shell Brand-geschoß n, -granate f ~ torch Brandfackel f
**incentive** Anreiz m, Ansporn m, Anstoß m; lohnanreizend
**inception** Anfang m, Ansprache f (rdo), (of metals) Einsetzen f
**inch** Zoll m ~ graduation Zollskalierung f ~ scale Zoll-stab m, -stock m ~ thread Zollsteigung f

**inching** (slow action) Druckabfangen *n* ~ **operation** Tippschaltung *f* ~ **service** Tippbetrieb *m* ~ **wheel** Handeinstellung *f*
**incide, to** ~ einfallen
**incidence** Anstellung *f*, Einfall *m* (phys.), Einfallen *n* ~ **of rays** Strahleneinfall *m* ~ **of rejects** Ausschußziffern *pl* ~ **of traffic over a period** Verteilung *f* des Verkehrs in einer bestimmten Zeit
**incidence,** ~ **bracing** Tiefenkreuzverspannung *f* ~ **indicator** Anstellwinkelanzeiger *m* ~ **range** Anstellbereich *m* ~ **setting former** Einstellwinkellehre *f* ~ **wire** Stielauskreuzung *f*, Tiefenkreuzdraht *m*
**incident** auftreffend (opt.), direkt (light) ~ **at small angle** flach auffallend oder auftreffend
**incident,** ~ **beam** Einfallsstrahl *m* ~ **light** auffallendes, eingestrahltes oder einfallendes Licht *n* ~ **light condenser** Auflichtkondensator *m* ~ **particle** einfallendes Teilchen *n* ~ **ray** einfallender Strahl *m* ~ **sound** einfallender Schall *m* (acoust.) ~ **wave** einfallende Welle *f*, Eingangswiderstand *m*
**incidental** beiläufig, zufällig ~ **action** Nebenklage *f* ~ **charge (or expense)** Nebenausgabe *f* ~ **computation** Nebenrechnung *f* ~ **light microscope** Auflichtmikroskop *n* ~ **proposition** Nebensatz *m* ~ **provision** Nebenbestimmung *f* ~ **resonance** Störresonanz *f* ~ **work** Nebenarbeit *f*
**incinerate, to** ~ einäschern, veraschen
**incinerating dish** Veraschungsschale *f*
**incineration** Veraschung *f*
**incinerator** Einäscherungs-, Kehricht-, Müllverbrennungs-ofen *m*
**incipience of firing (or of striking)** Einsetzen *n* der Zündung, Zündungseinsetzen *n*
**incipiency** Einsatz *m*
**incipient** anfangend, angehend, beginnend, eben beginnend, einleitend, träge; (explosives) Einleitung *f* ~ **beam return** Bildkippeneinsatz *m* ~ **combustion** träge Verbrennung *f* ~ **current** Anfangsstrom *m* ~ **failure** beginnender Bruch *m* ~ **(crack) flaw** Anriß *m* ~ **flyback** Zeilenkippeinsatz *m* ~ **fracture** Anbruch *m* ~ **frame flyback** Bildkippeinsatz *m* ~ **knocking** Klopfgrenze *f* ~ **plastic flow** Anlaufvorgang *m* des plastischen Fließens
**incised** eingekerbt
**incising** Einkerbung *f* (des Holzes)
**incision** Einschnitt *m*, Schnitt *m*
**incitation** Anregung *f*
**incite, to** ~ anreizen, anspornen, aufhetzen
**incitement** Anreizung *f*
**inclemency of the weather** Witterungsunbilden
**inclement** unfreundlich
**inclinable** kippbar, neigbar, schrägstellbar, umlegbar ~ **adjustment** Schrägeinstellbarkeit *f* ~ **(bascule) camera** Pendelkamera *f* ~ **prism** Kipp-Prisma *n* ~ **single-crank power press** Kurbelpresse *f* mit schrägstellbarem Körper ~ **stand** Stativ *n* mit Kippe
**inclination** Böschung *f*, Einfallen *n*, Fall *m*, Höhenrichtung *f* (Neigung), Inklination *f*, Lust *f*, Neigung *f*, Orientierung *f*, Schiefe *f*, Schräge *f*, Schrägstellung *f*, Senkung *f*, Steigung *f*
**inclination,** ~ **of axis** Achs(en)neigung *f* ~ **of the camera** Neigung *f* der Kamera ~ **of flight path**

Bahnneigung *f* ~ **of the grid bar** Roststabneigung *f* ~ **of letters** Schiefstehen *n* von Buchstaben ~ **of ray** Strahlneigung *f* ~ **of steering knuckle pivot** Spreizung *f* des Lenkzapfens ~ **of strata** Einfallen *n* der Schichten ~ **of strut** Stielneigung *f*, Strebenbiegung *f*
**inclination,** ~ **angle** Neigungswinkel *m* ~ **angle of flight path** Bahnneigungswinkel *m*, Flugbahnneigungswinkel *m* ~ **axis** Kippachse *f* ~ **balance** Neigungswaage *f* ~ **compass** Inklinationsbussole *f* ~ **joint (or hinge)** Kippgelenk *n* ~ **needle** Inklinationsnadel *f* ~ **weighing device** Neigungswägevorrichtung *f*
**incline, to** ~ abböschen, abschrägen, abweichen, auslenken (g/m), inklinieren, kippen, neigen
**incline** Abhang *m*, Anlauf *m*, Anzug *m*, Böschung *f*, Gefälle *n*, Rampe *f*, Steigung *f* **(length of)** ~ Schrägstrecke *f* **of runway** Gefällstrecke *f* ~ **housing** Auflaufergehäuse *n*
**inclined** abdachig, gekippt, geneigt, schief, schiefliegend, schräg ~ **to the left** linkssteigend ~ **to the right** rechtssteigend ~ **upward** Aufwärtsseilung *f*
**inclined,** ~ **apron** geneigt Sohle *f* in ~ **arrangement** schräg gelagert ~ **ascent** Schrägschuß *m* (rocket) ~ **ball mill** Schrägkugelmühle *f* ~ **belt conveyer** schrägliegender Bandförderer *m*, Schrägförderband *n* ~ **binocular tube** binokularer Schrägtubus *m* ~ **bore** schräge Bohrung *f* ~ **bottom shaft** schrägliegende Untermesserwelle *f* ~ **bracket** Schrägstütze *f* ~ **column** Schrägkolonne *f* ~ **conveyor** ansteigendes Förderband *n*, Schrägförderer *m* ~ **distance** Schrägentfernung *f* ~ **dock** Schrägrampe *f* ~ **drive** Schrägantrieb *m* ~ **elevator** Schrägaufzug *m*, schräger Aufzug *m* ~ **eyepiece** Schrägeinblick *m* ~ **eyepiece revolver** Okularrevolver *m* mit Schrägeinblick ~ **flat-belt-type conveyer** Schrägförderband *n* mit Flachband ~ **gallery** Schleppschacht *m* ~ **grate** schräger Rost *m*, Schrägrost *m* ~ **-grate-type gas producer** Gaserzeuger *m* oder Generator *m* mit Schrägrost ~ **hoist** Schrägaufzug *m* ~ **manometer** Schrägrohrmanometer *m* ~ **mother gallery** flach einfallende Förderstrecke *f* ~ **mounting** Schrägeinbau *m* ~ **outlet seat valve** Schrägsitzventil *n* ~ **pan conveyer** Schrägförderer *m* mit Plattenband ~ **path** Schrägsteg *m* ~ **pictures** Schrägbilder *m* ~ **pile** Druck- oder Zugpfahl *m*, Schrägpfahl *m* ~ **plane** Bremsberg *m*, Gefällebahn *f*, geneigte Fläche *f*, Neigungsebene *f*, schiefe Ebene *f* ~ **plane with rope** Seilebene *f* ~ **plane impact test** (for container testing) Prallprüfung *f* ~**-plane switch** (railroad) Kletterkreuzung *f*, -weiche *f* ~**-platform conveyer** schrägliegender Plattenbandförderer *m*, Schrägförderer *m* mit Plattenband ~ **position** Schiefstellung *f* ~ **position of tooth** Zahnschrägstellung *f* ~ **quoin** (print.) Schief-, Schräg-steg *m* ~ **reinforcement** abfallende oder geneigte Eiseneinlage *f* ~ **retort** geneigte Retorte *f* ~ **roll** Schrägwalze *f* ~**-seat valve** Schrägsitzventil *n* ~ **shaft** Schleppschacht *m*, donlägiger Schacht *m* ~ **slat conveyer** Schrägförderer *m* mit Plattenband ~ **slider bearing** Klotzlager *n* ~ **spark plug** schrägliegende Zündkerze *f* ~ **strut** Schrägstrebe *f* ~ **track** Schrägbahn *f* ~**-tube boiler** Schrägrohrkessel *m* ~**-tube manometer** Schräg-

rohrmanometer *n* ~ **U tube** geneigtes U-Rohr
~ **wharf** Schrägrampe *f*
**inclining** Neigen *n* ~ **cant(ing)** Kippung *f*
**inclinometer** Inklinations-kompaß *m*, -messer *m*,
-nadel *f*, Längsneigungsmesser *m*, Libelle *f*
(Fluginstr.), Markscheidergerät *n*, Neigungsmesser *m* (aviat.), Steigungsmesser *m* ~ **arc (or
bow)** Gradbogen *m* ~ **dial (or limb)** Gradbogen
*m*
**inclosing capsule** Verschlußkapsel *f*
**include, to** ~ beifügen, einbegreifen, einschließen, einsetzen, zwischenschalten **to** ~ **in** einrechnen **to** ~ **in parentheses** einklammern
**included** inbegriffen ~ **angle** einbeschriebener
Winkel *m* ~ **angle of thread** Flankenwinkel *m*
~ **grain** verwachsenes Korn *n*
**including** einbegriffen, einschließlich, zuzüglich
**(not)** ~ **packing** Verpackung *f* (nicht) einbegriffen
**inclusion** Einlagerung *f*, Einschluß *m* ~ **theorem**
Einschließungssatz *m*
**inclusions** Einschlüsse *pl* (Gestein)
**inclusive** einschließlich, enthaltend, umfassend,
umschließend ~ **of** einschließlich ~ **correction**
Gesamtausschaltung *f*
**incoherence** Inkohärenz *f*, Nichtübereinstimmung *f*
**incoherent rotation** nichtkohärenter Spin *m*
**incombustibility** Unverbrennbarkeit *f*
**incombustible** feuerfest, feuersicher, unverbrennbar, unverbrennlich ~ **constituent** unverbrenn-
licher Bestandteil *m* ~ **matter** unbrennbarer
Stoff *m*
**income** Einkommen *n*, Rente *f* ~ **tax** Einkom-
mensteuer *f*
**incoming** ankommend (teleph.), einfallend, eingehend, einlaufend ~ **audible current** ankommender Hörstrom *m* ~ **block** Anfangssperre *f*
(electr.) ~ **call** ankommendes Gespräch *n*
~**-call blocking** Anrufsperre *f* ~**-call signal light**
Anrufschauzeichen *n* ~ **circuit** ankommende
Leitung *f*, Eingangsleitung *f* ~ **control impulse**
V-Impuls *m* eines Impulsgebers ~ **current** ankommender Strom *m* ~ **dose** Einfalldosis *f* ~
**fields** einlaufende Felder *pl* ~ **junction** ankommende Verbindungsleitung *f* ~ **(wire) line** ankommende Leitung *f* ~ **one-way circuit** Leitung
*f* für ankommenden Verkehr ~ **oscillation** einfallende Schwingung *f* ~ **(current) panel** Strom-
zufuhrtafel *f* ~ **position** Ankunftsplatz *m*,
B-Platzschrank *m* ~ **potential** gerichtete Empfangsspannung *f* ~ **power** aufgenommene Leistung *f*, Empfangsleistung *f* ~ **register** Ge-
sprächsbuch *n* ~ **selector** Eingangswähler *m*
~**-signal level** Empfangsfeldstärke *f* ~ **spherical
wave** einlaufende Kugelwelle *f* ~ **track** Ein-
fahr-, Einfahrt-geleise *n* ~ **traffic** Ankunftsver-
kehr *m*, ankommender Verkehr *m* (teleph.) ~
**trunk** ankommende Fernleitung *f*
**incommensurable** inkommensurabel, ohne gemeinsame Maßeinheit *f*, unmeßbar (math.) ~
**number** Primzahl *f*
**incomparable** unvergleichlich
**incompatible** unverträglich
**incomplete** lückenhaft, unvollendet, unvollkommen, unvollständig ~ **call and release
control** Verbindungsweg-Kontrolle *f* ~ **combustion** unvollkommene oder unvollständige

Verbrennung *f* ~ **overfall** Grundwehr *n*
(hydraul.) ~ **transposition section** Schutzstrecke
*f* mit nicht ausgeglichenem Induktionsschutz
**incompletely,** ~ **burned gas** Schwelgas *n* ~ **dialed
call** unvollständige Verbindung *f*
**incompleteness** Lückenhaftigkeit *f*, Unvollstän-
digkeit *f*
**incomprehensible** unübersehbar
**incompressibility** Raumbeständigkeit *f*, Unzu-
sammendrückbarkeit *f*
**incongealable** ungefrierbar
**incongruity** Mißverhältnis *n*, Nichtübereinstimmung *f*
**inconsiderable** bedeutungslos, belanglos
**inconsistency** Unbeständigkeit *f*, Unvereinbar-
keit *f*, Wankelmut *m*, Widerspruch *m*
**inconsistent** unverträglich, widersprechend
**inconspicuous** unauffällig
**inconstancy** Inkonstanz *f*, Unbeständigkeit *f*
**inconstant** inkonstant, unbeständig, wankelhaft
**inconvenience** Unannehmlichkeit *f*, Unbequem-
lichkeit *f*, Unzuträglichkeit *f*
**inconvenient** lästig, ungelegen
**inconvertible** nichtumwandelbar
**incooler** Zwischenkühler
**incorporate, to** ~ angliedern, in etwas einbauen,
eingliedern, zu einer geschlossenen Einheit
machen, einlagern, einschließen, einverleiben,
hineinziehen, vermischen
**incorporated in an apparatus** eingebaut in einen
Apparat *m*
**incorporation** Eingliederung *f*, Einlagerung *f*,
Einschluß *m*
**incorrect** flasch, fehlerhaft, ungenau, unrichtig
~ **dimension** Abmaß *n* ~ **indication (or reading)**
Fehlanweisung *f*
**incorrectly** unsachgemäß ~ **centered front sight**
geklemmtes Korn *m* ~ **trimmed** flaschlastig
**incorrodible** unkorrodierbar
**incorrupted** unbeschädigt, unverdorben
**increase, to** ~ anlaufen, anschwellen, anwachsen, (of wind) aufbrisen, auflaufen, aufschlagen,
erhöhen, schwellen, steigen, steigern, (in size)
vergrößern, vermehren, verstärken, wachsen,
zunehmen **to** ~ **the angle of incidence** den Einfallswinkel vergrößern **to** ~ **distance between
aircraft in flight** sich absetzen **to** ~ **the gain** die
Verstärkung erhöhen **to** ~ **the power** die Spannung erhöhen **to** ~ **in proportion with the
square of** Anwachsen *n* mit dem Quadrat von
**to** ~ **speed** auftouren
**increase** Anwachsen *n*, Anwuchs *m*, Aufschlag
*m*, Erhöhung *f*, Steigen *n*, Steigerung *f*, Steigung *f*, Vergrößerung *f*, Zulage *f*, Zunahme *f*,
Zuschlag *m*, Zuwachs *m*
**increase,** ~ **of breadth** Verbreiterung *f* ~ **of
current** Stromzunahme *f* ~ **in density of the
subsoil** Verdichtung *f* des Baugrundes ~ **in
efficiency** Leistungssteigerung *f* ~ **of impact
strength by aging** Alterungskerbzähigkeit *f* ~
**in lattice spacing** Gitteraufweiterung *f* ~ **of lift**
Auftriebserhöhung *f* ~ **in load** Belastungser-
höhung *f*, Laststeigerung *f* ~ **of momentum**
Impulszunahme *f* ~ **in potential** Potentialan-
stieg *m* ~ **of potential** Potentialzunahme *f* ~ **of
pressure** Druckerhöhung *f*, Drucksteigerung *f*
~ **in price** Mehrpreis *m*, Steigen der Preise, Ver-
teuerung *f* ~ **of range** Reichweitensteigerung *f*

**~ in resistance** Widerstandszunahme *f* **~ of resistance** Widerstandserhöhung *f* **~ of salary** Gehaltszulage *f* **~ in speed** Geschwindigkeitserhöhung *f*, **-gewinn** *m* **~ in strength and rigidity** Verfestigung *f* **~ in temperature** Wärmezunahme *f* **~ in thickness of metal to allow for rust** Rostzuschlag *m* **~ in thrust** Schubgewinn *m* **~ in volume** Ausdehnung *f* (math.) Volumenvermehrung *f*, Volumenzunahme *f* **~ in weight** Gewichtszunahme *f*

**increased** erhöht, gesteigert, überhöht (geom) **~ alertness** erhöhte Aufmerksamkeit *f* **~ cant** (in a curve) Überhöhung *f* **~ delivery** Mehrförderung *f* **~ efficiency** Mehrleistung *f* **~ expenditure** Mehraufwand *m* **~ height** Überhöhung *f* **~ output** Mehrleistung *f*, verstärkte Förderung *f* **~ stepwise** stufenweise gesteigert **~ vigilance** erhöhte Aufmerksamkeit *f* **~ yield** Ausbeuteerhöhung *f*

**increasing, ~ of lime** Gedeihen *n* des Kalkes **~ a sentence** Strafverschärfung *f*

**increasing, ~ engine speed** steigende Motordrehzahl *f* **~ gear** Lagergetriebe *n* (jet) **~ load** zunehmende Belastung *f* **~ pressure at the center of the low-pressure area** Luftdruckauffüllung *f* **~ speed** Drehzahlerhöhung *f* **~ twist** Progressivdrall *m*, wachsender Drall *m*, (rifling) zunehmender Drall *m*

**increment** Anteil *m*, Anwuchs *m*, Differential *n*, Inkrement *n*, Steigerung *f*, Stufe *f*, Zunahme *f*, Zuwachs *m* **~ duty** Wertzuwachssteuer *f*

**incremental** zusätzlich, Zuwachs .... **~ area** Teilfläche *f* **~ computer** digitaler Integrator *m* **~ duplex** Additionsduplex *m* **~ impulse** Zuwachsimpuls *m* **~ permeability** zusätzliche Permeabilität *f* **~ thermocouple** Tendenz-Thermoelement *n* **~ value** Wertzuwachs *m* **~ velocity** Zusatzgeschwindigkeit *f*

**increments, by ~** stufenweise

**incrust, to ~** belegen, inkrustieren, überkrusten, überziehen, mit einer Kruste überziehen, verkleiden, verkrusten, (spinnerets) verstopfen, zuwachsen

**incrustate** ansetzen, inkrustiert, verkrustet

**incrustation** Anflug *m*, Ansatz *m*, Anwitterung *f*, Belag *m*, Inkrustierung *f*, dünne Kruste, Krustenbildung *f*, Überkrustung *f*, Überzug *m*, Verkrüstung *f*, (of spinneret) Wasserstein *m*, Verstopfung *f* **~ of boiler fur** Kesselsteinansatz **~ of scale** Wassersteinansatz *m* **~ heat** Sinterungshitze *f*

**incubation, ~ period** Latenzzeit *f* **~ time** Verzögerungszeit *f*

**incubator** Ausbrütapparat *m*, Brutofen *m*

**incumbent pressure of the ground** Firstendruck *m*

**incunable** Wiegendruck *m*

**incur, to ~** auf sich laden, übernehmen, sich etwas zuziehen **to ~ liability** Verbindlichkeit *f* eingehen

**incurable** unheilbar

**incurrend** eingegangen

**incus** amboßförmiger Kumulonimbus *m*

**indebted** schuldig, verschuldet **to be ~ to** schulden, verdanken

**indecomposability** Unzersetzbarkeit *f*

**indecomposable** unzerlegbar, unzersetzbar

**indefinite** unbenannte (math.), unbestimmt, undeutlich, unscharf, verschwommen **~ integral** unbestimmtes Integral *n* **~ number** unbenannte Zahl *f*

**indefinitely long** beliebig lange

**indefinition** Undeutlichkeit *f*

**indelible** unauslöschbare Farbe *f*, unauslöschlich, untilgbar, unvertilgbar **~ ink** unzerstörbare Farbe *f* **~ pencil** Tintenstift *m*

**indemnification** Abfindung *f*, Schadenersatz *m*, Schadloshaltung *f*

**indemnify, to ~** entschädigen, schadlos halten

**indemnity** Abfindungssumme *f*, Entschädigung *f*, Schadloshaltung *f*, Straflosigkeit *f*, Vergütung *f* **~ for property damage** Sachschadensatz *m* **~ claim** Ersatzanspruch *m*

**indene** Inden *n*

**indent, to ~** auszacken, dornen, eindrehen, einkeilen, einkerben, einschneiden, einzahnen, verzahnen, zähneln

**indent** Eindrehung *f*, Eindruck *m*, Einkerbung *f* **~ (ing)** Einkerben *n*

**indentation** Auszacken *n*, Beule *f*, Einbeulung *f*, Einbuchtung *f*, Einkerben *n*, Einpressung *f*, Eindrückung *f* (print.), Einschnitt *m*, Einziehen *n*, Kugelkaliber *n*, Vertiefung *f*, Zahnschnitt *m* (arch.) **~ of the coast line** Gestaltung *f* der Küste

**indentation, ~ cup** Eindruckkalotte *f* **~ depth** Eindrucktiefe *f* **~-depth dial gauge (or indicator)** Eindrucktiefenmeßuhr *f* **~-depth gauge** Eindrucktiefenmesser *m* **~ hardness** Eindruckhärte *f* **~ problem** Kerbproblem *n* **~ process** Eindringungsverfahren *n* **~ test** Eindringungs-, Eindruck-verfahren *n*

**indented** ausgezackt, gekerbt, gezahnt, verzahnt, zackig **~ beam** Zahnkolben *m* **~-built beam** verzahnter Balken *m* **~ connector** Kerbverbinder *m* **~ port** Eckpforte *f* **~ rules** Kerblinien *pl* **~ sill** Zahnschwelle *f*

**indenter** Prüfspitze *f*

**indenting, ~ hammer** Fallhammer *m* **~ tool** Eindruckstempel *m*, Einkerbungswerkzeug *n*

**indentor** Druckkörper *m* (rubber)

**indenture** (apprentice's) Lehrbrief *m*

**independence** Selbständigkeit *f*, Unabhängigkeit *f* **~ of frequency** Frequenzunabhängigkeit *f* **~ theorem** Unabhängigkeitssatz *m*

**independent** frei, freistehend, selbständig, selbstständig, unabhängig, verbandsfrei **~ of frequency** frequenzunabhängig **~ of voltage fluctuations** spannungsunabhängig

**independent, ~ adjustment** Einzelverstellung *f* **~ air cooling** Fremdbelüftung *f* **~-arc furnace** Strahlungsofen *m* **~ arc heating** indirekte Lichtbogenerhitzung *f* **~ axle** Schwingachse *f* **~ charging** Fremdladung *f* **~ circuit** unabhängiger geschlossener Weg *m* **~ coma of higher order** Komareste *pl* höherer Ordnung **~ drum hoist** Versteckhaspel *m* **~ excitation** Fremderregung *f* **~ fission yield** primäre Spaltausbeute *f* **~ front suspension** Vorderpendelachse *f* **~ manuel operation** Sprungschaltung *f* **~ sideband transmission** Übertragung *f* mit unabhängigen Seitenbändern **~ suspension** Einzelabfederung *f* (aviat.), Einzelaufhängung *f* **~ switch** (for opening or closing a circuit) Schalter *m* mit unabhängiger Schaltbewegung **~ time control** Gleichlauf *m*, lokale oder örtliche Synchronisierung *f* **~ time element** unabhängig

verzögerte Auslösung f ~ **variable** unabhängige Veränderliche f
**independently routed ship** Einzelfahrer m
**indestructibility** Unzerstörbarkeit f
**indestructible** unverwüstlich, unzerstörbar
**indeterminacy relation** Unbestimmtheitsrelation f
**indeterminate** unbekannt, unbestimmt ~ **standard** Unbestimmtheit f der Eichung
**indetermination** Unbestimmtheit f ~ **principle** Unbestimmtheits-, Ungewißheits-prinzip n
**indeterministic conception (or interpretation)** indeterministische Auffassung f
**index, to** ~ abecelich einreihen, anzeichnen, einteilen, registrieren, schalten, auf eine Teilmarke stellen, teilen
**index** Anhaltspunkt m, anzeigendes Organ n (Meßgerät), Anzeiger m, Beiwert m, Blatteinteilung f, Diopterlineal n, Einstellmarke f, Exponent m, Feststellstift m, Gradmesser m, Index m, Inhalts-angabe f, -verzeichnis n, Kartothek f, Kennmerkmal n, -ziffer f, Marke f, Maß n, Merkmal n, Nachweis m, Namenverzeichnis n, Register n, Seitenanzeiger m, Stellenzahl f, Tafel f, Teilmarke f, Verzeichnis n, Zeiger m
**index, ~ of co-operation** Kooperationsmodul m ~ **of figures** Figurenverzeichnis n ~ **of inertia** Trägheitsindex m ~ **of pH** Ph-Wert m ~ **of refraction** Brechungs-exponent m, -index m, -koeffizient m, -quotient m, -verhältnis n, -verhältniszahl f, -vermögen m, Refraktionsindex m ~ **of subjects** Realindex m
**index, ~ arrow** Kennpfeil m ~ **base** Teiluntersatz m ~ **bed** Leitschicht f ~ **board** Kartothekkarton m ~ **bolt** Rastenbolzen m ~ **card punches** Karteikerbzangen pl ~ **change gear** Teilwechselrad n ~ **circle** Teil-scheibe f, -trommel f ~ **correction** Indexkorrektion f, Instrumentalkorrektion f ~ **depression** Indexpunkt m (punktähnliche Vertiefung) ~ **disk** Indexscheibe f ~ **dot** Einstellmarke f ~ **drive** Teilgetriebe n ~ **drum clamping lever** Knebel m für Teilung ~ **error** Indexfehler m ~ **fossils** Leitfossilien pl ~ **fullautomatic turret screw machine** Indexrevolverautomat m ~ **gear** Teilrad m, Teilzahnrad n ~**-gear generator (or hobber)** Teilradwälzfräsmaschine f ~**-gear hob** Schaltradwälzfräser m ~ **graduation** Teilstriche pl ~**-head slide** Teilkopfschlitten m ~ **hole** Einstellbohrung f, Rastenloch n ~ **lever** Indexhebel m ~ **line** Ablesemarke f, Indexstrich m ~ **map** Übersichtsbild n ~ **map to a map series** Übersichtskarte f ~ **mark** Ablesemarke f, Index m bei Maßstäben, Indexstrich m, Tiefenzahl f ~ **number** Stellenziffer f ~ **numbers** Stellenzahl f ~ **path** Zeigerbahn f ~ **pin** Anzeigestift m ~ **plate** Indexscheibe f, (of telescopic sight) Markenplatte f, Rasterplatte f, Teil-platte f, -scheibe f, Zeigerwerk n ~ **point** Merkpunkt m ~ **property** Klassifizierungseigenschaft f ~ **ring** (telescopic sight) Teilring m ~ **value** Sollwert m ~ **wheel with notches** Rastenteilscheibe f ~ **worm shaft** Teilschneckenwelle f
**indexing** Bezifferung f, Einteilung f, Schaltung f, Teilansatz m ~ **of crystal faces and zones** Indizierung f von Flächen und Richtungen
**indexing, ~ arm** Teilhebel m ~ **attachment** Teilvorrichtung f ~ **bolt** Schaltzapfen m ~ **center**

Teilaufsatz m ~ **change gears** Teilwechselräder n, pl ~ **clutch** Teilkupplung f ~ **cycle** Teilvorgang m ~ **device** Teilgerät n ~ **disk** Schaltscheibe f ~ **gear** Schaltrad n ~ **head** Teilkopf m ~ **head change gears** Teilkopfwechselradsatz m ~ **head traverse screw** Teilkopfverschiebespindel f ~ **hole** Einstellmarke f ~ **mechanism** Schalt-antrieb m, -mechanismus m ~ **method** Teilverfahren n ~ **pin** Anzeigestift m, Indexbolzen m ~ **plate** Schaltscheibe f ~ **position** Schaltstellung f ~ **register** Indexregister n ~ **ring** Schaltscheibe f ~ **wheel** Schaltrad n
**India paper** Dünndruckpapier n
**Indian, ~ fiber** Indiafaser f ~ **red** Bergrot n ~ **yellow** Indischgelb n
**India-rubber** Gummi m, Kautschuk m ~ **cable** (vulcanized) Gummikabel n ~ **core cable** (vulcanized) Gummiabschlußkabel n ~ **cover** Gummihülle f ~ **linen** Gummiwäsche f ~ **wire** (vulcanized) Gummidraht m
**indican** Indikan n
**indicate, to** ~ andeuten, angeben, anzeichnen, anzeigen, aufzeichnen, besagen, bezeichnen, deuten, indizieren, nachweisen **to** ~ **optically (or visually)** sichtbar machen
**indicated** angezeigt, aufgezeichnet, indiziert ~ **by a dot-dash** strichpunktiert
**indicated, ~ air speed** angezeigte Eigengeschwindigkeit f Fahrtmesseranzeige f, angezeigte Fluggeschwindigkeit f ~ **altitude** aufgezeichnete Höhenleistung f ~ **angle** Anstell-, Einstellwinkel m ~ **angle of attack** angezeigter Anstell- oder Einstellwinkel m ~ **efficiency** indizierter Wirkungsgrad m ~ **horsepower** angezeigte Leistung f, aufgezeichnete Pferdekraft f, gemessene Pferdestärke f, indizierte Leistung f oder Pferdestärke f ~ **output** indizierte Leistung f ~ **static pressure** angezeigter statischer Druck m ~ **voltage** Anzeigespannung f ~ **work** indizierte Leistung f
**indicating, ~ apparatus** Indikator m, Nachweisgerät n, Weiser m ~ **(recording) device** Anzeigevorrichtung f ~ **disk** Laufkontrollrädchen n ~ **instrument** Ablese-, Anzeige-gerät n, Indikator m, Indikatorinstrument n, Zeigerinstrument n ~ **lamp** Signallampe f ~ **letter** Kennbuchstabe m ~ **light** Anzeigelampe f ~ **line** Meßstrich m ~ **meter** anzeigender Meßapparat m ~ **needle** Zeiger m ~ **peg** Hinweisstöpsel m, Hinweisungsstöpsel m ~ **plug** Passierrohr n ~ **plug gauge** Passimeter n ~ **pointer** Zeigerhebel m ~ **pressure gauge** Zeigermanometer n ~ **range** Anzeigebereich m ~ **scale** Anzeigeskala f ~ **screw-thread calipers** Gewindeschraublehre f mit Fühlhebel ~ **snap gauge** Rachenlehre f mit Maßanzeige durch Fühlhebel
**indication** Angabe f, Anhaltspunkt m, Anzeichen n, Anzeige f, Anzeigung f, Aufschluß m, Bestimmung f, Bezeichnung f, (engine trouble) Erkennungszeichen, Fingerzeig m, Hinweis m, Indizierung f, Kennzeichen n, Kennzeichnung f, Merkmal n, Nachweis m, Nachweisung f, Vorläufer m, Vorzeichen n, Weisung f, Zeichen n
**indication, ~ of the actual value** Istwertanzeige f ~ **of deviation** Abweichungsanzeige f ~ **of a lever position** Rückmeldung f ~ **of targets** Zielzuweisung f

indication, ~ signal Rückmeldung *f* ~ voltage
Anzeigespannung *f*
indications (oil) Aufschlüsse *pl*
indicative bezeichnend, kennzeichnend
indicator Ablesegerät *n*, (drop) Anrufzeichen *n*,
Anzeigegerät *n*, Anzeigeinstrument *n*, Anzeiger
*m*, Begleiter *m*, Folgezeigerempfänger *m*,
(test) Fühlhebel *m*, Indikator *m*, Indikatorinstrument *n*, Klappe *f*, Kommandozeiger *m*,
Maßbegrenzung *f*, Melder *m*, Melde-tafel *f*,
-zeichen *n*, Meßgerät *n*, Pegel *m*, Registriervorrichtung *f*, Richtungsanzeiger *m*, Schautafel *f*, Sichtgerät *n*, Signal *n*, Test *m*, Uhrzeiger
*m*, Zeiger *m*
indicator, ~ of circulation Kreislaufschreiber *m*
~ for earthen electric contact Erdschlußanzeiger *m* .~ for gyrostatic compass Tochterkompaß *m* ~ of impact Aufschlaganzeiger *m* ~ of
inclination Neigungsanzeiger *m* ~ of micrometric caliper Fühlhebel *m* ~ of swing Verkan-
tungsanzeiger *m* ~ on the top of a blast furnace
Gichtanzeiger *m*
indicator, ~ apparatus Anzeigevorrichtung *f* ~
armature Kontrollanker *m* ~ bar Anzeige-
stange *f* ~ bell Wecker *m* mit Fallscheibe ~
board Anzeigetafel *f*, Fallscheibenkasten *m*,
Tableau *n* ~ cabinet Meßgehäuse *n* ~ card
Diagrammpapier *n* für Indikator ~ cartridge
Deutpatrone *f* ~ case Anrufzeichenkästchen *n*
(teleph.) ~ casing Indikatorgehäuse *n* ~ chart
Indikatordiagramm *n* ~ clip Fanghaken *m*,
Skalenreiter *m* ~ cock Indikator-, Indizier-
hahn *m* ~ connection piece Indikatorstutzen *m*
~ control Bildschirmbediengerät *n* ~ diagram
Indikatordiagramm *n*, Leistungsbild *n* ~ dial
Anzeigerblatt ~ drive Indikatorantrieb *m* ~
element radioaktiver Indikator *m* ~ gate
Aktivierungsimpuls *m*, Anzeigetor *n* ~ gate
pulse Hellsteuerimpuls *m* (rdr) ~ gauge Anzeige-gerät *n*, -lehre *f*, -meßgerät *n*, -meßgerät-
lehre *f* ~ grid Anzeigegitter *n* ~ lag Anzeigeträgheit *f* ~ lamps Aufleuchtlampen *pl* ~ lead
Signalleitung *f* ~ mode switch Betriebsartenwahlschalter *m* ~ needle Index-, Merk-zeiger *m*
~ net Anweisernetz *n* ~ piston Indikatorkolben
*m* ~ plug screw Indikatorverschlußschraube *f*
~ point Zeigernase *f* ~ signal Schauzeichen *n* ~
slide Anzeigerschiene *f* ~ spring Anzeiger-,
Indikator-feder *f* ~ stylus Indikatorschreibstift
*m* ~ test indikatorische Untersuchung *f* ~
tube Anzeigeröhre *f*, Zählrohr *n*
indicatrix Indikatrix *f*
indices Achsenabschnitt *m*
indicial anzeigend ~ admittance (or conductance)
Kennleitwert *m* ~ equation determinierende
Gleichung *f* ~ impedance Kennwiderstand *m*
indictable klagbar
indictment Anklage *f*
indifferent gleichgültig, indifferent, neutral, teil-
nahmslos, unempfindlich
indigo, ~ in color indigofarbig ~ copper Kupferindigo *m* ~ print Blaudruck *m* ~ purple Phönizin *n* ~ vat Blauküße *f*
indirect indirekt, mittelbar ~ address Adresse
von Adresse *f* (info proc.), iterierte Adresse *f*
~-arc furnace Strahlungsofen *m* ~ arc heating
indirekte oder mittelbare Lichtbogenbeheizung
*f* ~ call Durchgangsverbindung *f* ~ control

Rückwärtsregelung *f* ~ cycle indirekter Kreislauf *m* (propeller) ~ excitation Umweganregung
*f* ~ fire Bogenschuß *m* ~-fired indirekt beheizt
~ firing indirekte Feuerung *f* ~ heating mittelbare Heizung *f* ~ laying indirektes Richten *n*,
Richtpunktverfahren *n* (artil.) ~ lighting indirekte Beleuchtung *f* ~ means Nebenweg *m* ~
measurement of length unechte (Längen)mes-
sung *f* ~ mode Indirektverfahren *n* (rdo nav.) ~
process indirektes Verfahren *n* ~ routing Benutzung *f* von Umwegen ~-routing system
System *n* mit mittelbar gesteuerter Wahl ~
scanning indirekte Abtastung *f* ~ support
operative Unterstützung *f* ~ tripping mittelbare
Auslösung *f* ~ wave Luftwegwelle *f*, indirekte
oder reflektierte Welle *f*, Rundwelle *f*
indirectly, ~ controlled system indirekte Regel-
strecke *f* ~ heated cathode Äquipotentialkathode *f*, indirekt geheizte Glühkathode *f* oder
Kathode *f* ~ heated tube (or valve) indirekt geheizte Röhre *f*
indiscernible unbestimmbar
indiscriminate unterschiedslos
indispensable unabkömmlich, unbedingt, uner-
läßlich
indisputable eindeutig ~ title nicht bestreitbarer
Anspruch *m*
indissoluble unlösbar
indistinct undeutlich, unklar, verschmiert, verschwommen ~ image unscharfes Bild *n*
indistinctness Undeutlichkeit *f*, Unklarheit *f*,
Verschwommenheit *f*
indistinguishable gleichartig
indium Indium *n* ~ chloride Indiumchlorid *n* ~
oxide Indiumoxyd *n* ~ sulphate Indiumsulfat *n*
individual eigen, einzeln, individuell; Einzel ....;
~ adjustment Einzelverstellung *f* ~ bank
breaker Vorstufenleistungsschalter *m* ~ blower
Einzelgebläse *n* ~-capacitor method Einzel-
kompensation *f* ~ change Eigenveränderung *f*
~ chassis Einzelchassis *n* ~ control Einzelantrieb *m* ~ crystallite Einzelkristallit *m* ~ cut
Einzelschnitt *m* ~ deflection (or deviation)
Einzeldeviation *f* ~ excitation Eigenerregung *f*
~ flying control Einzelsteuerung *f* ~ item Einzelheit *f* ~ job card Stückkarte *f* ~-line subscriber Einzelanschlußteilnehmer *m* ~ line switch
Vorwähler *m* ~ motor drive Einzelantrieb *m*
~-motor power factor improvement Einzel-
kompensation *f* (von Motoren) ~ needs Einzel-
fall *m* ~ operation Einzeldruckgang *m* ~
oscillation Einzelschwingung *f* ~ particle
model Einteilchenmodell *n* ~ parts of the
apparatus Baueinheiten *pl* ~ picture in a series
(or sequence) Einzelbild *n* einer Bildreihe ~
piece work Einzelakkord *m* ~ protection Einzel-
schutz *m* ~ release (bombing) Einzelabwurf *m*
~ results with a trend Einzelresultate *pl* mit
Gang ~ spot welding Einzelpunktschweißung *f*
~ stage Einzelschritt *m* ~ steering control Einzelsteuerung *f* ~ supercharger Einzelgebläse ~
~ suspension Einzelaufhängung *f* ~ tape divider
Einzelriemchenflorteilchen *n* ~ training Einzel-
ausbildung *f* ~ transport Einzeltransport *m* ~
traverse Einzelverstellung *f* ~ trunk group besonderes Leitungsbündel *n* ~ (or dead) weight
Eigenlast *f*
indivisible untrennbar, unzerlegbar

**indoctrination** Unterweisung *f*
**indole** Indol *n*
**indolent** faul, träge
**indoor,** ~ **antenna** Innen-, Zimmer-antenne *f* ~
**cable** Hausleiterkabel *n* ~**-cable terminator**
Kabelendverschluß *m* für Innenleitung ~
**firing equipment** Zimmerschießgerät *n* ~ **fittings**
Inneneinrichtung *f* ~ **flier** Zimmerflugmodell *n*
~ **furniture** Inneneinrichtung *f* ~ **isolating**
**switch** Trenner *m* ~ **lead distribution sleeves**
Bleiaufteilungsmuffen *pl* ~ **location (or moun-**
**ting)** Aufstellung *f* im Innenraum ~ **post in-**
**sulator** Innenraumstützer *m* ~ **service** Innendienst *m* ~ **shot** Innenaufnahme *f*
**indoors** inwendig; Innen...
**indraft** Ansaugluftströmung *f*, Einströmung *f*,
Saugluft *f*, Vorstrom *m* ~ **velocity** Saugluftge-
schwindigkeit *f*
**induce, to** ~ beeinflussen, einleiten, erzeugen,
induzieren, verleiten
**induced** angefacht, induziert ~ **charge** gebundene
oder induzierte Ladung *f* ~ **current** Induktionsstrom *m*, induzierter Strom *m*, Öffnungsinduktionsstrom *m*, Sekundärstrom *m* ~ **de-**
**tonation** Zündübertragung *f* ~**-detonation**
**charge** Folgeladung *f*, Übertragungskörper *m*
~ **draft** Induktionszug *m* ~ **drag** induzierter
Widerstand *m*, Randwiderstand *m* (aerodyn.)
~ **draught** Saugzug *m* ~**-draught fan** Saugkreis
*m*, Saugzuggebläse *n* ~**-draught plant** Saugzug
anlage *f* ~ **electricity** Induktionselektrizität *f* ~
**electromotive force** Induktionsspannung *f* ~
**magnet** induzierter Magnet *m* ~ **natural**
**radio-nuclides** induzierte natürliche Radionuklide *pl* ~ **noise** Induktionsgeräusch *n* ~
**pole** induzierter Pol *m* ~ **power** Sekundärlei-
stung *f* ~ **radioactivity** induzierte oder künstliche Radioaktivität *f* ~ **resistance** induzierter
Widerstand *m*, Randwiderstand *m*, Wirbel-
walze *f* ~**-single-circulation boiler** Zwangsdurchlaufkessel *m* ~ **voltage** induzierte Spannung *f*
**inducement** Anregung *f*, Veranlassung *f*
**inducer** Vorlaufrad *n*
**inducing** induzierend ~ **current** induzierender
Strom *m* ~ **magnet** induzierender Magnet *m* ~
**magnetic pole** induzierender Magnetpol *m* ~
**pole** induzierender Pol *m*
**inductance (coil)** Drosselspule *f*, Induktanz *f*,
induktiver Blindwiderstand *m*, induktive Reaktanz *f*, induktiver Widerstand *m*, Induktivität *f*
~ **per unit length** Induktivität *f* je Längeneinheit
**inductance,** ~ **balance** Induktivitätssymmetrie *f*
~ **box** Induktivitätskasten *m* für Meßzwecke ~
**bridge** Induktivitätsmeßbrücke *f* ~**-capacitance**
**coupling** Drosselkopplung *f* ~ **coil** Induktanz-,
Induktions-, Induktivitäts-, Selbstinduktionsspule *f* ~ **core** Spulenkern *m* ~ **coupling** magnetische oder induktive Kopplung *f* ~ **load** induktive Belastung *f* ~ **meter** Induktivitäts-
messer *m* ~ **rating (in henrys per 1,000 turns)**
A 1-Wert *m* ~ **reactance** Induktanz *f*, induktiver Blindwiderstand *m*, induktive Reaktanz *f*
~**-repeating amplifier** Verstärker *m* mit magnetischer Kopplung ~ **strain gauge** Induk-
tionsdehnungsmeßstreifen *m* ~ **unbalance** In-
duktivitätsunsymmetrie *f* ~ **variation** Abgleichverfahren *n*

**induction** Ansaugvorgang *m*, Einströmung *f*,
Einweisung *f*, Folgerung *f*, Induktion *f*, In-
duzierung *f*, Influenz *f*, Influenzwirkung *f*,
(fluorescence) Wechselwirkung *f* ~ **of duct**
Lufteintritt *m* in einem Düsenkühler ~ **in iron**
Eiseninduktion *f*
**induction,** ~ **accelerator** Resonanzbeschleuniger
*m* ~ **arrangement** Induktionsschutz *m* ~
**burner** Induktionsbrenner *m* ~ **channel** In-
duktionsrinne *f* ~ **coefficient** Induktionskoeffizient *m* ~ **coil** Funkeninduktor *m*, Induktionsapparat *m*, Induktionsspule *f*, Induktor *m*, Induktorium *n*, Sprechspule *f* ~ **coil of**
**operator's telephone set** Induktionsspule *f* der
Platzschaltung ~**-coil apparatus** Induktions-
gerät *n* ~**-coil interruptor** Induktorunterbrecher
*m* ~ **compass** Erdinduktor *m*, Induktions-
kompaß *m* ~ **current** Erreger-, Nebenschluß-
strom *m* ~**-current relay** Induktionsrelais *n* ~
**device** Induktionsschutz *m* ~ **electron accelera-**
**tor** Resonanzbeschleuniger *m* ~ **field** Induk-
tionsfeld *n* ~ **flowmeter** Induktionsströmungs-
messer *m* ~ **flux** Induktionsfluß *m* ~ **furnace**
Induktionsofen *m* ~ **fuse** Induktionszünder *m*
~ **hardening** Hartlagerhärtung *f*, induktives
Härten *n* ~**-heating apparatus** Gerät *n* mit
Induktionsheizung ~ **impedance** Drehregler *m*
~ **instrument** Induktionsinstrument *n* ~ **inter-**
**ference** Induktionsstörung *f* ~ **manifold** An-
saug-leitung *f*, -schacht *m*, Verteilerleitung *f*,
Verteilungsrohr *n*, verzweigte Ansaugleitung *f*
~ **meter** Induktions-meßgerät *n*, -zähler *m*,
L-Meßgerät *m* ~ **mixture** Brenngemisch *n* ~
**motor** Asynchronmotor *m*, Induktions-ma-
schine *f*, -motor *m* ~ **period** (experimenting
with engines) Einwirkungszeit *f*, Induktions-
dauer *f*, -periode *f* ~ **pipe** Ansaug-, Saug-rohr *n*
~ **port** Ansaugkanal *m* (Motorzylinder) ~
**pressure** Ladedruck *m* ~ **regulator** Dreh-regler
*m*, -transformator *m* ~ **scroll** Ansaugspirale *f* ~
**side** (left side of engine) Ansaugseite *f*, Einlaßseite *f*, Vergaserseite *f* ~ **spark** Induktions-
funke *m* ~ **stroke** Ansaugtakt *m*, Einlaßhub *m*,
(gas) Einsaugtakt *m* ~ **swirl** Wirbelbildung *f*
des Ladegemisches (aviat.) ~ **system** Einlaß-,
Einström-leitung *f*, Einströmsystem *n*, Induk-
tionsanlage *f*, Ladeleitung *f* ~**-system tem-**
**perature** Ladelufttemperatur *f* ~ **valve** Ein-
tritts-, Saug-ventil *n* ~ **voltage regulator** Dreh-
wandler *m*, regelbarer Zusatztransformator *m*
~ **winding** Erregerwicklung *f*
**inductive** induktiv; Induktions... ~ **ballast unit**
induktives Vorschaltgerät *n* (Lampen) ~
**capacity** Dielektrizitätskonstante *f*, Induk-
tions-kapazität *f*, -vermögen *n*, Verteilungs-
vermögen *n* (electr.) ~ **circuit** induktiver Stromkreis *m* ~ **coupling** induktive oder magnetische
Kopplung *f* ~ **current** induktiver Strom *m* ~
**displacement pick-up** induktiver Weggeber *m*
~ **disturbances** Übersprechen *n* ~ **effects bet-**
**ween high-voltage and low-voltage lines** gegenseitige Beeinflussung *f* von Starkstromleitun-
gen und Schwachstromleitungen ~ **feedback**
induktive Rückkopplung *f* ~ **heating** Induk-
tionsbeheizung *f* ~**-heating current** Induktions-
heizstrom *m* ~ **impedance** Schwungradwiderstand *m* ~ **interference** Induktionsstörung *f*,
induktorische Beeinflussung *f*, Seiteninduktion

*f* ~ **leaks in bridges** induktive Verluste *pl* in Brücken ~ **load** induktive Belastung *f* ~ **rail connection** Drosselstoß *m* ~ **reactance** Induktanz *f*, induktiver Blindwiderstand *m*, induktiver Widerstand, Widerdruck *m* ~ **resistance** Induktionswiderstand, induktiver Widerstand *m* ~ **shunt** induktiver oder magnetischer Nebenschluß *m* ~ **stirring mechanism (in induction furnace)** induktive Rührwirkung *f* ~ **train control** induktive Zugsicherung *f* ~ **transmitter** (induktiv) gekoppelter Sender *m* ~ **trouble** Induktionsstörung *f*, induktorische Beeinflussung *f* ~ **voltage drop** induktiver Spannungsabfall *m* ~ **winding** Induktionswicklung *f*
**inductively,** ~ **coupled** induktiv gekuppelt ~ **loaded** induktiv geladen
**inductivity** Dielektrizitätskonstante *f*, Induktanz *f*, Induktivität *f*
**inductor** Drossel *f*, Drossel-, Impedanz-spule *f*, Induktanzspule *f*, Induktionsapparat *m*, Induktionsspule *f*, Induktivitätsspule *f*, Induktor *m* ~ **alternator** Induktorgenerator *m* ~ **alternator with moving iron** Induktormaschine *f* ~ **alternator spider** Ständerkörper *m* ~ **pole** Induktorpol *m* ~ **speaker** elektromagnetischer Lautsprecher *m* ~ **station** Induktorstation *f* ~ **tap** Spulenanzapfung *f* ~**-type generator** Induktormaschine *f* ~**-type magneto with rotating magnet** Magnetzünder *m* mit umlaufendem Magnet ~ **wheel** Induktorrad *n*
**indurable** härtbar
**indurate, to** ~ verhärten
**indurated** hart geworden, verhärtet ~ **fabric** Hartgewebe *n*
**induration** Diagenese *f* (geol.), Hartwerden *n*, Verhärtung *f*
**industrial** Betriebs-; gewerblich, industriell, technisch ~ **agreement** Tarifvertrag *m* ~ **alternating currents** technische Wechselströme *pl* ~ **apparatus** Industriegerät *n* ~ **cardboard boxes** Industriekartonagen *pl* ~ **chemist** Betriebschemiker *m* ~ **combine** Kartell *n* ~ **cotton wool** Industriewatte *f* ~ **diamond** Industriediamant *m* ~ **engineer** Gewerbeingenieur *m* ~ **engineering** Betriebswissenschaft *f* ~ **exhibition** Gewerbeausstellung *f* ~ **firm** Industriefirma *f* ~ **frequency** technische Frequenz *f* ~ **fumes (or (smoke)** Hüttenrauch *m* ~ **furnaces** Industrieöfen *pl* ~ **installation** Industrieanlage *f* ~ **instances** Betriebsfälle *pl* ~ **legislation** Gewerbeordnung *f* ~ **mains** Industrienetz *n* ~ **maintenance** Werkinstandhaltung *f* ~ **material** technischer Werkstoff *m* ~ **measuring instruments** gewerbliche Meßgeräte *pl* ~ **plant** Industrie-anlage *f*, -werk *n* ~ **port** Industriehafen *m* ~ **power pack** Stromerzeugungsaggregat *n* ~ **printing** Werkdrucke *pl* ~ **process control** Verfahrensreglung *f* ~ **processing engineering** Verfahrenstechnik *f* ~ **product** Industrieerzeugnis *n* ~ **refractometer** Betriebsrefraktometer *n* ~ **standardization** Industrievereinheitlichung *f* ~ **truck** Hub-karren *m*, Hub-wagen *m* ~ **tube** Industrieröhre *f* ~ **utilization** gewerblicher Gebrauch *m* ~ **waste water** Industrieabwässer *pl* ~ **wood** Werkholz *n*
**industries fair** Industriemesse *f*
**industrious** arbeitsam, emsig, erwerbstätig, fleißig, geschäftig

**industry** Erwerb *m*, Erwerbstätigkeit *f*, Industrie *f*, Werktätigkeit *f* ~ **of finishing textiles fibers** Faserstoffveredlungsindustrie *f* ~ **using iron (or steel)** eisenverarbeitende Industrie *f*
**inedible** ungenießbar
**ineffective** unwirksam, wirkungslos ~ **blow** blinder Schlag *m*
**ineffectiveness** Unwirksamkeit *f*, Wirkungslosigkeit *f*
**inefficacy** Wirkungslosigkeit *f*
**inefficiency** schlechte Wirkung *f*, Unwirksamkeit *f*, Wirkungslosigkeit *f*
**inefficient** leistungsunfähig, unökonomisch, unwirksam, unwirtschaftlich, wirkungslos
**inelastic** starr, unelastisch ~ **cross-section** Querschnitt *m* unelastischer Streuung
**inequality** Ungleichheit *f*, Ungleichung *f*
**inert** (gases) edel, inaktiv, indert, indifferent, leblos, neutral, reaktionsträge, träge, unempfindlich, unentzündbar, unwirksam **with** ~ **filling** blindgeladen
**inert,** ~ **battery** Füllbatterie *f* ~ **bomb** Blindbombe *f* ~ **cell** Lagerelement *n* ~ **filling** Blindfüllung *f* ~ **gas** chemischträges Gas *n*, ~ **gas bleed control** Schubregelung *f* durch nicht brennbare Gase (rocket), Edel-, Inert-, Schutzgas *n* ~ **gas filled** edelgasgefüllt ~ **loaded** blindgeladen ~ **material** totes Material *n* ~ **missile** Leergeschoß *n*
**inertance** Inertanz *f*, Massenwirkung *f*
**inertia** Beharrungs-kraft *f*, -vermögen *n*, Ermüdung *f*, lebende Energie *f*, Massenkräfte *pl*, Trägheit *f*, Trägheitsvermögen *n* **circle of** ~ Trägheitskreis *m* ~ **to exposure to darkness** (photoelectric cells) Verdunk(e)lungsträgheit *f* ~ **of needle (or pointer)** Anzeigeträgheit *f*
**inertia,** ~ **disk valve** Scheibenventil *n* ~ **effect** Beharrungswirkung *f* ~ **force** Trägheitskraft *f*, Wuchtkraft *f* ~**-free switch** trägheitsloser Schalter *m* ~ **method** Auslaufverfahren *n* ~ **moment** Massenträgheitsmoment *n* ~ **nozzle** Anlaßdüse *f* ~ **revolution counter** Fliehpendeldrehzahlmesser *m* ~ **ring** Schleppring *m* ~ **starter** Schwerkraft-, Schwungrad-anlasser *m*, Trägheitsstarter *m* ~**(-type) starter** Schwungkraftanlasser *m* ~ **temperature** Beharrungstemperatur *f*
**inertiae, vis** ~ Beharrungsvermögen *n*
**inertial,** ~ **force** Trägheitskraft *f* ~ **frame** Inertialsystem *n* ~ **guidance** Trägheitsführung *f* ~ **mass** träge Masse *f* ~ **navigation** Trägheitsnavigation *f* ~ **reference integrating gyro (IRIG)** integrierender Trägheitskreisel *m* ~ **space** Inertialraum *m* ~ **waves** Trägheitswellen *pl*
**inertialess** trägheitsfrei, trägheitslos, verzögerungsfrei
**inertness** (of mechanical filter) Filtermassenwiderstand *m*, Massenwiderstand *m*, Trägheit *f*
**inexact** inexakt, ungenau
**inexactitude (or inexactness)** Nachlässigkeit *f*, Ungenauigkeit *f* ~ **of measurement** Meßungenauigkeit *f*
**inexcitability** Unanfachbarkeit *f*
**inexhaustible** unerschöpflich
**inexpansible** unausdehnbar
**inexpedient** unzweckmäßig
**inexpensive** billig ~ **controller** Kleingröße *m*

inexperience Unerfahrenheit *f*
inexperienced unerfahren ~ hand Handlanger *m*, Nichtfachmann *m*
inexploitable unbauwürdig
inexplosive unexplodierbar
inextensible unausdehnbar, undehnbar
inextinguishable unauslöschbar
inextricable unentwirrbar
infect, to ~ anstecken, verseuchen
infection Verseuchung *f*
infectious ansteckend ~ disease Infektionskrankheit *f*
in-feed Beschickung *f*, Zustellung *f* ~ and toolhead motion Zahntiefenbewegung *f* ~ connection Zulaufstutzen *m* ~ opening Einfüllöffnung *f* ~ pressure Eingangsdruck *m* ~ rate Einstechvorschub *m* ~ side Aufgabeseite *f*
infer, to ~ folgern to ~ from entnehmen
inference Folgerung *f*, Hinweis *m*, Rückschluß *m*
inferential meter Durchflußmesser *m*
inferior geringwertig, nieder to be ~ unterlegen sein of ~ make minderwertiges Erzeugnis *n*
inferior, ~ flow property at low temperatures schlechtere Kältefließfähigkeit *f* ~ ore Pochgänge *pl*
inferiority Minderwertigkeit *f*, Unterlegenheit *f*
inferred zero unterdrückter Nullpunkt *m*
infiltrate, to ~ durchsetzen, eindringen, einsickern, saigern, seigern
infiltrated air Falschluft *f*
infiltration Durchsickern *n*, Einfiltrierung *f*, Einsickern *n*, Einsickerung *f*, Einziehung *f*
infinite endlos, unendlich ~ aspect ratio unendliches Seitenverhältnis *n* ~ body Vollraum *m* ~ bus (system) starres Netz *n* ~ lattice unendliches Gitter *n* ~ line lange Leitung *f* ~ plate unendlich ausgedehnte Platte *f* ~ series unendliche Reihe *f* ~ slab reactor Reaktor *m* mit unendlichen Plattenabmessungen ~ span unendliche Spannweite *f* oder Flügelstreckung *f* (aerodyn.)
infinitely, ~ long unendlich lang ~ minute unendlich klein ~ safe geometry absolut sichere Geometrie *f* ~ variable stufenlos regelbar ~ variable output control stufenlose Leistungsregulierung *f* ~ variable speed transmission stufenloses Regelgetriebe *n*
infinitesimal unendlich klein ~ calculus Infinitesimalrechnung *f* ~ quantity Differential *n*, unendlich kleine Größe *f*
infinitive ratio unendliches Seitenverhältnis *n*
infinity Skalenendwert *m*, Überlaufzahl *f*, Unendliche *n*, Unendlichkeit *f*, Unendlichkeitsstelle *f*; unendlich viele ~ adjustment Unendlicheinstellung *f* (opt.) ~ focus Einstellung *f* (auf unendlich), konstante Brennweite *f* ~ plug Trennstöpsel *m*
inflame, to ~ entzünden
inflammability Entflammbarkeit *f*, Entzündbarkeit *f*, Entzündlichkeit *f*, Zündbarkeit *f*
inflammable brennbar, entflammbar, entzündbar, entzündlich, feuergefährlich, verbrennlich, zündbar not easily ~ schwer entzündlich ~ in contact with air luftentzündlich
inflammable, ~ constituent brennbarer Bestandteil *m* ~ metal Zündmetall *n*
inflammables Zündwaren *pl*
inflammation Entflammung *f*, Entzündung *f*

inflatable, ~ boat (or dinghy) Schlauchboot *n* ~ fin Wulstflosse *f*
inflate, to ~ aufblähen, aufblasen, (a ballute) Ballonschirm *m* aufblasen, aufpumpen, blähen
inflated aufgeblasen
inflating agent (or medium) Blähmittel *n*
inflation Aufblähung *f* (parachute), Aufblasen *n*, Füllung *f*, (of tire) Schlauchfüllung ~ of tires Aufpumpen *n* der Reifen
inflation, ~ balance Füllwaage *f* ~ deflection Druckdeformation *f* ~ device Aufblasevorrichtung *f* ~ manifold vielseitiges Aufblasen *n*, (balloon) Füllansatz *m* ~ net Füllnetz *n* ~ nipple (balloon) Füllansatz *m* ~ pressure Fülldruck *m* ~ scale Füllwaage *f* ~ sleeve Füllansatz *m*, Füllungsschlauch *m* ~ station (air pump) Füllplatz *m* ~ strength Schwellfestigkeit *f* ~ tube Füllschlauch *m* ~ valve (for tires) Ventileinsatz *m*
inflator hose Füllschlauch *m*
inflect, to ~ ablenken, beugen, biegen
inflecting Biegen *n*
inflection Ablenkung *f*, Beugen *n*, Beugung *f*, Einbiegung *f* (point of) ~ Biegung *f* ~ of voice Modulation *f* der Sprache *f* oder der Stimme *f* ~ point Momentennullpunkt *m*, (of curve) Winkelpunkt *m*
inflector Einlenkkondensator *m*
inflexibility Steifheit *f*, Unbiegsamkeit *f*
inflexible starr, steif, unbiegsam, ungefügig unnachgiebig
inflexion Knickpunkt *m*
inflict, to ~ auferlegen, (losses etc.) beibringen
inflight refueling Luftbetankung *f*
inflow Ansaugluft *f*, Einfluß *m*, Einströmung *f*, Zufluß *m* ~ current Zulaufstrom *m* ~ fittings Einlaufarmatur *f* ~ funnel Einlauftrichter *m*
influence, to ~ beeinflussen, influenzieren to be of ~ on hineinspielen
influence Beeinflußung *f*, Einfluß *m*, Einwirkung *f*, Influenz *f* ~ of mass Massenwirkung *f* ~ of slope of the target surface upon longitudinal dispersion Hangfaktor *m* ~ of temperature Temperatureinfluß *m*
influence, ~ fuse Annäherungszünder *m* ~ fuse squib Zündpille *f* für Annäherungszünder ~ line Einflußlinie *f* ~ machine Influenzmaschine *f*
influencing Beeinflussung *f*
influential einflußreich ~ reach (or sphere) Einflußbereich *m*
influx Einfluß *m*, Einlauf *m*, Zufluß *m*, Zustrom *m*, Zuzug *m* ~ of orders Auftragseingang *m*
inform, to ~ angeben, avisieren, benachrichtigen, berichten, nachweisen, unterrichten, unterweisen
informal formlos
informality Formfehler *m*, Formlosigkeit *f*
informant Gewährsmann *m*
information Angabe *f*, Anzeige *f*, Aufschluß *m*, Auskunft *f*, Avis *m*, Benachrichtigung *f*, Bericht *m*, Bescheid *m*, Impulshaushalt *m* (TV), Kunde *f*, Meldung *f*, Mitteilung *f*, Nachricht *f*, Nachweis *m*, Unterweisung *f*
information, ~ bureau Auskunftei *f*, Auskunftstelle *f* ~ call Ersuchen *n* um Auskunft ~ center Nachrichtenstelle *f* ~ content Informationsgehalt *n* ~ density Informationsdichte *f* ~

**desk** Auskunftsstelle *f* ~ **folder** Angebotsmappe *f* ~ **handling** nachrichtenbearbeitend ~ **operator** Auskunftsbeamter *m* ~ **pulse** Informationsimpuls *m* ~ **read-wire** Informationslesedraht *m* ~ **received from abroad** Auslandsnachricht *f* ~ **retrieval** Informationserschließung *f* ~ **trunk** Überweisungsleitung *f* ~ **unit/sec (bits)** Nachrichtenmenge *f* je Zeiteinheit (Nachrichtenfluß) ~ **write-wire** Informations-Schreibdraht *m*
**informational** aufklärend
**informative** aufschlußreich, belehrend, instruktiv
**informer** Anzeiger *m*
**infra,** ~**-acoustic telegraphy** Unterlagerungstelegrafie *f* ~**-audible** unterhörfrequent ~ **black condition** Ultraschwärzung *f* ~ **filter** Hochfrequenzsperrkette *f*, Kondensatorkette *f* ~**-gravity waves** Infraschwerewellen *pl* ~ **posed** unterlagert
**infrared** infrarot, überrot, ultrarot ~ **bands** Ultrarot-Banden *pl* ~ **block** Ultrarotsperre *f* ~ **emission** Ultrarotemission *f* ~ **focus calibration** Brennweitenkorrektion *f* für Ultrarot ~ **homing head** Infrarot-Steuerung *f* (Rakete) ~ **image converter** Infrarot-Bildwandlerröhre *f* ~ **position-finding set** Urortungsgerät *n* (rdr) ~ **range and direction detection** Infrarotortung *f* ~**-ray telephone** Lichtsprechgerät *n* ~ **rays** infrarote Strahlen *pl* ~ **transmittancy** Ultrarotdurchlässigkeit *f* ~ **viewer** Bildwandler *m*
**infra-Roentgen ray** Grenzstrahl *m*
**infrasonic** Infraschall.... *m*, Unterschall ...., untertonfrequent ~ **frequency** Unterhörfrequenz *f* ~ **wave** Unterschallwelle *f*
**infrequency** Seltenheit *f*
**infrequent** selten
**infringe, to** ~ beeinträchtigen, (patents) verletzen
**infringement** Beeinträchtigung *f*, Übertretung *f*, (of patent rights) Verletzung *f*, Zuwiderhandlung *f* ~ **of a patent** Patentverletzung *f*
**infuse, to** ~ aufgießen, einflößen
**infusibility** Unschmelzbarkeit *f*
**infusible** unschmelzbar
**infusion** Aufguß *m* (chem.), Eingießung *f*, Einguß *m* ~ **process** Aufgußverfahren *n*
**infusorial earth** Bergmehl *n*, Diatomeen-erde *f*, -pelit *m*, Infusorienerde *f*, Kieselgur *f*, Polierschiefer *m*
**ingate** Anschnitt *m*, Einguß-lauf *m*, -trichter *m*, Trichter-einlauf *m*, -lauf *m*, -zulauf *m* ~ **plot** Füllort *m*
**ingenious** erfinderisch, geistreich, scharfsinnig, sinnreich, sinnvoll
**ingenuity** Findigkeit *f*
**ingestion** Ansaugen *n* (Triebwerk), Einführung *f* von Stoffen
**ingoing charge** in den Zylinder eintretendes Gemisch *n*
**ingot** Barren *m*, Block *m*, Blöckchen *n*, Dackel *m*, Deul *m*, Einguß *m*, Eisenblock *m*, Gußblöck *m*, Luppe *f*, Rohblock *m*, Rohbramme *f*, Zain *m* **to make into** ~**s** zainen ~ **of gold** Goldbarren *m*
**ingot,** ~**-breaker press** Blockbrechpresse *f* ~ **butt** Restblock *m* ~ **charger** Blockeinsetzmaschine *f* ~**-charging car** Blockeinsetzwagen *m* ~**-charging crane** Blockeinsetzkran *m* ~**-conveying device** Blockvorrollvorrichtung *f* ~ **core** Blockkern *m* ~ **crane** Block-, Stripper-kran *m* ~ **crop**

**end** Blockkopf *m* ~**-drawing-out** Blockauszieher *m* ~ **gold** Stangengold *n* ~ **gripper** Blockgreifer *m* ~**-head heating device** Blockkopfbeheizung *f* ~ **iron** Block-, Fluß-eisen *n*, Flußstahl *m* ~ **metal** Rohmasseln *pl* ~ **mold** Block-, Gieß-form *f*, Kokille *f*, Kokillengußform *f*, Zainform *f* ~ **piping** Kerbstellen *pl*, Lunkerbildung *f* ~ **production** Blockerzeugung *f* ~ **pusher** Block-ausdrücker *m*, -drücker *m* ~ **reheating furnace** Blockwärmeofen *m* ~**-rolling mill** Blockwalze *f* ~**-roughing lathe** Blockschruppbank *f* ~ **shears** Blockschere *f* ~ **silver** Stangensilber *n* ~ **skin** Blockkruste *f* ~ **slab** Rohbramme *f* ~ **slab mold** Brammenform *f* ~**-slicing machine** Blockteilmaschine *f* ~ **steel** Blockeisen *n*, Flottstahl *m*, Fluß-eisen *n*, -stahl *m*, -stahl *m* mit hohem Kohlenstoffinhalt ~**-steel plate** Gußstahlblech *n* ~ **stirrup** Blockzange *f* ~ **stripper** Block-abstreifer *m*, -stripper *m* ~**-stripping crane** Blockabstreifkran *m* ~ **tilter** Block-hebetisch *m*, -kipper *m*, -wender *m* ~**-tipping device** Blockkipper *m* ~ **tongs** Blockzange *f* ~ **turning** Blockdreherei *f* ~ **withdrawing device** Blockauszieher *m*
**ingotism** Blockseigerung *f*
**ingrained** echt
**ingredient** Ansatz *m*, Bestandteil *m*, Einschluß *m*, Füllstoff *m*, Zugabe *f*, Zutat *f* ~ **of mixture** Mischungsbestandteil *m*
**ingredients,** Bestandteile *pl*, Gehalt *m*, Satz *m* ~ **for powder** Pulversatz *m*
**ingress, to** ~ eindringen, in eine Leitung eintreten
**ingress** Eindringen *n*, Eintritt *m* ~ **gate** Eingußkanal *m*, -lauf *m*, -trichter *m*, Gießtrichter *m*, Trichterlauf *m* ~ **reflector** Eintrittsreflektor *m* (Autokartograph)
**inhabit, to** ~ bewohnen
**inhabitable** bewohnbar
**inhabitants** Einwohnerschaft *f*
**inhaling valve** (gas mask) Einatemventil *n*
**inherent** anhaftend, eigentümlich, innewohnend, von Natur, original, untrennbar verbunden, zugehörig
**inherent,** ~ **characteristic** Eigencharakteristik *f* ~ **(or natural) color** Eigenfarbe *f* ~ **control** Eigenkontrolle *f* ~ **corrosion resistance** natürliche Widerstandsfähigkeit *f* gegen Korosion ~ **damping** Eigendämpfung *f* ~ **distortion factor** Eigenklirrfaktor *m* ~ **error** Anfangsfehler *m*, mitgeschleppter Fehler *m* ~ **feature** Eigentümlichkeit *f* ~ **feedback** innere Rückkopplung *f* ~ **film noise** Eigengeräusch *n* ~ **filtration** Eigenfilterung *f* ~ **fog** Grundschleier *m* ~ **frequency** Eigenfrequenz *f* ~ **instability** Eigeninstabilität *f* ~ **lag** Eigenverzögerung *f* ~ **loss of condenser (or capacitor)** Eigenverlust *m* des Kondensators ~ **regulation** relativer Spannungs(ab)fall *m* ~ **resistance** natürliche Widerstandsfähigkeit *f* ~ **stability** Eigen-, Form-stabilität *f* ~ **strain (or tension)** Eigenspannung *f* ~ **value** Eigenwert *m*
**inherently** von Natur aus ~ **rigid** eigensteif ~ **stable** eigen-, form-stabil
**inherit, to** ~ ererben
**inherited error** Anfangsfehler *m*, mitgeschleppter Fehler *m* (comput.)
**inhibit, to** ~ entgegenwirken, hemmen, unterbinden, zurückhalten

inhibit, ~ pulse Verbotsimpuls *m* ~ wire Blockier-
draht *m*
inhibited rotation (of molecules) behinderte
Rotation *f*
inhibiting input Verbotssignal *n*
inhibition Behinderung *f*, Hemmung *f*, Untersagung *f*, Verbot *n* ~ of plastic deformation
Fließbehinderung *f*
inhibitor Inhibitor *m*, Spar-beize *f*, -beizstoff *m*,
Verhütungs-, Verzögerungs-mittel *n*
inhibitory hemmend
inhomogeneity Inhomogenität *f*, Ungleichstoffig-
keit *f*
inhomogeneous inhomogen, uneinheitlich, un-
gleichstoffig
initial anfänglich, am Anfang stehend, Aus-
gangs-, Vorlauf *m* ~ acceptivity Nennaufnahme
*f* (electr.) ~ address Anfangsadresse *f* ~ adjustment Nulleinstellung *f* ~ amplification Anfangs-
verstärkung *f* ~ amplitude Anfangsausschlag *m*
~ blank Ausgangsrode *f* ~ body retention
Retentionsfaktor *m* ~ boiling point (petroleum)
Anfangspunkt *m*, Anfangssiedepunkt *m*, Beginn *m* des Siedens *n*, unterer Siedepunkt *m* ~
break-away torque (engin.) Losbrechdrehmo-
ment *n* ~ breakdown Anfangsdurchschlag *m* ~
call-up erster Anruf *m* ~ capacitance Anfangs-
kapazität *f* ~ capital Gründungskapital *n* ~
charge (blasting) Anfangsladung *f*, Hauptladung *f* ~ charge of secondary (or storage) cell
erste Ladung *f* eines Sammlers ~ climb An-
fangssteiggeschwindigkeit *f* ~ concentration
Ansatzkonzentration *f* ~ condition Anfangs-
bedingung *f*, -zustand *m* ~ cost(s) Anschaffungskosten *pl* ~ creep primäres Kriechen *n* ~
current Anfangsstrom *m* ~ daily production
Anfangstagesförderung *f* ~ data Eingangs-
werte *pl* ~ dimension Ausgangsmaß *n* ~ direction Ausgangsrichtung *f* ~ discharge caused by
rapid surge of ions Spritzentladung *f* ~ displacement Anfangsauslenkung *f* ~ disturbance
Anfangsstörung *f* ~ drift (of photocell) An-
fangsdrift *f* ~ equations Ausgangsgleichungen
*pl* ~ equipment Erstausstattung *f* ~ expenses
Abschlußkosten *pl* ~ force Initialkraft *f* ~
fusion Rauhschmelze *f* ~ gradient Anfangs-
gradient *m* ~ grid voltage Gittervorspannung
*f* ~ hardening Anziehen *n* (Mörtel) ~ hardness
Ausgangshärte *f* ~ heating Anheizen *n* (tube) ~
installation Ersteinbau *m* ~ inverse voltage
Sprungspannung *f* ~ ionizing event primäres
Ionisationsereignis *n* ~ issue Erstausstattung *f*
~ jet Vorstrahl *m* ~ lateral pressure ratio Ruhedruckziffer *f* ~ letter Anfangsbuchstabe *m* ~
levelling Grobhorizontierung *f* ~ lift coefficient
(or factor) Anfangsauftriebsbeiwert *m* ~ load
Anfangs-, Erst-belastung *f* ~ loading Vorbe-
lastung *f*, Vorlast *f* ~ material Ausgangsprodukt
*n* ~ meridian Anfangs-längenkreis *m*, -meridian
*n*; Null-meridian *m*, -mittagskreis *m* ~ moment
of stress Anfangsmoment *n* ~ ordinate Anfangs-
ordinate *f* ~ part Anfangsstück *n* ~ particle
Primärteilchen *n* ~ performance Anfangslei-
stung *f* ~ period Anfangsphase *f*, Gesprächseinheit *f* ~ permeability Anfangspermeabilität *f*
~ phase Auftakt *m* ~ pitch Anfangssteigerung *f*
~ point Ablaufpunkt *m* ~ position Anfangs-
lage *f*, -stellung *f*; Ausgangsstellung *f*, Nullage *f*

~ potential Vorspannung *f*, Zündspannung *f* ~
power Anhubmoment *n*, Anzugskraft *f* ~
pressure Anfangsdruck *m* ~ product Ausgangs-
erzeugnis *n*, Vorprodukt *n* ~ program Erst-
planung *f* ~ rate of climb Anfangssteigge-
schwindigkeit *f* ~ recombination anfängliche
Rekombination *f* ~ region Anfangsbereich *m*
~ revolutions per minute Anfangsdrehzahl *f* ~
salary Anfangsgehalt *m* ~ separatory cell
Grobflotator *m* ~ set Beginn *m* des Abbindens
~ shell Embryonalkammer *f* ~ side clearance
anfängliches Seitenspiel *n* (Kolbenring) ~
slackness Anfangs(lager)spiel *n* ~ slag An-
fangsschlacke *f* ~ slope Anfangssteigung *f* ~
sound receiver Schallabgangsempfänger *m* ~
sounding Anklingen *n* ~ speed Anfangsgeschwindigkeit *f* ~ stage Anfangsstadium *n* ~
state Anfangs-, Ausgangs-zustand *m* ~ state of
descent of a parachute Abfangsmoment *m* ~
steam pressure Frischdampfdruck *m* ~ step
Anfangsstufe *f* (rdo) ~ strain Anfangsbean-
spruchung *f* ~ strength Anfangs-, Ausrück-
stärke *f* ~ stress Anfangs-, Vor-belastung *f*,
Vorspannung *f* ~ stress in the spring Vorspannung *f* der Feder ~ stressing force Vorspann-
kraft *f* ~ striking energy to fracture Bruchfaktor
*m* ~ substance Ausgangsstoff *m* ~ sump Vor-
sumpf *m* ~ susceptibility Anfangs-aufnahme-
vermögen *n*, -suszeptibilität *f* ~ tank pressure
Tankvordruck *m* (g/m) ~ temperature Anstell-,
Bezugs-temperatur *f* ~ tension Anzugsmoment
*n* ~ tension in the spring Vorspannung *f* der
Feder ~ thrust Anfangsschub *m* ~ transient
Einschwingung *f* ~ twist Anfangsdrall *m* ~
value Anfangswert *m* ~-value method Anfangs-
wertmethode *f* ~ valve Vorventil *n* ~ velocity
Anfangsgeschwindigkeit *f* ~ voltage Anfangs-
spannung *f*, Spannung *f* am Anfang ~ vortex
Anfahrwirbel *m* ~ water pump Vorlaufpumpe *f*
~ wave (or spike) (cardiogram) Initialzacke *f* ~
wear Anfangsverschleiß *m* ~ whirl Vordrall *m* ~
x, y Anfangskoordinate X, Y
initially tensioned vorgespannt
initiate, to ~ ansetzen, beginnen, einführen, einleiten, in Gang bringen to ~ a call ein Gespräch
*n* einleiten
initiating Inbetriebsetzung *f* ~ data Anlaufs-
daten *pl* ~ power Initialkraft *f* ~ switch Schaltelement *n*
initiation Beginn *m*, Einleitung *f*, Einsatz *m*,
Initialzündung *f*, (of a procedure) Inmarschsetzung *f* ~ of firing Einsetzen *n* der Zündung,
Zündungseinsetzen *n* ~ of impulse Einschwingvorgang *m* ~ of striking Einsetzen *n* der Zündung, Zündungseinsetzen *n*
initiation flight Einweihungsflug *m*
initiator Knallaufsatz *m* (explosives), Zündsatz
*m*, Zündstoff *m*
inject, to ~ ausspritzen, (steam) blasen, einblasen, einpumpen, einspeisen (Spannung), einspritzen, impfen, injizieren, spritzen, tränken
to ~ cement mit Zement torkretieren
injected, ~ beam eingeschlossener Strahl *m* ~
spectrum Anfangsspektrum *n*
injecting powder Spritzpulver *n*
injection Einblasen *n*, Einführung *f*, Einspritzung *f*, Tränkung *f* ~ of fuel Brennstoffein-
spritzung *f*

**injection,** ~ **advance** Voreinspritzung *f* (Diesel-motor) ~ **advance lever timing control** Einspritzzeitverstellhebel *m* ~ **advance mechanism** Spritzversteller *m* ~ **air** Einblaseluft *f* ~-**air compressor** Einblase-luftkompressor *m*, -luftpumpe *f* ~-**air receiver** Einblasegefäß *n* ~-air reservoir Einblaseluftbehälter *m* ~ **can** Spritzkanne *f* ~ **cock** Einspritzhahn *m* ~ **condenser** Einspritzkondensator *m* ~ **conduit** Einblaseleitung *f* ~ **control** Füllungsführung *f*, Spritzverstellung *f* ~ **control hand lever** Einspritz-Handverstellhebel *m* ~ **cooling** Einspritzkühlung *f* ~ **die-casting** Spritzguß *m* ~ **equipment** Einspritzausrüstung *f* ~ **grid** Mischgitter *n* ~ **jet (or spray)** Einspritzstrahl *m* ~ **lag** Einspritzverzug *m* ~ **method of grouting** Injektionsverfahren *n* ~ **mold** Spritzgußform *f* ~ **mold for plastic material** Kunstharzspritzform *f* ~ **molding** Spritzguß(verfahren) *m* (*n*), Spritzling *m* ~-**molding compound** Spritzgußmasse *f* ~ **nozzle** Einblase-, Einspritz-düse *f* ~ **oil can** Spritzöler *m* ~ **order** Einspritzfolge *f* ~ **pin** Spritzzapfen *m* ~ **pipe** Anlaßgasleitungsrohr *n*, Einspritzrohr *n* ~ **plunger** Spritzkolben *m* ~ **port (for propellants into rocket motor)** Einspritzöffnung *f* ~ **press** Spritzgußpresse *f* ~ **pressure** Einspritzdruck *m* ~ **process** Einspritzvorgang *m*

**injection-pump** Einspritzpumpe *f* ~ **carrier** Einspritzpumpenträger *m* ~ **drive shaft** Einspritzpumpenantriebswelle *f* ~ **gear** Einspritzpumpenantriebsrad *n*

**injection,** ~ **set** Spritzaggregat *n* ~ **timer** Spritzzeitversteller *m* ~ **timing** Einspritzmoment *m* ~ **timing control** Einspritzzeitversteller *m* ~ **timing gear** Einspritzbeginnverstellung *f* ~-**type Diesel engine** kompressorlose Dieselmaschine *f* (Einspritzmotor) ~ **valve** Einspritzdüse *f*

**injector** Dampfstrahlgebläse *n*, Druckstrahlpumpe *f*, Einblasedüse *f*, Einspritzdüse *f*, Einspritzer *m*, Einspritzpumpe *f*, Injektor *m*, Spritzdüse *f*, Spritze *f*, (of steam) Strahlapparat *m*, Strahlenpumpe *f*, Strahlpumpe *f*

**injector,** ~ **bushing** Düsenmantel *m* ~ **cone** Injektordüse *f* ~ **grid** Mischgitter *n* ~ **nozzle** Injektordüse *f* ~ **pipe** Einspritzröhre *f* ~-**pump gear wheel** Einspritzpumpenrad *n* ~ **spring** Düsenfeder *f* ~ **valve** Injektorventil *n* ~ **yoke** Düsenhalterflansch *m*

**injunction** Einhaltsbefehl *m*, Untersagung *f*, Verbot *n* ~ **method** Beschwerdeverfahren *n* ~ **suit** Einstellungsklage *f* ~ **switch** Wähler *m* für ankommende Verbindungsleitungen

**injure, to** ~ beeinträchtigen, beschädigen, schaden, verletzen, verwunden

**injured** ramponiert, verletzt ~ **party** Geschädigter *m* ~ **person** Unfallverletzter *m*, Verwundeter *m* ~ **seriously** schwer verletzt

**injurious** schädlich

**injury** Beeinträchtigung *f*, Beleidigung *f*, Benachteiligung *f*, Beschädigung *f*, Nachteil *n*, Schade *m*, Schaden *m*, Verletzung *f*

**ink, to** ~ austauschen, (drawings) ausziehen, einfärben, Farbe auftragen, mit Tusche ausziehen **to** ~ **the form** Schwärze auftragen **to** ~ **(up)** einschwärzen

**ink** Farbe *f*, Tinte *f*, (india) Tusche *f* ~ **agitator** Farbrührwerk *n* ~ **block** Farbläufer *m*, Reib-

stein *m* ~ **blot** Tintenklecks *m* ~ **brush** Tuschpinsel *m* ~ **compound (or reducer)** Farbenzusatz *m* ~ **disk** Farbrädchen *n* ~ **distribution** Verreibung *f* der Farbe ~ **drums** Farbwalzen *pl* ~ **eraser** Tintengummi *n* ~ **feed** Farbzufuhr *f* ~ **feed regulation** Farbstreifenregulierung *f* ~ **feed roller adjustment** Farbregulierung *f* ~ **fountain** Farbkasten *m* ~-**grinding machine** Farbenreibmaschine *f* ~ **grinding table** Farbverreibtisch *m* ~ **lines** Linienblatt *m* ~ **medium** Farblöser *m* ~ **mill** Farbenverreibmaschine *f* ~ **pad** Farb-, Stempel-kissen *n* ~ **recorder** Linien-, Tinten-schreiber *m* ~ **reducer** Drucktinktur *f* ~ **regulating screw** Farbmesserstellschraube *f* ~ **reservoir** Farbbehälter *m* ~ **ribbon** Farbband *n* ~-**ribbon change** Farbbandwechsel *m* ~-**ribbon feed** Farbbandvorschub *m* ~-**ribbon reversal** Farbbandwechsel *m* ~ **roller** Auftrageröllchen *n*, Farben-, Farb-walze *f*, Schwärzrolle *f* ~ **(ing) roller** Farb-rolle *f*, -röllchen *n* ~ **slab** Farbtisch *m* ~ **spot** Klecks *m* ~ **stand** Schreibzeug *n* ~ **supply** Farbverreibungsversorgungsanlage *f* ~ **supply cut-out and reset** Farban-und-abstellung *f* ~ **supply shutoff** Farbabstellung *f* ~ **vapor recorder** Tintenstrahlschreiber *m* ~ **well** Farb-gefäß *n*, -kasten *m* ~ **wheel** Auftrageröllchen *n*, Farbrädchen *n* ~ **writer** Farbschreiber *m*

**inked ribbon** Farbband *n*

**inker** Auftragwalze *f*, Farbenwalze *f*, Farb-rolle *f*, -röllchen *n*, -schreiber *m*, -walze *f* (textiles), Morsefarbschreiber *m*, Tintenschreiber *m* ~ **drive gear segment** Zahnsegment *n* für Farbwerk

**inking** Einfärben *n*, Einfärbung *f*, Farbauftragung *f* ~ **by hand** Handeinfärbung *f*

**inking,** ~ **compass** Zirkel *m* mit Reißfeder ~ **cylinder** Auftragwalze *f* ~ **device** Farbwerk *n* ~ **pencil** Noppenstift *m* ~ **ribbons** Farbbänder *pl* ~ **roll** Gegendruckzylinder *m* ~ **roller** Auftragwalze *f*, Farbrädchen *n*, Farbwalze *f* (print.) ~ **rollers** Auftragrollen *pl* ~ **wheel** Farb-rolle *f*, -röllchen *n*

**inky** dunkel, tintig; tintenartig

**inlaid work** Einlegearbeit *f*

**inland** Binnenland *n*; inländisch, landeinwärts ~ **customer** Inlandabnehmer *m* **at** ~ **end** landseitig ~ **lowlands** Binnentiefland *n* ~ **market** Inlandmarkt *m* ~ **message** Inlandtelegramm *n* ~ **sea** Binnensee *m* ~ **shipping** Binnenschiffahrt *f* ~ **traffic** Binnenverkehr *m* ~ **trunk call** Ferngespräch *n* ~ **water(s)** Binnengewässer *n* ~ **waterway** Binnenwasserstraße *f*

**inlay, to** ~ furnieren **to** ~ **with black enamel** niellieren

**inlay** Einlage *f* ~ **fade** Schabloneneinblendung *f* ~ **shooting bar** Schußschiene *f*

**inlaying** Austäfelungsarbeit *f*, Getäfel *n*, Lambris *m*, Täfelwerk *n* ~ **apparatus** Durchschußapparat *m* (textiles)

**inlet** Bucht *f*, Einfahrt *f*, Einführen *n*, Einführung *f*, Einführungsöffnung *f*, Einfüllstutzen *m*, Eingang *m*, Einlaß *m*, Einlaßöffnung *f*, Einlauf *m* (v. Motoren), Einströmungsöffnung *f*, Eintritt *m*, Öffnung *f*, Zufluß *m*, Zufuhr *f*, Zuführung *f*, (in casting) Zulauf *m* ~ **and outlet** Zu- und Abflußleitung *f* ~ **and outlet element** Zulauf- und Ableitungselement *n* ~ **and outlet temperatures**

Einlaß- und Auslaßtemperatur *f* ~ **of scavenging air** Spüllufteintritt *m*

**inlet,** ~ **angle of impeller** Laufradeintrittswinkel *m* ~ **box** Einlaufbüchse *f* ~ **branch** Einlaßstutzen *m* ~ **cam** Einlaßnocken *m* ~**-cam roller lever** Einlaß-hebel *m*, -ventilverbindungshebel *m* ~ **camshaft** Einlaßsteuerwelle *f* ~ **chamber** Einströmkasten *m* ~ **connecting branch (or tube)** Eingußstutzen *m* ~ **connection (or nozzle)** Ansaugstutzen *m* ~ **cross section** Eintrittsquerschnitt *m* ~ **edge** Austritt-, Einlaß-kante *f* ~ **edge of blade** Eintrittskante *f* ~ **face** Einlaßfläche *f* ~ **fan** Vorsatzläufer *m* ~ **fitting** Füllansatz *m* ~ **fittings** Einlaufarmatur *f* ~ **flange** Flansch *m* für Einlaßöffnung ~ **funnel** Einführungspfeife *f* ~ **gate** Einguß-kanal *m*, -lauf *m* ~ **governer** Einlaßsteuerung *f* ~ **guide vane** Eintrittsleitschaufel *f* ~ **hole** Zulaufbohrung *f* ~ **limit** Einlaßfeld *n* ~ **manifold** Ansaugerohr *n* ~ **motor** Reihenmotor *m* mit hängenden Zylindern ~ **nozzle** Einströmdüse *f* ~ **opening** Einlaßöffnung *f* ~ **pipe** Einfluß-, Einlauf-, Einlaß-, Einströmungs-rohr *n*, Zulaufleitung *f*, Zulaufrohr *n*, Zuleitungsrohr *n* ~**-pipe connection** Eintrittsstutzen *m* ~ **piping** Einlaßleitung *f* ~ **plug** einlaßseitige Zündkerze *f* ~ **port** Einlaßkanal *m*, -öffnung *f*, -schlitz *m*, Einspritzöffnung *f*, Einströmschlitz *m* ~ **pressure** Admissionsdruck *m*, Ansaugspannung *f*, Einströmungsdruck *m*, Vordruck *m* ~ **seat** Einlaßsitz *m* ~ **side** Ansaug-, Einlaß-, Vergaser-seite *f* ~ **sliding plug** Einlaßschieber *m* ~ **sluice** Einfluß-, Jagd-schleuse *f* ~ **spot (or place)** Eintrittsstelle *f* ~ **strainer** Eingußsieb *n*, Einlaufseiher *m* ~ **strainer for insertion** Einlaufstechseiher *m* ~ **temperature** Eintritts-temperatur *f*, -wärmegrad *m* ~ **tube** Einfluß-, Einführungs-rohr *n* ~ **union** Druckrohr-anschluß *m*, -stutzen *m*

**inlet-valve** Ansaug-, Einlaß-, Einström-, Eintritts-ventil *n* ~ **cap** Einlaßventilverschraubung *f* ~ **cotter** Einlaßventilkeil *m* ~ **gear** Einllßsteuerung *f* ~ **stem** Einlaßventilspindel *f*

**inlet well** Gaseinpreßbohrung *f*

**in-line blending** Mischoperationen *pl* in Rohrleitungen

**innate** innewohnend

**inner** innen, inner ~ **ball-bearing race** Kugellagerinnenring *m* ~ **barrel** Futterrohr *n* ~ **basin** Innenbecken *n* ~ **brake** Innenbremse *f* ~ **bremsstrahlung** innere Bremsstrahlung *f* ~ **capsule** (explosives) Innenhütchen *n* ~ **chase** Innenrahmen *m* ~ **circumference** Innenumfang *m* ~ **collector ring jet** Innensammler *m* ~ **conductor** Innenleiter *m* ~ **cone** Innenkonus *m* ~ **contact** Innenkontakt *m* ~ **corner** einspringende Ecke *f* ~ **cover** Innenüberzug *m* (nicht abnehmbar) ~ **cover of manhole** Schmutzfänger *m* für Kabelschächte ~ **cowl** Innenhaube *f* ~ **crucible** Hintergestell *n* ~ **dam** Binnendeich *m* ~ **dock** Binnendock *n* ~ **duct ring jet** Innensammler *m* ~ **edge of rail** Fahrkante *f* der Schiene, Schieneninnenkante *f* ~ **electron** Rumpfelektron *n* ~ **electrons** kernnahe Elektronen *pl* ~ **envelope** Innenbelag *m* ~ **fabric strip (or band)** Grundgurt *m* ~ **fishplate with round holes** Innenlasche *f* mit runden Löchern ~ **fit** Inneneinpaß *m* ~ **form** Widerdruckform *f* ~ **gate (of the lock)** Oberhaupt *n* ~ **(or upper)**

**gates** Binnenhaupt *n* ~ **gearing** Innenzahnkranzantrieb *m* ~ **grid** Innengitter *n* ~ **harbor** Binnenhafen *m*, Innenbecken *n*, Innenhafen *m* ~ **infrasonic** innerer Infraschall *m* ~ **jib** Binnenklüver *m* ~ **leg** landseitige Stütze *f* (Kran) ~ **lining** Innenverkleidung *f* ~ **lining of a furnace** Schachtfutter *n* ~**-marker signal (IM)** Platzeinflugsignal *n* (aviat.) ~ **nozzle ring** Schaufelträger *m* ~ **orientation** (camera) innere Daten *pl*, innere Orientierung *f* ~ **partition** Zwischenwand *f* ~ **parts** Innenteile *pl* ~ **photoelectric effect** innerer Fotoeffekt *m* ~ **planking** (of ship) Wegerung *f* ~ **player** (piano) Einbau *m* ~ **point** Innenpunkt *m* (math.) ~**-pole armature** Innenpolanker *m* (electr.) ~**-pole frame** Innenpolgehäuse *n* ~ **quantum number** innere Quantenzahl *f* ~ **race (or raceway)** Innenlaufbahn *f* ~ **racer** innerer Laufring *m* (Triebwerk) ~ **ring wall of a furnace** Schachtfutter *n* ~ **rubber lining** Innengummi *n* ~ **seal ring** Innenseegerring *m* ~ **shell** Innenboden *m* ~**-shell electron** inneres Elektron *n* ~ **side** Innenseite *f* ~ **slide** Innenschieber *m* ~ **sole** Brandsohle *f* ~ **springs** Federeinlagen *pl* ~ **surfaces** innerer Rand *n* ~ **tank** Innenbehälter *m* ~ **tube** (of a gun) Futterrohr *n*, Kernrohr *n*, (tire) Luftschlauch *m*, Schlauch *m* ~ **tube of gun liner** Seelenrohr *n* ~ **wheel brake** Innenradbremse *f* ~ **work function** innere Austrittsarbeit *f*

**innocuous** unschädlich

**innovation** Neuerung *f*

**innumerable** unzählig

**inoculate, to** ~ einimpfen, impfen, okulieren

**inoculating pencil** Impfstift *m*

**inodorous** geruchlos

**inoffensive** harmlos

**inoperating contact** Ruhekontakt *m*

**inoperation** Nicht-in-Betrieb-Sein *n*, Stilliegen *n*

**inoperative** in Ruhe *f* befindlich, in der Ruhelage *f* oder Ruhestellung *f*, in der Stellung *f*; ruhend, unwirksam ~ **pass** (in rolling) blindes Kaliber *n*, totes Kaliber *n* ~ **transformer** ruhender Transformator *m*

**inorganic** anorganisch, unorganisch

**inoxidizable** nicht oxydierbar

**inphase** gleichphasig, konphas, phasengleich, phasenrichtig ~ **amplifier** Gleichtaktverstärker *m* ~ **coincidence** conphas ~ **component** Watt-, Wirk-komponente *f* ~ **component of the current** Wirkstromkomponente *f* ~ **component of the voltage** Wirkspannungskomponente *f* ~ **condition** Gleichphasigkeit *f* ~ **feedback** phasenfreie Rückkupplung *f* ~**-opposed** gegenphasisch ~ **opposition** gegenphasig ~ **recording** Gleichtaktaufnahme *f* (film) ~ **state** Phasengleichheit *f*

**in-process material** Zwischenprodukt *n*

**input** Aufnahme *f* (electr.), Aufwand *m*, Ausgabe *f*, Eingabespeicher *m* ~ **or output** Ein- oder Auslaß *m*, Eingabe *f* (data proc.), Eingang *m*, Eingangsformation *f*, eingespeiste Menge *f*, Eintrag *m*, Kraftbedarf *m*, Ladung *f*, aufgenommene Leistung *f* (of electric motors), zugeführte Leistung *f*, Leistungsaufnahme *f*, Spannungszuführung *f*, Speisung *f*, Wellenbestand *m*, Zufuhr *f*

**input,** ~ **active admittance** Eingangswirkleitwert *m* ~ **admittance** Eingangsadmittanz *f* ~ **alternating-current voltage** Eingangswechselstromspannung *f* ~ **amplifier** Eingangs-, Vor-verstärker *m* ~ **amplitude** Eingangsamplitude *f* ~ **area** Eingabespeicherbereich *m* ~ **attenuator** Eingangsabschwächer *m* ~ **axis** Meßachse *f* ~ **block** Eingangsspeicher *m* ~ **capacitance** Eingangskapazität *f* ~ **capacity** Gitterkreiskapazität *f* ~ **capacity disk** (klystron) Eintrittsblende *f* ~ **cavity buncher** Eingangshohlraum *m* ~ **circuit** Aufnahmestromkreis *m*, Eingangs-kreis *m*, -stromkreis *m*, Empfangsstromkreis *m*, Vorkreis *m* ~ **computer** Eingangssignalwandler *m* ~ **conductance** Eingangswirkleitwert *m* ~ **conductivity** Eingangsleitwert *m* ~ **connector** Eingangssignalanschluß *m* ~ **control** (soundwaves) Vorregler *m* ~ **current** ankommender Strom *m*, Eingangsstrom *m* ~ **damping** Eingangsbedämpfung *f* ~ **data** Ausgangsgrößen *pl* ~ **diaphragm disk** Eintrittsblende *f* ~ **equipment** Eingabegerät *n* ~ **filter** Filtereingang *m* ~ **gap** Eingangsspalt *m* ~ **grid capacity** Eingangsgitterkapazität *f* ~ **impedance** Eingangs-kreisimpedanz *f*, -scheinwiderstand *m*, -widerstand *m*, Gitterkreisimpedanz *f* ~**-impedance terminals** Eingangsscheinwiderstandsklemmen *pl* ~ **level** Eingangspegel *m* ~ **lever** Eingangshebel *m* ~ **linkage** Eingangsgestänge *m* ~ **monitoring** Bandkontrolle *f* (Mischpult) ~ **output media** Informationsträger *m* (tape rec.) ~ **potential** Eintrittspotential *n* ~ **preamplifier** Eingangsvorverstärker *m* ~ **quantity** Eingangsgröße *f* ~ **range** Eingangssignalbereich *n* ~ **rate** Eingangsgeschwindigkeit *f*, (inertial nav.) Größe *f* des Eingangssignals, ~ **reactance** Gitterkreisreaktanz *f* ~ **repeating coil** Vorübertrager *m* ~ **resistance** Eingangs-bedämpfung *f*, -widerstand *m* ~ **resistance of a thermionic valve** Gitterkreiswiderstand *m* ~ **routine** Lese-, Eingabe-programm *n* (info proc) ~ **section** Eingabespeicherbereich *m* ~ **selector** Eingangsumschalter *m* ~ **sensitivity** Eingangsempfindlichkeit *f* ~ **signal** Eingangssignal *n* ~ **stage** (of amplifier) Vorstufe *f* ~ **table** Funktionstrieb *m* ~ **tape** Eingabeband *n* ~ **terminal** Eingangsklemme *f* ~ **terminals** Speisepunkt *m* ~ **time constant** Eingangszeitkonstante *f* ~ **transformer** Eingangs-transformator *m*, -übertrager *m*, Vorübertrager *m* ~ **tube** Vorrats-, Vor-röhre *f* ~ **unit** Eingabeeinheit *f*, -werk *n* (comput.) ~ **value for receiver** Empfängervorröhre *f* ~ **variable** Eingangsgröße *f* ~ **voltage** Anfangs-, Eingangs-spannung *f* ~ **well** Einlaßsonde *f*

**inquire, to** ~ abfragen, anfragen, erfragen, sich erkundigen, nachfragen, Umfragen halten, untersuchen **to** ~ **into** nachforschen

**inquiry** Abfragen *n*, Anfrage *f*, Erhebung *f*, Erkundigung *f*, Ermittlung *f*, Umfrage *f*, Untersuchung *f* ~ **circuit** Überweisungsleitung *f* ~ **clerk** Auskunftsbeamter *m* ~ **docket** Auskunftsblatt *n* ~ **station** Abfragestelle *f*

**inrush** Einströmen *n*, Flut *f*, Zustrom *m* ~ **of air** Lufteinbruch *m* ~ **of water** Einbruch *m*

**inrush,** ~ **current** Einschaltstrom *m* ~ **load** Überlastung *f* durch Einschaltstromspitze

**inscribe, to** ~ einschreiben

**inscribed** einbeschrieben, eingeschrieben ~ **angle**

Peripheriewinkel *m* ~ **circle** eingeschriebener Kreis *m* ~ **circumference** einbeschriebene Kreislinie *f* ~ **polyhedron** einbeschriebenes Vieleck *n*

**inscriber** Kodiermaschine *f*

**inscription** Aufschrift *f*, Beschriftung *f* einer Zeichnung, Inschrift *f*

**insect lime** Raupenleim *m*

**insecticide,** ~ **plane used for crop dusting** staubstreuendes Flugzeug *n* ~ **powder** Insektenpulver *n*

**insecure** unsicher

**insensibility** Bewußtlosigkeit *f*, Unempfindlichkeit *f*

**insensible** unempfindlich ~ **for vibrations** Schwingungsfreiheit *f*

**insensitive** unempfindlich

**insensitiveness** Unempfindlichkeit *f*

**insensitivity** Unempfindlichkeit *f*

**inseparability** Untrennbarkeit *f*

**inseparable** untrennbar, unzertrennbar

**insert, to** ~ (in column) aufnehmen, einbauen, einbetten, einfügen, einführen, einlegen, einrücken, einschalten, einsetzen, einspannen, (lead) hineinschlagen, zwischenlegen, zwischenschalten **to** ~ **cables** einstecken **to** ~ **into one another** verschachteln **to** ~ **the packing** die Packung *f* einbringen **to** ~ **a plug** einen Stecker *m* oder einen Stöpsel *m* einsetzen, stöpseln **to** ~ **in resistance** Widerstand *m* einschalten **to** ~ **a rivet** einen Niet *m* einsetzen **to** ~ **tubes** aufschieben **to** ~ **in wrong jack** falsch stöpseln

**insert** Einbauplatte *f*, Einlage *f*, Einlaßstück *n*, Einsatz *m*, Einsatzstück *n*, Gewindeeinsatz *m* (Triebwerk), Haltestift *m*, Lochblende *f* ~ **in distributor** Zündverteilerzwischenstück *n*

**insert,** ~ **bit** Bohrer *m* mit Einsätzen ~ **coil form** Einsatzspulenrahmen *m* ~ **drawing** Teilblatt *n* ~ **gauge** Kaliberdorn *m* ~ **hole** Blendenöffnung *f* ~ **piece for casting mold** Gießformeinsatzstück *n* ~ **ring** Einsetzring *m* ~ **sheet** Einsteckbogen *m* ~ **switch** Einsatzschalter *m* ~ **thread** herausnehmbarer Lauffächenteil *m* der Reifenform ~ **unit** Steckeinheit *f* ~ **vessel** Einsatzkopf *m* (bei Ultrathermostat)

**insertable cutter** einsetzbare Schneidzunge *f*

**inserted** eingefügt, eingelassen, (advertisement) eingerückt, eingesetzt, geschaltet, zwischengeschaltet ~ **adapter** eingesetztes Paßstück *n* ~ **blade** eingesetztes Messer *n* ~**-blade end-milling cutter** Messerkopf mit eingesetzten Messern ~**-blade plain milling cutter** Walzenfräser *m* mit eingesetzten Messern ~ **ceiling** blinde Decke *f* ~ **cover** Einsatzdeckel *m* ~ **crank-pin** eingesetzter Kurbelzapfen *m* ~ **pillar** flacher Pilaster *m* ~ **ring** Eindrückring *m* ~ **tooth cutter** Mehrfachschneidenhalter *m*, Messerkopf *m*, ~**-tooth side-milling cutter** Scheibenfräser *m* mit eingesetzten Messern

**inserter** (tool) Eindreher *m*, Schraubenspundringeindreher *m*

**inserting** Einbringen *n*, Einstecken *n* ~ **crank** Einsteckkurbel *f* ~ **quoin** Einlegekeil *m*

**insertion** Ansatz *m*, Aufnahme *f*, Einbau *m*, Einfügen *n*, Einfügung *f*, Einführen *n*, Einführung *f*, Einlage *f*, Einrückung *f*, Einsatz *m*, Einschaltung *f*, Einsetzen *n*, Einspannung *f*, Zeitungsanzeige *f* ~ **of a pad** Einschaltung *f* einer Dämpfung

**insertion, ~ attenuation** Einfügedämpfung *f* **~ gain** Einfügungsgewinn *m* **~ handle** Aufsteckgriff *m* **~ leaf** Einschaltblatt *n* **~ loss** Einfügungs-, Einführungs-verlust *m* **~ mark** Einschaltungszeichen *n* **~ process** Einschaltvorgang *m* **~ step** Einsatzschale *f*

**inserts** (Geiger) Blenden *pl*, (from cemented carbide) Einlagen *pl* (aus gesintertem Hartmetall)

**inset** Einsatz *m*, Zwischenlage *f*; eingelassen **~ diaphragm** Einhängeblende *f* **~ mounted switch** eingebauter Schalter *m* **~ transmitter** Kapselmikrofon *n*

**insetter** Papierbahnrückeinführer *m* für Mehrfarbendruck

**inside** Innenseite *f*; innerhalb, inwendig **~ and outside chaser** Außenstrehler *m* **~ antenna** Zimmerantenne *f* **~ bark** Holzbast *m* **~ bottom of cartridge case** Zündglocke *f* **~ cable** Innenkabel *n* **~ caliper** Innenkaliber *n* **~-caliper gauge** Stichmaß *n* **~ calipers** Hohlzirkel *m*, Innen-fühlhebel *m*, -taster *m*, Lochtaster *m* **~ chaser** (or chasing) **tool** Innenstrehler *m* **~ compass** Lochzirkel *m* **~ control** (or drive) Innenlenkung *f* **~ crack** Innenriß *m* **~ cranks that necessitate a cranked axle** Innenkurbeln *pl* die Achskröpfung erfordern **~ diameter** Innendurchmesser *m*, lichter Durchmesser *m*, lichte Weite *f*, Loch-, Nenn-weite *f*, Rohr-seele *f*, -wand *f* **~ diameter of pipe** (or tube) Rohrkaliber *n* **~ dimension** Innenabmessungen *pl* **~ fittings** (or equipment) Innenausstattung *f* **~ flask jacket** Formkastenband *n* **~ gauge** Passierrohr *n*, Passimeter *n* **~ gearing** Innentreibradbetrieb *m* **~ gluing** Innenbeleimen *n* **~ journal** Achsinnenlagerstelle *f* **~ labor** Arbeit *f* (min.) **~ lead lining** Bleieinlage *f* **~ lead tube scrapers** Bleirohrausbohrer *m* **~ link** Innenglied *n* **~ measure** Innenmessung *f* **~ micrometer** Stichmaß *n* **~ micrometer calipers** Innenmikrometer-Stichmaß *n* **~ padding** Innenauspolsterung *f* **~ parts** Innenteile *pl* **~ plant** gesamte Fernsprecheinrichtung *f* **~ room** Innenraum *m* **~ space** Innenraum *m* **~ spider** (of loud speaker) Zentrierungsvorrichtung *f* **~ spraying period** Innenspritzzeit *f* **~ spring caliper** Innentaster *m* (Innenzirkel) **~ stroke** Binnengang *m* **~ taper** Innenkonus *m* **~ test indicator** Innenfühlhebel *m* **~ thread** Innengewinde *n* **~-thread calipers** Gewindeinnentaster *m* **~-threading tool** Innengewindeschneidestahl *m* **~ tube riveting** Rohrinnennietung *f* **~ turning equipment** Ausdrehvorrichtung *f* **~ wall** Innenwand *f* **~ width** lichte Weite *f*, Lichtweite *f* **~ wing** innerer Flügel *m* (Kurve) **~ wooden casing** Holzauskleidung *f*

**insight** Einblick *m*, Einsicht *f*, Erkenntnis *f*

**insignia** Abzeichen *n*, Hoheitszeichen *pl*

**insignificant** bedeutungslos, geringfügig, klein, nichtssagend, nicht von Belang *m*, unansehnlich, unbedeutend, unscheinbar

**insincere** versteckt

**insipid** fade, geschmacklos

**insist, to ~ on** bestehen auf

**insistant** beharrlich

**insolation** Besonnung *f*, Bestrahlung *f*, Einstrahlung *f*, Insolation *f*

**in-sole** Brand-, Einlage-sohle *f*

**insolubility** Unauflösbarkeit *f*, Unlöslichkeit *f*

**insoluble** nichtmischbar, unauflösbar, unauflöslich, unlöslich **~ in oil** ölfest **~ in spirit** sprit-unlöslich **~ in water** wasserunlöslich

**insolvable** nichtmischbar

**insolvency** Konkurs *m*, Zahlungsunfähigkeit *f*

**insolvent** zahlungsunfähig

**insonorous materials** schalltote Werkstoffe *pl*

**inspect, to ~** in Augenschein nehmen, beaufsichtigen, begehen, besichtigen, einsehen, inspizieren, mustern, nachprüfen, nachsehen, prüfen, säubern, überwachen, untersuchen, visitieren **to ~ for cracks** auf Risse prüfen **to ~ (the workings of) a mine** eine Grube befahren

**inspecting officer** Inspekteur *m*

**inspection** Appell *m*, Aufsicht *f*, Baukontrolle *f*, Beaufsichtigung *f*, Bemusterung *f*, Besichtigung *f*, Durchsicht *f*, Einsicht *f*, Inaugenscheinnahme *f*, Inspektion *f*, Inspizierung *f*, Kontrolle *f*, Nachprüfung *f*, Prüfung *f*, Revision *f*, Überwachung *f*, Untersuchung *f* **~ of equipment** Gerätedurchsicht *f* **~ of material** Materialprüfung *f* **~ of ordonance** Gerätedurchsicht *f*

**inspection, ~ agency** Überwachungsstelle *f* **~ control** Abnahmekontrolle *f* **~ cover** Schaulochdeckel *m* **~ crib** Prüfplatz *m* **~ department** Kontrollabteilung *f* **~ door** (or flap) Schauklappe *f* **~ flap** Handlochklappe *f* **~ gallery** Inspektionsstollen *m* **~ gauge** Abnahme-, Revisions-, Werkstattabnahme-lehre *f* **~ glass** Schauglas *n* **~ hole** Einguck *m*, Einschauloch *n* (aviat.), Schauloch *n* **~ lamp** Ableuchtlampe *f* **~ light** Prüflampe *f* **~ panel** Besichtigungspanel *n*, Schau-öffnung *f*, -tafel *f* **~ pit** Arbeits-, Ausbesserungs-, Reparatur-grube *f* **~ plate** Schauplatte *f* **~ port** Schaugang *m* **~ room** Schaukammer *f* **~ screen** Schaugitter *n* **~ shaft** Einsteig-, Inspektions-, Revisions-schacht *m* **~ sheet** Befundbogen *m*, Prüfungsprotokoll *n* **~ signature** Sichtvermerk *m* **~ slit** Sichtschlitz *m* **~ tour** Betriebsbegehung *f* **~ tube** Kontrollröhrchen *n* **~ tunnel** Inspektionsstollen *m* **~ window** Beobachtungsfenster *m*

**inspector** Abnahmebeamter *m*, Aufseher *m*, Bauaufseher *m*, Beschauer *m*, Grubenaufseher *m*, Grubenaufsichtsbeamter *m*, Inspekteur *m*, Inspektor *m*, Inspizient *m*, Intendant *m*, Kommissar *m*, Prüfer *m* **~ of products** Warenbeschauer *m* **~'s acceptance stamp** Abnahmestempel *m*

**inspectorate** Inspektion *f*, Inspektorat *n*

**inspectoscope** Kristallbeschauer *m*

**inspire, to ~** ansaugen, begeistern, ein- oder ausatmen

**inspired volume** Ansaugvolumen *n*

**inspissate, to ~** eindicken

**inspissation** Asphaltbildung *f*, Eindicke *f*

**instability** Haltlosigkeit *f*, Instabilität *f*, Labilität *f*, Unbeständigkeit *f*, Unruhe *f*, Unstabilität *f*, Ziehen *n*, Ziehvorgang *m* **~ factor** Unsicherheitsfaktor *m* **~ shower** Instabilitätsschauer *m*

**install, to ~** anlegen, anschließen, aufbauen, aufstellen, einbauen, einrichten, einsetzen, errichten, installieren, montieren, rüsten, verlegen

**installation** Anlage *f*, Anordnung *f*, Anschluß *m*, Anstellung *f*, Aufstellung *f*, Ausführung *f*,

Betriebsanlage *f*, Einbau *m*, Einbauanordnung *f*, Einrichtung *f*, Errichtung *f*, Installation *f*, Montage *f*, Montierung *f*, Neuanlage *f*, Rüstung *f*, Vorrichtung *f* ~ **of arms (or armament)** Waffeneinbau *m* ~ **for cement factories** Zementwerkseinrichtung *f* ~ **for the protection of rooms** Raumschutzanlage *f* (teleph.) ~ **of railroad safety appliances** Eisenbahnsicherungsanlage *f* ~ **of reactors** Einbau *m* von Reaktanzspulen ~ **for secret conversations** Geheimsprecheinrichtung *f* ~ **of a subscriber's telephone** Einrichtung *f* eines Teilnehmeranschlusses

**installation,** ~ **accessories shop** Betriebsausrüstung *f* ~ **charge** Einrichtungsgebühr *f* ~ **construction** Bauanweisung *f* ~ **diagram** Einbauschema *n* ~ **dimensions** Einbaumaße *pl* ~ **drawing** Einbauplan *m* ~ **error** Einbaufehler *m* ~ **folder** Einbaumappe *f* ~ **implement** Installationsgegenstand *m* ~ **jig** Einbaugroßvorrichtung *f* ~ **kit** Einbausatz *m* ~ **measurement** Baumaß *n* ~ **plan** Aufstellungsplan *m* ~ **position of motor** Einbaustellung *f* des Motors ~ **radio** Bild-, Empfangs-anlage *f* ~ **rig** Einbaugroßvorrichtung *f* ~ **signaling plant** Fernmeldeanlage *f* ~ **specification** Ausrüstungsvorschrift *f* ~ **switch** Kleinschalter *m* ~ **zone** Einbaubereich *m*

**installations** Anlagen *pl*, Einbauten *pl*
**installed** eingebaut, eingerichtet
**installer** Einrichter *m*
**installment** Akontozahlung *f*, Einzahlung *f*, Errichtung *f*, Rate *f*, Teilzahlung *f* **by** ~**s** ratenweise **to make a first** ~ anzahlen ~ **of a delivery** Teilsendung *f* ~ **payment** Akontozahlung *f*
**instance** Beispiel *n* **for** ~ beispielsweise
**instant** Augenblick *m*, Moment *m*, Zeitpunkt *m*; sofort, unverzüglich ~ **of exposure** Auslösemoment *m* ~ **of shutter release** Auslösemoment *m*
**instant,** ~ **coupling** Momentkupplung *f* ~ **mechanical throw-off** Momentausrückung *f* ~ **response of the engine to the throttle** sofortiges Ansprechen *n* des Motors auf Vollgasgeben ~ **return mirror** Rückschwenkspiegel *m* ~ **starting** Momentanstellung *f* ~ **stop** Schnellstop *m* (tape rec.) ~ **stopping** Momentabstellung *f*
**instantaneous** augenblicklich, blitzschnell, momentan, unverzögert, momentan ~ **assembly** direkter Kontakt *m* ~ **automatic gain control** (IAGC) unverzögerte automatische Verstärkerreglung *f* ~ **(spiral) axis** momentane Schraubungsachse *f* ~ **center** Momentanzentrum *n* ~ **center of rotation (or of motion)** Wälzpunkt *m* ~ **clamp** Momentklemme *f* ~ **coupling (hose)** Schnellkupplung *f* ~ **current** Augenblicks-, Momentan-strom *m* ~ **dazzle** Momentanblendung *f* ~ **engaging and release** augenblickliches Einrücken *n* und Ausrücken *n* ~ **excitation** Impulserregung *f* ~ **firing** Moment-, Schnellzündung *f* ~ **fuse** Knallzündschnur *f*, (detonator) Momentzünder *m*, Schnellzünder *m*, Schnellzündschnur *f* ~ **(high-sensitive) fuse** Augenblickszünder *m* ~**-fuse striker** Schlagstift *m* ~ **high rate resistor** Stoßbelastungswiderstand *m* ~ **ignition** Zünddurchschlag *m* ~ **lens** Momentobjektiv *n* ~ **load** Augenblicksbelastung *f* ~ **locking** Momentfeststellung *f* ~

**photograph** Augenblicks-, Moment-aufnahme *f*, Momentbild *n*, Schnellfotografie *f* ~ **photometer** Schnellfotometer *m* ~ **photomicrography** Momentmikrofotografie *f* ~ **power** Augenblicksleistung *f* ~ **power output** momentane Ausgangsleistung *f* ~ **pressure release system** Momententlastungsvorrichtung *f* ~ **release** Schnellauslösung *f* ~ **sampling** momentane Analyse *f* ~ **shutter** Momentverschluß *m*, (camera) Objektivverschluß *m* ~ **stand-by** Momentanreserve *f* ~ **stereo photograph** Augenblicksstereoaufnahme *f* ~ **trip** Schnellauslöser *m* ~ **value** Augenblicks-, Momentan-wert *m* ~ **voltage** Augenblicks-, Momentan-spannung *f*
**instantaneously** schlagartig
**instantaneousness** Augenblicksdauer *f*, Unverzüglichkeit *f*
**instantly** augenblicklich, sofort
**in-step** Spann *m*
**institute** Anstalt *f* ~ **of technology** technische Hochschule *f* ~ **(or legal) proceedings against...** gerichtliches Verfahren *n*
**institution** Anstalt *f*, Einrichtung *f*, Gründung *f*
**instruct, to** ~ angeben, anleiten, anlernen, anweisen, ausbilden, lernen, unterrichten, unterweisen
**instruction** Angabe *f*, Anleitung *f*, Anordnung *f*, Anweisung *f*, Auftrag *m*, Befehl *m* (info proc.), Behandlungsvorschrift *f*, Einweisung *f*, Lehre *f*, Richtlinie *f*, Schulung *f*, Unterricht *f*, Unterweisung *f*, Verhaltungsmaßregel *f*, Vorschrift *f* ~ **by correspondence** Fernunterricht *m* ~ **for hardening process** Härteanleitung *f* ~ **for maintenance** Wartungsvorschrift *f* ~ **for uniform factory control** Anweisung *f* für einheitliche Betriebsuntersuchungen
**instruction,** ~ **address** Befehlsadresse *f* ~ **book** Bedienungsanweisung *f* ~ **chart** Knopfkarte *f*, Unterweisertafel *f* ~ **film** Lehrfilm *m* ~ **group** Lehrgruppe *f* ~ **leaflet** Merkblatt *n* ~ **plate** Belehrungstafel *f*, Instruktionsschild *n* ~ **relative secrecy** Geheimhaltungsvorschrift *f* ~ **sequence** Befehlsfolge *f* ~ **tag** Anhänger *m* mit Gebrauchsanweisung ~ **word** Befehlswort *n* (comput.)
**instructions** Betriebsblatt *n*, Dienstvorschrift *f*, technische Daten oder Vorschriften *pl*, Verfügung *f*, Vorschrift *f* ~ **to be followed** Verhaltungsvorschrift *f* ~ **without force of order** Haltepunkt *m* ~ **to operator** Betriebsanweisung *f* ~ **for practices** Dienstanweisung *f* ~ **for setting up** (assembly) Montageanweisung *f* ~ **for working** Arbeitsanleitung *f*
**instructive** belehrend, instruktiv, lehrreich
**instructor** Ausbilder *m*, Lehrer *m*, Lehrgangsleiter *m*
**instrument** Apparat *m*, Einrichtung *f*, Gerät *n*, Instrument *n*, (controlling) Kontrollinstrument *n*, Vorrichtung *f*, Werkzeug *n*
**instrument,** ~ **for aerological measuring** aerologisches Meßgerät *n* ~ **to convert** Umwertegerät *n* ~ **of geodesy** Erdmeßinstrument *n* ~ **on the ground** Bodengerät *n* ~ **with locking device** Fallbügelinstrument *n* ~ **for measuring lengths** Längenmesser *m* ~ **for measuring lift and drag** Umlaufgerät *n* ~ **with optical pointer** Lichtmarkeninstrument *n* ~ **for photogrammetrical surveying** fotogrammetrisches Aufnahmegerät

*n* ~ **for ramming** Schlaginstrument *n* ~ **for setting out roads** Instrument *n* für Absteckungen im Straßenbau ~ **for straightening a bore** Bohrlochglätter *m* ~ **for surface mounting** Aufbaugerät *n* ~ **for taking bearings** Ortungsgerät *n* ~ **for testing tapers** Kegelmeßgerät *n* ~ **with tubular level** Libelleninstrument *n*
**instrument,** ~ **approach** Instrumentenanflug *m* ~ **auto transformer** Meßwandler *m* in Sparschaltung ~ **basket** Geräte-, Instrument-korb *m* ~ **bearing** Feinlagerung *f* ~ **board** Armaturen-, Geräte-, Instrumenten-brett *n*, Instrumenttafel *f* ~ **board light** Schalttafel-leuchte *f*, -licht *n* ~**-board thermometer** Fernthermometer *n* ~ **bulb** Instrumentbirne *f* ~ **cage** Geräte-, Instrument-korb *m* ~ **capsule** Meßgerätekapsel *f* (Rakete) ~ **cart** Apparatekarren *m* ~ **case** Besteck *n*, Instrumentengehäuse *n* ~ **center** Meßhaus *n* ~ **channel** Instrumentenrinne *f* (aviat.) ~ **compartment test set** Spitzenprüfgerät *n* (g/m) ~ **connection cable** Geräteanschlußkabel *n* ~ **cord** Anschlußschnur *f* (eines Meßgerätes) ~ **coverage** Instrumentenüberblick *m* ~ **cross-checking** Instrumentenvergleich *m* ~ **cubicle** Instrumentenschrank *m* ~ **dial** Zifferblatt *n* ~ **error** Instrumentenfehler *m* ~**-flight plan** Blindflugplan *m* ~ **flight rules** (IFR) Instrumenten-Flugregeln *pl* ~ **flying** Blind-fliegen *n*, -flug *m*, Instrumenten-fliegen *n*, -flug *m* ~ **flow diagram** Instrumentenfließschema *n* ~ **fuse** Geräteanschlußkabelschmelzeinsatz *m* ~ **installation** Instrumenteneinbau *m* ~ **interpretation** Anzeigedeutung *f* ~ **jack** Apparatklinke *f* ~ **landing** Blindflug-, Nebel-landung *f* ~**-landing line (or landing path)** Blindlandeachse *f* ~ **landing system (ILS)** Instrumenten-Lande-System *m* (ILS) ~ **layout** Instrumentenanordnung *f* ~ **lead** Apparatezuleitung *f* ~ **leads** Apparatzuleitungen *pl*, Meßleitung *f* ~ **light** Gerätelicht *n*, Instrumentenleuchte *f* ~ **movement** Maßwerk *n* ~ **name plate** Geräteschild *m* ~ **officer** Meßoffizier ~ **panel** Armaturenband *n*, Geräte-, Instrumenten-brett *n*, Einzelfeld *n* mit Meßinstrumenten ~ **panel assembly** Anpassungsteile *pl* ~ **panel equipment** Schaltbrettausrüstung *f* ~ **parameters** Gerätewerte *pl* ~ **piping** Instrumentenrohrleitungen *pl* ~ **plug** Instrumentenstecker *m* ~ **plug connection** Gerätesteckeranschluß *m* ~ **power bus** Instrumentensammelschiene *f* ~ **program** Geräteprogramm *n* ~ **quality** Instrumentengüte *f*, Meßgenauigkeit *f* ~ **range** Anzeigebereich *m*, Meßbereich *m* ~ **reading** Anzeigen *n* der Instrumente ~ **ready for use** Instrument *n* in gebrauchsfertig montiertem Zustand ~ **registering rise and fall of balloons** Windrädchen *n* ~ **ring light** Skalenbeleuchtung *f* ~ **room** Apparat-raum *m*, -saal *m* ~ **runway** Instrumenten-Start-und-Landebahn *f* ~ **schedule** Instrumentenliste *f* ~ **section position** Rechenstelle *f* ~ **shelf** Bedienkasten *n* ~ **shelter** Instrumentenhülle *f*, Instrumentschutz *m*, -umhüllung *f* ~ **table** Apparat-, Geräte-tisch *m* ~ **test board** Armaturenprüfbrett *n* ~ **time** Instrumentenzeit *f* ~ **transformer** Meßwandler *m* ~ **tuner** Instrumentenstimmer *m* ~ **turn** Flugwendung *f* mit Hilfe der Instrumente ~ **type** Gerätetyp *m*
**instrumental error** Instrumentenfehler *m*

**instrumentality** Apparatur *f*
**instrumentation** Instrumentausrüstung *f*, Instrumentierung *f*, Meßgeräteausrüstung *f* ~ **and control** Messen und Regeln *pl*
**insubmersible** unsinkbar
**insubordination** Achtungsverletzung *f*, Gehorsams-delikt *n*, -verweigerung *f*, Ungehorsam *m*
**insufferable** unerträglich
**insufficiency** Dürftigkeit *f*
**insufficient** mangelhaft, nicht ausreichend, ungenügend, unzureichend ~ **depth of focus** Tiefenunschärfe *f* ~ **focus** Unschärfe *f* ~ **lube oil** Schmierölmangel *m* ~ **modulation** Untersteuerung *f* ~ **oil pressure** Öldruckmangel *m*
**insufflate, to** ~ einblasen
**insufflator** Einblasegerät *n*, Gebläse *n*
**insulance** Isolationswiderstand *m*
**insulant** Isolierstoff *m*
**insulate, to** ~ abisolieren, isolieren, sperren **to** ~ **a line** eine Leitung isolieren
**insulated** isoliert ~ **conductor** isolierter Leiter *m* ~ **electrode** Zündstift *m* ~ **handle** isolierter Handhebel *m* ~ **pliers** isolierte Zange *f* ~ **plug ring** isolierter Stöpselring *m* ~ **system** ungeerdetes System *n* ~ **wall** wärmeundurchlässige Wand *f* ~ **wire** Gummi-ader *f*, -draht *m*, isolierter Draht *m*, isoliertes Kabel *n*
**insulating** isolierend ~ **anodic coating** Sperrschicht *f* ~ **asphalt** Pech *n* für Isolation ~ **base** Isoliersockel *m* ~ **board** Isolierpappe *f* ~ **bush** Isolierbüchse *f* ~ **bushing** Isolatormuffe *f* ~ **(protection) cap** Berührungsschutzkappe *f* ~ **capacitor** Blockkondensator *m* ~ **case** Isolierbecher *m* ~ **casing** Isoliermantel *m* ~ **chair** Isolierschemel *m* ~ **clamp** Isolier-klemme *f*, -platte *f* ~ **compound** Isoliermasse *f* ~ **condenser** Sperrkondensator *m* ~ **course** Dämmschicht *f* ~ **covering** Isolationshülle *f* ~ **effect** Dämmwirkung *f* ~ **fabric** Isoliergewebe *n* ~ **glove** Gummihandschuh *m* ~ **joint** Isolierstoß *m* ~ **lacquered tapes** Isolierlackbänder *pl* ~ **layer** Isolationsschicht *f*, Isolierzwischenlage *f*, isolierende Zwischenlage *f* ~ **ledge** Isolierleiste *f* ~ **line** Isolierleinen *n* ~ **machine** Spritzmaschine *f* zum Gummieren von Draht ~ **mast** Isoliersäule *f* (Rakete) ~ **mat** Isoliermatte *f* ~ **material** Bindemittel *n*, Isolationsmaterial *n*, Isolier-masse *f*, -material *n*, -mittel *n*, -stoff *m*, Masse *f* für Wärmeisolation, Wärme-schutzmasse *f*, -schutzmittel *n* ~ **mechanism** Isoliergerät *n* ~ **medium** Isoliermittel *n* ~ **oil** Isolieröl *n* ~ **paint** Isolationslack *m* ~ **paper** Isolierpapier *n* ~ **plug** Isolierstöpsel *m* ~ **property** Isolier-fähigkeit *f*, -vermögen *n* ~ **quality** Isolierfähigkeit *f* ~ **rod** Schaltstange *f* ~ **rubber mat** Gummiunterlage *f* ~ **separator** Isolierzwischenlage *f* ~ **sheath** Isolierbecher *m* ~ **sheats** Isolierschlauch *m* ~ **slab** Dämmplatte *f* ~ **sleeve** Isoliermuffe *f* ~ **stool** Isolierschemel *m* ~ **strength** Isolierfestigkeit *f* ~ **strip** Isolierstreifen *m* **(adhesive)** ~ **tape** Isolierband *n* ~**-tape-producing machine** Isolierbandherstellungsmaschine *f* ~ **tissues** Isolierseidenpapier *n* ~ **tube** Isolierrohr *n* ~ **tubing** Isolierschlauch *m* ~ **tubings without fabrics** gewebelose Isolierschläuche *pl* ~ **varnish** Isolierlack *m* ~ **wall board** Dämmwandplatte *f* ~ **washer** Isolierscheibe *f* ~ **wax** Paraffin *n* für elektrische Isolierung

**insulation** Abdämmung (acoust.) f, Abkleidung f, Dämmung f, Isolation f, (resistance) Isolationswiderstand m, Isolierschutz m, Isolierung f, Schutz m, Sperrung f, Wärme-schutzmasse f, -schutzmittel n ~ **for cold** Kälteschutzisolierung f ~ **between commitator segments** Lamellenisolation f ~ **against ground** Isolation f gegen Erde ~ **against loss of heat** Wärmeschutz m
**insulation,** ~ **beads** Isolierperlen pl ~ **blanket** Isoliermantel m ~ **clip** Isolationshaken m, Isolierbefestigung f ~ **covering** Isolierhülse f ~ **current** Isolationsstrom m ~ **failure (or fault)** Isolationsfehler m ~ **indicator** Isolationsprüfer m ~ **layer** Isolationslage f ~ **material** Isolationsstoff m, Isolierwerkstoff m ~ **material for pressed products** Isolierpreßmassen pl ~ **plate** (jet) Isolierungsplatte ~ **resistance** Durchgangswiderstand m ~ **rib** Isolierrippe f ~ **sheath** Isolierhülse f ~ **strength** Spannungsfestigkeit f ~ **stripping tool** Entisolierer m ~ **test** Isolations-messung f, -prüfung f ~ **tester** Isolations-messer m, -prüfer m
**insulator** (single or double petticoat) Doppelglocke f, (bushing) Durchführung f, Isolator m, Isoliermittel n, Nichtleiter m (**strain**) ~ Abspannisolator m **one-piece rod-type suspension** ~ Längsstabisolator m ~ **for cold** Kälteschutzisolierung f
**insulator,** ~ **bolt** Schraubenstütze f ~ **chain** Isolatorenkette f ~ **column** Isolatorensäule f ~ **groove** Drahtlager n der Doppelglocke oder des Isolators ~ **mounting** Isolatoreinspannung f ~ **pin** Abspannisolatorstütze f, gerade Stütze f, Isolatorstütze f, Konsole f ~ **radio** Isolierkörper m ~ **spacing** Isolierabstand m ~ **spindle** Isolatorstütze f ~**-type transformer** Topfwandler m
**insulatory** isolierend
**insuperable** unübersteigbar, unüberwindlich
**insurable** versicherbar ~ **value** Versicherungswert m
**insurance** Assekuranz f, Versicherung f, Versicherungswesen n ~ **on appurtenances (or equipment or hull)** Kaskoversicherung f ~ **of value** Wertversicherung f
**insurance,** ~ **company** Versicherungs-anstalt f, -gesellschaft f ~ **contract** Versicherungsvertrag m ~ **policy** Versicherungspolice f
**insure, to** ~ assekurieren, sichern, versichern
**insured** Versicherungsnehmer m; versichert ~ **letter** Wertbrief m
**insurer** Versicherer m
**insusceptible** unempfänglich, unempfindlich
**intact** intakt, unversehrt
**intaglio, to** ~ eingravieren, einschneiden
**intaglio** eingraviertes Bild n, Dunkelwasserzeichen n, Intaglioverfahren n, Schattiertwasserzeichen n ~ **printing** Tiefdruckverfahren n
**intake** (motors) Ansaugen n, Ansaugöffnung f, (structure) Einlaßbauwerk n, Einlaßöffnung f, Einführungsrohr n, (for motors, steam, water) Einlauf m, Einlauftrichter m, Einströmungsöffnung f, Eintritt m, Einziehstrecke f, Fang m, Wasserentnahme f (hydraul.), Werkkanal m, Zuführungsrohr n, Zulauf m, Zulauföffnung f
**intake,** ~ **adapter** Lufteintritt-Zwischenring m ~ **air** Ansaugluft f, Einziehstrom m, Frischluft f

~ **connection** (oil, etc.) Bordanschluß m ~ **cross-section** Ansaugquerschnitt m ~ **diffusory** Fangdiffusore f ~ **drag** Einlaufwiderstand m (aerodyn.) ~ **duct** Ansaugleitung f ~ **elbow** Ansaugkrümmer m ~ **fan** Vorsatzläufer m ~ **flange** Eintrittsstutzen m ~ **header** Einlaßführungskanal m ~ **manifold** Einström-, Ladeluft-, Saug-leitung f, Verteilungsrohr n ~ **manifold boost** Ladedruck m ~ **muffler** Ansauggeräuschdämpfer m ~ **order** Ineinnahmeverfügung f ~ **pipe** Ansaug-, Zulauf-rohr n ~ **port** Eintritts-, Gemischeintritts-öffnung f im Zylinder ~ **pressure** Ansaug-, Einlaß-, Lade-druck m ~ **roar silencer** Ansauggeräuschdämpfer m (Auto) ~ **side** Ansaugseite f ~ **stack** Ansaugstutzen m ~ **stator** Einlaufleitschaufel (jet) f ~ **stroke** Ansaugetakt m, Ansaughub m, (gas) Einsaugtakt m, Einström-, Lade-, Saug-hub m ~ **structure** Einlaufbauwerk n ~ **tube** Einlaßkanal m ~ **valve** Ansaug-, Einlaß-, Einström-ventil n ~**-valve chamber** Einlaßventilkammer f ~ **well** Einlaßbohrung f
**intangible** immateriell, unkörperlich
**intarsia** Intarsie f, Einlegearbeit f
**intarsiate** eingelegt
**integer** ganze Zahl f, ganzzahlig (info proc) ~**-slot winding** Ganzlochwicklung f
**integrability condition** Integrabilitätsbedingung f
**integrable** integrierbar
**integral** ein Ganzes bildend, Integral n (math.); einteilig, ganz, ganzzahlig, integrierend, vollständig ~ **action coefficient** I-Wirkungskoeffizient m ~ **action controller** indirekter Regler m, I-Regler m ~**-action time** Nachstellzeit f ~ **calculus** Integralrechnung f ~ **charge** ganzzahlige Ladung f ~ **control** Integralreglung f ~ **cylinder head** angegossener Zylinderkopf m ~ **domain** Integritätsbereich n ~ **dose** Integraldosis f ~ **electric starter** eingebauter elektrischer Anlasser m ~ **equation** Integralgleichung f ~ **heater thermocouple** Thermoelement m für eingebauten Heizkörper ~ **invariant** Integralinvariante f ~ **line breadth** gesamte Linienbreite f ~ **line width in solids** Linienbreite f in Festkörpern ~ **measurement** gesamte Messung f, Summenmessung f ~ **multiple** ganzes Vielfaches f ~ **number** ganze Zahl f ~ **quantity** ganzzahlig ~ **part** Einheit f ~ **rational function** ganze rationale Funktion f ~ **reflection** Gesamtreflexion f ~ **representation** quellenmäßige Darstellung f ~ **seriation** Integralzerlegung f ~ **sine function** Integralsinusfunktion f ~ **spin** ganzzahlige Ladung f ~ **taken over a surface** Oberflächenintegral n ~ **tank** Integralbehälter m ~ **theorem** Integralsatz m ~ **transmitter** Integralgeber m
**integrally cast** angegossen
**integralness** Ganzzahligkeit f
**integrand** Integrand m
**integrant** integrierend
**integraph** (for curves) Flächenmesser m
**integrate, to** ~ aufrechnen, ergänzen, integrieren **to** ~ **over a cycle** integrieren über eine Periode
**integrated** einbezogen ~ **data processing** zusammengefaßte Datenverarbeitung f ~ **frequency** Summenhäufigkeit f ~ **range** Gesamtweglänge f ~ **reflection** integrierte Reflexion f

**integrating** integrierend ~ **accelerometer** I-Gerät *n*, Innenschaltgerät *n* ~ **amplifier** summierender Verstärker *m* ~ **apparatus** Zählgerät *n* ~ **circuit** Integrationsglied *n*, Integrierschaltung *f* ~ **cube** Zählwürfel *m* ~ **detector** quantitativ arbeitender Detektor *m* ~ **divider** integrierender Frequenzteiler *m* ~ **dose meter** integrierendes Dosimeter *n* ~ **gear** Integrierwerk *n* ~ **indicator** integrierender Anzeiger *m* ~ **photometer** (lumenmeter) integrierendes Fotometer *m*
**integration** Ergänzung *f*, Integration *f* **for** ~ **of differences of synchronism** Auflaufen *m* der Gleichlauffehler
**integration,** ~ **constant** Integrationskonstante *f* ~ **device** Integrationsgerät *n* ~ **gyro** Integrationskreisel *m* ~ **period** Integrationszeit *f*
**integrator** Integrationsgerät *n*, Integrator *m*, Integriereinrichtung *f*, Planimeter *n*, Summierungsgerät *n* ~ **device of guided missile** I-Gerät *n* ~ **relay** Geräterelais *n* ~-**testing-device calibrator** Innenschaltuhr *f*
**integrity** Rechtschaffenheit *f*
**integument** Decke *f*
**intelligence** Auskunft *f*, Kunde *f*, Nachricht *f*, Verstand *m* ~ **bandwidth** Nachrichtenbandbreite *f* ~ **collator** Endauswerter *m*
**intelligibility** Sinnverständlichkeit *f*, Sprachdeutlichkeit *f*, Sprachverständigung *f*, Verständigung *f*, Verständlichkeit *f* ~ **of articulate sound (or of speech)** Sprachverständlichkeit *f* ~ **measurement** Silbenverständlichkeitsmessung *f*
**intelligible** deutlich, klar, vernehmlich, verständlich ~ **crosstalk** verständliches Nebensprechen *n*
**intemperateness** Atmosphärilien *pl*
**intended** berechnet, gewollt ~ **for** vorgesehen ~ **track** beabsichtigter Kurs *m* über Grund (navig.)
**intense** hochgradig, nachhaltig, rege, stark ~ **cooling plant** Tiefkühlanlage *f*
**intensification** Erhöhung *f*, Verstärkung *f* ~ **of fire** Feuersteigerung *f* ~ **modulation** Helligkeitsmodulation *f*
**intensifier** Druckerhöher *m*, Verstärker *m* ~ **electrode** Nachbeschleunigungselektrode *f* ~ **stage** Verstärkerstufe *f*
**intensify, to** ~ erhöhen, verstärken
**intensifying screen** Verstärker-folie *f*, -schirm *m*, Verstärkungsfolie *f*, (X-ray work) Verstärkungsschirm *m*
**intensitive time** Totzeit *f*
**intensitometer** Intensitometer *n*
**intensity** Grad *m*, Heftigkeit *f*, Helligkeit *f* (CRT), Intensität *f*, Stärke *f*, Stärkegrad *m*
**intensity,** ~ **of activation** Aktivierungsenergie *f* ~ **of the beam current** Strahlstromstärke *f* ~ **of bending stress** Biegespannungsgröße *f* ~ **of combustion** Brennkraft *f*, Verbrennungsintensität *f* ~ **of compressive stress** Druckspannungsgröße *f* ~ **of continuous spectrum** Intensität *f* des Bremsspektrums ~ **of draft** Zugstärke *f* ~ **of echo** Echointensität *f* ~ **of electric field** elektrische Feldstärke *f* ~ **of electron beam** Kathodenstrahlintensität *f* ~ **of field** Feld-intensität *f*, -stärke *f* ~ **of force** Kraftangriff *m* ~ **of heat** Hitzegrad *m*, Wärmeintensität *f* ~ **of illumination** Beleuchtungsstärke *f* ~ **of image current** Bildstromstärke *f* ~ **of ionization current** Ioni-

sationsstromstärke *f* ~ **of lighting** Belichtungsstärke *f* ~ **of magnetic field** (magnetische) Feldstärke *f*, Übertragungspegel *m* ~ **of magnetization** Intensität *f* der Magnetisierung, Magnetisierungsstärke *f*, Magnetismusmenge *f* ~ **of precipitation** Niederschlagsstärke *f* ~ **of radiation** Strahlungsintensität *f* ~ **of rain** Regendichtigkeit *f* ~ **of shearing stress** Schubspannungsgröße *f* ~ **of sound** Lautstärke *f* ~ **of tone** Tonstärke *f* ~ **of traffic** Verkehrsstärke *f*
**intensity,** ~ **balance** Intensitätsbilanz *f* ~ **control** Helligkeits-, Intensitäts-regelung *f* ~ **diaphragm** Helligkeitsblende *f* ~ **discrimination** Augenempfindlichkeit *f*, Kontrast-empfindlichkeit *f*, -empfindung *f* ~ **distribution** Intensitätsverteilung *f* ~ **modulation** Helligkeitsmodulation *f*, Intensitäts-beeinflussung *f*, -steuerung *f* ~ **modulation of control electrode** Aufhellungsorgan *n* ~ **modulation pulse** Helltastimpuls *m* (electron.) ~ **pedestal** Grundhelligkeit *f* ~ **per unit of intrinsic brilliance (or surface brilliance)** (of luminous surface) Leuchtdichte *f* ~ **range of a negative** Negativumfang *m* ~-**range limiter** Dynamikbegrenzer *m* (bei Verstärkeranlage) ~ **ratio** Helligkeitsverhältnis *n* ~ **setting** Helligkeitseinstellung *f* ~ **standard** Intensitätsmarke *f* ~ **steps (or grades)** Intensitätsabstufung *f* ~ **sum rule** Intensitätssummensatz *m* ~ **value** Helligkeitswert *m* (photo) ~ **voltage** Helligkeitsspannung *f*
**intensive** durchgreifend, intensiv, kräftig, (color) satt
**intensiveness** Intensität *f*
**intent** Vorhaben *n* (law) ~ **on** bedacht
**inter, to** ~ eingraben
**inter** (dar)unter, (da)zwischen, einander, gegenseitig, unter, Wechsel...; zwischen, Zwischen...
**interact, to** ~ aufeinander einwirken, zusammenwirken
**interacting from scattering** Wechselwirkung *f* zwischen Elektronenanordnungen aus Streuversuchen
**inter-action** gegenseitige Beeinflussung *f* oder Einwirkung *f*, (disturbance) Störung *f* durch Nebensprechen, Wechselwirkung *f*, Zusammenwirken *n* ~ **between electrons** Wechselwirkung *f* zwischen Elektronen- und Gitterwellen ~ **of radio waves** Luxemburg-Effekt *m*, Wechselwirkungseffekt *m*
**inter-action,** ~ **crosstalk coupling** Gesamtnebensprechkopplung *f* ~ **factor** Wechselwirkungsfaktor *m* ~ **gap** Wechselwirkungsspalt *m* ~ **impedance** Kopplungswiderstand *m* ~ **potential** Wechselwirkungspotential *m* ~ **space** Wechselwirkungsraum *m* ~ **splitting** Wechselwirkungsaufspaltung *f* ~ **terms** Wechselwirkungsglieder *pl*
**interaileron,** ~ **strut** Querruder-verbindungsstrebe *f* ~ **wire** Flügelverbindungsdraht *m*, Querverbindungsseil *n*
**inter-aircraft** Bord zu Bord ~ **command communication** Befehlsübertragung *f* zwischen Flugzeugen, Flugzeugbefehlsübertragung *f*, Kommandoübertragung *f* zwischen Flugzeugen ~ **voice communication** Sprechverbindung *f* zwischen Flugzeugen
**inter-atomic force** interatomare Kraft *f*
**interbedded, to be** ~ dazwischengelagert, wechselgelagert (geol.)

intercalate, to ~ zwischenlegen, zwischenschalten
intercalation Zwischenschaltung *f* ~ compound
Einlagerungsverbindung *f*
intercalculate, to ~ einfügen
intercardinal point Zwischenkardinalpunkt *m*
intercarrier, ~ noise suppression Rauschunterdrückung *f* ~ sound system Zwischenträgerver-
fahren *n*
intercede, to ~ dazwischentreten
intercellular zwischenzellig ~ substance Zwi-
schengewebesubstanz *f*
intercept, to ~ abfangen, abhören (teleph.), ablauschen, den Weg abschneiden, anhalten, auffangen, aufnehmen (rdo), peilen, unterwegs
abfangen to ~ a conversation eine Unterhaltung
abhören to ~ a length eine Länge abschneiden
to ~ a message eine Meldung abfangen to ~
water eine Anzapfung abzweigen
intercept Abschnitt *m* ~ length Schnittweite *f* ~
methode Mithörverfahren *n* ~ overcutting
Überschneidung *f* ~ post Horchstelle *f* ~
receiver Lauschempfänger *m* ~ search elektronische Suche *f* ~ set Lauschgerät *n* ~ station
Abhör-, Lausch-stelle *f*
intercepted abgehört
intercepting Bescheidansage *f* (teleph.) ~ cut
Durchstich *m* ~ ditch Abfanggraben *m* ~
(catching) effect abfangende Wirkung *f* ~ grid
Fanggitter *n* ~ screen Auffangschirm *m* ~
system Abfangsystem *n* ~ trunk Bescheid-,
Hinweis-leitung *f*
interception Abfangtaktik *f*, Abschneidung *f*,
Erfassung *f* (rdr), Unterbrechung *f*, Versper-
rung *f* ~ activity Abhörtätigkeit *f* ~ chaser
system Kollisionskursverfahren *n* ~ circuit
Haltestromkreis *m* ~ cord Sprechschnur *f* einer
B-Beamtin ~ course system Vorhalteverfahren
*n* ~ equipment Fangeinrichtung *f* (teleph.) ~
flight Abfangflug *m*, Sperrefliegen *n*, Sperrflug
*m* ~ service Abhördienst *m* ~ station Abhörstation *f* ~ trunk Überweisungsleitung *f*
interceptor Abfangjäger *m*, Abwehrjäger *m*,
Unterbrecherklappe *f*, (fighter) Verteidigungs-
jagdflugzeug *n*, Zusatzflügel *m* ~ of odors
Geruchsverschluß *m*
interceptor, ~ grid Fanggitter *n* ~ plane Zerstörer *m*
intercession Einsprache *f*
interchange, to ~ abwechseln, aufeinanderfolgen, austauschen, auswechseln, vertauschen,
wechseln
interchange Austausch *m*, Auswechs(e)lung *f*,
Vertauschung *f*, Wechseln ~ of current Strom-
austausch *m* ~ of heat Wärmeaustausch *m* ~
of power Kräfteaustausch *m* ~ of sites Platzwechsel *m*
interchange head lamp Austauschscheinwerfer *m*
interchangeability Austauschbarkeit *f*, austausch-
bare Fertigung *f*, Auswechselbarkeit *f*, Ver-
tauschbarkeit *f* ~ of types Buchstabenaus-
wechslung *f*
interchangeable austauschbar, auswechselbar,
wechselseitig ~ barrel Wechsellauf *m* ~ belt
pulley Wechselriemenscheibe *f* ~ carriage
wechselbarer Wagen *m* (typewriter) ~ coil set
auswechselbarer Spulensatz *m* ~ condenser
Wechselkondensor *m* ~ diaphragm Einsatz-
blende *f* (opt.) ~ gear Wechselrad *n* ~ inset

auswechselbarer Einsatz *m* ~ parts vollaus-
tauschbare Teile *pl* ~ power unit Austausch-
triebwerk *n* ~ welding and cutting torch
Wechselbrenner *m*
interchanger Austauscher *m* ~ air Luftaustauscher *m*
intercolumination Säulenweite *f*
intercom Befehlsübermittlungsanlage *f*, Bord-
verständigungsanlage *f* ~ system Gegensprechanlage *f*, Haussprechanlage *f*
intercommunicate, to ~ miteinander in Verbindung bringen, stehen, oder verkehren
intercommunication Eigenverständigung *f*, Linienführung *f*, Wechselverkehr *m*, Zwischen-
verkehr *m* ~ channel Funkquerverbindung *f* ~
plant Fernsprechreihenanlage *f* ~ system Bord-
verständigungsanlage *f*, Gegenverkehrssystem
*n*, Reihen-anlage *f*, -schalter *m*, Rundspruchanlage *f* ~ telephone Bordsprechgerät *n*, Bord-
telefon *n* ~ telephone plant Reihenanlagefernsprechreihen *pl*
intercompare, to ~ vergleichen
interconnect, to ~ verschränken, zusammenschalten
interconnected untereinander verbunden ~ network Maschennetz *n* ~ operation Verbundbetrieb *m*
interconnecting Verschränkung *f* ~ feeder
Kuppelleitung *f*, Zwischenschaltung *f* ~ plug
Durchverbindungsstecker *m*
interconnection gegenseitige Verbindung *f*, Ver-
maschung *f*, Zusammenschaltung *f* ~ box Verbindungsdose *f*
interconnector Kuppelleitung *f*, Umzündstutzen
*m* (Brennkammer), Verbindungsrohr *n* (Triebwerk), Zündrohr *n* (jet), Zündröhrchen *n* (jet)
intercontinental, ~ ballistic missile (ICBM)
interkontinentale Fernwaffe *f* ~ missile Inter-
kontinentalrakete *f*
intercooler Zwischenkühler *m*
inter-cooling Zwischenkühlung *f*
intercostal plate Interkostalplatte *f*
intercourse Verkehr *m*
intercrescence Verwachsung *f*
intercrest distance Wellenberglochweite *f*
intercropping Zwischenkultur *f*
intercrystalline interkristallin ~ brittleness interkristalline Brüchigkeit *f* ~ corrosion interkristalline Korrosion *f* ~ crack längs der
Kristallgrenzen verlaufender Riß *m*
intercylinder, ~ baffle Zylinderschaftleitblech *n*
~ baffle ring Zwischenzylinderleitring *m* ~
baffles Druckleitbleche *pl* ~ ring Zwischen-
zylinderring *m*
interdendritic interdendritisch
interdepartmental innerbetrieblich, intern
interdependence gegenseitige Abhängigkeit *f*,
Korrespondenz *f*
interdependent voneinander abhängig
interdict, to ~ ausschließen
interdict Einhaltsbefehl *m*
interdiction Untersagung *f*, Verbot *n*
inter-digit hunting time Freiwahlzeit *f*
interdigital, ~ magnetron Doppelkamm-Mag-
netron *n* ~ pause Zwischenimpulspause *f*
(Zeichenpause)
inter-district network zwischenamtliches Netz *n*
interdot flickering Zwischenpunktflimmern *n*

**interelectrode, ~ capacitance** innere Röhrenkapazität *f*, Röhrenkapazität *f*, Zwischenelektrodenkapazität *f* **~ capacity** Elektroden-, Innen-kapazität *f*, (of thermionic tube) innere Röhrenkapazität
**interest, to ~** interessieren
**interest** Anteil *m*, Anteilnahme *f*, Verzinsung *f*, Zins *m* **(rate of) ~** Zins-fuß *m*, -satz *m* **~-bearing security** zinstragendes Papier *n* **~ payable on arrears** Verzugszinsen *pl*
**interested** beteiligt **~ party** Interessent *m*
**interface** Abschnitt *m*, Berührungsfläche *f* Grenzfläche *f* (cryst.), Grenzschicht(e) *f*, Schnitt *m*, Sprungschicht *f*, Stirnseite *f* **~ conditions** Grenzflächenbedingungen *pl* **~ level** Zwischenniveau *n* **~ measurement** Trennschichtmessung *f* **~ normal** Grenzflächennormale *f* **~ plane** Trennungsfläche *f* **~ potential** Grenzflächenpotential *n* **~ resistance** Zwischenschichtwiderstand *m*
**interfacial, ~ angle** Flächenwinkel *m*, Grenzflächenwinkel *m* **~ relationships** Grenzflächenbeziehung *f* **~ tensiometer** Grenzschichtspannungsmesser *m* **~ tension** Grenzflächen-, Oberflächen-, Zwischenflächen-spannung *f*
**interfere, to ~** dazwischentreten, (with) eingreifen, (with) einmengen, sich einmischen, (with) einwirken, interferieren, intervenieren, in die Quere kommen **to ~ in a conversation** in ein Gespräch eingreifen
**interference** (X ray) Auslöschung *f*, (störende) Beeinflussung *f*, Betriebsstörung *f*, Eingreifen *n*, Einmischung *f*, Einspruch (Patent) *m*, Einwirkung *f*, Interferenz *f*, Pfiff *m* (TV), Störung *f*, Störung *f* durch andere Sender, Überlagerung *f*, Überlagerung *f* einer Störung, Übermaß *n* **~ by neighboring station in volume** Durchschlagen *n* vom Nachbarsender **~ from power lines** Störung *f* durch Starkstrom **~ from power system** Starkstromstörung *f* **~ with transmission** Sendestörung *f*
**interference, ~ action** Kollisionsverfahren *n* **~ area** Störgebiet *n* (rdr), Störungsgebiet *n* **~ band** Interferenzstreifen *m* **~ bearer** Störträger *m* **~ belt** Störgürtel *m* **~ blanker** Störaustastschaltung *f*, Störungssperre *f* **~ circuit** Störkreis *m* **~ coatings** Interferenzschichten *pl* **~ comparator** Interferenzkomparator *m* **~ current** Störstrom *m* **~ double refraction** Interferenzdoppelbrechung *f* **~ drag** Wechselwirkungswiderstand *m*, Widerstandskraft *f* **~ effect** Interferenzwirkung *f*, Störeffekt *m* **~ eliminating condenser** Störungsschutz *m* **~ elimination** Entstörung *f*, Störschutz *m*, Störungsbeseitigung *f* **~ elimination measuring** Entstörungsmessung *f* **~ eliminator** Entstörer *m*, Störschutz-anordnung *f*, -gerät *n* **~-eliminator kit** Störschutzpackung *f* **~ factor** Stör-, Störungs-faktor *m* **~-factor meter** Störungsfaktormesser *m* **~ field** Störfeld *n* **~ figure** Interferenz-aufnahme *f*, -bild *n* **~ figures** Achsenbilder *pl* **~ filter** Störschutzfilter *n* **~ fit** Festsitz *m*, Passung *f* mit Übermaß **~-free** störungsfrei **~ fringes** Interferenzstreifen *m* (opt.) **~ gravimeter** Interferenz-Gravimeter *n* **~-guard band** Frequenzabstand *m*, Schutzbereich *m* (d. Frequenzbandes) **~ inverter** Entstördiode *f* (TV) **~ level** Stör-höhe

*f*, -spiegel *m* **~ limiter** Störbegrenzer *m* **~ location** Störungssuche *f* **~ locator** Störsuchgerät *n* **~ measuring apparatus** Funkstörmeßgerät *n* **~ microphone** Rücksprachemikrofon *n* **~ note** Interferenzton *m* **~ pattern** Interferenzbild *n* (acoust.), Störstelle *f* **~ peak** Störspitze *f* **~ phenomenon** Interferenz-erscheinung *f*, -vorgang *m* **~ point** Schwingungsknoten *m* **~ prevention** Störungsverhinderung *f* **~ procedure** Kollisionsverfahren *n* **~ process** Störungsverfahren *n* **~ protection** Entstöranlage *f* **~ pulse** Störimpuls *m* **~ range** Verwirrungsgebiet *n* **~ relay** Störungsrelais *n* **~ search gear** Störsuchgerät *n* **~ signal** Störungszeichen *n* **~ spectrogram** Interferenzspektrogram *n* **~ spectroscope** Interferometer *n* **~ stage** (with variable air stratum) Interferenztischchen *n* **~ sump** (electricity-absorbing layer) Interferenzsumpf *m* **~ suppressiblity** Entstörbarkeit *f* **~ suppression (device)** Störsperre *f* **~ suppression socket** Entstörstutzen *m* **~ suppressor** Störabsperrer *m* **~ suppressor plate** Entstörplatte *f* **~ surface tester** Interferenzflächenprüfer *m* **~ threshold** Störschwelle *f* **~ tone** Interferenzton *m* **~ voltage** Störspannung *f* **~ wave** Stör-, Störungs-welle *f*
**interfering** störend, sich überlagernd **~ electromagnetic waves** störende elektromagnetische Wellen *pl* **~ frequency** Störfrequenz *f* **~ noise** Stör-, Unterbrechungs-geräusch *n* **~ tone** Störton *m* **~ transmitter** Störsender *m*
**interferometer** Interferenzialrefraktor *m*, Interferometer *n* **~ for liquids** Flüssigkeitsinterferometer *n*
**interferometric capacity** Interferenzfähigkeit *f*
**interferometry of electrons** Elektroneninterferometrie *f*
**interfin, ~ space** Rippenzwischenraum *m* **~ velocity** Geschwindigkeit *f* im Rippenzwischenraum
**interflow** Überströmen *n* (hydraul.)
**interfusion** Diffusion *f*
**intergranular** intergranular **~ corrosion** interkristalline Korrosion *f*, Korngrenzenkorrosion *f* **~ cracking** Korngrenzriß *m*
**intergrown** verwachsen
**intergrowth** Verwachsung *f*
**interim** Zwischenzeit *f* **~ addendum** vorläufiger Nachtrag *m* **~ cover note** vorläufige Deckungsmasse *f* **~ decision** Vorbescheid *m* **~ injunction** einstweilige Verfügung *f* **~ protection** einstweilige Verfügung *f* **~ solution** Interimslösung *f*
**interionic force fields** Kraftfelder *pl* zwischen den Ionen
**interior** Binnen..., Binnenland *n*, Innenraum *m*, Innenseite *f*, Innere *n* (math.); binnenländisch, inländisch, inner (er, e, es) **~ of the basket** Korbraum *m* **~ of a country** Hinterland *n* **~ of dry dock** Dockkammer *f* **~ of the earth** Erdinneres *n* **~ of fuselage** Rumpfinneres *n* **~ of the tank** Kesselinneres *n*
**interior, ~ angle** Innenwinkel *m* **~ architecture** Raumgestaltung *f* **~ ballistics** Innenballistik *f* **~ coating** Innenanstrich *m* **~ communication** Bordsprechanlage *f* **~ cooler unit** Kühlereinsatz *m* **~ corrosion protective oil** Innenkonservierungsöl *n* **~ drainage** Binnenentwässerung *f* **~ equipment** Inneneinrichtung *f* **~ height** lichte

Höhe *f*, Lichthöhe *f* ~ **panelling** Innenverkleidung *f* ~ **percussion cap** Innenhütchen *n* ~ **protection page** Vorsatzblatt *n* ~ **room** Innenraum *m* ~ **space** Innerraum *m* ~ **strutting and bracing** Innenverspannung *f* (aviat.) ~ **view** Innenansicht *f* ~ **water(s)** Binnengewässer *n*
**interlace, to** ~ durchflechten, eng verschlingen, ineinander verflechten, verschränken, verweben
**interlace** Antwortüberlappung *f* (sec. rdr.), Wechselabfrage *f* (sec. rdr.), Zwischenzeile *f*, Zwischenzeilenrhythmus *m* ~ **factor** Zeilensprungfaktor *m* ~ **method** Sprungzeilenverfahren *n* ~ **ratio** Zeilensprung-faktor *m*, -verhältnis *n* ~ **transmission** Zeilensprungendung *f*
**interlaced** ineinandergeschachtelt (Impulse) ~ **ground** verschränkter Grund *m* ~ **scanning** alternierende oder springende Zerlegung *f*, sprungweise Abtastung *f* nacheinanderfolgender Bildpunkte, Zeilensprung *m*, Zwischenzeilen-abtastung *f*, -verfahren *n* ~ scanning **field** Zeilensprungteilraster *m* ~ **scanning method** Zeilensprungmethode *f* ~ **scanning pattern** Zwischenzeilenraster *m*
**interlacing** (of railway or tramway lines) Gleisverschlingung *f*, verschränkte Numerierung *f*, Verschlingung *f*, Verschränkung *f*, Zeilensprung *m*, Zeilenverschiebung *f*, Zwischenzeilenverfahren *n* (TV) ~ **effect** Bindungseffekt *m*
**interlattice plane distance** Gitterebenen-, Netzebenen-abstand *m*
**interlay, to** ~ (da)zwischenlegen
**interlayer** Pufferlage *f*
**interleaf** Durchschußblatt *n* (print.), Zwischenlage *f*
**interleave, to** ~ ausschießen (print.), durchschießen (print.), einschichten
**interleaved**, ~ **scanning** alternierende oder springende Zerlegung *f*, Zwischenzeilenverfahren *n* ~ **scanning method** Zeilensprungmethode *f*
**interleaving**, ~ **device** Einschießvorrichtung *f* ~ **paper** Zwischenlagenpapier *n* ~ **sheet** Durchschußbogen *m*
**interlever** Zwischenhebel *m*
**interline, to** ~ einschalten, Zeilen durchschießen (print.)
**interline** Zwischenzeile *f* ~ **flicker** Zwischenlinien-, Zwischenzeilen-flimmer *m* ~ **rhythm** Zwischenzeilenrhythmus *m*
**interling** Zwischen-futter *n*, -schicht *f*
**interlink, to** ~ verketten
**interlinked** verkettet ~ **current** verketteter Strom *m* ~ **gas grid system** Gasverbundnetz *n* ~ **piece** Gleiszwischenstück *n* ~ **two-phase current** verketteter Zweiphasenwechselstrom *m* ~ **voltage** verkettete Spannung *f*, Verkettungsspannung *f*
**interlinking** Verkettung *f* ~ **of phase** Phasenverkettung *f* ~ **point** Verkettungspunkt *m*
**interlock, to** ~ eingreifen, einrücken, ineinandergreifen, kuppeln, verblocken, verriegeln, zusammenschließen
**interlock** Ineinandergreifen *n*, Sperrplatte *f*, Synchronisierungsorgan *n*, Verriegelung *f* ~ **board** Schalterwerk *n* (electr.) ~ **circuit** elektrische Welle *f* ~ **helical cutters** Satz *m* zum Planfräsen ~ **lever** Handradsperre *f* ~ **lining** Interlockfutterware *f* ~ **signals** abhängige oder

verriegelte Signale *pl* ~ **switch** Freigabeschutzschalter *m*
**interlocked** verfilzt (fibers), verriegelt, verzahnt, zusammengesetzt ~ **operation** verriegelte Verarbeitung *f* ~ **sinker cams** ineinandergreifende Platinenexzenter *pl*
**interlocking** Verblockung *f*, (device) Verriegelung *f*; ineinandergreifend ~ **contact** Verriegelungskontakt *m* ~ **device** Verblockungssystem *n* ~ **gear** Verschluß-einrichtung *f*, -gitter *n* ~ **installation** Stellwerksanlage *f* ~ **joint** Schloßverbindung *f* ~ **lever to prevent the engagement of more than one feed at a time** Sperrhebel *m* zur Verhütung des gleichzeitigen Einschaltens von mehr als einer Bewegung ~ **mechanism** Verschlußvorrichtung *f* ~ **post** elektrische Stellwerksanlage *f* ~ **relay** Stützrelais *n* ~ **roofing tile** Falzziegel *m* ~ **side milling cutter** Doppelscheibenfräser *m*, verstellbarer Scheibenfräser *f* ~ **solenoid** Verriegelungsbremsmagnet *m* ~ **steel sheathing** Stahlspundwand *f* mit Schloßverbindung ~ **switch system** Verriegelungsschaltung *f* ~ **system** Synchron-Halteeinrichtung *f*, Verriegelungssystem *n* (electr.) ~ **tile** Falzziegel *m* ~ **tower** Stellwerk *n*
**interlocution** Zwiegespräch *n*
**intermediary** Vermittlung *f*, Zwischenschaltung *f*; dazwischenliegend, intermediär ~ **blade** Zwischenflügel *m* ~ **header** Zwischensammler *m* ~ **member** Zwischenglied *n* ~ **plate** Einsatzplatte *f* ~ **product** Zwischenerzeugnis *n* ~ **stage** Übergangszustand *m* ~ **steps** Zwischenstufen *pl* ~ **transformer** Zwischenübertrager *m*
**intermediate** Einrollgewebe *f*, Zwischenglied *n*, Zwischenkern *m*; intermediär, zwischenliegend, zwischenstehend ~ **and extension piece** Zwischen- und Verlängerungsstück *n*
**intermediate**, ~ **accelerator** Zwischenbeschleuniger *m* ~ **air space** Luftzwischenraum *m* ~ **amplifier** Zwischenfrequenzverstärker *m*, Zwischenverstärker *m* ~ **anchoring** Zwischenverankerung *f* ~ **angle (or bracket)** Zwischenwinkel *m* ~ **annealing** Zwischenglühung *f* ~ **aperiodic circuit** aperiodischer Zwischenkreis *m* ~ **approach** Zwischenanflug *m* ~ **area** Zwischenfeld *n* ~ **asbestos member** Asbestzwischenstück *n* ~ **basis circuit** Zwischenbasisschaltung *f* ~ **bath** Zwischenbad *n* ~ **bearing** Zwischenlager *m* ~ **-beat frequency** Zwischenfrequenz *f* ~ **blade** Beruhigungsflügel *m*, Zwischenleitschaufel *f* ~ **block** Amboßeinsatz *m* ~ **bottom** Zwischenboden *m* ~ **brace** Zwischenquerstück *n* ~ **brake shaft** Bremszwischenwelle *f* ~ **cable** Kabelzwischenstück *n*, Zwischenkabel *n* ~ **case** Mittelding *n* ~ **cave** Multicurie-Zelle *f* ~ **ceiling** Zwischendecke *f* ~ **cell** Zwischenerhitzer *m* ~ **chain** Zwischenkette *f* ~ **chair** Zwischenstuhl *m* ~ **charge** Zwischen-ladung *f*, -zünder *m* ~ **check valve** Zwischenrückschlagventil *n* ~ **circuit** Übertragerkreis *m*, Verstärkerstufe *f*, Zwischenkreis *m* ~ **coat** Zwischenanstrich *m* ~ **color** Zwischenfarbe *f* ~ **compression** Zwischenkompression *f* ~ **connecting bay** Verbindungsgestell *n* ~ **construction** Zwischenausführung *f* ~ **container** Zwischenbehälter *m* ~ **contour** Zwischenkurve *f* ~ **control** Hauptgruppenkontrolle *f* ~ **cooler** Zwischenkühler *m* ~

**cooling** Zwischenkühlung f ~ **coupling** mittelstarke Kupplung f ~ **cut-off (or switch)** Zwischenausschalter m ~ **degree (or stage)** Mittelstufe f ~ **delaying position** Zwischenfeld n ~ **discharge station** Zwischenabwurfstelle f ~ **disk** Mittelscheibe f ~ **distributing frame** Zwischenverteiler m ~ **drive** Zwischengetriebe n ~ **drive shaft** Antriebswelle f (hydraul. pump) ~ **echo-suppressor** Zwischenecho-Unterdrücker m ~ **energy** Gebiet n der mittleren Energie ~ **exchange** Zwischen-amt n, -vermittlung f ~ **-film method** Zwischenfilm-methode f (TV), -verfahren n ~ **film television system** Filmfernsehsystem n ~ **-film transmitter** Fernsehzwischenfilmsender m, Zwischenfilm-geber m, -sender m ~ **flange** Zwischenflansch m ~ **fluorescence screen** Zwischenfluoreszenzschirm m ~ **focal length** mittlere Brennweite f ~ **frame** Zwischenspant n

**intermediate-frequency (IF)** Zwischenfrequenz f (ZF) ~ **amplification stage** Zwischenfrequenzverstärkerstufe f ~ **amplifier** Zwischenfrequenzverstärker m ~ **jamming** ZF-Stören n ~ **receiver** Superheterodyn-, Zwischenfrequenzempfänger m ~ **rectifier** Zwischenfrequenzgleichrichter m ~ **response ratio** ZF-Empfindlichkeitsverhältnis n ~ **section of radar equipment** Zwischenfrequenzteil m ~ **stage** Zwischenfrequenzstufe f ~ **transformer** zweikreisiger Bandfilter m, Zwischenfrequenztransformator m ~ **tube** Zwischenröhre f

**intermediate,** ~ **gate** Zwischentor n ~ **gear** Vorgelege n, Zwischen-schaltung f, -welle f ~ **gear shaft** Zwischenrad m mit Welle ~ **gear wheel** Zwischenrad n ~ **gearing** Zahnradvorgelege n, Zwischengetriebe n ~ **grade** Zwischenstufe f ~ **heating** Zwischenwärme f ~ **horizon** Zwischenhorizont m ~ **housing** Zwischengehäuse n ~ **image** Zwischenbild n ~ **-image modulation** Zwischenbildmodulation f ~ **implement** Zwischengeschirr n ~ **ions** Zwischenionen pl ~ **jamb** Mittelquaderpfeiler m, Mittelschaft m ~ **kind** Zwischensorte f ~ **landing** Zwischenlandung f ~ **layer** Beilauffaden m, Zwischen-lage f, -lager n ~ **line** Zwischenleitung f ~ **linkage** Zwischenglied n ~ **liquid** Zwischenflüssigkeit f ~ **load** Zwischenlast f ~ **longitudinal girder** Hilfslängsträger m ~ **mass** Zwischenmittelmasse f ~ **means** Zwischenorgane pl ~ **medium** Zwischen-körper m, -mittel n ~ **member** (of a series) Mittelglied n, Zwischenträger m ~ **neutron** mittelschnelles Neutron n ~ **nucleus** Zwischenkern m ~ **nut** Zwischenmutter f ~ **office** Zwischenamt n ~ **optic** Zwischenoptik f ~ **paper layer** Papierzwischenlage f ~ **part (or piece)** Zwischenstück n ~ **pillar** Zwischenpfeiler m ~ **plane** Mittelebene f ~ **plate** Zwischenplatte f ~ **plug-and-cable adapter assembly** Zwischenkupplung f ~ **point** Zwischenpunkt m ~ **position** Übergangsstellung f, Zwischen-lage f, -stellung f ~ **power amplifier** Leistungszwischenverstärker m ~ **-pressure stage** Mitteldruckstufe f ~ **printed recordings** Zwischendrucke pl ~ **produce** Zwischenerzeugnis n ~ **product** Halb-erzeugnis n, -fabrikat n, Vorprodukt n, Zwischen-erzeugnis n, -produkt n ~ **pump facilities** Zwischenpumpeneinrichtungen pl ~ **quality** Zwischensorte f ~ **range** Zwischen-

entfernung f ~ **reaction** Zwischenreaktion f ~ **reading** Zwischenablesung f ~ **receptacle** Zwischengefäß n ~ **reception** Zwischenempfang m ~ **relay** Zwischenrelais n ~ **repeater** Zwischenverstärker m ~ **rib** Nebenrippe f ~ **ring** Zwischenring m ~ **rinsing** Zwischenspülung f ~ **roll stand** Mittelgerüst n ~ **rolling mill** Mittelwalzwerk n ~ **rolling train** Mittelstraße f, Zwischenstrecke f ~ **selector** Gruppenwähler m ~ **service (or operation)** Kurzbetrieb m ~ **shade** Mittelnuance f, Zwischenton m ~ **shading value** (of wedge) Graukeilstufe f, Graustufe f ~ **shaft** Hilfs-, Lauf-, Neben-, Tunnel-, Vorgelege-, Zwischen-welle f ~ **size** Zwischenmaß n ~ **size reduction** Mittelzerkleinerung f ~ **sleeve** Zwischenhülse f ~ **softening** Zwischenglühung f ~ **sort** Zwischensorte f ~ **space** Lichte n ~ **span** Zwischenöffnung f ~ **spring leaf** Federzwischenlage f ~ **stage** Übergangszustand m, Zwischen-stufe f, -zustand m ~ **stand** Zwischengestell n ~ **station** Trennanstalt f (teleph.), Zwischen-amt n, -bahnhof m ~ **steam utilization** Zwischendampfentnahme f ~ **stop** Zwischenlandung f ~ **story** Halbgeschoß n ~ **strand of rolls** Mittelstrecke f ~ **stretching device** Zwischenspannvorrichtung f ~ **strut** Zwischenstiel m ~ **subcarrier** Zwischenhilfsträger m ~ **substance** Zwischen-körper m, -mittel n ~ **support** Mittel-, Zwischen-stütze f ~ **switchboard** Zwischenvermittlung f ~ **tank** Zwischenbehälter m ~ **tappet** Zwischenstößel m ~ **telephone set** Zwischenstelle f ~ **terminal (or connector)** Zwischenklemme f ~ **toll center** Zwischen-amt n, -anstalt f ~ **toll station** Zwischenanstalt f ~ **tonal value** Graustufe f ~ **tone** Zwischenton m ~ **town marker** Zwischenwegweiser m ~ **trainer (or training aircraft)** Übergangsflügzeug n ~ **transformer** Zwischentransformator m ~ **trestle** Sprengwerk n ~ **truss** Leergebinde n, Zwischengesperre n ~ **-type submarine cable** Flachseekabel n ~ **unit** Zwischenstück n ~ **value** Zwischenwert m ~ **vessel** Zwischengefäß n ~ **wall** Zwischenwand f ~ **wave** Übergangswelle f ~ **wave radio telephone** Grenzwellenfunktelefon n ~ **work** Zwischenwerk n ~ **zone** Zwischenzone f ~ **yield** Zwischennutzung f

**intermesh, to** ~ ineinandergreifen
**intermeshed scanning method** Zeilensprungmethode f
**intermeshing** Ineinandergreifen n
**intermetallic** intermetallisch
**intermingle, to** ~ beimischen, durchsetzen, durchwachsen, einmengen
**intermission** Pause f, Unterbrechung f
**intermit, to** ~ aussetzen, intermittieren, unterbrechen
**intermitted,** ~ **fillet welding** unterbrochene Kehlnahtschweißung f ~ **integration** absatzweise Summierung f ~ **load** kurzzeitige Belastung f
**intermittency** zeitweilige Unterbrechung f ~ **factor** Intermittenzfaktor m
**intermittent** pulsierende Belastung f; absatzweise (film), aussetzend, diskontinuierlich, instationär, intermittierend, kurzzeitig, nichtkontinuierlich, periodisch, pulsend, pulsierend, unstet, unstetig, unterbrochen, zeitweilig, zeitweilig aussetzend

**intermittent, ~ capacity** Kurzbetriebsleistung *f* **~ contact** intermittierender Kontakt *m*, zeitweise Leitungsberührung *f* **~ control (or feed)** Springschaltung *f* **~ current** intermittierender Strom *m*, Unterbrechungsstrom *m* **~ discharge** aussetzende Entladung *f* **~ disconnection** zeitweilige Unterbrechung *f* **~ drive** stoßweise beanspruchter Antrieb *m* **~ earth circuit** zeitweiser Erdschluß *m* **~ feed** schaltweise Bewegung *f*, Sprungvorschub *m* **~ feeding mechanism** Sprungschaltung *f* **~ film feed** bildweise Filmschaltung *f*, Filmbewegung *f*, Filmbildweise *f* **~ film motion** bildweise Filmschaltung *f* **~ film movement** bildweise Schaltung *f*, Filmbewegung *f*, Filmbildweise *f* **~ grinding** satzweise Vermahlung *f* **~ jet engine** intermittierendes Luftstrahltriebwerk *n*, Pulsstrahltriebwerk *n* **~ light** Blink-, Gleichfunkel-, Gleichtakt-, Wechsel-feuer *n* **~ line** gestrichelte Linie *f* **~ load** aussetzende Belastung *f* **~ mechanism** Schaltwerk *n* (film) **~ motion** Wechselbewegung *f* **~ movement** schaltweise Bewegung *f* **~ periodic load** periodisch aussetzender Betrieb *m* mit gleichbleibender Belastung **~ printer** Fensterkopiermaschine *f* **~ rating** Einschaltdauer *f* **~ reception** unterbrochener Empfang *m* **~ seam welding** Rollenschrittverfahren *m* (Schweißen) **~ service** aussetzender Betrieb *m* mit gleichbleibender Belastung **~ shock load** stoßweise Belastung *f* **~ sprocket** Transporttrommel *f* **~ stress** stoßweise Beanspruchung *f* **~ table** Vollherd *m* **~ table feed** Sprungtischschaltung *f*

**intermittently** absätzig, in Pausen, ruckweise, sprungweise

**intermitter** Pumpe *f* mit unterbrochener Leistung

**intermix, to** ~ beimischen, durchmengen, durchmischen, einmengen, einmischen

**intermixture** Durchmischung *f*, Einschlag *m*

**intermodulation** Erzeugung *f* von einem Differenzton, Erzeugung *f* von Differenztönen, Kreuzmodulation *f*, gegenseitige Modulation *f*, Zwischenmodulation *f* **~ distortion** Intermodulationsverzerrung *f* **~ frequency** Differenzton *n* **~ noise** Intermodulationsrauschen *n* **~ products** (in per cent) Intermodulationsfaktor *f*

**intermolecular** intermolekular, zwischenmolekular **~ chain** intermolekulare Kette *f* **~ force** zwischenmolekulare Kraft *f*

**internal** (built in) eingebaut, innen befindlich, inner(er, e, es), inwendig **~ administration** Hausverwaltung *f* **~ air lines** Inlandluftverkehrsstrecken *pl* **~-and face-grinding machine** Innen- und Stirnflächenschleifmaschine *f* **~ angle** innere (Mauer)Ecke *f* **~ annular shake** (in wood) Kern-, Ring-schäle *f* **~ antenna** Innen-, Zimmer-antenne *f* **~ audits** betriebseigene Prüfungen *pl* **~ band brake** Innenbandbremse *f* **~ base resistance** Basis-Innenwiderstand *m* **~ blackening** Kolbenschwärzung *f* **~ blower** Innengebläse *n* **~ bracing wire** Innenverspanndraht *m* **~ cable** Innenkabel *n* mit Vermittlungsstellen **~ caliper gauge** Bohrungs-, Loch-lehre *f* **~ cam operating propeller motion** Exzenter *m* zur Arretur **~ capacity** (innere) Röhrenkapazität *f*, (of coil) Windungskapazität *f* **~ chuck jaw** Innenspannbacke *f* **~ circuit** interne Leitung *f* **~ circulation** Werksumlauf *m*

**internal-combustion** Innenverbrennung *f* **~ engine** Brennkraftmaschine *f*, Explosionsmotor *m*, Innenverbrennungsmotor *m*, Verbrennungskraftmaschine *f* **~ engine with revolving piston** Drehkolbenmotor *m* (Wankelmotor) **~-engine installation** Verbrennungsmotorenanlage *f* **~ tractor** Rohöl-schlepper *m*, -traktor *m* **~ type** Innenverbrennungstyp *m*

**internal, ~ condensation** Inkohlung *f* **~ conductance** innerer Leitwert *m* **~ consumption** Eigenbedarf *m*, -verbrauch *m* **~ contact** Innenkontakt *m* **~ conversion spectrum** Umwandlungsspektrum *n* **~ copying** Innenkopieren *n* **~ cracking** Kernrissigkeit *f* **~ current** innerer Strom *m* **~ cylindrical gauge** Bohrungslehre *f*, Lehrdorn *m* **~ degree of freedom** innerer Freiheitsgrad *m* **~ detector of temperature** eingebautes elektrisches Thermometer *n* **~ diameter** Innendurchmesser *m*, Lichtweite *f*, Lumen *n* **~ diameter of pipe** innerer Rohrdurchmesser *m*, Rohrweite *f* **~ diameter of tube** Rohrweite *f* **~ disc** Innenlamelle *f* (Kupplung) **~ distribution** Hausverteilung *f* **~-drag bracing** Innenauskreuzung *f* (aviat.) **~ drag bracing fitting** Innenverspannungsbeschlag *m* **~ economies** Rationalisierung *f* innerbetrieblicher Art **~ economy** Binnenwirtschaft *f* **~ energy** innere Energie *f* **~ expanding brake** Innenausdehnungsbremse *f* **~ facing attachment** Innenplandreheinrichtung *f* **~ fault staff** Störungspersonal *n* für Amtsstörungen **~ faultsman** Störungssucher *m* im Innendienst **~ firebox** Flammenrohrkessel *m* **~-firebox boiler** Kessel *m* mit Innenfeuerung **~ firing** Innenfeuerung *f* **~ fissure** Innenriß *m* **~ fissuring** Innenrißbildung *f* **~ flue** durchgehendes Feuerrohr *n* **~-flue boiler** Feuerbuchs-, Flammenrohr-kessel *m* **~ fluid motion** Flüssigkeitsbewegung *f* im Inneren **~ focussing lens** Innenfokussier-, Schiebe-linse *f* **~ friction** Eigenreibung *f*, innere Reibung *f* **~ furnace boiler** Flammrohrkessel *m* **~ fuse** Innenzünder *m* **~ gauge** Lehrbolzen *m* **~-gauging optimeter** Optimeter *m* für Innenmessungen **~ gear** Innenverzahnung *f*, Innenzahnkranz *m* (einer Planscheibe), innere Verzahnung *f* **~ gear shaper** Stirnradinnenstoßmaschine *f* **~ gearing** Innenverzahnung *f* **~ gears** Hohlräder *pl*

**internal-grinding** Innen-schleifen *n*, -schliff *m* **~ of straight bores** Innenschliff *m* zylindrischer Büchsen **~ of taper bores** Innenschliff *m* konischer Rundflächen

**internal-grinding, ~ attachment** Innenschleifeinrichtung *f* **~ countershaft** Vorgelege *n* für die Innenschleifvorrichtung **~ fixture** Innenschleifvorrichtung *f* **~ head** Innenschleifkopf *m* **~ machine** Innenschleifmaschine *f* **~ spindle** Innenschleifspindel **~ work** Innenschleifarbeit *f*

**internal, ~ gripping (or stress)** Innenspannung *f* **~ hardening stress** Härtungsspannung *f* **~-head loss** Flüssigkeitswiderstand *m* **~ heating** Innenfeuerung *f* **~ idle time** interne Leerlaufzeit *f* **~ inductance** Eigeninduktivität *f* **~ infrasonic** innerer Infraschall *m* **~ input resistance** Gitterwiderstand *m* **~ intercommunication** (housetelephone) **line** Linienwählerleitung *f* **~ lapleakage** innere Undichtigkeit *f* **~-limit gauge** Grenz-lehrdorn *m*, -lochlehre *f*, Toleranzlochlehre $f$ ~ **loss** Eigenverlust $m$ ~ **measurement** Innenmessung $f$ ~ **measuring attachment** Innenmeßeinrichtung $f$ ~**-measuring instrument with spring-type rest** Innenmeßgerät $n$ mit federndem Anschlag ~ **milling attachment** Innenfräsapparat $m$ ~ **mixer with floating weight** Gummikneter $m$ ~ **modes** innere Eigenschaften $pl$ ~ **mold (or cast)** Steinkern $m$ ~ **momentum** inneres Moment $n$ ~ **operating ratio** innere Ausbeute $f$ ~ **perspective center** (of the lens) bildseitiger (hinterer) Hauptpunkt $m$ ~ **photoelectrical effect** innerer Fotoeffekt $m$ ~**-photography** oscillograph Innenfotografieoszillograf ~ **plant** Amts-, Innen-anlage $f$ ~**-pole dynamo** Innenpoldynamo $m$ ~ **pressure plate** Pendelfenster $n$ ~ **pressure test** Innendruckversuch $m$ ~ **quenching** Selbstlöschung $f$ ~ **rating** Typenleistung $f$ ~ **reading** Innenablesung $f$ ~ **recessing** Inneneinstechen $n$ ~**-reciprocating combustion engine** Ottomotor $m$ ~ **reinforcement** innere Versteifung $f$ ~ **reinforcing member** innere Versteifung $f$ (des Flügels) ~ **resistance** Innenwiderstand $m$, (of a cell or of a thermionic tube) innerer Widerstand $m$ ~ **resistance of a tube** Abschlußwiderstand $m$ einer Röhre ~ **resistance for zero grid bias** innerer Leistungswiderstand $m$ ~ **revenue** Akzise $f$ ~ **rubber shock absorber** Innenfederung $f$ ~ **safety ring** Innensicherung $m$ ~ **scale (or graduation)** Innenteilung $f$ ~ **service** Innendienst $m$ ~ **setting** unterer Rahmen $m$ ~ **shake** Innenriß $m$ ~ **spherical surface** Innenkugelfläche $f$ ~ **spiral insert** Wendeleinsatz $m$ ~ **springing** Innenfederung $f$ ~ **storage** eingebauter Speicher $m$, Innenspeicher $m$ ~ **strain** Eigenspannung $f$ ~ **stress** innere Spannung $f$ ~ **structure** Inneaufbau $m$ ~ **strutting** Innenverstrebung $f$ ~ **supercharger** Innenvorverdichter $m$ ~ **tap** Innengewinde $n$ ~ **tapering** Konusbohren $n$ ~ **teeth** Innenverzahnung $f$ ~ **tension** Eigenspannung $f$ ~ **tension of glass** Glasspannung $f$ ~ **thread** Innen-, Mutter-gewinde $n$ ~ **thread grinder** Innengewindeschleifmaschine $f$ ~ **thread measurement (or gauging)** Innenmessung $f$ ~**-thread-measuring instrument** Innenmeßgerät $n$ ~**-thread milling attachment** Innenfräsapparat $m$ ~ **threading** Innengewinde $n$ ~ **toothing** Innenverzahnung $f$ ~ **trouble** Eigenstörung $f$ ~ **upset** Innenstauchung $f$ ~ **vane** (rocket) Druckstück $n$ ~ **variables** innere Variable $f$ ~ **width** Lichtweite $f$ ~ **wiring** Amtsverdrahtung $f$, Innen-, Zimmer-leitung $f$ ~ **wrenching bolt** Innensechskantbolzen $m$

**internally,** ~ **controlled** innengesteuert ~ **heated rotary drier** Trommeltrockner $m$ mit Innenbeheizung ~ **sprung** innengefedert ~ **sprung starting wheel** innengefedertes Anlaufrad $n$

**international** international, zwischenstaatlich ~ **broadcasting station** internationaler Rundfunksender $m$ ~ **candle** internationale Kerze $f$ ~ **circuit** Auslandsleitung $f$, internationale Leitung $f$

**International Commission on Illumination** Internationale Beleuchtungskommission (CIE)

**international,** ~ **communication service** internationaler Fernmeldedienst $m$ ~ **date line** Linie $f$ des Datumwechsels ~ **electrical units** internationale elektrische Einheiten $pl$ ~

**exhibition** Weltausstellung $f$ ~ **law** Völkerrecht $n$ ~ **line** Auslandsleitung $f$ ~ **message** zwischenstaatliches Gespräch $n$

**International Normal Atmosphere (INA)** Norm-(al)-atmosphäre $f$

**international ohm** internationales Ohm $n$

**International Postal Union** Weltpostverein $m$

**international radio silence** internationale Funkstille $f$

**International Standard Atmosphere (ISA)** internationale Normalatmosphäre $f$

**international,** ~ **standard candle** Kerze $f$ ~ **telegraph circuit** Telegrafenleitung $f$ für den zwischenstaatlichen Verkehr ~ **telegraphic convention** Welttelegrafenvertrag $m$ ~ **telephone circuit** zwischenstaatliche Fernsprechleitung $f$ ~ **terminal office** Grenzausgangsanstalt $f$, zwischenstaatliches Fernamt $n$ ~ **trade** Welthandel $m$, Weltmarkt $m$ ~ **traffic** Weltverkehr $m$ ~ **transit exchange** zwischenstaatliches Durchgangsamt $n$

**internodal distance** Knotenabstand $m$

**internode** Schwingungsbauch $m$ ~ **of an oscillation** Bauch $m$ einer Schwingung ~ **of stationary oscillation (or wave)** Schwingungsschleife $f$

**internuclear distance** Kern-abstand $m$, -distanz $f$

**interocular distance** Augenabstand $m$

**interoffice,** ~ **slip (or tag)** Laufzettel $m$ ~ **trunk** Ortsverbindungsleitung $f$ ~**-trunk-cable plant** Kabelnetz $n$ mit Ortsverbindungsleitungen ~ **trunks** Amtsverbindungsleitung $f$

**interpenetrate, to** ~ durchdringen, durchwachsen

**interpenetration** (soldering) Diffusion $f$, Durchdringung $f$, Durchwachsung $f$ (cryst.)

**interphase** Phasengrenzschicht $f$, Zwischenphase $f$ ~ **transformer** Saugdrosselspule $f$

**interphone** Bord-sprechgerät $n$, -telefon $n$, -verständigungsanlage $f$, Eigenverständigungsanlage $f$ ~ **communication** Bordverständigung $f$

**interplanar spacing** Netzebenenabstände $pl$

**interplane,** ~ **bracing** Tiefenverspannung $f$ ~ **control** Flügelsteuerung $f$ ~ **strut** Flügelstiel $m$, Tragflächenstrebe $f$

**interplanetary aviation** Raumschiffahrt $f$

**interplay** Zusammenspiel $n$

**interpolar,** ~ **distance** Maulweite $f$ ~ **gap (or space)** Polzwischenraum $m$

**interpolate, to** ~ einschalten, interpolieren, zwischenschalten

**interpolated** geschaltet

**interpolation** Einschaltung $f$, Interpolation $f$, Zwischenschaltung $f$

**interpolator** Interpolator $m$

**interpole** Wende-, Zwischen-pol $m$ ~ **shim** Wendepolunterlegblech $n$ ~ **winding** Wendepolspule $f$

**interpolymer** Mischpolymerisat $n$

**interpose, to** ~ einfügen, intervenieren, zwischen-legen, -schalten, -setzen

**interposed** eingeschoben, zwischengeschaltet ~ **means (or medium)** Zwischen-körper $m$, -mittel $n$ ~ **optical means** Zwischenoptik $f$

**interposing** Zwischenschaltung $f$ ~ **alloy** Zwischenlegierung $f$

**interposition** Einfügen $n$, Einfügung $f$, Vermittlung $f$, Zwischenlage $f$, Zwischenschaltung $f$ ~ **circuit** Fernplatzrufleitung $f$ ~ **trunk** Verbin-

dungs-, Platzverbindungs-leitung *f* (teleph.) ~
**trunk at toll** Ferndienstleitung *f*
**interpret, to** ~ auslegen, auswerten, deuten, interpretieren
**interpretation** Ausdeutung *f*, Auslegen *n*, Auslegung *f*, Auswertung *f*, Deutung *f*, Erklärung *f*, Interpretation *f* ~ **of aerial photographs** Auswerten *n* von Luftbildern, Bildauswertung *f*
**interpretative subroutine** interpretatives Teilprogramm *n*
**interpreter** Dolmetscher *m*, Interpret *m*, interpretierendes Organ *n*, Interpretierprogramm *n*, Lochstreifenabtaster *m*, Zuordnungsprogramm *n* ~ **code** Interpretier-, Zuordner-kode *m*
**interpreting** Übertragung *f* der Lochstreifen in Klartext *m*
**interpretive,** ~ **routine** interpretatives Programm *m* ~ **system** Pseudoprogramm *n* (info proc)
**interpulse pause** Impulspause *f*
**interpupillary,** ~ **distance** Augenabstand *m* ~ **distance gauge** Augenabstandsmesser *m*
**interrack cable** Systemkabel *n*
**interrelation** Wechselbeziehung *f*, Zusammenhang *m*
**interrelationship** Beziehung *f*
**interrogate, to** ~ abfragen, ausfragen, verhören
**interrogating,** ~ **pulse** Abfrageimpuls *m* ~ **signal** Abfragesignal *n*
**interrogation** Abfrage *f*, Abfragung *f*, Verhör *n*, Verhörung *f*, Vernehmung *f* ~ **beamwidth coding** Abfragestrahlbreite *f*, Abfrageverschlüsselung *f* ~ **pulse** Abfrageimpuls *m* ~ **recurrence frequency** Abfragefolgefrequenz *f*
**interrogator** Abfragesender *m* ~ **responder (or responsor)** Frage/Antwort-Gerät *n* (secrdr)
**interrogatory** Fragestellung *f*
**interrupt, to** absetzen, aussetzen, einfallen, unterbrechen **to** ~ **a conversation** in ein Gespräch eingreifen **to** ~ **a line** eine Leitung auftrennen (electr.) oder auslösen
**interrupted** diskontinuierlich, durchbrochen, intermittierend, tönend, unterbrochen ~ **capacity** Aussetzleistung *f* ~ **circuit** unterbrochene Leitung *f* ~ **continuous wave** unterbrochene ungedämpfte Welle *f* ~-**continuous-wave transmission** Impulssenden *n* ~ **direct current** zerhackter Gleichstrom *m* ~ **earth circuit** zeitweiser Erdschluß *m* ~ **fire** unterbrochenes Feuer *n* ~ **hardening** unterbrochene Härtung *f* ~ **quenching (metal)** gebrochenes Härten *n* ~ **ringing** intermittierender oder selbsttätig wiederholter Ruf *m* ~-**screw-type breech- block** Schraubverschluß *m* ~ **signaling current** unterbrochener Tonfrequenzrufstrom *m* ~ **thread** unterbrochenes Gewinde *n* ~ **waves** periodisch zerhackte Wellen *pl*
**interrupter** Abreißhebel *m* (motor), Gleichstromunterbrecher *m*, elektrischer Hammer *m*, Schalter *m*, Stromunterbrecher *m*, Unterbrecher *m*, Unterbrecherdrossel *f* (jet) ~ **disk** Unterbrecherscheibe *f* ~ **lever** Unterbrecherhebel *m* ~ **pin** (fuse) Fliehbolzen *m* ~ **receiver** Unterbrecherempfänger *m* ~ **shaft** Schaltwelle *f* ~ **spring** (fuse) Fliehbackenfeder *f*, Unterbrecherfeder *f*
**interrupting,** ~ **capacity** Aussetzleistung *f* ~ **jack** Unterbrecherklinke *f* (TV)
**interruption** Absatz *m*, Pause *f*, Unterbrechung *f*
**interruption,** ~ **of the arc** Abreißen *n* des Licht-

bogens ~ **of cooling** Erstarrungsinterval *n* ~ **of current** Stromunterbrechung *f* ~ **of the gas flow** Brechung *f* der Gasströmung ~ **of line** Leitungsunterbrechung *f* ~ **of operation** Betriebsunterbrechung *f* ~ **of postal services** Postsperre *f* ~ **of speech** Dazwischensprechen *n* ~ **in supply** Zufuhrstockung *f* ~ **of traffic** Verkehrs-stockung *f*, -störung *f*
**interruption,** ~ **arc** Abreißbogen *m* ~ **cable** Notkabel *n* ~ **cord** Sprechschnur *f* einer B-Beamtin ~ **noise** Unterbrechungsgeräusch *n* ~ **signal** Unterbrechungszeichen *n* ~ **tone** Unterbrechungston *m*
**intersect, to** ~ durchkreuzen, durchschneiden, einschneiden, sich gabeln, kreuzen, rückwärtseinschneiden, schneiden, vorwärtseinschneiden **to** ~ **the ground** das Gebirge durchörten **to** ~ **lines of force** Kraftlinien *pl* schneiden
**intersected** durchschnitten, gekreuzt
**intersecting** Einschneiden *n* ~ **gills** Flachshechelmaschine *f*, (textiles) Glatthechel *m* ~ **plane** Schnittfläche *f*
**intersection** Anschnitt *m*, Durchhörterung *f* (min), Durchschlag *m*, Durchschneiden *n*, Durchschnitt *m*, Knotenpunkt *m*, Kreuz-feld *n*, -stelle *f*, Kreuzung *f*, Kreuzungspunkt *m*, Schnitt *m*, (of planes) Schnittlinie *f*, Stoßstelle *f*, Straßenkreuzung *f*, Vierung *f* (arch), Vorwärtseinschnitt *m* ~ **point of** ~ Kreuzungs-, Schnitt-punkt *m*, Überschneidung *f* ~ **of axes** Achsenkreuz *n* ~ **in the beam** Bündeleinschnürung *f* ~ **of curves** Kurvenschnittpunkt *m*
**intersection,** ~ **circle** Schnittkreis *m* (Bagger) ~ **error** Kreuzungsfehler *m* ~ **figures** Querschnittfiguren *pl* ~ **jog** Durchschneidungssprung *m* (cryst.) ~ **jog formation** Bildung *f* von Durchschneidungssprüngen ~ **line** Schnitt-gerade *f*, -linie *f* ~ **process** Schneideprozeß *m* ~ **table** Peiltischanlage *f* (rdo nav.) ~ **theorem** Schnittpunktsatz *m*
**intersector** Intersektionsschaltung *f*
**interspace** Zwischenraum *m*
**intersperse, to** ~ durchsetzen, einsprengen, einsprenkeln, einstreuen
**interspersed** (by sprinkling) eingesprengt
**interstage,** ~ **cooling** Zwischenstufenkühlung *f* ~ **coupling** Kopplung *f* zwischen Röhren ~ **fairing** Übergangsverkleidung *f* (Rakete) ~ **leak-off** Anzapfung *f* ~ **reservoir** Vorvakuumbehälter *m* ~ **transformer** Zwischen-transformator *m*, -übertrager *m*
**interstate** übergebietlich, zwischen den einzelnen Bundesstaaten, zwischenstaatlich ~ **air commerce** zwischenstaatlicher Luftverkehr *m*
**interstation noise suppression** Rauschunterdrückung *f*, Stummabstimmung *f*
**interstellar** interstellar, zwischen den Sternen ~ **craft** Weltraumschiff *n* ~ **flying** Raumfahrt *f* ~ **disturbance** außerirdische Störung *f* ~ **space** Weltenraum *m* ~-**space disturbances (or noise)** außerirdische Störungen *pl*
**interstice** Fuge *f*, Furche *f*, Hohlraum *m*, Lücke *f*, Masche *f*, Pore *f* (im Gewebe), Ritze *f*, Zwischenraum *m* ~-**synchronizing method** Lückensynchronisierungsverfahren *n*
**interstitial** zwischenräumlich ~ **atom** Zwischengitteratom *n* ~ **compounds** interstitielle Verbindungen *pl* ~ **distance** Fugenweite *f* ~ **impurity** Einlagerungsfremdatom *n* ~ **ion**

Zwischengitterion $n$ ~ **irradiation** interstitielle Bestrahlung $f$ ~ **material** Zwischenklemmungs-masse $f$ ~ **matter** Zwischenmasse $f$ ~ **mechanism** (of alloys) Zwischengitterplatzmechanismus $m$ ~ **particle** Zwischengitterteilchen $n$ ~ **position** Zwischengitterplatz $m$ ~ **space** Zwischenraum $m$

**interstratification** Zwischenlagerung $f$

**interstratified** durchzogen, zwischen-gelagert, -geschichtet ~ **material** Verwachsenes $n$

**interstratify, to** ~ dazwischenlagern, einspren-gen

**interstructure** Unterteilung $f$

**intersystem combination** Interkombinationen $pl$

**interthrough switch** Zwischenstellenumschalter $m$

**inter-tie** Wandriegel $m$

**inter-toll,** ~ **dialling** Fernwahl $f$ ~ **trunk** Fern-leitung $f$

**intertropical** intertropisch ~ **front** Gegend $f$ der Windstillen, Kalmengürtel $m$

**interturn,** ~ **capacitance** Windungskapazität $f$ ~ **continuity tester** Spulenwindungsschlußprüfer $m$ ~ **short-circuit tester** Spulenwindungsschluß-prüfer $m$, Windungsschlußprüfer $m$ ~ **voltage** Windungsspannung $f$

**intertwine, to** ~ verschlingen

**interurban,** ~ **bus** Landomnibus $n$ ~ **car** Vor-ortswagen $m$ ~ **railway** Überlandbahn $f$ ~ **traffic** Überlandverkehr $m$

**interval** Abstand $m$, Arbeits-, Betriebs-pause $f$, Getriebsfeld $n$ (mining), Intervall $n$, Pause $f$, Seitenabstand $m$, Spanne $f$, Spannweite $f$, Strecke $f$, Zeitperiode $f$, Zwischenraum $m$, Zwischenzeit $f$

**interval,** ~ **between graticule wires** Fadentanz $f$ ~ **of integration** Integrationsschritt $m$ ~ **between knubs** Noppenabstand $m$ ~ **between one thread** (of screw) and the next Gang auf Gang ~ **of oscillation** Oszillier-Intervall $m$ ~ **of retardation** Verzögerungsintervall $m$ ~ **of a shaft** Schacht-feld $n$, Schachtverzug $m$ ~ **between ships** Schiffs-abstand $m$ ~ **between spectral lines** Schrittweite $n$ ~ **between strips** Streifenabstand $m$ ~ **of time** Zeitabschnitt $m$ ~ **between two coils of a spring** Gang auf Gang ~ **between two scale lines of a graticule** Abstand $m$ zwischen zwei Teilstrichen einer Strichplatte

**interval,** ~ **decrease** Abstandsverringerung $f$ ~ **energy** wahre Energie $f$ ~ **error** Pausenfehler $m$ ~ **factor** Intervallfaktor $m$ ~-**recording disk** Kurzzeitmesserrad $n$ ~ **rule** G-Faktor $m$, Inter-vallregler $f$ ~-**selector circuit** Zeitdiskriminator-kreis $m$ ~ **signal** Pausenzeichen $n$ ~-**synchroniz-ing method** Lückensynchronisierungsverfahren $n$

**intervalometer** Bildfolgeregler $m$

**intervals** Lattenteilung $f$ ad ~ absatzweise ~ **of no speech** Gesprächspausen $pl$

**intervalve,** ~ **coupling (or linkage)** Kopplung $f$ zwischen zwei Röhren, Röhrenkopplung $f$ ~ **transformer** Zwischen-rohrtransformator $m$, -transformator $m$, -übertrager $m$

**intervene, to** ~ dazwischentreten, eingreifen, einspringen, intervenieren, vermitteln

**intervening** dazwischen-liegend, -tretend, zwi-schenliegend ~ **ground** Zwischengelände $n$ ~ **space** Zwischenraum $m$

**intervention** Dazwischenkunft $f$, Eingreifen $n$, Einmischung $f$, Vermittlung $f$

**interweave, to** ~ durchflechten, flechten, ver-flechten

**interwinding** zwischen den Wicklungen wirkend ~ **capacity of a transformer** Kapazität $f$ zwi-schen den Transformatorwicklungen

**interwing strut** Tragflächenstiel $m$

**interword space** Gruppen-, Wort-abstand $m$

**interwork, to** ~ zusammenarbeiten

**interwoven** durchflochten, geflochten

**intimate, to** ~ ankündigen, mitteilen

**intimate** eng, innig, intim, vertraulich ~ **adhesion** enge Anschmiegung $f$ ~ **contact** inniger Kon-takt $m$ ~ **mixture** Durchmischung $f$ ~ **threading** zügiger Gang $m$

**intimately,** ~ **connected** nahestehend ~ **mixed** innig gemischt ~ **related** nahestehend

**into,** ~ **each other (or one another)** ineinander

**intonation** Anklingen $n$, Einsatz $m$ (acoust.)

**intone, to** ~ intonieren

**intoxicated** betrunken

**intracavitary,** ~ **applicator** Körperhöhlenappli-kator $m$ ~ **irradation** intrakavitäre Bestrahlung $f$

**intracrystalline rupture** intrakristalliner Bruch $m$

**intrados** innere Bogenfläche $f$, Leibung $f$, innerer Rand $m$ ~ **of an arch** Bogenleibung $f$

**intramolecular** zwischenmolekular ~ **chain** innermolekular Kette $f$

**intrapearlitric** intraperlitisch

**intrashell transition** Übergang $m$ innerhalb einer Schale

**intrastate** innerhalb des Staates ~ **air commerce** Inlandsluftverkehr $m$

**intravital microscope** Intravitalmikroskop $n$

**intricate** beschwerlich, kompliziert, unüber-sehbar, verwickelt ~ **assembly** komplizierte An-ordnung $f$ ~ **casting** verwickelt gestaltetes Guß-stück $n$ ~ **sections** schwierige Formen $pl$ (des Werkstücks) ~ **shape** komplizierte Form $f$

**intrinsic** eigenleitend, eigentlich, innewohnend, wahr ~ **angular momentum** Eigendrehmoment $n$ ~ **brilliance** Flächen-helle $f$, -helligkeit $f$ ~ **coercive force** Induktionskoerzitivkraft $f$ ~ **conduction** Eigenleitung $f$ ~ **counter efficiency** intrinsike Empfindlichkeit $f$ des Zählers ~ **energy** innere oder wahre Energie $f$ ~ **geometry of a surface** innere Geometrie $f$ einer Fläche ~ **induction** Magnetisierungsintensität $f$ ~ **in-frasonic** innerer Infraschall $m$ ~ **light of eye** Augeneigenlicht $n$ ~ **light of retina** Augen-eigenlicht $n$, Eigenlicht $n$ des Auges ~ **luminous intensity** Leuchtdichte $f$ ~ **noise** Eigengeräusche $pl$ ~ **photoconduction** innerer Eigenfotoeffekt $m$ ~ **resistance** Kernwiderstand $m$ ~ **semi-conductor** Eigenhalbleiter $m$ ~ **temperature range** Eigenleitungstemperaturgebiet $n$ ~ **transconductance** Gegenwirkleitwert $m$ ~ **value** Wirklichkeitswert $m$ ~ **viscosity** Struktur-viskosität $f$

**intrinsically safe circuits** eigensichere Strom-kreise $pl$

**introduce, to** ~ einführen, einleiten, (value) ein-setzen, vorlagern, verlegen, vorstellen **to** ~ **customers** Kunden zuweisen **to** ~ **oneself** sich vorstellen

**introduced** (into) eingerückt

**introduction** Einführen *n*, Einführung *f*, Einführungsschreiben *n*, Vorbericht *m*, Vorstellung *f* ~ **to the specification** Beschreibungseinleitung *f* (patent)
**introductory page** Titelseite *f*
**intrude, to** ~ mit Gewalt *f* eindringen
**intruded body** Intrusionsmasse *f*
**intruder** Unbefugter *m*
**intrusion** Aufschaltung *f*, Eindringen *n* ~ **detector system** Raumschutzanlage *f*
**intrusive** eingedrungen, vordringlich ~ **rock** eingedrungenes Gestein *n*
**intumescence** Intumeszenz *f*
**inundate, to** ~ überfluten, überschwemmen
**inundated area** Ansumpfung *f*, Überschwemmungsgebiet *n*
**inundation** Ansumpfung *f*, Überflutung *f*, Überschwemmung *f* ~ **limit** Hochwasserbett *n*
**invade, to** ~ einfallen
**invalid** nichtig, ungültig
**invalidate, to** ~ aufheben, entkräftigen, ungültig machen
**invalidation** Nichtigkeitserklärung *f* ~ **suit** Nichtigkeitsverfahren *n*
**invalidity** Ungültigkeit *f* ~ **suit** (patents) Nichtigkeitsklage *f*
**invariability** Beständigkeit *f*, Unveränderlichkeit *f* ~ **with frequency** Frequenzunabhängigkeit *f*
**invariable** ausnahmslos, beständig, konstant, unveränderlich, unwandelbar ~-**coupling capacitor** Kopplungsblock *m*
**invariably** ohne Ausnahme *f*
**invariant** Invariante *f*, Unveränderliche *f*, gleichbleibend ~ **of magnetic rotation** magnetische Rotationsinvariante *f* ~ **subgroup** Normalteiler *m*
**invasion** Einfall *m*, Invasion *f*
**invention** Erfindung *f*, Fund *m*
**inventive** erfinderisch ~ **genius** Erfindergeist *m*
**inventor** Erfinder *m*
**inventory** Bestand *m*, Bestandbuch *n*, Bestandsaufnahme *f*, Inventar *n*, Lageraufnahme *f*, Lagerbestand *m*, Nachweis *m*, Verzeichnis *n* ~ **of equipment (or of supplies)** Bestandsliste *f* ~ **and material accounting** Inventur- und Materialabrechnung *f* ~ **verification** Inventurprüfung *f*
**inverse** entgegengesetzt, invers (math), reziprok, umgekehrt ~ **amplification factor** Durchgriff *m* ~ **current** entgegengesetzter Strom *m*, Rückstrom *m*, Rückstrom *m* der Anoden, Sperrstrom *m* ~ **electrode current** Kathodenstrom *m* ~ **(or negative) feedback circuit** Gegenkopplungsschaltung *f* ~ **function** reziproke Funktion *f*, Umkehrfunktion *f* ~ **grid current** positiver Gitterstrom *m* ~-**grid potential (or voltage)** Gegenspannung *f* ~-**grid transparency** Rückgriff *m* ~ **impedances** inverse Scheinwiderstände *pl* ~ **mapping** inverse Abbildung *f* ~ **photo-electric effect** inverser fotoelektrischer Effekt *m* ~-**rate curve** Osmondsche Kurve *f* ~ **rate curve** reziproke Geschwindigkeitskurve *f* ~ **ratio** umgekehrtes Verhältnis *n* ~ **relationship** spiegelbildliches Verhältnis *n* ~ **resistance** Widerstandsreziprok *n* ~ **sine** Arcus sinus ~ **speed-rate curve** Geschwindigkeitsreziproke *f* ~ **square law** Entfernungsquadratgesetz *n* ~ **time element** abhängig verzögerte Auslösung *f* ~ **time-limit release** abhängig verzögerter Aus-

löser *m* ~-**time relay** Invertzeitrelais *n* ~ **trigonometric function** zyklometrische Funktion *f* ~ **value** Kehrwert *m* ~ **voltage** Sperrspannung *f*
**inversely,** ~ **induced** entgegengesetzt induziert ~ **proportional** umgekehrt oder verkehrt proportional
**inversion** (image) Bildaufrichtung *f*, Figurenwechsel *m*, Inversion *f*, Umkehr *f*, Umkehrung *f*, Umschaltung *f*, Wärmeumkehr *f*, Wechsel *m* ~ **in space** Inversion *f* im Raum ~ **of syllables** Silbenumkehrung *f*
**inversion,** ~ **center** Symmetriezentrum *n* ~ **factor** Inversionsfaktor *m* ~ **formula** Umkehrformel *f* ~ **layer** Inversionszone *f* ~ **point** Umkehrpunkt *m* ~ **prism** (reverses in both axes of image) Doppelumkehrprisma *n*, doppelumkehrendes Prisma *n*, Umkehrprisma *n* ~ **rule** Umkehrungssatz *m* ~ **signal** Umschaltzeichen *n* ~ **telemeter** Invertentfernungsmesser *m*
**invert, to** ~ umkehren, umstellen, umstülpen, umwenden **to** ~ **rails** Schienen umlegen
**invertase** Invertin *n*
**inverted** überkippt, (image) umgekehrt, verkehrt ~ **arch** umgekehrter Bogen *m*, Erdbogen *m* ~ **brake cone** Bremskonus *m* ~-**coincidence range finder** Halbbild-, Kehrbild-entfernungsmesser *m* ~ **converter** Gleichstromwechselstromumformer *m*, Wechselrichter *m* ~ **cross talk** unverständliches Nebensprechen *n* ~ **cylinder** hängender Zylinder *m* ~ **engine** Hängemotor *m*, hängender oder umgekehrter Motor *m* ~ **field pulses** umgekehrte Teilbildimpulse *pl* ~ **flight** Rückenflug *m* ~-**flight carburetor** Rückenflugvergaser *m* ~ **fold** liegende oder überkippte Falte *f* ~ **guideways** Dachprismenführung *f* ~ **hour** Inhour *f* ~ **image** auf dem Kopf stehendes Bild *n*, umgekehrtes Bild *n*, Kehrbild *n* ~-**image range finder** Invertentfernungsmesser *m*, Kehrbildentfernungsmesser *m* ~ **in-line engine** Reihenmotor *m* mit hängenden Zylindern ~ **L-antenna** L-Antenne *f* ~ **load** Rückenlast *f* ~ **loop** Looping *m* aus der Rückenlage, Schleifenrückenflug *m* (aviat) ~ **mirage** Luftspiegelung *f* ~ **motor** Motor *m* mit hängenden Zylindern ~ **normal loop** Rückenflug *m* ~ **outside loop** Außenschleife *f* im Rückenflug ~ **parabolic superstructure** Parabolfachwerkträger *m* ~ **pendulum** umgekehrtes oder zeitgekehrtes Pendel *n* ~-**plan view** Ansicht *f* von unten ~ **position** Auskippstellung *f* **to put in** ~ **position** stülpen ~ **siphon** Dücker *m* ~ **speech** verschlüsselte Sprache *f* ~ **speech frequency** umgekehrte Sprachfrequenz *f* ~ **speech system** Telefonie *f* mit Frequenzumkehrung ~ **spin** Rückentrudeln *n* ~ **tooth chain with pin link** Zahnkette *f* mit Gleitgelenk ~-**type carburetor** Fallstromvergaser *m* ~-**type engine** Motor *m* mit hängenden Zylindern ~ **V** negative V-Stellung *f* (des Flügels) ~ **V twin engine** hängender Zweizylinder-V-Motor *m* ~-**V-type engine** A-Motor *m*, Motor *m* mit hängender V-Anordnung ~ **valve** Hängeventil *n*, hängendes Ventil *m*
**inverter** (thyratron) Inverter *m*, Sprachwende *f*, Umformer *m*, Umrichter *m*, Umwandler *m*, (static) Wechselrichter *m* ~ **electronics** Umkehrrohr *n* ~ **module** Umkehr-Modul *n* ~ **tube** Umkehrröhre *f*
**invertible** umstellbar

**inverting prism** doppelumkehrendes Prisma *n*, Umkehrprisma *n*
**invest, to** ~ abschließen, anlegen, einkreisen, investieren, umzingeln, zernieren **to** ~ **capital** Kapital *n* hineinstecken **to** ~ **money** Geld *n* anlegen **to** ~ **money in** Geld *n* hineinstecken in
**investigate, to** ~ erforschen, ergründen, forschen, nachforschen, prüfen, untersuchen, eine Untersuchung anstellen **to** ~ **by feel** abtasten
**investigation** Arbeit *f*, Erforschung *f*, Ermittlung *f*, Forschung *f*, Nachforschung *f*, Prüfung *f*, Untersuchung *f* ~ **of the accuracy** Genauigkeitsuntersuchung *f* ~ **of foundation** Baugrundforschung *f* ~ **of foundation soil** Baugrunduntersuchung *f* ~ **of soil** Bodenuntersuchung *f* ~ **of weakest points (or spots)** in a structural member Schwachstellenprüfung *f* ~ **bureau (or station)** Untersuchungsstelle *f*
**investigator** Erforscher *m*, Forscher *m*
**investment** Anlage *f*, Anlagekosten *pl*, Kapitalanlage *f*, Zernierung *f* ~ **of capital** Geldanlage *f* ~ **casting** Feingießtechnik *f*, Präzisionsformguß *m*
**inveterate** eingewurzelt, hartnäckig
**invigorate, to** ~ stärken
**invigoration** Stärkung *f*
**inviolability** Unverletzlichkeit *f*
**inviscid**, ~ **fluid** reibungsfeste Flüssigkeit *f* ~ **gas** reibungsfreies Gas *n*
**invisible** unsichtbar ~ **(actinic) radiation beyond the violet** Dunkelstrahlung *f*
**invoice, to** ~ fakturieren
**invoice** Faktura *f*, Lieferschein *m*, Rechnung *f* ~ **clerk** Fakturist *m*
**invoicing** Fakturieren *n*
**involute** Abwicklungskurve *f* (geom), Evolvente *f*; evolventisch, verwickelt ~ **of a circle** Kreisvolvente *f*
**involute**, ~ **gear** Zahnrad *m* mit Evolventenverzahnung ~ **gearing** Evolventenverzahnung *f* ~ **helical gear** Evolventenschrägzahnrad *n* ~ **helicoid surface** schraubenförmige Evolventenfläche *f* ~ **tooth** Evolventenzahn *m* ~ **tooth gear (or system)** Evolventenverzahnung *f* ~ **worm** Evolventenschnecke *f*
**involution** Potenzierung *f*, Zurückbildung *f*
**involve, to** ~ hineinziehen, potenzieren, in sich schließen, umhüllen, verflechten
**involved** kompliziert, verwickelt
**invulnerability** Unverletzlichkeit *f*
**inwall** Innenwand *f*, Kernmauerwerk *n*
**inward** einwärts, innerlich, Innen ... ~ **board** Eingangsplätze *pl*, ankommende Plätze *f* **(direct)** ~ **dialling** Durchwahl *f* zur Nebenstelle ~ **flange** Einhalsung *f* ~ **-flow turbine** Turbine *f* mit äußerer Beaufschlagung ~ **position** Ankunftsplatz *m*, B-Platz *m*, Verbindungsplatz *m*
**iodate** Jodat *n*, jod-saures Salz *n*
**iodic**, ~ **acid** Jodsäure *f* ~ **anhydride** Jodsäureanhydrid *n*
**iodide** Jodid *n*, Jodsalz *n*
**iodiferous** jodhaltig
**iodine** Jod *n* ~ **chloride** Chlorjod *m*, Jodchlrid *n* ~ **-gold test** Jod-Gold-Test *m* ~ **intake** Jodaufnahme *f* ~ **isotope** Jodisotop *n* ~ **monochloride** Chlorjod *m*, Eichfachchlorid *n*, Jodchlorür *n* ~ **number** Jodzahl *f* ~ **pentoxide** Jodpentoxyd *n* ~ **-potassium iodine activator** Jod-Jodkalium-

Verstärker *m* ~ **-potassium thiocyanate** Rhodanjodkalium *n* ~ **test** Jodprobe *f* ~ **trichloride** Jodchlorid *n* ~ **value** Jodwert *m*
**iodized**, ~ **paper** Jodpapier *n* ~ **starch** Jodstärke *f*
**iodizing** (of a negative) Jodieren *n*
**iodobenzene** Jodbenzol *n*
**iodoform** Jodoform *n*
**iodometric** jodometrisch
**iodopropionic acid** Jodpropionsäure *f*
**iodyrite** Jodsilber *n*
**iolite** Dichroit *m*
**ion** Ion *n*, positiver Träger *m* **having a common** ~ gleichionisch
**ion**, ~ **acceleration** Ionenbeschleunigung *f* ~ **accelerator** Ionenbeschleuniger *m* ~ **accepter** Ionenakzepter *m* ~ **avalanche** Trägerlawine *f* ~ **beam engine** Ionenstrahltriebwerk *n* ~ **burn** Ionenfleck *m* ~ **cluster** Ionenanhäufung *f* ~ **core** Ionenrumpf *m* ~ **counter** Ionenaspirationsapparat *m*, Ionisationsmesser *m* ~ **-counting** Ionenzählung *f* ~ **cross effect** Nullpunktfehler *m* ~ **current** Ionen(gitter)strom *m* ~ **density** Ionenkonzentration *f* ~ **drift velocity** Ionenwanderungsgeschwindigkeit *f* ~ **exchange** Ionenaustausch *m* ~ **gauge** Ionisationsmanometer *n* ~ **getter pump** Ionengetterpumpe *f* ~ **gun** Ionenquelle *f* ~ **-ion-recombination** Ion-Ion-Rekombination *f* ~ **meter** Ionisationsdosimeter *n* ~ **movement** Ionenbewegung *f* ~ **-pair yield** Ionenpaarausbeute *f* ~ **producing arc** Ionen-erzeugender Lichtbogen *m* ~ **receiver** Lauschempfänger *m* ~ **retarding field** Ionenbremsfeld *n* ~ **rocket** Ionenrakete *f* ~ **sheath** Ionenschicht *f* ~ **spot** Ionenfleck *m* (CRT) ~ **spreading** (i.e. the expanding Hofmeister series effect) Ionenspreizung *f* ~ **trajectory** Bahn *f* eines Ions ~ **trap** Ionen-fänger *m*, -falle *f*
**ionic** ionisch ~ **atmosphere** Ionenwolke *f* ~ **bond** Ionenbindung *f* ~ **cleavage** Ionenspaltung *f* ~ **concentration** Ionenkonzentration *f* ~ **current** Elektronenstrom *m* ~ **density** Ionendichte *f* ~ **dissociation** Dissoziation *f*, elektrolytische Zersetzung *f* ~ **equation** Ionengleichung *f* ~ **focusing** Gas-konzentration *f* (CRT), -konzentrierung *f* ~ **-heated cathode** Selbstaufheizkathode *f* ~ **increment** Ionenzuwachs *m* ~ **interference** Ionisationsstörung *f* ~ **medication** Ionentherapie *f* ~ **mobility** Ionenbeweglichkeit *f* ~ **quantimeter** Ionendosismesser *m* ~ **reaction** Ionenreaktion *f* ~ **state** Ionenzustand *m* ~ **strength** Ionenstärke *f* ~ **theory** Ionentheorie *f* ~ **tube** weiche Röhre *f* ~ **valve** Elektronen-röhre *f*, -ventil *n*, Ionenventil *n*, weiche Röhre *f* ~ **wind voltmeter** Ionenwindvoltmesser *f*
**ionical relay** Glimmrelais *n*
**ionicity** Anregungswahrscheinlichkeit *f*
**ionium** Ionium *n*
**ionizable** ionisierbar
**ionization** Ionenspaltung *f*, Ionisation *f*, Ionisierung *f* ~ **by collision (or impact)** Stoßionisation *f*
**ionization**, ~ **balance** Ionisationsgleichgewicht *n* ~ **chamber** Ionisationskammer *f* ~ **constant** Ionisationsgrad *n* ~ **cross section** Ionisierungsquerschnitt *m* ~ **current** Ionisationsstrom *m* ~ **dose meter** Ionisationsdosimeter *n* ~ **efficiency** Ionisierungsausbeute *f* ~ **gauge** Ionisationsmanometer *n*, Vakuummeter *n* ~ **potential**

Entstehungspotential *n*, Ionisations-, Ionisierungs-spannung *f* ~ **pulse** Impuls *m* durch Ionenbildung ~ **rate** Ionisationsgeschwindigkeit *f* ~ **state** Ionisationszustand *m* ~ **time** Aufbauzeit *f* ~ **track** Ionisierungsbahn *f* ~ **unit** Ionisierungsanordnung *f* ~ **voltage** Ionisations-, Ionisierungs-spannung *f*

**ionizator** Ionisator *m*

**ionize, to** ~ ionisieren

**ionized** gezündet, ionisiert ~ **gas detector** Glimmdetektor *m* ~ **impurity centers** geladene Störstellenreste *pl* ~ **layer** Ionisationsschicht *f*, ionisierte Schicht *f* ~ **spark gap** ionisierter Funkenzieher *m*

**ionizer** Ionisierungsmittel *n*

**ionizing,** ~ **electrode** Ionisierungselektrode *f* ~ **event** ionisierendes Ereignis *n* ~ **power** Ionisierungsvermögen *n*

**ionogenic** ionenbildend

**ionometer** Ionenstärkemesser *m*

**ionosphere** Ionosphäre *f*, Kennelly-Heaviside-Schicht *f*

**ionospheric,** ~ **cross-modulation** ionosphärische Kreuzmodulation *f* ~ **disturbance** ionosphärische Störung *f* ~ **scatter** ionosphärische Streuausbreitung *f* ~ **sounding** Ionosphärengeräusch *n* ~ **storm** ionosphärischer Sturm *m*

**iontophoresis** Iontophorese *f*

**iridesce, to** ~ irisieren

**iridescence** Farbenschiller *m*, Farbwechsel *m*, Irisation *f*, Irisieren *n*, Schillern *n*, Spektrum *n*

**iridescent** irisierend, regenbogenfarbig, schillernd, wechselfarbig ~ **with metallic color (or luster)** display schillerfarbig ~ **color** Schillerfarbe *f* ~ **printing** Irisdruck *m*

**iridium** Iridium *n*

**iridize, to** ~ irisieren

**iridizing** Changierung *f*

**irinite** Irinit *n*

**iris** Iris *f*, Regenbogenfarben *pl*, Regenbogenhaut *f* ~ **diaphragm** Iris-, Kreis-blende *f* ~ **ground** Irisfond *m* ~**-in** Abblenden *n* ~**-out** Aufblenden *n*

**irization** Spektrum *n*

**iron, to** ~ aufbügeln, bügeln, plätten **to** ~ **out** drosseln

**iron** Eisen *n*, Eisenpulver *n*, Flacheisen *n*, Gußschrott *m*; aus Eisen bestehend, eisern **(of)** ~ eisern **No. 2** ~ Paketierschweißstahl *m*, Paketstahl *m* **low-grade malleable** ~ **of easily drillable character** Bohrguß *m* ~ **of granular fracture** Korneisen *n* ~ **and lattice masts** Eisen- und Gittermaste *pl* ~ **of the pounder** Pocheisen *n* ~ **in the ship** Schiffseisen *n* ~ **and steel hot-rolling mills** Warmwalzwerke *pl* für Eisen und Stahl ~ **and steel industry** Hüttenindustrie *f* ~ **in voltage circuit** Spannungseisen *n* ~ **for wire** Drahteisen *n*

**iron,** ~ **algae** Eisenalgen *pl* ~ **analysis** Eisenuntersuchung *f* ~**-arc spark** Eisenlichtfunken *m* ~ **associates** Eisenbegleiter *m* ~ **bacteria** Eisenbakterien *pl* ~ **ballast tube** Eisenwasserstoffstromregulatorröhre *f* ~ **bark** Eisenholz *n* ~ **bear** Eisensau *f* (metal) ~**-bearing** eisenführend ~ **black** Eisenschwarz *n* ~ **block** Eisenklumpen *m*, Sau *f* ~ **bloom** Eisenluppe *f* ~ **boride** Eisenbor *n* ~ **borings** Eisenspäne *pl* ~ **box** Eisenmuffe *f* ~ **bracket** Eisenkonsole *f* ~

**bromide** Bromeisen *n* ~ **bronze** Eisenbrühe *f* ~ **buff shade** rostfarbige Nuance *f* ~ **cable** Eisendrahtseil *n* ~ **cancer** faulende Rostanfressung *f* ~ **cap** Eisenhaube *f* ~ **carbide** Kohleneisen *n*, Kohlenstoffeisen *n* ~**-carbon alloys** Eisenkohlenstofflegierungen *pl* ~ **carbonate** Eisenkarbonat *n*, kohlensau(e)res Eisen *n* ~ **carboxide** Eisenkohlenoxyd *n* ~ **case** Eisengehäuse *n* ~ **cast directly from the blast furnace** Hochofenguß *m* ~ **casting** Gußeisen *n* ~ **castings** Eisenguß *m* ~ **cell** Eisenelement *n* ~ **cement** Eisenkitt *m* ~ **chill** Kokille *f*, Kokilleneinlage *f*, Kühleisen *n* ~ **chips** Eisendrehspäne *pl*, Eisenspäne *pl* ~ **chloride** Chloreisen *n*, Eisendichlorid *n*, salzsaures Eisenoxyd *n* ~**-chrome black** Eisenchromschwarz *n* ~ **circuit** Eisen-kreis *m*, -weg *m*

**iron-clad** eisen-bewehrt, -umgeben ~ **dynamo** Panzerdynamo *m* ~ **electric motor** eingekapselter Elektromotor *m* ~ **magnet** Topfmagnet *m* ~ **magnetic lens** magnetische Linse *f* mit Eisenpanzer ~ **plate** Panzerplatte *f* ~ **spool** Topfkernspule *f* ~ **transformer** Mantel-, Panzertransformator *m*

**iron,** ~**-cladding** Panzerung *f* ~ **cladding** Panzer *m* ~ **clay** Eisenton *m* ~ **color** Eisenfarbe *f* ~**-colored** eisenfarbig ~ **column** Eisen-säule *f*, -ständer *m* ~ **companions** Eisenbegleiter *m* ~ **composition** Eisenlegierung *f* ~ **compound** Eisenverbindung *f* ~ **construction** Eisenbau *m* ~ **content** Eisengehalt *m* ~ **copper sulfide** Kupferkies *m*, -pyrit *m* ~ **core** Eisenkern *m* ~**-core(d) coil** Eisenkernspule *f* ~**-core deflecting yoke** gepanzerte, wenig streuende Spule *f* ~**-core deflecting yoke unit** wenig streuende gepanzerte Spule *f* ~**-core deflector unit** magnetische Linse *f* mit Eisenpanzer ~**-core inductance** Eisenkerninduktivität *f* ~**-core transformer** Eisentransformator *m*

**iron-cored** eisengeschlossen ~ **choke coil (or inductance)** Eisendrossel *f* ~ **coil** Eisen-, Massekern-, Neosidkern-spule *f* ~ **frame** Eisenrahmen *m* ~ **goniometer** Eisengoniometer *m*

**iron,** ~ **corner cramp** Eckschiene *f* ~ **cross section** Eisenquerschnitt *m* ~ **crucible** Eisentiegel *m* ~ **cup (with handle)** Eisenschale *f* (mit Stiel) ~ **cyanide** Zyaneisen *n* ~**-cyanogen compound** Eisenzyanverbindung *f*, Zyaneisenverbindung *f* ~**-cyanogen pigment** Eisenzyanfarbe *f* ~ **cylinder** Eisenflasche *f* ~ **deposit** Eisenniederschlag *m* ~ **dish (with handle)** Eisenschale *f* ~ **disulfide** Eisendisulfid, Eisenkies *m*, Zweifachschwefeleisen *n* ~ **drill** Bohreisen *n* ~ **driver** Eisenvollsetze *f* ~ **dross** Blachmahl *n*, Sinter *m* ~ **dust** Eisen-pulver *n*, -staub *m* ~**-dust coil** Eisenpulverkernspule *f* ~**-dust core** Eisenstaubkern *m* ~**-dust-core coil** Eisenstaubkern-, Massekern-, Staubkern-spule *f* ~**-encased** eisengekapselt ~**-filament ballast lamp** Eisenwasserstoffstromregulatorröhre *f*, Eisenwasserstoffstromregulatorröhre *f*, Lampe *f* beruhend auf Eisenwiderstand ~ **filings** Eisenfeilspäne *pl*, Eisenspäne *pl* ~ **flask** Eisenflasche *f* ~ **flattener** Eisenquetscher *m* ~ **flowers** Eisenblumen *pl* ~ **founder** Gießereimann *m* ~ **founding** Eisenschmelze *f* ~ **foundry** Eisengießerei *f*, Eisenschmelzhütte *f*, Stückofen *m* ~ **frame** Eisengerippe *n*, -rahmen *m* ~**-frame plane** Gestellhobel *m* ~ **framework** Eisenfachwerk *n* ~ **free coils** eisenfreie Spulen *pl* ~ **garnet** Eisengranat *m* ~ **gauze** Eisen-drahtnetz *n*, -gaze *f* ~ **girder** Eisenträger *m* ~ **glaze** Eisenglasur *f* ~-**gray** eisengrau ~ **guide** Führungseisen *n* ~ **hame** Kumteisen *n* ~-**handed** plump, schwerhändig ~ **handle** Trageisen *n* ~ **hook** (for drawing charcoal) Störhaken *m* ~ **hoop** Eisenband *n*, Schiene *f* ~-**hooped drum** Rollreifenfaß *n* ~-**hydrogen resistance** Eisenurdox-, Eisenwasserstoff-widerstand *m* ~ **hydroxide** Eisenhydroxyd *n* ~ **ink** Eisentinte *f* ~ **iodate** Eisenjodat *n* ~ **iodite** Eisendijodid *n* ~ **jack** eiserner Kamm *m* ~ **jacket** Eisenmantel *m* ~-**jacketed** mit einem Eisenmantel *m* versehen ~ **ladle** (with handle) Eisenschale *f* ~ **lamina** Eisenblech *n* ~ **lamination** Eisenblatt *n* ~-**lead matte** Bleieisenstein *m* ~ **like** eisen-ähnlich, -artig ~ **lining** Eisenfassung *f* ~ **liquid** Eisenflüssigkeit *f* ~ **liquor** Eisen-beize *f*, -brühe *f*, -grund *m*, -rostwasser *n*, Schwarzbeize *f* ~ **lode** Eisengang *m* ~-**logwood black** Eisenblauholzschwarz *n* ~ **loop** Eisenluppe *f* ~ **loss** Kernverlust *m* ~ **losses** Eisenverluste *pl* ~ **mandrel** Eisendorn *m* ~ **mass** Eisenkern *m* ~-**master** Hochofenbesitzer *m* ~ **material for telephone lines** Eisenbauzeug *n* für Telefonleitungen ~-**melting furnace** Gießereiofen *m* ~ **mica** Eisenglimmer *m* ~ (-ore) **mine** Eisengrube *f* ~ **mold** Eisenerde *f*, Kokille *f* ~ **mordant** Eisenbeize *f*, Eisenbrühe *f* ~ **mounting** Eisen-beschlag *m*, -fassung *f* ~-**nickel accumulator** Edison-, Eisennickel-sammler *m* ~-**nickel storage battery** Eisennickelakkumulator *m* ~ **nipper** Kabelarschäkel. *m* ~ **nitrate** salpetersaures Eisen(oxyd) *n* ~ **notch** Stich-auge *n*, -loch *n* ~ **ocher** Berggelb *n*, Eisenocker *m* ~ **ore** Eisen-erz *n*, -stein *m* ~-**ore calciner** Eisensteinröster *m* ~-**ore deposit** Eisenerzablagerung *f* ~ **oxide** Eisenoxyd *n*, Sauereisen *n* ~ **oxide layers** Eisenoxidschichten *pl* ~ **oxide paint** Eisenoxidfarbe *f* ~ **part** Eisenteil *m* ~ **particles** Sumpf *m* ~ **path** Eisenweg *m* ~ **perchloride** Eisenchlorid *n* ~ **pet** Absteck-pfahl *m*, -pflock *m* ~ **phosphide** Eisenphosphor *n*, Phosphoreisen *n* ~ **pig** Eisengans *f*, Eisenmassel *f*, Roheisengans *f* ~ **pin** Schließnagel *m* ~ **pipe** Eisen-rohr *n*, -röhre *f* ~-**pipe conduit** Eisenrohrstrang *m* ~ **plate** Eisenblech *n* ~-**plate box** Eisenblechmuffe *f* ~ **pole** Eisenmast *m* ~ **pot** Grapen *m* ~-**powder coil** Eisenpulverkernspule *f* ~-**powder core** Eisenmassekern *m* (teleph), Eisenpulverkern *m* ~ **precipitation** Eisenfällung *f* ~ **puddling** Eisenfrischerei *f* ~ **pyrite** Eisendisulfid *n* ~ **pyrites** Doppelschwefeleisen *m*, Eisenkies *m* ~ **rail** Schiene *f* ~ **refinery** Eisenfrischerei *f* ~-**refinery slag** Eisenfeinschlacke *f* ~ **refining** Eisenfrischerei *f* ~ **reinforcement** Eisen-armierung *f*, -einlage *f*, -gewebe *n* ~ **requirement** Eisenbedarf *m* ~ **ring** Eisenring *m* ~ **rod** Krücke *f*, Rundeisen *n* ~ **rod for stuffing collars** Polsterstange *f* ~ **roller** Eisenwalze *f* ~-**rolling mill** Eisenwalzwerk *n* ~ **roofing and siding** Eisentäfelung *f* ~ **rubber** Eisengummi *m* ~ **runner** Roheisenrinne *f* ~ **rust** Eisenrost *m* ~ **rust cement** Rostkitt *m* ~ **salt** Eisensalz *n* ~ **scale** Eisenhammerschlag *m* ~ **scrap** Brucheisen *n*, Eisenabfall *m* ~ **screen** Eisenpanzer *m* ~ **screw** Metallschraube *f* ~

**separation** Eisenabscheidung *f* ~ **series** Eisenreihe *f* ~ **set hammer** Eisenvollsetze *f* ~-**sheathed** eisenbewehrt ~ **shell** Eisenblechmantel *m* ~ **shield** Eisen-panzer *m*, -panzerung *f* ~-**shielded** eisengekapselt ~ **shoe of the stamper** Pocheisen *n* ~ **shot** Gußschrott *m* ~ **shutter** eiserner Schutzladen *m* ~ **slab** Bramme *f* ~ **slag** Eisenschlacke *f* ~ **sludge with clay content** tonhaltiger Eisenschlamm *m* ~ **smeltery** Eisenschmelze *f* ~ **smelting** Eisenschmelze *f*, Eisenverhüttung *f* ~ **sow** Eisensau *f* (metal) ~ **spot** Eisenfleck *m* ~-**spotted** eisenfleckig ~ **stain** Eisen-, Rost--fleck *m* ~-**stain remover** Rostfleckenwasser *n* ~-**stained** eisenfleckig ~ **standard** Eisen-säule *f*, -ständer *m* ~ **staple** Eisenbügel *m* ~ **stone** Eisenstein *m* ~ **stone china** Hartsteingut *n* ~ **stone concretion** (unechter) Alaunstein *m* ~ **structure** Stellungsanlage *f* ~ **stud** Anschweißende *n* ~ **sublimate** Eisensublimat *n* ~ **sulfate** Bergbutter *f*, Eisensulfat *n* ~ **sulfide** Eisensulfid *n* ~ **suspension tie** Aufhängeeisen *n* ~ **tank** Eisengefäß *n* ~ **tannage** Eisengerbung *f* ~-**tape winding** Eisenbandumspinnung *f*, Krarupbandumspinnung *f* ~ **taping** Eisenbandumspinnung *f*, Eisenumspinnung *f* ~-**tempering material** Eisenhärtemittel *n* ~ **tie** Schlauder *f*, Stichanker *m*, Zuganker *m* ~ **tongued** mit (eingesetzter) Eisenfeder *f* ~ **torus** Eisenring *m* ~ **tower** Stahlmast *m* ~ **tray** Eisengefäß *n* ~ **troughing** Eisenrinne *f* ~ **tube (or tubing)** Eisen-rohr *n*, -röhre *f* ~ **tungstate** Eisenwolframat *n* ~ **turnings** Eisendrehspäne *pl*, Eisenspäne *pl* ~ **valverate** Eisenvalerianat *n* ~ **varnish** Eisen-glasur *f*, -lack *m* ~ **vein** Eisengang *m* ~ **ware manufacture** Grobschmiede *f* ~ **waste** Eisenabbrand *m* ~-**whipped** eisenumsponnen ~-**whipped core** Krarupader *f* ~ **whipping** Eisen-drahtumspinnung *f*, -umspinnung *f*, -wicklung *f*, Krarupumspinnung *f* ~ **winding** Eisen-umspinnung *f*, -wicklung *f* ~ **window bar** Fenster-, Wind-eisen *n* ~ **wire** Eisen-draht *m*, -garn *n* ~ **wire core coil** Eisendrahtkernspule *f* ~-**wire mesh** Eisendrahtgeflecht *n* ~-**wire whipping** Eisendraht-, Krarupdraht-umspinnung *f* ~ **wood** Hartholz *n* ~ **work** eisernes Bauzeug *n*, Eisenbeschlag *m* ~ **works** Eisen-hütte *f*, -werk *n*, Hütte *f*, Hüttenwerk *n* ~ **wrapping** Eisenumspinnung *f* ~ **yellow** Eisengelb *n* ~ **yoke** Eisenbügel *m*, Eisenjoch *n*

**ironer** Bügelmaschine *f*

**ironing,** ~ **board** Bügelbrett *n* ~ **machine** Bügelmaschine *f*

**ironless** eisenfrei, eisenlos

**irons** Fesseln *pl*

**irradiance** (in lux or candle-meter units) Beleuchtung *f*

**irradiate, to** ~ belichten, bestrahlen, durchstrahlen

**irradiated cathode** bestrahlte Kathode *f*

**irradiation** Bestrahlung *f*, Durchstrahlung *f*, Überstrahlung *f* ~ **by solar rays** Sonnenbestrahlung *f* ~ **method** Durchstrahlungsmethode *f*

**irratic trouble** unregelmäßig auftretende Störung *f*

**irrational** gegensinnig, widersinnig ~ **flow** drallfreie Strömung *f* ~ **numbers** irrationale Zahlen *pl*

**irrationality** Irrationalität *f*
**irreducible** nichtreduzierbar **~ constants** akzessorische Parameter *pl*
**irregular** ordnungswidrig, regellos, regelwidrig, sprunghaft, ungleichförmig, ungleichmäßig, unpünktlich, unregelmäßig **~ arrangement** regellose Anordnung *f* **~ fracture** irreguläre Bruchfläche *f* **~ joint lines** unregelmäßige Teilungsebene *f* **~ lode** unterbrochener oder unregelmäßiger Gang *m* **~ oscillating motion** (of a locomotive) Schlingern *n* **~ running** unregelmäßiger Gang *m* **~ section** Fantasieprofil *n* **~ vein** Flader *m* **~ working** Rohgang *m* **~ working of the blast furnace** Rohgang *m* des Hochofens
**irregularity** Ungleichförmigkeit *f*, Unregelmäßigkeit *f*, Unruhe *f*
**irrelevant** belanglos, unsachgemäß
**irreparable** unersetzlich, unheilbar
**irreplaceable** unersetzlich
**irreproachable** einwandfrei
**irresistible** unwiderstehlich
**irrespective** ohne Rücksicht auf **~ of** unabhängig von
**irrespirable** unatembar
**irretrievable** unersetzlich
**irreversibility** Irreversibilität *f*, Unwiderruflichkeit *f*
**irreversible** nichtumkehrbar, nicht umkehrbar, selbsthemmend, selbstsperrend, unverrückbar **~ controls** irreversible Steuerung *f* **~ cycle** nicht umkehrbarer Vorgang *m* **~ process** nicht umkehrbarer Prozeß *m* **~ rotation** nicht umkehrbarer Drehsinn *m* **~ steering** selbstsperrende Lenkung *f*
**irrigate, to** **~** begießen, berieseln, bewässern, wässern
**irrigated land** bewässertes Land *n*
**irrigating shovels** Bewässerungsschaufeln *pl*
**irrigation** Berieselung *f*, Bewässerung *f*, Wässerung *f* **~ channel (or ditch)** Bewässerungsgraben *m* **~ field** Rieselfeld *n* **~ lock** Bewässerungsschleuse *f* **~ water** Rieselwasser *n* **~ works** Beriesenungsanlage *f*
**irritability** Reizbarkeit *f*
**irritant** Reiz *m*, (gas) Reizstoff *m* **~ and toxic smoke candle** Reiz- und Giftrauchkerze *f* **~ agent** reizender Kampfstoff *m* **~ gas** ätzender Kampfstoff *m*, Reizgas *n* **~ gas candle** Giftnebelkerze *f* **~-gas generator** Riechtopf *m* **~ smoke** Giftrauch *m* **~ smoke candle** Reizkerze *f*
**irritate, to** **~** reizen
**irritation** Reiz *m* **~ of the eyes** Tränenreiz *m*
**irrotational** drehungsfrei, rotationsfrei, strudellos, wirbelfrei **~ field** wirbelfreies Feld *n* **~ flow** drallfreie, wirbelfreie, oder wirbellose Strömung *f* **~ motion** Potentialströmung *f*
**irrotationality** Wirbelfreiheit *f*
**irruption of carbon dioxide** Kohlensäureausbruch *m*
**IR-wire** Informationsesedraht *m*
**isacoustic** von gleicher Lautstärke *f* **~ line** Isakuste *f*
**ischyetal lines** Ischyeten *pl*
**I scope** I-Schirm *m*
**I S-diagram** I-S-Diagramm *n*
**I section** Doppel-T-Querschnitt *m*, I-Profil *n*

**isenthalpic** von der gleichen Enthalpie *f*
**isentropic** isentrop **~ compressibility** adiabatische Kompressibilität *f* **~ cooling coefficient** Kühlungskoeffizient *m* bei konstanter Entropie
**ishikawaite** Ishikawait *m*
**I-signal (in-phase signal)** I-Signal *n* (phasenrichtiges Buntsignal)
**isinglass** Fischleim *m*, Hausenblase *f*, Marienglas *n*
**island** Eiland *n*, Insel *f* **~ deposit** inselartige Ablagerung *f* **~ effect** Inselbildung *f* **~ formation** Inselbildung *f*
**isle** Eiland *n*, Insel *f*
**islet** Holm *m*
**Isley flues** Isley-Kanäle *pl*
**isobar** Böenlinie *f*, Gleichdrucklinie *f*, Isobare *f* (meteor.), Isobar *n* (phys), Linie *f* gleichen Luftdrucks
**isobaric, ~ chart** Isobaren-, Luftdruck-karte *f* **~ control** Gleichdruckregelung *f* **~ corrections** isobare Korrekturen *pl* **~ drop** Druckgefälle *n* (meteor.) **~ pressure** gleichbleibender Druck *m* **~ slope** Druckgefälle *n* **~ space** Spin-Konfigurationsraum *m* **~ spin** Isotopenspin *m* **~ spin quantum number** Quantenzahl *f*
**isobarometrical racking apparatus** isobarometrischer Faßfüllapparat *m*
**isobath contours** Isobaten *pl*
**isobrontic chart (or map)** Isobrontenkarte *f*
**isobutyl** Isobutyl *n*
**isobutyric ether** Äthyl-i-butyrat *n*
**isocandela chart** Kurve *f* gleicher Lichtstärke
**isocenter** Fokalpunkt *m*, Isozentrum *n*, Metapol *m*
**isochasm** Linie *f* gleicher Ultrastrahlenintensität
**isochoric** isochor
**isochromatic** farbrichtig, gleichfarbig, isochromatisch
**isochrone determination** isochrone Ortsbestimmung *f*
**isochrones** Linien *pl* gleicher Zeiten
**isochronism** Gleichdauer *f*, gleichschneller Gang *m*, Isochronismus *m*
**isochronous** gleich schnell, gleichzeitig, isochron(isch) **~ curve** Isochrone *f*, Tautochrone *f* **~ scanning** isochrone Abtastung *f*
**isoclinal** von gleicher Inklination *f* oder Neigung *f* **~ fold** Isoklinalfalte *f* **~ line** Isokline *f* **~ lines** Linien *pl* gleicher Spannung oder Neigung **~ valley** Scheide *f*
**isocline** Isoklinale *f*, Isoklinalfalte *f* (geol)
**isoclines** Linien *pl* gleicher Spannung
**isoclinic** isoklinisch, von gleicher Neigung *f* **~ lines** isoklinische Linien *pl*
**isodiaphere** Isodiapher *n*
**isodose** Isodose *f*; gleichmäßig **~ chart** Isodosentafel *f* **~ contour** Isodosenkurve *f* **~ surface** Isodosenoberfläche *f*
**isodrome** Isodrom *n*
**isodynamia** isodynamischer Zustand *m*
**isodynamic** gleichkräftig **~ line** Isdyname *f* **~ lines** Linien *pl* mit gleicher Kraftwirkung
**isodynamics** Isodynamen *pl*
**isodynamostacy** isodynamischer Zustand *m*
**iso-electronic** isoelektronisch
**iso-energetic** isoenergisch
**isogon** Isogon *n*, Linie *f* gleicher Mißweisung

**isogonal** gleichwinklig
**isogonic** isogonisch ~ **chart** Isogonenkarte *f*, isogonische Karte *f* ~ **line** Isogone *f*, Linie *f* gleicher Mißweisung, Mißweisungsgleiche *f* ~ **lines** isogonische (gleichwinklige) Linien *pl*, Linien *pl* gleicher Winkel ~ **map** Isogonenkarte *f*
**isogram** Isogramm *m*
**isogrivs** Isogriven *pl* (Linien gleicher Gitternetzabweichung)
**isohydric** isohydrisch
**iso-intensity curve** Iso-Intensitätskurve *f*
**isokurtosis** Symmetrie *f* (der Häufigkeitskurve)
**isolate, to** ~ abschalten, absondern, herausziehen (data proc.), isolieren, scheiden, sperren, trennen **to** ~ **a line** eine Leitung isolieren
**isolated** einzeln, gesondert, isoliert, vereinzelt **to be** ~ freistehen ~ **basin** abflußlose Wanne *f* ~ **danger signal** Riffzeichen *n* ~ **mass** schwebendes Mittel ~ **pillar crane** freistehender Säulendrehkran *m*
**isolating,** ~ **links** Trennlaschen *pl* ~ **motion** Trennschub *m* ~ **pipe** Isolierröhre *f* ~ **resistor** Entkopplungs-, Trenn-widerstand *m* ~ **switch** Trennschalter *m* ~ **tube** Trennröhre *f* ~ **valve** Absperrventil *n*, Isolierröhre *f*
**isolation** Abdämmung *f*, Absonderung *f*, Dämmung *f*, Isolation *f*, Isolierung *f*, Schutz *m*, Sperrung *f*, Trennung *f* ~ **amplifier** Trennverstärker *m* ~ **procedure** Störungssuche *f* ~ **valve** Schleusenventil *n*
**isolator** Isolationsmittel *n*, Isolator *m*, Trenner *m*, Trennschalter *m* ~ **stage** Pufferstufe *f* ~ **valve** Absperrventil *n*
**isolux,** ~ **curve** Isoluxkurve *f* (Beleuchtungsgleiche) ~ **lines** Linien *pl* gleicher Beleuchtungsstärke
**isomer** Isomer *n*
**isometric** isomer, isomerisch ~ **acids** Isosäuren *pl*
**isomerism** Isomerie *f*
**isomerous** gleichteilig
**isometric** isometrisch (min.), maßgleich (math.), regulär
**isomorphism** Isomorphie *f*, Isomorphismus *m*
**isomorphous** gleichgestaltet, homöomorph, isomorph
**isophons** Kurven *pl* gleicher Lautstärke
**isophot** Isoluxdiagramm *n*, Isophot *n* ~ **curve** Kurve *f* gleicher Beleuchtungsstärke
**isophotic lines** Linien *pl* gleicher Beleuchtungsstärke
**isopleth** Band *n*
**isoporic focus** Isoporenfokus *m*
**isopotential,** ~ **cathode** indirekt geheizte Kathode *f* ~ **glide path** äquipotentieller Gleitweg *m* ~ **plane** Äquipotentialfläche *f* ~ **surface** Äquipotentialfläche *f*, Fläche *f* konstanten Potentials, konstante Potentialfläche *f*
**isoprene caoutchouc** Isoprenkautschuk *m*
**isopropyl,** ~ **alcohol** Isopropylalkohol *m* ~ **ether blend** Isopropyläthergemisch *n*
**isosceles** gleicheckig ~ **triangle** gleichschenkeliges Dreieck *n*
**isospic spin selection** Isospin-Auswahlregel *f*
**isostatic** isostatisch ~ **regulation** isodrome Regulierung *f* ~ **surface** Ausgleichfläche *f*
**isostatics** Isostatik *f*
**isosteres** isostere Verbindungen *pl*

**isostructural** isomorph
**isotenoscope** Isotenoskop *n*
**isotherm** Isotherme *f*, Wärmegleiche *f*
**isothermal** gleichtemperiert, gleichwarm, isotherm, isothermal, isothermisch ~ **annealing** Perlitisieren *n* ~ **change of state** (condition) isothermische Zustandsänderung *f* ~ **compression** isothermische Verdichtung *f* ~ **curve** Wärmegleiche *f* ~ **layer** isotherme Schicht *f* ~ ~ **lines** Isothermen *pl* ~ **region** atmosphärische Schicht *f* ~ **surface** Wärmegradfläche *f*
**isotone** Isoton *n*
**isotonic** isotonisch
**isotope** Isotop *n* ~ **and absorber slides** Präparat- und Filterschieber *m* ~ **container** Isotopengefäß *n* ~ **dating** Altersbestimmung *f* mit Hilfe von Isotopen ~ **dilution analysis** Isotopenverdünnungsanalyse *f* ~ **distribution** Isotopenverteilung *f* ~ **effect** Isotopeneffekt *m* ~ **shift** Isotopieverschiebung *f*
**isotopic** isotopisch ~ **abundance** relative Häufigkeit *f* ~ **abundance measurement** Isotopenhäufigkeitsmessung *f* ~ **composition** Isotopenverhältnis *n* ~ **dating** Datierung *f* mit Hilfe von Isotopen ~ **number** Neutronenüberschuß *m* ~ **spin quantum number** Isotopenspin-Quantenzahl *f*
**isotron** Isotron *n*
**isotropic** isotropisch ~ **antenna** isotroper Strahler *m* ~ **radiator** Kugelstrahler *m* ~ **turbulence** isotrope Turbulenz *f*
**isotropy** Isotropie *f*
**issuance** Ausgabe *f* ~ **of material** Materialausgabe *f*
**issue, to** ~ abziehen, ausfertigen, ausfließen, ausgeben, auslaufen, aussenden, ausströmen, austreten, entspringen, erteilen **to** ~ **from** ausgehen, entströmen **to** ~ **a decree** ein Urteil *n* fällen **to** ~ **orders** erlassen
**issue** (gas) Ableitung *f*, Abziehen *n*, Ausgabe *f*, Ausgang *m*, Ausstattung *f*, Ausströmung *f*, Austritt *m*, Emission *f*, (of a periodical) Heft *n*, Lieferung *f*, Nummer *f*, Streitfrage *f*, Streitpunkt *m* ~ **of bank notes** Notenausgabe *f* ~ **of equipment** Geräteanschlußkabelausgabe *f* ~ **of letters patent** Patenterteilung *f*
**issued** ausgestellt
**issuer of a license** Lizenzerteiler *m*
**issues, in** ~ lieferungsweise
**isthmus** Landenge *f*
**I strut** I-Stiel *m*
**itabirite** Itabirit *m*
**itabitite** Eisenglimmerschiefer *m*
**itacolumite** Gelenkquarz *m*
**italic** Kursiv-, Schräg-schrift *f*; kursiv
**italics, in** ~ in Schrägdruck *m*
**item** Gegenstand *m*, Posten *m* ~ **of apparatus** Apparatteil *m* ~ **of business** Beratungsgegenstand *m* ~ **of equipment** Gerätstück *n* ~ **counter** Postenzähler *m* ~ **No.** Positionsnummer *f* ~ **number** Sachnummer *f*
**itemize, to** ~ aufführen, einzeln aufführen, aufstellen, aufzählen, spezifizieren
**items covered by the contract** Lieferungsumfang *m*
**iterated** iteriert ~ **fission** Mehrfachspaltung *f* ~ **fission probability** asymptotische Spalterwartung *f* ~ **limit** sukzessiver Grenzübergang *m*

**iteration** wiederholtes Einsetzen *n*, Iterieren *n* (math.)

**iterative,** ~ **attenuation per section** Kettendämpfung *f* je Glied ~ **attenuation constant** Kenndämpfung *f* ~-**attenuation factor** Kettendämpfungsfaktor *m* ~ **frequency** Wiederholungsfrequenz *f* ~ **impedance** Kenndämpfungswiderstand *m*, Kettenwiderstand *m* ~ **instruction** Wiederholungsbefehl *m* ~ **network** Kettenleiter *m* ~ **phase constant** Kennwinkelmaß *n* ~-**phase constant** Kettenwinkelmaß *n* ~-**phase factor** Kettenphasenfaktor *m* ~-**propagation constant** Kettenübertragungsmaß *n* ~ **propagation factor** Übertragungsfaktor *m* eines Vier-

pols ~ **transfer coefficient** Koeffizient *m* für Kettenübertragung *f*

**itinerant** reisend ~ **traffic** Durchgangsverkehr *m*

**itinerary** Reiseweg *m* ~ **lever** Fahrstraßensignalschalter *m*

**ivory** Elfenbein *n*; elfenbeinern ~ **black** Spodium *n* ~ **cardboard** Elfenbeinkarton *m* ~ **nut** Steinnuß *f* ~ **tower** Elfenbeinturm *m*

**IW-wire** Informations-Schreibdraht *m*

**Izod,** ~ **impact-figures scale** Izod-Kerbzähigkeitsskala *f* ~ **test** Charpy-Schlagversuch *m*, Kerbschlagprüfung *f* ~ **value** Kerbschlagwert *m* nach dem Izod-Verfahren

# J

**jab, to** ~ picken, stechen
**jack, to** ~ schroppen **to ~ an aircraft** Flugzeug *n*
aufbocken **to ~ down** Holz *n* abhobeln **to ~
up** aufbocken, aufwinden, in die Höhe heben,
hochbocken, hochheben, hochwinden
**jack** Anhebevorrichtung *f*, Arbeitszylinder *m*,
Auslösungsknopf *m*, Bock *m*, Buchse *f*, Bug-
flagge *f*, Flaschenzug *m*, Fuchs *m*, Gestell *n*,
Gösch *m* (naut), Handwinde *f*, Hebe-baum *m*,
-bock *m*, -vorrichtung *f*, Hebel *m*, Hebelade *f*,
Heber *m*, Klinke *f*, Niederhalter *m* (Blechbiegemaschine), Pflock *m*, Steckdose *f*, (lamp)
Steckfassung *f*, Wagenwinde *f*, Winde *f*, (socket)
Zentralbüchse *f* ~ **(switch)** Klinkenstecker *m*
~ **for busy tone** Besetztklinke *f* ~ **and circle**
Winde *f* auf ringförmiger Zahnstange ~ **of a
repeater watch** Hämmerchen *n* einer Repetier-
uhr ~ **with two hooks** Doppelklauenwinde *f*
**jack,** ~ **adapter** Heberansatzstück *n* ~ **arrangement** Klinkenanordnung *f* ~ **barrel** Klinkenhülse *f* ~ **body** Klinkenkörper *m* ~ **box** Radio-
stöpselkasten *m* ~ **bush(ing)** Klinkenbuchse *f*
~ **button** Auslöserknopf *m* ~ **cam adjustment**
Einstellung *f* des Platinenexzenters ~ **car** Ein-
setzwagen *m* ~ **claw** Hebebockklaue *f* ~ **engine**
Hilfsmaschine *f* ~ **equipment** Klinkenausrü-
stung *f* ~ **field** Klinkenfeld *n* ~ **finder** Klinken-
sucher *m* ~ **float** Fußteller *m* ~ **frame** Spindel-
bank *f* ~ **hammer for wracking** Aufreißhammer
*m* ~ **head pit** blinder Schacht *m* ~ **holder**
Wagenheberhalter *m* ~ **key switch** Klinke *f*
~-**knife** Klappmesser *n* ~ **lamp** Stecklampe *f*
~ **latch** Federriegel *m* ~ **leg** Überlaufleitung *f*
~ **lift** Hubkarren *m* ~ **line** Büchsenleitung *f* ~
**multiple** Klinkenvielfachfeld *n* ~-**o'-lantern**
Irrlicht *n* ~ **panel** Klinken-brett *n*, -feld *n* ~
**plane** Schropp-, Rauh-hobel *m* ~ **planing**
Schrupphobeln *n* ~ **plug** Klinkenstöpsel *m* ~
**point** Aufbockpunkt *m* ~ **post** Kurbelwellen-
bock *m* ~ **rafter** Schifter *m*, Schift-, Walm-
sparren *m* ~ **retaining spring** Schraubenfeder-
ring *m* zum Halten der Einschließplatinen ~
**screw** Hebeschraube *f*, Schraubspindel *f* ~
**screw (winch)** Schraubenwinde *f* ~ **screw elevating mechanism** Schraubenspindelrichtmaschine
*f* ~**shaft** Übersetzungswelle *f*, Vorgelegewelle
*f*, hintere Zwischenachse *f* ~ **socket** Klinkenkörper *m* ~ **spring** Klinkenfeder *f* ~ **strip**
Klinkenstreifen *m* ~ **switch** Knebelschalter *m*
~ **switch spring** Klinkenspannung *f* ~ **switchboard** Klinkenumschalter *m* ~-**valve cone** Hub-
ventilkegel *m* ~-**wheels** Stützräder *pl*
**jacket, to** ~ einhüllen, einwickeln, umhüllen,
umkleiden, ummanteln, verschalen, mit einer
Hülse versehen
**jacket** Buchbinde *f*, Doppelwand *f*, Gehäuse *n*,
Glocke *f*, Hülle *f*, Hüllenrohr *n*, Hüllröhre *f*,
Hülse *f*, Jacke *f*, Mantel *m*, Mantelkühler *m*,
Rohrjacke *f*, Schutzmantel *m*, Umhüllung *f*,
Umkleidung *f*, Ummantelung *f*, (of a book)
Umschlag eines Buches, Wärmeisolierung *f*
~ **of compass bowl** Kesselmantel *m*

**jacket,** ~-**body machine** Jacken-maschine *f*,
-stuhl *m* ~ **box** Mantelboden *m* ~ **cooling**
Mantelkühlung *f* ~ **cradle** Jackenwiege *f* ~
**heating** Mantelheizung *f* ~ **rim** Mantelzarge *f*
~ **ring** Mantelring *m* ~ **tread band** Laufmantel
*m* ~-**type electrical heating zones** Beheizungs-
zonen *pl* durch Heizmanschetten ~-**ventilation**
Mantelkühlung *f* ~ **water** Zylinderwasser *n*
**jacketed,** ~ **barrel** Lauf *m* mit Oberrohr, Mantelrohr *n* ~ **bottom** Dampfboden *m* ~ **cylinder**
Zylinder *m* mit Dampfmantel ~ **gun** Mantel-
kanone *f* ~ **pipe** Heizmantelrohr *n* ~ **shelf drier**
Heizplattentrockner *m* ~ **vessel** Doppelkessel *m*
~ **wall** Doppelwandung *f*
**jacketing** Bekleidung *f*, Verkleidung, Verschalung *f*
**jacking** Verkleidung *f* ~ **the track** Gleisheben *n*
~ **up** Hochheben *n*
**jacking,** ~ **beam** Aufbockauflage *f* ~ **diagram**
Aufbockdiagramm *n* ~ **equipment** Aufwinde-
vorrichtung *f* ~ **pad** Aufbockansatz *m*
**jacks** Klinkenfeld *n*
**Jacob's ladder** Jakobs-, Strick-leiter *f*
**Jacquard,** ~ **card** Webekartenpappe *f* ~ **loom**
Webstuhl *m*
**jactation** ruckartige Bewegung *f*
**jade** Beilstein *m*, Jade *f*, Jadeit *m* (min), Neprit
*m*
**jag, to** ~ ausblatten, auszacken
**jag** Anschnitt *m*, Auszackung *f*, Kappung *f*,
Kerb *m*, Kerbe *f*, Zacke *f*, Zacken *m*, Zinken *m*
**Jäger,** ~ **hurdle** Jägerhorde *f* ~ **hurdle-type
washer** Jägerhordenwäscher *m* ~ **tube** Jäger-
röhrchen *n*
**jagged** ausgezackt, gezackt, kerbig, schartig,
zackig ~ **line** ausgezackte oder ausgezahnte
Linie *f*, (aus)gehackte oder gezackte Linie
**jagging board** Glauch-, Kehr-herd *m*
**jam, to** ~ bekneifen, blockieren (rdo), festfressen, -keilen, -klemmen, (tightly) sich festklemmen, kneifen, pressen, (sender) stören,
versperren, verstopfen **to ~ brakes** mit voller
Kraft bremsen **to ~ tight** sich festlaufen
**jam** Hemmung *f*, (firearms) Ladehemmung *f*,
Ladestörung *f* ~ **of the carriage** Schlitten-
hemmung *f* ~ **nut** Gegenmutter *f*, gesicherte
Mutter *f* ~ **pot cover** (airship) Ventilschutzhaube *f*
**jamb, to** ~ ecken
**jamb** Anschlagfläche *f*, Einfallung *f*, Gewände *n*,
Pfosten *m* ~ **lining of a door frame** Türfutter *n*
~**stone** Eckpfeiler *m*
**jamesonite** Querspießglanz *m*
**jammed, to be** ~ **tight** festsitzen
**jammer** Störsender *m*
**jamming** Fest-fressen *n*, -klemmen *n*; Klemmung
*f*, (of firearms) Ladehemmung *f*, Stör-einsatz
*m*, -geräusch *n*, -träger *m*, Störung *f*, Störung *f*
durch andere Sender ~ **of projectile in the bore**
Geschoßverkeilung *f* ~**station** Störer *m*, Störsender *m* ~ **transmitter** Störer *m* ~**wave** Stö-
rungswelle *f*

**jangly** klirr
**janitor** Hauswart *m*, Pförtner *m*
**japan** schwarzer Lack *m*
**Japan**, ~ **lacquer (or varnish)** Japanlack *m* ~
**wax** Japanwachs *n*, Pflanzentalg *m*
**Japanese** japanisch ~ **paper** Japanpapier *n*,
Seidenpapier *n*
**japanned** lackiert, mit Japanlack *m* überzogen
**japanning** Emaillelackierung *f*, Lackieren *n*
**jar, to** ~ erschüttern, knarren, rütteln, schnarren
**jar** Abtropfschale *f*, Anstoß *m*, Flasche *f*, Gefäß
*n*, Knarren *n*, Kratzen *n*, (noise) Kratzer *m*,
Krug *m*, Mißton *m*, Topf *m* ~ **conveyer** Fla-
schentransporteinrichtung *f* ~ **diffusion** Gefäß-
diffusion *f* ~ **latch** Fanghund *m* ~ **piston** Rüt-
telkolben *m* ~-**ram molding machine** Rüttelformmaschine *f*, Rüttelmaschine *f*, Rüttler *m*
~-**rammed mold** gerüttelte Form *f*
**jar-ramming** Rütteln *n* ~ **and-squeezing turneyer**
**draw machine** Rüttelpreßformmaschine *f* mit
Wendeplatte ~ **method** Rüttelverfahren *n* ~
**power draw-machine** Rüttelformmaschine *f* mit
Druckluftabhebung ~**power turnover machine**
Wendeplattenrüttler *m* mit selbsttätiger Wendevorrichtung ~ **process** Rüttelverfahren *n* ~
**roll-over molding machine** Umrollrüttelformmaschine *f* ~ **stripping machine** Rüttelformmaschine *f* mit Abstreifvorrichtung ~ **turnover**
**footdraw machine** Rüttelabformmaschine *f*
mit Fußhebelmodellabhebung ~ **turnover**
**molding machine** Wendeplattenrüttler *m* ~
**turnover pattern-drawing machine** Rüttelformmaschine *f* mit Wende- und Abhebeeinrichtung oder mit Abhebevorrichtung ~ **turnover**
**plate machine** Rüttelwendeformmaschine *f* ~
**jar,** ~ **ring** Konservenring *m* ~ **socket** Rutsch-
scherenfänger *m* ~-**squeeze flask-lift machine**
Rüttelpreßformmaschine *f* mit Stiftenabhebung ~-**squeeze machine** Rüttelpreßformmaschine *f* ~-**strip molding machine** Rüttelformmaschine *f* mit Abstreifvorrichtung
**jarrah** Dscharrah-Holz *n*, australisches Mahagoni *n*
**jarred mold** gerüttelte Form *f*
**jarring** Erschütterung *f*, Rütteln *n*, Schlagen *n*
eines Werkzeuges ~ **blow** Prellschlag *m* ~
**capacity** Rüttelleistung *f* ~**flask-lifting machine**
Rüttelformmaschine *f* mit Stiftenabhebung ~
**machine with air lift** Rüttelformmaschine *f* mit
Druckluftabhebung ~ **machine with hand-lever**
**lift** Rüttelformmaschine *f* mit Handhebel-
modellabhebung ~ **molding machine** Rüttelformmaschine *f*, Rüttelmaschine *f*, Rüttler *m*
~ **operation** Rüttelvorgang *m* ~ **piston** Rüttel-
kolben *m* ~ **plate** Rütteltisch *m* ~ **roll-over**
**pattern-drawing machine** Rüttelformmaschine *f*
mit Umroll- und Modellabhebeeinrichtung ~
**roll-over pattern-drawing molding machine**
Rüttelformmaschine *f* mit Wende- und Abhebeeinrichtung, Rüttelwendeformmaschine *f* mit
Abhebevorrichtung ~ **table** Rütteltisch *m* ~
**valve** Rüttelventil *n*
**jars** Bohrschere *f*, Rutschschere *f*
**jasper** Jaspis *m*
**Jato (jet-assisted takeoff)** Strahlstarthilfe *f*
**javelin** Kolonne *f* (aviat) ~-**shaped fuel rod**
speerförmiger Spaltstoffstab *m*
**jaw** Aufsatzbacke *f*, (holding fixture) Backe *f*,

Backen *m*; (testing machine) Einspannklemme
*f*, Fang-haken *m*, -schere *f*, -vorrichtung *f*;
Kiefer *m*, Klaue *f*, Klemmbacke *f*, Maul *n*,
(calipers) Schnabel *m*, Schuh *m*, Versegelung *f*
~ **of grab** Greiferbacke *f* ~ **for rifle bench (or**
**for rifling machine)** Ziehbacke *f* ~ **for wire rail**
**machine** Backe *f* an Drahtstiftmaschinen
**jaw,** ~ **arrangement** Klauenanordnung *f* ~ **blade**
Meßschenkel *m* ~ **bladeholder** Kimmenhalter
*m* ~ **bone** Kiefer *m* ~ **breaker** Backenbrecher
*m* ~ **chuck** Backen-, Klauen-futter *n* ~ **clutch**
**coupling** Ausdehnungs-, Zahn-kupplung *f* ~
**coupling** Klaukupplung *f* ~ **crusher** Backenbrecher *m*, -quetsche *f*, Steinbrecher *m* ~
**crushing** Vorzerkleinerung *f* ~ **distance** Backen-
abstand *m* ~ **liner** auswechselbares Backen-
futter *n* ~ **plate** Brechplatte *f* ~ **pliers** Schnabel-
zange *f* ~-**rolling machine** Rollbank *f* ~ **shaped**
klauenförmig ~ **slide** Tastbolzenhalter *m*
(beim Optimeter) ~-**type steady** Backensetzstock *m*
**jaws,** ~ **for jolting machine** Stauchbacken *pl* ~
**for knurling (or milling) machine** Ränderier-
backen *pl*
**jell, to** ~ verdicken
**jellied** dick, gallertartig, geronnen, verdickt
**jelly** fest werden, kristallisieren, verdicken;
Gallert *n*, Gallerte *f*, Gelee *n*, Leimgallerte *f*
~-**like** gallertähnlich ~-**like substance** Gallert-
masse *f*
**jenny** Laufkran *m*
**jeopardize** beeinträchtigen, gefährden
**jerk, to** ~ bewegen, rücken, stoßen, ziehen **to** ~
**up** aufschnellen
**jerk** ruckartige Bewegung *f*, Ruck *m*, Stoß *m* ~
**line** Stoß-, Zupf-seil *n*
**jerking stop** ruckweises Anhalten *n*, Ausschalten
*n*
**jerkingly** stoßweise
**jerks, by** ~ ruckweise
**jerky** ruckartig, sprunghaft
**jerry can** Benzinkanister *m* (slang)
**jet, to** ~ aus-strahlen, -stoßen, -spritzen, herausspritzen, mit Druckwasser spülen (einspritzen)
**jet** Ablauf *m*, Brennerdüse *f*, Durchflußrohr *n*,
Düse *f* (aus welcher der Strahl austritt), Düsenbohrung *f*, Gagat *n*, Gebläse *n*, Gießkopf *m*,
Guß *m*, Gußröhre *f*, Jet *m*, Läufer *m*, Laufradkanal *m*, Laufschaufel *f*, Pechkohle *f*, Strahl *m*
(Gas), Strahlrohr *n*; Strom *m*, tiefschwarz ~ **of**
**fire** Feuergarbe *f* ~ **of flame** Stichflamme *f* ~ **of**
**liquid fire** Feuer-, Flammen-strahl *m* ~ **of**
**water** Druckstrahl *m*
**jet,** ~ **agitator** Strahlmischer *m* ~ **aircraft** Luftfahrzeug *n* mit Strahlantrieb ~ **angle** Strahl-
winkel *m* ~ **apparatus** Strahlapparat *m* ~
**assisted take-off (JATO)** Strahlstarthilfe *f*
~-**black** kohl-, pech-, tief-schwarz ~ **blender**
Strahlmischer *m* ~ **bore** Strahlbohrung *f* ~
**boundary** Strahlgrenze *f* ~ **carburetor** Düsen-,
Zerstäubungs-vergaser *m* ~ **chamber** Zerstäu-
bungsgehäuse *n* ~ **compressor** Strahlkompres-
sor *m* ~ **condenser** Einspritz-, Strahl-konden-
sator *m* ~ **contraction** Strahlenschnitt *n* ~
**damper** Strahldämpfer *m* ~ **deflector** Strahlablenker *m* ~ **diffuser** Zerstrahler *m* ~ **disperser**
Düsenzerstäuber *m* ~ **dispersion** Strahlauflö-

şung *f* ~ **drive** Strahltrieb *m* ~-**edge source** Strahlschneidequelle *f* ~ **efficiency** Strahlwirkungsgrad *m* (aviat) ~ **engine** Strahl-motor *m*, -triebwerk *n* ~ **evator** (rocket) Strahlablenkring *m* ~ **exhaust stack reactor** Abgasrückstoßer *m* ~ **expansion** Strahlausbreitung *f* ~ **fighter** Strahljäger *m* ~ **flow** Strahlausfluß *m* ~ **gas turbine** Kompressortrommel *f* ~ **inclination in degrees** Strahlneigung *f* ~ **injection** Strahleinspritzung *f* ~ **injection-type engine** Strahleinspritzmotor *m* ~ **mill** Strahlmühle *f* ~ **mold** Spritzform *f* ~ **molding** Spritzverfahren *n* ~ **nacelle** Strahlgondel *f* ~ **needle** Düsennadel *f* ~ **needle valve** Düsennadel *f* ~ **nozzle** Ausstoß-, Schub-düse *f* ~ **oil-foam separator** Schmierstoffentschäumer *m* ~-**operated piston** Düsensteuerkolben *m* ~ **orifice** Düsen-austrittsfläche *f*, -öffnung *f* ~ **oscillograph** Kathodenstrahl *m* ~ **pipe** Strahlrohr *n* ~-**pitching apparatus** Pechspritzapparat *m* ~ **plane** Strahlebene *f* ~ **power plant** Strahltriebwerk *n* ~-**propelled aircraft** Strahlflugzeug *n* ~ **propulsion** Heißluftstrahl-, Reaktions-antrieb *m*, Rückstoßmotor *m*, Strahlantrieb *m*, Strahlvortrieb *m* ~-**propulsion engine** Heißluftstrahltriebwerk *n*, Strahlmotor *m*, Turbinenmotor *m* ~-**propulsion unit** Luftstrahltriebwerk *n* ~ **pump** Düsen-, Spritz-, Strahl-pumpe *f* ~ **radiator** Düsenkühler *m* ~ **reaction** Rückstoßkraft *f* ~ **regulator** Strahlregler *m* ~ **relay** (for submarine cables) Flüssigkeitsstrahlrelais *n* ~ **report (or sheet)** Düsenbericht *m* ~ **rinsing bowl** Strahlenspülmaschine *f* ~-**rudder hinge bracket** Strahlruderhaltebock *m* ~ **setting** Düseneinstellung *f* ~ **size** Düsenkaliber *n* ~ **spoiler** Strahlruder *n* (g/m) ~ **spraying** Strahlzerstäubung *f* ~ **spraying head** Düsenspritzkopf *m* ~ **stream** Strahlströmung *f* ~-**supercharger drive** Ladergetriebe *n* ~ **surface** Strahloberfläche *f* ~ **system** Düsensystem *n* ~ **thrust** Strahlschub *m* ~ **tone** Ausfluß-, Spalt-ton *m* ~ **tube** Düsenrohr *n* ~ **unit** Strahltriebwerk *n* ~ **vane** (rocket) Strahlruder *n* ~-**vane support** Druckstückhalterung *f* ~ **velocity** Strahlgeschwindigkeit *f* ~ **width** Strahlbreite *f*

**jetsam** Strand-gut *n*, -trift *m*

**jetting,** ~ **of a pile** Einspülen *n* eines Pfahles ~ **action** Spüleffekt *m* ~ **cap** Abfallhütchen *n*

**jettison, to** ~ ablassen (Treibstoff), absprengen, abstoßen, abwerfen, über Bord werfen

**jettison,** ~ **arrangement** Überbordwerfvorrichtung *f* ~ **gun point** Abstoßylinder *m*, Trennpunkt *m*

**jettisonable tank** Abwurftank *m*

**jettisoning** Schnellablaß *m*, Schnellablaßentleerer *m* (von Kraftstoff)

**jetty** Anlegebrücke *f*, Fluß-, Hafen-damm *m*, Kai *m*, Kofferleitdamm *m* (rock), Ladedamm *m*, Ladezunge *f*, Landebrücke *f*, Landungssteg *m*, Leitdamm *m*, Leitwerk *n*, Mole *f*, Pier *m*, Wellenbrecher *m* ~ **for fuel-oil bunkering** Tankanlage *f*

**jewel** Decklinse *f*, Edelstein *m*, Juwel *n*, Stein *m* (einer Uhr) ~ **bearing** Edelsteinlager *n* (Meßgerät) ~ **cup** Achathütchen *n* ~ **support** Lagersteinschraube *f*

**jeweled bearing** Lagerung *f* in Steinen, Spitzenlagerung *f*, Steinlager *n*

**jib** Arm *m* einer Schrämmaschine, Auslegearm *m*, Klüver *n*, Nasenkeil *m* ~ **arm** Trägerarm *m* ~ **arm of a crane** Dreharm *m* ~ **(arm) of derrick** Kranbalken *m* ~-**boom saddle** Sattel *m* des Klüverbaumes ~ **crane** Ausleger-, Brückendreh-kran *m* ~ **derricking** Auslegerantrieb *m* ~ **head** Auslegerkopf *m* (Kran) ~-**head sheave** Auslegerrolle *f*

**jibe, to** ~ sich drehen, giepen, Kurs ändern

**jig, to** ~ scheiden, separieren, setzen, waschen

**jig** Anreißvorrichtung *f*, Anschlag *m*, Aufspannvorrichtung *f*, (boring) Bohrfutter *m*, Bohrvorrichtung *f*, Einspannvorrichtung *f*, (textiles) Filzmaschine *f*, Futter *n*, Kaliber *n*, Montagebock *m*, Paßlehre *f*, Schablone *f*, Setzkasten *m*, Siebsetzmaschine *f*, Spannvorrichtung *f*, (in airpane assembly) Unterbauvorrichtung *f* (**assembly**) ~ Bauvorrichtung *f*

**jig,** ~ **borer** Lehrenbohrmaschine *f* ~ **boring** Lehren-, Schablonen-bohren *n* ~ **boring machine** Vorrichtungsbohrmaschine *f* ~ **department** Vorrichtungsbau *m* ~ **drill** Lehrenbohrwerk *n* ~ **drilling** Lehrenbohren *n* ~ **loom** Ausleger *m* ~ **point** Wiegemarke *f* ~-**saw** Folgeschnitt *m*, Dekopier-, Rahmenspalt-, Wipp-säge *f* ~-**saw characters** Skelettschablonen *pl* ~ **screen** Setzsieb *n* ~-**shaken tray** Rüttelrutsche *f* ~ **swing** Schwenkbereich *m* (Bagger) ~ **table** Schüttelrätter *m* ~ **work** Spannarbeiten *pl*

**jigger** Aufsetzkasten *m* (dyeing), Jigger *m*, Kopplungstransformator *m* (electr.), Töpferscheibe *f*

**jiggers** (in facsimile) Bildverzerrung *f*

**jigging** Sieb-setzarbeit *f*, -setzen *n*, (petroleum) Tanz *m*, Vorrichtungsbau *m* ~ **action** Setzvorgang *m* ~ **screen** Schwingsieb *n*

**jiggle, to** ~ rücken, rütteln, schaukeln, stoßen

**jiggle** Rütteln *n*, Wackeln *n*

**jiggling** stoßweise Hin- und Herbewegung *f* (film)

**jigs in lockmaking** Führungsschnitte *pl* (für die Schloßindustrie)

**jimmy** Brecheisen *n*

**jingle, to** ~ klimpern, klappern

**jingling** Klimpern *n*

**jitter** Flackern *n*, Flimmern *n*, Schaukeleffekt *m* (TV), Synchronisationsfehler *m*, Zittereffekt *m*, Zittern *n* (des Signals)

**job** Anstellung *f*, Arbeit *f*, Arbeitsplatz *m*, Arbeitsstück *n*, auszuführende Arbeit *f*, Stellung *f* ~ **control** Baukontrolle *f* ~ **crane** (for electrodes) Fahrsäule *f* ~ **lot** Ramschware *f* ~ **mark** Gedingstufe *f* ~ **office paper** Werkdruckpapier *n* ~ **order** Auftragszettel *m* ~ **printer** Akzidenzdrucker *m* ~ **time** Stückzeit *f* ~-**time recorder** Akkordzeitregistrierapparat *m* ~ **types** Akzidenzschrift *f* ~ **work** Akkordarbeit *f*

**jobber** Wiederverkäufer *m*, Zwischenhändler *m*

**jobbing** Akkordarbeit *f*, Stückwerk *n* ~ **compositor** Akzidenzsetzer *m* ~ **foundry** Handform-, Kunden-gießerei *f* ~ **fount** Akzidenzmaterial *n* ~ **house** Ersatzteilfirma *f* ~ **mill** Vorsturzwalzwerk *n* ~ **sheet** Mittelblech *n* ~ **sheet-rolling mill** Mittelblechwalzwerk *n* ~ **(or face) type** Akzidenzschrift *f* ~ **work** Akzidenzarbeit *f* (print)

**jobbings** Handform-, Kunden-guß *m*

**jockey** Reiter *m*, (wheel) Reiter-rädchen *n*, -röllchen *n* ~ **pulley** Führungsrolle *f* (Riemen),

Spannrolle *f* ~ **roller** Reiter-rädchen *n*, -röll-chen *n*, Spannrolle *f* (film) ~ **weight** Laufge-wicht *n*
**jog, to** ~ aufstoßen (paper), einkerben, glatt-stoßen
**jog** Anstoßen *n*, Rütteln *n*, Schütteln *n*, Unstetig-keit *f* einer Kurve ~ **formation** Sprungbildung *f* ~ **lathe** Profildrehbank *f*
**jogger** Bogengeradeleger *m* (print), Gerade-stoßer *m* ~ **piston** Kolben *m* zum Bogenschie-ber ~ **plate** Bogenschieberblech *n*
**jogging** Tastbetrieb *m*
**joggle, to** ~ falzen, rütteln, schütteln, ver-schränken, verzinken (aviat) **to** ~ **the control stick** den Steuerknüppel *m* leicht hin- und her bewegen **to** ~ **with(gear) teeth** verzahnen **to** ~ **two beams** zwei Balken *pl* miteinander ver-schränken
**joggle** Absetzung *f*, Auskerbung *f*, Falz *m*, Nut- und Federverband *m*, Schwalbenschwanz *m*, Verbindungspflock *m*, Versetzung *f*, Ver-zahnung *f*, Zapfen *m*, Zinken *m* ~ **beam** ver-zahnter Balken *m* ~ **truss** Hängewerksbinder *m*
**joggled** versetzt, verzahnt ~ **beam** verzahnter Balken *m*, verdübelter Balken *m* ~ **rivet** Bördelniete *f*
**jogless curve** kontinuierliche Kurve *f*
**joggling** Schütteln *n*, Stauchverschränkung *f* ~ **machine** Kröpfmaschine *f* ~ **roll** Stauchver-schränkungswalze *f*
**johannite** Uranvitriol *n*
**join, to** ~ (carp) abbinden, angrenzen, anhängen, sich anhängen, anschließen, anstücken, bei-treten, einfallen, fügen, kuppeln, an die Lei-tung anlegen, schalten, sich scharen, spleißen, stoßen, verbinden, vereinigen, verknüpfen, zu-sammenfügen
**join, to** ~ **by casting** angießen **to** ~ **closely** an-schmiegen **to** ~ **by cogging** aufkämmen **to** ~ **into line** sich einfädeln **to** ~ **like a loop** zur Schleife schalten **to** ~ **in multiple** vielfach schalten **to** ~ **onto** anarbeiten, anschmelzen **to** ~ **in parallel** parallel oder nebeneinander schalten **to** ~ **rafters together** schiften **to** ~ **in series** hintereinanderschalten, in Reihe schalten **to** ~ **by slit and tongue** (carp) zusammenscheren **to** ~ **a thread** einen Faden anspinnen **to** ~ **timbers by cogging** zwei Holzstücke verkäm-men **to** ~ **to** anlegen an **to** ~ **together** aneinan-der fügen **to** ~ **together without cement** zusam-mensprengen **to** ~ **by a triangular notch** auf-klauen **to** ~ **by twisting (or plaiting)** anflechten **to** ~ **up** einschalten **to** ~ **up with** sich anschlie-ßen
**join** Bindeglied *n*, Fuge *f*, Naht *f*, Verbindung(s-stelle) *f*
**joined** gefalzt ~ **surfaces** Anlagekante *f*
**joiner** Schreiner *m*, Tischler *m* ~ **plate** Tischler-platte *f*
**joiner's,** ~ **bench** Hobelbank *f* ~ **clamp** Schraub-zwinge *f* ~ **(or carpenter's) glue** Tischlerleim *m* ~ **shop** Schreinerei *f* ~ **work** Tischlerarbeit *f*
**joinery** Tischlerei *f* ~ **for buildings** Bauschreine-rei *f*, Bautischlerei *f* ~ **works** Tischlerarbeiten *pl*
**joining** Aneinanderlagerung *f*, Schalten *n*, Ver-bindung *f*, Vereinigungsstelle *f* ~ **of crystals** Vergraupelung *f* ~ **with key piece** Schurzwerk *n* ~ **a thread** Andrehen *n*

**joining,** ~ **line** Stoßlinie *f* ~ **means** Verbindungs-mittel *n* ~ **member** Stoßriegel *m* ~ **point** Ver-bindungspunkt *m* ~ **ring** Anschlußring *m* ~ **shackle** Verbindungsschäkel *m* ~ **stone** Satz-stein *m* ~ **surface** Verbindungsfläche *f* ~ **tub-bing of mine shaft** Anschlußstübbings *pl*
**joint, to** ~ ansplissen, anstücken, fugen, glie-dern, schäften, verspleißen, verzapfen **to** ~ **sta-ves** Dauben *pl* fügen **to** ~ **timbers** Holzstücke *pl* verbinden
**joint** gemeinsam, gemeinschaftlich, verbunden
**joint** Ablösungsfläche *f* (geol.), Abschluß *m*, Absonderungsfläche *f*, Anschluß *m*, Dichtung *f*, Eckpunkt *m*, Falte *f*, Falz *m* (Buchdecke), (foundry) Formlinie *f*, Fuge *f*, Gelenk *n*, Gelenkpunkt *m*, Gelenkstück *n*, Glied *n*, Kluft *f*, Knie *n*, Knotenpunkt *m*, Lötstelle *f*, Mörtel-fuge *f*, Naht *f*, Schäftung *f*, Scharnier *n*, Spalte *f*, Spleißstelle *f*, Stoß *m*, Talfuge *f*, Teilfuge *f*, Verbindung *f*, Verbindungsstelle *f* **(knuckle)** ~ Schloß *n* **(upright)** ~ Stoßfuge *f* ~ **of the bed** Lagerfuge *f* ~s **of bedded rocks** Schichtfugen *pl* ~ **by double rabbeting** mit Anschlag *m* und Überschlag *m* verbinden ~ **of facing** Fuge *f* der Verkleidung ~ **with inside-riveted fishplates for portable lines** Stoßverbindung *f* mit innen an-genieteten Stecklaschen ~ **and junction termi-nals** Einsatzklemmen *pl* ~ **of rods** Gestänge-schloß *n* ~ **of rupture** Bruchfuge *f* ~ **between the squeezing rollers** Quetschfuge *f* der Walzen ~ **of support** Stützengelenk *n* ~ **of tube** Röhren-stoß *m*
**joint,** ~ **abutment** Stoßstelle *f* ~ **area** Stoßstelle *f* ~ **bandage** Gelenkbinde *f* ~ **bend** Gelenk-krümmer *m* ~ **binding** Drahtverbindung *f* ~ **board** Aufstampf-, Leer-boden *m* ~ **bolt** Gelenkbolzen *m* ~ **box** Abzweigkasten *m*, Kabel-brunnen *m*, -muffe *f* ~-**box cover** Schachtabdeckung *f* ~ **capacity** gemeinsame Kapazität *f* ~ **capital** Aktienkapital *n* ~ **cast iron pipe** Muffengußrohr *n* ~ **communicator** gemeinsamer Fernmeldeverkehr *m* ~ **coupling** Anschlußstück *n* ~ **enterprise** gemeinschaft-liches Unternehmen *n* ~ **face** Dichtungsfläche *f*, Fugenseite *f*, Teilfläche *f*, Teilungsebene *f* ~ **face of molding box** Kastenteilfläche *f* ~ **file** Scharnierplatzfeile *f* ~ **flange** Gelenkflansch *m* ~ **(or hinge) frame** Scharnierband *n* ~ **grease** Dichtungs-, Hahn-fett *n* ~ **heating** Fugenhei-zung *f* (Klebstoff) ~ **hinge with peg (or pin)** Vorsteckerscharnierband *n* ~ **latch** Scharnier-klinke *f* ~ **leak** Verbindungsleck *m* ~ **line** Teil-fläche *f*, Teilungsebene *f* ~ **line of a mold** Formscheidung *f* ~ **measurement** gemeinsames Aufmaß *n* ~ **packing** Flanschen-dichtung *f*, (for pipes) -packung *f* ~-**party control** Auslö-sung *f* beim Einhängen des Hörers durch beide Teilnehmer, beiderseitige Auslösung *f* ~ **pin** Drehachse *f*, Gelenkbolzen *m* ~ **plate** An-schlußblech *n*, Gelenklasche *f*, Verbindungs-platte *f* ~ **point** Stoßstelle *f* ~ **products of a unified refinery operation** Kuppelprodukte *pl* ~ **pulley** Gelenkseilscheibe *f* ~ **resistance** kombinierter oder gemeinsamer Widerstand *m* ~ **responsibility** Solidarhaft *f* ~ **ring** Dichtungs-ring *m* ~ **shaped** gelenkförmig ~ **sleeper** Stoß-schwelle *f* ~ **spacing** Fugenabstand *m* ~-**stock company** Aktien-, Kommandit-gesellschaft *f*

~ **tie** Fugenschwelle *f* ~ **washer** Anschluß-
dichtung *f* ~ **weld** Verbindungsschweißung *f*
~ **welding** überlappte Schweißung *f* ~ **work**
Gemeinschaftsarbeit *f*
**jointed** gelenkig ~**-band conveyor** Gliederband-
förderer *m* ~ **cross shaft** Achse *f* mit Quer-
kardanwelle, Pendelachse *f*, Schwingachse *f*
~ **shank system** Scherensystem *n* (Walzeneinstellung) ~ **sighting rod** Gliedervisier
**jointer** Fügebank *f*, Fugeisen *n*, Fugkelle *f*,
Spleißer *m*, Vorschäler *m* **(cable)** ~ Kabellöter *m* ~'s **pliers** Lötzange *f* ~ **vise** Spleiß-
block *m*
**jointing** Abdichten *n*, Abdichtung *f*, Absonderung *f*, Dichten *n*, Dichtung *f*, Dichtungsmittel *n*, Fugen *n*, Gliederung *f*, Spleißung *f*,
Verbinden *n*, Verspleißung *f* ~ **chamber**
Kabel-, Kabellöt-, Löt-brunnen *m* ~ **clamp**
Drahtkluppe *f*, Kluppe *f* (electr.), Preßzange *f*
~ **compound** Dichtmasse *f* ~ **edge** Kleberand *m*
~ **machine** Fügemaschine *f* ~ **materials** Dicht-
stoffe *pl* ~ **pastes** Vorsatzstoffe *pl* ~ **plane** Teil-
fläche *f*, Teilungsebene *f* ~ **ring** Dichtungsband
*n* ~ **saw** Fügesäge *f* ~ **sleeve** Drahtverbindungs-
hülse *f*, Hülsenverbinder *m*, Übergangsform-
stück *n*, Verbindungshülse *f* ~ **solution** Dicht-
mittel *n* ~ **strip** Stoßschiene *f*
**jointless** fugenlos, nahtlos
**jointly** miteinander, sämtlich
**joints of bedded rocks** Schichtfugen *pl*
**joist** Balken *m*, Dielenbalken *m*, Formeisen-
träger *m*, Nebenbalken *m* ~ **of a landing place**
Podestbalken *m* ~ **rolling mill** Trägerwalzwerk
*n* ~ **rule** Schiene *f* ~ **transverse** Ausleger *m*
**joists** Gebälk *n*
**jolt, to** ~ rütteln, schütteln, stoßen **to** ~ **ram**
rütteln
**jolt** Ruck *m*, Rüttelstoß *m*, Stoß *m* ~ **blow**
Rüttel-schlag *m*, -stoß *m* ~ **cylinder** Rüttel-
zylinder *m* ~**-free** stoßfrei ~ **piston** Rüttel-
kolben *m* ~**-ramming machine** Rüttelformmaschine *f*, Rüttelmaschine *f*, Rüttler *m* ~ **roll-
over machine** Umrollrüttelformmaschine *f*
~ **roll-over squeeze pattern-draw machine**
Rüttelpreßformmaschine *f* mit Wende- und
Abhebeeinrichtung ~**-squeeze turnover machine**
Rüttelwendeformmaschine *f* mit Preßvorrich-
tung ~**-strip pattern-draw machine** Rüttelformmaschine *f* mit Durchzieh- und Abhebevorrichtung ~ **stripper** Rüttelformmaschine *f*
mit Abstreifvorrichtung ~**-stripper with hand-
lever lift** Rüttelformmaschine *f* mit Abstreif-
platte und Handhebelmodellaushebung ~
**table** Rütteltisch *m* ~ **turnover machine** Wendeplattenrüttler *m* ~ **valve** Rüttelventil *n*
**jolted** gestaucht
**jolter** Rüttelformmaschine *f*, Rüttelmaschine *f*,
Rüttler *m* ~ **with mold-lift attachment** Rüttelformmaschine *f* mit Abhebevorrichtung
**jolting** Rütteln *n* ~ **capacity** Rüttelleistung *f*
~ **furnace** Rüttelofen *m* ~ **molding machine**
Rüttel(form)maschine *f*, Rüttler *m* ~ **operation**
Rüttelvorgang *m* ~ **table** Rütteltisch *m*
**jolty** ruckartig
**jordan mill** Kegelstoffmühle *f*
**Joule losses with direct current** Gleichstromverluste *pl*
**Joule's,** ~ **heat loss** Joulescher Wärmeverlust *m*,

Stromwärmeverlust *m*, Ohmscher Verlust *m*,
~ **law** Joulesches Gesetz *m* ~ **law effect**
Joulesche Wärme *f*, Stromwärme *f*
**joulemeter** Joulezähler *m*
**journal** Buch *n*, (of an axle) Drehzapfen *m*,
Lagerzapfen *m*, Lauffläche *f*, Laufzapfen *m*,
Tagebuch *n*, (of shaft) Wellzapfen *m*, Zapfen *m*,
Zeitschrift *f* ~ **with collars** Halszapfen *m* ~ **of
crossbar** Querstückzapfen *m* ~ **of a roll** Walzenzapfen *m*
**journal-béaring** Achsen-, Achs-, Gleit-, Hals-
-lager *n*, hinteres Lager *n*, Trag-, Zapfen-,
Zylinder-lager *n* ~ **bushing** Gleitlagerbüchse *f*
~ **friction** Tragzapfenreibung *f* ~ **seal** Dichtungsring *m*
**journal,** ~ **bolt** Bundbolzen *m* ~ **box** Lagerbüchse *f*, Walzenlager *n* ~**-box guide** Achslagerführung *f* ~**-box oil** Achsenschmieröl *n*
~ **cross assembly** Zapfenkreuz *n* ~ **folding
machine** Zeitungsfalzmaschine *f* ~ **friction**
Zapfenreibung *f* ~ **gasket retainer** Dichtungshalter *m* ~ **grinding machine** Zapfenschleif-
maschine *f* ~ **load** (axial) Druck *m* ~ **oil**
Schmierzapfenöl *m* ~ **pin** Lagerzapfen *m*
~ **turning attachment** Achsdrehvorrichtung *f*
~ **voucher** Buchungsbeleg *m* ~ **web fillet**
Kurbelwangenausrundung *f* ~**-wheel spindle**
Achsschenkel *m*
**journaled on points** spitzengelagert
**journey, to** ~ reisen
**journey** Fahrt *f* ~ **there** Hinweg *m*
**joystick** Daumenhebel *m*
**J scope** J-Schirm *m*
**judder** Verwacklung *f*, Vibrieren *n* (Flugzeug)
**judge, to** ~ beurteilen, schätzen **to** ~ **by its
flavour** abschmecken
**judge** (competition) Kampfrichter *m*, Richter *m*
**judged by (or judging from)** beurteilt nach
**judgment** Ansicht *f*, Beurteilung *f*, Entscheidung
*f*, Erachten *n*, Ermessen *n*, Gutachter *n*,
Rechtsspruch *m*, Richterspruch *m*, Strafe *f*,
Urteil *n*, Urteilsspruch *m* **in my** ~ meines
Erachtens ~ **by default** Versäumnisurteil *n*
~ **test** Bewertungsprüfung *f*, überschlägliche
Prüfung *f*
**judicial** gerichtlich ~ **examination** Verhör *n*
**judiciary** gerichtlich, justiz **to bring** ~ **action**
gerichtlich belangen
**judicious** urteilsfähig
**jug** Kanne *f*, Krug *m*
**juice** Saft *m* ~ **boiler** Aufkocher *m* ~ **catcher**
Saftabscheider *m* ~ **density** Saftdichte *f*
~ **extraction** Saftgewinnung *f* ~ **filler** Saftvorfüller *m* ~ **gravity** Saftdichte *f* ~ **gutter**
Saftrinne *f* ~**(pre)heater** Saftvorwärmer *m*
~ **piping** Saftleitung *f* ~ **pump** Montejus *m*,
Saft-, Sirup-pumpe *f* ~ **purification agent** Saftreinigungsmittel *n* ~ **sampler** Saftprobenehmer
*m*
**jukebox** Musikautomat *m*
**jumble, to** ~ vermengen, verschlüsseln, verwürfeln
**jumble,** ~ **code** Verwürfelungsschlüssel *m*,
~ **perforator** Schlüsselstreifenlocher *m*
**jumbling part (or piece)** Zwischenstück *n*
**jump, to** ~ hüpfen, springen, stauchen **to** ~ **back**
zurückschnellen **to** ~ **in** einspringen **to** ~ **into
the groove** in die Rast einspringen **to** ~ **off** ab-

springen **to ~ over** überspringen **to ~ a pin** einen Bolzen aufsitzen lassen **to ~-weld** stumpf schweißen
**jump** Abgangsfehler *m* (artillery), Satz *m*, Sprung *m* **~ in brightness** Helligkeitssprung *m* (film) **~ in intensity** Intensitätssprung *m* **~ of temperature** Temperatursprung *m*
**jump, ~ characteristic** Sprungkennlinie *f* **~ feed** Springschaltung *f* **~-gap distributor** Verteiler *m* mit springenden Funken **~ order** Sprungbefehl *m* (info proc) **~ phenomenon** Kipperscheinung *f* **~ roughing mill** Vorsturzwalzwerk *n*
**jump-spark** überspringender Funke(n) *m*, Zündungsfunke *m* **~ gap** Spaltfunkenzünder *m* **~ ignition** Hochspannungszündung *f* **~ system** Funkenschirm *m*
**jump, ~ start** Sprungstart *m* **~ weld** Stumpfschweißen *n*, Stumpfschweißung *f* **~ welding** T-Schweißung *f*
**jumper, to ~** mit Schalterdraht *m* verbinden
**jumper** Bohrmeißel *m*, Kurzschluß-brücke *f*, -draht *m*, (wire) Schaltaderdraht *m*, Schaltdraht *m*, Stauchhammer *m*, Steinbohrer *m*, Überbrückungsdraht *m* **~ connection** Schaltdrahtverbindung *f* **~ cord with plug** Vermittlungsschnur *f* **~ plug** Verbindungsstecker *m* **~ ring** Tragring *m* **~ side** Verteilerseite *f* **~ wire** Rangierdraht (teleph.), Schalt-ader *f*, -draht *m*, Verteilerdraht *m* (teleph.)
**jumpering field of distribution frame** Schaltaderfeld *n* des Verteilers
**jumping** Bildtanzen *n* (film), Tanzeffekt *m*, Umspringen *n* **~ of the brushes** Hüpfen *n* der Bürsten **~ of image** mangelhaftes Stehen *n* des Bildes
**jumping, ~ apparatus** Puffervorrichtung *f* **~ arc amplifier** Sprungbogenverstärker *m* **~-off point** Abflugpunkt *m* **~-off position** Ausgangsstellung *f*
**junction** Abzweigung *f*, Anschluß *m*, Anschlußbahnhof *m*, Berührungspunkt *m*, Durchschlag *m* (min.), Einmündung *f* (Straße), Flächenkontakt *m*, Fuge *f*, Grenze *f*, Knotenpunkt *m*, Kontaktstelle *f* (in pyrometry), Kreuzung *f*, Lötstelle *f*, Netzknoten *m*, Stoßfuge *f*, Stoßpunkt *m*, Trennungsbahnhof *m*, Übergang *m*, (semiconductor) Übergangszone *f*, Verbindung *f*, Verbindungsleitung *f*, Verbindungspunkt *m*, Verbindungsstelle *f*, Vereinigung *f*, Vereinigungsstelle *f*, Vielfachanschluß *m*, Wegkreuzung *f* **in(coming) ~** ankommende Verbindungsleitung *f* **~ of lines** Gleisanschluß *m* **~ of lodes** Rammeln *n* von Gängen **~ of paths** Leitungsverbindung *f* **~ from private-branch exchange to exchange** Amtsleitung *f* **~ of side road with main road** Wegeinmündung *f* **~ of two map sheets** Zusammenstoß *m* zweier Kartenblätter **~ of veins** Scharung *f* von Gängen **~ of the wing with the fuselage** Anschluß *m* des Flügels an den Rumpf
**junction, ~ block** Anschlußklemmleiste *f* **~ board** Verbindungsleitungsschrank *m* **~ box** Abzweigdose *f*, Abzweigkasten *m*, Anschluß-dose *f*, -kasten *m* (electr.), -muffe *f*, Kabelkasten *m*, Kreuzdose *f*, Verbindungsdose *f*, Verbindungskasten *m*, Verteilerkasten *m* **~ box with plugs** Verteilerdose *f* (rdr) **~ box arranged in tiers**

Etagenabzweigkasten *m* **~ cable** Verbindungskabel *n*, Verbindungsleitungskabel *n* **~ canal** Sieltief *n*, Verbindungskanal *m* **~ center** Verbundamt *n* **~ circuit** Verbindungsleitung *f* **~ cross** Thermokreuzbrücke *f* **~ curve** Verbindungskurve *f* **~ diode** Flächengleichrichter *m* **~ finder** Verbindungsleitungssucher *m* **~ group** Leitungsbündel *n* **~ indicator** Verbindungsleitungsklappe *f* **~ jack** Verbindungsleitungsklinke *f* **~ line** Anschlußgleis *n*, Grenze *f*, Grenzlinie *f*, Umgrenzungslinie *f*, Verbindungsbahn *f*, Verbindungsleitung *f*, Zweiglinie *f* **~-line panel** Verbindungsleitungsfeld *n* (teleph.) **~ manhole** Abzweigkasten *m* **~ multiple** Verbindungsleitungsvielfachfeld *n* **~ network** Netzspinne *f*, Verbindungsleitungsnetz *n* **~ panel** Anschließ-, Anschluß-feld *n* **~ photocell** Flächenkontakt-Fotozelle *f* **~ plug** Verbindungsleitungsstöpsel *m* **~ point** Stoß-, Übergangs-stelle *f* **~ pole** Verzweigungsstützpunkt *m*, Übergangsstange *f* (electr.) **~ rail end** Anschlußende *n* der Schiene, Schienenanschlußstück *n* **~ rails** Verbindungsgleis *n* **~ ring** Sammellöse *f* **~ roller** Kreuzrolle *f* **~ selector** Amtswähler *m* **~ service** Nah-, Netzgruppen-, Verbindungsleitungs-verkehr *m* **~ signal** Vereinigungszeichen *n* **~ stage** Durchschaltebene *f* **~ switch box** Sammlerschalter *m* **~ switching position** Vermittlerplatz *m* **~ traffic** Bezirks-, Nachbarorts-, Nah-, Vororts-verkehr *m* **~ transistor** Flächentransistor *m* **~ trunk** Ortsverbindungsleitung *f* **~ working** Verbindungsleitungsbetrieb *m*
**junctions** Verbindung *f* zwischen den Ämtern
**juncture** Naht *f*, Stoßband *n*, Verbindungspunkt *m*
**jungle** Dschungel *m*, Urwald *m*
**junior cave** Zelle *f* für schwache Aktivitäten
**junk, to ~** beiseitelegen
**junk** Abfall-, Alt-eisen *n*, Altmaterial *n*; Dschunke *f* **~ basket** Kern- oder Sedimentrohr *n* (Kernbüchse) **~ dealer** Alteisenhändler *m* **~ goods** Ramschware *f* **~ head** Schieberzylinderkopf *m* **~ pile** Abfallhaufen *m*
**Jurassic** jurassisch **~ limestone** Jurakalk *m*
**Jurid lining** Juridbelag *m*
**juridical** rechtlich
**jurisdiction** Gerichtsstand *m*, Rechtszuständigkeit *f*
**jurisdictional area** Geltungsgebiet *n*
**jury** Schiedsgericht *n* **(prize) ~** Preisgericht *n* **~ of appeal** Berufungsgericht *n*
**jury, ~ man** Geschworener *m* **~ mast** Hilfsmast *m* **~ rudder** Notruder *n*
**just** angemessen, eben, genau, gerade, korrekt, richtig
**justification** Ausschluß *m* (print.), Justierung *f*, Zeilenlänge *f* **~ device** Ausschließapparat *m* **~ wedge** Ausschluß *m*, Ausschlußkeil *m*
**justify, to ~** abgleichen (electr.), justieren (print.), ausschließen
**justifying, ~ needle** Justiernadel *f* **~ pointer** (scale) Ausschlußzeiger *m*
**jut, to ~ out** ausfluchten, auskragen, hervorstehen, vorspringen
**jut** Vorsprung *m*
**jute** Flachs *m*, Jute *f* **~ bag** Jutesack *m* **~ board** Jutepappe *f* **~ covering** Juteumwicklung *f* **~**

**fiber** Jutefaser *f* ~ **filler** (cable making) Jutebeilauf *m* ~ **mill** Jutespinnerei *f* ~ **packing** Jutepackung *f* ~ **paper** Tauenpapier *n* ~**-served** mit Jute *f* umwickelt ~ **serving** Juteumwicklung *f* ~ **yarn** Flachs-, Jute-garn *n*

**jutting** ausspringend, hinausragend
**jutty** Erker *m*
**juxtapose, to** ~ nebeneinanderstellen
**juxtaposed to** angrenzend
**juxtaposition** Anlagerung *f*, Nebeneinanderstellung *f*, Nebensetzung *f*

# K

**K absorption limit** K-Absorptionskante *f*
**kahlerite** Kahlerit *m*
**kaleidophon** Tonschwingungsspiegel *m*
**kaleidoscope** Kaleidoskop *n*
**kaleidoscopic** kaleidoskopisch
**kamacite** Kamacit *m*
**kampometer** Kampometer *n*
**kaneite** Arsenikmangan *n*
**kangeroo-crane** Wippentlader *m*
**kaolin** Kaolin *n*, (geschlammte) Porzellanerde *f*,
Porzellanton *m* ~ **deposit** Kaolinlager *n*
**kaoliniferous sandstone** Kaolinsandstein *m*
**kaolinite** Kaolinit *m*
**kaolinization** Kaolinisierung *f*
**kaolinize, to** ~ in Kaolin *n* verwandeln
**Kaplan turbine** Kaplanturbine *f*
**kapok** Bombaxwolle *f*, Kapok *m*
**kappa-number** Kappazahl *f*
**Karman vortex street** Strömungsschatten *m*
**karst** Karst *m*
**kasolite** Kasolit *m*
**katabatic wind** Fallwind *m*
**katacaustic** (curve or surface) Katakaustik *f*
**kataphoresis** Kataphorese *f*
**Kater's pendulum** Reversionspendel *n*
**katharometer** Katharometer *n*
**Kauri pine** Dammarharz *n*, Kauriholz *n*
**kaurit,** ~ **glue** Kauritleim *m* ~ **gluing** Kaurit-
verleimung *f*
**K edge** K-Absorptionskante *f*
**kedge firing** Wurffeuerung *f*
**keel, to lay down the** ~ den Kiel legen
**keel** Kiel *m*, Längsträger *m* ~ **angle** Kielwinkel
*m* ~ **apex** Kielstoßpunkt *m* ~ **block** Kiel-,
Stapel-klotz *m*; (locks and dry docks) Kiel-
stapel *m* ~ **blocking** Kiellager *n* ~ **bolt** Kiel-
bolzen *m* ~ **bracing** Kielversteifung *f* ~ **corridor**
Lauf-brücke *f*, -gang *m* ~ **effect** Kielwirkung *f*
~ **fin** Kielflosse *f* ~ **framework (or structure)**
Kielgerüst *n* ~ **girder** Heckträger *m* ~ **line**
Kiellinie *f* ~ **scarf** Kiellaschung *f* ~ **shoe** After-
kiel *m* ~ **surface (or area)** Kielfläche *f*
**keelson** Kielschwein *n*
**keen, to** ~ schärfen
**keen** scharf, stark ~ **edge** scharfe Kante *f* ~-
**edged** mit scharfer Schneide *f* ~-**edged tool**
scharf schneidendes Werkzeug *n*
**Keene's cement** Estrichgips *m*
**keenness** Schärfe *f* ~ **of vision** Sehschärfe *f* des
Auges
**keep, to** ~ aufbewahren, behalten, beibehalten,
bewahren, dauern, halten, innehalten, unter-
halten **to** ~ **alive** (an application or patent) aufrechterhalten **to** ~ **apart** auseinander halten
**to** ~ **clear** klar halten **to** ~ **constant** gleichmäßig erhalten, konstanthalten **to** ~ **on course**
ansteuern **to** ~ **the course** Kurs halten **to** ~
**upon the course** Kurs *m* halten **to** ~ **the ground**
spuren **to** ~ **down** niederhalten **to** ~ **headway**
Fahrt *f* voraus halten, Geschwindigkeit *f*
halten **to** ~ **house** wirtschaften **to** ~ **invariable
(or stationary)** konstant erhalten **to** ~ **the**

**land aboard** unter Land *n* halten **to** ~ **up** erhalten, instandhalten **to** ~ **a mine in repair** ein
Bergwerk *n* bauhaft erhalten **to** ~ **out** Eintritt
*m* verboten **to** ~ **a relay excited** ein Relais *n*
halten **to** ~ **in reserve** vorbehalten **to** ~ **silent**
schweigen **to** ~ **step with** Schritt *m* halten
(mit) **to** ~ **in stock** auf Lager *n* halten **to** ~ **in
touch** Fühlung *f* halten **to** ~ **in the track** spuren
**to** ~ **watch** bewachen **to** ~ **off the waters by
timbering** die Gewässer *pl* durch Holzver-
dämmung abhalten, die Wasser *pl* verdämmen
**to** ~ **within** unterschreiten
**keep** Unterhaltungskosten *pl* ~-**alive electrode**
Halteanode *f*, Vorionisationselektrode *f* ~-
**alive voltage** Vorionisierungsspannung *f*
**keeper** Aufseher *m*, Kontermutter *f*, Magnetanker *m* (des Dauermagneten), Schiebe-
kopf *m*, Schieber *m*, Sperrung *f*, Tauschlaufe
*f*, Verschluß *m*, Wärter *m*, (of magnet) Zunge *f*
~ **nose** Schließkolbennase *f* ~ **pin** Mitnehmerbolzen *m* ~ **plates** Untergriffsleisten
*pl* ~ **support** Schließkolbenführung *f*
**keeping** Aufbewahrung *f*, Lagern *n* ~ **color**
farbehaltend ~ **dry** Trockenhaltung *f* ~
**spare parts** Ersatzteilhaltung *f* ~ **taut** Straff-
haltung *f*
**keeve** Bottich *m*
**keg** Faß *n*, Lägel *n*
**Keith,** ~-**line switch** Keithvorwähler *m* ~
**master switch** Gruppenwähler *m* (Keith Vorwähler)
**Keller,** ~ **arc furnace** Kellerlichtbogenofen
*m* ~-**control** Kellersteuerung *f*
**kelly** Mitnehmerstange *f* ~ **bushing** Mitnehmereinsatz *m*
**Kelly's rat hole** Loch *n* der Vierkantstange
**keloid** Keloid *n*
**Kelvin,** ~ **arrival curve** Thomsonkurve *f* ~ **balance** Stromwaage *f* ~ **bridge** Doppelbrücke *f*
~ **effect** Haut-effekt *m*, -wirkung *f*, Oberflächen-beschaffenheit *f*, -wirkung *f* ~ **scale**
Kelvinsche Skala *f* (Nullpunkt bei —273° C.)
**kennel** Fluter *n* ~ **of paving** Gassenrinne *f*
**Kennelly-Heaviside layer** Kennelly-Heaviside-
Schicht *f*
**kenotron** Gleichrichterelektronenröhre *f*, Hoch-
vakuum-gleichrichter *m*, -gleichrichterröhre
*f*, -glühkathodengleichrichterröhre *f*, Kenotron *n*, Zweielektrodenröhre *f*
**keps gear** Käpsvorrichtung *f*
**keratin** Hornsubstanz *f*, Keratin *n*
**keratometer** Keratometer *n*
**keraunophone** Fernblitzanzeiger *m*
**kerb** Spurbegrenzung *f*
**kerf** Brennschnittspalt *m*, Einschnitt *m*, Schnittbreite *f* einer Säge
**kermes mineral** Antimonzinnober *m*, Mineral-
kermes *m*
**kermisite** Antimonblende *f*, Pyrantimonit *m*,
Rohspießglanzerz *n*, Rotspießglanzerz *n*,
Spießglanz-blende *f*, -kermes *m*, -zinnober *m*
**kerned letter** unterschnittener Buchstabe *m*

**kernel** Kern *m*, (stable inner electron group) Rumpf *m*, Rumpfatom *m*, Wachstumszentrum *n* ~ **of an integral equation** Kern *m* ~ **of a roasted ore** Herz *n* eines gerösteten Erzes ~- **counting apparatus** Körnerzählapparat *m* ~ **electron group** Elektronenrumpf *m*

**kerogen** Kerogen *n*

**kerosene** Kerosin *n*, Leuchtöl *n*, Leucht-petroleum *n*, Paraffinöl *n*, Petroleum *n* ~ **engine** Petroleummotor *m* ~ **lamp** Petroleum-lampe *f* ~ **outfit** Petroleumanlage *f* ~ **tractor** Petroleum-traktor *m*, -schlepper *m*

**Kerr,** ~ **cell** Fremdlicht *n*, Kerr-kondensator *m*, -zelle *f* ~**-cell circuit** Kerrzellenschaltung *f* ~ **effect** Kerreffekt *n* ~ **magneto-optic apparatus** Kerr-Zelle *f*

**ketch** Ketsch *f*, Kutter *m*

**ketonic group** Ketongruppe *f*

**kettle, to** ~ verkochen

**kettle** Heizkessel *m*, Kessel *m*, Pfanne *f*, Waser-kessel *m* ~ **of the compass** Kompaß-büchse *f*, -kessel *m* ~ **casing** Kesselmantel *m* ~ **drum** Kesselpauke *f*, Pauke *f* ~ **drum wrench** Paukenstimmschlüssel *m* ~**-shaped** kesselförmig ~**-shaped deepening** Binge *f*, Pinge *f* ~ **soap containing large quantities of builders (or of fillers)** Sparkernseife *f*

**key, to** ~ festkeilen (mach.), kehlen, keilen, tasten, unterlegen to ~ **off** (a carrier) austasten (rdo) **to** ~ **on** aufkeilen **to** ~ **on or in** eintasten (rdo) **to** ~ **on the crank** die Kurbel aufkeilen **to** ~ **to the shaft** auf der Welle festkeilen **to** ~ **seat** nuten

**key** Anleitung *f*, Diebel *m*, Druckstöpsel *m*, Drucktaste *f*, Feder *f*, Fütterung *f*, Hellig-keitsumfang (TV) *m*, Keil *m*, (switching) Kippschalter *m*, (of wind instrument) Klappe *f*, Querkeil *m*, Schalter *m*, Schlüssel *m*, Setz-keil *m*, Splint *m*, (typebriting) Taste *f*, Taster *m*, Tonart *f*, (book-binding) Unterlage *f*, (river weirs) Verzahnung der Betonsohle, Vorsprung *m*, Zeichenerklärung *f*

**key,** ~ **of arch** Bogenscheitel *m*, Keil *m* des Gewölbes ~ **of an image** Bildcharakter *m* ~ **of joint** Fugenkeil *m* ~ **and key way** Nut *f* und Feder *f* ~ **for printing totals** Summendruck-taste *f* ~ **for signaling lamp** Blinktaster *m* ~ **and slot** Federkeil *m* und Nut *f* ~ **of a stopcock** Küken *n*, Lilie *f* ~ **to symbols** Zeichenerklärung *f*

**key,** ~ **action** Klappenmechanismus *m* ~ **address** Schlüsseladresse *f* ~ **alloy** Vorlegierung *f* ~ **arrangement** Keilanordnung *f* ~**-bar** Kombinationsstab *m* ~ **bar rotor** Keilstabläu-fer *m* ~ **bed** Keilfläche *f* ~ **bit** Schlüsselbart *m* ~ **blank** Rohschlüssel *m* ~ **block** Konturplatte *f*

**keyboard** Klaviatur *f*, Manual *n* (Orgel), Schlüsselfeld *n*, Tastatur *f* (Klavier), Tastenfeld *n*, Tastwerk *n* ~ **base** Sendersockel *m* ~ **eccentric** Klaviaturexcenter *m* ~ **lock** Tastensperre *f* ~ **perforator** Tastenlocher *m* ~ **printing telegraph** Tastenschnelltelegraf ~ **selection** Wahl *f* mittels Tastatur ~ **sender** Sender *m* mit Tastenfeld ~ **transmitter** Tastengeber *m* ~ **typewriter** Klaviaturschreibmaschine *f* ~ **unit** Sender *m*

**key,** ~ **bolt** Keilbolzen *m*, Schloßriegel *m* ~ **button** Taste *f*, Tastenknopf *m* ~ **button lifter** Tastenknopfheber *m* ~ **cabinet** Teilnehmer-handvermittlung *f* ~ **card punch** Handlocher *m* ~ **cartridge** Meß-kartusche *f*, -patrone *f* ~ **chatter** Tastschlag *m* ~ **chirp** Tastgeräusch *n*, (after key has been opened) Tastprellen *n* ~ **clerk** Prüfbeamter *m* ~ **click** Tast-klick *m*, (when closing key) -prellen *n*, Tast-schlag *m* ~ **clicking** (caused on closing key) Tastgeräusch *n* ~ **clicks** Geräuschtastung *f* ~ **comb** Tasten-kamm *m* ~**-controlled continuous wave** ge-tastete ungedämpfte Welle *f* ~ **cutter** Langloch-fräser *m* ~ **dowel** Tragdübel *m* ~ **driver** Keil-auszieher *m*, -treiber *m* ~ **equipment** lebens-wichtige Anlageteile *pl* ~ **face** (of nut) Schlüs-selfläche *f* (der Mutter) ~ **file** Spaltfeile *f* ~ **forcing screw** Abdruckschraube *f* ~ **form** Leit-form *f* ~ **fossil** Leitfossil *n* ~ **head** Keilnase *f*, Tastenknopf *m* ~ **hole** Schlüsselkranz *m* ~ **hole saw** Lochsäge *f* ~ **horn** Klappenhorn *n* ~ **industry** Schlüsselindustrie *f* ~ **joint** Schlüssel-gelenk *n* ~ **letter** Tastenaufschrift *f*

**key-lever** Schlüsselhebel *m*, Tasten-arm *m*, -hebel *m* ~ **end** Tastenhebelschwanz *m* ~ **guide** Tastenhebelführung *f* ~ **locking bar** Sperr-schiene *f*

**key,** ~ **list** Schlüsselliste *f* ~ **metering** Tasten-zählung *f* (telegr.) ~**-modulated continuous wave** getastete ungedämpfte Welle *f* ~ **Morse system** Handmorsesystem *n* ~ **note** Grundton *m* ~**-operated lock switch** schlüsselbetätigter Riegeltrenner *m* ~ **pair** Zähladernpaar *n* ~ **picture** Stimmungsbild *n* ~ **plate** Konturplatte *f* ~ **plug** Stöpsel *m* ~ **point** Ausgangs-, Schlüssel-, (fig.) Schwer-punkt *m* ~ **position** Schlüssel-stellung *f* ~ **profiling machine** Schlüsselkopier-maschine *f* ~ **proportioning** Keilberechnung *f* ~ **pulsing** Fernwahl *f* mit Nummernscheibe oder Tastensatz ~**-pulsing A position** A-Platz *m* mit Wahlzusatz ~**-pulsing position** Zahlen-geberplatz *m* ~ **punch** Tastenlocher *m* ~**-ready** schlüsselfertig ~ **recovery** Entzifferung *f* ~ **relay** Tastrelais ~ **ring** Schlüsselring *m* ~ **rod** Tastenstange *f*

**key-seat** Keil-nut *f*, -schlitz *m*, -sitz *m*, Nut *f*, Nute *f* ~ **cutter** Keilnutenfräser *m* ~ **milling cutter** Nutenfräser *m* ~ **milling machine** Nu-tenfräsmaschine *f* ~ **steel rule** Keilnutenlineal *n*

**key-seater** Keilnutenziehmaschine *f*

**key-seating** Keilnuten *n* mit ziehendem Schnitt fräsen; Splintlager *n* ~ **of parts** Nuten *n* von Werkstücken ~ **machine** Keilnutenziehmaschi-ne *f*, Nutenziehmaschine *f*

**key,** ~ **selector switch** Tastwahlschalter *m* ~ **sender** Sendeklavier *n*, Zahlengeber *m* (teleph.) ~**-sending position** Zahlengeberplatz *m* ~ **set** Tastatur *f*, Tasten-feld *n*, -satz *m* ~ **shaft** Keil-welle *f* ~ **shelf** Schaltergrundplatte *f*, Schlüssel-brett *n*, Umschaltergrundplatte *f* ~ **slide** Schlüs-selschieber *m* ~**-slot chisel** Keillochmeißel *m* ~ **socket** Fassung *f* mit Schalter ~ **speed** Hand-tempo *m* ~ **station** Muttersender *m* ~ **steel** Keilstahl *m*

**keystone** Bogenschluß *m*, Füllsplit *m* (Straßen-bau), Gewölbescheitel *m*, Schlußstein *m*, Tra-pezfehler *m*, Verschlußblock *m* (Schmelz-ofen) ~ **distortion** Schlußsteinverzerrung *f* ~ **effect** Trapezeffekt *m*

**key,** ～ **stroke** Tastenanschlag *m* ～ **substance** Ausgangsstoff *m* ～ **switch** Stromschlüssel *m* ～ **tape** Schlüssellochstreifen *m* ～ **thump** Tastgeräusch *n*, -klick *m*, -prellen *n*, -schlag *m*
**keyway** Keilnut *f*, Keilweg *m*, Kerbnute *f*, Mitnehmernut *f*, Nut *f*, Nute *f*, Schlüsselschlitz *m* ～ **and groove cutting** Nutenhobeln *n* ～ **in the hub** Nabennute *f*
**keyway,** ～**-broach (or cutter) bar** Keilnutenziehmesser *n* ～ **cutting** Keilnutenstoßen *n*, Nutenfräsarbeit *f* ～ **gauge** Keillehre *f*, Keilschablone *f* ～ **milling** Keilnutenfräsen *n*, Keilnuten *pl* fräsen ～ **planing machine** Nuten-hobelmaschine *f*, -stoßmaschine *f* ～ **tolerance** Abmaß *n* der Nutbreite
**key,** ～**-waying** Nutenfräsarbeit *f* ～ **wedge** Keilschlüssel *m*, Tangentialkeil *m* ～ **width** Keildicke *f* ～ **word** Schlüsselwort *n*, Stichwort *n* ～**-word index** Stichwörterverzeichnis *n* ～ **work** Klappenmechanismus *m* ～**-worked** handgetastet
**keyed** aufgekeilt, festgekeilt, getastet, verkeilt, versplintet ～ **advertisement** Kennzifferanzeige *f* ～ **blocks** mit Nut und Feder ineinander passende Blöcke *pl* ～ **CW jamming** getastete Dauerstrichstörung *f* ～ **harp** Klavierharfe *f* ～ **pointing** Kehlfuge *f* ～ **washer** Sicherungsscheibe *f*
**keyer** Radarmodulator *m*, Tastgerät *n* ～ **adapter** Faksimilemodulator *m* ～ **coupling** Querkeilverbindung *f*
**keying** Aufkeilung *f*; Funktasten *n*, Keilschluß *m*, Tastung *f* ～ **of tie of a lock gate** Keilvorrichtung *f* der Zugstange am Schleusentor
**keying,** ～ **apparatus** Tastgerät *n* (rdo) ～ **cable** Fernsteuerungskabel *n* ～ **chirps** Zirpen *n* (beim Tasten) ～ **circuit** Tast-kreis *m*, -leitung *f* ～ **device** Tasteinrichtung *f* (rdo), Taster *m* ～ **filter** Vorfilter *m* ～ **head** Tastkopf *m* ～ **line** Fernsteuerungsleitung *f* ～**-out** Austastung *f* ～ **rectifier** Tastgleichrichter *m* ～ **references** Bezugszahlen *pl* (in Zeichnungen) ～ **relay** Tast-, Taster-relais *n* ～ **scheme** Mischerschaltung *f*
**keyless,** ～ **ringing** selbsttätiger Anruf *m* (ohne Rufschlüssel) ～ **watch** Remontoiruhr *f*
**keys** Klaviatur *f*
**K factor** Multiplikationsfaktor *m*
**khaki** Khaki *n*
**khlopinite** Khlopinit *m*
**kick, to** ～ zurückstoßen **to** ～ **off** andrehen (einen Ausbruch veranlassen) **to** ～ **out** herausschlagen
**kick** Ausschlag *m*, Gegenwirkung *f*, Rückstoß *m*, Schlag *m*
**kickback, to** ～ rückwärts starten, zurückprallen
**kickback** Rücklauf *m* ～ **danger** Rückschlaggefahr *f* ～ **power supply** Rücklaufhochspannungsspeisung *f* ～**-proof** rückschlagsicher ～ **pulse** Rückschlagimpuls *m* ～ **safety device** Rückschlagsicherung *f*
**kick,** ～ **indication** (direction finding) Zuckanzeige *f* ～**-off pressure** Anfangsdruck *m* ～**-off valve** Aufschlag-, Start-ventil *n*, Startverschluß *m* ～**-out of cards** maschineller Ausstoß *m* von Karten ～ **plate** Trittleiste *f* ～ **reading** (direction finding) Zuckanzeige *f* ～ **sorter** Impulshöhenanalysator *m* ～ **starter** Kickstarter *m*, Tretanlasser *m*

**kicker** Schläger *m* ～ **light** Aufheller *m* (film)
**kicking,** ～ **coil** Drosselspule *f* ～ **process** Fettwalke *f*
**kid leather** Ziegenleder *n*
**kidney,** ～ **ore** Adlerstein *m*, roter Glaskopf, Nierenerz *n* ～**-shaped** nierenförmig, nierig
**kien oil** Kienöl *n*
**kier-boiling** Kesselkochung *f*
**kieselguhr** Infusorienerde *f*, Kieselgur *f* ～ **brick** Kieselgurstein *m* ～ **dosing unit** Kieselgurdosiergerät *n*
**kiesel kopal** Kieselkopal *n*
**kieserite** Kieserit *m*
**kill, to** ～ abstehen lassen, abtöten, abtun, aufheben, (melt) beruhigen, niedermachen, (flotation foam) niederschlagen, töten, unwirksam machen to ～ **a (power) circuit** eine Leitung *f* spannungslos machen **to** ～ **lime** Kalk *m* verlöschen
**killed** abgetötet ～ **spirits** Lötwasser *n* ～ **steel** beruhigter oder ruhiger Stahl *m* ～ **temperhardening automatic screw-machine steel** beruhigter Vergütungsautomatenstahl *m*
**killer** Verminderer *m* des Nachleuchtens ～ **circuit** fremdgesteuerte Austastschaltung *f*
**killing** Abstehenlassen *n*, Beruhigung *f*, (of a smelting bath) Ruhigwerden *n*, Tötung *f* ～ **of fluorescence** Fluoreszenzunterdrückung *f* ～ **period** (steel melting) Ausgarzeit *f*
**kiln, to** ～ ausdarren, dörren **to** ～**-dry** abdarren, darren, dörren
**kiln** Darre *f*, Darrofen *m*, Kaloriferendarre *f*, Kiln *m*, Ofen *m*, Schachtofen *m*, (for drying sleepers or ties) Schwellendörrofen *m*, Trokkendarre *f* ～ **with overlying beds** Etagenofen *m*
**kiln,** ～ **combined with separated hurdles** mit getrennten Horden kombinierte Darre *f* ～**-dried** ofentrocken ～**-dried wood** Darrholz *n* ～**-dry weight** Darregewicht *n* ～ **drying** Ofentrocknung *f* ～**-drying test** Darrprobe *f* ～**-floor clearer** Darreabräumer *m* ～ **plant** Darranlage *f* ～**-proof decalcomanias** einbrennbare Abziehbilder *pl* ～ **regulation** Darrordnung *f* ～ **sow (or warming tub)** Darrsau *f*
**kilo** Kilogramm *n* ～**-ampere** Kiloampere *n* ～ **curie** Kilocurie *f* ～**-cycle (Kc/s)** Kilohertz *n* **(KHz)** ～**-dyne** Kilodyn *n* ～**-electron-volt** Kilo-Elektronenvolt *n*
**kilogram** Kilogramm *n* ～**-calorie** große Kalorie, Kilogrammkalorie *f*, Kilokalorie *f* ～**-meter** Kilogrammeter *n*, Meterkilogramm *n* ～ **molecule** Kilogrammolekül *n*, Kilomol *n*
**kilohm** Kiloohm *n*
**kiloohm** Kiloohm *n*
**kilometer** Kilometer *m* ～ **divider (or circle)** Kilometerzirkel *m* ～**s flown** Flugkilometer *pl* **at . . .** ～ **intervals** in Abständen von . . . Kilometern *pl* ～**s per hours** Kilometerstunde *f*, Stundenkilometer *m*
**kilometric waves** Langwellen *pl*
**kilovar (kvar)** Blindkilowatt *n* ～ **control** Blindleistungsregelung *f* ～**-hour** Kilovarstunde *f* ～ **requirements** Blindleistungsbedarf *m*
**kilovolt** Kilovolt *n* ～**-ampere** Kilovoltampere *n* ～**-ampere reactive** Blindkilovoltampere *n*
**kilowatt** Kilowatt *n* ～**-hour** Kilowattstunde *f*
**kin** verwandt

kind Art *f*, Art und Weise, Gattung *f*, Marke *f*, Qualität *f*, Sorte *f*

kind, ~ of Bessel function Art *f* der Besselfunktion ~ of casting Gußart *f* ~ of charge Ladungsart *f* ~ of clay Tonart *f* ~ of coal Kohlensorte *f* ~ of current Strom-art *f*, -gattung *f* ~ of drive Antriebsart *f* ~ of fit Sitzart *f* ~ of goods Warengattung *f* ~ of lead Bleiart *f* ~ of loading Art *f* der Beanspruchung ~ of sand Sandsorte *f* ~ of stressing Beanspruchungsart *f* ~ of tire Bereifungsart *f*

kindle, to ~ anfachen, anmachen, anzünden, entzünden to ~ a fire Feuer *n* anmachen

kindling Anfachung *f*, Anheizen *n*, Entzündung *f* ~ point Entzündungspunkt *m*, Flammpunkt *m* ~ temperature Entzündungstemperatur *f*, Flammpunkt *m*

kindred Art *f*; artverwandt

kine Bildröhre *f* (TV)

kinematic viscosity Bewegungszähigkeit *f*

kinematical kinematisch

kinematics Bewegungslehre *f*, Getriebelehre *f*, Kinematik *f*, Zwangslauflehre *f*

kinemo drift indicator Kinemoabdrängungsmesser *m*

kinemoderivometer Kinemoabdrängungsmesser *m*

kinemometer Kinemometer *n*

kineophone Kineofon *n*

kineoptoscope Kineoptoskop *n*

kineplastic kineplastisch

kinescope Bildröhre *f*, Kineskop *n* ~ recording gefilmtes Fernsehprogramm *n*

kinestate Kinestat *n*

kinesthesia Kinesthesie *f*

kinesthetic kinesthetisch ~ hallucination Bewegungshalluzination *f* ~ sensitivity kinesthetisches Empfindungsvermögen *n*

kinetheodolitic photograph Kinetheodolitenaufnahme *f*

kinetic kinetisch ~-control system dynamische Reglung *f* ~ effect kinetische Wirkung *f* ~ energy Arbeitsvermögen *n*, Bewegungsenergie *f* kinetische Energie *f*, lebendige Kraft *f*, Strömungsenergie *f* ~ energy head Staudruck *m* ~ energy at the muzzle Mündungs-arbeit *f*, -wucht *f* ~ force lebendige Kraft *f* ~ magnitude Bewegungsgröße *f* ~ quantity Bewegungsgröße *f* ~ theory of gases kinetische Gastheorie *f*

kinetics Bewegungslehre *f*, Dynamik *f*, Getriebelehre *f*, Kinetik *f*

kinetoscope Betrachtungsapparat *m*

king block (Gießerei) Königsstein *m*

king-bolt Achszapfen *m*, Drehbolzen *m*, Hauptbolzen *m*, Königsbolzen *m*, Königsstange *f*, Königszapfen *m*

king-pin Achsschenkelbolzen *m*, (steeringknuckle pivot) Achszapfen *m*, Anlenkbolzen *m*, Bockdalben *m*, Königswelle *f*, Königszapfen *m* ~ angle Spreizung *f* (Vorderräder) ~ bearing Achsschenkellager *n* ~ bushing Achsschenkelbolzenbuchse *f* ~ inclination Achssturz *m*, Achszapfensturz *m* ~ side inclination angle Spreizwinkel *m*

king, ~ pillar Königszapfen *m* ~ pivot (or pin) Mittelzapfen *m* ~ post Außenspannturm *m*, Hängesäule *f*, Königsstück *n*, Stützträger *m* ~ strut Königsstück *n*

kingston valve Bodenventil *n*

kink, to ~ knicken, Kinken bilden (Seil)

kink Biegung *f*, (in chain or wire rope) Kink *m*, Klanke *f* (Seil), Knick *m*, Knickpunkt *m*, Knickstelle *f*, Knickung *f*, (in chain or wire rope) Knoten *m*, Schleife *f* ~ band density Knickbanddichte *f* ~ pair formation Sprungpaarbildung *f*

kinking Knicken *n*

Kipp, ~ oscillation Kippschwingung *f* ~ relay Röhrenwippe *f*

kips for clogs Pantinenkipse *f*

Kirchhoff's law Kirchhoffsches Gesetz *n*

kish Garschaum *m*

kit Ausstattung *f*, Bepackung *f*, hölzernes Gefäß *n*, Gepäck *n*, Werkzeugtasche *f*

kitchen Küche *f* ~ chopper Hackbeil *n* ~ dresser Küchenanrichte *f* ~ machine Küchenmaschine *f*

kitchenette Kleinküche *f*

kite Drache *m*, Drachen *m* ~ ascent Drachenaufstieg *m* ~ balloon Drachenballon *m* ~-control surface Kastenruder *n* ~ span Drachenspann *m* ~ start Drachenstart *m* ~ string Drachenschnur *f* ~ strop Drachenstropp *m* ~ sweep Drachenleine *f* ~ wing Drachenflügel *m*

Kjedahl flask Kjedahlkolben *m*

Kjellin furnace Kjellin-Ofen *m*

klaprotholite Kupferwismuterz *n*

klieg, ~ light Aufheller *m* ~ lights Sonne *f*

klirr factor Klirrfaktor *m* (acoust.)

klydonograph Klydonograf *m*

klystron Klystron *n*, Klystonröhre *f*

knack Kniff *m*, Kunstgriff *m*

knag Knorren *m* ~ in wood Knast *m* im Holz

knaggy knästig, knorrig

knapsack Rückengepäck *n*, Rucksack *m*, Tornister *m*

knead, to ~ (dough) einmachen, kneten, wirken to ~ the dough den Teig anmachen to ~ thoroughly durchkneten

kneadable knetbar, teigig, verformbar

kneader Kneter *m*, Knetmaschine *f*, Zerfaserer *m* ~ mixer Mischkneter *m*

kneading Knetung *f* ~ action komprimierende Wirkung *f* (der Reifen) ~-and-mixing machine Flügelknetmaschine *f* ~ blade Knetflügel *m* ~ mass (or material) Knetmasse *f* ~ mill Knet-apparat *m*, -mühle *f*, Mischmaschine *f* ~ process Knetprozeß *m* ~ trough Back-mulde *f*, -trog *m*

knee, to ~-halter koppeln

knee Knick *m*, Knickpunkt *m*, Knickstelle *f*, Knie *n*, Knierohr *n*, Konsol *n*, Kröpfung *f*, Kurbel *f*, Rohrknie *n*, Winkeltisch *m*, Zungenstuhl *m* (r.r.) ~ of characteristic oberer Kennlinienknick ~ of commutation Knie *n* der Kommutierungskurve

knee, ~-action shock absorber Hebelstoßdämpfer *m* ~-action suspension Kniegelenkfederung *f* ~-and-column-type milling machine Konsolfräsmaschine *f* ~ bar Kniestange *f* ~ boot Kniestiefel *m* ~ bracing Halbdiagonalverspannung *f*, K-Verband *m* ~ bracket Eckblech *n* ~ cap Kniescheibe *f* ~ clamp (for arm braces) Stützenfußhalter *m*, Traverse *f* ~ clamp lever Konsolklemmhebel *m* ~ drive Konsolantrieb *m* ~ elevating shaft Welle *f* für die Höhenstellung des Konsols ~ **joint** Kniegelenk *n* **~-joint bar** Kugelstange *f* ~ **lever** Kniehebel *m* ~ **pad** Knieschützer *m* ~ **pipe** Knierohr *n* ~ **roll** (saddle) Kniepausche *f* **~-shaped** knieförmig ~ **shot** mittlere Nahaufnahme *f* (film) ~ **support** Kniestütze *f* ~ **tap holder** gekröpfter Gewindebohrerhalter *m* ~ **timber** Knieholz *n* **~-type miller** Konsolfräsmaschine *f* ~ **valve** Knieventil *n*

**knife, to** ~ (Leder) beschneiden, rakeln, schneiden

**knife,** ~ **adjustment** Messer-anstellung *f*, -einstellung *f* ~ **attachment** Messereinrichtung *f* ~ **bar** Messerstange *f* ~ **barking machine** Schnellschäler *m* **~-blade contact** Messerkontakt *m* **~-blade switch** Klingelschalter *m* ~ **box base plate** Messerkastenunterplatte *f* ~ **clip** Messerhalter *m* ~ **contact** Messereinsatz *m* ~ **cover** Messerschützer *m* ~ **crusher** Messerbrecher *m* ~ **cutter** Vorschneider *m* ~ **disk** Messerscheibe *f* ~ **driving shaft** Messerwelle *f* ~ **drum (or roll)** Messerwalze *f*

**knife-edge** Messerschneide *f*, Schneide *f* ~ **bearing** Mittungsschneide *f*, Schneidlager *n*, Zentrierschneide *f* ~ **bearings** Schneidenlagerung *f* ~ **contact tip** Planmeßhütchen *n* mit Schneide ~ **effect** Messerschneideneffekt *m* ~ **lightning protector** Schneidenblitzableiter *m* ~ **nose pieces** Messerschnäbel *pl* ~ **pointer** Messerzeiger *m* ~ **relay** Ankerrelais *n*, Relais *n* mit Schneidenlagerung, Schneidenankerrelais *n* ~ **support** Schneidenlager *n* ~ **suspension** Schneide-, Schneiden-aufhängung *f*, Schneidenlagerung *f* ~ **test** Messerkantenversuch *m*

**knife,** ~ **file** Schneidfeile *f* ~ **folding drum** Messerfalztrommel *f* ~ **grinder** Messer-, Scheren-schleifer *m*, Schleifapparat *m* **~-grinding attachment** Messerschleifapparat *m* ~ **handle** Messerschale *f* **~-handle filer (or shaper)** Messerheftfeiler *m* ~ **leveler** Messerebener *m* ~ **motion** Messerführung *f* ~ **pick** Messerpicke *f* ~ **pruning saw** Messersäge *f* ~ **section** Klinge *f* ~ **seizing** Messerbändsel *n* ~ **shaft** Messerwelle *f* **~-shaft bearing** Messerwellenlager *n* **~-shaped lightning arrester** Schneidenblitzableiter *m* ~ **smith** Messerschmied *m* ~ **switch** Hebelschalter *m* **~(-blade) switch** Messerschalter *m* **~-switch prong** Kontaktmesser *n* ~ **tool** Drehmesser *m* ~ **weeder** Unkrautmesser *m*

**knifing,** ~ **the filler** Spachtelung *f* ~ **glaze** Spachtelmasse *f*

**knit, to** ~ stricken, wirken ~ **goods** Strickwaren *pl*

**knitting,** ~ **frame** Wirk-maschine *f*, -stuhl *m* ~ **industry** Strickereiindustrie *f* ~ **machine** Strickmaschine *f* ~ **machine with double mechanism** Doppelstrickmaschine *f* **~-machinery factory** Strickmaschinenfabrik *f* ~ **needle** Stricknadel *f* ~ **trade** Wirkerei *f* ~ **wool** Strickwolle *f* ~ **yarn** Strickgarn *n*

**knob** Blocktaste *f*, (on rolls etc.) Bombage *f*, Däumling *m*, Griff *m*, Hügel *m*, Isolierrolle *f*, Knagge *f*, Knauf *m*, Knollen *m*, Knopf *m*, Knoten *m*, Triebscheibe *f*, Warze *f* **~-headed** knopfartig

**knobbed** mit einem Knauf *m* oder mit einem Knopf *m* versehen, (as a roll on a textile-finishing machine) mit Bombagen *f* überzogen

**knobbled iron** Herdfrischeisen *n*

**knobbling** Herdfrischarbeit *f* ~ **fire** Feinfeuer *n*

**knobby** knästig, knollig

**knock, to** ~ klopfen, pochen, schlagen, stoßen **to** ~ **down** demontieren, umstoßen **to** ~ **loose** losschlagen **to** ~ **off** abhauen, abklopfen, abschlagen, abstoßen **to** ~ **out** ausschlagen, herausschlagen, herausstoßen **to** ~ **over** umwerfen **to** ~ **straight** geradeklopfen

**knock** Klopfen *n*, Schlag *m*, Stoß *m* ~ **of detonation** Zündschlag *m*

**knock,** **~-down form** Bauform *f* **~-down metal ~-free** klopffrei ~ **inhibitor** Klopfbremse *f* ~ **intensity** Klopfintensität *f*, Klopfstärke *f* ~ **limit** Klopfgrenze *f* ~ **meter** Klopfmeßgerät *n* **~-off action** Fadenwächter *m* **~-on electrons** angestoßene Elektronen *pl*

**knock-out** Ausbrechöffnung *f*, Ausheber *m* (metall.) ~ **bar** Ausstoßstift *m*, Auswerferstange *f* ~ **box** (petroleum) Auffang *m* für die Ausscheidung ~ **die** Stanzform *f* ~ **frame** Ausdrückrahmen *m* ~ **junction box** Scherbenwandverteilungsdose *f* ~ **key** Keiltreiber *m* ~ **pin** Auswerferbolzen *m* ~ **shop** Ausleerhalle *f* ~ **spindle** Steckachse *f*

**knock,** ~ **promoters** Klopfförderer *pl* ~ **proof** klopffest ~ **property** Klopfeigenschaft *f* ~ **rating** Bestimmung *f* der Klopffestigkeit, Klopfeinschätzung *f*, Oktanzahl *f* **~-reference fuel** Vergleichskraftstoff *m* zur Bestimmung der Klopffestigkeit ~ **stone** Schlagstein *m* ~ **suppressor** Klopfbremse *f* **~-testing method** Verfahren *n* zur Bestimmung der Klopffestigkeit ~ **wrench** Gestängerohrdrehschlüssel *m*

**knocked,** ~ **down** demontiert **~-on atom** Rückstoßatom *n*

**knocker** Klopfer *m*, Klöppel *m*, Türklopfer *m*

**knocking** klopfendes Geräusch *n*, Klopfen *n* (d. Motors) ~ **in the crank** Kurbelschlag *m* ~ **in the piping** Gestängeschläge *pl* ~ **of (car) bearings** Hämmern *n* von Lagern

**knocking,** ~ **behaviour** Klopfverhalten *n* **~-in device** Einstoßvorrichtung *f* ~ **inception depending on ignition moment** Klopfeinsatz *m* abhängig vom Zündzeitpunkt ~ **noise** Klopfgeräusch *n* **~-off** Abstoßen *n* **~-over bar** Abschlagschiene *f* **~-over cam (weaving)** Abschlagexcenter *n* **~-over comb (weaving)** Abschlagkamm *m* **~-over device (weaving)** Abschlageinrichtung *f* ~ **process** Klopfvorgang *m* ~ **tendency** Klopfneigung *f*

**knoll** Kuppe *f*

**Knorr air brake** Knorrbremse *f*

**knot, to** ~ knoten, (ver)knüpfen

**knot** (timber) Ast *m*, Knollen *m*, Knorren, Knötchen *n*, Knoten *m* (naut.), (wood) Knotenstück *n*, Schleife *f*, Seemeile *f*, (in rope) Stich *m* ~ **in wood** Ast(knoten) *m* im Holz ~ **catcher** Astfänger *m* ~ **crown** Kreuzknoten *m* ~ **load** Knotenpunktbelastung *f*

**knotless** astfrei, knorrenfrei

**knots** (in calcined ore) Erzkern *m* ~ **and ties** Bunde *pl* und Knoten *pl*

**knotted** geknotet, geknüpft ~ **bar iron** Knoteneisen *n* ~ **wire netting** Knotengeflecht *n* ~ **work** Makrameearbeit *f*

**knottiness** knotige Beschaffenheit *f*

**knotting machine** Knüpfmaschine *f*

knotty heikel, heikelig, knästig, knollig, knorrig, knotig
know-how Erfahrungen *pl*
knowledged bewandert
knub Noppe *f* ~ yarn Noppe *f*
knubbed knotig
knuckle Buckel *m* (min.), Gelenk *n*, Kardandrehzapfen *m*, Knie *n* ~ joint Gelenkverbinder *m*,
Kardandrehzapfen *m*, Kardangelenk *n*, Knie-,
Kreuz-gelenk *n* ~ lug on gun-carriage frame
Auflaufstück *n* ~ pin Anlenkbolzen *m*, Kuppelbolzen *m* ~ support Achszapfenlager *n* ~
thread Rundgewinde *n* ~ washer Kardan-
scheibe *f*
knurl, to ~ aufrauhen, kordeln, (an edge) kordieren, (coins) molletieren, randerieren, rändeln, riefeln, riffeln, zähneln
Knurl Kordelrad *n*, Kordelung *f*, Rad *n*, Rändelrad *n*, Ränderierrad *n*, Riffelung *f* ~ pitch
Kordel-, Rändel-teilung *f*
knurled gerändelt, gerändert, gereifelt, gerieft,
geriffelt ~ diopter ring Okularrändelring *m* ~
grip Griff *m* für Indexhebel ~ head Rändel *n*,
Rändelkopf *m* ~ head with diopters Dioptrien-
ring *m* ~-head screw Rändelschraube *f* ~
knob Kordelgriff *m*, Rändel-, Zier-knopf *m* ~
nut Kordelmutter *f*, gerillte Mutter *f* ~ ring
Rändelring *m* ~ screw Kordelschraube *f*
knurling Kordeln *n*, Kordelung *f*, Kordierung *f*
~ swing tool Rändelschwenkwerkzeug *n* ~
tool Händel *m*, Kordelapparat *m*, Kordier-,
Ränderier-werkzeug *n* ~ tool holder Rändel-

rollenhalter *m* ~ wheel Rändelrad *n*
kodurite Kodurit *m*
Koepe winding engine Koepe-Förderung *f*
Köhler, ~ (illuminating) principle Köhlerprinzip
*n* (Köhlersches Prinzip) ~ radiant-field principle Köhler-Leuchtfeldverfahren *n*
konimeter Staubmesser *m*
Koppers, ~ combination oven Koppers-Verbund-
ofen *m* ~ vertical-flue oven Koppers-Vertikal-
kammerofen *m*
Krarup, ~ cable Krarupkabel *n* ~ conductor
Krarupader *f* ~ loading Krarupisierung *f*
~ winding Krarupumspinnung *f*
krarupization Krarupisierung *f*
krarupize, to ~ krarupieren, krarupisieren
kremnitz white Kremserweiß *n*
Kruger cell Krügerelement *n*
krypton Krypton *n*
kryptoscope Kryptoskop *n*
kryptosterol Kryptosterin *n*
K truss Halbdiagonalverspannung *f*, K-Verband
*m*
Kullmann's communication system Kullmann-
Verfahren *n*
Kummer method of dismantframes Kummersche
Auslösung *f*
kyanising (of timber) Fäulnisverhütung *f*,
Kyanisierung *f*
kyanize, to ~ mit Sublimat tränken
kymograph Kymograf *m*, Schwingungsregistrier-
gerät *n*
kymography Kymografie *f*

# L

**L** Rohrbogen *m*
**label, to** ~ bekleben, beschriften, bezeichnen, bezetteln, etikettieren, signieren, mit einer Anschrift *f* versehen
**label** Aufklebezettel *m*, Aufschrift *f*, Beklebezettel *m*, Bezeichnung *f*, Bezeichnungsschild *n*, Etikett, Gütezeichen *n* (of quality), Kenngruppe *f*, Marke *f*, Markierung *f* (info proc.), Plättchen *n*, Schild *n*, Vorspann *m* (print.), Zettel *m* ~ **for sticking on** Tektur *f* ~ **coding** Kode-Markierung *f*
**labeling** Bezeichnung *f* ~ **machine** Etikettiermaschine *f*
**labile** gleitend, kippling, labil, schwankend, umfällig, unbeständig, unstet, unstetig ~ **oscillator** ferngesteuerter Oszillator *m*
**labor, to** ~ arbeiten
**labor** Anstrengung *f*, Arbeit *f*, schwere körperliche Arbeit *f*, Arbeiterschaft *f*, Werk *n* ~ **bureau** Arbeitsamt *n* ~ **camp** Arbeitslager *n* ~ **charges** Arbeitskosten *pl* ~ **conference** Arbeitstagung *f* ~ **cost** Arbeitskosten *pl* ~ **council** Arbeitsgemeinschaft *f* ~ **dispute** Arbeitskampf *m* ~ **efficiency** Arbeitsleistung *f* ~ **exchange** Arbeitsnachweis *m* ~ **market** Arbeitsmarkt *m* ~ **movement** Arbeitsbewegung *f* ~ **organization** Arbeitergemeinschaft *f* ~ **peace** Arbeiterfriede *f* ~ **plan** Arbeitsplan *m* ~ **pool** Arbeitseinsatz *m* ~**-saving** Arbeitsersparnis *f*; arbeitsparend ~ **supply** Arbeitseinsatz *m*
**laboratory** Erhitzungskammer *f*, Forschungsanstalt *f*, Laboratorium *n*, Offizin *f*, Prüfanstalt *f*, -raum *m*, Untersuchungsraum *m*, Versuchsraum *m*, Werkstatt *f*
**laboratory,** ~ **assistant** Laborant *m* ~ **crusher** Laboratoriumsmühle *f* ~ **inspector** Laborant *m* ~ **naphta** normales Benzin *n* für Laboratoriumsgebrauch *m* ~ **sample crushed to below 10 mm** Grobprobe *f* ~ **sample printing machine** Modelldruckmaschine *f* ~ **sole** Herd *m* ~ **test** Laboratoriums-prüfung *f*, -versuch *m* ~ **truck** Laboratoriumskraftwagen *m* ~ **worker** Laborant *m* ~ **yield** Laboratoriumsausbeute *f*
**laborer** Arbeiter *m*, Hilfsarbeiter *m*, Tagelöhner *m*
**laboring** werktätig
**laborious** umständlich
**labradorite** Labradorfeldspat *m*, Labradorit *m*
**labyrinth** Leitfläche *f* ~ **of loud-speaker** Lautsprechertonführung *f*
**labyrinth,** ~ **gland** Labyrinthdichtung *f* (turbine) ~ **gland point** Dichtungseinsatzspitze *f* ~ **packing** Labyrinthdichtung *f* ~ **seal** Labyrinthring *m*
**lac** Lack *m* ~ **varnish** Lackfirnis *m*
**laccinic** lacksauer
**lace, to** ~ (cables) abbinden, binden, fitzen, klöppeln, schnüren, umschnüren, verflechten **to** ~ **out** ausformen **to** ~ **together** zusammenschnüren **to** ~ **up** zuschnüren
**lace** Litze *f*, Schnur *f*, Tresse *f* ~**-and-trimming machine** Posamentmaschine *f* ~ **finishing**

Spitzenappretur *f* ~ **ground** Netzgrund *m* ~**-maker** Bortenwirker *m*, Posamentierer *m* ~ **man** Posamentier *m* ~ **paper** Spitzenpapier *n* ~ **warp fabric** durchbrochene Kettenware *f* ~ **work** durchbrochene Ware *f* ~ **working** Bortenwirkerei *f* ~ **workship** Bortenwirkerei *f*
**laced wiring harness** Kabelbaum *m*
**lacerate, to** ~ zerreißen
**lachrymator** augenangreifender Reizstoff *m*, Augenreizstoff *m*, Tränenstoff *m*
**lachrymatory** tränenreizend ~ **action** Tränenreiz *m*
**lacing** Anschlagleine *f*, Riemenverbinder *m*, Ringverbindung (aviat.) ~ **to bottom** Grundgurtverschnürung *f*
**lacing,** ~**-board** Kabelformbrett *n* ~**-out** Ausformen *n* ~ **twine (or cord)** gedrillte Schnur *f* ~ **wire** Bindedraht *m* ~ **yarn** Fitzgarn *n*
**lack, to** ~ entbehren, ermangeln, fehlen, mangeln
**lack** Mangel *m*, Unterschluß *m*
**lack,** ~ **of balance** Gleichgewichtsfehler *m* ~ **of compressed air** Druckluftmangel *m* ~ **of control** Steuerlosigkeit *f* ~ **of definition** (of a picture) Unschärfe *f* ~ **of depth of focus** Tiefenunschärfe *f* ~ **of equilibrium** Gleichgewichtsfehler *m* ~ **of focus** Unschärfe *f* ~ **of harmony** Unstimmigkeit *f* ~ **of iron** Eisenmangel *m* ~ **of picture definition** Bildunschärfe *f* ~ **of precision** Funktrübung *f* ~ **of provisions** Nahrungsmangel *m* ~ **of selectivity** Unselektivität *f* ~ **of sensitivity** Übersprecherscheinung *f* ~ **of space** Raummangel *m* ~ **of symmetry** mangelnde Symmetrie *f* ~ **of synchronism** Verschiebefehler *m* ~ **of uniformity** Ungleichförmigkeitsziffer *f* ~ **of work** Arbeitsmangel *m*
**lacking contrast** kontrastlos
**lacmoid** Lakmoid *n*
**lacquer, to** ~ firnissen, lacken, lackieren
**lacquer** Firnis *m*, Lack *m*, Lackfarbe *f*, Lackwaren *pl*, Schellack *m* ~ **for bronze** Bronzelack *m* ~ **for negatives** Negativlack *m* ~ **for tin cans** Blechdosenlack *m* ~ **for tin foil caps** Stanniolkapsellack *m*
**lacquer,** ~ **binding medium** Lackbindemittel *n* ~ **coating (or film)** Lackschicht *f* ~ **disk** Lackfolie *f* (phono) ~ **giving the effect of hammered metal** Hammerschlaglackierung *f* ~ **original** Lackfolienaufnahme *f* (photo) ~ **residues** Lackrückstände *pl* ~ **work** Lackierarbeit *f*
**lacquered cane** Glanzrohr *n*
**lacquering** Lackausmalung *f* ~ **stove** Lackierofen *m*
**lactate** Lactat *n*, Laktat *n*
**lacteal** milchig
**lacteous** milchartig, milchig
**lactic,** ~ **acid** Milchsäure *f* ~ **acid bacilli** Milchsäurebakterien *pl* ~ **fermentation** Milchsäuregärung *f*
**lactobutyrometer** Laktobutyrometer *n*
**lactodensimeter** Galaktometer *n*
**lactoglucose** Laktoglukose *f*

**lactometer** Galaktometer, Milchmesser *m*
**lactoscope** Galaktometer *n*
**lactose** Milchzucker *m*
**lacuna** (in the map) Lücke *f* in der Karte
**lacunarity** Lückenanteil *m*
**lacunary** lückenhaft
**lacustrine limestone** Seekalkstein *m*
**ladder** (way of mine) Fahrt *f*, Laufmasche *f* (text.), Leiter *f*, Niedergang *m*, Steigleiter *f* ~ **used for distant observation** Beobachtungsleiter *f*
**ladder,** ~ **attenuation** Dämpfungsleiter *f* in der Abzweigschaltung *f* ~ **beam** Leiterbaum *m* ~ **bridge** Leiterbrücke *f* ~ **chain** Hakenkette *f* ~ **dredge** Eimerbagger *m* ~ **hasp** Fahrhaspe *f* ~ **hook** Fahrthaken ~ **mounted on derrick** Auslegeleiter *f* ~ **network** symmetrisches Netzwerk *n* ~ **peg** Frosch *m*, Fröschel *n* ~ **rope** Schwenkseil *n* ~ **rung** Leitersprosse *f* ~ **shaft** Fahrschacht *m* ~ **stringer** Holm *m* ~**-type filter** kettenförmige Siebschaltung *f* ~ **way** Fahrschacht *m* ~ **wedge** Fröschel *n*
**laddered,** ~ **circuit** Brückenleitung *f* ~ **telephone circuit** Gitterleitung *f*
**lading** Beladung *f*, Schiffsladung *f*
**ladle, to** ~ auskellen **to** ~ **out** abschöpfen, ausschöpfen, schöpfen
**ladle** (brewing) Abschöpflöffel *m*, Gießpfanne *f*, Gußpfanne *f*, Kelle *f*, Pfanne *f*, Schöpflöffel *m* ~ **with tipping-gear appliance** Gießpfanne *f* mit Getriebekippvorrichtung *f*, Pfanne *f* mit Getriebekippvorrichtung *f*
**ladle,** ~ **analysis** Pfannen-, Schöpf-probe *f* ~ **bail** Gießpfannen-bügel *m*, -gehänge *n*, Pfannen--bügel *m*, -gehänge *n* ~ **barrow** Gießpfannenwagen *m*, Pfannenwagen *m* ~ **bowl** Gießpfannen-behälter *m*, -tiegel *m*, Pfannen--behälter *m*, -tiegel *m* ~ **brick** Pfannenstein *m* ~ **capacity** Gießpfanneninhalt *m*, Pfannen-inhalt *m* ~ **capacity inside lining** Gießpfannen-inhalt *m* nach der Ausschmierung *f* ~ **car** Gieß-, Roheisen-wagen *m* ~ **crab** Gießkatze *f* ~ **crane** Gießpfannenkran ~ **lining** Gießpfannen-, Pfannen-auskleidung *f* ~ **lip** Gießpfannen-ausguß *m*, -schnauze *f*, Pfannen--ausguß *m*, -schnauze *f* ~ **man** Pfannenmann *m* ~ **nozzle** Gießpfannenschnauze *f* ~ **plug** Gießpfropfen *m* ~ **pourer** Eingießer *m* ~ **sample** Schöpfprobe *f* ~ **shank** Gießpfannen-gabel *f*, -tragschere *f*, Pfannen-gabel *f*, -tragschere *f* ~ **skull** Pfannen-bär *m*, -kruste *f* ~ **test** Pfannen-, Schmiede-, Schöpf-probe *f* ~ **truck** Gießpfannenwagen *m*, Pfannenwagen *m*
**lag, to** ~ bremsen, nachbleiben, nacheilen, nachlaufen, verschieben, verzögern, zurückbleiben **to** ~ **behind** abhängen
**lag** Laufzeit *f*, Nacheilen *n*, Nacheilung *f*, Nachhängen *n*, Nachhinken *n*, Nachwirkung *f*, negative Phasenverschiebung *f* (electr.), Phasennacheilung *f*, Rücktrift *f* (aerodyn.), Trägheit *f*, Verschiebung *f*, Verspätung *f*, Verzögerung *f*, Verzugszeit *f* ~ **of air-intake valve** Nacheilen *n* des Einlaßventils *n* ~ **of the photoelectric cell** Nachhinken *n* der Zelle *f* ~ **of release** Auslösezeit *f*
**lag,** ~ **constant** Verzögerungskonstante *f* ~ **element** Verzögerungsglied *n* ~ **error** Schleppfehler *m* ~**-free** trägheitslos ~ **hinges** Schwenk-

**gelenke** *pl* ~ **screw** Ankerbolzen *m*, Ankerschraube *f*, Fundamentschraube *f*, Holzschraube *f* mit Rundkopf *m*, Rundkopf *m*, Schlüsselschraube *f*, Strebenschraube *f*
**lagged** ummantelt
**lagging** hemmend, nacheilend, phasen-verspätet, träge, verlangsamend
**lagging** Ausbau *m*, Auskleidung *f*, Auspolsterung *f*, Isolierschicht *f*, Nacheilen, Nacheilung *f*, Sperrwirkung *f*, Verkleidung *f*, Verschalung *f*, Verzögerung *f*, Verzug *m*, Zurückbleiben *n* ~ **of phase** Phasennacheilung *f*
**lagging,** ~ **coil** Nacheilspule *f* ~ **crankshaft** Schleppkurbel *f* ~ **current** (in der Phase) nacheilender Strom *m* ~ **material** Wärmeschutzmaterial *n*
**lagoon** Haff *n*, Lagune *f*, Strandsee *m* ~ **deposit** lagunare Ablagerung *f*
**laid,** ~ **flat** umgebrochen ~**-in float** Einlegeschlauch *m* ~**-in key** Einlegekeil *m* ~ **length** Einbaulänge *f* ~ **mold** Drahtform *f*, gerippte Form *f* ~ **open to public inspection** (patents) ausgelegt ~ **paper** geripptes Papier *n* ~ **rope** kabelweise geschlagenes Tau *n* ~**-up** abgerüstet ~**-up vessel** Abwrackschiff *n*
**laitence** Zementmilch *f*
**lake** Binnensee *m*, See *m* ~ **basin** Einmündung *f* ~ **iron ore** See-Erz *n* ~ **pigment** Pigmentfarbe *f*
**lambda limiting process** Lambda-Begrenzung *f*
**lamberito** Lamberit *m*
**Lambert's law** Lambertgesetz *n*, Lambert'sches Gesetz *n*
**lambskin** Lammfell *n*
**lamella** Blatt *m*, Blättchen *n*, Blech *n*, Lamelle *f*, Plättchen *n*, Platte *f*, Scheibe *f*, Streifen *m*
**lamellar** blättchenartig, blätt(e)rig, lamellar, lamellenartig, plattenförmig, schichtig, streifig, streifenförmig ~ **chisel** Lamellenmeißel *m* ~ **coupling** Blätterkupplung *f* ~ **pyrites** Blätterkies *m* ~ **shutter** Blättchen-, Lamellen-verschluß *m* ~ **structure** Schichtgefüge *n* ~ **tube** Lamellenröhre *f*
**lamelliform** blättchenartig
**lamina** Blatt *n* (Eisenkern), Blättchen *n*, Lage *f*, Lamelle *f*, dünnes Plättchen *n*, Platte *f*, Schicht *f* ~ **of wood** Dickte *f* ~**-planing (or thicknessing) machine** Dichtenhobelmaschine *f*
**laminar** aus dünnen parallelen Schichten *pl* bestehend, flächenförmig, flächenhaft, flächig, laminar, plattenförmig ~ **airfoil** Laminarprofil *n* ~ **boundary layer** laminare Grenzschicht *f* ~ **coil** flächenhafte Spule *f* ~ **flow** Bandströmung *f*, Fadenströmung *f*, plastisches Fließen *n*, Laminarströmung *f*, einfache Strömung *f*, gleichmäßige Strömung *f*, regelmäßige, schlichte, wirbelfreie oder wirbellose Strömung *f* ~ **flow in wake** glatter Abfluß *m* ~ **fracture** Schieferbruch *m* ~ **layer** wirbelfreie Grenzschicht *f*, laminare Schicht *f* ~ **path of current** flächenanziehungsartige Strombahn *f*
**laminarization** Laminarhaltung *f*
**laminate, to** ~ blättern, lamellieren, laminieren, walzen
**laminate** Kunststoff-, Plastik-folie *f*
**laminated** aus dünnen Folien *pl* bestehend, blättchenartig, blätterförmig, blätt(e)rig, feinstreifig, geblättert, geschichtet, lamellar, la-

mellenartig, lamelliert, plattenförmig, plattig, schichtig, streifig

**laminated,** ~ **board** mehrlagiger Karton *m* ~ **brush** Blätterbürste *f* ~ **core** Blätterkern *m*, Blechkern *m*, unterteilter Eisenkern *n*, geblätterter Kern *m*, Lamellenkern *m* ~ **covering** Furnierverschalung *f* ~ **fabric** Hartgewebe *n* ~ **fiber sheet** Preßspan *m* ~ **glass** Schicht-, Sicherheits-, Verbund-glas *n* ~ **hook** Lamellenhaken *m* ~ **iron** Dynamoblech *n*, geblättertes Eisen *n* ~ **iron core** Eisenblätterkern *m*, Eisenblechkern *m*, geblätterter Eisenkern *m* (electr.) ~ **iron core coil** Blätterkernspule *f*, Eisenblätterkernspule *f* ~ **leather coupling** Lederpaketkupplung *f* ~ **magnet** Lamellenmagnet *m* ~ **material (or plastic)** Schichtstoff *m* ~ **panel** Schichtstoffplatte *f* ~ **paper** Hartpapier *n* ~ **plastic** Schichtkunststoff *m* ~**-plated spring** Blattfederwerk *n* ~ **pole** Polpaket *n* ~ **radiator arrangement** Lamellenheizkörper *m* ~ **record** geschichtete Platte *f* (phono) ~ **sheet** Schichtstoffbahn *f* ~ **shell** Ring-geschoß *n*, -granate *f* ~ **shim** geschichtete Beilegescheibe *f* ~ **spring** Blattfeder *f* ~ **spring bending machine** Blattfederbiegemaschine *f* ~ **spring coupling** Federpaketkupplung *f* ~ **spring testing machine** Blattfederprüfmaschine *f* ~ **steel** Stahllamelle *f* ~ **structure** Blättchen-, Lamellar-, Streifengefüge *n* ~ **wood** geschichtetes (blättriges) Holz *n*, Lagenholz *n* ~ **wooden tube** Holzbandrohr *n* ~ **yoke** Blechpaket *n*

**lamination** blättrige Beschaffenheit *f*, (of magnetic core) Blatt *n*, Blätterung *f*, Doppelung *f*, Lamelle *f*, Lamellenstruktur *f*, Lamellierung *f*, Platte *f*, Schichtung *f*, Strecken *n*, Streckung *f*, Streifengefüge *n* ~ **of metal** Blech *n* ~ **of pole shoes** Pol(schuh)lamelle *f*

**lamination,** ~ **coating** Kaschieren *n* ~ **coupling** Lamellenkupplung *f* ~ **embedded in plastic material** einbakelierte Lamelle *f*

**laminiform** blättchenartig, plattenförmig

**laminography** Röntgenschichtverfahren *n*, Schichtenkunde *f*, Stratigrafie *f*

**lamp** Beleuchtungskörper *m*, Lampe *f*, Laterne *f*, Leuchte *f*, Licht *n* ~ **with crater light source** punktförmige Lampe *f* ~ **for interior lighting** Ableuchtlampe *f* ~ **with point-shaped light source** punktförmige Lampe *f* ~ **with solid carbons** Reinkohlenlampe *f* ~ **for time recording** Zeitmarkenlampe *f*

**lamp,** ~ **adary** Lampengestell *n* ~ **attachment** Beleuchtungsaufsatz *m* ~ **base** Lampensockel *m* ~ **black** Eisenschwarz *n*, Flatterruß *m*, Lampenruß *m*, Lampenschwarz *n*, Ruß *m* ~ **black bister** Rußfarbe *f* ~ **box** Lampenkasten *m* ~ **bracket** Lampen-arm *m*, -halter *m* ~**-burner case** Lampenbrennerkapsel *f* ~ **cage** Lampenkäfig *m* ~ **call** Glühlampenanruf *m* ~ **cap** Deckglas *n*, Decklinse *f*, Lampenklappe *f* ~ **capping cement** Sockelkleber *m* ~ **case (or chamber)** Lampengehäuse *n* ~ **chimney** Lampenzylinder *m* ~ **circuit** Lichtleitung *f* ~ **cleaner** Lampenputzer *m* ~**-control relay** Lampenrelais *n* ~ **cords** Fassungsadern *pl* ~ **current** Lichtstrom *m* ~ **detector** Lampensucher *m* (electr.), Suchlampe *f* ~ **equipment** Leuchte *f* (film) ~ **filament** leuchtender Faden *m* ~ **filling** Getter *n* ~ **fitting (or mount)** Lampenhalter *m* ~ **fuse** Licht-

sicherung *f* ~ **globe** Lampenkuppel *f* ~**-globe collar** Lampenkuppelfassung *f* ~**-globe cushion** Gummiring *m* ~**-holder** Fassung *f*, Lampenfassung *f* ~**-holder with cord grip** Lampenfassung *f* mit zentraler Drahteinführung ~**-holder for tubular incandescent lamps** Sofitten-Fassung *f* ~ **hood** Laternenkappe *f* ~ **housing** Lampen-gehäuse *n*, -kasten *m* ~ **indicator** Glühlampensignal *n* ~ **jack strip** Lampenstreifen *m* ~ **locking device** Festhaltevorrichtung *f* für Lampe ~ **oil** Brennöl *n*, Leuchtpetroleum *n*, Paraffinöl *n* ~ **panel** Lampenfeld *n* ~ **pole** Laternenpfahl *m* ~ **post** Lampengestell *n* ~ **press** Lampenfuß *m* ~ **reflector** Laternenspiegel *m* ~ **resistance** Lampenwiderstand *m* ~ **rim** Lampenkranz *m* ~ **ring** Lampenkranz *m* ~ **screen** Zellenbildschirm, Zellentafel *f* ~ **series resistance** Lampenvorwiderstand *m* ~ **shade** Lampenschirm *m*, Reflektor *m* ~ **shade holder** Schalenhalter *m* ~ **shield** Lampenkappe *f* ~ **signal** Blink-kennung *f*, Glühlampenanruf *m*, Lampensignal *n* ~**-signal apparatus** Blinkapparat *m*, -feuer *n* ~ **signal call** Glühlampenanruf *m* ~ **signaling** Lampensignalisierung *f* ~**-signaling apparatus** Blinkgerät *n* ~ **socket** Fassung *f*, Lampensockel *m* ~**-socket mounting** Lampenstreifen *m* ~ **soot** Blak *m* ~ **squash** Lampenfuß *m* ~ **strip** Lampenstreifen *m* ~ **switch-board** Glühlampenschrank *m* ~ **terminal** Horn *n* einer Röhre ~ **tripod** Lampenstativ *n* ~ **trolley** Lampenwagen *m* ~ **voltage** Lampenspannung *f* ~ **wick** Lampendocht *m*

**lamps, bark of** ~ Lampenfeld *m*

**lanarkite** Kohlenbleivitrolspat *m*

**Lancashire,** ~ **charcoal-hearth** Lancashire--Herdfrischprozeß *m* ~ **hearth** Lancashire-Frischherd *m*

**lance** Büchse *f*, Lanze *f*, Schlauchendstück *n*, Spieß *m* ~ **bucket** Lanzenschuh *m* ~ **cutting** Schneiden *n* mit der Sauerstoffpflanze ~ **pole** Baustange *f*

**lancet** Lanzette *f*, Laßeisen *n*

**land, to** ~ landen **to** ~ **a cable** ein Kabel *n* anlanden **to** ~ **slowly** langsam landen **to** ~ **on water** anwassern

**land** Gelände *n*, Grundbesitz *m*, hervorstehender Teil *m* zwischen zwei Nuten (mach.), Land *n*, Schallplattensteg *m*, tragende Fläche *f*, Steg *m* (electr.)

**land,** ~ **aerodrone** Landflugplatz *m* ~ **amalgamation** Feldbereinigung *f* ~ **area of bearing** tragende Fläche *f* ~ **breeze** Landwind *m* ~ **compass** Landkompaß *m* ~ **direction-finding station** Landpeilstation *f* ~ **fall** Küstenüberflug *m* ~ **fog** Landnebel *m* ~**-holder** Grundbesitzer *m*, Gutsbesitzer *m*, Landeigentümer *m*, Pächter *m* ~**-line** Landleitung *f*, Landlinie *f* ~**-line teletype circuit** Drahtfernschreibverbindung *f* ~**-locked port** eingeschlossener Hafen *m*

**land-mark** Ansegelungsmarke *f* (navig.), Bake *f*, Bezugspunkt *m*, Geländegegenstand *m*, Geländepunkt *m*, Gemarkung *f*, Landmarke *f*, Markstein *m*, Merkpunkt *m*, Ortungspunkt *m*, Richtbake *f* ~ **beacon** Landmarkenlicht *n*

**land,** ~ **measurement** Feldmessung *f* ~ **mine** Erd-, Fladder-, Land-, Luft-mine *f* ~ **owner** Grundeigentümer *m* ~ **plane** Landflugzeug *n* ~ **plaster** Feingips *m* ~ **reclaimed from sea by**

**means of dikes** Polder *m* ~ **register** Kataster *m* ~ **return** Bodenecho *n* (rdr) ~ **roller** Ackerwalze *f* ~ **slide** Bergrutsch *m*, Erdrutsch *m*, Erdsturz *m*, Grundlawine *f*, Lawine *f*, Murgang *m*, Rutschen *n* des Bodens *m*, Rutschung *f* ~**-station charges** Landgebühr *f* ~ **survey** Landvermessung *f* ~**-survey office** Landesstelle *f* ~ **surveying** Feldmessung *f*, Landmeßkunst *f*, Terrainaufnahme *f* ~ **surveyor** Feldmesser *m*, Geodät *m* ~ **tax** Grundsteuer *f* ~ **vehicle** Landfahrzeug *n* ~**-ward** landwärts ~**-ward board** Landbord *n* ~ **wind** ablandiger Wind *m*, Landwind *m*
**landing** Anschlagpunkt *m*, Füllort *m*, Landen *n*, Landung *f*, Landungsponton *n*, Leiterabsatz *m*, Podest *n*, Treppenboden *m* (staircase) ~ **with the airscrew stopped** Landung *f* mit stehender Schraube ~ **with firm footing** Vorplatz *m* mit befestigtem Boden *m* ~ **on nose** Kopfstand *m* ~ **of a shaft** Schachthängebank *f* ~ **with solid footing** Vorplatz *m* mit befestigtem Boden ~ **with stopped engine** Landen *n* mit stehendem Motor ~ **without use of brakes** Landen *n* ohne Bremsen
**landing,** ~ **aids** Landehilfen *pl* (aviat.) ~ **angle** Ausrollwinkel *m* ~ **area** Abdriftplatz *m*, Lande-bereich *m*, -platz *m*, -zone *f*; Landungs--gebiet *n*, -platz *m* ~ **barge** Landungsboot *n* ~ **beacon** Landefunkfeuer *n* ~ **boat** Landungsboot *n* ~ **brake** Landungsbremse *f* ~ **bridge** Landbrücke *f* ~ **buoy** Landungstonne *f* ~ **climb performance** Steigleistung *f* in der Landezustandsform ~ **competition** Landewettbewerb *m* ~ **craft** Landungsfahrzeug *n* ~ **crew** Landemannschaft *f* ~ **cross** Landekreuz *n* ~ **curve** Gleitweg *m* (aviat.) ~ **deceleration parachute** Landebremsschirm *m*, Landefallschirm *m* ~ **deck** Flugdeck *n* ~ **device** Landehilfe *f* ~ **direction indicator** Landrichtungsanzeiger *m* ~ **direction light** Landebahnfeuer *n* ~ **distance** Landestrecke *f* ~ **distance available** verfügbare Landestrecke *f* ~ **end** Landekopf *m* ~ **equipment** Lande-anlage *f*, -einrichtung *f* ~ **facilities** Landeeinrichtung *f* ~ **fee** Landegebühr *f* ~ **field** Landegelände *n*, Landeplatz *m*, Landungsfeld *n* ~**-field indicator** Flugzeiger *m*, Landungsfühler *m* ~ **flap** Landeklappe *f* ~**-flap-position indicator** Anzeiger *m* für Landeklappenausschlag, Landeklappenanzeiger *m* ~ **flap track** Landeklappenschiene *f* ~ **flare** Bordlandefackel *f* ~ **floodlight** Landebahnleuchte *f*, Landescheinwerfer *m* ~ **fuel weight** Kraftstoffgewicht *n* bei der Landung
**landing-gear** Fahrwerk *n*, Landungsgestell *n*, Rollwerk *n* ~ **adjustment** Fahrgestelleinstellung *f* ~ **axle** Fahrgestell-, Fahrwerk-achse *f* ~ **bracing** Fahrgestellauskreuzung *f* ~ **brake** Fahrgestellbremse *f* ~ **door** Fahrwerkklappe *f* ~ **fairing** Fahrgestellverkleidung *f* ~ **part** Fahrwerkteil *m* ~**-position indicator** Anzeigevorrichtung *f*, Fahrwerkanzeiger *m* ~ **release and stowage lever** Fahrgestelleinziehhebel *m* ~ **stirrup** Fahrgestellbügel *m* ~ **strut** Fahrgestell-schenkel *m*, -strebe *f*, Fahrwerkfederbein *n*, Flugzeugbein *n* ~ **strut socket** Fahrgestellstrebenschuh *m* ~ **truss** Fahrwerksbock *m* ~ **uplock** Fahrwerkeinfahrverriegelung *f* ~ **well flap** Fahrwerksklappe *f* ~ **winch** Fahrgestellwinde *f* ~ **yoke** Fahrwerk(s)brücke *f*

**landing,** ~ **gross weight** höchstzulässiges Landegewicht *n* ~ **groundroll** Landelaufstrecke *f* ~ **headlight** Landescheinwerfer *m* ~ **height recorder** Landehöhenschreiber *m* ~ **hook** Landespieß *m* ~ **hot** Landen *n* mit großer Schnelligkeit ~ **lamp** Landescheinwerfer *m* ~ **legs** Landegestell *n* (g/m) ~ **length** Landestrecke *f* ~ **light** Landescheinwerfer *m* ~ **load** Lande-last *f*, -gewicht *n* ~ **operation** Landungunternehmen *f* ~ **path** Landebahn *f* ~ **permit** Landeerlaubnis *f* ~ **pier** Landungssteg *m* ~ **place** Landungs-brücke *f*, -stelle *f* ~ **place of a pit** Schachtbühne *f* ~ **point** Aufsetzpunkt *m*, Flugzeugaufsetzpunkt *m*, Landeziel *n* ~ **position** Landestellung *f* ~ **procedure** Landungsvorgang *m* ~**-prohibited signal** Sperrzeichen *n* ~ **projector** Landungsscheinwerfer *m* ~ **quay** Landungskai *m* ~ **ramp** Landesteg *m*, Landungsklappe *f* ~ **report** Landemeldung *f* ~ **restriction** Landeverbot *n* ~ **rocket** Leuchtrakete *f* ~ **run** Auslaufstrecke *f*, Ausrollen *n* (aviat.), Landelauf *m*, Landelaufstrecke *f* ~ **run with brakes** Auslaufstrecke *f* mit Bremsen ~ **run with flaps** Auslaufstrecke *f* mit Landeklappen ~ **searchlight** Landungsscheinwerfer *m* ~ **sequence** Landefolge *f* ~ **sheet** Landungstuch *n* ~ **shock** Landestoß *m* ~ **signal** Landezeichen *n* ~ **ski assembly** Schneekufengestell *n* ~ **skids** Landekufen *pl* (g/m) ~ **slab** Laufplatte *f* ~ **smoke signal** Landungsrauchzeichen *n* ~ **speed** Ausrollgeschwindigkeit *f*, Landegeschwindigkeit *f* ~ **spot** Landepunkt *m*, Landestelle *f* ~ **stage** Abzugsbühne *f*, Anlandebrücke *f*, Dock *n*, Ladedamm *m*, Landbrücke *f*, Landebrücke *f*, Landungsbrücke *f* ~ **strip** Lande-bahn *f*, -platz *m*, -streifen *m* ~ **strut** Abfangstrebe *f* ~ **surface (or T)** Landekreuz *n*, Lande-T *n*, Landungs-T *n* ~ **tax** Landegebühr *f* ~ **tetradron** Landetetraeder *m* ~ **threshold** Landeschwelle *f* ~ **valve** Schlauchanschlußventil *n* ~ **weight** Landegewicht *n* ~ **wheel** Anlauf-, Lande-, Lauf-, Lauf-rad *n* eines Flugzeuges ~**-wheel axle** Laufradachse *f* ~**-wheel brake** Laufradbremse *f* ~ **wire** Gegendraht *m*, Gegenkabel *n* ~ **zone** Lande-raum *m*, -zone *f*
**lands,** Felder *pl* ~ **and grooves** Felder *pl* und Züge *pl*
**landscape** Gelände, Landschaft *f* ~ **development** Landschaftsgestaltung *f* ~ **strip** Grünstreifen *m* (Straße) ~ **treatment** Landschaftsgestaltung *f*
**landscaping** Landschaftsgestaltung *f*
**landside** landweitig
**lane** Fahrbahn *f*, Nullhyperbel *f* (rdr), Pfad *m*, Streifen *m* ~ **cut through forest** Durchhau *m* ~ **lay rope** Gleichschlagseil *n*
**lanscashire boiler** Flammrohr-Kessel *m*
**lantern** Drehling *m*, Käfig *m* (mach.), Laterne *f*, Leuchte *f* ~ **on post** Stocklaterne *f*
**lantern,** ~ **gear** Triebstockkranz *m* ~ **gear pinion** Triebstockkritzel *m* ~ **mask** Laternenbildermaske *f* ~ **picture** Diapositivbild *n* (film) ~ **pinion** Hohltrieb *m*
**lantern-slide** Diapositiv *n*, Durchsichtbild *n* ~ **binding strip** Laterneneinfaßleiste *f* ~ **box** Laternenbilderkasten *m* ~ **carrier** Schieberrahmen *m* ~ **projection** Diapositivprojektion *f* ~ **projection method** Stehbildverfahren *n*

lanthanide, ~ contraction Lanthanidenkontraktion *f* ~ elements Lanthaniden *pl* ~ series seltene Erden *pl*
lanthanite Lanthanit *m*
lanthanum Lanthan *n* ~ chloride Lanthanchlorid *n*
lanyard Abreißschnur *f*, Abzugsleine *f*, Abzugsstück *n*, Reißleine *f*, Taljecreep *n*, Zündschnur *f* ~ handle Abzugsgriff *m*
lanzet needle lanzettartige Nadel *f*
lap, to ~ glänzen, läppen, polieren, schmirgeln, überblenden, umhüllen, (round) umlappen, umwickeln to ~ in (valve) einschleifen to ~ over übergreifen, überlappen to ~-weld überlappt schweißen
lap Falte *f*, Falz *m*, Faserbandwinkel *m*, Läppscheibe *f*, Lappung *f*, Läppwerkzeug *n*, Naht *f*, Polierscheibe *f*, (on track) Runde *f*, Rundstrecke *f*, Schleifscheibe *f*, Überdeckung *f*, Überlappung *f*, Überwalzung *f*, Umwicklung *f*, Vorstoß *m*, (roll) Wattenwickel *m*, Wicklung *f* ~ of slide valve Deckfläche *f* ~ of the slide valve Schieberdeckung *f*
lap, ~ belt Bauchgurt *m* ~ creel Wickelträger *m* ~ dissolve weiche Überblendung *f* ~ dissolve shutter Überblendeinrichtung *f* ~-dissolving shutter Überblendungsblende *f* ~ formation Wickelbildung *f* ~ guard Wickelwächter *m* ~ joint überlappter Stoß *m*, überlappte Teilfuge *f*, Überlappung *f*, Überlappungsverbindung *f*, überlappte Verbindung *f* ~ lattice Überführungslattentuch *n* ~-pack parachute Schoßkissenfallschirm *m*, Schoßpackfallschirm *m* ~ riveting Überlappungsnietung *f* ~ seam weld überlappte Schweißung *f* ~ strap Bauchgurt *m* ~ tester Wickelprüfgerät *n* ~-time Auszugszeit *f* ~-type parachute Schoßkissenfallschirm *m* ~-weld überlappte Schweißung *f*, Überlappungsschweißung *f* ~-welded flachgeschweißt ~-welded tube überlappt geschweißtes Rohr *n* ~-welding Flach-, Übereinander-, Überlappungs-schweißung *f*, Schleifenwicklung *f* ~ winding Schleifenwicklung *f*
lapel microphone Knopflochmikrofon *n*
lapidary Edelsteinschneider *m*, Steinschleiferei *f*
lapis, ~ lazuli Azurstein *m*, Blaustein *m*, Lapislazuli *m*, Lazurstein *m* ~ style Lapisdruck *m*
Laplace, ~ domain Unterbereich *m* ~'s law Durchflutungsgesetz *n* ~ transform of function Bildfunktion
laplacian Gegenwölbung *f*
lapp Einsetzband *n*
lapped geläppt ~ armoring geschlossene Bewehrung *f* ~ bearing Paßlager *n* ~ finish Läppschliff *m*
lapper with inclined coil carrier Schrägspinner *m* (Papier)
lapping Bombage *f*, Läpparbeit *f*, Läppen *n*, Schleifen *n*, Trense *f*, Überlappen *n*, Umhüllung *f*, Umlappung *f*, Umwicklung *f* ~ abrasive Läppmittel *n* ~ bush Läppbüchse *f* ~ compound Läppaste *f* ~ engine Bandleitungsmaschine *f* ~-in Einschleifen *n* (Ventilsitz) ~ machine Läppmschine *f* ~ stick Läppdorn *m* ~ tool Läppwerkzeug *n* ~ wheel Läppscheibe *f*
lapse Abfall *m*, Ablauf *m*, Abnahme *f*, Verfall *m* ~ factor Schleppgröße *f* ~ rate Abweichungs-

rate *f*, senkrechtes Temperaturgefälle *n* (meteor.)
lapsing Verjährung *f*
larboard Backbord *n*
larch Lärche *f*
lard oil Specköl *n*
lardaceous fracture speckiger Bruch *m*
larder Fleischkammer *f*
large ausgedehnt, geräumig, groß, stark at ~ frei ~ in bore (or in diameter) großkslibrig in ~ pieces großstückig on a ~ scale im großen Stil *m*
large, ~ aircraft Großflugzeug *n* ~-angle-rainboundary Großwinkelkorngrenzen *pl* ~ angle scattering Streuung *f* in große Winkel
large-area Großraum ~ counter tube Großflächenzählrohr *n* ~ diaphragm Großflächenmembrane *f* ~ halogen counter tubes Großflächen-Halogen-Zählröhre *f* ~ methane flow counters Großflächen-Methandurchflußzähler *m*
large, ~ bit Breitmeißel *m* ~ body size Großkegel *m* ~-caliber machine gun überschweres Maschinengewehr *n* ~ calories große Kalorie *f*, Kilogrammkalorie *f*, Kilokalorie *f* ~-capacity cable hochpaariges Kabel *n* ~-capacity car Großraumwagen *m* ~-capacity condenser großer Kondensator *m* ~-celled großzellig ~ circulation Großauflage *f* ~ construction fixture jig (or fixture rig) Ausbaugroßvorrichtung *f* ~ construction jig Anbaugroßvorrichtung *f* ~ core Großkern *m* ~-current arc Hochstrombogen *m* ~ Diesel unit Großdiesel *m* ~ end of an ingot Blockfuß *m* ~-fibered großfaserig ~ glider Großsegler *m* ~ hammer Boßhammer *m* ~ kettle Kochkessel *m* ~-meshed großmaschig ~ observation telescope in special mount Mastfernrohr *n* ~ paper edition Luxus-, Pracht-ausgabe *f* ~-picture projection Großprojektion *f* ~ plant Großbetrieb *m* ~ range finding telescope Richtsäule *f* ~ river Strom *m* ~ scale in größerem Umfang *m*
large-scale ausgedehnt, großangelegt, umfangreich ~ experiment Großversuch *f* ~ indicator Großanzeige *f* ~ manufacture am laufenden Band *n* fabrizieren, serienmäßige Herstellung *f* ~ manufactured goods Massengüter *pl* ~ manufacturing operation Großbetrieb *m* ~ map Karte *f* im großen Maßstab ~ plant Großanlage *f* ~ production Großserienanfertigung *f*, Rihengroßanfertigung *f* ~ project site Großbaustelle *f* ~ serial production Großserienbau *m* ~ series production Großreihenfertigung *f*, Großserienbau *m* ~ test Großversuch *m* ~ use Großeinsatz *m* ~ workings Großabbau *m*, Großbauten *pl*
large, ~ screen picture Breitbildfilm *m* ~-size dynamo große Lichtmaschine *f* ~-size tapper Großgewindeschneider *m*
large-sized derbstückig, grobkörnig, großformatig, großkalibrig, großstückig ~ cable hochpaariges Kabel *n* ~ coke Großkoks *m* ~ furnace Großofen *m*
large, ~ sprinkler for irrigating Regnerdüse *f* ~-steam forgings manufacture Dampfhammerschmiede *f* ~ studio großer Ausführungsraum *m* ~ surface großflächig ~-surface cathode Flächenkathode *f* ~ text geneigte Mittelschrift *f* ~ tin can for shipping materials Hobbock *m*

~ **tongs** (glass mfg.) Hafenzange f ~ **trough** Beute f ~ **vessel of aluminum** Aluminiumgroß-gefäß n

**largeness** Weite f

**larger, to a** ~ **extent** in höherem Maße n

**largest,** ~ **flare** Trichtermundöffnung f, Trich-termündung f ~ **river** Hauptfluß m

**Larmor precession** Larmorpräzession f

**Larssen sheet pile** Larssenbohle f

**larva** Larve f

**laryngeal mirror** Kehlkopfspiegel m

**laryngophone** Kehlkopfmikrofon n

**laryngoscope** Kehlkopfspiegel m

**larynx** Kehle f

**Laser** (light amplification by stimulated emission of rediation) LASER m (Lichtverstärker m durch angeregte Strahlungsabgabe)

**lash, to** ~ abbinden, anbinden, durchlaschen, festmachen, festzurren, peitschen, verlaschen, zurren, zusammenschnüren

**lashing** Bändsel n, (rope) Bund m, Gerüsthalter m, Lasche f, Laschung f, Zurrung f ~ **cord** Befestigungsleine f ~ **point** Verzurrstelle f ~ **wedge** Schnurleiste f ~ **wire** Bindedraht m

**last, to** ~ anhalten, ausdauern, aushalten, dauern, währen

**last,** ~ **batch** Schlußpartie f ~ **brine pit** Kristall-isolationsbecken n ~ **day** Termin m ~ **diamond pass** Schlichtraute f ~ **explosion in the engine** Auspufftakt m ~ **finishing pass** Fertigschlicht-stich m ~ **groove** Polierkaliber n ~ **incidence of a configuration** letzte Inzidenz f einer Konfiguration f ~ **line** Ausgangszeile f (print.) ~ **pass** Endkaliber n, Fertigschlichtkaliber n ~ (**planishing**) **pass** Fertigpolierkaliber n **as a** ~ **resource** in letzter Instanz f ~ **runnings** Nachlauf m

**lasting** anhaltend, beständig, bleibend, dauerhaft, dauernd, haltbar, nachhaltig, stichhaltig ~ **effect** Dauerwirkung f ~ **finish** stehende Appretur f ~ **machine** Zwickmaschine f ~ **power** Dauerleistung f ~ **quality** Dauerhaftig-keit f

**latch, to** ~ einklinken, einschnappen, verriegeln

**latch** Anschlag m, Drücker m, Falle f, Falleisen n, Klinke f, Klinkwerk n, Rastklinke f, Rastnase f, Riegel m, Schließe f, Schnapper m, Schnepper m, Sperr-haken m, -hebel m, -klinke f, -zunge f

**latch,** ~ **blade** Schaft m der Zunge ~ **bolt** Schnäpper m ~ **clearing position** Einschließ-stellung f ~ **covering** Schloßhülle f ~ **lock** Federschloß n ~**-locked switch** verklinkter Schalter m ~ **mechanism with engaging roller** Schloß n mit Schließrolle f (Kühlschrank) ~ **member** Verriegelungsglied n ~**-needle machine** Zungen-nadelmaschine f ~ **opener** Zungenöffner m ~ **plate** Fangplatte f (text.) ~ **shaft** Fallenachse f ~ **slot** Lagerschlitz m für die Zunge im Nadel-schaft ~ **spoon** Löffel m am Ende der Zunge ~ **spring** Sperrhebelfeder f

**latched** eingeklinkt

**latching,** ~**-in** Verklinkung f ~ **switch (or relay)** Haftschalter m ~ **voltage** Haltespan-nung f

**late** (later) nachträglich, spät(er) ~ **bolter** Spät-schosser m ~ **combustion** Nachbrennen n ~ **effect** Spätwirkung f ~ **entry** Nachmeldung f

~ **gate former (or generator)** Spättorimpuls-generator m

**latency** Latenz f, Verborgenheit f, Wartezeit f (data proc.) ~ **time** Wartezeit f

**latent** aufgespeichert, gebunden, eingefroren, latent, ruhend, versteckt ~ **electronic image** gespeichertes Bild n ~ **energy** Arbeitsvermögen n, potentielle Kraft f ~ **heat** gebundene, latente oder verborgene Wärme f ~ **heat of conversion** Umwandlungswärme f ~ **heat of evaporation** latente Verdampfungswärme f ~ **heat of liquid** Flüssigkeitswärme f ~ **image** latentes Bild n ~ **image formation** latente Ab-bildung f ~ **period** Latenzzeit f ~ **stability** Querstabilität f ~ **steam** indirekter Dampf m ~ **tissue injury** latente Gewebeschädigung f

**lateral** lateral, quer, seitlich ~ **adjustment** Sei-teneinstellung f ~ **aisle** Quergang m ~ **amor-phosis** seitenständiges Wandlungsbild n (film) ~ **angular lead of target** Seitenvorhaltswinkel m ~ **appendage** Seiten-ansatz m, -arm m ~ **area** Seitenbezirk m ~ **arm** Seitenansatz m, Seiten-arm m ~ **axis** Holm-, (of rhombe crystal) Neben-, Quer-, Seiten-, Umlenk-achse f, Y-Achse f ~ **beam control** Leitstrahlseiten-führung f ~ **bearing surface** seitliche Führungs-fläche f ~ **blower** Seitenwandbläser m ~ **canal (or channel)** Lateral-, Seiten-kanal m ~ **com-pensation** Seitenaussteuerung f ~ **component of velocity** Seitengeschwindigkeit f ~ **contraction** Querkontraktion f, Querzusammenziehung f ~ **control** Quersteuerbarkeit f, Quersteuerung f (helicopter), Seitensteuerung f ~ **controls** Quersteuerung f ~ **corrugation** Querriffelung f ~ **cyclic pitch control** periodische Quersteue-rung f ~ **damping** Wendedämpfung f ~ **de-flection** Seitenverschiebung f ~ **deformation** seitliche Ausdehnung f oder Ausweichung f ~ **deviation** Seiten-abweichung f, -fehler m, -ver-schiebung f ~ **discharge** Nebenentladung f, Seitenstreuung f ~ **displacement** Seiten-ab-stand m, -verschiebung f ~ **displacement of an airplane** Flugzeugversetzung f ~ **drift** Seiten-trifft m ~ **drift landing** Schiebelandung f ~ **edge** Randkante f (cryst.), Seitenkante f ~ **effect** seitliche Wirkung f ~**electrode** Seiten-elektrode f ~ **elevation** Seiten-ansicht f, -auf-riß m, -riß m ~ **error** Seitenfehler m ~ **excurs-ion** seitliche Auslenkung f (phono) ~ **face** Randfläche f (cryst.), Seitenfläche f ~ **flexure** Knickung f ~ **flicker** (in panorama work) Schwimmen n ~ **flow** seitliche Ausweichung f ~ **force** seitliche Kraft f ~ **force axis** Seiten-kraftachse f ~ **force coefficient** Querkraftbei-wert m ~ **guidance** (of airplate) seitliche Führung f, Seitenführung f ~ **image shift** seit-liche Bildverschiebung f ~ **inversion** Seiten-umkehr f ~ **joint** Seitenfuge f ~ **jump** Seiten-abgangsfehler m ~ **lead** (distance correction for drift) Seitenvorhalt m ~ **lead plane** Seiten-länge f ~ **leveling** Quernivellierung f ~**limits** Seitengrenze f ~ **load** Seitenführung f ~ **magni-fying power** Quervergrößerung f (film) ~ **mo-raine** Seitenmoräne f ~ **motion** Seitwärtsbewe-gung f ~ **movement** Seitwärtsbewegung f ~ **observation** seitliche Beobachtung f ~ **oscillation** Querschwingung f ~ **overlap** seitliche Überdek-kung f ~ **(or horizontal) parallax** Seitenparal-

laxe *f* ~ **plane** Querebene *f* ~ **pull** Seitenzug *m*, seitlicher Zug *m* ~ **pushpull rod** Quersteuerschubstange *f* ~ **radiation** Nebenmaximum *n*, Seitenstrahlung *f* ~ **radiator** Ohrenkühler *m* ~ **radiator block** Kühlerteilblock *m*, Seitenblock *m* ~ **recording** Seitenschrift *f* (film) ~ **reinforcement** Querbewehrung *f* ~ **resection** Seitwärtseinschneiden *n* ~ **section** Querschnitt *m* ~ **sense** Breitenrichtung *f* ~ **separation** Seitenstaffelung *f* (aviat.) ~ **shift of image** seitliche Bildverschiebung *f* ~ **slides** Seitenschlitten *m* ~ **stability** Querstabilität *f* ~ **stay** Seitenanker *m* ~ **strength** Querfestigkeit *f* ~ **summit** Seiteneck *n* ~ **surface** Seitenfläche *f* ~ **sway** Seitenschwankung *f* ~ **thrust** (of turbines) Axialverschiebung *f* ~ **tilt** Nadirdistanz *f* quer zur Flugrichtung, Quer-kippung *f*, -neigung *f* ~ **torque tube** Querverbindungswelle *f* (Hubschrauber) ~ **track** Seitenrutsch *m* ~ **tracker** Entfernungsmeßmann *m* zur Seitenbeobachtung ~ **tracking telescope** Seitenrichtfernrohr *n* ~ **traffic** Querverkehr *m* ~ **trim** Quertrimmung *f*, Seitenlastigkeit *f* ~ **truss** Verschwertung *f* ~ **velocity** Schiebegeschwindigkeit *f* ~ **view** Seitenansicht *f* ~ **wave** Querschrift *f* ~ **wind bracing** Windverband *m*
**laterite** Laterit *m*
**latest development** letzter Stand *m* der Entwicklung
**latex** Milchsaft *m* ~ **cure** Latexvulkanisation *f*
**lath, to** ~ belatten, mit Latten *pl* verschalen oder einfassen
**lath** Abtreibepfahl *m*, Latte *f*, Pfahl *m*, Vortreibepfahl *m* (min.) ~**-and-plaster-wall** Fachwand *f* ~ **fence** Gitterzaun *m* ~**-frame** Lattengerüst *n* ~ **nail** Lattenspieker *m* ~**-wince (or winch)** Lattenhaspel *f* ~ **work** Lattengestell *n*
**lathe, to** ~ drechseln, auf der Drehbank *f* bearbeiten
**lathe** Drechselbank *f*, Drehbank *f*, Drehlade *f*, (am Webstuhl) Schlag *m* ~ **for bar work** Stangenmaschine *f* ~ **with draw-in attachment** Zangenspanndrehbank *f* ~ **for eccentric turning** Passigdrehbank *f* ~ **for facing axle ends** Achsspiegeldrehbank *f* ~ **for hollow rods** Gestängerohrdrehbank *f*
**lathe,** ~ **accessories** Drehbankzubehör *n*, Futter *n* ~**-and-planer tool grinder** Dreh- und Hobelstahlschleifmaschine *f* ~ **bearer** Einspannvorrichtung *f* ~ **bed** Bett *n*, Drehbankbett *n* ~ **bed extension** Drehbankbettverlängerung *f* ~ **carriage** Drehbanksupport *m* ~ **carrier center** Mitnehmerspitze *f* ~ **center** Drehbank-körner *pl*, -spitze *f*; Körnerspitze *f* ~ **chuck** Dreh-(bank)futter *n*, Spannfutter *n* ~ **dog** Drehbankherz *n*, Mitnehmer *m* ~ **faceplate** Drehplatte *f* ~ **fixture** Drehvorrichtung *f* ~ **hand** Dreher *m* ~ **headstock** Drehbankkopf *m* ~ **mandrel** Drehdorn *m* ~ **operator** Dreher *m* ~ **spindle** Drehbankspindel *f* ~**-testing tool machine** Meßsupport *m* ~ **tool** Drehling *m*, Drehmesser *m*, Drehstahl *m* ~**-tool holder** Drehstahlhalter *m*, Messerhaus *n* ~ **tools** Drehwerkzeuge *pl* ~ **work** Dreharbeit *f*
**lather, to** ~ schäumen
**lather** Schaum *m* ~ **mixture** Schaumgemisch *n*
**lathering power** Schaumfähigkeit *f*
**lathing** Ausschaltung *f*, Lattenbeschlag *m* ~ **hammer** Latthammer *m*

**laths** Ansteckpfähle *pl*
**latitude** Ausdehnung *f*, Breite *f*, geographische Breite, Höhe *f*, Spielraum *m*, Umfang *m* **in the** ~ **of** auf der Höhe *f* von ~ **of density** Schwärzungsumfang *m* ~ **in ecliptic** Breite *f* in der Ekliptik ~ **of exposure** Belichtungsspielraum *m* ~ **of fix** Besteckbreite *f* ~ **of a negative** Negativumfang *m*
**latitude,** ~ . . . **degree, north** (south) nördlicher (südlicher) Breitengrad *m* ~ **effect** (cosmic rays) Breiteneffekt *m* ~ **flying** Breitenflug *m* ~ **n° N.** nördliche Breite *f*, n° ~ **variation** Breitenschwankung *f*
**latitudinal metacenter** Quermetazentrum
**lattice, to** vergittern
**lattice** Gatter *n*, Gitter *n*, Linienraster *m*, Netz *n*, Raster *m* ~ **absorption** Gitterabsorption *f* ~ **box** Gitterschachtel *f* ~ **bridge** Gitterbrücke *f* ~**-cell model** Gitter-Zellen-Modell *n* ~ **coil** Korbbodenspule *f* ~ **conductivity** Gitterleitfähigkeit *f* ~ **constant** Gitterkonstante *f* ~ **construction** Gitterfachwerk *n* ~ **cooling stack** Lattengradierwerk *n* ~ **crystal** netzförmiges Kristall *m* ~ **defect** Gitterstörstelle *f* ~ **delivery** Lattenablieferungstisch *m* ~ **design** Gitterplanung *f* ~ **dislocation** Gitterverschiebung *f* ~ **dislocations** Gitterstörungen *pl* ~ **disorder** Fehlordnung *f*, Gitterfehlbau *m* ~ **distortion** Gitterverschiebung *f* ~ **distortions** Gitterstörungen *pl* ~ **door** Gittertür *f* ~ **electrons** Gitterelektronen *pl* ~ **energy** Gitterenergie *f* ~ **feed table** Zufuhrlattentisch *m* ~ **filter** Gitterfilter *n* ~ **flooring** Sprossenbelag *m* ~ **forces** Gitterkräfte *pl* ~**frame** Fachwerkgestell *n* (machin.), Gitterrahmen *m* ~ **gate** Fachwerkträger *m*, Gitterträger *m*, Gittertür *f* ~ **girder** Fachwerkträger *m* ~ **gland** Gitterträger *m* ~ **hole** Gitterloch *n* ~ **imperfections** Gitterstörungen *pl* ~ **jack** Scherenheber *m* ~ **jib** Fachwerkausleger *m* ~ **keelson** Gitterkielschwein *n* ~ **like polymerization** Vernetzung *f* ~ **mast** Gitter-, Streck-mast *m* (electr.) ~ **network** Vierpolkreuzglied *n* ~ **parameter** Gitterkonstante *f* ~ **pitch** Gitterabstand *m* ~ **place** Gitterplatz *m* ~ **plane** Gitterebene *f* (cryst.) ~ **plate** Lattenrostschale *f* ~ **point** Gitterplatz *m* ~ **point on a circular disk** Gitterpunkte *pl* auf einer Kreisscheibe ~**-point method** Gitterpunktmethode *f* ~ **pole** Gittermast *m* ~**(d) pole** Gitterständer *m* ~ **relaxation** Gitterrelaxation *f* ~ **running** Transportlattentuch *n* ~ **section** (of network or filter) Kreuzglied *n* ~ **skeleton** Gitterskelett *n* ~ **space** Gitterbereich *m* ~ **spacing** Gitterabstand *m* ~ **specific heat** spezifische Gitterwärme *f* ~ **spectrograph** Gitter-, Strich-spektrograf *m* ~ **structure** Netzstruktur *f*, Netzwerkstruktur *f* ~ **thermal conductivity** Gitterwärmeleitfähigkeit *f* ~ **tie** Fachwerkstrebe *f* ~ **tower** Gittermast *m* ~ **truss** Gitterbalken *m* ~**-type crane arm** gitterförmiger Ausleger *m* ~**-type filter** Mehrfachsieb *n* ~**-type lamppost** Gitterlichtmast *m* ~**-type network** Kreuzglied *n* ~ **unit** Einheits-, Elementar-zelle *f*; Gittereinheit *f* ~ **vacancy** Gitterleerstelle *f*, Gitterloch *n* ~ **vibration** Gitterschwingung *f* ~ **vibration superposition** Gitterschwingungsüberlagerung *f* ~ **void** Gitterlich *m* (cryst.) ~ **wet** Eintauchgitter *n* ~**-winch** Lattenwalze *f* ~ **work** Fach-

werk *n*, Gitterkonstruktion, Gitterwerk *n*, Lattenverschlag *m* ~ **work mast** Gittermast *m* ~**work skeleton** Gitterskelett *n* ~**-wound coil** Wabenspule *f*

**latticed** gitterartig, gitterförmig ~ **bar** Gitterstab *m* ~ **box** Gitterloge *f* ~ **brickwork** gitterförmiges Mauerwerk *n* ~ **girder** Gitterwerksträger ~ **girder bridge** Fachwerkbrücke *f* ~**-girder column** Gitterwerksäule *f* ~**-girder construction** Gitterwerk *n* ~**-girder pole** Gittermast *m*

**latticing** Vergitterung *f*

**Laue,** ~ **diagram** Lauediagramm *n* ~ **spots** Laueflecken *pl*, -punkte *pl*

**launch, to** ~ abschließen, ansetzen, katapultieren, auf die Reise *f* schicken, vom Stapel *m* (laufen) lassen, starten

**launch** Abschuß *m* (g/m), Stapellauf *m*, Start *m*, Vorhelling *f* ~ **by automobile tow** Autostart *m* ~ **by automobile winch** Autowindenstart *m* ~ **by an elastic cable** Gummiseilstart *m* ~ **by towing** Zugstart *m*

**launch,** ~ **cycle** Abfeuervorgang *m* ~ **ring** Abschußring *m* (g/m) ~ **site** Abschußort *m* (g/m)

**launching** Abschießen *n*, Abschuß *m*, Auslösung *f*, Ausrücken *n*, Ingangsetzen *n*, Start *m* (g/m) ~ **angle** Katapultwinkel *m* ~ **basin** Becken *n* ~ **carriage** Startwagen *m* ~ **cradle** Ablauf-apparat *m*, -gerüst *n*, -schlitten *m* ~ **crew** Startmannschaft *f* ~ **deck** (for ship planes) Startdeck *n* ~ **device** Startvorrichtung *f* ~ **elastic** Gummistartseil *n* ~ **flight** (of electrons) Ablauf *m* ~ **lugs** Führungsschuhe *pl* ~ **nose** Vorbauschnabel *m* ~ **platform** Abschußrampe *f* ~ **rack** Zielvorrichtung *f* ~ **rail** Abschußrinne *f* (g/m), Schleuderschiene *f* ~ **rope** Startseil *n* ~ **rotation** (of electrons) Ablauf *m* ~ **ship having a constant** geradlinige Ablaufbahn *f* ~ **silo** Startschacht *m* (g/m) ~ **site** Abschuß-base *f*, -stelle *f*, Raketengelände *n* ~ **switch (or holder)** Schießhalter *m* ~ **table** Abschußtisch *m* ~ **team** Abschußmannschaft *f* ~ **thrust** Startschub *m* ~ **trigger** Ablaufschlitten *m* ~ **tube** Lanzierrohr *n* ~ **ways** Ablaufbahn *f*, Gleitbalken *pl* ~ **weight** Ablaufgewicht *n*

**launder, to** ~ waschen

**launder** Gefluder *n*, (mining conduit) Gerinne *n*, Lutte *f*

**laundry** Waschanstalt *f*, Wäsche *f*, Wäscherei *f* ~ **bag** Wäschesack *m* ~ **plant** Wäschereianlage *f*

**lauric acid** Laurinsäure *f*

**Lauritsen electroscope** Lauritsen-Elektroskop *n*

**Lauth three-high plate mill** Lauth'sches Blechtrio *n*

**lava** Lava *f*

**lavatory** Abortanlage *f*, Stehabortanlage *f*, Toilette *f*, Wascheinrichtung *f* ~ **installation** Klosettanlage *f*

**lavender** Lavendel *m* ~ **copy (or print)** Lavenderkopie *f*

**law** Fundamentalsatz *m*, Gesetz *n*, Grundgesetz *n* (phys.), Recht *n*, Satz *m*, Verordnung *f*

**law,** ~ **of area** Flächensatz *m* ~ **of change** Zufallsgesetz *n* ~ **of conservation** Erhaltungsgesetz *n* ~ **of conservation of energy of matter** Gesetz *n* zur Erhaltung der Energie ~ **of conservation of momentum** Satz *m* vom Antrieb

~ **of debentures** Obligationsrecht *n* ~ **of diminishing returns** Gesetz *n* vom abnehmenden Ertragszuwachs ~ **of elasticity** Elastizitätsgesetz *n* ~ **of (energy) equipartition** Gleichverteilungssatz *m* ~ **of evidence** Beweisrecht *n* ~ **of exchange** Wechselrecht *n* ~ **of falling bodies** Fallgesetz *n* ~ **of gravity** Gravitations-, Massenanziehungs-gesetz *n* ~ **of hurricanes** Orkanregel *f* ~ **of imagery** Abbildungsgesetz *n* (von Abbe) ~ **of the index of remoteness** Entfernungsgesetz *n* ~ **of induction** Induktionsgesetz *n* ~ **of limitations** Verjährungsgesetz *n* ~ **of mass action** Massenwirkungsgesetz *n* ~ **of the mean** Mittelwertsatz *m* ~ **of motion** Bewegungsgesetze *pl* ~ **of partition** Teilungsgesetz *n* ~ **of prohability** Zufallsgesetz *n* ~ **of the propagation of error** Fehlerfortpflanzgesetz *n* ~ **of radiation** Entfernungsgesetz *n* (electron.) ~ **for receiving energy** Tiefempfangsgesetz *n* (acoust.) ~ **of reflection** Reflektionsgesetz *n* ~ **of refraction** Brechungsgesetz *n* ~ **of similarity** Ähnlichkeitsgesetz *n* ~ **of similtude** Ähnlichkeitsgesetz *n* ~ **of stages** Zustandsgesetz *n* ~ **of superannuation** Verjährungsgesetz *n* ~ **of symmetry** Symmetriegesetz *n* ~ **of thermodynamics** Wärmesatz *m* ~ **of the transmissibility of pressure** Druckfortpflanzungsgesetz *n*

**law,** ~ **costs** Prozeßkosten *pl* ~ **court** Gericht *n* ~ **governing prizes** (navy) Prisenrecht *n* ~ **interpretation** Gesetzauslegung *f* ~ **regime** Gültigkeitsbereich *m* ~ **suit** Prozeß *m*

**lawful** gerecht, gesetzlich, gesetzmäßig, rechtlich

**lawfullness** Gesetzmäßigkeit *f*, Gültigkeit *f*, Rechtmäßigkeit *f*

**lawn** Rasen *m*, Schleier *m*, Wiese *f* ~ **weaver** Feinleinwandweber *m*

**lawyer** Advokat *m*, Anwalt *m*, Rechtsanwalt *m*

**lay, to** ~ abflauen, anvisieren, (cable) auslegen, (cable) legen, verlegen, (cable) versenken **to** ~ **around** umlegen **to** ~ **away** versetzen **to** ~ **bare** bloßlegen **to** ~ **a bridge** eine Brücke *f* schlagen **to** ~ **a cable** ein Kabel *n* legen oder auslegen **to** ~ **cable-fashion** kabelweise schlagen **to** ~ **claim to** in Anspruch *m* nehmen **to** ~ **a course** (on a map or chart) einen Kurs *m* auf einer Karte absetzen **to** ~ **off a distance** Strecke *f* abschneiden **to** ~ **down** hinlegen, niederlegen, (ship) auf die Helling *f* strecken, (rules) festlegen **to** ~ **flat** abgleichen, einebnen **to** ~ **on form** die Form *f* einheben (print.) **to** ~ **the foundation** fundamentieren, gründen **to** ~ **down a gathering** Lagen *pl* machen **to** ~ **out on the ground** trassieren **to** ~ **in** (provisions) sich eindecken, einkellern **to** ~ **out money** auslegen **to** ~ **off** abbauen, abmustern, abtragen, (on a map or chart) absetzen, außer Arbeit *f* setzen **to** ~ **on** auftragen **to** ~ **on (or up)** auflegen **to** ~ **open** (a patent) auslegen, (metal) freilegen **to** ~ **out** abdocken, abstecken, aufreißen, auslegen, entwerfen, (a cable) stecken **to** ~ **over** umlegen **to** ~ **a pipeline** eine Rohrleitung legen **to** ~ **down as principle** als Grundsatz *m* aufstellen **to** ~ **reciprocally** gleichlaufend einstellen **to** ~ **a ridge** befirsten **to** ~ **out ridges** rigolen **to** ~ **down smoke** nebeln **to** ~ **a smoke screen** einnebeln **to** ~ **out a store** ein Lager *n* anlegen **to** ~ **out stores** Lager *pl* anlegen **to** ~ **stress on**

betonen **to ~ in swath** in Schwaden *pl* ablagen (agr.) **to ~ off a true bearing** eine rechtweisende Peilung *f* in die Karte tragen **to ~ up** einstellen
**lay** Drall *m* (naut.), Kabelschritt *m*, (of a cable) Schlag *m*, (length of) Schlaglänge *f*, (textiles) Webeinlage *f*; laienhaft **~ of a rope** (in ropemaking) Schlag *m* eines Seiles **~ of wire** Drahtlage *f*
**lay, ~ blasts** Tunnelschießen *n* **~ boy** Bogenableger *m* (print.) **~-by** Anlegestelle *f*, Ausweichstelle *f* **~ day (or days)** festgesetzte Zeit *f* zum Entladen eines Schiffes **~ edge** Anlagekante *f* **~-formation** Schichtenbildung *f* **~ gauge** Anlegemarke *f* **~ man** Laie *m* **~-out fee** Bekanntmachungsgebühr *f* **~ race** Schützenbahn *f* **~ shaft** Zwischenwelle *f* **~ turn** Schlag *m*
**layer, to ~** aufschichten
**layer** Bank *f*, Bett *n*, Flöz *n*, Horizont *m*, Lage *f*, Lager *n*, Schicht *f*, (of weld) Schweißlage *f*, Steckling *m* **in ~s** lagenweise, schichtenweise, schichtweise
**layer, ~ of adhesive rubber** Klebgummischicht *f* **~ of ballast (or metal)** Schotterlage *f* **~ of bead** Raupenlage *f* **~ of bitumastic** Bitumastikbelag *m* **~ of broken stones** Koffer *m* **~ of cable conductors** Adernlage *f* **~ of cement** Kittschicht *f* **~ of charge** Ladungsschicht *f* **~ of concrete** Betonschicht *f* **~ of cuttings** Spanschicht *f* **~ of disturbance** Störungsschicht *f* **~ of earth or stone** Erd- oder Gesteinschicht *f* **~ of fabric** Stoffschicht *f* **~ of glass-dust** Glasstaubschicht *f* **~ of gray paper** Fließpapierunterlage *f* **~ of grouting** Vergußschicht *f* **~ of mold** Humusschicht *f* **~ of oxide** Oxydschicht *f* **~ of puddled clay** Lettendamm *m* **~ of quicksand** Schwimmsandschicht *f* **~ of rock** Gesteinsschicht *f* **~ of rubble** Geröllschicht *f* **~ of rust** Rostschicht *f* **~ of separation** Trennungsschicht *f* **~ of slag** Schlackenkruste *f* **~ of sod** Rasenabdeckung *f* **~ of stability** Stabilitätsschicht *f* **~ of temperature** Inversionsschicht *f* **~ of vorticity** Wirbelschicht *f* **~ of warm air** warme Luftschicht *f* **~ of weather-worn material** Verwitterungsschicht *f* **~ of weld** Schweiße *f*, Schweißlage *f* **~ of wood** Dicke *f*
**layer, ~ blast** Kammerschießen *n* **~ charge** Schichtladung *f* **~ clouds** Schichtwolken *pl* **~ controller** Schichtregler *m* **~ distribution** Schichtverteilung *f* **~ filtration** Schichtenfiltration *f* **~ lattice** Schichtengitter *n* **~ line** Schichtlinie *f* **~-on** Anleger *m*, Einleger *m* **~-out** (dyeing) Abdocker *m* **~ resistance** Schichtwiderstand *m* **~ separation** Grenzschicht(e)ablösung *f* **~ winding** Lagenwicklung *f* **~-wound** lagenweise gewickelt
**layered structure** Schichtenbau *m*
**layering apparatus** Auslegeapparat *m*, Leger *m*
**laying** Auslegung *f*, (of cable) Legung *f*, (of cable) Verlegung *f* **~ of cable** Kabelauslegung *f*, Kabellegung *f* **~ for direction** Nehmen *n* der Seitenrichtung **~ down** (rules) Festlegung *f* **~ down an equation** Ansatz *m* einer Gleichung **~ of field cables** Feldkabelbau *m* **~ down a formula** Ansatz *m* einer Gleichung *f*, Gleichungsansatz *m* **~ of a gas obstacle** Begasung *f*, Geländevergiftung *f* **~ an oil-field cable** Verlegung *f* eines Ölkabels *n* **~ of pipes** Rohrleitungsführung *f*

**laying, ~-down of standard norms (or standard rules)** Normenaufstellung *f* **~-down gig** Verstreichrauhmaschine *f* **~ dressing** Richten *n* (metal) **~ gear** Richtgerät *n*, Richtvorrichtung *f* **~-gear housing** Richtgehäuse *n* **~-in** Stoffeinlage *f* **~ mechanism** Richtvorrichtung *f* **~-off position bearings (or lines)** Kursabsetzen *n* **~-off of work** Außerdienststellung *f* **~-off of works for repairs (or other purposes)** Außerbetriebsetzung *f* **~-on of colors** Farbenauftrag *m*, Farbenlage *f* **~-out** Absteckung *f* **~-out of a canal** Anlage *f* eines Kanals **~-out reel** Kabelwagen *m* **~ pieces reciprocally** Gleichlaufstellung *f* **~ pole** (ropemaker's) Drehpfahl *m* **~-up basin** Rückhaltbecken *n* **~-up machine** Vierer-Verseilmaschine *f* **~ various coils over one another** Übereinanderwickeln *n* mehrerer Schichten
**layout** Anlage *f*, Anordnung *f*, Arbeitsplanung *f*, Aufriß *m*, Ausstattung *f*, Entwurf *m*, Gruppierung *f*, Plan *m*, Zeichnung *f* **~ of the case** Kastenschema *n* **~ of cranes** Kranplan *m* **~ of equipment** Geräteanordnung *f* **~ of pipes** Rohrleitungsplan *m* **~ of a work on the ground** Abstecken *n* eines Werkes
**layout, ~ plan** Auslageplan *m*, Grundrißplan *m*, Situationsskizze *f* **~ sheet** Musterplan *m* **~ sketch** Anzeigeskizze *f*
**lazulite** Lasurspat *m*, Lazulit *m*
**lazy, ~ tongs** Scherenspreizer *m* **~ tuning** fühlbare Abstimmung *f*
**LB** (local battery) Ortsbatterie *f* (OB)
**LB-line** OB-Leitung *f* (teleph.)
**L-bar (or iron)** Stabeck *n*
**L-capture** L-Einfang *m*
**leach, to ~** aussüßen, laugen **to ~ out** auslaugen
**leach** Lauge *f* **~ line** Gording *f* **~ residue** Laugenrückstand *m* **~(ing) solution** Laugenlösung *f*
**leachable** laugefähig, laugungsfähig
**leached pulps** Auslaugerückstand *m*
**leaching** Auslaugung *f*, Laugerei *f*, Laugung *f*, (of metals) Tarnieren *n* **~ by agitation** Rührlaugung *f* **~ in place** Laugung *f* in situ
**leaching, ~ agent** Laugemittel *n* **~ barrel (or drum)** Laugetrommel *f* **~ installation** Laugeeinrichtung *f* **~ method** Laugeverfahren *n*, **~ plant** Auslaugerei *f*, Laugeneinrichtung *f*, Laugerei *f* **~ property** Laugbarkeit *f* **~ solution** Lauge *f* **~ tank** Laugebehälter *m*, Laugentank *m* **~ vat** Laugebehälter *m*
**lead, to ~** anführen, (joints) ausbleien, bleien, bleiern, führen, leiten, an der Spitze *f* stehen, plombieren, verbleien, voreilen, vorgehen, vorhalten **to (re) ~** mit einem neuen Bleimantel *m* versehen **to ~ back** zurückführen, rückführen, zurückleiten **to ~-coat** verbleien **to ~ down** herabführen **to ~ forward** vorführen **to ~-fuel** aufbleien **to ~ in** einführen, einleiten, zuführen, zuleiten **to ~ off** abführen, ableiten **to ~ on** anführen **to ~ through** durchführen **to ~ the web** aufführen (paper mfg.)
**lead** Ader *f*, Blei *n*, Durchführung *f*, Einklang *m*, Erzgang *m*, Gewindesteigung *f* (thread), Kabelverbindung *f*, Leitung *f*, Leitungskabel *n*, Linie *f*, Nine *f*, Schrägungswinkel *m*, Senkblei *n*, Spulen(draht)ende *n*, Steigung *f*, Voreilen *n*, Voreilung *f*, Vorsprung *f*, Zuführung *f* (electr.), Zuleitung *f*

lead, ~ of brushes Bürstenverschiebung *f*, Bürstenverstellung *f* ~ in continuous phase fein verteiltes Blei *n* ~ of the leadscrew Leitspindelsteigung *f* ~ for netting Netzblei *n* ~ and return Hin- und Rückführung *f* ~ of a screw Steigung *f* (Ganghöhe) einer Schraube *f* ~ of sound Vorlauf *m* ~ of sound recording Filmvorlauf *m* ~ of a spiral (or of a spire) Spiralsteigung *f* ~ to the trailer Anhängerleitung *f*

lead, ~ accumulator Bleisammler *m* ~ acetate Bleiazetat *n*, essigsaures Bleioxyd *n* ~ acid battery Säurebatterie *f* ~ activated bleiaktiviert ~ alloy Bleilegierung *f* ~ amalgam Bleiamalgam *n* ~ angle (of phases) Voreilungs-, Vorhalte-winkel *m* ~ anode Bleianode *f* ~-antimony alloy Antimonblei *n*, Bleiantimonlegierung *f* ~ aperture Durchlaßöffnung *f* ~ arsenate arsensaures Blei *n*, Bleiarsenik *m* ~ arsenide Bleiarsenik *m* ~ arsenite Bleiarsenit *n* ~ asbestos Bleiasbest *m* ~-ashes Blei-asche *f*, -schaum *m* ~ azide Bleiazid *n* ~ ball Bleikugel *f* ~ base Bleifuß *m* ~ basin Bleischale *f* ~-bath furnace Bleibadeofen *m* ~-bearing bleiführend ~ bismuth Bleiwismut *n* ~ blast furnace Bleischachtofen *m* ~ blocks Bleiquader *m* ~ borate glasses Bleiboratgläser *pl* ~ box Bleibüchse *f* ~ brick bleihaltiger Baustein *m* ~ bronze Bleibronze *f* ~ bullet Bleikugel *f* ~ burner Bleilöter *m* ~ burning autogene Schweißung *f* ~ cable Bleikabel *n* ~-cable jointer Bleikabellöter *m* ~-cable press Bleikabelpresse *f* ~ calx Bleikalk *m* ~ cam Leitkurve *f* ~ cap Bleikappe *f* ~ carbonate Bleikarbonat *n*, kohlensau(e)res Bleioxyd *n*, Cerusset *m* ~ case Bleibüchse *f* ~ cast Bleiabguß *m* ~ cathode Bleikathode *f* ~ caulking Bleiverstemmung *r* ~ caulking chisels Bleistemmer *pl* ~-chamber crystals Bleikammerkrystalle *pl* ~-chamber process Bleiprozeß *m* ~ charge Bleifüllung *f* ~ chloride Bleichlorid *n* ~ chromate Bleichromat *n*, Bleigelb *n*, Chromgelb *n*, chromsaures Bleioxyd *n* ~ clamp Bleibacke *f* ~-coated verbleit ~ coating Bleiüberzug *m*, Verbleien *n* ~ colic Bleikolik *f* ~-colored bleifarben, -fabig ~ compound Bleiverbindung *f* ~ computer Auswanderungsmesser *m* ~ content Bleigehalt *m* ~ core Bleikern *m* ~ covered bleiausgekleidet ~-covered cable Bleirohrkabel *n*, Kabel *n* mit Bleimantel ~-covered rubber cable Gummibleikabel *n* ~-covered twin-(four) wire cable zweiadriges (vieradriges) Bleirohrkabel *n* ~ covering Bleimantel *m* ~-covering plant Verbleiungsanlage *f* ~ crust Bleiansatz *m* ~ deposit Bleischlamm *m* ~ die Bleigesenk *n* ~ dioxide Bleidioxyd *n*, -perioxyd *n*, Plumbioxyd *n* ~ dish Bleischale *f* ~ dross Blei-abgang *m*, -asche *f*, -krätze *f*, -schaum *m* ~ dust Bleistaub *m* ~-dust storage cell Bleistaubsammler *m* ~ engraving Bleischnitt *m* ~ error Steigungsfehler *m* ~ extraction Bleigewinnung *f* ~ extrusion press Bleistrangpresse *f* ~ feature Leitartikel *m* ~ figure Vorhaltemaß *n* ~ fluosilicate Kieselfluorblei *n*, kieselfluorwasserstoffsaures Blei *n* ~ foil Bleifolie *f* ~ fume Blei-dampf *m*, -rauch *m* ~ (safety) fuse Bleisicherung *f* ~ glass Bleiglas *n* ~ glaze Bleiglasur *f* ~-grave bleigrau ~ gravel Bleigrieß *m* ~ grid Bleigitter *n* ~ hardening Abschrecken *n* in Blei, Blei-

härtung *f* ~ hydroxide Bleioxydhydrat *n* ~ hyposulfite unterschwefligsaures Blei *n*

lead-in Ableitung *f* (antenna), Ausführung *f*, Durchführung *f*, Einführen *n*, Einführung *f*, Einführungsdraht *m*, Zufuhr *f* ~ of an overhead line oberirdische Leitungsführung *f*

lead-in, ~ cable Einführungskabel *n* ~ groove Einlaufrille *f* (phono) ~ insulator Einführungs-, End-isolator *m* ~ lights Leitfeuer *pl* (aviat.) ~ outdoor bushing Freiluftdurchführung *f* ~ pole Einführungsgestänge *n* (electr.) ~ spiral Einlaufrille *f* (phono) ~ tube Einführungsstutzen *m* ~ wire Antennenzuleitung *f*, Einführungsdraht *m*, Leitungseinführung *f*, Spulen(draht)ende *n*, Stromzuführung *f*, Zuleitungsdraht *m*

lead, ~ inductance Leitungsinduktivität *f*, Zuleitungsinduktivität *f* ~-insulated cable bleiummanteltes Kabel *n* ~ iodate Bleijodat *n* ~ iodide Jodblei *n* ~ jacket Bleimantel *m* ~ jaw socket Bleibacke *f* ~-lag relation between picture and sound Bild-Ton-Abstand *m* ~-like bleiartig, bleiisch ~ line Lotleine *f* ~-lined bleiausgefüttert, mit Blei *n* ausgeschlagen ~-lined pipe verbleites Rohr *n* ~ lining (accumulators) Bleiausschlag *m*, Bleiseele *f*, Bleiverkleidung *f*, Verbleien *n*, Verbleiung *f* ~ linoleate Leinölsaures Bleioxyd *n* ~ loaded steel balance weight Gegengewicht *n* aus Stahl mit Bleizusatzgewicht ~ magazine Minenkammer *f* ~ mine Blei-bergwerk *n*, -grube *f*, -mine *f* ~ mold Bleiform *f* ~-molding matrix Bleimatrize *f* ~ molybdate Bleimolybdat *n* ~ monoxide (gelbe) Bleiglätte *f*, Bleioxyd *n*, Glätte *f* ~ nitrate Bleisalpeter *n*, salpetersaures Blei(oxyd) *n* ~ obtained from dross Abzugsbrei *n* ~ oleate Bleioleat *n* ~ ore Bleierz *n* ~-ore roasting Bleierzröstung *f* ~-out groove Auslaufrille *f* ~ output Bleiproduktion *f* ~-over groove Überleitrille *f* (phono) ~ oxalate oxalsaures Blei *n* ~ oxide Bleikalk *m*, Bleioxyd *n* ~ oxychloride Kasseler Gelb *n* ~ packing Bleidichtung *f* ~ packing ring Bleischeibe *f* ~ palmitate Bleipalmitat *n* ~ paper Bleipapier *n* ~ pea Bleierzstück *n* ~ pencil Blei-feder *f*, -stift *m* ~-pencil sharpener Bleistiftfräser *n* ~ peroxide Bleidioxyd *n*, Bleiperoxyd *n*, Bleisuperoxyd *n* ~ phosphide Phosphorblei *n* ~ phosphite Bleiphosphit ~ pig Blei-barren *m*, -block *m*, -mulde *f* ~ pin Durchführungsstift *m* ~ pipe Blei-rohr *n*, -röhre *f* ~-pipe press Bleirohrpresse *f* ~ plaster Bleipflaster *n* ~ plate Bleiplatte *f* ~-plate test (explosives) Bleiplattenprobe *f* ~-plating plant Verbleiungsanlage *f* ~ point Vorhaltepunkt *m* ~ poisoning Bleikrankheit *f* ~ pot Vortiegel *m* ~-powder mills Bleistaubmühlen *pl* ~ precipitate Bleiniederschlag *m* ~ press Bleipresse *f* ~ production Bleiproduktion *f* ~-pursuit course (or path) Zielverfolgungskurs *m* ~ quenching Abschrecken *n* in Blei ~ rasp Bleiraspel *f* (grobe Feile) ~ rate Auswanderungszeit *f* ~ refinery Bleiraffinerie *f* ~ refining Bleiraffination *f* ~ refining plant Bleihütte *f*, Bleiraffinerie *f* ~ regulus Bleikönig *m* ~ rein Führerzügel *m* ~ response (fuel) Mischfähigkeit *f* ~ response of a fuel Bleiempfindlichkeit *f* eines Kraftstoffes ~-roasting process Bleiröstprozeß *m* ~ salicylate Bleisalizylat *n* ~ salt Bleioxydsalz *n*,

Bleisalz *n* ~ **scoria** Bleiabgang *m* ~ **screen**
Bleischirm *m* ~**screen opaque to radiations** für
Strahlung undurchlässiger Bleischirm *m* ~
**screw** Bewegungsschraube *f*, Führungsschraube
*f*, Leit-, Schrauben-spindel *f* ~**-screw inspection**
Leitspindelkontrolle *f* ~**-screw reversal** Drehrichtungsumkehr *f* ~ **seal** (for a meter) Bleiplombe *f*, Plombe *f*, Plombenverschluß *m*
~**-sealed** plombiert ~**-sealing pliers** Plombenzange *f* ~ **selenide** Klaustalit *m* ~ **sheath
stripping machine** Bleimantelabstreifmaschine *f*
~**-sheathed** mit einem Bleimantel *m* umpreßt
~**-sheathed cable** bleiumhülltes Kabel *n* ~ **shielding** Bleiabschirmung *f* ~ **shot** Bleischrot *n* ~ **sight**
Vorhaltevisier *n* ~**silicate** Bleiglas *n* ~ **silicofluoride** siliziumfluorwasserstoffsaures Blei
*n* ~**-silver telluride** Tellursilberblei *n* ~ **skim**
Abstrichblei *n* ~ **slag** Bleischlacke *f* ~**sleeve**
Bleimuffe *f* ~ **slime** Bleischlich *m* ~ **sludge**
Bleischlamm *m* ~ **smeltery** Bleihütte *f* ~ **smelting** Bleiarbeit *f*, Bleiverhüttung *f* ~**-smelting
furnace** Bleiofen *m* ~**-smelting hearth** Blei-
schmelzherd *m* ~**smoke** Bleirauch *m* ~ **solder**
Bleilot *n*, Lötblei *n* ~ **soldering** Blei-löterei *f*,
-lötung *f* ~ **soldering process** Bleilötverfahren *n*
~ **spar** Bleispat *m* ~ **spatter (or sponge)** Bleischwamm *m* ~ **spring** Bleibügel *m* ~ **stearate**
Bleistearat *n* ~ **storage battery** Bleiakkumulator *m* ~ **storage container** Isolopenbehälter
*m* ~**-strip fuse** Blechsicherung *f* ~ **subacetate**
Bleiessig *m* ~ **suboxide** Bleiasche *f*, Bleioxydul
*n* ~ **sulfate** Anglesit *m*, Bleisulfat *n* ~**-sulfate**
Bleivitriol *m* ~ **sulfate pigment** Milch-, Mühl-
häuser-weiß *n* ~ **sulfide detector** Bleiglanz-
detektor *m* ~ **sulfide type semiconductor** Blei-
sulfidtyp-Halbleiter *m* ~ **sulfocyanate** Blei-
rhodonid *n* ~ **sulfuric acid cell** Bleisammler *m*
~ **supply** Zulauf *m* ~ **telluride** Tellurblei *n* ~
**tempering** Anlassen *n* in Blei ~ **tester** Stei-
gungsprüfer *m* ~ **tetra-acetate** Bleitetra-Azetat
*n* ~ **tetraethyl** Bleitetra-Äthyl *n* ~**-through
capacitor** Durchführungskondensator *m* ~
**time** Auswanderungszeit *f* ~**-tin solder** Blei-
zinnlot *n* ~ **titanate** titansaures Bleioxyd *n* ~
**track** Ausziehzeichen *n* ~ **tube** Lotröhre *f* ~
**tube drawn through iron pipe as liner for hot
sea-water lines of large diameter** Bleiseele *f* ~
**tungstate** Bleiwolframat *n*, Scheel-bleierz *n*,
-bleispat *m*, Stolzit *m*, Wolframbleierz *n*,
wolframsaures Blei *n* ~**-up strap** Rückenriemen
*m* ~ **value** Vorhaltemaß *n* ~ **vapor** Bleidampf
*m* ~ **vein** Blei-ader *f*, -gang *m* ~**vitrol** Anglesit
*m*, Bleivitrol *m* ~**water** Bleiwasser *n* ~ **weight**
Blei-ballast *m*, -gewicht *n* ~ **weight in the
bottom of the compass bowl** Bleiboden *m* ~ **wire**
Bleidraht *m*, Kabel *n*, Leitungsdraht *m*, Verbindungsleitung *f*, Zuführungsdraht *m*, Zuleitung *f*, Zünderdraht *m* ~ **yield** Bleiausbrin-
gen *n*

**leaded** durchschossen, (fuel) gebleit ~ **fuel** ge-
bleiter Kraftstoff *m* ~**-in bolt** eingebleiter
Bolzen *m* ~ **joint** Bleigefüge *n* ~ **matter** durch-
schossener Satz *m* ~ **zinc** Mischoxyd *n*

**leaden** bleiern ~ **dividing box of cable** Bleiauf-
teilungsmuffe *f* ~ **end box of cable** Bleischluß-
muffe *f* ~ **jaw** Bleibacke *f*

**leader** Ader *f*, Draht *m*, Führer *m*, Gewindeleit-
patrone *f*, (driving gear) Haupttriebrad *n*,

Leitartikel *m*, Leiter *m*, Trum *n*, Vorpolier-
kabel *n*, Vorschlichtkaliber *n*, Vorspann *m*
(film) ~ **cable** Erd-, Leit-, Leitungs-kabel *n* ~
**end** Papiervorlauf *m* ~ **field change** Leitkanal-
feldänderung *f* ~ **head** Rinnenkasten *m* ~**-
stroke discharge** strahlartige Entladung *f*
~**-tape** Vorband *n* (tape rec.), Vorspannband
*n*

**leaders** Geviertpunkt *m*, Leitepunkte *pl*
**leading** Bleiverschluß *m*, Führung *f*, (calking)
Plombierung *f*, Verbleiung *f*, Voreilen *n*, Voreilung *f*; maßgebend, vorangehend, vorausgehend, voreilend ~ **(45 degrees)** voreilend
(um 45 °) ~ **of phase** Voreilung *f*
**leading, ~ airplane of a formation** Führerflug-
zeug *n* ~ **article** Leitartikel *m* ~ **cable** Einführungsleitung *f* ~ **channel** Leitkanal *m* ~ **coil
section** Spulenfeld *n* ~ **cord** Zuleitungskabel
*n* ~ **current** voreilender Strom *m* ~ **dimensions**
Hauptabmessungen *pl* ~ **down** Herabführung *f*
~ **echelon** Führungsstaffel *f*
**leading-edge** (entering) Anlaßkante *f*, Anlauf-
seite *f*, Anlegekante *f* (film), Aufstrich *m*
(Impuls), Flügeleintrittskante *f* (aviat.), Pro-
filvorderkante *f*, Steuer-, Vorder-kante *f*,
(Impuls) Vorderflanke *f*, Vorderteil *n* ~ **control
horn** Blattnasen-Ausgleichshorn *n* ~ **flap**
vordere Flügelklappe *f*, Nasenklappe *f* ~
**radiator** Nasenkühler *m* ~ **radius** Nasen-
haubenradius *m* ~ **skin** Außenhautleitkante *f*
~ **stiff against torsion** drehsteife Flügelnase *f*
~ **strip** Formrippe *f*, Nasenleiste *f*, Stirnleiste *f*
**leading, ~-end resistance** Stirnwiderstand *m* ~
**form** Leitform *f* ~ **fossil** Leitfossil *n* ~ **grate**
Leit-, Trift-rechen ~ **hand** Vorarbeiter *m*
**leading-in** Leitungseinführung *f* ~ **cable** Zuführungskabel *n* ~ **double bracket** Durchführungsisolator *m*, Einführungsdoppelstütze
*f* (electr.) ~ **line** Zufuhrgleis *n* ~ **manhole**
Einführungsbrunnen *m* ~ **pin** Durchführungsstift *m* ~ **wire** Antennenzuleitung *f*
**leading, ~ light** (single) Bündelfeuer *n*, (marine)
Leitfeuer *n*, Richtfeuer *n*, Strahlfeuer *n* ~ **line**
Deckpeilung *f* ~**-out** Ausführung *f* ~**-out cable**
Ausführungskabel *n* (electr.), Durchführungs-
kabel *n* ~ **particulars** Hauptmerkmale *pl* ~
**pass** Vorpolierstich *m* ~ **pile** Richtungspfahl *m*
~ **pole tip** auflaufende Polkante *f* ~ **pulley**
Lenkrolle *f* ~ **role** Hauptrolle *f* ~ **ship** Spitzen-
schiff *n* ~ **species** gangbarste Sorte *f* ~ **spindle**
Führungs-, Leit-spindel *f* ~ **spring** Voreilfeder
*f* ~**-through tape** Führungsbänder *pl* (paper
mfg) ~ **tip** (of airplane wing) Leitkante *f* ~
**valve** Voreilventil *n* ~ **wind edge** Flügelnase *f*
~ **zero** führende Nullen *pl* (info proc.)
**leads** (terminals) Anschlußleiter *m*, Durchschuß *m*

**leady** bleiartig, bleiern, bleiisch, bleireich, ~
**matte** Bleistein *m* ~ **spelter** bleiisches Boden-
zink *n*
**leaf, to** ~ blättern, (bronze) schwimmen
**leaf** Aufziehklappe *f*, Blatt *n*, Flügel *m*, Folie *f*,
Klappe *f*, Lamelle *f*, (table) Platte *f*, (driving
wheel) Zahn *m*
**leaf, ~ of door** Türflügel *m* ~ **of fan-shaped gate**
Fächerflügel *m* ~ **on flap valve** Scharnierventil
*n* ~ **of the gauge loom** Gazeschaft *f* ~ **of a
pinion** Triebstock *m* eines Getriebes ~ **of**

**sluice gate** Schleusentorflügel *m* ~ **of a spring** Federblatt *n*

**leaf,** ~ **catcher** Krautfänger *m* ~ **electrometer (or electroscope)** Blattelektrometer *m* ~ **fiber** Blattfaser *f* ~ **gold** Blatt-, Buch-, Schlag-gold *n* ~ **prunning (or laminated)** blätterförmig ~ **sight** Klapp-, Kurven-visier *n* ~ **spring** Blatt-, Flach-feder *g* ~ **spring gaiters** Federschutzgamaschen *pl* ~**-spring tail skid** Blattfederspron *m* ~**-type quadrant sight** Quadrantenvisier *n* ~ **valve** (referring to the leaf) Federplattenventil *n*, Klappenventil *n* ~ **wood** Laubholz *n*

**leaflet** Broschüre *f*, Faltblatt *n*, Merkblatt *n* ~ **dropping** Flugblattabwurf *m*

**leafy** belaubt, laubreich ~ **wood** Laubwald *m*

**leak, to** ~ absintern, auslaufen, durchlässig sein, (of current) kriechen, laufen, lecken, leck sein, leck werden, rinnen, sichern, streuen (electr.), tröpfeln, undicht sein **to** ~ **in (or into)** eindringen **to** ~ **load** mit Ableitung *f* belasten **to** ~ **off** abfließen, (current) ableiten, **to** ~ **out** durchsickern **to** ~ **through** durchlecken

**leak** Ableitung *f*, Abzweig *m*, Dämpfung *f* (electron.), Entweichung *f*, Leck *n*, Loch *n*, Nebenschließung *f*, Nebenschluß *m*, Rinne *f*, undichte Stelle *f*, Undichtigkeit *f* **in** ~ in Brücke geschaltet **to have sprung a** ~ leck sein **the** ~ **has been stopped** das Leck *n* hat sich zugezogen ~ **in an electric current** Abteil *n*

**leak,** ~ **circuit** Mitlesestromkreis *m* ~ **coil** Abzweigwiderstand *m*, Querspule *f* ~ **conductance** Ableitung *f*, Leitungskonstante *f* ~**(age) current** Ableitungs-, Leck-strom *m* ~ **detector** Leckstellensucher *m*, Lecksucher *m* ~ **detector head** Lecksuchsonde *f* ~ **detector tube** Lecksuchröhre *f* ~ **hole** Durchlaßöffnung *f* ~ **impedance** Parallel-, Quer-impedanz *f* ~ **inductance** Querinduktivität *f* ~ **instrument** Mitleseapparat *m* ~ **load** Querspulenbelastung *f* ~**-loaded** mit erhöhter Ableitung *f* belastet, mit Querspulen *pl* belastet ~ **load(ing)** Belastung *f* der Ableitung ~ **loading** Ladung *f* mit erhöhter Ableitung, Querspulenbelastung *f* ~ **loss** Ableitverlust *m* ~**-off nipple stud** Leckölrückleitungsnippel *m* ~ **oil discharge** Leckölabfluß *m* ~**-oil gallery** Leckölbohrung *f* ~**-oil pipe** Leckölleitung *f* ~ **path** (for grid current) Abfluß *m* ~**-proof** dicht, tropfsicher ~ **proofness** Lecksicherheit *f* ~ **resistance** Abzweig-, Nebenschließungs-, Nebenschluß-widerstand *m* ~**(age) resistance** Ableitungswiderstand *m* ~**(shunt) resistance** Querwiderstand *m* ~**-retardant fuel tank** selbstabdichtender Kraftstoffbehälter *m*

**leakage** Abfließen *n* (des Ladungsbildes), (of current) Ableitung *f*, Dämpfung *f* (rdo), Durchsickerung *f*, Eindringen *n*, Hopkinsonscher Streufaktor *m*, Leck *n*, Leckage *f*, (of a liquid) Lecken *n*, Leck-schaden *f*, -verlust *m*, -werden *n*, Nebenschließung *f*, Nebenschluß *m*, Schwund *m*, Schwunderscheinung *f*, Streuung *f*, Stromverlust *m*, Verlust *m*

**leakage, to test for** ~ auf Dichtigkeit *f* prüfen ~ **on one phase** Phasenverschluß *m* ~ **of water** Verlauf des Wassers *n*, Wasserleck *n*

**leakage,** ~ **area between cylinder walls and baffles** freier Durchtrittsquerschnitt *m* zwischen Druckleitblechen und Zylindern ~ **current**

Anzapf-, Kriech-, Leck-strom, vagabundierender Strom *m*, Verluststrom *m*, (dielectr.) Vorstromentladung *f* ~ **error** Kriechstromfehler *m* ~ **factor** Verlustfaktor *m* ~ **field** Streufeld *n* ~ **flux** Streufluß *m* ~ **impedance** Streuimpedanz *f* ~ **indicator** Erdschlußprüfer *m* ~**-indicator panel** Fehleranzeigefeld *n* ~ **inductance** Streukapazität *f* ~ **light** falsches Licht *n* ~ **loss** Leckverlust *m* ~ **meter** Ableitungsmesser *m* ~ **oil stop** Leckölsperre *f* ~ **path** Nebenschließungsweg *m* ~ **pipe** Leckleitung *f* ~ **pressure** Dichtdruck *m* ~**-proof** durchschlagsicher ~ **protection** Fehlerschutz *m* ~ **radiation** Sickerstrahlung *f* ~ **reactance transformer** Streutransformator *m* **(high)** ~ **reactance transformer** Streuflußtransformator *m* ~ **resistance** Ableitwiderstand *m* ~ **spectrum** Verlustspektrum *n* ~ **steam** Leckdampf *m* ~ **test** Dichtheitprüfung *f* ~ **tester** Erdschlußprüfer *m* ~ **yoke** Streujoch *n*

**leakance** dielektrische Leitfähigkeit *f*, Leitungskonstante *f*, Nebenanschluß *m* ~ **per unit length** Ableitung *f* je Längeneinheit ~ **current** Ableitungs-, Leck-strom *m* ~ **loss** Ableitungsdämpfung *f*

**leaker-off** Anzapf *m*

**leakiness** Durchlässigkeit *f*, Undichtheit *f*, Undichtigkeit *f*

**leaking** Gießen *n*, Schwitzen *n*; leckend ~ **fuel** Leckbrennstoff *m* ~ **gasoline** Tropfbenzin *n*

**leaky** mit Ableitung *f* behaftet, durchlässig, leck, mit einem Nebenschluß *m* behaftet, undicht ~ **capacity** Verlustkapazität *f* ~ **condenser** unvollkommener Kondensator *m* ~ **dielectric** unvollkommener Isolator *m* ~**-grid detection** Audiongleichrichtung *f* ~ **insulation** unvollkommener Isolator *m* ~**-waveguide antenna** Hohlleiterschlitzantenne *f*

**lean, to** ~ lehnen, neigen **to** ~ **against** anlehnen **to** ~ **over** krängen

**lean arm,** dürr, mager ~ **anthracitic coal** Sandkohle *f* ~ **best power mixture** stärkstes Feingemisch *n* (von Brennstoff und Luft) ~ **clay** Magerungsmittel *n* ~ **coal** Magerkohle *f*, magere Kohle *f* ~ **concrete** Mager-, Sparbeton *m* ~ **concrete used to fill space (or used because of weight)** Füllbeton *m* ~ **gas** armes oder geringwertiges Gas *n*, Schwachgas *n* ~ **material** Magerungsmittel *n* ~ **mixture** Feingemisch *n*, mageres oder armes Gemisch *n*, Spargemisch *n* ~**-to roof** Flugdach *n*, Halbdach *n* ~**-to tent** Einerzelt *n* ~**-to trassed strut** Bockstütze *f* ~**-type** gemeine Schrift *f*

**leaning** (against) angelehnt

**leap, to** ~ **at** anspringen

**leap,** ~ **frog test** Durchprüfung *f* (comput.) ~ **probability** Sprungwahrscheinlichkeit *f* ~ **year** Schaltjahr *n*

**leaper** Folger *m*

**learn, to** ~ lernen **to** ~ **anew** umlernen **to** ~ **by listening** abhorchen

**learned** gebildet, gelehrt

**learner** Anfänger *m*, Anlernling *m*, Lehrling *m*

**learning** Lernen *n*, Wissenschaft *f*

**lease, to** ~ mieten, pachten, vermieten, verpachten

**lease** (weaving) Kreuz *n*, Pacht *f*, (weaving) Schrank *m* ~ **bar** Fitzenstock *m* ~**-hold** Erb-

zins-, Pacht-vertrag *m* ~ **peg (or pin)** Kreuznagel *m* ~ **pins** Fitzenstock *m* ~ **rod** Fitzenstock *m*, Kreuzrute *f*
**leased,** ~ **circuit** Mietleitung *f* ~**-line contract** Mietvertrag *m* für eine Leitung ~ **wire** Mietleitung *f*
**leasing machine** Fadenkreuzeinlesemaschine *f*
**least** geringst-, kleinst-, mindest- (e, er, es) ~ **action** kleinste Wirkung *f* ~ **circle of aberration** Kreis *m* der geringsten Aberration ~ **common multiple** das kleinste gemeinsame Vielfache *n* ~ **constraint (or resistance)** geringster Zwang *m* ~ **squares** kleinste Quadrate *pl* ~ **squares method** Fehlerquadratmethode *f* ~**-squares principle** Prinzip *n* der kleinsten Quadrate
**leather** Leder *n* ~ **from central part of hide** Kernleder *n*
**leather,** ~ **apron** Lederschürze *f* ~ **bag** Ledertasche *f* ~ **ball** Druckerballen *m* ~ **beating machine** Lederklopfmaschine *f* ~ **bellows** Lederbalg *m* ~ **belt** Lederriemen *m*, Ledertreibriemen *m* ~**-belt dressing** Lederriemenfett *n* ~ **binding** Ledereinfassung *f* ~ **board** (paper) Lederpappe *f* ~ **body** Lederaufbau *m* ~ **boot lace** Lederschnürsenkel *m* ~**-bound** ledergebunden ~ **brief case** Ledermappe *f* ~ **bucket** Ledermanschette *f* ~ **cap** Lederkappe *f* ~ **case** Ledertasche *f* ~**-cement** Riemenkitt *m* ~ **combination suit** Lederüberanzug *m* ~**-covered** beledert ~ **(collar) cuff** Ledermanschette *f* ~ **cuffs** Lederharzstulpen *pl* ~ **cup** Lederstulp *m* ~ **diaphragm** Ledermembran *f* ~ **dresser** Lederzurichter *m* ~ **dressing** Lederpflegemittel *n* ~ **edge** Lederschutzecke *f* ~ **fitting** Lederbeschlag *m* ~**-flap valve** Lederklappenventil *n* ~ **fluffing machine** Lederbimsmaschine *f* ~ **gasket** Dampfkolbenpackung *f*, Lederring *m* ~ **gauntlet with bent fingers** (for wire laying etc.) Krummfingerlederhandschuh *m* ~ **grease** Lederfett ~ **headgear** Lederbügel *m* (am Kopfhörer) ~ **industry** Lederindustrie *f* ~ **knife** Messer *n* für Lederbearbeitung *f* ~ **lace** Nähriemen *m* ~ **legging** Ledergamasche *f* ~ **oil** Lederöl *n* ~ **packer cap** Liderungsklappe *f* ~ **packing** Leder-dichtung *f*, -klappe *f*, -liderung *f*, -packung *f* ~ **portfolio** Ledermappe *f* ~ **pressing roller** Lederwalze *f* ~ **puttee** Beinleder *n* ~ **ring** Lederring *m* ~ **shim** Lederbeilage *f* ~ **shoelace** Lederschnürsenkel *m* ~**-skiving and -splitting machine** Lederschärf- und Spaltmaschine *f* ~ **slide** Lederschieber *m* ~ **strap** Leder-, Näh-riemen *m* ~ **strip** Lederstreifen *m* ~ **tightening** Lederdichtung *f* ~ **upholstery** Lederpolsterung *f* ~ **washer** Leder-dichtung *f*, -ring *m*, -scheibe *f*, unterlegscheibe *f* ~ **work** Lederarbeit *f* ~ **wrapper** Ledermappe *f*
**leatherette** Kunstleder *n*, Kunstlederpapier *n*
**leathering of a piston** Dichtung *f* eines Kolbens
**leatheroid paper** Kunstleder *n*, Lederimitation *f*
**leave, to** ~ abgeben, abtürmen, abziehen, austreten, belassen, überlassen, verlassen **to** ~ **blank** freilassen **to** ~ **cable slack in manholes** Kabelzuschlag *m* (Vorratslänge) in den Schächten *pl* lassen **to** ~ **a formation** ausscheren **to** ~ **an instrument on** ein Meßgerät *n* eingeschaltet lassen **to** ~ **a line "live"** eine Leitung *f* unter Spannung lassen **to** ~ **off** einstellen **to** ~ **out**

**auslassen to** ~ **port** aus dem Hafen *m* auslaufen **to** ~ **untouched** aussparen
**leave** Urlaub *m* **on** ~ auf Urlaub *m*
**leaving-out (sound) base** Schallüberschlagsbasis *f*
**Leblanc,** ~ **exciter** Leblancscher Phasenschieber *m* ~ **system** Leblancsche Schaltung *f*
**Lecher,** ~ **line** Lecherleitung *f*
**Léclanche cell** Leclanche-, Salmiak-element *n*
**lectern** Lesepult *n*, Pult *n* ~**-shaped** pultförmig
**lecture** (university) Vorlesung *f*, Vortrag *m* ~ **demonstration** Vorleseversuch *m* ~ **experiment** Schauversuch *m* ~ **reprint** Vortragsabdruck *m* ~ **room** Hörsaal *m*
**ledge** Absatz *m*, Anschlag *m*, Ionisierungsbank *f*, Lager *n*, Leiste *f*, vorspringender Rand *m*, Sims *m* ~ **of plane** Hobelanschlag *m*
**ledger** (trestle) Fußlatte *f*, Hauptbuch *n*, Knagge *f*, Leiste *f*
**Leduc current** Leducstrom *m*
**lee** Lee *n*, Schwert *n*, Windschatten *m* ~ **of a wave** Leeseite *f* der Welle ~ **side** Leeseite *f*
**lees presser** Treberpresser *m*
**leeward** Lee, Leeseite *f*; leewärts, windabwärts **to be to** ~ unter Wind *m* liegen ~ **bank** leeseitiges Ufer *n* ~ **drop of a wave** Leeabhang *m* ~ **eddy** Leewirbel *m*
**leeway** Abdrängung *f*, Abdrift *f*, Abtrieb *m*, Abtrift *f*, Vertreiben *n*
**left** link, links **on (or to) the** ~ links ~ **in the open air** im Freien *n* bleiben **from** ~ **to right** rechtsläufig
**left,** ~ **averted photographs** linksverschwenkte Aufnahmen *pl* ~ **half of the girder** linke Riegelhälfte *f*
**left-hand,** ~ **(side)** links ~ **action** Linksgang *m* ~ **(or port) curve** Linkskurve *f* (aviat.) ~ **cutter** linksschneidender Fräser *m* ~ **deviation** Linksabweichung *f* ~ **diamond-point tool** linker Schruppstahl *m* mit Rhomboidspitze ~ **drill** linksschneidend ~ **drive** Linksantrieb *m* ~ **engine** linksläufiger Motor *m* ~ **helix** Linksschraube *f* ~ **lay** Linksdrall *m* ~ **(or port) loop** Linkskurve *f* (aviat.) ~ **notched** linksgenutet ~ **post** linker Stiel *m* ~ **rope** linksgeschlagenes Tau *n* ~ **rotation** Linksdrehung *f*, linkslaufend ~ **rule** Linkehandregel *f* ~ **screw** linksgängige Schraube *f* ~ **side tool** linker Seitenstahl *m* ~ **siding tool** linksseitiger Abflachstahl *m* ~ **sloped girder** linker Schrägstab *m* ~ **spiral** Linksspirale *f* ~ **switch** Linksweiche *f* ~ **thread** (of screw) linksgängiges Gewinde *n*, Linksgewinde *n* ~ **threading attachment** Gewindeschneideeinrichtung *f* für Linksgewinde ~ **transmission** Linksantrieb *m* ~ **twist** Linksdrall *m* ~ **zero** seitlicher Nullpunkt *m*
**left-handed** linksgängig ~ **arbor** Linkser *m* ~ **polarization** Linksdrehung *f* ~ **quartz** Linksquarz *m* ~ **screw** linksgängige Schraube *f*
**lefthandedness** Linkshändigkeit *f*
**left,** ~ **regular lay cable** Kabel *n* mit Linksdrall **the** ~**(-hand) side** die linke Seite *f* ~ **side of engine** Einlaßseite *f* ~ **side of an equation** linke Seite *f* einer Gleichung ~**-sided** linksseitig ~**-side engine** linksseitiger Motor *m* ~ **turn** Links-drehung *f*, -kurve *f* ~**-turning** linksgängig ~ **wind** linksgängiges Wind *m*
**leg** Bein *n*, Pfeil *m*, Quadrant *m*, Schenkel *m*, Steg *m*, Stütze *f*, (of traffic pattern) Teil *n*, (of

angle) Wange *f*, Zweig *m* ~ **of the compasses**
Zirkelbein *n* ~ **of the crane** Brückenfuß *m* ~
**of a curve** Ast *m* einer Kurve ~ **of tripod** Ge-
stellbein *n*

**leg,** ~ **difference coefficient** Querwegübersetzung
*f* ~ **length of weld** Schweißnahthöhe *f* ~ **pipe**
Kniestück *n* ~ **section** Stegteil *m* ~ **spring**
Schenkelfeder *f* ~ **strap** Beingurt *m*

**legal** gesetzlich, gesetzmäßig, gültig, rechtlich,
rechtmäßig, rechtsgültig, statthaft **to bring** ~
**action** gerichtlich belangen

**legal,** ~ **address** Wohnsitz *m* ~ **advisor** Sachwalter *m*, Syndikus *m* ~ **claim** Rechtsanspruch
*m* ~ **contest** Rechtsstreitigkeit *f* ~ **expenses**
Gerichtskosten *pl* of ~ **force** rechtskräftig ~
**ground** Rechtsgrund *m* ~ **impediment** gesetzliches Hindernis *n* ~ **loophole** Gesetzeslücke *f*
~ **partner** Beteiligte(r) ~ **procedure** Rechtsweg *m*
~ **proceedings** Gerichtsverfahren *n* ~ **protection
of registered designs** Gebrauchsmusterschutz *m*
~ **punishment** gerichtliche Strafe *f* ~ **reason**
Rechtsgrund *m* ~ **records** Gerichtsakten *pl* ~
**redress** Rechtsmittel *n* ~ **requirement** gesetzliche Bestimmung *f* ~ **successor** Rechtsnachfolger *m* ~ **tender** Zahlungsmittel *n* ~ **time**
Einheitszeit *f*, gesetzliche Uhrzeit *f* ~ **title**
Rechtsanspruch *m*

**legality** Gesetzmäßigkeit *f*, Gültigkeit *f*, Rechtmäßigkeit *f*

**legalize, to** ~ beurkunden, rechtskräftig machen

**legalized,** ~ **copy** beglaubigte oder rechts-
gültige Abschrift *f* ~ **copy of assignment**
Rechtsnachfolgerungserklärung *f*

**legend** Aufschrift *f*, Beschriftung *f* einer Be-
zeichnungserklärung *f* (map), Inschrift *f*, (list
of symbols) Schaltzeichen *n*, Umschrift *f*,
Zeichenerklärung *f*

**Legendre polynoms** Legendresche Polynome *f*

**legged** schenklig

**legging frame** Längenstuhl *m*

**leggings** Gamaschen *pl*

**legibility** Ablesemöglichkeit *f*, Güte *f* (telegraph.
signal), Lesbarkeit *f*

**legible** ablesbar, lesbar, leserlich

**legitimacy** Gesetzmäßigkeit *f*, Rechtmäßigkeit *f*

**lehr** Kühlofen *m* ~ **bachite** Selenquecksilberblei
*n* ~ **stacker** (glass) Einstoßvorrichtung *f*

**leisure hour** Freistunde *f*

**LEM (lunar excursion module)** Mondlandefahrzeug *n*

**lemma** Lemma *n*, lexikografisches Stichwort *n*

**lemniscate** Lemniskate *f*, (curve of characteristic)
Lemniskatenkennlinie *f* (rdo), Schleifenlinie *f*
(geom.)

**Lenard tube** Kathodenstrahlröhre *f*, Lenard-
röhre *f*

**lend** ausleihen, auslohnen, leihen, verleihen
**to** ~ **a hand** Vorschub *m* leisten

**lender** Abnehmer *m*, Borger *m*

**lending** Ausleihen *n*, Verleihen *n* ~ **circuit** Ferndienstleitung *f* ~ **condition** Leihbedingung *f* ~
**position** Durchgangsplatz *m*

**length** Feldeslänge *f*, Länge *f*, Quantität *f*
(metr.), Wegstrecke *f*

**length,** ~ **of action** Ausladung *f* ~ **of airplane**
Flugzeuglänge *f* ~ **of any chain** Kettenlänge *f*
~ **of barrel** Rohrlänge *f*, (sparkling plug)
Schaftlänge *f*, (in calibers) Seelenlänge *f* ~ **of**

**brake path** Bremsweg *m* ~ **of break** Ausschalt-
strecke *f* ~ **of cable** Kabelstrecke *f* ~ **between
centers** Spitzenabstand *m* ~ **of chord** Sehnen-
länge *f* ~ **of conversation** Gesprächsdauer *f* ~
~ **of cordite sticks** Pulverlänge *f* ~ **of crane
runway** Kranbahnlänge *f* ~ **of a cut** Schnittlänge *f* ~ **of end section** Auslauflänge *f* ~ **of
engagement** Einschraublänge *f* ~ **of filament**
Fadenlänge *f* ~ **of glow column** Glimmfaden-
länge *f* ~ **of grip** Klemmlänge *f* ~ **of length
under head** Bolzenlänge *f* ~ **of inclination**
Länge *f* der Hochführung ~ **of jaws** Schna-
bellänge *f* ~ **of jet** Strahlreichweite *f* ~ **of
jib** Ausladung *f* ~ **of the land chain** Kettenlänge *f*, -maß *n* ~ **of landing run** Auslauflänge *f*, Ausrollstrecke *f* (beim Landen)
~ **of lay** Drallage *f* ~ **of levelling off** Ausschwe-
belänge *f* ~ **of lines of force** Kraftlinienlänge *f*
~ **of measurement** Meßlänge *f* ~ **of operation**
Betriebsdauer *f* ~ **of operation spark** Betriebs-
funkenlänge *f* ~ **of overhang** Ausladung *f* ~
of **path** Weg-länge *f*, -strecke *f* ~ **of pipe** Rohrlänge *f*, -schuß *m* ~ **of pitch by inches** Länge *f*
einer Zahnteilung in Zoll ~ **of ricochet** Ab-
prallweite *f* ~ **of rifling** Drallage *f* ~ **of shank
chucked** Einspannlänge *f* ~ **of skids** Länge *f* der
Kufen (des kufenförmigen Unterbaues) ~ **of
slot** Schlitzlänge *f* ~ **of span** Spannlänge *f* ~ **of
starting spark** Anlaßfunkenlänge *f* ~ **of strike**
streichende Länge *f* ~ **of stroke** Hubhöhe *f*,
Zylinderhub *m* ~ **of stroke adjustment** Hubbegrenzung *f* ~ **of support** Auflageläne *f* ~ **of
surface** Auflagelänge *f* ~ **of take-off run** Länge
*f* der Anlaufstrecke ~ **of thread** Gewindelänge
*f* ~ **of threaded end of spark plug** Zünderkerzenreichweite *f* ~ **of travel** Schalthub *m*
(electr.) ~ **to be turned** Drehlänge *f* ~ **of twist**
Drallage *f*, (of cables) Schlaglänge *f* ~ **of an
undulation** Wellenlänge *f* ~ **of wire for pins**
Nadellänge *f*

**length,** ~ **dimensions** Längenmaß *n* ~ **done in a
day** Tagesvortrieb *m* ~ **feed** Längsvorschub *m*
~ **gauging** Längenmessung *f* ~ **keel blocks**
Länge *f* auf Kielstapeln ~ **measuring** Längenmessung *f* ~ **measuring (or gauging) instrument**
Längsmeßinstrument *n* ~ **parameter** Längen-
parameter *n* ~ **path** Bahnlänge *f* ~**-preserving**
Längentreue *f* ~ **preserving mapping** längentreue Abbildung *f* ~ **quotient** Längen-quotient
*m*, -streuung *f* ~ **shear** Längsschub *m* ~ **stop**
Längsanschlag *m* ~ **subjected to bending** Biege-
länge *f* ~**-to-diameter ratio** Längedurch-
schnittsverhältnis *n* ~ **zone** Längsstreuung *f*

**lengthen, to** ~ anschuhen, ausrecken, strecken,
verlängern **to** ~ **range** das Feuer *n* vorverlegen
oder vorwärts verlegen, vorverlegen

**lengthened** breit, verlängert ~ **dots** breite Morsepunkte *pl*

**lengthening** Strecken *n*, Streckung *f*, Verlängerung *f* ~ **bar** Verlängerungsstück *n* ~ **bar of
compass** Zirkelverlängerung *f* ~ **coil** Verlänge-
rungsspule *f* ~ **piece** Anstückung *f*, Verlängerungsstück *n* ~ **pipe** Aufsatzrohr *n* ~ **rod** Ver-
längerungsstange *f*

**lengthways perforater tape** Längslochstreifen *m*

**lengthwise** der Länge *f* nach; längs ~ **adjustable
anvil** quer verstellbarer Meßtisch *m* ~ **bubble
(or level)** Längslibelle *f* ~ **dimension** Längsab-

messung *f* ~ **folding mechanism** Längsfalzein-richtung *f* ~ **graduation** (scale) Längsteilung *f* ~ **linse** Längsstrich *m* ~ **seat** Längssitz *m*
**lengthy** langandauernd, langwierig, weitläufig
**lens** Brillenglas *n*, Glaslinse *f*, Linse *f*, Objektiv *n* ~ **aberration** Linsenfehler *m* ~ **adapter** Objektivring *m* ~ **adjustment** Objektivverstellung *f* ~ **antenna** Linsenantenne *f* ~ **aperture** Blende *f*, Objektivöffnung *f* ~ **arrangement** Linsenanordnung *f* ~ **assembly** Linsenzusammenstellung *f* ~ **attachment** Objektivfassung *f*, Vorhänger *m* ~ **barrel** Linsenfassung *f*, Objektivfassung *f* ~ **bias** Linsenvorspannung *f* ~ **board** Objektivbrett *n* ~ **cap** Objektivdeckel *m* ~ **carrier** Objektivträger *m*, Standarte *f* ~ **combination** Linsen-folge *f*, -satz *m*, -zusammenstellung *f*, Objektivsatz *m* ~ **concentration** Linsenkonzentration f ~ **cone** Objektivdeckel *m* ~ **cooler** Linsenkühler *m* ~ **cover** Objektivdeckel *m* ~ **coverage** Gesichtsfeld *n* (film) ~ **current** Linsenstrom *m* ~ **curvature** Linsenkrümmung *f* ~ **defect** Linsenfehler *m* ~ **disk scanner** (scanning device) Linsenkranzabtaster *m* ~ **distortion** Objektivverzeichnung *f*, Verzeichnung *f* einer Linse ~ **drum scanner** Linsenkranzabtaster *m* ~ **effect** Dreherwirkung *f* ~ **element** Einzellinse *f* ~ **equation** Linsengleichung *f* ~ **extension** Bodenauszug *m* ~ **flange** Anschraubring *m* ~ **floodlight** Linsenscheinwerfer *m* ~ **guard** Objektivschutzdeckel *m* ~ **head** Objektiv-kopf, -stutzen *m* ~ **holder** Linsen-kreuz *n*, -träger *m* ~ **hood** Gegenlichtblende *f* ~-**like** linsenförmig ~ **making** Linsenschleifen *n* ~-**meter** Linsenmeßgerät *n* ~ **mount** Objektiv-, Linsen-fassung *f* ~ **mounting** Fassung *f* eines Objektivs ~ **opening** Lichtöffnung *f* (Scheinwerfer) ~ **panel** Objektivbrett *n* ~ **plane** Objektivebene *f* ~ **port** Kabinenfenster *n* (film) ~ **rim** Fensterfassung *f* (Scheinwerfer) ~ **ring** Objektivring *m* ~ **scanning disk** Linsenscheibe *f* ~ **screen** Gegenlichtblende *f*, Linsenschirm *m* ~-**shaped** linsenförmig ~ **shield** Sonnenblende *f* (film) ~ **sleeve** Objektivhülse *f* ~ **slide** Objektivschlitten(stück) *n* ~ **socket** Deckelring *m* (Scheibenfassung) ~ **speed** Lichtstärke *f* des Objektivs ~ **stand** Lupenständer *m* ~ **stereoscope** Linsenstereoskop *n* ~ **system** Linsenzusammensetzung *f* ~ **tube** Linsenstutzen *m* ~ **turret** Objektivrevolver *m*, (of camera) Revolverkopf *m* ~-**type machine gun mount** Linsenlafette *f* ~ **vertex** Glas-, Lupen-scheitel *m* ~ **vesicle** Linsenbläschen *n* ~ **voltage** (or potential) Linsenspannung *f*
**lensed disk** Linsenscheibe *f*
**lenses, set of** ~ Linsensatz *m* ~ **are bedewed,** (or fogged) die Gläser *pl* laufen an
**lensometer** Linsenprüfer *m*
**lenticular** länglichrund, linsen-ähnlich, -förmig ~ **apparatus** Linsenapparat *m* ~ **color-printing method** Linsenrasterfarbverfahren *n* ~ **color screen film** Linsenrasterfilm *m* ~ **screen** Linsenraster *m* ~ **vein** Linsengang *m* ~ **vug** linsenförmige Druse *f*
**lenticulated** linsenähnlich ~ **color-printing method** Linsenrasterfarbverfahren *n* ~ **film** Linsenrasterfilm *m*
**lenticulation** Rasterung *f*
**lentiform** linsenförmig ~ **compensator** Linsen-

rohrkompensator *m*
**lentil** (Gestein) Linse *f* (geol.)
**Lentz valve gear** Lentzsteuerung *f*
**Lenz's law** Lenzsches Gesetz *n*
**lepidolite** Lepidolith *m*, Lithiumglimmer *m*
**leptometer** Leptometer *n*
**lepton** Lepton *n* ~ **conservation** Leptonerhaltung *f*
**lesion** Verletzung *f*
**less** geringer, minder, ohne ~ **and** ~ immer weniger **more or** ~ mehr oder weniger
**lessee** Berg(e)pächter *m*, Bergeunternehmer *m*, Pächter *m*, Pachtmieter *m*
**lessen, to** ~ abnehmen, herabmindern, mindern, schmälern, verkleinern, vermindern, verringern **to** ~ **by planing** dünnhobeln
**lessening** Verringerung *f*
**lesser** minder
**let, to** ~ erlauben, gestatten, vermieten **to** ~ (have) lassen **to** ~ **alone** absehen (von) **to** ~ **come down** sacken lassen **to** ~ **down** abklappen, ablegen, (of an airplane) heruntergehen, herunterlassen, hinablassen, niederkurbeln **to** ~ **down in hinges** vorschlagbar **to** ~ **fly** entsenden **to** ~ **go** ablaufen lassen, abwerfen, ausrauschen lassen, schießen lassen **to** ~ **in** einlassen **to** ~ **loose** loslassen **to** ~ **mine water drain off** das Grubenwasser *n* versickern lassen **to** ~ **the motor run itself** in den Motor *m* einlaufen lassen **to** ~ **out** auslassen, herauslassen **to** ~ **run in** einlaufen lassen **to** ~ **stand** (cool) abstehen lassen **to** ~ **through** durchlassen, hindurchlassen
**let,** ~-**down vessel** Entspannungsgefäß *n* ~-**off stripper** Gewebevorrollapparat *m*
**lethal** tödlich ~ **gas** Giftgas *n*
**letter, to** ~ beschriften, Maß *n* eintragen **to** ~ **a drawing** eine Zeichnung *f* beschriften
**letter** Brief *m*, Buchstabe *m*, Schreiben *n*, Type *f*, Urkunde *f* **by** ~ brieflich
**letter,** ~ **of allotment** Zuteilungsanzeige *f* ~ **of application** Bewerbungsschreiben *n* ~ **of credit** Akkreditiv *n*, Gutschrift *f*, Kreditbrief *m* ~ **of hypothecation** Verpfändungsurkunde *f* ~ **of indemnity** Ausfallbürgschaft, Garantieschein *m* ~ **of renunciation** Verzichterklärung *f*
**letter,** ~ **blank** Buchstaben-abstand *m*, -blank *n*, -weiß *n* ~-**blank key** Buchstabenblanktaste *f* ~ **box** Briefkasten *m* ~ **brush** Abziehbürste *f* ~-**card** Briefkarte *f* ~ **case** Setzkasten *m* ~-**counting device** Buchstabenzählvorrichtung *f* ~ **drum** Buchstabentrommel *f* ~ **engraving** Letternstich *m* ~ **facing machine** Briefaufstellmaschine *f* ~-**facing system** Sortieranlage *f* (f. Briefe) ~ **feed** Buchstabenvorschub *m* ~-**file** Briefordner *m* ~ **nail** (electric line or lead) Bezeichnungsnagel *m* ~ **prefix** Kennbuchstabr *f* ~ **press** Druckerpresse *f* ~ **punch** Locher *m* ~ **requesting payment** Mahnbrief *m* ~ **rolling machine** Schrifteinwalzmaschine *f* ~ **scale** Briefwaage *f* ~ **shift** (signal) Buchstaben-umschaltung *f*, -wechsel *m* ~ **shoot** Rohrpost *f* ~ **sorter** Briefsortieranlage *f* ~ **space** Buchstaben-abstand *m*, -blank *n*, -weiß *n* ~ **stamps** Schlagbuchstaben *pl* ~ **weight** Briefbeschwerer *m*
**lettering** Bezifferung *f*, Buchstabenbezeichnung *f*, Schrift *f* ~ **punch** Buchstabenstempel *m*

**letters,** ~ **of reference** Zeichenerklärung *f* ~ **case** Buchstabenreihe *f* ~ **patent** Erfindungspatent *n*, Patent-brief *m*, -schrift *f* ~ **position** Zeichenstellung *f* ~ **shift** Buchstabeninversion *f*

**letting,** ~-**on** Anlaß *m* ~-**out the jib** Ausfahren *n* des Auslegers ~ **run through** hindurchströmen lassen

**leucite** Leucit *m*, Leuzit *m* ~-**bearing** leucitführend

**leuco,** ~ **base** Leukobase *f*

**leucoester compound** Leukoesterverbindung *f*

**leucometer** Leukometer *n*

**leucopyrite** Arseneisen *n*

**levee** Deich *m*, Wall *m*

**level, to** ~ abebnen, abgleichen, abkanten, abschrägen, aufrichten, ausgleichen, ausrichten, ausstreichen, begradigen, ebnen, egalisieren, einebnen, einkippen, einwiegen, horizontieren, nivellieren, planen, planieren, (Bleche) richten **to** ~ **balance** gleichmachen **to** ~ **down** gratfrei machen **to** ~ **fibers** abfasern **to** ~ **off** das Flugzeug *n* abfangen, in Horizontalflug *m* übergehen, flach werden, verflachen **to** ~ **up** erhöhen, nach oben ausgleichen **on a** ~ **with** in der Höhe *m* von

**level** eben, flach, platt, söhlig, waagerecht **not** ~ uneben

**level** Abbausohle *f*, Ausschlag *m*, Bausohle *f* (min.), Damm *m*, Dekade *f*, Ebene *f*, Etage *f*, Fläche *f*, (instrument) Grundwaage *f*, (sluice) Haltung *f*, Höhe *f* (geogr.), Höhenlinie *f*, Höhenregion *f* (teleph.), Horizontale *f*, Libelle *f*, Niveau *n*, Niveaumessung *f*, Nivellierinstrument *n*, Pegel *m*, Richtlatte *f*, Richtscheit *n*, (drift) Sohlenstrecke *f*, Spiegel *m*, Stand *m*, Waagerechte *f*, (bubble) Wasserwaage *f*

**level,** ~ **of brightness** Helligkeitsniveau *n* ~ **of convection** Konvektionsstand *m* ~ **of a crystal** Abflachung ~ **for delayed action mechanism** Hebel *m* für Vorlaufwerk ~ **of drive** Belastung *f*, Quarzlastung *f* (cryst.) ~ **of free convention** freier Konvektionsstand *m* ~ **of liquid** Flüssigkeitsstand *m* ~ **of radiation** Strahlungsspiegel *m* ~ **of reflection** Reflexionshöhe *f* ~ **of synchronization** Synchronisierpegel *m* ~ **of upper pool** Stauspiegel *m* ~ **of white modulation** Pegel *m* der Weißmodulation ~ **of work** Objektivhöhe *f*

**level,** ~ **adjustment** Pegeleinstellung *f* ~ **base plate** ebene Grundplatte *f* ~ **bedded rubble** Bruchsteinmauerwerk *n* (mit Steinlagen) ~ **blotch effects** glatte Böden *pl* ~ **board** Setzplatte *f* ~ **broading** Niveauverbreiterung *f* ~ **characteristic** Brodelstörung *f*, flache Kennlinie *f* ~ **chart** Höhenlinien-darstellung *f*, -diagramm *n*, Niveaukarte *f* ~ **clinometer** Richtbogen *m* ~ **compensator** Pegelausgleicher *m* ~-**compound excitation** Flachverbunderregung *f* ~ **control** Pegelregler *m*, Pegelreglung *f* ~ **crossing** Bahn-, Niveau-übergang *m*, Überführung *f* in gleicher Weghöhe ~-**crossing signal** Warnsignal *n* an Wegübergängen ~ **density** Termdichte *f* ~ **deviation** Pegelabweichung *f* ~ **diagram** Höhenlinien-darstellung *f*, -diagramm *n*, Niveaukarte *f*, (transmission) Pegellinie *f*, (transmission) Pegelschaulinie *f* ~-**difference indicator** Höhendifferenzmelder *m* ~ **displacement** Termbeeinflussung *f* ~ **distribution**

Energieniveauverteilung *f* ~ **drift** Feldortstrecke *f* ~ **drive** Kegelantrieb *m* ~**equalizer** Pegelausgleich *m* ~ **flight** Horizontalflug *m*, Normalfluglage *f* ~ **fluctuation** Pegelschwankung *f* ~ **flying speed** Horizontalfluggeschwindigkeit *f* ~ **free vein** gelöster Gang *m* ~ **gangway** Gezeug-, Sohlen-strecke *f* ~ **gauge** Standglas *n* ~ **housing** Libellengehäuse *n* ~ **hunting** Wahl *f* über verschiedene Höhenschritte ~ **indicator** Pegel-, Stand-anzeiger *m* ~ **jack** Hebewinde *f* ~ **landing** waagerechte Landung *f* ~ **line** Höhenlinie *f* ~ **luffing crane** Einziehkran *m*, Wippdrehkran *m* ~ **luffing deck crane** Bordwippkran *m* ~ **luffing jib** einziehbarer Ausleger *m* ~ **mark** Pegelmarke *f* ~ **marker** Höhenmarkierung *f* ~ **measurements** Pegelmessung *f* ~ **measuring set-up** Pegelmeßplatz *m* ~ **meter** (sound recording) Aussteuerungsanzeiger *m* ~ **mirror** Libellenspiegel *m* ~ **multiple** Dekadenvielfach *n*, Höhenschrittvielfach(feld) *n* ~ **numbers** Termanzahl *f* ~ **oscillator** Pegelsender *m* ~ **plane** Planie *f* ~ **point** Fallpunkt *m* ~ **quadrant** Winkellibelle *f* (mit Mikroskop) ~ **recorder** Intensitätsmeßgerät *n*, Pegelschreiber *m* ~ **regulating valve** Niveauregelventil *n* ~ **regulation** Pegelung *f* ~ **road (or section)** waagerechte Strecke *f* (r.r.) ~ **scheme** Niveauschema *n* ~ **setting** Horizontierung *f* ~ **shade** glatte Färbung *f* ~ **spacing** Niveauabstand *m* ~ **structures** Termstrukturen *pl* ~ **surface** ebene Fläche *f*, Sohlenoberfläche *f* ~ **systematic** Termsystematik *f* ~ **tangent railroad** ebene Straße *f* (gerade Strecke) ~ **testing equipment** Pegelprüfer *m* ~ **transportation** Waagerechtförderung *f* ~ **(ing) tube** Niveauröhre *f* ~ **width** Breite *f* des Energieniveaus

**leveled** abgeebnet, planiert ~ **position** Horizontierung *f*

**leveler** Dosenlibelle *f*, Einebner *m*, Planierer *m*, (in sheet metal rolling) Richt-maschine *f*, -bank *f*

**leveling** Abflachung *f*, (removal of ground) Abgleichung *f*, Ausrichten *n*, Einebnen *n*, Einebnung *f*, Einpegelung *f*, Einspielen *n* der Libelle *f*, Einwägung *f*, Höhenbestimmung *f*, Horizontierung *f*, Nivellieren *n*, Nivellierung *f*, Planierung *f*, Verlauf *m* ~ **of ground** Ebnung *f* des Bodens ~ **by means of barometer** barometrische Höhenmessung *f* ~ **a site** Planieren *n* von Gelände

**leveling,** ~ **agent** Egalisierer *m* ~ **blade** Ausstreichmesser *m* ~ **block** Richtplatte *f* ~ **board** Richt-latte *f*, -scheit *n*, Visierlatte *f* ~ **bottle** Hub-, Niveau-flasche *f* ~ **cam** Rückstreifschloßteil *m* ~ **concrete** Gefällbeton *m* ~ **device** Horizontierungs-, Nivellier-vorrichtung *f* ~ **doctor** Verstreichrakel *n* ~ **dyestuff** Egalisierfarbstoff *m* ~ **head** Horizontierkopf *m* ~ **indicator** Dosenlibelle *f* ~ **instrument** Fernglaslibelle *f*, Nivellierinstrument *n*, Richtzeug *n* ~ **lever** Geradeführungshebel *m* ~ **lug** Nivellierauflage *f* ~ **machine** Planiermaschine *f* ~ **mechanism** Einkipptrieb *m*, Horizontierung *f* ~ **mill** (autopilot) Nachdrehgetriebe *n* ~ **operation** Flächenanziehungsaufnahmen *pl* ~ **plate** Richtplatte *f* ~ **pole** Nivellierplatte *f* ~ **rod** Nivellierkreuz *n*, Planierstange *f* ~ **rule** Nivelliermaßstab *m* ~ **screw** Einstell-, Nivellier-,

Stell-schraube f ~ staff Nivellierlatte f ~
torque motor Aufrichtmotor m ~ wedge
Nivellierkeil m
levels of the ground taken along a line Längen-
nivellelement n
lever Anker m (watch), Bedienungshandgriff m,
Brechstange f, Drehstock m, Förderwerk n,
Hanspake f, Hebebaum m, Hebel m, Hebel-
system n, Heber m mit Klöppel
lever, ~ for adjusting brushes Bürstenstellhebel m
~ with equal arms gleicharmiger Hebel m ~ for
operating radiator shutter Kühlerklappenhebel
m ~ with rolling contact Wälz(ungs)hebel m
~ for run-out test Rundlaufhebel m ~ for
shifting friction-disk shaft Hebel m zur Verschiebung der Riemenscheibenwelle ~ for
slabs Plattenhebel m ~ for tilt control Stell-
hebel m (print.) ~ for vibrator roller Hebwal-
zenbewegung f
lever, ~ action Hebelwirkung f ~-adjusted
catwhisker crystal detector Hebeldetektor m ~
advantage Übersetzung f des Hebels ~ apparatus Hebelvorrichtung f ~ arm Hebelstange f ~
arm for recording stability curves of ships
Stabilitätshebearm m ~ arms Hebelarme pl ~
arrangement Hebelanordnung f ~-arresting
device Hebelfangvorrichtung f ~ axle stop
Hebelachsenanschlag m ~-bar Hebelgestänge n
~ bearing Kulissenarm m ~ bearing plate
Hebellagerplatte f ~-bottle Hebelflasche f ~
brake Hebelbremse f ~ breechmechanism
Hebelverschluß m ~ chain Hebelkette f, Spann-
und Hebelkette f ~ control Hebelsteuerung f ~
dolly Nietwippe f ~ escapement (clockwork)
Ankerhemmung f ~-expanding device Hebel-
verlängerungseinrichtung f ~ fastener Hebelverschluß m ~ foot Hebelschuh m ~ guide
Hebelführung f ~ handle Hebelgriff m ~ head
Hebelknopf m ~ jack Hebe-, Hebel-, Lastwinde f, Winde f ~ key Hebeltaste f ~ knob
Hebelknopf m ~ lighting switch Drehlichtschal-
ter m ~ micrometer screw Hebelarmmikro-
meterschraube f ~-operated mandrel Hebel-
dornpresse f ~ operated recorder Hebelanzeige-
vorrichtung f ~-operation Zangenspannein-
richtung f ~ path Hebelweg m ~ pin Hebelstift
m ~ position Hebelstellung f ~ press Hebel-
presse f ~ pump Hebelpumpe f, Schlagbrunnen
m ~ punch Hebel-, Hebelloch-stanze f ~
punching machine Hebelstanzmaschine f ~
ratchet Aufziehhebel m ~ ratio Übersetzung f
~-relation diagram Hebelbeziehung f ~ release
Hebelauslassung f ~ rod Hebelstange f ~
safety valve Hebelsicherheitsventil n ~ scale
Hebelwaage f ~ set keyboard Hebelstellwerk n
~-shaped hebelförmig ~ shears Alligator-,
Hebel-schere f ~ shifter Hebelschalter m ~
stand Stellbock m ~ starter Hebelanlasser m
~-stick Keilverschluß m ~ stop Hebelanschlag
m ~ support Hebellager n ~ switch Handschal-
ter m, Hebel-schalter m, -umschalter m, Kur-
bel-ausschalter m, -schalter m, -umschalter m
~ system Antriebsgestänge n, Gestänge n,
Gestängeantrieb m, Hebelwerk n ~-testing
machine Prüfmaschine f mit Laufgewichtswaage ~ track Hebelbahn m ~ transmission
Hebelübersetzung f ~ travel Hebelweg m ~
tumbler Zuhaltung f ~-type (or screwed) bottle

Hebel- oder Schraubflasche f ~-type Brinell
machine Brinell-Presse f mit Hebelwaage ~
valve Hebelventil m ~ watch Ankeruhr f
leverage Hebel-kraft f, -übersetzung f, -verhält-
nis n, -wirkung f ~ of load Lastarm m ~ ratio
Hebelverhältnis n
leverless hebellos
levigate, to ~ dekantieren, pulvern, schlämmen,
zerreiben
levigating Schlämmung f
levigation Abschlemmen n, Dekantieren n
levogyrate crystal Linkswurz m
levorotation Linksdrehung f
levorotatory linksdrehend
levulose Fruchtzucker m
levy, to ~ erhaben to ~ a charge eine Gebühr f
erheben
lewis Kropflich n ~ bolt Ankerschraube f, Stein-
dolle f ~ pins Steinwolf m
lewisite Chlorvinyldichlorarsin
leyden jar Leydener Flasche f
L. F. (low frequency) Niederfrequenz f (N. F.) ~
acceleration NF-Beschleunigungsgeber m ~
beat oscillator NF-Schwebungsoszillator m
L-head-type engine L-förmiger Verbrennungsraum m
liability Ausgesetztsein n, Ausstand m, Ersatz-
pflicht f, Feuergefährlichkeit f, Haftbarkeit f,
Haftpflicht f, Haftung f, Passivum n, Verbindlichkeit f ~ to detonation Klopfneigung f ~ to
punishment Strafbarkeit f ~ to make restitution
Erstattungspflicht f ~ on third party insurance
Haftpflichtversicherung f
liability insurance Haftpflichtversicherung f
liable ausgesetzt, ersatzpflichtig, haftbar, unterworfen, verantwortlich to be ~ haften to be ~
to unterliegen to be ~ to indemnity entschädi-
gungspflichtig sein ~ to break-down störanfällig
~ to break off in splinters splintrissig ~ for
damages schadenersatzpflichtig ~ to duty ab-
gabenpflichtig ~ to postage portopflichtig ~
to recourse regreßpflichtig
liason Fühlungsnahme f, Verbindung f ~ airplane Verbindungsflugzeug n ~ communication
Nachrichtenverbindung f ~ net Verbindungs-
netz n
liberate, to ~ abscheiden, ausscheiden, befreien,
entbinden, entwickeln, freigeben, freimachen,
loslösen
liberated befreit to be ~ auftreten
liberating area Auslöserfläche f
liberation Abscheidung f, Befreiung f, Entbindung f, Freigabe f ~ of atomic energy Energie-
lieferung f ~ of energy Energiefreisetzung f ~
of gas Gasausbruch m
libethenite Phosphorkupfererz n
librarian Bibliothekar m
library Bibliothek f, Bücherei f ~ subroutine
Bibliotheksprogramm n (info proc.)
license Ausweis m, Bewilligung f, Erlaubnis f,
(amtliche) Genehmigung f, Konzession f,
Lizenz f, (permit) Schein m, Steuerausweis m,
Unbedenklichkeitsvermerk m, Verleihungsurkunde f, Zulassung f, Zulassungsschein m ~
to export Ausfuhrbewilligung f ~ for a miner to
quit Abkehrschein m
license, ~ fee Lizenzgebühr f ~ folder Aus-
weishalter m ~ plate Kennzeichen-, Nummern-

schild *n* ~ **plate illumination** Kennzeichen-
leuchte *f*
**licensed** eingetragen ~ **under patents** Lizenzbau
*m* auf Grund von Patentrechten ~ **engineer**
Diplomingenieur *m* ~ **pilot** zugelassener Luftfahrzeugführer *m*
**licensee** Erlaubnis-, Konzessions-, Lizenz-in-
haber *m*, Lizenznehmer *m*
**licenser** Lizenzerteiler *m*
**lichen** Flechte *f*
**lick, to** ~ lecken ~ **(stone)** Leckstein *m*
**licking** Lecken *n*; leckend ~ **clamp** Klemmhebel
*m*
**licorice manufacture** Lakritzenfabrik *f*
**lid** (box) Abdeckplatte *f*, Deckel *m*, Kappe *f*,
Klappe *f*, Verschlußdeckel *m* ~ **of carburetor**
Vergaserdeckel *m*
**lid,** ~ **arrester** Deckelraste *f* (tape rec.) ~ **catch**
Deckelverriegelung *f* ~ **knob** Deckelknauf *m* ~
**layer** Deckelaufleger *m* ~-**locking device**
Deckelverschluß *m* ~ **oil cup** Deckelöler *m*
**lie, to** ~ liegen **to** ~ **against** anliegen **to** ~ **ahead**
bevorstehen **to** ~ **at anchor** vor Anker *m* liegen
**to** ~ **at the bottom of the sea** auf dem Grund *m*
des Meeres liegen **to** ~ **near** naheliegen
**lie** Lage *f*; Lüge *f* ~ **to the rear** Rückwärtslenkung *f* ~ **detector** Lügendetektor *m*
**Lieben valve** Liebenröhre *f*
**Liebig condenser** Liebig-Kühler *m*
**lien** Eigentumsvorbehalt *m*
**lieu, in** ~ **of** an Stelle *f*
**life** Leben *n*, Brenndauer *f*, (of a patent) Gültigkeit *f*, (durability) Haltbarkeit *f*, (span) Lebensdauer *f* **for** ~ lebenslänglich; (tool) Standzeit *f*
~ **of lining** Futterhaltbarkeit *f* ~ **of a patent**
Patentdauer *f*
**life,** ~ **annuity** Leibrente *f* ~ **assurance** Lebens-
versicherung *f* ~ **belt** Rettungsring *m* ~ **boat**
Rettungsboot *n* ~ **boat wireless set** Rettungs-
bootfunkeinrichtung *f* ~ **buoy** Rettungsboje *f* ~
**center** Mitlaufkörnerspitze *f* ~ **expectancy**
Lebenserwartung *f* ~ **guards** Rettungsmannschaft *f* ~-**guard service** (at river crossings)
Rettungsdienst *m* ~ **insurance** Lebensversiche-
rung *f* ~ **interest** Nutznießung *f* ~ **jacket**
Schwimmweste *f* ~ **line** Manntau *n*, Rettungs-
leine *f*, (parachute) Sicherungsleine, Signalleine
*f* ~ **net** Sprungtuch *n* ~ **part** spannungführen-
der Teil *m* ~ **preserver** Rettungsgürtel *m*,
Schwimmweste *f* ~-**restoring apparatus** Wieder-
belebungsapparat *m*
**life-saving,** ~ **boat** (at river crossings) Rettungsfahrzeug *n* ~ **equipment** Rettungs-gerät *n*,
-mittel *n* ~ **jacket** Schwimmweste *f* ~ **raft**
Floßboot *n*
**life test** Lebensdauer *f* (von Röhren)
**life-time** Leben *n*, Lebensdauer *f* ~ **of an agree-**
**ment** Vertragsdauer *f* ~ **grease lubricated bear-**
**ing** Lager *n* mit Dauerschirmung
**lifeless** leblos, tot, unbelebt
**lifelike** naturgetreu
**lift, to** ~ anheben, (color) auffärben, aufnehmen,
aufziehen, (the teleph.) aushängen, erheben,
fortheben, heben, hochheben, in die Höhe *f*
heben, (clear of something) lüften, roden,
steigen, (artillery fire) vorverlegen, wältigen
**lift, to** ~ **clear of** (pawl) ausheben **to** ~ **a form**
**into the press** einschießen **to** ~ **off** abheben

(aviat.), aushaben **to** ~ **off round crusts of**
**copper** resettieren **to** ~ **out** ausheben, heraus-
heben, herausnehmen **to** ~ **out the form** die
Form *f* aushaben **to** ~ **over** überheben **to** ~ **up**
abheben (aviat.), anlüften, aufheben, aufklap-
pen, hochwinden, hochziehen
**lift** Abbauhöhe *f*, Anhub *m*, Auftrieb *m* (aviat.),
Aufzug *m*, Aushub *m*, Fahrstuhl *m*, Förderhöhe *f*, -werk *n*, -zug *m*, Hebedaumen *m*,
Heben *n*, Hebung *f*, Hub *m*, Nocke *f*, Stauhöhe
*f*, Tragfähigkeit *f*
**lift,** ~ **of cam** Nockenhub *m* ~ **and delivery pump**
Saug- und Druckpumpe *f* ~ **per hour** Stundenleistung *f* ~ **of key** Tastenhub *m* ~ **of lifting cog**
Wellendaumen *m* ~ **of a lock** Höhe *f* des
Schleusenabfalls *m* ~ **of sleeve** Muffenhub *m* ~
**of tail unit** Leitwerksauftrieb *m* ~ **per unit area**
Flächenanziehungsbelastung *f*
**lift,** ~ **arm** Hubhebel *m* ~-**augmenting** auftriebserhöhend ~ **axis** Auftriebsachse *f* ~ **balance**
Auftriebswaage *f* ~-**boosting** auftriebserhöhend
~ **brake** Fallbremse *f* ~ **bridge** Fahrstuhlbrücke
*f*, Hubbrücke *f* ~ **car** Aufzugskabine *f* ~
**change** Auftriebsänderung *f* ~ **coefficient**
Auftriebs-koeffizient *m*, -wert *m*, -zahl *f*
(aviat.) ~ **component** Komponente *f* senkrecht
zur Bewegungsrichtung ~ **control cable** Auf-
zugsteuerleitung *f* ~ **conveyer** Höhenförderer
*m* ~ **distribution** Auftriebsverteilung *f* ~-**drag**
**ratio** Gleitzahl *f*, Gleit- und Widerstandsbei-
wert *m* ~ **engine** Hubtriebwerk *n* ~ **factor**
Auftriebsbeizahl *f* ~ **force formula** Hubkraft-
formel *f* ~ **formula** Auftriebsformel *f* ~ **frame**
Förderturm *m* ~ **hammer** Aufwerfhammer *m*
~ **hoisting cable** (or rope) Aufzugseil *n* ~-**in-**
**creasing** auftriebserhöhend ~-**increasing device**
auftriebserhöhende Einrichtung *f* ~ **indication**
Auftriebsbestimmung *f* ~ **indicator** Auftriebs-
anzeiger *m* ~-**off thrust** Anfangsschub *m* ~-**off**
**type parachute** Abhebefallschirm *m* ~-**over**
**arrangement** Überhebevorrichtung *f* ~ **parabola**
Auftriebsparabel *f* ~ **pod** Hubgondel *f* ~ **pump**
Hebpumpe *f* ~ **rotor** Hubrotor *m* ~ **shaft** Hub-
welle *f* ~ **strut** auftrieberzeugende Strebe *f*,
I-Strebe *f* ~ **thrust** Hub/Längsstrich/Schub-
Mantelstromtriebwerk *n* ~ **truck** Hub-karren
*m*, -stapler *m*, -wagen *m* (elevating platform) ~
**truss** Tragwand *f* ~ **valve** Druckventil *n* ~
**variation** Auftriebsänderung *f* ~ **vector** Schub-
vektorneigung *f* ~ **vortex** Auftrieb(s)wirbel *m*
~ **wall** Fallmauer *f* (lock) ~ **wire** Abfangkabel
*n*, Haupttragseil *n* (aviat.), Tragdraht *m*, Tragkabel *n*
**lifted** gehoben, (lacquer) gerollt ~ **off** aufgehoben
**lifter** Ausrückhebel *m*, Gehänge *n*, Heber *m*,
Hebestück *n*, Hublader *m*, Regulierstößel *m*,
Rodemaschine *f*, Wuchtbaum *m* ~ **of arbor**
Wellendaumen *m*
**lifter,** ~ **action** Baskühlverschuß *m* ~ **propeller**
Tragschraube *f*
**lifting** Abheben *n*, Abhebung *f*, Aufheben *n*,
Ausheben *n*, Hub *m*, Rupfen *n* ~ **and forcing**
**pump** Hub- und Druckpumpe *f* ~ **and lowering**
Höhenstellung *f* ~ **and lowering arrangement**
Hebe- und Senkvorrichtung *f* ~ **the rack** Frei-
legen *n* der Gleise ~ **by suction** Saugförderung *f*
**lifting,** ~ **apparatus** Hebe-gerät *n*, -vorrichtung *f*,
-zeug *n* ~ **appliance** Hebezeug *n* ~ **armature**

Hebemagnetanker *m* ~ **arrangement** Förderwerk *n* ~ **bar** Hebezwinge *f* ~ **bolt** Kranbolzen *m* ~ **cable** Hebeseil *n* ~ **cam** Hebedaumen *m* ~ **capacity** Hebekraft *f*, Hub-kraft *m*, -leistung *f*, -vermögen *n*, Lasthebekraft *f*, Lasthub *m*, Tragvermögen *n* ~ **chain** Hubkette *f*, Zugstange *f* ~ **cog** Nocke *f* ~ **contrivance** Aushebevorrichtung *f* ~ **cylinder** Hubzylinder *m* ~ **device** Abhebe-, Aushebe-, Hebe-vorrichtung *f* ~ **device for rolls** Walzenaushebevorrichtung *f* ~ **element** Huborgan *n* ~ **equipment** Hebezeug *n* ~ **eye** Aufhänge-, Heiß-öse *f* ~ **eyebolt** Heißöse *f* ~ **flank** Auflaufkurve *f* (Nocken) ~ **force** Auftriebs-, Hub-, Trag-kraft *f* ~ **frame** Hebegerüst *n* ~ **gas** (balloon) Füllgas *n*, Traggas *n* ~ **gear** Aufhängevorrichtung *f*, Windwerk *n* ~ **handle** Handgriff *m*, Trageisen *n* ~ **hook** Haken *m* zum Aufheben ~ **jack** Hebebock *m*, Hebewinde *f*, (screwform) Kopfwinde *f*, Lastwinde *f*, Spindel *f*, Wagenheber *m*, Winde *f*, Windenbock *m* ~ **jack with anchor guard plate** Windenbock *m* mit Schürze ~ **jet** Absauger *m* ~ **line** tragende Linie *f*, Verbindungsleine *f* ~ **lug** Eisenbeschlag *m*, Haken *m*, Heißbeschlag *m*, Öhr *n*, Tragzapfen *m*
**lifting-magnet** Hebeelektro-, Hebe-, Hub-, Lasthebe-, Last-magnet *m* ~ **with movable pole shoes** Lastmagnet *m* mit beweglichen Polen ~ **circuit** Hubstromkreis *m* ~-**type crane** Magnetkran *m*
**lifting,** ~ **mechanism** Hebemechanismus *m*, Hubstange *f* ~ **medium** Huborgan *n* ~ **moment** Übermoment *n* ~ **motion** Hubbewegung *f* ~ **motor** *m* zum Aufheben ~ **movement** Hubbewegung *f* ~-**over** Überheben *n* ~ **pad** Anhebeblock *m* ~ **pawl** Hebeklinke *f* ~ **pin** (or post) Abhebestift *m* ~ **plate** Aushebeplatte *f* ~ **platform** Hebe-tisch *m*, -trog *m*, Walztisch *m* ~ **power** Hebekraft *f*, Hubkraft *f*, Tragfähigkeit *f*, Zugkraft *f* ~ **pressure** (valve of oil engine) Anlaufdruck *m*, (pumps) Förderdruck *m* ~ **pump** Hebepumpe *f* ~ **ramp** Hebebühne *f* ~ **ring** Huböse *f*, Ring *m* zum Aufheben ~ **rod** Abhebesäule *f*, Königs-, Zug-stange *f* ~ **screw** Abdrück-, Hebe-schraube *f*, Hubschnecke *f* ~ **set** Hubsatz *m* ~ **ship** Hebeschiff *n* (für Unterseeboote) ~ **sling** Aufhängejoch *n* ~ **speed** Hubgeschwindigkeit *f* ~ **spindle** Hubspindel *f* ~ **spring** Hebefeder *f* ~ **stage** Hebebühne *f* ~ **strap** Handgriff *m* ~ **stud** Hebestift *m* ~ **surface** Flügeldruck *m*, tragende Fläche *f*, Tragfläche *f* (aviat.) ~ **table** Hebe-tisch *m*, -trog *m* ~ **tackle** Flaschenzug *m* zum Aufheben, Hebevorrichtung *f*, Hißzeug *n* ~ **template** Anlaufschablone *f* ~ **through** Hubrinne *f* ~ **time** Hubzeit *f* ~ **unit** Hebezeug *n* ~-**water swivel** Hebewasserwirbel *m* ~ **wheel** Hubrad *n* ~ **yoke** Aufhängejoch *n*
**light, to** ~ anbrennen (anzünden), anmachen, anzünden, beleuchten, bestrahlen, entzünden, leuchten **to** ~ **with beacon lights** befeuern **to** ~ **a fire** Feuer *n* anmachen **to** ~ **up** aufleuchten, erhellen
**light** blaß, gering, von geringem Gewicht *n*, hell, hellgetönt, (spezifisch) leicht, leicht flüchtig, licht, (wind)schwach, unbedeutend **with** ~ **draught** leichtzügig **of** ~-**gauge wire** schwachdrähtig **of** ~ **precipitation** niederschlagarm

**light** Beleuchtung *f*, Feuer *n*, Funke *m*, Lampe *f*, Leuchte *f*, Leucht-feuer *n*, -turm *m*, Licht *n*, Licht-öffnung *f*, -quelle *f*, Schein *m*, Tageslicht *n* ~ **with flashes at regular intervals** Einzelblitzfeuer *n* ~ **of the night sky** Licht *n* des Nachthimmels *m* ~ **and power station** Lichtzentrale *f* ~ **from vertical sources** Auflicht *n*
**light,** ~-**absorbing** lichtschluckend ~ **absorption** Lichtschwächung *f* ~ **adaption** Hellanpassung *f* ~-**admission window** Lichtwerbung *f* ~ **aging** Lichtalterung *f* ~ **air** leiser Luftzug *m* ~ **aircraft** Leichtflugzeug *n* ~ **alloy** Leichtmetallegierung *f*; leichtlegiert ~ **aperture** Lichtspalt *m* ~ **arc** Lichtbrücke *f* ~ **armor** leichte Armierung *f*, leichte Bewehrung *f* ~ **band** Lichtlinie *f* ~ **barrier** Lichtschranke *f* ~ **beacon** Feuer-, Leucht-bake *f*
**light-beam** Beleuchtungsbalken *m* (Scheinwerfer), Licht-bündel *n*, -büschel *n*, -streifen *m* ~ **of intensive intrinsic brilliancy** blendender Lichtstrahl *m*
**light-beam,** ~ **depth finder** Lichtstrahltiefenlot *n* ~ **diaphragm** Lichttubusblende *f* ~ **pickup** elektrooptischer Tonabnehmer *m* ~ **pointer** Lichtzeiger *m* ~ **projector** (for determining ceiling of clouds) Wolkenhöhenmeßscheinwerfer *m* ~ **transmitter** Lichtstrahlabtastsender *m*
**light,** ~ **benzol** Leichtbenzol *n* ~ **border** Lichtrand *m* ~ **buoy** Leuchtboje *f* ~ **car** Kleinkraftwagen *m*, Kleinwagen *m* ~ **characteristic** Lichtstärkenkennlinie *f*, (of a glow lamp) optische Kennlinie *f* ~ **chopper** Lichtsirene *f*, Lochscheibengenerator *m* ~ (-beam) **chopper** Lichtzerhacker *m* ~-**circuit antenna** Lichtnetzantenne *f* ~-**collection angle** Lichtaufnahmewinkel *m* ~ **coloration** Lichtfarbe *f* ~-**colored** hellfarbig ~ **column** Glimmsäule *f* ~ **communication** (or **connection**) Lichtverbindung *f* ~ **construction** Leichtbau *m* ~ **contact arm** kleiner (schmaler) Meßbügel *m* ~-**control engineering** Lichtsteuerkennlinie *f* ~-**control means** Lichtregler *m* ~ **cracks** Lichtrisse *pl* ~ **cross** Lichtkreuz *n* ~-**current engineering** Schwachstromtechnik *f* ~ **curtain** Lichtvorhang *m* ~ **cut-off** Fallklappe *f* (film) ~-**dark lines** Hellinien *pl* ~-**dark range** Helldunkelintervall *n* ~ **density** Lichtdichte *f* ~-**dependent resistor** lichtabhängiger Widerstand *m* ~ **development** Lichtentwicklung *f* ~ **diffraction method** Lichtbeugungsmethode *f* ~-**diffuse** lichtzerstreuend ~-**diffusing glass** Opalglas *n* ~ **diffusion** diffusierendes Gewebe *n* (film), Licht-durchlässigkeit *f*, -schwächung *f* ~ **dimming** Lichtschwächung *f* ~ **dispenser** Lichtspender *m* ~-**dispersing** (or **diffusing**) lichtstreuend ~-**dispersive** lichtzerstreuend ~ **duct** Lichtkanal *m* ~-**duty design** leichte Bauweise *f* ~-**duty parts** gering beanspruchte Teile *pl* ~ **effect** Lichterscheinung *f*, Lichtgebilde *n* ~ **efficiency** Lichtausbeute *f* ~ **efficiency of fluorescent material** Lichtausbeute *f* der Leuchtstoffe ~ **electrician** Beleuchter *m* ~ **element** leichtes Element *n* ~ **emission** Lichtstrahlung *f* ~-**emissive** lichtgebend ~-**emitting** lichterzeugend ~ **engineering** Feinmechanik *f*, Lichttechnik *f* ~ **entry** (or **opening**) Lichteinlaß *m* ~ **etching** Hellätzerei *f* ~ **feed** Auslauf *m* mit Schauglas ~ **fillet** leichte Kehlnaht *f* ~ **filter**

Licht-, Strahlen-filter *n* ~-filter trough Vorsatz-kuvette *f* ~ fire Leuchtfeuer *n* ~ flap Leucht-rakete *f* ~ flash Lichtblitz *m* ~ fluctuation Lichtschwankung *f* ~ flux (in lumen units) Lichtstrom *m* ~ fog falsches Licht *n* ~ fuel Leichtkraftstoff *m*, Leichtöl *n* ~-fuel injection valve Leichtöleinspritzventil *n* ~ funnel Licht-schutztrichter *m* ~ gain Lichtgewinn *m* ~-gap testing Lichtspaltprüfung *f* ~ gate lid (or cover) Lichtklappe *f* ~ gathering Lichtsammeln *n* ~-gathering power Helligkeit *f* (eines Fern-rohrs) ~-gauge sheet dünnes Blech *n* ~-gauge sheet steel welding Dünnblechschweißung *f* ~ gauge standard outlet pipe LNA-Rohr *n* (Leichtes Normal-Abflußrohr) ~ generator Lichtanlasser *m* ~-giving lichtgebend ~ guard Lichtabschluß *m* ~ gun Lichtkanone *f*, Signal-scheinwerfer *m* ~ halve Leuchtfeld *n* (tape rec.) ~ hood Lichtschutzkappe *f* (photo) ~ hours verkehrsschwache Zeit *f* ~ house Feuerturm *m*, Hafenfeuer *n* ~ house tender Seezeichendampfer *m* ~ increment Lichtgewinn *m* ~ installation Lichtanlage *f*

light-intensity Lichtstärke *f* ~ control Regulie-rung *f* der Lichtintensität ~ distribution Licht-stärkeverteilung *f* ~ drop Lichtschwächung *f* ~ variation Lichtintensitätsschwankung *f* ~ variations Wechsellicht *n*

light, ~ intersection Lichtschnitt *m* ~ leak Licht-einfall *m* ~ line Leuchtstrich *m*, dünne Linie *f* ~-line antenna Lichtnetzantenne *f* ~ list Leucht-feuerverzeichnis *n* ~ loading Leichtbespulung *f*, leichte Pupinisierung *f* ~ loss Lichtschwächung *f* ~ mark deflection read Lichtzeigerablesevor-richtung *f* ~ material Leichtstoff *m* ~-measuring apparatus (or meter) Luxmesser *m* ~ message Leuchtmeldung *f*

light-metal, ~ alloy Leichtmetall-Legierung *f* ~ casting Leichtmetallguß *m* ~ cover Leicht-metall-deckel *m*, -überzug *m* ~ framework Leichtmetallgerippe *n* ~ packing Leichtmetall-dichtkörper *m* ~ piston Leichtmetallkolben *m* ~ working Leichtmetallbearbeitung *f*

light, ~ metals Avionallegierungen *pl* ~ micro-phone lichtelektrische oder lichtempfindliche Zelle *f* ~ microscope Lichtmikroskop *n* ~ modulation Lichtsteuerung *f* ~ modulator Licht-regler *m*, -relais *n*, -steuergerät *n* ~ motorcycle Klein-, Leicht-kraftrad *n* ~ oil Leichtöl *n*, dünnflüssiges oder leichtflüssiges Öl *n* ~ oil recovery Leichtölgewinnung *f* ~ opening Lichtspalt *m* ~-optical microscope Lichtmikroskop *n* ~ optics Lichtoptik *f* ~ passage Austrittsöffnung *f* ~-path Meß-, Strahlen-weg *m* ~ pattern Lichtspur *f* ~ period Aufleuchtdauer *f*, Leuchtdauer *f* ~ pile driver Bär *m* ~ pilot Stichflamme *f* ~ plane Leicht-flugzeug *n* ~ plant Lichtanlage *f* ~-plate rolling mill Mittelblechwalzwerk *n* ~ pointer Projek-tionspfeil *m* ~ port Lichtpforte *f* ~ positions Hellstellungen *pl* ~-positive lichtpositiv ~ probe Lichtsonde *f*, materielose Sonde *f* ~ probe techniques Lichtsondentechnik *f* ~ pro-ducer Lichterreger *m* ~-producing lichterzeu-gend ~-proof lichtundurchlässig ~-proof con-necting funnel Lichtverschluß *m* ~-proof connecting sleeve Lichtmanschette ~ proofness Lichtdichte *f*, Lichtdichtigkeit *f* ~ protection

flap Lichtschutzklappe *f* ~ quantum Licht-quantum *n*, Photon *n* ~ radiation Lichtaus-strahlung *f* ~ railway Sekundärbahn *f* ~ range Lichtweite *f* (opt.) ~-range regulator Leucht-weiterregler *m* ~ ray Lichtstrahl *m* ~-ray ben-ding Lichtablenkung *f* ~-ray indicator Licht-strahlanzeiger *m* ~-reactive lichtempfindlich ~-reactive cell licht-empfindliche Zelle *f* ~ reduction Lichtschwächung *f* ~ reflectance (or reflexivity) Lichtreflexionsvermögen *n* ~ relay Fremdlicht *n*, Licht-modulator *m*, -regler *m*, -relais *n*, -steuerungseinrichtung *f*, -ventil *n* ~ relay depending on Faraday effect elektro-optisches Lichtrelais *n* ~ relay of the glow tube selbstleuchtender Lichtmodulator *m* ~ re-sistance Lichtwiderstand *m* ~ rocket Leucht-rakete *f* ~ scanning Lichtsteuerung *f* ~ scatter Lichtstreuung *f* ~ scattering lichtzerstreuend ~ scattering apparatus Streulichtmeßgerät *n* ~ screen Lichtschirm *m* ~ screen for side light Blendschirm *m* für Positionslicht ~ screening cap Lichtschutzkappe *f* ~ screening sleeve Lichtschutzhülse *f*

light-section kleines Profil ~ iron Kleineisen *n* ~ method Lichtschnittverfahren *n* ~ rolling mill Feineisenstraße *f* ~ steel Feineisen *n*

light, ~ sector Leuchtraum *m* ~ sensation Licht-empfindung *f* ~ sense Lichtsinn *m* ~-sensitive substance Schichtkathode *f* ~-sensitive surface Fotoschicht *f* ~-sensitive tube Fotozelle *f* ~ sensitivity Helligkeitsempfindlichkeit *f* ~ shield Lichtblende *f* ~ ship Feuerschiff *n* ~ shot Nebenlicht *n* ~ signal Blinkkennung *f*, Fanal *n*, Leuchtzeichen *n*, Lichtsignal *n* ~-signal call system Lichtrufanlage *f* ~-signal message Lichtspruch *m* ~ signaling Lichttelegrafie *f* ~-signaling system Lichtrufanlage *f* ~ siren Licht-hupe *f*, -sirene *f* ~-sized kleinstückig ~ slit Lichtspalt *m* ~ socket Licht-anschlußdose *f*, -steckdose *f* ~-socket antenna Lichtnetzan-tenne *f* ~-socket plug Netzstecker *m* ~ softener Lichtdämpfer *m* (film) ~-sound apparatus Lichttongerät *n* ~-source efficiency Licht-quellen-ausbeute *f*, -stärke *f* ~-source image Lichtquellenbild *n* ~ sorce slit Lichtquellen-spalt *m* ~ spar Leichtspat *m*, Lenzin *n*

light-spot Leuchtfleck *m*, Licht-fleck *m*, -punkt *m* ~ galvanometer Lichtmarkengalvanometer *n* ~ instrument Lichtzeigerinstrument *n* ~ meter Lichtzeigerinstrument *n* ~ scanner Licht-punktabtaster *m*

light, ~ station Lichtstelle *f* ~ steel structures Stahlleichtbau *m* ~ stop Blende *f*, Lichtblende *f* ~ storage Lichtspeicherung *f* ~ stream con-verter (or transformer) Lichtstromumformer *m* ~ stream varier Lichtstromumformer *m* ~-struck plates Lichtschleier *m* (photo) ~ subduing Lichtdämpfung *f* ~ sum Lichtsumme *f* ~ surface peat Rasentorf *m* ~ tight licht-dicht, -undurchlässig ~ tones helle Boldstellen *pl* ~ tracing Lichtspur *f* ~-tracing establishment Lichtpausanstalt *f* ~ track Lichtspur *f* ~ transmission Lichtfortpflanzung *f* ~-trans-missive lichtdurchlässig ~ transmissivity Licht-freundlichkeit *f* ~ transmittance Lichtdurch-lässigkeit *f* ~ transmittancy (specific) Licht-durchlässigkeit *f*, Lichtfreundlichkeit *f* ~-trans-mitting lichtdurchlässig ~ transmitting power

Lichtdurchlässigkeit *f* ~ **trap** Lichtfalle *f*, -schleuse *f* ~ **treatment** Fototherapie *f* ~ **truck** Lichtwagen *m* (film) ~ **tunnel** (of sensitometer) Lichtschacht *m* ~ **value** Lichtempfindlichkeitszahl *f* ~ **valve** Lichtsteuerungseinrichtung *f*, Lichtventil *n* ~ **variation** Lichtschwankung *f* ~ **vibration** Lichtschwingung *f* ~ **visor** Lichtvisier *m* ~ **volatile** (fuel) leicht flüchtig ~ **volume** Lichtstrahlung *f* ~ **wave** Lichtwelle *f* ~-**wave interference** Lichtinterferenz *f*

**lightweight,** ~ **alloy** Leichtlegierung *f* ~ **building board (or slab)** Leichtbauplatte *f* ~ **construction** Leichtbauweise *f* ~ **Diesel rail car** Leichtmetall-Dieseltriebwagen *m* ~ **oil** dünnflüssiges Öl *n*, Leichtöl *n* ~ **steel construction** Stahlleichtbau *m* ~ **steel structural members** Stahlleichtbauteile *pl*

**light,** ~ **weld** leichte Schweißnaht *f* ~ **window** Lichtfenster *n* ~ **wire** Lichtdraht *m* ~ **wiring** Lichtleitung *f* ~ **year** Lichtjahr *n* ~ **yield** Lichtausbeute *f* ~ **yield of fluorescent material** Lichtausbeute *f* der Leuchtstoffe ~ **zone** Leuchtzone *f*

**lighted,** ~ **buoy** Leuchttonne *f* ~ **field (or area)** Lichterscheinung *f*

**lighten, to** ~ erleichtern, sich aufklären **to** ~ **a ship** ein Schiff leichtern

**lightened,** ~ **silver dross** Bleisack *m* ~ **web** Erleichterungssteg *m*

**lightening,** ~ **of silver** Silberblick *m* ~ **arrester** Blitzableiter *m* ~ **hole** (metal. construction) Erleichterungsloch *n*

**lighter** Feueranzünder *m*, Hubprahm *f*, Kahn *m*, Lastkahn *m*, Leichter *m*, Leichterschiff *n*, Lichter *m*, Prahm *m*, Schlepper *m*, Schleppkahn *m*, Zünder *m*

**lighter,** ~ **than air** leichter als Luft ~ **bar** (paper) Aufhelfung *f*, Hebelade *f* ~-**than-air-aircraft** Aerostat *m*, Luftfahrzeug *n* leichter als Luft

**lighting** (of lights) Anzünden *n*, Aufleuchten *n*, Befeuerung *f*, Beleuchten *n*, Beleuchtung *f*, Leuchten *n*, Zünden *n* ~ **for form and delivery** Form- und Ablagebeleuchtung *f* ~ **of instruments** Instrumentbeleuchtung *f* ~ **of lamp** Leuchten *n* der Lampe ~ **of squares** Platzbeleuchtung *f*

**lighting,** ~ **activity** Leuchttätigkeit *f* ~ **aids** Befeuerungshilfen *pl* ~ **aperture** Beleuchtungsfenster *n* ~ **attachment** Maschinenleuchte *f* ~ **cable** Lichtkabel *n* ~ **circuit** Licht-leitung *f*, -netz *n* ~ **circuit socket** Lichtleitungssteckdose *f* ~ **condition for searchlight** Leuchtbedingung *f* ~ **conduit** Lichtleitungsrohr *n* ~ **current** Lichtstrom *m* ~ **dynamo** Licht-maschine *f*, -zündmaschine *f* für Kraftfahrzeuge ~ **engineering** Beleuchtungstechnik *f* ~ **equipment** Beleuchtungsanlage *m* ~ **fittings** Beleuchtungsarmaturen *pl* ~ **fixture** Beleuchtungs-apparat *m*, -körper *m* ~ **gas** Leuchtgas *n* ~ **generator** Impulsgenerator *m* ~ **hour** Brennstunde *f* ~ **load** Lichtnetzleitungsbelastung *f* ~ **mains** (Haupt)lichtleitung *f*, Lichtnetz *n* ~ **oil burner** Ölzündbrenner *m* ~ **operator** Beleuchter *m* ~ **panel** Beleuchtungs-, Lichtschalt-tafel *f* ~ **part** Lichtteil *m* ~ **purpose** Beleuchtungszweck *m* ~ **reflector** Beleuchtungsschirm *m* ~ **set** Lichtaggregat *n* ~ **switch** Beleuchtungs-, Lichtschalter *m* ~ **system** Beleuchtung *f*, Beleuch-

tungsanlage *f* ~ **transformer** Beleuchtungstransformator *m* ~ **unit** Beleuchtungskörper *m* ~-**up time** Brennstunde *f*

**lightly,** ~ **armored** leichtgepanzert ~ **loaded** leicht belastet

**lightness** geringes Gewicht *n*, Leichtigkeit *f*

**lightning** Blitz *m* ~ **arrester** Blitzschutz *m*, Fritterschutz *m*, Überspannungsableiter *m*, (on mast bases) Überspannungssicherung *f* ~-**arrester collar** Blitzableiterschelle *f* ~ **bolt** Blitzschlag *m* ~ **call** Blitzgespräch *n* ~ **circuit receiver** Netzanschlußempfänger *m* ~ **conductor** Ableitstange *f*, Blitz-ableiter *m*, -kabel *n* ~ **conductor for poles** Mastblitzableiter *m* ~-**conductor cable** Blitzableiterseil *n* ~ **discharge** Blitzentladung *f*, Blitzschlag *m* **(most powerful)** ~ **discharge** Hauptstrahl *m* ~ **expectancy** Blitzanfälligkeit *f* ~ **flash** Blitz-schlag *m*, -strahl *m* ~-**flash cluster** Blitzbündel *n* ~ **foil** Leuchtfolie *f* ~ **fuses and protector blocks** Blitzableiter *m* mit Sicherung ~ **like** blitzartig ~ **protector** Blitzschutz *m* ~ **rod** Blitzableiter *m*, Fangstange *f* des Blitzableiters, Stangenblitzableiter *m* **the** ~ **strikes** der Blitz schlägt ein ~ **stroke** Blitzschlag *m* ~ **switch** Blitzschutzschalter *m*

**lightproofness** Lichtdichtigkeit *f*

**lighttightness** Lichtdichtigkeit *f*

**lignate, to** ~ ausseigern

**ligneous** holzartig, holzig, lignitartig ~ **asbestos** Holzasbest *m* ~ **beet** holzige Rübe *f* ~ **coal** Braunkohle *f* ~ **stucco** Holzstück *m*

**lignicidal** holzzerstörend

**lignification** Verholzung *f*

**lignify, to** ~ verholzen

**lignin** Holzfaserstoff *m*, Holzstoff *m*

**lignite** Bergkohle *f*, Braunkohle *f*, Lignit *m* ~ **beds** Braunkohlenvorkommen *n* ~ **breeze** Braunkohlenklein *n* ~ **briquette** Braunkohlenbrikett *n* ~-**carbonization plant** Braunkohlenschwelerei *f* ~ **coking** Braunkohlenschwelung *f* ~ **dressing plant** Braunkohlenaufbereitungsanlage *f* ~ **field** Braunkohlenlager *n* ~ **firing** Braunkohlenfeuerung *f* ~ **low-temperature carbonization power plant** Braunkohlenschwelkraftwerk *n* ~ **mining** Braunkohlentiefbau *m* ~-**mining machine** Braunkohlengrubenmaschine *f* ~ **pitch** Braunkohlenpech *n* ~ **power plant** Braunkohlenkraftwerk *n* ~ **tar** Braunkohlenteer *m* ~-**tar oil** Braunkohlenteeröl *n*

**lignitic,** ~ **carbonization coke** Grude *f* ~ **coal** Wachsfirniskohle *f* ~ **earth** Erdkohle *f*

**lignocellulose** Holzfaserstoff *m*

**lignoleate** Holzölsäure *f*

**lignone** Holzstoff *m* (chem.)

**lignose** Lignose *f*

**lignum vitae** Guajak-, Poch-holz *n* ~ **bearing** Pochholzlager *n*

**ligroine** Ligroin *n*

**like,** ~ **flour (or meal)** mehlartig ~ **peat** torfartig ~ **a plate** tellerförmig ~ **pole** entgegengesetzter oder gleichnamiger Pol *m* ~ **steel** stählern ~ **a telescope** posaunenartig verschiebbar ~ **water** wasserartig

**likelihood** Anschein *m*, Wahrscheinlichkeit *f* **in all** ~ voraussichtlich

**likeness** Abbild *n*, Ähnlichkeit *f*, Gleichheit *f*

**Lilienfeld silk** Kupferstreckseide *f*

**limb** Arm *m*, Bogen *m*, Glied *n* ~ **of a magnet** Magnetkern *m* ~ **of syncline** Muldenflügel *m* ~ **of a transformer** Säule *f*, Schenkel *m* ~ **darkening** Randverdunklung *f*
**limbed** schenklig
**limber** biegsam, elastisch, geschmeidig
**limber** Lafette *f*, Protze *f*, Vorderwagen *m* ~ **axle** Protzachse *f* ~ **box** Protzkasten *m* ~ **frame** Protzgestell *n* ~ **hook** Protz-, Schlepphaken *m* ~ **position** Protzenstellung *f*
**limbs, having two** ~ zweischenklig
**lime, to** ~ äschern, kalken, (Lederherstellung) kälken, scheiden **to** ~-**coat** kälken
**lime** Kalk *m*, Kalziumoxyd *n*, Linde *f* (Lindenbaum) ~ **alkalinity** Kalkalkalität *f* ~-**base grease** kalkverseiftes Fett *n* ~ **basin** Kalkloch *n* ~ **beater** Kalk-hacke *f*, -krücke *f* ~ **blast** Kalkschattenfleck *m* ~ **boil** (in the Martin process) Schlackenfrischreaktion *f* ~-**bond** kalkgebunden ~ **burning** Kalkbrennen *n* ~-**burning kiln** Kalkbrennofen *m* ~-**cast** Kalkbewurf *m* ~-**combining capacity** Verbindungs-vermögen *n* oder -fähigkeit *f* mit Kalk ~ **concrete** Kalkbeton *m* ~ **cream** Schwödebrei *m* ~ **crucible** Kalktiegel *m* ~-**cutting machine for chalky soil** Kalkstechmaschine *f* für Wiesenkalk ~ **dust** Kalkstaub *m* ~ **ground** Kalkuntergrund *m* ~ **harmotome** Anorthit *m* ~ **kiln** Kalkbrennofen *m*, Kalkofen *m* ~-**kiln gas** Kalkofengas *n* ~ **liquor** Äscher-, Kalk-brühe *f* ~-**loam brick** Kalksandziegel *m* ~ **marl** Kalkmergel *m* ~ **milk** Kalk-milch *f*, -tünche *f* ~ **mordant** Kalkbeize *f* ~ **mortar** Kalkmörtel *m* ~ **ore** Kalkerz *n* ~ **paste** Kalkbrei *n* ~ **pit** Äscher *m*, Kalkgrube *f*, -loch, Schwödgrube *f* ~ **powder** Düngekalk *m*, Kalkpulver *n* ~ **precipitation** Kalkausscheidung *f* ~ **process** Äscherverfahren *n* ~ **pyrolignite** Graukalk *m* ~ **rake** Kalk-haken *m*, -krücke *f*, Rudel *n*, Rührkrücke *f* ~ **requirement** Kalkbedarf *m* ~ **roasting** Verblaseröstung *f* ~ **sandstone** Kalksandstein *m* ~ **saturator** Kalksättiger *m* ~ **scum** Kalk-abscheidung *f*, -schaum *m* ~ **set** (of blast furnaces) Kalkelend *n* ~ **silica** Kalksilikat *n* ~ **slag** Kalkschlacke *f* ~ **slaked in air** abgestandener, abgestorbener oder verwitterter Kalk *m* ~ **slaking** Kalklöschung *f* ~ **slaking drum** Kalklöschtrommel *f* ~-**slaking plant** Löschanlage *f* für Kalk
**lime-soap,** ~ **dispersing properties** Kalkseifendispersvermögen *n* ~ **precipitation** Kalkseifenausflockung *f* ~ **preventing properties** Kalkseifenschutzvermögen *n*
**lime,** ~ **solution** Kalkmilch *f* ~ **sower (or spreader)** Kalkstreuer *m* ~ **stains** Kalkschäden *pl*
**limestone** Eisenbitterkalk *m*, Kalkstein *m* ~ **for flux** Zuschlagkalkstein *m* ~ **addition** Kalksteinzuschlag *m* ~ **bin** Kalksilo *m* ~ **breaker** Kalkbrecher *m* ~ **brick** Kalkziegel *m* ~ **concrete facing** Muschelkalkvorsatzbeton *m* ~ **flux** Flußspat *m* ~ **pit** Kalksteingrube *f* ~ **processing** Kalksteinverarbeitung *f* ~ **quarry** Kalkbruch *m*
**lime,** ~ **store** Kalklager *n* ~ **sucrate** Zuckerkalk *m* ~ **tree bast** Lindenbast *m* ~ **uranite autunite** Kalkuranglimmer *f*
**limewash, to** ~ kalken, weißen
**limewash** Kalktünche *f* ~ **whiting** Weißkalk *m*

**lime,** ~ **water** Kalk-milch *f*, -wasser *n* ~ **wood charcoal** Lindenholzkohle *f*
**limed,** ~ **juice** gekalkter Saft *m* ~ **resin** abgekalktes Harz *n*
**limer** Kalksprenger *m*
**liminal** schwellig ~ **value of intensity of a color** Farbschwellenwert *m*
**liming** Scheidung *f* ~ **of valves** Öffnungsdauer *f* bei Ventilen ~ **drum** Äschertrommel *f* ~ **paper** Beklebepapier *n* (print.) ~ **tank** Scheidepfanne *f* ~ **wheel** Drehkalke *f*
**limit, to** ~ begrenzen, beschränken **to** ~ **amplitude** abkappen **to** ~ **the field** ausblenden
**limit** Begrenzung *f*, Bereich *m*, Ende *n*, Grenze *f*, Ziel *n*
**limit,** ~ **of adhesion** Adhäsions-, Klebe-grenze *f* ~ **of age-hardening (or aging)** Aushärtungsgrenze *f* ~ **of audibility** Hörbarkeitsgrenze *f* ~ **of brittleness** Alterungsgrenze *f* ~ **of capacity** Leistungsgrenze *f* ~ **of current** (intensity) Grenzstromstärke *f*, Höchststromstärke *f* ~ **of definition** Schärfenfeldgrenze *f* ~ **of deflection** (wheels) Ausschlagbegrenzung *f* ~ **of density** Festigkeitsgrenze *f* ~ **of detection** Nachweisgrenze *f* ~ **for disruption of flow** Abreißgrenze *f* ~ **of elasticity** Elastizitätsgrenze *f*, zulässige Spannkraft *f* ~ **of endurance** Kaltverfestigungsgrenze *f* ~ **of error** Fehlergrenze *f* ~ **of expansion** zulässige Ausdehnung *f* ~ **of haze** Dunstgrenze *f* ~ **of illumination** Beleuchtungsgrenze *f* ~ **of integration** Integrationsgrenzen *pl* ~ **of measurement** Meßgrenze *f* ~ **of mist** Dunstgrenze *f* ~ **of the normal bed at low water** (summer) Niedrigwasser-grenze *f*, -linie *f* ~ **of the normal major bed** (winter) Hochwassergrenze *f*, -linie *f* ~ **of perceptibility** Wahrnehmbarkeitsgrenze *f* ~ **of power range** Leistungsgrenze *f* ~ **of proportionality** Gleichmaßgrenze *f* ~ **of resistance** Festigkeitsgrenze *f* ~ **of resolution** Auflösungsgrenze *f* ~ **of scattering** Streugrenze *f* ~ **of sensitivity** Empfindlichkeitsgrenze *f* ~ **of sharpness** Schärfenfeldgrenze *f* ~ **of stability** Festigkeitsgrenze *f* ~ **of strain-hardening** Kaltverfestigungsgrenze *f* ~ **of stress** Beanspruchungsgrenze *f* ~ **of territorial sovereignty** Hoheitsgrenze *f* ~ **of tolerance** Erträglichkeitsgrenze *f* ~ **of work-hardening** Kaltverfestigungsgrenze *f*
**limit,** ~ **bar** Begrenzungsstab *m* ~ **bar ball** Visierkugel *f* ~ **characteristics** Grenzcharakteristik *f* ~ **conception** Grenzbegriff *m* ~ **distance** Grenzentfernung *f* ~ **field** Grenzfeld *n* ~ **gauge** Grenzladeprofil *n*, Grenzlehre *f* ~ **(snap) gauge** Grenzrachenlehre *f* ~ **gauges for outside maximum and minimum** Doppelringlehre *f* für Grenzmessung der Dicke ~ **hydrocarbon** Grenzkohlenwasserstoff *m* ~ **lamp** Begrenzungslampe *f* ~ **load** äußerstes Gewicht *n* ~ **loads** sichere Lasten *pl* ~ **mark** Grenzstrich *m* ~ **moment** Tragmoment *n* ~ **point** Grenzpunkt *m* ~ **pointer** Sollwertzeiger *m* ~ **screw** Begrenzungsschraube *f* ~ **size** Grenzmaß *n* ~ **snap gauge** Toleranztasterlehre *f* ~ **stop** Anschlag *m* ~ **stop mechanism** Endabstellvorrichtung *f* ~ **stop switch** Endabstellschalter *m* ~ **switch** Auflaufkontakt *m*, Begrenzungsschalter *m*, Einlagenschalter *m*, End-ausschalter *m*, -begrenzungsschalter *m*, -lagenschalter *m*, -schalter *m*,

-taster *m*, -umschalter *m*, Grenzschalter *m* ~ **switch drive** Schaltwalzenantrieb *m* ~ **thread gauges** Grenzgewindelehren *pl* ~-**type singularity** Grenzliniensingularität *f* ~ **value** Schwellenwert *m* ~ **value switch unit** Grenzwertschaltereinheit *f*

**limitation** Bedingung *f*, Begrenzung *f*, Beschränktheit *f*, Beschränkung *f*, Einengung *f*, Einschränkung *f*, Frist *f*, Grenze *f*

**limitation,** ~ **of the draft** Begrenzung *f* des Tiefganges ~ **of frequency band** Frequenzbegrenzung *f* ~ **of groove** Kaliberbegrenzung *f* ~ **of the normal width** Begrenzung *f* der Normalbreite ~ **of radio traffic** Funkbeschränkung *f* ~ **on speed** Drehzahlbegrenzung *f*

**limitation stop** Begrenzungsanschlag *m*

**limited** bedingt, begrenzt, beschränkt ~-**area smoke screen** Kleinvernebelung *f* ~-**liability company** Gesellschaft *f* mit *f* mit beschränkter Haftung ~ **partnership** Kommanditgesellschaft *f* ~ **radio silence** Funkbeschränkung *f* ~ **reverse impedance** endliche Sperrschicht *f* ~ **solubility** beschränkte Löslichkeit *f* ~ **variation** beschränkte Gesamtschwankung *f* ~ **view** beschränkte Fernsicht *f*

**limiter** (Amplituden) Begrenzer *m* ~ **characteristic** Begrenzerkennlinie *f* ~ **diode** Begrenzerdiode *f* ~ **valve** Begrenzerröhre *f*

**limiting** begrenzend, einschränkend ~ **amplifiableness** Verstärkbarkeitsgrenze *f* ~ **angle** (of refractometer) Grenzwinkel *m* ~ **angle of elevation** Höhenrichtbereich *m* ~ **angle of a refractometer** Refraktometergrenzwinkel *m* ~ **angle of rolling** Greif-, Walzen-winkel *m* ~ **aperture** Ausblendmittel *n* ~ **area** Grenzgebiet *n* ~ **bar** Begrenzungsschiene *f* ~ **bolt** Grenzriegel *m* ~ **boundery** Grenzbedingung *f* ~ **case** Grenzfall *m* ~ **characteristic value** Grenzkennwert *m* ~ **circle** Grenzkreis *m* ~ **coil** Begrenzungsdrossel *f* ~ **condition** Randbedingung *f* ~ **creep strength** Kriechgrenze *f* ~ **creep stress** Dauerstandfestigkeit *f*, Kriechgrenzenspannung *f* ~ **current** Grenzstrom *m* ~ **curve** Grenzkurve *f* ~ **date line** Datumgrenze *f* ~ **deflection** Endausschlag *m* ~ **design** Grenzleistungskonstruktion *f* ~ **device** Begrenzer *m* ~ **diameter** (of dielectric line) Grenzdurchmesser *m* ~ **dynamic current** dynamischer Grenzstrom *m* ~ **factor** Begrenzungsfaktor *m* ~ **flange** Anschlagflansch *m* ~ **frequency** Grenzfrequenz *f* ~ **gradient** Grenzgefälle *n* ~ **height** Grenzhöhe *f* ~ **lamp** Begrenzungsleuchte *f* ~ **line** Grenzlinie *f* ~ **load** Grenzbelastung *f* ~ **means** Begrenzungsanschlag *m* ~ **mobility** Grenzbeweglichkeit *f* ~ **molar conductance** Grenzwert *m* der molaren Leitfähigkeit ~ **number of load alternations** Grenzlastwechselzahl *f* ~ **output** Grenzleistung *f* ~ **point** Häufungspunkt *f* ~ **position** Englage *f* ~ **quantity** Einflußgröße *f* ~ **rabbet** Hubbegrenzungsstück *n* ~ **range of stress** Beanspruchungsgrenze *f*, Dauerfestigkeit *f* ~ **regulator** Grenzregler *m* ~ **resistance** Begrenzungs-, Grenzwiderstand *m* ~ **resolution factor** auflösungsbegrenzender Faktor *m* ~ **resolving power** Grenzauflösungsvermögen *n* ~ **screen** Durchgangssieb *n* ~ **set** Grenzmenge *f* ~ **size** Grenzmaß *n* ~ **slope** Grenzgefälle *n* ~ **stage** Begren-

zerstufe *f* ~ **stop** Anschlagstück *n*, Zonenblende *f* ~ **stress** Grenzspannung *f*, höchstzulässige Spannung *f* (Festigkeit) ~ **surface of stress** Streckungsgrenzfläche *f* ~ **value** Grenz-, Schwellen-wert *m* ~ **value of stimulus** Reizschwelle *f* ~ **value of stress cycle endured** Grenzlastspielzahl *f* ~ **viscosity** Grenzviskosität *f* ~ **wave length** Grenzwellenlänge *f*

**limitless** grenzenlos, unbegrenzt

**limits** Grenzmaße *pl* ~ **of accuracy** Genauigkeitsgrenzen *pl* ~ **of applicability** Anwendungsgrenzen *pl* ~ **of validity** Gültigkeitsgrenzen *pl*

**limnimeter** Seepegelmesser *m*

**limonite** Brauneisenmulm *m*, Braunerz *n*, Eisenrahm *m*, brauner Glaskopf *m*, Limonit *m*, Quellerz *n*

**limousine** geschlossener Wagen *m*

**limp** schlaff

**limpid** durchsichtig

**limping** Hinken *n*

**limpness** Schlaffheit *f*

**limy** kalk-artig, -haltig, kalkig ~ **condition** Bildflachheit *f*, -flauheit *f*; flau

**linac** linearer Beschleuniger *m*

**linarite** Bleilasur *f*, Kupferbleivitriol *n*

**linch,** ~ **hoop** Achsring *m* ~ **pin** Achsen-nagel *m*, -stift *m*, Achsnagel *m*, Lünse *f*, Radbolzen *m*, Radsplint *m*, Splint *m*, Vorstecker *m*, Vorsteckstift *m* ~ **washer** Lünscheibe *f*

**Lindemann glass** Lindemannglas *n*

**linden,** ~ **tree** Linde *f* ~ **wood** Lindenholz *n*

**line, to** ~ abfüttern, aufreißen, ausfüttern, ausgießen, auskleiden, (a box with zinc) ausschlagen, ausstampfen, bekleiden, dublieren, füttern, in eine Linie bringen, linieren, polstern, schraffieren, überziehen, umkleiden, verkleiden, verzimmern, zustellen **to** ~ **a bearing** ein Lager ausfüttern **to** ~ **with brick** ausmauern **to** ~ **with lead** ausbleien **to** ~ **out** abvisieren, eine Linie ausfluchten **to** ~ **stuff** (or timber) Holz abschnüren **to** ~ **with turf** stocken **to** ~ **up** auskleiden, ausstampfen, zum Flüchten bringen, zurechtmachen

**line** Achse *f*, Bildzeile *f* (rdo), Fach *n*, Flucht *f*, (of diffraction grating) Furche *f*, Leine *f*, Leitungszug *m* (teleph.), Linie *f*, (of a graph) Linienzug *m*, Reihe *f*, Schnur *f*, Seil *n*, Serie *f*, Stellung *f*, (of drive shafts) Strang *m*, Strecke *f*, Strich *m*, Strick *m*, Stromkreis *m*, Tau *n*, Trasse *f*, Zeile *f* **(straight)** ~ Gerade *f* **in** ~ in Dwarslinie *f*, in Reihe *f* geschaltet **the** ~ **is short-circuited** die Leitung ist kurzgeschlossen oder kurzverbunden

**line,** ~ **of action** Angriffs-, Eingriffs-, (Zahnrad) Kraft-, Wirkungs-linie *f* ~ **of advance** Vormarschstraße *f* ~ **of aim** Schußlinie *f* ~ **of anchors** Ankerlinie *f* ~ **of application** Angriffslinie *f*, (of a force) Wirkungskreis *m* ~ **of approach** Annäherungsweg *m* ~ **of arrest** Aufhaltelinie *f* ~ **of attack** Angriffslinie *f* ~ **of bearing** Peillinie *f*, Staffel *f* ~ **of breach** Trennabschnitt *m* ~ **of business** Geschäftskreis *m* ~ **of collimation** Kolimationsachse *f*, Sehlinie *f*, Ziellinie *f*, Zielstrahl *m* ~ **of communication** Nachschubstraße *f*, Verbindungslinie *f* ~ **of concrete dugouts** Bunkerlinie *f* ~ **of confinement** Kettenlinie *f* ~ **of contact** Berührungsfläche *f*, -linie *f*, Eingriffslinie *f* (Zahnrad) ~

**of the course** Kurslinie *f* ~ **of curvature** Krümmungslinie *f* ~ **of deflection** Durchbiegungslinie *f* ~ **of demarcation** Demarkationslinie *f*, Trennungs-linie *f*, -strich *m* ~ **of departure** Ausgangslinie *f*, Frontausgangsstellung *f*, Schußlinie *f* ~ **of development** Entwicklungslinie *f* ~ **of dip** Deklinationslinie *f* ~ **of direction** Baufluchtlinie *f* ~ **of displacement** Verschiebungslinie *f* ~ **of droplets** Perlschnur *f* **in** ~ **of duty** aus dienstlicher Veranlassung *f*, in Ausübung *f* des Dienstes ~ **of equal amplitude (of waves)** Hubhöhenlinie *f* ~ **of equal anomaly** Linien *pl* des gleichen Wertes der Unregelmäßigkeit ~ **of equal barometric pressure** Linie *f* gleichen Luftdrucks ~ **of equal rise** Hubhöhenlinie *f* ~ **of fire** Schußrichtung ~ **of flight** Fluglinie *f* ~ **of flux** Feldlinienvektor *m*, Vektorlinie *f* ~ **of force** Hilfskreis *m*. Kraftlinie *f* ~ **of future position** Schußlinie *f* ~ **with heavy gradients** Hügellandstrecke *f* ~ **with hook-shaped brackets** Linie *f* mit Hakenstützen ~ **of impression** Drucklinie *f* ~ **of inclination** Gefällinie *f* ~ **of induction** Erregungslinie *f* ~ **of intersection** Anschneide-, Durchschnitts-, Schnitt-linie *f* ~ **of junction** Anheftungslinie *f* ~ **of levels** Nivellementszug *m* ~ **of light wave of oscilloscope** Lichtstrich *m* ~ **of no loss** verlustlose Leitung *f* ~ **of magnetic force** magnetische Kraftlinie *f* ~ **of map grid** senkrechte Gitterlinie *f* ~ **of most rapid flow** Stromstrich *m* ~ **with numerous curves** kurvenreiche Linienführung *f* ~ **of observed position** Meßortungslinie *f* ~ **of operations** Operationslinie *f* ~ **of partition between adjoining grid sections** Gittertrennungslinie *f* ~ **of pegs** Pfahlreihe *f* ~ **of pipes** Rohrkolonne *f* ~ **of position (LOP)** Positionslinie *f*, Standlinie *f* (nav.) ~ **of present position** Visierlinie *f* ~ **of pressure** (in masonry analysis) Stützlinie *f* ~ **of print** Druckzeile *f* ~ **of production** Produktionsgang *m* ~ **of projection** Wurflinie *f* ~ **of range on map** Kartenentfernungslinie *f* ~ **of regard** Betrachtungsrichtung *f* ~ **of repeater bays** Bucht *f*, Gestellreihe *f* ~ **of resistance** Widerstandslinie *f* ~ **of retirement (or retreat)** Rückzugsrichtung *f* ~ **of rods** Gestängehöhe *f* ~ **of seepage** Durchströmungslinie *f* ~ **of segregation** (in an ingot) Seigerungszone *f* ~ **of self-intersection** Durchdringungslinien *pl* ~ **of sight** Blicklinie *f*, Schußlinie *f*, Sehlinie *f*, quasioptische Sicht *f*, Sichtweite *f*, Visierachse *f*, Visierrichtung *f*, Visur *f*, Ziellinie *f* ~ **of sight checking gear** Fotojustiergerät *n* ~ **of sight range** optische Reichweite *f* ~ **of sight system** Zieldeckungsmethode *f* ~ **of simultaneous high tide** Flutstundenlinie *f* ~ **of skirmishers** Feuerkette *f* ~ **of slide (or slip)** Gleitlinie *f* ~ **of solidification** Erstarrungslinie *f* ~ **of stakes** Pfahlreihe *f* ~ **of stress** Streckungsgrenzfläche *f* ~ **of striction** Kehllinie *f* ~ **of strike** Streichlinie *f* ~ **of supply** Nachschubstraße *f* ~ **of teeth** Zackenreihe *f* ~ **of thrust** Vortriebsachse *f* ~ **of total force** totale Kraftlinie *f* ~ **of trail** Spurlinie *f* ~ **of train** Spur *f* ~ **of vision** Blicklinie *f*, Blickrichtung *f*, Sehlinie *f* ~ **of weld** Schweißfuge *f*, -naht *f* ~ **of zero drift** Steuerstrich *m* ~ **of zero pressure** Nulldrucklinie *f*

**line,** ~ **abatis** Baumverhau *m* ~ **adding key** Zei-

lenaddiertaste *f* ~ **advance** Zeilenabstand *m* ~ **amplifier** Leitungsverstärker *m* ~**-amplifying coil** Pupinspule *f* ~ **amplitude control** Zeilenbreitenregler *m* ~ **angle** Fortpflanzungs-größe *f*, -maß *n* ~ **arm** Längs-arm *m*, -zweig *m* ~ **array** Dipolzeile *f* ~**-at-a-time printer** Zeilendrucker *m* ~ **attenuation** Leitungsdämpfung *f* ~ **balance** Leitungsausgleich *m* ~ **bank** A/B-Kontaktbank *f* (electr.), Leitungskontaktbank *f* ~ **bar** Leitungsschiene *f* ~ **barrage** Sperriegel *m* ~ **bend** Zeilenentzerrung *f* ~ **blanking** Zeilenaustastung *f* ~ **block** Strichätzung *f* ~ **break** Umbruch *m* der Linie ~**-break relay** Drahtbruchrelais *n* ~ **breaker** Linienunterbrecher *m* ~ **broadening by particle size** Linienverbreiterung *f* durch Teilchengröße ~ **brush** Kontaktarm *m* für eine Sprechleitung ~ **calling current** Linienrufstrom *m* ~ **capacity** Leitungskapazität *f* ~ **carried on brackets** Linie *f* mit Hakenstutzen ~ **carrying very high voltage** Höchstspannungsleitung *f* ~**(s) casting machine** Zeilengießmaschine *f* ~ **change** Leitungsumschaltung *f* ~**-change impulse** Zeilenwechselimpuls *m* ~ **characteristic** Leitungseigenschaft *f* ~ **chart** Stabliniensystem *n* ~ **choke** Liniendrossel *f* ~ **circuit** Leitungsschaltung *f* ~ **clear** Freisignal *n* ~ **coil** Leitungsspule *f* ~ **communication** Netzverkehr *m* ~ **commutator** Linienumschalter *m* ~ **concentrator** Leitungsdurchschalter *m* (telegr.) ~ **condition** Leitungszustand *m* ~ **connecting element** Leitungsanschlußelement *n* ~ **connecting points of greatest depth along stream course** Talweg *m* ~ **connection** Netz-anode *f*, -verkehr *m* ~ **connector** Leitungskupplung *f* ~ **consolidation** Linienverstärkung *f* ~ **constant** Grundwert *m*, Konstante *f* je Längeneinheit, Leitungs-eigenschaft *f*, -konstante *f* ~ **construction** Leitungsbau *m* ~**-construction service** Telegrafenbaudienst *m* ~**-construction unit** Leitungsbaueinheit *f* ~ **contact** Berührung *f* längs einer Linie ~ **contact bank** a-b-Kontaktsatz *m*, Leitungskontaktsatz *m* ~ **contrast micrometer** Strichkontrastmikrometer *n* ~**-controlled** leitungsgesteuert ~**-controlled oscillator** leitungsgesteuerter Oszillator *m* ~ **copying** Zeilendrehen *n* ~ **cord** Netzschnur *f* ~ **costs** Leitungskosten *pl* ~ **crawl** Zeilenflimmern *n* ~ **crew** Entstörungstrupp *m* ~ **cross** Fadenkreuz *n* ~ **crossing** Leitungskreuzung *f* ~ **current** Leitungs-, Linien-, Netzstrom *m* ~**-current thermal relay** Hauptstromthermorelais *n* ~ **curve** Linienzug *m* ~ **cycle** Horizontalwechsel *m* ~**-deflecting apparatus (or unit)** Zeilenablenkgerät *n* ~**-deflecting coil** Zeilenlenkspule *f* ~**-deflecting tube** Zeilenablenkröhre *f* ~ **density** Längendichte *f* ~ **density of electrons** Liniendichte *f* der Elektronen ~ **diameter** Grenzdurchmesser *m* ~ **diapositive** Strichdiapositiv *n* ~**-dipole** Liniendipol *n* ~ **dismantling** Leitungsabbau *m* ~ **division staff** Strichlatte *f* ~ **drawing** Strich-, Umriß-zeichnung *f* ~ **drop** Anrufklappe *f*, Leitungsabfall *m* ~ **element** Leitungsstück *n* ~ **emitted by neutral atoms** Neutralatomlinie *f* ~ **end switch** Kopfschienenweiche *f* ~ **equipment** Streckenausrüstung *f* ~ **equivalent** Leitungsäquivalent *n* ~ **equivalent in miles of standard cable** Leitungsäquivalent *n* in Meilen Standardkabel ~ **etching** (on zinc) Strichätzung f ~ **extent** (in length) Linienstrecke f ~ **failure** Leitungsfehler m ~ **fault** Leitungs-fehler m, -störung f ~ **fault locator** Fehlerortungsgerät n ~ **faults** Massenstörung f ~ **feed** Zeilenvorschub m ~ **feed lever** Abdeckhebel m ~**-feed magnet** Zeilenmagnet m ~ **finder** Leitungs-, Linien-sucher m ~ **finder with allotter switch** Anrufsucher m mit Vorwähler (ein Anrufsucher und ein Vorwähler) ~ **flickering** Zeilenflimmern n ~ **flyback** Zeilenrücklauf m ~ **flyback pulse** Rücklaufimpuls m ~ **focus** (of X rays) Strichfokus m ~ **force** Linienkraft f ~ **formation** Reihenflug m ~**-free circuit** Rückblockungsstromkreis m ~ **frequency** Netz-, Zeilen-frequenz f, Zeilenwechselfrequenz f, -zahl f, (number) Zeilenzahl f ~**-frequency blanking impulse** Zeilenaustastimpuls m ~ **gap** Zeilensynchronisierungslücke f ~ **grating** Linienraster m ~ **height** Zeilenhöhe f ~ **hum** Netzton m ~ **hydrophone** lineares Hydrofon n ~ **identification circuit** Teilnehmerfeststellung f ~ **image** Zeilenbild n ~ **(-synchronizing) impulse** Zeilen-impuls m, -stoß m ~**-impulse amplifier** Zeilenverstärker m ~**-impulse method** Zeilenstoßverfahren n ~ **index** Zeilenindex m ~ **inductance** Leitungsinduktivität f ~ **inspection** periodische Kontrolle f ~ **insulator** Krückenisolator m ~ **integral** (of a vector) Linienintegral n, Liniensumme f ~ **intercept** Lauschempfänger m ~ **interception** Lauschdienst m ~ **interlace transmission** Springzeilensendung f ~ **interval** Linien-abstand m, -intervall n, Zeilensynchronisierungslücke f ~ **jack** Anschluß-, Leitungs-klinke f ~ **lag** Laufzeit f des Stroms über eine Leitung, Leitungsverzögerung f ~ **lamp** Anrufglühlampe f, Anruflampe f ~ **limitating device** Postenbegrenzer m ~ **load** Linien-, Strecken-last f ~ **loop resistance** Schleifenwiderstand m ~ **loss** Leitungs-dämpfung f, -verlust m ~ **losses** Betriebsdämpfung f ~ **maintenance** Feldinstandsetzung f ~ **man** Bahnwärter m, Leitungsmann m, Störungssucher m, Streckenarbeiter m **(section)** ~ **man** Leitungsaufseher m ~ **man's assistant** Hilfswärter m ~ **man's climbers** Appareils pl ~ **man's tool bag** vollständige Gerätetasche f für Störungssucher ~ **mark** Markierungsstrich m ~ **marker** Schauzeichen n ~ **material** Bauzeug n, Linienmaterial n ~ **method** (paper test) Strichmethode f ~ **monitoring tube** Zeilenkontrollröhre f ~ **multiplexing equipment** Wählersternanschluß m ~ **negative** Strichnegativ n ~ **noise** Leitungsgeräusch n ~**-offset scan** Zeilenverschiebung f ~ **output** zeilenfrequente Kippspannung f (TV) ~ **output pentode** Horizontal-Endpentode f ~ **overlap** Zeilenüberdeckung f ~**-overlap photograph** Reihenbild n ~ **pattern** Linienbild n ~ **period** Zeilenintervall n ~ **photo** Strichaufnahme f ~ **pipe** Leitungsrohr n ~ **pitch** Teilungsfehler m, Zeilenteilung f ~ **plate** Futterstück n ~ **plomb** Schnurlot n ~ **points** Bahnpunkte pl ~ **pole** Baustange f ~ **positions** Fernamt n ~ **pressure** Leitungsdruck m ~ **printer** Zeilendrucker m (info proc.) ~ **production** Straßenfertigung f ~**-production system** Durchlauf-, Fließ-fertigung f ~ **pulse** Zeilenimpuls m ~

**pulsing** Laufzeitkettentastung f ~ **radiation** Linienstrahlung f ~ **radio** Drahtfunk m, Drahtrundfunk m, Hochfrequenzverbindung f (über Leitungen) ~ **ranger** Doppelwindelprisma n ~ **record** Liniennachweis n ~ **relay** Anruf-, Leitung-, Linien-relais n, Rufmagnet m ~**-relay rack** Leitungsrelaisgestell n ~ **repeater** Leitungsverstärker m ~ **repeating coil** Überträgerpaar n ~ **residual current** Erdrückstrom m ~ **residual equalizer** Restverzerrungskorrektor m ~ **resistance** Fern-, Längs-, Leitungswiderstand m ~ **retrace (or return)** Zeilenrücklauf m ~ **reversal** Linienumkehr f ~**-reversal method** Linienumkehrmethode f ~ **riveting** Reihennietung f ~ **saw-tooth** Zeilensägezahn m ~**-scale micrometer** Strichmikrometer n ~ **scan** horizontale Ablenkung f, Horizontalablenkung f, Zeilenablenkung f, Zeilenbewegung f (TV) ~**-scan coil** Zeilenspule f ~**-scanning mirror drum** Zeilenspiegelrad n ~**-scanning pattern** Zeilenrester m ~ **scheme** Linienbild n ~ **scratches** Kratzergeräusche pl ~ **screen** Linienraster m, Netznegativ n (photo), Schraffürplatte f ~ **segment** Strecke f ~ **selector** Leitungswähler m ~**-selector switch** Linienwähler m ~ **separation** Leitungsabstand m ~ **sequence** Zeilenzug m ~ **sequential system** Linienfolgesystem n ~ **series** Zeilenzug m
**line-shaft** Antriebswelle f, Hauptwelle f, Transmissionsstrang m, Wllenleitung f ~ **bearing** Transmissionslager n ~ **drive chain** Transmissionstreibkette f ~ **equipment** Triebwerkanlage f ~ **erection** Wellenmontage f
**line,** ~ **shafting** Wellenstränge pl ~ **shape** Linienform f ~**-shunt admittance** Scheinleitwert m einer Leitung gegen Erde oder zwischen den Drähten einer Doppelleitung ~**-shunt conductance** Leitungskonstante f ~ **side of the main distributing frame** Leitungsseite f des Hauptverteilers ~ **signal** Anrufssignal n, Anrufzeichen n, Linienstrom m, Rufzeichen n ~ **simulator** künstliche Leitung f ~ **sizes** Maße pl der Leitungen ~ **source** Linienquelle f
**line-space** Zeilenabstand m ~ **adjusting mechanism** Zeileneinstellvorrichtung f ~ **lever** Zeilenschalthebel m ~ **mechanism** Zeilenschaltwerk n ~ **regulator** Zeilenabstandskala m, Zeilensteller f
**line,** ~ **spacing** Zeilenschaltung f ~**-spacing lever** Zeileneinstellhebel m ~**-spacing mechanism** (typewriter) Zeilenfortschaltmechanismus m ~ **spectrum** Linienspektrum n ~ **speed** Telegrafen-, Startkontroll-geschwindigkeit f (aviat.) ~ **splitting** Linienaufspaltung f ~ **spring** (of jack) Leitungsfeder f ~ **squall** Bö f mit breiter Front, Front-, Linien-, Reihen-bö f ~ **standard** Strichmaß n, Zeilennorm f ~ **stress** Linienspannung f ~ **stretcher** (ausziehbare) Leitung f, Posaune f
**line-sweep** Zeilenablenkung f, Zeilenvorschubbewegung f ~ **apparatus** Zeilenkippgerät n ~ **coil** Zeilenkippspule f, Zeilenspule f ~ **period** Zeilenkipperiode f ~ **tube** Zeilenkippröhre f
**line,** ~ **switch** Hauptschalter m, Vorwähler m ~**-switch spring** Leitungsklinke f ~ **switchboard** Klinken-, Linien-umschalter m, Vorwählergestell n ~ **synchronism** Zeilengleichlauf m ~ **synchronizing impulse** Zeilengleichlaufstoß m ~

synchronizing pulse Horizontalwechsel *m*
~-synchronizing signal Zeilenimpuls *m*, Zeilen-
stoß *m*, Zeilensynchronisierimpuls *m*, Zeilen-
synchronisierungsimpuls *m*, Zeilensynchroni-
sierzeichen *n* ~ system Leitungsnetz *n* ~
tapping Lauschdienst *m* ~ terminal Leitungsklemme *f* ~ test Leitungsprobe *f* ~ test set
Feldprüfgerät *n* ~ tester Leistungsprüfer *m*
~-testing apparatus Abfrageapparat *m* ~-test-
ing vehicle Leitungsmeßkraftwagen *m* ~
theory Leitungstheorie *f* ~-throwing apparatus
Leinenwurfapparat *m* ~-throwing gun Wurf-
leinenkanone *f* ~ tilt Zeilenentzerrung *f* ~
time Leitungs-, Linien-zeit *f* ~ time base
Zeilenkipper *m* ~-time-base impulse Zeilen-
kippschwingung *f* ~-to-~ verkettet ~ train
sweep Zeilenzug *m* ~ transformer Zeilentrans-
formator *m* ~-transformer coil Leitungsüber-
träger *m* ~ translation Zeilenumsetzung *f* ~
transmission Leitungsübertragung *f* ~ transparency Strichdiapositiv *n* ~ trap (for powerline carrier) Trägerfrequenzsperre *f* ~ traversal
Bildzeilendurchlauf *m* ~-traversing motion
Zeilenvorschubbewegung *f* ~ turn Windungslinie *f* ~ unit Anschlußeinheit *f* ~-up Bestük-
kung *f*, Ordnung *f*, Vorbereitung *f* ~ voltage
Netzspannung *f*, verkettete Spannung *f*
~-voltage generator Zeilengenerator *m* ~
walker Leitungsaufseher *m* ~ width Zeilen-
breite *f* ~ winding Leitungswicklung *f* ~ wiper
a/b-Arm *m* ~ wipers a/b-Bürsten *pl* ~ wire
Blockleitung *f*, Leitungsdraht *m* ~ wires
Linienleitung *f* ~ wiring diagram Verdrah-
tungsschema *n*
**lineage structure** Verzweigungsstruktur *f*
**lineal** direkt, geradlinig ~ scale length Skalenlänge *f*
**linear** fortschreitend, geradlinig, linear, linien-
förmig ~ acceleration lineare Beschleunigung *f*
~ accelerator Linearbeschleuniger *m* ~ amplification lineare Verstärkung *f* ~ amplifier
linearer Verstärker *m* ~ amplifier plug-in unit
Linearverstärkereinschub *m* ~ array Dipol-
reihe *f* ~ attenuation lineare Dämpfung *f* ~ ball
bearing lineares Kugellager *n* ~ contraction of
cross-sectional area Querkürzung *f* ~ course of
gradation Gradationsverlauf *m* ~ dependence
lineare Abhängigkeit *f* ~ detail (or element)
lineare Einzelheit *f* ~ detection lineare Gleichrichtung *f* ~ detector linearer Detektor *m* ~
dimension lineare Größe *f* ~ distortion Amplitudenverzerrung *f*, lineare Verzerrung *f* ~
drawing Umrißzeichnung *f* ~ drift zeitlich
linearkriechende Abweichung *f* ~ electrical
constants Leitungsparameter *pl* ~ electrode
lineare Elektrode *f* ~ electron accelerator
Elektronenlinearbeschleuniger *m* ~ element
Strecken-element *n*, -teilchen *n* (math.) ~
elongation Längsdehnung *f* ~ equation Gleichung *f* ersten Grades ~ expansion Längen-
ausdehnung *f* ~ extension Dehnungslänge *f* ~
extrapolation distance linearer Extrapolationsabstand *m* ~ figure Liniengebilde *n* ~ framework lineare Methode *f* ~ function lineare
Funktion *f* ~ graph lineares Netz *n* ~ height of
burst Flugweite *f*, Restflugweite *f* ~ high-
frequency amplifier linearer Hochfrequenzverstärker *m* ~ increase of velocity of ascending

(or upward) currents Aufwindgeschwindigkeit *f*,
lineare Zunahme *f* des Aufwindes oder der
Aufwindgeschwindigkeit ~ ionic chain lineare
Ionenkette *f* ~ measure Längenmaß *n*, Linear-
maß *n*, -maßstab *m*, Linienmaß *n*, laufendes
Maß *n* ~ measurement Längenmessung *f* ~
meter laufender Meter *m* ~ momentum
linearer Impuls *m* ~ motion eindimensionale
Bewegung *f* ~ perspective Linienperspektive *f*
~ pinch discharge fadenförmige Entladung *f* ~
pitch Achsteilung *f* ~ power amplifier linearer
Leistungsverstärker *m* ~ program part gerades
Programmstück *n* ~ projection Perspektivität *f*
~ range lineare Reichweite *f* ~ rectifier linearer
Detektor *m* oder Gleichrichter *m* ~ response
lineares Ansprechen *n* ~ rolling speed horizontale Walzgeschwindigkeit *f* ~ scan lineare Abtastung *f* ~ scanning zeilenweises Abtasten *n* ~
~ (or normal) section Normalstück *n* ~ slip
wirklicher Schlupf *m* ~-speed date computer
lineares Kommandogerät *n* ~ stopping power
lineares Bremsvermögen *n* ~ strain Schubzahl
*f* ~ stress distribution lineare Spannungsverteilung *f* ~ sweep lineare Abtastung *f* ~ taper
linearveränderlicher Widerstand *m* ~ time base
lineare Zeitbasis *f* ~ transducer linearer Transduktor *m* ~ travel of target Auswanderungsstrecke *f*
**linearity** Geradlinigkeit *f*, Linearität *f* ~ of
response lineares Ansprechen *n* ~ control Linearitätsregelung *f*
**linearization** Linearisierung *f*
**linearize, to** ~ linearisieren
**linearized equations of observation** linearisierte
Beobachtungsgleichungen *pl*
**linearizing,** ~ action Linearisierung *f* ~ resistance
Linearisierungswiderstand *m*
**linearly** vom ersten Grad ~ dependent linear abhängig ~ polarized wave linear polarisierte
Welle *f* ~ viscous fluid linear-viskose Flüssigkeit *f*
**lineated** längsgestrichelt
**lined** ausgefüttert, ausgekleidet, gefüttert, überzogen ~ borehole verrohrtes Bohrloch *n* ~ foil
kaschierte Folie *f* ~ packing materials kaschierte
Verpackungsmaterialien *pl* ~-up vorgeschaltet
**linen** Laken *n*, Leinen *m*, Leinwand *f*, (cloth)
Linnen *n*, (blind) Tuch *n*, Wäscheleinen *n*,
Weißwaren *pl*
**linen,** ~ cap for screw bungs Dichtungsläppchen
*n* für Spunde ~ cloth Leinenstoff *m*, Leinöl-
wand *f*, Leinwand *f* ~ fabric Leinengewebe *n* ~
finish Leinenprägung *f* ~ flap Leinwandkappe *f*
~ goods Weißwaren *pl* ~ grinding blade
Leinenschleifblatt *n* ~ hosiery Leinstrickerei
*f* ~-ironing establishment Plättanstalt *f* ~-mak-
ing machine Wäschefabrikationsmaschine *f*
~ pulp Leinenhalbstoff *m* ~ sheet Laken *n* ~
tester Fadenzähler *m* ~ texture Leinengewebe
*n* ~ woof Leinengewebe *n* ~ wove paper Leinen-
velinpapier *n*
**liner** Einlage *f*, Einlegerohr *n* (artil.), Einsteck-
lauf *m*, Führungsröhre *f*, Füllstück *n*, Futter *n*,
Futterrohr *n* (artil.), Futterstück *n*, Gehäuseoberteil *n*, Innenbüchse *f*, Keil *m*, Kolbenlauf-
mantel *m* (mach.), Laufbuchse *f*, Liniendampf-
er *m*, Linienflugzeug *n*, Produktionsrohre *pl*
(gelocht), Rezipientenbüchse *f*, Schärfeinlage *f*,

Stoffeinlage *f* ~ **of eccentric strap** Exzentereinlage *f*, -futter *n*
**liner,** ~ **artillery** Kernrohr *n* ~ **flange** Laufbuchsenflansche *m* ~ **locking screw** Laufbuchsenhalteschraube *f* ~ **pipe** Einsatzrohr *n*
**liners** Lagerbüchse *f*
**lines** Linienführung *f* ~ **for the bottom limit** Begrenzungslinien *pl* ~ **of communication** rückwärtige Verbindungen *pl* ~ **of dip** Fallinien *pl* ~ **of equal phase relations** Isophasen *pl* (rdo) ~ **of ferrite** Ferritstreifen *pl* ~ **of force** Kraftlinien *pl* ~ **of force on a model** Stromlinienaufzeichner *m* ~ **of intersection of two surfaces of second order** Schnittkurven *pl* zweier Flächen zweiter Ordnung ~ **of simultaneous high tides** Fluthöhenlinie *f* ~ **of steepest gradient** Fallinien *pl* ~ **of stress** Fließfiguren *pl*, Streckfigur *f*
**lines,** ~ **connecting collimation points** Bildmarkenverbindungsgeraden *pl* ~ **distance** Strichabstand *m* ~ **emitted by ionized atoms** Ionenatomlinien *pl* ~ **outlets** Anschlußmöglichkeiten *pl* ~ **per square cm** Gauß *n*
**linger, to** ~ verweilen
**lingerie factory** Wäschefabrik *f*
**lingering period** Verweilzeit *f*
**linguist** Sprachkundiger *m*
**lining** Ausfüllung *f*, Ausfütterung *f*, Auskleidung *f*, Auspolsterung *f*, Ausschlag *m*, Ausschlagen *n*, Ausstampfung *f*, Bekleidung *f*, Belag *m*, Einlage *f*, (rocket combustion chamber) Feuerhaut *f*, Futter *n*, (clothing) Futterstoff *m*, Fütterung *f* (print.), Garnierung *f*, Isolationsschicht *f*, Isolierung *f*, (book binding) Kapitalband *n*, Kaschierung *f*, Tübbings *pl*, Überzug *m*, Unterfutter *n*, Unterlage *f*, Verkleidung *f*, Zustellung *f*
**lining,** ~ **of the bearing** Lager-ausguß *m*, -futter *n* ~ **of the bed** (of the bottom) Befestigung *f* des Kanalbettes ~ **a borehole** Füttern *n* eines Bohrloches ~ **of canal bed** Kanalsohle *f* ~ **with flocks** Verflocken *n* ~ **with metal** Metallisierung *f* ~ **of a shaft** Schachtbekleidung *f* ~ **up for shooting pictures** Bildaufnahmevorbereitung *f*, Vorbereitung *f* für Bildaufnahmen ~ **with slabs** (or with tables) Plattenverblendung *f* ~ **and trimming** Zutat *f* ~ **of the slot** Auslegung *f* der Nut ~ **of a well** Brunnenbrüstung *f*, -einfassung *f*, -mantel *m*
**lining,** ~ **calender** Kaschierkalender *m* ~ **device** Kaschiereinrichtung *f* ~ **frame** Scherengerüst *n* ~ **mass** Zustellungsmasse *f* ~ **material** Futtermasse *f*, Fütterungsstoff *m* ~ **paper** Kaschierpapier *n*, Makulatur *f*, Überzugspapier *n* ~ **plate** Futterblech *n* ~ **rail** Randschiene *f* ~ **split** Futterspalt *m* ~ **tissues** Futterseidenpapier *n* ~ **tube** Futterrohr *n* ~ **wall** Verkleidungsmauer *f*
**link, to** ~ ketteln, ketten, verbinden, verketten
**to** ~ **up (with)** anschließen
**link** (connecting member of) Bindeglied *n*, Gelenk *n*, Glied *n*, Hebebügel *m*, Kettenglieder *pl*, Krampe *f*, Kulisse *f*, Lenkstange *f*, Masche *f*, Meßkettenglied *n*, (chain) Schake *f*, Schäkel *m*, Schlinge *f*, Schmelzeinsatz *m*, Verbindungsdraht *m*, -leitung *f*, -stück *n*, Zwischenstück *f*
**link,** ~ **baffle** Gelenkwand *f* ~ **belt** Glieder-kette *f*, -riemen *m* ~**-belt chain drive** Zahnkettentrieb

*m* ~ **block** Gleitklotz *m* ~ **box** Gelenkmuffe *f* ~ **bracket** Kulissen-, Schwingen-lager *n* ~ **bush** Schakenbuchse *f* ~ **chain** Gelenkkette *f*, Schäkel *n* ~ **circuit** Zwischenkreis *m* ~ **column** Pendelsäule *f* ~**-grinding and copying machine** Kulissenschleif- und Kopiermaschine *f* ~ **hook** Gelenkglückshaken *m* (min.) ~ **joint** Gabelgelenk *n* ~**-loading machine** Gurtfüller *m* ~ **motion** (Kurvenzeichen) Gelenksystem *n*, Kulissensteuerung *f* ~ **motion part** Steuerungsteil *m* ~ **pin** Anlenkbolzen *m*, (chain) Kettenbolzen *m*, Nebenpleuel-kopf *m*, -zapfen *m* ~ **plate** Verbindungsplatte *f* ~ **polygon** Seilpolygon *n* ~ **quadrangle** Gelenkviereck *n* ~ **quadrilateral bellcrank throttle control** Gelenkviereck *n* ~ **radius** Anlenkradius *m* ~ **reversing motion** umstellbare Kulissenbewegung *f* ~ **rod** Anlenkpleuel *n*, Gelenkstange *f*, Neben-pleuel *n*, -stange *f* ~**-rod assembly** Pleuelstern *m* ~ **rod end** Nebenpleuelzapfenkopf *m* ~**-rod pin** Gelenkschaftbolzen *m* ~ **support** Pendelstütze *f* ~ **tongs** Kettenzange *f* ~ **trainer** Drehscheiben (flug)lehrapparat *m* ~ **transmitter** Zwischensender *m* ~ **width** (of a chain) Gliederbreite *f* ~ **work** Gelenksystem *n*, Gliederwerk *n*
**linkage** Bindung *f*, Flußverbindung *f*, Gestänge *n*, Gliederwerk *n*, Kopplung *f* (electr.), (magnetic) Windungsfluß *m* ~ **for tracing a plane** Ebenführung *f* ~ **for tracing a straight line** Geradführung *f*
**linkage,** ~ **force** Bindungskraft *f* ~ **formula** Wertigkeitsformel *f* ~ **hole** Gestängeloch *n* ~ **lever** Betätigungshebel *m*
**linkages** Gelenkmechanismen *pl*
**linked** gelenkig, verbunden, verkettet
**linker left hand turning tool** Drehstahl *m*
**linking** Scharung *f*, Verkettung *f* ~ **bar** Kettelschiene *f* ~ **labor to suit structure** Arbeitsgliederung *f* nach Eignungsstruktur ~ **machine** Kettelmaschine *f* ~ **station** Zwischenstelle *f*
**linnaeite** Linneit *m*
**Linnet hole** Fuchs *m*
**linoleate** Leinölsäure *f*
**linoleic acid** Linolsäure *f*
**linoleum** Linoleum *n* ~**-engraving** Linolschnitt *m* ~**-engraving tool** Linolschnittgerät *n*
**linotype** Linotype *f*
**linseed** Leinsaat *f* ~ **mucilage** Leinsamenschleim *m* ~ **oil** Baum-, Lein-öl *n* ~ **oil for varnish** Lackleinöl *n* ~ **oil size** Leinölschlichte *f* ~**-oil vehicle** Leinölbindemittel *n*
**lint doctor** Gegenrakel *f* einer Walzendruckmaschine
**lintel** Fenster-sturz *m*, -träger *m*, -überlage *f*, Riegel *m* ~ **beam over door opening** Türsturz *m* ~ **ring** Tragkranz *m*
**lip** Ausflußschnauze *f*, Ausguß *m*, Lippe *f*, Nase *f*, Ösenblatt *n*, Schnauze *f*, Schneppe *f*, Tülle *f*, (baffle) Überlaufkante *f* ~ **of a casting ladle** Ausflußmündung *f*, Ausgußöffnung *f* (einer Gießpfanne)
**lip,** ~ **angle** Keil-, Meißel-, Schleif-, Spitzen-, (of cutter) Zuschärf-winkel *m* ~**-clearance angle** Schnittkantenwinkel *m* ~ **drilling** Kantenbohrung *f* ~ **nozzle** Zapfendüse *f* mit Ablenkfläche ~ **pour ladle** Gießpfanne *f* mit Ausgußschneuze oder mit Schnauzenausguß ~**-poured ladle** Kipp-Pfanne *f* ~ **pouring** Schnabel-

gießen *n* ~ **rest** Lippenstütze *f* ~ **surface** obere Schneidfläche *f*
**lipolysis** Fettspaltung *f*
**lipolytic** fettspaltend
**Lipowitz alloy** Lipowitzmetall *n*
**liquate, to** ~ abseigern, absintern, ausseigern, (copper) darren, entmischen, seigern **to** ~ **again (or subsequently)** nachseigern
**liquated copper** Darrkupfer *n*
**liquating** Seigern *n* ~ **furnace** Seigerofen *m* ~ **kettle** Seigerkessel *m*
**liquation** Abschmelzung *f*, Ausseigerung *f*, Entmischung *f*, Saigerung *f*, Seigerung *f* ~ **apparatus** Seigerapparat *m* ~ **cake** Seigerstück *n* ~ **hearth** Darre *f*, Darrofen *m*, Seigerherd *m* ~ **lead** Seigerblei *n* ~ **residue** Seigerungsrückstand *m* ~ **works** Seiger-hütte *f*, -werk *n*
**liquefacient** Verflüssigungsmittel *n*
**liquefaction** Verflüssigung *f* ~ **failure** Rutschung *f* durch Verflüssigung
**liquefied natural gas** Flüssigerdgas *n*
**liquefier** Verflüssigungsapparat *m*
**liquefy, to** ~ auflösen, flüssig machen, flüssig werden, schmelzen, verflüssigen, zerlassen
**liquefying** Verflüssigung *f* ~ **plant** Verflüssigungsanlage *f*
**liquescent** flüssigwerdend
**liquid** Flüssigkeit *f*; fließend, flüssig, tropfbar ~ **of crystallization** Kristallflüssigkeit *f* ~ **to be filtered** Unfiltrat *n* ~ **at high temperature (or to melting temperature)** feuerflüssig ~ **at rest** ruhende Flüssigkeit *f*
**liquid,** ~ **accumulator** Flüssigkeitskraftspeicher *m* ~ **addition** flüssiger Zusatz *m* ~ **air** flüssige Luft *f* ~ **ammonia** Salmiakgeist *m* ~**-and gas-spraying apparatus on a plane** Aus- und Abgießgerät *n* ~ **asphalt** flüssiges Pech *n* ~ **barometer** Flüssigkeitsbarometer *n* ~ **bath** Flüssigkeitsbad *n* ~ **binder** Naßbinder *m* ~ **body** Fließkörper *m* ~ **bottled motor fuel** Flüssiggas *n* ~ **cell** Elektrolytzelle *f* ~ **consonant** Halbvokal *m* ~ **contraction** Lunkern *n*, Lunkerung *f* ~ **coolant** flüssiges Kühlmittel *n* ~**-cooled** flüssigkeitsgekühlt ~ **cooler** Flüssigkeitskühler *m* ~ **cooling** Flüssigkeits-kühler *m*, -kühlung *f* ~ **Flüssigkühlung *f* ~ **core** flüssiger Erdkern *m* ~ **cycle** Flüssigkeitskreislauf *m* ~ **cyclones** Hydrozyklone *pl* ~ **damping** Flüssigkeitsdämpfung *f* ~ **density meter** Flüssigkeitsdichtemesser *m* ~ **drop model** Tröpfchenmodell *n* ~ **extinguisher** Flüssigkeitsfeuerlöscher *m* ~ **filled** Füllgut *n* ~ **filtrate** flüssiger Ablauf *m* (Filter) ~**-flow counter** Flüssigkeitsdurchflußzählrohr *n* ~**-flow limit tests** Fließgrenzenversuche *pl* ~**-flow measuring instrument** Flüssigkeitsdurchflußmesser *m* ~ **friction** Flüssigkeitsreibung *f* ~ **fuel** flüssiger Brennstoff *m* ~**-fuel nozzle** Flüssigkeitszerstäuber *m* ~ **fuse** Flüssigkeitszünder *m* ~ **glucose** Stärkesirup *m* ~ **head** Flüssigkeitshöhe *f* ~ **incendiary bomb** Flüssigkeitsbrandbombe *f* ~**-inglass thermometer** Flüssigkeitsthermometer *n* ~ **jet** Flüssigkeitsstrahl *m* ~**-jet microphone** Flüssigkeitsstrahlmikrofon *n* ~**-jet transmitter** Flüssigkeitsstrahlmikrofon *n* ~ **layer** Flüssigkeitsschicht *f*
**liquid-level** Flüssigkeitsspiegel *m*, Libelle *f* ~ **controller** Flüssigkeitsstandregler *m* ~ **displacement float** Verdrängungskörper *m* ~

**gauge** Flüssigkeitsstandmesser *m* ~ **gauge glass** Flüssigkeitsmeßglas *n* ~ **indicator** Flüssigkeitsstandanzeiger *m* ~ **measuring instrument** Flüssigkeitsstandmesser *m* ~ **recorder** Flüssigkeitsstandschreiber *m* ~ **regulator** Flüssigkeitsstandregler *m*
**liquid,** ~ **leveling switch** Flüssigkeitsvorratsmesser *m*, Libellenschalter *m* ~ **limit** Fließgrenze *f* ~ **manometer** Druckmesser *m*, Flüssigkeitsdruckmesser *m* ~ **manure** Jauche *f* ~ **measure** Flüssigkeitsmaß *n* ~ **metal coolant** Kühlmittel *n* auf verflüssigtem Metall ~ **metal vapor** Flüssigmetalldampf *m* ~ **meter** Flüssigkeitsmesser *m*, Hohlmaß *n* ~**-oxygen explosive** Sprengluft *f* ~ **phase** Flüssigkeitsphase *f* ~ **pitch** Harzpech *n* ~**-pressure gauge** Flüssigkeitsdruckmesser *m* ~ **pressure relief** Flüssigkeitsentlastung *f* ~ **prism** Flüssigkeitsprisma *n* ~**-propellant rocket** Flüssigkeitsrakete *f* ~ **propeller anti-icing** Blattbenetzung *f* ~ **quantity measuring instruments** Flüssigkeitsvorratsmesser *m* ~ **radiator** Flüssigkeitskühler *m* ~ **reflection** Flüssigkeitsbrechung *f* ~ **remnant** Flüssigkeitsrest *m* ~ **resin** Tallöl *n* ~ **resistance** Flüssigkeitswiderstand *m* ~**-reversing and** ~**starting resistance** Flüssigkeitsumkehranlaßwiderstand *m* ~ **seal** Sperrflüssigkeit *f* ~ **seal pump** Wasserringpumpe *f* ~**-seal trap** hydraulischer Verschluß *m* ~ **shrinkage** flüssige Schwindung *f* ~ **shut-off (or stop) valve** Flüssigkeitsabsperrventil *n* ~ **siphon** Flüssigkeitsheber *m* ~ **slide valve** Flüssigkeitsschieber *m* ~ **spraying** Spritzflüssigkeit *f* ~ **spring** Ölfederstrebe *f* ~ **squirting-out** Übersprudeln *n* ~ **starter** Flüssigkeitsanlasser *m* ~ **starting resistance** Flüssigkeitsanlaßwiderstand *m* ~ **sugar** flüssige Raffinade *f* ~**-type compass** Flüssigkeitskompaß *m*
**liquidate, to** ~ liquidieren
**liquidation** Abwicklung *f*, Auflösung *f*
**liquidensitometer** Flüssigkeitsdichtemesser *m*
**liquidity index** Fließ-index *m*, -zahl *f*
**liquidogenic** flüssigkeitsbildend
**liquidometer** flüssiger Brennstoffmesser *m*, Liquidometer *n*
**liquidor foaming** Überschäumen *n*
**liquor** Deckel *m*, (dye-works) Flotte *f*, Flüssigkeit *f*, Lauge *f* ~ **for swelling hides** Schwellbeize *f*
**liquor,** ~ **circulation** Laugenüberführung *f* ~ **discharge** abgeschleuderte Lauge *f* ~ **emitted in discharging** Abgasbrüden *n* ~ **gauge** Aerometer *n*
**liroconite** Linsenerz *n*, Lirokonit *m*
**Lisle thread** Florgarn *n*
**Lissajous,** ~ **curves** Lissajousfiguren *pl* ~ **figures** Lissajousfiguren *pl*, Lissajous'sche Schwingungsfiguren *pl*
**list, to** ~ aufführen, aufzählen, krängen, registrieren, überliegen, verzeichnen **to** ~ **a ship** einem Schiff Schlagseite *f* geben
**list** Aufstellung *f*, Liste *f*, Nachweis *m*, Nachweisung *f*, Neigung *f*, Plättchen *n*, Rand *m*, Rolle *f*, Saum *m*, Schlagseite *f*, Tabelle *f*, Tafel *f*, Überhängen *n*, Verzeichnis *n*, Zusammenstellung *f*
**list,** ~ **of contents** Inhaltsverzeichnis *n* ~ **of descriptions** Namenverzeichnis *n* ~ **of goods**

Warenverzeichnis *n* ~ **of names** Namenverzeichnis *n* ~ **of queries** Fragebogen *m* ~ **of square parts** Ersatzteilliste *f*

**list,** ~ **control** Schreibsteuerung *f* ~ **pot** Abwerfpfanne *f* ~ **price** Listenpreis *m* ~ **printing** Listenschreiben *n* ~ **speed** Einzelgang *m*

**listen, to** ~ anhören, horchen, lauschen, **to** ~ **in** abhören, ablauschen, hineinhören, lauschen, mithören **to** ~ **to** abhorchen

**listener** Horcher *m*, Hörer *m*, Zuhörer *m* ~ **echo** Hörecho *n*

**listening** Abhörbarkeit *f*, Abhörverfahren *n*, Hören *n* ~ **attachment** Hörsatz *m* ~ **circuit** Abhörstromlauf *m* ~ **coil** Hörspule *f* ~ **connection** Höranschluß *m* ~ **device** Abhorchvorrichtung *f* ~ **equipment** Lauschgerät *n* ~ **gear** Abhörapparat *m*, Horchgerät *n* ~**-in** Aufschalten *n* (teleph.) ~**-in box** Abfragekasten *m* ~**-in equipment for inquiry purposes** Anfrageapparat *m* (teleph.) ~ **jack** Mithörklinke *f* ~ **key** Hörtaste *f*, Mithör-schalter *m*, -schlüssel *m*, -taste *f* ~ **position** Hörstellung *f*, Mithörstellung *f* ~ **post** Abhörstelle *f*, Höranlage *f*, Horch-posten *m*, -stelle *f* ~ **relay** Mithörrelais *n* ~ **room** (submarine) Horchraum *m* ~ **service** Horch-, Lausch-dienst *m* ~**-set for Morse practice** Höranlage *f* ~ **station** Abhörstation *f* ~ **trench** Horchsappe *f* ~ **trumpet** Horchtrichter *m* ~ **watch** Empfangsbereitschaft *f*, Hörbereitschaft *f*, Horchwache *f*, Hörwacht *f*

**lister** Listendruckvorrichtung *f* ~ **attachment** Listerapparat *m* ~ **bottom** Pflugkörper *m* für Reihenhäufler ~ **cultivator** Dammkulturhackmaschine *f* ~ **planter** Dammkulturpflug *m*

**listing** Krängung *f* ~ **calculator** selbstschreibende Rechenmaschine *f*

**liter** Liter *n*

**literal** wörtlich

**literature references** Schriftmaterial *n*

**litharge** Abstrich *m*, (gelbe) Bleiglätte *f*, (gelbes) Bleioxyd *n*, Glätte *f* ~ **channel** Glättgasse *f*

**lithia** Lithiumoxyd *n* ~ **mica** Lithion-, Lithiumglimmer *m*

**lithium** Lithium *n* ~ **citrate** Lithiumzitrat *n* ~ **hydroxide** Lithionhydrat *n* ~ **loaded** lithiumgetränkt ~ **nitride** Stickstofflithium *n* ~ **oxide** Lithion *n* ~ **schist** Lepidolith *m*

**lithochromic print** Ölbilderdruck *m*

**lithochromy** Farbendruck *m*

**lithograph, to** ~ lithografieren

**lithograph** Lithografie *f*, Steindruck *m*

**lithographer** Lithograf *m*

**lithographic,** ~ **(al)** lithografisch ~ **press** Steindruckschnellpresse *f* ~ **print** Steinabdruck *m* ~ **printing** Steindruckerei *f* ~ **printing roller** Steindruckwalze *f* ~ **sheet** Lithografenblech *n* ~ **stone dresser** Steinschleifer *m*

**lithography** Steindruck *m*, Steindruckerei *f*

**lithologic unit** Gesteinseinheit *f*

**lithology** Gesteinskunde *f*, Lithologie *f*

**lithomarge** Steinmark *n*, Wundererde *f* ~ **containing iron** Eisensteinmark *n*

**lithopaper** Steindruckpapier *n*

**lithoplate graining machine** Offsetplattenkörnmaschine *f*

**lithopone** Lithopone *f*, Schwefelzinkweiß *n*

**litho-printing** Flachdruck *m*

**lithosphere** Lithosphäre *f*

**lithotomy** Steinschnitt *m*

**litigation** Rechtsstreit *m*

**litmus** Lackmus *m*, Lakmus *m* ~ **lichen** Lackmusflechte *f* ~ **paper** Lackmuspapier *n* ~ **solution** Lackmusflechtentinktur *f*

**litter** Krankentrage *f*, Papierabfall *m*, Streu *f*, Tragbahre *f*

**little** klein, wenig ~ **end bearing** Kolbenzapfenlager *n* ~ **knob** Knöllchen *n* ~ **knot in silk yarn** Garnknötchen *n* ~ **loops in weaving** Schleifchen *pl* im Gewebe ~ **marking pole** Piket *n* ~ **node** Knöllchen *n* ~ **pocket** Täschchen *n* ~ **seam** hammer Abbindhammer *m* ~ **ship** Schiffchen *n* ~ **stick** Hölzchen *n* ~ **thumb** Däumling *m*

**littleness** Geringfügigkeit *f*, Kleinheit *f*

**littoral,** ~ **current** Uferströmung *f* ~ **deposit** Küstenablagerung *f*

**litz,** ~ **coil** Litzenspule *f* ~ **wire** Litzendraht *m*

**live, to** ~ leben, geladen, spannungsführend, stromführend, wohnen ~ **axle** drehende Achse *f*, direkte Antriebsachse *f*, Differentialachse *f*, Triebachse *f* ~ **broadcast** Originalsendung *f* ~ **center** umlaufende Aufnahmespitze *f*, laufende oder mitlaufende Spitze *f* ~ **end** Reflexionswand *f* ~ **hole** Gang *m* mit hochwertigen Erzen ~ **load** Arbeits-, Betriebs-belastung *f*, bewegliche, stoßweise oder wechselnde Belastung *f*, bewegliche Last *f*, Nutzlast *f*, Verkehrslast *f* ~ **parts** Spannung führende Teile *pl* ~ **rail** Stromschiene *f* ~ **roll** Arbeitswalze *f*

**live-roller** getriebene Rolle *f* ~ **bed** Rutschbahn *f* ~ **feed bed** Auflaufrollgang *m* ~ **(-type) feeding table** Zufuhrrollgang *m* ~**-loading bed** Verladerollgang *m*

**live,** ~ **shell** scharfgeladene Granate *f*, Vollgeschoß *n* ~ **skid** Plateaukarre *f* ~**-spindle driving pulley** Mitnehmerspindelantriebsscheibe *f* ~**-spindle locking pin** Mitnehmerspindelfeststellung *f*

**live-steam** direkter oder offener Dampf *m*, Frischdampf *m*, Kesseldampf *m* ~ **inlet** Frischdampfeinlaß *m* ~ **line** Frischdampfleitung *f* ~ **nozzle** Frischdampfdüse *f* ~ **piping** Frischdampfleitung *f* ~ **valve** Frischdampfventil *n*

**live,** ~ **studio** nachhallreiches Studio *n* ~ **switchboard** Klinkenumschalter *m* ~ **test** scharfe Prüfung *f* ~ **transmission** Originalsendung *f* ~ **wire** geladener, spannungsführender oder stromführender Draht *m*, unter elektrischer Spannung stehender Draht *m*, stromführender Leiter *m*, stromdurchflossene Leitung *f*

**liveliness** Lebendigkeit *f*, Lebhaftigkeit *f*

**lively** lebendig, lebhaft ~ **shade** frischer Farbton *m*

**liveness** Halligkeit *f*, Lebendigkeit *f*, Raumhalleffekt *m* (acoust.)

**liver, to** ~ (paint) eindicken

**liver stone** Hepatit *m*

**livered** kurz, stockig

**livering** (paint) Eindickung *f*

**livid** blaugrau

**living** Auskommen *n*, Leben *n*; lebend, lebendig ~ **room** Wohnzimmer *n* ~ **space** Wirtschaftsraum *m*, Wohnraum *m*

**lixiviant** Laugungsmittel *n*

**lixiviate, to** ~ auslaugen, extrahieren

**lixiviated** ausgelaugt ~ **tan** Gerberlohe *f*

**lixiviation process** Laugeverfahren *n*

Ljungström, ~ turbine Gegenlaufturbine *f* ~ type turbine Ljungström-Turbine *f*
Llewellyn formula Llewellynformel *f*
L network L-Glied *n*
load, to ~ aufladen, beanspruchen, beladen, belasten, bepacken, beschicken, beschweren, chargieren, durchladen, einladen, einlegen, füllen, laden, speisen, verladen to ~ a cable bespulen, gleichförmig belasten to ~ continuously krarupisieren to ~ (a line) continuously krarupieren to ~ the dark slides Platten in Kassetten einlegen to ~ inductively induktiv belasten to ~ up beladen to ~ the valve das Ventil belasten
load Beanspruchung *f*, Belastbarkeit *f*, Belastung *f*, Bepackung *f*, (of men or machinery) Betriebsdienstleistung *f*, Bürde *f*, Druck *m*, Fuhre *f*, Kraft *f*, Lade *f*, Ladung *f*, Last *f*, Leistung *f*, Schüttlast *f*, Tragkraft *f*, Traglast *f*, Verbraucher *m* no ~ (in specifications) unbelastet ~ (ing) Fracht *f*
load, ~ of ammunition Munitionsausstattung *f* ~ on axle Achsenbelastung *f* ~ acting at the bottom von unten belastet ~ on bracket Konsolbelastung *f* ~ on cantilever Konsolbelastung *f*, Kragarmlast *f* ~ of the cathode Ladung *f* der Kathode ~ to be dropped by the magnet Fallkugel *f* ~ per horsepower Last *f* pro Pferdestärke ~ acting on the inside Innenbelastung *f* ~ at nodal point Knotenpunktbelastung *f* ~ in pounds per square foot of wing area Flächenanziehungsbelastung *f* ~ on soil Bodendruck *m* ~ on spring Federbelastung *f* ~ acting at the top von oben belastet ~ per unit area of (propeller) blade surface Blattflächenbelastung *f* ~ per unit of length Längeneinheitslast *f*
load, ~ angle Polradwinkel *m* ~ application Lastspiel *n* ~ area Belastungs-, Last-fläche *f* ~ axis Lastachse *f*
load-bearing, ~ assembly Tragwerk *n* ~ capacity Tragfähigkeit *f*, Widerstandsfähigkeit *f* ~ corrugated plate Trägerwellblech *n* ~ member tragender Bauteil *m* ~ slab Flächentragwerk *n*
load, ~ brake Lastbremse *f* ~ cable Zugseil *n* ~ capacity Belastungs-fähigkeit *f*, -möglichkeit *f*, Ladegewicht *n*, Rollbahnbelastungswert *m* (aviat.), Tragfähigkeit *f* ~ (ing) capacity Ladefähigkeit *f* ~-carrying equipment Förderanlage *f* ~ cell Kraftmeßdose *f* (electr.) ~ -center unit substation Transformator-Schwerpunktsstation *f* ~ chain Lastkette *f* ~-chain guide Lastkettenführung *f* ~-chain sheave Lastkettenrolle *f* ~ change Belastungswechsel *m* ~ changes Belastungsgänge *pl* ~ characteristic curve Belastungskennlinie *f* ~ charge Tracht *f* ~ circuit Entnahme-, Verbraucher-kreis *m* ~-circuit efficiency Belastungskreisausbeute *f* ~ coefficient identical with the cross-line distances Belastungsglieder *pl* ~ coil Pupinspule *f* ~ compensation Belastungsausgleich *m* ~ condition Belastungszustand *m* ~ conditions of network Belastungsverhältnisse *pl* des Netzes ~ configuration Außenlastenanordnung *f* ~-controlled feed(ing) quantity lastabhängig regelbare Fördermenge *f* ~ counting Leistungszählung *f* ~ current Ruhestrom *m* ~ curve Belastungskurve *f* ~ cycle Lastspielzahl *f*,

Lastwechsel *m* ~ cycle counter Lastspielzähler *m* ~ cycles Belastungsgänge *f* ~ decrease Lastabwurf *m* ~-deflection diagram Durchbiegungsdiagramm *n* ~-deformation diagram in compression Spannungsdruckdiagramm *n* ~ diagram Belastungsdiagramm *n* ~ dispatcher Lastverteiler *m* ~ dispatcher center Lastverteilerstelle *f* ~ dispatching Lastverteilung *f* ~ dispatching plant Lastverteileranlage *f* (electr.) ~ distributed over a length Streckenlast *f* ~-distributing station Lastverteilungsstelle *f* ~ distribution Belastungs-, Ladungs-, Last-verteilung *f* ~-distribution switch Mischwähler *m* ~ disturbance Laststörung *f* ~ divider Lastverteiler *m* ~ draft Tiefgang *m* des beladenen Schiffes ~ driver Einschreibtreiber *m* ~ duration Belastungsdauer *f* ~ equalization Belastungsausgleich *m* ~-extension diagram Spannungsdehnungs-, Zerreiß-diagramm *n* ~ face Ladefläche *f* ~ factor Belastungs-faktor *m*, -grad *m*, -koeffizient *m*, -zahl *f*, -ziffer *f*, Lastfaktor *m*, -vielfaches *n* ~ fluctuation Belastungsschwankung *f* ~ gauge shipping profile Ladeprofil *n* ~ grade Belastungsstufe *f* ~-graph Belastungskurve ~ hook Lasthaken *m* ~ impedance Endimpedanz *f* ~ increase Spannungserhöhung *f* ~ increment Laststufe *f*, stufenweiser Zuwachs *m* (stumpf angefügt) ~-independent d-c eingeprägter Gleichstrom *m* ~ index Belastungsverhältnis *n* ~ indicator Aussteuerungskontrollgerät *n*, Belastungsanzeiger *m* ~ indicator monitor Ladeanzeigeleuchte *f* ~ isolator Lastanpassungsglied *n*, Lastschalter *m* ~ level Lastamplitude *f* ~ life Lebensdauer *f* bei voller Belastung ~ lifting member Lastaufnahmemittel *m* ~ limit Belastungsgrenze *f* ~ limitations Beladungs-beschränkungen *pl*, -grenzen *pl* ~ line Arbeitlinie *f*, Lademarke *f*, Tiefladelinie *f*, Widerstandsgerade *f* ~ peak Belastungsspitze *f* ~ plate Lastplatte *f* ~ platform Lade-bock *m*, -gestell *n* ~ pressure Lastdruck *m* ~ range Belastungsbereich *m*, Laststufe *f* ~ rating Belastungs-zahl *f*, -ziffer *f* ~ ratio control Lastregelung *f* (electr.) ~ regulating spannungsregelnd ~ regulation Leistungsregelung *f* ~ repetitions Belastungsgänge *pl* ~ resistance Außen-, Belastungs-, Verbraucher-widerstand *m* ~ resistance of a tube Abschlußwiderstand *m* einer Röhre ~ resistor Belastungswiderstand *m* ~ reversal Lastwechsel *m* ~ ring Sammelöse *f* ~ rope Lastseil *n* ~ selector switch Belastungswahlschalter *m* ~-settlement curve Lastsenkungslinie *f* ~-settlement diagram Lastsenkungsdiagramm *n* ~ sheave Lastweilscheibe *f* ~ slab Lastplatte *f* ~ spacing Spulenentfernung *f* ~ spectra Belastungsfolgen *pl* ~ station Beladepunkt *m* ~-stone Gangerz *n*, Magnet-eisenerz *n*, -eisenstein *m*, natürlicher Magnet *m* ~ surge Belastungsstoß *m* ~ tension Belastungsspannung *f* ~ terms (of an equation) Belastungsglieder *pl* ~ test Belastungs-, Last-probe *f* ~ test up to breaking Belastungsprobe *f* bis zum Bruch *m* ~ triangle Belastungsdreieck *n* ~ unit Ladeeinheit *f* ~ unit surface Flächenanziehungsbelastung *f* ~ value Belastungswert *m* ~ variation Belastungsschwankung *f* ~ voltage Belastungsspannung *f* ~ water line Ladewasserlinie *f* ~ work Fundgrube *f*

**loaded** beladen, belastet, geladen **when** ~ im Lastfall *m* ~ **with blanks** blindgeladen ~ **in shear** auf Scherung *f* beansprucht

**loaded,** ~ **antenna** belastete Antenne *f* ~ **cable** belastetes Kabel *n*, Kabel *n* mit erhöhter Induktivität ~ **cable** (continously) Krarupkabel *n* ~ **circuit** belastete Leitung *f*, Leitung *f* mit erhöhter Induktivität ~ **grinding wheel** verschmierte oder zugesetzte Schleifscheibe *f* ~ **long-distance telephone cable** pupinisiertes Fernsprechkabel *n* ~ **mixings** gefüllte Mischungen *pl* ~ **Q** Lastgüte *f* ~ **stock** gefüllte Mischung *f* ~ **transmission line** bespulte Leitung *f* ~ **valve** Belastungsventil *n* ~ **weight** Dienst-, Fluggewicht *n*

**loader** Einfüller *m*, Lader *m*, Verlader *m*, Verladevorrichtung *f* ~**'s seat** Ladesitz *m*

**loading** Aufladen *n*, Auskolkung *f*, Beanspruchung *f*, Beladen *n*, Belasten *n*, Belastung *f* (aerodyn.), Beschickung *f*, Beschwerung *f*, Einladen *n*, Energiebedarf *m*, Füllrumpf *m*, (of firearms) Ladehemmung *f*, Laden *n*, Ladung *f*, Last *f*, (coil) Pupinisierung *f*, (operation) Verladen *n*, Verladung *f*, Verpackung *f*

**loading,** ~ **of cable by series coils** Tageslichtfüllung *f* ~ **of the frame** Rahmenlast *f* ~ **of the paper** Papierbeschwerung *f* ~ **of ships** Beladung *f* der Schiffe ~ **per square meter** Belastung *f* je Quadratmeter ~ **per unit area** Flächeneinheitsbelastung *f*

**loading,** ~ **action** Ladevorgang *m* ~ **agents** Zusätze *pl* ~ **agents on fillers** Appreturkörper *m* (paper) ~**-and unloading wharf** Beladestelle *f* ~ **aperture** Ladeloch *n* ~ **arrangement** Verladeeinrichtung *f* ~ **bay** Beladerampe *f* ~ **belt** Verladeband *n* ~ **berth** Schiffsliegeplatz *m* ~ **bin** Füllrumpf *m* ~ **bomb rack** Anhängevorrichtung *f* ~ **bridge** Ladebrücke *f* ~ **(or handling) bridge** Ladesteg *m*, Verladebrücke *f* ~ **bunker** Verladetasche *f* ~ **capacity** Belastbarkeit *f*, Belastungsvermögen *n*, Ladekapizität *f* (electr.), Ladevermögen *n* ~ **capacity of the magazine** Fassungsvermögen *n* der Kassette ~ **case** Belastungsfall *m* ~ **chamber** Füllraum *m* ~ **chute** Verlade-rinne *f*, -rutsche *f* ~ **chutes of an orehandling bridge crane** Beladeschurren *pl*

**loading-coil** abgestimmte Spule *f*, Antennenverlängerungsspule *f*, Belastungsspule *f*, Pupinspule *f* (teleph.) ~ **case** Spulenkasten *m* ~ **pot** Pupinspulenkasten *m* ~ **section** Ladungsabschnitt *m* ~ **spacing** Spulenabstand *m* ~ **unit** Spulensatz *m*

**loading,** ~ **condenser** Sammelkondensator *m* ~ **conditions** Belastungszustände *pl*, Lastfälle *pl* ~ **crane** Batterie-, Lade-, Verlade-kran *m* ~ **curve** Belastungskurve *f* ~ **data** Ladedaten *pl* ~ **detail** Aufladekommando *n*, Ladetrupp *m* ~ **device** Belast-, Lade-vorrichtung *f*, Verladeeinrichtung *f* ~ **diagram** Beladeplan *m* ~ **direction** Belastungsrichtung *f* ~ **disc** Belastungsscheibchen *n* ~ **equipment** Ladevorrichtung *f* ~ **factor** Belastungsmaß *n* (rdr), (explosives) Verspannungsfaktor *m* ~ **frame** (photoelasticity) Belastvorrichtung *f*, Ladegestell *n* ~ **frame with box** Ladegestell *n* mit Aufsatz ~ **funnel** Hosenrutsche *f* ~ **gauge** Begrenzung *f* der Ladung, Lademaß *n*, Meßrahmen *m* ~ **grab** Verladegreifer *m* ~ **hatch**

**Ladeluke** *f* ~ **hopper** Fülltrichter *m*, Sammeltasche *f*, Schüttrumpf *m* ~ **(charging) hopper** Ladetrichter *m* ~ **inductance** Belastungsspule *f* ~ **key** Ladekurbel *f* ~ **lag** Ladeverzug *m*, Ladeverzugszeit *f* ~ **lever** Ladehebel *m* ~ **line** Belastungs-, Lade-linie *f* (shipping), Lademarke *f* (shipping) ~ **machine** Lademaschine *f* ~ **manifest** Ladeliste *f* ~ **material** Beschwerungs-, Füll-mittel *n* ~**-mechanism cam** Winkelhebel *m* ~ **mechanism lever** Patronenhebel *m* ~ **movement** Ladegriff *m* ~ **operations** Ladetätigkeit *f* ~ **place** Halteplatz *m* ~ **plan** Spulenplan *m* ~ **plant** Verlade-anlage *f*, -vorrichtung *f* ~ **plate** (explosives) Ladelöffel *m* ~ **platform** Ausladebrücke *f*, Belade-, Lade-bühne *f*, Laderampe *f* ~ **pocket** Ladetasche *f* ~ **point** Belastungsgrenze *f*, Lade-platz *m*, -stelle *f* ~ **port** Ladungs-, Verlade-hafen *m* ~ **position** Ladestellung *f* ~ **pot for overhead line** Spulenkasten *m* für Freileitungen ~ **process** Ladevorgang *m* ~ **rack** Verladerampe *f* ~ **ramp** Ladebrücke *f*, -podest *m*, -rampe *f*, Verladerampe *f* ~ **ramp of a ferry ship** Anfahrtrampe *f* ~ **recess** Ladeloch *n* ~ **resistance** Ballast-, Belastungs-widerstand *m* ~ **resistor** Belastungswiderstand *m* ~ **safe** (fuse) ladesicher ~ **scheme** Bespulungsplan *m* ~ **section** Belastungs-, Spulen-abschnitt *m*, Spulenfeld *n* ~ **sequence** Belastungsfolgen *pl* ~ **siding** Ladegleis *n* ~ **spindle** Ladespindel *f* ~ **station** Aufgabe-, Belade-, Lade-station *f* (electr.), Verladebahnhof *m* ~ **surface** Ladefläche *f* ~ **table** Beladeplan *m*, Lottabelle *f*, Verladeplan *m* ~ **tax** Einschiffungsgebühr *f* ~ **tension roller** Belastungsrolle *f* ~ **test** Belastungsprüfung *f*, Probebelastung *f* ~ **time** Beladezeit *f*, Lade-verzug *m*, -verzugszeit *f*, -zeit *f* ~ **track** Ladegleis *n* ~ **tray** Lade-mulde *f*, -schale *f*, -schwenger *m*, -tisch *m* ~ **trough** Einflußrinne *f*, Lademulde *f* ~ **tube** Lade-büchse *f*, -rohr *n* ~ **(coil) unit** Spulensatz *m*, Verladungseinheit *f* ~ **weight** Belastungs-, Last-gewicht *n* ~ **wharf** Ladepodest *m* ~ **winch** Ladewinde *f*

**loam** Lehm *m*, Letten *m*, Tonmergel *m*, Ziegelton *m* ~ **beater** Lehmmesser *n* ~ **board** Musterbrett *n* ~ **brick** Lehmziegel *m* ~**-cast roll** Lehmgußwalze *f* ~ **casting** Lehmguß *m* ~ **core** Lehmkern *m* ~ **mold** Lehmform *f* ~ **molding** Lehmformen *n* ~**-molding shop** Lehmformerei *f* ~ **mortar** Lehmmörtel *m* ~ **pit** Lehmgrube *f* ~ **sand** Lehmsand *m* ~ **seal** Lehmdichtung *f*

**loamy** lehmig, fettig ~ **marl** Lehmmergel *m* ~ **sand** fetter Formsand *m*

**loan, to** ~ leihen **to** ~ **money on security** lombardieren

**loan** Anleihe *f* **as a** ~ leihweise ~ **business** Lombardgeschäft *n* ~**-capital** Anleihekapital *n*

**lob-type blower** Kapselgebläse *n*

**lobate** lappig

**lobe** (as of a cam) Erhöhung *f*, Flügel *m*, Keule *f*, Lappen *m*, Lobe *f*, Maul *n* des Steuersackes, Strahlungskeule *f* (electron), (on cam) Überhöhung *f*, Vorsprung *m* eines Nockens, Zipfel *m* (electron.)

**lobe,** ~ **frequency** Keulenumtastfrequenz *f* ~ **plate** Grundplatte *f* ~ **splitting** (antenna) Keulenaufgliederung *f* ~ **switching** Keulenumschaltung *f*, Zipfelumschaltung *f* ~**-switching**

device Umtastgerät *n* ~-**switching line section**
Umtastleitung *f* ~-**switching system** Phasenumtastgerät *n* ~ **width** Strahlungslappenbreite *f*
**lobed** lappig ~ **cross section** gelappter oder
lappiger Querschnitt *m*
**lobelike condenser** posaunenartige Leitung *f*
**lobing** Aufzipfelung *f* (antenna), Leitstrahldrehung *f* (rdr)
**lobster back (or bend)** Segmentbogen *m*
**local** begrenzt, eingeengt, nah, örtlich ~ **action**
Selbstentladung *f* ~-**and long-distance tele-**
**phone** Fernamt *n* ~ **announcing device** Orts-
ansagegerät *n* ~ **apparent time** Ortszeit *f* ~ **area**
Orts-bezirk *m*, -gebiet *n* ~ **attraction** (magnetic
needle) lokale Anziehung *f* ~ **automatic**
**exchange** selbsttätiges Ortsamt *n* ~ **average**
örtliches Mittel *n*
**local-battery (LB)** Ortsbatterie *f* (OB), Schrank-
batterie *f* ~ **telephone set** Fernsprechapparat *m*
mit Ortsbatterie, (magneto and buzzer calling)
Fernsprechapparat *m* mit Induktor- und
Summeranruf ~ **telephone set with battery**
**ringing** Fernsprechapparat *m* mit Batterieanruf
(Ortsbatterie) ~ **working** Ortsbatteriebetrieb *m*
**local,** ~ **boundary** Platzgrenze *f* ~ **buckling**
Beulen *n* ~ **buckling stress** Beulspannung *f* ~
**business** Platzgeschäft *n* ~ **busy** ortsbesetzt ~
**busy condition** Ortsbesetztsein *n* ~ **call** Anruf *m*
im Ortsverkehr, Fernanruf *m*, Ortsanruf *m*,
-gespräch *n*, -ruf *m*, -verbindung *f* ~ **carrier**
örtlich überlagerte Trägerfrequenz *f* ~ **cell**
Lokalelement *n* ~ **central office** Ortsamt *n* ~
**channel** Gemeinschaftswelle *f* ~ **circuit** Orts-
stromkreis *m* ~-**circuit line** Nebenanschlußleitung *f* ~ **conditions** örtliche Gegebenheiten
*pl* ~ **contact bank** c-Kontaktsatz ~ **control**
Bedienung *f* am Gerät, örtliche Betätigung *f*,
Nah-, Orts-bedienung *f* ~ **control zone** Nahver-
kehrsbezirk *m* ~ **conversation** Ortsverbindung *f*
~ **correction** Zeichenkorrektur *f* am Empfänger ~ **corrosion** örtliche Korrosion *f* ~
**current** Ortsstrom *m* ~ **custom** Ortsgebrauch *m*
~ **declination** Ortsmißweisung *f* ~ **diurnal**
**variation** örtlicher Tagesgang *m* ~ **earthquake**
Ortsbeben *n* ~ **electrical disturbances** Lokal-
störungen *pl* ~ **exchange** Orts-amt *n*, -ver-
mittlungsstelle *f* ~ **flight** örtlicher Flug *m* ~
**frequency** Hilfsfrequenz *f* ~ **gravity** örtliche
Schwerkraft *f* ~ **group switch** Ortsgruppenum-
schalter *m* ~ **hardening** Brennerhärtungsver-
fahren *n* ~ **hour angle** örtlicher Stundenwinkel
~ **indicator** Ortsanzeigegerät *n* ~ **inker** Farbschreiber *m* mit vorgeschaltetem Relais ~
**inquiry** Rückfrage *f* ~ **inspection** Ortsbesichti-
gung *f* ~ **interference** Lokalstörung *f* ~ **inter-**
**office trunking** Betrieb *m* über Ortsverbindungsleitungen (teleph.) ~ **level** örtliche
Tangentialebene *f* ~ **line** Orts-leitung *f*, -linie *f*
~ **line receiving allowance** zulässige Dämpfung
*f* der Teilnehmerleitung auf der Empfangsseite
~ **line sending allowance** zulässige Dämpfung *f*
der Teilnehmerleitung auf der Sendeseite ~
**link** Innenverbindungsleitung *f* (P.A.B.X.) ~
**load-distributing station** Lokallastverteilungs-
stelle *f* ~ **loop** Teilnehmer-leitung *f*, -schleife *f*
~ **magnetic attraction** örtliche magnetische
Anziehung *f* ~ **mains** Ortsnetz *n* ~ **mean time**
mittlere Ortszeit *f* ~ **message** Ortsgespräch *n*

~ **observation** Nahbeobachtung *f* ~ **office code**
**number** Ortskennzahl *f* (teleph.) ~ **operation**
Ortsbesprechung *f* ~ **operator** Ortsbeamtin *f* ~
**oscillation frequency** Überlagerungsfrequenz *f*
~ **oscillator** Empfangsüberlagerer *m*, Orts-
generator *m*, örtlicher Überlagerer *m* ~ **oscilla-**
**tor unit** Schweigesenderüberlagerer *m* ~ **paral-**
**lax on the horizon** Waagerechtparallaxe *f* ~
**party release relay** Trennrelais (TR) *n* ~ **pitting**
örtliche Korrosion *f* ~ **plant** Ortsnetz *n* ~
**position** Ortsplatz *m* ~ **railway** Lokalbahn *f* ~
**rate zone in large cities** Ortsverkehrsbereich *m*
~ **reception** Ortsempfang *m* ~ **record** Mitlese-
text *m* ~ **relay** Ortsrelais *n* ~-**remote control**
Nahsteuerung *f* ~ **repair shop** Ortsausbesse-
rungsanstalt *f* ~ **resistance** örtlicher Widerstand *m* ~ **ring** Ortsring *m* ~ **sawtooth** Steuer-
sägezahn *m* ~ **send circuit** Sendeortskreis *m* ~
**sender** Ortssender *m* ~ **sensitivity** örtliche
Empfindlichkeit *f* ~ **service** Nahverkehr *m* ~
**service area** Ortsverkehrsbereich *m* ~ **sidereal**
**time** Ortssternzeit *f* ~ **smoke screening** Kleinvernebelung *f* ~ **station** Nebenstelle *f* ~-**station**
**reception** Ortsempfang *m* ~ **subscriber** Orts-
teilnehmer *m* ~ **supervision** Überwachung *f* am
Ursprungsort ~ **surcharge** Ortszuschlag *m* ~
**swelling of water** Stauhöhe *f* ~ **synchronization**
lokal geregeltes Synchronisieren *n* ~ **synchro-**
**nizing** lokal geregelter Gleichklang *m* ~
**system network** Ortsnetzbau *m* ~ **telephone**
**exchange** Fernsprechamt *n* ~ **telephone net-**
**work** Ortsnetz *m* ~ **telephone traffic** Ortsfern-
sprechverkehr *m* ~ **tidal diagram** Flutkurve *f* ~
**time** Ortszeit *f* ~ **traffic** Nah-, Orts-verkehr *m*
~ **train** Personenzug *m* ~ **transmitted syn-**
**chronization control** örtliche, übertragene
Gleichlaufregelung *f* ~ **usage** Platzgebrauch *m*
~ **variation** Ortsgang *m* ~ **vertical** Ortsvertikale
*f* ~ **weather forecast** lokale Wetterprognose *f*
**locality** Lage *f*, Örtlichkeit *f* ~ **allowance** Orts-
zulage *f*
**localization** Begrenzung *f*, Eingrenzen *n*, Eingrenzung *f*, Festlegung *f*, Lagebestimmung *f*,
Ortsbestimmung *f* ~ **of a fault** Fehlerortsbe-
stimmung *f*, Störungseingrenzung *f* ~ **of heat**
Wärmestau *m* ~ **of trouble** Eingrenzen *n* von
Fehlern, Eingrenzung *f* von Fehlern
**localization,** ~ **test** Eingrenzungs-, Fehlerortsmessung *f* ~ **theorem** Lokalisierungssatz *m*
**localize, to** ~ eingrenzen, (a fault) feststellen,
lokalisieren, auf einen bestimmten Ort zurückführen
**localized,** ~ **corrosion** örtliche Korrosion *f* ~
**selective corrosion** Lochfraß *m* ~ **stress**
Spitzenbeanspruchung *f*
**localizer (transmitter)** Landekurssender *m* (ILS)
**localizing trouble** Fehlereingrenzung *f*
**locate, to** ~ auffinden, aufsuchen, eingrenzen,
(by direction finder) einpeilen, einstellen, festlegen, Fehler feststellen, lokalisieren, orten
**to** ~ **a claim** Mutung *f* einlegen **to** ~ **faults**
Fehler *pl* eingrenzen **to** ~ **by guard rings**
mittels Sicherungsringen *pl* sichern **to** ~ **by**
**intersections** anschneiden **to** ~ (a point) **by**
**surveying** einmessen **to** ~ **a working part** ein
Werkstück *n* in die Vorrichtung aufnehmen
**locating,** ~ **a bore** Bohrpunktbestimmung *f* ~
**bolt bearing** Aufhängerbolzenlager *n* ~ **cross**

Hilfskreuz n ~ **device** Ortungsgerät n ~ **fence** Auflageschiene f ~ **hole** Halteloch n ~ **key** Führungsnase f ~ **mark** Strichmarke f ~ **nose** Führungsnase f (aviat.) ~ **pin** Heftbolzen m ~ **screw** Heftschraube f ~ **stop** Anschlaglẹiste f
**location** Anbringung f in einer bestimmten Lage, Aufnahme f eines Werkstückes in einer Vorrichtung , Aufnahme-gelände n (film) -stelle f (film), Aufstellpunkt m, Eingrenzen n, Eingrenzung f, Feststellung f, Lage f, Lageort m, Linienführung f (r.r.), Lokalisierung f, Ortung f, Platz m, Speicher-stelle f, -platz m (comput.), Stelle f, Trassierung f
**location,** ~ **of the center of gravity** Schwerpunktslage f ~ **of claims** Muten n ~ **of defect (or error) in casting** Fehlstelle f ~ **of a fault** Fehlort m (rdr) ~ **of instruction** Befehlsadresse f ~ **of spar** Holmlage f
**location,** ~ **finding** Standort-bestimmung f, -peilung f ~ **hole** Aufnahmeloch n ~ **indicator** Ortsnamenkennung f (aviat.) ~ **light** Ansteuerungsfunkfeuer n ~ **method** Eingrenzungsverfahren n ~ **order** Befehlsadresse f (info proc.) ~ **truck (or unit)** Tonwagen m
**locator beacon** Platzfunkfeuer n (aviat.)
**lock, to** ~ abschließen, absperren, anhalten, arretieren, bremsen, eingreifen, einklinken, einrasten, festhalten, festlegen, feststellen, fixieren, hemmen, schließen, spannen, sperren, verklemmen, verriegeln, versperren, (up) verschließen, zuschließen **to** ~ **up capital** Kapital n festlegen **to** ~ **in** einschließen **to** ~ **on** anhängen, (sich) aufschalten (rdr), (target) festhalten **to** ~ **out** aussperren **to** ~ **in position** verriegeln (Objektiv) **to** ~ **wire** mit Draht m sichern
**lock** Bremsvorrichtung f, Feststellung f, Gewehrschloß n, Hemmschuh m, Ineinandergreifen n, Riegel m, Schleuse f, Schleusenkammer f, Schließvorrichtung f, Schloß n, Sicherung f, Sperrer m, Stauung f, Stockung f, Verschluß m, Verstopfung f **(control)** ~ Schiffsschleusung f, Sperrschloß n ~ **(ing)** Sperre f ~ **with cheek gates** Drempelschleuse f ~ **with circular chamber** Kesselschleuse f, Trommelschleuse f ~ **with double opposed gates** doppelte Kammerschleuse f ~ **with falling latch** Fallschloß n ~ **with high lift** Schachtschleuse f ~ **for long tows (or trains) of barges** Schleppzugschleuse f ~ **to prevent table from being moved** Sicherung f für den Stillstand des Tisches ~ **for safety gear** Reißklinke f ~ **in which water is reused** Sparschleuse f
**lock,** ~-**and-block system** Blockung f ~ **axis line** Schleusenachse f ~ **bar** Sperrschiene f ~ **bolt** (of gun) Riegel m, Sperriegel m ~ **box** (post office) Fach n, Schloßkasten m ~ **case** Schloßkasten m ~ **catch** Verschlußfanghebel m ~ **center line** Schleusenachse f ~ **chamber** Schleusenkammer f ~ **cylinder** Schließzylinder m ~ **field** Verriegelungsfeldwicklung f ~ **filling** Schleusenfüllung f ~ **floor** Schleusen-boden m, -sohle f ~ **frame** Verschlußschlitten m ~ **gate** Schleusentor n ~ **hasp** Schloßzuhaltung f ~ **hatch** Stellfalle f ~ **hook** Schließhaken m ~ **housing** Schloßgehäuse n ~-**in amplifier** Synchrondetektor m ~-**in detector** Phasendetektor m ~-**in range** Einspring-, Mitnahme-bereich m ~ **keeper** Schleusenwärter m ~ **knob** Einrast-,

Verschluß-knopf m ~ **lift** Schleusen-einsatz m, -fall m ~ **(ing) magnet** Sperrmagnet m ~ **nut** Band-, Gegen-mutter f, gesicherte Mutter f, Klemm-, Konter-mutter f, Schloß-schraube f, selbstsichernde Mutter f, Sicherheits-, Stell-, Verschluß-mutter f ~-**on** Anhängen n, Aufschaltung f (rdr), Zielfesthalten n ~-**on voltage** Anhängespannung f, Aufschaltspannung f (rdr)
**lockout** Aussparung f, Aussperrung f ~ **of workers by the employer** Aussperrung f der Arbeiter durch den Unternehmer ~ **relay** Sperrelais n (SP) ~ **valve** Doppelschloßventil n
**lock,** ~ **part** Schloßteil n ~ **pawl** Sperrklinke f ~ **pin** Spannstift m ~ **plate** Schloßblech n ~ **ring** (of magnifier) Verschlußring f ~ **rod** Zurrstange f ~ **screw** Lochschraube f ~-**seam sockets for cathodes** Falzen pl aus Blech, gefalzte Hülsen pl ~ **shaft** Schloßkurbel f ~ **smith** Schlosser m ~ **smith's shop** Wagenschlosserei f ~-**spar support** Bocksprengwerk n ~ **spigot** Verriegelungszapfen m ~ **spring** Riegel-, Verschluß-feder f ~ **staple** Verriegelungseinrichtung f ~-**stitch** Steppstich m ~ **support** Schloßhalter m, Zurrlager n ~-**up** Verschließen n, Verschluß m ~-**up bar** Schließbalken m ~-**up the round plates** Einspannen n der Rundplatten ~ **washer** Federring m, Gegen-dichtung f, -mutter f, Mutterunterlegscheibe f, Verschlußscheibe f, (internally toothed) Zahnscheibe f ~ **wire** Drahtschluß m
**lockable** (in position) feststellbar, schließbar ~ **joining member** arretierbarer Stoßriegel m
**lockage** Durchschleusen n, Schleusensystem n
**locked** blockiert, eingerückt (mach.), fest, festgehalten, geschlossen, unlösbar
**lock,** ~ **armor** geschlossene Armierung f oder Bewehrung f ~ **basin** Binnenhafen m, Innenbecken n ~ **cable (or rope)** vollschlächtiges Seil n ~ **controls** blockierte Steuerung f ~ **groove** Auslaufrille f (phono) ~ **keyboard** Tastenfeld n mit Sperre **in a** ~ **manner** zwangsläufig ~ **nut** gesicherte Mutter f ~ **oscillator** Mitnahmeoszillator m ~-**out** ausgesperrt ~ **position** Zurrstellung f ~-**rotor torque** (motor) Anzugsmoment n ~ **synchronism (or synchronization)** Synchronisierzwang m ~ **teeth** gesperrte Verzahnung f ~-**up** verschlossen ~-**up stress** innere Verspannung f ~-**wire rope** Drahtverschlußkabel n
**locker** Back f, Kammer f, Kleiderschrank m, Schließfach n, Schrank m, Werkzeugspind n
**lockering rollers** Flachshechelmaschine f
**locking** Blockierung f, Festlegemittel n, Verblockung f, Verriegelung f ~ **by center-punching** Sicherung f durch Körnerschlag ~ **in (or of) circuit** Mitnahme f ~ **of the form** Formenschließen n ~ **in position** Lagensicherung f
**locking,** ~ **apparatus** Verschluß m ~ **ball** Sperrkugel f ~ **bar** Riegel m, Sperrstange f ~ **battery** Haltebatterie f ~ **bolt** (torpedo) Arretier-, Halte-, Riegel-, Sicherungs-, Verschluß-bolzen m ~ **bracket** Feststellbock m ~ **cam** Einzahnrad n (film), Schließnocken m ~ **cap** Verschlußkappe f ~ **circuit** Haltekreis m (Relais) ~ **clamp** Verschuß-klammer f, -klaue f ~ **claw** Arretiergreifer m (film) ~ **cog** Festhaltezahn m ~ **collar** Lagerring m ~ **cone** Verschlußkegel m

**~ contact** Anschlußkontakt *m* **~ device** Abstell-, Festhalte-, Feststell-vorrichtung *f*, Feststellung *f* **(inter) ~ device** Sperrvorrichtung *f* **~ disk** Arretierungsscheibe *f* **~ drum** Sperrtrommel *f* **~ frame** Stellwerk *n*, Zurrbrücke *f* **~ furniture** Schließbeschlag *m* **~ gear** Festsetzvorrichtung *f*, Gesperre *n* **~ handle** Griffklemme *f*, (traveling lock) Zurrgriff *m* **~ hook** Verschluß-, Zurr-haken *m* **~ key** Gegenschlüssel *m*, Schalter *m* mit festen Stellungen **~ lever** Sperr-, Zurr-hebel *m* **~ lever shaft** Sperrhebelwelle *f* **~ load** Verriegelungskraft *f* **~ lug** Verriegelungsnase *f* **~ magnet** Verschlußmagnet *m* **~ means** Rastenarretierung *f* **~ mechanism** Gesperre *n*, Sperrgetriebe *n*, Sperrwerk *n*, Verriegelung *f*, Verschlußmechanismus *m* **~ notch** Sperrnute *f* **~ nut** Knebelgriff *m*, Kreuzlochschraube *f* **~ pawl** Feststellklinke *f*, Sperrdaumen *m* **~ piece** Verschlußstück *n* **~ pin** Feststellstöpsel *m*, Riegelbolzen *m*, Stützsäule *f*, Verschlußbolzen *m*, Vorsteckstift *m*, Zurrbolzen *m* **~-pin hole** Splintloch *n* **~ plate** Federscheibe *f*, Sicherungsblech *n* **~ plug** Verschlußpfropfen *m* **~ port** Verriegelungsanschluß *m* **~ pressure** Formschließkraft *f* **~ pulse** Sperrimpuls *m*, -puls *m* **~ pushbutton key** Taste *f* mit fester Stellung **~ pushbutton with magnetic release** Taste *f* mit selbsttätiger Auslösung **~ ratchet** Sperrzahn *m* **~ relay** Halte-, Sperr-, Verriegler-relais *n* **~ ring** Klemm-, Spann-, Sperr-, Verbindungs-ring *m*, Verschraubungsstück *n* **~ ring of the handle bar** Lenkstangenklemmring *m* **~ screw** Abschließ-, (on gear) Fang-, (on gear) Grenz-, Verschluß-schraube *f* **~ segment** Sperrsegment *n* **~ sheet-iron disk** Blechverschluß *m* **~ signal** Sperrzeichen *n* **~ sleeve** Sicherungshülse *f*, Verschlußmuffe *f* **~-slide** Sicherungsschieber *m* **~ socket** Feststellbuchse *f* **~ spring** Schließfeder *f* **~ strip** Sicherungsblech *n* **~ switch** Achsenschalter *m* **~ tab** Verriegelungszunge *f* **~ tongue** Sperrzunge *f* **~ tooth** Sperrzahn *m* **~-type button** Taste *f* mit Rastung (feste Taste) **~ washer** Sicherungsscheibe *f* **~ wheel** Arretierrad *n* **~ wire** Sicherungsdraht *m*

**locomotion** Ortsveränderung *f*

**locomotive** Lokomotive *f*; bewegungsfähig **~ with cable drum** Lokomotive *f* mit abrollbarem Leitungskabel **~ with hauling drum** Lokomotive *f* mit Seilwinde

**locomotive, ~ boiler** Lokomotivkessel *m* **~ crank axle** Lokomotivkurbelachse *f* **~ firebox** Lokomotiv-feuerkiste *f*, -stahlkessel *m* **~ furnace** Feuerbüchse *f* **~ heaver** Lokomotivhebebock *m* **~ jackscrew** Flaschenwinde *f* **~ running without cars** einzeln fahrende Lokomotive *f* **~ shifting crane** Lokomotivversatzkran *m* **~ traverser** Lokomotivscheibebühne *f* **~ works** Lokomotivwerk *n*

**locular** fächerig

**locus** geometrischer Ort *m*, Ort *m* **~ diagram** Ortsdiagramm *n*

**locust tree** Robinie *f*

**lode** Ader *f*, Bergader *f*, Berg-, Erz-gang *m*, Gang *m*, Trum *n* **the ~ changes its course** der Gang ändert das Streichen **one ~ is faulted by the other** ein Gang *m* verwirft einen anderen **the ~ grows** der Gang wird mächtiger **~ of medium**

**dip** flachfallender Gang *m*, flachfallendes Land *n* (geol.) **~ of steep dip** tonnlägiger Gang *m* (geol.)

**lode, ~ forming a network** Netzgang *m* **~ miner** Ganghauer *m* **~ reef** Erzgang *m* **~ seam** Erzader *f* **~ star** Polarstern *m* **~ stone** Erzgang *m* **~ wall** Gangwand *f*

**lodge** Hütte *f*, Loge *f*, Portierloge *f*

**lodgement** Sitz *m*

**lodger** Hausbewohner *m*

**lodging** Obdach *n* **~ knee** Winkelknie *n*

**loess** Briz *m*, Löß *m*

**loft** Dachboden *m*, Schnürboden *m*, Speicher *m* **~ board** Strakbrett *n* **~ door** Bodentür *f* **~ floor** großer Zeichentisch *m*

**loftiness** Hoheit *f*

**lofting, ~ pattern (or template)** Zuschneideschablone *f*

**loftsman** Aufhänger *m*, Trockner *m* (paper mfg.)

**lofty** himmelanstrebend, hoch

**log, to ~** eichen, loggen **to ~ a station** feststellen

**log** Balken *m*, (railway track, pulley) Block *m*, Bohr-bericht *m*, -profil *n*, -rapport *m*, -tagesbericht *m*, Buch *n*, Holzklotz *m*, Klotz *m*, Log *n*, Logbucheintragung *f*, Scheit *m* & *n* **~ of pressure altitude** Logarithmus *m* der Druckhöhe **~ of total pressure** Logarithmus *m* des Gesamtdruckes

**logbook** Bautage-, Bord-, Eintrage-, Flug-, Funktage-, Log-buch *n*, (of motor) Lebenslaufakte *f* (des Motors)

**log, ~ bunker** Holzbunker *m* **~ calipers** Gabelmaße *pl* **~ coordinates** doppellogarithmische Koordinaten *pl* **~ crib** Balkenstapel *m* **~ frame saw** Gattersäge *f* **~ glass** Logglas *n* **~ line** Logleine *f* **~ raft** Balkenfloß *n* **~ roller** Aufbanker *m* **~ rules** Forstgabelmaße *pl* **~ runner** Baumstammkufe *f* **~ saw with quick-grip feed saddle gear** (band saw) Dielenbandsäge *f* mit schnellgreifendem Sattelgetriebe **~ sheet** Versuchsprotokoll *n* **~ transmitter** Loggeber *m* **~ wall** Blockwand *f*

**logarithm** Logarithmus *m* **~ to the base ten** Zehnerlogarithmus *m* **~ action** Logarithmierung *f*

**logarithmic, (al)** logarithmisch **~ condenser** Mittellonienkondensator *m* **~ curve** logarithmische Kurve *f* **~ decreement** Schwingungsdekrement *n* **~ horn** Exponentialtrichter *m* **~ (cross-section) paper** Logarithmenpapier *n* **~ scale** logarithmischer Maßstab *m* **~ spiral** logarithmische Spirale *f*

**logatom** Logatom *n*, Silbe *f* **~ articulation** Silbenverständlichkeit *f*

**logger** Mitschreiber *m*

**logging** Eintragung *f*, Holz-fällung *f*, -transport *m* **~ the results of tests** Versuchaufschreibung *f*

**logic** Folgerichtigkeit *f*, Logik *f*; binär, boolesch, funktionell (data proc.) **~ of the computer** Rechnerlogik *f*

**logic, ~ circuit** Logikschaltung *f* **~ element** Verknüpfungsglied *n* (comput.) **~ variable** Schaltvariabel *f* (comput.)

**logical** folgerichtig, logisch, sinngemäß, sinnvoll **~ circuit** logische Schaltung *f* **~ comparison** logischer Vergleich *m* **~ design** Rechnerlogik *f* **~ diagram** Funktionsplan *m* **~ element** Ver-

knüpfungsglied *n* ~ **shift** binäres Schieben *n* ~
~ **value** Wahrheitswert *m*
**logistics** Logistik *f*
**logotype matrix** Logotypenmatrize *f*
**logwood** Blau-, Block-, Blut-, Kampeche-holz *n*
~ **black** Blauholzschwarz *n*
**loiter** Warte-, Spar-flug *m*
**loktal**, ~-**base tube** Preßbodenröhre *f*, quetsch-
fußfreie Röhre *f* ~ **tube** fußlose Röhre *f*
**löllingite** Glanzarsenikkies *m*, Löllingit *n*
**lone electron** Einzelelektronensystem *n*
**long** lang, länglich **for a** ~ **time** langfristig **of** ~
**wave length** langwellig
**long**, ~-**and luminous-flaming semi-bituminous**
**nut coal of high rank** Flammnußkohle *f* ~-**arc**
**lamp** Langbogenlampe *f* ~-**base method** Lang-
basisverfahren *n* ~-**base range** Langbasisent-
fernungsmesser *m* ~ **beam** Langholz *n* ~ **bend**
Schwenkung *f* ~ **bolt** Ankerbolzen *m* ~ **borer**
Abbohrer *m*, Langbohrer *m* ~-**burning torch**
Dauerbrandfackel *f* ~ **channel bar of chassis**
**frame** Längsträger *m* ~ **course** großer Federweg *m* ~ **cross** Mittelsteg *m* ~ **dash** Dauerstrich
*m* ~-**delay echo** Langzeitecho *n*
**long-distance** Fern-, Lang-, Weit-strecke *f* ~
**navigation** Langstreckennavigation *f* ~ **broad-**
**cast** Fernfunk *m* ~ **cable** Fernkabel *n* ~ **cable**
**system** Fernkabelnetz *n* ~ **call** Fern-anruf *m*,
-ruf *m*, Weitverkehrsverbindung *f* ~ **call signal**
Fernanrufzeichen *n* ~ **circuit** Fernamtsschaltung *f*, Fernleitung *f* ~ **communication and**
**travel** Fernverkehr *m* ~ **communications center**
Fernmeldebetriebszentrale *f* ~ **connection**
Weitverkehrsverbindung *f* ~ **connector** Fernleitungswähler *m* ~ **conversation** Fernverbindung *f* ~ **cruise** Dauerfahrt *f* ~-**delay fuse** Zünder *m* mit langer Verzögerung ~ **direct-wire**
**telephone call** Durchgangsfernspruch *m* ~
**driving** Fernverkehr *m* ~ **drop** Fernklappe *f* ~
**flight** Fernflug *m*, Langstreckenflug *m* ~ **gas-**
**supply plant** Ferngasversorgungsanlage *f* ~
**headlight** Fernlicht *n* (Scheinwerfer) ~ **heating**
**system** Fernheizungsanlage *f* ~ **indicator** Fern-
zeiger *m* ~ **jack lamp** Fernklinkenlampe *f* ~
**jack lines** Fernklinkenleitungen *pl* ~ **line** Fernleitung *f*, -linie *f* ~ **mains** Fernleitungsrohre *pl*
~ **operator** Fernamts-, Fern-beamtin *f* ~
**power station** Überlandzentrale *f* ~ **propagation**
**of radio wave** Fernausbreitung *f* ~ **radio station**
Großfunkstelle *f*, Radiogroßstation *f* ~ **receiv-**
**ing equipment** Weitverkehrsempfangsanlage *f* ~
**reception** Fernempfang *m* ~ **reconnaissance**
**plane** Fernaufklärer *m* ~ **reconnoitering** Fern-
aufklärung *f* ~ **record** (gliding) Entfernungs-
rekord *m*, Streckenrekord *m* ~ **selection** Fernwahl *f* ~ **signal station** Großfunkstelle *f* ~
**station** Großfunkstation *f*, Großstation *f* ~
**supply** Fernversorgung *f* ~ **supply work** Fern-
kraftwerk *n* ~ **switch** Fernnachwähler *m* ~
**switchboard** Fernamt *n*, Fernschrank *m* ~
**table** Ferntisch *m* ~ **telephone administration**
Durchgangsverwaltung *f* ~ **telephone cable**
Fernkabel *n* ~ **telephone call** Ferngespräch *n*
~ **telephone exchange** Durchgangsvermittlung *f*,
Überlandzentrale *f* ~ **telephone service** Fern-
sprechweitverkehr *m* ~ **traffic** Fern-, Weit-
verkehr *m* ~ **trial** Dauer-fahrt *f*, -probe *f* ~
**type printer** Ferndrucker *m*

**long**, ~-**duration test** Dauerversuch *m* ~ **echo**
Langzeitecho *n* ~-**eye auger** Stangenbohrer ~
(or wide) **face pinion** Zahnradwelle *f* ~-**fibered**
**pulp** Langschliff *m* ~ **filter press** langgebaute
Filterpresse *f* ~-**flame coal** langflammige Kohle
*f* ~-**flame gas coal** Gasflammkohle *f* ~-**flaming**
langflammig ~-**flaming coal** langflammige
Kohle *f* ~-**flaming run-of-the-mine gas coal**
Gasflammförderkohle *f* ~ **flashes** Blitzfeuer *n*
~-**focus lens** (or **objective**) langbrennweitige
Linse *f* ~-**focus telescope** langbrennweitiges
Fernrohr *n* ~ **fuse cord** Zeitzündschnur *f* ~
**grinding** Längsschleifen *n* ~-**handed tool** Fernbedienungsgerät *n* ~ **handle shovel** Lochschaufel *f* ~-**haul toll circuit** Weitverkehrs-
leitung *f* ~-**hole cutter** Langfräser *m*, Langlochfräser *m* ~ **jumper** Abbohrer *m* (min.),
Endbohrer *m* ~ **key** Untertaste *f* ~ **knurled nut**
hohe Rändelmutter *f* ~ **lay** große Schlaglänge
*f* (cable) ~ **lease** Erbpacht *f* ~ **length** langer
Abschnitt *m* ~ **letters** geschwänzte Schrift *f* ~
**life** große Standzeit *f* ~-**life mold** Dauerform *f*
~-**life pentode** Weitverkehrspentode *f* ~-**lines**
**company** Fernleitungsgesellschaft *f* ~ **link**
Langzugdraht *m* ~-**link chain** langgliedrige
Kette *f* ~-**linked** langgliedrig ~-**lived** langlebig
~-**loop apparatus** Langmascher *m* ~ **luminous-**
**flaming semi-bituminous coal of high rank**
Flammkohle *f* ~ **lunge** Ausfall *m* ~ **measure**
Längenmaß *n* ~-**membered** langgliedrig ~-**neck-**
**ed** langhalsig ~ **needle** Spieß *m* ~-**nose pliers**
Röhrenzange *f* ~-**oil-type-varnish** fetter Öllack
*m* ~ **pane of a roof** Langseite *f* eines Daches
~-**period recording** Dauerregistrierung *f* ~-**peri-**
**od waves** langperiodische Wellen *pl* ~ **per-**
**sistence screen** Nachleuchtschirm *m* (CRT)
~-**pitch winding** Wicklung *f* mit verlängertem
Schritt ~ **plane** Füge-, Füg-hobel *m* ~-**playing**
**tape** Langspielband *n* ~ **primer** Korpus *f* ~
**pulse** Langimpuls *m* ~ **ramp** Längerampe *f*
**long-range** Langstrecken..., große Reichweite *f*;
weitreichend **at** ~ auf große Entfernung *f* ~
**communications tube** Weitverkehrsröhre *f* ~
**focus** Ferneinstellung *f* ~ **mail plane** Lang-
streckenpostflugzeug *n* ~ **navigation** Langstreckennavigation *f* ~ **order** Fernordnung *f* ~
**ordering** Fernordnungsgrad *m* ~ **plane** Lang-
streckenflugzeug *n* ~ **radar** Weitbereichsradar
*n* ~ **radio beacon** (station) Fernfunkfeuer *n* ~
**service** Langstreckendienst *m* ~ **transmission**
Fernübertragung *f* ~ **transmitter** Fernsender *m*
**long**, ~ **saw** Spaltsäge *f* ~-**scale meter** Weitwinkelinstrument *n* ~-**service valve** Langleberöhre
*f* ~-**shore current** küstenparallele Strömung *f*
~-**shoreman** Hafenarbeiter *m* ~-**skeined** lang-
strähnig ~-**skirted piston** langschäftiger Kolben
*m* ~ **slot milling machine** Langnutenfräsma-
schine *f* ~ **spar** Langholz *n* ~ **splice** Langs-
spleißung *f* ~-**stapled** langschnürig ~ **strip** (or
**continuous**) **footing** Streifenfundament *n*
~-**stroke** langhubig ~ **taper** flacher Kegel *m*
**long-term** langfristig ~ **average value** langzei-
tiger Mittelwert *m* ~ **constancy** Langzeitkon-
stanz *f* ~ **run** Dauerexperiment *n*
**long**, ~-**thread milling** Langgewindefräsen *n* ~
**thrust** Ausfall *m* ~-**time alternating-strength**
**stress** Dauerwechselfestigkeit *f* ~ **time creep**
**test** Prüfung *f* der Dauerstandfestigkeit ~ **ton**

Langtonne *f* ~ **tube effect** Rohrlängeneffekt *m*
~**-turning automatic** Langdrehautomat *m*
~**-turning lathe** Langdrehbank *f* ~ **wall** Lang-
frontbau *m* ~**-wall working on small veins**
Krummhälzerarbeit *f* (min.)
**long-wave** lange Welle *f*, Langwelle *f*; langwellen..., langwellig ~ **band** Langwellenbereich
*m* ~ **coil** Langwellenspule *f* ~ **direction finder**
**(or finding station)** Langwellenpeiler *m* ~
**transmitter** Langwellensender *m* ~ **transmitting**
**set** Langwellensenderanlage *f*
**long,** ~ **wire antenna** Langdrahtantenne *f* ~
**wood** Langholz *n*
**longer** länger **no** ~ **in force** hinfällig
**longeron** Hauptträger *m*, Holm *m*, Längs-holm
*m*, -träger *m*, Rumpflängsholm *m* ~ **of fuselage**
Rumpfholm *m* ~ **support** Rumpflängsleiste *f*
**longevity** Lebensdauer *f* ~ **pay** Besoldungszulage
*f* nach dem Dienstalter
**longitude** geographische Länge *f* ~ **in ecliptic**
Länge *f* in der Ekliptik ~ **of fix** Bestecklänge *f*
~ **of vertex** Scheitellänge *f*
**longitude minute** Längenminute *f*
**longitudinal** Längs.., längs verlaufend, longitudinal ~**-advance clutch** Längs-vor-Kupp-
lung *f* ~ **arch** Längengurt *m* ~ **axis** Längsachse
*f* ~ **axis of the spar** Holmlängsachse *f* ~ **ball**
**bearing** Kugellängslager *n* ~ **bar** Längsbe-
wehrung *f* ~ **batten** Längsleiste *f* ~ **beam**
Längsbalken *m* ~ **bond (or bracing)** Längsver-
band *m* ~**-brake-clutch** Längs-Bremse-Kupp-
lung *f* ~ **bulkhead** Längsschott *n* ~ **carrier**
**cable** Längstragseil *n* ~ **conductivity** Längsleit-
fähigkeit *f* ~ **controls** Höhensteuerung *f* ~
**copying** Kopierdrehen *n* in Längsrichtung ~
**crack** Längsriß *m* ~ **cross bracing** Tiefenverspannung *f* ~ **cross section** Längsschnitt *m* ~
**culvert** Umlaufkanal *m* ~ **dead limit** Längen-
begrenzung *f* ~ **dihedral angle** Schränkungs-
winkel *m* ~ **dike** Längs-, Parallel-werk *n* ~
**direction** Längsrichtung *f* ~ **dispersion** Länge-,
Längen-streuung *f* ~ **displacement** Längsverschiebung *f* ~ **divergence** Längsdivergenz *f* ~
**drum** Längstrommel *f* (boiler) ~ **equalizer**
Längsausgleichhebel *m* ~ **error** Längenfehler
*m* ~ **feed** Langvorschub *m* ~ **(traverse) feed**
Längszug *m* ~ **feed rack** Langzugzahnstange *f*
~ **feed stop** Längsanschlag *m* ~ **field** Längsfeld
*n* ~ **force** Axialkraft *f*, Längs-kraft *f*, -schiffs-
kraft *f* ~ **forming (or copying) attachment**
Längskopiervorrichtung *f* ~ **fracture** Längen-
bruch *m* ~ **frame** Längsspant *n* ~ **friction**
**device** Längsreiber *m* ~ **girder** Längsträger *m*
~ **grinder** Längsreiber *m* ~ **groove** Längsnut *f*
~ **iron mass** Längsschiffeisenmasse *f* ~ **level**
Längslibelle *f* ~ **leveling** Längsnivellierung *f* ~
**loop** Längsschleife *f* ~ **lurch** (aircraft) Aufbäumen *n* ~ **magnetization** Längsmagnetisie-
rung *f* ~ **magnification** Tiefenvergrößerung *f* ~
**measuring attachment** Längenmeßeinrichtung *f*
~ **member of frame** Längsträger *m* des Rahmens
~ **metacenter** Längenmetazentrum *n* ~
**milling machine with copying (or profiling)**
**device** Langkopierfräsmaschine *f* ~ **moment**
Längsmoment *n* ~ **motion head** Längsbewe-
gungstrieb *m* ~ **movement** Längsverschiebung *f*
~ **oscillation** Längsschwingung *f* ~ **outrigger**
Längsträger *m* ~ **overlap** Längsüberdeckung *f*

~ **pitch** Längsteilung *f* ~**-ply grain** Längsfurnier
*n* ~ **pole** Längsschiffspol *m* ~ **power traverse of**
**the table** Längsselbstgang *m* des Tisches ~
**profiling apparatus** Plankurvenfräsapparat *m* ~
**push-pull rod** Längssteuerschubstelle *f* ~ **rapid**
**return** Längsseilrücklauf *m* ~ **rapid traverse**
Längsschnellverstellung *f* ~ **recess** Längsaus-
sparung *f* ~**-return** „Längs-Zurück" ~ **runner**
**of carriage body** Längsträger *m* der Karosserie
~ **scale** Längenmaßstab *m* ~ **seam** Längsnaht *f*
~ **section** Längen-durchschnitt *m*, -profil *n*,
-schnitt *m*, Längs-durchschnitt *m*, -schnitt *m* ~
**shear** Längsschubkraft *f* ~ **sheet cutter** Längsschneider *m* ~ **shrinkage** Längsschrumpfung *f*
~ **slope** Längsneigung *f* ~ **slot** Längsnut *f* ~
**spacing** Längsteilung *f* ~ **spring** Längsfeder *f* ~
**stability** Längsstabilität *f* ~ **stay** Linienanker *m*
~ **stick positioner** Längstrimmeinrichtung *f* ~
**stiffening** Längsversteifung *f* ~ **stop** Längsanschlag *m* ~ **streak (or stria)** Längsstreifen *m*
~ **stringer** Längs-achse *f*, -profil *n* ~ **stripe**
Längsstreifen *m* ~ **support** Längsversteifung *f*
~ **tie** Langschwelle *f* (r.r.) ~ **travel of table**
Tischlängsbewegung *f* ~ **trim** Höhentrimmung
*f*, Längslästigkeit *f*, Längstrimm *m* ~ **valve**
**assembly** Planventilblock *m* ~ **vibration** Längs-,
Longitudinal-schwingung *f* ~ **vibration of a**
**crystal** Kristallschwingung *f* ~ **vibrations**
Dehnungsschwingungen *pl* ~ **view** Längs-
ansicht *f*, -aufriß *m*, -schnitt *m* ~ **wave** longi-
tudinale Welle *f* ~ **weld** Längsschweißung *f* ~
**wiring** Längsverspannung *f*
**longitudinally** längs, der Länge nach ~ **ad-**
**justable** längsverstellbar ~ **slit** längsgeschlitzt ~
**stiffened** längssteifig
**loofah** Luffa *f*
**look, to** ~ blicken, schauen, sehen **to** ~ **after**
nachsehen **to** ~ **at** sich besehen **to** ~ **at closely**
in Augenschein nehmen **to** ~ **for** nachsehen,
suchen **to** ~ **hot** heiß aussehen **to** ~ **into** einsehen **to** ~ **up the logarithm of** logarithmieren
**to** ~ **over** durchsehen, nachsehen, überscheuen
**to** ~ **upward** aufblicken
**look** Aussehen *n*, Blick *m* ~ **box** Ablaufglocke *f*,
Schauglas *n* ~ **out!** Obacht!
**lookout** Ausguck *m*, Ausgucksmann *m*, Auslug
*m*, Beobachter *m*, Wart *m* ~ **on the bridge**
Brückenwache *f* **on the** ~ **for** auf der Suche
**lookout,** ~ **post** Ausguckposten *m*, Warnstelle *f*
~ **station** (on ship) Ausguckposten *m* ~ **tower**
Wartturm *m*
**loom** Webstuhl *m* ~ **of an oar** Ruderschaft *m*
**loom,** ~ **beam** Weberbaum *m* ~ **fixer** Webstuhlvorrichter *m* ~ **fixing** Webstuhlvorrichtung *f* ~
**state** stuhlroher Zustand *m* (Gewebe)
**looming** Fata Morgana *f*, Luftspiegelung *f* nach
oben
**loop, to** ~ kettein, zur Schleife schalten, (a line)
schleifen, umführen, umschlingen **to** ~ **two**
**circuits** zwei Leistungen zur Schleife schalten
**to** ~ **into the connection** einschleifen (Verbindung) **to** ~ **by hand** (textiles) anschlagen **to** ~ **in**
(an office) einschleifen **to** ~ **a line** eine Leitung
schleifen
**loop** Armgerüst *n*, Bauch *m*, Bügel *m*, Dackel *m*,
Deul *m*, Lagebügel *m*, Lastbügel *m*, (circuit)
Leitungsschleife *f*, Looping *n* (aviat.), Luppe *f*,
Öhr *n*, Öse *f*, Rahmen *m*, Rahmenantenne *f*

(mit einer Windung), Ringbahn *f*, Rohrleitung *f*, Sackbahn *f*, Schlaufe *f*, senkrechte Schleife *f*, Schleifen-berührung *f*, -flug *m*, Schlinge *f*, (in rolling) Schräge *f*, Zyklus *m* (info proc.) **on the** ~ in Schleifenschaltung *f*, Schwingungsschleife *f*, Seil *n*, Überschlag *m* in der Luft (aviat.), Wellenbauch *m*, Windung *f* ~ **of oscillation** Schwingungsbauch *m*
**loop,** ~**-and hook** Aufsetzband *n*, Hakenband *n*, Kegelband *n* ~**-and-trunk layout** Netzplan *m* ~ **antenna** Empfängerrahmen *m*, Lichtantennenbauch *m*, Rahmenantenne *f*, Schleifenantenne *f* ~ **antenna circuit tuning** Rahmenkreisabstimmung *f* ~**-bending test** Schlaufprobe *f* ~ **breadth** Schleifenbreite *f* ~ **channels** Umführungen *pl* ~ **(ed) circuit** Doppelleitung *f* ~ **connection** Schleifenschaltung *f* ~ **coupling** Schleifenkupplung *f* ~ **current** Schleifenstrom *m* ~ **decremeter** Schleifdämpfungsmesser *m* ~ **dialling** Schleifenwahl *f* ~ **drier (festoon drier)** Schleifentrockner *m* ~ **drive** Peilantrieb *m* ~ **formation** Ösenbildung *f* ~ **forming** Schlingenbildung *f* ~ **frame** Rahmenluftleiter *m* ~ **gain** Umlaufverstärkung *f* ~ **galvanometer** Schleifengalvanometer *n* ~ **glass dome** durchsichtige Rahmenantennenkuppel *f* ~ **hanger** Aufhängebügel *m* ~ **hole** Schartenschlitz *m*, Schießschlitz *m*, -scharte *f* ~ **hole cover** Schartenschild *n* ~ **line** Ringstrecke *f* ~ **line (or main)** Schleifenleitung *f* ~ **maker** Bortenwirker *m*, Posamentier *m* ~ **(resistance) measurement** Schleifenmessung *f* ~ **mill rolling** Umlaufführen *n* ~ **navigation** Funkeigenpeilung *f* ~ **oscillograph** Schleifenoszillograf *m* ~ **receiver** Rahmenempfänger *m* ~ **reception** Rahmenempfang *m* ~ **rectangularity** Rechteckformgüte *f* ~ **regeneration** Umlaufverstärkung *f* ~**-reversing method** Flimmerpeilung *f* ~ **resistance** Doppelleitungs-, (conductor) Schleifenwiderstand *m* ~ **ringing** Durchrufen *n* in Schleifenschaltung ~ **scavenging** Schleifenspülung *f* ~ **short** Schleifenschluß *m* ~ **setter** Schleifenbildner *m* ~ **strength** Knickbruch-, Knoten-festigkeit *f* ~ **test** Erddraht-, Erdfehler-schleifenmessung *f*, Schleifenverfahren *n* ~ **(-resistance) test** Schleifenmessung *f* ~ **thread guide** Schlaufenfadenführer *m* ~**-tip terminal** Kabelschuh *m* ~ **traction eye** Zugöse *f* ~ **troughs** Umführungen *pl* ~ **tunnel** Kehrtunnel *m* ~ **twist** Schlingenzwirn *m* ~**-type clamp** Rohrschelle *f* ~ **value** Schleifenwert *m* ~ **wheel** Maschenleger *m* ~ **wire** Schließungsdraht *m* ~**-wire antenna** Rahmen-, Schleifenantenne *f*
**looped** zur Schleife verbunden ~ **baling wire** Ösendrähte *pl* für Ballenumschnürung ~ **circuit** Doppelleitung *f* ~ **fabric** Schlingengewebe *n* ~ **pipe** Trompetenrohr *n*
**looping** Installation *f* von Parallelleitungen, Kettelmaschine *f*, Looping *n* (aviat.), Maschenbildung *f*, Schleifen-, Schlingen-bildung *f*
**looping,** ~ **angle** Umschlingungswinkel *m* ~**-in** Durchschleifen *n* ~ **mill** Wechselduo *n* ~**-mill train** Drahtstrecke *f* ~**-type rolling mill** Umsteckwalzwerk *n* ~ **wheel** Maschenrad *n*
**loose, to** ~ (current) abfallen, lockern, lösen **to be** ~ wackeln **to become** ~ loskommen
**loose** frei (chem.), locker, los, lose, schlaff, un-

gebunden, wacklig ~ **axle** Blindachse *f*, Blindwelle *f* ~ **bushing** Leerlaufbuchse *f* ~ **cable** lose verseiltes Kabel *n* ~ **cap** abnehmbare Kappe *f* ~ **change gear** Aufsteckwechselrad *n* ~ **coiler** Einrollmaschine *f* ~ **color** nicht farbechte Farbe *f* ~ **connection** Wackelkontakt *m* ~ **cord and plugs** lose Stöpselschnut *f* ~ **coupling** lose Kopplung *f* ~**-fill packing** Stopfpackung *f* ~ **fit** Grobpassung *f* ~ **flange** loser Bord *m* oder Flansch *m* ~ **framing** weite Umrahmung *f* ~**-ground lister planter** Listenpflanzer *m* für losen Boden ~**-ground lister plow** Dammkulturpflug *m* für losen Boden ~ **heel breasting machine** Absatzfrontbeschneidemaschine *f* ~ **hub** Nabenmittelstück *n* ~**-jointed post** Gelenkpfosten *m* ~ **leaf binder** Schnellhefter *m* ~ **ledger** Loseblatthauptbuch *n* ~ **liner** auswechselbares Futterrohr *n* ~ **material** Schüttgut *n* ~ **pattern** Naturmodell *n* ~ **pin** einsteckbarer Bolzen *m* ~ **pointer** Schleppzeiger *m* ~ **pulley** Leer-, Los-scheibe *f*, bewegliche Rolle *f* ~**-pulley drive** Losscheibenantrieb *m* ~ **pulp** Zellstofflocken *pl* ~ **road** grundlose Straße *f* ~ **sand** loser Sand *m* ~ **selfsealing liner** (brewing) Zylinderfutter *n* ~ **shaft** Blindwelle *f* ~ **soil** lockerer Boden *m* ~ **sticks** Reisig *n* ~ **weight** Schüttgewicht *n*
**loosely** locker, lose, ungenau ~ **braided cordage** lose geklöppeltes Seil *n* ~ **coupled** losegekoppelt ~ **woven goods** undichte Gewebe *pl*
**loosen, to** abbinden, abheften, ablösen, (screw, nut) abschrauben, lockern, lockerwerden, lösen, loslösen, nachlassen **to** ~ **the loops** (textiles) locker arbeiten **to** ~ **up the structure** das Gefüge auflockern **to** ~ **up** auflockern **to** ~ **the wedge** loskeilen **to** ~ **wool by arsenic** Wolle abgiften
**looseness** Lockerheit *f*, Schlaffheit *f*, Spannungslosigkeit *f*, Ungenauigkeit *f*
**loosening** Abbinden *n*, Auflockerung *f*, Auflösung *f*, Lockerung *f*, Lockerwerden *n*, Lösen *n*, Lösung *f* ~ **of tread** Loslösen *n* der Lauffläche
**lop, to** ~ (wood) zöpfen **to** ~ **off** (peaks or crests) abkappen, kappen, köpfen
**loper** Folger *m*
**lopped tie** gekappte Schwelle *f*
**lopper** Amplitudenabschneider *m*
**lopping off of amplitude crests ('peaks') by limiter means** Abschneiden *n* der Amplitudenspitzen
**lorry** Lastwagen *m* (UK) ~ **loading crane** Ladeschwinge *f* ~**-trailer combination** Lastzug *m*
**lose, to** ~ abgeben, einbüßen, (a chance) entfallen, verlieren **to** ~ **color** bleichen, verblassen **to** ~ **consciousness** Bewußtsein *n* verlieren **to** ~ **flying speed** sinken **to** ~ **height** Höhe *f* verlieren **to** ~ **qualification** die Befähigung einbüßen **to** ~ **selectivity** verflachen **to** ~ **shape** deformieren **to** ~ **sight of** aus dem Auge *n* verlieren **to** ~ **weight** abnehmen
**losing the horizon** Kimmfehler *m*
**loss** Abfall *m*, Abgang *m*, Ansatz *m*, Ausfall *m*, (of energy) Dämpfung *f*, Einbuße *f*, Nachteil *m*, Verbrauch *m*, Verlust *m* **by** ~ **of** unter Abgang von
**loss,** ~ **by absorption of radiant heat** Einstrahlungsverlust *m* ~ **of accuracy** Genauigkeitsver-

lust *m* (info proc.) **~ of altitude** Höhenverlust *m*
**~ in bends** Krümmverlust *m* **~ of charge** La-
deverlust *m* **~ of charge method** Entladungs-
methode *f* **~ by circulation** Spülverlust *m* **~ of
civil rights** Verlust *m* der bürgerlichen Ehren-
rechte **~ of coal** Kohlenverlust *m* **~ of contrast**
(in picture) Aufhellung *f* schwarzer Stellen **~
of control** Selbständigmachen *n*, Verlust *m* der
Steuerbarkeit (aircraft) **~ in cooling water**
Kühlwasserverlust *m* **~ by cupellation** Kapellen-
zug *m* **~ due to cutting (or tool) action** Schnitt-
verlust *m* **~ through diffusion** Diffusionsverlust
*m* **~ around the edges** Randabfall *m* **~ in
efficiency** Leistungsverlust *m* **~ of energy** Abgang *m*, Energie-verbrauch *m*, -verlust *m*,
Sprachverlust *m* **~ by evaporation** Verdun-
stungsverlust *m* **~ by grinding** Schleifverlust *m*
**~ of head** Druckhöhenverlust *m*, Verlust *m* an
Geschwindigkeitshöhe **~ of heat** Wärme-abführung *f*, -abgabe *f*, -entwicklung *f* **~ on
ignition** Glühverlust *m* **~ in intelligibility** Ver-
ständlichkeitsverlust *m* **~ of interest** Zinsver-
lust *m* **~ at a junction** Spiegelungsverlust *m*,
Übergangsverlust *m* **~ due to leakage** Dichtig-
keitsverlust(beiwert) *m* **~ of light** Lichtverlust
*m* **~ of manganese** Manganabbrand *m* **~ of
official position** Amtsverlust *m* **~ in output**
Förderausfall *m* **~ of pay** Besoldungseinbuße *f*
**~ by percolation** Versickerungsverlust *m* **~ by
percussions** Stoßverlust *m* **~ in power** Stromverlust *m* **~ of power** Energieverlust *m*, Kraft-
verlust *m*, Leistungsabfall *m* **~ of pressure**
Widerstandshöhe *f* **~ by radiation** Strahlungsverlust *m* **~ of radiation of heat** Wärmestrah-
lungsverlust *m* **~ of rank** Rangverlust *m* **~ by
roasting** Röstverlust *m* **~ from roasting** Röstverlust *m* **~ of selectivity by parallel interal
resistance** Pseudodämpfung *f* **~ of silicon**
Siliziumabbrand *m* **~ of speed** Geschwindig-
keitsabfall *m* **~ in suppressed band (or range)**
Sperrdämpfung *f* **~ of synchronism** Gleichlauf-
verlust *m* **~ of thrust** Schubverlust *m* **~ of time**
Stockung *f*, Zeitverlust *m* **~ in the transmission
range** Lockdämpfung *f* **~ of velocity** Geschwindigkeitsverlust *m* **~ in volume** Volumenverlust
*m* **~ of volume** Mengenverlust *m* **~ in weight**
Gewichtsverlust *m* **~ of weight on (or by)
roasting** Röstabgang *m* **~ of yield** Ertragsausfall
*m*
**loss, ~ allocation** Dämpfungsaufteilung *f* **~
angle** Dämpfungswinkel *m*, (of a condenser)
Verlustwinkel *m* **~ current to earth** Erdschlußstrom *m* **~ damping** Verlustdämpfung *f* **~
factor** Betriebsdämpfung *f* **~-free** verlustfrei **~
record on packages** Gebindeverlustübersicht *f*
**~ resistance** Dämpfungs-, Verlust-widerstand
*m* **~-summation method** Einzelverlustverfahren
*n*
**lost** verloren **~ call** Fehlanruf *m* **~ container**
Einweggebinde *n* **~ head** verlorenes Gefälle *n*,
verlorener Kopf *m* **~ motion** toter Gang *m*,
Hubverlust *m*, Leergang *m*, totes Spiel *n* **~
motion of a screw** toter Gang *m* einer Schraube
**~ time** Zeitverlust *m* **~ work** Arbeitsverlust *m*
**lot** größere Anzahl *f*, Förderabgabe *f*, Grundstück *n*, Lieferung *f*, Los *n*, Parzelle *f*, Posten *n*
**lots** Menge *f* **in ~** (surv.) parzellarisch
**loud** (color) auffallend, lärmend, laut, stark **~**

**blast horn** Starktonhorn *n* **~ color** schreiende
Farbe *f* **~ hailer** Megafon *n*, Richtlautsprecher
*m* hoher Leistung **~ note buzzer** Starktonsum-
mer *m*
**loud-speaker** Lautsprecher *m*, Schallsender *m*,
Trichter *m*, Tonfunkensender *m* **~ with folded
and corrugated horn** Riffelfaltenlautsprecher *m*
**loudspeaker, ~ car** Lautsprecherwagen *m* **~
chassis** Lautsprecherhaltevorrichtung *f* **~
clusters** Tonkorb *m* **~ cone** Lautsprechtrichter
*m* **~ horn** Lautsprechtrichter *m* **~ operating
mechanism (or motor) element** Schalltreibwerk
*n* **~ report** Höranzeige *f* **~ screen** Schallwand *f*
**~ service** Lautsprecherbetrieb *m* **~ telephone
set** Teilnehmerendverstärker *m* **~ truck** Lautsprecherwagen *m* **~ trumpet** Lautsprechtrichter
*m* **~ tube** Endröhre *f* **~ unit** Lautsprecherdose *f*
**~ voice coil** Schwingspule *f*
**loud-speaking telephone** Lautfernsprecher *m*
**loudness** Lautheit *f*, Lautstärke *f* **~ contour of ear**
Empfindungsgrenze *f* des Ohres, Ohrenemp-
findungsgrenze *f* **~ contours** Isofonen *pl*
(acoust.) **~ level** Lautheits-, Lautstärke-pegel
*m* **~-level meter** Lautstärkemesser *m* **~ unit**
Lautstärkenstufe *f*
**louver** Ausschnitt *m*, Jalousie *f*, Luftschlitz *m*
(der Motorhaube). Schalloch *n* (am Glockenturm), Schallöffnung *f* **~ of a fan** Abdeck-
jalousie *f* (d. Gebläses)
**louver, ~ boards** Jalousie-bretter *pl*, -laden *pl*,
Luken *pl* (paper mfg.) **~ damper** Jalousieklappe
*f* **~-shutter** Jalousieverschluß *m* **~ turret** Dachreiter *m*
**louvered** mit Kühlschlitzen *pl* versehen **~ slide**
Schlitzwand *f*
**louvers** Haubenausschnitte *pl*
**Love's waves** Love-Wellen *pl*
**low** gering, minderwertig, nieder, niedrig, nie-
drigstehend, platt, tief, tiefhängend
**low, of ~ capacity** kapazitätsarm **~ in carbon**
kohlenstoffarm, niedriggekohlt **of ~ content**
geringhaltig **of ~ humidity** wasserarm **of ~
inductance** induktionsarm **of ~ light intensity**
lichtschwach **~ in manganese** manganarm **of
~ noise** geräuscharm **of ~ noise level** rausch-
arm **of ~ permeability** schwerdurchlässig **~ in
phosphorus** phosphorarm **of ~ pressure** tief **~
in silicon** siliziumarm **of ~ standard** geringhaltig **of ~ transmittance** schwerdurchlässig
**low** erster Gang *m* (auto), Tief *n*, Tiefdruckgebiet *n* **~ absolute pressure per unit surface**
Hochvakuum *n* **~ aerobatics** Kunstflug *m* in
niedriger Höhe **~-alloy steel** niedrig legierter
Stahl *m* **~-alloyed** schwachlegiert **~-angle
fading antenna** Nahschwundantenne *f* **~-angle
fire** Feuer *n* in der unteren Winkelgruppe **~
angle shot** Tiefwinkelaufnahme *f* (film) **~
antenna** Niedrigantenne *f* **~-ash coal** aschen-
arme Kohle *f* **~ assay** Versuch *m* mit schwacher
Konzentration **~-atomic** niederatomig **~
boiler** niedrig siedender Stoff *m* **~-boiling** leicht-
siedend, tiefsiedend **~-boiling oil** leicht sieden-
des Öl *n* **~-built chassis** Niederrahmenfahr-
gestell *n* **~ butt needle** Tieffußnadel *f* **~-capa-
citance winding** Korbwicklung *f* **~ capacity**
von geringem Aufnahmevermögen
**low-carbon** niedriggekohlt **~ core** Kern *m* mit
niedrigem Kohlenstoffgehalt (Einsatzhärtung)

~ **free-cutting steel** Schnellautomatenweich-stahl *m* ~ **steel** Schmiedeeisen *n*, niedrigge-kohlter Stahl *m*, Stahl *m* mit niederem Kohlen-stoffgehalt ~-**steel grades** Flußstahlsorten *pl*
**low,** ~ **ceiling** niedrige Wolkenuntergrenze *f* ~ **center of gravity** Haftung *f* am Erdboden ~ **cloud** tiefhängende Wolke *f* ~-**compression engine** niedrigverdichtender Motor *m* ~-**con-sumption** von geringem Verbrauch *m* ~ **consumption** geringer Verbrauch *m* ~-**con-sumption cathode** Sparkathode *f* ~ **country** Niederung *f* ~-**coupling** kopplungsarm ~ **current fuse** Schwachstromsicherung *f* ~-**cur-rent protector** Feinspannungsschutz *m* (telegr.) ~-**definition television** Fernsehen *n* geringer Güte ~-**dosage counter tube** Niederdosiszähl-rohr *n* ~-**down pump** (double-acting hand-force pump) Tiefbrunnenpumpe *f* ~-**drag cowling** aerodynamisch günstige Haubenverkleidung *f* ~-**energy neutron** Neutron *n* niedriger Energie ~ **energy tail** (emission band) langwelliger Aus-läufer *m* ~ **explosive** Sprengstoff *m* mit ge-ringer Brisanz ~ **fermentation** Untergärung *f* ~-**fermentation yeast** Unterhefe *f* ~-**flux reactor** Reaktor *m* mit niedrigem Neutronen-fluß ~-**flying aircraft** Tiefflieger *m* ~-**freezing mixture** schwergefrierbare Mischung *f*
**low-frequency (LF)** niederfrequent, Langwellen... ~ **amplification** Niederfrequenz-, Tonfre-quenz-verstärkung *f* ~ **amplification stage** Niederfrequenzverstärkungsstufe *f* ~ **amplifier** Niederfrequenz-, Tonfrequenz-verstärker *m* ~ **cable** Niederspannungsleitung *f* ~ **choke** Niederfrequenzdrossel *f* ~ **cone loud-speaker** Tiefen-, Tiefton-konus *m* ~ **current** Nieder-frequenzstrom *m* ~ **dialing** Unterlagerungs-fernwahl *f* ~ **disturbance** Niederfrequenz-störung *f* ~ **effect of local emission den-sity** Funkeleffekt *m* (electron) ~ **filter** Nieder-frequenzsiebkette *f* ~ **furnace** Niederfrequenz-ofen *m* ~ **load alternator** Lastwechselgerät *n* ~ **muffler** Hochtondurchlasser *m* ~ **notes** Tiefen *pl* ~ **oscillator** Unterbandoszillator *m* ~ **ringer** Niederfrequenzrufsatz *m* ~ **signaling** Zeichen-gebung *f* mit Niederfrequenz ~ **signaling cur-rent** niederfrequenter Rufstrom *m* ~ **stage** Niederfrequenzstufe *f* ~ **transformer** Nieder-frequenz-transformator *m*, -übertrager *m* ~ **tube** Niederfrequenzröhre *f* ~ **wave** lange Welle *f*
**low,** ~-**frequent** niederperiodig ~ **gear** An-fahrtsgang *m*, niedriger oder unterer Gang *m*, hohe Untersetzung *f*
**low-grade** geringhaltig, geringwertig, minder-wertig ~ **lead obtained by the Carinthian process** Preßblei *n* ~ **massecuite** Nachproduktfüllmasse *f* ~ **matte** armer Stein *m* ~ **ore** minderhaltiges Erz *n* ~ **stock** minderwertige Mischung *f* ~ **strike** Nachproduktsud *m* ~ **vacuum pan** Nach-produktvakuum *n* ~ **varnish** magerer Lack *m*
**low,** ~ **ground** Niederung *f* ~-**head screw** Flach-kopfschraube *f* ~ **impedance triode** Triode *f* mit niedriger Impedanz ~-**inertia** trägheitsarm ~-**inertia manograph** Druckschreiber *m* von geringer Trägheit der beweglichen Teile ~ **intensity carbon service** Reinkohlenbetrieb *m* ~ **intensity lighting system** Niederleistungsbe-feuerung *f* ~-**key picture** dunkles Stimmungs-

bild *n* ~-**lag plant** verzögerungsarme Regel-strecke *f* ~ **land** Tiefebene *f* ~ **(est) level** Tief-stand *m*
**low-level,** ~ **continuous signaling** Ruhestrom-systemwahl *f* ~ **counter** Zählrohr *n* für schwache Intensität ~ **flight** Nieder-, Tief-flug *m* ~ **freight-jettisoning device** Tieflastenabwurfgerät *n* ~ **indicator** Leerstandanzeiger *m* ~ **modu-lation** Modulation *f* auf niedrigem Pegel
**low,** ~ **lever blower** Bodenlager *m* ~-**lift platform truck** Niederhubkarren *m* ~ **load** Schwachlast *f* ~-**loss cable** verlustarmes Kabel *n* ~-**loss con-denser** Kondensator *m* mit geringen dielektri-schen Verlusten ~-**loss RF lacquer** Hochfre-quenzlack *m* (verlustarm) ~-**loss tuning con-denser** verlustarmer Abstimmkondensator *m* ~ **losses** Dämpfungsarmut *f* ~ **luminosity picture** lichtschwaches Bild *n* ~-**lying** tiefliegend ~-**lying bog** Flachmoor *n* ~-**lying coast** Flach-küste *f* ~ **machine-(or mill-)finished** schwach geglättet (paper mfg.) ~ **magnetic fields** schwache Magnetfelder *pl* ~-**manganese** man-ganarm ~-**melting** tiefschmelzend ~-**melting-point alloy** niedrigschmelzende Legierung *f* ~ **minimum cyclone** Zyklone *f* ~-**modulation (sound) track** schwach ausgesteuerter Ton-streifen *m* ~-**molecular intermediate products** niedermolekulare Zwischenprodukte *pl* ~ **noise in picture background due to thermal effects** Störgrieß *m* ~-**noise mixer** Empfindlich-keitsmischer *m* ~-**noise tube** Tiefpaß *m* ~ **number of revolutions** niedrige Drehzahl *f* ~-**oblique convergent photograph** (konvergente) Steilaufnahme *f* ~-**oblique photograph** Steil-aufnahme *f* ~ **performance** Leistung *f* bei nie-driger Drehzahl ~-**phosphorous pig iron** phosphorsaures Roheisen *n* ~ **pile delivery** Kleinstapelausleger *m* ~ **pile feeder** Aufsetz-bogenleger *m* ~ **pitch** kleiner Blatteinstellwinkel *m* ~-**pitch notes** Tiefen *pl* ~-**pitch sound** tiefer Ton *m* ~ **pitched** tieftönend, schwach geneigt ~ **position** Tieflage *f* ~ **position of center of gravity** Tieflage *f* des Schwerpunktes
**low-power** von schwacher Leistung *f* ~ **electro-lytic capacitor** Niedervoltelektrolytkondensator *m* ~-**objective** schwaches Objektiv *n* ~ **range** Reaktorbetrieb *m* mit Hilfsquelle ~ **reactor** schwachbelasteter Reaktor *m*
**low-powered** schwach, schwachmotorig
**low-pressure** Nieder-druck *m*, -spannung *f*, Tiefdruck *m*, Unterdruck *m* (meteor.) ~ **air system** Niederdruckluftsystem *n* ~ **area** Nie-derdruck-, Tiefdruck- (aviat.), Niederdruck-gebiet *n* ~ **atomizer** Niederdruckzerstäuber *m* ~ **blower** Niederdruck-gebläse *n*, -luftgebläse *n* ~ **burner** Niederdruckbrenner *m* ~ **chamber** Unterdruckkammer *f* ~ **conveyance** Nieder-druckförderung *f* ~ **cylinder** Niederdruck-zylinder *m* ~ **end** Niederdruckseite *f* ~ **fan** Niederdrucklüfter *m* ~ **gas storage** Gasnieder-druckspeicherung *f* ~ **gauge** Unterdruckmesser *m* ~ **helium** verdünntes Helium *n* ~ **hose** Niederdruckschlauch *m* ~ **indication** Unter-druckanzeige *f* ~ **piston** Niederdruckkolben *m* ~ **return pipe** Abflußrohr *n* des verbrauchten Druckwassers ~ **sensing line** Niederdruck-geberleitung *f* ~ **side** Saugseite *f* (Kühlma-schine) ~ **stage** Niederdruckstufe *f* ~ **steam**

Niederdruckdampf *m* ~ **steam generator**
Niederdruckdampferzeuger *m* ~ **tire** Nieder-
druck-, Tiefdruck-reifen *m* ~ **torch** Injektor-
brenner *m* ~ **transport** Niederdruckförderung *f*
~ **turbine** Niederdruckturbine *f* ~ **vacuum tank**
Unterdruckkessel *m* ~ **wind tunnel (or channel)**
Unterdruckkanal *m*
**low,** ~ **propeller pitch** niedrige Luftschraubensteigung *f* ~**-quality goods** Ausschuß *m* ~ **range**
unterer Bereich *m* ~**-range prints** Kleinum-
fangkopien *pl* ~**-rank sub-bituminous coal** gemeine Braunkohle *f* ~ **red** dunkle Rotglut *f*
~**-resistance** niederohmig ~**-resistance direction** (rectifier) Durchgangs-, Durchlaß-richtung
*f*, (of a dry or oxide rectifier) Flußrichtung *f*
~**-resistance trap** Falle *f* mit niedrigem Widerstand ~ **rudder** Tiefenruder *n* ~**-set type electrode holder** Tieffassung *f* ~**-side valve of
pressure gauge** saugseitiges Ventil *n* am Manometer ~**-silicon pig iron** niedrigsiliziertes Roheisen *n* ~ **slung** niedrig aufgehängt ~**-speed**
langsamlaufend ~**-speed buck** Bocken *n* bei
zu niedriger Drehzahl ~**-speed mill** langsam-
laufende Mühle *f* ~**-speed shooting** Unterdrehen
*n* ~**-speed shooting for highspeed projection**
Zeitraffaufnahmeverfahren *n* ~ **spot** Bodensenkung *f* ~**-stage modulation** Vorstufen *n* ~
**stand** Bodengestell *n* ~**-supercharged engine** Motor *m* mit geringer Aufladung ~ **supercharged
engine** Bodenmotor *m* ~ **surface** Unterseite *f*
(of wing) ~**-synchronizing impulse** Bildsynchronisierimpuls *m* ~ **temperature** Tieftemperatur *f*
**low-temperature,** ~ **brine** Tiefkühlsohle *f* ~ **carbonization** Schwelen *n*, Schwelung *f*, Tief-ver-
gasung *f*, -verkokung *f*, Urdestillation *f*, Urverkokung *f*, Verschwelung *f* ~ **carbonization
of coal** Kohlenschwelung *f* ~ **carbonizer**
Schweler *m* ~ **carbonizing furnace** Schwelofen
*m* ~ **carbonizing plant** Schwelanlage *f* ~ **coal
tar** Steinkohlenschwelteer *m* ~ **coil** Tiefkühl-
spule *f* ~ **coke** Schwel-, Ur-koks *m* ~ **coking**
Urverkokung *f* ~ **cooling** Tiefkühlung *f* ~
**cooling chest** Kühltruhe *f* ~ **distillation** Tiefvergasung *f*, Urdestillation *f*, Urverkokung *f* ~
**process** Tiefkälteverfahren *n* ~ **purification
plant** Kältereinigungskala *f* ~ **retort coal gas**
Schwelgas *n* ~ **tar** Tief-, Ur-teer *m*
**low-tension** Niederspannung *f* ~ **blasting machine** Glühzünder *m* ~ **current** niedergespann-
ter Strom *m* ~ **ignition cable** Niederspannungs-
zündleitung *f* ~ **line** Niederspannungs-leitung
*f*, -transformator *m* ~ **magneto with igniter**
Niederspannungsmagnet *m* oder Abreißzündmagnet *m* mit Zündapparat ~ **side** Nieder-
spannungsseite *f* ~ **transformer** Niederspan-
nungstransformator *m* ~ **voltage** Unterspannung *f* on ~**-voltage side** unterspannungsseitig
**low,** ~ **test sprinkler** normalgrädiger Sprinkler
*m* ~ **tide** Ebbe *f*, Niedrigwasser *n*, Tidenfall *m*
~ **tone** tiefer Ton *m* ~ **viscosity** Leichtflüssigkeit *f* ~ **visibility** beschränkte Sicht *f*
**low-voltage** Niederspannung *f* ~ **circuit** Nieder-
spannungskreis *m* ~ **current** Schwachstrom *m*
~ **generator** Niederspannungsgenerator *m* ~
**microscope lamp (or bulb)** Niedervoltmikro-
skopierlampe *f* ~ **relays** Schwachstromrelais
*pl* ~ **terminal junction** Klemmenanschluß *m*
für Schwachstrom

**low,** ~**-warp tapestry** tiefschäftige Tapete *f* ~
**water** Nieder-, Niedrig-wasser *n* ~ **water of
ordinary neap tide** Nippniedrigwasser *n* ~ **water
of spring tides** Springniedrigwasser *n* (concrete)
**of** ~ **water content** Erdfeuchte *f* (Betonmasse) ~
**water level** Niedrigwasserspiegel *m* ~ **water
mark** Niedrigwassermarke *f* ~ **weir** Grundwehr *n* ~**-wing cantilever monoplane** freitragender Tiefdecker *m* ~**-wing model** Tiefdeckerflug-
modell *n* ~**-wing monoplane** Ein-, Tief-decker *m*
~ **wire entanglement** Stacheldrahtzaun *m*
**lower, to** ~ abnehmen, absenken, abwärtsbewegen, (flaps or undercarriage) ausfahren,
auskurbeln, (boat) aussetzen, einlassen, erniedrigen, fieren, hängen, hangen, herablassen,
herabsetzen, herunter-drücken, -fieren, hinablassen, kleiner machen, niederholen, niederlassen, niedriger machen, reduzieren, senken,
sinken lassen, tiefstellen, vermindern, verringern, (boat) zu Wasser fieren **to be** ~
(surveying) niedriger gelegen sein **to** ~ **the
antenna** Antenne *f* auskurbeln **to** ~ **a boat** ein
Boot *n* aussetzen **to** ~ **the cage one level** den
Förderkorb *m* um eine Etage senken **to** ~ **the
colors** die Flagge niederholen **to** ~ **the pressure**
abspannen (Strom) **to** ~ **in price** verbilligen **to**
~ **a seaplane** ein Seeflugzeug *n* aussetzen **to** ~
**target** Ziel *n* verschwinden lassen **to** ~ **the
undercarriage** das Fahrwerk ausfahren
**lower** Filterrohr *n*, Unter-; niedriger, unter ~
**aileron** Querruder *n*, Unterflügel *m* ~ **band**
Unterring *m* ~ **band spring** Unterringfeder *f*
~ **bed** Unterlager *n* ~ **(upper) bed drive rack**
Rollradzahnstange *f* ~ **bell** Unterglocke *f* ~
**bench** Unterbank *f* (min.) ~ **bend** unteres Knie
*n* ~ **bend of the characteristic** unterer Kennlinienknick *m* ~ **bend of valve characteristic**
unterer Knick *m* der Röhrenkennlinie ~ **blade**
Untermesser *m* ~ **block** Unterflasche *f* ~ **bush
of the axle box** Unterteil *m* der Schmierbüchse
(r.r.) ~ **camber** untere Krümmung *f* ~ **carriage
of a crane** Unterwagen *m* ~ **carrier roll** Unter-
zylinder *m* ~ **case** Unterkasten *m*, Zeichenfeld
*n* ~**-case letter** Minuskel *f* ~ **chassis of a crane**
Unterwagen *m* ~ **chord** (of truss or girder)
Untergurt *m* ~ **component of a composite aircraft** Unterflugzeug *n* ~ **course** Unterlauf *m* ~
**court** (of law) Untergericht *n* ~ **crank mechanism** Schubkurbelgetriebe *n* ~ **cross girder**
unterer Querträger *m* ~ **cut** Unterschnitt *m* ~
**cycle** Niederdruckkreis *m* ~ **deck** Banjerdeck *n*,
unteres Deck, Zwischendeck *n* ~ **deviation**
unteres Abmaß *n* ~ **die** Gesenkunterteil *n*,
Untergesenk *n*, Unterstanze *f* ~ **drag brace**
Knickstrebenschenkel *m* ~ **edge of beams**
Balkenunterkante *f* ~ **finishing hurdle** Ab-
darrhorde *f* ~ **flange** Fuß *m* der Schiene ~
**floor** Sturzbett *n* ~ **furnace** Unterofen *m* ~
**gates** Außenhaut *f* ~ **(or outer) gates** (of lock)
Unterhaupt *n* ~ **guide roll** Siebleitwalze *f*
(paper mfg.) ~ **gun carriage** Unterlafette *f*
(artil.) ~ **half nut** Mutterschloßunterteil *n* ~
**heating** Unterhitze *f* ~ **leg** Saugarm *m* ~ **level**
vertiefte Sohle *f* ~ **limit** Untergrenze *f* ~**-limit-
ing filter** Kondensatorkette *f* ~ **longeron**
Rumpfunterholm *m* ~ **most** unterst ~ **parhelion** Untersonne *f* ~ **part** Unterteil *n* ~ **part
of crankcase** Motorwanne *f*

**Lower Permian (shale) sandstone** Rotliegendes *n*
**lower,** ~ **pintle casting** Zapfenlager *n* ~ **plate of
transit** Limbus *m*, Teilkreis *m* ~ **platen** unterer
Pressentisch *m* ~ **platform** untere Bühne *f* ~
**pool elevation** Unterwasserspiegel *m*, Wasserstand *m* der unteren Haltung ~ **port** Unter-
pforte *f* ~ **portion of the combustion unit** Heiz-
behälterunterteil *n* ~ **power** schwache Vergrößerung *f* ~ **production costs** wirtschaftlichere
Herstellung *f* ~ **pump box** Pumpenherz *n* ~
**reaches of a stream** Unterstrom *m* ~ **rib
flange** Rippenuntergurt *m* ~ **river** Mündungs-
gebiet *n*, Unterlauf *m* ~ **roll** Matrizenwalze *f* ~
**shaft** Unterschacht *m* (of furnace) ~ **shed** Tief-
fach *n* ~ **shield** Unterschild *n* ~-**shield bracket**
Unterschildträger *m* ~ **side of the lode** Bett-
schicht *f* ~ **side band** unteres Seitenband *n* ~
**side band position** Kehrlage *f* ~ **slide** Gleitschiene *f* ~ **slides magazine** unteres Magazin *n*
~ **spar of engine mounting** Motorlagerunter-
holm *m* ~ **spring** Unterfeder *f* ~ **stone** (in mills)
Grundstein *m* ~ **story** Untergeschoß *n* ~
**stratum** Unterschicht *f* ~ **studding sail** Fock-
leesegel *n* ~ **surface** Profilbauch *m* ~ **tensile
limit of the fatigue cycle** Zugursprungfestigkeit
*f* ~ **thread** Unterfaden *m* ~ **torque link** unterer
Teil *m* der Federbeinschere ~ **transom** Unterriegel *m* (electr.) ~ **variation** unteres Abmaß *n*
~ **wind** Unterwind *m* ~ **window** Dachluke *f* ~
**wing** Unterdeck *n*, Unterflügel *m* ~-**wing span**
Spannweite *f* des Unterflügels ~ **wing spar**
Unterflügelholm *m*
**lowerable flap** Klappe *f* zum Niederlassen
**lowered** gesunken ~ **antenna** ausgekurbelte Antenne *f* ~ **wicket** niedergelegte Schütztafel *f*
**lowering,** ~ **of ground water** (well-point method)
Grundwasserabsenkung *f* ~ **of the level by the
wind** Absenkung *f* des Wasserspiegels ~ **of price**
Preisermäßigung *f* ~ **of the water by the effect
of wind** Windsenkung *f* des Wasserspiegels ~
**of the water level** Senkung *f* des Wasserstandes
~ **of the water table** Absenkung *f* des Grund-
wassers *n*, Grundwasser-absenkung *f*, -ver-
senkung *f*
**lowering,** ~ (**switch**) **cam** Abzugsdreieck *n* ~
**chain** Einlaßkette *f* ~ **delay** Senkverzögerung *f*
~ **device** Niederlaß-, Senk-vorrichtung *f* ~
**direction** im Senksinn *m* ~ **distance** Senkweg *m*
~ **handle** Herablaßgriff *m* ~ **lever** Herablaßhebel *m* ~ **mechanism** Absenkvorrichtung *f* ~
**motion** Senkbewegung *f* ~ **movement** Niedergang *m*, Niedergehen *n*, Senkbewegung *f* ~
**position** Senkstellung *f* (hydraul.) ~ **speed** Senk-
geschwindigkeit *f* (hydraul.) ~ **spindle** Herab-
laßstange *f* ~ **spring** Herablaßfeder *f*
**lowest** geringst, mindest, unterst ~ **bid** das billig-
ste Angebot ~ **isobathic line** Tiefenlinie *f* ~
**pool elevation** niedrigstes Unterwasser *n* ~
**price** äußerster Preis *m* ~ **rock holding oil**
Farewell-Sand *m* ~ **stratum** Bodenschicht *f* ~
**threshold of audibility** untere Hörschwelle *f* ~
**upper pool elevation** niedrigstes Oberwasser *n*,
niedrigster Wasserspiegel *m* ~ **usable high
frequency (LUHR)** Grenzfrequenz *f* ~ **useful
frequency (LUF)** niedrigste brauchbare Frequenz *f*

**loxodrome** Kursgleiche *f*, Loxodrome *f*, Schieflaufende *f*
**loxodromic** loxodromisch, schräglaufend, schief-
laufend ~ **spiral** Dwarslinie *f*
**lozenge** Pastille *f*, Plätzchen *n*, Raute *f*, Rhombus *m*, Tablette *f* ~ **riveting** verjüngte Nietung
*f* ~ **section** Spießkantenprofil *n*
**lozenged glas** Rautenglas *n*
**lozengelike arrangement** (**or position**) Rauten-
stellung *f*
**L scope** L-Schirm *m*
**L-shaped bar** Stabeck *n*
**L strap** Winkelschiene *f*
**lubber line** Anliegestrich *m* (compass), Steuerstrich *m*, Windstrich *m*
**lube discharge filter** Schmierölfilter *m*
**lube-oil,** ~ **cooler** Schmierölkühler *m* ~ **gear
pump** Schmierölzahnradpumpe *f* ~ **service
truch** Ölbetankungswagen *m*
**lube pump** Schmierölpumpe *f*
**lubricant** Fett *n*, Gleitmittel *m*, Schmiere *f*,
Schmier-flüssigkeit *f*, -material *n*, -mittel *n*,
-öl *n*, -stoff *m* ~ **additive blender** Schmiermittel-
zusatzregler *m* ~ **consumption** Schmierstoff-
verbrauch *m* ~ **film** Schmierölhaut *f* ~ **inlet**
Schmiermitteleinlaß *m* ~ **preheater** Schmier-
stofferwärmungsgerät *n* ~ **pressure** Schmier-
stoffdruck *m* ~ **supply** Schmierstoffversorgung
*f* ~ **tank** Schmierstoff-, Schneidöl-behälter *m*
**lubricate, to** ~ einfetten, einschmieren, fetten,
Fett *n* einspritzen, ölen, schmieren **to** ~ **the
bearing** das Lager schmieren
**lubricated** geschmiert
**lubricating** Abschmieren *n* ~ **action** Schmierwir-
kung *f* ~ **cup** Schmiergefäß *n* ~ **device** Schmiervorrichtung *f* ~ **film** Schmierfilm *m* ~ **grease**
Abschmierfett *n*, Schmiere *f*, Schmiermaterial *n*
~ **liquid** Schmierflüssigkeit *f* ~ **oil** Maschinenöl *n*, Schmiere *f*, Schmier-flüssigkeit *f*, -material
*n*, -öl *n* ~ **oil collector** Schmierölsammeltank *m*
~-**oil distillate** Schmieröldestillat *n* ~-**oil filter**
Schmierölfilter *n* ~-**oil jet body** Schmieröldüse
*f* ~-**oil pump** Schmierölpumpe *f* ~ **paste**
Schmiermasse *f* ~ **power** Schmierfähigkeit *f* ~
**pump** Ölschleuderring *m* ~ **sheet** Abschmier-
blech *n* ~ **system** Schmier-stoffanlage *f*, -system
*n* ~ **value** Schmierwert *m*
**lubrication** Befetten *n*, Einfettung *f*, Fettung *f*,
Ölförderung *f*, Ölung *f*, Ölversorgung *f*,
Schmieren *n*, Schmierung *f* ~ **chart** Schmier-
plan *m* ~ **connection** Schmierstutzen *m* ~
**diagram** Schmier-plan *m*, -schema *n* ~ **pipe**
Öl-, Schmier-rohr *n* ~ **piping** Schmierleitung *f*
~ **points to the cylinders** Zylinderschmierstellen
*pl* ~ **system** Schmiermittelkreislauf *m*, Schmie-
rungssystem *n*
**lubricative** ölend, schmierend ~ **quality** Öligkeit *f*
**lubricator** Druckschmierapparat *m*, Fettbüchse *f*,
Öler *m*, Schmierer *m*, Schmier-apparat *m*,
-vase *f*, -vorrichtung *f* ~ **for top head** Deckel-
schmiergefäß *n* ~ **disc** Ölförderscheibe *f* ~
**wheel** Ölerantriebsrad *n*
**lubricity** Fettigkeit *f*, Schlüpfrigkeit *f*, Schmierfähigkeit *f*
**lucid** durchsichtig, klar
**lucite** Lucit *n*, Plexiglas *n*
**lucrative** gewinnbringend, rentabel

**ludlum,** ~ **direct-arc non-conducting hearth furnace** Ludlum-unmittelbarer Lichtbogenofen *m* ohne Herdbeheizung ~ **series-arc furnace** Ludlum-Mehrelektrodenlichtbogenofen *m*
**Ludolph's altitude-reduction instrument (or protractor)** Ludolphsches Auswertgerät *n*
**Ludwik's cone-hardness test** Ludwiks Kegeldruckpresse *f*
**Lufberry circle** Lufberry-Kreis *m*
**luff** Luv *n*
**luffa** Luffa *f*, Loofah *f* ~ **fiber** Luffafaser *f*
**luffer board** Schall-ader *f* (acoust.), -brett *n* (acoust.)
**luffing** Wippen *n* ~ **beetle** Messingkäfer *m* ~ **crane** Wippkran *m* ~ **drum** Einziehtrommel *f* ~ **gear** Einziehwerk *n* ~ **gear block** Blockeinziehwerk *n* (Kran) ~ **jib** verstellbarer Ausleger *m* ~ **movement** Wippbewegung *f*
**lug** Anguß *m*, Ansatz *m*, Anschlag *m*, Aufhängeöse *f*, Aufhängungsohr *n*, Auge *n*, Bock *m*, Fahne *f*, Henkel *m*, Kabelhalter *m*, Knagge *f*, Lappen *m*, (parachute) Lascheröse *f*, Nase *f*, Öse *f*, Pfanne *f*, Rute *f*, Stift *m*, Vorstoß *m*, Warze *f*, Zapfen *m*, Zapfenlager *n*, Zinken *m* (connector) ~ **Kabelschuh** *m* ~ **of a battery plate** Fahne *f* ~ **on bedplate** (for fixing) Ankernocken *pl* ~ **of a collar** Nase *f* eines Ringes ~ **of a molding box** Formkastenlappen *m* ~ **of a ring** Nase *f* eines Ringes
**lug,** ~ **bolt** Bolzen *m* mit rechtwinklig eingebogenem (Anker-) Ende ~ **cam** Anschlagnocken *m* ~ **crimper** Kabelschuhkerbzange *f* ~ **eye** Schenkelauge *n* ~ **face** Polschuh *m* ~ **head** Lappen-, Ohren-kopf *m* ~ **hole** Lappenbohrung *f* ~ **strip** Klemmleiste *f*
**luggage** Gepäck *n*
**lugger** Galliot *f*, Logger *m*
**lugging capacity of an engine** Durchzugmoment *n*
**lukewarm** lau, lauwarm
**lull, to** ~ abflauen, flau werden, sich legen
**lull** Flaute *f*, Stille *f*, Windstille *f*
**lumber** (structural) Bauholz *n*, Holz *n*, Nutzholz *n* ~ **for derrick** Bohrturmgebälk *n* ~ **man** Holzhauer *m* ~ **preservation** Holzimprägnierung *f* ~ **wagon** Leiterwagen *n* ~ **yard** Holz-lager *n*, -platz *m*, Zimmerplatz *m*
**lumen** Lumen *n* (Einheit des Lichtstromes) ~ **-hour** internationale Lumenstunde *f* ~ **meter** Lichtstrom-, Lumen-messer *m*
**luminance** Helligkeit *f* (fotometrische Größe), Helligkeit *f* (als physikalische Größe), Leuchtdichte *f*, gemessene Leuchtkraft *f* oder Lichtstärke *f* ~ **information** Helle-Information *f* (TV) ~ **scale** Leuchtdichteskala *f*
**luminant** Beleuchtungsmittel *n*, Leuchtkörper *m*; glänzend, leuchtend
**luminary** Himmelskörper *m*, leuchtender Körper *m*, Leuchtkörper *m*
**luminesce, to** ~ fluoreszieren, leuchten, lumineszieren
**luminescence** Leuchten *n*, Lumineszenz *f* ~ **excitation** Lumineszenzanregung *f*
**luminescent** leuchtend, lumineszierend ~ **digital indicator** leuchtender Zahlenindikator *m* ~ **lamp** Leuchtstofflampe *f* ~ (screen or lamp) **materials** Phosphore *pl* ~ **screen** Leuchtschirm *m* ~ **substance** Leuchtmasse *f* ~ **tube** Leuchtstofflampe *f*

**luminosity** Glanz *m*, Helle *f*, Helligkeit *f*, Leuchtdauer *f*, -fähigkeit *f*, -kraft *f*, Lichtstärke *f* ~ **curve** Leuchtstärkenverteilungskurve *f*, Lichtverteilungsfläche *f* ~ **factor** Hellempfindlichkeit *f* ~ **factor of a monochromatic radiation** fotometrisches Strahlungsäquivalent *n* ~ **response of eye** Augenlichtempfindlichkeit *f*
**luminous** leuchtend, licht, lichtgebend, mit Nachtbeleuchtung *f*, nachtleuchtend, selbstleuchtend ~ **advertising** Lichtreklame *f* ~ **annunciator** Leuchtzahlenfeld *n* (teleph.) ~ **beam** Licht-bündel *n*, -büschel *n* ~ **brightness of a surface** Flächen-helle *f*, -helligkeit *f* ~ **center of gravity** Beleuchtungsschwerpunkt *m* ~ **-circuit diagram** Leuchtschaltbild *n* ~ **column** Leuchtfaden *m*, Lichtsäule *f* ~ **(-dial) compass** Leuchtkompaß *m* ~ **cone** Lichtkegel *m* ~ **current discharge** Glimmentladung *f* ~ **density** Leuchtdichte *f* ~ **density of screen** Bildwandleuchtdichte *f* ~ **dial** Leuchtzifferblatt *n* ~ **discharge current** Glimmentladungsstrom *m* ~ **-discharge lamp** Glimmentladungslampe *f*, Leuchtröhre *f* ~ **-discharge tube** Leuchtröhre *f* ~ **dust** Lichtstaub *m* ~ **edge** magischer Rahmen *m* ~ **effect** Leucht-, Licht-wirkung *f* ~ **efficiency** Leuchtwirksamkeit *f* ~ **energy** Licht-, Strahlungs-energie *f* ~ **envelope** Leuchthülle *f* ~ **figure** Leuchtziffer *f* ~ **flame** aufleuchtende Flamme *f* ~ **flux** (in lumen units) Lichtstrom *m*, Stromdichte *f* ~ **-flux density** Lichtstromdichte *f* ~ **glow** Leuchtglimmen *n* ~ **groundwind indicator** Leuchtgrundwindzeiger *m* ~ **haze** Lichtnebel *m* ~ **-headed press button** Leuchttaste *f* ~ **indicator board** Lichttableau *n* ~ **indicator panel with remote-control and telemetering devices** Leuchtschaltbild *n* mit Fernsteuerung und Fernmessung ~ **instrument** nachtleuchtendes oder selbstleuchtendes Instrument *n* ~ **intensity** (of a point source) Lichtstärke *f* ~ **key** Leuchttaste *f* ~ **lamp** Leuchtstofflampe *f* ~ **lettered text** Leuchtschrift *f* ~ **number-indicator board** Nummernlichttableau *n* ~ **paint** Fluoreszenz-, Leucht-farbe *f*, Radiumbelag *m* ~ **pattern** Leuchtgebilde *n* ~ **phenomenon (or light) produced during detonation** Detonationsleuchter *m* ~ **power** Leuchtkraft *f* ~ **quartz** Leucht-quarz *m*, -resonator *m* ~ **ray** Lichtstrahl *m* ~ **resonator** Leuchtresonator *m* ~ **screen** Leuchtschirm *m* ~ **sheath** Leuchthülle *f* ~ **sight** Leuchtvisier *n* ~ **signal** Feuer *n*, Lichtsignal *n*, -zeichen *n* ~ **source** Lichtquelle *f* ~ **sources** Flächenlampe *f* ~ **spectrum** sichtbares Spektrum *n* ~ **spot** Leuchtbild *n* ~ **-spot ring condenser** Leuchtbildkondensor *m* ~ **standard** Normallichtquelle *f* ~ **stimulus** Lichtreiz *m* ~ **streamer** Leuchtstreifen *m* ~ **streamers** Entladungsleuchtstreifen *m* ~ **tube** Leuchtstofflampe *f*
**Lummer-Brodhun contrast (cube) photometer** Lummer-Brodhun'sches Fotometer *n*
**lump, to** ~ konzentrieren **to** ~ **-load** punktförmig belasten, pupinisieren, mit Querspulen *pl* belasten **to** ~ **-weld** stumpfschweißen
**lump** Batzen *m*, Brocken *m*, Dackel *m*, Deul *m*, Klumpen *m*, Knollen *m*, Külbchen *n* (glass mfg.), Luppe *f*, Masse *f*, Menge *f*, Posten *m* (glass mfg.), Stück *n*, Stückchen *n*, Wolf *m*
**lump,** ~ **of coke** Koksstück *n* **a pig** ~ **of different**

**metals** Gans *f* ~ **of ice** Eisklumpen *m* ~ **of lead** Bleibrocken *pl* ~ **of ore** Erzbrocken *m* ~ **of tin ore** Zinnstufe *f*
**lump,** ~ **coal** Füllkohle *f*, grobstückige Kohle *f*, Knorpel *m*, Stückkohle *f* ~ **coke** Stückkoks *m* ~ **freight** Pauschalfracht *f* ~**-loaded** pupinisiert ~**-loaded cable** punktförmig belastetes Kabel *n* ~**-loaded circuit** Leitung *f* mit punktförmiger Ladung, punktförmig belastete Leitung *f*, Pupinleitung *f* ~**-loaded open circuit** Pupinfreileitung *f* ~**-loading** Pupinisation *f* ~ **(ed) loading** punktförmig verteilte Ladung *f* ~ **ore** Stückerz *n* ~ **peat** Brocken-, Krümel-torf *m* ~ **slag** Stückschlacke *f* ~ **sugar** Stückenzucker *m* ~ **sum** Pauschalsumme *f* ~**-sum contract** Pauschalakkord *m* ~**-sum payment** einmalige Abfindung *f* ~**-sum settlement** Kapitalabfindung *f*
**lumped** punktförmig verteilt ~ **capacity** punktförmig (verteilte) Kapazität *f* oder Ladung *f*, punktförmig verteilter Widerstand *m* ~ **characteristic** Schwinglinie *f* ~ **circuit element** konzentriertes Schaltelement *n* ~ **inductance** punktförmig verteilte Induktivität *f* ~ **leak load(ing)** Belastung *f* mit Querspulen ~ **(series) load** punktförmige Belastung *f* ~ **load(ing)** punktförmig verteilte Ladung *f* ~ **loading** punktförmige Ladung *f*, Pupinisierung *f* ~ **parameter** räumlich konzentrierter Parameter *m* ~ **resistance** punktförmig verteilter Widerstand *m* ~**-series loading** Belastung *f* durch Reihenspulen
**lumpiness** klumpige Beschaffenheit *f*, Stückigkeit *f*
**lumps** Erzkern *m* (in calcined ore) **in** ~ punktförmig, punktförmig verteilt, stückig
**lumpy** ballig, derbstückig, großstückig, klumpig, stückig
**lunar** lunarisch ~ **aurora** Mondhof *m* ~ **caustic** Höllenstein *m* ~ **crater** Mondkrater *m* ~ **distance** Mondentfernung *f* ~ **eclipse** Mondfinsternis *f* ~ **excursion module (LEM)** Mondlandefahrzeug *n* ~ **halo** Mond-halo *n*, -ring *m* ~ **month** Lunarmonat *m* ~ **node** Mondknotenpunkt *m* ~ **observation** Monddistanzbeobachtung *f* ~ **orbit** Mondumlaufbahn *f* ~ **probe** Mondraumsonde *f* ~ **year** Mondjahr *n*
**lunate** mondförmig
**lunation** Mondumlauf *m*
**lune** Halbmond *m*, Kreiszweieck *n*

**lunge, to** ~ ausholen
**lunge** Vorstoß *m*
**luni-solar** lunisolar
**lurch** Gierung *f*, Neigung *f*, Rollen *n*, Schlingern *n*
**lürman,** ~ **front (or** ~**'s closed front)** Lürmannsche Schlackenform *f*
**luster** Glanz *m*, Lüster *m*, Politur *f*, Schein *m*, Schimmer *m* ~ **finish** Glanzappretur *f* ~ **shrinking machine** Preßglanzdekatiermaschine *f*
**lustering agent** Glanzmittel *n*
**lusterless** glanzlos, matt, stumpf ~ **paint** Mattanstrich *m*
**lustrous** perlglänzend ~ **carbon** Glanzkohle *f* ~ **yarn** Glanzgarn *n*
**lustrum** Jahrfünft *n*
**lutation** Kitten *n*
**lute, to** ~ abstreichen, dichten, lutieren, verkitten, verkleben, verschmieren, zusammenkitten
**lute** Dichtungsmittel *n*, Gummiring *m*, Kitt *m*, Lutierungsmittel *n*
**lutecium** Lutetium *n*
**luthern** Dachluke *f*, Gaupe *f*, Mansardenfenster *n*
**luting** Dichtung *f*, Kitten *n*, Verkittung *f*, Verschmierung *f* ~ **with clay** Lehmschmierung *f* ~ **agent** Lutierungsmittel *n* ~ **material** Dichtungsmaterial *n*, -stoff *m*
**lux** (unit of light) Lux *n* ~ **meter** Beleuchtungsmesser *m*, -stärkefotometer *n*, -stärkemesser *m*
**Luxembourg effect** Dämpfungsmodulation *f*
**lycopodium** Lykopodium *n* ~ **spores** Bärlappsamen *m*
**Lydian stone** schwarzer Kieselschiefer *m*
**lye** Alkalilauge *f*, Ätznatron *n*, Beuche *f* ~ **brush** Laugenbürste *f* ~**-dissolving tank** Laugenauflösebehälter *m* ~ **graduating tank** Laugenbereitungsbehälter *m* ~ **pot** Laugenfänger *m* ~**-proof** laugenfest ~ **recovery plants** Laugenrückgewinnungsanlagen *pl* ~ **siphon** Laugenheber *m* ~ **sludge** alkalischer Schlamm *m* ~ **solution** Lauge *f* ~ **vat** Laugenkessel *m*
**lying** liegend ~ **in drydock** auf Stapel liegen ~ **within** innenliegend
**lymph** Lymphe *f*
**lyre** Lyra *f* ~ **crystallizer** Lyramaische *f*
**lysoform** Lysoform *n*

# M

**macadam** Chaussierung *f*, Makadam *n* & *m*, Schotter *m*, Steinschotter *m* ~ **pavement** Kieselschlag *m* ~ **road** Makadamweg *m*, Schotterbahn *f* ~ **surface** Makadamdecke *f*
**macadamization** Chaussierung *f*
**macadamize, to** ~ chaussieren, makadamisieren, mit Steinschotter *m* belegen
**macadamized road** Schotterstraße *f*
**macadamizing** Kieselschlag *m*
**macerate, to** ~ abbeizen, ätzen, auswässern, einweichen, mazerieren
**macerate molding** Schnitzelpreßstoff *m*
**macerating** Ausziehen *n*
**maceration** Ausziehen *n*, Einweichung *f*, Mazeration *f*
**macerator** Einweichbütte *f*
**Mach, ~ bands** Machsche Streifen *pl* ~ **cone** Machscher Kegel *m* ~ **cutout** transsonische Ausblendung *f* ~ **front** Machscher Kegel *m* ~ **hold** Mach-Haltesignal *n* ~ **line** Machsche Welle *f* ~ **meter** Machmeter *n* ~ **number** Machsche Zahl *f*, Machzahl *f* ~ **quadrangle** Machsches Viereck *n* ~ **region** Mach-Gebiet *n* ~ **stem** Machscher Kegel *m* ~ **unit** Mach-Einheit *f*
**machete** Hackmesser *n*
**machinability** Bearbeitbarkeit *f*, Bearbeitungsfähigkeit *f*, Schnittbearbeitbarkeit *f*, Verarbeitungsfähigkeit *f*, Zerspanbarkeit *f*
**machinable** bearbeitbar, feilbar, verarbeitbar, verspanbar
**machine, to** ~ abspannen, bearbeiten, mit Maschinen *pl* bearbeiten oder herstellen, maschinell bearbeiten oder herstellen, nacharbeiten, nachbearbeiten, schleudern, spanabheben, spanabhebend bearbeiten, verarbeiten **to ~ all over** allseitig bearbeiten **to ~-lap** maschinenläppen **to ~ together** zusammen bearbeiten
**machine** Apparat *m*, Arbeitsmaschine *f*, Getriebe *n*, Maschine *f*, Mechanismus *m*, Triebwerk *n*, Vorrichtung *f*
**machine, ~ for bending casings** Mantelbeugemaschine *f* ~ **for breaking up pavement** Straßenaufreißer *m* ~ **for burnishing axle journals** Achsschenkelprägepoliermaschine *f* ~ **for centering and countersinking nuts** Zentrier- und Muttersenkmaschine *f* ~ **for cutting out** Zuschneidemaschine *f* ~ **for drawing off** Abfüllbock *m* ~ **for full-width acid impregnation** Breitsäureeinrichtung *f* ~ **with inherent self excitation** läufererregte Kommutatormaschine *f* ~ **for inspecting cloths** Repassiermaschine *f* ~ **for making handles** Griffeherstellungsmaschine *f* ~ **for making rules** Maßstabmaschine *f* ~ **for making tin boxes** Dosenfertigungsmaschine *f* ~ **for making washers** Unterlagscheibenherstellungsmaschine *f* ~ **for making window-sash bars** Kreuzsprossenmaschine *f* ~ **for mining lignite** Braunkohlenbruchschlitzmaschine *f* ~ **for rasping colors, paints, and varnishes to perfect fineness** Feinzerreibemaschine *f* für Farben und Lacke ~ **for reproduction of drawings** Zeichnungskopiermaschine *f* ~ **for**

**rolling mills** Walzwerksmaschine *f* ~ **for simultaneous testing of several specimens** Vielprobenmaschine *f* ~ **for static-endurance tests** Dauerversuchsmaschine *f* für ruhende Beanspruchung ~ **for testing durability arranged for alternate loads** Wechselbelastungsdauerprüfmaschine *f* ~ **for testing wheels** Zahnräderlaufprüfmaschine *f* ~ **for vertical stripes** Langstreifenmaschine *f* ~ **for the working of armor plates** Panzerplattenbearbeitungsmaschine *f*
**machine, ~ audit** maschinelle Überprüfung *f* ~ **available time** Maschinenwirkzeit *f* ~ **bed** Maschinenbett *n* ~ **bits** Maschinenbohrer *pl* ~ **bolt** Maschinenbolzen *m* ~ **broke** Papiermaschinenausschuß *m* ~ **builder** Maschinenbauer *m* ~ **building industry** Maschinenbau *m* ~ **cast** Maschinenguß *m* ~ **casting** Maschinenguß *m* ~ **chest** Rührbütte *f*, Stoffbütte (paper mfg.), Stoffkiste *f* ~ **code** Maschinenkode *m* ~ **composition** Maschinensatz *m* ~ **compositor** Maschinensetzer *m* (print.) ~ **construction** Maschinenbau *m* ~ **control** maschinelle Steuerung *f* ~-**control room** Zentrale *f* für die Bedienung der Maschine(n) ~-**controlling cabin** Bedienung *f* der Maschinen, Zentrale *f* für die Bedienung ~ **cords** Maschinenschnüre *pl* ~ **current** Maschinenstrom *m* ~ **cutter** Maschinenmesser *n* ~ **cycle** Maschinen-periode *f*, -zyklus *m*, Rechnerperiode *f* ~ **designer** Maschinenkonstrukteur *m* ~ **direction** Papierlaufrichtung *f* ~ **drive** Maschinenantrieb *m*, mechanischer Antrieb *m* ~ **driven** mit Maschinenantrieb *m* ~-**driven part** maschinenangetriebener Teil *m* ~ **driving gear** Gangwerk *n* ~ **education** Rechnertraining *n* ~ **element** Maschinenelement *n* ~ **embroidery** Kurbelei *f* ~ **equation** Maschinenrechnergleichung *f* ~ **fault** Maschinendefekt *m* ~-**finished** maschinen-fertig, -glatt ~ **folding** Maschinenfalzung *f* ~ **frame** Maschinengerippe *n* ~ **giving northings and eastings** Hoch- und Seitenwertgeber *m* ~ **glaze cylinder** Glättzylinder *m* ~ **glazed** einseitig glatt, satiniert ~ **gray** maschinengrau *n* ~ **guard** Maschinengitter *n* ~ **gun** Maschinengewehr *n* ~ **handle** Ballengriff *m* ~ **hoses** Maschinenschläuche *pl* ~ **impression** Maschinendruck *m* ~ **inertia constant** Anlaufzeit *f* ~ **key ringing** halbselbsttätiger Ruf *m* ~ **lamp** Gelenkluchte *f* ~ **language** Maschinensprache *f* ~-**lapped finish** Maschinenläppschliff *m* ~-**lapping** Maschinenläppen *n* ~ **load** Maschinenbelastung *f* ~-**made** maschinell hergestellt, maschinengearbeitet ~-**made grey and colored cardboard** Speltpappe *f* ~ **master tap** Maschinenbackenbohrer *m* ~ **member** Maschinenteil *m* ~ **mixing** maschinelle Mischung *f*, Maschinenmischung *f* ~-**nut tap** Maschinenmuttergewindebohrer *m* ~ **oar** Rührflügel *m* ~ **oil** Maschinenöl *n* ~ **part** Maschinenteil *m* ~ **pistol** Maschinenpistole *f* ~ **plotting** Maschinen-auswertung *f*, -kartierung *f* ~ **programming** Maschinenprogrammierung *f* ~ **proof** Maschinen-abzug *m*, -bogen *m* ~ **pulp**

**board** Maschinenholzpappe *f* ~ **reamer** Maschinenreibahle *f* ~ **retouching** Maschinenretusche *f* ~ **ringing** Fünfsekundenruf *m*, selbsttätiger Ruf *m* ~ **riveting** Stanznietung *f* ~ **room** Maschinensaal *m* ~ **room overseer** Obermaschinenmeister *m* ~ **ruler** Linierer *m*, Liniermaschine *f* ~ **ruler thread** Liniermaschinenfaden *m* ~ **scrap** Maschinenschrott *m* ~ **screw** Maschinenschraube *f* ~ **set** maschinengesetzt ~ **setting of photographs** Einpassen *n* am Bildkartiergerät ~ **sewer** Maschinenhefter *m* ~ **shaft** Maschinenwelle *f* ~ **shop** Maschinenwerkstatt *f*, mechanische Werkstatt *f* ~-**sized** oberflächengeleimt ~ **spindle** Maschinenspindel *f* ~ **steel** Maschinenbaustahl *m* ~ **stitcher** Maschinenhefter *m* ~ **stitching** Kurbelei *f* ~ **switching system** Maschinenwählersystem *m* ~ **tap** Maschinenbohrer *m* ~ **telegraph** Maschinentelegraf *m* ~ **tinging** Anruf *m* mit Maschinenstrom, selbsttätiger Anruf (mit Maschinenstrom), Fünfsekundenruf *m*, periodischer oder selbsttätiger Ruf *m* ~ **tool** Maschinenwerkzeug *n*, Werkzeugmaschine *f* ~-**tool builder** Werkzeugmaschinenbauer *m* ~-**tool construction (or industry)** Werkzeugmaschinenbau *m* ~ **tool microscope** Einbaumikroskop *n* für Werkzeugmaschinen ~ **transmitter** Maschinensender *m* ~ **trouble** Maschinenstörung *f* ~ **type** Maschinen-ausführung *f*, -bezeichnung *f* ~ **unit** Maschinenanlage *f*, -einheit *f* ~-**value** Maschinenwert *m* ~ **vice** Maschinenschraubstock *m* ~ **wire** Langsieb *n*, Maschinendraht *m*, endloses Sieb *n*, Siebband *n* ~ **word** Maschinen-, Rechnerwort *n* ~ **work** Maschinenarbeit *f*

**machined** bearbeitet ~ **all over** allseitig bearbeitet ~ **from the solid** aus dem Vollen herausgearbeitet ~ **together** in einem Arbeitsgang *m* zueinander passend bearbeitet

**machined,** ~ **blank** vorgearbeiteter Werkstoff *m* ~ **nut** bearbeitete Mutter *f* ~ **surface** bearbeitete Fläche *f* ~ **washer** blanke Scheibe *f*

**machinery** Betriebsanlage *f*, Getriebe *n*, Maschinenpark *m*, Maschinerie *f*, Mechanismus *m* ~ **for constructing a permanent way for railroad** Eisenbahnoberbaumaschine *f*

**machinery,** ~ **allowance** verlorener Kopf *m* ~ **casting** Maschinenteil *m* ~ **equipment** Maschinenausrüstung *f* ~ **house** Maschinenhaus *n* ~ **materials** Betriebsmittel *pl*

**machining** Bearbeiten *n*, Bearbeitung *f*, maschinelle oder mechanische Bearbeitung *f*, spangebende oder spanabhebende Bearbeitung *f*, Verarbeitung *f*, Weiterverarbeitung *f*, Zerspannung *f*; spanabhebend

**machining,** ~ **allowance** (for extra thickness) Bearbeitungs-zugabe *f*, -zuschlag *m* ~ **alloy stock** Automatenlegierung *f* ~ **costs** Bearbeitungskosten *pl* ~ **operation** Bearbeitungs-, Zerspannungs-vorgang *m* ~ **possibility** Bearbeitungsmöglichkeit *f* ~ **property** Bearbeitbarkeit *f* ~ **quality** Bearbeitungsgüte *f* ~ **requirements** Aufgabenstellung *f* ~ **step** Bearbeitungsstufe *f* ~ **symbols** Bearbeitungszeichen *n* ~ **time** Hauptzeit *f*

**machinist** gelernter Arbeiter *m* des Maschinenbaues, Maschinen-arbeiter *m*, -bauer *m*, -führer *m*, -schlosser *m*, -wärter *m*, Maschinist *m* ~'**s**

**hammer** Schlosserhammer *m* ~'**s tools** Schlosserwerkzeug *n*

**mackerel sky** Schäfchen-himmel *m*, -wolken *pl*

**mackintosh** Gummimantel *m*

**mackle, to** ~ duplieren (print.)

**mackled sheets** Ausschußpapier *n*, Makulator *f*

**mackling** Dublieren *n* (print.)

**Macquisten(tube) process** Macquistenprozeß *m*

**macro,** ~-**analysis** Makro-analyse *f*, -untersuchung *f* ~-**axis** Makroachse *f* (cryst.) ~-**crystalline** makrokristallin ~-**equipment** Makro-Einrichtung *f* ~-**etching** Grobätzung *f*

**macrography** Makrografie *f*

**macrogravimetric** makrogravimetrisch

**macro,** ~-**imprinting** Makrodruck *m* ~-**incident light stage** Makro-Auflichttisch *m*

**macrometer** Makrometer *n*, Spiegelentfernungsmesser *m*

**macro,** ~-**molecule** Makromolekül *n* ~-**molecules** großmolekulare Stoffe *pl* ~-**program** Makroprogramm *n* ~-**projection** Makroprojektion *f* ~-**rheology** Makrorheologie *f*

**macroscopic** makroskopisch, das grobe Gefüge betreffend ~ **state** makroskopischer Zustand *m*

**macroscopy** Grobstrukturuntersuchung *f*

**macro,** ~-**stage** Makrotisch *m* ~-**stage illumination** Makro-Beleuchtung *f* ~-**structure** Grobgefüge *n*, Groß-gefüge *n*, -struktur *f*, Gußgefüge *n*, -struktur *f*, Makrostruktur *f* ~-**structure of an ingot** Blockgefüge *n* ~-**structure analysis** Grobstrukturanalyse *f* ~-**turbulence** Makroturbulenz *f*

**madder, to** ~ krappen

**madder** Färberröte *f*, Krapp *m* ~ **lake** Krapplack *m*

**made** hergestellt, zurückgelegt ~ **to precision limit** passungsgenau gefertigt ~ **to be** ~ **red-hot** ausgeglüht werden ~ **of a single piece** aus einem Stück *n* gefertigt ~ **to specifications** nach Vorschrift *f* gebaut

**made,** ~ **ground** Auffüllung *f* ~ **ready** bereitgestellt

**maelstrom** Strudel *m*

**mafic** ferromanganhaltig

**magazine** Behältnis *n*, Filmvorratsspule *f*, Kassettenhalter *m*, Lager *n*, Magazin *n*, Munitionsraum *m*, Niederlage *f*, Spulenträger *m*, Vorratsraum *m*, Zeitschrift *f* ~ **of woven fabrics** Stoffkassette *f*

**magazine,** ~ **arm** Spulenarm *m* ~ **attachment** Kassettenansatz *m* ~ **case** Kassettenkoffer *m* ~ **catch** Magazinhalter *m* ~ **chamber** (gun) Magazingehäuse *n* ~ **clip** Ladestreifen *m* ~ **compartment** Magazinkammer *f* ~ **coupling** Kassettenkupplung *f* ~ **cover** Kassettendeckel *m* ~ **creel** Magazin-Zettelgatter *n* ~ **drive** Kassetten-antrieb *m*, -laufwerk *n* ~ **driving pin** Kassettenschlüssel *m* ~ **drum** Filmabwickler *m* ~ **feed** Magazinspeisung *f* ~ **feed attachment** Magazinzuführung *f* ~ **feed equipment** Zuführungsmagazin *n* ~-**feed mechanism** Mehrladevorrichtung *f* ~ **floor plate** Kastenboden *m* ~ **frame** Magazinhalter *m* ~ **hand lamp** Kabellampe *f* ~ **holder** Magazinhalter *m* ~ **latch** Magazinsperre *f* ~ **load** Kassettenfüllung *f* ~ **lock** Kassettenriegel *m* ~ **mechanism** Mehrladeeinrichtung *f* ~ **opening** Magazineintritt *m* ~ **release button (or locking gear)**

Magazinverriegelung *f* ~ **retaining pin** Magazinhaltehebel *m* ~ **rifle** Mehrlader *m* ~ **roll** Vorratsrolle *f* ~ **seating frame** Kassetteneinsatz *m* ~ **slide** Magazinschlitten *m* ~ **spring-base stud** (pistol) Federkopf *m* ~ **stop** Magazinsperre *f* ~ **support** Spulenarm *m* ~ **valve** Kassettenschlitz *m*

**magic,** ~ **eye** Abstimmanzeigerröhre *f*, Anzeigeröhre *f*, magisches Auge *n* ~ **lantern** Laterna magica *f* ~ **number** magische Zahl *f* ~ **tee** Hohlleiterdifferentialübertrager *m*

**magma** Magma *n*, aufgemaischter Zucker *m* ~ **base** Grundmasse *f* ~ **reservoir** magmatischer Herd *m* ~ **tank** (crystallizer) Auffangbarke *f*

**magmatic intrusion** Magmenintrusion *f*

**magnaflux, to** ~ auf magnetischem Wege zerstörungsfrei auf Risse prüfen

**magnaflux** magnetische Oberflächenprüfung *f* ~ **inspection means** Fehlerscheibchensucher *m* ~ **inspection method** Eisenpulververfahren *n* ~ **inspection test** Magnaflux-Prüfung *f*

**magnalite** Magnalit *m*

**magnalium** (aluminum-magnesium alloy) Magnalium *n*

**magnascope** Vorsatzlinse *f*

**magna wheel** Magnarad *n*

**magnesia** Bittererde *f*, Magnesia *f*, Talkerde *f* ~ **alum** Magnesia-Alaun *m* ~ **brick** Magnesiastein *m* ~ **mordant** Magnesia-Beize *f* ~ **powder** Magnesia-Pulver *n* ~ **stone** Magnesiastein *m*

**magnesian** magnesiahaltig ~ **limestone** Dolomit *m*, Magnesiakalk *m* ~ **slate** Talkschiefer *m*

**magnesite** Bitterspat *m*, Magnesit *m*, Talk-spat *m* ~ **bottom** Magnesitherd *m* ~ **brick** Magnesitziegel *m* ~ **lining** Magnesit-auskleidung *f*, -zustellung *f*

**magnesium** Magnesium *n* ~ **antimonide** Magnesium-Antimonid *n* ~ **arsenate** Magnesiumarseniat *n* ~**-base alloy** Magnesiumlegierung *f* ~ **carbonate** Bitterspat *m* ~ **chloride** Chlormagnesium *n* ~ **content** Magnesiumgehalt *m* ~ **exychloride** Steinholz *n* ~ **flare** Magnesiumfackel *f* ~ **formate** Magnesiumformiat *n* ~ **hydroxide** Magnesiumoxydhydrat *n* ~ **hypophosphite** Magnesiumhypophosphit *n* ~ **light** Blitz-lampe *f*, -licht *n*, Magnesiumlicht *n* ~ **limestones** Bitter-kalk *m*, -kalkspat *m* ~ **oxide** Bittererde *f*, gebrannte Magnesia *f*, Magnesiausta *n*, Magnesiumoxyd *n* ~ **perborate** Magnesiumperborat *n* ~ **peroxide** Magnesiumhyperoxyd *n* ~ **plate** Magnesiumklischee *n* ~ **powder burner** Blitzlichtlampe *f* ~ **silicide** Siliziummagnesium *n* ~ **sulfate** Bittersalz *n*, Epsom-Salz *n*, Magnesiumsulfat *n*, schwefelsaures Magnesium *n* ~ **thiocyanate** Magnesiumrhodanid *n* ~ **torch** römische Kerze *f*, Magnesiumfackel *f* ~ **wire** Magnesiumdraht *m*

**magnet** Magnet *m* ~ **with anisotropic properties** Magnet *m* mit Vorzugsrichtung ~ **and motor driven shaking device** Magnet- und Motorrüttler *m*

**magnet,** ~ **bracket carrier** Magnetträger *m* ~ **carrier** Magnetschelle *f* ~ **case** Magnetgehäuse *n* ~ **coil** Feld-spule *f*, -wicklung *f*; Magnetspule *f*, -wicklung *f* ~ **coil voltage** Spulenbetätigungsspannung *f* ~ **construction** Magnetkonstruktion *f* ~ **core** Magnetkern *m* ~ **couple** Magnetpaar *n* ~ **cover** Magnetgehäusedeckel

*m* ~ **crab** Magnetkatzbahn *f* ~ **frame** Magnetkörper *m* ~ **galvanometer** D'Arsonval-Galvanometer *n*, Drehspulgalvanometer *n* ~ **gear** Magnetbetrieb *m* ~ **holder** Magnet-schelle *f*, -träger *m* ~ **housing** Magnetgehäuse *n* ~ **leg** Magnetschenkel *m* ~ **limb** Magnetschenkel *m* ~ **mirror** Magnetspiegel *m* ~ **separator** Magnetabschneider *m* ~ **service** Magnetbetrieb *m* ~ **short circuiting device** Magnetkurzschlußvorrichtung *f* ~ **space** Luftspalt *m* ~ **steel** Magnetstahl *m* ~ **support** Magnetträger *m* ~**-testing apparatus** Magnetprüfer *m* ~ **wheel** Magnet-, Pol-rad *n* ~ **yoke** Magnetjoch *n*

**magnetic** magnetisch, (as opposed to true) mißweisend ~ **after-effect** magnetische Nachwirkung *f* ~ **aging** zeitliche Desakkomodation *f* ~ **alloy** magnetische Legierung *f* ~ **alternating induction** magnetische Wechselinduktion *f* ~ **amplification** magnetische Verstärkung *f* ~ **amplifier** Transduktor *m*, magnetischer Verstärker *m* ~ **anomaly** magnetische Unregelmäßigkeit *f* ~ **antitank hollow charge** Hafthohlladung *f* ~**-armature speaker** elektromagnetischer Lautsprecher *m* ~ **attraction** magnetische Anziehungskraft *f*, Polsteinkraft *f* ~ **axis** magnetische Achse *f*, Magnetachse *f* ~ **azimuth** Kurswinkel *m*, Nadelzahl *f* ~ **balance** Feldwaage *f*, Magnetwaage *f*, magnetische Waage *f* ~ **ball-joint valve** Kugelschliff-Magnetventil *n* ~ **bearing** magnetische oder mißweisende Peilung *f* ~ **bias** magnetische Vorspannung *f* ~ **biasing** Vormagnetisierung *f* ~ **block** Magnetblock *m* ~ **blow** magnetische Blaswirkung *f* ~ **blow-out** magnetische Bogenbeeinflussung *f* ~ **blow-out coil** Funkenlöscher*f*, Schmelz-spule *f* ~ **brake** Magnetbremse *f* ~ **button** Magnettaste *f* ~ **capacity** magnetische Kapazität *f* ~ **changeover switch** Magnetumschalter *m* ~ **charge** magnetische Belegung *f* ~ **chuck** magnetische Aufspannvorrichtung *f*, Magnetspann-futter *n*, -platte *f*, -schraube *f* ~ **circuit** magnetischer Kreis *m*, Magnetkreis *m* ~ **circuit of regulator** Reglermagnetkreis *m* ~ **clutch** magnetische Kupplung *f*, Magnetkupplung *f* ~ **cobber** Magnetscheider *m* ~ **coil** Erregerwickelung *f*, Magnetisierungsspule *f* ~ **compass** (Magnet)-Kompaß *m*, Bussole *f* ~ **conductance** (or conductivity) magnetische Leitfähigkeit *f* ~ **conveyor** Magnetförderer *m* ~**-core** Massekern *m* ~**-core memory** Magnetkernspeicher *m* ~**-core storage** Magnetkern-speicher *m*, -speicherung *f* ~ **counter** Zählmagnet *m* (Zählrelais) ~ **coupling** induktive oder magnetische Kopplung *f* ~ **course** mißweisender Kurs *m* ~ **course to steer** mißweisender Zielkurs *m* ~ **crack detector** magnetische Anrißsucher *m*, Magnafluxgerät *n* ~ **cross flux** magnetische Streuung *f* ~ **current** magnetischer Fluß *m* ~ **cutter** magnetischer Schreiber *m*, Tonschreiber *m* (phono) ~ **cycle** magnetischer Kreislauf *m* oder Prozeß *m* ~ **declination** magnetische Deklination *f*, Nadelabweichung *f*, Mißweisung *f* ~ **deflection** magnetische Ablenkung *f* ~ **delay-line** magnetische Verzögerungsleitung *f* ~ **density** magnetische Dichtigkeit *f* ~ **detector** Erdfeldsonde *f*, Magnetdetektor *m* ~ **deviation** magnetische Abweichung *f* ~ **dip** Inklination *f* ~ **dipole** magnetischer Dipol *m* ~

**dipole moment** magnetisches Moment *n* **~ -disk storage** Magnetplattenspeicher *m* **~ dispersion** magnetische Streuung *f* **~ disturbance** magnetische Störung *f* **~ dram plug** Magnetstopfen *m* **~ drive** magnetische Kupplung *f* **~ drum** Magnettrommel *f* **~ fatigue** magnetische Nachwirkung *f* **~ field** magnetische Belegung *f*, Magnetfeld *n* **~ field of earth** magnetisches Erdfeld *f* **~ field coil** Magnetfeldspule *f* **~-field strength** magnetische Feldstärke *f* **transverse ~ field** transversales Magnetfeld *n* **~ figure** Kraftlinienbild *n* **~ flux** magnetischer Fluß *m*, Magnetfluß *m*, magnetische Induktion *f*, magnetischer Induktionsfluß *m*; Kraftfluß *m*, Kraftlinienfluß *m* **~ flux density** magnetische Flußdichte *f* **~ flux path** magnetischer Kraftlinienweg *m* **~ flux reversal** Kraftflußwechsel *m* **~ focusing** magnetische Fokussierung *f* **~ force** magnetische Feldstärke *f* oder Kraft *f* **~ forming** Magnetformverfahren *n* **~ frame** Magnetgestell *n* **~ friction** magnetische Reibung *f* **~ grader** Magnetscheider *m* **~ grate** Magnetabscheider *m* **~ gyro** magnetischer Kreisel *m* (mit magnetisch gelagertem Läufer) **~ heading** mißweisender Steuerkurs *m* **~ hum** magnetischer Brumm *m* **~ hysteresis** magnetische Hysterese *f* oder Hysteresis *f*, (magnetische) Reibungsarbeit *f* **~ inclination** Inklination *f*, Inklinationswinkel *m* **~ induction** magnetische Belegung *f* oder Induktion *f*, Magnetinduktion *f* **~ induction gyro** Magnetinduktionskreisel *m* **~ inductive capacity** magnetische Durchdringbarkeit *f*, Permeabilität *f* **~ inductivity** magnetische Durchdringbarkeit *f* oder Durchlässigkeit *f*, Permeabilität *f* **~ intensity** magnetische Feldstärke *f* oder Kraft *f* **~ interrupter** Magnetsummer *m* **~ iron** weiches Eisen *n* **~ iron ore** Ferro-Oxyd *n* **~ iron oxide** Eisenoxyduloxyd *n* **~ key** Tastrelais *n* **~ lag** magnetische Rücktrift *f* **~ laminated shunt** Eisenblechsäule *f* **~ latitude** magnetische Breite *f* **~ latitude effect** magnetischer Breiteneffekt *m* **(electro) ~ leak** magnetischer Nebenschluß *m* **~ leakage** magnetische Streuung *f* **~ lens** elektromagnetische Linse *f* **~ line of force** Kraftlinie *f* **~ links** magnetische Glieder *pl* **~ marker** Schreibmagnet *m* **~ memory** Magnetspeicher *m*, magnetischer Speicher *m* **~ meridian** magnetische Nord-Südrichtung *f* **~ microscope** Elektronenmikroskop *n* mit magnetischen Linsen **~ modulator** Steuer-, Tast-, Telefonie-drossel *f* **~ moment** magnetisches Moment *n* **~ moment density** Intensität *f* des magnetischen Moments **~ needle** Kompaß-, Magnet-nadel *f* **~ north** mißweisend Nord, magnetische Nordrichtung *f* **~ ore** magnetisches Erz *n* **~ particle inspection (or test)** Magnetpulverprüfung *f* **~ permeability** magnetischer Induktionskoeffizient *m*, magnetische Leitfähigkeit *f* **~ permeance** magnetischer Leitwert *m* **~ perturbation** magnetische Störung *f* **~ pick-up** elektromagnetischer Tonabnehmer *m* oder Geber *m* **~ picture tracing** Magnetbildverfahren *n* **~ plate memory** Magnetplattenspeicher *m* **~ plated wire** plattierter Magnetondraht *m* **~ plug** Abreißstecker *m*, Magnetstopfen *m* **~ plug box** Abreißsteckdose *f* **~ plug ignition** Magnetkerzenzündung *f* **~ polarization** magnetische Vor-

spannung *f* **~ pole** Magnetpol *m* **~ pole strength** Magnetpolstärke *f*, magnetische Polstärke *f* **~ potential** magnetische Spannung *f* **~ potential of coil** Spulendurchflutung *f* **~ potentiometer** Kompensator-Feldstärkemesser *m* **~ powder coupling (or clutch)** Magnetpulverkupplung *f* **~ power** Polsteinkraft *f* **~ powerclutch** magnetische Kopplung *f* **~ preamplifier to adjust the mixing ratio** Magnet-Vorverstärker *m* für Gemischregelung **~ printing** Kopiereffekt *m* (tape rec.) **~ properties** magnetische Eigenschaften *pl* **~ property** Magnetisierbarkeit *f* **~ pulley separator** Magnetabscheidetrommel *f* **~ pyrites** Magnetkies *m* **~ pyrometer** Galvanopyrometer *n* **~ radiation** magnetische Strahlung *f* **~ recorder** Magnetton-gerät *n*, -schreiber *m*, magnetischer Schreiber *m* **~ recording head** Magnettonkopf *m* **~ recording medium** Magnettonmaterial *n* **~ relay timer** magnetisches Zeitrelais *n* **~ reluctance of iron** magnetischer Eisenwiderstand *m* **~ residual loss** magnetische Nachwirkung *f* **~ resistance** Reluktanz *f*, magnetischer Widerstand **~ resistance of iron** magnetischer Eisenwiderstand *m* **~ resonance accelerator** Resonanzbeschleuniger *m* **~ retraction device** Magnetrückzugseinrichtung *f* **~ return path** magnetische Rückleitung *f*, magnetischer Rückschluß *m* **~ reversal** Ummagnetisierung *f* **~ reversing type controller** Schützenumkehrsteuerung *f* **~ rigidity** magnetische Steifigkeit *f* **~ ripple** magnetischer Brumm *m* **~ rotation** magneto-optische Drehung *f*, Magnetrückzug *m* **~ rotation spectrum** Magnetorotationsspektrum *n* **~ rotor** Magnetläufer *m* **~ scanner** Bildabtasterröhre *f* mit mechanischer Blende **~ screen** magnetischer Schirm *m* **~ separation** magnetische Scheidung *f* **~ separator** Magnet-ausscheider *m*, -scheider *m* **~ sheet** Magnettonfolie *f* **~ shell** magnetisches Blatt *n*, magnetische Doppelschicht *f*, Magnetschale *f* **~ shield** magnetischer Schirm *m* **~ shielding** magnetische Schirmung *f* **~ shunt** Eisenschluß *m*, magnetischer Nebenschluß *m* oder Schluß *m* **~ slide coupling** Magnetgleitkupplung *f* **~ solenoid valve** Magnetkreuzventil *n* **~ sound recorder** Magnetofon *n*, magnetischer Tonaufzeichner *m* **~ sound-recording equipment** Magnettonanlage *f* **~ sound-recording head** Magnetisierungskopf *m* **~ sound-recording method** Magnettonverfahren *n* **~ sound reproducer** Magnettonabtastgerät *n* **~ spark plug** Magnetkerze *f* **~ specimen holder** Magnethalter *m* **~ spectrograph** Geschwindigkeitsspektrograf *m*, magnetischer Spektrograf *m* **~ standard** Standardmagnet *m* **~ steel-tape recording** Stahlbandaufnahme *f* **~ strap for surface regulator** Magnetbügel *m* für Aufbauregler **~ stray flux** magnetischer Streufluß *m* **~ substance** magnetische Masse *f* **~ surface** Anzugfläche *f* **~ survey** magnetisches Schürfen *n* **~ susceptibility** magnetische Aufnahme *f*, magnetisches Aufnahmevermögen *n*, magnetische Einstreuung *f*, magnetische Empfindlichkeit *f*, magnetische Kapazität *f* **~ switch** Magnetschalter *m*, magnetisch betätigter Schalter *m*, Schütz *m* **~ tachometer** magnetischer Umdrehungsmesser *m* **~ tape** magnetisches Band *n*, Magnetband *n*,

Magnetofonband *n*, Magnetton-band *n*, -folie *f*, Tonband *n* ~-tape memory Magnetbandspeicher *m* ~-tape reader Magnetbandleser *m* ~-tape recorder Magnetofon *n*, Tonbandgerät *n* ~ test coil magnetische Prüfspule *f* ~ transition temperature magnetische Umwandlungstemperatur *f* ~ transmission Magnetübertragung *f* ~ trip (on circuit breaker) magnetische Auslösung *f* ~ tube of force magnetische Kraftröhre *f* ~ type controller Schützensteuerung *f* ~ variation magnetische Änderung *f* oder Deklination *f*, Mißweisung *f*, Ortsmißweisung *f* ~ viscosity magnetische Nachwirkung *f* ~ visual signal sichtbares Zeichen *n* ~ voltage stabilizer magnetischer Spannungsgleichhalter *m* ~ wire Magnettondraht *m* ~ wire recorder Drahtmagnetofon *n*

**magnetically,** ~ **hard** magnetisch hart ~ **soft** magnetisch weich

**magnetism** Anziehungskraft *f*, Magnetismus *m*

**magnetite** Magnet-eisenstein *m*, -stein *m* ~ **dense-medium** Magnetittrübe *f*

**magnetizability** Magnetisier-barkeit *f*, -fähigkeit *f*, Magnetisierungs-fähigkeit *f*, -koeffizient *m*, Suszeptibilität *f*

**magnetizable** magnetisier-bar, -fähig

**magnetization** (mechanical) Aufmagnetisierung *f*, Magnetisierung *f*, Vormagnetisierung *f* ~ **of a super-conductor** Magnetisierung *f* eines Supraleiters ~ **by touch** Strichmagnetisierung *f*

**magnetization,** ~ **curve** Magnetisierungskurve *f* ~ **cycle** Magnetisierungszyklus *m* ~ **vector** Magnetisierungsvektor *m*

**magnetize, to** ~ magnetisieren, magnetisch machen

**magnetized** magnetisiert ~ **fagot** Magnetbündel *n* ~ **sample** magnetische Probe *f*

**magnetizing** magnetisierend ~ **coil** Erregerwicklung *f*, Feldspule *f*, Feldwicklung *f*, magnetisierende Wicklung *f*, Magnet-spule *f*, -wicklung *f* ~ **current** Magnetisierungsstrom *m* ~ **device** Magnetisiervorrichtung *f* ~ **force** magnetische Feldstärke *f*, Magnetisierungs-kraft *f*, -stärke *f* ~ **jaw** Magnetisierbacke *f*

**magneto** (generator) Kurbelinduktor *m*, Magnetapparat *m*, -induktor *m*, -zündapparat *m*, -zünder *m*, magnetelektrische Maschine *f*, Zündapparat *m* ~ **with rotating sleeve** Magnetzünder *m* mit Umlaufhülse (ignition) ~ Zündmagnet *m*

**magneto,** ~ **bearing** Magnetlager *n* ~-**bell** Magnetwecker *m*, Wechselstrom-glocke *f*, -wecker *m* ~ **block** Magnetblock *m* ~ **booster** Anlaßmagnet *m* ~ **breaker box** Unterbrechergehäuse *n* ~ **caloric** magnetokalorisch ~ **carbon** Magnetkolben *m* ~ **central office** Vermittlungsamt *n* mit Induktoranruf ~ **coupling** Magnetkupplung *f* ~ **distributor** Verteiler *m* des Magnetapparates ~ **drive** Magnetantrieb *m* ~ **dynamo** Magnetdynamo *m* ~ **elastic** magnetoelastisch ~ **electric** magnetelektrisch ~ **electric instrument** Drehspulinstrument *n* ~ **exchange area** Netz *n* mit Ortsbatteriebetrieb ~-**firealarm system** Induktorfeuermeldesystem *n* ~-**flywheel ignition** Schwungradmagnetzündung *f* ~-**gear** Magnetrad *n* ~-**gear distance piece** Abstandsstück *n* zum Magnetrad ~-**gear**

**driving shaft** Magnetradantriebswelle *f* ~ **generator** Kurbelinduktor *m*, Zündlichtmaschine *f* (dynamo)

**magnetogram** Magnetogramm *n*

**magneto,** ~-**hydrodynamic** Magnetohydrodynamik *f* ~ **ignition** Magnetzündung *f* ~ **ignition switch** Magnetzünderschalter *m* ~-**ionic double refraction** magneto-ionische Doppelbrechung *f* ~ **lighter for gas appliances** Magnetzünder *m* für Gasgeräte

**magnetometer** Magnetometer *n*

**magneto,** ~-**metrical** magnetometrisch ~-**motive force** magnetometrische Kraft *f*, magnetomotorische Kraft *f*, Umlaufspannung *f* ~-**motive force meter** Spannungsjoch *n* ~-**optical rotation** magneto-optische Drehung *f*

**magnetophone** Magnetbandspieler *m*, magnetischer Tonaufzeichner *m*

**magneto resistance** magnetische Widerstandsänderung *f*

**magnetoscope** Magnetoskop *n*

**magentoscopical** magnetoskopisch

**magneto,** ~ **spindle** Zündmagnetwelle *f* ~ **spindle oil cup** Magnetspindelölfangschale *f* ~ **starter** Magnetanlasser *m* ~ **starting ignition** Magnetanlaßzündung *f*

**magnetostatic** magnetostatisch ~ **oscillator** magnetostatischer Oszillator *m*

**magnetostatics** Magnetostatik *f*

**magnetostriction** Magnetostriktion *f* ~ **delay line** magnetostriktives Laufzeitglied *n* ~ **oscillator** Magnetostriktions-generator *m*, -oszillator *m*, magnetostriktiver Oszillator *m* ~ **transceiver** Magnetostriktionsübertrager *m*

**magnetostrictive** magnetostriktiv

**magneto,** ~-**strop** Magnetbügel *m* ~-**switchboard** Klappenschrank *m* für Induktoranruf ~-**switchboard exchange** OB-Vermittlung(sstelle) *f* ~-**system** Magnetsystem *n* ~-**telephone set** Fernsprechapparat *m* mit Induktoranruf ~ **telephone station** Induktorapparat *m* ~-**tellurics** Magneto-Tellurik *f* ~ **timing** Magnetregelung *f*, Magnetzünder-, Zündpunkt-einstellung *f* ~ **timing scale** Magneteinstellskala *f* ~ **wrench** Magnetschlüssel *m*

**magneton** Magneton *n* ~ **number** Magnetonzahl *f*

**magnetron** Magnetfeldröhre *f*, Magnetron *n* ~ **amplifier** Magnetronverstärker *m* ~ **arcing** Überschläge *pl* im Magnetron ~ **cut-off** Nullanodenstromeffekt *m* ~ **oscillation** Magnetronschwingung *f* ~ **oscillator** Magnetrongenerator *m* ~ **pulling** Magnetronlastverstimmung *f* ~ **pulsing circuit** Magnetronkreis *m* ~ **pushing** Fehltastung *f* des Magnetrons

**magnets with anisotropic properties** Magnete *pl* mit Vorzugsrichtung

**magnettor** magnetischer Modulator *m*

**magnification** Überhöhung *f*, Vergrößerung *f*, Vergrößerungsvermögen *n*, Verstärkung *f*, Verstärkungscharakteristik *f* ~ **of the circuit** Resonanzschärfe *f*, Überspannungsverhältnis *n* ~ **in diameter** lineare Vergrößerung *f* ~ **on the focusing screen** Mattscheibenvergrößerung ~ **change revolver** Revolver *m* zum Vergrößerungswechsel

**magnification,** ~ **factor** Q-Faktor *m*, Widerstandsverhältnis *n* ~ **meter** Q-Messer *m* ~ **step** Vergrößerungsstufe *f*

magnified verstärkt
magnifier Lupe *f*, Vergrößerer *m*, Verstärker *m*
~ reading Lupenablesung *f* ~ stand Lupen-
gestell *n*, -ständer *m*, -stativ *n* ~ unit Einzellupe
*f* ~ work Lupenaufnahme *f*
magnify, to ~ vergrößern, verstärken
magnifying, ~ and diminishing mirror Vergrö-
ßerungs- und Verkleinerungsspiegel *m*
magnifying, ~ glass Ableselupe *f*, Lupe *f*, Vergrößerungsglas *n* ~ lens Lupe *f*, Vergrößerungsglas *n*, -linse *f* ~ lens for photographs Foto-
grafielupe *f* ~ lens for precious stones Steinlupe
*f* ~ power Vergrößerungs-kraft *f*, -stärke *f*,
-vermögen *n*
magnitude Ausschlag *m*, Betrag *m*, Größe *f*,
Größen-klasse *f*, -ordnung *f*, -zahl *f*, Helligkeit
*f*, Höhe *f*, Mächtigkeit *f*
magnitude, ~ of cloudiness Bewölkungs-grad *m*,
-größe *f* ~ of the command given Aufschaltgröße *f* (g/m) ~ of error Fehlergröße *f* ~ of
oscillation Schwingungsgröße *f* ~ and phase
Betrag *m* und Phase *f* ~ of resistance Widerstandskraft *f* ~ of vector Tensor *m* des Vektors
magnitude, ~ diagram Größenklassen-Diagramm
*n* ~ ratio Amplitudenverhältnis *n*, Betrag *m*
Magnus effect Magnuseffekt *m* (ballistics)
mahogany Akajou-, Amaranth-holz *n*, Mahagoni
*m*, Mahagoniholz *n* ~ brown mahagonibraun
~ lacquer Mahagonilack *m*
maiden flight Erst-, Jungfern-flug *m*
mail, to ~ absenden, aufgeben
mail Briefpost *f*, (letters) Post *f* ~ bag Brief-
beutel *m*, Postsack *m* ~ box Brief-, Post-kasten
*m* ~ bus Postomnibus *m* ~ compartment
Fracht-, Post-abteil *n* ~ delivery Postzustellung
*f* ~ flight Postflug *m* ~ heald (textiles) Draht-
ösenlitze *f* ~ load Postladung *f* ~-order house
Versandhaus *n* ~ plane Postflugzeug *n* ~ room
Postabteil *n* ~ service Postdienst *m* ~ slot
Briefeinwurf *m* ~ steamer Postdampfer *m*
mailer postfertige Verpackung *f*
mailing, ~ label address Paketadresse *f* ~ tube
Versandrolle *f*
mails Postsachen *pl*
maim, to ~ verstümmeln
main Hauptrohr *n*, Leitung *f*, Röhrengang *m*,
Sammelrohr, Verbrauchsleitung *f*; hauptsächlich ~ A. C. bus system Hauptwechselstrom-
schienensystem *n* ~ adit Hauptschlüsselstollen
*m* ~ adit for draining a mine Erbstollen *m* ~
adjusting lever Hauptverstellhebel *m* ~ adjustment Hauptregelung *f* ~ air intake Hauptlufteinlaß *m* ~ aisle Mittel-, Haupt-gang *m*
(Lager) ~ amplifier Hauptverstärker *m* ~
anchor Hauptanker *m* ~-and-tail-rope haulage
Doppelseilförderung *f* ~ anode Hauptanode *f*
~ arch Mittelöffnung *f* ~ assembly jig Haupt-
spannvorrichtung *f* ~ axis of inertia Hauptträgheitsachse *f* ~ balance beam Laufwerk-
schwinge *f* ~ ballast pump Hauptlenzpumpe *f*
~ ballast tank Hauptballasttank *m*, Haupt-
tauchtank *m* ~ bang Auslöse-, Sende-, Start-,
Steuer-impuls *m* ~ bar Hauptbewehrung *f* ~
barrel (carburetor) Hauptluft-düse *f*, -trichter
*m* ~ base Hauptliegehafen *m* ~ basic material
Hauptausgangsmaterial *n* ~ basis Hauptgrund-
lage *f* ~ beam Binderlängsbalken *m*, Fernlicht
*n*, Haupt-balken *m*, -kettbaum *m*, -träger *m* ~

beam adjustment Fernlichteinstellung *f* ~ beam
filament Fernlichtfaden *m* ~ beam indicator
light Fernlichtanzeigelampe *f* ~-beam killing
Hauptstrahlunterdrückung *f* ~ bearer plate
Hauptlagerplatte *f* ~ bearing Hauptantriebs-
lager *n*, Haupt-, Rahmen-, Wellen-lager *n*,
Hauptunterbau *m* ~ bearing of a crankshaft
Hauptlager *n* der Kurbelwelle ~ bearing shell
Grundlagerschale *f* ~ bearing support Grund-
lagerstühle *pl* ~ bell Unterglocke *f* ~ blast line
Hauptwindleitung *f* ~ body Hauptfallschirm *m*
~ boiler Hauptkessel *m* ~ boiler blow-off valve
Kesselablaßventil *n* ~ boom Giekbaum *m* ~
bottom festes Gestein *n* unter alluvialer Ablagerung ~ brace Großbrambraß *f* ~ bracing
Hauptverspannung *f* ~ breadth (ship) Flanken-
ausbauchung *f* ~ building Hauptgebäude *n* ~
bulb Hauptlampe *f* ~ bulkhead Haupt-spant *n*,
-schnitt *m* ~ burner pressure point Manometerstutzen *m* für Brennerdruck ~ bus bar Haupt-
sammelschiene *f* ~ cable Haupt-kabel *n*,
-linienkabel *n*, -strangkabel *n* ~ cam Haupt-
exzenter *m* ~ canal (in sewerage system) Vorfluter *m* ~ car (or gondola) Hauptgondel *f* ~
carrier Hauptträger *m* ~ center Knotenamt *n* ~
central office Knotenamt *n* ~ chains Großrüst *f*
~ change in form Hauptformänderung *f* ~
channel große Rüste *f* ~ channels Großrüst *f*
~ charge Haupt-, Kern-ladung *f* ~ check valve
Hauptabsperrventil *n* ~ circuit Hauptstromkreis *m*, Kreislaufsystem *n*, Stammleitung *f* ~
clutch Hauptkupplung *f* ~ cogwheel Hauptzahnrad *n* ~ column Haupt-pfeiler *m*, -säule *f*,
Königssäule *f* ~ combustion chamber Haupt-
verbrennungsraum *m* ~ connecting device Vorschaltgerät *n* ~ connecting rod Haupttriebs-,
Königs-stange *f* ~ contactor Hauptschütz *m* ~
contract Hauptvertrag *m* ~ control surface
Haupt(steuer)ruder *n* ~ control valve Haupt-
steuerventil *n* ~ couple Binderbalken *m* ~ crab
Hauptkatze *f* ~ current Hauptstrom *m* (electr.),
(river) Hauptströmung *f* ~ current filter Haupt-
stromfilter *m* ~ current relay Primärrelais *n* ~
cutting edge (für Schneidestahl oder Fräszahn)
Hauptschneide *f* ~ cylinder Haupt-trommel *f*,
-walze *f*, -zylinder *m*, Tambour *m* ~ data-con-
version instrument Zentralumwertegerät *n* ~
deck Hauptdeck *n* ~ deformation Hauptformänderung *f* ~ depot Hauptniederlage *f* ~
difference Hauptunterschied *m* ~ dike Haupt-
deich *m*, -schütz *n* ~ dimensions Hauptabmessungen *pl* ~ direction Hauptrichtung *f* ~
directorate Hauptausschuß *m* ~ distributing
frame Hauptverteiler *m* ~ district exchange
Bezirksamt *n* ~ ditch (in sewerage system) Vorfluter *m* ~ drainage canal Hauptentwässerungs-
graben *m* ~ drive clutch lever Hauptantriebs-
kupplungshebel *m* ~ drive shaft Hauptantriebs-,
Zwischen-welle *f* ~ drive wheel Hauptantriebs-
rad *n* ~ driving bevel gear Spindelkegelrad *n* ~
driving gear Hauptantriebs-achse *f*, -rad *n* ~
driving pinion Hauptplanzugritzel *n* ~ driving
spring Hauptantriebsfeder *f* ~ drop Leitungsabfall *m* ~ drum Hauptwalze *f* ~ drum wheel
Haupttrommelrad *n* ~ eccentric Grundexzenter
*n* ~ effort Hauptangriff *m* ~ emplacement
Hauptstellung *f* ~ engine Haupt-maschine *f*,
-motor *m* ~ engine drive Hauptmotorenantrieb

*m* ~-engine test Vollmotorenversuch *m* ~ essential condition Hauptbedingung *f* ~ exchange Haupt(fernsprech)amt *n*, Hauptvermittlungsstelle *f*, Vermittlungszentrale *f* ~ factor Hauptmoment *n* ~ feature Hauptmerkmal *n* ~-feed check valve Hauptspeiseventil *n* ~ feeder Hauptspeiseleitung *f* (electr.) ~-feeder trench (sluice) Hauptspeisegraben *m* ~ field Hauptfeld *n* ~ field compensating coil Hauptfeldausgleichspule *f* ~ field magnetic switch coil Hauptfeldmagnetschalterspule *f* ~ field winding Haupterregerwicklung *f* ~ file Grundregister *n* ~ filter Hauptfilter *n* ~ flash Hauptstrahl *m* ~ float Hauptschwimmer *m* ~ floodgate Hauptschütz *n* ~ flooding valve Hauptanblaseventil *n* ~ floor Hauptgeschoß *n* ~ flow Grund-, Haupt-strömung *f* ~ flue (boiler) Hauptzug *m* ~ formants Hauptformanten *pl* ~ foundation Fundamentsohle *f* ~ frame Hauptspant *n*, -spantschnitt *m*, -verteilergestell *n*, Richtspant *n* ~ framework Hauptrahmen *m* ~ framing Hauptgestell *n* ~ fuel tank Kraftstoffhaupttank *m* ~ fuse Hauptsicherung *f* ~ fuse alarm lamp Hauptalarmlampe *f* (HA) ~ gallery Haupt-arm *m*, -stollen *m* ~ gangway Gezeugstrecke *f*, Lauf *m*, Sohlenstrecke *f* ~ gap Hauptentladungsstrecke *f* ~ gas valve (balloon) Ballonventil *n* ~ gasket Hauptdichtung *f* ~ gear Hauptantrieb *m* ~ gearing Haupträdergetriebe *n* ~ girder Binderbalken *m*, Haupt-langträger *m*, -träger *m* ~ governor (of a turbine) Hauptregler *m* ~ hatch Großluke *f* ~ hauling line Fessel-kabel *n*, -seil *n* ~ headings Hauptgruppen *pl* ~ highway Hauptverkehrsstraße *f* ~ hoisting tackle Haupthubwerk *n* ~ hood beam (windmill) Fügbalken *m* ~ hook Haupthaken *m* ~ housing Grundsonde *f* ~ ignition unit Hauptzündgerät *n* ~ inlet Hauptzuleitung *f* ~ inlet dike (or sluice) Haupteinlaufschütze *f* ~ item Hauptsache *f* ~ jet Hauptdüse *f*, -strahl *m* ~ lamp Hauptlampe *f* ~ land Festland *n* ~ landing gear Hauptfahrwerk *n* ~ lead sleeve for multiple joint Verzweigungsmuffe *f* aus Blei ~ leg of position-vector between guided missile and cutoff ground site Hauptweg *m* ~-leg coefficient Hauptwegübersetzung *f* ~ level Hauptstrecke *f* ~ limit switch Grenzschalter *m* ~ line Haupt-bahn *f*, -fluglinie *f*, -leitung *f*, -linie *f*, Vollbahn *f* ~ line of resistance Hauptkampflinie *f* (mil.), Hauptwiderstandsstellung *f* ~ line current Ablaßleitung *f* ~-line telephone Verbindungsrichtung *f* ~ load-distributing station for entire network Hauptlastverteilungsstelle *f* für das gesamte Netz ~ lobe Hauptkeule *f* ~ lobe of radiation Hauptstrahlzipfel *m* ~ longitudinal beam (or girder) Hauptlängsbalken *m* ~ lubricating oil filter Hauptstromölfilter *n* ~ lubricating oil line Hauptverteilerleitung *f* ~ lubricating oil pipe Hauptschmierleitung *f* ~ machine (Scherbius sets) Vordermaschine *f* ~-magnet fuse Hauptmagnetzünder *m* ~ meterological station Hauptwetterwarte *f* ~ mooring cable Hauptfesselkabel *n* ~ mooring line Haupthalteleine *f* ~ mooring-mast line Hauptverankerungsmastseil *n* ~ motor Hauptantriebsmotor *m* ~ object Hauptzug *m* ~ oblique tie Zugdiagonale *f* ~ office Haupt-amt *n*, -verwaltung *f*, Vollamt *n* ~ oil

gallery Hauptölbohrung *f* ~ oil pipe Hauptölleitung *f* ~ operating platform Hauptbedienungsstand *m* ~ oscillation Grundschwingung *f* ~ oscillation generator (in independent-drive system) Hauptrohr *n* ~ oscillator Hauptsender *m* ~ oscillator stage nachgeschaltete gesteuerte Stufe *f* ~ oxygen valve A-Entlufterhauptventil *n* ~ pantograph Hauptstromabnehmer *m* ~ part Hauptteil *m* ~ patent Hauptpatent *n* ~ phase Hauptphase *f* ~ pinion Rollrad *n* ~ pipe Haupt-leitung *f*, -rohr *n* ~ pipes Stammleitung *f* ~ piston Arbeitskolben *m* ~ pit Hauptschachtbetrieb *m* ~ plane Tragwerk *n* ~ plane structure Tragwerk *n* ~ plate Hauptfederbett *n* ~ plotting station Hauptmeßstelle *f* ~ plug-out connection (signal) Ausnehmehauptanschluß *m* ~ point Hauptsache *f* ~ point of penetration Haupteinbruchsstelle *f* ~ pole dynamo sheet Hauptpoldynamoblech *n* ~ pole shim Hauptpolunterlegblech *n* ~ poop Binderbalken *m* ~ position Hauptstellung *f* ~ power plant Hauptmaschinenanlage *f* ~ presentation Oszillograf *m* ~ presentation unit Anzeigelehre *f* ~ problem Kernfrage *f* ~ proceedings Hauptverhandlung *f* ~ product Haupterzeugnis *n* ~ pulley Antriebs-, Hauptriemen-scheibe *f* ~ pulse Sendeimpuls *m* (rdr) ~ pump spears Hauptpumpengestänge *n* ~ purpose Hauptzweck *m* ~ quantum number Hauptquantenzahl *f* ~ radiation direction Hauptstrahlrichtung *f* ~ rail of a siding Hauptschiene *f* einer Weiche ~ railroad station Hauptbahnhof *m* ~ range tube Übersichtsrohr *n* (electron.) ~ reduction gear Hauptvorgelege *n* ~ reed Hauptriet *n* ~ regulator (of a turbine) Hauptregler *m* ~-regulator board Hauptreglertafel *f* ~ reinforcement Hauptbewehrung *f* ~ reinforcing bar Tragstab *m* ~ (-line) relay Linienrelais *n* ~ repeater station Hauptverstärkeramt *n* ~ reserve Hauptreserve *f* ~ reservoir Hauptbehälter *m* ~ resonance frequency Hauptresonanzfrequenz *f* ~ restoration relay Wiedereinschaltrelais *n* ~ rib (wing) Hauptrippe *f* ~ river Haupt-fluß *m*, -strom *m* ~ road Chaussee *f*, Haupt-förderstrecke *f*, -straße *f*, -strecke *f*, Heerstraße *f* ~ rod Schachtgestänge *n* ~ rope Schwenk-, Schwung-seil *n* ~ rotor Hauptrotor *m* ~ rotor blade Hauptrotorblatt *n* ~ rotor blade grip bearing Hauptblattanschlußlager *n* ~ rotor hub Hauptkopf *m* ~ route Haupt-kabel *n*, -linie *f* ~ royal Großroyal *n* ~ runner Hauptlauf *m* ~ runway Hauptpiste *f* ~ sail (of parachute) Hauptfallschirm *m* ~ scavenging Hauptspülung *f* ~ separation screen Hauptseparationstrommel *f* (min.) ~ set Fernsprechhauptanschluß *m*, Teilnehmer-Hauptanschluß *m* ~ sewer Hauptkanal *m* ~ shaft Antriebs-, Getriebeantriebsförderwelle *f*, Empfängerachse (teletype), Haupt-achse *f*, -schacht *m*, -welle *f*, Motorwelle *f* ~ shaft bearing Hauptantriebslager *n* ~-shaft mounted generator Wellengenerator *m* ~-shaft spacing gear Vorschubschaltbuchse *f* ~ shifting Hauptschaltung *f* ~ side Hauptgang *m* ~ signal frame Hauptsignalrahmen *m* ~ sill Haupt-schwelle *f*, -sohle *f* ~ sleeve for multiple joint Verzweigungsmuffe *f* ~ source Hauptquelle *f* ~ span (Brücke) Hauptöffnung ~ spar Hauptholm *m*, Hauptträger *m* (aviat.), Stand-

holm *m* ~-spar flange Hauptholmgurt *m* ~
spears Hauptgestänge *n* ~ spindle Arbeits-,
Haupt-spindel *f* ~ spindle speeds Hauptspindel-
drehzahlen *pl* ~ spring Schlag-, Trieb-feder *f* ~
spring drum Zugfederhaus *n* ~ sprocket Belichtungszahntrommel *f* ~ standard Haupt-
ständer *m* ~ starting air valve Hauptanfahr-
ventil *n* ~ station Hauptstelle *f* (subscriber's) ~
station Fernsprechhauptanschluß *m* ~-station
intersection method Hauptschnittverfahren *n* ~
steamplant valve Hochdruckventil *n* (g/m) ~
step on a flotation gear (or stage) Hauptstufe *f*
~ stop Haupttakt *m* ~ stream Grundströmung
*f*, Hauptfluß *m* ~ strut Haupt-stab *m*, -stiel *m*
~ supporting surface Haupt-haltfläche *f*,
-tragfläche *f* ~ switch Haupt-betriebschalter *m*,
-schalter *m*, Netzumschalter *m* ~ switch-board
Haupt-anschluß *m*, -schalttafel *f* ~ switch
station Hauptschaltwarte *f* ~ switching, Hauptschaltung *f* ~ system Leitungs-anlage *f*, -netz *n*
~ tank Haupt-behälter *m*, -tank *m* ~ tank vent
(torpedo) Haupttauchzellenventil *n* ~ telephone station Teilnehmer-Hauptanschluß *m* ~
thrust bearing Hauptdrucklager *n* ~ tide
Stammtide *f* ~ tilting axis Hauptkippachse *f*
(photo) ~ timber of a truss frame Spannriegel
*m* eines Hängewerkes ~ tool slide Hauptwerk-
zeugschlitten ~ toothed wheel Hauptzahnrad *n*
~ top Großmars *m* ~ track Haupt-bahn *f*,
-geleise *n* ~ transformer Netzspeisetransforma-
tor *m* ~ transmission gear Hauptgetriebe *n* ~
transmitter Hauptsender *m* ~ traverse Haupt-
ring *m* ~ trunk line Hauptstrang *m* ~ truss
Binderbalken *m* ~ tuning button Hauptab-
stimmknopf *m* ~ types of combat Hauptge-
fechtsarten *pl* ~ undercarriage Hauptfahrge-
stell *n* ~ valve Haupt-schütz *n*, -ventil *n* ~
vanishing point Hauptverschwindungspunkt *m*
~-velocity component Hauptgeschwindigkeit *f*
~ ventilating pipe Hauptlüftungsrohr *n* ~
voltage Netzspannung *f* ~ voltage fluctuations
Netzspannungsschwankungen *pl* ~ wave hin-
laufende Welle *f* ~ wave guide Haupthohlleiter
*m* ~ weather station Hauptwetterwarte *f* ~
wheel Hauptrad *n* ~ (driving) wheel Stirntreib-
rad *n* ~ winding Hauptwicklung *f* ~ wing
Hauptflügel *m* ~ zone center Hauptfernamt *m*
mains Stammleitung *f*, Starkstromnetz *n* ~ and
high-voltage apparatus Netz- und Hochspan-
nungs-Geräte *pl*
mains, ~ antenna Lichtantenne *f*, Lichtnetzan-
tenne *f* ~ connecting plug Starkstromstecker *m*
~ connection Netzanschluß *m* ~ connection
unit Netzanschlußgerät *n* ~ contactor Netz-
schütz *m* ~ control Netzkontrolle *f* ~ current
supply Netzstromversorgung *f* ~ cutout Netzausschalter *m* ~-driven receiver Lichtnetz-,
Netzanschluß-, Netz-empfänger *m* ~ fed (or
operated) netzgespeist ~ feed unit Netzspei-
sungsgerät *n* (rdo) ~ filter Siebkreis *m* ~ final
stage Netzendstufe *f* ~ fluctuation Netzunruhe *f*
~ frequency Netzfrequenz *f* ~ fuse Netzsiche-
rung *f* ~ heating unit Netzgerätheizung *f* ~ hold
Netzsynchronisierung *f* ~ hum Netzton *m* ~
input Netz-eingang *m*, -einspeisung *f* ~ interruption Netzstörung *f* ~ line hum Netzbrummen
*n* ~ noise Netzgeräusch *n* ~-operated netz-
betrieben ~-operated receiver Netzempfänger

*m* ~ output Netzendstufe *f* ~ plug-in unit Netzgeräteeinschub *m* ~-power bell Starkstrom-
glocke *f* ~ power supply Bordnetz *n*, Netzbe-
trieb *m* ~-powered netzgespeist ~ receiver
Empfänger *m* für Gleichstrom (Netzanschluß),
Empfänger *m* für Wechselstrombetrieb, Empfänger *m* für Wechselstrom(-Netzanschluß),
Lichtnetzempfänger *m*, Netzanschlußempfänger *m*, Netzempfänger *m* ~ restoration relay
Wiedereinschaltrelais *n* ~ ripple Netzunruhe *f*
~ section Netzzuführungseinschub *m* ~ set
Lichtnetz-, Netzanschluß-, Netz-empfänger *m*
~ supply (public electric supply system) Netzanschluß *m*, -versorgung *f*, Vollnetzanschluß *m*
~ switch Netzschalter *m* ~ switchboard Netz-
schalttafel *f* (rdo) ~ synchronization Netzsynchronisierung *f* ~-synchronized netzsynchronisiert ~ tapping panel Netzspannungs-
wähler *m* ~ transformer Netz-trafo *m*, -trans-
formator *m* ~ transmitter Netzsender *m* (rdo)
~ unit Netzteil *m* ~-unit adapter Netzgerät-
einsatz *m* ~ voltage Netzspannung *f* ~ voltage
fluctuations Netzspannungsschwankungen *pl* ~
voltage stabilizer Netzspannungsstabilisierung *f*
maintain, to ~ aufrechterhalten, behaupten, beibehalten, erhalten, hemmen, instandhalten,
pflegen, unterhalten, wahren, warten to ~
constant konstant halten to ~ contact Fühlung
*f* halten to ~ course and speed durchhalten to ~
equal pressure Druck *m* konstant halten to ~
speed Fahrt halten to ~ stationary anhalten,
arretieren, bremsen, festhalten, feststellen to ~
the trajectory die Bahn einhalten
maintainable speed Dauergeschwindigkeit *f*
maintained contact push button Dauerkontakt-
schalter *m*
maintaining constant Konstanthaltung *f*
maintenance Aufrechterhalten *n*, Aufrechterhaltung *f*, Erhaltung *f*, Instandhaltung *f*, Instandsetzungsarbeiten *pl*, Nacharbeit *f*, (work)
Pflege *f*, Unterhalt *m*, Unterhaltung *f*, (costs)
Unterhaltungskosten *pl*, Wahrung *f*, Wartung *f*
maintenance, ~ of the depth Erhaltung *f* oder
Unterhaltung *f* der Tiefen ~ of an installation
Unterhaltung *f* einer Anlage ~ of law and order
Aufrechterhaltung *f* der öffentlichen Ordnung
~ of the navigable channel Unterhaltung *f* des
Fahrwassers ~ of service Ausübung *f* des Dienstes, Dienstverrichtung *f*, Verrichtung *f* des
Dienstes
maintenance, ~ book Wartungsbuch *n* ~ charge
Erhaltungskosten *pl*, (periodic) Instandhal-
tungsgebühr *f*, Unterhaltungskosten *pl* ~ check
Wartungs-kontrolle *f*, -überprüfung *f* ~ circuit
Festhaltestromkreis *m* ~ cost Instandhaltungs-,
Wartungs-kosten *pl* ~ costs Unterhaltungskosten *pl* ~ department Leitungsüberwachungs-
dienst *m* ~ echelon Instandsetzungsstaffel *f*
~-free wartungsfrei ~ instructions Betriebsanweisung *f* ~ log book Wartungsbuch *n* ~
manual Service-Anleitung *f*, Wartungshand-
buch *n* ~ measurement Instandhaltungsmessung
*f* ~ party Werkstatttrupp *m* ~ personnel War-
tungspersonal *n* ~ pressure (balloon) Prall-
haltung *f* ~ program Überwachungsplan *m* ~
release Wartungsabnahmeschein *m* ~ routine
laufende Wartung *f* ~ section Gefechtstroß *m*,
Waffenmeisterei *f* ~ service Meßdienst *m* ~

**stand** Arbeitsbühne *f* ~ **standard** Erhaltungszustand *m* ~ **test** Überwachungsliste *f* ~ **testing schedule** Meßplan *m* ~ **truck** Instandsetzungskraft-, Werkstatt-wagen *m* ~ **true bearing** konstante wahre Peilung *f* ~ **unit** Instandsetzungsabteilung *f* ~ **work** Unterhaltungsarbeit *f* ~ **(routine) work** Instandhaltungsarbeiten *pl*, Wartungsarbeit *f*
**maize-flaking mill** Maisflockenstuhl *m*
**majolica** Majolika *f*
**major** Dur *n*, Major *m*; hauptsächlich, Haupt..., vorwiegend ~ **apex face** Hauptspitzenfläche *f* (eines Quarzkristalls) ~ **axis** große Achse *f*, Haupt-, Längs-achse *f* ~ **axis of airplane** Flugzeuglängsachse *f* ~ **bed** Hochwasserbett *n* ~ **commitment** Großeinsatz *m* ~ **constituents** Hauptgemengteile *pl* (geol.) ~ **cycle** Hauptperiode *f* ~ **diameter** Außengewindedurchmesser *m* ~ **feedback** Hauptrückführung *f* ~ **graduations** Hauptskalenteilungen *pl* ~ **lobe** Haupt-keule *f* (rdr), -lappen *m*, (of space pattern) -zipfel *m* ~ **part** Majorität *f* ~ **principal stress** größere Hauptspannung *f* ~ **resonant speed** überkritische Geschwindigkeit *f* ~ **septum** Hauptsepten *pl* (geol.) ~ **switch** großer Wähler *m* ~ **thread diameter** Gewindeaußendurchmesser *m*
**Majorana particle** Majorana-Teilchen *n*
**majorant** Majorante *f*
**majority** Majorität *f*, Mehrheit *f*, Mehrzahl *f*, Überzahl *f*, (in numbers) Zahlenüberlegenheit *f* ~ **of votes** Stimmenmehrheit *f*
**majority,** ~ **carrier** Majoritäts-Mehrheitsträger *m* ~ **carrier contact** Majoritätsträgerkontakt *m* ~ **charge carrier** Majoritätsladungsträger *m* ~ **emitter** Majoritäts- oder Mehrheitsemitter *m*
**make, to** ~ anbringen, anfertigen, ausmachen, bereiten, bewerkstelligen, bilden, erzeugen, fabrizieren, fertigen, machen, verfertigen
**make, to** ~ **an accounting** abrechnen **to** ~ **an agreement** einen Vertrag *m* abschließen **to** ~ **(a circuit) alive** (eine Leitung) unter Spannung *f* setzen **to** ~ **an allowance** ablassen, Rabatt *m* bewilligen oder geben **to** ~ **an allowance on a charge** einen Gebührennachlaß *m* gewähren **to** ~ **amends** entschädigen **to** ~ **apparent** verdeutlichen **to** ~ **an appointment** sich verabreden **to** ~ **with appropriateness** sachgemäß herstellen **to** ~ **arcuate** wölben **to** ~ **arrangements** Anstalten *pl* machen **to** ~ **available** bereitstellen, herausbringen, nutzbar machen, zur Verfügung stellen **to** ~ **into bars** zainen **to** ~ **bricks** ziegeln **to** ~ **bull** (foundry) sich beruhigen lassen **to** ~ **a carbon copy** durchpausen **to** ~ **the chimes of a cask** ein Faß *n* gergeln oder gargeln **to** ~ **a circuit** einen Stromkreis *m* herstellen oder schließen **to** ~ **a claim** erheben, geltend machen (Anspruch) **to** ~ **clear** verdeutlichen **to** ~ **cloudy** verschleiern **to** ~ **a complete revolution** durchdrehen **to** ~ **concrete** festmachen **to** ~ **conditions** Bedingungen *pl* aufstellen **to** ~ **connection** Kontakt *m* machen **to** ~ **conspicious** hervorheben **to** ~ **constant** stabilisieren **to** ~ **contact-** Fühlung *f* aufnehmen, Kontakt *m* machen, einen Kontakt *m* öffnen, Verbindung *f* aufnehmen **to** ~ **a contact (or connection)** eine Verbindung *f* herstellen **to** ~ **contact with the water** (airplane) anwassern **to** ~ **a contract**

einen Vertrag *m* abschließen, verdingen, vereinbaren **to** ~ **a crash landing** bruchlanden **to** ~ **dead** stromlos machen **to** ~ **a detour** umgehen **to** ~ **effective** in Wirkung *f* setzen **to** ~ **an emergency landing** notlanden **to** ~ **an emergency landing on water** notwassern **to** ~ **an end-to-end test** Streckenmessung *f* machen **to** ~ **an estimate** überrechnen **to** ~ **even** ausgehen, gleichmachen **to** ~ **evident** ausweisen, verdeutlichen **to** ~ **a fair (or final) copy** ins Reine *n* schreiben **to** ~ **fast** (ein Tau) anschlagen, festigen, festmachen **to** ~ **fast to shore** sich am Lande *n* vertauen **to** ~ **final** besiegeln **to** ~ **a finishing cut** sauber fertig arbeiten **to** ~ **a fire** anfeuern, heizen **to** ~ **firm** festigen, festmachen, verfestigen **to** ~ **a first cut** vorschneiden **to** ~ **flush with** bündig machen **to** ~ **for** ansteuern **to** ~ **a forced landing** notlanden **to** ~ **good** ersetzen, vergüten **to** ~ **good a faulty circuit** eine fehlerhafte Leitung *f* ersetzen **to** ~ **good a loss** Verlust *m* ausgleichen (wettmachen) **to** ~ **headway** Boden *m* gewinnen **to** ~ **heavy** erschweren **to** ~ **impermeable** dicht machen **to** ~ **indistinct** verschmieren **to** ~ **inquiries** Umfrage *f* halten **to** ~ **insensitive** desensibilisieren **to** ~ **intersections** anschneiden **to** ~ **known** bekanntmachen, offenbaren, verkünden **to** ~ **the land** auf den Strand *m* kommen, stranden **to** ~ **lean** magern **to** ~ **legal** rechtskräftig machen **to** ~ **lighter** aufhellen **to** ~ **the line busy** einen Wähler *m* sperren **to** ~ **live** (of bombs) scharfstellen **to** ~ **a loop test** Schleifenmessung *f* machen **to** ~ **the most of** ausnutzen **to** ~ **a motion** beantragen **to** ~ **a noise like a projectile** fauchen **to** ~ **nonappealable** rechtskräftig machen **to** ~ **nonluminous** entleuchten **to** ~ **normal** stabilisieren **to** ~ **a note of** vormerken **to** ~ **oblate** abflachen **to** ~ **off** abtürmen **to** ~ **an offer** einen Antrag stellen **to** ~ **opaque by grinding** mattschleifen, matt schleifen **to** ~ **out** abstecken, (report) ausfertigen **to** ~ **partitious** unterabteilen **to** ~ **into a paste** aufschlämmen **to** ~ **poor** magern **to** ~ **port** in einen Hafen *m* einlaufen **to** ~ **preparations** Anstalten *pl* machen **to** ~ **a position inaccessible** einen Platz *m* sperren **to** ~ **prominent** hervorheben **to** ~ **propaganda** werben **to** ~ **public** bekanntgeben **to** ~ **pulse** Stromschließungsstoß *m* (teleph.) **to** ~ **ready** bereiten, bereitmachen, klar machen, zubereiten, zurichten **to** ~ **a river navigable** einen Fluß *m* schiffbar machen **to** ~ **a rough outline** flüchtig entwerfen **to** ~ **a rustling sound** knirschen **to** ~ **secondary rank** unterordnen **to** ~ **shiny** blänken **to** ~ **slag** Schlacke *f* bilden **to** ~ **smaller** kleiner machen **to** ~ **smooth** spachteln **to** ~ **a straight line** eine gerade Linie *f* bilden **to** ~ **a straightaway test** eine Streckenmessung *f* machen **to** ~ **sure** feststellen, sichern **to** ~ **sure of a thing** sich einer Sache *f* vergewissern **to** ~ **terms** Bedingungen *pl* aufstellen **to** ~ **tight** abdichten **to** ~ **a traverse survey** polygonisieren **to** ~ **turbid** verschleiern **to** ~ **uniform** gleichmachen, stabilisieren **to** ~ **up (with, to)** auffüllen, aufstellen, ausarbeiten, zusammensetzen **to** ~ **up for** nachholen, entzerren **to** ~ **(or fill) up to the mark** auffüllen zur Marke *f* **to** ~ **up a charge** gattieren **to** ~ **up the margin** Format *n* machen **to** ~ **up the page** mit der Seite *f* ausgehen lassen (print.) **to** ~ **use (of)**

bedienen **to ~ useful** nutzbar machen **to ~ visible** sichtbar machen **to ~ water-repellent** hydrophobieren **to ~ worse** verschlechtern
**make** Arbeit *f,* Bearbeitung *f,* Erkennung *f,* Erzeugnis *n,* Fabrikat *n,* (contact) Schließen *n,* (contact) Schließung *f,* Ware *f*
**make-and-break** (of circuit) Schließen *n* und Unterbrechen *n,* Unterbrecher *m* ~ **cam** Zündnocken *m* ~ **contact** Ruhe- und Arbeits-, Öffnerschließ-, Umschalte-, Wechsel-kontakt *m* ~ **current** zeitweilig unterbrochener Strom *m* ~ **device** Unterbrecherscheibe *f* ~ **ignition** Abreißzündung *f* ~ **keying method** Wechseltastverfahren *n* ~ **magneto** Abreißzündmagnet *n* ~ **magneto with igniter** Niederspannungsmagnet *m* oder Abreißzündmagnet *m* mit Zündapparat ~ **mechanism** Abreißvorrichtung *f* ~ **rotary table** Schraub- und Abschraubdrehtisch *m* ~ **switch** Stromschlüssel *m,* Zünd- und Löscheinrichtung *f* ~ **switch method** Wechseltastverfahren *n* ~ **transients** Ein- und Ausschwingen *n* ~ **type magneto** Abreißmagnetzünder *m*
**make, ~-and-hold** Abrufauftrag *m* (paper mfg.) ~**-before-break** Folgekontakt *m* ~**-before-break contact** Folge-Arbeits-Ruhekontakt *m,* Folgekontakte *pl* (Schließen vor Öffnen), Schleppkontakt *m* ~**-busy arrangement** Belegungsvorrichtung *f,* Besetzteinrichtung *f* ~ **contact** Arbeitskontakt *m,* Schließen *m* ~**-contact unit** Arbeitskontakt *m* ~ **impulse** Schließungsimpuls *m,* Stromschließungsstoß *m* ~ **pulse** Schließungsimpuls *m* ~ **ready** Zurichtung *f* (print.) ~ **setting** Ansprech-Einstellung *f*
**makeshift** Aushilfe *f,* Aushilfsmittel *n,* Behelf *m,* Notbehelf *m,* Notmittel *n*; provisorisch ~ **antenna** Behelfsantenne *f* ~ **construction** Behelfskonstruktion *f* ~ **ferry** Behelfsfähre *f* ~ **landing state** Behelfslandebrücke *f* ~ **tools** Behelfswerkzeug *n*
**make transient** Ausgleichsvorgang *m* bei Stromschließung (Stromunterbrechung)
**make-up** Aufbau *m,* Aufmachung *f,* Aussehen *n,* Schminke *f,* Umbruch *m* (print.) ~ **of a book** Buchausstattung *f* ~ **of charge** Einsatzverhältnisse *pl*
**make-up, ~ medium** Frisch-, Zusatz-trübe *f* ~ **piece** Fasson-, Paß-rohr *n* ~ **pieces** Futterstücke *pl* ~ **railroad yard** Aufstellbahnhof *m* ~ **section** Mettage *f* ~ **track** Aufstell-, Ordnungsgleis *n* ~ **water** Ersatzwasser *n,* Zusatzwasser *n,* Zusatz-Wasserspeisung *f*
**maker** Hersteller *m,* Macher *m,* Schöpfer *m,* Werkbesitzer *m* ~**'s name plate** Fabrik-, Firmen-schild *n*
**making** Anfertigung *f,* Bau *m,* Fabrikation *f,* Verfertigung *f* ~ **an accounting** Rechnungslegung *f* ~ **a boring** Bohrung *f* ~ **out quantities** Mengenberechnung *f* ~ **a record upon** Beschriften *n* ~ **a sentence more severe** Strafverschärfung *f* ~ **a sound track upon** Beschriften *n* ~ **for uniformity** Vereinheitlichung *f*
**making, ~ contact** Arbeitskontakt *m* ~ **fast** Festlegung *f* ~ **firm** Verfestigung *f* ~ **independent** Selbständigmachen *n* ~ **regenerative coupling closer** Anziehen *m* der Rückkupplung ~ **tape** Klebstreifen *m* ~**-up price** Liquidationspreis *m* ~ **use of clouds** Wolkenausnutzung *f* ~ **useless** Unbrauchbarmachen *n*

**malacca cane** Bambusrohr *n*
**malachite** Atlaserz *n,* Kupferspat *m,* Malachit *m* ~ **green** Malachitgrün *n*
**malacolite** Malakolith *m*
**maladjustment** Falscheinstellung *f*
**malanders** Mauke *f*
**maldonite** Wismutgold *n*
**male** männlich ~ **collar** Patrize *f* ~ **connecting nipple** Verschlußkegel *m* ~ **contact connector strip** Steckerleiste *f* ~ **coupling** Einsteckkupplungsteil *m,* Kupplungsteil *m* mit Außengewinde ~ **die** Oberschnitt *m* ~ **fishing trip** Fangzapfen *m* ~ **ground joint** Kernschliff *m* ~ **half** Gegenstückhälfte *f* ~ **insulator plate** Steckplatte *f* ~ **mold** Patrize *f* ~ **part of ground-in joint** Kernschliff *m* ~ **piece** hervorstehender Teil *m* ~ **screed** Gegenprofil *n* ~ **screw** Schrauben-Preßspindel *f* ~ **screw thread** Außengewinde *n* ~ **thread** Außen-, Bolzen-gewinde *n*
**malformation** Fehlöffnung *f* (parachute), Mißbildung *f*
**malfunction** Fehl-leistung *f,* (parachute) -öffnung *f,* Funktionsstörung *f* (info proc), Maschinenfehler *m,* Störung *f*
**malfunctioning** schlechtes Arbeiten *n* (eines Gerätes)
**malic, ~ acid** Apfelsäure *f* ~ **ether** Apfeläther *m*
**malignant neoplasm** bösartiges Neoplasma *n*
**malinowskite** Nickelfahlerz *n*
**malleability** Dehnbarkeit *f,* Formbarkeit *f,* Geschmeidigkeit *f,* Hämmerbarkeit *f,* Kaltbildsamkeit *f,* Kaltverformbarkeit *f,* Schmiedbarkeit *f,* Verformbarkeit *f,* Verwalzbarkeit *f* ~ **when cold** Kalt-hämmerbarkeit *f,* -schmiedbarkeit *f*
**malleable** dehnbar, geschmeidig, hämmerbar, kaltverformbar, schmiedbar, streckbar **not ~** nicht hämmerbar
**malleable, ~ alloy** Knetlegierung *f* ~ **aluminum alloy** Aluminiumknetlegierung *f* ~**-annealing furnace** Temperglühofen *m* ~ **cast iron** Glüh-, Halb-stahl *m*; Temper-eisen *n,* -guß *m,* -gußeisen *n,* -stahlguß *m* ~ **casting** Tempergußstück *n* ~ **chain** gegossene Kette *f* ~ **founder** Tempergießer *m* ~ **hard cast iron** weißes Gußeisen *n* ~ **hard iron** Temperrohguß *m* ~ **ingot iron** Flußschmiedeeisen *n* ~ **iron** schmiedbares Eisen *n,* schmiedbarer Guß *m,* Schmiedeeisen *n,* Temperguß *m* ~ **(cast) iron** Weichguß *m* ~**-iron foundry** Tempergießerei *f* ~ **metal** Schmiedemetall *n* ~ **pig iron** Temperroheisen *n* ~ **scrap** Temperschrott *m* ~ **(iron) scrap** Tempergußbruch *m* ~ **white iron** Temperrohguß *m*
**malleableize, to ~** glühfrischen
**malleableizing** Glühfrischen *n,* Tempern *n* ~ **oven** Temperofen *m*
**mallet** Fäustel *n,* Hammer *m,* Handklöppel *m,* Holz-hammer *m,* -schlegel *m,* Klöpfel *m,* Klopfholz *n,* Klöppel *m* des Wheatstonelochers, Kolben *m,* Mokerhammer *m,* Schlägel *m,* Schlegel *m* (wooden) ~ Holzschlägel *m*
**malonic acid** Malonsäure *f*
**Malsi table** Malsitisch *m*
**malt, to ~** malzen
**malt** Malz *n* ~**-bruising plant** Malzschroterei *f* ~ **detrition apparatus** Abreibevorrichtung *f* für Malz, Malzabreibeapparat *m* ~ **drier** Malzdarre *f* ~**-drying kiln** Malzdarre *f* ~ **floor** Malztenne *f*

**Maltese, ~ cross** Malteserkreuz *n* **~ cross gear** Maltesergetriebe *n* **~-cross transmission** Malteserkreuzgetriebe *n*
**maltha** Berg-, Erd-teer *m*, Maltha *m*, Mineralteer *m*
**malthouse** Mälzerei *f*, Malzhaus *n*
**malting** Mälzen *n*
**maltose determination** Maltosebestimmung *f*
**maltreat, to ~** mißhandeln
**maltreatment** Mißhandlung *f*
**mammato-cumulus** Mammatokumulus *m*
**mammilated** warzenförmig
**mammography** Mammografie *f*
**mammoth, ~ antenna** Mammutantenne *f* **~ battery** Mammutbatterie *f*
**man, to ~** bemannen, besetzen
**man** Mann *m*, Mensch *m* **~ on the left (or right)** Nebenmann *m*
**man, ~ cage** Personenförderkorb *m* **~-carrying kite** bemannter Drachen *m* **~-eater** Beißer *m* **~ head** Mannloch *n*
**manhole** Durchlaßöffnung *f*, Einsteig-loch *n*, -schacht *m*, (of an engine) Fahrloch *n*, Kabelschacht *m*, Mannloch *n*, (of a water conduit) Revisionsloch *n*, Schacht *m* **~ of a water channel** Revisionsschacht *m*
**manhole, (cable) ~** Kabelbrunnen *m* **~ cover** Mannloch-deckel *m*, Schacht-abdeckung *f*, -deckel *m* **~-cover hook** Hebevorrichtung *f* für Schachtdeckel **~ culvert** Einsteigöffnung *f* **~ davit** klappbares Mannloch *n*, Mannlochdeckelbehälter *m* **~ door** Mannloch-verschluß *m*, -deckel *m* **~ end plate** Mannlochboden *m* **~ frame** Deckrahmen *m* **~ guard** Absperrgestell *n* **~ hook** Plattenhebel *m* **~ junction box** Schachtkabelkasten *m* **~ lock** Mannlochverschluß *m* **~ opening** Einsteigöffnung *f* **~ ring** Mannlochversteifung *f* **~ saddle** Mannlochbügel *m* **~ steps** Einsteigeisen *n* (für Kabelschachte) **~ wall** Schachtwand *m*
**man, ~-hour** Arbeitsstunde *f* **~ kind** Menschheit *f* **~-made fibers** Chemiefasern *pl* **~-made interference (or static)** nichtatmosphärische Störungen *pl* **~ power** menschliche Arbeitskraft *f*, -leistung *f*, Personalbestand *m* **~ power level** Arbeiterstand *m* **~ rope** Manntau *n*, Schwungseil *n* **~ trolley** Führerstandslaufkatze *f* **~ trolley transporter** Verladebrücke *f* mit Führerstandslaufkatze **~ trolley unloader** Katzbrücke *f* **~ winding** Seilfahrt *f* (min.)
**manacle** Hand-fessel *f*, -schelle *f*
**manage, to ~** bewirtschaften, dirigieren, disponieren, führen, handhaben, hantieren, leiten, verfahren, verwalten, werken, wirtschaften **to ~ with blade** beschaufeln **to ~ a firm** einer Firma vorstehen
**manageability** Bearbeitbarkeit *f*
**manageable** handlich, lenksam, steuerbar
**manageableness** Wendigkeit *f*
**management** Betrieb *m*, Bewirtschaftung *f*, Direktion *f*, Disposition *f*, Geschäfts-führung *f*, -leitung *f*, Leitung *f*, Verwaltung *f*, Werkleitung *f*, Wirtschaftsführung *f* **~ of a workshop** Betriebsregelung *f*
**manager** Betriebsleiter *m*, Bewirtschafter *m*, Intendant *m*, Leiter *m*, Verwalter *m* **(general) ~** Betriebsführer *m*

**managing** betriebsführend **~ administration** geschäftsführende Verwaltung *f* **~ committee** Vorstand *m* **~ director** Betriebsdirektor *m*, Geschäftsführer *m* **~ partner** geschäftsführender Teilhaber *m*
**Manchester, ~ plate** Rosettenplatte *f* **~ velvet** Manchester *m*
**Manchon stretcher** Manchonspanner *m*
**mandarining** Mandarinagearbeit *f*
**mandated territory** Mandatsgebiet *n*
**mandator** Auftraggeber *m*, Mandant *m*, Vollmachtgeber *m*
**mandatory** vorschreibend **~ proxy** Bevollmächtigter *m*
**mandrel** Aufspanndorn *m*, Docke *f*, Doppelkeilhaue *f*, Dorn *m*, Drehbankspindel *f*, Drehstift *m*, Gewindepatrone *f*, Lochdorn *m*, Patrone *f*, Richtdorn *m*, Spindel *f*, Stanzdorn *m*, Stopfen *m*, Wickelturm *m* **~ for shaping machine** Fräsdorn *m*
**mandrel, ~ bar** Dornstab *m* **~ coiling test** Dornumwickelprobe *f* **~ diameter** Dorndurchmesser *m* **~ dressing** Dornaufbereitung *f* **~ plug (or point)** Walzstopfen *m* **~ press** Dornpresse *f* **~ rod** Dornstange *f* **~ socket** Fangbirne *f* **~ stripper** Dornausziehvorrichtung *f* **~ upsetting press** Dornstauchpresse *f*
**manganese** Glasermagnesia *f*, Mangan *n* **~ (-ore)** Braunstein *m* **~ acetate** essigsaures Manganoxydul *n* **~ alum** Manganalaun *m* **~ ammonium** Manganbadmethode *f* **~ bister shades** Manganbistertöne *pl* **~ carbide** Mangankarbid *n* **~ composition** Mangankitt *m* **~ content** Mangangehalt *m* **~ dioxide** Mangansuperoxyd *n* **~ dioxide cylinder** Braunsteinzylinder *m* **~ dioxide sack (in mask)** Braunsteinbeutel *m* **~ nickel silver** Manganneusilber *n* **~ ore** Manganerz *n* **~ oxide** Manganoxyd *n* **~ perchloride** Manganperchlorid *n* **~ peroxide** künstlicher Braunstein *n*, Mangansuperoxyd *n*, Pyrolusit *m* **~ salt** Mangansalz *n* **~ sesquioxide** Manganoxyd *n* **~-steel frog** Manganstahlherzstück *n* **~-steel rail** Manganstahlschiene *f* **~ tetroxide** Manganoxyduloxyd *n*
**manganic, ~ acid** Mangansäure *f* **~ aluminate** Manganialuminat *n* **~ compound** Manganverbindung *f* **~ perchloride** Manganperchlorid *n* **~ phosphate** Manganiphosphat *n* **~ sulfate** Mangansulfat *n*
**manganiferous** manganhaltig, manganreich
**manganin** Manganin *n*
**manganite** Braunstein *m*, Graumanganerz *n*, graues Manganerz *n*, Manganit *m*
**manganous, ~ aluminate** Manganoaluminat *n* **~ chloride** Manganchlorür *n* **~ hydroxide** Manganhydroxydul *n*, Manganohydroxyd *n* **~ oxide** Manganoxydul *n* **~ sulfide** Mangansulfür *n*
**manger** Krippe *f*
**mangle, to ~** mangeln, zerstückeln
**mangle** Mangel *f*, Wäscherolle *f* **~ gear** Triebstockverzahnung *f* **~ rack motion** Doppelrechen-, Mangelrad-bewegung *f*
**mangler** Hack-Mangelmaschine *f*, Mangler *m*
**manifest, to ~** betätigen, bezeigen, offenbaren
**manifest** durchsichtig, offenbar **~ of cargo** Ladungsmanifest *n*
**manifestation** Erscheinung *f*, Erscheinungsform *f*, Offenbarung *f* **~ of force** Kraftäußerung *f*

**manifold, to** ~ hektofieren, vervielfachen, vervielfältigen

**manifold** Ansaugrohr *n* mit Verzweigungen, Auspuffröhre *f* zum Auspufftopf, Ladeleitung *f*, Leitung *f*, verzweigtes Leitungssystem *n*, Rohr-abzweigung *f*, -leitung *f*, -krone *f*, -spinne *f*, -verteiler *m*, Sammelleitung *f*, Verteiler *m*, Verteilerrohr *n*, Verteilungskopf *m*; differenziert, mannig-fach, -faltig, mehrfach, viel-fach, -förmig, -gestaltig, -seitig ~ **of electronic states** Summe *f* der Elektronenzustände

**manifold,** ~ **conduit** Leitungsrohr *n* ~ **density method** Bestimmung *f* der Leistung auf Grund des Ladedrucks ~ **girt** Rohrverteilergrube *f* ~ **inlet** Einlaßleitung *f* ~ **pen** Durchschreibefeder *f* ~ **plug** Vielfachstecker *m* ~ **pressure** Ladeluft *f*, (intake) Lade-, Leitungs-, Startlade-druck *m* ~**-pressure control** Ladedruckregler *m*, Leitungsdruckkontrolle *f* ~**-pressure gauge** Lade-, Leitungs-druckmesser *m* ~**-pressure limiting regulator** Ladedruck-begrenzer *m*, -grenzregler *m* ~ **printing** Umdruck *m* ~ **requirements** vielseitige Verwendbarkeit *f* ~ **writer** grafischer Vervielfältigungsapparat *m*

**manifolding,** ~ **of pipeline system** Ausbildung *f* und Anordnung *f* eines verzweigten Rohrleitungssystems ~ **work** Vervielfältigungsarbeit *f*

**Manila,** ~ **hemp** Gewebepisang *m*, Manilahanf *m* ~ **paper** Bast-, Hart- Manila-papier *n* ~ **rope** Hanfseil *n*, Hanf-, Manila-tau *n*

**manipulability** Bedienbarkeit *f*

**manipulable** bedienbar, behandelbar

**manipulate, to** ~ bedienen, behandeln, betätigen, handhaben, hantieren, manipulieren, mit Verstand behandeln, tasten, werken

**manipulated** betätigt, gestaltet, manipuliert ~ **variable** Stellgröße *f*

**manipulating,** ~ **key** (Morse)Taste *f* ~ **stand** Hilfs-Operationsstativ *n*

**manipulation** Behandlung *f*, Betätigung *f*, Handgriff *m*, Handhabung *f*, Hantierung *f*, (of slag) Schlackenarbeit *f* ~ **of arms** Waffenhandhabung *f*

**manipulator** Blockwender *m*, Handgeber *m*, Kanter *m*, Kant- und Verschiebevorrichtung *f*, Objekttisch *m*, mechanischer Schalttisch *m*, Wender *m*, (in rolling) Wendevorrichtung *f* ~ **for turning sheets** Blechwender *m* ~ **gear** Wendegehänge *n*

**manner** Art *f*, Art *f* und Weise *f*, Manier *f*, Maß *n*, Stil *m*, Weg *m*, Weise *f*

**manner,** ~ **of craftsmen** Handwerksbrauch *m* ~ **of formation** Bildungsweise *f* ~ **of loading** Belastungsfall *m* ~ **of mixing** Mischweise *f* ~ **of notation** Bezeichnungsweise *f* ~ **of representation** Darstellungsweise *f* ~ **of treatment** Behandlungsweise *f*

**Mannesmann,** ~ **process** Pilgerschrittverfahren *n* ~ **roll-piercing process** Mannesmannschrägwalzverfahren *n* ~ **rolling process** Schrägwalzverfahren *n*

**Mannheim gold** Goldkupfer *n*, Kupfergold *n*, Mannheimer Gold *n*

**manning** Bemannung *f*

**manocryometer** Druckmesser- und Gefrierpunktmesser *m*, gefrierpunktmessendes Manometer *m*

**manoeuver, to** ~ manövrieren **to** ~ **out of position** herausmanöverieren

**manoeuver** Andrehvermögen *n*, Bedienbarkeit *f*, Beschleunigungsvermögen *n*, Lenkbarkeit *f*, Manöver *n*, Steuer-barkeit *f*, -fähigkeit *f*, Übung *f*, Wendigkeit *f*, Wendigkeitsvermögen *n*

**manoeuverable** steuerbar, wendig

**manoeuvrability,** ~ **chain** Bedienungskette *f* ~ **device** Rangieranlage *f* ~ **line** Hubseil *n* ~ **load factor** Manöverbelastungsfaktor *m* ~ **spider** Halteleinenbund *m* ~ **valve** Manöverventil *n* ~ **winch** Bedienungs-, Rangier-winde *f*

**manoeuvring,** ~ **area** Rollfeld *n* (aviat.) ~ **brake** Fahrbremse *f* bei Fördermaschine, Manövrierbremse *f* ~ **speed** Geschwindigkeit *f* für Ruderbetätigung

**manograph** Druckschreiber *m*, Manograf *m*, Registriermanometer *n*

**manometer** Dampfmesser *m*, Druckanzeiger *m*, Manometer *n* ~ **capsule** Manometerdosensatz *m* ~ **elastic pipe (or tube)** Manometerfederrohr *n* ~ **hand** Manometerzeiger *m* ~ **indicator** Manometerzeiger *m* ~ **pressure** durch das Manometer angezeigter Druck *m* ~ **vacuum gauge** Manovakuummeter *n*

**manometric** manometrisch ~ **head** manometrische Förderhöhe *f* ~**-lifting height** manometrische Förderhöhe *f* ~ **pressure** manometrischer Förderdruck *m*

**manometrically** manometrisch

**mansard roof** Mansardendach *n*

**mantissa** Mantisse *f* (info proc)

**mantle** Formmantel *m*, (incandescent) Glühstrumpf *m*, Mantel *m* ~ **of a furnace** Ofenstock *m* ~ **of wall** Mauermantel *m*

**mantle,** ~ **carrier** Glühkörperhalter *m* ~ **column** (of blast furnace) Tragkranzsäule *f* ~ **corbel** Mantelknagge *f* ~ **guard** Glühkörperschutzkorb *m* ~ **piece of plaster** Gipsmantelbelag *m* ~ **ring** Tragkranz *m* ~ **stack of a wall** Stirnseite *f* einer Mauer ~ **suspender** Glühkörperhalter *m*

**mantlet,** ~ **for cylindrical mount** Walzenblende *f* ~ **of flexible ball mount** Kugelblende *f*

**manual** schriftliche Anleitung *f*, Hand..., Hand-, Hilfs-buch *n*, Leitfaden *m*, Manual *n*, Vorschrift *f*; handbetriebsmäßig, handtätig, manuell ~ **and test start** Hand- und Probestart *m*

**manual,** ~ **alarm box** Druckknopfmelder *m* ~ **automatic** Handverstellungsautomatik *f* ~**-automatic regulator** Handschnellregler *m* ~ **central office** Handamt *n*, Vermittlungsamt *n* mit Handbetrieb ~ **charger** Handdurchlader *m* ~ **control** willkürliche Steuerung *f* ~ **control for damping signal** Dämpfungswahl *f* ~ **control wheel** Richthandrad *n* ~ **controller** handgesteuerter Apparat *m*, Handsteuerschalter *m* ~ **data input** manuelle Dateneingabe *f* ~ **disengagement** Handauslösung *f* ~ **drive for switches** Handantrieb *m* für Schaltgeräte ~ **exchange** Handamt *n*, Handvermittlung *f*, Handvermittlungsamt *n* (teleph.) ~ **fire engine** Handspritze *f* ~ **fuse-setter** Zünderstellschlüssel *m* ~ **highpressure valve** Hochdruckhandabsperrventil *n* (g/m) ~ **labor** Handarbeit *f* ~ **laborer** Handarbeiter *m* ~ **motor** Handdrehmaschine *f* ~ **operation** Hand-antrieb *m*, -bedienung *f*,

-betrieb *m*, -gebrauch *m* ~ **operation of rams** Stößelbetätigung *f* ~ **override device** manueller Notauslöser *m* ~ **regulator** Handregler *m* ~ **release** Handauslösung *f* ~ **release lever** Handauslösehebel *m* ~ **retransmission** Weitergabe *f* mit der Hand ~ **ringing** handbetätigter Ruf *m* ~ **sample changer** Handprobenwechsler *m* ~ **selection** Handwählschalter *m* (teleph.) ~ **speed** Hand-Tastgeschwindigkeit *f* ~ **start set** Handanlaßaggregat *n* ~ **switchboard** Handamtsschrank *m*, Umschalteschrank *m* eines Handamts ~ **switching** Handschaltung *f* ~ **switching for built-up connections with no delay (or demand) service** Handbetrieb *m* in Durchgangsämtern bei Sofortverkehr ~ **switching system** Handschaltverfahren *n* ~ **switchroom** Betriebsraum *m* ~ **system** Handsystem *n* ~ **telephone area** Ortsnetz *n* mit Handbetrieb ~ **telephone set** Fernsprechapparat *m* für Handbetrieb ~-**telephone system** Fernsprechnetz *n* mit Handbetrieb ~ **timing** Handverstellung *f* ~ **timing lever** Handverstellhebel *m* ~ **transmission** Handgeben *n* ~ **use** Handgebrauch *m* ~ **volume control** Handlautstärkeregler *m* ~ **welding** Handschweißung *f* ~ **worker** Handwerker *m* ~ **working** Handbetrieb *m*

**manually,** ~ **controllable propeller** handbetätigter Verstellpropeller *m* ~-**given cutoff signal** Handbrennschluß *m* (g/m) ~-**operated brake** Feststellbremse *f* ~-**operated circuit breaker** Handausschalter *m* ~-**operated conveyer** Handfahrgerät *n* ~-**operated fuel pump** Kraftstoffhandpumpe *f* ~-**operated gear** von Hand betätigte Durchdrehungsvorrichtung *f* ~-**operated injection-pump testing equipment** Handeinspritzpumpen-Prüfvorrichtung *f* ~-**operated mixture control** mit der Hand regulierter Mischer *m* ~-**operated parachute** Fallschirm *m* mit Handauslösung, manueller Fallschirm *m*, Handfallschirm *m* ~ **started machine ringing** halbselbsttätiger Ruf *m*

**manufacture, to** ~ anfertigen, ausführen, darstellen, fabriksmäßig herstellen, fabrizieren, fertigen, herstellen, machen, verarbeiten, verfertigen **to** ~ **on a conveyer belt** am laufenden Band fabrizieren

**manufacture** Anfertigung *f*, Ausbildung *f*, Bau *m*, Bearbeitung *f*, Bereitung *f*, Darstellung *f*, Erzeugnis *n*, Erzeugung *f*, Fabrikat *n*, Fabrikation *f*, Fertigung *f*, Herstellung *f*, Verfertigung *f*

**manufacture,** ~ **of apparatus** Apparatebau *m* ~ **of armor** Panzerfertigung *f* ~ **of dies** Gesenkbau *m* ~ **of electric steel** Elektrostahlschmelzen *n* ~ **of gas** Gasmachen *n* ~ **of inks** Farbenfabrik *f* ~ **on a large scale** Großfabrikation *f* ~ **of mild steel** Flußstahlerzeugung *f* ~ **of pig iron** Roheisendarstellung *f* ~ **of plastics** Kunststoffherstellung *f* ~ **of soft steel** Flußstahlerzeugung *f* ~ **of steel** Stahlherstellung *f* ~ **of tin** Zinnproduktion *f* ~ **of tubing** Röhrenfabrikation *f* ~ **of welded-steel tubing** Schweißrohrherstellung *f*

**manufactured,** ~ **gas** künstliches Gas *n*, Vergasergas *n* ~ **goods** Fabrikwaren *pl*, Fertigerzeugnis *n* ~ **head** Setzkopf *m*

**manufacturer** Erzeuger *m*, Fabrikant *m*, Hersteller *m*, Lieferer *m*, Macher *m*, Werkbesitzer

*m* ~ **of knit goods** Strickwarenfabrikant *m* ~ **of machinery** Maschinenbauer *m*

**manufacturer's mark** Fabrikzeichen *n*

**manufacturing,** ~ **in series** Serienfertigung *f*

**manufacturing,** ~ **affiliate** Konzernwerk *n* ~ **arrangement** Fabrikationsaufstellung *f* ~ **costs** Produktionskosten *pl* ~ **defect** Fertigungsfehler *m* ~ **department** Fabrikationsabteilung *f* ~ **district** Industrie-bezirk *n*, -gebiet *n* ~ **efficiency** Produktionsleistung *f* ~ **engineer** Betriebs-ingenieur *m*, -wirtschaftler *m* ~ **engineering department** Werkeinrichtungsabteilung *f* ~ **expenses** Gestehungskosten *pl* ~ **firm** Hersteller-firma *f*, -werk *n* ~ **flaw** Fabrikationsfehler *m* ~ **grinding** Produktionsschleifarbeit ~ **industries** Fabrikationsbetriebe *pl* ~ **lathe** Produktionsdrehbank *f* ~ **length** Fabrikationslänge *f* ~ **loss** Betriebsverlust *m*, Schrottanfall *m* ~ **method** Herstellungs-gang *m*, -methode *f*, -weise *f* ~ **operation** Bearbeitungsverfahren *n*, Fabrikationsvorgang *m* ~ **permit** Nachbaurecht *n* ~ **plant** Erzeugerwerk *n*, Fabrik *f*, Fabrikanlage *f*, Herstellungswerk *n*, Werk *n* ~ **practice** Betriebspraxis *f* ~ **preparation** Fertigungsvorbereitung *f* ~ **process** Arbeitsverfahren *n*, Fabrikations-gang *m*, -prozeß *m*, Herstellungsverfahren *n*, Produktionsgang *m* ~ **program** Fabrikations-, Herstellungs-programm *n*, Fertigungsfolge *f* ~ **right** Herstellungsrecht *n* ~ **schedule** Fabrikationsprogramm *n* ~ **tolerance** Fertigungstoleranz *f* ~ **town** Industriestadt *f*

**manure, to** ~ düngen

**manure** Dung *m*, Düngemittel *n*, Dünger *m*, Mist *m* ~ **distributor (or drill)** Düngerstreumaschine *f* ~ **dump** Jauchegrube *f* ~-**manufacturing plant** Düngemittelfertigungsanlage *f* ~ **spreader** Düngersteuer *m*, Stalldüngerstreuer *m*

**manurial requirement** Düngerbedürfnis *n*

**manuring lime** Düngekalk *m*

**manuscript** Druckvorlage *f*, Handschrift *f*, Urschrift *f* ~ **copy** Abschrift *f* mit der Hand ~ **cover** Manuskriptmappe *f*

**many,** ~-**bodied forces** Mehrkörperkräfte *pl* ~-**bodied vaporizer** Mehrkörperverdämpfer *m* ~-**body problem** Mehrkörperproblem *n* **in** ~ **cases** vielfach ~-**colored** vielfarbig ~-**copy typewriter** Schreibmaschinenschriftsetzmaschine *f* **with** ~ **faces (or sides)** flächenreich ~-**line spectrum** Viellinienspektrum *n* ~-**lined system** Vielliniensystem *n* ~-**sided** mehrkantig, vielseitig ~-**time formalism** Mehrzeitformalismus *m* ~-**valued** vieldeutig **of** ~ **years** langjährig

**map, to** ~ abbilden (math.), ausarbeiten, entwerfen, kartographieren, landkartenmäßig erfassen, planen **to** ~ **a graph** eine Kurve zeichnen **to** ~ **lines of force** Stromlinien *pl* aufzeichnen

**map** Abbildung *f*, Karte *f*, Plan *m* **(topographic)** ~ Landkarte *f*

**map,** ~ **of the immediate environment** Umgebungskarte *f* ~ **of line** Streckenplan *m* (teleph.) ~ **of mine** Gruben-bild *n*, -riß *m* ~ **of network** Leitungsplan *m* ~ **of stock** Bestandskarte *f* (forest)

**map,** ~ **carrier** Kartenhalter *m* ~ **case** Kartenbehälter *m*, -schutzhülle *f*, -tasche *f* ~-**chart** Küstenkarte *f* ~ **chest** Karten-schrank *m*,

-spind *m* ~ **compilation (or completion)** Karten-vervollständigung *f* ~ **composed of aerial photographs** Bildlagekarte *f* ~ **content** Karten-inhalt *m* ~ **coordinate** Quadratzahl *f* ~**-coordinates system** Kartenkoordinatensystem *n* ~**-correction book** Kartenverbesserungsbuch *n* ~ **course** Kartenkurs *m* ~ **cover** Kartenschutz-hülle *f* ~ **distance** Kartenentfernungslinie *f* ~ **drawing** Kartenherstellung *f*, Kartierung *f* ~ **equipment** Kartenausstattung *f* ~ **grid** Karten-gitternetz *n*, -netz *n* ~ **holder** Kartenhalter *m* ~ **level** Kartenniveau *n* ~**-making** Planzeichnen *n* ~ **making institute** kartografische Anstalt *f* ~**-mounting board** Kartenbrett *n* ~ **outfit** Kar-tenausrüstung *f* ~ **paper** Karten-, Landkarten-druck-, Plandruck-papier *n* ~ **plane** Karten-ebene *f* ~ **plotting** Kartenherstellung *f*, Kar-tierung *f*, Planzeichnen *n* ~**-plotting apparatus** Kartierungsgerät *n* ~ **plumb point** Kartennadir *m* ~ **point quadrilateral** Kartenpunktviereck *n* ~ **position** Planstelle *f* ~**-printing office (or plant)** Kartendruckerei *f* ~ **projection** Karten-entwurf *m*, -projektion *f* ~ **protractor** Karten-winkelmesser *m*, Meßkreis *m* ~**-protractor set** Plansektor *m* ~ **quadrangle** Kartenblatt *n* ~ **range** Kartenentfernung *f* ~ **reading** Karten-lesen *n* ~**-reading scale** Planzeiger *m* ~ **recti-fication** Kartenberichtigung *f* ~ **room** Karten-kammer *f* ~ **scale** Kartenmaßstab *m* ~ **scale table** Meßtischblatt *n* ~ **section** Kartenaus-schnitt *m* ~ **sector** Plansektor *m* ~ **sheet** Kar-tenblatt *n* ~ **showing altitudes** Höhenkarte *f* ~ **sketch** Kartenskizze *f* ~ **square** Grad-abteilung *f*, -feld *n*, Planquadrat *n* ~ **sticker** Landkarten-aufzieher *m* ~ **table** Kartentisch *m* ~ **vernier of map protractor** Neugradteilung *f* ~ **work atlas** Kartenwerk *n*
**maple** Ahorn *n* ~ **sirup** Ahornsirup *m* ~ **sugar** Ahornzucker *m* ~ **wood** Ahornholz *n*
**mapping** Abbildung *f*, Aufnahme *f* des Geländes, Kartenzeichnen *n*, Kartierung *f* (rdr), Karto-grafie *f* ~ **a curve** Kurvenzeichen *n* ~ **by means of central prospective** Abbildung *f* durch Zen-tralperspektive ~ **by means of parellel normals or tangent planes** Abbildung *f* durch parallele Normalen bzw. Tangentialebenen ~ **from photographs** Bildkartierung *f* ~ **of the projective plane into itself** Abbildung *f* der projektiven Ebene auf sich ~ **of regions** Abbildung *f* von Gebieten ~ **of the torus into itself** Abbildung *f* des Torus auf sich
**mapping,** ~ **camera** Meßbildkamera *f*, Reihen-meßkamera *f* ~ **group** Reihenbildzug *m* ~ **scale** Kartiereinrichtung *f* ~ **section** Karten-batterie *f*
**mar, to** ~ verderben
**mar,** ~**-proof** beständig, kratzfest, reibfest ~**-resistance** Kratzfestigkeit *f*
**marble, to** ~ adern, marmorieren
**marble** Marmor *m*, Murmel *f* ~ **cement** Alaun-gips *m* ~**-like** marmorartig ~ **paper** Marmor-papier *n* ~ **plate** Marmorplatte *f* ~ **quarry** Marmorbruch *m* ~ **slab** Marmor-platte *f*, -tafel *f*
**marbled** geädert, gefleckt, marmoriert ~ **edge** Marmorschnitt *m* ~ **paper** Marmorpapier *n* ~ **paper made by tipping** getuftes Marmorpapier *n*

**marbling** Aderung *f*, Geäder *n*, Marmorierung *f* ~ **color** Marmorierfarbe *f*
**marcasite** Blätterkies *m*, Grauseisenerz *n*, Mar-kasit *m*, Strahl-, Vitriol-, Wasser-, Weich-kies *m*
**mare** Stute *f* ~**'s-tail** Federwolke *f*, Windbaum *m* ~**'s-tail clouds** Haarwolken *pl* ~**'s-tails** Pferdeschwanzwolken *pl*
**marengo effect** Marengoeffekt *m*
**margaric acid** Margarinsäure *f*
**margarite** Bergglimmer *m*, Margarit *m*, Perl-glimmer *m*, -schnur *f*
**margin, to** ~ auf einen Unterschied einstellen **to** ~ **an "n" milliampere** auf eine Ansprech-stromstärke *f* von "n" MA einstellen **to** ~ **a relay to pull up at ...** das Relais auf einen Grenzstrom von ... einstellen
**margin** Abweichung *f*, Damm *m* (phono), Durchschuß *m*, Rand *m*, Spielraum *m*, Ufer *n*
**margin,** ~ **of an apparatus** Spielraum *m* eines Apparates ~ **of commutation** Sicherheitswinkel *m* ~ **of lift** Aufstiegsspielraum *m* ~ **of manu-facture** Einpaßzugabe *f* ~ **of the plate** Platten-rand *m* ~ **of power** Leistungsreserve *f* ~ **of profit** Gewinn-marge *f*, -spanne *f* ~ **of safety** Sicherheit *f*, Sicherheits-grad *m*, -koeffizient *m*, -spielraum *m*, Sicherungszuschlag *m* ~ **of screen** Schirmrand *m* ~ **of stability** Abstand *m* vom Pfeifpunkt
**margin,** ~ **control lever** Randstellerhebel *m* ~ **release** Tabelle *f* ~ **release mechanism** Rand-auslösung *f* ~ **scale** Randstellerskala *f* oder Zeilenstab *m* ~ **sheet** Unterlegebogen *m* (print.) ~ **stop (left)** Anfangsrandsteller *m*, **(right)** End-, Schluß-randsteller *m*, Randsteller *m* ~ **stop mechanism** Randsperre *f* ~ **stop rack** Randstellerzahnstange *f* ~ **stops** Randstellung *f*
**marginal,** ~ **angle** Randwinkel *m* ~ **annotation** Rand-anmerkung *f*, -bemerkung *f* ~ **check(ing)** Grenzwert-, Randwert-prüfung *f*, Toleranz-prüfung *f* ~ **current** Grenzstrom *m* ~ **definition** Randschärfe *f* ~ **effect** Randeffekt *m* ~ **embossing** Randprägung *f* (film) ~ **force** Rand-kraft *f* ~ **gloss** Seitenanmerkung *f* ~ **groove** leere Rille *f* ~ **inscription** (coin) Umschrift *f* ~ **layer** Grenzschicht *f* ~ **note** Anmerkung *f*, Rand-anmerkung *f*, -bemerkung *f*, -glosse *f*, Seitenanmerkung *f* ~ **oil** Quetschöl *n* ~ **opera-tion** Grenzstrombetrieb *m* ~**-operation relay** Grenzstromrelais *n* ~ **parabola** Randparabel *f* (aviat.) ~ **point** Randpunkt *m* ~ **property** An-liegergrundstück *n* ~ **punched card** Randloch-karte *f* ~ **ray** Randstrahl *m* ~ **release** Randaus-lösung *f* ~ **sketch** Randzeichnung *f* ~ **stop** Rand-hemmung *f*, -steller *m* ~ **track** Randspur *f* (tape rec.) ~ **vibration** Grenzschwingung *f* ~ **vortex** Randwirbel *m* ~ **zone** verstickte Rand-zone *f*
**marginalia** Rand-anmerkung *f*, -bemerkung *f*
**Marietta process** Mariettaverfahren *n*
**marigraph** Flutzeiger *m*
**marine** Marine *f*; ozeanisch, See... ~ **airport** Wasserflughafen *m* ~ **auxiliary set** Bordhilfs-aggregat *n* ~ **boiler** Schiffskessel *m* ~ **cable** Tiefseekabel *n* (sub) ~ **cable** Flußkabel *n* ~ **chart** Seekarte *f* ~ **climate** ozeanisches Klima *n*, Seeklima *n* ~ **connection box** Schiffsanschluß-dose *f* ~ **current** Meeresströmung *f* ~ **deposit** Meerablagerung *f* ~ **Diesel (engine)** Schiffs-

diesel *m* ~ **distribution box** Schiffsverteilerdose
*f* ~ **emergency set** Bordnotaggregat *n* ~ **engine**
Schiffsdampfmaschine *f*, Schiffs-maschine *f*,
-motor *m* ~ **fan** Schiffs-gebläse *n*, -ventilator *m*
~ **gear** Schleppgeschirr *n* ~ **glue** Marineleim *m*
~ **green** seegrün ~ **handbook** Seehandbuch *n* ~
**indicator** Schiffssignal *n* ~ **insurance** Seetrans-
portversicherung *f*, Seeversicherung *f* ~ **jack**
Schiffsklinkenstecker *m* ~ **junction box**
Schiffsabzweigdose *f* ~ **lighting unit** Schiffs-
beleuchtungsanlage *f* ~ **lights** Seefahrtfeuer *pl*
~ **meteorology** Meereswetterkunde *f* ~ **oil**
Marineöl *n* ~ **oil engine** Schiffsölmaschine *f* ~
ore See-Erz *n* ~ **overhead light** Schiffshänge-
lampe *f* ~ **phosphorescence** Seelicht *n* ~ **portable**
**light** tragbares Schiffslicht *n* ~ **propulsion**
Schiffsantrieb *m* ~ **radio** Seefunk *m* ~ **radio**
**service** Seefunkdienst *m* ~ **service** Bordbetrieb
*m* ~ **supplies** Bootszubehörteile *pl* ~ **telephone**
**equipment** Schiffstelefonanlage *f* ~ **transfer**
**switch** Schiffsumschalter *m* ~ **(or ship) trans-**
**mitter** Schiffssender *m* ~**-type steam engine**
stehende Dampfmaschine *f*
**mariner** Schiffer *m*, Seefahrer *m* ~**'s compass**
Schiffskompaß *m*
**maritime** zur See gehörig ~ **canal** Seekanal *m* ~
**law** See-gesetz *n*, -recht *n* ~ **meteorological**
**office** Seewetteramt *n* ~ **navigation** Seeschiff-
fahrt *f* ~ **radio beacon** Navigationsfeuer *n* ~
**right** Seerecht *n* ~ **signals** Seezeichenwesen *n*
**mark, to** ~ ankern, anmerken, anstreichen,
anzeichnen, bezeichnen, einzeichnen, kennzeichnen, marken, markieren, merken, signalisieren, stempeln, unterstreichen, vermarken,
vermerken, Zeichenstrom geben **to** ~ **by**
**beacons (or stakes)** abbaken (das Fahrwasser)
**to** ~ **with beacon lights** befeuern **to** ~ **with**
**ciphers** beziffern **to** ~ **engaged** als besetzt kennzeichnen, Besetztzeichen *pl* belegen **to** ~ **goods**
**with prices** anpreisen **to** ~ **off** abgrenzen, abklatschen, abtragen, abzeichnen **to** ~ **on** eintragen (in einen Plan) **to** ~ **out** anreißen auszeichnen; Linienführung *f* festlegen
**mark** Abdruck *m*, Abzeichen *n*, Bezeichnung *f*,
Eindruck *m*, Eindruck eines Schlagstempels,
Kennzeichen *n*, Kennzeichnung *f*, Kerbe *f*,
Mal *n*, (coin) Mark *f*, Marke *f*, Markierung *f*,
Markierungsstrich *m*, Maser *f*, Merkmal *n*,
Morsezeichen *n*, (boat) Peilboot *n*, Signum *n*,
Sorte *f*, Stempel *m*, Strich *m*, Zeichen *n*, Ziel *n*
**(guide)** ~ Merkzeichen *n*, Strom-, Trenn-
schritt *m* (electr.) ~ **of amplification** Verstär-
kungsziffer *f* ~ **of suspension** Gedankenstrich
*m* ~ **of test ball** Eindruckkalotte *f*
**mark,** ~ **frame** Meßrahmen *m* ~ **made with a hot**
**iron** Brandzeichen *n* ~ **post** Stangenseezeichen
*m* ~ **scraper** Anreißnadel *f* ~ **sense position**
Zeichenlochstelle *f* ~ **sensing** Zeichenlochung *f*
~**-sensing punch card** Zeichenlochkarte *f*
~**-signal pulse** stromerfüllter Zeichenschritt *m*
~**-up** Aufschlag *m*
**marked** ausgeprägt, ausgezeichnet, bemerkenswert, durchschlagend, erkennbar, markant,
stark ~ **dependence of amplitude upon fre-**
**quency** starker Frequenzgang *m* der Amplitude
~ **pair** Zähladernpaar *n* ~ **wire** Zählader *f*
**marker** Anreißer *m*, Anzeiger *m*, Indikator *m*,
Kabelmerkstein *m*, Markier-apparat *m*, -ein-

richtung *f*, -stein *m*, Markierer *m*, Markierung
*f*, Markierungsvorrichtung *f*, Meßmeister *m*,
Pfosten *m*, Stellmarke *f* ~ **beacon** Markierungs-
bake *f*, -funkfeuer *n*, -sender *m* ~**-beacon signal**
Kennung *f* ~ **buoy** Anzeigerboje *f*, Seemarke *f*
~ **cable** Kabelbezeichnung *f* ~ **connecting**
**relay** Anschalterelais *n* für Markierer ~ **genera-**
**tor** Frequenzmarkierungsgenerator *m* ~ **gun**
Lichtspucker *m* ~ **light** Abstandslicht *n* ~
**receiver** Markierungsfunkfeuer-Empfänger *m*
~ **slide** Schreibschlitten *m* ~ **track** Markier-
bahn *f* (info proc)
**market** Absatz-gebiet *n*, -markt *m*, Geschäftslage *f*, Markt *m*, (condition) Marktlage *f* ~
**analysis** Markt-analyse *f*, -forschung *f* ~ **con-**
**ditions** Marktgeschehen *n* ~ **grade** Handels-
sorte *f* ~ **outlet** Absatzmarkt *m* ~ **place** Markt-
platz *m* ~ **quotations** Marktkurs *m*
**marketable** absatzfähig, absetzbar, gangbar,
gängig, handelsgängig, konkurrenzfähig, lieferbar, markt-fähig, -gängig, verkäuflich
**marketing,** ~ **expenses** Verkaufskosten *pl* ~
**margin** Vertriebsspanne *f* ~ **organization** Ver-
triebsorganisation *f* ~ **report** Marktbericht *m*
~ **service department** Verkaufsdienst *m*
**marking** Absteckung *f*, Abstempeln *n*, Arbeits-
zustand *m*, Beschilderung *f*, Bezeichnung *f*,
Feldereinschnitt *m*, Gepräge *n*, Geschoßfelder-
einschnitt *m*, Kennzeichnung *f*, Markierung *f*,
Markung *f*, Vermarkung *f* ~ **of conductor**
Leitungsbezeichnung *f* ~ **by the electro-mark-**
**ing method** Beschriften *n* im Elektrosignierver-
fahren ~ **on metal** Stichelwirkung *f* ~ **of**
**minimum beam** Minimumstrahlkennzeichnung
*f* ~ **of surveyed points** Vermarkung *f* ~ **of time**
Synchronisierzeitmarke *f*, Zeit-marke *f*, -maß *n*
**marking,** ~ **aids** Markierungshilfen *pl* ~**-and**
**laying-out tools** Anreiß- und Prüfwerkzeuge *pl*
~ **apparatus** Anzeigevorrichtung *f*, Markier-
apparat *m* ~ **awl** Reißspitze *f* ~ **battery** Zei-
chenbatterie *f* ~ **(or spacing) bias** Überwiegen *n*
nach der Zeichen(Trenn-)seite ~ **connecting**
**relay** Markierer-Anschalterelais *n* ~ **contact**
Zeichenkontakt *m* ~ **crayon** Signierkreide *f* ~
**current** Arbeits-, Zeichenstrom *m*, Zeichen-
sendung *f* ~ **device** Anreißgerät *n*, Markierein-
richtung *f* ~ **disk** Anzeigerdeckung *f*, Kelle *f*,
Ziel-kelle *f*, -löffel *m* ~ **engaged** Belegen *n* ~
**gauge** Anreißlehre *f*, Parallel-maß *n*, -reißer *m*
~ **identification** Kenntlich-machen *n*, -machung
*f* ~ **indicator** Markenstück *n* ~ **ink** Signier-
farbe *f*, -tinte *f*, Stempelfarbe *f* ~ **iron** Brennstempel *m* ~ **keysender** Markierzahlengeber *m*
~ **lamp** Begrenzungsleuchte *f* ~ **machine** Stem-
pelmaschine *f* ~**-off** Abschnürung *f* ~**-off**
**board** Reißboden *f* ~**-out** Einfluchten *n* ~**-out**
**table** Maßplatte *f* ~ **pencil** Signierstift *m* ~ **per-**
**centage** Arbeitsprozentsatz *m* ~ **plate** Anreiß-,
Markierplatte *f* ~ **pointer** feststehender Zeiger *m*
~ **post** Kabelmerkstein *m*, Markierpfahl *m* ~
**punch** Schlagstempel *m*, Schreibstift *m* ~ **side**
Zeichenseite *f* ~ **signal** Strom-, Trenn-schritt *m*
~**-staff** Markierstab *m* ~ **stick** Markstab *m* ~
**stop** Arbeitsschiene *f* (electr.), Zeichenkontakt
*m* ~ **strip** Bezeichnungsstreifen *m* ~ **surface** Begrenzungsfläche *f* der Markierung ~ **tool** Anreißnadel *f*, Vorreißer *m* ~ **value** Zeichenstrom-
röhre *f* ~ **wave** Arbeits-, Zeichen-welle *f*

**markings** Marken *pl* (Flächenzeichen)
**marks, the** ~ **split** die Zeichen *pl* brechen
**marl, to** ~ bekleiden, (Tau) marlen, mergeln
**marl** Mergel *m* ~ **clay** Tonmergel *m* ~ **pellet** Tongalle *f* ~ **pit** Mergelgrube *f* ~ **slate** Mergelschiefer *m*
**marline** Marlleine *f*, Marling *f* ~ **spike** Marleisen *n*, -pfriem *m*, Spleiß-eisen *n*, -nadel *f*, Splißholz *n* ~ **tie** Bindung *f* (des Luftkabels am Tragseil)
**marling-spike** Marlspieker *m*
**marly** mergelartig ~ **limestone** Mergelkalk *m* ~ **sandstone** Mergelsandstein *m*
**marmatite** Eisenzinkblende *f*
**maroon** kastanienbraun
**marquee** Markise *f*
**marquetry** Holzeinlegearbeit *f* ~ **cutting** Holzausschnittarbeit *f*
**married print** Musterkopie *f* (film)
**marrow** Kern *m*, Mark *n*, Wesentlichste *n*
**Mars yellow** Marsgelb *n*
**marsh** Marsch *f*, Moor *n*, Niederung *f*, Sumpf *m* ~ **drying** Feldentwässerung *f* ~ **gas** Erd-, Gruben-, Sumpf-gas *n* ~ **land** sumpfiges Gelände *n*, Sumpfgebiet *n*
**marshalled** vorgeordnet
**marshaller** Einwinker *m* (Flughafen)
**marshalling signal** Einwinkzeichen *n*
**marshes** Polder *m*
**marshy** sumpfig ~ **areas** sumpfige Gebiete *pl* ~ **land** Marschboden *m* ~ **soil** Marsch-, Moorboden *m* ~ **thicket** Röhricht *n*
**martempering** Warmbadhärten
**Martens, ~ mirror and scale extensometer** Spiegelfeinmeßgerät *n* von Martens ~ **surface-scratching test** Martens-Ritzprobe *f*
**martensite** Martensit *m* ~ **formation** Martensitbildung *f* ~ **point** Martensitpunkt *m*
**Martin, ~ clinker ejector** Martin-Entschlacker *m* (Siemens-) ~ **furnace** Martin-ofen *m*, -prozeß *m*, -verfahren *n* ~ **steel** Martinstahl *m*
**martite** Eisenmohr *m*
**marver, to** ~ marbeln
**marver** Marbeltisch *m*
**maser (microwave amplification by simulated emission of radiation)** Maser *m*
**mash, to** ~ einmaischen, maischen, quetschen, zerdrücken, (zu Brei) zerquetschen **to** ~ **in** einteigen **to** ~ **off** abmaischen
**mash** Brei *m*, Maische *f* ~ **cooler** Maischekühler *m* ~ **tub** Maischpfanne *f* ~ **tun** Braubottich *m*
**masher** (sugar working) Breiapparat *m*
**mashing, ~ apparatus** Maischapparat *m* ~ **in** Einteigen *n* ~ **machine** Schüttel-maschine *f*, -vorrichtung *f*
**mask, to** ~ abdecken (aviat.), maskieren, überdecken, verdecken, verhüllen, verkleiden **to** ~ **with apertured diaphragm (or stop)** ausblenden **to** ~ **off** abblenden **to** ~ **the signal** Signal *n* überlagern oder überdecken
**mask** Abdeckblende *f*, Abdeckung *f*, Abschirmung *f*, Bildröhrenmaske *f*, Deckung *f*, Diaphragm *n*, Maske *f* ~ **shot** Maskenaufnahme *f* (film)
**masked** gedeckt, verdeckt ~ **battery** gedeckte Batterie *f* ~ **headlight** Tarnscheinwerfer *m* ~ **inlet valve** Schirmventil *n* ~ **position** versteckte Stellung *f* ~ **side** Abdeckseite *f* (film)

**masking** Maskierung *f*, Schwundeffekt *m*, Tarnung *f*, Verdeckung *f*, Verhüllen *n*, Verschiebung *f*, Verschleierung *f* ~ **asphalt** Abdeckasphalt *m* ~ **audiogram** Maskierungs-Verdeckungs-Audiogramm *n* (acoust.) ~ **blade** Abdeckflügel *m* ~ **disk** Wechselblende *f* ~ **effect** Verdeckungserscheinung *f* ~ **frames** Filmmasken *pl* ~ **lacquer** Abdecklack *m* ~ **level** Verdeckungspegel *m* ~ **means** Ausblendmittel *n* ~ **paint** Abdeckpaste *f* ~ **plate** Abdeckplatte *f* ~ **process** Rasterverfahren *n* ~ **pulse** Austast-, Wegtast-impuls *m* ~ **recording** Abdeckaufzeichnung *f* ~ **tape** Abdeckband *n*
**mason, to** ~ mauern
**mason** Maurer *m*, Steinmetz *m* ~**'s broom (or brush)** Annetzer *m*, Netzpinsel *m* ~**'s chisel** Steinmeißel *m* ~**'s level** Setzwaage *f*
**masonry** Ausmauerung *f*, Gemäuer *n*, Mauer-arbeit *f*, -werk *n* ~ **apron (or downstream)** gemauertes Sturzbett *n* ~ **dam** Staumauer *f* ~ **wall** Steinmauer *f*
**mass, to** ~ anhäufen, massieren **to** ~**-produce** serienmäßig herstellen
**mass** Haufen *m*, Klumpen *m*, Masse *f*, Menge *f*
**mass, ~ of the aircraft** Flugzeugmasse *f* ~ **of cold air** Kaltluftmasse *f* ~ **of the electron** Elektronenmasse *f*, Masse *f* des Elektrons ~ **of flywheel** Schwungradmasse *f* ~ **of rock** Fluh *f* ~ **of tail unit** Leitwerkmasse *f* ~ **of the universe** Masse *f* des Weltalls ~ **of warm air** Wärmeluftmasse *f*
**mass, ~ absorption** Massenabsorption *f* ~ **absorption coefficient** Massenabsorptions-, Massenschwankungs-koeffizient *m* ~ **action** Massenwirkung *f* ~ **action law** Massenwirkungsgesetz (M.W.G.) *n* ~ **air flow** durch einen Querschnitt hindurchströmende Luftmenge *f* ~ **assignment** Massenbestimmung *f* ~ **attenuation coefficient** Massenschwächungskoeffizient *m* ~ **balance** Gewichtsausgleich *m*, Massengleichgewicht *n* ~ **balance weight** Ausgleichsgewicht *n* ~ **balancing** Masse(n)ausgleich *m* ~ **beater** Massenschläger *m* ~ **central point** Massenmittelpunkt *m*, -schwerpunkt *m* ~ **coefficient of absorption** Massenabsorptionskoeffizient *m* ~ **color** deckende Farbe *f* ~ **concrete** Massenbeton *m* ~ **conversion velocity** Massenumsetzungsgeschwindigkeit *f* ~ **conveyance** Massenförderung *f* ~ **coverage** Massenbelegung *f* ~ **defect** Massendefekt *m* ~ **deficiency** Massendefekt *m* (phys.) ~ **density** Dichtheit *f* oder Dichtigkeit *f* der Masse, Massendichte *f* ~**-discharge piston** Beschleunigerpumpenkolben *m* ~ **distribution** Massenverteilung *f* ~ **edition** Massenauflage *f* ~ **effect** Massen-effekt *m*, -wirkung *f* ~**-energy equivalence** Masse-Energie-Äquivalenz *f* ~**-energy total** Summe *f* von Masse und Energie ~ **equilibrium** Massenausgleich *m* ~ **equivalent** Massenäquivalent *n* ~ **estimation** Massenschätzung *f* ~ **extinction coefficient** Massenschwächungskoeffizient *m* ~ **flow** durch einen Querschnitt hindurchströmende (Luft-)Menge *f*, Massenstrom *m*, Massenströmungsdichte *f*, Mengenstrom *m* ~ **flow control valve** Durchflußregler *m* ~ **force** Trägheitskraft *f* ~ **formula** Massenformel *f* ~ **fraction** Massenverhältnis *n* ~ **inertia** Beharrungsvermögen *n* der Masse ~ **influence** Wandstärkenempfindlichkeit *f* ~**-lock-**

ing Massenverriegelung *f* ~ **moment of inertia**
Massenträgheit *f* ~ **motion** Massenbewegung *f*
~ **movement** Massenbewegung *f* ~ **number**
Massenzahl *f* ~ **point** Massenpunkt *m* ~ **point
mechanics** Massenpunktmechanik *f* ~**-produced
part** Massenteil *n* ~ **producer** Massenhersteller
*m* ~**-product goods** Massengut *n* ~ **production**
am laufenden Band, Groß-erzeugung *f*, -ferti-
gung *f*, -serienfertigung *f*, Massen-anfertigung
*f*, -erzeugung *f*, -fabrikation *f*, -fertigung *f*,
-herstellung *f*, -produkt *n*, Menge-reihenferti-
gung *f*, serienmäßige Herstellung *f*, Serien-
fabrikation *f* ~ **radiator** Massenstrahler *m* ~
**range** Reichweite *f* eines Teilchens ~ **ratio**
Massenverhältnis *n* ~ **reactance** Massenreak-
tanz *f* (acoust.) ~ **renormalization** Massen-
renormierung *f* ~ **resulting from a reaction**
Reaktionsmasse *f* ~ **scale** Massenskala *f*
(phys.) ~**-scattering coefficient** Massenstreu-
koeffizient *m* ~ **selection** Massenauslese *f* ~
**separation** Massentrennung *f* ~ **spectrograph**
Massen-spektrograf *m*, -spektrogramm *n* ~
**spectrometer** Massenspektrometer *n* ~ **spectrum** Massenspektrum *n* ~ **stopping power**
Massenbremsvermögen *n* ~ **susceptibility**
Massensuszeptibilität *f* ~ **synchrometer** Mas-
sensynchrometer *m* ~ **testing** Massenunter-
suchung *f* ~ **transfer** Massenübergang *m* ~
**transfer coefficient** Massenübergangszahl *f*
~**-type plate** Masseplatte *f*, pastierte Platte *f* ~
**unbalance** Unwucht *f* ~ **velocity** Massenge-
schwindigkeit *f* ~ **welding fixture (or jig)**
Schweißgroßvorrichtung *f*
**massage, to** ~ massieren
**massaging implement** Frottierartikel *m*
**massecuite** Füllmasse *f* ~ **chute** Füllmasseschure
*f* ~**-emptying gate** Füllmasseschieber *m* ~
**screw conveyer** Füllmasseschnecke *f* ~ **slide
discharge valve** Füllmasseschieberventil *n* ~
**tank** Füllmassekasten *m* ~ **wagon** Brei-, Füll-
masse-kutsche *f*
**massicot** Blei-gelb *n*, -oxyd *n*, Neugelb *n*, rote
Bleiglätte *f*
**massif** Gebirgsmasse *f*
**massing** Anhäufung *f*, Massenbildung *f*
**massive** derb, dicht, massig, massiv, schwer,
voll, vollwandig ~ **block** Klotz *m* ~ **coil** Hauptspule *f* ~ **one-piece box section** kastenförmiger
Hohlkörper *m* aus einem Stück
**mast** Mast *m*, Mastbaum *m*, (crane) Säule *f*,
Turm *m* ~ **set in concrete** einbetonierter Mast
*m* ~ **of derrick** Kransäule *f*
**mast,** ~ **base** Mastfuß *m* ~ **beacon** Mastbefeue-
rung *f* ~ **buoy** Spierentonne *f* ~ **cart** Mast-
karren *m* ~ **cleat** Mastklampe *f* ~ **foundation**
Mastfundament *n* ~ **guide** Mastkoker *m*
**masthead** Mast-spitze *f*, -topp *m* ~ **cap** Esels-
haupt *n* ~ **lantern** Focklaterne *f* ~ **light** Focklaterne *f*, Topplicht *n*
**mast,** ~ **hoop** Mastbügel *m* ~ **line** Mast-leine *f*,
-seil *n* ~ **link** Mastenankopplung *f* ~ **man** Gast
*m* ~ **mooring** Mastverankerung *f* ~**-mooring
gear** Mastfesselgeschirr *n* ~ **mooring point**
Mastfesselpunkt *m* ~ **prop for careening** Mast-
stütze *f* ~ **rapping** Mastschlagen *m* ~ **step**
Mastspur *f* ~ **support** Masthalterung *f* ~
**thwart** Mastducht *f* ~ **trunk** Mastkoker *m* ~**-type
antenna** Funkmast *m* ~ **wedge** Mastkeil *m*

**master, to** ~ bewältigen, meistern, überwinden
**master** Dienstherr *m*, Lehr-herr *m*, -meister *m*,
Meister *m*, Werkmeister *m* ~ **amplifier** Summenverstärker *m* (Mischpult) ~**-and link-rod
assembly** Gruppe *f* von Hauptpleueln und
Nebenpleueln, Pleuelstern *m* ~ **antenna system**
Gemeinschaftsantennenanlage *f* ~ **attenuator**
Summenregler *m* ~ **batch** Grundmischung *f* ~
**battery** Leitbatterie *f* (rdr) ~ **battery cutoff
switch** Netzausschalter *m* ~ **binder** Sammelpro-
spekt *m* ~ **blade** Abdeckflügel *m* (film) ~ **blank**
Musterkegel *m* ~ **cam** Führungsschablone *f*
einer Kopiereinrichtung, Meisternocke *f*, Ur-
nocke *f* ~ **cam disc** Befehlsscheibe *f* ~ **card**
Leit-, Matrizen-, Stamm-karte *f* (Lochkarte) ~
**carrier** Meisterwellenträger *m* ~ **caution light**
Hauptwarnleuchtschild *n* ~ **clock** Haupt-,
Mutter-, Zentral-uhr *f*; Takt-, Zeit-geber *m* ~
**compass** Mutterkompaß *m* ~ **compass card**
Mutterrose *f* ~ **component** Meisterwelle *f* ~
**condenser** Hauptkondensator *m* ~ **connecting
rod** Hauptpleuel-, Hauptschub-stange *f* ~
**contactor** Hauptschütz *m* ~ **control** Regiepult
*n* (film), Summenregler *m*, Zentralpult *n*
~**-control carburetor** Vergaser *m* mit vollkommen selbsttätiger Regelung ~ **control link**
(helicopter) Hauptsteuerhebel *m* ~ **control
valve** Hauptkontrollschieber *m* ~ **controller**
Führungsregler *m*, Meisterschalter *m* ~ **copy**
Musterkopie *f* ~ **copyholder** Tischbett *n* ~
**cylinder** (brake) Druckzylinder *m* ~ **dimension**
Passungsmaß *n* ~ **dividing gear** Urteilrad *n* ~
**drill jig** Urbohrvorrichtung *f* ~**-excited
oscillator (or transmitter)** fremderregter Sender
*m* ~**-excited tube** fremdgesteuerte Röhre *f* ~
**file** Hauptkartei *f*, Stammaufzeichnung *f*,
Stammband *n* (data proc.), Stammkartei *f* ~
**flask** Musterformkasten *m* ~ **frequency** Grund-
frequenz *f* ~ **frequency meter** integrierender
Frequenzmesser *m* ~ **gate valve** Hauptkontrollschieber *m* ~ **gauge** Einstell-, Gegen-, Haupt-,
Kontroll-, Normal-lehre *f*, (check gauge)
Prüflehre *f*, Ur-form *f*, -lehre
*f*, -maß *n*, Vorrichtungslehre *f* ~**-generator**
Muttergenerator *m* ~ **gyro compass** Kreisel-
kompaßmutteranlage *f* ~ **gyroscope** Kreisel-
mutter *f* ~**-impulse generator frequency** Haupt-
taktgeberfrequenz *f* ~ **indicator** Haupt-,
Mutter-anzeigegerät *n* (Kompaß) ~ **instrument**
Normalinstrument *n* ~ **key** Haupt-schlüssel *m*,
-taste *f* ~ **level** Kontrollibelle *f* ~ **lode** Haupt-
ader *f* ~ **magnetic compass** Kurskreisel *m* ~
**mechanic** Betriebs-, Werk-meister *m* ~ **monitor**
Endkontroll-gerät *n*, -vorrichtung *f*, Endmoni-
tor *m* ~ **negative** Matrize *f*, Meister-, Original-
negativ *n* ~ **office** Hauptamt *n* ~ **operation
card** Arbeitsganghauptkarte *f* ~ **oscillator**
Hauptoszillator *m*, Steuer-generator *m*, -kreis
*m*, -oszillator *m*, -röhre *f*, -sender *m*, -stufen-
röhre *f* ~ **oscillator circuit** Steuerkreisteil *m*
~**-oscillator control stage** Steuerkreisabstim-
mungsstufe *f* ~**-oscillator tuning** Steuerkreis-
abstimmung *f* ~ **panel** Kommandotafel *f* ~
**pattern** Muster-, Mutter-, Original-modell *n*,
Urform *f*, Urschablone *f* ~ **piece** Meisterstück *n* ~ **pilot lamp** Hauptüberwachungslampe
*f* ~ **plan** Bebauungs-, Gesamt-plan *m* ~ **plant**
Mutteranlage *f* ~ **plate** Kopierschablone *f*,

Schablone *f* für Gesenkfräser ~ **positive** Meister-, Original-, Zwischen-positiv *n* ~ **power-switch** Netzhauptschalter *m* ~ **print** Musterkopie *f* (film) ~ **program** Haupt-Leitprogramm *n* (info proc) ~ **pulse generator** Hauptimpulsgeber *m* ~ **record** Stamm-aufzeichnung *f*, -band *n*, -kartei *f* (data proc.) ~ **reference system** Ureichkreis *m* ~ **regulator** Hauptregler *m* ~ **release key** Hauptauslösetaste *f* ~ **(connecting) rod** Hauptpleuel *n* ~ **rod assembly** Pleuelstern *m* ~ **rod cylinder** Hauptzylinder *m* beim Sternmotor ~ **scale** Mutterskala *f* ~ **searchlight** Leitscheinwerfer *m* ~ **shot** Gesamtaufnahme *f* (film) ~**-slave manipulator** Fernbedienungsgerät *n* ~ **sound negative** Mustertonnegativ *n* ~ **sound track** Mustertonspur *f* ~ **spline** breiterer Keilzahn *m* in sonst gleichmäßiger Teilung der Kerbverzahnung ~ **spring** Hauptfeder *f* ~ **spring leaf** oberstes Federbett *n* ~ **stage stabilized by quartz crystal** fremdgesteuerte, quarzgesteuerte Steuerstufe *f* ~ **station** Hauptstation *f* (rdo), Leitstelle *f*, Mithör- und Sprechapparat *m* (teleph.), Muttersender *m* ~ **steel pattern** Tauchform *f* ~ **stroke** Meisterstück *n* ~ **switch** Befehls-, Hauptbetriebs-, Netz-, Steuer-schalter *m*, Gruppenwähler *m*, Vorwählerantrieb *m*, kleiner Wähler *m* ~ **switching unit** Ausgangsbildwähler *m* ~ **table** Tischbett *n* ~ **tap** Backen-, Normal-, Original-bohrer *m* ~ **tape** Hauptband *n*, Lochstreifenmater *f*, Mutterlochstreifen *m*, Stammband *n* ~ **tape-load** Stammband-Ladeprogramm *n* ~ **telephone transmission reference system** Ureichkreis *m* ~ **template** Urschablone *f* ~ **test gauge** Urlehre *f* ~ **test pump** Eichpumpe *f* ~ **thyratron** Befehlsrohr *n* ~ **touch** letzter Schliff *m* ~ **transmitter** Haupt-, Mutter-sender *m*, Steuersender *m* in Ortungsgeräten ~ **trigger (or pulse)** Hauptsteuerimpuls *m* ~ **unit** Haupteinheit *f* ~ **vibrator** Autotrembleur *m*
**masterly** meisterhaft
**master's certificate** Schifferpatent *n*
**mastery** Meisterschaft *f* ~ **battery C main switch** Netzausschalter *m*
**mastic** Kitt *m*, Mastik *m*, Mastix-harz *n*, -zement *m*, Spachtel-kitt *m*, -masse *f* ~ **asphalt** Gußasphalt *m* ~ **asphalt flooring** Gußasphaltestrich *m* ~ **cement** Kittzement *m*, Steinkitt *m*, Zementmastix *m* ~ **varnish** Mastixfirnis *f*
**masticable** kaubar
**masticate, to** ~ zerkleinern, zerkneten
**mastication** Kneten *n*
**masticator** Kneter *m*, Knetmaschine *f*, Schneidekessel *m*
**masting sheers** Mastenkran *m*
**mat, to** ~ mattieren, verfilzen, verflechten
**mat** (rush or cane) Bastmatte *f*, Decke *f*, (brewery) Greifhaufen *m*, Matte *f*; matt ~ **of rush** Binsenmatte *f*
**mat,** ~ **color** Mattfarbe *f* ~ **dike** Mattendamm *m* ~ **finish** Mattzurichtung *f* ~ **foundation** Flachfundation *f* ~ **glaze** Mattglasur *f* ~ **impression** Mattdruck *m* ~ **lighting** Mattscheibe *f* ~ **molding** Mattenpreßverfahren *n* ~ **pickle** Matt-, Mattier-brenne *f* ~**-surface paper** Mattkunstdruckpapier *n* ~ **varnish** Mattlack *m*
**match, to** ~ anpassen, aufeinander passen, aufeinanderpassen, (electric current) aufschleifen,

aussteuern, einhalten, entsprechen, sich gleichen, koinzidieren, passen, passend machen, passend verbinden, schwefeln, im Ton treffen, vergleichen, zusammenpassen
**match** Gegenstück *n*, Lunte *f* (min.), Streichholz *n*, Zündholz *n*, Zündhölzchen *n* ~ **box** Zündholzschachtel *f* ~ **box paste** Reibzündmasse *f* für Streichhölzchen ~ **cardboard** Zündholzkarton *m* ~ **cord** Feuerzeuglunte *f* ~**-head** Zündholzmasse *f* ~ **manufacturing** Zündholzherstellung *f* ~ **plane** Spundhobel *m* ~ **plate** Formplatte *f*, (doppelseitige) Modellplatte *f*, zweiseitiges Modell *n* ~ **splint** Zündholzdraht *m*
**matched** gleich kommen, gleich sein ~ **cable** abgestimmtes Kabel *n* ~ **circuits** Anpassungskreise *pl* ~ **crystal diodes** (matched for uniformity) angepaßte Richtleiter *pl* ~ **die molding** Zweiformpressen *n* ~ **doublet** ausgelesene Dublette *f* ~ **junction** angepaßtes Verzweigungsglied *n* ~ **line** angepaßte Leitung *f* ~ **receiver** angepaßter Empfänger *m* ~ **repeating coils for phantom circuits** Übertragerpaar *n* für Viererleitung und Nachbildung ~ **termination** angepaßter Abschlußwiderstand *m*, reflexionsfreier Abschluß *m* ~ **transformers** Gabelschaltung *f* ~ **transmission line** angepaßte Leitung *f* ~ **waveguide** angepaßter Hohlleiter *m*
**matching** Angleichung *f*, Anpassung *f*, Koinzidenz *f*, Schwefeln *n* von Fässern, Vergleich *m*, Vergleichung *f* ~ **attenuation** Anpassungsdämpfung *f* ~ **bar (or stub)** Anpassungsstück *n* ~ **circuit** Anpassungskreis *m* ~ **condition** Anpassungsverhältnis *n* ~ **edge** Anschlag *m* ~ **equipment** Abgleichmittel *n* ~ **field** Vergleichsfeld *n* ~ **hole** Paßloch *n* ~ **load** Anpassungslast *f* ~ **part** Anschlußteil *m*, Paßstelle *f* ~ **pillar** Anpassungsstift *m* ~ **plane cutters** Nuthobeleisen *pl* ~ **plate** Anpassungsblende *f* ~ **plug** Gegenstecker *m* ~ **repeating coils** Gabelschaltung *f* ~ **resistance** Anpassungswiderstand *m* ~ **strip** Anpassungsstreifen *m* ~ **stub** Abgleichspindel *f* (rdr, electr.), Abstimmpfeife *f*, Anpaßstichleitung *f*, Anpassungsblindschwanz *m* ~ **surfaces** Anlagekante *f* ~ **transformer** Anpassungs-, Stichleitungs-transformator *m* (impedance) ~ **transformer** Anpaßübertrager *m* ~ **unit** Anpassungsgerät *n* (rdr)
**matchless** konkurrenzlos, unvergleichlich
**mate, to** ~ eingreifen, in Eingriff stehen, entsprechen, (gears) kämmen
**mated-surface rusting** Passungsrost *m*
**material** Artikel *m*, Ausrüstung *f*, Gut *n*, Masse *f*, Material *n*, Materie *f*, Stoff *m*, Werkstoff *m*, Zeug *n*; körperlich, materiefest, materiell, stofflich
**material,** ~ **to be admixed** Zusatzstoff *m* ~ **to be annealed** Glühfrischgut *n* ~ **to be blasted** Blasgut *n* ~ **for construction** Ausführungsmaterial *n*, Baustoff *m* ~ **to be conveyed** Fördergut *n* ~ **to be crushed** Mahlgut *n* ~ **for deadening** Schalldämpfungsmittel *n* ~ **to be discharged** Umschlagsgut *n* ~ **for disk (or jaw) crushing** Brechgut *n* ~ **to be dried** Trockengut *n* ~ **of which the frame is built** Rahmenbaustoff *m* ~ **to be ground** Mahlgut *n*, Reibgut *n* ~ **for linings and stiffenings** Futter- und Einlagestoff *m* ~ **to be milled** Mahlgut *n* ~ **for mine equipment** Ein-

richtungsgegenstand *m* für Gruben ~ **to be
mixed** Mischgut *n* ~ **for permanent road** Ober-
baustoff *m* ~ **for protection against gas** Gas-
schutzmittel *n* ~ **to be pulverized** Feinmahlgut *n*
~ **to be roasted** Röstgut *n* ~ **under test** Meßgut
*n* ~ **for track work** Oberbaumaterial *n*
**material,** ~ **accompanying iron** Eisenbegleiter *m*
~ **accounting** Materialabrechnung *f* ~ **assets**
Sachwerte *pl* ~ **axis** Stoffachse *f* ~ **balance**
Materialienverhältnis *n*, Stoffausgleich *m* ~
**buckling** Materialwölbung *f* ~ **category** Stoffgliederung *f* ~ **cathode** körperliche Kathode *f*
~ **cement** Bindemittel *n* ~ **chamber** Material-
abteil *n* ~ **code** Materialschlüssel *m* ~ **constituents** stofflicher Bestand *m* ~ **consumption**
Werkstoffverbrauch *m* ~ **coordinates** materie-
feste Koordinaten *pl* ~ **cost** Materialkosten *pl*
~ **damage** Sachschaden *m* ~ **defect** Materialfehler *m* ~ **delivered** Fördergut *n* ~ **derivative**
massenfeste oder substantielle Ableitung *f* ~
**detaching work** spanabhebende Arbeit *f* ~
**economy** Materialausnutzung *f* ~ **examination**
Materialuntersuchung *f* ~ **form** substantielle
Form *f* ~ **gauge** Begrenzung *f* der Fahrzeuge
**material-handling,** ~ **crane** Überlade-, Verlade-
kran *m* ~ **equipment** Beförderungs-, Transportmittel *n* ~ **ramp** Verladerampe *f* ~ **system**
Transportsystem *n*
**material,** ~ **inventory** Materialvorrat *m* ~ **load**
Materialbelastung *f* ~ **made of wooden latticework** Holzstabgewebe *n* ~ **man** Werkstoffver-
walter *m* ~ **particle** Materieteilchen *n* ~ **proof**
Sachbeweis *m* ~ **requisition** Materialanforderung *f*, Werkstoffbestellung *f* ~ **retained on
trash rack** Rechengut *n* ~ **rise in the output**
Materialscherung *f* ~ **specification** Güte-,
Güter-vorschrift *f* ~ **status** Materialbestand *m*
~ **stocks** Material-bestände *pl*, -vorräte *pl* ~
**stop** Werkstoffanschlag *m* ~ **stores** Magazin-
vorräte *pl* ~ **stress** Materialbeanspruchung *f* ~
**substitute** Werkstoffumstellung *f* ~ **supplies**
Sachlieferung *f*
**material-testing,** ~ **apparatus** Material-, Werk-
stoff-prüfgerät *n* ~ **device** Stoffprüfeinrichtung
*f* ~ **laboratory** Materialprüfungsanstalt *f* ~
**machine** Material-, Werkstoff-prüfmaschine *f*
~ **sample** Materialprobe *f*
**materialization** Verstofflichung *f*
**materially intrinsic** wesentlich
**materials** Stoffe *pl* ~ **for construction of air-
planes** Flugzeugbaubedarf *m* ~ **in transit**
Durchlaufmagazinvorräte *pl*
**materials,** ~ **category number** Stoffgliederungs-
ziffer *f* ~ **control engineer** Materialüberwa-
chungsingenieur *m* ~ **research** Werkstoff-For-
schung *f* ~ **testing** Material-, Werkstoffprüfung
*f* ~ **testing institute** Materialprüfungsanstalt *f*
**material** Material *n*
**mathematical** mathematisch, rechnerisch ~
**check** mathematische Kontrolle *f* ~ **composition** mathematischer Satz *m* ~ **concepts** mathematische Begriffsbildung *f* ~ **constant** Beiwert
*m* ~ **evidence** rechnerischer Nachweis *m* ~
**geography** mathematische Erdkunde *f* ~ **interrelationship** Gesetzmäßigkeit *f* ~ **investigation**
rechnerische Erfassung *f* ~ **logic** formale Logik
*f*, Symbollogik *f* ~ **point** gedachter, ideeller

oder imaginärer Punkt *m* ~ **representation**
mathematische Darstellung *f* ~ **rigor** mathematische Strenge *f*
**mathematically,** ~ **exact** rechnerisch genau bestimmt ~ **transformed diagram** Anamorpho-
gramm *n*
**mathematician** Mathematiker *m*
**mathematics** Größenlehre *f*, Mathematik *f*
**matildite** Silberwismutglanz *m*
**mating** Eingreifen *n*; zusammen-gehörig, -pas-
send ~ **allowance** Paarungsabmaß *n* ~ **clutch**
Gegenklaue *f* ~ **dog** Gegenklaue *f* ~ **edge** Anschlag *m* ~ **features** Paßelemente *pl* ~ **gauge**
Gegenlehre *f* ~ **plug** zugehöriger Stecker *m* ~
**plug gauge** Gegenlehrdorn *m* ~ **profile** Gegenprofil *n* ~ **size** Paarungsmaß *n* ~ **surface** Gegen-
führungsfläche *f* ~ **surfaces** aufeinanderliegende
oder aufeinander arbeitende Flächen *pl* ~
**tooth profile** Gegenflanke *f* ~ **wheel** Gegenrad *n*
**matrices** Matern *pl*, Matrizen *pl* ~ **for type
setting machines** Setzmaschinenmatrizen *pl*
**matriculate, to** ~ aufnehmen
**matrix, to** ~ einschließen
**matrix** Bergart *f*, Bindemittel *n* nach Abbinden,
Form *f*, Gangart *f*, Gefüge *n*, (casting) Gieß-
mutter *f*, Grund-gefüge *n*, -masse *f*, -material *n*,
-stock *m*, Hohlstempel *m*, Lochteil *m* einer
Stanze, Magma *n*, Mater *f*, Matrize *f*, Matrix *f*,
Mutterkaliber *n*, Nonne *f*, Präge-form *f*,
-stock *m*, Preßform *f*, Schriftmutter *f*, Stanzform *f*, Stempel *m*, Untergesenk *n*, Unterstanze *f*, Zwischenmasse *f* ~ **whose elements are
matrices** Übermatrix *f* ~ **for large body size**
Großkegelmatrize *f* ~ **of a transformation**
Transformationsmatrix *f*
**matrix,** ~ **adjusting plate** Matrizenrichtlineal *n*
~ **algebra** Matrix-Algebra *f* ~ **board** Matrizen-
pappe *f* ~ **box** Matrizenbehälter *m* ~ **case** Matrizenrahmen *m* ~ **centering lever** Matrizen-
zentrierhebel *m* ~ **element** Matrix-, Matrizen-
element *n* ~ **elevator** Matrizenheber *m* ~ **embossing press** Matrizenprägepresse *f* ~ **encoder**
Kodierschaltung *f* ~ **formulation of quantum
mechanics** Matrizenform *f* der Quantenmechanik ~ **function** Matrizenfunktion *f* ~ **guide**
Matrizengleitschiene *f* ~ **guide wire** Matrizen-
führung *f* ~ **justifying machine** Maternjustier-
maschine *f* ~ **material** Einbettungsmaterial *n*
~ **mechanics** Matrizenmechanik *f* ~ **memory**
Matrixspeicher *m* ~**-molding press** Matrizenprägepresse *f* ~ **notation** Matrixschreibweise *f*
~ **polynomical** Matrizenpolynom *n* ~ **positioning jaw** Matrizenbacke *f* ~ **release finger** Ma-
trizenhaken *m* ~ **service** Maternkorrespondenz
*f* ~ **storage** Matrizenspeicher *m* ~ **striking press**
Matrizenprägepresse *f*, Schlag-, Stock-presse *f*
**matte** Glanzlosigkeit *f*, Lech *m*, Maske *f* (film),
Mattheit *f*, Stein *m* ~ **calendered** matt satiniert
~ **dip** Mattbrenne *f* ~ **fall** Steinfall *m* ~ **finished**
glanzlos ~ **formation** Steinbildung *f* ~ **lacquer**
Mattlack *m* ~ **rolls** bewegliche Masken *pl*
(film) ~ **smelting** Steinschmelzen *n*
**matted** mattgeschliffen, mattiert, verfilzt
**matter, to** ~ darauf ankommen, von Bedeutung
*f* sein, daran gelegen sein
**matter** Angelegenheit *f*, Bestandteil *m*, Ding *n*,
Masse *f*, Material , *n*, Materie *f*, Sache *f*,
Schriftsatz *m*, Stoff *m*, Substanz *f*, Wesen *n* ~

**at bar** Streitfrage *f* ~ **in dispute** Streitfall *m* ~ **of experience** Erfahrungs-sache *f*, -tatsache *f* **in a** ~ **of fact way** sachlich ~ **of practice** Erfahrungssache *f* **as a** ~ **of principle** im Grunde genommen **as a** ~ **of routine** von Amts wegen

**matter,** ~ **constants** Materialkonstanten *pl* ~ **track** Schnürspur *f* (film) ~ **transport** Materialtransport *m* ~ **wave** Materiewelle *f*

**matting** Abdeckaufzeichnung *f*, Bodenbelag *m*, Mattfläche *f*, Mattieren *n*, Mattierung *f*, Sparterie *f*, Tonstreifenverschmälerung *f* ~ **filter** Anschwemmfilter *n* ~ **frame** Mattenrahmen *m* ~ **salt** Mattsalz *n* ~ **varnish** Mattierungslack *m*

**mattock** Berg-, Schräg-, Spitz-hammer *m*, Breithacke *f*, Breit-, Erd-, Flach-, Letten-haue *f*, Haue *f*, Haueisen *n* (min.), Knappeisen *n*, Zweispitze *f*

**mattress** (for foundations) Faschine *f* ~ **of fascine work** Senkstück *n* ~ **pegged to the ground** Deck-matte *f*, -werk *n*

**maturation** Ausreifung *f* ~ **of a call** Bereitstellung *f* eines Gesprächs

**mature, to** ~ ablagern, altern, reifen

**matured** abgelagert ~ **construction (or design)** ausgereifte Konstruktion *f* ~ **paper** reifes, abgelagertes Papier *n*

**maturing** Altern *n*, Ausreifung *f*, Nachreifung *f* (film)

**maturity** End-, Schluß-alter *n*, Reife *f*, (for spinning) Spinnreife *f*, Verfall *m*

**maul** Handramme *f*, Holz-, Zuschlag-hammer *m*, (double headed) Maker *m*, Schlägel *m*, Stampfer *m*

**maximum** (value) Gipfelwert *m*, Höchst-grad *n*, -maß *n*, Maximum *n*; größt, höchst, maximal, optimal, voll ~ **on either side** Nebenmaximum *n* **at** ~ **speed** volle Drehzahl *f* ~ **of the year** Jahreshöchstmaß *n*

**maximum,** ~ **alarm** Maximalmelder *m* (telegr.) ~ **all-out level power** Notleistung *f* ~ **allowable airspeed indicator** Sicherheitsfahrtmesser *m* ~ **allowable error** zulässige Fehlertoleranz *f* ~ **allowable working pressure** zulässiger Arbeitsdruck *m* ~ **amplification** Verstärkbarkeitsgrenze *f* ~ **amplitude** Maximalamplitude *f* ~ **amplitude of the oscillation** Höchstausschlag *m* der Schwingung ~ **angle of turn** Lenkeinschlag *m* ~ **angular limits** maximaler Ausschlag *m* (Steuerorgane) ~ **aperiodic oscillation** Grenzschwingung *f* ~ **area** Maximalfeld *n* ~ **assymmetric three-phase short-circuit current** dreiphasiger Stoßkurzschlußstrom *m* ~ **automatic device** Maximalautomat *m* (electr.) ~ **boost** Höchstladedruck *m* ~ **boost altitude** Volldruckhöhe *f* ~ **capacity of a line** Aufnahmefähigkeit *f* oder Höchstbelastung *f* einer Leine ~ **case** Höchstfall *m* ~ **ceiling** Flughöhe *f* ~ **charging current** Höchstladestrom *m* ~ **clearance** Größtspiel *n* ~ **coefficient of lift** Höchstauftriebsbeiwert *m* ~ **combustion pressure** Zündhöchstdruck *m* ~ **combustion velocity** maximale Brenngeschwindigkeit *f* ~ **continuous power** höchste Dauerleistung *f* ~ **continuous rating** höchste Dauer(nenn)leistung *f* ~ **cross section** Nullspant *m* ~ **cruising power** höchstzulässige Dauerleistung *f* ~ **current** Höchst-, Maximal-strom *m* ~ **current control switch** Strom-

stärkeumschalter *m* ~ **current impulse** Höchststromimpuls *m* ~ **current intensity** Höchststromstärke *f* ~ **current relay** Höchststromrelais *n* ~ **cut-out** Höchst-, Maximal-ausschalter *m*, Überstromschalter *m* ~ **deflection** Biegungspfeil *m* ~ **deflection of the oscillation** Höchstausschlag *m* der Schwingung ~ **delivery** Vollförderung *f* ~ **demand** Spitzenbelastung *f* ~ **demand indicator** Maximumzähler *m*, Zähler *m* mit Höchstverbrauchsangabe oder mit Maximumzeiger ~ -**demand meter** Höchstverbrauchsmesser *m* ~ -**demand recorder** Zähler *m* mit schreibendem Höchstverbrauchsanzeiger ~ **deviation** maximaler Ausschlag *m*, Spitzenhub *m* ~ **diameter of the work piece** größter Werkstückdurchmesser *m* ~ **dimension** Größtmaß *n* ~ **displacement** Weite *f* der Ausschwingung ~ **distance spindle to table** Ausladung *f* ~ **duration of call** Höchstdauer *f* des Gespräches ~ **economy cruising power with a weak mixture** größtzulässige Reisedauerleistung *f* mit armem Gemisch ~ **efficiency** Höchstleistung *f* ~ **elevation** Dachwinkel *m* ~ **emergency power** Nothöchstleistung *f* ~ **endurance** Höchstflugdauer *f* ~ **energy** obere Grenzenergie *f* ~ **engine overspeed** höchste Motorüberdrehzahl *f* ~ **error** Maximalfehler *m* ~ **except take-off power** höchste längere Zeit *f* entnehmbare Leistung ~ **feed** Vollförderung *f* ~ **flexibility of usefulness** vielseitige Anwendbarkeit *f* ~ **flight range** Reichweitenflug *m* ~ **freezing point** Maximalgefrierpunkt *m* ~ **frequency** Höchstfrequenz *f* ~ **gas pressure** Höchstgasdruck *m* ~ **governed speed** höchste geregelte Geschwindigkeit *f* ~ **horsepower** Höchst-, Maximum-pferdekraft *f* ~ **impuls-indicator** Höchstwertzeiger *m* ~ **indicator pointer** Schleppanzeiger *m* ~ **input voltage** Empfindungsvermögen *n* ~ **intensity** Maximaleffekt *m* ~ -**intensity method** Höchstwertverfahren *n* ~ **interference** Größtübermaß *n* ~ **length** Maximallänge *f* ~ **level** Vollaussteuerung *f* ~ **lift** Höchstauftrieb *m* (aviat.) ~ **load** Belastungs-grenze *f*, -spitze *f*, Bruch-belastung *f*, -last *f*, Grenzlast *f*, größte Zuladung *f*, Höchst-belastung *f*, -last *f*, maximale Beladung *f* oder Belastung *f*, Reißkraft *f*, Zerreißfestigkeit *f* ~ **magnetizing force** maximale Magnetisierungskraft *f* ~ **material condition** (oberer) Abmaßzustand *m* ~ **mean camber** Höchstdurchschnittskrümmung *f*, Maximumdurchschnittswölbung *f* ~ **modulation frequency** Modulationshöchstfrequenz *f* ~ **moment** größtes Moment *n*, Größtmoment *n* ~ **noise zone** Störnebel *m* ~ **number** Höchstzahl *f* ~ **number of electrons in shell** Höchstbesetzungszahl *f* ~ **operating pressure** Höchstbetriebsdruck *m* ~ **ordinate of trajectory** Gipfelpunkt *m* ~ **output** Höchstleistung *f*, Maximalausbeute *f*, maximale Ausgangsleistung *f*, Spitzenleistung *f* ~ **overshoot** Überschwingweite *f* ~ **overspeed** höchste Überdrehzahl *f* ~ **peak inverse plate voltage** maximale Sperrspannung *f* (Spitzenwert) ~ **period** Höchstzeit *f* ~ **permissible** höchstzulässige ~ **permissible elevation of reservoir water surface** Eichpfahl *m* ~ **permissible water level in reservoir** Stauziel *n* ~ **picture signal** Oberstrich *m* (TV) ~ **pointer** Schleppzeiger *m* ~ **position**

Höchststellung *f* ~ **possible rate of change of controller** Stellgeschwindigkeit *f* ~ **power** äußerste Kraft *f*, Höchstleistung *f* ~ **power altitude** Höchstleistungsvolldruckhöhe *f*, Volldruckhöhe *f* für Höchstleistung, Volleistungshöhe *f* ~ **power for climb** Nennleistung *f* (Flugmotor) ~**-power rating** Höchstkurzleistung *f* ~ **pressure** Höchst-, Maximal-druck *m* ~ **pressure steam pipe** Höchstdruckdampfleitung *f* ~ **propeller governed speed** höchste geregelte Propellerdrehzahl *f* ~ **propeller overspeed** höchste Propellerüberdrehzahl *f* ~ **radius** Außenradius *m* ~ **range** Aktionsbereich *m*, (of aircraft) Gesamt-flugstrecke *f*, -schußweite *f*; Höchst-reichweite *f*, -schußweite *f*, maximale Reichweite *f* ~ **range of draft** Gesamtstreckweite *f* ~ **range of gain variation** Regelgrad *m* ~ **rating** Grenzwert *m* ~ **reception** Empfangshöchststärke *f* ~ **recommended cruising power** höchste empfohlene Reiseleistung *f* ~ **response** breitestes Maximum *n* ~ **retention time** Speicherdauer *f* ~ **revolutions** Höchst-drehzahl *f*, -umdrehungen *pl* ~ **safe value** höchstzulässiger Wert *m* ~ **scale reading** Skalenendwert *m* ~ **scale value** Maximalskalenwert *m* ~ **short-circuit current** Stoßkurzschlußstrom *m* ~ **signal (strength) method** (direction finding) Maximalpeilung *f* ~ **sparking distance** Höchstschlagweite *f* des Funkens ~ **speed** Größtgeschwindigkeit *f*, Höchst-drehzahl *f*, -fahrt *f*, -geschwindigkeit *f*, Nenndrehzahl *f* ~ **speed governing** Endregelung *f* ~ **speed governor** Enddrehzahl-, Höchstgeschwindigkeits-, Vollastregler *m* ~ **speed range** Enddrehzahlbereich *m* ~ **speed spring** Enddrehzahlfeder *f* ~ **stress** Höchst-beanspruchung *f*, -last *f*, -spannung *f*, Maximumspannung *f*, Reißkraft *f*, Spitzenbeanspruchung *f*, Zerreißfestigkeit *f* ~ **stress limit** Oberspannung *f* der Dauerfestigkeit ~ **system deviation** höchstzulässige Frequenzabweichung *f* ~ **take-off power** höchste Startleistung *f* ~ **take-off weight** Starthöchstgewicht *n* ~ **taxying weight** Rollhöchstgewicht *n* ~ **temperature** Temperaturmaximum *n*, Wärmehöchstgrad *m* ~ **temperature rise** maximale Temperaturerhöhung *f* ~ **tensile load** Reißlast *f* ~ **tensile strength** Höchstzugfestigkeit *f* ~ **tension** Höchstzug *m* ~ **tension stress** Höchstzugspannung *f* ~ **time** Höchstzeit *f* ~ **torque** Kippmoment *n* (mech.) ~**-trajectory ordinate** Gipfelhöhe *f* ~ **turning diameter** Drehdurchmesserbereich *m* ~ **turning length** Drehlängenbereich *m* ~ **usable frequency** Grenzfrequenz *f*, höchste brauchbare Weitverkehrsfrequenz *f* ~ **useful load** Höchstzuladung *f* ~ **valence** maximale Valenz *f*, Maximalwertigkeit *f* ~ **value** Größ-, Höchst-, Maximal-, Scheitel-, Spitzenwert *m* ~ **vane deflection** Ruderanschlag *m* ~ **velocity** Höchstströmungsgeschwindigkeit *f* ~ **velocity of slipstream** Schraubenstrahlhöchstgeschwindigkeit *f* ~ **vertical ceiling** Gipfel-, Schuß-höhe *f* ~ **voltage** Höchst-, Maximalspannung *f* ~ **volume** Vollaussteuerung *f* (tape rec.) ~ **wavelength** Höchstwellenlänge *f* ~ **weak-mixture power** höchste Sparleistung *f* ~ **weight** Höchstgewicht *n* ~ **working pressure** Genehmigungsdruck *m* (Kessel) ~ **zero fuel weight** Leertankhöchstgewicht *n*

**Maxwell** Maxwell *n* ~ **(statistical) distribution** Maxwellverteilung *f* ~ **earth** Maxwellerde *f* ~'s **equations** Maxwellsche Gleichungen *pl* ~ **series formula** Maxwellsche Reihenformel *f* ~ **top** Maxwellsche Spitze *f* ~ **turn** Maxwellsche Windung *f*, Windungslinie *f* ~ **view** Maxwellsches Gesichtsfeld *n*
**meadow** Flur *f*, Weide *f*, Wiese *f* ~ **green** Wiesengrün *n* ~ **ore** Ortstein *m*
**meager** mager ~ **chalk** Graukalk *m*
**meal** Essen *n*, Kost *f*, Mahlzeit *f*, Mehl *n* ~ **hopper** Mehlbunker *n* ~**-powder composition** rascher Satz *m* ~ **time** Arbeitspause *f*
**mean, to** ~ bedeuten, meinen
**mean** Durchschnitt *m*, (number) Durchschnittszahl *f*, Mittel...; durchschnittlich, mittel, mittelmäßig, mittler(er, e, es) ~ **of separation** Scheidungsmittel *n* (chem.)
**mean,** ~ **aerodynamic chord** Ersatzflügeltiefe *f*, mittlere aerodynamische Tiefe *f* ~ **annual temperature** Jahresmittel *n* der Temperatur, Temperaturjahresmittel *n* ~ **approximation** Approximation *f* im Mittel ~ **brightness value of picture** Mittelbildhelligkeit *f* ~ **camber** Durchschnittskrümmung *f*, Krümmung *f* des Skeletts ~ **camber line** Skelettlinie *f* ~ **central fields** gemittelte Zentralfelder *pl* ~ **chord** mittlere Flügeltiefe *f* ~ **chord of a wing** mittlere Tiefe *f* ~ **day** Tag *m* und Nacht *f* (aviat.) ~ **deviation** mittlerer Fehler *m* ~ **draft** mittlerer Tiefgang *m* ~ **effective pressure** Mitteldruck *m*, mittlerer Arbeitsdruck *m* oder Nutzdruck *m*, Nutzmitteldruck *m*, (of brake) mittlerer Wirkungsdruck *m* ~ **error** mittlerer Fehler *m* ~ **flow** Hauptströmung *f* ~ **flow velocity** mittlere Durchflußgeschwindigkeit *f* ~ **flying condition** mittlerer Flugzustand *m* ~ **free path** mittlere freie Weglänge *f* ~ **free time** mittlere freie Zeit *f*, mittlere Stoßzeit *f* ~ **frequency** Durchschnittsfrequenz *f* ~ **frequency of speech** mittlere Sprachfrequenz *f* ~ **geometric chord** mittlere geometrische Flügel- oder Profiltiefe *f* ~ **geometric distance** mittlerer geometrischer Abstand *m* ~ **hemispherical candle power** halbräumliche Lichtstärke *f* ~ **high water (level)** Mittelhochwasser *n* ~ **horizontal candle power** mittlere räumliche oder horizontale Lichtstärke *f* ~ **hydraulic depth** mittlerer Profilradius *m* ~ **impulse indicator** Mittelwertzeiger *m* ~ **inclination of bank** mittlere Böschungsneigung *f* ~ **indicated pressure** indizierter mittlerer Druck *m* ~ **latitude** Mittelbreite *f* ~ **latitude triangle** Mittelbreitendreieck *n* ~ **length** mittlere Länge *f* ~ **life (time)** mittlere Lebensdauer *f* ~ **line** Mediane *f*, Mittellinie *f*, mittlere Linie *f* ~ **longitudes** mittlere Längen *pl* ~ **low water (level)** Mittelniedrigwasser *n* ~ **noon** Mittag *m*, mittlerer Tag *m* ~ **path of lines of force** mittlerer Kraftlinienweg *m* ~ **pitch** mittlere Steigung *f* ~ **power** mittlere Leistung *f* ~ **power level** mittlerer Pegel *m* ~ **pressure** Mitteldruck *m*, mittlerer Druck *m* ~ **proportional** mittlere Proportionale *f* ~ **pulse time** mittlere Impulsdauer *f* ~ **radius** Mittelradius *m* ~ **range** mittlere Reichweite *f* oder Weglänge *f* ~ **rate of circulation (or flow)** mittlere Durchflußgeschwindigkeit *f* ~ **rotation tensor** Tensor *m* der mittleren Drehung ~ **scale** mittlerer Maß-

stab m ~ sea level (MSL) mittlerer Meeresspiegel m, Normal Null (NN) ~ shading component Grundhelligkeit f ~ solar day mittlerer
Sonnentag m ~ solar time mittlere Sonnenzeit
f ~ speed mittlere oder durchschnittliche Geschwindigkeit f ~ spherical candle power mittlere räumliche Lichtstärke f ~ square deviation
mittlere quadratische Abweichung f ~ square
value mittlerer Quadratwert m ~ stage Mittelwasser n ~ stress Mittelspannung f ~ sun
mittlere Sonne ~ temperature mittlere Temperatur f ~ temperature difference mittlerer
Temperaturunterschied m ~ tensile strain
Bruchlast f ~ tensile strength Formänderungs-
festigkeit f ~ terms mittlere Glieder pl (Proportion) ~ thermal mittlere Thermische f ~
thrust Tagesschub m ~ time Mittelzeit f ~
(solar) time mittlere (Sonnen)Zeit f ~ trajectory mittlere Flugbahn f ~ traveling speed
Reisegeschwindigkeit f ~ upper hemispherical
candle power obere halbräumliche Lichtstärke
f ~ value Durchschnitts-, Mittel-wert m
(math.), mittlere Größe f ~ value of a periodic
quantity Mittelwert m einer periodischen Größe
~-value deviation Mittelwertabweichung f
~-value indication Mittelwertanzeige f ~-value
theorem Mittelwertsatz m ~ velocity Durchschnittsgeschwindigkeit f ~ velocity of flow
mittlere Durchflußgeschwindigkeit f ~ wall
Mittelmauer f ~ water level Mittelwasser n,
mittlerer Wasserstand m ~ wing chord mittlere
Flügeltiefe f ~ yield Durchschnittsertrag m
meander Mäander m ~ belt Mäanderband n ~-
-shaped mäanderförmig ~ strip Mäanderband n
meandering gewunden, schlangenförmig
meaning Begriffbestimmung f
means Geldmittel pl, Mittel n, mittlere Glieder
pl (Proportion), Vermittlung f by ~ of an Hand
von, durch, mittels by that ~ dadurch
~ of attack Angriffsmittel pl, Aufschlußmittel
n (chem.) ~ of communication Verkehrsmittel
n ~ of conveyance Transportmittel n ~ for
deadening Schalldämpfungsmittel n ~ of detonation (explosive) Zündmittel n ~ of illumination Beleuchtungsmittel n ~ of lighting Leucht-
mittel ~ to mask Strahlabblender m ~ of
measurement Meßmittel n ~ of protection Ab-
deckmasse f ~ of separation Trennungsmittel n
~ of supply movement Nachschubmittel n ~ of
transportation Beförderungsmittel n, Fördermittel n, Transportmittel n ~ of wire communication Drahtnachrichtenmittel n
measurable abmeßbar, meßbar, meßfähig ~
quantities meßbare Größen pl ~ variable Meß-
größe f
measurably changeable meßbar veränderlich
measurand Fernmeßwert m, Meß-größe f, -wert
m
measure, to ~ ab-, aus-, be-, nach-, ver-messen,
abtasten, eichen, messen, peilen to ~ accurately
feinmessen to ~ the depth of a shaft with a
plumb line einen Schacht m abseigern to ~
distance with compasses die Entfernung f abgreifen (Karte) to ~ over Übermaß n geben to
~ the perpendicular height absiegern to ~ with
planimeter planimetrieren to ~ short Untermaß
n geben to ~ throughout durchmessen to ~ up
to erfüllen

measure Ausmaß n, Faktor m, Flöz n (geol.),
Grad m (chem.), Kolumnen-, Zeilen-breite f
(print.), Lager n, Maß n, Maßeinheit f (math.),
Maßnahme f, Maßregel f, Maßstab m, Maßstock m, Mensur f, Messen n, Meß-gefäß n,
-instrument n
measure, ~ of acceleration Beschleunigungsmaß
n ~ of altitude Höhenmaß n ~ to bring to terms
Beugemittel n ~ of capacity Hohlmaß n ~ of
comparison Vergleichsmaß n ~ of consistency
of the adjustment Konsistenzmaß n der Ausgleichsrechnung ~ of deformation Deforma-
tionsmaß n ~ of deviation Schwankungs-breite
f, -maß n ~ of distinctness Schärfe f ~ of ductility Zähigkeitsmaß n ~ of elevation Höhenmaß
n ~ of hardness Härte-maß n, -maßstab m ~ of
heat Wärmemaß n ~ of length Längsmaßein-
heit f ~ of settling Sackmaß n ~ of stability
Stabilitätsmaß n ~ of temperature Temperaturmaßeinheit f ~ of value Wertmaßstab m ~ of
wear and tear Verschleißmaß n
measure, ~ column Kolumnenbreite f ~ criterion Maßstabzahl f ~ indicator Formatzeiger
m ~ scale Maßskala f ~ taken by the eye
Augenmaß n ~ theoretical approach maß-
theoretische Behandlung f ~ theory Maß-
theorie f
measured bemessen, gemessen, regelmäßig ~
from A von A aus gemessen ~ with psophometric filter geräuschbewertet in ~ quantities
dosiert
measured, ~ (or actual) difference Istdifferenz f
~ distance Stoppstrecke f ~ feedback gemessene
Rückkopplung f ~ quantity Maßgröße f ~ rate
Einzel-, Zeit-gebühr f ~-rate subscriber Ge-
sprächsgebührenteilnehmer m ~ result Meßwert m ~ trajectory (of projectile) Meßstrecke f
~ value Meß-wert m, -zahl f ~ variable Meß-
größe f
measurement Ab-, Auf-, Aus-maß n, Ab-, Aus-,
Ver-messung f, Dimension f, Messen n, Meß-
ergebnis n, -methode f, Messung f
measurement, ~ of the activity Messung f der
Aktivität ~ of air currents Ventilationsmessung
f ~ of angle Winkelmessung f ~ of areas Flächenmessung f ~ of compression Druckkraftmessung f ~ of consumption Verbrauchsmes-
sung f ~ of current velocity Geschwindigkeitsmesser m ~ of density Dichtemeßanlage f ~ in
dimensions Fadenmaß n ~ of discharge Wasser-
mengenmessung f ~ of discharges Messung f
der Abflußmengen ~ of distance at earth's
surface Erdstreckenmessung ~ of efficiency
Bestimmung f des Wirkungsgrades ~ of
elevation Höhenmessung f ~ of emission
Emissionsmessung f ~ in fathoms Fadenmaß n
~ in feet Fußmaß n ~ of hardness Härtemessung f ~ of hardness by corrosion Korrosions-
härtemessung f ~ of illumination Beleuchtungs-
messung f ~ of ion concentration Ionenmessung
f ~ of the mark Größe f des Stempelaufdruckes
~ of photographs taken with multiple cameras
Mehrbildmessung f ~ of power Kraftmessung f
~ of pressure Druckmessung f ~ of radiant
energy Strahlenmessung f ~ in radians (or
radius) (of an angle) Bogenmaß n ~ of the
radioactivity Radioaktivitätsmessung f ~ of
resistance Widerstandsmessung f ~ of strip

**thickness** Banddickenmessung f ~ **of taper** Kegelmessung f ~ **of tension** Spannungsmessung f ~ **and test equipment** Maß- und Prüfeinrichtungen pl ~ **of thrust** Druckkraftmessung f ~ **of ventilation** Ventilationsmessung f **measurement** ~ **accuracy** Meßgenauigkeit f ~ **amplifier** Meßverstärker m ~ **bridge** Widerstandbrücke f ~ **checking** Nachmeldung f ~ **component** Instrumenteneinzelteil n ~ **concept** Messungsbegriff m ~ **control** Bestimmungskontrolle f ~ **device** Meßanordnung f ~ **energy** Energieverbrauch m ~ **equipment** Maßapparatur f ~ **lengthway** Längsmessung f ~ **mechanism** Meßmechanik f ~ **possibility** Meßmöglichkeit f ~ **range** zuverlässiger Meßbereich m ~ **result** Meß-ergebnis n, -wert m ~ **skeleton chase** (innere) Rahmenweite f des Sparschließrahmens ~ **standard chase** (innere) Rahmenweite f (print.) ~ **system** Meßeinrichtung f ~ **transmission** Meßdurchführung f

**measures** Gebirge n ~ **of the day** Lichtenmaß n **measuring** Abmessung f, Mensur f (chem.), Messen n, Messung f, Meß-gerät n, -vorgang m, -vorrichtung f, Nachmessung f, Vermessen Vermessung f

**measuring,** ~ **an arc with dividers** Kreisbogenabsteckung f ~ **the attachment probability** Messung f der Anlagerungswahrscheinlichkeit ~ **of base** Bodenstärkenmessung f ~ **by chain** Kettenmessen n ~ **by conjunction** Anschlußmessung f ~ **the dose values** Dosisleistungsmessung f ~ **of horizontal angles** Horizontalwinkelmessung f ~ **of outside circumference** Außenrundmessung f ~ **of powder temperature** Pulvertemperaturmessung f ~ **of the reverberation** Nachhallmessung f ~ **of revolutions** Drehzahlmessung f ~ **by sight** Augenmaß n ~ **and signal equipment** Meß- und Signaleinrichtung f ~ **in space** Raummessung f ~ **by stadia** Lattenmessung f

**measuring,** ~ **agent** Meßorgan n ~ **amplifier** Meßverstärker m ~ **angle** Meßwinkel m ~ **anvil** Meßamboß m ~ **aperture** Stauscheibe f ~ **apparatus** Ausmeßgerät n, Meßapparat m ~ **appliance** Meßvorrichtung f, Vermessungsgerät n ~ **arm** Meßzweig m ~ **arrangement** Meßanordnung f ~ **bin** Meßbunker m ~ **bowl** Meßgefäß n ~ **bridge** Meßbrücke f ~ **carriage** Meßwagen m ~ **chain** Lachter-, Meß-kette f ~ **chamber** Meßkammer f ~ **channel** Meß-gerinne n, -kanal m ~ **circuit** Meß-kreis m, -schaltung f ~ **coil** Meßspule f ~ **condenser** Meßkondensator m ~ **content in liters** Ausliterung f ~ **converter** Meßwertumformer m ~ **cord** Meßschnur f ~ **counter** Zähluhr f ~ **cup** Meßbasis f ~ **current** Meßstrom m ~ **cylinder** Standglas n, Meßzylinder m ~ **day** Gedingeabnahme f ~ **detector** Meßdemodulator m ~ **device** Meßeinrichtung f, Messer m, Meßinstrument n, -vorrichtung f ~ **device for oscillation (or acceleration)** Schwingbeschleunigungsmeßgerät n ~ **device for sand** Sandabmeßgerät n ~ **dial** Rundmaßskala f ~ **diaphragm** Meßblende f ~ **dimensions** Meßgeometrie f ~ **direction** Meßrichtung f ~ **disk** Meßscheibe f ~ **drum** Meßtrommel f ~ **eccentricity** Schlagmessung f ~ **element** Meß-organ n, -werk n ~ **equipment** Meß-anordnung f, -platz m ~ **error** Meßfehler

m ~ **face** Meßfläche f ~ **fault** Meßfehler m ~ **flask** Meß-flasche f, -kolben m ~ **fluid** Meßflüssigkeit f ~ **gauge** Meßlehre f ~ **glass** Mensurglas n, Meß-glas n, -zylinder m ~ **grid** Meßgitter n ~ **head** Meß-kopf m, -sonde f ~ **hopper** Meßtrichter m ~ **hose** Meßschlauch m ~ **inset** Meßeinsatz m ~ **instrument** Messer m, Meß-apparat m, -gerät n, -instrument n, -vorrichtung f, -werkzeug n; Vermessungsgerät n, Zeiger m ~ **instrument with alternating-current power supply** netzbetriebenes Meßgerät n ~ **instrument for cylindrical work** Bolzenmeßgerät n ~ **instrument engineering** Meßgerätebau m ~**-jack strip** Meßklinkenstreifen m ~ **jaw** Meß-einsatz m, -schnabel m ~ **junction** Meßstelle f ~ **length** Längenmessung f ~ **level** Meßpegel m ~ **line** Eichstrich m, Meßmarke f, Meß-, Prüf-leitung f ~ **loop** Meßschleife f ~ **machine** Ausmeßgerät n, Dubliermeß- und Wickelmaschine f, Meßmaschine f ~ **magnifier** Meßlupe f ~ **map range with dividers (or compass)** Abgreifen n der Entfernung ~ **mark** Bildfeld-, Einstell-, Meß-, Wander-, Zielmarke f ~ **mark scale** Meßmarkenskala f ~ **means** Meßglied n, Meßwerk n (des Reglers) ~ **mechanism** Meßmechanismus m ~ **method** Meß-methode f, -verfahren n ~ **methodology** Meßmethodik f ~ **microphone** Meßmikrofon n ~ **microscope** Meßmikroskop n ~**-off** Abmessung f, Abtragung f (math.) ~ **office** Meßamt n ~ **orifice** Meßblende f, Stauscheibe f ~ **panel** Messerfeld n ~ **passage** Meßdurchlauf m ~ **peg** Meßstift m ~ **photocell** Meßzelle f ~ **pin** Meßschnabel m, Meßstift m, Tastbolzen m ~ **plug** Meßstöpsel m ~ **point** Meßort m (autom. control), Meß-punkt m, -stelle f, -zapfen m, Tastpunkt m ~ **point accuracy** Meßgenauigkeit f ~ **point amplifier** Meßverstärker m ~ **point distance** Meßabstand m ~ **point fluorescent screen** Meßleuchtschirm m ~ **point resistor** Meßwiderstand m ~ **point selector switch** Meßstellenumschaltung f ~ **point unit** Meßstelleneinheit f ~ **pressure** Meßdruck m ~ **principle** Meßprinzip n ~ **probe** Meßsonde f ~ **process** Meßvorgang ~ **pump** messende Pumpe f, Meßpumpe f ~ **range** Meßbereich m (phys.) ~ **range switch** Meßbereichschalter m ~ **record (or sheet)** Meßprotokoll n ~ **reel** Meßleinenhaspel f ~ **resistor** Meßwiderstand m ~ **rod (for charge descent)** Gichtmesser m (meteor.), Meß-, Peil-stab m ~ **rod for the descent of the charges** Gichtmesser m ~ **roll** Meßwalze f ~ **round (for powder temperature)** Meßkartusche f ~ **screw** Meß-schraube f, -spindel f ~ **seabottom slopes** Messen n von Neigungen im Meeresboden ~ **section** Meß-strecke f, -trupp m ~ **set-up** Meßanordnung f ~ **spanner (or key)** Meßschlüssel m ~ **spark-gap** Meßfunkenstrecke f ~ **spindle sleeve** Meßspinole f ~ **spring** Meßfeder f ~ **staff** Meßstange f ~ **stage** Meßtisch m ~ **stand** Meßstativ n ~ **stereoscope** Meßstereoskop n ~ **stick** Peilstab m ~ **surface** Meßfläche f ~ **table** Meßtisch m ~ **tank** Meßbehälter m, -gefäß f ~ **tape** Bandmaß n, Meßleine f, -schnur f ~ **technique** Meßtechnik f ~ **temperature** Meßtemperatur f ~ **time** Meßdauer f ~ **tool** Meß-gerät n, -werkzeug n ~ **transducer** Meßumformer m ~ **transformer**

Meß-transformator *m*, -übertrager *m* (teleph.), -wandler *m* ~ **transmitter** Meßsender *m* ~ **trolley** Meßwagen *m* ~ **tube** Meß-glas *n*, -röhre *f* ~ **value** Meßzahl *f* ~ **vessel** Maßgefäß *n* ~ **(test) voltage** Meßspannung *f* ~ **wedge** Meßkeil *m* ~ **weir** Meßwehr *n* ~ **wheel** Meßrad *n* ~ **wire** Meßkabel *n*

**meat** Fleisch *n* ~ **charcoal** Fleischkohle *f* ~ **chopper** Fleischhackmaschine *f* ~-**chopper knife** Fleischhackmesser *n* ~ **cleaver** Hackmesser *n* ~ **grinder** Fleischhackmaschine *f* ~-**packing plant** Schlachthof *m* ~ **safe** Vorratsschrank *m*

**mechanic** Handarbeiter *m*, Monteur *m*, Schlosser *m* ~ **auger** Tiefbohrapparat *m* ~'**s bench lathe** Feinmechaniker-Tischdrehbank *f* ~'**s drilling machine** Feinmechaniker-Bohrmaschine *f* ~'**s seat** Maschinistensitz *m* ~'**s tool bag** Gerätetasche *f* für Störungssucher ~'**s tools** Feinmechanikerwerkzeuge *pl*

**mechanical** handwerksmäßig, maschinell, maschinenmäßig, mechanisch, schablonenmäßig ~ **advance and return movement** maschineller Vor- und Rücklauf *m* ~ **advantage** Arbeitsgewinn *m* ~ **advantage shifter** verstellbare Steuerungsuntersetzung *f* ~ **analysis** Kornanalyse *f*, -verteilungsbestimmung *f* ~ **arbor press** mechanische Richtpresse *f* ~ **assembly technique (MAT)** Baukastenprinzip *n* ~ **atomizer burner** Druckzerstäuber *m* ~ **atomizing oil burner** Ölbrenner *m* für mechanische Zerstäubung ~ **balance** Auswuchtungsverhältnis *n*, Massenausgleich *m* ~ **bench** Werkstattbank *f* ~ **bottle stopper** Verschluß *m* für Flaschen ~ **caging lever** mechanischer Feststellhebel *m* ~ **centering** mechanische Zentrierung *f* ~ **centrifugal tachometer** Fliehpendeltachometer *n* ~ **charger** Beschickungsanlage *f* ~ **cokepusher ram** Koks-ausdrückmaschine *f*, -ausstoßvorrichtung *f* ~ **compliance** mechanische Kapazität *f* ~ **contact** Kopfkontakt *m* ~ **control** meachanische Prüfung *f*, mechanisch wirkender oder mechanischer Regler *m* ~ **convection** mechanische Konvektion *f* ~ **conveying and handling** Fördertechnik *f* ~ **copying attachment** Formlineal *m* ~ **de-icer** mechanische Konstruktion *f* ~ **device** maschinelle Einrichtung *f* ~ **diaphragm** mechanische Meßblende *f* ~ **differential** Differentialgetriebe *n* ~ **disengagement** Maschinenauslösung *f* ~ **dividing device** Dividierbetrieb *m* ~ **dodge** Handwerkskniff *m*, Kunstgriff *m* ~ **drawing** Maschinenzeichnen *n* ~ **drive** mechanischer Antrieb *m* ~ **effect** Nutzwert *m* ~ **efficiency** mechanischer Wirkungsgrad *m*, Nutz-effekt *m*, -leistung *f* ~ **engineer** Maschinen-bauer *m*, -baumeister *m*, -ingenieur *m* ~ **engineering** Maschinen-bau *m*, -wesen *n* ~ **engineering institute** Maschinenbauanstalt *f* ~ **equipment** maschinelle Einrichtung *f* ~ **equivalent of heat** Arbeitswert *m* der Wärmeeinheit, mechanischer Hitzegegenwert *m*, Wärme-arbeitswert *m*, -gleichwert *m* ~ **filter** Schutzsieb *n*, Schwebstoffilter *n* ~ **floor clearer** Darresel *m* ~ **followup mechanism** Rückführgestänge *n* ~ **force** mechanische Kraft *f* ~ **fuse** mechanischer Zünder *m* ~ **gas seller** Gasautomat *m* ~ **impedance** mechanische Impedanz *f*, mechanischer Scheinwiderstand *m* ~ **impurity** me-

chanische Verunreinigung *f* ~ **injury** mechanische Beschädigung *f* ~ **instability** mechanische Unbeständigkeit *f* ~ **integrating device** Integriertrieb *m* ~ **knack** Handwerkskniff *m*, Kunstgriff *m* ~ **lift** mechanischer Auftrieb *m* ~ **make and break** mechanischer Abreißmechanismus *m* ~ **mass** mechanische Masse *f* ~ **mixing** maschinelle Mischung *f*, Maschinenmischung *f* ~ **mortar** Mörsermühle *f* ~ **multiplying** Multipliziertrieb *m* ~ **music instrument** Musikwerk *n* ~ **notched-racktype cooling bed** Morgankühlbett *n* mit gezahnten Kipprinnen ~ **oil-atomizer burner** Öldruckzerstäuber *m* ~ **operation** Maschinenbetrieb *m* ~-**optical composition of the image** mechanisch-optische Bildzusammensetzung *f* ~-**optical image scanning (dissection)** mechanisch-optische Bildzerlegung *f* ~ **outfit** Apparatur *m* ~ **passivity** mechanische Passivität *f* ~ **pick** Abbauhammer *m* ~ **pilot** automatische Kurssteuerung *f* ~ **pivoting** mechanisches Schwenken *n* ~ **plans** Installationspläne *pl* ~ **poker** Stochapparat *m* ~ **power** mechanische Kraft *f*, (output) Maschinenleistung *f* ~ **press** mechanische Presse *f* ~ **properties** mechanische Eigenschaften *pl* ~ **properties strength** Festigkeitseigenschaften *pl* ~ **property** Gütewert *m* ~ **puddling furnace** Schaukelofen *m* ~ **pulp** Holzschliff *m* ~ **pusher** Ausdrückmaschine *f* ~ **reactance** mechanische Reaktanz *f* ~ **recording head** Schneid-dose *f*, -kopf *m*, Schneider *m* (phono) ~ **rectifier** Kontaktgleichrichter *m*, mechanischer Gleichrichter *m* ~ **register** mechanisches Zählwerk *n* ~ **reproducer** Schalldose *f*, Tonabnehmer *m* ~ **resistance (or impedance)** mechanische Dämpfung *f* ~ **scanning** mechanische Abtastung *f* ~ **seller** Verkaufsautomat *m* ~ **service shop** Betriebsschlosserei *f* ~ **shop** mechanische Werkstätte *f* ~ **sight** mechanisches Visier *n* ~ **skill in puttering** Bastelbegabung *f* ~ **sleeve** Buchse *f* ~ **stabilizer** mechanische Kippsicherung *f* ~ **stage** Kreuztisch *m* ~ **"step-up"** Übersetzung *f* ~ **stoker** Ladevorrichtung *f* zur Bekohlung von Feuerungen, Schürmaschine *f*, Wanderrost *m* ~ **strain** mechanische Beanspruchung *f*, mechanischer Zug *m* ~ **stress** mechanische Spannung *f* ~ **system** Mechanik *f* ~ **time fuse** Uhrwerkzünder *m*, mechanischer Zeitzünder *m* ~ **towing** Windenschlepp *m* ~ **treatment** Weiterverarbeitung *f* ~ **trick** Handwerkskniff *m* ~ **twin** Zwillingslamelle *f* ~ **upcurrent** Hangwind *m* ~ **ventilation** Zwangslüftung *f* ~ **wood pulp** geschliffenes Holz *n*, Holzschliff *m* ~ **working** Umformung *f* ~ **workshop** mechanische Werkstatt *f* ~ **zero** mechanischer Nullpunkt *m*

**mechanically** schablonenartig ~ **biased** mechanisch vorgespannt ~ **controlled** mechanisch betätigt ~ **controlled variable-pitch propeller** mechanisch verstellbare Schraube *f* ~ **operated** mechanisch betätigt ~ **operated admission valve** gesteuertes Einlaßventil *n* ~ **operated flaps** mechanisch betätigte Klappen *pl* ~ **operated switch** Betätigungsschalter *m* ~ **operated valve** gesteuertes Ventil *n* ~ **operating instruments** mechanisch arbeitende Geräte *pl* ~ **polished** mechanisch bearbeitet (Oberfläche) ~ **retractable undercarriage** mechanisch einziehbares Fahrgestell *n*

mechanics Bewegungslehre *f*, Mechanik *f* ~ of
aircraft Luftfahrzeugmechanik *f* ~ of fits
Passungsmechanik *f* ~ of flight Flugmechanik
*f* ~ of mold making Formtechnik *f* ~ of a
particle Punktmechanik *f* ~ of similitude
Ähnlichkeitsmechanik *f*
mechanism Anlage *f*, Apparatur *f*, Ausrüstung *f*,
Laufwerk *n*, Mechanik *f*, Mechanismus *m*,
Vorrichtung *f*, Werk *n*, Zusammenhang *m* ~ of
conduction Leitungsmechanismus *m* ~ of the
conductivity Leitungsmechanismus *m* ~ of
deformation Deformationsmechanismus *m* ~
for delivery pile Ablagestapelmechanismus *m* ~
of flow Strömungsvorgang *m* ~ for releasing
tow cable from sailplane Ausklinkvorrichtung *f*
~ unit mechanischer Teil *m* (Servo)
mechanization Mechanisierung *f*
mechanize, to ~ mechanisieren
mechanized mechanisiert
mechano-caloric effect mechanisch-kalorischer
Effekt *m*
medal Auszeichnung *f*, Medaille *f*, Münze *f*,
Plakette *f*
meddle, to ~ sich einmischen
medial moraine Mittelmoräne *f*
median die Mitte *f* einnehmend, Mittelwert *m*,
Zentralwert *m* (statist.) ~ lethal time mittlere
letale Zeit *f* ~ line Mediane *f* ~ perpendicular
Mittel-lot *n*, -senkrechte *f* ~ plane Median-,
Mittel-ebene *f* ~ surface Mittelebene *f* ~ value
Medianwert *m*
mediate, to ~ vermitteln
mediate mittelbar ~ strut Mittelstrebe *f*
mediation Schlichtung *f*, Vermittlung *f*, Zwischenschaltung *f*
mediator Unterhändler *m*, Vermittler *m*
medical ärztlich ~ attendance ärztliche Behandlung *f* ~ electrology Elektromedizin *f* ~
electronics Elektromedizin *f* ~ examination
ärztliche Untersuchung *f* ~ ionization Ionentherapie *f* ~ kit Sanitätskasten *m* ~ record
Krankenblatt *n* ~ service Gesundheits-, Sani-
täts-dienst *m*
medicinal arzneilich ~ capsule Arzneikapsel *f* ~
oil Öl *n* für medizinischen Gebrauch ~ plant
Arzneipflanze *f* ~ wax Paraffin *n* für Arznei-
gebrauch
medicine Arznei *f*, Arzneimittel *n*, Medizin *f* ~
chest Arznei-kasten *m*, -kiste *f*, Sanitäts-, Verbands-kasten *m*
mediocrity Mittelmäßigkeit *f*
medium Durchschnitt *m*, Medianpapier *n*
(print.); mittelmäßig, Medium *n*, Mitte *f*,
Mittel *n*, Trübe *f*, Vermittlung *f*
medium, ~ carbon steel Stahl *m* mit mittlerem
Kohlenstoffgehalt ~ circuit Trübeumlauf *m* ~
close-up mittlere Nahaufnahme *f* ~ coarse
mittelgrob ~ cut Mittelschnitt *m* ~ cutting
mower bar Mittelschnittmähbalken *m* ~
delivery Halbförderung *f* ~ density mitteldicht
~ detonation mittelstarkes Klopfen *n* ~ diameter mittelstark ~ distillates Mitteldestillate
*pl* ~-faced halbfett (print.) ~ fine mittelfein ~
fit Gewindepassung *f* (mittel bis fein), Mittel-
passung *f* ~ force fit Edelgleit-, Fest-sitz *m* ~
frequency (MF) Mittelfrequenz *f*, Mittel-
wellenbereich *m*, mittlere Frequenz *f* ~-fre-
quency wave mittlere Welle *f*, Rundfunkwelle *f*

~ (super-charger) gear erster Ladergang *m* (für
mäßige Aufladung) ~ gliding turn mittlere
Gleitflugkurve *f* ~ granulation Mittelbruch *m*
~ gun sight gestrichenes Korn *n* ~ hard halb-,
mittel-hart ~ heavily loaded mittelstark belastet
~ heavy mittel-schwer, -stark ~-heavy-loaded
mittelstark pupinisiert ~-heavy loading mittel-
schwere Bespulung *f* oder Pupinisierung *f* ~
heavy petrol Mittelbenzin *n* ~-heavy-weight oil
mittleres Schmieröl *n* ~-high vacuum range
Feinvakuum *n* ~ indicated and medium
effective pressure mittlerer indizierter und
effektiver Druck *m* ~ loading mittlere Bespulung *f* ~-long shot mittlere Fernaufnahme *f*
(film), Totale *f* ~ oil Mittelöl *n*, mittelschweres
Öl *n* ~ oil varnish halbfetter Öllack *m* ~ paper
Medianpapier *n* ~ plate Mittelblech *n* ~ pole
mittelstarke Stange *f* ~ power mittelstark ~
pressure Mitteldruck *m* ~ pressure cylinder
Mitteldruck-flasche *f*, -zylinder *m* (Kälteanlage) ~-pressure part Mitteldruckteil *m*
~-pressure tire Mitteldruckreifen *m* ~ range
mittlere Entfernung *f*, Mittelbereichs..., Mittel-
strecken... ~ range aircraft Mittelstreckenflug-
zeug *n* ~ range radar Mittelbereichsradar *n* ~
rule halbfette Linie *f* ~ section mill Mittelstraße
*f* ~ sheet Mittelblech *n* ~ shot Halbtotale *f*,
Mittelaufnahme *f* (film) ~ sight gestrichenes
Korn *n* ~ signal area Bereich *m* mittlerer Feldstärke *f* ~ size Mittelgröße *f* ~-size enterprise
Mittelbetrieb *m* ~-sized mittelgroß ~-sized
force mittlere Führung *f* ~-sized plant Mittelbetrieb *m* ~-sized square nail Formstift *m* ~
soft halbschlicht ~ supercharged engine Mittel-
druckladermotor *m* ~ supercharged motor
mäßig aufgeladener Motor *m* ~ supercharger
ratio Laderübersetzung *f* für mäßige Aufladung
(mittlere Nennhöhe, erster Ladergang) ~-thick
profile halbdickes oder mittleres Profil *n* ~ tone
Zwischenton *m* ~ (climbing) turn mittlerer
(gezogener) Kurvenflug *m* ~ wave Mittelwelle
*f*, mittlere Welle ~ wave range Mittelwellen-
bereich *m* ~-wave receiver Mittelwellenemp-
fänger *m* ~-wave transmitter Mittelwellensen-
der *m* ~-weight loading mittlere Bespulung *f*
medullary, ~ ray Markstrahl *m* ~ tube Mark-
röhre *f*
meet, to ~ (by thrust or push) aneinanderstoßen,
antreffen, begegnen, sich berühren, einhalten,
entgegenkommen, entsprechen, genügen, sich
scharen, treffen, zusammentreffen to ~ with
approval Anklang finden to ~ the demand der
Forderung *f* entsprechen to ~ demands genügen to ~ the requirements den Anforderungen
*pl* entsprechen to ~ at station abholen to ~ a
test einer Prüfung *f* entsprechen to ~ violently
aufeinanderprallen
meet Wettbewerb *m*
meeting Begegnung *f*, Fuge *f*, Sitzung *f*, Stoß *m*
(arch.), Tagung *f*, Veranstaltung *f*, Versammlung *f*, Wechselstelle *f*, Zusammentreffen
*n* ~ of the cages Wechselort *m* (min.) ~
of the share-holders Gesellschafterversamm-
lung *f*
meeting, ~ point Treffpunkt *m*, Vereinigungsstelle *f* ~ post Stemmsäule *f*
mega, ~ cycle per second (MC/s) Megahertz *n*
(MHz) ~ dyne Megadyn *n* ~-electron-volt Me-

gaelektronenvolt *n* ~ **farad** Megafarad *n* ~ **meter** Megameter *n*

**megampere** Megampere *n*

**mega,** ~ **phone** Lautsprecher *m*, Megafon *n*, Sprachrohr *n* ~ **seism** Weltbeben *n* ~ **volt** Megavolt *n* ~ **voltage therapy** Hochvolttherapie *f*

**megerg** Megaerg *n*, Megerg *n*

**megger** Isolationsmesser *m*, Megohmmeter *n*

**megohm** Megohm *n* ~ **meter** Megohmmeter *n*

**Meidinger cell** Meidinger-Element *n*

**Meissner circuit** Röhrenschaltung *f* mit magnetischer Rückkopplung

**Méker burner** Mékerbrenner *m*

**mel** Mel *n*

**melaconite** Kupferschwärze *f*, Schwarzkupfererz *n*

**melalite insulator** Melalithisolator *m*

**melange, to** ~ mischen (Farben, Garne, Wolle)

**mélange** Mischung *f*, Vermengung *f*

**melanging** Melangieren *n*

**melanite** Melanit *m*, schwarzer Granat *m*

**melanterite** Melanterit *m*, natürliches Eisenvitriol *n*

**melaphyre** Melaphyr *m*

**melassigenic** melassebildend

**meldometer** Meldometer *n*, Mineralien-Schmelzpuntmesser *m*

**melinite** Melinit *m*

**mellite** Mellit *m*

**mellitic acid** Honigsäure *f*

**mellow** mürbe, teigig, weich

**mellowness** Auflösung *f* ~ **of the soil** Bodengare *f*

**melonite** Tellurnickel *n*

**melt, to** ~ (fuse) ansprechen, auflösen, auftauen, ausschmelzen, durchbrennen (Sicherung), durchschmelzen, einbinden, einschmelzen, erschmelzen, flüssig machen, flüssig werden, schmelzen, verschmelzen, zu Wasser werden, zerfließen, zergehen, zerlassen, zerrinnen **to** ~ **around** umschmelzen **to** ~ **away** wegschmelzen **to** ~ **down** niederschmelzen, weichfeuern **to** ~ **fuse on** anschmelzen **to** ~ **off** abschmelzen, wegschmelzen **to** ~ **on** aufschmelzen **to** ~ **out** ausschmelzen **to** ~ **with reducing flame** reduzierend schmelzen **to** ~ **together** zusammenschmelzen

**melt** Rohschmelze *f* (metall.), Schmelz *m*, Schmelze *f*, Schmelzfluß *m*, Schmelzgang *m*, Schmelzung *f* ~ **computation** Schmelzdurchrechnung *f* ~**-down test** Einschmelzprobe *f*

**meltable** schmelzbar

**melted,** ~ **addition** flüssiger Zusatz *m* ~ **ashes** Ochras *m*, Pottaschefluß *m* ~ **asphalt** Gußasphalt *m* ~ **quartz** geschmolzener Quarz *m* ~ **silicate solution** Silikatschmelzlösung *f* ~ **snow and ice** Schmelzwasser *n* ~ **sugar** aufgelöster Zucker *m*

**melter** Auflöse-, Einschmelz-pfanne *f*, Gießer *m*, Schmelzarbeiter *m*, Schmelzer *m*, Schmelzgefäß *n*, -ofen *m*, -tiegel *m*

**melting** (away) Abschmelzen *n*, Ansprechen *n*, Durchbrennen *n*, Durchschmelzen *n*, Durchschmelzung *f*, Einschmelzen, Erschmelzung *f* ~ **of a fuse** Abschmelzen *n* einer Sicherung ~ **on** Aufschmelzen *n* ~ **without oxidation** Schmelzen *n* ohne Oxydation ~ **together** Kontaktschmoren *n*, Kontaktzusammenschmoren

*n* ~ **under a white slag** Schmelzen *n* ohne Oxydation

**melting,** ~ **aggregate** Schmelzaggregat *n* ~ **apparatus** Schmelzapparat *m* ~ **band** Schmelzband *n* ~ **bath** Schmelz *m*, Schmelzbad *n* ~ **capacity** Schmelzleistung *f* ~ **chamber** Schmelzraum *m* ~ **charge** Schmelzgut *n* ~ **condition** Schmelzbedingung *f* ~ **cone** Schmelzkegel *m* ~ **crucible** Schmelz-gefäß *n*, -tiegel *m* ~ **crucible tongs** Schmelztiegelzange *f* ~ **curve** Schmelzkurve *f* ~**-down power** (**or efficiency**) Schmelzleistung *f* ~ **equipment** Schmelzeinrichtung *f* ~ **flashover** schmelzbarer Selfbogen *m* ~ **furnace** Schmelzofen *m* ~ **hearth** Schmelzgefäß *n* ~ **heat** Schmelz-hitze *f*, -wärme *f* ~**-hole furnace** Unterflurofen *m* ~ **installation** Schmelzeinrichtung *f* ~ **kettle for glass** Glasschmelzhafen *m* ~ **ladle** Schmelzlöffel *m* ~ **loss** Abbrand *m*, Kupolofenabbrand *m* ~ **medium** Schmelzaggregat *n* ~ **operation** Schmelzarbeit *f*, Schmelzbetrieb *m*, Schmelzungsgang *m* ~ **pan** Auflöse-, Einschmelz-pfanne *f* ~ **parameter** Schmelzparameter *n* ~ **phenomenon** Schmelzerscheinung *f* ~ **plant** Schmelzanlage *f* ~ **point** Fließpunkt *m*, Schmelz-grad *m*, -punkt *m*, -wärmegrad *m* ~**-point curve** Schmelzkurve *f* ~ **point depression** Schmelzpunktserniedrigung *f* ~ **pot** Schmelz-hafen *m*, -tiegel *m*, -topf *m*, Tiegel *m* ~ **pot lever** Gießtropfrollenhebel *m* ~ **practice** Schmelzbetrieb *m* ~ **pressure** Schmelzdruck *m* ~ **process** Schmelz-arbeit *f*, -prozeß *m*, -verfahren *n* (electr.) ~ **rack** Anschmelzrost *m* (print.) ~ **shop** Schmelzhalle *f* ~ **stock** Beschickung *f*, Beschickungs-gut *n*, -material *n*, Schmelzgut *n* ~**-stock column** Schmelzsäule *f* ~ **tank** Schmelztiegel *m* ~ **temperature** Schmelztemperatur *f* ~ **unit** Schmelzaggregat *n* ~ **zone** Schmelzzone *f*

**member** Angehöriger *m*, Bauteil *m* (aviat.), Einzelteil *m*, Flächenspiere *f* (aviat.), Glied *n*, Körper *m*, Mitglied *n*, Stab *m*, Teil *m* **any point of the** ~ beliebiger Stabpunkt *m* ~ **of frame** Rahmenträger *m*

**member,** ~ **measuring the departure of the condition from its prescribed value** Meßglied *n* ~ **sensing the condition to be controlled** Fühlglied *n*

**membership** Mitgliedschaft *f*

**membrane** Diaphragm *n*, Federplatte *f*, Haut *f*, Luftschwinger *m*, Membran(e) *f*, Schwingplättchen *n*, Schwingwand *f* ~ **dome** Membrandom *m* ~ **duct** Membranleiter *m* ~ **gauge** Dämpfungsmanometer *n* ~ **stress** Membranspannungszustand *m* ~ **valve** membranbelastetes Ventil *n* ~ **waterproofing** Außenhautdichtung *f*

**memo** Merkzettel *m*

**memorandum** Aktenvermerk *m*, Anmerkung *f*, Aufzeichnung *f*, Merkblatt *n*

**memorial** Denkmal *n*, Schriftsatz *m* ~ **grove** Ehrenhain *m*

**memory** Speicher *m* (data proc.) ~ **address register** Speicheradreßregister *n* ~ **drum** Speichertrommel *f* ~ **effect** Nachwirkungseffekt *m* ~ **solenoid** Speicherbremsspule *f* ~ **tube** Speicherröhre *f* in Rechenmaschinen ~ **unit** Speicher *m*

**menace, to** ~ drohen

menace Drohung *f*, Gefahr *f*
menacing drohend
mend, to ~ ausbessern, flicken, reparieren, repassieren, wiederherstellen
menders Blech 2. Güte
Mendheim (chamber-type) kiln Mendheim-Ofen *m*
mending Stopfen *n* ~ machine Stopfmaschine *f*
menilite Leberopal *m*
meniscal meniskenförmig, mondförmig ~ lens mondförmiges Glas *n*
menisciform menisken-artig (-förmig)
meniscoid menisken-artig (-förmig)
meniscular formations Berührungskapillarradikale *pl*
meniscus konvexkonkave Linse *f*, Kuppe *f*, Linse *f* mit sichelförmigem Querschnitt, Meniskus *m* (phys.), Möndchen *n* ~ edge Meniskuskante *f* ~ lens Meniskenglas *n*, Meniskuslinse *f* ~ spectacle glasses durchbogene Brillengläser *pl*
mensuration Meßbestimmung *f*, Meßkunst *f*, Vermessung *f* ~ of earthwork Massenberechnung *f*
mental, ~ calculations (or computation) Kopfrechnen *n* ~ picture Vorstellung *f*
mention, to ~ anführen, bemerken, erwähnen
mention Erwähnung *f*
méplat abgeflachter Teil *m* der Patrone
Mercator's, ~ chart Merkatorkarte *f*, wachsende Karte *f* ~ map Merkatorkarte *f* ~ projection Merkator-karte *f*, -kartenprojektion *f*
mercatorial merkatorisch ~ bearing merkatorische Peilung *f* ~ flying nach der Merkatorkarte *f* fliegen
mercerization Merzerisierung *f*
mercerized merzerisiert ~ cotton merzerisierte Baumwolle *f* ~ yarn Glanzgarn *n*
mercerizing, ~ in the grey Rohmerzerisation *f* ~ process Merzerisierverfahren *n*
merchandise account Handelsgüter *pl*, Waren *pl*, Warenkonto *n*
merchant Großkaufmann *m* ~ bar Paket-, Paketierschweiß-, Raffinier-stahl *m* ~ bar iron Handels-eisen *n*, -stabeisen *n* ~ bar rolling mill Stabstahlwalzwerk *n* ~ copper Handelskupfer *n* ~ (marine) flag Handelsflagge *f* ~ fleet Handelsflotte *f* ~ iron Grobeisen *n* ~ lead Frischblei *n* ~ man Handels-, Kauffahrteischiff *n* ~ marine Handels-marine *f*, -schiffahrt *f* ~ mill Kunstmühle *f*, Universaleisenwalzwerk *n* ~ (iron) mill Handelseisenwalzwerk *n* ~-mill train Stabstraße *f* ~ rollers Reckwalzwerk *n* ~ rolls Grobeisen-, Reck-walzwerk *n* ~ shipping Handelsschiffahrt *f* ~ tonnage Handelsschiffsraum *m* ~ vessel Handelsschiff *n*
merchantable käuflich
mercurial merkurhaltig, quecksilberhaltig ~ air pump Quecksilberluftpumpe *f* ~ barometer Quecksilberbarometer *n* ~ gray copper ore Graugültigerz *n* ~ horn ore Quecksilberhornerz *n* ~ ointment graue Salbe *f*
mercuric, ~ arsenite Quecksilberarsenitoxyd *n* ~ bromide Merkuribromid *n* ~ chloride Ätzsublimat *n*, Merkurchlorid *n*, Quecksilberchlorid *n* ~ chloride spray Sublimat *n* ~ chromate Quecksilberchromatoxyd *n* ~ compound Merkuriverbindung *f* ~ cyanide Quecksilber-

zyanid *n*, Queckzyanid *n*, Zyanquecksilberoxyd *n* ~ iodide Jodquecksilberoxyd *n*, Quecksilberjodid *n* ~ nitrate Quecksilberoxydnitrat *n*, salpetersaures Quecksilberoxyd *n* ~ oxide Queckoxyd *n*, Quecksilberoxyd *n* ~ oxycyanide Quecksilberoxyzyanid *n* ~ sulfide Quecksulfid *n* ~ sulphate Quecksilbervitriol *n* & *m* ~ sulphocyanate Pharaoschlange *f*, Quecksilberrhodanid *n* ~ sulphocyanide Merkurisulfozyanid *n* ~ thiocyanate Merkurirhodanid *n*
mercurochrome Merkurochrom *n*
mercurous quecksilbrig ~ acetate Quecksilberazetatoxydul *n* ~ azide Stickstoffquecksilberoxydul *n* ~ bromide Merkurobromid *n* ~ chloride Calomel *n*, Kalomel *n*, Merkurchlorür *n*, Quecksilberchlorür *n* ~ cyanide Quecksilberzyanür *n* ~ iodide Jodquecksilberoxydul *n*, Quecksilberjodür *n* ~ nitrate Queckoxydulnitrat *n*, Quecksilberoxydulnitrat *n*, salpetersaures Quecksilberoxydul *n* ~ (azide) nitride Stickstoffkalomel *n* ~ oxalate Quecksilberoxalatoxydul *n* ~ oxide Quecksilberoxydul *n* ~ sulfate Quecksilbersulfatoxydul *n* ~ sulfide Quecksilbersulfür *n* ~ sulphocyanide Quecksilberrhodanür *n* ~ tannate Quecksilberrhodanür *n*
mercury Merkur *m*, Quecksilber *n* ~ air pump Quecksilberpumpe *f* ~ alloy Amalgam *n*, Quecksilberlegierung *f* ~ ammoniumchloride Merkuriammoniumchlorid *n* ~ arc Quecksilber-lichtbogen *m*, -kurve *f* ~-arc convertor Quecksilberdampfstromrichter *m* ~-arc rectifier Quecksilberdampfgleichrichter *m* ~ bag Quecksilbersack *m* ~ barometer Quecksilberbarometer *n*, -luftdruckmesser *m*, -wärmegradmesser *m* ~ bath Quecksilberbehälter *m* ~ bowl Quecksilberkessel *m* ~ box Quecksilberdose *f* ~ break Quecksilberstahlunterbrecher *m* ~ bulb Quecksilberkessel *m* ~ cathode Quecksilberkathode *f* ~ cell Quecksilberelement *n* ~ chloride Merkur-, Quecksilber-chlorid *n* ~ circuit breaker Quecksilberschalter *m* ~ column Quecksilber-faden *m*, -säule *f* ~ contact Quecksilberkontakt *m* ~-contact relay Relais *n* mit Quecksilberkontakten ~ contact tube Quecksilber-kontaktröhre *f*, -schaltröhre *f* ~ contact tubes Quecksilberschalter *m* ~ contactor Quecksilberschütz *m* ~ contacts relay Relais *n* mit Quecksilberkontakten ~ content Quecksilbergehalt *m* ~ converter Quecksilbergleichrichter *m* ~ cup Quecksilbernäpfchen *n* ~ cyanide Quecksilberzyanid *n* ~ delay line Quecksilberlaufzeitglied *n* ~ diffusion pump Quecksilber-Diffusionspumpe *f* ~ ejector Quecksilber-Dampfstrahlpumpe *f* ~ fulminate Knallquecksilber *n* ~ gauge Quecksilbermanometer *n* ~ halide Halogenquecksilber *n*, Quecksilberhalogen *n* ~ iodide Quecksilberjodid *n* ~ jet interrupter Quecksilberstrahlunterbrecher *m* ~ lamp Quecksilberlichtlampe *f* ~ maximum pressure burner Quecksilber-Höchstdruckbrenner *m* ~ meniscus Quecksilberkuppe *f* ~ mirror Quecksilberspiegel *m* ~ mordant Quecksilberbeize *f* ~ motor meter Quecksilberumlaufzähler *m* ~ nitride Stickstoffquecksilber *n* ~ oleate Quecksilberoleat *n* ~ ore Quickerz *n* ~ oxide Quecksilberoxyd *n* ~ phenolate Phenolqueck-

silber n ~ **pool** (in discharge tube) Quantum-vorrat m, Quecksilbermasse f ~-**pool cathode** Quecksilberkathode f ~ **potassium cyanide** Zyanquecksilberkalium n ~ **process** Queck-silberverfahren n ~ **ring balance** Quecksilber-ringmeßwaage f ~ **salicylate** Quecksilber-salizylat n ~ **selenide** Selenquecksilber n ~ **solution** Quickwasser n ~ **stem** Quecksilber-faden m ~ **storage** Quecksilberspeicher m ~ **sump** Quecksilbermasse f ~ **switch** Quecksilber-kippschalter m, -umschalter m, Schaltröhre f ~ **switch breaker** Quecksilberschalter m ~ **switching tube** Quecksilberschaltrohr n ~ **tank** Quecksilberspeicher m ~ **thread** Quecksilber-faden m ~ **trap** Quecksilberfalle f ~ **trough** Quecksilberdose f ~-**type differential gauge** Schwimmermanometer n

**mercury-vapor** Quecksilberdampf m ~-**filled** quecksilberdampfgefüllt ~ **high pressure lamp** Quecksilber-Höchstdrucklampe f ~ **lamp** Quecksilberdampflampe f ~ **rectifier** Queck-silberdampfgleichrichter m ~ **rectifier tube** Quecksilberdampfgleichrichterröhre f ~ **stream** Quecksilberdampfstrom m ~ **tube** (with thyratron) Quecksilberdampfröhre f (mit Glühkathode), Thyrotronröhre f ~ **tube with control grid** Quecksilberdampflampe f mit Steuerelektrode ~ **tube of the ignitron type (or of the thyratron type)** Quecksilberdampflampe f mit Steuergitter

**mercury watthour-meter** Quecksilberleistungs-messer m

**merge, to** ~ aufgehen, aufgehen in, einmischen, fusionieren, mischen, verschmelzen

**merger** Fusion f, Fusionierung f, Verschmelzung f, Zusammen-legung f, -schluß m

**merging** (of pictures) Verschmelzung f ~ **of light impressions** Verschmelzen n der Lichtein-drücke ~ **of traffic** Mischen n des Verkehrs

**meridian** Längenkreis m, Meridian m, Mittags-kreis m, Stundenkurve f ~ **of the earth** Erd-meridian m ~ **of Greenwich** Meridian m von Greenwich, Nullmeridian m ~ **of longitude** Erdmittagskreis m, Längenkreis m

**meridian,** ~ **altitude** Breiten-, Meridian-höhe f ~ **ball (or body)** Meridiankugel f ~ **figure** Meri-dianfigur f ~ **gyroscope** Meridiankreisel m ~ **line** Meridiankurve f, Mittagslinie f ~ **line on chart** Kartenmeridian m ~ **plane** Meridianebene f ~ **ray** Meridianstrahl m ~ **section** Meridian-streifen m

**meridional** meridional, gegen Süden m gerichtet ~ **flux** Meridionaltransport m ~ **focal line** meridionale Brennlinie f ~ **plane** erster Haupt-schnitt m (opt.), Meridionalebene f ~ **ray** Meridionalstrahl m ~ **section** Meridionalteil m

**merit, to** ~ verdienen

**merit** Güte f, Wertigkeit f ~ **of picture** Bildgüte f, Filmwirkung f ~ **rating** positive Bewertung f

**Merton** (straight-line rabble) **furnace** Merton-Ofen m

**mesa transistor** Mesatransistor m

**mesh, to** ~ eingreifen, in Eingriff m stehen, ein-rücken, einspuren, kämmen to ~ **into** abwalzen

**mesh** Bild-element n, -punkt m (of screen), Dreieckglied n, Eingriff m, Geflecht n, Gitter n, Glied n, Masche f, (triangle or delta connection) Polygonschaltung f, Raster m, Sieb n, (size) Siebfeinheit f, Vieleckschaltung f, (of tooth) Zahneingriff m **in** ~ (engaged) eingekuppelt **out of** ~ ausgekuppelt ~ **of a network** Ketten-leitermasche f ~ **of sieve** Siebmasche f ~ **of texture** Gewebemasche f

**mesh,** ~ **amplifier** Netz-verstärker m, -verviel-facher m ~ **aperture** Maschenweite f ~-**connec-ted** in Dreieckschaltung f ~ **connection** Delta-, Dreieck-schaltung f ~ **gauge** Siebskala f ~ **grating** Siebrost m ~ **method** schrittweise Näherung f ~ **multiplier** Gittervervielfältiger m, Netzverstärker m, Netzvervielfacher m ~ **screen (or sieve)** Maschensieb n ~ **size** Ma-schenweite f ~ **voltage** kleinste verkettete Spannung f ~ **width** Maschenweite f ~ **work** Netzwerk n, Stahldrahtgeflecht n

**meshed** maschig, netzartig ~ **anode** Maschen-anode f ~ **bag hydro-extractor** Netzzentrifuge f ~ **network** Maschen-netz n, -werk n ~ **struc-ture** Verbundkonstruktion f ~-**system breaker** Maschennetzschalter m

**meshing** Eingriff m, (of teeth) Ineinandergreifen n ~ **direction** Einspurrichtung f ~ **drive** Ein-spurtrieb m ~ **gear** Gegenrad n ~ **lever** Ein-spurhebel m ~ **magnetic switch** Einspurmagnet-schalter m ~ **movement** Einspurbewegung f

**mesitesite** Mesitinspat m

**mesitylene** Mesitylen n

**meson** Meson n, Mesotron n ~ **detection** Mesonennachweis m ~ **field** Mesonenfeld n ~ **pair term** Mesonpaarterm m ~ **production** Mesonenerzeugung f ~ **theory** Mesotron-theorie f

**mesonic,** ~ **atom** mesonisches Atom n ~ **levels** Meson-Niveaus pl, Mesonterme pl

**mesotergum** Rhachis m, Spindelachse f

**mesothorium** Mesothorium n

**mesotron** schweres Elektron n ~ **track end** Mesonenbahnende n

**mess** Messe f; Unordnung f

**message** Benachrichtigung f, Botschaft f, Fern-schreiben n, Fernspruch m, Funkspruch m, Gespräch n, Meldung f, Mitteilung f, Nach-richt f, Rundspruch m, Spruch m (mil.), Tele-gramm n ~ **bag** Abwurf- (aviat.), Melde-tasche f ~ **blank** Melde-blatt n, -karte f, Tele-gramm-formular n, -vordruck m, -vordrucks-blatt n ~ **book** Meldeblock f, Melde-, Pendel-heft n ~ **call** ausgeführte Verbindung f ~ **center** Fernmeldezentrale f ~ **desk** Telegramm-pult n ~ **disseminator** Hinweis-Ansagegerät n ~-**dropping center** Meldeabwurfstelle f (aviat.) ~-**dropping ground** Abwurfstelle f ~ **fee** Ge-sprächsgebühr f ~ **form** Telegramm-vordruck m, -vordrucksblatt n ~ **heading** Übermittlungs-vorsatz m ~-**interception net** Abhörnetz n ~ **minute** Belegungsminute f ~ **pickup** Aufhaken n (aviat.) ~ **rate** Einzel-gebühr f, -gesprächs-gebühr f ~-**rate subscriber** Teilnehmer m mit Einzelgebührenanschluß ~-**rate subscription** Einzelgebührenanschluß m ~ **register** Ge-sprächszähler m, mechanischer Schreiber m ~ **storage unit** Nachrichtenspeicher m ~ **trans-mitted by hand signals (or signaling disk)** Wink-spruch m ~ **unit** Gesprächseinheit f

**messenger** Besteller m, Bote m, Botschafter m, Laufbursche m, Läufer m, Meldeübermittler m, Telegrammbesteller m ~ **cable** Kabelring f

~-cable strand Kabeltragseil n ~ call Gespräch n mit Herbeiruf ~ relay service Stafette f ~ route Meldeweg m ~ trip Botengang m ~ wire Tragseil n ~ wire for overhead cables Luftkabeltragseil n ~ wire clamp Tragseil-klemme f, -schelle f

metabolic process Stoffwechselvorgang m

metabolism Stoffwechsel m, Umwandlung f ~ balance Stoffwechselbilanz f

metabolon Folgeprodukt n, radioaktives Folgeprodukt n oder Produkt n

metaboric acid Metaborsäure f

metacenter Aufrichtemoment m, Metazentrum n

metacentric metazentrisch ~ height metazentrische Höhe f, Metazenterhöhe f

metachromotype Abziehbild n

metadyne Metadyne f ~ drive Amplidynsystem n ~ dynamo Zwischenbürstenmaschine f ~-position Mittelstellung f ~-silicate Metasilikat n

metal, to ~ mit Metall n bedecken to ~-braid metallumspinnen

metal Beschotterung f, Eisenbahnschienen pl, Geleise n, (road) Kiesfüllung, König m, Korn n, Kupferstein m, Lech m, Metall n, Regulus m, Schotter m ~ to be melted Schmelzgut n for ~ surface protection atramentieren (Phosphat)

metal, ~ abrasion Metallabrieb m ~ arc Metall-lichtbogen m ~ arc weld Schweißarbeit f ~ arc welding Bogenschweißung f ~ architectural box Metallbaukasten m ~-armed metallarmiert ~ armoring Bewehrung f ~ assay Metallprobe f ~ backed metallhinterlegt (Leuchtschirm) ~ background Metallhintergrund m ~ backing Metall-hinterlegung f, -spiegel m ~ bagging twine Metallsackbinde f ~ ball Metallkugel f ~ bar Metall-barren m, -stange f ~ base Metallfuß m, -sockel m ~ bath Metallbad n ~ bead Metallkügelchen n ~ bellow Federungskörper m ~ bellows on delivery side austrittseitiger Federungskörper m ~ bending Metallbiegen n ~ block Metallblock m ~ borings Metallspäne pl ~ box Metallbüchse f ~ braid Metallgeflecht n ~-braided cable metallumklöppeltes oder metallumsponnenes Kabel n ~ bronzing Metallbronzieren n ~ carrier Metallträger m ~ (lic) case Metallgehäuse n ~-cased im Metallgehäuse n ~-cased capacitor Becherkondensator m ~-cased conductor Panzeraderleitung f ~ casing Metallfassung f ~ casting Metallgußstück n ~ cementation process Metallschutzverfahren n ~ ceramic metallkeramisch; Keramik-Metallgemisch n ~ ceramic valve Metallkeramikröhre f ~ charge Metallbeschickung f ~ chassis Empfängerblechgestell n ~ chips Metallspäne pl ~-clad metall-gekapselt, -gepanzert, -geschützt ~ clamp Metallschelle f ~ cleaning powder Metallreinigungspulver n ~ clip Metallschelle f ~ cloth Metallgewebe n ~ coat Metallüberzug m ~ coated metallbeschlagen ~ coated on glass Metall n auf Glas aufgedampft ~ coating Metall-auflage f, -auskleidung f, -beschlag m, -überziehung f ~ (lic) coating Metall-anstrich m, -überzug m ~-coating process Metallschutzverfahren n ~ color Metallfarbe f ~ cone Metallkonus m (Bildröhre) ~ cone kinescope Bildröhre f mit Konus-

teil aus Metall ~ construction Metall-ausführung f, -bauweise f ~ constructional kit Metallbaukasten m ~ container Blech-behälter m, -emballage f, -kanister m ~ content Metallgehalt n ~ conveyer Metallförderer m ~ core Metallseele f ~ cover Metall-deckel m, -kappe f ~-covered yarn Metallgespinst n ~ covering Metall-bekleidung f, -beplankung f, -ummantelung f, -verschalung f ~ cross slide Metallkreuzschlitten m ~ cutter Metallausschneider m ~ cutting Gewindeschneiden n, spanabhebend Bearbeitung f ~-cutting spanabhebende Metallbearbeitung f ~-cutting machine Fräse f, spanabhebende Maschine f ~-cutting-machine operator Fräser m ~-cutting material Schneidmetall n ~ cylinder Metallzylinder m ~ dark slide Metallkassette f ~ deposit Metallbelag m ~ detector Metallmeldegerät n ~ diaphragm Plattenfeder f ~ drain Gußgerinne n ~ drum Trommel f ~ dust Metallstaub m ~ embosser Metallpräger m ~ enamel Metallglas n ~ enameling manufacture Emaillierwerk n ~-enclosed apparatus gekapselter Apparat m ~ envelope Metall-hülle f, -ummantelung f ~ extraction Metallgewinnung f ~ eyelet Metallöse f ~ facing Metallverkleidung f ~ fancy goods Metallgalanteriewaren pl ~ fastening Beschlag m (Zellenbau) ~ fatigue Metallmüdigkeit f ~ ferrules Metallkappen pl ~ filings Metallschliff m ~ film Metallschicht f ~ fitting Metall-beschlag m (am Flugzeugrumpf), -fassung f ~ fittings Beschläge pl ~ float Metallschwimmer m ~ foil Metall-blatt n, -folie f ~ form Stahlschale f ~ forming Metallformgebung f ~ foundry Metallgießerei f ~ fragment Metall-brocken m, -stück n ~ framework Metallgerippe n ~ framework of a wing Flügelmetallgerüst n ~ furnace Metallumschmelzofen m ~ fuselage Metallrumpf m ~ gauge Blechlehre f, Metall-manometer n, -schablone f ~ gauze Drahtgitter n, Metallgewebe n ~-glass seal Metallglaskitt m ~ glaze Metallglas n ~ globe frame Metallrahmen m für Beleuchtungskuppel ~ headgear Metallbügel m (am Kopfhörer) ~ heat insulator Wärmeschutzblech n ~ helve Stirnhammer m ~ hinged falling gate eisernes Klapptor n ~ hood Metallkasten m ~ (lic) hose Metallschlauch m ~ hull Metallbootskörper m ~ impurity Begleitmetall n ~ insert Metall-einlage f, -einsatz m ~ label Metallschildchen n ~ lamina (or lamination) Metallblatt n ~ lath Rabitz m ~ lattice body (or fuselage) Metallfachwerkrumpf m ~ leaf Metallblättchen n ~ level Metallhöhe f ~-lined hose Schlauch m mit Metalleinlage ~ lining Metallbelag m ~ loss Abbrand m, Metallverlust m ~ magazine Metallkassette f ~ mass of the engine Maschinenmasse f ~ master Matrize f, Meisternegativ n, Negativ n (phono) ~ mesh Streckmetall n ~ mesh road Drahtstraße f ~ microscope Metallmikroskop n ~ migration Feinwanderung f ~ mist Metallnebel m ~ mixer Roheisenmischer m ~ model plane Metallflugmodell n ~ mold casting Metallformguß m ~ mount Bleifuß m ~ mounting Metallfassung f ~ notch Abstich-loch n, -öffnung f ~ nut Metallmutter f ~ object cast in a mold gegossener Rohling m ~ packing Metallpackung f ~ (lic) paint Metallanstrich m ~ (lic) paper Metallpapier n ~ part Metallteil m ~ particles Metallrieb m ~ passage to die (of die-casting machine) Spritzkanal m ~ pattern Metall-modell n, -schablone f ~ plane Metall-flugzeug n ~ plate Grobblech n, Metall-blech n, -platte f, -schild n, -tafel f ~-plate products Blecherzeugnisse pl ~-plated plywood Panzer-holz n ~ plating Metall-auftrag m, -überzug m ~-plating process Metallschutzverfahren n ~ plug Metallstopfen m ~ poisoning Metallver-giftung f ~ polish Metall-poliermittel n, -schliff m ~ pontoon Metallschwimmer m ~ press for making round bars Stangenpresse f ~ primer Metallgrundierung f ~ probe Metallfühler m ~ propeller Metalluftschraube f ~-protecting cage Drahtgeflecht n ~ protection cap for tripod Stativschuh m ~ puncher Metallausschneider m ~ purifying medium Metall-Läuterungsmittel n ~ purse Metallbörse f ~ quoin Metallschließ-zeug n ~ radiography Röntgenmetallografie f ~ rectifier Metall-, Trocken-gleichrichter m ~ rectifier of reduced size Gleichrichterpille f ~ removal Spanabhebung f ~ removal rate Zer-spannungsleistung f ~ removing capacity Span-leistung f (Schruppen) ~-removing process Ver-spannungsvorgang m ~ research Metallfor-schung f ~ reservoir Metallvorrat m ~ residues Metallabrieb m ~ rib Metallrippe f ~ rim edge Blechrand m ~ ring Metallring m ~ saw Metall-säge f ~ saw file Metallsägefeile f ~ scalpel Stahlmesserchen n ~ science Metallkunde f ~ scrap Metallabfall m ~ scrap dealer Schrott-händler m ~ screen Blechlochsieb n, Metall-schirm m ~ screw Metallschraube f ~ seal Metall-dichtung f, -plombe f ~ section Metall-schnitt m ~ shade Metallschirm m ~ shavings Metallspäne pl ~ shearing Metallschneiden m ~-sheathed metallgeschützt ~-sheathed con-ductor Panzeraderleitung f ~ sheathing Metall-bekleidung f, -ummantelung f, -verschalung f ~ sheet Metall-blech n, -platte f ~ sheet with good bend properties Falzblech n ~ sheet covering Mantelblech n ~ sheet poster Blechplakat n ~ shielding Kapselung f bei Zündapparat (magneto), Metallabschirmung f ~ sieve Metallsieb n ~ single crystal Metalleinkristall m ~ skin Blechbeplankung f ~-skinned metall-beplankt ~ sleeve Metallhülse f ~ slitting machine Trennschleifmaschine f ~-slitting saw Metallkreissäge f ~ smelting Steinarbeit f ~ socket of a candlestick Metalleuchterdille f ~ spar Metallholm m ~ spray method Metall-spritzmethode f ~ sprayed zinc coating Flamm-spritz-Verzinkung f ~ spraying Spritzmetalli-sieren n ~ spraying process Metallspritzver-fahren ~ spring Sprungfeder f ~ stamper Metallausschneider m ~ stamping press Metall-präge f ~ stencil Metallschablone f ~ straps eiserner Bügel m ~ strip Metall-leiste f, -streifen m ~ structure Metallbau m ~ surface Metall-schliff m ~ sweep Anhaublech n (Mähma-schine) ~ tag Blechetikett n ~ tape Metallband n ~ test (ing) Metallprobe f ~-to-~ brake Metallbackenbremse f ~-to-~ joint metallische Abdichtung f ~-to-salt contact Metallsalzkon-takt m ~ tube Metall-hülse f, -röhre f (electron.) ~ tube press Metallrohrpresse f ~ tubing Me-

tallröhren pl ~ turbulence Wirbelströmung f ~ turnings Metallspäne pl ~ twist (usually of woven copper or brass wire) Tress m ~ ware Metallwaren pl ~ waste Metallabfall m ~ wearing plate Verschleißblech n ~ wing Metall-flügel m ~-wire pot cleaner Metalldrahttopf-reiniger m

**metalwork** Bronzearbeit f, Eisenwerk n, Metall-warenfabrik f, Schmiedearbeit f ~ of applied art kunstgewerbliche Metallarbeiten pl

**metalworker and welder** Schlosser m und Schwei-ßer m

**metal, ~-working** Metall-bearbeitung f, -bear-beitungstechnik f, -verarbeitung f, spanlose Bearbeitung f ~-working industry metallver-arbeitende Industrie f ~-working lathe Dreh-bank f für Metallbearbeitung ~-works Metall-hütte f ~ zapon lacquer Metallzaponlack m

**metaled** beschottert

**metalize, to** ~ mit Metall n imprägnieren, metal-lisieren, verspiegeln

**metalized,** ~ paper capacitor Metallpapierkon-densator m ~ screen metallisierter Schirm m ~ strip Silberfolie f (tape rec.)

**metalizing** Metallisieren n ~ with masks Masken-bedampfung f ~ alloy Auftropflegierung f ~ technique Aufdampftechnik f

**metallic** aus Metall n, metallähnlich, metallen, metallisch, metallisch klingend ~ air conduit Metallutte f ~ arc Metalldichtbogen m ~-arc welding Metallichtbogenschweißung f ~ arsenic Scherbenkobalt m ~ automatic telephone system Schleifensystem n ~ azide Metallazid n ~ bond metallische Bindung f ~ bonded metall-gebunden ~ calcium Kalziummetall n ~ carbide Metallkarbid n ~ cartridge Mantel-geschoß n ~ circuit doppeldrähtige Leitung f, Leitungsschleife f ~ circuit against earth current Erdstromschleife f ~ cistern Metall-gefäß n ~ composite set Simultaneinrichtung f für Doppelleitung ~ compound Metallverbin-dung f ~ cord Metallschnur f ~ cyanide Zyan-metall n ~ effect Metalleffekt m ~ electron emitter elektronenemittierendes Metall n ~ filament Metallfaden m ~-filament lamp Me-tallfadenlampe f ~ fuel element Metall-Spalt-stoffelement n ~ fusion metallische Verbindung f ~ gold Goldmetall n ~ ink Bronzedruckfarbe f ~ link belt Stahlgurt m ~ luster varnish Me-talleffektlack m ~ nitride Stickstoffmetall n ~-oxide cathode Metalloxydkathode f ~ pack-ing Metalldichtung f, metallische Dichtung f, Metallstopfbüchsenpackung f ~ packing ring Metalldichtungsring m ~ polar duplex Doppel-stromgegensprechen n auf Doppelleitungen ~ polar duplex telegraphy Gleichstromunter-lagerungstelegrafie f ~ powder Hilfsmetallstaub m ~ printing Bronzedruck m ~ pulley Metall-scheibe f ~ rectifier Sperrschichtgleichrichter m ~ rectifier stack Stapelgleichrichter m ~ recti-fier stack assembly Gleichrichtergruppe f ~ reed Metallblatt n ~ reflector Metallspiegel m ~ reservoir Metallgefäß n ~ return metallische Rückleitung f ~ rhodium Rhodiummetall n ~ ribbon Metallband n ~ salt solution Metallsalz-lösung f ~ screen Gegenelektrode f, metalli-scher Schleier m, Metallsieb n ~ sealing strip Metallstreifenverschluß m ~ signaling system

Schleifensystem *n* ~ **silicide** Siliziummetall *n* ~
**silver** Silbermetall *n* ~ **soap** Metallseife *f* ~
**sodium** Natriummetall *n* ~ **starting rheostat**
Metallanlaßwiderstand *m* ~ **sulfide** Metall-
sulfid *n* ~ **teasel** Metallkratze *f* ~ **telephone**
**circuit** Fernsprech-amtdoppelleitung *f*, -dop-
pelleitung *f* ~ **thermometer** Metallthermometer
*n* ~ **tin** Zinnmetall *n* ~ **vapor** Metalldampf *m* ~
**vein** Metallader *f* ~ **zirconium** Zirkoniummetall *n*
**metalliferous** metallhaltig ~ **group** Metallgruppe
*f*
**metalline** metallähnlich
**metallization** Aufspritzen *n* flüssigen Metalls
auf eine Oberfläche mittels Preßluft; Metallbelag *n*, Metallspritzverfahren *n*, Spritzverfahren *n* ~ **of a tube** Metallbelag *m*
**metallo-chemistry** Metallochemie *f*
**metallochromy** (metal coloring) Metallochromie
*f*
**metallograph** Gefüge-aufnahme *f*, -bild *n*,
metallografische Aufnahme *f*
**metallographer** Metallograf *m*
**metallographic(al)** metallografisch
**metallography** Gefügelehre *f*, Metall-beschrei-
bung *f*, -kunde *f*, Metallografie *f*
**metalloid** Begleitelement *n*, Metalloid *n*; metallartig ~ **cutting tool** Hartmetallwerkzeug *n* ~
**properties** halbmetallische Eigenschaften *pl*
**metallophone** Metallofon *n*
**metallurgic** metallurgisch ~ **hard lead** Hütten-
hartblei *n* ~ **lead** Hüttenblei *n* ~ **merchant lead**
Hüttenweichblei *n* ~ **raw zinc** Hüttenrohzink *n*
**metallurgical** metallurgisch, Hütten... ~ **coke**
Groß-, Hochofen-, Hütten-koks *m* ~ **engineer**
Hüttenmann *m* ~ **engineering** Hütten-technik
*f*, -wesen *n* ~ **furnace** hüttenmännischer Ofen
*m* ~ **knowledge** Hüttenkunde *f* ~ **lime** Hütten-
kalk *m* ~ **oil** Öl *n* für die Metallindustrie ~
**operations** Verhüttung *f* ~ **plant** Hüttenbetrieb
*m*, Metallhütte *f* ~ **process** Hüttenprozeß *m* ~
**product** Hüttenerzeugnis *n* ~ **works** Hüttenwerk *n*
**metallurgist** Hüttenfachmann *m*, Metallurge *m*
**metallurgy** Hütten-betrieb *m*, -kunde *f*, -wesen
*n*, Metallurgie *f* ~ **of copper** Kupferverhüttung
*f* ~ **of iron** Eisenhüttenwesen *n* ~ **of quicksilver**
Quecksilberhüttenwesen *n*
**metals** "sound off" Metallgeräusche *pl*
**metamagnetic** metamagnetisch
**metameric** metamer
**metamerism** Metamerie *f*
**metamorphic** metamorph ~ **rock** metamorphisches Gestein *n*
**metamorphism** Metamorphie *f*, Metamorphis-
mus *m*, Umprägung *f* (geol.), Umwandlungs-
fähigkeit *f*
**metamorphosis** Metamorphose *f*, Verwandlung *f*
**metanilic acid** Metanilsäure *f*
**metapectin** Metapektin *n*
**metaphosphoric acid** Metaphosphorsäure *f*
**metaphosphorous** metaphosphorig
**metapole** Fokalpunkt *m* (photo), Metapol *m*
**metaposition** Metastellung *f*
**metascope** Bildwandler *m*
**metasomatic** metasomatisch
**metastability** Metastabilität *f*
**metastable** bedingt, halbbeständig, metastabil ~

**atomic state** metastabiler Energiezustand *m* ~
**currents** metastabile Ströme *pl* ~ **equilibrium**
metastabiles Gleichgewicht *n* ~ **level** metastabiles Niveau *n* ~ **nuclei** metastabile Kerne *pl* ~
**phase** begrenzt beständige Phase *f* ~ **state**
metastabiler Zustand *m*
**metastannic acid** Metazinnsäure *f*
**metastatic electron** metastatisches Elektron *n*
**metathesis** Umsetzung *f* (chem.)
**metatitanic acid** Metatitansäure *f*
**metatungstic acid** Metawolframsäure *f*
**meteor** Feuerkugel *f*, Meteor *m*, Sternschnuppe *f*
~ **shower** Meteor-, Sternschnuppen-schwarm
*m* ~ **trail** Meteorschweif *m*
**meteoric**, ~ **iron** Meteoreisen *n* ~ **light** Meteor-
licht *n* ~ **steel** Meteorstahl *m* ~ **water** Niederschlagswasser *n*
**meteorite** Meteorit *m*
**meteorograph** Meteorograf *m*
**meteorological** meteorologisch ~ **airplane** Wet-
terflugzeug *n* ~ **apparatus** Absonderungsappa-
rat *m* ~ **balloon** Pilotballon *m* ~ **code** Wetter-
kode *m* ~ **conditions** Wetter-lage *f*, -verhält-
nisse *pl* ~ **disturbance** meteorologische Störung
*f*, Wetterstörung *f* ~ **effect** Witterungseinfluß
*m* ~ **element** meteorologischer Bestandteil *m* ~
**factor** Witterungseinfluß *m* ~ **limit** Wetterscheide *f* ~ **measurement** Wettermessung *f* ~
**message** Wetterfunk-meldung *f*, -spruch *m* ~
**minima** Wettermindestbedingungen *f* *pl* ~
**observation** Wetterbeobachtung *f* ~ **report**
Wetter-beratung *f*, -bericht *m* ~ **service** Wetterdienst *m* ~**-service instruments** Wetterdienst-
geräte *pl* ~ **station** Flugwetterwarte *f*, Wetter-
dienststelle *f*, Wetterwarte *f* ~ **tropics** meteorologische Wendekreise *pl* ~ **weather service**
Flugwetter-beratung *f*, -dienst *m*
**meteorologist** Meteorologe *m*, Wetterbeobachter *m*
**meteorology** Meteorologie *f*, Wetterkunde *f*
**meteors** Sternschuppen *pl*
**meter, to** ~ messen, zählen, zumessen **to** ~ **a call**
ein Gespräch *n* zählen
**meter** integrierendes Meßgerät *n*, Kontroll-
indexinstrument *n*, Literzähler *m*, Meßapparat
*m*, Messer *m*, Meßinstrument *n*, Meter *n*,
Zähler *m*, Zählwerk *n*, Zeiger *m* **(conversation)**
~ Gesprächszähler *m* **(water)** ~ **with mechanism**
**and indicator operating wet** Naßläufer *m* ~ **for**
**oil** Messer *m* für Öl
**meter,** ~**-ampere** Meterampere *n* ~**-amperes**
Leistungsfähigkeit *f* (eines Strahlers) ~ **armature**
Zähleranker *m* ~ **base** Zählergrundplatte *f* ~
**battery** Zählerbatterie *f* ~ **board** Zählertafel *f*
~ **box** Zählergehäuse *n* ~**-calibrating equip-**
**ment** Zähler-eicheinrichtung *f*, -prüfeinrich-
tung *f* ~ **candle** Meterkerze *f* ~**-candle-second**
Sekundenmeterkerze *f* ~ **case** Zählergehäuse *n*
~ **change-over clock** Zählerschaltuhr *f* ~ **con-**
**stant** Zählerkonstante *f* ~ **cover** Zählerkappe *f*
~ **creeping** Zählerleerlauf *m* ~ **frame** Systemträger *m* ~ **head** Zählkopf *m* ~ **housing** Zähler-
werkgehäuse *n* ~ **indicator** Zählerkontroll-
zeichen *n* ~ **key** Zähltaste *f* ~ **lamp** Zähler-
kontroll-, Zählerüberwachungs-lampe *f*
(teleph.) ~ **limit** Metergrenze *f* ~ **mounting**
**limits** nutzbare Grenzmaße *pl* ~ **oil** Meß-
apparateöl *n* ~ **pulse** Zählimpuls *m* (teleph.)

(service) ~ rack Zählergestell n ~ reading Meß-
uhranzeige f, Zählerablesung f ~ rectifier Maß-
gleichrichter m ~-relay Kontaktgeberzähler m,
Zählerrelais n ~ support Tragrahmen m ~ test
bench Uhrenmeßstand m ~ wire Zählader f
metering Abmessen, Messung f (Wasser, Gas
usw.), Zählung f, Zumessen n (call) ~ Gesprächszählung f ~ characteristics Zumeßkenn-
werte pl ~ diaphragm Meßmembran f ~ element Meßglied n ~ orifice Meßdüse f ~ pump
Dosierpumpe f ~ relay Zählimpulsrelais n ~
section Homogenisierzone f ~ stud Probe-
klemme f für Zähler ~ tank Dosiertank m ~
valve Dosier-, Nadel-, Zumeß-ventil n
meters per second (m/s) Metersekunde f, Se-
kundemeter pl
methane Erdgas n, Grubengas n, leichtes Kohlenwasserstoffgas n, Methan n, Sumpfgas n ~
flow counter Methan-Durchflußzähler m
methanol Methanol n ~ formation Methanolbil-
dung f
method Ansatz m, Art f, Art f und Weise f, Ausführungsart f, Methode f, Prozeß m, System n,
Verfahren n, Weg m, Weise f
method, ~ of achieving the optimum of production
Bestverfahren n ~ of adjustment Einpaßverfahren n ~ of ad-measurement Ausmeßver-
fahren n ~ of aerophysical measurement
aerophysikalisches Meßverfahren n ~ of
aggregate motion Wegsummenverfahren n
(Fernschreiber) ~ of allotting woods to a number of periods Fachwerkbauart f ~ by altitude
Höhen-methode f, -verfahren n ~ of analysis
Bestimmungsmethode f ~ of application An-
wendungsweise f ~ of approximation Annäherungsverfahren n, Näherungsmethode f
~ of assignment Zuordnungsverfahren n ~ of
attack Angriffs-art f, -verfahren n, Art f der
Inangriffnahme ~ of balancing Abgleichverfahren n, Nachbildungsverfahren n (teleph.)
~ based on minimum deviation Minimumver-
fahren n (opt.) ~ of calculation Berechnungs-
weise f ~ of calibration Eichverfahren n ~ of
catapulting gliders by motorcar and cable Autoschlepp m ~ of characteristics Charakteristi-
kenverfahren pl ~ of connecting successive
photographs Methode f des Anschlusses von
Folgebildern ~ of construction Bauweise f ~ of
continuity Stetigkeitsmethode f ~ of continuous
addition Addiermethode f ~ of cooling Erkal-
tungsmethode f ~ of correct associations Tref-
fermethode f ~ of destination Bestimmungsmethode f ~ used to determine alkali in glass
manufacturing Autoklavenverfahren n ~ of
determining content of austenite, ferrite, and
carbide by acid etching in conjunction with color
changes Anlaßätzverfahren n ~ of differences
Differenzenrechnung f ~ of dimension similitude consideration Dimensionsbetrachtung f ~
of discharging Entlademethode f ~ of distribution Verteilungs-art f, -methode f ~ of
dressing Aufbereitungsverfahren n ~ by
eigths Achtelverfahren n ~ of electric images
Spiegelbildmethode f ~ of employment An-
wendungsart f ~ of evaluation of aerial photographs Papierstreifenverfahren n ~ of excitation Anregungsmethode f ~ of fire Feuerart f,
Feuerordnung f, Schießverfahren n ~ of firing

Feuerführung f, Feuerungsverfahren n ~ of gas
attack Blasverfahren n ~ of gripping Einspan-
nungsweise f ~ of heating Feuerungsverfahren
n ~ of ignition Einspritzverfahren n (Diesel) ~
of images Spiegelungs-methode f, -prinzip n,
-verfahren n ~ for impermeabilizing Dichtungsmittel n ~ of independent image pairs Methode
f der unabhängigen Bildpaare ~ of infinite
continued fractions Kettenbruchmethode f ~
of injection molding Spritzverfahren n ~ of
investigation Untersuchungsmethode f ~ of
isolation Isolierungsmethode f ~ of iteration
Iterationsverfahren n ~ of least squares Methode f der kleinsten Quadrate ~ of line construction Leitungsbauweise f ~ of making
charges Tarifpolitik f ~ of manufacture Gewinnungsverfahren n ~ of measurement Ausmeß-
verfahren n ~ of mining Gewinnungsverfahren
n ~ of mixed construction of aircraft Gemisch-
bauweise f ~ of modulation Modulationsverfahren n ~ of molding without flasks Formweise
f ~ of molecular orbitals Molekülbahnmethode
f ~ of notation Bezeichnungsweise f ~ of
obtaining Darstellungsverfahren n ~ of offsets
Seitwärtsabschnitt m ~ of operation Arbeits-
prozeß m, -verfahren n, -weise f, Betriebs-art f,
-weise f ~ of phase contrast Phasenkontrast-
verfahren n ~ of plotting fatigue test results
Versuchsauswertung f ~ of preparation Darstellungsweise f ~ of preparing Aufbereitungsverfahren n ~ of production Erzeugungsart f ~
of proof Beweisführung f ~ of the pyramid
Pyramidenverfahren n ~ of random sampling
Stichprobenkontrolle f ~ of recording Ver-
buchungsmethode f ~ of relieving the pressure
Entlastungsart f ~ of reproduction (or duplication) Reproduktionsverfahren n ~ of
residues Restmethode f ~ of revaluation used
angewandte Methode f der Wiederbewertung
~ of reversals Inversionsmethode f ~ of right
and wrong cases Konstanzmethode f ~ of
sample taking Bemusterungsmethode f ~ of
separation Trennungsmethode f ~ of settlement
Methode f der Begleichung ~ of sighting
Richtverfahren n ~ of slides Gleitflächen-
methode f ~ of smelting Schmelzverfahren n ~
of sparging Gußführung f ~ of stoking Feuerungsverfahren n ~ of stress analysis Be-
rechnungsmethode f ~ of surveying Vermes-
sungsart f ~ of taking bearings Peilverfahren n
~ of test Versuchsdurchführung f ~ of testing
Prüfungsart f, Versuchs-anordnung f, -methode
f ~ of translation of impulses Impulsumrechnungsverfahren n ~ of travel of the flames
Flammenführung f ~ of variable area Schwarz-
weißverfahren n ~ of welding Schweißverfahren
n ~ of wood preservation using zinc chloride
Burnett-Verfahren n ~ of working Arbeitsmethode f
methodic(al) methodisch
methodical ordnungsgemäß, planmäßig, plan-
voll, regelrecht
methodology Methodik f, Methodenlehre f
methods, ~ of multiplexing Multiplexmethoden
pl ~ of radio communication Funkverkehrs-
formen pl
methyl Methyl n ~ alcohol Methylalkohol m,
Vergällungsholzgeist m ~ benzene Toluol n ~

cellulose Colloresin n ~ hydrochloride salzsaures Methylamin n
methylate, to ~ methylieren
methylated spirit denaturierter Alkohol m
methylation Methylierung f
methylene, ~ chloride Dichlormethan n ~
iodide Dijodmethan n
methylic ether Holzäther m
meticulous pedantisch genau
metonic cycle metonischer Zyklus m
metric, (~al) metrisch, Maß n und Gewicht n
betreffend; Metrik f, Periodik f, Taktlehre f
with ~ measure scale mit metrischer Teilung f
metric, ~ calculation Dezimalrechnung f ~
determination Maßbestimmung f ~ graduation
Millimeterteilung f ~ horsepower Pferdestärke
f ~ measure Dezimal-, Meter-maß n ~ measure
model Millimeterausführung f ~ scale Meter-
maßstab m ~ system metrisches System n ~
system of measurement das metrische Maß-
und Gewichtssystem n ~ thread metrisches
Gewinde n, Millimetergewinde n ~ ton Tonne
f ~ ton per day Tagestonne f ~ ton per hour
Stundentonne f ~ ton per month Monatstonne
f ~ ton per year Jahrestonne f ~ waves Meterwellen pl, Ultrakurzwellen pl (UKW)
metrical, ~ count metrische Nummer f ~ entropy metrische Entropie f ~ measure metrisches Maß n
metrology Maß- und Gewichtskunde f, Meßtechnik f, Metrologie f
metronome Metronom n, Taktgeber m
metropolis Weltstadt f
metropolitan, ~ area Großraum-, Großstadt-
gebiet n ~ railroad Stadtbahn f ~ railway
Stadtbahn f ~ region Großraumgebiet n
mev Megaelektronenvolt n
mezzanine Halbgeschoß n, Zwischen-boden m,
-stock m ~ floor Zwischengeschoß n
mezzotint Halbtonbild n
mica Eisenglas n, Felsglimmer m, Frauenglas n,
Glimmer m, Katzen-glimmer m, -silber n ~
capacitor Glättungs-, Glimmer-kondensator m
~ (dielectric) condenser Glimmerkondensator
m ~ cover Glimmer-dach n, -deckel m ~-foil
machine Mikafoliummaschine f ~ insert
Glimmereinlage f ~ insulator Glimmerscheibe
f ~ mine Glimmerbruch m ~ plate Glimmerplatte f ~ powder Glimmerpulver n ~ punching
machine Glimmerstanzmaschine f ~ schist
Glimmerschiefer m ~ sheet Glimmerplatte f ~
slate Glimmerschiefer m, Phyllit m, Ton-
glimmerschiefer m ~ spangle Glimmerflitter m
~ spark plug Glimmer-kerze f, -zündkerze f ~
strip Glimmerstreifen m ~ substrate Glimmer-
unterschicht f ~ top Glimmer-dach n, -deckel
m ~ washer Glimmerscheibe f
micaceous glimmer-artig, -haltig ~ copper Kup-
ferglimmer m ~ form of cerussite Bleiglimmer
m ~ iron ore Eisenglimmer m ~ rock Glimmer-
gestein n ~ sand Flitter-, Glimmersand m
micanite paper Mikanitpapier n
micellar structure Mizellenaufbau m
micelle Micell n
micro, ~ adjustment device Feineinstellung f ~
ammeter Mikroamperemeter n ~ ampere
Mikroampere n ~ analysis Mikroanalyse f ~
balance Mikrowaage f ~ bar Mikrobar n ~

barograph Mikrobarograf m ~ beam technique
Feinstrahlmethode f ~ burner Mikrobrenner
m ~ caliper square Mikrometerschublehre f
~-cell Mikroküvette f ~ chemical mikro-
chemisch ~ chemical analysis Mikroanalyse f
~ chemistry Mikrochemie f ~ chronograph
Mikrozeitmeßgerät n ~ chronometer Behm-
Zeitmesser m, Kurzzeitmesser m ~ chronometer wheel Kurzzeitmesserrad n ~ cosmic salt
Phosphorsalz n ~ coulomb Mikrocoulomb n
~ cracks Mikrorisse pl ~ crystal Mikrokristall
m ~ crystalline mikrokristallin ~ crystallography Mikrokristallografie f ~ curie Mikro-
curie f ~ densitometer Mikrodichtemesser m
~ densitometer record Mikrodensogramm n
~ farad Mikrofarad n ~ film Mikrofilm m ~
filter Feinstfilter m, Molekülsieb n ~ finish
Fein-bearbeitung f, -schleifen n ~ flaw Haarriß
m ~ fuse Feinsicherung f ~ graph Mikrofotografie f, Schliffbild n ~ graphic mikro-
granophyrisch (geol.) ~ graphic(al) mikro-
grafisch ~ graphy Mikrografie f ~ hardness
tester (microsclerometer) Mikrohärteprüfer m
~ hardness testing Mikro-härtemessung f,
-härteprüfung f ~ henry Mikrohenry n ~ inch
Mikrozoll m ~ indication Feinanzeige f ~
indicator Mikroanzeiger m ~ houle Mikrojoule
n ~ lathe Feinstdrehbank f ~ limit control
Dickenmessung f ~ manipulator Feinregelungs-
organ n ~ manometer Feindruck-manometer n,
-messer m, Mikro-druckmesser m, -mano-
meter n ~ melting point apparatus Mikro-
schmelzpunktapparat m
micrometer Dicken-, Fein-messer m, Mikrometer n, Schrauben-, Schraub-lehre f (screw-
thread) ~ Schraubenmikrometer n ~ adjusting
dial Mikrometereinstellskala f ~ adjustment
(electron microscope) Einstelltubus m, Feinst-
einstellung f, Mikrometer-, Schrauben-ver-
stellung f ~ caliper Bügelschraublehre f, Fein-
meß-lehre f, Mikrometer n, Mikrometerschraube f, -taster m Schraublehre f ~-caliper
gauge Mikrometerschraubenlehre f ~ caliper
gauge Feinmeßschraublehre f ~ calipers Feinmeßschraublehre f, Präzisionsstichmaß n,
Schraubenmikrometer n ~ calipers with dial
indicator Uhrschraublehre f ~ casing Mikro-
meterkasten m ~ collar Skalentrommel f ~
control Mikrometerschraubeneinstellung f ~
depth gauge Feinmeßtiefenlehre f, Tiefenmikro-
meter n ~ drum Mikrometertrommel f ~
eyepiece Meßokular n ~ friction thimble Fühl-
ratsche f ~ gap Mikrometer-Funkenstrecke f ~
gauge Feinmeßlehre f, Mikrometerschraube f,
Schraubenlehre f ~-gauge screw Feinmeß-
schraube f ~ head Schraublehre f ohne Bügel,
Triebscheibe f ~ knob Rändel n, Teiltrommel f
~ magnifier Mikrometerlupe f ~ ring Teilring
m ~ scale Feineinstell-, Mikrometer-skala f
~-scale knob Handrädchen n ~ screw (gauge)
Feinstellschraube f, Meßschraube f, Mikrometerschraube f ~-screw extensiometer Mikro-
metertaster m für Längenmessung ~ screw
gauge Feinschraublehre f ~ setting Feinstein-
stellung f, Teilringzahl f ~ slide Feinmeßschie-
ber m ~ slide gauge Mikrometerschublehre f ~
slit Mikrometerspalt m ~ spark-gap Mikrometer-Funkenstrecke f ~ stage Objekttisch m

~ **stand** Mikrometerhalter *m* (kleiner Ständer) ~ **(-motion) table** Meßtisch *m* ~ **value (or reading)** Mikrometerwert *m*
**micrometric** mikrometrisch ~ **measurement** Feinmessung *f* ~ **spark** Mikrometerfunkenstrecke *f* ~ **spark discharger (or gap)** Funkenmikrometer *n*
**micrometrically movable slide** mikrometrisch verstellbarer Blendschieber *m*
**micrometry** Mikrometrie *f*
**micro,** ~ **microcurie** Mikro-Mikro-Curie *f* ~ **microfarad** Mikromikrofarad *n*, Picafarad *n* ~ **microfarad capacitance** Anodengitterkapazität *f* ~ **miniaturization** Mikroschalttechnik *f* ~ **organism** Kleinstlebewesen *pl*, Mikroorganismus *m*
**microphone** (piezo-electric) Kristallmikrofon *n*, Mikrofon *n*, Schallempfänger *m* ~ **with spherical mouthpiece** Mikrofon *n* mit Kugeleinsprache
**microphone,** ~ **amplifier** Mikrofonverstärker *m* ~ **battery** Mikrofonelement *n* ~ **blanket** Mikrofonkappe *f* ~ **boiling** Mikrofonbrodeln *n* ~ **boom** Mikrofon-antenne *f*, -galgen *m*, -stange *f* ~ **button** Mikrofonknopf *m* ~ **cap** Kugeleinsprache *f* ~ **circuit** Mikrofonstromkreis *m* ~ **concentrator** Schallsammler *m* ~ **connection** Mikrofonanschluß *m* ~ **current** Mikrofonstrom *m* ~ **current supply** Mikrofonspeisung *f* ~ **diaphragm** Mikrofonmembran *f* ~ **effect** Klingneigung *f*, Mikrofon-effekt *m*, -kreis *m* ~ **hiss** Zeichen *n* des Mikrofons ~ **howler (or hummer)** Mikrofonsummer *m* ~ **mouthpiece** Mikrofontrichter *m* ~ **noise** Mikrofon-geräusch *n*, -schmoren *n* ~ **outrigger** Mikrofonantenne *f* ~ **position** Mikrofonanordnung *f* ~ **preamplifier** Mikrofonverstärker *m* ~ **range** Mikrofonkreis *m* ~ **relay** Mikrofonschütz *n* ~ **relay circuit** Mikrofonrelaisschaltung *f* ~ **responsiveness** Ansprechkonstante *f* ~ **socket** Kupplungsdose *f* (tape rec.) ~ **supply circuit** Mikrofonspeisestrom *m* ~ **susceptibility** mikrofonische Einstreuung *f* ~ **switch** Sprechtaste *f* ~ **transformer** Mikrofon-transformator *m*, -überträger *m* ~ **transmitter** Mikrofonsender *m* ~ **voice frequency currents (vf-currents)** (Mikrofon) Sprechströme *pl* ~ **voltage** Mikrofonspannung *f*
**microphonic,** ~ **(al)** mikrofonisch ~ **carbon** Mikrofonkohle *f* ~ **effect** Mikrofoneffekt *f* ~ **noise** Klingfähigkeit *f*, Mikrofonie *f*, Röhrenklingen *n* ~ **troubles (or noise)** Mikrofoniestörung *f* ~ **tube** mikrofonische Röhre *f*
**microphonics** Röhrenklingen *n* ~ **test** Mikrofonie-Prüfung *f*
**microphonism** Mikrofoneffekt *m*
**microphony** Mikrofoneffekt *m*, Mikrofonie *f*
**micro,** ~ **photograph** Kleinbild *n*, Mikro-aufnahme *f*, -fotografie *f* ~ **photographic** fotomikrografisch ~ **photography** Kleinbild-aufnahme *f*, -fotografie *f*, Mikrofotografie *f* ~ **photometer** Mikrofotometer *n* ~ **photoscope** Mikrofotoskop *n* ~ **physics** Mikrophysik *f* ~ **pipettes** Mikropipetten *pl* ~ **planar** Mikroplanar *n* ~ **porous** mikroporös ~ **porphyritic** mikroporphyritisch ~ **-projection equipment** Mikroprojektionseinrichtung *f* ~ **pyrometer** Mikropyrometer *n* ~ **radiography** Mikroradio-

grafie *f* ~ **radiometer** Mikroradiometer *n* ~ **ray** Mikro-strahl *m*, -welle *f*, Zwergwelle *f* ~ **rays** Barkhausen-Kurzwellen *pl*, Dezimeterwellen *pl*, Deziwellen *pl* ~ **-rheology** Mikrorheologie *f* ~ **rivet shaver** Präzisions-Nietkopffräser *m* ~ **scale** Mikroskala *f* ~ **sclerometric measurement** Mikrohärteprüfung *f*
**microscope** Feinsehrohr *n*, Mikroskop *n* ~ **for evaluating negatives** Negativmeßmikroskop *n* ~ **with lines** Strichmikroskop *n* ~ **for use on machine tools** Einbaumikroskop *n*
**microscope,** ~ **arc lamp** Mikro(skopier)bogenlampe *f* ~ **attachment** Mikroskopaufsatz *m* ~ **eyepiece for reading the circle** Kreisablesemikroskop *n* ~ **fluorescence lamp** Fluoreszenzleuchte *f* ~ **image** mikroskopisches Bild *n* ~ **lamp** Mikroskopier-lampe *f*, -leuchte *f* ~ **lens** Mikroskopobjektiv *n* ~ **mounting** Mikroskopaufbau *m*, -aufhängung *f* ~ **objective changer** Objektivschlitten *m* ~ **projection** Mikroprojektion *f* ~ **slide** Objektivträger *m* ~ **spot arc lamp (or bulb)** Mikroskopierpunktlichtlampe *f* ~ **stage** Mikroskop-stativ *n*, -tisch *m*, -träger *m*, Objekttisch *m* ~ **tube** Mikroskoptubus *m*
**microscopic,** ~ **(al)** mikroskopisch ~ **examination** Mikroskopuntersuchung *f* ~ **field** Gesichtsfeld *n* ~ **inspection** mikroskopische Untersuchung *f* ~ **instruments** Mikrogeräte *pl* ~ **optics** Mikroskopie *f* ~ **sharpness** gestochene Schärfe *f* (photo) ~ **state** mikroskopischer Zustand *m* ~ **structure** Kleinstgebilde *n*
**microscopist** Mikroskopiker *m*
**microscopy** Feinstrukturuntersuchung *f*, Mikroskopie *f*
**micro,** ~ **second** Mikrosekunde *f* ~ **-section** Dünn-, Metall-schliff *m* ~ **section surface** Schliff *m* ~ **seism** Mikroseismus *m* ~ **seisms** mikroseismische Bewegungen *pl* ~ **-set** Mikroeinstellung *f* ~ **slip** Feingleitung *f* ~ **spectroscopic camera** Mikroskopspektralkamera *f* ~ **spectroscopic eyepiece** Mikrospektralokular *n* ~ **-stirring effect** Mikrorühreffekt *m* ~ **-strain** Mikrodehnung *f* ~ **strip** Mikrostreifenleiter *m*, Mikrostrip *m*, Streifenleitung *f* ~ **-strip-line** Bandleitung *f* (Mikrowelle) ~ **structure** Feingefüge *n*, Feinstruktur *f*, Kleingefüge *n*, Mikrostruktur *f* ~ **suction filter** Mikronutsche *f* ~ **supporting effect** Mikrostützwirkung *f* ~ **switch** Mikroschalter *m* ~ **switch plunger** Mikroschalterstöpsel *m* ~ **tasimeter** Mikrotasimeter *n* ~ **technique** Mikrotechnik *f*
**microtelephone** Empfängersenderkombination *f*, Handapparat *m*, Handfernsprecher *m*, Mikrotelefon *n* ~ **combination** Sprechhörer *m* ~ **transmitter** Mikrotelefon *n*
**micro,** ~ **tome** Mikrotom *n* ~ **tube** Mikroröhre *f* ~ **turbulence** Mikroturbulenz *f* ~ **turner** Raster *m*
**microvolt** Mikrovolt *n* ~ **meter** Mikrovoltmeter *n* ~ **sensitivity** Mikrovoltempfindlichkeit *f*
**microwatt** Mikrowatt *n*
**microwave** Dezimeterwelle *f*, Mikrostrahl *m*, Mikrowelle *f*, Ultrakurzwelle *f*, Zwergwelle *f* ~ **break-down** Mikrowellendurchschlag *m* ~ **cavity** Mikrowellenhohlraum *m* ~ **discharges** Mikrowellenentladungen *pl* ~ **energy** Mikrowellenenergie *f* ~ **generator** Mikrowellengenerator *m* ~ **highway** Richtfunkstrecke *f* ~ **inter-**

**ferometry** Mikrowelleninterferometrie *f* ~ **link** Mikrowellen-Richtfunkverbindung *f* ~ **measurement** Mikrowellenmessung *f* ~ **propagation** Mikrowellenausbreitung *f* ~ **radio link** Mikrowellen-Relaisstrecke *f* ~ **region** Mikrowellengebiet *n* ~ **relay link** Mikrowellenstrecke *f* ~**-relay system** Richtfunk-Relaiskette *f* ~ **spectra** Mikrowellenspektren *pl*
**microhm** Mikrohm *n*, Mikroohm *n*, Mikrosiemens *n*
**micron** Mikromillimeter *n*, Mikron *n*
**micronizer** Feinstmahlvorrichtung *f*
**micronometer** Mikronometer *n*
**micrurgy** Mikrurgie *f*
**mid** mittler-er, -e, -es, Mittel... ~**-band width** Halbbandbreite *f* (electron.) ~**-cut** (of a file) Halbschlichthieb *m* ~**-faced rule** halbfette Linie *f* ~**-frequency** mittlere Frequenz *f* ~**-line condenser** Mittellinienkondensator *m* ~ **load impedance** Impedanz *f* einer mit halber Spule beginnenden Leitung ~**-meridian** Mittelmeridian *m* ~ **plane** mittlere Ebene *f* ~**-point** Meß-, Wechsel-punkt *m* ~**-point range** Meßentfernung *f* ~ **position** Mittel-lage *f*, -stellung *f* ~ **proportion** Mittelteil *m* ~ **range** mittlerer Bereich *m* ~**-section** Mittelteil *m* (d. Rumpfes)
**mid-series** Kettenglied *n* zweiter Art ~ **filter termination** Filterabschluß *m* durch ein halbes Längsglied ~ **terminated filter** in einem halben Längsglied endendes Filter *n* ~**-terminated network** mit einem halben Längsglied endender Kettenleiter *m* ~ **termination** Abschluß *m* eines Filters durch ein halbes Längsglied; Kettenglied zweiter Art ~ **(or mid-shunt) termination** Abschluß *m* durch ein halbes Längs (Quer-) glied ~ **(or mid-shunt) termination of a network** Kettenleiterabschluß *m* durch ein halbes Längsglied (Querglied)
**midship,** ~ **frame** Mittelspant *n* ~ **section** Richtspant *n* ~**s section** Haupt-schnitt *m*, -spant *n*
**mid-shunt** Kettenglied *n* erster Art ~ **filter termination** Filterabschluß *m* durch ein halbes Querglied ~**-terminated network** mit einem halben Querglied endender Kettenleiter *m* ~ **termination** Abschluß *m* durch halbes Querglied; Kettenglied *n* erster Art
**mid,** ~**-square method** Mittelquadratenmethode *f* ~ **(-point) tap** Mittelanzapfung *f* ~**-vertical** Mittelsenkrechte *f* ~**-wall case** Schachtscheider *m* ~ **watch** Mittelwache *f* ~**-wind monoplane** Halbhockdecker *m* ~**-wing monoplane** Mittel-, Schulter-decker *m*
**middle,** Mitte *f*, Mittel..., Zwischenkarton *m* (paper mfg.) ~ **of band** Bandmitte *f* ~ **of the beam** Balkenmitte *f* ~ **of the girder** Riegelmitte *f*
**middle,** ~ **aisle** Mittelschiff *n* ~ **bowl** Mittelwalze *f* ~ **breaker** Mittelbrechpflug *m* ~ **breaker bottom** Pflugkörper *m* für Mittelbrecher ~ **connecting rod** Mittelscharnier *n* ~**-contact basin** Mittelkontaktgefäß *n* ~ **course** Mittellauf *m* ~ **crosspiece** Mittelriegel *m* ~ **diaphragm** Mittelblende *f* ~ **grain jig** Mittelkornmaschine *f* ~ **horizontal (line)** Mittelwaagrechte *f* ~ **horizontal section** Mittelquerschnitt *m* ~ **jib** Mittelklüver *m* ~**-latitude flying** mittlerer Breitenflug *m* ~ **line of tooth** Zahnmittenlinie *f* ~**man** Zwischenhändler *m* ~ **marker** Haupteinflugzeichen *n* (ILS) ~ **mill** Mittelstraße *f* ~

**part of casing** Gehäusemittelteil *m* ~ **portion of rib** Rippenmittelstück *n* ~ **position of eccentric** Exzentermittellage *f* ~ **price** Mittelkurs *m* ~ **roll** Mittelwalze *f* ~**-shot water wheel** mittelschlächtiges Wasserrad *n* ~**-shot wheel** Kropfrad *n* ~ **strut** Mittelstiel *m* ~ **thread** Mittelfaden *m* ~ **traverse** Mittelriegel *m* ~ **wheel** Mittelrad *n*
**middlingness** Mittelmäßigkeit *f*
**middlings** Mittelgut *n*, Mittelprodukt *n*
**midget** Miniatur..., Kleinst... ~ **molder** Zwergspritzgußmaschine *f* ~ **receiver** Midgetempfänger *m* ~ **set** Kleinstempfänger *m* ~ **size** Kleinstformat *n* ~ **tape** Taschentonbandgerät *n* ~ **wave** Zwergwelle *f*
**migrate, to** ~ wandern
**migration** Auswanderung *f*, Bewegung *f*, Wandern *n*, Wanderung *f* ~ **of gravel bars** Wanderung *f* der Kiesbänke ~ **of ions** Ionen-bewegung *f*, -wanderung *f* ~ **of line position** Zeilenverlagerung *f*
**migration,** ~ **area** Migrationsgebiet *n*, überstrichene Fläche *f* ~ **length** Migrationslänge *f* ~ **tendency** Wanderungstendenz *f* (Farbe) ~ **velocity** Wanderungsgeschwindigkeit *f*
**mike** (= **microphone**) Kurzform für Mikrofon *n* ~ **blanket** Mikrofonkappe *f* ~ **boiling** Mikrofonbrodeln *n*
**mil** Mil *n* (1/6400 des Kreisumfanges)
**mild** gemäßigt, lau, mild, sanft, weich ~ **steel** Fluß-eisen *n*, -stahl *m*, niedriggekohlter (Fluß-) Stahl *m*, Schweißstahl *m*, Weicheisen *n*, weicher Stahl *m*, Weichstahl *m* ~**-steel pipe** Flußeisenrohr *n* ~**-steel plate flanged for fireboxes** Feuerblech *n* aus Flußeisen ~**-steel sheet** Flußeisenblech *n*
**mildew** Mehltau *m*, Meltau *m*, Moder *m*, Schimmel *m*, Stockfleck *m*
**mile** Meile *f* ~ **of standard cable** Meile *f* Standardkabel ~ **stone** Mark-, Meilen-stein *m*
**mileage** Beförderungsweite *f*, Meilen-geld *n*, -länge *f*, zurückgelegte Meilenzahl *f* ~ **meter** Kilometerzähler *m* ~ **recorder** Wegmesser *m* ~ **subsidy** Meilenbeisteuer *f*
**mileometer** Meilenzähler *m*
**military crest** Böschungskante *f*
**milk** Milch *f* ~ **of lime** Naßkalk *m*
**milk,** ~ **can** Milchkanne *f* ~ **churn washer** Milchkannenwascher *m* ~ **cooler and aerator** Milchkühler *m* und Milchlüfter *m* ~ **(trickling) cooler with direct evaporation** (Berieselungs-) Milchkühler *m* für unmittelbare Verdampfung ~**-cooling plant** Milchkühlanlage *f* ~ **drip period** Milchaustropfzeit *f* ~ **glass (or porcelain)** Milchglas *n* ~ **glass plate (disk, sheet, screen or pane)** Milchglasscheibe *f* ~ **glass scale** Milchglasskala *f* ~ **glass window** Milchglasfenster *n* ~ **level** Milchspiegel *m* ~**-of-lime grit** Kalkgrieß *m* ~**-of-lime-grit separator** Kalkmilchgrießabscheider *m* ~ **storage** Milchkammer *f* ~ **sugar** Milchzucker *m* ~ **weighing machine** Milchwaage *f*
**milked steel for automatic lathes** Automatenweichstahl *m*
**milkiness** Trübung *f*
**milky** milchig
**mill, to** ~ abschleifen, ausmahlen, (glass, metal) fräsen, mit Fräser *m* bohren, kordeln, (an

edge) kordieren, (coins) kräuseln, mahlen, (soap) pilieren, rändeln, rändern, riefeln, vermahlen, walken, walzen, zähneln, zermahlen **to ~-groove** Nut *f* fräsen **to ~ keyway** Nut *f* oder Nute *f* fräsen **to ~ off** abfräsen **to ~ out (or to ~ ream)** ausfräsen **to ~ threads** Gewinde *pl* fräsen

**mill** Anlage *f*, Betrieb *m*, Druckwalze *f* (print.), Druckwerk *n*, Fabrik *f*, Hammerwerk *n*, Hüttenwerk *n*, Mühle *f*, Prägwerk *n*, (grating or grinding) Reibmaschine *f*, Spinnerei *f*, Walzwerk *n*, Werk *n*

**mill, ~ for fluorite** Flußspatmühle *f* **~ and grinding stones** Mahl- und Mühlsteine *pl* **~ for groats** Graupenmühle *f* **~ for rolling breakdown** Vorsturzwalzwerk *n* **~ for rolling center-disk-type wheels** Scheibenradwalzwerk *n* **~ for rolling circular shapes** Scheibenwalzwerk *n* **~ for rolling light sections** Feineisenwalzwerk *n* **~ for rolling sections** Profile-eisenwalzwerk *n*, -walzwerk *n* **~ for rolling shapes** Kaliberwalzwerk *n* **~ with smooth rolls** Glattwalzenstuhl *m*

**mill, ~ accident** Betriebsunfall *m* **~ bar** Plattine *f* **~ barrel** Rommel *f*, Roll-, Rummel-faß *n* **~ board** geformte Pappe *f*, Graupappe *f*, Zellstoffpappe *f* **~ capacity** Mühlenleistung *f* **~ chute** Rolloch *n* **~ clack** Anschlageholz *n* **~ construction** Mühlenbau *m* **~-cut** (paper) unbeschnitten **~-cutter pick** Mühlpicke *f* **~ efficiency** Mühlenleistung *f* **~-finished** maschinenglatt **~ floor** Hütten-flur *m*, -sohle *f* **~ foreman** Pochsteiger *m* **~ grinding** Müllerei *f* **~ iron** Puddelroheisen *n* **~ limit** handelsübliche Toleranz *f* **~ limits** Walztoleranz *f* **~ line** Walzenreihe *f* **~ pick** Mühlpicke *f* **~ pond** Mühlteich *m* **~ race** Fließbett *n*, Flutgang *m* **~ saw blades** Gattersägeblätter *pl* **~ scale** Glühspan *m*, Hammerschlag *m*, Walzenschlacke *f*, Walz-haut *f*, -sinter *m*, -zunder *m* **~ sifter** Mühlensichter *m*

**mill-stone** Mühlstein *m* **~ adjusting gear** Steinstellung *f* **~ arrangement** Mahlgang *m* **~ grit** Mühlsandstein *n* **~ hammer** Mühlsteinpicke *f* **~ rock** Mühlkalkstein *m* **~ runner** Mahlstein *m*

**mill, ~ tandem** Verbundwalzwerk *n* **~ train** Walzenstraße *f*, Walz-straße *f*, -strecke *f* **~ trough** Flutgang *m* **~ wheel** Mühlrad *n* **~ work** Tischlerarbeiten *pl* **~ works** Hütte *f*

**millable** pillierfähig (soap)

**milled** gefräst, gekordelt, gemahlt, gerändelt, gerändert, gereifelt, gerieft, geriffelt, gewalzt, (ground) zerflockt **~ asbestos** zerflockter Asbest *m* **~ distance setting ring** Rändelring *m* für die Entfernungseinstellung **~ edge** Hochkante *f*, Kräuselung *f* (einer Münze), gekerbter Rand *m* (film), geriffelter oder geriefter Rand *m* **~ edge of a coin** Rändelung *f* einer Münze **~ file** Fräserfeile *f* **~ free from backlash** spielfrei gefräst **~ groove** Fräsnute *f* **~ head** Drehknopf *m*, Rändel-knopf *m*, -kopf *m*, Triebknopf *m* **~-head screw** Rändelschraube *f* **~ knob** Kordel-, Rändel-knopf *m*, Rändel *n* **~ nut** Kordel-, Zier-mutter *f* **~ out** ausgefräst **~ reamed** ausgefräst **~ recess** Einfräsung *f* **~ roller** Kordelrolle *f* **~ rubber** gewalzter Gummi *m* **~ screw** Rändel-knopf *m*, -schraube *f* **~ slot** Einfräsung *f* **~ soap** pilierte

Seife *f* **~ surface** Fräsbild *n* **~ thread** gefrästes Gewinde *n* **~ tooth** gefräster Zahn *m*

**millefiori** Millefiori *pl*

**miller** Fräser *m*, Fräsmaschine *f*, Müller *m*

**millerite** Haarkies *m*, Millerit *m*, Nickel-kies *m*, -sulfid *n*

**milli, ~ ammeter** Milliamperemeter *n* **~ ampere** Milliampere *n* **~ amperemeter** Milliamperemeter *n* **~ ampere-second** Milliamperesekunde *f* **~ ampere-second meter** Dosismesser *m* **~ bar** Millibar *n* **~ curie-hour** Millicurie-Stunde *f* **~ gram-hour** Milligrammelement-Stunde *f* **~ henry** Millihenry *n* **~ lambert** Millilambert *n* **~ mass unit** Millimaßeinheit *f*

**millimeter, ~ line** Millimeterzeile *f* **~ measure** Millimetermaß *n* **~ scale** Millimeter-einteilung *f*, -skala *f*

**milli, ~ metric waves** Millimeterwellen *pl* **~ roentgen** Milliroentgen *n* **~ second** Millisekunde *f* **~ volt** Millivolt *n* **~ voltage** Millivoltspannung *f* **~ voltmeter** Millivoltmeter *n* **~ watt** Milliwatt *n*

**milliard** Milliarde *f*

**milling** Brechen *n*, Fräsen *n*, Mahlen *n*, Rändeln *n*, Walken *n*, Walzen *n* **~ to break down the structure** Quälen *n*, Totmahlen *n* **~ in the stocks** Trockenwalken *n*

**milling, ~ apparatus** Mahlvorrichtung *f* **~ arbor** Fräserzapfen *m* **~ attachment** Fräsvorrichtung *f* **~ carriage** Frässchlitten *m* **~ crown** Fräskrone *f* **~ cutter** Fingerfräser *m*, Fräse *f*, Fräser *m*, Fräs-vorrichtung *f*, -werkzeug *n* **~-cutter backed for clearance** hinterdrehter Fräser *m* **~-cutter driving gear** Fräserantriebsrad *n* **~-cutter relief-grinding attachment** Fräserhinterschleifeinrichtung *f* **~ cycle** Mahlgang *m* **~ file** Fräser-, Turbo-feile *f* **~ fixture** Fräsvorrichtung *f* **~ head** Fräserzapfen *m*, Fräskopf *m* **~ jig** Fräservorrichtung *f* **~ length** Fräslänge *f* **~ machine** Fräs-bank *f*, -maschine *f*, Walke *f* **~ machine with adjustable housings** Fräsmaschine *f* mit ausfahrbaren Ständern **~ machine with slotting attachment** Fräsmaschine *f* mit Stoßeinrichtung **~ machine attachment** Fräsmaschinenzusatzausrüstung *f* **~ methods** Fräsarten *pl* **~ operation** Fräsvorgang *m* **~ ore** Pocherz *n* **~ path** Fräsweg *m* **~ pattern** Kopierfrässchablone *f* **~ plant** Mahl-anlage *f*, -werk *n*, Müllerei *f* **~ process** Mahlverfahren *n* **~ properties** Verwalzbarkeit *f* **~ shoe** Fräs-, Zahn-schuh *m* **~ shop** Fräserei *f*

**milling-spindle** Frässpindel *f* **~ column** Frässpindelständer *m* **~ head** Frässpindel-kopf *m*, -stock *m* **~ mounting** Frässpindellagerung *f* **~ nose (or end)** Frässpindelnase *f* **~ slide** Frässpindelschlitten *m* **~ speed** Frässpindeldrehzahl *f*

**milling, ~ star** Putzstern *m* **~ support** Frässchlitten *m* **~ template** Kopierfrässchablone *f* **~ tool** Fräse *f*, Fräser *m*, Fräswerkzeug *n* **~ travel (or traverse)** Fräsweg *m* **~ wheel** Schlagrädchen *n* **~ work** Fräsarbeit *f*, Fräserei *f*

**millings** Frässpäne *pl*

**million electron-volt** Megaelektronenvolt *n* (MeV)

**millionth** Millionstel *n*

**mimeograph, to ~** vervielfältigen

**mimeograph ink** Druckfarbe *f*

mimeographed, ~ copy vervielfältigte Kopie *f*, Vervielfältigung *f* ~ reproduction Umdruck *m*
mimetite Arsenikbleispat *m*, Bleiblüte *f*, Grünbleierz *n*, Mimetesit *m*, Traubenblei *n*
mimic, to ~ nachahmen
mimic, ~ (connection) diagram Blindschaltbild *n* ~ diagram control Anlagenbildsteuerung *f* ~ network Netzbild *n*
minator Eisenfänger *m* (Entaschungsanlage)
mincing Hacken *n*, Zerkleinern *n*, zerkleinernd ~ knife Wiegemesser *n* ~ machine Fleisch-wolf *m*, -zerkleinerungsmaschine *f*, Hackmaschine *f*
mine, to ~ abbauen, fördern, minieren, verminen
mine Bergwerk *n*, Erzgrube *f*, Grube *f*, Grubenanlage *f*, Mine *f*, Zeche *f* ~ of energy Energiemine *f*
mine, ~ adits Förderstollen *pl* ~ cage Förderkorb *m* ~ captain Obersteiger *m* ~ car Grubenwagen *m*, Hunt *m*, Kasten-, Mulden-kipper *m* ~ carrier Minenträger *m* ~ chamber Minenhalle *f* (min.), Minenkammer *f*, Ofen *m*, Sprengkammer *f* ~ chamber in bridges Minenanlage *f* ~ chute Ausleger *m* ~ coal Zechenkohle *f* ~ crater Minen-, Spreng-trichter *m* ~ damp Grubenwetter *n* ~ detecting set elektronisches Minensuchgerät *n* ~ disaster Bergwerksunglück *n* ~ door Grubentür *f* ~ effect Minenwirkung *f* ~ exploder Explosionszünder *m* ~ fan Wettermaschine *f* ~ field Minen-feld *n*, -sperre *f* ~-field detonating net Knallnetz *n* ~-field float Sprengboje *f* ~ filling Berg *m*, Versatz *m* ~ fire Grubenbrand *m* ~ gallery Minenstollen *m* ~-laying cruiser Minenkreuzer *m* ~-leveling staff Grubennivellierlatte *f* ~-locating device Minensuchgerät *n* ~ locomotive Grubenlokomotive *f* ~ manager Grubendirektor *m* ~ not working Feiertag *m* ~ office Berg-amt *n*, -büro *n* ~ openings Grubenräume *pl* ~ ore Roherz *n* ~ owner Gewerke *m*, Grubeneigner *m* ~ priming machine Minenzündmaschine *f* ~ projector Minenwerfer *m* ~ prop Gruben-holz *n*, -stempel *m* ~ pump Grubenpumpe *f* ~ railroad Schiebbahn *f* ~ resistance Grubenwiderstand *m* ~ slack Fördergruskohle *f*, Grubenklein *n* ~-smalls separating house Grubenkleinwäsche *f* ~ splinter Minensplitter *n* ~ survey Grubenmessung *f* ~-survey instrument Markscheidinstrument *n* ~ surveying Markscheidewesen *n*, Markscheidung *f* ~-surveying appliances Markscheiderzeug *n* ~-surveying instrument Markscheidergerät *n* ~ surveyor Markscheider *n* ~ sweeper Minen-suchboot *n*, -sucher *m*; Räumboot *n* ~ telephone system Grubenfernsprechanlage *f* ~ tramway Grubenbahn *f* ~ tubbing Grubenverschalung *f* ~ ventilating fan Grubenlüfter *m* ~ waste Berge *pl*, Grubenabfall *m* ~ water Grubenwasser *n* ~ winder Grubenfördermaschine *f* ~ worker Bergmann *m*
mined tin Bergzinn *n*
miner Bergarbeiter *m*, Bergmann *m*, Bergwerksarbeiter *m*, Grubenarbeiter *m*, Hauer *m*, Knappe *m*, Kohlenbergmann *m*
miner's, ~ ax Grubenbeil *n*, Kaukamm *m* ~ box Gezähnekiste *f* ~ dial Diopterkompaß *m* ~ hammer Bohrfäustel *n* ~ hammer for striking

the borer Handschlägel *m* ~ hoe Minenkratze *f* ~ house coal Hausbrand *m* ~ lamp Grubenlampe *f*, Grubenlicht *n* ~ level Gradbogen *m* ~ scraper Minenkratze *f* ~ tram Förderbahn *f* ~ trough Bergtrog *m* ~ truck Förder-hund *m*, -karren *m*, Grubenhund *m* ~ wedge Fimmel *m*, Pfändekeil *m*
mineral Mineral *n*; mineralisch ~ acid Mineralsäure *f* ~ blue Wolframblau *n* ~ charcoal Faserkohle *f* ~ coal fossile Kohle *f*, Steinkohle *f* ~ color Erdfarbe *f* ~ concentration Erzkonzentrat *n* ~ constituent Mineralbestandteil *m* ~ deposit Minerallager *n* ~ dust Gesteinstaub *m* ~ green Berggrün *n* ~-grinding plant Mineralienmahlanlage *f* ~ kingdom Mineralreich *n* ~ lard oil Schneidöl *n* ~ lines Gesteinslinien *pl* ~ lubricant Mineralschmierstoff *m* ~ matter mineralische Bestandteile *pl* ~ oil Mineralöl *n*, Naphta *n/f*, Petroleum *n*, Vulkanöl *n* ~-oil refinery Mineralölraffinerie *f* ~ particle Mineralteilchen *n* ~-pigment quarry Farberdgewinnung *f* ~ pigments Pigmentfarben *pl* ~ pitch Asphalt *m*, Asphaltstein *m*, Berg-, Erd-, Mineral-pech *n* ~ resin Kopalin *m*, mineralisches Harz *n* ~ resource (or wealth) Bodenschätze *pl* ~ salt Steinsalz *n* ~ soil Mineralboden *m* ~ spirit Petroleumsolvent *n* ~ spirits Lackbenzin *n*, Ligroin *n*, Lösungsmittel *n*, Schwerbenzin *n* ~ spring Mineralbrunnen *m* ~ spring charged with carbon dioxide Säuerling *m* ~ tallow Mineraltalg *m* ~ tallow (or wax) Berg-, Erd-talg *m*, Mineralfettwachs *n* ~ tar Berg-, Erd-, Mineral-teer *n* ~ vein Mineralgang *m* ~ water Mineralwasser *n* ~-water spring Mineralquelle *f* ~ wax Berg-, Mineral-wachs *n* ~ wool Schlackenwolle *f* ~-wool filter Schlackenwollfilter *n*
mineralizable vererzbar
mineralization Inkohlung *f*, Verzehrung *f*, Versteinerung *f*
mineralize, to ~ vererzen
mineralized carbon getränkte Kohle *f*
mineralizer Mineralbildner *m*, Vererzungsmittel *n*
mineralizing agent Versteinerungsmittel *n*
mineralogical mineralogisch ~ occurrence bergbauliches Vorkommen *n*
mineralogist Mineraloge *m*
mineralogy Gesteins-kunde *f*, -lehre *f*, Mineralienkunde *f*, Mineralogie *f*, Steinkunde *f*
minette Minetteeisenstein *m* ~ ore Minette *f*
mingle, to ~ mischen, vermengen
mingler Maischtrog *m*
mingling Vermengung *f*
miniature Miniatur *f* ~ bearing Kleinstkugellager *n* ~ camera Kleinbildkamera *f* ~ camera photography Kleinbildfotografie *f* ~ cell Kleinküvette *f* ~ contactor Kleinschütz *m* ~ direction finder Kleinstpeilgerät *n* ~ Edison screwcap Mignonsockel *m* ~ edition Miniaturausgabe *f* ~ filament bulb Glühlämpchen *n* ~ film adapter Kleinbildansatz *m* ~ lamp Klein-, Zwerg-lampe *f* ~ machine Kleinstmaschine *f* ~ microscope Taschenmikroskop *n* ~ photo Kleinbild *n* ~ receiver Taschenempfänger *m* ~ size Kleinbildformat *n*
miniaturization Miniaturisierung *f*
minim Grundstrich *m* (Kalligrafie)

**minimal** minimal ~ **change** Minimalveränderung f ~ **curves** Minimalkurven pl ~**-latency coding** Bestzeitprogrammierung f ~ **theorems** Minimalsätze pl

**minimeter** Minimeter n ~ **instrument for testing the distance between pinholes and piston head** Minimetergerät n zum Messen der Kompressionshöhe

**minimize, to** ~ auf das kleinste Maß, auf das Kleinstmaß oder auf das Mindestmaß zurückführen, herabdrücken, herabmindern, herunterdrücken, verringern

**minimum** Geringst-maß n, -wert m, kleinster Absolutwert m (math.), kleinste Größe f, Mindest-maß n, -wert m, Minimum n, niedrigsten Meßwert m registrierend, Tiefdruckgebiet n (meteor.); geringst, kleinst, mindest, tief

**minimum,** ~ **access** Schnellprogrammierung f ~ **access programming** Minimalsuchzeitprogrammieren n ~ **access routine** Bestzeitprogramm n ~ **amplitude** Minimalamplitude f ~ **axial length** Mindestachsiallänge f ~ **back tension** Rückstellfeder-Restkraft f ~ **barrel elevation** Mindestrohrerhöhung f ~ **band** Mindestbiegung f ~ **blow-out pressure** Mindestzerplatzdruck m ~ **body diameter** Dehnschaftsdurchmesser m ~ **characteristic** Minimaleigenschaft f ~ **clearance** Kleinst-, Mindest-spiel n, Mindestabstand m ~ **content** Niedrigstgehalt m ~ **control speed** Mindestgeschwindigkeit f mit gewahrter Steuerbarkeit ~ **cross section** Mindestquerschnitt m ~ **current** Mindeststrom m ~ **current intensity** Mindeststromstärke f ~ **cut-out** Minimalausschalter m ~ **deflection** Mindestausschlag m ~ **degree of error** (direction finder) Minimumwanderung f ~ **delay** Mindestverzögerung f ~ **design weight** Entwurfsminimumgewicht n ~ **dimension** Kleinstmaß n ~ **distance** Minimalabstand m ~ **elastic limit** Mindeststreckgrenze f ~ **elevation** Deckungswinkel m ~ **energy** Energieminimum n ~ **equivalent** Mindeststrestdämpfung f ~ **feed** Mindestförderung f ~ **flow at low water** kleinstes Niedrigwasser n ~ **flow resistance** geringster Strömungswiderstand m ~ **flying speed** Mindestfluggeschwindigkeit f ~ **gap** Engstspalt m ~ **gliding angle** kleinster Gleitwinkel m, Minimumgleitwinkel m ~ **gliding ratio** Kleinstgleitzahl f ~ **height** Mindesthöhe f über Grund ~ **horizontal range** Kartenentfernung f zum Wechselpunkt ~(**-voltage**) **indicator** Mindest-, Minimum-wertanzeiger m ~ **interference** Kleinstübermaß n ~ **inventory** Mindestbestand m ~ **ionization** Minimalionisation f ~ **latency** Mindestlatenz f ~ **length** Kleinstmaß n ~ **liftdrag ratio** Kleinstgleitzahl f ~ **limiting gauge** Restmelder m ~ **magnification** Mindestvergrößerung f ~ **material condition** unterer Anmaßzustand m ~ **net loss** Mindestrestdämpfung f ~ **number of cycles** Mindestlastwechselzahl f ~ **number of theoretical plates** theoretische Mindestbodenzahl f ~ **obstacle clearance** Mindesthindernisfreiheit f ~ **operating current** Fehl-, Mindestansprech-strom m ~ **operating pressure** Mindestbetriebsdruck m ~ **output** Minimalleistung f ~ **percentage** Mindestgehalt m (ore) ~**-phase network** Minimumphasennetzwerk n ~ **potential** Mindestvoltgeschwindigkeit f ~ **potential gradient** Mindestspannungsgradient m ~ **pressure** Druckminimum n, Mindest-, Minimal-druck m ~ **pressure of response** Ansprechdruck m ~ **pressure drop** Minimumdruckgefälle n ~ **pressure point** Unterdruckspitze f ~**-range clinometer** Deckungswinkelmesser m ~ **rate** Mindestgesprächsgebühr f ~ **reception** Empfangsminimum n ~ **resistance at end position** Endspringwert m ~ **resistance at starting position** Anfangsanschlagwert m ~ **safe height (or level)** Sicherheitsmindesthöhe f über Grund ~ **scale value** Minimalskalenwert m ~**-signal direction-finding method** Minimumpeilung f ~ **size** Minimal-format n, -größe f ~ **speed** Geringst-, Mindest-geschwindigkeit f ~ **speed of descent** Mindestsinkgeschwindigkeit f ~ **speed governor** Leerlaufregler m ~ **speed prescribed** vorgeschriebene Mindestgeschwindigkeit f ~ **standard of living** Armutsgrenze f ~ **starting voltage** Mindestanlaßspannung f ~ **steady flight speed** stationäre Mindest(flug)geschwindigkeit f ~ **storage capacity** Mindestspeichervermögen n ~ **stress limit** Unterspannung f der Dauerfestigkeit ~ **suction quantity** Mindestabsaugmenge f ~ **take-off output** Kurzleistung f ~ **temperature** Niedrigstwärmegrad m, Temperaturminimum n, Wärmeniedrigstgrad m ~ **tensile strength** Mindestzugfestigkeit f ~ **tension** Minimalspannung f ~ **triggering level** Auslöseschwelle f ~ **tripping current** Auslösegrenzstrom m ~ **value** Kleinst-, Mindest-, Minimal-, Niedrigst-, Tal-, Tiefst-wert m ~ **vertical clearance** vertikaler Mindestabstand m (von Bodenhindernissen) ~ **voltage amplifier** Nullverstärker m ~ **wave antenna arrangement** Grenzwellenantennenordnung f ~ **wave length** Grenzwellenlänge f, kürzeste Wellenlänge f ~ **wave receiver** Grenzwellenempfangsgerät n ~ **weight** Mindest-, Passier-gewicht n (allowable) ~ **working current** Ansprechstrom m

**mining** Bergarbeit f, Bergbau m, Bergbauliche n, Grubenbau m ~ **with filling (or stowing)** Abbau m mit Bergeversatz ~ **with numerous shafts** Reihenschachtbetrieb m

**mining,** ~ **accountant** Grubenrechnungsführer m ~ **bell** Grubenwecker m ~ **camp** Bergarbeitersiedlung f ~ **car** Grubenförderwagen m ~ **claim** Grubenfeld n ~ **company** Bergbaugesellschaft f, Bergkompanie f, Bergwerksgesellschaft f, Gewerkschaft f ~ **dial** Markscheideinstrument n ~ **drill** Grubenbohrer m ~ **engineer** Bergwerksingenieur m ~ **expert** Bergbausachverständiger m ~ **hoist** Grubenkabelwinde f ~ **industry** Montanindustrie f ~ **inspector** Bergrevierbeamter m ~ **installation** Bergwerksanlage f ~ **law** Berg-gesetz n, -recht n ~ **locomotive** Schleppzeug n ~ **machine that cuts and breaks coal in low-level lignite mines** Bruchschlitzmaschine f ~ **method** Abbauverfahren n, unterirdische Gewinnungsmethode f ~ **region** Abbaufeld n ~ **shares** Bergwerksaktien pl ~ **spoon** Raumlöffel m ~ **stocks** Bergwerksaktien pl ~ **subsidence** Bodensenkung f ~ **survey instrument** Markscheidgerät n ~ **theodolite** Markscheideinstrument n ~ **tools** Miniergerät n ~ **working** Gewinnungsmethode f

**minion** Kolonel *f* (print.), Mignon *f*
**minium** Blei-mennige *f*, -oxyduloxyd *n*, -plumbat *n*, -rot *n*, tetroxyd *n*, -zinnober *m*, Minium *n*, rotes Bleioxyd *n*
**minor** klein, unbedeutend, untergeordnet; Hilfs . . ., Neben . . ., Unter . . . **~ of original determinant** Unterdeterminante *f*
**minor,** **~ apex face** Nebenspitzenfläche *f* (cryst.) **~ axis** kleine Achse *f*, Halb-, Neben--achse *f* (math., techn.) **~ bed** Niedrigwasserbett *n* **~ case** Leichtkranker *m* **~ control** Untergruppenkontrolle *f* **~ control plot of rhomboidal figures** Rautenkette *f* **~ cycle** Klein--periode *f*, -zyklus *m*, Nebenperiode *f* **~ determinant** Unterdeterminante *f* **~ diameter of screw** Kerndurchmesser *m* der Schraube oder des Schraubengewindes **~ drive** Spiegelantrieb *m* **~ ear** Nebenzipfel *m* **~ exchange** Neben-, Unter-, Verbund-amt *n* **~ graduation** Nebenskalenteilungen *pl* **~ lobe** Neben-keule *f*, -lappen *m*, -zipfel *m* (rdr) **~ resonant speed** unterkritische Geschwindigkeit *f* **~ septa** Nebensepten *pl* **~ switch** kleiner Wähler *m* **~ thread diameter** Gewindebolzenkern-, Kern-durchmesser *m*
**minority** Minderheit *f*, Minderzahl *f* **~ carrier** Minderheits-, Neben-träger *m* **~ charge carrier** Minoritätsladungsträger *m* **~ emitter** Minderheitsemitter *m*
**minos,** **~ cycle** Kleinperiode *f* **~ plate condenser** Minosplattenkondensator *m*
**mint, to** **~** ausmünzen, ausprägen, münzen
**mint** Münze *f*, Münz-amt *n*, -anstalt *f*, -stätte *f*, -werk *n* **~ mill** Münzstreckwerk *n*
**mintage** Münzrecht *n*, Prägschatz *m*
**minting** Münzkunst *f* **~ of money** Münzen *n*
**minuend** Minuendus *m*
**minus** minus, negativ, weniger **~ deviation** Minusabweichung *f* **~ gauge** Minuslehre *f* **~-plus wiper** a/b-Arm *m* **~ sign** Minuszeichen *n*
**minuscle** Minuskel *f* (print.)
**minute** Minute *f*, Sitzungsbericht *m*, Zeitspanne *f*; minuziös, sehr klein, sorgfältig, unbedeutend **per ~** in der Minute *f*, minütlich **~ of arc** Kreisbogenminute *f* **~ of latitude** Breitenminute *f*
**minute,** **~ card** Minutenrose *f* **~ crystal** Kriställchen *n* **~ distribution** Feinverteilung *f* **~ hand** Minutenzeiger *m* **~ switch** Minutenausschalter *m* **~ wheel** (watchmaking) großes Bodenrad *n*, Mittelrad *n*
**minuteness** Kleinheit *f*
**minutes** Besprechungsniederschrift *f*, Protokoll *n* **~ of proceedings** Aufzeichnung *f*, Errichtungsprotokoll *n*
**minutiae** Einzelheit *f*
**miocene** Miocän *n*
**mirage** Fata Morgana *f*, Luftspiegelung *f*, Spiegelung *f* (rdo) **~ landing** Glattwasserlandung *f*
**mirbane oil** Mirbanöl *n*
**mire** Schlamm *m*, Sumpf *m*
**mirror, to** **~** ab-, wider-spiegeln **to ~-finish** hochglanzpolieren
**mirror** Reflektor *m*, Spiegel *m* **~ apparatus** Spiegelapparat *m* **~ arc-lamp** Spiegellampe *f* **~ balancing technique** Spiegel-Nullmethode *f*

**~-coating** Spiegelbelag *m*, Verspiegelung *f* (Scheinwerfer) **~ comparator** Spiegellehre *f* **~ compass** Spiegel-bussole *f*, -kompaß *m* **~ device** Einsatz *m* für getrennt aufgestellte Leuchten **~ diameter** Spiegeldurchmesser *m* **~ drum** Spiegelrad *n* **~ effect** Spiegelwirkung *f* **~-effect eliminator** Störungsgetriebe *n* **~ (and scale) extensometer** Spiegelapparat *m* **~ finish** Spiegelschliff *m* **~-finished surface** Hochglanzfläche *f* **~ frame** Spiegelrahmen *m* **~ galvanometer** Spiegelgalvanometer *n* **~ glass** Spiegelglas *n* **~ helix** Spiegelschraube *f* **~ image** Spiegelbild *n* **~-image firing** Spiegelbildschießen *n* **~ image function** Spiegelbildfunktion *f* **~-image negative** Glasnegativ *n* **~ instrument** Lichtzeiger-, Spiegel-instrument *n* **~ inversion** Spiegelverkehrung *f* **~ inverted** spiegelverkehrt **~ like** spiegelig **~ like polished surface** spiegelblanke Oberfläche *f* **~ like reflection** spiegelnde Reflektion *f* **~ nuclei** Spiegelkerne *pl* **~ nuclides** Spiegelnuklide *pl* **~ oscillograph** Spiegeloszillograf *m* **~-picture condition** spiegelbildlich **~ plane** Spiegelebene *f* **~ point** Spiegelpunkt *m* **~ position finder** Spiegelortungsgerät *n* **~-quicksilvering** Spiegelbelegung *f* **~ recorder** Spiegelschreiber *m* **~-reflection echoes** Mehrfach-, Umweg-echos *pl* **~ reflector** Spiegelreflektor *m* **~ rim** Spiegelkranz *m* **~ screw** (scanner) Spiegelschraube *f* **~ sight** Spiegelvisier *n* **~ speaker** Sprachgalvanometer *n* **~ support** Spiegelhalter *m* **~ symmetry** Spiegel-gleichheit *f*, -symmetrie *f* **~ wheel** Spiegelrad *n* **~ writing** Spiegelschrift *f*
**mirrored surface** Spiegelfläche *f*
**misadjustment** Fehleinstellung *f*
**misaligned** nichtfluchtend
**misalignment** (of image or track) falsche Ausrichtung *f*, Fehlabgleichung *f*, Fehlausrichtung *f*, Fluchtungsfehler *m*, schlechte Ausrichtung *f*, Verlagerung *f* **~ of shaft** Wellenversetzung *f*
**miscalculate, to** **~** falsch berechnen
**miscalculation** Fehlrechnung *f*, Rechen-, Rechnungs-fehler *m*, Schätzungsirrtum *m*
**miscarriage of justice** Rechtsbeugung *f*
**miscarry, to** **~** fehlschlagen, mißglücken, mißlingen
**miscellaneous,** **~ charges** verschiedene Kosten *pl* **~ folding** Gemischtfalzung *f*
**misch metal** Mischmetall *n*
**miscibility** Mischbarkeit *f* **~ gap** Mischungslücke *f*
**miscible** mischbar
**mis-classify, to** **~** fehlklassieren
**misconclusion** Fehlschluß *m*
**misconduct, to** **~** mißleiten
**miscount, to** **~** falsch berechnen oder kalkulieren, verzählen
**miscount** Fehlkalkulation *f*, Rechenfehler *m*
**misdirect, to** **~** irre-führen, -leiten, mißleiten
**miser rods** Gestänge *n* mit Blattschloßverbindung
**miserable** elend, erbärmlich, kümmerlich, (business) trübe
**misfire, to** **~** abblitzen, fehlzünden, versagen
**misfire** Aussetzer *m*, Fehlschluß *m*, (ignition) Fehlzündung *f*, Versager *m*, Zündungsausfall *m*

**misfiring** Aussetzen *n* der Zündkerze, fehlerhaftes Arbeiten *n* (d. Magnetrons), Fehlzündung *f*, Zündaussetzer *m*, Zündungsversager *m*
**misfit, to** ~ schlecht passen
**misfitting** unpassend ~ **slip** unpassender Streifen *m*
**misframing** falsche Bildnachstellung *f*, falsche Bildstricheinstellung *f*
**misguide, to** ~ mißleiten
**mishap** Panne *f*, Unfall *m*, Zwischenfall *m*
**misinterpretation** falsche Auslegung *f*
**mislabeling** falsche Etikettierung *f*
**mislay, to** ~ verlegen
**mislaying** Verlegung *f*
**mislead, to** ~ irreführen, vortäuschen
**mismanagement** Mißwirtschaft *f*
**mismatch** Fehlanpassung *f*, Konturabweichung *f* (mech.) ~ **factor** Reflexions-koeffizient *m*, -verlustfaktor *m* ~ **protection** Impedanzschutz *m*
**mismatched impedance** Falschanpassung *f* des Widerstandes
**mismatching** Fehlanpassung *f* ~ **of impedance** Falschanpassung *f* des Widerstandes ~ **of impedances** Widerstandsfalschanpassung *f* ~ **factor** Falsch-, Fehl-anpassungsfaktor *m*
**misorientation** Orientierungsfehler *m*
**misphased** falschphasig, phasenfalsch
**mispick** Webfehler *m*
**mispickel** Arsenkies *m*, Mispickel *m*
**misplace, to** ~ verlegen, versetzen
**misplaced** unpassend, versetzt ~ **core** versetzter Kern *m*
**misplacing** Versetzung *f*
**misplug, to** ~ falsch stöpseln
**misprint, to** ~ fehldrucken, verdrucken
**misprint** Druckfehler *m*, Erratum *n*
**misreading** Fehlanweisung *f*
**misregistration** Fehlüberdeckung *f* (der Farben auf dem Bildschirm)
**misroute, to** ~ fehlleiten
**misrouted message** falsche Kenngruppe *f*, fehlgeleiteter Spruch *m*
**misrouting** Fehlleitung *f*, Leitfehler *m*
**misrun** mangelhaft ausgelaufene Form *f*, nicht ausgelaufenes Gußstück *n* ~ **casting** Fehlgußstück *n*
**miss, to** ~ aussetzen, fehlen, verfehlen, vermissen **to** ~**-fire** abbrennen **to** ~ **one's way** fehlgehen **to** ~ **the target** das Ziel *n* verfehlen
**miss** Aussetzer *m*, Fehlgänger *m*, Fehlschuß *m* ~ **distance** Trefferablage *f* ~**-fire** Fehlfeuerung *f* ~ **vector** Fehlervektor *m*
**missal** Missale *n*
**missed approach** Fehlanflug *m*
**missent message** fehlgesendete Meldung *f*
**misshaped** mißgestaltet
**missile** Geschoß *n*, Lenkwaffe *f*, Schleuderwaffe *f*, Wurfgeschoß *n* a ~ **that does not follow the desired path** Kurvenklemmer *m*
**missile,** ~**-angle** Aggregatlagerwinkel *m* ~ **case** Raketengehäuse *n* ~ **fuel** Raketentreibstoff *m* ~ **guide beam receiver** Leitstrahlempfänger *m* ~ **gyro battery** Bordbatterie *f* ~**-mounted on board** bordseitig ~**-mounted recording instrument** Bordregistrierung *f* ~**-mounted terminal board** Bordsammelschiene *f* ~ **oscillation** Geschoßpendelung *f* ~ **target indicating radar** Zielerfassungsradar *n*

**missiles** Raketen-, Wurf- oder Schleuder-waffen *pl*
**missing** Vorbeigang *m*, Vorbeischießen *n* ~ **of the ignition (misfiring)** Aussetzen *n* der Zündung ~ **line** Nullinie *f* ~ **part** Fehlteil *m*
**mission** Aufgabe *f*, Auftrag *m* ~ **accomplished** Auftrag erledigt
**misstep** Fehltritt *m*
**mist, to** ~ beschlagen
**mist** Bergnebel *m*, Bodennebel *m*, Dunst *m*, Dunstbildung *f*, (leichter) Nebel *m*, Nebeldunst *m*, Schleier *m*, Staubregen *m* ~ **due to water vapor** Dunstnebel *m*
**mistake, to** ~ verwechseln
**mistake** Fehler *m*, Fehlgriff *m*, Irrtum *m*, Versehen *n*, Verwechselung *f* ~ **in charging** Fehler *m* in der Gebührenberechnung
**mistaken** irrig **to be** ~ irren ~ **indication blank** Fehlanzeige *f*
**mistuning** Falschabstimmung *f*, falsche oder schlechte Abstimmung *f*, Verstimmung *f*
**misty** diesig, dunstig, nebelig, schleierig **to be** ~ duften
**misuse, to** ~ mißbrauchen, schlecht behandeln
**misuse** Mißbrauch *m*
**miter (mitre)** Achtelschlag *m*, Gehre *f*, Gehren *m*, Gehrung *f*, Gierung *f* ~**-box** Gehrungsstoßlade *f*, Gehr-, Kröpp-, Schneid-lade *f* ~**-box saw** Gehrungssäge *f* ~ **clamps** Gehrungszwingen *pl* ~**-cutting guide** Gehrungsanschlag *m* ~**-cutting machine** Gehrungsstanzmaschine *f* ~**-cutting shears** Gehrungsschere *f* ~**-dovetail** Gehrungszinke *f* ~ **fillet weld** Vollkehlnaht *f* ~ **gate** Stemmtor *n* ~ **gear** Kegelrad *n* (mit fünfundvierziggradiger Achsenstellung), Winkel-getriebe *n*, -rad *n* ~ **(-wheel) gearing** Kegelrädergetriebe *n* ~ **joint** Gehr-, Gehrungs-fuge *f*, Gehrstoß *m*, Stoß *m* auf Gehrung, stumpfe Gehrung *f* ~ **joint by wood split** Zusammenschlitzen *n* ~ **knife** Gehrungsstichel *m* ~ **line** Kropf-grat *m*, -kante *f* ~ **plane** Gehrungshobel *m* ~ **post** Schlag-, (sluice) Stemm-säule *f* ~ **rule** Gehrmaß *n*, Schmiegewinkel *m* ~ **saw** Rücksäge *f* ~ **sill** Drempel *m*, (gate stop) Schlagschwelle *f* ~ **sill of a canal lock** Blankscheit *m* ~ **sill's foundation** Drempel *m* ~ **square** Achtelwinkellineal *n*, Achtwinkelmaß *n*, festes Gehrdreieck *n*, Gehrungswinkel *m* ~ **valve** Kegelventil *n* ~ **weld** Gehrungsschweißung *f* ~**-wheel gearing** Kegelradgetriebe *n*, Winkelgetriebe *n*
**mitered joint** Gierung *f*
**mitering machine** Gehrungsstoßlade *f*
**mitigate, to** ~ abschwächen, lindern, mäßigen, mildern
**mitigating circumstances** Milderungsgründe *pl*
**mitigation** Mäßigung *f* ~ **of induction by means of transpositions** Induktionsschutz *m* ~ **of sentence** Urteilsmilderung *f*
**mitogenic radiation** mitogenetische Strahlung *f*
**mitre:** see **miter**
**mitred** ~ **dovetail** versenkte Zinke *f* ~ **quad** Gehrungsquadrat *n* ~ **rules** Gehrungslinien *pl*
**mitring cutter** Gehrungsschneider *m*
**mitted quad** Eckquadrat *n*
**mitten** Fausthandschuh *m*
**mix, to** ~ anmachen, anmengen, anrichten, anrühren, ansetzen, durchführen, durchsetzen,

einmaischen, einmischen, melieren, mengen, mischen, überblenden, vermengen, vermischen, verrühren, verschneiden, versetzen, zusetzen **to ~ a charge** gattieren **to ~ ores and additions for a furnace charge** möllern **to ~ signals** aufschalten (g/m) **to ~ thoroughly** durchmischen **to ~ up** aufmischen, verwechseln **to ~ with** zumischen
**mix** Gattierung *f*, Mischung *f*
**mixable** mischbar
**mixed** gemischt, gescheckt, meliert, scheckig **of ~ color** mischfarbig
**mixed, ~ acid for nitration** Mischsäure *f* **~-base crude oil** Hybridrohöl *n* **~-base notation** Gemischtbasisschreibweise *f* **~ cargo** gemischte Ladung *f* **~ combustion** gemischte Verbrennung *f* **~ construction** Mischbau *m* **~ crystal** Mischkristall *n* **~-crystal alloy** Mischkristalllegierung *f* **~ crystals** Austenit *n* **~ cycle** gemischte Verbrennung *f* **~-cycle engine** Semidieselmotor *m* **~ ensemble** gemischte Gesamtheit *f* **~ fabric** Mischgewebe *n* **~ field illumination** Mischfeldbeleuchtung *f* **~ flow compressor** Axialradial-Verdichter *m*, Diagonalverdichter *m* **~-flow pump** Schraubenradpumpe *f* **~ gangues** Geschütte *n* **~ gas** Mischgas *n* **~ gas pipe** Mischgasflöte *f* **~ gas producer** Mischgaserzeuger *m* **~-grain steel** Stahl *m* mit unterschiedlicher Korngröße **~ impeller** Mischflußschnellrührer *m* **~-in-situ cement** örtlich hergestellter Zement *m* **~ load** Mischlast *f* **~ melt** gemischter Schmelzfluß *n* **~ metal** Mischmetall *n* **~ molecule** Mischmolekül *n* **~ nitration acid** Mischsatz *m* **~ nuclei** Mischkerne *pl* **~ number** gemischter Bruch *m* **~ octane numbers** Mischoktanzahl *f* **~ ore** Mischerz *n* **~ oxide** Mischoxyd *n* **~ particles** Gemengteilchen *pl* **~ polyamide** Mipolid *n* **~-pressure turbine** gemischtbeaufschlagte Turbine *f* **~ semi-automatic advance** gemischte Verstellung *f* **~ semi-conductor** gemischter Halbleiter *m* **~ service** gemischter Betrieb *m* oder Verkehr *m* (teleph.) **~ setting** Mischsatz *m* **~ sliver** Mischzug *m* **~ stand** Mischbestand *m* **~ system** gemischter Betrieb *m* **~ weighting** gemischte Beschwerung *f* **~ woods** Mischwald *m*
**mixer** Kristall-Mischstufe *f*, Misch-apparat *n*, -glied *n*, -maschine *f*, -pult *n*, -stufe *f*, -vorrichtung (acoust.), -walze *f*, Mischer *m*, Mixer *m*, Mörtelmaschine *f*, Regler (studio), Rührer *m* **~ of main receiver** Mischkopf *m* zum Senderüberlagerer **~ with orifice for the entraining medium** Strahldüsenmischer *m*
**mixer, ~ bus** Mischgerät *n* **~ console** Mischpult *n* **~ duplexer** Hohlleiterweiche *f* und Mischstufe *f* **~-first detector stage** Mischstufe *f* **~ head** Mischkopf *m* (electr.) **~ metal** Mischereisen *n* **~ (or fader) pad** Widerstandsbahn *f* (d. Reglers) **~ platform** Mischerbühne *f* **~ stage** Misch-, Überlagerungs-stufe *f* **~ tube** Modulationsröhre *f* **~-tube scheme** Mischerschaltung *f* **~ unit** Mischgerät *n* **~ valve** Mischventil *n*
**mixing** (of gases) Bindung *f*, Gattierung *f*, Homogenisierung *f* (des Rückstandes), Versatz *m*, Versetzung *f* **~ of composition** (fireworks) Ansetzen *n* des Satzes **~ by gravity**

Mischung *f* durch Schwerkraft **~ by hand** Handmischung *f* **~ of material** Materialdurchmischung *f*
**mixing, ~ action** Mischwirkung *f* **~ air** Luftbeimischung *f* **~ amplifier** Überblendverstärker *m* **~ amplifier conversion gain** Mischverstärker *m* **~ apparatus** Mischapparat *m* **~ baffle** Leitblech *n* mit absichtlich erzeugter Wirbelbildung **~ booth** Mischer *m* **~ bottle** Mischflasche *f* **~ chamber** (of carburetor) Mischgehäuse *n*, Mischkammer *f*, Zerstäubungsraum *m* **~ cone** (injector) Fangdüse *f* **~ control** Mischregler *m* **~ desk** Misch-pult *n*, -tisch *m*, Regiepult *n* **~ device** Mischvorrichtung *f* **~ drum** Misch-behälter *m*, -trommel *f* **~ efficiency** Homogenisierungsausbeute *f* **~ flask** Mischflasche *f* **~ floor** Schichtboden *m* **~ hexode** Mischhexode *f*, Sechspolmischröhre *f* **~ installation** Mischanlage *f* **~ jet** Rührgebläse *n* **~ length** Mischungslänge *f* **~ lever** Übersetzhebel *m* **~ loud-speaker** Mischlautsprecher *m* **~ machine** Misch-maschine *f*, -trommel *f* (f. Beton) **~ method** Mischungsweg *m* **~ mill** Mischwalzwerk *n* **~ motion** Mischbewegung *f* **~ nozzle** (injectors) Druckdüse *f*, (steam jet) Fangdüse *f*, Mischdüse *f* **~ pan** Mischpfanne *f* **~ panel** Misch-brett *n*, -tafel *f* **~ place** Schichtboden *m* **~ plant crane** Mischerkran *m* **~ pot** Grapen *m* **~ preheater** Mischvorwärmer *m* **~ process** Mischungs-, Misch-vorgang *m* **~ property** Mischfähigkeit *f* **~ proportion** Mischungsverhältnis *n* **~ ratio** Mischungsverhältnis *n* **~ ratio for angular velocity signal** Geschwindigkeitsaufschaltung *f* (g/m) **~ roll** Mischwalze *f* **~ room** Mischraum *m* (film) **~ screw** Mischschnecke *f* **~ section** Misch-stufe *f*, -teil *m* **~ shed** Beschickungsboden *m* (metall.) **~ space** Zerstäubungsraum *m* **~ stage** Misch-schaltung *f*, -stufe *f*, Summierstufe *f* (TV) **~ table** Regiepult *n* **~ tank** Mischbehälter *m* **~ time** Mischdauer *f* **~ transformer** Mischtransformer *m* **~ trough** Mischtrog *m*, Rührkübel *m* **~ tube** Misch-röhre *f*, -rohr *n* **~ unit** Mischgerät *n* **~ valve** Misch-röhre *f*, -ventil *n* **~ vat** Anmachebottich *m* **~ vessel** Ansatzbehälter *m*, Misch-behälter *m*, -gefäß *n* **~ water** Anmachwasser *n* **~ wing** Mischflügel *m* **~ zone** Mischzone *f*
**mixture** Ansatz *m*, Beimengung *f*, Beimischung *f*, Gattierung *f*, Gemenge *n*, Gemisch *n*, Luft-Gasgemisch *n*, meliertes Tuch *n*, Mischlösung *f* (combination), Mischung *f*, Mixtur *f*, Vermengen *n*, Vermischung *f*
**mixture, ~ of air masses** Luftmischung *f* **~ of fuel vapor** Brennstoffdampfgemisch *n* **~ of gold and graphite** Goldgraphit *m* **~ of isotopes** Isotopengemisch *n* **~ of ores** Möllerung *f* (natural) **~ of sand and gravel** Kiessand *m* **~ of tallow and sal ammoniac** Lötfett *n*
**mixture, ~ admission** Gemisch-zuführung *f*, -zutritt *m* **~ chamber** Mischraum *m* **~ component** Mischfarbstoff *m* **~ control** Arm-Reich-Schalter *m*, Arm-Reich-Schaltung *f*, Gemisch-regelung *f*, -regulierung *f*, -steuerung *f* **~-control device** Düsenkorrektor *m* **~ controls** Gemischregler *m* **~ corrector** Gemischregler *m* **~ felt** Melangefilz *m* **~ formation** Gemischbildung *f* **~ hot spot** Gemischbeheizer *m* **~**

inlet Gemischzutritt *m* ~ **limit for the appearance of explosion** Gemischgrenze *f* für Eintreten der Explosion ~ **method** Mischungsmethode *f* ~ **ratio** Mischungsverhältnis *n* ~ **ratio for complete combustion** theoretisches Gemischverhältnis *n* ~ **setting** Gemischeinstellung *f* ~ **strength** Gemisch-anreicherungsgrad *m*, -wert *m*, -zusammensetzung *f* ~ **supply** Gemischzuführung *f* ~ **temperature** Mischungstemperatur *f*

**mixup** Durcheinander *n*

**mizzen,** ~ **mast** Besan-, Kreuz-mast *m* ~ **top** Besanmars *m* ~ **top-gallant mast** Besanbramstenge *f*

**moat** Diamant *m*; Festungs-, Wall-, Wasser--graben *m*

**mobile** bewegbar, beweglich, fahrbahr, leicht--beweglich, -flüssig, mobil, schnell, verlegbar ~ **crane** Autokran *m*, Elektrokrankarren *m* ~ **radio service** beweglicher Funkdienst *m*, fahrbare Funkverbindung *f* ~ **radio station** bewegliche Funkstelle *f* ~ **surface (radio) station** bewegliche Bodenfunkstelle *f* ~ **transformer** Wandertrafo *m* ~ **transmitter** fahrbarer Sender *m*

**mobility** Beweglichkeit *f*, Leichtflüssigkeit *f*, Mobilität *f*, Wendigkeit *f* ~ **of liberated carriers** Beweglichkeit *f* von freien Trägern ~ **in winter** Winterbeweglichkeit *f*

**mobility,** ~ **coefficient** Beweglichkeitskoeffizient *m* ~**-controlled** beweglichkeitsgeregelt ~ **tube** Bewegungsrohr *n*

**mobilometer** Mobilometer *n*

**Möbius** (electrolytic silver-refining) ~ **net** Möbiusnetz *n* ~ **process** Möbiusprozeß *m*

**mock,** ~ **fog** falscher Nebel *m* ~ **seam** falsche Naht *f* ~**-up** Attrappe *f*, hölzernes Probemodell *n* in natürlicher Größe (aviat.) ~ **window** blindes Fenster *n*

**mode** Art und Weise, Betriebsart *f* (IFF), Methode *f*, Modus *m*, Schwingungstyp *m*, (major or minor) Tonart *f*, unwahrscheinlichster Verteilungswert *m* (statist.), Weg *m*, Weise *f*

**mode,** ~ **of action** Verfahren *n*, Wirkungsweise *f* ~ **of anchorage** Verankerungsart *f* ~ **of formation** Bildungsweise *f* ~ **of frequency values** Modus *m* der Häufigkeitswerte ~ **of operation** Betriebsart *f*, Wirkungsweise *f* ~ **of origin** Entstehungsbedingung *f* ~ **of oscillation** Eigenbewegung *f*, Schwingungsweise *f* ~ **of payment** Zahlungs-art *f*, -weise *f* ~ **of procedure** Prozeß *m* ~ **of propagation** Ausbreitungsform *f* ~ **of resonance** Resonanzform *f* ~ **of suppression** Abschirmungsmethode *f*, Entstörungsart *f* ~ **of vibration** Schwingungs-art *f*, -form *f*

**mode,** ~ **changer** Wellenformwandler *m*, Wellentyp-Transformator *m* ~ **filter** Wellentypfilter *m* ~ **filter slot** Schlitzblende *f*, Schlitz-Wellenfilter *n* ~ **selector** Betriebsartenwahlschalter *m* ~ **separation** Frequenzabstand *m* ~ **shift** Frequenzinkonstanz *f*, Frequenzverlauf *m*, Schwingbereichänderung *f* ~ **skip** (magnetron) Fehlzündung *f* ~ **transducer** Tonwandler *m*

**model, to** ~ abbilden, modellieren

**model** Ausführung *f*, Bau-größe *f*, -weise *f*, Gerät *n*, Konstruktion *f*, Modell *n*, Muster *n*, Norm *f*, Schablone *f*, Schema *n*, Typ *m*, Type *f*, Vorbild *n*; mustergültig, vorbildlich ~ **of**

**nucleus** Einteilchenmodell *n* ~ **of solid** Festkörpermodell *n* ~ **with tubular fuselage** Modell *n* mit Rohrrumpf

**model,** ~**-aircraft competition** Flugmodellwettbewerb *m* ~ **airplane** Flugzeugmodell *n* ~ **airplane flying** Flugmodellsport *m* ~ **airway** Versuchsluftstraße *f* ~ **aviation** Modellflugwesen *n* ~ **basin** Probierbassin *n*, Schleppbehälter *m* (aviat.) ~ **basin test** Bassinversuch *m* ~**-basin towing testing** Schleppversuchswesen *n* ~ **builder** Modellbauer *m* ~ **construction kit** Modellbaukasten *m* ~ **designation** Baumusterbezeichnung *f*, Typenbezeichnung *f* ~ **display** Musterdekoration *f* ~ **drawing** Modellzeichnung *f* ~ **flying meeting** Modellflugveranstaltung *f* ~ **laws** Modellgesetze *pl* ~ **missile** Modellaggregat *n* ~ **number** Baumusterbezeichnung *f*, Typennummer *f* ~ **(aircraft)school** Modellbauschule *f* ~ **shop** Versuchswerkstatt *f* ~ **size** Modellgröße *f* ~ **test** Modellversuch *m* ~ **test datum** Modellversuchswert *m* ~ **value** Modellversuchswert *m* ~ **warpage** Modellverbiegung *f*

**modeler** Modellierer *m*, Mustermacher *m*

**modeling** Formgebung *f*, Formung *f* ~ **board** Musterbrett *n* ~ **clay** Modellier-masse *f*, -ton *m* ~ **shop** Fräswerkstatt *f* ~ **wax** Modellierwachs *n*

**moderate** gemäßigt, mäßig, mittelmäßig ~ **in price** preiswert ~ **focal length** mittlere Brennweite *f*

**moderated reactor** moderierter Reaktor *m*

**moderately,** ~ **chilled iron** mildharter Guß *m* ~ **live room** gedämpfter, mittelgedämpfter Raum *m*

**moderating,** ~ **glass** Blendglas *n* ~ **glass revolver** Blendglaskappe *f* ~ **ratio** Bremsverhältnis *n*

**moderation** Bremsung *f*, Mäßigkeit *f*, Mäßigung *f* ~ **of neutrons** Bremsung *f* von Neutronen

**moderator** Brems-stoff *m*, -substanz *f* ~**-coolant** Kühlbremssubstanz *f* ~ **lattice** Bremsstoffgitter *n*

**modern** (hoch)modern, neuzeitlich ~ **engineering** neuzeitliche Technik *f*

**modernize, to** ~ modernisieren

**modes** Eigenzustände *pl* ~ **of anchorage** Verankerungsarten *pl* ~ **of oscillation of quartz** Schwingungsarten *pl* von Quarzen ~ **of waves** Eigenschwingungen *pl*

**modest** anspruchslos, bescheiden ~ **cost** mäßigster Preis *m*

**modification** Abänderung *f*, Abart *f*, Abwandlung *f*, Änderung *f*, Einschränkung *f*, Modifikation *f*, Neugestaltung *f*, Umformung *f*, Umwandlung *f*, Vered(e)lung *f* ~ **of delay** Änderung *f* der Wartezeit ~ **of quality** Qualitätsregulierung *f* ~ **of rules** Änderung *f* der Verfahrensordnung

**modification,** ~ **plate** Formänderungsschild *n* ~ **test** Änderungsprüfung *f*

**modified,** ~ **arrangement** Abwandlung *f* ~ **binary code** zyklischbinärer Kode *m* ~ **bridge** variante Brücke *f*

**modifier** Modifiziermittel *n*, Umsteuergröße *f* ~ **register** Indexregister *n*

**modify, to** ~ abändern, abwandeln, einschränken, modifizieren, umändern, umgestalten, umsteuern, umwandeln

**modulability** Aussteuerfähigkeit *f*, Modulierbarkeit *f*
**modulable** steuerbar
**modular,** ~ **construction** Baukastenweise *f*, Kassettenbauweise *f* ~ **design** Modulbauweise *f* ~ **equation** Modulargleichung *f* ~ **function** Modulfunktion *f*
**modulate, to** ~ abdämpfen, abstimmen, aufschalten, aussteuern, modeln, modulieren
**modulate on** aufmoduliert
**modulated** abgestimmt, abgestuft, angepaßt, moduliert (rdo), reguliert ~ **at audible frequencies** tonüberlagert
**modulated,** ~ **amplifier** modulierter Verstärker *m*, Modulationsstufe *f* ~ **antenna** abgestimmte Antenne *f* ~ **carrier current** modulierter Trägerstrom *m* ~ **continuous wave** (M.C.W.) modulierte ungedämpfte Welle *f* ~ **continuous-wave transmission** Tonsenden *n* ~ **current** gemodelter Strom *m* ~ **light source** Lichtquelle *f* ~-**light speech equipment** Lichtsprecher *m*, Lichtsprechgerät *n* ~ **reception** Tonempfang *m* ~ **transmitter** modulierter Sender *m* ~ **wave** modulierte Welle *f*
**modulating** Modelung *f*, Steuerung *f* ~ **choke** Steuerdrossel *f* ~ **circuit** Modulatorschaltung *f* ~ **coefficient** Aufschaltwert *m* ~ **current** Modulationsstrom *m* ~ **electrode** Lichtsteuerelektrode *f*, Steuerblende *f*, Wehneltzylinder *m* ~ **frequency** Modulationsfrequenz *f* ~ **method** Modulationsverfahren *n* ~ **motor** Klappenantriebsmotor *m* ~ **piston** Regelschieber *m* ~ **tube** Beeinflussungsröhre *f* ~ **valve** Modulations-, Modulator-röhre *f* ~ **voltage** Durchflußregelventil *n* (Drosselventil), Modulationsspannung *f* ~ **wave** modulierende Welle *f*
**modulation** Abstimmung *f*, Aufschaltung *f*, Aussteuerung *f*, Beeinflussung *f*, Modelung *f*, Modulation *f*, Modulierung *f*, Stufenleiter *m*, Tasten *n* ~ **at audible frequency** Tonüberlagerung *f* ~ **of beam intensity** Strahlintensitätskontrolle *f* ~ **of a carrier current** Modulation *f* eines Trägerstroms ~ **of the cathode ray** Modulation *f* des Kathodenstrahls ~ **to dark condition** Modulation *f* oder Modulierung *f* auf Dunkel, Steuerung *f* auf Dunkel ~ **of the electron beam** Modulation *f* eines Elektronenstrahles ~ **by key action** Tastung *f* ~ **to light condition** Hellsteuerung *f*, Modulation *f* oder Steuerung *f* auf hell ~ **in opposition** Gegenmodulation *f* ~ **by voice** Sprachbeeinflussung *f* ~ **by voice action with magnetic modulator** Besprechung *f* mit Eisendrossel
**modulation,** ~ **amplifier** Modulationsverstärker *m* ~ **amplitude** Modulationsamplitude *f* ~ **capability** Höchstmodulationsgrad *m*, Modulierbarkeit *f* ~ **carrier** Modulationsträger *m* ~ **characteristic** Modulationskennlinie *f* ~ **choke** Modulationsdrossel *f* ~ **circuit** Modulationsanordnung *f* ~ **control** Modulations-messer *m*, -regler *m* ~ **depth** Modulationsgrad *m* ~ **device** Modulationseinrichtung *f* ~ **due to iron saturation** Drosselmodulation *f* ~ **element** Modulationszelle *f* ~ **eliminator** Modulationsunterdrücker *m* ~ **envelope** Modulationshüllkurve *f* ~ **factor** Aufschaltwert *m*, Modulationsgrad *m* ~ **factor meter** Modulationsmesser *m* ~ **frequency** Kennungs- und

Modulationsfrequenz *f* ~ **frequency harmonic characteristic** Verzerrungscharakteristik *f* der Harmonischen ~-**frequency intermodulation distortion characteristic** Intermodulationsverzerrungscharakteristik *f* ~ **grid** Steuerblende *f* ~ **index** Modulationsgrad *m* ~ **measurement** Modulationsmessung *f* ~ **meter** Aussteuerungsmesser *m* (acoust.), Modulationskontrolle *f* ~ **monitor** Modulationsüberwachungsgerät *n* ~ **noise** Modulationsrauschen *n* ~ **note** Modelton *m* ~ **percentage** prozentuelle Aussteuerung *f* ~ **power** Modulationsleistung *f* ~ **rate** Telegrafiergeschwindigkeit *f* ~ **reactor** Modulations-, Steuer-drossel *f* ~ **shield** Steuerblende *f* ~ **stage** Modulationsstufe *f* ~-**suppression ratio** Modulationsunterdrückungsgrad *m* ~ **track** ausgesteuerter Tonstreifen *m* ~ **tube** Taströhre *f* (electr.) ~ **unit** Bedienungs-, Modulations--gerät *n* ~ **voltage** Tastspannung *f*
**modulator** Modler *m*, Modulator *m* ~ **blocking pulse** Ausgangssperrimpuls *m* (rdr) ~ **cell** Modulatorzelle *f* ~-**driver** Modulator-Treiberstufe *f* ~ **electrode** Steuer-, Wehnelt-elektrode *f* ~ **output pulse** Ausgangstastimpuls *m* (rdr) ~ **stage** Modulatorstufe *f* (rdr) ~ **tube** Modulationsverstärker-, Modulator-, Steuer-röhre *f* ~ **unit** Steuergerät *n* ~ **valve** Modulations-, Modulator-röhre *f* ~ **valve (or tube)** Beeinflussungsröhre *f*
**module** Glied *n* (schaltungstechnisch), Maß--einheit *f*, -stab *m*, Model *m*, Modul *m*, Verhältniszahl *f* ~ **range** Modulbereich *m* ~ **unit** Bau-gruppe *f*, -stein *m*
**modulo-n-check** Querrestkontrolle *f*
**modulus** konstanter Koeffizient *m*, Modul *m*, Verdichtsteife *f* ~ **in gears** Querteilung (Kennzahl) ~ **with gears** Durchmesserteilung *f*, Kennzahl *f*
**modulus,** ~ **of compression** Elastizitätsmodul *n* für Druck, Zusammendrückungsmodul *m* ~ **of cubic compressibility** K-Modulus *m* ~ **of decay** Abkling- (acoust.), Zerfalls-modul *m* ~ **of direct elasticity** Elastizitätsmodul *m* für Zug ~ **of distribution** Verteilungsmodul *m* ~ **of elasticity** Dehnmaß *n*, Dehnsteife *f*, Dehnungsmodulus *m*, Elastizitäts-maß *n*, -modul *m*, -zahl *f*, -ziffer *f*, E-modul *m*, Gleitmodulus *m*, (shearing) Schubelastizitätsmodul *m*, Steifigkeitszahl *f* ~ **of elasticity for (or in) shear** Elastizitätsmodul *m* für Schub, Schub--modul *m*, -steife *f* ~ **of elasticity for (or in) tension** Elastizitätsmodul *m* für Zug, Zug-elastizitätsmodul *m* ~ **of extension** Dehnungs--koeffizient *m*, -modulus *m* ~ **of resistance** Festigkeits-koeffizient *m*, -zahl *f*, -ziffer *f* ~ **of rigidity** Elastizitäts-, Schub-, Starrheits-, Torosions-modul *m* ~ **of rupture** Biegespannung *f*, Bruch-, Zerreiß-modul *m* ~ **of shear** Elastizitätsmodul *m* für Schub, Gleitmaß *n*, Schub-modul *m*, -steife *f* ~ **of shearing** Schermodul *m*, Schubelastizitätsmodul *m* ~ **of stiffness** Elastizitätsmodul *m* ~ **of stretch** Reckmodul *m* ~ **of subgrade reaction** Bettungsziffer *f* ~ **of torsional shear** Drehmodul *m*, Schubmodul *m* für Verdrehung, Torsionsmodulus *m*, Verdrehungsmodul *m* ~ **of transmission** Transparenzschwankungen *pl* ~ **of transverse elasticity in shear** Gleitmaß *n* ~ **of vector** Betrag *m* des Vektors

**modumite** Tesseralkies *m*
**Mohr's,** ~ **envelope** Mohrsche Hüllkurve *f* ~
**salt** Mohrsches Salz *n*
**Mohs' scale** Mohs' Härteskala *f*
**moiety** Halbscheid *n*
**moire** Moiré *n*, Wasserglanz *m*; geflammt, gewässert, moiriert ~ **métallique** Perlmuttblech *n* ~ **pattern of noise** moiréartiges Geräusch *n*
**moist** dampfhaltig, feucht, naß ~**-adiabatic lapse** feuchtadiabatisch ~ **air** feuchte Luft *f* ~ **curing** Feuchthaltung *f* (concrete)
**moisten, to** ~ anfeuchten, annässen, annetzen, aufweichen, befeuchten, begießen, benetzen, bewässern, feuchten, nässen, netzen **to** ~ **slightly** anwässern **to** ~ **thoroughly** durch-feuchten, -nässen
**moistener** Anfeuchtapparat *m*
**moistening,** ~ **chamber** Befeuchtungskasten *m* ~ **plant** Befeuchtungsanlage *f* ~ **power** Benetzungsfähigkeit *f*
**noistness** Feuchte *f*, Feuchtigkeit *f*
**moist-o-graph** Luftfeuchtigkeitsmesser *m*
**moisture** Feuchte *f*, Feuchtigkeit *f*, Feuchtigkeitswasser *n*, Nässe *f* ~**-attracting** wasseranziehend ~ **chamber** Speichelkammer *f* ~ **content** Feuchtigkeits-gehalt *m*, -grad *m*, Wassergehalt *m* ~ **determination** Feuchtigkeitsbestimmung *f* ~ **indicator** Feuchtigkeitsanzeiger *m* ~ **measuring device** Feuchtigkeitsmeßgerät *n* ~ **meter** Feuchtigkeitsgradmesser *m* ~ **particle** Dunstteilchen *n* ~ **percentage** Feuchtigkeitsgehalt *m* ~ **pick-up** Feuchtigkeitsaufnahme *f* ~**-proof** feuchtigkeits-fest, -sicher ~**-repellent** hydrophob ~**-resistant (or resisting)** feuchtigkeitsfest ~ **sensitive** feuchtigkeitsempfindlich ~ **separator** Feuchtigkeitsabscheider *m* ~ **trap** Trockenpatrone *f*
**molar** körperlich, massig, massiv, molar ~ **boiling-point elevation** molare Siedepunktserhöhung *f* ~ **heat** Molwärme *f* ~ **magnitude** Molgröße *f* ~ **ratio** Molen-, Mol-verhältnis *n* ~ **rotation** Molekularrotation *f* ~ **state** Molzustand *m* ~ **weight** Molar-, Molekular-, Mol-gewicht *n* ~ **volume** Molvolumen *n*
**molarity** Molarität *f*
**molasses** Melasse *f*, Scheideschlamm *m*, Sirup *m* ~ **proof stick** Melasseprobestecher *m* ~ **pulp** melassierte Schnitzel *pl* ~ **residue** Melasseschlempe *f* ~ **slop (or vinasse)** Melasseschlempe *f* ~ **wash** Melassedecke *f*
**mold, to** ~ abformen, abmallen, bemallen, bosseln, formen, gestalten, gießen, modellieren, pressen, profilieren, schimmeln, stocken, verschimmeln
**mold** Abdruck *m* (geol.), Abguß *m*, Druckfutter *n*, Einguß *m*, Form *f*, (briquette press) Formbuchse *f*, Formschablone *f*, Gesenk *n*, Gießmodell *m*, (foundry) Gießereiform *f*, Guß-material *n*, -mutter *f*, -schale *f*, -stück *n*, Hartgußform *f*, Hohl-form *f*, -kehle *f* (arch.), -stempel *m*, Kokille *f*, Kopfspindel *f*, (shipbuildg.) Mall *n*, Matrize *f*, Modell *n*, Nonne *f*, Prägeform *f*, Preßform *f*, Riester *m*, Schablone *f*, Schimmel *m*, Stanzform *f*, Verschalung *f* ~ **for block-making** Gießform *f* ~ **for composition inking rollers** Gießhülse *f* ~ **for initials** Initialinstrument *n*
**mold,** ~ **board** Formkasten-, Pflugstreich-,

Streich-brett *n* ~ **board plow** Riesterpflug *m* ~ **box dumper** Formkastenausleerer *m* ~ **candle** gegossenes Licht *n* ~ **castings** Formguß *m* ~ **catch** Formenhalter *m* ~ **cistern** Form--back *n* (sugar mfg.), -trog *m* ~ **clamping** Formenschluß *m* ~ **closure** Schließen *n* der Form ~ **cure** Formheizung *f* ~ **disc** Gießrad *n* ~ **drying oven** Formen-, Form-trockenofen *m* ~ **engraving machine** Profilgraviermaschine *f* ~ **face** Form-fläche *f*, -wand *f* ~**-filling capacity** Formfüllungsvermögen *n* ~**-finishing smoke black** Formschlichte *f* ~ **fungus** Schimmelpilz *m* ~ **growth** Schimmelbildung *f* ~**-handling crane** Kokillenkran *m* ~ **holder** Glasformenschieber *m* ~ **impression** Matrize *f* ~ **insert** Gesenkeinsatz *m* ~ **lubricant** Gleitmittel *n* ~**-made paper** Maschinenbüttenpapier *n* ~ **maker** Formengießer *m* ~ **maker's vice** Gesenkmacherschraubstock *m* ~ **mark** (on glassware) Formnaht *f* ~ **misalignment** schlechte Zentrierung *f* ~ **opening** Formöffnungshub *m* ~ **part** Formhälfte *f* ~ **release agent** Trennmittel *n* ~ **release medium** Entformungsmittel *n* ~ **setter** Kokillenmann *m* ~ **shaft** Gießformachse *f* ~ **shell** Gießschale *f* ~ **side pieces** Gießformseitenteile *pl* ~ **spindle** Formspindel *f* ~ **table** Formtisch *m* ~ **wiper** Gießformwischer *m*
**moldability** Bildsamkeit *f*
**moldable** bildsam, formbar, verformbar ~ **insulating substance** Isoliermasse *f* ~ **material** Preß-masse *f*, -stoff *m*
**molded,** ~ **from gypsum** aus Gips gegossen **capability of being** ~ (by pressure and/or heat application) Bildsamkeit *f*
**molded,** ~ **article** Formling *m* ~ **articles** Preßmasse *f* ~ **bakelite** Preßstoff *m* ~ **base** Preßmassefuß *m* ~ **blank** Formling *m* ~ **board** Modell-, Zieh-pappe *f* ~ **body** Formstück *n* ~ **breadth** (maritime signals) Breite *f* über Spant ~ **brick** Form-, Profil-ziegel *m* ~ **candle** gegossenes Licht *n* ~ **cardboard** Ziehkarton *m* ~**-coil core** Spulenmark *n* ~ **core** Form-, Masse-, Staub-kern *m* ~ **fiber board** Hartfaserplatte *f* ~ **insulation** Formstück *n* (Isolierform), gepreßte Isoliermasse *f*, Isolierformstück *n* ~ **laminated plastics** Schichtpreßstoff *m* ~ **part** Formpreßstück *n* ~ **peat** Streichtorf *m* ~ **piece** Preßteil *n* ~ **plastic** Preßmasse *f* ~ **plastics** Preßstoff *m* ~ **plywood** Formpreßholz *n* ~ **rubber** Profilgummi *n*
**molder, to** ~ **away** vermodern
**molder** Former *m*, Formgießer *m*, Formmaschine *f*, Gießer *m*, Modellierer *m*, Muttergalvano *n* (print.)
**molder's,** ~ **brush** Formerpinsel *m*, Gußputzbürste *f* ~ **ladle** Handpfanne *f* ~ **nail** Kernnagel *m* ~ **pitch** Formerpech *n* ~ **punch** Formermesser *n* ~ **riddle** Formersieb *n* ~ **rule** Schwindmaß *n* ~ **sieve** Formersieb *n* ~ **tools** Formerwerkzeug *n*
**molding** (inside flask jacket) Band *n*, Borde *f*, Formarbeit *f*, Formen *n*, Formgebung *f*, Fries *m*, (wood) Füllungsleiste *f*, Gesims *n*, Kehlschneiden *n*, Kehlung *f*, Leiste *f*, Modellieren *n*, Preßprofil *n*, Sims *m*, Simswerk *n*, Zierleiste *f* ~ **with clay sheets** Schwartenformerei *f* ~ **of fillet plane** Leistenhobel *m*

~ **at the foot (or at the head) of a window** Fenstergesims *n* ~ **with pins** Nagelleiste *f*
**molding,** ~ **appliance** Formeinrichtung *f* ~ **art** Formereitechnik *f* ~ **batch** Formmasse *f* ~ **bench** Formbank *f* ~ **blank** geharzte Pappe *f* ~ **board** Formbrett *n* ~ **box** Form-flasche *f*, -kasten *m*, -presse *f*, Kasten *m* ~**-box part** Formkastenteil *n* ~**-box pin** Formkastenstift *m* ~ **composition** Preßmasse *f* ~ **cycle** Preßdauer *f* ~ **edge** Mallkante *f* ~ **equipment** Form-anlage *f*, -einrichtung *f* ~ **flask** Form-flasche *f*, -kasten *m*, Gußflasche *f* ~ **floor** (founding) Form-platz *m*, -raum *m* ~ **hall** Formhalle *f* ~**-in lead** Bleiprägung *f* ~ **job** Formarbeit *f* ~ **loft** Mallboden *m* ~ **machine** Formmaschine *f* ~ **machine with power lift** Formmaschine *f* mit selbsttätiger Modellaushebung ~ **machine with runout table** Formmaschine *f* mit Formkastenwagen ~ **machine with swing-out-type squeezing head** Formmaschine *f* mit ausschwingbarem Preßhaupt ~ **machine for use with snap flasks** Formmaschine *f* für kastenlosen Guß ~ **machine-practice (or -work)** Formmaschinenbetrieb *m*, Maschinenformerei *f* ~ **machinery** Formmaschinenanlage *f* ~ **material** Formerstoff *m* ~ **method** Formweise *f* ~ **mixture** Preßmischung *f* ~ **operation** Formarbeit *f*, Formerei *f* ~**-out board** Musterbrett *n* ~ **pin** Formerstift *m* ~ **plane** Profilhobel *m* (carp.) ~ **plane iron** Kehleisen *n*, Kehlhobeleisen *n* ~ **powder** Preßpulver *n* ~ **practice** Formerei *f* ~ **preparation** Preßmasse *f* ~ **press** Kunststoffpresse *f* ~ **pressure** Preßdruck *m* ~ **process** Form-arbeit *f*, -verfahren *n* ~ **sand** Form-, Geis-, Gieß-, Gießerei-sand *m* ~ **sand for iron** Gießereiformmasse *f* ~**-sand core** Formsandkern *m* ~**-sand quarry (or pit)** Formsandlager *n* ~ **sand testing** Formsandprüfung *f* ~ **shop** Formerei *f* ~ **shrinkage** Formenschwindmaß *n* ~ **technique** Formtechnik *f* ~ **time** Stehzeit *f* ~ **wax** Bossierwachs *n*
**molds** Stockflecke *pl* ~ **in bottlemaking** Flaschenform *f*
**moldy** muffig, schimmelig, stockig
**mole** Fluß-, Hafen-damm *m*, Grammolekül *n*, Mol *n*, Mole *f*, Molekülmasse *f*, Pier *m* ~ **fraction** Molenbruch *m* ~ **ratio** Mol(en)verhältnis *n*
**molecular** molar, molekular ~ **adhesion** Moleküladhäsion *f* ~ **amplifier** Molekularverstärker *m* ~ **attraction** Molekularanziehung *f* ~ **beam** Molekularstrahl *m* ~ **bond** molekulare Bindung *f* ~ **cluster** Molekülaggregat *n* ~ **clusters** Molekülhaufen *pl* ~ **concentration** Molarität *f*, Normalität *f* ~ **diagram** Molekülabbildung *f* ~ **diameter** Molekulardurchmesser *m* ~ **energy** Molekularkraft *f* ~ **excitation** Molekülanregung *f* ~ **field approximation** Näherung *f* des Molekularfeldes ~ **force** Molekularkraft *f* ~ **friction** Molekularreibung *f* ~ **gauge** Moleкulardruckmanometer *n* ~ **heat** Molarwärme *f* ~ **label** molekülarer Indikator *m* ~ **lattice** Molekülgitter *n* ~ **magnet** Elementarmagnet *m* ~ **mass** Molekulargewicht *n* ~ **method** Atomstrahlmethode *f* ~ **modes** Schwingungsfreiheitsgrade *pl* im Molekül ~ **motion** Molekularbewegung *f* ~ **number** Molekülzahl *f* ~ **ray** Molekülstrahl *m* ~ **refraction** Lichtbrechungs-

vermögen *n* ~ **scattering** Molekülstreuung *f* ~ **scattering law** Molekularstreugesetz *n* ~ **stopping power** molekulares Bremsvermögen *n* ~ **velocity** Molekülgeschwindigkeit *f* ~ **volume** Molarvolumen *n* ~ **weight** Molekulargewicht *n*
**molecularity** Molekularität *f*
**molecule** Massenteilchen *n*, Molekel *n*, Molekül *n* ~ **disintegrating impact** molekularzertrümmernde Stoßkraft *f*
**moleskin** Englischleder *n*, Feuchtfilz *m*, Feuchtwalzenstoff *m*, Moleskin *m*, (molleton) Wischwalzenstoff *m*
**molette of turbine shaft** Wellenstern *m*
**mollify, to** ~ feuchten
**mollusk** Weichtier *n*
**molten** erschmolzen, feuerflüssig, flüssig, geschmolzen, schmelzflüssig ~ **ash chamber** Schmelzkammer *f* ~ **bath** Schmelzbad *n* ~ **metal** flüssiges Metall *n*, Metallschmelze *f*
**molybdeniferous** molybdänhaltig
**molybdenite** Molybdänglanz *m*, Molybdänit *m*, Molybdänkies *m*, Wasserblei *n*
**molybdenum** Molybdän *n* ~ **ore** Molybdänerz *n* ~ **oxide** Molybdänoxyd *n*
**molybdic,** ~ **acid** Molybdänsäure *f* ~ **anhydride** Molybdänsäurenanhydrid *n* ~ **ocher** Wasserbleiocker *m*
**molybdite** Wasserbleiocker *m*
**moment** Augenblick *m*, (of inertia) Beharrungsvermögen *n*, Moment *n*, Zeitpunkt *m*
**moment,** ~ **of acceleration** Beschleunigungsmoment *n* ~ **of bolt and nut tension** Anzugsmoment *n* ~ **of flexure** Biegemoment *n* ~ **of force** Kraftmoment *n* ~ **of a force** statisches Moment *n* ~ **of gravitational force** Schwerkraftmoment *n* ~ **of gyration** Schwungmoment *n* ~ **of ignition** Zündmoment *m* ~ **of image change** Bildkipp-Punkt *m* ~ **of inertia** Beharrungs-, Massen-, Schwung-, Trägheitsmoment *n*, Rotationsträgheit *f* ~ **of inertia of the column** Stielträgheitsmoment *n* ~ **of inertia of the member** Stabträgheitsmoment *n* ~ **of injection** Spritzzeitpunkt *m* ~ **at the joint** Eckmoment *n* ~ **of momentum** Drall *m* (phys.), Drehimpuls *m*, Impuls-, Winkeldreh-moment *n*, Moment *n* der Bewegungsgröße ~ **of nose heaviness** Kopflastmoment *n* ~ **about the nucleus** Kernpunktmoment *n* ~ **of overhead** Kragmoment *n* ~ **on the pin joint (or on the support)** Einspannungsmoment *m* ~ **of propeller thrust** Schrauben-kraftmoment *n*, -zugmoment *n* ~ **of reaction** Reaktionsmoment *m* ~ **of resistance** Trag-, Widerstands-moment *n* ~ **of resultant air force** Luftkraftmoment *n* ~ **of rotation** Dreh-, Drehungs-, Torsions-, Verdrehungs-moment *n* ~ **of stop** Einhalteaugenblick *m* ~ **of tail unit** Leitwerkmoment *n*
**moment,** ~ **area** Momentenfläche *f* ~ **arm** Hebel *m* ~ **coefficient** Momentbeiwert *m* ~ **curve** Momentenlinie *f*, Momentkurve *f* ~ **diagram** Momenten-diagramm *n*, -schaubild *n* ~ **distribution method** Momentenausgleichsverfahren *n* ~ **ellipsoid** Momentenellipsoid *n* ~ **equation** Aggregatgleichung *f* (g/m) ~ **line** Momentenlinie *f* ~ **planimeter** Integrimeter *n* ~ **pole** Momenten-pol *m*, -punkt *m* ~ **process** Momentprozeß *m* ~**-resisting** biegungsfest
**momental vector** Momentvektor *m*

**momentary** augeblicklich, kurzzeitig ~ **contact** Momentkontakt *m* ~ **contactor** Momentkontaktgeber *m* ~ **deviation** vorübergehende Regelabweichung *f* ~ **load** Augenblicksbelastung *f* ~ **running** Momentbetrieb *m* ~ **target** Augenblicksziel *n* ~ **trigger switch** Andrückkippschalter *m* ~ **value** Augenblickswert *m*

**moments,** ~ **of force** Massenkraftmomente *pl* ~ **of stress** Momente *pl* der Spannungskomponenten

**momentum** Angriffskraft *f*, Antrieb *m*, bewegende Kraft *f*, Bewegungsgröße *f*, (dynamics) Impuls *m*, mechanisches Moment *n*, Moment *n*, Momentum *n*, Schwung *m*, Triebkraft *f* ~ **of a body** Bewegungsmoment *n* ~ **and energy exchange** Impuls- und Energieaustausch *m* ~ **of a sound wave** Impuls *m* einer Schallwelle

**momentum,** ~ **conservation** Erhaltungssatz *m* ~ **conservation law** Impulserhaltungssatz *m* ~ **density** Impulsdichte *f* ~ **determination** Impulsbestimmung *f* ~ **distribution** Impulsverteilung *f* ~ **flow tensor** Impulstransporttensor *m* ~ **integral** Impulsintegral *n* ~ **measurement** Impulsmessung *f* ~ **production** Impulsproduktion *f* ~ **space** Geschwindigkeitsraum *m* ~ **space representation** Impulsraumdarstellung *f* ~ **starter** Schwungkraftanlasser *m* ~ **tensor** Impulstensor *m* ~ **theorem** Impuls-, Momenten-satz *m* ~ **thickness** Impulsdicke *f* ~ **transfer** Impulsübertragung *f*, Momenttransport *m* ~ **transfer collision** Stoß *m* mit Impulsübertragung ~ **transfer equation** Impulstransportgleichung *f* ~**-transportation theory** Impulstransporttheorie *f*

**monatomic** einatomig, monoatomar ~ **solids** monoatomare Festkörper *pl*

**monaural** einkanalig, einohrig, monaural ~ **listening** einohriges oder monaurales Hören *n* ~ **reception** monauraler Empfang *m*

**monazite** Monazit *m*, Turnerit *m* ~ **sand** Monazitsand *m*

**Mond gas** Mondgas *n*

**Monel,** ~ **metal** Monelmetall *n* ~ **process** Monelverfahren *n*

**monetary standard** Münzfuß *m*

**money** Geld *n* ~ **advanced** Vorschuß *m* ~ **allowance** Geldabfindung *f* ~ **changer** Wechsler *m* ~**-changer's office** Wechselstube *f* ~ **chest** Kasse *f* ~ **due** Geldforderung *f* ~ **market** Kapitalmarkt *m* ~ **matter** Geldangelegenheit *f* ~ **order** Geldanweisung *f* ~ **paid in settlement** (of claims etc.) Abfindungssumme *f* ~ **rate** Geldkurs *m* ~ **telegram** telegrafische Postanweisung *f* ~ **turnover** Geldumsatz *m*

**Monier,** ~**'s arch** Moniergewölbe *n* ~**'s building** Monierbau *m* ~ **slab** Monierplatte *f* ~**'s vault** Moniergewölbe *n*

**monitor, to** ~ abhören, kontrollieren, mithören, überwachen (rdr.), in eine Verbindung *f* eintreten **to** ~ **a circuit** in eine Leitungs *f* (un)verstärkt eintreten, in einer Leitung *f* (un)verstärkt mithören

**monitor** Abhorch-, Abhör-gerät *n*, Kontroll-empfänger *m*, -gerät *n*, Mithöreinrichtung *f*, Monitor *m*, Prüfrohr *n*, Warngerät *m*, Wendestrahlrohr *n*, Überwacher *m*, Überwachungseinrichtung *f*, (pressure, flame monitor) Wächter *m* ~ **button** Mithörtaste *f* (teleph.)

~ **console** Überwachungstisch *m* ~ **cord** Überwachungsschnur *f* ~ **desk** Beschwerdestelle *f* ~ **dredging** Spülen *n* ~ **earphone** Kontroll-Kopfhörer *m* ~ **jack** Mithör-, Überwachungs-klinke *f* ~ **man** Tonmeister *m* (film) ~ **part of roof** Dachaufbau *m* ~ **picture** Fernsehbild *n* ~ **picture tube** Kontrollwiedergaberöhre *f* ~ **position** Kontroll-, Prüf-platz *m* ~ **rack** Überwachungsgestell *n* ~ **receiver** Abhörempfänger *m* ~ **reel** Kontrollwalze *f* ~ **room** Abhörraum *m* (film, acoust.) ~ **screen** Sichtgerät *n* ~ **socket** Meßbüchse *f* ~ **speaker** Kontroll-Lautsprecher *m* ~ **transmitter** Kontrollsender *m*

**monitored** abgehört

**monitoring** Aktivitätskontrolle *f*, (of sound recording or sound tracks) Aufzeichnungskontrolle *f*, Kontrolle *f*, Überwachung *f* ~ **amplifier** Abhör-, Kontroll-, Mithör-, Überwachungs-verstärker *m* ~ **antenna** Kontrollantenne *f* ~ **board** Mithörschrank *m* ~ **box** Abhörbox *f* ~ **circuit** Mithöreinrichtung *f*, Mithör-, Prüf-schaltung *f*, Überwachungskreis *m* ~ **coil** Mithörübertrager *m* ~ **desk** Aufsichts-, Misch-tisch *m*, Misch-, Regler-pult *n* ~ **device** Mithöreinrichtung *f* ~ **device for radio** Mithörer *m* ~ **discriminator** Kontrolldiskriminator *m* ~ **element** Ausgangssignalwandler *m* ~ **facility** Mithörvorrichtung *f* ~ **feedback** stabilisierende Rückführung *f* ~ **installation** Mithöreinrichtung *f* ~ **instruments** Schutzgeräte *pl* ~ **jack** Mithörklinke *f* ~ **key** Mithör-schalter *m*, -taste *f*, Überwachungs-schalter *m*, -schlüssel *m* ~ **loudspeaker** Abhör-, Kontroll-lautsprecher *m* ~ **meter** Überwachungsinstrument *n* ~ **operator** Tonmeister *m* ~ **panel** Misch-brett *n*, -tafel *f* ~ **picture** Sucher-, Kontroll-bild *n* (TV) ~ **position indicator** Antwortgeber *m* ~ **print** Kontrolldruck *m* ~ **racks** Kontrollgestelle *pl* ~ **receiver** Kontrollempfänger *m* ~ **reception** Kontrollempfang *m* ~ **system** Überwachungsanlage *f* ~ **tube** Kontrollröhre *f*

**monitron** Alarmgeber *m*

**monkey** Affe *m*, Fall-bär *m*, -hammer *m*, -klotz *m*, -werk *n*, Hund *m*, Rammbär *m*, Ramme *f* ~ **chatter** Nachbarkanalstörung *f*, Seitenbandinterferenz *f* ~ **spanner** Universalschlüssel *m* ~ **wrench** englischer Schraubenschlüssel *m*, Engländer *m*, Franzose *m*, Mutterschlüssel *m*, Schraubenschlüssel, Universalschlüssel *m*, Universalschraubenschlüssel *m*, verstellbarer Schraubenschlüssel *m*, Vierkantschlüssel *m*

**monkshood** Eisenhut *m*

**mono-acid** einsäurig

**mono-atomic** einatomar, einatomig ~ **layer** monoatomare oder einatomige Schicht *f* ~ **liquids** einatomige Flüssigkeiten *pl*

**monobasic** einbasisch ~ **acid** einbasische Säure *f*

**monoblock,** ~ **barrel** Vollrohr *n* ~ **casting** Blockgußstück *n*

**mono,** ~ **cellular** einzellig ~ **centric** monozentrisch ~ **chloracetic acid** Monochloressigsäure *f* ~ **chloride (protochloride) of sulfur** Einfachchlorschwefel *m* ~ **chord** Monochord *n*

**monochromatic** einfarbig, monochromatisch ~ **illuminator** einfarbige Leuchtstrahlquelle *f* ~ **light** einfarbiges oder monochromatisches

Licht *n* ~ **objective** Monochromat *n* ~ **pyrometer** Glühfadenpyrometer *m* ~ **sensitivity** monochromatische Empfindlichkeit *f*
**mono,** ~ **chromatism** Monochromasie *f* ~ **chromatization** Monochromatfilter *f* ~ **chromator** einfarbige Lichtquelle *f*, Monochromator *m*, Registrierschreiber *m*
**monochrome** einfarbig, monochrom ~ **light** einfarbiges Licht *n* ~ **printing** Einfarbendruck *m* ~ **signal** Schwarz-Weiß-Signal *n* (TV) ~ **transmission** Monochromübertragung *f*
**mono,** ~ **chromic** monochrom ~ **clinal** monoklinal, nur in einer Richtung geneigt ~ **cline** Monoklinale *f* ~ **clinic** monoklin, monoklinisch
**monocoque** Wickelrumpf *m* ~ **body** Holzwickelrumpf *m* ~ **construction** Schalen-bau *m*, -bauweise *f*, -konstruktion *f* ~ **fuselage** Schalen-bauweise *f*, -rumpf *m*, Walfischrumpf *m*, wickelfournierter Rumpf *m* ~ **metal fuselage** Metallschalenrumpf *m* ~ **type** Schalenbauweise *f* ~ **wing** Schalenflügel *m*
**mono,** ~ **cord** Einschnur(schaltung) *f* ~ **cryometer** Druck- und Gefrierpunktmesser *m* ~-**crystal** Einkristall *m* ~ **crystal filament** Einkristallfaden *m* ~ **crystals** Einkristalle *pl*
**monocular** einäugig, monokular ~ **vision** einäugiges Sehen *n*
**mono,** ~ **cyclic** monozyklisch ~ **dimetric** monodimetrisch ~ **energetic** monoenergetisch ~ **energy** einheitliche Energie *f*
**monogram** Monogramm *n*
**monograph** Einzeldarstellung *f*
**mono,** ~ **gyrocompass** Einkreiselkompaß *m* ~ **hydric alcohol** einwertiger Alkohol *m* ~ **iodoethane** Äthyljodid *n* ~ **layer** Einfachschicht *f*, monolekularer Film *m* ~ **lithic conduit** monolithischer Kabelkanal *m* ~ **lithic foundations** Blockfundamente *pl* ~ **meter** Monometer *m* ~ **metric** monometrisch
**monomial** eingliedrige Zahlengröße *f*
**mono,** ~ **molecular** monomolekular, von der Dicke *f* eines Moleküls ~ **molecular layer** Einfachschicht *f*, monomolekularer Film *m* ~ **morphic** monomorphisch ~ **nuclear** einkernig ~ **pack method** Monopackverfahren *n*
**monophase** einphasig ~ **alternomotor** Einphasenwechselstrommotor *m* ~ **current** Einphasenstrom *m* ~ **equilibrium** einphasiges Gleichgewicht *n*
**monophonic** einstimmig
**monoplane,** ~ **of low-wingtype** Tiefdeckerflügel *m* ~ **fuselage** Rumpfeindecker *m* ~ **model** Eindeckerflugmodell *n* ~ **star model** Stabeindecker *m*
**monopole wave** Monopolwelle *f*
**monopolize, to** ~ monopolisieren
**monopoly** Alleinhandel *m*, Monopol *n*
**mono,** ~ **propellant** (rocket) Einzeltreibstoff *m* ~ **pulse method** Einpulsverfahren *n* (electron.) ~ **pulse operation** Monopulsbetrieb *m*
**monorail** Deckenschiene *f*, Einschienenbahn *f*; eingleisig ~ **bucket crab running on an overhead track** Einschienengreiferkatze *f*, auf Hochbahn *f* ~ **conveyer** Einschienenförderbahn *f*, Hängebahn *f* ~ **crab** Einschienenlaufkatze *f* ~ **hoist** Einschienenkran *m* ~ **motor hoist** Einbahnkatze *f* (Winde) ~ **railway** ein-

schienige Schwebebahn *f* ~ **track** einschienige Schwebebahn *f*, Einschienenlaufbahn *f* ~ **traveling crab** Einschienenhängekatze *f*
**mono,** ~ **saccharide** Monose *f* ~ **silicate** Singulosilikat *n* ~ **silicate slag** Singulosilikatschlacke *f*
**monoscope** Monoskop *n*, Testbildröhre *f*
**mono,** ~ **spar** einholmig ~ **stable** monostabil ~ **stable multivibrator** Flip-Flop-Generator *m*, monostabile Kippschaltung *f* ~ **static range-finder** Einstandentfernungsmesser *m* ~ **symmetric** monosymmetrisch ~ **telephone** Monotelefon *n*
**monotone** eintönig, monoton
**monotonicity theorem** Monotonitätssatz *m*
**monotonous** einförmig ~ **system of linear equations** monoton lineares Gleichungssystem *n*
**monotonously decreasing** monoton abnehmend
**monotorial device** Mitleser *m* (teleph.)
**monotrimetric** monotrimetrisch
**monotron** Fernsehröhrenprüfer *m*, Prüfröhre *f*
**mono,** ~-**tube boiler** Einröhrenkessel *m* ~ **type** Monotype *f*, Monotypie *f* ~ **type caster** Monotypegießmaschine *f* ~ **type keyboard** Monotypetaster *m* ~ **valent** einwertig ~ **variant** einfachfrei, monovariant ~ **variant equilibrium** einfachfreies oder monovariantes Gleichgewicht *n*
**monsoon** Monsun *m*
**montage** Montage *f*
**monte-jus** Druckbehälter *m*, Druckbirne *f*, Montejus *m*
**monthly** monatlich ~ **allowance** monatlicher Etat *m* ~ **report** Monatsbericht *m* ~ **statement** Monatsauszug *m* ~ **stock summary** monatliche Bestandsübersicht *f*
**montmorillonite** Montmorillonit *n*
**monument** Denkmal *n*, (surveying) Punkt *m*
**moon,** ~ **motion mechanism** Mondgetriebe *n* ~ **ring** Mondring *m*, Wasserstreifen *m* ~ **stone** Abart *f* des Feldspates
**moor, to** ~ (Schiff) festmachen, verankern, vermuren, vertäuen **to** ~ **a balloon (an aircraft)** einen Ballon (Flugzeug) verankern
**moor** Hochmoor *n*, Moor *n* ~ **land** Moorboden *m*
**moorage** Liegeplatz *m*, Vertäuung *f*
**moored** vertäut ~ **mine** Ankermine *f*, Ankertaumine *f*
**mooring** Ankergerät *n*, Ankern *n*, Festmachen *n*, Mooring *f*, Verankerung *f*, (of a boat) Vermooren *n* ~ **area** Liegeplatz *m* ~ **band** Verankerungsgurt *m* ~ **bitts** Ankerbeting *f* ~ **bollard** Haltepfahl *m*, Vertauungspoller *m* ~ **buoy** Ankerboje *f*, Festmache-boje *f*, -tonne *f*, Festmacherboje *f* ~ **cable** Fesseltau *n*, Verankerungskabel *n* ~-**cable strain indicator** Fesselkraftanzeiger *m* ~ **cone** Fessel-, Verankerungs-kegel *m* ~ **drag** Heckbeschwerung *f* bei der Verankerung ~ **equipment** Festmachevorrichtungen *pl* ~ **gear** Ankermaterial *n*, Festmachevorrichtungen *pl*, Verankerungsvorrichtung *f*, Vertäuung *f* ~ **harness** Halte-, Verankerungs-netz *n* ~ **line** Ankerseil *n*, Fesselkabel *n*, Festmacher *m*, Halteleine *f* ~ **loop** senkrechte Schleife *f* für den Sandsack ~ **mast** Anker-, Verankerungs-, Vertäu-mast *m* ~ **pile** Festmache-, Streich-pfahl *m* ~ **point** Fessel-Ver-

ankerungspunkt *m* ~ **post** Haltepfahl *m*,
Poller *m*, Vertäupfahl *m* ~ **posts** Dalben *pl*,
Dallen *m* ~ **ring** Anker-, Boots-ring *m*, Fest-
mache-bügel *m*, -ring *m*, Haltekreuz *n* ~-**ring
bolt** Schiffsring *m* ~ **rod** Zugstange *f* ~ **rope**
Ankertau *n*, Halteleine *f*, Verankerungstau *n*
~-**rope eyelet** Anlegeseilöse *f* ~ **screw** Schraubenanker *m* ~ **sinker** Ankerstein *m* ~ **tower**
Ankermast-, Verankerungs-turm *m*
**moorings** Hafenanker *m*
**moot** strittig
**mop, to** ~ aufwischen **to** ~ **up** aufräumen, ausheben, auskämmen, ausräumen
**mop** Büschel *m*, Dweil *n*, Mop *m*, Putzlappen *m*,
Quaste *f*, Schwabbel *m*, Schwabber *m*, Wisch-
lappen *m* ~ **polishing** Schwabbeln *n*
**mopping,** ~ **up** Aufräumung *f* ~-**up action** Säu-
berungsaktion *f*
**moraine** Moräne *f*, Moränenschutt *m*
**moratorium** Aufschub *m*, Frist *f*
**mordant, to** ~ beizen
**mordant** Ätzbeize *f* (print.), Ätzwasser *n*, Beize
*f*, Beizmittel *n*; ätzend, beizend ~ **action**
Beizkraft *f* ~ **dyestuff** Beizenfarbstoff *m* ~-
-**padded style** Beizenklotzartikel *m* ~ **printed
style** Beizendruckartikel *m* ~ **printing** Beizen-
druck *m* ~ **steam color** Beizendampffarbe *f*
~ **style** Beizartikel *m* ~ **toning** Beiztönung *f*
**mordanted bottom** Beizgrund *m*
**mordanting,** ~ **bath (or liquor)** Beizflotte *f*
**more-stage demagnetization** Mehrstufenentmag-
netisierung *f*
**Morgan,** ~ **(continuous-rod) mill** Morganstraße
*f* ~ **pipe reel** Morganhaspel *f*
**morning** Morgen *m* ~ **fog** Morgennebel *m* ~
**reckoning** Vormittagsbesteck *n* ~ **shift** Früh-
schicht *f* ~ **(routine) test** (regelmäßige) Früh-
messung *f* ~ **watch** Morgenwache *f*
**morocco (leather)** Saffian *m* ~ **paper** Marocain-
papier *n*
**morphologic extinction** morphologische Aus-
löschgesetze *pl*
**morphological changes due to effects of gravita-
tion** Baromorphose *f*
**morphometry** Morphometrie *f*
**morphotropic** morphotropisch
**morphotropy** Morphotropie *f*
**Morris tube** Abkommrohr *n*
**Morse,** ~ **apparatus** Morse-apparat *m*, -schrei-
ber *m* ~ **code** Morse-alphabet *n*, -schrift *f*,
Seekabelalphabet *n* ~ **corruption in reception**
Horchfehler *m* ~ **dash** Morsestrich *m* ~ **error in
reception** Horchfehler *m* ~ **key** Morsetaste *f* ~
**key connector** Tastanschluß *m* ~ **printer** Morse-
schreiber *m* ~ **receiver** Farbschreiber *m*, Morse-
empfänger *m* ~ **receiving apparatus** Klopfer *m*
~ **signal** Morsezeichen *n* ~ **slip** Morsestreifen
*m* ~ **system** Morsesystem *n* ~ **taper** Morse-
konus *m* ~ **taper gauge** Morsekegellehre *f* ~
**taper shank** Morsekegel *m*
**mortality rate** Sterblichkeitsziffer *f*
**mortar** Minenwerfer *m*, Mörser *m*, Mörtel *m*,
Pochladen *m*, Reibschale *f* **(powder)** ~ Pulver-
mörser *m* ~ **of overburnt lime grains** Krump-
mörtel *m* ~ **of wax** Wachskitt *m*
**mortar,** ~ **block** Pochklotz *m* ~ **box** Pochtrog *m*
~ **carriage** Tauchlafette *f* ~ **funnel** Kalkrutsche
*f* ~ **joint** Mörtelfuge *f* ~ **lime** Luftmörtel *m*

~ **mill** Mörtelmühle *f* ~ **pillar** Mörtelständer *m*
~ **sand** Bausand *m* ~ **test** Mörserprobe *f* ~
**worker** Fugenausstreicher *m*
**mortgage, to** ~ verpfänden
**mortgage** Grundpfand *n*, Hypothek *f* ~ **bond (or
deed)** Pfandbrief *m* ~ **free** lastenfrei
**mortise, to** ~ einlassen, einscheren, einstemmen,
einzapfen, nuten, verschwalben, verzapfen, zu-
sammenfügen **to** ~ **a tenon** einen Zapfen *f* ein-
lochen
**mortise** Einschnitt *m*, Falz *m*, Fuge *f*, Nut *f*,
Nute *f*, Stemmloch *n*, Zapfenloch *n* ~-**and-
-tenon joint** Zapfenverbindung *f* ~ **ax** Stich-,
Stoß-axt *f* ~ **bolt** Einsteckriegelschloß *n*,
Zapfennagel *m* ~ **chisel** Lochbeitel *m* ~ **chisel
with socket** Düllochbeitel *m* ~ **dead lock** Ein-
steckschloß *n* ~ **gauge** Zapfenstreichmaß *n*
~ **joint** Verzapfung *f* ~ **knob latch** Einsteck-
riegel *m* ~ **wheel** Zahnrad *n* mit Winkelzähnen
**mortiser** Stemmaschine *f*, Verzapfer *m*, Zapfenlochmaschine *f*
**mortising** Stemmlochen *n* ~ **and saw chains**
Fräs- und Sägeketten *pl* ~ **machine** Stemmaschine *f* (Holz), Zapfenlochmaschine *f*
**mosaic** Bildzusammenstellung *f* (Luftaufnahme),
Mosaik *f*, Mosaikelektrode *f* (TV), Mosaik-
schirm *m*, **(work)** Musivarbeit *f*, **(in photo
cathode tube)** Raster *m*, Reihenbild *n* ~ **block**
Mosaikblock *m* ~ **camera** Reihenbildner *m*
~ **(photo sensitized) coat** gerasterte Schicht *f*
~ **disease** Gelbstreifenkrankheit *f* ~ **electrode**
Mosaikelektrode *f* ~ **gold** Katzen-, Mosaik-,
Musiv-gold *n* ~ **paving** Mosaikpflaster *n* ~
**plate** Mosaik *f*, Mosaikplatte *f* (TV) ~ **screen**
(photosensitized) gerasterte Schicht *f*, ge-
rasterter Schirm *m*, Raster-kathode *f*, -schirm *m*
~ **structure** Fehlbauerscheinung *f*, Mosaik-
struktur *f* ~ **telegraphy** Mosaikverfahren *n* ~
**type storage iconoscope** Ladungsspeicherbild-
abtaster *m*
**Moseley's law** Moseleysches Gesetz *n*
**mosquito net** Moskitonetz *n*
**moss** Moos *n* ~ **agate** Mokkastein *m*, Moos-
achat *m* ~-**rubber articles** Moosgummi-
Artikel *m*
**mossy,** ~ **lead** Bleischwamm *m* ~ **tin metal**
Zinnwolle *f* ~ **zinc metal** Zinkwolle *f*
**most** meist(e, er, es), größt(e, er, es) ~ **effective
best-** ~ **exposed part of a dike** Schardeich *m*
~ **favorable** bestwert, günstig, optimum ~
**finely dispersed** feinstdispers ~ **meticulous
cleanliness** peinlichste Sauberkeit *f* ~ **practicable type** gangbarste Sorte *f* ~ **probable
value** wahrscheinlichster Wert *m* ~ **promising**
vielversprechend ~ **rarefied packing of spheres**
dünnste Kugellagerung *f* ~ **unfavorable** un-
günstigst
**mote (weaving)** Knötchen *n*, Stäubchen *n*
**mother** Essigmutter *f*, Mutter *f* ~ **disk** Mutter *f*
~ **gate** Hauptfördersohle *f*
**Mother Hubbard bit** Mutter Hubbard Meißel *m*
**mother,** ~ **liquor** Mutter-lauge *f*, -magma *n*
~ **lode** Hauptgang *m* ~ **lye** Mutterlauge *f* ~
**matrix** Mutter *f* ~-**of-pearl** Perlmutter *f*;
perlmutterartig ~ **record** Meisterpositiv *n*
~ **rock** Urgestein *n* ~ **substance** Stammsub-
stanz *f* ~ **vat** Mutterfaß *n* ~ **water** Mutter-
lauge *f*

motion Antrag *m*, Antrieb *m*, Begriff *m*, Bewegung *f*, Gang *m*, Gang-, Geh-werk *n*, Getriebe *n*, Kreuzkopf *m*, Lauf *m*, Querhaupt *n*, Räderwerk *n*, Regung *f*, Zug *m*

motion, ~ of aircraft Flugzeugbewegung *f* ~ in altitude (or altitude motion) Höhenbewegung *f* ~ of the center of gravity Schwerpunktsbewegung *f* ~ in a circle (or orbit) Kreisbahnbewegung *f* ~ of crank Kurbelbetrieb *m* ~ of a line of a sight Visierlinienbewegung *f* ~ of the nucleus Kernmitbewegung *f* ~ in pitch um die Querachse *f* ~ of the plane into itself Bewegung *f* der Ebene in sich ~ in right ascension Stundenbewegung *f* ~ of the sea Seegang *m* ~ of a silent chain Gang *m* einer geräuschlosen Kette ~ of the wind Windbewegung *f*

motion, ~ balance Wegvergleich *m* ~ box Triebkasten *m* ~ class Bewegungsklasse *f* ~ contactor Triebwerksschütz *m* ~ drive Bewegungsantrieb *m* ~ driving mechanism Antriebs-, Trieb-mechanismus *m* ~ element Bewegungsorgan *n* ~ (or manipulating) head Bedienungsknopf *m* ~ indicator Laufkontrolle *f* ~ knob Bewegungsknopf *m* ~ means Bewegungsorgan *n* ~ mechanism Bewegungsorgan *n* ~ phenomenon Bewegungserscheinung *f* ~ pickup Weg-aufnehmer *m*, -geber *m*

motion-picture Film *m* ~ camera Filmaufnahmekamera *f* ~ equipment Tonfilmanlage *f* ~ lens (or objective) Kinoaufnahmeobjektiv *n* ~ operator Kameramann *m*, Vorführer *m* ~ projector Filmvorführapparat *m* ~ screen Kinoleinwand *f* ~ studio Filmstudio *n* ~ technology Kinomatografie *f* ~ theater Kino *n* ~ theodolite Kinotheodolit *n*

motion, ~ pictures Reihenbilder *pl* ~ study Bewegungsstudie *f* ~ type Bewegungstype *f*

motional, ~ action Bewegungsvorgang *m* ~ energy Bewegungsenergie *f* ~ field Bewegungsfeld *n* ~ impedance Bewegungsimpedanz *f*, kinetische Impedanz *f* (acoust.), kinetischer Scheinwiderstand *m* (electron.) ~ magnitude (or quantity) Bewegungsgröße *f* ~ transient Bewegungsübergang *m*

motionless bewegungslos, regungslos, unbeweglich, still

motive Antrieb *m*, Beweggrund *m*, Motiv *n*, Veranlassung *f*; bewegend, treibend ~ energy Bewegungsenergie *f* ~ fluid Treibmittel *n* ~ force Bewegungskraft *f* ~ lock (door) Blindschloß *n* ~ power Antrieb *m*, Betriebskraft *f*, Bewegungsantrieb *m*, Bewegungskraft *f*, Getriebe *n*, Stoßkraft *f*, Treibkraft *f*, Triebkraft *f* ~ (moving) power Antriebskraft *f* ~ substance Treibmittel *n*

motivity Bewegungs-fähigkeit *f*, -kraft *f*

motley bunt, scheckig, ungleich, verschiedenartig

motor antreibendes Element *n*, Antriebsmaschine *f*, Elektromotor *m*, Kraftmaschine *f*, Kraftwagen *m*, Motor *m*, Motorfahrzeug *n*, Triebkraft *f*, Verbrennungsmotor *m*; antreibend, bewegend by ~ motorisch ~ with commutable poles polumschaltbarer Motor *m* ~ of the external-rotor type Außenläufermotor *m* ~ for independent electric drive Einbaumotor *m* ~ in position of depth of immersion Motorherstel-

lung *f* bei Eintauchtiefe ~ with reciprocating movement Schwingmotor *m* ~ with separate cylinders Motor *m* mit einzelstehenden Zylindern ~ with smooth speed regulation Motor *m* mit weicher Drehzahl ~ with swivel bearing Pendelmotor *m* ~ of a unit Aggregatmotor *m*

motor, ~ ambulance Krankentransportwagen *m* ~ apparatus Bewegungsapparat *m* ~ ataxia Bewegungsataxie *f* ~ balloon Fesselballon *m*, motorisierter Ballon *m*, ~ barge Motorlastkahn *m* ~ bearing Motorträger *m* ~ block Zylinderblock *m* ~ board Laufwerkplatte *f* (acoust.) ~ boat Motor-boot *n*, -schiff *n* ~ boating Blubbern *n* (rdo), Motorbrummen *n*, Pumpen *n* des Empfängers *n*, Selbsterregung *f* in tiefer Frequenz, Surrerscheinung *f* ~ bracket Konsole *f*, Motorgrundplatte *f* ~ brake Motorbremse *f* ~ bus Autobus *m*, Kraftomnibus *m* ~ cade Wagenkolonne *f*

motorcar Kraftwagen *m*, Motorwagen *m* ~ for winch launching Windenauto *n* ~ crane Autoaufbaukran *m* ~ dealer Kraftfahrzeughändler *m* ~ engine Kraftfahrzeugmotor *m* ~ jack Autowinde *f* ~ polish Autopoliermittel *n* ~ repair crane Autoinstandsetzungskran *m* ~ service Kraftfahrzeugbetrieb *m*

motor, ~ centrifuge Motorschleuder *f* ~ circuit contactor Motorschutzschalter *m* ~-circuit switch Motorschalter *m* ~ coach Omnibus *m*, Triebwagen *m* ~ continuous duty condenser Motordauerlaufkondensator *m* ~ control Motorüberwachung *f* ~-control unit Statoranschlußgerät *n* ~-controller Drehzahlregler *m*, Motorschalter *m*, Motorsteuerschalter *m* ~ converter Drehgleichstromumformer *m*, Drehstromgleichstromumformer *m*, Umformeraggregat *n* ~-converter dynamotor Einankerumformer *m* (Drehstrom-Gleichstrom) ~ cover Motorschutzkasten *m* ~ cue Motorstartmarke *f* (film) ~ cultivator Motorhackmaschine *f* ~ cultivator with rotary knives Bodenfräse *f* ~ current Motorstrom *m* ~ cycle Krad *n*, Kraft-, Motor-rad *n* ~ cycle tube Kraftradschlauch *m* ~ cyclist Kradfahrer *m*, Motorradfahrer *m* ~ drive Motorantrieb *m* ~-drive electric verifier Motorlochprüfer *m* ~-drive make and break Motorunterbrecher *m* ~-drive punch Motormagnetlocher *m*

motor-driven mit Motorantrieb *m* ~ airplane model Antriebsmodell *n* ~ air pump Motorluftpumpe *f* ~ chain saw Motorkettensäge *f* ~ feed water pump Elektrospeisepumpe *f* ~ overhead trolley conveyer Elektrohängebahn *f* ~ pump Motorpumpe *f*, Pumpe *f* mit elektrischem Antrieb ~ sounding machine Motorlotmaschine *f* ~ sweeper Motorkehrmaschine *f* ~ truck Motorkarren *m* ~ unit Triebwageneinheit *f*

motor, ~ element Antriebelement *n*, (of loud-speaker) Triebelement *n*, Triebwerk *n* ~ engine Kraftmaschine *f* ~ extension ladder Drehleiter *f*, Motordrehleiter *f* ~ failure Motordefekt *m* ~ fan elektrischer Fächer *m* ~ fire engine Motorspritze *f* ~ fitter Autoschlosser *m* ~ flange Motorflansch *m* ~ frame Motorgestell *n* ~ freight car train Lastkraftwagenzug *m* ~ fuel Motorbetriebsstoff *m* ~ fuse Motorsicherung *f* ~ gasoline Motorbenzin *n* ~ generator Maschinensatz *m*, Motorgenerator *m*, Umformer *m*, Umformersatz *m* ~-generator panel Motor-generatortafel *f* ~-generator set Motordynamo *m*, Umformergruppe *f* ~ grab Motorgreifer *m* ~ head Motor-haube *f*, -haubenblech *n* ~ installation Triebwerkseinbau *m* ~ instrument Triebwerksüberwachungsinstrument *n* ~ launch Barkasse *f*

motorless motorlos

motor, ~ lorry Kraftkarren *m* ~ lubricant Autoschmieröl *n* ~ man Wagenführer *m* ~-march column Autokolonne *f* ~ means Triebmittel *n* ~ mechanic Motorenwart *m* ~ meter Motor-, Umlauf-zähler *m* ~ oil Motor-, Motoren-öl *n* ~ oil without additives Motoröl *n* ohne Zusätze ~ oils and greases Autoschmierstoffe *pl* ~ operating capacitor Betriebskondensator *m* ~ operator Motorerregerteil *n* ~ park Kraftfahrpark *m* ~ part Motorteil *n* ~-performance gauge Triebwerksgerät *n* ~-protecting release Motorschutzauslöser *m* ~-protection Motorschutzschalter *m* ~ pulley Motorriemenscheibe *f* ~ rating Motornennleistung *f* ~ reversing contactor Umkehrschütz *m* ~ room Motorenraum *m* ~ safety switch Motorschutzschalter *m* ~ scooter Motorroller *m* ~ scrapers Motorschürfwagen *m* ~ shaft Motorwelle *f* ~ shaft end Motorwellenstumpf *m* ~ ship Motorschiff *n* ~ show Automobilausstellung *f* ~ socket Motorhülse *f* ~ specifications Motordaten *pl* ~ starter Anlasser *m* ~ starting condenser Motoranlaufkondensator *m* ~-starting switch Motoranlaßschalter *m* ~ support Motorträger *m* ~ sweeper Straßenkehrmaschine *f* ~ switch armature Motorwippe *f* ~ symbolism Motorrechnung *f* ~ tank truck Motorkesselwagen *m* ~ tanker Tankmotorschiff *n* ~ terminal board Motorklemmbrett *n* ~ terminal box Motorklemmenkasten *m* ~ torque Motordrehmoment *n* ~ tractor Schlepper *m*, Traktor *m*, Zugmaschine *f* ~ traffic Auto-, Kraft-verkehr *m* ~ transmitter Motorgeber *m* ~ transport Kraftwagentransport *m* ~-transport supply Kraftfahrversorgung *f* ~ trolley hoist elektrische Laufwinde *f* ~ truck Kraftkarren *m*, Motorlastwagen *m* ~ tube (or inner tube) Autoschlauch *m* ~-type relay Relais *n* mit Spuleneinrichtung ~-type steam engine stehende Dampfmaschine *f* ~ uniselector Motorwähler *m* ~ unit Motoreinheit *f*

motor-vehicle Kraftfahrzeug *n*, Kraftwagen *m*, Motor-fahrzeug *n*, -wagen *m* ~ accident Kraftfahrunfall *m* ~ column Autokolonne *f* ~ driver Kraftfahrzeugführer *m* ~ inertia starter Wagenschwungkraftanlasser *m*

motor, ~ winch Motorwinde *f* ~ winch for cable Kabelkraftwinde *f*

motoring Anlaufen *n* des Motors

motorist Kraftfahrer *m*

motorized motorisiert

mottle, to ~ flecken, marmorieren, sprenkeln, tüpfeln

mottle Marmormuster *n*, Sprenkelung *f*

mottled bunt, (cast iron etc.) fladerig, (cast iron) geädert, gefleckt, (cast iron) gemasert, gesprenkelt, halbiert, meliert, runzelig, scheckig ~ with pig iron Kerneisen *n*

mottled, ~ effect Melliereffekt *m* ~ iron hal-

biertes Eisen *n* ~ paper Faserpapier *n*, gemasertes Papier *n* ~ pig iron Forelleneisen *n*, (stark) halbiertes Roheisen *n* ~ sandstone Bunt-, Tiger-sandstein *m*

mottling Aderung *f*, Fleckung *f*, Sprenkelung *f*, Tüpfelung *f* ~ colors Melierfarben *pl*

mould: see mold

mound Aufwurf *m*, Erdaufwurf *m*, Hügel *m*, Meiler *m*, Meilerofen *m*, Wall *m*

mount, to ~ Abzüge aufkleben, anbringen, aufbauen, aufklotzen, aufpflanzen, aufsitzen, aufspannen, aufstellen, (pictures) aufziehen, befestigen, besteigen, einbauen, einfädeln, emporkommen, errichten, ersteigen, montieren, rüsten, steigen, zusammenbauen, in passender Weise zusammensetzen to ~ flexibly onto fuselage gelenkig am Flugzeuggerippe *n* lagern to ~ a map Karte *f* aufziehen to ~ by parbuckle aufschroten

mount Aufhängevorrichtung *f*, Aufsatz *m*, Berg *m*, Einbauvorrichtung *f*, Fassung *f* eines Objektives, Gehäuse *n*, Gestell *n*, Lafette *f* (artil.), Metallbeschlag *m* (aviat.), (microscope) Objektglas *n*, Objektträger *m*, Träger *m*, Unterlagsteg *m*

mount, ~ adapter Lafettenaufsatzstück *n* ~-arm assembly Lagerbrücke *f* ~ base Grundrahmen *m* ~ eclipse Vignettierung *f*

mountable case Aufsatzkasten *m*

mountain Berg *m* ~ and valley breezes Berg- und Talwind *m* ~ barometer Höhenbarometer *n* ~ breeze Bergwind *m* ~ building gebirgsbildend ~ butter Bergbutter *f* ~ chain Höhenzug *m* ~ creep Bergsturz *m* ~ crest Gebirgskamm *m* ~ declivity Bergabhang *m* ~ effect Abschattungsverlust *m*, Gebirgseffekt *m* ~ flesh (asbestos) Bergfleisch *n* ~ hollow Bergkessel *m* ~ irrigation Rückenbau *m* ~ limestone Kohlenkalkstein *m* ~ pass Gebirgs-joch *n*, -paß *m* ~ pasturage Alp *n* ~ peak Berggipfel *m* ~ pine Bergkiefer *f* ~ railway Gebirgsbahn *f* ~ range Gebirge *n* ~ ridge Gebirgsrücken *m* ~ sickness Höhenkrankheit *f* ~ side Bergwand *f*, Gebirgshang *m* ~ slope Berghang *m* ~ spar Vorgebirge *n* ~ stream Bergstrom *m*, Gebirgsfluß *m* ~ summit Berggipfel *m* ~ tallow Bergfett *n* ~ terrain Gebirgsgelände *n* ~ thunderstorm Reliefgewitter *n* ~ top Berggipfel *m* ~ torrent Bergstrom *m*, Gebirgsfluß *n* ~ trail Höhenweg *m*

mountainous bergig, gebirgig ~ ground Gebirgsgelände *n*

mountains Gebirge *n* ~ formed of disrupted folds Bruchfaltengebirge *n* ~ formed of folds Faltengebirge *n* ~ formed of overthrust (or recumbent) folds Deckfaltengebirge *n* ~ formed by plateau-forming movements Schollengebirge *n*

mounted aufgestellt, aufgezogen (Bild), eingebaut, montiert, zusammengebaut ~ on an apparatus eingebaut in einen Apparat ~ in ... bearings ... fach gelagert ~ in an elevated position hoch gelegen ~ on gimbals kardanisch gelagert ~ in rockers schwingend gelagert

mounted, ~ externally außen angebracht ~ lens gefaßtes Linsensystem *n* ~ lines zusammengesetzte Gleise *pl* ~ pattern plate Modellplatte *f* mit angeheftetem Modell

mounter Webstuhlvorrichter *m*

mounting Anbringung *f*, Armatur *f*, Aufbau *m*, (of wheel) Auflaufen *n* (des Rades), Aufsatz *m*, Aufstellung *f*, Einbau *m*, Einbettungsmedium *n*, Einrichtung *f*, Fassung *f*, Futter *n*, Halterung *f*, (gun) Lafettierung *f*, Montage *f*, Montierung *f*, Rüstung *f*, Streifen *m*, (strip) Tragleiste *f*, (plate) Tragplatte *f*, Wehgeschirr *n*; aufsteigend

mounting, ~ for cars Waggonbeschlagteil *m* ~ and demounting presses Auf- und Abziehpressen *pl* ~ and dismantling Auf- und Abziehen *n* ~ of the screen Lage *f* der Blende ~ for weaving mills Webegeschirr *n*

mounting, ~ arm Befestigungsarm *m* ~ attachment Befestigungsanschluß *m* ~ base Befestigungsplatte *f* ~ board Aufziehkarton *m* ~ bracket Auflagewinkel *m*, Befestigungs-platte *f*, -schelle *f*, -winkel *m* ~ brackets Aufhängung *f* ~ cell Füllfassung *f* (opt.) ~ device Aufklotzapparat *m* ~ drawing Montagezeichnung *f* ~ fixture Einbauvorrichtung *f* ~ flange Anbau-, Befestigungs-, Montage-flansch *m*, Einbauplatte *f* (für Sternmotoren) ~ frame Aufhänge-(rdr), Einbau-rahmen *m* ~ glasses Montagegläser *pl* ~ hole Befestigungsloch *n* ~ hole separation Lochabstand *m* ~ instruction Montageanleitung *f* ~ iron Montiereisen *n* ~ ladder Aufstiegsleiter *m* ~ master workparts Musterwerkstücke *pl* ~ material Befestigungsmaterial *n* ~ neck Aufsteckhals *m* ~ pad Anbaukonsole *f*, bearbeitende Anbaufläche *f* ~ paper tympan Aufzugpapier *n* ~ pillar Steigleitung *f* ~ plate Einbauplatte *f* für Sternmotoren, Grund-, Montage-platte *f* ~ position Aufstellung *f* ~ rail Befestigungsschiene *f* ~ ring Anschraubring *m* (print.), Mitnehmer *m* (lens) ~ section of airplane engine Gehäuseteil *n* zur Befestigung des Flugmotors am Traggerüst ~ sleeve Einbauhülse *f* ~ socket Aufsteckfuß *m* (Scheinwerfer) ~ strap Befestigungslasche *f* ~ stud Befestigungsbolzen *m*, Stiftschraube *f* zur Befestigung ~ support Halterung *f* ~ system Halterungssystem *n* ~ test Saughöheprüfung *f* ~ trunnion Federbeinlager *n* ~ wall Befestigungswand *f*

mountings Garnitur *f*

mouse Schubdüseneinsatz *m*, Zugleine *f* mit Gewicht ~ gray mausgrau ~ mill Motor *m* des Heberschreiters (telegr.) ~ trap Fangklappe *f*, (petroleum) Klappventil *n*

Mousse rubber Kautschukschaum *m*

mouth Ausfluß *m*, (of a river) Auslauf *m*, Ausmündung *f* (print.), (waterway) Einfahrt *f*, Gicht *f*, (of converter) Helm *m*, Mund *m*, Mündung *f*, Öffnung *f*, Schnauze *f*

mouth, ~ of an adit Stollenmundloch *n* ~ of a bore Bohrlochmund *m* ~ of furnace Füllöffnung *f* ~ of a furnace Einsatzöffnung *f*, Floßloch *n* ~ of fuse setter Stellbecher *m* ~ of a hook Hakenweite *f* ~ of horn Trichter-mund *m*, -mündung *f*, -öffnung *f* ~ of a retort Retortmündung *f* ~ of river Ein-, Fluß-mündung *f* ~ of a shaft Hängebank *f* ~ of tongs Zangenmaul *n* ~ of Venturi tube Düsenendfläche *f*

mouth, ~ area of nozzle Mündungsfläche *f* ~ correction Mündungskorrektion *f* ~ diameter Mündungsdurchmesser *m* ~ drum Maultrommel *f* ~ gauge Mundschraube *f* ~ lock (para-

chute) Öffnungssicherung *f* ~ opening of horn Lautsprechertrichtermundöffnung *f* ~ organ Mundharmonika *f* ~ piece Ansatz *m*, (mask) Atemmundstück *n*, Einspracheöffnung *f*, Mundstück *n*, Sprechtrichter *m*, Tülle *f* ~ piece of microphone Mikrofonbecher *m* ~ screen Gichtmantel *m*

movability Beweglichkeit *f*, Fahrbarkeit *f*, Verschiebbarkeit *f*

movable bewegbar, beweglich, fahrbar, lose, ortsveränderlich, schwenkbar, verschiebbar, verstellbar ~ abatis Schleppverhau *n* ~ arms bewegliche Waffen *pl* ~ beam Wiege-balken *m*, -träger *m* ~ block Losblock *m* ~ bollard beweglicher Poller *m* ~ bridge bewegliche Brücke *f* ~ (aircraft-) camera mounting Kameragestell *n* für beweglichen Einbau (in das Flugzeug) ~ carbon rod verstellbarer Kohlestab *m* ~ crane Rollkran *m* ~ dam bewegliches Wehr *n* ~ disk bewegliche Scheibe *f* ~-disk relay Scheibenrelais *n* ~ drive verstellbarer Antrieb *m* ~ index Laufmarke *f* ~ installation fahrbare Einrichtung *f* ~ jib-type pillar crane Hammerwippkran *m* ~ key Matritzenanschlag *m* ~ kidney nierenförmiger dynamischer Schwingungsdämpfer *m*, pendelndes Gegengewicht *n* ~ leakage yoke Streujochschieber *m* ~ mounting beweglicher Einbau *m*, (gun) Schwenkbügel *m* ~ plate Anhängeplatte *f* (print.) ~ pressure-gauge connection drehbare Manometeraufnahme *f* ~ pulley Losrolle *f* ~ rail bewegliches Auflager *n* (Schiene) ~ rose fahrbares Gießrohr *n* ~ spindle end Spindelzapfen *m* ~ sprayer fahrbares Gießrohr *n* ~ stanchion Klappstütze *f* ~ stop verschiebbarer Anschlag *m* ~ surface Steuerfläche *f* ~ table Einschiebetisch *m* ~ thread of the warp Polfaden *m* ~ tripod Fahrstativ *n* ~ vertically der Höhe *f* nach verstellbar

movables bewegliche Anlagewerte *pl* oder Güter *pl*

move, to ~ ausziehen, beantragen, bewegen, fahren, fortbewegen, gehen, rücken, rühren, treiben, sich verlagern, an einen anderen Ort verlegen, verschieben, versetzen, verstellen, ziehen to ~ about abtasten to ~ along anfahren to ~ away wegziehen to ~ back zurückbewegen to ~ a cable auslegen (ein Kabel) to ~ in a circle kreisen to ~ down absenken, absinken, niedergehen to ~ downward abwärtsbewegen to ~ in either direction auf- oder abfahren to ~ forward vorrücken to ~ with a lever hebeln to ~ near(er) aneinander rücken to ~ off abrücken to ~ in or out of range in oder außer Wirkungsbereich *m* kommen to ~ and redeposit soil by means of water umschlämmen to ~ through hindurchströmen to ~ to and fro hin- und herbewegen, schaukeln to ~ up aufrücken, aufsteigen, aufwärtsbewegen, ausrücken, vorziehen

move Bewegung *f*, Zug *m* ~ of declination (or of deflection) Ausweichbewegung *f*

movement Antriebsmechanismus *m*, Ausschlag *m*, Bewegung *f*, Fortbewegung *f*, Lauf *m*, Regung *f*, Verschiebung *f*

movement, ~ of the aileron Querruderausschlag *m* ~ of the air Luftbewegung *f* ~ at an angle Schrägfahrt *f* ~ in blank Rohuhrwerk *n* ~ of boulders Geschiebeführung *f* ~ of the bubble

Blasenbewegung *f* ~ **of control surface** Ruderausschlag *m* ~ **of the counter** Zählertransport *m* ~ **in a curved line** krummlinige Bewegung *f* ~ **in depth** Parallaxenbewegung *f*, Tiefenbewegung *f* (film) ~ **of frame** Bildschaltung *f* (film) ~ **of the guide bars** Legeschienenschwung *m* ~ **of gyroscope** Kreiselbewegung *f* ~ **of pendulum** Pendelbewegung *f* ~ **of phonograph** Sprechmaschinenlaufwerk *n* ~ **of soil** Grundbruch *m* ~ **of stroke** Hubbewegung *f* ~ **of supplies** Nachschub-bewegung *f*, -verkehr *m* ~ **of the water** Wasserbewegung *f*
**movement,** ~ **area** Bewegungsfläche *f* (airport) ~ **blur** Bewegungsunschärfe *f* ~ **control mechanism** Steuermechanismus *m* ~ **free from play** spielfreier Gang *m* ~ **parallel to front table** Querfahrt *f* ~ **plan** Bewegungsübersicht *f* ~ **table** Transportbefehl *m* ~ **timetable** Bewegungsübersicht *f* **to-and-fro** ~ hin- und hergehende Bewegung *f*
**mover** Antriebsmotor *m*, Motor *m*, Trieb-kraft *f*, -werk *n*, Zug *m*
**moving** Rührung *f*, Versetzung *f*; bewegend, bewegt, wandernd ~ **in a closed cycle** geschlossen kreisend ~ **in and out** Ein- und Ausschwenken *n* ~ **into position** Instellunggehen *n* ~ **to and fro** hin- und hergehend ~ **up and down** Auf- und Abwärtsbewegung *f*
**moving,** ~ **arm** Dreharm *m* ~ **beam radiation** Bewegungsbestrahlung *f* ~**-belt production** Bandarbeit *f* ~ **blade** Laufschaufel *f* ~ **charge** bewegliche Ladung *f*
**moving-coil** bewegliche oder drehbare Spule *f*, Rotor-, Schwing-, Tauch-spule *f* ~ **ammeter** Drehspul-, Drehspulen-strommesser *m* ~ **galvanometer** D'Arsonval-Galvanometer *n*, Drehspulengalvanometer *n* ~ **indicating instrument** Drehspulanzeigeinstrument *n* ~ **instrument** Drehspul-instrument *n*, -meßwerk *n* ~ **loudspeaker** elektrodynamischer Lautsprecher *m* ~ **measuring system** Drehspulmeßwerk *n* ~ **meter** Drehspulmeßinstrument *n* ~ **microphone** dynamisches oder elektrodynamisches Mikrofon *n*, Tauchspulenmikrofon *n* ~ **pick-up** Drehspultonabnehmer *m* ~ **pointer galvanometer** Drehspulzeigergalvanometer *n* ~ **recorder** Drehspulschreiber *m* ~ **relay** Drehspul--relais *n*, -schütz *m* (electr.) ~ **speaker** (elektro)dynamischer Lautsprecher *m*, Tauchspulenlautsprecher *m* ~ **telewriter** Drehspulschnellschreiber *m* ~ **torquer** Stellspule *f* ~ **vibrator (or oscillograph)** Spulenschwinger *m*
**moving,** ~**-conductor receiver** elektrodynamischer Fernhörer *m* ~ **contact** bewegliche Kontaktbrücke *f* ~ **contact spring** bewegliche Kontaktfeder *f* ~ **cylinder** Ventilkörper *m* ~ **element** bewegliches Organ *n* ~ **expenses** Umzugskosten *pl* ~ **ferry** Trajekt *n* ~ **field** bewegliches oder wanderndes Feld *n*, Wanderfeld *n* ~ **force** bewegende Kraft *f* ~ **frame** drehbare Rahmenantenne *f*, drehbarer Rahmen *m*, Drehrahmen *m* ~ **gear** Fahrvorrichtung *f* ~ **grid** beweglicher Raster *m* ~ **ground** Gebirgsbewegung *f*
**moving-iron,** ~ **ammeter** Weichblei-, Weicheisen-strommesser *m* ~ **amperemeter** elektromagnetisches Amperemeter *n* ~ **instrument** Dreheisen-, Weichblei-, Weicheisen-instrument *n* ~ **loud-speaker** induktordynamischer Laut-

sprecher *m* ~ **measuring instrument** Weicheisenmeßgerät *n* ~ **microphone** elektromagnetisches Mikrofon ~ **receiver** elektromagnetischer Fernhörer *m* mit Eisenanker ~ **speaker** elektromagnetischer Lautsprecher *m* ~ **type of ammeter (or of voltmeter)** Dreheiseneinheitsmeßwerk *n* ~ **voltmeter** Dreheisen-, Weicheisenspannungsanzeiger *m*
**moving,** ~ **joint** Gelenk-stelle *f*, -verbindung *f* ~ **load** bewegliche Last *f* ~ **loud speaker** Tauchspulenlautsprecher *m* ~**-magnet galvanometer** Nadelgalvanometer *n* ~**-magnet instrument** Drehmagnetinstrument *n* ~**-needle galvanometer** Drehmagnetinstrument *n* ~ **part** Antriebsaggregat *n* (of the plant), beweglicher Teil *m*, Bewegungselement *n* ~ **parts** arbeitende Maschinenteile *pl*, Triebwerk *n* ~ **period** Bewegungsintervall *n* (film), Bildwechselzeit *f* (film), Schaltperiode *f* ~ **pictures camera** Filmkamera *f* ~ **plate of the variable condenser** Rotor-paket *n*, -platte *f* ~ **section** beweglicher Teil *m* ~ **sheets** laufende Bahnen *pl* ~ **system** bewegliches System *m* ~ **tape transmitter** Gleitbandsender *m* ~ **target** bewegliches oder fahrendes Ziel *n* ~ **target indicator (MTI)** Festzeichenlöscher *m* (rdr) ~ **the track** Gleisverlegung *f* ~ **trihedral** begleitender Dreikant *m* ~ **trihedral of a curve in space** begleitender Dreikant *m* einer Raumkurve ~**-vane instrument** Weicheiseninstrument *n* ~**-vehicle communication** Fahrzeugfunk *m* ~ **vertically and revolving** Auslegerdrehscheibenkran *m* ~ **water** Aufschlagwasser *n* ~ **wave** fortschreitende Welle *f*, Wanderwelle *f*
**mow, to** ~ mähen **to** ~ **down** niedermähen
**mower** Grasmäher *m*, Grasschneider *m*, Mäher *m*, Mähmaschine *f* ~ **cutter bar** Mähbalken *m* ~ **cutter bar lift** Mähbalkenaufzug *m* ~ **drive** Mähantrieb *m* ~ **drive housing** Mähantriebsgehäuse *n* ~ **lever** Mähhebel *m*
**M-quadrat** Schließgeviert *n*
**mucic acid** Schleimsäure *f*
**mucilage** Gummilösung *f*, Klebstoff *m*, Leim *m*, Schleim *m*
**mucilaginous** klebrig, schleimhaltig, schleimig
**muck** Berge *pl*, Schlick *m* ~ **bar** Luppenstab *m*, Rohschiene *f* ~**-bar faggot** Luppeneisenpaket *n* ~**-bar pile** Luppeneisenpaket *n*, paketierte Luppenstäbe *pl* oder Luppeneisenstäbe *pl*, Rohschienenpaket *n* ~ **bar pile** Schweißeisenpaket *n*, Schweißpaket *n* ~**-bar piling** Paketierung *f* ~ **bars** Luppenstäbe *pl* ~ **iton** Luppeneisen *n*, Rohschiene *f* ~ **mill** Luppen-, Puddeleisen--walzwerk *n* ~ **pile** Paket *n* ~ **rolling mill** Puddeleisenwalzwerk *n*
**mucker** Lader *m*, Schaufler *m*, Verlader *m* ~**'s car (or truck)** Minenhund *m*
**mucous** schleimartig, zähschleimig ~ **membrane** Schleimhaut *f*
**mud, to** ~ **off** mit Schlamm *m* verdichten
**mud** Morast *m*, Schlamm *m*, Schlammgrund *m*, Schlick *m*, Schmutz *m*, Trübe *f* ~ **analysis** Schlammanalyse *f* ~ **bit** Schlammeißel *m* ~ **bottom** Schlickgrund *m* ~ **box** Reinigungs-, Schlamm-kasten *m* ~ **capping** Auflegeschuß *m* ~ **cock** Schlamm-ablaßhahn *m*, -hahn *m* ~ **collector** Schlammsammler *m* ~ **container** Schlammbecher *m* ~ **conveyer** Schlammbahn *f*

~-discharging valve Schlammablaßventil ~
drain Schlammablaß *m* ~ dredger Schlamm-
bagger *m* ~ exhauster Schlammsauger *m* ~
exhauster for cleaning sewers Kanalschlamm-
sauger *m* ~ flat Sumpfebene *f*
**mudgard** Kotflügel *m*, Schmutzfänger *m*, Schutzblech *n* ~ skirt Kotflügelschürze *f* ~ stay Kot-
strebe *f* ~ struts Schutzblechstreben *pl*
**mud,** ~ hole Bodenluke *f* (boilers) ~ hook Gleit-
schutzgreifhaken *m* ~ jacking Schlamm-
-injektion *f*, -einpressung *f* ~-laden fluid
Schlammflüssigkeit *f* ~ ladle Schlammlöffel *m*
~ lighter Baggerprahm *m* ~ mixer Schlamm-
Mischapparat *m* ~ pan Schlammtopf *m* ~ pit
Schlammloch *n* ~ pump Schlamm-löffel *m*,
-pumpe *f* ~ pump plunger Schlammlöffel-
kolben *m* ~ pump valve Schlammpumpen-
ventil *n* ~ residues Schlammrückstände *pl* ~
scraper Schlammkatze *f* ~ settling pond
Schlammteich *m* ~ sill Schlammschwelle *f* ~
slinging Dreckschleudern *pl* ~ socket Schlammbüchse *f* mit Klappventil ~ stone Tonstein *m*
~ tube Schlammröhre *f* ~ volcano Salse *f*,
Schlamm-sprudel *m*, -vulkan *m*
**muddle, to** ~ verfahren
**muddy** lehmig, schlammartig, schlammhaltig,
schlammig, schmutzig, trübe ~ ground Moddergrund *m* ~ soil Schlammboden *m* ~ structure
Schlammgefüge *n*
**muff** Flanschenstück *n*, Muffe *f*, Stutzen *m*,
(glassworks) Walze *f*, Zylinder *m* ~ coupling
Kupplungsbuchse *f*
**muffing** Schalldämpfung *f*
**muffle, to** ~ einwickeln, Schall dämpfen, umhüllen
**muffle** Auspufftopf *m*, Flaschenzug *m*, Muffel *f*,
Rollkloben *m*, Schalldämpfer *m* ~ brazing
Feuerlöten *n* ~ furnace Muffelofen *m* ~-type
lehr Kanalmuffel *f*
**muffler** Auspuff-dichtung *f*, -topf *m*, Schalldämpfer *m* ~ explosion Knallen *n* im Auspufftopf
**mug** Becher *m*, Kanne *f*
**muggy** feuchtwarm
**mulch paper** Bodenpappe *f*
**mule** Maulesel *m*, Maultier *n*, Schlepper *m*, Mule-
maschine *f* (spinning), Selfaktor *m*, Traktor *m*,
Treidel-(Förder-)lokomotive *f*, Wagenspinner
*m* ~ fitter Webereimechaniker *m* ~ pulley
Riemenspannrolle *f* ~ spinner Mulespinner *m*
~-stand Riemenleiter *m* ~ track Saumpfad *m*
~-type header Eselsohrumkehrstück *n*, Rücklaufbüchse *f* mit Konsolen
**muley sawer** Blockbandsäger *m*
**mull** Mull *m*
**muller** Läufer *m*, Läufer *m* einer Erzmühle ~
mixer Eirich-Mischer *m* (Mischkneter)
**Müllerian fibers** Müllersche Streifen *pl*
**mulling,** ~ agent (leather industry) Dunstmittel
*n* ~ effect Dunstwirkung *f*
**mullion** Mittelfries *m*
**mullock** Berge *pl*
**multi,** ~-address circuit Sammelverbindung *f*
~-angular vieleckig ~-anode rectifier Mehr-
anodenventil *n* ~-arm(ed) mehrarmig ~-axial
mehrachsig ~-band antenna Allwellenantenne
*f* ~ band drive Mehrbänderbetrieb *m*
**multibank,** ~ engine Vielreihenmotor *m* ~ radial

engine Mehrreihen-, Reihen-, Viel-sternmotor
*m* ~ radial two-stroke engine Mehrreihenzwei-
taktsternmotor *m*
**multi,** ~-barrel mehrläufig ~-bay biplane Mehr-
stieler *m* ~-bayed building mehrschiffiges Gebäude *n* ~-branched mehrfach verzweigt
~-cavity magnetron Vielkammermagnetron *n*
~-cell horn Fasettentrichter *m* ~-cellular mehrzellig, multizellular, vielzellig ~-cellular horn
mehrzellige Trichter *m* (acoust.) ~ cellular
pump Differentialpumpe *f* ~ cellular voltmeter
Multizellularvoltmeter *n* ~ cellular wing viel-
zelliger Flügel *m* ~ chamber kiln Mehrkammer-
ofen *n*
**multi-channel** Mehrkanal *m*; mehrkanalig, mehrwegig ~ carrier telephone system Trägerstrom-
vielfachtelefonie *f* ~-carrier telephone system
Einrichtung *f* für Mehrfachfernsprechen mit
hochfrequenten Trägerströmen ~ data recorder
Mehrkanal-Datenschreiber *m* ~ distributor
Mehrfachverteiler *m*, mehrwegiger Verteiler *m*
~ electric pulse communication system Mehr-
kanalimpulsnachrichtenanlage *f* ~ pulse spectrometer Vielkanalimpuls-Spektrometer *m* ~
recorder Vielspur-Magnetband-Gerät *n* ~
speaker Lautsprecherkombination *f* ~ stereophonic sound reproduction Mehrkanal-Lichtton-
Stereofonie *f* ~ telegraph Mehrfachtelegraf *m*,
mehrwegiger Telegraf *m* ~ television Mehr-
kanalfernsehen *n* ~ transmission mehrwegige
Übertragung *f*, Vielkanalübertragung *f* ~
voice telegraphy Vielfachtonfrequenztelegrafie *f*
**multi,** ~-circuit switch Serienschalter *m* ~-coin
pay telephone Münzfernsprecher *m* für verschiedene Geldsorten
**multicolor** mehr-, viel-farbig ~ effect Vielfarben-
effekt *m* ~ flexographic printer Mehrfarben-Ani-
lindruckmaschine *f* ~ gravure printing Mehrfar-
bentiefdruck *m* ~ printing Mehrfarbendruck *m*
~ printing card Vielfarbendruckkarton *m* ~ recorder Mehrfarbenschreiber *m* ~ rotary press
Mehrfarbenrotationspresse *f* ~ temperature recorder Mehrfarbentemperaturschreiber *m*
**multi,** ~-corored mehr-, viel-farbig ~-component
alloy Mehrstofflegierung *f* ~-component signal
zusammengesetztes Signal *n* ~-conductor cable
mehradriges Kabel *n* ~-contact dimmer Bühnenlichtregulator *m* ~-contact switch Stufenschalter *m* ~-control Vielfachkontrolle *f* ~-
coordinate drive-field combination Mehrkoor-
dinaten-Schreibfeldkombination *f* ~-core viel-
adrig ~-core cable Mehrleiterkabel *n*, vieladriges Kabel *n*, Vielfachkabel *n* ~-core cable
end Aderendhülse *f* ~-cut lathe Mehrschneid-,
Vielschnitt-drehbank *f* ~-cyclon dust filter
Multiklonentstaubung *f* ~-cylinder Mehr-
zylinder *m* ~-cylinder engine Mehrzylinder-
maschine *f*, Vollmotor *m* ~-dimensional mehr-
dimensional ~-directional motion of liquid
mehrdimensionale Flüssigkeitsbewegung *f* ~-
disc brake Lamellenbremse *f* ~-electrode tube
Mehrelektrodenröhre *f* ~-engine mehr-, vielmotorig ~-engine plane Mehrmotorenflugzeug
*n* ~-engined airplane mehrmotoriges Flugzeug
*n* ~-flame torch Mehrfachbrenner *m* ~-flow
turbine Mehrstromturbine *f* ~ fold Vielfache *f*
~-form vieldeutig, viegestaltig ~-frequency
heterodyne generator Schwebungsnummer *f*

~-frequency transmitter Mehrfrequenzsender
*m* ~-fuel Mehrstoffbetrieb *m* ~-fuel ability Viel-
stoffeignung *f* ~-fuel engine Motor *m* für Mehrstoffbetrieb *m* ~-gap mit mehreren Zwischen-
räumen *pl* ~-grid mixing valve Mehrgitter-
mischröhre *f* ~-group model Vielgruppen-
modell *n* ~-gun tube Mehrstrahlkathoden-
strahlröhre *f* ~-headed cable machine mehr-
spulige Kabelmaschine *f* ~-hearth roaster (or
roasting furnace) mehrherdiger Röstofen *m*
~-hole nozzle Mehrlochdüse *f* ~-hop path
Zickzackweg *m* (rdo) ~-impression mold Mehrfachform *f* ~-jet burner Mehrstrahlbrenner *m*
~-lamp fire Viellampenfeuer *n* ~-lateral molding machine mehrseitige Fräsmaschine *f* ~-
lateral sound recording Vielfachzackenschrift *f*
~-lateral sound track Mehrfachzackenschrift *f*
multi-layer mehrlagig, mehrschichtig; Mehr-
schichten *pl* ~ of plastic covering Mehrschich-
ten-Kunststoffbelag *m*
multi-layer, ~ cathode (of a phototube) zusammengesetzte Fotokathode *f* ~ coil Mehrlagenspule *f*, mehrlagige Spule *f* ~ filter Mehrschichtenfilter *m* ~ glass Verbundglas *n* ~
hydraulic press Etagenpresse *f* ~ insulating
cover Mehrschichtisolierhülle *f* ~ winding
mehrlagige Wicklung *f*
multi, ~-layers Mehrfachschichten *pl* ~-length
number Zahl *f* veränderlicher Länge ~-lobe
radiation pattern mehrblättriges Strahlungsdiagramm *n* ~-lobed mehrlappig ~-loop control
system vermaschter Regelkreis *m* ~-man crew
mehrköpfige Besatzung *f* ~-manned station
mehrköpfiges Stationspersonal *n*
multimesh mehrgliedrig ~ filter mehrgliedrige
Siebkette *f* ~ filter circuit mehrgliedriges Siebgebilde *n* ~ network mehrgliedriger oder vielgliedriger Kettenleiter *m*
multi, ~-meter Multimeter *m*, Universalprüfer
*m*, Vielfachmeßgerät *n* ~-mode propagation
Vieltypenausbreitung *f* ~-motor vielmotorig
~-motor drive Mehrmotorenantrieb *m* ~-
motored mehrmotorig ~-mutube Exponentialröhre *f*, Röhre *f* mit variablem Verstärkungsfaktor, Röhre *f* mit veränderlichem Durchgriff
~ needle compass card (or rose) Mehrnadelrose
*f* ~ noded mehrknotig ~ nominal Polynom *n*
~ nominal installation Aggregat *n* ~ nozzle
motor vieldüsiger Motor *m* ~ nuclear vielkernig ~ office (exchange) Fernsprechnetz *n*
mit mehreren Vermittlungsämtern, Ortsfern-
sprechanlage *f* mit mehreren Ämtern ~ office
exchange area Netzabschnitt *m*, Ortsnetz *n* mit
mehreren Vermittlungsämtern ~ office system
Verteileramt *n* ~ pactor Multipactorleitwert *m*
~ pair vielpaarig ~ pair cable vielpaariges
Kabel *n* ~ part mehrteilig ~ particle wave function Mehrteilchenwellenfunktion *f* ~ partite
gespalten, vielteilig
multipass, ~ copying Mehrschnittkopieren *n* ~
device Mehrschnitteinrichtung *f* ~ drier Mehr-
fachbahnentrockner *m* ~ evaporator Röhren-
kesselverdampfer *m* ~ heat exchanger Viel-
stufenwärmeaustauscher *m* ~ welding Mehrlagenschweißung *f*
multi-path, ~ effect Echo *n* ~ fading Mehrwege-
schwund *m* ~ transmission mehrwegige Übertragung *f*

multi, ~-phase mehrphasig ~-phase current
Durch-, Mehrphasen-strom *m* ~-piston feed
Mehrkolbenförderung *f* ~-plane Mehr-, Viel-
decker *m*, Vielebene *f* ~-plate air filter Lamel-
lenluftfilter *n* ~-plate clutch Lamellenkupplung
*f* ~-plate friction clutch Mehrscheibentrocken-
kupplung *f* ~-platen hydraulic press hydraulische Etagenpresse *f* ~-play back Umkopieren
*n* (tape rec.)
multiple, to ~ vielfach schalten
multiple Mehrfaches *n*, Parallel-anordnung *f*,
-schaltung *f*, Vielfaches *n*, Vielfachfeld *n*, Viel-
fachstreuung *f*; mehrfach, mehrgängig, mehr-
köpfig, mehrzügig, parallel geschaltet, vielfach
(jack) ~ Klinkenfeld *n* to have ~ echoes hallen
~ of image Vielspiegelung *f*
multiple, ~ adress Mehrfach-adresse *f*, -anschrift
*f*, Sammelmeldung *f* ~ alloy steel mehrfach
legierter Stahl *m* ~ antenna Mehrfachantenne
*f* ~-arc connection gemischte Schaltung *f*
~-arch dam Gewölbereihensperre *f* ~ arrangement Nebeneinanderschalten *n* ~-axle drive
Mehrachs-, Mehrfach-antrieb *m* ~ balloon
cable Seilschürze *f* ~-band receiver Allwellenempfänger *m* ~ bank contact Vielfachkontakt-
feld *n* ~ beads Mehrlagenraupen *pl* (Schweißung) ~-beam interferometry Vielstrahl-Inter-
ferometer *n* ~ belt mehrfacher Riemen *m*
~-belt Flächendrahthindernis *n* ~-belt drive
Mehrbänderbetrieb *m* ~-blade drags Mehrfach-
planierschleppe *f* ~-blade saw frame Bund-,
Voll-gatter *n* ~ blow test Mehrfachschlagprobe
*f* ~ branching Vielfachverzweigung *f* ~ burner
mehrflammiger Brenner *m* ~ cable System-,
Vielfachfeld-, Vielfach-kabel *n*, vieladriges
Kabel *n* ~ cabling Vielfachverkabelung *f*
~-call indicator Reihenrufanzeiger *m* ~ camera
Mehrfach-, Verbund-, Vielfach-kamera *f* ~-
cell flotation machine Mehrzellenflotator *m*
~ charge gestreckte Ladung *f* (explosives) ~
charged mehrfach geladen ~-clay conduit
mehrzügiges Tonformstück *n* ~ coatings
Mehrfachüberzüge *pl* ~ collisions Vielfachstöße
*pl* ~-color dotted-line recorder Mehrfarben-
punktschreiber *m* ~ combinations for flush
mounting Mehrfachunterputz-Kombinationen
*pl* ~ condenser Kondensatorwanne *f*, Mehr-
gang-, Mehrfach-kondensator *m* ~-conductor
cable (or lead) mehradriges Kabel *n* ~-cone
classifier Luttenapparat *m* ~-connected vielfachgeschaltet ~ connection Mehrfach-, Neben-
einander-, Parallel-, Verzweigungs-schaltung *f*,
Vielfach-anschluß *m*, -schaltung *f* ~ connection
of cable wires Multiplexverteilung *f* von Kabeladern ~ connection box Vielfachanschlußdose
*f* ~ contact Mehrfach-, Vielfach-kontakt *m* ~
contact plug Mehrfachstecker *m* ~ container
Blockkasten *m* ~ converter Mehrfachstrom-
richter *m* ~-cornered lathe Mehrkantblock-
drehbank *f* ~ crosstalk Mehrfachnebensprechen
*n* ~ current generator Mehrstromgenerator *m*
~ cutter Mehrschneider *m* ~-cutter lathe Viel-
stahldrehbank *f* ~ decay Aufzweigung *f*,
Dualzerfall *m* ~ delay discriminator Vielfach-
verzögerungsdiskriminator *m* ~ die Mehr-
fachgesenk *n* ~-die plunger press Mehrstempel-
presse *f* ~-die press Stufenpresse *f* ~-die
punching Vielstempelstanzen *pl* ~-disk brake

Lamellen-, Mehrfachscheiben-bremse f ~-disk clutch Mehrscheibenkupplung f ~-disk steel clutch Stahllamellenkupplung f ~ distribution box Vielfachdose f ~-domed vielkuppelig ~ double-edged variable-width track Vieldoppelzackenschrift f ~-drill press Gelenkspindelbohrmaschine f ~ drilling machine Vielspindelbohrmaschine f ~ duct (cable) Mehrlochkanal m ~ duct concrete block mehrzügiges Zementformstück n ~-duct conduit Kabelkanal m mit mehreren Einzelrohrzügen, Mehrlochkanal m, mehrzügige Rohrpost f ~ earthing point Erdungsschleife f ~ echo Mehrfach-, Vielfachecho n ~-effect evaporator Mehrfachverdampfungs-, Mehrkörperverdampf-, Vielkörperverdampf-apparat m ~-electrode furnace Mehrelektrodenofen m ~-electrode spot weld Vielfachpunktschweißung f ~ exploitation (of el.circuits) Mehrfachausnutzung f ~ exposure Mehrfachbelichtung f ~ face joint Vielflächenverbindung f ~ feed Mehrfachanlage f ~-feed nozzle Mehrfachmengendüse f ~ field Vielfachfeld n ~-filament lamp mehrfädige Lampe f ~-flight screw conveyor Mehrfachschneckenförderer m ~ form Mehrfachformular n ~ form cutter Gewinderillenfräser m ~ fuse Mehrfach-, Mehr-zünder m ~-graph Mehrfachschaubild n ~ grating Beugungsgitter n (acoust.) ~-grid tube Mehrgitterrohr n ~-grid valve Mehrgitterröhre f ~ gyro-inclinometer Mehrfachkreiselneigungsmesser m ~-hearth roaster Plattenofen m ~ image Mehrfachbild n (TV) ~ inlet head Einlaßkopf m ~ integration Mehrfachintegration f ~ interlacing Mehrfachzeilensprung m ~ ionization Mehrfachionisation f ~ jack Verbindungs-, Vielfach-klinke f ~-jack field Vielfachklinkenfeld n ~-jet carburetor Mehrdüsenvergaser m ~-jet gear Mehrfachdüsensatz m ~ joint Kabelmuffenverzweigung f ~ junction Knotenpunktbahnhof m ~ labelling mehrfache Markierung f ~ lead cable joint Verteilbleimuffe f ~ lens camera Mehrfach-, Viellinsen-kamera f ~-lens combination Mehrlinsenobjektiv n, mehrteiliges Objektiv n ~ lens system mehrgliedriges Linsensystem n ~-lever cash register Mehrzählerhebelkontrollkasse f ~ line printing Mehrzeilenschreibung f ~ line read selection Mehrzeilenabfühlsteuerung f ~ manometer Mehrfach-druckmesser m, -manometer m ~ marking Mehrfachmarkierung f ~ metering Mehrfachzählung f ~ modulation Mehrfachmodulation f ~-modulation beacon mehrstrahliges Funkfeuer n ~-needle telegraph Mehrnadeltelegraf m ~ nozzle Vielfachdüse f ~ nozzle carburetor Registervergaser m ~ operation Mehrfach-, Vielfach-betrieb m ~-operator welding machine Mehrfachschweißmaschine f ~ outlines along edges of objects Vielfachkonturen pl ~ paddle agitator Mehrfachbalkenrührer m ~ parallel winding mehrgängige Parallelwicklung f ~-part mehrteilig ~-part flask mehrteiliger Formkasten m ~-part shaft Gelenkwelle f ~-pass drier Wandertrockner m ~ peg Hinweisstöpsel m ~ photogrammetric camera Mehrfachmeßkamera f ~-plate brake Lamellenbremse f ~ plug Vielfachstecker m ~-ply belt mehrfacher Riemen m

~-point switch mehrteiliger Schalter m, Rastenschalter m ~ pole switch Vielfachschalter m ~ positioning switch Meßstellenumschalter m ~ prints Mehrfachkopien pl ~ production Reihen-, Serien-herstellung f ~ pump Mehrfachpumpe f ~ punch Mehrfachstanze f ~-punching attachment Mehrfachstanzeinrichtung f, Mehrstempellochapparat m ~ purpose Mehrzweck m ~-purpose plane Mehrzweckflugzeug n ~ purpose pliers Kombi-, Mehrfach-zange f ~ purpose set Mehrzweckstudio n (film) ~-purpose tester Universalprüfer m ~ receiver connection to antenna Antennenringleitung f ~ rectifier circuit Mehrfach-Gleichrichterschaltung f ~ rectilinear punching machine Vielstempelreihenlochmaschine f ~ reel machine Mehrfachrollenherstellungsmaschine f ~ reflection Flatterwiderhall m ~-reflection echo Umwegecho n (rdr) ~ reflection path Zickzackweg m ~ regeneration Mehrfachrückkopplung f ~ registration Mehrfachzählung f ~-request-line apparatus Mehrfachanschlußapparat m (teleph.) ~ resonant circuit Drossel-, Parallelresonanz-, Resonanz-kreis m (Parallelresonanz) ~ retarder Vielfachverzögerer m ~-row mehrreihig ~ scanning Mehrkanalmethode f, Vielfachabtastung f ~ scattering Mehrfach-, Vielfach-streuung f ~-seat valve Stufenventil n ~ seizure Mehrfachbelegung f ~-series camera Mehrfachreihen-bildner m, -kamera f ~-shear riveting mehrschnittige Nietung f ~ signal Mehrfachzeichen n ~ slip Mehrfachgleitung f ~ slotter Mehrfachschlitzmaschine f ~ sound track Mehrfachtonspur f ~ span mehrfeldrig ~ spark discharger Vielfachfunkenstrecke f ~ spark gap Mehrfach-, Reihen-, Serien-funkenstrecke f, unterteilte Funkenstrecke f ~ spectral term Mehrfachterm m ~-speed windshield wiper Mehrgangwischer m

multiple-spindle, ~ automatic chucking machine Mehrspindelautomat m ~ bar-type automatic lathe Vielspindelstangenautomat m ~ chucking-type automatic lathe Vielspindelfutterautomat m ~ drill Mehrspindelbohrmaschine f ~ fourway precision boring machine Vierwegemehrspindel-Feinbohrmaschine f ~ head Vielspindelkopf m

multiple, ~ spinning nozzle Spinnbrause b ~ spline shafts Vielkeilwellen pl ~-splined driving shaft Antriebskeilwelle f ~ splining Mehrfachnutung f ~ spooling machine Fachmaschine f ~-stage mehrstufig ~-stage boiler Etagenkessel m ~-stage cementing stufenweise Zementierung m ~-stage fan mehrstufiger Ventilator m ~ stage tests Stufenversuche pl ~-stand rolling mill mehrgerüstiges Walzwerk n ~ step device Anschlagkreuz n ~ step pulley Stufenscheibe f ~ storied mehrstöckig ~-story frame Stockwerkrahmen m ~-story furnace Mehretagenofen m ~-strand chain mehrsträngige Kette f ~ supension Vielfachaufhängung f ~ switch Mehrfachschalter m, Umschalter m ~ switchboard Vielfach-schrank m, -umschalter m ~ tariff meter Mehrfachtarifzähler f ~ telephony Mehrfachfernsprechen n ~ thread mehrgängiges Gewinde n ~ thread milling cutters Gewinderillenfräser m ~-threaded (screw)

mehrgängig **~-throw crankshaft** mehrfach gekröpfte Kurbelwelle *f* **~-throw switch** Kellog-
schalter *m* **~ tile** mehrzügiges Tonformstück *n*
oder Formstück *n* **~-tile conduit** mehrzügiges
Tonformstück *n* **~ tile duct** (cable laying) mehrzügiges Formstück *n* **~-tone transmitter** Viel-
tonsender *m* **~-tool attachment** Umschlag-
werkzeug *n* **~-tool carrier** Mehrstahlsupport *m*
**~-tool operation** Mehrstahlarbeit *f* **~-track**
mehrgleisig **~ transmission** Mehrfachsenden *n*,
Vielfachübermittlung *f* **~ transmission on lines**
Mehrfachbetrieb *m* auf Leitungen (teleph.) **~**
**tripping** mehrfaches Anschlagdrehen *n* **~ tube**
Mehrfachröhre *f* **~-tuned antenna** mehrfach
abgestimmte Antenne *f* **~-tuned manometer**
Vielfachmanometer *n* **~ tuner** Vielfachabstimm-
gerät *n*, -vorrichtung *f*, vielfache Abstimmvorrichtung *f* **~ turnbuckle** Sammelspannschloß *n*
**~ turning** Mehrfachdrehverfahren *n* **~ twin**
**cable** Dieselhorst Martin-Kabel *n*, Mehrfach-
zwillingskabel *n*, Vielfachzwillingskabel *n*, vielpaariges Kabel *n*, Viererkabel *n*, viererverseiltes
Kabel *n* **~-twin formation** Dieselhorst Martin-
Verfeilung *f*, Vielfachzwillingsverseilung *f* **~**
**twin quad** D. M.-Vierer *m* **~ twinning** Wiederholungszwillinge *pl* (geol.) **~-type typewriter**
Vieltypenschreibmaschine *f*

**multiple-unit, ~ condenser** Abzweig-, Mehrfachkondensator *m* **~ control** Vielfachsteuerung *f*,
Zugsteuerung *f* **~ steerable antenna** Antenne *f*
mit schwenkbarer Charakteristik, Musa-
Antenne *f* **~ steerable antenna reception** Musa-
Empfang *m* **~ train** Triebwagen-, Trieb-zug *m*
**~ tube (or valve)** Verbundröhre *f*

**multiple, ~ V gear** Keilrädergetriebe *n* **~ valve**
Mehrfach-, Verbund-röhre *f* **~-way duct** Kabelformstück *n*, mehrzügiger Kanal *m* **~-way**
**stopcock** Vielwegehahn *m* **~-way switch** mehrwegiger Umschalter *m* **~-way telegraph** Mehrfachtelegraf *m* **~ wheel landing gear** Mehrfach-
fahrwerk *n* **~-wire** mehrdrähtig **~-wire antenna**
mehrdrähtige Antenne *f* **~ wires** Vielfachverdrahtung *f* **~-word acess system** Entmagneti-
sierungsfaktor *m*

**multipled** vielfachgeschaltet

**multiplet** Linienkomplex *m*, Multiplette *f*, zusammengesetzte Linie *f*

**multiplex, to ~** in Mehrfachschaltung betrieben,
mehrere Signale gleichzeitig senden (über Draht
oder Funk)

**multiplex** Multiplexverbindung *f* **~ code trans-**
**mission** Multiplexübertrag *f* von kodierten
Sprüchen **~ distributor** Mehrfachverteiler *m*
**~ lap** mehrgängige Parallelwicklung *f* **~ signal**
**system** Mehrfachnachrichtensystem *n* **~ tele-**
**graph** Mehrfachtelegraf *m* **~ telegraphy** Mehr-
fachtelegrafie *f* mit Verteilern, Vielfachtele-
grafie *f* **~ transmission** Vielfachverkehr *m* **~**
**working** Multiplexbetrieb *m*

**multiplexer** Mehrfachkoppler *m*, Mehrwegschalter *m*

**multiplexing equipment** Mehrfachanschlüsse *pl*

**multiplicand** Multiplikand *m* **~ register** (MD)
Multiplikant-Divisorregister *n* (info proc.)

**multiplication** (gears) Getriebeübersetzung *f*,
Multiplication *f*, Vervielfachung *f* **~ of charge**
Ladungsvervielfachung *f* **~ of power** Kraftver-
vielfachung *f*

**multiplication, ~ constant** Multiplizierungsfak-
tor *m* **~ factor** Vervielfältigungsfaktor *m* **~**
**sign** Mal-, Multiplikations-zeichen *n*

**multiplicative, ~ constant** Multiplikationskon-
stante *f* **~ mixture** multiplikative Mischung *f*

**multiplicity** Mannigfaltigkeit *f*, Mehrheit *f*,
Mehrzahl *f*, Vielfalt *m*, Vielfältigkeit *f*, Zählig-
keit *f* **~ factor** Formfactor *m*

**multiplier** Faktor *m*, Multiplikator *m*, Multipli-
ziereinrichtung *f* (data proc.), Verstärker *m*,
Vervielfacher *m*, Voltmeterverschaltwiderstand
*m* **~ coil** Multiplikatorrahmen *m* **~ photocell**
Fotozelle *f* mit Sekundärelektronenvervielfacher **~ photoelectric cell** Vervielfacher-
fotozelle *f* **~ quotientsregister** Multiplikator-
Quotientregister *n* **~ register** Multiplikator-
register *n* (info proc.) **~ tube** Vervielfältigungs-
röhre *f*

**multipling** Vielfachschaltung *f*

**multiply, to ~** multiplizieren, vermehren, vervielfachen, vervielfältigen, verstärken, vielfach schalten, vielschichtig **to ~ through** erweitern

**multiplying on the expression (or quantity) in**
**brackets** Ausmultiplizieren *n* der Klammer

**multiplying, ~ gauge** Feindruckmesser *m* **~ lever**
**extensometer** Rollenapparat *m* **~ manometer**
Feindruckmesser *m* **~ power** (of a galvanometer
shunt) Erweiterungs-, Empfindlichkeits-faktor
*m* **~ stage** Vervielfachungsstufe *f* **~ voltages**
Spannungsvervielfachung *f*

**multipoint, ~ gas water heater** Gasdurchlauf-
erhitzer *m* **~ ignition** Mehrfunkenzündung *f* **~**
**plug** Doppelstecker *m* **~ recorder** Mehrpunkt-
schreiber *m* **~ switch** mehrwegiger Schalter *m*,
Stufenschalter *m*, vielstufiger Schalter *m* **~**
**tap-switch** Vielfachschalter *m*

**multipolar** mehrpolig, vielpolig **~ switch** mit getrennten Polen *pl*

**multipolarity** Multipol-ordnung *f*, -strahlung *f*

**multipole, ~ expansion** Multipolentwicklung *f* **~**
**transition probabilities** Multipolstrahlungsüber-
gänge *pl*

**multiport injector** Mehrlocheinspritzdüse *f*

**multiposition, ~ propeller** Einstellschraube *f* **~**
**switch** Vielstellenschalter *m* **~ switchboard**
Mehrplatzvermittlung *f*

**multiprong plug socket case** Kastenhalterung *f*
(g/m)

**multipurpose, ~ appliance** Vielfachgerät *n* **~ com-**
**puter** Mehrzweckrechner *m* **~ patrol plane**
Mehrzweck-Seeflugzeug *n* **~ tube** Mehrfach-,
Verbund-röhre *f*

**multi, ~-range ammeter** umschaltbares Amperemeter *n* **~-range receiver** Allwellenempfänger
*m* **~-rate meter** Mehrtarifzähler *m* **~-reduction**
**unit drive** mehrstufiges Blockgetriebe *n* **~-rib**
**reinforcing bars** Rippenstahl *m* **~-roller** Multi-
roller *m* **~-rotation** Mehrdrehung *f* **~-run**
**weld** Mehrlagenschweißung *f* **~-screw extrusion**
Mehrschnecken-Preßverfahren *n* **~-seated valve**
Etagenventil *b* **~-seater** Mehrsitzer *m*, mehr-
sitziges Flugzeug *n* **~ seating control** Mehrkan-
tensteuerung *f* **~ section** mehrgliedrig **~ section**
**filter** mehrgliedrige Siebkette *f* **~ section net-**
**work** mehrgliedriger oder vielgliedriger Kettenleiter *m* **~ section tube** Mehrfach-, Verbund-
röhre *f* **~ sectional** mehrteilig **~-segment**

**magnetron** Vielschlitzmagnetron $n$ ~ **shape key plug switchboard** Steckschlüsselschrank $m$ ~-**sheet filter** Mehrschichtenfilter $n$ ~ **shell condenser** Elementenverflüssiger $m$ ~ **(way) socket and plug device** Steckersäule $f$

**multi-spar** mehrholmig ~ **construction** mehrholmige Bauform $f$ ~ **wing** mehrholmiger Flügel $m$, Vielholmflügel $m$

**multi,** ~-**speed motor** Motor $m$ mit (mehreren) Drehzahlstufen ~-**speed transmission** Mehrganggetriebe $n$ ~-**spiral scanning disk** Mehrfachspirallochscheibe $f$ ~-**spline shaft** Sternkeilwelle $f$ ~ **splined fit** Mehrkeilsitz $m$ ~ **splined shaft** Mehrkeilwelle $f$ ~ **spot-welding machine** Vielpunktschweißmaschine $f$

**multi-stage** mehrstufig, vielstufig ~ **amplification** mehrstufige Verstärkung $f$ ~ **amplifier** Kaskaden-, Mehrfach-verstärker $m$, mehrstufiger Verstärker $m$ ~ **blower** mehrstufiges Gebläse $n$ ~ **centrifugal pump** vielzellige Zentrifugalpumpe $f$ ~ **copying** Folgekopieren $n$ ~ **diffusion unit** mehrstufiges Diffusionsgerät $n$ ~ **fatigue test** Mehrstufendauerschwingversuch $m$ ~ **method** Folgekopierverfahren $n$ ~ **pump** Differentialpumpe $f$

**multi,** ~-**step action** Mehrstellenwirkung $f$ ~ **storied** mehrstöckig ~-**storied building** Hochhaus $n$ ~-**strand** mehrlitzig ~-**strand cable** Mehrfachkabel $n$ ~-**strut** mehrstielig ~-**tester** Multimeter $m$, Vielfachmeß- oder Prüfgerät $n$ ~-**tone** Heulton $m$ ~-**tone signal** Mehrklangsignal $n$ ~-**toolholder** Mehrfachstahlhalter $m$

**multi-tubular,** ~ **boiler** Vielröhrenkessel $m$ ~ **boiler with removable firebox and smoke tubes** ausziehbarer Flammrohrsiederohrkessel $m$

**multitude** Menge $f$

**multi,** ~-**unit machine** Mehreinheitenrechner $m$ ~-**use steel rule** Mehrzweckstahllineal $n$ ~ **valent** mehr-, viel-wertig ~-**valued displacement** mehrdeutige Verrückung $f$ ~ **valve amplifier** Mehrröhrenverstärker $m$ ~ **valve arrangement** Mehrventilanordnung $f$ ~ **vector** Multivektor $m$ ~ **vibrator** Multivibrator $m$, Vielfachschwingungserzeuger $m$, Zungenfrequenzmesser $m$ ~ **wave property** Mehrwelligkeit $f$

**multi-way** mehrwegig ~ **circuit** Zeitungsleitung $f$ mit mehreren Empfangsstellen ~ **switch** Mehrwegumschalter $m$, Vielfachschalter $m$

**multi-wire** mehradrig ~ **cable** mehradriges Kabel $n$ ~ **counter** Vieldrahtzählrohr $n$ ~-**triatic antenna** Dreieckanfenne $f$

**municipal** städtisch ~ **airport** Stadtflughafen $m$, städtischer Flughafen $m$ ~ **area** Weichbild $n$ ~ **authorities** Ortsbehörde $f$ ~ **district** Weichbild $n$ ~ **planning** Stadtplanung $f$ ~ **public works** städtische Werke $pl$ ~ **undertaking** kommunaler Betrieb $m$

**municipality** Gemeinde $f$

**muon,** ~ **capture** Myoneneinfang $m$ ~ **pair production** Myonenpaarerzeugung $f$

**mural** mauer-, wand-artig, steil ~ **decoration** Wanddekoration $f$

**muriate** Chlorstrontium $n$

**muriatic,** ~ **acid** Salzsäure $f$ ~ **ether** Salzäther $m$

**muricalcite** Bitter-kalk $m$, -spat $m$

**murkiness** Lufttrübung $f$, Trübung $f$

**murky** trübe

**murmur, to** ~ rauschen

**Murray telegraph** Murray's Reihentelegraf $m$

**musa antenna** Mehrfachrautenantenne $f$

**muscle** Muskel $m$ ~ **diaphragm** Stellmembran $f$ ~ **power** Muskelkraft $f$ ~-**power flight** Muskelflug $m$ ~-**power plane** Muskelkraftflugzeug $n$

**muscovite** Kaliglimmer $m$, Muskovit $m$, weißer Glimmer $m$

**mush area** Interferenzgebiet $n$, Störgebiet $n$

**mushing** Fliegen $n$ mit halbüberzogenem Flugzustand

**mushroom** Pilz $m$, pilzförmig ~ **anchor** Pilzanker $m$ ~ **contact** Pilzkontakt $m$ ~ **control valve** Tellersteuerventil $n$ ~ **cowl** Windhutze $f$ mit Rundeinlaß ~ **form** Muschelanordnung $f$ ~ **head** (of breech) Verschlußkopf $m$ ~ **head** (obturator) Liderungskopf $m$, Pilz $m$ ~ **insulator** Pilzisolator $m$ ~ **loud-speaker** Pilzlautsprecher $m$ ~ **mixer** schrägstehender pilzförmiger Mischer $m$ ~ **parachute** Pilzfallschirm $m$ ~ **piston** Pilzkolben $m$ ~ **rock** Pilzfelsen $m$ ~ **strainer** (pump) Saugkorb $m$, Seiherkasten $m$ ~ **tappet** Pilzstößel $m$ ~-**type air filter** Pilz-Luftfilter $m$ ~-**type follower** Pilzstößel $m$ ~-**type valve** Forter-Ventil $n$, Pilz-, Teller-ventil $n$, Verstellpilz $m$ ~ **valve** Kegelventil $n$, Muschelschieber $m$, pilzartiges Sicherheitsventil $n$, pilzförmiges Ventil $n$, Ringventil $n$

**mushy** blasig ~ **controls** weiche Steuerruder $pl$

**music** Musik $f$ ~ **circuit** Musikleitung $f$ ~ **engraver** Notenstecher $m$ ~ **pilot** Musiklotse $m$ ~ **printing** Notendruck $m$ ~ **ruling pen** Rastral $n$ ~ **stand** Notenständer $m$ ~ **wire** Drahtsaite $f$, Klavierdraht $m$

**musical** musikalisch ~ **automaton** Musikwert $m$ ~ **chimes** Glockenspiel $n$ ~ **frequency** Tonfrequenz $f$ ~ **generator** Tongenerator $m$ ~ **instrument** Musikinstrument $n$ ~ **instrument with direct percussion** Musikinstrument $n$ mit direktem Schlag ~ **note** musikalischer Ton $m$ ~ (**spark**) **signal** tönendes Signal $n$ ~ **sound** Klang $m$ ~ **spark** tönender Funke(n) $m$ ~ **spark gap** tönende Funkenstrecke $f$ ~-**spark transmitter** tönender Funkensender $m$, tönender Sender $m$, Tonfunksender $m$ ~ **tone** musikalischer Ton $m$

**musk** Moschus $m$

**musket** Flinte $f$

**musketry** Schießwesen $n$ ~ **manual** Schießvorschrift $f$ ~ **training** Schießausbildung $f$

**muslin bandage** Mullbinde $f$

**musty** dumpfig, muffig

**mutant** Mutant $m$ ~ **gene** mutierendes Gen $n$

**mutarotation** Umkehrung $f$ oder Wechsel $m$ des Drehungsvermögens

**mutation,** ~ **of energy** Energieumfang $m$ ~ **stop** (organ) Kombinationsstimme $f$, gemischte Stimme $f$ ~ **stops** (organ) gemischte Stimme $f$

**mutator** Ignitronstromtorkombination $f$, Stromrichter $m$, Umwandler $m$, Wechselrichter $m$

**mute, to** ~ (musical instrument) dämpfen (sound) schwächen

**mute** (musical instrument) Dämpfer $m$, Sourdine $f$ ~ **antenna** Antennenersatzstromkreis $m$, Ersatzantenne $f$, künstliche oder verstimmte Antenne $f$

**mutilate, to** ~ beschädigen, entstellen, verstümmeln

**mutilated selection** unvollständige Wahl $f$

**multilation** Beschädigung *f*, Entstellung *f*, Verstümmelung *f* ~ **of speech** Abschneiden *n* von Silben und Worten ~ **of syllables** Silbenabschneidung *f* ~ **of words** Wortabschneidung *f* (teleph.)

**muting** Dämpfung *f* ~ **switch** Schalter *m* für Stummabstimmung

**mutoscope** Mutoskop *n*

**mutual** gegenseitig, gemeinschaftlich, wechselseitig ~ **action** gegenseitige Beeinflussung *f*, Wechselwirkung *f* ~ **capacitance** Gegenkapazität *f* ~ **capacitance of a pair of a phantom** Betriebskapazität *f* einer Doppel- oder Viererleitung ~ **capacity** Betriebskapazität *f*, Gegenkapazität *f*, gegenseitige Kapazität *f* ~ **capacity of a pair of a phantom** Betriebskapazität *f* einer Doppel- oder Viererleitung ~ **characteristic** Charakteristik *f* der Röhre, Transimpedanz *f* ~ **conductance** gegenseitige Leitfähigkeit *f*, (tube) Steilheit *f*, Transponierungssteilheit ~ **conductance of tube** Röhrensteilheit *f* ~ **debt** Gegenschuld *f* ~ **displacement of the key pair** Versatz *m* der Keilpaare ~ **impedance** gegenseitige Impedanz *f* ~ **inductance** Wechselinduktion *f* ~-**inductance coupling** induktive Kopplung *f* ~ **induction** Gegeninduktion *f*, Gegeninduktivität *f*, gegenseitige Induktion *f* ~ **induction bridge** gegeninduktive Brücke *f* ~ **inductive coupling** gegen-induktive Kopplung *f* ~ **inductivity** Gegeninduktion *f* ~ **influence** gegenseitige Einwirkung *f* ~ **interaction** Wechselwirkung *f* ~ **interference** gegenseitige Beeinflussung *f* ~-**potential energy** Wechselwirkungsenergie *f* ~ **relation(ship)** Wechselbeziehung *f* ~ **slope** (of the characteristic curve

of a valve) Betriebssteilheit *f*, Umwandlungssteilheit *f* (electron.) ~ **support** Zusammenarbeit *f*

**mutually** beiderseits, gegenseitig **to cancel** ~ (math., physics, electron.) sich gegenseitig aufheben ~ **circumscribed quadrangles** wechselweitig umschriebene Vierecke *pl* ~ **exclusive** gegenseitig ausschließen ~ **perpendicular** zweiseitig senkrecht

**mu-tube (variable)** Röhre *f* mit veränderlichem Durchgriff

**muzzle** Mund *m*, Mündung *f*, Rohrmündung *f* ~ **of charging valve** Füllventilsitz *m* ~ **and sight cover** (rifle) Mündungsschoner *m*

**muzzle,** ~ **blast** Mündungsknall *m* ~-**blast smoce** Mündungsrauch *m* ~ **brake** Mündungsbremse *f* ~ **cap (or cover)** Mündungskappe *f* ~ **energy** Mündungs-energie *f*, -wucht *f* ~ **flash** Mündungsfeuer *n* ~ **gas pressure** Mündungsgasdruck *m* ~ **guide** Rohrklaue *f* (artil.) ~ **hoop** Schnellband *n* ~ **loader** Vorderlader *m* ~ **plug** Mündungsschoner *m* ~ **sound** Mündungsknall *m* ~ **velocity** Anfangsgeschwindigkeit *f* des Geschosses, (of guns) Austrittsgeschwindigkeit *f*, Geschoßanfangsgeschwindigkeit *f*, Grundstufe, Mündungsgeschwindigkeit *f*

**mycalex** Mikalex *n*

**mylonite** Mylonit *n*

**myopia** Kurzsichtigkeit *f*, Myopie *f*

**myopic** kurzsichtig ~ **eye** kurzsichtiges oder nahsichtiges Auge *n*

**myriametric waves** Längst-, Myriameter-wellen *pl* (VLF)

**myriare** Quadratkilometer

**myristic acid** Myristinsäure *f*

# N

**nab** Schließblech *n*

**nacelle** Ballonkorb *m*, Flugzeugrumpf *m*, Gondel *f*, Luftschiffgondel *f* ~ **swivelling device** Gondelschwenkvorrichtung *f*

**nacreous paper** Perlmutterpapier *n*

**n-address electronic computer** n-Adressen-Rechner *m*

**nadir** Fußpunkt *m*, Nadir *m* ~ **of the image** Bildlotpunkt *m*, Bildnadir *m*

**nadir,** ~ **angle** Nadirwinkel *m* ~ **distance** Nadirdistanz *f* ~ **plumbing** Nadirlotung *f* ~ **ward** nadirwärts

**nagelfluh** Nagelfluh *f* (geol.) ~ **quarry** Nagelfluhsteingrube *f*

**nagyagite** Blätter-erz *n*, -tellur *m*, Nagyagit *m*, Tellurglanz *m*

**nail, to** ~ annageln, benageln, festnageln, nageln, vernageln **to** ~ **a lock** (to a door) ein Schloß *n* anschlagen

**nail** Formerstift *m*, Nagel *m*, Spieker *m*, Stift *m* ~ **bore** Nageldocke *f* ~ **catcher** Nagelfänger *m* ~ **claw (or drawer)** Kuhfuß *m* ~ **hammer** Klauenhammer *m* ~ **head** Nagelkopf *m* ~ **iron** Nageleisen *n* ~ **maker** Nagelschmied *m* ~ **mold** Nageldocke *f* ~ **pass** Nagelkaliber *n* ~**-polish** Nagellack *m* ~ **puller** Nagel-heber *m*, -zange *f*, -zieher *m* ~ **rod** Nageleisen *n* ~ **set** Durchschlag *m*, Handdurchschläger *m* ~ **smith** Nagelschmied *m* ~ **test** (for blasting caps) Nagelprobe *f* ~ **tool** Nageldocke *f*

**nailable** nagelbar

**nailer** Nagelschmied *m*

**nailing** Nageln *n*, Nagelung *f*, Spikerung *f*

**nails,** ~ **with convex heads** Gurtstifte *pl* ~ **with large heads** Breitkopfstifte *pl*

**naked** blank, bloß, entblößt, nackt **with the** ~ **eye** frei mit dem Auge *n* ~ **eye** unbewaffnetes Auge *n* ~ **wire** blanker Draht *m*

**name, to** ~ anführen, bezeichnen, erwähnen, nennen

**name** Bezeichnung *f*, Name *m*, Titel *m* ~ **plate** Motor-, Namen-, Typen-schild *n* ~ **plate on plaited wirework** Drahtluftschild *n*

**naming** Benennung *f*

**nap, to** ~ aufrauhen, (fabrics) rauhen

**nap** (textiles) Flor *m*, Haardecke *f*, Pole *f* ~**-lifting apparatus** Velourhebeapparat *m* ~ **pattern** Noppenmuster *n* ~ **warp** Oberkette *f*, (textiles) Polkette *f*

**nape** Genick *n*, Nacken *m*

**naphta** Benzin *n*, Erdöl *n*, Kreoson *n*, Leucht-öl *n*, -petroleum *n*, Naphta *n*, Schwerbenzin *n*, Steinmetzöl *n* ~ **pitch** Naphtapech *n* ~ **solution** Benzinlösung *f*

**naphtacene** Naphthazen *n*

**naphthalene** Naphthalin *n* ~ **series** Naphthalinreihe *f*

**naphthalenesulphonic acid** Naphthalinsulfosäure *f*

**naphthene** Naphthen *n*

**naphthenic** naphtenisch ~ **soap** Naphtenseife *f*

**naphthol** Naphthol *n*, Oxynaphthalin *n* ~ **feeding addition** Naphtolnachsatz *m* ~ **green** Naphtholgrün *n* ~ **yellow** Naphtholgelb *n*

**naphtholate printing process** Naphtolatdruckverfahren *n*

**naphtholize, to** ~ naphtholieren

**naphthyl** Naphthyl *n*

**naphthylamine** Naphthylamin *n* ~ **hydrochloride** Naphthylaminchlorhydrat *n*

**naphtolating** Grundieren *n* (Naphtholieren)

**Napierian,** ~ **logarithm** Napierscher Logarithmus *m* ~ **logarithms** hyperbolische Logarithmen *pl*

**nappe** Decke *f*, Mantel *m*, Strahl *m*, Überfall *m*

**napped** (fabrics) gerauht

**narrow, to** ~ begrenzen, beschränken, einengen, einschränken, verengen, verjüngen, vermindern, verringern

**narrow** Enge *f*, Engpaß *m*; beschränkt, eng, knapp, räumlich schmal **with a** ~ **internal diameter** englumig

**narrow,** ~ **-angle projector** Normalwinkelprojektor *m* ~**-beam absorption** Kleinfeldabsorption *f* ~ **bit** Schmalmeißel *m* ~ **channel** Kille *f*, Meerenge *f* ~ **construction** Schmalbauart *f* (Batterie) ~ **dismension** nichtkritische Abmessung *f*, schmale Seite *f* ~ **edge** Schmalkante *f* ~ **film** Kleinfilm *m*, Schmalfilm *m* ~ **furniture** Schmalschließstege *pl* ~ **gauge** Schmalspur *f*; schmalspurig ~ **gauge field railroad** Förderbahn *f* **gauge field railway** Feldbahn *f* ~ **gauge railroad** Feldeisenbahn *f*, Kleinbahn *f*, Schmalspurbahn *f* ~ **gauge railroad car** Schmalspurwagen *m* ~ **gauge track** Schmalspurgleis *n* ~ **guide** Schmalführung *f* ~ **inlet of horn** Trichtermundstück *n* ~**-meshed** engmaschig ~**-minded** engherzig ~**-mouthed** enghalsig ~**-mouthed bottle** Enghalsflasche *f* ~**-mouthed shovel** schmale Schaufel *f* ~**-necked** enghalsig ~ **pass** Engpaß *m*, Hohlweg *m* ~ **passage** Durchschlupf *m* ~ **pitch cut** Tiefschnitt *m* ~ **place** Pfeilerdurchhieb *m* ~ **shower** schmaler Schauer *m* ~ **side** Schmalkante *f*, Schmalseite *f* ~**-spaced** engspaltig ~**-test flume** Cuvette *f*, Küvette *f* ~**-track seeder** schmalspurige Säemaschine *f*

**narrowed** verengt ~ **goods** geminderte Ware *f*

**narrowing** Einengung *f*, Minderung *f*, Verschmälerung *f* ~ **of a river** Verengung *f* eines Flusses

**narrowing** ~ **chain** Deckkette *f* ~ **down** Verengung *f* ~ **machine** Mindermaschine *f* ~ **the tube** Rohrverengung *f*

**narrowly defined electron beam** sehr fein ausgeblendeter Elektronenstrahl *m*

**narrowness** Enge *f*, Schmalheit *f*

**narrows** Enge *f*, Meeresenge *f*

**nasal,** ~ **consonant** Nasalkonsonant *m* ~ **speculum** Nasenspiegel *m*

**nascent** im Entstehen *n* begriffen, freiwerdend, naszierend ~ **state** Entstehungszustand *m*

**nasturan** Pechblende *f*

**nasty weather** schlechtes Wetter *n*

**nation-wide dialling network** Landesfernwahlnetz n

**national,** ~ **grid** Energiennetz n ~ **guard frequency** nationale Wachfrequenz f ~ **insignia** Nationalitätenzeichen n ~ **park** Naturschutzgebiet n ~ **prosperity** Volkswohlstand m

**nationality** Staatsangehörigkeit f, Staatszugehörigkeit f ~ **mark** Staatszugehörigkeitszeichen n ~ **marking** Nationalitätenzeichen n

**nationalization** Nationalisierung f

**nationalize, to** ~ verstaatlichen

**native** Eingeborener m, Einheimischer m; bodenständig, einheimisch, gebürtig, (ores, metals) gediegen, heimisch, inländisch, natürlich ~ **aluminum sulphate** Alugen n ~ **amalgam** Amalgamsilber n ~ **arsenic** Fliegenstein m, Scherbenkobalt m ~ **borax** Tinkal m ~ **cinnabar** Bergzinnober m ~ **copper** Bergkupfer n, gediegenes Kupfer n ~ **gold** Freigold n, gediegenes Gold n ~ **lead** gediegenes Blei n ~ **manufacture** Heimaterzeugnis n ~ **mellitate of alumina** Mellit m ~ **ore** einheimisches Erz n ~ **paraffin** Erdwachs n ~ **platinum** Polyxen n ~ **rubber** eingeborener Kautschuk m ~ **salpeter** Gayerde f, Gaysalpeter m ~ **sulphate of barium** Baryt m

**natrolite** Natrolith m, Radiolit m

**natron** Natron n

**natural** natur-bedingt, -gemäß, -getreu, natürlich, üblich, unbearbeitet **with a** ~ **grain** naturgenarbt

**natural,** ~ **action** Naturerscheinung f ~ **aggregate** Naturaggregat n ~ **bed** Flußbett n ~ **blocking layer** physikalische Sperrschicht f ~ **capacitance** Eigenkapazität f ~ **characteristic** natürliche Beschaffenheit f ~ **circulation cooling** Thermosiphonkühlung f ~ **clay** natürliche Bleicherde f ~ **color** Naturfarbe f ~ **cooling** Selbstkühlung f ~ **crack** (in wood) Luft-, Sonnen-riß m ~ **crack in wood** Waldriß m ~ **detector** Kontaktgleichrichter m ~ **divergence** Eigenstreuung f ~ **draft** natürlicher Zug m ~ **drying** Lufttrocknung f ~ **earth currents** (natürliche) tellurische Ströme pl ~ **earth potentials** natürliche Erdbodenpotentiale f, Potentiale f des Erdbodens ~ **frequency** Eigen-, Resonanz-frequenz f ~ **frequency in bending** Biegeeigenschwingung f ~ **gas** Erd-, Gruben-, Leicht-, Natur-gas n ~-**gas firing** Naturgasfeuerung f ~ **gas rights** Erdgasschürfrechte pl ~-**gas source** Erd-, Natur-gasquelle f ~-**gas well** Erdgasquelle f ~ **gasoline** Erdgas n, natürliche Treibstoff m ~ **grading** (chromatic scala) empfindungsgemäße Farbstufe f, empfindungsgerechte Abstufung f ~ **(or subjective) grading (or spacing)** empfindungsgerechte Abstufung f ~ **ground** verwachsener Boden m ~ **ground level** gewachsener Boden m ~ **hardness** Naturhärte f ~ **inductance** natürliche Induktivität f ~ **lag** Eigenverzögerung f ~ **leak** Eigenverlust m ~ **limit of stress** Ursprungsfestigkeit f ~ **logarithm** Napierscher Logarithmus m ~ **magnet** natürlicher Magnet m ~ **manganese dioxide** Braunstein m ~ **motion** Eigenschwingung f ~ **negative feedback** Eigen(gegen)kopplung f ~ **neutralization** wechselseitige Neutralisierung f ~ **number** natürliche Zahl f ~ **obstacle** Geländehindernis n ~ **oil seepage** Fettloch n ~ **order** natürliche (An)Ordnung f ~ **oscillation** Eigen-,

Grund-schwingung f ~ **parameter of a curve** natürlicher Parameter m einer Kurve ~ **period** (of oscillation) Eigenperiode f, Eigenschwingungszahl f ~ **period of oscillation** Eigenschwingungsperiode f ~ **period frequency** Eigenschwingung f ~ **phenomenon** Naturerscheinung f ~ **philosophy** Naturlehre f ~ **process** Naturerscheinung f ~ **protection** Geländeschutz m ~ **radionuclides** natürliche Radionuklide pl ~ **recources** Naturschätze pl ~ **rock** asphalthaltiges Gestein n ~ **rubber** Naturgumme m, Naturkautschuk m ~ **scale** Achse f der natürlichen Zahlen (math.) ~ **science** Naturlehre f, Naturwissenschaft f ~ **shade** naturfarbiger Ton m ~ Zahlen (math.) ~ **science** Naturlehre f, Naturwissenschaft f ~ **shade** naturfarbiger Ton m ~ **size** natürliche oder wirkliche Größe f ~ **slope** natürliche Böschung f ~ **slope of the earth,** (or of the soil) natürliche Erdböschung f ~ **soil** gewachsener Boden m, Mutterboden m ~ **spacing** (chromatic scale) empfindungsgemäße Farbstube f, empfindungsgerechte Abstufung f ~ **stability** Formstabilität f ~ **(dynamic) stability** Eigenstabilität f ~ **steel** Renn-, Roh-, Schmelz-, Wolfs-stahl m ~ **stone** Naturwerkstein m ~ **tendency** Naturanlage f ~ **time** Zeitkonstante f ~-**uranium reactor** Reaktor m mit natürlichem Uran m ~ **vibration** Eigenschwingung f ~ **vision** natürliches Sehen n ~ **volume** Eigenvolumen n ~ **water** Rohwasser n ~ **wave length** Eigenwelle f, Eigenwellenlänge f

**naturalize, to** ~ naturalisieren

**naturally,** ~ **aspirated** selbstsaugend ~ **aspirated engine** selbstsaugender Motor m ~ **hard** naturhart ~ **occurring talc** anstehender Talk m ~ **varnished** naturfarben lackiert

**nature** (natürliche) Beschaffenheit f, Natur f, Verlauf m, Wesen n ~ **of the bottom of sea** die Art f des Grundes ~ **of a charge** Ladungssinn m ~ **of grain** Körnungsbeschaffenheit f ~ **of the ground** (or soil) Bodenbeschaffenheit f ~ **of the terrain** Geländebeschaffenheit f

**nature brown** Braunholzpapier n

**naumannite** Naumannit m, Selensilberglanz m

**nautical** nautisch ~ **chart** Seekarte f ~ **gear** Schiffsgerät n ~ **map** Seekarte f ~ **mile** nautische Meile f, Seemeile f ~ **navigation** Schiffsortung f ~ **sign** Schiffahrts-signal n, -zeichen n ~ **sign on post** Stangenseezeichen n ~ **signal** Schiffahrts-signal f, -zeichen n ~ **surveying** nautische Vermessung f

**nautics** Nautik f, Seewesen n

**nautophone** Luftschallsender m, Nautofon n

**naval,** ~ **academy** Marineakademie f ~ **administration** Marineverwaltungsamt n ~ **agreement** Flottenabkommen f ~ **airship** Marineluftschiff n ~ **architect** Schiffsbauingenieur m ~ **armament** Seerüstung f ~ **aviation** Marineflugwesen n ~ **base** Ausfallhafen m, Flottenstützpunkt m, Kriegshafen m, Marinestation f, Seefestung f, Seestützpunkt m ~ **battle** Seeschlacht f ~ **coast artillery** Marineartillerie f ~ **control service** Hafenüberwachung f ~ **detachment** Flottenabteilung f ~ **district** Flottenstützpunkt m, Marinestation f ~ **dockyard** Arsenal n, Seearsenal n ~ **engineer** Marine-, Schiffsbau-ingenieur m ~ **engineer officer** Marineingenieur-

offizier *m* ~ **flier** Seeflieger *m* ~ **home base**
Flottenstation *f* ~ **meteorological service** Mari-
newetterdienst *m* ~ **observatory** Seewarte *f* ~
**officer** Marine-, See-offizier *m* ~ **plane** Marine-,
See-flugzeug *n* ~ **radio station** Marinefunkstelle
*f* ~ **signal station** Marinesignalstelle *f* ~ **transport** Seetransport *m* ~ **vessel** Kriegsschiff *n* ~
**warfare** Seekrieg *m*
**nave** Haupt-, Mittel-schiff *n* (arch.), (Rad)Nabe *f*
~ **(plate) bolt** Nabelbolzen *m* ~ **borer** Nabenbohrer *m* ~ **box** Nabenbüchse *f* ~ **disk** Nabenscheibe *f*
**navigable** fahrbar, schiffbar ~ **for rafts** flößbar
**navigable,** ~ **balloon** Lenkballon *m* ~ **pass** Schiff-
fahrtsöffnung *f* ~ **waterway** Schiffahrtsstraße *f*
**navigate, to** ~ fahren, franzen (slang), lenken,
navigieren, orten, steuern **to** ~ **by soundings**
nach Lotungen *pl* navigieren
**navigating compartment** Navigationsraum *m*
**navigation** Führung *f*, Nautik *f*, Navigation *f*,
Ortung *f*, Schiffahrt *f*, Schiffahrtskunde *f*,
Schiffsführung *f*, Seefahrt *f* ~ **chart** Navigationskarte *f* ~ **compass** Navigations-, Orter-
kompaß *m* ~ **guide** Bake *f* ~ **instrument** Navi-
gationsinstrument *n* ~ **light** Kennlicht *n*,
Positions-lampe *f*, -licht *n*, Stellungslicht *n* ~
**lights** Navigations-lampen *pl*, -leuchten *pl*,
-lichter *pl* ~ **map** Navigationskarte *f* ~ **officer**
Navigationsoffizier *m* ~ **service** Hafendienst *m*
~ **sign (or signal)** Schiffahrts-signal *n*, -zeichen
*n* ~ **table** Steuertafel *f*
**navigational,** ~ **aid** Navigationshilfe *f* ~ **radar**
Navigationsradar *n* ~ **radio beacon** Navigationsfeuer *n* ~ **smoke buoy** Dauerrauchpeil-
boje *f* ~ **star** Navigationsstern *m*
**navigator** Navigationsoffizier *m*, Orter *m*, Seefahrer *m* ~'**s chart table** Ortungskartentisch *m*
~'**s compartment** Orterraum *m* ~'**s compass**
Peilkompaß *m* ~'**s table** Ortertisch *m*
**navvy** Exkavator *m*
**navy** Flotte *f*, Kriegsmarine *f*, Marine *f* ~ **aircraft** Marineflugzeug *n* ~ **yard** Marinewerft *f*
**n-channel tape** n-Spurenband *n*
**N-conductor cord** N-adrige Schnur *f*
**"n"-degrees,** ~ **absolute** N-Grad absolut ~
**centigrade** N Grad Celsius ~ **Kelvin** N Grad
absolut
**neap tide** Nippflug *f*, Nipptide *f*, taube Flut *f*,
Taubetide *f*
**near** bei, nahe, neben ~ **(it)** daneben **to be** ~
naheliegen ~ **the ground (or the earth's surface)** bodennah, erdennah
**near,** ~-**by illustration** nebenstehende Abbildung
*f* ~-**by interference** Nahstörung *f* ~ **echo** Nah-
echo *n* ~-**end cross talk** Nebensprechen *n* am
Anfang ~-**end crosstalk attenuation** Nebensprechdämpfung *f* ~ **fading** Nahschwund *m*
~-**mesh material** Grenzkorn *n* ~ **miss** Fastzu-
sammenstoß *m* (aviat.), Fehl-, Nah-, Neben-
treffer *m* ~ **point** Nahe-, Nah-punkt *m* ~ **portion** Nahteil *m* ~ **side** Sattelseite *f*
**nearsighted** kurzsichtig ~ **eye** nahsichtiges Auge *n*
**near,** ~-**singing condition** Pfeifneigung *f* ~ **vision**
Nahsehen *n* (Nahsicht) ~ **zone** Nahzone *f*
**nearest** nächst ~ **neighbor coupling** Kopplung *f*
nächster Nachbarn ~ **neighbor interaction**
Wechselwirkung *f* ~ **size** nächstgrößte(r)

**nearing metal** Antifriktionsmetall *n*
**nearness** Nähe *f*
**neat** knapp, ordentlich, sauber **to be** ~ **in appearance** gut aussehen
**neat,** ~ **cement** Zementbrei *m* ~'**s-foot oil** Klau-
enfett *n*, Klauenöl *n*, Knochenöl *n*, Rinder-
klauenöl *n* ~ **grouping** klare oder übersichtliche Anordnung *f* ~'**s leather** Rindsleder *n*
**nebula** Nebel *m* (astron.), Nebel . . ., Nebula . . .
~ **of Andromeda** Andromedanebel *m* (astron.)
~ **projector** Nebelapparat *m*
**nebular** nebel-ähnlich, -artig ~ **lines** Nebellinien
*pl* ~ **shell** Nebelhülle *f*
**necessary** erforderlich, nötig, notwendig, zwangsläufig ~ **article (or object)** Bedarfsgegenstand
*m* ~ **power** Kraftbeanspruchung *f*
**necessitate, to** ~ erfordern, nötigen
**necessity** Not *f*, Notwendigkeit *f*
**neck, to** ~ zusammenschnüren **to** ~ **out** aushalsen
**neck** Ansatz *m*, Füllstutzen *m*, Genick *n*, Hals
*m*, Hohlkehle *f*, (of rolling mill) Laufzapfen *m*,
Nacken *m*, Wellenhals *m*, Zapfen *m*
**neck,** ~ **of axle** Achsstummel *m* ~ **of a Bessemer
converter** Hals *m* einer Bessemerbirne ~ **of bulb**
Kolbenhals *m* ~ **of cartridge** Hülsenhals *m* ~ **of
the float** Schwimmerhals *m* ~ **of hatchet** Kehle
*f* der Axt ~ **of land** Erdzunge *f* ~ **of a retort**
Retortenhals *m* ~ **of a roll** Walzen-, Walz-
zapfen *m* ~ **of a shaft** Hals *m* einer Welle ~ **of
telescope** Fernrohrhals *m* ~ **of tube** Röhren-
hals *m* ~ **of a vessel** Halsverengerung *f* eines
Gefäßes
**neck,** ~ **collar journal** Halszapfen *m* ~ **groove**
Drahtlager *n*, Halskerbe *f*, (of insulators) Hals-
rille *f* ~-**groove binding** Halsbindung *f* (electr.)
~ **halter** Halshalfter *m* ~ **halter strap** Hals-
halfterriemen *m* ~ **journal** Halslager *n* ~ **lace**
Halsband *n* ~ **radius** Radius *m* einer Hohlkehle
~ **strap** Hals-koppel *f*, -riemen *m*, (mask)
Nackenband *n*, (harness) Nackenriemen *m*
~**tie** Binder *m*, Halsbindung *f*, Kragenbinde *f*
~ **yoke** Brustholz *n*, Halskoppel *f*, Nacken-
joch *n*
**necked-down** eingeschnürt ~ **bolt** abgesetzte
Schraube *f*, Dehn-, Taillen-schraube *f* ~ **portion** mit einer Einschnürung *f* versehen, einge-
schnürter Teil *m* (aviat.) ~ **portion of a pipe** Einschnürung *f* (eines Ventilschaftes)
**necking** Einschnürung *f*, Querschnitts-verminde-
rung *f*, -verringerung *f*, Querzusammenziehung
*f*, Verengung *f*, Zusammenschnürung *f* ~
**down** Kerbe *f* ~ **operation** Aushalsungsarbeit *f*
~ **tool** Aushalsestahl *m*
**nectarine** Nektarine *f*
**need, to** ~ bedürfen, bebötigen, brauchen, nötig
haben
**need** Bedarf *m*, Bedürfnis *n*, Dürftigkeit *f*,
Mangel *m*, Not *f* **in case of** ~ erforderlichen-
falls ~ **for raw materials** Rohstoffhunger *m* **in**
~ **of repair** ausbesserungsbedüftig ~ **for replacement** Ersatzteilbedarf *m* ~ **for space**
Raumnot *f*
**needle, to** ~ durchstechen, nadelförmig kristallisieren, nähen
**needle** Grammofonnadel *f*, (building) horizontales Rüstholz *n*, Magnetnadel *f*, Nadel *f*,
(engraving) Radiernadel *f*, Räumnadel *f*

(min.), (weaving) Rietnadel *f*, Stößel *m*, Ventil-nadel *f*, Zeiger *m*, Zunge *f*,

**needle,** ~ **beam** Balkennadel *f* ~ **(roller) bearing** Nadellager *n* ~ **bearing bush** Nadellagerbüchse *f* ~**-bearing support-roller** Nadellagerstütz-rollen *pl* ~ **board** Nadelbett *n* ~ **bottom** Nadel-boden *m* ~ **bushing (or sleeve)** Nadelhülse *f* ~ **case** Nadelbüchse *f* ~ **caster** Nadelsetzer *m* ~**-controlled (valve)** nadelgesteuert ~ **counter** Spitzenzähler *m* ~ **dam** Nadelverschluß *m* ~ **die** Nadelmatrize *f* ~ **drag** Rückstellkraft *f* (d. Nadel) ~ **dressing pliers** Nadelrichtzange *f* ~ **effect** Spitzenwirkung *f* ~ **electrode** Spitzen-elektrode *f* ~ **eye** Nadelöhr *n* ~ **file** Fitzfeile *f* ~ **filler of embroidering machines** Stickmaschi-neneinfädlerin *f* ~ **force** Auflagedruck *m* (phono), Nadeldruck *m* (phono) ~ **fracture** na-delförmiges Gefüge *n* ~ **friction** Nadelreibung *f* ~ **galvanometer** Nadelgalvanometer *n* ~ **gap** Nadelfunkenstrecke *f* ~**-gap discharge** Spritz-entladung *f* ~ **gauge** Nadelmaß *n* ~ **hammer** Nadelhammer *m* ~ **induction** Nadelinduktion *f* ~ **ironstone** Nadeleisenstein *m* ~ **lamina** Nadeleisenstein *m*, Nadellamelle *f* ~ **lift control** Nadelhubregulierung *f* ~ **like** nadel-artig, nadelförmig, nadelig ~ **like crystal** langstrahliges Kristall *n* ~ **lubricator** Stift-schmierapparat *m* ~ **ore** Belonit *m* ~**-point (corona)-discharge** strahlartige Entladung *f* ~ **point file** Nadelfeile *f* ~**-point spark gap** Nadel-funkenstrecke *f* ~**-point streamer discharge** strahlartige Entladung *f* ~ **point velocity** Nadel-spitzengeschwindigkeit *f* ~ **proof** nadelfest ~**-recorded and needle reproduced sound** Nadel-ton *m* ~ **roller-cages** Nadelkäfige *pl* ~**-scratch filter** Nadelgeräuschfilter *m* ~ **seat** Nadelsitz *m* ~**-shaped** nadelförmig ~ **slide** Nadelschiene *f* ~ **splitter** Nadelaufreiher *m* ~ **telegraph** Nadel-telegraf *m* ~ **thread take-up** Nadelfadenzug *m* ~ **throw** Nadelausschlag *m* ~ **valve** Nadelkappe *f*, Nadelventil *n*, Ventilnadel *f* ~ **valve spray** geschlossene Zapfendüse *f* ~**-valve spray nozzle with pilot pin** geschlossene Zapfendüse *f* (Ein-spritzdüse) ~ **weir** Nadelwehr *n* ~ **width** Zeiger-breite *f* ~ **work** Hand-, Näh-arbeit *f*, Näherei *f* *f* ~ **worker** Handarbeiter *m*

**needless** entbehrlich
**needling** Aufklebepunktur *f* (print.)
**needy** notdürftig
**negate, to** ~ **an effect** eine Wirkung *f* aufheben, ins Gegenteil *n* verkehren
**negation** Verneinung *f*
**negative, to** ~ negieren
**negative** Matrize *f*, Minuszeichen *n* (math.), Negativ *n*, negativer Pol *m* (electr.), negative Zahl *f*; negativ, (test) ohne Befund *m*
**negative, without contrasts** flaues Negativ *n* ~ **with too strong contrasts** hartes Negativ *n*
**negative,** ~ **acceleration** negative Beschleunigung *f*, Verzögerung *f* ~ **angle of site** negativer Gelän-dewinkel *m* ~**-angle circuit** Bremskreis *m* ~**-anode potential** (in electronoscillation tube) Bremsspannung *f* ~ **aperture** Negativebene *f* ~ **area** negatives Gebiet *n*, Stromeintrittszone *f* ~ **bias** negative Vorspannung *f* ~ **biasing of grid** Gitterblockierung *f* ~ **booster** Zusatzmaschine *f* in Gegenschaltung ~ **boosting transformer** Saugtransformator *m* ~ **buoyancy** Untertrieb

*m* ~ **charge** Dielektrizitätskonstante *f* ~ **collar** Patrize *f* ~ **conductance** negativer Leitwert *m* ~ **contact** Minuskontakt *m* ~ **curvature** negative Krümmung *f* ~ **density** Negativschwärzung *f* ~ **(pole) deposit** Minusbelag *m* ~ **dihedral** nega-tive V-Form *f* ~ **dispersion** negative Dispersion *f* ~ **distortion** kissenförmige Verzeichnung *f* ~ **drag** im Vortrieb umgekehrter Luftwiderstand *m* ~ **electricity** Minuselektrizität *f* ~ **electrode** Kathode *f*, negative Elektrode *f* ~ **electron** Negatron *n* ~ **feedback** Gegenkopplung *f*, negative Rückkopplung *f* oder Rückführung *f* ~ **follow-up** Gegenkopplung *f* ~ **gamma** Kopier-, Negativ-gamma *n* ~ **glow(light)** nega-tives Glimmlicht *n* ~ **grid current** negativer Gitterstrom *m* oder Ionengitterstrom *m* ~ **holder** Negativhalter *m* ~ **image** negatives Bild *n* ~ **ion** Sauerstoffion '*n* ~**-ion vacancy** Ionen-loch *n* ~ **lead** Minusleitung *f* ~ **lens** negatives System *n*, Zerstreuungslinse *f* ~ **material** Negativmaterial *n* ~ **modulation** Modulation *f* auf dunkel, Steuerung *f* auf dunkel ~ **picture** Negativbild *n* ~ **plane** Negativebene *f* ~ **plate** Minusplatte *f* ~**-plate circuit** Bremskreis *m* ~ **polarity of transmission** negative Polarität *f* der Sendung ~ **pole** Minuspol *m* (electr.) ~ **pressure** Unterdruck *m*, Zug *m* ~ **pressure in the atomizer** Unterdruck *m* im Zerstäuber ~ **pressure relief** Unterdruckausgleich *m* ~ **print** negatives Bild *n*, Negativpause *f* ~ **proton** Antiproton *n* ~ **reactance** Kapazitätsreaktanz *f*, kapazitive Reaktanz *f*, kapazitiver Blindwiderstand *m* ~ **reaction** negative Rückkopplung *f* ~**-reading microscope** Negativmeßmikroskop *n* ~ **re-generation** negative Rückkopplung *f* ~ **reply** abschlägige Antwort *f* ~ **report** Fehlanzeige *f* ~ **resistance** Kapazitanz *f*, negativer Wider-stand *m* ~**-sequence field impedance** Gegen-impedanz *f* ~ **slope** Böschung *f* im Abtrag ~ **source** Senke *f* ~ **space charge** Bremswirkung *f* der negativen Raumladung, Eigenabstoßung *f* der Elektronenwolke ~ **stagger** Staffelung *f* nach hinten ~ **storage** Negativaufbewahrung *f* ~ **sweep** Vorpeilung *f* ~ **transconductance generator (or oscillator)** Bremsfeldgenerator *m* ~ **transmission** negative Übertragung *f*, Über-tragung *f* von Dias mit negativer Modulation ~ **unvaried portion of luminous glow** (in glow tube) Rumpf *m* ~ **valence** negative Wertigkeit *f* ~ **viewing microscope** Betrachtungsmikroskop *n* für Negative ~ **wire** an Spannung liegende Ader *f* ~ **work due to back pressure** Gegenarbeit *f*

**negatively,** ~ **biased detector** Anodengleichrich-ter *m*, Richtverstärker *m* ~ **electric** negativ-elektrisch ~ **rake wing** Rückklappflügel *m*
**negatoscope** Negativschaukasten *m*
**negatron** Negatron *n*
**neglect, to** ~ außer Acht *f* lassen, außerachtlassen, verabsäumen, vernachlässigen
**neglect** Vernachlässigung *f*, Versäumnis *n*
**neglected** unberücksichtigt
**neglectful** nachlässig
**negligence** Fahrlässigkeit *f*, Nachlässigkeit *f*
**negligent** fahrlässig, nachlässig, säumig
**negligible** vernachlässigbar
**negotiability** (of securities) Übertragbarkeit *f*
**negotiable** einlösbar, übertragbar (handelsfähig)

**negotiate, to** ~ verhandeln **to** ~ **a loan** eine Anleihe *f* abschließen
**negotiation** Verhandlung *f*
**negotiator** Unterhändler *m*
**negro head rubber** Negerköpfe *pl*
**neighbor** Nachbar *m* ~ **atom** Nachbaratom *n*
**neighborhood** Nachbarschaft *f*, Umgebung *f* ~ **traffic** Nachbarortsverkehr *m*
**neighboring** angrenzend, benachbart ~ **earthquake** Nahbeben *n* ~ **position** Nachbarstellung *f* ~ **station's break-through (or swamp signals)** Durchschlagen *n* vom Nachbarsender ~ **turns** benachbarte Windungen *pl*
**Noecomian** Neokom *n*
**neodymium** Neodym *n* ~ **chloride** Neodymchlorid *n* ~ **content** Neodymgehalt *m*
**neolite** Neolith *m*
**neolithic** neolithisch
**neon** Edelgas *n*, Neon *n* ~ **advertising** Lichtreklame *f* ~ **arc lamp** selbstleuchtender Licht-modulator *m* ~ **ground light** Neonbodenlicht *n* ~ **indicator** Glimmlichtanzeigeröhre *f* ~ **lamp** Glimmlampe *f*, Neon-glimmlampe *f*, -lampe *f*, -röhre *f* ~ **lamp with plate-shaped cathode** Flächenglimmlampe *f* ~ **light** Neonlicht *n* ~ **liquefier** Neonverflüssigungsmaschine *f* ~ **sign** Neonschild *n* ~ **stabilizer** Glimmstabilisatorröhre *f* ~ **time base** Glimmsummer *m* ~ **tube** Glimmlampe *f*, Neon-glimmlampe *f*, -lampe *f*, -röhre *f* ~ **tube rectifier** Leuchtröhrengleichrichter *m* ~**-tube volume indicator** Aussteuerungskontrolle *f* mit Neonröhre, Neonröhren-aussteuerungskontrolle *f*, -lautstärkemesser *m* ~**-tube wires** Leuchtröhrenleitungen *pl* ~ **tuning indicator** Glimmabstimmanzeiger *m* ~ **wire** Neonleitung *f*
**neoplasm** Neoplasma *n*, Neubildung *f*
**neoplastic coil** neugebildete Zelle *f*
**neoprene** Neopren *n*
**neotype** Neotyp *m*
**Neozoic group** neozoische Gruppe *f*
**neper** Neper *n*
**nepheline** Elaolith *m*, Fettstein *m*
**nephelite** Dawyn *m*, Fettstein *m*, Nephelin *m*
**nephelogy** Wolkenkunde *f*
**nephelometer** Nebel-, Trübungs-messer *m*
**nephelometric analysis** (turbidimetric analysis) nephelometrische Analyse *f*
**nephelometry** Nebelmessung *f*
**nepheloscope** Wolken-spiegel *m*, -zugmesser *m*
**nephoscope** Nephoskop *n*
**nephrite** Beilstein *m*, Nephrit *m*
**neptunium series** Neptuniumreihe *f*
**Nernst,** ~ **lamp** Nernst-brenner *m*, -lampe *f* ~ **lamp of a microscope** Mikronernstlampe *f* ~ **needle** Nernststift *m*
**neroli oil** Neroliöl *n*
**nerve** Nerv *m* ~ **of a vault** Gewölberippe *f* ~ **structure** adrige Struktur *f*
**nest, to** ~ sich einnisten, verschachteln
**nest** Nest *n*, Satz *m* ~ **of boiler tubes** Rohrbündel *n* ~ **of screens** Siebnest *n*
**nesting,** ~ **strap** Paßleiste *f* ~ **strip** Paßstreifen *m*
**net** Leitungsnetz *n*, Netz *n*, netzartiges Gewebe *n*, Straßennetz *n*; netto ~ **of slip lines** Gleitliniennetz *n* ~ **of triangulation** Dreiecksnetz *n*, trigonometrisches Netz *n*
**net,** ~ **action** resultierende Kraft *f* ~ **assets**

Betriebsvermögen *n* ~ **attenuation** Restdämpfung *f* ~ **balance** Nettosaldo *n* ~ **brake horsepower** an der Welle verfügbare reine Nutzleistung *f*, Wellenleistung *f* ~ **control station** Leitfunkstelle *f* ~ **cost** Selbstkosten *pl* ~ **cutter** Netzschere *f* ~ **density** Reindichte *f* ~ **dredger** Sackbagger *m* ~ **efficiency** Nutz-effekt *m*, -leistung *f*, -wirkung *f* ~ **energy gain** effektiver Energiegewinn *m* ~ **equalitation** Netzausgleich *m* ~ **export** Reinausfuhr *f* ~ **flux** wirksamer Fluß *m* ~ **force** resultierende Kraft *f* ~ **gain** Reineinnahme *f* ~ **gradient of climb** Netto-Steigwinkel *m* ~ **group dial technique** Netzgruppentechnik *f* (teleph.) ~ **height** Netto-Höhe *f* ~ **increase** Nettozunahme *f* ~ **layer (or -laying vessel)** Netzleger *m* ~ **like** netzartig ~ **load** Nutzlast *f* ~ **loss** Restdämpfung *f*
**net-loss,** ~ **frequency measurements** Restdämpfungsmesser *f* bei mehreren Frequenzen ~ **measurement** Restdämpfungsmessung *f* ~ **variation with amplitude** Dämpfungsänderung *f* mit der übertragenen Leistung, Spannungsabhängigkeit *f* der Dämpfung
**net,** ~ **masonry** Netzverband *m* ~ **pattern** Gitter-muster *n* ~ **pay** Nettoverdienst *m* ~ **proceeds (or profit)** Reinertrag *m* ~ **receipts** Betriebsüberschüsse *pl* ~ **result** Enderfolg *m* ~ **shaped** netzförmig, siebförmig ~**-shaped electrode** Netzelektrode *f* ~ **tonnage** Handelsschiffsraum *m* ~ **transmission equivalent** Restdämpfung *f* ~ **weight** Eigen-, Leer-, Netto-, Rein-gewicht *n*
**network** Filet *n*, Funkverbindung *f*, Geflecht *n*, Kette *f*, Kettenleiter *m*, Kettenleiter *m* erster Art, Leitungsnetz *n* (electr.), Maschenwerk *n*, Netz *n*, Netzwerk *n*, räumliches Fachwerk *n*, Sendernetz *n* (rdo), Sieb *n*, Siebgebilde *n*, Süll *m*, Vernetzung *f*, Verteilungsnetz *n* (electr.), Zellenstruktur *f*
**network,** ~ **of air routes** Luftstreckennetz *n* ~ **of conductors** Stromverzweigung *f* ~ **of lines** Gleisnetz *n* ~ **of lodes** Gangverzweigung *f* ~ **of streams and torrents** Fluß-netz *n*, -system *n*
**network,** ~ **analysis** Netzwerkanalyse *f* ~ **analyzer** Netzmodell, Netzwerkgleichungslöser *m* ~ **calling machine** Netzrufmaschine *f* (teleph.) ~ **communication** Netzverkehr *m* ~ **concept** Netzwerkvorstellung *f* ~ **design** Riffelungsmuster *n* ~ **diagram** Leitungsskizze *f*, Linienkarte *f* ~ **hum** Netzton *m* ~ **layout** Netzgestaltung *f* ~ **level** Netzebene *f* ~ **map** Netzplan *m* ~ **mesh** Kettenglied *n*, Kettenleiterglied *n* ~**-modifying ions** netzwerkändernde Ionen *pl* ~ **parameter** Netzwerkparameter *n* ~ **plan** Netzplan *m* ~ **protector** Maschennetzschalter *m* ~ **(balancing) rack** Nachbildungsgestell *n* (teleph.) ~ **radio communication** Sternverkehr *m* ~ **side** Netzseite *f* ~ **structure** Netzwerkstruktur *f* ~ **system** Leitungsnetz *n* ~ **system of wires** Drahtnetz *n* ~ **terminated at midshunt (position)** in halb Querglied endender Kettenleiter *m* ~**-type structure** netzwerkartiges Gefüge *n* ~ **yield** Netto-ertrag *m*, -gewinn *m*
**netted** maschig, netzförmig
**netting** Filet *n*, Gewebe *n*, Netz *n*, Netzbildung *f*, Netzknüpfen *n* ~ **machine** Netzknüpfmaschine *f* ~ **wire** Webedraht *m*
**nettle** Knittels *pl*, Nessel *f* ~ **cloth** Nessel *f*, Nesselgewebe *n*

**nettling** Spleißen *n*
**neutral** indifferent, neutral, träge **the ~ Mittelleiter** *m* **to set to ~** neutral einstellen **in ~ equilibrium** statisch indifferent
**neutral, ~ adjustment** Neutralstellung *f* **~ air layer** indifferentes Gleichgewicht *n*, neutrale Luftschicht *f* **~ axis** Neutralachse *f*, neutrale Faser *f* (Festigkeitslehre), Null-achse *f*, -linie *f*, Schwerpunktachse *f* **~ body** Vergleichskörper *m* **~ conductor** Nulleiter *m* **~ current** Einfachstrom *m* **~ density filter** Abschwächungsfilter *n* **~ equilibrium** indifferentes Gleichgewicht *n*; indifferent statisch **~ feeder** Nulleitung *f* **~ fiber** neutrale Achse *f*, Neutralfiber *f*, Null-achse *f*, -linie *f* **~ filter** neutral absorbierender Körper *m* **~ glass** Neutralglas *n* **~ goods** Neutralladung *f* **~ hydrocarbon** Neutralkohlenwasserstoff *m* **~ lard** Flaumfett *n* (chem.) **~ leach** neutraler Prozeß *m* **~ line** Knoten-, Mittel-linie *f*, Neutrale *f* **~ line of a magnet** Indifferenzstelle *f* eines Magnetes **~ lines** (of a machine with commutator) neutrale Zone *f* **~ material** totes Material *n* **~ medium** neutral (grau, selektiv) absorbierender Körper *m* **~ oil** unviskoses Öl *n* **~ plane** Mittelfläche *f*, Nullzone *f* (electr.) **~ point** Indifferenz-, Null-, Stern-punkt *m* **~-point resistance** Nullpunktwiderstand *m* **~ position** Abschlußstellung *f*, Mittelage *f*, Nulllage *f*, -stellung *f* **~ process** neutraler Prozeß *m* **~ red** Neutralrot *n* **~ relay** polarisiertes Relais *n* mit mittlerer Ruhestellung des Ankers, Relais *n* mit drei Stellungen **~ screen** neutraler Schirm *m* **~ solution** neutrale Lösung *f* **~ state** jungfräulicher Zustand *m* **~ terminal** Sternpunktklemme *f* **~ tint** Neutraltinte *f* **~ value** Ruhewert *m* **~ (gray) wedge** Goldberg-, Graukeil *m* **~ wire** Mittel-, Null-leiter *m* **~ zone** Indifferenzzone *f*, neutrale Zone f
**neutrality** (of a gimbal or gyro mounting, phys.) Indifferenz *f*, Neutralität *f*, Neutralstellung *f*
**neutralization** Aufhebung *f*, Auskopplung *f*, Entkopplung *f*, Entkupplung *f*, Lahmlegung *f*, Neutralisierung *f* **~ of sound waves** Ausgleichen *n* von Schallwellen, Schallwellenausgleich *m* **~ number** Neutralisations-, Säure-zahl *f*
**neutralize, to ~** abstumpfen, anschärfen, eine Wirkung *f* aufheben, auskoppeln, entkoppeln, lähmen, lahmlegen, neutralisieren, niederhalten, symmetrieren, unschädlich machen, wiederaufheben
**neutralized, ~ by a counterweight** ausgeglichen durch Gegengewicht *n* **~ white** aufgefärbtes Weiß *n*
**neutralizing** wirkungsaufhebend **~ capacitor** Mehrfachdrehkondensator *m* **~ condenser** Neutralisationskondensator *m*, Neutrodon *n*, Neotrodynkondensator *m* **~ controls** neutralisierende Steuerung *f*
**neutrally, ~ adjusted** neutraleingestellt **~ mounted** (as a compass mounted on gimbels) indifferent aufgehängt
**neutretto** neutrales Meson *n*, Neutretto *n*
**neutrino** neutrales Elementarteilchen *n*, Neutrino *n* **~ momentum** Neutrino-Impuls *m*
**neutro-apparatus** Neutrogerät *n*
**neutrodyne, to ~** neutrodynisieren
**neutrodyne** Neutrodyn *n* **~ circuit** Neutrodynschaltung *f* **~ circuit organization** Neutralisationsschaltung *f* **~ (hookup)** Neutrodynschaltung *f* **~ receiver** Neutrodynempfänger *m*
**neutron** Neutron *n* **~ absorber** Neutronenabsorptionsmittel *n* **~ bombardment** Beschießung *f* mit Neutronen **~ capture** Neutroneneinfang *m* **~-capture cross-section** Einfangsquerschnitt *m* für Neutronen **~ cross section** Neutronenquerschnitt *m* **~ damage** Neutroneneinwirkung *f* **~ degradation** Neutronen-bremsung *f*, -verlangsamung *f* **~ diffraction pattern** Neutronenbeugungsaufnahme *f* **~ diffractometer** Neutronenbeugungsgerät *n* **~ flux distribution** Verteilung *f* der Gewinnrücklage des Neutronenflusses **~ hardening** Zunahme *f* der Neutronenenergie **~-induced reaction** neutroneninduzierte Reaktion *f* **~ irradiation** Neutronenbestrahlung *f* **~ monitor** Neutronenüberwachungsgerät *n* **~ monitoring** Neutronenflußmessung *f* **~ reaction** neutroneninduzierte Reaktion *f* **~ reflection** Neutronenspiegelung *f* **~ robber** Neutronenfänger *m* **~ thermopile** Neutronenthermsäule *f* **~ velocity selector** Neutronengeschwindigkeitsselektor *m*
**never exceed speed** zulässige Höchstgeschwindigkeit *f*
**new** frisch, neu **~ achromat** Neuchromat *n* **~ assembly** (of an engine) Erstzusammenbau *m* **~ building** Neubau *m* **~ design** Neukonstruktion *f* **~ dimensions system** Mineaturisierung *f* **~ edition** Neu-ausgabe *f*, -druck *m* **~ field** Neuland *n* **~ lining** Neuzustellung *f* **~ order** Neuordnung *f* **~ paint** Übermalung *f* **~ point** Neupunkt *m* **~ recording** Neuaufnahme *f* **~ red-aluminate process** Neurot-Aluminatverfahren *n* **~ red oiling** Neurot-Ölung *f* **~ sand** Neusand *m* **~ works** Neuanlage *f*
**newel** Eckpfeiler *m*, Seele *f*, Spindel *f* **~ post** Pfosten *m*
**newly, ~ formed** neugebildet **~ mixed sand** Frischsand *m* **~ won land** Neuland *n*
**newness** Neuheit *f*
**news** Bericht *m*, Botschaft *f*, Kunde *f*, Nachricht *f*, Neuigkeit *f* **~ agency** Nachrichtenagentur *f* **~ circuit** Zeitungsleitung *f* **~ film** Reportagefilm *m* **~ message** Presse-, Zeitungs-telegramm *n*
**newspaper** Zeitung *f* **~ article** Zeitungsartikel *m* **~ clipping** Zeitungsausschnitt *m* **~ folder** Zeitungspfalzerin *f* **~ printing** Zeitungsdruck *m* **~ printing blankets** Zeitungsdrucktücher *pl* **~ printing plant** Zeitungsdruckerei *f* **~ rotary machine** Zeitungsrotationsdruckmaschine *f* **~ service plane** Zeitungsflugzeug *n* **~ work** Zeitungsdruck *m*
**newsphoto service** Bilderdienst *m*
**newsprint** Zeitungsdruckpapier *n* **~ color** Zeitungsdruckfarbe *f* **~ paper** Zeitungspapier *n*
**news, ~ reel** Reportagefilm *m* **~-reporting work** Reportage *f* **~ stand** Zeitungskiosk *m* **~-system communications** Nachrichtenwesen *n* **~ work** Zeitungsdienst *m*
**Newton's, ~ color rings** Newtonsche Ringe *pl* **~ law** Newtonsches Gravitationsgesetz *n* **~ second law** mechanische Grundgleichung *f*
**next** folgend, nächst (e, er, es)
**N-fixture** N-förmiges Gestänge *n*
**nib** Düseneinsatz *m*, Kerngröße *f*, Schnabel *m*, Schreibfeder *f*, Spitze *f*, Teil *m* eines Kombinationsschlüssels **~ point** Federspitze *f*

**nibbed** anspitzen ~ **spring leaf** Federblatt *n* mit Längsnut und Rippe

**nibble, to** ~ anfressen, ausnagen, beknabbern

**nibble machine** Knabberschere *f* (Nibbelmaschine)

**nibbler** Knabberer *m*, Nagemaschine *f*, Stanzschere *f*

**nibbling** Knabbern *n* ~ **machine** Ausbau-, Dekupier-maschine *f*

**niccolite** Antimonarsennickel *n*, Arseniknickel *n*, Arsennickel *m*, Nickelin *n*, Rotnickelkies *m*

**niche** Auskehlung *f*, Einbuchtung *f*, Fuchsloch *n*, Halbkuppel *f*, Nische *f*

**Nicholl's chart** Nichollsches Diagramm *n*

**nichrome** Chromnickel *n* ~ **tube** Chromnickelrohr *n* ~ **wire** Chromnickeldraht *m*

**nick, to** ~ einkeilen, einkerben, (in roughening rollers) kerben **to** ~ **out** ausfurchen, furchen

**nick** Dolle *f*, Einkerbung *f*, Einschnürung *f*, Schere *f* **(screw)** ~ Schraubenschlitz *m* ~ **bend test** Kerbbiegeprobe *f*

**nicked** eingekerbt, schartig ~ **sinker cam** eingearbeiteter Platinexzenter *m*

**nickel, to** ~ vernickeln **to** ~**-plate** vernickeln

**nickel** Nickel *n* ~**-alkali storage battery** Nife-Akkumulator *m* ~**-alkaline cell** Nickelsammler *m* ~ **alloy** Nickellegierung *f* ~ **aluminum** Nickelaluminium *n* ~ **arsenide** Arsennickel *m* ~**-bearing** nickelführend ~ **bloom** Nickelblüte *f* ~**-cadmium battery** Nickel-Kadmium-Sammler *m* ~ **carbonyl** Kohlenoxydnickel *m* & *n*, Nikkelkohlenoxyd *n* ~ **coating** Nickelüberzug *m* ~ **coin** Nickelmünze *f* ~ **content** Nickelgehalt *m* ~ **converting** Nickelfeinsteinverarbeitung *f* ~ **crucible** Nickeltiegel *m* ~ **cube** Nickelwürfel *m* ~ **cubes** Würfelnickel *n* ~ **delay-line** Nickelverzögerungsleitung *f* ~ **deposit** Nickelniederschlag *m* ~ **electro** Nickelgalvano *n* ~ **electrolyte** Nickellauge *f* ~ **fittings** vernickelte Armaturen *pl* ~ **foil** Nickelpapier *n* ~ **green** Nickelblüte *f* ~ **ingots** Blocknickel *m* ~ **iron** Nickeleisen *n* ~**-like** nickelartig ~ **lining** Nickelauskleidung *f* ~ **matrix** Nickelmater *f* ~ **matte** Nickelstein *m* ~**-matte refining** Nickelfeinsteinverarbeitung *f* ~ **ocher** Nickelocker *m* ~ **ore** Nickelerz *n* ~ **oxalate** oxalsaures Nickel *n* ~ **pan** Nickelschabe *f* ~ **pans** Nickelspäne *pl* ~ **pellets** Kugelnickel *n* ~ **plate** Nickelblech *n* ~**-plate(d)** vernickelt ~**-plated gauges** vernickelte Skalen *pl* ~**-plating** Nickelüberzug *m*, Vernickeln *n*, Vernickelung *f* ~ **recovery** Nickelgewinnung *f* ~ **rondelles** Rondellennickel *m* ~ **sheet** Nickelblech *n* ~ **shot** Granaliennickel *n*, Nickelgranalien *pl* ~ **silver** Neusilber *n* ~ **speiss** Nickelspeise *f* ~ **steel** Nickelflußeisen *n*, Nickelstahl *m* ~**-steel casing** Nickelstahlguß *m* ~ **stibine** Antimonnickelglanz *m* ~ **sulfate** Nickelvitriol *n* ~ **sulfide** Nikkelsulfid *n* ~ **telluride** Tellurnickel *n* ~ **trimmings** Nickelbeschlag *m* ~ **wire** Nickeldraht *m*

**nickeliferous** nickel-führend, -haltig ~ **pyrrhotite deposit** Nickelmagnetkieslagerstätte *f*

**nickeline** Nicklin *n*

**nickeling** Vernickeln *n*

**nickelous** nickelhaltig, nickelig ~ **bromide** Nikkelbromür *n* ~ **chloride** Nickelchlorür *n* ~ **compound** Nickelverbindung *f* ~ **cyanide** Nikkelzyanür *n* ~ **hydroxide** Nickelhydroxydul *n* ~ **oxide** Nickeloxydul *n* ~ **sulfide** Nickelsulfür *n*

**nicking** Einkerben *n*, Einstechstahl *m* ~ **buddle** Klaubherd *m* ~ **tool** Einstechstahl *m*

**nicol prism** Nikolsches Prisma *n*

**nig, to** ~ **an ashlar** einen Stein *m* scharrieren

**night,** ~ **air mail** Nachtluftpost *f* ~ **air traffic** Nachtluftverkehr *m* ~ **alarm** Nachtwecker *m* ~**-alarm key** Schalter *m* für Nachtwecker ~ **alarm switch** Nachtschalter *m* ~ **bell** Nachtwecker *m*, Schalter *m* für Nachtwecker ~ **binoculars** Dunkelsuchgerät *n* ~**-blind** nachtblind ~ **blindness** Nachtblindheit *f* ~ **bolt** Nachtriegel *m* ~ **concentration position** Nachtplatz *m* ~ **concentrator** Nachtzentralschalter *m* ~ **depository** Nachtschalter *m* (Bank) ~**-driving regulator** Nachtmarschgerät *n* ~ **duty** Nachtdienst *m* ~ **effect** Dämmerungseffekt *m*, Nacht-effekt *m*, -wirkung *f* ~ **extension** Pikettstelle *f* ~ **flight (or flying)** Nachtflug *m* ~**-flying chart** Nachtflugkarte *f* ~**-flying equipment** Nachtflugausrüstung *f* ~ **gliding** Nachtsegelflug *m* ~ **goniometer** Nachtwinkelmesser *m* ~ **illumination** Nachtbeleuchtung *f* ~ **landing** Nachtlandung *f* ~ **load** Nachtbelastung *f* ~**-lodging allowance** Übernachtungsgeld *n* ~ **marking** Nachtmarkierung *f* ~ **operations** Nachtflugbetrieb *m* ~ **operator** Nachtdienstbeamtin *f* ~ **photograph** Nachtaufnahme *f* ~ **(air) photograph** Nachtluftaufnahme *f* ~ **pilot** Nachtflieger *m* ~ **position** Nachtplatz *m* ~ **power** Nachtstrom *m* ~ **rate** Nachtgebühr *f* ~ **screen** Nachtblende *f* ~ **service** Nachtdienst *m* ~ **service button** Nachttaste *f* ~ **service connection** (as applied to P.B.X.'s) Dauerverbindung *f*, Nachtverbindung *f* (teleph.) ~ **service position** Nachtdienstplatz *m* ~ **service relay** Nachtrelais *n* ~ **shift** Nacht-arbeit *f*, -schicht *f* ~ **signal indicating ship's course to following unit** Nachtfahrtanzeiger *m* ~ **switching relay** Nachtrelais *n* ~ **telegraph letter** Brieftelegramm *n* ~ **tracer** Glimmspur *f* ~ **trunk position** Nachtfernschrank *m* ~ **vision** Nacht-sehvermögen *n*, -sicht *f* ~ **watch** Nachtwache *f* ~ **watchman** Nachtwächter *m* ~ **wave** Nachtwelle *f* ~ **wind** Nachtwind *m* ~ **work** Nachtarbeit *f*

**nil** Null *f*; nichts, nichtvorhanden ~ **balance** Nullabgleich *m* ~ **report** Fehlanzeige *f* ~ **return** Fehlanzeige *f*

**nimble** gewandt

**nimbostratus cloud** Nimbostratuswolke *f*

**nimbus cloud** Nimbuswolke *f*

**niobate** Niobat *n*

**niobic,** ~ **acid** Niobsäure *f* ~ **anhydride** Niobsäureanhydrid *n* ~ **(pent)oxide** Niobpentoxyd *n*

**niobium** Niob *n*, Niobium *n*

**nip, to** ~ abzwicken, klemmen, kneifen, zwicken **to** ~ **in the bud** im Keime *m* ersticken **to** ~ **off** abkneifen

**nip** Abkneifen *n*, Abzwicken *n*, Knick *m*, Walzenspalt *m* ~ **of the press roller** Klemmpunkt *m* der Preßwalze ~ **roll** Haltewalze *f*

**Nipkow,** ~ **channel transmission** Nipkowkanalübertragung *f* ~ **disk** Kreislochscheibe *f*, Lochscheibe *f*, Lochspirale *f*, Nipkowscheibe *f*

**nipper,** ~ **for needle bending** Nadelbiegezange *f* ~ **of a pile driver** Auslösungshaken *m* ~ **frames** (textiles) Ausheberahmen *pl*

**nippers** Draht-abschneider *m*, -schere *f*, -zange *f*, Kneifzange *f*, Pinzette *f*, Teufelsklaue *f*, Zange *f*
**nipping cut** Quetschschnitt *m*
**nipple** Ansatz *m*, Anschlagzapfen *m*, Anschluß-stück *n*, Düse *f*, Einbaustutzen *m* (aviat.), Gußwarze *f*, Nippel *m*, Pimpel *m*, Rohrstutzen *m*, Sauger *m*, Speichen-, Schmier-nippel *m*, Stutzen *m*, Warze *f* ~ **of spring buckle** Federbundzapfen *m*
**nipple, ~ joint** Nippelverbindung *f* ~ **key** Nippelschlüssel *m* ~ **threading machine** Nippelgewindeschneidemaschine *f*
**niter cake** Natriumkuchen *m*
**nitol** (nitric acid) alkoholische Salpetersäure *f*
**niton** Radon *n* ~ (**or nitra**) **bulb** Nitrabirne *f* ~ **lamp** Nitralampe *f*
**nitralloy, ~ barrel** Nitralloynabe *f* ~ **steel** Nitrierstahl *m*
**nitraniline** Nitranilin *n*
**nitraphoto lamp** Nitrafotolampe *f*
**nitrate, to** ~ nitrieren
**nitrate** Nitrat *n*, salpetersaures Salz *n* ~ **of** salpetersauer, salpetrigsauer
**nitrate, ~ discharge** Nitratätze *f* ~ **dope** nitrozelluloser Spannlack *m* ~ **explosive** Sprengsalpeter *m* ~ **mordant** Nitratbeize *f* ~ **plant** Nitratanlage *f*
**nitrated** salpetersauer ~ **case** Nitrierschicht *f*
**nitrating, ~ acid** Nitriersäure *f* ~ **apparatus** Nitrierapparat *m* ~ **mixture** Nitriergemisch *n*
**nitration** Nitration *f*, Nitrierung *f* ~ **case** Nitrierschicht *f* ~ **furnace** Nitrierofen *m* ~ **process** Nitrierhärteverfahren *n*
**nitre cake** (sodium bisulfate) Weinsteinpräparat *n*
**nitric** salpetersauer ~ **acid** Salpetersäure *f*, Scheidewasser *n* ~ **anhydride** Salpetersäureanhydrid *n* ~ **ether** Salpeteräther *m* ~ **oxide** Stickstoff-dioxyd *n*, -oxyd *n*
**nitridation** Nitrierung *f*, Nitrierungsgrad *m*
**nitride, to** ~ nitrieren (metal.)
**nitride** Nitrid *n* ~ **case** Nitrierschicht *f* ~ **hardening** Nitrierhärtung *f*
**nitrided** nitriert ~ **edge** verstickte Randzone *f* ~ **steel** Verstickstahl *m*
**nitriding** Nitrierhärten *n*, Nitrierung *f* ~ **action** Nitrierwirkung *f* ~ **box** Nitrierkasten *m* ~ **equipment** Nitrieranlage *f* ~ **furnace** Nitrierofen *m* ~ **process** Nitrier-härteverfahren *n*, -verfahren *n* ~ **steel** Nitrierstahl *m* ~ **time** Nitrierdauer *f*
**nitrifiable** nitrierbar
**nitrification** Nitrierung *f*, Salpeterbildung *f*
**nitrify, to** ~ nitrieren
**nitrite** salpetrigsaures Salz *n*
**nitro** Nitro . . ., Salpeter . . . ~ **benzene** Nitrobenzol *n* ~ **carbon** Stickkohlenstoff *m* ~ **cellulose** Nitrozellulose *f* ~ **cellulose dope** Nitrozelluloselack *m* ~ **cellulose filler mass** Nitrospachtel *f* ~ **cellulose finishes** Collodiumdeckfarben *f pl* ~ **cellulose flake powder** Nitrozelluloseblättchenpulver *n* ~-**cotton** nitrierte Wolle *f*
**nitrogen** Azot *n*, (gas) Stickgas *n*, Stickluft *f*, Stickstoff *m* ~ **carbide** Stickkohlenstoff *m* ~ **chloride** Chlorstickstoff *m* ~ **compound** Stickstoffverbindung *f* ~ **dioxide** Stickdioxyd *n*, Stickstoffdioxyd *n*, Untersalpetersäure *f* ~

**fixation** Bindung *f* des atmosphärischen Stickstoffes ~ **halide** Stickstoffhalogen *n* ~-**hardening** Nitrierhärten *n* ~ **monoxide** Stickstoffoxydul *n* ~ **plant** Stickstoffanlage *f* ~ **tetroxyde** Stickstofftetroxyd *n*
**nitrogenous** stickstoffhaltig
**nitroglucose** Knallzucker *m*
**nitroglycerin** Glonoin *n*, Nitroglyzerin *n*, Sprengöl *n* ~ **flake powder** Nitroglyzerinblättchenpulver *n* ~ **powder** Nitroglyzerinpulver *n*
**nitrolic acid** Nitrolsäure *f*
**nitrometer** Nitrometer *n*, Stickstoffmesser *m*
**nitromethane** Nitromethan *n*
**nitron** Nitron *n*
**nitronaphthalene** Nitronaphthalin *n*
**nitrophenol** Nitrophenol *n*
**nitrosobenzene** Nitrosobenzol *n*
**nitrososulfonic acid** Nitrososulfosäure *f*
**nitrosulfuric acid** Nitroschwefelsäure *f*
**nitrotoluene** Nitrotoluol *n*
**nitrous** salpeterhaltig, salpetrig, salpetrigsauer ~ **acid** Salpetrigsäure *f* ~ **oxide** Stickoxydul *n*, Stickstoffoxydul *n* ~ **oxide gas** Lachgas *n* ~ **vapor** Stickstoff-dampf *m* ~ **vitriol** Nitrose *f*
**nitro varnishes** Nitrolacke *f*
**nitroxylene** Nitroxylol *n*
**no-alloy** ohne Zusatz *m*
**noble metal** Edelmetall *n*
**no-bond resistance** Hyperkonjugation *f*
**no-current pulse** stromlose Zeichenschrift *f*
**no-delay, ~ operation** Schnellverkehr *m* ohne direkten Anruf *m*, Sofortverkehr *m* ~ **telephone** (**or traffic**) **exchange** Schnellverkehrsamt *n* ~ **telephone** (**or traffic**) **network** Schnellverkehrsnetz *n* ~ **trunk junction** Schnellverkehrsseitenamtsleitung *f* ~ **working** A- und B-Betrieb *m* im Fernverkehr
**no-drag slot** widerstandslose Flügelnute *f*
**no-feed** (**or zero feed**) Nullförderung *f*
**no-flicker frequency** Flimmerfrequenz *f*
,,**no-hang-up**'' **service** Sofortverkehr *m*
**no-impulse setting** Kein-Stromschritt-Stellung *f*
**no-lift** Nullauftrieb *m* ~ **line** Nullauftriebslinie *f*
,,**no lines**'' alle Leitungen *pl* besetzt
**no-load** Leerlauf *m* ~ **capacity** Eigenarbeit *f* ~ **characteristic** Leerlauf-charakteristik *f*, -kennlinie *f*, statische Kennlinie *f* (electr.) ~ **condition** Nullzustand *m* ~ **consumption** Leerlaufverbrauch *m* ~ **current** Leerlaufstrom *m* ~ **cutout** Nullausschalter *m* ~ **friction** Leergang-, Leerlauf-reibung *f* ~ **impedance** Leerlauf-impedanz *f*, -widerstand *m* ~ **kilovar requirements** Leerlaufblindleistungsbedarf *m* ~ **loss** Leerlaufverlust *m* ~ **position** Leerstellung *f* ~ **range of motor** (**or engine**) Motorleerlaufbereich *m* ~ **relay** Minimalrelais *n* ~ **release** Nullauslösung *f*, Nullstromauslösung *f* ~ **speed** Leerlaufdrehzahl *f* ~ **stroke** Leerhub *m* ~ **test** Leerversuch *m* ~ **time** Leerzeit *f* (teleph.) ~ **valve** Freilaß-Leerlauf-Ventil *n* ~ **voltage** Leerlauf-, Zündspannung *f* ~ **work** Leerlaufarbeit *f*
**no-man's land** Niemandsland *n*
**no-modulation lighting** Ruhebelichtung *f*, Ruhelicht *n*
**no parking** Parkverbot *n*
**no-reply call** unbeantworteter Ruf *m*
**no-rub wax** gleitsicheres Wachs *n*

**no-signal,** ~ **anode plate-cathode voltage** Anodenruhespannung f ~ **current** Anodenruhestrom m ~ **noise voltage** Ruhegeräuschspannung f ~ **potential** Ruhespannung f ~ **value** Ruhewert m

**no-slip,** ~ **condition** Haftbedingung f ~ **point** Fließscheide f

**no-sound,** ~ **density** Ruheschwärzung f ~ **illumination** Ruhelicht n ~ **lighting** Ruhebelichtung f

**no-such-number-tone generator** Hinweistonerzeuger m

**no trunks** alle Leitungen pl besetzt

**no-volt release** Nullspannungsauslösung f

**no-voltage,** ~ **circuit breaker** Null-, Nullspannungs-ausschalter m ~ **relay** Nullspannungsrelais n, Spannungsrelais n ~ **release** Ruhrstromauslöser m ~ **tripping breaker** Unterspannungsschalter m

**no-wind position** windlose Stellung f, Windstillepunkt m

**nob in wood** Knast m im Holz

**nobbing** Zängearbeit f, Zängen n

**noble** edel(metall) ~ **metal** Edelmetall n, edles Metall n ~ **opal** Edelopal m

**noctovision** Fernsehen n im Dunkeln, Nachtsehen n ~ **scanner** Strahlabtaster m

**noctovisor scan** Abtastung f nacheinanderfolgender Bildpunkte mit unsichtbaren ultraroten Strahlen, Ultrarotabtastung f

**nod** Wink m

**nodal,** ~ **curve** Knotenlinie f, Knotenpunktkurve f ~ **line** Knotenlinie f ~ **plane** Knotenebene f (opt.), Knotenfläche f ~ **point** (lens) Knotenpunkt m, (lens) Objektivknotenpunkt m, Schwingungsknoten m (phys.) (of undulating waves) Wellenknoten m ~-**point fitting** Knotenpunktbeschlag m ~ **surface** Knotenfläche f, -ebene f (opt.)

**node** Knollen m, Knoten m, Knotenpunkt m, Objektivknotenpunkt m, Verzweigungspunkt m ~ **factor** Knotenfaktor m ~ **theorem** Knotenregel f

**nodes of vibration** Bewegungsknoten pl

**nodular** knotig, kugelartig, kugelig, warzig ~ **cementite** kugeliger Zementit m ~ **powder** Fadenpulver n

**nodule** Druse f, Klümpchen n, Knöllchen n, Knollen m, Knötchen n, Knoten m, Niere f **in** ~s niedrig

**nodulous cementite** körniger Zementit m

**noil** Kämmling m

**noils,** ~ **of carded waste silk** Seidenwerg n ~ **dyeing** Kämmlingswollfärberei f

**noise** Brummen n, Geräusch n, Krach m, Lärm m, Rauschen n, Schall m, Summen n, Störung f ~ **abatement** Geräuschbekämpfung f, Lärm-abwehr f, -bekämpfung f, Schallabwehr f ~ **amplifier** Regelverstärker m ~ **analyzer** Geräuschanalysator m ~ **audiogram** Geräusch-audiogramm n ~ **bandwidth** Rauschbandbreite f ~ **behind the signal** Modulationsrauschen n ~ **box** Nasenkappe f ~-**bucking** Störgeräuschkompensation f ~ **caused by listening in** Geräusch n durch Mithören ~ **component** Schrotanteil m (TV) ~ **current** Diodengleichstrom m ~ **diode** Hochtastrauschdiode f, Rauschdiode f (electron.) ~ **decrease** Rauschabnahme f ~ **due to transients** Geräusch n durch Einschwing-

vorgänge ~ **eliminator** Glättungsdrossel f ~ **filter** Störschutz m ~ **gate** automatische Geräuschbeseitigung f, Krachtöter m, Störsperre f ~ **guards** Lärmschutzmittel pl (acoust.) ~ **height** Lärmspitze f ~ **increase** Rauchzunahme f ~ **input** Eingangsrauschen n ~ **intensity** Geräuschstärke f ~ **killer** Drosselsatz m, Geräuschvernichter m, Knallschutzgerät n (rdo) ~ **level** Fremdspannungsabstand m, Geräusch-pegel m, -spiegel m, Rauschpegel m, Stör-amplitude f, -höhe f, -lautstärke f, -niveau n, -pegel m, -spiegel m ~ **limiter** Rausch-, Störbegrenzer m ~ **lock** Rauschsperre f ~ **matching** Rauschanpassung f ~ **measure** Kettenrauschzahl f (electr. tube) ~-**measuring set** Geräuschmesser m ~ **meter** Geräuschmesser m, Geräuschspannungs-messer m, -zeiger m ~ **output** Rausch-, Stör-ausgang m ~ **output power** Ausgangsrauschleistung f ~ **parameters** Rauschkennwerte pl ~ **pattern** Geräuschmuster n ~ **potential** Rauschspannung f ~ **power** Rauschleistung f ~-**propagating** (structures not acoustically sound) hellhörig ~ **reduction circuit** Geräuschunterdrückungsstromkreis m ~-**reduction stop** Reintonblende f ~ **resistance** Rauschwiderstand m ~ **standard** Geräuschnormal n ~ **standard temperature** Rauschbezugstemperatur f ~ **suppression** Entstörung f, (automatische) Geräuschbeseitigung f, Schallabwehr f, Stör-schutz m, -packung f ~ **suppressor** Geräuschsperre f, Rauschunterdrückungsschaltung f ~-**transmission impairment** Minderung f der Übertragungsgüte durch Leitungsgeräusche ~ **transmitter** Rauschsender m (rdo) ~ **tuning** Rauschabstimmung f ~ **twoport** Rauschvierpol m (electr. tube) ~ **unit** gebräuchliches Geräuschmaß n ~ **voltage** Geräusch-, Rausch-spannung f

**noiseless** geräuscharm, geräuschlos, lautlos ~ **commutation** geräuschlose Kommutierung f ~ **film method** Klartonverfahren n ~ **recording** (on film) Reintonverfahren n ~ **running** (engine) geräuschloser Gang m, ruhiger Gang m ~ **running of an engine** ruhiger Lauf m des Motors

**noiselessness** Geräuschlosigkeit f, Rauscharmut f

**noisy** geräuschvoll, klirrend, laut ~ **blacks** gestörtes Schwarz n ~ **circuit** geräuschvolle Leitung f, Leitung f mit Geräusch oder mit Kraftgeräuschen ~ **running** (engine) geräuschvoller Gang m, unruhiger Gang m ~ **valve** stark rauschende Röhre f

**nomenclature** Benennung f, Bezeichnung f, Namenverzeichnis n, Nomenklatur f

**nominal,** ~ **acceptivity** Nennaufnahme f ~ **allowance** Nennmaß n ~ **black signal** Testsignal n für Schwarz ~ **capacitance** Nennkapazität f ~ **capacity** Nenn-, Normal-leistung f ~ **circuit voltage** maximal zulässige Betriebsspannung f ~ **conductor (cross) section** Leiternennquerschnitt m ~ **cross section** Nennquerschnitt m ~ **cutoff frequency of a loaded line** Nennwert m der Grenzfrequenz einer bespulten Leitung ~ **diameter** Nenndurchmesser m, Nennweite f, Solldurchmesser m ~ **diameter of case** Gehäusenenndurchmesser m ~ **easy axis** Sollvorzugsachse f ~ **feedback ratio** nominales Rückkopplungsverhältnis n ~ **frequency** Nennfrequenz f ~ **gauge size** Lehrensollmaß n ~ **line**

**width** Zeilennennlänge *f* ~ **load** Nennbelastung *f* ~ **margin** Nennspielraum *m* ~ **maximum circuit** fiktiver Bezugskreis *m* ~ **measure** Nennmaß *n* ~ **output** Nennleistungsdurchgang *m* ~**-power** Nominalkraft *f* ~ **pressure** Nenndruck *m* ~ **range** Nennreichweite *f* ~ **rating** Regelstärke *f* ~ **ratio error** Nennübersetzungsfehler *m* ~ **size** Nenndurchmesser *m*, (of a screw) Nennmaß *n* ~ **stress at fracture** Nennlast *f* beim Bruch ~ **torsional moment** Nenndrehmoment *n* ~ **transformation ratio** Nennübersetzungsverhältnis *n* ~ **value** Nennwert *m*, Nominalbetrag *m*, Sollwert *m* ~ **value setter** Sollwerteinsteller *m* ~ **voltage** Nennspannung *f* ~ **volume** Nenngasinhalt *m* ~ **white signal** Testsignal *n* für Weiß ~ **width** Nennweite *f*

**nominally transverse word current** Nennwort-Querstromstärke *f*

**nomogram** Gaußsches Verteilernetz *n*, Nomogramm *n*

**nomograph** Fluchtlinientafel *f*, Gaußsches Verteilungsnetz *n*, Nomogramm *n*

**nomography** Nomografie *f*

**non-abrasive quality** Abriebfestigkeit *f*

**non-absorbent** hydrophob

**non-absorbing** schallhart ~ **horn wall** schallharte Trichterwand *f*

**non-acceptance** Nichtannahme *f*

**non-acid Manila paper** säurefreies Manilapapier *n*

**non-acknowledgment** Nichtanerkennung *f*

**non-additivity correction** nicht additive Korrektur *f*

**non-adhesive** nichthaftend

**non-adjacent** nichtbenachbart

**non-affected** unangreifbar

**nonagon** Neuneck *n*

**non-alcoholic** alkoholfrei

**nonane** Nonan *n*

**non-animated picture** Stehbild *n*

**non-aperiodic indication (or reading)** ungedämpfte Anzeige *f* (instrument)

**non-apertured** kompakt

**non-appearance** Nichterscheinen *n*

**non-aqueous** nichtwässerig

**non-arcing** funkenfrei (metal) ~ **property** Lichtbogensicherheit *f*

**non-astatic** richtungsempfindlich

**non-attendance** Nichterscheinen *n*

**nonavalent** neunwertig

**non-axial trolley** Rollenstromabnehmer *m* für seitlichen Fahrdraht

**nonbaking coal** magere Kohle *f*, Magerkohle *f*, Sandkohle *f*

**nonballistic reading** gedämpfte Anzeige *f*

**nonbituminous coal** magere Kohle *f*, Magerkohle *f*

**nonbleeding** nicht auslaufend

**nonblistered steel** blasenloser Stahl *m*

**nonbounded** nichtbeschränkt

**nonbranched** unverzweigt

**nonbranching** verzweigungslos

**nonbreakable** unzerbrechlich ~ **skid** Gleitschutz *m*

**nonbreeding material** nichtbrütendes Material *n*

**nonbrowning glass** stabilisiertes Glas *n*

**nonbuckled** wölbungsfrei

**noncaking,** ~ **coal** Sand-, Sinter-kohle *f* ~ **condition** (of materials) Rieseln *n*

**noncapacitive** kapazitätsfrei

**noncapsizable** nichtkenterbar

**noncarbonaceous** kohlefrei

**noncarbonate hardness** Nichtkarbonathärte *f*

**noncenterable** nicht zentrierbar

**noncentering** nicht zentrierbar

**noncentral** nichtzentral

**noncentrosymmetric** nichtzentralsymmetrisch

**nonchromatic** chromatisch korrigiert

**noncircular** unrund

**noncirculating water cooling** Durchlaufkühlung *f*

**nonclinkering coal** schlackenreine Kohle *f*

**nonclouding windscreen** Klarsichtscheiben *pl*

**noncohesive** nicht bindig (kohäsionslos)

**noncolinear forces** Kräftepaar *n*

**noncombustible** unbrennbar, unentflammbar

**noncommissioned** nicht bevollmächtigt

**noncompetitive** außer Wettbewerb *m*, konkurrenzlos

**noncompliance** Nichteinhaltung *f*

**noncompulsory** fakultativ

**nonconcentric** exzentrisch

**noncondensing,** ~ **engine** Auspuffmaschine *f* ~ **operation** Gegendruckbetrieb *m*

**nonconducting** nichtleitend ~ **material** Wärmeschutz-masse *f*, -mittel *n*

**nonconductive** nichtleitend ~ **finish** nichtleitender Überzug *m*

**nonconductor** Nichtleiter *m*

**nonconforming** unstimmig

**nonconservation** Nichterhaltung *f*

**nonconservative motions** nichtkonservative Bewegungen *pl* (Klettern)

**noncontact,** ~ **gauging** berührungsloses Messen *n* ~ **piston** Kurzschlußschieber *m*

**noncontinuous** nichtkontinuierlich

**noncontrasty,** ~ **condition** Flachheit *f* (photo) ~ **image (or picture)** flaues Bild *n* ~ **quality** Flachheit *f* (photo) ~ **quality of a picture** Bild-flachheit *f*, -flauheit *f*

**noncontrollable** nichtreproduzierbar

**noncorrodibility** Rost-festigkeit *f*, -sicherheit *f*

**noncorrodible** korrosionssicher, rostbeständig, rostsicher, unzerstörbar

**noncorroding** korrosionsfrei, nichtrostend ~ **steel** rostbeständiger Stahl *m*

**noncorrosible** korrosionsbeständig

**noncorrosion property** Rostsicherheit *f*

**noncorrosive** korrosionsfrei, nichtkorrosiv, unangreifbar, unzerstörbar ~ **grease** säurefreies Fett *n* ~ **steel** rostfreier Stahl *m*

**noncrossing rule** Gesetz *n* der Nichtüberkreuzung

**noncrucible furnace** tiegelloser Ofen *m*

**noncrystalline** bildlos, gestaltlos, nichtkristallisch, unkristallinisch

**noncutting** spanlos ~ **shaping** spanlose Verformung *f*

**nondazzling** blendungsfrei

**nondecimal base** Nicht-dezimalsystem *n*, -zehnersystem *n*

**nondeformability** Form-beständigkeit *f*, -festigkeit *f*

**nondegenerate** nichtentartet

**nondelay fuse** Aufschlagzünder *m* ohne Verzögerung

**nondelivery post office** Abholungspostamt *n*
**nondestructive** zerstörungsfrei ~ **readout** löschungsfreier Lesevorgang *m* ~ **readout memory** löschungsfreier Lesespeicher *m* ~ **test** nichtzerstörende Prüfung *f*, Widerstandsprobe *f* ~ **testing** zerstörungsfreie Werkstoffprüfung *f*
**nondiaphanous** lichtundurchlässig, undurchsichtig
**nondimensional** unbenannt ~ **figure** unbenannte Zahl *f* ~ **quantity** absoluter Zahlenwert *m*, dimensionslose Größe *f* ~ **ratio** absoluter Zahlenwert *m*, Zahlenverhältnis *n* in Form einer unbenannten Zahl
**nondimensionalized** dimensionslos
**nondirectional** richtwirkungsfrei, richtungsunempfindlich, ungerichtet ~ **antenna** Rundstrahler *m*, ungerichtete Antenne *f* ~ **beacon (NDB)** ungerichtetes Funkfeuer *n* ~ **microphone** nichtgerichtetes Mikrofon *n*, Raummikrofon *n*, ungerichtetes Mikrofon *n* ~ **transmission** Raumstrahlen *n*
**nondirective** ungerichtet ~ **antenna** Rundstrahlantenne *f*, ungerichtete Antenne *f*
**nondisintegrability** Nichtzerfallbarkeit *f*
**nondisintegrable** nicht zerfallbar
**nondissipative** energielos, leistungslos, ohne Energieverbrauch *m*, verlustlos ~ **line** ideale Leitung *f* (teleph.) ~ **stub** Blindleitung *f* (zur Anpassung), Hohlleiter-Anpassungsglied *n*, verlustfreie Stichleitung *f*
**nondissociated** nichtdissoziiert
**nondistorting** verzerrungsfrei
**nondiving** überschlag(s)sicher
**nondrying oil** nicht eintrocknendes Öl *n*
**nonductile** unausdehnbar, undehnbar, unstreckbar, unziehbar
**non-eccentric valve gear** Gelenkschiebersteuerung *f*
**non-echoing** rückschallfrei
**non-eddying** strudellos
**non-effective** dienstunfähig
**non-elastic neutron cross sections** Neutronenquerschnitte *pl* bei nichtelastischer Streuung
**non-electrolyte** Nichtelektrolyt *n*
**non-equivocal** vieldeutig
**non-erasable storage** nichtlöschbare Speicherung *f*
**non-euclidian** nichteuklidisch
**non-exploding reservoir** explosionssicheres Gefäß *n*
**non-extensible tube** nichtausziehbarer Tubus *m*
**non-exuding gelatin** nichttropfende Gelatine *f*
**nonfading** lichtecht
**nonferrous** nicht eisenhaltig ~ **copper** eisenfreier Kupferdraht *m* ~ **forging** Metallschmiedestück *n* ~ **metal** Buntmetall *n*, nicht eisenhaltiges Metall *n*, Nichteisenmetall *n* ~-**metal alloy** Nichteisenmetallegierung *f* ~ **metal strips** Buntmetallbänder *pl* ~ **metallurgy** Metallhüttenkunde *f* ~ **metals** Buntmetalle *pl*
**nonfibrous** nichtfaserig
**nonfilamentary coil** flächenhafte Spule *f*
**nonfission absorption** Absorption *f* ohne Spaltung
**nonfissionability** Nichtzerfallbarkeit *f*
**nonflam film** nichtentzündbarer Film *m*
**nonflashing** rückschlaglöschend

**nonflexible connection for glider towing** Starrschlepp *m*
**nonflicker shutter** flimmerfreie Verschlußblende *f*
**non-flow condition** Nullförderung *f* (hydr. pump)
**non-flying scale model** nichtfliegendes maßstäbliches Modell *n*
**nonforfeiture** Unverfallbarkeit *f* ~ **values** Garantiewerte *pl*
**nonfraying** nicht ausfasernd
**nonfreezable** frostsicher
**nonfreezing** kältebeständig ~ **lubricant** Frostschutzfett *n* ~ **mixture** Frostschutzmittel *n*
**non-frequency-critical** nicht frequenzempfindlich
**nonfulfilment** Nichterfüllung *f*
**nonfully development state** Anfachungsstadium *n*
**non-gagged** ohne Knebel *m*
**nongaseous coal** gasarme Kohle *f*
**nongasifiable** nichtvergasbar
**nongenuine** unecht
**nongrid** unstarr
**nongrowth cast iron** nicht dilatierendes Gußeisen *n*
**nongumming** harzfrei, nicht harzend
**nonhalating** lichthoffrei, lichthofsicher
**nonhalation plate** lichthoffreie Platte *f*
**nonheat treatable** thermisch nicht vergütbar
**nonheatable** nichtheizbar
**nonheterodyne reception** Suchempfang *m*
**nonholonomic** nicht-holonom
**nonhorizontal** schräg
**nonhydrocarbon impurities** nichtkohlenwasserstoffhaltige Unreinheiten *pl*
**nonhygroscopic** feuchtigkeitssicher, hydrophob, unhygroskopisch, wasserabstoßend
**non-idle spinning machine** Doppelspulspinnmaschine *f*
**non-ignitable** unentflammbar
**non-illuminated** unbeleuchtet
**non-immersion objective** Trockenlinse *f*
**non-impregnated sleeper** Rohschwelle *f*
**non-inclinable stand** Stativ *n* ohne Kippe
**non-inductive** (free of inductance) induktionsfrei ~ **circuit** nicht induktiver Stromkreis *m* ~ **resistance** induktionsfreier Widerstand *m* ~ **winding** bifilare oder induktionsfreie Wicklung *f*
**non-inductively wound** bifilar
**non-inflammable** flammsicher, unbrennbar, unentflammbar
**non-intelligible crosstalk** unverständliches Nebensprechen *n*
**non-interacting** rückwirkungsfrei
**non-interchangeability** Unvertauschbarkeit *f*
**non-interchangeable** nichtaustauschbar, unauswechselbar, unvertauschbar, unverwechselbar ~ **plug** unverwechselbarer Stecker *m* ~ **wall socket** unverwechselbare Anschlußdose *f*
**non-intermittent** fortlaufend ~ **feed (or motion) of film (or picture strip)** stetiger Bildwechsel *m* ~ **voltage** Dauerspannung *f*
**non-intersecting** ungekreuzt
**non-ion** Nichtion *n*
**non-ionized** nichtionisiert
**non-isotropic** anisotrop
**nonius** Nonius *m* ~ **zero** Noniusnullpunkt *m*
**nonkinking** drallfrei

**nonknocking surface ignition** klopffreie Oberflächenzündung *f*
**nonlagging** nachwirkungsfrei
**nonleaded gasoline** unverbleites Benzin *n*
**nonlinear** nichtlinear, ungerade, ung(e)radlinig ~ **characteristic** nicht lineare Kennlinie *f* ~ **conductance** (cryst.) asymmetrische Leitfähigkeit *f* ~ **device** nicht lineare Vorrichtung *f* ~ **distortion coefficient** Klirrfaktor *m* (einer Verstärkerröhre), Klirrverzerrung *f*, nichtlineare oder ungradlinige Verzerrung *f* ~ **harmonic-distortion coefficient (or factor)** Klirrfaktor *m* ~ **potentiometer** nichtlinearer Spannungsteiler *m* ~ **scale** nichtlineare Skala *f*
**nonlinearity** Nicht-, Un-linearität *f*, Ung(e)radlinigkeit *f*
**nonlisting cycle** Sammelgang *m*
**nonloaded** unbelastet, unbespult, ungeladen ~ **Q** Leerlaufgütefaktor *m* ~ **(submarine)telegraph cable** Thomsonkabel *n*
**nonlocal theory** nichtlokale Theorie *f*
**nonlocalized** nichtlokalisiert
**nonlocking,** ~ **key** nicht festlegbarer Schalter *m* ~ **type button** Taste *f* ohne Rastung (lose Taste)
**nonmagnetic** nicht-, un-magnetisch ~ **steel** unmagnetischer Stahl *m*
**nonmalleable** unstreckbar
**nonmarring** kratzfest
**nonmeasuring pump** nichtmessende Pumpe *f*
**nonmetal** Ametall *n*, Nichtmetall *n* ~ **shot (alloy)** Granalien *pl*
**non-metalized theatre screen** nicht metallisierte Leinwand *f*
**nonmetallic** nichtmetallisch ~ **sheathed wires and cables** Mantelleitungen *pl*
**nonmetering** Zählunterdrückung *f* ~ **relay** Zählverhinderungsrelais *n*
**nonmicrophonic** antimikrofonisch, klingfrei
**nonmobile radio station** Funkfeststation *f*
**nonmovable overload** ruhende Überlast *f*
**non-obligatory** fakultativ
**non-observance** Nichtbeobachtung *f*
**non-operate current** Fehlstrom *m* (v. Relais)
**non-operative** in Ruhe *f* befindlich ~ **time** Totzeit *f* (bei Maschinen)
**non-oscillable** schalltot
**non-oscillating** schwingungsfrei ~ **condition** schwingungsfreier Zustand *m* ~ **phenomenon** schwingungsfreier Vorgang *m*
**non-oscillatory,** ~ **circuit** nichtschwingungsfähiger Kreis *m* ~ **discharge** aperiodische Entladung *f*
**non-oxidizable** unoxydierbar
**non-oxidizing** nichtoxydierend ~ **annealing and hardening furnace** Blankhärte- und Glühofen *m*
**nonpaying mine** Zubußzeche *f*
**nonpayment** Nichtzahlung *f*
**nonperformance** Nichterfüllung *f*
**nonperiodic** unperiodisch ~ **action (or phenomenon)** unperiodischer Vorgang *m* ~ **temperature changes** nichtperiodische Temperaturunterschiede *pl*
**nonpermanent** nichtständig ~ **magnet** fremderregter Magnet *m* ~ **runway** neutrale Bahn *f*
**nonpersistence** Flüchtigkeit *f*
**nonpersistent,** ~ **chemical agent (or gas)** flüchtiger Kampfstoff *m* ~ **chemical particle** Schwebstoffteilchen *n*

**nonphantom(ed) circuit** nicht viererfähige Leitung *f*
**nonplanar network** nicht ebenes Netzwerk *n*
**nonplastic film** Flachfilm *m*
**nonpoisonous (non-toxic)** ungiftig
**nonpolarized** neutral, unpolarisiert ~ **relay** neutrales, nicht polarisiertes oder unpolarisiertes Relais *n*
**nonpolarizing** unpolarisierbar
**nonporous** blasenfrei ~ **fabric** luftundurchlässiges Gewebe *n*
**nonpower driver** ohne eigenen Antrieb *m*
**nonprecedence call** Gespräch *n* ohne Vorrang
**nonpreferential orientation** nichtbevorzugte Orientierung *f*
**nonpressure tunnel** Freispiegelstollen *m*
**nonpressurized** drucklos, nicht druckbelüftet, unter Außendruck *m* stehend ~ **water** nicht drückendes Wasser *n*
**nonproductive,** ~ **lead mine** unergiebige Bleimine *f* ~ **operations** Routineoperationen *pl* ~ **time** Nebenzeit *f*
**nonprofit insurance** Versicherung *f* ohne Gewinnanteil
**nonquadded cable** doppeladriges oder paarverseiltes Kabel *n*
**nonquantized system** klassisches oder nicht quantisiertes System *n*
**nonradiative** strahlungslos
**nonrandom orientation** bevorzugte Orientierung *f*
**nonreacting** rückstoßfrei
**nonreactive** Wirkwiderstand *m*; induktionsfrei, phasenfrei, reaktionslos, unempfindlich, winkelfrei ~ **artificial cable** künstliches Kabel *n* (aus reinen Widerständen gebildet) ~ **coupling** reaktionslose Kopplung *f* ~ **resistance** Dämpfungswiderstand *m*, reiner Ohmscher Widerstand *m*, Verlustwiderstand *m*
**nonrecoil(ing)** rückstoßfrei
**nonrecurrent** einmalig, unperiodisch ~ **action** einmaliges Ereignis *n*, einmaliger Vorgang *m*, unperiodischer Vorgang *m* ~ **event** einmaliges Ereignis *n* ~ **phenomenon** unperiodischer Vorgang *m*
**nonreflected sound** nichtrückkehrende Schallstärke *f*
**nonreflecting** reflexions-, rückschall-frei ~ **diaphragm** atmende Membran(e) *f* ~ **termination** angepaßter Abschlußwiderstand *m*, reflexionsfreier Abschluß *m*
**nonreflection attenuation** Anpassungsdämpfung *f*, Reflexionsverlust *m*
**nonrefractory** schmelzbar
**nonregenerative** rückkuppelungsfrei
**nonregistering** Zählunterdrückung *f*
**nonrelativistic** unrelativistisch
**nonrelief type** Negativschrift *f*
**nonremovable** unverlierbar
**nonrepeatered toll circuit** Fernleitung *f* ohne Verstärker
**nonrepetitive** einmalig
**nonreproducible** nichtreproduzierbar
**nonresinifying** nichtverharzend
**nonresinous** harzfrei
**nonresisting** unecht
**nonresonant** außer Resonanz *f* befindlich, nicht abgestimmt, nicht mitschwingend, nicht resonierend, schalltot ~ **line** (mit) Wellenwider-

stand *m* abgeschlossene Leitung *f* (aperiodische Leitung)
**nonresounding** rückschallfrei
**nonreturn** nicht umkehrbar ~ **flap** Rückschlag-klappe *f* ~ **valve** Rückhaltklappe *f*, Rückschlag -klappe *f*, -ventil *n*, Rückströmventil *n*
**nonreversed** seitenrichtig
**nonreversibility of the thread** Selbsthemmung *f* oder Selbstsperrung *f* des Gewindes
**nonreversing lock** Rückdrehsperre *f*
**nonrevolving** nichtdrehbar
**nonrigid** unstarr ~ **airship** unstarres Luftschiff *n* ~ **dirigible** Pralluftschiff *n*, unstarres Luftschiff *n* ~ **drive** elastischer Antrieb *m* ~ **molecule** unstarres Molekel *n*, unstarres Molekül *n* ~ **noncircular cone** (with curved radiating surface) Falzmembran *f*
**nonrigidity** Unstarrheit *f*
**nonrising stem** (valve) mit innenliegendem Spindelgewinde *n*
**nonrotating** (rope) drall-arm, -frei
**nonrub polish** glänzend auftrocknende Politur *f*
**nonrubberized** ungummiert
**nonrusting** rostsicher
**nonsalient pole generator** Generator *m* mit gleichförmigem Luftspalt
**nonscaling** zunderbeständig ~ **property** Zunderbeständigkeit *f*
**nonselective** nicht selektiv, wahllos ~ **pitch** Luftschraube *f* mit elektrisch kontrollierter Steigung
**non-self-maintained discharge** unselbständige Entladung *f*
**non-self-sustained discharge** unselbständige Entladung *f*
**non-self-sustaining** unselbständig
**nonsensical** gegensinnig
**nonsensitive indicator** Grobzeiger *m*
**nonsequential** sprunghaft
**nonsequentially** sprungweise
**nonsharp** unscharf
**nonsiliceous** siliziumfrei
**nonsinous** ungewellt (Teilkammer)
**nonskid** (tires) profiliert, rutschsicher ~ **chain** Gleitschutz-, Schnee-kette *f* ~ **cover of tire** Gleitschutzmantel *m* ~ **device** Gleitschutzvorrichtung *f* ~ **equipment** Gleitschutzmittel *pl* ~ **pattern** Laufflächendessin *n* ~ **property** Griffigkeit *f* (Straße) ~ **surfacing** rutschsicherer Belag *m* ~ **thread** Gleitschutz *m* ~ **tire** Gleitschutzreifen *m*, nichtschleudernder Reifen *m*
**nonslip** schiebefest, (in drive) zwangläufig ~ **drive** schlupffreier Antrieb *m* ~ **finish** Schiebefestappretur *f*, schiebefeste Ausrüstung *f* ~ **plate** Gleitschutzplatte *f*
**nonslipping** gleitfrei
**nonsoldering** unlötbar
**nonsparking** funkenfrei; Funkenlosigkeit *f* (electr.)
**nonspherical** asphärisch, nichtspärisch ~ **aplanatic focusing lens** aplanatische asphärische Ophthalmoskoplinse *f*
**nonspinning** trudelsicher ~ **airplane** trudelsicheres Flugzeug *n* ~ **wire rope** dralloses Kabel *n*
**nonsplintering** nicht splitternd
**nonspontaneous** unselbständig ~ **transition** erzwungener Übergang *m*
**nonsqueaking** nichtquietschend
**nonstandard(ized)** nichtgenormt

**nonsteady** (flow) instationär, nichtstationär
**nonsterioscopic,** ~ **film** Flachfilm *m* ~ **vision** einäugiges Sehen *n*
**nonstochiometric compound** geordnete Mischphase *f*
**nonstop** aufenthaltslos ~ **flight** Dauer-, Ohnehalt-flug *m*, Flug *m* ohne Halt ~ **train** Durchgangszug *m*, durchgehender Zug *m*
**nonstrain aging steel** reckalterungsbeständiger Stahl *m*
**nonsuit, to** ~ abweisen (eine Klage)
**nonsuperconducting** nichtsupraleitend
**nonsymmetric** nichtsymmetrisch
**nonsymmetrical profile** gewölbtes Profil *n*
**nonsynchronous** asynchron ~ **pulsation-weld timer** Asynchronschwingungsregler *m* ~ **rotating spark gap** Asynchronfunkenstrecke *f*
**nonsyndicated** verbandsfrei
**nontacky** nicht klebrig
**nonterminating** unendlich
**nonthermal** nichtthermisch
**nontoxid** giftfrei
**nontracking quality** Kriechstrombeständigkeit *f*
**nontransferrable** unübertragbar
**nontransparency** Glanzlosigkeit *f*
**nontransparent** undurchsichtig
**nontransposed metallic curcuit** ungekreuzte Doppelleitung *f*
**nontwisting** drallfrei, nicht verdreht ~ **wire rope** torsionsfreies Drahtseil *n*
**non-typing mechanical functions** nichtschreibende Arbeitsgänge *pl* (teletype)
**non-uniform** nicht stationär, uneinheitlich, ungleich, ungleich-artig, -förmig, -mäßig, unstet, unstetig ~ **frequency response** Frequenzverzerrung *f*
**non-uniformity** Ungleich-förmigkeit *f*, -mäßigkeit *f*, Unstetigkeit *f* ~ **coefficient** Ungleichförmigkeitsziffer *f*
**non-use** Nichtbenutzung *f*
**nonvalent** nullwertig
**nonvalid** ungültig
**nonvariable,** ~ **capacitor (or condenser)** Festkondensator *m* ~ **head torch** Einzelbrenner *m* ~ **head welding torch** Einzelschweißbrenner *m*
**nonvariant** nichtvariant ~ **equilibrium** nonvariantes oder unfreies Gleichgewicht *n*
**nonvertical** wirbel-frei, -los
**nonvibratile** erschütterungsfrei
**nonviscous** reibungslos, unviskos
**nonvitreous inclusion** Steinchen *n* (in glass)
**nonvocal consonant** stimmloser Konsonant *m*
**nonvolatile** nicht flüchtig, nichtvergasbar ~ **storage** Dauerspeicher *m*, leistungsloser Speicher *m*
**nonwoody tissue** nichtverholztes Gewebe *n*
**nonworkable** unverformbar
**nonwoven** ungewebt ~ **fabric** Faservlies *n*
**nonyielding** unausgiebig
**noose** Fallstrick *m*, Schlaufe *f*, Schleife *f*, Schlinge *f*
**noria** Bewässerungsgrad *n*, Entwässerungsmühle *f*, Eimer-, Paternosterwerk *n*
**norm** (used for comparison) Bezugssystem *n*, Norm *f*
**normable** normierbar
**normal** (line) Einfallslot *n*; (line) Normale *f*, Senkrechte *f*, gewöhnlich, normal, norment-

sprechend, normgerecht, normrecht, regelmäßig, regelrecht, in der Ruhestellung *f*, senkrecht, in der Stellung *f*, in Stellung *f*, üblich,
vorschriftsmäßig
**normal, to set at** ~ normalschalten **from** ~
**flying position** aus der Normallage *f* ~ **of a**
**plane curve** Normale *f* einer ebenen Kurve ~ **of**
**a plane surface** Normale *f* einer Fläche ~ **to**
**point of contact** Berührungsnormale *f* ~ **to sur-**
**face** Oberflächennormale *f*
**normal,** ~ **accelerometer** Normalbeschleuni-
gungsmesser *m* ~ **adjustment** Normaleinstel-
lung *f* ~ **axis** Z-achse *f* ~ **background** normaler
Nulleffekt *m* ~ **band** normale Spurengruppe *f*
~ **barrage** Notfeuer *n* ~ **boost** Normaldruck *m*
~ **calomel electrode** (of Ostwald electrode) Nor-
malkalomelektrode *f* ~ **capacity** Regelleistung
*f* ~ **case** Normalfall *m* ~ **charge** Gebrauchs-,
Verbrauchs-ladung *f* ~ **class** Achsenkreuz *n* ~
**component of force** Normalkraft *f* ~ **component**
**of force on control surface** Rudernormalkraft *f*
~ **concentration** Normalkonzentration *f* ~
**contact** Ruhekontakt *m*, Wählerruhekontakt *m*
~ **cut** X-Schnitt *m* (cryst.) ~ **divider** Normalteiler *m* ~ **effect in gravity measurements** normaler Einfluß *m* in gravimetrischen Messungen
~ **efficiency** Nennleistung *f* ~ **electrode potential**
**of a metal** Normalpotential *n* eines Metalls ~
**energy level** Normalzustand *m* ~ **equation** Nor-
malgleichung *f* ~ **fault** gewöhnliche Verwerfung *f* (geol.) ~ **field of a mine where coal was**
**recently disclosed by borings** neugemutetes Nor-
malfeld *n* einer Grubenanlage ~ **flying position**
Normallage *f* ~ **force** Axial-, Normal-, Senk-
recht-kraft *f* ~-**force coefficient** Pfeilkrafttreibwert *m* ~ **frequency** Vergleichsfrequenz *f* ~
**frequency auxiliary transmitter** Vermessungs-
sender *m* ~ **gauge** Regelspur *f* ~-**gauge railway**
Normalspurbahn *f* ~ **head turbine** Normal-
gefällerad *n* ~ **horsepower** Normalpferdestärke
*f* ~ **hydrogen electrode** Normalwasserstoff-
elektrode *f* ~ **iconoscope** Bildspeicherröhre *f* ~
**junction working** A-B-Betrieb *m* ~ **law of**
**errors** Fehlernetz *n* ~ **law curve** normale Wahrscheinlichkeitskurve *f* ~ **lift** Normalauftrieb *m*
~ **line** Lot *n*, Plusstrang *m* ~ **liquid** Normal-
flüssigkeit *f* ~ **load** Nennlast *f*, Normalbelastung *f*, Normallast *f*, Regelbelastung *f* ~-**load**
**case** Regelbelastungsfall *m* ~ **mode** Normal-
schwingung *f* ~ **moisture content** Normalwas-
sergehalt *m* ~ **operating current** Regelstrom *m*
~ **operation** Normalbetrieb *m* ~ **output** Nenn-,
Regel-leistung *f* ~ **permeability** gewöhnliche
Permeabilität *f* ~ **photoelectric emission** normaler Fotoeffekt *m* ~ **photograph** Normalauf-
nahme *f* ~ **plane of a curve in space** Normal-
ebene *f* einer Raumkurve ~ **plane pitch** Nor-
malmodul *n* ~ **pool elevation** gewöhnlicher
Stau *m* ~ **position** Ausgangs-, Grund-, Normal-,
mal-, Null-, Regel-stellung *f*, Ruhelage *f*,
Ruhestellung *f* ~ **power** Dauerleistung *f* (Triebwerk) ~ **pressure** Normal-, Wand-druck *m* ~
**profile** Normalprofil *n* ~ **radiation** Zentralstrahl *m* ~ **rating** Dauerleistung *f* ~ **running**
**potential** Brennspannung *f* ~ **scale** Normal-
skala *f* ~ **scheme of erection** Normalaufstellung
*f* ~ **section of a surface** Normalschnitt *m* einer
Fläche ~ **sighted** normalsichtig, rechtsichtig

~-**sighted eye** rechtsichtiges Auge *n* ~ **size** Regelgröße *m* ~ **solution** Normal-flüssigkeit *f*,
-lösung *f* ~ **speed** Betriebsdrehzahl *f* ~ **standard**
**cell** Weston-Normalelement *n* ~ **stereogram**
Normalstereogramm *m* ~ **strength** Normalstärke *f* ~ **stress** Normalbeanspruchung *f*, (in
testing) Normalspannung *f* ~ **surface** oberflächengerichtet ~ **temperature** Bezugstemperatur *f* ~ **thermometer** Normalthermometer *n*
~ **treshold of feeling** normale Schmerzschwelle
*f* (acoust.) ~ **turn** normaler Kurvenflug *m* ~
**twisting** Normaldrall *m* ~-**type ignition** Ein-
fachzündung *f* ~ **upspacing** Zeilentransport *m*
~ **value** Nennwert *m* ~ **velocity** Vertikalgeschwindigkeit *f* ~ **voltage divider** Spannungsteiler *m* ~ **width** Normalbreite *f*
**normalization** Normenaufstellung *f*, Normierung
*f*, Vereinheitlichung *f*
**normalize, to** ~ ausglühen, normalglühen, (in
heat-treating) normalisieren, spannungsfrei
glühen (aviat.)
**normalized** normiert, weich geglüht ~ **admittance**
relativer Leitwert *m* ~ **eigen-functions** normier-
te Eigenfunktionen *pl* ~ **plateau slope** normalisierte Plateauneigung *f*
**normalizing** Ausglühen *n*, Glühen *n* mit nachfolgender Abkühlung an der Luft, (annealing
followed by air cooling) Normalglühen *n*, Normalisieren *n* ~ **furnace** Normalisierofen *m* ~
**plant** Normalisierungsanlage *f*
**normally** für gewöhnlich, im rechten Winkel *m*
~ **aspirated engine** Motor *m* ohne Lader, selbstansaugender Motor *m* ~ **closed contact** Öffner
*m*, Ruhekontakt *m* des Relais ~ **open contact**
Arbeitskontakt *m*, Schließer *m* ~ **pulled up** an-
gezogen
**normatron** Normatron *n*
**Norrish pattern** Norrish-Schema *n*
**north** Nord *m*, Norden *m*, Nordpunkt *m* ~
**direction** (magnetic, grid, true) Nordrichtung *f*
**northeast** Nordosten *m* ~ **monsoon** Nordost-
monsun *m* ~ **trade wind** Nordostpassat *m*
**north,** ~ **magnetic** Mitternachtsgang *m* ~-**magnetic**
nordmagnetisch ~ **magnetic pole** Nordpol *m* ~
**point** Nordpunkt *m* ~ **pole** Nordpol *m* ~-**seek-**
ing nordsuchend ~-**seeking gyro** nordsuchender Kreisel *m* ~-**seeking pole** nordsuchender
Pol *m* ~-**south component** Hochwertkompo-
nente *f* ~ **star** Polarstern *m* ~ **ward** nördlich,
nordwärts ~ **west** Nordwesten *m* ~ **wester**
Nordwester *m* ~ **wind** Nordwind *m*
**northerly** nördlich ~ **turning error** Norddreh-
fehler *m*, nördlicher Querneigungsfehler *m*
**northern** nördlich ~ **column** Nordpfeiler *m* ~
**hemisphere** Nord-halbkugel *f*, -hemisphäre *f*,
nördliche Halbkugel *f* ~ **latitude** nördliche Breite *f* ~ **lights** Nordlicht *n* ~ **lode** Mitternachtsgang *m* ~ **region** Norden *m*
**Norway,** ~ **iron** Holzkohleneisen *n* ~ **spruce**
Balsamtanne *f*
**Norwegian harrow** Rollenegge *f*
**nose, to** ~ **over** (Flugzeug am Boden) überschlagen
**nose** Arbeitsende *n* (eines Drehstahles), Bug *m*
(aviat.), Helm-rohr *n*, -schnabel *m*, Kanzel *f*,
Mündung *f*, Nase *f*, (of fuselage) Rumpfnase *f*,
Rüssel *m*, Schnabel *m*, Schnauze *f*, Schneid-

kopf *m*, Spitze *f*, Vorsprung *m*, (of a turner's chisel)

**nose,** ~ **of airplane** Flugzeugstirn *f* ~ **of a Bessemer converter** Hals *m* einer Bessemerbirne ~ **fuselage** Rumpfspitze *f* ~ **of the gear case** Vorderteil *n* des Getriebegehäuses ~ **of projectile** Geschoßspitze *f* ~ **of shell** Kopfteil *m* ~ **of ship** Gallion *n* ~ **of tuyere** Formrüssel *m* ~ **of wing** Flügelnase *f*

**nose,** ~**-and-axle-suspended motor** Tatzlagermotor *m* ~ **band** Nasenriemen *m* ~ **box** Nasenkasten *m* ~ **cap** Bugkappe *f* (Flugzeug), Kopf *m* ~ **clamp** Nasenklammer *f* ~ **clip** (mask) Nasenklemme *f* ~ **clutch** Mitnehmer *m* ~ **cowl** Nasenhaube *f* ~ **cowling** Stirn-gehäuse *n*, -kappe *f* ~ **dive** kopflastige Fluglage *f*, Kopfsturz *m*, Sturzflug *m* ~ **fan** Buggebläse *n* (Flugzeug) ~ **fuse** Kopfzünder *m* ~**-fuse shell** Kopfgranate *f* ~**-gun station** Bugstand *m* ~ **heaviness** Kopflastigkeit *f*, Vorderlastigkeit *f* ~ **heavy** buglastig, kopflastig (aviat.), vorderlastig ~ **high** überzogene Stellung *f* (aviat.), in überzogener Stellung *f* ~ **key** Hakenkeil *m*, Nasenhaubenkeil *m* ~ **landing gear** Bugfahrwerk *n* ~ **landing gear steering** Bugradlenkung *f* ~**-over** Kopfstand *m*, Überschlag *m* (aviat.); kopfüber (aviat.) ~ **panel** Kopfbahn *f* ~ **piece** (goggles) Nasensteg *m*, Nasenstück *n*, Vorderteil *n* ~ **piece of the fuse** Zündrüssel *m* ~ **pipe** Düse *f* ~ **radar** Bugradargerät *n* ~ **radiator** Brust-, Bug-kühler *m* ~ **rib** Nasenrippe *f* ~ **rod** Zünderabstandsrohr *n* ~ **section** Rumpf-bug *m*, -vorderteil *n* ~ **section of a motor** Getriebeteil *m* des Motors ~ **sill** Endquerschwelle *f* ~**-slot cowling** Nasenspaltschlitzhaube *f* ~ **spinner** Nabenhaube *f* ~**-suspension motor** Tatzenmotor *m* ~ **turret** vordere Kanzel *f* ~**-up** schwanzlastige Fluglage *f* ~ **wave of a projectile** Bugwelle *f* eines Geschosses, Geschoßbugwelle *f* ~ **wheel** Laufrad *n* unter der Flugzeugstirn, Spornrad *n* ~ **wheel landing gear** Bugradfahrwerk *n* ~ **wheel turntable** Bugraddrehscheibe *f* ~ **window** Bugfenster *n*

**nostril** Nüster *f*

**not,** ~ **circuit** Nichtschaltung *f* ~ **crocking** abrußecht ~ **decodable** unentzifferbar ~ **discharged** ungekündigt ~ **easily inflammable** schwerentzündlich ~ **edged** ungerandet ~ **entitled** nichtberechtigt ~ **functioning antenna** verstimmte Antenne *f* ~**-go gauge** Ausschußlehre *f* ~**-go screw plug member** Anschlußgewindelehrzapfen *m* ~**-go setting gauge** Anschlußeinstellehre *f* ~**-go side** Ausschußseite *f* (einer Lehre) ~ **grounded** massefrei (rdo) ~ **including packing** Verpackung *f* nicht einbegriffen ~ **level** uneben ~ **malleable** unhämmerbar ~ **proper** uneigentlich ~ **ready** unfertig ~ **real** uneigentlich ~ **resonated** nicht abgestimmt ~ **run-in** nicht eingelaufen ~ **running dry** unversiegbar ~ **used** ungebraucht ~ **volatile** schwerflüchtig

**notable** bemerkenswert, namhaft

**notarial fee** Notariatsgebühr *f*

**notary** Notar *m* ~**'s office** Notariat *n*

**notation** Aufzeichnung *f*, Beifügung *f*, Bemerkunf *f*, Bezeichnung *f*, Bezeichnungssystem *n*, Buchstabenbezeichnung *f* (math. formula), Eintragung *f*, Schreibweise *f*, Vermerk *m*, Zahlendarstellung *f* (data proc.) ~ **on map** Karteneintragung *f*

**notch, to** ~ (in milling machine) ausfräsen, ausklinken, ausscharten, auszacken, einbühnen, einkerben, einschneiden, mit Einschnitten *pl* versehen, falzen, kerben, nuten, zähneln **to** ~ **a cask** ein Faß *n* gargeln oder gergeln

**notch** Anschnitt *m*, Auskämmung *f* (Zimmerei), Ausschnitt *m*, Aussparung *f*, Einfeilung *f*, Einschnitt *m*, Falz *m*, Feilhieb *m*, Gargel *f*, Kamm *m*, Kappung *f*, Keep *f*, Kerb *m*, Kerbe *f*, Kimme *f*, Nut *f*, Nute *f*, Raste *f*, Rastzahn *m*, Scharte *f*, Schere *f*, Schlitz *m*, Visiereinschnitt *m* (r.r.) ~ **of rear sight** Kimme *f* ~ **with root radius** Rundkerb *m*

**notch,** ~**-and-bead** Kimme *f* und Korn *n* ~**-and-corn** Kimme- und Korn-Visier *n* ~ **arm** Rastenhebel *m* ~ **bending test** Kerbschlag-probe *f*, -versuch *m* ~**-brittleness** Kerbsprödigkeit *f* ~**-diplexer** Fernsehfilterweiche *f* ~ **effect** Kerbwirkung *f* ~ **fatigue strength** Kerbdauerfestigkeit *f* ~ **filter** Kerbfilter *n* ~ **gun** Stopfmaschine *f* ~ **hook** Einklinghaken *m* ~ **impact-bending test** Kerbschlagbiegeversuch *m* ~ **impact resistance** Kerbzähigkeit *f* ~ **impact strength** Kerbschlagzähigkeit *f*, Kerbzähigkeit *f* ~ **impact tenacity** Kerbschlagzähigkeit *f* ~**-impact test** Stoß- und Schlagbeanspruchung *f* ~ **plate** V-Meßwehr *n* ~ **sensitiveness** Kerbempfindlichkeit *f* ~ **sensitivity** Kerbschlagempfindlichkeit *f* ~ **stick** kurzer Holzpflock *m* zur Ermittlung der Förderung ~ **tenacity** Kerbzähigkeit *f* ~ **toughness** Kerbfestigkeit *f*, Kerbzähigkeit *f* ~ **wheel** Zählrad *n*

**notchable** rastbar

**notched** eingekerbt, gezinkt, mit Nuten *pl* versehen ~ **bar** (weaving) Platine *f* ~ **bar bend test** Kerbbiegeprobe *f* ~ **bar bending test** Einkerbbiegeversuch *m* ~ **bar impact bending test** Kerbschlagbiegeversuch *m* ~ **bar impact-bending test (or specimen)** Kerbschlagbiegeprobe *f* ~ **bar impact-endurance test** Dauerkerbschlagversuch *m* ~ **bar impact test** Kerbschlag-probe *f*, -versuch *m*, Kerbzähigkeits-probe *f*, -prüfung *f* ~ **bar iron** Zaineisen *n* ~ **bar pull test** Kerbzugprobe *f* ~ **bar strength** Kerbschlagfestigkeit *f* ~ **bar tensile test** Kerbzugprobe *f* ~ **bar toughness** Kerbzähigkeit *f* ~ **bars** Gleitschiene *f* ~ **bolt** Brechscherbolzen *m* ~ **conical pin** Kegelkerbstift *m* ~ **disk** Klinkenscheibe *f*, Schaltrad *n* ~ **edge** (of plates) Außenkerbe *f* ~**-in** eingerastet ~ **line** (aus)gehackte oder ausgezackte (ausgezahnte) Linie *f* ~ **locking quadrant** Feststellsegment *n* ~ **nail** Kerbnagel *m* ~ **nut** Rastenmutter *f* ~ **(taper)pin** Kerbstift *m* ~ **quadrant** gekerbter Quadrant *m* ~ **ring** Nutenring *m* ~ **segment** Rastbogen *m* ~ **specimen** Probestab *m* für Kerbschlagversuche ~ **star** Raststern *m* ~ **stick** Kerbholz *n* (min.) ~ **taper pin** Kegelkerbstift *m* ~ **test specimen** gekerbter Probestab *m* ~ **trowel** Schabkelle *f*

**notcher** Gargelkamm *m*, Markierungsvorrichtung *f* (filen)

**notching** Einkerben *n*, Einkerbung *f*, Kerbe *f*, Verblattung *f*, Verkämmung *f* ~ **of the fishplate** Laschenausklinkung *f*

**notching,** ~ **block** Kerbblock *m* ~ **chisel** Kammeisen *n* ~ **cutter** Falzfräser *m* ~ **filter** Saugfilter

*n* ~ **iron** Flammeisen *n* ~ **machine** Ausklink-, Einklink-maschine *f* (bookbindery), Einsäge-maschine *f*, Nutenstanze *f*, Stanzpresse *f* ~ **press** Nutenstanzautomat *m* ~ **saw** Kerbsäge *f* ~ **test** Kerbprüfung *f* ~ **tool** Kimmhobel *m*, Kröse *f*
**notchy** schartig
**note, to** ~ anmerken, aufzeichnen, bemerken, bezeichnen, (from) ersehen, notieren, vermerken **to** ~ **beforehand** vormerken
**note** Anmerkung *f*, Bemerkung *f*, Mitteilung *f*, Note *f*, Notiz *f*, Ton *m*, Tonhöhe *f*, Tonzeichen *n*, Vermerk *m*, Zettel *m* ~ **of hand** Schuld-brief *m*, -schein *m* ~ **with one mode of vibration** Reinton *m* ~ **of modulation** Modulationston *m* ~ **and wave of tuning** Abstimmton *m* ~ **and wave tuning** Abstimmung *f* von Tonhöhe und Welle
**note,** ~ **amplifier** Hörfrequenz-, Niederfrequenz-, Tonfrequenz-, Ton-verstärker *m* ~ **book** Notiz-, Taschen-buch *n* ~ **circulation** Notenumlauf *m* ~ **magnification** Tonverstärkung *f* ~ **magnifier** Tonverstärker *m* ~ **pad** Notiz-, Schreib-block *m* ~ **paper** Briefpapier *n* ~ **tuning** Tonabstimmung *f*
**noted for** bekannt wegen
**noteworthy** bemerkenswert
**notice, to** ~ anzeichnen, bemerken, merken, wahrnehmen
**notice** Ankündigung *f*, Anzeige *f*, Aufkündigung *f*, Bekanntmachung *f*, Benachrichtigung *f*, Kündigung *f*, Meldung *f*, Nachricht *f*, Notiz *f*, Vorladung *f* ~ **for airmen (NOTAM)** Nachrichten *pl* für Luftfahrer (NfL)
**notice,** ~ **of allowance** Erteilungsbericht *m* ~ **of default** Inverzugsetzung *f* ~ **of departure** Abmeldung *f* ~ **of sale by auction** Versteigerungsbekanntmachung *f*
**notice,** ~ **board** Anschlagtafel *f* ~ **showing the totale trade tax basis amount** Gewerbebemeßbescheid *m*
**noticeability** Wahrnehmbarkeit *f*
**noticeable** bemerkenswert, kennbar, merkbar, merklich, wahrnehmbar
**notification** Ansage *f*, Anzeige *f*, Aufzeichnung *f*, Bekanntmachung *f*, Vorladung *f* ~ **in advance of incoming call** Vorbereitung *f* einer Fernverbindung ~ **within a given period** befristete Anzeige *f*
**notify, to** ~ avisieren, förmlich anzeigen, kundgeben
**notion** Einfall *m*, Vorstellung *f* ~ **screw** Bewegungsschraube *f*
**notorious** allgemein bekannt, offenkundig
**noumeite** Numeait *n*
**noun** Hauptwort *n*, Substantiv *n*
**novation** Novation *f*
**novel** neu
**novelty** Neuigkeit *f*
**novice** Anfänger *f*, Neuling *m*
**noxious** schädlich ~ **clearance** schädlicher Raum *m*
**nozzle** Ansatzrohr *n*, Ausgleichdüse *f*, Ausguß *m*, Ausströmöffnung *f*, Düse *f*, Gasdüse *f*, Helm-rohr *n*, -schnabel *m*, Lötspitze *f*, Mundstück *n*, Mündung *f*, Muschel *f*, Nase *f*, Rohrpoststutzen *m*, Rüssel *m*, Schlauch-mundstück *n*, -stutzen *m*, Schnabel *m*, Schnauze *f*, Schneppe *f*, Stutzen *m*, Tülle *f*, Wasserstrahlmundstück

*n*, Wellenleiteröffnung *f* (antenna), Zerstäuber *m*
**nozzle,** ~ **of a blowpipe** Lötrohrspitze *f* ~ **of charging pipe** Füllhändel *m* ~ **of a ladle** (Gieß)-Pfannen-ausguß *m*, -schnauze *f* ~ **with pilot pin** geschlossene Zapfendüse *f*
**nozzle,** ~ **adapter** Düsen-ansatz *m*, -paßstück *n* ~ **angle** Anstellwinkel *m* der Düse ~ **area control** Schubdüsenregelung *f* ~ **arrangement** Düsenaustritt *m* ~ **blower** Gebläsestutzen *m* ~ **body** Düsenkörper *m* ~ **box** Düsen-kammer *f*, -kasten *m*, Turbineneintrittsgehäuse *n* ~ **brick** Lochstein *m* ~ **cap** Verschlußkappe *f* für Zapfhahn ~ **casing** Düsenkasten *m* ~ **coefficient** Düsenbeiwert *m* ~ **cone** Schubdüsenkörper *m* ~ **connection** Düsenstock *m* ~ **-control-shaft jet** Düsennadelverstellwelle *f* ~ **diaphragm** Düsenzwischenboden *m* ~ **duct** Düsenkanal *m* ~ **efficiency** Düsenwirkungsgrad *m* ~ **ejector** Düsenauswerfer *m* ~ **exhaust box** Düsenauspufftopf *m* ~ **exhaust muffler** Düsenauspufftopf *m* ~ **feeding to carburetor venturi** Mischblockschnabel *m* ~ **fixing** Düsenfixierung *f* ~ **flap** Düsen-klappe *f*, -zunge *f* ~ **gauge** Düsenlehre *f* ~ **group** Düsengruppe *f* ~ **guide vane** (gas turbines) Laufradeinleitschaufel *f*, Leitschaufel *f* ~ **head** Düsenkörper *m* ~ **holder** Düsen-halter *m*, -rohr *n* ~ **holder body** Düsenhalterkörper *m* ~ **holder flange profile** Düsenhalterflanschprofil *n* ~ **hole** Düsenbohrung *f* ~ **inlet** Düseneintritt *m* ~ **jet** Düsenstrahl *m* ~ **jet vane** Leitschaufel *f* ~ **magnifying torch** Düsenleuchtlupe *f* ~ **meter** Düsenmesser *m* ~ **-mixing burner** Kreuzstrombrenner *m* ~ **needle** Düsennadel *f* ~ **opening pressure** Düsenöffnungsdruck *m* ~ **outlet area** Düsenmündungsfläche *f* ~ **pin** Düsenzapfen *m*, Strahlendüsennadel *f* ~ **pintle** Düsenpilz *m* ~ **plate** Düsenplatte *f* ~ **(or jet) pressure** Düsendruck *m* ~ **-propulsion aircraft** Düsenmaschine *f* ~ **pulverizer** Strahlprallmühle *f* ~ **pump** Schubdüsenverstellpumpe *f* ~ **radiator** Düsenkühler *m* ~ **-reconditioning equipment** Düsennacharbeitsgerät *n* ~ **regulator** Düsenregler *m* ~ **ring** Düsenplatte *f*, (turbines) Düsenring *m*, Leitkranz *m* ~ **screw** Düsenschraube *f* ~ **scroll** Düsenkranz *m* ~ **shaft** Strahlendüsennadel *f* ~ **slot** Düsenschlitz *m* ~ **spring** Düsenfeder *f* ~ **tester** Düsenprüfvorrichtung *f* ~ **test(ing)** Düsenprüfstand *m* ~ **throat** Düsenhals *m* ~ **tip** Düsen-kopf *m*, -spitze *f*, Schlauch-mundstück *n*, -tülle *f*, Strahlrohrmundstück *n* ~ **type** stenter Düsenbelüftung *f* ~ **valve** Düsenventil *m*
**n-point switch** Wähler *m* für Richtungen
**N-pole** N-förmiges Gestänge *n*
**NP-sections** Deutsche Normal-Profile *pl*
**n-quadrate** Halbgeviert *n* (print.)
**N-strut** N-Stiel *m*, N-Strebe *f*
**nth order** nte Ordnung
**n-type,** ~ **conductivity** N-Leitung *f*, Überschußleitung *f* ~ **semiconductor** n-Leiter *m*, Überschußleiter *m*
**nub twist** Knotenzwirn *m*
**nuclear** atomgetrieben, kernförmig, kerntechnisch, nuklear ~ **action** Keimwirkung *f* ~ **alignment** Kernausrichtung *f* ~ **burst** Kernspaltung *f* ~ **cascade process** Kaskadenkernreaktion *f* ~ **chain reaction** nukleare Ketten-

reaktion f ~ **charge** Kernladung f ~ **charge number** Atom-zahl f, -ziffer f, Kernladungszahl f ~ **concentration** Kerngehalt m ~ **disintegration** Kern-spaltung f, -zerfall m, -zertrümmerung f ~-**disintegration process** Kernumänderungsprozeß m ~ **division** Kernteilung f ~ **electron** Kernelektron n ~ **energy** Atom-, Kern-energie f ~ **engineering** Kernenergietechnik f ~ **excitation** Kernanregung f ~ **fission** Kern-spaltung f, -zertrümmerung f ~ **fission fragment** Kernspaltungsfragment n ~ **fluid** Kernbindungsfluidum n ~ **fuel** Atomtreibstoff m, Spaltmaterial n ~ **fusion** Kernverschmelzung f ~ **gyro** Kernkreisel m ~ **inflammability** Kernreaktionsfähigkeit f ~ **interaction** Kernwechselwirkung f ~ **isobar** Kernisobare f ~ **isomerism** Kernisomerie f ~ **magic moment** magnetisches Kernmoment m ~ **magnetic alignment** magnetische Kernausrichtung f ~ **magneton** Kernmagneton n ~ **matter** Kernmaterie f ~ **migration** Kernübertritt m ~ **moment** Kernmoment n ~ **opacity** Opazität f von Atomkernen ~ **paramagnetism** Kernparamagnetismus m ~ **particle** Kernteilchen n, Nukleon n ~-**penetration factor** Potenzschwelle f ~-**penetration function** Kerndurchdringung f ~ **phase shift** Kernprozeß-Streuphase f ~ **photodisintegration** Kernfotoeffekt m ~ **physics** Kernphysik f ~ **polymerism** Kernpolymerie f ~ **power plant** Atomkraftwerk n ~ **procession** Präzession f des Kerns ~ **reactor** Atomreaktor m ~ **spin** Kernspin m ~ **stain** Kernfarbe f ~ **star** Zertrümmerungsstern m ~ **structure** Kernaufbau m ~ **submarine** Atom-U-Boot n ~ **theory** Kerntheorie f ~ **transformation** Kernumwandlung f ~ **transmutation** Atomzertrümmerung f, Kernumwandlung f

**nucleated** kernreich

**nucleation** Bildung f von Kristallisationskernen, Kernbildung f, Kristallkernbildung f, Kristallisationskernbildung f

**nucleic acid** Nukleinsäure f

**nucleolus** Kernkörperchen n

**nucleon** Kernteilchen pl, Nukleon n ~ **number** Massenzahl f

**nucleonic field** Nukleonfeld n

**nucleonics** Atomtechnik f, Kernphysik f, Kerntechnologie f, Nukleonik f

**nucleous** kernhaltig

**nucleus** Atomkern m, Keim m (cryst.), Keimling m, Kern m, Kernschatten m (opt.), Mittelpunkt m, Nukleus m, Wachstumszentrum n ~ **of a crystal** Kristallkern m ~ **of crystallization** Kristallisations-kern m, -zentrum n ~ **of an integral equation** Kern m

**nucleus,** ~ **denudation** Keimbloßlegung f ~ **formation** Keimbildung f ~ **isolation** Keimisolierung f

**nuclide** Nuklid n

**nugget** natürlicher Klumpen m von gediegenem Metall ~ **of lime** Kalknest n

**nuisance** Belästigung f ~ **area** Verwirrungsgebiet n ~ **value** Störwert m (gegen Dritte)

**null** (Funkpeilung) Minimum n, Null f, Nullpunkt m (point), Peilnull f, toter Punkt m (Empfangsgeräte); leer, nichtig, nicht vorhanden, ungültig ~ **astatic magnetometer** Ausgleichmagnetometer n ~ **balance** Nullabgleich m ~-**balance amplifier** Nullstromverstärker m

~ **balance device** Nullungsglied n ~ **circle** Nullkreis m ~ **coil magnetometer** Magnetometer m mit Abgleichspule ~ **cone** Lichtkegel m ~ **cut** Nullschnitt m ~ **detector** Abgleichdetektor m, Nullindikator m ~ **indicator** Nullinstrument n ~ **method** Null-methode f, -verfahren n ~ **off-set** astatische Regelung f ~-**phase modulation** Nullphasenmodulation f ~ **point** Anschlussung f (teleph.), Auslöschung f, Erlöschung f, Nullpunktfehler m der Teilkreise, Schwingungsknoten m ~ **signal** Nullzacke f (rdr) ~ **stress** Nulltensor m der Spannung

**nullification** Annulierung f, Streichung f

**nullify, to** ~ annullieren, aufheben, ungültig machen, vernichten

**nullity** Ungültigkeit f ~ **of a patent** Ungültigkeit f eines Patents

**nullode** elektrodenlose Röhre f, Nullode f

**number, to** ~ benummern, beziffern, numerieren, rechnen, zählen **to** ~ **consecutively** fortlaufend numerieren **to** ~ **continuously** durchlaufend numerieren

**number** Anzahl f, (of yarn) Feinheitsnummer f, (of a periodical) Heft n, Meßzahl f, Nummer f, Reihe f, Zahl f, Zahlwert m, Ziffer f **in** ~**s** heftweise, lieferungsweise

**number,** ~ **of alternations** Polwechselzahl f ~ **of ampere turns** Amperewindungszahl f, A-W Zahl f (Amperewindungszahl) ~ **of atoms** Atom-zahl f, -ziffer f ~ **of blades** Blattzahl f ~ **in brackets** eingeklammerte Zahl f ~ **of bursts** Stoßzahl f ~ **centers** Kernzahl f ~ **of collisions** Stoßzahl f ~ **of components** Stoffzahl f ~ **of conductors** (in the armature) Leiterzahl f ~ **of connections** Schaltzahl f ~ **of cycles to failure** (Bruch)lastspielzahl f ~ **of cycles per second** Periodenzahl f ~ **of degrees** Gradzahl f ~ **of digits** Stellenzahl f ~ **of divisions** Teilzahl f ~ **of dust nuclei** Stäubchenzahl f ~ **of the equation** Gleichungsnummer f ~ **of faces** Flächenzahl f ~ **of feed rates** Drehzahlenbereich m ~ **of feeders** Maschinensatz m ~ **of frames per second** Bildwechselfrequenz f ~ **of the frog** Herzwinkel m (r.r.) ~ **of gears** Gangzahl f ~ **of graduations** Grad-, Teilstrich-zahl f ~ **of grid wires** Stegzahl f ~ **of illuminations of a picture point** Helligkeitswechsel m ~ **of image elements** Bildpunktzahl f ~ **of impacts** Trefferzahl f ~ **of lines** Zeilenzahl f ~ **of lines of force** Kraftlinienzahl f ~ **of links** (of a chain) Gliederzahl f ~ **of load alternations** Lastwechselzahl f ~ **of load cycles at which fracture occurs** Bruchlastwechselzahlen pl ~ **of men in the crew** Besatzungsstärke f ~ **of nuclei** Kernzahl f ~ **of parts in a batch** Anzahl f von Teilen, die gleichzeitig bearbeitet werden ~ **of periods** Periodenzahl f ~ **of phase** Phasenzahl f ~ **of the photograph** Bildnummer f ~ **of picks** Schußzahl f ~ **of the picture** Bildnummer f ~ **of picture elements** Bildpunkt-, Raster-zahl f ~ **of planes** Tragdockanzahl f ~ **of poles** Polzahl f ~ **of propeller revolutions** Schraubendrehzahl f ~ **of reflections** Reflexionszahl f ~ **of revolutions** Drehzahl f, Randintegral n, Touren-, Umdrehungs-, Umlauf-zahl f ~ **of rings** Ringzahl f ~ **of rounds** Schlußzahl f ~ **of shocks** Trefferzahl f ~ **of slots** Nutenzahl f ~ **of starts** Gängigkeit f ~ **of starts of a hob (or worm)** Gangzahl f eines Wälzfräsers oder einer

Schnecke ~ **of stays** Stegzahl *f* ~ **of strands** Flechtungszahl *f* ~ **of strokes** (of a piston, etc.) Hubzahl *f* ~ **of strokes of rocker arm** Schwinghebelhubzahl *f* ~ **of strokes per second** Anschläge *pl* je Sekunde ~ **of successive images** Bildwechselzahl *f* (TV) ~ **of supports** Stegzahl *f* ~ **of transfer units** Zahl *f* der Übergangszellen ~ **of turns** Dreh-wert *m*, -wertigkeit *f*, Touren-, Umlauf-, Windungs-zahl *f* ~ **of watts** Wattzahl *f*

**number,** ~ **barrel mechanism** Zahlenrollenwerk *n* ~ **checking arrangement** Fangeinrichtung *f*, Halteeinrichtung *f* ~ **density** (of plasma) Teilchendichte *f* ~ **detector** Nummernsucher *m* ~ **dialling** Nummernwahl *f* (teleph.) ~ **domain** Zahlenbereich *m* ~ **index of the vector group** Schaltgruppenziffer *f* ~**-indicating system** Nummergeber *m* ~ **nail** Bezeichnungsnagel *m* ~ **peg** Nummernpfahl *m* ~ **perforating-machine** Zahlenlochmaschine *f* ~ **period** Zahlenperiode *f* ~ **plate** Nummern-scheibe *f*, -schild *n* ~ **plate light** Nummernlicht *n* ~ **plate lighting** Nummernbeleuchtung *f* ~ **printer** Zahlendruckvorrichtung *f* ~ **representation** Zahlendarstellung *f* ~ **sleeve** Nummernhülse *f* ~ **tape** Zahlen-band *n*, -lochstreifen *m* ~ **triplet** Zahlentripel *n* ~ **type bar** Zifferntypenhebel *m* ~**-unobtainable tone** Störungssignal *n*, Summerzeichen *n* zur Anzeige unausführbarer Verbindung ~ **wheel** Nummernrad *n*

**numbered,** ~ **addendum** numerierter Zusatz *m* ~ **consecutively** fortlaufend numeriert

**numberer** Nummerwerk *n*

**numbering** Benummerung *f*, Bezifferung *f*, Numerieren *n*, Numerierung *f*, Nummernbezeichnung *f*, Nummerngebung *f*, Nummerung *f* ~ **of cable conductors** (color scheme) Adernzahlfolge *f*

**numbering,** ~ **axis** Numerierachse *f* ~ **box** Druckkopf *m*, Numerierwerk *n* ~ **frame (or box)** Numerierrahmen *m* ~ **machine** Nummerndruckwerk *n* ~**-machine operator** Maschinenpaginierer *m* ~ **nail** Bezeichnungsnagel *m* (Leitung) ~ **stamp** Zahlenstempel *m* ~ **systems** Numerierungsarten *pl* ~ **wheel** Paginierrad *n* ~ **wheel engraving device** Typenradgravierbock *m*

**numberless** unzählig

**numeral** Nummer *f*, numerisches Zeichen *n*, Zahl *f*, Zahlzeichen *n*, Ziffer *f*; numerisch

**numeration** Bezifferung *f*

**numerator** Zähler *m*

**numeric(al)** numerisch

**numerical** numerisch, rechnungsmäßig, zählbar, zahlenmäßig, der Zahl *f* nach ~ **aperture of objective** numerische Objektivöffnung *f* ~ **calculation** zahlenmäßige Berechnung *f* ~ **check** Probe *f* (info proc.) ~ **code** Kennziffer *f*, numerischer Kode *m* ~ **coding** numerische Kodierung *f* ~ **comparison** zahlenmäßiger Ver-

gleich *m* ~ **control (system)** Datensteuerung *f*, numerische Steuerung *f* ~ **criterion** Maßstabzahl *f* ~ **date** Zahlenangabe *f* ~ **digits** Ziffernzeichen *pl* ~ **example** Zahlenbeispiel *n* ~ **impulse** Ziffernfeldimpuls *m* ~ **increment** Ziffernschritt *m* ~ **index of the vector group** Schaltgruppenziffer *f* ~ **order** Zahlenfolge *f* ~ **positioning control** numerische Stellensteuerung *f* ~ **punching** Ziffernlochung *f* ~ **quantitiy** Zahlengröße *f* ~ **rating** numerische Bewertung *f* ~ **result of blank test** Blindwert *m* ~ **selection** erzwungene Wahl *f* ~ **superiority** Überzahl *f* ~ **switch** Nummerwähler *m* ~ **table (or tabulation)** Zahlentafel *f* ~ **value** Meßzahl *f*, Zahlenwert *m* ~ **work** Rechenarbeit *f*

**numerically evaluable** zahlenerfaßbar, zahlenmäßig erfaßbar

**numerology** Zahlenlehre

**numerous** zahlreich

**numismatic** numismatisch

**nummulitic limestone** Nummulitenkalk *m*

**n-unit** N-Einheit *f*

**nursery** Gärtnerei *f*, Pflanzenschule *f*, (hort.) Schonung *f* ~ **plant** Zierpflanze *f*

**nut** Düsenmutter *f*, Frosch *m*, Gewindering *m*, Mutter *f*, Nuß *f*, (for keyed end) Nutmutter *f*, Radnabenmutter *f*, Überwurfmutter *f* **(bolt)** ~ Schraubenmutter *f* **to do up a** ~ eine Mutter *f* aufschrauben ~ **with fiber insert** Mutter *f* mit Fibereinlage ~ **on hub** Nabenbolzenmutter *f* ~ **with round spigot** Halsmutter *f* ~ **with slipped thread** ausgeleiertes Gewinde *n* ~ **of a swivel** Nuß *f* eines Wirbels ~ **with two holes** Zweilochmutter *f*

**nut,** ~**-beveling machine** Mutternabkantmaschine *f* ~ **channel** Mutternleiste *f* ~ **coal** Nußkohle *f* ~ **coke** Klein-, Knabbel-koks *m* ~ **key** Mutterschlüssel *m* ~**-loading pocket** Nußtasche *f* ~ **lock** Mutterverschluß *m*, Schraubensicherung *f* ~ **locking** Muttersicherung *f* ~ **machine** Mutterfräsmaschine *f* ~ **plate** Nabenmutter *f* ~ **punch** Mutternstempel *m* ~ **runner** (elektromotorisch betriebenes) Mutteraufschraubgerät *n* ~ **setter** Mutterfestziehmaschine *f* ~ **size** Nußkörnung *f* ~ **tap** Mutterbohrer *m*, Muttergewindebohrer *m* ~ **tapper** Muttergewindeschneider *m* ~ **thread** Muttergewinde *n* ~ **washer** Mutterscheibe *f*

**nutating horn** kreisender Hornstrahler *m* (rdr)

**nutation** Nutation *f*

**nutator** Nutationsvorrichtung *f*

**nutrient,** ~ **content** Nährstoffgehalt *m* ~ **medium** Nährboden *m* ~ **transmission** Stoffleitung *f*

**nutriment** Nähr-, Nahrungs-mittel *n*

**nutrition** Ernährung *f*

**nutritious** nahrhaft

**nutritive,** ~ **medium** Nährboden *m* ~ **(or alimentary) substance** Ernährungsstoff *m* ~ **substratum** Nährboden *m* ~ **value** Nährwert *m*

# O

oak Eiche *f* ~ tanning Eichenlohgrubengerbung *f* ~ wood Eichenholz *n* ~ wood for caskmaking Eichenfaßholz *n*
oaken eichen
oakum Abdichtfaser *f*, Hede *f*, Putzwolle *f*, Werg *n* ~ picker Wergzupfer *m*
oar Krücke *f*, Riemen *m*, Ruder *n* ~ lock Ruderklappe *f*
oat crusher Haferquetsche *f*
oatmeal Hafer-mehl *n*, -grütze *f* ~ ingrains Rauhfaserpapier ~ paper Holzmehlpapier *n*
object, to ~ reklamieren, weigern to ~ to aussetzen, beanstanden
object Gegenstand *m*, Objekt *n*, Tatbestand *m*, Ziel *n* ~ of attraction Gegenstand *m* der Anziehung ~ that catches the wind windfangender Gegenstand *m*, Windhindernis *n* ~ of invention Anmeldungs-, Erfindungs-gegenstand *m* ~ of measurement Meßobjekt *n* an ~'s own shadow Eigenschatten *m* ~ being shot Aufnahmeobjekt *n* on the ~ side dingseitig ~ to be transmitted by video signals Übertragungsgegenstand *n*
object, ~ air lock Objektschleuse *f* ~ chamber Objektkammer *f* ~ clip Objektklemme *f* ~ distance Dingweite *f*, Gegenstands-, Start-weite *f* ~ drawing Zeichnen *n* nach Vorlage ~ finder Objektsucher *m* ~ fouling Objektverschmutzung *f* ~ glass Gegenstandsglas *n*, Objektiv *n* ~ glass of microscope Mikroskopobjektiv *n* ~-glass aperture Objektivöffnung *f* ~-glass carrier Objektivträger *m* ~-glass collar Objektivring *m* ~-image distance Objektbildabstand *m* ~ lesson Anschauungsunterricht *m*, praktisches Beispiel *n* ~ marker Objektmarkierapparat *m* ~ plane Dingebene *f* ~ point Ding-, Gegenstand-, Objektiv-punkt *m* ~ preserving objekttreu ~ program maschinenübersetztes Programm *m* ~ register Objektkartei *f* ~ slide Objektträger *m* ~ sluice Objektschleuse *f* ~ space Ding-, Objekt-raum *m* ~ staff Nivellierstab *m*, Zielstange *f* ~ stage Objekt-bühne *f*, -tisch *m* ~ stand Objekttisch *m* ~ support Objektträger *m* ~-support lamina (or slide) Objektträgerplättchen *n* ~ table Objektbühne *f* ~ teaching Anschauungsunterricht *m*
objectification Objektivierung *f*
objection Beanstandung *f*, Bedenken *n*, Einrede *f*, Einspruch *m*, Einwand *m*, Einwendung *f*, Rechtseinwand *m*, Weigerung *f*, Widerspruch *m*
objectionable nicht einwandfrei, unerwünscht
objective angestrebtes Ziel, Angriffsziel *n*, (lens) Dingglas *n*, Gefechtsziel *n*, Objektiv *n*, Operationsziel *n*, Ziel *n*, Zielpunkt *m*, Zweck *m*; objektiv, sachlich, unvoreingenommen ~ composed of several lenses mehrlinsiges Objektiv *n* ~ with correction mount Objektivfassung *f* mit Korrektur ~ of great lighttransmitting capacity lichtstarkes Objektiv *n*
objective, ~ adapter Objektivzwischenstück *n* ~ angular field objektives Gesichtsfeld *n* ~

barrel Objektivfassung *f* ~ board Objektivbrett *n* ~ carriage Objektivträgerwagen *m* ~ centering device Objektivzentriervorrichtung *f* ~ coil Objektivspule *f* ~ diopter Objektivdiopter *m* ~ field of view objektives Gesichtsfeld *n* ~ glass Linse *f* ~ guard glass Objektivschutzglas *n* ~ lens Ausblick *m*, Objektivlinse *f* ~-lens socket Ausblickstutzen *m* ~ mount Objektivfassung *f* ~ panel Objektivbrett *n* ~ plane Objektivebene *f* ~ point Aufschlag-,, Treff-punkt *m* ~ pole shoes Objektivpolschuhe *pl* ~ prism Vorsatzprisma *n* ~-prism housing Ausblickprismengehäuse *n* ~ revolver Objektrevolver *m* (Winkel) ~ revolver desk Objektivscheibe *f* ~ revolver stop Rast *f* für den Objektivrevolver ~ scale ring Objektivsockel *m* ~ slide Objektivschlittenstück *n* ~ stop (or diaphragm) Objektivblende *f* ~ variable Hilfsregelgröße *f*
objectivity Objektivität *f*, Sachlichkeit *f*
oblate, to ~ abplatten
oblate, ~ ellipsoid of revolution abgeplattetes Rotationsellipsoid *n* ~ spheroid an den Polen abgeflachte Kugel *f*, sphäroide Gestalt *f*
oblateness (at poles) Abflachung *f*, Abflachung *f* der Kugel, Abplattung *f*
obligated verpflichtet
obligation Bindung *f*, Pflicht *f*, Verbindlichkeit *f*, Verpflichtung *f* ~ to make good any loss by breakage Ersatzverbindlichkeit *f* für Bruch ~ to pay Zahlungspflicht *f*
obligatory verbindlich ~ preventive treatment Pflichtschutzbehandlung *f* ~ reporting Meldepflicht *f*
obliging verbindlich
oblique flach auffallend, flach auftreffend, quer, schief, schräg-laufend, -liegend ~ aerial photography Schrägaufnahme *f* ~ angle schiefer Winkel *m* ~-angled schiefwinklig ~ axial schiefachsig ~-bored stopper schräg gebohrter Stopfen *m* ~ (mapping) camera Schrägbildkamera *f* ~-convergent photographs vertikalkonvergente Aufnahmen *pl* ~ coordinates schiefwinkliges Achsenkreuz *n* ~ crank axle schrägschenk(e)lige Kropfachswelle *f* ~ cut schräge Befestigungsschelle *f*, schräg geschnitten ~ drilled duct Schrägbohrung *f* ~ edge Schrägschnitt *m* ~ exposure Geneigtaufnahme *f* ~ extruder head Schrägspritzkopf *m* ~ head Schräg- oder Querspritzkopf *m* ~ incidence schiefer oder streifender Einfall *m*, Schrägeinfall *m* ~ incidence angle schräger Auftreffwinkel *m* ~-incidence transmission Übertragung *f* unter schrägem Einfall ~ lamination falsche Schieferung *f* ~ light schiefes, schräg einfallendes Licht *n* ~ parallelogram schiefwinkliges Parallelogramm *n* ~ photograph Schrägaufnahme *f* ~ pull Schrägzug *m* (Drahtseil) ~ ray flach auffallender oder schräger Stahl *m* ~ strike Querschläger *m* ~ stroke Teilstrich *m* ~ suspension rod Zugdiagonale *f* ~ triangle schiefwinkeliges Dreieck *n*

**obliquely, ~ bedded** schrägschichtig **~ incident light** schiefes, schräg einfallendes Licht *n* **~ positioned** schiefliegend

**obliqueness** Schiefe *f*, Schiefheit *f*, Unregelmäßigkeit *f*

**obliquity** Neigung *f*, Neigungsgrad *m*, Schiefe *f*, Schräge *f*, Schrägstellung *f* **~ of geartooth profile** Eingriffswinkel *m* der Zahnflanke **~ of the wheels** Radeinschlag *m*

**obliterate, to** ~ auslöschen, ausradieren, ausstreichen, auswischen, tilgen, verschmieren, verstümmeln, verwaschen, verwischen

**obliterating, ~ field** magnetisches Löschfeld *n* **~ magnet** Löschmagnet *m* **~ pole piece** Auslöschmagnet *n*

**obliteration** Tilgung *f*, Wortabschneidung *f*, Wortverstümmelung *f* **~ of speech** Abschneiden *n* von Silben und Worten **~ of syllables** Silbenabschneidung *f* **~ magnet** Auslöschmagnet *n*

**oblong** langgestreckt, länglich, (of reading glass) viereckig **~ hole** Langloch *n*, Längsloch *n* **~ perforation** Schlitzlochung *f* **~ reading glass** viereckiges Leseglas *n* **~ size** Langformat *n* **~ slot** Langloch *n*

**obnoxious odor** Geruchsbelästigung *f*

**oboe system** Bumerangverfahren *n*

**obscuration** Verdunk(e)lung *f*

**obscure, to** ~ abdunkeln, verdunkeln, verwischen

**obscure** dunkel, düster, finster, ungeklärt **~ radiation** Dunkelstrahlung *f* **~ situation** ungeklärte Lage *f*

**obscuring period** Dunkelpause *f* (film)

**obscurity** Dunkelheit *f*, Finsternis *f*

**observability** Wahrnehmbarkeit *f*

**observable** wahrnehmbar

**observating report** Überwachungsblatt *n*

**observation** Ablesen *n*, Bemerkung *f*, Beobachtung *f*, Betrachtung *f*, Erfahrung *f*, Sichtung *f*, Überwachung *f*, Wahrnehmung *f* **~ in darkness** Dunkelmethode *f* **~ of the sun** Sonnenbeobachtung *f*

**observation, ~ aperture** Durchblicköffnung *f* **~ area** Beobachtungsraum *m*, Überwachungsbezirk *m* **~ balloon** Fesselballon *m* **~ base** Betrachtungsbasis *f* **~ basket** (balloon) Beobachtungskorb *m* **~ desk** Dienstüberwachungsplatz *m*, Überwachungsstelle *f* (teleph.) **~ hole** Guck-, Schau-loch *n* **~ instrument** Beobachtungsgerät *n* **~ kite** Beobachtungsballon *m* **~ panel** Beobachtungs-haube *f*, -feld *n* (aviat.) **~ parallax** Betrachtungsparallaxe *f* **~ plane** Aufklärungsflugzeug *n* **~ platform** Ausguckplattform *f* **~ point** Beobachtungsstelle *f*, Meßstelle *f* **~ port** Sehklappe *f* **~ post** Ausguck-, Beobachtungs-posten *m* **~ scale tube** Orientierungsskala *f* **~ slit** Seh-schlitz *m*, -spalt *m* **~ telescope** Beobachtungsfernrohr *n* **~ tower** Aussichtsturm *m*, Beobachtungswarte *f* **~ tube** (in electron microscope) Beobachtungsrohr *n* **~ window** Beobachtungsfenster *n*

**observational equation** Fehlergleichung *f*

**observatory** Observatorium *n*, Sternwarte *f*

**observe, to** ~ aufklären, bemerken, beobachten, innehalten, merken, überwachen, wahrnehmen

**observed, ~ angular height** Meßhöhenwinkel *m* (zwischen Meßortungslinie und Horizontale) **~ output** tatsächlich gemessene Leistung *f* **~ point** Meßkartenpunkt *m* **~ radio bearing** roher Funkazimut *m*, rohe Funkpeilung *f* **~ result** Beobachtungsergebnis *n*

**observer** Beobachter *m*, Beschauer *m*, Zuschauer *m* **~ on land** Landbeobachter *m* **~'s cockpit** Beobachterraum *m* **~'s horizon** Vermessungshorizont *m* **~ navigator** Flugzeugbeobachter *m*

**observing position of a photograph** (positive position) Betrachtungs-, Positions-stellung *f*

**obsidian** Glas-achat *m*, -lava *f*, Obsidian *m*

**obsolescence** Veralten *n*

**obsolescent** veraltet

**obsolete** alt, außer Gebrauch gekommen, veraltet **~ model (or type)** veraltetes Baumuster *n*

**obstacle** Behinderung *f*, Hindernis *n*, Hinderung *f*, Sperre *f* **~ clearance limit (OCL)** Hindernisfreigrenze *f* **~ clearance surface (OCS)** Hindernisfreifläche *f* **~ construction unit** Sperrverband *m* **~ course** Hindernis-bahn *f*, -strecke *f* **~ course for riding (practice)** Sprunggarten *m* **~ field** Sperrfeld *n* **~ light** Begrenzungsleuchte *f* **~ line** Sperrlinie *f* **~ pit** Wolfsgrube *f* **~ trench** Fallgrube *f*

**obstacles in depth** Sperrfeld *n*

**obstinacy** Sprödigkeit *f*

**obstinate** hartnäckig, spröde, widerspenstig

**obstruct, to** ~ behindern, hemmen, legen, versetzen, versperren, verstopfen, zusetzen

**obstructing** Versetzung *f*

**obstruction** Ansatz *m*, Hemmnis *n*, Hindernis *n* (aviat.), Sperrung *f*, Stauung *f*, Stockung *f*, Störung *f*, Verstopfung *f* **~ in blast furnace** Feststauen *n*

**obstruction, ~ clearance** Hindernisfreiheit *f* **~ clearing** Hindernisbeseitigung *f* **~ lighting** Hindernisbefeuerung *f* **~ lights** Hindernisfeuer *n* **~ marking** Hinderniskennzeichnung *f*

**obtain, to** ~ anfallen, erhalten, erreichen, erwerben, erzielen **to ~ a bearing on** anschneiden

**obtainable** erreichbar

**obtained** gewonnen **to be ~** anfallen **~ wedge** (foundry) tatsächliche Keilbreite *f*

**obtaining** vordringlich

**obturate, to** ~ abdichten, lidern, verschließen

**obturation** Dichtung *f*, Liderung *f*, Verschließung *f*, Verstopfung *f* **~ by means of the cartridge case** Hülsenliderung *f* **~ pad** Liderungsplatte *f*

**obturator** Dichtungsmittel *n*, Liderung *f*, Schließ-, Verschluß-vorrichtung *f* **~ head** Liderungskopf *m* (artil.), Pilz *m* **~ modulation** Ausblendsteuerung *f* **~ pad seat** Liderungslager *n* (artil.) **~ ring** gegen Verbrennungsgase abdichtender Kolbenring *m* mit L-förmigem Querschnitt, Liderungsring *m*

**obtuse** stumpf **~ angle** stumpfer Winkel *m* **with ~ angle** schiefwinklig **~-angled** stumpfwinklig **~ bisectrix** stumpfe oder zweite Mittellinie *f* **~ corner** verbrochene Ecke *f*

**obverse** Vorderseite *f*; umgekehrt

**obviate, to** ~ einer Sache abhelfen oder zuvorkommen, umgehen, verhüten, vorbeugen, zuvorkommen

**obvious** augenscheinlich, handgreiflich, naheliegend, offenbar, offenkundig **to be ~** auf der Hand liegen, naheliegen **~ reason** erkennbarer Grund *m*

**occasion, to** ~ verursachen

occasion Anlaß *m*, günstiger Zeitpunkt *m* as ~
demands nach Bedarf *m*
occasional beiläufig, gelegentlich
occasionally verschiedentlich
occlude, to ~ absorbieren (an der Oberfläche),
einschließen, mitführen, okkludieren, verschließen
occluded, ~ gas eingeschlossenes oder okklu-
diertes Gas *n* ~ hydrogen eingeschlossener
Wasserstoff *m*
occlusion Absorption *f* (durch die Oberfläche)
(chem.), Aufsaugung *f*, Einlagerung *f*, Einschließung *f*, Einschluß *m*, Okklusion *f*, Verstopfung *f*, Zusammenstoß *m* von Warm- und
Kaltfront ~ of gas Gas-aufnahme *f*, -ein-
schluß *m* ~ of gases Gasblasenbildung *f*
occult, to ~ (lens) abdecken, abblenden, abdunkeln
occult rays Strahlabblender *m*
occultate, to ~ verdunkeln (astr.), (sich) ver-
finstern
occultation Sternbedeckung *f*, Verdeckung *f*,
Verfinsterung *f* (astr.)
occulter Blinklichtapparat *n*
occulting light unterbrochenes Feuer *n*
occupancy Besitzergreifung *f*
occupant Insasse *m*
occupation Beruf *m*, Beschäftigung *f*, Besetzung
*f*, Einnahme *f*, Erwerb *m*, Fach *n*, Gewerbe *n*,
Inbesitznahme *f* ~ efficiency Wirkungsgrad *m*
~ hygiene Arbeitshygiene *f* ~ numbers Be-
setzungszahlen *pl* ~ tax Gewerbesteuer *f* ~
time Belegungszeit *f*
occupational, ~ disease Berufs-krankheit *f*,
-noxe *f* ~ eczema Berufsekzem *n* ~ hazard
Berufsnoxe *f*
occupied besetzt to be ~ sich beschäftigen ~
position besetzter Platz *m*
occupy, to ~ belegen, beschäftigen, besetzen,
bewohnen, einnehmen to ~ oneself with sich
befassen mit to ~ a position beziehen
occur, to ~ aufsetzen, auftreten, einfallen, eintreten, geschehen, passieren, sich abspielen,
sich ereignen, (accidentally) unterlaufen, vorkommen, zustande kommen to ~ in an equation in eine Gleichung *f* eingehen
occurred aufgetreten
occurrence Anlaß *m*, Auftreten *n*, Begebenheit *f*,
Ereignis *n*, Geschehnis *n*, Vorfall *m*, Vorgang
*m*, Vorkommen *n*, Vorkommnis *n*, zufälliges
Auftreten *n* (einer Erscheinung), Zustande-
kommen *n* ~ in floors stockförmiges Vorkommen *n* (min.)
occurring auftretend
ocean Meer *n*, Ozean *m*, See *f*, Weltmeer *n* ~
cable Ozeankabel *n* ~ chart Ozeankarte *f* ~
flight Ozeanflug *m* ~ floor Meeresboden *m*
~-going steamship Hochseeschleppdampfer *m*
~-going tugboat Hochseeschlepper *m* ~ liner
Ozeandampfer *m* ~ map Ozeankarte *f* ~ station Ozeanstation *f* ~ swells Ozeanseen *pl* ~
trip Ozeanfahrt *f* ~ wave Meereswelle *f*
oceanic ozeanisch
oceanographic section Ozeanografische Abteilung
ocelit Ocelit *n* ~ rod Ocelitstab *m*
ocher (or ochre) Berggelb *n*, Menning *m*, Ocker
*m*

ocherous ocker-artig, -farben, -haltig
ochery brown iron ore Brauneisenerz *n*
octa-acetyl sucrose Octaacetylsaccharose *f*
octagon Achteck *n*, Achtkant *n* ~ bar Acht-
kantstange *f* ~ column Achtecksäule *f* ~ ram
Achtkantstößel *m* ~ stock Achtkantmaterial *n*
~ tool arm Achtkantschlitten *m*
octagonal acht-eckig, -kantig, -seitig ~ error
achtelkreisiger Peilfehler *m* ~ pile achteckiger
Pfahl *m* ~ pillar (or column) achteckiger Pfeiler
*m*
octagons Achtkanteisen *n*
octahedral achtflächig, achtseitig, oktaedrisch
~ shear stress Oktaederfläche *f* in Schubspannung
octahedrite Anatas *m*
octahedron Achtflach *n*, Achtflächner *m*, Achtkant *n*, Oktaeder *n*
octal oktal (info proc.) ~ base Oktal-fassung *f*,
-sockel *m* ~ digit Oktalziffer *f* ~ notation
oktale Schreibweise *f*
octanal, ~ component of error Oktanfehlerkom-
ponente *f* ~ error Oktanfehler *m*
octane Oktan *n* ~ control Oktaneinstellung *f* ~
control lever Oktanverstellhebel *m* ~ number
Klopffestigkeitsgrad *m*, Oktanzahl *f* ~ number
requirement Oktananspruch *m* ~ pointer
Oktanzeiger *m* ~ rating Klopffestigkeits-grad
*m*, -wert *m*, Oktan-einschätzung *f*, -wert *m*,
-zahl *f* ~ value Oktanzahl *f*
octant Achtelkreis *m*, Oktant *m*
octavalent achtwertig
octave Oktave *f* ~ analyzer Oktavsieb *n* ~
coupler (organ) Oktavkoppel *f*
octavo volume Oktavband *m*
octet ring Achterschraube *f*
octode Achtpolröhre *f*, Okthode *f*
octoferric carbide Achtelkohleneisen *n*
octogon Achteck *n*
octoid system Oktoidenverzahnung *f*
octonary notation oktale Zahlendarstellung *f*
octulpe achtfach ~-phantom circuit Zweiund-
dreißiger Leitung *f*
octuplex telegraph Achtfachtelegraf *m*
octupole Oktupol *m* ~ excitation Oktopolan-
regung *f* ~ transitions Oktopolübergänge *pl*
ocular Okular *n*, Schauglas *n*; augenscheinlich,
sichtbar ~ diopter Okulardiopter *m* ~ head
Okularkopf *m* ~ jacket Okularhülse *f* ~ lens
Okularlinse *f*
O.D. (out of order) fehlerhaft
odd einzeln, seltsam, sonderbar, über eine
runde Zahl hinausgehend, (integer) ungerade,
ungleich, ungrad an ~ piece of pairs Einzelteil
von Paaren of ~ valence ungeradwertig
odd, ~-even check Gerade-Ungerade-Kontrolle
*f* ~-even nucleus Ungerade-Gerade Kern *m* ~
function ungerade Funktion *f* ~ harmonic ungerades Vielfaches *n* (TV) ~ harmonics ungerade (ungeradzahlige) Harmonische *f*, ungerade
Oberharmonische *f* ~-line interlace ungerad-
zähliger Teilraster *m* ~-line interlaced scan
ungeradzahliger Raster *m* ~-line scanning ungerader Zwischenzeilenraster *m* ~ mass number
nuclei Kerne *pl* ungerader Massenzahl ~ multiple ungerades Vielfaches *n* ~ number ungerade Zahl *f* ~-numbered ungeradzahlig ~-odd
nucleus Ungerade-Ungerade-Kern *m* ~ page

Rekte-, Vorder-seite *f* ~ **particles** Einzelteil-
chen *pl* ~ **parts** einzelne Stücke *pl* ~ **side** Formboden *m* ~ **term of atom** ungerader Atomkern
*m* ~ **valent groups** ungeradwertige Gruppen *pl*
**oddness** Ungeradheit *f*
**odiometer** Odiometer *n*
**odometer** Entfernungsmesser *m*, Kilometerzähler *m*, Meßrad *n*, Odometer *n*, Wegmesser
*m* ~ **drive** Kilometerzählerantrieb *m*
**odometrical method** Wegmeßverfahren *n*
**odor** Duft *m*, Geruch *m* ~ **seal** Geruchsverschluß *m*
**odoriferous matter** Riechstoff *m*
**odorless** geruchlos
**odorometer** Odorometer *n*
**oenometer** Weinmesser *m*
**O factor** Kreisgüte *f*
**off** ab, (on a switchboard) aus, hinweg, (the
port of) auf der Höhe von, (on a switch) ohne,
querab ~ **axis** achsenentfernt ~**-center** außermittig ~**-center dipole** Exenterdipol *m* ~-
**-center plan display** exzentrisches Schirmbild *n*
~**-center position** Abweichen *n* ~ **color** Fehl-
farbe *f* ~**-contact** Ruhekontakt *m* ~**-course**
**computor** automatischer Kursanzeiger *m* ~
**course correction** Kursverbesserung *f* ~**-course**
**position**), Ziehablage *f* ~**-cut** Nebenbahn *f*
(print.), Papierabschnitt *m* ~**-cuts** Ausschuß *m*
~ **duty** dienstfrei ~**-end stand** Endgestell *n* ~
**-field landing** Außenladung *f* (aviat.) ~**-flag**
Warnzeichen *n* (an Instrumenten) ~**-gauge**
Sondermaß *n* ~**-gauge materials** abnormaler
Werkstoff *m* ~**-grade iron** Ausfall-, Spezial-,
Übergangsrohr-eisen *n* ~**-grade rubber** Kautschuk *m* niederer Sorte ~ **hand** freihändig, von
der Hand ~ **heat** Ausfallschmelzung *f* ~**-limit**
Sperrabschnitt *m* ~ **line** indirekt ~**-load** unbelastet ~**-net station** netzfremde Fernmeldestelle *f* ~ **normal** in Arbeitsstellung *f* befindlich
~**-normal check** Grenzwertkontrolle *f* ~**-normal**
**contact** Arbeits-, Kopf-kontakt *m* ~**-normal**
**memory** Störwertspeicher *f* ~**-normal position**
Arbeitsstellung *f* ~**-normal spring** (of dial)
Nebenschlußfeder *f* ~**-on switch** Ein-Aus-
-Schalter *m* ~**-peak** Belastungstal *n* ~**-peak**
**hour** Sperrstunde *f* ~ **period** Sperrseite *f*
~**-pitch note** Mißton *m* ~ **position** Ausschalt-,
Null-stellung *f* ~**-premises extension station**
Außenmeldung *f* ~**-resonance condition**
Falschabstimmung *f* ~**-resonance tuning** un-
genaue Abstimmung *f* ~ **season** Nachsaison *f*,
tote Saison *f* ~ **side** Handseite *f* ~**-size** Abmaß
*n* ~**-size condition** Maßabweichung *f* ~**-stand-**
**ard goods** Partieware *f* ~**-the-air recording** Aufnahme *f* direkt vom Empfänger ~**-the-highway**
**manoeuvrability** Geländegängigkeit *f* ~**-the-**
**-record** inoffiziell ~**-time** Nebenzeit *f* ~**-track**
**belt switch** Schieflaufschalter *m* ~**-tuned po-**
**sition** schlechte Abstimmung *f* ~ **warehouse** ab
Lager *n* ~**-white** grauweiß, mattweiß
**offend, to** ~ Anstoß geben, beleidigen, kränken,
verletzen
**offense** Beleidigung *f*, Straftat *f*, Übertretung *f*,
Verletzung *f*
**offensive** Angriff *m*, Offensive *f* **of** ~ **smell**
übelriechend
**offer, to** ~ anbieten, offerieren, vorbringen **to**
~ **a call** Verbindung *f* anbieten **to** ~ **evidence**

Beweis *m* antreten **to** ~ **a prize** einen Preis *m*
aussetzen **to** ~ **a resistance** Widerstand *m* entgegensetzen **to** ~ **for sale** auf den Markt *m*
bringen
**offer** Angebot *n* **to make an** ~ einen Antrag *m*
stellen
**offering** Anerbieten *n*, Angebot *n* ~ **signal** Auf-
schaltezeichen *n*
**office** Amt *n*, Amtsgebäude *n*, Anstalt *f*, Büro *n*,
Funktion *f*, Geschäftszimmer *n*, Kontor *n*,
Schreibstube *f*, (of company) Wohnsitz *m*
~ **of allocations** Steuerungsstelle *f* **other** ~ **in**
**the circuit** Gegenamt *n* (teleph.) ~ **of destina-**
**tion** Bestimmungsanstalt *f* ~ **for the distribution**
**of orders** Auftragslenkungsbüro *n* ~ **with**
**extension lines** Amt *m* mit Verlängerungslei-
tungen ~ **of origin** Absendestelle *f* ~ **of supply**
**services** Nachschubdienststelle *f*
**office,** ~ **action** Prüfungsbescheid *m* ~ **boy** Laufbursche *m* ~ **busy hour** Hauptverkehrsstunde *f*
des Amtes ~ **cable** Amtskabel *n*, Innenkabel für
Vermittlungsstellen (teleph.), Zimmerleitung *f*
~ **clip** Büronadel *f* ~ **code** Amts-bezeichnung *f*,
-namen *pl*, -schlüssel *m*, Leitbuchstaben *pl*
~**-code register** Amtsnamen-, Leitbuchstaben-
-speicher *m* ~**-code system** Amtsbezeichnungs-
system *n* ~ **equipment** Amtseinrichtung *f* ~ **key**
Amtstaste *f* ~ **prefix** Amtsbezeichnung *f* ~
**record** Mitlesestreifen *m* ~ **requisites** Büro-
bedarf *m* ~ **routine** Amtstätigkeit *f* ~ **selector**
Amtswähler *m* ~ **staff** Büropersonal *n* ~
**wiring** Amtsverkabelung *f*, Zimmerleitung *f*
~ **work** Zimmerarbeit *f*
**officer,** ~ **in charge** leitender Beamter *m* ~**'s**
**mess** Kasino *n*
**official** Beamter; amtlich, dienstlich, formell,
offiziell **in** ~ **course** von Amtswegen
**official,** ~ **approval** dienstliche Genehmigung *f*
~ **authorization** dienstliche Genehmigung *f*
~ **call** Dienstanruf *m* ~ **capacity** dienstliche
Stellung *f* ~ **car** Dienstkraftfahrzeug *n* ~
**channels** Dienstweg *m*, Instanzenweg *m* ~
**civil-service employee** Beamter *m* ~ **communi-**
**cation** dienstlicher Verkehr *m* ~ **confirmation**
amtliche Bestätigung *f* ~ **confirmatory order**
bestätigte Bestellung *f* ~ **copy** Ausfertigung *f*
~ **correspondence** Dienstschreiben *n* ~ **docu-**
**ment** Handakt *m* ~ **duty** Dienstpflicht *f* ~
**envelope** Dienstumschlag *m* ~ **explanation of**
**the law** Gesetzesbegründung *f* ~ **gazette (or**
**journal)** Amtsblatt *n* ~ **leaflet** Handakt *m* ~
**message** Dienstgespräch *n* ~ **order** Befehl *m*
in Dienstsachen ~ **pamphlet** Handakt *m* ~
**power** Amtsgewalt *f* ~ **record** offizieller Rekord *m* ~ **registration (designation or number)**
amtliches Kennzeichen *n* ~ **residence** Amtswoh-
nung *f* ~ **seal** Dienstsiegel *n* ~ **secret** Amtsge-
heimnis *n* ~ **stamp** Dienststempel *m* ~ **survey**
Erhebung *f* ~ **telephone** Dienstanschluß *m*
~ **title** Diensttitel *m* ~ **traffic** Dienstverkehr *m*
~ **weighing station** Waageamt *n*
**officially sanctioned** behördlich zugelassen
**offing** von der Küste fort nach dem Horizont
gelegener Teil *n* der See **in the** ~ draußen
**offlet** Abzugsrohr *n*
**offprint** Separat-, Sonder-druck *m*
**offset** Abbiegung *f*, Absatz *m*, Abziehen *n*,
Abzug *m* (lithogr.), Abzweigung *f*, Anrechnung

*f*, Ausgleich *m*, Ausweichung *f*, Hervorstehen *n* (aviat.), Kompensation *f*, Kröpfung *f*, Offsetdruck *m* (print.), (bleibende) Regelabweichung *f* (aut. contr.), Schränkung *f*, Verrechnung *f*, (of front axle) Vorlauf *m*, Zweigleitung *f* (electr.); abgebogen, abgesetzt (aviat.), gekröpft, versetzt ~ **of step** Abstufung *f*, Abtreppung *f*

**offset, ~ angle** Reibungswinkel *m* (acoust.) ~ **base** vorspringende Grundschicht *f* ~ **bent chisel** gekröpftes Balleisen *n* ~ **blanket** Offsettuch *n* ~ **carrier** versetzter Träger *m* (TV) ~ **carrier system** versetztes Trägerwellensystem *n* ~ **cartridge paper** Gummidruckpapier *n* ~**-course computer** automatischer Kursanzeiger *m* ~ **crank drive** schiefer Kurbeltrieb *m* ~ **crankshaft** desaxierte Kurbelwelle *f* ~ **cylinder** Absatzzylinder *m* ~ **deep printing process** Offsettiefverfahren *n* ~ **demand** Gegenforderung *f* ~ **frequency simplex** frequenzabgesetzter Simplexbetrieb *m* ~ **high-speed press** Offsetschnellpresse *f* ~ **holder** gekröpfter Elektrodenhalter *m* ~ **horn** exzentrischer Hornstrahler *m* (rdr) ~ **machine** Gummidruckmaschine *f* ~ **molding** Offsetverfahren *n* ~ **plate** Unterlagplatte *f* ~ **plunger edge** abgesetzte Kolbenkante *f* ~ **position of aiming circle in axial-lateral conduct of fire** Richtstelle *f* ~ **printing** Druck *m* auf Gummi, Offsetdruck *m* ~ **printing machine** Gummiwalzendruckmaschine *f* ~ **pulse** Verschiebungsimpuls *m* (rdr) ~ **ratio** Abweichungsverhältnis *n* ~ **reproduction** Umdruck *m* ~ **roller** Gummidruckwalze *f* ~ **rotary press** Offsetrotationsmaschine *f* ~ **screen** Offsetraster *m* ~ **screw driver** Winkelschraubenzieher *m* ~ **sheet** Durchschußbogen *m* ~**-signal method (TV)** Signalverschiebungsverfahren *n* ~ **spin antenna** exzentrisch angeordnete rotierende Antenne *f* ~ **thrust** versetzter Schub *m* ~ **tool** Drehstrahl *m* (gekröpfter) ~ **well** außerhalb der Antiklinie gelegener Schacht *m*

**offsetting well** Grenz-, Nachbar-bohrloch *n*

**offshoot** Auskäufer *m*, Ausstülpung *f*

**offshore** ablandig ~ **breeze (or wind)** Landwind *m*

**offtake** (tube) Ableitung *f*, Abzug *m*, Abzugs--kanal *m*, -rohr *n*, -röhre *f*, Aufsatz *m* ~ **gate** Entnahmeverschluß *m* ~ **order** Abnahmeauftrag *m*

**ogee** Kehlleiste *f*, steigender Karnies *m* (arch.)

**ogival** Spitzbogen *m*

**ogive** Bogenspitze *f*, Spitzbogen *m*, Tellerfläche *f* ~ **of shell** Geschoßspitze *f* ~ **washer** spitzbogige Unterlagscheibe *f*

**ogligist iron ore** Glanzeisenstein *m*

**ohm** Ohm *n* ~ **meter** Ohmmeter *n*, Widerstandsmesser *m* ~ **resistance** induktionsfreier Widerstand *m*, Wirkwiderstand *m*

**ohmic** ohmsch (e, er, es), ohmisch ~ **drop of voltage** ohmscher Spannungsabfall *m* ~ **heating** Joule-Effekt *m* ~ **international resistance** Innenwiderstand *m* im Arbeitspunkt *m* ~ **loss** Joulescher Wärmeverlust *m*, Ohmscher Verlust *m*, Stromwärmeverlust *m* ~ **resistance** Dämpfungswiderstand *m*, induktionsfreier Widerstand *m*, Luftwiderstand *m*, Ohmscher Widerstand *m*, Verlustwiderstand *m* ~ **re-**

**sistance of a tube filament** Fadenwiderstand *m*

**Ohms Law** Ohmsches Gesetz *n*

**oil, to ~** einfetten, einölen, einschmieren, fetten, ölen, schmieren, das Lager schmieren **to ~-coat** ölbefeuchten **to ~-temper** in Öl *n* härten

**oil** Öl *n*, Schmierstoff *m* ~ **of bitter almonds** Bittermandelöl *n* ~ **for switches** Schalteröl *n* ~ **of turpentine** Terpentinöl *n* ~ **of vitriol manufacture** Vitriolhütte *f*

**oil, ~ atomization** Ölzerstäubung *f* ~ **atomizer** Öldüse *f* ~ **backstreaming** Ölrücktritt *m* ~ **baffle** Ölprallblech *n*, (Triebwerk) Schmierstoffabweiser *m* ~ **barrel** Ölfaß *n* ~ **base** Ölbasis *f* ~ **base for varnish** Anlegeöl *n*, Anschließöl *n* ~ **base for varnish on which to apply gold leaf** Blattgoldgrundöl *n*, Goldanlegeöl *n*, Mixtion *f* ~ **bath** Ölbad *n* ~**-bath air filter** Ölbadluftfilter *m* ~**-bearing** öldurchtränkt, ölführend ~**-bearing series** öldurchtränkte oder ölführende Schichtenfolge *f* ~ **blast circuit breaker** Ölflammofen *m* ~ **bomb** Ölbombe *f* ~ **box** Ölbüchse *f* ~ **brake** Öldruckbremse *f* ~**-break switch** Öltrennschalter *m* ~ **breather** Ölansaugventil *n* ~ **burner** Ölbrenner *m*, Ölfeuerung *f* ~ **burning** Ölfeuerung *f* ~**-burning furnace** Ölflammofen *m* ~**-burning installation** Ölfeuerungsanlage *f* ~ **cable head** Ölendverschluß *m* ~ **cake** Ölkuchen *m* ~ **can** Ausbauchung *f* (aviat.), Ausbuchtung *f* (aviat.), Blechflasche *f*, Ölkännchen *n*, Ölkanister *m*, Ölkanne *f*, Schmierkanne *f* ~ **(or lubricating) can** Schmierölkanne *f* ~ **cap** Helmöler *m* ~ **carbon** Öhlkohle *f* ~**-carbureted** ölkarburiert ~ **carburetion** Ölkarburierung *f* ~ **catch** Öltropfblech *n* ~ **catcher** Ölsammler *m*, Ölschöpfer *m* ~ **cellar** Schmierölbehälter *m* ~ **centrifugal catch** Ölschleudernase *f* ~ **centrifuge** Ölzentrifuge *f* ~ **chamber** Ölkammer *f* ~ **change** Ölwechsel *m* ~ **charge** Ölfüllung *f* ~ **circuit breaker** Ölschalter *m* ~**-circulating device (or system)** Schmierölfördereinrichtung *f* ~**-circulating tank** Ölkreislaufbehälter *m* ~ **circulation** Ölumlauf *m* ~ **cloth** Ölleinen *n*, Öltuch *n*, Wachstuch *n* ~ **cloth-making machine** Wettertuchherstellungsmaschine *f* ~ **cloud** Ölnebel *m* ~ **coating** Ölüberzug *m* ~ **cock** Ölhahn *m* ~ **coil** Ölspule *f* ~ **coke** Petroleum-, Petrol-koks *m* ~**-collector tray** Ölfangmulde *f* ~ **color** Ölfarbe *f* ~ **company** Petroleumgesellschaft *f* ~ **(dielectric) condenser** Ölkondensator *m* ~ **conduit** Ölführung *f* ~ **connection** Ölleitung *f* ~ **conservator** Ölausdehnungsgefäß *n* ~ **consumption** Ölverbrauch *m* ~ **container** Ölfangbehälter *m*, Ölkanister *m*, Ölsammelgefäß *n* ~ **control ring** Ölabstreif-, Ölhalte-ring *m*, Ölzurückhaltungsring *m* ~**-controlling valve** Ölregelventil *n* ~ **cooler** Ölkühler *m* ~ **cooling** Öhlkühlung *f* ~ **cover** Öldeckel *m*, Ölerdeckel *m*, Ölklappe *f* ~ **cup** Ölbüchse *f*, Öler *m*, Ölfänger *m*, Ölschale *f*, Öltasse *f*, Ölvase *f*, Schmier-apparat *m*, -büchse *f*, Schmierer *m*, Schmiervase *f*, Schwimmergefäß *n*, Tropfschale *f* ~ **cushion** Ölkissen *m* ~ **damping** Öldämpfung *f* ~ **dashpot** Bremszylinder *m* mit Ölfüllung, Öldämpfung *f*, Ölkatarakt *m* ~ **deflector** Ölstaubblech *n* ~**-delivery pipe** Ölzufuhrleitung *f* ~ **deposit** Ölkruste *f* ~ **derrick**

Derrick m, Ölbohrturm m ~ **dilution** Ölver-
dünnung f ~ **dipstick** Öl-meßstab m, -peilstab
m, -prüfstock m ~ **dispenser** Ölverteiler m
~**-display tube** Schauglas n für Öl ~ **distribution duct** Ölschöpfbohrung f ~ **drain** Ölablaß
m ~ **drain banjo** ringförmiges Rohranschlußstück n ~**-drain plug** Ölablaßpfropfen m ~
**drain pump** Ölrücksaugpumpe f ~ **drain screw**
Ölablaß-, Ölentleerungs-schraube f ~**-drain
valve** Ölablaßhahn m ~**-drainage container** Ölfangbehälter m ~ **drainage hole** Ölablauföff-
nung f ~ **dregs** Öldraß m ~ **drip rail** Ölfang m
~**-dripping plate** Öltropfblech n ~**-drop method**
Öltröpfchenmethode f ~ **duct** Ölkanal m ~
**ejector booster pump** Öldampfstrahlpumpe f
~ **enamel** Öllack m ~ **engine** Ölmaschine f,
Ölmotor m ~ **face** Ölspiegel m ~**-fed** ölge-
speist ~ **feed** Öleintritt m, Ölzufluß m, Ölzu-
fuhr f, Ölzuführung f, Triebölförderung f ~
**feed by gravity** Fallöl n ~**-feed pump** Öldruckpumpe f ~ **feeder** Selbstöler m, Spritzkännchen
n ~ **feeding groove** Ölfördernute f ~ **felt pad**
Ölfilz m ~ **field** Ölfeld n ~**-filled cable** Öl-
kabel n ~ **filler** Öleinfüllstutzen m, Ölfüllstutzen m ~**-filler inlet** Öleinfüllung f ~**-filler
pipe** Öleinguß m ~**-filler plug** Öleinfüllver-
schraubung f ~ **filler screw** Öleinfüllschraube f
~ **filler tube** Öleinfüllstutzen m ~ **filling filter**
Öleinfilter m ~ **filling neck** Öleinfüllung f ~
**filling socket** Öleinfüllstutzen m ~ **film** Öl-
film m, Ölhaut f, Ölüberzug m ~ **filter** Öl-
filter m, Ölreinigungsapparat m, Ölseiher m,
Ölseparator m ~ **filter case** Ölfiltergehäuse n
~ **filter gasket** Ölfilterdichtung f ~**-filter head**
Ölfilterkopf m ~ **filtering plant** Ölreinigungs-
anlage f ~ **filtration** Ölfiltration f ~ **fire** Ölfeuerung f ~**-fired** ölgefeuert, ölgeheizt ~ **fired
(air) furnace** Ölflammofen m ~**-fired space
heating** Raumbeheizung f mit Ölfeuerung ~
**firing** Ölfeuerung f ~ **flinger ring** Ölschleuderring m ~**-flooding** Ölberieselung f ~ **flotation
process** Ölschwemmverfahren n ~ **flow** Ölzufluß m ~ **foam** Ölschaum m ~ **foots** Ölsatz
m ~ **fountain** Ölfontäne f ~ **fuel** Brenn-, Heiz-
-öl n ~**-fuel pump** Heizölförderpumpe f ~
**funnel** Ölfüllstutzen m ~ **gallery** langer Ölkanal m ~ **gas** Fett-, Öl-gas n ~**-gas manufacture** Ölgaserzeugnis f ~**-gas tar** Ölgasteer
m ~ **gauge** Ölstands-anzeiger m, -messer m
~ **gauge glass (or window)** Ölschauglas n ~
**gear** Ölgetriebe n ~ **groove** Ölnute f ~**-groove
bushing** Leerlaufbuchse f mit Ölschmiernuten,
Ölrinne f, Schmiernute f ~ **groove milling
cutter** Ölnutenfräser m ~ **guard** Öldichtung f
~ **gun** Ölspritze f, Schmierölfördereinrichtung f
~**-hardened** ölgehärtet ~ **hardening** Ölhärtung f
~**-hardening steel** Ölhärtungsstahl m ~ **heater**
Ölheizung f, Ölofen m ~ **hole** Öl-, Schmier-
-loch n, Schmierstelle f ~ **hose** Rohpetroleum-
schlauch m ~**-hydraulic device** Öldruckvorrich-
tung f ~**-immersed** ölgekühlt ~ **immersed
contactor** Ölschütz m ~**-immersed transformer**
Öltransformator m ~**-immersion system** Öl-
tauchsystem n ~ **injection** Öleinspritzung f
~ **injection device** Einspritzvorrichtung f für
Öl, Öleinspritzvorrichtung f ~ **inlet** Einlaßöffnung f für Öl, Öleinlaß m, Öleintritt m ~
**intake pipe** Ölsaugrohr n ~ **jack** Ölwinde f ~

**jacketed carburetor** Vergaser m mit ölgespei-
stem Heizmantel ~ **jet** Öldüse f ~ **jet breaker**
Ölströmungsschalter m ~ **jet cap** Öldüsen-
kappe f ~ **jet mount** Öldüsenhalterung f ~ **jet
nozzle** Öldüse f ~ **joint** Ölabdichtung f ~ **lamp**
Öllaterne f ~ **lands** Erdölgelände n ~ **layer**
Ölschicht f ~ **lead** Ölführung f ~ **leather** fett-
gares oder sämiges Leder n ~ **lees** Öltrester pl
~ **lens** ölhaltige Sandlinse f ~ **level** Öl-spiegel
m, -stand m ~ **level check** Ölstandskontrolle f
~ **level screw** Ölstandprüfschraube f ~ **level
gauge** Ölstandsanzeiger m ~ **level gauge cock**
Ölstandhahn m ~ **level gauge glass** Ölstands-
glas n ~ **level indicator** Ölstandskontrolle f
~ **level inspection window** Ölstandschauglas n
~ **level plug** Ölstand-pfropfen m, -schraube f
~ **level screw** Ölkontrollschraube f ~ **line** Ölleitung f ~ **linen diagonal tape** Ölleinendiago-
nalband n ~ **lines** Ölzufluß m ~ **lubrication**
Ölschmierung f ~ **measuring stick** Ölmeß-
stab m ~ **meter** Ölmesser m ~ **mill** Öhlmühle f
~ **mist** Öl-nebel m, -staub m ~ **mist lubrication**
Ölbeschmierung f ~ **mixture** Ölmischung f
~ **mordant** Ölbeize f ~ **nipple** Schmiernippel m
~ **obtained by destructive distillation** Brandöl n
~ **operated** ölbetätigt ~ **outlet** Öl-abfluß m,
-austritt m ~ **overflow cup** Überlaufschale f ~
**overflow pipe** Ölüberlaufrohr n ~ **pad bearing**
Ölkissenlager m ~ **paint** Ölfarbe f ~ **paint(ing)**
Ölanstrich m ~ **pan** Öl-fänger m, -fangschale
f, -wanne f ~ **pan drain-plug** Ölablaßstopfen m
~ **pan gasket** Ölwannendichtung f ~ **paper**
Ölpapier f ~ **passage** Ölkanal m ~ **pipe** Öl-,
Schmier-rohr n ~ **piping** Schmierstoffleitung f
~ **pit** Ölloch n ~ **plug** Ölstöpsel m ~ **pocket
bearing** Ölkammerlager n ~ **pool** Ölfeld n, Öl-
lagerstätte f ~**-poor breaker** ölarmer Schalter
m ~ **preheater** Ölvorwärmer m ~ **press** Öllade
f, Ölpresse f
**oil-pressure** Öldruck m ~ **box** Öldruckdose f ~
~ **brake** Öldruckluftbremse f ~ **discharge**
Druckölabfluß m ~ **gauge** Öldruck-anzeiger m,
-messer m, Ölmanometer n, Schmierstoff-
druckmesser m ~ **gauge pipe** Ölmanometer-
leitung f ~ **lead** Druckölzuleitung f ~ **line**
Schmierstoffdruckleitung f ~ **pipe** Öldruck-
leitung f ~ **pump** Ölförderpumpe f ~ **supply**
Druckölzufluß m ~ **valve** Druckölventil n
~ **winch** Druckölgewinde f
**oil,** ~ **primer** Ölspachtelgrund m ~ **printing** Öldruck m (photo) ~ **probe tester** Ölstichprober
m ~ **production** Ölgewinnung f ~**-proof** öldicht
~ **protection** Ölschutz m ~ **pump** Ölpumpe f
~ **pump body** Ölpumpengehäuse n ~ **pump
casing** Ölpumpengehäuse n ~ **pump drive gear**
Zahnradölpumpenantriebsrad n ~ **pump drive
pinion** Ölpumpenantriebsritzel n ~ **pump drive
shaft** Ölpumpenantriebswelle f ~ **pump gear**
Ölpumpenantriebsrad n ~ **pump idler gear** getriebenes Rad n der Zahnradölpumpe ~
**pump screen** Ölsaugkorb m ~ **pump shaft**
Schmierpumpenwelle f ~**-pumping outlift** Öl-
pumpenanlage f ~ **purifier** Ölreiniger m ~
**purifying apparatus** Ölreinigungsapparat m
~ **radiator** Ölkühler m ~ **rating** Öleinschät-
zung f ~ **reconditioning setup** Ölrückgewin-
nungsanlage f ~ **recovered from waste oils and
fats** Altöl n ~ **recovery plant** Ölrückgewinnung

*f* ~ **rectifier** Ölreiniger *m* ~ **refinery** Öl-raffinerie *f*, -ölraffinieranlage *f* ~ **refining plant** Ölraffinieranlage *f* ~**-relay** ölgesteuert ~**-relay steam pressure reducing valve** Dampfdruckregelventil *n* mit Ölsteuerung ~ **relief valve** Ölüberdruckventil *n* ~ **reservoir** Ölbehälter *m* ~ **reservoir bearing** Ölkammerlager *n* ~ **residue** Ölrückstand *m* ~ **resistant** ölbeständig ~ **restrictor** Öldrosseldüse *f* ~ **retainer** Ölabdichtungsring *m*, Ölfangblech *n*, Ölstaublech *n* ~ **retainer plate** Ölspritzblech *n* ~ **retainer ring** Ölabdichtungsring *m* ~ **retaining packing** Lagerölabdichtung *f* ~ **retention** Abdichtung *f* gegen Ölaustritt ~ **return line** Ölrücklauf *m* ~ **return orifice** Ölrücklaufdüse *f* (Kühlschrank) ~ **return pipe** Ölrücklauf-leitung *f*, -rohr *n* ~ **return (feed) thread** Ölrückförderschnecke *f* ~ **rights** Ölausbeutungsrechte *pl* ~ **ring** Öl-, Ölstell-, Schmier-, Spritz-ring *m* ~**-ring bearing** Ringschmierlager *n* ~ **riser pipe** Ölsteigrohr *n* ~ **run** Ölnute *f*, Schmierloch *n* ~ **safety lamp** Ölsicherheitslampe *f* ~ **sample** Ölprobe *f* ~ **sand** ölhältiger Sand *m* oder Sandstein *m*, Ölsand *m* ~ **saver** Ölretter *m* ~ **scavenge** Ölrückförderung *f* ~ **schist** Ölschiefer *m* ~ **scoop** Ölschöpfer *m* ~ **scooping nose** Ölschöpfnase *f* ~ **scraper** Ölabstreifer *m* ~ **scraper ring** Schmierstoffabstreifring *m* ~ **screen** Ölsieb *n* ~ **seal** Ölabdichtung *f*, Simmerring *m* ~ **seal of crankcase** Ölfangring *m* ~ **seal high vacuum pump** Ölluftpumpe *f* ~ **seal tightening** Ölabdichtung *f* ~ **sediment** Ölsatz *m* ~ **seepage** Ölanzeichen *n* ~ **separation** Ölabscheidung *f* ~ **separator** Ölabscheider *m* ~ **separator for steam** Dampfentöler *m* ~ **service tank** Brennölbehälter *m* ~ **servo-operated** servoölbetätigt ~ **servo-piston** ölhydraulischer Servokolben *m* ~ **shale** Ölschiefer *m*, schieferiges Ölgestein *n* ~ **shield** Ölschutz *m* ~ **shock absorber** Ölstoßdämpfer *m* ~ **shock absorption** Öldruckfederung *f*, Ölstoßdämpfung *f* ~ **show** Ölspur *f* ~ **sight glass** Schauglas *n* für Öl ~ **silk** Öl-seide *f*, -taft *m* ~ **skin** Ölanzug *m*, Ölleinwand *f*, Öltuch *n*, Öltuch *n* ~ **skin coat** Ölmantel *m* ~ **skin jacket** Öljacke *f* ~ **skin trousers** Ölhose *f* ~ **slick** Ölschlick *m* ~ **slinger** Ölschleuder-, Ölspritz-ring *m* ~ **sludge** Ölschlamm *m* ~ **sludge separation** Ölschlammabsonderung *f* ~ **sluice valve** Ölschieber *m* ~ **smoke** Öl-qualm *m*, -rauch *m*, -staub *m* ~ **soap** Ölseife *f* ~ **spray** Öl-nebel *m*, -staub *m*, Spritzöl *n* ~ **spray lubrication** Ölbeschmierung *f* ~ **sprayer** Ölzerstäuber *m* ~ **spring** Mineralölquelle *f* ~ **squirt** Ölspritze *f* ~ **stone** Abzieh-, Ölabzieh-, Öl-, Schleif-, Wetz-stein *m* ~ **storage** Öllagerung *f* ~ **storage cabinet** Ölkabinett *n* ~ **storage depot** Öllager *n* ~ **strainer** Öl-seiher *m*, -sieb *n* ~ **submerged** in Öl *n* liegend, ölüberlagert ~ **substitute** Ölersatz *m* ~ **sump** Ölsammelnapf *m*, Ölwanne *f* ~ **supply** Ölvorrat *m*, Ölzuführung *f* ~ **supply nozzle** Ölzuführungsdüse *f* ~ **supply pump** Öldruckförderpumpe *f* ~ **supply valve** Ölzulaufventil *n* ~ **switch** Ölschalter *m* ~ **tanned** fettgar ~ **tank** Heizöltank *m*, Ölbehälter *m*, Ölkasten *m*, Öltank *m*, Schmierstoffbehälter *m* ~ **tank vent** Öffnung *f* des Ölbehälters ~ **tanker** Öltankschiff *n*, Tanker *m* ~ **tawed** ölgar ~ **temperature**

**gauge** Ölthermometer *n* ~ **temperature regulator** Öltemperaturregler *m* ~ **test** Ölprobe *f* ~ **tester** Ölprüfer *m* ~ **testing machine** Ölprüfmaschine *f* ~ **thermometer** Ölthermometer *n* ~ **thrower** Ölschleuderer *m*, Ölschleuderring *m* ~ **thrower funnel** Ölabschirmtrichter *m* ~ **tight** öldicht ~ **trace** Ölspur *f* ~ **transfer pump system** Schmierstoffumpumpanlage *f* ~ **(cooled) transformer** Öltransformator *m* ~ **trap** Ölabscheider *m* ~ **tray** Ölfangblech *n* ~ **treated** ölgar ~ **tube** Ölzuführungsrohr *n* ~ **vapor** Öldampf *m* ~ **vapor exhaust** Öldunstableiter *m* ~ **vapor vent** Öldunstfilter *n* ~ **varnish** Leinölfirnis *m*, Öllack *m*, Ölüberzug *m* ~ **varnish paint** Öllackfarbe *f* ~ **varnish vehicle** Firnisbindemittel *n* ~ **water** Lagerstättenwasser *n*, Ölnute *f*, Ölrinne *f*, Schmiernute *f* ~ **wax stain** Ölwachsbeize *f* ~ **well** Erdölbohrung *f*, Öl-bohrloch *n*, -bohrung *f*, -kammer *f*, -quelle *f*, -sammelnapf *m*, -tasche *f* ~ **well casing** Bohrkolonne *f* ~ **well derrick** Bohrturm *m* für Erdölbohrung ~ **wetted** ölbenetzt ~ **wick** Öl-, Schmier-docht *m* ~ **wiper** Ölabstreicher *m*, Ölabstreifer *m*, Schmierfänger *m* ~ **zone** Petroleumzone *f*

**oiled** ölgetränkt ~ **board** geölte Deckpappe *f* ~ **paper** Firnispapier *n* ~ **silk** Ölseide *f*

**oiler** (in flotation) Kollektor *m*, Öler *m*, Ölkanne *f*, Öltropfer *m*, Schmierer *m*, Schmierkanne *f*

**oiliness** Festigkeit *f*, ölige Beschaffenheit *f*, Öligkeit *f*, Schmierfähigkeit *f* ~ **dope** Ölzusatz *m*

**oiling** Einfettung *f*, Fettung *f*, Ölschmierung *f*, Ölung *f*, Schmierung *f* ~ **circuit** Ölumlauf *m* ~ **hole** Schmierloch *n* ~ **pad** Schmierkissen *n* ~ **point** Ölstelle *f* ~ **pump** Ölschleuderring *m* ~ **regulations** Schmierungsvorschriften *pl* einer Maschine ~ **screw** Ölschraube *f* ~ **system** Ölversorgung *f* ~**-up** Verölen *n*

**oilite** selbstschmierende Bronze *f*

**oilless** öllos ~ **bushing** Leerlaufbüchse *f* ohne Ölumlauf ~ **unit switchgear cubicle** Trockenschaltschrank *m*

**oillike** ölig

**oily** fett, fettig, ölhaltig, ölig, schlüpfrig, schmierig ~ **lacquer** fetter Lack *m* ~ **rag** Öllappen *m*

**ointment** Salbe *f*, Schmierstoff *m*

**okay-signal** Fertigsignal *n* (min.)

**okonite** Okonit *n*

**old** alt ~ **age insurance** Altersversicherung *f* ~ **books** Altbücher *pl* ~ **brass** Altmessing *n* ~ **excavation** alter Mann *m* ~**-fashioned** alt, altmodisch, rückständig ~ **floor** Althaufen *m* ~ **ground** alter Mann *m* ~ **ham coupling** Kreuzklauenkupplung *f* ~ **heap** Althaufen *m* ~ **holdings** Altbesitz *m* ~ **metal** Altmetall *m*, Abwrackmetall *n* ~ **red ground** Altrotgrundierung *f* ~ **red-oil-preparation** Altölpräparation *f* ~ **red oiling** Altrotölung *f* ~ **river bed** Altwasser *n* ~ **sand** Altsand *m* ~ **tan liquor** Treibbrühe *f* ~ **timer** Altholz *n*

**oleate** Oleat *n* ~ **of** ölsauer

**olefiant** ölbildend ~ **gas** Äthylen *n*, Äthylenverbindung *f*, ölbildendes Gas *n*, Ölgas *n*

**olefin** Olefin *n*

**oleic acid** Elainsäure *f*, Oleinsäure *f*, Ölsäure *f*

**oleiferous** ölhaltig
**olein** Olein *n* ~ **emulsion (or lubricant)** Oleinein-
schmelze *f*
**oleo,** ~ **gear** Ölstoßfänger *m* ~ **leg** Ölfeder-bein
*n,* -strebe *f* ~-**pneumatic brake** Ölluftdruck-
bremse *f* ~-**pneumatic shock absorber** Ölluft-
stoßdämpfer *m* ~-**resin** Ölharz *n* ~-**resin still**
Harzdestillieranlage *f* ~-**resinous varnish** Öl-
harzlack *m* ~-**strut** Federbein *n,* Ölfeder-bein
*n,* -strebe *f,* Ölluftfeder-bein *n,* -strebe *f,* Öl-
strebe *f* ~-**strut filling valve** Federbeinfüll-
ventil *n* ~-**strut trunnion** Ölstrebenbügel *m*
**oleographic paper** Öldruckpapier *n*
**oleography** Öldruck *m*
**oleometer** Oleometer *n,* Öl-messer *m,* -waage *f*
**oleum** Oleum *n*
**olfactometer** Geruchmesser *m*
**olfactory,** ~ **bulb** Riechkolben *m* ~ **test** Riechprobe *f*
**olibanum** Oliban *n*
**oligist iron mine** Glanzeisenerzgrube *f*
**oligiste iron ore** Eisenglanz *m*
**oligocene** Oligozän *n*
**oligoclase** Oligoklas *m*
**oligodynamic** oligodynamisch
**olive** Olive *f* ~ **drab** olivgrün ~ **oil** Olivenöl *n*
~ **oil black soap** Olivenölschmierseife *f* ~ **oil**
**fatty acid** Olivenölfettsäure *f*
**olivenite** Olivenerz *n,* Olivenit *n*
**olivine** Olivin *m*
**ombrograph** Regenschreiber *m*
**ombrometer** Regenmesser *m*
**omegatron** Massenspektrometer *n,* Omegatron *n*
**omission** Auslassung *f,* Unterlassung *f,* Vernachlässigung *f,* Wegfall *m,* Weglassung *f*
**omit, to** ~ auslassen, übersehen, überspringen,
unterlassen, verabsäumen, vernachlässigen
**omnibearing,** ~ **indicator (OBI)** automatischer
Azimutanzeiger *m* ~ **selector** Kurswähler *m*
(navig.)
**omnibus** Omnibus *m* ~ **circuit** Gemeinschaftsleitung *f,* Omnibusleitung *f*
**omnidirectional,** ~ **all-wave receiving antenna**
Allwellen-Rundfunkempfangsantenne *f* ~ **an-**
**tenna** Rundstrahlantenne *f,* Rundstrahler *m*
~ **exponential sound radiator** Pilzlautsprecher *m*
~ **light** Rundstrahllicht *n* ~ **microphone** All-
richtungsmikrofon *n* ~ **radio range** Drehfunkfeuer *n*
**omnigraph** Omnigraf *m*
**omnirange** Drehfunkfeuer *n*
**on** im Betrieb ~-**call channel** Reservekanal *m*
**oncoming** einfallend, entgegenkommend, heran-
kommend, sich nähernd ~ **wave** ankommende
oder einfallende Welle *f*
**on,** ~-**course signal** Aufkurssignal *n* ~-**course**
**zone** Ausflugsektor *m* ~-**line data reduction** mit
laufender Datenverarbeitung *f,* Parallelreduk-
tion *f* ~-**load speed** Lastdrehzahl *f* ~-**load tap**
**changer** Lastschalter *m* ~-**net station** netzeigene
Fernmeldestelle *f*
**on-off** auf-zu, ein-aus ~ **control** Helldunkelsteuerung *f,* Schwarzweißsteuerung *f,* Zweipunktregelung *f* ~ **controller** Zweipunktregler *m* ~
**course** Schwarzweißsteuerung *f* ~ **course-**
**indication control** Ja-Nein-Steuerung *f* ~
**keying** Ein-Aus-Tastung *f* ~ **ratio** Öffnungsverhältnis *n* ~ **type of servo** Ein-Aus-Schaltung *f*

**on,** ~ **period** Flußzeit *f* ~ **position** Einschaltstellung *f* ~ **position of switch** Schaltstellung *f*
~ **site requirements** die im Werk erforderlichen
Bedingungen *pl* ~-**the-air-monitor** Endkontrol-
gerät *n* ~-**the-job-training** Ausbildung *f* am
Arbeitsplatz ~-**the-job-training period** Einar-
beitungszeit *f* ~ **the-spot-broadcast** Rundfunk-
reportage *f*
**once, at** ~ alsbaldig, einmal, sofort, sogleich,
auf der Stelle **for** ~ ausnahmsweise ~ **through**
Zwangsdurchlauf *m* ~ **through cooling** Zwangs-
durchlaufkühlung *f* ~ **through operation** direkter Durchsatz *m*
**ondograph** Ondograf *m,* Wellenlinienschreiber *m*
**ondometer** Frequenz-, Wellen-messer *m*
**ondoscope** Glimmlichtoszilloskop *n* .
**one** ein ~ **after another** nacheinander ~ **and a**
**half times** anderthalbfach ~ **into the other** ineinander **in** ~ **row** einreihig ~ **at a time** einmal
**one,** ~-**address instruction** Einadreßbefehl *m*
~-**and multi-spindle machine** Ein- und Mehr-
spindelmaschine *f* ~-**armed** einarmig ~-**armed**
**lever** einarmiger Hebel *m* ~-**bath dyeing process**
Einbadfärbeverfahren *n* ~-**chamber brake** Ein-
kammerbremse *f* ~-**circuit set** Einkreisempfänger *m,* Einkreiser *m* ~-**cylinder** einzylindrig
~-**digit** einstellig ~-**dimensional disorder** eindimensionale Fehlordnung *f* ~-**dimensional**
**flow** Fadenströmung *f* ~-**dimensional group**
einfaltige Gruppe *f* ~-**dip aniline black** Ein-
bandanilinschwarz *n* ~-**figure** einstellig ~-
-**filament bulb** Einfadenglühlampe *f* ~-**floored**
einhordig ~-**floored kiln** Darre *f* mit einer
Horde, Einhordendarre *f* ~-**flue boiler** Ein-
flammrohrkessel *m* ~-**fold** einfach, einzeln ~-
-**fourth wave transformer** Transformationslei-
tung *f* ~-**group model** Eingruppenmodell *n*
~ **half of scissors telescope** Halbschere *f* ~
**half full size** halbe natürliche Größe *f* ~-**hand**
**operation** Einhandbedienung *f* ~-**hole mounting**
Einlochbefestigung *f* ~-**hole nozzle** Einlochdüse *f* ~-**horse** Einspänner *m* ~-
-**horse (hauling) wagon** einspänniger Transportwagen *m* ~-**hour-rating-speed** Stundendreh-
zahl ~-**hour-rating-torque** Stundendrehmoment
*n* ~ **hundredth part** Hundertstel *n* ~-**in arrears**
Restant *m* ~-**knob coil adjustment (or tuning)**
Eingriffsspulenwechsel *m* ~-**knob coil changing**
Einspulentrommel *f* ~-**knob tuning** Eingriffab-
stimmung *f,* Einknopfbedienung *f* ~-**layer**
**coiling** Einschichtwicklung *f* ~-**layer filter** Ein-
fachfilter *n* ~-**layer insulating cover** Einschicht-
isolierhülle *f* ~-**line business** Spezialgeschäft *n*
~-**line diagram** Einlinienschaltbild *n* ~-**lobed**
einhöckrig ~ **off item** Einzelfertigung *f* ~ **output**
Einerausgabe *f* ~-**parameter family of curves** ein-
parametrische Familienscharen *pl* ~-**parameter**
**family of lines** Geradenbüschel *n* ~-**part** einteilig
~ **phase** einphasig ~-**phase grounding** Phasenerd-
anschluß *m* ~-**piece** aus einem Stück, einteilig ~-
-**piece crankshaft** Kurbelwelle *f* aus einem Stück
~-**piece gun barrel** Vollrohr *n* ~-**piece rod-type**
**insulator** Langstabilisator *m* ~-**piece wheel**
Vollrad *n* ~-**piece wing** durchgehender Flügel
*m* ~-**plus-one instruction** Eins-plus-Eins-Befehl
*m* ~-**point** Achtelpetit *n* (print.) ~-**point**
**emergency-cell switch (or end-cell switch)** Ein-
fachzellenschalter *m* ~-**point space** Einpunkt-

spatium n ~-pole plug einpoliger Stecker m ~
pole switch einpoliger Schalter m unter Verputz
~ pressure stage Gleichdruckstufe f ~-shot
camera Einbelichtungs-, Spaltbild-, Techni-
color-kamera f ~-shot lubrication Zentralschmierung f ~-shot multivibrator mono-
stabiler Multivibrator m ~-shuttle einschützig
~-side finish einseitige Appretur f ~-sided
einseitig ~-sided bit Exzentermeißel m, schiefer
Bohrmeißel ~-sided cutter Einflankenwerkzeug
n ~-sided rope eintrümmiges Seil n ~ sided
surfaces einseitige Flächen pl ~ sixtieth of one
angular degree Winkelminute f ~-stage amplifier einstufiger Verstärker m ~-stage blower
Einstufengebläse n ~-state Einszustand m
~-story (building) einstöckig ~-strand rope
eintrümmiges Seil n ~ stroked octave einge-
strichene Oktave f ~ tenth normal solution
Zehntelnormallösung f ~-throw crankshaft
einfach gekröpfte Kurbelwelle f ~-time tape
Schlüssellochstreifen m (teletype) ~-to-~-cor-
relation eineindeutige Zuordnung f ~-to-~-cor-
respondence umkehrbare Eindeutigkeit f ~-
-to-~-salts einwertige Salze pl ~-to-~-se-
quential doppeleindeutig ~-to-partial-select
ratio Verhältnis n Einersignal/Teilselektions-
signal ~-to-zero ratio Verhältnis n Einersignal/
Nullsignal ~-two-three system Eins-zwei-drei-
System n ~-valued function eindeutige Funktion f ~-valve receiving set Einröhrengerät n
~-valve repeater Einrohrverstärker m ~-
-voiced einstimmig
one-way einwegig, in einer Richtung f ~ circuit
einseitig oder in einer Richtung betriebene
Leitung f für gerichteten Verkehr ~ communication einseitiger Fernmeldeverkehr m ~ conductor Hinleiter m ~ current flow Richtstrom m
~ street Einbahnstraße f ~ traffic Einbahnver-
kehr m, einseitiger Verkehr m ~ working
Arbeiten pl in einer Richtung (einseitiges
Arbeiten)
one, ~ wheel landing gear Einradfahrgestell n
~ wing heavy querlastig (aviat.) ~ wire circuit
Einfachleitung f
onion Leuchtrakete f ~ skin Florpost f, Luft-
postpapier n ~ skin and copy paper Flor- und
Durchschlagpost f
onlooker Zuschauer m
onofrite Quecksilberglanz m
onset Anfang m, Ansatz m, Beginn m ~ of
impulse Einschwingvorgang m ~ of ionization
Einsetzen n der Ionisierung ~ of a sound
Klangeinsatz m ~ of steady corona Einsetzen n
des Dauerkoronadurchbruchs
onset time (of a tone) Anklingzeit f
onsetter Anschläger m (min.)
onsetting carriage Einstoßschlitten m
onus Last f
onyx Onyx m
oölite Oolith m, Rogenstein m
oölitic oolitisch ~ brown iron ore Minette f ~
iron mine Rogen(eisen)steingrube f ~ ironstone
Eisenoolith m, Likrokonit m (min.) ~ limestone Oolithkalk m, schlickiger Kalk m ~
limonite Linsenerz n ~ ore Bohnerz n ~ period
Juraformation f
ooze, to ~ durchsickern, sickern to ~ away
versickern to ~ in einsickern to ~ out aus-

sickern, heraussickern to ~ out of herausströmen to ~ through durchschwitzen
ooze Beizbrühe f, Lohbrühe f, Schlamm m,
Schlammgrund m, Schlick m
oozed leather lohgares Leder n
oozing Beizbrühe f ~ out Ausschwitzung f
oozy schlammig ~ botton Moddergrund m
opacifier Trübungsmittel n (glass mfg.)
opacimeter Trübungsmesser m
opacity Deckkraft n, Farbdeckfähigkeit f,
Glanzlosigkeit f, Lasurfähigkeit f, Lichtun-
durchlässigkeit f, Trübung f ~ of a lens Linsentrübung f ~ tester Lichtdurchlässigkeitsprüfer
m
opal Opal m; getrübt ~ finish Opalausrüstung f
~ glass Milch-, Opal-, Trüb-glas n ~ glass
globe Milchglasglocke f ~ glass window Milchglasfenster n ~ lamp Opallampe f
opalesce, to ~ bunt schillern, opalisieren
opalescence Farben-spiel n, -wechsel m, Opa-
leszenz f, Opalisieren n, Schillern n
opalescent opalisierend, schillernd ~ glass
Milchglas n ~ liquid schillernde Flüssigkeit f
opaline opalartig
opaque deckend (color), lichtundurchlässig,
nicht durchscheinend, opak, trübe, undurchlässig, undurchsichtig ~ to infrared ultrarot-
undurchlässig
opaque, ~ illuminator Opakilluminator m ~
meal Kontrast-einlauf m, -füllung f, -mahlzeit
f ~ pigment Deckfarbe f ~ reflector Opakilluminator m ~ rubber Bleigummi m
opaqueness Deckkraft f (color), Lichtundurch-
lässigkeit f, Undurchsichtigkeit f
open, to ~ anstechen, aufgehen, aushalsen, aufhauen, (connections) auflösen, aufreißen, aufschlagen, aufschließen, einbrechen, eine Röhre
öffnen, einleiten, eröffnen, erschließen, entsperren, klappen, münden, öffnen, unterbrechen
open, to ~ after bending nach dem Biegen zurück-
federn to ~ a block zurückblocken (r.r.)
to ~ with caustic aufätzen to ~ a chamber
einen Abbau m aufhauen to ~ a circuit den
Stromkreis öffnen to ~ the circuit ausschalten
to ~ a credit akkreditieren to ~ the draft den
Durchlaß m öffnen to ~ up an engine Drossel f
öffnen to ~ up the engine Gas n geben to ~
fire Feuer n eröffnen oder einschalten, das
Schießen beginnen to ~ fire on unter Feuer n
nehmen to ~ up on hinges aufklappen to ~ a
mine den Betrieb m einer Grube eröffnen to
~ the mold die Gußform losbrechen, entfor-
men to ~ out aufreiben, aufweiten, dornen,
nach außen erweitern to ~ the parachute den
Fallschirm m öffnen to ~ out a pipe ein Rohr n
aufweiten to ~ to public inspection auflegen
(patents) to ~ a switch einen Schalter m öffnen
to ~ the throttle Gas n geben to ~ the throttle
wide Vollgas n gegeben to ~ to traffic freigegeben für den Verkehr m to ~ up auflockern,
aufmachen, sich öffnen to ~ wide aufsperren
open Drahtbruch m (teleph line); frei, geöffnet,
locker, offen, öffentlich in the ~ über Tag
into the ~ ins Freie
open-air Freiluft f ~ drying lufttrocknend ~
exhibition space Freigelände n (Messe) ~
ground Freigelände n (Messe) ~ ionization

**chamber** offene Luftionisationskammer *f* ~
**piping** Freileitung *f* ~ **plant** Freiluftanlage *f*
~ **theater** Freilichtbühne *f* ~ **transformer plant**
Freilufttransformatorenanlage *f* ~ **transmission line** Freileitung *f* ~ **wall duct** Freiluftdurchführung *f* ~ **weathering test** Freiluftbewitterung *f* ~ **wind tunnel** Freiluftwindkanal *m*
**open,** ~ **antenna** Außenantenne *f* ~ **armor** offene
Bewehrung *f* ~ **boat** offenes Boot *n* ~ **burning
coal** Gasflammkohle *f* ~ **cable** offenes Seil *n*
~ **cast mining** Abräumarbeiten *pl* ~ **cast working** Übertagebetrieb *m* ~ **channel** Werkgraben
*m* ~ **center control** Mittelpunktausbreitung *f*
~ **center press wheel** Druckrad *n* mit offener
Mitte, Druckrolle *f* mit offener Mitte, ~ **circuit** unterbrochener Kreis *m*, offener Stromkreis *m*
**open-circuit,** ~ **admittance** Leerlaufadmittanz *f*
~ **alarm system** Alarmeinrichtung *f* mit Arbeitsstrom ~ **connection** Arbeitsstromschaltung
*f* ~ **fault** Kabelunterbrechung *f* ~ **impedance**
Leerlauf-impedanz *f*, -widerstand *m* ~ **monitor** Leitungsbruchwächter *m* ~ **operation**
Arbeitsstrombetrieb *m* ~ **signalling** Arbeitsstromverfahren *n* ~ **stable** leerlaufstabil ~
**system** Arbeitsstromsystem *n* ~ **ventilation**
Frischluftkühlung *f* ~ **voltage** Leerlaufspannung *f* ~ **working** Arbeitsstrombetrieb *m*
**open,** ~ **circuited** geöffnet ~ **clamp** Stadel *f*
~ **core** offener Kern *m* ~ **core transformer**
Transformator *m* mit offenem Eisenkern ~
**country** freies Gelände *n* ~ **cup** offener Tiegel
*m* ~ **cup flash point** Flammpunkt *m* im offenen
Tiegel ~ **cure** Vulkanisation *f* durch trockene
Hitze ~**cut** offener Einschnitt *m* ~ **cycle**
offener Kreislauf *m* ~ **die** Preßbacke *f* ~
**diggings** Gräberei *f*, Pingenbau *m* ~ **drainage**
Oberflächenentwässerung *f* ~ **end** leerlaufend,
offen; Austrittsöffnung *f* ~ **end barometer**
Baroskop *n* ~ **end impedance** Leerlaufimpedanz
*f* ~ **end ratchet wrench** Ratschen-Maulschlüssel
*m* ~ **end wrench** Gabelschraubenschlüssel *m*
~ **ended circuit** am Ende offener Stromkreis *m*
~ **ended line** offene Leitung *f* ~ **ends of bridge**
Brückenspitze *f* ~ **exhaust** freier Auspuff *m*
~ **feeder** Stichleitung *f* ~ **fire kiln** Rauchdarre *f*
~ **flank** offene Flanke *f* ~ **flow** freie Eruption *f*
~ **flow test** Messung *f* bei offenem Bohrloch
~ **flow well** Springer *m* (Erdöl) ~ **flux structure**
Eigenschaft *f* des offenen Kraftflusses ~ **form
counting chamber** offene Zählkammer *f* ~
**framework** räumliches Fachwerk *n* ~ **front
eccentric press** Einständerexzenterpresse *f* ~
**grained** großluckig ~ **grained structure** grobes
Gefüge *n* ~ **grid** weitmaschiges Gitter ~ **groove**
offenes Kaliber *n*
**open-hearth,** ~ **cinder** Martinofenschlacke *f* ~
**furnace** Hochofen *m* mit offener Brust, Martionofen *m*, Siemens-Martin-Ofen *m* ~ **iron**
Martinflußeisen *n* ~ **pig(iron)** Martinroheisen
*n*, Roheisen *n*, Siemens-Martin-Ofen-Roheisen
*n* ~ **pig iron** Stahlroheisen *n* ~ **plant** Martinofen-anlage *f*, -betrieb *m* ~ **practice** Martinofenbetrieb *m*, Siemens-Martin-Betrieb *m* ~
**process** Herdofen-, Martin-prozeß *m*, Martinverfahren *n*, Siemens-Martin-Ofen-Prozeß *m*,
Siemens-Martin-Ofen-Verfahren *n* ~ **refining**
Herdfrischen *n* ~ **steel** Herdfrischstahl *m*,

Martinflußstahl *m*, Martinstahl *m*, Puddeleisen *n*, Siemens-Eisen *n*, Siemens-Flußstahl *m*
~ **(furnace) steel** Siemens-Flußeisen *n* ~ **steel
furnace** Martinstahlofen *m* ~ **steelworks** Martinstahlhütte *f*
**open,** ~ **jet wind channel** Freistrahltunnel *m* ~
**jet wind tunnel** Freistrahlwindkanal *m* ~
**language** offene Sprache *f* ~ **line** Freileitung *f*.
oberirdische Linie *f*, freie Strecke *f* (r.r.) ~
**line balancing network** Freileitungsnachbildung *f* ~ **line construction** Freileitungsbau *m*
~ **link chain** Glieder-, Schaken-kette *f* ~ **lode**
Anbruch *m* ~ **loop control** einfache offene
Steuerung *f* ~ **loop control system** rückführungslose Steuerung *f* ~ **matter** Satz *m* mit
viel Ausschuß (print.) ~ **meshed grid** weitmaschiges Gitter ~ **mold cast** Offenguß *m*
~ **nozzle** offene Düse *f* ~ **oscillating circuit**
offener Schwingungskreis *m* ~ **oscillator** offener Schwinger *m* ~ **pass** offenes Kaliber *n* ~
**phase protection** (relay) Leiterbruchschutz *m*
~ **pilot's cockpit** offener Führersitz *m* ~ **pit
mining** Tagebau *m* ~ **pit ore mining** Erztagebau
*m* ~ **position** Öffnungsstellung *f* ~ **pressure**
Ausflußdruck *m* ~ **radiative circuit** offener
Schwingungskreis *m* ~ **rectangular kiln** Stadel *f*
~ **resistance** unendlicher Widerstand *m* ~ **road**
freie Strecke *f* ~ **roadstead** Außenreede *f* ~
**sand** offener Sand *m* ~ **sand casting** Herdguß *m*
~ **sand mold** Herdform *f*, Herdgußform *f*,
offene Form *f* ~ **sand molding** Herdformerei *f*
~ **sea** offener See *f* ~ **seam tubing** offenes Rohr
*n* mit einer Naht ~ **sheet delivery** Planoausleger *m* ~ **shop** betriebseigene Programmierung
*f* ~ **side grinding machine** Einständerschleifmaschine *f* ~ **side planer** Einständerhobelmaschine *f* ~ **sight for heavy guns for close range
finding** Nahfeuervisier *n* ~ **sight searching
telescope** Sucherfernrohr *n* ~ **size** Maulweite *f*
(of wrench) ~ **slot coil** Schlitzspule *f* ~ **space**
Freifläche *f* ~ **stamp mill** Klotze *f* ~ **starter**
Anlasser *m* (offen) ~ **steam cure** Vulkanisation
*f* mit direktem Dampf ~ **stope** versatzlose Abbaufirste *pl* ~ **stope with pillar** Abbau *m* in
regelmäßigen Abständen ~ **storage** Lagerung *f*
im Freien ~ **surface condenser** Rieselkondensator *m* ~ **surface cooler** Rieselkühler *m* ~
**surface evaporator** Rieselverdampfer *m* ~ **tail**
Gitterschwanz *m* ~ **tank steeping** Trogtränkung *f* (Holz) ~ **terrain** deckungsloses Gelände *n* ~ **throttle** Vollgas *n* ~ **top** offen; ohne
Verdeck *n* ~**-type motor** offener Motor *m* ~
**undulating ground** flachwelliges Gelände *n* ~
**weaving** loses Gewebe *n* ~ **wind tunnel** Freistrahlwindkanal *m* ~ **wire** oberirdische Linie *f*
~ **wire circuit** Einfachleitung *f*, oberirdische
Linie *f* ~ **wire line** Freileitungslinie *f*, Luftleitung *f* ~ **wiring** offene Leitungsführung *f* ~
**work** durchbrochene Arbeit *f*, Tagebau *m*
(min.) ~ **work fabric** durchbrochenes Gewebe *n*
~ **work mining** Tagebau *m* ~ **worked** durchbrochen ~ **working** Tagebaubetrieb *m* ~
**workings** über Tag (min.)
**openable** zu öffnen
**opener** Öffner *m*, Reißwolf *m*
**opening** Abbaustrecke *f*, Anbruch *m* (min.),
Aufhauen *n*, Aufschließung *f*, Auslösung *f*,
Ausschnitt *m*, Aussparung *f*, Ausweitung *f*,

Durch-bruch *m*, -fahrt *f*, -laß *m*, Erweiterung *f*,
freie Stelle *f*, Freigabe *f*, Hohlraum *m*, Lichtung *f*, Loch *n*, (of a connection) Lösung *f*,
Lücke *f*, Mund *m*, Niederlegung *f*, Öffnen *n*,
Öffnung *f*, Schurf *m*, Spalt *m*, Spannweite *f*,
Unterbrechung *f*, Weite *f*, Zugang *m*
**opening,** ~ **up of accounts** Aufstellung *f* der
Rechnungen ~ **to admit light** Lichtöffnung *f*
~ **for axle-box guide** Achslagerausschnitt *m*
(r.r.) ~ **of a circuit** Ausschalten *n* ~ **of a
circuit breaker** Auslösung *f* ~ **and forewinning**
Aus- und Vorrichtungsarbeiten *pl* ~ **of groove**
Kaliberöffnung *f* ~ **out a hole** Auftreiben *n*
eines Loches ~ **for light** Lichtloch *n*, Öffnung *f*
~ **of the mine shaft** Grubenweite *f* ~ **of passage**
Durchgangsöffnung *f* ~ **between rolls** Walzenspalte *f* ~ **in the roof** Dachluke *f* ~ **for the
sale** Absatzmöglichkeit *f* ~ **of scavenging port**
Spülschlitzöffnung *f* ~ **of the screw spanner**
Maulweite *f* des Schraubenschlüssels ~ **of
telephone service** Eröffnung *f* von Sprechbeziehungen, Zulassung *f* ~ **(door) for withdrawing ashes** Aschenausziehöffnung *f*
**opening,** ~ **bit** Räum-, Reib-ahle *f*, Schneidbohrer *m* ~ **bridge** bewegliche Brücke *f* ~ **cam for
swinging grippers** Exzenter *m* für Vorgreiferbewegung ~ **capacity** (of a switch) Unterbrechungsvermögen *n* ~ **contact** Öffnungskontakt
*m* ~ **feed** Maulweite *f* ~ **fire** Feueraufnahme *f*
~ **force** Öffnungskraft *f* ~ **indicator** Stellungsanzeiger *m* ~ **pin** Öffnungsstift *m* ~ **pipe** Einführungsisolator *m* ~ **rail** Ausbreiter *m* ~ **ram**
Rückdrückkolben *m* ~ **rate** Abnäherung *f*
(artil.), (exchange) Anfangskurs *m* ~ **section**
(swing bridge) Durchlaß *m* ~ **shock** Entfaltungs-, Öffnungs-stoß *m* ~**-shock indicator**
(parachute) Entfaltungsstoßschreiber *m* ~
**signal** Aufforderungszeichen *n* ~ **(or initial)
spark** Eröffnungsfunke *m* ~ **speed** Entfaltungsgeschwindigkeit *f* ( Fallschirm) ~ **stress**
Sprengspannung *f* ~ **time** (parachute) Entfaltungszeit *f* ~ **wedge** Bresche *f* ~ **width** Spaltbreite *f*
**operable** bedienbar, betriebsfähig, durchführbar
**operand** Rechengröße *f*
**operate, to** ~ antreiben, arbeiten, (magnet) ansprechen, bedienen, betätigen, betreiben, funktionieren, handhaben, hantieren, schalten, werken, wirken **to** ~ **on an automatic system** nach
einem selbsttätigen System *n* betreiben **to** ~
**above capacity** überlasten **to** ~ **on a closed
circuit** mit Ruhestrom *m* betreiben **to** ~ **only
when engine is clear** Betätigung *f* nur wenn
Maschine frei **to** ~ **to the full capacity** voll
arbeitsfähig **to** ~ **mechanically** zwangsläufig
arbeiten **to** ~ **on open circuit** mit Arbeitsstrom
*m* betreiben **to** ~ **a relay** ein Relais *n* ansprechen
lassen **to** ~ **a winch** spillen
**operate,** ~ **lag** (of relay) Anzugsverzögerung *f*
~ **switch** Betriebsschalter *m* ~ **time** Ansprechzeit *f*
**operated** betätigt ~ **by cables** durch Kabel *n* betätigt ~ **by means of tie rods** durch Gestänge *n*
betätigt ~ **position** Arbeits-lage *f*, -stellung *f*
**operating** Arbeiten *n*; ansprechend, in Betrieb
befindlich, wirkend **not** ~ ausgeschaltet sein ~
**and design pressures and temperatures** Betriebs- und Solldrücke und -temperaturen *pl*

~ **by hand** Hantierung *f* ~ **of switches** Weichenstellung *f*
**operating,** ~ **alternating current** Altstrombetrieb *m* ~ **angle** Schaltwinkel *m* ~ **aperture**
Arbeitsöffnung *f* ~ **arm** Arm *m* zum Operieren
~ **ball** Schaltkugel *f* ~ **bar** Betätigungsstange *f*
~ **cam** (lubricator) Arbeitshubrad *n* ~ **capacity
of valve** Ventilgängigkeit *f* ~ **characteristic**
Laufeigenschaft *f* ~ **characteristics** Betriebs
-kennzahlen *pl*, -merkmale *pl* ~ **circuit** Arbeitsstromkreis *m* ~ **code** Betriebsvorschrift *f*
~ **coil** Arbeitsbetriebsspule *f* ~ **coil winding**
Betriebsspulenwicklung *f* ~ **condition** Betriebszustand *m* ~ **condition(s)** Betriebs-bedingungen
*pl*, -daten *pl* ~ **contact** Betriebskontakt *m* ~
**cost** Arbeitskosten *pl*, Verarbeitungskosten *pl*
~ **crank** Antriebskurbel *f* ~ **crew** Bedienungsmannschaft *f* ~ **current** Ansprechstrom *m*, (minimum) Betriebsstrom *m* ~ **curve** Arbeitskennlinie *f* ~ **cycle** Arbeitszyklus *m*, Förderweg *m*
~ **cylinder** Betätigungs-, Stell-zylinder *m* ~
**data** Betriebs-daten *pl*, -werte *pl* ~ **datum** Bestimmungsstück *n* ~ **device** Bedienungs-, Betriebs-einrichtung *f* ~ **difficulty** Betriebsschwierigkeit *f* ~ **direct current** Allstrombetrieb *m* ~ **efficiency** Betriebsleistung *f* ~
**expense** Betriebskosten *pl* ~ **expenses** Betriebsausgaben *pl* ~ **experience** Betriebserfahrung *f* ~ **explosion** Einsatzexplosion *f* ~
**facilities** Betriebseinrichtung *f* ~ **features**
Arbeitselemente *pl* ~ **field strength** Betriebsstärke *f* ~ **figures** Betriebskennzahlen *pl* ~
**force** Arbeitskolonne *f*, Betriebspersonal *n*
~ **fork** Schließgabel *f* ~ **frequency** Betriebsfrequenz *f* ~ **grippers** Greiferbewegung *f* ~
**handle** Bedienungshebel *m* ~ **handwheel** Bedienungsrad *n* ~ **installation** Betriebseinrichtung *f* ~ **instructions** Betriebsanleitung *f*, Gebrauchsanweisung *f* ~ **key** Funktionstaste *f*
(tape rec.), Taste *f* ~ **knob** Bedienungsknopf *m*
~ **(water) level** Betriebswasserstand *m* ~ **lever**
Antriebs-, Steuer-hebel *m* ~ **lever fulcrum**
Druckhebelböckchen *n* ~ **lever shaft** Druckhebelachse *f* ~ **lever spring** Schenkelfeder *f* ~
**limitations** betriebliche Begrenzungen *pl* ~
**line** Arbeitskurve *f* ~ **load** Betriebsbeanspruchung *f* ~ **locus** Arbeitspunkt *m* ~
**machinery** Antriebsvorrichtung *f*, Bewegungsmechanismus *m* ~ **magnet** Stell
-magnet *m*, -zylinder *m* ~ **maintenance** nach
Bedarf durchgeführte Instandsetzung *f* ~ **man**
Betriebsmann *m* ~ **mechanism** Bewegungsvorrichtung *f*, Steuergerät *n* ~ **noise temperature**
System-Rauschtemperatur *f* ~ **panel** Bedienungsplatte *f* ~ **photocell** Tonlampe *f*, Tonlichtlampe *f* ~ **pin** Abstellstift *m* ~ **piston**
Regel-, Schalt-kolben *m* ~ **piston sleeve** Zündachse *f* ~ **platform** Manövrierstand *m* ~ **plunger** (pump) Arbeitskolben *m*, Auslösebolzen *m*
~ **point** Ruhepunkt *m* ~ **position** Arbeits-platz
*m*, -stellung *f* ~ **practice** Betriebspraxis *f* ~
**practices** Betriebsregeln *pl* ~ **pressure** Arbeitsdruck *m*, Betriebs-druck *m*, -spannung *f* ~ **pump**
Arbeitspumpe *f* ~ **range** Brennbereich *m*, Fahrbereich *n* ~ **ratio** effektiver Betriebsfaktor *m*
~ **regulations** Betriebsordnung *f* ~ **reliability
(or safety)** Betriebssicherheit *f* ~ **result** Betriebsergebnis *n* ~ **rhythm** Arbeitsrhythmus *m*

**~ rod** Stoßstange f **~ room** Betriebsraum m, Vorführraum m **~ rules** Betriebsregeln pl **~ screw** Steuerschraube f **~ screw mounted on ball bearings** Schraube f auf Kugellager **~ shaft** Schloßkurbel f **~ signal** Verkehrsabkürzung f **~ site** Betriebsstelle f **~ skill** Handfestigkeit f **~ span** Reichweite f **~ speed** (relay) Ansprechgeschwindigkeit f, Arbeits-, Betriebs-geschwindigkeit f, Betriebsdrehzahl f **~ staff** Betriebspersonal n **~ statement** Betriebsbericht m **~ station** Bedienungsstand m **~ stop** Arbeitsanschlag m **~ stoppage** Betriebsstillstand m **~ supervision** betriebliche Überwachung f **~ switch** Bedienungsschalter m (electr.), Betriebs-artenschalter m, -schalter m **~ temperature** Betriebstemperatur f **~ tension** Betriebsdruck m **~ time** Ansprechzeit f, Arbeitszeit f, Herstellungsdauer f, Schaltzeit f **~ transmitter at high modulation percentage** Durchmodulierung f des Senders **~ trouble** Betriebsstörung f **~ unit** Bedienungsgerät n **~ voltage** Betriebs-, Biegungs-, Brenn-spannung f **~ wave lengths** Betriebswellenlängen pl **~ weight** Betriebsleergewicht n **~ winding** Erregerspule f

**operation** Ansprechen n, Anziehung f, Arbeit f, Arbeiten n, Arbeitsvorgang m, Bedienung f, Betätigung f, Betrieb m, Funktion f, Handhabung f, Operation f, Trieb m, Unternehmen n, Unternehmung f, Verfahren n, Vorgang m, Werkstufe f, Wirken n, Wirkungsweise f, Zusammenbau m

**operation, in ~** eingeschaltet **in one ~** in einem Arbeitsgang m **to be in ~** in Gang sein, in Betrieb sein **to be out of ~** kaltliegen **~ under constant pressure** Gleichdruckbetrieb m **~ by hand lever** Handhebelbetrieb m **~ to maintain the edges true** (in rolling) Stauchdruck m **~ of measurement** Meßtätigkeit f **~ to one side** einseitiges Arbeiten n **~ at reduced pressure** Unterdruckbetrieb m **~ of transmitter station at high modulation percentage** Senderdurchmodulierung f

**operation, ~ details** Bedienungsvorrichtung f **~ efficiency** Arbeitsleistung f **~ governed motor** Arbeitsreglermotor m **~ level** Schalthebel m **~ life** Laufdauer f **~ number** Arbeitsgang-, Operations-nummer f **~ period** Betriebszeit f **~ record** Arbeitsfolgenplan m **~ sheet** Bearbeitungsplan m **~ test** Betriebsprüfung f **~ time** Operationszeit f **~ unit** Einheitbedienungsgerät n

**operational** betrieblich, einsatzbereit, operativ; Arbeits . . ., Betriebs . . ., Funktions . . ., **~ aerodrome** Einsatz(flug)hafen m **~ altitude** Betriebshöhe f **~ amplifier** Rechenverstärker m **~ calculus** Operatorenrechnung f **~ chart** Schaubild n **~ check** Funktionsprüfung f **~ code** Operationsbefehl m **~ control** Leitung f der Flugdurchführung **~ error** Betriebsfehler m **~ fault** Laufstörung f **~ flexibility** Anpassungsfähigkeit f im Betriebsverhalten **~ height** Gebrauchshöhe f **~ layout** Arbeitsanweisungen pl **~ life** Betriebslebensdauer f **~ method** Arbeitsmethode f **~ objective** Operationsobjekt n, operative Planung f, operatives Ziel n **~ possibility** Einsatzmöglichkeit f **~ range** Anwendungsfeld n **~ readiness of plotters**

Auswertebereitschaft f **~ research** Ökonometrie f **~ snag** betrieblicher Mangel m **~ standard** Arbeitsbereitschaft f **~ summing amplifier** Summenverstärker m **~ test** Funktionserprobung f **~ time sequence chart** Arbeitsablaufdiagramm n **~ unit** Bearbeitungsorgan n **~ write-up** Arbeitsanweisungen pl

**operationally, ~ fit** einsatzreif, serienreif **~ secure** betriebssicher

**operations** Arbeitsweise f **~ manual** Betriebshandbuch n **~ map** Operationskarte f **~ office** Betriebsamt n **~ order** Operationsbefehl m **~ plan** Operationsplan m **~ scheduling** Arbeitsvorbereitung f **~ section** Betriebs-, Operations-abteilung f **~ staff** Führungsstab m **~ unit** Betriebseinheit f

**operative** arbeitend, betrieblich, betriebsfähig, gebrauchsfähig, wirksam **~ attenuation** Betriebsdämpfung f **~ characteristics** Arbeitsweise f **~ condition** Arbeitsbedingung f **~ missile** Einsatzaggregat n **~ position** Arbeitslage f **~ use** betriebsmäßige Verwendung f

**operator** Arbeiter m, Beamter m, Beamtin f, Bedienungsmann m, Betriebsführer m, Luftfahrzeughalter m, Maschinenwärter m, Taster m **~ in charge** Apparataufsicht f, Gruppenführer m

**operator, ~ aid system** Abwerfsystem n **~ calculus** Operatorkalkül n **~ controlling sound volume** Tonsteuermann m **~ dialing working** Wählbetrieb m durch die Beamtin **~ manipulation** Operatorrechnung f

**operator's, ~ cabin** Kontrollbude f **~ circuit** Abfragestromkreis m **~ control cab(in)** Führerkabine f **~ head set** Abfrageapparat m **~ jack** Anschalte-, Anschalt-, Doppelanschalt-, Doppel-, Einschalt-klinke f für Abfrageapparate (teleph.) **~ key equipment** Beamtenschalterausrüstung f **~ load during busiest hours** Belastung f einer Beamtin in der Hauptverkehrsstunde **~ permit** Führerschein m **~ platform** Führerstand m **~ plug** Anschaltstöpsel m **~ position** Arbeits-, Schrank-platz m **~ set** Abfragegarnitur f **~ (telephone) set** Sprecheinrichtung f, Sprechgarnitur f **~ speaking circuit** Platzschaltung f **~ team** Dienstgruppe f **~ telephone** Abfragegarnitur f **~ telephone set** Abfrage-apparat m, -einrichtung f, -gehäuse n **~ telephone-set jack** Klinke f zur Einschaltung des Sprechzeuges **~ telephone set plug** Stecker m

**ophitic texture** Intersertalstruktur f

**ophtalmology** Augenlehre f

**ophtalmometer** Augenmesser m

**ophtalmoscope** Augenspiegel m, Netzhautspiegel m

**opinion** Anschauung f, Ansicht f, Erachten n, Gutdünken n, Meinung f, Urteil n (expert) **~** Gutachten n **adjustment of ~s** Gleichschaltung f

**opponent** Gegenpartei f, Gegner m

**opportune** zeitgemäß, zweckmäßig

**opportunity** Gelegenheit f

**oppose, to ~** entgegen-halten, -setzen, -stellen, -wirken, gegenarbeiten, gegenüberstellen, widersetzen, -streben, -streiten, sich sträuben, Widerstand leisten

**opposed** entgegen-gesetzt, -gerichtet, gegensätzlich, gegenüberliegend **~ to** gegen

**opposer,** ~ **bearing bush** Gegenlagerbüchse *f* ~ **contact holder** Gegenträger *m* ~ **current** entgegengesetzter Strom *m* ~ **cylinder engine** Boxermotor *m* ~ **cylinder-type engine** Boxermotor *m* ~ **horizontally engines working on the same crankshaft** Gegenzwilling *m* ~ **ions** Gegenionen *pl* ~ **piston** gegenläufiger Kolben *m* ~**-piston Diesel engine** Doppelkolben-Dieselmotor *m* ~**-piston engine** Doppelkolbenmaschine *f*, Gegenkolbenmotor *m* ~ **piston two--stroke engine** Zweitakt-Boxermotor *m* ~**-piston-type engine** Motor *m* mit gegenläufigen Kolben

**opposing** gegeneinander geschaltet, gegensinnig, widerstrebend ~ **in series** gegensinnig in Reihe geschaltet

**opposing,** ~ **electromotive force** gegenelektromotorische Kraft *f* ~ **field** Gegenfeld *n* ~ **force** Gegenkraft *f* ~ **lane** Gegenspur *f* (Straße) ~ **moment** Rückkehrmoment *n* ~ **spring** Gegenfeder *f* ~ **torque** Rückkehrmoment *n* ~ **winding** Gegenwicklung *f*, Gegenwindung *f*

**opposite** Gegenteil *n*; entgegengesetzt, gegenseitig, gegenüber, gegenüberstehend, (of poles) ungleichnamig, widersprechend ~ **in direction** rücklaufend, widersinnig **in** ~ **sense** widersinnig **on the** ~ **tack** über den andern Bug ~ **to** angemessen, entgegengesetzt gerichtet, passend

**opposite,** ~ **angle** Gegen-, Scheitel-winkel *m* ~ **charge** entgegengesetzte Ladung *f* ~ **contact** Gegenkontakt *m* ~ **control** entgegengesetzte Steuerung *f* ~ **course** Gegenkurs *m* ~ **cranks** gegenläufige Kurbeln *pl* ~ **direction** Gegenrichtung *f* ~**-end derrick** Gegenausleger *m* ~ **extreme limit** entgegengesetzter Grenzwert *m* ~ **forces** Kräftepaar *n* ~ **illustration** nebenstehende Abbildung *f* ~ **number** Gegenfunkstelle *f* ~ **party** Gegenpartei *f* ~ **pistons** gegenläufige Kolben *pl* ~ **polarity** Gegenpolung *f* ~ **pole** ungleichnamiger Pol *m* ~ **poles** entgegengesetzte Pole *pl* ~ **signs** entgegengesetzte Vorzeichen *pl* ~ **station** (photogrammetry) Gegenstandspunkt *m*

**oppositely,** ~ **charged** ungleichnamig elektrisch ~ **directed** gegenläufig ~ **electrified** ungleichnamig elektrisch ~ **oriented** gegenläufig

**opposition** Einspruch *m* (patent), Entgegensetzung *f*, Gegenläufigkeit *f*, Gegenphase *f*, Gegensatz *m*, Gegen-schein *m*, -stellung *f* (astr.), Kontrast *m*, Phasenverschiebung *f* um 90 Grad, Widerstand *m* **in** ~ entgegengesetzt gerichtet ~ **of phase** entgegengesetzte Phase *f* ~ **of phases** Gegenläufigkeit *f* ~ **of polarity** Gegenpolung *f*

**opposition,** ~ **duplex** Duplex *m* in Gegenspannung ~ **method** Gegenschaltungsmethode *f*, halbpotentiometrische Methode *f*, Kompensationsverfahren *n* ~ **period** Einspruchsfrist *f* ~ **proceedings** Einspruchsverfahren *n*

**oppress, to** bedrücken, beengen

**oppressed** bedrückt

**oppression** Bedrückung *f*, Niederdrückung *f*

**oppressive** drückend ~ **air** mattes Wetter *n* ~ **heat** Schwüle *f*

**optic,** ~ **angle** Achsen-, Gesichts-winkel *m* ~ **axis** optische Achse *f*, Sehachse *f* ~ **light filter** Graufilter *n* (TV), Grau(glas)scheibe *f*, Neutralfilter *n* ~ **nerve** Sehnerv *m*

**optic(al)** optisch

**optical,** ~ **apparatus** Leuchte *f* ~ **axis of a lens** Linsenachse *f* ~ **balance (or equalizing)** optischer Ausgleich *m* ~ **barrel** Optiktubus *m* ~ **bench** optische Bank *f* (photo) ~ **center of the plate** Achsendurchstoßpunkt *m* mit der Platte ~ **center of the registering frame** optischer Rahmenmittelpunkt *m* ~ **density** Schwärzung *f* (acoust.) ~ **direction** optische Richtung *f* ~ **direction finding** Sichtpeilung *f* ~ **distance** Sichtweite *f* ~ **distance measurement** optische Streckenmessung *f* ~ **efficiency** optischer Wirkungsgrad *m* ~ **electron** Leuchtelektron *n* ~ **exaltation** optische Berechnungsüberschreitung *f* ~ **flat** Ebene *f*, Planglas *n* ~ **flat gauge** Glasprüfmaß *n* ~ **flats** Glasendmasse *f* ~ **focus** optischer Brennpunkt *m* ~ **gauge** optisches Maß *n* ~ **glass** Linsenglas *n* ~ **grinding art (or artifice)** optische Schleifkunst *f* ~ **illusion** Augen-, Licht-täuschung *f* ~ **indicator** Lichtzeigerinstrument *n*, Optimeter *n* ~ **inhomogeneities** Schlieren *pl* ~ **isotope shift** Isotopieverschiebung *f* ~ **lens converter** Bildwandler *m* ~ **level quadrant** Winkellibelle *f* ~ **lever** optischer Zeiger *m* ~ **light filter** Reflexschutzfilter *n* ~ **magnifying extensometer** Spiegelapparat *m* ~ **microscope** Lichtmikroskop *n* ~ **mode** optischer Schwingungsfreiheitsgrad *m* ~ **panel** Linsenfach *n*, Scheinwerferfeld *n* ~ **parallels** parallele Glasprüfmasse *f* ~ **path difference** optische Wegdifferenz *f* ~ **pattern** Lichtband *n* ~ **performance** Fernrohrleistung *f*, (Bildkartiergeräte) optische Leistung *f* ~ **phenomenon** Lichterscheinung *f* ~ **power** optische Leistung *f* ~ **print of altered dimensions** Umkopie *f* ~ **printing** Umkopieren *n* ~ **protractor level** Windellibelle *f* mit Mikroskop *n* ~ **pyrometer** Glühfadenpyrometer *m/n*, Teilstrahlungspyrometer *n* ~ **range** Sichtweite *f* ~**-reduction print** Umkopie *f* ~ **rotation** Drehpolarisation *f*, optische Drehung *f*, Rotationspolarisation *f* ~ **rotary dispersion** optische Rotationsdispersion *f* (ORD) ~ **rotary power** optische Drehung *f* ~ **scanning** optische Abtastung *f* ~**-screw--thread-measuring machine** Gewindemeßkomparator *m* ~ **sight** optisches Zielgerät *n* ~ **signal** Schauzeichen *n*, sichtbares Signal *n* ~ **soundhead** Lichttongerät *n* ~ **sound-on-film** Lichtton *m* ~ **sound recorder** optischer Tonschreiber *m* ~ **sound reproducer** elektro-optischer Abtaster *m* ~ **square level** Spiegelkreuzlibelle *f* ~ **stimulation** Ausleuchten *n* ~ **stimulus** Sehreiz *m* ~ **system** Abtastoptik *f*, Optik *f*, (interchangeable lens barrels) Optikauswechslung *f* ~ **system for the projection** Projektionsspiegelsystem *n* ~ **telegraph** optischer Telegraf *n* ~ **telemetry** optische Streckenmessung *f* ~ **thread tool gauge** Drehbankmikroskop *n* ~ **train** optisches System *n* ~ **transmission amplifier** Spiegelübertragungsverstärker *m* ~ **view finder** optischer Sucher *m* ~ **visibility** Durchsicht *f* ~ **wedge** optische Kammblende *f*

**optically,** ~ **absorbing (or absorptive)** lichtschluckend ~ **empty water** Ultrawasser *n* ~ **permeable** optisch durchlässig ~ **uniaxial** optisch-einachsig

**optician** Optiker *m*

**optics** Licht-kunde *f*, -lehre *f*, -wissenschaft *f*, Optik *f*, Sehlehre *f* ~ **of stress and strain photoelectricity** Spannungsoptik *f*
**optimal performance** Höchstleistung *f*
**optimally codes program** Bestzeitprogramm *n* (info proc.)
**optimeter** Optimeter *n* ~ **tube** Optimeterfernrohr *n*
**optimizing** zugängliche Anordnung *f*
**optimum** Bestwert *m*; günstigst, optimal, optimum ~ **angle of incidence** optimaler Anstellwinkel *m* ~ **bunching** optimale Ballung *f* ~ **coding** Bestzeit-, Schnell-programmierung *f* ~ **coupling** optimale oder kritische Kopplung *f* ~ **efficiency** Bestwirkungsgrad *m* ~ **form** günstigste Form *f* ~ **value** Bestwert *m*, optimaler Wert *m* ~ **working frequency (OWF)** günstige Verkehrsfrequenz *f*
**option** Bezugsrecht *n*, Option *f*, Wahl *f* ~ **agreement** Optionsvertrag *m* ~ **switch** Fakultativschalter *m*
**optional** fakultativ, freigestellt, wahlweise ~ **clause** Fakultativklausel *f* ~ **equipment** Sondereinrichtung *f* ~ **feature** wahlweise Zusatzeinrichtung *f* ~ **pilotage** Lotsenfreiheit *f* ~ **rating** wahlweise Leistung *f*
**optionally** eventuell
**optometer** Optometer *n*, Sehweitebestimmer *m*
**optometry** Optometrie *f*, Sehkraftmessung *f*
**optophone** optischer Fernsprecher *m* ~ **terminal** Lichtsprechstelle *f*
**optophysical** physikalisch-optisch
**or** oder ~ **-circuit** inklusives Oder *n*, Oder-Glied *n*, Oder-Schaltung *f* ~ **else** sonst ~ **else circuit** exklusives Oder *n*, Oderschaltung *f* ~ **gate** Oder-Glied *n*, Oder-Schaltung *f*, Oder-Tor *n* ~ **-module** Oder-Glied *n*, Oder-Modul *n* ~ **operator** Oder-Zeichen *n*
**oral** mündlich ~ **hearing** mündliche Verhandlung *f*
**orange** Apfelsine *f*, Orange *f*; orange-farben, -farbig ~ **lead** Bleisafran *m* ~ **lily** Feuerlilie *f* ~ **peel effect** (of laquers) Apfelsinenschalen--effekt *m*, -erscheinung *f* ~ **peel grab** Polypgreifer *m*
**orb** Ball *m*, Kugel *f*
**orbed** gerundet, kugelförmig, rund
**orbicular** kugel-, ring-, scheiben-förmig ~ **diorite** Kugeldiorit *m* ~ **structure** Augentextur *f* ·
**orbit, to** ~ die Erde umkreisen
**orbit** Bahn *f*, Kreis-, Umlauf-bahn *f* ~ **of electrons** Elektronenkreis *m* ~ **of a planet** Planetenbahn *f*
**orbit,** ~ **inclination** Bahnhöhenneigung *f* ~ **shift coil** Ablenkspule *f* ~ **time of electron** Elektronenlaufzeit *f*
**orbital,** ~ **angular momentum operator** Bahndrehimpulsoperator *m* ~ **cavity** Augenhöhle *f* ~ **electron capture** Einfang *m* eines Hüllenelektrons ~ **electrons** Bahn-, Hüllen-elektronen *pl*, kernferne oder kreisende Elektronen *pl* ~ **flight speed** Umlaufbahngeschwindigkeit *f* ~ **loop** Bahnschlinge *f* ~ **moment** Bahnimpuls *m* ~ **motion** Kreis-, Orbital-bewegung *f* ~ **movement** Kreisbahnbewegung *f* ~ **path** Orbitalbahn *f* ~ **plane** Umlaufebene *f* ~ **quantum number** azimutale Quantenzahl *f*, Bahndrehimpulsquantenzahl *f* ~ **road** ringförmige Um-

gehungsstraße *f* ~ **stability** Bahnstabilität *f* ~ **stabilization** Bahnstabilisierung *f* ~ **state** Bahnzustand *m* ~ **valence** Bahnvalenz *f* ~ **wave function** Wellenbahnfunktion *f*
**orbited period** Umlaufperiode *f*
**orbiting** Warteschleife *f* ~ **altitude** Bahnhöhe *f* ~ **time** Umlaufzeit *f*
**orchard,** ~ **gang plow** mehrschariger Obstgartenpflug *m* ~ **harrow** Obstgartenegge *f*
**order, to** ~ anordnen, anweisen, beauftragen, befehlen, bestellen, beziehen, gebieten, heißen, kommandieren, ordnen, verordnen **to** ~ **beforehand** vorausbestellen
**order** Anordnung *f*, Anweisung *f*, Auftrag *m*, Befehl *m*, (of a court of law) Beschluß *m*, Bestellung *f*, Erlaß *m*, Folge *f*, Grad *m* (einer Gleichung), (of magnitude) Größenordnung *f*, Klasse *f*, Kommando *n*, Mandat *n*, Order *f*, Orderscheck *m*, Ordnung *f*, Rang *m*, Reihe *f*, Reihenfolge *f*, schriftliche Anweisung *f*, Sorte *f*, System *n*, Verfügung *f*, Verhaltungsmaßregel *f*, Verordnung *f*, Weisung *f*, Zustand *m*
**order, in** ~ angebracht, betriebsklar **in bad** ~ in schlechtem Zustand **in due** ~ ordnungsmäßig **in open** ~ aufgelöst **by** ~ **of** im Auftrag **out of** ~ in Unordnung ~ **of accuracy** Genauigkeitsordnung *f* ~ **to blast** Sprengbefehl *m* ~ **of calls** Reihenfolge *f* der Gespräche ~ **of drop** Fallfolge *f* ~ **for evidence** Beweisbeschluß *m* ~ **on hand** Auftragsbestand *m* ~ **to launch** Startbefehl *m* ~ **of magnitude** Größenordnung *f* ~ **of merit** Reihenfolge *f* der Wichtigkeit ~ **to pay** Zahlungsbefehl *m* ~ **of precedence** Vorrangordnung *f* ~ **of priority** Rangfolge *f* ~ **of start** Startordnung *f*
**order,** ~ **-and security regulations** Ordnungs- und Sicherheitsbestimmungen *pl* ~ **blank** Bezugsanweisung *f* ~ **channel** Dienstkanal *m* ~ **code** Befehlskode *m* ~ **counter** Befehlszähler *m* ~ **department** Auftragsabteilung *f* ~ **determination** Ordnungsbestimmung *f* ~ **-disorder** Ordnung-Unordnung *f* ~ **-disorder transformation** Ordnungs-Unordnungs-Umwandlung *f* ~ **-disorder-transition** Ordnungs-Unordnungs--Umwandlung *f* ~ **form** Bestell-formular *n*, -schein *m*, -vordruck *m*, -zettel *m*, Bezugsanweisung *f* ~ **number** Anforderungszeichen *n*, Auftragsnummer *f* ~ **parameter** Ordnungsparameter *m* ~ **place** Kommandostelle *f* ~ **register** Befehlsregister *m* ~ **sheet** Bestell--schein *n*, -vordruck *m* ~ **slip** Auftragszettel *m* ~ **-sorter** Ordnungstrenner *m* ~ **symbol** Anforderungszeichen *n*
**order-wire** Dienstleitung *f* ~ **button** Diensttaste *f* ~ **button strip** Diensttastenstreifen *m* ~ **circuit** Dienstleitung *f* ~ **distributor** selbsttätiger Dienstleitungsverteiler *m* ~ **junction** Verbindungsleitung *f* für Dienstleitungsbetrieb ~ **key** Dienstleitungstaste *f* ~ **lamp** Ferndienstlampe *f* ~ **operation** Dienstleitungsbetrieb *m* ~ **panel** Dienstleitungsfeld *n* ~ **system (or trunking)** Dienstleitungsbetrieb *m*
**ordered** bezogen ~ **arrangement** geordneter Zustand *m* ~ **scattering** Braggsche Streuung *f* ~ **state** geordneter Zustand *m*
**ordering,** ~ **domain** Ordnungsdomäne *f* ~ **No.** Bestell-Nummer *f* ~ **reference** Bestellzeichen *n* ~ **transition** Ordnungsumwandlung *f*

orderly geordnet, methodisch, ordentlich, ordnungsgemäß ~ **arrangement (or state)** geordneter Zustand *m*
**ordinal index (or number)** Ordnungszahl *f*
**ordinance** Bestimmungen *pl*
**ordinary** alltäglich, einfach, üblich ~ **bath** Vollbad *n* ~ **call** Privatgespräch *n* ~ **freight** Frachtgut *n* ~ **gradient** Durchschnittsneigung *f* ~ **iron** gewöhnliches Eisen *n* ~ **key** Längskeil *m* ~ **light containing red rays** gewöhnliches rothaltiges Licht *n* ~ **microphone** Besprechungsmikrofon *n* ~ **milling** Gegenlauffräsen *n* ~ **occupancy sprinkler system** Sprinkleranlage *f* für normale Risiken ~ **photograph** ruhendes Bild *n* ~ **(or normal) process** Normalprozeß *m* ~ **share** Stammaktie *f*
**ordinate** Ordinate *f* ~ **of the image** Bildordinate *f* ~ **of trajectory** (ballistics) Flughöhe *f*
**ordinate,** ~ **axis** Ordinatenachse *f* ~**-controlled** ordinatengesteuert ~ **line** Achse *f* (math.)
**ordinance** Bestückung *f* ~ **map** Generalstabskarte *f*
**ore, to** ~ **down** Erz *n* zugeben
**ore** Erz *n* ~ **and cargo chain-bucket conveyers** Becherwerke *pl* ~ **for crushing** Walzerz *n* ~ **in large lumps** Stockerz *n* ~ **rough from the mine** Grubenerz *n* ~ **in sight** anstehendes Erz *n* oder Gestein *n*
**ore,** ~ **assaying** Erzprobe *f* ~ **barrow** Erzkarre *f* ~**-bearing** erzenthaltend, erzhaltig ~**-bearing rock** erzführendes Gestein *n* ~ **bed** Erz-lager *m*, -stätte *f* ~ **bin** Erz-behälter *m*, -bunker *m*, -füllrumpf *m* ~**-bin gate** Erztaschenverschluß *m* ~ **body** Erzkörper *m* ~ **boil** Erzfrischreaktion *f* ~**-boring machine** Erzbohrmaschine *f* ~ **breaker** Erzbrecher *m* ~**-breaking plant** Erzzerkleinerungsanlage *f* ~ **briquetting** Erzbrikettierung *f* ~**-briquetting press** Erzbrikettpresse *f* ~ **bucket** Erzkübel *m* ~ **bucket-handling crane** Erzkübelkran *m* ~ **bunker** Erz-behälter *m*, -bunker *m* ~ **burden** Erzmöller *m* ~ **burdening** Erzmöllerung *f* ~ **cargo** Erzladung *f* ~ **charge** Erz-charge *f*, -gicht *f* ~ **chimney** Erzschlauch *m* ~ **chimneys** Schläuche *pl* von Erz ~ **chute** Erzrutsche *f* ~ **column** Erzsäule *f* ~ **compartment** Masche *f* ~ **crusher** Erz-brecher *m*, -pocher *m* ~**-crusher plant** Erzzerkleinerungsanlage *f* ~**-crushing shoe** Pochschuh *m* (min.) ~ **deposit** Erz-feld *n*, -lager *m*, -lagerstätte *f*, -vorkommen *n* ~**-dressing plant** Erzaufbereitungsanlage *f* ~**-drying kiln** Erzröstofen *m* ~**-dumping yard** Erzplatz *m* ~ **dust** Mulm *m* ~ **extractor** Eisenabschneider *m* ~ **fines** Erzklein *m* ~ **formation** Erzformation *f* ~ **forming** erzbildend ~ **furnace** Rohofen *m* ~ **grab** Erz-greifer *m*, -messer *n* ~ **hammer** Pochschlegel *m* ~**-handling equipment (or machinery)** Erzverladeanlage *f* ~ **heap** Erzhaufen *m* ~ **jigger** Siebsetzarbeit *f*, Siebsetzer *m* ~**-leaching** Erzlaugung *f* ~ **leaching plant (or practice)** Erzlaugerei *f* ~ **load** Erzladung *f* ~ **loading** Erzverladung *f* ~**-loading bridge** Erzverladebrücke *f* ~ **lode** Erzader *f* ~ **lump** Erzstück *n* ~ **matrix** Erzmutter *f* ~ **mine** Erzbergwerk *n* ~ **mining** Erzbergbau *m* ~ **mixture** Erzgemisch *n* ~ **modules** Erznieren *pl* ~ **picker** Erzklauber *m* ~**-picking belt** Erzleseband *n* ~ **pile** Erzhaufen *m* ~ **pipe** Erzschlauch *m* ~ **pipes** Schläuche *pl* von Erz ~ **pocket** Erznest *n* ~**-pocket gate** Erztaschenverschluß

*m* ~ **process** Erzverfahren *n* ~ **puddling** Erzpuddeln *n* ~ **pulp** Erztrübe *f* ~ **pulverizer** Erzmühle *f* ~ **purchase** Erzkauf *m* ~ **refining** Erzsonderung *f* ~ **roasting** Erzröstung *f* ~**-roasting thorn** Röstdorn *m* ~ **separation** Erzsortierung *f* ~ **separator** Erz-scheider *m*, -sieb *n* ~ **sheet** derbes Erztrumm *n* ~**-shipping quay** Erzkai *m* ~ **shoot** Erzfall *m* ~ **shovel** Erzschaufel *f* ~ **slag** Erz-, Roh-schlacke *f* ~ **slime** Erzschlamm *m*, Poch-satz *m*, -schlamm *m*, -schlich *m* ~ **sludge (or slurry)** Erzschlamm *m* ~ **smelting** Erz-schmelzen *n*, -verarbeitung *f*, Roharbeit *f*, Rohschmelzen *n* ~**-smelting furnace** Erzschmelzofen *m*, Rohofen *m* ~ **sorting** Erzsortierung *f* ~ **stamper** Erzpocher *m* ~**-storage bunker (or picket)** Erztasche *f* ~**-storage pocket** Erzbehälter *m* ~ **thickener** Erzeindicker *m* ~ **traffic** Erzhandel *m* ~ **transporter** Erzverladebrücke *f* ~ **treatment** Erzverhüttung *f* ~ **valuation** Erzbewertung *f* ~ **vein** Erz-ader *f*, -gang *m* ~ **wash** Erzschlamm *m* ~ **washer** Erzschlämmer *m* ~ **washing** Erzwaschen *n* ~**-washing machine** Erzwaschmaschine *f* ~**-washing plant** Erzwäscherei *f* ~**-washing room** Erzwäsche *f* ~ **working** Erzverarbeitung *f* ~ **yard** Erzlagerplatz *m*, Erzplatz *m* ~ **yielding** Erzgeben *n*
**Orford receiver (or settler)** Orford-Vorherd *m*
**organ** Organ *n*, Orgel *f* ~ **with depth sensibility** Tiefenaufnahmeorgan *n* ~ **of vision** Sehnervapparat *m* ~**-pipe metal** Orgelmetall *n*
**organic** organisch ~ **arrangement** Eingliederung *f* ~ **carbon cycle** Kreislauf *m* des Kohlenstoffs ~ **chemistry** organische Chemie *f* ~ **laquer foil** Zwischenfolie *f*
**organization** Bildung *f*, Einrichtung *f*, Gebilde *n*, Gestaltung *f*, Gliederung *f*, Organisation *f*, Zusammenschluß *m* ~ **in width** Breitengliederung *f*
**organizational staff** Organisationsstab *m*
**organize, to** ~ einrichten, gliedern, organisieren
**organogenic sedimentary rock** organogenes Sedimentgestein *n*
**organometallic** organometallisch ~ **compound** Organometall *n*
**orient, to** ~ ausrichten, orientieren, orten
**orientable** schwenkbar ~ **tail wheel** schwenkbares Spornrad *n*
**oriental** östlich
**orientation** Ausrichtung *f*, Lagerung *f* (cryst.), Orientierung *f*, Ortsbestimmung *f*, Ortung *f*, Phasenunterschied *m* zwischen Geber und Empfänger, räumliche Lage (cryst.), Sichzurechtfinden *n* ~ **by map** Geländeorientierung *f* ~ **of a map** Einrichten *n* einer Karte ~ **of striae** Fiederstreifung *f*
**orientation,** ~ **chart** Ortungskarte *f* ~ **instrument** Ortungsgerät *n* ~**-responsive (or -sensitive)** orientierungsabhängig
**orientational distortion** Bildzerdrehung *f*
**oriented,** ~ **body** orientierter Körper *m* ~ **surface** orientiertes Flächenstück *n* ~ **voids** orientierte Leerstellen *pl*
**orienting** Adressenzuweis *m* (info proc.), Adressierung *f* (info proc.) ~ **a map** (die) Richtung der Landkarte bestimmen ~ **line** Orientierungslinie *f* ~ **point** Anlegepunkt *m* ~ **straight-edge** Ziellineal *n*

**orifice** Ausflußöffnung *f*, Ausmündung *f*, Austrittsöffnung *f*, Blende (in nozzles, screen), Drosselbohrung *f*, kalibrierte Bohrung *f*, Düse *f*, Maßblende *f* (meas.), Meduse *f*, Muffenloch *n*, Mund *m*, Mundloch *n*, Mundstück *n*, Mündung *f*, Öffnung *f*, Schützenöffnung *f*, Spritzdüse *f* (jet nozzle) **~ of conduit** Kanalmundstück *n*

**orifice, ~ assembly** Meß-blende *f*, -flansch *m*, Rückschlagdrossel *f* **~ correction** Mündungskorrektion *f* **~ gauge** Ringkammerflansch *m* **~ meter** Standscheibenmesser *m* **~ plate** Drossel-düse *f*, -scheibe *f*, Loch-platte *f*, -scheibe *f*, Meßblende *f*, Staurand *m* **~ plate in pipe lines** kreisförmige Blendenscheibe *f* in Rohrleitungen *f* **~ pressure** Mündungsdruck *m* **~ type blender** Düsenmischertyp *m*

**origanum oil** Dostenöl *n*

**origin** (metamorphic process) Abfolge *f*, Anfang *m*, Anfangspunkt *m*, Ausgangspunkt *m*, Beginn *m*, Entstehung *f*, Herkunft *f*, Null (im Koordinatensystem), Nullpunkt *m*, Pol *m* (geom.), Ursprung *m*

**origin, ~ of coordinates** Koordinaten-anfangspunkt *m*, -ursprung *m* **~ of coordinator** Scheitel *m* **~ of curve** Kurvenanfang *m* **~ of force** Angriffspunkt *m* der Kraft **~ of generation (or production)** Entstehungsort *m* **~ of a river** Flußquelle *f*

**origin distortion** Nullpunkt-anomalie *f*, -fehler *m*

**original** Original *n*, Urbild *n*; original, ursprünglich **in ~** urschriftlich

**original, ~ cause** Entstehungsursache *f* **~ cellulose** Ausgangszellulose *f* **~ constituent** Ausgangsprodukt *n* **~ copy** Vorlage *f* **~ creation** Neuschöpfung *f* **~ current** Anfangsstrom *m* **~ datum** Ausgangswert *m* **~ document** Urschrift *f* **~ ground surface** natürliche Oberfläche *f* des Bodens **~ head** Setzkopf *m* **~ jurisdiction** Gerichtsbarkeit *f* erster Instanz **~ material** Ausgangsmaterial *n* **~ (or base) metal** Ausgangswerkstoff *m* **~ model** Urbaumuster *n* **~ patent** Hauptpatent *n* **~ position** Ausgangslage *f* **~ prototype** Originalbild *n* **~ scale** Originalmaßstab *m* **~ specimen** Ursprungstyp *m* **~ subject copy** Ursprungsbild *n* **~ tap** Normal-, Original-bohrer *m* **~ test** Originalarbeit *f* **~ type** Ursprungstyp *m* **~ unit** Stammtruppenteil *m* **~ volume** Eigenvolumen *n*

**originality** Neuheit *f*, Originalität *f*, Ursprünglichkeit *f* **~ case** Neuigkeitswert *m* der Erfindung

**originate, to ~** (a report) abfassen, ausgehen, beginnen, entstehen, herrühren **to ~ a call** ein Gespräch *n* einleiten

**originating** Einleitung *f* **~ exchange** Anmeldeanstalt *f* (teleph.) **~ firm** Herstellwerk *f* **~ office** Ursprungsanstalt *f* **~ station** Aufgabenstelle *f* **~ toll center** Abgangsanstalt *f*, Überweisungsfernamt *n*, Ursprungsanstalt *f* **~ traffic** ausgehender Verkehr *m*, Ursprungsverkehr *m*

**origination office (or toll center)** Abgangsanstalt *f*

**originator** Aufgeber *m*, Urheber *m*

**orlop (deck)** Raumdeck *n*

**ornament** Aufsatz *m*, Putz *m*, Verzierung *f* **~ for trellis fences** Gitterornament *n*

**ornamental** dekorativ, schmückend, zierend **~ base** Zierfuß *m* **~ castings** Ornamentguß *m* **~ iron** Ziereisen *n* **~ rule** Abschnittlinie *f* **~ sheet** Zierblech *n* **~ sleeve** Ziersockel *m* **~ thread** Verzierungsfaden *m* **~ type** Grotesk-, Zier-schrift *f* **~ wood** Edelholz *n*

**ornamented hinge** Zierband *n*

**ornithopter** Schlagflügler *m*, Schwingenflügler *m*

**orogenic disturbance** Gebirgsstörung *f*

**orographic** orografisch **~ influences** Einfluß *m* von Gebirgen **~ rain** orografischer Regen *m* **~ thunderstorm** Reliefgewitter *n*

**orography** Gebirgskunde *f*

**orometer** Höhenbarometer *n*

**orpiment** Arsenblende *f*, Arsentrisulfid *n*, Auripigment *n*, gelbe Arsenblende *f*, gelbes Arsensulfid *n*, Operment *n*

**orsat gas analysis** Orsat-Rauchmesser *m*

**ortho** gerade, korrekt, recht **~ and parahydrogen** Ortho- und Parawasserstoff *m* **~ acid** Orthosäure *f* **~ basic** orthobasisch **~ carbonic acid** Orthokohlensäure *f*

**orthochromatic** farbenempfindlich, farbenrichtig (photo), farbrichtig, farbtonrichtig, gleichfarbig, orthochromatisch **~ filter** tonrichtiger Filter *m* **~ processing** Orthochromatisierung *f*

**orthochronous** orthochron

**orthoclase** monokliner Kalifeldspat *m*, Orthoklas *m*

**orthodiagraphy** Orthodiagrafie *f*, Orthoröntgenografie *f*

**orthodox** althergebracht, orthodox, streng **~ design** übliche Bauart *f*

**orthodromes** raumgeradlinige Peilstrahlen *pl*

**orthodromic** raumgeradlinig

**orthodromy** Orthodrome *f*

**orthogonal** orthogonal, rechtwinklig, winkelrecht **~ axes** senkrechtes Achsenkreuz *n* **~ biplane** rechtwinkliger Doppeldecker *m* **~ directions** aufeinander senkrechte Richtungen *pl* **~ families of curves** orthogonale Kurvenscharen *pl* **~ parallel projection** orthogonale Parallelprojektion *f* **~ system** dem Kern zugeordnetes Orthogonalsystem *n* **~ system of surfaces** orthogonale Flächensysteme *pl*

**orthogonality** Orthogonalität *f* **~ of error** Fehlerorthogonalität *f*

**orthogonalization** Orthogonalisierungsverfahren *n*

**orthogonalized** orthogonalisiert

**orthogonally** im rechten Winkel

**orthogonic** winkeltreu **~ point** winkeltreuer Punkt *m*

**orthographic** orthografisch

**orthography** Orthografie *f*, Rechtschreibung *f*

**orthohydrogen** Orthowasserstoff *m*

**orthomorphic projection** Merkatorkartenprojektion *f*

**orthonormal system** normiertes Orthogonalsystem *n*

**orthophonic** klanggetreu, lautgetreu **~ reproduction** getreue Wiedergabe *f* (of loud speaker)

**orthophtalic acid** Orthophthalsäure *f*

**orthopolar method** Orthopolarenverfahren *n*

**orthopositronium** Orthopositronium *n*

**orthopter** Ornithopter *n*, Schwingen-flieger *m*, -flugzeug *n*

**orthoptic** Diopter *n*, Dioptrie *f*

orthorhombic ein-und-ein-achsig (cryst.), rhombisch
orthoscopic orthoskopisch, tiefenrichtig, verzeichnungsfrei ~ eyepiece verzerrungsfreies
Okular n ~ image unverzeichnetes oder unverzerrtes Bild n
orthotest Orthotest m ~ bore gauge Bohrungsorthotest m
orthotoluidine Orthotoluidin n
orthotropy Orthotropie f
Ortinghaus clutch Ortinghauskupplung f
oscillate, to ~ auspendeln, hin- und herschwingen, hochfrequente Schwingungen erzeugen
pl (electr.), oszillieren, pendeln, schaukeln,
schwanken, schwingen, vibrieren to ~ about an
average value um einen Mittelwert m schwingen
to ~ in resonance (sympathy, or unison) mitschwingen
oscillating oszillierend, pendelnd, schwingend,
vibrierend ~ of electrons Elektronenpendelung f
oscillating, ~ arc Lichtbogengenerator m ~ arm
Pendelarm m ~ armature Schwinganker m ~
armature-type magneto Pendelmagnetzünder m
~ back and forth Vor- und Rückwärtskippen n
~ beacon Schwebungsfeuer n ~ characteristic
Schwingkennlinie f ~ circuit Schwingungsgebilde, -kreis m ~ condenser-amplifier Schwingkondensator-Verstärker m ~ cone Schwingtrichter m ~ conveyor Schüttelrutsche f ~
crystal Dreh-, Schwenk-kristall m ~ current
oszillierender Strom m ~ cylinder schwingender
Zylinder m ~ device Schaukelapparat m ~
direction indicator Pendelwinker m ~ discharge
oszillatorische, oszillierende oder schwingende
Entladung f, Schwingentladung f ~ disk
Schwingteller m, Schwingungsscheibe f ~ drive
Pendelantrieb m ~-electron scheme with
reflecting electrode Bremsfeldschaltung f ~
evener Abstreichhacker m ~ exposure Schwenkaufnahme f ~ feeding apparatus Schüttelspeiseapparat m ~ field schwingendes Feld n ~ flashovers mehrpolige Überschläge pl ~ flat stripping comb schwingender Kupphacker m ~
funnel Schwingtrichter m ~ klystron Oszillator-
Klystron n ~ lapping Schwingungsläppen n ~
lever hin- und hergehender Hebel m, Schwenker m (am Filter) ~ limit Schwinggrenze f ~
meter oszillierender Zähler m, Schwingungsspulenzähler m ~ mirror Kippspiegel m ~
motion Hin- und Herbewegung f, Schwanken n
~ movement hin- und hergehende Bewegung f
~ process Schwingen n ~ property Schwingungseigenschaft f ~ pump Flügelpumpe f ~ quality
Schwingungseigenschaft f ~ quantity Schwingungsgröße f ~ riddle Rüttelsieb n ~ roll
schwingende Walze f ~ saw Kapssäge f ~ shaft
pendelnde Welle f ~ shoot Schwingtransportrinne f ~ shutter schwingender Verschluß m ~
starter Pendelanlasser m ~ steering axle Pendellenkachse f ~ strainer Schüttelsortierer m ~
stretcher Schaukelbreithalter m ~ table
Schüttelherd m ~ transmitting circuit schwingender Sendungskreis m, Sendererregerkreis m
~ traverse motion Wackelchangierung f ~
trough washer Trogschwingwascher m ~ tube
Schwingröhre f ~ valve Generatorröhre f ~ vat
Pendelbütte f ~ wet-end of the paper machine
schwingende Siebpartie f

oscillation Ausschub m, Drehachse f, Eigenschwingung f, (in superheterodyne) Hilfsfrequenz f, Kristallquerschwingung f, Oszillation
f, Pendelung f, Schwankung f, Schwenkung f,
Schwingung f, Schwung m, Vibration f, Welle f
oscillation, ~ of air Luftschwingung f ~ about a
center Drehschwingung f ~ of the center of
gravity Schwerpunktschwingung f ~ of compass
needle Nadelschwankung f ~ of gas globe Gasblasenschwingung f ~ of a pendulum Pendelschwingung f ~ about zero Nulleinspielung f
oscillation, ~ amplitude limit Grenzamplitude f
~ choke Schwingungsdämpfer m ~ condition
Anfachungsbedingung f ~ detector Schwingaudion n, Schwingungs-, Wellen-anzeiger m,
Wellendetektor m ~ energy Schwingungsenergie
f ~ frequency Eigenschwingung f, Schwingungszahl f ~ generator Schwingungserzeuger m ~
hysteresis Ziehvorgänge pl ~ hysteresis phenomena Ziehen n, Zieherscheinung f ~ indicator
Schwingungsanzeiger m ~ lubricator oszillierender Öler m, Pendelschmierer m ~ mechanism
Schwingungsmechanismus m ~ method Schwingungsverfahren n ~ nodal (or node) point
Schwingungsknoten m ~ time Schwingungsperiode f ~-torsion test Drehschwingungsversuch m ~ transformer Schwingungs-transformator m, -transformer m ~ valve Generatorröhre f
~ valve detector Ventilröhrendetektor m
(teleph.)
oscillations, ~ in coupled circuits Kopplungsschwingungen pl ~ of higher order Leitbahnschwingungen pl (electron.) ~ increasing in
amplitude aufklingende Schwingungen pl
oscillator Anrichter m, Oszillator m, Schwinger
m, Schwingungs-erreger, -erzeuger m, Sender
m, Summer m, Tonsender m (heterodyne) ~
Überlagerer m ~ for measuring and signaling
purposes Wechselstromerzeuger m für Meß-
und Rufzwecke
oscillator, ~ alignment Oszillatorabgleich m ~
calibration Oszillatoreichung f ~ circuit
Schwungkreis m ~ drift Frequenzabweichung
f ~ frequency Oszillatorfrequenz f ~ lead
Summerleitung f ~ stage Quarzstufe f ~ tube
Generator-, Oszillator-, Schwingungserzeugerröhre f ~ valve Generator-, Schwing-röhre f
oscillatory oszillierend, schwingend, schwingungsfähig ~ circuit Energieschaltung f,
Oszillatorkreis m, Schwingkreis m, Schwingungs-gebilde n, -kreis m ~-circuit potential
(or voltage) Schwingkreisspannung f ~ current
schwingender Strom m ~ discharge oszillatorische oder oszil lerende Entladung f,
Schwingentladung f ~ field schnelles Wechselfeld n, schwingendes Feld n ~ form Schwingungsform f ~ formations Schwingungsgebilde
n ~ mirror Schwingspiegel m ~ modes Schwingsystem n ~ motion schwingende Bewegung f ~
movements of electrons Elektronentanz m ~
power Schwingleistung f ~ resistance Wellenwiderstand m ~ yaw Gierschwingung f
oscillion Triodenoszillator m
oscillogram Oszillogramm n, Wellenbild n
oscillograph Kurvenschreiber m, Oszillograf m,
Schallaufnahmegerät n, Schwingungs-messer
m, -schreiber m, Wellen-aufzeichner m,
-schreiber m ~ curve Oszillogramm n ~ loop

Oszillografenschleife *f* ~ **record** Oszillografen-
ausnahme *f* ~ **tube** Oszillografenröhre *f* ~
**vibrator** Oszillografenschleife *f*
**oscillographic** oszillografisch ~ **relay** Oszillo-
grafenrelais *n*
**oscillography,** ~ **angle** Laufzeitwinkel *m* ~
**figure** Laufzeitfigur *f*
**oscillometer** Oszillometer *n*
**oscilloscope** Glimmlichtoszilloskop *n*, Kathoden-
strahlröhre *f*, Oszilloskop *n*
**osculating,** ~ **circle** Oskulations-, Schmiegungs-
kreis *m* ~ **curve** oszillierende Kurve *f* ~ **parabola** Schmiegungsparabel *f* ~ **plane** Schmie-
gungsebene *f* (math.) ~ **sphere** Schmiegungs-
kugel *f*
**osculation** Anschmiegung *f*, Berührung *f*, Berührung *f* zweiter Ordnung (math.), Oskulation
*f* **point of** ~ Berührungspunkt *m*
**osculatory** oskulierend
**O-seal** Rundschnurring *m*
**osier** Korbweide *f* ~ **furniture** Korbmöbel *pl*
~**-goods manufacture** Korbflechterei *f* ~ **plaiter**
Weidenflechter *m*.
**osmi-iridium** Osmiridium *n*
**osmic,** ~ **acid** Osmiumsäure *f* ~ **anhydride**
Osmiumtetroxyd *n*
**osmious oxide** Osmiumoxydul *n*
**osmium** Osmium *n* ~ **alloy** Osmiumlegierung *f* ~
**tetroxide** Osmiumtetroxyd *n*, Überosmium-
säure *f*
**osmogene** Osmoseapparat *m*
**osmology** Osmologie *f*, Wissenschaft *f* von den
Riechstoffen
**osmometer** Osmometer *n*, osmotischer Druckmesser *m*
**osmondite** Osmondit *m*
**osmonditic** osmonditisch
**osmo-regulator** Osmoregeneriervorrichtung *f*
**osmosis** Diffusionsverfahren *n*, Durchdringung *f*,
Osmose *f* ~ **process** Osmoseentzuckerung *f*
**osmotic** osmotisch ~ **equilibrium** osmotisches
Druckgleichgewicht *n* ~ **pressure** osmotischer
Druck *m*
**osophone** Knochenhörer *m*
**ossein plant** Osseinanlage *f*
**osseous** knöchern, knochig
**osteogenic** knochenbildend, osteogen
**Ostwald electrode** Normalkalomelelektrode *f*
**Otto,** ~ **carburetor engine** Ottomotor *m* ~**-cycle**
Viertaktprozeß *m* ~ **methode (or procedure)**
Ottoverfahren *n*
**ounce** Unze *f*
**oust, to** ~ austreiben, entfernen
**out, to be** ~ ausgeschaltet sein ~ **of alignment**
dejustiert (mech.), nicht ausgerichtet (mech.),
nicht fluchtend (mech.), nicht eingelaufen,
verstimmt (telecom.) ~ **of balance** unausgeglichen
~ **of balance condition** unausgeglichener Zustand *m* ~ **of balance exciter** Fliehkrafterreger
*m* ~ **of balance force** Unwuchtkraft *f* ~ **of
balance moment** Schwerpunktmoment *m* ~ **of
band signaling** systemeigenes Signalisieren *n* ~
**of breakdown** Entstörung *f* ~ **of date** verjährt,
unzeitgemäß ~ **of focus** dekonzentriert ~ **of
focus picture** Außerfokusbild *n* ~ **of frame
condition** falsche Bildnachstellung *f* oder Bildstricheinstellung *f*, Verschiebefehler *m* ~ **of line**
aus der Richtung *f*, ungerade ~ **of one's sphere**

außer Bereich ~ **of order** gestört ~ **of order tone**
Gestörtsummerzeichen *n*, Störungssignal *n* ~
**of phase** außer Phase *f*, falschphasig, phasenverschoben, versetzt ~ **of phase component**
Zweiphasenkomponente *f* ~ **of phase condition** verschobene Phase *f* ~ **of phase current**
phasenverschobener Strom *m* ~ **of phase wave**
phasenverschobene Welle *f* ~ **of print** vergriffen
~ **of range** außerhalb der Schußweite *f*, schuß-
frei ~ **of reach** außer Bereich *m* ~ **of repair**
baufällig ~ **of round** Schlag *m* einer Welle; unrund ~ **of the running** außer Wettbewerb *m* ~
**of service record** Statistik *f* über die Unbenutz-
barkeit ~ **of step protection** (relay) Pendelschutz
*m* ~ **of trim** auswärts ~ **of trim** vertrimmt ~ **of
true** nicht rundlaufend, Schlag *m*, mit Schlag *m*
behaftet, Schlag *m* einer Welle ~ **of tune** verstimmt ~ **of tune antenna** verstimmte Antenne *f*
~ **of water** Austauchung *f*
**outbalance, to** ~ überwiegen
**outbalance** überwiegend
**out-band signaling** Systemwahl *f*
**outbid, to** ~ überbieten
**outbidding** Mehrgebot *n*
**outboard** außenbords, äußerer; Abflug... ~
**bearing** Außenborlager *m* ~ **fin** Endscheibe *f*
(am Leitwerk) ~ **float** (seitlicher) Stützschwimmer *m*, (aviat.) ~ **hinge fitting** Außenbord-
scharnier *n* ~ **motor** Außenbordmotor *m* ~
**part of landing gear** Auslegerfahrgestell *n* ~
**stabilizing float** außenbord stabilisierender
Schwimmer *m* ~ **strut** Endstiel *m* ~ **thrust
bearing** Außenendlager *n* ~ **wing** Außenflügel
*m*
**outbound tracking** Kursabflug *m* (nav.)
**outbreak** Ausbruch *m*
**outburst of carbon dioxide** Kohlensäureausbruch
*m*
**outcome** Ausgang *f*, Befund *m*, Ergebnis *n*,
Resultat *n*
**outcrop, to** ~ anstehen, ausbeißen, ausstreichen,
zutage treten (geol.)
**outcrop** Anstehendes *n*, Aufschluß *m*, Ausbiß *m*
(min.), Ausstreichen *n* ~ **coal** Ausbiß *m* eines
Kohlenflößes
**outdated** überholt, veraltet
**outdoor** außenseitig, draußen, im Freien ~ **antenna** Außen-, Hoch-antenne *f* ~**-apparatus**
Freiluftapparatur *f* ~ **boiler** Freiluftkessel *m* ~
**durability** Außenbeständigkeit *f* ~ **equipment**
Außenanlage *f* ~ **exposure** Freiluftlagerung *f* ~
**gantry crane** Hofbockkran *m* ~ **insulator** Frei-
leitungs-, Freiluft-isolator *m* ~ **location** Aufstellung *f* im Freien ~ **metering device** Außen-
meßvorrichtung *f* ~ **pickup** Außenübertragung
*f* ~ **picture** Außenaufnahme *f* ~ **plant** Freiluftanlage *f* ~ **scene** Außenaufnahme *f* ~ **shooting**
Freilichtaufnahme *f* ~ **sub-station** Freiluftanlage *f* (Umspannanlage) ~ **transformer station**
Freiluftumspannwerk *n*
**outer** Außenleiter *m*; außer, äußer ~ **air** Außenluft *f* ~ **arm** Reichweite *f* ~ **ball-bearing race**
Kugellageraußenring *m* ~ **band brake** Außenbandbremse *f* ~ **barrel** Außenbohrrohr *n*,
Pumpengehäuse *n* ~ **basin** Vorhafen *m*, Vortiegel *m* ~ **bearing** Außenlager *n* ~ **bell jar**
Außenglocke *f* ~ **border** Außenrand *m* ~
**bracket** äußerer Bock *m* ~ **brake** Außenbremse

f ~ **cable filler** Außenzwickel m ~ **casing** Ofenstock m, Rauchschacht m, Tragmantel m ~
**chamber** Vorraum m ~ **coating** Außenbelag
m ~ **column** Mantelrohr n (Autolenkung)
~ **cone** Außenkegel m ~ **contour** Umriß-
form f ~ **cover** Außen-belag m, -hülle f, Ballonhülle f, Reifenmantel m, Tragkörper m ~
**covering** Außenbelag m, -haut f ~ **down** Vordüne f ~ **duct** Außensammler m ~ **electrons**
Elektronenhülle f ~ **(level) electrons** kernferne
Elektronen pl ~ **end support** Spindelgegenlager
n ~ **field** Umfeld n ~ **fortification** Außenfort n
~ **gates** Außenhaut f ~ **globe** Außenglocke f ~
**grid** Außengitter n ~ **harbor** Außenhafen m,
freier Hafen m, Vorhafen m ~ **hull** Außenrumpf
rumpf m, der äußere Rumpf ~ **jacket** Außen-
mantel m ~ **jib** Außenklüver m ~ **layer** Rand-
schicht f ~ **layer of rolled piece** Walzhaut f ~
**leg (or limb)** Außenschenkel m ~ **liner** Außen-
mantel m ~ **main** Außenleiter ~ **main plane**
Außenflügel m ~ **marker (beacon)** Voreinflugzeichen n ~ **marking signal** Vorsignal n ~
**orientation** äußere Daten pl, äußere Orientierung f ~ **post** Außenstiel m ~ **pylon** Außenspannturm m ~ **race** Außen(lauf)ring m ~ **race**
**spacer** äußerer Abstandsring m ~ **race(way)**
Außenlaufbahn f ~ **ring** Begrenzungsring m
(Haspel) ~ **seal ring** Außenseegerring m ~ **shell**
(missiles) Außenboden m ~**-shell electron**
kernfernes Elektron n, Valenzelektron n ~
**sidewall surface** Mantelaußenoberfläche f ~
**skin** Außenhaut f (des Drahtes) ~ **sleeve on**
**engine (or slide)** Außenschieber m ~ **space**
Weltraum m ~ **stack of a wall** Stirnseite f einer
Mauer ~ **stack of wall** Mauermantel m ~ **steam**
**tube** Sperrohr n ~ **steering column** Stützrohr n
(Auto) ~ **stock support** Stangenhalter m ~
**string** Außenwange f (Treppe) ~ **strut** Außen-
strebe f ~ **support** Gegenlager n ~ **surface**
Außenfläche f, äußerer Rand m ~ **surface of**
**rolled piece** Walzhaut f ~ **temperature** Außenwärme f ~ **trench** Vorgraben m ~ **tube** äußeres
Rohr n ~ **wall** Außenwand f, Umfassungsmauer f ~ **wing panel** Außenflügel m ~ **work**
**function** äußere Austrittsarbeit f ~ **works**
Außenwerke pl ~ **zone** Randschicht f

**outfall** Abflußleitung f ~ **drain** Vorflutleitung f
**outfit** Anlage f, Apparat f, Ausrüstung f, Ausstattung f, Bepackung f, Einrichtung f, Vorrichtung f, Werkzeug n
**outflank, to** ~ flankieren, überflügeln, umfassen
**outflanking movement** Umgehungsbewegung f
**outflow** Abgang m, Ablauf m, Ausfluß m, Ausgang m, Auslauf m, Auspuff m, Ausströmung f
~ **method** Ausströmverfahren n ~ **port lantern**
Ausströmlaterne f ~ **tube** Ausflußrohr n ~
**valve** Ausströmventil n
**outflowing** ausfließend
**outgasing** Entgasung f (of electr. tubes), Gasaustreibung f
**outgoing** abgehend (current), ausscheidend, zurückgehend ~ **cable** Kabelabgang m ~ **call** abgehendes Gespräch n ~ **circuit** abgehende Leitung f ~ **conductor** Hinleiter m ~ **connection**
Hinverbindung f ~ **current** abgehender Strom
m ~ **data** Ausgangstrommel f ~ **direction** abgehende Richtung f ~ **flight** Hinflug m ~
**junction** abgehende Ortsverbindungsleitung f ~

**lead** Hinleiter m ~ **line** abgehende Leitung f ~
**one-way circuit** Leitung f für abgehenden Verkehr ~ **outlet** abgehende Verbindung f ~
**position** Abgangsplatz m (teleph.), A-Platz m ~
**selector** Ausgangswähler m ~ **spherical wave**
auslaufende Kugelwelle f ~ **stream** Ebbstrom
m ~ **subject copy** Sendebild n ~ **traffic** Abgangs-
verkehr m, abgehender Verkehr m ~ **trunk** abgehende Ortsverbindungsleitung f, abgehende
Verbindung f ~ **trunk multiple** Verbindungs-
leitungsbündel pl für alle Abgangsplätze ~
**tube** Senderohr n (Rohrpost) ~ **wire** Hinleiter m
**outgrowth** Auswuchs m
**outhaul** Ausholer m
**outlast, to** ~ länger dauern als, überdauern
**outlay** Ausgabe f, Auslage f
**outlet** Abflußöffnung f, Abfuhr f, Abführung f
(print.), Ablaß m, Ablaßöffnung f, Ablauf m,
Ableitung f, Absatz m, Absatz-gebiet n, -weg
m, Abzug m, Abzugs-öffnung f, -rohr n, -röhre
f, Anschlußpunkt m, Ausfluß m, Ausflußkanal m, -öffnung f, Ausfüllöffnung f, Ausgang m, Auguß m, Auslaß m, Auslaßschleuse
f, Auslauf m, Ausmündung f, Auspuff m, Aus-
schlitz m, Austritt m, Durchlaß m, Notschott n,
Spülloch n, Ventil n ~ **of a canal** (into a river)
Ausmündung f eines Kanals ~s **from discharge**
**culverts** Austrittöffnung f ~ **of the valve** Ventil-
austrittseite f ~ **of well** Bohrlochkopf m
**outlet,** ~ **branch** Auslaßstutzen m ~ **cam** Auslaßnocken m ~ **chamber** Ablaufraum m ~ **channel**
Abführungskanal m ~ **connecting piece** Ausblasstutzen m ~ **connection** Auslauf-, Auslaßstutzen m ~ **edge** Austrittskante f ~ **end** Austrittseite f ~ **flange** Austrittsstutzen m ~ **flap**
Auslaßklappe f ~ **funnel** Auslauftrichter m ~
**gas** Gichtgas n ~ **nozzle** Ausfluß-, Ausgußschnauze f, Austrittsdüse f ~ **opening** Auslauf-
öffnung f, Ausström-, Ausströmungs-öffnung f
~ **orifice** Auslaßöffnung f ~ **passage** Auslaßkanal m ~ **pipe** Ableitungs-, Ab-, Ausfahrungs-,
Ausgangs-, Auslauf-, Fall-rohr n ~ **pipe (or**
**tube)** Abflußrohr n ~ **pipe connection** Austrittsstutzen m ~ **port** Ausströmöffnung f, Austritts-
schlitz m ~ **pressure** Mündungsarbeitdruck m
~ **side** Austrittseite f ~ **slide** Auslaßschieber m
~ **sluice** Fluchtschleuse f ~ **socket** Netzsteck-
dose f ~ **structure** Auslaufbauwerk n ~
**temperature** Austrittswärmegrad m ~ **tower**
Entnahmeturm m ~ **transformer** Ausgangs-
transformator m, -übertrager m, Nachübertrager m ~ **tube** Ablaß-, Ablauf-, Ausström-
mungs-rohr n ~ **valve** Ablaß-, Abzugs-, Aus-
blase-ventil n, Auslaßseite f, Auslaß-, Aus-
ström-, Austritts-, Druck-, Entweichungs-ventil
n ~ **well** Auspreßsonde f
**outliers** Ausreißer pl (quality control)
**outline, to** ~ angeben, aufzeichnen, entwerfen, in
großen Zügen entwerfen, markieren, skizzieren,
umreißen
**outline** Abriß m, Aufzeichnung f, Außenlinie f,
Beschreibung f, Einhüllende f, Entwurf m,
Gerippe n, Grundlinie f, Grundriß m, Grundzug m, Linienführung f, Richtlinie f, Skizze f,
Umriß m, Umrißlinie f ~ **of rib** Rippen-profil n,
-umriß m
**outline,** ~ **drawing** Maßzeichnung f, Umriß-
skizze f mit Einbaumaßen ~ **illustration** Um-

rißzeichnung *f* ~ **letter** Konturschrift *f* ~ **plan** schematischer Plan *m*

**outlive, to** ~ länger dauern als, überdauern

**outlook** Ausblick *m*, Aussicht *f*, Sicht *f*

**outlying,** ~ **department** Außenverwaltung *f* ~ **lands** Vorgelände *n* ~ **post (or station)** Außenstelle *f*

**outnumber, to** ~ überwiegen, an Zahl übertreffen

**outphasing modulation** Chireixmodulation *f*

**outpost** Außenstelle *f* ~ **observer** Vorwarner *m* ~ **well** Außenbohrung *f*

**output** Abgabe *f*, Arbeitsleistung *f*, Ausbeute *f*, Ausbringen *n*, Ausbringung *f*, Ausgabe *f* (data proc.), Ausgang *m*, Ausgangsklemme *f*, Ausstoß *m*, Betriebsleistung *f*, Empfindlichkeit *f*, Endamplitude *f*, entnommene Leistung *f*, Erzeugungsmenge *f*, Fertigung *f*, Fördergut *f* (min.), Förderquantum *n*, Förderung *f*, Gewinnung *f*, Leistung *f*, Leistungsabgabe *f*, Nutzeffekt *m*, Produktion *f*, Verdampfleistung *f*, (in watts) Wirkleistung *f*, Wirkungsgrad *m*

**output,** ~ **of castings** Gußerzeugung *f* ~ **of coke** Kokausbringen *n* ~ **from master-oscillator control tube** Steuerkreisabstimmungsröhre *f* ~ **from master oscillator stage** Steuerkreisabstimmungsleistung *f* ~ **of plotting maps** Auswerteleistung *f*, Kartierungsleistung *f* ~ **of a pump** Druckleistung *f* einer Pumpe ~ **of pure noble metal per ton of ore** Edelmetallgehalt *m* pro Tonne ~ **for the recorder** Schreiberausgang *m* ~ **per unit** Einzelarbeit *f* ~ **per unit of displacement** Hubraumleistung *f*,

**output,** ~ **admittance** Ausgangs-admittanz *f*, -leitwert *m* ~ **alternating current** Ausgangswechselstrom *m* ~ **amplifier** Endverstärker *m* ~-**and input voltage** Ausgangs- und Eingangsspannung *f* ~ **anode** Sammelanode *f* ~ **aperture** Ausgangsspalt *m* ~ **area** Ausgabespeicherbereich *m* ~ **axis** (gyro) Abgriffachse *f* ~-**balance meter** Ausgangsspannungsmesser *m* ~ **block** Ausgabespeicher *m* ~ **cabinet** Endgestell *n* ~ **capacity** Erzeugungsleistung *f*, Leistungsvermögen *n* ~ **capacity per unit time** Leistung *f* je Zeiteinheit **circuit** ~ Ausgangs-kreis *m*, -stromkreis *m*, Entnahmekreis *m*, Nutzkreis *m*, Verbrauchszweig *m* ~ **conductance** Ausgangsleitwert *m* ~ **current** Ausgangs-, End-strom *m* ~ **electrode** Abnahmeelektrode *f*, Anode *f* einer Elektronenröhre, Entnahmeelektrode *f*, Sammelelektrode *f* ~ **end** Kraftabgabeseite *f* ~ **equipment** Ausgabegerät *n* ~ **field** Auskoppelfeld *n* ~ **gap** Auskoppel-raum *m*, -spalt *m* ~ **governor** Kraftversteller *m* ~ **hub** Sendebuchse *f* ~ **impedance** Ausgangs-impedanz *f*, -kreisimpedanz *f*, -scheinwiderstand *m* (rdo), -widerstand *m*, Quellwiderstand *m* ~ **limiter** Abkapper *m*, Verstärkungsbegrenzer *m* ~ **load circuit** Verbraucherkreis *m* ~ **meter** Leistungsmesser *m* ~ **monitor** Ausgangskontrollgerät *n* ~ **monitoring** akustische Ausgangskontrolle *f* ~ **noise voltage** Ausgangsstörspannung *f* ~ **part** Ausgangsteil *m* ~ **pentode** Endpenthode *f*, Fünfpolendröhre *f*, Schutzgitter(end)röhre *f* ~ **power** Ausgangs-energie *f*, -leistung *f* ~ **pulse** Ausgangsimpuls *m* ~ **quantity** Endgröße *f* ~ **rating** Leistungsrate *f* ~ **recorder** Leistungsschreiber *m* ~ **resistance** innerer Widerstand *m*

~ **resistor** Ausgangswiderstand *m* ~ **resonator** Ausgangsresonator *m* ~ **routine** Ausgabeprogramm *n* (info proc.) ~ **section** Ausgabespeicherbereich *m* ~-**selector-lever** Leistungswählhebel *m* ~ **shaft** Kraftabgabewelle *f* ~ **side** Arbeitsseite *f* ~ **signal** Ausgangssignal *n* ~ **socket** Ausgangsbuchse *f* ~ **stage** Ausgangs-, End-, Leistungs-stufe *f* ~ **suppressor** Verstärkungs-begrenzer *m*, -unterdrücker *m* ~ **tape** Ausgabeband *m* **terminal noise** Ausgangsklemme *f*, entnehmbare Klemme *f* ~ **terminals** Entnahme-klemmen *pl*, -punkt *m* ~ **transformer** Ausgangs-beträger *m*, -transformator *m*, -übertrager *m*, (repeater) Nachübertrager *m* ~ **triode** Endtriode *f* ~ **tube** Endröhre *f* ~ **unit** Ausgabe-einheit *f*, -werk *n* ~ **value** Resultatwert *m* ~ **values** Endwerte *pl* ~ **valve** Endröhre *f* ~ **variable** Ausgangsgröße *f* ~ **voltage** Ausgangs-, End-spannung *f* ~ **voltage furnished by rectifier** Relaisspannung *f* ~-**voltage meter** Ausgangsspannungsmesser *m* ~ **windings** Ausgangswicklungen *pl*

**outrigger** Ausleger *m*, (for direction finder) Braßbaum *m*, Holm *m*, Spreize *f* ~ **base** Bettung *f* ~ **drive** Auslagergetriebe *n* ~ **gear** Auslagergetriebe *n* ~ **lashing** (platform, lock) Holmzurrung *f* ~ **leveling handle** (platform) Horizontiergriff *m* ~ **spring** Auslegerfeder *f* ~ **tail** Gitterschwanz *m* ~ **tail boom** Gitterschwanzträger *m*

**outriggers** (gun platform) Lafettenkreuz *n*

**outrigging** ausladend

**outrun, to** ~ überholen

**outsell, to** ~ unterbieten

**outset** Anfang *m* **from the** ~ von vornherein

**outshining** (of spectral lines) Überstrahlen *n*

**outshoot, to** ~ hervorsprießen

**outshrink, to** ~ ausschrumpfen

**outside** Außenseite *f*; außen oder außerhalb befindlich, außerhalb, auswendig, draußen, werkfremd

**outside,** ~ **air** Außenluft *f* ~-**air temperature** Außenlufttemperatur *f* ~-**air temperature gauge** Außenluftthermometer *n* ~ **antenna** Außenantenne *f* ~ **appearance** äußere Form *f* ~ **box** Außenlager *n* ~ **cable** Außenkabel *n* ~ **caliber** Außenkaliber *n* ~ **calipers** Außentaster *m* ~ **camshaft** fremde Nockenwelle *f* ~ **chaser** Außensträhler *m* ~ **coating** Außenanstrich *m* ~ **crank** Außenkurbel *f* ~ **cylinder** (locomotive) Außenzylinder *m* ~ **diameter** Außendurchmesser *m* ~ **energy** Hilfsenergie *f* ~ **extension** außenliegende Nebenstelle *f* ~ **feeding attachment** Außenvorschubeinrichtung *f* ~ **fiber** Randfaser *f* ~ **gauging** Außenmessung *f* ~ **gear ring** Außenzahnkranz *m* ~ **girder** Außenträger *m* ~ **interference** Frequenzstörung *f* ~ **light** Fremdlicht *n* ~ **line of the tooth** Kopflinie *f* des Zahnes ~ **link** Außenglied *n* ~ **lip** Lippe *f* inklusiv *m* ~ **made spark plugs** Fremdkerzen *pl* ~ **make** Fremdfabrikat *n* ~ **measurement** Außenmaß *n* ~ **plant** Leitungsnetz *n* ~ **protection tube** äußeres Schutzrohr *n* ~ **reel support** äußere Haspelstütze *f* ~ **register** Außen-einpaß *m*, -zentriersitz *m* ~ **roll** Außen-rolle *f*, -überschlag *m* ~-**roller chain** Kette *f* mit seitlichen Laufrollen ~ **rotor** Außenläufer *m* ~ **screw thread** Außengewinde *n* ~ **storage** Freilagerung

*f* ~ **stringer of a bridge** Ölträger *m* ~ **taper**
Außenkegel *m* (Keillochbohrer) ~ **temperature**
Außenlufttemperatur *f*, Außenwärme *f*
~**-thread calipers** Gewindeaußentaster *m* ~
**vibrator** Außenrüttler *m* ~ **wall** Außenwand *f*
~ **way** Bettführung *f* (Drehbank) ~ **wheel**
Außenrad *n*

**outskirts** Außenbezirk *m* ~ **of a village** Dorf-
rand *m*

**outspeed, to** ~ an Geschwindigkeit *f* übertreffen,
an Schnelligkeit *f* übertreffen, überholen

**outstanding** außerordentlich, hervorragend **to
be** ~ (claims) ausstehen

**outstanding,** ~ **debt** Ausstand *m* ~ **debts** Außenstände *pl*, ausstehende Gelder *pl* ~ **feature**
Hauptzug *m* ~ **liabilities** Außenstände *pl*

**outstrip, to** ~ überholen

**outward** außen, außer, äußerlich, auswärts ~
**and inward** hin und zurück

**outward,** ~ **appearance** Äußere *n* ~ **bend of bank**
Ufervorsprung *m* ~ **board** abgehende Plätze
*pl*, Ausgangsplätze *pl* ~ **bound** auslaufend ~
**course** ablaufender Kurs *m*, Gegenkurs *m* ~
**dialling** Durchwahl *f* in abgehender Richtung
~**-flanged** ausgehalst ~**-flow turbine** Turbine *f*
mit innerer Beaufschlagung ~ **grid** Außengitter *n* ~**-inward operator** (toll) A- oder B-
Beamtin *f* ~ **keel-grooving** Kiel-sponung *f*,
-spündung *f* ~ **run of the carriage** Wagenaus-
fahrt *f*

**outwash** Sand *m* ~ **plain** Kies-, Sand-feld *n*

**outweigh, to** ~ aufheben, aufwiegen, ausgleichen,
überwiegen

**outwork, to** ~ ausarbeiten

**outwork** Außenarbeit *f*, Kronwerk *n*

**oval** Ellipse *f*, Oval *n*; eiförmig, eirund, länglichrund, oval ~ **disk meter** Ovalradzähler *m* ~
**drawing pass** Streck-kaliber *n*, -oval *n* ~
**fillister head screw** Linsenschraube *f* ~ **flange**
ovaler Flansch *m* ~ **flat-head screw** Linsensenk-
schraube *f* ~ **groove** Ovalkaliber *n* ~ **head
countersunk rivet** Linsenkniet *m* ~**-head screw**
Halbrundkopf-, Linsensenk-schraube *f* ~ **pass**
Ovalstich *m* ~ **point** (spring) Flachspitze *f* einer
Feder ~**-shaped** ovalförmig ~**-shaped body**
ovaler Rumpf *m* ~**-shaped roughing pass**
Streck-oval *n*, -kaliber *n* ~ **tube (or tubing)**
Ovalröhrchen *n* ~ **winch** Ovalhaspel *f*

**ovalization** Unrundwerden *n* ~ **of cylinder** Unrundwerden *n* des Zylinders

**oven** Backofen *m*, (baking) Brennofen *m*, Kochherd *m*, Ofen *m* ~ **battery** Ofenblock *m* ~
**brickwork** Ofenmauerwerk *n* ~ **builder** Ofenbauer *m* ~ **chamber section** Thermostat *m* ~
**charge** Ofenfüllung *f* ~ **construction** Ofenkonstruktion *f* ~**-dried** ofentrocken ~**-dry** absolut
trocken, knochentrocken ~**-dry weight** Darregewicht *n* ~ **drying** Ofentrocknung *f* ~ **floor**
Kammer-, Ofen-sohle *f* ~ **operator** Öfner *m* ~
**plant** Ofenanlage *f* ~ **rake** Ofengabel *f* ~ **sole**
Ofensohle *f*

**over** über **to go** ~ übergehen, übertreten
~ **against** querüber ~ (**or undercrossing**)
**station** Kreuzungsbahnhof *m* ~ **or under-
recoveries** Über- oder Unterdeckung *f* ~
**the telephone** telefonisch ~**s** überzählige
Teile *pl* (aviat.)

**overacidify, to** ~ übersäuern

**overaging** zu starkes oder zu langes Altern *n*
(Werkstoff), übervergütet

**overall** Arbeits-anzug *m*, -kittel *m*, -kleidung *f*,
Kombinationsanzug *m*, Overall *m*, Schutzanzug *m*

**overall** alles deckend, gesamt, total ~ **aerodynamic efficiency** aerodynamischer Gesamtwirkungsgrad *m* ~ **amplification** Gesamtverstärkung *f* ~ **attenuation** Restdämpfung *f* ~
**backlash** Gesamtflankenspiel *n* ~ **balance**
Gesamtsaldo *n* ~ **block diagram** Gesamtblock-
schaltbild *n* ~ **coefficient** Wärmedurchgangszahl *f* ~ **construction width** Baubreite *f* ~ **contrast ratio** Gesamtkontrast *m*, Gradationsver-
hältnis *n* ~ **diameter** Spitzendurchmesser *m* ~
**dimension** Außenmaß *n* ~ **dimension(s)** gesamte oder größte Abmessung *f* ~ **dimensions**
Raumbedarf *m* ~ **efficiency** Ausnutzungsfaktor
*m*, gesamter Wirkungsgrad *m*, Gesamt-leistung
*f*, -wirkungsgrad *m*, Totalnutzeffekt *m* ~ **enrichment** globaler Anreicherungsfaktor *m* ~
**equivalent** Restdämpfung *f* ~ **expenses** Allgemeinkosten *pl* ~ **formula** Bruttoformel *f* ~
**gain** Gesamtverstärkung *f* ~ **height** Bauhöhe *f*,
ganze Höhe *f* ~ **interference** Sammelentstörung
*f* ~ **length** Ausmaß *n*, äußerste Länge *f*, Baulänge *f*, Gesamtlänge *f*, Länge *f* über alles ~
**length of a boat (or vessel)** Schiffslänge *f* ~
**loss** Betriebsdämpfung *f* ~**-luminosity** Gesamt-
leuchtkraft *f* ~ **net loss** Restdämpfungsver-
zerrung *f* ~ **net-loss measurements** Restdämpfungsmessung *f* zwischen den Leitungs-
enden ~ **noise factor** mittlerer Rauschfaktor *m*
~ **projector height** Gesamthöhe *f* des Projektors
~ **rate** Gesamtsatz *m* ~ **sensitivity** Gesamtempfindlichkeit *f* ~ **situation** Gesamtlage *f* ~
**span** gesamte Breite *f* ~ **spread** gesamte Spannweite *f* ~ **system performance** Arbeitsreich-
weite *f* ~ **thickness** Gesamtmächtigkeit *f* ~
**transmission equivalent (or loss)** Restdämpfung
*f* ~ **transmission test** Dämpfungsmesser *m*,
Restdämpfungsmessung *f* ~ **transmission time**
Betriebslaufzeit *f* ~ **treatment time** Gesamtbe-
handlungszeit *f* ~ **voltage** Gesamtspannung *f*
~ **width** Gesamtbreite *f*

**overarch, to** ~ überwölben

**overarch** Obergurt *m*

**overarm** Gegenhalter *m* (Normalzubehör) ~
**pilot wheel** Überarmhandkreuz *n*

**overbalance, to** ~ überausgleichen

**overbalanced** Überausgleich *m*

**overbating** Überbeizung *f*

**overbeating** Totmahlen *n*

**overbias** Übersteuerung

**overbid, to** ~ überbieten

**overblow, to** ~ überblasen

**overblown steel** übergarer Stahl *m*

**overboard** überbord ~**-ambient line** Außendruck-
meßleitung *f* ~ **steam-vent** Dampfablaßleitung *f*

**overbunching** überkritische Ballungen *pl*

**overburden, to** ~ überbürden, überlasten

**overburden** Abraum *m*, Deckgebirge *n* (min.),
obere Schicht (min.), Obergestein *n* (min.)

**overburding** Überbelastung *f*

**overburn, to** ~ totbrennen **to** ~ **lime** Kalk *m*
totbrennen

**overburned particle of limestone** Kalk-kern *m*,
krumpe *f*

**overbusy** überbelastet
**overcast** bedeckt, bewölkt, düster, überwendliche Naht *f* ~ **day** trüber Tag *m* ~ **sky** bedeckter Himmel *m*
**overcenter arrangement** Endlagensperre *f*
**overcharge, to** ~ überladen
**overcharge** Aufschlag *m*, Überladung *f*, Überlastung *f*
**overcharging** Prellerei *f*, Überladung *f*
**overcoat shell** Mantel *m*
**overcome, to** ~ bewältigen, bezwingen, übermannen, überwältigen, überwinden **to** ~ **difficulties** Schwierigkeiten *pl* beseitigen
**overcoming** Überwinden *n*, Überwindung *f*
**overcompensate, to** ~ überkompensieren
**overcompensation** Überbesserung *f*, Überkompensierung *f*, Überkorrektur *f*, Überregulierung *f*
**overcompound, to** ~ überzusammensetzen
**overcompound excitation** Überverbunderregung *f*
**overcompounded dynamo** Überverbunddynamo *m*
**overcompressed** überverdichtet
**overcompression, ~ impeller** Nachschaltgebläse *n* ~ **ratio** Überkompressionsverhältnis *n*
**overcontrol, to** ~ (das Flugzeug) übersteuern
**overcooling** Überkühlung *f*
**overcorrect, to** ~ überkorrigieren
**overcorrection** Überbesserung *f*, Überkorrektur *f*, Überkorrigieren *n*
**overcouple, to** ~ überkoppeln
**overcritical** überkritisch
**overcure** Übervulkanisation *f*
**overcurrent** Überstrom *m* ~ **relay** Höchststrom-, Überlastungs-relais *n* ~ **release** Arbeitsstrom-, Überstrom-auslöser *m* ~ **trip-out** Überstromschalter *m*
**overcut** Überspringen *n* (acoust.)
**overcutting coalcutter** hochschneidende Schrämm-Maschine *f*
**overdam, to** ~ überstauen
**overdamming** Überstauung *f*
**overdamped** zu stark gedämpft
**overdamping** überkritische Dämpfung *f*
**overdense** zu dicht
**overdepth** Übertiefe *f*
**overdeveloped** überentwickelt
**overdie** Oberstempel *m*
**overdistill, to** ~ überdestillieren
**overdone** überentwickelt, (flax, hemp) überrottet
**overdoped fuel** Kraftstoff *m* mit übermäßig hohem Bleigehalt
**overdraft** überdisposition *f*
**overdrawn** übertrieben
**overdried** übertrocken
**overdrive, to** ~ übersteuern
**overdrive** Schnellgang *m*, Schnellgang-, Schongang-getriebe *n* ~ **gear** Ferngang-, Schnellganggetriebe *n*
**overdriven amplifier** übersteuerter Verstärker *m*
**overdue** rückständig, überfällig
**overdye, to** ~ überfärben
**overedge** überkant
**overedging** Känteln *n*
**overestimated** zu hoch veranschlagt
**overetch, to** ~ tiefätzen, überätzen
**over-excitation** Überregung *f*, Übersteuerung *f*

**overexcite, to** ~ überregen
**overexcited** übersteuert
**overexert, to** ~ überanstrengen
**overexpose, to** ~ überbelichten
**over-exposed** überexponiert
**overexposed print** verbranntes Bild *n*
**overexposure** Überbelichtung *f*, Überlichtung *f*
**overfall** Überfall *m*
**overfeed, to** ~ überfüttern
**overfeed stoker** Überschubfeuerung *f*
**overfill, to** ~ überfüllen
**overfilled metal** (welding) Perlgrat *n*
**overfilling** Überfüllung *f*
**overfinned** grätig
**overflow, to** ~ durchdrehen, überfließen, überlaufen, überströmen, übertreten
**overflow** Erguß *m*, Überlauf *m*, Überlaufen *n*, Überlaufrinne *f* ~ **from a mold** (plastics) Grat *m* ~ **of ore** Erztrübe *f*
**overflow, ~ baffle** Überlaufkante *f* ~ **cap** Überströmverschluß *m* ~ **card** Überlaufkarte *f* ~ **chamber** Überlaufraum *m* ~ **cock** Überlaufhahn *m* ~ **coupling** Überlaufverschraubung *f* ~ **damper** Überlaufschieber *m* ~ **dike** Überlaufdeich *m* ~ **drain water** Abwasser *n* ~ **edge** Überlaufkante *f* ~ **ejection** Übertragungsvorschub *m* ~ **exchange** Spitzenamt *n* ~ **flange** Überlaufflansch *m* ~ **flask** Auslaufflasche *f* ~ **groove** Stoffabflußnute *f* ~ **gutter** Überlaufrinne *f* ~ **level** (dam) Stauhöhe *f* ~ **lip** Überflußkante *f* ~ **meter** Belegungszähler *m* ~**-oil line** Leckölleitung *f* ~ **piece** (Pumpe) Überstromstück *n* ~ **pipe** Abfall-Lute *f*, Abwasserleitung *f*, Ausguß-rohr *n*, -röhre *f*, Leitblech *n*, Überfallrohr *n*, Überlauf-leitung *f*, -rohr *n*, Überschleusleitung *f* ~**-pipe valve** Abflußrohrventil *n* ~ **port** Überlaufbohrung *f* ~ **position** Überlaufplatz *m* ~**-prevention valve** Überfüllschutzventil *n* ~**-protective clutch** Überlaufsicherheitskupplung *f* ~ **register** Zähler *m* für die Besetzt- und Verlustfälle ~ **regulator** Überlaufschieber *m* ~ **reservoir** Überlaufgefäß *n* ~**-type ball mill** Überlaufmühle *f* ~ **valve** Entlastungsklappe *f*, Überströmventil *n* ~ **vessel** Übersteiggefäß *n* ~ **water** Überschußwasser *n* ~ **weir** Überfallwehr *n*, Wehrüberlauf *m* ~ **well** (die-casting) Luftsack *m*, Überlauf *m*
**overflowing** Überlaufen *n* ~ **water** Überwasser *n*
**overflux relay** Abschaltrelais *n*
**overfold** (fault) liegende Falte *f*
**overfolded rocks** Überfaltungsgebirge *n*
**overfolding** Überfaltung *f* (min.)
**overform, to** ~ nachformen
**overfreight** Überfracht *f*
**overglaze, to** ~ überglasen
**overglaze** Überglasung *f*
**overglazing** Überglasung *f*
**overglove** Überziehhandschuh *m*
**overgrinding** Totmahlen *n*
**overgrown** bedeckt, bewachsen ~ **soil** gewachsener Boden
**overgrowth** Überwachsung *f*
**overhand, ~ stope** Firstenstoß *m* ~ **stoping** Firstenbau *m*
**overhang, to** ~ überhängen, überkragen
**overhang** Ausladung *f* (aviat.), Freilänge *f*, Überhang *m*, Überhängen *n*, Überkragung *f*, Vorsprung *m* ~ **of arm** Armausladung *f*

overhang, ~ angle of approach Fahrzeugüberhang *m* ~ beam (of a rotating crane) Ausleger *m* ~ effect Nachwirkung *f* eines Einflusses ~ leakage flux Stirnstreuung *f* ~ strut Auslegerstiel *m*, -strebe *f*

overhanging hervorstehend, überhängend, überstehend ~ beam Freiträger *m* ~ screw press with friction drive einarmige Spindelpresse *f* mit Reibungsantrieb ~ shaft überhängende Welle *f* ~ side überhängender Stoß *m* (min.) ~ stairs freitragende Treppe *f* ~ wall überhängender Stoß *m*

overhaul Überholung *f* ~ handbook Überholungsanweisung *f*

overhauled instandgesetzt

overhauling Prüfung *f*, Überholung *f*, Überholungsarbeit *f* ~ balance of control surface Hornausgleich *m* (aerodyn.)

overhead kopfüber laufend, oben, obengesteuert, oberirdisch, über der Erde befindlich, über Kopf laufend ~ beam Deckenbalken *m* ~ bin Hochbehälter *m* ~ bridge obere Bedienungsbrücke *f* ~ bunkers Dachbunker *m* ~ cable Luftkabel *n* ~ camshaft obenliegende oder überhängende Nockenwelle *f* ~ carrier rail Fahrbahnträger *m* ~ chain-and-trolley conveyer Hängebahn *f* ~ chute Füllrohr *n* ~ communication line Fernmeldefreileitung *f* ~ conducter rail Hänge-Stromschiene *f* ~ contact hoop Stromabnehmerbügel *m* ~ conveyor Hängeförderband *n* ~ conveyer trolley Hängebahnlaufkatze *f* ~ cost allgemeine Unkosten *pl*, Gemeinkosten *pl* ~ cover Eindeckung *f*, Überdeckung *f* ~ (traveling) crane Laufkran *m* ~ crossing Kreuzungsklemme *f*, Wegüberführung *f* ~ distribution point Freileitungsverteilungspunkt *m* ~ door Falttür *f* (Garage) ~ drive Oberantrieb *m*, oberhalb gelagerter Antrieb *m* ~ driving gear Antriebsdeckenvorgelege *n* ~ earthing conductor Erd-kabel *n*, -seil *n* ~ electric cable Luftkabel *n* ~ expenses allgemeine (Geschäfts-)Unkosten *pl*, Generalunkosten *pl* ~-fire table Tiefenfeuertafel *f* ~-flange runway Oberflanschlaufwerk *n* ~ hand travelling crane Handkran *m* ~ highway Brücken-, Hochstraße *f* ~ insulator Außenisolator *m* ~ irrigation Beregnung *f* ~ lighting senkrechte Beleuchtung *f* ~ line Freileitung *f*, Freileitungslinie *f*, Freiluftleitung *f*, Luftleitung *f*, oberirdische Leitung *f* oder Linie *f*, Oberleitung *f* (electr.) ~ line connectors Freileitungsabzweigklemmen *pl* ~-line construction Hochbau *m* von Leitungen ~-line fittings Freileitungsarmaturen *pl* ~-line insulator Fahrleitungsisolator *m* ~ line insulator for maximum voltage Freileitungshöchstspannungsisolator *m* ~ line isolator Freileitungstrennschalter *m* ~ line material for streetcars Straßenbahnoberleitungsmaterial *n* ~ lubricant Obenschmieröl *n* ~ pilot bar Oberführungsstange *f* ~ pilot sleeve Oberführungsbuchse *f* ~ plumb Firstlot *n* ~ product Obendestillat *n* ~ radiation heating system Deckenstrahlungsheizung *f* ~ railway Hochbahn *f* ~ reservoir Hochbehälter *m* ~ roadway Brücken-, Hoch-straße *f* ~ room Höhenraum *m* ~ rope railway Seilhängebahn *f* ~ runway Fahrbahn *f* ~ scoop Oberlicht *n*, Sofittenlampe *f* ~ shaft hochgebaute Welle *f* ~ shafting Deckentransmission *f* ~ single-rail track Einschienenhochbahn *f* ~ space Ausbauhöhe *f* ~ stoping Firstenbau *m* ~ structure (at road interception) Überführung *f* ~ suspension Aufhängung *f*, (for airships) Luftschiffsaufhängung *f* ~ switch panel Deckenschalttafel *f* (im Flugzeug) ~ tank Hochbehälter *m* ~ telpher with remote control Laufkatze *m* (mit elektr. Hebe- und Fahrvorrichtung mit Zugseilbetätigung) ~ track Hochbahn *f* ~ transmission Deckentransmission *f* ~ transmission gear Deckenvorgelege *n* ~ transmission line Freileitung *f* ~ transportation Deckentransport *m* ~ travelling balance Hängebahnwaage *f* ~ travelling crane Laufkran *m* ~ travelling crane with grab Greiferlaufkran *m* ~ tray Kabelrost *m* ~ trolley Deckenlaufkatze *f*, Hängebahn *f* ~-trolley-conveyer rail Hängebahn-geleise *n*, -schiene *f* ~-trolley equipment Hängebahnanlage *f* ~-trolley rail switch Hängebahnweiche *f* ~-trolley track Hängebahnstrang *m* ~-trolley-track scales Hängebahnwaage *f* ~-type hängend angeordnet (Ventile) ~-underground system ober- und unterirdisches Netz *n* ~ valve hängendes oder obengesteuertes Ventil *n* ~ weld Schweißung *f* über Kopf ~ welding Überkopfschweißung *f* ~ wire Hochleitung *f*

overhear, to ~ abhorchen, mithören

overhearing Mitsprechen *n*

overheat, to ~ heißlaufen, überheizen, überhitzen, zu stark beheizen

overheated heißgelaufen, überhitzt ~ steam ungesättigter Dampf *m* ~ steel überhitzter Stahl *m*

overheating Überheizung *f*, Überhitzung *f*

overhouse structure Dachgestänge *n*

overhung fliegend angeordnet ~ arrangement (of rockets) fliegende Anordnung *f* ~ crank Stirnkurbel *f* ~ cylinder überhängender Zylinder *m* ~ flywheel fliegendes Schwungrad *n* ~ impeller fliegender Kreisel *m* (Kompressor) (in) ~ position fliegend angeordnet ~ screw freitragende Schraube *f* ~ spring Stützfeder *f* ~ starter pinion freifliegendes Anlasserritzel *m* ~ type of roll überhängende Walze *f*

overland, ~ cable oberirdisches Kabel *n* ~ longdistance cable Überlandfernkabel *n* ~ transportation Landtransport *m*

overlap, to ~ sich teilweise decken, überblatten, überdecken, übereinandergreifen (TV), übergreifen, überkreuzen, überlappen, überschneiden, verlaschen

overlap Überdeckung *f*, Übergreifen *n*, Überlappung *f*, Überschneidung *f*, Schutzstrecke *f* ~ of wave trains Überlappen *n* der Wellenzüge

overlap, ~ control Überdeckungsregler *m* ~ indicator Überdeckungsanzeiger *m* ~ seam Überlappungssaum *m* ~ welding Überlappschweißung *f*

overlapped überlappt

overlapping Überdeckung *f*, Übereinandergreifen *n*, Überlagerung *f*, Überlappen *n*, Überlappung *f*, Überschneiden *n*, Überschneidung *f*, Verlaschung *f*; übergreifend, überkant ~ of influence Überschiebung *f* ~ of lines Zeilenüberlappung *f* ~ of vibrations Überlagerung *f* von Schwingungen

**overlapping,** ~ **joint** überlappter Stoß *m* ~ **multiple** gestaffelte Vielfachschaltung *f*, Teilvielfachfeld *n* ~ **roller conveyer** übereinander angeordnete Rollenbahn *f* ~ **seam** Kappnaht *f*
**overlay, to** ~ überlagern, überziehen
**overlay** Auflage *f*, (harness) Belag *m*, Deckpause *f*, Planpause *f*, Transparentauflage *f*, Überzug *m* ~ **board** Relieffolie *f* ~ **shelf** Abraum *m*
**overlie, to** ~ überlagern
**overlift, to** ~ überheben
**overlighting** Szenenbeleuchtung *f* (film)
**overlimed** überäschert
**overload, to** ~ überladen, überlasten
**overload** Mehrbelastung *f*, Überbeanspruchung *f*, Überladung *f*, Überlast *f*, Überlastung *f*, (of microphone) Überschreien *n*, Überspannung *f* ~ **and undervoltage protection** Spule *f* für maximale Stromstärke und minimale Spannung
**overload,** ~ **breaker** Kurzschlußbügel *m* ~ **capacity** Überlastbarkeit *f*, Überlastungsfähigkeit *f* ~ **chopper** Übersteuerungsabschneider *m* ~ **circuit** Überlastungsschutzkreis *m* ~ **circuit breaker** Maximal-, Überstrom-ausschalter *m* ~ **clutch** Drehmomentbegrenzungskupplung *f*, Kupplung *f* zur Sicherung gegen Überlastung ~ **current** Überlastungsstrom *m* ~ **cutout** Überlastschalter *m* ~ **device** Überströmvorrichtung *f* ~ **indicator** Übersteuerungs-messer *m*, -zeiger *m* ~ **level** Belastungsgrenze *f* (acoust.) ~ **limiter** Übersteuerungsschutz *m* ~ **margin** Überlastbarkeitsbereich *m* ~ **point** (of recorder) Übersteuerungspunkt *m* ~ **protection** Überlastungs-, Überstrom-schutz *m* ~ **protective device** Überstromschutz *m* ~ **relay** Maximalspannungs-, Überstrom-relais *n* ~ **release** Arbeitsstromauslöser *m* ~ **relief wave** Überlastungsschutz *m* ~ **safety device** Überlastungssicherung *f* ~ **safety switch** Überspannungsschutzschalter *m* ~ **security** Überlastungsschutz *m* ~ **sensing control** Überlastungsselbstschalter *m* ~ **switch** Höchststromschalter *m*, Kurzschlußbügel *m* ~ **tariff** Überverbrauchstarif *m* ~ **test** Überlastungsprüfung *f*, Überlastversuch *m* ~ **valve** Überdruckventil *n*
**overloadable** überlastbar
**overloaded** überbelastet
**overloading** Überbelastung *f*, Überladung *f*, Überlast *f*, (of tubes) Übersteuerung *f*
**overlook, to** ~ nicht bemerken, übersehen
**over-lubricated** überölt
**overlying** darübergelagert, überstehend ~ **rock** Dachgestein *n* ~ **rocks** Deckgebirge *n*
**overmagnification** Übervergrößerung *f*
**overmatch** Überanpassung *f*
**overmatching of impedance** Widerstandsüberanpassung *f*
**overmature** überständig
**overmeasure** Übermaß *n*
**overmilling** Totmahlen *n*
**overmodulate, to** ~ übermodeln, übermodulieren, übersteuern
**overmodulated** übersteuert ~ **modulation** Übermodelung *f*, Überschreien *n*, (in film recording) Übersteuerung *f* ~ **oxidation** Überoxydation *f*, Überoxydierung *f*
**overpass for pedestrians** Fußgängerüberführung *f*
**overlumped** überschwellt
**overpoling** Überpolen *n*

**overpotential** Überpotential *n*
**overpredefecation** Übervorscheidung *f*
**overpress, to** ~ verdrücken
**overpressure** Überdruck *m*, Überspannung *f* ~ **grease gun** Überdruckpresse *f* ~ **pipe** Überdruckleitung *f*
**overprint, to** ~ einkopieren
**overprint** Aufdruck *m*, Überdruck *m*, Überschuß *m* ~ **color** überfallende Farbe *f*
**overprinted part** Überdruckstelle *f*
**overproduction** Über-erzeugung *f*, -produktion *f*
**overproof spirit** Spiritus *m* über Normalstärke
**overpunch** Überlochung *f*
**overrate, to** ~ übermessen, überschätzen, zu hoch veranschlagen
**overreach, to** ~ überholen
**overregulate, to** ~ überregeln
**override, to** ~ abhetzen, im Lauf überholen, steuern (picture), durch Übergehen einer Begrenzung ausschalten (aviat.), überholen, übersteuern
**override,** ~ **action** Übersteuerungsvorgang *m* ~ **brake** Auflaufbremse *f* ~ **stop** Überspringanschlag *m* ~ **switch** Überdruckschalter *m*
**overriding** Übersteuern *n* ~ **of noise** Geräuschverdeckung *f*
**overriding,** ~ **clutch** Überholkupplung *f* ~ **control** Sicherung *f* gegen Übersteuern ~ **pulse** aufgedrängter Impuls *m*
**overroad stay** Überweganker *m*
**overrun, to** ~ überfluten, überlaufen, umbrechen (print.)
**overrun** Über-fließen *n*, -laufen *n*, -schwemmen *n*, Übermaß *n*, Überschuß *m* ~ **brake** Auflaufbremse *f* ~ **control** Filmnachlaufregler *m* ~ **length** Durchrutschweg *m* ~ **safety device** Überlaufsicherung *f*
**overrunning** Freilauf *m*, Überkoppelung *f*, (a lamp) Überlastung *f*, Überlauf *m*, überspannter Betrieb *m*, Überspannung *f* ~ **of a lamp** Lampenüberlastung *f*
**overrunning,** ~ **clutch** Freilaufkupplung *f*, Rollenfreilauf *m* (für Anlasser) ~ **sprag clutch** Freilaufkupplung *f* ~ **switch** Grenzschalter *m* ~ **torque** Überholdrehmoment *n*
**oversaturate, to** ~ übersättigen
**oversaturation** Übersaturation *f*
**oversea,** ~ **plane** Überseeflugzeug *n* ~ **trade** Überseehandel *m*
**overseas** Übersee *f*, über See ~ **broadcasting** Hochseerundfunk *m*
**oversee, to** ~ übersehen
**overseer** Aufseher *m*, Druckereifaktor *m*, Polier *m*, Vorarbeiter *m*
**oversensitive** überempfindlich
**oversewing machine** Überwendlichnähmaschine *f*
**overshoes** eisentfernender Gummimechanismus *m* (aviat.)
**overshoot, to** ~ überschwingen, sich verrechnen, übers Ziel fliegen, zu weit bewegen
**overshoot** Durchschlag *m*, Überreichweite *f*, Überschuß *m*, Überschwingen *n* (Meßinstrument) ~ **(ing)** (of the indicator or pointer) zu weite Bewegung *f* ~ **period** Überschwingdauer *f* ~ **ratio** Überschwingfaktor *m*
**overshooting** Überschießen *n*
**overshot** Gestängekeilfänger *m*, Glocke *f*, Überschreitung *f*, Überschwingung *f* (TV); ober-

schlächtig ~ **haystacker** Heustapler *m* (Überkopfwurfsystem) mit Abladevorrichtung, Überwurfheustapler *m* ~ **(water) wheel** oberschlächtiges Wasserrad *n*
**oversight** Unterlassung *f*
**oversize, to make** ~ überdimensionieren
**oversize** (tire) Übergröße *f* (pieces) Überkorn, Übermaß *n* ~ **engine** überbemessener Motor *m* ~ **track shoe for operations in deep snow** Winterkettenglied *n*
**oversized** überdimensioniert
**overspeed, to** ~ (the engine) überdrehen
**overspeed** mit zu hoher Drehzahl *f*, Überdrehzahl *f* ~ **condition** Übergeschwindigkeitszustand *m* ~ **governor** Ausschaltvorrichtung *f* bei Übergeschwindigkeit, Geschwindigkeitssicherheitsausschalter *m* ~ **indicator** Drehzahlwarngerät *n* ~ **limiter** Übergeschwindigkeitsbegrenzer *m* ~ **monitor** Drehzahlwächter *m* ~ **protection** Überdrehzahlschutz *m* ~ **safety governor** Drehzahlregler *m* mit Sicherung gegen Überdrehzahl
**overspeeding of a machine** Durchgehen *n* einer Maschine
**overstatement** Übertreibung *f*
**overstrain, to** ~ überlasten, zerschmieden
**overstrain** bleibende Formänderung *f*, Überlastung *f* ~**(ing)** Überbeanspruchung *f* ~ **test** Überlastversuch *m*
**overstrained** überspannt ~ **state** überspannter Zustand *m*
**overstrength** fester als erforderlich
**overstress, to** ~ überbeanspruchen, überbiegen, überspannen
**overstress** Überbelastung *f*
**overstressing** (static) Überbeanspruchung *f*
**overstretch, to** ~ überstrecken
**overstrung** kreuzsäitig
**oversubscribed** überzeichnet
**oversupply** Überproduktion *f*
**overswing** (of volume indicator) Überanzeige *f*, (of compass needle) Überschwingung *f*
**overt** offen
**overtake, to** ~ (an aircraft) abhängen, einholen, überholen
**overtaking** Überholen *n*, Überholung *f*,
**overtax** Überlastung *f* ~ **tension** Überspannung *f*
**overthrow, to** ~ stürzen, über den Haufen werfen, umstoßen, umstürzen, zu weit ausschlagen, zu weit bewegen
**overthrowing** zu weites Ausschlagen *n* (des Zeigers)
**overthrown** gestürzt
**overthrust** Überkippung *f*, Überschiebung *f* ~ **of folds** Faltungsüberschiebung *f*
**overthrust** ~ **mass** Überschiebungsmasse *f* (geol.) ~ **transverse to the strike** Querüberschiebung *f* (geol.)
**overtime** Überstunde *f*
**overtone** (harmonische) Oberschwingung *f*, Oberton *m*, Oberwelle *f* ~ **absorption band** Oberband *n* ~ **band** harmonisches Band *n* (opt.)
**overtravel, to** ~ zu große Steuerwege *pl* ausführen, (control) übersteuern
**overtravelling** Überfahren *n* der Endstelle
**overturn, to** ~ kentern, überkippen, umkanten, umkehren, umstülpen, umstürzen

**overturn** Umkippen *n*, Umstürzen *n*
**overturned** überkippt, umgekippt ~ **fold** liegende Falte *f*
**overturning** Umfallen *n*, Umkanten *n* ~ **of wall** Kippen *n* der Mauer
**overturning**, ~ **bending moment** Kipp- und Knickmoment ~ **(or tipping) moment** Kippmoment *m* ~ **skip** Gefäß *n* mit Kopfkippung
**overtype armature** unterständiger Anker *m*
**overventilation** Überentlüftung *f*
**overvoltage** Überbeanspruchung *f* ~ **diverter** Überspannungsableiter *m* ~ **operation** überspannter Betrieb *m* ~ **protection** Überspannungsschutz *m* ~ **trip-out** Überspannungsausschalter *m*
**overvolted** überspannt ~ **discharge** überspannte Entladung *f* ~ **operation** überspannter Betrieb *m*
**overvolting** Überlastung *f*
**overweight** Mehr-, Über-gewicht *n*
**overwind, to** ~ überdrehen
**overwinding** Überfahren *n* der Endstelle
**overwing refueling** Oberflügelbetankung *f*
**overwire measurement** Maßwert *m* (für Dreidrahtmessung)
**overwriting error** Überschreibungsfehler *m*
**ovoid** eiförmiger Körper *m*
**owe, to** ~ schulden **to** ~ **to** verdanken
**owing to** angesichts, beruhen auf, durch, herrührend
**owl-light** Abenddämmerung *f*
**own, to** ~ besitzen, haben
**own** Eigen *n*, Eigentum *n* **one's** ~ eigen **on one's** ~ **authority** eigenmächtig
**own**, ~ **manufactured** selbsthergestellte Produkte *pl* ~ **plan** (estimate of situation) eigene Absicht *f* ~ **produced oil** selbstgefördertes Öl *n* ~ **speed** Flugzeugeigengeschwindigkeit *f* ~ **speed compensation** Eigengeschwindigkeitsausgleich *m* ~ **use** Eigenverbrauch *m* ~ **weight** Eigengewicht *n*
**owner** (of building under construction) Bauherr *m*, Besitzer *m*, Eigentümer *m*, Grundbesitzer *m*, Halter *m* ~ **of mines** Bergeigentümer *n*
**ownerless** herrenlos
**ownership** Besitz *m*, Besitzstand *m*, Eigentumsrecht *n*
**oxalic acid** Äthandisäure *f*, Kleesäure *f*, Oxalsäure *f*
**ox**, ~ **bow** Altwasser *n* ~ **cart tongue** Ochsendeichsel *f* ~ **hitch** Ochsenanspannvorrichtung *f*
**oxidant** Sauerstoffträger *m* (rocket)
**oxidation** Anwitterung *f*, Frischen *n*, Frischung *f*, Oxidierung *f*, Verbrennung *f* ~ **under high pressure** Druckoxydation *f*
**oxidation**, ~ **bleaching** Sauerstoffbleiche *f* ~ **discharges** Oxydationsätzen *n* ~ **film** Oxydhaut *f* ~ **inhibitor** Antioxydationsmittel *n* ~ **number** Oxydationsziffer *f* ~ **period** Frischperiode *f* ~ **process** Oxydationsvorgang *m* ~ **semi-conductor** Oxydationshalbleiter *m* ~ **stability** Oxydationsbeständigkeit *f* ~ **state** Oxydationszustand *m* ~ **tint** Ablaß-, Anlauf-farbe *f* ~ **zone** Verbrennungszone *f*
**oxidative slagging** Verschlackung *f*
**oxide** Oxyd *n*, oxydisch ~ **(oxygen compound)** Sauerstoffverbindung *f* (Oxyd) ~ **cathode** Oxydfaden *m* ~ **ceramic** oxydkeramisch ~

coat Oxydhaut *f* ~-coated oxydüberzogen
~-coated cathode Oxydkathode *f* ~-coated
filament vacuum tube Oxydkathodenröhre *f* ~
coating Oxyd-beschlag *m*, -schicht *f* ~ filament
Oxydfaden *m* ~ film Oberflächenoxydschicht *f*,
Oxyd-belag *m*, -häutchen *n*, -schicht *f* ~
inclusion Oxydeinschluß *m* ~ paint Rostfarbe *f*
~ paste Oxydpaste *f* ~ rectifier of reduced size
Gleichrichterpille *f* ~ skin Oxydhaut *f*
**oxidic** oxydhaltig
**oxidizability** Oxydierbarkeit *f*
**oxidizable** oxydationsfähig, oxydierbar
**oxidize, to** ~ beschlagen, frischen, oxydieren,
rosten **to** ~ **off** aboxydieren, fortoxydieren,
herausoxydieren
**oxidized** sauerstoffhaltig ~ **asphalt** oxidiertes
Pech *n* ~ **films** oxydierte Flächen *pl*
**oxidizer** Sauerstoffträger *m* ~ **added to rocket**
**fuel** Oxydierstoff *m*
**oxidizing,** ~ **action** Oxydationswirkung *f* ~
**agent** Oxydationsmittel *n* ~ **flame** Oxydations-
flamme *f*, oxydierende Flamme *f* ~ **frame** Luftbahn *f* ~ **period** Frischperiode *f* ~ **process**
Frischungs-prozeß *m*, -verfahren *n* ~ **reaction**
Frischwirkung *f* ~ **slag** Einschmelz-, Frisch-,
Oxydations-schlacke *f*
**oxy-acetylene** Azetylensauerstoffgas *n*; autogen
~ **blowpipe** Azetylensauerstoffbrenner *m* ~
**cutter** autogener Schweißapparat *m* oder
Schneidapparat *m* ~ **cutting** oxyazetyle-
nisches Schneidbrennen *n*, Sauerstoffschneiden
*n* ~ **cutting torch** Azetylensauerstoffschneid-
brenner *m* ~ **torch** Autogenschweiß-, Azety-
lensauerstoff-brenner *m* ~ **weld** Sauerstoff-
azetylenschweißung *f* ~ **welding** autogene
Schweißung *f*, Autogenschweißen *n*, Azety-
lensauerstoffschweißung *f*, Gasschmelzschweißung *f*, Oxyazetylenschweißung *f*
**oxybitumen** Oxybitumen *n*
**oxycalorimeter** Sauerstoffkalorimeter *n*
**oxychloride** Oxydchlorid *n*
**oxydated oil** geblasenes Öl *n*
**oxydations state** Oxydationszustand *m*
**oxygen** (rockets) A-Stoff *m*, Sauerstoff *m* ~ **in**
**cylinders** Flaschensauerstoff *m*
**oxygen,** ~ **ageing** Sauerstoffalterung *f* ~ **appara-**
**tus** Sauerstoffgerät *n* ~ **apparatus for high-**
**altitude flying** Höhen-atmer *m*, -atmungsgerät
*m* ~ **assimilability** Sauerstoffbindfähigkeit *f* ~
**blinker** Sauerstoffwächter *m* ~ **bottle** Sauer-
stoffflasche *f* ~ **breathing apparatus** Isoliergerät
*n* ~-**breathing apparatus for high-altitude flying**
Sauerstoffhöhenatmer *m* ~ **carrier** Sauerstoffträger *m* ~ **charging truck** Sauerstoffumfüll-
wagen *m* ~-**containing** sauerstoffhaltig ~
**content** Sauerstoffgehalt *m* ~ **cylinder** Sauer-

stoffflasche *f* ~ **deficiency** Sauerstoffmangel *m*
~ **deseaming** Sauerstoffhobeln *n* ~ **displace-**
**ment** Sauerstoffverschiebung *f* ~ **equipment**
Atemanlage *f* ~ **feed** Sauerstoffzuführung *f* ~
**feed line** Langrohrleitung *f* (g/m) ~ **filler valve**
Betankungsventil *n* ~ **fuel sprayer** A-Zerstäu-
ber *m* ~ **fuel tanker** A-Stoffwagen *m* ~-**hydro-**
**gen combustion** Knallgasverbrennung *f* ~-**indi-**
**cator gauge** Behälterstandsonde *f* ~ **inhaling**
**apparatus** Sauerstoffrettungsapparat *m* ~
**mask** Atem-, Sauerstoff-makse *f* ~ **mixture**
Sauerstoffgemisch *n* ~ **pipe** A-Rohr *n* (aviat.) ~
**plant** Sauerstoff(-gewinnungs)anlage *f* ~ **pole**
Sauerstoffpol *m* ~ **pressure gauge** Sauerstoff-
druckschlauch *m* ~-**pressure regulation pipe**
A-Belüftungsanlage *f* ~ **producer** Sauerstoff-
erzeugungsanlage *f* ~ **pump** A-Stoff-Pumpe *f*
~-**pump impeller** A-Stoff-Pumpenrad *n* ~-**pump**
**shaft** A-Stoff-Pumpenwelle *f* ~ **quenching**
Sauerstofflöschung *f* ~ **regulator** Sauerstoff-
druckminderer *m* ~ **relationship** Sauerstoffver-
wandtschaft *f* ~ **replenisher valve** Abfüllventil
*n* (g/m) ~ **respirator** Sauerstoffschutzgerät *n* ~
**supply** Sauerstoffvorrat *m* ~ **supply bleeder**
Sauerstoffentleerer *m* ~ **tank** A-Stoff-Tank *m*,
Sauerstoffbehälter *m* ~ **tank pressure control**
**line** Tankdruckregelleitung *f* ~ **topping-up con-**
**nection** A-Stoff-Nachtankkupplung *f* ~ **topping-**
**up nipple** A-Stoff-Nachtankstutzen *m* ~ **valve**
Sauerstoffventil *n* ~ **vent valve** A-Entlüfter *m*
**oxygenate, to** ~ oxygenieren
**oxygenation** Oxydierung *f*, Oxygenierung *f*
**oxygenic acid** Sauerstoffsäure *f*
**oxygenize, to** ~ oxygenieren
**oxyhydrogen** Knallgas *n* ~ **blowpipe** Knallgas-
gebläse *n* ~ **flame** Knallgasflamme *f* ~ **welding**
hydrooxygene Schweißung *f*, Sauerstoffschwei-
ßung *f*, Wassersauerstoffschweißung *f*
**oxysebacic acid** Oxyfettsäure *f*
**oxysiloxene** Oxysiloxen *n*
**oxysulphuric acid** Oxyschwefelsäure *f*
**oyster** Auster *f*
**ozocerite** Bergfett *n*, Erdwachs *n*, Mineralwachs
*n*, Ozokerit *m* ~ **paraffin** Bergtalg *m*
**ozone** Ozon *n* ~-**annihilation** Ozonzerstörung *f*
~ **dissipation** Ozonzersetzung *f* ~ **formation**
Ozonbildung *f* ~-**paper** Ozonreagenzpapier *n*
~-**proof** Ozonfest *n* ~ **radiosonde** Ozonradio-
sonde *f* ~ **spectra** Ozonkontinua *pl* ~ **variations**
Ozonschwankungen *pl*
**ozoned oxygen** Ozonsauerstoff *m*
**ozoner** Ozonisator *m*
**ozonifureous** ozonhaltig
**ozonize, to** ozonisieren
**ozonometer** Ozonmesser *m*
**ozonosphere sound** Ozonsphärenschall *m*

# P

**pace, to** ~ abgehen, abschreiten, Schritt halten (mit) **to** ~ **a distance** eine Strecke *f* abschreiten
**pace** Schritt *m*, Tempo *n*, Tritt *m* ~**-length** Schrittlänge *f* ~**-maker** Schrittmacher *m*
**pachymeter** Dickemesser *m*, Pachymeter *n*, Sicherungsgrenzenmesser *m*, Tastzirkel *m*
**pacing off** Begehung *f*
**pack, to** ~ abdichten, dichten, eindosen, einpacken, konservieren, (with leather) lidern, packen, schichten, stauen, stopfen, verdichten, verpacken, zusammenpressen **to** ~ **the carbons of electrodes** die Elektroden *pl* abdichten **to** ~ **up** einpacken, unterstopfen
**pack** Ballen *m*, Bündel *n*, Gepäck *n*, Konservierungsmethode *f*, (of dogs) Koppel *f*, Pack *m*, Packen *m*, Packung *f*, Paket *n*, (for apparatus) Rahmen *m*, Rückengebäck *n*, (for carrying cable reel on back) Rückentrage *f* Sturz *m* (metall.) Tornister *m*, Verpackungssack *m* (Fallschirm)
**pack,** ~ **animal** Last-, Pack-tier *n* ~ **carburizing** Pulverementieren *n* ~ **cloth** Pack-tuch *n*, -leinwand *f* ~ **cord** Binde-garn, -zwirn *m* ~ **fong** Neusilber *n* ~ **frame** Rucksackgestell *n* (für Gebläse) ~ **hardening process** Einsatz-, Kasteneinsatz-verfahren *n* ~ **horse** Hand-, Pack-, Saum-pferd *n* ~ **house** Konservenfabrik *f*, Lagerhaus *n* ~ **ice** Eisbank *f*, Packeis *n* ~ **radio set** Tornisterfunkapparat *m* ~ **receiver radio** Tornisterempfänger *m* ~ **rolling** Paketwalzung *f* ~ **rope** Packstrick *m* ~ **saddle** Packsattel, Tragesattel *m* ~ **stowing** Berg(e)versatz *m* ~ **strap** Packriemen *m* ~ **telephone** Fernsprechtornister *m* ~ **thread** Bindfaden *m*, Kordel *f* ~ **twine** Bindfaden *m*, Packzwirn *m* ~**-type filter** Packungstypfilter *m* ~**-type water purification unit** Tornisterfiltergerät *n* ~ **wall** Bergmauer *f*
**package, to** ~ packen, paketieren, verpacken
**package** Bündel *n*, Emballage *f*, Ladungsstück *n*, Packung *f*, Paket *n*, Verpacken *n*, Verpackung *f* ~ **conveyer cradle** Kreuzspultransport *m*, Spulrahmen *m* ~ **diameter** Spulendurchmesser *m* ~ **dyeing** Packsystem *n* ~ **holding rod** Spulenhalterstange *f* ~ **movement statement** Gebindebewegungsmeldung *f* ~ **reactor** transportabler Reaktor *m*
**packaged magnetron** betriebsfertige Magnetfeldröhre *f*
**packages** Kolli *pl*
**packaging** Einzelverpackung *f*, Verpacken *n* ~ **fabrics** Verpackungsgewebe *n* ~ **industry** Verpackungsmaschinenbau *m* ~ **installation** Verpackungsanlage *f* ~ **line** Verpackungsstraße *f* ~ **machine** Verpackungsmaschine *f*
**packed, to be** ~ paketieren, verpacken
**packed** abgefüllt ~ **column** Füllkörperkolonne *f* ~ **joint** Trennstelle *f* mit Dichtpackung ~ **stocks** Bestände *pl* an verpackten Waren ~ **tower** Füllkörpersäule *f*
**packer** Packer *m*, Stopfbüchsenpackung *f*, Stopfhacke *f*, Verpacker *m*, Versatzarbeiter *m* ~ **of ties** Schwellenstopfer *m*

**packet** kleines Paket *n*, Päckchen *n* ~ **bed** Festbett *n* ~ **dyes** Päckchenfarben *pl*
**packing** Dichtpackung *f*, Dichtung *f*, Dichtungsmittel *n*, Einfüllstoff *m*, Emballage *f*, Feststampfen, Filmabsatz *m*, Füllung *f*, Futter *n*, Futterstück *n*, Liderung *f*, (tightening) Liderungsdruck *m*, Luderung *f*, Öldichtungsring, Packen *n*, Packmaterial *n*, Packung *f*, Stampfen *n*, Stoffeinlage *f*, Tempermittel *n*, Verdichtung *f*, Verpacken *n*, Verpackung *f*, Zusammenbacken *n*
**packing,** ~ **of circles** Kreislagerung *f* ~ **about an explosive charge** Besatz *m* ~ **of ice** Eisstopfung *f* ~ **of a piston** Dichtung *f* eines Kolbens *f* ~ **ranging the permanent way** Heben *n* und Richten *n* des Geleises ~ **of the spheres** Kugellagerung *f*, **-packung** *f* ~ **up** Aufhäufung *f*
**packing,** ~ **awl** Pack-ahle *f*, -pfriem *m* ~ **base** Versandbock *m* ~ **block** Beschlagklotz *m* ~ **board** Packpappe *f* ~ **bolt** Liderungs-, Packkungs-bolzen *m* ~ **box** Stopfbüchse *f* ~ **case** Packkiste *f*, Transportkoffer *m*, Verpackungskiste *f* ~ **chamber** Dichtungskammer *f* ~ **clamp** Aufzugschiene *f* ~ **cord** Dichtungsschnur *f*, Verschnürungsleine *f* ~ **cover** Verpackungsplane *f* ~ **density** Dichte *f* (comput.), Informations-, Packungs-dichte *f* ~ **depth** (of tower) Rieselhöhe *f* ~ **disk** Verdichtungsscheibe *f* ~ **fluid** Sperrflüssigkeit *f* ~ **fraction** Packungsanteil *m* ~ **gland** Stopfbüchsenpackung *f* ~ **house** Konservenfabrik *f*, Warenlager *n* ~ **invoice** Packzettel *m* ~ **list** Begleitschein *m* ~ **loss** Massendefekt *m* ~ **machine** Formstampf-, Pack-maschine *f* ~ **material** Dichtungs-material *n*, -stoff *m*, Packungsmaterial *n*, Verpackungsmittel *n* ~ **method** Verpackungsmethode *f* ~ **paper** Einschlag-, Pack-papier *n* ~ **performance** Stopfarbeit *f* ~ **phenomenon** Packungserscheinung *f* ~ **piece** Beilageblech *n* ~ **plate (or shim)** Unterlegblech *n* ~ **point** Einpackstelle *f* ~ **pot** Einsatztopf *m* ~ **press** Bündel-, Pack-, Paketierpresse *f* ~ **ring** Buchsring *m*, Dichtungs-ring *m*, -scheibe *f*, Liderungsring *m*, Manschette *f* **(cylinder)** ~ **ring** Zylinderdichtungsring *m* ~**-ring press** Dichtungsfadenandrückmaschine *f* ~ **rod** Stopfstange *f* ~ **room** Packraum *m* ~ **sand** Füllsand *m* ~ **sheet** Pack-leinwand *f*, -tuch *n*, Dichtungsplatte *f* ~ **shim** Unterlagblech *n* ~ **sleeve** Dichtungsbuchse *f*, Rillenbüchse *f* ~ **slip** Packzettel *m* ~ **stand** Versandbock *m* ~ **stocks** Bestände *pl* an Verpackungen ~ **strip** Dichtleiste *f* ~ **stud** Liderungsbolzen *m* ~ **test** Einbettprobe *f* ~ **tube** Verpackungsrohr *n* ~ **washer** Dichtscheibe *f*, Liderungsdeckel *m*, Stopfdichtung *f*, Unterlegscheibe *f* ~ **weight** Umschließungsgewicht *n* ~ **worm** Packungszieher *m*
**pad, to** ~ ausfüttern, klotzen (Färberei), polstern
**pad** Abschußrampe *f* (rocket), Bohrfutter *n*, Dämpfungsglied *n*, Flausch *m*, Kissen *n*, Konsole *f* (für Hilfsgeräte), Mehrzweckhandstück *n*, Polster *n*, Puffer *m*, Verlängerungs-

leitung *f* (electr.), Watte *f* ~ **of cotton wool** Wattebausch *m*

**pad,** ~ **control signal** Dämpfungsschaltkennzeichen *n* ~ **embedding** Einbettung *f* (Polster) ~ **ground** Verklotzung *f* ~ **having distortion** verzerrende Verlängerungsleitung *f* ~ **lubricator** Polsterschmiervorrichtung *f* ~ **roll** Druckrolle *f* (film feed) ~ **roller** (without sprockets) Leitrolle *f* ~ **saw** Blatt-, Fuchsschwanz-säge *f* ~ **thrust bearing** Blocklager *n*

**padded** gepolstert, unterpolstert ~ **face mask** wattierte Maske *f*

**padding** Abfederung *f*, Polster *n*, Polsterung *f*, Stopfen *n* ~ **capacitor** Serientrimmer *m*, Verkürzungskondensator *m* ~ **condenser** Antennenverkürzungskondensator *m* ~ **machine** Farbmaschine *f* ~ **mangle** Imprägniermaschine *f* ~ **recipe** Klotz-ansatz *m*, -vorschrift *f* ~ **rubber sheets** Polstergummiplatte *f*

**paddle, to** ~ paddeln

**paddle** Brechstange *f*, Kratze *f*, Rühr-scheit *n*, -spatel *m*, -stab *m*, Schaufel *f* ~ (s) Büttkrück *m*, Streichbrett *n* (Fallschirm)

**paddle,** ~ **balance** Nebenausgleich *m* ~ **blade** Schaufelprofil *n* ~**-bladetype mixing machine** Flügelmischmaschine *f* ~ **box** (of a ship) Radkasten *m* ~ **drier** Schaufeltrockner *m* ~ **dyeing machine** Schaufelradfärbemaschine *f* ~ **fan** Gemischverteiler *m*, Quirl *m* ~ **hole** Freiarche *f* ~ **lime** Haspeläscher *m* ~ **liquor** Haspelfarbe *f* ~ **mixer** Paddelrührer *m* ~ **steamer** Raddampfer *m* ~ **tanning** Haspelgerbung *f* ~ **valve** Abschlußventil *n* (hydraul.), Kulissenschütz *n*, Registerschütz *n* ~ **wheel** Schaufelrad *n* ~**-wheel airplane** Flügelrad-, Schaufelrad-flugzeug *n*

**paddler staff** Raumlöffel *m*

**paddling door** Arbeitstür *f*

**paddock** Baggerstich *m*, Koppel *f*

**padlock** Hänge-, Vorhänge-, Vorlege-schloß *n* ~ **maker** Vorhängeschloßmacher *m*

**padwalk** Laufleiste *f*

**page, to** ~ foliieren (a book), paginieren, Seitenvorschub ausführen (up)

**page** Blatt *n*, Folio *n*, Kolumne *f*, Seite *f* ~ **copy** Ausbindeschnur *f*, Blattausfertigung *f* ~ **feed** Blatt-, Seiten-vorschub *m* ~ **gauge** Kolumnenmaß *n* ~ **printer** Blatt-, Seiten-drucker *m* ~ **printing** Blattdruck *m* ~ **(tape) printing** Abdruck *m* auf Blättern (Streifen) ~ **printing apparatus** Blattdrucker *m* ~**-printing telegraph** Blattdruck-, Seitendruck-telegraf *m* ~ **rule** Kolumnenmaß *n* ~ **string** Kolumnenschnur *f* ~ **teleprinter** Blattfernschreiber *m*

**paging** Paginierung *f*, Seitennumerierung *f* ~ **apparatus** Paginierapparat *m* ~ **machine** Paginiermaschine *f* ~**-up** Blattvorschub *m*

**paid,** ~ **call** gebührenpflichtiger Anruf *m* ~**-in capital** eingezahltes Kapital *n* ~**-out antenna** ausgekurbelte Antenne *f* ~ **stamp** Quittungsstempel *m* ~ **time** Gebührenminuten *f* ~ **time ratio** Hundertsatz *m* der Gebührenminuten je Stunde (teleph.) ~**-up capital** eingezahltes Kapital *n*

**pail** Eimer *m*, Kübel *m*

**painstaking** sorgfältig

**paint, to** ~ anstreichen, ausstreichen, bestreichen, bestreichen mit, färben, lacken, malen, pinseln, streichen, verstreichen

**paint** Anstrich-farbe *f*, -stoff *m*, Bildspur *f* (rdr), Farbe *f*, Ölfarbe *f*, Schminke *f*, Tünche *f* ~ **made of bituminous mastic** Bitumasticanstrich *m* ~ **and lacquer industries** Farben- und Lackindustrie *f* ~ **for protecting iron** Eisenschutzfarbe *f*

**paint,** ~ **booth** Lackbad *n* ~ **box** Farb-, Mal-, Tusch-kasten *m* ~ **brush** Maler-, Tusch-pinsel *m* ~ **burning tip** Farbabbrennmundstück *n* ~ **chemist** Farbenchemiker ~ **coat** Farbüberzug *m* ~**-coating installation** Farbanlage *f* ~ **drying** Lacktrocknung *f* ~ **grinder** Farben-mühle *f* (mach.), -reibmaschine *f* ~ **mist** Farbnebel *m* ~ **mixer** Farbmischmaschine *f* ~ **remover** Abbeizmittel *n*, Farbenabbeizmittel *n* ~ **resin** Lackharz *m* ~ **roller mill** Farbreibemaschine *f* ~ **shop** Streichanlage *f* ~ **spray gun** Farbspritzpistole *f* ~ **spraying apparatus** Farbspritzgerät *n* ~ **spraying booth** Farbspritzkabine *f* ~**-spraying system** Lackierspritzverfahren *n* ~**-storage equipment** Lagerausrüstung *f* für Farben ~ **thinner** Lösungsmittel *n* für Anstriche, Solvent *n*

**painter** Anstreicher *m*; Fangleine *f*; Maler *m* ~ **on china** Porzellanmaler *m* ~**'s-enamel** Maleremail *f* ~**'s shop** Malerwerkstatt *f* ~**'s varnish** Malerfirnis *m*

**painting** Farbanstrich *m*, Spritzlackieren *n* ~ **with bronze** Bronzieren *n* ~ **appliances** Anstrichgeräte *pl*

**pair** Paar *n*; doppeladrig (of), zweiadrig

**pair,** ~ **of antennas** Antennenpaar *n*, Zwillingsantenne *f* ~ **of bevel gears** (or wheels) Kegelräderpaar *n* ~ **of brushes** Bürstenpaar *n* ~ **of cables** Adernpaar *n* ~ **of cam** Exzenterpaar *n* ~ **of compasses** Zirkel *m* ~ **of cords** Schnurpaar *n* ~ **of curved scissors** gebogene Schere *f* ~ **of equations** Gleichungspaar *n* ~ **of eyepieces** (or **oculars**) Okularpaar *n* ~ **of fluorescent tubes** Tandemleuchtstofflampen *pl* ~ **of helical bevel gears** schrägverzahntes Kegelradpaar *n* ~ **of jacks** Zwillingsklinke *f* ~ **of lineman's climbers** Steigeisen *n* ~ **of particles** Teilchenpaar *n* ~ **of pawls** Doppelsperrklinke *f* ~ **of pincers** Beißzange *f* ~ **of planes** Ebenpaar *n* ~ **of plates** Plattenpaar *n* ~ **of plugs** Zwillingsstecker *m* ~ **of poles** Polpaar *n* ~ **of rolls** Walzenpaar *n* ~ **of shafts** Gabeldeichsel *f* ~ **of split bearings** (or **split brasses**) Lagerschalenpaar *n* (mach.) ~ **of spur gears** Stirnräderpaar *n* ~ **of stereoscopic pictures** stereoskopisches Bildpaar *n* ~ **of struts** Strebenpaar *n* ~ **of switch blades** Zungenpaar *n* ~ **of teeth** Zahnpaar *n* ~ **of test cords** Prüfschnurpaar *n* ~ **of values** Wertepaar *n* ~ **of wedges** Keilpaar *n* ~ **of wheels** Satz *m*

**pair,** ~ **annihilation** Paarvernichtung *f* ~**-creation** Paarbildung *f* ~ **damping** Mesonpaartermdämpfung *f* ~ **distribution function** Zweiteilchen-Verteilungsfunktion *f* ~ **formation** Paarbildung *f* ~ **furnace** Doppel-, Platinenwärme-(metall.), Plattenwärm-ofen ~ **production absorption** Absorption *f* durch Paarbildung ~**-to-~ capacity** Viererkapazität *f* ~ **twisting** (cable) Paarverseilung *f*

**paired** gepaart ~ (or **nonquadded**) **cable** paarverseiltes Kabel *n* ~ **compensating eyepieces** Kompensationsokularpaar *n* ~ **echoes** Echopaare *pl* ~ **eyepieces** (or **oculars**) Okularpaar *n*

~ **lattices** zugeordnetes Gitter *n* ~ **lenses** Linsenpaar *n*
**pairing** Paar-bildung *f*, -erzeugung *f*, Paarigkeit *f* ~ **of lines** Paarigerscheinen *n* von Linien
**pairings** Abgang *m*
**pairs, in** ~ paarweise ~ **cabled in quadpair formation** Doppelsternvierer *m* (teleph.)
**palatable** genießbar
**pale, to** ~ erblassen
**pale** abgeblaßt, blaß, bleich, hell, licht **of** ~ **fracture** weißbrüchig
**pale,** ~ **blue** hell-, licht-blau ~ **gold** Bleichgold *n* ~ **oil** helles Öl *n* ~ **yellow** falb
**paled** mit Pfählen *pl* versehen
**paleness** Blässe *f*
**paleolithic** paläolithisch
**paleontology** Paläontologie *f*
**paleozoic** paläozoisch
**palette knife** Farbspachtel *m*
**paling** Pfahlzaun *m*, Staket *n*, Verpfählung *f*
**palisade, to** ~ einfählen
**palisade** Palisade *f*, Pfahlwerk *n*, Staket *n*
**pall** Gabelkreuz *n*
**palladium** Palladium *n* ~ **alloy** Palladiumlegierung *f* ~ **black** Palladiummohr *m* ~ **hydride** Palladiumwasserstoff *m*
**palladous,** ~ **bromide** Palladiumbromür *n* ~ **chloride** Palladiumchlorür *n*, Palladochlorid *n* ~ **iodide** Palladiumjodür *n* ~ **nitrate** Palladiumoxydulnitrat *n* ~ **oxide** Palladiumoxydul ~ **salt** Palladiumoxydulsalz *n*
**pallet** (of harmonium) Klappe *f*, Palette *f*, Stapelplatte *f*, Töpferscheibe *f* ~ **of escape·ment** Hemmungslappen *m*
**pallial sinus** Mantelsinus *m*
**pallid** abgeblaßt, farblos
**palm, to** ~ **in** palmen
**palm** Palme *f* ~ **of the hand** Handfläche *f* ~ **butter** Palmfett *n* ~ **oil** Palmkernöl *n*, Palmöl *n* ~ **rest** Ballenlage *f* ~ **wax** Palmenwachs *m*
**palmitic acid** Cetyl-, Palmin-, Palmin-säure *f*
**palmitin** Palmitin *n*
**palnut lock** Palmuttersicherung *f*
**palpable** greifbar
**palpate, to** ~ abtasten, betasten, ertasten
**palpation** Betastung *f*
**pamphlet** Druckheft *n*, Flugblatt *n*, Flugschrift *f*, Merkblatt *n*, Schrift *f*
**pan** Angelring *m*, Becken *n*, Frosch *m* (print.), Gefäß *n*, Kessel *m*, Mulde *f*, Napf *m*, Pfanne *f*, Schale *f*, Schüssel *f*, Tiegel *m*, Trog *m*, Türangelpfanne *f* **(separating)** ~ Sichertrog *m* ~ **for dressing goldbearing sands** Rührwäsche *f* ~ **of the rhumb card** Kompaßpfanne *f* ~ **for roasting limestone** Steinhärtekessel *m*
**pan,** ~ **amalgamation** Kessel-, Pfannen-amalgamation *f* ~ **arrest** Schalenarretierung *f* ~ **base** Schalenfuß *m* ~ **boiling** Verkochen *n* ~-**boiling control apparatus** Kochkontrollapparat *m* ~ **bottom** Kollergangsteller *m*, Läuferplatte *f* ~ **car** Muldenwagen *m* ~ **conveyer** Pfannentransporteur *m* ~ **drain plug** Ölablaßpfropfen *m* ~ **grinder** Koller-gang *m*, -mühle *f*, Misch-koller *m*, -gang *m* ~ **grinding** Kollern *n* ~ **handle** Pfannenstiel *m* ~ **head (or cheese head) screw** Flachkopf-, Flachzylinder-, Kegelkopf-schraube *f* ~ **impermeable to light**

lichtdichte Kassette *f* ~ **mixer** Tellermischer *m* ~ **riveting** Flachkopfnietung *f*
**pancake, to** ~ durchsacken lassen, sacklanden (aviat.)
**pancake,** ~ **coil** flache Schlangenröhre *f* oder Spiralrohrschlange *f*, Flachspule *f*, quadratische Flachspule *f*, Scheibenspule *f* ~ **landing** Durchsacklandung *f*, Landung *f* mit Durchsacken ~ **loudspeaker** Flachlautsprecher *m* ~-**type pick-off coil** flache Abnahmespule *f*
**pancaking** Geschwindigkeitsverlust *m*, Überziehen *n* (aviat.)
**panchromatic** panchromatisch ~ **film** Panfilm *m*
**pancratic condenser** pankratischer Kondenser *m*
**pane** Feld *n*, (of a hammer) Finne *f*, Platte *f*, (of a hammer) Scheibe *f*, (of a glas) Tafel *f* ~ **of a wall** Mauerfeld *n*
**panel, to** ~ Feld *n* (eines Linienfeldes) in Felder unterteilen, in Felder unterteilen, panellieren, täfeln
**panel** Abbaufeld *n*, Abteilung *f*, Bauabteilung *f*, Beplankung *f*, Brett *n*, Fachwerkfeld *n*, Feld *n*, Feld *n* einer Täfelung, Frontplatte *f*, Füllung *f* (einer Täfelung), Hüllenbahn *f*, Jochfeld *n*, Paneel *n*, Platte *f*, Schalt-feld *n*, -kasten *m*, -tafel *f*, Scheibe *f*, Spant *n*, Stoffbahn *f* (Fallschirm), Tafel *f*, Tafelfeld *n*, Verkleidung *f*, Vorderseite *f*, Wand *f* ~ **of relay sets** Gemeinschaftsübertragung *f* ~ **of spar** Holmfeld *n* ~ **of a trussed beam** Feld *n* eines Fachwerkträgers
**panel,** ~ **annunciator** Signallampentafel *f* ~ **board** Füllbrett *n* ~-**call-indicator operation** Handbetrieb *m* mit mittelbar gesteuertem Nummernanzeiger ~ **door** Füllungstür *f* ~ **heating** Deckenheizung *f* ~ **illumination** Instrumentenbeleuchtung *f* ~ **lamp** Frontplattenlampe *f* ~ **light switch** Stirnwandlampenschalter *m* ~ **meter** Schalttafelinstrument *n* ~ **planing machine** Dickenhobel *m* ~ **point** Knotenpunkt *m* (statics) ~-**point load** Knoten-last *f*, -punkt-last *f* (statics) ~ **selector** Flachwähler *m* ~ **stock** Blechtafel *f* ~ **switch (or -type selector)** Stangenwähler *m* ~ **work** Pfeilerabbau *m*
**paneled** getäfelt ~ **ceiling** getäfelte Decke *f*
**panelling** Fachwerk *n*, Felderbau *m*, Getäfel *n*, Lambris *m*, Täfelwerk *n*, Felderbau *m*, Getäfel *n*, Lambris *m*, Täfelwerk *n*, Verkleidung *f*, Wandbekleidung *f*
**panels** Gestell *n*
**panning** Kameraschwenkung *f* ~-**out** Auswaschen *n* ~ **shot** Panoramabild *n*
**panorama** Rundbild *n*, Rundbildaufnahme *f* (photo) ~ **camera** Rundbildkamera *f* ~ **radar** (for early warning) Rundumsuchgerät *n*
**panoramic,** ~ **camera** Panorama-, Rundblickkamera *f* ~ **display** Rundumdarstellung *f* (rdr) ~ **exposure** Rundbild-, Rundblick-aufnahme *f* ~ **monitor** Frequenzspektrograf *m* ~ **photograph** Panorama-, Rundblick-aufnahme *f* ~ **picture** Panorama *n*, Panorama-, Raum-, Rund-bild *n* ~ **radar** Weitwinkelradar *n* ~ **receiver** Panoramaempfänger *m* (rdr) ~ **reception** Panoramaempfang *m* ~ **sight** Rundblickfernrohr *n* ~ **sketch** Ansichtsskizze *f* ~ **telescope** Rundblickfernrohr *n* ~ **teletype** Bildfernschreiber *m* ~ **tripod** Verfolgungsstativ *n* ~ **view** Panorama-, Rund-bild *n*, Rundblickaufnahme *f*

**pantelephone** Pantelefon *n*, verzerrungsfreier Fernhörer *m*
**pantile** Fittichziegel *m*
**pantograph, to** ~ pantografieren
**pantograph** Allzeichner *m*, Gleitbügel *m*, Pantograf *m*, Scherenstromabnehmer *m*, Storchschnabel *m* ~ **isolating switch** Greifertrenner *m* ~ **isolator** Scherentrenner *m* ~ **main joint** Pantografenhauptlager *n* ~ **support** Pantografenträger *m*
**pantometer** Pantometer *n*, Winkelmesser *m*
**pantoscopic spectacles** Bifokalgläser *pl*
**pap** Kleister *m*
**paper, to** ~ in Papier einwickeln, tapezieren **to** ~ **the pins** die Stecknadeln *pl* einbriefen
**paper** Abhandlung *f*, Bericht *m*, Papier *n*, Vortrag *m* **in** ~ **cover** (binding) geheftet ~ **for embossing** Prägepapier *n* ~ **for filter process** Filterpressenpapier *n* ~ **for insulating material** Isolierrohrpapier *n* ~ **for posters** Plakatpapier *n* ~ **from Schrenz** Schrenzpapier *n* ~ **for slate pencils** Griffelpapier *n* ~ **(made) from wood pulp** holzhaltiges Papier *n*
**paper,** ~ **bag** (for tamping) Papierschlauch *m*, Papiersack *m*, Tüte *f* ~**-bag effect** (burst of rocket tanks) Knalltüteneffekt *m* ~ **bail** Papierbügel *m* ~ **blank** Papierblatt *n*, Vordruck *m* ~ **blinds** Papier-rolläden *pl*, -rollos *pl* ~ **block condenser** Papierblock *m* ~ **board** Pappdeckel *m*, Pappe *f* ~ **board for boxes** Kartonagenpappe *f* ~ **board stock** Graupappe *f* ~ **bobbin** Papierspule *f* ~**-bound** broschiert ~ **cable** Papierkabel *n* ~ **capacitor** Papier-, Wickelblock-kondensator *m* ~ **carriage** Papierschlitten *m* ~ **carrier** Papierunterlage *f* ~**-carrying cylinder of the indicator** Indikatortrommel *f* ~ **clip** Aktenklammer *f*, Büronadel *f* ~ **coal** Blatt-, Blätter-, Papier-kohle *f* ~ **coating machine** Papierstreichmaschine *f* ~ **condenser** Becher-, Papierwickel-kondensator *m*, Wickel-block *m*, -kondensator *m* ~ **converting** Papierverarbeitung *f* ~ **core cable** Papierkabel *n* ~**-covered model** papierbespanntes Modell *n* ~ **covered wire** papierisolierter Draht *m* ~ **covering** Papierbespannung *f* ~ **credit** Wechselkredit *m* ~ **cutter** Papier-ausschläger *m*, -schneidemaschine *f* ~ **cutting** Papierstanzen *n* ~**-cutting knife** Papierschneidemesser *n* ~ **cylinder of an embossing (or goffering) machine** Gegenwalze *f* der Gaufriermaschine ~ **deformation** Papierschaden *m* ~ **drawer** Streifenlade *f* ~ **drill** Papierbohrer *m* ~ **drum** Papiertrommel *f* ~ **dust filter** Papierstaubsieb *n* ~ **element** Filterpapiereinsatz *m* ~ **embossing press** Papierstempelpresse *f* ~ **envelope** Papierhülle *f* ~ **etiquette** Papierschild *n* ~ **fancy goods** Papierausstattung *f* ~ **fastener** Büronadel *f* ~ **feed** Papier-nachschub *m*, -vorschub *m* ~ **feed magnet** Papierfortschubmagnet *m* ~ **feed rollers** Papierandruck-rollen *n*, -walzen *n*, Papier-führungswalze *f*, -transportwalzen *n* ~ **feeding** Papiervorschub *m* ~**-feeding cam** Streifenvorschubdaumen *m* ~**-feeding device** Streifenvorschubeinrichtung *f* ~**-feeding lever** Papierführungshebel *m* ~ **felt** Papiermacherfilz *m* ~ **file** Akte *f* ~ **filter element** Papierfiltereinsatz *m* ~ **finger** Papierführungsfinger *m* ~ **finishing** Papier-ausrüstung *f*, -veredelung *f* ~

**gasket** Papierdichtung *f* ~ **gauge** Piknometer *n* ~ **gauging contact tip** Meßhütchen *n* für Papiermessungen ~ **glazer** Glätter *m* ~**-glazing works** Papiersatinieranstalt *f* ~ **goods** Papierwaren *pl* ~ **guide** Papierführung *f* ~ **guide roll** Papierleitwalze *f* ~ **guide wheel** Papierführungsrad *n* ~ **half stuff** Papierhalbstoff *m* ~**-hanger's work** Tapezierarbeit *f* ~ **hanging device** Papieraufhängevorrichtung *f* ~ **holder** Bogenhalter *m* ~**-holder bail** Papier-bügel *m*, -halteschiene *f* ~**-insulated cable** Papierkabel *n* ~**-insulated enameled wire** Lackpapierdraht *m* ~ **insulated lead covered cables** Papierbleikabel *n* ~**-insulated wire** papierisolierter Draht *m* ~ **jointing tube** Papierröhrchen *n* ~ **knife** Papiermesser *n* ~ **lapping machine** Papierumwicklungsmaschine *f* ~ **layer** Papierlage *f* ~ **linen** Papierwäsche *f* ~**-machine oil** Papiermaschinenöl *n* ~ **machine winders** Papiermaschinenaufwickler *m* ~ **mangle** Papierkalander *m* ~ **mill** Papiermühle *f* ~ **model** Papierflugmodell *n* ~ **mold** Papiermatrize *f* ~ **pattern** Schmittmuster *n* ~ **pleating machine** Plissiermaschine *f* ~ **pressing plant** Spanpreßanlage *f* ~ **pulp** Ganzzeug *n*, Papier-brei *m*, -masse *f* ~ **reel cutter** Papierrollenschneidemaschine *f* ~ **refuse** Papierausschuß *m* ~ **release** Papierlöser *m* ~ **release lever** Papierauslösehebel *m* ~ **resistance tester** Papierwiderstandsprüfgerät *n* ~ **rod** Papierrandrückstange *f* (typewriter) ~ **roll** Papierrolle *f* ~ **rubbing** Papierabdruck *m* ~ **shredder** Papierwolf *m*, -zerfaserer *m* ~ **sign** Papierschild *n* ~ **sleeve** Papier-hülse *f*, -röhrchen *n* ~ **slip** Papierstreifen *m* ~**-staining machine** Fonciermaschine *f* ~ **stamp** Papierstampfe *f* ~ **stencil** Papierschablone *f* ~ **stereo process** Papiersterotypie *f* ~ **streamer** Papierschlange *f* ~ **stretch** Papierdehnung *f*, -spannung *f* ~ **string** Papierkordel *f* ~ **strip feed** Papierstreifentransport *m* ~ **table** Papierauflegeblech *n* ~ **tape** Papier-band *n*, -streifen *m* ~ **tape guide** Papierführer *m* (telet.) ~**-tearing device** Papierabreißvorrichtung *f* ~ **terminal cable** Papierabschlußkabel *n* ~ **texture** Papiergewebe *n* ~ **thickness gauge** Papierstärkemesser *m* ~ **throw** Papiervorschub *m* ~ **transport wheel complete** Transportträger *m* ~ **tube** Papier-hülse *f*, -rohr *n* ~ **web** Papierstreifenrolle *f* ~ **weight** Papierbeschwerer *m* ~ **winding device** Papieraufwicklung *f* ~ **wrapper** Streifband *m* ~ **wrapping** Papier-umhüllung *f*, -umwicklung *f* ~ **wrapping machine** Umwicklungsmaschine *f*
**papering,** ~ **arrangement** Spanapparat *m* ~**-in** Einspannen *n* ~ **lift** Spänebühne *f* ~**-out** Ausspannen *n* ~ **pump** Spanpreßpumpe *f*
**papers** Ausweis *m*, Schriftstück *n*
**papier-mâché** Papier-masché *n*, -teig *m*, Pappmasse *f*
**pappy** breiartig, breiig
**papyraceous** papierartig, papieren, papyrusartig, aus Papyrus angefertigt ~ **lignite** Papierkohle *f*
**par** Gleichheit *f*, pari **at** ~ paritätisch **on a** ~ gleichwertig ~ **of exchange** Wechselpari *n*
**par,** ~**-axis rays** achsnahe Strahlen *pl* ~ **value** Pariwert *m* ~ **weighting** Parierschwerung *f* (Seide)
**parabola** Kegel(schnitt)linie *f*, Parabel *f*

**parabolic** parabelförmig, parabolisch ~ **curvature** parabolische Krümmung f ~ **detection** quadratische Gleichrichtung f ~ **glass reflector** Glasparabolspiegel m ~ **line** parabolische Kurve f ~ **load** Parabelbelastung f ~ **mirror** Parabolreflektor m, -spiegel m, parabolischer Spiegel m ~ **projector** Parabelscheinwerfer m ~ **reflector** Parabol-reflektor m, -spiegel m, parabolischer Reflektor m ~ **shape** Parabolgestalt f
**parabolical** kegellining, parabolisch
**paraboloid of revolution** Rotationsparaboloid n
**paraboloid,** ~ **dish** (antenna) Reflektorschale f ~ **generated by rotation** Umdrehungsparaboloid n
**parabrake, to** ~ mit dem Fallschirm m bremsen ~ **reflector** Parabolspiegel m ~ **shape** Glockenfläche f
**paracentric** parazentrisch
**parachor of fuels** Parachor m von Treibstoffen
**parachute, to** ~ mit dem Fallschirm m abspringen
**parachute** Fallschirm m, Fangvorrichtung f, Stoßsicherung f the ~ **opens** der Fallschirm geht auf
**parachute,** ~ **brake** Fallgleiter m ~ **canopy** Fallschirm-dach n, -hülle f, -kappe f ~ **cartridge** Fallschirmpatrone f ~ **cartridge for measuring wind velocity** Fallschirmpatrone f für Windmessung ~ **cords** Fallschirmleinen pl, Fangleine f ~ **descent** Fallschirmabsprung m ~ **equipment** Fallschirmgerät n ~ **fabric** Fallschirmstoff m ~ **flare** Fallschirm-feuer n, -leuchte f, Leuchtfallschirm m, schwebende Kugel f ~**-flare cartridge** Fallschirmleuchtpatrone f ~ **flare signal** Fallschirmrakete f ~ **harness** Fallschirmgeschirr n, -gurt m, -gurtwerk n, Gurtwerk n, Haltegurt m ~ **jump** Fallschirmabsprung m ~ **jumper** Fallschirmspringer m ~ **jumping** Fallschirmspringen n ~ **lobe** Zusatzfallschirm m ~ **opening** Falltür f ~ **pack** Fallschirm-pack m, -verpackungssack m ~ **release** Fallschirmauslösung f ~ **rigger** Fallschirmwart m ~ **rip cord** Fallschirmreißleine f ~ **rocket signal** Fallschirmrakete f ~ **seat** Fallschirmsitz m ~ **static line** Reißleine f (zw. Fallschirm und Flugzeug) ~**-testing device** Entfaltungsstoßschreiber m ~ **tower** Fallschirm(sprung)turm m, Sprungturm m ~ **vent** Fallschirmluftloch n
**parachutist** Fallschirm-jäger m, -springer m
**paracresol** Parakresol n
**paradioxybenzene** Hydrochinon n
**parados** Rücken-deckung f, -wehr n
**paraffin, to** ~ paraffinieren
**paraffin** Paraffin n ~**-base crude oil** paraffinisches Rohöl n ~ **bath** Paraffinbad n ~ **coating** Paraffinanstrich m ~ **crystallizer** Paraffinkristallisator m ~ **distillate** Paraffinöldestillat n ~ **oil** Paraffinöl ~**-resistant color** paraffinfeste Farbe f ~ **scale** Paraffinschuppe f ~ **series** Paraffinreihe f ~ **spraying apparatus** Paraffinsprühapparat m ~**-waxed paper** Paraffinrohpapier n ~**-waxed tissue paper** Paraffinrohseidenpapier n
**paraffined,** ~ **paper** paraffiniertes Papier n ~ **wire** Wachsdraht m
**paraffinic** paraffinisch ~ **hydrocarbons** Paraffinkohlenwasserstoffe pl
**paraformaldehyde** Paraformaldehyd n

**paraglider** Gleitschirm m
**paragon** Textschrift f
**paragraph** Absatz m, Abschnitt m
**para-helium** Parhelium n
**para-hydrogen** Parawasserstoff m
**paraldehyde** Paraldehyd n
**paralketon** Paralketon n
**parallactic** parallaktisch ~ **angle** parallaktischer Winkel m ~ **axis** parallaktische Achse f ~ **displacement of coordinates** Koordinatenparallaxe f ~ **movement** Parallaxenbewegung f
**parallax** Parallaxe f, parallaktische Verschiebung f ~ **of coordinates** Koordinatenparallaxe f
**parallax,** ~ **arm** Parallaxenlineal n ~ **corrector** Parallonrechner m ~ **error** Parallaxenfehler m ~**-free mirror** parallaxfreier Spiegel m ~ **offset mechanism** Verbesserungsgerät n ~ **panoramogram** Parallax-Panoramogramm n ~ **rule (or ruler)** Parallaxenlineal n ~ **slide** Parallaxenschlitten m ~ **stereogram** Parallax-Stereogramm n

**parallel** Breitenkreis m, Nebeneinander-, Parallel-schaltung f; gleichgerichtet, parallel, Parallele f, Parallelität f **in** ~ nebeneinander, parallelgeschaltet **out of** ~ Außertrittfallen n ~ **along the path** bahnparallel **a** ~ **to the axis of tilt** Bildwaagerechte f ~ **of departure** (declination) Abweichungsparallele f ~ **of latitude** Breiten-grad m, -kreis m, -parallel n, Parallelkreis m (geogr.) ~ **to the path** bahnparallel ~ **to the principal line** Bildsenkrechte f ~ **in space** raumparallel **in** ~ **with** parallel zu
**parallel,** ~ **access** parallelle Ein- und Ausgabe f ~ **access to four words** Vierwortparallelzugriff m ~ **adjustment** Parallelverstellung f ~ **aiding** gleichsinnig, parallel ~ **alignment** Parallelstellung f ~ **arithmetic unit** Parallelrecheneinheit f ~ **arrangement** Nebeneinanderschaltung f ~ **balance** Summenkontrolle f ~ **bedding** Parallelstruktur f ~ **capacitance** Parallelkapazität f ~ **capacitor** Parallelkondensator m ~ **cascade action** Parallel-Kaskadenverhalten n ~ **circle** Parallelkreis m ~ **circuit** Nebeneinanderschaltung f, Parallel-kreis m, -schaltung f ~ **computer** Simultanrechner m ~**-connected** gleichsinnigparallel ~ **connection** Nebeneinanderschalten (electr.), Nebeneinander-, Parallel-, Zweig-schaltung f ~ **construction** Parallel-bauart f, -führung f ~ **cut** X-Schnitt m (cryst.) ~ **design** Parallelbauart f ~ **digital computer** Paralleldigitalrechner m ~ **displacement** Verpflanzung f ~ **equalizer** Querentzerrer m ~ **experiment** Parallelversuch m ~ **fabric** Parallelstoff m ~ **faced shut-off valve** Parallelabsperrschieber m ~**-faced sluice valve** Parallelabsperrschieber m ~ **feeding** Parallelspeisung f ~**-flange girder** Parallelflanschträger m ~ **flat nose pieces** Planschnäbel pl ~ **flight** Parallelflug m ~ **flow** Parallel-strom m, -umlauf m, Schichtenströmung f, wirbelfreie Strömung f ~ **flow cataract condenser** Parallelstromkataraktkondensator m ~**-flow condenser** Gleichstromkondensator m ~**-flow-type valve** Gleichstromventil n ~ **fold** Parallelbruch m ~ **forces** Kräftepaar n ~ **gliding planes** Parallelverschiebung f ~ **guidance (or guide)** Parallelführung f ~ **guide segment** Gradführungssegment n ~ **hand vise** Parallelfeilkloben m ~ **hook-up** Nebeneinander-

schaltung f ~ inverter Hauptstromwechselrichter m ~ jack Parallelimpedanzklinke f ~ joint (ing) sleeve Abzweigmuffe f ~ key Flachkeil m ~ laminated parallelgeschichtet ~ light (rays) gerichtetes Licht n ~ line Gleichlaufende f ~ lines parallele Geraden pl ~ measuring Parallelmessung f ~ motion Parallelbewegung f ~ mounter Anschliffpresse f ~-moving target querfahrendes Ziel n ~ multiple jack Vielfachparallelklinke f ~ ohm method Parallelohmmethode f ~ operation Parallelbetrieb m ~ orientation Parallelstellung f ~-planing machine Langhobelmaschine f, Tangentialhobelmaschine f ~ plate counter Parallelplattenzähler m ~ plate guide planparalleler Wellenleiter m ~ projection Parallel-projektion f, -riß m ~-ray bundle Parallelstrahlenbündel n ~ resistance Parallelwiderstand m ~ resonance Anti-, Nebenschluß-, Parallel-, Spannungs-, Sperr-resonanz f, Sperrkreiswirkung f, Stromresonanz f ~ resonance coupling Sperrkreiskopplung f ~ resonant circuit Drossel-, Parallelresonanz-, Resonanz-, Stromresonanz-kreis m ~ row atomizer plate Parallelzeilendüse f ~ ruler Parallellineal n ~ scribe Parallelreißer m ~-series connection gemischte Schaltung f ~ sheaf Gleichlaufstellung f (artil.) ~-sided parallelwandig ~ single test Vergleichs-Einzelversuch m ~-slide steam stop valve Dampfabsperrschieber m ~-slide stop valve Absperrschieber m ~ slideway Parallelgleitführung f ~ spark gap Funkenstrecke f im Nebenschluß ~ storage Parallelspeicher m ~ straight edge Parallaxenlineal n ~ stretch Parallelstrecke f ~-stroke (or straight milling) Zeilenfräsen n ~ surface Parallelfläche f ~ system of distribution System n mit Parallelschaltung ~ thread zylindrisches Gewinde f ~ trace of lines Parallelenzug m ~-tube pole zylindrische Rohrpost f ~ vise Parallelschraub(en)stock m ~-wire line Paralleldrahtleitung f ~-wire stop line Paralleldraht-Sperrleitung f ~-wire system Gitter n ~ wiring Gruppierung f im Parallelogramm
paralleled parallelgeschaltet
parallelepiped Parallelpipedon n ~ block (masonry) Quader m
parallelepipedon Parallelpipedon n
paralleling Parallelschaltung f ~ of a synchronous machine Parallelschalten n ~ device Synchronisiereinrichtung f
parallelism gleichlaufende Näherung f, Parallelismus m, Parallelität f, Parallelverlauf m, Symbasis f
parallelogram Parallelogramm n ~ of forces Kräfteparallelogramm n ~ of velocities Parallelogramm n der Geschwindigkeiten ~ linkage system Parallelogrammgestänge n
parallels, ~ of declination Deklinationsparallelen pl ~ of latitude Breitengrade pl ~ for levelling up Richtschienen pl
paralysis Sperrung f (rdr)
paralyze, to ~ lähmen, niederhalten, unwirksam machen
paramagnetic paramagnetisch ~ alloy paramagnetische Legierung f ~ substance paramagnetischer Stoff m
paramagnetism Paramagnetismus m

parameter Achsenabschnitt m, Beiwert m, Bestimmungsstück n, Größe f, Kenn-größe f, -wert m, Konstante f, Meß-größe f, -wert m, Parameter n, Zahlenangabe f ~ of gap length distribution Parameter m der Lückenverteilung ~ of a thermionic valve charakteristische Größe f der Röhre, charakteristische Röhrenkonstante f
parameter, ~ setting instructions Substitionsteil n ~ setting orders Substitionsteil m (info proc.)
parametric, ~ amplifier Reaktanzverstärker m ~ representation Parameterdarstellung f
parametrization technique Parametrisierungstechnik f
parametron Parametron n
paramorphic paramorph
paramorphosis Paramorphose f
paramp parametrischer Verstärker m
paranthelion Nebengegensonne f
parapet Aufwurf m, Bank f, Brust-mauer m, -wehr f, -werk n, Brüstung f, Erdaufwurf m, Fensterbrüstung f, Gewehrauflage f, Grabenwehr f, Schutzgeländer n, Wehr f ~ wall Brustwehr f
paraphase, ~ amplifier Gegentakt-, Kalhodenverstärker m ~ stage Phasenumkehrstufe f
paraphrase, to ~ abwandeln, gliedern, umschreiben
paraphrase Einschaltung f (im Text), kürzende oder vereinfachende Wiedergabe f in anderer Fassung, Umschreibung f
paraposition Parastellung f
parapositronium Parapositronium n
para-red Pararot n
pararosolic acid Pararosolsäure f
paraselene Nebenmond m
parasite Schädling m, Schmarotzer m ~ drag Profilwiderstand m, schädlicher Widerstand m (aerodyn.) ~ resistance schädlicher Widerstand m
parasitic parasitär, störend ~ antenna strahlungsgekoppelte Antenne f ~ capture parasitärer Neutroneneinfang m ~ coupling Strahlungskopplung f (antenna) ~ current Außen-, Fremd-strom m ~ diminution Verbrauchsverschlechterung f ~ disturbance elektromagnetische Störungen pl ~ drag schädlicher Luftwiderstand (aviat.), schädlicher Widerstand m ~ element strahlungsgekoppeltes Antennenelement n ~ harmonic Störoberwelle f ~ lobe Nebenschleife f ~ oscillation Streuschwingung f, Störwelle f, wilde Schwingung f ~ reflector strahlungserregter oder strahlungsgekoppelter Reflektor m ~ resistance schädlicher Widerstand m (aviat.) ~ wave Störungs-, Stör-welle f
parasitically excited antenna strahlungsgekoppelte Antenne f
parasol Parasol n, Sonnendach n ~ monoplane Schirmeindecker m ~-type antenna Gitterantenne f ~ wing of monoplane Parasol m
parastate Parazustand m
paratoluidine Paratoluidin n
paravane Minenabweiser m
paraxial achsparallel, parachsial ~ field Längsfeld n ~ orbits Paraxialbahnen pl ~ ray achsenparalleler Strahl, Nullstrahl m
parboil, to ~ abbrühen (cloth), ankochen
parbuckle, to ~ aufschroten

parcel, to ~ parzellieren
parcel Ballen *m*, Bündel *n*, Lieferung *f*, Paket *n*, Parzelle *f*, Posten *m*, Stückgut *n* ~ of land Grundstück *n*
parcel, ~ closing stamp Paketverschlußmarke *f* ~ rack Gepäcknetz *n* ~ sent by book post Kreuzbandsendung *f* ~ tag Kollianhänger *m*
parcels, in ~ parzellarisch ~ porterage Eingutbeförderung *f*
parch, to ~ dörren, sengen
parched gedörrt, geröstet
parchment Pergament *n* ~ base paper Pergamentrohpapier *n* ~ paper Papierpergament *n*, Pergamentpapier *n*
parchmentability Pergamentierfähigkeit
parchmentize, to ~ pergamentieren
pare, to ~ abschälen, schaben, schälen to ~ quarry stones Bruchsteine *pl* abschalen
pare-flash Gegenfunk *m*
pared veneer Schälfurnier *n*
parenchyma schwammige Zellensubstanz *f*
parent ursprünglich ~ body Stammkörper *m* ~ firm Ursprungsfirma *f* ~ lattice Hauptgitter *n* ~ material Ortstoff *m*, Urstoff *m* ~ nucleus Mutterkern *m* ~ nuclide Ausgangsnuklid *n* ~ patent Stammpatent *n* ~ plant Stamm(-baum)werk *n* ~ rock Ausgangsgestein *n*, Urgestein *n* ~ specification Hauptpatent *n* ~ state Ausgangszustand *m* (eines Zerfalls) ~ substance Grundkörper *m*, Muttersubstanz *f*, Stammkörper *m*
parentage Abstammung *f*, Urheberschaft *f*
parentheses runde Klammern *pl* in ~ das in Klammern Gesetzte
parfocal oil immersion abgeglichene Ölimmersion *f*
parfocalized objective Objektivabgleich *m*
parhelion Gegen-, Neben-sonne *f*, Parhelion *n*
paring Hobelspan *m* ~ ax Schälaxt *f* ~ knife Schäl-, Schärfen-messer *n*, Wirkeisen *n* ~ plow Schälpflug *m* ~ tool Krummeißel *m*, Schnitzer *m*
parings (of leaf gold) Schabin *n*, Schabsel *n*, Schnitzel *pl* ~ of leaf gold Goldabfall *m*
Paris, ~ green Deck-, Lackier-, Mitis-, Neu-, Original-, Patent-, Staub-, Staubhaltig-, Wiesen-grün *n* ~ green and a little chrome yellow Papageigrün *n*
parison (glass mfg.) Külbchen *n*, Külbel *m*, Meßform *f*
parity Gleichheit *f*, Parität *f* ~ bit Paritätsbit *n* ~ check (infro proc.) Paritätsprüfung *f* ~ conversation Paritätserhaltung *f* ~ digit Paritätsziffer *f*
parley Unterhandlung *f*, Unterredung *f*
park, to ~ abstellen, deponieren (Material), lagern, parken, unterstellen
park Park *m*, Park-anlage *f*, -gelände *n*, -platz *m*
parka Anorak *n*
parked abgestellt
parker(ize), to ~ parken (Rostschutz)
Parker, ~'s cement Patentzement *m* (rustproofing) process Parker-Verfahren *n*
parkerizing Parkerisieren *n*, Phosphorisierung *f*, Rostschutzverfahren *n*
parking, no ~ Parkverbot *n* ~ apron Abstellvorfeld *n* ~ area Abstell-, Park-platz *m* ~ brake Feststell-, Park-bremse *f* (aviat.) ~ brake for an airplane Bremse *f* für ein Flugzeug ~ charge

Unterkunftsgebühr *f* ~ light Park-, Stand-licht *n* ~ lot Parkplatz *m* ~ orbit Wartebahn *f* ~ place Abstell-, Halte-platz *m*
parkmeter Parkiermesser *m*, Parkmeter *n*, Park(zeit)uhr *f*
parlor rifle Zimmerstutzen *m*
parquet Parkett *n*
parquetry Parkett *n*, Parkettafeln *pl*, Täfelung *f* ~ abrasive paper Parkettschleifpapier *n*
parry Fangstoß *m*, Parieren *n*
pars pro mille Promille *f*
parsonite Parsonsit *n*
part, to ~ abstechen, brechen, sich lösen, scheiden, teilen, trennen, zerreißen
part Anteil *m*, Beitrag *m*, Bruchteil *m*, Glied *n*, Organ *n*, Teil *m* in ~ auszugsweise, teilweise to make a ~ of eingliedern ~ of apparatus Apparatteil *m* ~ of a circle ring Kreisringstück *n* ~ of construction Einbaustück *n* ~ of the front Frontabschnitt *m* ~ of the hair Scheitel *m* ~ of installation material Installationsteil *m* ~ of load Lastanteil *m* ~ by measure Maßteil *m* ~ of a mold Formteil *m* as ~ payment abschlägig ~ of prospected volume subject to current (or displacement lines) Aufschlußraum *m* any ~ put (or set) on Aufsatz *m* ~ of scale Skalateil *m* ~ of series of lines forming a band Bandenzweig *m* ~ of stress Spannungsanteil *m* ~ of structure Einbaustück *n* ~ of trajectory with pull-out effect Abfangbahn *f* ~ of underframe Untergestellteil *n* ~ by volume Raumteil *n* ~ by weight Gewichtsteil *n*
part, ~-checking department Teileprüferei *f* ~ chill roll Halbhartwalze *f* ~ cut Vorschnitt *m* ~ distributor Einzelhändler *m* ~ drawing Detailskizze *f*, -zeichnung *f* ~ load Teillast *f* ~ number Teilenummer *f* ~-open throttle teilweise geöffnete Drossel *f* ~ owner Miteigentümer *m*, Teilhaber *m* ~ payment Abschlagzahlung *f*, Anzahlung *f*, Teilzahlung *f* ~ pressure Teildruck *m* ~ printing Teildruck *m* ~-sectioned drawing Teilschnittzeichnung *f* ~-sectional view teilweise Ansicht *f* in Schnittdarstellung, teilweise im Schnitt dargestellte Ansicht *f* ~ subject to wear Verschleißteil *m* ~ throttle Sparlauf *m*, Teillast *f* ~-time employment Halbtagsbeschäftigung *f*
partake, to ~ sich beteiligen to ~ in a competition an einem Wettbewerb *m* teilnehmen
parted casing gebrochene oder gerissene Verrohrung *f*
partial Teilstrom *m*, Teilton *m* (acoust.); partiell, teilweise ~ acceleration Teilchenbeschleunigung *f* ~ acceptance Teilakzept *n* ~ admission impulse turbine beaufschlagte Gleichdruckturbine *f* ~ alienation Teilenteignung *f* ~ blackout geschränkte Beleuchtung *f* ~ bulkhead Halbschott *n* ~ capacity Teilkapazität *f* ~ carbonization Halbverkokung *f*, Tiefvergasung *f*, (of coal, peat, etc.) ~ carbonization of coal (peat, etc.) Halbkokung *f* ~ carbonization at a low temperature without admission of air Urverschwelung *f* ~ carbonizer Schwelzylinder *m* ~ carry Teilübertrag *m* ~ cementation Einsatzerhärtung *f* ~ color (images) Farbauszüge *pl* ~ combustion chamber Teilbrennkammer *f* ~ cross sections Partialquerschnitte *pl* ~ damage Teilbeschädigung *f* ~ delivery Teilförderung *f* ~ detonation

**of charge** Sitzenbleiben *n* eines Teiles einer Ladung beim Sprengen ~ **discharge** Partialfunke *m* ~ **dislocation** Teilversetzung *f* (unvollständige Versetzung) ~ **drag** Teilwiderstand *m* ~ **entropy** Partialentropie *f* ~ **error** Teilfehler *m* ~ **expropriation** Teilenteignung *f* ~ **extension** Teilausbau *m* ~ **feed** Teilförderung *f* ~**-flow lubrication-oil filter** Nebenstromschmierölfilter *m* ~ **force** Teilkraft *f* ~ **fraction** Teilbruch *m* ~ **fraction rule** Partialbruchregel *f* ~ **frequency** Teilfrequenz *f* ~ **function** Zustandssumme *f* ~ **image** Teilbild *n* ~ **liquefaction** Teilverflüssigung *f* ~ **load** Teillast *f* ~**-load region** Teillastgebiet *n* ~ **miscibility** Teilmischbarkeit *f* ~ **multiple** Teilvielfachfeld *n* ~ **node** angenäherter Knotenpunkt *m* ~ **note** Teilton *m* ~ **overhaul** Teilüberholung *f* ~ **overtone** Teilton *m* ~ **packing** Rippenversatz *m* ~ **pitch** Teil(wicklungs)schritt *m* ~ **pitch at the commutator** Teilwicklungsschritt *m* ~ **pressure** Partialdruck *m*, Teilspannung *f* ~ **process** Teilversorgung *m* ~**-pyritic process** Halbpyritschmelzen *n* ~**-read pulse** Teilleseimpuls *m* ~ **report** Teilbericht *m* ~ **restoring time** Teilnachwirkzeit *f*, (echo suppression) Teilsperrzeit *f* ~ **result** Teilerfolg *m* ~ **scan** Teilabtastung *f* ~**-scan impulse** Zeilenfolgeimpuls *m* ~ **scanning pattern** Teilraster *m* ~**-select output** teilweise selektive Ausgabe *f* ~ **shadow** Halbschatten *m* ~ **short circuit** teilweiser Kurzschluß *m* ~ **solidification** Teilerstarrung *f* ~ **solubility** beschränkte Löslichkeit *f* ~ **sphere** Kugelschale *f* ~ **splice** Teilstoß *m* ~ **sprinkler system** Teilsprinkleranlage *f* ~ **stowing** Rippenversatz *m* ~ **success** Teilerfolg *m* ~ **suppression of high frequencies** Benachteiligung *f* der hohen Frequenzen ~ **thread angle** Teilflankenwinkel *m* ~ **tide** Partialtide *f* ~ **tone reversal** teilweise Tonumkehrung *f* ~ **vacuum** Unterdruck *m*, Vorvakuum *n* ~ **view** Teilansicht *f* (aviat.) ~ **voltage** Teilspannung *f* ~ **volume** Volumteil *m* ~ **vortex** Teilwirbel *m* ~ **wave** Kopplungs-, Kupplungs-, Teil-welle *f* ~ **wave solutions** Partialwellenlösungen *pl* ~ **width** partielle Niveaubreite *f* ~**-write pulse** Teilschreibimpuls *m*

**partially** auszugsweise ~ **carbonized lignite** Braunkohlenhalbkoks *m* ~ **controlled (sketched) mosaic** Bildplanskizze (aviat.) ~ **occupied band** teilweise besetztes Energieband *n* ~ **throttled** gedrosselt, mit verminderter Drosselöffnung *f*

**participant** Teilnehmer *m*

**participate, to** ~ mitarbeiten, sich beteiligen, teilhaben, teilnehmen

**participating preference share** gewinnbeteiligte Vorzugsaktie *f*

**participation** Beteiligung *f*, Mitwirkung *f*, Teilnahme *f*

**particle** Körperchen *n*, Korpuskel *n*, Massenpunkt *m*, Partikel *f*, Spur *f*, Stoffteil *m*, Stoffteilchen *n*, Teilchen *n* ~ **accelerator** Teilchenbeschleuniger *m* ~ **aspect** Teilchenbild *n* ~ **charge** Teilchenladung *f* ~ **density** Teilchendichte *f* ~ **detection** Teilchennachweis *m* ~ **displacement** Schallausschlag *m* ~ **flow in configuration space** Teilchenstrom *m* im Konfigurationsraum ~ **gyro** Kern-, Partikel-kreisel *m* ~ **impact** Teilchenstoß *m* ~ **patch** Teilchen-

**bahn** *f* ~ **radiation** Korpuskularstrahlung *f* ~ **size** Körnung *f*, Teilchengröße *f* ~**-size distribution** Korngrößenverteilung *f* ~ **size reduction** Zerkleinerungsvorgang *m* ~ **sizes** Partikelgrößen *pl* ~ **state** Teilchenzustand *m* ~ **static** Störung *f* durch Partikeln ~ **velocity** Maßeteilchen-, Teilchen-geschwindigkeit *f* **particles** Schwebeteilchen *n* ~ **in suspension** Schwebeteilchen *n* ~**-structure** Teilchenstruktur *f*

**particular** Einzelheit *f*, Umstand *m*; ausgeprägt, ausgezeichnet, besonder, besonders ~ **solution** partikuläre Lösung *f* (von Diff. Gleichg.)

**particularity** Besonderheit *f*, Genauigkeit *f*, besonderer Umstand *m*

**particularize, to** ~ ausführlich angeben

**particulars** Angaben *pl*, Näheres *n*

**particulate solid** Feststoffpartikel *n*

**particulates** Makroteilchen *n*

**parting** Ablösung *f*, Bruch *m*, Formsand (Gießerei) *m*, Riß *m*, Scheiden *n*, Scheidung *f*, Teilfuge *f*, Trennung *f* ~ **of a mold** Formscheidung *f* ~ **of the rolls** Öffnung *f* des Kalibers

**parting,** ~ **case** Schachtscheider *m* ~ **compound** Trennmittel *n* ~ **device** Scheidevorrichtung *f* ~ **force** Trennungsebene *f* ~ **furnace** Scheideofen *m* ~ **glass** Scheidegefäß *n* ~ **kettle** Scheidekessel *m* ~ **line** Teil-fuge *f*, -linie *f*, Trennfuge *f*, Trennlinie *f*, Trennungslinie *f* ~ **medium** Einstaubmittel *n* ~**-off** Trennschleifen *n* ~**-off operations** Abstecharbeiten *pl* ~ **place** Trennstelle *f* ~ **plane** Trennungsfläche *f* ~ **sand** Streusand *m*, trockener Formsand *m* oder Sand *m* ~ **silver** Scheidesilber *n* ~ **test** Trennversuch *f* ~ **tool** Abstechwerkzeug *n* (aviat.), Einstichstahl *m* ~ **wall** Scheidewand *f* ~ **work** Scheideanstalt *f*

**partings of material in fatigue** Ausrisse *pl*

**partition, to** ~ abfachen, ab-, auf-teilen, verschlagen **to** ~ **by means of bulkheads** abschotten

**partition** Back *f*, Feld *n*, Querwand *f*, Regal *n*, (wall) Scheidewand *f*, Schott *n*, Teilung *f*, Unterteilung *f*, Verschlag *m*, Verteilung *f*, Wand *f*, Wandung *f*, Zwischen-boden *m*, -wand *f* ~ **of boards** Bretterverschlag *m*

**partition,** ~ **coefficient** Konzentrationsverhältnis *n* ~ **constant** Verteilungskonstante *f* ~ **function** Verteilungsgesetz *n* ~ **noise** Stromverteilungs-rauschen *n* ~ **opening** Schottöffnung *f* ~ **plate** Verteilungsplatte *f* ~ **port** Schottöffnung *f* ~ **ratio** (tube) Stromübernahmeverhältnis *n* ~ **timber** Einstrich *m*, Scheidewand *f* ~ **wall** Scheidemauer *f*, Trennungswand *f*, Versatzmauer *f* ~ **wall of xylolite** Xylolithwand *f*

**partitioned** zerteilt ~ **radiator** Scheidenkühler *m*

**partitioning** Unterteilung *f*

**partly,** ~ **automatic** teilautomatisch ~ **automatic rectifier** teilautomatisches Entzerrungsgerät *n* ~ **double preselection** teilweise doppelte Vorwahl *f* ~ **finished article** Zwischenerzeugnis *n* ~ **formed bottle** Külbel *m* ~ **machined** teilweise bearbeitet ~ **motorized** teilmotorisiert ~ **peat and partly mineral** (soil) anmoorig ~ **printed sheets** kopflose Zeitung *f* ~ **satisfied** halbgesättigt

**partner** Teilhaber *m* ~ **in a firm** Mitinhaber *m* ~ **in a triple collision** Dreierstoßpartner *m*

**partnership** Mitbeteiligung *f*, Partnerschaft *f*, Teilhaberschaft *f* ～ **with limited liability** Gesellschaft *f* mit beschränkter Haftung (G.m. b.H.) ～ **contract** Gesellschaftsvertrag *m*

**parts, in** ～ heftweise ～ **for bicycles** Fahrradbestandteile *pl* ～ **of breech mechanism** Schloßteile *pl* ～ **in contact with liquid** flüssigkeitsberührte Teile *pl* ～ **of the drive** Antriebsteile *pl* ～ **of the plant** Anlageteile *pl*

**parts,** ～ **catalog(ue)** Ersatzteilkatalog *m* ～ **lists** Bestandteillisten *pl*

**party** gerufener Teil *m*, Teilhaber *m*, Teilnehmer *m* ～ **to contract** Vertragspartei *f*

**party,** ～ **antenna** Gemeinschaftsantenne *f* ～ **concerned (or interested or involved)** Beteiligter *m* ～ **line** Gemeinschaft-, Gesellschafts-anschluß *m*, Neben-, Sammel-anschluß *m* ～-**line system** Gruppenstellensystem *n* (teleph.) ～ **reception** Blockempfang *m* ～-**wall** Zwischenmauer *f*

**Pascal's,** ～ **law** Pascalsches Gesetz *n* ～ **straight line** Pascalsche Gerade *f*

**pass, to** ～ aufkommen, durchfließen, durchlassen, durchstreichen, fällen, gehen, hinausgehen, hindurchlassen, hinwegstreichen (across), streichen, strömen, überholen, überschreiten, verbringen, vergehen (time), verlaufen, vorbei-fahren, -streichen, -streifen, -ziehen

**pass, to** ～ **along** entlangstreichen (of gases), weitergeben **to** ～ **around** umspülen **to** ～ **away** fortschaffen **to** ～ **back** (in rolling) rückführen **to** ～ **a booking** eine Gegenanmeldung *f* übermitteln **to** ～ **by** verstreichen, vorbeikommen **to** ～ **a call** eine Gesprächsanmeldung *f* übermitteln **to** ～ **a call again** eine Gesprächsanmeldung *f* weitermelden oder weitergeben **to** ～ **to the debit** anrechnen **to** ～ **an examination** bestehen **to** ～ **in** einleiten, fließen **to** ～ **in and out** durchstreichen **to** ～ **inspection** von Kontrolle *f* abgenommen werden, von Kontrolle *f* abnehmen **to** ～ **into** einströmen, fließen **to** ～ **judgment** absprechen **to** ～ **judgment against** verurteilen **to** ～ **off** abgehen, ableiten, abströmen, abziehen (gas), entweichen, fortschaffen **to** ～ **on** weiterleiten **to** ～ **out** ausströmen **to** ～ **over** aushändigen, durcheilen, leiten, überführen, übergehen, überleiten **to** ～ **in review** defilieren **to** ～ **through** durch-fahren, -fallen, -heilen, -gehen, (wire) -führen, -laufen, -leiten, -schreiten, -sinken,-strömen, durchziehen, hindurch-gehen, -laufen, -streichen, -strömen **to** ～ **through the muffle** einbrennen **to** ～ **up** ansteigen, aufsteigen, emporsteigen **to** ～ **a vessel through a lock** ein Schiff *n* durchschleusen **to** ～ **the word** durchrufen, durchsagen **to** ～ **work over to** Arbeit *f* weitergeben an

**pass** Absinken *n*, Ausweis *m*, Ausweiskarte *f*, Durchgang *m*, Durchgangsschein *m*, Durchlaßschein *m*, (in rolling) Einstich *m*, Erlaubnisschein *m*, Fadeneinzug *m*, Freipaß *m* (for admission), Furche *f*, Gang *m*, Geschirreinzug *m*, Kaliber *n*, Paß *m*, Passage *f*, (mountain) Sattel *m*, Schafteinzug *m*, Schweißlage *f*, Spur *f*, Stich *m*, Urlaubsschein *m*, Walzenfurche *f*, Walzenkaliber *n* Walzstich *m* (iron and steel mfg.), (of boiler) Zug *m* ～ **of journey** Hinlauf *m* ～ **of operation** Arbeitsgang *m* ～ **of welding material** Schweißlage *f*

**pass,** ～ **band** Durchlässigkeitsbereich *m* (eines Filters) ～ **direction** Durchlaßrichtung *f* ～ **key** Nachschlüssel *m* ～-**out branch** Entnahmestutzen *m* ～-**out operation** Kondensationsbetrieb *m* ～-**out point** Anzapfpunkt *m* ～-**out steam** Entnahmedampf *m* ～-**out turbine** Anzapfturbine *f* ～ **range** Durchlässigkeitsbereich *m* eines Filters ～ **template** Blech-, Kaliber-, Walz(en)-schablone *f* ～-**template drawing** Kaliberschablonen-, Kalibrierungs-zeichnung *f* ～ **through capacitor** Durchschleifungskondensator *m* ～ **wave** Durchlaßfrequenz *f*

**passable** befahrbar, brauchbar, fahrbar, gangbar, leidlich

**passage** Ausflußkanal *m*, Durchfahrt *f*, Durchfluß *m*, Durchgang *m*, Durchlaß *m*, Durchzug *m*, Gang *m*, Steg *m*, Stelle *f*, Stelle in einem Buch, Stromübergang *m*, Überfahrt *f*, Übergang *m*, Umführung *f*, Verbindungsgang *m*, Weg *m* **(air)** ～ Luftweg *m* **(of beta rays)** Fensterdurchlässigkeit *f* ～ **of cards through machine** Kartendurchlauf *m* ～ **of carriage** Schlittendurchgang *m* ～ **of current** Stromdurchgang *m* ～ **of the goods** Warendurchgang *m* ～ **of heat** Wärmeübergang *m* ～ **to the limit** Grenzübergang *m* (chem.) ～ **of a line through a town** Linienführung *f* durch eine Ortschaft ～ **for the plug** Aussparung *f* für die Zündkerze, Zündkerzenöffnung *f* ～ **for steam** Dampfkanal *m* ～ **through** (of a film) Durchlauf *m*

**passage,** ～ **grid controllance** Durchgriff *m* ～ **section** Durchgangsprofil *n* ～ **width** Durchgangsweite *f*

**passed** geprüft ～ **through** hindurchgeführt

**passenger** Fahr-, Flug-gast *m*, Passagier *m* **to be a** ～ mitfahren

**passenger,** ～ **airplane** Personenflugzeug *n* ～ **automobile** Personenwagen *m* ～ **berth** Fahrhafen *m* ～ **cabin** Fluggast-kabine *f*, -raum *m* ～ **car** Automobil *n*, Personenkraftwagen *m*, Verkehrsfahrzeug *n* ～ **coach** Personenwagen *m* ～-**communication apparatus** Zugsignalvorrichtung *f* ～ **ferry** Personenfähre *f* ～ **flight** Passagierflug *m* ～ **flying** Personenluftverkehr *m* ～ **flying boat** Verkehrsflugboot *n* ～ **hall** Schalterhalle *f* ～ **hop** kurzer Passagierflug *m* ～ **kilometer** Passagierkilometer *m* ～ **landing** Landungsponton *n* ～ **list** Passagierliste *f* ～ **locomotive** Personenzuglokomotive *f* ～ **plane** Passagier-, Verkehrs-flugzeug *n* ～ **ramp** Flugsteig *m* ～ **rate** Passagiersatz *m* ～ **seat** Gastsitz *m* ～ **service** Beförderung *f* von Personen, Personenbeförderung *f* ～ **ship** Passagierschiff *n* ～ **station** Personenbahnhof *m* ～ **steamer** Passagierdampfer *m* ～ **traffic** Personenbeförderung *f*, -verkehr *m* ～ **train** Personenzug *m* ～ **transport** Personenbeförderung *f* ～ **vehicle** Personenwagen *m*

**passimeter** Fahrkatendruckmaschine *f*

**passing** hinausgehend, vorübergehend ～ **through** durchgehend ～ **through iron** Eisenschluß *m* ～ **the spectrum in review** Durchmusterung *f* des Spektrums ～ **through the velocity-range of sound** Schalldurchgang *m*

**passing,** ～ **axle** durchgehende Achse *f* ～ **below** Unterschreitung *f* ～ **color** Übergangsfarbe *f* (chem.) ～ **flight** Vorbeiflug *m* ～ **light** Backbord-Positionslampe *f* ～-**over** Übergang *m* ～

**place** Ausweicheplatz *m*, Ausweichstelle *f*, Ausweichung *f*, Kreuzungsstelle *f* ~ **siding** Überholungsgleis *n* ~ **signal** Überholmelder *m* ~ **tenon** durchgehender Zapfen *m* ~ **threads into combs** Blattstechen *n*

**passivate, to** ~ (surface against corrosion) Oberfläche *f* zum Schutz gegen Korrosion unangreifbar machen

**passivation** Passivierung *f*

**passive** passiv ~ **air defense** passiver Luftschutz *m* ~ **balance of trade** passive Handelsbilanz *f* ~ **resistance** Übergangswiderstand *m* ~ **transducer** passiver Wandler *m*

**passivity** Passivität *f*

**passometer** Passometer *n*, Schrittmesser *m*

**passport** Paß *m*

**paste, to** ~ auf-, be-kleben, (pictures) aufziehen, kleistern, kleben, pappen, pastieren, verkleben, verkleistern (up) **to** ~ **on** ankleben, bekleben **to** ~ **up** aufkleben **to** ~ **upon copies** Abzüge *pl* aufkleben

**paste** Abguß *m*, Brei *m*, Klebstoff *m*, Kleister *m*, Leim *m*, Papp *m*, Paste *f*, Teig *m*, (ceramics) Tonmasse *f*, Weichenpfahl *m* ~ **atomizer** Pastenzerstäubungstrockner *m*

**pasteboard** Karton *m*, Pappdeckel *m*, Pappe *f*, Pappenart *f* ~ **box** Karton *m*, Kartonage *f* ~**-box stabbing machine** Kartonagenheftmaschine *f* ~ **card** Pappkarte *f* ~ **casting** Pappengußmaschine *f* ~ **converting** Pappebearbeitung *f* ~ **cutter** Pappschere *f* ~**-making machine** Pappenmaschine *f* ~ **work** Papparbeit *f*

**paste,** ~ **brush** Kleisterstreichpinsel *m* ~ **cathode** Pastekathode *f* ~ **former** Pastenverformer *m* ~ **forming dehydrator** Verformungseinrichtung *f* ~ **knife** Pappmesser *n* ~ **paint** Pastenfarbe *f* ~ **pot** Kleistertopf *m* ~ **resin** Harzpaste *f* ~ **resist** Pappreserve *f* ~ **resist printed goods** Pappdruckware *f* ~ **soap** Leimseife *f* ~ **tube** Quetschtube *f* ~**-up** Umbruch *m*

**pasted** geklebt ~ **cathode** Pastekathode *f* ~ **plate** geschmierte oder pastierte Platte *f*, Masseplatte *f* ~ **squares** Massefelder *pl*

**pastel** Pastell *n*, Pastell-stift *m*, -ton *m*; pastellfarbig ~ **crayon** Pastellstift *m* ~ **work** Pastellmalerei *f*

**pastern** Fessel *f*

**pasteurize, to** ~ pasteurisieren

**pastilles** Massefelder *pl*

**pasting** Anreiben *n*, Verleimung *f* ~ **of silks** Verkleisterung *f* von Seide

**pasting,** ~ **agent** Anteigmittel *n* ~**-on** Einkleben *n* ~ **plate** Klebeschild *n* ~ **seam** Klebenaht *f*

**pasty** breiartig, breiig, klebrig, pappig, pastenartig, pastös, teigartig, teigig ~ **iron** teigiges Eisen *n*

**patch, to** ~ ausbessern, ausflicken, flecken, flicken, überlochen (info proc.), zusammenschalten **to** ~ **up** bosseln

**patch** Anwurf *m*, Befestigungspflaster *n*, Butz *m*, Butze *f*, Butzen *m*, Fleck *m*, (road) Flickstelle *f*, Gänsefuß *m*, Heftpflaster *n*, Korrekturbefehl (comput.), Lappen *m*, Pflaster *n* (for cordage, of airship) ~ **(ing)** Flicken *n* ~ **for screw bungs** Dichtungsläppchen *n* für Spunde

**patch,** ~ **area** Fleckengebiet *n* ~ **board** Schalttafel *f* (comput.) ~ **cord** Rangiersteckerschnur

*f* ~ **effect** Fleckeffekt *m* ~**-line base** Flickstandlinie *f* (photo) ~ **panel** Klinkenfeld *n* ~ **system** Lappenbefestigung *f* ~ **theory** Fleckentheorie *f* ~ **work** Flickarbeit *f*

**patched** geflickt ~ **slate** Fleckschiefer *m*

**patching** Ausbesserung *f*, Flickerei *f*, Rangierring *f* (electr.), Schleifen *n* (von Leitungen) **(re)** ~ **cord** Umschalteschnur *f*, Vermittlungsschnur *f* ~ **material** Flickmasse *f* ~ **work** Flickarbeit *f*

**patent, to** ~ patentieren

**patent** Erfinderpatent *n*, Patent *n*, Patentschrift *f* **to be** ~ auf der Hand liegen, naheliegen ~ **applied for** angemeldetes Patent *n* ~ **has been granted** Patent *n* ist erteilt worden ~ **of improvement** Verbesserungspatent *n* ~ **for invention** Erfindungspatent *n* ~ **in issue** Klagepatent *n* ~ **and printed copies issued** ausgegeben ~ **in suit** Klagepatent *n* ~ **of utilization** Gebrauchsmusterschutz *m*

**patent,** ~ **action** Patent-klage *f*, -streit *m* ~ **agent** Patentanwalt *m* ~ **allowance** Patenterteilung *f* ~ **anticipation** Patentvorwegnahme *f* ~ **anticipatory reference** Patentvorwegnahme *f* ~ **appeal** Patenteinspruch *m* ~ **application** Patentanmeldung *f*, -gesuch *n* ~ **attorney** Patentanwalt *m* ~ **closure** Patentverschluß *m* ~ **code** Patentgesetz *n* ~ **fastening** Patentverschluß *m* ~ **fee** Patentgebühr *f* ~ **fuel** Preßkohle *f* ~ **gazette** Patentblatt *n* ~ **grant** Patenterteilung *f* ~ **grid** Patentrost *n* ~ **infringement suit** Patentverletzungsverfahren *n* ~ **infringing** patentverletzend ~ **law** Patentgesetz *n* ~ **lead pencil** Füllbleistift *m* ~ **leather** Glanleder *n* ~ **leather effect** Lackeffekt *m* (auf Leder) ~ **letter** Patenturkunde *f* ~ **litigation** Patentstreit *m* ~ **lock** Patentverschluß *m* ~ **log** Patentlog *n* ~ **medicine** Arzneifertigware *f* ~ **office** Patentamt *n* ~**-office drawing** Patentzeichnung *f* ~ **office examination procedure** Prüfungsverfahren *n* ~**-office notification (or official notice)** amtlicher Bescheid *m* ~ **pending** angemeldetes Patent *n* ~**-renewal fee** Patentjahres-, Patentverlängerungs-gebühr *f* ~ **right** Patentrecht *n* ~ **rights** Schutzrechte *pl* ~ **rolls** Patent-register *n*, -rolle *f* ~ **slip** Aufschlepphelling *f* ~ **specification** Patent-beschreibung *f*, -schrift *f* ~ **statutes** Patentgesetz *n* ~ **suit** Patentklage *f* ~ **survey** Patentuntersuchung *f* ~ **tax** Patentgebühr *f* ~**-welded casing** patentgeschweißte, verschraubte Bohrröhre *f* ~ **yellow** Montpelliergelb *n*

**patentability** Patentierbarkeit *f*

**patentable** patentfähig, patentierbar

**patented** gesetzlich geschützt, patentiert ~ **at home and abroad** im In- und Ausland *n* patentiert

**patented,** ~ **procedure** gesetzlich geschütztes Verfahren *n* ~ **provisionally** (trademark or model filed) eingetragenes Gebrauchsmuster *n* ~ **steel wire** Patentdraht *m*

**patentee** Patentinhaber *m*

**patenting** Bleihärten *n*, Patentieren *n*, Perlitisieren *n* (wire)

**paternoster,** ~ **bailing work** Paternosterschöpfwerk *n* ~ **work** Noria *f*

**path** Bahn *f*, Gehweg *m*, Laufbahn *f*, Pfad *m*, Verlagerung *f*, Verlauf *m*, Weg *m*

path, ~ of armature winding Ankerzweig *m* ~ of
blade element Luftschraubengang *m* ~ of contact Eingriffslinie *f* (Zahnrad) ~ of contact of
rolling Walzbahn *f* ~ of current Stromweg *m*
~ of the depression Depressionsbahn *f* ~ of the
flames Flammenweg *m* ~ of flight Flugweg *m*,
Zielweg *m* ~ of integration Integrationsweg *m*
~ of interference wave Störungswellenweg *m*
~ of light ray (beam) Lichtstrahlengang *m* ~ of
lightning Blitzbahn *f* ~ of lines of force Kraftlinienverlauf *m*, -weg *m* ~ of liquid Flüssigkeitsbahn *f* ~ of motion Bewegungsbahn *f* ~ of
percolation Sickerweg *m* ~ of pilot balloon
Pillotballon-bahn *f*, -kurve *f*, -linie *f* ~ of projectile Flugbahn *f*, Schußkanal *m* ~ of propagation Ausbreitungsweg *m* ~ of rays Strahlengang *m*, Strahlverlauf *m* ~ of rest Ruheweg *m*
~ of spark Funkenbahn *f* ~ of stylus Schreibstiftweg *m* ~ of waves Wellenpfad *m* ~ of wind
Windbahn *f* ~ of winding Ankerzweig *m* ~ in
woods Holzweg *m* ~ of work Arbeitsweg *m*
path, ~ angle Bahnneigung *f*, Bahntangentenneigungswinkel *m* ~ available for actual filling
Füllweg *m* ~ closed through iron Eisenschluß *m*
~ computation Bahnvermessung *f* ~-controlled
weggesteuert ~ cut through a forest Durchhau
*m* ~ difference Gangunterschied *m* ~ length
Schichtlänge *f* ~-length difference Wegdifferenz
*f* ~ line Bahnkurve *f* ~ normal bahnsenkrecht
~-reversal principle Bahnumkehrprinzip *n* ~
way Laufsteg *m*
pathless unwegsam
pathogenic krankheitserregend
patina Altersfärbung *f*, Edelrost *m*, Patina *f*
patinate, to ~ patinieren (to coat with patina)
patrol Patrouille *f*, Runde *f*, Spähtrupp *m*, Streife
*f* ~ airplane Späh-, Wach-flugzeug *n* ~ boat
Bewacher *m*, Patrouillenschiff *n*, Vorpostenboot *n* ~ vessel Wachboot *n*
patrolling Begehen *n*, Begehung *f* ~ vessel Bewachungsfahrzeug *n*
patron Lehrherr *m*
patronage Schirmherrschaft *f*
patten Latsche *f* ~ of rail Fuß *m* der Schiene
pattern, to ~ ausführen, kopieren, nachbilden
pattern Anreißschablone *f*, Art *f*, Ausführung *f*,
Dessin *n*, Drehbrett *n*, Entwurf *m*, Exemplar *n*,
Lehre *f*, Modell *n*, Muster *n*, Patrone *f*, Probe *f*,
Schablone *f*, Schema *n*, Streichmodell *n*, Verlauf *m*, Vorbild *n*, Vorlage *f* ~ of coordination
Koordinatenbild *n* ~ with great intricacy of
detail verwickelt gestaltetes Modell *n* ~ of
intricate external shape verwickelt gestaltetes
Modell *n* ~ of motion Bewegungsbild *n* ~ of
proof firing Beschußbild *n*
pattern, ~ box Musterkästchen *n* ~ bracket elevating screw Schablonentischspindel *m* ~
bracket screw Schablonenbockspindel *m* ~
card Muster-brief *m*, -karte *f* ~-card winding
machine Musterkartenwickelmaschine *f* ~
chain Schaltkette *f* ~ chain drum Kettenrad *n*
~ cut Musterschliff *m* ~ cutting of chilled castiron rolls Riffeln *n* von Hartgußwalzen ~
design Modellentwurf *m* ~ disc housing Trommelböckchen *n* ~ distortion Ablenkfehler *m*
(CRT) ~ draw Modell-abhebung *f*, -abhub *m*,
-aushebung *f* ~ draw mechanism Abhebevorrichtung *f* ~-draw turn-over molding machine

Abhebeformmaschine *f* mit Wendeplatte ~
drawing Modellaushebung *f* ~-drawing contrivance Modellabhebevorrichtung *f* ~-drawing
operation Modellabhebung *f* ~-drawing piston
Modellabhebekolben *m* ~ drum Mustertrommel *f* ~ folder Patronenfalzerin *f* ~ frame
Modellrahmen *m* ~ gear Formmodellrad *m* ~
lever Musterhebel *m* ~ lift Modell-abhub *m*,
-aushub *m* ~ lifting Modellaushebung *f* ~
maker Modell-macher *m*, -schlosser *m* ~
maker's rule Schwindmaß *n* ~ making Modellherstellung *f* ~ making practice Formtechnik *f*
~ making shop Modellbauwerkstätte *f* ~
modulation Bildmodulation *f* ~ molding Formen *n* nach Modell, Formerei *f* ~ movement
register Schieberegister *n* ~ paper Malblatt *n*,
Patronenpapier *n* ~ plate Form-, Modellplatte *f* ~ plate with half patterns mounted on
both sides Relief-, Wende-platte *f* ~-plate
making Modellplattenherstellung *f* ~ practice
Modelltechnik *f* ~ printing Dessindruck *m* ~
selector Musterwähler *m* (am Maßgerät) ~
selector lever Musterstopperhebel *m* ~ set
Modellgarnitur *f* ~ sheet gemusterter Deckbogen *m*, Mustertafel *f* ~ shop Fräs-, Modellwerkstatt *f* ~ signal generator Bildmustergenerator *m* ~ stationary stehende Figur *f*
(CRT) ~ storage Modellboden *m* ~ stud
Schaltstift *m* ~ table Modelltisch *m* ~ turnover
arrangement (or device) Wendevorrichtung *f* ~
wheel Formmodellrad *m*
patterning mechanism Mustereinrichtung *f*
pattinsonize, to ~ pattinsonieren
paucity geringe Anzahl *f*
Pauli exclusion principle Pauli-Fermi-Prinzip *n*
pause, to ~ innehalten, eine Pause einlegen
pause Absatz *m*, Anhalt *m*, Anstand *m*, Haltezeit *f*, Pause *f*
pave, to ~ auskleiden, pflastern, stampfen, verkleiden to ~ the way bahnen
paved gepflastert ~ road gepflasterte Straße *f* ~
runway befestigte Startbahn *f* oder Start- und
Landebahn *f* ~ slope gepflasterte Böschung *f*
~ taxiway befestigte Rollbahn *f*
pavement Belag *m*, Decke *f*, Fußbodenbelag *m*,
Pflaster *n*, Pflasterung *f*, Straßenpflaster *n* ~
of paving tiles Plattenbelag *m* ~ of small
cobblestones in interlocking pattern Kleinpflaster *n*
pavement, ~ breaker Aufreißhammer *m* ~
distress Deckenschaden *m* (Straße) ~ work
Pflasterarbeit *f*
paver Dammsetzer *m*, Steinsetzer *m* (floor) ~
Plattenleger *m*
paving Abpflasterung *f*, Belag *m*, Deckplattenbelag *m*, Fußbodenbelag *m*, Pflaster *n*, (a road)
Überschütten *n* ~ of a bank Böschungspflaster
*n* ~ of the elliptic plane Pflasterung *f* der elliptischen Ebene ~ of the Euclidean plane Pflasterung *f* der euklidischen Ebene ~ of the Euclidean
space Pflasterung *f* des euklidischen Raums
paving, ~ asphalt Pech *n* für das Pflastern der
Straßen ~ brick Pflaster-stein *m*, -ziegel *m* ~
coat Straßendecke *f* ~ material Pflasterungsmaterial *n* ~ stone Pflasterstein *m*, Uferpflaster
*n* ~ tamp Erdstampfer *m* ~ tile Bodenziegel *m*
pavior Dammsetzer *m*
paw Klaue *f*, Pfote *f*, Tatze *f*

**pawl** Absperrklaue *f*, Klaue *f*, Klinke *f*, Mitnehmerklinke *f*, Rastklinke *f*, Schaltklinke *f*, (locking) Sperrhaken *m* Sperrhebel *m*, (locking) Sperrkegel *m*, (holding) Sperrklinke *f*, Vorstecker *m* ~ **with roller** Rollenklinke *f*
**pawl,** ~ **carrier** Klinkenträger *m* ~ **clip chain** Tasterklappenkette *f* ~ **coupling** Klinkenapparat *m*, -kupplung *f* ~ **drive** Sperrklinkenantrieb *m* ~ **holder** Klinkenhalter *m* ~ **latch voltage** Klinkenspannung *f* ~ **operating rod** Klinkenstange *f* ~ **pin** Klinkenbolzen *m* ~ **release** Klinkenauslösung *f* ~ **spring** Sperrhakenfeder *f* ~ **stop** Rücklaufsperre *f* ~ **stud** Sperrhakenstift *m* ~ **wheel** Sperrad *n*
**pawn, to** ~ verpfänden
**P.A.X.** (**private automatic exchange**) automatische Hauszentrale *f*
**pay, to** ~ bezahlen, herhalten, honorieren, pechen, zahlen **to** ~ **on account** anzahlen **to** ~ **additionally** zuzahlen **to** ~ **attention (to)** achtgeben, aufpassen **to** ~ **out a cable** ein Kabel abrollen **to** ~ **collect (or later)** nachbezahlen **to** ~ **duty** versteuern, verzollen **to** ~ **in full** ausbezahlen **to** ~ **out gradually** wegfieren (a cable) **to** ~ **in** einzahlen **to** ~ **out the log line** das Log werfen **to** ~ **off** abbezahlen, abmustern, auslohnen, durchlaufen (tape rec.) **to** ~ **an official call** einen dienstlichen Besuch *m* machen **to** ~ **out** abrollen, auskurbeln (antenna), auslegen, ausstechen (cable), ausstecken (cable), fieren (cable) **to** ~ **the surcharge** zuzahlen **to** ~ **wages** löhnen **to** ~ **out wire** Draht *m* abwickeln **to** ~ **out wire (or rope) from a reel** abfieren
**pay** Arbeitslohn *m*, Besoldung *f*, Gehalt *n*, Lohn *m*, Löhnung *f* ~-**and allowance administration** Besoldungswesen *n* ~-**as-you-enter bus** Einmannwagen *m* ~ **call** Löhnungsappel *m* ~ **day** Zahltag *m* ~ **dirt** goldführendes Erdreich *n* ~ **grade** Wehrsoldgruppe *f* ~ **gravel** Pinge *f* ~ **hop** zahlender Kurzflug *m* ~ **law** Besoldungsgesetz *n* ~ **load** zahlende Last *f* oder Nutzlast *f* ~-**load structure fuel weight ratio** Nutzlastaufbaukraftstoffverhältnis *n* (g/m) ~ **load volume** Nutzraumvolumen *n* (g/m) ~-**off spool** Ablauftrommel *f* ~ **ore** abbauwürdiges Erz *n* ~-**out** Rentabilität *f* ~-**out reel** Drahthaspel *f* ~ **regulation** Besoldungsvorschrift *f*
**payroll** Lohn-, Löhnungs-liste *f* ~ **accounting** Lohnabrechnung *f* ~ **master card** Matrizenlohnkarte *f* ~ **scale** Lohnverrechnung *f* ~ **tax** Lohnsummensteuer *f*
**pay,** ~ **sand** rentabler Sand *m* ~ **slip** Lohnzettel *m* ~ **station** Fernsprechamtautomat *m*, Münzfernsprecher *m*
**payable** bauwürdig, fällig, zahlbar ~ **to bearer** zahlbar an Überbringer *m* ~ **deposit** abbauwürdige Mächtigkeit *f* ~ **places** ergiebige Ausbeutungsplätze *pl*
**payed-out yardage** ausgefahrene Meterzahl *f*
**payee** Zahlungsempfänger *m*
**paying** zahlend ~ **of additional fare** Nachlösung *f* ~-**off** Entlohnung *f* ~-**off pendant** Heimatswimpel *m* ~ **out the line** Auslegen *n* der Leitung ~-**out brake** Ausstechbremse *f* ~ **out of cable** Ablaufen *n* des Kabels ~-**out drum** Auslegetrommel *f* ~-**out (reel) machine** Auslegemaschine *f*

**payment** Auszahlung *f*, Besoldung *f*, Bezahlung *f*, Einzahlung *f*, Zahlung *f* ~ **on account** Akontozahlung *f* ~ **to account** Zahlung *f* für Rechnung ~ **in advance** Vorauszahlung *f*, Vorschußzahlung *f* ~ **by anticipation** Vorauszahlung *f* ~ **in cash** Barzahlung *f* ~ **after a certain date** Nachtragszahlung *f* ~ **of dividend** Dividendenausschüttung *f* ~ **of duty** Verzollung *f* ~ **in installment** Abschlagzahlung *f* ~ **in installments** Ratenzahlung *f* ~ **by job** Gedingelohn *m* ~ **in kind** Naturalleistung *f* ~ **made on account** à conto Zahlung *f* ~ **on sugar-content basis** Bezahlung *f* nach dem Zuckergehalt ~ **on tonnage basis** Bezahlung *f* nach dem Gewicht ~ **by weight basis** Bezahlung *f* nach dem Gewicht
**pea** Erbse *f* ~ **attachment** Erbsensäeapparat *m* ~ **coal** Erbsen-, Erbs-, Grieß-, Klein-, Perlkohle *f* ~ **coke** Perlkoks *m* ~ **gravel** Kies *m* von Erbsengröße ~ **lamp** Zwerglampe *f* ~ **ore** Bohnerz *n* ~ **size** Erbsengröße *f*
**peach wood** Pfirsichholz *n*
**peacock coal** Glanzkohle *f*
**peak** Berg *m* (einer Kurve), Erhebung *f*, Gipfel *m*, Höhepunkt *m*, Kuppe *f*, (of a curve) Maximum *n*, Scheitel *m* einer Welle, (value) Scheitelwert *m*, Spitze *f*, (of microphone) Überschreien *n*; höchst ~ **of the curve** Kurvenscheitelpunkt *m* ~ **of preparedness** Schlagfertigkeit *f* ~ **of resistance** Widerstandhöhe *f* ~ **of traffic** Verkehrssritze *f* ~ **of a wave** Wellengipfel *m*
**peak,** ~ **anode current** Anodenspitzenstrom *m* ~ **anode forward voltage** Scheitelwert *m* der Anoden(vorwärts)spannung ~-**anode-voltage radio** Anodenspitzenspannung *f* ~ **capacity** Spitzenleistung *f* ~ **carrier amplitude** Oberstrich *m* ~ **cathode current** Kathodenspitzenstrom *m* ~ **cathode fault current** maximaler Kurzschlußstrom *m* ~ **clipper** Amplitudenbegrenzer *m* ~ **current** Höchst-, Scheitel-, Spitzen-strom *m* ~-**current generating station** Spitzenkraftwerk *n* ~ **demand (or discharge)** Spitzenentnahme *f* ~ **effective noise voltage** Spitzenstörspannung *f* ~ **efficiency** Gipfelleistung *f* ~-**envelope power** Spitzenleistung *f* ~ **factor** Scheitelfaktor *m* ~ **flux density** Spitzenflußdichte *f* ~ **forward anode-voltage** Anodenspitzenspannung *f* ~ **halyard** Gaffelfall *n* ~ **indicator** Höchstwert-, Maximumwert-anzeiger *m*; Spitzen-spannungszeiger *m*, -wertzeiger *m* ~ **intensity of light** maximale Lichtstärke *f* ~ **inverse anode voltage** Spitzenwert *m* der Sperrspannung ~ **inverse voltage** Sperrspannungsscheitelwert *m* ~ **limiter** Amplituden-abschneider *m*, -begrenzer *m* ~ **line voltage** Spitzenleitungsspannung *f* ~ **load** Hochstabilbelastung *f*, Höchst-, Spitzen-belastung *f* ~-**load operation** Spitzenlastfahren *n* ~-**load power station** Spitzenkraftwerk *n* ~ **load service** Spitzenlastdeckung *f* ~ **lopper** Amplitudenbegrenzer *m* ~ **magnetic field** Scheitelwert *m* ~ **magnetizing force** Spitzenmagnetisierungskraft *f* ~ **output** Spitzenleistung *f* ~ **overpressure** Spitzenüberdruck *m* ~ **performance** Arbeitshöhe *f* ~ **performance altitude** Arbeitsgipfelhöhe *f* ~ **power** Höchstleistung *f*, Spitzenleistung *f* ~ **power of transmitter** Antennenkreisleistung *f* ~ **power of the transmitter** Telegrafieleistung *f* ~ **power**

output Spitzenausgangsleistung *f* ~ **pressure**
Spitzendruck *m* ~ **pressure meter** Spitzendruck-
messer *m* ~ **program(me)meter** Pegelmesser *m*
~ **pulse amplitude** Amplitudenspitzenwert *m*
~-**reading (level) indicator** Spitzenanzeiger *m*
~ **reverse voltage** Umkehrspannung *f* ~ **reverse**
**voltage point** Umkehrpunkt *m* ~ **rider** Spitzen-
demodulator *m* ~-**shaving** Spitzenausgleich *m*
~ **sideband power (PSP)** Seitenbandspitzen-
leistung *f* ~ **signal voltage** Spitzennutzspannung
*f* ~ **sound pressure** maximaler Schalldruck *m*
~-**to-peak value** Spitzen-Spitzenwert *m* ~-**to-**
**valley ratio** Einsattlung *f* (TV) ~-**to-zero** Spitze-
zu-Null *f* ~ **torque** Spitzendrehmoment *n* ~
**traffic** Spitzenverkehr *m* ~ **value** Höchst-,
Spitzen-wert *m* ~ **velocity** Scheitelgeschwindig-
keit *f* ~ **voltage** Gipfel-, Scheitel-, Spitzen-
spannung *f* ~ **voltmeter** Impuls-, Scheitelwert-
spannungs-messer *m*
**peaked** scharf ~ **curve** spitze Kurve *f*
**peaker** Differenzierschaltung *f* ~ **strip** Differen-
zierspule *f*
**peaking,** ~ **circuit** Differenzierkreis *m* ~ **coil**
Entzerrerdrossel *f*, Entzerrspule *f* (TV) ~
**resistor** Impulswiderstand *m*
**peaks and dips** Spitzen *pl* und Einbrüche *pl*
**peal** Schall *m*
**pear** Birne *f* ~-**key** Birntaster *m* ~-**shaped**
birnenförmig ~-**shaped lamp (or bulb)** Birnen-
lampe *f* ~-**shaped lifting eye** Bügel *m*, Ladebügel
*m*, Lastbügel *m*, Schlaufe *f* ~ **switch** Birnen-
(aus)schalter *m* ~ **wood** Birnbaumholz *n*
**pearl** Perle *f* ~ **ash** Aschensalz *n*, Perlasche *f*
~-**beaded screen** Glasperlenleinwand *f* ~ **lamp**
innenmattierte Lampe *f* ~ **long stripes pattern**
Perllängsstreifenmusterung *f* ~ **ore** Perlerz *n* ~
**printing** Perldruck *m* ~ **sinter** Kiesel-, Perl-
sinter *m* ~ **spar** Braunkalk *m*, Perlspat *m* ~
**white** Permanentweiß *n*, Schwerspat *m* (paper
mfg.)
**pearlite** Austenit *n* ~ **area** Perlitinsel *f*
**pearlitic** perlitisch
**pearlitize, to** ~ perlitisieren (metall.)
**pearloid** perlitähnlich
**pearly** perlmutterartig ~ **constituent** Perlit *m* ~
**(zinc) gray** Perlgrau *n*
**peat** Heidetorf *m*, Moorboden *m*, Torf *m* ~
**ashes** Torfasche *f* ~ **backfilling** Torfpackung *f*
~ **bog** Torf-lager *n*, -moor *n* ~ **carcoal** Torf-
kohle *f* ~ **charring** Torfverkohlung *f* ~ **deposit**
Fehn ~ **digger** Torfstecher *m* **dug** ~ Handtorf
*m* ~ **dust** Torf-mehl *n*, -staub *m* ~ **excavator**
Torfbagger *m* ~ **fiber** Torffaser *f* ~ **litter** Torf-
mull *m*, -streu *f* ~ **locality** Fehn *n* ~ **moor** Torfmoor *n* ~ **slab** Torfplatte *f* ~ **sod** Torfsoden *m*
~(y) **soil** Torfboden *m* ~ **tar** Torfteer *m*
**peaty** torfartig
**pebble** (rayon mfg.) Gaufre *m*, Geröll *n*, Kiesel
*m*, Kieselstein *m* ~ **manganese** Braunstein *m*,
Mangansuperoxyd *n* ~ **mill** Kugelmühle *f* ~
**pavement** Kleinpflaster *n*
**pebbling** (in a sprayed film) Spritznarben *pl*
**pebbly** kieselig, körnig ~ **bottom** Kieselgrund *m*
~ **ground** Kegelgrund *m*, Keigrund *m*
**pecker** Abfühlnadel *f* ~ **block** Prallstück *n*
**pectic acid** Gallertsäure *f*
**pectine solution** Pektinlösung *f*
**peculiar** ausgeprägt, besonders, eigen, eigentüm-

lich, merkwürdig, seltsam, sonderbar ~ **flavor**
**(or taste)** Beigeschmack *m*
**peculiarity** Besonderheit *f*, Charakter *m*, Eigenheit *f*, Eigenschaft *f*, Eigentümlichkeit *f*
**pecuniary difficulties** Geldverlegenheit *f*
**pedagogical** erzieherisch
**pedal** Fuß-auflage *f*, -hebel *m*, -punkt *m* (math.),
Hebel *m*, Kurbeltritt *m*, Kurve *f*, Pedal *n*,
Regeltritt *m*, Tretkurbel *f*, Trittbrett *n* ~
**bracket** Fußhebellagerblock *m* ~ **circuit** Stromkreis *m* des Schienenstromschließers ~ **control**
Pedalsteuerung *f* ~ **control circuit** Fußregel-
schalter *m* ~ **disk** Fußscheibe *f* ~ **dynamo**
Tretdynamo *m* ~ **engagement** Fußeinrückung *f*
~ **fully depressed** den Hebel *m* ganz durch-
treten ~ **generator** Tretgenerator *m* (rdo) ~
**lever shaft** Fußhebelwelle *f* ~ **locus (or pedal**
**locus curve)** Fußpunktkurve *f*
**pedal-operated** fußbedient ~ **generating unit**
Tretsatz *m* ~ **generator** Tretmotor *m* ~ **machine**
Tretmaschine *f*
**pedal,** ~ **pin** Pedalwelle *f* ~ **pressure switch** Fuß-
druckknopftaste *f* ~ **shaft** Fußhebelwelle *f* ~
**switch** Fußschaltung *f* ~ **treadle** Fußhebel *m*
~-**type brake valve** Trittplattenbremsventil *n*
~-**type clutch valve** Trittplattenkupplungsventil
*n*
**pedestal** Auflager *n*, Bock *m*, Bügelgleitbacke *f*,
Fuß *m*, Fußgestell *n*, Fußgestell *n* einer Säule,
Fußstück *n*, gerade Achslagerführung *f*, Lager
*n*, Lagerbock *m*, (of latern) Laternenfuß *m*,
Mastfuß *m*, (of latern) Mauersockel *m*, Postament *n*, Rechteckkomponente *f*, Säulenfuß *m*,
Sockel *m*, Sockelimpuls *m*, Schwarzwertpegel
*m*, Ständer *m*, Untergestell *n*, Untersatz *m*,
Vorraum *m* ~ **of compass rose** Rosensäule *f*
**pedestal,** ~ **bearing** Bock-, Deckel-, Rumpf-,
Steh-lager *n* ~ **chain mortiser** Ständerketten-
fräsmaschine *f* ~ **height** Sprunghöhe *f* ~ **level**
Schwarzwertpegel *m* (TV), Sprunghöhe *f* ~
**(desk) telephone station** Ständerfernsprecher *m*
~ **tie bar (or binder)** Achsgabelsteg *m*
**pedestrian** Fußgänger *m* ~ **crossing** Fußgänger-
überweg *m* ~ **island** Fußgängerschutzinsel *f* ~
**subway** Fußgängerunterführung *f*
**pedicle,** ~ **foramen** Stielforamen *n* ~ **valve** Stiel-
klappe *f*
**pedometer** Passimeter *n*, Schritt-messer *m*,
-zähler *m*
**pedrail** Rad-gürtel *m*, -schienenkette *f*
**peel, to** ~ ablösen, abschaben, ausscheren
(aviat.), blättern, schälen, sich abschälen
(aviat.) **to** ~ **away** schülpen **to** ~ **off** abblättern,
abbröckeln, abplatzen, abschälen, abspringen,
abstreifen, (in patches) abschülpen
**peeler** Abschäler *m*, Schälgerät *n* ~ **centrifuge**
Schälzentrifuge *f*
**peeling** (of glazes) Abrollen *n* ~ **device** Schäl-
vorrichtung *f* ~ **knife** Schälmesser *n* ~ **machine**
Schälmaschine *f* ~-**off** Abblätterung *f*, Abschälen *n*, Abstreifen *n* ~ **shim** geschichtete
Beilegescheibe *f* aus dünnen Folien ~ **test**
Schäl- und Abhebeprüfung *f*
**peen, to** ~ etwas mit leichtem Hammerschlag
gestalten, mit der Finne hämmern, walzen **to**
~ **over** (aviat.) überwalzen, verstemmen, Werkstoff *m* über eine Kante stemmen

**peen** Finne *f*, (of a hammer) Pinne *f* ~ **hammer** Flachfinnhammer *m*
**peening** Hämmern *n* ~ **test** Hammerschlagprobe *f*
**peep,** ~ **chamber** Guckkasten *m* ~ **hole** Beobachtungsfenster *n*, Guckloch *n*, Schau-loch *n*, -öffnung *f*, Sehschlitz *m*, Sehspalt *m* ~ **sight** Lochdiopter *n*
**peg, to** ~ anstiften, dübeln, stiften, zusammendübeln **to** ~ **out** abpfählen, abstecken **to** ~ **out a line** eine Linie *f* abpfählen **to** ~ **out the line** die Linie abstecken oder auspflocken
**peg** Dorn *m*, Dübel *m*, hölzerner Nagel *m*, Holzpflock *m*, Keil *m*, Knagge *f*, Markierpfahl *m*, Mitnehmer *m*, Nase *f*, Pfahl *m*, Pflock *m*, Pfropf *m*, Picket *n*, Pinne *f*, Schließe *f*, Splint *m*, Stab *m*, Stift *m*, Stopfen *m*, Stöpsel *m*, (in string instrument) Wirbel *m*, Zapfen *m*, Zwecke *f* ~ **for drawing boards** Heftzwecke *f*
**peg,** ~ **count** Stichzählung *f*, Zählung *f* durch Stichproben **~-count meter** Leistungszähler *m* **~-count register** Belegungszähler *m* **~-count summary** Leistungs-, Verkehrs-zählung *f*, Zahlergebnis *n* ~ **drum threshing machine** Stiftendreschmaschine *f* ~ **feather key** Zapfenfeder *f* ~ **gate** (in casting) Stoßfang *m* ~ **hole** Zapfenloch *n* ~ **ladder** Stangen-, Stock-leiter *f* **~-legging** (oil drilling) Hinken *n* ~ **rail** Leitpflock *m* **~-tooth harrow** biegsame Zapfenegge *f*, verstellbare Ackeregge *f* ~ **wood** Nagelklotz *m*
**pegging** Bodenpunkt *m*, Verdübelung *f*
**pegmatite** Pegmatit *m* ~ **vein** Pegmatitgang *m*
**pegmatolite** Pegmatolit *m*
**pellet, to** ~ tablettieren
**pellet** Kugel *f*, Kügelchen *n*, Pille *f*, Würfel *m*
**pelletize, to** ~ (ore) pelletisieren
**pelletizing** Tablettenherstellung *f*
**pellets** stückige Kontaktkörper *pl* (für ortfeste Katalysatoren)
**pellicle** Haut *f* (dünn und zäh)
**pellucid film stock** glasklarer Film *m*
**pelt** Blut-, Schlacht-wolle *f*
**Peltier effect** Peltiereffekt *m*
**pelting rain** Platzregen *m*
**Pelton,** ~ **turbine** Peltonturbine *f* ~ **turbine bucket** Becherausschnitt *m* ~ **water wheel** Peltonrad *n* ~ **wheel** Freistrahlturbine *f*
**pen, to** ~ aufschreiben, verfassen
**pen** Box *f*, Damm *m*, Hürde *f*, (Schreib)Feder *f*, Schreib-mine *f*, -röhrchen *n*, Talsperre *f*, (of a recorder), Verschlag *m* ~ **with turned up point** Löffelschreibfeder *f*
**pen,** **~-and-ink drawing** Federzeichnung *f* ~ **arm** Schreibhebel *m* ~ **arm guide** Schreibhebelführung *f* ~ **holder** Federhalter *m* ~ **knife** Federmesser *n* ~ **lever** Schreibfederhebel *m* ~ **point** Stahlfeder *f* ~ **recorder** Schreibgerät *n* **~-shaped dosimeter** Füllhalterdosimeter *m*
**penal,** ~ **authority** Strafgewalt *f* ~ **code** Strafgesetz *n* ~ **law** Strafrecht *n* ~ **regulation** Strafbestimmung *f* ~ **tax** Strafsteuer *f*
**penalty** Erschwerung *f*, Strafbestimmung *f*, Strafe *f*, Straf-geld *n*, -punkt *m*, -summe *f*, ungünstiger Einfluß *m* ~ **for breach of contract** Vertragsstrafe *f* ~ **for delayed delivery** Verzugsstrafe *f* ~ **for nonfulfillment of a contract** Konventionalstrafe *f*

**penaltry test** Zusatztest *m*
**pencil** Schreibstift *m*, Stift *m*, (Strahlen)-Bündel *n*, -Büschel *m*
**pencil,** ~ **of light** Licht-bündel *n*, -kegel *m* ~ **for making carbon copies** Durchschreibestift *m* ~ **of nonuniform section** Knotenstrahl *m* ~ **of parallel light** paralleles Lichtstrahlenbüschel *m* ~ **of rays** Büschel *m*, Licht-bündel *n*, -büschel *n*, Parallelstrahlenbündel *n*, Strahlbündel *n*, Strahlen-bündel *n*, -büschel *n* ~ **of sound (rays)** Schallstrahlbündel
**pencil,** ~ **beam** Nadelstrahl *m*, scharfgebündelter Strahl *m* **~-beam antenna** Nadelstrahl-, Schmalbündel-antenne *f* ~ **blue** Kasten-, Schilder-blau *n* ~ **bow** Bleistifteinsatz *m* ~ **box** Federkasten *m* ~ **deflection** Strahlablenkung *f* ~ **drawing** Bleistift-skizze *f*, -zeichnung *f*, Blei-, Hand-zeichnung *f* ~ **eraser** Bleistiftgummi *m* ~ **extension** Bleistiftverlängerer *m* **~-head brush** Pinselkopfbürste *f* ~ **lead** Bleistiftmine *f* ~ **leverage** Strahlhebelarm *m* **~-line** strichartig ~ **marking** Verweisungsmarkierung *f* ~ **point** Bleistifteinsatz *m* ~ **pointer** Bleistiftspitzer *m* ~ **rays discharge** Fadenstrahlen *pl* **~-shaped brush** Pinselbürste *f* ~ **sharpener** Bleistiftspitzer *m* ~ **sketch** Bleistiftskizze *f* ~ **slabs** Bleistiftbrettchen *n* ~ **slate** Griffel-, Zeichen-schiefer *m* ~ **transmitter** Walzenmikrofon *m*
**pendant** Gegenstück *n*, Gehänge *n*, Gehäuseknopf *m*, Pendant *n*, Ständer *m* ~ **speaker arrangement** Schallampel *f* ~ **sprinkler** hängender Sprinkler *m* ~ **switch** Schnurschalter *m*
**pendants** Pendel *n*
**Pendel telegraph** Pendeltelegraf *m*
**pendent** hängend
**pending** laufend, schwebend, vorstehend **to be** ~ herabhängen, schweben ~ **patent** angemeldetes oder schwebendes Patent *n*
**pendular region** Ringbereich *m*
**pendulate, to** ~ pendeln
**pendulosity** Pendel-eigenschaft *f*, -wirkung *f*
**pendulous** pendelartig ~ **float** Schwebependel *n* ~ **gyro** Pendelkreisel *m* ~ **vane** Aufrichtpendel *n*
**pendulously suspended** pendelnd aufgehängt
**pendulum** Pendel *n*, Perpendikel *n*, Schwunggewicht *n* ~ **altimeter** Pendelhöhenmesser *m* ~ **anemometer** Pendelwindmesser *m* ~ **balance** Pendelwaage *f* ~ **bearings** Pendellagerung *f* ~ **bob** Pendellinse *f* ~ **bucket conveyer** Pendelbecherwerk *n* ~ **change-over switch** Pendelumschalter *m* ~ **coal chute** Pendelschurre *f* ~ **contact** Pendelkontakt *m* ~ **counterweight** pendelndes Gegengewicht *n* ~ **cross beam** Pendelstauer *m* ~ **danamometer** Neigungswaage *f*, Pendelmanometer *n* ~ **emplacement** Pendellagerung *f* ~ **governor** (conical) Pendelregler *m* ~ **hammer** Pendel(schlag)hammer *m* ~ **hydroextractor** Pendelzentrifuge *f* **~-impact testing machine** Pendelhammerschlagwerk *n*, Pendelschlagwerk *n* ~ **inclinometer** Pendelneigungsmesser *m* ~ **lever** Pendelheber *m* ~ **magnet** Pendelmagnet *m* **~-meter** Aronzähler *m* ~ **mill** Pendelmühle *f* ~ **motion of electrons** Elektronen-pendelung *f*, -pendelschwingungen *pl*, -tanz *m*, -tanzschwingungen *pl* ~ **multiplier** Pendelvervielfacher *m* ~ **recording gauge** Anschütz-Punkter *m* ~ **saw** Pendelsäge *f* **~self-**

interrupter Pendelselbstunterbrecher *m* ~
sextant Pendelsextant *m* ~ **stability** Pendel-
stabilität *f* ~ **startstop telegraph** Pendeltelegraf
*m* ~ **switch gear** Pendelschaltwerk *n* ~ **transmission dynamometer** Pendelkraftmesser *m* ~
**tup** Pendelschlagwerk *n* ~**-type barrel valve**
Pendelentlüfter *m* ~**-type dynamometer** Pendelmanometer *m* (-prüfmaschine) ~**-type impact
testing machine** Pendelhammer *m* ~ **weight**
Pendelgewicht *n*
**peneplain** Rumpffläche *f*
**penetrability** Durchdringfähigkeit *f*, Durchdringungsvermögen *n*, Durchlässigkeit *f*
**penetrable** durchdringbar, durchlässig, durchschlagbar ~ **by air** luftdurchlässig
**penetrant** durchdringend ~ **test** Penetrations-
prüfung *f*
**penetrate, to** ~ durchbohren, durchdringen,
durchschlagen, durchsetzen, durchstoßen, (in
depth) durchteufen, einbrennen, eindringen,
hineindringen
**penetrating** durchdringend ~ **component** harte
Komponente *f* ~ **component of cosmic rays**
harte Komponente *f* der Höhenstrahlung ~
**light** Nebenlicht *n* (photo) ~ **point** Prüfspitze *f*
~ **power** Durchschlagskraft *f* ~ **properties** Ein-
färbevermögen *n* (Druckfärbevermögen) ~
**radiation** harte Strahlung *f* ~ **twins** Durchdrin-
gungszwillinge *pl*
**penetration** Durch-bruch *m*, -dringen *n*, -drin-
gung *f*, -schlagen *n*, (of a projectile) Durch-
schlagsleistung *f*, (in welding) Einbrand *m*,
Einbruch *m*, Eindringung *f*, (of water) Innen-
tränkung *f*, Scharfblick *m*, Spritzweite *f*,
Strahllänge *f* (aviat.) ~ **of armor plate** Panzer-
durchschlag *m* ~ **by assault** Durchstoß *m* ~ **by
burning** Einbrand *m* ~ **in depth** Tiefeinbruch *m*
~ **of fuel gases** Durchschlagen *n* der Brenngase
~ **of ignition** Zünddurchschlag *m* ~ **in (into)**
Eindringen *n* ~ **of potential barries** Tunneleffekt
*m* ~ **of prints** Durchdruck *m* ~ **by rays** Durchleuchtung *f*, Durchstrahlung *f*
**penetration, ~ coefficient** (magnification) Durchgriff *m* **depth on mean free path** Eindringtiefe *f*
von der mittleren freien Weglänge ~ **factor**
Durchgriff *m* ~ **frequency** Durchdringungs-,
Durchgangs-frequenz *f* ~ **hardening** Durchhärtung *f* ~ **law** Eindringgesetz *n* ~ **method**
Durchstoßverfahren *n* (aviat.) ~ **notch** Ein-
brandkerbe *f* ~ **probability** Durchlässigkeits-
koeffizient *m* ~ **record** Eindringujngsprotokoll
*n* ~ **spike** Landedorn *m* ~ **tension** Haftspannung *f* ~ **test** Penetrationsversuch *m* ~ **twins**
Durchwachungszwillinge *pl* (geol.) ~**-weld**
Bindung *f*
**penetrative** durchgreifend ~ **ability** Durchdringungsvermögen *n* ~ **power** Geschoßwirkung *f*
**penetrativeness** Eindringungsfähigkeit *f*
**penetrator** Prüfspitze *f*
**penetrometer** Eindruck(tiefen)-, Penetrations-,
Qualitäts-, Röntgenstrahlenhärte-, Strahlen-
härte-messer *m*
**penetrons** schwere Elektronen *pl*
**peninsula** Halbinsel *f*
**penitentiary** Strafanstalt *f*, Zuchthaus *n*
**pennant** Fahnenband *n*, Ständer *m*, Wimpel
*m*

**penning vacuum gauge** Penning-Vakuummeter *m*
**penny** Pfennig *m* ~ **in-the-slot machine** Einwurf-
automat *m* ~ **royal oil** Poleyöl *n*
**penstock** Druckrohr *n*, Kniestück *n*, Schütze *f*,
Stollen *m*, Turbinendruckleitung *f*
**pentaerythritol,** ~ **tetranitrate** Nitropentaery-
thrit *n*, Pentrit *n* ~ **tetranitratenitroglycerin
mixtures** Pentrinit *n*
**pentagon** (räumliches vollständiges) Fünfeck *n*
~ **turret** Fünfkantenrevolverkopf *m*
**pentagonal** fünf-eckig, -kantig ~ **prism** Penta-
gonalprisma *n*
**pentagrid** Siebenpolröhre *f* ~ **converter** Misch-
heptode *f* ~ **value** Heptode *f* ~ **valve** Pentagrid-,
Siebenpol-röhre *f*
**pentahedral** fünfflächig
**pentahedron** Fünfflach *n*
**pentane lamp** Pentanlampe *f*
**pentaprism** Dachkantprisma *n*
**pentatomic** fünfatomig
**pentavalent** fünfwertig
**pentene** Penten *n*
**penthouse** Schauer *m*, Schirmdach *n*, Schutzdach *n*, Wohnung *f* auf dem Dach ~ **combustion chamber** dachförmiger Verbrennungsraum *m*
**pentlandite** Eisennickelkies *m*, Pentlandit *m*
**pentode** Audionkraft *f*, Bremsgitter-, Fünf-
elektroden-, Fünfpol-röhre *f*, Pentode *f*,
Schutzgitterfanggitterröhre *f* ~ **gun** Pentoden-
system *n* ~ **tube (or valve)** Fünfpolröhre *f*
**pentroof** Halbdach *n*
**penumbra** Halbschatten *m* ~ **boundary** Halb-
schattengrenze *f*
**penumbral** halbdunkel ~ **blur** Halbschatten-
schleier *m* ~ **compensation** Halbschattenkom-
pensation *f*
**penury** Mangel *m*
**pepple** Chagrin *n*
**peppling roller** Chagrinierrolle *f*
**peptize, to** ~ verteilen (chem.)
**peptonate** Peptonat *n*
**peptonization** Stärkeabbau *m* ~ **temperature** Abbautemperatur *f*
**per** durch, für, mit, pro ~ **degree** Gradmaß *n* ~
**diem payment** Tagessatz *m* ~ **man hour** pro
Mann und Stunde ~ **unit of time** in der Zeiteinheit *f* ~ **unit time** zeitlich
**peracetic acid** Peressigsäure *f*
**peracid** Persäure *f*
**perambulator** Meßrad *n*
**perbenzoic acid** Perbenzoesäure *f*
**percale** Perkal *n*, Perkalstoff *m*
**perceivable** wahrnehmbar
**perceive, to** ~ auffassen, bemerken, empfinden,
sehen, wahrnehmen
**percent (or per cent)** Hunderstel *n*, Prozent *n*;
prozentig, pro Hundert, vom Hundert ~ **by
volume** Volum(en)prozent *n*
**percent,** ~ **consonant articulation** Konsonanten-
verständlichkeit *f* ~ **content** Prozentgehalt *m* ~
**drift** Abweichungsgrad *m* ~ **hearing** prozentuales Hörvermögen *n* ~ **linearity** Prozent-
linearisierung *f* ~ **ripple** Welligkeitsgrad *m* ~
**ripple voltage** Brummspannungsverhältnis *n*
~ **time out** Ausbesserungszeitfaktor *m* ~ **time
on reserve** Reservezeitfaktor *m* ~ **vowel articulation** Vokalverständlichkeit *f*

**percentage** Anteilszahl *f*, Betrag *m*, Gehalt *m*, Hundertsatz *m*, Maß *n*, Prozent-gehalt *m*, -satz *m*, Vomhundertanteil *m*, -satz *m*
**percentage,** ~ **of carbon** Kohlengehalt *m* ~ **of completion** Hundertsatz *m* der ausgeführten Anmeldungen ~ **of delayed calls** Verzögerungsziffer *f* ~ **of drag power** Leistungsanteil *m* in Prozenten für Luftwiderstand ~ **of inclination** prozentuale Steigerung *f* ~ **of inspection** Stichprobenprüfung *f* mit bestimmtem Hundertsatz ~ **of load** Teillast *f* ~ **of lost calls** Verlustziffer *f* ~ **of modulation** Aussteuerungskoeffizient *m*, Beeinflussungsfaktor *m*, Hub *m*, Modelungsgrad *m*, Modulations-grad *m*, -tiefe *f*, -ziffer *f*, prozentuale Modulation *f* ~ **of modulation factor** Aussteuerungsgrad *m* ~ **of moisture** Wassergehalt *m* ~ **of ozone in the air** Ozongehalt *m* ~ **of passing sieve** Siebdurchlaß *m* in von Hundert ~ **of reaction** Reaktionsgrad *m* ~ **of resin** Harzgehalt *m* ~ **of scrap obtained** Schrottentfall *m* ~ **of share** Anteilziffer *f* ~ **of smoke nuclei in the air** Rauchgehalt *m* ~ **of solids** Trockensubstanzgehalt *m* ~ **of voids in the rubble** Verhältnis *n* der Hohlräume ~ **by volume** Raum-hundertstel *n*, -prozent *n* ~ **by weight** Gewichtsprozent *n*
**percentage** prozentuell ~ **change** prozentuale Änderung *f* ~ **depth dose** prozentuale Tiefendosis *f* ~ **drag power** prozentueller Anteil *m* der zur Überwindung des Luftwiderstandes erforderlichen Leistung ~ **error** Fehlerprozentsatz *m* ~ **figures** Prozentangabe *f* ~ **inclination** prozentuale Steigung *f* ~ **increase** prozentuelle Zunahme *f* ~ **loss** Verlustprozentsatz *m* ~ **modulation** prozentuale Aussteuerung *f* ~ **polarimeter** Prozentpolarimeter *n* ~ **retained** zurückgehaltener Prozentgehalt *m* ~ **synchronization** Synchronisierungsnutzeffekt *m* ~ **table of rejects** Ausschlußkurve *f*
**percental** prozentig
**percentual** prozentig ~ **break down** prozentuales Aufteilen *n*
**perceptibility** Wahrnehmbarkeit *f*
**perceptible** feststellbar, fühlbar, merkbar, merklich, wahrnehmbar ~ **bearing wear** feststellbarer Lagerverschleiß *m*
**perception** Empfindung *f*, Erkennen *n*, Wahrnehmung *f* ~ **of depth** Tiefenwahrnehmung *f* ~ **of direction** Richtungsempfindung *f* ~ **of a light** (maritime signals) Ausmachen *n* ~ **of relief** Tiefenwahrnehmung *f* ~ **of sound** Klangempfindung *f*
**perceptive faculty** Auffassungsgabe *f*
**perceptual** gegenständlich
**perch** Lang-, Lenk-baum *m*, Meßstange *f*, Pricke *f*, Rute *f*, Stange *f* ~ **plate** Langbaumblech *n*
**perched water table** gespannter Grundwasserspiegel *m*
**perchlorate** übersaures Salz *n*
**perchloric acid** Überchlorsäure *f*
**perchloride** Sesquichlorid *n*
**perchromic acid** Überchromsäure *f*
**percolate, to** ~ aussickern, durchseihen, durchsickern, durchsintern, sickern
**percolating** Rieselung *f*, Sickerung *f* ~ **filter** Tropfkörper *m* ~ **leach** Sickerlaugung *f* ~ **tank** Sickertank *m* ~ **water(s)** Sickerwasser *n*

**percolation** Durchseihung *f*, Durchsickerung *f*, Filtrierung *f*, Perkolation *f*, Sickerung *f* ~ **basin** Versickerungsbecken *n* ~ **range** Sickerungsstrecke *f* ~ **well** Versitzgrube *f*
**percolator** Filtrier-beutel *m*, -trichter *m*, Perkolator *m*, Seihetrichter *m*
**percuss, to** ~ beklopfen
**percussed chord (or string)** geschlagene Saite *f*
**percussion** Aufschlag *m*, Erschütterung *f*, Schlag *m*, (schlagartiger) Stoß *m* ~ **cap** Zündhütchen *n* ~ **contact** Erschütterungskontakt *m* ~ **drill** Stauch-, Stoß-bohrer *m* ~ **(rope) drill** Schlagbohrer *m* ~ **drilling** Schlagbohren *n* ~ **feeder** Stoßaufgabevorrichtung *f* ~ **fire** Aufschlagschießen *n* ~ **frame** Stoßherd *m* ~ **fuse** Anschlag-, Aufschlag-, Doppel-, Fall-, Granat-, Schlag-zünder *m* ~ **ignition** Aufschlagzündung *f* ~ **instrument** Schlag-instrument *n*, -zeug *n* ~ **jig** Stauchsetzmaschine *f* ~ **lever** Stoßhebel *m* ~ **piston** Schlagkolben *m* ~ **pneumatic tool** Druckluftschlagwerkzeug *n* ~ **press** Spindelschlagpresse *f* ~ **pressure** Stoßdruck *m* ~ **primer** (fuse) Pulverkorn *n*, Schlagpatrone *f*, -zündschraube *f* ~ **priming** Schlagzündung *f* ~ **riveting** Schlagnietung *f* ~ **riveting machine** Schlagnietmaschine *f* ~ **shell** Aufschlaggranate *f* ~ **sieve** Stoßsieb *n* ~ **table** Stoßherd *m* ~ **test** Schlag-probe *f*, -versuch *m* ~ **wave** Stoßwelle *f* ~ **welding** Schlagschweißung *f*
**percussive power** Stoßkraft *f*
**perfect, to** ~ ausbilden, veredeln, vervollkommen, widerdrucken
**perfect** betriebsfähig, fehlerfrei, fehlerlos, fertig, ideal, vollendet, vollkommen, vollwertig ~ **and imperfect polarization** vollkommene und unvollkommene Polarisation *f* **of** ~ **conductance** ideal leitend
**perfect,** ~ **circuit** betriebsfähige Leitung *f* ~ **combustion** vollkommene Verbrennung *f* ~ **configuration** geordneter Zustand *m* ~ **diffuser** vollkommen streuender Körper *m* ~ **elasticity** vollkommene Elastizität *f* ~ **fluid** reibungsfreie Flüssigkeit *f* ~ **gas** ideales Gas *n* ~ **impression** Reindruck *m* ~ **interval** reines Intervall *n* ~ **junction** unlösbare Verbindung *f* ~ **match** mustertreu ~ **paid time** Gebührenzeit *f*
**perfecting** Widerdruck *m* ~ **cylinder** Widerdruckzylinder *m* ~ **engine** Kegelmühle *f*, Knetmaschine *f*, (paper mfg.) Zerfaserungsmaschine *f* ~ **form** Widerdruckform *f* ~ **machine** Komplettmaschine *f* ~ **sheet** Widerdruckbogen *m*
**perfection** Ausbildung *f*, Vervollkommnung *f*, Vollendung *f*, Vollkommenheit *f*
**perfometer** Impulstastgerätschalter *m*
**perforate, to** ~ durchbohren, durchlochen, durchlöchern, durchschlagen, lochen, perforieren, stanzen, vorreißen
**perforated** durchbohrt, durchbrochen, durchlöchert, gelocht, löcherig ~ **(brick or stone)** **block** Lochstein *m* ~ **bottom** Nadel-, Siebboden *m* ~ **bottom plate** Siebboden *m* ~ **brick** Lochstein *m* ~ **cap** siebartige Kappe *f* ~ **-card system** Hollerithsystem *n* ~ **casing** Filterverrohrung *f* ~ **code** Lochschrift *f* ~ **disk** Kreisloch-, Loch-scheibe *f* ~ **-disk powder** Würfelpulver *n* ~ **filter plate** Filtriersieb *n* ~ **iron plate** siebartig durchlochtes Eisenblech *n* ~ **metal sheets** Lochblendenauflagen *pl* ~ **paper tape**

gelochter Papierstreifen *m* ~ **pipe** Berieselungs-
rohr *n* ~ **pipe casing** Casing *n* mit durchlöchertem Rohr ~ **plate** gelochtes Blech *n*, Zerstäuberlochplatte *f* ~ **plate sieve** Blechlochsieb *n* ~
**portion of casing of drainage well** (sometimes
slotted instead of perforated) Filterrohr *n* ~
**powder** Röhrenpulver *n* ~ **receive tape** Emp-
fangslochstreifen *m* ~ **screen** durchlochte Leinwand *f* (film) ~ **seat** Lochsitz *m* ~ **send tape**
Sendelochstreifen *m* ~ **sheet** gelochtes Blech *n*,
Gitterblech *n* ~ **slip** Lochstreifen *m* ~ **tank**
Tank *m* mit Lochung ~ **tape** Lochstreifen *m* ~
**wing flap** durchlöcherter Hilfsflügel *m*
**perforating,** ~ **gauge** Perforierkamm *m* ~
**machine** Bohrmaschine *f*, Perforier- und Aus-
zackmaschine *f* ~ **poking section** Stanzstelle *f*
(für Lochstreifen) ~ **press** Lochpresse *f*
**perforation** Durch-bohrung *f*, -löcherung *f*,
-lochung *f*, Einstich *m*, Loch *n*, Lochen *n*,
Lochnaht *f*, Lochung *f*, Perforation *f*, Perfo-
rierung *f* ~ **hole** Randloch *n* ~ **pitch** Lochabstand *m* (film) ~ **placed in diagonals** (X-ray)
diagonal stehende Lochung *f* ~ **point** Durchstoßpunkt *m* ~ **sprocket hole** Perforation *f*
**perforator** Casing *n* mit durchlöchertem Rohr,
Locher *m*, Lochzange *f*, Perforiermaschine *f*,
Stanzer *m*, Streifenlocher *m*
**perforce-centering** Zwangszentrierung *f*
**perform, to** ~ abwickeln, anrichten, arbeiten,
ausführen, ausrichten, bewerkstelligen, durchführen, erfüllen, leisten, tun, verrichten, wirken to ~ **a system** ein Verfahren ausüben oder
benutzen (Patent)
**performance** Abwicklung *f*, Arbeiten *n*, Arbeitsleistung *f*, -weise *f*, Aufführung *f*, Ausführung *f*,
Betrieb *m*, Durchführung *f*, Leistung *f*, Leistungsfähigkeit *f*, (of a play) Vorstellung *f*,
Werk *n*, Wirken *n*, Wirkungsweise *f* ~ **of**
**aircraft engine** Leistung *f* eines Flugmotors ~
**and capability** Leistungsfähigkeit *f* ~ **per head**
Produktivität *f* (Leistung pro Kopf ~ **with**
**respect to direct current** Gleichstromverhalten *n*
~ **of the survey** Aufnahmeleistung *f*
**performance,** ~ **bond** Haftrückstand *m* ~ **calcu-**
**lation** Leistungs(be)rechnung *f* ~ **characteristic**
Arbeitskennlinie *f*, Leistungs-angabe *f*, -kenn-
wert *m*, -kennzahl *f* ~ **characteristics** Arbeits-
kenngrößen *pl*, Leistungsdaten *pl* ~ **chart**
Arbeitsdiagramm *n* ~ **coefficient** Leistungsbei-
wert *m* ~ **curve** Leistungskurve *f* ~ **data** Lei-
stungsgrößen *pl* ~ **figure** Leistungszahl *f* ~
**figures not given** Leistungen *pl* nicht angegeben
~-**graph** Leistungskurve *f* ~ **load** Leistungsbelastung *f* ~ **margin** Leistungsspielraum *m* ~
**meter** Genauigkeitsmesser *m* ~ **monitor** Kontrollapparat *m* ~ **rating** Leistungsgrad *m* ~
**requirements** Leistungsanforderungen *pl* ~
**sailplane** Leistungssegler *m* ~ **switch** Betriebs-
artenschalter *m* ~ **test** Bewährung *f* ~ **tester**
Leistungsprüfer *m* ~ **testing** Leistungs-fähig-
keitsprobe *f*, -prüfung *f* ~-**type glider** Leistungs-
segelflugzeug *n*, Schwingenflieger *m*, Segelflugzeug *n* ~ **value** Endwert *m*
**performed winding** Formspulenwicklung *f*
**perfume** Wohlgeruch *m*
**perfumed** wohlriechend
**pergamoid** Pergamoid *n*
**perhydrol** Perhydrol *n*

**periastron** Sternnähe *f* (astr.)
**pericline** Albit *m*
**peridot** Peridot *m*
**peridotite** Peridotit *n*
**perigee** Erdnähe *f*, Perigäum *n*
**perihelion** Sonnennähe *f*
**perikon detector** Perikondetektor *m*
**peril** Gefahr *f*
**perilous** lebensgefährlich
**perimeter** Peripherie *f*, Umfang *m*, Umkreis *m*
~ **cabling** Ringleitungsnetz *n* ~ **track** Ringstraße *f*, Rollfeldringstraße *f*
**perimorph** Umhüllungspseudomorphose *f*
**perimorphous crystal** Kristallumhüllung *f*
**perineal electrode** Damm-, Perinial-elektrode *f*
**period** Abschnitt *m*, Dauer *f*, Jahrgang *m*,
Periode *f*, Periodendauer *f*, Punkt *m*, (of
oscillation) Schwingungsdauer *f*, Umlaufzeit *f*,
Wiederkehr *f*, Zeit *f*, Zeitabschnitt *m*, Zeitraum
*m*

**period,** ~ **of acceleration** Beschleunigungszeit *f*
~ **of action** Wirkungszeit *f* ~ **of admission** Ein-
laßperiode *f* ~ **of beat** Schwebungsdauer *f* ~ **of**
**blowing** Windperiode *f* ~ **of change** Halbzeit
*f* ~ **of compression** Verdichtungsperiode *f* ~ **of**
**contact** Eingriffdauer *f* ~ **during which call is**
**active** Gültigkeitsdauer *f* einer Gesprächsanmeldung ~ **of duty** Dienstschicht *f* ~ **of em-**
**bargo** Sperrfrist *f* ~ **of engagement** Eingriffs-
dauer *f* ~ **of evolution** Entwicklungsperiode *f* ~
**of exhaust** Auspuffperiode *f* ~ **for giving**
**grounds** (law) Begründungsfrist *f* ~ **of hail**
Hagelstadium *m* ~ **of half life** Halbzeit *f* ~ **of**
**idleness** Stilliegezeit *f* ~ **of illumination** Aufhellungsdauer *f* ~ **of inactivity** Stilliegezeit *f* ~
**of inoperation** Stillstandsperiode *f* ~ **of issue**
Ausgabefrist *f* ~ **of nitration** Nitrierdauer *f* ~
**of notice of termination** (e.g. of a contract)
Kündigungsfrist *f* ~ **of oscillation** Schwingungsperiode *f* ~ **of the oscillation** Schwingungszeit *f*
~ **of radioactive element** Halbwertszeit *f* ~ **of**
**reconstruction** Aufbauzeit *f* ~ **of restitution**
Wiederherstellungsperiode *f* ~ **per second** Herz-
einheit *n*, Periode *f* pro Sekunde ~ **of service**
Betriebs-periode *f*, -zeit *f* ~ **of starting engine**
Anlaufzeit *f* ~ **of throw-out** (switch) Auslösungszeit *f* ~ **of time** Frist *f*, Zeitdauer *f* ~ **of use** Ver-
wendungsdauer *f* ~ **of validity** Gültigkeitsdauer *f* ~ **of validity of a prearranged call**
Gültigkeitsdauer *f* einer Voranmeldung ~ **of a**
**wave** Wellenperiode *f*
**period,** ~ **meter** Periodenmeter *n* ~ **range**
Periodenmeßgerät *n* ~ **time contract** Zeitkarte-
kontrakt *m*
**periodate** Perjodat *n*
**periodic** periodisch, regelmäßig wiederkehrend,
zeitweilig ~ **acid** Perjodsäure *f* ~ **boundary**
**condition** Randbedingung *f* der Periodizität
~ **components** (of a wave). Teilschwingung *f* ~
**electromotive force** periodische elektromotorische Kraft *f* ~ **inspection** periodische Untersuchung *f* ~ **magnetic field** Wechselfeld *n* ~
**orbits** periodische Bahnen *pl* ~ **pass** periodisches
Kaliber *n* ~ **permanent magnets** permanent-
magnetische Linsenkette *f* ~ **pulse-train**
periodische Impulsgruppe *f* ~ **quality (or**
**quantity)** periodische Größe *f* ~ **service** Dauerbetrieb *m* mit periodisch veränderlicher Be-

lastung ~ **time** Schwingungsdauer *f* ~-**trigger-type receiver** Hilfsfrequenzrückkupplungsempfänger *m*, Pendelrück-kopplungsempfänger *m*, -**koppler** *m*, Superregenerativempfänger *m*, Überrückkupplungsempfänger *m* ~ **uniform motion** gleichförmige wiederkehrende Bewegung *f* ~ **volume** Periodizitätsvolumen *n*
**periodical** periodisch, regelmäßig erscheinend; (Fach)Zeitschrift *f* ~ **bulk charges to users** periodische Gesamtbelastung *f* an die Verbraucher ~ **flame effect** Maserung *f* ~ **motion** periodische Bewegung *f* ~ **overhaul** periodische oder regelmäßige Überholung *f* ~ **unevenness** periodische Ungleichmäßigkeit *f* ~ **uniform motion** gleichförmig wiederkehrende Bewegung *f* ~ **variation** periodische Ungleichmäßigkeit *f*
**periodically recurrent structure** Kettenleiter *m*
**periodicals** Zeitschriftenliteratur *f*
**periodicity** Frequenz *f*, Periodenzahl *f*, Periodizität *f*, regelmäßige Wiederkehr *f* ~ **condition** Periodizitätsbedingung *f*
**periodometer** harmonischer Analysator *m*
**periods of disuse** Stillstandszeiten *pl*
**periosteum** Periost *n*
**peripheral** an der Außenseite *f* befindlich, peripherisch ~ **air seal** äußerer Luftdichtungsring *m* ~ **area** Umfangsfläche *f* ~ **change** randliche Umwandlung *f* ~ **circuitry** Schaltsystem *n* der Außenanlagen ~ **direction** Umfangsrichtung *f* ~ **discharge** Peripherieaustrag *m* ~ **drive** Randantrieb *m* ~ **electrons** kernferne Elektronen *pl* ~ **equipment** Zusatzgeräte *pl* ~ **force** Umfang(s)kraft *f* ~ **line** Umgrenzungslinie *f* ~ **ray** Randstrahl *m* ~ **region** Peripherie *f* ~ **speed** Umfangsgeschwindigkeit *f* ~ **speed of wheel** Laufwert *m* ~ **tension** Umfangsspannung *f* ~ **velocity** Umfangs-, Umkreis-geschwindigkeit *f*
**periphere** Peripherie *f*
**peripheric** am Umfang *m* peripherisch, befindlich
**peripherical electron** kernfernes Elektron *n*, Valenzelektron *n*
**periphery** Außenfläche *f* eines Körpers, Außenseiter *f*, Kreisumfang *m*, Peripherie *f*, Rand *m*, Umfang *m*, Umkreis *m*, Umriß *m*
**periplanatic** periplanatisch
**periscope** Ausguck *m*, Beobachtungsspiegel *m*, Geländespiegel *m*, Periskop *n*, Scherenfernrohr *n*, Sehrohr *n*, Unterseebootfernrohr *n*, Winkelspiegel *m* ~ **depth** Sehrohrtiefe *f*
**periscopic** konvex-konkav, periskopisch
**perish, to** ~ eingehen, verderben, verlorengehen
**perishable** leicht verderblich, verderblich, vergänglich ~ **goods** verderbliche Artikel *pl*
**perished** eingegangen
**peristyle** Säulengang *m*
**peritectic** Peritektikum *n*; peritektisch
**perlaceous** perlmutterartig
**perlite** Perlit *m*, Perlstein *m*
**permafrost** Dauerfrost *m*
**permalloy** Permalloy *n* ~ **film** Permalloyfilm *m*
**permanence** Beharrungszustand *m*, Beständigkeit *f*, Dauer *f*, Dauerhaftigkeit *f*, Fortdauer *f* ~ **of regulation** Dauerhaftigkeit *f* der Regulierung ~ **principle** Permanentsatz, Summen-regel *f*, -satz *m*
**permanent** andauernd, anhaltend, beständig, bleibend, bodenständig, dauerhaft, dauernd, durchgehend, fest, fortdauernd, (of dyes) halt-

bar, laufend, nachhaltig, ortsfest, permanent, ständig **of** ~ **shape** formbeständig
**permanent,** ~ **blip** Festzeichen *n* (rdr) ~ **breaking load** Dauerstandbruchbelastung *f* ~ **call** Dauer-belegung *f*, -brenner *m* ~ **contactor** Dauerkontaktgeber *m* ~ **current** Dauerstrom *m* ~ **deformation** dauernde Formänderung *f* oder Deformation *f* ~ **direct current** Dauergleichstrom *m* ~ **disposal** Dauerabfallbeseitigung *f* ~ **distortion** bleibende Formänderung *f* ~ **echo** Fest-marke *f*, -zacke *f*, -zeichen *n* (rdr) ~ **echo cancellation** Festzeichenunterdrückung *f* (rdr) ~ **effect** Dauerwirkung *f* ~ **extension** bleibende Dehnung *f* ~ **feed** Dauervorschub *m* ~ **ferrying operations** Dauerfärbetrieb *m* ~ **gas** Permanentgas *n* ~ **hardness** konstante Härte *f* (des Wassers) ~ **implant** Dauerimplantation *f* ~ **installation** fixe Anlage *f* ~ **joint** nicht lösbare Verbindung *f* ~ **junction** unlösbare Verbindung *f* ~ **light** Dauerfeuer *n* ~ **load** ständige Belastung *f* ~ **locking pin** Kegelstift *m* ~ **loop** Schleifenberührung *f* ~ **magnet** Dauer-, Permanent-magnet *m* ~-**magnet dynamic loud speaker** permanentdynamischer Lautsprecher *m* ~-**magnet steel** Dauermagnetstahl *m* ~-**magnet-type moving-coil speaker** Permanentdynamik *f*, permanentdynamischer Lautsprecher *m* ~ **magnetization** Dauermagnetisierung *f* ~ **maximum carrier (or current)** Oberstrich *m* ~ **memory** Dauerspeicher *m*, energieunabhängiger Speicher *m* ~ **mold** Dauerform *f* ~-**mold casting** Dauerformguß, Kokillenguß *m* ~ **molding machine** Dauerformmaschine *f* ~ **note** Dauer-strich *m*, -ton *m* ~ **offset** bleibende Regelabweichung *f* ~ **operating pressure** Dauerbetriebsbeanspruchung *f* ~ **position** Dauerstellung *f*, ständige Stellung *f* ~ **post** Lebensstellung *f* ~ **set** bleibende Dehnung *f*, bleibende Durchbiegung *f*, bleibende Formänderung *f*, bleibende oder permanente Veränderung *f*, bleibende Verformung *f* ~ **set-up** Dauereinstellung *f* ~ **signal** Dauerbrenner *m*, Dauerton *m*, Fehlanruf *m* ~ **slide culture** Dauerpräparat *n* ~ **sound** Dauerton *m* ~ **staff** Stamm-personal *n*, -truppe *f* ~ **state** Dauerzustand *m* ~ **steel magnet** Dauerstrahlmagnet *m* ~ **storage** Dauerspeicher *m*, energieunabhängiger Speicher *m* ~ **store** Totspeicher *m* (info proc.) ~ **target** Standziel *n* ~ **telephone lines** Dauerlinien *pl* ~ **telephone-pole line signal** Dauergestänge *n* ~ **tone** Dauerton *m* ~ **way** Geleise *n*, Oberbau *m* ~-**way bolt** Oberbauschraube *f* ~ **way construction** Eisenbahnbautechnik *f* ~-**way material** Oberbaumaterial *n* ~ **weather station** ortsfeste Wetterwarte *f* ~ **white** Bariumsulfat *n*, Barytweiß *n*, Permanentweiß *n* ~ **wiring** feste Verdrahtung *f*
**permanently,** ~ **deformed** verformt bleibend ~ **drooping voltage characteristic** statisch dauernde Ungleichförmigkeit *f* ~ **plastic** verformbar bleibend
**permanganate** Permanganat *n*
**permanganic acid** Übermangansäure *f*
**permatron** Röhre *f* mit magnetischer Steuerung
**permeability** Durchdring-barkeit *f*, -lichkeit *f*, Durchdringungsvermögen *n*, Durchlässigkeit *f*, magnetische Durchlässigkeit *f* oder Durchdringbarkeit *f*, magnetische Leitfähigkeit,

Permeabilität *f*, Undichtheit *f*, Undichtigkeit *f*, (water) Wasserdurchlässigkeit *f* ~ **to air** Luftdurchlässigkeit *f* ~ **to gas** Gasdurchlässigkeit *f*, Gaspermeabilität *f* ~ **at low magnetizing forces** Anfangspermeabilität *f*, Permeabilität *f* bei kleinen Feldstärken

**permeability,** ~ **bridge** magnetische Brücke *f* ~ **model** Durchlässigkeitsmodell *n* ~ **test** Durchlässigkeitsversuch *m* ~ **tuner core** Permeabilitätsabstimmung *f*

**permeable** durchdringbar, durchdringlich, durchlässig, permeabel, undicht ~ **to gas** gasdurchlässig ~ **to light rays** lichtdurchlässig

**permeable,** ~ **formation** durchlässige Schicht *f* ~ **material** filterfähiger Werkstoff *m*

**permeameter** Diffusionsgerät *n*, Durchlässigkeitsmesser *m*, Permeabilitätsabstimmungsmesser *m*, Permeameter *n*

**permeance** magnetische Leitfähigkeit *f*, magnetischer Leitwert *m*

**permeate, to** ~ durchdringen, durchsetzen

**permeating,** ~ **efficiency of the voice** Durchschlagskraft *f* der Stimme ~ **light** durchfallendes Licht *n*

**permeation** Durchdringung *f*

**Permian series** Permisches Gestein *n*

**permissible** zulässig ~ **current-carrying capacities of insulated conductors** Belastungstabelle *f* für isolierte Leitungen ~ **dynamite** Wetterdynamit *n* ~ **limits** Toleranz *f*, Zulaß *m* ~ **load** Regellast *f*, zulässige Beanspruchung *f* ~ **signal distortion** zulässige Zeichenverzerrung *f* ~ **speed** Fahrgeschwindigkeit *f* ~ **stress** zulässige Beanspruchung *f* ~ **variation** Fehlergrenze *f*, Zulaß *m*, zulässige Abweichung *f*

**permission** Bewilligung *f*, Einwilligung *f*, Erlaubnis *f*, Genehmigung *f*, Vergünstigung *f*, Zustimmung *f* ~ **to fire** Feuererlaubnis *f* ~ **to land** Landeerlaubnis *f* ~ **to mine** Bergbaufreiheit *f* ~ **to pass** Vorfahrterlaubnis *f* ~ **to print** Druckgenehmigung *f* ~ **to take off** Starterlaubnis *f*

**permissive** fakultativ, zulässig ~ **block** bedingte Raumfolge *f* ~ **(down) signal** bedingtes Haltesignal *n*

**permit, to** ~ bewilligen, erlauben, genehmigen, gestatten, vergünstigen, zugeben, zulassen

**permit** Ausweis *m*, (for transit) Durchgangsschein *m*, Durchlaßschein *m*, Erlaubnis *f*, Erlaubnisschein *m*, Paß *m*, Steuerausschlag *m*, Unbedenklichkeitsvermerk *m*, Zollabfertigung *f* ~ **to export** Ausfuhrbewilligung *f* ~ **license** Freibrief *m*

**permitted** erlaubt, statthaft ~ **energy zone (or level)** erlaubtes Energieniveau *n* ~ **quantum state** erlaubtes Energieniveau *n* ~ **transition** erlaubter Übergang *m*

**permittivity** Dielektrizitätskonstante *f*

**permolded pile** geformter Pfahl *m*

**permutation** Versetzung *f*, Vertauschung *f* ~ **bar** Zuteilungsscheibe *f* ~ **code switching system** Wahl *f* mittels Tastatur ~ **disk** Zuteilungsscheibe *f* ~ **plate** Zuteilungskamm *m* ~ **table** Permutations-tabelle, -tafel *f*

**permute, to** ~ permutieren, vertauschen

**peroxidation** Überoxydation *f*, Überoxydierung *f*

**peroxide** Peroxyd *n*, Superoxyd *n*, Überoxyd *n* ~ **of magnesium** Magnesiumsuperoxyd *n*

**peroxide,** ~ **bleach** Oxydationsbleiche *f* ~ **injection nozzle** T-Stoff-Spritzkopf *m* ~ **vent valve** T-Stoff-Entlüfter *m*

**peroxidize, to** ~ peroxydieren, übersäuern

**perpend** Vollbinder *m*

**perpendicular** Einfallslot *n*, Perpendikel *n*; im Lot, lotrecht, perpendikulär, rechtwinklig, scheitelrecht, seiger, senkrecht ~ **(line)** Lot *n* ~ **to the trajectory plane** Bahnnormale *f*

**perpendicular,** ~**-and base adjoining hypotenuse** Katheten *pl* ~ **conveyance** Seigerförderung *f* ~ **line** normale Linie *f*, Senkrechte *f* ~ **magnetization** Quermagnetisierung *f* ~ **position** Senkrechtstellung *f* ~ **shaft** Seigerschacht *m* ~ **throw** seigere Sprunghöhe *f* ~ **tolerance** Rechtwinkligkeitsfehlertoleranz *f*

**perpendicularly incident light** gerade einfallendes Licht *n*

**perpetration** Ausführung *f*, Begehung *f*

**perpetrator** Täter *m*

**perpetual** fortwährend, stetig, unaufhörlich ~ **calendar** Dauer-, Umsteck-kalender *m* ~ **inventory** permanente Inventur *f* ~ **inventory file** laufende Bestandskartei *f*

**perplex, to** ~ verfitzen, verwirren

**perpropionic acid** Perpropionsäure *f*

**perquisites** Nebeneinnahmen *pl*

**perrate, to** ~ auszacken

**Perrins' rolling process** Pilgerschritt-, Schrägwalz-verfahren *n*

**perron** Freitreppe *f*

**perrotine printing** Perrotindruck *m*

**persecute, to** ~ verfolgen

**persecution** Verfolgung *f*

**perseverance** Beständigkeit *f*

**persevere, to** ~ beharren

**persevering** nachhaltig

**persist, to** ~ anhalten, beharren, verharren

**persistence** Beharrung *f*, Festigkeit *f*, Nachleuchtdauer *f* (CRT), Nachwirkung *f*, Persistenz *f*, Wirkungsdauer *f* ~ **characteristic** Nachglüh-, Nachleucht-charakteristik *f* ~ **phenomenon** Nachleuchterscheinung *f* ~ **screen** Nachleuchtschirm *m*

**persistence,** ~ **of the fluorescent screen** Nachleuchtdauer *f* des Fluoreszenzschirmes ~ **of image spot** Bildpunktdauer *f* ~ **of plastic flow** Nachfließen *n* ~ **of the spot** (cathode-ray oscillograph) Auflösungsvermögen *n* ~ **of vision** Augenträgheit *f*, Nachbildwirkung *f*, Persistenz *f* des Netzhauteindruckes, Phasendrehung *f*, Visionspersistenz *f*

**persistent** anhaltend, seßhaft, stetig ~ **current** Dauerstrom *m* ~ **oscillations** ungedämpfte Schwingungen *pl* ~ **screen** Speicherschirm *m* (TV) ~ **spectrum** Grundspektrum *n* ~ **wave** ungedämpfte Schwingungen *pl*, ungedämpfte Welle *f*

**personal** persönlich ~ **damage** Personenschaden *m* ~ **data** Personaldaten *pl* ~ **data sheet** Personalbogen *m* ~ **equation** persönliche Gleichung *f* ~ **error** Beobachtungsfehler *m* ~ **hygiene** Körperpflege *f* ~ **management** Menschenführung *f* ~ **monitor** Individual-dosimeter *n*, -monitor *m* ~ **paging system** Personenrufanlage *f* ~ **record** Lebenslauf *m*, persönliche Führung *f*

**personality** Persönlichkeit *f*

**personally liable partner** persönlich haftender Gesellschafter *m*
**personnel** Belegschaft *f*, Besatzung *f*, Besetzung *f*, Menschenbestand *m*, Personal *n*, Truppe *f* ~ **card** Personalkarte *f* ~ **manager** Leiter *m* der Personalabteilung ~ **monitoring** individuelle Überwachung *f* ~ **section** Personalabteilung *f*
**perspective** Aussicht *f*, Bildweite *f*, Perspektive *f*; perspektivisch ~ **axis** Achse *f* der Perspektivität ~ **center** (of the lens) Objektivhauptpunkt *m*, Zentrum *n* der Perspektive ~ **drawing** Fernzeichnung *f*, perspektivische Zeichnung *f* ~ **glass** Taschenfernrohr *n* ~ **representation** Sichtbild *n* (rdr) ~ **view** Fernsicht *f*, perspektivische Ansicht *f*, Schaubild *n* ~ **visibility** Fernsicht *f*
**perspectivity** Perspektivität *f*
**perspectograph** Perspektograf *m*
**perspicacious** durchdringend, scharf-sichtig, -sinnig, verständlich
**perspiration** Ausdünstung *f* ~ **corrosion** Schwitzwasserkorrosion *f*
**perspire, to** ~ schwitzen
**perspiring** schweißtriefend
**persulphuric acid** Perschwefelsäure *f*, Überschwefelsäure *f*
**pertain, to** ~ **to** betreffen
**pertaining** ~ **to commercial customs** handelsüblich ~ **to deflection balance system** wegschlüssig ~ **to exterior ballistics** außerballistisch ~ **to flying** fliegerisch ~ **to glass technology** glastechnisch ~ **to ice** eisklüftig ~ **to industrial engineering** betriebswissenschaftlich ~ **to interior ballistics** innerballistisch ~ **to mechanical stress causing optical phenomena** spannungsoptisch ~ **to ordnance engineering (or mechanics)** waffentechnisch ~ **to quanta (or the quantum) theory** quantenhaft
**pertinacious** beharrlich
**pertinax** Pertinax *n* ~ **preset control** Pertinaxschnittschraube *f* ~ **screw-driver rod** Pertinaxstäbchen *n*
**pertinent** gehörig, passend, sachlich, zweckdienlich
**perturbance theory** Störungstheorie *f*
**perturbation** Perturbation *f*, Schwankung *f*, Störung *f* ~ **calculation** Störungsrechnung *f* ~ **character** Störungscharakter *m* ~ **equation** Störungsdifferentialgleichung *f* ~ **function** Störungsfunktion *f* ~ **insensitive physical properties** störungsunempfindliche physikalische Eigenschaften *pl* ~ **method** Störungsverfahren *n* ~ **sensitive** störungsempfindlich
**perturbed** gestört ~ **correlation** gestörte Korrelation *f* ~ **motion** Störbewegung *f*
**perturbometer** Perturbometer *n*
**perusal** Durchsicht *f*, Einsicht, Prüfung *f*
**peruse, to** ~ durch-lesen, -sehen, einsehen
**Peruvian,** ~ **balsam** Perubalsam *m* ~ **bark** Chinarinde *f*
**pervade, to** ~ durch-dringen, -setzen
**perveance** Perveanz *f*, Raumladungskonstante *f*
**pervious** durch-dringlich, -lässig; undicht ~ **to light rays** lichtdurchlässig ~ **to water** wasserdurchlässig ~ **subsoil** durchlässiger Baugrund *m*
**perviousness** (sand) Durchlässigkeit *f*, Undichtheit *f*, Undichtigkeit *f*
**Peschel insulating tube** Peschelrohr *n*

**pest** (plant) Pflanzenschädling *m* ~ **control** Schädlingsbekämpfung *f*
**pestle** Mörser-keule *f*, -stempel *m*, Pistill *n*, Schlägel *m*, Stampfe *f*, Stempel *m*, Stößel *m*, Stößer *m*
**petcock** Kompressions-, Kondenswasser-, Probier-, Wasserablaß-hahn *m* ~ **for pumps** Pumpenprobhahn *m*
**Petersen coil** Erdschlußlöschspule *f*
**petition, to** ~ beantragen, einreichen **to** ~ **for an audience** um Gehör *n* ersuchen
**petition** Bitte *f*, Bittschrift *f*, (patents) Eingabe *f*, Gesuch *n* ~ **on appeal** Berufungsschrift *f* ~ **for time (or respite)** Stundungsgesuch *n*
**petitioner** Antragsteller *m*
**Petri dish** Petrischale *f*
**petrifaction** Petrefakt *n*, Versteinerung *f*, Versteinung *f*
**petrified** fossil, versteinert
**petrify, to** ~ versteinern
**petrographic** petrografisch
**petrography** Gesteins-beschreibung *f*, -kunde *f*, Petrografie *f*, Steinbeschreibung *f*
**petrol** Benzin *n*, Treibstoff *m* ~ **engine** Ottomotor *m* ~ **feed** Benzinförderung *f* ~ **injection** Benzineinspritzung *f* ~ **lubrication** Gemischschmierung *f* (Oberschmierung) ~ **tank body** Benzintankaufbau *m*
**petrolatum** Petrolatum *n*, Vaseline *f* (für Kabel)
**petrolene** Petrolen *n*
**petroleum** Bergöl *n*, (mineral oil) Erdöl *n*, Felsöl *n*, Mineralöl *n*, Petroleum *n*, Steinmetzöl *n* ~ **by-product industries** Industrie *f* der Erdölnebenprodukte ~ **chemistry** Brennstoffchemie *f* ~ **coke** Petroleum-, Petrol-koks *m* ~**-cracking industry** Krackindustrie *f* ~ **engine** Petroleummotor *m* ~ **ether** Gasolin *n*, Petrol-, Petroleumäther *m*, Rhigolen *n* ~ **furnace** Petroleumofen *m* ~ **hydrocarbon** Benzinkohlenwasserstoff *m* ~ **jelly** natürliche Vaseline *f* ~ **product** Petroleumprodukt *n* ~ **refinery** Petroleumraffinerie *f* ~ **refinery plants** Erdölraffinerieanlagen *pl* ~ **research** Erdölforschung *f* ~ **set-off device** Petrolör *f* ~ **sluice valve** Petrolschieber *m* ~ **spirit** Ligroin *n*, Masut *m* ~ **spring** Petroleumquelle *f* ~ **vapor** Petroleumdampf *m* ~ **well** Erdölquelle *f*
**petrolic acid** Petrolsäure *f*
**petroliferous** petroleumhaltig
**petrological** gesteinkundlich
**petrous portion of temporal bone** Felsenbein *n*
**petticoat** Isolierglocke *f* ~ **insulator** Doppelglokkenisolator *m*
**petty** klein
**pewter** Weißmetall *n* **(hard)** ~ Hartzinn *n* **of** ~ zinnen, zinnern
**Pfauter hobber** Pfauter-Verzahnungsmaschine *f*
**phacometer** Phakometer *n*
**phanotron** Phanotron *n*, Quecksilberdampfflampe *f* mit Heizkathode, ungesteuerte Gleichrichterröhre *f*
**phantastron** Kippschaltung *f* mit einmaliger Ablenkspannung, Linearsägezahn-Schaltung *f* Phantastron *n*
**phantom, to** ~ zum Phantomkreis *m* schalten, zum Vierer *m* schalten
**phantom** Trugbild *n*, Vierer *m* ~ **antenna** Antennenersatzstromkreis *m*, Ersatzantenne *f*,

künstliche Antenne f ~ **cable** Kabel n mit Viererverseilung, viererverseiltes Kabel n ~ **capacity** Viererschleifenkapazität f ~ **circuit** Doppelsprechschaltung f, Phantom-kreis m, -leitung f, Simultanverbindung f, viererfähige Leitung f Vierer-kreis m, -leitung f, -verbindung f ~ **-circuit loading coil** Viererspule f ~**-circuit operation** Viererbetrieb m ~**-circuit repeat coil** Viererkreisübertrager m ~ **coil** Phantom-, Vierer-spule f ~**-coil set** Viererspulensatz m ~ **connection** Phantomkreis-, Vierer-schaltung f ~ **frequency selector plug** Tarnfrequenzwähler (g/m) ~**-loaded** viererpupinisiert ~ **loading** Vierer-belastung f, -pupinisierung f ~ **pair** Viererleitung f ~ **repeating coil** Doppelsprechringübertrager m ~ **signal** Irrzacke f, Irrzeichen n ~ **telephone connection** Doppelsprechschaltung f ~ **telephone operation** Doppelsprechbetrieb m ~**-to-side unbalance** Mitsprechkopplung f ~ **transposition** Platzwechsel m ~ **view** Phantombild n

**phantoming** Doppelsprechen n, Phantombildung f, Viererbildung f

**pharmaceutical,** ~ **balance** Tarierwaage f ~ **industry** pharmazeutische Industrie f ~ **oil** Öl n für medizinischen Gebrauch

**pharmacolite** Arsenblüte f, Pharmakolith m

**pharmacy** Apotheke f, Drogerie f

**phase, to** ~ in die richtige Phase f bringen (Wechselstrom), in Phasen einteilen, synchronisieren

**phase** Erscheinungsform f, Periode f, Phase f, Stand m, Stufe f, Zustand m **in** ~ (on film) Gleichtakt m in Phase f mit **to be out of** ~ in der Phase f verschoben sein **in** ~ **coincidence** konphas **in** ~ **opposition** phasenverschoben **in** ~ **quadrature** um 90 Grad phasenverschoben ~ **of progress** Bewegungsstufe f **in locked** ~ **relation** phasenstarr **in proper** ~ **relation** phasenmäßig ~ **of a sine wave** Phasenwinkel m ~ **of a sinusoidal quantity** Phase(nwinkel) m

**phase,** ~ **adjustment** Phasen-einstellung f, -regelung f, -umschaltung f ~ **advance** Phasenvoreilung f ~ **advancer** Phasen-schieber m, -vorschieber m ~ **advancing** Phasenverschiebung f ~ **amplitude distortion** Phasen/Amplituden-Verzerrung f ~ **angle** Phasen-unterschied m, -verschiebungswinkel m, Phasen-winkel m, Verlustwinkel m, Winkelmaß n ~ **angle of a condenser** dielektrischer Verlustwinkel m ~ **angle-curve** Winkelmaßkurve f ~ **angle difference** Dämpfungswinkel m, (of a condenser) dielektrischer Verlustwinkel m, (of condenser) Verlustfaktor m, Verlustziffer f ~ **angle error** Winkelfehler m ~ **angle indicator** Pendelzeiger m ~ **angle shift** Phasensprung m ~ **average** Phasenmittel n ~ **balance** Phasen-abgleichvorrichtung f, -bilanz f ~ **balance relay** Phasenunterbrechungsrelais n ~ **balancer** Symmetriereinrichtung f ~ **bandwidth** Bereich m der Phasenlinearität ~ **belt (or spread)** Zonenbreite f ~ **boundary** Phasengrenzschicht f ~ **change** Gefüge-neubildung f, -umwandlung f, Phasenänderung f, -variation f ~**-change coefficient** Phasenkonstante f ~ **changer** Blindleistunsgsmaschine f, Phasen-abgleichvorrichtung f, -schieber m, Wendephase f ~ **coincidence** Phasengleichheit f ~ **comparator** Phasenvergleicher

m ~ **comparsion** Phasenvergleich m ~ **compass** Phasenkompaß m ~**-compensating transformer** Meßwandler m mit Kunstschaltung ~ **compensation** Phasen-ausgleich m, -entzerrung f ~ **compensator** Phasen-entzerrer m, -entzerrungskette f ~ **constant** Phasen-konstante f, -maß n, Winkelkonstante f (unit) ~ **constant** Winkelmaß n ~ **constant per section** Kettenwinkelmaß n je Glied ~ **contrast images** Phasenkontrastbilder pl ~ **control** Phasenregelung f ~ **control relay** Phasenüberwachungsrelais n ~ **converter** Blindleistungsmaschine f, Phasenumformer m ~ **corrected** phasenentzerrt ~**-corrected horn** phasenkorrigierter Hornstrahler m, Planwellenhorn n ~ **correction** Phasen-ausgleich m, -entzerrung f ~ **corrector** Phasenentzerrer m ~ **curve** Phasenkurve f ~ **decrement** Phasenabnahme f ~ **delay** Phasenverzögerung f ~ **delay error** Laufzeitfehler m ~ **detector** Phasendetektor m ~ **deviation** Phasenhub m ~ **diagram** Zustandsdiagramm n ~ **difference** Gangunterschied m, Phasen-differenz f, -unterschied m, -unterschiedsverschiebung f (angle of) ~ **difference** Phasenwinkel m ~ **discriminator** Phasendiskriminator m ~**-displaced current** phasenverschobener Strom m ~ **displacement** Gangverschiebung f, Phasen-kehrschleife f, -unterschied m, -unterschiedverschiebung f, -verschiebung f ~**-displacement angle** Fehlwinkel m ~ **distortion** Laufzeitverzerrung f, Phasen-fehler m, -verzerrung f ~ **distortion index** Phasenverzerrungsgrad m ~ **dot** Tüpfel m (TV) ~ **equalization** Laufzeitausgleich m (TV) ~ **equalizer** Phasenausgleicher m ~ **factor** Phasenfaktor m ~ **fading** Phasenschwund m ~ **fluctuation** Phasenhub m ~ **focusing** Antreten n zum Tanz, (in beams tubes) Elektronengruppierung f mit der Grundfrequenz Intrittkommen n, Phasenaussortierung f, -fokus m ~**-frequency characteristic** Phasengang m ~ **grating** Phasengitter n ~ **grouping** Phasengruppenbildung f ~ **indicator** Phasen-anzeiger m, -indikator m, -zeiger m ~ **integral** Phasenintegral n ~ **interfaces** Phasengrenzen pl ~ **inversion** Phasenumkehr f ~ **inverter** Phasenumkehrer m ~ **jitter** Phasengeräusch n ~ **lag** Nacheilung f, Phasennacheilung f, -verzögerung f ~ **lagger** Phasenrückdreher m ~ **lagging** Phasenvergrößerung f ~ **lamp** Phasenlampe f ~ **lead** Phasenvor(aus)eilung f ~ **leading** Phasenvoreilung f ~ **leg** Phasennacheilung f ~ **linearity** Phasenlinearität f ~**-locked** phasenstarr ~ **margin** Phasenspielraum m ~ **meter** Phasenmesser m, Phasometer n ~**-modulated transmitter** phasenmodulierter Sender m ~ **modulation** Phasen-modelung f, -modulation f ~ **opposition** Gegenphase f, Gegenphasigkeit f, Phasenverschiebung f von 180 Grad ~ **quadrature** Phasenverschiebung f um 90 Grad, Quadratur f ~ **recorder** Phasenschreiber m ~ **reference** Phasenbezug m ~ **regulation** Phaseneinstellung f ~ **relation** Phasenbeziehung f ~ **relationship** Phasen-beziehüng f, -lage f, -verhältnis n ~ **resistance** Ausgleichwiderstand m ~ **resonance** Phasenresonanz f ~ **retardation** Phasenverzögerung f ~ **reversal** Phasenumkehr f ~**-reversing loop** Phasenumkehrschleife f ~ **reverter** Phasenumkehrer m ~**-reverter stage** Phasenumkehrstufe f ~**-rigid**

**control** phasenstarre Steuerung *f* ~ **rotation**
Phasen(ver)drehung *f* ~ **rule** Phasenregel *f*
~ **select switch** Phasenwählschalter *m* ~ **selective rectifier** phasenabhängiger Gleichrichter *m*
~ **sensitive rectifier** phasenempfindlicher
Gleichrichter *m* ~ **sequence indicator** Drehfeldrichtungsanzeiger *m* ~ **shift** Phasen-dreh-
ung *f*, -regelung *f*, -verzerrung *f* ~ **shift(ing)**
Phasenverschiebung *f* ~ **shift analysis** Phasen-
winkelanalyse *f* ~ **shift angle** Phasendrehungs-
winkel *m* (-verschiebungswinkel) ~-**shift constant** Wellenlängenkonstante *f* ~ **shift control**
Phasenschieberregelung *f* ~ **shifter** Phasen-abgleichvorrichtung *f*, -rückdreher *m*, -schieber
*m*, Wellenschieber *m* ~ **shifting** Phasenverdreh-
ung *f* ~-**shifting section** Verzögerungsglied *n*
~-**shifting transformer** Phasen-schiebertrans-
formator *m*, -transformator *m* ~ **space** Phasenraum *m* (math.) ~ **splitter** Phasenteiler *m* ~
**splitting** Phasen-spaltung *f*, -teilung *f* (TV)
~-**splitting device** Phasenteiler *m* ~-**splitting
tube circuit** Kathodynschaltung *f* ~ **spreading**
Phasenverschmierung *f* ~ **step** Phasenstufe *f*
~ **swinging** Pendeln *n* des Rotors, Pendel-
schwingungen *pl* ~ **switch** Phasenablösung *f* ~
**theory** Phasentheorie *f* ~-**to-ground voltage**
Phasenspannung *f* ~-**to-phase voltage** verkettete Spannung *f* ~ **transition** Phasenübergang
*m* ~ **variation** Phasenhub *m* ~ **variations** Pha-
senschwankung *f* ~ **velocity** Phasen-, Wellen-
geschwindigkeit *f* ~ **(group) velocity** Gruppenlaufzeit *f* ~ **voltage** Phasenspannung *f* ~ **voltmeter** Phasenvoltmeter *n* ~ **wave** Elektronen-,
Materie-, Phasen-welle *f* ~-**white** Phasierung *f*
auf Weiß ~ **winding** Phasenwicklung *f* ~ **zero
passage** Phasennulldurchgang *m*
**phaser** Phasenschieber *m*
**phases, in** ~ abschnittsweise
**phasing** Ausgleich *m*, Bildnachstellung *f*, Phaseneinstellung *f* ~ **of combustion** Verbrennungsverlauf *m* ~ **of picture** Bildstricheinstellung *f* ~
**of shutter** Blendennachstellung *f* ~
**phasing** ~ **section** Anpaßstück *n* (Leitung zur
Phasenverschiebung) ~ **signal** Phasengabe *f* ~
**signals** Gleichlauf-stöße *pl*, -zeichen *pl* ~ **system**
Phasenumtastgerät *n*
**phasotron** Synchrotron *n*
**phasotropy** Phasotropie *f*
**phenacite** Phenazit *m*
**phenanthrene** Phenantren *n*
**phenic acid** Phensäure *f*
**phenol** Karbol *n*, Karbolsäure *f*, Phenol *n*, Phenylsäure *f* ~ **extraction** Phenolextraktion *f* ~
**fiber** bakelisierter Faserstoff *m*, Phenolfiber *f*
~-**soluble** phenollöslich
**phenolic,** ~ **compounds** Phenolmasse *f* ~ **resin**
Phenolharz *n*
**phenology** Phänologie *f*
**phenoloid** phenolartig
**phenolphthalein** Phenolphthalein *n* ~ **(test) paper**
Polpapier *n*
**phenocryst** Einsprengling *m*
**phenolsulphonic acid** Phenolsulfonsäure *f*
**phenomena,** ~ **of friction** Reibungserscheinun-
gen *pl* ~ **of inertia** Trägheitserscheinungen *pl*
~ **of interference** Interferenzerscheinungen *pl*
~ **of scattering** Streuungsvorgänge *pl* ~ **of
wear** Abnutzungserscheinungen *pl*

**phenomenological** phänomenologisch
**phenomenon** Ereignis *n*, Erscheinung *f*, Geschehnis *n*, Phänomen *n*, Vorgang *m* ~ **of diffraction**
Beugungserscheinung *f* (phys.) ~ **of discharge**
Entladungserscheinung *f* ~ **of flow** Fließer-
scheinung *f* ~ **of motion (or of movement)** Bewegungsvorgang *m* ~ **of oscillations** Schwin-
gungsvorgang *m* ~ **of reverberation** Nachhall-
erscheinung *f*
**phenomenon,** ~ **involving foam** Schaumschwimmaufbereitung *f* ~ **occurring in etching** Ätzer-
scheinung *f*
**phenyl** Phenyl *n*
**phenylacetic,** ~ **acid** Phenylessig *m*, Phenylessig-
säure *f* ~ **ether** Phenylessigäther *m*
**phenylacetylene** Phenylacetylen *n*
**phenylamine** Anilin *n*
**phenylate, to** ~ phenylieren
**phenyl(iso) cyanide** Benzonitril *n*
**phenylene blue** Phenylenblau *n*
**phenyl,** ~ **ethylene** Phenyläthylen *n* ~ **ethylic**
phenyläthyl ~ **formic acid** Phenylameisensäure
*f*
**phenylic acid** Phenylsäure *f*
**phenyl,** ~ **hydrazine hydrochloride** Phenylhydra-
zinchlorhydrat *n* ~ **hydride** Benzol *n*, Phenylwasserstoff *m* ~ **sulphonic acid** Benzolsulfonsäure *f* ~ **sulphuric acid** Phenylschwefelsäure *f*
**phial** Fläschchen *n*
**phlobaphene** Rindenfarbstoff *m*
**phloroglucinol** Phloroglucin *n*
**pH-meter** pH-Messer *m*
**phonautograph** Langwellenaufzeichner *m*
**phone, to** ~ anrufen
**phone** Fernhörer *m*, Kopfhörer *m* ~ **jack** Kopf-
hörerbuchse *f* ~ **plug** Kopfhörerstecker *m*
**phonegram** zugesprochenes Telegramm *n*
**phonetic** lautlich ~ **alphabet** Buchstabentafel *f*,
Buchstabiertafel *f* ~ **error** Hörfehler *m* ~ **symbol** Lautzeichen *n*
**phonetics** Lautlehre *f*, Lautsystem *n*, Phonetik *f*
**phonic** phonisch ~ **wheel** phonisches Rad *n*
**phonodeik** Schallwellenaufzeichner *m*, Wellen-
aufzeichnungsgerät *n*
**phonoelectrocardioscope** Phono-Elektrokardio-
graf *m*
**phonogram** zugesprochenes Telegramm *n* ~
**circuit** Telegrafenleitung *f* mit Sprechbetrieb ~
**section** Fernsprechtelegrammaufnahme *f*
**phonograph** Fonograf *m*, Plattenspieler *m*,
Sprechmaschine *f* ~ **amplifier** Nadeltonver-
stärker *m* ~ **needle** Abspiel-, Abtast-nadel *f* ~
**pickup** Tonabnehmer *m*, Wiedergabedose *f* ~
**record** Fonografenplatte *f*, Meßschallplatte *f*,
Schallplatte *f* ~ **turntable** Fonografen-dreh-
platte *f*, -plattenteller *m*
**phonographic,** ~ **pickup** Adaptor *m* ~ **shooting**
Aufnahme *f* ~ **taking** (of pictures) Aufnahme *f*
**phonolite** Klingstein *m*, Phonolith *m*
**phonometer** Schallstärkenmesser *m*
**phonon** Schallquant *m* ~ **entropy** Schallquanten-
Entropie *f*
**phonoscope** Phonoskop *n*
**phorometer** Phorometer *n*
**phosgene (gas)** Chlorkohlenoxyd *n*, Phosgen *n*
~ **gas** Phosgengas *n*
**phosgenite** Bleihorn-erz *n*, -spat *m*, Chlorblei-
spat *m*

**phosphate** Phosphat *n* **~ of iron** Eisenblau *n* **~ of lime** Kalziumphosphat *n*
**phosphatic** phosphorhältig
**phosphide** Phosphid *n*
**phosphite** Phosphormetall *m*
**phosphomolybdic acid** Phosphormolybdänsäure *f*
**phosphor** Leuchtschirmsubstanz *f*, Selbstleuchter *m* **~ bronze** Phosphorbronze *f* **~-bronze wire** Phosphorbronzedraht *m* **~ dot** Leuchtstoffpunkt *m* (TV) **~ dot trio** Leuchtstoffpunktdreier *m* (Tüpfeldreier) **~ screen** Leuchtschirm *m* **~ strip** Leuchtstoffstreifen *m* (TV)
**phosphorate, to** **~** phosphorisieren
**phosphorated** phosphorhaltig
**phosphoresce, to** **~** nachstrahlen, phosphoreszieren
**phosphorescence** Nachglühen *n*, Nachleuchten *n*, Phosphoreszenz *f* **~ of picture** Bildnachleuchten *n* **~ of the sea** Meeresleuchten *n* **~ period** Leuchtdauer *f*
**phosphorescent** phosphoreszierend **~ compound** Leuchtmasse *f* **~ paint** Leuchtfarbe *f* **~ picture point** nachleuchtender Bildpunkt *m* **~ screen** Fluoreszenzschirm *m* **~ substance** Leuchtmasse *f*
**phosphoretic** phosphorig
**phosphoric** phosphorhältig, phosphorig, phosphorisch **~ anhydride** Phosphorsäureanhydrid *n* **~ pig iron** Phosphorroheisen *n*
**phosphorite** Phosphorit *m*
**phosphorize, to** **~** phosphorisieren
**phosphormanganese** Phosphormangan *n*
**phosphorogen** Leuchtkraftverstärker *m*, Phoaphorogen *n*
**phosphorogenic** Leuchtenergie *f* erzeugend, phosphoreszenzerzeugend
**phosphoroscope** Fluoreszenzmesser *m*
**phosphorous** phosphorig **~ acid** phosphorige Säure *f* **~ composition attendant** Phosphorteigbereiter *m* **~ pentoxide** Phosphorpentoxyd *n*
**phosphors** Phosphore *pl*, Schirmluminophore *pl*
**phosphorus** Phosphor *m* (**containing**) **~** phosphorhältig **~ bromide** Bromphosphor *m*, Phosphorbromid *n* **~ chloride** Phosphor-chlorid *n*, -chlorür *n* **~ content** Phosphorgehalt *m* **~ di-iodide** Phosphorjodür *n* **~ drop** Entphosphorung *f* **~ iodide** Phosphorjodid *n* **~-like** phosphorartig **~ oxychloride** Phosphoroxychlorid *n*, Phosphorylchlorid *n* **~ pentabromide** Phosphorbromid *n* **~ pentachloride** Phosphorchlorid *n* **~ segregation** Phosphorsegregierung *f* **~ tribromide** Phosphorbromür *n* **~ trichloride** Phosphorchlorür *n* **~ tri-iodide** Phosphor(tri)jodid *n*
**phosphotungstic acid** Phosphorwolframsäure *f*
**phot** Phot *n*
**photistor** lichtempfindlicher Transistor *m*
**photo, ~ audio generator** Lichtsirene *f*, Lichttongenerator *m* **~-barrier cell** Sperrschicht-Fotozelle *f* **~ cartograph** Fotokartograf *m* **~ cathode** Fotokathode *f*
**photocell** Fotoelement *n*, Fotozelle *f*, Lichtelement *n*, lichtelektrischer Wandler *m*, lichtempfindliche Zelle *f*, Lichtzelle *f* **~ amplifier** Fotostromverstärker *m*, Fotozellenverstärker *m* **~ current** Fotostrom *m* **~ resistance** lichtelektrischer Widerstand *m* **~ supply unit** Fotozellennetzgerät *n* **~ window** Belichtungsfenster *n*

**photo, ~ cells and light relays** Fernsehwandlerorgane *pl* **~ centric orbit** Bahn *f* des Lichtzentrums **~ ceramic** fotokeramisch **~ chemical** fotochemisch, lichtchemisch **~ chemical color printing** Lichtfarbendruck *m* **~-chemistry** Fotochemie *f* **~ chromation** Heliochromie *f* **~ coated with lightsensitive layer** geschichtete Kathode *f* **~ compensating eyepiece** Foto-Kompensokular *n* **~-composing machine** Lichtsetzmaschine *f* **~-composition** Fotosatz *m*, Lichtsetzen *n* **~-conducting cell** Widerstandszelle *f* **~ conducting phosphors** fosphorleitende Leuchtstoffe *pl* **~ conduction sensitivity** Fotoleitungsempfindlichkeit *f* **~ conductive cell** Widerstandszelle *f* **~-conductive effect** innerer lichtelektrischer Effekt *m* **~ conductive transducer** lichtleitender Wandler *m* **~ conductivity** lichtelektrische Leitfähigkeit *f* **~ conductor** Licht-, Foto-leiter *m* **~-copying apparatus** Fotokopiermaschine *f* **~-current** Fotostrom *m* **~ current decay** Fotostromabfall *m* **~ current stimulation** Fotostromanregung *f* **~ densitometry** Fotographische Dichtemessung *f* **~ detachment** Fotoablösung *f* **~ dielectric effect** fotodielektrischer Effekt *m* **~ diffusion voltage** Fotodiffusionsspannung *f* **~ diode** Fotodiode *f* **~ disintegration** Fotokernspaltung *f* **~ disintegration of beryllium** Fotoeffekt *m* an Beryllium **~ dissociation** Fotodissoziationskontinua *pl* **~ elastic** polarisationsoptisch, spannungsoptisch **~-elastic method** spannungsoptisches Verfahren *n* **~ elasticity** Spannungsoptik *f*
**photo-electric** fotoelektrisch, lichtelektrisch **~ cathode** Fotokathode *f* **~ cell** fotoelektrische Zelle *f*, Fotoelement *n*, Fotozelle *f*, lichtelektrische Zelle *f*, Lichtelement *n* **~-cell amplifier** Fotozellenverstärker *m* **~ colorimeter** lichtelektrisches Kolorimeter *n* **~ control equipment** lichtelektrische Steueranlagen *pl* **~ current** Fotostrom *m*, lichtelektrischer Strom *m* **~ densitometer** fotoelektrischer Schwärzungsmesser *m* **~ door opener** lichtelektrischer Türöffner *m* **~ effect** Fotoeffekt *m*, fotoelektrische Wirkung *f*, lichtelektrischer Effekt *m*, lichtelektrische Wirkung *f*, Sperrschichtfotoeffekt *m* **~ electron-multiplier tube** Vervielfacherzelle *f* **~ emission** äußerer lichtelektrischer Effekt *m*, äußerer Fotoeffekt *m*, Hallwachseffekt *m*, lichtelektrische Ausbeute *f*, lichtelektrische Elektronenemission *f* **~ emissivity** fotoelektrische Ausbeute, fotoelektrisches Emissionsvermögen *n* **~ eye** elektrisches Auge *n* **~ fatigue** Ermüdungserscheinung *f* (beim Fotoeffekt) **~ film** lichtelektrische Schicht *f* **~ glossmeter** fotoelektrischer Glanzmesser *m* **~ intrusion detector** fotoelektrische Einbruchsicherung *f* **~ layer** lichtelektrische Schicht *f* **~ plethysmograph** fotoelektrischer Plethysmograf *m* **~ pyrometer** fotoelektrisches Pyrometer *n* **~ reflection meter** Weißgradmesser *m* **~ reflection photometer** Elektroremissionsfotometer *n* **~ reflectometer** Weißgradmesser *m* **~ scanning** Fotozellenabtastung *f* **~ scieroscope** fotoelektrischer Härtemesser *m* **~-sound set** Lichttongerät *n* **~ surface** Fotoschicht *f* **~ threshold** Schwellenfrequenz *f* **~ threshold frequency** fotoelektrische Schwellenfrequenz *f* **~ tone generator** Lichtsirene *f*, Licht-

tongenerator *m* ~ **tube** Fotoelement *n*, Fotozelle *f*, lichtelektrische Zelle *f*, Lichtelement *n* ~ **yield** fotoelektrische Ausbeute *f*, fotoelektrisches Emissionsvermögen *n*

**photo,** ~ **electricity** Fotoelektrizität *f*, Lichtelektrizität *f* ~ **electrolytic cell** Elektrolytzelle *f* ~ **electromagnetic voltage** Foto-Hall-Spannung *f* ~ **electromotive force** fotoelektromotorische Kraft *f* ~ **electron** Fotoelektron *n*, (electron micros) Leuchtelektron *n*

**photo-emissive,** ~ **cell** Fotozelle *f* mit äußerem lichtelektrischem Effekt ~ **effect** äußerer Fotoeffekt *m*, äußerer lichtelektrischer Effekt *m*, lichtelektrische Ausbeute *f*, lichtelektrische Elektronenemission *f* ~ **gasfilled cell** Glimm-, Vorglimmlicht-zelle *f*

**photo,** ~ **engraving** Fotogravüre *f*, fotomechanische Vervielfältigung *f*, Lichtdruck *m* ~**-excitation cross section** Fotoanregungsquerschnitt *m* ~ **eyepiece** Fotookular *n* ~ **finish apparatus** Zielfotogerät *n* ~ **fission** Fotospaltung *f* ~ **flash bomb** Blitzlichtbombe *f* ~**-fluorograph** Röntgenschirmbildgerät *n* ~**-galvanomagnetic effect** fotogalvanomagnetischer Effekt *m*

**photogenic** Leuchtenergie *f* erzeugend, licht-entwickelnd, -erzeugend, -gebend ~ **subject (or matter)** Bildmaterial *n*

**photo** ~**-glow tube** Glimmzelle *f* ~**-goniometer** Bildmeßtheodolit *n*, Bildtheodolit *m*, Doppelbildtheodolit *m* ~**-gram** Fotogramm *n*, Meßbild *n*

**photogrammetric,** ~ **apparatus** Bildmeßgeräte *pl* ~ **camera** Meßbildkamera *f* ~ **camera for single photographs** Einfachmeßkamera *f* ~ **camera on stand** Stativmeßkamera *f* ~ **fixing of position** fotogrammetrische Ortsbestimmung *f* ~ **flight** Meßflug *m* ~ **instrument** Luftbildvermessungsinstrument *n* ~ **plotting instrument** fotogrammetrisches Auswertegerät *n* ~ **position finding** fotogrammetrische Ortsbestimmung *f* ~ **stereocamera** Stereomeßkammer *f* ~ **survey** fotogrammetrische Aufnahme *f*

**photogrammetry** Bild-messung *f*, -meßwesen *n*, -verfahrenmesser *m*, Fotogrammetrie *f*, Lichtbildzeichnung *f*, Meßbildverfahren *n* ~ **by intersection** Einschneidefotogrammetrie *f*

**photograph, to** ~ aufnehmen, fotographieren

**photograph** Abbildung *f*, Abdruck *m*, Aufnahme *f*, Bildnis *n*, Fotografie *f*, Fotogramme *n*, fotographische Aufnahme *f*, Licht-bild *n*, -bildaufnahme *f*, -druck *m* ~ **taken from the air** Lichtbild aus der Luft aufgenommen ~ **with approximately horizontal axis** Horizontalaufnahme *f* ~ **of the fragmentation of a shell casing** Splitterbild *n* einer Ladungshülle ~ **of structual defects** Gefügeabbildung *f*

**photograph printing** Lichtdruck *m*

**photographer** Fotograf *m*, Lichtbildner *m*

**photographic** fotografisch ~ **apparatus** Lichtbildgerät *n* ~ **attachment** Fotozusatzgerät *n* ~ **basin** Lichtbildschale *f* ~ **compass** Lichtbildkompaß *m* ~ **density measurement** fotographische Dichtemessung *f* ~ **emulsion** lichtempfindliche Schicht *f* ~ **equipment** Lichtbildausrüstung *f* ~ **extension** Anschluß *m* aufeinanderfolgender Aufnahmen, Bildanschluß *m*, Folgebildanschluß *m* ~ **field lens** Aufnahmeobjektiv

*n* ~ **flight** Bildflug *m* ~ **glass scala** Glasdiapositiv *n* ~ **image** Fotoablichtung *f*, fotographische Aufnahme, Lichtbild *n* ~ **intersection** Einschneidefotogrammetrie *f* ~ **layout drawing** fotographische Zeichnung *f* ~ **map** Bildkarte *f*, Kartenaufnahme *f*, Luftbildkarte *f* ~ **mapping** fotographische Kartenaufnahme *f* ~ **mission** Lichtbildauftrag *m* ~ **observer** fotographisches Registriergerät *n* ~ **overlay** Lichtbildpause *f* ~ **pair** (aviat. photo) Raummeßbilder *pl* ~ **paper** lichtempfindliches Papier *n* ~ **plan** Bildplan *m* ~ **plane** Fotoflugzeug *n* ~ **printing** Lichtpausverfahren *n* ~ **processing apparatus** Bildbearbeitungsgerät *n* ~ **reconnaissance** Bildaufklärung *f*, Bilderkundung *f*, Lichtbilderkundung *f*, Luftbildaufklärung *f* ~ **record** Fotogramm *n* ~ **recorder system** Bildmeßverfahren *n* ~ **report** Bildmeldung *f* ~ **reversal** Umkehrerscheinung *f* ~ **shooting** Aufnahme *f* ~ **shutter** Fotoverschluß *m* ~ **sketch** Bildskizze *f* ~ **solution** fotographische Lösung *f* ~ **sound-film head** Lichttonansatz *m* ~ **sound-film recording** Lichttonaufnahme *f* ~ **sound pick-up head** Lichttonabtastkopf *m* ~ **sound recording** Lichtton *m* ~ **sound reproducer** elektro-optischer Abtaster *m* ~ **sound track** Lichttonschicht *f* ~ **station** Bildstelle *f* ~ **strip** Luftbildreihe *f* ~ **survey** Bilderkundung *f* ~ **surveying** Fotogrammetrie *f*, Meßbildkunst *f* ~ **surveying apparatus** Fotogrammeter *n* (Gerät) ~ **taking** (of pictures) Aufnahme *f* ~ **tracing** Lichtpause *f* ~ **transmission by electromechanical (or electrochemical) means** elektromechanische und elektrochemische Bildübertragung *f* ~ **travel ghost** Bildziehen *n* ~ **unit** Luftbildabteilung *f* ~ **wood-engraving** Fotoxylografie *f* ~ **writing reception** Lichtschreiberempfang *m*

**photographical recorder** fotografischer Empfänger *m*, Lichtschreiber *m*

**photographically recordable** fotografisch erreichbar

**photographs** Aufnahmen *pl* ~ **faithful to minute details** Aufnahmen *pl* mit großer Auflösung

**photography** Fotografie *f*, Lichtbild *n*, Lichtbild-aufnahme *f*, -kunst *f*, -wesen *n* ~ **of lightning** Blitzfotografie *f*

**photograving** Lichtdruckverfahren *n*

**photogravure** Foto-, Printo-gravüre *f*, Kupferlichtdruck *m*, -tiefdruck *m* ~ **color printing** Farbentiefruck *m* ~ **printing machine** Tiefdruckmaschine *f*

**photo,** ~ **impact** Quantenstoß *m* ~ **interpreter** Bildauswerfer *m* ~ **ionisation efficiency** Fotoionisierungsausbeute *f* ~**litho** Fotolithografie *f* ~**lithographer** Fotolithograf *m* ~**lithography** Licht-druck *m*, -steindruck *m* ~ **loft template** Fotostrakschablone *f*

**photology** Licht-lehre *f*, -wissenschaft *f*

**photoluminescence** Fotolumineszenz *f*

**photolyctic cell** Elektrolytzelle *f*

**photomacrography** Makrofotografie *f*

**photo-mechanical printing** Lichtdruck *m*

**photomeson** Fotomeson *n*

**photometer** Fotometer *n*, Licht-messer *m*, -stärkemesser *m*, -strommesser *m* ~ **deflection** Fotometerausschlag *m* ~ **head** Fotometeraufsatz *m* ~ **light entry opening** Fotometeröffnung *f* ~ **screen** Fotometerschirm *m* ~ **test** Fotometerprobe *f*

**photometric** fotometrisch ~ **brightness** gemessene Leuchtkraft *f*, Helligkeit *f* als physikalische Größe, Lichtstärke *f* ~ **evaluation** Fotometrierung *f* ~ **integrator** Topffotometer *n* ~ **recording** Fotometrierung *f* ~ **screens** fotometrische Tafeln ~ **standard** Normallampe *f* ~ **unit** Lichtmaßeinheit *f* ~ **wedge** Streifenkeil *m*
**photometrograph** fotometrische Aufnahme *f*
**photometry** Fotometrie *f*, Leuchtkraftbestimmung *f*, Licht-messung *f*, -stärkemessung *f*
**photo,** ~ **microgram** Mikrofotogramm *n* ~ **micrograph** Mikrofotografie *f*, Schliffbild *n* ~**micrographer** Mikrofotograf *m* ~ **micrographic** mikrofotografisch ~**micrographic** mikrofotografisch ~**micrography** Kleinbildfotografie *f*, Kleinlichtbildkunst *f*, Mikrofotografie *f* ~**microscopy** Mikrofotografie *f* ~**missive effect** Hallwachseffekt *m* ~ **mosaic** Fotomosaik *n* ~**mount** Fotorahmen *m*
**photomultiplier** Fotoverstärker *m* ~ **counter** Szintillationszähler *m* ~ **tube** Fotovervielfachungsröhre *f*, Fotozelle *f*
**photon** Photon *n*, Licht-korpuskel *n*, -quant *m*, Torland *n* (opt.) ~ **absorption** Photonenabsorption *f*
**photoneutron** Fotoneutron *n*
**photonuclear,** ~ **interaction** Kernfoto-Wechselwirkung *f* ~ **reaction** Kernfotoeffekt *m*
**photo,** ~ **nucleons** Fotonukleonen *pl* ~**optics** Lichtoptik *f* ~**parallel** Bildparallele *f* ~**peak efficiency** Fotomaximumausbeute *f* ~**philic (or philous)** lichtfreundlich ~**phobia** Lichtscheu(e) *f* ~**phobic** lichtscheu ~**phone** Fotofon *n* ~**physics** Fotophysik *f* ~**planigraph** Fotoplanzeichner *m*
**photoprint** Abzug *m* von einem Negativ, Kopie *f*, Lichtdruckätzung *f* ~ **copy** Fotogramm *n*
**photoprinting** fotografischer Zeichendruck *m* ~ **paper** Lichtpauspapier *n* ~ **telegraph** Telegraf *m* mit fotografischem Zeichendruck
**photo,** ~ **production of deuterons** Fotoauslösung *f* von Deuteronen ~ **radio transmission** Bildfunk *m* ~**restistance** lichtelektrischer Widerstand *m* ~**resistance cell** Widerstandszelle *f* ~**resistor** lichtelektrische Widerstandszelle *f* ~**-scintigraph** Fotoszintigraf *m* ~**sensitive surface** Fotoschicht *f*, lichtempfindliche Oberfläche *f* ~**sensitivity** Lichtempfindlichkeit *f*
**photosensitize, to** ~ lichtempfinglich machen
**photosensitized mosaic plate** Rasterplatte *f*
**photosensitized mosaic** gerasterte Schicht *f*
**photosphere** Leucht-halle *f*, -hülle *f*
**photostat, to** ~ fotokopieren
**photostat** Fotokopie *f*, Lichtpause *f* ~ **equipment** Nachbildungsgerät *n*
**photostatic copy** Ablichtung *f*
**photo,** ~**sulphide of iron** Einfachschwefeleisen *n* ~**synthesis** Fotosynthese *f* ~**telegram** Bildtelegramm *n* ~**telegraphy** Bildübertragungsgerät *n*, Faksimiletelegrafie *f*, Fernbildschrift *f*, Fernfotografie *f* ~**telephony** Lichttelefonie *f* ~**theodolite** Fototheodolit *m* ~**timer** fotoelektrischer Belichtungsautomat *m* ~ **topography** Erdbildmessung *f* ~**transistor** Kristalllichtverstärker *m* ~**tropism** (phototropy) Fototropie *f* ~**tube** Foto-elektronenröhre *f*, -element *m*, -zelle *f*, Lichtelement *n* ~**type** Lichtdruck *m* ~**valve** Fotozelle *f* ~**varistor** Foto-

varistor *m* ~**visual technique** fotovisuelle Methode *f*
**photovoltaic** fotoelektrisch, lichtelektrisch ~ **barrier-layer cell with posterior metallic layer** Hinterwandzelle *f* ~ **cell** Elektrolyt-, Sperrschicht-zelle *f* ~ **effect** Becquereleffekt *m*, (photovoltage) Fotospannung *f*, Fotovolteffekt *m* ~ **photocell** Sperrschichtfotozelle *f* ~ **transducer** Fotovoltwandler *m*
**phrase** Ausdrucksweise *f*, Redensart *f*, Stil *m* ~ **in parenthesis** Einschaltung *f* ~ **intelligibility** Satzverständlichkeit *f*
**phreatic,** ~ **line** Durchströmungslinie *f*, Linie *f* der Sättigung, Sickerlinie *f* ~ **nappe** Grundwasserspiegel *m* ~ **surface** Sickerfläche *f* ~ **water** Grundwasser *n*
**phthalic acid** Phthalsäure *f*
**phthalide** Phthalid *n*
**phugoid,** ~ **motion** Dauerschwingung *f*, Phugoidbewegung *f* ~ **oscillation** phugoide Schwingung *f*
**pH-value** pH-Wert *m*
**phylum** Stamm *m*
**physical** körperlich, physikalisch, physisch ~ **ability to stand high altitude** Höhenfestigkeit *f* ~ **appearance** Erscheinungsform *f*, Habitus *m* ~ **cathode** körperliche Kathode *f* ~**-chemical** physikalisch-chemisch ~ **circuit** Stammkreis *m* ~ **condition** Aggregat-, Form-, Gesundheitszustand *m* ~ **connection** galvanische Verbindung *f* ~ **constants** Kenndaten *pl* ~ **defects** Materialmängel *pl* ~ **evidence** physikalische Definition *f* ~ **examination** ärztliche oder körperliche Untersuchung *f* ~ **impossibility** absolute Unmöglichkeit *f* ~ **inventory** körperliche Bestandsaufnahme *f* ~ **line** Stammleitung *f* ~ **mass unit** Atommasseneinheit *f* ~ **optics** Lichtoptik *f* ~ **photometer** objektives Fotometer *n* ~ **properties** Festigkeitseigenschaften *pl* ~ **scale** physikalische Skala *f* ~ **test** physikalische Prüfung *f* ~ **tracer** physikalischer Indikator *m* ~ **verification** physische oder effektive Bestandsaufnahme *f* ~ **weather analysis** physische Wetteranalyse *f*
**physically,** ~ **realizable** physikalisch sinnvoll ~ **separated** galvanisch getrennt
**physician** Arzt *m*
**physicist** Naturforscher *m*, Physiker *m*
**physico,** ~**-chemical** physikalischchemisch ~**-metallurgical** physikalischmetallurgisch
**physics** Naturlehre *f*, Physik *f*
**physiological** physiologisch ~ **research pertaining to aviation** Flugmedizin *f*
**physiology** Physiologie *f*
**physique** Körperbeschaffenheit *f*
**phyto-sanitary certificate** Pflanzenschutzzeugnis *n*
**pians linge** Scharnierband *n*
**pica** Cicero *f*, Pica *f* (print.)
**pick, to** ~ auszupfen, belesen, einschießen, kletten, noppen, scheiden, (Weberei) werfen to ~ **apart** analysieren, zerreißen to ~ **off** abgreifen to ~ **out** ausersehen, aushalten, ausklauben, auslesen, aussortieren, aussuchen, klauben, lesen to ~ **up** abtasten, anleuchten, (a signal) auffangen, auffassen, aufholen, auflesen, aufnehmen, (energy from an electron beam) entnehmen, erfassen, herausreißen, auf Null ab-

nehmen, (as of an engine) wieder aufholen **to ~ up a buoy** an die Boje gehen **to ~ up a cable** ein Kabel aufnehmen **to ~ up speed** auftouren **to ~ the wool** die Wolle plüsen

**pick** Erweiterungskeilhaue *f*, Flachhaue *f*, (one end flat-edged) Flachspitze *f*, Hacke *f*, Keilhaue *f*, Kreuzhacke *f*, Messerpicke *f*, Picke *f*, Pickel *m*, Pickhacke *f*, Pickhammer *m*, Preßlufthammer *m*, (textiles) Schützenschlag *m*, Spitzenhaue *f*, Spitz-hacke *f*, -haue *f*, -krampen *m*, (tamping) Stopfhacke *f*, Zweispitze *f*

**pick-a-back, ~ airplane** Huckepackflugzeug *n* **~ operation** Reitersitz *m* **~ towing** Tragschlepp *m*

**pickax** Beilpicke *f*, Breithacke *f*, Erdhaue *f*, (one end flat-edged) Flachspitze *f*, Gerinnhaue *f*, Haue *f*, Keilhaue *f*, Kreuzhacke *f*, Picke *f*, Pickel *m*, Spitz-hacke *f*, -haue *f*, -krampen *m*

**pick, ~ breaker** Gußzerkleinerer *m* **~ carrier** Pickenträger *m* (min.) **~ counter** Schußzähler *m* **~ hammer** Brech-, Klaub-hammer *m* **~ handle** Pickengriff *m* **~ lock** Dietrich *m*, Nachschlüssel *m* **~ mattock** Brechhammer *m*, Klapphacke *f* **~-off** Abgriff *m* (gyro), Abstreifvorrichtung *f* **~-off attachment** Greifeinrichtung *f* **~-off coil** Abnahmespule *f* **~-off gears** Wechselscheiben *pl* **~ pincers** Noppeisen *n* **~ point** Hauespitze *f*

**pick-up, ~ on the cutting edge** Aufbauschneide *f* (Drehstahl) **~ from external electrical sources** Einstreuung *f* durch Störungen fremder Stromquellen (rdo) **new ~ of film** Frischaufnahme *f* **~ of motion** Bewegungsempfänger *m* **~ of production** Beschleunigungsmöglichkeit *f*

**pick-up, ~ amplifier** Schallplattenverstärker *m* **~ box** Druckmeßdose *f* **~ bush** Schalldosenbuchse *f* **~ camera** Aufnahmekamera *f*, Bildgeber *m* **~ carrier** Greiferwagen *m* **~ cartridge** Tonabnehmerkopf *m* **~ characteristic** Abtasterkennlinie *f* **~ coil** Prüfspule *f* **~ compensation** Schallplattenentzerrer *m* **~ condenser** Fühlerkondensator *m* **~ device** Tonabnehmereinrichtung *f* **~ factor** Aufnahmefaktor *m* **~ grab** Fangzange *f* **~ groove** Haltekerbe *f* **~ head** Tonkopf *m* **~ hook** Aufnahmehaken *m* **~ key** Tonabnehmertaste *f* (tape rec) **~ light** Leitscheinwerfer *m* **~ link transmitter** Kameraverbindungssender *m* **~ magnet** Aufnahmemagnet *m* **~ means** Fühlhebel *m* **~ method** Aufnahmeverfahren *n* **~ microphone** Lauschmikrofon *n* **~ microphonic device** Abhorchgerät *n* **~ needle with sphericrl point** (soundengraved film) Kugelspitzenfühlnagel *f* **~ performance** (of antenna) Aufnahmefähigkeit *f* **~ plate** Signalplatte *f* **~ point** Aufnahmeort *m* (TV) **~ rate** Abtastgeschwindigkeit *f* (phono) **~ roll** Aufnahmewalze *f* **~ sound box** Abtastdose *f* **~ system** Abtastsystem *n*, Seilpost *f* **~ tongs** Aufhebezange *f* **~ traffic** Zusteigeverkehr *m* **~ transmitter** Aufnahme-, Lausch-mikrofon *n* **~ truck**

schneller Motorlastwagen *m* **~ tube** Abtast-, Aufnahme-röhre *f* **~ unit** Abtastdose *f* **~ value** Ansprechwert *m* **~ velocity** Abtastgeschwindigkeit *f* **~ winding** Geberwicklung *f*

**picked** ausgesucht, mit der Spitzhaue bearbeitet **~ cotton** gezupfte Baumwolle *f* **~ore** Scheideerz *n* **~ up** abgehört

**picker** Ankerpfahl *m* (electr), (stick) Bergeisen *n*, Bohrnadel *f*, Klauber *m*, Klaubhammer *m*, Klaubjunge *m*, Schlagarm *m*, (weaving) Treiber *m*, (weaving) Webvogel *m* **~ cams** Schlagexzenter *n* **~ knife** Greifmesser *m* **~-stick blow** (textiles) Schützenschlagvorrichtung *f*

**picket, to ~** verankern

**picket** Absteck-pfahl *m*, -pflock *m*, (surveying) Kettenhalter *m*, (surveying) Kettenstab *m*, Pfahl *m*, Pfeil *m*, Pflock *m*, Pfosten *m*, Piket *n*, Posten *m*, Stange *f* **~ post** Lagerpfahl *m*

**picketing-line attachments** Fesselungsleine *f*

**pickets** Buhnenpfähle *pl*

**picking** Abpalen *n*, Auswicken *n*, Entketten *f*, Scheidung *f* **~ belt** (conveyers and crushers) Förderband *n*, Klaubeband *n* **~ belt conveyer** Leseband *n* **~ drum** Läutertrommel *f* **~ machine** (hair and hides) Enthaar- und Glattmaschine *f* **~-out slate** Steinausklaubung *f* **~ plant** Sortieranlage *f* **~ plates of loom** Antriebsplatten *pl* der Schlagarme **~ shaft** bestiftete Schlagwelle *f* **~ table** Klaub(e)-, Lese-, Wander-tisch *m* **~ table for line assembly** Wandertisch *m* für fließende Fertigung **~-up** Aufnehmen *n* **~-up of motor** Auftourenkommen *n* des Motors **~-up target** Einsteuerung *f* (aviat.)

**pickings** Klaubeberge *pl*

**pickle, to ~** abbeizen, abbrennen, beizen, dekapieren, entzundern, pickeln **to ~ brass** Messing *n* abtrennen

**pickle** Beize *f* **~ action** Pickelwirkung *f*

**pickled** (goods) gepökelt **~ sheet metal** Mattblech *n* **~ sheets** dekapierte Bleche *pl*

**pickling** Abbeizen *n*, Beizbehandlung *f*, Beizen *n*; beizend **~ agent** Abbeizmittel *n* **~ basket** Beizkorb *m* **~ bath** Beizbad *n* **~ compound** Beizzusatz *m* **~ fluid** Beizflüssigkeit *f* **~ house** Beizerei *f* **~ plant** Beizanlage *f* **~ solution** Beizlösung *f*

**picks** Mühl- und Messerpicken *pl*

**picofarad** Pikofarad *n*

**picot edge** (textiles) Zackenrand *n*

**picotite** Pikotit *m*

**picric acid** Bitter-, Piktin-säure *f*

**picrosmine** Pikrosmin *m*

**pictorial** bildhaft **~ character** Bildcharakter *m* **~ depth** Bildtiefe *f* **~ feature** Bildbericht *m* **~ service** Bilddienst *m*

**picture** Abbildung *f*, Aufnahme *f*, Bild *n*, Bildnis *n*, Fotografie *f*, Schnittzeichnung *f*, Zeichnung *f* **~ of low brightness** lichtschwaches Bild *n* **~ of proper shading** tönungsrichtiges Bild *n* **~ with proper shading values and contrast** tönungsreiches Bild *n* **~ rich in detail and contrast** detailreiches Bild *n*

**picture, ~ amplifier** Bildverstärker *m* **~ amplifier cubicle** Bildaufzeichnungsverstärker *m* **~ analyzer** Bildzerleger *m* **~ area** Bildfeld *n* **~ aspect ratio** Bildseitenverhältnis *n* **~ background** Bildhintergrund *m* **~ black** Schwarzpegel *m* **~ carrier frequency** Bildträgerfrequenz *f* ~-change-over Bildüberblendungsvorrichtung *f* ~ **charge** Bildladung *f* ~ **check print** Probekopie *f* (film) ~ **chief** Bildmeister *m* ~content Bildinhalt *m* ~control Bildsteuerung *f* (TV) ~ **coordinate** Bildkoordinate *f* ~-creation Bilderzeugung *f* ~ **current** Bildstrom *m* ~ **cycle** Bild-sprung *m*, -wechsel *m*, wechselzeit *f* (film) ~ **cylinder** Bildwalze *f* ~ **definition** Bild-auflösung *f*, -schärfe *f* ~ **delineation** Bild-aufzeichnung *f*, -beschreibung *f* ~ **delineator** Bildzusammensetzvorrichtung *f* ~ **detail** Bildausschnitt *m* ~ **director** Bildmeister *m* ~ **dissector** Bild(feld)zerleger *m* ~ **distortion** Bild-fehler *m*, -verzerrung *f* ~ **element** Bild-element *n*, -punkt *m*, Raster-element *n*, -punkt *m* ~ **embossing** Bildprägung *f* ~ **explorer** Bildzerleger *m* ~-exploring means Bildfeldzerleger *m* (TV) ~ **eye** Bilderöse *f* ~ **feed roller** Bildtransportrolle *f* ~ **field** Bildfläche *f* ~ **field illumination** Bildfeldausleuchtung *f* ~ **format** Bildformat *n* ~-forming tube on infrared spotter Bildwandler *m* ~ **frame** Rahmen *m* ~-frame cabane Rahmenspannturm *m* ~ **frame girder** Viereckträger *m* ~-frame strut Rahmentiefenkreuz *n* ~ **frame-type fracture** Bilderrahmenbruch *m* ~ **frequency** Bild-frequenz *f*, -modulations-frequenz *f*, -punktfrequenz *f*, -wechselfrequenz *f*, -zahl *f* ~ **gate** Abtaststelle *f*, Bildtaststelle *f* (film), Bildfenster *n* ~ **gate lenses** Bildfensterlinsen *pl* ~ **impulses** Bildschwingungen *pl* ~-interval regulator Bildfolgeregler *m* ~ **level** Bildpunkt *m* ~ **line** Bildzeile *f* ~ **lock** Bildhalt *m* ~ **measuring** Bildmessung *f* ~-measuring apparatus Bildmeßgeräte *pl* ~ **mirroring** Spiegelung *f* ~ **modulation** Bildmodulation *f* ~ **molding** Bilderleisten *pl* ~ **monitor** Bildmonitor *m* (TV) ~ **output** Bildzeitbasis *f* ~ **panel** Bildertafel *f* ~ **pashing** Bildstrichverstellung *f* ~-plan Bildplan *m* ~ **point** Bildpunkt *m*, Raster-element *n*, -punkt *m* ~ **portions of composition (or of different nature)** verschieden gegliederte Bildteile *pl* ~ **portions of dissimilar classification (or composition, make-up, nature, organization)** Bildteile *pl* verschiedener Gliederung, verschieden gegliederte Bildteile *pl* ~ **portions of organization** verschieden gegliederte Bildteile *pl* ~ **projector** Bildprojektor *m* ~ **quality** Bildgüte *f* ~ **raster** Bildraster *m* ~ **ratio** Bildformat *n*, Seitenverhältnis *n* ~ **receiver** Bildempfänger *m* ~ **receiving office (or station)** Bildempfangsstelle *f* ~ **reconstructor** Bildempfänger *m* ~ **recording equipment** Bildaufzeichnungsanlage *f* ~ **recreation** Bild-aufzeichnung *f*, -beschreibung *f*, Nachschrift *f* ~ **recreator** Bildzusammensetzvorrichtung *f* ~ **recreator aperture** Zusammensetzblende *f* ~ **repetition frequency** Bildsprung *f* ~ **reproducing device** Fernsehempfänger *m* ~ **reproducing tube** Bildschreibröhre *f* ~ **reproduction means** Bildzusammensetzvorrichtung *f* ~ **resolution** Bildauflösung *f* ~ **rotate control** Regelung *f* der Bilddrehung ~ **scale** Abbildungsmaßstab *m* ~ **scan** Bildablenkung *f*, Bildabtastung *f* ~ **scanner** Bildfeldzerleger *m*, (at receiving end) Bildpunktverteiler *m*, Bildstelle *f*, Bildzerleger *m*, Bildzusammensetzvorrichtung *f* ~ **screen** Bildfenster *n*, Bildschirm *m* ~ **sequence** Reihenbildaufname *f* ~ **shape** Bildformat *n*, Seitenverhältnis *n* ~

**sheet** Bilderbogen *m* ~ **signal** Bild-, Fernsehsignal *n* ~ **signal input** Bildsignaleingang *m* ~ **signal modulation** Bildmodulation *f* ~ **signals** Bildschwingungen *pl* ~ **strip** Bildzeile *f* ~ **synchronized** bildsynchron ~ **synchronizing impulse (or signal)** Bildimpuls *m*, Bildsynchronisierimpuls *m* ~ **synthesis** Bildaufbau *m* ~ **telegraph apparatus** Bildübertragungsgerät *n* ~ **telegraphy** Bildtelegrafie *f*, Faksimiletelegrafie *f*, Fernbildschrift *f*, Fernfotografie *f* ~ **timebase oscillation** Bildkippschwingung *f* ~ **tracing** Beschreibung *f* von Bildern auf Schirm, Bild-aufzeichnung *f*, -beschreibung *f*, -erzeugung *f* ~ **transmission** Bildfernübertragung *f*, Bildübertragung *f* ~ **transmission by electromechanical (or electrochemical) means** elektromechanische und elektrochemische Bildübertragung *f* ~ **transmitter** Bildübertrager *m* ~ **trap** Bildsperre *f* (TV) ~ **traversing** Zeilenschaltung *f* ~ **tube** Bildröhre *f* (TV) ~ **unit** Bildgerät *n* ~ **unit area** Bildpunkt *m* ~ **viewer** Bildlupe *f* ~ **viewing tube** Bildempfänger *m* ~ **voltage** Bildspannung *f* ~ **white** Weißpegel *m*
**pictures** Bildreihe *f* (film)
**piebald** gescheckt
**piece, to** ~ anstücken, ausbessern **to** ~ **together** anstückeln
**piece** Ballen *m*, Brocken *m*, Faß *m*, Külbchen *n*, Posten *m*, Scherbe *f*, Schrot *m*, Schrott *m*, Stück *n*, Teil *m* **by** ~ stückweise
**piece,** ~ **of equipment** Gerätstück *n* ~ **of furniture** Möbel *n* ~ **of installation** Einbaustück *n* ~ **made of a single** aus einem einzigen Stück *n* gefertigt **new reading an adjusting the** ~ Festlegezahl *f* ~ **in one** einstückig, einteilig ~ **of rule** Linienstück *n*
**piece,** ~ **calculation** Stückrechnung *f* ~ **cost** Stückkosten *pl* ~ **cutter** Stückschneidemaschine *f* ~ **doubler** Warendoppler *m* ~ **dyer** Stückfärber *f* ~ **end printing machine** Endleistendruckmaschine *f* ~ **end sewing machine** Kuppelmaschine *f* ~ **good service** Stückgutbetrieb *m* ~ **goods** Stück-gut *n*, -ware *f* ~ **goods service** Lasthakenbetrieb *m* ~ **hand** Paket-, Stücksetzer *m* ~-meal stückweise ~ **part** Einzelteil *n* ~ **per-hour rate** Akkordsatz *m* ~ **production costs** Herstellungsstückkosten *pl* ~ **scouring** (textiles) Entfettung *f* im Stück ~ **smooth** stückweise glatt (math.) ~ **wages** Akkord *m* ~
**piece-work** Akkordarbeit *f*, Gedinge *n*, Gedingearbeit *f*, Paketsatz *m*, Stücksatz *m*, Stückwerk *n* ~ **rates** Akkordlohn *m* ~ **wage** Gedingelohn *m*
**piece-worker** Stücklohnarbeiter *m*
**piecer** Fadenanleger *m* (textiles) ~ **and doffer** Anspinner *m*
**piedmontite** Manganepidot *m*
**pier** Anlegestelle *f*, Brückenpfeiler *m*, Dock *n*, Fluß-, Hafen-damm *m*, Hafenhaupt *n*, Kai *m*, Ladeplatz *m*, Ladezunge *f*, Landebrücke *f*, Landungs-brücke *f*, -damm *m*, Mole *f*, Pfeiler *m*, Pier *m*, Strebepfeiler *m*, Strompfeiler *m*, Wellenbrecher *m*, Widerlager *n* ~ **of blocks** Blockschichten *pl*
**pier,** ~ **arch** freistehender Längengurt *m* ~ **head** (in a pier used to unload ships) Molenkopf *m* ~ **nose** (in birdge piers) Pfeiler-kopf *m*, -vorkopf *m*

**pierce, to** ~ anstechen, bohren, durch-bohren, -brechen, -dringen, -löchern, -stechen, -stoßen, eindringen, lochen, stechen, vorstechen
**pierced** durch-brochen, -löchert ~ **steel plank (PSP)** Stahlstartbahn *f*
**piercer** Lochdorn *m*, Stopfen *m* ~ **rod** Lochdornstange *f*
**piercing** durchgehend, scharf (acoust.) ~ **die** Lochwerkzeug *n* ~ **effect** (explosives) Durchschlagskarft *f* ~ **mandrel** Lochdorn *m* ~ **mill** Stopfenwalzwerk *n* ~ **point** Durchstoßpunkt *m* ~ **point connection** (Zündanker) Stechspitzenanschluß *m* ~ **test** Durchschlagprobe *f*
**piezo-electric** druck-, piezo-elektrisch ~ **crystal** piezoelektrischer Kristall *m* oder Quarz *m*, Quarzsender *m*, Rechtsquarz *m* ~ **effect** Piezoeffekt *m* ~ **gauge** Druckmeßdose *f*, Piezoindikator *m* ~ **loudspeaker** Kristallautsprecher *m*, piezoelektrischer Lautsprecher *m* ~ **manometer** piezoelektrischer Druckmesser *m* ~ **microphone** Kristallmikrofon *n*, piezoelektrisches Mikrofon *n* ~ **oscillator** piezoelektrischer Oszillator *m* ~ **quartz** piezoelektrischer Quarz *m* ~ **receiver** Kristallfernhöhrer *m* ~ **resonator** piezoelektrischer Resonator *m*, Piezo-, Quarz-resonator *m* ~ **strain gauge** piezoelektrischer Dehnungsmeßstreifen *m*
**piezoelectricity** Druck-, Piezo-elektrizität *f*
**piezoid** fertiger Quarzkristall *m*
**piezometer** Druckmesser *m*, (for water pressure) Druckmeßgerät *n*, Flüssigkeitsdruckmesser *m*
**piezometric head** piezometrisches Gefälle *n*
**piezotropy** Piezotropie *f*
**pig** Block *m*, Blöckchen *n*, Blockeisen *n*, Eisenklumpen *m*, Floß *n* (metall.), Flosse *f* (metall.), Gans *f* (metall.), Massel *f*, Mulde *f*, Roheisen *n*, Roheisenstück *n*, Schwein *n*, Zain *m* ~ **of lead** Bleibarren *m*
**pig,** ~**-and-ore process** Erzschmelzen *n*, Erzstahlverfahren *n*, Roheisenerzschmelzverfahren *n*, Siemens-Prozeß *m* ~**-and-scrap method** Schrott-martinieren *n*, -verfahren *n* ~**-and--scrap process** Siemens-Martin-Ofen-Prozeß, Siemens-Martin-Ofen-Verfahren *n* ~ **bed** Gieß-, Massel-bett *n* ~ **bed dressing machine** Gießbettaufbereitungsmaschine *f* ~ **bed pattern** Masselkamm-Modell *n* ~ **boiling** Kochpuddeln *n*, (of iron) Fettpuddeln *n*, Puddeln *n*, Schlakken-frischen *n*, -puddeln *n* ~ **breaker** Fallwerk *n*, Masselbrecher *m* ~ **casting machine** Masselgießmaschine *f* ~ **casting yard** Gießhalle *f* ~ **driver** Ramme *f* ~ **handling crane** Masseltransportkran *m* ~ **iron** Block-, Dreck-, Guß-, Massel-, Roh-, Schweine-eisen *n* ~ **iron for making steel** Rohstahleisen *n* ~ **iron for refining** Frischerei-, Herdfrisch-roheisen *n*
**pig-iron,** ~ **barrow** Roheisenkarren *m* ~ **brand** Roheisen-gattung *f*, -marke *f* ~ **charge** Roheiseneinsatz *m* ~ **grade** Roheisen-marke *f*, -sorte *f* ~ **handling bridge** Masselverladebrücke *f* ~ **output** Roheisen-ausbringen *n*, -erzeugungsmenge *f* ~ **production** Roheisenerzeugung *f* ~ **purifying process** Roheisenfrischverfahren *n* ~ **truck** Roheisenkarren *m*
**pig,** ~ **lead** Bleigans *f*, Blockblei *n*, Muldenblei *n* ~ **machine** Massel-, Roheisen-gießmaschine *f* ~ **mold** Flossenbett *n*, Masselform *f*, Mulde *f* ~ **nickel** Blocknickel *m* ~ **process** Roheisenver-

fahren *n* ~ **strand** Masselstrang *m* ~ **string** Masselstrang *m* ~ **tails** Anschlußdrähte *pl* (bei Transistoren)
**pigeon** Taube *f* ~ **hole** Ablagefach *n*, Fach *n* ~ **loft** Brieftaubenstelle *f*, Schlag *m*, Taubenschlag *m*
**pigging** Roheisenzugabe *f*
**"piggyback"-case** „Huckepack"-Gehäuse *n*
**pigment** Farbe *f*, Farb-körper *m*, -stoff *m*, Pigment *n*, Pigmentfarbstoff *m* ~ **gilding** Beizvergoldung *f*
**pigmented** gefärbt
**pigs** Roheisenmasseln *pl*
**pigsty timbering** Ausbau *m* mittels Holzpfeilern
**pigtail** Ausgleichswindung *f* (aviat.)
**pike** Kerbhaue *f*
**pikeman** Hauer *m*
**pilaster** Pilaster *m*, Wandpfeiler *m*
**pile, to** ~ aufschütten, auspfählen, häufeln, packetieren, paketieren, stapeln, zusammensetzen **to** ~ **in layers** anschichten **to** ~ **up** anhäufen, aufhäufen, aufschichten, aufstapeln, schichten, zusammenhäufen **to** ~ **and weld the steel** den Stahl *m* gerben
**pile** Batterie *f*, Bock *m* (bridge), Brenner *m*, galvanische Säule, Garbe *f*, Haardecke *f*, Haufen *m*, Kreuz *n* (print.), Meiler *m*, Paket *n*, Pfahl *m*, (in fabrics) Pol *m*, Pole *f*, Reaktor *m*, Säule *f* (electr.), Schicht *f*, Schweiß(eisen)paket *n*, Stapel *m*, Stoß *m*, Stützpfahl *m*, Uranbrenner *m*
**pile,** ~ **of arms** Gewehrpyramide *f* ~ **of documents** Aktenstoß *m* ~ **to be driven** Rammpfahl *m* ~ **with drawing engine** Pfahlausheber *m* ~ **of rubble** Trümmerhaufen *m* ~ **of scrab** Schrottberg *m* ~ **of substructure** Grundpfahl *m* **the** ~ **sits on the ground** der Pfahl sitz auf
**pile,** ~ **battery** Anodenbatterie *f* ~ **bent** Pfahlgründung *f*, -joch *n* ~ **bents** Joch *n* ~ **block** Afterramme *f*, Ramm-jungfer *f*, knecht *m* ~ **board** Stapeltisch *m* ~ **body** Pfahllänge *f* ~ ~ **bridge** Jochbrücke *f*, Pfahljochbrücke *f* ~ **brushing device** Strichbürsteinrichtung *f* ~ **cap** Pfahlrost *m* ~ **charring** Meiler-verfahren *n*, -verkohlung *f* ~ **cluster** Pfahlrost *m* ~ **coil** Mehrlagenspule *f* ~ **coking** Meilerverkokung *f* ~ **cover** Schlaghaube *f* ~ **delivery** Druckstapler *m*, Stapelausleger *m* ~ **driver** Fallwerk *n*, Ramme *f*, Rammklotz *m* ~ **driver earth tamper** Explosionsramme *f* ~ **driver raft** Rammfähre *f* ~ **driver staging** Rammbühne *f* ~ **driving** Auslängen *n*, Pfählen *n*, Pfahlschlagen *n*, Rammarbeit *f*, Rammen *n*, Verpfählung *f* ~ **driving engine** Rammvorrichtung *f* ~ **driving frame** Pfahlführungsrahmen *m* ~ **driving plant** Pfahlrammanlage *f* ~ **extraction drawing** Ausziehen *n* eines Pfahles ~ **extractor** Pfahlausziehungsmaschine *f* ~ **fabric** Sammetstoff *m*
**fabrics** Polgewebe *n* ~ **factor** Überhöhungsfaktor *m* ~ **foundation** Pfahl-gründung *f*, -rostfundament *m*, -werk *n* ~ **guide** Raumträger *m* ~ **hammer** Ramme *f* ~ **height** Einsatzhöhe *f* ~ **lifting apparatus** Velourhebeapparat *m* ~ **loading test** Probebelastung *f* eines Pfahles ~ **lock** (dam) Sturzwehr *n* ~ **lowering controls** Stapeltischschaltung *f* ~ **operations** Betrieb *m* eines Reaktors ~ **oscillator** Reaktorenoszillator *m* ~ **point** Pfahlspitze *f* ~ **poisoning** Reak-

torverseuchung f ~ **reactor** Reaktor m ~ **reinforcement** Pfahlbewehrung f ~ **road** Bohlenbahn f ~ **shearing cutter** Polschermaschine f ~ **toe** Pfahlspitze f ~ **tube** Florschlauch m ~ **warp** Oberkette f, Polkette f ~ **wood lag** Schichtholzbelagbrettchen n ~ **work** Pfahlbau m, Pfählung f ~ **wound coil** Scheibenwicklung f
**piled** gestapelt ~ **fendering** Leitwerk n ~ **up** aufgehäuft
**piling** Längsbrett n, Paketierung f, Paketierverfahren n, Pfählen n, Pfahlschlagen n, Pfählung f, Rammarbeit f ~ **for buffer wall of a pier** Gordungspfahl m ~ **through loose rock** Abtreibearbeit f in losem Gebirge ~ **through quicksand** Getriebearbeit f in losem Gebirge ~ **up** Anhäufung f, (wood) Schichtung f ~ **up of water by any means** Anstau m
**piling,** ~ **elevator** Aufstauelevator m ~ **frame** Ramme f ~ **pole** Rammpfahl m ~ **stores** Vorratshaufen m ~ **winch** Aufhängehaspel f
**pillar** Bergepfeiler m, Pfeiler m, Pfosten m, Säule f, Spindel f, Ständer m, Ständersäule f, Streckenpfeiler m, Stütze f, Tragsäule f
**pillar,** ~ **of a blast furnace** Vierpaß m eines Hochofens ~ **of coal** Kohlenpfeiler m ~ **of ground** Grundstreckenpfeiler m ~ **of lamps** Flächenlampe f, Lampenaggregat n, Mehrfachstrahler m ~ **of luminous sources** Flächenleuchte f ~ **of ore** Erzpfeiler m ~ **of a telescope** Stativsäule f eines Fernrohres
**pillar,** ~ **bracket** Säulenarm m ~ **buoy** Spierentonne f ~ **crane** feststehender Verladekran m ~ **dredger** Säulenschwenkbagger m ~ **file** dickflache Feile f ~ **hydrant** Überflurhydrant m ~ **lamp** Säulenlampe f ~**-like** säulenförmig ~ **pier** Pilarwalzwerk n ~ **sample** Säulenprobe f ~ **station point** Pfeilerpunkt m ~ **test box** Untersuchungssäule f
**pillbox** Hohlleiter m mit parallelen Platten (microwaves) ~ **antenna** Segmentantenne f
**pillion** Soziussitz m ~ **rider** Beifahrer m
**pillow** Lager n, Lagerschale f, Pfanne f, Polster n, Zapfenlager n ~ **block** Lagerbock m, Stehlager m ~ **block bearing** Stehlager m ~ **block bolt** Lagerbockbolzen m ~ **block cap** Stehlageroberschale f ~ **block frame** Stehlagergehäuse n ~ **block liner** Stehlagerunterschale f ~ **bush** Lagerfutter n, Pfanne f ~ **jointing** Wollsackabsonderung f (geol.) ~ **lace** geklöppelte Kante f ~ **shaped scanning pattern** kissenförmiger Raster m ~**-type speaker** Kissenlautsprecher m
**pilot, to** ~ fliegen, führen, leiten, lenken, lotsen, steuern
**pilot** Auslöseeinrichtung f, Betätigungselement n, Flugzeug-, Luftfahrzeug-führer m Führer m, Führungs-stift m, -zapfen m (Ventil), Leiter m, Lotse m, Pilot m, Spitze f, Zapfen m ~ **at controls** Pilot m am Steuer
**pilot,** ~ **alarm** Kontrollwecker m ~ **balloon** Pilotballon m ~ **bar** Führungsdorn m ~ **beam shunt (or switch factor)** Leitstrahlaufschaltgröße f ~ **bell** Kontrollwecker m ~ **bit** Führungsmeißel m ~ **boat** Lotsenboot n ~ **bulb** Signalglimmlampe f ~ **burner** Sparbrenner m ~ **bush for boring bar** Bohrstangenführung f ~ **cable** Leit-, Lotsen-kabel n ~ **carrier** Pegelstrom m ~ **charges** Lotsengebühr f ~ **circuit**

maßgebende Leitung f, Pilotkreis m ~ **claw** Justiergreifer m (film) ~ **current** Leit-, Pegelstrom m ~ **ejector seat** Schleudersitz m ~ **exitation** Eigenerregung f ~ **exiter** Hilfserreger m ~ **flag** Lotsenflagge f ~ **flame** Lockflamme f, Zünd-flämmchen n, -flamme f ~ **frequency** Hilfs-, Steuer-frequenz f ~ **glow plug** Glimmkontrolle f ~ **hole** Führungsloch n ~ **house** Kartenhaus n ~ **ignition marking** Zeichen f für Zündstellung ~ **in-command** verantwortlicher Luftfahrzeugführer m ~ **indicator** Gruppenmeldezeichen n, Leitsignal n, Platzlampe f ~ **jack** Überwachungsklinke f ~ **jet** Führungs-, Vor-düse f ~ **lamp** Kontroll-, Melde-, Platz-, Signal-, Überwachungs-, Warn-lampe f ~ **lamp connector relay** Signallampeneinschaltrelais n ~ **lamp switching key** Schalter m für die Platzlampe ~ **light** Anzeigeleuchte f, Sparflamme f, Zünd-flämmchen n, -flamme f ~ **operated** ferngesteuert, hilfsgesteuert, indirekt wirkend ~ **oscillator** Steuer-kreis m, -röhre f, -sender m ~ **pair** Zähladernpaar n ~ **parachute** Hilfs-(fall)schirm m, Zusatzfallschirm m ~ **pin** Greiferstift m, Haltezahn m, Justierstift m, Sperrgreifer m ~ **pin nozzle** Zapfendüse f ~ **plant** Einrichtung f für halbindustrielle Versuche, Versuchsanlage f ~ **plant scale** Großversuchsmaßstab m ~ **pressure** Vorsteuerdruck m ~ **projection welding** Zweistufenwarzenschweißen n ~ **pulse** Auslöse-, Start-impuls m ~ **relay** Melde-, Platzlampen-, Steuer-, Überwachungsrelais n ~ **service** Lotsenwesen n ~ **signal** Gruppen-leitsignal n, -meldezeichen n, Steuerzeichen n, Überwachungs-lampe f, -zeichen n, ~ **spark** Zündfunker m ~ **speaker** Abhörlautsprecher m ~ **station** Lotsenstation f ~ **tape** Leitlochstreifen m ~ **-tone process** Pilot-Tonverfahren n ~ **tunnel** Richtstollen m ~ **valve** Schalt-, Steuer-ventil n, Steuerwelle f, Vorfülleinrichtung f, Vorsteuerventil n ~ **wave** Steuerwelle f ~ **wheel** Handkreuz n ~ **wire** Hilfsleiter m, Meßdraht m, Prüfdraht m, Prüfleitung f, Reglerleitung f, (circuit) Steuerleitung f, Zählader f ~ **wire regulator** Reglerleistung f
**pilotage** Steuerung f
**piloted thyratron** Folgerohr n
**piloting** Flugzeugleitung f, Führung f, Lotsen n ~ **of an airplane** Flugzeugführung f
**pilotless** führerlos, unbemannt
**pilot's,** ~ **cabin** Führerkabine f ~ **certificate** Flugzeugführerschein m ~ **cockpit** Führerkanzel f, -raum m, -stand m, Pilotenraum m ~ **compass** Führerkompaß m ~ **intercom** Eigenverständigung f des Piloten ~ **license** Flugzeugführerschein m ~ **school** Flugzeugführerschule f ~ **seat** Führersitz m ~ **throttle lever** Gashebel m am Führersitz (aviat.), Pilothebel m
**pimple** Warze f ~ **metal** (copper) Pimpelstein m
**pin, to** ~ anflöcken, anstiften, heften, stecken, verbinden, verdübeln, verstiften **to** ~ **down** fesseln **to** ~ **to** aufstiften
**pin** Bolzen m, Diebel m, Dorn m, Frosch m, Klöppel m, Knagge f, Niet m, Nocke f, Pflock m, Pinne f, Sperrkegel m, Sperrstift m, Spurzapfen m, Stecknadel f, Stift m, Stütze f, Stützer m, Stützerführung f (electr.), Tragzapfen m, Vorstecker m, Warze f, Zapfen m, Zwecke f

**pin, ~ and arc indicator** Winkelindikator *m* ~ **in bellcrank** Bolzen *m* im Winkelhebel ~ **of coils** Ringe *pl* an der Stange ~ **with cotter** Bolzen mit Splint ~ **of drilling bit** Meißelansatz *m* ~ **with eyebolt** Angelring *m* ~ **of forked (knuckle) joint** Gabelgelenkbolzen *m* ~ **with head and hole for splint pin** Bolzen *m* mit Kopf und Splintloch ~ **with round eye** Türangelpfanne *f* ~ **with thread** Gewindezapfen *m* ~ **of the tube base** Steckerstift *m* eines Röhrensockels ~ **of a tube socket** Sockelstift *m*

**pin, ~ bar** Wagenrunge *f* ~ **barrel** Stiftbüchse *f* ~ **bearing** Nadel-, Stift-lager *n* (hinge) ~ **beater mill** Schlagstiftmühle *f* ~ **board programming** Stecknadelprogrammierung *f* ~ **bolt** Federbolzen *m* ~ **break shovel** Sicherheitsschare *f* ~ **chain** Gall'sche Kette *f*, Nietbolzenkette *f* ~ **clutch** Drehkeilkupplung *f* ~ **connection lock for stepwise setting** Absteck-Arretierung *f* für stufenweise Rampeneinstellung ~ **connector** Verbindungsstecker *m* ~ **coupling** Bolzenkupplung *f* ~ **coupling head** Stiftkupplungskopf *m* ~ **cracks** (kleine) Gebirgsspalten *pl* ~ **cushion distortion** Kissenverzeichnung *f* ~ **drill** Zapfenbohrer *m*, -reißer *m* ~ **electrode** Stiftelektrode *f* ~-**ended beam** eingespannter Balken *m* ~-**ended support** Pendelstütze *f* ~-**feathered head** Tellerkopf *m* ~-**feed platen device** Stachelwalzenführung *f* ~-**feed wheel** Sternrad *n* ~ **gear** Schaftverzahnung *f* ~ **gear-ring** Triebstockring *m* ~ **guide** Führung *f*, Stift(en)führung *f* ~ **head** Nieten-, Stecknadel-kopf *m* ~ **hinge saw** Fischbandsäge *f* ~ **holder** Führung *f*, Stiftenführung *f*

**pinhole** Feinlunker *m*, Gaspore *f*, Loch *n* (opt.), Madenloch *n*, Nadelloch *n*, Okularöffnung f. Splintloch *n*, Vakuole *f*, Zapfenloch *n* ~ **camera** Lochkamera *f* ~ **image** Lochbild *n* ~ **plug** (of converter) Nadelboden *m* ~ **sight** Lochdiopter *n* ~ **stencil** Lochmarke *f*

**pinholes** Durchschlagstellen *pl* (electr.)

**pin, ~ insulator** Knüppel-, Stütz-isolator *m* ~ **jack** Buchse *f* für Bananenstecker ~ **joint** Bolzengelenk *n* ~ **jointed** (statics) gelenkig gelagert ~ **jointed beam** Auslegerbalken *m*, Gerberträger *m* ~ **length** Nadellänge *f* ~ **lever movement** (Uhr) Stiftankerwerk *n* ~ **movement** Greiferantrieb *f* (film) ~ **paper** rostfreies Nadelpapier *n* ~ **plate** Nadelleiste *f* ~-**point** Nadelspitze *f*; punktförmig ~-**point gating** Nadelpunktanguß *m* ~-**point** Punktziel *n* ~-**pointed disk** Stachelscheibe *f* ~-**pointing of a target** Zielfestlegung *f* ~ **punch** Splintauszieher *m* ~ **punches** Splinttreiber *pl* ~ **rack** Nagelbank *f* ~ **riveting** Stiftnietung *f* ~ **safety device** Stiftsicherung *f* ~ **spanner** Stiftschlüssel *m* ~ **stenter** Anschlagmaschine *f*, Nadelspannrahmen *m* ~ **support** Stiftlagerung *f* ~ **suspension** Stiftlagerung *f* ~-**type insulator** Isolator *m* mit gerader Stütze ~-**type locking system** Bolzenverriegelung *f* ~-**type nozzle** Nadeldüse *f* ~ **vice** Feil-, Stift-kolben *m*, Stielfeilkloben *m* ~ **way of the grain** Hirnseite *f* ~ **weir** Nadelwehr *n* ~ **wheel** Stiftrad *n* ~ **wire** Nietstiftdraht *m*, Splintdraht *m* ~ **wrench** Steckdorn *m*

**pinacoid** Pinakoid *n*

**pinacoidal** pinakoidisch

**pincers** Beiß-, Feder-zange *f*, Kloben *m*, Kneifzange *f*, Pinzette *f*, Punkturzange *f*, Zange *f*, Zwackeisen *n*, Zwickzange *f*

**pincette** Pinzette *f*

**pinch, to** ~ (a wire or hose) abzwicken, (sich) einklemmen, klemmen, kneifen, quetschen, zwicken **to** ~ **off** abkneifen **to** ~ **out** ausgehen, auskeilen, sich erschöpfen

**pinch** Einklemmen *n*, Kneifen *n*, Zwicken *n* ~ **of a lamp** (or of a tube) Quetschfuß *m*

**pinch, ~ bar** Brech-eisen *n*, -stange *f*, Klauenmaul *n*, Rückstange *f* ~ **cock** Quetschhahn *m* ~**effect** Einschnürungs- (electr.), Klemm-, Pinch-effekt *m* ~-**proof tube** nichteinklemmbarer Schlauch *m* ~ **rolls** Ausführungswalzen *pl* ~ **stopper** Griffstöpsel *m* ~-**type** Locheisentype *f* ~ **welder** Punktschweißzange *f*

**pinched** eingeklemmt, verdrückt, zusammengepreßt ~ **base** Quetschfuß *m* (electr.) ~ **gas** eingeschnürte Gassäule *f*

**pincher** Brechstange *f*, Kneifzange *f*

**pinching** Blockierung *f* ~ **nut** Gegen-, Klemmmutter *f* ~ **weight** Klemmgewicht *n*

**pine** Föhre *f*, Kiefer *f* ~ **needle** Fichtennadel *f* ~ **needle wool** Waldwolle *f* ~ **oil** Kieferöl *n*, Silbertannenöl *n* ~ **oil tar** Teeröl *n* ~ **plank** Kieferplatte *f* ~ **resin** Fichtenharz *n* ~ **tar** Kienteer *m* ~ **tree array** Tannenbaumanordnung *f* ~ **wood** Kiefer(n)-, Kien-, Nadel-holz *n*

**piney tallow** Pineytalg *m*

**ping** Sonarimpuls *m*

**pinion** Antriebs-kegelrad *n*, -rad *n*, -ritzel *n*, -zahnrad *n*, Drehling *m*, Flügelspitze *f*, Getriebe *n*, Kammwalze *f*, Langzugritzel *n*, Ritzel *n*, Spindel *f*, treibendes Rad *n*, Treibrad *n*, Trieb *m*, Triebel *n*, Triebling *m*, Vorgelege *n* (rim) Zahnrad *n*, Zahnritzel *n* ~ **of report** Schneckenzapfen *m* ~ **of slewing gear** Drehwerkritzel *n* ~ **and spur-gear drive** Ritzel- und Stirnradantrieb *m*

**pinion, ~ advance** Ritzelvorschub *m* ~ **box** Getriebekasten *m* ~ **brake** Getriebebremse *f* ~ **cage** Planetenträger *m* ~ **drive shaft** Ritzelantriebswelle *f* ~ **free wheeling** Ritzelfreilauf *m* ~ **gauge** Triebmaß *n* (Uhrmacherei) ~ **grease** Getriebe-, Zahnrad-fett *m* ~ **head** (in microscope) Triebknopf *m* ~ **head for focusing eyepiece** Okularfokussierung *f* ~ **helical advance** Ritzelverschraubung *f* ~ **housing** Kammwalzgerüst *n* ~ **neck** Kammwalzenzapfen *m* ~ **nut** Ritzelmutter *f* ~ **srew fastening** Ritzelverschraubung *f* ~ **shaft** Ritzel-, Trieb-welle *f* ~ **side** Ritzelseite *f* ~ **spindle** Zahnradwelle *f* ~ **steel** Triebstahl *m* ~ **switch** Ritzelschalter *m* ~ **travel** Getriebeschubweg *m* ~-**type cutter** Radform-, Stoß-messer *m*

**pink, to** ~ durch-bohren, -stechen, (Motor) klopfen

**pink** Heftel *m*; blaßrot, rosa

**pinked cotton tape** gezackter Baumwollstreifen *m*

**pinking** Klopfen *n*, leichtes Klopfen *n* des Motors ~ **iron** Auszackeisen *n* ~ **punch** Zackenausstecher *m*

**pinnace** Pinasse *f*

**pinned** aufgestiftet, genagelt ~ **on shaft** Vorsteckachse *f*

**pinned, ~ disk mill** Stiftmühle *f* **~ fitting** Verstiftung *f* **~ head** Tellerkopf *m*

**pinning** Benadelung *f* **~ down** Fesselung *f*

**pins for tipping plates** Stoßplattenstifte *pl*

**pintle** Achsnagel *m*, Drehbolzen *m*, (of rudder) Fingerling *m*, Kupplungsbolzen *m*, Spurzapfenlager *n*, Zapfen *m*, Zughaken *m* **~ cotter** Steckbolzensicherung *f* **~ hook** Anhängeröse *f* **~ injection valve** Zapfendüse *f* **~ stone** Torpfannenstein *m*

**Pintsch regulator** Pintschregler *m*

**pion decay** Pionenzerfall *m*

**pioneer** Pionier

**pioneering** bahnbrechend **~ experiment** Fundamentalversuch *m*

**pip** Echozeichen *n*, Höcker *m*, kurzer Impuls *m*, Pip *m* (rdr), Pumpspitze *f*

**pipe, to ~** (durch Rohr oder Rohrleitung) leiten, lunkern, weiterleiten **to ~ gas** Gas *n* durch Rohrleitungen pumpen **to ~ into** zuleiten

**pipe** Abflußrohr *n*, Erderplatte *f*, Gaspore *f*, Hohlstange *f*, (primärer) Lunker *m*, Lutte *f*, Pfeife *f*, Rohr *n*, Röhre *f*, Rohrleitung *f*, (in a casting) Saugtrichter *m*, Schlauch *m*, Trichterlunker *m* **~ without branch** unverzweigte Leitung **~ for filtrate outflow** Filtratabführrohr *n* **~ for oil under pressure** Rohr *n* für Drucköl **~ with single bordered end** umgebördeltes Rohr

**pipe, ~ angle** (Rohr) Winkel-stück *n*, -verschraubung *f* **~ arrangement** Rohranordnung *f* **~ assembly** Rohrbündel *n* **~ bending** Rohrbiegen *n* **~ bending machine** Rohrbiegmaschine *f* **~ bending tongs** Rohrbiegezange *f* **~ breakage** Rohrbruch *m* **~ bridge crossing** Überführung *f* mit Rohrbrücke **~ burst** Rohrbruch *m* **~ cased shaft** Rohrschacht *m* **~ casing** Rohrbandagierung *f* **~ casing head** Ausgußkopf *m* **~ castings** Rohrguß *m* **~ chase** Installationszelle *f* **~ chocking** Rohrverstopfung *f* **~ clamp** Röhrenklammer *f*, Rohrsattel *m*, Rohrschelle *f* **~ clamp fitting** Rohrverbindungsschelle *f* **~ clay** Tonröhre *f* **~ clip** Rohrschelle *f* **~ clogging** Rohrverstopfung *f* **~ closer** Rohrverschluß *m* **~ coil** Rohrschlange *f*, Spiralrohr *n* **~ conduit** Rohrnetz *n* **~ connection** Anschlußstutzen *m*, Rohr-anschluß *m*, -verbindung *f* **~ connection diagram** Rohrleitungsschaltbild *n* **~ coupling** Rohrverbindung *f* **~ cramp** Rohrhaken *m* **~ crown** Rohrscheitel *m* **~ culvert** Rohrdurchlaß *m* **~ culvert through a leave** Rohrsiel *n* **~ cutter** Abschneider *m*, Rohrabschneider *m*, Rohrschneider *m* **~ diagram** Rohrleitungsschema *n* **~ discharge** Rohrentladung *f* **~ ditch** Röhrenverleggraben *m* **~ dog** Rohrfänger *m* **~ drain** Entwässerungsrohr *n* **~ drilling appliance** Anbohrvorrichtung *f* **~ duct** Rohrdurchführung *f* **~ elbow** Kniestück *n*, Rohrknie *n*, Rohrkrümmer *m* **~ elevator** Gestängeelevator *m* **~ elimination** Lunkerverhütung *f* **~ eliminator** Lunkerverhütungsmittel *n* **~ end** Leitungsende *n* **~ end opener** Bördelgerät *n* für Rohrenden **~ end preparing machine** Rohr-Abstech- und Fräs-Maschine *f* **~ eradicator** Lunkerverhütungsmittel *n* **~ expanding and flanging machine** Rohrstauch- und muffenmaschine *f* **~ failure** Rohrbruch *m* **~ filter** Röhrenfilter *m* **~ fitter** Rohrschlosser *m* **~ fitting** Röhrenverbindung *f*, Rohrleitungsstück *n* **~ fittings** Rohr-

**fittings** *pl* **~ flange** Rohrflansch *m* **~ flanging tool** Bördelgerät *n* für Rohrenden **~ flow** Rohrströmung *f* **~ flow resistance coefficient** Rohrwiderstandsziffer *f* **~ formation** Tütenbildung *f* **~ fracture** Rohrbruch *m* **~ friction** Rohrreibung *f* **~ grip** Greifzange *f* **~ guide** Röhrenführung *f* **~ hanger** -sattel *m*, Rohr-schelle *f* **~ hangers** Rohrleitungsaufhängungen *pl* **~ hob** Strehlbohrer *m* für Rohrgewinde **~ hook** Rohrhaken *m* **~ housing** Rohrleitungsgehäuse *n* **~ joint** Rohrmuffe *f*, Rohrverbindung *f*, Verbindungs-rohr *n*, -röhre *f* **~ jumping press** Röhrenstauchpresse *f* **~ knee** Rohrkrümmer *m* **~ laying** Röhrenlegung *f*, Rohrlegung *f*, Rohrmontage *f* **~ laying hook** Wegehaken *m* (f. Rohre) **~ laying jacks** Rohrverlegböcke *pl* **~ layout** Installationsplan *m* **~ leading** Leitungsrohr *n* **~ lifter** Rohrheber *m*

**pipeline** konzentrisches Kabel *n*, (konzentrische) Leitung *f*, Leitungsrohr *n*, Ölleitungsrohr *n*, Röhren-fahrt *f*, -fernleitung *f*, -leitung *f*, Rohrfernleitung *f*, -kanal *m*, -leitung *f*, Rohrstrang *m* **~ company** Ölleitungsgesellschaft *f* **~ compensators** Kompensations-bogen *pl* oder -schleifen *pl* **~ construction** Rohrleitungsbau *m* **~ coupling** Rohrleitungskupplung *f* **~ facilities** Leitungen *pl* (Leitungsnetze) **~ mixer** Durchflußmischer *m* **~ project** Ölleitungsprojekt *n* **~ pump** Gasleitungspumpe *f* **~ strainer** Schmutzabschneider *m* **~ under pressure** Druckleitung *f*

**pipelines** Dampfrohrleitung *f*

**pipe, ~ lining** Rohrauskleidung *f* **~ manifold** Rohrverteiler *m* **~ molding machine** Rohrformmaschine *f* **~ nail** Kernnagel *m* **~ nipple** Rohrnippel *m* **~ nut** Rohrmutter *f* **~ outlet** Kanalöffnung *f* **~ packer** Verrohrungsliderung *f* **~ passage through a levee** Rohrsiel *n* **~ plan** Rohrleitungsplan *m* **~ pusher** Erdbohrer *m*, Gestängedrucker *m* **~ reducer** Rohrreduktionsstück *n* **~ reducing machine** Rohrreduziermaschine *f* **~ reel** Drahthaspel *f* mit rotierendem Führungsrohr **~ relaying** Rohrverlegung *f* **~ riser** Standrohr *n* **~ roller** Rohrwalzer *m* **~ rolling mill** Röhrenwalzwerk *n* **~ run** Rohrleitung *f* **~ saddle** Anzweigschelle *f*, Rohr-sattel *m*, -schelle *f* **~ scraper** Rohrkratzer *m* **~ seal** Rohrverschluß *m* **~ section** Rohrabschnitt *m* **~ slicing lathe** Rohrabstechbank *f* **~ socket** Röhrenmuffe *f*, Rohr-flansch *m*, -muffe *f* **~ solder** Rohr-, Röhren-lot *n* **~ steel** Schweißstahl *m* **~ still** Röhrendestillationsofen *m*, Röhrenofen *m* **~ stock** Kluppe *f*, Rohrgewinde *n* **~ stock die** Kluppenbacke *f* **~ stocktap** Backenbohrer *m* **~ stocks** Rohrschneidkluppe *f* **~ straightener** Rohrausrichtvorrichtung *f* **~ strap** Rohrschelle *f* **~ string** Rohrstrang *m* **~ support** Röhrenträger *m*, Rohrhalter *m* **~ swage** Verrohrungsbirne *f* **~ switch** Rohrschalter *m* **~ system** Rohrleitungsnetz *n* **~ system with shut-off device** absperrbarer Leitungsstrang *m*, Leitungsstrang *m* mit Absperrvorrichtung **~ tap** Rohrgewinde-bohrer *m*, -schneider *m* **~ tee** T-Stück *n* **~ testing apparatus** Röhrenprüfapparat *m* **~ thermocouple** Rohrthermoelement *n* **~ thread** Rohrgewinde *n* **~ thread protector** Rohrgewindeschutzkappe *f*, Schutzmuffe *f* **~ threading machine** Gewindedrehmaschine *f*, Rohrgewindeschneidemaschine *f* **~**

**threading tool** Schneidkluppe f ~ **tongs** Rohrzange f ~ **union** Rohrverbindungsstück n ~ **valve** Tellerventil n ~ **vise** Rohrschraubstock m ~ **way** Rohrdurchbruch m (durch Unterzüge, etc.) ~ **weaver** Schlauchweber m ~ **welding** Rohrschweißung f ~ **wrapping** Rohrbandagierung f ~ **wrench** Rohr-schlüssel m, -zange f
**piped** lunkerig ~ **program** Drahtfunk m
**pipeless** lunkerfrei, lunkerlos
**pipette, to** ~ pipettieren **to** ~ **out** herauspipettieren, mit Pipette entnehmen
**pipette** Heber m, Pipette f, Saug-, Tropf-röhrchen n, Stech-heber m, -kolben m, Tropfglas n ~ **holder (or stand)** Pipettenständer m
**piping** Lunker m (metall), Lunkerbildung f, Lunkern n, Lunkerung f, Rohr n, Rohranlage f, Rohr-bildung f, -legung f, -leitung f, -montage f, -netz n, Röhren-anlage f, -leitung f, Saugen n, Trichterlunkerbildung f (metall.), Tütenbildung f (in ingot or casting), Zugfalten n (paper mfg.), Zweiwachs m ~ **of the exhaust gases** Auspufführung f
**piping,** ~ **action** Rohrblock m ~ **bent to form right angles** rechtwinklig gebogene Röhren pl ~ **convection** Leitung f ~ **(or pipe line) drawings** Rohrleitungszeichnungen pl ~ **system** Rohrleitungssystem n ~ **unit** Rohranlage f
**pique** Pikee n
**pirate listener** Schwarzhörer m
**pirn, to** ~ spulen
**pirn** Einschußspule f, Garnrolle f, Kötzer m, Schleifspule f ~ **core** Spulenkern m
**pirning layout** Spulaggregat n
**pisolite** Erbsenstein m, Schalenkalk m
**pisolitic** pisolitisch
**pissasphalt** Bergteer m, Pissasphalt m
**pistol** Pistole f ~ **with safety on** gesicherte Pistole f
**pistol,** ~ **grip** Handstütze f, Pistolengriff m ~ **holster** Pistolentasche f ~ **shot silencer** Knallschutzgerät n ~ **stock** Griff-schale f, -stück n ~ **target** Pistolenscheibe f ~-**type nozzle** Pistolenhahn m
**piston** Druck-kolben m, -stempel m, Gleitventil n, Kolben m, Stempel m, Zug m ~ **with metallic packing** Metalliderungskolben m
**piston,** ~ **acceleration** Kolbenbeschleunigung f ~ **area** Kolbenfläche f ~ **attenuator** Hohlleiterspannungsteiler m ~ **baller** Kolbenlöffel m ~ **barrel** Kolbenkörper m ~-**bearing big end of connecting rod** Kurbelwellenkopf m ~-**bearing surface** Kolbentragfläche f ~ **block** Luftkolbenblock m ~-**block base** Luftventilblock m ~ **blow** Durchblasen n des Kolbens ~ **blower** Kolben-gebläse n, -vorverdichter m ~ **body (or shaft)** Kolbenschaft m (Bremse) ~ **bore** Kolbenbohrung f ~ **boss** Kolbennabe f ~ **boss bushing (or box)** Kolbenbüchse f ~ **button** Sicherungspilz m für Kolbenbolzen ~ **casting** Kolbenguß m ~ **compressor** Kolben-gebläse n, -vorverdichter m ~ **control** Kolbenschieberregler m ~ **controlled** kolbengesteuert ~ **crown** Kolben-boden m, -krone f ~ **dead center** Kolbentotlage f ~ **descent** Kolbenniedergang m ~ **diaphragm** als Ganzes schwingende Membran(e) f, Kolbenmembran f ~ **displacement** Hubraum m, Hubvolumen n, beim Kolbenhub verdrängter Raum m, Kolbenverdrängung f, Zylinder-hub-

volumen n, -inhalt m, -volumen n ~ **drill** Kolbenbohrmaschine f ~ **drive** Kolbenantrieb m ~ **engine** Kolbenmotor m ~ **feeder** Kolbenaufgabevorrichtung f ~ **ferrule** Kolbenband n ~ **foghorn** Kolbensirene f ~ **follower bolt** Kolbendeckelschraube f ~ **fracture** Kolbenbruch m ~ **guide** Kolbenführung f ~ **head** Hubraum m, Kolben-boden m, -kopf m, -körper m ~ **head ring** Kolbenring m ~ **hole** Kolbenbohrung f ~ **indicator** Kolbenindikator m ~ **jig** Kolbensetzmaschine f ~ **knock** Klopfen n des Kolbens ~ **land** Kolbensteg m ~ **lever group** Luftkolbenhebelblock m ~ **life** Kolbenleistung f ~ **load** Kolbenbelastung f ~ **lock** Kolbenschloß n ~ **lubrication** Kolbenschmierung f ~ **main ram** Preßkolben m ~ **manometer** Druckwaage f, Kolbenmanometer n ~ **motion** Kolbenschub m ~ **oil pump** Kolbenölpumpe f ~ **packing** Kolben-dichtung f, -liderung f, -packung f ~ **packing leather** Kolbenstulp m ~ **passage** Durchgang m des Kolbens ~-**phone** Pistonphon n ~ **pin** Kolben-bolzen m, -zapfen m ~ **pin bore** Kolbenbolzenbohrung f ~ **pin bushing** Kolbenbolzen-buchse f, -lager n ~ **pin pliers** Kolbenbolzenzange pl ~ **pin plug** Kolbenbolzenpilz m ~ **ported engine** kolbengesteuerter Motor m, Motor m mit kolbengesteuerten Ein- und Auslaßschlitzen, Motor m mit Schlitzsteuerung ~ **position time diagram** Kolbenwegdruckdiagram n ~ **pressure** Kolbendruck m ~ **pump** Kolbenpumpe f ~ **return spring** Kolbenrückholfeder f ~ **ring** Kolben(dicht)ring m
**piston-ring,** ~ **blow-by** Durchblasen n des Kolbens, Kolbengasdurchlaß m ~ **casting** Kolbenringguß m ~ **clamp** Spannband n für Kolbenringeinbau ~ **face grinding machine** Kolbenringschleifmaschine f ~ **flutter** Flattern n der Kolbenringe ~ **gap** Stoß m oder Stoßfuge f des Kolbenringes ~ **groove** Kolbenring-nut n, -riefe f ~ **lock** Dichtungs-, Kolbenring-schloß n ~ **slot** Kolbenringschlitz m ~ **sticking** Kolbenringstecken n ~-**type oil seal** Dichtring m für Öl (kolbenringähnlich), kolbenringähnlicher Öldichtring m
**piston,** ~ **rod** Kolbenstange f, Pleuel n ~ **rod collar** Kolbenstangenbund m ~ **rod guide** Kolbenstangenführung f ~ **rod nut** Kolbenstangenmutter f ~ **rod taper** Kolbenstangenkegel m ~ **sampler** Kolbenprobenehmer m ~ **servo-valve** Regelschieber m ~ **shock-absorber** Kolbenstoßdämpfer m ~ **skirt** Kolben-gleitbahn f, -mantel m, -unterteil m ~ **slap** Kippen n des Kolbens, Kolbenkippen n ~ **speed** Kolbengeschwindigkeit f ~ **spring** Kolbenfeder f ~ **stroke** Kohlen-hub m, -takt m, -weg m ~ **stroke indicator** Kolbenhubanzeiger m ~ **stroke volume** Kolbenhubraum m ~ **supercharger** Kolben-gebläse n, -vorverdichter m ~ **surface** Kolbenfläche f ~ **swept volume** Hubvolumen n ~-**top land** oberster Kolbensteg m ~ **travel** Kolbenweg m ~ **tube** Rohrkolben m, Stempelrohr n ~ **turning lathe** Kolbendrehbank f ~-**type compressor** Kolbenkompressor m ~-**type die-casting machine** Kolbenmaschine f ~-**type steam engine** Kolbendampfmaschine f ~ **valve** Kolbenschieberventil n, Kolbenventil n, Ventilkolben m ~ **valve chamber** Kolbenschieberkammer f ~ **valve control** Kolben-

schiebersteuerung *f* ~ **valve liner** Kolben-
schieberbüchse *f* ~ **working in opposite direction**
gegenläufiger Kolben *m* ~ **wrench** Kolben-
schlüssel *m*
**pit, to** ~ anfressen, angreifen, einfressen, eingraben, zerfressen
**pit** (foundry) Abstichherd *m*, Bergwerk *n*,
Damm-, Erd-grube *f*, Einsenkung *f*, Gesenk *n*,
(foundry work) Gießgraben *m*, Grube *f*, (of
surfaces) Grübchen *n*, (cable) Kabelbrunnen
*m*, Kiesgrube *f*, Kohlenschacht *m*, Loch *n*,
Mine *f*, (of surfaces) Narbe *f*, (from corrosion)
Pockholzlager *n*, Rostgrübchen *n*, Schacht *m*,
Schlackengrube *f*, Vertiefung *f*, Zeche *f* ~ **for
setting off blasting charges** Sprenggrube *f* ~ **of
traverser** Schiebebühnengrube *f* ~ **of turntable**
Drehscheibengrube *f*
**pit,** ~ **arch** Grubenausbau *m* ~ **bottom** Füllort
*m* ~ **burning** Grubenverkohlung *f* ~ **caisson**
Schachtkasten *m* ~ **coal** Steinkohle *f* ~ **coal
plant (or power station)** Steinkohlenkraftwerk
*n* ~ **cover** Grubenabdeckung *f* ~ **fire** Grubenbrand *m*, Unterflur(tiegel)ofen *m* ~ **foreman**
Hängebankarbeiter *m* ~ **frame** Zimmerwerk *n*
eines Schachtes ~ **fueling** Hydrantenbetankung
*f* ~ **furnace** Schachtofen *m* ~ **gate locking
device** Schachttürverschluß *m* ~ **gear** Förderschacht *m* ~ **gravel** Grubenkies *m* ~ **guide** Leitstange *f* ~ **head** Blindschacht *m* ~ **head baths**
Waschkaue *f* (min.) ~ **head-gear** Schachtgerüst
*n* ~ **head winch** Förderhaspel *m* ~ **heating
furnace** Tiefofen *m* ~ **life-saving station** Gruben-
rettungsstation *f* ~ **made by floodwater** Kolk *m*
~**-man** Bergmann *m* ~ **man shaft** Lenkwelle *f*
~ **masonry** Senkmauerung *f* ~ **mouth** Schacht-
öffnung *f* ~ **prop** Gruben-, Stempel-holz *n* ~
**rope** Förder-, Schacht-seil *n* ~ **rope pulleys** För-
derseilscheiben *pl* ~ **sample** Pfannen-, Schmier-,
Schöpf-probe *f* ~ **sand** Grubensand *m* ~ **saw**
Baum-, Treck-, Trumm-säge *f* ~**-shaft** H.P.
Schachtpferdekraft *f* ~ **silo** Grubensilo *m* ~
**sinking** Schachtabteufen *n* ~ **tannage** Gruben-
gerbung *f* ~ **timber** Schalholz *n* ~ **wagon** Grubenwagen *m* ~ **water** Grubenwasser *n* ~ **wood**
Grubenholz *n* ~ **work** Wasserhaltung *f*
**pitch, to** ~ (brewing) anstellen, befestigen, be-
pichen, bocken (naut.) (ship) eintauschen, ineinandergreifen, (um die Querachse) neigen,
(aviat) pechen, (ship) stampfen, steigen, teeren,
verpechen, verpichen, **to** ~ **down** abkippen, absacken **to** ~ **a tent** ein Zelt *n* aufschlagen
**pitch** Abstand *m*, regelmäßiger Abstand *m*
(aviat), (propeller) Blattsteigung *f*, (distance or
degree) Distanz *f*, Drall *m*, Einfall *m*, Einfallen *n*, (of thread) Gang *m*, (of a spiral, thread)
Ganghöhe *f*, Gefälle *n*, (of a screw) Gewindesteigung *f*, Grundton *m*, Längsneigung *f*, (of
cam compensator) Leitkurve *f*, Lochabstand *m*
(acoust.), Neigung *f*, Pech *n*, (of screw) Spin-
delsteigung *f*, (of screw) Steighöhe *f*, (of screw
thread) Steigung *f*, Steigungsmaß *n*, Stufe *f*,
Teer *m*, Teilung *f*, (of sound sensation) Ton-
helligkeit *f*, Tonhöhe *f*, (of turns) Windungs-
ganghöhe *f*, Zahnteilung *f*, Zeilenabstand *m*
**pitch,** ~ **along an arc** Bogenabstand *m* ~ **of
blades** Blatteinstellung *f* ~ **of the control edge**
Steuerkantensteigung *f* ~ **of female thread**
Steigung *f* des Muttergewindes ~ **of helical**

**spring** Steigung *f* der Federwindung ~ **of
intermittent fillet welding** Teilung *f* unterbrochener Kehlnahtschweißung ~ **of lead screw**
Leitspindelsteigerung *f* ~ **of rivets** Nietentfer-
nung *f* ~ **of the rivets** Nietteilung *f* ~ **of a screw**
Gang *m* eines Gewindes ~ **of a screw thread**
Ganghöhe *f* der Schrauben oder des Schraubengewindes, Gewindegang *m* ~ **of the signal note**
Zeichentonhöhe *f* ~ **of sleepers** Schwellen-
teilung *f* ~ **of spin in detonations** Ganghöhe *f*
des Spins bei Detonationen ~ **of a spiral (or
spire)** Spiralsteigung *f* ~ **of spring** Federweg *m*
~ **of thread** Gewindeteilung *f* ~ **of tone** Tonhöhe *f* ~ **of turns** Ganghöhe *f* der Windungen
~ **of weld** Abstand *m* zwischen Schweißnähten
**pitch,** ~ **accent** Tonakzent *m* ~ **altitude** Inklinationswinkel *m* ~ **angle** Steigungswinkel *m* ~
**arm** Blattverstellarm *m* ~ **attitude** Inklinationswinkel *m* (rdr), Längsneigung *f* (aviat) ~
**attitude vane** Anstellwinkelfühler *m* ~ **axis**
Nickachse *f*, Querachse *f* ~ **black** pechschwarz
~**blende** Eisenblende *f*, Gummierz *n*, Pech
-blende *f*, -uran *n*, Uranpech-, -blende *f*, -erz *n*
~ **cam** Steigungskurve *f* ~ **change bearing** Blatt-
verstellager *n* ~ **change control** Blattverstelłein-
richtung *f* (Blattsteuerung) ~ **change drum**
Blattverstelltrommel *f* ~ **change indicator** Ver-
stellanzeiger *m* ~ **changing range** Steigungsver-
stellbereich *m* ~ **circle** (applied to pinions),
Grund-, Teil-, Wälz-kreis *m*, Zahnteilbahn *f*
~ **circle of worm** Teilkreis *m* der Schnecke ~
**circle diameter** Teilkreisdurchmesser *m* ~ **circle
radius** Teilkreisradius *m* ~ **coal** Pech(stein)kohle
*f* ~ **coat** Pechüberzug *m* ~ **coke** Pechkoks *m*
~ **cone** Grund-, Teil-kegel *m* ~ **cone angle** Kegelwinkel *m* der Walzfläche ~ **cone distance** Teil-
kegellänge *f* ~ **control** Blattwinkelverstellung *f*,
Lattenführung *f* ~ **control lever** Blattsteuerungs-
hebel *m* ~ **control mechanism** Blattwinkel-
verstellvorrichtung *f* ~ **diameter** Teilkreisdurchmesser *m* ~ **diameter of tapped hole** Flankendurchmesser *m* des Muttergewindes ~
**diameter ratio** relative Steigung *f* (Propeller)
~ **discrimination threshold** Tonhöhenunter-
schiedsschwelle *f* ~ **error** Längsneigungs-, Stei-
gungs-fehler *m* ~ **factor** Polbedeckungsfaktor
*m* ~ **fin** Höhenruder *n* ~ **fluctuation** Tonhöhen-
schwankung *f* ~ **fork** Heugabel *f* ~ **formation**
Verpichung *f* ~ **horn** Blattsteuerungshorn *n* ~
**indicator** Längsneigungsmesser *m*, Steigungs-
anzeiger *m*, Stellungsanzeiger *m* für die Luftschraubeneinstellung ~ **interval** Tonhöhen-
verhältnis *n* ~ **line** Teilkreis *m*, Zahnteilbahn *f*
~ **line of groove** Kaliberlinie *f* ~ **line of spur
gear** Teilbahn *f* des Stirnrades ~ **line contact**
Teilkreiskontakt *m* ~ **line speed** Umfangsgeschwindigkeit *f* ~ **link** Blattverstellstange *f*
~ **measuring** Teilprüfgerät *n* ~ **operation cylinder**
Verstellzylinder *m* ~ **peat** Specktorf *m* ~ **pine**
Föhrenholz *n* ~ **pipe** Stimmflöte *f*, Stimmpfeife
*f*, Tonhöhenvergleicher *m* ~ **point** Wälzpunkt
*m* ~ **range** Modul *n*, Verstellbereich *m* ~ **rate**
Nickwinkelgeschwindigkeit *f* ~ **ratio** Teilungsverhältnis *n* ~ **ratio of a propeller** Steigungs-
verhältnis *n* ~ **residue** Pechrückstand *m* ~
**setting** Steigungseinstellung *f* ~ **setting mechanism** Steigungsverstellvorrichtung *f* ~ **speed of
a propeller** Geschwindigkeit *f* bei Schlüpfung

Null ~ **spraying apparatus** Pechspritzapparat *m* ~ **stone** Pechstein *m* ~ **thread** Gewindeanstieg *m* ~ **tool** Pechschale *f* (zum Schleifen von Linsen) ~ **variation indicator** Tonhöhenschwankungsmesser *m*

**pitched** gängig ~ **blade** schräggestellte Schaufel *f* ~ **curd soap** abgesetzte oder geschliffene Kernseife *f* ~ **face** Außenfläche *f* ~ **roof** Steildach *n* ~ **slope** gepflasterte Böschung *f*

**pitcher** Ablader *m*, Auflader *m*, Brecheisen *n*, Hacke *f*, Haue *f*, Kanne *f*, Krug *m*, Pflasterstein *m*

**pitching** Abdeckung *f*, Aufstellen *n*, Ausschleifen *n*, Errichten *n*, Pflasterung *f*, Schwankung *f*, (Schiff) Stampfen *n*, Steinpackung *f*

**pitching,** ~ **angle** (of ship) Stampfwinkel *m* ~ **apparatus** Pechapparat *m* ~ **borer** Meißelbohrer *m* ~ **indicator** Kippzeiger *m* ~ **machine** Pechmaschine *f* ~ **moment** Kipp-, Längs-, Stampfmoment *n* ~ **moment coefficient** Längsmomentenbeiwert *m* ~ **moment compensator** Nickmomentausgleicher *m* ~ **motion** Galoppieren *n*, Querschwingung *f* (aviat.), Stampfbewegung *f* (aviat.), Stoßbewegung *f* ~ **poll pick** Hammerspitzhaue f (min.) ~ **temperature** Anstelltemperatur *f* ~ **tool** Geradhängevorrichtung *f*

**pitchy** pechartig

**pitfall** Falle *f*

**pith** Mark *n* ~ **of rattan** Peddig *n* ~ **ball** Holundermarkkugel *f* ~ **helmet** Tropenhelm *m*

**pitot,** ~ **boom** Staurohrausleger *m* ~ **gauge** Pitotmanometer *n* ~ **head** Stau-düse *f*, -rohr *n* ~-**head fresh-air heating** Staufrischluftheizung *f* ~-**head heating** Staudüsen-, Staurohr-heizung *f* ~ **meter** Pitotmesser *m* ~ **pressure connection** Staudruckverbindung *f* ~ **static airspeed indicator** Fahrtmesser *m* (mit Staurohr), Staudruckmesser *m* ~ **static head** Staurohrkopf *m* ~ **static tube** Pitot's statisches Saugrohr *n* ~ **tube** Luftansaugemeter *n*, Pitotmanometer *n*, Pitotröhre *f*, Pitotsches Rohr *n*, Saugrohr *n*, Staudruckmesser *m*, Staudüse *f*, Staurohr *n*, Stauröhre *f* ~ **venturi airspeed indicator** Düsenluftstrommesser *m*

**pits, having** ~ durchlöchert, höhlig

**pitted** löcherig, narbig, wabenartig, zellähnlich ~ **contact** ausgefressener Kontakt *m* ~ **structure** zellenartige Struktur *f*

**pitticite** Arseneisensinter *m*, Eisen-pecherz *n*, -sinter *m*

**pitting** Abhauen *n*, Abnutzung *f*, Anfraß *m*, Annagung *f*, Einfressung *f*, grübchenartige Anfressung *f*, Grübchen-bildung *f*, -korrosion *f*, Kornen *n* (metall), Kraterbildung *f*, Lochfraß *m*, lochfraßähnliche Zerstörung *f*, (corrosion) Metallpickelbildung *f*, Narbenbildung *f*, örtliche Anfressung *f*, (of metal) Pickelbildung *f*, Rostanfressung *f*, Verschleiß *m* ~ **corrosion** grübchenartige Rostanfressung *f*

**pituitary body** Hypophyse *f*

**pivot, to** ~ anlenken, drehbar lagern, einschwenken, schwenken, schwingen, in Zapfen lagern **to** ~ **sideways** seitlich einschwenken

**pivot** Angel *f*, Anlenkzapfen *m*, Dreh-achse *f*, -punkt *m*, -zapfen *m*, Fallenachse *f*, Gickel *m*, Spindel *f*, Spitze *f*, Spurzapfen *m*, Stellzapfen *m*, Stützzapfen *m*, Tragzapfen *m*, Wellenzapfen

*m*, Zapfen *m* ~ **of compass** Kompaßspinne *f* ~ **of a hinge** Angelzapfen *m* einer Türangel

**pivot,** ~-**and-bearing arrangement** Steckhülseneinrichtung *f* ~-**and-bearing joint** Steckhülsenverbindung *f* ~ **axis** (of an accelerometer) Lagerung *f* ~ **axle cultivator** Hackmaschine *f* mit beweglichen Achsen ~ **bearing** Dreh-, Fuß-, Kipp-, Pendel-, Spur-, Zapfen-lager *n* ~ **bolt** Drehbolzen *m* ~ **bridge** Drehbrücke *f* ~ **broach** Zapfenreibahle *f* ~ **coiler** Pinolenwickler *m* ~ **collar** Zapfenkragen *m* ~ **coupling** Schwenkantrieb *m* ~ **end pressure** Zapfenkantenpressung *f* ~ **friction** Lager-, Pinnen-reibung *f* ~ **hole** Zapfenloch *n* ~ **inclination** Achs-(zapfen)sturz *m* ~ **jewel** Spitzenlagerung *f* ~ **journal** Spur-, Stütz-zapfen *m* ~ **leaf gate** Drehschützklappe *f* ~ **length** Zapfenlänge *f* ~ **lever for rear brake** Bremszwischenhebel *m* ~ **lever attachment** bewegliche Hebelvorrichtung *f* ~ **mounting** Drehgestell *n*, Pivot-, Sockel-lafette *f* ~ **nozzle** Zapfendüse *f* ~ **pin** (Unterdruckklappe) Achse *f*, Drehbolzen *m*, Kipp-, Lagerzapfen *m*, Zapfennadel *f* ~ **point** Aufsattelungspunkt *m* (Sattelschlepper), Lagerspitze *f* ~ **point set-screw** Druckschraube *f* mit Zapfen ~ **pole cultivator** Hackmaschine *f* mit beweglichem Rahmen ~ **post** Pinnenträger *m*, Trägerstift *m* ~ **pressure** Zapfendruck *m* ~ **stone** Pfannenstein *m*, Torpfannenstein *m* ~ **stud** Bolzen *m*, Seilrollenbolzen *m*, Türkloben *m* ~ **suspension** Spitzenaufhängung *f* ~ **thrust** Zapfendruck *m* ~ **tongue attachment** bewegliche Deichselvorrichtung *f*

**pivotal** zentral ~ **bearing** Steckhülse *f* ~ **pin** Hebeldrehpunkt *m* ~ **point** Anlenkungspunkt *f* ~ **shaft** Hebeldrehpunkt *m*

**pivotally attached** drehbar gelagert oder angeordnet

**pivoted** angelenkt (drehbar), drehbar angelenkt, eingesetzt oder gelagert ~ **armature** (loud speaker) Drehanker *m* ~ **bogie steering** Drehschemellenkung *f* ~ **bucket conveyer** Pendelbecherwerk *n* ~ **carriage** (of a machine tool) Drehschlitten *m* ~ **leg** drehbare Stütze *f* ~ **lever** drehbar gelagerter Hebel *m* ~-**type bar** Zapfentypenhebel *m* ~-**type trailer coupling** drehbare Anhängerkupplung *f*

**pivoting** Drehung *f*, Zapfenlagerung *f* ~ **arm** Drehbalken *m* ~ **grooving saws** schwenkbare Nutsägen *pl* ~ **point in launching** Aufschwimmen *n* ~ **sash** Wendeflügel *m*

**pix carrier** Bildträger *m* (TV)

**placard, to** ~ bekleben

**placard** Anschlagzettel *m*, Aushang *m*, Patent *n*, Plakat *n*, Zettel *m* ~ **speed** zulässige Höchstgeschwindigkeit *f* ~ **sticker** Zettelankleber *m*

**place, to** ~ anbringen, einbringen, legen, plazieren, schlichten, setzen, stellen, tun, unterbringen **to** ~ **across** in Brücke *m* schalten **to** ~ **alongside** nebeneinanderstellen **to** ~ **aside** wegstellen **to** ~ **on a base** kaschieren **to** ~ **in a box** ausbüchsen, ausbuchsen **to** ~ **a butt strap over the joint** überlaschen **to** ~ **the ceiling** wegern **to** ~ **to the credit** kreditieren **to** ~ **at disposal** zur Verfügung stellen **to** ~ **on dollies** schwenkbar aufstellen **to** ~ **on edge** hochkant stellen **to** ~ **on file** den Akten *pl* einverleiben **to** ~ **in layers** in Schichten *pl* einbringen **to** ~ **in matrix** einbetten to ~ on auflegen to ~ in position ausrichten (fluchten), einsetzen to ~ in readiness zurechtlegen, -stellen to ~ side by side nebeneinanderstellen to ~ in a sloping position schräglegen to ~ on a support (or on a surface) kaschieren to ~ a toll call ein Gespräch *n* anmelden to ~ underneath unterlegen

**place** Ort *m*, Ortschaft *f*, Platz *m*, Posten *m*, Raum *m*, Stelle *f*, Stellung *f*

**place,** ~ **of application** Angriffspunkt *m* ~ **of appointment** Sammel-ort *m*, -platz *m* ~ **of consumption** Verbrauchsstelle *f* ~ **of contact** Berührungsstelle *f* ~ **of control** Kommandostelle *f* ~ **of departure** Aufstiegsort *m* ~ **of discharge** Bodenkippe *f* ~ **of discovery** Fundort *m* ~ **of embarkation** Landungsbrücke *f* ~ **of error** Fehlerstelle *f* ~ **of governor attachment** Reglerangriff *m* ~ **of installation** Aufstellungsplatz *m* ~ **where pressures (or samples) are taken** Entnahmestelle *f* ~ **of publication** Erscheinungsort *m* ~ **of reshipment** Umschlagplatz *m* ~ **of residence** Aufenthaltspunkt *m*, Wohnort *m* ~ **of temporary stop** Aufenthaltspunkt *m* ~ **of transfer** Übergangs-punkt *m*, -stelle ~ **of transition** Übergangsstelle *f*

**place,** ~ **brick** Weichbrand *m* ~ **indicator** Bandzählwerk *n* ~ **mark** Lattenpunkt *m* ~ **name** Ortsname *m*

**placed,** ~ **below (or underneath)** unterliegend ~ **side by side** nebeneinander angeordnet

**placer** Goldseife *f* ~ **bed (or deposit)** Erzseife *f* ~ **gold** Seifen-, Wasch-gold *n*

**places, in** ~ stellenweise

**placing** Anbringung *f*, (concrete) Verlegen *n* ~ **of the compass card (or rose)** Rosenlagerung *f* ~ **out of operation** Außerbetriebsetzung *f* ~ **an order** Auftragserteilung *f* ~ **of an order** Vergebung *f* eines Auftrages ~ **and removing** Ein- und Auslegen *n* (der Walzen)

**plagioclase** Plagioklas *m*

**plain** Ebene *f*, Fläche *f*, Flachland *n*; anschaulich, deutlich, eindeutig, einfach gestaltet, flach, glatt, glattgeschoren, ohne Zubehör, platt, schlicht

**plain,** ~ **of denudation** Abrasionsfläche *f* **in** ~ **language** in offener Sprache *f*

**plain,** ~ **antenna** Einfach-, Linear-antenne *f* ~ **back rest** einfacher Setzstock *m* ~ **bar** Flachstab *m* ~ **bearing** Gleit-, Zapfen-lager *n* ~ **bearing race** Gleitlagerschale *f* ~ **bolt** glatter, nicht abgesetzter Bolzen *m* ~ **carbon steel** unlegierter Kohlenstoffstahl *m* ~ **color** Grundierer *m*, Unifarbe *f* ~ **connecting rod** einfache Schubstange *f* ~ **continuous (nonmosaic) photocathode** zusammenhängende Fotokathode *f* ~ **crane truck ladle** Krangießpfanne *f* mit Wagen- und Handkippvorrichtung ~ **cylindrial plug gauge** Normallehrdorn *m* ~ **cylindrical ring gauge** Normallehrring *m* ~ **dead center** feste Körnerspitze *f* ~ **disk** Flachscheibe *f* ~ **division** Flächenteilung *f* ~ **dyeing** Glattfärberei *f* ~ **edge** Ansetzblatt *n* ~ **-end tubes** Glattendrohre *pl* ~ **fabric** einfacher Stoff *m* ~ **feather** Feder *f* in Nut mit planen Enden, gewöhnliche und geradstirnige Keilnutenfeder *f* (mach.) ~ **-feed table** regulärer Einleger *m*, Tisch *m* mit Handeinlage ~ **film** Flachfilm *m* ~ **film base** unbeschichteter Blankfilm *m* ~ **fit** Schlichtpassung *f* ~ **flap** einfache

Flügelklappe *f*, Wölbungsklappe *f* ~ **forceps** anatomische Pinzette *f* (med.) ~ **girder** Vollwandträger *m* ~ **girding machine** Rundlaufmaschine *f* ~ **glass** glattes Glas *n* ~ **grain drill** einfache Getreidedrillmaschine *f* ~ **grinding** Rundschleifen *n* ~ **grinding maschine** Rundschleifmaschine *f* ~ **grinding machine driven by a handwheel** Handradschleifmaschine *f* ~ **harmonic motion** reine sinusförmige Bewegung *f* ~ **horizontal milling machine** Einfachwagerechtfräsmaschine *f* ~ **joint** Leimfuge *f* ~ **jointed floor** gefugter Dielen(fuß)boden *m* ~ **language** offene Sprache *f* ~ **limit gauge** Rundpassungslehre *f* ~ **milling cutter** einfacher Fräser *m*, Walzen(stirn)fräser *m* ~ **milling cutter with nicked teeth** Walzenfräser *m* mit Spanbrechernuten ~ **mirror photocathode** zusammenhängende Fotokathode *f* ~ **nut** Universalschraubenmutter *f* ~ **oil circuit breaker** vorstufenloser Ölschalter *m* ~ **orifice flange** Meßflansch *m* (einteilig) ~ **pattern** glattes Muster *n* ~ **pole** rohe oder unzubereitete Stange *f* ~ **resistance** induktionsfreier Widerstand *m* ~ **rod** nackte Elektrode *f* ~ **roll** glatte Walze *f*, Glattwalze *f*, Walze *f* mit glatten Ballen ~ **running fit** Schlichtlaufsitz *m* ~ **scale** natürlicher Maßstab *m* ~ **shades** Unitöne *pl* ~ **sliding fit** Schlichtgleitsitz *m* ~ **solid mandrel** (einfacher) Drehdorn *m* ~ **spark discharger** feste Funkenstrecke *f* ~ **spiral mill** Walzenspiralfräser *n* ~ **steel** unlegierter Stahl *m* ~ **striped goods** (textiles) glatte Ringelware *f* ~ **superrefined steel** unlegierter Edelstahl *m* ~ **tile** Biberschwanz *m*, Flachziegel *m* ~ **tool post (or toolholder)** einfacher Support *m* ~ **turning** Langdreharbeit *f* ~ **veneer** einfacher Schraubstock *m* ~ **washer** glatte Unterlegscheibe *f*, Scheibe *f* für Zylinder- und Halbrundschrauben

**plaint** Klageschrift *f*

**plaintiff** Kläger *m*

**plait, to** ~ falten, flechten, plattieren, verflechten to ~ **around** beflechten

**plait** Falte *f*, (cords, jute) Flechte *f*, (cords, jute) Geflecht *n*, Zopf *m* ~ **point** Faltpunkt *m*

**plaited** geflochten ~ **tress** Haarflechte *f* ~ **tubular goods** (textiles) plattierte Schlauchware *f*

**plaiting** Falten *n*, Verschnüren *n* ~ **fold** Lagenbruch *m* ~ **machine** Dockenmaschine *f*, Flechtung *f*, Umspinnmaschine *f* ~ **material** Flechtstoff *m* ~ **shovel** Legeschaufel *f* ~ **thread** Plattierfaden *m*

**plan, to** ~ aushobeln, auslegen, beabsichtigen, entwerfen, konstruieren, planen, projektieren

**plan** Abriß *m*, Aufsicht *f*, Aufstellung *f*, Draufsicht *f*, Entwurf *m*, Grundplan *m*, Lageplan *m*, Plan *m*, Projekt *n*, Riß *m*, Schema *n*, Übersicht *f*, Übersichtsskizze *f*, Vorsatz *m*, Zeichnung *f*

**plan,** ~ **of cable layout** Kabellageplan *m* ~ **of the concession** Mutungsriß *m* **in** ~ **form** im Grundriß *m* ~ **for housebuilding** Bebauungsplan *m* ~ **of operations** (management) Betriebsplan *m* ~ **and side elevation** Grund- und Seitenriß *m* ~ **of study** Studienplan *m*

**plan,** ~ **form** Grundriß *m* ~ **paper** Karten-, Landkartendruck-, Plandruck-papier *n* ~ **position indicator (PPI)** Planbildanzeiger *m*, Rundsichtgerät *n* (rdr) ~ **showing position** Lageplan *m*, Situationsplan *m* ~ **sifting machine** Plan-

sichter *m* ~ **taper** Tiefenverjüngung *f* ~ **verification** Planung *f* ~ **view** (top) Draufsicht *f*, Grundrißansicht *f*
**planar** planar ~ **diode** planparallele Diode *f* ~ **implant** flache Implantation *f* ~ **network** ebenes Netzwerk *n* ~ **sense** planparallele Richtung *f* ~ **symmetric** eben symmetrisch
**planchet** Münzplatte *f*
**planchets** Küvetten *pl*
**Planck's,** ~ **constant** Plancksche Konstante *f*, Planck's Wirkungsquantum *n* ~ **law of radiation** Plancksches Strahlungsnetz *n*
**plane, to** ~ abgleichen, abhobeln, abrichten, ausbeulen, behobeln, ebnen, einebnen, einsetzen, glätten, hobeln, planen, planieren, pritschen, schlichten **to** ~ **down** glattklopfen **to** ~ **off** abhobeln, abschrubben, schroppen **to** ~ **over** überschlichten **to** ~ **a switch tongue** eine Weichenzunge *f* abhobeln (r.r.) **to** ~ **off timber** Holz *n* abhobeln
**plane** Ebene *f*, (hand or machine planes) Fausthobel *m*, ebene Fläche *f*, Fläche *f*, Gehrungsstichel *m*, Gerademesser *n*, Hobel *m*, Kiste *f*; eben, flach, schlicht
**plane,** ~ **of axes** Achsenebene *f* (opt.) ~ **of buoyancy** Schwimmebene *f* ~ **of complex numbers** Zahlenebene *f* ~ **of cross section** Querschnittsebene *f* ~ **of deflection lead** Seitenvorhaltsebene *f* ~ **of departure** Schußebene *f* (artil.) ~ **of division** Teilungs-ebene *f*, -fläche *f* ~ **and elevational measurements** Flächen-anziehungsberechnung *f*, -bestimmung *f* ~ **of failure** Bruchfläche *f* ~ **of the film** Filmebene *f* ~ **of flexure** Biegeebene *f* ~ **of flotation** Schwimmebene *f* ~ **of forces** Kraftebene *f* ~ **of fracture** Bruchebene *f* ~ **of frame** Rahmenebene *f* ~ **of the ground glass** Mattscheibenebene *f* ~ **with handle** Hobel *m* mit Nase ~ **of incidence** Einfallsebene *f* ~ **at infinity** unendlich ferne Ebene *f* ~ **of the instrument** Instrumentenebene *f* ~ **of motion** Bewegungsebene *f* ~ **of oscillation** Schwingungsebene *f* ~ **of the paper** Bildebene *f* ~ **of the plate** Plattenort *m* ~ **of polarization** Polarisationsebene *f* ~ **of position** Ortungsebene *f* ~ **of present position** Meß-, Visier-ebene *f* ~ **of projection** Bild-, Projektions-ebene *f* ~ **of the real intermediate image** Ort *m* des reellen Zwischenbildes ~ **of reference** Bezugs-, Einstell-ebene *f* ~ **of reflection** Reflexionsebene *f* ~ **of refraction** Brechungsebene *f* ~ **of ribbon** totes Feld *n* ~ **of rotation** Rotationsebene *f*, Umdrehungsfläche *f*, ~ **of rotation propeller** Luftschraubenebene *f* ~ **of separation** Trennfuge *f* ~ **of shaft support** Wellenauflagerebene *f* ~ **of shear** Scherfläche *f* ~ **of sighting** Richtfläche *f*, Visierebene *f* ~ **of site** Bauebene *f* ~ **of standard trajectory** Sollschußebene *f* ~ **of stratification** Schichtebene *f* **having a** ~ **surface** ebenflächig ~ **of surface** Verstellflosse *f* ~ **of symmetry** Ebenmaßfläche *f*, Symmetrie-achse *f* -ebene *f* ~ **of transposition** Verschiebungsplan *m* ~ **of union** Vereinigungsebene *f* ~ **of vision** Blickebene *f*
**plane,** ~ **altitude** Flughöhe *f* ~ **angle** Flächenwinkel *m*, flächiger Winkel *m* ~ **antenna** Flächenantenne *f* ~ **arrangement** ebene Anordnung *f* ~ **(load)-bearing structure** Flächentragwerk *n* ~ **bending-test** Planbiegeversuch *m*

~ **bolter** Flachbeutel *m* ~**-bottom hammer** Setzhammer *m* ~ **bracing** Ebeneverspannung *f* ~ **carbon bearing** Kohleplanlager *n* ~ **cell** Planküvette *f* ~ **chart** Flächenkarte *f*, gleichgradige Karte *f* ~**-concave mirror** Hohlplanspiegel *m* ~ **configuration** ebene Konfiguration *f* ~ **contact tip** Meßhütchen *n* mit Planfläche, Planmeßhütchen *n* ~ **curve** einfach gekrümmte Linie *f* ~ **curves** ebene Kurve *f* ~ **cutters** Hobeleisen *pl* ~ **effect** Flächengitterwirkung *f* (cryst.) ~ **failure surface** ebene Gleitfläche *f* ~ **flow** ebene Strömung *f* ~ **flying** Flächenfliegen *n* ~ **frame system** ebenes Rahmensystem *n* ~ **geometry** Geometrie *f* der Ebene ~ **glass disk** Planglas *n* ~ **grate** Plan-, Stangen-rost *m* ~ **grating** Plangitter *n* ~ **height** Flughöhe *f* ~ **horn** Hobelnase *f* ~ **imagery** Bildfeldebnung *f* ~ **implant** flache Implantation *f* ~ **iron** Hobel-eisen *n*, -messer *n* ~ **knife** Schlichtmesser *n* ~ **layer** Halbleiterschicht *f* ~ **lens** ebene Glasscheibe *f* ~ **milling miller** Langfräsmaschine *f* ~ **mirror** Planspiegel *m* ~ **model** Flugmodell *n* ~ **motion** ebene Bewegung *f* ~ **net of a polyhedron** ebenes Netz *n* eines Polyeders ~ **optics** Planoptik *f* ~ **parallel** planparallel ~ **-parallel color filter** Meßfarbfilter *m* ~ **parallel plate** Planplatte *f* ~ **parallel plate micrometer** Planplattenmikrometer *n* ~ **parallelism** Planparallelität *f* ~ **point lattice** ebenes Punktgitter *n* ~ **polarization** lineare Polarisation *f* ~**-polarized** eben polarisiert (Wellenstrahlung), geradlinig oder linear polarisiert ~ **polarized light** linear polarisiertes Licht *n* ~ **polarized wave** linear polarisierte Welle *f* ~ **projection of a sphere** Planiglob *n* ~ **resection** ebener Rückwärtsschnitt *m*, Rückwärtseinschnitt *m* in der Ebene ~ **seal** Plandichtung *f* ~ **sealing surface** Plandichtfläche *f* ~ **soap film (or membrane)** Seifenlamelle *f* ~ **source** flache Quelle *f* ~ **stock** Hobelkasten *m* ~ **strain** ebene Deformation *f* ~ **surface** Ebene *f*, ebene Fläche *f*, Planfläche *f* ~ **surfaces (tools)** Flachseiten *pl* ~ **symmetry** Flächensymmetrie *f* ~ **table** Kartenunterlage *f*, Mensel *f*, Meßtisch *m*, (survey) Peiltafel *f*, Planunterlage *f* ~**-table apparatus** Peilvorrichtung *f* ~**-table photogrammetry** Meßtischfotogrammetrie *f* ~ **-table sheet** Meßtischblatt *n* ~**-table surveying** Meßtischaufnahme *f* ~ **tree** Platane *f* ~ **wave** ebene Welle *f* ~ **wave front** ebene Wellenfront *f*
**planed,** ~ **board** gehobeltes Brett *n* ~ **tooth** gehobelter Zahn *m*
**planeness** Ebenheit *f*, Flachheit *f*
**planer** Fein-, Glatt-säge *f*, Flächenschaber *m*, Hobelmaschine *f*, Klopfholz *n* (print.), Planbank *f* ~ **drive** Hobelmaschinenantrieb *m* ~ **tool** Hobelstahl *m* ~**-type grinding machine** Doppelständerschleifmaschine *f* ~**-type horizontal-spindle grinding machine** Langflächenschleifmaschine *f* mit Horizontalschleifkopf ~**-type milling machine** Planfräsmaschine *f*
**planet** Planet *m*, Wandelstern *m* ~ **mixing arm** Planetrührer *m* ~ **pinion** Planeten(zahn)rad *n* ~ **stirrer** Planetenrührwerk *n* ~ **wheel** Umlaufrad *n*
**planetary** planetarisch ~ **circulation** planetarische Strömung *f* ~ **electrons** kernferne Elektronen *pl* ~ **frame work** Planetarium-, Planetengerüst *n* ~ **gear** Differentialrad *n*, Umlauf-

getriebe *n* ~ **gearing** Planetengetriebe *n* ~ **miller** Planetenfräsmaschine *f* ~ **milling** Pendelfräsen *n* ~ **paddle mixer** Planetenrührwerk *n* ~ **pinion** Sternrad *n* ~ **plunge milling** Pendeltauchfräsen *n* ~ **reduction gear** planetarisches Untersetzungsgetriebe *n* ~**-type reduction gear** Planetenraduntersetzungsgetriebe *n* ~ **wheel** Umlaufrad *n*
**planigraph devices** Schichtbildgeräte *pl*
**planigraphy** Schichtaufnahmeverfahren *n*
**planimeter** Flächen(inhalts)messer *m*, Planimeter *n*
**planimetering** Flächenanziehungsausmessung *f*, Planimetrierung *f*
**planimetric,** ~ **arm** Grundrißlineal *n* (Stereo Autograf) ~ **map** Grundrißkarte *f* ~ **measurement** Lagemessung *f* ~ **position** Grundrißlage *f* ~ **ration** (propeller) Völligkeitsgrad *m* ~ **survey** Grundkataster *m*, Grundrißaufnahme *f*
**planimetry** Flächenmessung *f*, Geometrie *f* der Ebene, Lageplan *m*, Planimetrie *f*, Situationsplan *m* ~ **and elevation** Lage *f* und Höhe *f*
**planing** Hobeln *n*, Planieren *n*, Verflachung *f* ~ **by generating** Walzhobeln *n*
**planing,** ~ **bench** Hobelbank *f* ~ **bottom** (seaplanes) Gleitboden *m* ~ **cut** Planschnitt *m* ~ **file** Bestoßfeile *f* ~ **fin** Gleitflosse *f* ~ **fixture** Hobelvorrichtung *f* ~ **jig** Planiervorrichtung *f* ~ **knife** Hobelmesser *n*, Schabeisen *n* ~ **machine** Abricht-, Hobel-maschine *f*, spanabhebende Maschine *f* ~**-machine table** Hobelmaschinenschlitten *m* ~ **mill** Hobelwerk *n* ~ **piece** Planierstück *n* ~ **resistance** (of seaplanes) Gleitwiderstand *m* ~ **surface** Gleit-fläche (cryst.), -platte *f*, Hebelfläche *f* ~ **tool** Hobel *m*, Hobel-meißel *m*, -stahl *m*, Stichel *m* ~ **works** Hobelwerk *n*
**planish, to** ~ ausbeulen, dressieren, fertigschlichten, glänzen, nachwalzen, planieren, polieren, schlichten **to** ~ **extra-bright** hochglanzpolieren
**planished sheet** hochglanzpoliertes Blech *n*
**planisher** Poliergerüst *n*, Schlicht-, Vorpolier-, Vorschlicht-kaliber *n*
**planishing** Fertig-polieren *n*, -schlichten *n* ~ **hammer** Plätt-, Schlicht-hammer *m* ~ **knife** Geradeisen *n* ~ **pass** Polier-kaliber *n*, -stich *m*, Schlicht-, Vorpolier-stich *m* ~ **stand** Poliergerüst *n*
**planisphere** Weltkarte *f*
**planispiral** flachgewunden
**planisymmetric** planisymmetrisch
**plank, to** ~ mit Brettern abdecken, dielen, einwalken, planken, verkleiden, verschalen, verzimmern **to** ~ **over** bebohlen **to** ~ **up** fertigwalken *n*
**plank** (bridge) Belag *m*, (thick) Bohle *f*, Brett *n*, Büttenbrett *n*, Diele *f*, Fichtenholz *n*, Gautschbrett *n*, Münzplatte *f*, Planke *f*, Schrötling *m*, Schwarte *f*
**plank,** ~ **bed** Pritsche *f* ~ **butt** Plankenstoß *m* ~ **crib** Bretterstapel *m* ~ **flattener** Plätthammer *m* ~ **floor** Brettertafel *f* ~ **flooring** Bohlenbelag *m*, Bretterteppich *m* ~ **log** Sägeblock *m* ~ **nail** Spundnagel *m* ~ **partition** Schalwand *f* ~ **pile** Spundpfahl *m* ~ **pole** Schwartenpfahl *m* ~ **revetment** Bretterbekleidung *f* ~ **road** Bohlenweg *m* ~ **scaffolding on a pontoon** Aufrüstung *f* ~**-sheer** Schandeck *n* ~ **timber** Sägeblock *m* ~ **wall** Bretterbahn *f*
**planked** gewalkt

**planking** Belag *m*, Beplankung *f*, Bohlenbelag *m*, Verkleidung *f*, Verschalung *f* ~ **fixture** Beplankungsgroßvorrichtung *f*
**plankton-microscope** Planktonmikroskop *n*
**planned economy** Planwirtschaft *f*
**planning** Entwerfen *n*, Planen *n*, Planung *f* ~ **mark** Planungskennzeichen *n* ~ **office** Planungsstelle *f* ~ **stage** Planung *f*
**plano,** ~**-concave** plankonkav ~**-convex** plankonvex ~**-convex lens** plankonvexe Linse *f* ~**-film inclosure for plateholder** Planfilm-Einlage *f* für Plattenkassette
**planographic process** Flachdruckverfahren *n*
**planomiller** Langtischfräsmaschine *f*
**plant, to** ~ pflanzen **to** ~ **trees** beholzen
**plant** Aggregat *n*, Anlage *f*, Betrieb *m*, Betriebs-anlage *f*, -einrichtung *f*, Fabrik *f*, Gewächs *n*, Maschinenanlage *f*, Pflanze *f*, Werk *n*, Werkstatt *f* **of the** ~ werkseigen ~ **with distribution boxes** Schaltadernetz *n*
**plant,** ~ **breeding** Pflanzenzüchtung *f* ~ **cane** Pflanzrohr *n*, Stecklingsrohr *n* ~ **conditions** Betriebsbedingungen *pl*, Betriebsverhältnisse *pl* ~ **design** Auslegung *f* der Anlage (oder Konstruktion) ~ **ecology** botanische Wanderungen *pl* ~ **engineer** Betriebsingenieur *m* ~ **equipment** Betriebseinrichtung *f*, Werksausrüstung *f* ~ **extension** Erweiterungsausbau *m* ~ **layout** Auslegung *f* der Anlage (oder Konstruktion) ~ **maintenance** Betriebsinstandhaltung *f* ~ **management** Betriebsleitung *f* ~ **operation** Aggregatbetrieb *m* ~ **protective** Pflanzenschutzmittel *n* ~ **public-address system** Betriebslautsprecheranlage *f* ~ **(investment) risks** Anlagerisiko *n* ~ **supervision** Betriebsüberwachung *f* ~ **switchboard** Stationsschalttafel *f*
**plantation** Pflanzung *f*
**planter** Pflanzer *m*
**Planté-type plate** Großoberflächenplatte *f*
**planting** Anpflanzung *f* ~ **of beacons** Bebakung *f*, Betonung *f* ~ **attachment** Pflanzenvorrichtung *f*
**plasma** Plasma *n* ~ **accelerator** Plasmabeschleuniger *m* ~ **balance** Plasmagleichgewicht *n* ~ **beam** Plasmastrahl *m* ~ **boundary** Plasmagrenzfläche *f* ~ **drive** Plasmatriebwerk *n* ~ **interaction** Plasmawechselwirkung *f* ~**-jet boring** Flammenstrahlbohrer *m* ~ **oscillations** Plasmaschwingungen *pl* ~ **sphere** Plasmakugel *f*
**plasmagene** plasmagen
**plasmus of gas discharge** Gasentladungsplasmen *pl*
**plaster, to** ~ bepflastern, bewerfen, gipsen, vergipsen, verputzen, verschmieren **to** ~ **the sound floor** den Fehlboden *m* verfüllen **to** ~ **a wall** elne Mauer *f* abputzen
**plaster** Bewurf *m*, Gipsmörtel *m*, Gipsstein *m*, Mörtel *m*, Pflaster *n*, Putz *m*, Verputz *m* ~ **for facing** Edelputz *m* ~ **of Paris** gebrannter Gips *m*, Gips *m*, Stuckgips *m*
**plaster,** ~ **burning** Gipsbrennerei *f* ~ **cast** Gipsabdruck *m*, -abguß *m*, -verband *m* ~**-cast saw** Gipsverbandsäge *f* ~ **cement** Gipskitt *m* ~ **floor** Gipsestrich *m* ~ **flooring** Gipsestrich(fuß)boden *m* ~ **hatchet** Tünchhacke *f* ~ **lagging** Gipsmantel *m* ~ **mold** Gipsform *f* ~**-mold casting** Gipsformguß *m* ~ **mortar** Stuckmörtel *m* ~ **paints** Fassadenfarben *pl* ~ **slab** Gipsdiele *f* ~ **stucco** Gipsputz *m* ~ **work** Verputz *m*

**plasterer** Gipser *m*
**plastering** Gips-arbeit *f*, -bewurf *m*, -stein *m*, Putz *m*, Rapputz *m*, Stuckarbeit *f*, Wandbewurf *m* ~ **material** Verkittungsmaterial *n*
**plastic** Kunst-harz *n*, -produkt *n*, -stoff *m*, plastisches Material *n*; bildnerisch, erhaben, formänderungsfähig, formbar, knetbar, plastisch, teigig, verformbar
**plastic,** ~ **alloy** Knetlegierung *f* ~ **art** Plastik *f* ~ **audition** zweiohriges Hören *n* ~ **bulletproof glass** Glaspanzer *m* ~ **clay** bildfähiger Ton *m* ~**-coated paper** kunststoffbeschichtetes Papier *m* ~ **coating** Kunststoffüberzug *m* ~ **cold working** Kaltverformung *f* ~ **compliance** Plastizitätskoeffizient *m* ~ **compound** Preßstoff *m* ~ **deformation** bildsames Gleiten *n*, plastische Verformung *f* ~ **effect** Bildplastik *f*, Plastik *f*, (of picture) Tiefenwirkung *f* ~**-fabric bearing** Preßstofflager *n* ~ **film** plastischer Film *m* ~ **floor** Kunstharzboden *m* ~ **flow** Dehnung *f*, plastische Ausweichung *f* ~ **flow persistence** elastische oder plastische Nachwirkung *f* ~ **foil** Kunststoffolie *f* ~ **foil capacitor** Kunststofffolienkondensator *m* ~ **force** Bildungskraft *f* ~ **gear** Kunststoffrad *n* ~ **hearing** zweiohriges Hören *n* ~ **lens** Bakelit(licht)scheibe *f*, Plastiklinse *f* ~ **limit** Ausroll-, Plastizitätsgrenze *f* ~ **material** Formerstoff *m*, Preß-masse *f*, -stoff *m* ~ **mixture** plastische Mischung *f* ~ **molding material** Preßmasse *f* ~ **paint** plastische Farbe *f* ~ **picture** Preßmasse *f*, Raumbild *n* ~ **press** Presse *f* für thermoplastische Massen ~ **process** Halbtrockenverfahren *n* ~ **quality** Geschmeidigkeitseigenschaft *f* ~ **range** Fließbereich *m*, Plastizitätsgebiet *n* ~ **refractory clay** Klebsand *m* ~ **resin** Kunststoff *m* ~**-rigid boundary** plastisch-starre Grenze *f* ~ **shear** Abgleitung *f* ~ **slideway** Preßstoffgleitbahn *f* ~ **strain** bleibende Formänderung *f* ~ **stress-strain curve** Verfestigungskurve *f* ~ **tube** Isolierschlauch *m* ~ **welding** Preßschweißung *f* ~ **welding technique** Kunststoffschweißverfahren *n* ~ **wood** Knetholz *n* ~ **workability** Formveränderungsvermögen *n* ~ **working** plastische Verformung *f* ~ **working condition** Verformungsverhältnis *n*
**plasticine** Plastilin *n*
**plasticity** Bildsamkeit *f*, Formänderungsvermögen *n*, Formbarkeit *f*, Geschmeidigkeit *f*, Körperlichkeit *f*, Plastizität *f*, (of pictures) Relief *n*, Verformbarkeit *f* ~ **of image** Tiefenwirkung *f* ~ **index** Plastizitäts-zahl *f*, -ziffer *f* ~ **needle** Prüfnadel *f*
**plasticize, to** ~ erweichen, plastizieren, weichen
**plasticizer** Plastizierer *m*, Plastiziermittel *n*, Weichmacher *m*, Weichmachungsmittel *n*
**plasticizing** Plastizierung *f* ~ **efficiency** weichmachende Wirkung *f*
**plastics** Kunststoffe *pl*, Preßstoffe *pl* ~ **foil** Kunststoff-Folie *f* ~ **industry** Kunstharzindustrie *f* ~ **linings** Kunststoffüberzüge *pl* ~ **suspension** Kunststoffsuspension *f* ~ **tyre** Kunststoffbandage *f*
**plastometer** Plastizitätsmesser *m*, Plastometer *n*
**plate, to** ~ auskleiden, platieren, plattieren, überziehen **to** ~ **with cobalt** verkobalten **to** ~ **out** ausbreiten, ausplatten

**plate** Abbildung *f*, Anode *f*, Belegung *f*, Blatt *n*, Blech *n*, dickes Blech, (distilling column) Boden *m*, Fußteller *m*, Grobblech *n*, Lamelle *f* einer Gürtung, Nockenbahn *f*, Platine *f*, Platte *f*, Plattine *f*, (disk) Scheibe *f*, Schild *n*, Streifen *m*, Sturz *m*, Tafel *f*, Teller *m*, Tellerzinn *n*, Zacke *f*, Zacken *m*, Zunge *f*
**plate,** ~ **for accessories** Zubehörblech *n* ~ **of clay** Tonplatte *f* ~ **of a condenser** Belegung *f* oder Platte *f* eines Kondensators ~ **of earthenware** Tonplatte *f* ~ **free from halation** lichthoffreie Platte *f* (photo) ~ **of main anode** Hauptanode *f* in einer Vakuumröhre ~ **of refined copper** Garscheibe *f* ~ **to be rolled** Walzplatte *f* ~ **for safes** Geldschrankblech *n* ~ **for the skin** Hautblech *n* ~ **for tilting sucker bar** Schild *n* für Saugerkippung ~ **with wedgeshaped groove** Keilnutplatte *f*
**plate,** ~ **air preheater** Taschenlufterhitzer *m* ~ **alternating-current potential (or voltage)** Anodenwechselspannung *f* ~ **amalgamation** Platten-, Pochwerk-amalgamation *f* ~ **anemometer** Flügelradwindmesser *m* ~ **atomizer** Plattenzerstäuber *m* ~ **axes** Bildachsen *pl* ~ **bar** Platine *f* ~ **battery** Anoden-, Bord-batterie *f* ~ **bearing (or bed)** Plattenlagerung *f* ~ **bearing test** Plattenbelastungsversuch *m* ~**-bending machine (or rolls)** Blechbiegemaschine *f* ~ **by-pass capacitor** Anodenfußkondensator *m* ~ **cam** Kurven-Steuerscheibe *f* ~ **camera** Plattenkamera *f* ~ **caps** Anodenkappen *pl* ~ **carrier** Bildträger *m* (TV), Einlegerahmen *m*, Plattenhalter *m*, -träger *m* ~**-carrying box car (or cat)** Plattenhalterwagen *m* ~ **caster** Lenkrolle *f* mit Blattzapfen ~ **cell** Tellerzelle *f* ~ **center** Plattenmittelpunkt *m* ~ **change** Plattenwechsel *m* ~ **characteristic** Anodenstromkennlinie *f*, Bremsröhre *f* ~ **circuit** Anoden(strom)kreis *m* ~**-circuit detector** Anodengleichrichter *m*, Richtverstärker *m* ~**-circuit modulation** Anodenbesprechung *f* ~ **clamp** Blechpratze *f* ~ **clamping bar** Plattenspannschiene *f* ~ **closer** Blechschlußring *m* ~ **closing tool** Niet-antreiber *m*, -zieher *m* ~ **clutch** Einscheiben-, Platten-, Scheibenkupplung *f* ~ **coil** Anoden(kreis)spule *f* ~ **column** Plattenturm *m* ~ **condenser** Plattenkondensator *m* ~ **conductance** Anodenleitwert *m* ~ **conjunction** Plattenschluß *m* ~ **contour** Plattenschnitt *m* ~ **conveyer** Platten-band *m*, -förderer *m* ~ **coordinate** Bild-, Platten-koordinate *f* ~ **coordinate system** Bildkoordinatensystem *n* ~ **coupling** Flanschen-, Scheibenkupplung *f* ~ **covering** Blechbelag *m* ~ **covering the balance** Unruhedeckplatte *f* ~ **current** Anoden(gleich)strom *m*, Batteriestrom *m*
**plate-current,** ~ **curve** Anodenabsinkkurve *f* ~ **excursion** Anodenstromaussteuerung *f* ~ **grid voltage characteristic** Charakteristik *f* der Röhre, Anodenstrom-Gitterspannungskennlinie *f*, normale Kennlinie *f* ~ **measuring switch** Anodenstromschalter *m* ~**-plate-voltage characteristic** Anodenstrom-Anodenspannungskennlinie *f* ~ **rectification** Anodengleichung *f* ~ **variation** Anodenstromänderung *f*
**plate,** ~ **cut to size** beschnittene Platte *f* ~**-cutting attachment (or machine)** Blechschere *f* ~ **detection** Anodengleichrichtung *f* ~ **dimension**

Plattengröße f ~ **direct current** Anodengleich-
richterspannung f, Anodengleichstrom m ~
**direct current potential (or voltage)** Anodengleichspannung f ~ **dissipation** Anodenverlustleistung f ~ **distance** Kameralänge f ~ **dryer**
Tellertrockner m ~**-drying rack** Plattentrocken-
gestell n ~**-edge planing machine** Blechbesäum-
maschine f, Blechkantenhobelmaschine f ~
**efficiency** Anodenwirkungsgrad m ~ **electrode**
Anode f einer Elektronenröhre ~ **electrometer**
Plattenelektrometer n ~ **end** Tellerboden m ~
**-feed mechanism** Plattenfördermechanismus m
~ **feedback** Anodenrückwirkung f ~ **feedback
coupling** Anodenrückkopplung f ~ **feeder** Tel-
leraufgabeapparat m ~**-filament circuit** Aus-
gangsstromkreis m ~ **filter** Zellenfilter n ~
**fitting** Plattenbeschlag m ~**-flanging machine**
Blechbördelmaschine f ~ **form spring** Federung
f mit Quer- und Längsfedern ~ **free from halo**
lichthoffreie Platte f ~ **frame** Plattenrahmen m
~ **fuses** Streifensicherungen pl ~ **gauge** Blechlehre f ~ **girder** Blech-balken m, -träger m, vollwandiger Träger m ~ **gland** Blechträger m ~
**glass** Fensterglas n, Scheibenglas n, Spiegel-glas
n, -platte f, -scheibe f ~**-glass pane** Spiegelglas-
platte f ~ **glass production** Flachglaserzeugung
f ~ **gravure** Plattentiefdruck m ~ **grid** (of
storage batteries) Gitter n ~ **grinding and
polishing machine** Blechschleif- und -polier-
maschine f ~ **holder** Blechschlußring m, Kassette f, Platten-halter m, -kassette f, -magazin n
~ **jet** Plattenanguß m ~**-joggling machine** Plat-
tenkröpfmaschine f ~ **keel** Plattenkiel m ~ **key**
Legeschlüssel m ~ **keying** Anodenspannungstastung f ~ **layer** Oberbauarbeiter m ~ **laying**
Gleisverlegung f ~ **lightning arrester** Plattenblitzableiter m ~ **lightning rod** Grobblitzablei-
ter m ~ **liner** Plattenfüllstück n ~**-link rigging**
Laschenfedergehänge n ~ **load** Anodenarbeits-
widerstand m, Anodenbelastung f ~ **loading
test** Plattenbelastung f mit Platte ~ **locking
device** Plattenzuhaltungen pl (Sicherheitsschloß)
~ **lug** Plattenfahne f ~ **magazine** Plattenkassette f ~ **maker** Systemmacher m ~ **measuring**
Bildmessung f ~**-measuring apparatus** Bildmeßgerät n ~**-meshing-type capacitor** Plattenkondensator m ~**-metal chute** Blechrinne f ~
**mill** Blechwalzwerk n, Grobblech-walzwerk n,
-straße f ~**-mill roll** Flacheisenwalze f ~**-mill
stand** Blechgerüst n ~**-mill train** Blech-straße f,
-strecke f, Grobstraße f ~ **modulation** Anodenbesprechung f, -modulation f, Formen n nach
Modell, Modellformerei f, Modulation f durch
Änderung der Anodenspannung ~**-molding
shop** Modellformerei f ~ **mounting** Plattenauf-
spannung f ~ **multiplier** Plattenvervielfacher m
~ **neon lamp** Neonflächenglimmlampe f (TV)
~ **neon light** Flächenglimmlampe f ~ **nut**
Anniet-, Platten-mutter f ~ **oil** Druckerfirnis
m ~ **orifice** Staurand m ~ **painter** Blechstrei-
cher m ~ **panel of aluminium alloys (or of metal)**
Blechtafel f ~ **pattern molding** Modellformerei f
~ **penotrometer** Härtemesser m mit Platte ~
**plane** Plattenebene f ~ **planers** Blechkantenhobelmaschine f ~ **plumb point** Bildlotpunkt m,
Bildnadier m ~ **potential** Anodenspannung f ~
**products** Grobblecherzeugnisse pl ~ **protector**
Plattenblitzableiter m ~**-punching machine**

Blechlochmaschine f ~ **racks** Plattenlager n ~
**rail** Flach-, Platt-schiene f ~ **reaction** Anodenrückwirkung f ~ **resistance** Anodenwiderstand
m, innerer Widerstand m ~ **return** Anodenrück-
leitung f ~ **roll** Blechwalze f, Grobblechwalze f
~ **rolling** Flacheisenwalzen n ~**-rolling mill**
Blechwalzwerk n, Grobblechwalzwerk n ~ **rope
connecting tackle** Laschenzwischengeschirr n ~
**section** Plattenschnitt m ~**-shape cutter** Teller-
fräser m ~**-shaped** plattenförmig ~**-shaped base**
Tellerfuß m ~**-shaped cathode** Flächenkathode
f ~**-shaped foot** Tellerfuß m ~**-shaped press**
Tellerfuß m ~ **shearer** (wool) Plättchenma-
schinenarbeiter m ~**-shearing attachment (or
machine)** Blechschere f ~ **shears** Metallschere f,
Tafelschere f ~ **sheath** Plattenträger m ~ **shield**
Anodenschutznetz n ~ **singeing machine** Plat-
tensenge f ~ **size** Platten-format n, -größe f ~
**slab** Bramme f, Platine f ~**-slab shears** Platinenschere f ~ **sounder** Klangplattenklopfer m ~
**specimen** Flachstabprobe f ~ **spring** Blatt-,
Platten-, Scheiben-feder f, tellerförmige Feder f
~**-spring manometer** Plattenfederdruckmesser
m ~ **store** Grobblechlager n ~**-straightening
machine (or roll)** Blechrichtmaschine f ~
**-straightening rolls** Grobblechmaschine f ~
**supply** Anodenspeisung f ~ **terminal** Anoden-
klemme f ~ **test** (explosives) Plattenbeschuß m
~ **testing device** Druckstockprüfapparat m ~
**-tinning furnace** Blechverzinnungsofen m ~**-to
filament circuit** Anodenkreis m ~ **tower** Plattenturm m ~ **transition-time angle** Anodenlaufzeit-
winkel m ~**-type apparatus** Plattenapparat m
~**-type filter** Spaltebenekristallfilter m ~**-type
(oil) filter** Spaltfilter n ~ **valve** Plattenventil n ~
**voltage** Anodenspannung f ~**-voltage characteristic** Anodenspannungskennlinie f ~**-voltage
excursion** Anodenspannungsaussteuerung f ~
**-voltage keying** Anodentastung f ~**-voltage
winding** Anodenspannungswicklung f ~ **voltages** Anodenseite f ~ **volts** Anodenspannung f
~ **warning** Warnungstafel f ~ **washer** Ankerplatte f ~ **wave** Plattenwelle f ~ **welding** Blech-
weißung f ~ **work** Blecharbeit f ~ **-working
machine** Blechbearbeitungsmaschine f
**plateau** Blachfeld n, Hochebene f, Plateau n,
Platte f ~ **seat** Furniersitz m ~ **slope** Plateauneigung f ~ **structure** Netzwerkstruktur f
**plated** überzogen, (electroplated) plattiert ~
**with metal** blechbeschlagen
**plated,** ~ **coat** Plattierungsschicht f ~ **fabric**
Plattierware f ~ **seat** Furniersitz m ~ **structure**
Netzwerkstruktur f ~ **tubular goods** plattierte
Schlauchware f
**plateless valve** anodenlose Röhre f
**platelet** Plättchen n, Scheibchen n, Thrombozyt n
**platelike** lamellar, platten-artig, -förmig, schichtig, steifig
**platen** Druck-rolle f, -tafel f, -walze f (teletype),
-zylinder m, (Presse) Holm m, Platte f, Tiegel m,
Walze f ~ **area** Aufspannfläche f ~ **feed clutch
lever** Tischvorschubkupplungshebel m ~ **feed
screw** Tischvorschubspindel f ~ **knob** Handrädchen n, Walzendrehknopf m ~ **machine** Flach-
druckmaschine f ~ **machine minder** Tiegel-
drucker m ~ **press** Tiegeldruckpresse f ~ **release
mechanism** Walzen-freilauf m, -löser m ~
**superheater** Schottüberhitzer m

**plates** Mittel- und Grobbleche *pl* ~ **(in stamping or deep-drawing quality)** Bleche *pl* (in Stanz- und Tiefziehgüte) ~ **with projections** Warzenbleche *pl* ~ **and sheets** Bleche *pl*
**platform** Altan *m*, Auffahrt *f*, Auftritt *m*, Bahnsteig *m*, Bedienungsstand *m*, (gun) Bettung *f*, Brücke *f*, Bühne *f*, Kanzel *f*, Ladeplatz *m*, Ladestelle *f*, Längerampe *f*, Laufbühne *f*, Platte *f*, Plattenband *n*, Plattform *f*, Podest *n*, Rampe *f*, Schußbühne *f*, Steuer-bühne *f*, -tisch *m*, Zulage *f* **(timber)** ~ Schwellenrost *m* ~ **for the operator** Bedienungsbühne *f*
**platform,** ~ **balance** Brückenwaage *f* ~ **body** Pritschenaufbau *m* ~ **conveyer** Platten(band)förderer *m*, Wandertisch *m* ~ **conveyer for handling cased goods** Kistentransportanlage *f* ~ **detail** Bettungsstaffel *f* ~ **extension** Plattenformverlängerung *f* ~ **flooring** Bühnenbelag *m*, Brückenboden *m* ~ **flooring plate** Bühnenbelagblech *n* ~ **grate** Plattformrost *m* ~ **hand truck** Tafelwagen *m* ~ **production** Brückenfertigung *f* ~ **roof** Terrassendach *n* ~ **scales** Tafelwaage *f* ~ **spring** Federung *f* mit Quer- und Längsfedern ~ **support** Laufbrettträger *m* ~ **truck** Plateaukarre *f*, Plattformwagen *m* ~ **underpass** Bahnunterführung *f*
**plating** Belag *m*, Belegung *f*, Beplattung *f*, Blechbekleidung *f*, Blechhaut *f*, Metall-auflage *f*, -überzug *m*, Plattieren *n*, (of metal) Plattierung *f*, Überziehen *n*, Überzug *m*, Versilberung *f* ~ **material** Plattierwerkstoff *m* ~ **mill** Plattierwalzwerk *n* ~-**out** Ausplatten *n* ~-**out test** Ausbreiteprobe *f* ~ **process** Galvanisierprozeß *m*, Plattierverfahren *n*
**platinic,** ~ **ammonium chloride** Platinsalmiak *m* ~ **chloride** Platinchlorid *n* ~ **compound** Platinoxydverbindung *f* ~ **salt** Platinsalz *n*
**platinicyanic acid** Platinzyanwasserstoffsäure *f*
**platiniferous** platinhaltig
**platinization** Platinierung *f*
**platinize, to** ~ platinieren
**platinized charcoal** Platinkohle *f*
**platinizing** Platinisieren *n*
**platino-cyanide** Platinzyanür *n*
**platinoid** Platinoid *n*
**platinotype paper** Platinpapier *n*
**platinous** platinhaltig ~ **barium cyanide** Platinbariumzyanür *n* ~ **bromide** Platinbromür *n* ~ **chloride** Platinchlorür *n*, Platinochlorid *n* ~ **compound** Platinoxydulverbindung *f* ~ **cyanide** Platinzyanür *n* ~ **hydroxide** Platinhydroxydul *n* ~ **magnesium cyanide** Platinmagnesiumzyanür *n* ~ **oxide** Platinoxydul *n* ~ **potassium chloride** Platinkaliumchlorür *n* ~ **sodium chloride** Platinnatriumchlorür *n* ~ **sulphide** Platinsulfür *n*
**platinum** Platin *n* ~ **apparatus** Platingerät *n* ~ **black** Platin-mohr *m*, -schwarz *n* ~ **boat** Platinschiffchen *n* ~ **coil** Platinspirale *f* ~ **cone** Platin-kegel *m*, -konus *m* ~ **contact bead** Platinkontaktperle *f* ~ **contact piece** Platinkontakt *m* ~ **content** Platingehalt *m* ~ **crucible** Platintiegel *m* ~ **dish** Platinschale *f* ~ **foil** Platin-blech *n*, -folie *f* ~ **fusion** Platinverbindung *f* ~ **gauze** Platindrahtnetz *n* ~ **gray** Platingrau *n* ~ **group** Platinreihe *f* ~ **lamp** Platinfadenlampe *f* ~ **oxide** Platinoxyd *n* **to** ~-**plate** (ver)platinieren ~ **plate** Platinblech *n* ~-**plated** platiniert ~ **plating** Verplatinierung *f* ~-~-**rhodium couple**

Platinplatinrhodiumelement *n* ~ **point** Platinkontakt *m*, -punkt *m*, -spritze *f* ~ **residue** Platinrückstand *m* ~ **resistance thermometer** Platinthermometer *n* ~ **rhodium** Platinrhodium *n* ~ **screw** Platinschraube *f* ~ **series** Platinreihe *f* ~ **spatula** Platinspatel *m* ~ **sponge** Platinschwamm *m* ~ **still** Platinblase *f* ~ **thermocouple** Platinelement *n* ~-**tipped screw** mit Platinspritze versehene Schraube *f* ~ **vessel** Platingefäß *n* ~ **ware** Platingerätschaft *f* ~ **waste** Platinabfall *m* ~ **wire** Platindraht *m* ~-**wire loop** Platindrahtöse *f*
**platonic polyhedra** platonische Körper *pl*
**platoon** Zug *m*
**platter** Aufnahmeplatte *f*
**Plattner process** Plattner-Prozeß *m*
**plattnerite** Schwerbleierz *n*
**platynite** Platinit *n*
**platymeter** Platymeter *m*
**play, to** ~ laufen, sich leicht hin- und herbewegen, spielen **to** ~ **back** abspielen (phono) **to come into** ~ in Funktion *f* treten **to** ~ **in** gängig machen
**play** Ausdehnungsspiel *n*, Gang *m*, Schauspiel *n*, Spiel *n*, Spiel *n* eines Pochstempels, Spielraum *m* **(end)** ~ toter Gang *m*
**play,** ~ **of bit** Meißelspiel *n* ~ **between contacts** Kontaktspielraum *m* ~ **of forces** Kraftspiel *n* ~ **of gears** Spiel *n* der Getrieberäder ~ **of piston** Kolbenspiel *n* ~ **of the piston** Spiel *n* des Kolbens ~ **of spring** Federspiel *n*, Schwingung *f* der Feder ~ **of tongue** Anker-hub *m*, -spiel *n*
**playback** Abspielen *n* (tape rec.), Rückspielen *n*, Wiedergabe *f* ~ **equalization** Wiedergabeentzerrung *f* ~ **head** Abspiel-, Hör-, Wiedergabekopf *m* ~ **loudspeaker** Hintergrundlautsprecher *m* ~ **strength** Abspielstärke *f* ~ **unit** Wiedergabegerät *n*
**playground** Spielplatz *m*
**play-pipe** Strahlrohr *n*
**plea** Einspruch *m*, Einwendung *f*, Rechtfertigungsgrund *m*, Rechtseinwand *m*, Verteidigungsschrift *f* **on the** ~ unter dem Vorwand *m* ~ **of nullity** Nichtigkeits-beschwerde *f*, -klage *f*
**plead, to** ~ antworten, vorbringen
**pleading** Verhandlung *f*
**pleasing appearance** Formschönheit *f*
**pleat, to** ~ falten, plissieren
**pleat** Falte *f*, Flechtung *f*, Sieke *f*
**pleated,** ~ **diaphragm** Riffelmembrane *f* ~ **filter** Faltenfilter *n* ~ **hose (or tube)** Faltenschlauch *m*
**pleating** Faltung *f* ~ **machine** Faltenlegemaschine *f*, Flechtung *f*
**pledge, to** ~ verpfänden
**pledge** Bürgschaft *f*
**pledget** Meißel *m*
**plenary** uneingeschränkt, voll(ständig) ~ **perspective** Füllperspektive *f* ~ **session** Plenarsitzung *f*
**Plendell's adapter** (on dive bombers) automatische Abfangvorrichtung *f*
**plenipotentiary** Bevollmächtigter *m*
**plentiful** ergiebig, in Hülle und Fülle, reichhaltig, reichlich
**plentifulness** Fülle *f*, Überfluß *m*
**plenty** Fülle *f*, Menge *f*

**plenum** Anfüllung *f*, ausgefüllter Raum *m*; voll ~ **chamber** Beruhigungsraum *m*, Luftkammer *f*, Luftspeicherraum *m*, Trockenkammer *f*, Verdichtervorkammer *f* ~ **chamber** Flächenstrahler *m* (BEF) ~ **system** Durchventilationssystem *n*

**pleochroic** mehrfarbig ~ **hale** mehrfarbiger Lichthof *m*, pleochroitischer Halo *m*

**plessite** Plessit *m*

**plethysmograph** Blutfarbemesser *m*

**plexiglass** Plexiglas *n* ~ **light conductor** Plexiglaslichtleiter *m*

**pliability** Biegbarkeit *f*, Biegsamkeit *f*, Biegungsvermögen *n*, Geschmeidigkeit *f*

**pliable** biegbar, biegsam, biegungsfähig, gelenkig, mürbe, nachgiebig ~ **glass window** Plexiglasfenster *n* ~ **sheeting** biegsames Plattenmaterial *n*

**pliableness** Biegsamkeit *f*, Geschmeidigkeit *f*, Weichheit *f*

**pliant** biegsam, gefügig, schmiegsam, weich ~ **rule** Klappmaßstab *m*, Kurvenlineal *n*

**plication** Faltung *f* einer Lagerstätte

**plied yarn** gefachtes Garn *n*

**pliers** Beißzange *f*, Draht-kluppe *f*, -zange *f*, Klemme *f*, Kluppe *f*, Spannzange *f*, Zange *f*, Ziehzange *f*

**plies of fabric** Gewebeeinlage *f*

**Plimsoll mark** Lademarke *f*

**plinth** Fußleiste *f*, Säulenplatte *f*, Sockel *m*, Sockelplatte *f* (arch.) ~ **brick** Plinthziegel *m*

**pliodynatron** Doppelgitterröhre *f* (in Schutznetzschaltung), Pl(e)iodynatron *n*

**pliotron** Hochvakuumgitterröhre *f*, Pl(e)iotron *n*

**plodder** (soap mfg.) Peloteuse *f*, Strangpresse *f*

**plot, to** ~ (with a compass) abgreifen, (on a map or chart) absetzen, (a curve) anlegen, anreißen, (curves) auftragen, aufreißen, aufzeichnen, auswerten, (curve) bestimmen, (a curve) darstellen (als Kurve), einzeichnen, festlegen, hinzeichnen, orten, planen, zeichnen

**plot, to** ~ **against** auftragen gegen, Kurvenwerte *pl* auftragen (über) **to** ~ **a curve** eine Kurve *f* zeichnen, erstellen oder bilden **to** ~ **curves** Kurven *pl* aufnehmen **to** ~ **lines of force** Stromlinien *pl* aufzeichnen **to** ~ **a map** kartieren **to** ~ **off** abtragen **to** ~ **on (a plan)** eintragen (in einen Plan) **to** ~ **the route of a pipeline** die Trasse freilegen **to** ~ **tubes** Stromlinien *pl* aufzeichnen **to** ~ **a value against another** einen Wert *m* in Abhängigkeit von einem anderen darstellen **to** ~ **versus** Kurvenwerte *pl* auftragen über

**plot** Anschlag *m*, Auftragung *f*, (sketch) Aufzeichnung *f*, Bauplan *m*, Diagramm *n*, geortetes Ziel *n*, grafische Darstellung *f*, Grundriß *m*, Grundstück *n*, Plan *m*, Zielort *m* ~ **of ground located upstream** Oberlieger *m* ~ **of a mine** Markscheideriß *m* ~ **limits** Anlagegrenzen *pl*

**plottable** auswertbar

**plotted** abgetragen (math.), grafisch ~ **against** über ~ **hits** (firing) Trefferbild *n* ~ **point** Aufpunkt *m*

**plotter** Auswerter *m*, Kennlinienschreiber *m*, Kurvenschreiber *m*, Planzeichner *m*

**plotting** (a course) Abstecken *n*, (curves, graphs) Aufnahme *f*, Aufnehmen *n*, (of a map) Auf-

tragung *f*, Aufzeichnen *n*, (of photogrammetric data) Auswertung *f*, Entwerfen *n*, (a curve) Darstellung *f*, (of course) Kursermittlung *f*, Mitkoppeln *n* (nav.), Planzeichnen *n*, ~ **from aerial photographs** Bildauswertung *f* ~ **of curves** Auswertung *f* ~ **of maps (or plans) from photograps** Bildkartierung *f* ~ **by polar coordinates** Polverfahren *n* ~ **of target course** Zielwegaufschreibung *f* ~ **of timing-curve** Verstellinienaufnahme *f*

**plotting,** ~ **apparatus** Auswertegerät *n*, Bildkartiergerät *n* ~ **apparatus for pairs of photographs** Bildkartiergerät *n* für Bildpaare ~ **board** (flash ranging) Flächenmeßplan *m*, Planzeiger *m* ~ **center** Meßstelle *f* ~ **equipment** Plangerät *n* ~ **field configuration (or map)** (of electrostatic field) Feldmessung *f* ~ **interval** Meßpause *f* (rdr) ~ **lens** Auswerteobjektiv *n* ~ **method** Auswerteverfahren *n* ~ **paper** Millimeter-, Zeichen-papier *n* ~ **position bearings (or lines)** Kursabsetzen *n* ~ **scale** Auswerte-, Kartierungs-, Zeichen-maßstab *m* ~ **station** Auswertestelle *f* ~ **table** Auswertetisch *m*

**plough: see plow**

**plow, to** ~ ackern, durchfurchen, furchen, pflügen **to** ~ **in (or under)** unterpflügen **to** ~ **and tongue together** verspunden **to** ~ **up** umpflügen

**plow** Beschneidehobel *m*, Filmabheber *m*, Falz-, Kehl-, Nut-hobel *m*, Pflug *m*, Stromabnehmer *m* für unterirdische Stromschiene (r.r.), Zuschneidehobel *m* ~ **bean** Grindel *m* ~ **harrow** Grubberegge *f* ~ **knife** Beschneidehobel *m* ~ **plane** Spundhobel *m* ~ **share** Pflugmesser *n* ~ **tail** Pflugsterz *m* ~ **wire** Pflugdraht *m*

**plowed,** ~ **and tongued door** gespundete Tür *f* ~ **field (or land)** Sturzacker *m*

**plowing** Nut *f*, Rinne *f* ~ **machine** Auflockerungsmaschine *f*, Maschinenpflug *m* ~ **-up of the roller** Anreißen *n* der Druckwalze

**pluck, to** ~ anreißen, (out) ausreißen, (a string or chord) reißen, zupfen **to** ~ **the wool** die Wolle *f* plüsen

**plucked,** ~ **chord (or string)** gezupfte Saite *f* ~ **wool** Blut-, Schlacht-wolle *f*

**plucker** (one who plucks out) Ausreißer *m*

**plucking** Abpalen *n* ~ **instrument** Zupfinstrument *n* ~ **machine** Enthaar- und Glattmaschine *f* ~ **woman** (textiles) Ausrupferin *f*

**plug, to** ~ durchkneten, stöpseln, verbleien, verspunden, verstemmen, verstopfen, zupfropfen **to** ~ **a hole** (aus)-spünden **to** ~ **in** den Stöpsel *m* in eine Klinke einführen, einen Stöpsel *m* einsetzen, einschalten, einstöpseln, mit Stöpseln *pl* einschalten, stöpseln, vorstekken **to** ~ **up** abstöpseln, mit Stöpseln *pl* anschalten oder einschalten, zustöpseln

**plug** Abschlußschraube *f*, Absperr-glied *n*, -organ *n*, -teil *n*, -vorrichtung *f*, (switch) Abstecker *m*, Bausch *m*, Boden *m*, Döbel *n*, Gießstöpsel *m*, Kegel *m*, kegelförmige Messerwalze *f*, Klotz *m*, (gauge) Lochdorn *m*, Losboden *m*, Normalbohrer *m*, Pflock *m*, Pfropfen *m*, Quellkuppe *f*, Schaltstöpsel *m*, Schlußpfropfen *m*, Spund *m*, Stecker *m*, (contact) Steckkontakt *m*, Stift *m*, Stopfholz *m*, Stöpsel *m* (electr.), Verschluß-kappe *f*, -stopfen *m*, Zapfen *m*, Zündeinrichtung *f*

**plug,** ~ **of a cock** Hahn-kegel *m,* -küken *n,* -schlüssel *m* ~ **with earth contact** Schukostecker *m* ~ **and jack** Stecker *m* und Büchse *f* ~ **for operator's headset** Abschlußstöpsel *m* des Sprechzeuges ~ **with protection collar** Stecker *m* mit Schutzkragen ~ **on sleeve wire** Stöpselleitung *f* ~ **of a stopcock** Küken *n,* Lilie *f* ~ **of a tap** Lilie *f* eines Faßhahns *m* ~ **with three current consumption places** Drillingstecker *m* ~ **of three-way cock** Dreiweghahnküken *n* ~ **of the trunk junction circuit** Vorhaltestöpsel *m* ~ **of two-way cock** Durchgangshahnküken *n* ~ **of a valve** Ventilküken *n*

**plug,** ~ **adapter** Gewindefassung *f* für Zündkerzen ~ **board** Schalttafel *f* ~ **body** Kerzengehäuse *n,* Stöpselkörper *m* ~ **bolt** Gewindebolzen *m* ~ **boreholes** Hahnkükenbohrungen *pl* ~ **box** Klemmenkasten *m,* Steckdose *f* ~ **bridge contact plate of plug** Stöpselbrücke *f* ~ **cap** Stöpselkopf *m* ~ **capacitance box** Stöpselkondensator *m* ~ **cartridge** Steckpatrone *f* ~-**cartridge fuse** Steckpatronensicherung *f* ~ **cock** Reiberhahn *m* ~ **commutator** Stöpsel-umschalter *m,* -wähler *m* ~ **connection** Steckeranschluß *m* ~ **connector** Kerzenstecker *m* ~ **contact** Stöpselkontakt *m* ~ **contact pressure** Kükenanpressung *f* ~ **cord** Stöpselschnur *f* ~ **couple** Zündkerzenthermoelement *n* ~ **cover** Stöpselhülse *f* ~-**cover thread** Stöpselhülsengewinde *n* ~ **crystal** steckbarer Quarz *m* ~-**cut-out** Stöpselsicherung *f* ~ **device** Steckvorrichtung *f* ~-**ended cord** Stöpselschnur *f* ~ **eye** Aufstecköse *f* ~ **face** Kerzengesicht *n* ~ **feeder** (on a glass furnace) Postenspeiser *m* ~ **fitting** Zapfenbeschlag *m* ~ **fuse** Patronenstöpsel *m,* Stöpselsicherung *f* ~ **ga(u)ge** Bolzen *m,* Grenzlehre *f,* Innengrenzlehre *f,* Kaliberbolzen *m,* Lehrdorn *m,* Lochlehre *f* ~ **handle** Stöpselgriff *m* ~ **head** Stöpselkopf *m* ~ **hole** Stöpselloch *n* ~ **housing** Kerzengehäuse *n*

**plug-in** Einschub *m,* steckbar ~ **amplifier** Kasettenverstärker *m* ~-**and socket control** Steckregler *m* ~ **assembly** Einsteckteil *m* ~ **chassis** Einschub *m* mit Steckverbindung ~ **circuit card** Einsteckschaltkarte *f* ~ **coil** Aufsteck-, Einsteck-, Steck-spule *f* ~ **connection** Aufstecker *m* (electr.) ~ **crystal** Steckquarz *m* ~ **design** Einschubbauweise *f* ~ **indicator** Anzeigeeinschub *m* ~ **inlet strainer** Einsteckseiher *m* ~ **input** Eingangseinschub *m* ~ **inserts** Steckeinheiten *pl* ~ **keyboard perforator** einstöpselbarer Tastenlocher *m* ~ **linear amplifier** Linienverstärkereinschub *m* ~ **mains section** Netzzuführungseinschub *m* ~ **measuring unit** Meßeinschub *m* ~ **ocular** Steckokular ~ **power pack** Netzgeräteeinschub *m* ~ **rheostat** ausstöpselbarer Rheostat *m* ~ **single -channel discriminator** Einkanaldiskriminatoreinschub *m* ~ **steel conduits** Stahlsteckröhren *pl* ~ **switch** Steckumschalter *m* ~ **time printer** Zeitdruckereinschub *m* ~ **timer** Steuereinschub *m* ~ **two-channel discriminator** Zweikanaldiskriminatoreinschub *m* ~ **unit** Einschubsteckeinheit *f,* Einsteckteil *m,* Steckeinheit *f*

**plug,** ~ **inductance box** Stöpselinduktivität *f* ~ **inductor** Steckspule *f* ~ **insulator** Kerzenstein *m* ~ **junction** Steckanschluß *m* ~ **mill** Knetmischer *m,* Schwedenstraße *f,* Stopfenwalzwerk *n*

~ **neck** Stöpselhals *m* ~ **point** Steckkontakt *m* ~-**protector pan** Aufbewahrungskasten *m* für Stöpsel ~-**ramming machine** Bodenstampfmaschine *f* ~ **receptable** Anschluß-steckdose *f,* -steckkontakt *m* ~ **resistance box** Stöpselwiderstand *m* ~-**restored shutter** Rückstellklappe *f* ~ **ring** Stöpsel-hals *m,* -ring *m* ~ **sand blasting** Kerzensandstrahlung *f* ~ **screw** Absperrschraube *f* ~ **seat** Stöpselsitzplatte *f* (teleph.) ~-**seat switch** Stöpselsitzschalter *m* ~ **selector** Stöpsellinienwähler *m* ~ **shelf** Stöpselbrett *n* ~ **sleeve** Stöpsel-hals, -körper *m* ~ **socket** Ansteckdose *f,* Dose *f,* Steckbuchse *f,* Steckerbuchse *f,* -hülse *f* ~ **switch** Klinkenstecker *m,* Stöpsel(um)schalter *m* ~ **system** Steckersystem *n* ~ **tap** (for screwing) Gewindenachbohrer *m,* Mittelschneider *m* ~ **tester** Zündkerzenprüfvorrichtung *f* ~ **tip** Stöpselspitze *f* ~-**type cutout** Patronensicherung *f* ~ **valve** Hahnschieber *m,* Kegelventil *m* ~ **weld** Nietschweißung *f* ~ **weld(ing)** Lochschweißung *f*

**plugged** zugestopft ~ **line** verstopfte Rohrleitung *f* ~ **well** versiegelte Quelle *f* (Bohrloch)

**plugging** Bremsen *n* durch Gegenstrom, Füllmaterial *n,* schalldichte Zwischenwand *f* (arch.), Zustopfen *n,* Zustöpseln *n* ~ **up a spring** Dichten *n* einer Quelle

**plugging,** ~ **chisel** Ausstechmeißel *m,* Durchbrechmeißel *m* ~ **unit** Gerätsblock *m* ~-**up device** Sperrvorrichtung *f*

**plum-tree wood** Zwetschgenbaumholz *n*

**plumb, to** ~ ab-, aus-loten, abseigern, bleien, einloten, loten, plombieren, sondieren, verbleien, mit Blei *n* verlöten **to** ~-**line** einloten

**plumb** Bleigewicht *n,* Lot *n,* Senkblei *n;* lotrecht, scheitelrecht, seiger, senkrecht ~ **(line)** Senkrechte *f* **out of** ~ Abweichung *f* vom Lot *n*

**plumb,** ~ **bob** Brahme *f,* Brahne *f,* Lot *n,* Senkblei *n,* Senkel *m,* Senklot *n,* Setzwaage *f* ~ **drop rod** Pendelfallstab *m* ~ **line** Bleilot *n,* Lot *n,* Lot-linie *f,* -richtung *f,* -riß *m,* -schnur *f,* senkrechte Richtung *f* ~-**line** altimeter Pendelhöhenmesser *m* ~-**line deflection** Lotabweichung *f* ~-**line hook** Lothaken *m* ~-**line level** Lotwaage *f* ~-**line position** Senkrechtstellung *f* ~-**line slider** Lotschnurschieber *m* ~ **point** Nadirpunkt *m* ~-**point triangulation** Nadirpunkttriangulation *f* ~ **rule** Schrotwaage *f*

**plumbago** Glanzerz *n,* Graphit *n,* Pottlot *n,* Reißblei *n,* Wasserblei *n* ~ **crucible** Graphittiegel *m*

**plumbeous** bleiartig

**plumber** Bleigießer *m,* Bleilöter *m,* Installateur *m,* Klempner *m,* Rohrleger *m,* Spengler *m* ~ **bearing** Schwungradlager *n*

**plumber's,** ~ **lead cutter** Bleischere *f* ~ **pliers** Rohrzange *f* ~ **solder** Lötmörtel *m* ~ **wiped joint** Lötwulst *m* (der Bleimuffe), Plombe *f*

**plumbic** bleihaltig ~ **acid** Bleisäure *f* ~ **compound** Plumbidverbindung *f* ~ **oxide** Plumbioxyd *n*

**plumbiferous** bleiführend, bleihaltig

**plumbing** Ab-, Aus-loten *n,* Bleiarbeit *f,* Bleigießerarbeit *f,* Installationsarbeiten *pl,* Loten *n,* Plombierung *f,* Rohrlegearbeit *f,* Gas-, Rohr-leitung *f* ~ **material** (vacuum work) Dichtungsstoff *m* ~ **rod** Lotstab *m*

**plumbo, ~ gummite** Gummi(blei)spat *m* **~ resinite** Bleigummi *m*
**plumbous, ~ compound** Plumboverbindung *f* **~ salt** Plumbosalz *n*
**plume** Feder *f*, künstliches Echo *n*, Wassersäule *f*
**plumed** gefiedert
**plummer, ~ bearing** Rumpflager *n* **~ block** Lagerblock *m*, Stehlager *n*
**plummet** Lot *n*, Reißschnur *f*, Richtblei *n*, Schrot-, Senk-waage *f*, Senker *m*
**plump** plump
**plunge, to ~** einsenken, eintauchen, stürzen, tauchen, versenken **to ~ down** herabstürzen
**plunge** Eintauchen *n*, Sturz *m*, Stürzen *n* **~ of the axle** Achseneinfallen *n*
**plunge, ~ bath** Tauchbad *n* **~ battery** Tauchbatterie *f* **~-cut grinding** Einstechschleifen *n* **~-cut internal grinder** Einstechinnenschleifmaschine *f* **~-cutting** Einstechen *n* **~-cutting operations** Einstecharbeiten *pl* **~ cylinder** Tauchzylinder *m* **~-milling** Tauchfräsen *n*
**plunger** Abstimmkolben *m*, Ausstoßdorn *m*, Druck-kolben *m*, -stange *f* -stöpsel *m*, Kolben *m*, (Tauchstange) Krücke *f*, Meßbolzen *m*, Mönchskolben *m*, Pimpel *m*, Plunger *m*, Plungerkolben *m*, Riegelbolzen *m*, Schwingspule *f*, Stempel *m*, Stößel *m*, Taucherkolben *m*, Tauch -bolzen *m*, -kern *m*, -platte *f*, -spule *f*, -stange *f*, -zylinder *m*, Ventil *n*, Ventilkolben *m*, Verschlußklinke *f* **~ armature stud** Druckknopf *m*, Pimpel *m* **~ barrel** Arbeitsrohr *n* **~ bucket** Plungerkolben *m* **~ cam** Mitnehmerriegel *m* **~ case** Kolbenrohr *n* **~ crib** Kolbenbrücke *f* **~ disk** Kolbenscheibe *f* **~ driving vane** (injection pump) Kolbenmitnehmer *m* **~ follower** Kolbenscheibe *f* **~ housing** Mitnehmergehäuse *n* **~ initial position** Stößelausgangsstellung *f* **~ jig** Kolbensetzmaschine *f* **~ leather** Kolbenleder *n* **~ lift** Tauchkolbenheber *m* **~ lock** Druckknopfsperre *f* **~ lug** Kolbenfahne *f* **~ movement** Schlägerschaltung *f* **~ piston** Plunger-, Tauch-kolben *m* **~ piston with sealing collar** Manschettenkolben *m* **~ poppet** Kolbenkegel *m* **~ position** Kolbenstellung *f* (bei der Pumpe) **~ pump** Kolben-, Plunger-pumpe *f* **~ relay** Taucherrelais *n* **~-rod nut** Kolbenmutter *f* **~ roller** Mitnehmerrolle *f* **~ spring** Kolbenfeder *f* **~ stud** Kolbenstutzen *m* **~ switch** Stechschalter *m* **~ tip** Meßbolzen *m* **~ tube** Stempelrohr *n* **~ type** Kolbensetzmaschine *f* **~-type air compressor** Tauchkolbenluftpresser *m* **~-type armature** Tauchanker *m* **~-type die casting machine** Kolbenspritzmaschine *f* **~ valve** Kolben-, Pumpenkolben-ventil *n* **~ working barrel** Pumpenkolbenrohr *n*
**plunging** Eintauchen *n* **~ fire** Senkfeuer *n* **~ siphon** Steckheber *m*
**plural scattering** Mehrfachstreuung *f*
**plurality** Mehrheit *f*, Mehrzahl, Vielzahl *f*
**plus** plus, zuzüglich **~ and minus limits** Plus- und Minusabmaß *n* **~ circuit** Viererkreis *m*, Viererverbindung *f* **~ gauge** Pluslehre *f* **~ hubs** Additionskontrollbuchsen *pl* **~ sign** Pluszeichen *n*
**plush copper** Chalkotrichit, Kupferfedererz *n*
**plutonium fission** Plutoniumspaltung *f*
**plutonyl ions** Plutonylionen *pl*
**pluviograph** Regenschreiber *m*

**pluviometer** Pluviometer *n*, Regenmesser *m*
**pluviometry** Regenmessung *f*
**ply, to ~ between** den Verkehr *m* vermitteln zwischen **to ~ yarn** Garn *n* fachen
**ply** (coat, fold, thickness) Falte *f*, Gewebeanlage *f* (bei Reifen), Schicht *f* **two- or three ~** zwei- oder dreisträhnig **~ metal** Sperrmetall *n*
**plywood** Furnierblatt *n*, Schicht-, Sperr-holz *n* **~ board** Sperrholzplatte *f* **~ covering** Sperrholzbeplankung *f* **~ fairing** Sperrholzverkleidung *f* **~ framework** (or hull) **structure** Sperrholzgerippe *n* **~ fuselage** (of aircraft) Sperrholzrumpf *m* **~ layer** Furnierblatt *n* **~ panel** Sperrholzplatte *f* **~ paper** Holzfurnierpapier *n* **~ plate** Sperrholzplatte *f* **~ rib** Sperrholzrippe *f* **~ sheet** Furnierholz *n* **~ shell** Sperrholzverschalung *f* **~ shelter** Sperrholzzelt *n* **~ slab formwork** Sperrholzschalung *f*
**p-n boundary** pn-Übergang *m*, Störstelleninversionszone *f*
**pneumatic** durch Druckluft *f* betätigt, pneumatisch, preßluftangetrieben **~ air brake** Knorrbremse *f* **~-air plant** Saugluftanlage *f* **~ assault boat** Behelfsfloßsack *m*, Floßsack *m* **~ attachment** pneumatischer Zusatzantrieb *m* **~ bailing** Luftdruckspülung *f* **~ bed** Schlauchbett *n* **~ boat** Schlauchboot *n* **~ brake** Luft(druck)-bremse *f* **~ braking circuit** Druckluftbremskreis *m* **~ buffer** Luftpuffer *m* **~ bumper** Gondelpuffer *m* **~ caisson** Druckkasten *m* **~ catch** Druckluftfangvorrichtung *f* **~ charger** Druckluftdurchlader *m* **~ chipping hammer** Druckluftmeißel *m* **~ chisel** Preßluftmeißel *m* **~ chuck** druckluft-betätigtes Spannfutter *n*, Druckluftspannfutter *n* **~ chucking** Preßluftspannung *f* **~ clipper** Preßluftmeißel *m* **~ coal pick hammer** Preßluftabbauhammer *m* **~ control** Druckluft-regelung *f*, -regler *m*, -steuerung *f* **~ pneumatic(r) Regler** *m*, Regelung *f* **~ controller with an electrical measuring element** elektropneumatischer Regler *m* **~ conveyer** Rohrpostanlage *f* **~ conveying** pneumatische Förderung *f* **~ conveying drier** Rohr-, Strom-trockner *m* **~ conveying plant for stone powder** Druckluftgesteinstaubförderanlage *f* **~ conveying system** Staubpumpensystem *n* **~ cushioning** Luftfederung *f* **~ dashpot** Tauchtasse *f* **~ de-icer** Druckluftenteiser *m* **~ descaling apparatus** Preßluftabklopfapparat *m* **~ dispatch** Fernrohrpost *f* **~ drill** Druckluftbohrer *m* **~ drive** Druckluftantrieb *m* **~ elevator** Getreidesaugaufzug *m* **~ feed mechanism** Druckluftvorschubeinrichtung *f* **~ feeding arrangement** pneumatische Bogenführung *f* (print.) **~ flushing** Luftdruckspülung *f* **~ governor** Membranregler *m* **~ hammer** Drucklufthammer *m*, Luft(druck)-hammer *m* **~ horn** Drucklufthorn *n* **~ ingot-dressing shop** pneumatische Blockrüsterei *f* **~ jack** Hubwinde *f* **~ lever switch** Drucklufthebelschalter *m* **~ lifting gear** Drucklufthebezug *f* **~ loud-speaker** Druckhammer-, Preßluft-lautsprecher *m* **~ motor** Druckluftmotor *m* **~ packing** Blasversatz *m* **~ pick** Pickhammer *m* **~ pile driver** Druckluftramme *f* **~ piston** Preßluftkolben *m* **~ positioning relay** Einstellrelais *n* **~ post capsule** Rohrpostkapsel *f* **~ pressure** Luftdruck *m* **~ pressure control device** pneumatische Regeleinrichtung *f* **~**

**pressure switch** Druckschalter *m* ~ **process** Bessemer-Verfahren *n* ~ **process of foundation** Gründung *f* mit Preßluft, Preßluftgründung *f* ~ **pump** Luftpumpe *f* ~ **raft** Schlauchboot *n* ~ **rammer** Preßluftstampfer *m* ~ **rapper** Druckluft-, Preßluft-klopfapparat *m* ~ **recuperator cylinder** Luftzylinder *m* ~ **riddle** Preßluftsiebmaschine *f* ~ **riddle sifter (or riddler)** Schüttelsieb *n* mit Druckluftantrieb ~ **riveter** Druckluftniethammer *m*, pneumatischer Niethammer *m* ~ **riveting** Druckluftnietung *f* ~ **sample irradiation system** Reaktorrohrpostanlage *f* ~ **scaling hammer** Preßluftabklopfer *m* ~ **servobrake system** Druckluft-Servobremsanlage *f* ~ **shaft sinking** Druckabteufung *f* ~ **shock absorber** Luftstoßdämpfer *m* ~ **shock absorption** Luftfederung *f* ~ **shutter** Ballverschluß *m* (photo) ~ **single-chamber brake energizer** Drucklufteinkammerbremsverstärker *m* ~ **size analysis** Sichtanalyse *f* ~ **splitting hammer** Luftkeilhammer *m* ~ **starting device** Preßluftanlasser *m* ~ **stowage** Druckluftversatz *m* ~ **stower** Druckluftversatzmaschine *f* ~ **stowing** Blasversatz *m* ~ **stowing pipes** Blasversatzrohre *pl* ~ **suspension** lastabhängige Luftfederung *f* ~ **ticket carrier** Zettelrohrpost *f* ~-**ticket-distribution position** Rohrpostzettelverteiler *m* ~ **tire** Luftreifen *m* ~ **tired roller** Walze *f* mit Luftbereifung ~ **tires** Luftbereifung *f* ~ **tool** Druckluftwerkzeug *n* ~-**tool oil** Lufthammeröl *n* ~ **trench mortar** Luftmörser *m* ~ **trough** (in gas drying) pneumatische Wanne *f* ~ **tube** Rohrpost *f*

**pneumatic-tube,** ~ **carrier** Rohrpostbüchse *f* ~ **dispatching station** Rohrpostsendestelle *f* ~ **flarer** Preßluft-Kelchmaschine *f* ~ **plant** Rohrpostanlage *f* ~ **receiver** Rohrpostempfänger *m* ~ **receiving station** Rohrpostempfangstelle *f* ~ **system** Zettelrohrpost *f* ~ **systems (PTS) with flat carriers** Flachrohrpostanlagen *pl* ~ **ticket distributor** Rohrpost *f* für die· Beförderung der Gesprächsblätter oder Gesprächszettel

**pneumatic,** ~ **tubes** Rohrpostanlage *f* ~ **warning unit** Druckluftwarngerät *m* ~ **wheel brake** Luftdruckradbremse *f* ~ **winch** Drucklufthaspel *m* ~ **window wiper** Druckluftwischer *m* ~ **windscreen (or windshield) wiper** Druckluftscheibenwischer *m*

**pneumatically controlled** druckluftgesteuert

**pneumatometer** Pneumometer *n*

**poach, to** ~ aufschlagen, mischen

**poacher** Stabilisator *m*

**poaching engine** Bleichholländer *m*

**pocket, to** ~ einkesseln

**pocket** Ablagefach *n*, Behälter *m*, Fangraum *m*, Gaspore *f*, Loch *n*, (of ore) Nest *n*, Niere *f*, Pore *f*, Sack *m*, Tasche *f* ~ **of ore** Butz *m*, Butze *f*, Butzen *m*, Putzen *m* ~ **of resistance** Widerstands-kessel *m*, -nest *n* ~ **for a spring (or for a valve stem)** Federkammer *f*

**pocket,** ~ **ammeter** Taschenstrommesser *m* ~ **book** Taschenbuch *n* ~ **chamber** Taschendosimeter *n* ~ **compass** Taschen-bussole *f*, -kompaß *m* ~ **counter** Taschenzähler *m* ~ **drill** Sackbohrer *m* ~ **edition** Taschenausgabe *f* ~-**knife blade** Taschenmesserklinge *f* ~ **lamp** Taschenlampe *f* ~ **lighter** Taschenfeuerzeug *n* ~ **lumin-**

ous **magnifier** Taschenleuchthupe *f* ~ **magnifier** Taschenlupe *f* ~ **meter** (folding) Taschenmeßinstrument *n* ~ **rule** zusammenlegbarer Maßstab *m* ~-**size** Taschenformat *n* ~ **spring** Überfallquelle *f* ~ **spring balance** Taschenfederwaage *f* ~ **tape recorder** Taschentonbandgerät *n* ~ **telescope** Taschenfernrohr *n* ~ **tracer** Taschenprüfgerät *n*

**pocketing of valve seats** Einschlagen *n* der Ventilsitze (durch das Hämmern des Ventils)

**pockets** Ablagefächer *pl* **in** ~ nierig

**pocking** Narbung *f*

**pockwood** Pockholz *n*

**podal line** Fußpunktlinie *f*

**podded engine** Motor *m* in Gondel

**podometer** Meßrad *n*

**podsel** Grauerde *f*

**Pohl commutator** Pohlsche Wippe *f*

**Pohl's diagram** Dauerfestigkeitsschaubild *n* nach Pohl

**point, to** ~ anschneiden, anspitzen, im Gesenk *n* anspitzen, anzielen (einen Gegenstand), aufrichten, einstellen (optics), punktieren, richten, stellen, verjüngen, visieren, zielen, zuschärfen, zuspitzen **to** ~ **up joints** die Fugen *pl* glattstreichen **to** ~ **out** angeben, anweisen, darlegen **to** ~ **precisely (roughly)** fein (grob) einstellen **to** ~ **with steel** stählen **to** ~ **to** richten auf **to** ~ **up** (brickwork) ausfugen, (masonry) verfugen

**point** Erhebung *f*, Körper *m*, Landspitze *f*, Ort *m*, Punkt *m*, Punktur *f*, Spitze *f*, Stelle *f*, (of compass) Strich *m*, Zuspitzung *f* **at any** ~ an einer beliebigen Stelle *f*

**point,** ~ **of accumulation** Häufungspunkt *m* ~ **of action** Angriffspunkt *m* (der Kraft) ~ **of aim** Abkommen *n*, Haltepunkt *m* ~ **of air burst** Luftsprengpunkt *m* ~ **of angle** Scheitel *m* ~ **of application** (of attack) Angriffspunkt *m*, Angriffsstelle *f*, Wirkungslinie *f* ~ **of attachment** Angriffs-, Ansatz-, Anschluß-punkt *m* (Rumpf) ~ **of attack** Angriffsstelle *f* ~ **of attenuation** Dämpfungspol *m* ~ **of beam origin** Strahlenpunkt *m* ~ **of blade welded to rail** angeschweißtes Zugende *n* ~ **of bolt** Bolzenende *n* ~ **of break-through** Durchbruchstelle *f* ~ **of burst** Brennpunkt *m*, Einschlagstelle *f*, Sprengpunkt *m* ~ **of calescense** Haltepunkt *m*, Kaleszenzpunkt *m* ~ **of change** Sprungstelle *f*, Umschlagspunkt *m* ~ **of charge** Brechpunkt *m* ~ **on a circle** Kreißpunkt *m* ~ **of comparison** Vergleichspunkt *m* ~ **of compass** Himmelsgegend *f* ~ **of the compass** Kompaßstrich *m*, Zirkelspitze *f* ~ **of concentration** Schwerpunkt *m* ~ **of concentration of the inductive load** Blindverbrauchsschwerpunkt *m* ~ **of condensation** Taupunkt *m* ~ **of congelation** Stockpunkt *m* ~ **of connection** Verbindungsstelle *f* ~ **of contact** Anknüpfungspunkt *m*, Anstoß *m*, Berührungs-punkt *m*, -stelle *f*, Kontaktstelle *f* ~ **of contraflexure** Wendepunkt *f* ~ **of control** Kontrollpunkt *m* ~ **of a crossing** Herzspitze *f* ~ **of decalescence** Kaleszenzpunkt *m* ~ **of delivery** Luftaustritt *m* ~ **of departure** Abflugort *m*, Ausgangsort *m* ~ **of deployment** Entfaltungspunkt *m* ~ **of determination** Stelle *f* der Bestimmtheit ~ **of development** Entfaltungspunkt *m* ~ **of discontinuity** Sprungstelle *f*, Unstetigkeitsstelle *f*, Wirbelpunkt *m* ~ **of discovery**

Fundpunkt *m* ~ of the earth Erdpunkt ~ of
embarkation Verladestelle *f* ~ of engagement
Angriffspunkt *m* ~ of equilibrium Gleichgewichtslage *f* ~ of equivalence Äquivalenzpunkt
*m* ~ of fall Fallpunkt *m* ~ of the feeler Tasterspitze *f* ~ where film is fed onto drum Einlauf-
punkt *m* auf die Rolle ~ where film strip leaves
drum Auslauf *m* von der Rolle ~ of firing pin
Schlag(bolzen)spitze *f* ~ of force Angriffspunkt
*m* der Kraft ~ of fuse Zünderspitze *f* ~ on the
horizon Horizontalpunkt *m* ~ of ignition
Flammpunkt *m* ~ of image change Bildkipp-
punkt *m* ~ of the image perpendicular Bildlotpunkt *m*, Bildnadir *m* ~ of impact Artillerie-
einschlag *m*, Aufschlagpunkt *m*, Auftreffpunkt
*m*, Geschoß-aufschlag *m*, -einschlag *m*, Stoß-
(mittel)punkt *m*, Treffpunkt *m* ~ of incipient
current flow at grid Gitterstrompunkt *m* ~ of
incipient grid-current flow Gitter(strom)einsatz-
punkt *m* ~ of inflection Umkehr-, Wende-punkt
*m* (math.) ~ of insertion Ansatzpunkt *m*, Ein-
baustelle *f* ~ of instability Unstetigkeitsstelle *f*
~ of interpretation Auslegepunkt *m* ~ of
interruption Unterbrechungsstelle *f* ~ of intersection Durchschlagspunkt *m*, (of a line with a
plane) Durchstoßpunkt *m* ~ of intersection of
the collimating axes Rahmenkreuzpunkt *m*
~ of intersection of diagonals Diagonalenschnitt-
punkt *m* ~ of intersection of plane of target and
plane of sighting Durchpunkt *m* ~ of intersection of principal line with horizon trace Hori-
zonthauptpunkt *m* ~ of intersection of tangents
Tangentenschnittpunkt *m* ~ of introduction
Einführungsstelle *f* ~ of inversion Umwende-
punkt *m* ~ of law Rechtspunkt *m* ~ of light
Lichtpunkt *m* ~ of lubrication Schmierstelle *f*
~ of magnetic transformation magnetischer
Umwandlungspunkt *m* ~ of maximum bending
höchster Drehungspunkt *m* ~ of maximum load
Bruchgrenze *f* ~ of maximum rainfall during a
rainstorm Regenkern *m* ~ of minor control Paß-
punkt *m* ~ of neutral stability Indifferenzpunkt
*m* ~ of no return Umkehrgrenzpunkt *m* (aviat.)
~ of observation Beobachtungspunkt *m* ~ of
opening (of parachute) Entfaltungspunkt *m* ~
of orientation Ablaufpunkt *m* ~ of origin Ur-
sprungspunkt *m* ~ of origin of coordinates
Koordinatenausgangspunkt *m* ~ of oscillation
Berührungspunkt *m* ~ of penetration Durchschlagspunkt *m*, Einbruchstelle *f* ~ of pouring
Gießstelle *f* ~ of projectile Geschoßspitze *f* ~ of
projectile ogive Bogenspitze *f* ~ of puncture
Durchschlagsstelle *f* ~ of recalescence Rekales-
zenzpunkt *m* ~ of reckoning Besteckort *m* ~ of
reference Grundlinie *f* ~ of regression Haltepunkt *m* ~ of resistance Widerstandsnest *n* ~
of revolution (or rotation) Drehpunkt *m* ~ of
rising (of sum) Aufgangspunkt *m* ~ of rupture
Sprengstelle *f* ~ of self-oscillation Schwingungseinsatzpunkt *m* ~ of separation Abreißungs-,
Spaltungs-, (aerodyn.) Trennungs-, Übergangs-
-punkt ~ of separation out Ausscheidungspunkt
*m* ~ of setting Untergangspunkt *m* ~ of shallow
depth between reversed curves in river course
Furt *f* ~ of shedding Ablösungsstelle *f* ~ of
sight Augenpunkt *m* ~ of on the streamline
Stromlinienpunkt *m* ~ of sudden irregularity
Sprungstelle *f* ~ of support Auflagepunkt, Ein-

spannstelle *f*, Kämpfer *m*, Unterstützungspunkt *m* ~ of suspension Aufhängepunkt *m*,
Unterstützungspunkt *m* ~ of the switch Wei-
chenspitze *f* ~ of tangency Berührungspunkt *m*
~ of time Zeitpunkt *m* ~ of transfer Übergabe-
stelle *f* ~ of transition Umwandlungspunkt *m* ~
of triangulation Netzpunkt *m* ~ of unsteadiness
Sprungstelle *f*, Unstetigkeitsstelle *f* ~ of view
Augenmerk *n*, Blickpunkt *m*, Gesichtspunkt *m*,
Standpunkt *m* ~ in the winding cycle Aufzug-
stellung *f* ~ of zero moment Momentennullpunkt *m*
point, ~-and-edge table Hilfseinrichtung *f* zum
Prüfen von Parallelendmaßen ~ (ed) angle
Spitzenwinkel *m* ~ appearance of distant light
punktförmiges Sehen *n* ~ bearing pile Stand-
pfahl *m* ~-blank range Kernschußweite *f* ~
blank shot Fleck-, Kern-schuß *m* ~-by-~ punktweise ~-by-~ evaluation punktweise Auswertung *f* ~-by-~ measuring punktweise Ausmessung *f* ~ cathode Spitzenkathode *f* ~ charge
system Punktladungssystem *n* ~ contact Spit-
zenkontakt *m* ~ contact photodiode Punkt-
kontaktfotodiode *f* ~-contact rectifier Spitzen-
gleichrichter *m* ~-contact transistor Spitzen-
transistor *m* ~ control Weichungssteuerung *f*
~ controls operated from vehicles Weichenstell-
vorrichtungen *pl* ~ counter tube Spitzenzähl-
rohr *n* ~-cylinder Nadelzylinder *m* ~ defect
Punktfehlordnung *f* ~-designation grid Punkt-
tafel *f* ~ determination along an orienting line
Einschneiden *n* auf Richtlinie ~ detonating fuse
Kopfzünder *m* ~ detonator chamber Kopfkam-
mer *f* ~ discharge Spitzenentladung *f* ~ dot (or
granule) raster Punktraster *m* ~ drag Punkt-,
Spitzen-widerstand *m* ~ east Ostpunkt *m* ~
effect Spitzeneffekt *m* ~-electrode method
Methode *f* der punktförmigen Elektroden ~
-emission effect Punktwirkung *f* ~ (-shaped)
emitter punktförmige Quelle *f* ~ exposure
(slotted shutter) Punktbelichtungszeit *f* (photo)
~-focal glasses punktuell abbildende Brillengläser *pl* ~ focal imagery punktuelle Abbildung
*f* ~-focal vision punktuelle Abbildung *f* ~ focus
punktförmiger Brennpunkt *m* ~-for-~ punktweise ~-force Einzelkraft *f* ~ gap taster Kon-
taktabstandprüfer *m* (Magnetzünder) ~ gauge
Spitzenlehre *f* ~ group Punkt-, Raum-gruppe *f*
~ hit Kopftreffer *m* ~ hole Punkturloch *n* ~
image Bildpunkt *m*, punktförmige Abbildung *f*
~ imperfection Punktfehlstelle *f* ~ indicator
Weichensignal *n* ~ insertion (or interpolation)
Punkteinschaltung *f* ~-junction transistor Spit-
zenflächentransistor *m* ~ lamp Punkt(licht)-
lampe *f* ~ lattice Punktgitter *n* ~ lattice in the
plane ebenes Punktgitter *n* ~ lattice in space
räumliches Punktgitter *n* ~ load Einzel-,
Punkt-last *f* ~ locking Weichensicherung *f*
~ motor Weichenantriebsmotor *m* ~ neon
lamp Punktglimmlampe *f* ~ object Ding-
punkt *m*, Gegenstandspunkt *m* (photo) ~
paper Linien-, Muster-, Tupf-papier *n* ~
paper design Patrone *f* ~ position Punkt-
lage *f* ~ rail Leitzunge *f* ~ resistance Spit-
zenwiderstand *m* ~ scatterer punktförmiger Streuer *m* ~ singularity Punktsingularität *f*
~ size Kegelstärke *f* ~ slope method Polygon-
zugverfahren *n* ~ (-shaped) source punktför-

mige Quelle *f* ~ **source of light** Punkt(licht)-
lampe *f* ~**-source cathode** Punktkathode *f* ~
**sources** Punktquelle *f;* Punktlichtquelle *f* ~ **spot
focus** Punktschärfe *f* ~ **spur** Punktspur *f* ~
**support** Spitzenlagerung *f* ~ **surface transformation** Punktflächentransformation *f* ~ **suspension** Spitzen-aufhängung *f*, -lagerung *f* ~ **target**
Einzelziel *n*
**point-to-point,** ~ **circuit** Standverbindung *f* ~
**communication** Direkt- (rdo), Punkt-zu-Punkt-,
Strecken-verbindung *f* ~ **density** jeweilige
Schwärzung *f* ~ **focussing** Punktabbildung *f* ~
**gap** Spitze-Spitze-Zündstrecke *f* ~ **radiotelephone link** Sprechfunkbrücke *f* ~ **service**
Direktverkehr *m*, Standverbindung *f* (teletype)
**point,** ~ **vertically beneath a given point** Lotpunkt
*m* ~ **west** Westpunkt *m*
**pointed** (at) gerichtet sein, spitz, spitzig, zugespitzt ~ **arch** Spitzbogen *m* ~ **bar** Spieß *m* ~
**bits** Spitzmeißel *m* ~ **bullet** Spitzgeschoß *n* ~
**chisel** Spitz-eisen *n*, -meißel *m* ~ **countersink**
Senkkolben *m* ~ **(end) drill** Spitzbohrer *m* ~
**ear stencil** Spitzenmarke *f* ~ **filament** Spitzenkathode *f* ~ **groove** Spitzbogenkaliber *m* ~ **gun
sight** gestrichenes Korn *n* ~ **hammer** Anspitz-
hammer *m* ~ **hammer work** Spitzarbeiten *pl* ~
**hand reamers** Aufreiber *m* ~ **journal** Spitzzapfen *m* ~ **lightning protector** Spitzenblitzableiter
*m* ~ **mark** Spitzenmarke *f*, (stencil) Zacken-
marke *f* ~ **pin** Tastspitze *f* ~ **pliers** Spitzzange *f*
~ **screw** Stachelschraube *f* ~ **shovel** zugespitzte
Schaufel *f* ~ **spark gap** Spitzenfunkenstrecke *f*
~ **spigot** Ansatzspitze *f* ~ **stencil** Spitzenmarke
*f* ~ **stone chisel** spitzer Steinmeißel *m* ~ **tool**
Spitzstichel *m* ~ **tube** Spitzlutte *f* ~ **weapon**
Stichwaffe *f*
**pointer** Ablesemarke *f*, Anzeiger *m*, Fingerzeig
*m*, Folgezeiger *m*, Kettengabel *f*, Nadel *f*, Stellmarke *f*, Vorzeiger *m*, Weiser *m*, Wink *m*, (of
balance) Zunge *f* ~ **of autopilot** Steuerrose *f* ~
**of directionfinder** Peilzeiger *m* **the** ~ **moves** der
Zeiger schlägt aus ~ **on selsyn** Kommandozeiger *m*
**pointer,** ~ **base** Zeigerlagerung *f* ~ **eyepiece**
Zeigerokular *m* ~ **frequency meter** Zeigerfre-
quenzmesser *m* ~ **galvanometer** Zeigergalvano-
meter *n* ~ **indicator** Gegenzeiger *m* ~ **instrument** Anzeigegerät *n*, Zeigerinstrument *n* ~
**reading** Zeigerablesung *f* ~**-shaped knob** Knebelkopf *m* ~ **support** Zeigerlagerung *f* ~ **telegraph** Zeigertelegraf *m* ~ **tip** Zeigerspitze *f* ~
**typewriter** Zeigerschreibmaschine *f*
**pointing** Fugenausfüllung *f*, Voreinteilung *f*, Vor-
punktieren *n*, Zeigen *n*, Zuspitzen *n* ~ **of a tap**
Gewindebohreranschnitt *m* ~ **up the joints** Fu-
genverstreichen *n*
**pointing,** ~ **engine** Anspitzmaschine *f* ~ **mechanism** Folgezeigereinrichtung *f* ~ **sill** Schlagschwelle *f* (einer Schleuse) ~ **stand** Anspitzwerk
*n* (Kabel) ~ **tool** Anspitzwerkzeug *n* ~
**upstream** stromaufwärts gerichtet
**points** Hörner *pl*, Tastflächen *pl* ~ **of emergence
of the axes** Achsenaustrittspunkte *pl* ~ **and
straight lines at infinity** unendlich ferne Punkte
*pl* und Geraden *pl*
**pointwise** punktweis
**poise, to** ~ im Gleichgewicht *n* erhalten, im
Gleichgewicht *n* sein

**poise** Gleichgewicht *n*, Schwebezustand *m* ~
**unit of viscosity** Zähigkeitseinheit *f*
**poised, to be** ~ im Gleichgewicht sein
**poison, to** ~ vergiften, verseuchen
**poison** Gift *n*, (afterglow) Verminderer *m* des
Nachleuchtens ~ **gas** Giftgas *n*, Kampfgas *n*,
Kampfstoff *m* ~ **tower** Gift-fang *m*, -turm *m*
**poisoned by gas** gasvergiftet
**poisoning** Vergiften *n*, Vergiftung *f*, Verseuchen *n*
**poisonous toxic** gifthaltig
**Poisson,** ~ **bracket** Klammerausdruck *m* ~ **'s
law** Poissonsches Gesetz *n* ~ **'s number** Form-
änderungszahl *f*, Querzahl *f* ~ **'s ratio** Kontraktionskoeffizient *m*, Poissonsche Zahl *f*,
Poissonzahl *f*, Querdehnungsziffer *f*, Quer-
kontraktions-koeffizient *m*, -ziffer *f*
**poke, to** ~ schüren, stochern **to** ~ **the fire** das
Feuer *n* anschüren **to** ~ **tuyères** Düsen *pl*
putzen
**poke,** ~ **hole** Schürloch *n* ~ **welding** Schweißung
*f* mit Spreizelektrode
**poker** Feuerhaken *m*, Schür-eisen *n*, -haken *m*,
-zeug *n*, Spieß *m*, Stocheisen *n*, Stocher *m* ~
**bar** Stocheisen *n* ~ **tip** Stochstangenspitze *f*
**poking** Stocharbeit *f* ~ **bar** Schürhaken *m*, Stocheisen *n*, Stocher *m* ~ **door** Schürloch *n*, Stocher-
öffnung *f*
**polar** gepolt, polar, polarisch, polarisiert, polig
~ **adjustment** Polhöhenverstellung *f*, Verstellung *f* in Polhöhe (teleph.) ~ **air** arktische Kaltluft *f*, arktische Luft *f*, Polarluft *f* ~ **altitude**
Polhöhe *f* ~ **altitude adjustment** Polhöhenverstellung *f* ~ **altitude correction** Polhöhenkor-
rektion *f* ~**-altitude motor** Polhöhen(bewe-
gungs)motor *m* ~ **altitude reading** Polhöhen-
ablesung *f* ~ **Atlantic air mass** atlantische Polarluftmasse *f* ~ **aurora** Polarlicht *n* ~ **axis** Polar-,
Stunden-achse *f* ~ **blackout** Polarverdunkelung
*f* ~ **characteristic** Polarcharakteristik *f* ~ **chart**
Polarkarte *f* ~ **circle** Polarkreis *m* ~ **continental
air mass** kontinentale Polarluftmasse *f* ~ **coordinate** Polarkoordinate *f* ~**-coordinate tube**
Polarkoordinatenröhre *f* ~ **current** Polarstrom
*m* ~ **curve** Polare *f* (aerodyn.) ~ **diagram** Polardiagramm *n* ~ **distance** polare Luftlinienentfernung *f*, Poldistanz *f* ~ **duplex system** Dop-
pelstromgegensprechsystem *n* ~ **eddy** Polarwirbel *m* ~ **frequency response locus** Ortskurve
*f* ~ **front** Polarfront *f* ~ **line** Polstrahl *m* ~ **low**
polarischer Tiefdruck *m* ~ **maritime air** polare
Seeluftmasse *f* ~ **moment of inertia** Flächenträgheitsmoment *n* (acoust.) ~ **Pacific air mass**
Polarluftmasse *f* über dem Stillen Ozean ~
**pattern** Polardiagramm *n* ~ **planimeter** Polar-
planimeter *n* ~ **points** Polfigur *f* ~ **semi-axis**
Polarhalbmesser *m* ~ **surface** Polfläche *f* ~
**surface of light distribution** Lichtverteilungs-
körper *m* ~ **triangle** polarisches Dreieck *n* ~
**vortex** Polarwirbel *m* ~ **wandering** Polarver-
schiebung *f*
**polarimeter** Polarimeter *n*, Polariskop *n*
**polariscope** Polarisationsapparat *m*, Polariskop
*n*
**polaristrobometer** Polaristrobometer *n*
**polarity** Polarität *f*, Polrichtung *f* ~**-finder glow
tube** Polsuchglimmlampe *f* ~ **indicator** Pol-
sucher *m*, Polsuchglimmlampe *f*, Polungswei-
ser *m*, Stromrichtungsanzeiger *m*, Umpolung *f*

**polarizability** Polarisierbarkeit *f* ~ **tensor** Polarisierbarkeitstensor *m*
**polarizable** polarisierbar
**polarization** Polarisation, Polarisierung *f*, Vormagnetisierung *f* ~ **of a medium** Polarisation *f* eines Mediums *n*
**polarization,** ~ **capacitance** Polarisationskapazität *f* ~ **cell** Polarisationszelle *f* ~ **charge** Polarisationsladung *f* ~ **current** Polarisationsstrom *m* ~ **decoupling** Polarisationsentkopplung *f* ~ **detection** Polarisationsnachweis *m* ~ **ellipse** Polarisationsellipse *f* ~ **error** Drehung *f* der Polarisationsebene, Nachteffekt *m*, Polarisationseffekt *m* ~ **fading** Polarisations-fading *n*, -schwund *m* ~ **phenomenon** Polarisationserscheinung *f* ~ **spectrophotometer** Polarisationsspektralfotometer *n* ~ **state** Polarisationszustand *m* ~ **unit vector** Einheitsvektor *m* der Polarisation ~ **voltage** Polarisationsspannung *f*
**polarize, to** ~ polarisieren, vormagnetisieren
**polarized** gepolt, polarisiert ~ **bell** polarisierter Wecker *m* ~ **light** polarisiertes Licht *n* ~ **radiation** polarisierte Strahlung *f* ~ **relay** gepoltes oder polarisiertes Relais *n* ~ **ringer** polarisierter Wecker *m* ~ **wave** polarisierte Welle *f*
**polarizer** Polarisationsprisma *n*, Polarisator *m*
**polarizing,** ~ **apparatus** Polarisationsapparat *m* ~ **attachment** Polarisationseffekt *m* ~ **constant** Polarisationskonstante *f* ~ **current** Magnetisierungsstrom *m*, Polarisationsstrom *m* ~ **filter** Bernotar *m* ~ **marks** Polarisierungsbezeichnung *f* ~ **neutrons** Neutronenpolarisator *m* ~ **nicol** Polarisatornikol *n* ~ **potential** Vorspannung *f* ~ **power** polarisierende Wirkung *f* ~ **prism** Halbschattenpolarisator *m*
**polarogram** Polarogramm *n*
**polarograph** Polarograf *m*
**polarography** Polarografie *f*
**polaroid** Polarisator *m* ~ **glasses** polarisierte Brille *f* ~ **material** polaroides Material *n*
**pole, to** ~ polen, rühren **to** ~ **copper** Kupfer *m* polen
**pole** Bohrstange *f*, Deichsel *f*, Langbaum *m*, Lenkbaum *m*, Mast *m*, Pfahl *m*, Pol *m*, Rute *f*, Ständer *m*, Stange *f*, Tragmast *m* ~ **with bracket** Auslegermast *m* ~ **with cross arms** Abzweiggestänge *n* ~ **of the face** Flächenpol *m* ~ **of inertia** Trägheitspol *m* ~ **with line stays** Linienfestpunkt *m* ~ **for overhead lines** Leitungsmast *m* ~ **with socle** angeschuhte Stange *f* ~ **and stay** verankerte Stange *f*
**pole,** ~ **arc** Polbedeckung *f*, Polbogen *m* ~ **arm** Querträger *m* ~ **armature** Sternanker *m* ~ **auger** Löffelbohrer *m* ~ **bore** Polbohrung *f* ~ **boring apparatus** Holzgestängebohrapparat *m* ~ **box** Polgehäuse *n* ~ **brace** Riegel *m* ~ **butt** Stangenende *n* ~ **-butt reinforcement** Stockschutz *m* ~ **casing** Polgehäuse *n* ~ **chain** Deichselkette *f* ~ **change motor** Motor *m* mit umschaltbarer Polzahl ~ **change-over switch** Polwechselschalter *m*, -wendeschalter *m* ~ **-changeable** polumschaltbar ~ **changer** Polwechsler *m* ~ **changing** Polumschaltung *f* ~ **-changing spring** Polwechselfeder *f* ~ **changing switch** Polumschalter *m* ~ **charge** geballte Ladung *f* ~ **clamp** Gestängebündel *n*, Stangenstuhl *m* ~ **clearance** freier Polabstand *m*, Polabstand *m*, Pollücke *f*, Polzwischenraum *m* ~ **climbers**

Steigestütze *f* ~ **coil** Polspule *f* ~ **core of a magnet** Magnetkern *m* ~ **cribbing** Schwelle *f*, unterer Querriegel *m* ~ **diagram** Stangenbild *n* ~ **diagram in local telephone plants** Gruppierungsbild *n* für Ortsfernsprechnetze ~**-diagram book** Heft *n* mit Stützpunktbildern ~ **dislocation** Polversetzung *f* ~ **distance** Pol-, Stangen-, Stützpunkt-abstand *m* ~ **drill** Stützenbohrung *f*, Stützenbohrer *m* (electr.) ~ **earth wire** Blitzschutzdraht *m* ~ **effect** Poleffekt *m* ~ **end** (wagon) Deichselkopf *m* ~ **face** Pol(schuh)fläche *f* ~ **fender** Prellpfahl *m* ~ **finder** Polanzeiger *m* ~**-finding paper** Polklemme-, Polreagenz-, Polsuch-papier *n* ~ **fingers** Gerüstgabel *f*, Gestängerechen *m* ~ **fittings** Gestängeausrüstung *f*, Stangenausrüstung *f* ~ **float** Stangen-, Stock-schwimmer *m* ~ **flux** Polfluß *m* ~ **footing** Mast-, Stangen-fuß *m* ~ **foundation** Mastfundament *n* ~ **gap** Polspalt *m* ~ **hardware** Mastarmatur *f* ~ **hole** Stangenloch *n* ~ **induction** Polinduktion *f* ~ **leg** Polschenkel *m* ~ **lightning arrester** Stangenblitzableiter *m* ~ **line** Gestängeleitung *f*, Stangenlinie *f* ~**-location chart** Stangenbild *n* ~ **lug** Polfahne *f* ~ **mast** Polankermast *m* ~ **paper** Polklemmepapier *n* ~ **pedestal** Stangenfuß *m* ~ **piece** (of an electric furnace) Polplatte *f*, Polschuh *m* des Magnets ~ **piece face** Polschuhfläche *f* ~ **pin** (wagon) Bolzen *m*, Deichselbolzen *m* ~ **pitch** Polbogen *m*, Polteilung *f* ~ **plate** Dachschwelle *f*, Fußrahmen *m*, Polplatte *f*, Sparrensohle *f*, Stoßscheibe *f* ~ **prop** (wagon) Deichselstütze *f* ~ **rail** Polschiene *f* ~ **reinforcement** Stangenverstärkung *f* ~ **reverser** Polumschalter *m* ~ **reversing device** Polwechseleinrichtung *f* ~ **reversing switch** Polwendeschalter *m* ~ **ring with winding** Polring *m* mit Wicklung ~ **roof(ing)** Stangenabdachung *f* ~ **set in concrete** einbetonierter Mast *m* ~**-setting derrick** Kranbaum *m* zum Aufrichten der Stangen ~ **shoe** Pol-, Stangen-schuh *m* ~ **spoke** Polzahn *m* ~ **step** Polschritt *m* (electr.), Steigeisen *n* ~ **steps** Steigestütze *f* ~ **store** Stangenlagerplatz *m* ~ **strap** Mastschelle *f* ~**-strap bit** Laschenbohrer *m* ~ **strutted in pyramidal form** Bockgestänge *n* ~ **switch** Mastschalter *m* ~ **swivel** Gestängedrehkopf *m* ~ **system** Hängesystem *n* ~ **test box** Stangenuntersuchungskasten *m*, Untersuchungsstange *f* ~ **test paper** Polreagenzpapier *n* ~ **tester** Zuwachsbohrer *m* ~ **timber** Stangenholz *n* ~ **tip** Polhorn *n*, Polranf *m*, Polschuhrand *m* ~ **tooth** Polzahn *m* ~ **top end** Stangenspitze *f* ~ **tow** (wagon) Deichselschlepp *n* ~ **trestle** Stangenbock *m* ~ **wrench** Halt-, Stangenschlüssel *m* ~ **wrencher** Verschrauber *m*
**poled** gepolt, gerichtet
**poleless** polschuhlos
**poles,** ~ **and zeros** Pole *pl* und Nullstellen *pl* ~ **capable of being reversed** umpolbar ~ **statics** Stangenstatik *f* (electr.)
**polhode** Polhodie *f*
**polhody** Polhodie-Kegel *m*
**polianite** Polianit *m*
**police** Polizei *f*; polizeilich ~**-man** Polizist *m*, Schutzmann *m* ~ **radio service** Polizeifunkdienst *m* ~ **signal system** Polizeirufanlage *f* ~ **station** Polizeiamt *m*, Polizeiwache *f*
**policy** Police *f*, Politik *f*

**poling** Kappenverzug *m*, Polen *n*, Umpolung *f*, (of copper) Umrühren *n* ~ **board** Brustholz *n*, Spießbrett *n* ~ **boards** Schaltbretter *pl*, Streichwand *f* ~**-boards sheathing** Stulpwand *f* aus Bohlen ~ **switch** Kopplungswechsler *m*, Umkehrschalter *m* ~ **tough pitch** Hammergarmachen *n*

**polish, to** ~ abschleifen, abschmirgeln, ausputzen, blank machen, blankputzen, glänzen, glanzschleifen, glätten, polieren, putzen, schleifen, schlichten, (leather) wichsen **to** ~ **the commutator** den Kollektor *m* abschmirgeln **to** ~- **-grind** polierschleifen **to** ~ **with pumice** bimsen

**polish** Glanz *m*, Glänze *f*, Glätte *f*, Politur *f*, Schliff *m* ~ **for floors** Bohnermasse *f* ~ **rod** glatte Bohrstange *f*

**polishability** Schleifbarkeit *f*

**polishable** polierfähig

**polished** blank, glatt, poliert ~ **with emery** abgeschmirgelt

**polished,** ~ **blue sheet** blaublankes Blech *n* ~ **face** polierte Ansichtsfläche *f* ~ **lacquer** Schleiflack *m* ~**-metal surface** Metallschliff *m* ~ **plate glass** Spiegelglas *n* ~ **section** Anschliff *m*, Metallschliff *m*, Schliffbild *n* ~ **sheet material** Glanzblech *n* ~ **specimen** Schliffpräparat *n* ~ **steel** polierter Stahl *m* ~ **surface** Spiegel *m* ~ **washer** blanke Scheibe *f*

**polisher** Glätter *m*, (Buchbinderei) Glättzahn *m*, Kaliber-Schlichtbohrer *m*, Polierer *m*, Polierbürste *f*, -maschine *f*, -mittel *n*, -scheibe *f*, -stahl *m*, Schleifer *m* ~ **of convex glasses** Rundglasschleifer *m* ~'**s frame** Polierbock *m*

**polishing** Fertig-polieren, -schlichten *n*, Glätten *n*, Hochglanzgebung *f*, Polieren *n*, Scheuern *n*, Schleifen *n*, Schmirgeln *n* ~ **bit** Kaliber-, Schlicht-bohrer *m* ~ **cloth** Polier-filz *m*, -tuch *n* ~ **composition** Polier-kaliber *n*, -masse *f* ~ **disc** Polierscheibe *f* ~ **hammer** Glanzhammer *m* ~ **iron** Glätteisen *n*, Polierstahl *m* ~ **lacquer** Polier-, Schleif-lack *m* ~ **liquid** Polierflüssigkeit *f* ~ **material** Glänze *f*, Poliermittel *n* ~ **medium** Poliermittel *n* ~ **mop** Politurballen *m* ~ **motor** Poliermotor *m* ~ **oil** Polier-, Putz-öl *n* ~ **paper** Polierpapier *n* ~ **paste** Glanz-, Polier-, Schleifpaste *f*, Wichse *f* ~ **plant** Polieranlage *f* ~ **powder** Polier-mittel *n*, -pulver *n*, Schleifpulver *n* ~ **property** Schleiffähigkeit *f* ~ **red** Polierrot *n* ~ **roll** Polierwalze *f* ~ **room** Poliererei *f* ~ **slate** Polierschiefer *m*, Weststein *m* ~ **stand** Poliergerüst *n* ~ **stick** Glätt-holz *n*, -schiene *f* ~ **surface** Polierfläche *f* ~ **tool** Filzstock *m*, Glatt-bein *n*, -heft *n*, -stahl *m*, -stange *f*, -zahn *m*, Poliereisen *n* ~ **varnish** Glanzfirnis *m*, Schleiflack *m* ~ **wheel** Glanzscheibe *f*, Polierscheibe *f*, Schleifrad *n*, Schwabbelrad *n*

**political** politisch ~ **economy** Volkswirtschaft *f*, (science) Volkswirtschaftslehre *f*

**politics** Politik *f*

**poll** Gestänge *n*

**pollute, to** ~ verunreinigen

**pollution** Verschmutzung *f*, Verunreinigung *f* ~ **of the air** Luftverunreinigung *f*

**poly** mehr . . ., poly . . ., viel . . . ~**-acid** mehrsäurig ~**-atomic** mehratomig ~**-basic** mehrbasisch ~**-basite** Eugenglanz *m* ~**-cellular** mehrzellig ~**-chromatic** vielfarbig ~**-chrome** mehrfarbig ~**-chromy** Mehrfarbigkeit *f*, Vielfarben-druck *m* ~**-conic** vielkegelig ~**-crase** Euxenit *n*, Polycrasit *n* ~**-crystal** Mehr-, Viel-kristall *n* ~**-crystalline** aus mehreren Kristallen *pl* bestehend, polikritallin ~**-cyclic** vielkernig

**polydirectional,** ~ **antenna** Rundstrahlantenne *f* ~ **microphone** Raummikrofon *n*

**polyene** (butadiene, etc.) Polyene *n*

**polyenergetic** polyenergetische Strahlung *f*

**polyester film** Polyesterband *n*

**polyethylene** Polyäthylin *n*

**polyganization** Polyganisierung *f*

**polygon** Polygon, Vieleck *n* ~ **of flexure** Biegungspolygon *n*

**polygon,** ~ **circumscribed about a circle** Tangentenvieleck *n* ~ **connection** Vieleckschaltung *f* ~ **inscribed in a circle** Sehnenvieleck *n* ~ **inscribed into and circumscribed about itself** sich selbst ein- und umschriebenes Polygon *n* ~ **mapping** Polygonabbildung *f* ~ **measurement** Polygonmessung *f*

**polygonal** polygonisch, vieleckig ~ **angle** Brechungswinkel *m* bei Polygonzugmessung *f* ~ **bar** Mehrkantstab *m* ~ **bow-string girder** Bogensehnenträger *m* ~ **coil** vieleckige Spule *f* ~ **course** Polygonzug *m* ~ **ingot mold** Vieleckkokille *f* ~ **lines** gebrochene Linienzüge *pl* ~ **trace (or front)** polygonaler Grundriß *m*

**polygoniometry** Polygonmessung *f*

**polygonization** Polygonisierung *f*

**polygraphy** polygrafisches Gewerbe *n*

**polyhalite** Polyhalit *m*

**polyhedra** Polyeder *n*

**polyhedral** flächenreich, polyedrisch, vielflächig ~ **angle** polyedrischer Winkel *m* ~ **mirror (or scanner)** Vielkantscheibe *f* ~ **prism** Mehrkantprisma *n* ~ **structure** Polyederstruktur *f* ~ **surface** Vielflächenmantel *m*, Vielflächnermantel *m*

**polyhedron** Polyeder *n*, Rautenglas *n* (opt.), Vielfach *n*

**polymer** polymere Körper *pl* ~ **chains** Polymerenketten *pl*

**polymeric** polymer

**polymeride** Polymerisat *n*

**polymerization** Polymerie *f*, Polymerisation *f*

**polimerize, to** ~ polymerisieren

**polymers** Polymerisate *pl*

**polymeter** Polymeter *n*

**polymorphic** polymorph, polymorphisch

**polymorphism** Polymorphie *f*, Polymorphismus *m*, Vielgestaltigkeit *f*

**polymorphous** polymorph, polymorphisch

**polymorphy** Polymorphie *f*, Polymorphismus *m*

**polynominyl** mehrgliedrige Größe *f*, Polynom *n*; vielgliedrig (math.), vielseitig ~ **installation** Aggregat *n* ~ **ring** Polynomring *m*

**poly,** ~**-nuclear** mehrkernig, vielkernig ~**-nucleate** vielkernig

**polyode valve** Mehrpolröhre *f*

**polyphase** mehrphasig, mehrpolig, mehrwellig, vielphasig ~ **aggregate** mehrphasiges Aggregat *n* ~ **alternating current** Drehstrom *m* ~ **current** Mehrphasen(wechsel)strom *m* ~ **induction motor** Drehstrommotor *m* ~ **motor** Mehrphasenmotor *m* ~ **system** Mehrphasensystem *n* ~ **watthourmeter** Vielphasenwattstundenzähler *m* ~ **wattmeter** Vielphasen-Wattmeter *n*

**polyphonic** (of organ) mehr-, viel-stimmig

**polyphonous** vielstimmig
**polypole cable coupler** Vielfachstecker *m*
**poly,** **~-propylene sheet** Polypropylenfolie *f*
**~-saccaride** Polyose *f* **~-somatic** polysomatisch
**~-styrene (or styrole)** Polystyrol *n* **~-sulfide**
Polysulfid *n* **~-symmetry** Polysymmetrie *f*
**~-synthetic** polysynthetisch
**polytechnic** polytechnisch **~(al) institute** Poly-
technikum *n*
**polytropic** politropisch **~ change of state** poly-
tropische Zustandsänderung *f* **~ exponent**
Exponent *m* der Polytrope **~ line** Polytrope *f*
**polyvalence** Vielwertigkeit *f*
**polyvalent** vielwertig
**polyvinyl chloride fiber** PeCe-Faser *f*
**pomade** Salbe *f*
**pommel** Krispelholz *n*, Sattelkopf *m*
**Pompeian red** Pompejanischrot *n*
**poncelet** Poncelet *n*
**Poncelet wheel** Ponceletrad *n*
**pond, to** **~** Wasser anstauen
**pond** Becken *n*, Teich *n* **~ plug** Ablaßschütz *n*
**~ water** Teichwasser *n*
**ponderomotive,** **~ force** ponderomotorische
Kraft *f* **~ law** Bewegungsgleichung *f*
**ponding** (a fill) Schlämmen *n*
**poniard** Dolch *m*
**pontonier** Pontonier *m*
**pontoon** Brücken-boot *n*, -schiff *n*, Hebeprahm
*m*, Ponton *m*, Prahm *m*, Pralan *m*, Schwimmer
*m*, Schwimmerflugzeug *m* **~ balk** Scherbalken
*m* **~ bridge** Klappbootbrücke *f*, Pontonbrücke
*f*, Schiffsbrücke *f* **~ carrier** Hacket *m*, Ponton-
wagen *m* **~ ferry** Pontonfähre *f* **~ shears**
schwimmender Mastkran *m* **~ truck** Ponton-
wagen *m*
**pony,** **~ engine** Rangierlokomotive *f* **~ masher**
Maischapparat *m* **~ mixer** Ponymischer
**~ motor** Anwurfmotor *m* **~ packer** Zwerg-
packer *m* (min.) **~ rougher** Vorsteckgerüst *n*
**~ roughing** Vorstrecken *n* **~ roughing strand** (of
rolls) Vorstreckgerüst *n* **~ truck** Drehgestell *n*,
Laufwagen *m*
**pool** (in a river) Auskolkung *f*, Becken *n*, Interessengemeinschaft *f*, Kanalhaltung *f*, (in a
river) Kolk *m*, Kuhle *f*, Lache *f*, Ölfeld *n*, Pfütze
*f*, Schmelzbad *n*, Sumpf *m*, Teich *m*, Tümpel *m*
**~ of factories** Arbeitsgemeinschaft *f* **~ of
molten metal** Schmelzkrater *m*
**pool,** **~ cathode** flüssige Kathode *f* **~ elevation**
Stauspiegel *m*, Wasserstand *m* **~ freight** Pari-
tätsfracht *f* **~-rectifier** Entladungsgefäß *n* mit
flüssiger Kathode, Quecksilberdampfgleichrichter *m*, Ventil *n* mit flüssiger Kathode
**pooled air traffic** Poolverkehr *m* (aviat.)
**poop** Heck *n*, Kampanie *f* **~ deck** Achterdeck *n*
**~ lantern** Achterlaterne *f*, Hecklaterne *f*
**poor** arm, (of gases, vacuum, etc.) flau, (of gases
vacuum) geringwertig, notdürftig, schlecht **~ in
inertia** trägheitsarm **~ in seams** flözarm
**poor,** **~ balance** Unwucht *f* **~(-quality) clay**
Schluff *m* **~ coal** geringwertige Kohle *f*
**~ conductor** Halbleiter *m*, schlechter Leiter *m*
**~ contact** schlechter Kontakt *m* **~ dielectric**
schlechtes Dielektrikum *n* **~ drafting** Verzeichnung *f* **~ gas** armes oder geringwertiges
Gas *n*, Schwachgas *n* **~ geometry** ungünstige
Geometrie *f* **~ lead mine** unergiebige Bleimine *f*

**~ lignite of recent growth** junge geringwertige
Braunkohle *f* **~ lime** Magerkalk *m* **~ lode**
tauber Gang *m* **~ mixture** mageres Gemisch *n*
**~ ore** Pocherz *n*, Scheidegang *m* **~ slag** rohe
Frischschlacke *f* **~ transmission** mangelhafte
oder schlechte Verständigung *f*, Schwierigkeit
*f* in der Verständigung **~ visibility** schlechte
Sicht *f*
**poorly** lediglich **~ defined** unscharf
**pop, to** **~ off** abspringen
**pop** Knall *m*, Pistole *f* **~-hole** Knäpperlich *n*
**~ rivet** Dornniet *m*, Knallniete *f*, Schußniete *f*
**poplar** Pappel *f*, (wood) Pappelholz *n*
**poppet** Holzstütze *f*, Ventilkegel *m* **~ assembly**
Ventil-kegel *m*, -teller *m* **~ ball** Riegelkugel *f*
**~ spring** Dockenfeder ' **~ valve** Einschieber-
steuerung *f*, pilzförmiges Ventil *n*, Ringventil *n*,
Rohrventil *n*, Schnüffelventil *n*, Spindelventil *n*,
Tellerventil *n* **~ valve cylinder** Ventilzylinder *m*
**~-valve engine** Motor *m* mit Tellerventilen
**populace** Bevölkerung *f*
**populate, to** **~** besetzen
**population** Bevölkerung *f*, Einwohnerschaft *f*
**populous** volkreich
**porcelain** Porzellan *n*, porzellanartig **~ boat**
Porzellanschiffchen *n* **~ body** Porzellanmasse *f*
**~ cement** Porzellankitt *m* **~ clay** Kaolin *n*,
Porzellan-erde *f*, -ton *m* **~ cup** Isolierglocke *f*
**~ die** Porzellanmatrize *f* **~ earth** Steinmark *n*
**~ evaporating basin** Porzellandampfschale *f*
**~ eyelet** Porzellanöse *f* **~ funnel** Nutsche *f*
**~ glass** Milchglas *n* **~ glaze** Porzellanglasur *f*
**~ insulator** Lüsterklemme *f*, Porzellanisolator
*m* **~ marble** Porzellanmärbel *f* **~ pot-eye**
Porzellanring *m* **~ spark plug** Glimmzünd-
kerze *f* **~ suction funnel** Glasnutsche *f* **~ tube**
Einführungspfeife *f* aus Porzellan *n* (teleph.),
Porzellanrohr *n*
**porcellanic** porzellanartig
**porch** Portal *n*, Schwarzschulter *f* (TV) **~ paint**
Sockelfarbe *f*
**porcupine** Igel *m* (Spinnerei), Nadel-, Kamm-
-walze *f*
**pore** Loch *n*, Pore *f*, winzige Öffnung *f* **~ conductivity** Porenleitfähigkeit *f* **~ content**
Porengehalt *m* **~ cross section** Porenquerschnitt
*m* **~ damming** porenschließend **~ filler** Vorlack *m* **~-fluid anelasticity** Porenflüssigkeits-
elastizität *f* **~ pressure dissipation** Poren-
wasserdruckabnahme *f* **~ sealer** Sperrschicht *f*
**~ section** Porenquerschnitt *m* **~ space** Hohlraum *m*, Porenraum *m* **~ volume** Porenraum *m*
**~ water** Porenwasser *n* **~ water head** Poren-
wasserdruckhöhe *f* **~-water pressure** Poren-
wasserdruck *m*
**porometer** Porometer *n*
**porosimeter** Porosimeter *n*, Wasserdurchlässigkeitsprüfer *m*
**porosity** Durchdringbarkeit *f*, Durchlässigkeit *f*,
Feinlunkerung *f*, Lunkerung *f*, Porigkeit *f*,
Porosität *f*, Undichtheit *f*, Undichtigkeit *f* **~
density** Porendichte *f*
**porous** blasig, löcherig, locker, lückig, luckig,
porig, porös, schaumig, schwammig **~ barrier** poröse Wand *f* **~ brick** (from clay and
gravel) Schwemmstein *m* **~ cast iron (or casting)**
undichter Guß *m* **~ ceramic** Tongut *n* **~ iron**
luckiges Eisen *n* **~ point** Lunkerstelle *f* **~ pot**

poröses Gefäß *n*, poröse Zelle *f*, Tonzelle *f*
~-pot method Methode *f* mit unpolarisierbaren
Elektroden ~ reactor Reaktor *m* mit porösem
Material ~ spot Lunkerstelle *f*, schwarze Stelle
*f*, Stich *m* ~ white pig Weichfluß *m*
**porphyritic** porphyrisch ~ structure porphyri-
sches Gefüge *n*
**porphyry** Porphyr *m* ~ roller Porphyrwalze *f*
**porpoise, to** ~ hüpfen, tauchstampfen
**porpoise** Aufsummen *n*, Landung *f*, in Stampfbewegung *f* (aviat.) ~ oil Meerschweinöl
*n*
**porpoising** Stampfen *n*, (seaplane) Tauchbewegung *f*
**Porro's,** ~ photogoniometric method Bildmessung *f* nach Porro ~ principle Porrosches
Prinzip *n*
**port, to** ~ tragen
**port** Backbord *n*, Durchgangsöffnung *f*, Durchlaß *m*, Durchströmungskanal *m* (mach.),
Hafen *m*, Kopf *m*, Mündung *f*, Ofenkopf *m*,
Öffnung *f*, Schlitz *m*, Steueröffnung *f*; links
on the ~ beam querab Backborf *n* on the ~
bow Backbord *n* voraus
**port,** ~ of call Anflughafen *m*, Anlaufhafen,
Anlegehafen *m* ~ of delivery Löschplatz *m*
~ of departure Abflughafen *m*, Abgangshafen
*m* ~ of destination Bestimmungshafen *m*,
Zielhafen *m* ~ of distress Nothafen *m* ~ of
embarkation Verladehafen *m* ~ of entry Ein-
fuhrhafen *m* ~ of loading Verschiffungshafen
*m* ~ of registry Heimathafen *m* ~ of transit
Transithafen *m* ~ of transshipment Umschlag-
hafen *m*
**port,** ~ aft Backbord *n* achteraus ~ authority
Hafenabgabeamt *n* ~ charges Hafen-abgaben
*pl*, -gebühren *pl* ~ control Hafenüberwachung
*f* ~ control officer Hafenoffizier *m* ~-controlled
schlitzgesteuert ~ corrector Backbordkorrektor
*m* ~ cover Schildklappe *f* ~ covered and uncovered by the piston kolbengesteuerter Schlitz
*m* ~ (left hand) curve Backbordkurve *f* ~
dolphin Backborddalbe *f* ~ dues Hafenge-
bühren *pl* ~ end Heizkopf *m* (Ofen) ~ engine
Backbord-maschine *f*, -motor *m* ~ entrance
Hafen-einfahrt *f*, -mündung *f* ~ equipment (or
facilities) Hafenausrüstung *f* ~ inner engine
der innere Backbordmotor *m* ~ installation
Hafenanlage *f* ~ light Backbord(seiten)licht *n*,
Hafenfeuer *n* ~ pale Backborddalbe *f* ~ radar
Hafenradar *m* ~scavenging Schlitzspülung *f* ~
scavenging engine Schlitzspülmaschine *f* ~
senior officer Hafenkommandant *m* ~ side
Auspuffseite *f*, Backbord *n* ~ signal Backbord-
zeichen *n* ~ vessel Hafenfahrzeug *n*
**portability** Fahrbarkeit *f*, Tragbarkeit *f*
**portable** beweglich, fahrbar, förderbar, ortsveränderlich, tragbar, transportabel, umsetzbar, verlegbar, versetzbar ~ barrel drainer
tragbare Faßentleerungsvorrichtung *f* ~ blinker
Blinktornister *m* (semi) ~ boiler Lokomobil-
kessel *m* ~ bridge Kolonnenbrücke *f* ~ bridge
unit K-Brückengerät *n* ~ conveyor fahrbarer
Förderer *m* ~ decontamination duster Hand-
streutrommel *f* ~ drill Handbohrmaschine *f* ~
dust sampler Staubprobensammler *m* ~ electric drill Elektrohandbohrmaschine *f* ~ engine
fahrbarer Motor *m*, Lokomobil *n* ~ footbridge

Schnellbrücke *f* ~ forge Feldschmiede *f* ~
form Kofferform *f* ~ generating unit Kleinmaschinensatz *m* ~ grinder Handschleif-
maschine *f* ~ ionization chamber tragbare
Ionisationskammer *f* ~ lamp Handlampe *f*
(contractor's) ~line verlegbares Gleis *n* ~ line
crossing tragbare Wegübergang *m* ~ line
section Gleisjoch *n* ~ loader Gleitbandlade-
wagen *m* ~ long-wave radio telephone Langwellenkleinfunkapparat *m* ~ metering unit
bewegliches Meßgerät *n* ~ microphone bewegliches entfesseltes Mikrofon *n* ~ mixer An-
klemmrührer *m* ~ motor Fahrmotor *m* ~ nozzle
tragbarer Zapfhahn ~ obstacle bewegliches
Hindernis *n*, Schnellhindernis *n* ~ obstacle for
speedy obstruction Schnellsperre *f* ~ plant fahrbare Anlage *f* ~ plummet Taschenlot *n* ~
position Fahrbereitung *f* ~ power pump Trag-
kraftspitze *f* ~ projector Wandergerät *n* ~
projector equipment Kofferkino *n* ~ propeller
mixer Ekatorührer *m* ~ pump fahrbare Benzinpumpe *f* ~ radio Feldfunksprecher *m* ~
radio set Tornisterfunkgerät *n* ~ radio telephone
Kleinfunkapparat *m* ~ radio transmitter Klein-
sendegerät *n* ~-radio-transmitting station
Kleinfunkstelle *f* ~ railway Feldbahn *f* ~ ramp
Rampengerät *n* ~ receiver Tornisterempfänger
*m* ~ reel for heavy field cable Rückentrage *f* für
schweres Feldkabel ~ searchlight Hand-
scheinwerfer *m* ~ set Anhängegerät *n*, Koffer-
gerät *n* (rdo) ~ (telephone) set Tischapparat *m*,
tragbarer Apparat *m* ~ short-wave apparatus
Kurzwellen(klein)funkapparat *m* ~ signal lamp
K-Blink *m* ~ singlephase low-voltage transformer tragbarer Einphasen-Kleinspannungs-
Transformator *m* ~ storage accumulator (or
cell) tragbarer Sammler *m* ~ switchboard Feldklappenschrank *m*, Klappenschrank *n* ~
telephone Fernsprechtornister *m* ~ telephone
set (or station) Streckenfernsprecher *m* ~
teletype Feldfernschreiber *m* ~ test set Prüf-
gehäuse *n*, Prüfgerät *n* (tragbar oder fahrbar)
~ TV service set Fernseh-Servicekoffer *m* ~
unit dust collectors Kleinentstauber *m* ~ wire
entanglement Schnelldrahthindernis *n* ~ wireless set Landungsfunkstelle *f*
**portal** Pforte *f* ~ crane Portalkran *m* ~ revolving crane Vollportaldrehkran *m* ~ tower Portal-
mast *m*
**ported piston valve** Steuerkolben *m*
**porter** Lastträger *m* ~ bar Dachwippe *f*
**portfolio** Akten-mappe *f*, -tasche *f*, Briefmappe
*f*, Mappe *f*, Markttasche *f*, Zeichenmappe *f*
**porthole** Bullauge *n*, Durchlaßöffnung *f*, Einschnitt *m* ~ light Ochsenauge *n*
**portico** überdachter Vorbau *m*
**portion** Anteil *m*, (of a curve) Bereich *m*, Los *n*,
Menge *f*, Portion *f*, Quantum *n*, Quote *f*,
Stück *n*, Teil *m*, Teilgebiet *n*, Wegabschnitt *m*
~ which is cut out to produce segment Aus-
schliff *m* (bei Bifokalgläsern) ~ of line Leitungsstück *n* ~ passing through screen Durchfall *m* ~ of the scene Objektausschnitt *m* ~ of
the trajectory of rockets after combustion cutoff
antriebsfreie Bahn *f*
**portions, in** ~ portionweise
**portray, to** ~ abbilden
**portrayal** Beschreibung *f*, Schilderung *f*

ports covered and uncovered by the working piston kolbengesteuerte Schlitze *pl* (Zweitakt)
pose, to ~ vorlegen
posing the problem Problemstellung *f*
position, to ~ einstellen, in die richtige Lage *f* bringen, in Stellung *f* bringen
position Anstellung *f*, Besteck *n*, Existenz *f*, Lage *f*, Ort *m*, Posten *m*, Situation *f*, Stand *m*, Standort *m*, Stelle *f*, Stellung *f*, Zustand *m* on ~ Einschaltrelaisstellung *f*
position, ~ according to dead reckoning gekoppelter Standort *m* ~ of advance ignition Frühzündungsanlage *f* ~ of the azimuthal circle Kreisstellung *f* (eines Theodolts) ~ of the band pass of a filter Lochlage *f* eines Filters *n* ~ of the center of gravity Schwerpunktslage *f* ~ of the coating Schichtlage *f* ~ of control surface Ruder-lage *f*, -stellung *f* ~ of crank Kurbelstellung *f* ~ of the cylinders Zylinderanordnung *f* ~ of device Einbaulage *f* ~ on dominant height Höhenstellung *f* ~ of fix Besteckort *m* ~ of grating resulting in a one-order spectrum Einordnungstellung *f* ~ in the horizontal plane Meßdreieck *n* ~ of the lead Lotstand *m* ~ and level Lage *f* und Höhe *f* ~ of load Laststellung *f* ~ of maximum signal Peilmaximum *n* ~ of minimum signal (in direction finder) Peilminimum *n* ~ of the picture frame bars Bildstrichlage *f* ~ of reckoning Besteckort *m* ~ at (of) rest Ruhelage *f* ~ of spar Holmlage *f* ~ of the sun Sonnenstand *m* ~ of symmetry Ruhestellung *f* ~ as teacher Lehramt *n* ~ of test points Lage *f* der Meßpunkte ~ of tool during form dressing Stahlstellung *f* beim Profilieren ~ of transmission range Filterlochlage *f* ~ of the transmission range of a filter Lochlage *f* eines Filters ~ of trolley Katzenstellung *f*, Laufkatzenstellung *f* ~ invertical sense Höhenlage *f* ~ of weld Schweißlage *f* ~ of welding Schweißführung *f* ~ of the zero (or graduation) Nullpunktlage *f*
position, ~ booking Platzbelegung *f* ~ busy relay Platzbesetztrelais *n* ~ circle Positionskreis *m* ~ control Wegregelung *f* ~ coupling Platzzusammenschaltung *f* ~ coupling key Verbindungsschalter *m* ~ determination by neutron diffraction Ortsbestimmung *f* durch Neutronenbeugung ~ distributor Meldeverteiler *m* ~ equipment Platzausrüstung *f* ~ filar micrometer Positionsfadenmikrometer *m* ~ finder Abstandsmesser *m* ~ finding Funkortung *f*, Ortsbestimmung *f*, Schiffsortung *f*, Standortbestimmung *f*, -peilung *f* ~-finding apparatus Ortungsgerät *n* ~ fix Standort *m* ~ fixing Navigation *f*, Ortsbestimmung *f* ~ function Ortsfunktion *f* ~ group Lagegruppe *f* ~ grouping key Verbindungsschalter *m* ~ gyro Lagekreisel *m* ~ indicator Bandzählwerk *n* (tape rec.), Positionszeiger *m* ~ lever for ram stroke Drehhebel *m* für den Stößelhub ~ light Positionslampe *f*, -laterne *f* ~ lights Positionslichter *pl* ~ line Standlinie *f* ~-line bearing Standlinienpeilung *f*, Standortbestimmung *f* ~ line triangle Standliniendreieck *n* ~ map Stellungskarte *f* ~ mark Lagezeichen *n* ~ measurement Lage-, Orts-messung *f* ~ message Positionsmeldung *f* ~ meter Platzzähler *m* ~ number Stellenzahl *f* ~-peg-count register Leistungszähler *m* ~

pilot lamp Platzlampe *f* ~ repeating device Stellungsrückmelder *m* ~-repeating means Rückmelder *m* ~ report Lagebericht *m*, Standortmeldung *f* (aviat.) ~selector Stellungsschalter *m* ~ shunt factor Lageaufschaltgröße *f* ~ sketch Umgebungskarte *f* ~ space Konfigurationsraum *m* ~ switch factor Lageaufschaltgröße *f* ~-switching key Platzumschalter *m* ~ telemeter positionsindizierender Fernmesser *m* ~ tracker Zeichenplatte *f* ~ vector Fahrstrahl *m*, Ortsvektor *m*, Radiusvektor *m* ~ wiring Platzbelegung *f*
positional, ~ accuracy Positionsgenauigkeit *f* ~ information Standortdaten *pl* ~ notation Stellenschreibweise *f* (comput.), Zahlensystem *n* ~ representation Stellenschreibweise *f* ~ tolerance Lagetolerenz *f*
positioned so as to fill gaps auf Lücke *f* stehen
positioner Stellwerk *n* ~ part mechanischer Teil *m* (eines Servos)
positioning, ~ for approach Radarführung *f* zum Anflug ~ of landing gear Fahrgestellstellung *f*
positioning, ~ action Lage-regelung *f*, -verhalten *n* ~ coil Verschiebespule *f* ~ control Lageeinstellung *f* ~ device Einsteller *m* ~ device for leads Vorhalte-Einstellung *f* ~ pin Anschlagstift *m*, Greiferstift *m* ~ screw Einstellschnecke *f*
positions transfer print entry Stellenschreibsteuerung *f*
positive Abzug *m* (photo), Positiv *n*, positive Eigenschaft *f*, positives Bild *n* (photo), Schwarzweißfoto *n*; positiv, vorgeschrieben, zwangsläufig ~ and negative booster Zusatzmaschine *f* für Zu- und Gegenschaltung ~ and negative carbons Plus- und Minuskohlenstift *m* ~ and negative electricity positive und negative Elektrizität *f* ~ and negative stagger Rückwärts- und Vorwärtsstaffelung *f*
positive, ~ accelerator Sauganode *f* ~ air pressure Luftüberdruck *m* ~ area positives Gebiet *n*, Stromaustrittszone *f* ~ bias positive Vorspannung *f* ~ bias potential Rückkupplungsspannung *f* ~ blower Kapselgebläse *n* ~ booster Spannungserhöher *m* ~ brush Plusbürste *f* ~ buoyancy dynamischer Auftrieb *m* ~ carbon core Pluskohlenkrater *m* ~ charge positive Ladung *f* ~ collars Matrizen *pl* ~ column positive Lichtsäule *f* ~ contact Pluskontakt *m* ~ control stufenlose Regulierung *f* ~ coupling Kraftverbindung *f*, (kraft- und) formschlüssige Verbindung *f* zweier Teile ~ curvature Plusbelag *m* ~ dihedral positive V-Stellung *f* (aerodyn.) ~ direction Wirkungsrichtung *f* ~ dispersion positive Dispersion *f* ~-displacement pump Pumpe *f* der Verdrängebauart, Verdrängerpumpe *f* ~ distortion positive Verzerrung *f*, Tonnenfehler *m*, tonnenförmige Verzeichnung *f* ~ drive schlupffreier Antrieb *m*, Vortrieb *m*, zwangsläufiger Antrieb *m* ~ electricity Glaselektrizität *f* ~ electrode Anode *f*, positive Elektrode *f* ~ electron Positron *n* ~ electrons positive Elektronen *pl* ~ feedback Mitkopplung *f*, Rückkopplung *f* ~ field Absaugfeld *n*, Saugfeld *n* ~ gas pressure Gasüberdruck *m* ~ glow light positives Glimmlicht *n* ~ grid Zuggitter *n* ~-grid circuit scheme Bremsfeldschaltung *f* ~-grid detector Bremsfeldaudion *n* ~-grid

oscillator Bremsfeldgenerator *m* ∼ **hole** positiv geladene Lücke *f* (cryst.) ∼ **infinitely variable gear** zwangsläufiges und stufenlos regelbares Getriebe *n* ∼ **ion beam** Strahl *m* positiver Ionen ∼ **lead** Plusleitung *f* ∼ **lens** Sammellinse *f* ∼ **lift** dynamischer Auftrieb *m* ∼ **line** Plusleitung *f* ∼ **locking of bolt** formschlüssige Sicherung *f* der Schraube ∼ **modulation** Hellmodulation *f*, Modulation *f* auf hell, Steuerung *f* auf hell, subtraktive Modulation *f* ∼ **movement** zwangsläufige Bewegung *f* ∼ **plate** (of storage battery) Plusplatte *f* ∼ **polarity of transmission** positive Polarität *f* der Sendung ∼ **pole** Pluspol *m* ∼ **potential** (of photocell) Saugspannung *f* ∼ **pressure** Überdruck *m* ∼**-pressure cabin** positive Druckkabine *f* ∼ **print** Positivpause *f* ∼ **quench potential** Rückkupplungsspannung *f* ∼ **quench voltage** Pendelrückkopplungsspannung *f* ∼ **ray** Kanalstrahl *m* ∼ **rays** Anodenstrahlen *pl* ∼ **reactance** Induktanz *f*, induktive Reaktanz *f*, induktiver Blindwiderstand *m* ∼ **regeneration** Mitkopplung *f*, positive Rückkupplung *f* ∼ **release** (on parachutes) Zwangsauslösung *f* ∼ **rotation** positiver Drehungssinn *m* ∼ **seat** arretierter Sitz *m* (Ventil) ∼ **sign** positives Vorzeichen *n* ∼ **slope** Böschung *f* im Auftrag, Vorderhang *m* ∼ **stagger** Staffelung *f*, Staffelung *f* nach vorne ∼ **stock** Positivmaterial *n* ∼ **supply** Anodenspannungsquelle *f* ∼ **sweepback** Tragflügelpfeilform *f* ∼ **threading** (of gear) zügiger Gang *m* ∼ **transmission** positive Übertragung *f* ∼ **transparency** Glaslichtbild *n* ∼ **valence** positive Wertigkeit *f* ∼ **wire** (A) an der Erde liegende Ader *f* ∼ **yawing moment** Giermoment *n*

**positively**, ∼ **actuated** zwangsläufig, zwangsläufig betätigt ∼ **connected** kraftschlüssig ∼ **controlled valve** gesteuertes Ventil *n*

**positiveness (or certainty) of measurement** Meßsicherheit *f*

**positron** Positron *n* ∼**-decay** Positronenzerfall *m* ∼**-emission** Positronenzerfall *m*

**positronium formation** Positroniumbildung *f*

**positrons** positive Elektronen *pl*

**possess, to** ∼ besitzen

**possession** Besitz *m* **act of taking** ∼ Inbesitznahme *f*

**possessor** Besitzer *m*

**possibilities of variation** Variationsmöglichkeiten *pl*

**possibility** Möglichkeit *f* ∼ **of adjustment** Regelmöglichkeit *f* ∼ **of auxiliary operation** Anrufhilfe *f* ∼ **to connect** Anschlußmöglichkeit *f* ∼ **of delivery** Liefermöglichkeit *f* ∼ **of mapping one surface into another** Abbildbarkeit *f* zweier Flächen (geom.) ∼ **of reading** Ablesemöglichkeit *f* ∼ **of use** Verwendungsmöglichkeit *f*

**possible** angängig, eventuell, möglich ∼ **definition** Auflösungsvermögen *n*

**post, to** ∼ absenden, anschlagen, aufgeben, aufstellen, eintragen

**post** Anstellung *f*, Bausch *m* (paper mfg.), Bolzen *m*, Existenz *f*, Halter *m*, Pausch *m*, Pfahl *m*, Pfeiler *m*, Pfosten *m*, Pilar *m*, Post *f*, Posten *m*, Puscht *m*, Stab *m*, Ständer *m*, Ständersäule *f*, Stange *f*, Stelle *f*, Stellung *f*, Stempel *m*, Steven *m*, Stiel *m*, Stoß *m* (paper) mfg.), Strebe *f*, Streckenpfeiler *m* **with one** ∼ ein-

hüftig **with** ∼**s of equal height** mit gleich hohen Stielen *pl*

**post,** ∼ **accelerating anode** Nachbeschleunigungsanode *f* ∼ **acceleration** Nachbeschleunigung *f* ∼ **assessment** Nachveranlagung *f* ∼ **bracket** Säulenarm *m* ∼ **card** Postkarte *f* ∼ **chain mortisers** Wandkettenfräsmaschinen *pl* ∼ **crane** Säulenkran *m* ∼ **dam** Nadelwehr *n* ∼ **dating** Nachdatierung *f* ∼ **defecation** Nachscheidung *f* ∼ **defecation juice** Nachscheidesaft *m* ∼**-deflection accelerating electrode** Nachbeschleunigungselektrode *f* ∼ **deflection accelerator tube** Nachbeschleunigungsröhre *f* ∼ **diffusion alloyed** diffundiert legiert (Transistor) ∼ **drill** Säulenbohrmaschine *f* ∼ **drill stand** Wandbohrständer *m* ∼ **edition** Nachausgabe *f* ∼ **emphasis** Deakzentuierung *f* (acoust.) ∼**-equalization** Entzerrung *f* (acoust.) ∼**-equalizing pulse** Nachtrabant *m* (TV) ∼ **exposure** Nachbelichtung *f* ∼**-fermentation** Nachgärung *f* ∼ **hanger** Säulen-armlager *n*, -konsollager *n*, -träger *n* ∼ **hole** Stangenloch *n* ∼**-hole auger** Erweiterungsbohrer *m* ∼**-hole drilling machine** Stangenbohrmaschine *f* ∼**-ignition** nach dem normalen Zündfunken *m* ∼ **jig** Säulengestell *n* ∼ **lanterns** Lichtmastaufsätze *pl* ∼ **liming** Nachscheidung *f* ∼ **magnification** Nachvergrößerung *f* ∼**-man** Briefträger *m*, Postbote *m* ∼**-mark** Aufgabestempel *m*, Poststempel *m* ∼ **mortem** Obduktion *f* ∼**-mortem report** Fundbericht *m* ∼ **obstacle** Pfahlsperre *f* ∼ **office** Postamt *n*

**post-office,** ∼ **authorities** Postdirektion *f* ∼ **box** Brief-, Post-, Schließ-fach *n* ∼ **earthing point** Amtserde *f* ∼ **exchange connection** Amtsanschluß *m* ∼ **line** Amtsleitung *f* ∼ **money order** Postanweisung *f*, Postnachnahme *f* ∼ **standard relay** Postrelais *n*

**post,** ∼ **record** Nachsynchronisierung *f* (film) ∼**-record tape equalizer** Wiedergabeentzerrer *m* ∼**-scoring** Vertonung *f* ∼**-script** Nachschrift *f*, Zusatz *m* ∼**-selection** Nachwahl *f* ∼ **squeezer** Säulenpreßformmaschine *f* ∼**-synchronization** Nachsynchronisierung *f* ∼ **truss** Hängesäulenwerk *n* ∼**-type molding machine** Säulenformmaschine *f*

**postage** Porto *n* ∼ **due** Strafporto *n* ∼ **meter machine** Frankiermaschine *f* ∼ **paid** portofrei ∼ **stamp** Postwertzeichen *n* ∼**-stamp guards** Briefmarkenwalze *f* ∼ **stamp printing** Briefmarkendruck *m*

**postagram** Kurzbrief *m*

**postal** postalisch ∼ **address** Wohnsitz *m* ∼ **check account** Postscheckkonto *n* ∼ **costums** Postzoll *m* ∼ **delivery** Postzustellung *f* ∼ **plane** Postflugzeug *n* ∼ **receipt** Posteinlieferungsschein *m* ∼ **savings account** Postsparkasse *f* ∼ **union** Postverband *m*

**poster** Affiche *f*, Anschlag *m*, Anschlagzettel *m*, Aushang *m*, Plakat *n* ∼ **advertising** Plakatwerbung *f* ∼ **cardboard** Plakatkarton *m* ∼ **edgings** Plakatleisten *pl*

**posterior,** ∼ **end** Hinterrand *m*, Schloßrand *m* ∼ **focal plane** hintere Brennebene *f* ∼ **lobe** Hinterlappen *m* ∼ **part** Hinterteil ∼ **projection** a-p-Projektion *f*

**posteriori, a** ∼ aus der Erfahrung *f* folgernd

**posteriorly** hinten

**posters** Affichen-, Anschlag-, Plakat-papier *n*

**posting** Aufgeben *n*, Anschlagen *n*, Beförderung *f*, Buchung *f*, Eintragung *f*, Übertragung *f* ~ **fluid** Umdruckflüssigkeit *f*

**postpone, to** ~ aufschieben, einstellen, nachsetzen, verlegen, verschieben, verschleppen

**postponed** aufgeschoben

**postponement** Aufschiebung *f*, Aufschub *m*, Verlegung *f*, Vertagung *f*

**postulate, to** ~ eine Forderung *f* stellen, voraussetzen

**postulate** Bedingung *f*, Postulat *n*, Voraussetzung *f*

**posture** Anordnung *f*, Aufstellung *f*, Lage *f*

**pot, to** ~ vergießen **to** ~ **galvanize** feuerverzinken

**pot** Büchse *f*, Gefäß *n*, Kanne *f*, Kübel *m*, (Lautsprecher) Magnettopf *m*, Pfanne *f*, (Plastikverfahren) Spritztopf *m*, Tiegel *m*, Topf *m* ~-**annealed** kastengeglüht ~ **annealing** Kastenglühung *f*, Topf-glühen *n*, -glühverfahren *n* ~ **annealing furnace** Kasten-, Topf-glühofen *m* ~ **capacitor** Topfkondensator *m* ~-**core coil** Topfspule *f* ~ **experiment** Gefäßversuch *m* ~ **fire** Unterflurtiegelofen *m* ~ **furnace** (for melting glass) Büttenofen *m*, Gefäßofen *m*, (glass mfg.) Hafenofen *m*, Kesselofen *m* ~ **galvanization** Feuerverzinkung *f* ~ **galvanizing** Feuerverzinken *n*, Feuerverzinkung *f* ~-**head** (cable) Abschlußmuffe *f* ~-**head insulator** Einführungsisolator *m* mit Vergußkammer, Überführungsisolator *m* mit Vergußkammer ~-**head jointing sleeve** Abschlußmuffe *f* ~-**head tail** Bleirohrkabel *n* zwischen Überführungskasten und Freileitung ~ **head terminal** Kabelendmuffe *f* ~ **hole** ausgewaschene Rundung *f* im Flußbett, Gletschertopf *m*, Strudelkessel *m* ~ **hook** Topfhaken *m* ~ **magnet** Magnettopf *m*, Mantelmagnet *m* ~ **mold** Tiegelhohlform *f* ~ **mouth wiper** Gießmundwischer *m* ~ **retort** Ausbrenntopf *m* ~ **roaster** Röstkonverter *m* ~ **roasting** Verblaserröstung *f* ~-**shaped magnet** Topfmagnet *m* ~ (**bucket**) **spinning process** Zentrifugenspinnen *n* ~ **stone** Topfstein *m* ~ **throat** Gießhals *m* ~ **time** (Klebstoff) Topfzeit *f*

**potable** trinkbar ~ **water** Trinkwasser *n*

**potash** Aschensalz *n*, Kali *n*, Pottasche *f* ~ **alum** Kalium-alaun *n*, -aluminiumsulfat *n* ~ **case** (mask) Kalipatrone *f* ~-**cobalt glass** Blaufarbenglas *n* ~ **fertilizer** Kalidünger *m* ~ **fertilizer salt** Kalidüngesalz *n* ~-**free** kalifrei ~ **fusion** Kalischmelze *f* ~ **iron alum** Kalieisenalaun *m* ~ **lye** Kalilauge *f* ~ **melt** Kalischmelze *f* ~ **mica** Kaliglimmer *m*, weißer Glimmer *m* ~ **mine** Kalibergwerk *n* ~ **mold** Kaliform *f* ~ **olive oil soap** Kaliolivenölseife *f* ~ **salt** Kaliumsalz *n* ~ **solution** Kalilösung *f* ~ **water glass** Kaliwasserglas *n*

**potassic** kalihaltig

**potassium** Kalium *n* ~ **acetate** essigsaures Kali *n* ~ **aluminate** Tonerdekali *n* ~ **aluminium sulphate** Kalialaun *m*, Kalium-alaun *m*, -aluminiumsulfat *n* ~ **aurichloride** Kaliumgoldchlorid *n* ~ **auricyanide** Zyangoldkalium *n* ~ **aurobromide** Goldkalium-, Kaliumgold-bromür *n* ~ **aurochloride** Goldkaliumchlorür *n* ~ **aurocyanide** Aurokaliumzyanid *n*, Goldkaliumzyanür *n*, Kaliumgoldzyanür *n*, Zyangoldkalium *n* ~ **binoxalate** Kleesalz *n* ~ **bitartrate**

**doppeltweinsaures Kali** *n* ~ **borate** borsaures Kalium *n* ~ **bromate** bromsaures Kali *n* ~ **bromide** Bromkalium *n* ~ **carbide** Kohlenstoffkalium *n* ~ **carbonate** kohlensau(e)res Kali *n*, Pottasche *f* ~ **carboxide** Kohlenoxydkalium *n* ~ **cartride** Kalipatrone *f* ~ **cell** Kaliumzelle *f* ~ **chloride** Chlorkali *n*, Chlorkalium *n* ~ **chlorplatinate** Kaliumplatinchlorid *n* ~ **chromate** gelbes Chromkali *n* ~ **citrate** zitronensaures Kali *n* ~ **cobalticyanide** Kobaltizyankalium *n* ~ **cobaltinitrate** Kobaltikaliumnitrat *n* ~ **cobaltocyanide** Kobaltozyankalium *n* ~ **container** Kaliumbehälter *m* kali *n*, -kalium *n* ~ **dichromate** Chromkali *n*, rotes Chromkali *n* ~ **ferrate** Eisenkalium *n* ~ **ferricyanide** Ferrizyankalium *n*, Kaliumeisenzyanid *n*, rotes Blutlaugensalz *n*, Rotkali *n* ~ **ferrisulphate** Kalium-eisenalaun *m*, -ferrisulfat *n* ~ **ferrocyanide** Blausalz *n*, blausaures Kali *n*, Blutlaugensalz *n*, Cyaneisenkalium *n*, eisenblausaures Kali *n*, Eisenzyankalium *n*, gelbes Blutlaugensalz *n*, Gelbkali *n*, Kaliumeisenzyanür *n*, Zyaneisenkalium *n* ~ **fluoride** Fluorkalium *n* ~ **fluosilicate** Kieselfluorkalium *n* ~ **formate** Kaliumformiat *n* ~ **hydroxide** Ätzkali *n*, Kali *n*, Kalilauge *f*, Kalium-hydrat *n*, -hydroxyd *n* ~ **hypochlorite** Chlorkali *n* ~ **iodate** jod-saures Kali *n* ~ **iodide** Jodkali *n*, Jodkalium *n* ~ **iridichloride** Kaliumiridiumchlorid *n* ~ **metabisulphite** (**or meta sulphate**) Kaliummetabisulfat *n* ~ **nitrate** Kalisalpeter *m* ~ **perchlorate** über(chlor)saures Kali *n* ~ **permanganate** Chamäleon *n*, Kaliumpermanganat *n*, übermangansaures Kali *n* ~ **permanganate solution** Chamäleonlösung *f* ~ **persulphate** über schwefelsau(e)res Kali *n* ~ **phenolate** Phenolkalium *n* ~ **phosphate** phosphorsaures Kali *n* ~ **platinchloride** Kaliumplatin-chlorid *n*, -chlorür *n* ~ **platinocyanide** Kaliumplatinzyanür *n* ~ **selenocyanate** Selenzyankalium *n* ~ **silicate** Kaliwasserglas *n*, kieselsaures Kali *n* ~ **sodium carbonate** Kaliumnatriumkarbonat *n* ~ **sodium tartrate** Kaliumnatriumtartrat *n*, weinsau(e)res Kaliumnatrium *n* ~ **stearate** stearinsaures Kalium *n* ~ **sulphide** Schwefelkalium *n* ~ **sulphocyanate** Kalium-rhodanat- *n*, -rhodanid *n*, -sulfozyanid *n*, Rhodankali *n*, Schwefelzyankalium *n*, Sulfozyankalium *n* ~ **tellurate** tellursaures Kalium *n* ~ **thiocyanate** Kalium--rhodanat *n*, -rhodanid *n*, Rhodan-kali *n*, -kalium *n*, Thiozyankalium *n* ~ **tungstate** Kaliumwolframat *n* ~ **valerate** valeriansau(e)res Kalium *n* ~ **vandate** vanadinsaures Kalium *n* ~ **xanthogenate** xanthogensaures Kalium *n*

**potato** Kartoffel *f* ~ **digger** Kartoffelgräber *m* ~ **hiller** Kartoffelhäufer *m* ~ **masher** Kartoffelpresse *f* ~ **planter** Kartoffelpflanzmaschine *f*

**potch, to** ~ aufschlagen, (paper mfg.) mischen

**potching** Bleichen *n* (der Papiermasse) ~ **stick** Rührscheit *n*

**potency** Potenz *f*

**potent** einflußreich, stark überzeugend

**potential** Hilfsquellen *pl*, Möglichkeit *f*, Potential *n*, (difference) Spannung *f*, Verspannung *f*; möglich, potentiell ~ **antinode** Spannungs-bauch *m*, -gegenknoten *m* ~ **barrier** Potential-berg *m*, -hügel *m*, -schwelle *f* ~ **causing quadrantal errors** Trübungsspannung *f* ~ **circ-**

uit Spannungskreis *m* ~ **correction** vorübergehende Regelabweichung *f* ~ **difference** Ausgleichdruck *m*, Potential-differenz *f*, -unterschied *m*, Spannungs-abfall *m*, -differenz *f*, -unterschied *m* ~**-difference ripple** Spannungswellen *pl* ~ **discharge** Spannungsstoß *m* ~ **distribution** Potentialverteilung *f* ~ **divider** Spannungsschalter *m* ~ **drop** Potential-abfall *m*, -sprung *m*, -verringerung *f*, Spannungsabfall *m*, -abnahme *f*, -gefälle *n*, -verlust *m* ~ **energy** Arbeits-inhalt *m*, -vermögen *n*, elektrische Energie *f*, Energie *f* der Lage ~ **equation** Potentialgleichung *f*, Potenzreihe *f* ~ **fall** räumlicher Spannungsabfall *m*, Spannungs-abnahme *f*, -gefälle *n* ~ **flow** Potentialströmung *f* ~ **function** Potential *n* ~ **gradient** Gradient *m* des Potentials, Potentialgradient *m*, Potenzgefälle *n*, Spannungsgradient *m* ~ **heat** potentielle Wärme *f* ~ **hill** Potentialberg *m* ~ **hole** Potential-kasten *m*, -mulde *f* ~ **impulse** Spannungsstoß *m* ~ **leakage** Potentialstreuung *f* ~ **liberation** Potentialemission *f* ~ **line** Druck-, Niveau-, Potential-linie *f* ~ **loop** Spannungsbauch *m* ~ **motion** wirbelfreie Bewegung *f* ~ **nodal point** Spannungsknotenpunkt *m* ~ **node** Spannungsknoten *m* ~ **peak** Potentialhügel *m* ~ **position** Potentiallage *f* ~ **production** Produktionsmöglichkeit *f* ~**-ratio method** Methode *f* der Verhältnisse der Potentiale ~ **rise** Spannungsanstieg *m* ~ **scattering** Potentialstreuung *f* ~ **stabilizer** Spannungsglättung *f* ~ **stray** Potentialstreuung *f* ~ **surface** Potentialfläche *f* ~ **theorem** Potentialsatz *m* ~ **transformer** Spannungstransformator *m* ~ **trough** Potentialnapf *m* ~ **value** Potentialwert *m* ~ **vortex** Potentialwirbel *m* ~ **well** Potential-kasten *m* -mulde *f*

**potentiality** Möglichkeit *f*

**potentiograph** registrierender Spannungsteiler *m*

**potentiometer** Drehwiderstand, Kompensator *m*, Potentiometer *n*, regelbarer Widerstand *m*, Schwächungswiderstand *m*, Spannungsteiler *m* (rdo), Widerstandskette *f* ~ **arm** Potentiometerschleifer *m* ~ **arrangement** Potentiometeranordnung *f* ~ **circuit** Potentiometerkreis *m* ~ **controller** Potentiometerregler *m* ~ **indicator** Potentiometeranzeigeinstrument *n* ~ **pick-off** Potentiometergeber *m* ~ **pyrometer** Potentiometerpyrometer *n* ~ **ratio** Aufteilungsverhältnis *n* ~ **recorder** Potentiometerschreiber *m* ~ **step** Stellung *f* des Reglungswiderstandes, Stufe *f* des Spannungsteilers ~ **stud** Stufe *f* des Spannungsteilers

**potentiometric** potentiometrisch ~ **oxidation method** oxydometrische Methode *f* ~ **recorder** Kompensationsschreiber *m* ~**-reduction method** reduktometrische Methode *f*

**Potier's,** ~ **coefficient of equivalence** Umrechnungsfaktor *m* nach Potier ~ **electromotive force** EMK *f* des Luftspaltfeldes ~ **reactance** Potierreaktanz *f*

**potter** Hafner *m*, Töpfer *m* ~**'s clay** Letten *m*, Töpfer-erde *f*, -ton *m* ~**'s lathe** Töpferscheibe *f* ~**'s ore** Glasurerz *n* ~**'s wheel** Töpferscheibe *f*

**pottery** Mufflerie *f*, Steingut *n*, Tonwaren *pl*, Töpferei *f*, Töpferware *f* ~ **mine** Kohleneisenstein *m* ~ **plaster** Formgips *m*

**potting** Einbetten *n*, Potten *n*, Vergießen *n*

~ **compound** Vergußmasse *f* ~ **resin** Vergußmasse *f*

**pouch** Beutel *m*, Sack *m*, Täschchen *n*, Tasche *f*

**pouchlike** sackartig

**Poulsen arc** Poulsenscher Lichtbogen *m*

**poultice** Umschlag *m* (med.) ~ **head cylinder** Zylinder *m* mit flachem Verbrennungsraum

**pounce, to** ~ mit Bimsstein *m* oder Sandpapier *n* bearbeiten, durchpausen, hämmern, prägen, schleifen

**pounce** Bimsstein-, Glätt-, Paus-pulver *n* ~**-box dredger** Streubüchse *f* ~ **paper** Pauspapier *n*

**pouncing** Glätten *n* (mit Bimsstein oder Sandpapier)

**pound, to** ~ pochen, pulvern, rammen, stampfen, stoßen, zerstampfen, (to pieces) zerstoßen

**pound** Pfund *n* ~ **per square inch** Pfund *n* per Quadratzoll

**pounded** eingeschlagen, (von aufeinander arbeitenden Teilen) gestoßen ~ **ore** Rausch *m*

**pounding** (of valve seats) Einschlagen *n*, Schlagen *n*, Stampfen *n*, Zerstoßen *n* ~ **hammer** Pochhammer *m* ~ **machine** Stampfwerk *n* ~ **trough** Pochladen *m*

**pour, to** ~ abgießen, abkippen, ausströmen, ausströmen lassen, strömen **to** ~ **away** abgießen **to** ~ **cold** kaltvergießen **to** ~ **on end** abgießen (stehend) **to** ~ **horizontally** waagrecht gießen oder abgießen **to** ~ **in** einfüllen, eingießen, einschütten, zugießen **to** ~ **into** hineingießen **to** ~ **metal** ergießen **to** ~ **off** abgießen **to** ~ **on** angießen, zuschütten **to** ~ **out** ausgießen, auskippen, ausschütten **to** ~ **out against** anschütten **to** ~ **over** übergießen **to** ~ **through** durchgießen **to** ~ **together** zusammengießen **to** ~ **from the top** fallend gießen **to** ~ **to the top** vollgießen **to** ~ **upon** aufgießen **to** ~ **water again** Wasser nachgießen

**pour** Einguß *m*, Fließen *n*, Metallguß *m*, Rinnen *n*, (in founding) ~**-in hole** Füllöffnung *f* ~ **point** Gieß-, Stock-punkt *m* ~ **test** Bestimmung *f* des Stockpunktes, Gießprobe *f*, Tropfprobe *f*

**pourability** Gießarbeit *f*, Vergießbarkeit *f*

**pourable compound** Gußmasse *f*

**pourer** (foundry) Formeneingießer *m*, Gießer *m*

**pouring** Gießen *n*, Guß *m*, Vergießen *n* ~ **of malleable (cast) iron** Glühstahlguß *m* ~ **out** Ausguß *m* ~ **takes place every . . . hours** alle . . . Stunden *pl* wird zum Abstich geschritten

~**-basin** Einguß-sumpf *m*, -tümpel *m*, Gießtümpel *m*, Sumpf *m*, (in casting) Tümpel *m* ~ **bed** Gießbett *n* ~ **concrete by chute** Einbringen *n* des Betons mittels Rutschen ~ **concrete from walkway** Einbringen *n* des Betons von einer Laufbühne ~ **conveyer** Gießband *n* ~ **crew** Gießkolonne *f* ~ **cup** Gießlöffel *m* ~ **gang** Gießkolonne *f* ~ **gate** Einguß *m*, Einguß-mündung *f*, -trichter *m*, Gießtrichter *m* ~ **head** Gußzapfen *m* ~**-in** Eingießung *f*, Einguß *m* ~ **ingate** Gußtrichter *m* ~**-in hole** Gußöffnung *f* ~ **jar** Gießtopf *m* ~ **ladle** Gießpfanne *f* ~ **level** Abstichsohle *f* ~ **lip** Ausguß-, Gieß-schnauze *f* ~ **mouth** Ausgußschnauze *f* ~ **nozzle** Ausguß-, Gieß-schnauze *f* ~**-off** Abguß *m* ~**-on** Anguß *m* ~ **outfit** Gießausrüstung *f*, Gießeinrichtung *f* ~ **platform** Gießbühne *f* ~ **position** Auskipp-, Gieß-

stellung f ~ **pot** Gießtopf m ~ **practice** Gieß-
technik f ~ **rod reel** Drahthaspel f mit Drehung
des Drahtringes im Aufnahmekorb ~ **snout**
Ausguß-rohr n, -röhre f ~ **spout** (of a crucible)
Ausgußlippe f, Ausguß-, Gieß-rinne f ~ **tech-
nique** Gießtechnik f ~ **temperature** Gieß-,
Vergieß-temperatur f ~ **test** Aufguß-, Auslauf-,
Fließ-, Gieß-probe f ~ **time** Gießzeit f

**powder, to** ~ bestäuben, bestreuen, einpudern,
einstauben, feinmahlen, pudern, pulvern, pulverisieren, stäuben, zerstoßen **to** ~ **with magnesia** einweißen

**powder** Mehl n, Puder m, Pulver n, Staub m
~ **anodes** Pulverglühanoden pl ~ **bag** Kartuschbeutel m, Kartusche f, (foundry) Streubeutel m
~ **camera** Pulverkamera f ~ **casting** Gießling m
~ **catapult** Sprengstoffschleuder f ~ **cathode**
Sinterkathode f ~ **chamber** Kartusch-, Pulver-
-raum m ~ **charge** Füllpulver n, Pulverladung f
~ **charge of cubical shape** Würfelkohle f
~ **coherer** Pulverfritter m ~ **core** Pulver-kern m,
-seele ~ **factory** Pulverfabrik f ~ **fouling of
barrel** Pulverrückstand m ~ **fuse** Pulverzünder
m ~ **grain** Pulverblättchen n (Stück des zer-
schnittenen Pulverbandes) **~-impulse catapult**
pulvergetriebener Katapult m ~ **man** Spreng-
stoffverwalter m ~ **metallurgy** Metallkeramik f
~ **mill** Pulverfabrik f ~ **outfit** Bestäubungs-
gerät n ~ **pattern** Pulverdiagramm n, Staubfiguren pl ~ **rocket** Pulverrakete f ~ **spray
apparatus** Puderapparat m ~ **strand** Pulver-
band m, -streifen m ~ **temperature** Pulvertem-
peratur f **~-temperature thermometer** Pulver-
thermometer n ~ **train** Brandsatz m, (fuse)
Pulversatz m **~-train ignition** Pulverbrennzün-
dung f **~-train ring** (for projectiles) Satzring m
~ **transmitter** Pulvermikrofon n **~-weighing
machine** Dosiermaschine f

**powdered** feingemahlen, gepulvert, pulverisiert,
staubförmig ~ **asbestos** Asbestmehl n ~ **brown
coal** Braunkohlenstaub m ~ **charcoal** Holzkohlenpulver n ~ **coal** Kohlen-mehl n, -staub
m, staubförmige Kohle f, Staubkohle f

**powdered-coal,** ~ **combustion** Kohlenstaubver-
brennung f ~ **firing** Brennstaub-, Kohlen-
staub-feuerung f ~ **flame** Kohlenstaubflamme
f ~ **operation** Kohlenstaubbetrieb m ~ **storage
bin** Kohlenstaubbehälter m

**powdered,** ~ **coke** Koksmehl n ~ **coloring** Staubfarbe f ~ **fiber** Faserpulver n ~ **fuel** Brennstaub m **~-fuel preparation plant** Staubaufbe-
reitungsanlage f ~ **glass** Glasstaub m ~ **ground
quartz** Quarzmehl n **~-iron core (or coil)**
Massekern m ~ **lignite** Braunkohlenstaub m ~
**melis** Melispuder n ~ **peat** Torf-mehl n, -staub
m ~ **semi-coke** Halbkoksstaub m ~ **slag** Zerfallschlacke f ~ **tin** Zinnpulver n

**powdering** Einstauben n, Feinmahlen n, Stäuben
n, Zerreiben n

**powdery** puderig, pulverartig, pulverförmig,
pulverig, staubartig, staubig ~ **snow** Pulver-
schnee m ~ **tin dross** Zinnpuder n ~ **tin skimming** Zinnpuder n

**power, to** ~ antreiben, potenzieren

**power** Antriebskraft f, Arbeit f, Arbeits-effekt m,
-kraft f, -leistung f, Befugnis f, Brennstärke f
(opt.), Effekt m, Energie f, Exponent m, Gewalt f, Kraft f, Kräfte pl, Leistung f, (brake)

Leistungsfähigkeit f, Leistungsvermögen n,
Macht f, Mächtigkeit f, Mandat n, Motor,
Ordnung f (math.), Potenz f, Trieb m, Ver-
größerungskraft f (opt.), Vermögen n

**power,** ~ **of accomodation** Akkomodations-
fähigkeit f, -kraft f ~ **to act** (law) Legitimation
f ~ **at altitude** Höhenleistung f ~ **of analysis**
Auflösungsvermögen n ~ **in antenna circuit**
Antennenkreisleistung f ~ **of appreciation of
differences in depth** Tiefenunterscheidungsver-
mögen n ~ **of apprehension** Auffassungsgabe ~
**of attorney** Vollmacht f ~ **of a control device**
Verstellkraft f einer Regelvorrichtung ~ **of
decision** Entschlußkraft f ~ **of expansion** Dehn-
kraft f ~ **of gravitation** Anziehungskraft f ~ **at
the ground** Bodennähendruck m ~ **of movement** Regungskraft f ~ **of observation** Beobach-
tungsgabe f ~ **off** mit abgestelltem Motor m
~ **on** mit arbeitendem Motor m ~ **of operating
piston** Kraft f am Schaltkolben, am Schalt-
kolben verfügbare Kraft f ~ **of radiating light**
Lichtausstrahlungsvermögen n ~ **of of radio-
transmitter** Sonderleistung f ~ **of rebound (or
recoil)** Springkraft f ~ **of refraction** Brechkraft
f ~ **of resistance** Aufhaltekraft f, Widerstandskraft f ~ **of resisting heat** Feuerbeständigkeit f
~ **of selection** Unterscheidungsvermögen n ~
**of set** Mächtigkeit f einer Menge (math.) ~ **of
sight** Sehvermögen n ~ **of stereoscopic vision**
räumliches Sehvermögen n ~ **of ten** Zehnerpotenz f ~ **of a transmitter** Senderleistung f ~
**per unit of displacement** Hubraum-, Literleistung f ~ **per unit area** Leistung f pro Ober-
flächenheit ~ **per unit surface** Flächenleistung f
~ **of vision** Sehvermögen n

**power,** ~ **absorption** Leistungs-aufnahme f, -be-
darf m **~-actuated** kraftbetrieben ~ **agitator**
Rührwerk n mit Motorantrieb ~ **amplification**
Leistungsverstärkung f **~-amplification rate**
Verstärkungsgrad m **~-amplification ratio** Lei-
stungsverstärkungsverhältnis n, Verstärkungs-
verhältnis n ~ **amplifier** Endverstärker m,
Kraftverstärker m, Leistungs-stufe f, -ver-
stärker m, Nachverstärker m **~-amplifier stage**
Endverstärkerstufe f **~-amplifier tube (or valve)**
Großverstärkerröhre f ~ **anode** Großleistungs-
anode f ~ **approach** Anflug m mit dem Motor
noch an **~-area ratio** Flächenleistung f ~ **arm**
Kraftarm m **~-assisted control system** Servo-
Steueranlage f ~ **attenuation** Leistungsverlust
m ~ **available** verfügbare Leistung f ~ **available
for continous cruising** verfügbare Dauerleistung
f (aviat.) ~ **balance** Energieausgleich m
**~-balancing jig (or rig)** Auswuchtgroßvor-
richtung f ~ **bell** Starkstromwecker m ~
**(switch) board** Kraftschalttafel f ~ **bogie** Drehgestell n (einer Maschine) ~ **brake** Biegestanze
f, Bist m, Servobremse f ~ **break** Abkantpresse
f ~ **breed reactor** Leistungsbrutreaktor m ~
**breeder** Leistungsbrutreaktor m ~ **cable** Stark-
stromkabel n ~ **canal** Fluder m ~ **characteristic
curve (or line)** Leistungskennlinie f ~ **circuit**
Arbeitsstromkreis m, Kraftleitung f, Stark-
stromleitung f ~ **climb** Antriebsanstieg m ~
**coefficient** Leistungsbedarfszahl f ~ **component**
Wattkomponente f, Wirk(spannungs)kompo-
nente f ~ **condenser** Kondensator m für große
Leistung **~-connection plug** Kraftsteckdose f

~ **constant** Leistungskonstante *f* ~ **consumption** Energieverbrauch *m*, Kraft-bedarf *m*, -verbrauch *m*, Leistungs-aufnahme *f*, -bedarf *m*, -entnahme *f* ~ **control** Drosselventil *n*, Leistungsregelung *f*, (apparatus) Leistungsregler *m* ~ **control assembly** Kraftsteuergerät *n* ~ **control handle** Steuerhebel *m* zum Heben und Senken des Querbalkens ~ **control lever** Leistungs-Gashebel *m* ~ **control rod** Leistungsregelstab *m* ~ **converter** Umformer *m* ~ **cord** Netzkabel *n*, -schnur *f* ~ **corn sheller** Maiskolbenschäler *m* mit Kraftbetrieb ~ **corrected to standard** INA-Leistung *f*, Leistung *f* nach Internationaler Normalatmosphäre ~ **craft** Kraftfahrzeug *n* ~ **cross-feed** selbsttätiger Planzug *m* ~ **cross-traverse** Querselbstgang *m* ~ **current** Starkstrom *m* ~ **curve** Leistungskurve *f* ~ **delivery** Leistungsabgabe *f* ~ **demand** Energiebedarf *m*, Kraftanforderung *f* ~ **density** Leistungsdichte *f* ~ **detector** Kraftdetektor *m* ~ **device adjusting the final control element** Stellmotor *m* ~ **diffraction** Kristallpulver *n* ~ **dispatcher** Schaltwart *m* ~ **dissipation** Leistungsbedarf *m*, -verbrauch *m*, -zerstreuung *f* ~ **distribution panel** Stromverteilertafel *f* ~ **distribution system** Stromverteilung *f* ~ **dive** Vollgassturzflug *m* ~ **divider** Leistungsteiler *m* ~ **drain** Stromverbrauch *m* ~ **drift** Leistungsabweichung *f* ~ **drill** elektrische Bohrmaschine *f*, Standbohrmaschine *f* ~ **drive** Kraftantrieb *m* ~ **drive chain** Transmissionstreibkette *f* ~**-drive transmission** Verteilergetriebe *n*
**power-driven** betrieben mit Motor *m*, kraftangetrieben, kraftbetrieben, mit eigenem Kraftantrieb *m* ~ **elevating lever** Hebel *m* für den Vertikalselbstgang ~ **machine for constructional engineering** Eisenbearbeitungsmaschine *f* für Kraftbetrieb ~ **model** Motorflugmodell *n* ~ **system** System *n* mit Maschinenwählern oder Motorwählern ~ **winch** Motorwinde *f*
**power,** ~ **economy** Energiewirtschaft *f* ~ **efficiency** Kraftleistung *f*, Leitungsverbrauch *m* ~ **elevation of table** Selbstgang *m* für die Höhenstellung des Tisches ~ **end of the engine** Kraftabgabeseite *f* des Motors ~ **engineering** Starkstromtechnik *f* ~ **equalizer** Echofalle *f* ~ **equation** Arbeitsgleichung *f* ~ **equipment** Stromlieferungs-, Stromversorgungs-anlage *f* ~ **exerting thread** Kraftgewinde *n* ~ **expanded in cooling** Leistungsaufwand *m* für Kühlzwecke ~ **expenditure** Leistungsaufwand *m* ~ **exponent** Potenzexponent *m* ~ **factor** Leistungsfaktor *m*, Phasenmeter *m*, Verlustwinkel *m* ~ **factor of grid-plate capacity** Gitteranodenkapazität *f* ~**-factor correcting capacitor** Phasenschieberkondensator *m* ~**-factor improvement** Blindstromkompensation *f* ~**-factor indicator** Leistungsfaktorzeiger *m* ~**-factor meter** Leistungsfaktor-messer *m*, -zeiger *m*, Phasenmesser *m* ~ **failure** Starkstromstörung *f* ~ **failure flag** Ausfall-Warnzeichen *n* ~**-failure point** Leistungsausfallpunkt *m* ~ **feed** Kraftvorschub *m* ~**-feed lock** Selbstgangeindrückrad *n* ~**-feeding attachments** Vorschubapparate *pl* ~ **flight** Kraft-, Motor-flug *m* ~ **frequency** Netzfrequenz *f* ~ **fuel** Betriebsstoff *m* ~ **function** Kraftfunktion *f* ~ **gain** Leistungsverstärkung *f* ~ **gas** Dowson-, Kraft-, Saug-, Treib-gas *n* ~**-gas cell** Triebgas-

zelle *f* ~ **generation** Krafterzeugung *f* ~ **generator** Stoßgenerator *m* ~ **glider** Motorsegler *m* ~ **gluing fixture** Leimgroßvorrichtung *f* ~ **grid-current detector** Kraftaudion *n* ~**-grid detector** Audionkraft *f*, Kraftaudion *f* ~ **ground wire** geerdeter Nulleiter *m* ~ **hack saw** mechanische Bogensäge *f* ~ **hammer** Kraft-, Maschinen-hammer *m*, mechanischer Hammer *m* ~ **hammers** Fallwerk *n* ~**-handling capacity** (transistor) Belastbarkeit *f*, Tonfrequenzbelastung *f* ~**-handling capacity of capacitor** Kondensatorleistung *f* ~**-harnessing equipment** Energiegewinnungseinrichtung *f* ~ **hay (press) baler** Motorheupresse *f* ~ **holder** Bevollmächtigter *m* ~ **house** Krafthaus *n*, Kraftstation *f*, Maschinenhaus *n* ~**-house crane** Maschinenhauskran *m* ~ **increment** Leistungsaufwand *m* ~**-indicator** Leistungszeiger *m* ~ **indicator relay** Unterspannungsrelais *n* ~**-inducted noise** induzierte Geräusche *pl*, Starkstromgeräusch *n* ~**-induction noise** induziertes Geräusch *n*, Starkstromgeräusch *n* ~ **input** aufgenommene Leistung *f*, Leistungsaufwand *m*, zugeführte Leistung *f* ~ **installation** Kraftbetrieb *m* ~ **isolating switch** Leistungstrennschalter *m* ~ **jack** Batterieklinke *f* ~ **jet** Anreicherungsdüse *f* für Volleistung, Spritzdüse *f*, Treibstrahl *m* ~ **knitting loom** Strickmaschine *f* ~ **landing** Landung *f* mit noch nicht abgestorbenem Motor ~ **lathe** Hochleistungsdrehbank *f* 3/2 ~ **law** 3/2-Gesetz *n* ~ **lead** Speiseleitung *f*, Starkstromzuführung *f*, Stromversorgung *f* ~ **lead-in** Anzapfleitung *f* ~ **leads** Stromzuführungsleitung *f* ~ **let-down** Sinkflug *m* mit gedrosseltem Triebwerk ~ **level** Leistungspegel *m* ~ **lift** Elektrohub *m*, Krafttheber *m* ~**-lift tractor disk plow** Schlepperscheibenpflug *m* mit Kraftaushebung ~**-lift tractor drill** Schlepperdrillmaschine *f* mit Kraftaushebung ~**-lift tractor plow** Schlepperpflug *m* mit Kraftaushebung ~**-lift truck** Elektrohubwagen *m* ~ **line** Haupt-, Kraft-, Starkstrom-, Überland-leitung *f* ~**-line carrier** Trägerfrequenzübertragung *f* auf Hochspannungsleitungen (TFH) ~**-line crossing** Starkstromkreuzung *f* ~**-line filter** Netzfilter *m* & *n* ~**-line hum** Netzbrummen *n* ~ **loading** Belastung *f* pro Pferdestärke, Leistungsbelastung *f* ~ **loom** mechanischer Webstuhl *m* ~ **loss** Energie-, Kraft-, Leistungs-verlust *m*, Verlustleistung *f* ~ **loudspeaker** Groß(flächen)lautsprecher *m* ~ **magnet** Kraftmagnet *m* ~ **magnification** Leistungsverstärkung *f* ~ **main** Kraftleitung *f* ~ **margin** Leistungsüberschuß *m* ~ **measurement** Leistungsmessung *f* ~**-measuring device** Leistungsmeßgerät *n* ~ **meter** Leistungsmeßgerät *n* ~ **mixture** Kraftmischung *f* ~ **number** Lichtstärke *f* ~ **off** antriebslos; Leerlauf *m*, leerlaufender Motor *m* ~ **on** laufender Motor *m*, Vollgas *n* ~**-operated** kraftbetätigt ~**-operated air chuck** kraftbetätigtes Preßluftspannfutter *n* ~**-operated calender** Kraftkalander *m* ~**-operated chuck** Kraftspannfutter *n* ~**-operated molding machine** Kraftformmaschine *f* ~ **oscillator** Kraftschwingröhre *f* ~ **output** abgegebene Leistung *f*, Ausgangs-, Nenn-, Sende-leistung *f*, Kraftausbeute *f*, Leistungsentnahme *f* ~ **output limitation** Leistungsbegrenzung *f* ~ **output per liter** Liter-

leistung *f* **~-output stage** Kraftendstufe *f* ~
**overlap** Kraftüberlappung *f* ~ **pack** Netzanschlußgerät *n*, -gerät *n*, -teil *m* ~ **package**
Antriebsaggregat *n* ~ **panel** Kraftschalttafel *f*
~ **pentode** (cathode grid connected with control
grid rather than with cathode) Doppelsteuerröhre *f* ~ **pile** Leistungsreaktor *m* ~ **piping**
Hochdruckrohrleitung *f* ~ **plane** Motorflug-
zeug *n* ~ **plant** Kraftanlage *f*, Kraftwerk *n*,
Maschinensatz *m*, Starkstromanlage *f*, Stoßanlage *f*, Triebwerk *n*, Triebswerk(s)anlage *f* ~
**plant with facilities for pumping to storage pond**
Speicherwerk *n* ~ **plant without storage**
**facilities** Laufwerk *n* **~-plant control** Trieb-
werksbediengestänge *n* **~-plant gas** Rauchgas *n*
~ **plug** Kraftstromstecker *m* ~ **press** Presse *f* ~
**product** Potenzprodukt *n* ~ **production** Krafterzeugung *f* **~-propelled vehicle** Fahrzeug *n* ~
**pump** angetriebene Pumpe *f*, Motorpumpe *f* ~
**punching and shearing machinery** Loch- und
Schermaschinen *pl* für Kraftbetrieb ~ **radiated**
**in main direction** gerichtete Leistung *f* (bei
Antennen) ~ **rail** Arbeitsschiene *f* ~ **ram**
**magneto** Rammenmagnetzünder *m* ~ **range**
Leistungsbereich *n* ~ **ranges** Dimensionierung
*f* ~ **rating** Kraftbemessung *f*, Krafteinschätzung
*f*, Nennleistung *f* ~ **ratio** Leistungsverhältnis *n*
~ **reactor** Leistungsreaktor *m* ~ **rectifier** Netzgleichrichter *m* ~ **rectifying valve** Gleichrichterröhre *f*, Hochleistungsgleichrichterröhre *f*
**~-reference level** Leistungsbedarfspegel *m* ~
**relay** Netz-, Schalt-relais *n*, Steuerschütz *m* ~
**required** Kraftbedarf *m* ~ **required for take-off**
Abflugleistung *f* ~ **required across the test**
Leistungsbedarf *m* während des Versuches ~
**requirement** Arbeits-, Leistungs-bedarf *m* ~
**requirement during test** Leistungsbedarf *m*
während des Versuches ~ **requirements** Energiebedarf *m*, Kraftbedarf *m* ~ **reserve** Kraftreserve
*f* ~ **resistor** Hochleistungswiderstand *m* ~
**riddle sifter (or riddler)** Schüttelsiebmaschine *f*
~ **ring** Kran *m* mit Gasmaschinen- oder Diesel-
maschinenantrieb ~ **ringing** Anruf *m* mit
Maschinenstrom ~ **saving** Kraftersparnis *f* ~
**saw** Kraft-, Maschinen-säge *f* ~ **section** Lei-
stungsteil *m* ~ **sensitivity** Leistungsempfind-
lichkeit *f* ~ **series** Potenzreihe *f* ~ **series**
**expansion** Potenzreihenentwicklung *f* ~ **set**
Maschinensatz *m* ~ **shears** Schnittstanze *f* ~
**shovel** Löffelbagger *m* ~ **slush pump** Kraftspül-
pumpe *f* ~ **socket** Kraftstecker *m* ~ **source**
Stromquelle *f* ~ **spectrum** Leistungs-, Potenz-
spektrum *n* ~ **spin** Trudeln *n* mit noch nicht
erstorbenem Motor (aviat.) ~ **squeezer** Preßformmaschine *f* für Luftdruckbetrieb, Preßformmaschine *f* mit selbsttätigem Abhub
**~-squeezing molding machine** Preßformmaschine *f* mit selbsttätigem Abhub **~-squeezing**
**turnover machine** Preßformmaschine *f* mit
Wendeplatte ~ **stage** Kraftstufe *f*, nachgeschaltete Stufe *f*, Steuerstufe *f* ~ **stall** überzogener Flug *m* mit Motor an **~-stall landing**
Landung *f* mit noch nicht erstorbenem Motor
~ **standing wave ratio** Stehwellenverhältnis *n* ~
**station** Elektrizitätswerk *n*, Eltwerk *n*, Energie-
zentrale *f*, Kraft-station *f*, -werk *n*, -zentrale *f*,
Maschinenhaus *n*, Stromerzeugungsanlage *f*
~ **steering** Servolenkung *f* **~-stitching fixture**

Heftgroßvorrichtung *f* ~ **stroke** Arbeits-hub *m*,
-takt *m*, Ausdehnungshub *m*, Expansionshub
*m*, Krafthub *m*
**power-supply** Energievorrat *m*, Kraftquelle *f*,
Kraftversorgung *f*, Leistungszufuhr *f*, Netz *n*,
Netzanschluß *m*, (mains-operated) Netzteil *m*,
Speise *f*, Stoßanlage *f*, Strom-führung *f*, -ver-
sorgung *f*, -zuführung *f* ~ **assembly** Netzgerät *n*
~ **connection** Kraftanschluß *m* ~ **plate** Netzzu-
führungsplatte *f* ~ **plug** Netzstecker *m* ~
**plug-in unit** Netzzuführungseinschub *m* ~
**synchronization** Netzsynchronisierung *f* ~
**system** Starkstromnetz *n* ~ **trailer** Stromver-
sorgungswagen *m* ~ **unit** Netz(anschluß)gerät *n*
~ **variation** Stromversorgungsschwankung *f*
~ **voltage** Netzspannung *f* ~ **voltage from mains**
Netzwechselspannung *f* ~ **voltage switch** Netz-
spannungsumschalter *m*
**power,** ~ **sweeper** Kehrmaschine *f* ~ **switch**
Hauptschalter *m* ~ **switch machine** Weichenantrieb *m* (electr.) ~ **synchro** Synchronomotor
*m* ~ **system** Kraftanlage *f* ~ **take-off** Außenan-
trieb *m*, Zapfwellenantrieb *m* ~ **take-off guard**
Zapfwellenschutzdeckel *m* ~ **take-off rain**
**cover** Zapfwellenschutz *m* ~ **take-off shaft** Antriebswelle *f* ~ **take-up reel** Antriebshaspel *f* ~
**test run** Probelauf *m* bei Vollast, Vollprobelauf
*m* ~ **tire pump** Motorluftpumpe *f* ~ **train**
**assemblies** Kraftübertragungsaggregate *pl* ~
**transformer** Leistungs-transformator *m*, -über-
trager *m*, Netztrafo *m*, Netztransformator *m* ~
**transmission** Energie-, Kraft-, Leistungs-über-
tragung *f* **~-transmission cable** Kraftkabel *n*
**~-transmission plant** Transmissionsanlage *f*
**~-transmission system** Kraftübertragungsan-
lage *f* ~ **transmitter** Kraftumlenker *m* **~-trans-**
**mitting chain** Transmissionstreibkette *f* ~
**traverse** Kraftverstellung *f*, selbsttätiger Vorschub *m* ~ **tube** End-, Großleistungs-, Kraft-,
Sende-röhre *f* ~ **(amplifier) tube** Leistungs-
röhre *f* **(high)** ~ **tube** Hochleistungsröhre *f*
**~-tube rack** Senderöhrengestell *n* ~ **unit** Kraftwerk *n*, Leistungseinheit *f*, Maschinensatz *m*,
Triebwerkeinheit *f* **~-unit mounting** Umformer-
aufhängerahmen *m* ~ **valve** Großleistungs-,
Kraft-, Leistungs-röhre *f* ~ **Venturi tube** Kraft-
saugrohr *n*, Kraft-Venturitube *f* ~ **waste** Kraft-
verschwendung *f* ~ **water** Druck-, Preß-wasser
*n* ~ **weight ratio (lb/lb)** Schubgewicht *n* (Kg/Kp)
~ **winch** Kraftwinde *f* ~ **winding** (of transducer)
Arbeitswicklung *f*, Leistungswicklung *f*
**powered,** ~ **aileron** kraftbetätigtes Querruder *n*
~ **flying** Motorfliegerei *f* ~ **glider** Motorgleiter
*m* ~ **phase** Brenndauer *f* (g/m) ~ **wing** Trieb-
flügel *m*
**powerful** gewaltig, kräftig, leistungsfähig, mächtig, stark, wirksam, wuchtig ~ **wide-aperture**
**lens** lichtstarke Linse *f*
**powerless** kraftlos, machtlos
**Poynting,** ~ **vector** Poyntingscher Vektor *m*
**~'s vector** Poyntingscher Vektor *m*, Strahlungsvektor *m*
**pozzuolana** Pozzolan *n*, Pozzolanerde *f* ~ **cement**
Pozzolanzement *m*
**PPI** (plan position indicator) Rundsichtradar-
gerät *n*
**practicability** Anwendbarkeit *f*, Ausführbarkeit
*f*, Brauchbarkeit *f*, Durchführbarkeit *f*

**practicable** ausführbar, durchführbar, fahrbar, gangbar, praktisch, wirksam **most ~ type** gangbarste Sorte *f*
**practical** angewandt, anstellig, anwendbar, konstruktiv, praktisch, praktisch verwendbar, zweckmäßig **~ amplification** Nutzanwendung *f*, Verstärkbarkeitsgrenze *f* **~ ceiling** Betriebs-, Gebrauchs-gipfelhöhe *f*, praktische Gipfelhöhe *f*, Versuchsgipfelhöhe *f* **~ data** Erfahrungstatsache *f* **~ embodiment** (of an invention) Verwirklichungsform *f* **~ experience** Betriebserfahrung *f*, Sachkenntnis *f*, Sachkunde *f* **~ facts** Erfahrungssache *f* **~ man** Praktiker *m* **~ operating diagram** Wirkschema *n* **~ range** praktische Reichweite *f* **~ research** Zweckforschung *f* **~ science** Realwissenschaft *f* **~ stage** (fully developed) Betriebsreife *f* **~ system of measurement** praktisches Meßsystem *n* **~ test** praktischer Versuch *m* **~ unit** praktische Einheit *f* **~ units system** praktisches Unitätssystem *n* **~ working diagram** Wirkschema *n*
**practice, to ~** ausüben, handeln, praktizieren, verfahren **to ~ one's profession** Beruf *m* ausüben
**practice** Anwendung *f*, Arbeitsweise *f*, Ausübung *f*, Betrieb *m*, Betriebsweise *f*, Brauch *m*, Erfahrung *f*, Gebrauch *m*, Praxis *f*, Verfahren *n*, Wesen *n* **~ buzzer** Übungssummer *m* **~ device** Übungsgerät *n* **~ field** Übungsanlage *f* **~ fire** Abkomm-, Schul-schießen *n* **~ firing** Übungsschießen
**practiced** ausgeübt, praktiziert
**practices** Betriebsvorschrift *f*
**Prandtl group** Prantlsche Zahl *f*
**praseodymium** Praseodym *n*
**pre-accelerator** Vorbeschleuniger *m*
**preadjusted** vorjustiert
**preadjustment of homing device** Voreinweisung *f* (g/m)
**preadvice call** Voranmeldungsgespräch *n*
**preamble** Einleitung *f*, Kopf *m* eines Telegramms, Telegrammkopf *m*, Vorwort *n* **~ on radio signal (or message)** Funkspruchkopf *m* **~ of specification** Einleitung(s)formel *f* zur Beschreibung
**preamplification** Vorverstärkung *f*
**preamplifier** Vorverstärker *m* **~ stage** Vorverstärkerstufe *f*
**prearranged call** (with operator) Gespräch *n* mit Voranmeldung
**prearranging worm** Vorordnerschnecke *f*
**préavis, ~ call** Gespräch *n* mit Voranmeldung **~ fee** Zuschlaggebühr *f* für Voranmeldung
**prebody, to ~** vorverdicken
**precalculated** vorberechnet **~ position** Sollage *f*
**precast, to ~** vorfabrizieren
**precast, ~-concrete pile** Betonfertigpfahl *m* **~ construction** Fertigbauweise *f* **~ floor** Fertigdecke *f* **~ reinforced concrete (structural) unit** Stahlbetonfertigteil *m* **~ slab** Fertigbetonplatte *f* **~ slab unit** Kassettenplatte *f*
**precaution** Vorkehrung *f*, Vorsicht *f*, (measure) Vorsichtmaßnahme *f* **~ against skidding** Gleitschutz *m*
**precautionary, ~ measure** Vorkehrung *f*, Vorsichtsmaßregel *f* **~ steps** sicherstellende Vorkehrungen *pl*
**precautions, to take ~** vorbauen, vorkehren

**precede, to ~** voraufgehen, vorgeben
**precedence** Dringlichkeitsstufe *f*, Vorrang *m* **~ call** Vorranggespräch *n* **~ indicator** Voranganzeiger *m*
**preceding** vorangehend, vorausgehend, vorstehend **~ page** umseitig **~ pass** Vorkaliber *n* **~ stage** Vorstufe *f*
**precept** Begriff *m*
**precess, to ~** präzedieren
**precessability** Präzedierbarkeit *f*
**precession, ~ of a gyro** Präzession *f* eines Kreisels **~ (in gyroscope)** Abwanderung *f* **~ of sound** Vorlauf *m* **~ of sound recording** Filmvorlauf *m*
**precession, ~ amplifier** Präzessionsverstärker *m* **~ dial** Präzessionszifferblatt *n* **~ oscillation of a gyropendulum** Präzessionsschwingung *f* des Kreisels **~ motor** Aufrichtmotor *m*
**prechamber, ~ compression ignition engine** Vorkammer-Dieselmotor *m*
**precharge** Vorbeladung *f*
**precharged (or preloaded) plate** (battery) vorgeladene oder formierte Platte *f*
**precheck switch** Vorprüfschalter *m*
**precinct** Bereich *m*, Bezirk *m*, Revier *n*, Weichbild *n*
**precious** kostbar, (gems or stones) edel, wertvoll **~ metal** edles Metall *n* **~-metal assaying** Edelmetallprobe *f* **~-metal collector** Edelmetallsammler *m* **~-metal contacts** Edelkontakte *pl* **~-metal refiner** Edelmetallscheider *m* **~ opal** Edelopal *m*
**preciousness** Kostbarkeit *f*
**precipice** Abgrund *m*, Fluh *f*, Gehänge *n*, steiler Abfall *m*
**precipitability** Fällbarkeit *f*
**precipitable** abscheidbar, fällbar
**precipitant** Ausscheidungsmittel *n*, Fällmittel *n*, Fällungsmittel *n*, Niederschlagsmittel *n*
**precipitate, to ~** abscheiden, absetzen, ausfallen, ausflocken, ausscheiden, sich auscheiden, fällen, niederfallen, niederschlagen, präzipitieren, setzen, steil abfallen, übereilen **to ~ in crystal form** soggen
**precipitate** abgeschiedener Stoff *m*, Abscheidung *f*, Abscheidungsstoff *m*, Ausfall *m*, Ausfällen *n*, Ausscheidung *f*, Fällprodukt *n*, Niederschlag *m*, Präzipitat *n* **~ of iron** Eisenniederschlag *m* **~ of lime** Kalkniederschlag *m* **~ of lime-soap** Kalkseifenniederschlag *m*
**precipitated, to be ~** niederfallen **~ chalk** Schlämmkreide *f* **~ lead** Bleiniederschlag *m*
**precipitating** Fällung *f* **~ agent** Fällungs-, Niederschlags-mittel *n* **~ reagent** Fällungsreagenz *f* **~ vessel** Fällkessel *m*, Niederschlagsgefäß *n*
**precipitation** Ausfall *m*, Ausfällen *n*, Ausfällung *f*, Auslagerung *f*, Ausscheidung *f*, Fällung *f*, Nieder-schlag *m*, -schlagsmenge *f*, -schlagung *f*, Präzipitation *f*, Regen *m*, Regenfall *m*, steiler Abfall *m*, Übereilung *f* **~ by electrolysis** elektrolytische Fällung *f* **~ of excess copper** Kupferaushärtung *f*
**precipitation, ~ box** Fäll-, Niederschlags-kasten *m* **~ effect** Absetzeffekt *m* **~ forecast** Niederschlagsprognose *f* **~ fractination** Fällungsfraktionierung *f* **~ hardening** Ausscheidungshärtung *f*, Metallwärmebehandlung *f* (aviat.), Seigerungshärtung *f* **~ heat treatment** künst-

liche Alterung f ~ **number** Ausfällungszahl f,
Niederschlagsziffer f ~ **process** Niederschlags-
arbeit f ~ **static** Funkströmung f durch Nieder-
schläge ~ **tank** Absatz-, Abscheide-gefäß n ~
**vat** Zementierfaß n
**precipitator** Ausfällapparat m, Niederschlags-
apparat m
**precipitous** abschüssig, plötzlich, schroff
**precipitron** Ausfällapparat m (electron.)
**précis** Kurzbeschreibung f
**precise** bestimmt, genau ~ **fitting** paßgerecht ~
**hole pitch** genaue Lochtasterteilung f ~ **mechanical work** feinmechanische Arbeit f ~
**tuning** scharfe Abstimmung f
**precision** Bestimmtheit f, Exaktheit f, Genauigkeit f, Präzision f, Schärfe f ~ **of test** Meßgenauigkeit f
**precision,** ~ **adjustment** Feineinstellung f
~-**adjustment drive** Feinstelltrieb m ~-**adjustment fire** genaues Einschießen n ~ **altimeter**
Feinhöhenmesser m ~ **approach radar (PAR)**
Präzisions-Anflug-Radar n (PAR) ~ **balance**
chemische Waage f, Präzisionswaage f ~
**bombing** gezielter Bombenwurf m ~ **boring
machine** Feinbohrblock m, Präzisionsbohr-
maschine f ~ **caliper** Präzisionskluppe f ~
**castings** Genauigkeitsguß m ~ **center** Präzisions-
körnerspitze f ~ **chuck** Präzisionsfutter n ~
**contact** Feinkontakt m ~ **control valve** Fein-
steuerventil n ~ **cross-winding** Präzisionskreuz-
spulung f ~ **cut** Genauigkeitsschnitt m ~ **drill**
Feinbohrmaschine f ~ **drilling** Feinbohren n ~
**engineering** Feinwerktechnik f ~ **forging** Prä-
zisionsschmieden n ~ **gauge block** End-,
Parallel-maß n ~ **gear** Genauigkeitsverzahnung
f ~ **gearing** Präzisionsgetriebe n ~ **grinding**
Feinschleifen n ~ **high-voltage plug-in unit**
Präzisionshochspannungseinschub m ~ **hoist**
Feinhubsteuerung f ~ **hoisting gear** Feinhub-
werk n ~ **honing** Feinziehschleifen n ~ **instrument** Präzisionsinstrument n ~-**instrument oil**
Instrumentöl n ~ **landing** Ziellandung f ~
**lathe** Feinmechanikerdrehbank f ~ **lens** Meßobjektiv n ~ **leveling** Präzisionsnivellement n
~ **limit** Genauigkeitsgrenze f ~ **load regulator**
Feinsteuerdruckhalter m ~ **measurement** Feinmessung f ~ **measuring instruments** Feinmeß-
geräte pl ~ **mechanics** Feinmechanik f ~ **metal
plateholder** Präzisionsmetallkassette f ~ **milling
machine** Feinstfräsmaschine f ~ **plug gauge**
Prüflehrdorn m ~ **plug resistance** Präzisions-
stöpselwiderstand m ~ **production boring
machine** Produktionsfeinbohrmaschine f ~ **ray
condenser** Feinstrahlkondensor m ~ **regulator**
Feinregler m ~ **resistance for measurement
purposes** Meßwiderstand m ~ **revolving stage**
Präzisionsdrehtisch m ~ **round steel** Präzisions-
rundstahl m ~ **scale** Feinskala f ~ **selector
switch** Meßstellenumschalter m ~ **slit** Mikrometerspalt m ~ **slitter** Präzisionsritzapparat m
~ **slotter** Präzisionsschlitzstanze f ~ **spin** Prä-
zisionsrudeln n ~ **sweep** Feinmeßbasis f ~
**tool** Genauigkeits-, Präzisions-werkzeug n ~
**trigger** Präzisionsauslöseimpuls m ~ **turn** Prä-
zisions-kurve f, -waage f ~ **turning** Feindrehen
n ~ **turning lathe** Feindrehbank f ~ **turning
operation** Präzisionsdrehvorgang m ~-**type
centrifugal governor** Fliehkraftfeinregler m

~-**wire-drawing works** Genaudrahtzieherei f ~
**work** Genauigkeitsarbeit f
**preclarification** Vorklärung f
**pre-clarified** vorgeklärt
**preclassification** Vorklassierung f
**pre-cleaner** Vorreiniger m
**preclude, to** ~ ausschließen, verhindern, vorbeugen
**preclusion period** Präklusivfrist f
**precoat filter** Hilfsschichtfilter n
**precombustion** Vorverbrennung f (Dieselmotor)
~ **chamber** Vorkammer f, Vorverbrennungs-
raum m (Dieselmotor) ~ **chamber engine** Vor-
kammermaschine f
**pre-compounded** vorimprägniert
**precompression** Vorverdichtung f ~ **in the
boundary layer** Grenzschichtbelüftung f
**precondition** Vorbedingung f
**preconsolidation** Vorverdichtung f ~ **load** Vor-
belastung f
**precool, to** ~ vorkühlen
**precooler** Raumkühler m, Vorkühler m
**pre-crushed material** vorgebrochenes Gut n
**pre-curing** Vorvulkanisation f
**precursor** (Generator) Unterschwingen n, Vorläufer m
**predecessor** Vorderglied n, Vorgänger m, Vorläufer m
**predefecate, to** ~ vorscheiden
**predefecation** Vorscheidung f ~ **tank** Vorscheide-
pfanne f
**pre-deflected spring** vorgespannte Feder f
**predetermination** Voraus-berechnung f, -be-
stimmung f
**predetermine, to** ~ vorausberechnen, vorherbe-
stimmen
**predetermined** vorbestimmt; Sollbruchstelle f ~
**moment** bestimmter Zeitpunkt m
**predetonation** vorzeitige Explosion f oder Detonation f ~ **path** Prädetonationsweg m
**predicate, to** ~ die Ansicht f vertreten, behaupten
**predictable** bestimmbar
**predicted,** ~ **fire** Planschießen n ~ **position** Vorhaltepunkt m ~ **position system** Vorhalteverfahren n
**predicting interval** Vorhaltestrecke f
**prediction** Voraussage f ~ **computer** Vorhalt-
rechengerät n ~ **mirror** Vorhaltspiegel m ~'s
theoretischer Wert m
**predictor** Artilleriekommandogerät n Kommandogerät n, Voraussager m, Zielrechen-
maschine f
**predig, to** ~ vorroden
**predischarge** Vorentladung f ~ **track** Vorent-
ladungskanal m
**predisposition** Anlage f
**predissociation** Prädissoziation f
**predistorter** Vorverzerrer m (Modulation)
**predistribution cylinder** großer Reibzylinder m
**predock** Vorbereitungsdock n
**predominance** Übergewicht n
**predominant** hervortretend, überwiegend, vorwiegend
**predomination** Vorherrschen n
**pre-dry, to** ~ vortrocknen
**pre-edition** Vorausgabe f
**pre-eminence** Vorrang m

**pre-emphasis** Akzentuierung *f* (acoust.) ~ **accentuation** Vorverzerrung *f*
**pre-equalization** Vorverzerrung *f*
**pre-establish, to** ~ vorher festsetzen
**pre-evaporation** Vorverdampfung *f*
**pre-evaporator** Vorverdampfer *m*
**pre-exhaust** Vorauspuff *m*
**pre-exposure** Vorbelichtung *f*
**prefab** Fertighaus *n*
**prefabricate, to** ~ vorfertigen
**prefabricated** vorfabriziert, zusammensetzbar ~ **house** Fertighaus *n* ~ **piece** Fertigbauteil *m* ~ **pipe** vorgearbeitete Röhre *f* ~ **walk-in cooler** zerlegbare Kühlzelle *f*
**pre-fabrication** Vorfertigung *f*
**preface, to** ~ einleiten
**preface** Kopf *m* eines Telegramms, Telegrammkopf *m*, Vorrede *f*, Vorwort *n*
**prefer, to** ~ bevorzugen, vorziehen **to** ~ **a claim (or point)** Anspruch *m* geltend machen
**preference** Bevorzugung *f* ~ **share** Vorzugsaktie *f*
**preferential,** ~ **rate** Vorzugspreis *m* ~ **recombination** bevorzugte Wiedervereinigung *f* ~ **tariff** Vorzugszoll *m* ~ **treatment** Vorzugsbehandlung *f*
**preferred** bevorzugt ~ **configuration** geordneter Zustand *m* ~ **crystallographic axis orientation** bevorzugte Kristallachsenrichtung *f* ~ **frame** Vorzugssystem *n* ~ **number** Normungszahl *f* ~ **numbers** irrationale Zahlen *pl* ~ **orientation** bevorzugte Richtung *f* ~ **share** Vorzugsaktie *f* ~ **value** Vorzugswert *m*
**prefilter** Vorfilter *n*
**prefiltered** vorgereinigt
**prefiltering section** Vorfilterteil *n*
**prefiltration** Vorfiltration *f*
**pre-firing** Voraktivierungszeit *f*, Vorfeuerung *f*
**prefix** Kennziffer *f*, Vorimpuls *m* ~ **of the metric system** Vorsatz *m* im metrischen System ~ **evaluator** Netzgruppenschalter *m*
**preflame,** ~ **ignition** Vorreaktion *f* im Brenngemisch ~ **reactions** Vorreaktion *f*
**preflight,** ~ **check** Abflugprobe *f* (rdo) ~ **inspection** Prüfung *f* auf Flugklarheit
**prefocus** Vorkonzentration *f*
**prefocusing** Anfangskonzentration *f*
**prefogging** Vorbelichtung *f* (photo)
**prefolded filter** Faltenfilter *n*
**prefolding** Vorfalzen *n*
**preform process** Vorformverfahren *n*
**preformed** vorgeformt ~ **precipitate** vorgebildeter Niederschlag *m* ~ **winding** Schablonenwicklung *f*
**preforming,** ~ **attachment** Vordralleinrichtung *f* ~ **press** Tablettiermaschine *f* ~ **scaffold (or stand)** Vorprofilierungsgerüst *n*
**pre-glow(ing) current** Vorglühstrom *m*
**pregnant solution** Mutterlösung *f*
**pre-grain, to** ~ vorkörnen
**preheat, to** ~ vorwärmen
**preheated** erhitzt, vorgewärmt
**preheater** Anwärmeapparat *m*, (boiler) Röhrenvorwärmer *m*, Vorerhitzer *m*, Vorwärmer *m* ~ **of mixture** Gemischvorwärmer *m* ~ **using flue gases** Rauchgasvorwärmer *m*
**preheating** Anheizen *n*, Vorerhitzung *f*, Vorwärmung *f* ~ **of air** Luftvorwärmung *f* ~ **of the blast** (smelting) Erwärmung *f* des Gebläse-

windes ~ **of gas mixture** Gemischvorwärmung *f*
**preheating,** ~ **bench** Vorwärmtisch *m* ~ **chamber** Vorwärmkammer *f* ~ **furnace** Vorwärmeofen *m* ~ **period** Vorglühwiderstand *m* ~ **plant** Vorwärmanlage *f* ~ **relay** Vorglührelais *n* ~ **resistance** Vorglühwiderstand *m* ~ **space** Vorwärmeraum *m* ~ **time** Anheizzeit *f* ~ **torch** Anwärme-, Heiz-brenner *m* ~ **tube** Vorwärmeröhre *f* ~ **zone** Vorwärmezone *f*
**prehnite** Kupholit *m*
**pre-ignition** Frühzündung *f*, Frühzündungsdruck *m*, Kerzenzündung *f*, vorzeitige Zündung *f*, Vorzündung *f* ~ **Diesel** Vorkammer-Diesel *m*
**pre-impulse sender** (generator) Vorimpulserzeuger *m*
**pre-ionization** Selbstionisation *f* ~ **potential** Anregungsspannung *f*
**pre-ionized** vorionisiert ~ **track** vorionisierte Spur *f*
**prejudice, to** ~ beeinträchtigen, verletzen
**prejudice** Beeinträchtigung *f*, Nachteil *m*, Voreingenommenheit *f*, vorgefaßte Meinung *f*, Vorurteil *m*
**preknock pulse** Vorimpuls *m*
**prelimed juice** vorgeschiedener Saft *m*
**preliminaries** Präliminarien *pl*, Vorbereitung *f*
**preliminary** vorläufig ~ **action** Vorbescheid *m*, Vorprüfungsverfahren *n* ~ **analysis** Vorprobe *f* ~ **blast** Vorblasen *n* ~ **blowing** Vorblasen *n* ~ **breaking** Vorbrechen *n*, Vorzerkleinerung *f* ~ **call** vorläufiger Anruf *m* ~ **charge** Vorbeladung *f* ~ **cleaning** Vorreinigung *f* ~ **coating process** Verstreichverfahren *n* (Klebstoff) ~ **container** Vorbehälter *m* ~ **crusher** Vorzerkleinerungsmühle *f* ~ **design** Vorentwurf *m* ~ **discussion** Rücksprache *f* ~ **drying** Vorentwässerung *f*, Vortrocknung *f* ~ **drying chamber** Vortrocknungskammer *f* ~ **dust extraction** Grobentstaubung *f* ~ **dust separator** Staubvorabscheider *m* ~ **engineering for the project** Ausarbeiten *n* des Vorprojektes ~ **examination** Vorprüfungsverfahren *n* ~ **expenses** Aufwand *m* für Vorbereitungsarbeiten ~ **experiment** Testversuch *m* ~ **fat-liquoring** Vorlickerung *f* ~ **feed valve** Vorfüllventil *n* ~ **filter** Vorabscheider *m*, Vorfilter *m*, Vorreiniger *m* ~ **finishing** Vorappretur *f* ~ **fire refining** Vorraffination *f* im Schmelzfluß ~ **focusing** Vorkonzentration *f* ~ **forging** Vorschmieden *n* ~ **frequency divider stage** Voruntersetzereinheit *f* ~ **fuel valve** B-stoffvorventil *n* ~ **grinder** Vorzerkleinerungsmühle *f* ~ **grinding** Vormahlen *n* ~ **heating** Vorglühen *n*, Vorwärmung *f* ~ **inspection** Vorrevision *f* ~ **investigation** Voruntersuchung *f* ~ **leaching** Vorlaugung *f* ~ **operation** Vorarbeit *f* ~ **peace** Präliminarfrieden *m* ~ **pickle** Vorbrenne *f* (metall.) ~ **preparation** Rohaufbereitung *f* ~ **proceedings (or procedure)** Vorverfahren *n* ~ **(or first) proof** Vorkorrektur *f* ~ **purification** Vorreinigung *f* ~ **reading** indirekte Ablesung *f* ~ **reduction** Vorreduktion *f* ~ **rinsing** Vorspülung *f* ~ **roasting** Vorröstung *f* ~ **search** Vorprüfungsverfahren *n* ~ **separation** Vorabscheidung *f* ~ **shaft** Vorschacht *m* ~ **shaping** Vorprofilieren *n* ~ **spectrum** Vorspektrum *n* ~ **stage** Anfahrstufe *f* (g/m) ~ **stop** Vortakt *m* ~ **study** Vorstudie *f* ~ **tensioned spring**

vorgespannte Feder *f* ~ **test** Einschmelzprobe *f*,
Vorprobe *f*, Vorversuch *m* ~ **treatment** Vorbehandlung *f* ~ **trial** Vorversuch *m* ~ **visit** Instruktionsreise *f* ~ **warning position** Vorwarnstellung *f* ~ **washing** Vorwäsche *f* ~ **work** Vorarbeit *f*
**preliming** Vorscheidung *f*
**pre-loaded** vorgespannt, vorimprägniert
**prelubricated ball bearing** Kugellager *n* mit Fettkammer für Dauerschmierung
**prelubricating pump** Vorschmierpumpe *f*
**premature** frühzeitig, verfrüht, vorzeitig; (torpedo) Frühzünder *m* ~ **burst** (of shell) Frühkrepierer *m*, Frühsprenger *m* ~ **disconnection** vorzeitige Trennung *f* oder Unterbrechung *f* ~ **explosion** vorzeitiges Losgehen *n* ~ **ignition** Frühzündung *f*, Vorzündung *f* ~ **release** vorzeitige Auflösung *f* ~ **shot** Frühzündung *f* ~ **wear** frühzeitige Abnutzung *f*
**prematurely bursting shell** Frühzerspringer *m*
**prematureness** Früh-, Vor-zeitigkeit *f*
**premeditated violation of duty** vorsätzliche Verletzung *f* einer Dienstpflicht
**premeditation** Vorsatz *m*
**premelting furnace** Vorschmelzofen *m*
**pre-milled** vorgefräst
**premise, to** ~ vorausschicken
**premise** Voraussetzung *f*
**premises** Grund *m* und Boden *m*, Grundstück *n*
**premium** Agio *n*, Prämie *f*, Preis *m*, Zugabe *f* ~ **blend** Super *n* (Benzin) ~ **bonus system** Prämienlohnsystem *n* ~ **installment (or rate)** Prämienrate *f* ~ **notice writing** Prämienrechnungsschreibung *f*
**premix, to** ~ (viscose) anteigen
**premixed** vorgemischt
**premixing** Vormischung *f*
**pre-modifications** Vorwegänderungen *pl* (info proc.)
**premodulation** Vormodelierung *f*
**pre-onset corona streamer** Korona *f* der Zündung
**preoscillation current** Anschwingstrom *m*
**prepaid** franko (frei) **not** ~ unfrankiert, unfrei
**preparation** Aufbereitung *f*, Ausarbeitung *f*, Avivage *f*, Bearbeitung, Bereitstellung *f*, Bereitung *f*, Darstellung *f*, Präparat *n*, Rüstung *f*, Verfertigung *f*, Vorarbeit *f*, Vorbereitung *f*, Vorrichtung *f*, Zubereitung *f*, Zustellung *f*
**preparation,** ~ **for flight** Flugvorbereitung *f* ~ **for flotation** Schaumschwimmaufbereitung *f* ~ **of fuel** Brennstoffaufbereitung *f* ~ **of inventories** Aufstellung *f* der Inventarlisten ~ **for mass production** Serienreifmachung *f* ~ **of molding sand** Formsandaufbereitung *f* ~ **of molds** (founding) Formherstellung *f* ~ **of ores** Erzaufbereitung *f* ~ **of poles** Stangenzubereitung *f* ~ **of a position** Einrichten *n* einer Stellung ~ **of production facilities** Betriebsmittelvorbereitung *f* ~ **of site** Vorbereitung *f* des Bauplatzes ~ **of site for deposit of fill** vorgeschriebenes Profil *n* ~ **of a stock vat** Stammküpenansatz *m* ~ **of the weft** Vorbereitung *f* des Einschlags
**preparation,** ~ **concentrate** Aufbereitungskonzentrat *n* ~ **plant** Aufbereitungsanlage *f* ~ **tube** Sammelglas *n* ~ **needle** Präpariernadel *f*
**preparative column** präparative Säule *f*

**preparatory** vorbereitend ~ **and final processing** Vor- und Nachbehandeln *n* ~ **pass** Vorbereitungskaliber *n* ~ **training** Vorbereitung *f* ~ **treatment** Vorbehandlung *f* ~ **work** Vorarbeit *f*, Vorrichtungsarbeit *f*
**prepare, to** ~ anfertigen, anmachen, anrichten, ansetzen, aufbereiten, ausarbeiten, bereiten, bereitmachen, bereitstellen, darstellen, einrichten, fertigen, herrichten, herstellen, präparieren, rüsten, verfertigen, (a sample) vorbereiten, zubereiten, zurechtmachen, zurichten, (a furnace) zustellen **one who** ~**s** Aufbereiter *m* **to** ~ **a bath** ein Bad *n* ansetzen **to** ~ **for firing** entsichern **to** ~ **the tape** Streifen *m* herstellen **to** ~ **the way** bahnen
**prepared** aufbereitet, bereit, fertiggemacht, gebrauchsfertig, hergestellt, vorgearbeitet, zubereitet ~ **for long-distance call** fernvorbereitet (apparatus) ~ **to sweep (or to bring in)** (Gerät) klar zum Einnehmen *n*
**prepared,** ~ **chalk** Schlämmkreide *f* ~ **charcoal** Glühstoff *m*
**preparedness** Bereitschaft *f*
**preparer** Vorspinner *m*
**preparing,** ~ **of composition** (pyrotechnics) Ansetzen *n* des Satzes
**preparing,** ~ **vat** Anmachebottich *m* ~ **vessel** Ansatzbottich *m*
**pre-patch board** Steckbrett *n*
**prepay, to** ~ frankieren, vorausbezahlen
**prepayment** Voraus(be)zahlung *f* ~ **electricity meter** Elektrizitätsselbstverkäufer *m* ~ **meter** Automatenzähler *m*, Münzzähler *m*
**prephotoglow region** Vorglimmlichtgebiet *n*
**preplasticising method** Vorplastifiziermethode *f*
**pre-plumbed** vorabgestimmt ~ **system** starre Koaxialleitung *f*, starrer Hohlleiterzug *m*
**pre-polymer** Vorpolymerisat *n*
**preponderance** Übergewicht *n*, Überlegenheit *f*, Überwiegen *n*, Vorherrschen *n*
**preponderant** überwiegend, vorwiegend
**preponderate, to** ~ vorherrschen, vorwiegen, überwiegen
**prepreg process** Vorimprägnierverfahren *n*
**preprint** Vorabdruck *m*
**pre-processing** Vorverarbeitung *f* ~ **of steel material** Eisenvorbereitung *f* ~ **shop** Vorbetrieb *m*
**preproduction model** Vorserienmodell *n*
**pre-punch, to** ~ vorlochen
**prepunched cards** vorgelochte Karten *pl*
**prerarefy, to** ~ vorverdünnen
**pre-reaction** Vorreaktion *f*
**pre-record tape equalizer** Aufnahmeentzerrer *m*
**pre-reduce, to** ~ vorreduzieren
**pre-reduction** Vorreduktion *f* ~ **paste** Vorreduktionssatz *m*
**pre-refining** Vorraffination *f*
**prerequisite** Voraussetzung *f*, Vorbedingung *f*
**pre-roast, to** ~ vorrösten
**pre-roast** Vorröstung *f*
**preroasting furnace** Vorröstofen *m*
**prerogative** Vorrecht *n*
**presbyopic** altersichtig ~ **eye** altersichtiges Auge *n*
**prescoring** Vorsynchronisieren *n*
**prescribe, to** ~ verordnen, verschreiben, vorschreiben

**prescribed** vorgeschrieben, vorschriftsgemäß, vorschriftsmäßig ~ **course** Sollkurs *m* ~ **form** Formblatt *n*; vorgegeben (math.)
**prescription** Rezept *n*, Verordnung *f*, Vorschrift *f* ~ **to be observed in operating** (ship, plane, plant, etc.) Betriebsvorschriften *pl*
**preselect, to** ~ vorwählen
**pre-selectable** vorwählbar
**pre-selected,** ~ **frequency** Rastfrequenz *f* ~ **pulse count** Impulszahlvorwahl *f*, Zeitvorwahl *f*
**preselecting,** ~ **the depth** Spantiefenvorwahl *f* ~ **control** Vorwählschaltung *f* ~ **rotary line switch** Drehwähler *m* als Vorwähler verwendet
**preselection** Vorselektion *f* (rdo), Vorwahl *f* ~ **of aperture** Blendenvorwahl *f* ~ **diaphragm** Vorwählblende *f* ~ **pulse switch** Vorwahlimpulsschaltereinheit *f*
**preselective,** ~ **gear** Vorwählgetriebe *n* ~ **transmission** Vorwählgetriebe *n*
**preselector** Bandfilter *n*, Hochfrequenzvorstufe *f*, kleiner Wähler *m*, Rufordner *m*, Vorselektionskreis *m*, Vorwähler *m* ~ **for single lens** Blendenvorwahleinrichtung *f*
**preselector,** ~ **key control** Tastenanwahlsteuerung *f* ~ **means** Primärabstimmung *f* ~ **mechanism** Vorwählschalteinrichtung *f*
**presence** Anwesenheit *f*, Gegenwart *f*, Vorhandensein *n*, Vorkommen *n*
**present, to** ~ einreichen, präsentieren, vorführen **to** ~ **oneself** sich stellen
**present** anwesend, da, laufend, vorliegend **at** ~ augenblicklich, vorderhand **those** ~ **die** Anwesenden *pl* **to the** ~ bisher **up to the** ~ bislang
**present,** ~ **age** Neuzeit *f* ~ **angle of sight** Meßhöhenwinkel *m* ~ **bearing** Meßseitenwinkel *m* ~**-day value** Gegenwartswert *m* ~ **position** Anschlag *m*, Gegnerpunkt *m*, Meßdreieck *n* ~**-position data** Ortungswerte *pl* ~ **position in the horizontal plane** Meßpunkt *m* in der Kartenebene ~ **position of target** Meßpunkt *m* ~**-position triangle defined by zero** Meßdreieck *n* ~ **slant range** Abschußentfernung *f* ~ **state of technology** heutiger Stand *m* der Technik ~ **time** Gegenwart *f*
**presentation** Anzeige *f*, Bilddarstellung *f* (rdr), Bildeindruck *m* (TV), Form *f* des Echos (rdr) ~ **screen** Meßtisch *m*
**presented** eingereicht
**preservation** Aufbewahrung *f*, Erhaltung *f*, Haltbarmachung *f*, Imprägnierung *f*, Konservierung *f*, Tränkung *f* ~ **of lumber** Holzkonservierung *f* ~ **of meat by cold storage** Fleischkühlanlage *f* ~ **of wood** Holzkonservierung *f*, Holzzubereitung *f*
**preservation rectifier** Konservierungsgleichrichter *m*
**preservative** Anstrichstoff *m*, Konservierungsmittel *n*, Schutzerdungsmittel *n*, Schutzpräparat *n*; schützend ~ **(agent)** Schutzmittel *n* ~ **coat** Schutzanstrich *m* ~ **coating** Schutzüberzug *m* ~ **treatment** Schutzbehandlung *f*
**preserve, to** ~ aufbewahren, aufheben, bewahren, eindauern, einhalten, (fruit) einmachen, einwecken, erhalten, haltbar machen, konservieren, wahren **to** ~ **contact** eine Verbindung *f* halten
**preserve** Dauerware *f*, Konserve *f*

**preserved** (goods) eingemacht, gepökelt, gesalzen ~ **food** Konserve *f*
**preserving** Haltbarmachen *n*; erhaltend ~ **angles unaltered** winkeltreu ~ **film (or layer)** Konservierungsschicht *f* ~ **rectifier** Konservierungsgleichrichter *m*
**preset, to** ~ abschirmen, vorwählen
**preset** festgesetzt ~ **apparatus** Apparat *m* mit Voreinstellung ~ **breaking point** Sollbruchstelle *f* ~ **count** Impulsvorwahl *f* ~ **count and time** Impuls- und Zeitvorwahl *f* ~ **course** Programmsteuerung *f* ~ **disk** Einstellrad *n* ~ **exciter lamp** Vorfokus-Erregerlampe *f* ~ **frequency** gerastete Frequenz *f* ~ **guidance** Programmlenkung *f* ~ **knob** Einstellknopf *m* ~ **parameter** vorgegebener Parameter *m*, Vorwegparameter *m* (info proc.) ~ **pulse rate** Impulszahlvorwahl *f* ~ **pulse switch unit** Vorwahlimpulsschalter *m* ~ **row counter** Vorwählreihenzähler *m* ~ **stability** Voreinstellstabilität *f* ~ **time** Zeitvorwahl *f* ~ **trim-tab position** Vorwählschaltung *f*
**presetting** Rückstellung *f*
**preshoot** Vorschwinger *m* (TV)
**preshrink, to** ~ einlaufen lassen, sanforisieren
**preshrinkage** Sanforisieren *n*
**presidency** Vorsitz *m*
**presignal** Vorsignal *n* ~ **and release signal** Vor- und Auslösesignal *n*
**presinter, to** ~ vorsintern
**pre-slit** vorgeschlitzt
**pre-sort, to** ~ vorsortieren
**pre-spooling sprockets** Vorwickelzahntrommeln *pl*
**press, to** ~ drücken, eindrücken, einpressen, pressen, streben, walken, zusammen-drücken, -pressen **to** ~ **against** anpressen **to** ~ **around** (herum)pressen **to** ~ **backward** zurückdrücken **to** ~ **down** herab-, herunter-, nieder-drücken, niederstauchen **to** ~**-forge** preßschmieden **to** ~ **forward** vordringen **to** ~ **to gauge** auf Maß drücken **to** ~ **hollow** hohlpressen **to** ~ **home** festpressen **to** ~ **on** andrücken, (book cover) anreiben, aufpressen, (barrel) aufschrumpfen **to** ~ **out** abpressen, ausdrücken, auspressen **to** ~**-pack** packen **to** ~ **on a pedal** latschen **to** ~ **and shrink into** warmschrumpfen **to** ~ **shut** zudrücken **to** ~ **together** zusammenpressen **to** ~ **through** durch-drücken, -pressen **to** ~ **the trigger** durchkrümmen
**press** Beanspruchung *f*, Druck *m*, Drucken *n*, Druckerei *f*, Druckerpresse *f*, Feuchtstein *m* (print.) **the** ~ Zeitungswesen *n* **to be in the** ~ sich im Druck *m* befinden, Fruchtpresse *f*, Fuß *m*, Kelter *f*, (lamp) Lampenstempel *m*, Quetschfuß *m* (Elektronenröhre), Schrein *m*, Stempel *m*
**press,** ~ **for dismounting bands from laminated springs** Federbundabziehpresse *f* ~ **for glazing** (textiles) Glanzpresse *f* ~ **for illustration printing** Bilddruckpresse *f* ~ **for printing slips** Fahnenpresse *f* (print.) ~ **that removes burr** Abgratpresse *f* ~ **for removing roller shells** Walzenabziehpresse *f* ~ **for stamping monograms** Monogrammprägepresse *f* ~ **for sugar strips** Streifenpresse *f* ~ **of a tube** Röhrenstempel *m* ~ **of a valve** Quetschfuß *m* einer Röhre, Röhrenstempel *m* ~ **for zigzag feed** Zickzackpresse *f*

press, ~-and-blow operation (for forming glass-ware) Kippverfahren *n* ~ bar Preßbengel *m* ~ bench Preßbank *f* (paper mfg.) ~ blanket Drucktuch *n* ~ board Preßbrett *n*, Preßspan *m*, Stanzpappe *f* ~ board strips Preßspanstreifen *m* ~ body Pressen-gestell *n*, -körper *m* ~-button Druckknopf *m*, Einnietdruckknopf *m* ~-button housing Mitnehmerkasten *m* ~-button key Druckknopf *m*, Taste *f* mit fester Stellung ~-button plug Drucktastenaggregat *n* (tape rec.) ~-button receiver set Druckknopfempfänger *m* ~-button switch Tastendruckschalter *m* ~ cake Filter-, Preß-, Schlamm-kuchen *m*, Schneide-schlemm *m* ~ casing Preßkoffer *m* ~-cast method Preßgußverfahren *n* ~ casting Warm-preßguß *m* ~ chamber Preßkoffer *m* ~ clamp Druckklemme *f* (electr.) ~ conference Presse-besprechung *f* ~ containing pressboards Span-presse *f* ~ control Pressensteuerung *f* ~ control desk Pressensteuerstand *m* ~ control panel Flursteuerstand *m* der Presse ~-cube Preß-würfel *m* ~ cure Preßvulkanisation *f* ~-cured article Preßartikel *m* ~ cylinder Preßzylinder *m* ~ die Präge-, Preß-stempel *m* ~-drawing Preßziehen *n* ~ drive and main switch panel Antrieb *m* mit Schaltkasten ~ filter Druck-, Preß-filter *n* ~ fit Paßsitz *m*, Preßsitz *m*, Schrumpfsitzpassung *f* ~ frame Pressen-gestell *n*, -körper *m* ~ gate Längsbauten *pl* ~ jack Preßbengel *m* ~ key Taste *f* ~ lever Preß-baum *m*, -hebel *m* ~ man Pressenarbeiter *m*, Presser *m* (glass mfg.), Zubereiter *m* (print.) ~ mandrel Preßdorn *m* ~ manufacture Pressenbau *m* ~ mark Signatur *f* ~ message Presse-, Zeitungs-telegramm *n* ~ mold Preßfutter *n* ~ molding machine Preßformmaschine *f* ~ molding machine with stationary yoke Preßformma-schine *f* mit feststehendem Querhaupt ~ molding machine for use with snap flasks Preß-formmaschine *f* für kastenlosen Guß ~-off cam Absprengschloß *n* ~ pad Niederhalter *m* ~ pan insulation material Preßspan *m* ~ photo-grapher Bildberichterstatter *m* ~ pin Preß-stempel *m* ~ plate (for manufacture of blasting caps) Löffel *m*, Preßtisch *m* ~ plunger Preß-kolben *m* ~ power Preßdruck *m* ~ proof letzte Korrektur *f* (print.), Maschinenkorrektur *f*, Presseabzug *m* ~ ram Preßstempel *m* ~ release Veröffentlichung *f* ~ rod Druckstelze *f* ~ roll Preßwalze *f* ~ rolled board Wickelpappe *f* ~ room Druckerei *f* (print.), Maschinensaal *m* ~-spahn Preßspan *m* ~ spoon Preßdaumen *m* ~ stick Preßbengel *m* ~ stone Preßfundament *n*, Setzstein *m* ~ sweating process Preßschwitzver-fahren *n* ~-to-open valve federbelastetes Ablaß-ventil *n* ~-to-talk switch Sprechtaste *f* ~-to-test indicator light Anzeigeleuchte *f* mit Taster ~-type core machine Kernpresse *f* ~ valve federbelastetes Ablaßventil *n* ~ vulcanization Vulkanisation *f* mit direktem Dampf in der Presse ~ work Druckarbeit *f* ~-wheel attach-ment Druckrollenvorrichtung *f* ~ working of metals Stanzereitechnik *f* ~ yoke Preßtraverse *f*
pressable preßbar
pressed (sheet-metal) gepreßt ~ to shape geprägt
pressed ~ aluminium Preßaluminium *n* ~-and-shrunk warm eingeschrumpft ~ article Preß-körper *m*, Preßling *m* ~ bale Preßballen *m* ~

brass Druckmessing *n* ~ brick Preßziegel *m* ~ coal Brikett *n* ~ distillate Preßöl *n* ~ flanges aufgepreßte Briden *pl* ~ girder Preßträger *m* ~ glass base Preßglassockel *m* ~ glass lens Preß-glasscheibe *f* ~ material Preßstoff *m* ~ metal part Metalldruckteil *n* ~ mica Preßglimmer *m* ~ object Preßkörper *m*, Preßling *m* ~ paper disc Preßpapierscheibe *f* ~ parts of asbestos cement Asbestzementpreßteile *pl* ~ peat Preß-torf *m* ~ piece Preßstück *n* ~ pump (sugar mfg.) Preßling *m* ~ raffinade Preßraffinade *f* ~-rubber article Gummipreßformartikel *m* ~ steel Preßstahl *m* ~ thread gedrücktes Gewinde *n* (sheet metal)
presser (book binding) Abpreßmaschine *f*, Bandschläger *m*, Drucker *m* (print.), (Buch u. Stoffdruck) Druckwalze *f*, Formen-, Tuch-presser(in) *m* & *f*, Preßapparat *m*, Presse *f*, Preßfinger *m* (spinning mach.), Pressur *m*, Quetsche *f*
presser, ~ bar Nadelpresse *f* ~ board Formblock *m*, Preßklotz *m*, (in molding) Preßplatte *f* ~ foot Steppfuß *m* ~ frame Preßleier *m*, Spindel-bank *f* ~ head Preßhaupt *n* ~ plate (textiles) Preßblech *n* ~ roll (film feed) Druckrolle *f* ~ (water) tank Druckluftwasserkessel *m*
pressing Drücken *n*, Glätten *n*, Plattenabdruck *m*; dringend, dringlich, drückend, pressend, Pressen *n*, Preß-erzeugnis *n*, -platte *f* (phono), -stück *n*, -teil *m*, Preßling *m*, Satinieren *n*, Stanzen *n* (paper mfg.)
pressing, ~ bag Preßbeutel *m* ~ board Brand-pappe *f*, Preß-, Gautsch-brett *n* ~ brush An-drückbürste *f* ~ cover einseitiger Umschlag *m* ~ die Preßwerkzeug *n* ~ effect Preßdruck *m* ~ excentric Andrückexzenter *m* ~ head of casing Futterrohrkopf *m* ~-in mandrel Eindrückdorn *m* ~ iron roll Plattwalze *f* ~ lever Andrückhebel *m* ~ method Preßverfahren *n* ~ mill Preßmühle *f* ~ nipple Drucknippel *m* ~ operation Einpreß-vorgang *m*, Preßarbeit *f* ~ plant Preßanlage *f* ~ process Preßverfahren *n* ~ rod Druckstange *f* ~ roller Preßwalzwerk *n* ~ screw Druck-schraube *f* ~ set Presseneinsatz *m* ~ table Preßtisch *m* ~ tool Preßwerkzeug *n* ~ wheel Abdrehrädchen *f*
pressings Preßerzeugnis *n*
pressure Andruck *m*, Anpressen *n*, Druck *m* (me-*f* (electr.) ~ of arch Druckstoß *m* des Bogens ~ below atmospheric Unterdruck *m* ~ of axle Achsdruck *m* ~ in the battery Drücken *n* der Diffusionsbatterie ~ between Binnendruck *m* ~ of burst Zerplatzdruck *m* ~ of business Ge-teor., phys., techn.), Druck-amplitude *f*, -kraft *f*, Drücken *n*, Keltern *n*, Pressen *n*, Spannung schäftsdrang *m* ~ at center Mittendruck *m* ~ of (or from) the chip Schnittdruck *m* ~ in excess of atmospheric pressure Überdruck *m* ~ of fluidity Fließdruck *m* ~ on foundation soil Baugrundpressung *f* ~ of geartooth profile Ein-griffswinkel *m* der Zahnflanke ~ from head-race side Oberwasserdruck *m* ~ at the jet exit Strahlaustrittsdruck *m* ~ on the joint Liederungs-druck *m* ~ on soil Bodenpressung *f* ~ on the support Auflagerungsdruck *m* ~ from a system of levers Hebeldruckeinrichtung *f* ~ in terms of millimeters of mercury Druck *m* in mm Queck-silbersäule, Quecksilbersäulendruck *m* in Mil-

limetern ~ **of traffic** Verkehrszufluß *m* ~ **per unit area** Flächendruck *m* ~ **per unit of area** Flächenpressung *f* ~ **in units of water height** Wassersäule *f* (WS) ~ **and vacuum vent valve** Sicherheitsventil *n* für Druck und Vakuum **pressure,** ~ **accumulator** Druckakkumulator *m*, Drucksammler *m* ~ **adjusting device** Druckstellvorrichtung *f* ~ **adjusting handle** Druckstellbügel *m* ~ **adjusting screw** Druckeinstellschraube *f* ~ **adjustment** Druckeinstellung *f* ~ **air** Druckluft *f* ~ **air supply** Preßluftzufuhr *f* ~ **airship** Pralluftschiff *n* ~ **alarm** Druck-, Gaszellen-alarm *m* ~ **altimeter** barometrischer Höhenmesser *m* ~ **altitude** Barometer-, Druckhöhe *f*, barometrische Höhe *f*, Prallhöhe *f* ~ **altitude gradient** Druckhöhengefälle *n* ~ **amplitude** Druckamplitude *f* ~ **angle** Eingriffswinkel *m* (metall), Zahneingriffswinkel *m* ~ **apparatus** Druckapparat *m* ~ **area** Druckgebiet *n*, Luftdruckgebiet *n* ~ **arm** Druckkolben *m* ~ **baffles** Druckleitbleche *pl* ~ **balance** Druckausgleich *m* ~**-balancing device** Druckausgleicheinrichtung *f* ~ **bar** Druckstange *f*, Drucksteg *m* (piano) ~**-bar controller** Druckbügelregler *m* ~ **block** Preßguß *m* ~ **blower** Zylindergebläse *n* ~**-board mine** Brettstück-, Druckbrett-mine *f* ~ **boiler** Druckkessel *m* ~ **box** Druck-dose *f*, -gefäß *n* ~ **broadening** (gas spectrum) Druckverbreiterung *f* ~ **bulb** Druckzwiebel *f* ~ **bulkhead** Druckschott *n* ~ **cabin** unter Druck gesetzte Kabine *f* beim Höhenflug, Druck-, Höhen-kabine *f* ~ **calibration** Druckeichung *f* ~ **cam** Druckkurve *f* ~ **capsule** Meßuhr *f* ~**-cast carton (or cardboard)** Pappenguß *m* ~ **casting(s)** Preßguß *m* ~ **cell** Druck-gefäß *n*, -kammer *f*, Meßdose *f* ~ **center** Druck-mitte *f*, -punkt *m* ~ **chamber** Druckkammer *f* ~**-chamber conditioning** Druckkammerprüfungsmethode *f* ~**-chamber loudspeaker** Druckkammerlautsprecher *m* ~ **clip** Druckstück *n* ~ **coil** Spannungsspule *f* ~ **column** Drucksäule *f* ~ **compensating valve** Druckausgleichventil *n* ~ **compensator** Druckregler *m* ~ **cone** Druck-, Spannungs-kegel *m* ~ **connection** Druckstutzen *m* ~ **contour** Luftdruckhöhenlinie *f* (meteor) ~ **control** Druckkontrolle *f* ~ **control line** Steuerdruckleitung *f* ~**-control switch** Niederdruckkontaktmanometer *n* ~**-control valve** Druckluftanschluß-, Drucksteuer-ventil *n*, Sicherheitskolben *m* ~ **cooker** Druck-, Schnell-kocher *m* ~ **cooling** Druckbelüftungs -kühlung *f*, -kühlverfahren *n*, Zwangsbelüftung *f* ~ **core** Druckkern *m* ~ **cover** Anpreßdeckel *m* ~ **crank-end bearing** Druckkurbellager *n* ~ **cup** Druckkappe *f* ~ **curve** Druckkurve *f*, -linie *f*, Spannungslinie *f* ~ **cylinder** Druckzylinder *m* ~ **cylinder casing** Druckzylindergehäuse *n* ~ **cylinder guiding** Druckzylinderführung *f* ~ **cylinder head** Druckzylinderkopf *m* ~ **de-aerator** Druckentgaser *m* ~ **decrease** Druckminderung *f* ~ **defined** Druckausgleich *m* ~ **demand regulator** Bedarfsregler *m* ~ **density relation** Druck-Dichte-Beziehung *f* ~ **dependence** Druckabhängigkeit *f* ~**-dependent** druckabhängig ~ **diagram** Druckfigur *f* ~ **diaphragm** Überdruckmembran *f* ~ **die-cast alloy** D-Liegerung *f* ~ **difference** Überdruck *m* ~**-difference gauge** Druckunterschiedsmesser

*m* ~**-difference transducer** Druckdifferenzgeber *m* ~ **digester** Druckkocher *m* ~**-displacement phase shift** Druckverschiebungs-Phasendifferenz *f* ~ **distillate** rohes Krackbenzin *n* ~**-distillate bottoms** Redestillationsrückstand *m* von Krackbenzin ~ **distribution** Druck-verlauf *m*, -verteilung *f* ~**-distribution meter** Druckverteilungsmesser *m* ~ **drag** Druckwiderstand *m* ~ **drilling** Unterdruckbohren *n* ~ **drop** Druckgefälle *n*, -minderung *f*, -sprung *m*, -verlust *m*, -verminderung *f*, Entspannung *f*, Spannungsabfall *m* ~ **drop between receiver and cylinder** Spannungsabfall *m* zwischen Aufnehmer und Zylinder ~ **drum** Baumtrommel *f* ~ **effect** Druckwirkung *f* ~ **elbow** Druckkrümer *m* ~ **element** Druck-dose *f*, -meßzelle *f* ~ **engine** Druckwerk *n* ~ **equalizer** Stoßfänger *m* ~**-equalizing reservoir** Druckausgleichbehälter *m* ~**-equalizing valve** Gleichdruckventil *n* ~ **estimates** Druckabschätzungen *pl* ~ **exerted by a rivet on inside of rivet hole** Lochleibung *f* ~ **fan** blasender Ventilator *m* ~**-fashioning method** Preßprozeß *m* ~**-fed carburetor** Vergaser *m* mit Druckförderung ~ **feed** Druck-förderung *f*, -speisung *f* ~**-feed lubrication** Druckölung *f*, Preßölschmierung *f* ~**-feed pump** Öldruckpumpe *f* ~ **film** Druckhaut *f* ~ **filter** Druckfilter *m* ~ **firing device** Druckzünder *m* ~ **flange** Druckflansch *m* ~ **flap** Luftdruckausgleich-hutze *f*, -klappe *f* ~ **float** Druckschwimmer *m* ~ **fluid** Druckflüssigkeit *f*, Preßflüssigkeit *f* ~ **foot** Druckstempel *m*, Meßtaster *m* ~ **foot of a sewing machine** Stoffdruckerfuß *m* ~ **force** Druckkraft *f* ~ **front** Stoßfront *f* ~ **fuelling** Druckbetankung *f* (g/m) ~ **gas** Preßgas *n* ~ **gas connection** Druckgasanschluß *m* ~ **gas fitting** Druckgasarmatur *f* ~ **gas plant** Druckgasanlage *f* ~**-gas producer** Druckgasgenerator *m*, geblasener Gaserzeuger *m* oder Generator *m* ~ **gate** Druckfenster *m* (film) ~ **gauge** Druckanzeiger *m*, -anzeigerohr *n*, -kraftmesser *m*, -lehre *f*, -manometer *m*, -messer *m*, Federlehre *f*, Kraftmeßdose *f*, Manometer *n*, Manometerzeiger *m*

**pressure-gauge (or -gage),** ~ **case** Manometergehäuse *n* ~ **connection** Druckmesseranschluß *m* ~ **pointer** Manometerzeiger *m* ~ **reading** Manometerstand *m* ~ **spring** Manometerfeder *f* ~ **stopcock with air discharge** Manometerabsperrhahn *m* mit Entlüftung ~ **testing machine** Manometerprüfapparat *m* ~ **throttles** Beruhigungsdrossel *pl* für die Druckmesser ~ **tube** Manometerstutzen *m* ~ **valve** Druckmesser-, Manometer-ventil *n*

**pressure,** ~ **gradient** barometrisches oder spezifisches Druckgefälle *n*, Druckgradient *m*, Drucksteigung *f*, Luftdrucksteigung *f* ~**-gradient microphone** Bewegungsmikrofon *n*, Druckgradienten -empfänger *m*, -mikrofon *m*, Geschwindigkeitsmikrofon *n* ~ **grease fitting** Druckschmierknopf *m* ~ **guide** Andrückschiene *f*, Gleitschuh *m*, Kufe *f* ~ **half coupling** Druckhälfte *f* ~ **head** Druck-gefälle *n*, -höhe *f* (water), -knopf, -öffnung *f*, -säule *f*, Förderhöhe *f*, Staudruck *m* ~**-head indicator** Staudruckmesser *m* ~**-head speed indicator** Staudruckfahrtmesser *m* ~**-head speed recorder** Staudruckfahrtschreiber *m* ~**-head table** Staudrucktafel *f* ~ **height** Prallhöhe *f* ~ **height gradient** Druckhöhengefälle *n* (meteor) ~ **hole** Druckbohrung *f* ~ **hose** Druckschlauch *m* ~ **hull** Druckkörper *m* ~ **hull suction connection** Bordlenzanschluß *m* ~ **igniter** Druckzünder *m* ~ **indication** Druckanzeige *f* ~ **indicator** Druckindikator *m* ~ **inlet** Vorverdichterdruckstutzen *m* ~-**inlet nipple** Druckstutzen *pl* ~ **injection of cement grout** Zementeinpressung *f* ~ **jump** Drucksprung ~ **lever** Druck-, Preß-hebel *m* ~ **limit** Druckgrenze *f* ~ **line** Druckleitung *f*, Eingriffslinie *f* (Zahnrad) ~ **line of telescopic jack** Hubraumdruckleitung *f* (g/m) ~ **load** Druckbeanspruchung *f* ~ **locking** (of bolt) kraftschlüssige Sicherung *f* ~ **loss** Druck-minderung *f*, -verlust *m*, -verminderung *f* ~ **lubrication** Druckumlaufschmierung *f* ~ **lubricator** Preßöler *m* ~ **manometer** Druckmanometer *n* ~ **mark** Druckstelle *f* (Färberei) ~ **measuring point with plug screw** Manometerstutzen *m* mit Zylinderschraube ~ **mechanism** Anpreßvorrichtung *f* ~ **medium** Druckmittel *n* ~ **meter** Druckmesser *m*, -meßgerät *n* ~ **microphone** Druckmikrofon *n*, (sound) Schalldruckmikrofon *n* ~ **modulus** Kompressibilität *f* ~ **mounting** Druckhalterung *f* (cryst.) ~ **nipple knurl** Drucknippelrändel *n* ~ **nozzle** Druck-düse *f*, -meßrohr *n*, Stau-düse *f*, -rohr *n* ~ **oil** Druck-, Preß-öl *n* ~ **oil pump** (telemotor) Druckölpumpe *f* ~ **oil set** Druckölerzeugungsanlage *f* ~ **oil unit** hydraulischer Regler, ölhydraulische Regelvorrichtung *f* ~-**operated alcohol valve** B-Stoffvorventil *n* ~-**operated circuit closer** Druckschalter *m* ~-**operated valve** Vorventil *n* ~ **orifice** Drucköffnung *f* ~ **pad** Andrückschiene *f*, Druckkufe *f*, Gleitschuh *m*, Kufe *f*, Preßkissen *n* (film) ~-**pattern navigation** barometrische Navigation *f* ~ **peak** Stoßdruckspitze *f* ~ **period** Druckdauer *f* ~ **perpendicular to the road exerted by a vehicle** Bahndruck *m* ~ **perturbation** Druckstörung *f* ~ **pick-off** Druckgeber *m* ~ **pickup** Druckmeßdose *f* ~ **pile** Druckpfahl *m* ~ **pin** Druckstift *m* ~ **pipe** Druckrohr *n*, (from supercharger) Ladeluftleitung *f* ~ **pipe connection (or socket)** Druckrohrstutzen *m* ~ **pipe elbow** Druckrohrwindelanschluß *m* ~-**pipe line** Druckrohrbahn *f* ~-**plank mine** Druckbohlenmine *f* ~ **plate** Anpreßteller *m*, Druck-fenster *m* (film), -platte *f*, -stück *n*, Preßplatte *f* ~ **point** Druck-punkt *m*, -stelle *f* ~ **polishing** Druckpolieren *n* ~ **probe** (in article or tilted weir plates) Druckaufnahmeröhrchen *n*, Entnahmerohr *n*, Entnahmeröhrchen *n* ~ **process** Preßdauer *f* ~-**proof** druckfest ~ **pulley** Andrückrolle *f* ~ **pump** Druck-pumpe *f*, -werk *n*, Hub-, Preß-pumpe *f* ~ **range** Druckgebiet *n* ~ **ratio** Druck-quotient *m*, -verhältnis *n* ~ **reading** Druckanzeige *f* ~ **receiving valve** Druckluftempfänger *m* (Rohrpost) ~ **recording** Druckregistrierung *f* ~ **recovery** Druckrückgewinn *m* ~ **reducer** Druck-minderer *m*, -reduktionsapparat *m*, -regler *m* ~-**reducing valve** Druckminderer *m*, Druckverminderungs-, Reduzierventil *n* ~ **reducing valve with diaphragm** Membrandruckreduzierventil *n* ~-**reducing valve regulator** Druckminderventil *n* ~-**reduction device** Verdichtungsminderer *m* ~ **reduction valve** Druckminderventil *n* ~ **refueling** Druckbetan-

kung *f* ~ **region** Staugebiet *n* ~-**regulating bolt** Druckstellbolzer *m* ~ **regulating device** Druckreguliervorrichtung *f* ~-**regulating valve** Drossel *f*, Druck-regler *m*, -regulierventil *n* ~ **regulator** Druck-regler *m*, -verminderungsventil *n* ~ **regulator pipe** Belüftungsleitung *f* ~ **release** Druckauslösung *f*, Entlüftung *f*, Entspannung *f* ~-**release valve** Druckverminderungsventil *n* ~ **relief** Druckentlastung *f* ~ **relief pipe** Entlüftungsrohr *n* ~ **relief plug** Einschraubstück *n* (Zylinderkopf) ~-**relief valve** Druckminder-, Überdruck-, Sicherheits-ventil *n* ~-**relief vent** Luftauslaßventil *n*, Überdruckaustritt *m* ~-**relieving device** Entlastungseinrichtung *f* ~ **reservoir** Druckspeicher *m* ~ **resistance** Druckhaltung *f* ~ **resistant container** druckfester Behälter *m* ~-**resistant type** druckfeste Sonderausführung *f* ~-**responsive** druckabhängig ~-**rigid airship** Druckstarrluftschiff *n* ~ **rim** Druckrand *m* ~ **ring** (piston) Gasring *m*, Preßring *m* ~ **rise** Druckanstieg *m* ~ **roller** Druckwalze *f* ~ **rolls for rail-straightening machines** Druckrollen *pl* für Geleiserückmaschinen ~-**sampling device** Druckmeßsonde *f* ~ **scala** Luftdruckstufe *f* ~ **screw** Hebe-, Preß-schraube *f* ~ **screw joint (or connection)** Druckverschraubung *f* ~ **screws** Preß- und Hebevorrichtung *f* ~ **sensing device** Druckfühler *m* ~ **sensitive tape** Klebestreifen *m*, Trockenklebeband *n* (tape rec.) ~ **sensitivity** Druckempfindlichkeit *f* ~-**shaping method** Preßprozeß *m* ~ **shift** (spectral lines) Druckverschiebung *f* ~ **shoe** Druckschiene *f* (film) ~ **shoulder** Druckschulter *f* ~ **shutter** Spannschütze *f* ~ **side** Druckseite *f* ~ **side of a valve** Eintrittsseite *f*, Förderseite *f* ~ **sleeve** Druckhülse *f* ~ **solenoid** Ausdruckmagnet *m* (tape rec.) ~ **space** Druckräume *pl* ~ **speed indicator** Staufahrtmesser *m* ~ **sphere** Staukugel *f* ~ **spray gun** Druckzerstäuber *m* ~ **spread** Druckausbreitung *f* ~ **spring** Druckfeder *f* ~ **stage** Druckstufe *f*, Luftdruckstufe *f* ~ **stamp** Druckstempel *m* ~ **steaming** Dämpfen *n* unter Druck ~ **step** Luftdruckstufe *f* ~ **step -up means** Druckerhöher *m* ~ **stirrup** Druckbügel *m* ~ **stroke** Preßhub *m* ~ **suit** Druckanzug *m* (Höhenflug) ~ **surface** Druckseite *f* ~ **surge** Druckstoß *m* ~ **survey rake** Harkensonde *f* ~ **switch** Druckschalter *m*, Kontaktmanometer *n*, Rubidkontakt *m* ~ **system** Drucksystem *n* ~ **tank** Druckbehälter *m*, (of a sandblast-tank machine) Druckluftkessel *m*, Drucktank *m* ~ **tap** Druckhahn *m* ~ **test** Druckversuch *m*, Härteprobe *f* ~-**testing chamber** Unterdruckkammer *f* ~ **testing set** Abpreßvorrichtung *f* ~-**tight** druckdicht ~-**transmittance ratio** Druckübersetzungsverhältnis *n* ~ **tube** Druck-anzeiger *m*, -anzeigerohr *n*, -meßrohr *n*, -rohr *n*, Stau-, Pitot-rohr *n* ~-**tube anemometer** Staurohrwindmesser *m* ~-**tube joint** Ablaßstutzen *m* ~ **tubing** Druckschlauch *m*

**pressure-type, ~ baffle** Druckbelüftungsblech *n*, Leitblech *n* ~ **capacitor** Hochdruckkondensator *m* ~ **carburetor** Druckvergaser *f* ~ **deflector** Druckbelüftungsblech *n*, Leitblech *n* ~ **hose sand-blast tank machine** Druckstrahlgebläse *n* ~ **sand-blast cabinet** Sandstrahldruckapparat *m* ~ **sand-blast machine** Drucksand-

strahlgebläse *n* ~ **sand-blast unit** Sandstrahlgebläse *n* nach dem Drucksystem
**pressure,** ~ **unit** Druckerzeugungsanlage *f* ~ **vacuum gauge** Manovakuummesser *m* ~ **valve** Druckventil *n* ~**-valve spring** Druckventilfeder *f* ~ **vane** Druckflügel *m* ~ **vessel** Druck-behälter *m*, **-gefäß** *n* ~ **viscosity** Druckviskosität *f* ~**-void ratio curve** Druckporenzifferdiagramm *n* ~ **warning unit** Alarmmanometer *n*, Druckalarmeinrichtung *f* ~ **wash** Außenwasch *n*, Druckwäsche *f* ~ **water** Druckwasser *n* ~ **water-cooling system** Wasserpumpkühlung *f* ~**-water tank** Druckwasserkessel *m* ~ **wave** Druckwelle *f* ~ **welding** Preßschweißung *f* ~ **within** Binnendruck *m* ~ **worm** Druckschnecke *f*
**pressureless** drucklos
**pressurization** Druckbelüftung *f* ~ **port** Druckbelüftungsanschluß *m* ~ **system** Druckbelüftungssystem *n*
**pressurize, to** ~ druckbelüften, unter Druck setzen
**pressurized** druckdicht ~ **alarm (boiler)** Abblasedruck *m* ~ **cabin** Druckkabine *f* ~ **casing** Druckgefäß *n*, druckfeste Umhüllung *f* ~ **component** unter Überdruck gesetzter Bauteil *m* ~ **suit for emergency jumps** Rettungsdruckanzug *m* ~ **water** drückendes Wasser *n* ~ **water reactor** Druckwasserreaktor *m*
**pressurizing** Druckerzeugung *f* ~ **pipe** Staudruckrohr *n* ~ **system** Druckanlage *f* ~ **valve** Druckaufbau-, Druckbelüftungs-, Vorspannventil *n*
**pre-stall buffeting** Schütteln *n* vor dem Abreißen (der Strömung)
**Prestone** Äthylenglykol *n*, Preston *n* ~ **cooling** Glykolkühlung *f*, Heißkühlung *f*
**prestore, to** ~ vorspeichern, vorzuweisen
**pre-strain** Vorverformung *f*
**prestress, to** ~ vorbelasten, vorspannen, Vorspannung *f* geben
**prestressed** vorbeansprucht, vorbelastet, vorgespannt ~ **concrete** Spannbeton *m*
**prestressing** Bauschinger-Effekt *m*, Festigkeitssteigerung *f* des Werkstoffes durch Vorbeanspruchung, Vorspannung *f* ~ **reinforcement** Spannarmierung *f* ~ **wire** Vorspannungsdraht *m*
**pre-stroke measuring instrument** Vorhubmeßgerät *n*
**presume, to** ~ annehmen
**presumption** Annahme *f*, Voraussetzung *f*
**presumptive loss** Verschollenheit *f*
**pretend, to** ~ als Vorwand *m* brauchen, beanspruchen, vorgeben
**pretense, under the** ~ unter dem Vorwand *m*
**pretention** Vorspannung *f*
**pretest** Testversuch *m*
**pretrajectory condition (or time)** Vorweg *m*
**pretreat, to** ~ vorbehandeln, vorrichten
**pretreatment** Vorbearbeitung *f*, Vorbehandlung *f*
**pretrigger** Vorimpuls *m*
**prevail, to** ~ in Geltung sein, herrschen, die Oberhand haben, vorherrschen
**prevailing** Vorherrschen *n*; (vor)herrschend ~ **direction of wind** Hauptwindrichtung *f* ~ **direction of wind movement** vorherrschende Windrichtung *f* ~ **westerlies** vorherrschende west-

liche Winde *pl* ~ **wind** vorherrschender Wind *m*
**prevalent** das Übergewicht *n* habend, überwiegend, vorherrschend, weitverbreitet
**prevent, to** ~ abbiegen, abhalten, hindern, steuern, verhindern, verhüten, vorbeugen, (from) zurückhalten, zuvorkommen, einer Sache *f* zuvorkommen
**preventable** abwendbar
**preventative** Verhütungsmittel *n*
**preventer** Schutzmaßnahme *f* ~ **brace** Borgbrasse *f* ~ **mechanism** Rückhaltvorrichtung *f* ~ **pin** Arretierstift *m*
**preventing** Unterbindung *f*
**prevention** Verhinderung *f*, Verhütung *f* ~ **against spring fracture** Federbruchsicherung *f*
**preventive** Verhinderungsmittel *n* ~ **maintenance** Schutzwartung *f*, vorbeugende Wartung *f* ~ **measure** Schutz-, Verhütungs-maßnahme *f* ~ **measures** sicherstellende Vorkehrungen *pl*
**preverberate, to** ~ vorhalten
**preview monitor** Kamerakontrollgerät *n*
**previous** bisher, früher, vorausgehend, vorherig **on the** ~ **page** vorseitig ~ **examination** Vorprüfung *f* ~ **history** Vorgeschichte *f* ~ **pass** (in rolling) Vorkaliber *n*
**previously cut** vorgeschnitten
**prevulcanization** Anvulkanisation *f*
**pre-vulcanize, to** ~ vorvulkanisieren
**prewar airplane** Vorkriegsflugzeug *n*
**preweigh, to** ~ vorwiegen
**preweld, to** ~ vorschweißen
**pre-wetting** Anfeuchten *f* ~ **vat for wetting** Netzbehälter *m* zum Vornetzen
**price** Ansatz *m*, Preis *m*, Satz *m* ~ **of admission** Eintrittspreis *m* ~ **of a single copy** Einzelpreis *m*
**price,** ~**-controlled** preisgebunden ~ **cutting** Preisdrückung *f* ~ **list** Preisverzeichnis *n* ~ **raise** Kostenaufbau *m* ~ **reduction** Preis-ermäßigung *f*, -senkung *f* ~ **storage** Preisspeicher *m* ~ **tag** Preisschild *n*
**prick, to** ~ stechen to ~ **out** (forest) ausplentern
**prick** Ahle *f*, Körner *m*, Pfriem *m*, Pricke *f*, Stichel *m* ~ **of a needle** Nadelstich *m*
**prick,** ~ **mark** Nadelsicht *m* ~ **pin** Pikiernadel *f*
**pricked** gestochen ~ **drawing** Pause *f*
**pricker** Ahle *f*, Hülsenzieher *m*, Loch-, Stech-, Steig-eisen *n*, Räumnadel *f*
**pricking** Lochen *n*, Lochung *f*, Punktieren *n*, Stechen *n* ~ **machine** Schablonenstech-, (textile) Stupfel-, Tüpfel-maschine *f* ~ **out** (of a nipple) Durchstechen *n* ~ **pin** Punktiernadel *f* ~ **pin with vise** Pikiernadel *f* mit Halter
**prim** Aufwinderöhre *f*
**primary** Hauptleitung *f*, Hauptsache *f*, Primärstromkreis *m*, Primärwicklung *f*; direkt, induziert (electr.), primär, unmittelbar, ursprünglich ~ **AC bus** Primär-Wechselstromschiene *f* ~ **aerodynamic characteristic** primäre aerodynamische Charaktereigenschaften *pl* (Kennwerte) ~ **aiming point** Hauptrichtungspunkt *m* ~ **air** Primärluft *f* ~ **air line** Primärluftleitung *f* ~ **air pressure** Primärluftdruck *m* ~ **air supply** Erstluftzufuhr *f* ~**-aluminum pig** Hüttenaluminium *n* ~ **amplifier** Grundverstärker *m* (rdo) ~ **axis** Haupt-, Primär-achse *f* ~ **battery** Primärbatterie *f* ~ **beam cone** Hauptstrahlkegel *m* ~ **bow** Hauptbogen *m* ~ **cell** elektrochemische Zelle *f*, galvanisches Element *n*,

Primärelement n ~ **circuit** Erstkreis m, Hauptstromkreis m, Primär-kreis m, -schaltung f, -stromkreis m ~ **classifying screen** Vorklassiersieb n ~ **coil** Hauptspule f, Induktor m, Niederspannspule f, Primärspule f ~ **color** Grund-, Stamm-farbe f ~ **color signal** Primärfarbensignal n (TV) ~ **condition** Grundbedingung f ~ **control surface** Hauptsteuerfläche f, primäre Steuerfläche f ~ **controls** Hauptsteuerung f ~ **cooler** Luft-, Vor-kühler m ~ **cooling** Vorkühlung f ~ **crack** Primärriß m ~ **cracking** Vorkracken n ~ **crusher** erster Steinklopfer m ~ **crushing** Vorbrechen n, Vorzerkleinerung f ~ **crystallite** Primärkristallit m ~ **crystallization** primäre Kristallisation f, Primärkristallisation f ~ **current** Erst-, Primär-strom m ~ **DC bus** Primär-Gleichstromschiene f ~ **deposit** Kontaktlagerstätte f, primäre Lagerstätte f ~ **detector** primärer Fühler m ~ **dynamo** Primärgenerator m ~ **electron** Primärelektron n ~ **element** Meßfühler m, Meßwertgeber m ~ **emission** Primäremission f ~ **etching** Primärätzung f ~ **feedback** Hauptrückführung f ~ **fission yield** primäre Spaltausbeute f ~ **flooding tank** Voranschwemmbehälter m ~ **flow** Primärstrom m ~ **force** Grundkenntniskraft f ~ **forces** Hauptkräfte pl ~ **form** Grundform f ~ **frequency** Arbeits-, Haupt-frequenz f ~ **fuel tank** Brennstofftagesbehälter m ~ **furnace** Vorfrischofen m ~ **head** Setzkopf m ~ **ignition lead** Hauptzündung f ~ **inductance** Primärinduktion f ~ **industry** Grundstoffindustrie f ~ **instability** primäre Instabilität f ~ **ionization process** Primärvorgang m der Ionisierung ~ **layer** Grundschicht f ~ **lead** Primärleitung f ~ **line** Hauptlinie f ~ **line switch** erster Anrufsucher m, erster Vorwähler m ~ **luminous standard** Lichteinheit f ~ **manifold** Primärverteilerleitung f ~ **material** Urstoff m ~ **meaning** Grundbedeutung f ~ **member** primäres Glied n ~ **natural radio-nuclides** primäre natürliche Nuklide pl ~ **nuclear particle** Elementarteilchen n ~ **ore minerals** aszendente Erzmineralien pl ~ **plane** erster Hauptschnitt m ~ **power input** Primäraufnahme f ~ **power supply** Hauptstromversorgung f ~ **propelling charge** Hauptladung f ~ **pulse** Primärimpuls m ~ **purification** Vorfrischen n ~ **radar** Primärradar n ~ **radiation** Primärstrahlung f ~ **radiator** Primärstrahler m ~ **reactor** Primär-brenner m, -reaktor m ~ **receiver** Einkreisempfänger m, Primärempfänger m ~ **reception** Einkreisempfang m ~ **requirement** Haupterfordernis n ~ **resistance** Primärwiderstand m ~ **road** Straße f erster Ordnung ~ **rock** Urgebirge n ~ **route** Regelweg m ~ **shaft** Eingangs-, Primär-welle f ~ **side** Primärseite f ~ **skip zone** primäre stille Zone f ~ **slimes** primäre Schlämme f ~ **speed** Grunddrehzahl f ~ **standard** Urmaß n ~ **standard weight** Urgewicht n ~ **starter** Gehäuseanlaßwiderstand m ~ **station** Hauptfunkstelle f ~ **steam** Primärdampf m ~ **step grate** Treppenvorrost m ~ **still** Vordestiller m ~ **target** Hauptziel n ~ **terminal** Primärklemme f ~ **trainer** Schulflugzeug n für Anfänger ~ **training glider** Anfängersegelflugzeug n ~ **traverse station** trigonometrischer Punkt m ~ **treatment** Vorbehandlung f ~ **triangulation** Haupttriangulation f ~ **turns** Primärwicklungen pl ~-**type glider** primäres Gleitflugzeug n ~ **unit** Primärkomponente f (TV) ~ **use** hauptsächliche Verwendung f ~ **valence chain** Hauptvalenzkette f ~ **vertex refraction** Grundscheitelbrechwert m ~ **voltage** Primärspannung f ~ **winding** Erstwicklung f, Niederspannspule f, Primärseite f, -wicklung f ~ **wire** Primär-draht m, -leitung f ~-**zone casing** Ringblech n (jet)

**prime, to** ~ anheben, (pump) anstechen, (pump) durch Auffüllen n zum Ansaugen vorbereiten, (Kraftstoff) einspritzen, füllen, grundieren, instruieren, (water) mitreißen, (Bad) schärfen, spachteln, (Kessel, Motor) spucken, untermalen, vorbereiten, zünden **to** ~ **with hot oil** heißes Öl n in die Lagerstellen des Motors vor der Inbetriebsetzung einspritzen **to** ~ **a pump** eine Pumpe f anlassen

**prime** Auslese f, Strich m (math.); erstklassig, unbedingt ~ **coat** Grund(ier)anstrich m ~ **conductor** Hauptleiter m, Primärleitung f ~ **cost** Anschaffungswert m, Einkaufspreis m, Gestehungspreis m, Selbstkostenpreis m ~ **factor** Primfaktor m ~ **lard oil** unbehandeltes Speck öl n ~ **membrane** Grundierüberzug m ~ **meridian** der erste Meridian m ~ **mover** Antriebsmaschine f, Geschützschlapper m, Kraftmaschine f, Schlepper m, Zugmaschine f ~ **mover for the pump** Pumpenmotor m ~-**mover truck** Zugkraftwagen m ~ **movers** Antriebsmaschine f, Getriebeturbine f ~ **number** Grund- (math.), Primzahl f

**primed** zündfertig ~ **battery** aufgeladene Batterie f ~ **symbol** Formelzeichen n mit Strichindex, mit Strichindex versehenes Formelzeichen n

**primer** Anlaßkraftstoff m, Einspritzvorrichtung f, Firnis m, Grundier-farbe f, -masse f, Initiale f, Lack m, Leitfaden m, Porenfüller m, Spachtelmasse f, (protective coatings) Zünd-apparat m, -hütchen n, -mittel n, -nadel f, -vorrichtung f, Zündung f ~ **of bomb (or shell)** initiator Initialzünder m **any** ~ **that is sprayed on** Spritzspachtel f

**primer,** ~ **agent** Anlaßmittel n ~ **anvil** Amboß m ~ **cap** Sprengkapsel f ~ **capsule** Zündpille f ~ **charge** Übertragungsladung f, Zündladung f ~ **composition** Zündsatz m ~ **cup** Zündhütchenhülse f ~ **detonation** Initialzündung f ~ **pellet** Zündpille f ~ **pump** Anlaßeinspritzpumpe f ~ **tubing** Anlaßeinspritzleitungen pl, Anlaßrohrleitung f ~ **valve** Anlaß-, Zünd-ventil n

**primes** Blech n erster Güte

**priming** Füllen n, Grund m, Grundierung f, Gründung f, Vorspannung f, Zündung f ~ **of the pump** Anfüllen n der Pumpe ~ **of a shell** Geschoßzündung f

**priming,** ~ **cartridge** Schlagpatrone f ~ **charge** Beiladung f, Eingangszündung f ~ **coat** Grund(ier)anstrich m ~ **cock** Kompressions-, Verdichtungs-, Zisch-hahn m ~ **color** Grundfarbe f ~ **composition** Initialzündsatz m ~ **exposure** Vorbelichtung f ~-**grid fuse element** Anlaßeinspritzanlage f ~-**grid** fuse element Zündsteg m ~-**grid voltage** Gitterverschiebungsspannung f, Gittervorspannung f ~ **illumination** Grundbeleuchtung f, Vorbeleuchtung f, Vorbelichtung f, Vorlicht n ~ **(bias)** illumination Fotozellenvorlicht *n* ~(-water)level Säugungswasserstand *m* ~ **lever** Vorpumphebel *m* ~ **material** Spachtelmasse *f* ~ **paint** Grundier-, Vorstreich-farbe *f* ~ **pipe** Anlaßgasleitungsrohr *n* ~ **plug** Füllschraube *f*, Primärstöpsel *m* (electr.) ~ **powder** Zündpulver *n* ~ **pump** Benzin-, Einspritz-pumpe *f*, Zischhahn *m* ~ **reservoir** Saugkessel *m* ~ **rods** Startregulierungsgestänge *n* ~ **speed** Vorspannungsgrad *m* ~ **spindle** Vorpumpbolzen *m* ~ **substance** Zündmittel *n* ~ **tube** Friktionszünder *m* ~ **valve** Anfüll-, Anstech-ventil *n*

**primitive** primitiv, ursprünglich ~ **material** Ausgangsmaterial *n* ~ **rock** Grundgestein *n*

**primordial,** ~**atom** Uratom *n* ~ **specimen** Ursprungstyp *m* ~ **stage** Primordialstufe *f* (geol.) ~ **type** Ursprungstyp *m*

**Prince,** ~**'s metal** Prinzmetall *n* ~ **Rupert's metal** Prinzmetall *n*, Rupertsmetall *n*

**principal** Auftraggeber *m*, Chef *m*, Grundkapital *n*, Hauptgebälk *n* (arch.), Stützbalken *m*; führend, hauptsächlich, Vollmachtgeber *m*, Vorgesetzter *m*

**principal,** ~ **abode** Hauptniederlassung *f* ~ **adjustment** Hauptregel *f* ~ **angular movement** Hauptausschlag *m* ~ **axis** Hauptachse *f* ~ **axis of inertia** Hauptträgheitsachse *f* ~ **axis of vision** Hauptsehrichtung *f* ~ **azimuth** Hauptazimut *m* ~ **beam** Binderbalken *m*, (tie beam) Hauptschwelle *f* ~ **business** Hauptaufgabe *f* ~ **channel filter** Hauptbandfilter *n* ~ **chord of an airfoil** Hauptsehne *f* ~ **circuit** Grundschaltung *f* ~ **clock** Hauptuhr *f* ~ **collimating mark** Hauptmeßmarke *f* ~ **co-ordinate system** Hauptachsensystem *n* ~ **curvature terms** Hauptkrümmungsfunktionen *pl* ~ **curvatures of a surface** Hauptkrümmungen *pl* einer Fläche ~ **deflection** Lagenausschlag *m* ~ **deflection movement** Hauptausschlag *m* ~ **diagonal** Hauptdiagonale *f* ~ **direction** Krümmungsrichtung *f* ~ **direction of vision** Hauptsehrichtung *f* ~ **distance** Bildweite *f* ~ **distance of the photograph** Aufnahmebildweite *f* ~ **ditch** Hauptzuggraben *m* ~ **duty** Haupttätigkeit *f* ~ **eduction canal** Hauptabflußgraben *m* ~ **float** Hauptschwimmer *m* ~ **girder** Binderbalken *m* ~ **horizontal plane** Haupthorizontalebene *f* ~ **indices of refraction** Hauptbrechungsindizes *pl* ~ **ingredient in a mineral-oil base** Hauptbestandteil *m* einer Mineralölmischung ~ **invariants** Hauptinvarianten *pl* ~ **item** Kernstück *n* ~ **lateral deflection angle** Gesamthaltwinkel *m* ~ **line** Bildhaupt-senkrechte *f*, -vertikale *f*, (spectroscopy) Hauptlinie *f* ~ **(vertical) line** Hauptsenkrechte *f* (photo) ~ **line of sight** Hauptblickrichtung *f* ~ **magnetic susceptibility** magnetische Hauptsuszeptibilität *f* ~ **market** Hauptabsatzgebiet *n* ~ **meridian** Hauptschnitt *m* ~ **minor** Hauptminor *n* ~ **moment supporting planes** Haupttragflügelmoment *n* ~ **moments** Hauptträgheitsmomente *pl* ~ **motive** Hauptantrieb *m* ~ **normal of a curve** Hauptnormale *f* einer Kurve ~ **normal vector** Hauptnormalenvektor *m* ~ **objective** Hauptziel *n* ~ **path** Hauptbahn *f* ~ **performer** Hauptrollendarsteller *m* ~ **piece** Hauptstück *n* ~ **plan** Hauptebene *f*, Hauptlotebene *f* (photo), Übersichtsplan *m* ~ **(vertical) plane** Hauptvertikalebene *f*

(photo) ~ **plane of the lens** Hauptebene *f* des Objektives ~ **point** Einheits-, Haupt-, Plattenhaupt-punkt *m* ~ **point of registering frame** Rahmenhauptpunkt *m* ~ **point on image** Bildhauptpunkt *m* ~**-point refeaction** Hauptpunktbrechwert *m* ~**-point triangulation** Hauptpunkttriangulation *f* ~ **quantum number** Hauptquantenzahl *f* ~ **radius of curvature** Hauptkrümmungshalbmesser *m* ~ **rafter** Bindersparren *m* ~ **railway** Hauptbahn *f* ~ **ratings** Hauptdaten *pl* ~ **ray** Hauptstrahl *m* ~ **section** Hauptschnitt *m* ~ **sluice** Hauptschütz *n* ~ **spectrum** Hauptspektrum *m* ~ **story** Hauptgeschoß *n* ~ **strains and stresses** Hauptzüge und -drücke *pl* ~ **street** Verkehrsader *f* ~ **stress** Hauptspannung *f* ~ **stress-ratio** Hauptspannungsverhältnis *n* ~ **stretches** Hauptdehnungen *pl* ~ **stretchings** Hauptstreckungen *pl* ~ **support** Haupt-pfeiler *m*, -stütze *f* ~ **surface** Hauptebene *f* ~ **tangent curves** Haupttangentenkurven *pl* ~ **theorem** Hauptsatz *m* ~ **trajectories** Haupttrajektorien *pl* ~ **truss** Hauptbinder *m* ~ **value** Hauptwert *m* ~ **vertical deflection angle** Reglerwinkel *m* ~ **wing** Hauptflügel *m* ~ **wing rib** Hauptflügelrippe *f* ~ **workshop** Zentralwerkstatt *f*

**principle** Element *n*, Gesetz *n*, Gesichtspunkt *m*, Grundsatz *m*, Prinzip *n*, Satz *m*, Verfahren *n* (basic) ~ Grundbegriff *m* **in** ~ im wesentlichen

**principle,** ~ **of averages** Mittelwertsatz *m* ~ **of conservation of area** Flächenerhaltungsgesetz *n* ~ **of continuity** Kontinuitätsgleichnug *f* ~ **of design** Konstruktionsprinzip *n* ~ **of distribution** Verteilungssatz *m* ~ **of (energy) equipartition** Gleichverteilungssatz *m* ~ **of heat balance** Wärmeausgleichsprinzip *n* ~ **of immobility** Stationarität *f* ~ **of invariance** Invarianzprinzip *n* ~ **of law** Rechtsgrundsatz *m* ~ **of least constraint** Prinzip *n* des kleinsten (geringsten) Zwanges ~ **of least work** Grundsatz *m* der geringsten Formveränderung ~ **of regeneration** Regenerativprinzip *n* ~ **of the stationary state** Stationarität *f* ~ **of unit construction** Bauprinzip *n*

**principle indetermination** Unbestimmtheitsbeziehung *f*

**principles** Grundsätze *pl*

**print, to** ~ abdrucken, abziehen, drucken, kopieren (photo) **to** ~ **over (or on)** bedrucken **to** ~ **through** durchkopieren **to** ~ **with the type** Abdruck *m* mit der Schrift **to** ~ **waste** makulieren

**print** Abzug *m*, Bericht *m*, Druck *m*, Druckschrift *f*, Kernauge *f*, Kopie *f*, Pause *f* ~**(ing)** Abdruck *m* ~ **of test ball** Eindruckkalotte *f*

**print,** ~ **cutter** Formschneider *m* ~**-cutter's spoon-bit chisel** Gravierschaufelbeitelchen *n* ~ **grader** Lichtbestimmer m (film) ~ **selection common exit** Schreibsteuerungsausgang *m* ~ **sprayers** Druckbestäuber *m* ~ **style** Druckartikel *m* ~ **through** Kopiereffekt *m* (tape rec.) ~ **trimmer** Beschneideglas *n*

**printable** druckfähig

**printed,** ~ **in bold** fettgedruckt, bedruckt ~ **on** bedruckt

**printed,** ~ **advertising posters** Werbedrucke *pl* ~ **airmail matter** Luftpostdrucksache *f* ~ **characters** Druckbuchstaben *pl*, Druck-

schrift *f* ~ **circuit** gedruckte Schaltung *f* ~ **data** gedruckte Daten *pl* ~ **guide** schriftliche Anleitung *f* ~ **instructions** Gebrauchsanweisung *f* ~ **matters** Drucksache *f* ~ **page** Druckseite *f* ~ **sheet** Druckbogen *m* ~**-side-up delivery** Gleitausleger *m* ~ **styles** Druckware *f* ~ **wire board** (P.W.B.) Montageplatte *f* mit gedruckter Schaltung ~ **work** Druckschrift *f*
**printer** Buchdrucker *m*, Drucker *m*, Farbschreiber *m*, Kopiermaschine *f*, Übersetzer *m* ~ **and typesetter** Druck- und Setzmaschine *f* ~ **factor** Lichttonwert *m* (film)
**printergram** Printergramm *n*
**printer's,** ~ **blanket** Druckdecke *f* ~ **broaches** Buchdruckerahlen *pl* ~ **emblem** Buchdruckerwappen *n* ~ **flower** Randverzierung *f* ~ **mark** Druckermarke *f* ~ **roller composition** Walzenmasse *f* ~**-screen** Druckraster *m* ~ **supply** Druckereibedarf *m* ~ **supply house** Druckereifachgeschäft *n*
**printing** Buchdruck *m*, Vervielfältigung *f* ~ **from aluminium plates** Aluminiumdruck *m* ~ **on back** Widerdruck *m* ~ **in black** Schwarzdruck *m* ~ **with a deep edge plate** Tiefdruck *m* (film) ~ **and function cam** Druckkurve *f* ~ **of half-tone engravings (or etchings)** Autotypiedruck ~ **of illustrations** Illustrationsdruck *m* ~ **of illustrative matter** Bilderdruck *m* ~ **from plates** Plattendruck *m* ~ **in relievo** Reliefdruck *m* ~ **of warp** (textiles) Kettendruck *m* ~ **of wool fabrics** Wolldruckerei *f*
**printing,** ~ **apparatus** Druckapparat *m*, Drucker *m* ~ **area** Druckfeld *n* ~ **arrangement** Druckeinrichtung *f* ~ **auxiliary** Druckereihilfsprodukt *n* ~ **bail** Druckbügel *m* (teletype) ~ **bail arm** Druckarm *m* ~ **bail track** Druckfalle(nschine) *f* (teletype) ~ **blankets** Schnellpressendrucktücher *pl* ~ **block** Cliché *n*, Druckbrett *n*, -form *f*, -stock *m*, Handdruckmodell *n* ~ **cam** Druckdaumen *m* ~ **capacity** Druckleistung *f* ~ **carrier** Druckträger *m* ~ **circuit magnet** Druckschaltungsmagnet *m* ~ **clutch throwout lever** Drucksachensperre *f* ~ **color** Druckansatz *m* ~ **costs** Druckkosten *pl* ~ **craft** Buchdruckerkunst *f* ~ **cylinder** Druckwalze *f* ~ **density** Kopierschwärzung *f* ~ **device** Druckorgan *n* (teletype), Druckvorrichtung *f* ~ **disk** Schreibscheibe *f* ~ **drum** Belichtungsrolle *f*, Kopierstelle *f* ~ **equipment** Buchdruckereieinrichtung *f* ~ **factory** Druckerei *f* ~ **felt** Druckfilz *m* ~ **form** Druckform *f* ~ **frame** Einlage *f* (photo), Kopierrahmen *m* ~ **frame cover** Kopierrahmendecke *f* ~ **hammer** Druck-, Typen-hammer *m* ~ **head** Druckkopf *m* ~ **indicator** Schreib--anzeige *f*, -anzeiger *m* ~ **ink** Druckerschwärze *f*, Druck-farbe *f*, -tinte *f*, Vervielfältigungsfarbe *f* ~ **ink oil** Druckerschwärzeöl *n* ~ **job** Druckarbeit *f* ~ **keyboard perforator** druckender Tastaturlocher *m* ~ **letter** Letter *f* ~ **lever** Druckhebel *m* ~ **loss** Kopierverlust *m* (film) ~ **machine** Lichtpausapparat *m* ~ **machine web** Maschinenband *n* für Druckereien ~ **magnet** Druckmagnet *m* ~ **manager** Druckereileiter *m* ~ **mechanism** Druckwerk *n* ~ **medium** Druckhilfsmittel *n* ~ **office** Buchdruckerei *f*, Druckerei *f*, Offizin *f* ~**-on of dissolved caustic** Laugenaufdruck *m* ~**(-out) paper** Auskopierpapier *n* ~ **papers** (blue or white prints) Blaupau-

senpapier *n* ~ **perforator** druckender Handlocher *m* ~ **periodicals** Zeitschriftendruck *m* ~ **plane** Kopierfenster *n* ~ **plant** Druckereibetrieb *m*, Kopierwerk *n* ~ **plate** Druck-form *f*, -platte *f* ~ **point** Typenabdruckstelle *f* ~ **position** Schreibstelle *f* ~ **press** Druckerpresse *f* ~ **process** Druckverfahren *n* ~ **progress** Druckabwicklung *f* ~ **property** Druckfähigkeit *f* ~ **reader** registrierender Schreiber *m* ~ **receipt on sales slips** Quittungsdruck *m* ~**-receiving apparatus** Druckempfänger *m* ~ **reception** Druckempfang *m* ~ **recipe** Druckvorschrift *f* ~ **relay** Druckrelais *m* ~ **reperforator** Druckerempfanglocher *m* ~ **roller** Auftrag-, Druck-walze *f* ~ **roller composition** Buchdruckwalzenmasse *f* ~ **sequence** Druckfolge *f* ~ **shaft** Druckachse *f* ~ **shop** Kopier-anstalt *f*, -werk *n* ~ **shop auxiliary** Druckerei-Hilfsmaschine *f* ~**-shop detachment** Druckereitrupp *m* ~ **specimen** Druckmuster *n* ~ **speed** Druckgeschwindigkeit *f* ~ **sprocket** Belichtungszahntrommel *f* ~ **stamp for block printing** Druckstempel *f* für Holzdruck ~ **station** Kopierfenster *n* ~ **studio** Kopierwerk *n* ~ **surface** Druckfläche *f* ~ **table** Drucktisch *m* ~ **tape** Druckband *n* ~ **telegraph** Drucktelegraf *m*, Ferndrucker *m*, Typendrucker *m*, Typendrucktelegraf *m* ~ **timer** Zeltdrucker *m* ~ **trade** Buchgewerbe *n* ~ **trade expert** Druckfachmann *m* ~ **trip magnet** Auslösemagnetdruck *m*, Druckauslösemagnet *m* ~**-type** Letter *f* ~ **unit** Druckeinheit *f* ~ **unit of three rollers** dreiwalziges Eindruckwerk *n* ~ **waste** Makulaturdruck *m* ~ **width** Druckbreite *f* ~ **worm** Schreibspirale *f*
**prints** Boden *m*, Drucke *pl* ~ **of sharp outlines** scharfstehende Drucke *pl* ~ **free from specks** punktfreie oder stippenfreie Drucke *pl*
**prior** früher, vorausgehend ~ **to locking** vor der Verriegelung *f*
**prior,** ~ **delivery** Vorlieferung *f* ~ **heat treatment** Wärmevergangenheit *f* ~ **right** Vorzugsrecht *n* ~ **use** offenkundige Vorbenutzung, öffentliche Vorbenutzung *f*, Vorbenutzung *f*
**priority** Berechtigungsschein *m*, Dringlichkeit *f*, (degree of) Dringlichkeitsgrad *m*, Priorität *f*, Vorfahrtsrecht *n*, (for machines) Vormerkschein *m*, Vorrang *m*, Vorrecht *n* ~ **call** vorberechtigter Anruf *m* ~ **circuit** Vorrangsschaltung *f* ~ **claim** Prioritätsbeanspruchung *f* ~ **line circuit** berechtigter Teilnehmer *m* (Teilnehmerschaltung) ~ **list** Dringlichkeitsliste *f* ~ **metal** Sparmetall *n* ~ **number** Dringlichkeitsstufe *f* ~ **prefix** Rangvermerk *m* (von Meldungen) ~ **proof** Prioritätsbeleg *m* ~ **rating** Dringlichkeitsstufe *f*, (building trade) Kennummer *f* ~ **regulations** Vorfahrtregelung *f* ~ **right** (patents) Prioritätsrecht *n* ~ **routine** Prioritätsspektroskop *n* ~ **schedule** Dringlichkeitsliste *f* ~ **select switch** Vorwählschalter *m* ~ **term** Prioritätsfrist *f*
**prism** Glasring *m*, Prisma *m*, Säule *f*, Spektralfarben *pl*, Spektrum *n* ~ **of light flint** Leichtflintprisma *n* ~ **of the second order** Deuteroprisma *n* ~ **of the third order** Tritoprisma *n*
**prism,** ~ **antenna** Reusenantenne *f* ~ **binoculars** Prismaglas *n* ~ **cam** Prismennocke *f* ~ **casing** Prismenkasten *m* ~ **edge** Prismenkante *f* ~ **face**

Prismenfläche *f* ~ **glas** Prismenglas *n* ~ **head**
Keilkopf *m* ~ **illuminator** Prismenilluminator
*m* ~ **lens** Keilglas *n* ~**-like** prismaähnlich ~**-
shaped** prismaförmig, prismatisch, prismen-
förmig ~**-shaped running surfaces** Prismalauf-
flächen *pl* ~ **strength** Prismenfestigkeit *f* ~
**support** Prismenstuhl *m* ~ **surface** Prismenfläche *f* ~ **width** Prismenbreite *f*
**prismatic** prismaförmig, prismatisch, prismen-
förmig, säulenartig ~ **arsenate of copper** Phar-
makochalzit *m* ~ **axis** Säulenachse *f* ~ **bed** Pris-
menbett *n* ~ **colors** Beugungsfarben *pl* ~ **compass** Marschkompaß *m* ~ **finder** Prismensucher
*m* ~ **guide** Prismenführung *f* ~ **light** Cellonlam-
pe *f* ~ **powder** Würfelpulver *n* ~ **seeker** Prismen-
mensucher *m* ~ **shape** Prismengestalt *f* ~ **spectrum** Prismenspektrum *n* ~ **square** Winkelprisma *n* ~ **structure** Faltwerk *n*, Säulenform *f* ~
**surveying instrument** Prismenkreuz *f* ~ **telescope** Prismenfernrohr *n* ~ **wedge** Winkelblock
*m*
**prismoid** säulenförmig
**prismoidal** prismoid -artig, -förmig
**prisometer** Prisometer *n*
**privacy** Geheimhaltung *f*
**private** privat ~ **address system** Gegensprechanlage *f* ~ **airplane** Zivilflugzeug *n* ~ **automatic
branch exchange** Selbstanschlußnebenstellen-
zentrale *f*, selbsttätige Teilnehmerzentrale *f* ~
**automatic exchange (P.A.X.)** automatische
Hauszentrale *f*, Selbstanschluß-nebenstellen-
anlage *f*, -privatzentrale *f*, selbsttätige Privat-
fernsprechzentrale *f* ~ **bank** Prüfkontaktbank
*f* ~ **(contact) bank** c-Kontaktsatz *m* ~**-branch
exchange (P.B.X.)** Fernsprechnebenstellen-
anlage *f*, Klappenschrank *m*, Nebenstellen-
zentrale *f*, Sammelanschluß *m* (teleph.),
Teilnehmerzentrale *f*, Vielfachanschluß *m*
~**-branch-exchange final selector** Mehrfach-
leitungswähler *m* ~**-branch-exchange installation for private lines** Nebenstellenanlage *f* für
Amts- und Hausverkehr ~**-branch-exchange
junction group** Mehrfachanschlußbündel *n* ~**-
branch-exchange switchboard** Nebenstellenanlage *f* ~ **connection** Privatanschluß *m* ~ **correspondence** privater Verkehr *m* ~ **corridor** geheimer Gang, Nebengang *m* ~ **exchange** Pri-
vatfernsprechzentrale *f* ~ **experimental station**
private Versuchsfunkstelle *f* (rdo) ~ **manual
branch exchange** Privatnebenstellenanlage *f*
mit Handbetrieb ~ **message** Privattelegramm
*n* ~ **pilot** Privatflieger *m* ~**-pilot certificate** Be-
scheinigung *f* eines Privatfliegers ~ **sidings** An-
schlußgeleise *n*, Privatanschlußbahn *f* ~ **station**
Haus-, Privat-stelle *f* ~ **telegraph plant** Privat-
telegrafenanlage *f* ~ **telegraph wire** Privattele-
grafenleitung *f* ~**-telephone plant** Privatfern-
sprechanlage *f* ~ **telephone station** Privatfern-
sprechstelle *f* ~ **telephone wire** Privatfernsprech-
leitung *f* ~ **wiper** Ader *f* zum Stöpselkörper, c-
Ader *f*, c-Arm *m*, c-Bürste *f*, c-Leitung *f*, Prüfkontaktarm *m* ~ **wire** Prüfdraht *m* ~ **wire
circuit** Mietleitung *f* ~**-wire service** Betrieb *m*
auf den Privatleitungen
**privately owned capital** Eigenkapital *n*
**privation** Mangel *m*
**privilege** Befugnis *f*, Gerechtsame *f*, Vergünstigung *f*, Vorrecht *n* ~ **of the author** Urheber-

recht *n* ~ **formerly granted to brew beer** Frei-
brauen *n* ~ **to mine** Berggerechtsame *f*
**privileged,** ~ **direction** Vorzugsrichtung *f* ~
**orientation** bevorzugte Orientierung *f* ~ **work**
bevorrechtigte Anlage *f* (teleph)
**privy tube for mines** Abortkübel *m* für Bergwerke
**prize** Prise *f*, (navy) Seebeute *f* ~ **court** Prisenge-
richt *n* ~ **cup** Ehrenpreis *m*
**pro,** ~ **forma invoice** Proformarechnung *f* ~ **rata**
anteilig
**proactinides** Proaktinide *pl*
**probability** Wahrscheinlichkeit *f*, Zufälligkeit *f*
~ **of collision** Stoßwahrscheinlichkeit *f* ~ **of
delay** Verzögerungsziffer *f* ~ **of detection** Nach-
weiswahrscheinlichkeit *f* ~ **of engagement**
Wahrscheinlichkeit *f* des Besetztseins ~ **of
hitting** Trefferwahrscheinlichkeit *f* ~ **of loss**
Verlustziffer *f* ~ **of the ionisation** Ionisations-
wahrscheinlichkeit *f*
**probability,** ~ **amplitude** Wahrscheinlichkeits-
amplitude *f* ~ **calculus** Wahrscheinlichkeits-
rechnung *f* ~ **curve** Wahrscheinlichkeitskurve
*f* ~ **density** Wahrscheinlichkeitsdichte *f* ~
**diagram** Bodentreffbild *n* ~ **distribution** Wahr-
scheinlichkeitsverteilung *f* ~ **frequency function**
Wahrscheinlichkeits-Häufigkeitsfunktion *f* ~
**index** Wahrscheinlichkeitsindex *m* ~ **law**
Wahrscheinlichkeitsgesetz *n* ~ **theory** Wahr-
scheinlichkeitsrechnung *f*
**probable** augenscheinlich, voraussichtlich, wahr-
scheinlich **most** ~ **value** wahrscheinlichster
Wert *m*
**probable,** ~ **error** wahrscheinlicher Fehler *m* ~
**extent of the deposit** mutmaßliche Erstreckung
*f* der Lagerstätte
**probation, (time of)** ~ Probezeit *f*
**probe, to** ~ abtasten, eindringen, prüfen, sondieren
**probe** Fühler *m*, Geber *m*, Koppel-schleife *f*,
-stift *m* (waveguide), Meßkopf *m*, Prüfung *f*,
Senknadel *f*, Sonde *f*, Sondierung *f*, Stechheber
*m*, Sucher *m*, Testkopf *m*, Versuchsrakete *f* ~
**advance spindle** Sondentransportspindel *f* ~
**characteristics** Sondencharakteristiken *pl* ~
**coupling** Tast-(Sucher)Kupplung *f* ~**electrode**
Suchelektrode *f* ~**ended knife** geknöpftes Messer *n* ~**-holding device** Sondenhalterung *f* ~
**proof of concrete** Betonbeschuß *m* ~ **scanning**
Sondenabtastung *f* ~ **scanning tube** Sondenab-
taströhre *f* ~ **technique** Sondentechnik *f* ~ **test**
Sondierversuch *m* ~ **tip** Tastenspitze *f* ~ **unit**
Sonde *f* mit Verstärker
**probing (for depth)** Lotung *f*, Tasten *n*, Untersuchung *f* ~ **a bore** Sondierungsbohrung *f*
**probing,** ~ **auger** Sondierbohrer *m* ~ **head** Son-
despitze *f* ~ **lever** Fühlhebel *m* ~ **rod** Sondier-
nadel *f* ~ **unit** Tastkopf *m*
**problem** Aufgabe *f*, Frage *f*, Problem *n* ~ **of
solid bodies** Festkörperproblem *n*
**problematic** problematisch, ungewiß
**problems of disgression** Ausweichprobleme *pl*
**procedural** Verfahrenssache *f*
**procedure** Arbeits-prozeß *m*, -weise *f*, Behandlungsweise *f*, Gang *m*, Gesetzmäßigkeit *f*, Prozedur *f*, Verfahren *n*, Versuchsanordnung *f*,
Vorgang *m*, Vorgehen *n*
**procedure,** ~ **message** Betriebsspruch *m* ~ **sign**
Funkabkürzung *f* ~ **turn** Verfahrenskurve *f*
(aviat) ~ **word** Betriebswort *n*

proceed, to ~ fortschreiten, gehen, verlaufen to ~ at blank speed einsteuern to ~ in open order mit vergrößertem Abstand *m* fahren

proceed, ~-to-dial signal Wählbereitschaftszeichen *n* ~-to-send signal Sendefreigabesignal *n*

proceeding Fortschreiten *n*, Prozeß *m*, Verhandlung *f*, Vorgang *m* ~ in the field of stability (of ships) Stabilitätsbelange *m* ~ wave sich ausbreitende Welle *f*

proceedings Abhandlungen *pl*, Sitzungsbericht *m* ~ by default Versäumnisverfahren *n*

proceeds Ertrag *m*, Gewinn *m* ~ from goods Warenerlös *m* ~ from investments Einnahmen *pl* aus Beteiligungen ~ of sum of net basic costs Stichsummenertrag *m*

proceeds, ~ figure Erlöszahl *f* ~ realized by a sale Erlös *m*

process, to ~ (for further disposition) abfertigen, bearbeiten, fertigen, herstellen, unterwerfen, verarbeiten, etwas einem Verfahren *n* to ~ a material einen Werkstoff behandeln

process Arbeitsvorgang *m*, Art *f*, Gang *m*, Klischeeherstellung *f* print; Methode *f*, Prozeß *m*, Übereinanderkopieren (photo), Verfahren *n*, Verlauf *m*, Vorgang *m*,

process, ~ of aftertreating with sulfate of copper Nachkupferungsverfahren *n* ~ of brazing Hartlöten *n*, Lötarbeit *f* ~ of calculation Rechengang *m* ~ of combustion Verbrennungsvorgang *m* in the ~ of being constructed im Bau begriffen ~ of conversion Umwandlungsverfahren *n* ~ of detection Detektionsprozeß *m* to be in the ~ of dissolution in Auflösung begriffen sein ~ for dressing pictures Bildzurichteverfahren *n* ~ of evolution Kubikwurzelziehen *n* ~ of flow Strömungsvorgang *m* ~ of formation Bildungsvorgang *m*, Entstehungsvorgang *m* ~ of fuel conversion Brennstoffumwandlungsverfahren *n* ~ of gasification Vergasungsverfahren *n* ~ of gold refining Goldscheideverfahren *n* ~ of hardening Hartmachen *n* ~ of hardsoldering Hartlöten *n* ~ of irradiation Bestrahlung *f* ~ of manufacture Herstellungsgang *m* ~ of marking out a line by means of surveying rods Abpfählen *n* der Linie ~ of printing Druckvorgang *m* ~ of production Werdegang *m* ~ for the production of a lake Verlackungsverfahren *n* ~ of program Programmverlauf *m* ~ of setting Abbindeverlauf *m*, Erhärtungsverlauf *m* ~ of soldering Lötarbeit *f* ~ of solution Lösungsvorgang *m*

process, ~ annealing Ausglühen *n* ~-block making Netzätzung *f* ~ card Laufkarte *f* ~ control Fertigungssteuerung *f*, Verfahrensregelung *f* ~ design Verfahrensplan *m* ~ diagram Betriebsschaubild *n* ~ engraver Chemigraf *m* ~ engraving Chemigrafie *f* ~ fields Prozeßfelder *pl* ~ flow-sheet Prozeßfließschema *n* (diagram), Verfahrensstammbaum *m* ~ function Prozeßbelegung *f* ~ lens Reproduktionsobjektiv *n* ~ piping Rohrleitungen *pl* für das Verfahren *n* ~ piping diagram Prozeßrohrleitungsschema *n* ~ planning Arbeitsuntersuchung *f* ~ plate fotomechanische Platte *f* ~ reaction rate Änderungsgeschwindigkeit *f* ~ sheet Aufbereitungsverfahren *n* ~ stages Prozeßschritte *pl* ~ steam Betriebs-, Fabrikations-dampf *m* ~ tank Verfahrenstank *m* ~ unit Verarbeitungsanlage *f* ~ vapor fractometer Betriebskontrollfraktome-

ter *m* ~ variable Meßgröße *f* ~ work Reproduktionsaufnahme *f*

processable behandelbar

processing Bearbeitung *f*, Behandlung *f*, fabrikationsmäßige Herstellung *f*, Verarbeitung *f*, Verfahren *n* (film) ~ of any construction material Werkstoffverfeinerung *f* ~ of waste materials Verarbeitung *f* von Abfuhrstoffen

processing, ~ agent Vered(e)lungsmittel *n* ~ industry Vered(e)lungsindustrie *f* ~ line Bearbeitungs-, Fertigungs-straße *f* ~ liquor Vered(e)lungsbad *n* ~ period Bearbeitungszeit *f* ~ plant Verarbeiter *m* ~ roughness Bearbeitungsrauhigkeit *f* ~ scars Bearbeitungsnarben *pl* ~ shop Bearbeitungsbetrieb *m* ~ stage Verarbeitungsstufe *f* ~ steam condensate Fabrikkondensat *n* ~ water Industrie-, Nutz-wasser *n*

procession Programm *n* (g/m), Prozession *f*, Zug *m*

proclaim, to ~ ankündigen, kundgeben, (a judgment) verkünden

proclamation Ankündigung *f*, Bekanntmachung *f*, Proklamation *f*

procreation Zeugung *f*

Proctor needle Prüfnadel *f*

procurable erhältlich

procuration Prokura *f*

procure, to ~ anschaffen, beschaffen, besorgen, erzielen, herbeischaffen, verschaffen, zulegen

procurement Beschaffung *f*, Besorgung *f* ~ department Bestellungsabteilung *f* ~ office Beschaffungsamt *n* ~ service agency Einkaufs-, büro *n*

procuring of water Wassergewinnung *f*

prod, to ~ stacheln, stechen

prod-type terminals Dornleitungsschuhe *pl*

prodding stick Stechstock *m*

produce, to ~ anfallen, anrichten, ausbringen, darstellen, erbringen, erzeugen, fabrizieren, gewinnen, herstellen, hervorbringen, machen, produzieren, (geom) verlängern, verursachen, vorführen, wirken, züchten to ~ on a large scale am laufenden Band fabrizieren to ~ a magnetic field ein Magnetfeld *n* erzeugen to ~ in quantity serienmäßig herstellen to ~ output Ausbeute *f* liefern to ~ rosette copper rosettieren to ~ by selfgenerating method (milling machine) nach dem Abwälzverfahren *n* herstellen

produce Ausbeute *f*, Erzeugnis *n*, Leistung *f*, Produkt *n*

producer Erzeuger *m*, Generator *m*, Hersteller *m*, Produzent *m* ~ base Generatorsockel *m* ~ body Generatorschacht *m* ~ gas Gaserzeugergas *n*, Generatorgas *n*, Luft(generator)gas *n*, Sauggas *n*, Siemens-Gas *n*, Steinkohlengeneratorgas *n* ~-gas engine Sauggasmotor *m* ~-gas flame Generatorgasflamme *f* ~-gas fuel Gasogen *n* ~-gas operation Generatorgasbetrieb *m* ~-gas plant Generatorgasanlage *f* ~-gas tar Gasteer *m* ~ shell Generatormantel *m*

producing erzeugend ~ light lichtentwickelnd

product Ergebnis *n*, Erzeugnis *n*, Fabrikat *n*, Produkt *n*, Ware *f*

product, ~ used in acetylene lamps Beagid *n* ~ of amperes by volts Produkt *n* aus Ampere und Volt ~ of broiling, (or calcining) Röstprodukt *n* ~ of combustion Verbrennungs-ergebnis *n*, -erzeugnis *n*, -produkt *n* ~ of decomposition

Zerfallsprodukt *n* ~ **of disintegration** Verwitterungsgebilde *n*, Zerfallgebilde *n* ~ **of disunion** Spaltprodukt *n* ~ **of inertia** Deviationsmoment *n*, Trägheitsprodukt *n*, Zentrifugalmoment *n* ~ **of injection molding** Spritzgut *n* ~ **of power loading and root of wing loading** Bauzahl *f* ~ **of roasting** Röst-erzeugnis *n*, -produkt *n* ~ **of separation** Spaltprodukt *n*, Trennungserzeugnis *n* ~ **of transformation** Umwandlungsprodukt *n*

**product,** ~ **demodulator** Produktdemodulator *m* ~ **density** Produktdichte *f* ~ **loss analysis statement** Verlustübersicht *f* bei Produkten ~ **modulator** idealer Modulator *m* ~ **molded from insulating material** Spritzteil *m* aus Isoliermaterial ~ **movements** Mengenbewegungen *pl* ~ **nucleus** erzeugter Kern *m*, Folgekern *m*, Nachfolgekern *m* ~ **patent** Stoffpatent *n* ~ **representation** Produktdarstellung *f* ~ **space** Warenlagerraum *m* ~ **stocks** Warenbestände *pl*

**production** Ausbringen *n*, Ausbringung *f*, Ausführung *f*, Betriebsleistung *f*, Darstellung *f*, Erzeugung *f*, Fabrikation *f*, Fertigung *f*, Gewinnung *f*, Herstellung *f*, Hervorbringung *f*, Produktion *f*, Werk *n*, Zeugung *f* **to be in** ~ serienmäßig hergestellt werden

**production,** ~ **of castings** Gußerzeugung *f* ~ **of cores** Kernherstellung *f* ~ **of fibers** Fasergewinnung *f* ~ **of heat** Wärmebildung *f* ~ **of ingot steel in an open hearth** Flußstahlerzeugung *f* auf offenem Herd ~ **of maps from aerial photographs** Fotogrammetrie *f* ~ **of nitrogen** Stickstoffgewinnung *f* ~ **of oscillations** Schwingungserzeugung *f*, -vorgang *m* ~ **of oxygen** Sauerstoffgewinnung *f* ~ **of pharmaceutical products** Arzneimittelerzeugung *f* ~ **of pure gas** Reindarstellung *f* des Gases ~ **of small work** Werkstück *n* ~ **of sparks** Funkenbildung *f* ~ **of stress at rest** Ruhespannung *f* ~ **of water gas** Wassergaserzeugung *f*, -herstellung *f*

**production,** ~ **airplane** serienmäßiges Flugzeug *n*, Serienflugzeug *n* ~ **analyzer** Auswertgerät *n* ~ **bottleneck** Fertigungsengpaß *m* ~ **capacity** Fertigungskapazität *f*, Produktionsfähigkeit *f* ~ **certificate** Produktionsbescheinigung *f* ~ **coefficient** Nachlieferungskoeffizient *m* ~ **control office** Betriebswirtschaftsstelle *f* ~ **cost** Produktionskosten *pl* ~ **costs** Baukosten *pl*, Erzeugungskosten *pl*, Gestehungspreis *m* ~ **department** Betriebsabteilung *f* ~ **drop** Leistungsknick *m* ~ **engine** Serienmotor *m* ~ **engineer** Fertigungsfachmann *m* ~ **engineering** Fertigungsplanung *f* ~ **engineering department** Abteilung *f* für Fertigungsplanung ~ **equipment** Fabrikationsanlage *f* ~ **gauge** Arl (Arbeitslehre), Arbeitslehre *f* ~ **goods** Massenware *f* ~ **improvement** Leistungsertüchtigung *f* ~ **lag** Leistungsminderung *f* ~ **lathe** Fabrikationsdrehbank *f* ~ **line** Fertigungsstraße *f*, Förderband *n* ~ **-line fabrication (or manufacture)** Bandfabrikation *f* ~ **manager** Betriebsdirektor *m* ~ **method** Produktionsverfahren *n* ~ **model** Baureihe *f* ~ **order** Fertigungsauftrag *m* ~ **part** Produktionsteil *m*, Werkstück *n* ~ **plan** Erzeugungsvorhaben *n* ~ **planning** Arbeits-planung *f*, -vorbereitung *f* ~ **process** Bau-, Fertigungs-, Produktions-programm *n*, Herstellungsverfahren *n* ~ **range** Baureihe *f* ~**rate** Produktions-

ziffer *f* ~ **requirement** Fertigungsaufwand *m* ~ **research** Betriebsforschung *f* ~**schedule** Arbeitsplan *m*, Produktionsplan *m*, Schmelzprogramm *n* ~ **scheduling** Arbeitsplanung *f* ~ **sheet** Arbeitsplan *m* ~ **stage** Fertigungs-, Serien-reife *f* ~ **supervision** Betriebsüberwachung *f* ~ **supervisor (or manager)** Produktionsleiter *m* ~ **target** Produktionsziel *n* ~ **test** Fabrikmessung *f* ~ **testing** Erprobung *f* der Erzeugnisse der Reihenfertigung, Fertigungsreife *f*, Reihenprüfung *f*, Serienprüfung *f* ~ **testing of engines** Prüfung *f* der Serienmotoren ~ **time** Fertigungszeit *f* ~**-type milling machine** Produktionsfräsmaschine *f* ~ **units** Fabrikationsbauten *pl* ~ **wages** Leistungslohn *m*

**productive** einbringend, ergiebig, ertragsfähig, produktiv, rentabel ~ **bed** abbauwürdige Schicht *f* ~ **capacity** Ertragsfähigkeit *f*, Mengen-, Produktions-leistung *f* ~ **coal formation** abbauwürdige Steinkohlenlagerung *f* ~ **mine** Ausbeutezeche *f* ~ **time** Hauptzeit *f*

**productiveness** Ausgiebigkeit *f*, Ergiebigkeit *f*, Rentabilität *f*, Tragfähigkeit *f*

**productivity** Ergiebigkeit *f*, Leistungsfähigkeit *f*, Produktivität *f*, Rentabilität *f* ~ **situation** Ertragslage *f*

**products,** ~ **of combustion** Feuergase *pl* ~ **of inertia** Diviationsmomente *pl* ~ **for peacetime industries** Friedens-erzeugnisse *pl*, -material *n*, -werkstoff *m* ~ **pressed from insulating material** Preßteile *pl* aus Isoliermaterial

**proeutectic** proeutektisch

**proeutectoid** proeutektoid

**profess, to** ~ Beruf ausüben

**profession** Beruf *m*, Fach *n*, Gewerbe *n*, Handwerk *n*, Kunst *f*, Stand *m*

**professional** Sachbearbeiter *m*; berufsmäßig, gewerbsmäßig ~ **colleague** Fachgenosse *m* ~ **disease** Berufskrankheit *f* ~ **draftsman** Berufszeichner *m* ~ **education** Fachbildung *f* ~ **engineering** fachtechnisch ~ **expert** fachmäßig ~ **hand camera** Berufshandkamera *f* ~ **knowledge** Fachkenntnis *f* ~ **man** Fachmann *m* ~ **photographer** Fachfotograf *m* ~ **pilot** Berufsflugzeugführer *m* ~ **school** Lehranstalt *f* ~ **service unit** Fachabteilung *f* ~ **speciality** Fachsparte *f* ~ **terminology** Fachsprache *f* ~ **training** Berufs-ausbildung *f*, -erziehung *f* ~ **world** Fachwelt *f*

**professorship** Lehramt *n*, Professur *f*

**proffer, to** ~ anbieten, antragen

**proficient** bewandert, fertig

**profilated** profiliert ~ **brick** Formziegel *m*

**profile, to** ~ fassondrehen, fassonieren, kopierfräsen, profilieren to ~**-mill** profilfräsen

**profile** Durchschnitt *m*, Geländedurchschnitt *m*, Kontur *f*, Längsprofil *n*, Profil *n*, Profilform *f*, Profilieren *n*; Querschnitt *m*, Seiten-ansicht *f*, -riß *m* ~ **of a line** Linienprofil *n* ~ **whose mean line is an arc of circle** Kreisbogenprofil *n* ~ **of rip** Rippen-umriß *m*, -profil *n* ~ **of a roll** Walzenprofil *n* ~ **of surface** Bodenprofil *n* ~ **across valley (cross section)** Querprofil *n* des Tales ~ **of welding rod** Schweißdrahtform *f*

**profile,** ~ **coefficient** Profilbeiwert *m* ~ **coordinates measured from mean line** Dickenbelegung *f* ~ **cords** Profilschnüre *pl* ~**-correcting factor** Profilverschiebungsfaktor *m* ~ **curve** Querschnittslinie *f* ~ **cutter** Fassonfräser *m*, Form-fräser *m*, Fräser *m* mit gefrästen Zähnen, Profil-fräser *m*, -zieher *m* ~drag Profilwiderstand *m* ~ **drag area** Profilwiderstandsfläche *f* ~ **drag coefficient** Profilwiderstandsbeiwert *m* ~ **form** Profillehre *f* ~ **gauge** Formlehre *f* ~ **gliding coefficient** Profilgleitzahl *f* ~ **ground** Zahnflanken *pl* geschliffen ~ **iron** Fasson-, Form-, Profil-eisen *n* ~ **line** Schnittlinie *f* ~ **map** Profilkarte *f* ~ **mill cutter** Profilwalze *f* ~ **milling** Umrißfräsen *n* ~ **milling machine** Schablonenfräsmaschine *f* ~**profile offset** Profilverschiebung *f* ~ **press for metal rods** Metallstrangpresse *f* ~ **recording** Profilregistrierung *f* ~ **resistance** Profilwiderstand *m* ~ **stamp** (surface pressing) Profilstempel *m* ~ **strut** Profilrohrstiel *m*, Profilrohrstrebe *f* ~ **thickness** Profildurchmesser *m* ~ **tool steel** Formstahl *m* ~ **tracer** Umrißfühler *m* ~ **tracer attachment** Umrißkopiereinrichtung *f* ~ **tracing** Umrißtastung *f* ~ **turned piece** Formdrehteil *m*, geformter Drehteil *m* ~-**type chart** Reliefkarte *f* ~ **washer** Profildichtung *f*

**profiled** geformt, gerillt ~ **bar** profilierter Stab *m* ~ **strand** Profillitze *f* ~ **top** profiltragendes Oberteil *n* ~ **wire** Fassondraht *m*

**profiler** Kopier-, Nachform-fräsmaschine *f*

**profiling** Formgebung *f*, Profilgestaltung *f*, Profilieren *n*, Profilierung *f* ~ **attachment** Kopiervorrichtung *f* ~ **cutters** Fassonmesser *n* ~ **job** Kopierarbeit *f* ~ **machine** Nachformfräsmaschine *f* ~ **lathe** Fassondrehbank *f* ~ **point** Profilierungspunkt *m* ~ **pressure** Kopierdruck *m* ~ **slide** Schablonensupport *m*

**profilometer** Profilometer *n*, Profil-, Rauheitsprüfer *m*

**profit, to** ~ benutzen, gewinnen

**profit** Ausbeute *f*, Bruttoertrag *m*, Ertrag *m*, Erwerb *m*, Gewinn *m*, Nutzen *m*, Profit *m*, Reinertrag *m*, Verdienst *m* ~ **and loss statement** Gewinn- und Verlustrechnung *f* ~ **margin** Gewinnspanne *f* ~ **sharing** Gewinnbeteiligung *f*

**profitability statement** Rentabilitätsaufstellung *f*

**profitable** ergiebig, fett, gewinnbringend, nutzbringend, nützlich, vorteilhaft **to be** ~ sich lohnen ~ **ore** abbauwürdiges Erz *n*

**profitableness** Einträglichkeit *f*, Rentabilität *f*

**profound** tief, tiefschürfend

**profoundity** Tiefe *f*

**profuse** überfließend, verschwenderisch

**progenitor** Vorgänger *m*

**prognosis** Vorhersage *f*

**prognostic chart** Vorhersagekarte *f* (meteor)

**prognostics** Prognose *f*

**program, to** ~ programmieren

**program** (material) Darbietung *f*, Erzeugungsbereich *n*, Plan *m*, Programm *n* ~ **of experimental work** Versuchsprogramm *n* ~ **of routine maintenance** Überwachungsplan *m*

**program, ~ checking** Prüfungsprogramm *n* ~ **circuit** Rundfunk-leitung *f*, -übertragungsverbindung *f* ~ **circuit loading** sehr leichte Bespulung *f* für Musikübertragung ~ **control** Folgeregelung *f* ~ **counter** Befehlsadreßregister *n*, Behelfszähler *m*, Steuerbefehlsspeicher *m* ~ **cylinder** Programmwalze *f* ~ **dated** programmgespeichert ~ **desk** Regiepult *m* ~ **device** Programmsteuerung *f* ~ **display** Programmabbildung *f* ~ **distribution system** Fernausbreitung *f*,

Rundfunkvermittlung *f* ~ **flow chart (or diagram)** Programmablaufplan *m* ~ **library** Befehls-bibliothek *f*, Programm-bibliothek *f* ~ **parameter** Jeweilsparameter *m* (info proc) ~ **register** Befehlsregister *n*, Programm-, Steuerspeicher *m* ~ **repeater** Rundspruchverstärker *m* ~ **scaler** Programmzähler *m* ~-**sensitive error** programmbedingtes Versagen *n* ~ **step** Programmschritt *m* ~ **stop-switch** Programmstoppschalter *m* ~ **switch** Programmschalter *m* ~ **tape** Programm-band *n*, -lochstreifen *m* ~ **transmission** Rundfunkübertragung *f* ~ **transmission over wires** Drahtrundspruch *m* ~ **transmitter** Programmgeber *m*

**programmatic** programmgemäß

**programmed, ~ check(ing)** programmierte Prüfung *f* ~ **control** Zeitplanregelung *f*

**programmer** Programmierer *m*, Programmplaner *m*

**programming** Datensteuerung *f* ~ **plug-in unit** Programmiereinschub *m*

**progress, to** ~ fortschreiten, vorrücken

**progress** Ablauf *m*, Fortschritt *m*, Raumgewinn *m* **progress** Ablauf *m*, Fortschritt *m*, Raumgewinn *m*, Verlauf *m*, Zug *m* ~ **of elongation** Dehnungsverlauf *m* ~ **of fatigue** Ermüdungsverlauf *m* ~ **in an invention** Erfindungsschritt *m* ~ **with ordinary flame** Fortschreiten *n* bei gewöhnlicher Flamme (min.) ~ **in penetration** Eindringen *n* der Sonde

**progress, ~ flow chart (or sheet)** Arbeitslaufkarte *f* ~ **reports** Fortschrittsberichte *pl*

**progression** Abstufung *f*, Folgeordnung *f*, Fortschreiten *n*, fortschreitende Bewegung *f*, fortschreitende Reihe *f* (math.), Progression *n*, Reihe *f* (math.), Reihenansatz *m* (math.), Reihenentwicklung *f*, senkrechte Reihe *f*, Stufenleiter *m*, Stufensprung *m* (einer Drehbank) ~ **by stages (or degrees)** Stufengang *m*

**progressive** fortschreitend, gestaffelt, progressiv, stufenweise, (allmählich) vorrückend ~ **assembly** Fließ-arbeit *f*, -bandmontage *f*, fließende Fertigung *f*, fließender Zusammenbau *m* ~ **assembly-conveyer machinery** Fließarbeitseinrichtung *f* ~ **assembly line** Fließband *n* ~ **burning powder** rauchloses Pulver *n* ~ **control of traffic lights** grüne Welle *f* ~ **cut** Schälschnitt *m* ~ **development** Reihenentwicklung *f* ~ **diminution** fortschreitende Abnahme *f* ~ **failure** fortschreitender Bruch *m* ~ **manufacture** Fließarbeit *f* ~ **motion** fortschreitende Bewegung *f* ~ **position** aussichtsreiche Stelle *f* ~ **rate** Staffelung *f* ~ **ratio** Stufensprung *m* ~ **scanning** unmittelbar aufeinanderfolgende Abtastung *f*, Zeitfolgeverfahren *n* ~ **set of photographs** Bildsatz *m* ~ **settlement** fortschreitende Setzung *f* ~ **smoke screening** Nebelwalze *f* ~ **totals** Staffelsummen *pl* ~ **wave** fortschreitende Welle *f* ~ **welding** Vorwärtsschweißung *f*

**progressively burning powder** Kernpulver *n*

**prohibit, to** ~ verbieten

**prohibited** verboten ~ **area** Luftsperrgebiet *n*, verbotene Zone *f* ~ **zone** Sperrgebiet *n*

**prohibiting equipment** Verhinderungsschaltung *f*

**prohibition** Sperre *f*, Untersagung *f*, Verbot *n* ~ **of importation** Einfuhrverbot *n* ~ **of payment** Zahlungsverbot *n*, ~ **of trade** (with foreign countries) Grenzsperre *f*

**prohibitive** ausschließend ~ **cost** untragbare Kosten *pl*

**prohibitory** schützend ~ **duty** Sperrzoll *m*

**project, to** ~ ausladen, entwerfen, heraustragen, hervortragen, konstruieren, loten, planen, projektieren, projizieren, überhängen, vorführen (film), vorspringen, werfen, wiedergeben (film) **to** ~ **beyond** übertragen **to** ~ **(a screen)** ausleuchten

**project** Bauentwurf *m*, Entwurf *m*, Plan *m*, Planung *f*, Projekt *n*, Unternehmen *n*, Vorhaben *n* ~ **engineer** ausführender Ingenieur *m*, (design) Entwurfsingenieur *m* ~ **manager** Bauleiter *m* ~ **site** Baustelle *f*

**projected** projiziert ~ **area** Entwurfsteil *m* ~ **area in contact with rolls** gedrückte Fläche *f* ~ **area of piston** projizierte Kolbenoberfläche *f* ~ **propeller-blade area** projizierter Flächeninhalt *m* des Luftschraubenblattes, projizierte Schraubenflügelfläche *f* ~ **scale** Projektionsaussteuerungsanzeiger *m*

**projectile** Geschoß *n*, Granate *f*, Projektil *f* ~ **for signal pistols** Wurfkörper *m*

**projectile,** ~ **base** Geschoßboden *m* ~ **charge** Geschoßfüllung *f* ~ **force** Wurfkraft *f* ~ **jacket** Geschoßmantel *m* ~ **lathe** Geschoßdrehbank *f* ~ **motion** Wurfbewegung *f* ~ **time** Geschoßflugzeit *f*

**projecting** abstehend, ausladend, ausspringend, erhaben, (beam) überstehend, vorragend, vorspringend, vorstehend ~ **arm of the driving end bearing** Auslegearm *m* des Antriebslagers ~ **base** vorspringende Grundschicht *f* ~ **beam** Projektionsstrahl *m* ~ **camera** Projektionskammer *f* ~ **edge** vorspringender Wulst *m* (r.r.) ~ **length (of bolt)** Überstand *m* ~ **lens** Projektions-linse *f*, -objektiv *n* ~ **lug** vorspringender Lappen *m* ~ **shaft** überhängende Welle *f* ~ **stones** vorspringende Bruchsteine *pl*

**projection** Abbildung *f*, Ablichtung *f*, (anode) Ansatz *m*, Auskragung *f*, Ausladung *f*, Auslauf *m*, Erhebung *f*, (of pipe, etc) Halsansatz *m*, Kimme *f*, Nase *f*, Projektion *f*, Risalit *m*, (geom.) Riß *m*, Überhängen *n*, Vorsprung *m*, Vorstoß *m*, Warze *f*, Wurf *m* ~ **of a cam** Vorsprung *m* eines Nockens ~ **without distortion of range** streckentreue Abbildung *f* ~ **of flight path** Flugbahnprojektion *f* ~ **of meridian** Meridianfigur *f*

**projection,** ~ **aperture** (film projector) Fenster *n*, Projektionsfenster *n* ~ **apparatus** Laufbildwerfer *m*, Projektions-einrichtung *f*, -vorrichtung *f* ~ **booth** Vorführungs-kabine *f* (film), -raum *m* (film) ~ **coil** Projektionsspule *f* ~ **distance** Projektionsentfernung *f* (film) ~ **equipment** Projektionseinrichtung *f* ~ **eyepiece** Projektionsokular *n* ~ **factor** Tonkopfleistung *f* ~ **gate** Projektionsfenster *n* ~ **lamp** Bildwerferlampe *m* ~ **optics** Projektionsoptik *f* ~ **printer (or printing) machine** optische Kopiermaschine *f* ~ **procedure** Projektionsverfahren *n* ~ **receiver** Projektionsempfänger *m* ~ **room** Vorführungs-kabine *f*, -raum *m* (film) ~ **scale** Kinoskala *f* ~ **screen** Bildschirm *m*, Bildwand *f*, Leinwand *f*, Projektionsschirm *m* ~ **screen loudspeaker** Bühnenlautsprecher *m* ~ **screen pointer** Projektionspfeil *m* ~ **system** Projektions-einrichtung *f*, -vorrichtung *f* ~ **table**

**running on gimbal bearings** Projektionstisch *m* (kardanisch gelagert) ~ **television** Projektionsverfahren *n* ~ **tube** Leuchtschirmrohr *n*, Projektionsröhre *f* ~-**type television receiver** Projektionsbildschreibröhre *f* ~ **video picture** (of large size) Fernbildgroßaufnahme *f* ~ **weld** Dellennaht *f*, Warzenpunktschweißung *f* ~ **welder** Buckelschweißmaschine *f* ~ **welding** Warzenschweißen *n*

**projectionist** Vorführer *m*

**projective** projektiv ~ **mapping** projektive Abbildung *f* ~ **plane** projektive Ebene *f* ~ **scanning field** projektives Punktfeld *n*

**projectivity** Projektivität *f*

**projector** Bildwerfer *m*, Lichtbildwerfer *m*, Lichtwerfer *m*, Projektions-apparat *m*, -gerät *n*, Projektor *m*, Sammelsystem *n*, Scheinwerfer *m*, Vorführgerät *n* (photo), Wurfgerät *n* ~ **with lenses** Linsenscheinwerfer *m*

**projector,** ~ **frame** Werferrahmen *m* ~ **lamp** Scheinwerfer- und Bildwerferlampe *f* ~ **lens** Scheinwerferlinse *f* ~ **rake** Projektorneigung *f* ~ **signal** Granatsignal *n* ~ **surface** (of loudspeaker) Lautsprecherabstrahlfläche *f*

**projectors** Darstellungslinien *pl*

**projecture** Risalit *m*

**prolate ellipsoid** gestrecktes Rotationsellipsoid *n*

**prolific** erfindungsreich, fruchtbar

**prolixity** Weitschweifigkeit *f*

**prologue** Vorspannprogramm *n*

**prolong, to** ~ erneuern, verlängern (pipe organ)

**prolong** Abstrakte *f*

**prolongation** Ansatz *m*, Aufschub *m*, Fristgewährung *f*, Fristung *f* (min.), Prolongation *f*, Stillstandsfrist *f* (min.), Verlängerung *f* ~ **of the axis of the bore** verlängerte Seelenachse *f* ~ **of a bill** (of exchange) Wechselverlängerung *f* ~ **of delay** Verlängerung *f* der Wartezeit

**prolonged** prolongiert, verlängert ~ **alternating loading** Dauerwechselbeanspruchung *f* ~ **alternating-stress strength** Dauerwechselfestigkeit *f* ~ **effect** Andauern *n* einer Wirkung ~ **test** Langzeitversuch *m*

**promenade deck** Promenadendeck *n*

**prominence** Erhöhung *f*, Kuppe *f*, Vorsprung *m*

**prominent** hervortretend **to be** ~ vorspringen

**prominently marked** deutlich bezeichnend

**promise, to** ~ versprechen, zusichern

**promise** Versprechen *n*, Zusage *f*

**promising** aussichtsreich, erfolgversprechend, hoffnungsvoll **most** ~ vielversprechend

**promissory note** Schuldbrief *m*, Schuldschein *m*

**promontory** Vorgebirge *n*

**promote, to** ~ aufrücken lassen, befördern, fördern, im Rang befördern, Vorschub leisten, vorwärts bringen

**promoter** Förderer *m*, Gründer *m*

**promotin**<sub></sub> **swelling** quellungsfördernd

**promotion** Beförderung *f*, Erhebung *f* ~ **by selection** Beförderung *f* auf Grund der Auswahl ~ **by seniority** Beförderung *f* nach dem Dienstalter

**promotion,** ~ **expenses** Gründungskosten *pl* ~ **list** Beförderungsliste *f* ~ **money** Gründungskosten *pl*

**promotive effort (or endeavor)** Förderungsbestreben *n*

**prompt, to** ~ anregen, antreiben, eingehen, soufflieren, zu etwas treiben, vorsagen

**prompt** prompt, (execution of orders) pünktlich, schlagartig ~ **fission neutrons** prompte Neutronen *pl* ~ **gamma** prompte Gammastrahlung *f* ~ **note** Mahnzettel *m*

**prompter** Souffleurkasten *m*

**promptly** rechtzeitig

**promptness** Raschheit *f*

**prone** liegend **to be** ~ **to** neigen zu

**prong** Forke *f*, Gabel *f*, Klaue *f*, Spitze *f*, Stachel *m*, Zacke *f*, Zacken *m*, Zinke *f*, Zinken *m* ~ **of jack** Kontaktstift *m* ~ **of the tube base** Steckerstift *m* eines Röhrensockels ~ **of a tuning fork** Arm *m* einer Stimmgabel, Stimmgabelzinken *m*, Zinke *f* einer Stimmgabel

**prong,** ~ **hoe** Karst *m* ~ **templet** Aufsteckschablone *f*

**pronged** gezinkt, zackenförmig

**pronounce, to** ~ ansprechen als, (deutlich) aussprechen

**pronouncement of sentence** Verurteilung *f*

**pronunciation** Aussprache *f*

**Prony brake** Bremszaum *m*, Pronischer Zaum *m*, Pronybremse *f*

**proof, to** ~ prüfen, dicht oder undurchlässig machen

**proof** Beleg *m*, Beweis *m*, Beweis-führung *f*, -stück *n*, Erweis *m*, Fahnenkorrektur *f* (print.), Nachweis *m*, Nachweisung *f*, Probe *f*, Probe--abdruck *m*, -abzug *m*, -druck *m*, Prüfabzug *m*, Vordruck *m*; beständig, echt, fest, undurchlässig, widerstandsfähig ~ **of content** Inhaltsnachweis *m* ~ **by evidence** Zeugenbeweis *m* ~ **of identity** Ausweis *m*, Legitimation *f* ~ **of stability against lifting and tilting** Standsicherheitsnachweis *m* gegen Abheben und Umkippen ~ **of title** Nachweis *m* der Empfangsberechtigung

**proof,** ~ **certificate** Prüfungszeugnis *n* ~ **charge** Versuchsladung *f* ~ **cock** Probehahn *m* ~ **copy** Andruck *m* ~ **impression** Probe-abzug *m*, -druck *m* ~ **load** Probe-, Versuchs-last *f* ~ **mark** Korrektur *f*, Probestempel *m*, Stempelplatte *f* ~ **paper** Abziehpapier *n* ~ **plane** Prüf--körper *m* (electr.), -platte *f* (electr.) ~--**reading** Korrekturlesen *n* ~ **sheet** Korrektur--abzug *m*, -bogen *m*, Probe-abdruck *m*, -bogen *m* ~ **spirit** Normalweingeist *m* ~ **stick** (sugar) Probestecher *m* ~ **stress** Prüfbeanspruchung *f*, Prüfspannung *f*, Streckfließgrenze *f*, Streckgrenze *f* ~ **vinegar** Normalessig *m* ~ **voltage** Prüfspannung *f* ~ **weight** Probe-, Versuchs-last *f*

**proofed sheets** Bogenkorrektur *f*

**proofing** Abziehen *n* (print.), Dichtung *f*, Dichtungsmittel *n*, (material) Durchtränken *n* ~ **layer** (gas, water) Dichtungsschicht *f* ~ **machine** Andruckmaschine *f* ~ **paper** Korrekturpapier *n* ~ **press for rubber plates** Gummiandruckpresse *f*

**prop, to** ~ abfangen, abspreizen, (mine) absteifen, abstützen, aufbocken, stützen, unterstützen, versteifen, verstreben **to** ~ **up** aufbocken

**prop** Absteifung *f*, Balkenstütze *f*, Bolzen *m*, (Hebel) Dreh-, Stütz-punkt *m*, Druckstrebe *f*, Gestängebelastung *f*, Knagge *f*, Pfahl *m*, Rohrstrebe *f*, Schore *f*, Stapelstütze *f*, Steife *f*, Stempel *m*, Stütz-balken *m*, -säule *f*, Stütze *f*, Verstrebung *f*, Widerhalt *m* **back** ~ Strebe *f* ~ **of a lever** Hebelunterlage *f* ~-**withdrawer** Raubwinde *f*

**propaganda** Propaganda *f*, Werbung *f* ~ **drawing** Werbezeichnung *f* ~ **using light effects** Lichtwerbung *f* ~ **work** Werbearbeit *f*

**propagate, to** ~ (waves) ausbreiten, fortpflanzen, verbreiten

**propagated error** mitlaufender Fehler *m*

**propagating wave** Raumwelle *f*, sich ausbreitende oder auslaufende Welle *f*

**propagation** Ausbreitung *f*, Fortpflanzung *f*, Übertragung *f*, Verbreitung *f* ~ **of error** Fehlerfortpflanzung *f* ~ **of glide** Gleitungsausbreitung *f* ~ **of heat** Wärme-ausbreitung *f*, -fortpflanzung *f* ~ **beyond the horizon** Überreichweite *f* ~ **of pressure (wave)** Druckausbreitung *f* ~ **of sound** Schall-ausbreitung *f*, -fortpflanzung *f* ~ **of waves** Fortpflanzung *f* (Verbreitung) elektrischer Wellen

**propagation,** ~ **action** Ausbreitungsvorgang *m* ~ **characteristic** Ausbreitungserscheinung *f* ~ **charge** Übertragungsladung *f* ~ **coefficient** Ausbreitungskonstante *f* ~ **conditions** Ausbreitungs-bedingungen *pl*, -verhältnisse *pl* ~ **constant** Ausbreitungs-konstante *f*, Fortpflanzungsgröße *f*, (spezifische) Fortpflanzungskonstante *f*, (spezifisches) Übertragungsmaß *n* ~ **constant per section** Übertragungsmaß *n* je Glied ~ **constant per unit length** Fortpflanzung *f* je Längeneinheit (teleph.) ~ **direction** Fortpflanzungsrichtung *f* ~ **distance** Überschlagslänge *f* ~ **factor** Übertragungsfaktor *m* einer glatten (homogenen) Leitung ~ **phenomenon** Ausbreitungsvorgang *m* ~ **speed** Durchschlagsgeschwindigkeit *f* ~ **time** Laufzeit *f* ~ **time distortions** Laufzeitverzerrungen *pl* ~ **velocity** Ausbreitungsgeschwindigkeit *f*

**propagational movement** Fortschreitungsbewegung *f*

**propane** Propan *n*

**propanol** Propanol *n*

**propel, to** ~ antreiben, treiben, vorwärtstreiben **to** ~ **around** herumschleudern

**propellant** Kraftstoff *m*, Treib-ladung *f*, -mittel *n*, -stoff *m* (g/m) ~ **charge** Treib-ladung *f*, -mittel *n*, -satz *m*, -stoffzuladung *f* ~ **control valve** Treibstoffregelventil *n* ~ **cut-off** Brennschluß *m* (g/m) ~ **efficiency** Vortriebswirkungsgrad *m* ~ **gas** Treibgas *n* ~ **grain** Festtreibstoff *m* ~ **ingredient** Treibstoffkomponente *f* ~ **injector** Einspritzsystem *n* (g/m) ~ **mixing ratio test** Abstimmversuch *m* (g/m) ~ **requirement** Treibstoffbedarf *m*

**propelled swinging crane** Drehlaufkran *m*

**propeller** Antreiber *m* (aviat.), Antriebs-gerät *n*, -maschine *f*, Luftschraube *f*, Propeller *m*, Schiffsschraube *f*, Schraube *f*, Treiber *m*, Treibschraube *f* **(full-feathering)** ~ Luftschraube *f* ~ **with adjustable blades** Luftschraube *f* mit verstellbaren Blättern ~ **in feathering position** Luftschraube *f* in Segelstellung

**propeller,** ~ **anti-icer** Luftschraubenvereisungsschutz *m* ~ **area** Luftschraubenflächeninhalt *m* ~ **axis** Luftschraubenachse *f* ~ **balancing**

**machine** Propellerwaage *f* **~-balancing stand** Gleichgewichtswaage *f* für die Luftschraube **~ bar** Luftschrauben-eisen *n*, -stab *m* **~ bearing** Luftschraubenlagerung *f* **~ blade** Luftschrauben-, Propeller-blatt *n*, Propeller-, Schrauben-flügel *m*
**propeller-blade, ~ angle** Luftschraubenblattwinkel *m* **~ area** Luftschraubenblattflächeninhalt *m* **~ back** Saugseite *f* des Luftschraubenflügels **~ beat** Propellerflügelschlag *m* **~ element theory** Luftschraubenblattgrundtheorie *f* **~ face** Druckseite *f* des Luftschraubenflügels **~ flexural deflection** Schraubenblattdurchbiegung *f* **~ rake** Blattneigung *f* **~ section** Konstruktionsblattschnitt *m*, Querschnitt *m* des Luftschraubenflügels **~ shank** Luftschraubenflügelschaft *m*, Schaft *m* des Luftschraubenblattes **~ solidity** Schraubenflügelsolidität *f* **~ speed** Schraubenblattgeschwindigkeit *f* **~ tip** Luftschraubenblattspitze *f* **~ width** Luftschraubenblattbreite *f*
**propeller, ~ blast** Luftschrauben-windstoß *m*, -strahl *m* **~ blower** Luftschraublüfter *m* **~ bolt** Schraubenbolzen *m* **~ boss** Luftschraubennabenwulst *m* **~ brake** Luftschraubenbremse *f* **~ cap** Propellerhaube *f* **~ cavitation** Luftschraubenaushöhlung *f* **~ chamber ratio** Luftschraubenwölbungsverhältnis *n* **~ characteristic curve** Luftschraubenkennlinie *f* **~ clearance** Bodenfreiheit *f* der Luftschraube, Propellerblattspitzenfreiheit *f* **~ cone** Propellerzapfen *m* **~ covering** Luftschraubenüberzug *m* **~ disc** Luftschraubenkreis *m*, Schraubenkreis *m* **~ disc area** Kreisfläche *f*, Luftschraubenkreisfläche *f*, Schrauben-kreisfläche *f*, -flächeninhalt *m* **~ disc area loading** Schraubenführungskreisflächenbelastung *f* **~-disc loading** Schraubenkreisbelastung *f* **~ drive** Luftschraubenantrieb *m* **~ driving face** Druckseite *f* des Luftschraubenflügels **~ efficiency** Luftschraubenleistung *f*, Schraubenwirkungsgrad *m* **~ end** Andrehseite *f*, Luftschraubenseite *f* **~ etching** Luftschraubenätzung *f* **~ fan** Schraubenlüfter *m* **~ flange hub** Luftschraubenflansch *m* **~ flutter** Luftschraubenschwankung *f* **~ fracture** Propellerbruch *m* **~ gear** Treibapparat *m* **~ governor** Luftschraubensteigungsregler *m* **~ gust** Propellerbö *f* **~ hub** Luftschrauben-nabe *f*, -nabenstock *n*, Propellernabe *f*, Schraubennabe *f* **~ hub attachment flange** Luftschraubenanbauflansch *m* **~ hub barrel** Nabenschaft *m* **~-hub cone** Luftschraubennabenkonus *m*, Zentrierkonus *m* **~ hub flange** Nabenteller *m* **~ hub lock nut** Nabenschraube *f* **~ interference** Luftschraubeninterferenz *f*, Strahlstörung *f* **~ load curve** Schaubild *n* der Luftschraubenleistung **~ master blade** Musterschraubenblatt *n* **~ mixer** Planeten-, Schrauben-rührer *m*, Propellermischer *m* **~ modulus** Fortschrittsgrad *m* der Luftschraube **~ noise** Luftschrauben-, Propeller-geräusch *n* **~ nosepiece** Luftschraubenlager *n* **~ nozzle** Schubdüse *f* **~ nut** Luftschraubenbefestigungsmutter *f* **~ output** Vortriebsleistung *f* **~ path** Luftschraubengang *m* **~ performance** Schraubenleistung *f* **~ pitch** Luftschrauben-, Propeller-, Schrauben-steigung *f*
**propeller-pitch, ~ change** Luftschraubenverstel-

lung *f* **~ control** Luftschraubenregelung *f* **~ controls** Propeller-steuerung *f*, -verstelleinrichtung *f* **~ gauge** Luftschraubenuhr *f* **~ indexing** Einstellung *f* der Luftschraubensteigung, Luftschraubeneinstellung *f* **~ indicator** Luftschraubensteigungs-anzeiger *m*, -messer *m*, Luftschraubenstellungsanzeiger *m* **~ reversing** Verstellen *n* des Propellers auf Bremsstellung
**propeller, ~ plane** (of rotation) Schraubenebene *f* **~ post** Schraubensteven *m* **~ protection** Luftschraubenschutz *m* **~ pump** Schraubenschaufler *m* **~ race** Luftschrauben-brunnen *m*, -tunnel *m*, -wind *m* **~ radius** Luftschraubenradius *m* **~ reduction gear** Luftschraubenuntersetzungsgetriebe *n* **~ regulator valve** Ölumsteuerventil *n* für Zweistellungsschraube **~ root** Luftschraubenwurzel *f* **~ r.p.m.** Luftschraubendrehzahl *f* **~ rupture** Flügelbruch *m* **~ section** Luftschrauben(durch)schnitt *m* **~ shaft** Antriebs-, Propeller-, Schrauben-welle *f* **~ shaft gear spider** Luftschraubenwellengetriebekreuz *n* **~ shaping machine** Luftschraubenfräsmaschine *f* **~ shroud** Manteltiefe *f* **~ slip** Schlupf *m* der Luftschraube **~ slip ratio** Luftschraubenschlupfverhältnis *n* **~ slip stream** Luftschraubenstrahl *m*, Schraubenstrahl *m* **~ speed** Schraubendrehzahl *f* **~-speed control** Luftschraubendrehzahlregler *m* **~ spinner** Luftschraubennabenhaube *f* **~ structural clearance** Abstand *m* des Propellers von Flugzeugteilen **~ strut** Schraubenstrebe *f* **~ test stand** Propellerprüfstand *m* **~ thrust** Druckseite *f* des Luftschraubenflügels, Luftschrauben-schub *m*, -zug *m*, Propellerschub *m*, Schrauben-druck *m*, -kraft *f*, -schub *m*, -zugkraft *f* **~ thrust block** Luftschraubenschublager *n* **~ thrust in ratio to pitch** Zug-und-Steigung *f* **~ thrust bearing** Drucklager *n* zur Aufnahme des Luftschraubenschubes, Luftschraubenlager *n* für Schubaufnahme, Luftschraubenschublager *n* **~ thrust line** Schubleistung *f* des Schraubenstrahls **~ tip** Blattspitze *f* der Luftschraube **~-tip failure** Luftschraubenblattspitzenschaden *m* **~-tip radius** Luftschraubenhalbmesser *m* **~ torque** Drehmoment *n* (aviat.), Drehplatte *f*, Luftschraubendrehmoment *n* **~ trailer** Luftschraubenförderwagen *m* **~ tube** Kardanrohr *n* **~ unit** Propelleraggregat *n* **~ wash** Schraubenstrahl *m* **~ whirling test** Schleuderprüfung *f* **~ width ratio** Schraubenflügelweitenverhältnis *n*
**propelling, ~ action** Wirbelwirkung *f* **~ cartridge for rifle grenade** Gewehrkartusche *f* **~ charge** Ladung *f*, Pulvertreibladung *f*, Wurfladung *f* **~ charge of powder** Pulvertreibladung *f* **~-charge container** Kartusche *f* **~-charge increment** Teilladung *f* **~ equipment** Fahrgerät *n* **~ force** Trieb-, Vortriebs-, Zug-kraft *f* **~ gas valves and fittings** Treibgasarmaturen *pl* **~ machinery** Antriebsmaschinerie *f* **~ means** Treiber *m* **~ nozzle** Rückstoßdüse *f* **~ pawl** Stoßklinke *f* **~ piston** Strömungskolben *m* **~ power** Treibkraft *f* **~ pressure** Rückdruck *m*
**propensity** Empfindungsdauer *f*
**proper** angemessen, eigen, eigentlich, geeignet, gehörig, passend, sachgemäß, schicklich, tauglich, zugehörig, zweckmäßig **not ~** uneigentlich **~ to felt** (textiles) verfilzungsfähig **in ~ phase relation** phasenrichtig

proper, ~ authority zuständige Stelle *f* ~ energy Eigenenergie *f* ~ energy density Energieeigendichte *f* ~ feed Normalvorschub *m* ~ filling einwandfreie Abfüllung *f* ~ fraction echter Bruch *m* (math.) ~ frequency Eigenfrequenz *f* ~ function Eigenfunktion *f* ~ functioning Betriebsbereitschaft *f* ~ mass Ruhmasse *f* ~ motion wahre Eigenbewegung *f* ~ nipotent element eigentlich nipotentes Element *n* ~ number Eigenwert *m* ~ orthogonal eigentliche Orthogonale *f* ~ rate Eigengangsgeschwindigkeit *f* ~ setting of springs richtige Einstellung *f* der Federn ~ state Eigenzustand *m* ~ states of the field of matter Eigenschwingungen *pl* des Materialfeldes ~ symmetries Eigensymmetrien *pl* ~ time Eigenzeit *f* ~ value problem echtes Randwertproblem *n* ~ variables Eigengrößen *pl* ~ vector Eigenvektor *m*

properties Dekors *pl* ~ of symmetry Symmetrieeigenschaften *pl*

property Besonderheit *f*, Eigenheit *f*, Eigenschaft *f*, Eigentum *n*, Eigentümlichkeit *f*, Gut *n* ~ of absorbing solvents Lösungsmittelaufnahmefähigkeit *f* ~ of dying through Durchfärbevermögen *n* ~ of being an integer Ganzzahligkeit *f* ~ of material Werkstoffeigenschaft *f*

property, ~ damage Sachbeschädigung *f*, Sachschaden *m* ~ damage insurance Haftpflichtversicherung *f* ~ disposal Materialverwaltung *f* ~ divisible among the creditors Teilungsmasse *f* ~ insurance Sachversicherung *f* ~ owner Landeigentümer *m* ~ right Eigentumsrecht *n* ~ spinning Spinnbarkeit *f*

Properzi rod Properzi-Walzdraht *m*

prophase Prophase *f*

prophylactic verhütend, vorbeugend ~ antidote Gegenmittel *f*

propinquity Nähe *f*

propionic acid Propionsäure *f*

propolis Bienenharz *m*, Kitt-, Kleb-, Vor-wachs *n*

proportion, to ~ anpassen, bemessen, proportionieren

proportion Ausmaß *n*, Eben-, Gleich-maß *n*, Größe *f*, (of ingredients) Mengenverhältnis *n*, Proportion *f*, Umfang *m*, Verhältnis *n*, Verhältnisgleichung *f*

proportion, in ~ anteilig ~ of acid Säureverhältnis *n* ~ of ingredients Mengungsverhältnis *n* ~ of load Lastanteil *m* ~ of mix Mengungsverhältnis *n* ~ of a square quadratisches Verhältnis *n* in ~ to the square (of) im quadratischen Verhältnis *n* ~ of stress Spannungsanteil *m* in ~ to nach Maßgabe *f* von ~ of voids Porenanteil *m* ~ by volume Raum-, Volum-verhältnis *n* ~ by weight Gewichtsverhältnis *n*

proportional proportional, verhältnismäßig ~ action statisches Verhalten *n* ~ action controller Proportionalregler *m* (P-Regler) ~ band Proportionalbereich *f* ~ compass Proportional-, Reduktions-zirkel *m* ~ control factor Proportionalitätsfaktor *m* ~ controller gain Proportionalitätsfaktor *m* ~ counter proportionaler Zähler *m* ~ counter tube proportionales Zählrohr *n* ~ divider Reduktionszirkel *m* ~ limit verhältnismäßige Grenze *f*, Proportionalitätsgrenze *f* ~ offset P-Abweichung *f* (aut. contr.) ~ plus derivative action

controller PD-Regler *m* (aut. contr.) ~ plus integral action controller PI-Regler *m* ~ plus rate action controller PD-Regler *m* ~ plus reset action controller Proportional-Integral-Regler *m* (PI-Regler), Regler *m* mit Angleichung ~ position action Proportionalregelung *f* ~ reset (control) action (nachgehende) Rückführung *f* ~ reset control Integralregelung *f* ~ test bar Proportionalstab *m*

proportionality Proportionalität *f*, Verhältnismäßigkeit *f* ~ factor Verhältniszahl *f*

proportionate anteilig, proportional, verhältnismäßig ~ to angemessen

proportioner Mischanlage *f*

proportioning Anpassung *f*, Bemessung *f*, Berechnung *f* ~ device Mischungsregler *m* ~ distributor Zumeßverteiler *m* ~ elements Mischungsregelglieder *pl* ~ plant Dosieranlage *f* ~ plunger Zumeßkolben *m* ~ screw Dosierschnecke *f* ~ valve Zumeßventil *n*

proportions Bemessung *f*

proposal Anregung *f*, Antrag *m*, Vorschlag *m* ~ for settlement Vermittlungsvorschlag *m* ~ form Antragsformular *n*

propose, to ~ anregen, beantragen, vorlegen, vorschlagen

proposed cable scheme Kabelplanung *f*

proposition Antrag *m*, Entwurf *m*, Lehrsatz *m* (math.), Planung *f*, Projekt *n*, Satz *m*, Vorschlag *m*

propped up gestützt

propping Versteifung *f*, Verstreben *n*, Verstrebung *f* ~ of a beam Absteifung *f* eines Balkens

proprietary, ~ article (Monopol) Erzeugnis *n* ~ right Eigentumsrecht *n*

proprietor Eigentümer *m*, Inhaber *m*

proprietorship Eigentumsrecht *n*

props Stollenholz *n*, Stützbalken *m*

propshore Firstenstempel *m*

propulsion Antrieb *m*, Antriebskraft *f*, Fortbewegung *f*, Stoß-, Vorwärts-bewegung *f*, Vortrieb *m* ~ efficiency Vortriebs-wirkung *f*, -wirkungsgrad *m* ~ machine Vortriebsmaschine *f* ~ nozzle Rückstoß-, Vortriebs-düse *f* ~ position Schubstellung *f* ~ system Marschantrieb *m* (g/m) ~ unit for continuous operation Langzeit-Triebwerk *n* ~-unit contractor Getriebeschaltschütz *m* (g/m) ~-unit governor Triebwerksregler *m* ~ unit instrument panel Triebwerksgeräte-brett *n*, -tafel *f*

propulsive, ~ efficiency Antriebswirkungsgrad *m* ~ effort Vortriebserzeugung *f* ~ force Vortriebskraft *f* ~ jet Antriebsstrahl *m* ~ means Vortriebseinrichtung *f* ~ output Antriebsleistung *f* ~ unit Triebwerk *n*, Vortriebseinheit *f*

propyl alcohol Propylalkohol *m*

propylamine Propylamin *n*

propylene Propylen *n* ~ hydride Propan *n*

prorate, to ~ einschränken

proscenium Proszenium *n*

prosecute, to ~ fortführen

prosecution Fortsetzung *f*, (of a patent application) Prüfverfahren *n*

prosign Funkbetriebszeichen *n*

prospect, to ~ prospektieren, schürfen to ~ on common account auf gemeinschaftliche Rechnung *f* schürfen

**prospect** Ausblick *m*, Aussicht *f*, Erwartung *f* ~ **drilling** Schürfbohrung *f* ~ **hole** Schürfloch *n* **prospecting** Aufsuchung *f*, Lager-, Schürfstätten-suche *f*, Prospektieren *n*, Schürfen *n*, Schürfung *f* ~ **bore** Untersuchungsbohrung *f* ~ **operations** Schürfbetrieb *m* ~ **shaft** Schürf-, Versuchs-schacht *m* ~ **work** Ausrichtungs-, Schürf-arbeit *f*

**prospective** angehend, aussichtsreich, in Aussicht stehend, voraussichtlich ~ **buyer** Interessent *m*

**prospector** Schürfer *m*

**prospectus** Ankündigung *f*, Broschüre *f*, Preisliste *f*, Prospekt *m*, Werbe-druckschrift *f*, -schreiben *n*

**prosperous** schwunghaft

**prostrate** auf der Erde hingestreckt

**protect, to** ~ abschirmen, abwehren, armieren, beschützen, decken, schonen, schützen, in Schutz nehmen, sichern, umdecken **to** ~ **from draft** vor Zugluft *m* schützen

**protected** geschützt ~ **against corrosion** korrosionsgeschützt ~ **against high tension** hochspannungsgeschützt ~ **against jets of water** spritzwassergeschützt ~ **by patent from** patentiert vom ~ **by registration** zeichenrechtlich geschützt

**protected, to** ~ **frequency** geschützte Frequenz *f* ~ **front sight** Mantelkorn *n* ~ **switchgear** Schutzschalter *m*

**protecting,** ~ **box** äußere Gewindeschutzmuffe *f* ~ **cap** Schutzkappe *f* ~ **(grounded) cap** Schutzerdungs-haube *f*, -kappe *f* ~ **clothes** Schutz-anzug *m*, -kleidung *f* ~ **cover** Schutz-abdeckung *f*, -kappe *f* ~ **edge** Schutzrand *m* ~ **the gearing** das Getriebe schonend ~ **grounding network** Schutzerdungsnetz *n* ~ **screen** Abschlußblech *n*, Fangkorb *m*, Schutzsieb *n* ~ **sheet** Schutzblech *n* ~ **strip** Stoßleiste *f* ~ **tube for pushrod** Stoßstangenschutzrohr *n* ~ **varnish** Deckfirnis *m* ~ **wall** Futter-, Schutz-mauer *f* ~ **walls** Dammschutz *m* ~ **wire helix** Schutzerdungsspirale *f*

**protection** Befestigung *f*, Beschirmung *f*, Beschützung *f*, Bewahrung *f*, Deckung *f*, Schirmwirkung *f*, Sicherung *f*, Schutz *m*, Verteidigung *f*, Wehr *f*

**protection,** ~ **against accidental contact** Berührungsschutz *m* bei Motoren ~ **of the bank** Uferschutz *m* ~ **against cold** Kälteschutz *m* ~ **against corrosion** Korrosionsschutz *m* ~ **of dunes by planting** Dünenbepflanzung *f* ~ **of dunes by tree planting** Dünenaufforstung *f* ~ **against dust** Staubschutz *m* ~ **against electric--shock hazard** Berührungsschutz *m*, Berührungsspannungsschutz *m* ~ **from freezing** Gefrierschutz *m* ~ **of fuel deposits** Behälterschutz *m* ~ **from glare** Blendschutz *m* ~ **against ice** Eisschutz *m* ~ **of industrial property** gewerblicher Rechtsschutz *m* ~ **against lightning** Blitzschutz *m* ~ **of pole butt** Stockschutz *m* ~ **against projectile fragments** Schutz *m* gegen Splitterwirkung ~ **against radiation** Strahlungsschutz *m* ~ **of river bank** Uferschutzwerk *n* ~ **by smoke screening** Nebelschutz *m* ~ **of sol** Solschutz *m* ~ **of the stock** Fußschutz *m* ~ **of toothed wheels** Abdeckung *f* der Zahnräder ~ **against torsion** Verdrehsicherung *f* ~ **of**

**trade-marks** Markenschutz *m* ~ **against vibration** Erschütterungsschutz *m* ~ **against wear and tear** Verschleißschutz *m* ~ **against weather** Witterungsschutz *m* ~ **by zinc** Zinkschutz *m*

**protection,** ~ **bag** Schutzbalg *m* ~ **box** Schutzkapsel *f* ~ **contact** Schukostecker *m* ~ **covering** Schutzbelag *m* ~ **enamel finish** Schutzlackierung *f* ~ **fence** Sicherheits-Leitplanke *f* (Straße) ~ **grease** Schutzfett *n* ~ **grill** Steinschlag-, Vorsatz-gitter *n* ~ **ground** Schutzerde *f* ~ **hole** Vorbohrloch *n* ~ **housing** Schutzgehäuse *n* ~ **mounts** Schutzfassung *f* ~ **plate** Schutzleiste *f* ~ **shell** Schlagpanzer *m* (Hochofen) ~ **sleeve** Schutzhülse *f* ~ **stone** Uferpflaster *n* ~ **stone below dam** Sturzbett *n* aus geschütteten Steinen ~ **strip** Schutzleiste *f* ~ **survey** Strahlenschutz-überwachung *f* ~ **switch with resistance** Vorstufenschalter *m* ~ **timber** Holzschutz *m* ~ **tuning** Abschirmung *f*

**protective** schützend ~ **action** Schutzwirkung *f* ~ **agent** Schutzmittel *n* ~ **apparatus for high tension** Hochspannungsschutz-apparat *m*, -vorrichtung *f* ~ **apron** Bleigummischutzschürze *f* ~ **arrangement surrounding the bulb** strahlenundurchlässiger Röhrenschutzmantel *m* ~ **awning** Gasplane *f* ~ **barrage** Feuersperre *f* ~ **belt of trees** Waldschutzstreifen *m* ~ **cable** Blitzschutzkabel *n* ~ **cap** (for thread cutting) innere Gewindeschutzmuffe *f*, Schutzmütze *f* ~ **capacitor for overvoltages** Überspannungsschutzkondensator *m* ~ **casing** Schutz-mantel *m*, -ummantelung *f* ~ **choke** (coil) Schutzdrosselspule *f* ~ **circuit** Schutzschaltung *f* ~ **clothing** Gasbekleidung *f*, Körperschutzmittel *n*, Schutzbekleidung *f* ~ **coat** Deckmittel *n*, Schutzmantel *m* ~ **coat of paint** Schutzfarbenanstrich *m* ~ **coating** (etching) Ätzgrund *m*, Schutz-anstrich *m*, -überzug *m*, -umkleidung *f* ~ **coloration** farbiger Anstrich *m* ~ **conductor** Schutzleiter *m* ~ **container** Schutzkasten *m* ~ **cover** Schutz-abdeckung *f*, -mantel *m* ~ **cover of the focal plane frame** Meßrahmenschutzdeckel *m* ~ **covering** Schutz-hülle *f*, -überzug *m*, -verkleidung *f* ~ **covering of cables** Kabelschutzbekleidung *f* ~ **device against exhaled air** Atemschutzgerät *n* ~ **device for blind-flying formations** Rammschutz *m* ~ **dome** Windschutzhaube *f* ~ **earth** Schutzerdung *f* ~ **edge rail** Bandschutzschiene *f* ~ **effect** Schutzwirkung *f* ~ **envelope** Schutzumschlag *m* ~ **fence** Fang-, Schutz-zaun *m* ~ **film** Schutzüberzug *m* ~ **fitting** Schutzbeschlag *m* ~ **flank** Sicherungsflügel *m* ~ **functions relay** Relaisschutzfunktionen *pl* ~ **fuse** Spannungssicherung *f* ~ **gap** Schutzzwischenraum *m* ~ **gas** Schutzgas *n* ~ **gauze** Schutzgaze *f* ~ **glass** Schutzglas *n* ~ **glasses** Schutzbrille *f* ~ **globe** Schutzglocke *f* ~ **glove** Schutzhandschuh *m* ~ **grounding device** Schutzerdungsvorrichtung *f* ~ **grounding means** Schutzerdungsmaßnahme *f* ~ **grounding resistance** Schutzerdungswiderstand *m* ~ **head leader** Startband *n* (film) ~ **hedge** Fangzaun *m* ~ **helmet** Schutzhelm *m* ~ **helmet against dust** (or sandblast) Staubschutzhelm *m* ~ **hoarding** Schutzzaun *m* ~ **horn** Schutzhorn *n* ~ **housing** Schutzumkleidung *f*, Verkleidungsgehäuse *n* ~ **layer** Deck-, Schutz-schicht *f* ~ **lead-glass** Schutzbleiglas *n* ~ **mask** Gesichts-, Schutz-

-maske f ~ mat Gasschutzdecke f ~ measure Schutzmaßnahme f ~ metal covering Metallschutzschlauch m ~ metal ribbon Metallschutzband n ~ ointment Gasschutzsalbe f ~ paint Schutzfärbung f ~ pipe Schutzrohr n ~ process Schutzverfahren n ~ reactance coil Strombegrenzungsdrossel f ~ relay Schutzrelais n ~ resistor Schutzwiderstand m ~ rubber sleeve Knickschutz m ~ screen Schutz-schirm m, -wand f, Strahlenschutz m ~ screen for overturn structure Schutzschirm m für den Kopfübersturz ~ shell Schutzmantel m ~ shield Schutzschild n ~ sleeve Schutz-hülle f, -muffe f ~ spark gap Schutzfunkenstrecke f ~ spring Schutzfeder f ~ strap Vorsteckriemen m ~ suit undurchlässiger Schutzanzug m ~ switchgear against contact voltage Berührungsschutzschalter m ~ system Schutzsystem n ~ tariff Schutzzoll m ~ tube Futter-, Verkleidungs-rohr n ~ use Schutzanwendung f ~ wall Schutzwand f ~ wing Sicherungsflügel m ~ wire Drahthindernis n ~ works Schutzanlagen pl

protector Gestängeschoner m, Schoner m, Sicherungskästchen n ~ against ice Eisschutz m

protector, ~ box Sicherungskasten m ~ frame Sicherungsgestell n ~ ground Schutz-, Sicherungs-erdung f ~ rack Sicherungsgestell n ~ sleeve Schutzmuffe f ~ strip Blitzableiterstreifen m

protectorate Schutzgebiet n

protein Protein n ~ material Eiweißkörper m ~ metabolism Eiweißstoffwechsel m

proteolysis Eiweißabbau m

protest, to ~ against beanstanden

protest Einsprache f, Einspruch m, Protest m, Widerspruch m

proto, ~-bitumen Protobitumen n ~-chloride Chlorür n ~-chloride of copper Kupferchlorid n

protocol Niederschrift f

protogenic primär

protogine Talkgranit m

proton Proton n ~ binding energy Protonenbindungsenergie f ~ bombardment Protonenbeschuß m ~ capture Protoneneinfang m ~ collisions Protonenstöße pl ~-electron mass ratio Protonelektron-Massenverhältnis n ~ precession frequency Protonenpräzessionsfrequenz f ~-proton chain Proton-Proton-Kettenreaktion f ~ radiation Protonenstrahlung f ~ range Protonenreichweite f ~ recoil chamber Protonenrückstoßkammer f ~ resonance Protonenresonanz f ~ scattering Protonenstreuung f ~ spin Protonenspin n

prototype Ausgangsbaumuster n, Erstausführung f, Prototyp m, Urbaumuster n, Urbild n, Urmuster n, Ursprungstyp m, Vorbild n ~ aircraft Musterflugzeug n ~ analysis Vergleich m mit Prototyp

protoxide Oxydul n ~ of iron Eisenmonoxyd n, Ferrooxyd n

protract, to ~ auftragen (geom.) to ~ a survey eine Aufmessung f auftragen

protracted langwierig, nachhaltig ~ level Winkellibelle f (mit Mikroskop) ~ treatment Protrahierung f

protractor Anlegegoniometer n, Auswert(e)gerät n, Auswertungsinstrument n, Gehrmaß n, Gradbogen m, Kursmesser m, Schmiege f, Strahlen-

zieher m, Transporteur m, Winkelmesser m, Winkelmeß-gerät n, -instrument n, Zirkel m ~ and scale Zielgevierttafel f ~ ocular head Winkelmeßokularkopf m

protrude, to ~ herausragen, heraustreten, hervortreten, vorspringen

protruding ausspringend, überstehend ~ lug (fuse) Zünderstutzen m

protrusion Vorsprung m, Vorwölbung f

protuberance Ausstülpung f, Auswuchs m, Höcker m

proustite Arseniksilberblende f, lichtes Rotgültigerz n, Proustit m, Silberblende f

provable erweisbar

prove, to ~ bestätigen, beweisen, (ordnance) erproben, nachweisen, die Probe machen (math.), prüfen to ~ correct zutreffen to ~ a failure fehlschlagen to ~ good bewähren to ~ unavailing scheitern to ~ useful bewähren

proven territory bekanntes Gebiet n

provide, to ~ anschaffen, ausfüllen, ausstaffieren, beschaffen, besorgen, versehen, versorgen to ~ with choke coal verdrosseln to ~ for bereithalten, vorsehen to ~ with a gutter einkehlen to ~ with a lead coating ausbleien to ~ with lid (or cover) deckeln to ~ oneself with sich eindecken to ~ with rails verschienen to ~ a resist coat (in metal etching) abdecken to ~ with ausstatten, versehen to ~ with winding bespulen

provided vorgesehen ~ by the budget etatmäßig ~ for versorgt ~ with ausgerüstet oder ausgestattet mit

providing Beschaffung f

province Gebiet n

proving Erprobung f, Prüfung f ~ of fitness Bewährung f ~ ground Prüffeld n, Versuchsfeld n ~ load Prüflast f

provision, to ~ verproviantieren

provision Bereitstellung f, Beschaffung f, Einrichtung f, Maßnahme f, Proviant m, Reserven pl, Verfügung f, Verordnung f, Vorrat m, Vorschrift f ~ of control signals Abgabe f der Steuersignale ~ for loading means (or for sinding coil) Bespulung f

provisional behelfsmäßig, provisorisch, vorläufig ~ agreement Vorvertrag m ~ decree einstweilige Verfügung f ~ dike Pinnplanke f ~ estimate Voranschlag m ~ injunction einstweilige Verfügung f ~ map Ersatzkarte f ~ plant Notanlage f ~ specification vorläufige Patentbeschreibung f

provisions Nahrungsmittel n, Proviant m ~ of law Rechtsvorschrift f ~ concerning the protection of submarine cables Kabelschutzrecht n ~ for taxes account Rückstellungen pl für Steuern

provisions engineering Nährmitteltechnik f

prow Bug m ~ problem Bugwellenproblem n

proximate analysis Immediat-, Kurz-analyse f

proximity geringer Abstand m, Nähe f ~ effect Eigenkapazitätseffekt m, Nahwirkung f ~ effect term Nahwirkungsglied n ~ field meter Nahfeldmesser m ~ fuse Abstand-, Annäherungs-, Influenz-zünder m ~ fuse tube Röhre f für Distanzzünder

proxy Bevollmächtigter m, Stellvertreter m

prune, to ~ ausästen, (electric lines) beschneiden

**pruner** Baumschere *f*
**pruning** Aufpfropfung *f*, Ausästen *n*, Beschneiden *n* ~ **hook** Haumesser *n*, Sichel *f* ~ **knife** Gartenmesser *n*, Saß *m* ~ **rod with shears** Baumschere *f* ~ **saw** Fuchsschwanz *m*, Pfropfmesser *n* ~ **shears** Baumschere *f*
**Prussian blue** Berlinerblau *n*, Ferriferrozyanid *n*, Preußischblau *n*
**prussiate** Blutlaugensalz *n*
**prussic acid** Blausäure *f*, Zyanwasserstoffsäure *f*, Hydrozyansäure *f*
**pry, to** ~ **open** aufstemmen **to** ~ **over** verstemmen
**pry pole** Hebeladefuß *m*
**pseudo** scheinbar, vorgetäuscht, unecht ~**-adiabatic** pseudoadiabatisch ~**-code** Pseudobefehl *m* ~**-crystalline** pseudokristallin ~**-dichroism** Pseudodichroismus *m* ~**-fading** Pseudoschwund *m* ~**-image** Pseudobild *n* ~**-instruction** Pseudobefehl *m*, symbolischer Befehl *m* ~**-lineal motion** pseudolineale Bewegung *f* ~**-malachite** Phosphorkupfer *n* ~**-molecule formation** scheinbare Molekülbildung *f*
**pseudomorph crystal** Afterkristall *m*
**pseudomorphic** pseudomorph
**pseudomorphosis** Pseudomorphose *f*
**pseudomorphy** Pseudomorphie *f*
**pseudonym** Deckname *m*, Pseudonym *n*
**pseudo,** ~ **periodic quantity** scheinbar periodische Größe *f* ~**-plane motion** pseudo-ebene Bewegung *f* ~**-random result** Pseudostreuergebnis *n* ~**-random sequence** Pseudostreufolge *f* ~ **scalar quantity** pseudoskalare Größe *f*
**pseudoscopic** tiefenverkehrt ~ **space image** tiefenverkehrtes Raumbild *n*
**pseudo,** ~**-sphere** Pseudosphäre *f* ~ **stereoscopic effect** stroboskopischer Effekt *m* (film) ~**-stress** Pseudospannung *f* ~ **tensor** Pseudotensor *m* ~**-thermostatics** Pseudothermostatik *f*
**psilomelane** Hartmanganerz *n*, Psilomelan *n*, Schwarzbraunstein *m*, schwarzer Glaskopf *m*
**psophometer** Geräuschmesser *m*, Geräuschspannungsmesser *m*
**psophometric,** ~ **electromotive force** Geräusch *n* elektromotorischer Kraft *f*, Geräusch-EMK *f* ~ **filter** Ohrkurvenfilter *m* ~ **potential difference** Geräuschspannung *f* ~ **power** Rauschleistung *f* ~ **voltage** Geräuschspannung *f*
**psychogalvanometer** Psychogalvanometer *n*
**psychological,** ~ **examination** Eignungsuntersuchung *f* ~ **test** Eignungsprüfung *f*
**psychologically unique yellow** farbtongleiches reines Gelb *n*, reines farbengleiches Gelb *n*
**psychrometer** Feuchtemesser *m*, Feuchtigkeitsmesser *m*, Psychrometer *n*
**p-type,** ~ **conduction** Defekt-, Mangel-halbleitung *f* ~ **conductivity** Löcherleitung *f*, P-Leitung *f* ~ **semiconductor** p-Typ-Halbleiter *m*
**public** Benutzer *pl*, Öffentlichkeit *f*; offenkundig, öffentlich **of** ~ **utility** gemeinnützig
**public,** ~ **address amplifier** Rundfunkgroßverstärker *m* ~ **address loudspeaker** Großflächen-, Groß-, Hochleistungs-lautsprecher *m* ~**-address loud-speaker system** Großschallübertragungsanlage *f* ~**-address system** Großlautsprecheranlage *f*, Lautsprecher *m*, Rundspruchanlage *f*, Schallübertragungseinrichtung *f* ~ **call office** Münzfernsprecher *m* ~ **correspondence** öffentlicher Verkehr *m* ~ **electricity supply** Elektrowirtschaft *f* ~ **exchange** öffentliches Fernsprechamt *n* ~ **health service** Gesundheitsdienst *m* ~ **institute** Körperschaft *f* des öffentlichen Rechts ~ **law** öffentliches Recht *n* ~ **notice** Aufgebot *n*, Aufgebotsverfahren *n* ~ **notification** Aufgebotsverfahren *n* ~ **regulations** behördliche Vorschriften *pl* ~ **road** öffentliche Straße *f*, öffentlicher Weg *m*, Verkehrsweg *m* ~ **subscriber** Teilnehmer *m* ~ **telephone** öffentliche Fernsprechstelle *f* ~ **telephone station** öffentliche Fernsprechzelle *f* ~**-telephone--station agent** Inhaber *m* oder Verwalter *m* einer öffentlichen Sprechstelle ~ **use** öffentlicher Verkehr *m*, öffentliche oder offenkundige Vorbenutzung *f* ~ **works** Regiebetrieb *m*
**publication** Auslegung *f*, Bekanntmachung *f*, Druckschrift *f*, Veröffentlichung *f*, Werbeschrift *f* ~ **date** Erscheinungsdatum *n*
**publicity** Bekanntgabe *f*, Öffentlichkeit *f*, Propaganda *f*, Werbung *f* ~ **letter** Werbebrief *m*
**publish, to** ~ anzeigen, auslegen, bekanntmachen, herausgeben, (notice) kundgeben, (books) verlegen
**publisher** Herausgeber *m*, Verleger *m*
**publishers' and job bookbinders** Verlags- und Lohnbuchbindereien *pl*
**publishing house** Verlag *m*, Verlags-buchhandlung *f*, -haus *n*
**puck** Andruckrolle *f* (tape rec.)
**pucker, to** ~ fälteln **to** ~ **up** krumpeln
**pucker** schlecht ausgeführte Naht *f*
**puckered** faltig, runzelig ~ **vent** elastische Scheitelöffnung *f*
**puckering** Faltenwerfen *n*
**pudding stone** Flintkonglomerat *n*, Puddingstein *m*
**puddle, to** ~ puddeln, rühren, verpuddeln
**puddle** Lehmschlag *m*, Tonschlag *m* ~ **ball** Puddel-, Roh-luppe *f* ~**(d) ball** Luppe *f* ~**-bar pile** Schienenpaket *n* ~**-bar roll** Rohschienenwalze *f* ~ **(puddling) cinder** Puddelschlacke *f* ~ **iron** Puddel-, Schweiß-eisen *n* ~ **mill** Luppenwalzwerk *n* ~ **roll** Puddelwalze *f* ~ **rolling mill** Puddeleisenwalzwerk *n* ~**(d) steel** Puddelstahl *m*
**puddled,** ~ **with clay** mit Stampfmasse *f* abgedichtet
**puddled,** ~ **bar** Luppenstab *m* ~**-bar-iron pile** paketierte Luppeneisenstäbe *pl* ~ **bars** Luppenstäbe *pl* ~ **iron** Luppeneisen *n*, Puddel-eisen *n*, -stahl *m*, Rohschiene *f*, Schweißeisen *n* ~ **steel** Schweißstahl *m* ~**-steel ball** Stahldeul *m*
**puddler** Puddelarbeiter *m*, Puddler *m*, Puddelofen *m* ~ **bar** Puddelstab *m* ~**'s paddle** Puddelspitze *f*
**puddling** (of the weld) Aufkochen *n*, (in ironworks) Frischen *n*, Ofenfrischerei *f*, Puddeln *n*, Puddelverfahren *n*, Rührfrischen *n*, (of iron) Umrühren *n* ~ **of fibrous iron** Sehnepuddeln *n* ~ **of fine-grained iron** Feinkornpuddeln *n*, Kornpuddeln *n* ~ **of foundation walls** Verfüllung *f* der Grundmauern
**puddling,** ~ **basin** Puddelherd *m* ~ **furnace** Eisenfrischflamm-, Puddel-ofen *m* ~**-furnace bed** Puddelbett *n*, Puddelsohle *f* ~ **hearth** Puddelherd *m* ~ **operation** Puddelbetrieb *m* ~ **process** Puddelverfahren *n* ~ **roll** Zängwalze *f* ~ **slag**

Dörner-, Puddel-schlacke *f* ~ **works** Puddelwerk *n*

**puff, to** ~ **(off)** ausschießen, verpuffen **to** ~ **up** sich bauschen

**puff,** ~ **of steam** Entweichen *n* des Dampfes

**puffing out** Ausschließen *n*

**puffy** kurz, stockig

**pug, to** ~ kneten

**pug,** ~ **and kneading machine** Kneter *m*, Knetmaschine *f* ~ **mill** Knetfaß *n*, Koller-gang *m*, -mühle *f*, Misch-koller *m*, -gang *m*, Schlägermühle *f* ~**-mill runner** Kollerläufer *m*

**pugging** (rough plastering) Tragdecke *f*, Tragwerk *n* ~ **mill** Mörtelmaschine *f*

**pull, to** ~ anziehen, ausziehen, recken, reißen, (rip cord) wegziehen, zerren, ziehen, zupfen **to** ~ **along** mitreißen, mitziehen, schleppen **to** ~ **alongside** längsseits gehen **to** ~ **apart** aufreißen, auseinanderziehen, zerreißen **to** ~ **(a bolt) apart by means of simultaneous shearing and tensile stresses** abwürgen **to** ~ **back** zurückziehen **to** ~ **out of a dive** (aviat.) abfangen, aus dem Sturzflug abfangen, durchziehen **to** ~ **down** abbrechen, abreißen, niederreißen **to** ~ **the form to pieces** (print.) die Form ausschlachten **to** ~ **in** einziehen **to** ~ **up an „n" milliampere** auf eine Ansprechstromstärke von „n" MA einstellen **to** ~ **off** einen Druck abziehen **to** ~ **out** abfangen, das Flugzeug ziehen, herausreißen, hochziehen, Höhensteuer geben **to** ~ **over** überheben, überstreifen **to** ~ **a proof** abziehen **to** ~ **a proof off a sheet** einen Korrekturbogen *m* abziehen **to** ~ **off a proof** einen Probedruck abziehen **to** ~ **out the reed** den Hanf pellen **to** ~ **round** (dial) aufziehen **to** ~ **round the dial** die Nummernscheibe aufziehen **to** ~ **in slips** in Fahnen abziehen **to** ~ **in step** mitziehen **to** ~ **through** (propeller) durchdrehen, hindurchziehen **to** ~ **the trigger** abdrücken **to** ~ **up** (aviat.) abfangen, ansprechen, (relay or armature) anziehen, aufhissen, aufziehen, erregt werden, heranziehen, hochziehen **to** ~ **the wool** die Wolle plüsen **to** ~ **work off a sheet** einen Korrekturbogen *m* abziehen

**pull** Griff *m*, Ziehen *n*, Ziehkraft *f*, Zug *m*, Zugkraft *f* ~ **of wire** Drahtzug *m*

**pull,** ~**-and-push** zweizügig ~ **bar guide** Führungskamm *m* der Zugstäbe ~ **bar-type magnetic** (für Gasgerät) Stößelmagnetzünder *m* ~ **box** Anschlußkasten *m* ~ **contact** Zugknopf *m*, Zugkontakt *m* ~ **cord** Zieh-, Zug-schnur *f* ~**-down claw mechanism** Greifermechanismus *m* ~**-down sprocket** Vorwickel *m*, Vorwickel-rolle *f*, -trommel *f* ~ **electrode** Zugelektrode *f* ~ **guide** Ziehmarken *pl* ~ **head** Ziehknopf *m* (Räummaschine) ~ **igniter** Zugzünder *m* ~**-in** Mitziehen *n*, Mitzieherscheinung *f* ~**-in range** Fangbereich *m*, Mitziehbereich *m* ~**-in step** Mitnahme *f* ~**-in step range** Mitnahmebereich *m* ~ **knob** Zug-knopf *m*, -kontakt *m* ~ **magnets for A.C.** Zugmagnete *pl* für Wechselstrom ~**-off roll** Abrollwalze *f*, Rückbeförderungsrolle *f*, Rücklaufrolle *f* ~**-off spring** Abziehfeder *f* ~**-out** Abfangen *n* ~**-out and push-in type** (switch) Ausziehtype *m* ~**-out device** Abfangvorrichtung *f* ~**-out fuse** Ausziehsicherung *f* (electr.), Einsatzsicherung *f*, Einstecksicherung

*f* ~**-out point** Kippunkt *m* ~**-out radius** Abfangradius *m* ~**-out strength** Zugkraft *f* (beim Ziehen der Rohrfahrt) ~**-out torque** (synchronous-motor practice) Kippmoment *n* ~**-out type crucible furnace** Unterflurtiegelofen *m* ~**-over arrangement** Überhebevorrichtung *f* ~**-over gauge** Rachen-, Überstreif-lehre *f* ~**-over mill** Duowalzwerk *n* mit Anordnung zum Überheben des Walzgutes ~**-pressure igniter** Zugdruckzünder *m* ~ **ring** Zugring *m* ~ **rod** Zugstange *f* ~ **switch** Zugschalter *m* ~**-test** Reckprobe *f*, Zugprüfung *f* ~**-test machine** Zerreiß-, Zugprüfungs-maschine *f* ~**-through winding** Durchzieherwicklung *f*, Fädelwicklung *f* ~**-type starting switch** Anlaßzugschalter *m* ~**-type switch** Schubschalter *m* ~**-up** Hochziehen *n* ~ **wire** (cord, cable) Handzug *m*

**pulled** gezogen *d* ~ **apart** auseinandergezogen

**puller** Abziehvorrichtung *f* (aviat.), Ausziehvorrichtung *f* ~ **propeller** Zugschraube *f* ~ **screw** Abdrück-, Abzieh-schraube *f*

**pulley** Block *m*, Blockrolle *f*, Flaschenzug *m*, Kloben *m*, Riemenscheibe *f*, Riemscheibe *f*, Rolle *f*, Rollkolben *m*, Scheibe *f*, Seilscheibe *f*, Übersetzungsrad *n* **(guide)** ~ Leitrolle *f*

**pulley,** ~ **axle** Scheibenachse *f* (Riemenscheibe) ~ **bearing** Rollager *n* ~**-block** (Hebezug) Flasche *f*, Rollen-block *m*, -kloben *m*, -zug *m*, Rollkloben *m* ~ **bushing** Riemenscheibenbüchse *f* ~**-case** Seilflasche *f* ~ **chain** Flaschenzug-, Kolben-kette *f* ~ **chain hoist** Kettenflaschenzug *m* ~ **crown** Riemenscheibenwölbung *f* ~ **drive** Riemenscheibenantrieb *m*, (cone) Stufenscheibenantrieb *m* ~ **drive for raising crossrail** Riemenantrieb *m* für den Querbalken ~ **frame** Flaschenzuggehäuse *n*, Klobengehäuse *n* einer Rolle, Kopfgerüst *n* ~ **groove contour** Rillenprofil *n* einer Seilscheibe ~ **guide** Riemenleiter *m* ~ **hub** Riemenscheibennabe *f* ~ **ratio** Scheibenübersetzung *f* ~ **rim** Riemenscheibenkranz *m*, Scheibenkranz *m* ~ **screw** Lenkschraube *f* ~ **stand** Riemenscheibenlagerarm *m* ~ **weight** Rollen-, Schnur-gewicht *n* ~ **winder** Haspelversetzer *m*

**pulleys, set of** ~ Flaschenzug *m*

**pulling** Rohrziehen *n*, Zerrung *f*, Ziehen *n*, Ziehvorgänge *pl* ~ **down** Abbruch *m* ~ **in** Einziehen *n*, Mitnahmeerscheinung *f* ~ **in transversal direction** Querschlepper *m* ~ **into tune** Mitnahme *f*, Mitnahmeerscheinung *f*, Mitziehen *n*, Mitzieherscheinung *f* ~ **over** Überheben *n* ~ **the valve** Ventilziehen *n* ~ **on whites** Nachziehen *n*

**pulling,** ~ **action** Zugwirkung *f* ~ **capacity** Räumkraft *f* (Räummaschine) ~ **chuck** Abziehfutter *n* ~ **figure** Belastungsverstimmung *f*, Frequenzziehwert *m* ~ **forward** Vorziehen *n* ~ **handle** Zuggriff *m* ~ **head** Ziehwerk *n* ~ **head for chains** Einspannkopf *m* für Ketten ~ **hook** Ziehhaken *m* ~ **hook for tension springs** Einspannvorrichtung *f* für Zugfedern ~**-in line** Zugseil *n* ~ **power** Durchzugskraft *f* (des Riemens) ~ **ram** Ziehwerk *n* ~ **strand** Zugtrumm *n* ~ **test** Grundbuchuntersuchung *f* ~**-through** Durchzug *m* ~ **tools** Ausholen *n* des Bohrgeräts ~**-up** Abfangen *n*

**pully block** Flaschenzug *m*

**pulmotor** Sauerstoffrettungsapparat *m*

**pulp, to** ~ zermahlen, zermalmen
**pulp** ausgelaugte Diffusionsschnitzel *pl*, Brei *m*, Glanzzeug *n*, Mark *n*, Masse *f*, Papierbrei *m*, Pulpe *f*, Schlamm *m*, Schlich *n* (min.), Trübe *f*, Wascherz *n*, (paper) Zellstoff *m* ~ **finely divided by means of a rasp** geschliffener Rübenbrei *m*
**pulp,** ~ **board** Zellstoffpappe *f* ~ **catcher** Faser-, Stoff-fänger *m* ~ **collector** Stoffänger *m* ~ **dilution** Trübeverdünnung *f* ~ **elevator** Trübeelevator *m* ~ **engine** Holländer *m* ~ **factory** Holzschleiferei *f* ~ **feed pipe** Trübezuführungsrohr *n* ~ **machine** Entwässerungsmaschine *f* ~ **machinery for paper** Drescher *m* ~ **making** Holzschliffbereitung *f* ~ **meter** Zeugregler *m* ~ **press** (for paper) Büttenpresse *f* ~ **product** Saugling *m* ~ **screenings** Siebrückstand *m* ~ **screw conveyer** Schnitzelschnecke *f* ~ **sheet** Pappe *f* ~ **silo** Baggergrube *f*, Schnitzel--einmietung *f*, -einsäuerung *f*, -sumpf *m* ~ **sizing** Büttenleimung *f* ~ **strainer** Knotenfänger *m*, Zeugsichter *m* ~ **stream** Trübestrom *m* ~ **thickener** Schlammeindicker *m*, Trübeverdicker *m*
**pulper** (sugar working) Breiapparat *m*, Papierknetmaschine *f*, Pulper *m*, Rübenbreimaschine *f*
**pulping rolls** (first process) Auspreßwalzen *pl*
**pulpit** Kanzel *f*, Steuer-kanzel *f*, -tisch *m*
**pulpy** breiartig, breiig
**pulsate, to** ~ pulsen, pulsieren, schwingen, vibrieren
**pulsating** Pulsieren *n*; intermittierend, pulsierend, stoßweise, wellig ~ **current** pulsierender Strom *m*, Wellenstrom *m*, welliger Strom *m* ~ **effect** Stoßwirkung *f* ~ **fatigue strength under bending stresses** Biegeschwellfestigkeit *f* ~ **jet** intermittierendes Gerät *n* ~ **load** pulsierende oder stoßweise Belastung *f* ~ **load pressure** Wechseldruck *m* ~ **potential** wellige Gleichspannung *f* ~ **stress** stoßweise Beanspruchung *f* ~ **turbojet** Turbinenluftstaugerät *n* ~ **unit** intermittierendes Gerät *n* ~ **voltage** pulsierende Spannung *f* ~ **wave** gleichstromüberlagerte oder pulsierende Welle *f*
**pulsation** Eckfrequenz *f*, Flattern *n*, Kreisfrequenz *f*, Pulsation *f*, Pulsieren *n*, Schwingung *f*, Winkelgeschwindigkeit *f* ~ **of current** Strompulsation *f*
**pulsation,** ~ **choke** Pulsationsdrossel *f* ~ **factor** Welligkeit *f* ~ **welding** Vibrationsschweißung *f*
**pulsations** Stromstoßreihe *f*
**pulsator** Pulsator *m*, Pulser *m*, Schüttelmaschine *f* (min.), Taktgeber *m* ~ **classifier** Pulsatorsetzmaschine *f* ~ **machine** Zugdruckmaschine *f*
**pulse** Entladungsstoß *m*, Impuls *m*, Pulsieren *n*, Schwingung *f*, Stoß *m*, Stromschritt *m* ~ **of light** Lichtblitz *m*
**pulse,** ~ **action** Impulswahl *f* ~ **amplitude** Impulshöreinheit *f* ~ **bandwidth** Breite *f* des Impulsspektrums ~ **cam** Impulsträger *m* ~ **carrier** Impulsträger *m* ~ **clipper** Impulsabtrennstufe *f* (TV) ~ **coding** Impulsverschlüsselung *f* ~ **counting instrument** Impulszählwerk *n* ~ **decay time** Impulsabfallzeit *f* ~ **depth** Impulshöhe *f* ~ **direction finder** Impulspeiler *m* ~ **distortion** Impulsverzerrung *f* ~ **droop** negative Dachschräge *f* ~ **duration modulation** Impulslängenmodulation *f* ~ **duration ratio** Impulstastver-

hältnis *n* ~ **duty factor** Impulsleistungsverhältnis *n* ~ **emission** pulsierende Emission *f* ~ **excitation** Impulstastung *f*, Stoßerregung *f* ~ **forming stage** Impulsbildungsstufe *f* ~ **frequency modulator** (or **generator**) Impulsgerät *n* (rdr), Impulstempomodulator *m* ~ **height distribution** Impulshöhenverteilung *f* ~ **height spectrum** Impulsgrößenspektrum *n* ~ **interleaving** Impulsverflechtung *f* ~ **interrogation** Impulsabfragung *f* ~ **ionization chamber** zählende Ionisationskammer *f* ~ **jet drive** Pulsdüsenantrieb *m* ~-**jet engine** pulsierender Staustrahlmotor *m* ~ **jitter** Impulsinstabilität *f* ~ **lengthener** Impulsumformer *n* ~ **limiting rate** Impulsbegrenzungsmaß *m* ~ **listening** Impulshöreinheit *f* ~ **machine** Maschinenzahlengeber *m* ~ **mark generator unit** Impulsmarkengebereinheit *f* ~ **mode** Impulsart *f* ~-**modulated radar** impulsmoduliertes Radar *n* ~-**modulated waves** impulsmodulierte Wellen *pl* (rdr) ~ **modulator** Impulsgerät *n* (rdr) ~ **motor** Argusrohr *n*, intermittierendes Luftstrahltriebwerk *n* ~ **multiplex** Impulsverschlüsselung *f* ~ **phase modulation** Pulsphasenmodulation *f* ~ **position modulation** Impulslagemodulation *f* ~ **power** Impulsleistung *f* (rdr) ~ **rate** Impulsrate *f* ~ **ratio** Impuls-, Stromstoß-verhältnis *n* ~ **recurrence frequency (PRF)** Impulsfolgefrequenz *f*, Tastfrequenz *f* ~ **repeater** Impuls-, Stromstoß-übertrager *m* ~ **repetition frequency (PRF)** Impulsfolgefrequenz (rdr) ~ **repetition rate** Impulsfolgefrequenz *f* ~ **rise time** Pulsanstiegzeit *f* ~ **slope modulation** Impulssteilheitmodulation *f* (rdr) ~ **spacing** Impulsabstand *m* ~ **spectrometer** Impulsspektrometer *n* ~ **spring** Nummernschalter-Kontaktfeder *f* ~ **springs** (of dial) Stromstoßfedern *pl* ~ **stretcher** Impulskorrektor *m* ~ **stretching** Impulsschwanz *m* (Nachleuchtschleppe) ~-**switching condition** Impulsschaltbedingung *f* ~ **tilt** Dachschräge *f* (des Impulses) ~ **timing** Laufzeit *f* (rdr) ~ **train** Impuls-folge *f*, -reihe *f*, -zug *m* ~ **width** Impulsbreite *f* (rdr)
**pulsed,** ~ **column** pulsierende Säule *f* ~ **electron beam** pulsierender Elektronenstrahl *m* ~ **voltage** Stromimpuls *m*
**pulsimeter** Pulsmesser *m*
**pulsing** Impulsgabe *f*, Stromstoßgabe *f* (im) ~ **cam** Stromstoßnocke *f* ~ **circuit** Impulsstromkreis *m* ~ **keysender** Speicherzahlengeber *m* ~ **keysender pilot lamp** Speicherzahlengeberlampe *f* ~ **ratio** (of diode) Tastverhältnis *n* ~ **relay** Stromstoßempfangsrelais *n* ~ **signal** Zifferzeichen *n*
**pulsometer** Dampfdruckpumpe *f*, Pulsometer *n*, Wasserheber *m*
**pulverizable** pulverisierbar, zerstäubbar
**pulverization** Feinstmahlung *f*, Mahlung *f*, Pulverisierung *f*, Pulverung *f*, Walzen *n*, Zerreibung *f*, Zerstäubung *f*
**pulverize, to** ~ feinmahlen, mahlen, zu Pulver stoßen, pulverisieren, pulvern, vermahlen, walzen, zerkleinern, zermahlen, zerpulvern, zerreiben, zerstäuben, zerstoßen
**pulverized** feingepulvert ~ **coal** feingemahlene Kohle *f*, Kohlen-mehl *n*, -stab *m* ~ **coal bunker** Kohlenstaubbunker *m* ~-**coal burner** Kohlenstaubbrenner *m* ~ **coal-fired** stabkoh-

lengefeuert ~-coal firing Kohlenstaubfeuerung
f ~-coal flame Kohlenstaubflamme f ~-coal
plant Brennstaubanlage f ~-coal practice
Staubbetrieb m ~-coal storage bin Kohlenstaubbunker m ~ firing system Mühlenfeue-
rung f ~ fuel Brennstaub m ~ fuel supply (in
firing) Staubzufuhr f ~ ore Pochmehl n
pulverizer Mühle f, Pulverisierungsmühle f,
Zerkleinerer m, Zerkleinerungsmaschine f,
Zerstäuber m ~ mill Mahlanlage f
pulverizing Feinmahlen n, Vermahlung f ~
action Mahlwirkung f ~ element Mahl-körper
m, -organ n ~ equipment Mahl-anlage f, -werk
n ~ machine Maschine f des Pilieraggregats ~
plant Mahlanlage f, Müllerei f, Zerkleinerungsanlage f ~ practice Mahltechnik f ~ process
Mahlvorgang m ~ test Mahlversuch m
pulverous pulverartig, pulverig
pulverulent feinpulverig, pulverartig, pulverförmig, pulverig, pulvrig, staubartig ~ brown
coal mulmige Braunkohle f
pulvinar kissenförmig
pulviniform kissenförmig
pumice, to ~ abbimsen, mit Bimsstein abreiben
pumice Bimsstein m ~ concrete Bimsbeton m ~
paper Bimssteinpapier n ~ powder Bims-staub
m, -steinmehl n ~ slabs Bimsdieler pl ~ soap
Bimssteinseife f ~ stone Bims-, Schwemm-stein
m ~ substitute Bimssteinersatz m
pumiceous bimsstein-ähnlich, -artig
pump, to ~ aus-, herauf-, leer-pumpen, pumpen
to ~ down abpumpen to ~ out auspumpen,
lenzen to ~ up aufblasen, aufpumpen, herauf-
pumpen
pump Paternoster n, Pumpe f the ~ draws die
Pumpe zieht the ~ knocks die Pumpe schlägt
~ for clean water Reinwasserpumpe f ~ for
introducing impregnating fluid Tränkmittel-För-
derpumpe f ~ with solid piston Scheibenkolben-
pumpe f ~ for untreated water Rohwasser-
pumpe f
pump, ~ action of piston rings Pumpen n der Kolbenringe ~ assembly Pumpenmaschinensatz m
~ barrel Kolbenstiefel m, Pumpenzylinder m ~
bit Schrotbohrer m ~ bucket Pumpenkolben m
mit Klappe ~ camshaft Pumpennockenwelle f
~ casing Pumpengehäuse n ~ caster Pumpen-
gießwerk n ~ chain Pumpenkette f ~ chamber
Pumpenraum m ~ circulation system Umpump-
aggregat n ~ cistern Pumpenkasten m ~ cover
Pumpenkörper m ~ crank pin Bolzen m zur
Pumpe ~ curves Pumpenkennlinien pl (Förderhöhe) ~ cylinder Pumpenzylinder m ~ defect
Pumpenbeschädigung f ~-down time Auspump-
zeit f ~ drill Rennspindel f ~ drive Pumpen-
antrieb m ~ drive gear treibendes Pumpenrad n
~ drive shaft Pumpenantriebswelle f ~ driver
Pumpenflügel m ~ element Pumpenelement n
~-equipped ballon flask Ballonabfüller m ~-fed
power station Pumpspeicherwerk n ~ feeding
Pumpenförderung f ~ fittings Pumpenarmatur
m ~ fluid Treibmittel n ~ fluid filling Treib-
mittelfüllung f ~ fluid vapor Treibmitteldampf
m ~ gasoline Pumpenbenzin n ~ gear Pumpen-
beschlag m ~ governor Pumpenregler m ~
grease Pumpenfett n ~ handle Brunnenschwengel m, Pumpenschwengel m ~ head Kesselboden m, Pumpenkopf m ~ holder Pumpen-

halter m ~ house Pumpenhaus n ~ idler gear
getriebenes Pumpenrad n ~ impeller Pumpenrad n ~-impeller blade (or vane) Pumpenflügel
m ~ installation Umpumpanlage f ~ intake
Saugstutzen m ~ jack Pumpengestell n,
Pump-vorgelege n, -vorrichtung f ~ kettle
Pumpenkorb m ~ lever Pumpenhebel m ~ lift
Pumpenhub m ~ mechanism Pumpentriebwerk
n ~ nozzle Saugstutzen m ~ packing Pumpen-
liderung f ~ pinion Pumpenantriebsrad n ~
piping Umpumpleitung f ~ piston Pumpen-kol-
ben m, -stempel m ~ plate Pumpenplatte f ~
plunger Plunger m, Tauchkolben m ~ plunger
drive Pumpenstempelantrieb m ~ primer lever
Anpumphebel m ~ rig Transmissionspumpe f
~ rods Pumpengestänge n ~ room Pumpen-
kammer f, -raum m ~ shaft Pumpenwelle f ~
spears Schachtgestänge n ~ spindle Pumpenwelle f ~ staff Pumpenstock m ~ stand Pump-
stand m ~ station Pumpwerk n ~ stock Pumpenkasten m ~ storage station Pumpspeicherwerk n ~ strainer Pumpenkorb m ~ stroke
Pumpenhub m ~ suction pipe Pumpensauglei-
tungsrohr n ~ sump Pumpensumpf m ~ tappet
Pumpenstößel m ~ telecontrol Pumpenfern-
steuerung f ~ trailer Pumpenanhänger m ~
unit Einzelpumpenelement n ~ well Pumpen-
brunnen m, -sod m ~ wheel Pumprad n ~ work
Pumpwerk n
pumpable pumpfähig
pumpage Pumpwirkung f
pumped, ~ out abgepumpt, ausgepumpt ~
concrete gepumpter Beton m ~ rectifier Ventil
n mit Vakuumhaltung ~ storage (at a hydro-
electric plant) Pumpspeicher m
pumping Aufpumpen n, Pumpen n (beim Barometer) ~ the material from a barge Entleerung f
einer Baggerschute
pumping, ~ automation Pumpautomatik f ~
capacity Pumpleistung f ~-down time Aus-
pumpzeit f ~ engine Pump-, Wasserhaltungsmaschine f ~ facilities Pumpeinrichtung f ~
gear Pumpenwerk n, Pumpgetriebe n ~ installation Pumpanlage f ~ jack Pumpen-bock m,
-winde f ~ lead Pumpstengel m ~ line Seil n für
Bohrpumpenantrieb ~ losses Ausscheide- und
Ladeverluste pl ~ machinery Pumpanlage f ~
plant Pump(en)anlage f, Schutensauger m ~
speed Sauggeschwindigkeit f ~ station Pumpstation f
pumps for elevator service Schöpfwerke pl
pun, to ~ feststampfen
punch, to ~ ankernen, ankörnern, (holes) bohren, durch-bohren, -lochen, -löchern, -schlagen,
knabbern, (ticket) knipsen, lochen, stanzen to
~-mark ankörnen to ~ out ausstanzen
punch Ausschlageisen n, Dorn m, (drift) Durchschlag m, Durchtreiber m, Gesenke n, Körner
m, Loch-eisen n, -stempel m, -zange f, Mönch
m, Nageltreiber m, Oberstanze f, Oberstempel
m, Patrize f, Patrone f, Prägestempel m, Punze
f, Rändel n, Schlag-locher m, -werkzeug n,
Schneidstempel m, Stampfe f, Stanze f, Stanz-
eisen n, -presse f, -stempel m, Stemmer m,
Stoßdorn m, Treib-eisen n, -stahl m, -werkzeug
n, Zentrierstift m ~ for expanding Dorn m zum
Ausdornen ~ for making fine grooves (or
flutes) Lupferpunzen m

**punch,** ~ **bail** Stanzbügel *m* ~ **bail link** Stanz-bügelzwischenstück *n* ~ **block** Stanzblock *m* ~**-card** Lochkarte *f* ~**-card register** Lochkarten-kartei *f* ~ **clutch** Locherkupplung *f* ~ **cutter** Stempelschneider *m* ~ **die** Kümpelstempel *m* ~ **die assembly** Lochermatrizeneinheit *f* ~ **drawer** Kluppe *f* ~ **drift** Durchschläger *m* ~ **hammer** Dorn-, Durchschlag-hammer *m* ~ **knives** Stanzmesser *pl* ~ **leg** Lochkaliber *n* (film) ~ **lever** Stanzhebel *m* ~ **(ing) magnet** Stanzmagnet *m* ~ **mark** Ankörnung *f*, Körner *m* ~**-marked** angekörnt, durch Körnerschlag *m* gekennzeichnet ~ **marks** Körnermarkierung *f* ~ **pliers** Lochzange *f* ~ **pliers tube** Lochhülse *f* für Lochzangen ~ **position** Lochstelle *f* ~ **press** Lochstanze *f*, Lochungspresse *f*, Stanz-maschine *f*, -presse *f* ~ **type** Locheisen-Type *f* ~ **verifier** Lochprüfer *m*
**punched** durchbohrt, gepresst, gestanzt ~ **card** Lochkarte *f* ~**-card reader** Lochkartenabtaster *m* ~ **holes** Lochung *f* (film) ~ **laminations** ge-stanzte Bleche *pl* ~**-out round** rund ausgestanzt ~**-plate screen** Siebblech *n* ~ **screen** Sieb *m* von gelochtem Blech ~ **side** Stanzkante *f* ~ **tape** Lochstreifen *m*
**puncheon** Futterholz *n*
**puncher** Locher *m*, Markierungsvorrichtung *f*
**punching** Knabbern *n*, Lochen *n*, Lochung *f*, Putzen *n*, Stanzen *n*, Stempeln *n* ~**-and-eye-letting machine** Loch- und Öseneinsetzmaschine *f* ~ **capacity** Stanzleistung *f* ~ **device** Stanzein-richtung *f* ~ **die** Lochkaliber *n*, Stanzmatrize *f* ~ **effect** Durchschlagswirkung *f* ~ **form** Aus-stanzstück *n* ~ **handle** Klöppel *m* des Wheat-stonelochers ~ **machine** Ausstanz-, Dessinier-, Durchstoß-, Loch-, Stech-, (textiles) Stupfel-, Vorstech-maschine *f*, Lochstanze *f* ~ **material for folding boxes** Faltstanzmaterial *n* ~ **method** Stanzverfahren *n* ~**-out** Ausstanzen *n* ~ **platens** Stanztiegel *m* ~ **press** Lochpresse *f* ~ **recorder** Lochschreiber *m* ~ **relay** Stanzrelais *n* ~ **shear** Stanzschere *f* ~ **test** Lochprobe *f*, Lochungs-versuch *m*, Lochversuch *m* ~ **tool** Stanz-matrize *f*, -werkzeug *n* ~ **tool for workmen** Lochhand-werkzeug *n* ~ **tool steel** Schnittstahl *m*, Stanz-stahl *m* ~ **tools of steel strip** Bandstahlschnitte *pl* ~ **wire** Stanzdraht *m*
**punchings** Stanzabfälle *pl*
**punctiform** punktförmig ~ **lamp** Punkt(licht)-lampe *f* ~ **light source** punktförmige Lichtquelle *f*
**punctiliousness** peinliche Genauigkeit *f* oder Pünktlichkeit *f*
**punctual** prompt, pünktlich, rechtzeitig
**punctuate, to** ~ punktieren
**punctuation mark** Interpunktions-, Satz-zeichen *n*
**puncture, to** ~ bohren, durch-bohren, -stechen, einstechen
**puncture** Durch-bruch *m*, -schlag *m*, Einstich *m*, Loch *n*, (tire) Panne *f*, Reifenpanne *f*, Stich *m* ~ **mark** Lochmarke *f* ~**-proof** nagelsicher, stich-fest, undurchlochbar ~**-proofness** Spannungs-sicherheit *f* ~ **resistance** (paper) Durchschlags-kraft *f* ~ **-resisting** punktursicher ~ **strength** Spannungsfestigkeit *f* **having high** ~ **strength** hochspannungssicher ~ **voltage** Durchschlags-spannung *f*

**punctured** durch-bohrt, -brochen, -löchert, (electr) durchgeschlagen *m*, löcherig ~ **carbu-retor float** undichter Vergaserschwimmer *m* ~ **tire** geplatzter Reifen *f*
**puncturing** Durchschlagen *n*
**punget** beißend, scharf
**punk** Glühzündstück *n* ~ **wire** (for lighting fuses or explosives) Glühkörper *m*
**punning** Feststampfen *n*
**pupil** Augenstern *m*, (of eye) Lichtloch *n*, Pupille *f* (med), Schüler *m*
**Pupin coil** Pupinspule *f* ~ **for open lines** Pupin-freileitungsapparat *m*
**pupinization** Bespulung *f*, Pupinisierung *f*, Spu-lenbelastung *f* ~ **section** Ladungsabschnitt *m*, Spulenfeld *n*
**pupinize, to** ~ bespulen, pupinisieren
**purchasable** käuflich
**purchase, to** ~ abkaufen, anschaffen, beziehen, einkaufen, kaufen
**purchase** Anschaffung *f*, Flaschenzug *m*, Hebel-kraft *f*, -wirkung *f*, Hebevorrichtung *f*, Kauf *m* ~ **of current** Fremdbezug *m* von Strom, Strom-bezug *m*
**purchase,** ~ **cost** Ankaufspreis *m*, Erwerbungs-kosten *pl* ~ **inspection gauge** Abnahmelehre *f* für Besteller ~ **order number** Bestellliste *f* ~ **price** Anschaffungspreis *m*, Einkaufspreis *m*, Er-werbungskosten *pl*, Kaufsumme *f* ~ **request** Bedarfsanforderung *f*
**purchaser** Besteller *m*, Erwerber *m*
**purchases** Gien *n*
**purchasing** Ankauf *m* **on** ~ bei Abnahme *f* von ~ **cost** Anlagekosten *pl* ~ **department** Bestel-lungsabteilung *f*, Einkaufs-abteilung *f*, -dienst *m* ~ **executive** Einkäufer *m* ~ **manager** Ein-kaufsdirektor *m* ~ **power** Kaufkraft *f*
**pure** echt, gediegen, lauter, makellos, rein ~ **aerodynamic form** aerodynamisch günstige Form *f* oder Formgebung *f* ~**-air-rich-mixture sandwich system** geschichtete Ladung *f* aus rei-ner Luft und reichem Gemisch ~ **aluminum** Reinaluminium *n* ~ **anthracene** Reinanthrazen *n* ~ **bismuth** Reinwismut *n* ~ **cantilever** voll-freitragend ~**-cantilever wing** vollfreitragender Flügel *m* ~ **clay** Tonerde *f* ~ **cod oil** Blanktran *m* ~**-color filter** tonrichtiger Filter *m* ~ **culture** Reinkultur *f* ~ **gold** gediegenes Gold *n* ~ **iron** reines Eisen *n* ~ **jet** Einstromtriebwerk *n* ~**-lumped inductance** punktförmige Induktanz *f* ~ **mathematics** reine Mathematik *f* ~ **note** reiner Ton *m* ~ **rubber tape** reines Paraband *n* ~ **scrap** Edelschrott *m* ~ **sine electromotive force** rein sinusförmige elektromotorische Kraft *f* ~ **toluol** Reintuluol *n* ~ **tone** reiner Ton *m* ~ **undamped wave** Reinwasser *n* ~ **yeast** Rein-zuchthefe *f* ~**-yeasting machine** Hefereinzucht-apparat *m*
**purge, to** ~ klären, läutern, reinigen
**purge cock** Ausstoßhahn *m*
**purging** Ablassen *n*, (sugar) Decken *n*, Reinigung *f* ~ **of distilled gas by rinsing gas** Spülgasschwe-lung *f*
**purging,** ~ **cock** Durchblasehahn *m*, Reinigungs-hahn *m*, Schlammhahn *m* ~ **sirup** Deckablauf *m* ~ **valve** Schlammventil *n*
**purification** Abläuterung *f*, Aufbereitung *f*, Frischen *n*, Frischung *f*, Klärung *f*, Läuterung

*f*, Reinigen *n*, Reinigung *f*, Vered(e)lung *f*, Windfrischen *n* **dry ~** Trockenreinigung *f* ~ **process** Frischherdverfahren *n*, Reinigungsprozeß *m*, -verfahren *n*

**purified** gereinigt, veredelt ~ **semolina** geputzter Grieß *m* ~ **steel** Frischstahl *m*

**purifier** Reiniger *m*, Reinigungs-apparat *m*, -mittel *n* ~ **with automatic brush clearing of the bolt sheet** (flourmilling) Flachsiebeputzmaschine *f* ~ **of semolina** Grießputzmaschine *f*

**purify, to** ~ abläutern, abschlämmen, feinen, frischen, läutern, reinigen, veredeln, waschen

**purifying** Frischarbeit *f*, Veredeln *n*, Windfrischen *n* ~ **agent** Reinigungsmittel *n* ~ **drum** Läutertrommel *f* ~ **furnace** Frischofen *m* ~ **material** Reinigermasse *f* ~ **method** Frischungsprozeß *m*, -verfahren *n* ~ **plant** Reinigungsanlage *f* ~ **process** Vered(e)lungs-, Windfrischverfahren *n* ~ **reaction** Frischwirkung *f* ~ **tank** Reinigungsbottich *m*

**purify** Reinheit *f* ~ **of the gas** Gasreinheit *f* ~ **of tone** Tonreinheit *f*

**purify, ~ coil** Farbreinheits- (TV), Sätte-spule *f* (TV) ~ **criterion** Reinheitskriterium *n* ~ **drop** Reinheitsabfall *m* ~ **indicator** Reinheitsmesser *m*

**purifying tower** Reinigungsturm *m*

**purlin** Dachpfette *f*, Dachrahmen *m*, Pfette *f*

**purling cut** Schälschnitt *m*

**purple** purpurrot ~ **of Cassius** Goldpurpur *m*

**purple, ~ light** Purpurlicht *n* ~ **ore** Kiesabbrand *m*, Purpurerz *n* ~ **oxide of iron** Oxydrot *n* ~ **willow** Purpurweide *f*

**purpose** Bestimmung *f*, Verwendungszweck *m*, Zweck *m* ~ **of exposure to rays** (of illumination or of irradiation) Bestrahlungszweck *m*

**purposely** absichtlich, vorsätzlich

**purpura** Purpurschnecke *f*

**pursue, to** ~ (a discussion) anknüpfen an etwas, hetzen, nachsetzen, nachstellen, nachstoßen, verfolgen

**pursuit** Fortsetzung *f*, Nachdrängen *n*, Verfolgung *f* ~ **curve** Hunde-, Verfolgungs-kurve *f* (aviat)

**push, to** ~ antreiben, drücken, durchdrücken, schieben, stoßen **to ~ against** anschieben **to ~ along** rücken **to ~ back** aufrollen, verdrängen, zurückdrängen **to ~ down** herunterdrücken, herunterstoßen, nach unten stoßen **to ~ forward** nachstoßen, weiterschieben **to ~ in** einstoßen **to ~ off** abstoßen **to ~ on** aufschieben **to ~ open (or out)** ausdrücken, ausstoßen, herausdrücken, herausstoßen **to ~ over** umstoßen **to ~ shut** zuschieben, zustoßen **to ~ through** durchstoßen **to ~ up** aufstoßen, hochtreiben **to ~ one's way through** sich Bahn brechen **to ~ on the works** auslängen

**push** (key) Druckknopf *m*, Schub *m*, Stoß *m*, Vorschub *m*, Vorstoß *m* ~ **of an arch** Seitenschub *m* eines Bogens

**push, ~-and-pull** Zug *m* und Druck *m* ~**-and-pull-jack** Schraubenschlittenwinde *f* ~ **bench** Stoßbank *f* ~ **binder** Stoßbinder *m* ~ **brace** Stange *f* mit Strebe ~ **bracing** Verstrebung *f* von Gestängen ~ **broach** Räumnadel *f* zum Stoßen ~ **button** Druck-knopf *m*, Kontaktknopf *m*, Zugknopf *m* ~**-button for hand tripping** Ausschaltknopf *m* für Handauslösung

**push-button, ~ control** Druckknopf-schaltung *f*, -steuerung *f*, Knopfsteuerung *f* ~**-controlled variable-speed motor** druckknopfgesteuerter Regelmotor *m* ~ **key** Druckknopf *m*, Taste *f* mit fester Stellung ~ **mushroom head** Druckknopfschalter *m* mit pilzförmigem Kopf ~ -**operated** druckknopfgesteuert ~**-operated rocker starter** Pendelanlasser *m* mit elektromagnetischer Einrückung ~**-operated starter** Anlasser *m* mit Druckknopfbetätigung ~ **panel** Druckknopftafel *f* ~ **receiver set** Druckknopfempfänger *m* ~ **relay** Tastenrelais *n* ~ **single circuit** einpoliger Druckknopfschalter *m* ~ **starter** Druckknopfstarter *m* ~ **switch** Druckknopf-, Tastenschalter *m* ~ **switch-board** Druckknopftafel *f* ~ **tuning** Druckknopf-, Drucktasten-abstimmung *f*

**push, ~ cam** Stoßdaumen *m* ~ **cart** Steinkarren *m* ~ **collet chocking** Druckspannung *f* ~**-cut milling** Fräsen *n* im Gegenlauf ~ **feeder** Schubaufgabevorrichtung *f* ~ **fit** Schiebesitz *m* ~ **igniter** Druckzünder *m* ~**-in fuse** Ausziehsicherung *f* (electr), Einsatzsicherung *f* (electr), Einstecksicherung *f* ~ **key** Schaltschlüssel *m*, Schubtaste *f*, Taste *f* ~**-off** Abstoß *m* ~**-out collet** auf Druck wirkende Spannpatrone *f* ~**-out door** abwerfbare Tür *f* ~**-out star** Ausschubstern *m* ~**-out tube** Spannrohr *n* ~ **pick** Spitzspaten *m* ~ **plate** Abkratzer *m*, Abschaber *m* ~ **plates** Schublaschen *pl* ~ **plug** Steckblende *f* ~ **pole for wagons** Wagenschieber *m*

**push-pull** abwechselnd in entgegengesetzter Richtung arbeitend, Gegentakt *m* ~ **amplifier** Auswahl-, Druckzug-, Gegentakt-verstärker *m*, Verstärker *m* mit zwei gegeneinander geschalteten Röhren ~ **amplifier circuit** Gegentaktverstärkerschaltung *f* ~ **arrangement** Gegentakt *m*, Gegentakt-anordnung *f*, -schaltung *f* ~ **breaker** Gegentaktschalter *m* ~ **carbon microphone** Doppelkohlemikrofon *n* ~ **carbon transmitter** Kohlenmikrofon *n* ~ **circuit** Druckzuschaltung *f* (electr) ~ **communication** Gegentaktverkehr *m* ~ **connection** Gegentaktschaltung *f* ~ **detection circuit** Gegentaktdetektionsschaltung *f* ~ **detector** Gegentaktgleichrichter *m* ~ **energization** Gegentakterregung *f* ~ **grid** Gegentaktaudion *n* ~ **input circuit** Gegentakteingangskreis *m* ~ **microphone** Gegentaktmikrofon *n* ~ **modulation** Gegen-modulation *f*, -steuerung *f* ~ **modulation stage** Modulationsgegentaktstufe *f* ~ **modulator** Gegentaktmodulator *m* ~ **operation** Gegentaktarbeiten *n* ~ **oscillator** Gegentaktgenerator *m* ~ **output** Gegentaktausgang *m* ~ **output control** Gegentaktendröhre *f* ~ **pentode** Gegentaktendpentode *f* ~ **power stage** Gegentaktendstufe *f* ~ **recording** Gegentaktaufzeichnung *f* (film) ~ **reproduction** Gegentaktaufzeichnung *f* ~ **stage** Gegentaktstufe *f* ~ **switch** Gegentaktschaltung *f* ~ **time base** Gegentaktkippgerät *n* ~ **tract** Gegentaktschrift *f* ~ **transducer** Kippschallerzeuger *m* ~ **transformer** Gegentakttransformator *m* ~ **transmitter** Doppel(kohle)mikrofon *n*, Druckzugmikrofon *n* ~ **valve** Gegentaktrohr *n*

**push-push** Gleichtakt *m* (rdo) ~ **circuit** Push-Push-Schaltung *f* (rdo)

**push, ~ rod** Gestänge *n*, Schub-rohr *n*, -stange *f*, Stößel *m*, (valve) Stößelstange *f*, Stößer *m*,

Stoßstange *f*, Ventilstoßstange *f* ~ **rod and rocker arm** Stoßstange *f* und Kipphebel *m* **~-rod ball end** Kugelkopf *m* der Stoßstange **~-rod (chamber) cover** Stoßstangenkammerdeckel *m* **~-rod cover tube** Stoßstangenverkleidungsrohr *n* **~-rod enclosure** Stoßstangenverkleidung *f* **~-rod type actuation typewriter** Stoßstangenmaschine *f* **~-switch** Anstoßschalter *m* **~-to-talk button** Sprechtaste *f* **~-to-talk switch** Sprechtaste *f* ~ **truck with trays** Hordenwagen *m* **~-up** Stauchung *f*

**pusher** Drücker *m*, (in rolling) Einstoßvorrichtung *f*, Schieber *m*, Stoßvorrichtung *f*, Treiber *m*, Wagenschieber *m* ~ **airplane** Druckpropellerflugzeug *n*, Druckschrauber *m* ~ **airsrew** Druckluftschraube *f* ~ **(or ejector) bar** Ausstoßstange *f* ~ **(or ejector) blade** Ausstoßplatte *f* ~ **centrifuge** Schubzentrifuge *f* ~ **control** Stoßhubregelung *f* ~ **rod** Schubstange *f* ~ **screw** Treibschraube *f* ~ **side** Maschinenseite *f* **~-type airplane** Flugzeug *n* mit Druckschraube, Schubtypflugzeug *n* **~-type engine** Schiebermotor *m* **~-type furnace** Durchlauf-, Stoß-ofen *m* **~-type propeller** Druckschraube *f*

**pushing** Schieben *n*, Stoßen *n* ~ **arm** Ausdrückstange *f* ~ **back** Zurückschieben *n* ~ **blade** Mitnehmerschaufel *f* ~ **device** Stoßvorrichtung *f* ~ **figure** Stromverstimmungsmaß *m* ~ **force** Schubkraft *f* ~ **forward of rails in a longitudinal direction** Verschieben *n* des Geleises in der Längsrichtung ~ **machine** Ausdrückmaschine *f* **~-off** Abstoßung *f* ~ **ram** Ausdrückstange *f* ~ **rod** Stößel *m* ~ **side** Maschinenseite *f* ~ **trough** Schubförderrinne *f*, Stoßmulde *f*

**put, to** ~ legen, an die Leitung anlegen, setzen, stellen

**put, to** ~ **across** schränken **to** ~ **into action** in Gang *m* setzen **to** ~ **out of action** ausscheiden, außer Betrieb setzen, außer Gefecht setzen, kampfunfähig machen **to** ~ **an additional locomotive to** eine Lokomotive *f* vorspannen **to** ~ **ashore** an die Leitung anlegen **to** ~ **aside** wegstellen **to** ~ **aside the roasted ore** den Rost *m* fortsetzen **to** ~ **back** zurückstellen **to** ~ **in best shape in** Höchstform *f* bringen **to** ~ **in a bid** submittieren **to** ~ **braiding around** umklöppeln **to** ~ **(or set) bricks in the kiln** den Brand *m* einfahren **to** ~ **through a call** eine Verbindung *f* herstellen **to** ~ **in circuit** einschalten **to** ~ **a circuit regular** eine Leitung wieder normal schalten **to** ~ **out of circuit** außer Strom *m* setzen **to** ~ **into circulation** girieren **to** ~ **out of commission** stillegen **to** ~ **distance between** abhängen **to** ~ **down** absetzen, niederlegen, niedermachen **to** ~ **on draught** ausschenken **to** ~ **to earth** erden **to** ~ **into effect (or execution)** zur Ausführung *f* bringen **to** ~ **an end to** beheben **to** ~ **an end to a grant** die Verleihung aufheben **to** ~ **in a garage** Wagen *m* einstellen **to** ~ **in a gear** einen Gang *m* einrücken **to** ~ **into gear** einschalten **to** ~ **into a gear** einen Gang *m* einrücken **to** ~ **out of gear** außer Betrieb setzen **to** ~ **the hood down** das Verdeck öffnen **to** ~ **the hood up** das Verdeck aufklappen, aufschlagen **to** ~ **in** einfüllen, einlegen, einsetzen, einstellen **to** ~ **inside out** umstülpen **to** ~ **out of joint** verrenken **to** ~ **on left rudder** links Seitensteuer *n* geben **to** ~ **on the market** auf den Markt *m*

bringen **to** ~ **a motion** einen Antrag stellen **to** ~ **into motion** in Bewegung *f* setzen **to** ~ **the nose down** drücken (avist) **to** ~ **off** vertrösten **to** ~ **on** anlassen, aufbringen, aufsetzen, auftragen, geben, überziehen **to** ~ **into operation** inbetriebnehmen **to** ~ **out of operation** außer Betrieb setzen, außer Gang setzen **to** ~ **up opposition** Widerstand *m* entgegensetzen **to** ~ **out** abschalten, ausfahren, auslöschen, ausmachen, ausschalten, aussetzen, löschen **to** ~ **over** überziehen **to** ~ **in the piston** den Kolben *m* einbringen **to** ~ **(around something) by pressure** umpressen **to** ~ **in readiness** bereitstellen **to** ~ **up for sale** versteigern **to** ~ **to sea** auslaufen **to** ~ **into service** in Betrieb setzen **to** ~ **out of service** außer Betrieb setzen, außer Gang setzen, blockieren, stillegen **to** ~ **up a smoke screen** sich einnebeln **to** ~ **on speed** andrücken **to** ~ **stickers on** bezetteln **to** ~ **under stress** spannen **to** ~ **up the target** das Ziel aufsitzen lassen **to** ~ **to the test** auf die Probe stellen **to** ~ **through** anschalten, durchschalten, (a call) durchverbinden, verbinden **to** ~ **up** aufstecken, aufstellen, einstellen, hochheben **to** ~ **up with** belassen **to** ~ **upright** hochstellen **to** ~ **upside down** umstülpen

**putlog** Gerüststange *f*, Rüstbaum *m*, Schußriegel *m*

**putrefaction** Fäulnis *f*, Verwesung *f*

**putrefactive** fäulniswirkend

**putrefied** verfault

**putrefy, to** ~ faulen, verwesen

**putter** Fördermann *m*, Karrenläufer *m*

**putting,** ~ **forward** Vorbringen *n* ~ **into operation** Inbetriebsetzung *f*, Ingangsetzen ~ **in readiness** Bereitstellung *f*

**putting,** ~ **apparatus into operation** Inbetriebnahme *f* ~ **out of commission** Außerdienststellung *f* ~ **under the command of** Unterstellung *f* **~-down machine for transporting barren rock** Absetzmaschine *f* für den Abraumbetrieb **~-on device for ingots** Blockauflegevorrichtung *f* ~ **-on device for slabs** Brammenauflegevorrichtung *f* ~ **waste land into forest** Aufforsten *n* von Ödland

**putty, to** ~ einkitten, kitten, verkitten, zusammenkitten

**putty** Dichtungsmasse *f*, (for castings) Füllmasse *f*, Glaserkitt *m*, Glaserkitter *m*, Kitt *m*, Spachtel *m* **(oil)** ~ Ölkitt *m* ~ **of emery** Schmirgelasche *f*

**putty,** ~ **coat** Spachtelschicht *f* ~ **knife** Kittmesser *n*, Spachtel *m* ~ **powder** Polierasche *f*

**puttying groove** Kittfalz *m*

**puzzle** Geheimschloß *n*, Rätsel *n* ~ **lock** Kombinations-, Vexier-schloß *n*

**puzzling** ausgeklügelt

**puzzolana concrete** Traßbeton *m*

**P-wiper** Prüfkontaktarm *m*

**pycnite** Stangenstein *m*

**pycnometer** Pyknometer *n*

**pylon** Mast *m*, Meßmarke *f*, Pfeiler *m*, Pfosten *m*, Pylon *m*, (viereckiger) Turm *m*, Wendeturm *m* ~ **bracing** Spannturmverspannung *f*, Tragrohr *n* ~ **drop tank** Abwurftank *m* ~ **eights** (tower) Abspannmast *m*, Wendeturm *m*

**pyramid** Pyramide *f* ~ **antenna** Reusenantenne *f* ~ **column** Pyramidenständer *m* ~ **stand** Pyra-

midenstativ *n* ~ **switchboard** Pyramidenschrank *m* (teleph)
**pyramidal** pyramidenförmig, pyramidisch ~ **horn** Reusenstrahler *m* ~ **point** Pyramidenspitze *f* ~ **separator** Spitzkasten *m* ~ **strut** Pyramidenstiel *m*, -strebe *f* ~ **tract** Pyramidenbahn *f*
**pyrargyrite** Antimonsilberblende *f*, dunkles Rotgültigerz *n*, Pyrargyrit *m*, Rotgültig *n*, Rubinblende *f*, Silberblende *f*
**pyrex** Hartglas *n*
**pyrgeometer** Pyrgeometer *n*
**pyrheliometer** Sonnen-strahlungsmesser *m*, -wärmemesser *m*, Strahlungsthermometer *n*
**pyridine** Pyridin *n* ~ **base** Pyridinbase *f*
**pyrite** Grundstein *m* (min.), Pyrit *m*, Schwefelkies *m* ~ **cinders** Pyritabbrände *pl* ~ **detector** Pyritdetektor *m* ~ **furnace (or oven)** Pyritofen *m* ~ **smelting** Pyritschmelzen *n*
**pyrites** Kies *m* ~ **kiln** Fließofen *m*
**pyritic** kiesähnlich, pyritisch ~ **(al)** pyritartig ~ **ore** Pyriterz *n* ~ **smelting** pyritisches Schmelzen *n*, Pyritschmelzen *n* ~ **smelting furnace** Pyritschmelzofen *n*
**pyritiferous** kieshaltig, pyrithaltig
**pyritous** kiesähnlich ~ **impurity** Beimengung *f*
**pyroacid** Brenzsäure *f*
**pyrobitumen** Pyrobitumen *n*
**pyrocatechin** Brenzkatechin *n*, Pyrokatechin *n*
**pyrocatechol** Brenzkatechin *n*
**pyrochemical** pyrochemisch
**pyrocrystalline** pyrokristallin
**pyroelectric(al)** pyroelektrisch
**pyroelectric conductor** Heißleiter *m*
**pyroelectricity** Pyroelektrizität *f*, Wärme-Elektrizität *f*
**pyrogallic acid** Brenzgallussäure *f*, Pyrogallöl *n*, Pyrogallussäure *f*
**pyrogallol** Brenzsäure *f*, Pyrogallöl
**pyrogen-free** pyrogenfrei
**pyrogenic** pyrogen
**pyrogens** Pyrogenen *pl*
**pyrognomic** pyrognomisch

**pyroligneous acid** Holzessig *m*
**pyrolignite** Holzessig *m* ~ **lead** holzessigsaures Blei *n*
**pyrolusite** Braunstein *m*, Graubraunstein *m*, graues Manganerz *n*, Graumangan *n*, Mangansuperoxyd *n*, Pyrolusit *m*, Weich-braunstein *m*, -mangan *n*, -manganerz *n*
**pyrolytic** pyrolytisch
**pyrometamorphose** Pyrometamorphose *f*
**pyrometer** Brand-, Glut-, Hitz(e)grad-, Hitzestrahlungs-, Hitz-messer *m*, Pyrometer *n*, Strahlungs-, Temperatur-messer *m*, Wärmemeßgerät *n* ~ **effect** Verbrennungstemperatur *f* ~ **equipment** Pyrometeranlage *f* ~ **fire end** Pyrometereintauchende *n* ~ **lead** Pyrometerdraht *m* ~ **protecting tube** Pyrometerschutzrohr *n*
**pyrometric(al)** pyrometrisch, wärmetechnisch ~ **cone** Brenn-, Schmelz-kegel *m* ~ **effect** Heizwert *m* ~ **scale** Temperatur-, Wärme-skala *f*
**pyrometry** Parometrie *f*
**pyromorphite** Braunbleierz *n*, Phosphorblei *n*, Pyromorphit *m*, Traubenblei *n*
**pyromorphous** pyromorph
**pyrope** böhmischer Granat *m*, Karfunkelstein *m*
**pyrophoric** luftentzündlich ~ **alloy** pyrophore Legierung *f* ~ **property** Entzündlichkeit *f*
**pyrophorus** Luftzünder *m*, Pyrophor *m*
**pyrophosphate** pyrophosphorsaures Salz *n*
**pyrophosphoric acid** Pyrophosphorsäure *f*
**pyroscope** Pyroskop *n*
**pyrostilpnite** Feuerblende *f*
**pyrosulfuric acid** Pyroschwefelsäure *f*
**pyrosulfurous acid** pyroschweflige Säure *f*
**pyrotechnic** feuerungstechnisch ~ **composition** Feuerwerkskörper *m* ~ **pistol** Leuchtpistole *f* ~ **projector** Leuchtraketengestell *n*
**pyrotechnics** Feuerwerkerei *f*, Pyrotechnik *f*
**pyroxylin(e)** Kollodiumwolle *f* ~ **lacquer** Nitrozelluloselack *m*
**pyrrhotite** Leber-eisenerz *n*, -kies *m*, Magnetkies *m*, Magnetopyrit *m*, Pyrrhotin *m*

# Q

**Q external** Lastgüte *f*
**Q factor** Blindwiderstandsverhältnis *n*, Güte-
faktor *m*, Q-Faktor *m*, Überspannungsfaktor
*m*
**Q loaded** Gesamtgüte *f*
**Q series** Jacobi (Reihen q) *pl*
**quack** Pfuscher *m*
**quad, to** ~ zum Vierer *m* verseilen
**quad** Adervierer *m* (electr.), Ausschließquer-
stück *n* (print), Quadruplextelegraf *m* (electr.),
(spiralfour) Sternvierer *m*, Vierer *m*, Viererbündel *n*, -kabel *n*, -seil *n* ~ **cable** Vierer-kabel
*n*, -seil *n* ~ **formation** Viererverseilung *f* ~**-pair
cable** Sternkabel *n* ~ **stripper** Blindspannungs-
unterdrücker *m*
**quadded** viererverseilt ~ **cable** Kabel *n* mit Vie-
rerseilen, viererverseiltes Kabel *n*
**quadding machine** Viererverseilmaschine *f*
**quadrangle** Viereck *n*. ~ **formed by collimation
points** Bildmarkenviereck *n* ~ **formed by points
of collimation** Markenviereck *n*
**quadrangular** quadratisch, viereckig ~ **housing**
Viereckgehäuse *n*
**quadrant** Gleitbogen *m*, Henry *n*, Kreisquadrant
*m*, kreissegmentförmige Skala *f*, Quadrant *m*,
Schaltbogen *m*, Viertelkreis *m* (einer Signal-
zone), Winkelmesser *m* ~ **of a circular arch**
Bogenviertel *n*
**quadrant, ~ angle** Erhebungswinkel *m* ~ **angle
of departure (or start)** Abgangswinkel *m* ~
**angle of elevation** Gesamterhöhung *f* ~ **compass**
Quadrantzirkel *m* ~ **electrometer** Gradbogen-,
Quadranten-elektrometer *n* ~ **elevation** Erhö-
hungswinkel *m*, Gesamt(rohr)erhöhung *f*, Ge-
samtschußrohrerhöhung *f*, Rohrerhöhung *f*,
Schußrohrerhöhung *f*, Zielhöhenwinkel *m* ~
**-elevation drum** Rohrerhöhungstrommel *f* ~
**gear** Umsteckrad *n* ~ **lever** Zahnsegmenthebel
*m* ~ **plate** Quadrantenfläche *f* ~ **scale** Gradbogen *m* ~ **seat** Winkelmesserebene *f* ~ **sight**
Libellenaufsatz *m*
**quadrantal** viertelkreisförmig, viertelkreisig ~
**compass error** Viertelfehler *m* ~ **corrector**
Kugel-, Quadrantal-korrektor *m* ~ **cruising
levels** Quadrantenflughöhen *pl* (aviat) ~ **deviation** Quadrantalausschlag *m* ~ **error** Beschik-
kung *f*, (due to structural parts or metal) Bordeffekt *m*, D-Wert *m* (aviat.), Funkbeschickung
*f*, Funkfehlweisung *f* ~**-error clearing (or
compensating) loop** Funkbeschickungsdraht-
schleife *f* ~**-error compensator (or corrector)**
Funkbeschicker *m* ~ **errors** Nachteffekt *m* ~
**point** Quadrantenpunkt *m* ~ **points** Viertel-
kreispunkte *pl* ~ **zone** Kugelzone *f*
**quadrat** breites Spatium (print), Geviert *n*,
Quadrat *n*
**quadrate** quadratisch; Quadrat . . ., Viereck . . .
**quadrated drum** Teilring *m*
**quadratic** quadratförmig, quadratisch ~ **equation** Gleichung *f* zweiten Grades, quadratische
Gleichung *f* ~ **form** quadratische Form *f* ~
**mean** quadratischer Mittelwert *m*

**quadrature** Inhaltsbestimmung *f* einer Fläche,
Phasenverschiebung *f* um 90 Grad, Quadratur
*f*, Vierung *f* in ~ **to each other** senkrecht zueinander
**quadrature, ~-axis component of a magnetomo-
tive force (or of a current)** Querdurchflutung *f*
(einer Synchronmaschine) ~**-axis synchronous
impedance** synchrone Querimpedanz *f* ~ **component** Blind-, Quadratur-komponente *f* ~
**component of the current** Blindstromkomponente *f* ~ **component of voltage** Blindspannungskomponente *f* ~ **error** Blindkomponente *f* ~
**field** wanderndes Feld *n* aus zwei senkrechten
Komponenten ~ **filter** Blindspannungsunter-
drücker *m* ~ **stripper** Blindspannungsunter-
drücker *m* ~ **voltage** Blindspannung *f*
**quadrennial** vierjährlich
**quadric** quadratische Form *f*
**quadricycle** Vierrad *n*
**quadriform optical apparatus** vierfache Leuchte *f*
**quadrilateral** Rechteck *n*, Viereck *n*, Vierseit *n*;
rechteckig
**quadrilaterally inclined** vierseitig geneigt
**quadrille paper** kariertes oder rautiertes Papier *n*
**quadrillion** Quadrillion *f*
**quadrinomial** Quadrinom *n*
**quadripartition** Vierteilung *f*
**quadripole** Vierpol *m* ~ **attenuation factor** Vierpoldämpfungsfaktor *m* ~ **propagation factor**
Übertragungsfaktor *m* eines Vierpols
**quadrivalence** Vierwertigkeit *f*
**quadrivalent** vierwertig
**quadropol moment** Quadrupolmoment *n*
**quadruplane** Vierdecker *m*
**quadruple, to** ~ vervierfachen
**quadruple** vierfach, vierfältig, vierzählig ~**-effect
evaporator** Vierkörperverdampfapparat *m* ~
**expansion engine** Vierfachexpansionsmaschine
*f* ~ **multiplex apparatus** Vierfachapparat *m* ~
**pair** achterverseilt ~**-pair cable** achterverseiltes
Kabel *n*, Vierfachzwillingskabel *n* ~ **phantom
circuit** Sechzehnerleitung *f* ~ **pole** Vierfach-
gestänge *n* ~**-prism quartz spectograph** Vier-
prismenquarzspektrograf *m* ~ **recorder** Vierfachschreiber *m* ~ **register** vierfacher Meteorograf *m* ~ **scanning** vierfaches Zwischenzeilenverfahren *n* ~**-screw liner** Vierschraubendamp-
fer *m* ~**-serial airsurvey camera (or photogrammetric camera)** Vierfachreihenmaßkamera *f* ~
**-spiral scanning disk** Vierfachspirallochscheibe
*f* ~ **system** Quadruplexverfahren *n* ~ **telegraph
(system)** Vierfachtelegraf *m* ~ **twin** Achter *m*,
Achterbündel *n*, Vierfachzwilling *m* ~ **twin
circuit** Achterschaltung *f*
**quadrupler** Vervierfacher *m*
**quadruplet** Quadruplett *n*, Vierergruppe *f*
**quadruplex** Quadruplextelegraf *m*; vierfach, vierfältig ~ **system** Doppelgegen-sprechen, -sprech-
system *n*, Vierfachbetrieb *m* ~ **telegraphy** Doppelgegensprechbetrieb *m*, Vierfachtelegrafie *f*
~ **working** Doppelgegensprechbetrieb *m* (teleph.), Quadruplexbetrieb *m*

**quadrupole** Quadrupol *m* ~ **constant** Vierpolkonstante *f* ~ **coupling terms** Quadrupolkopplungsterme *pl* ~ **distortion** Quadrupolverzerrung *f* ~ **moment** Quadrupolmoment *n*
**quail shot** Flintenschrot *n*
**quake, to** ~ beben, zittern
**quake center** Erdbebenherd *m*
**qualification** Befähigung *f*, Bewertung *f*, Charakter *m*, Qualifikation *f*, Qualifizierung *f*
**qualificatory** einschränkend, befähigend, qualifizierend
**qualified** berechtigt **to be** ~ sich eignen
**qualify, to** ~ befähigen, qualifizieren
**qualifying,** ~ **equation** Bestimmungsgleichung *f* ~ **period** (of instruction) Probezeit *f* ~ **reference** Fähigkeitsnachweis *m*
**qualimeter** Röntgenstrahlenhärtemesser *m*, (for X-rays) Strahlenhärtemesser *m*
**qualitative** nach der Beschaffenheit, qualitativ ~ **analysis** qualitative nicht mengenmäßig bestimmte Analyse *f* ~ **governing** Füllungsregelung *f*
**quality** Beschaffenheit *f*, Charakter *m*, Eigenschaft *f*, Fähigkeit *f*, Färbung *f* des Schalles, Güte *f*, Gütestufe *f*, Klasse *f*, Marke *f*, Qualität *f*, Sorte *f* **in** ~ qualitativ **of even** ~ ebenbürtig
**quality,** ~ **of the casting** Gußqualität *f* ~ **of coefficient** Gütegrad *m* ~ **of control** Regelgüte *f* ~ **in grinding** Schleifgüte *f* ~ **of a hard spring** federharte Eigenschaft *f* ~ **of the image** Bildschärfe *f* ~ **of picture** Bildgüte *f*, Filmwirkung *f* ~ **of reception** Verständigung *f* ~ **of reproduction** Wiedergabe-güte *f*, -qualität *f* ~ **of service** Betriebsgüte *f* ~ **of a sound** Klangfarbe *f* ~ **of transmission** Übertragungsgüte *f* ~ **of vapor** Dampffeuchtigkeit *f*
**quality,** ~ **article** Qualitätsware *f* ~ **control** Güteregelung *f* ~ **criterion** Gütemaßstab *m* ~ **equal to sample** gleiche Qualität *f* ~ **factor** Blindwiderstandverhältnis *n*, Gehaltswert *m*, Gütefaktor *m*, Gütezahl *f*, Phasenwinkel *m*, Wirkwiderstandverhältnis *n* ~ (or Q) **factor** Gütewert *m* ~ **features** Güteeigenschaften *pl* ~ **grade of materials** Werkstoffgütegrad *m* ~ **grading** Güteklasse *f* ~ **index** Güteziffer *f* ~ **production** Wertarbeit *f* ~ **reserve** Qualitätsreserve *f* ~ **specification** Güte-, Qualitäts-vorschrift *f*
**quantal** Zentner *m*
**quantified system analysis** quantitative Systemanalyse *f*
**quantimeter** Dosismesser *m*
**quantitative** nach der Menge, mengenmäßig, quantitativ ~ **analysis** Maßanalyse *f*, Mengenbestimmung *f*, quantitative Messung *f* ~ **composition** Mengenverhältnis *n* ~ **determination** Mengen-, Quantitäts-bestimmung *f* ~ **governing** Gemischregelung *f* ~ **measurement** quantitative Messung *f* ~ **relationship** Mengenverhältnis *n*
**quantitatively regulated** dosiert
**quantities** Größen *pl* ~ **of charge** Größenordnungen *pl* der Ladungen (im Blitz) ~ **by volume** Mengen *pl* nach Volumen
**quantity** Anzahl *f*, Größe *f*, Masse *f*, Menge *f*, Posten *m*, Quantität *f*, Quantum *n*
**quantity,** ~ **of air** Luftmenge *f* ~ **of air supplied** zugeführte Luftmenge *f* ~ **of circulating water**

durchlaufende Wassermenge *f* ~ **of color** Farbmenge *f* ~ **of comparative standard** Menge *f* im Vergleichsnormal ~ **of deposit** Ablagerungsmenge *f* ~ **of filling** Füllungsmenge *f* ~ **of flow** Durchflußmenge *f* ~ **of heat** Wärme-betrag *m*, -menge *f* ~ **of heat exchanged** ausgetauschte Wärmemenge *f* ~ **of light** Lichtmenge *f* ~ **of magnetization** Magnetismusmenge *f* ~ **of metal removed** Spanmenge *f* ~ **of motion** Moment *n* ~ **of oil in engine necessary for operation** Betriebsölfüllung *f* ~ **of radiation** Strahlungsmenge *f* ~ **of scavenging air required** Spülaufwand *m* ~ **of turbid matters** Trübstoffgehalt *m* ~ **of water** Wassermenge *f* ~ **of work** Arbeitsmenge *f*
**quantity,** ~ **adjustment lever** (accelerator) Mengenverstellhebel *m* ~ **balance** Betragsabgleich *m* ~ **being transported** Fördermenge *f* ~ **brewed** Gebräu *n* ~ **control** Quantitätsregulierung *f* ~ **delivered** Liefermenge *f* ~ **delivered per hour** stündliche Fördermenge *f* ~ **delivered per second** sekundliche Fördermenge *f* oder Durchflußmenge *f* ~ **discharged** Ausflußmenge *f* ~ **discount** Mengenrabatt *m* ~ **efficiciency** Wirkungsgrad *m* nach Menge und nach Energie ~ **gauge** Vorratsmesser *m* ~ **measuring fuse** Rohrbruchventil *n* ~ **meter** Ampèrestundenzähler *m* ~ **metered** Meßgröße *f* ~ **plate** Schild *n* mit Mengenbezeichnung *f* ~ **production** Massen-erzeugung *f*, -fabrikation *f*, -herstellung *f*, Serien-fertigung *f*, -herstellung *f*, serienmäßige Herstellung *f* ~ **production of prints** Massenkopie *f* ~ **recorder** Mengenschreiber *m* ~ **stop** Anschlagraste *f*, Einstellanschlag *m* ~ **stop rod** Anschlagstange *f* ~ **stops** Einstellrasten *pl* ~ **transmitter** Vorratsgeber *m*
**quantivalence** äquivalente Menge *f*
**quantivalent ratio** Gewichtsverhältnis *n*
**quantization** Quantelung *f* (phys.), Quantisierung *f* ~ **of lattice waves** Quantisierung *f* der Gitterwellen
**quantization,** ~ **distortion** Quantisierungs-rauschen *n*, -verzerrung *f* ~ **level** Quantisierungspegel *m*
**quantize, to** ~ quanteln
**quantized** gequantelt ~ **pulse modulation** Deltamodulation *f*, quantisierte Impulsmodulation *f* ~ **system** quantisiertes System *n*
**quantizer** Größenwandler *m*, Umwandler *m* in Digitalschreibweise
**quantizing** Quantisierung *f*
**quantometer** Quantimeter *n*, Quantometer *n*, Spektralapparat *m*
**quantum** Massenpunkt *m*, Quantum *n* ~ **condition** Quantenbedingung *f* ~ **efficiency** Quanten-, Quantum-ausbeute *f* ~ **jump** Quantensprung *m*, Quantumübergang *m* ~ **leakage** Tunneleffekt *m* ~ **leap** Quantensprung *m*, Quantumübergang *m* ~ **limit** Grenzwellenlänge *f* ~ **mechanical** quantisch ~ **mechanical system** Quantenmechanik *f* ~ **mechanics** Quantenmechanik *f* ~ **numbers of levels** Quantenzahlen *pl* für Terme ~ **optics** Quantenoptik *f* ~ **state** Energie-niveau *n*, -stufe *f*, Quantumzustand *m* ~ **theory** Quantentheorie *f* ~ **transition** Quantensprung *m*, Quantumübergang *m* ~ **weight** Quantumgewicht *n* ~ **yield** Quanten-, Quantum-ausbeute *f*

**quarantine** Quarantäne *f*, Sperre *f* ~ **flag** Quarantäneflagge *f* ~ **port** Quarantänehafen *m*
**quarry, to** ~ abbauen **to** ~ **out** ausbrechen
**quarry** Grube *f*, Halde *f*, Quaderstein *m* (window pane) quadratisches Fach *n*, Steinbruch *m* ~ **bed** Steinbruchlager *n* ~ **face** Steinbruch *m* ~ **man** Steinbrecher *m* ~ **picks** Steinspaltmesser *n* ~ **rocks** Bruchgestein *n* ~ **rubbish** Steinbruchabraum *m* ~ **stone** Bruch-, Werk-stein *m* **to ax a** ~ **stone** einen Bruchstein *m* bossieren **to set the** ~ **stones horizontal** die Werksteine einwiegen ~ **tile** Fußbodenplatte *f* ~ **water** Bergfeuchtigkeit *f*
**quartation** Quartierung *f* (**in**) ~ Scheidung *f* durch die Quart
**quarter, to** ~ beherbergen, unterbringen, vierteilen, vierteln
**quarter** Revier *n*, (district) Stadtviertel *n*, Stollenholz *n*, Unterkunft *f*, Viertel *n* ~ **of the heavens** Himmelsgegend *f* ~ **of year** Quartal *n*
**quarter,** ~ **bend** Krümmer *m* ~ **circle** Abrundung *f*, Viertelkreis *m* (aviat.) ~ **cone filling** kegelförmige Anschüttung *f* ~-**deck** Achter-, Hinter-, Quarter-deck *n* ~ **grain** Spiegelholz *n* ~ **leather binding** Halblederband *m* ~-**life span** Viertelwertzeit *f* ~ **load** Viertellast *f* ~ **period** Viertelperiode *f* ~-**phase** vierphasig ~ **point** Viertelpunkt *m* ~ **round** Anzug *m* ~-**sawed veneer** Querfurnier *n* ~-**sawn** geviert ~-**sized** viertelgeleimt ~ **turn** Vierteldrehung *f* ~-**turn belt** halbgeschränkter Riemen *m*, Halbkreuzriemen *m* ~-**twisted** halbgekreuzt, viertelgeschränkt ~-**wave antenna** Viertelwellenantenne *f* ~-**wave attenuator** Lambdaviertel-Dämpfungsglied *n* ~ **wavelength** Viertelwellenlänge *f* ~ **wavelength line** Lambda-Viertel-Leitung *f*, Viertelwellenleitung *f* ~ **wavelength plate** Halbschattenplatte *f* ~ **wave plate** Viertelwellenlängenplättchen *n* (opt.) ~ **wave vertical antenna** Einpolluftleiter *m*
**quartering** Sparrenholz *n*, einen rechten Winkel bildend, rechtwinklige Verbindung *f*, Vierteilen *n* ~ **area** Unterbringungsgebiet *n*, Unterkunftsbezirk *m* ~ **gauge** Nachprüfvorrichtung *f* ~ **machine** Kurbelzapfensitzbohrmaschine *f*
**quarterly** vierteljährlich, quartalsweise ~ **review** Vierteljahresschrift *f*
**quarternary formation** Quartärformation *f*
**quartile** Quadratur *f*
**quarto** Quart *n*, Quartformat *n* ~ **volume** Quartband *m*
**quartz** Bergkristall *m*, Quarz *m* ~ **cell** Quarzkammer *f* ~ **control** Quarzsteuerung *f* (rdo) ~ **controlled transmitter** quarzgesteuerter Sender *m* ~ **crystal** Quarz-druckelement *n*, -kristall *m* ~ **crystal-controlled oscillator** Quarzoszillator *m* ~ **crystal resonator** Quarzschwinger *m* ~ **crystal unit** Quarzkristallgerät *n* ~ **delay-line** Quarzverzögerungsleitung *f* ~ **deposit** Quarzlager *n* ~ **detector** Quarzempfänger *m* ~ **fiber dose meter** Quarzfadendosismesser *m* ~-**fiber manometer** Quarzfadenmanometer *m* ~ **filament** Quarzfaden *m* ~ **glass** Quarzglas *n* ~ **gravel** Quarzkiesel *m* ~ **indicator** Quarzindikator *m* ~ **measure** Quarzlager *n* ~ **mica rock** Quarzglimmerfels *m* ~ **monochromator** Quarzmonochromator *m* ~ **normal** Quarznormal *n* ~ **optical system** Quarzoptik *f* ~ **oscillator** Quarzoszillator *m* ~ **plate** Quarzplatte *f* ~ **porphyry** Euritporphyr *m*, Quarzporphyr *m* ~ **prism** compensator Quarzkeilkompensator *m* ~-**reference oscillator** quarzgesteuerter Bezugsgenerator *m* ~ **resonator** Quarzresonator *m* ~ **rock** Quarzifels *m* ~ **rod** Quarzstab *m* ~ **sand** Quarzsand *m* ~ **schist** Quarzschiefer *m* ~-**stabilized transmitter** Quarzsender *m* ~ **stage** Quarzstufe *f* ~ **thread** Quarzfaden *m* ~ **transducer** Quarzschallgenerator *m* ~-**treated** bequarzt ~ **tube lamp** Quarzbrenner *m* ~ **wavemeter** Quarzwellenmesser *m* ~ **wedge** Quarzkeil *m*
**quartziferous** quarzhaltig, quarzig
**quartzite** Quarzitfels *m* ~ **rock** Quarzfels *m*, Quarzitgestein *n* ~ **stone** Quarzitgestein *n*
**quartzose (or quartzous)** quarz-ähnlich, -artig, -haltig, quarzig
**quartzy** quarzhaltig, quarzig
**quasi** ähnlich, etwas gleichend ~-**arc process** Flußschweißen *n* ~-**chemical** quasichemisch ~ **continuum of levels** Termkontinuum *n* ~ **detonation** Quasidetonation *f* ~-**elastic aether** quasi-elastischer Äther *m* ~ **equilibrium** Quasigleichgewicht *n* ~-**equilibrium distribution** Quasigleichgewichtsverteilung *f* ~-**ergodic hypothesis** Quasi-Ergodenhypothese *f* ~-**geostrophic approximation** quasigeostrofische Näherung *f* ~-**infinite** quasi-unendlich ~-**infinite line** unendlich(lange) Leitung *f* ~-**linear** quasilinear ~ **molecule** Quasimolekül *n* ~-**momentum** Quasi-impuls *m* ~-**optic wave** ultrakurze Welle *f* ~ **optical wave** Mikro-strahl *m*, -welle *f* ~ **optical waves** Barkhausen Kurzwellen *pl*, Dezimeterwellen *pl*, Deziwellen *pl* ~ **solenoidal** Quasisolenoidal ~-**static** quasistatisch ~-**stationary** quasistationär ~-**stationary states** quasistationäre Zustände *pl* ~-**steady flow** quasistationäre Strömung *f* ~-**viscous flow** quasiviskoses Fließen *n*
**quassia** Bitterholz *n* ~ **wood** Quassiaholz *n*
**quaternary** geviert, quatär, quaternär ~ **alloy** quaternäre Legierung *f* ~ **notation** quaternäre Schreibweise *f* ~ **steel** Quaternärstahl *m* ~ **system** Vierstoffsystem *n*
**quaternion** Quaternion *n*
**quaver, to** ~ vibrieren
**quay** Bollwerk *n*, Hafendamm *m*, Hafenmole *f*, Kai *m*, Kaje *f*, Ladeplatz *m*, Quaianlage *f*, Schiffslände *f*, Ufermauer *f* ~ **bulkhead** Kaimauer *f* ~ **crane** Quai-, Hafen-kran *m* ~ **edge** Kaieinfassung *f* ~ **shed** Kaischuppen *m*
**quayside,** ~ **appliances** Umschlagvorrichtung *f* ~ **conveyor** Kai-, Förder-, Kai-band *n*
**quay wall** Kaieinfassung *f*, Kaimauer *f*
**quayage** Kai-anlagen *pl*, -gelände *n*
**quebracho wood** Quebrachoholz *n*
**quebrachotannic acid** Quebrachogerbsäure *f*
**queen post** Hängesäule *f*
**quench, to** ~ abbrausen, abdämpfen, abkühlen, ablöschen (coke), abschrecken (metall), löschen (electr.), vergüten (steel) **to** ~-**age** (Leichtmetall) vergüten **to** ~ **ashes** Asche *f* ablöschen **to** ~ **interruptedly** gebrochen härten
**quench** Abschreckung *f* (metall), Härtung *f* ~ **action** Dämpfungswirkung *f* ~-**aging** Abschreckalterung *f* ~ **effect** Löschwirkung *f* ~ **frequency** Pendelfrequenz *f* ~ **gap** Löschfunkenstrecke *f*

~ **generator (or oscillator)** Pendelgenerator *m*
(rdr) ~ **oil** Kühlöl *n* ~ **oscillation** Unterbre-
chungs-, Unterdrückungs-schwingung *f* ~
**period** Dämpfungsperiode *f* ~ **pump** (petroleum)
Abkühlpumpe *f*
**quenchant** Abschreckflüssigkeit *f*
**quenched** abgeschreckt, gehärtet ~ **charcoal**
Löschkohle *f* ~-**gap transmitter** Löschfunksen-
der *m* ~ **glass** abgeschrecktes oder gelöschtes
Glas *n* ~ **spark** gelöschter Funken *m*, Lösch-
funken *m* ~ **spark gap** Lösch-, Ton-funken-
strecke *f* ~-**spark resistance value** Funken-
löschwiderstandswert *m* ~ **spark system** Ton-
funkensystem *n* ~-**spark transmitter** Lösch-
funkensender *m* ~ **specimen (or test piece)** vergütete Probe *f*
**quencher** Löscher *m*
**quenching** Abkühlung *f*, (coke) Ablöschen *n*, Ablöschung *f*, Abschrecken *n*, Abschreckung *f*, (of
fluorescence) Ausleuchtung *f* und Tilgung *f*,
Härten *n*, Tilgung *f* ~ **of coke** Kokskühlung *f*,
Kokslöschen *n* ~ **of fluorescence** Fluoreszenzunterdrückung *f* ~ **of lines** Löschung *f* von
Spektrallinien ~ **of the orbital moments** Auslöschung *f* der Bahndrehimpulse ~ **of oscillations** Schwingungsabreißen *n* ~ **in water** Was-
serabschreckung *f*
**quenching,** ~ **action** Löschwirkung *f* ~ **agent** Ab-
löschmittel *n* ~ **bath** Abschreck-, Härte-bad *n*
~ **bench** Löschrampe *f* ~ **brush** Löschquaste *f*
~ **choke** Löschdrossel *f* ~ **circuit** Lösch-kreis
*m*, -schaltung *f*, Löschungskreis *m* ~ **compound**
Abschreckmittel *n* ~ **condenser** Löschkonden-
sator *m* ~ **crack** Härteriß *m* ~ **device** Lösch-
vorrichtung *f* ~ **drum** Löschtrommel *f* ~ **effect**
Abschreckwirkung *f* ~ **frequency** Hilfs-, Pendelfrequenz *f* ~ **gap** Löschfunkenstrecke *f* ~ **hood**
Löschgestell *n* ~ **installation** Abschreckeinrich-
tung *f* ~ **medium** Ablösch-, Abschreck-mittel *n*
~ **operation** Löschvorgang *m* ~ **oscillator** Pen-
delfrequenzgenerator *m* ~ **point** Abschreck-
punkt *m* ~ **press** Härtepresse *f* ~ **process** Ab-
schreckvorgang *m* ~ **property** Löscheigenschaft
*f* ~ **resistor** Löschwiderstand *m* ~ **solution** Abschreckmittel *n* ~ **stress** Abschreck-, Härtungs-
spannung *f* ~ **tank** Abschreck-behälter *m*,
-bottich *m* ~ **tower** Löschturm *m* ~ **water**
Löschwasser *n*
**quercitannic acid** Eichengerbsäure *f*
**quercitine** Quercitin *n*
**quercitron** Querzitron *n*
**query** Frage *f*, Nachforschung *f*
**quest** Nachforschen *n*
**question, to** ~ ausfragen, verhören
**question** Frage *f* ~ **of economics** Wirtschaftlich-
keitsfrage *f* ~ **at issue** Streit-frage *f*, -punkt *m*
~ **mark** Fragezeichen *n*
**questionable** fraglich
**questioning** Abfragung *f*, Befragung *f*, Fragestellung *f*, Verhör *n*
**questionnaire** Formular *n*, Fragebogen *m*
**quick** geschwind, rasch, schleunig, schnell ~ **as
lightning** blitzschnell
**quick,** ~ **access storage** zugriffzeitfreie Speicherung *f* ~ **access store** Schnellspeicher *m* ~
**acting** schnell wirkend
**quick-acting,** ~ **carrier** Schnellspannmitnehmer
*m* ~ **clamping device (or socket)** Schnellspann-

vorrichtung *f* ~ **controls** Schnellschaltung *f* ~
**couplings** Schnellschlußkupplungen *pl* ~ **disengagement** Schnellausrückung *f* ~ **driver**
Schnellspannmitnehmer *m* ~ **eccentric** Schnell-
spannexzenter *m* ~ **engagement** Schnellein-
rückung *f* ~ **gate valve** Schnellschlußventil *n* ~
**regulator** Schnellregler *m* ~ **screw** Schnellein-
spannschraubstock *m* ~ **tailstock** Schnellspann-
reitstock *m* ~ **vise** Schnellspannschraubstock
*m*
**quick-action** Arbeiten *n* ohne Verzögerung, sofortige Anziehung *f* ~ **and release** schnelles
Ein- und Ausschalten *n*
**quick-action,** ~ **adjustment** Schnellverstellung *f*
~ **brake valve** Schnellbremsventil *n* ~ **drier**
Schnelltrockenapparat *m* ~ **engine-stopping
device** Schnellausschalter *m* eines Motors ~
**fuse** Flinksicherung *f* ~ **release** Schnellausschalter *m* ~ **switch** Moment-, Schnapp-schal-
ter *m* ~ **vise** Schnellspanner *m*
**quick,** ~-**adjusting lever** Schnelleinstellhebel *m*
~-**adjusting monkey wrench** Schnellspannschraubenschlüssel *m* ~-**adjusting screw spanner
wrench** Schnellspannschraubenschlüssel *m* ~
**analysis** Schnellanalyse *f* ~ **ash** (sugar refining)
Flugasche *f* ~-**ash outlet** Flugaschenabzug *m*
~-**ash separators** Flugascheabscheider *m* ~
**bleaching** Schnellbleiche *f* ~ **blow** Schnellschlag
*m* ~ **break** Momentunterbrechung *f* (electr.) ~
**break action** Momentschaltung *f* ~-**break fuse**
Hochleistungssicherung *f* ~-**break switch**
Augenblicks-, Moment-, Schnapp-, Schnell-,
Sprung-schalter *m* ~ **breaking** Schnell-aus-
lösung *f*, -ausrückung *f*
**quick-change,** ~ **adaptor** Schnellwechselauf-
nahme *f* ~ **chuck** Schnellwechselfutter *n* ~
**engine lathe** Schnellwechseldrehbank *f* ~ **gear
box** Schnellwechselräderkasten *m* ~ **gear drive**
Norton-Antrieb *m*, Schnellwechselgetriebe *n* ~
**gear mechanism** Norton-Getriebe *n*, Schnellwechselgetriebe *n* ~ **holder** Schnellwechselhal-
ter *m* ~ **tool block** Schnellwechselstahlhalter *m*
~ **toolpost** Schnellwechselstahlhalter *m* ~ **unit**
Schnellwechseleinheit *f*
**quick,** ~ **chuck** Schnellfutter *n* ~ **clay** Fließton
*m* ~-**connector-type parachute** schnell anknüpf-
barer Fallschirm *m* ~ **cooling** Schnellkühlung *f*
~ **coupling** Schnellkupplung *f* ~-**deadbeat
instrument** Meßgerät *n* mit Dämpfung, schnell
zur Ruhe kommendes Meßgerät *n* ~ **de-energizing** Schnellentregung *f* ~ **delivery table
elevating mechanism** Schnellbug *m* für den Aus-
lege-Stapeltisch ~-**disconnect (plug)** Schnell-
trennkupplung *f* ~-**drying** schnelltrocknend
~ **drying oil** ätherisches Öl *n* ~ **excitation**
Schnellerregung *f* ~-**feathering propeller** Luftschraube *f* mit Schnellverstellung (in Segelstellung) ~ **fire** Schnellfeuer *n* ~-**firing boiler**
Schnellheizkessel *m* ~-**firing gun** Schnellfeuer-
geschütz *n*, -kanone *f*, Schnelladekanone *f* ~
**flashing light** Funkelfeuer *n* ~-**freezing microtome** Gefriermikrotom *n* ~ **fuse** empfindlicher
Zünder *m*, Schnellzünder *m* ~-**gripping** Schnell-
spannung *f* ~ **lime** Ätzkalk *m*, ätzender, gebrannter, lebender oder ungelöschter Kalk *m*,
Löschkalk *m* ~ -**make-and-break-switch** Mo-
mentschalter *m* ~-**match machine** Zündschnur-
maschine *f* ~ **motion** Schnellgang *m* ~-**motion**

**position** Eilgangsstellung *f* ~**-motion safety trip** Eilgang-Sicherheitsschaltung *f* ~**-motion screw** Grobbewegungsschraube *f* ~ **opening valve** Schnellschlußschieber *m*, Schnellverschlußventil *n* ~**-operating relay** schnell ansprechendes Relais *n* ~ **operation** sofortige Anziehung *f* ~ **print** Schnellkopie *f* ~ **procedure** Schnellmethode *f* ~ **recharging** Schnelladung *f* (einer Batterie) ~ **release** Schnellablaß *m*, sofortige Trennung *f* oder Unterbrechung *f*

**quick-release** schnell lösend ~ **box** Zentralschloß *n* ~ **cable** Auslöseseil *n* ~ **catch** Schnellverschlußvortreiber *m* ~ **connection** Schnelltrennstelle *f* ~ **coupling** Reißkupplung *f* ~ **harness** (parachute) Schnellablegegurt *m* ~ **hook** Sicherungshaken *m* ~ **lever** Schnellauslösehebel *m*, Schnellbremshebel *m* ~ **valve** Schnellablaßventil *n*

**quick,** ~ **releasing** schnellösend ~ **return lever** Hebel *m* für den schnellen Rückgang, Rückganghebel *m* ~**-return traverse** Eilbewegung *f*, Eilgang *m*, Eilrückgang *m* ~ **return valve** Schnellrücklaufventil *n* ~ **returns** schneller Umsatz *m* ~ **rewind** rascher Rücklauf *m* ~**-rotating shaft** schnellaufende Welle *f* ~ **sand** Fließ-, Flott-, Flug-, Schwimm-, Trieb-sand *m* ~ **set-hedged wall** Knick *m* ~**-setting cement** rasch abbindender Zement *m*, Schnellbinder *m* ~ **shut-off** Schnellschluß *m* ~ **silver** Quecksilber *n* ~ **silver mine** Quecksilbergrube *f* ~ **silver ore** Quecksilbererz *n* ~ **snap connector** Karabinerhaken *m* ~ **solving** schnellösend ~ **start** Sprungstart *m* ~ **starting snap magnet** Schnellstartschnappermagnet *m* ~**-stop device** Schnellbremse *f* ~ **table motion** Tischeilgang *m* ~ **test** Vorprobe *f* ~ **time** Gleichschritt *m* ~ **traverse motor** Eilgangmotor *m* ~ **traverse winder** Kreuzspulmaschine *f*

**quickness** Geschwindigkeit *f*, Raschheit *f*, Schnelligkeit *f*

**quiescent** ruhend ~ **air gap** Ruheluftspalt *m* ~ **automatic volume control** Stummabstimmung *f* ~ **carrier** unterdrückter Träger *m* ~ **condition** Ruhearbeitspunkt *m* ~ **current** Ruhestrom *m* ~ **grid** Gitterruhe *f* ~ **period (steel melting)** Ausgarzeit *f* ~ **point** (of a valve) Arbeitspunkt *m*, Ruhearbeitspunkt *m* ~ **potential** Arbeitspunkt-, Ruhe-potential *n* ~ **push pull** Ausgleichsschaltung *f* ohne Gleichstromkomponente, B-Verstärker *m* ~ **push-pull switch** Ausgleichsschaltung *f* ~ **transmitter system** Sender *m* mit unterdrückter Trägerwelle ~ **value** Ruhewert *m*

**quiet, to** ~ beruhigen **to** ~ **down** sich legen

**quiet** Ruhe *f*, Stille *f*; beruhigt, geräuschlos, lautlos, ruhig, still, stillschweigend ~ **automatic volume control** Stummabstimmung *f* ~ **circuit** geräuschfreie Leitung *f* ~ **flow** Fließströmung *f* ~ **running** ruhiger Gang *m* ~ **shade** dezente Farbe *f* ~ **steel** ruhiger Stahl *m* ~ (**automatic-valve-control) tuning** leise Abstimmung *f*

**quieting** Beruhigung *f* ~ **method** Geräuschdämp-

fungsverfahren *n* ~ **sensitivity** Empfindlichkeitsschwelle *f*

**quietly working** geräuscharm arbeitend

**quietness** Geräuschlosigkeit *f*, (of running motor) Laufruhe *f*

**quill, to** ~ aufspulen, aufwinden

**quill** Federpose *f*, hohles Wellenstück *n*, Hohlwelle *f*, Spindel *f*, Spule *f* ~ **bearing** Hohlwellenlager *n* ~ **bit** Hohlbohrer *m* ~**-drive motor** Motor *m* mit Hohlwelle ~ **feather** Kielfeder *f* ~ **gear** Federachse *f* ~**-gear drive** Federachsantrieb *m*, Federtopfantrieb *m* ~ **shaft** (Triebwerk) Drehstab *m*, Hohlwelle *f* ~ **winding machine** Schlußspulmaschine *f*

**quillai bark** Quillayarinde *f*

**quilling** Rüsche *f*

**quilt, to** ~ (textiles) abheften, steppen

**quilting** (textiles) Pikee *n*

**quinary notation** quinäre Schreibweise *f*

**quince** Quitte *f*

**quincunx order** schachbrettförmige Aufstellung *f*

**quinhydrone** Chinhydron *n* ~ **electrode** Chinhydronelektrode *f*

**quinic acid** Chinasäure *f*

**quinine** Chinin *n*

**quinoline** Chinolin *n*

**quinolinic acid** Chinolinsäure *f*

**quinone** Chinon *n*

**quinopyridine** Chinopyridin *n*

**quinoxaline** Chinoxalin *n*

**quintessence** Quintessenz *f*

**quintuple** das Fünffache *n*; fünffach ~ **coupling** Fünffachkupplung *f* ~**-effect evaporator** Fünfkörperverdampfapparat *m* ~ **harmonics** fünffache Oberharmonische *f* ~ **point** Fünffachpunkt *m*

**quire** Lage *f* (Buchbinderei)

**quirk, to** ~ spitzkehlen (arch)

**quirk** spitze Kehlung *f*; spitz gekehlt ~ **head** Einziehung *f*, Halskehle *f* ~ **ogee iron** Karnieseisen *n*

**quit, to** ~ aussetzen, quittieren, verlassen

**quiver, to** ~ beben, zittern

**quiver** Beben *n*

**quoin, to** ~ einkeilen, schließen (print), verkeilen

**quoin** Ecke *f*, Eck-, Kropf-, Winkel-stein *m*; Schließ-keil *m*, -zeug *n* ~ **brick** Eck-, Keil-ziegel *m* ~ **chase** Keilrahmen *m* ~ **key** Formenschlüssel *m* ~ **post** Dreh-, Wend-säule *f* ~ **screw** Keilschraube *f*

**quoins of miter** Drempelquader *m*

**quota** Anteil *m*, Beteiligungszahl *f*, Kontingent *n*, Quote *f*, Steg *m* (print) ~ **restriction** Quoteneinschränkung *f* ~ **system** Kontingentierung *f*

**quotation** Anführung *f*, (of price) Angebot *n*, Ansatz *m*, Aufführung *f*, Notierung *f*, (of prices) Preisangabe *f* ~ **sheet** Angebotsblatt *n*

**quote, to** ~ anführen, (price) ansetzen, berechnen, notieren, Preisangebot *n* machen, quotisieren

**quote mark** Abführung *f* (print)

**quotient** Ableitung *f*, Quotient *m*, Teilungszahl *f* ~ **meter** Quotientenmesser *m*

# R

rabbet, to ~ einfügen, falzen, fugen

rabbet Anschlag *m*, Falz *m*, Fuge *f*, Nute *f*, Nutverbindung *f*, Stoßreitel *m* ~ of keel Kiel-sponung *f*, -spündung *f*

rabbet, ~ cutter Falzfräser *m* ~ iron Kröseleisen *n* ~ ledge Schlagleiste *f* ~ pipe Falzrohr *n* ~ plane Falz-, Sims-hobel *m*

rabbeted, ~ corner pile Bundpfahl *m* mit Nuten ~ encased lock eingelassenes Schloß *n*

rabbit Anlage *f* für rasche Bestrahlung, Rohrpostbüchse *f*

rabble, to ~ durchkrählen, kratzen, rühren, schummeln, schüren to ~ out auskratzen, herauskratzen

rabble Feuerkrücke *f*, Haken *m*, Krähle *f*, Kratze *f*, Kratzer *m*, Krücke *f*, Pack *n*, Röstschaufel *f*, Rotte *f*, Rührhaken *m*, Rührstab *m* ~ arm Krähl-arm, -behälter *m* ~ carriage Krählwagen *m* ~ furnace Krählofen *m*, Rührofen *m* ~ rake Krählofen *m*

rabbling Krählung *f*, Schummeln *n* ~ hole Arbeits-öffnung *f*, -tür *f*, Schummelloch *n* ~ mechanism Rühreinrichtung *f* ~ operation Krählarbeit *f*

R.A.C. rating Steuer-PS *f*

race, to ~ (gear, motor) durchdrehen, durchgehen, rennen to ~ (or run) up abbremsen (engine)

race (of gun) Bettungsring *n*, Lauf *m*, (Kugellager) Laufbahn *f*, Lauf-kranz *m*, -ring *m*, -spur *f*, (weaving) Schützenbahn *f*, Strömung *f*, Wettbewerb *m*, Wettfahrt *f* ~ circular Laufkranz *m* (Laufring einer Kuppel) ~ mate optisch inaktiver Körper *m* oder Quarz *m*, optisch inaktives Quarzkristall *n* ~ track Rennstrecke *f* ~ way Führungsschiene *f*, Kabel-leiste *f*, -rost *m*, Laufbahn *f*, Laufring *m* ~ way of balls Kugellaufbahn *f*

racemic, ~ quartz optisch inaktiver Quarz *m* ~ substance optisch unaktiver Körper *m*

racemization Inaktivierung *f* (opt.), optische Inaktivierung *f*, optische Kristallinaktivierung *f*, optische Quarzinaktivierung *f*

racemized material optisch inaktiver Körper *m*

racer Rennwagen *m*

rachet, ~ handle Ratschenschlüssel *m* ~ pawl Gegen-, Schalt-klinke *f*

rachis Fruchtspindel *f*

racing, ~ of a machine Durchgehen *n* einer Maschine ~ of the propeller Blindschlagen *n*

racing, ~ car Rennwagen *m* ~ plane Rennflugzeug *n* ~ plug Rennkerze *f* ~ seaplane Wasserrennflugzeug *n* ~-type magneto Rennmagnetzünder *m* ~ yacht Rennjacht *f*

rack, to ~ abfüllen, ausrecken, dehnen, recken, steckkühlen to ~ down niederkurbeln to ~ up hochkurbeln

rack Aufhänger *m*, (battery) Batteriegestell *n*, Bock *m* (print.), Etagere *f*, Gestell *n*, Gerüst *n*, Gitter *n*, Kehrrad *n*, Rahmen *m*, Ratschenstange *f*, Reck-bank *f*, -vorrichtung *f*, Spanner *m*, Sperrstange *f*, Stellage *f*, Stockleiter *m*,

Traggerüst *n*, Zahnlatte *f*, Zahnstange *f*

rack, ~ of amplifier Bunker *m* für Verstärker ~ for collecting logs transported by steam Triftrechen *m* ~ with helical teeth Zahnstange *f* mit schrägen Zähnen ~ of an inlet (hydroelectric plant) Rechen *m* ~ and pinion Stangentriebwerk *n*, Zahn *m* und Trieb *m*, (gear) Zahnstange *f* und Ritzel *m*, Zahnstangentrieb *m* ~ and pinion for tilting the frame Zahnstangengetriebe *n* zur Verstellung des Körpers

rack, ~-and-pinion drive Zahnstangenantrieb *m*, Zahntrieb *m* ~-and-pinion gear Zahnstangengetriebe *m* ~-and-pinion press Zahnstangenpresse *f* ~ attachment Rechenansatz *m* ~ bar of the jack Windenstock *m* ~-bar-control Zahnstangensteuerung *f* ~ bit Rollmeißel *m* ~ brace Bohrknarre *f* ~ carrier Führungshebel *m* ~ construction Gestellkonstruktion *f* ~ development Rahmenentwicklung *f* ~ drying apparatus Hordentrockenapparat *m* ~ gear Stangentriebwerk *n*, Zahnstangenrad ~ gear(ing) Zahnstangentrieb *m* ~ hoisting gear Zahnstangenhubeinrichtung *f* ~ indexing attachment Zahnstangenteilvorrichtung *f* ~ jack Zahnstangenwagenwinde *f* ~ lashing Schleuderbund *m* ~ negative Rahmennegative *f* ~ pinion Zahnstangenritzel *n* ~ rail Zahnschiebe *f* ~ railway Zahnradbahn *f* ~ rigging Rödeltau *n* ~ section Gestein-, Rahmen-einheit *f* ~-shaped cutter Hobelkamm *m*, Zahnstangenwerkzeug *f* ~ shelving Fächergestell *n* ~ spacing attachment Zahnstangenteilvorrichtung *f* ~ steering Zahnstangenlenkung *f* ~ stick Rödel *m* ~ tool Kammstahl *m* ~ tooth cutter Zahnstangenfräser *m* ~-type core oven Kerntrockenofen *m* mit ausziehbaren Trockenzellen ~ wagon Leiterwagen *m* ~ wheel Schieber-, Sperr-rad *n* ~ wiring Gestellverdrahtung *f*

racked ausgezogen, gezogen, gereckt

racker Abfüllapparat *m*, (brewing) Abzieher *m*

racking Abfüllen *n*, Abziehen *n*, (by framing device) Bildbühneneinstellung *f*, Bildnachstellung *f* (film, TV), Recken *n*, Rödeln *n*, Strecken *n* ~ apparatus Abfüll-apparat *m*, -vorrichtung *f*, (brewing) Abziehapparat *m* ~ back treppenförmiges Absetzen *n* ~ bar Versatzleiste *f* ~ bench (or block) Abfüll-bock *m*, -ständer *m* ~ clamp Rödelzange *f* ~ gear Versatzzahnrad *n* ~ gut (brewing) Abfüll(arm)-schlauch *m* ~ hose (or pipe) Abfüllschlauch *m* ~ pawl arm Klinkenträger *n* ~ plant (brewing) Abfüllanlage *f* ~ room (brewing) Abziehraum *m* ~ square Abfüllbütte *f* ~ stud Versatzbolzen ~ table Planherd *m* ~ wedge Rödelkeil *m*

racks for bottles (tools, etc) Flaschengestell *n*

racon (radar beacon) Radarbake *f*

rad (radiation dose unit) Rad *n* (Einheit der Strahlungsdosis)

radar Ablenkzeichen *n*, Bildfeld *n*, Funkmeßanlage *f*, -gerät *n*, -ortungsgerät *n*, -technik *f*, Radar *n*, Radargerät *n* ~ altimeter Radarhöhenmesser *m* ~ attenuation Radardämpfung *f* ~

**beacon (racon)** Radarbacke *f* ~ **bearing** Funkpeilung *f* ~ **blip** Zacke *f*, Zacken *m* ~ **branch** Abzweigung *f* ~ **calibration** Radareichung *f* ~ **clutter** Radarstörflecke *pl* (auf dem Bildschirm) ~ **contact** Radarerfassung *f* ~ **controller** Radarlotse *m* ~ **cooling blower** Radarkühlgebläse *n* ~ **coverage** Radarbereich *m* ~ **dipole** Sendendipol *n* ~ **display** Radarwiedergabe *f* (auf dem Bildschirm), Schirmbild *n* ~ **display drum** Radardatenspeichertrommel *f* ~ **echo** Radarecho *n* ~ **equation** Radargleichung *f* ~ **equilibrator** Ausgleichen *n* ~ **equipment** Funkmeßgerät *n*, Radargerät *n* ~ **indicator** Radarbildschirm *m* ~ **jammer** Funkmeßstörsender *m* ~ **jamming** Radarstörung *f* ~ **link** Radarbildübertragung *f* ~ **lobe** Sichtkeule *f* ~ **locator** Radarortungsempfänger *m* ~ **mechanic** Radarmechaniker *m* ~ **monitoring** Radarüberwachung *f* ~ **mosaics** Radar-Mosaikbilder *pl* ~ **navigation** Radarnavigation *f* ~ **picket** Radarwarngerät *n* ~ **pickup horn** Hornantenne *f* ~ **plotting** Radarzeichnung *f* ~ **point** Arbeitspunkt *m* ~ **probing** Radarsondierung *f* ~ **ranging equipment** Radarentfernungsmeßgerät *n* ~ **relay** Radarbildübertragung *f* ~ **repeater** Tochtersichtgerät *n* ~ **scan pattern** Radarabtastschema *n* ~ **scope** Radarbildschirm *m* ~ **screen** Radarschirm *m* ~ **selector switch** Radarwählschalter *m* ~ **speed meter** Radargeschwindigkeitsmesser *m* ~ **station** Funkmeßturm *m*, Radarstation *f* ~ **target** Funkmeßziel *n*, Radarziel *n* ~ **tower** Funkmeß-, Radar-turm *m* ~ **track** Radarbildspur *f* ~ **watch** Radarüberwachung *f*
**radial (bearing)** Querlager *n*, Radial *m*, Radialtäfelchen *n*; rad-, speichen-, stern-förmig, strahlig ~ **acceleration** Fliehkraftbeschleunigung *f* ~ **admission** Radialbeaufschlagung *f* ~ **arm switch** Kurbelschalter *m* ~ **armature** Sternanker *m* ~ **arrangement of cylinders** Zylinderstern *m* ~ **axle-box housing** Radialachslagergehäuse *n* ~ **ball bearing** Radial-, Ringrillen-lager *n* ~ **bearing** Querlager *n* ~ **bore** radiale Bohrung *f* ~ **boring machine** Auslegerbohrmaschine *f* ~ **brick** Schachtziegel *m* ~ **center** Radialzentrum *n* ~ **clearance** Radialluft *f* ~ **crystallization** Stengelkristallisation *f* ~ **current transmitter** Sternmikrofon *n* ~ **cut** Radialschnitt *m* ~ **cylinder compressor (or supercharger)** Kreiskolbengebläse *n* ~ **cylinder supercharger** Kreiskolbengebläse *n* ~ **distribution** offene Verteilung *f* ~ **drill** Radialbohrmaschine *f* ~ **drilling machine** Kranbohrmaschine *f*, Radialbohrmaschine *f* ~ **drive shaft** Radialwelle *f* ~ **eigenfunctions** Radialeigenfunktionen *pl* ~ **end axle** kurvenbewegliche Endachse *f*, kurvenbewegliche Hinterachse *f* (r.r.) ~ **engine** Radialmotor *m*, sternförmiger Motor *m*, Stern(umlauf)motor *m* ~ **expansion** Selbstschrumpfung *f* ~-**flow compressor** (jet) Radialkompressor *m* ~ **flow in velocity space** Radialstrom *m* im Geschwindigkeitsraum ~-**flow turbine** Radialturbine *f* ~ **force** Radialkraft *f* ~ **gate** Segmenttor *n* ~ **gear** Lenksteuerung *f* ~ **grating** Radialfilter *n* ~ **integrals and transition probabilities** Radialintegrale *pl* und Übergangswahrscheinlichkeiten *pl* ~ **load** Radialbelastung *f*, Umfangslast *f* ~ **method** Radialmethode *f* ~-**milling cutter** Radialfräser *m* ~ **modes** Radialschwin-

gungen *pl* ~ **motor** Sternmotor *m* ~ **network** Sternnetz *n* ~ **oil seal ring** Radialdichtring *m* ~ **packing** Radialdichtung *f* ~ **plane** Radialebene *f* ~ **raceway** Radiallaufring *m* ~ **range** Ausladung *f* ~ **row atomizer plate** Radialzeilendüse *f* ~ **seal** Radialdichtring *m* ~ **section** Radialschnitt *m* ~ **serrations** Radial-, Stirn-verzahnung *f* ~ **slide** Meßwagen *m* ~ **speed** Radialgeschwindigkeit *f* ~ **spoke** Radialspeiche *f* **in ~ symmetry** radialsymmetrisch ~ **thrust** Radialdruck *m* ~ **thrust bearing** Lager *n* für axiale und radiale Belastung ~ **tolerance** Radialluft *f* (mech.) ~ **tool holder** Radialstahlhalter *m* ~ **triangulator** Radialtriangulator *m* ~-**type** radiale Bauart *f* ~ **valve** Segmentventil *n* ~ **velocity** Radialgeschwindigkeit *f* ~ **wiring** Radialverspannung *f*
**radially, ~-arranged type bars** sternartig gelagerte Typenhebel *pl* ~ **laminated core** Radialkernbauart *f* (eines Transformators) ~ **symmetric** radialsymmetrisch
**radian** Bogeneinheit *f*, Einheitswinkel *m*, Kreisgrad *m*, Radian *n*, Radiant *m*, Radiationspunkt *m*, Winkelgrad *m* ~ **frequency** Winkelfrequenz *f* ~ **measure** Bogenmaß *n*
**radiance** Helligkeit *f* als physikalische Größe, spezifische Lichtausstrahlung *f*, Strahlungswiderstand *m* ~ **of a surface** Oberflächenstrahlung *f*
**radiant** Lichtquelle *f* (phys.), Radiant *m*; strahlend, strahlenförmig angeordnet, Radiationspunkt *m*, Strahl *m* (math.), Strahlungspunkt *m* ~ **area** Strahlfläche *f* ~ **body** Leuchtkörper *m* ~ **disk** Leucht-scheibe *f*, -scheibchen *n* ~ **energy** Strahlungsenergie *f* ~-**field diaphragm** Leuchtfeldblende *f* ~ **field stop** Leuchtfeldblende *f* ~ **flux** Strahlungsfluß *m* ~ **flux density** Strahlungsdichte *f* ~ **heat** strahlende Wärme *f*, Strahlungswärme *f*, Wärmestrahlung *f* ~ **intensity** Strahlungs-intensität *f*, -stromdichte *f* ~ **point** Strahlpunkt *m* ~ **power** Strahlungsvermögen *m* ~ **section** Strahlungswärmezone *f* ~ **surface** (loudspeaker) Abstrahlfläche *f* ~ **superheater** Strahlungsüberhitzer *m* ~ **type of furnace** Strahlungstype *f* ~ **wall tube** Strahlungswandröhre *f*, Wandröhre *f*
**radiate, to** ~ abgeben, abstrahlen, aussenden, ausstrahlen, elektromagnetische Wellen *pl* aussenden, radialangeordnet, radieren, strahlen, strahlig, verbreiten, widerstrahlen to ~ **back** zurückstrahlen to ~ **heat** Wärme *f* abgeben to ~ **(ultra-)sound waves to act (or impinge) upon** Beschallung *f* to ~ **through** durchstrahlen
**radiated** ausgestrahlt, strahlenförmig, ~-**crystalline** kristallinisch strahlig ~ **energy** Strahlungsenergie *f* ~ **power** ausgestrahlte Leistung *f*, Strahlungsleistung *f*
**radiating** ausstrahlend, strahlend ~ **bridge** Fächerbrücke *f* ~ **capacity** Ausstrahlungs-, Emissions-vermögen *n* ~ **circuit** Strahler-, Strahlungs-kreis *m* ~ **fin (or flange)** Kühlrippe *f* ~ **guide** Hohlleiterantenne *f* ~ **power** Ausstrahlungsvermögen *n*, Strahlungsvermögen *n* ~ **sheaves** radialstrahlig ~ **sound source** strahlende Schallquelle *f* ~ **structure** strahliges Gefüge *n* ~ **surface** Ausstrahlungsfläche *f*, (of a cooler) Emissionsfläche *f*, (of loudspeaker) Lautsprecherabstrahlfläche *f* ~ **system** Luftdraht-

gebilde *n*, Strahler *m*, Strahler-gebilde *n* ~ **vane** Kühlrippe *f*

**radiation** Abkühlung *f* (aviat.), Anstrahlung *f*, Ausbreitung *f*, Ausstrahlung, Ausstrahlungsrichtung *f*, Bestrahlung *f*, Energieausstrahlung *f*, Kühlung *f*, Strahl *m*, Strahlung *f* ~ **in a continuous spectrum** Bremsstrahlung *f* ~ **of energy** Energieausstrahlung *f* ~ **of heat** Wärme(aus)-strahlung *f* ~ **of light** Lichtstrahlung *f* ~ **of the oscillations generated in the oscillator** Ausstrahlung *f* der Oszillatorschwingungen ~ **into space** Ausstrahlung *f* in den Raum

**radiation,** ~ **annealing** Bestrahlungserholung *f* ~ **background** umgebende Strahlung *f* ~ **balance meter** Strahlungsbilanzmesser *m* ~ **characteristic** Feldstärkendiagramm *n*, Strahlungs-charakteristik *f*, -diagramm *n*, -kennlinie *f* ~ **characteristic impedance** Strahlungswiderstand *m* ~ **cone** Strahlungskegel *m* ~ **constant** Leistungsfähigkeit *f* eines Strahlers, Meteramperezahl *f* ~ **coupled reflector** strahlungserregter oder strahlungsgekoppelter Reflektor *m* ~ **coupling** Strahlungskopplung *f* ~ **damage** Strahlungsschäden *pl* ~ **damping** Strahlungsdämpfung *f* ~ **density** Strahlungsdichte ~ **detecting instruments** Strahlungsdetektoren *pl* ~ **detector** Deckenfühler *m* ~ **dose meter** Strahlungsmengenmesser *m* ~ **effect** Fernwirkung *f* ~ **efficiency** Strahlungswirkungsgrad *m* ~ **efficiency of an antenna** Wirkungsgrad *m* einer Antenne ~ **elimination** Strahlungsauslöschung *f* ~**emissive surface** Strahlfläche *f* ~ **exposure limits** Strahlendosisgrenzen *pl* ~ **factor** Meteramperezahl *f* geteilt durch die Wellenlänge ~ **field** Strahlungsfeld *n* ~ **fog** Strahlungsdunst *m* ~ **furnace** Strahlungsofen *m* ~ **hazard** Strahlengefahr *f* ~ **heat** Strahlungswärme *f* ~ **height** Effektivhöhe *f*, effektive Höhe *f*, Strahl(ungs)-höhe *f* ~ **intensity meter** Röntgenwertmesser *m* ~ **load** Strahlungsbelastung *f* ~ **lobe** Strahlungslappen *m* ~ **loss** Strahlungsverlust *m* ~ **losses** Ausstrahlungsverluste *pl* ~ **maze** Strahlenschleuse *f* ~ **method of plotting** (using polar coordinates from a central station) Rayonieren *n* ~ **monitoring** Strahlungsüberwachung *f* ~ **monitoring instrument** Strahlenwarngerät *n* (Strahlenschutzgerät) ~ **pattern** Feldstärkendiagramm *n*, Richtcharakteristik *f* (Antenne), Strahlungs-charakteristik *f*, -diagramm *n*, -kennlinie *f* ~ **potential** Resonanzspannung *f*, Strahlungspotential *n* ~ **pressure** Strahlungsdruck *m* ~ **principle** Strahlungsprinzip *n* ~ **-proof** strahlungsbeständig ~ **property** Strahlungseigenschaft *f* ~ **protecting wall safe** Wandtresor *m* mit Strahlenschutz ~ **pyrometer** Ardometer *n*, Strahlungspyrometer *n* ~ **reaction** Strahlungsgegendruck *m* ~ **receiver** Strahlungsempfänger *m* ~ **resistance** Strahlungsdämpfung *f* ~ **resistance impedance** Strahlungswiderstand *m* ~ **responsive (or** ~**-sensivity) pickup** Strahlungsempfänger *m* ~**-sensivity surface** Empfängerfläche *f* ~ **surface** (loudspeaker) Abstrahlfläche *f* ~ **temperature** Strahlungstemperatur *f*

**radiative,** ~ **equilibrium** Strahlungsgleichgewicht *n* ~ **stopping** Bremsstrahlen *pl*

**radiator** Abstrahler *m*, Heizkörper *m*, Heizregister *n*, Kalorifer *m*, (auto) Kühler *m*, Radiator

*m*, Raumstrahlantenne *f*, Rippenheizkörper *m*, Sendeantenne *f*, Strahler *m*, Strahlkörper *m* ~ **box** Kühler-gehäuse *n*, -körper *m*, -rahmen *m* ~ **bracket** Fuß *m* des Kühlers *m*, Kühler-fuß *m*, -halter *m* ~ **bumper rod** Kühlerschutzbügel *m* ~ **cap** Kühler-kappe *f*, -verschluß *m*, -verschlußkappe *f*, -verschraubung *f* ~ **coil** Kühl(er)schlange *f*, Schlangenkühler *m* ~ **control valve** Kühlerventil *n* ~ **core** Kühler-block *m*, -netz *n* ~ **cover** Kühler-abdeckung *f*, -haube *f*, -schutzhaube *f* ~ **cowl** Kühlerablaßschraube *f* ~ **cowling** Kühlerverkleidung *f* ~ **damper** Kühlluftregler *m* ~ **draw-off-plug** Kühlerablaßschraube *f* ~ **element** Kühlerpaket *n* ~ **emblem** Kühlerzeichen *n* ~ **fan** Kühlerventilator *m* ~ **filler** Kühlereinguß *m* ~**-filler cap** Kühlereinfüllstutzen *m* ~ **flap** Kühler-abdeckung *f*, -kappe *f*, -klappe *f* ~ **frame** Kühler-gehäuse *n*, -körper *m*, -rahmen *m* ~ **grid** Kühlerschutzgitter *n* ~ **grill** Heizkörperverschalung *f* ~ **guard** Kühlerschutz *m* ~ **guard ring** Kühlerschutzring *m* ~ **header tank** Kühlwassertank *m* ~ **inlet pipe** Kühlereinlaufstutzen *m* ~ **lamination** Kühlerlamelle *f* ~ **muff** Kühlerschutzdecke *f* ~ **nozzle** Kühlerdüse *f* ~ **pipe** Heiz-, Kühl-, Zentralheizungs-rohr *n* ~ **piping** Kühlrohrleitung *f* ~ **rib** Kühlerlamelle *f* ~ **segment** Kühlerlamelle *f* ~ **shell** Kühlerrahmen ~ **shutter** Kühler-abdeckung *f*, -kappe *f*, -klappe *f*, -kulisse *f*, Kühlluftregler *m* ~**-shutter control** Kühlerkulissenschaltung *f* ~ **sill** Kühlerquerträger *m* ~ **spreader** Ablenkplatte *f* des Kühlers ~ **standard** Kühlerfuß *m* ~ **stay** Kühlerstrebe *f* ~ **strainer** Kühlersieb *n* ~ **strip** Kühlerbänder *pl* ~ **strut** Kühlerverstrebung *f* ~ **support** Kühlerfuß *m* ~ **surface** Strahlfläche *f* ~ **thermometer** Kühlerthermometer *n* ~ **tube** Kühlrohr *n* ~ **vent pipe** Kühlerentlüftung *f* ~ **water hose** Kühlwasserverbindungsschlauch *m*

**radical** Radikal *n*, (of molecule) Rest *m*; durchgreifend, Wurzelzeichen *n*

**radically** grundlegend, von Grund auf, radikal ~ **new** bahnbrechend

**radii** Radien *pl*

**radio, to** ~ durch-geben, -sagen, funken, senden, übertragen **to** ~ **control** fernsteuern

**radio** Funkbeobachtungsgerät *n*, Funkbetrieb *m*, Radio- oder Rundfunk-apparat *m*, -empfänger *m*, -gerät *n*, Radium *n*; drahtlos, funktechnisch, radioaktiv, Röntgen *n*, Rundfunk *m* **by (or via)** ~ auf dem Funkweg *m*

**radio abbreviation** Funkabkürzung *f*

**radio-active** radioaktiv ~ **carbon** Radiokohlenstoff *m* ~ **chain** radioaktive Zerfallsreihe *f*, Zerfallfolge *f* ~ **contamination** radioaktive Verseuchung *f* ~ **deposit** radioaktiver Niederschlag *m* ~ **equilibrium** radioaktives Gleichgewicht *n* ~ **fall-out** radioaktiver Niederschlag ~ **source** radioaktiver Strahler *m* ~ **tracer** radioaktiver Indikator *m* ~ **wastes** radioaktive Abfallstoffe *pl*

**radioactivity** Radioaktivität *f*

**radio,** ~ **advertising** Funk-reklame *f*, -werbung *f* ~ **aid** Funkhilfe *f* ~ **aid to air navigation** Funknavigationshilfe *f* ~**-aiming device** Funkzielgerät *n* ~ **alert** Funkbereitschaft *f* ~ **alignment** Funkrichtweisung *f* ~ **alternator** Hochfrequenzgenerator *m* ~ **altimeter** Funkgerät *n*, Funkhöhen-, Radiohöhen-messer *m* ~ **amateur** Funkbastler *m*, Radioamateur *m* ~ **assay** Messung *f* der Radioaktivität *f* ~**-autocontrol** Funkselbstbesteuerung *f* ~ **autograph** Strahlungsfotografie *f* ~ **balance** Strahlungswärmemesser *m* ~ **balloon** Radiosonde *f* ~ **beacon** Funk-bake *f*, -feuer *n* ~ **beam** Funk-, Leit-, Richt-strahl *m* ~ **bearing** Funkpeilung *f* ~ **bearing beam** Peilstrahl *m* ~ **bearing chart** Funkortungskarte *f* ~ **bearing installation** Peilvorrichtung *f* ~ **bearing service** Peildienst ~**-bearing station** Funkpeilstelle *f*, Peilfunkstelle *f* ~ **brevity** Funkverkehrsabkürzung *f* ~ **buoy** Funkboje *f* ~ **cabin** Funker-kabine *f*, -raum *m* ~ **cable** Hochfrequenzkabel *n*, -leiter *m* ~ **call** Funk-ruf *m*, -zeichen *n* ~ **call sign** Funkrufzeichen *n* ~ **car** Funkkraftwagen *m*, Funkwagen *m* ~**-carrier frequency** Funkträgerfrequenz *f* ~ **case** Funkkiste *f* ~ **chopper** Ticker *m* ~ **circuit** Funkverbindung *f* ~**-code board** Signaltafel *f* ~ **code message** geschlüsselter Funkspruch *m* ~**-code table** Funk-, Signal-tafel *f* ~ **coding** Funkverschleierung *f* ~ **coil** Rundfunkspule *f* ~ **command** Funkbefehl *m* ~ **communication** drahtlose Verbindung *f*, Funk-nachricht, -verbindung *f*, -verkehr *m*, -verkehrsverbindung *f*, -wesen *n* ~**-communication diagram** Funkskizze *f* ~ **compartment** Funk-kabine *f*, Geräteraum *m* ~ **compass** Bordpeiler *m*, Funk-, Radio-kompaß *m* ~ **compass calibration** Beschickung *f* ~ **control** Fernlenkung *f*, Funksteuerung *f* ~**-controlled** drahtlos gesteuert, funkgesteuert ~ **controlled glider model** ferngesteuertes Segelflugmodell *n* ~ **conversation** Funkgespräch *n* ~ **deception** Funktäuschung *f* ~ **decoy set** Funktäuschungsgerät *n* ~ **dermatitis** Hauptbeschädigung *f* ~ **detachment** Funk-staffel *f*, -trupp *m* ~ **detection** Funkerfassung *f*, -warnung *f* ~ **dial** Radioskala *f* ~ **diffusion** Drahtfunk *m* ~**-directing frequency** Leitfrequenz *f* ~ **direction-finding** Anpeilen *n*, drahtlose Ortsbestimmung *f*, Funkpeilung *f*, Peilung *f*, Richtungs-bestimmung *f*, -ermittlung *f* ~ **direction-finding apparatus** Funkpeilgerät *n* ~ **direction-finding chart** Funkpeilungskarte *f*, Karte *f* der Funkpeilung ~ **direction-finding equipment** Bordpeilgerät *n* ~ **direction-finding post** Funkpeilstelle *f* ~ **direction-finding service** Funkpeildienst *m* ~ **direction-finding station** Funkpeilstelle *f* ~ **direction-finder** Funkpeileinrichtung *f*, Radioortungsgerät *n* ~**-direction-finder indicator** Funkpeilanzeigegerät *n* ~ **discipline** Funkdisziplin *f* ~ **distress signal** Funknotsignal *n* ~ **distribution** Fernprogramm *n* ~ **distribution system** Rundfunkvermittlung *f* ~ **duplex service** drahtloser Duplexverkehr *m* ~ **(signal of wave) echo** Wellendurchbiegung *f* ~ **electric** licht-, radio-elektrisch ~**-electric wave** Funkwelle *f* ~ **element** radioaktives Element *n* ~ **engineer** Funk-, Radio-techniker *m* ~ **engineering** Funk-technik *f*, -wesen *n*, Hochfrequenztechnik *f* ~ **equipment** Funk-ausrüstung *f*, -einrichtungen *pl*, -gerät *n* ~ **equipment of plane (or vessel)** Bordfunkeinrichtung *f* ~ **facilities** Funkeinrichtungen *pl* ~ **facility chart** Funkpeilungskarte *f* ~ **fade-out** Funkpeilschwund *m* ~ **field intensity** Hochfrequenzfeldstärke *f* ~ **fix** Funkbesteck *n*, Standort *m* (durch Funk bestimmt) ~ **flying** Funkfliegen *n*

**radiofrequency (RF, r.f.)** Funk-frequenz *f*, -schwingungszahl *f*, HF, Hochfrequenz *f*, hohe Frequenz *f*, Radiofrequenz *f* ~ **amplification** Hochfrequenzverstärkung *f* ~ **amplification stage** Hochfrequenzverstärkungsstufe *f* ~ **amplifier** Hochfrequenzverstärker *m*, regelbare Hochfrequenz *f* ~ **choke** Hochfrequenzdrossel *f* ~ **coil** Hochfrequenzspule *f* ~ **current** Hochfrequenzstrom *m* ~ **detector** erster Detektor *m* ~ **field** Hochfrequenzfeld *n* ~ **oscillator stage** erste (zweite) Schwingstufe *f* ~ **preselection stage** Hochfrequenzvorstufungsröhre *f* ~ **pulse** pulsmodulierter Träger *m* ~ **range** Funkfrequenzbereich *m* ~ **rectifier** Hochfrequenzgleichrichter *m* ~ **resistance** Hochfrequenzwiderstand *m* ~ **section of radar equipment** Abstimmteil *m* ~ **stage** Hochfrequenzstufe *f* ~ **suppression equipment** Radioentstöreinrichtung *f* ~ **television signal** hochfrequentes Fernsignal *n* ~ **transformer** Hochfrequenztransformator *f* ~ **transparent** hochfrequenzdurchlassend ~ **unit** Ultrateil *m*

**radiogenic** radiogen, strahlungserzeugt ~ **heat** radiogene Wärme *f*

**radiogoniometer** Funkkompaß *m*, Funkpeiler *m*, Radiogoniometer *n* ~ **antenna** Radiogoniometerantenne *f*

**radiogoniometry** Funkpeiltechnik *f*, Radiogoniometrie *f*

**radiogram** Blitzfunk *m*, Drahtung *f*, Empfänger *m* für Schallplattenwiedergabe, Funk-meldung *f*, -telegramm *n*, Radiotelegramm *n*

**radiograph, to** ~ röntgenisieren

**radiograph** Radiogramm *n*, Röntgen-aufnahme *f*, -bild *n*, -spektrogramm *n*

**radiographic** radiografisch, röntgenografisch ~ **couch** Lagerungstisch *m* ~ **effects** röntgenografische Wirkungen *pl* ~ **equipment** Röntgenausrüstung *f* ~ **examination of metals** radiografische Metalluntersuchung *f* ~ **exposure unit** Röntgenbelichtungseinheit *f* ~ **inspection** Röntgen-prüfung *f*, -untersuchung *f*

**radiography** Durchstrahlung *f*, Radiografie *f*, Röntgen-aufnahmeverfahren *n*, -kunde *f*, -prüfung *f*, -untersuchung *f*, Röntgenografie *f*

**radio, ~ guidance** Funkkommandoführung *f*, rechtweisende Kursweisung *f* ~ **guiding** Radiowarte *f* ~ **ham** Radioamateur *m* ~ **head-phone** Radiokopfhörer *m* ~ **homing beacon** Zielflugfunkfeuer *n* ~ **index** Funklade *f* ~ **installation** Funkanlage *f* ~ **intelligence** Funkaufklärung *f* ~ **intercept service** Funkhorchdienst *m*, Horchortung *f* ~**-intercept station** Funkhorchstelle *f* ~ **interception** Funkaufklärung *f* ~ **interference** Funkstörung *f*

**radio-interference, ~ control** Entstörung *f* ~ **control equipment** Funkentstörmittel *n* ~ **equipment suppression** Entstörungsmittel *n* ~ **screening parts** Abschirmteile *pl* ~ **suppression** Funkentstörung *f*, Störschutzmittel *n* ~ **suppression capacitor** Entstörungskondensator *m* ~ **suppression reacter** (choke) Entstörungsdrossel *f* ~ **suppression resistor** Störschutzwiderstand *m* ~ **suppressor** Entstörungsvorschaltgerät *n*, Störschutzvorsatzgerät *n*

**radio, ~ interferometer** Radiointerferometer *n* ~ **interphone system** Bordsprechanlage *f* ~**-isotope** radioaktives Isotop *n* ~ **jammer** Störsender *m*

~ **jamming** Funkstörung f ~ **jamming station** Störfunkstelle f ~ **kit** Radioblei n ~ **letter-telegram** Funkbrief m ~ **license** Rundfunkgebühren pl ~ **line** Funklinie f ~ **link** Funk-brücke f, -linie f, -sprechverbindung f, -verkehrslinie f, Richtfunkstrecke f ~ **link system** Richtfunkanlage f ~ **link terminal** Richtverbindungsendstelle f ~**-listening receiver** Funkhorchempfänger m ~**-listening service** Funkhorchdienst m ~ **location** Funkortung f ~ **location service** Funkortungsdienst m ~ **locator** Abhörgerät n, Echolotung f (rdr) ~ **log** Funk-platte f, -tagebuch n, -verkehrsbuch n

**radiological** radiologisch, röntgenstrahlenkundlich ~ **physics** Physik f der Radiologie

**radiologist** Radiologe m, Röntgenologe m

**radiology** Radiologie f, Röntgen-kunde f, -lehre f, -prüfung f, Röntgenologie f

**radio,** ~ **loop** Ösen(rahmen)antenne f, Rahmenantenne f ~ **luminescence** Radiolumineszenz f ~**-maintenance (work)shop** Funkwerkstatt f ~ **marker (beacon)** Markierungsfunkfeuer m ~ **mast** Antennen-, Funk-mast m, Funkturm m ~ **mechanic** Funkwart m ~ **message** Funk-meldung f, -nachricht f, -spruch m ~ **message in clear** offener Funkspruch m ~**-message transmission** Funkspruchvermittlung f ~ **metallography** Röntgenmetallografie f ~ **meteorograph** Radiosonde f ~ **meteorographic** sonde Radiosonde f für Wetterforschung ~ **meteorological service** Funkwetterdienst m ~ **meter** Lichtmühle f, Radiometer m, Strahlen-messer m, -prüfer m, Strahlungsmesser m ~ **meter vane** Radiometerflügel m ~ **meter-vane effect** Lichtdruck m ~ **metric** radiometrisch ~ **metric gauge** Strahlungsmanometer m ~ **metry** Strahlungsmessung f ~ **micrometer** Radiomikrometer n ~ **mobile unit** Rundfunkübertragungswagen m ~ **monitor** Funksendemonitor m ~**-monitoring service** Mithördienst m ~ **navigation** Funkortung f ~ **navigational aid** Funknavigationshilfe f ~**-net diagram** Funkskizze f ~ **network** Fernmelde-, Funk-, Nachrichten-, Rundfunknetz n ~**-network chart** Funkverkehrskarte f ~**-network communication** Kreisverkehr m ~ **noise field intensity** Störfeldstärke f ~ **noise filter** Störschutzfilter n ~ **noise suppressor** Störschutz m ~ **officer** Funk-beamter, -offizier m ~ **operation** Funkbetrieb m ~**-operation station** Funkbetriebsstelle f ~ **operator** Bordfunker m, Funker m ~ **operator's compartment** Funkerraum m ~ **operator's office** Funkerkabine f ~ **operator's table** Funkertisch ~**-optical range** radiooptische Reichweite f ~ **order** Funkbefehl m ~ **outfit** Funkgerät n ~ **outlet** Radiosteckdose f

**radiopacity** Röntgensichtbarkeit f, Strahlendurchlässigkeit f

**radiopaque** röntgenstrahlungsundurchlässig, strahlenundurchlässig, undurchlässig für Röntgenstrahlen

**radio,** ~ **patrol car** Funkstreifenwagen m ~ **pattern** Strahlungscharakteristik f ~ **phony** Rundfunk m ~**-photo luminescence** Radiofotolumineszenz f ~ **photogram** Funkbild n, Radiofotogramm n ~ **photograph transmission** Bildfunk m ~ **photography** Röntgenfotografie f ~ **phototelegraphy** Bildtelegrafie f ~ **picture** Funkbild n

~**-picture set** Bildsendeempfangsanlage f ~**-picture transmitter** Bildfunkanlage f ~ **pilot balloon** Wettersonde f ~ **pilotage** rechtweisende Kursweisung f ~ **plant** Funkanlage f ~ **pocket** abgeschirmte Stelle f, Empfangsloch n, Funkschatten m ~ **position finding** Funkortung f ~ **procedure** Funkverfahren n ~ **prospection** Funkmutung f ~ **pulse gear** Netzanschlußgerät n ~ **range** Funk-bereich m, -feuer n, -reichweite f, Kursfunkbake f, Leitfunksender m ~ **receiver** Empfänger m, Empfangsgerät n, Funkempfänger m, Radio-empfänger m, -gerät n ~ **reception** Funk-, Radio-empfang m ~ **reference line** Funkstandlinie f ~ **relay** Draht(rund)funk m, Funkübertragung f ~**-relay exchange** Rundfunkvermittlungsanlage f ~**-relay station** Überleitamt n, Vermittlungsfunkstelle f ~**-relay system** Richtfunksystem n ~ **repairman** Funkgerätemechaniker m ~ **repeating** Ballsenden n ~**-repeating station** Ballsendestelle f ~ **resistance** Strahlungsfestigkeit f ~ **room** Funkraum m

**radioscope** Durchleuchtungsapparat m ~ **box** Durchleuchtungskasten m ~ **table** Durchleuchtungstisch m

**radioscopy** Durchleuchtung f, Radioskopie f, Röntgendurchleuchtung f

**radio,** ~ **screened** funkentstört ~ **secrecy** Funkgeheimnis n ~ **secret** Funkgeheimnis n ~ **section** Funktrupp m ~ **sensitivity** Strahlenempfindlichkeit f ~ **service** Funkbetrieb m ~ **service between permanent stations** fester Funkdienst m ~ **set** Funkgerät n, Rundfunkempfänger m ~ **shadow** abgeschirmte Stelle f, Empfangsloch n, Funkschatten m ~ **shielded** abgeschirmt ~ **shielding** Störschutz m ~ **signal** Funkspruch m, Funksignal n ~**-signal communication** funktelegrafische Verbindung f ~**-signal cutoff** elektrische Abschaltung f ~**-signal number sheet (or signal unit journal)** Funkspruchübersicht f ~ **silence** Funk-stille f, -verbot n ~ **sonde** Pilotballonaufstieggerät n, Radiosonde f, Wettermeßinstrument n mit funktelegrafischer Übertragung f ~**-sounding** Radiosondierung f ~**-spectrometer** Radiospektrometer n ~ **spectrum** Funkwellenspektrum n ~ **station** Fuge, Funk-empfänger m, -gerät n, -station f, -stelle f, Leitstelle f ~ **station of destination** Bestimmungsfunkstelle f ~ **studio** Rundfunkaufnahmeraum m ~ **substation** Nebenfunkstelle f ~ **synoptic weather message** Wettersammelfunkspruch m ~ **system** Funknetz n ~ **technician** Funktechniker m, Funkwart m ~ **technics** Funktechnik f ~ **technology** Funkwesen n ~**-telecontrol** Funkfernsteuerung f ~ **telegraph** Tastfunkgerät n ~ **telegraph plant** Funktelegrafenanlage f ~ **telegraph transmitter** Telegrafiesender m

**radiotelegraphic** funk-, radio-telegrafisch ~ **connection** funktelegrafische Verbindung f ~ **transmitter** Funktelegrafiesender m ~ **valve** Empfängerröhre f, Gitterröhre f, Glühkathodendetektor m

**radio,** ~ **telegraphy** drahtlose Telegrafie f, Funktasten n, -telegrafie f, Radiotelegrafie f ~**-telemeteorographic (or telemetric) instruments** Wettermeßinstrumente pl mit funkentelegrafischer Fernübertragung

**radiotelephone, to** ~ funksprechen, funktelefonisch übermitteln
**radiotelephone** Funk-sprechgerät *n*, -telefon *n* ~ **circuit** Funkfernsprechverbindung *f*, Funksprecher-stromkreis *m*, -verbindung *f* ~ **communication** Funksprechen *n* ~ **installation** Funkfernsprechanlage *f* ~ **message** Funkspruch *m* ~ **push button** Telefonietaste *f* ~ **set** Funkfernsprechanlage *f* ~ **system** Funksprechanlage *f* ~ **technique** Funkfernsprechbetrieb *m* ~ **transmitter** Telefoniesender *m*
**radiotelephonic transmitter** Funktelefoniesender *m*
**radiotelephony** drahtloses Fernsprechen *n*, drahtlose Telefonie *f*, Funk(fern)sprechen *n*, Funktelefonie *f*, Radiotelefonie *f* ~ **communication** Sprechfunkverkehr *m*, Telefoniebetrieb *m* ~ **discipline** Sprechdisziplin *f* ~ **network** Sprechfunknetz *n* ~ **operator** Fernsprechbetriebsfunker *m* ~ **set** Funksprechergerät *n*, Telefongerät *n* ~ **traffic** Sprechverkehr *m*
**radio,** ~ **teleprinter** Funkfernschreibmaschine *f* ~ **telescope** Radioteleskop *n* ~ **teletype** (RATT) Funkfernschreiber *m* ~ **tellurium** Radiotellur *n* ~ **tester** Funkprüfer *m* ~-**testing flight** Funkbetriebsflug *m* ~-**testing set** Funkprüfer *m* ~ **theodolite** Höhenwinkelpeiler *m*, Radiotheodolit *m* ~ **therapy** Röntgenbestrahlung *f*, Röntgen-, Strahlen-therapie *f* ~-**thermoluminescence** Radiothermolumineszenz *f* ~ **thermy** Hochfrequenzdiathermie *f*, Kurzwellenbehandlung *f* ~ **time signal** Funkzeitzeichen *n* ~ **timing apparatus** Zielkontrollapparat *m* ~ **tower** Antennenmast *m*, Funkmast *m*, Funkturm *m* ~ **traffic** Funkbetrieb *m*, Funkverkehr *m*, Tastverkehr *m* ~ **transmission** Funksenden *n*, Funksendung *f* ~ **transmission of pictures** Bildfunk *m*, Bildtelegrafie *f* ~-**transmitter** Funkgeber *m*, Funksender *m*, Netzanschlußgerät *n*, Röhrensender *m* ~-**transmitting center** Funksendezentrale *f* ~ **truck** Funk(gerätekraft)wagen *m* ~ **tube** Elektronen-, Radio-röhre *f* ~ **tube getter** Getter *m* ~-**tube performance chart** Röhrenkennlinie *f* ~-**type supervisory control** Sternverkehr *m* ~ **valve** Radioröhre *f* ~ **vision** drahtloses Fernsehen *n*, Funkfernsehen *n* ~ **vision studio** Fernsehsenderaum *m* ~ **wagon** Funkfahrzeug *n*, Funkwagen *m* ~ **wave** Funkwelle *f*, elektromagnetische Welle *f*, Radiowelle *f* ~-**wave deviation** Funkfehlweisung *f* ~-**wave frequency** Funkfrequenz *f* ~-**wave spectrum** Wellenspektrum *n* ~ **waves** Hertzsche Wellen *pl* ~ **weather-broadcasting station** Funkwetterwarte *f* ~ **weather service** Wettersendedienst *m* ~ **wind flight** Radiosonde *f*
**radium** Radium *n* ~ **coating** Radiumbelag *m* ~ **disintegration** Radiumzerfall *m* ~ **emanation** Radon *n* ~ **excretion** Radiumausstoßungsverfahren *n* ~ **iodide** Radiumjodid *n* ~ **mold** Radiummoulage *f* ~ **packing table** Radiumpacktisch *m* ~ **paint** Radiumbelag *m* ~ **plaque** Plattenträger *m* ~ **rays** Radiumstrahlen *pl* ~ **secondary ducts** Radonfolgeprodukte *pl* ~ **seed** Radiumpillare *f* ~ **sulfate** Radiumsulfat *n*
**radius** Aktionsradius *m*, Auslenkung *f*, Bereich *m*, Drehungsmesser *m*, Exzentrizität *f*, Furche *f*, Halbmesser *m*, Hub *m*, Kreiselhalbmesser *m*, Krümmung *f*, Radius *m*, Rayon *m*, Reichweite

*f*, Strahl *m* (geom.), Strahlung *f* (geom.), Wirkungskreis *m* **within a** ~ **of** . . . im Umkreis *m* von . . .
**radius,** ~ **of action** Aktions-bereich *m*, -radius *m*, Arbeitsfeld *n*, Eindringtiefe *f*, Einflußbereich *m*, Fahrbereich *m*, Flugbereich *m*, Reichweite *f*, Wirkungs-bereich *m*, -kreis *m* ~ **of action of a machine** Dampfstrecke *f* ~ **of axle turn** Achsenverschränkbarkeit *f* ~ **of bend** Kurvenradius *m* ~ **of center of gravity** Scherpunkt-halbmesser *m*, -radius *m* ~ **of crown** Wölbungsradius *m* ~ **of curvature** Abrundungs-halbmesser *m*, -radius *m*, Biegungshalbmesser *m*, Krümmungs-halbmesser *m*, -radius *m* ~ **of curvature of a corner** Eckenradius *m* ~ **of curvature at flange** Krempenübertragung *f* ~ **of curve** Kurvenradius *m* ~ **of development of spiral** (sinoid) Abwicklungshalbmesser *m* ~ **of effect** Wirkungsbereich *m* ~ **of flight** Flugbereich *m* ~ **of gyration** Kreiselradius *m*, Trägheits-halbmesser *m*, -radius *m* ~ **under the head** (fillet) Kopfübergangsradius *m* ~ **of influence** Einflußzone *f*, Reichweite *f* ~ **of jib** Ausladung *f* ~ **of notch** Kerbhalbmesser *m* ~ **of pipe bend** Rohrkrümmung *f* ~ **of service area** Reichweite *f* eines Senders ~ **of shock** Erschütterungshalbmesser *m* ~ **of sphericity** Abrundungsradius *m* ~ **of star** Gestirnhalbmesser *m* ~ **of tailspin** Trudelradius *m* ~ **of travel** Ausladung *f* ~ **of turn** Drehkreis *m*, Kurvenradius *m*, Wendekreis *m* ~ **of turn of driving axles** Verschränkbarkeit *f* der Achsen ~ **of vortex core** Wirbelkernhalbmesser *m* ~ **of wheel path** Lenkradius *m*
**radius,** ~ **bar** Führungskrummzapfen *m* ~ **cross-cut saw machine** Balanziersäge *f* ~ **gauge** Radienlehre *f* ~ **link** Gleitführung *f* ~ **rod** Einziehstrebe *f*, Gegenlenker *m*, (automobile) Hinterachsenschubstange *f*, Knickstrebe *f*, Leitstange *f*, Radiusstange *f*, Schwenkstrebe *f* ~ **strut** Knickstrebe *f* ~ **vector** Fahrstrahl *m*, Polstrahl *m*, Radiusvektor *m*
**radiused,** ~ **flange** gewölbter Flansch *m* ~ **nose pieces** Halbrundschnäbel *pl*
**radiusing** Verrunden *n*
**radix** Basis *f* (math.), Grundzahl *f* ~ **notation** Basisschreibweise *f*, Stellenwertschreibung *f* ~ **point** Radixpunkt *m*
**radome** Radar-haube *f*, -kuppel *f*, -nase *f*
**radon** Radon *n* ~ **seed** Radonhohlnadel *f*
**raff** Abfälle *pl*
**raffinate** Raffinat *n*, Raffinationsprodukt *m* ~ **layer** raffinierte Schicht *f*
**raft** Floß *n*, Fundamentrahmen *m*, Pfahlrost *m*, Rettungsfloß *m* ~ **of casks** Tonnenfloß *n*
**raft,** ~ **channel** Floßgasse *f* ~ **foundation** Wannen-fundation *f*, -gründung *f* ~ **pass** Floßgasse *f* ~ **port** Ladepforte *f* ~ **release** Schlauchbootauslösung *f* ~-**release lever** Schlauchbootauswurfhebel *m* ~ **wood** Flößholz *n*
**rafter** Dachsparren *m*, Sparren *m*, Sparrenholz *n*, Sprosse *f* ~ **beam (or girder)** Sparrenstab *m* ~ **nail** Floßnagel *m* ~ **plate** Sparrensohle *f* ~ **timbering** Sparrzimmerung *f* ~ **trimmer** Schifter *m*
**rafters** Balkenwerk *n*
**raftsman** Flößer *m*
**rag, to** ~ aufrauhen, kerben

**rag** Fetzen *m*, Lappen *m* ~ **board** Hadernkarton *m* ~ **boiler** Hadernkocher *m* ~ **bolt** Hackbolzen *m*, Hakenbolzen *m*, Steinschraube *f* ~ **bucket** Lumpenbutte *f*, Zuber *m* (paper mfg.) ~ **cleaner** Hadernstäuber *m* ~-**cutting knife** Lumpenschneidemesser *m* ~ **cylinder driver** Halbholländerführer *m* ~ **duster** Lumpenstäuber *m* ~ **engine** Holländer *m*, Stoffmühle *f* (paper mfg.) ~ **head** Kerbkopf *m* ~ **machine knife** Holländermesser *m* ~ **paper** Hadernpapier *n* ~-**pulp** Hadern-brei *m*, -halbstoff *m* ~ **pump** Kettenpumpe *f* ~-**tearing machine** Lumpenwolf *m* ~ **tissue paper** Hadernseidenpapier *n* ~ **tub** Lumpenbutte *f* ~ **tube** Zuber *m* (paper mfg.) ~ **work** Mauerwerk *n* in Polygonverband
**ragged** geschuppt, (in outline) gezackt, hadrig, zackig, zerfetzt
**ragging** Bettsetzen *n* (Aufbereitung), (roll surface) Einhauen *n*, Einkerben *n*, Hau *m* ~ **hammer** Berg-, Groß-fäustel *n*
**raggins** Erzstücke *pl*, die sich auf dem Siebboden abgesetzt haben
**rags** (paper) Hadern *pl*, Lumpen *pl*, (waste) Putzwolle *f* ~ **impregnated with gold solution** Goldzunder *m*
**rail, to** ~ gittern, schienen **to** ~ **off** absperren, einzäunen
**rail** Eisenbahnschiene *f* (r.r.), Eisenschiene *f*, Fahrschiene *f*, Geländer *n*, Geleise *n*, Gleis *n*, Quer-balken *m*, -stange *f*, Reling *f*, Schiene *f* ~ **of a bay work** Wandriegel *m* ~ **with bottom-discharge tubs** Standbahn *f* mit Bodenentladern
**rail,** ~ **barrier** Schienen-hindernis *m*, -sperre *f* ~ **base** Fuß *m* der Schiene, Schienenfuß *m* ~ **bender** Schienenbeuger *m* ~ **bending machine** Schienenbiegmaschine *f* ~ **bond** Schienen-stoß *m*, -verbinder *m* ~ **bottom** Schienenunterkante *f* ~ **brake** Bremszeug *n*, (electromagnetic) Gleisbremse *f*, Gleisbremszeug ~ **breakage** Schienenbruch *m* ~ **bridge** Gleisbrücke *f* ~ **bus** Triebwagen *m* ~ **car** Schienenfahrzeug *n*, Triebwagen *m* ~ **chair** Schienen-lager *n*, -stuhl *m* ~ **changing machine** Schienenverlegemaschine *f* ~ **clamp** Schienenzange *f* ~ **connection** Bahnanschluß *m* ~ **contact** Gleis-, Schienen-kontakt *m* ~ **contact-making device of mercury** Quecksilberschienenkontakt *m* ~ **conveyer system** Schiebebühnenanlage *f* ~ **end** Schienenende *n* ~ **foot** Schienenfuß *m* ~ **fork** Schienentraggabel *f* ~ **gauge** Umgrenzung *f* der Stromschiene ~ **gauge template** Spurlehre *f* ~ **guard** Bahnräumer *m* ~-**guided** gleisgebunden ~ **head** Auslade-bahnhof *m*, -spitze *f*, Eisenbahnendprodukt *n*, Endbahnhof *m*, Endstation *f*, Kopfbahnhof *m*, -station *f*, Schienenkopf *m* ~ -**hoisting engine** Gleishebewinde *f* ~ **joint** Schienen-stoß *m*, -verbinder *m* ~ **layer** Schienendecke *f* ~ **layout** Schienenaggregat *n* ~ **lifter** Gleis-heber *m*, -traghaken *m* ~-**lifting jack** Gleishebewinde *f*, Schienenheber *m* ~-**mill train** (in rolling) Schienenstraße *f* ~ **motorcar** Triebwagen *m* ~ -**mounted** gleisgebunden ~ **overlap** Schienenauszug *m* ~ **pass** Schienenkaliber *n* ~ **post** Geländer-docke *f*, -pfosten *m* ~ **return** Schienenrückleitung *f* ~ **ring** Schienenring *m*
**railroad** Eisenbahn *f*, Eisenbahn-linie *f*, -strecke *f*, Schienenweg *m* ~ **allocator** Zugverteiler *m* ~

**ballast** Eisenbahnschotter *m* ~ **bedding** Gleisbettung *f* ~ **bridge** Eisenbahnbrücke *f* ~ **buildings** Bahnanlagen *pl* ~ **car** Eisenbahnwagen *m*, Waggon *m* ~ **central station** Eisenbahnzentrale *f* ~ **coating** Eisenbahnanstrich *m* ~ **collection point** Sammelbahnhof *m* ~ **communication** Bahnverbindung *f* ~-**company** Eisenbahngesellschaft *f* ~ **connection** Eisenbahn-anschluß *m*, -verbindung *f* ~ **contruction** Eisenbahnbau *m* ~ **coupling** Eisenbahn-, Wagen-kupplung *f* ~ **crossing** Gleiskreuzung *f* ~ **earthworks** Eisenbahndamm *m*, Eisenbahnunterbau *m* ~ **ferry** Eisenbahnfähre *f* ~ **installation** (buildings and property) Eisenbahnanlage *f* ~ **installations** Bahnanlagen *pl* ~ **journal grease** Fett *n* zur Schmierung des Achsenlagers ~ **junction** Eisenbahnanschluß *m*, Eisenbahnknotenpunkt *m*, Verkehrsknotenpunkt *m* ~ **layout** Gleiseanlage *f* ~ **line** Bahnverbindung *f* ~-**line junction** Gleisanschluß *m* ~ **management** Eisenbahnverwaltung *f* ~ **material shops (or works)** Eisenbahnwerke *pl*, Eisenbahnwerkstätte *f* ~ **net** Bahnnetz *n* ~ **network** Gleisnetz *n* ~ **oil** Schmieröl *n* für die Achsenbüchsen ~ **operation** Eisenbahnbetrieb *m* ~ **siding** Eisenbahnanschluß *m* ~ **signal tower** Eisenbahnstellwerk *n* ~ **station** Bahnhof *m* ~ **supply-collecting depot** Sammelbahnhof *m* ~ **supply line** Eisenbahnnachschublinie *f* ~ **system** Bahnnetz *n* ~ **tank station** Eisenbahntankstelle *f* ~ **terminal** Eisenbahnendpunkt *m* ~ **track** Eisenbahngleis *n*, Oberbau *m* ~ **tracks** Gleisstränge *pl* ~ **traffic** Eisenbahnbetrieb *m*, -verkehr *m*, Zugverkehr *m* ~-**transport timetable** Eisenbahntransportfolge *f* ~ **transportation** Bahntransport *m*, Eisenbahnbeförderung *f* ~ **yard** Eisenbahnhof *m*, Geleis(e)anlage *f*
**railroading** Eisenbahnwesen *n*
**rail,** ~ **rolling mill** Eisenbahnschienenwalzwerk *n*, Schienenwalzwerk *n* ~ **rope** Geländerleine *f* ~ **scales** Gleiswaage *f* ~ **section** Schienenprofil *n* ~-**shifting machine** Gleisrückmaschine *f* ~ **sluer** Schienenrücker *m* ~ **square** Schienenwinkel *m* ~ **steel** Schienenstahl *m* ~-**straightening machine** Schienenricht-bank *f*, -maschine *f* ~ **surfacing** Schienenauftrag *m* ~ **tongs** Gleisheber *m*, Schienenzange *f* ~ **track** Gleisweg *m* ~ **vehicle** Schienenfahrzeug *n*
**railway** Eisenbahn *f* ~ **car with chute for sluicing the soil from the car into the fill** Spülkippe *f* ~ **carriage** Eisenbahnwaggon *m* ~ **connection** Anschlußgleis *n* ~ **coupling** Wagenkupplung *f* ~ **curve** Kurvenlineal *n* ~ **cutting** Eisenbahneinschnitt *m* ~ **equipment** Bahnausrüstungsteile *pl* ~ **ferry** Trajektschiff *n* ~ **gate** Gatterbalken *m*, Schranke *f*, Wegschranke *f*, Zugschranke *f* ~ **gauge** Spurweite *f* ~ **guide** Kursbuch *n* ~ **heating member** Zugheizkörper *m* ~ **implements** Streckenwerkzeug *n* ~ **junction** Gleisverzweigung *f* ~ **line** Bahnstrecke *f*, Eisenbahn-linie *f*, -strecke *f* ~ **loading gauge** Lichtraumprofil *n* ~ **property** Bahneigentum *n* ~ **rates** Bahnfrachtsätze *pl* ~ **repairshop** Bahnbetriebswerkstatt *f* (BBW) ~ **safety installations** Eisenbahnsicherungswesen *n* ~ **siding** Anschlußgleis *n* ~ **signaling system** Eisenbahnsignalanlagen *pl* ~ **staff** Zugpersonal *n* ~ **(electrical) switchgear** Bahnschaltanlage *f* ~ **tariff** Bahnfrachttarif *m*

~ **terminal** Abgangsbahnhof *m* ~ **territory** Eisenbahngelände *n* ~ **tire** Eisenbahnradreifen *m* ~ **tools** Streckenwerkzeug *n* ~ **track** Eisenbahn-linie *f*, -strecke *f* ~ **trolley** Draisine *f* ~ **yard** Güterbahnhof
**rail web** Schienensteg *m*
**railing** Brüstung *f*, Geländer *n*, Geländer-holz *n*, -stange *f*, Grießholm *m*, Staket *n*, Treppengeländer *n* ~ **of balustrade** Erkerbrüstung *f*
**railing,** ~ **bar** Schrankenstange *f* ~ **post** Geländerstütze *f*
**railings** Staketen *pl* (rdr)
**railless** gleislos
**rain, to** ~ regnen
**rain** Beregnung *f*, Regen *m* ~ **band** Regen-bande *f*, -linie *f* ~ **bow** Regenbogen *m* ~ **bow-colored** regenbogenfarbig ~ **bow ground** Irisfond *m* ~ **bow wheel** Farbenrad *n* ~ **cap** Regen-haube *f*, -kappe *f* ~ **chamber** Regenkammer *f* ~ **chart** Regenkarte *f* ~ **cloud** Regenwolke *f* ~ **clutter** Regenstörung *f* (rdr) ~ **drop** Regentropfen *m* ~ **effect** Regeneffekt *m* ~**-exposure-test specimen** Beregnungsprobe *f* ~ **fall** Niederschlag *m*, Niederschlagsmenge *f*, Regenfall *m*, Wasserhaushalt *m* ~ **fall intensity** Regendichte *f* ~ **fall recorder** Regenschreiber *m* ~ **gauge** Regenmesser *m* ~ **gutter** Regenröhre *f* ~ **hood** Regenkappe *f* ~ **making** Regenmachen *n* ~ **measurement** Regenmessung *f* ~ **pipe** Dachröhre *f*, Traufe *f* ~ **print** Regenabdruck *m* ~**-proof** regen-, wasser-dicht ~ **remover nozzle** Regenabweiserdüse *f* ~ **shadow** Regenschatten *m* ~ **-shield cap** (waterproof cover) Regenschutzkappe *f* ~ **shutter** Regenblende *f* ~ **spell** Regenperiode *f* ~ **squall** Regenbö *f* ~ **static** Regenstatik *f* ~ **test** Beregnungsversuch *m*, Regenversuch *m* ~ **trap** Regenverschluß *m* ~**-type condenser** Regenkondensator *m* ~ **visor** Regenschutzrohr *n*
**rainwater** Niederschlags-, Regen-wasser *n* ~ **deflector** Regenleiste *f* ~ **head** Rinnenkasten *m*
**raining plant** Regenanlage *f*
**rainless** regenarm, regenlos ~ **side** Regenschatten *m*
**rainy** niederschlagsreich, regenreich, regnerisch ~ **day** Regentag *m* ~ **film** verregneter Film *m* ~ **period** Regenzeit *f* ~ **side** Regenseite *f* ~ **weather** Regenwetter *n* ~ **year** Regenjahr *n*
**raisable** hebbar
**raise, to** ~ abheben, (a question) anschneiden, aufheben, (steam) aufmachen, aufnehmen, aufrichten, (dust etc.) aufwirbeln, erheben, erhöhen, (steam) erzeugen, fördern, heben, heranziehen, hissen, hochheben, hochstellen, hochwinden, (clear of something) lüften, rauhen, steigern, ziehen, züchten
**raise, to** ~ **a claim** reklamieren **to** ~ **a claim (or point)** einen Anspruch *m* geltend machen **to** ~ **a dam** überstauen **to** ~ **high** hochtreiben **to** ~ **to a higher power** potenzieren **to** ~ **the hood** das Verdeck *n* aufklappen oder aufschlagen **to** ~ **an opposition** Einspruch *m* erheben **to** ~ **ore** Erze *pl* gewinnen **to** ~ **to the nth power** in die nte Potenz *f* erheben **to** ~ **the question** nahelegen **to** ~ **to second power** quadrieren **to** ~ **steam** Dampf *m* aufmachen **to** ~ **to third power** kubieren **to** ~ **trusses of a bridge** eine Brücke *f* einbringen **to** ~ **up** hochziehen **to** ~ **upward** auf-

wärtsbewegen **to** ~ **a wall** eine Mauer *f* aufhöhen **to** ~ **with a windlass** aufhaspeln
**raise** Aufschäumen *n*, Aufwallen *n*, Steigerung *f* ~ **in salary** Gehaltsaufbesserung *f*
**raised** erhaben, gehämmert, gehoben, getrieben ~ **arch** Spitzbogen *m* ~ **beach** Strandstraße *f* ~ **characters** erhabene Schrift *f* ~ **countersunk head wails** Stemmnägel *pl* ~ **design** erhabenes Muster *n* ~ **end** Federwulst *m* ~ **face** Leistenfläche *f* ~ **fascine road** Faschinendamm *m* ~ **grain** Flader *m* ~ **ground floor** Hochpaterre *n* ~ **manhole** (engine) Fahrstuzen *m* ~ **pad** vorgezogene Anbaufläche *f* ~ **screw** hochköpfige Schraube *f* ~ **style** Rauhartikel *m* ~ **thread surface printing** Schleifdruck *m* ~**-type** erhabene Type *f* ~ **vein** Flader *m* ~ **wicket** aufgerichtete Schütztafel *f* ~ **wing** (of building) Eckflügel *m* ~ **work** (in relief) getriebene Arbeit *f*
**raising** Abhebung *f*, Abschließen *n*, Aufrichten *n*, Aufrichtung *f*, Ausheben *n*, Förderung *f*, Heben *n*, Hebung *f*, Hochstellen *n*, Lüften *n*, Richten *n*
**raising,** ~ **of an area by filling** Landaufhöhung *f* (durch Spülbetrieb) ~ **a dam** Aufkadung *f* eines Deiches ~ **the dust** Staubaufwirbelung *f* ~ **of the frames** Spantenaufrichtung *f* ~ **of the level** (by the wind) Aufreibung *f* des Wasserspiegels ~ **of rods** Gestängeziehen *n* ~ **to third power** Kubizierung *f* ~ **of water** Wasserförderung *f* ~ **of the water level** Wasserspiegelerhöhung *f* ~ **of a wreck** Beseitigen *n* eines Wracks *n*
**raising,** ~**-and lowering mechanism** Absenkeinrichtung *f* ~ **apparatus** Aufsatzform *f* ~ **board** Krispelholz *n* ~ **gig** Verstreichrauhmaschine *f* ~ **hammer** ovaler Stahlhammer *m* ~ **knife** Böttchermesser *m* ~ **lever** Vorwählhebel *m* ~ **machine** Hebe-, Rauh-maschine *f* ~ **pipe** Steigrohr *n* ~ **screw** Hebeschraube *f* ~**-up** Rohaufbrechen *n* ~ **waste** Rauhflocken *pl*
**rake, to** ~ aufkratzen, (fire) bestreichen, kratzen, schüren **to** ~ **off** abkrücken, putzen **to** ~ **out** ausbrechen, auskratzen, ausräumen, ausziehen, herauskratzen, ziehen **to** ~ **out cinder** ausschlacken, Schlacke *f* abziehen **to** ~ **up the ground** den Boden *m* aufgraben **to** ~ **out slag** (or cinder) abschlacken, ausschlacken, entschlacken, Schlacke *f* abziehen **to** ~ **together** zusammenscharren
**rake** Abstreicher, Ausschießen *n* des Stevens, Gabelrechen *m*, Harke *f*, Krähle *f*, Krammeisen *n*, Kratze *f*, Krücke *f*, Maischgitter *n*, (of a pole) Neigung *f* des auf Zug gestellten Gestänges, Ofenbrücke *f*, Rechen *m*, Rechenflügel *m*, Roststab *m*, Rühr-haken *m*, -scheit *n*, Scharre *f*, Stocheisen *n*, Überhang *m*, (for removing slag) Ziehhaken *m* ~ **of a leviathan** Leviathanrechen *m* ~ **of a tool** Ansatzwinkel *m* ~ **of wing tip** Flügelumriß *m*
**rake,** ~ **angle** (Räumnadel) Freiwinkel *m*, Neigungs-, Span-winkel *m* ~ **conveyer** Förderrechen *m* ~ **handle** Rechenstiel *m* ~ **stacker** Rechenstecker *m* zum Heuaufschichten, Strohgebläse *n* ~ **stand** Rechenständer *m* ~ **tooth** Kratzenzinke *f*
**raker** Räum-löffel *m*, -zahn *m*
**rakes** Gezäh(e) *n*
**raking** Enfilierung *f* ~ **molding** ansteigendes Gesims *n* ~ **operation** Krählarbeit *f* ~**-out** Aus-

sparung f ~-out door (or opening) Ausziehöffnung f ~ pile Schrägpfahl m ~ shoring Abstrebung f (Gebäude) ~ strut Druckstrebe f
rally Sammelpunkt m
ram, to ~ abrammen, (a shell) ansetzen, ausstampfen, besetzen, einstampfen, feststampfen, rammen, stampfen, stoßen to ~ down the earth die Erde f rammen to ~ an engine das Ansaugsystem des Motors unter Saugdruck setzen to ~ by hand handstampfen to ~ home einrasten to ~ in einstoßen to ~ up a mold eine Gußform f aufstampfen to ~ up aufstampfen, umstampfen, vollstampfen
ram (of a press) Bär m, Druckkolben m, Fallblock n, -hammer m, Kolben m, Kropfeisen n, Mönchskolben m, Plunger m, (of a press) Prägebär m, Preß-kolben m, -stempel m, Ramm-bär m, -bock m, -klotz m, Ramme f, Stampfe f, Stau m (aerodyn.), Stempel m, Stößel m, Stoßfänger m, Stoßvorrichtung f, Sturmbock m, Tauchkolben m, Werkzeugstößel m, Wuchtbaum m, Widder m (hydraul.)
ram, ~ adjuster Stößeleinstellhebel m ~ air (Klimaanlage) Frischluft f, Stauluft f ~ air induction Frischluftzufuhr f ~ air inlet Staulufteintritt m (des Ballonschirmes) ~ air turbine Staulufturbine f ~ angle Stößelwinkel m ~ charge Staudruckleitung f ~ charger Staudruckventil n ~ crane Fallwerkskran m ~ displacing device (presse) Stempelverschiebung f ~ drive motor Stößelmotor m ~ effect Auftreffwucht f ~ flushing Spülstampfe f ~ guide Stößelführung f ~ head Schleuderkopf m ~ head slide Hobelkopfschlitten m ~'s hook Doppelthaken m (Kran) ~'s horn Geröllhaken m
ramjet Lorinantrieb m, Manteltiefe f, Staustrahltriebwerk n ~ airplane Staudüsenflugzeug n ~ engine Staustrahltriebwerk n ~ flight Stoßstrahlflug m ~ power plant Lorintriebwerk n ~ propulsion Staudüsenantrieb m ~ wall Düsenwand f
ram, ~ movement Stößelbewegung f ~ point Staupunkt m ~ pressure Stau-, Preß-druck m ~ pressure due to flight Flugstau m ~ receiver port Staufangbohrung f ~ reciprocating motor Motor m für die Stößelbewegung ~ scoop Stauhutze f ~ shaft Rammwelle f ~ slide Stößelschlitten m ~ socket Rammwelle f ~ stroke Stößelhub m ~ temperature Stautemperatur f ~ tube Stauhutze f ~ way Stößelgleitbahn f ~ weight Bärgewicht n
ramie covered wire Ramé-isolierter Draht m
ramification Abzweigung f, Verzweigung f
ramify, to ~ ramifizieren, verzweigen
rammed (in) gerammt ~ bottom lining Stampfherd m ~ concrete gestampfter Beton m ~ layer Stampfschicht f ~ lining Stampfauskleidung f ~-loam construction Lehmstampfbau m
rammer Ansetzereinrichtung f, Bär m, Bläuel m, Fallbär m, Hund m, Ladestock m, Rammbär m, Ramme f, Schlägel m, Stampfe f, Stampfer m, Stößel m, Stößer m ~ foot Stampffuß m ~ head Setzkolben m ~ heads Stoßköpfe pl
ramming Ausstampfung f, Druckvermehrung f für die Luftstromgeschwindigkeit f, Einstampfen n, Stampfen n ~ of the air Luftstauung f ~ of profile castings Stampfen n von Betonteilen

ramming, ~ air intake Luftsaugrohr n für die Luftgeschwindigkeit des Schraubenstrahlen ~ apparatus Rammapparat m ~ block Formblock m, Preßklotz m ~ board Preßplatte f ~ effect Stampfwirkung f ~ head Schleuderkopf m ~ machine Schlag-, Stampf-maschine f ~ mass Stampfmasse f, Stößel m ~ plane Rammflugzeug n ~ pressure Staudruck m ~ rate (in sand slinging) Schleudergeschwindigkeit f
ramp Ansteigung f, Auffahrt m, Böschung f, Flugsteig m, Gleitschiene f, Ladeplatz m, Rampe f, schiefe Ebene f (Abbau) ~ for climbing Auslaufzunge f
ramp, ~ carrier Rampenwagen m ~ connector Rampenverbinder m ~ drive for infinitely variable ramp setting Rampenantrieb m für stufenlose Rampeneinstellung ~ floodlights Vorfeldscheinwerfer m ~ girder Rampenträger m ~ length Rampenlänge f ~ service Vorfelddienst m (airport) ~ shaft ballig gedrehte Welle f ~ sight Sattelkorn n
rampant überhandnehmend, üppig wuchernd
rampart Wall m
ramshackle baufällig
rancid ranzig
rancidness Ranzigkeit f
random zufällig, zufallsverteilt at ~ beliebig, nach Belieben, blindlings, aufs Geratewohl, unsystematisch, wahllos, willkürlich with ~ orientation regellos geordnet
random, ~ access direkter oder beliebiger Zugriff m ~ access programming stochastische Programmierung f ~ access storage RAM-Speicherung f ~-access storage system Speichersystem n mit direktem Zugriff ~ choice Stichwahl f ~ coincidence Zufallskoinzidenz f ~ drift rate Geschwindigkeit f der Zufallsauswanderung (gyro.) ~ encounter zufälliger Stoß m ~ impurities statistisch verteilte Fremdatome pl ~ light statistische Bewegung f ~ local variation regellose örtliche Abweichung f ~ noise Grundgeräusch n ~ number Zufallszahl f ~ orientation nichtbevorzugte oder regellose Orientierung f ~ phase Zufallsphase f ~ problem Regellosigkeitsproblem n ~ process Grundprozeß m ~ rubble unregelmäßiges Bruchsteinmauerwerk n ~ sample Stichprobe f ~ sampling Bewertungsprüfung f ~ sensivity mittlere Empfindlichkeit f ( acoust.) ~ sequence Zufallsfolge f ~ stone foundation Gründung f auf Steinschüttung f ~ test Stichprobe f ~ variable statistische Variable f, Zufallsvariable f ~ variation willkürliche Schwankung f ~ velocity ungerichtete Geschwindigkeit f ~ walk Brownsche Bewegung f
randomization willkürliche Verteilung f
randomize, to ~ willkürlich verteilen
randomizer Zufallsmaschine f
randomly, ~ distributed willkürlich verteilt ~ oriented nicht bevorzugte Orientierung f
range, to ~ (cable) aufschießen, einreihen, (of a gun) einschießen, sich erstrecken, reißen, schwanken to ~ out a line (eine) Linie f ausfluchten to ~ low tief liegen to ~ in a pole einen Stab m einfluchten to ~ straight lines with a telescope mit dem Fernrohr n die Geraden abstecken

**range** Abstand *m* (rdr), Aktionsradius, Ausdehnung *f*, (Schallmessung) Basislinie *f*, Baufluchtlinie *f*, Bereich *m*, Eindringtiefe *f*, Entfernung *f*, Flugweite *f*, Herd *m*, Intervall *n*, Kochherd *m*, Meß-, Ortungs-bereich *m*, Reichweite *f*, Reihe *f*, Schießplatz *m*, Schußentfernung *f*, Serienprogramm *n*, Spanne *f*, Spielraum *m*, Strecke *f*, Streuungsbreite *f*, Tragweite *f*, Umfang *m*, Umkreis *m*, Weite *f*, Wurfweite *f*, (phys.), Zeitabstand *m*, Zwischenraum *m*

**range,** ~ **of action** Aktionsreichweite *f* (aviat.), Fahrstrecke *f*, Flugbereich *m* ~ **of adjustment** Einstell-, Regel-, Verstell-bereich *m* ~ **for alternating stresses** Wechselbereich *m* ~ **of angular deflection** (of control surface) Anschlagwinkelbereich *m* ~ **of applicability** Anwendungsgrenzen *pl* ~ **of application** Anwendungsbereich *m* ~ **of an atom** Atombezirk *m* ~ **of audibility** Hörbereich *m* ~ **of authority** Kompetenz *f* ~ **of boring column head** Weihnachtsbaum *m* ~ **of brake** Bremsbereich *m* ~ **of brittleness** Brüchigkeitsgebiet *n* ~ **of capacity** Leistungsbereich *m* ~ **of characteristics and uses** (of practical applications) Anwendungsgebiet *n* ~ **of comfort** Fahrbereich *m* ~ **of concentration** Konzentrationsbereich *m* ~ **between contacts** Kontaktspielraum *m* ~ **of corpuscular emission** Strahlenreichweite *f* ~ **at cruising speed** Reichweite *f* bei Reisegeschwindigkeit ~ **of damage** Bereich *m* der Dauerbruchgefahr ~ **of deflection of control surface** Ausschlagbereich *m* ~ **of deformation** gesamte Formänderung *f* ~ **of density** Schwärzungsumfang *m* ~ **of effectiveness** Sicherheitsabstand *m* ~ **of exposure** Belichtungsspielraum *m* ~ **of feed** Vorschubbereich *m* ~ **of flight** Flugbereich *m* ~ **of flying** Aktionsradius *m* ~ **of forced oscillation** Mitnahmebereich *m* ~ **of free transmission** Durchlässigkeitsbereich *m* ~ **of gear sizes** Abmessungsbereich *m* ~ **of grain sizes** Körnung *f* ~ **of heat treatment** Gebiet *n* der Vergütung ~ **of hills** Höhenzug *m*, Hügelkette *f* ~ **of incipient current flow** Anlaufstromgebiet *n* ~ **of influence** Einflußhöhe *f* ~ **of investigation** Untersuchungsbereich *m* ~ **of lays** Drallbereich *m*, Schlaglängenbereich *m* ~ **of light oscillation** Lichtumfang *m* ~ **of linearity** linearer Bereich *m* ~ **of loading** Beanspruchungsbereich *m* ~ **of magnification** Vergrößerungsbereich *m* ~ **to mask** Gipfelentfernung *f* ~ **of measurement** Meßstrecke *f* ~ **of modulation** Aussteuerbereich *m* ~ **of motion** Verstellmöglichkeit *f* ~ **of movement** Spielraum *m* ~ **of nominal tension** Nennspannungsbereich *m* ~ **of ovens** Ofenbatterie *f* ~ **below point where overload begins** Aussteuerungsbereich *m* ~ **of positive pressure** Überdruckgebiet *n* (phys.) ~ **of primary absorption** Eindringtiefe *f* der primären Absorption ~ **for pulsating (or fluctuating) compressive stresses** (Druck)schwellbereich *m* ~ **for pulsating (or fluctuating) tensile stresses** Zugschwellbereich *m* ~ **of pure sensation** Gebiet *n* des reinen Gefühles ~ **of ratios** Übersetzungsbereich *m* ~ **of rays** Strahlenreichweite *f* ~ **of receiver** Empfängerbereich *m* ~ **of reception** Empfangsbereich *m* ~ **of regulation** Regelbereich *m* ~ **of sensitiveness** Empfindlichkeits-

bereich *m* ~ **of setting** Einstell-bereich *m*, -weite *f* ~ **of shades** Nuancenskala *f* ~ **of sight** Sichtweite *f* ~ **of sizes** Kornbereich *m* ~ **of spring** Federspiel *n*, Schwingung *f* der Feder ~ **of stability** Stabilitätsbereich *m* ~ **of the starting current** Anlaufstromgebiet *n* ~ **in still air** Flug *m* in ruhiger Luft *f* ~ **of stress** Belastungsbereich *m* ~ **of target** Treffpunktentfernung *f* ~ **of temperature** Temperaturgang *m*, Wärmegang *m* ~ **of temperature regulation** Temperaturverstellbereich *m* ~ **of tests** Versuchsgrundlage *f* ~ **of tide** Flutgröße *f*, Flutintervall *n* ~ **of tolerance** Toleranzbereich *m* ~ **of transmission** Reichweite *f*, Sendebereich *m* ~ **of transmittancy** Bereich *m* der Durchlässigkeit ~ **of transmitter** Senderbereich *m* ~ **of tune** Tonbereich *m* ~ **of type** Baumusterprogramm *n* ~ **of use** Verwendungsmöglichkeit *f* ~ **of variation** Schwankungsbereich *m* ~ **veins** Gangzug *m* ~ **of vision** Blickfeld *n* **within** ~ **of vision** Sichtweite *f* ~ **of welding** Schweißbereich *m*

**range,** ~ **adjustment** Einschießen *n* nach der Länge ~ **angle** (aerial bombing) Bombenabwurfwinkel *m* ~ **attachment** Meßzusatz *m* ~ **averager** Entfernungsmittler *m* ~ **bearing display** Höhenschirm *m* (rdr) ~ **calibration** Entfernungseichung *f* ~ **capacity** Meßbereich *m* ~ **card** Punktplan *m*, Zielpunktkarte *f* ~ **circuit** Entfernungsmeßschalter *m*, Meßstrang *m* ~ **clock** Entfernungs-anzeiger *m*, -uhr *f* ~ **coefficient** Reichenbeiwert *m* ~ **correction** Entfernungs-änderung *f*, -vorhalt *m*, Längenbesserung *f* ~ **corrector** Auswanderungsmesser *m* ~ **cursor** E-Meßmarke *f* ~ **determination** Entfernungs-ermittlung *f*, -messung *f* ~ **deviation** Höhenabweichung *f*, Längenabweichung *f* ~ **dial** Entfernungsskala *f* ~ **difference** Entfernungs-unterschied *m*, -vorhalt *m* ~ **discrimination** Abstandsunterscheidung *f* (rdr) ~ **dispersion** mittlere Längenstreuung *f* ~ **dispersion on horizontal target** Tiefenstreuung *f* (artil.) ~ **dispersion on vertical target** Höhenstreuung *f* ~ **display writer** Bildfernschreiber *m* (electron.) ~ **distribution** Reichweitenverteilung *f* ~ **drum** Entfernungs-, Teil-, Über-trommel *f* ~ **effect** Reichweiteeffekt *m* ~ **energy** Reichweiteenergie *f* ~ **error** Entfernungsfehler *m* ~-**estimating instrument** Entfernungsschätzgerät *m* ~ **estimation** Entfernungsschätzen *n* ~ **estimator** Entfernungsspinne *f* ~ **finder** E-meßgerät *n*, Entfernungs-gerät *n*, -messer *m*, -meßgerät *n*, -meßkontrollanlage *f*, Fernmesser *m*, Fernmeßgerät *n*, Meßwalze *f*, Tenemeter *m* ~-**finder with split (or deviced) image** Gegenbildentfernungsmesser *m* ~-**finder operator** E-messer *m*, Entfernungsmeßmann *m*, Fernmesser *m* ~-**finder platform** Entfernungsmeß-stand *m*, -stelle *f* ~-**finder station** Entfernungsmeßstand *m* ~ **finding** Entfernungsmessung *f* ~-**finding apparatus** Entfernungsmesser *m*, Telemeter *n* ~ **finding post** Meßstand *m* ~ **gate** Entfernungstor *n*, -torschaltung *f*, ~-**height marker** Abstandsschirm *m* (rdr) ~ **indicator** Entfernungsmarke *f* ~-**indicator unit** Entfernungsanzeigegerät *n* ~ **knob** Entfernungstrommel *f* ~ **lead** Entfernungsvorhalt *m* ~ **light** Landerichtungsfeuer *n*, Richtfeuer *n* ~ **marker generator** Markierungsgenerator *m* ~-**marker pip** Hellmark *f*

~ **measurement** Abstandsmessung *f* (rdr), Reichweitenmessung *f* ~ **outlet** Steckdose *f* für Elektroherd ~ **piping** Verteilleitungen *pl* ~ **pole** Visierstab *m* ~ **rate** Entfernungsunterschied *m* ~ **reading** Entfernungsablage *f* ~ **reducing wedge** Vorschlagkeil *m* ~ **reduction** Reichweiteverkürzung *f* ~ **resolution** Entfernungsauflösungsvermögen *n* ~ **ruler** Entfernungslineal *n* (Vermessung) ~ **scale** Entfernungsskala *f*, -teilung *f*, Meßskala *f* ~ **selector** Bereichsumschalter *m* ~ **sensing** Festlegen *n* der Längenabweichung ~**-setting wheel** (torpedoes) Laufstreckenmarkenrad *n* ~ **slant** Entfernungsvorhalt *m* ~ **spacer** Abstandhalter *m* ~ **span** Bereichsumfang *m* ~ **spider** Entfernungsspinne *f* ~ **spotting** Festlegen *n* der Längenabweichung *f* ~ **straggling** Reichweitenstreuung *f* ~ **sweep** Entfernungsmeßbasis *f* ~ **switch** Wellen(um)-schalter *m* ~ **table** Schußtafel *f*, Zielkarte *f* ~ **taker** Entfernungsschätzer *m*, Meßmann *m* ~ **transmission unit** Abstandgeber *m* (rdr) ~ **unit with moving targets** Scheibenzuganlage *f* ~ **view finder** Meßsucher *m* ~ **wind component** Längswindkomponente *f*

**ranged line** ausgefluchtete Linie *f*

**ranging** Meßbeobachtung *f*, Meßverfahren *n*, Richtungs- und Entfernungsbestimmung *f*, Skalenkontrolle *f* ~ **curves** Kurvenabsteckung *f* ~ **delay** Meßkette *f* ~ **device** Visiervorrichtung *f* ~ **fire** Erschießen *n* der Entfernung ~ **fuse** Zielzünder *m* ~ **magnifier** Visierlupe *f* ~ **rod** Fluchtstab *m* ~ **section** Meßtrupp *m* ~ **unit** Entfernungsübersichtsgerät *n*, Meßkette *f* (rdr)

**rank** Dienstgrad *m*, Glied *n*, Grad *m*, Ordnung *f*, (of switches) Rang *m*, Rangordnung *f*, Rangstufe *f*, Reihe *f*, Sorte *f*, Stand *m*, Stellung *f*, Stufe *f* ~ **and file** Reih *f* und Glied ~ **of piles** Pfahlwand *f*

**rap, to** ~ (molding) abklopfen, klopfen, losklopfen, losschlagen **to** ~ **the pattern** ein Modell *n* losklopfen (aus der Form)

**rape** Raps *m*; (Weinkelterei) Treber *m*, Trester *m* ~ **cake** Rapskuchen *m* ~ **oil** Rüböl *n* ~ **seed** Rübsamen *m* ~ **wine** Tresterwein *m*

**rapid** Stromschnelle *f*; geschwind, schnell ~ **action pliers** Blitzzange *f* ~**-action screw press** Schnellspindelpresse *f* ~ **ager** Schnelldämpfer *m* ~ **analysis** Schnellanalyse *f* ~ **band** Schnellband *n* ~ **charging** Schnelladung *f* ~ **charging unit** Schnellladegerät *n* ~**-closing mechanism** Schnelleinschaltvorrichtung *f* ~ **combustion** lebhafte Verbrennung *f* ~ **cooling system** Schnellkühlung *f* ~ **deceleration** Schnellbremsung *f* ~ **developer** Rapidentwickler *n* (photo) ~ **discharger** Schnellentladewagen *m* ~ **drier** Schnelltrockner *m* ~ **face advancement** Streckenschnellvortrieb *m* ~ **fall-off** schneller Abfall *m* ~ **firing** Schnellzündung *f* ~**-freezing compartment** Schnellgefrierfach *n* ~**-hardening cement** schnell erhärtender Zement *m* ~ **interrupter** Schnellunterbrecher *m* ~ **lens** lichtstarkes Objektiv *n* ~ **letter files** Schnellhefter *m* ~ **-locking device** Schnellverschluß *m* ~ **lowering brake** Schnellsenkbremse *f* ~ **memory** Schnellspeicher *m* ~ **method** Abkürzungsverfahren *n*, Schnellmethode *f* ~ **percussion table** Schnellstoßherd *m* ~ **phase change** Phasensprung *m* ~ **plate locking-up device** Plattenschnellspann-

vorrichtung *f* ~ **power traverse** Eilbewegung *f*, Eilgang *m*, Eilrückgang *m*, Schnellgang *m* ~ **power traverse in all directions of movement** Eilgang *m* in allen Bewegungsrichtungen ~ **power traverse shaft** Eilgangwelle *f* ~ **preparation** kurze Rüstzeit *f* ~ **production lathe** Schnelldrehbank *f* ~ **release valve** Schnellentlastungsventil *n* ~ **retting** Schnellrette *f* (Flachs) ~**-return motion** Eilbewegung *f*, Eilgang *m*, Eilrück-gang *m*, -lauf *m* ~ **storage** Schnellspeicher *m* ~ **starting** Schnellanlauf *m* ~ **striking** Schnellzündung *f* ~ **test** Schnellprobe *f* ~**-testing method** Kurzprüfverfahren *n* ~ **three-side trimmer with one knife only** Einmesser-Schnelldreischneider *m* ~ **transmitter** Schnellgeber *m* ~ **travel** Schnellgang *m* ~ **traverse** Eilbewegung *f* (Maschine) ~ **traverse pump** Eilgangpumpe *f* ~ **traverse rates** Eilganggeschwindigkeit *f* ~ **traverse valve** Eilgangventil *n* ~ **wind lever** Schnellaufzug *m*

**rapidity** Schnelligkeit *f* ~ **of action** Arbeitsgeschwindigkeit *f* ~ **of firing** Feuergeschwindigkeit *f*, Schußfolge *f* ~ **of lens** Lichtkraft *f* ~ **of a lens** Objektivlichtkraft *f* ~ **of measurement** Meßgeschwindigkeit *f* ~ **of modulation** Modelgeschwindigkeit *f*

**rapier** Schläger *m*

**rapper** (molding) Abklopfapparat *m*, Signalhammer *m*, Türklopfer *m*

**rapping** Klopfen *n*, Losklopfen *n*; (molding) ~ **device** Abklopfvorrichtung *f* ~ **plate** Aushebeplatte *f*

**rare** (air) dünn, (gases) edel, selten ~ **earth** Edelerde *f*, seltene Erde *f* ~ **gas** Edelgas *n*, inertes Gas *n* ~**-gas lightning arrester** Edelgassicherung *f* ~**-gas photoelectric cell** Edelgasfotozelle *f* ~**-gas rectifier** Edelgas(glühkathoden)gleichrichter *m* ~**-gas tube** Edelgasröhre *f* ~ **medium** dünnes Mittel *n* ~ **spark** Knall-, Knarr-funken *m*

**rarefaction** Verdünnung *f* ~ **of the air** Luftverdünnung *f* ~ **wave** Verdünnungswelle *f*

**rarefiable** verdünnbar

**rarefied** verdünnt ~ **air** verdünnte Luft *f* ~ **gas** verdünntes Gas *n* **most** ~ **packing of spheres** dünnste Kugellagerung *f*

**rarefy, to** ~ (gases) verdünnen

**rarity** Rarität *f*, Seltenheit *f*

**rash result** Ausschlag *m*

**rashness** Übereilung *f*

**rasp, to** ~ feilen, raspeln, sich reiben, schaben

**rasp** Grobfeile *f*, Raspel *f*, Reibe *f*, Reibeisen *n* ~ **cylinder** Sägewalze *f*

**rasping** kratzend, raspelnd ~ **sound** krächzender Ton *m*

**rasps** Raspelspäne *pl*

**raster** Linienraster *m*, Raster *m* ~ **definition (or detail)** Rasterfeinheit *f* ~ **scan microscope** Rastermikroskop *n* ~ **scanning** Rasterabtastung *f* (TV) =**-screen aperture** Rasterblende *f* ~ **size** Rasterabmessung *f*

**rat** Ratte *f* ~ **hole** Rattenloch *n* ~ **tail** Rattenschwanz *m* ~**-tailed file** Rattenschwanz *m*, Rundfeile *f*

**ratch** gezahnte Stange *f*, Zahnstange *f* mit Sperrzähnen

**ratchet** Feile *f*, Gesperr *n*, Klinke *f*, Knarre *f*, Ratsche *f*, Rechen *m*, Sperre *f*, Sperrhaken *m*, Sperrklinke *f* ~ **and pawl** Zahngesperre *n* ~

**-and-pawl mechanism** Ratschenschaltung f ~
**action** Kippvorgang m ~ **antenna** Sägezahnantenne f ~ **bar feed** Stangenratschenvorschub
m ~ **box end wrench** Ringratschenschlüssel m
~ **brace** Bohr-knarre f, -ratsche f, -winde f,
Ratschbohrer m ~ **brake** Klemm-gesperre n,
-sitz m, Klemmenspannung f ~ **casing** Sperr-
gehäuse n ~ **circuit scheme** Blinkschaltung f ~
**coupling** Sperrklinken-, Zahnrad-kupplung f ~
**cutter** Knarrenfräser m ~ **device** Sperrklinkenvorrichtung f ~ **die plate** Knarrenkluppe f ~
**disk** Klinkenscheibe f ~ **dog** Ratschenhebel m
~ **drill** Bohrknarre f, Bohrratsche f, Knarren-
bohrer m, Ratschbohrer m, Ratsche f ~ **drive**
Ratschengetriebe n **~-drive mechanism** Schalt-
klinkengetriebe n ~ **drum** Sperrzahnkranz m ~
**gear** Gesperre n, Sperrklinkenvorrichtung f
**~-gear mechanism** Zahnbogenrichtmaschine f
~ **handle** Knarrenhebel m ~ **head** Ratschen-
kopf m ~ **hob** Knarren-, Sperrad-wälzfräser m
~ **jack** Sperrad-heber m, -winde f, Wagenheber
m ~ **knob** Klinkengriff m ~ **lever** Klinkhebel
m ~ **nut** Sperradschraube f ~ **oscillation** Inter-
mittenz-, Kipp-, Relaxations-schwingung f ~
**pawl** Schaltratsche f ~ **pipe stock** Knarrenkluppe f ~ **screwstock** Ratschenkluppe f ~
**setting** weitergeschaltete Sperrzähnezahl f ~
**shaft** Fortschaltachse f ~ **spanner** Ratschenschlüssel m ~ **spring** federnder Zahn m, Klinkenfeder f ~ **stock** Knarre f ~ **stop** Gefühls-
ratsche f **~-stop control** Ausgleicher m ~
**tightening** Ratschenanpressung f ~ **time base**
Sperrzeitbasis f ~ **tooth** Schaltzahn m **~-type
drive** Sperrgetriebe n **~-type sprag** Sperradberg-
stütze f ~ **voltage** Sägezahnspannung f ~ **wheel**
Klinkrad n, Schalt-rad n, -werk n, Sperr-kranz
m, -rad n, -scheibe f, Steigrad n, Zahnscheibe f
~ **wrench** Radschlüssel m, Ratschenhebel m,
Windenbohrknarre f
**rate, to** ~ abschätzen, anschlagen, bemessen,
beurteilen, einschätzen, rechnen, schätzen,
taxieren, veranschlagen, vermessen, zuordnen
**rate** Abgabe f, Ansatz m, Berechnung f, Betrag
m, (engine) Fahrtstufe f, Ganggeschwindigkeit
f, Gebühr f, Grad m, Maß n, Maßstab m,
Rang m, Satz m, Steuer f, Takt m, Tarif m,
(of subscription) Teilnehmergebühr f, Verhältnis n **at the** ~ **(or fee) of** unter Abgabe von
**rate,** ~ **of acceleration** Änderungsgeschwindigkeit f ~ **of air flow** Luftdurch-flußmenge f, -satz
m ~ **of attack** Angriffsgeschwindigkeit f ~ **of
blowing** Blasgeschwindigkeit f ~ **of breaking**
Zerreißgeschwindigkeit f ~ **of change** Änderungsgeschwindigkeit f, Markierungsdifferenz
f, zeitliche Änderung f ~ **of change of angle of
sight** Höhengeschwindigkeit f ~ **of change of
leg difference** Quergeschwindigkeit f ~ **of
change of range** Entfernungsgeschwindigkeit f
~ **of charge** (or discharge) Aufladegeschwindig-
keit f, Entladegeschwindigkeit f, Ladefähigkeit
f ~ **of charges** Gebührensatz m ~ **of charging**
Durchsatzzeit f ~ **of charging time** Satzzeit f
~ **of climb** Anstieg m, Steiggeschwindigkeit f,
Steigzeit f ~ **of climb indicator** Steiggeschwin-
digkeits-anzeiger m, -messer m, Variometer n
~ **of climb meter** Steigleistungsmesser m ~ **of
closure** Annäherungsgeschwindigkeit f ~ **of
coking** Verkokungsgeschwindigkeit f ~ **of**

**combustion** Brenn-, Verbrennungs-geschwindig-
keit f ~ **of concentration** Konzentrationsgrad m
~ **of consumption per second** Sekundenver-
brauchszahl f ~ **of convergence** Konvergenz-
geschwindigkeit f, Stärke f der Konvergenz ~ **of
cooling** Abkühlungsgeschwindigkeit f ~ **of corrosion** Korrosions-geschwindigkeit f, -verlauf
m ~ **of crystalline growth** Kristallisationsge-
schwindigkeit f ~ **of decay** Abklingfaktor m
(acoust.) ~ **of decrease** Abfallwert m ~ **of
decrease in temperature** Wärmegefälle n ~ **of
deflection** Durchbiegungsgeschwindigkeit f ~
**of deformation tensor** Deformationsgeschwin-
digkeitstensor m ~ **of deposition** Abscheidungs-
Auftragege-schwindigkeit f ~ **of depreciation**
Abnutzungsrate f, Entwertungssatz m ~ **of
descent** Sinkgeschwindigkeit f (aviat.) ~ **of die
casting** Einspritzgeschwindigkeit f ~ **of
discharge** Entladegeschwindigkeit f ~ **of dissolving** Lösegeschwindigkeit f ~ **of driving**
Durchsatzgeschwindigkeit f ~ **of elevation**
Höhenrichtgeschwindigkeit f ~ **of energy**
Energiemenge f pro Zeiteinheit ~ **of evacuation**
(of vacuum pump) Sauggeschwindigkeit f ~ **of
exchange** Geld-, Wechsel-kurs m ~ **of feed** Zu-
teilungsgeschwindigkeit f ~ **of filtration** Fil-
trationsgeschwindigkeit f ~ **of fire** Feuer-folge
f, -geschwindigkeit f, Gruppen-, Schuß-folge f
~ **of flash** Blitzfolge f ~ **of flow** Durchflußmenge f, Durchströmungsgeschwindigkeit f,
Flüssigkeitsdurchsatz m, Zuflußmenge f ~ **of
flow meter** Wasserflußmesser m ~ **of freight**
Fracht f ~ **of fuel consumption** Betriebsstoff-,
Kraftstoff-verbrauchssatz m ~ **of gasification**
Vergasungsgeschwindigkeit f ~ **of growth**
Wachstumsgeschwindigkeit f ~ **of handling**
Umschlagsleistung f ~ **of heat transfer** Wärme-
durchsatz m ~ **of heating** Erhitzungsgeschwindigkeit f ~ **of impulse** Impulsfrequenz f ~ **of
incorporation** Einlagerungsrate f (cryst.) ~ **of
inflow** Einfließgeschwindigkeit f ~ **of injection**
Einspritzgeschwindigkeit f ~ **of insulation
breakdown** Durchschlagsgeschwindigkeit f ~ **of
interest** Diskontsatz m ~ **of load application**
Belastungsgeschwindigkeit f ~ **of melting**
Schmelz-geschwindigkeit f, -leistung f ~ **of
opening** Öffnungszeit f ~ **of outflow** Ausfluß-
geschwindigkeit f ~ **of perforation** Lochungs-
geschwindigkeit f ~ **of placing concrete** Ver-
legungs- oder Verarbeitungsgeschwindigkeit f
des Betons ~ **of postage** Portosatz m ~ **of
pouring** Gießgeschwindigkeit f ~ **of power input**
Leistungsaufnahme f ~ **of power loading** (in
propellers) Leistungsbelastungsgrad m ~ **of
pressure rise** Druckanstiegsgeschwindigkeit f
~ **of production** Arbeitsgeschwindigkeit f, Er-
zeugungssatz m ~ **of pulse repetition** Impuls-
folgegrad m ~ **of reading** Lesegeschwindigkeit
f ~ **of reduction** Reduktionsgeschwindigkeit f
~ **of retention** Einbindungsgrad m ~ **of revolution** Drehgeschwindigkeit f ~ **revolutions**
Drehzahl f ~ **of roll** Drehgeschwindigkeit f um
die Längsachse ~ **of self-regulation** Selbstrege-
lungsgeschwindigkeit f ~ **of sending** Gebetempo
n ~ **of setting** (Zement) Abbindegeschwindig-
keit f ~ **of shear** Scherungskoeffizient m ~ **of
shrinkage** Schrumpfungsgeschwindigkeit f ~ **of
side-slipping** Seitenabrutschgeschwindigkeit f

~ of side slipping Seitengeschwindigkeit *f* ~ of speed Streckengeschwindigkeit *f* ~ of speed of diffusion Diffusionsgeschwindigkeit *f* ~ of steaming Verdampfungsziffer *f* ~ of strain Deformationsgeschwindigkeit *f* ~ of stretch-(ing) Streckgeschwindigkeit *f* ~ of thrust loading Schubbelastungsgrad *m* ~ of transformation Umwandlungsgeschwindigkeit *f* ~ of travel Durchgangs-, Durchsatz-, Fortbewegung-geschwindigkeit *f* ~ of traverse Seitenrichtgeschwindigkeit *f* ~ of vertical descent Sinkgeschwindigkeit *f* ~ of wages Lohnsatz *m* ~ of wear and tear Verschleißmaß *n* ~ of withdrawal Ausscheidewahrscheinlichkeit *f* ~ of work Arbeitsleistung *f*
**rate,** ~ **action** differenzierender Einfluß *m*, Regelung *f*, Vorhaltwirkung *f* ~ **change** Geschwindigkeitswechsel *m*, Stückzeitänderung *f* ~ **change of couple** Dralländerung *f* ~ **controller** Mengenmesser *m* ~ **curve** Geschwindigkeitskurve *f* ~ **discount** Mengenrabatt *m* ~**-firing voltage** Ansprech-, Auslöse-spannung *f* ~ **grown transistors** (transistors) Ziehkristalle *pl* ~ **gyro** Meß-Wende-Kreisel *m* ~ **meter** Gebührenzähler *m*, Impuls-frequenzmeter *m*, -zähler *m*, Zählgeschwindigkeitsmesser *m* ~ **notification** Gebührenansage *f* ~ **registration** Gebührenerfassung *f* ~ **structure** Tarifart *f* ~ **table** Drehtisch *m* ~ **time** Vorhaltezeit *f* bei D-Wirkung ~**-type servomotor** geschwindigkeitsgesteuerter Servomotor *m*
**rateable value** steuerpflichtiger Wert *m*
**rated** abgeschätzt ~ **for (or at)** bemessen
**rated,** ~ **altitude** Gleichdruck-, Nennleistungs-, Nenn-höhe *f* ~ **boost** Nennladedruck *m* ~ **breaking point** Sollbruchstelle *f* ~ **burden** (instruments) Nenn-bürde *f*, -leistung *f* ~ **capacity** Nennleistung *f* ~ **cross section** Nennquerschnitt *m* ~ **current** Nennstrom *m*, zulässige Stromstärke *f* ~ **diagram** Leistungsdiagramm *n* ~ **duty** Nennbetriebsart *f* ~ **electrode dissipation** Nennverlustleistung *f* ~ **fatigue limit** Gestaltfestigkeit *f* ~ **feedback impulse** dosierter Rückimpuls *m* ~ **frequency** Nennfrequenz *f* ~ **full-load speed** Normaldrehzahl *f* ~ **height** Nennhöhe *f*, Nennleistungshöhe *f* ~ **horsepower** Nennleistung *f*, Pferdestärke *f* ~ **impedance** Nennbürde *f* ~ **load** Nennlast *f* ~ **manifold pressure** Nennladedruck *m* ~ **output** Solleistung *f* ~ **phase angle** Nennwinkelfehler *m* ~ **phase displacement (or angle)** Nennwinkelfehler *m* ~ **power** Nennleistung *f* ~ **power supply** Nennleistung *f* ~ **power value** Nennleistungswert *m* ~ **pressure** Nennladedruck *m* ~ **primary current** primärer Nennstrom *m* ~ **primary voltage** primäre Nennspannung *f* ~ **quantity** Nenngröße *f* ~ **range** Nennreichweite *f* ~ **revolutions** gewöhnliche Drehzahl *f* ~ **RPM (r.p.m.)** Solldrehzahl *f* ~ **secondary current** sekundärer Nennstrom *m* ~ **secondary voltage** sekundäre Nennspannung *f* ~ **speed** Nenn-, Soll-drehzahl *f*, Nenngeschwindigkeit *f* ~ **temperature** Solltemperatur *f* ~ **temperature range** Nenntemperaturbereich *m* ~ **temperature-rise current** maximal zulässiger Betriebsstrom *m* ~ **thrust** Entwurfschub *m* (g/m) ~ **value** Nenn-, Soll-wert *m* ~ **voltage** (of a cable) Nennspannung *f*

**rates** Gebührnisse *pl*
**ratification of treaty** Vertragsbestätigung *f*
**ratify, to** ~ ratifizieren, rechtskräftig machen
**rating** Auslegung *f* (motor), Beanspruchung *f*, Bemessung *f*, Berechnung *f*, Bewertung *f*, Dienstgrad *m*, Dienstgradbezeichnung *f*, Einschätzung *f*, genaue Bezeichnung *f* (Pflichtenblatt), Größenbestimmung *f*, Leistungsbilanz *f*, Nenn-betrieb *m*, -leistung *f*, Pflichten-blatt *n*, -heft *n*, Rechnung *f*, Schätzung *f* ~ **up in age** Alterserhöhung *f* ~ **of an apparatus** Betriebsverhältnisse *pl* ~ **of machine** Betriebsverhältnisse *pl* ~ **of a meter** Nennleistung *f* eines Zählers
**rating,** ~ **chart** Leistungsdiagramm *n* ~ **curve of discharges** Linie *f* der registrierten Durchflüsse ~ **curve for gauge** (discharge plotted against gauge reading) Pegelschlüssel *m* ~ **formula** Steuerformel *f* ~ **horsepower** Steuerleistung *f* ~ **plate** (on machines) Datenschild *n*, Leistungsschild *n* ~ **sheets** Betriebsbeanspruchungslisten *pl*
**ratio** Verhältnis *n*, Verhältniszahl *f*
**ratio,** ~ **of amplitudes** Amplitudenverhältnis *n* ~ **of attenuation** Dämpfungsverhältnis *n* ~ **between the channels** Stärkeverhältnis *n* der Kanäle ~ **of curvature** Krümmungsverhältnis *n* ~ **of deviation** Kursabweichungsverhältnis *n* ~ **of distance of pressure center from mouth of Venturi to caliber** Druckpunktabstand *m* ~ **of division** Teilverhältnis *n* ~ **of energy** Energieverhältnis *n* ~ **of enlargement** Umgrößerungsverhältnis *n* ~ **of expansion** Ausdehnungsverhältnis *n*, (in a nozzle) Erweiterungsverhältnis *n* ~ **of the flow velocity to the velocity of sound** Machsche Zahl *f* ~ **of forward peripheral speed of propeller** Fortschrittsgrad *m* der Luftschraube ~ **of friction** Reibungsverhältnis *n* ~ **of horizontal breadth to vertical height of a slope** Abdachverhältnis *n* einer Böschung ~ **of intensities** Helligkeitsverhältnis *n* ~ **of length** Längen-quotient *m*, -streuung *f* ~ **of lens aperture** Öffnungsverhältnis *n* eines Objektivs, relative Öffnung *f* eines Objektivs ~ **of lift to drag** Gleitverhältnis *n* ~ **of load** Belastungsverhältnis *n* ~ **of luminous flux to area of element of surface in lux units** Beleuchtungsstärke *f* ~ **of magnification** Umgrößerungsverhältnis *n* ~ **of maximum and effective pressure** Volldruckverhältnis *n* ~ **of minimum stress to maximum stress** Spannungsverhältnis *n*, Verhältnis *n* von Unterspannung zur Oberspannung ~ **of net weight of missile to thrust** Schubbelastung *f* ~ **of rest** Ruhegrad *m* ~ **of rise to span** (of arch) Pfeilverhältnis *n* ~ **of similitude** Ähnlichkeitsverhältnis *n* ~ **of slenderness** Schlankheits-grad *m*, -verhältnis *n* ~ **of soil pressure to settlement** Bettungsziffer *f* ~ **of speed reduction** Übersetzungsverhältnis *n* ~ **of tensile strength to weight density** Reißfestigkeit *f* ~ **of tension to thrust** Zugdruckverhältnis *n* ~ **of transformation** Übersetzungsverhältnis *n* ~ **of a transformer** Verhältnis *n* der Spannungen und Ströme ~ **of transmission times** Laufzeitenverhältnis *n* ~ **of two** Zweierteiler *m* ~ **of weight to cylinder displacement** Hubraumgewicht *f* ~ **of weight to piston displacement** Hubraumgewicht *n* ~ **of the windings** Windungsverhältnis *n*

**ratio,** ~ **arm** Abgleichzweig *m* (Brückenschaltung) ~ **arms** feste Brückenarme *pl* ~ **arms of bridge** Brückenverhältnisarme *pl* ~ **bridge circuit** Vergleichsbrückenschaltung *f* ~ **controller** Verhältnisregler *m* ~ **detector** Verhältnisdetektor *m* ~ **engine horse-power to frontal area** Leistungsdichte *f* ~ **error** Übersetzungsfehler *m* (Meßwandler) ~ **indicator** Verhältnisanzeiger *m* ~ **meter** Quotientenmesser *m*, Verhältniszeiger *m* ~ **number** Maßstabszahl *f* (Verhältniszahl) ~-**type telemeter** Verhältnis-Telemeter *n*

**ration, to** ~ rationieren

**rational** rational, rationell, sinngemäß, vernunftgemäß, verstandesmäßig ~ **formula** Konstitutionsformel *f* ~ **fraction** rationaler Bruch *m* ~ **integral** ganzrational ~ **number** aufgehende Zahl *f* ~ **quantity** aufgehende Zahl *f* ~ **slip** kreisförmige Gleitung *f*

**rationalization** Rationalisierung *f*

**rationalize, to** ~ rationalisieren, wirtschaftlich gestalten

**rationed material** Sparstoff *m*

**rationing** Bewirtschaftung *f*

**rations in kind** Naturalverpflegung *f*

**RATO** (rocket-assisted take-off) Start *m* mit Hilfsrakete

**rattle** Klappern *n*, Prasseln *n*, Rasseln *n*, Ratsche *f*, Schnarre *f* ~ **factor** Klirrfaktor *m* ~ **proof** klapperfrei

**rattler** Trommelmühle *f* ~ **attrition test** (in a drum mill) Mahlversuch *m* ~ **star** Putzstern *m* ~ **test** Schüttelversuch *m*, Trommel-probe *f*, -versuch *m*

**rattling** Klappern *n*, Rasseln *n* ~ **of the coupling** Scheppern *n* der Kupplung ~ **of valve** Schnarren *n* des Ventils

**rattling noise** Klirrgeräusch *n*

**raucous** rauh

**ravage, to** ~ verheeren

**ravehook** Nahthaken *m*, Werghaken *m*

**ravel, to** ~ auf-fasern, -lösen, ausfransen, verflechten

**ravelin** Ravelin *n*

**ravine** Bergschlucht *f*, Hohlweg *m*, Klamm *f*, Mulde *f*, Schlucht *f*, Tobel *m*

**raving** rasend

**raw** rauh, roh, unbearbeitet, unreif, unzubereitet ~ **bismuth** Rohwismut *n* ~ **brickwork** Ziegelrohbau *m* ~ **casting** Ausgangsguß *m* ~ **castings** Rohguß *m* ~ **coal** Förder-, Grob-, Roh-kohle *f* ~ **coal screen** Förderkohlensieb *n* ~ **copper matte** Kupferrohstein *m* ~ **figures** grobe Werte *pl* ~ **flax** Roh-leinen *n*, -stoff *m* ~ **glycerin** Rohglycerin *n* ~ **helium** Rohhelium *n* ~ **hide** Rohhaut *f*, ungegerbtes Fell *n* oder Leder *n* ~-**hide gear** Rohhautgetriebe *n* ~ **hide hammer** Rohhauthammer *m* ~ **hide pinion** Rohhaut-rad *n*, -ritzel *n* ~ **hide suspender** Rohhauthänger *m* ~ **ingot** Rohblock *m* ~-**juice pulp catcher** Rohsaftpülpefänger *m* ~ **kaolin** Rohkaolin *n* ~ **lead** Werkblei *n* ~ **lignite** Rohbraunkohle *f* ~ **linen** Roh-leinen *n*, -stoff *m* ~-**machined material** zwischenbearbeitbares Schlichtmaterial *n* ~ **magnesite** Rohmagnesit *m* ~ **massecuite** Rohzuckerfüllmasse *f* ~ **material** Ausgangsmaterial *n*, -stoff *m*, -werkstoff *m*, Grundstoff *m*, Rohmaterial *n*, Rohstoff *m*, Vorstoff *m*,

Zutat *f* ~ **material delivery** Rohstoffaufgabe *f* ~ **material entrance** Rohstoffeinlauf *m* ~ **material production** Urproduction *f* ~ **materials** Werkstoff *m* ~ **matter (or mass)** Rohmasse *f* ~ **ore** Frischerz *n*, Haufwerk *n*, Roherz *n* ~ **petroleum** Naphtha *n* & *f* ~ **product** Ausgangsprodukt *n* ~ **radar data** rohe Radardaten *pl* ~ **rubber** Rohgummi *m* ~ **silk** Grez-, Roh-seide *f* ~ **slag** Rohschlacke *f* ~ **smelting** Rohschmelzen *n* ~ **state** Rohzustand *m* ~ **steel** Rohstahl *m* ~ **stock** Rohfilm *m*, Rohmaterial *n* ~ **stocks** Rohstoff *m* ~-**sugar liquor** Raffineriekläre *f* ~-**sugar mixer** Affiniermaische *f* ~ **tape** Leerband *n* ~ **tin** Werkzinn *n* ~ **unwound silk** Florettseide *f* ~ **video** Rohvideo *n* ~ **viscose** Rohviskose *f* ~ **water** Frischwasser *n* ~ **wire** Blankdraht *m* ~ **zinc** Werkzink *n*

**ray** Funke *m*, Garbe *f*, Leitstrahl *m*, Lichtstrahl *m*, Spur *f*, Strahl *m*, Strahlung *f* ~ **through a point image** Bildstrahl *m* ~ **of sound** Schallstrahl *m*

**ray,** ~ **anisotropy** Strahlungsanisotropie *f* ~ **beam** Funkbüschel *n* ~ **cell** Strahlenzelle *f* ~ **center** Radialzentrum *n* ~ **deflection** Strahlablenkung *f* ~ **deflector** Strahlablenker *m* ~ **filter** Farben-, Licht-, Strahlen-filter *n* ~ **incident at oblique angle, at small angle, at small oblique angle** flach auffallender Strahl *m* ~ **mark** Radarbake *f* ~ **path** Bahnverlauf *m*, Strahlenbahn *f* ~ **producing system** Strahlerzeugungssystem *n* ~ **proofing** Strahlenschutz *m* ~ **radiation** Reststrahl *m* ~ **radiator system** Strahlerzeugungssystem *n* ~ **refraction** Strahlablenkung *f* ~-**shaped** strahlenförmig ~ **tracing** Bahnverlauf *m*, Strahlengang *m*, Strahlverlauf *m*

**rayed** strahlenförmig

**Rayleigh disk** Rayleigh-Scheibe *f*, Schallreaktionsrad *n*

**rayless** dunkel, strahlenlos, unerleuchtet

**raylet** kleiner Strahl *m*

**rayon** Kunstseide *f*, Rayon *m* ~ **fiber** Stapelfaser *f* ~ **staple** Zellwolle *f* ~ **tire cord** Reifencordrayon *n* ~ **velvet** Kunstseidensamt *m* ~-**yarn processing** Kunstseideverarbeitung *f*

**rays, (actinic)** ~ Strahlen *pl* ~ **caused by particle retardation** Bremsstrahlen *pl*

**raze, to** ~ abtragen, kratzen, schleifen

**razor** Rasier-apparat *m*, -messer *n* ~ **blade** Rasiermesser *n* ~ **edge** scharfer Rand *m* ~ **paste** Streichriemenpaste *f* ~ **strop** Abzieh-, Streichriemen *m*

**re** (on documents and letters) Betreff *m* **in** ~ betreffend

**re-absorb, to** ~ resorbieren, wiederaufsaugen

**re-absorber** Resorber *m*

**re-absorption** Resorption *f*, Wiederbenetzen *n*

**reach, to** ~ (by climbing) erklimmen, (high speeds, pressures) erreichen, erstrecken, reichen **to** ~ **an agreement with** sich abfinden mit **to** ~ **full growth** auswaschen **to** ~ **a height of . . . meters** Höhe von . . . Metern erreichen **to** ~ **out** auslaugen

**reach** Abschnitt *m*, Ausdehnung *f*, Ausladung *f*, Bereich *m* (of a beam, gun, girder), Erstreckung *f*, Griff *m*, Kupplungsdeichsel *f*, Lang-, Lenkbaum *m*, Riech-, Sicht-weite *f*, Spielraum *m* **within** ~ in Reichweite ~ **of call** Rufweite *f* (teleph.) ~ **of plug** Zündkerzenreichweite *f* ~

of river Flußhaltung $f$ ~ of a river Haltung $f$ eines Flusses ~ of the river Flußabschnitt $m$ ~ of screw Einschraublänge $f$ ~ of talk Sprachweite $f$ ~ of threaded end of spark plug Zünderkerzenreichweite $f$
reached by flying erflogen
reacher Aufgeber $m$, (textiles) Fadenanreicher $m$
reacidulate, to ~ nachsäuern
react, to ~ ansprechen, eine Reaktion eingehen, einwirken, gegenwirken, reagieren, rückwirken, umsetzen
reactance Blindwiderstand $m$ (electr.), Drossel $f$, Gegenwirkung $f$, Gesamtwiderstand $m$, Reaktanz $f$, Wider-druck $m$, -stoß $m$ ~ amplifier Rückkopplungsverstärker $m$ ~ characteristic Blindwiderstandskennlinie $f$ ~ coil Rückkupplungsspule $f$ ~ current Blindstrom $m$ ~ drop of voltage Spannungsabfall $m$ durch Blindwiderstand ~ modulator Blindmodulator $m$ ~ regulator veränderliche Drosselspule $f$ ~ resistance blinder Widerstand $m$ ~ tube Blindröhre $f$ ~ tube modulator Blindröhrenmodulator $m$, Reaktanzmodulatorröhre $f$ ~ valve Reaktanzröhre $f$ ~ voltage Blindspannung $f$
reactant Verbrennungshilfsstoff $m$
reacting, ~ force Gegenkraft $f$ ~ spring Gegenfeder $f$
reaction Einwirkung $f$, Gegen-druck $m$, -wirkung $f$, Kernreaktion $f$, Reaktion $f$, Rück-druck $m$ (phys.), -kopplung $f$, -prall $m$, -schlag $m$, -stoß $m$, -wirkung $f$, sekundäre Wirkung $f$, Stützdruck $m$, Um-satz $m$, -setzung $f$, -wandlung $f$, Vorgang $m$, Widerdruck $m$, ~ at the abutment Auflagerreaktion $f$ ~ in the front Reaktion $f$ in der Front ~ and heating velocity Reaktions- und Erhitzungsgeschwindigkeit $f$ ~ of oxyhydrogen gas Knallgasreaktion $f$ ~ of support Stützwiderstand $m$ ~ on (or of) support Auflagerkraft $f$ ~ of supports Auflagerreaktion $f$, Stützknagge $f$
reaction, ~ agent Reaktionsmittel $n$ ~ alternator Reaktionsmaschine $f$ ~ amplifier Rückkopplungsverstärker $m$ ~ avalanche Reaktionslawine $f$ ~ chamber Reaktionskammer $f$ ~ channels Reaktionswege $pl$ ~ circuit Rückkopplungskreis $m$ ~ coefficient Rückkopplungs-faktor $m$, -grad $m$ ~ coil Drosselspule $f$, Rückkopplungsspule $f$ ~ control Rückkopplungseinstellung $f$ ~ coupling Rückkopp(e)lung $f$ ~ drive Rückstoßantrieb $m$ ~ engine Rückstoßtriebwerk $n$ ~ field Rückwirkungsfeld $n$ ~ force Rückstoßkraft $f$ ~ formula Rückkupplungsformel $f$ ~ inhibitor Reaktionsbremse $f$ ~ gas Reaktionsgas $n$ ~ kinetics Reaktionskinetik $f$ ~ lump Reaktionsstein $m$ ~ mass Reaktionsmasse $f$ ~ material Rauchstoff $m$ (g/m) ~ mechanism Reaktionsmechanismus $m$ ~ method Rückkopplungsmethode $f$ ~ nozzle Ausstoßdüse $f$ ~ oscillation Rückkopplungsschwingung $f$ ~ pattern Reaktionsschema $n$ ~ point Auflagepunkt $m$ ~ pressure Rückdruck $m$ ~ principle Rückkopplungsprinzip $n$ ~ product Reaktionsprodukt $n$ ~ products in hydrocarbon combustion Reaktionsprodukt $n$ bei der Kohlenwasserstoffverbrennung ~ propulsion Reaktionsantrieb $m$ ~-propulsion jet Rückstoßdüse $f$ ~ rate Reaktions-, Umsetzungs-geschwindigkeit $f$ ~ rim Reaktions-hof $m$, -rand $m$ ~ scale

Rückkopplungsskala $f$ ~ spring Abreißfeder $f$ ~ stage Überdruckstufe $f$ ~ suppressor Rückkopplungssperre $f$ ~ test Reaktionsversuch $m$ ~ time (aut. contr.) Anlaufzeit $f$, Reaktionszeit $f$, Schrecksekunde $f$, Wirkzeit $f$ ~-time apparatus Reaktionszeitmesser $m$ ~ tower Reaktionsturm $m$ ~ transformer Rückkupplungstransformator $m$ ~ turbine Reaktions-, Überdruck-turbine $f$ ~ value Anlaufwert $m$ ~ valve Reaktionsventil $n$ ~ velocity Reaktionsgeschwindigkeit $f$ ~ vessel Reaktionsgefäß $n$ ~ wheel Gegenwirkungsrad $n$, Segnersches Rad $n$
reactionary rückständig
reactionless reaktionslos
reactivate, to ~ wiederbeleben
reactivated tungsten filament wiederbelebter Wolframfaden $m$
reactivation Neuaktivierung $f$, Reaktivierung $f$, Wiederbelebung $f$ ~ of thoriated cathode Wiederbelebung $f$ der thorierten Kathode
reactive empfindlich, gegenwirkend, mit Reaktanz, reaktionsfreudig, reaktiv, rückwirkend ~ admittance Rückwirkungsleitwert $m$ ~ attenuator nicht absorbierender oder verlustfreier Abschwächer $m$ ~ circuit mit Blindwiderstand behafteter Stromkreis $m$ ~ coil Drosselspule $f$ ~ component Blindkomponente $f$, wattlose Komponente $f$ ~ component of the current Blindstromkomponente $f$ ~ component of voltage Blindspannungskomponente $f$ ~ coupling Rückkupplung $f$ ~ current Blind-, Quer-strom $m$, wattloser Strom $f$ ~ effect Rückwirkung $f$ ~ energy meter Blindverbrauchszähler $m$ ~ jet Treibstrahl $m$ ~ kilovolt-ampere Blindkilowatt $n$ ~ load Blindbelastung $f$ ~ load adjuster Blindlasteinsteller $m$ ~ load resistance Blindlastwiderstand $m$ ~ means Reagenzien $pl$ ~ mixture reaktionsfähiges Gemisch $n$ ~ network (filter theory) verlustlose Schaltung $f$ ~ paper Reaktionspapier $n$ ~ power Blindleistung $f$ ~ resistance gerichteter Widerstand $m$ ~ volt-ampere Blindvoltampere $n$, Var $n$ ~ volt-ampere hour meter Blindverbrauchszähler $m$ ~ voltage Blindspannung $f$
reactivity Reaktionsfähigkeit $f$
reactor Drossel(spule) $f$, Induktanz-, Kernreaktor $m$ (phys.), Reaktanz-, Reaktions-spule $f$, Reaktionsgefäß $n$, Umwandlungsanlage $f$ ~ blanket Reaktorbrutmantel $m$ ~ carriage Transportwagen $m$ ~ core Spaltraum $m$ ~ debris radioaktiver Abfall $m$ ~ design Reaktorberechnung $f$ ~ engineering Reaktor-Ingenieurwesen $n$ ~ hardware Reaktorvorrichtung $f$ ~ period Reaktorzeitkonstante $f$ ~ pressure vessel Reaktordruckbehälter $m$ ~ room Reaktanzenraum $m$ ~ shell Reaktorhülle $f$ ~ sphere Reaktorkuppel $f$
read, to ~ (an instrument on a graph) abgreifen, abtasten, lesen
read, to ~ an angle einen Winkel $m$ ablesen to ~ off the angle den Winkelwert $m$ ablesen to ~ in eingeben (data proc.), schreiben to ~ (off or from) a map (or chart) eine Karte $f$ ablesen to ~-out ablesen, ausgeben (data proc.) to ~ through auslesen, durchlesen to ~ a value by an instrument einen Meßwert $m$ von einem Maßgerät ablesen, ein Meßgerät $n$ ablesen

**read,** ~ **address line** Aufrufleitung *f* ~**-around ratio** Zugriffzahl *f* ~**-disturbing** Lesestörung *f* ~**-in program** Eingabe-, Lese-programm *n* (data proc.)

**read-out** Auslesen *n*, Lesevorgang *m* ~ **clock** Fallblattuhr *f* ~ **rate** Ausgabegeschwindigkeit *f*

**read,** ~ **pulse** Leseimpuls *m* ~ **time** Lesedauer *f* ~ **winding** Lesewicklung *f*

**readability** Ablesemöglichkeit *f*, Lautstärke *f*, Verständlichkeit *f* (rdo)

**readable** ablesbar

**re-adapt, to** ~ umstellen

**re-address, to** ~ readressieren

**reader** Abtaster *m* ~'s **mark** Korrekturzeichen *n*

**readied** gebrauchsfertig

**readily** gern, leicht, prompt ~ **meltable** schnellflüssig ~ **visible** übersichtlich ~ **workable** gut bearbeitbar

**readiness** Bereit-schaft *f*, -willigkeit *f*, in Bereitschaft *f* ~ **of combustion** Anbrennfreudigkeit *f* ~ **for immediate operation** Betriebsbereitschaft *f* ~ **to locate by sound** Schallauswertebereitschaft *f* ~ **of medical detachment** Sanitätsbereitschaft *f* ~ **to transmit** Sendebereitschaft *f* ~ **for use** Betriebsbereitschaft *f*

**reading,** Ablesen *n*, Ablesung *f* (Instrument), Anzeige *f*, Ausschlag *m*, Lesen *n*, Maßangabe *f*, Meßwert *m*, Stand *m* ~(s) **at a distance** Fernlesen *n* ~ **by microscope** Mikroskopablesung *f* ~ **by mirror (or by reflection)** Spiegelablesung *f* ~ **in scale division** Ablesung *f* in Skalenteilen ~ **of a telegram by sound** Aufnahme *f* eines Telegramms nach dem Gehör

**reading,** ~ **brush** untere Bürste *f* ~ **condenser** Maxwell-anordnung *f*, -schaltung *f* ~ **device** Ablesevorrichtung *f* ~ **eyeglass** Nahklemmer *m* ~ **flip-flop** Abgreifkippschaltung *f* ~ **glass** Ableselupe *f*, Leseglas *n*, Vergrößerungsglas *n* ~ **index** Ableseindex *m* ~ **indication** Ablesbarkeit *f* ~ **line** Ablesestrich *m* ~ **magnifier** Ablese-lupe *f* ~ **microscope** Ablesemikroskop *n* ~ **precision** Ablesegenauigkeit *f* ~ **spectacles** Nahbrille *f* ~ **speed** Lesegeschwindigkeit *f* ~ **station** Abfühlstation *f* ~ **system** Lesesystem *n* ~ **table** Lesetisch *m* ~ **telescope** Ablesefernrohr *n* ~ **window** Ablesefenster *n* ~ **zenith eyepiece** Ablesezenithokular *n*

**readings on the sight** Visiereinteilung *f*

**readjust, to** ~ aufsetzen (tape rec.), nachregeln, nachrichten, neu einstellen, umstellen, wiedereinstellen

**readjusting** Neujustierung *f* ~ **the electrodes by bending** Nachbiegen *n* der Elektroden

**readjustment** Nach-justierung *f*, -regelung *f*, Neueinstellung *f* ~ **of coil (or frame)** (direction finder) Nachführung *f* des Rahmens

**ready** bereit, bereitwillig, fertig, reif, (for firing) zündfertig **for** ~ **access** leicht zugänglich, zugänglich **made** ~ bereitgestellt **not** ~ unfertig **to be** ~ in Schuß sein

**ready,** ~ **for action** Feuerbereitschaft *f*; feuerbereit, schußfertig ~ **to cast** vergießbar ~ **for driving** fahrfertig ~ **for fault indication** alarmbereit ~ **to fire** schußbereit ~ **to fly** flugklar ~ **for grinding** schleiffertig ~ **for installation** anschlußfertig (print.) ~ **for loading** ladebereit ~ **to move** fahrbereit ~ **for occupation** schlüssel-

fertig (Haus) ~ **to operate** betriebsfertig ~ **for operation** Dauerbetriebsbereitschaft *f*; betriebsfertig, -klar ~ **for pouring** gießfertig ~ **for press** druckreif ~ **for printing** druckfertig ~ **for the road** betriebsfertig ~ **to sail** seeklar, segelklar ~ **for service** betriebsfertig, dienstbereit ~ **for shipment** seemäßig, versandfähig ~ **to start** startbereit ~ **for take-off** startklar ~ **for use** betriebsfertig, gebrauchs-fähig, -fertig **not** ~ **for use** unklar ~ **for work** betriebsbereit

**ready,** ~ **fitted** fertig einmontiert ~**-fixed fuse** Fertigzünder *m* ~**-ground for analysis** analysenfein ~**-made** gebrauchsfertig ~**-made paints** angemachte Farben *pl* ~ **mixed paint** anstrichfertige Farbe *f* ~ **reading** schnelles Ablesen *n* ~ **reckoner** Rechenknecht *m* ~ **sale** schneller Umsatz *m* ~ **sensitized paper** Dauerpapier *n* ~ **state** Beharrungszustand *m*

**readying** Vorbereitung *f*

**reagent** Agens *n*, Reagens *n*, Reagenslösung *f*, Reaktionsmittel *n*, Reaktiv *n*

**reagents** Reagenzien *pl*

**real** echt, natürlich, reell, sachlich, tatsächlich, wirklich ~ **circuit** Stammleitung *f* ~ **component** reelle Komponente *f*, Wirk-komponente *f*, -teil *m*, -wert *m* ~ **domain** reelles Gebiet *n* ~ **estate** Grundeigentum *n* ~**-estate broker** Häusermakler *m* ~ **figures** absolute Zahlen *pl* ~ **horizon** natürlicher Horizont *m* ~ **line** wirkliche Leitung *f* ~ **measure** Sollmaß *n* ~ **number** reelle Zahl *f* ~ **part** Realteil *m* ~ **part of the refractive index** Brechungsindexrealteil *m* ~ **power** Wirkleistung *f* ~ **resistance** Wirkwiderstand *m* ~ **reversed image** reelles, umgekehrtes Bild *n* ~ **sailing flight** reiner Segelflug *m* ~ **shell** Schildpatt *n* ~ **size** Sollmaß *n* ~ **solubility** wirkliche Lösung *f* ~**-time** Echtzeit . . ., Realzeit . . . ~**-time operation** Echtzeitbetrieb *m*, schritthaltende Datenverarbeitung *f*

**realgar** Bergrot *n*, Rauschrot *n*, Realgar *n*, rote Arsenblende *f*, rotes Arsensulfid *n*

**re-align, to** ~ nachrichten

**realism of reproduction** Natürlichkeit *f* der Wiedergabe, Wiedergabenatürlichkeit *f*

**realistic** realistisch, sachlich, wirklichkeitsnah ~ **properties** (in architectural acoustics) Lebendigkeit *f*, Naturtreue *f*

**reality** Sachlichkeit *f*, Tatsächlichkeit *f*, Wirklichkeit *f*

**realizable** nutzbar

**realization** Durchführung *f*, Realisation *f*, Realisierung *f*, Vergegenwärtigung *f*, Verwirklichung *f* ~ **of space** Darstellung *f* des Raumes ~ **function** Realisierungsfunktion *f*

**realize, to** ~ abstoßen, erkennen, sich vorstellen, vergegenwärtigen

**realtor** Häusermakler *m*

**ream, to** ~ abschrägen, aus-reiben, -schärfen, erweitern, nach-bohren, -nehmen, reiben **to** ~ **a hole** Loch *n* auftreiben **to** ~ **out** aufreiben, nachbüchsen **to** ~ **up** aufweiten

**ream** Ries *n* ~ **wrappers** Riespapier *n*

**reamed** aufgerieben, erweitert ~ **bolt** eingepaßte Schraube *f*, Paßbolzen *m*, Paßschraube *f* ~ **hole** Paßloch *n*

**reamer** Aufdornwerkzeug *n*, Auftreiber *m*, Bohrlochausräumer *m*, Bohrwerkzeug *n*, Erweiterungsbohrer *m*, Freibohrer *m*, Freischneidebohrer *m*, Gewindeschneidemaschine *f*, Loch-strecker *m*, Nachnahmebohrer *m*, Nachnehmer *m*, Nachräumer *m*, Räumer *m*, Räum-, Reib-ahle *f*, Vollbohrer *m* ~ **bit** Nachschneidebohrer *m*

**reaming** Aufräumarbeit *f*, Aufreiben *n*, Gewinde-schneiden *n*, Nachschneiden *n*, Schleifen *n*; aufreiben ~-**and scraping machine** Aufreibe-und Ausreibemaschine *f* ~ **attachment** Auf-reibvorrichtung *f* ~ **auger** Nachbüchse *f* ~ **bit** Nachschneide-büchse *f*, -meißel *m*, Reibahle *f* ~ **diameter** Aufreibedurchmesser *m* ~ **fixture** Reibvorrichtung *f* ~ **instrument** Erweiterungs-instrument *n* ~ **press** Aufweitepresse *f* ~ **tool** Reibwerkzeug *n*

**reamplify, to** ~ wiederverstärken

**reams** Glasschlieren *pl*

**re-anneal, to** ~ nachglühen

**reap, to** ~ ernten

**reaper** Garbenbinder *m*, Mähmaschine *f* ~ **for wheat** Getreidemäher *m*

**reaping,** ~ **attachment** Getriebeableger *m*, Hand-ablage *f* ~ **hook** Sichel *f*

**rear, to** ~ aufstellen, bauen, errichten

**rear** Antriebseite *f*, Heck *n*, Hintergrund *m*, Rücken *m*, Rückseite *f*, Westseite *f* (aviat.); hinter **in the** ~ hinten **to the** ~ nach hinten, rückwärts ~ **of the plug-in units** Einschubrück-seite *f* ~ **of receiver well** Kreuzteil *n*

**rear,** ~ **anchoring clevis (or stirrup)** Hinterbügel *m* ~ **angle** Rückenwinkel *m* ~ **arch of saddle** Hinterzwießel *m* ~ **area** Hintergelände *n* ~ **armor** Heckpanzer *m* ~ **axle** Hinter-achse *f*, -brücke *f*, -radachse *f*

**rear-axle,** ~ **beam** Hinterachs-brücke *f*, -körper *m* ~ **brake** Hinterachsbremse *f* ~ **brake cylinder** Hinterachsbremszylinder *m* ~ **brake shaft** Hin-terachsbremswelle *f* ~ **carrier** Hinterachsträger *m* ~ **differential casing** Autobrücke *f* ~ **drive** Hinterachsantrieb *m* ~ **driving shaft** Hinter-achsantriebswelle *f* ~ **gear** Hinterachstrieb-werk *n* ~ **housing** Hinterachs-brücke *f*, -gehäuse *n*, -mittelgehäuse *n* ~ **housing cover** Hinterachs-gehäusedeckel *m* ~ **load** Hinterachsdruck *m* ~ **part** Hinterachsteil *n* ~ **pressure** Hinterachs-druck *m* ~ **shaft** Hinterachs(trieb)welle *f* ~ **thrust** Hinterachsschub *m* ~ **tie bar** Hinter-achsstrebe *f* ~ **tie rod** Hinterachsunterzug *m* ~ **tube** Hinterachsrohr *n*

**rear,** ~ **bearing** Hinterlager *n* ~ **bourrelet** Füh-rungswulst *f* ~ **brake** hintere Bremse *f* ~ **build-ing** Hintergebäude *n* ~ **bumper** hinterer Ab-weiser *m* ~ **cockpit** Rumpfkanzel *f* ~ **connec-tion** rückseitiger Anschluß *m* ~ **(or stern) control surface** Heckruder *n* ~ **cover** hinterer Abschlußdeckel *m* ~ **crankcase section** Ap-parateträger *m* ~ **door** Bodenverschluß *m* ~ **drive** rückwärtiger oder rückwärts liegender Antrieb *m* ~ **dumper** Hinterkipper *m* ~ **dump-ing** Hinterkippung *f* ~-**echelon train** Kolonnen-staffel *f* ~ **edge** Hinterkante *f* ~ **element lens mount** Hinterlinsenring *m* ~ **elevation** hintere Ansicht *f* ~ **end** rückläufiges Ende *n* ~ **end of assembly line** Bandende *n* ~ **end of the sheet** Bogenende *n* ~ **end plate** Hinterplatte *f* ~ **engine** Heckmotor *m* ~-**engine drive** Heck-antrieb *m* ~-**engine driven** heckangetrieben ~-**engine motor car** Heckmotorwagen *m* ~

**entrance** Hinteneinstieg *m* ~ **flange** Naben-mittelstück *n* ~ **focus** bildseitiger oder hinterer Bildbrennpunkt *m* ~ **frame** Hinterrahmen *m* ~ **guide** hintere Führung *f* ~ **gun turret** Heck-gefechtsstand *m*, Rumpfkanzel *f* ~ **handle attachment** hintere Griffvorrichtung *f* ~ **hub** hintere Nabe *f* ~ **jack** hintere Stütze *f* ~ **joint** Hintergelenk *n* **(pistol)** ~ **lamp** Rück-lampe *f*, -laterne *f* ~-**lamp bracket** Halter *m* für die Schlußlaterne ~ **landing gear** Heckfahrwerk *n* ~ **landing gear leg (or strut)** Fahrgestellhinter-strebe *f* ~ **lay standard** Rückenständer *m* ~ **light** Hecklicht *n*, Rückscheinwerfer *m* ~ **light changeover switch** Schlußlichtumschalter *m* ~ **machine gun** hinteres Maschinengewehr *n* ~ **main bearing of crankshaft** hinteres Haupt-lager *n* der Kurbelwelle

**rearmost** hinterst (e, er, es)

**rear,** ~ **nodal point** Bildknotenpunkt *m*, (hin-terer) Knotenpunkt *m* ~ **organs** Aftergegend *f* ~ **outrigger (gun platform)** Hinterholm *m* ~ **panel of the apparatus** Geräterückseite *f* ~ **panel door** rückwärtige Tür *f* ~ **part** Hinterteil *n* ~ **part of the body (or fuselage)** Rumpfhinter-stück *n* ~ **plate** hintere Scheibe *f* ~ **point (of a march column)** Nachspitze *f* ~ **portion of rib** Rippenendstück *n* ~ **position** Auffangstellung *f* ~ **position of engine** Hecklage *f* des Motors ~-**position control valve** nachgeschaltetes Steuer-gerät *n* ~ **projection** Durchprojektion *f* (film) ~ **propeller** Hinterschraube *f* ~ **radiator** Heck-kühler *m* (aviat.) ~ **rest** Hintersupport *m* ~ **scanning** Rückwärtsabtastung *f* (film) ~ **seat** Rücksitz *m* ~ **set trigger** Rückstecher *m* ~ **ship** Schlußschiff *n* ~ **side** Hinterseite *f* ~ **sight** Visierkimme *f* ~ **sight on gun** Kimme *f* ~-**sight base spring** Visierfeder *f* ~-**sight joint pin** Visier-(halte)stift *m* ~-**sight notch** Visierkimme *f* ~-**sight slide** Aufsatzschieber *m* ~ **spar** (antiair-craft gun) Hinterholm *m* ~ **spring** Hinterfeder *f* ~-**spring bracket** Hinterfederbock *m* ~-**spring shackle** Hinterfederlasche *f* ~-**steering system** Rückwärtslenkung *f* ~-**step on-float gear** Ne-benstufe *f* ~ **support** Hinterstütze *f* ~ **suspension** Hinterrad-aufhängung *f*, -federung *f* ~ **tip frame support** Auflagesattel *m* (g/m) ~ **tire** Hinterreifen *m* ~-**to-front ratio** Vorwärts-zu-Rückwärts-Verhältnis *n* ~ **tool post** hinterer Stahlhalter *m* ~ **trail** Hinterstütze *f* ~ **truck frame** Laufgestellrahmen *m* ~ **undercarriage strut** hintere Fahrgestellstrebe *f* ~ **view** hintere Ansicht *f*, Rück(an)sicht *f*, Sicht *f* nach hinten ~-**view mirror** Bordspiegel *m* (aviat.), Rück(blick)spiegel *m* ~ **viewing (of pic-tures)** Durchsichtsbetrachtung *f* ~-**vision mirror** Rückspiegel *m* ~ **wheel** Hinterrad *n*

**rear-wheel,** ~ **attachment** Hinterradeinrichtung *f* ~ **brake** Hinterradbremse *f* ~ **brake cylinder** Hinterradbremswelle *f* ~ **brake linkage** Hinter-radbremsgestänge *n* ~ **drive** Hinter-antrieb *m*, -radantrieb *m* ~ **hub** Hinterradnabe *f* ~ **pair** Hinterradsatz *m* ~ **spring suspension** Hinter-radfederung *f*

**rear window** Rückfenster *n*

**rearing** Kultur *f* ~ **form** Aufsatzform *f*

**re-arrange, to** ~ ordnen, um-formen, -gruppie-ren, -lagern, -ordnen, -wandeln

**re-arrangement** Disproportionierung *f*, Neugestaltung *f*, Neu-, Um-ordnung *f*, Umlagerung *f* **~ collisions** Umordnungsstöße *pl*
**re-arranging, ~ machinery in a plant** Betriebsumstellung *f* **~ operator** Anordnungsoperator *m*
**re-arrest, to ~** wiederanhalten
**rearward** rückliegend, rückwärtig **~ movement** Rückwärtslauf *m* **~ point** Rückhaltepunkt *m* **~ position** Rücklage *f*, rückwärtige Stellung *f* **~ turns (or return) counter** Rückwickelzähler *m*
**reason, to ~** folgern, vorbringen
**reason** Begründung *f*, Grund *m*, Überlegung *f*, Ursache *f* **~ code** Begründungskode *m*
**reasonable** angemessen, annehmbar, maßvoll, naheliegend, verstandesmäßig, verständig **~ assumption** Bestimmtheit *f* **~ diligence** angemessene Sorgfalt *f* **~ price** annehmbarer Preis *m*
**reasoning** Schlußfolgerung *f*
**reasons for exclusion** Ausschließungsgründe *pl*
**re-assemble, to ~** wiederzusammenbauen
**re-assembly** Einbau *m*, Rückmontage *f*, Wiederzusammenbau *m*
**re-assess, to ~** aufwerten
**re-assignment** Wiederabtretung *f*
**Réaumur scale** Reaumur-Skala *f*
**reaving of rope** Seilanordnung *f*
**rebag, to ~** umsacken
**rebalance, to ~** neu ausgleichen, neu nachbilden, neunachbilden
**rebalancing** neues Ausgleichen *n* **~ potentiometer** Abgleichpotentiometer *n*
**rebate, to ~** (rails) überblatten, vergüten **to ~ a charge** eine Gebühr *f* erstatten
**rebate** Ermäßigung *f*, Nachlaß *m*, Preisabzug *m*, Preisnachlaß *m*, Rabatt *m*, Rückvergütung *f*, Vergünstigung *f*, Vergütung *f*
**rebated joint** (half lap) Falzfuge *f*, (half lap) Fugeneinschnitt *m*, überfalzte Fuge
**rebating cutter** (woodwork) Falzfräser *m*
**rebellion** Erhebung *f*
**rebend, to ~** nach-biegen, -spannen
**rebending of the spring** Nachspannen *n* der Feder
**reboiler** Aufkochen *m*, Aufwärmen *m*
**reboiling** Umkochen *n*
**rebore, to ~** ausschleifen, nachbohren
**reboring** Nachbohren *n* **~ equipment** Ausbohrvorrichtung *f*
**rebound, to ~** abprallen, abspringen, aufspringen, prallen, zurückprallen, zurückspringen
**rebound** Abprallen *n*, Aufprall *m*, Aufschlag *m*, Rückprall *m*, Rückschlag *m*, Rücksprung *m*, Rückstoß *m*, Zurückspringen *n*; (book) neugebunden **~ of ball** Kugelrücksprung *m*
**rebound, ~ clip** Federklammer *f* **~ coefficient** Schwellbeiwert *m* **~ curve** Entlastungs-, Schwell-kurve *f*, Schwellast *f* **~ electrode** Rückprallelektrode *f* **~ hardness** Rückprall-, Rücksprung-, Skleroskop-, Sprung-härte *f* **~ impulse** Rückprallimpuls *m* **~ nozzle** Pralldüse *f* **~ process** Rücksprungverfahren *n* (metall.)
**rebroadcast** Rundfunkübertragung *f* **~ station** Ballsender *m*, Ballsendestation *f* **~ transmitter** Ballsender *m*, Ballsendestation *f*, Relais-sender *m*, -station *f*, Übertragungsfunkstelle *f*, Zwischensender *m*
**rebroadcasting** Ballsenden *n*, Relaisübertragung *f*

**rebromination** Rebromierung *f*
**rebuff** Abweisung *f*, abschlägige Antwort *f*, Rückstoß *m*
**rebuild, to ~** umbauen, umkonstruieren, wieder aufbauen, wiederaufbauen, wiederherstellen
**rebuild** Umbau *m*, Wiederaufbau *m*
**rebuilt** umgebaut, wiederaufgebaut
**rebush, to ~ a bearing** Lagerbuchse *f* auswechseln
**rebuttal** Widerlegung *f*, Zurückweisung *f*
**rebutting evidence (or testimony)** direkter Gegenbeweis *m*
**recalculated** umgerechnet
**recalculation** Neuberechnung *f*, Umrechnung *f*
**recalescence (point)** Haltepunkt *m* **~ curve** Abkühlungskurve *f*
**recalibrate, to ~** nacheichen
**recalibration** Nacheichung *f*, Neueichung *f* **~ shop** Kalibrierwerkstatt *f*
**recalk, to ~** nachstemmen
**recall, to ~** ab(be)rufen, aufkündigen, sich erinnern, wieder anrufen, den Anruf wiederholen, zurückrufen **to ~ from circulation** außer Kurs setzen **to ~ operator** Flackerzeichen *n* geben
**recall** Abruf *m*, Aufkündigung *f*, Rückruf *m* **~ signal** Nachruf *m*
**recalling, ~ fork** Geradesteller *m* **~ key** Rufschalter *m*
**recapitulation** Zusammenfassung *f*
**recapping** Reifenerneuerung *f*
**recapture, to ~** wieder nehmen
**recarbon, to ~** die Kohlen *pl* wechseln
**recarboning** Kohlenwechsel *m*
**recarburization** Aufkohlung *f*, Rückkohlung *f*
**recarburize, to ~** aufkohlen, kohlen, rückkohlen
**recarburizer** kohlender Zusatz *m*
**recarburizing addition** kohlender Zusatz *m*
**recast, to ~** um-formen, -gestalten, -gießen, -schmelzen
**recast** Um-arbeitung *f*, -guß *m*, -prägung *f*, -schmelzung *f* **~ alloy** Umschmelzlegierung *f* **~ material** Umgußkörper *m*
**recasting** Umguß *m*, Umschmelzung *f*, Wiedereinschmelzen *n*
**recede, to ~** zurückfließen
**receding, ~ leg of target course** Auswanderungsstrecke *f* **~ target** gehendes Ziel *n*
**receipt, to ~** quittieren
**receipt** Annahme *f*, Eingang *m*, Einnahme *f*, Einnahmeschein *m*, Empfang *m*, Empfangsbescheinigung *f*, -schein *m*, Erledigungsschein *m*, Quittung *f*, Quittungsschein *m* **~ of delivery** Aufgabeschein *m* **~ of goods** Wareneingang *m* **~ of a telephone call** Eingang *m* eines Anrufes **receipt, ~ book** Quittungsbuch *n* **~ stamp** Quittungsstempel *m* **~ voucher** Empfangsbescheinigung *f*
**receipts** Zugang *m* (am Lager)
**receive, to ~** abnehmen, annehmen, aufnehmen, bekommen, einnehmen, empfangen, erhalten, fangen, nehmen **to ~ hits** Treffer *pl* erhalten **to ~ a shock** einen Schlag *m* erhalten
**received, as ~** im Anlieferungszustand *m* (money) **~ on account** Rechnungseinzahlung *f*
**received, ~ field strength** Empfangsfeldstärke *f* **~ power** Empfangsleistung *f* **~ wave** Empfangswelle *f*

**receiver** Abnehmer *m*, Auffangsgefäß *n*, Aufnahme *f*, Aufnahmekasten *m*, Aufnehmer *m*, Auspuffsammler *m*, Empfänger *m*, Empfangsapparat *m*, -gerät *n*, Folgezeigerempfänger *m*, Füllbecken *n*, (of air pump) Glocke *f*, Hörerkapsel *f*, -muschel *f*, Kasten *m*, Mischer *m*, Sammel-flasche *f*, -gefäß *n*, -raum *m*, Überstromkanal *m*, Verbraucher *m*, Vorherd *m*, Vorlage *f* ~ **with automatic tuning means** Druckknopfempfänger *m* ~ **with feedback** Empfänger *m* mit Rückkopplung ~ **for multiple tubes** Mehrfachröhrenempfänger *m* (teleph.)

**receiver,** ~**-adjustment device** Empfangssteller *m* ~**-amplifier radio** Empfangsverstärker *m* ~ **antenna** Empfangsteil *m*, Empfängerantenne *f* ~**-antenna switch** Antennenausschaltgerät *n* ~ **arrangement** Empfangsvorrichtung *f* ~ **band filter** Rundfunkbandfilter *m* (rdo) ~ **cabinet** Empfängergehäuse *n* ~ **cap** Höhrermuschel *f*, Schallöffnung *f* ~ **case** Fernhörerkapsel *f*, Hörergehäuse *n* ~ **chassis** Empfangsapparatur *f* ~ **choke radio** Empfängerverdrosselung *f* ~ **circuit** Empfänger-reis *m*, -schaltung *f*, Entnahmestromkreis *m*, Hörerstromkreis *m*, Verbraucherkreis *m* ~ **connections** Empfängerschaltung *f* ~ **converter** Empfängerumformer *m* ~ **cover** Hülsenbrücke *f* ~ **cup** Hörmuschel *f* ~ **decrement** Empfängerdekrement *n* ~ **earpiece** Hörermuschel *f*, Schallöffnung *f* ~ **end** Entnahme-, Verbraucher-seite *f* ~ **floor plate** Kastenboden *m* ~ **gating** Empfängertastung *f*, Entblockierung *f* ~ **ground** (connection) Empfindungserde *f* ~ **magnet with graduated scale** Empfangsmagnet *m* mit Teilkreis ~ **mixer** Roheisenmischer *m* ~ **motor** Empfängermotor *m* ~ **mounting frame** Empfängerrahmen *m* ~ **noise** Empfängerrauschen *n* ~ **radiation** Empfänger-Eigenstrahlung *f* ~ **rectifier** Empfangsgleichrichter *m* (rdo) ~ **response time** Trägheit *f* eines Empfängers ~ **shell** Hörergehäuse *n* ~ **standard** Fernhörernormal *n* ~ **station for a picture telegraph installation** Empfangseinrichtung *f* einer Bildtelegrafenanlage *f* ~ **top** Kastendeckel *m* ~ **-transmitter amplifier** Mikrofonrelais *n* ~ **tuning** Empfängerabstimmung *f* ~ **vibrator unit** Empfängerwellenerreger *m*

**receiving** aufnehmend ~ **amplification** Empfangsverstärkung *f* ~ **amplifier** Empfängerverstärker *m* ~**-and milling machine** Aspirationsreinigungsmaschine *f* ~**-and transmitting** Aufführung *f* ~ **antenna** Ampfangsantenne *f* ~ **apparatus for red-filter light-signal equipment** Rotfilterempfangsgerät *n* ~ **band filter pass** Empfangsbandfilter *n* ~ **basket** Auffangkasten *m* ~ **board (or screen)** Empfängertafel *f* ~ **box** Ablegekasten *m* ~ **brushes** Empfangsbürsten *pl* ~ **buoy** Aufnahmebehälter *m* ~ **case** Sammelkasten *m* ~ **circuit** Empfangsschaltung *f* ~ **condition** Empfangsstellung *f* ~ **conditions** Empfangsverhältnisse *pl* ~ **current** ankommender Strom ~ **device** Empfangseinrichtung *f* ~ **disk** Empfangsscheibe *f* ~ **distributor** Empfangs(ver)teiler *m* ~ **drier** Vortrockner *m* ~ **drum** Aufnahmebehälter *m* ~ **end** Aufnahme-, Belade-station *f*, (of a line) Empfangsende *f*, Empfangsseite *f* ~ **energy** Empfangsenergie *f* ~ **equipment** Empfangsapparatur *f* ~ **flask**

Auffangkolben *m* ~ **frequency** Empfangs-frequenz *f*, -welle *f* ~ **ground** Empfangserde *f* ~ **guide** Einlaßführung *f* ~ **head end** Abnahmestation *f* ~ **hole** Aufnahmeloch *n* ~ **hopper** Schüttrumpf *m* ~ **lens** Empfangslinse *f* ~ **level** Empfangspegel *m* ~ **line** Empfangsleitung *f* ~ **loop** Empfangsrahmen *m* ~ **loop loss** Empfangsverluste *pl* ~ **lug** Aufnahmedorn *m* ~ **measurement** Empfangsmessung *f* ~ **office** Annahmestelle *f* ~ **operator** Empfangsbeamter *m* ~ **perforator** Empfangslocher *m*, Lochstreifenempfänger *m* ~ **pit** Einwurfgrube *f* ~ **place** Aufnahmestelle *f* (telegr.) ~ **platform** Verladerampe *f* ~ **point** Aufpunkt *m* (nav.) ~ **range** Hörreichweite *f* ~ **reflector** Empfangsspiegel *m* ~ **relay** Aufnahme-, Empfangs-relais *n* ~ **ring** Aufnahme-, Empfangs-ring *m* ~ **roller** Auflaufwalze *f* ~ **segment** Aufnahmesegment *n*, Empfangssegment *n* ~ **selector** Empfangs-verteiler *m*, -wahlschalter *m* ~ **set** Empfänger *m*, Empfangs-anlage *f*, -gerät *n*, -satz *m*, Funkenempfänger *m* ~ **shaft** Empfängerachse *f* ~ **silo** Empfangssilo *n* ~ **slit** Eintrittasphalt *m* ~ **station** Abnahmestelle *f*, Empfangs-station *f*, -zentrale *f*, (electronic current) Empfangsstelle *f*, Großfunkempfangsstelle *f* ~ **stream** Vorfluter *m* ~ **system** Empfänger *m*, Empfangs--anlage *f*, -system *n* ~ **tank** (sugar) Absetzbecken *n*, Sammelkasten *m* ~ **tape** Empfangsstreifen *m* ~ **tray** Aufgaberost *m* ~ **tube (or valve)** Empfangsröhre *f* ~ **vessel** empfangender Tank *m* ~ **voltage** Empfangsspannung *f* ~ **wire** Auffangdraht *m* (antenna)

**recent** frisch, kürzlich, neu, neu entstanden ~ **and prehistoric** Alluvial(bildung) *f*, Alluvium *n*

**re-center, to** ~ nachzentrieren

**recently** kürzlich, neuerdings, neulich ~ **built** neukonstruiert ~ **eroded valley** Runse *f* ~ **opened cast working** neuaufgeschlossener Tagebau *m*

**recentness** Frische *f*

**receptacle** Aufnahme *f* für Stecker, Behälter *m*, Empfangskasten *m* (electr.), Fach *n*, Fassung *f* (electr.), Gefäß *n*, Sammel-flasche *f*, -gefäß *n*, -raum *m*, Steckdose *f* ~ **for beer working out of the bunhole** Gärbecken *n*

**receptacle,** ~ **furnace** Gefäßofen *m* ~ **latch** Kupplungsriegel *m*

**reception** Audienz *f*, Aufnahme *f*, Aufnehmen *n*, Einnahme *f*, Empfang *m* ~ **by buzzer (or sounder)** Aufnahme *f* nach dem Gehör, Hörempfang *m* ~ **by tape** Bandempfang *m* ~ **with vibrating relay** Vibrationsempfang *m*

**reception,** ~ **amplification** Empfangsverstärkung *f* ~ **amplifier** Empfängerverstärker *m* ~ **bolt** Aufnahmebolzen *m* ~ **interference** Empfangsstörung *f* ~ **measurement** Empfangsmessung *f* ~ **range** Eingangsspannung *f* ~**-selection switch** Empfangswahlschalter *m* ~ **test** Empfangsversuch *m* ~ **test set** Stationsprüfer *m* ~ **zone** Empfangszone *f* ~ **zone beyond the skip distance** Fernempfangszone *f*

**receptivity** Aufnahme-fähigkeit *f*, -vermögen *n*, Empfänglichkeit *f* ~ **of stimulus** Reizaufnahme *f*

**recess, to** ~ mit einem Absatz versehen, (in milling machine) ausfräsen, aussparen, eindrehen, einstechen, hinterstechen, mit Vertiefung versehen

**recess** Absatz *m*, Ansatz *m*, Aufweitung *f*, Ausdrehung *f*, Ausnehmung *f*, Aussparung *f*, Einbuchtung *f*, Eindrehung *f*, Eindeckung *f*, Einschnitt *m*, Einschnürung *f*, Falz *m*, Höhlung *f*, Kammer *f*, Kreisnut *f*, Nische *f*, Pause *f*, Restperiode *f*, Seitenfalz *m*, Vertiefung *f*, (in wall) Wandaussparung *f* ~ **in the floor** Sohlenfalz *m*
**recess,** ~ **cutter** Dosensenker *m* ~ **head** Einziehung *f*, Halskette *f*
**recessed** abgesetzt, ausgebuchtet ~ **pointing** vertiefte Fuge *f* ~ **sprinkler** versenkter Sprinkler *m* ~ **square** Innenvierkant *n*
**recessing** Ein-, Hinter-stechen *n* ~ **feed** Einstechvorschub *m* ~**-off operations** Einstecharbeiten *pl* ~ **swing tool** Hinterstechschwenkwerkzeug *n* ~ **tool** Einstechstahl *m*, Hinterstechwerkzeug *n* ~ **tools for internal slots** Einstechwerkzeuge *pl* für Innennuten
**recession** Rückschritt *m* ~ **gas generator** Verdrängungsentwickler *m*
**recessive** zurückgehend
**recharge, to** ~ aufladen, nachladen, nachsetzen, wiederaufladen, wieder aufladen, wiederbeschicken, wieder laden
**recharge** Wiederaufladung *f* ~ **of battery** Erholung *f* des Akku ~ **of a secondary cell** Nachladung *f* eines Sammlers
**recharging phenomenon** Nachladeerscheinung *f*
**recheck, to** ~ nachmessen
**recheck of coil (or frame)** Rahmennachstimmung *f*
**rechecking,** ~ **collet** Umspann-patrone *f*, -zange *f* ~ **device** Umspann-futter *n*, -vorrichtung *f*
**rechuck, to** ~ wiedereinspannen
**re-chuck** Umspannen *n* (Drehbank)
**recinoleic (recinolic) acid** Rizinusölsäure *f*
**recipe** Mischungsvorschrift *f*, Rezept *n*
**recipient** Empfänger *m*
**reciprocal** Gegenstück *n*, Kehrwert *m*; gegenseitig, reziprok, umgekehrt, wechselseitig
**reciprocal,** ~ **of amplification factor** Durchgriff *m* ~ **of process** Anlaufwert *m* ~ **of process gain** Ausgleichsgrad *m* ~ **of steady-state plant** Ausgleichsgrad *m* ~ **of steady-state process gain** Ausgleichswert *m*
**reciprocal,** ~ **action** Wechselwirkung *f* ~ **bearing** gegenseitige Peilung *f* ~ **conversion** Wechselumsetzung *f* ~ **depression** (of curve) Nachlaufdelle *f* ~ **debt** Gegenschuld *f* ~ **effect** Wechselwirkung *f* ~ **integration** Kehrwertintegration *f* ~ **interstate air traffic** Poolverkehr *m* ~ **lattice** reziprokes Gitter *n* ~ **laying** Gleichlaufverfahren *n* ~ **milling** Pendelfräser *n* ~ **network** duales Netzwerk *n* ~ **orientation** gegenseitige Orientierung *f* ~ **pole** Gegenpol *m* ~ **projection** Reziprokalprojektion *f* ~ **relation** Wechselverhältnis *n* ~ **relationship** Wechselbeziehung *f* ~ **replacement** gegenseitige Ersetzung *f* ~ **service** Gegendienst *m* ~ **theorem** Reziprozitätssatz *m* ~ **transducer** reziproker Wandler *m* (acoust.) ~ **value** Kehrwert *m*, reziproker Wert *m*
**reciprocally** wechselweise
**reciprocate, to** ~ erwidern, hin- und herbewegen, hin- und hergehen, (to and fro) schwingen, in Wechselwirkung stehen
**reciprocating** abwechselnd, gegenwirkend ~ **compressor** Kolbenverdichter *m* ~ **cutter generating methods** Abwälzhobelverfahren *n* ~

**cylinder** Reibzylinder *m* ~ **engine** Kolben-maschine *f*, -motor *m* ~ **filing machine** Hubfeilmaschine *f* ~ **flat-die thread-rolling machine** Gewindewalzmaschine *f* mit Flachbacken ~ **grid** beweglicher Raster *m* ~ **hearth** Schwingherd *m* ~ **molding-sand riddle** Formsandschüttelsiebmaschine *f* ~ **motion** Hin- und Herbewegung *f*, (movement) hin- und hergehende Bewegung, Vor- und Rückgang *m* ~ **motor compressor** Elektrotauchkolbenverdichter *m* ~ **parts** reziproke Teile *pl* ~ **plate feeder** Förderschwinge *f*, Schwingförderrinne *f*, Wuchtförderer *m* ~ **plate feeding** Wuchtförderung *f* ~ **pump** Kolben-, Kurbel-pumpe *f* ~ **rolling process** Pilgerschrittverfahren *n* ~ **saw** Gattersäge *f* ~ **secondary-electron multiplier** Pendelvervielfacher *m* ~ **trough** Schüttelrinne *f* ~ **trumpet** Drehtrichter *m* ~**-type supercharger** hin- und hergehender Vorverdichtertypus *m* ~ **windscreen (or windshield-)wiper motor** Pendelwischmotor *m*
**reciprocation** Wechselwirkung *f*
**reciprocator** Kolbenmaschine *f*
**reciprocity** Gegenseitigkeit *f*, Reziprozität *f*, Wechselseitigkeit *f* ~ **failure** Abweichung *f* vom Reziprozitätsgesetz ~ **principle** Reziprozitätsprinzip *n*, Umkehrungssatz *m* ~ **property** Reziprozitätseigenschaft *f* ~ **relation** Reziprozitätsbeziehung *f* ~ **theorem** Reziprozitätssatz *m*
**recirculated air** Umluft *f*
**recirculating furnace** Luftumwälzofen *m*
**reckless** rücksichtslos
**reckon, to** ~ in Ansatz bringen, berechnen, einrechnen, mit einem Umstand rechnen **to** ~ **up** aufrechnen **to** ~ **with** sich abfinden mit
**reckoned longitude** Bestecklänge *f*
**reckoner** Zählerin *f*
**reckoning** Berechnung *f*, (ship's position by dead reckoning) Besteckrechnung *f*, Gegenrechnung *f*, Rechnung *f*
**reclaim, to** ~ (land) abgewinnen, regenerieren, wiedergewinnen, zurückgewinnen
**reclaimable waste** Abfallstoff *m*
**reclaimed,** ~ **land** Koog *m* ~ **lubricating oil** regeneriertes Schmieröl *n* ~ **product** Regenerat *n* ~ **rubber** (by regeneration) wiedergewonnener Gummi *m*
**reclaiming** Regenerierung *f* ~ **of solid matter** Feststoffgewinnung *f* ~ **tank** Rücklaufbehälter *m*
**reclamation** Gewinnung *f*, Rückgewinnung *f*
**reclamp, to** ~ wiedereinspannen
**recline, to** ~ anlehnen, stützen
**reclining,** ~ **position** Rückwärtslage *f* ~ **seat** Sitz *m* mit verstellbarer Rückenlehne ~ **spectacle** Liegebrille *f*
**reclose, to** ~ (a switch) wiedereinschalten
**reclose blocking** Wiedereinschaltsperre *f*
**recocking spring** Wiederspannfeder *f*
**recognition** Anerkennung *f*, Erkennung *f* ~ **chart** Kennblatt *n* ~ **light** Kennlicht *n* ~ **mark** Kennzeichen *n* ~ **signal** Erkennungssignal *n*, Unterscheidungssignal *n* ~**-signal disk** Kennungsscheibe *f*
**recognize, to** ~ anerkennen, erkennen
**recognized** anerkannt
**recoil, to** ~ zurück-laufen, -prallen, -stoßen

**recoil** Gegenwirkung *f*, Lafettenrücklauf *m*, Prallstock *m*, Prellklotz *m*, Reitel *m*, Rück-lauf *m*, -prall *m*, -schlag *m*, -stoß *m*, Stoß *m*, Stoß-reitel *m* ~ **atom** Rückstoßatom *n* ~ **block** Puffer *m*, Rückschlagklotz *m* ~ **booster** Rückstoßverstärker *m* ~ **brake** Rohr(rücklauf)bremse *f*, Rückbremse *f*, Schußbremse *f* ~ **brake cylinder** Bremszylinder *m* (artil.) ~ **buffer** Bremsvorrichtung *f*, (wedge) Hemmkeil *m*, Rückbremse *f*, Rückstoßdämpfer *m*, ~ **chamber** Rückstoßhülse *f* ~ **cushioning device** Rückschlagdämpfung *f* ~ **cylinder** Rücklauf-, Verdränger-zylinder *m* ~**-cylinder cap** Zylinderkappe *f* ~ **detection** Rückstoßnachweis *m* ~ **electron** Rückstoßelektron *n* ~ **escapement** zurückspringende Hemmung *f* ~ **guard** Abweiser *m* ~ **indicator** Rücklaufmesser *m* ~**-indicator slide** Rücklaufschieber *m* ~ **key** Prellkeil *m* ~ **liquid** Bremsflüssigkeit *f* ~ **lug** Hülsenzapfen *m*, Lagerbock *m* ~ **mechanism** Rücklauf-einrichtung *f*, -stellvorrichtung *f* ~ **nucleus** Rückstoßnukleon *n* ~**-operated gun** Rückstoßlader *m* ~ **piston rod** Regelstange *f* ~ **potential** Rückstoßpotential *n* ~ **rod** (typewriter) Prellanschlag *m* ~ **spring** Rücklauf-, Rückstoß-, Vorhol-feder *f* ~ **streamer** Rückstoß *m* ~ **track** Rückstoßbahn *f* ~**-type gun** Rohrrücklaufgeschütz *n* ~ **valve** Rücklaufventil *n* ~ **vector** Rückstoßvektor *m*
**recoilless** rückstoßfrei
**recoin, to** ~ umprägen
**recollect, to** ~ sich erinnern
**recombination** (of ions in gas) Molisierung *f*, Rekombination *f*, (of ions in gas) Wiedervereinigung *f* ~ **center** Rekombinationszentrum *n* ~ **coefficient** Wiedervereinigungskoeffizient *m* ~ **continuum** Rekombinationskontinuum *n* ~ **radiation** Rekombinationsstrahlung *f* ~ **rate** Rekombinationswahrscheinlichkeit *f* ~ **velocity** Rekombinationsgeschwindigkeit *f*
**recombiner** Rekombinationsapparat *m*
**recommend, to** ~ anpreisen, befürworten, empfehlen
**recommendation** Antrag *m*, Beratung *f*, Empfehlung *f*
**recommission, to** ~ wieder in Dienst stellen
**recommissioning** Wiederzulassung *f*
**recompact, to** ~ wiederverdichten
**recompensation** Rückerstattung *f*
**recomposition** Wiedergesetztes *n* (print.)
**recompression** Wiederzusammendrückung *f*
**recomputation** Umrechnung *f*
**recompute, to** ~ nachrechnen
**reconcile, to** aussöhnen, in Einklang *m* bringen
**reconciliation** Abstimmung *f*, Vergleichsangabe *f*
**recondition, to** ~ aufarbeiten, instandsetzen, nacharbeiten, überholen
**reconditioning** Aufarbeitung *f*, Instandsetzung *f*, (motor) Überholen *n*, Wiederinstandsetzung *f* ~ **plant** Neubearbeitungsanlage *f*, (reclaiming sediment in petroleum industry) Schlammaufbereitungsanlage *f*
**reconduct, to** ~ zurückleiten
**reconducting belt** Rückführungsband *n*
**reconduction band** Zusammenführungsband *n*
**reconnaissance** Aufklärung *f*, Erkundung *f*, Rekognoszierung *f* ~ **satellite** Aufklärungssatellit *m*
**reconnect, to** ~ wiederverbinden

**reconnection,** ~ **delay relay** Rückschaltverzögerungsrelais *n* ~ **lock** Wiedereinschaltsperre *f*
**reconnoiter, to** ~ aufklären, ausspähen, erkunden, streifen
**reconnoitering** Auskundschaftung *f*, Rekognoszierung *f* ~ **prism** Erkundungsprisma *n*
**reconstruct, to** ~ umbauen, wieder-aufbauen, -herstellen
**reconstruction** Aufbau *m*, (of a facsimile picture) Bildwiederaufbau *m*, Neukonstruktion *f*, Umbau *m*, -gestaltung *f*, Wieder-aufbau *m*, -herstellung *f* ~ **of image** Bild-punktverteilung *f*, -synthese *f*, -wiedergabe *f*, -zusammensetzung *f* ~ **of the image** Bilderzeugung *f*
**reconvene, to** ~ zurückführen
**reconversion** Rückverwandlung *f*
**reconverter** Tonrückumsetzer *m*
**reconvey, to** ~ zurückführen
**recooler** Nachkühler *m*
**recooling** Rückkühlung *f* ~ **installation** Rückkühlanlage *f* ~ **plant** Rückkühlanlage *f*
**record, to** ~ anzeigen, (sound) aufnehmen, aufschreiben, aufzeichnen, beurkunden, buchen, eintragen, hinterlegen, registrieren, (graphically) schreiben, vermerken, (on a meter) zählen **to** ~ **a call on the meter** ein Gespräch zählen **to** ~ **a diffraction pattern** Beugungsbild *n* aufnehmen **to** ~ **by instruments** selbstschreiben **to** ~ **photographically** (on record sheet) photografisch festhalten
**record** Akte *f*, Anzeige *f*, Aufzeichnung *f*, Beleg *m*, Bestleistung *f*, Eintragung *f*, Nachweis *m*, Nachweisung *f*, Registrierstreifen *m*, Registrierung *f*, Rekord *m*, Schallplatte *f*, Schrift *f*, Tonband *n*, Zeitstudie *f* ~ **of joints** Lötstellennachweis *m* ~ **of transactions** Bewegungsbuch *n*
**record,** ~ **accomplishment (or achievement)** Spitzenleistung *f* ~ **attempt** Rekordversuch *m* ~ **book** Kennbuch *n* ~ **changer** Plattenwechsler *m* ~ **chart** Registrierstreifen *m* ~ **circuit** Meldeleitung *f* ~ **cutter** Plattenschneider *m* ~ **(disk) diameter** Foliendurchmesser *m* ~ **distance** Streckenrekord *m* ~ **gap** Satzzwischenraum *m* ~ **head** Speicherkopf *m* ~ **ink** Dokumententinte *f* ~ **layout** Aufzeichnungsanordnung *f* ~ **level indicator** Aussteuerungskontrolle *f* (tape rec.) ~ **library** Plattenarchiv *n* ~ **mark(er)** Satzmarke *f* ~ **means (or tape)** Aufzeichnungsträger *m* (in der Bildtelegrafie) ~ **noise** Plattengeräusch *n*, Schallplattengeräusch *n* ~ **office** Archivstelle *f* ~ **(table) operator** Meldebeamter *m* ~ **operator's line** Meldeleitung *f* ~ **paper** Registrierpapier *n* ~ **performance** Spitzenleistung *f* ~**-playback head** Aufnahme- und Wiedergabekopf *m* ~ **playback unit** Aufnahme- und Wiedergabegerät *n* ~ **player** Plattenspieler *m* ~ **position** Meldeplatz *m* ~ **relay** Melderelais *n* ~ **section** Anmeldeabteilung *f*, Meldeamt *n* ~ **section with selectors** Meldeverteileramt *n* ~ **sheet** Empfänger-, Schreib-fläche *f*, Strahlungsempfänger *m* ~ **storage** Registeraufbewahrung *f* ~ **support** (sound on film, disk) Aufzeichnungsträger *m* ~ **surface** Schreibfläche *f*, Strahlungsempfänger *m* ~ **system** Registriersystem *n* ~ **table** Anmelde-, Melde-tisch *m* ~ **time** Rekordzeit *f* ~ **traffic** Meldeverkehr *m* ~**-transfer operator** Beamtin *f* am Anmeldespitzenplatz, Meldespitzenplatzbeamtin *f* ~**-transfer position**

Anmeldespitzenplatz *m*, Meldespitzenplatz *m*
**recorded, ~ copy** (in facsimile) Empfangsbild *n*
**~ image** Registrierbild *n*
**recorder** Anzeiger *m*, Auf-nahme *f*, -nahmegerät
*n*, -schreiber *m*, -zeichnung *f*, Bildschreiber *m*,
Meldebeamter *m*, Meßschreiber *m*, Registrierapparat *m*, -einrichtung *f*, -gerät *n*, -instrument
*n*, -vorrichtung *f*, Rekorder *m*, Schreiber *m*,
(automatic) Ton-aufnahme *f*, -wiedergabegerät
(automatic) Selbstschreiber *m*, Ton-aufnahme
*f*, -wiedergabegerät *n*, Zähler *m* **~ for tele-
hydrobarometer** Wasserstandsregistrierapparat
*m*
**recorder, ~ chart** Diagramm-, Registrier-streifen
*m* **~ connection** Schreiberanschluß *m* **~ head**
Schreibkopf *m* **~ lamp** Aufnahmelampe *f* **~
lamp with constricted neon arc** Lichtspritze *f* **~
paper** Registrierpapier *n* **~ pen** Registrierfeder
*f* **~ reception** Schreibempfang *m* **~ tape** Papier-
band *n* **~ unit** Schreibereinschub *m*
**recorder's feed rate** Schreibervorschubgeschwin-
digkeit *m*
**recording** Anmeldung *f*, Aufnahme *f*, Aufnehmen
*n*, Aufschreibung *f*, (of calls) Aufzeichnung der
Gesprächsanmeldungen, Band *n*, (on film) Beschriftung einer Zeichnung, Protokollierung *f*,
Registrierung *f*; schreibend **~ of calls** Aufzeichnung *f* der Gesprächsanmeldungen **~ of
deeds** Grundbuchwesen *n*
**recording, ~ accelerometer** Beschleunigungs-
schreiber *m* **~ airspeed indicator** Fahrtschrei-
ber *m* **~ altimeter** Höhenschreiber *m* **~ ammeter**
registrierender Strommesser *m*, Registrier-
strommesser *m*, Schreibstrommesser *m* **~ amplifier** Aufnahmeverstärker *m* **~-and diagram paper**
Registrier- und Diagrammpapier *n* **~ anemometer** Registrieranemometer *n*, Windregistrierapparat *m*, Windschreiber *m* **~ apparatus** Aufzeichnungs-schlitz *m*, -spalt *m* **~ apparatus** Auf-
zeichnenvorrichtung *f*, Registrierapparat *m*,
Schreiber *m* **~ arm** Schreibhebel *m* **~ arrangement** Registrieranlage *f* **~ barometer** registrierendes Barometer *n* **~ blank** Auswertevordruck
*m* **~ board** Meldeamt *n* **~ bridge** Aufnahme-
brücke *f* **~ buoy** Meßboje *f* **~ cabinet** Registrier-
schrank *m* **~ center** Schallauswertstelle *f* **~
chamber** (in electron miscroscope) Aufzeich-
nungskammer *f* **~ characteristic** Schneidekurve
*f* **~ chart** Registrierpapier *n* **~ circuit** Meldeleitung *f* **~ curve** Registrierkurve *f* **~ cylinder**
Schreibzylinder *m* **~ densimeter** Luftdichte-
schreiber *m* **~ device** Anzeigeinstrument *n*,
Registriereinrichtung *f* **~ equalizer** Aufsprech-
entzerrer *m* (tape rec.) **~ disk** Zählscheibe *f* **~
driftmeter** Derivograf *m* **~ drum** Beleuchtungs-
trommel *f*, Belichtungstrommel *f* (film),
Schreibtrommel *f* **~ furnace-filling counter** registrierender Gichtenzähler *m* **~ galvanometer**
Registriergalvanometer *n* **~ head** Aufnahme-,
Ton-kopf *m* **~ head output voltage** Ausgangs-
spannung *f* des Aufzeichnungskopfes **~ head
pole gap** Aufzeichnungskopf-Polspalt *m* **~
image of the counter** Zählerabbildung *f* **~
instrument** Aufnahmeapparat *m*, Registrier-ap-
parat *m*, -instrument *n*, Schreib-gerät *n*, -instru-
ment *n*, -werk *n*, Selbstschreiber *m* **~ lamp** Registrierlampe *f* **~ level indication** Aussteuerungsanzeiger *m*, Aussteuerungskontrolle *f* (tape

rec.) **~ level meter** Aussteuerungsanzeige *f* **~
magnet** Aufnahmemagnet *m* **~ manometer**
Dampfzähler *m* **~ mechanism** Meßschreiber *m*,
Registrier-gerät *n*, -mechanismus *m*, -vorrich-
tung *f*, schreibendes Meßgerät *n*, Schreib-werk
*n*, -zeug *n* **~ medium** Aufzeichnungsträger *m*
(in der Bildtelegrafie), Schallträger *m* **~
objective** Aufzeichnungsobjektiv *n* **~ operator**
Anmeldebeamtin *f*, Meldebeamter *m* **~ paper**
Registrierpapier *n* **~ pen** Schreiber *m* **~ pendulum without natural frequency** Pendel *n* ohne
Eigenschwingung **~ plug-in unit** Registrierein-
schub *m* **~ point** Bildempfangsstelle *f* **~ position**
Meldeplatz *m* **~ process** Anzeigeverfahren *n* **~
range** Aufzeichnungsumfang *m* **~ receiver**
Schreibempfänger *m* **~ roll** Tonrolle *f* **~ room**
Aufnahmeraum *n* **~ scale** Zeichenmaßstab *m*
**~ section** Anmeldestelle *f*, Melde-schrank *m*,
-stelle *f* **~ signal indicator** registrierender Signal-
anzeiger *m* **~ slit** Aufzeichnungs-schlitz *m*,
-spalt *m*, Lichttonspalt *m* **~ sound head** Sprechkopf *m* **~ spot** Schreibfleck *m* **~ station** Melde-
stelle *f* **~ stylus** Schneidestichel *m* **~ tachometer**
Drehzahlenschreiber *m* **~ tape** Band *n* **~
theodolite** Kino-, Registrier-theodolit *m* **~
thermometer** aufzeichnendes Thermometer *n*,
Schreibthermometer *m* **~ time** Laufdauer *f* **~
torsiometer** Verdrehungsschreiber *m* **~ torsion
meter** Drehschwingungsschreiber *m* **~ track**
Magnetspur *f* **~ transmission measuring set**
Pegelschreiber *m* **~ trunk** Anmelde-, Melde-lei-
tung *f* **~ tube** Aufzeichnung *f*, Aufzeichnungs-
röhre *f* **~ tube monitor** Bildröhre *f* **~ unit** Aufnahmegerät *n* **~ voltameter** schreibendes Voltameter *n* **~ voltmeter** schreibender Voltmesser *m*
**~ watt- and varmeter** Wirk- und Blindleistungs-
Schreiber *m* **~ wattmeter** Leistungsschreiber *m*,
registrierendes Wattmeter *n* **~ weigher** selbstschreibende Waage *f* **~ window** Ausschnitt *m*
**recordings of proceedings** Aufzeichnung *f*
**recordist** Tonkameramann *m*, Tonmeister *m*
**records** Protokoll *n*, Schriftmaterial *n* **as the ~
prove** nachweislich
**recountersink, to ~** nachsenken
**recoupling effect** Rückkupplungserscheinung *f*
**recourse** Regreßrecht *n*, Zuflucht *f*
**recover, to ~** aufarbeiten, beitreiben, bergen,
genesen, nachholen, regenerieren, rückgewin-
nen, sich erholen, wiedergewinnen, zurück-er-
halten, -gewinnen
**recoverable** eintreibbar, wiedergewinnbar, zu-
rückgewinnbar
**recovered** hergestellt, regeneriert **~ energy** Rückfederung *f*
**recoveries** Einnahmen *pl*
**recovering** (residues) Abfangen *n*
**recovery** Anschluß *m*, Aufschwung *m*, Ausbeute
*f*, Ausbringen *n*, Ausbringung *f*, Beitreibung *f*,
Bergen *n*, Besserung *f*, Erholung *f*, Genesung *f*,
Gewinnung *f*, Herstellung *f*, Regenerierung *f*,
Rück-bildung *f*, -gewinnung *f*, Vergütung *f*,
Wieder-erlangung *f*, -gewinnung *f* **~ of the airship** Einholen *n* des Luftschiffes **~ of distillate**
Destillatgewinnung *f* **~ of helium** Heliumgewinnung *f* **~ of a line** Abbruch *m* einer Linie **~
of solvents** Lösungsmittelrückgewinnung *f*
**recovery, ~ annealing** Erholungsglühen *n* **~ (or
salvage) column** Bergekolonne *f* **~ curve** Aus-

gleichsvorgang *m*, Regelverlauf *m* ~ **device** Bergungsgerät *n* ~ **equipment** Berg(e)gerät *n* ~ **flap** Abfangklappe *f* ~ **grate** Fangrost *m* ~ **installation** Rückgewinnungsanlage *f* ~ **parachute** Tragschirm *m* (g/m) ~ **party** Abschleppkommando *n* ~ **plant** Wiedergewinnungsanlage *f* ~ **processes** Erholungserscheinungen *pl* ~ **rate** Erholungsgeschwindigkeit *f* ~ **service** Abschlepp-, Berge-dienst *m* ~ **temperature** Erholungstemperatur *f* ~ **time** Ausgleichs-, Entionisierungs-, Erholungs-, Nachhol- (rdr), Rückkehr-zeit *f*, innere Totzeit *f* ~ **vehicle** wrecker Abschleppfahrzeug *n* ~ **voltage** wiederkehrende Spannung *f*, Wiederkehrspannung *f* ~-**work** Abschleppdienst *m*

**recrank, to** ~ nachkurbeln

**recreation** Neubildung *f* ~ **on screen** Beschreibung *f* von Bildern auf dem Schirm ~ **room** Erholungsraum *m*

**recruit, to** ~ anwerben, ausheben

**recrystallization** Neukristallisation *f*, Rekristallisation *f*, Rückkristallisation *f*, Umkristallisation *f*, Wiederkristallisierung *f* ~ **due to cold working between stages of heat-treatment** Reckalterung *f* ~ **hardening** (of aluminum alloys) Aushärtung *f*

**recrystallize, to** ~ wiederkristallisieren

**recrystallizing** Umkristallisieren *n*

**rectangle** Rechteck *n*, Viereck *n* ~ **formed by points of collimation** Markenviereck *n*

**rectangled** recht-eckig, -wink(e)lig

**rectangular** rechteckförmig, rechteckig, rechtwinklig, viereckig, winkelrecht **with** ~ **winding section** mit rechteckigem Wicklungsquerschnitt

**rectangular,** ~ **array** rechteckiges Gitter ~ **(steel) bars** Flacheisen *n* ~ **cathode** Rechteckkathode *f* ~ **coil** rechteckige Spule *f* ~ **connector plug** Rechteckstecker *m* ~ **construction** winkelrechte Ausführung *f* ~ **coordinates** rechtwinkliges Achsenkreuz *n* ~ **course** rechtwinkliger Kurs *m* ~ **cross section** rechteckiger Querschnitt *m* ~ **curve** Rechteckkurve *f* ~ **flange** Rechteckflansch *m* ~ **frame** Rechteckrahmen *m* ~ **parallelepipedon** Rechtkant *n* ~ **picture tube** Rechteckröhre *f* ~ **piece** Soden *m* ~ **piece of peat** Torfsoden *m* ~ **plate** Rechteckplatte *f* ~ **prism** Winkelprisma *n* ~ **pulse** Rechteckimpuls *m* ~ **scan** Rechteckabtastung *f* ~ **signal** Rechtecksignal *n* ~ **swiveling table** Langschwenktisch *m* ~ **table** Plantisch *m* ~ **tank** rechteckiger Tank *m* ~ **timber** Kanteln *n* ~ **tube** Rechteckrohr *n* ~-**type checkerwork** Ziegelfachwerk *n* ~ **voltage** Rechteckspannung *f* ~ **wave** Rechteckwelle *f* ~ **waveguide** Rechteckhohlleiter *m* ~ **wing** rechteckiger Flügel *m* ~ **wire** rechteckiger Leiter *m* ~ **wiring** Gruppierung *f* im Rechteck

**rectifiable** rektifizierbar

**rectification** Abziehen *n*, Ausgleich *m*, Berichtigung *f*, (image) Bildaufrichtung *f*, Demodulation *f*, Entmodelung *f*, Entmodulierung *f*, Entzerrung *f* (photo), Gleichrichtung *f*, Läuterung *f*, Nacharbeit *f*, Reindarstellung *f*, Rektifikation *f*, Rektifizierung *f*, Richtigstellung *f*, Wiedergebrauchsfähigmachen *n* ~ **of aerial photographs** Lichtbildentzerrung *f* ~ **of error of the moment** Ausschalten *n* der besonderen und Witterungseinflüsse ~ **of the intermediate**

**frequency** Gleichrichtung *f* der Zwischenfrequenz ~ **of wire** Ausrecken *n* des Drahtes

**rectification,** ~ **apparatus** Rektifizierapparat *m* ~ **characteristic** Gleichrichtungscharakteristik *f*, Richtkennlinie *f* ~ **constant** Richtkonstante *f* ~ **effect** Richteffekt *m* ~ **efficiency** Gleichrichterwirkungsgrad *m* ~ **factor** Richtfaktor *m*

**rectified** gleichgerichtet, rein ~ **airspeed** berichtigte Fluggeschwindigkeit *f* ~ **alcohol** gereinigter Branntwein *m* ~ **alternating current** gleichgerichteter Wechselstrom *m* ~ **current** gleichgerichteter Strom *m*, Richtstrom *m* ~ **image** aufgerichtetes oder aufrechtes Bild *n* ~ **tension** Gleichspannung *f*

**rectifier** Angleicher *m*, Entzerrer *m*, Entzerrungsgerät *n*, Gleichrichter *m*, Rektifizierer *m*, Rektifiziergerät *n*, ruhender Umformer *m*, Strom(gleich)richter *m*, Ventil *n*, Wechselrichter *m* ~ **for power supply from power mains** Gleichrichter *m* für Netzteil

**rectifier,** ~ **(or tube) arrangement** Gleichrichter- oder Röhrenanordnung *f* ~ **bulb** Gleichrichterkolben *m* ~ **cell** Gleichrichterzelle *f*, (operating with a blocking layer) Sperrschichtzelle *f* ~ **current** gleichgerichteter Strom *m* ~ **effect** Richtwiderstand *m* ~ **filter reactor** Netzdrossel *f* ~ **instrument** Gleichrichterinstrument *n* ~ **photocell** Absperrschicht-, Halbleiter-, Sperrschicht-fotozelle *f* ~ **plant** Gleichrichteranlage *f* ~ **railway** Gleichrichterbahn *f* ~ **resistance** Gleichrichterwiderstand *m* ~ **stack** Gleichrichter-satz *m*, -säule *f* ~ **strengthener** Richtverstärker *m* ~ **sub-station** Gleichrichterunterwerk *n* ~ **triode** Audion *n* ~ **tube** Gleichrichterröhre *f* ~ **unit** Stromrichtergefäß *n* ~ **valve** Gleichrichter(elektronen)röhre *f*, Ventilröhre *f*

**rectify, to** ~ (current) abgleichen, abwickeln (math.), abziehen, ändern, ausbessern, berichtigen, demodulieren, destillieren, entzerren, geradleiten, gleichen, gleichrichten, korrigieren, läutern, redestillieren, rektifizieren, umdestillieren, umkehren **to** ~ **trouble** Fehler *pl* beheben

**rectifying** Entzerrung *f*, Rektifizierung *f*, Richtigstellung *f*; gleichrichtend ~ **action** Gleichrichterwirkung *f*, gleichrichtende Wirkung *f* ~ **apparatus** Entzerrungsgerät *n* ~ **audion** Detektorröhre *f* ~ **camera** (aerial photo) Entzerrungsgerät *n*, Entzerrungskamera *f* ~ **circuit** Gleichrichterschaltung *f* ~ **coil** Ausgleichsspule *f* ~ **column** Rektifikationssäule *f* ~ **crystal** Gleichrichterkristall *m* ~ **current** ausgleichender Strom *m* ~ **effect** Sperrwirkung *f* ~ **film** Sperrschicht *f* ~ **installation** Gleichrichteranlage *f* ~ **interval** Gleichrichterintervall *n* ~ **photogrammetry** Entzerrungsfotogrammetrie *f* ~ **plane** rektifizierende Ebene *f* ~ **process** Entzerrungsverfahren *n* ~ **repeater** entzerrende Übertragung *f* ~ **strengthener** Richtverstärker *m* ~ **tube** Detektorröhre *f*, Gleichrichter-lampe *f*, -röhre *f*, Ventilröhre *f* ~ **valve** Detektor-, Gleichrichter-, Ventil-röhre *f* ~ **voltage** regulierbare Anodenspannung *f* ~ **voltmeter** Gleichrichtervoltmesser *m* ~ **wash heater** (for a rum still) Flüssigkeitsrektifizierheizer *m*

**rectilineal** gradlinig

**rectilinear** geradeaus, geradlinig ~ **flight level** Geradeausflug *m* ~ **punching machine** Reihenlochmaschine *f* ~ **pressure spread** geradlinige Druckausbreitung *f* ~ **scanning** Streifenabtastung *f* (TV)

**recumbent fold** überkippte Falte *f*

**recuperate, to** ~ auffrischen, sich erholen, wiedergewinnen, zurückgewinnen

**recuperation** Auffrischung *f*, Erholung *f*, (of current or energy) Rückgewinnung *f*, Zurückgewinnung *f* ~ **of current** Stromrückgewinnung *f*

**recuperative,** ~ **capacity** Erholungsfähigkeit *f* ~ **furnace (or oven)** Rekuperativofen *m*

**recuperator** Abgasspeicherofen *m*, Gegenstromluft- und Gasvorwärmer *m*, Rekuperator *m*, Verschlußvorholer *m*, Vorholeinrichtung *f*, Vorholer *m*, Winderhitzer *m* ~ **cylinder** Vorholzylinder *m* ~ **piston** Verdränger-, Vorholerkolben *m* ~ **spring** Vorholfeder *f* ~ **valve** Vorlaufventil *n*

**recur, to** ~ zurücklaufen

**recurrence** Wiederauftreten *n*, Wiederholung *f* ~ **formula** Rekursionsformel *f* ~ **theorem** Wiederkehrsatz *m*

**recurrent** periodisch, (periodically) regelmäßig wiederkehrend, rücklaufend, rückläufig, (periodically or not) wiederkehrend ~ **cold snap** Kälterückfall *m* ~ **flow** Rückstrom *m* ~ **network** Vierpolkette *f* ~ **structure** Teilvierpol *m*

**recurring** wiederholend ~ **issue** wiederkehrende Ausgabe *f* ~ **series** zurücklaufende Reihe *f* (math.)

**recursion** Rück-schlag *m*, -stoß *m* ~ **formula** Rekursionsformel *f*

**recursive,** rekursiv, Rekursiv . . .

**recurvature of storm** Umbiegen *n* der Sturmbahn

**recut, to** ~ nachpolieren, wiederaufarbeiten **to** ~ **a file** Feile *f* aufhauen

**recutting** Nach-, Neu-schnitt *m* ~ **of used files** Aufhauen *n* von stumpfen Feilen

**recycle, to** ~ rückführen

**recycle** Umlauf *m* ~ **gas** (refinery) Kreislaufgas *n* ~ **ratio** Umlaufverhältnis *n* ~ **stock** Umlaufprodukt *n*

**recycled,** ~ **fuel** aufgearbeitetes Spaltmaterial *n* ~ **heavy medium** Rücklauftrübe *f*

**recycling** Umlauf *m*

**red** rot ~**-abstracting filter** rotarmer Filter *m* ~ **adder** Rotbeimischer *m* ~ **antimony ore** Pyrantimonit *m* ~ **antimony sulfide** Antimonzinnober *m* ~ **arsenic sulfide** Realgar *n*, rote Arsenblende *f* ~ **beech** Rotbuche *f*, Rotbuchenholz *n* ~ **brass** Rotguß *m*, Rotgußmessing *n*, Tombak *m* ~ **bronze** Rotmessing *n* ~ **cast (or tint)** Rotstich *m* ~ **cedar** rote virginische Zeder *f* ~ **chalk** Rötel *m* ~ **component** Röte *f* ~ **copper** Kupferrot *n*, (ore) Kuprit *m*, Rotkupfer *n* ~ **copper foundry** Rotgießerei *f* ~ **copper ore** Kupferrot, Rotkupfererz *n* ~ **ferric oxide** Rotoxyd *n* ~ **filter** Rotfilter *m* ~**-filter light-signal equipment** Rotfiltergerät *n* ~ **fir** Föhre *f* ~ **fog** dichroitischer oder zweifarbiger Schleier *m*, Rotschleier *m* (photo) ~**-free light** Rotfreilicht *n* ~ **hardness** Rot-glühhärte *f*, -gluthärte *f*, -warmhärte *f*, Warmschneidhaltigkeit *f* ~**-heat** rotglühend ~ **heat** Rotglut *f*, (of incandescence) Rotglühhitze *f*, Rothitze *f* ~ **hematite** Roteisenstein *m*, roter Glaskopf *m* ~**-hot** glühend, rotglühend, rotwarm **to be made** ~**-hot** ausgeglüht werden

~**-hot iron** Glüheisen *n* ~**-illuminated** rotleuchtend ~ **indigo carmine** Phönizin *n* ~ **iron ore** Blutstein *m*, Roteisen-erz *n*, -stein *m* ~ **iron oxide** Crocus Martis *n* ~ **iron oxide paint** Roteisenrostfarbe *f* ~ **lead** Blei-mennige *f*, -oxyd *n*, -rot *n*, -zinnober *m*, Mennig *m* ~**-lead chromate** Chromrot *n* ~**-lead coating** Bleimennige-Grundierung *f* ~**-lead ore** Rotblei-erz *n*, -spat *m* ~**-lead oxide** Bleiplumbat *n*, Bleitetroxyd *n*, Mennige *f*, Minium *n*, rotes Bleioxyd *n* ~**-lead putty** Mennigekitt *m* ~**-lead substitute** Mennigeersatz *m* ~ **light** rotes Licht *n* ~ **liquor** Rotbeize *f* ~ **marl** Keupermergel *m* ~ **mercuric iodide** Jodzinnober *m* ~ **mercuric oxide** rotes Präzipitat *n* ~ **metal box** Rotgußbüches *f* ~ **metal in rolls** Rolltombak *m* ~ **mine stone** Blutstein *m*, Hämatit *m* ~ **mordant** Rotbeize *f* ~ **mud** Rotschlamm *m* ~ **ocher** Berggelb *n*, Eisenmennige *f*, Goldocker *m*, Roteisenocker *n* ~ **osier tree** Purpurweide *f* ~ **oxide of copper** Kuprit *m* ~ **oxide of zinc** Rotzinkerz *n* ~ **pencil** Rotstift *m* ~ **precipitate** rotes Präzipitat *n* ~ **print** Rotpause *f* ~ **rain** Blutregen *m* ~ **record** (color film) Rotauszug *m* ~ **resist** Rotreserve *f* ~ **sandstone** Rotsandstein *m* ~**-sensitive** rotempfindlich ~ **shift** Rotverschiebung *f* ~**-shortness** Rotbruch *m*, Rotbrüchigkeit *f* ~ **silver ore** Rotgültigerz *n* ~ **slag ironstone** Blutstein *m*, Hämatit *m* ~ **tape** bürokratisch ~**-tape operations** organisatorische Operationen *pl* (info proc.) ~ **trisulphide of antimony** Kermes *m* ~ **zinc ore** Rotzinkerz *n* ~ **zone of intersection** Schnittlinie *f* in der roten Zone

**redan** Redan *n*

**redden, to** ~ röten

**reddish** rötlich

**redeem, to** ~ abzahlen, wiederherstellen

**redeemable debenture** Schuldverschreibung *f*

**redefinition** Neudefinition *f*

**redemption** Amortisation *f*, Auslösung *f*, Rückzahlung *f* ~ **of shares** Aktieneinziehung *f*

**redeposition** Wiederausfallung *f* (geol.)

**redesign, to** ~ umbauen, umkonstruieren

**redesign** Neukonstruktion *f*, Umgestaltung *f*

**rediffusion** Draht(rund)funk *m*

**redimension, to** ~ neu bemessen

**redirect, to** ~ zurück-leiten, -senden

**redissolve, to** ~ wiederauflösen

**redistill, to** ~ umdestillieren, umschmelzen

**redistillation** Redestillation *f*, Rückdestillation *f*, Umdestillation *f*

**redistribute, to** ~ von neuem verteilen, wiederverteilen

**redistribution** Wiederverteilung *f* ~ **of forces** Umgliedern *n*

**redness** Rotglut *f*, Rötung *f*

**redoubt** Redoute *f*, Schanze *f*, Werk *n*

**redraft, to** ~ nochmals entwerfen

**redrawing** Weiterziehen *n* ~ **wire** Vorziehdraht *m*

**redress, to** ~ ausbessern, berichtigen, gleichrichten, nachpolieren, picken, pillen, schärfen

**redress** Abhilfe *f*

**redressed current** gleichgerichteter Strom *m*

**redressing** Nachdrehen *n*

**reduce, to** ~ abnehmen, mit einem Absatz versehen, abschwächen, abspannen, abwärts transformieren, (a fraction) aufheben, beschränken, bezwingen, einschränken, (dia-

meter) einziehen, ermäßigen, (temperature) erniedrigen, frischen, herab-mindern, -setzen, herunter-drücken, -setzen, kleinieren, nachlassen, niederkämpfen, reduzieren, schmälern, umrechnen, untersetzen, vereinfachen (math.), verjüngen, verkleinern, vermindern, verringern, zurückführen **to ~ by** reduzieren um **to ~ by boiling** eindicken **to ~ cost** einsparen **to ~ the damping** entdämpfen **to ~ a drawing** eine Zeichnung *f* verjüngen **to ~ the flow to a trickle** den Durchfluß *m* bis zum Tropfen reduzieren **to ~ a fraction to lowest terms** ausheben **to ~ the gain** die Verstärkung herabsetzen **to ~ out** herausreduzieren **to ~ plasticity** (of ceramics) magern **to ~ to practice** in die Praxis überführen **to ~ to a practical form** überführen **~ in price** verbilligen **to ~ the quantity** abbrechen (an Menge) **to ~ in rank** degradieren **to ~ to scrap** verschrotten **to ~ to slime** aufschlämmen **to ~ the strength** verarmen **to ~ to** reduzieren auf
**reduced** abgesetzt, verkleinert **of ~ length (or style)** kurzgefaßt **on a ~ scale** in verkleinertem Maßstab *m*
**reduced, ~ admittance** relativer Leitwert *m* **~ crude** Rückstand *m* der ersten Rohöldestillation **~ detail contrast** Detailkontrast-Verringerung *f* **~ equations** reduzierende Gleichung *f* **~ image** verkleinertes Bild *n* **~ nickel oxide** Nickelschwamm *m* **~ nucleon width** reduzierte Nukleonenbreite *f* **~ output** Minderleistung *f* **~ photostat copy** verkleinerte Fotokopie *f* **~ pressure** Unterdruck *m* **~ print** Coupüre *f* **~ profile drag area** verminderte Profilwiderstandsfläche *f* **~ rate** ermäßigte Gebühr *f* **~ scale** verjüngter oder verkleinerter Maßstab *m* **~-scale rule** verkleinerter Maßstab *m* **~ structure** Grobstruktur *f* **~ tension** Spannungsnachlaß *m* **~ width** reduzierte Linienbreite *f*
**reducer** Abschwächer *m*, Reduktionsmittel *n*, Reduzierstück *n*, Übergangsstück *n*, Untersetzungsgetriebe *n*, Verjüngungsrohrstutzen *m* **~ sleeve** Einsatzbüchse *f* **~ solution** Abschwächungslösung *f*
**reducers** Übergänge *pl*
**reducibility** Reduktions-fähigkeit *f*, -vermögen *n*, Reduzierbarkeit *f*
**reducible** reduzierbar, zurückführbar **~ circuit** reduzibler geschlossener Weg *m*
**reducing** reduzierend **~ of noise(s)** Abdämpfung *f* der Geräusche **~ of the tidal capacity of an inlet** Abflußvermögen *n*
**reducing, ~ action** Reduktionswirkung *f* **~ agent** Abschwächer *m*, Reduktions-mittel *n*, -stoff *m* **~ bush** Reduktionsmuffe *f* **~ bushing** Reduzierbüchse *f*, Reduziernippel *m* **~ capacity** Reduktionsfähigkeit *f* **~ chuck** Reduzierfutter *n* **~ coupling** Reduktions-kupplung *f*, -muffe *f* **~ coupling box** Absatzmuffe *f* **~ curve** Übergangsbogen *m* **~ elbow** Reduzierwinkelstück *n* **~ first costs** Verringerung *f* der Anschaffungskosten **~ fitting** Übergangsstück *n* **~ flame** reduzierende Flamme *f* **~ flange** Übergangsflansche *m* **~ gas** Reduktionsgas *n* **~ gear** Hubverminderer *m*, Reduzier-getriebe *n*, -vorgelege *n*, Vorgelege *n* für Tourenreduzierung **~ gear ratio** Untersetzungsverhältnis *n*

**~ gearing** Verzögerungskopplung *f* **~ installation** Reduziereinrichtung *f* **~ medium** Reduktionsmittel *n* **~ mill** Reduzierwalzwerk *n* **~ nipple** Verjüngungsnippel *m* **~ pass** Reduzierzug *m* **~ piece** Reduzierstück *n* **~ pipe** Rohrstutzen *m*, Übergangsrohr *n* **~ power** Reduktions-kraft *f*, -vermögen *n* **~ process** Reduktionsvorgang *m* **~ resistance** Abschwächungswiderstand *m* **~ retort** Verdampfungskolben *m* **~ scale** Verjüngungsmaßstab *m* **~ set** Reduzierstation *f* **~ slag** Reduktionsschlacke *f* **~ sleeve** Reduktionsmuffe *f* **~ smelting process** reduzierender Schachtofenprozeß *m* **~ socket** Reduzier-einsatz *m*, -stück *n*, Reduktions-, Übergangs-muffe *f* **~ steel** Reduzierstahl *m* **~ T** Reduktions-T-Stück *n*, (flanged) reduziertes T-Stück *n*, (flanged) Verengungs-T-Stück *n* **~ tachymeter (or tacheometer)** Reduktionstachymeter *n* **~ transformer** Abwärtstransformator *m* **~ valve** Druckminderungsventil *n* **~ (the) voltage to zero** Nullung *f*
**reduction** Abflachung *f*, Abminderung *f*, (of traffic) Abnahme *f*, Absatz *m*, Abschlag *m*, Abschwächung *f* (photo), Auflösung *f* von Gleichungen, Beschränkung *f*, (in thickness) Dickenabnahme *f*, Einschränkung *f*, (paper) Einstampfen *n*, Ermäßigung *f*, fotografischer Abzug *m* (print.), Frischen *n*, Herabsetzung *f*, Konzentration *f*, Minderung *f*, Niederkämpfung *f*, (amount of) Querschnittsabnahme *f*, Reduktion *f*, Reduzierung *f*, Rückführung *f*, Schmälerung *f*, Übersetzung *f*, Umgrößerung *f*, Vereinfachung *f*, Verjüngung *f*, Verkleinerung *f*, Verminderung *f*, Verringerung *f*, Verwandlung *f*, Ziehabnahme *f*, (amount of) Ziehgrad *m*, Zugabnahme *f*, Zurückführung *f*
**reduction, ~ in amplitude** Amplitudenabnahme *f* **~ of area** Abnahme *f*, Einschnürung *f*, Durchschnittsverminderung *f* (electron.), (cros-section) Querschnittsverminderung *f*, Querschnittszusammenziehung *f* **~ of axle speed** Achsuntersetzung *f* **~ in bulk** Reduktion *f* von großen Mengen **~ of charge** Gebührenermäßigung *f* **~ in charges** Gebührenermäßigung *f* **~ of chord** Tiefenverringerung *f* **~ in the cross-section of the exhaust port** Querschnittsverengung *f* des Auspuffschlitzes **~ of the course** Kursverwandeln *n* **~ of cross-sectional area** Querschnittsabnahme *f* **~ on edge** Stauchstich *m* **~ to equal temperature** Beschickung *f* auf gleiche Temperatur **~ of feedback** Rückkupplungsunterdrückung *f* **~ of flux** Lichtstromdrosselung *f* **~ by formula** Formelwertung *f* **~ of gloss** Glanzverminderung *f* **~ of intensity** Abschwächung **~ in landing speed** Verringerung *f* der Landegeschwindigkeit **~ of litharge to lead** Glättfrischen *n* **~ in load** Lastsenkung *f* **~ of man-hours** Arbeitsstreckung *f* **~ of masses** Reduktion *f* der Massen **~ of ore** Erzreduktion *f* **~ per pass** Einzel-, Stich-abnahme *f* **~ to practice of a patent** Benutzung *f* eines Patentes **~ in proceeds** Erlösminderung *f* **~ of range** Wellenvernichtung *f* **~ of resistance** Widerstandsverringerung *f* **~ of speed** Untersetzung *f* **~ of strength** Verarmung *f* **~ of stress** Spannungsermäßigung *f* **~ of structure** Grobstruktur *f* **~**

of time element Zeitraffung *f* ~ of tolerance
Toleranzeinengung *f* ~ of traffic Rückgang *m*
des Verkehrs, Verkehrsrückgang *m* ~ of
waves Abschwächung des Seegangs ~ in
weight Gewichtsverringerung *f* ~ with white
Weiß-verdünnung *f*, -verschnitt *m*
**reduction,** ~ **coal** Reduktionskohle *f* ~ **coeffi-**
**cient** effektiver Absorptionskoeffizient *m*,
Herabsetzungsbeiwert *m* ~ **cone** Einsatzfutter
*m* ~ **crucible** Reduktionstiegel *m*, Regulusofen
*m* ~ **discharge** Reduktionssätze *pl* ~ **factor**
Reduktionsfaktor *m*, Umrechnungsfaktor *m* ~
**furnace** Reduktionsofen *m* ~ **gas** Reduktionsgas *n* ~ **gear** Übersetzungsmotor *m*, Untersetzungsgetriebe *n*, Vorgelege *n*, Zahnrad *n* für
Geschwindigkeitsabstufung ~ **gear housing**
Hinterachstrichter *m* ~ **gear train** Übersetzungsgetriebe *n* ~ **gear(ing)** Reduktionsgetriebe *n*, reduzierende Übersetzung *f*, Zahnradvorgelege *n* ~ **instrument** Auswert(e)gerät
*n*, Auswertungsinstrument *n* ~ **jaw** Reduzier-
backe *f* ~ **paste** Verschnittansatz *m* ~ **point**
Reduktionsziffer *f* ~ **pot** Elektrolysiszelle *f* ~
**print** Verkleinerungskopie *f* ~ **process** Nieder-
schlagsarbeit *f* ~**-proof** reduktionsbeständig ~
**ratio** Reduktionsverhältnis *n*, Untersetzungsverhältnis *n* ~ **reverbatory** Regulusofen *m* ~
**schedule** Abnahmereihe *f* ~ **scheme** Reduktions-
schema *n* ~ **semiconductor** Reduktionshalb-
leiter *m* ~ **slag** Reduktionsschlacke *f* ~ **smelt-**
**ing** Reduktionsschmelzen *n* ~ **table** Umrech-
nungs-tabelle *f*, -tafel *f* ~ **works** Reduktions-
hütte *f*
**reduit** Kernpunkt *m*, Reduit *n*
**redundance** Weitschweifigkeit *f*
**redundancy** Redundanz *f* ~ **in determination**
Überbestimmung *f* (math.) ~ **check** Redun-
danzkontrolle *f*
**redundant** redundant, überfließend, im Überfluß vorhanden, überzählig ~ **check** redundante
Prüfung *f* ~ **digit** Kontrollziffer *f* ~ **member**
überzähliger Stab *m*
**reduplicate, to** ~ reduplizieren, verdoppeln
**redwood** rotes Sandelholz, Rotholz *n*
**re-dye, to** ~ umfärben
**re-dyeing process** Umfärbeverfahren *n*
**reed** Blattfeder *f*, Halm *m*, Mauerrohr *n*, Rohr
*n*, (of clarinet) Rohrblatt *n*, Schilf *n*, Schilfrohr
*n*, Schwingfeder *f*, (of horn) Zunge *f* ~ **that**
**vibrates in both directions past neutral point**
durchschlagende Zunge *f* ~ **for weaving looms**
Weberkamm *m*
**reed,** ~ **armature arrangement to accentuate low**
**frequencies** (in loudspeaker) Tiefenkompensa-
tion *f* ~ **binding** Blattbinden *n* ~ **board** Stimmplatte *f* ~ **hook** (textiles) Blattstecher *n*,
Einziehmesser *n* ~ **hummer** Zungensummer
*m* ~ **maker** Webeblattsetzer *m* ~ **mat** Rohr-
matte *f* ~ **mounting** Blatteinsetzen *n* ~ **organ**
**pipe** Orgelzungenpfeife *f* ~ **pipe** Rohr-,
Zungen-pfeife *f* ~ **plane iron with square and**
**cove** Doppelkarnieseisen *n* ~ **plate** (of accordion) Stimmplatte *f* ~ **tachometer** Zungenfrequenzmesser *m* ~ **telephone receiver** Zun-
gentelefon *n* ~**-type frequency meter** Zungenfrequenzmesser *m*
**reeding** Berohrung *f*, Riffelung *f*
**reef, to** ~ reffen

**reef** Ader *f* (min.), Fels *m*, Felsen *m*, Felsenriff *n*,
Flöz *n*, Klippe *f*, (Quarz) Gang *m*, Reff *n*, Riff
*n* ~**-building corals** riffbildende Korallen *pl* ~
**knot** Kreuz-, Weber-knoten *m* ~ **shoal** See-
klippe *f*
**reel, to** ~ auf-rollen, -spulen, friemeln, haspeln,
schwanken, spulen, weifen **to** ~ **in** Antenne *f*
einziehen, aufhaspeln **to** ~ **off** abhaspeln, abspulen, abweifen, abwickeln **to** ~ **out** auskurbeln **to** ~ **up** aufhaspeln
**reel** Ablaufhaspel *m*, Auflaufhaspel *m*, Auf-
spuler *m*, Bandspule *f* (tape rec.), Bobine *f*,
Drahthaspel *f*, Filmkern *m*, Garnweife *f*,
Garnwinde *f*, Gurttrommel *f*, Haspel *f*, Kabeltrage *f*, Rolle *f*, Scheibe *f*, Spule *f*, Trommel *f*,
Weife *f*, Winde *f* ~ **of cotton** Garnrolle *f* ~ **and**
**wire** Haspel *f* und Draht *m* .
**reel,** ~ **antenna** Kurbelantenne *f* ~ **bearings**
Papierrollenlagerung *f* ~ **chain** Haspelkette *f*
~ **drive** Papierrollenantrieb *m* ~**-end signal**
Aktendalarm *m* (film) ~ **flange** Seitenwand *f*
der Kabeltrommel ~ **handle** Apspuler *m* ~
**insulator** Isolierrolle *f* ~ **jack** Trommelwinde *f*
~**-lifting device** Rolleneinhebevorrichtung *f* ~
**paper** endloses Papier *n* ~ **protector** Spindelblitzableiter *m* ~ **slitting and rewinding ma-**
**chine** Rollenschneid- und -wickelmaschine *f* ~
**star** Rollenstern *m* (print.) ~ **stump** Rollenrest
*m* (print.) ~ **wheel** Haspelrad *n*
**reeled,** ~ **card for automatic machines** Automa-
tenrollenkarton *m* ~ **foil mechanism** Rollfolien-
einrichtung *f* ~**-in antenna** eingefahrene Antenne *f* ~**-out antenna** ausgekurbelte Antenne *f*
**reeler** Glättwalzwerk *n*, Haspeler *m*
**reeling** Haspelarbeit *f*, Weifen *n* ~ **and winding**
**machine** Haspel- und Wickelmaschine *f* ~
**frames and reels** (unwinding) Abzugspulma-
schine *f*, Garnhaspel *f* ~ **friction** Aufwickel-
friktion *f* ~ **machine** Aufroll-, Aufwickel-,
Friemel-, Haspel-maschine *f*, Friemelwalz-
werk, (of tubes) Glättwalzwerk *n* ~ **mill** Glätt-
straße *f* ~ **room** Haspelei *f* ~**-up drum** Aufroll-
trommel *f*
**reels and discs for recording instruments** Dia-
grammrollen und -scheiben *pl*
**re-embark, to** ~ wieder einschiffen
**re-enactment** Wiederinkraftsetzung *f*
**re-engagement of the winding unit** Einschaltung *f*
der Spule
**re-engaging** Wiedereinrücken *n*
**re-enrichment** Neuanreicherung *f*
**re-enter, to** ~ (cloth print.) eindrucken, (of an
angle) einspringen
**re-entering angle** einspringender Winkel *m*,
Rentrant *m*
**re-entrant** einspringender Winkel *m*, Mulde *f*;
mehrfach zusammenhängend ~ **angle** einspringende Ecke *f* ~ **cavity resonator** mehrfach
zusammenhängender Hohlraumresonator *m* ~
**horn** gefalteter Trichter *m* ~ **jet** rückkehrender
Strahl *m* ~ **press (or squash)** umgebördelter
Röhrenfuß *m*, (of a tube) umgekehrter Fuß *m*,
umgestülpter Fuß *m*
**re-entry** Wiedereintritt *m* ~ **body** Wiederein-
trittskörper *m* (g/m) ~ **capsule** Wiedereintritts-
körper *m* (g/m) ~ **head** Nasenkörper *m* (g/m) ~
**oscillation** Einschwingen *n* ~ **velocity** Einsturz-
geschwindigkeit *f* (g/m)

re-equipment Umbau *m*
re-establish, to ~ entzerren (Luftbildaufnahmen), retablieren, wiederherstellen
re-establishing force Rückstellkraft *f*
re-establishment Wiederherstellung *f*
re-evaluation Aufwertung *f*
reeve, to ~ (a wire line) aufziehen, durchscheren, einscheren
re-exportation Wiederausführ *f*
reface, to ~ nachschleifen
refer, to ~ beziehen, überweisen, zurückführen to ~ to anspielen auf, sich beziehen auf, vergleichen, verweisen
referee Schiedsrichter *m* ~ test Schiedsversuch *m*
reference Anspielung *f*, Betreff *m*, Beziehung *f*, Bezug *m*, Dienstzeugnis *n*, Festpunkt *m*, Hinweis *m*, Inhaltsnachweis *m*, Nachweis *m*, Rückfrage *f*, Rücksicht *f*, (bibliographical) Schrifttumsstelle *f*, Verweisung *f* by ~ to an Hand von with ~ (to) bezüglich with ~ to betreffend, mit Bezug auf ~ to literature Schrifttums-hinweis *m*, -nachweis *m* ~ to notes Notenhinweis *m*
reference, ~ address Bezugsadresse *f* (data proc.) ~ altitude Bezugshöhe *f* ~ angle Meßwinkel *m* ~ axis Bezugs-achse *f*, -linie *f* ~ beam attenuator Vergleichsstrahlabschwächer *m* ~ black level Vergleichsschwarzpegel *m* (TV) ~ body Vergleichskörper *m* ~ book Nachschlagebuch *n* ~ bursts Bezugsimpulsgruppe *f* ~ card index (or system) Sichtkartei *f* ~ character Bezugszeichen *n* (bei Formelgrößen) ~ circuit Bezugs-(strom)kreis *m*, Vergleichsleitung *f* ~ code Bezugsordnung *f* (aerodyn.) ~ condenser Normalkondensator *m* ~ coordinate Bezugskoordinate *f* ~ core Bezugskern *m* ~ data Bezugsdaten *pl* ~ datum Bezugsebene *f* ~ designation Bezugszeichen *n* (bei Formalgrößen) ~ dimension Kontrollmaß *n* (nicht toleriert) ~ disk Meßscheibe *f* ~ drawing Hinweiszeichnung *f* ~ electrode Bezugselektrode *f* ~ ellipsoid Referenzellipsoid *n* ~ equivalent Bezugsdämpfung *f* eines Übertragungssystems ~ equivalent of sidetone Rückhördämpfung *f* ~ equivalent of the transmitter Sendebezugsdämpfung *f* ~ file Nachschlageverzeichnis *n* ~ frame Meßrahmen *m* ~ frequency Bezugsfrequenz *f* ~ fuel Bezugskraftstoff *m*, Musterbrennstoff *m*, Verweisungsbrennstoff *m* ~ gauge Einstellmaß *n*, Prüflehre *f* ~ height Bezugshöhe *f* ~ hook Bezugshaken *m* ~ humidity Bezugsfeuchtigkeit *f* ~ index Handregister *n* ~ input (servo) Führungsgröße *f*, Sollwert *m* ~ input elements Vergleicher *m* ~ instrument Etalonapparat *m*, Vergleichsapparat *m* ~ jet Bezugsdüse *f* für Eichzwecke, Eichdüse *f* ~ junction Vergleichsstelle *f* ~ letter Betrachtungszeichen *n*, Verweisungsbuchstabe *m* ~ level Bezugsebene *f* ~ library Hand-, Stand-bücherei *f* ~ line Bezugs-, Null-, Vergleichs-linie *f*, Justiermarke *f* ~ liquid Vergleichsflüssigkeit *f* ~ lobe Bezugsstrahlungskeule *f* (of antenna) ~ magnitude Bezugsgröße *f* ~ mark Einstell-, Rahmenmeßmarke *f*, (on documents) Geschäftszeichen *n*, Hinweiszeichen *n* ~ noise Bezugsrauschen *n* ~ number Aktenzeichen *n* ~ numeral Betrachtungszeichen *n* ~ performance Bezugsbeute *f* ~ pilot Steuerwelle *f* ~ plane of the fuselage

Rumpfbezugsebene *f* ~ point Ablesemarke *f*, Anhaltspunkt *m*, Anlegepunkt *m*, Bezugspunkt *m*, Bodenpunkt *m*, Festlegepunkt *m*, Geländepunkt *m*, Grundrichtungspunkt *m*, Hilfsziel *n*, Merkpunkt *m*, Ortungspunkt *m*, Wechselpunkt *m* ~-point reading Grundzahl *f* ~ pointer Aufsatzzeiger *m* ~ quantity Bezugsgröße *f* ~ range Bezugsentfernung *f* ~ record Bezugsdaten *pl* ~ sample Vergleichsprobe *f* ~ signal Bezugsschwingung *f* ~ source Bezugsquelle *f* ~ square (map) Meldequadrat *n* ~ standard Bezugsnormal *n*, Etalonapparat *m* ~ symbol (numeral or letter) Bezugszeichen *n* ~ system Bezugssystem *n* ~ telephone system Fernsprechnormalsystem *n* ~ telephonic power Bezugslautstärke *f* ~ temperature Bezugstemperatur *f* ~ time Bezugszeit *f* ~ tone (= 1000 c/s) Normalton *m* (= 1000 Hz) ~ trigger Bezugsauslöseimpuls *m* ~ value Bezugswert *m* ~ voltage Vergleichsspannung *f* ~ voltage level Bezugsspannung *f* (Bezugspotential) ~ volume Bezugspegel *m* (acoust.) ~ white level Vergleichsweißpegel *m* (TV) ~ work Nachschlagewerk *n*
references Hinweise *pl*, Literaturverzeichnis *n*, Quellenangaben *pl*
referred to bezogen (auf)
reffle, to ~ auslosen
refill, to ~ wiederauffüllen
refill Nach-, Neu-füllung *f* ~ fan Umfüllventilator *m* ~ pump Auffüllpumpe *f*
refilling Baugrube *f*, Gasnachfüllen *n*, Nachguß *m* ~ of rolls Neubekleiden *n* von Walzen
refilter, to ~ nachfiltern
refinability Raffinierbarkeit *f*
refine, to ~ abdörren (meteor.), abläutern, (with mercury) abquicken, abscheiden, affinieren, (metal) anfrischen, aufbereiten, (gaseous or liquid) ausgaren, (tin) sich bauschen, feinen, flößen, frischen, garen, (metal) gerben, läutern, pauschen, raffinieren, reinigen, (copper) spleißen, treiben, veredeln, verfeinern, verfrischen, vergüten, zerrennen to ~ copper Kupfer *n* hammergar machen to ~ highly hochtreiben to ~ metal gallern to ~ petroleum Petroleum *n* raffinieren to ~ steel Stahl *m* gar machen to ~ tin Zinn *n* gattern
refined fein, gar, gereinigt, raffiniert ~ alloy steel Edelstahl *m* ~ brine gradierte Sole *f* ~ copper Fein-, Gar-kupfer *n*, hammergares Kupfer *n*, Raffinatkupfer *n* ~ garnierite (for gaining copper) Konzentrationsstein *m* ~ garnierite for gaining copper Feinstein *m* ~ gold Brandgold *n* ~ gold containing silver Blickgold *n* ~ iron Fein-, Frisch(feuer)-, Garn-eisen *n*, Gerbstahl *m*, Paketierschweißstahl *m*, Paketstahl *m*, Raffinierstahl *m* ~ lead Arm-, Frisch-, Raffinat-, Weich-blei *n* ~ liquor Raffinadekläre *f* ~ material Feingut *n* ~ metal Fein-, Frisch-metall *n* ~ mineral oil Mineralölraffinat *n* ~ pig iron geläutertes Roheisen *n* ~ product Raffinat *n* ~ products Weißprodukte *pl* ~ pure silver Blicksilber *n* ~ silver Brandsilber *n* ~ speiss Raffinierspeise *f* (metall) ~ spelter (zinc) Raffinadzink *n* ~ state Gare *f* ~ steel Frisch-, Raffinier-stahl *m* ~ steel ingots Edelstahlblöcke *pl* ~ sugar Raffinade *f* ~ wrought iron abgeschweißtes Eisen *n* ~ zone Nachbar-, Übergangs-zone *f*

**refinement** (as ingenious detail) Feinheit *f*, Frischung *f*, Konstruktionsfeinheit *f*, Raffinationsbehandlung *f*, Verbesserung *f*, Vered(e)lung *f*, Vered(e)lungsbehandlung *f*, Verfeinerung *f*, Vergütungsbehandlung *f*, Vollkommenheit *f* ~ **of structure** Gefügeverfeinerung *f*

**refiner** Abscheider *m*, Frischer *m*, Raffineur *m*, Treiber *m*, Vorfrischofen *m*, Zerrenner *m* ~ **bar** Raffineurmesser *n* ~'s **differential** Raffinierspanne *f*

**refineries proceeds** Werksergebnis *n*

**refinery** Affinierung *f*, Destillierbetrieb *m*, Einmalschmelzerei *f*, Eisenhütte *f*, Feinfeuer *n*, Frischerei *f*, Frisch-feuer *n*, -herd *m*, -hütte *f*, -werk *n*, Raffinationsanlage *f*, Raffinerie *f*, Raffinieranlage *f*, Raffinierung *f*, Scheideanstalt *f*, Schmelzfeuer *n*, Siederei *f* ~ **cinder** Frischlacke *f*, Herdfrischschlacke *f* ~ **cinders** Eisen(frisch)schlacke *f* ~ **gas** Raffineriegas *n* ~ **gas tail** Raffinerieabgas *n* ~ **molasses** Raffineriemelasse *f* ~ **operation** Herdfrischarbeit *f* ~ **plant** Raffinationsanlage *f* ~ **process** Frischherdverfahren *n*, Herdfrisch-arbeit *f*, -prozeß *m*, Vered(e)lungsverfahren *n* ~ **slag** Feinschlacke *f*, Herdfrischschlacke *f*

**refining** Abläuterung *f*, Abscheidung *f*, Affinierung *f*, Farbe *f*, Feinen *n*, (of steel) Feinern *n*, Feintreiben *n*, Feinung *f*, Frischarbeit *f*, Frischen *n*, Frischung *f*, Läutern *n*, Läuterung *f*, Raffination *f*, Raffinierung *f*, Veredeln *n* ~ **of optical glass** Glas-, Linsen-vergütung *f* ~ **of waste** Krätzfrischen *n* ~ **the steel** Frischen *n* des Eisens

**refining**, ~ **action** Raffinationswirkung *f* ~ **assay** Garnprobe *f* ~ **battery** Läuterbatterie *f* ~ **cupel** Abtreibkapelle *f* (metall) ~ **cupellation** Treibprozeß *m* ~ **effect** Raffinationswirkung *f* ~ **fire** Frischfeuer *n*, Frischherd *m*, Reinigungsfeuer *n*, Zerrennfeuer *n* ~ **furnace** Abwerfofen *m*, Feinofen *m*, Frisch(ungs)ofen *m*, Läuterofen *m*, Raffinier-feuer *n*, -herd *m*, -ofen *m*, Treibofen *m* ~ **hearth** Abtriebeherd *m*, Feinherd *m*, Frisch-feuer *n*, -herd *m*, Garherd *m*, (for copper) Rosettenherd *m* ~ **hearth of furnace** Treibherd *m* ~ **kiln** Läuterofen *m* ~ **means** Läuterungsmittel *n* ~ **melt** Abtreibschmelze *f* ~ **metals** Garen *n* der Metalle ~ **method** Scheideverfahren *n* ~ **operation** Raffinierarbeit *f* ~ **pan** Abwerfpfanne *f*, Schlackenscherbe *f*, -scherben *m* ~ **period** Feinungsperiode *f* ~ **plant** Raffinationsanlage *f*, Raffinieranlage *f* ~ **plate** Pauschherd *m* ~ **procedure** Vered(e)lungsverfahren *n* ~ **process** Frischungsprozeß *m*, -verfahren *n*, Gargang *m*, Raffinations-prozeß *m*, -verfahren *n*, -vorgang *m*, Raffinierverfahren *n*, Reinigungs-prozeß *m*, -verfahren *n*, Vered(e)lungs-verfahren *n*, -vorgang *m* ~ **protection** Verarbeitungsschutz *m* ~ **puddling** Feinpuddeln *n* (lowgrade oxides) Armoxyde *pl*, Poldreck *m* ~ **slag** Garschlacke *f* ~-**slag bed** Garschlackenboden *m* ~ **still** Raffinieranlage *f* ~ **treatment** Raffinations-, Vered(e)lungs-, Vergütungs-behandlung *f*

**refinish, to** ~ aufpolieren, nacharbeiten, nachpolieren, wiederaufarbeiten

**refinishing operation** Nacharbeit *f*

**refit, to** ~ instandsetzen, reparieren

**refit** Neuausstattung *f*, Wiederinstandsetzung *f*

**reflect, to** ~ abprallen, abspiegeln, abstrahlen, nachdenken, prellen, reflektieren, rückstrahlen, rückwerfen, spiegeln, wider-spiegeln, -strahlen, wieder-spiegeln, -strahlen, zurückwerfen **to** ~ **into** hineinspiegeln

**reflectance** Reflexions-faktor *m*, -stärke *f*, -vermögen *n*, Spiegelungsfaktor *m* ~ **measurement** Reflexionsmessung *f* ~ **spectrometry** Remission *f*

**reflected** reflektiert, (signal through transformer) übersetzt, zurück-gestrahlt, -geworfen **to be** ~ prallen, spiegeln, zurückprallen **to be** ~ **(or delineated)** sich abbilden

**reflecter**, ~ **beam** Widerstrahl *m* ~ **binary code** zyklischbinärer Kode *m* ~ **color** Aufsichtsfarbe *f* ~ **face type** Spiegelschrift *f* ~ **(ghost) image** Spiegelbild *n* ~ **image pick-up** (a type of practice firing) Spiegelbildaufnahme *f* ~-**light principle of scanning** Reflexionsabtastung *f* ~ **power** reflektierte Leistung *f* ~ **pressure** Reflexdruck *m* ~ **ray** Abstrahl *m*, Gegenstrahl *m*, Reflexstrahl *m*, reflektierter Strahl *m*, rückgestrahlter Lichtstrahl *m* ~ **ray angle** Abstrahlwinkel *m* ~ **together** zusammengespiegelt ~ **wave** Echowelle *f*, gespiegelte Welle *f*, Höhenwelle *f*, Raumwelle *f*, reflektierte Welle *f*, rücklaufende Welle *f*, rückschreitende Welle *f*, Rückstrahl *m*, Rundwelle *f*, Spiegelwelle *f*

**reflecting** spiegelnd ~ **ability** Rückstrahlungsvermögen *n* ~ **area** Abstrahlfläche *f* ~ **circle** Spiegelkreis *m* ~ **compass** Spiegel-bussole *f*, -kompaß *m* ~ **effect** Spiegelwirkung *f* ~ **electrode** (in positive-grid tube) Brems-, Bremsfeld-, Prall-elektrode *f* ~ **field** (vacuum-tube, engin.) Bremsfeld *n* ~ **film** Spiegelbelag *m* ~ **force** rücktreibende Kraft *f* ~ **galvanometer** Spiegelgalvanometer *n* ~ **goniometer** Reflexionsgoniometer *n* ~ **impactor** Prallelektrode *f* ~ **instrument** Spiegelinstrument *n* ~ **link** Lenkspiegel *m* ~ **plate** Prall-, Spiegel-platte *f* ~ **power** Reflexionsvermögen *n* ~ **prism** Reflexionsprisma *n* ~ **projector** Episkop *n* ~ **screen** Aufhellschirm *m* (film) ~ **sight** Spiegelvisier *n* ~ **square** Winkelspiegel *m* ~ **stereoscope** Spiegelstereoskop *n* ~ **strain indicator** Spiegelapparat *m* ~ **surface** Abstrahlfläche *f*, Spiegel *m* ~ **tail lamp** Rückstrahlschlußleuchte *f* ~ **telescope** Spiegelteleskop *n* ~ **volume indicator** Lichtzeigersteuerinstrument *n* (Mischpult) ~ **wattmeter** Reflexioswattmeter *n*

**reflection** Abbild *n*, Abglanz *m*, (optics) Abprallung *f*, Absprung (phys.), Abstrahlung *f*, Auswirkung *f*, Betrachtung *f*, (in magnetron) Bremsung *f* der Elektronen, Gegenstrahl *m*, Reflektion *f*, Reflexion *f*, Rückstahl *m*, Rückstrahlung *f*, Rückwurf *m*, Spiegelung *f*, Überlegung *f*, Widerschein *m*, Zurückwerfen *n*, Zurückwerfung *f* ~ **of electrons** Elektronenbremsung *f* ~ **of glass surfaces** Glasflächenspiegelung *f* ~ **of light** Lichtreflexion *f* ~ **with respect to a circle** Spiegelung *f* am Kreis ~ **with respect to a line to the hyperbolic plane** Umklappen *n* der hyperbolischen Ebene ~ **of seismic wave** reflektierte seismische Welle *f* ~ **of sound** Reflexion *f* des Schalls ~ **by water drops** Spiegelung *f* an Tropfen

**reflection, ~ abundance** Reflexreichtum *m* ~
**attenuation** Reflexionsverlust *m* ~ **coefficient**
Reflexionsfaktor *m* ~ **density** Aussichtsdichte
*f* ~ **diffraction** streifende Beugung *f* ~ **factor**
Falschanpassungs-, Fehlanpassungs-, Refle-
xions-, Spiegelungs-faktor *m* ~ **halo** Reflexionslichthof *m* ~ **invariance** Spiegelungsin-
varianz *f* ~ **loss** Spiegelungsverlust *m* ~
**measurement** Echomessung *f* (TV), Remissions-
messung *f* ~ **measuring chamber** Rückstrahl-
meßkammer *f* ~**-measuring set** Rückfluß-
dämpfungsmesser *m* ~ **method** Rückstrahlverfahren *n* ~ **peak** Rückstrahlzacke *f* ~ **plane**
Spiegelebene *f* ~ **plate** (in wind tunnel)
Symmetriewand *f* ~ **plotter** Reflexzeichenvor-
richtung *f* ~ **polarizer** Reflexionspolarisator *m*
~ **sounding** Echolotung *f* ~ **target** reflektierende
Treffplatte *f*
**reflective** reflektierend, zurückstrahlend ~
**optics** Projektoroptik *f* ~ **paint** Rückstrahl-
farbe *f* ~ **power** Reflektions-kraft *f*, -vermögen
*n*, Spiegelungsvermögen *n*
**reflectometer** Fehlanpassungszeiger *m*, Rück-
strahlungsmesser *m*
**reflectometry** Reflexionsmessung *f*
**reflector** Katzenauge *n*, Lichtspiegel *m*, Lichtwerfer *m*, Reflektor *m*, Rückstrahler *m*, Rückwerfer *m*, Scheinwerfer *m*, Spiegel *m*, Spiegelteleskop *n*, Spiegler *m* ~**-adjusting device** Spie-
gelversteller *m* ~ **aperture** Spiegelöffnung *f* ~
**arc lamp** Spiegelbogenlampe *f* ~ **back** Spiegel-
rücken *m* ~ **bearing** Spiegellager *n* (Scheinwerfer) ~ **framework** Spiegelgerüst *n* ~ **dial**
Spiegelskala *f* ~ **diameter** Spiegeldurchmesser
*m* ~ **floodlight** Spiegelscheinwerfer *m* ~
**glasses** Reflektorengläser *pl* ~ **grinding machine**
Spiegelausschleifmaschine *f* ~ **head** Ausblicks-
kopf *m* (eines Sehrohrs) ~ **luminaires** Spiegel-
leuchten *pl* ~ **mounting** Spiegellager *n* (des
Scheinwerfers) ~ **plate** Reflexscheibe *f* ~ **ring**
Spiegelhals *m* (Lampenfassung ) ~ **savings**
Reflektorersparnis *f* ~ **screen** Spiegelfeld *n* ~
**setting device** Spiegeleinstell-Vorrichtung *f* ~
**sheet** Brennspiegel *m* ~ **sight** Reflex-optik *f*,
-visier *n* ~ **space** Reflexionsraum *m* ~ **support**
Spiegelträger *m* ~ **surface** Spiegelfläche *f* ~
**tracker** Zeichenplatte *f* ~ **tube** Spiegel-rohr *n*,
-tubus *m* ~**-tube clip** Spiegelrohrhalter *m*
~**-type cross-level** (panoramic telescope) Spiegelkreuzlibelle *f* ~ **voltage** Reflektorspannung *f*
**reflex** Reflexbewegung *f* ~ **amplification** gleichzeitige Hoch- und Niederfrequenzverstärkung
*f*, Reflexverstärkung *f* ~ **arc** Reflexbogen *m* ~
**baffle** reflektierende Schallwand *f* (acoust.) ~
**bunching** Ballung *f* im Bremsfeld ~ **camera**
Spiegelreflexkamera *f* ~ **circuit** Reflexschaltung *f* ~ **drift sight** Spiegeltriftrohr *n* ~ **halation**
Reflexionslichthof *m* ~ **klystron** Reflexions-
klystron *n*, -laufzeitröhre *f*, Spiegeltriftrohr *n* ~
**print** Rückstrahlkopie *f* (photo) **(regenerative)**
~ **receiver** Reflexempfänger *m* ~ **reception**
Röhrenempfang *m* mit gleichzeitiger Hoch-
und Niederfrequenzverstärkung ~**-reducing**
(antireflecting) reflexmindernd ~ **reflector**
Rückstrahler *m* ~ **sight** Reflexvisier *n*
~**-through-the lens camera** Spiegelreflexkamera
*f*
**reflexed section** Widderhornprofil *n*

**reflexion: see reflection**
**refloating a wreck** Heben *n* eines Wracks
**reflux, to** ~ zurück-steigen, -strömen
**reflux** Rückfluß *m* (chem.), Rücklauf *m*, Rückstrom *m*, Zurückfließen *n* **with** ~ rückfließend
(chem.)
**reflux, ~ condenser** Rückflußkühler *m*, Rück-
lauf-kondensator *m*, -verdichter *m* ~ **drum**
Rücklaufbehälter *m* ~ **ratio** Rücklauf *m*, Rück-
strömgrad *m* ~**-type condenser** Gegenstromkondensator *m*
**re-focus, to** ~ nachfokussieren
**refocusing** Nachkonzentration *f*
**reforestation** Aufforsten *n* von Ödland
**reform (re-form), to** ~ neugestalten, um-bilden,
-gestalten, zurückbilden
**reform** Reform *f*, Verbesserung *f* ~ **ability** Umformbarkeit *f* ~ **action** Rückbildung *f*
**reformed gasoline** reformiertes Benzin *n*
**reforming** Reformierung *f*, Umformung *f*
**refound, to** ~ umgießen, umschmelzen
**refounding** Umschmelzung *f*
**refract, to** ~ biegen, brechen, zerlegen **to** ~
**sound waves** Schallwellen *pl* ablenken
**refracted** Brechungswelle *f* ~ **ray** Brechungs-
strahl *m*, gebrochener Strahl *m*
**refracting** brechend, lichtbrechend ~ **medium**
brechendes Medium *n* ~ **system** Brechungs-
system *n*
**refraction** Beugen *n*, Beugung *f*, Brechung *f*
(opt.), Brechungserscheinung *f*, Refraktion *f*,
Zurückwerfung *f* ~ **by droplets** Brechung *f* an
Tröpfchen ~ **of light** Lichtbrechung *f* ~ **of rays**
Strahl(en)brechung *f* ~ **of a traveling wave**
Brechung *f* einer Welle ~ **of ultrasound waves**
Brechung *f* von Ultraschallwellen ~ **of waves**
Wellenrefraktion *f*
**refraction, ~ constant** Brechungskonstante *f* ~
**loss** Brechungsverlust *m* (acoust.) ~ **meter**
Brechungsmesser *m* (seismic) ~ **method** seismische Brechungsmethode *f* ~ **unit** Refrak-
tionssäule *f*
**refractionometer** Refraktionometer *n*
**refractive** lichtbrechend ~ **index** Brechungszahl
*f* ~ **medium** brechendes Mittel *n* ~ **modulus**
Brechungsmodul *m* ~ **potential** Brechungsan-
teil *m* des Potentials ~ **power** Brechungsver-
mögen *n*, Lichtbrechungsvermögen *n* (opt.)
**refractivity** Brechungs-exponent *m*, -vermögen *n*,
Lichtbrechungskörper *m*
**refractometer** Brechungsmesser *m*, Strahlen-
brechungsmesser *m* ~ **prism** Meßprisma *n*
**refractometry** Brechungsindexbestimmung *f*
**refractor** Lichtbrechungskörper *m*, Refraktor *m*
**refractoriness** Feuer-beständigkeit *f*, -festigkeit *f*,
Zähflüssigkeit *f* ~ **under load** Druckfeuerbe-
ständigkeit *f*
**refractory** feuerbeständig, feuerfest, hitzebeständig, refraktär, schwer schmelzbar, strengflüssig, widersetzlich, widerspenstig, widerstandsfähig ~ **to fire** feuersicher, kaltbläsig
(metall)
**refractory, ~ building material** feuerfester Baustoff *m* ~ **cement** feuerfester Zement *m* ~ **clay**
Schamotteton *m* ~ **glaze** strengflüssige Glasur *f*
~ **gold ore** vererztes Gold *n* ~ **lining** feuerfestes
Futter *n* ~ **materials** Baustoffe *pl* ~ **mortar**
feuerfester Mörtel *m*, Feuerzement *m* ~ **ore**

strengflüssiges Erz *n* ~ **products** feuerfeste Materialien *pl* ~ **property (or quality)** Feuerbeständigkeit *f* ~ **sand** Glühsand *m* ~ **stone** feuerfester Stein *m*
**refrain, to** ~ sich enthalten **to** ~ **from** Abstand nehmen von, unterlassen
**refreezing** Regelation *f*
**refresh, to** ~ auffrischen, (battery) aufladen, erneuern
**refresher instruction** Wiederholungsunterricht *m*
**refreshment** Erfrischung *f*, Labung *f*
**refrigerant** Abkühlungsmittel *n*, Kühlmittel *n* ~**-metering device** Drosselorgan *n* (Kühlmaschine)
**refrigerate, to** ~ abkühlen, gefrieren, zum Gefrieren bringen
**refrigerated** gekühlt ~ **display case** Kühlvitrine *f* ~ **(mercury vapor) trap** Ausfrierfalle *f* ~ **wind tunnel** gekühlter Windtunnel *m*
**refrigerating,** ~ **capacity** Kälteleistung *f* ~ **chamber** Kühl-kammer *f*, -raum *m* ~ **effect** Kühlwirkung *f* ~ **machine** Kälte-, Kühlmaschine *f* ~ **machinery** Eis- und Kühlmaschine *f* ~ **pipe** Kühlrohr *n* ~ **plant** Gefrieranlage *f*, Kühlwerk *n* ~ **storage space capacity** Kühlrauminhalt *m*
**refrigeration** Abkühlung *f*, Gefrierung *f*, Kältetechnik *f*, Kühlung *f* ~ **plant** Gefrier-, Kühlanlage *f* ~ **unit** Kälteerzeugungsanlage *f*, Kühlaggregat *n*
**refrigerator** Gefrier-maschine *f*, -schrank *m*, Kältemaschine *f*, Kühl-apparat *m*, -faß *n*, -maschine *f*, -schrank *m*, Kühler *m*, Refrigerant *m* ~ **capacity** Kühlvermögen *n* der Kühlanlage ~ **car** Kühlwagen *m* ~ **carriage** Kühlwaggon *m* (r. r.) ~ **device** Kühlvorrichtung *f* ~ **package** Klimagerätesatz *m* ~ **screen** Kühlschrankblende *f* ~ **truck** Kühlwagen *m*
**refringent** lichtbrechend
**reftone** Normalton *m* (acoust.)
**refuel, to** ~ auftanken, (Kraftstoffbehälter *m*) nachfüllen, (in flight) nachtanken, tanken
**refuel,** ~ **connection** Nachtankkupplung *f* ~ **manifold (line)** Betankungsverteiler *m*
**refueling** Auftanken *n*, Betriebsstoffergänzung *f*, Brennstoffaufnahme *f* ~**-and-catapult-launching ship** schwimmender Flugstützpunkt *m* ~ **circuit** Betankungsstromkreis *m* ~ **device** Auftankvorrichtung *f* ~ **during flight** Lufttanken *n* ~ **proke** Betankungsausleger *m* ~ **unit** Tankvorrichtung *f*
**refuge** Fußgängerschutzinsel *f*, Hafen *m*, Schutz *m*
**refund, to** ~ ersetzen, vergüten, zurückerstatten **to** ~ **a charge** Gebühr *f* erstatten
**refund** Rück-vergütung *f*, -zahlung *f*
**refusal** Ablehnung *f*, abschlägige Antwort *f*, abschläger Bescheid *m*, Abweisung *f*, Beanstandung *f*, Versagen *n*, Verweigerung *f*, Weigerung *f* ~ **of acceptance** Annahmeverweigerung *f* ~ **to obey orders** Widersetzlichkeit *f* ~ **to work** Arbeitsverweigerung *f*
**refusal speed** Startabbruchgeschwindigkeit *f* (aviat.)
**re-fuse, to** ~ umschmelzen, wiederzusammenschmelzen
**refuse, to** ~ ablehnen, abweisen, ausschlagen, versagen, verweigern, weigern, wiederschmel-

zen, zurückweisen **to** ~ **acceptance** beanstanden **to** ~ **a call** eine Verbindung *f* ablehnen, verweigern
**refuse** Abfall *m*, Abschaum *m*, Ausschuß *m*, Berge *pl*, Gekrätz *n*, Kehricht *m*, Müll *n*, Rückstand *m*, Schund *m*, Schutt *m* ~ **coke** Koksabfall *m* ~**-collecting plant** Abfallsammelanlage *f* ~ **collection** Müllabfuhr *f* ~ **destructor** Müllverbrennungsanlage *f* ~**-disposal plant** Abfallvernichtungsanlage *f* **(house-)** ~ **dressing** Müllverwertung *f* ~ **exhauster** Müllsauger *m* ~ **gas lime** Grünkalk *m* ~ **gasification** Bergevergasung *f* ~ **gate** Austragschieber *m* ~ **incinerator** Müllverbrennungsanlage *f* ~ **lighter** Abfallprahm *m* ~ **oil** Nachöl *n* ~ **paper** Altpapier *n*, Papiermaschinenausschuß *m* ~**-removing cart** Straßenkehrrichtabfuhrwagen *m* ~ **sewage plant** Abfallvernichtungsanlage *f* ~ **spout** Abfallrinne *f* ~ **wagon** Müllast(kraft)wagen *m*
**refused call** abgelehntes Gespräch *n* (teleph.)
**refutation** Widerlegung *f*
**regain, to** ~ wiedergewinnen
**regard, to** ~ betrachten
**regard** Achtung *f*, Berücksichtigung *f*, Betreff *m*, Beziehung *f*, Hinsicht *f*, Hochachtung *f*, Rücksicht *f* **in this** ~ in dieser Beziehung, in dieser Hinsicht **in** ~ **to** in Hinsicht auf
**regarding** betreffend, hinsichtlich
**regardless** ohne Rücksicht auf, unbeschadet
**regelate, to** ~ wiedergefrieren
**regelation** Regelation *f*, Wiedergefrieren *n*
**regenerable** regenerierbar ~ **cell** aufladbares Element *n*
**regenerate, to** ~ auffrischen, entzerren, erneuern, regenerieren, rückkoppeln, verjüngen, wiederbelegen, wieder einschreiben (data proc.)
**regenerated** zurückgewonnen ~ **cellulose** Zellwolle *f* ~ **dense medium** eingedickte Trübe *f* ~ **energy** zurückgewonnene Energie *f* ~ **fuel** aufgearbeitetes Spaltmaterial *n* ~ **material** Gefrisch *n* (rubber or oil), Regenerat *n* ~ **rubber** Regeneratgummi *m*
**regeneration** Auffrischen *n*, Dämpfungsverminderung *f*, Entdämpfung *f*, Entzerrung *f* (rdo), Nachladung *f*, Neubildung *f*, Regeneration *f*, Regenerierung *f*, Rück-bildung *f*, -gewinnung *f*, -kopplung *f*, -kopplungsverstärkung *f*, -wirkung *f*, Verjüngung *f*, Wieder-belebung *f*, -gewinnung *f*, Wiederbrauchbarmachen *n* (chem.) ~ **of current** Stromrückgewinnung *f*
**regeneration,** ~ **crucible furnace** Tiegelofen *m* mit Umschaltfeuerung ~ **distortion** Rückkupplungsverzerrung *f* ~ **mark** Regenerationsmarke *f* ~ **period** Regenerierungsintervall *n*
**regenerative** rückkoppelnd ~ **amplification** Anfachung *f*, Dämpfungs-reduktion *f*, -verminderung *f*, Entdämpfung *f*, Rückkopplungsverstärker *m* ~ **amplifier** rückgekoppelter Verstärker *m* ~ **braking** Nutzbremsung *f* ~ **chamber** Kammer *f*, Regenerativkammer *f* ~ **circuit** Rückkopplungs-kreis *m*, -schaltung *f* ~ **coefficient of coupling** Rückkuppelungsfaktor *m* ~ **coupling** Rückkupplung *f* ~ **cycle** Zwischendruckverfahren *n* ~ **detector** Rückkopplungsdetektor *m* ~ **exchange operation** (in gas liquefaction) Umschaltwechselbetrieb *m* ~ **feedback** Mitkopplung *f*, Rückkopplung *f* ~ **firing** Regenerativ-, Speicher-, Wärmespeicher-

feuerung *f* ~ **furnace** Abgasspeicherofen *m*,
Regenerativ-feuerung *f*, -ofen *m*, Speicherofen
*m*, Umschaltfeuerung *f*, Wärmespeicherofen *m*
~ **gas furnace** Regenerativ-flammofen *m*, -gas-
ofen *m*, Umschaltgasfeuerung *f* ~ **grid-current
detector** rückgekuppeltes Audion *n*, Rück-
kupplungsaudion *n* ~ **hearth** Regenerativ-
feuerung *f* ~ **open-hearth furnace** Regenerativofen *m* ~ **oscillator** Rückkopplungsgenerator *m*
~ **oven** Regenerativofen *m* ~ **process** Regene-
rierungsverfahren *n* ~ **reactor** Produktions-
reaktor *m* ~ **receiver** Rückkopplungsempfänger
*m* ~ **reception** Rückkopplungsempfang *m* ~
**repeater** entzerrende Übertragung *f*, Tele-
grafenübertragung *f* mit Berichtigung der
Zeichenform ~ **rocket motor** Wiedergewin-
nungsmotor *m* ~ **set** Rückkopplungsempfänger
*m* ~ **system** Umschaltsystem *n* ~ **valve detector**
Rückkopplungsaudion *n*
**regenerator** Regenerativfeuerung *f*, Regenerator
*m*, Umschaltflammofen *m*, Wärmespeicher *m*,
Winderhitzer *m*, Winderhitzungsapparat *m* ~
**coil** Vorwärmerschlange *f*
**regime** Betriebsbedingungen *pl*
**regimen** Beharrungszustand *m*
**region** Distrikt *m*, Gebiet *n*, Gebietsteil *m*,
Gegend *f*, Gelände *n*, Land *n*, Ort *m*, Umgebung *f*, Zone *f*
**region,** ~ **of ascending currents** Aufwindfeld *n*
~ **of change of form** Formänderungsbereich *m*
~ **of combustion instability** Verbrennungsbe-
reichunbeständigkeit *f* ~ **of convergence** Kon-
vergenzgebiet *n*, Körper *m* ~ **of deformation of
form** Formänderungsbereich *m* ~ **of descending
currents** Abwindfeld *n* ~ **of disturbance**
Schüttergebiet *n*, Störungs-bereich *m*, -gebiet *n*
~ **of downward currents** Abwindfeld *n* ~ **of
flow** Strömungsgebiet *n* ~ **of incipient current
flow** Anlaufstromgebiet *n* ~ **of interference**
Verwaschungszone *f* ~ **of radiation** Strahlen-
bereich *m* ~ **of reduction** Reduktionszone *f* ~
**of saturation** Sättigungsgebiet *n* ~ **of silence**
Auslöschzone *f*, Totraum *m* ~ **of stability** Stabilitätsbereich *m* ~ **of steady flow** Beruhigungsstrecke *f* ~ **of upward currents** Aufwindfeld *n*
**region sheltered from the (prevailing) winds**
Windschatten *m*
**regional** gebietsweise, räumlich begrenzt ~
**broadcasting station** Leitsender *m* ~ **chart**
regionale Gebietskarte ~ **frontier** Binnengrenze
*f* ~ **guard frequency** regionale Wachfrequenz *f*
~ **manager** Bezirksleiter *m* ~ **metamorphism**
Belastungsmetamorphose *f*
**register, to** ~ anzeigen, aufeinanderpassen
(aviat.), aufführen, aufspeichern, aufzeichnen,
buchen, einrasten, einschreiben, eintragen,
hinterlegen, registrieren, schreiben, (one part
with another) übereinstimmen, verzeichnen,
**to** ~ **hits** Treffer *pl* verzeichnen **to** ~ **by instruments** selbstschreiben **to** ~ **a trade mark** Fabrikzeichen *n* anmelden
**register** Anschlag *m*, Eintragebuch *n*, Impuls-
speicher *m*, Lüftungsschieber *m*, Nummern-
speicher *m*, Raststellung *f*, Register *n*, Rolle *f*,
Seitenanzeiger *m*, (digit-storing) Stromstoß-
speicher *m*, Verzeichnis *n*, Zähler *m*, Zählrelais
*n*, Zählwerk *n* ~ **of dues and taxes** Heberolle *f*
~ **of land ownership** Grundbuch *n*

**register,** ~ **difference** Paßdifferenz *f* (print.) ~
**factor** Zählfaktor *m* ~ **form** Paßform *f* ~ **key**
Zähltaste *f* ~ **length** Register-kapazität *f*,
-länge *f* ~ **number** Rollennummer *f* ~ **pilot
lamp** Zahlüberwachungslampe *f* ~ **punching**
Ausrichtlochung *f* ~ **reading** Zählerablesung *f*
~ **rotation** zyklische Versetzung *f* ~ **sheet** Zurichtebogen *m* (print.) ~ **surveying** Grundbuch-
vermessung *f* ~ **tube** Einpaßrohr *n*
**registered** eingeschrieben, eingetragen ~ **address**
Telegrammkurzanschrift *f* ~ **capital** eingetragenes Kapital *n* ~ **design** Gebrauchsmuster
*n* ~ **letter** eingeschriebener Brief *m* ~ **name** gesetzlich geschützter Name *m* ~ **office** Hauptniederlassung *f* ~ **pattern** Gebrauchsmuster *n*,
Musterschutz *m* ~ **trade-mark** eingetragene
Schutzmarke *f*, eingetragenes Warenzeichen *n*
**registering** Zählung *f* ~ **balloon** Ballonsonde *f*,
Registrier-, Stich-ballon *m* ~ **bar** Anschlagleiste *f* ~ **frame** Anlegerahmen *m*, Rahmen *m*
~**-frame center** Rahmenmittelpunkt *m* ~ **ledge**
Auflageleiste *f* ~ **mark** Kennmarke *f*, Kennzeichen *n* ~ **(recording) mechanism** Anzeigemechanismus *m* ~ **rim** Auflageleiste *f* ~
**shoulder** Auflageleiste *f* ~ **sight** Peilgerät *n* ~
**stop** Anschlagleiste *f* ~ **straight-edge** Anlege-
leiste *f* ~ **strap** Anschlagleiste *f*
**registrar** Protokollführer *m*
**registration** Einschreiben *n*, Eintragung *f* ~ **of a
call** Gesprächsanmeldung *f* ~ **of film** Film-
justierung *f*, Filmregistrierhaltigkeit *f*
**registration,** ~ **certificate** Eintragungsschein *m*
~ **fire** Kontrollschießen *n* ~ **marks** Zulassungs-
zeichen *n* ~ **office** Anmeldeplatz *m*, Meldeplatz
*m* ~ **pin** Greiferstift *m*, Justierstift *m* Justierstift *m* (film, print.) ~ **stability** Registrierhaltig-
keit *f*
**registrogram** Registrogramm *n*
**registry** Deckung *f* ~ **fee** Anmelde-, Eintragungs-
gebühr *f*
**reglet** Linie *f* (print.) ~ **casting mold** Regletten-
gießform *f*
**regrain, to** ~ nach-krispeln, -körnen
**regression** Rückbildung *f*
**regressive** rückgängig ~ **light** regredientes Licht
*n* ~ **mutation** rückläufige Knotendrehung *f* ~
**wave** rückschreitende Welle *f*
**regrind, to** ~ nachschleifen
**regrind** Nachschliff *m*
**regrinding** Feinzerkleinerung *f*, Nachschleifen *n*
**regroup, to** ~ umlagern
**regrouping** Umgliedern *n*
**regulable** nachstellbar, regulierbar ~ **grid voltage** Gittereinstellung *f*
**regular** aktiv, isometrisch, regelmäßig, regelrecht, regulär, serienmäßig, stufig ~ **air service**
regelmäßige Luftverbindung *f* ~ **air traffic**
regelmäßiger Luftverkehr *m* ~ **alignment of
letters** Zeilengeradheit *f* ~ **battery connection**
Verbandanode *f* ~ **equipment** serienmäßig eingebaute Ausrüstung *f* ~ **geometrical outline**
regelmäßige Fläche *f* ~ **hexagonal nut** Sechskantmutter *f* ~ **lay rope** Kreuzschlagseil *n* ~
**octagon** regelmäßiges Achteck *n* ~ **point of a
curve** regulärer Kurvenpunkt *m* ~ **point of a
surface** regulärer Flächenpunkt *m* ~ **polygon**
regelmäßiges Vieleck *n* ~ **reflection** gerichtete
Reflexion, Rückspiegelung *f*, spiegelnde Rück-

strahlung *f*, Spiegelreflexion *f* ~ **site** Gitterplatz *m* ~ **system of points** reguläres Punkt-
system *n* ~ **train** fahrplanmäßiger Zug *m*
**regularity** Gleich-förmigkeit *f*, -mäßigkeit *f*,
Regelmäßigkeit *f*, Reglung *f* ~ **of features**
Gesetzmäßigkeit *f* ~ **attenuation** Rückflußdämpfung *f*
**regularization** Vereinheitlichung *f*
**regularize, to** ~ stabilisieren
**regularly, not** ~ **prescribed** unvorschriftsmäßig
**regulate, to** ~ anrichten, einregeln, einregulieren, einstellen, nachregulieren, nachstellen,
normalisieren, normieren, ordnen, regeln, regulieren, stellen, steuern **to** ~ **down (or in a downward direction)** herunter -regeln, -steuern
**regulated** ausgeglichen ~**-filament current** heizabgestimmter Strom *m* ~ **power supply** span-
nungsregelndes Netzgerät *n* ~ **quantity** (servo)
Stellgröße *f*
**regulating** Regulieren *n*, Regulierung *f* ~ **of
performance** Regulierung *f* der Leistung
**regulating,** ~ **apparatus** Regelapparatur *f* ~
**attenuator** Regelglied *n* ~ **cathode** Regulierwiderstand *m* ~ **characteristic** Regelkennlinie *f*
~ **cock** Abstell-, Regulier-, Stell-hahn *m* ~
**controls** Reguliergestänge *n* ~ **damper** Einstellschieber *m* (Rauchgase) ~ **device** Regeleinrichtung *f*, Regulierungsvorrichtung *f* ~ **dikes**
Längsbauten *pl* ~ **equipment** Regelungsvor-
richtung *f* ~ **feed valve** Schieberdruckregler *m*
~ **flap** Regulierklappe *f* ~ **float valve** Schwim-
merregelventil *n* ~ **gate** Regulierungsschieber
*m* ~ **gear** Regulierapparat *m* ~ **inductor** Regeldrossel *f* ~ **line** Reglerlinie *f* ~ **machine** (Scherbius set) Belastungs- oder Antriebsmaschine *f*
~ **member** Regelglied *n* ~ **nut (screw)** Einstellmutter *f* ~ **point** Reglerpunkt *m* ~ **power** Verstellkraft *f* ~ **pulley** Regelscheibe *f* ~ **quantity**
Stellstrom *m* ~ **range** Regelbereich *m* ~ **ratio**
Stellgrad *m* ~ **resistance** Regulier-, Über-
schalt-widerstand *m* ~ **rod** Streichstange *f* ~
**screw** Stell(ring)schraube *f* ~ **sector** Einstellsektor *m*, Regelsegment *n* ~ **shaft** Reglerwelle
*f* ~ **siphon** Ablaßdüker *m* (hydraul.) ~ **speed**
Stellgeschwindigkeit *f* (autom. contr.) ~**-speed
gear** Reguliervorgelege *n* ~ **starter** Regelanlasser *m* ~ **switch** Regelschalter *m*, Regulier-
schalter *m*, -widerstand *m* ~ **tap** Regulierge-
windebohrer *m* ~ **transformer** Reguliertrans-
formator *m* ~ **unit** Stellglied *n* ~ **valve** Regelventil *n*, Regulier(ungs)ventil *n*, Sicherheits-
klappe *f* ~ **value of voltage regulator** Regulier-
wert *m* des Spannungsreglers ~ **voltage (or
tension)** Regulierspannung *f* ~ **wire resistances**
Drahtstellwiderstände *pl*
**regulation** Anleitung *f*, Anordnung *f*, Bestimmung *f*, Einstellung *f*, Einweisung *f*, Kontrolle
*f*, Ordnung *f*, Regel *f*, Reglung *f*, Regulierung
*f*, Steuerung *f*, Verordnung *f*, Verstellung *f*,
Vorschrift *f*
**regulation,** ~ **of consumption** Verbrauchslenkung
*f* ~ **of electrodes** Elektrodenregelung *f* ~ **of fire**
Feuerregelung *f* ~ **of flames** Flammenführung
*f* ~ **of the Gauging Office** eichbehördliche Vorschrift *f* ~ **of a governor in percent** Statik *f* eines
Reglers in Prozent ~ **of light intensity** Hellig-
keitsregulierung *f* ~ **of light volume** Helligkeits-
regulierung *f* ~ **of the liquor level** Laugen-

standsregulierung *f* ~ **of output** Ausgangs-
energieregelung *f* ~ **of pressure** Druck-reglung
*f*, -regulierung *f* ~ **of the radiator** Kühlerreglung *f* ~ **of scarce supplies** Bewirtschaftung *f*
~ **in steps** stufenweise Regelung *f* ~ **of torrents**
Wildbachverbauung *f* ~ **by valves** Abschützung
*f*
**regulation,** ~ **barrage** regelnder Abschluß *m* ~
**curve** Regelkurve *f* ~ **dam** regelnder Abschluß
*m* ~ **device of the hydraulic gears** Getrieberegel-
vorrichtung *f* ~ **field cap** Einheitsmütze *f* ~
**light** Regellicht *n* ~ **parachute** Dienstfallschirm
*m* ~ **ratio (servo)** Stellgrad *m* ~ **step** Regelstufe
*f* ~ **table** Spannungstafel *f*
**regulations** Betriebsvorschrift *f*, Dienstvorschrift *f* ~ **for gauging weights and measures**
Eichordnung *f* ~ **having the force of law** gesetzliche Regelungen *pl* ~ **for plant operation**
Betriebsbestimmungen *pl*
**regulations,** ~ **governing prizes** (nav.) Prisen-
ordnung *f* ~ **governing rates of allowances (or
pay scale)** Tarifordnung *f* ~ **relating to exchange** Wechselordnung *f*
**regulator** Gang-, Motor-regler *m*, Regel-ein-
richtung *f*, -gerät *n*, -vorrichtung *f*, Regler *m*,
Regulator *m*, Regulier-schraube *f*, -vorrichtung
*f*, Steuerung *f*, Ventil *n*, (Lokomotive) Zentri-
fugalregulator *m* ~ **for altitude** Höhenregler *m*
~ **with constant angular vane velocity** Vollge-
schwindigkeitsregler *m* ~ **with constant vane
angle** Hartlagenregler *m* ~ **for cross-feed** Stellvorrichtung *f* für den Quervorschub ~ **and
cutout switch** Reglerschalter *m* ~ **of effective air
pressure** Luftüberdruckregler *m* ~ **for electrolysis circuits** Badstromregler *m* für Elektrolyse
~ **for fuel quantity injected** Füllungsregler *m* ~
**for power feed** Selbstgangeindrücker *m* ~ **for
vertical feed** Stellvorrichtung *f* für den Auf- und
Niedergang
**regulator,** ~ **armature** Regleranker *m* ~ **base**
Reglersockel *m* ~ **box** Reglerkasten *m* ~
**changeover switch** Reglerumschalter *m* ~
**characteristic (curve)** Reglerkennlinie *f* ~ **core**
Reglerkern *m* ~ **current coil** Reglerstromspule *f*
~ **force** Reglerkraft *f* ~ **housing** Reglergehäuse
*n* ~ **lay-out** Reglerauslegung *f* ~ **spring** Einstell-
feder *f* ~ **tube** Kontrollrohr *n* ~ **valve** Reglerventil *n* ~ **valve rod** Zugstange *f* des Reglers ~
**voltage coil** Reglerspannungsspule *f* ~ **wheel**
Stellrad *n*
**reguline** kubisch (min.) regulär, regulinisch
(chem.) ~ **metal** kompaktes Metall *n*
**regulus** Lech *m* (chem.), Metallkorn *n*, Regulus
*m* ~ **of gold** Goldkönig *m* ~ **furnace** Regulusofen *m*
**rehabilitation** Wiedereinsetzung *f* in frühere
Rechte
**re-handling point** Umschlagplatz *m*
**rehang, to** ~ umhängen
**rehearsal** Probe *f* ~ **hall** Probesaal *m* ~
**reheat, to** ~ anlassen, nachhitzen, tempern,
wieder-anwärmen, -erhitzen, -erwärmen
**reheat** Nachbrenner *m*, Nachverbrennung *f* ~
**boiler** Abwärmekessel *m* ~ **effect** (jet) Aufheiz-
wirkung *f* ~ **superheater** Rücküberhitzer *m*
**reheater** Abgasvorwärmer *m*, Nacherhitzer *m*
**reheating** Wiedererhitzung *f* ~ **furnace** Glüh-,
Nachwärme-, Schweiß-, Wärm-ofen *m* ~**-fur-**

nace cinder Wärmofenschlacke f ~-furnace slag
Schweißofenschlacke f ~ hearth Schweißfeuer
m
**re-ignite, to** ~ sich wieder entzünden, wieder
zünden
**re-igniter chamber** (rocket) Wiederzündbrenn-
kammer f
**re-ignition** Neuzündung f, Wiederzündung f
**reimburse, to** ~ wiedererstatten
**reimbursement** Abfindung f, Entschädigung f,
Nachnahme f, Rückzahlung f ~ **of expenses**
Spesennachnahme f ~ **ratio** Vergütungssatz m
**rein** Zaum m ~ **and saddle girth** Zügel- und Sattelgurt m
**Reinartz circuit** Reinartzschaltung f
**reinforce, to** ~ absteifen, abstützen, (concrete)
armieren, aussteifen, bewehren, kräftigen,
nähren, verstärken, versteifen, verstreben **to** ~
**at the bearing** am Auflager n verstärken **to** ~
**with concrete** betonieren
**reinforce** Materialverstärkung f, Verstärkungs-
stück n
**reinforced** armiert, verstärkt **not** ~ unverstärkt
~ **with concrete** betoniert
**reinforced concertina roll** Drahtwalze f
**reinforced-concrete** armierter oder bewehrter
Beton m, Eisenbeton m ~ **bridge** Eisenbeton-
brücke f ~ **construction** Eisenbetonbau m ~
**facing** Verkleidung f von Eisenbeton ~ **floor**
Massivdecke f ~ **pile** Eisenbetonpfahl ~ **pipe**
Spannbetonröhre f ~ **pole** Eisenbeton-mast m,
-stange f ~ **support** untere Verkleidung f von
Eisenbeton ~ **turret** Eisenbetonkuppel f
**reinforced** ~ **core wall** Kernmauer f von Eisenbeton ~ **floor** bewehrte Sohle f ~ **glass** verstärk-
tes Glas n ~ **papers** Papyrolin n ~ **parallel-
truss superstructure** Parallelfachwerkträger m
~ **plaster** Pflaster n von Eisenbeton ~ **plywood**
Blocksperrholz n ~ **seam** Wulstnaht f ~ **weld**
volle Schweißnaht f, Wulstnaht f
**reinforcement** Absteifung f, Armierung f, Befestigung f, Verstärkung f, Versteifung f, Verstrebung f, Wulst m **(steel)** ~ **(in concrete)**
Bewehrung f ~ **of gradation** Gradationsstei-
gerung f ~ **of pole butt** Stockschutz m
**reinforcement,** ~ **plate** Verstärkungsblech n ~
**weld** Schweißwulst f ~ **wire mesh** Baustahlge-
webe n
**reinforcing** Aussteifung f, Bewehrung f ~ **of wire
netting** Bewehrung f von Drahtgeflecht
**reinforcing,** ~ **bar(s)** (reinforced-concrete construction) Bewehrungseisen n ~ **beam** Stützbalken m ~ **crease (or fin)** Sicke f ~ **iron** (for
concrete) Betoneisen n ~ **pad** Verstärkungs-
sohle f ~ **plate** Verstärkungsplatte f ~ **pleat**
Sicke f ~ **rib** Stütz-, Verstärkungs-rippe f ~
**ring** Verstärkungsring m ~ **seam** Sicke f ~
**spring** Verstärkungsfeder f ~ **steel** Eisen- oder
Stahleinlage f ~ **strengthener** (dentistry) Ver-
stärkungseinlage f ~ **strip** Verstärkungsprofil n
~ **web** Verstärkungssteg m
**reins** Schenkel m, Zügel m
**re-inserted** wiedereingefügt
**re-insertion** Wiedereinführung f ~ **of carrier**
Trägerzusatz m
**re-inspect, to** ~ nachkontrollieren
**re-install, to** ~ wiedereinbauen
**re-instate, to** ~ wiederinstandsetzen

**re-insurance** Rückversicherung f
**re-integration** Rückgliederung f
**re-introduce, to** ~ wiedereinführen
**re-issue, to** ~ wiederherausgeben
**re-issue** Neuausgabe f, Wiederanmeldung f ~
**patent** Abänderungspatent n
**re-iterable** wiederholbar
**re-iterate, to** ~ repetieren, (Theodolit) wiederholen
**re-iteration** Verstellung f, Widerdruck m, Wiederholung f
**reject, to** ~ ablehnen, abschlagen, abweisen,
ausmustern, als Ausschuß ausscheiden, auswerfen, beanstanden, verweigern, verwerfen,
wegwerfen, zurückweisen
**reject,** ~ **card** ausgeworfene Karte f ~ **counter**
Schrottschautisch m ~ **pocket** Restablagefach
n
**rejected,** ~ **as unfit** ausgemustert ~ **material** Ausschuß m
**rejecter** Stößer m
**rejection** Ablehnung f, Abnahmeverweigerung f,
Abweisung f, Ausschuß m, Ausstoßung f, Aus-
werfung f, Beanstandung f, Verwerfung f,
Weigerung f, Zurückweisung f ~ **band** Sperr-
band n ~ **factor** Schwächungsfaktor m
**rejector** Ausstoßer m, Rückwerfer m, Sperre f ~
**circuit** Ausstoßer m, Drossel-kreis m, -satz m,
Parallelresonanzkreis m, Sperr-filter n, -kreis m,
-sieb n, -stromkreis m, Wellenschlucker m
~-**circuit air coil** Sperrkreisluftspule f
**rejects** Abgang m, Ausschuß m
**rejoinder** (in an appeal) Beschwerdeeinrede f,
Erwiderung f, Gegen-, Rück-antwort f
**rejoiner** Gegenschrift f
**rejuvenate, to** ~ regenerieren, verjüngen
**rejuvenation** Neuaktivierung f, Verjüngung f ~
**of thoriated cathode** Wiederbelebung f der
thorierten Kathode
**relapping** Ventileinschleifen n
**relapse** Rückfahrt f
**relatch, to** ~ wiederverriegeln, wieder verriegeln
**relatching** Wiederverriegelung f
**relate, to** ~ berichten, beziehen, erzählen **to** ~
**to** Bezug haben auf, in Beziehung setzen
**related** anverwandt, verwandt ~ **to** bezogen auf
~ **color** bezogene Farbe f ~ **companies** Schwe-
sterunternehmen n
**relating** (thereto) diesbezüglich ~ **to** mit Bezug
auf ~ **to business** geschäftlich
**relation** Ansatz m, (ship) Bezeichnung f, Beziehung f, Bezug m, Verbindung f, Verhältnis n,
Verwandschaft f, Zusammenhang m **to be in** ~
**with** in Beziehung stehen ~ **between coefficients** Koeffizientenbeziehung f ~ **between constants** Konstantenbeziehung f ~ **between
pressure and density** Adiabatengleichung f ~ **of
the inclination** Neigungsverhältnis n ~ **of levers**
Hebeverhältnis n
**relationship** Abhängigkeit f, Beziehung f, Verhältnis n, Verwandtschaft f ~ **between energy
and mass** Energiemassenbeziehung f ~ **between
gauge readings at different stations** Pegelbe-
ziehung f
**relative** bezüglich, relativ ~ **to** sich beziehend
auf, bezogen auf, bezüglich ~ **to high frequency**
hochfrequenzmäßig ~ **to vibration** schwin-
gungstechnisch

**relative** Verwandtes *n*, verwandtes Derivat *n* (chem.) ~ **address** relative Adresse *f* ~ **amounts** Mengungsverhältnis *n* ~ **aperture** Öffnungsverhältnis *n*, relative Öffnung *f* ~ **aperture of the lens** Öffnungsverhältnis *n* eines Objektivs, relative Öffnung *f* eines Objektivs ~ **bearing** relative Peilung *f*, wahre Seitenrichtung *f* ~ **deformation** Formänderung *f* je Meßlängeneinheit *f* ~ **density** Dichteverhältnis *n* ~ **efficiency** relative Dämpfung *f*, relativer Nutzeffekt *m* ~ **efficiency of biplane wings** relative Leistungsfähigkeit *f* der Doppeldeckerflügel ~ **elevation** Höhenzahl *f*, Meereshöhenzahl *f* ~ **flying height** Flughöhe *f* über Grund, relative Flughöhe *f* ~ **frequency** Häufigkeitskoeffizient *m*, relative Häufigkeit *f* ~ **harmonic content** Oberschwingungsgehalt *m* ~ **humidity** relative Feuchtigkeit *f* ~ **inclinometer** relativer Neigungsmesser *m* ~ **intelligibility** relative Sinnverständlichkeit *f* ~ **interfering effect** relative Störung *f* ~ **level** Bezugspegel *m* ~ **luminosity curve** spektrale Hellempfindlichkeitskurve *f* ~ **luminosity factor** spektrale Hellempfindlichkeit *f* ~ **moisture** relative Feuchtigkeit *f* ~ **motion** Relativbewegung *f*, relative Bewegung *f* ~ **path** Relativweg *m* ~ **period** Begegnungsperiode *f* ~ **permeability** relative Permeabilität *f* ~ **plateau slope** relative Plateauneignung *f* ~ **refractive index** relativer Brechungsindex *m* ~ **retardation** Wegdifferenz *f* ~ **sensitivity** relative Empfindlichkeit *f* ~ **sizes** Größenverhältnis *n* ~ **spectral curve** Kurve *f* der spektralen Energieverteilung *f* ~ **speed** Relativgeschwindigkeit *f* ~ **stopping power** relatives Bremsvermögen *n* ~ **value** Bezugswert *m* ~ **velocity** bezogene Geschwindigkeit *f*, Relativgeschwindigkeit *f* ~ **velocity based on smooth flow** relative Geschwindigkeit *f* bezogen auf ungestörte Strömung ~ **voltage drop** relativer Spannungs(ab)fall *m* ~ **voltage response of the exciter** (Nenn-)Erregungsgeschwindigkeit *f* ~ **voltage rise** relative Spannungssteigerung *f* ~ **wind** Fahrtwind *m* ~ **wind for flying** Flugwind *m* ~ **wind direction** relative Anströmrichtung *f*

**relatively** beziehungsweise **(any)** ~ **immovable point (or surface) sustaining pressure** Gegenlager *n*

**relativistic** relativistisch

**relativity** Relativität *f* ~ **principle** Relativitätsprinzip *n*

**relax, to** ~ abspannen, auflockern, entspannen lockern, nachlassen **to** ~ **a spring** eine Feder *f* entspannen

**relaxation** Entspannung *f* ~ **circuit** Kippkreis *m*, Röhrenkippschaltung *f* ~ **distance** Abbremsungslänge *f* ~ **frequency** Kippfrequenz *f* ~ **generator** Kippschwingungsgenerator *m* ~ **inverter** Kippschwingungswandler *m*, Wechselrichter *m* ~ **mode** Relaxationslösung *f* ~ **oscillation** Intermittenz-, Kipp-, Relaxationsschwingung *f* ~ **-oscillation connection** Kippschwingungsschaltung *f* ~ **oscillatons** Kippschwingungen *pl* ~ **oscillator** Kipp-gerät *n*, -schaltung *f*, -schwingschaltung *f*, Kippschwingungs-gerät *n*, -oszillator *m* ~ **part** Relaxationsanteil *m* ~ **period** Kippschwingungsdauer *f* ~ **scanning** Kippablenkung *f* ~ **time** Einstellzeit *f* ~ **time of ionic atmosphere**

Relaxationszeit *f* der Ionenwolke ~ **vibration** Kippschwingung *f* ~ **wave** Intermittenzschwingung *f*

**relaxed,** ~ **eye** entspanntes oder fernakkomodiertes Auge *n* ~ **state of accomodation** Akkomodationsruhe *f*

**relaxing** (of springs) Entspannung *f* ~ **of adaptation** Akkomodationsentspannung *f* ~ **of a spring** Entspannen *n* einer Feder

**relaxing media** relaxierende Körper *pl*

**relay, to** ~ ansprechen, mit Relais *n* übertragen, übertragen, (paving) umlegen, weitergeben **to** ~ **a cable** Kabel *n* verlegen **to** ~ **completely** vollständig umlegen **to** ~ **a message** Spruch *m* weiterleiten **to** ~ **a roof** Dach *n* umdecken

**relay** Abfallverzögerung *f* (electr.), Hilfstriebwerk *n*, Kraftflied *n*, Relais *n*, Schütz *n*, Weitergabe *f* **non metering** ~ Zählverminderungsrelais *n* **quick operating** ~ schnell ansprechendes Relais *n* ~ **and breaker panel** Relais- und Automatentafel *f* ~ **of clearing section** Abrückrelais *n* ~ **with holding winding** Relais *n* mit Haltewicklung *n* ~ **with magnetic shunt** Relais *n* mit magnetischem Nebenschluß *n* ~ **of runners** Meldekette *f* ~ **with sequence action** Stufenrelais *n* **the** ~ **is sticking** das Relais klebt

**relais,** ~ **armature** Relaisanker *m* ~ **automatic telephone system** Relaissystem *n* ~ **bay** Relaisgestell *n* ~ **blinker signaling** Blinken *n* über eine Zwischenstelle ~ **board** Relaistafel *f* ~ **box** Relaiskasten *m* ~ **chain** Relaiskette *f* ~ **charge** Schlagladung *f* ~ **chatter** Prellerscheinung *f*, Relaisflattern *n* ~ **coil** Relaisspule *f* (electr.) ~ **contact** Relaiskontakt *m* ~ **controlling local bell circuits** Weckerrelais *n* ~ **core** Relaiskern *m* ~ **equipment** Verteilungsanlage *f* ~ **finder** Relaisanrufsucher *m* ~ **group** Relaissatz *m* ~ **interrupter** Relaisunterbrecher *m* ~ **key** Tastrelais *n* ~ **line finder** Relais-Anrufsucher *m* ~ **-operated controller** Regler *m* mit Hilfsenergie *m* ~ **operating quantity** Relaisfunktion *f* ~ **phase sequence** Relaisphasenfolge *f* ~ **pick-up voltage** Relaisansprechspannung *f* ~ **point** Relaiskontakt *m*, Umschlagstelle *f* ~ **post** Zwischenstelle *f* ~ **preselector** Relaisvorwähler *m* ~ **rack** Relaisgestell *n* ~ **repeater** Relaisübertragung *f* ~ **releasing time** Relaisabfallzeit *f* ~ **responsive to kilovars** Blindleistungsrelais *n* ~ **screen** Relaisschirm *n* ~ **sender** Relaiszahlengeber *m* ~ **set** Relaissatz *m* ~ **station** Relais-sender *m*, -station *f*, Übertragungsfunkstelle *f*, Zwischensender *m* ~ **switch** Schaltschütz *n* ~ **time delay** Schaltzeit *f* ~ **-type** relaisbetätigt ~ **-type echo suppressor** unstetig arbeitende Echosperre *f* ~ **unaffected by alternating current** gegen Wechselstrom unempfindliches Relais *n* ~ **unit** Relaissatz *m* ~ **valve** Luftsteuerrelais *n* ~ **winding attachment** Relaiswickelzusatz *m* ~ **windings** Bespulung *f* eines Relais', Bewicklung *f* ~ **-working diagram** Erregungsdiagramm *n*

**relayed** mit Relais *n* ausgerüstet ~ **ringing** Rufen *n* mit Durchrufrelais

**relaying,** ~ **sounder** Übertragungsklopfer *m* ~ **station** Ballsender *m*, Ballsendestation *f*

**releasable** lösbar

**release, to** ~ abfallen, ablaufen lassen, abspannen, abwerfen, auflösen, ausklinken, auskuppeln, auslassen, (shutter) auslösen, ausrasten (tape rec.), ausrücken, ausschalten, ausschnappen, entbinden, entlassen, entlasten, entriegeln, freigeben, (a model) herausbringen, lösen, loslassen, loslösen, slippen, trennen, (a relay) vollständig abfallen, zurückstellen **to** ~ **bombs** Bomben *pl* abwerfen **to** ~ **the brake** die Bremse lösen **to** ~ **capital** Kapital *n* flüssig machen **to** ~ **carrier pigeons** Brieftauben *pl* auslassen **to** ~ **the catch** entriegeln **to** ~ **a connection** eine Verbindung *f* aufheben **to** ~ **the dial** die Nummernscheibe ablaufen lassen **to** ~ **gas** abblasen **to** ~ **the parachute** den Fallschirm *m* öffnen **to** ~ **by pressure lever** abdrücken **to** ~ **the safety arm** entsichern **to** ~ **the safety catch** entsichern **to** ~ **the tension** entspannen **to** ~ **the towing cable** das Schleppseil ausklinken **to** ~ **the trigger** betätigen

**release** Abfallen *n* (des Relais), (armature) Abwerfen *n*, Abwurf *m*, Auflösen *n*, Auflösung *f*, Ausklinkung *f*, Auslöse *f*, (lever) Auslöser *m*, Auslösung *f*, Ausrücken *n*, Befreiung *f*, Entbindung *f*, Entkuppelung *f*, Entlassung *f*, Entlastung *f*, Entriegelung *f*, Erledigungsschein *m*, Freigabe *f*, Rückstellung *f*

**release,** ~ **of the coupling** Auslösung *f* (Ausrückung) der Kupplung ~ **of energy** Auslösung *f* der Energie ~ **of film** Filmvertrieb *m* ~ **of heat** Wärmeentbindung *f* ~ **of oscillations** (increment of oscillation) Anstoßen *n* der Schwingungen ~ **of parachute** Fallschirmausstoß *m* ~ **of pressure** Druckentlastung *f* ~ **and retarding equipment** (for shunting yard ridges) Ablauf- und Bremsanlagen *pl* ~ **of the shift key** (typewriter) Freigabe *f* der Umschaltetaste ~ **of smoke** Nebelabblasen *n* ~ **of space bands** Keilauslösung *f* ~ **and spacing magnet** Auslösemagnet und Fortschub *m*

**release,** ~ **bar** Auslösestange *f* ~ **bearing** Ausrücklager *n* ~ **bearing for clutch** Ausrückmuffe *f* ~ **bearing sleeve** Schleifbügel *m* ~ **button** Auslöseknopf *m* ~ **button for magnetic buttons** Auslösetaste *f* für Magnettasten ~ **cam** Auslösenocke *f* ~ **catch** Ausllsewippe *f* ~ **clutch** Rutschkupplung *f* ~ **contact** Ausllsekontakt *m* ~ **cord** Aufziehleine *f*, (parachute) Aufzugleine *f* ~ **device** (typewriter) Abstellvorrichtung *f* ~ **disk** Auslösescheibe *f* ~ **drive** Getriebeausrücken *n* ~ **(emitting) electrode** Auslöseelektrode *f* ~ **gear** Ausklinkvorrichtung *f*, Auslöseeinrichtung *f*, Entlastungsgetriebe *n*, Freilaß *m*, Klinkwerk *n* ~ **guard signal** Auslösungszeichen *n* ~ **handle** Auslösegriff *m* ~ **hook** Starthaken *m* ~ **key** Auslöse-, Lösch-taste *f*, Schalter *m* für die Platzlampe ~ **knob** Auslöseknopf *m* ~ **lag** (of relay) Abfallverzögerung *f* ~ **lever** Auslösehebel *m* ~ **magnet** Auslösemagnet *m*, Rückstell(elektro)magnet *m* ~ **mechanism** Abdrückvorrichtung *f*, Abfeuerungseinrichtung *f*, Auslöseeinrichtung *f* ~ **pin** Ausrückbolzen *m* ~ **point** Abschußpunkt *m* (rocket), (bomb) Ausklinkpunkt *m*, Auslösepunkt *m* ~ **position** Lösestellung *f* ~ **print** Ansicht-, Massen-kopie *f* ~ **printer** Kopieranlage *f* ~ **relay** Auslöserelais *n* ~ **retardation** Auslöseverzögerung *f* ~ **rod** Entriegelungsstange *f* ~ **screw** Entlüftungs-schraube *f* ~ **shaft cam** Preßrollen-Nocken *m* ~ **sleeve** Auslösemuffe *f* ~ **spring** Auslöse-, Rückzugs-feder *f* ~ **switch box** Abwurf-, Auslöse-schaltkasten *m* ~ **system** Ausklinkvorrichtung *f* ~ **trigger** Ausrückklinke *f* ~ **valve** Entlastungs-, Sicherheits-, Verminderungsventil *n* ~ **velocity** Auslaßgeschwindigkeit *f*

**released by censor** freigegeben

**releasing** Abfallen *n* (des Relais), Auslösung *f*; Vertrieb *m* ~ **and circulating overshot** lösbarer Zirkulations-Overshot *m* ~ **of electrons** Elektronenaustritt *m*

**releasing,** ~ **axle** Auslösewelle *f* ~ **cam** Auslösedaumen *m* ~ **current** Auslöse-, Rückstellstrom *m* ~ **device** Auslösevorrichtung *f* ~ **key** Freigabetaste *f* ~ **lever** Ausrücker *m* ~ **lug** (or lobe) Auslösenase *f* ~ **magnet** Lösungsmagnet *m* ~ **relay** Auslöserelais *n* ~ **rod** Auslöseschubstange *f* ~ **socket** lösbare Fangglocke *f* ~ **time** Abfallzeit *f*

**relegate, to** ~ beiordnen

**relentless** unnachgiebig

**relevant** anwendbar, Bezug haben, passend, sachdienlich, sachgemäß

**reliability** (of operation) Betriebssicherheit *f*, Kreditfähigkeit *f*, Zuverlässigkeit *f* ~ **of control** Schaltsicherheit *f* ~ **in service** Betriebszuverlässigkeit *f*

**reliability test** Zuverlässigkeits-probe *f*, -prüfung *f*

**reliable** (in operation) betriebssicher, einwellig, kreditwürdig, sicher, verläßlich, vertrauenswürdig, zuverlässig ~ **working** Betriebssicherheit *f*

**relic** Rest *m*, Überbleibsel *n*

**relief, to** ~**-mill** hinterfräsen

**relief** Abhilfe *f*, Ablösung *f*, Aussparung *f*, Austausch *m*, Entlastung *f*, Entsatz *m*, Entsetzung *f*, Erleichterung *f*, Hilfe *f*, (eccentric) Hinterdrehung *f*, Hochbild *n*, Höhendarstellung *f*, Plastik *f*, (draw) Relief *n*, Unterstützung *f*; erhaben ~ **of strain** (on pilot) Bedienungsentlastung *f*

**relief,** ~ **angle** Hinter-drehwinkel *m*, -stellwinkel *m* ~ **cock** (air) Entlüftungshahn *m*, Entlüftungsventil *n*, Zischhahn *m* ~ **curve** Hinterdrehkurve *f* ~ **effect** Bildplastik *f*, (pictures) Plastik *f* ~ **embossing** Mosaikdruck *m* ~ **embossing machine** Hochprägemaschine *f* ~ **engraving** Wetzgravur *f* ~ **etching** Hochätzung *f* ~ **grinding** Hinterschleifen *n* ~ **grinding attachment** Hinterschleifeinrichtung *f* ~**-grinding wheel** Hinterschleifscheibe *f* ~ **groove** Freistrich *m* ~**-ground worm hob** hinterschliffener Walzenfräser *m* ~ **hose** Entlüftungsschlauch *m* ~ **lever** Rückmeldehebel *m* ~ **map** Höhenkarte *f*, Reliefkarte *f*, Schichtenplan *m* ~ **mapping** Stereoplastik *f* ~ **milling device** Relieffräseinrichtung *f* (Autokartograf) ~ **notch** Entlastungskerbe *f* ~ **period** Erholungspause *f* ~ **pipe** Überdruckleitung *f* ~ **polishing** Reliefpolieren *n* ~ **port** Überlauföffnung *f* ~ **ports** Entlastungslöcher *pl* ~ **pressure valve** Überdruckventil *n* ~ **print** Hoch-, Präge-druck *m* ~ **printing** (embossing) Blinddruck *m* ~**-printing machine** Reliefdruckmaschine *f* ~ **road** Umgehungsstraße *f* ~ **sluices** Notschütz *n* ~ **spring** Entlastungsfeder *f* ~ **station** Aushilfsstelle *f* ~

**support** (of embossed film) Reliefträger *m* ~
**surface** Bodenprofil *n* ~ **valve** Ablaufventil *n*,
Entlüftungshahn *m*, Entspannungsventil *n*,
Rückflußventil *n*, Rückschlag-klappe *f*, -ventil
*n*, Schnüffelventil *n*, Stoßventil *n*, Überströmventil *n* ~-**valve jet** Druckbegrenzungsventil *n*
~ **wave** Entlastungsventil *n* ~ **well** Entlastungs-
brunnen *m* ~ **work** Notstandsarbeit *f*
**relieve, to** ~ abfallen, (in service) ablösen, auslösen, befreien, degagieren, eine Feder *f* entspannen, entheben, entlassen, entlasten, entsetzen, erleichtern, (eccentrically) hinterdrehen
**to** ~ **a circuit of some of its traffic** eine Leitung
*f* entlasten **to** ~ **eccentrically** hinterschleifen
/ **to** ~ **a spring** entspannen **to** ~ **of strain (or
tension)** abspannen
**relieve,** ~ **distortion** Entlastungsverdrehung *f*
~ **stresses** Warmbehandlung *f* (entspannen)
**relieved-tooth milling cutters** hinterdrehte Fräser *pl*
**reliever** Ablöser *m*
**relieving** Hinterarbeiten *n* ~ **of a spring** Entspannen *n* einer Feder
**relieving,** ~ **anode** Hilfsanode *f* ~ **arch** Ent-
lastungsbogen *m* ~ **attachment** Hinterdrehkurve *f* ~ **force** Entlastungskraft *f* ~ **gear** Ent-
lastungs-, Feststell-vorrichtung *f* ~ **lathe** Hinterdrehbank *f* ~ **piston** (turbine) Ausgleichkolben *m*, Entlastungskolben ~ **plate** Entlastungsplatte *f* ~ **test** Hinterdrehsupport *m* ~
**timbers** Auswechseln *n* (von Hölzern) ~ **work**
Hinterdreharbeit *f*
**relievo-engraving** Hochschätzung *f*
**relight, to** ~ luftanlassen, wieder anzünden,
wieder erleuchten
**reline, to** ~ ausbessern, ausflicken, flicken
**reliner** neue Stoffeinlage *f* eines reparierten Reifens
**relining** Ausbesserung *f*, Ausflicken *n*, (of
furnace) Besatz *m*, Flicken *n*
**relinquish, to** ~ aufgeben, ausliefern, verlassen,
verzichten
**relish, to** ~ Geschmack finden an
**relish** Geschmack *m*
**reload, to** ~ umladen, wiederbeschicken
**reload** Umladen *n* ~ **charges** Umlade-gebühr *f*,
-spesen *pl* ~ **point** Umschlagstelle *f*
**relocation** Verlegung *f*
**relubricate, to** ~ nachschmieren
**reluctance** magnetischer Widerstand *m*, Reluktanz *f*, Widerstand *m*
**reluctivity** Reluktivität *f*, reziproker Wert *m* der
Permeabilität, spezifischer magnetischer Widerstand *m*
**rely, to** ~ **on** sich stützen auf
**remain, to** ~ bleiben, haften, verweilen **to** ~ **in
gear** in Eingriff *m* bleiben **to** ~ **in a listing
condition** mit Schlagseite *f* liegenbleiben **to** ~
**over (or behind)** zurückbleiben **to** ~ **true**
gelten **to** ~ **workable** (Asphalt) bearbeitbar
bleiben
**remainder** Residuum *n*, Rest *m*, Rückstand *m*,
Saldo *m*, Überrest *m*
**remaining** überschüssig, übrig, verbleibend ~
**behind** zurückbleibend ~ **power (or term)** Rest-
glied *n* ~ **value method** Restwertmethode *f* ~
**velocity** Vekt, (ballistics) Änderungsgeschwindigkeit *f*

**remains** Rest *m*, Trümmer *pl*, Überbleibsel *n* ~
**of a furnace** Ofenansätze *pl*
**remake, to** ~ wiedermachen
**remaking** Wiederherstellung *f*
**remanence** Remanenz *f*, (of magnetism) Rest-
magnetismus *m*, (of magnetism) Restkraft *f*
**remanent** remanent ~ **deviation** Rest-ablenkung
*f*, -deviation *f* ~ **magnetism** remanenter Magnetismus *m*
**remanufactured wool** Kunstwolle *f*
**remark, to** ~ bemerken
**remark** Anmerkung *f*, Beachtung *f*, Bemerkung
*f*, Vermerk *m*
**remarkable** beachtlich, bemerkenswert, eigenartig, erheblich, kennbar, merkwürdig, sehens-
wert, wesentlich ~ **evenness** Gleichmäßigkeit *f*
**re-marking** Ausbesserung *f*
**Rembrandt illumination** Hinterbeleuchtung *f*
**remeasure, to** ~ übermessen
**remedial,** ~ **action** Abhilfe *f* ~ **measure** Hilfs-
maßnahme *f*
**remedy, to** ~ einer Sache abhelfen, Mangel *m*
oder Schaden *m* beheben
**remedy** Abhilfe *f*, Arzneimittel *n*, Hilfe *f*, Hilfsmittel *n*, Mittel *n*
**remelt, to** ~ umschmelzen, wieder(ein)schmelzen
**remelt,** ~ **metal** Umschmelzmetall *n* ~ **sirup (or
liquor)** Einschmelzsirup *m* ~ **strike** Sud *m* von
aufgelöstem Zucker ~ **sugar** Einwurfzucker *m*
**remelting** Umschmelzen *n*, Umschmelzung *f*,
Wiedereinschmelzen *n* ~ **furnace** Umschmelz-
ofen *m* ~ **pressure die-cast alloy** Umschmelz-
druckgußlegierung *f* ~ **process** Umschmelz-
verfahren *n*
**remesh, to** ~ wieder in Eingriff gelangen
**remetal, to** ~ **a bearing** ein Lager neu ausgießen
**remetalling bearings** Ausgießen *n* der Lager
**remilling** Nachdrehen *n*
**remind, to** ~ erinnern, mahnen
**reminder** Erinnerungsschreiben *n*, Mahnung *n*
**reminiscence** Erinnerung *f*
**reminiscent** die Erinnerung betreffend
**remission** Erlaß *m* ~ **of fees (or rates)** Gebührenermäßigung *f* ~ **of sentence** Ermäßigung *f* der
Strafe
**remit, to** ~ einsenden, erlassen, remittieren,
überweisen
**remittance** Anweisung *f*, Beschaffung *f*, Geld-
sendung *f*, Rimesse *f*, Übersendung *f* ~ **statement** Überweisungsmitteilung *f*
**remitter** Absender *m*, Remittent *m*
**remnant** Rest *m*, Überbleibsel *n*
**remodel, to** ~ umbauen, umbilden, umformen,
umgestalten, ummodeln
**remodeling** Knetung *f*, Umformung *f*, Ummode-
lung *f*
**remodulation** Ummodelierung *f*
**remold, to** ~ neu formen, umformen
**remolding,** ~ **gain** Störungsgewinn *m* ~ **loss**
Störungsverlust *m*
**remonstrate, to** ~ Einwände erheben, protestie-
ren, remonstrieren
**remote** abseitsliegend, fern ~ **from the ground
or the earth's surface** erdenfern
**remote,** ~-**action equipment** Fernwirkeinrichtung
*f* ~ **action release** Fernausschaltung *f* ~ **compass** Fernkompaß *m* ~ **control** Fern-antrieb *m*,
-bedienung *f*, -einstellung *f*, -lenkung *f*, -schal-

tung f, -steuerung f, -taster m, -tastung f, -trieb m, Folgesteuerung f, Tastung f ～ **control of guns** Waffenfernbedienung f

**remote-control,** ～ **action** (by selsyn-type motor) Fernsteuerung f ～ **activation** Fernantriebsteuerung f ～ **airplane** ferngesteuertes Flugzeug n ～ **anemometer** Fernanemometer m, Fernwindmeßanlage f ～ **apparatus** Fernbediengerät n ～ **beam** Fernleitungswelle f ～ **change-over switch** Verholumschalter m ～ **desk** Fernbedienpult n ～ **device** Fernsteuereinrichtung f ～ **filling equipment** Distanzfüllgerät n ～-**gear** Fernregelung f, Fernsteuerung f ～ **head** Geber m eines Fernanzeigegerätes ～-**key radio** Ferntastgerät n ～ **panel** Fernbedieneinsatz m, Fernbedienungstafel f ～ **plane** Fernlenkflugzeug n ～ **plant** Fernsteueranlage f ～ **receiver** Fernlenkkommandoempfänger m ～ **shaft** Fernleitungswelle f ～ **starter** Fernanlasser m ～ **steering gear** elektrische Rudersteuerung f ～ **switch** Fernschalter m ～ **system** Fernsteueranlage f ～ **testing equipment** elektromagnetisch gesteuerte Untersuchungseinrichtung f ～ **unit** Fernbesprechgerät n ～ **valve** Fernsteuerungsventil n

**remote-controlled** fernbedient, ferngelenkt, ferngetastet ～ **master switch** Ferntrennschalter m ～ **sweeping device** Fernräumgerät n

**remote,** ～ **controller** Fernsteuergerät n ～ **cut-off tube** Exponentialröhre f, Regel-pentode f, -röhre f ～ **drive** Fernantrieb m ～ **effect** Fernwirkung f ～ **effective zone** Fernwirkzone f ～-**firing device** Fernzündgerät n ～ **group switch** Ferngruppenumschalter m ～-**guided rocket** ferngesteuerte Rakete f ～ **gun control** Fernrichten n des Geschützes ～ **handling equipment** Fernbedienungsgerät n ～ **handling tongs** Ferngreifer m ～ **indicating dial thermometer** Zeigerfernthermometer n ～-**indicating instrument** Fernmeßgerät n mit Fernablesung, Meßgerät n mit Fernablesung oder Fernanzeige ～ **indicator** Fernanzeigegerät n ～ **keying** Ferntastung f ～ **level indicator** Fernmelder m für den Flüssigkeitsstand ～ **measurement** Fernmessung f ～-**operating water-level transmitting plant** Wasserstandsfernmeldeanlage f ～ **pipetting device** Pipettiervorrichtung f ～-**position indicator** Stellungsgeber m ～ **pressure transducer** Ferndruckgeber m ～ **radiation field** Fernfeld n ～ **reading** Fernablesung f ～ **reading thermometer connection** Fernthermometeranschluß m ～-**reading water-level indicator** Anzeigegerät n der Wasserstandsfernmeldeanlage ～ **recorder** Fernzähler m ～-**recording system** Fernschreibwerk n eines Meßgerätes ～ **regulation** Fernregulierung f ～ **setting** Fernsteuerung f ～ **signaling** Fernmeldung f ～ **signaling plant** Fernmeldeanlage f ～ **speed regulator** Ferndrehzahlverstellvorrichtung f ～ **starting connection** Fernstartanschluß m ～ **starting relay** Fernanlaßrelais n ～ **stopping relay** Fernabstellrelais n ～ **summation meter** Zähler m für Summenfernmessung (Durchfluß) ～ **supervision** Fernüberwachung f ～ **switch** Fernschaltapparat m ～-**synchronization** Fremdsynchronisierung f ～ **transducer** Ferndruckgeber m ～ **transmitter** Fernsender m ～ **welding control** Fernschweißsteuerung f

**remotely,** ～ **controlled** fernbetätigt, ferngesteuert ～ **broadcast transmitter** Rundfunkzwischensender m ～ **switch** Fernschalter m

**remount** Remonte f ～ **commission** Remonte f ～ **school** Remonteschule f

**remounting (or reinstallation)** Wiederanbau m

**removable** abhebbar, abnehmbar, absetzbar, auswechselbar, ausziehbar, entnehmbar, herausnehmbar, verschiebbar, versetzbar ～ **by washing** auswaschbar

**removable,** ～ **bottom** Losboden m ～ **core bit** ausziehbares Kernrohr n ～ **cover** abnehmbarer Deckel m ～ **cowling** abnehmbare Haubenverkleidung f ～ **cutter** verschiebbare Schneidzunge f ～ **dual control** ausbaubare Doppelsteuerung f ～ **handle** Einsteckgriff m ～ **hood** abnehmbares Verdeck n ～ **inking apparatus** abfahrbares Farbwerk n ～ **inner boiler** herausnehmbarer Einsatzkessel m ～ **liner** auswechselbares Futterrohr n oder Seelenrohr n, auswechselbare Zylinderbüchse f, Futterrohr n ～ **plug** Abnehmboden m ～ **sheath** Abhebezylinder m ～ **shoulder stock** Anschlagkolben m ～ **strainer** auswechselbarer Schmutzabscheider m ～ **top** abnehmbares Verdeck n ～ **vane** bewegliche Schaufel f (Turbine)

**removal** Abfuhr f, Abheben n, Abhebung f, Absetzen n, Abtransport m, Aufnehmen n, Aushub m, Ausscheidung f, Ausschlag m, Ausziehen n, Beseitigung f, Entfernung f, Gewältigung f, Fortrückung f, Fortschaffung f, (of clinker) Fortschlagen n, Verdrängung f, Verlegung f, Wältigung f

**removal,** ～ **of the annealed material** Ausfahren n des Glühgutes ～ **of ashes** Aschen-abfuhr f, -austragung f, -entfernung f ～ **of the bur** Gratentfernung f ～ **of a cable** Kabelaufnehme f ～ **of cinder** Entschlackung f, Schlackenziehen n ～ **of coagulated material** Entschleimung n ～ **of empty cars** Ausfahrbewegung f (Waggonkipper) ～ **of fat** Entfettung f ～ **of faults** Fehlerbeseitigung f ～ **of fences** Entgitterung f ～ **of fibers by suction** Faserabsaugung f ～ **of ground** Ebnung f des Bodens ～ **of heat** Wärmeabführung f ～ **of iron** Enteisenung f ～ **by ligature** Abbindung f ～ **of the load** Entlastung f ～ **of metal by cutting tool** Zerspanung f ～ **of the receiver** Abheben n des Hörers ～ **of rubbish** Aschenabfuhr f ～ **of sand** Entsandung f ～ **of slag** Abschlackung f, Entschlackung f ～ **of slime** Entschleimung f ～ **of solids** Feststoffabscheidung f ～ **of stress** Entspannung f ～ **of subscriber** Verlegung f eines Fernsprechanschlusses ～ **by suction** Absaugung f ～ **of sulphur** Schwefelentfernung f ～ **of water** Wasserentziehung f ～ **of a wreck** Beseitigen n eines Wracks

**remove, to** ～ abbauen, abheben, (tire) abmontieren, abnehmen, abräumen, abraumen, abstellen, abtragen, aufnehmen, aufräumen, ausbauen, ausheben, ausräumen, ausscheiden, ausziehen, beheben, beseitigen, entfernen, (a load) entlasten, entnehmen, entziehen, fortheben, heraus-heben, -nehmen, -ziehen, räumen, umziehen, verdrängen, verhauen, verschieben, vertreiben, wegnehmen

**remove, to** ～ **the (etching or pickling) acid** poltern **to** ～ **with acid** abätzen **to** ～ **with aqua**

**fortis** mit Scheidewasser *n* abbeizen **to ~ air** entlüften **to ~ arsenic** desarsenizieren, entarsenizieren **to ~ bacteria** entkeimen **to ~ a bridge** abbrücken **to ~ burrs** entgraten **to ~ camouflage** enttarnen **to ~ a circuit from service for trouble investigation** eine Leitung *f* auf Anruf legen **to ~ copper by drossing** entkupfern **to ~ with corrosives** abbeizen **to ~ by corrosive means** abfressen **to ~ from the crankshaft** von der Kurbelwelle *f* abziehen **to ~ difficulties** Schwierigkeiten *pl* beseitigen **to ~ the dross** abschäumen **to ~ dust** entstauben **to ~ fat** abfetten, abschäumen **to ~ faults** Fehler *pl* beseitigen **to ~the fibers** entfasern **to ~ fins** entgraten **to ~ the flooring** den Fußboden *m* aufreißen **to ~ forms** ausrüsten, (from concrete) ausschalen **to ~ iron** enteisnen **to ~ lead** entbleien (metall.) **to ~ the leaves** abblatten **to ~ material from** aussparen **to ~ metal by cutting** zerspanen **to ~ by mopping** austupfen **to ~ the oil** entölen **to ~ with pipette** herauspipettieren **to ~ a pump** eine Pumpe *f* abbauen **to ~ the receiver** den Hörer *m* abhängen, abheben, abnehmen **to ~ a remedy of law** Rechtsmittel *n* aufheben **to ~ and remount** aus- und einbauen **to ~ resin from a tree** einen Baum *m* abharzen **to ~ retort residues** räumen **to ~ ridges** entgraten **to ~ rivets** entnieten **to ~ rust** entrosten **to ~ scaffolding** ausrüsten **to ~ slag** ausrosten, ausschlacken, Schlacke *f* abziehen **to ~ out slag** entschlacken **to ~ slime** entschleimen **to ~the sludge** entschlammen **to ~ the spent grains** austrebern **to ~ by sponging** austupfen **to ~ with a spoon** auslöffeln **to ~ spots** abflecken **to ~ stress** entspannen **to ~ by suction** absaugen **to ~ surface of molten metal** abschöpfen **to ~ in turning** abdrechseln **to ~ varnish** ablackieren

**removed** abgesetzt

**removing** Abnehmen *n*, Ausbau *m*, Verschiebung *f* **when ~ ground (or shelf)** beim Ausbau *m* der Bank **~ of kinks** Recken *n* des Drahtes **~ of stocks from vulnerable places** Auslagerung *f*

**removing, ~ capacity** Abtragsvermögen *n* **~ gates from castings** Abkneifen *n* von Gußtrichtern **~ plants** Absetzanlagen *pl*

**remunerate, to ~** entschädigen, vergüten

**remuneration** Entgelt *n*, Honorar *n*, Vergütung *f*

**remunerative** einträglich

**renail, to ~** umnageln

**renascence** Wiedererstarken *n*

**rend, to ~** einreißen

**render, to ~** ausschmelzen, liefern **to ~ audible** hörbar machen **to ~ an award (a decision, a judgment, a sentence, a verdict)** ein Urteil *n* fällen **to ~ conducting** leitend machen **to ~ conductive** eine Röhre *f* öffnen **to ~ difficult** erschweren **to ~ fit** gebrauchsfähig machen **to ~ harmless** unschädlich machen **to ~ ineffective** lahmlegen **to ~ inoffensive** ungefährlich machen **to ~ inoperative** unwirksam machen **to ~ necessary** erfordern **to ~ operative** auslösen, wirksam machen **to ~ pliable** assouplieren (Seide) **to ~ possible** ermöglichen **to ~ safe** entschärfen **to ~ soluble** aufschließen **to ~ supple** assouplieren (Seide) **to ~ visible (or**

**perceptible)** sichtbar machen **to ~ visually perceptible** sichtbar machen

**rendering** Ausschweißen *n*, (Mauerei) Berappen *n*, Rohbewurf *m* **~ astative** Astasierung *f* **~ balancing the sweep voltages** Symmetrierung *f* der Wippspannungen **~ impervious** Abdichtung *f*, Dichtung *f* **~ novelty negative** neuheitsschädlich **~ operative** Inbetriebsetzung *f* **~ symmetrical the sweep voltages** Symmetrierung *f* der Wippspannungen

**rendezvous** Sammelpunkt *m*, Treffpunkt *m*

**rendition** Übersetzung *f*, künstlerische Wiedergabe *f*

**renew, to ~** erneuern, ersetzen, verjüngen, (a bill of exchange) verlängern **to ~ the air** auslüften **to ~ a bill** einen Wechsel *m* verlängern

**renewable** auswechselbar, ersetzbar **~ sill** aufgesetzte Schwelle *f*

**renewal** Auffrischung *f*, Auswechs(e)lung *f*, Erneuerung *f*, Ersatz *m*, Ersetzung *f* **~ of a bill of exchange** Prolongation *f* eines Wechsels **~ of option** Optionserneuerung *f* **~ of a patent** Patentverlängerung *f*

**renewal, ~ commission** Inkassoprovision *f* **~ cost** Erneuerungskosten *pl* **~ fee** Erneuerungs-, Verlängerungs-gebühr *f* **~ notice** Prämienrechnung *f*

**renewed acknowledgement of indebtedness** Neuerungsvertrag *m*

**renewer** (a plane which renews target marking) Erneuerer *m*

**renewing** Auffrischen *n*

**reniform** nierenförmig, nierig

**renormalization** Renormalisierung *f*, Umnormierung *f*

**renounce, to ~** aufgeben, verzichten, Verzicht leisten

**renovate, to ~** erneuern, renovieren

**renovation** Erneuerung *f*, Wiederinstandsetzung *f*

**rent, to ~** heuern, mieten, vermieten

**rent** Miete *f*, Pacht *f*, Spalte *f* **~ of a mine** Berg(e)pachtzins *m*

**rental** Miete *f*, Pachteinnahme *f* **~ allowance** Wohnungsgeldzuschuß *m* **~ fee** Mietgebühr *f* **~ tariff** Mietgebühr *f*

**rented wire** Mietleitung *f*

**renunciation** Verzicht *m*, Verzichtleistung *f*

**re-occupation** Wiederinbesitznahme *f*

**re-open, to ~** wieder in Betrieb *m* setzen

**re-opening** (of a legal case) Wiederaufnahme *f*, Wiedereröffnung *f*

**reoperate, to ~** wieder in Betrieb *m* setzen

**reorganization** Neugestaltung *f*, Umbildung *f*, Umgliedern *n*

**reorganize, to ~** neugestalten, reorganisieren, sanieren, umbilden, umgestalten, umgliedern, umorganisieren

**reorientation** Umlenkung *f*, Umorientierung *f*

**reoxidation** Reoxydation *f*, Rückoxydation *f*, Wiederoxydation *f*

**repack, to ~** nachstopfen, umpacken **to ~ the ties** (sleepers) das Geleise *n* nachstopfen

**repaint, to ~** wieder verstreichen

**repair, to ~** ausbessern, ausflicken, flicken, in Stand setzen, instandsetzen, reparieren, wiederherstellen, wiederinstandsetzen, wieder instandsetzen

**repair** Ausbesserung *f*, Ausflicken *n*, Instandsetzung *f*, Nacharbeit *f*, Reparatur *f*, Wiederherstellung *f* ~ **and maintenance** Instandsetzung *f* und Unterhaltung *f* ~ **of motor vehicles** Instandsetzung *f* von Kraftfahrzeugen
**repair,** ~ **car** Ausbesserungskraftfahrzeug *n* ~ **contract** Instandsetzungsauftrag *m* ~ **costs** Ausbesserungskosten *pl* ~ **gang** Bau-, Instandsetzungs-, Störungs-trupp *m* ~ **hangar** Werfthalle *f* ~ **kit** Reparaturkasten *m* ~ **links** Notglieder *pl* ~ **outfit** Flick-kasten *m*, -zeug *n* ~ **part** Ersatzteil *m* ~ **service** Betriebsüberwachung *f*, Störungsdienst *m* ~ **ship** Mutter-, Reparatur-schiff *n* ~ **shop** Flickstube *f*, Handwerkerstube *f*, Instandsetzungswerkstatt *f*, Reparatur-betrieb *m*, -werkstatt *f* ~ **slip** Fehlerzettel *m* ~ **squad** Wiederherstellungstrupp *m* ~ **timber** Pfändholz *n* ~ **tools** Reparaturwerkstatt *f* ~ **truck** Instandsetzungskraftwagen *m* ~ **welding** Reparaturschweißung *f* ~ **work** Ausbesserungsarbeit *f*, Flickarbeit *f*, Instandsetzung *f*, Instandsetzungsarbeiten *pl*, Wiederherstellungsarbeit *f* ~ **yard** Instandsetzungswerft *f*
**repaired, to be** ~ **easily** leicht zu reparieren
**repairing** Flicken *n* ~ **kit** Reparaturausrüstung *f* ~ **mesh** Reparaturglied *n* ~ **outfit** Reparaturausrüstung *f*
**repairs of tools** Werkzeuginstandsetzung *f*
**reparability** Instandsetzungsfähigkeit *f*
**reparable** ausbesserungsfähig, instandsetzungsfähig, reparaturfähig
**reparation** Ersatzleistung *f*, Reparation *f*, Rückerstattung *f*
**repay, to** ~ zurückzahlen
**repayable** rückzahlbar
**repayment** Rückzahlung *f*
**repeak, to** ~ wieder auf Maximum stellen (meas., rdr)
**repeal, to** ~ widerrufen
**repeal** Abschaffung *f*
**repeat, to** ~ repetieren (Theodolit), weitergeben, wiederholen **to** ~ **a message** vermitteln **to** ~ **an order** nachbestellen **to** ~ **verbally** abrufen
**repeat** Wiederholung *f* ~ **in the design** Rapport *m* ~ **of design** Musterrapport *m*
**repeat,** ~ **determination** Doppelbestimmung *f* ~ **key** (typewriter) Wiederholtaste *f* ~ **order** Nachbestellung *f* ~ **signal** Wiederholungszeichen *n* ~ **wheel** Rapportrad *n*
**repeatability** Reproduzierbarkeit *f*
**repeatable** wiederholbar
**repeated** mehrfach, nochmalig, wiederholt ~ **bending-stress strength** Biegeschwingungsfestigkeit *f* ~ **bending-stress test** Dauerbiegeversuch *m* ~ **-bending-stress testing machine** Biegeschwingungsmaschine *f* ~ **-direct-stress test** Dauerzugversuch *m* ~ **direct-stress testing machine** Dauerversuchsmaschine *f* für Zugbeanspruchung ~ **dynamic-stress testing machine** Dauerversuchsmaschine *f* für pulsierende Beanspruchung ~ **flexing** Biegungswechsel *m* ~ **flexural stress** Dauerbiegebeanspruchung *f*
**repeated-impact,** ~ **bending strength** Dauerschlagbiegefestigkeit *f* ~ **bending test** Dauerschlagbiegeversuch *m* ~ **energy** Dauerschlagarbeit *f* ~ **strength** Dauerschlagfestigkeit *f* ~ **tension test** Dauerschlagzugversuch *m* ~ **test**

Dauerschlag-, Schlagdauer-versuch *m* ~ **testing machine** Dauerschlagwerk *n*
**repeated,** ~ **observation** Reihenbeobachtung *f* ~ **solidification** wiederholte Erstarrung *f* ~ **stress** Wechselfestigkeit *f* ~ **stress failure** Ermüdungsbruch *m* ~ **-stress test** Dauerversuch *m* mit pulsierender Beanspruchung *m*, Schwingungsversuch *m*, Wechselfestigkeitsversuch *m* ~ **-stress testing machine** Dauerversuchsmaschine *f* ~ **stresses** dynamische Beanspruchung *f* ~ **-tension test** Dauerzugversuch *m* ~ **-tension testing machine** Dauerversuchsmaschine *f* für Zugbeanspruchung ~ **-torsion test** Dauerverdrehungs-, Drehschwingungs-versuch *m* ~ **torsional-test specimen** Drehschwingungsprobe *f* ~ **transverse-stress** Dauerschlagwerk *n* ~ **transverse-stress strength** Dauerschlagbiegefestigkeit *f* ~ **transverse-stress test** Dauerschlagbiegeversuch *m* ~ **twinning** Wiederholungszwillinge *pl* (geol.) ~ **-until-acknowledged signal** Zeichen *m* mit Wiederholung bis zur Bestätigung
**repeatedly, by** ~ **taking the mean** durch mehrfache Mittelung *f* (math.)
**repeater** Schlingenführung *f*, Tastrelais *n*, Übertrager *m*, (in rolling) Umführer *m*, Umführung *f*, (in rolling) Umlaufführung *f*, (in telephone line) Verstärker *m* ~ **arrangement** Weitergabeanordnung *f* ~ **bay** Verstärker-bucht *f*, -bunker *m*, -gestellreihe *f* ~ **circuit** Verstärkerschaltung *f* ~ **compass** Kreiseltochter *f* ~ **equipment** Übertragungseinrichtung *f* ~ **gain** Entdämpfung *f*, Verstärkungsgrad *m*, (equivalent) Verstärkungsmaß *n* ~ **gain measurements** Verstärkungsgradmessung *f* ~ **gain measuring set** Verstärkungsmessung *f* ~ **indicator** Tochter-anzeigegerät *n*, -kompaß *m* ~ **insertion on automatic longdistance tandem exchanges** Einschaltung *f* von Verstärkern bei den Durchgangsfernämtern mit Wahlbetrieb ~ **motor** Tochtermotor *m* ~ **one-way** Einwegeverstärker *m* (telephonic) ~ **operation** Verstärkerbetrieb *m* ~ **place** Verstärkerplatz *m* ~ **rack** Verstärker-bucht *f*, -bunker *m*, -gestell *n* ~ **room** Verstärkersaal *m* ~ **section** Verstärker--abschnitt *m*, -feld *n* ~ **spacing** Verstärkerabstand *m* ~ **station** Relaisstelle *f*, Telegrafenübertragungsamt *n*, Übertrageramt *n*, Übertragungsamt *n*, Verstärke(r)amt *n*, Zwischensender *m* ~ **station test desk** Meßschrank *m* für Verstärkerämter ~ **test rack** Übertragerprüfgestell *n* ~ **theodolite** Repetitionstheodolit *m* ~ **transmitter** Nebensender *m* ~ **unit** Verstärkereinheit *f*, (basic) Verstärkersatz *m*
**repeating,** ~ **of the calling lamp** Anrufwiederholung *f* (teleph.) ~ **a signal** Rückmeldung *f*
**repeating,** ~ **amplifier** Kaskadenverstärker *m* ~ **center** Knotenamt *n* ~ **-center system** Knotenamtsystem *n*, Zonensystem *n* ~ **circle** Wiederholungskreis *m* ~ **coil** Fernsprech-, Leitungs-, Ring-übertrager *m*, (matched) Gabelschaltung *f*, Übertrager *m* (teleph.), Übertragerspule *f* ~ **-coil rack** Übertragergestell *n* ~ **-coil unit** Übertragerkästchen *n* ~ **gear** Kopierwerk *n* ~ **installation** Telegrafenübertrager *m* ~ **lever** Rückmeldehebel *m* ~ **method** Wiederholungsverfahren *n* ~ **register** Speicher *m* mit Übertragung ~ **relay** Übertragungs-, Weiter-

geber-relais *n* ~ **rifle** Mehrlader *m* ~ **signal** Rückmeldesignal *n* ~ **station** Gegenfunkstelle *f*, Übertrageramt *n*, Zwischensender *m* ~ **telegraph station** Telegrafenübertragungsamt *n* ~ **transformer** Kopplungstransformator *m*

**repel, to** ~ abschlagen, abstoßen, abweisen **to** ~ **mutually** sich gegenseitig abstoßen

**repellent** abweisend

**repeller** (clystron) Reflektoranode *f*

**repelling** abstoßend ~ **action of electron charge** Bremswirkung *f* der negativen Raumladung, Eigenabstoßung *f* der Elektronenwolke ~ **force** rücktreibende Kraft *f*

**repercussion** Rückprall *m*, Rückwirkung *f*

**repercussive spring** Rückdruckfeder *f*

**reperforating attachment** Empfangslocher *m*

**reperforation** Empfangslochung *f*

**reperforator** Empfangslocher *m*, Lochstreifenempfänger *m*

**repetition** Wiederholung *f*

**repetition,** ~ **of bending stress** Dauerbiegebeanspruchung *f* ~ **of dynamic stress** Dauerschwingungsbeanspruchung *f*, pulsierende Dauerbeanspruchung *f* ~ **of particulars of call** Vergleichung *f* der Angaben einer Gesprächsanmeldung ~ **of pictures** Bildwechsel *n* ~ **of stress** Dauerbeanspruchung *f* ~ **of torsional stress** Drehschwingungsbeanspruchung *f*

**repetition,** ~ **angle measurement** repetitionsweise Winkelmessung *f* ~ **border** Einfassungsstück *n* ~ **clamp catch** Klemmhebel *m* für Repetitionsklemme ~ **equivalent** Verkehrsgüte *f* ~ **frequency** Wiederholungsfrequenz *f* ~ **measurement** Repetitionsmessung *f* ~ **rate** Bildfrequenz *f* (TV), Rückfragehäufigkeit *f*, Zähl-, Zeitgeber-frequenz *f* ~ **rate of pulses** Impulsfrequenz *f* ~ **theodolite** Repetitionstheodolit *m* ~ **work** Reihen-arbeit *f*, -fertigung *f*, Serien-arbeit *f*, -fertigung *f*

**repetitive,** ~ **distribution** identische Verteilung *f* ~ **error** Wiederholungsfehler *m* ~ **instruction** Wiederholungsbefehl *m* ~ **stress** Schwellbelastung *f* ~**-stress test** Dauerversuch *m*

**rephosphorization** Rückphosphorisierung *f*

**replace, to** ~ auswechseln, einhängen, entheben, ersetzen, rückstellen, zurückbringen **to** ~ **the receiver** abhängen, den Hörer auflegen

**replace** Zurück-bringen *n*, -stellen *n*

**replaceability** Auswechslungsfähigkeit *f*

**replaceable** auswechselbar, ersetzbar ~ **graticule** Vorsteckraster *m* (CRT) ~ **inset** auswechselbarer Einsatz *m*

**replacement** Auflegen *n*, Auswechseln *n*, Auswechs(e)lung *f*, Einhängen *n*, Ergänzungsbedarf *m*, Ersatz *m*, Ersatz-lieferung *f*, -teil *m*, Ersetzung *f*, Nachschub *m*, Rückstellung *f* ~ **of air** Lufterneuerung *f*

**replacement,** ~ **body** Verdrängungskörper *m* ~ **calendar block** Kalenderersatzblock *m* ~ **center** Ergänzungsstelle *f* ~ **cost** Wiederbeschaffungs-kosten *pl*, -wert *m* ~ **diagram** Ersatzschaltbild *n* ~ **engine** Ersatzmotor *m* ~ **guide** Wegweiser *m* ~ **materials** Austauschwerkstoff *m* ~ **part** Ersatz-, Reserve-teil *n* ~ **parts manufacture** Ersatzteilfertigung *f* ~ **scheme** Ersatzschaltbild *n* ~ **supply** Ersatzbeschaffung *f* ~ **tube** Ersatzröhre *f*

**replacing** Auswechseln *n* ~ **electron** Ersatzelektron *n* ~ **telephone receiver** Einhängen *n* des Hörers

**replay** Wiedergabe *f*

**replenish, to** ~ auffüllen, ergänzen, nachfüllen

**replenisher** Auffrischer *m*

**replenishing,** ~ **chamber** Nachfüllkammer *f* ~ **cup** Einfülltopf *m* ~ **device** Nachtankvorrichtung *f* ~ **line** Nachtankleitung *f* ~ **liquor** Speiseflotte *f*

**replenishment** Anreicherung *f*, Auffüllung *f*, Bestandsergänzung *f*, Ersatz *m*

**replete** angefüllt

**replica** Gegenbild *n*, Nachbild *n*, Nachbildung *f*, (of curves) Spiegelung *f* ~ **method** Abdruckverfahren *n* ~ **process** Hautabdruckverfahren *n* ~ **specimen** Reproduktionsproben *pl*

**replication** (of plaintiff to defendant's answer) Erwiderung *f*

**reply, to** ~ antworten, erwidern

**reply** Antwort *f*, Bescheid *m*, Erwiderung *f*, Gegensignal *n*, Klagebeantwortung *f* **there is no** ~ Teilnehmer *m* antwortet nicht ~ **is requested** um Antwort *f* wird gebeten

**reply,** ~ **bell** Rückmeldeläutewerk *n* ~ **code** Antwortkode *m* (sec. rdr.) ~ **coupon** Antwortschein *m* ~ **hit** Antworttreffer *m* ~ **lever** Rückmeldehebel *m* ~ **pulses** Antwortimpulse *pl*

**repolish, to** ~ nachpolieren

**report, to** ~ angeben, anmelden, austragen, bekanntgeben, berichten, Bericht erstatten, melden, ruhen (lagern) **to** ~ **one's arrival** sich anmelden **to** ~ **faulty** gestellmelden (teleph.) **to** ~ **for work** den Dienst *m* antreten

**report** Angabe *f*, Anzeige *f*, Bericht *m*, Berichterstattung *f*, Knall *m*, Meldung *f*, Mitteilung *f*, Nachricht *f*, Protokoll *n*, Schuß *m* **to make a** ~ Bericht *m* erstatten ~ **on examination** Prüfungsbescheid *m* ~ **of execution of orders** Vollzugsmeldung *f* ~ **of a gun** Abschuß *m*

**report,** ~ **charge** Benachrichtigungsgebühr *f*, Vorbereitungsgebühr *f* ~ **produced by shell wave** Geschoßknall *m*

**reported** angezeigt

**reporter** Berichter *m*

**reporting** Berichterstattung *f*, Gutachten *n* ~ **of aircraft** Flugmeldung *f*

**reporting,** ~ **center** Meldeplatz *m* ~ **point** Meldepunkt *m* (aviat.) ~ **trunk telephone calls by telegraph order wire with buzzer (or sounder)** Summermeldedienst *m*

**repose, to** ~ lagern, liegen

**repose** Ruhe *f* ~ **escapement** ruhende Hemmung *f*

**reposing on an elastic half plane** auf elastisch nachgiebiger Bettung *f*

**repository** Ablage *f*

**represent, to** ~ darstellen, vertreten **to** ~ **a firm** einer Firma *f* vorstehen **to** ~ **graphically** grafisch darstellen, in Kurvenform *f* darstellen **to** ~ **by symbols** versinnbildlichen **to** ~ **vectorially** vektoriell darstellen

**representable** darstellbar

**representation** Abbildung *f*, Agentur *f*, Anzeige *f* (electron.), Darstellung *f*, Darstellungsweise *f*, Vertretung *f*, Vorstellung *f* ~ **of test results** Versuchsauswertung *f* ~ **theory** Darstellungstheorie *f* representative Abgeordneter *m*, Beauftragter *m*, (business) Geschäftsträger *m*, Stellvertreter *m*; darstellend **to be ~ of** darstellen

representative, **~ agent** Vertreter *m* **~ circle** Bildkreis *m* **~ ensemble** repräsentatives Ensemble *n* **~ fraction** Maßstab *m* **~ fuel** Vergleichskraftstoff *m* **~ observation** darstellende Beobachtung *f* **~ scale** darstellender Maßstab *m* **~ value** Eigenwert *m*

represented, **to be ~** vertreten sein **~ as rolled out** abgewickelt gezeichnet

representing darstellend

re-press, **to ~** erdrücken, nachpressen

repressuring Drainage *f* mittels Preßluftinjektion *f*, Wiederunterdrucksetzung *f*

reprint, **to ~** nachdrucken, neudrucken

reprint Abdruck *m*, Nachdruck *m*, Separatabdruck *m*, Sonderabdruck *m*, Umdruck *m* **to ~ a book** ein Buch *n* neu auflegen

reprinting Wiederabdruck *m* **~ ink** Umdruckfarbe *f* **~ press** Umdruckpresse *f* **~ process** Nachdruckverfahren *n*

reprisal Repressalie *f*, Vergeltung *f* **~ weapon** Vergeltungswaffe *f*

reproach Vorwurf *m*

reprocessing Auffrischen *n* **~ loss** Aufarbeitungsverlust *m*

reproduce, **to ~** (by printing or impression) abdrucken, doppeln (data proc.), nachbilden, reproduzieren, (an image or picture) schreiben, vervielfältigen, wiedergeben **to ~ by autograph** pantografieren **to ~ by photography** umfotografieren

reproducer Abspielgerät *n*, Geber *m* (acoust.), Kartendoppler *m*, (sound) Lautsprecher *m*, Schallsender *m*, Tonkopf *m* **~ aperture** Tonschlitz *m* **~ head** Lichttonansatz *m* **~ slit** Tonschlitz *m*

reproducibility Reproduzierbarkeit *f* **~ from run to run** losmäßige Herstellungsgleichmäßigkeit *f*

reproducible reproduzierbar, wiederholbar **~ copy** pausfähige Kopie *f*

reproducing Vervielfältigen *n* **~ apparatus** Wiedergabeapparatur *f* **~ camera** Nachbildungskamera *f* **~ lathe** Schablonendrehbank *f* **~ method** Abdruckmethode *f* **~ stone** Wiedergabestein *m* **~ stylus** Abspielnadel *f*

reproduction (print.) Abdruck *m*, Abspielen *n*, Abzug *m*, Kopie *f*, Nachbildung *f*, Reproduk-

reproduction (print.) Abdruck *m*, Abspielen *n*, Abzug *m* (photo), Kopie *f*, Nachbildung *f*, Reproduktion *f*, Vervielfältigung *f*, Wiedergabe *f* (electr.), Wiederhervorbringung *f* **~ with high acoustic frequencies** hohle Wiedergabe *f* **~ of image** Bilderzeugung *f*, Bildpunktverteilung *f*, Bildsynthese *f*, Bildwiedergabe *f*, Bildzusammensetzung *f* **~ of scanning** Wiederabtastung *f* **~ of type faces** Abdruck *m* der Typen

reproduction, **~ channel** Wiedergabekanal *m* **~ developer** Reproentwickler *m* **~ equalizer** Wiedergabeentzerrer *m* **~ equipment** (camera) Reproduktionsapparat *m* **~ (or multiplication) factor** Vervielfachungsfaktor *m* **~ printing process** Reduzierkopierverfahren *n* **~ process** Umdruckverfahren *n* **~ ratio** Reproduktionsverhältnis *n* **~ room** Reproduktionsraum *m*

reproductive abbildend **~ accuracy** Nachfahrgenauigkeit *f*

reprojection equipment Rückprojektionsanlage *f*

reprove, **to ~** mißbilligen, tadeln

republish, **to ~ a book** ein Buch *n* neu auflegen

repudiate, **to ~** verwerfen

repudiation Verwerfung *f*

repugnance Abneigung *f*

repugnant widerlich, widerstrebend

repulp, **to ~** (paper) einstampfen

repulse, **to ~** abschlagen, abstoßen, abwehren, abweisen, zurückschlagen, zurücktreiben

repulse force Abstoßungskraft *f*

repulsing zurückschlagend

repulsion Abstoßung *f*, Abstoßungs-kraft *f*, -typ *m*, Repulsion *f*, Rückprall *m*, Rückschlag *m*, Rückstoß *m*, Zurückweisung *f* **~ motor** Repulsionsmotor *m* **~ potential** Abstoßungspotential *n*

repulsory abstoßend

repunch, **to ~** neu stanzen, nochmals stanzen

repurchase, **to ~** zurückkaufen

repurchase Rückkauf *m*

repusher of loop wheel Maschenlegerrückstreifer *m*

reputable namhaft

reputation Ansehen *n*, Ruf *m*

request, **to ~** ansuchen, bitten, erbitten, fordern, Rückfrage halten, rückfragen **to ~ equipment** Gerät *n* anfordern **to ~ a pilot** um einen Lotsen *m* bitten

request Anforderung *f*, Anfrage *f*, Anliegen *n*, Ansuchen *n*, Ansuchung *f*, Antrag *m*, Aufforderung *f*, Bitte *f*, briefliche Anforderung *f*, Ersuchen *n*, Ersuchung *f*, Forderung *f*, Gesuch *n* **at the ~ of** auf Veranlassung von **~ for a bearing** Ortungsanforderung *f* **~ for deferment** Zurückstellungsantrag *m* **~ for a respite** Stundungsgesuch *n*

request, **~ form** Anforderungsformular *n* **~ stop** Bedarfshaltestelle *f* **~ time** Anmeldezeit *f*

requested, **as ~** wunschgemäß

require, **to ~** anfordern, in Anspruch nehmen, beanspruchen, bedürfen, erfordern, fördern, nötig haben, verlangen

require Anspruch *m*

required benötigt, erforderlich, gefordert **as ~** nach Bedarf *m*

required, **~ accuracy** verlangte Genauigkeit *f* **~ distance** erforderliche Strecke *f* (beim Start) **~-elevation dial** Sollgeberhöhe *f* (rdr) **~ floor space** Platzbedarf *m* **~ forepressure** benötigtes Vorvakuum *n* **~ output** Leistungsbedarf *m* **~ pressure** Kraftaufwand *m* **~ range** Sollschußweite *f* **~ strength** Etatsstärke *f*, Sollbestand *m* **~ subscriber** angerufener oder verlangter Teilnehmer *m* **~ value** Soll *n*, Sollwert *m*

requirement Anforderung *f*, Bedarf *m*, Bedarfsgegenstand *m*, Bedingung *f*, Bedürfnis *n*, Erfordernis *n*, Forderung *f*, Soll *n* **~ for ignition** Zündzubehör *n* **~ of material** Stoffaufwand *m* **~ for weaving** Webereibedarfsartikel *m*

requirements Anforderungen *pl*, Ergänzungsbedarf *m*, Inanspruchnahme *f*, Nachfrage *f* **~ of the filling technique** abfülltechnische Erfordernisse *pl*

requiring beanspruchend

requisite Erfordernis *n*; erforderlich, notwendig

**requisition, to** ~ anfordern, beanspruchen
**requisition** Anforderung *f*, Beitreibung *f*, Bestellschein *m*, Bestellung *f*, Requisition *f* ~ **for supplies** Materialanforderung *f*
**requisition,** ~ **form** Entnahmeschein *m* ~ **law** Naturalleistungsgesetz *n* ~ **number** Anforderungszeichen *n*, Bestellnummer *f* ~ **receipt** Beitreibungsschein *m* ~ **symbol** Anforder(ungs)-zeichen *n*
**requisitioned accuracy** verlangte Genauigkeit *f*
**requisitioning** Beitreibung *f*, Heranziehung *f*
**re-radiate, to** ~ wiederausstrahlen
**re-radiation** Ballsenden *n*, Wiederausstrahlung *f*
**re-radiator** Rückstrahler *m*
**rerail, to** ~ aufgleisen
**rerailing** Aufgleisung *f*
**re-ranging** Umstellung *f* des Meßbereichs
**re-record, to** ~ umspielen (phono)
**re-recording** Überspielen *n*, Umschreiben *n* ~ **of film** Frischaufnahme *f* ~ **of sound track** Nachvertonung *f*
**re-reel (or re-wind), to** ~ umhaspeln
**re-reeling machine** Umrollmaschine *f*
**re-regulate, to** ~ neu einregeln
**re-regulation** Neueinregelung *f*
**re-ring, to** ~ abrufen, den Ruf wiederholen lassen
**reroll, to** ~ nachwalzen
**reroute, to** ~ (den Verkehr) umleiten
**rerouting** Umlenkung *f*, Verlegung *f*
**rerun, to** ~ nochmals laufen lassen, nochmals senden, wiederholen (comput.)
**rerun** Wiederholung *f*, Wiederholungsprüfung *f* ~ **point** Wiederholpunkt *m* (comput.) ~ **program** Wiederholprogramm *n*
**rerunning stills** Anlagen *pl* zum Feinfraktionieren (chem.)
**resack, to** ~ umsacken
**resale** Wieder-veräußerung *f*, -verkauf *m*
**rescind, to** ~ rückgängig machen, widerrufen
**rescrape, to** ~ nachschaben
**rescrew, to** ~ wiederaufschrauben
**rescue, to** ~ bergen, retten
**rescue** Rettung *f* ~ **and refuge chamber** Rettungs- und Sicherheitskammer *f* ~ **apparatus** Rettungsapparat *m* ~ **brigade** Rettungstruppe *f* ~ **chamber** Rettungskammer *f* ~ **hoist** Bergungswinde *f* ~ **operations** Bergungsarbeiten *pl* ~ **party** Rettungsmannschaft *f* ~ **squad** Rettungstruppe *f* ~ **vehicle** Rettungsfahrzeug *n* ~ **vessel** Bergungsschiff *n* ~ **work** Hilfeleistung *f*
**rescuing operation** Rettungsarbeit *f*
**research, to** ~ erforschen, forschen, wissenschaftlich arbeiten
**research** Forschung *f*, Forschungsarbeit *f*, Nachforschung *f*, Untersuchung *f*, Versuch *m* ~ **of the probalitiy of large groups** (statistics) Großzahluntersuchungen *pl*
**research,** ~ **airfield** Forschungsflughafen *m* ~ **bureau** Forschungsamt *n* ~ **control** Forschungsführung *f* ~ **department** Forschungsabteilung *f* ~ **directorate** Forschungsamt *n* ~ **division** Entwicklungsabteilung *f* ~ **domain** Forschungsgebiet *n* ~ **engineer** Forschungs-, Versuchs-ingenieur *m* ~ **experiment** Versuchung *f* ~ **field** Forschungsgebiet *n* ~ **institution** Untersuchungsanstalt *f* ~ **laboratory** Versuchslaboratorium *n* ~ **method** Untersuchungsmethode *f*

~ **paper** Forschungsbericht *m* ~ **reactor** Forschungs-, Versuchs-reaktor *m* ~ **report** Forschungsbericht *m* ~ **rocket** Forschungsrakete *f* ~ **work** Forschungsarbeit *f*
**researcher** Bearbeiter *m*, Forscher *m*
**reseating** Einschleifen *n*
**réseau** Netz *n*, Reseau *m*
**resect, to** ~ rückwärts einschneiden
**resecting** Einschneiden *n*
**resection** Rückwärtseinschnitt *m* ~ **in space** Pyramidenverfahren *n*, Rückwärtseinschnitt *m* im Raum
**resemblance** Ähnlichkeit *f*
**resemble, to** ~ gleichen
**resembling** ähnlich ~ **lye** laugenartig
**reservation** Einschränkung *f*, Reservat *n*, Reservation *f*, Rückstellung *f*, Schutzgebiet *n*, Vorbehalt *m*, Vorbestellung *f*
**reservations, with** ~ vorbehaltlich
**reserve, to** ~ ausscheiden, hinterhalten, offenhalten, reservieren, vorbehalten **to** ~ **right of recourse** sich den Regreßanspruch *m* vorbehalten
**reserve** Aufnahme *f*, Bestand *m*, Deckpapp *m*, Ersatz *m*, Reserve *f*, (in calico printing) Reservierungsmittel *n*, Schutzpapp *m*, Zurückhaltung *f* ~ **aircraft** Reserveflugzeug *n* ~ **antenna** Ersatzantenne *f* ~ **buoyancy** Hilfsschwimmkraft *f* ~ **depot** Ersatzmagazin *n* ~ **factor** Reserve-faktor *m*, -zusatz *m* ~ **-fuel valve** Reservebrennstoffventil *n* ~ **fund** Sicherungsrücklage *f* ~ **layer** Reserverandschicht *f* ~ **liability** Nachschußpflicht *f* ~ **pair** Reserveader *f* ~ **part** Ersatzteil *m* ~ **power** Bewetterung *f*, Energiereserve *f*, Kraft-reserve *f*, -überschuß *m* ~ **pumping** Umpumpen *n* ~ **semiconductor** Assoziationshalbleiter *m* ~ **set of inserts** Reserveeinsatz *m* ~ **source of power** Leistungsreserve *f* ~ **sparking power** Zündleistungsreserve *f* ~ **tank** (for fuel) Hilfsbehälter *m*, Reservebehälter *f* ~ **unit** Ersatzeinheit *f* ~ **wire** Reserveader *f*
**reserved** gehalten ~ **for special use** zweckgebunden ~ **seat** Sperrsitz *m*
**reservedly** unter Vorbehalt
**reserves** Ergänzung *f*, Nachschub *m*, unausgerichtetes Feld *n*
**reservoir** Bassin *n*, Becken *n*, Behälter *m*, Kessel *m*, Klärbottich *m*, Sammel-behälter *m*, -brunnen *m*, -gebiet *n*, -gefäß *n*, -tank *m*, Speicher *m*, Staubecken *m*, Stauteich *m*, Stauweiher *m*, Tank *m*, Wasserbehälter *m* ~ **of supply** Speisereservoir *n*
**reservoir,** ~ **bleed** Behälterentlüftung *f* ~ **capacitor** Ladekondensator *m* ~ **capacity** Lade- und Siebkapazität *f* ~ **condenser** Speicher-, Vorrats-kondensator *m* ~ **cupola** Kupolofen *m* mit erweitertem Herd ~ **drain** Behälterablaß *m* ~ **keeper** Wasserturmwärter *m* ~ **ladle** Mischpfanne *f* ~ **rock** Speichergestein *n*
**reset, to** ~ auslösen, löschen, nachrichten, nachstellen, nullen, Rückführdaumen führen, rückstellen, versetzen, wiedereinstellen, zurückstellen, zurücksetzen
**reset** Wiedereinsetzen *n* ~ **action** Integralwirkung *f* ~ **-action controller** I-Regler *m*, integral wirkender Regler *m* ~ **coil** Einschaltspule *f* ~ **contact** Löschkontakt *m* ~ **control circuit**

Rückstellregelungskreis *m* ~ **device** Rückhol-
vorrichtung *f* ~ **feedback** nachgebende Rückführung *f* ~ **flux level** Rückstellflußpegel *m* ~
**key** Berichtigungstaste *f* ~ **pulse** Löschungs-,
Nullungs-impuls *m* ~ **solenoid** Rückstellspule *f*
~ **time** Nachgebe-, Rückstell-zeit *f* ~ **winding**
Wiederherstellungswicklung *f*
**resetting** Auslösung *f*, Nach-justierung *f*, -setzen
*n*, Neusatz *m*, Rückstellung *f*, Umstellung *f* ~
**of coil** (direction finder) Nachführung *f* des
Rahmens ~ **of coil tuning** Rahmennachstimmung *f* ~ **of frame** (direction finder) Nachführung *f* des Rahmens ~ **of frame tuning** Rahmennachstimmung *f*
**resetting,** ~ **cam** Auslöse-, Rückführ-daumen *m*
~ **counter** Nullstellenzähler *m* ~ **half-cycle**
Rückstellhalbzyklus *m* ~ **key** Rückstelltaste *f*
~ **magnet** Rückstellmagnet *m* ~ **mechanism**
Rücksetzeinrichtung *f* ~ **member** Rückstell-
glied *n* ~ **time** Nachstell-, Umricht-zeit *f* ~
**tool** Nachstellwerkzeug *n*
**reshape, to** ~ regenerieren
**resharpen, to** ~ nachschärfen, nachschleifen
**resharping** Nachschärfen *n*
**reshooting** Nachschießen *n*
**reside, to** ~ sich aufhalten, wohnen
**residence** Aufenthaltsort *m*, Hof *m*, Wohnsitz *m*,
Wohnung *f* ~ **telephone** Wohnungsanschluß *m*
**resident** Anlieger *m*; ortsansässig ~ **engineer**
Bauingenieur *m* ~ **inspector** Bauaufsichtsin-
spektor *f* ~ **physician** Assistenzarzt *m*
**residential** ansässig, wohnhaft ~ **construction**
Wohnungsbau *m* ~ **district** Wohngegend *f* ~
**housing** Wohnungsbau *m* ~ **section** Wohn-
viertel *n*
**residing at** ansässig
**residual** remanent, residuell, restlich, rückständig, übrigbleibend, zurückbleibend ~
**aberration** Restabweichung *f* ~ **activity** Rest-
aktivität *f* ~ **affinity** Affinitätsrest *m* ~ **air**
Restluft *f* ~ **alkalinity** Endalkalität *f* ~ **charge**
Ladungsrückstand *m* ~ **charge of condenser**
Kondensatorrückstand *m* ~ **charge phenomenon**
Nachladeerscheinung *f* ~ **chromatic aberration**
Farbenreste *pl* ~ **clay** Gehänge-lehm *m*, -ton *m*
~ **clay soil** toniger Rückstandsboden *m* ~ **combustion gas** Feuergasrest *m* ~ **compressive stress**
Druckeigenspannung *f* ~ **current** Reststrom *m*
*m* ~ **current state** Anlaufstromgebiet *n* ~
**deflection** Nullpunktfehler *m*, residueller Ausschlag *m* ~ **deviation** Rest-ablenkung *f*,
-deviation *f* ~ **distortion** Eigenklirrdämpfung *f*
~ **drag area** Restwiderstandsfläche *f* ~ **electrons** Elektronen-, Rest-plasma *n* ~ **error**
Restfehler *m* ~ **exhaust gas** Abgasrest *m* ~ **flux
density** Restflußdichte *f* ~ **force** Restkraft *f*
~ **frequency modulation** Frequenzmodulations-
rest *m* ~ **fuel** Rückstandsöl *n* ~ **gas** Gasrück-
stand *m*, Restgas *n* ~ **hardness** (of water) Rest-
härte *f* ~ **heat** Restschmelze *f* ~ **image** Bilder-
residuum *m*, Nachleuchtbild *n* ~ **impulse** Reststromstoß *m* ~ **intensity** Restintensität *f* ~
**ionization** Restionisation *f* ~ **liquid** Rest-
-flüssigkeit *f*, -lösung *f* ~ **liquor** Restlauge *f* ~
**loss** magnetische Nachwirkung *f*, Nachwirkungsverlust *m* ~ **magnetism** magnetische Remanenz, remanenter oder zurückbleibender
Magnetismus *m*, Remanenz *f* ~ **magnetization**

remanente Magnetisierung *f* ~ **modulation**
Trägerrauschpegel *m* ~ **moisture** Restfeuchtig-
keit *f* ~ **naphtha** Naphtharückstand *m* ~
**nucleus** Rest-keim *m*, -kern *m* ~ **oil** Rückstandsöl *n* ~ **parasitic coupling within the attenuator**
Übersprechdämpfung *f* innerhalb der Eichleitung ~ **pore pressure** zurückbleibender Po-
renwasserdruck *m* ~ **products** Abfall-, Rück-
stands-erzeugnisse *pl* ~ **pulse** Reststromstoß *m*
~ **radiation** Reststrahlung *f* ~ **range** Restreich-
weite *f* ~ **ray** Reststrahl *m* ~ **resistance** Rest-
widerstand *m* ~ **shrinkage** Restkrümpfung *f* ~
**soil** Eluvialboden *m* ~ **stop** Klebestift *m* ~
**stress** Eigen-, Rest-, Vor-spannung *f* ~ **time
constant** Eigenzeitkonstante *f* ~ **tolerance** Abweichung *f* ~ **twist** Restdrall *m* ~ **voltage** Rest-
spannung *f* ~ **water** Wasserreste *pl*
**residuary** rückständig ~ **acid** Abfallsäure *f* ~
**product** (weiter verwertbares) Abfallerzeugnis
*n*
**residue** Residuum *n*, Rückstand *m*, Überbleibsel *n*, Überrest *m* ~ **from combustion** Verbren-
nungsrückstand *m* ~ **on evaporation** Abdämpf-
rückstand *m* ~ **on ignition** Glührückstand *m*
~ **of moisture** Feuchtigkeitsrest *m*
**residue,** ~ **bagasse** Bagasse *f* ~ **class** Restklasse *f*
~ **class ring** Restklassierung *f* ~ **heat** Rest-
wärme *f*
**residues,** ~ **of coal washing** Nachwaschkohle *f*
~ **of a furnace** Ofenansätze *pl*
**residuum** Geschür *n*, Hüttenafter *n*, Residuum *n*,
Überbleibsel *n* ~ **of weathered rocks** Verwitterungsrückstand *m*
**resign, to** ~ abdanken, aufgeben
**resignation** Aufgabe *f*, Entlassungsgesuch *n*,
Verzicht *m*, Verzichtleistung *f*
**resilience** Abprall *m* (aviat.), Abprallen *n*, Beweglichkeit *f*, Durchfederung *f*, Elastizität *f*, Federungsvermögen *n*, Kerbschlag-wert *m*, -wi-
derstand *m*, Rückfederung *f*, Rückprall *m*,
Verformungsarbeit *f*, Weichheit *f*, Zurückspringen *n* ~ **of a spring** Federkraft *f*
**resiliency** Abprallen *n*, Elastizität *f*, federharte
Eigenschaft *f*, Prallkraft *f*, Prellkraft *f*, Rückstellkraft *f*, Springkraft *f*
**resilient** elastisch, erschütterungsfrei, federnd,
geschmeidig, zurückspringend ~ **coupling**
elastische Kupplung *f*, Federkupplung *f* ~
**cushioning** Schwingmetall *n* ~ **gasket material**
nachgiebiges Dichtungsmaterial *n* ~ **inter-
ponent members** elastische Betätigungsglieder
*pl* ~ **log support** Federbeinstütze *f* ~ **support**
erschütterungsfreie Unterlage *f*
**resin, to** ~ mit Harz tränken, imgrägnieren
**resin** Dammarharz *m*, Harz *n*, Kolophon(ium) *n*
~**-bed separation** Ionenaustausch *m* (Harz)
~ **blende** Zinkharz *n* ~**-coated** (nails) geharzt
~**-coated sand** harzumhüllter Sand *n* ~**-cored
solder** Röhrenlotzinn *n* ~ **drier** Harzsikkative *n*
~**-free** harzfrei ~ **gas load isolator** Harzgas-
leistungsschalter *m* ~**-gas protector tube**
Harzgasableiter *m* ~**-like** harzähnlich ~ **milk**
Harzlösung *f*, Leimmilch *f* ~ **oil** Harzöl *n* ~
**powder** Harzpulver *n* ~ **resist** Harzreserve *f* ~
**size** Harzmilch *f* ~**-sized** harzgeleimt ~ **soap**
Harzkernseife *f* ~ **solder** Harzlot *n* ~ **varnish**
Kolophoniumlack *m* ~ **wool** Kunstharzwolle *f*
**resinate, to** ~ harzen

**resinate** Harzsalz *n* ~ **of** harzsauer ~ **boiled oil** Resinatfirnis *f*
**resinferous** harzhaltig
**resinic** harzhaltig, harzig ~ **acid** Harzsäure *f* ~ **body** Harzkörper *m*
**resinification** Verharzung *f* ~ **of the oil** Harzbildung *n* in Öl
**resinified, to become** ~ harzig werden
**resinify, to** ~ verharzen
**resining** verharzen
**resinous** harzhaltig, harzig ~ **earth** Erdharz *n* ~ **electricity** Harzelektrizität *f* ~ **exudate on outside of trees** Baumharz *n* ~ **luster** Wachsglanz *m* ~ **quartz** Fettquarz *m* ~ **substance** Harz-körper *m*, -stoff *m* ~ **tar** Harzteer *m* ~ **wood** Kineholz *n*
**resist, to** ~ beständig sein gegen . . ., widerstehen
**resist** Abdecklack *m* (photo.), Ätzgrund *m* (print.), Deck-mittel *n*, -lack *m*, Reservierungsmittel *n*, Schutzmasse *f* ~ **against raising** Rauhreserve *f*
**resist,** ~ **paste** Deckpapp *m*, Schutzpapp *m* ~ **print** Reservedruck *m* (Erzeugnis) ~**-printed** mit Reserve *f* bedruckt ~**-printed goods** Reservierungsartikel *m* ~ **printing** Deckpappdruck *m*, Reservage-Druck *m* ~ **printing paste** Reservedruckfarbe *f* ~ **style** Reservage-Artikel *m*, Schutzbeizendruck *m*
**resistance** Abschubwiderstand *m*, Auflehnung *f*, Bedämpfung *f*, (of a machine) Beharrungszustand *m*, Beständigkeit *f*, Blindwiderstand *m*, Festigkeit *f* (mech.), Gegenwehr *f*, Gleichstromwiderstand *m*, (to buckling) Knickerscheinung *f*, Resistanz *f*, Stauwirkung *f*, Wehr *f*, Widerhalt *m* (electr.), Widerstand *m*, Widerstands-fähigkeit *f*, -vermögen *n*
**resistance,** ~ **to abrasion** Abnutzungswiderstand *m*, Abreibfestigkeit *f*, Querstabilität *f*, Reibfestigkeit *f*, Verschleißfestigkeit *f* ~ **to adverse climatic conditions** Klimafestigkeit *f* ~ **to aging** Alterungs-beständigkeit *f*, -widerstand *m* ~ **of air flow** Luftdurchströmungswiderstand *m* ~ **to atmospheric conditions** Witterungsbeständigkeit *f* ~ **of the axis** Achswiderstand *m* ~ **to bending** Biegungs-festigkeit *f*, -widerstand *m*, Durchbiegungswiderstand *m* ~ **to bending of cable** Seilsteifigkeit *f* ~ **to bending strain** Biegefestigkeit *f* ~ **to buckling (or bulging)** Knickwiderstand *m* ~ **to compression** Druckfestigkeit *f*, Zerdrückfestigkeit *f* ~ **to conduction** Leitungswiderstand *m* ~ **of contact** Berührungswiderstand *m* ~ **to corrosion** Anfressungsbeständigkeit *f*, Korrosionsbeständigkeit *f*, Rostbeständigkeit *f*, Rostwiderstand *m* ~ **to creasing** Bauschelastizität *f* ~ **to creep** Dauerstandfestigkeit *f* ~ **to creepage** Kriechfestigkeit *f* ~ **to crinkling** Knickwiderstand *m* ~ **to crushing** Druckfestigkeit *f*, Zerdrückfestigkeit *f* ~ **in curves** Kurvenwiderstand *m* ~ **to cutting** Schnittwiderstand *m*, Zerspannungsfestigkeit *f* ~ **to deformation** Deformationswiderstand *m*, Formveränderungswiderstand *m*, (of material) Wanderwiderstand *m* ~ **to displacement** Verschiebewiderstand *m* ~ **of driving gear** Triebwerkwiderstand *m* ~ **of earth plate** Ausbreitungswiderstand *m* ~ **to elongation** Streckfestigkeit *f* ~ **to erosion** Anfressungsbeständigkeit *f* ~ **to expansion and contraction** Raumbe-

ständigkeit *f* ~ **to flexing** Scherfestigkeit *f* ~ **to flow** Fließwiderstand *m*, Strömungswiderstand *m*, Wanderwiderstand *m* ~ **to the flow of current** Leitungswiderstand *m* ~ **to fouling** Verschmutzungsunempfindlichkeit *f* ~ **to frictional swivel movement** Reibungsschwenkwiderstand *m* ~ **to frost** Frostbeständigkeit *f* ~ **due to gradients** Steigungswiderstand *m* ~ **to grinding** Mahlwiderstand *m* ~ **to gunfire** Beschußfestigkeit *f* ~ **to heat** Wärmebeständigkeit *f*, Warmfestigkeit *f* ~ **to impact** Schlag-festigkeit *f*, -widerstand *m* ~ **of iron** Eisenwiderstand *m*, magnetischer Eisenwiderstand *m* ~ **to lateral wave action** Querstabilität *f* ~ **to lime** Kalkbeständigkeit *f* ~ **per loop mile** Widerstand *m* je Meile Doppelleitung ~ **of the material** Festigkeit *f* des Werkstoffes ~ **to motion** Fahrwiderstand *m*, Verschiebewiderstand *m* ~ **to overstress** Überbelastbarkeit *f* ~ **to passage of water** Wasserdurchflußwiderstand *m* ~ **to polish** Schleifhärte *f* ~ **to penetration** Beschlußfestigkeit *f* ~ **to pick test** Rupffestigkeit *f* ~ **to plucking** Rupffestigkeit *f* ~ **to pressure** Druckfestigkeit *f* ~ **to rolling** Fahrwiderstand *m*, Rollwiderstand *m* ~ **to the rolling of a car** Fahrwiderstand *m* ~ **to rupture** Bruchfestigkeit *f* ~ **to scaling** Zunderbeständigkeit *f* ~ **to shear** Schubwiderstand *m* ~ **to shock** Schlag-festigkeit *f*, -widerstand *m*, Stoßfestigkeit *f* ~ **to sinking** Eindringungswiderstand *m* ~ **to slagging** Verschlackungsbeständigkeit *f* ~ **to sliding** Gleitsicherheit *f*, Standsicherheit *f* gegen Drehen ~ **during speech** Schwingwiderstand *m* (bei Sprechkapsel) ~ **to sterilization** Sterilisierfähigkeit *f* ~ **to stretching** Streckfestigkeit *f* ~ **to suction** Saugfestigkeit *f*, Saugwiderstand *m* ~ **to sudden changes of temperature** Temperaturwechselbeständigkeit *f* ~ **to swelling** Quellwiderstand *m* ~ **to swing** Schwenkwiderstand *m* ~ **to switching transients** Schaltfestigkeit *f* ~ **to tarnish** Anlaufbeständigkeit *f* ~ **to taxiing** Fahrwiderstand *m*, Rollwiderstand *m* ~ **to tearing** Reißfestigkeit *f* ~ **to torsion** Torsionswiderstand *m* ~ **to tropical conditions** Tropenfestigkeit *f* ~ **to turning** Ablenkungswiderstand *m* ~ **to twisting** Verdrehungsfestigkeit *f* ~ **per unit length** Widerstand *m* je Längeneinheit ~ **to vibration** Schwingungswiderstand *m* ~ **to wear** Abnutzungs-beständigkeit *f*, -widerstand *m*, Verschleißfestigkeit *f* ~ **against wear and tear** Verschleißhärte *f* ~ **to wear and tear** Abnutzungsfestigkeit *f* ~ **to weathering** Widerstandsfähigkeit *f* gegen atmosphärische Korrosion ~ **with zero phase angle** winkelfreier Widerstand *m*
**resistance,** ~ **alloy** Widerstandslegierung *f* ~ **amplifier** Widerstandsverstärker *m* ~**-arc furnace** Lichtbogenwiderstandsofen *m* ~ **balance** Widerstands-ausgleich *m*, -symmetrie *f* ~ **ballast** Widerstandsballast *m* ~ **box** Widerstands-büchse *f*, -kasten *m* ~ **box with plugs** Stöpselwiderstand *m* ~ **breaking** Reißfestigkeit *f* ~**-capacity arrangement** Widerstandskapazitätsanordnung *f* ~**-capacity-coupled** kapazitätswiderstandsgekuppelt ~ **cell** Mikrofonkapsel *f* ~**-choke coupling** Widerstandsdrosselkupplung *f* ~ **coefficient** Widerstands-beiwert *m*, -koeffizient *m* ~ **coil** Widerstands-

spule *f* ~ **component** Widerstandskomponente *f* ~**-coupled** widerstandsgekoppelt ~**-coupled amplifier** Widerstandsverstärker *m* ~ **coupling** galvanische Kopplung *f*, Widerstandskopplung *f* ~ **decade** Widerstandsdekade *f* ~**-dependent** widerstandsabhängig ~ **derivation** Widerstandsableitung *f* ~ **drag** schädlicher Widerstand *m* ~ **drop** ohmscher Spannungsabfall *m* ~ **due to sizing** Leimfestigkeit *f* ~ **formula** Widerstandsformel *f* **(electric)** ~ **furnace** Widerstandsofen *m* ~ **head** Widerstandshöhe *f* der Flüssigkeitsbewegung, (in filter wells) Widerstandshöhe *f* ~ **heating** Widerstands--beheizung *f*, -erhitzung *f*, -heizung *f* ~**-heating apparatus** Gerät *n* mit Widerstandsheizung ~ **holder** Widerstandshalter *m* (teleph.) ~ **indicator** Anzeigewiderstand *m* ~ **instrument** Widerstandsinstrument *n* ~ **lamp** Widerstandslampe *f* ~ **lattice** Widerstandsnetz *n* ~ **loss** Joulescher Wärmeverlust *m*, Ohmscher Verlust *m*, Stromwärmeverlust *m*, Widerstands-dämpfung *f*, -verlust *m* ~ **manometer** Widerstandsmanometer *n* ~ **measurer** Widerstandsbrücke *f* ~ **metal** Widerstandswerkstoff *m* ~ **offered by earth to being displaced** Erdwiderstand *m* ~ **operator** Widerstandsoperator *m* ~ **opposition to electric flow** Durchgangswiderstand *m* ~ **reciprocal** Widerstandsreziprok *n* ~ **relation** Widerstandsbeheizung *f* ~**-repeating amplifier** Verstärker *m* mit magnetischer Kopplung ~ **standard** Widerstandsnormal *n* ~ **step** Widerstandsstufe *f* ~ **strain gauge** Widerstandsdehnungsmeßstreifen *m* ~ **strengthener** Widerstandsverstärker *m* ~ **tapping** Widerstandsabgriff *m* ~ **test** Widerstandsmessung *f* ~ **thermometer** Widerstands-thermometer *n*, -wärmegradmesser *m* ~ **thermometry** Widerstandsthermometrie *f* ~ **tuning** Abstimmung *f* durch veränderlichen Widerstand ~ **unit** Doseneinheit *f*, Heizwiderstand *m* ~ **value** Widerstandswert *m* ~ **vance** Dämpfungsplättchen *n* ~ **variation** Widerstands-änderung *f*, -schwankung *f* ~ **voltmeter** Widerstandvoltmeter *n* ~ **welding** Widerstandsschweißung *f* ~ **welding process** Widerstandsschweißverfahren *n* ~ **winding** Ohmwicklung *f* ~ **windings** Widerstandswendel *m* ~ **wire** Widerstandsdraht *m* ~ **wires and tapes** Heiz- und Widerstandsdrähte und -bänder *pl*

**resistanceless** widerstandslos ~ **circuit** widerstandsloser Kreis *m*

**resistant** beständig, widerstandsfähig

**resistant,** ~ **to alkalies** laugenbeständig ~ **to bending** biegesteif, biegungssteif ~ **to compression** druckfest ~ **to corrosion** korrosionssicher ~ **to fracture** bruch-fest, -sicher ~ **to leaching solutions** laugebeständig ~ **to metallic salts** metallsalzbeständig ~ **to rupture** bruch-fest, -sicher ~ **to water** wasserfest

**resistant,** ~ **basic lining** haltbares basisches Futter *n* ~ **target** widerstandsfähiges Ziel *n*

**resistible** reservierbar, widerstandsfähig

**resisting** widerstandsfähig ~ **agent** Reservemittel *n* ~ **breaking** bruchfest ~ **force** Widerhalt *m* ~ **moment** Biegungsfestigkeit *f*, Rückkehrmoment *n* ~ **power** Widerstandsvermögen *n* ~ **torque** Rückkehrmoment *n*

**resistive,** ~ **circuit** aus Ohmschen Widerständen

bestehender Stromkreis *m* ~ **coupling** Widerstandskopplung *f* ~ **element** Widerstandsglied *n* ~ **lining** Widerstandsbelag *m* ~ **load** Ohmsche Belastung *f* ~ **wire strain** Dehnungsmeßstreifen *m*

**resistively** widerstandsbegrenzt

**resistivity** Resistivität *f*, spezifischer Widerstand *m*, Widerstandsfähigkeit *f* ~ **to heat** Wärmedurchgangswiderstand *m* ~ **to transfer** Übergangswiderstandsfähigkeit *f*

**resistivity,** ~ **minimum** Widerstandsminimum *n* ~ **survey** Widerstandsmessung *f*

**resistor** Baretter *m*, Rheostat *m*, Vorschaltwiderstand *m*, Widerstand *m*, Widerstand(s)-leister *m*, Widerstands-apparat *m*, -gerät *n* ~ **of extremely high ohmic value** Höchstohmwiderstand *m*

**resistor,** ~ **element** Heizwiderstand *m* ~ **hearth furnace** herdbeheizter Ofen *m* ~ **noise** Widerstandsrauschen *n* ~ **spark plug** Widerstandszündkerze *f* ~**-step indicator lamp** Widerstandsstufen-Anzeigelampe *f* ~ **suppressor** Entstörwiderstand *m* ~ **tube** Widerstandszylinder *m* ~**-type interference** Widerstandsentstörung *f*

**resists,** ~ **under pads** Reserven *pl* unter Klotzungen ~ **under steam colors** Reserven *pl* unter Dampffarben

**re-size, to** ~ nachleimen

**resolute** entschlossen

**resolution** Auflösung *f* (TV), Auflösungsvermögen *n* (rdr), Beschluß *m*, Entschluß *m*, Rasterfeinheit *f*, Schärfe *f*, Spaltung *f*, Widerstandsinkrement *n*, Zerlegung *f* ~ **on border** Bildkantenschärfe *f* ~ **of a force** Kraftzerlegung *f* ~ **of forces** Kraftzerlegung *f*, Zerlegung *f* von Kräften ~ **of a picture** Kantenschärfe *f*, Rasterung *f* ~ **of radiation** Zerlegung *f* der Strahlung

**resolution,** ~ **border** Randschärfe *f* (TV) ~ **pattern** Testbild *n* ~ **time** Auflösungszeit *f*

**resolvability** Lösbarkeit *f*

**resolvable** auflösbar

**resolve, to** ~ beschließen, sich entschließen, trennen, zerlegen **to** ~ **in(to)** auflösen

**resolved** zerlegbar ~ **line patterns** aufgelöste Linienbilder *pl*

**resolvent** Resolvente *f*

**resolver** Auflöser *m*, Koordinatenwandler *m*, Rechengerät *n*, Zerleger *m*

**resolving,** ~ **distance** auflösbarer Abstand *m* ~ **power** Abbildungsvermögen *n*, auflösende Kraft *f*, (of telescopes, photographic and shortwave telescope) Auflösungsvermögen *n*, Leistungsfähigkeit *f*, (opt.) Schärfe *f*, (of eye) Sehschärfe *f*, (of eye) Sehvermögen *n* ~ **power of eye** Sehschärfe *f* des Auges ~ **property** Auflösungsvermögen *n* ~ **time** Auflösungszeit *f*, Totzeit *f*

**resonance** Abstimmung *f*, Gleichklang *m*, Mitschwingen *n*, Nachhall *m*, Resonanz *f*, Tönung *f*, Widerhall *m*, Wieder-hall *m*, -klingen *n* **in** ~ Abspannung *f* (rdo), in Resonanz *f* befindlich **in** ~ **with** in Resonanz *f*

**resonance,** ~ **amplifier** Resonanzverstärker *m* ~ **bridge** Resonanzbrücke *f* ~ **capture** Resonanz--absorption *f*, -einfang *m* ~ **cavity** Resonanzhohlraum *n* ~ **chamber** Resonanzhohlraum *m* ~ **characteristic** Schwingkennlinie *f* ~ **circuit** Resonanzschaltung *f* ~ **cross-section** Resonanzwirkungsquerschnitt *m* ~ **curve** Antennen-abspannung *f*, -kurve *f*, Resonanz-kurve *f*, -verlauf *m* ~ **discontinuity phenomena** Ziehvorgänge *pl* ~ **effect** Resonanzwirkung *f* ~ **escape probability** Resonanz-Durchlaßwahrscheinlichkeit *f* ~ **formants** Hallformanten *pl* ~ **frequency** Resonanzfrequenz *f* ~**-frequency meter** Resonanzfrequenzmesser *m* ~ **indicator** Amplituden-anzeiger *m*, -glimmröhre *f*, Resonanz-anzeiger *m*, -röhre *f* ~**-instability phenomena** Ziehvorgänge *pl* ~ **level** Resonanzniveau *n* ~ **matching** Resonanzanpassung *f* ~ **method** Resonanzmethode *f* ~ **peak** Resonanzspitze *f* ~ **phenomenon** Resonanzerscheinung *f* ~ **point** Resonanzstelle *f* ~ **potential** Resonanzspannung *f*, Strahlungspotential *n* ~**-protraction phenomena** Ziehvorgänge *pl* ~ **radiation** (fluorescence) Resonanzfluoreszenz *f*, Resonanzstrahlung *f* ~ **relay** Resonanzrelais *n* ~ **resistance** Resonanzwiderstand *m* ~ **rise** Resonanzüberhöhung *f* ~ **scattering** Resonanzstreuung *f* ~ **screen** Resonanzsieb *n* ~ **shape** Resonanzkurvenform *f* ~ **spark gap** tönende Funkenstrecke *f* ~ **test** Klangprobe *f* ~ **transformer** abgestimmter Transformator *m*, Resonanztransformator *m* ~ **trap** Resonanzeinfang *m* ~ **wave meter** Resonanzwellenmesser *m*

**resonant** frequenzabhängig, mitschwingend, in Resonanz (befindlich) ~ **aerothermo-dynamic duct** Verpuffungsstrahlrohr *n* ~ **bubble** Resonanzblase *f* (acoust.) ~ **cavity in klystron chamber** Resonanzkammer *f* ~ **cavity X-ray tube** Hohlraumresonatorröhre *f* ~ **circuit** Resonanzkreis *m*, Schwingkreis *m*, (parallel) Schwungradkreis *m* ~**-circuit impedance** Schwingkreiswiderstand *m* ~**circuit inductance** Schwingkreisinduktivität *f* ~**-circuit resistance** Schwingkreiswiderstand *m* ~**circuit combination** Resonanzgebilde *n* ~ **condition** Resonanz--bedingung *n*, -zustand *m* ~ **conductor** mitschwingender Leiter *m* ~ **diaphragm** Anpassungsblende *f* ~ **effect** Resonanzwirkung *f* ~ **frequency** Resonanzfrequenz *f* ~ **line** in Resonanz befindliche Leitung *f* ~ **mode** Resonanzmodus *m* ~ **range** Resonanz-bereich *m*, -lage *f* ~ **rise** Aufschaukeln *n*, Hinaufpendeln *n* ~ **shunt** Resonanznebenschluß *m* ~ **vibration** Resonanzschwingung *f*

**resonate, to** ~ mitklingen, mitschwingen, resonieren, auf Resonanz abstimmen, in Resonanz bringen, in Resonanz sein

**resonated, not** ~ nicht abgestimmt

**resonating** in Resonanz *f* ~ **circuit** Resonanzkreis *m* ~ **method** Resonanzmethode *f*

**resonator** Funkenanzeiger *m*, Resonanzkreis *m*, Resonator *m* ~ **grid** Hohlraumgitter *n* ~ **mode** Kreismodus *m*

**resonoscope** Resonoskop *n*

**resorb, to** ~ aufsaugen, resorbieren

**resorcinol** Resorzin *n*, Resorzinöl *n*

**resorption** Aufnahme *f*, Resorption *f*

**resound, to** ~ erschallen, ertönen, mittönen, widerhallen

**resource** Mittel *n*

**resources** Erwerbsmittel *pl*, Geldmittel *pl*, Zufluß *m*

**resowing** Nachbestellung *f*

**respective** betreffend, einschlägig

**respectively** beziehungsweise

**respirable** atembar

**respiration** Atemholen *n*, Atmung *f*, Respiration *f*

**respirator** (oxygen) Atemgerät *n*, Atemschutzapparat *m*, Atmungs-apparat *m*, -gerät *n*, -maske *f*, Lungenschützer *m* ~ **bag** Atemsack *m* ~ **made of rubber** Gummimaske *f* ~ **penetrant** Maskenbrecher *m*

**respire, to** ~ atmen

**respite** Aufschub *m*, Bedenkzeit *f*, Frist *f*, Fristung *f*, Fristverlängerung *f*, Nachfrist *f*, Stillstandsfrist *f*, Stundungsfrist *f* **(term of)** ~ Stundung *f* ~ **of payment** Zahlungsaufschub *m*

**resplendent** schimmernd, widerstrahlend

**respond, to** ~ anschlagen, antworten **to** ~ **to** ansprechen auf **to** ~ **to operate** zum Ansprechen *n* bringen (relay)

**respond** Ansprechen *n* eines Relais

**respondability** Ansprechvermögen *n*

**responder** Antwortsender *m*, Anzeiger *m*, Detektor *m* ~ **beacon** Antwort-, Ansprech-funkfeuer *n*, Antwortsenderbake *f* (sec rdr)

**response** Anregung *f*, Ansprechen *n* (Regler), Ansprechvermögen *n*, Antwort *f*, (characteristic) Gang *m*, Reizbeantwortung *f*, Rückantwort *f*, Schalleistung *f*, Transponderantwort *f* (sec rdr), Verhalten *n*, Verlauf *m*, Widerschall *m* ~ **of ear to signal strength** Empfindungslautstärke *f* ~ **of pupil** Pupillenspiel *n*

**response,** ~ **characteristic** Empfindlichkeitskurve *f* ~ **curve** Empfindlichkeits-, Lichtempfindlichkeits-kurve *f* ~ **equalization** Frequenzentzerrung *f* ~ **factor** Ansprechwert *m* (aut. contr.) ~ **functions** Ansprechwahrscheinlichkeiten *pl* ~ **lag** Ansprechzeit *f* ~ **pulse shape** Pulsformungsfaktor *m* ~ **signal** Antwortsignal *n* ~ **stage** Ansprechstufe *f* ~ **synchro** Rückmelder *m* ~ **time** Einstell-, Reaktions-zeit *f*, Trägheit *f*

**responsibility** Haftbarkeit *f*, Haftpflicht *f*, Verantwortlichkeit *f*, Verantwortung *f* ~ **for consequences** Erfolgshaftung *f*

**responsible** haftbar, verantwortlich **to be** ~ haften

**responsive** empfindlich **to be** ~ gehorchen ~ **to** ansprechend

**responsiveness** Beruhigungs-dämpfung *f*, -fähigkeit *f*, Empfänglichkeit *f*, Empfindlichkeit *f*, Stabilisationsvermögen *n*

**responsor** Antwortempfänger *m* (sec rdr)

**rest, to** ~ lagern, rasten, ruhen **to** ~ **against** anliegen an **to** ~ **on** auflegen, beruhen auf, stützen

**rest** Auflage *f*, Lehne *f*, Pause *f*, Rast *f*, Rest *m*, Ruhe *f*, Schlitten *m*, Support *m* **at** ~ Ruhe *f*, in der Ruhe-lage *f*, -stellung *f* **to be at** ~ sich , in der Ruhelage *f* befinden ~ **of the magma** Restlösung *f*

**rest,** ~ **beam** Stützbalken *m* ~ **block** Auflegekasten *m* ~ **contact** Ruhekontakt *m* ~ **contact relay** Ruhekontaktrelais *n* ~ **mass density** Ruhemasse *f*, Ruhemassendichte *f* ~**-mass energy** Ruheenergie *f* ~ **position** Ruhestellung *f* ~ **resistance** Ruhewiderstand *m* ~ **resonance** Ruheresonanz *f* ~ **room** Aufenthaltsraum *m*

**re-stake, to** ~ nachstollen

**re-start, to** ~ wideranfangen, wideranlassen (engine)
**rested** gerastet
**resting** Abstehenlassen *n*; ruhend ~ **contact** Ruhekontakt *m* ~ **frequency** Ruheträgerfrequenz *f*, ummodulierte Trägerfrequenz *f* ~ **place of a pit** Schachtbühne *f* ~ **position** Ruhe-lage *f*, -stellung *f*
**restitute, to** ~ wiedererstatten
**restitution**Entschädigung *f*, Entzerrung *f* (photo), Herstellung *f*, Wieder-erstattung *f*, -gabe *f* ~ **of aerial photographs** Entzerrung *f* von Flugaufnahmen ~ **from air photographs** Luftbildauswertung *f*
**restitution,** ~ **coefficient** Rückkehr(ungs)koeffizient *m* ~ **delay** Wiedergabeverzögerung *f* ~ **element** Wiedergabeelement *n*
**restless** unruhig
**re-stone, to** ~ nachwetzen (mit Ölstein), mit dem Ölstein nachschleifen
**restoration** Erneuerung *f*, Instandsetzung *f*, Rekonstruktion *f*, Wieder-einführung *f*, -einrichtung *f*, -herstellung *f* ~ **of balance ( or of equilibrium)** Gleichgewichtsherstellung *f*
**restoration constant** Rückstellungskonstante *f*
**restorative** Belebungsmittel *n*
**restore, to** ~ ergänzen, erneuern, erstatten, nullen, picken, pillen, rekonstruieren, restaurieren, schärfen, umlagern, wiederherstellen, zurück-erstatten, -führen, -setzen, -stellen **to** ~ **a circuit to service** eine Leitung *f* wieder in Betrieb geben **to** ~ **the circuit** den (Fern)hörer *m* anhängen **to** ~ **normal** rücklaufen **to** ~ **the receiver** den Hörer *m* auflegen **to** ~ **the shutter** die Klappe *f* aufrichten
**restore** Abfallen *n* (d. Relais) ~**-to-normal switch** Apparat *m* mit Ruhestellung
**restored** zurückgewonnen ~ **energy** zurückgewonnene Energie *f* ~ **voltage** wiederkehrende Spannung *f*
**restoring** Auffrischen *n*, Auflegen *n*, Einhängen *n* ~ **action** Gegenkupplung *f* ~ **amplifier** Fesselverstärker *m* ~ **force** Rückstellkraft *f* ~ **mechanism** Gegenkupplung *f*, Rückführung *f* ~ **moment** aufrichtendes Moment *n*, Gleichgewichtskraft *f*, rückdrehendes Moment *n* (aviat.) Rück-drehmoment *n*, -kehrmoment *n*, -stellmoment *n*, Wiederherstellungskraft *f* ~ **moment coefficient** Rückstellmomentbeiwert *m* ~ **moment field** Luftmomentenfeld *n*, Rückstellfeld *n* ~ **moment theorem** Rückstellgesetz *n* ~ **spring** Abreißfeder *f*, Rück-führfeder *f*, -stellfeder *f*, -zugsfeder *f* ~ **time** Rückstellzeit *f* ~ **torque** Rückstellung *f* ~ **torque gradient** Richtmomentgradient *m*
**restraighten, to** ~ nachrichten
**restrain, to** ~ abhalten, anhalten, arretieren, behindern, bremsen, Einhalt tun, einhalten, festhalten, feststellen, hemmen
**restrained** gefaßt, unfrei ~ **to the ground** erdgebunden
**restrainer** Spar-beize *f*, -beizstoff *m*, Verzögerer *m*
**restraining** Abdämmung *f*, Abseilung *f* ~ **bath** Verzögerungsbad *n* ~ **coil** Haltespule *f* ~ **force** hemmende Kraft *f* ~ **substance** Hemmungskörper *m*
**restraint** Behinderung *f*, Beschränkung *f*, Ein-

schränkung *f*, (at ends) Einspannung *f*, Fesselung *f* (gyro), Hemmung *f*, Untersagung *f*, Zwang *m* ~ **of trade** Wettbewerbsverbots-Klausel *f*
**restraint forces** (in structure) Zwangkräfte *pl*
**restrict, to** ~ (to) beschränken, einschränken
**restricted** begrenzt, beschränkt, eingeschränkt **of** ~ **dimensions** gedrängt
**restricted,** ~ **area** Flugbeschränkungsgebiet *n* ~ **discharge** (ground effect machine) Abströmbegrenzung *f* ~ **flow condition** Teilförderung *f* (hydr.) ~ **guidance** zwangsläufige Führung *f* ~**-hour maximum indicator** Teilzeit-Höchstverbrauchsmesser *m* ~ **internal rotation** Rotationsbehinderung *f* ~ **lighting** Abblendung *f* ~ **line** nicht berechtigte Leitung *f* ~ **number control** Rufnummernsperre *f* ~ **space** Platzmangel *m* ~ **tariff** Pauschaltarif *m*
**restriction** Bedingung *f*, Beschränkung *f*, Eindämmung *f*, Einschränkung *f*, Maßnahme *f*, Verbot *n* ~ **of new constructions** Neubaubeschränkung *f* ~ **of obstructions** Hindernisbeschränkung *f*
**restrictive** maßgebend
**restrictor** Drossel *f*, Drosselstelle *f*, Durchflußbegrenzer *m*, Meßblende *f* ~ **and dump valve** Drossel- und Ablaßventil *n* ~ **correctors** Meßblendenkorrektoren *pl*
**restriking** Wiederzündung *f* ~ **voltage** Wiederzündspannung *f*
**resublime, to** ~ resublimieren
**resublimed** doppelt sublimiert
**resulphurization** Rückschwefelung *f*
**result, to** ~ anfallen, ausschlagen, entstehen, (from) erfolgen, ergeben, sich ergeben, folgen, resultieren **to** ~ **in** erzeugen, verursachen, wirken
**result** Ausfall *m*, Ausgang *m*, Ausschlag *m*, Auswirkung *f*, Befund *m*, Erfolg *m*, Ergebnis *n*, Folge *f*, Leistung *f*, Produkt *n*, Resultat *n*, Wirkung *f* ~ **of analysis** Analysenbefund *m* ~ **of experiment** Versuchsergebnis *n* ~ **of measurement** Meßergebnis *n* ~ **of the operation** Rechnungsergebnis *n* ~ **of research** Forschungsergebnis *n* ~ **of trial** Versuchsergebnis *n* ~ **of weathering** Verwitterungsprodukt *n*
**resultant** Ergebniskraft *f* (phys.), Gesamte *n*, Mittelkraft *f*, Resultante *f*, Resultierende *f*; resultierend, zusammengesetzt ~ **of lift** Auftriebresultierende *f*
**resultant,** ~ **action** resultierende Kraft *f* ~ **air force** Luftkraftresultierende *f* ~ **current** resultierender Strom *m* ~ **deviation** Gesamt-ablenkung *f*, -deviation *f* ~ **fire** Erfolgsfeuer *n* ~ **force** Gesamtkraft *f*, Mittelkraft *f*, resultierende Kraft *f* ~ **measurement** Meßergebnis *n* ~ **pitch of a winding** resultierender Wicklungsschritt *m* ~ **resistance** resultierender Widerstand *m* ~ **steering signal** Gesamtausschaltung *f* ~ **value** Endwert *m* ~ **wind force** resultierende Windkraft *f*
**resulting** entstehend ~ **from** gegründet auf ~ **moment** Mittelmoment *n*
**resume, to** ~ wieder-anfangen, -aufnehmen
**résumé** Zusammenfassung *f*
**resumption** Wiederaufnahme *f*, (of a law suit) Wiedereröffnung *f*
**resupply, to** ~ wiederherstellen

**resurrection** Auferstehung *f*
**resurvey** Neuaufnahme *f*, neue Vermessung *f*
**retail** Kleinverkauf *m* ~ **bookseller** Sortimenter *m* ~ **book trade** Buchhandel *m* ~ **business** Einzelhandel *m* ~ **price** Ladenpreis *m*, Preis *m* im Kleinhandel, Wiederverkaufspreis *m* ~ **sale** (**or selling**) Stückverkauf *m* ~ **trade** Kleinhandel *m*
**retailer** Kleinhändler *m*, Wiederverkäufer *m*
**retain, to** ~ aufstauen, behalten, bei(be)halten, binden, festhalten, zurück(be)halten **to** ~ **height** Höhe *f* behalten (aviat.) **to** ~ **the water** das Wasser anstauen
**retained, to be** ~ (of a quality) erhalten bleiben ~ **sample** Rückstellmuster *n* ~ **water level** Stauspiegel *m*
**retainer** Aufnahmehülse *f*, zur Befestigung dienender Bauteil *m*, Haltebügel *m*, (Kugellager) Käfig *m*, Knebel *m*, (Rollenlager) Laufrille *f* ~ **lock** Seegersicherung *f* ~ **ring** (of ball bearing) Scheibenkäfig *m*, Seeger-, Spreng-ring *m* ~ **spring** Klemmfeder *f*
**retaining** Abfangen *n*, Verankerung *f* ~ **angle strap** Kabelhaltewinkel *m* ~ **bolt** Schlüsselbolzen *m* ~ **bracket** Halteklammer *f* ~ **button** Halteklammer *f*, Halteknopf *m* ~ **circuit** Haltestromkreis *m* ~ **clip** Halte-bügel *m*, -klammer *f* ~ **current** Haltestrom *m* ~ **dam** Staudamm *m* (der Spülfläche) ~ **effect** zurückhaltende Wirkung *f* ~ **hoop** Arbeitsreif *m* ~ **level** Abschlußdamm *m* (der Spülfläche) ~ **lid plate for perforated tapes** Abfühlklappe *f* (teletype) ~ **lug** Halterriegel *m* ~ **nut** Befestigungs-, Halte-mutter *f*, Sicherungsring *m*, Überwurfmutter *f* ~ **pin** Auswerfer-, Schlüssel-bolzen *m*, Vorsteckstift *m* ~ **plate** Flacheisenverankerung *f* ~ **pressure heating system** Staudruckheizung *f* ~ **ring** Außen-, Haken-spreng-, Halte-, Klammer-, Sicherungs-, Spreng-, Übersteck-ring *m*, Bandagierung *f* ~ **rings** Feldklappen *pl* ~ **rod** Spannstange *f* ~ **screw** Verbohrschraube *f* ~ **slot** Nutenführung *f* ~ **spindle wedge plate** Keil *m* mit Spindelangriff ~ **spring** Halte-, Sicherungs-, Sperr-feder *f* ~ **strap** Kabelhalteschelle *f* ~ **strip** (**or bar**) Halteleiste *f* ~ **valve** Fußventil *n*, Standventil *n* ~ **wall** Futtermauer *f*, Stützmauer *f* ~ **washer** Sicherungsscheibe *f*
**retake, to** ~ wieder nehmen
**retake** Nachaufnahme *f*, Wieder(holungs)aufnahme *f*
**retaliate, to** ~ erwidern
**retaliation** Erwiderung *f*, Vergeltung *f*
**retaming bolt** Mitnehmerbolzen *m*
**retamp, to** ~ nachstopfen **to** ~ **the ties** (**or sleepers**) das Geleise *n* nachstopfen
**retard, to** ~ abbremsen, hemmen, hint(an)-halten, hinterhalten, nachstellen, retardieren, verlangsamen, verzögern, zurückhalten
**retard** Retardation *f*, Verzögerung *f* ~ **circuit** Verzögerungskreis *m* ~ **coil** Drossel-, Induktanz-spule *f* ~ **mechanism** Hemmwerk *n*
**retardation** Abnehmen *n* der Geschwindigkeit, Bremsung *f*, Bremswirkung *f*, Hemmung *f*, Nacheilung *f*, Nachgang *m*, negative Beschleunigung *f*, Rückverlegung *f*, Verlangsamung *f*, Verspätung *f*, Verzögerung *f* ~ **of boiling** Siedeverzug *m* ~ **of the deflection** Verzögerung *f*

des Ausschlages ~ **of ebullition** Siedeverzug *m* ~ **of electrons** Bremsung *f* der Elektronen, Elektronenbremsung *f*
**retardation,** ~ **angle** Verzögerungswinkel *m* ~ **coil** Abflachungsdrossel *f*, Drossel *f*, Drossel-, Induktions-spule *f*, Verzögerungs-spule *f*, -widerstand *m* ~ **method** Auslaufverfahren *n* ~ **pressure** Verzögerungsdruck *m*
**retarded** verzögert ~-**action method** Zeitlupenverfahren *n* ~ **burst** Spätzerspringer *m* ~ **combustion** verlangsamte oder verzögerte Verbrennung *f* ~ **control system** (servo) Regelsystem *n* mit Verzögerungsgliedern ~-**field triode** Bremsröhre *f* ~ **ignition** Nach-, Spät-zündung *f*, verzögerte Zündung *f* ~ **ignition position** Spätzündlage *f* ~ **indication** Anzeigeverzögerung *f* ~ **injection** Späteinspritzung *f*
**retarder** Verzögerer *m*, Gleisbremse *f* (r.r.)
**retarding** Abbremsung *f* ~ **of the ignition** Verschleppen *n* des Zündzeitpunktes
**retarding,** ~ **action** Sperrwirkung *f* ~ **additions** (**or additives**) hemmende Zusätze *pl* (chem.) ~ **circuit** Bremsanordnung *f* ~ **conveyor** Bremsförderer *m* ~-**curve obstacle** Hemmkurvenhindernis *n* ~ **device** Verzögerungseinrichtung *f* ~ **electrode** Bremselektrode *f* ~ **field** Bremsfeld *n*
**retarding-field,** ~ **circuit scheme** Bremsfeldschaltung *f* ~ **electrode** Bremselektrode *f*, Bremsfeldelektrode *f* ~ **oscillation** Bremsfeldschwingung *f* ~ **oscillator** Bremsfeldgenerator *m* ~ **tube** Bremsröhre *f*
**retarding,** ~ **force** Verzögerungskraft *f* ~ **grid** Bremsgitter *n* ~ **mechanism** Verzögerungsvorrichtung *f* ~ **parachute** Bremsschirm *m* ~ **potential method** Gegenfeldmethode *f* ~ **power** Bremsvermögen *n* ~ **scraper conveyor** Bremskratzbänder *pl* ~ **torque** (of a meter) Bremsmoment *n* ~ **valve** Vorlaufventil *n*
**retention** Beibehaltung *f*, Rückstand *m* ~ **of hardness** Anlaßbeständigkeit *f*
**retention,** ~ **bolt** Haltebolzen *m* ~ **mixer** Speichertankmischer *m* ~ **money** Sicherheitssumme *f* ~ **pin** Arretierstift *m* ~ **range** Halte-, Synchronisierungs-bereich *m* ~ **time** Verweilzeit *f* ~ **valve** Rückventil *n*
**retentive** festhaltend
**retentivity** (of adsorbents) Festhaltungsvermögen *n*, Koerzitivkraft *f*, Remanenz *f*
**retest, to** ~ nochmals untersuchen
**retest** Nachprüfungsversuch *m*, Wiederholungsversuch *m*
**rethreading dies** Nachschneideisen *pl*
**reticence** Zurückhaltung *f*
**reticle** Fadenkreuz *n*, Gradnetz *n*, Netz *n*, Strich(kreuz)platte *f* ~ **of squares** Rasterquadrat *n*
**reticle,** ~ **glass** Fadenkreuzlupe *f* ~ **image** Leuchtkreuz *n* ~ **lens** (**or magnifier**) Fadenkreuzlupe *f*
**reticular** netzartig, netzförmig ~ **structure** Netzstruktur *f*
**reticulate** faserig, netzartig, verworren
**reticulated** maschig, runzelig ~ **bond** Netzverband *m* ~ **glass** Faden-, Spitzen-glas *n* ~ **work** Netzverband *m*, Strickwerk *n*
**reticulation** Geflecht *n*, Netz *n*, Netz-bildung *f*, -struktur *f*, Vernetzung *f*

reticule Fadenkreuz *n* ~ **support** Fadenstück *n* (opt.)

**reticulum** Netzwerk *n*

**retie, to** ~ umbinden

**retiform** gitter-, netz-förmig, maschig

**retighten, to** ~ nachspannen, (screw) wieder anziehen

**retightening** Nachspannen *n*

**retimbering** Auswechseln *n* (der Zimmerung)

**retina** Netzhaut *f*

**retinal,** ~ **illumination** Netzhautbeleuchtung *f* ~ **image** Netzhautbild *n* ~ **persistence** Netzhautträgheit *f* ~ **photometer** Lichtsinnprüfer *m*

**retinoscope** Augenhintergrundspiegelung *f*, Retinoskopie *f*

**retinue** Gefolge *n*, Gefolgschaft *f*

**retire, to** ~ ausscheiden, sich verabschieden, zurück-gehen, -treten, sich zurückziehen

**re-tire, to** ~ auffrischen

**retired** außer Dienst *m*, im Ruhestand *m*

**retirement** außer Dienst *m*, Außerdienststellung *f*, Austritt *m*, Rückzug *m*, Ruhestand *m*

**re-tiring of tire casing** Neubelegen *n* der Laufdecke

**retort** Destillierblase *f*, Glühfrischkolben *m*, Kolben *m*, Muffel *f*, Retorte *f* ~ **for low- -temperature carbonization (or distillation)** Schwelretorte *f*

**retort,** ~ **bulb** Bauch *m* ~ **carbon** Retort(en)-kohle *f* ~ **carbonization** Retortenverkohlung *f* ~ **chamber** Retortenkammer *f* ~ **channel** Retortenmulde *f* ~ **coke** Gas-, Retorten-koks *m* ~**-coke plant** Gasanstalt *f* ~ **coking** Retorten- -verkokung *f*, -verkohlung *f* ~ **condenser** Retortenvorstoß *m* ~**-contact method (or process)** Retortenkontaktverfahren *n* ~ **furnace** Gefäß-, Muffel-, Retorten-ofen *m* ~ **gasification** Retortenvergasung *f* ~ **head** Retortenhelm *m* ~ **helm** Retortenhelm *m* ~ **oven** Retortenofen *m* ~**-oven coke** Gaskoks *m* ~ **press** Muffelpresse *f* ~ **process** Muffel-prozeß *m*, -verfahren *n* ~ **residue** (in zinc recovery) Räumasche *f*, Retortenrückstand *m* ~ **slide** Retortenschieber *m*

**retouch, to** ~ retuschieren

**retouch** Nachbesserung *f*, Retusche *f*, Überarbeitung *f*

**retouching** Retuschieren *n* ~ **on metal** Metallretusche *f* ~ **work** Retuschieren *n*

**retrace** Rückführung *f* (TV), Rücklauf *m* ~ **time** Rücklaufzeit *f*

**retract, to** ~ abfallen, einziehen, zurück-schlagen, -ziehen **to** ~ **the antenna wire** Antenne *f* einziehen **to** ~ **the undercarriage** Fahrgestell *n* einziehen

**retract line** Einfahrleitung *f*

**retractable** aufziehbar, einschwenkbar, einziehbar, zurückziehbar ~ **antenna mast** einziehbarer Antennenmast *m* ~ **core barrels** zurückziehbares Kernrohr *n* zur Kerngewinnung ~ **engine** einziehbarer Motor *m* ~ **float** einziehbarer Schwimmer *m* ~ **gear** einziehbare Vorrichtung *f* ~ **landing gear** einziehbares Fahrwerk *n* ~ **landing light** einziehbarer Landescheinwerfer *m* ~ **radiator** ausziehbarer oder einziehbarer Kühler *m* ~ **soot blower** Stoßbläser *m* (Kessel) ~ **spring** Zugfeder *f* ~ **tail wheel** einziehbares Spornrad *n* ~ **turret** Senkturm *m* ~ **undercarriage** einziehbares Fahrge-

stell ~ **wheel** einziehbares Rad *n* ~ **wing cam** rückziehbarer Hilfssenker *m*

**retracted** eingezogen ~ **antenna** eingefahrene Antenne *f* ~ **landing gear** eingefahrenes Fahrwerk *n* ~ **position** Einziehstellung *f* ~ **undercarriage** eingefahrenes Fahrgestell *n*

**retractile** einziehbar ~ **spring** Abreißfeder *f*, Rück-führfeder *f*, -zugsfeder *f*

**retractility** Rückstellkraft *f*

**retracting** ausschwenkbar ~ **link** Einziehstreben *n* ~ **mechanism** Einziehvorrichtung *f* ~ **roller** Abzugsrolle *f* ~ **spring** Abreißfeder *f*, Rückführfeder *f* ~ **strut** Einziehstrebe *f*

**retraction** (of landing gear, etc) Einfahrt *f*, Zurückziehung *f* ~ **of landing gear** Einziehen *n* des Fahrwerks

**retraction,** ~ **chamber** Rückzugsraum *m* ~ **cylinder** Arbeitszylinder *m* ~ **line** Rückzugsleitung *f* ~ **piston** Rückzugskolben *m*

**retractive shift** rückwärts tendierende Schaltbewegung *f*

**retractor** Wundhaken *m*, Wundrandhalter *m* ~ **spring** Rückholfeder *f* ~ **strut** Retraktorstrebe *f*

**retrain, to** ~ umstellen

**retraining** Umschulung *f*

**retransfer, to** ~ rückübertragen

**retransfer** Rückübertragung *f*

**retransition to ground state** Zurückfallen *n* in Grundzustand

**retranslate, to** ~ rückübersetzen, rückverwandeln

**retranslation** Rückübersetzung *f*

**retransmission** Nachrichtenweitergabe *f*, Umtelegrafierung *f*, Weiter-beförderung *f*, -gabe *f*, -sendung *f* ~ **relay** Weiterrufrelais *n*

**retransmit, to** ~ umtelegrafieren, weiter-befördern, -geben, -senden

**retransmitter** Weitergeber *m*, Zwischensender *m*

**retransmitting** Umtelegrafierung *f* ~ **arrangement** Weitergabeanordnung *f* ~ **installation** speichernde Übertragung *f* ~ **station** Blinkrelaisvermittlung *f*

**retreaded tires** runderneuerte Reifen *pl*

**retreading** Erneuerung *f* der Lauffläche ~ **of tire casing** Neubelegen *n* der Laufdecke ~ **mold** Form *f* für die Laufflächenerneuerung von Reifen

**retreat, to** ~ begradigen, zurück-gehen, -stellen, -weichen

**retreating,** ~ **blade stall** Abreißen *n* der Strömung an rücklaufendem Blatt ~ **part** (building) Hinterflügel *m*

**retreatment** Nachbehandlung *f*

**retrenchment** Kürzung *f*

**retrieve, to** ~ apportieren, wieder-auffinden, -erlangen

**retrieving spring** Rückschlagfeder *f*

**retrimming** Umtrimmen *n*

**retro-act, to** ~ rückwirken

**retro-action** Rückkopplung *f*, Rückwirkung *f*

**retro-active** rückgekoppelt, rückwirkend ~ **amplification** Rückkopplungsverstärkung *f* ~ **audion** Rückkopplungsaudion *n* ~ **circuit** Rückkopplungsschaltung *f* ~ **control action** Rückwärtsregelung *f* ~ **receiving circuit** Rückkopplungsempfangsschaltung *f* ~ **reception** Rückkopplungsempfang *m*

**retro-actor** Rückkopplungsröhre *f*

**retrodirective reflection** rückweisende Reflexion *f*
**retrograde, to** ~ zurücklaufen
**retrograde** rückgängig, rücklaufend, rückläufig, rückwärtsschreitend ~ **motion** Rückwärtsbewegung *f* ~ **ray** Umkehrstrahl *m*
**retrogression** Rückbewegung *f*, Rückgang *m*, rückläufige Bewegung *f* ~ **of alkalinity** Alkalitätsrückgang *m*
**retrogressive** rückgängig, rückgreifend, rückständig ~ **slide** rückwärtsschreitender Rutsch *m* ~ **wave** rücklaufende Welle *f* ~ **welding** Rückwärtsschweißung *f*
**retro-rocket** Bremsrakete *f* (g/m)
**retrospect** Rückblick *m* ~ **mirror** Rückblickspiegel *m*
**retrospective** rückschauend
**retterey** Rösterei *f*
**retting** Röste *f*, Röstung *f* ~ **pit** Einweichgrube *f*, (ropemaking) Flachsröste *f* ~ **vat** Röst-kufe *f*, -küpe *f*
**return, to** ~ (dial switch) ablaufen, (dial switch) ablaufen lassen, erstatten, (to normal) rücklaufen, umkehren, wieder-erstatten, -kehren, zurück-führen, -kehren, -leiten **to** ~-**cool** rückkühlen **to** ~ **to normal** in die Grundstellung zurückführen (zurückkehren), in die Ruhelage zurückkehren **to** ~ **to the working position** in Bereitschaftsstellung *f* bringen
**return** Aufstellung *f*, Ausziehstrom *m*, Biegung *f*, Gegensatz *m*, Gewinn *m*, Kröpfung *f*, Krümmung *f*, Rück-erstattung *f*, -fahrt *f*, -förderer *m*, -führung *f*, -gabe *f*, -gang *m*, -kehr *f*, -lauf *m*, -leitung *f*, -lieferung *f*, -schlag *m*, statistischer Ausweis *m*, Übersicht *f*, Windung *f*, Zurückgehen *n*
**return,** ~ **of cold weather** Kälterückfall *m* ~ **of dial** Rücklauf *m* der Nummernscheibe **in** ~ **direction** rücklaufend ~ **to ground state** Zurückfallen *n* in Grundzustand ~ **of the key** Zurückschlagen *n* der Taste **by** ~ **mail** postwendend ~ **of pencil** Kippen *n* ~ **of premium** Prämiengewähr *f* ~ **of ram** Stößelrückkehr *f* ~ **to service** Wiederinbetriebnahme *f* ~ **to the starting point** Rückkehr *f* zum Startpunkt ~ **to suction point** Rücklauf *m* zur Saugstelle ~ **of time-base potential to zero** Kippspannungsrücklauf *m* ~ **of time base to zero** Rücklauf *m* der Kippspannung
**return,** ~ **address** Rückkehradresse *f* (info proc.) ~ **air** Rückluft *f*, Umluft *f* (Klimaanlage) ~ **air damper** Umluftschieber *m* ~ **air pipe** Rückluft-leitung *f*, -rohr *n* ~ **air valve** Rückluftventil *n* ~ **arrangement** Rücklaufeinrichtung *f* ~ **belt** Rücktrum *m* (Förderanlagen) ~ **bend** (pipe) Doppelkniestück *n*, Umkehrbogen *m* ~-**bend plug** Umkehrstückstöpsel *m* ~ **blower** Rückfördergebläse *n* ~ **button** Rücklaufknopf *m* ~ **cam** Rückführnocken *m*, Rückzugskurve *f* ~ **channel** Rückströmkanal *m* ~ **chute** Rücklaufrinne *f* ~ **circuit** Herleitung *f*, Rück-führung *f*, -leitung *f* ~ **condenser** Rückflußkühler *m* ~ **conductor** Rück-leiter *m*, -leitung *f* ~ **conveyance** Rückförderung *f* ~ **coupling** Rückkupplung *f* ~ **crank** Gegenkurbel *f* ~ **current** Rückstrom *m*, Rückströmung *f* ~-**current cutout** Rückstromausschalter *m* ~-**current juice beater** Rückstromsaft-

vorwärmer *m* ~-**current relay** Rückstromschütz *n* ~ **driving pulley** Riemenscheibe *f* für den Rücklauf ~ **echo** Echorückkunft *f* ~-**feed equipment** Rücksaugeinrichtung *f* ~-**feed locking mechanism** Rücklaufsperre *f* ~-**feed spring** Rückspulfeder *f* ~ **feeder** Rückleitung *f* mit besonderem Leiter ~ **feeder cable** Rückspeisekabel *n* ~ **flight** Rückflug *m*, Rückmarsch *m* ~ **flow** Rückstrom *m* ~-**flow cooler** Rückflußkühler *m* ~-**flow hydraulic accumulator** Rückstrom-Druckspeicher *m* ~-**flow system** Gegenflußsystem *n* ~-**flow wind tunnel** geschlossener Windkanal *m* ~ **force system** Kraftrückleitung *f* ~ **form** Rückgabeschein *m* ~ **gallery** Rücklaufbohrung *f* ~ **gear** Rücklaufgetriebe *n* ~ **grilles** Rückengitter *n* ~ **idler** Unterbandrolle *f* ~ **interval** Rücklaufzeit *f* (rdr) ~ **journey** Rück-fahrt *f*, -gang *m*, -reise *f* ~ **lead** Rückführung *f*, Rückleiter *m* ~ **lever** Rückstellhebel *m* ~ **light** Antwortsignal *n* ~ **lighting stroke** rücklaufender Blitz *m* ~ **line** Rückführleitung *f* ~ **loss** Fehlerdämpfung *f* ~-**loss measuring set** Nachbildungsmesser *m* ~ **mechanism** Rückführungsgestänge *n* ~ **motion** Heimlauf *m*, Rückgang *m*, Rücksendung *f* ~ **movement** Rücklauf *m*, Rückwärtslauf *m* ~ **oil** Rücklauföl *n* ~ **pass** (in rolling) Rückgang *m*, Rücklauf *m*, (in rolling) Rückwärtskaliber *n*, (in boiler) Umschaltung *f* ~ **passage** Überstromkanal *m* ~ **path** Rück-führung *f*, -leitung *f*, -weg *m* ~ **pipe** Absaug-, Rücklauf-leitung *f* ~ **pole piece** (of a relay) Joch *n* ~ **pressure** Rücklaufdruck *m* ~-**pressure valve** rückläufiges Druckventil *n* ~ **pulley** Übertragungs-, Umleitungs-rolle *f* ~ **pump** Rückförderpumpe *f* ~ **push button** Rücklaufdruckknopf *m* ~ **remittance** Gegenrimesse *f* ~ **scoop** Rücklaufschaufel *f* ~ **scrap** (in foundry) Kreislaufmaterial *n* ~ **screen** Rücklaufsieb *n* ~ **screw (or screw conveyer)** Rückführschnecke *f* ~ **shaft** Ausziehschacht *m* ~ **shock** kalter Schlag *m* ~ **signal** Rücksignal *n* ~ **slag** Retourschlacke *f* ~ **speed** Rücklaufzeit *f* ~ **spiral** Kehrwandel *m* ~ **spring** Gegenfeder *f*, Rückhol-, Rückstell-feder *f* ~ **strand** Leertrum *n* ~ **stroke** Abwärtshub *m*, Heimlauf *m*, kalter Schlag *m*, Leerlaufhub *m*, Rückbewegung *f*, Rückgang *m*, Rückhub *m*, Rückbewegung *f*, **switch** Rückstellschalter *m* ~ **ticket** Rückfahrtkarte *f* ~ **time** (of a dial switch) Ablaufzeit *f* ~ **trace** Rücklauf *m*, Rücklaufspur *f* (CRT) ~-**trace elimination** Rücklaufverdunkelung *f* ~ **track curve** Umkehrschleife *f* ~ **transportation** Rückbeförderung *f* ~ **travel** Rücklauf *m* ~-**travel mechanism** Rückgangmechanismus *m* ~-**type steam trap** Kondenswasserrückleiter *m* ~ **valve** Rückflußventil *n* ~ **voltage** Rückflußspannung *f* ~ **wall** (building) Flügelmauer *f* ~ **warrant** Rückkehrschein *m* ~ **winding counter** Rückwickelzähler *m* ~ **wire** metallische Rückleitung *f*, Nulleiter *m*, Rückleiter *m*, Stromrückleitung *f* ~ **worm** Rückführschnecke *f*
**returnable** rückdrehbar ~ **drums** (container) Leihgebinde *n*
**returned shipment** Retour-, Rück-sendung *f*
**returning** Nachdrehen *n*, Umlenkung *f* ~ **device** Nachdrehvorrichtung *f*

**returns** Erwerb *m*, Geldumsatz *m*, Rück-fälle *pl*, -sendungen *pl*
**reunion** Wiedervereinigung *f*
**re-usability** Wiederverwendungsmöglichkeit *f*
**re-usable, ~ energy** Energieüberfluß *m* **~ shuttering** Dauerschalung *f* **~ waste heat** Energieüberfluß *m*
**re-use** Wieder-gebrauch *m*, -verwendung *f*
**re-used** wiederverwendet **~ wool** Reißwolle *f*
**re-utilization** Wiederverwendung *f* **~ of waste water** Abwasserrücknahme *f*, Rücknahme *f* der Abwässer
**revalorize, to ~** aufwerten
**revamping** Erneuerung *f*
**reveal step** bündige (Tür)Stufe *f*
**revelation wrapper** Sichtpackung *f*
**revenue** Einkünfte *pl*, Ertrag *m*, Rente *f* **~ cutter** Zollkutter *m* **~ expenditure** budgetierte Kosten *pl* **~ stamp** Banderole *f*
**reverberate, to ~** hallen, nachhallen, zurückstrahlen, (a ray) zurückwerfen
**reverberating** tönend **~ chamber (or enclosure)** Hallraum *m* **~ roof** Kuppel *f*
**reverberation** Echo *n*, Gegenschein *m*, Nachhall *m*, Nachhallen *n*, Widerhall *m*, Wiederhall *m*, Zurück-strahlung *f*, -werfen *n* **~ chamber** Hohlraum *m* **~ enclosure** Nachthall-kammer *f*, -raum *m* **~ period** Widerhallzeit *f* **~ phenomenon** Nachhallvorgang *m* **~ room (or space)** Nachhall-kammer *f*, -raum *m* **~ time** Nachhall-dauer *f*, -zeit *f*, Widerhallzeit *f*
**reverberatory** zurückstrahlend **~ calciner** (ore roasting) Fortschauflungsofen *m*, Krählofen *m* **~ furnace** Reverberier-, Strahlungs-ofen *m* **~-furnace practice** Flammofenbetrieb *m* **~ hearth** Flammofenherd *m* **~ hearth furnace** Herdflammofen *m* **~ puddling furnace** Flamm(en)ofen *m* **~ roaster** (ore roasting) Fortschauflungsofen *m*, Krählofen *m* **~ smelting** Flammofen-arbeit *f*, -schmelzen *n* **~ tin furnace** Zinnflammofen *m*
**reverify, to ~** nachkontrollieren
**reversal** Bewegungsumkehr *f*, Richtungsumkehr *f*, Rücklauf *m*, Stornierung *f*, Umkehr *f*, Umkehrung *f*, (of motion) Umklappen *n*, Umlegen *n*, Umschaltung *f*, Umschlagen *n*, Umstellung *f*, Umsteuerung *f*, Wechsel *m*, Wendung *f* (electr.); umkehrend
**reversal, ~ of air** Luftumsteuerung *f* **~ of the beams of light** Umkehrung *f* der Strahlen **~ of current** Strom-umkehr *f*, -wendung *f* **~ of damping** Entdämpfung *f* **~ of direction** Richtungswechsel *m* **~ of the magnetic needle** Umschlagen *n* der Magnetnadel **~ of motion** Laufumschaltung *f* **~ of pencil** (in scanning) Kippen *n* **~ of polarity** Polwechsel *m*, Umpolarisierung *f* **~ of rotation** Drehrichtungsumkehr *f* **~ of stress** Belastungs-, Last-wechsel *m* **~ of stroke** Hubwechsel *m* **~ of table** Tischumsteuerung *f*
**reversal, ~ development** Umkehrentwicklung *f* (photo.) **~ dip** Linienumkehr *f* **~ error** Umkehrspanne *f* **~ indication** Stornierungsvermerk *m* **~ period** Umschaltperiode *f* **~ point** Umkehrpunkt *m* **~ prevention** Rücklaufhemmung *f* **~ processing** Umkehrentwicklung *f* **~ stage** Umkehrstufe *f*

**reversals** Wechselzeichen *pl* (telegr.)
**reverse, to ~** (polarity connections, etc.) kommutieren, reversieren, rückwärts fahren, umkehren, umlegen, umpolen, umschalten, umsetzen, umstellen, umsteuern, umwenden **to ~ the magnetism** unmagnetisieren **to ~ polarity** umpolen **to ~ the relay** das Relais *n* umlegen
**reverse** Gegenteil *n*, Rückschlag *m*, Rückseite *f*, Rückwärtsgang *m*, Umkehrung *f*, Umschaltung *f*, Umsteuerung *f*; (order) umgekehrt, rückläufig **in ~ order** versetzt angeordnet **~ of ram** Stößelumkehr *f*
**reverse, ~ absorption edge (or limit)** Umkehrabsorptionskante *f* **~ action** Rücklauf *m* (film) **~-battery metering** Zählung *f* durch Stromumkehr **~-bending test** Hin- und Herbiege-probe *f*, -versuch *m* **~ blocking direction** Sperrichtung *f* **~ brake** Rückbremse *f* **~ caponier** Reverskaponniere *f* **~ circulation** Gegen-, Konter-spulung *f* **~ current** Gegen-, Rück-strom *m* **~ current circuit-breaker (or cut-out)** Rückstromschalter *m* **~-current relay** Gegenstrom-, Rückstellungs-, Rückstrom-relais *n* **~-current switch** Rückstromausschalter *m* **~ current trip** Rückstromauslöscher *m* **~ curve** Gegenkrümmung *f* **~ direction** Umkehrrichtung *f* **~ domain** unmagnetisierte Domäne *f* **~ double-gear wheel** Rücklaufdoppelrad *n* **~ driving lamp switch** Rückfahrschalter *m* **~ fault** Überkippung *f* **~ feedback** Gegenkupplung *f* **~-flow cooling** Kühlung *f* durch Rückbelüftung, Rückbelüftungskühlung *f* **~ gear** Rückwärtsgang *m*, Stirnradwendegetriebe *n* **~ gear shaft** Rücklauf-bolzen *m*, -welle *f* **~ gradient** Gegenneigung *f* **~ grid current** negativer oder positiver Gitterstrom *m*, Rückgitterstrom *m* **~-grid potential (or voltage)** Gegenspannung *f* **~ idler gear device** Rücklaufsicherung *f* **~ leakage current** Sperrstrom *m* **~ line** Minusstrang *m* **~ locking relay** Rückstrom-Sperrelais *n* **~ mechanism for adjustable-thrust nozzle** Wendebremsschütz *n* **~ moment** Rückwärtsbewegung *f* **~ motion** Gegenlauf *m*, Rück-(wärts)gang *m* **~-motion belt drive** Umkehrpunkt *m* **~ movement** Rückwärtsfahrt *f* **~ order of taking the photograph** Umkehrung *f* des Aufnahmevorganges **~ pedal** Fußhebel *m* für Rückwärtsfahrt **~ pinion** Rücklaufzahnrad *n* **~ pitch** Brems(steigungsein)stellung *f* **~ pitch propeller** Bremspropeller *m*, (Verstell-)Propeller *m* mit einstellbarer Bremsstellung **~ polarity** Gegenpolung *f* **~-power protection** Rückleistungsschutz *m* **~ rapid motion** Umkehrgang *m* **~-reduction gear** Wendeuntersetzungsgetriebe *n* **~ rotation** Gegenrakel *f* **~ saturation current** (in transistor) Kollektorreststrom *m* **~ scavenging** Umkehrspülung *f* **~ shaft** Kehr-, Rücklauf-welle *f* **~ side** Kehrseite *f*, Verso *n* **~ slope** Hinterhang *m* **~ slope position** Hinterhangstellung *f* **~ speed** Rückwärtsgang *m* **~ stop** Rückdrehsicherung *f* **~ strain** Rückwärtsverformung *f* **~ taper ream** konisch erweiterte Bohrung *f* **~ thrust** Bremsschub *m* **~ torsion** Retorsion *f* **~ valve** Umsteuerklappe *f* **~ voltage** Sperrspannung *f* **~ wheel** Rücklaufrad *n* **~ winding** Rücklaufwicklung *f*

**reversed** gekantet, umgebogen, umgefalzt, umgekehrt, zurück-gebogen, -laufend ~ **aileron control** Umkehr *f* der Querruderwirkung ~ **bending strength** Biegewechselfestigkeit *f* ~ **block** Negativklischee *n* ~ **controls** umgekehrtes Steuerwerk *m* ~ **current** Gegenstrom *m* ~ **curve** Kontrakurve *f* ~ **double loop** Kehrdoppelwendel *n* ~ **Elliot-type axle** Faustachse *f* ~ **fatigue strength** Dauerbiegefestigkeit *f* ~ **fault** Überschiebungswechsel *m*, widersinnige Verwerfung *f* (geol.) ~ **feed grates** Rückschubroste *f* ~ **fold** überkippte Falte *f* ~ **frame** Gegenspant *n* ~ **image** umgekehrtes Bild *n* ~ **impulsing** rückwärtige Stromstoßgabe *f* ~ **installation (or mounting)** umgekehrter Einbau *m* ~ **position** Umkehrstellung *f* ~ **reprint** Gegenumdruck *m* ~ **sign** entgegengesetztes Vorzeichen *n* ~ **signal** umgekehrtes Zeichen *n* ~ **signals** umgekehrte Schrift *f* ~ **spiral** Kehrwendel *n*

**reversement** verdrehter Überschlag *m*

**reverser** Richtungswender *m*, Stromwechsler *m*, Umkehrer *m* (electr.), Umschalter *m*, Wender *m*

**reversibility** Reversibilität *f*, Umkehrbarkeit *f*, Umsteuerbarkeit *f* ~ **paradox** Umkehreinwand *m* ~ **principle** Umkehrungsprinzip *n*

**reversible** drehbar, durchschlagbar, (reaction) hin und zurück, reversibel, reversierbar, rückläufig, umdrehbar, umkehrbar, umlegbar, umschaltbar, umstellbar, umsteuerbar, umwendbar, wendbar ~ **cam-shaft** umsteuerbare Nockenwelle *f* ~ **cell** umkehrbares Element *n* ~ **chain** umkehrbare Kette *f* ~ **clamping device** Wendespannvorrichtung *f* ~ **cloth** Doppelgewebe *n* ~ **connection** Umkehrschaltung *f* ~ **cycle** umkehrbarer Kreisprozeß *m* oder Kreisvorgang *m* ~ **delivery** umsteuerbarer Ausleger *m* (print.) ~ **Diesel engine** umsteuerbarer Dieselmotor *m* ~ **double lube oil filter** Doppelschmierölfilter *m* ~ **finder** umlegbarer Klappsucher *m* ~ **machine** Reversiermaschine *f* ~ **motion** reversible Bewegung *f* ~ **motor** reversierbarer oder umsteuerbarer Motor *m*, Wendemotor *m* ~ **pattern plate** Reversiermodellplatte *f* ~ **pendulum** Reversionspendel *n* ~ **permeability** reversible oder umkehrbare Permeabilität *f* ~ **permeability factor** Rückgangsziffer *f* ~ **pitch propeller** Bremspropeller *m*, Luftschraube *f* mit umkehrbarer Steigung ~ **printing units** umsteuerbare Druckwerke *pl* ~ **process** umkehrbarer Vorgang *m* ~ **propeller** Drehflügel-propeller *m*, -schraube *f*, umkehrbare Luftschraube *f*, Wendeschraube *f* ~ **pump** wechselseitig wirkende Pumpe *f* ~ **readout** reversibler Lesevorgang *m* ~ **segment** Umkehrsegment *n* ~ **shear** Umkehrschere *f* ~ **spirit level** Umkehr-, Wende-libelle *f* ~ **telescope** umlegbares Fernrohr *n*

**reversing** (machine) Umkehrung *f*, Umschalten *n*, (machine) Umsteuern *n*, Umsteuerung *f*; zweizügig ~ **of battery terminals** Kreuzung *f* der Batteriepole ~ **of the flow of current** Stromeinrichtungsumkehr *f* ~ **of sense of rotation** Drehrichtungsumkehr *f* ~ **and starting resistance** Umkehranlaßwiderstand *m* ~ **of the winding** Umklemmen *n* der Wicklung

**reversing,** ~ **air valve** Luftwechselklappe *f* ~ **apparatus** Umsetzvorrichtung *f* ~ **arrangement** Umsteuervorrichtung *f* ~ **bar** Umsetzschubstange *f* ~ **bath** Umkehrbad *n* ~ **blade** Umkehrschaufel *f* ~ **blooming mill** Reversierblockwalzwerk *n*, Umkehrblockwalzwerk *n* ~ **blooming--mill train** Reversierblockstraße *f* ~ **box** Wechselrichtungskasten *m* ~ **bucket** Umkehrschaufel *f* ~ **clutch** Umkehrkupplung *f* ~ **cogging mill** Reversierblockwalzwerk *n*, Umkehrblockwalzwerk *n* ~ **cogging-mill train** Reversierblockstraße *f* ~ **contact breaker** Pendelunterbrecher *m* ~ **cycle** Umschaltperiode *f* ~ **danger** Rücklaufgefahr *f* ~ **device** Umsteuervorrichtung *f* ~ **disc plow** Scheibenwendepflug *m* ~ **disc shutter** Drehscheibenverschluß *m* mit Umkehrscheiben, Umkehrscheibenverschluß *m* ~ **engine** Umkehrmaschine *f* ~ **friction clutch** Umkehrreibkupplung *f* ~ **gas valve** Gaswechselklappe *f* ~ **gear** Reversier-getriebe *n*, -vorrichtung *f*, Rückwärtsgang *m*, Umkehrgetriebe *n*, Umsteuergetriebe *n*, Umsteuerung *f*, Wende-getriebe *n*, -kupplung *f* ~ **handle** Richtungs-, Umsteuer-hebel *m* ~ **key** Gegensprechtaste *f*, Stromwender *m*, Umkehrtaste *f*, Wende(pol)schalter *m* ~ **lamp** Rückfahrscheinwerfer *m* ~ **latch** Tischumsteuerhebel *m* ~ **layer** umkehrende Schicht *f* ~ **lever** Umstell(ungs)hebel *m*, Umsteuerhebel *m* für vor- und rückwärts ~ **light** Rückfahrleuchte *f* ~ **mechanism** Umkehr-, Umschalt-mechanismus *m* ~ **mill** Reversier-, Umkehr-walzwerk *n* ~ **mill strand** Reversierstrecke *f* ~ **mill train** Reversier-, Umkehr-straße *f* ~ **motion** Rückwärtsantrieb *n*, Rückwirkung *f* ~ **motor box (or housing)** Wendemotorkasten *m* ~ **nozzle** Umlenkungsdüse *f* ~ **pin** Ablaufstift *m* ~ **piston** Umsteuerkolben *m* ~ **pitch propeller** Bremspropeller *m* ~ **plate mill** Reversierblechwalzwerk *n* ~ **plate train** Reversierblechstraße *f* ~ **point** Wendepunkt *m* ~ **point of shear action** Belastungsscheide *f* ~ **pole** Wendepol *m* ~ **process** (of regenerator) Wechselbetrieb *m* ~ **propeller installation** Umsteuerschraubenanlage *f* ~ **pump** Manöverpumpe *f* ~ **rolling mill** Kehr-, Reversier-, Umkehr-walzwerk *n* ~ **screw** Umsteuerschraube *f* ~ **shaft** Umsteuerwelle *f* ~ **shape mill** Reversierprofileisenwalzwerk *n* ~ **sheave** Umkehrrolle *f* ~ **slide** Umsteuerschreiber *m*, Wendesupport *m* ~ **starter** Umkehranlasser *m* ~ **stator blades** Umkehrkranz *m* ~ **strand of rolls for roughing** Reversierstreckgerüst *n* ~ **strip mill** Umkehrbandwalzwerk *n* ~ **switch** Fahrtrichtungsschalter *m*, Stromwender *m*, Umkehr-anlaßwiderstand *m*, -schalter *m*, Umschalter *m*, Wendeschalter *m* ~**-switch cylinder (or drum)** Umschaltwalze *f* ~**-switch method** Wechseltastverfahren *n* ~ **system** Umsteuerschraube *f* ~ **trip** Tischumsteuerhebel *m* ~**-type regenerator** Regenerator *m* mit Wechselbetrieb ~ **unit** Wendegetriebe *n* ~ **valve** Reversier-, Umschalt-ventil *n*, Umstell-klappe *f*, -ventil *n*, Umsteuerventil *n*, Wechselklappe *f* ~**-valve sleeve** Umschaltventileinsatz *m* ~ **weft** Kehrschuß *m* ~ **winding** Umschaltwicklung *f*

**reversion** Anwartschaft *f*, Rückfall *m*, Umkehrung *f*, Umpolung *f*, Umsteuerung *f*, Zurück-

gehen *n* ~ **level** Reversionsnivellierlibelle *f* ~
**rolling mill** Umkehrwalzwerk *n*
**reversive control** Rückkontrolle *f*
**revert, to** ~ umkehren, umwenden
**revertible** umkehrbar
**reverting,** ~ **call** Anruf *m* zwischen Teilnehmern
einer Gesellschaftsleitung, Rückruf *m* auf eigene
Leitung, Rückruf *m* auf die Gesellschaftsleitung ~ **call switch** Rückrufwähler *m*
**revertive,** ~ **blocking** rückwärtige Sperrung *f*
~ **control** rückwärtige Stromstoßgabe *f*, Wähler
*m* mit Rückimpulsgabevorgang ~ **impulses**
rückwärtige Stromstöße *pl* ~ **impulsing (or**
**pulsing)** rückwärtige Stromstoßgabe *f* ~ **pulsing**
**circuit** Stromkreis *m* für rückwärtige Stromstoßgabe ~ **pulses** rückwärtige Stromstoßgabe
*f* ~**-signal means** Antwortgeber *m*, Rückmelder
*m* ~**-signal panel** Rückmeldefeld *n*, Spiegelfeld
*n*
**revet, to** ~ abpflastern, bekleiden, füttern, verkleiden
**revetment** Befestigung *f*, Bekleidung *f*, Futtermauer *f*, Steinverkleidung *f*, Verschalung *f* ~ **of the banks** Ufer-bekleidung *f*,
-deckwerk *n*, -schutz *m* ~ **of reinforced gunite**
verstärkter Torkretputz *m* ~ **of screened rock**
Verkleidung *f* von gesiebten Steinen ~ **of**
**upstream face (or slope)** Verkleidung *f* der
wasserseitigen Böschung
**revetting** Verschalung *f* ~ **knob** Vorhalter *m*
**review, to** ~ durchmustern, durchsehen
**review** Bericht *m*, Besichtigung *f*, Durchsicht *f*,
Musterung *f*, Nachprüfung *f*, Rezension *f*,
Rückschau *f*, Übersicht *f*, Zeitschrift *f*, zusammenfassender Bericht *m* ~ **copy** Bespre-
chungsexemplar *n* ~ **room** Mustervorführungs-
raum *m*
**reviewer** Bearbeiter *m*, Referent *m*
**reviewing** Prüfung *f*
**revise, to** ~ durchsehen, revidieren
**revise** Korrekturabzug *m*, Revisionsbogen *m*
**revised,** ~ **proof** Korrekturabzug *m* ~ **version**
Neubearbeitung *f*
**reviser** Bearbeiter *m*, Korrektor *m*, Revisor *m*
**revision** Durchsicht *f*, Nachprüfung *f*, Revision
*f*, Überprüfung *f* ~ **of accounts** Prüfung *f* der
Rechnungen ~ **of test** Gegenprobe *f*
**revitalize** wiederbeleben
**revival** Wiederbelebung *f* ~ **in business** Geschäfts-
belebung *f*
**revive, to** ~ auffrischen, wiederbeleben
**revivification** Wiederbelebung *f*
**revivify, to** ~ wiederbeleben
**revivifying kiln** Wiederbelebungsofen *m*
**reviving** Auffrischen *n*
**revocable** widerruflich
**revocation** Aufhebung *f*, Widerruf *m*, Zurücknahme *f*
**revoke, to** ~ aufheben, rückgängig machen
**revolution** Doppelhub *m*, Drehbewegung *f*,
Drehung *f*, Revolution *f*, Rotation *f*, Tour *f*,
Umdrehung *f*, Umlauf *m*, Umschwung *m*, Umwälzung *f* ~ **of camshaft** Steuerwellendrehung *f*
~ **in counterclockwise direction** Linksdrehung *f*
**to make a full** ~ durchdrehen ~ **per minute**
Umlauf *m*
**revolution,** ~ **axis** Drehachse *f* ~ **counter** Dreh-,
Rotations-, Stich-, Touren-, Umdrehungs-,

Umlauf-zähler *m* ~ **indicator** Drehzahlmesser
*m* ~ **mark** Umdrehungsspur *f*
**revolutionize, to** ~ bahnbrechend sein, umwälzen
**revolutions,** ~ **per minute** Drehzahl *f* per Minute,
Umdrehungen *pl* in der Minute ~**-per-minute**
**reading** Drehzahlanzeige *f*
**revolve, to** ~ drehen, sich drehen, kreisen, rotieren, rundlaufen, schwenken, umdrehen,
umkreisen, umlaufen, umwälzen
**revolver** Revolver *m* ~ **camera** Revolverkamera *f*
~ **loom** Revolverstuhl *m*
**revolving** Dreh . . .; sich drehend, kreisend,
schwenkbar, umlaufend ~ **and centering me-**
**chanical (compound) stage** dreh- und zentrier-
barer Kreuztisch *m* ~ **and grading facilities**
Umwälz- und Sichteranlagen *pl* ~ **armored**
**turret** Panzerdrehturm *m* ~**-barrel sandblast**
**machine** Drehtrommelsandstrahlgebläse *n*,
Sandstrahlgebläse *n* mit umlaufender Trommel,
Sandstrahlputztrommel *f* ~ **bath** Karussellbad
*n* ~ **beacon** Blinkfeuer *n*, Drehlichtsignal *n* ~
**beacon light** Drehlichtscheinwerfer *m* ~ **belt**
**punches** Revolverlochzangen *pl* ~ **blade** Rund-
messer *n* ~ **blower** Drehbläser *m* ~ **bobbin** Abrollspule *f* ~ **breech** Drehtrommel *f* ~ **bucket**
Drehkübel *m* ~ **can** Drehtopf *m* ~ **case** dreh-
bares Regal *n* ~ **cement drier** Zementrotier-
ofen *m* ~ **chair** Drehstuhl *m* ~ **chute** Dreh-
schnurre *f* ~ **clearer cloth** umlaufendes Putz-
tuch *n* ~ **coil** drehbare Spule *f* ~ **contact maker**
Kollektor *m* ~ **cover** Aufklappdeckel *m* ~ **crane**
Dreh-, Titan-kran *m* ~ **cross-cut saw** rotierende
Ablängsäge *f* ~ **crystal** Drehkristall *m* ~ **cutter**
**head** Revolverbohrkopf *m* ~ **cylinder** umlaufender Zylinder *m* ~**-cylinder roaster** Trommel-
ofen *m* ~ **cylindrical furnace** Drehrohr-,
Trommel-ofen *m* ~ **device** Umlaufgerät *n* ~
**diaphragm** Revolverblende *f* ~ **endless wire** umlaufendes Langsieb *n* ~ **eyepiece head** Okularrevolver *m* ~ **fatigue-testing machine** Umlauf-
biegemaschine *f* ~ **field** Innenpolfeld *n*, Pol-
rad *n* ~ **field generator** Innenpolgenerator *m*
~ **filter** Trommel-filter *m*, -gebläse *n* ~ **flat**
Laufdeckel *m* ~ **flat card** Drehscheibenkarte *f*
~ **frame** Getriebequerhaupt *n* ~ **front sight**
Sternkorn *m* ~ **furnace** Drehofen *m* ~ **grate**
Drehrost *m* ~**-grate producer** Drehrostgene-
rator *m* ~ **harrow** Rollenegge *f* ~ **head** Drehknopf *m*, Lauftrommel *f* ~ **jib** drehbarer Ausleger *m* ~ **knob** Drehknopf *m* ~ **light** Blink-,
Dreh-feuer *n* ~ **mechanism** Umschaltvorrichtung *f* ~ **mirror** Drehspiegel *m*, rotierender
Spiegel *m* ~ **nosepiece** Objektiv-revolver *m*,
-wechsel *m* ~ **objective changer** Objektivrevolver *m* ~ **one-legged gantry crane** schwenkbarer
Halbportalkran *m* ~ **platform** Drehtisch *m* ~
**pot (for glass-blowing machines)** Drehwanne *f*
~ **press** Revolverpresse *f* ~ **puddling furnace**
Drehpuddelofen *m* ~**-rate gas producer** Drehrostgaserzeuger *m* ~ **reverberatory furnace**
Drehflammofen *m* ~ **rolls** drehbare Walzen *pl*
~ **screen** Drehsieb *n* (min.), Separations-trom-
mel *f*, Sortiertrommel *f*, Trommelsieb *n* ~**-**
**-screen drum** Siebzylinder *m* ~ **searchlight**
Drehlichtscheinwerfer *m* ~**-shackle suspension**
Drehbügelaufhängung *f* ~ **shears** Rollschermesser *n* ~ **shutter** Rolladen *m* ~ **singer** Wal-
zensengmaschine *f* ~ **slide plate** Ringschieber

*m* ~ **slide rest** Kugelsupport *m* ~ **storm** Wirbelsturm *m* ~ **switch** Drehschalter *m*, umlaufender Schalter *m* ~ **table** Drehherd *m* ~ **template dial** Revolverstrichplatte *f* ~ **tippler** Kreiselwipper *m* ~ **tube** Spinnröhrchen *n* ~ **tubular furnace** Drehrohrofen *m* ~ **turret** Drehkuppel *f*, Drehturm *m* ~**-type hangar** Drehhalle *f* ~ **wagon crane** Wagendrehkran *m* ~ **worm** Förderschnecke *f*

**revulsion** Umschwung *m*

**reward** Belohnung *f*, Lohn *m*

**rewarm, to** ~ wiedererwärmen

**rewash** Nachwaschen *n* ~ **box** Nachsetzkasten *m*

**rewashed middlings** Nachwaschmittel *n*

**rewashing machine** Nachsetzmaschine *f*

**rewind, to** ~ neu wickeln, neuwickeln, rückspulen, umspulen, umwickeln, eine Spule umwickeln

**rewind** Bandrückspulen *n* ~ **collar** Hülse *f* (photo) ~ **control switch** Rangierschalter *m* ~ **fork** Spulenmitnehmer *m* ~ **knob** Rückwickelknopf *m* ~ **mechanism** Rückwicklung *f* ~ **motor** Rückwickelmotor *m* (tape rec.) ~ **period** Umspielzeit *f* (eines Bandes) ~ **release** Rücklaufentkupplung *f* ~ **shaft** Führungsstift *m*

**rewinder** Umroller *m* (film), Umspuler *m*, Umwickler *m* (electr.) ~ **machine** Umroller *m* (paper mfg.)

**rewinding** Neuwicklung *f*, Umbäumen *n*, Umspulen *n* ~ **machine** Umhaspelmaschine *f*, Umroller *m* (paper mfg.), Umrollmaschine *f* ~ **roller** Aufwickelwalze *f*

**reword, to** ~ anders formulieren, neu abfassen

**rework, to** ~ nach(be)arbeiten, überarbeiten, wiederaufarbeiten

**rework** Nachbearbeitung *f*

**reworking** Überarbeitung *f*

**rewrite, to** ~ regenerieren, umschreiben, wieder einschreiben (data proc.)

**rewriting** Reportage *f*, Umschreiben *n*

**Reynolds number** Reynolds'sche Zahl *f*

**R.F., r.f. (radio frequency)** Hochfrequenz ... (HF) ~ **choke** Hochfrequenzdrossel *f* ~ **connection** HF-Anschluß *m* ~ **filtering of power leads** Netzverriegelung *f* ~ **head** HF-Teil *m* ~ **line** HF-Leitung *f* ~ **plumbing** festverlegtes Wellenleitersystem *n* ~ **service oscillator** Empfänger *m* ~ **standard signal generator** HF-Standard-Meßgenerator *m* ~ **test set** HF-Prüfgerät *n* ~ **transmission** HF-Übertragung *f*

**rhabdomancy** Untersuchung *f* mit der Wünschelrute

**Rhenish** rheinisch ~ **furnace** rheinischer Ofen *m*

**rheograph** Rheograf *m*

**rheologic property** Formänderungsfähigkeit *f*

**rheological,** ~ **behaviour** rheologisches Verhalten *n* ~ **properties** Fließeigenschaften *pl*

**rheology** Fließ-kunde *f*, -lehre *f*, Wissenschaft *f* der Verformung unter Fließen

**rheometer** Galvanometer *n*, Rheometer *n*

**rheonomic** rheonom

**rheopexy** Verfestigung *f*

**rheostat** Drehwiderstand *m*, elektrischer Widerstand *m*, Magnetfeldregler *m*, regelbarer Widerstand *m*, Regelwiderstand *m*, Reglerwiderstand *m*, Rheostat *m*, Stromregler *m*, Widerstand *m* ~ **resistance** Vorschaltwiderstand *m*

**rheostatic braking** Widerstandsbremsung *f*

**rheostriction** Pincheffekt *m*, Rheostriktion *f*

**rheotan** Rheotan *n*

**rheotron** Rheotron *n* ~ **accelerator** Resonanzbeschleuniger *m*

**rho ratio** Kontraktionskoeffizient *m*

**rhodium** Rhodium *n* ~ **metal** Rhodiummetall *n*

**rhodochrosite** Dialogit *m*, Manganspat *m*, roter Braunstein *m*

**rhodonite** Eisenrhodonit *m*, Kieselmangan *n*, Mangan-kiesel *m*, -stein *m*, Rhodonit *m*, Rotbraunsteinerz *n*, Rotspat *m*

**rhomb** Rhombus *m* ~ **spar** Bitter-, Rauten-spat *m*

**rhombic** rautenförmig, rhombenförmig ~ **antenna** Rauten-, Rhombus-antenne *f* ~ **dodecahedron** Granat-, Rhomben-dodekaeder *n* ~ **line portion of a great circle** Orthodrome *f* ~ **pane** Rautenglas *n*

**rhombohedral** rhomboedrisch

**rhombohedron** Rautenflach *n*, Rhomboeder *n*

**rhomboid** Rhomboeder *n*, Rhomboid *n*, schiefwinkliges Parallelogramm *n*; rautenförmig, rhomboidisch ~ **chain** Rautenkette *f* ~ **point** Rautenpunkt *m*

**rhomboidal** rhomboidförmig, rhomboidisch

**rhombus** Raute *f*, Rautenfläche *f*, Rhombus *m* ~ **design (or pattern)** Rautenmuster *n*

**rhometer** Rhometer *n*

**rhomboidal arseniate of copper** Tamasit *m*

**rhumb** Kompaßstrich *m*, Windstrich *m* ~ **line** Dwarslinie *f*, Kursgleiche *f*, (straight course) Loxodrome *f*, Schieflaufende *f*, Schräge *f*

**rhumbatron** Hohlraumresonator *m*

**Rhumkorff coil** Rhumkorffscher Apparat *m*

**rhyacolite** Eisspat *m*

**rhysimeter** Rhysimeter *n*

**rhythm** Arbeitstakt *m*, Rhythmus *m*, Schrittmaß *n*, Takt *m*, Zeitmaß *n* **in** ~ im Rhythmus *m*

**rhythmic,** ~ **buzzer** Unterbrechersummer *m* ~ **light** Taktfeuer *m* ~ **variation** Pendel *n*

**rhythmometer** Rhythmusmesser *m*

**rib, to** ~ rippen, durch Rippen verstärken

**rib** Bein *n*, Feder *f*, Gewölberippe *f*, Laufschiene *f*, Leiste *f*, Rippe *f*, Spant *m*, Speiche *f*, Spiere *f* (aviat.), Stab *m*, Stange *f*, Strebe *f*, Verstärkungsring *m* ~ **of a radiator** Heiz-, Kühlrippe *f* ~ **of a semicircle** Gerüstrippe *f*

**rib,** ~ **band** Rödelträger *m* ~ **barrier** Sicherheitspfeiler *m* ~ **flange** Deckleiste *f*, Rippen-gurt *m*, -gurtung *f* (parachute) ~ **frame** Rippenspant *n* ~ **plate** Federblattrippe *f* ~**-roof knife** Dachrippenmesser *n* ~ **seam** Rippennaht *f* ~ **stiffening** falsche Rippe *f* ~ **stitching** Rippen-nähen *n*, -naht *f* ~ **width** Rippenbreite *f*

**ribbed** (pavement) geriffelt, gerippt, mit Rippen versehen, verrippt ~ **bolt** Rippenschraube *f* ~ **ceiling** gerippte Decke *f* ~ **cross section of frame** Rippenquerschnitt *m* ~ **funnel** Rippentrichter *m* ~ **glass** Riffelgals *n* ~ **guide in breechblock** Führungsleiste *f* ~ **heating pipe** Rippenheizrohr *n* ~ **insulator** Rippenisolator *m* ~ **nail** Kerbnagel *m* ~ **radiator** Lamellen-, Streifen-kühler *m* ~ **spring leaf** Rippenfeder *f* ~ **tank** Rippengefäß *n* ~ **tube** Lamellenrohr *n*

**ribbing** Rippenwerk *n*, Verrippung *f*

**ribbon** Band *n*, Borte *f*, Farbband *n*, Latte *f*, Leiste *f*, Maßband *n*, Metallband *n*, Strang *m* (Spinnerei), Streifen *m* ~ **of wadding** Watteband *n*

ribbon, ~ agate Bandachat *m* ~ antenna Bandantenne *f* ~ blender Gegenstrommischer *m* ~
brake parachute Bänderbremsschirm *m* ~
building Reihenbau *m* ~(-shaped) cable Band-,
Flach-kabel *n* ~ carrying an electric current
stromdurchflossenes Band *n* ~ chip Bandspan
*m* ~ coil Bandspule *f* ~ conductor Bandleitung *f*
~ container Farbbanddose *f* ~ control mechanism Bandhub *m* ~ feed Farbbandtransport *m*
~ flat coils Flachbandteilspulen *pl* ~ folder
Bandwickler *m* ~ guide Farbbandführung *f* ~
lap Faserbandwinkel *m* ~ lift mechanism Farb-
bandhub *m* ~ lightning Bandblitz *m* ~ lockout
bar Bandhubausschalthebel *m* (teletype) ~
loom (textiles) Bandstuhl *m* ~ loom gaiter
Bandeinrichter *m* ~ loudspeaker Bandsprecher
*m* ~ microphone Bändchenmikrofon *n*, Bandmikrofon *n* ~ mixer Band(schnecken)mischer
*m* ~ oscillator lever Farbbandhubhebel *m* ~
reverse mechanism Bandschaltung *f* ~ reverse
shaft Bandumkehrachse *f* ~ saw Bandsäge *f* ~
spools Fahrbandspulen *pl* ~-wrapped stahlband-
armiert, -bewehrt

riblet Hilfsrippe *f*

rice Reis *m* ~ fiber Reisfaser *f* ~ paper Reis-
papier *n* ~ straw Reisstroh *n* ~ tresher Reis-
dreschmaschine *f*

rich (mine) edel, fett, (land) fruchtbar, (in ore)
haltig, hältig, reich

rich, ~ in alumina tonerdereich ~ in carbon koh-
lenstoffreich ~ in copper kupferreich ~ in gas
gasreich ~ in gold goldreich ~ in graphite
graphitreich ~ in hydrogen wasserstoffreich ~
in iron eisenreich ~ in iron oxide eisenoxydreich
~ in lead bleireich ~ in lines linienreich ~ in
losses verlustreich ~ in manganese manganreich
~ in oxide oxydreich ~ in oxygen sauerstoff-
reich ~ in silver silberreich ~ in tin zinnreich
~ in tone volltönend

rich, ~ binding Prachteinband *m* ~ coal Fett-
kohle *f* ~ concentrations hochhaltige Anreiche-
rungen *pl* ~ finery cinder Schwal *m* ~ gas hochwertiges Gas *n*, Reichgas *n* ~ lead Reichblei *n*
~ mixture fettes Gemisch *n*, Gemenge *n*, fette,
reiche oder zementreiche Mischung *f* ~ ore
Edelerz *n* ~ silver ore Formerz *n* ~ slag gare
Frischschlacke *f*, Reichschlacke *f*

riches Reichtum *m*

ricinoleic acid Risinol-, Rizinol-säure *f*

ricochet, to ~ abprallen

ricochet Abprallen *n*, Abpraller *m*, Indirekt-
treffer *m*, Prall-, Prell-schuß *m*, Querschläger
*m*, Rücksprung *m* ~ burst Abpraller *m* ~
effect Abprallerwirkung *f* ~ fire Abpraller-
schießen *n*

ricocheting Kugelrücksprung *m*

rid, to ~ befreien, entziehen, freimachen, losmachen, reinigen to get ~ of abschaffen, los
werden

riddle, to ~ absieben, durchsieben

riddle Drahtzieh-brett *n*, -platte *f*, Rätter *m*,
Schüttelsieb *n*, Siebrahmen *m* ~ sifter with
return chute Schüttelsiebmaschine *f* mit Rücklaufrinne

riddled, ~ like a sieve durchsiebt

riddler Schüttelsieb *n*

riddling Absieben *n*, Absiebung *f*, Siebwäsche
*f*

riddlings Abfallasche *f*, Abgang *m* (min.), Durch-
fallkohle *f*, Rostdurchfall *m*

ride, to ~ ausfahren, entlanggleiten, fahren, sich
durch Schwerkraft fortbewegen to ~ along mitfahren to ~ at anchor zu Anker liegen to ~ a
bicycle radfahren to ~ on abrollen, abwalzen
to ~ over hinweglaufen über (relay) to ~ on the
vents mit geschlossenen Ventilklappen *pl* und
offenen Flutventilen *pl* fahren

ride, ~ index Wertziffer *f* ~ plow Furchenpflug
*m*

rideau Geländewelle *f*

rider (piano) Brücke *f*, kleines Begleitflöz *n*,
Lauf-, Reiter-gewicht *n* (Waage), Lenkschemel
*m*, Reiter *m*, (Kartei) Reiterchen *n*, Salband *n*
(min.), Wendeschemel *m* ~ bench Halteteil *m*
(Halteschiene) ~ carrier Reiterlineal *n* ~ plate
Binnenspant *m* ~ roller Reiterwalze *f* ~ weight
Laufgewicht *n*

ridge, to ~ a house ein Dach *n* verfirsten to ~
a roof befirsten

ridge Bergrücken *m*, Dachfirst *m*, Einfädelschlitz *m* (tape rec.), Falzriegel *m*, First *m*,
Gebirgskamm *m*, Grad *m*, Grat *m*, Hoch-
druckrücken *m* (meteor), Höhen-rippe *f*,
-rücken *m*, Kamm *m*, Kante *f*, Leiste *f*, Riff *n*,
Riffel *m*, Rücken *m*, Rückgrat *n*, rückgrat-
artige Erhebung, Wulst *f* ~ between furrows
Furchenrain *m* ~ of a roof Dachspitze *f*, First-
haube *f*, -linie *f*

ridge, ~ beam (buildings or cars) Firstbalken *m*
~ buster Dammglätter *m* ~ circuit Firstleitung
*f* ~ covering Verfirstung *f* ~ cultivation Kamm-
bau *m* ~ knife Dachrippenmesser *n* ~ lead
Firstpfette *f*, Sattelblech *n* ~ line Kamm-,
Rücken-linie *f* ~ piece Wolfrähm *m* ~ piece of
a house Dachzinne *f* ~ plate Sattelblech *n* ~
plough Furchenzieher *m*, Häufelpflug *m*
~ purlin Firstpfette *f* ~ rib Scheitelrippe *f* ~
rod Firststange *f* ~ slices Dachrippenschnitzel
*n* ~ soaring Hangsegeln *n* ~ tile First-, Grat-zie-
gel *m* ~ turret Dachreiter *m* ~ waveguide Steg-
hohlleiter *m* (microw.)

ridged, ~ coping Firstabdeckung *f* ~ end Federwulst *m* ~ roof Satteldach *n*, zweihängiges
(zweihängiges) Dach *n*

ridgeless stufenlos, wulstlos ~ seam wulstlose
Naht *f* ~ seam-welding wulstlose Nahtschweißung *f*

ridges granite steiles Felsenufer *n*

ridging Verfirstung *f* ~ plate Firstblech *n*

riding Fahren *n* ~ attachment Sitzvorrichtung *f*
~ bed Lenk-, Wende-schemel *m* ~ beet puller
Rübenheber mit Führersitz ~ bolster Lenk-,
Wende-schemel *m* ~ chain Schwimmkette *f* ~
comfort Fahrbequemlichkeit *f* ~ cotton planter
Baumwolldrillmaschine *f* mit Sitz, Baumwollpflanzer *m* mit Führersitz ~ cultivator Hackmaschine *f* mit (Führer-) Sitz ~ disk gang plow
Mehrscheibenpflug *m* mit Führersitz ~ disk
plow mehrschariger Pflug *m* mit Sitz, Scheibenpflug *m* mit Führersitz ~ gate Gleistor *n* ~
light Anker-lampe *f*, -leuchte *f*, -licht *n* ~
plow Pflug *m* mit Sitz ~ system Reitersystem
*n*

Riedel starter Riedelanlasser *m*

Riemann's mapping theorem Riemannscher Ab-
bildungssatz *m*

**riffle** Durchlaß *m*, Riefelung *f*, Rille *f*, Rinne *f*, Stromschnelle *f* ~ **calender** Riffelkalander *m* ~ **sampler** Riffelprobenteiler *m* ~ **sheet iron** geriffeltes Blech *n* ~ **surface table** Rillenherd *m* ~ **tube** Rippenrohr *n*
**riffled tube** Rippenrohr *n*
**riffler** Lochfeile *f*, Raspel *f*, Riffelpfeile *f*
**rifle, to** ~ ausreifen
**rifle** Büchse *f*, Flinte *f*, Gewehr *n* ~ **barrel** Gewehrlauf *m* ~ **range** Schieß-platz *m*, -stand *m* ~ **telescope** Zielfernrohr *n*
**rifling** (gun) Drall *m*; Riefung *f*, Zug *m* ~ **with increasing twist** Keilzüge *pl* ~ **rod** Ziehstange *f*
**rift** Ritze *f* ~ **valley** Graben *m*, Senkungsgraben *m*
**rifted lead** Riffelblei *n*
**rig, to** ~ aufrüsten, auftakeln, betakeln, luftverlasten, takeln **to** ~ **for diving** zum Tauchen *n* bereitmachen **to** ~ **for surface** zum Auftauchen *n* bereitmachen **to** ~ **up** (eine Anlage) aufbauen
**rig** Anlage *f*, Aufbau *m*, Ausrüstung *f*, Gestell *n*, Rüstung *f*, Takelung *f* ~ **bar** Einstell-Lehre *f* ~ **builder** Turmmontagearbeiter *m* ~ **irons** Bohrkranteile *pl*, mechanischer Teil *m* einer Seilmaschine ~ **pin** Einstellstift *m*
**rigger** Bordmonteur *m*, (for airplane) Rüster *m*
**riggers angle** Einfallswinkel *m*
**rigging** Abspannung *f*, Aufrüsten *n*, Betakelung *f*, (balloon) Geleine *n*, Gut *n*, Leinenwerk *n*, Montieren *n*, Rüsten *n*, Rüstungsverfahren *n*, Takelage *f*, Takelung *f*, Verspannung *f*, Vertakelung *f*, Zusammenbau *m* in einer Vorrichtung (Flugzeugbau), Zutakelung *f* ~ **angle of incidence** Einstellwinkel (aerodyn) ~ **band on captive balloon** Ballongurt *m* ~ **datum line** Lehren-, Rüst-, Verspannungs-bezugslinie *f* ~ **diagram** Verspannungsschema *n* ~ **gear** Takelage *f* ~ **hatch** Montageluke *f* ~ **line** Fangleine *f* ~ **machine** Dubliermeß- und Wickelmaschine *f* ~ **patch** Gänsefuß *m* ~ **position** Aufrüstungsposition *f*, Montierungshaltung *f* ~ **screw** Wantschraube *f* ~ **system** Leinensystem *n*, Takelage *f*
**right, to** ~ aufrichten **to** ~ **path of flight** aufrichten **to** ~ **a plane** herumschlagen
**right** Anrecht *n*, Anspruch *m*, Berechtigung *f*, Bezugsrecht *n*, Gerechtsame *f*, Recht *n*; recht, rechts, richtig ~ **to mine** Berggerechtsame *f*
**right, at** ~ **angles** quer, rechtwinklig senkrecht zueinander ~ **of appeal** Berufungsrecht *n* ~ **of capture** Prisenrecht *n* ~ **to a concession** Mutungsrecht *n* ~ **of the discover or the finder** Finderrecht *n* ~ **of mortgagee** Pfandrecht *n* ~ **of pasture** Abtrift *f* ~ **by the plummet** im Lot *n* ~ **of preemption** Vorkaufsrecht *n* ~ **of prior use** Vorbenutzungsrecht *n* ~ **of prospecting** Schürfrecht *n* ~ **of redemption** Rückkaufsrecht *n* ~ **of refusal** Vorkaufsrecht *n* ~ **of repurchase** Rückkaufsrecht *n* ~ **of way** Vortritts-, Vorflugsrecht *n* ~ **of way signal** Einfahrtszeichen *n*
**right-angle** rechtwink(e)lig, rechter Winkel *m* ~ **bend** Rohrbogen *m* ~ **connector** Winkelstecker *m* ~ **friction gear** Planscheibengetriebe *n* ~ **joint** Winkelstoß *m* ~ **speed reducer** Winkeluntersetzungsgetriebe *n* ~ **unit** Kegelradgetriebe *n* ~ **valve** Eckventil *n*
**right-angled** rechtwink(e)lig (math.) ~ **brace** Nachstellwinkel *m* ~ **triangle** rechtwinkliges Dreieck *n*

**right,** ~ **ascension** gerade Aufsteigung *f* (des Gestirns), Rektaszension *f* ~ **ascension circle** Stundenkreis *m* ~ **cone** rechter Kegel *m* ~ **crown** Kreuzknoten *m* ~ **cylinder** gerader Zylinder *m* ~ **half of the girder** rechte Riegelhälfte *f*
**right-hand** rechts-drehend, -laufend, -schneidend ~ **drive** Rechtslenkung *f* ~ **engine** rechtsläufiger Motor *m* ~ **frog** Rechtsweiche *f* ~ **helix** Rechtsschraube *f* ~ **lay** Rechtsdrall *m* ~ **(left-hand) member of an equation** rechte (linke) Seite *f* einer Gleichung ~ **post** rechter Stiel *m* ~ **rule** Dreifingerregel *f* ~ **scala** Skalenplättchen *n* ~ **screw** rechtsgängige Schraube *f* ~ **side tool** rechter Seitenstahl *m* ~ **siding tool** rechtsseitiger Abflachstahl *m* ~ **sloped girder** rechter Schrägstab *m* ~ **switch** Rechts-schalter *m*, -weiche *f* ~ **thread** Rechtsgewinde *n* ~ **(steel) tool** rechtseitiger Stahl *m* ~ **traffic** Rechtsverkehr *m* ~ **transmission** Rechtsantrieb *m* ~ **turning tool** rechter Drehstahl *m*, Rechtsstahl *m* ~ **turnoff** Rechtsweiche *f* ~ **twist** (of a cable) Rechtsdrall *m* ~-**wound** rechts gewunden
**right-handed** rechtshändig, von rechts nach links arbeitend ~ **curve** Rechtskurve *f* ~ **helix** Rechtsspirale *f* ~ **motion** Rechtslauf *m* ~ **notched** rechtsgenutet ~ **polarization** Rechtsdrehung *f* ~ **quartz** Rechtsquarz *m* ~ **rotation** Rechtsrotation *f* ~ **srew** Rechtsschraube *f*, rechtswendige Schraube *f*
**right,** ~ **height** Handhöhe *f* ~ **helicoid** Wendelfläche *f* ~ **lay rope** Rechtsschlagseil *n* ~ **knot** Kreuzknoten *m* ~ **prism** gerades Prisma *n* ~ **proportion** Gleichmaß *n* ~ **raising cam** Einsatzplättchen *n* ~ **section** Normalschnitt *m* ~ **side** (of paper) Oberseite *f* ~-**side drive** Rechtsantrieb *m* ~ **side of an equation** rechte Seite *f* einer Gleichung ~-**side engine** rechtsseitiger Motor *m* ~-**side round iron** Hohlkehleneisen *n* ~ **spur gearing** Stirn-räderwerk *n*, -radgetriebe *n* ~-**to-left twist** Rechts- auf Linksdraht *n* ~-**(angled) triangle** rechtwinkliges Dreieck *n* ~ **turn** Rechtskurve *f*
**righteous** gerecht
**rightfulness** Rechtmäßigkeit *f*
**righting,** ~ **of aircraft** Abfangen *n* ~ **capacity** (of a ship) Aufrichtvermögen *n* ~ **moment** aufrichtendes Moment *n*, Rückdrehmoment *n*
**rightness** Richtigkeit *f*
**rights,** ~ **of an author** Urheberrecht *n* ~ **of an inventor** Urheberrecht *n*
**rigid** fest, formhaltend, kräftig, prall, schallhart, stabil, standfest, standsicher, stark, starr, steif, stetig, straff, stramm, streng, unbiegsam, unwandelbar ~ **(ly) built** stark gebaut ~ **in phase** phasenstarr
**rigid,** ~ **airship** Starrluftschiff *n* ~ **axle** Starrachse *f* ~ **body** starrer Körper *m* ~-**body displacement** Starrkörper-Verschiebung *f* ~ **coupling** Starrkoppeln *n* ~ **drawgear** feste Zugvorrichtung *f* ~ **draw hook** fester Zughaken *m* ~ **fastening** feste Verbindung *f* ~ **foundation** starre Grundlage *f* ~ **frame** Rahmenbinder *m*, Steifrahmen *m* ~ **frame with two hinged supports** Zweigelenkrahmen *m* ~ **horn wall** schallharte Trichterwand *f* ~ **motion** starre Bewegung *f* ~ **parts equipped with springs** abgefederte starre Teile *pl* ~ **pavement** Hartbelag *m*, starre Decke *f* ~ **pinion drive** starrer Ritzeltrieb *m* ~-**plastic**

**approximation** starr-plastische Näherung *f* ~
**rail** Anschlagschiene *f*, (of the switch) Stock-
schiene *f* ~ **stay** Rundeisenanker *m* ~ **steel rule**
Stahllineal *n* ~ **support** Dach- und Mauerstän-
der *m* ~ **suspension** starre Aufhängung *f*
**rigidity** Bieg(e)festigkeit *f*, Dauerstandfestigkeit
*f*, Festigkeit *f*, Hartnäckigkeit *f*, Stabilität *f*,
Stand-festigkeit *f*, -sicherheit, *f*, Starrheit *f*,
Steife *f*, Steifheit *f*, Steifigkeit *f*, Unbeugsam-
keit *f*, Unnachgiebigkeit *f* ~ **in space** Beharrungsvermögen *n* im Raum
**rigidity,** ~ **modulus** Drillungs-, Gleit-modulus *m*,
Scherungs-modul *m*, -modulus *m*, Schiebungsmodulus *m*, Torsionsmodulus *m* ~ **number** Steifigkeitszahl *f*
**rigidly,** ~ **connected** starrschlüssig ~ **locked in
phase** phasenstarr verkoppelt ~ **mounted** fest
montiert ~ **mounted counterweight** festes Gegengewicht *n* ~ **terminated tube** hartgeschlos-
senes Rohr *n*
**rigor** Härte *f*
**rigorous** streng ~ **proof** strenger Beweis *m* ~
**solution** strenge Lösung *f*
**rigorously correct** mathematisch genau
**rill** Furche *f* ~ **stoping** Firstenbau *m* mit geneigter Firste
**rilles srew** Rändelschraube *f*
**rim, to** ~ bördeln, rändeln **to** ~ **a wheel** ein Rad
*n* befelgen
**rim** Außenzone *f*, Bord *m*, Bordrand *m*, (of tire
or wheel) Felge *f*, (of wheel) Felgenkranz *m*,
Kante *f*, (of wheel) Radkranz *m*, Rand *m*, Rand-
fassung *f*, Randzone *f*, Reifen *m*, Schiene *f*,
(face) Spurkranz *m* **(gear)** ~ Kranz *m* **(wheel)**
~ Radreifen *m* **to fit the** ~ **on** die Felge aufziehen
**rim,** ~ **of the basket** Korbrand *m* ~ **of bearing
surface** Tragerand *m* ~ **of cartridge case** Bodenrand *m* ~ **of the compass card (or rose)** Rosenrand *m* ~ **of cutaway** Sichtausschnittbogen *m*
~ **for head lamp** Verschlußdeckel *m* mit
Dichtplatte für Scheinwerfer ~ **of toothed
wheel** Zahnradkranz *m* ~ **of the valve disk**
Ventiltellerrand *m* ~ **of the wheel** Unterreifen
*m*
**rim,** ~ **angle** Randwinkel *m* ~ **band** Wulstband
*n* ~ **bolt** Felgenbolzen *m* ~ **clutch** Bandkupplung *f* ~ **collar** Kranzwulst *m* ~ **drive** Bortelantrieb *m* ~ **edge** Felgenrand *m* ~**-fire cartridge**
Randfeuerpatrone *f* ~ **joint** Kranzstoß *m* ~
**knob lock** Einsteckkastenschloß *n* ~ **padding**
Randpolsterung *f* ~ **pattern** Kranzmodell *n* ~
**ring** Felgenring *m* ~ **roll** Spurkranzwalze *f* ~
**size** Felgengröße *f* ~ **sizing machine** Felgen-
rundstauchmaschine *f* ~**-straddling turbine
bucket** Reiterschaufel *f* ~ **syncline** Randmulde
*f* ~ **tape** Felgenband *n* ~ **tool** Felgen-abzieh-
hebel *m*, -montiereisen *n* ~ **truing anvil** Felgen-
richtamboß *m* ~**-type** felgenartig ~ **wedge** Fel-
genkeil *m* ~ **zone** Randschicht *f*
**rime, to** ~ bereifen
**rime** Anraum *m*, Duftanhang *m*, Haarfrost *m*,
Reif *m*
**rimless,** ~ **holder** (glasses) randloses Gestell *n* ~
**lens** ungefaßtes Glas *n*, ungefaßte Linse *f*
**rimmed,** ~ **letter** musierte Schrift ~ **steel** unberuhigter Stahl *m* ~ **steel ingot** unruhig vergos-
sener Block *m*

**rimming,** ~ **action** Bildung *f*, Kragenbildung *f*
~ **steel** unberuhigter Stahl *m*, unberuhigtes
Flußeisen *n*
**rimy** bereift
**rind** Baumrinde *f*, Schwarte *f* ~ **gall** Holzkropf *m*
**ring, to** ~ beringen, läuten, rufen, schallen,
wecken **to** ~ **back** zurückrufen (teleph.) **to** ~
**the exchange** das Amt anrufen **to** ~ **off** abklin-
geln, abläuten, abrufen **to** ~ **out** ausquetschen
**to** ~ **through** durchrufen **to** ~ **up** anklingeln,
anläuten
**ring** (jet) Außensammler *m*, Bügel *m*, Glied *n*,
(of a tube) Klingen *n*, Läuten *n*, Öse *f*, Rad-
kranz *m*, Reif(en) *m*, Ring *m*, Ruf *m*, Schall *m*,
Verstellring *m*, (of fuse) Zünderteller *m*, ~ **of
an anchor** Rohrring *m* ~ **of bell clapper** Glok-
kenring *m* ~ **of case** Gehäusebügel *m* ~ **of
compass bowl** Kesselring *m* ~ **for dredging bolt**
Baggerbolzenring *m* ~ **for fishing socket** Kern-
fangring *m* ~ **of forts** Festungsgürtel *m* ~ **of
glaze** Glasurring *m* ~ **on pliers** Zangenring *m*
~ **of plug** Stöpselhals *m* ~ **with spring** Kapsel
*f* mit Feder (Rückzugfeder) ~ **of wire** Drahtring *m*
**ring,** ~**-and-bead sight** Kreiskorn *n* ~**-and-ex-
pander** Radialdichtung *f*, Simmering *m* ~**-and-
speak-key** Rufumschalter *m* ~ **angle** Ringeck
*n* ~ **anode** Lochscheibenanode *f* ~ **armature**
Ringanker *m* ~ **around** kreisförmige Zielan-
zeigeverbreiterung *f* (sec. rdr.) ~**-back kay**
Rückruf-schalter *m*, -taste *f* ~**-back signal**
Nach-, Rück-ruf *m* ~ **balance** Ringwaage *f*
~**-band tool** Spannband *n* für Kolbenringeinbau ~ **barrage** Ringsperre *f* ~ **brushes** Maler-
ringpinsel *m* ~ **buckle** Ringschnallverschluß *m*
~ **burner** Heizkranz *m*, Kranzbrenner *m*, Kronenbrenner *m* (chem.) ~ **cell** Ringküvette *f* ~
**channel** Abfangkanal *m* (hydraul) ~ **chuck**
Ringfutter *n* ~ **circuit** Ringverzweigung *f* ~**-
compression coupling** Hülsenkupplung *f* ~
**conduit** Ringleitung *f* ~ **counter** Ringzähler *m*
~ **cowl(ing)** Ringhaube *f* ~ **cradle** Jackenwiege
*f* ~ **die** Mahlring *m*, Mahlteiler *m* ~ **discharge**
Ringentladung *f* ~**-down operation** Anrufbetrieb *m* ~**-down operation of a no-delay or on-
demand basis** Sofortverkehr *m* mit Anrufbetrieb ~ **end stocks** Ringkluppen *pl* ~ **engine**
Zugramme *f* ~ **envelope** Hüllring *m* ~ **extractor**
Ringzieher *m* ~ **fastener** Befestigungsring *m* ~
**fitting** Ringstutzen *m* ~ **formation** Ringbildung
*f*, Ringlichkeit *f* ~**-forward signal** Vorwärtsruf
*m* ~**-gap mixing nozzle** Ringspaltmischdüse *f*
~ **gauge** Einstellehrring *m*, Kaliber-lehre *f*,
-ring *m*, Lehrring *m*, Ring-kaliber *n*, -maß *n*,
-messer *m* (aviat) ~ **gauges** Dornringe *pl* ~
**gear** Drehkranz *m*, Glockenrad *m*, (Umlaufgetriebe) Zahnkranz *m* ~ **gearing** Ringführung *f*
~ **girder** (Rohrleitg.) Kragenversteifung *f* ~
**grating** Ringgitter *n* ~ **groove** Kolbennut *n* ~
**head** Ringknopf *m* ~ **land** Ringsteg *m* ~ **lap**
Läppring *m* ~ **lock** Ringschloß *m* ~ **lubricated
bearing** Schmierringlager *n* ~ **lug** Ringöse *f*
**magnet system** Linsenkette *f* ~ **main** Ringleitung *f* ~ **micrometer** Ringmikrometer *n* ~ **mode
filter** ringförmiges Wellenfilter *n* ~ **mount** Drehkranz *m*, Drehring *m* ~ **neck** umgelegter Hals
*m* ~ **nozzle** Ringdüse *f* ~ **nut** Ringmutter *f* ~**-
off** Abläuten *n* (teleph.) ~**-off signal** Abläut-

zeichen *n* ~-oil bearing Lager *m* mit Schmierring ~ oiler Ringöler *m* ~ oiling Ringschmierung *f*
ring-oiling, ~ bearing Ringschmierlager *n* ~ drop hanger Ringschmierhängelager *n* ~ pillow block Ringschmierstehlager *n* ~ post hanger Ringschmiersäulenarmlager *n* ~ quill bearing Ringschmierhohlwellenlager *n*
ring, ~ ore Ringelerz *n* ~ oscillator ringförmig angeordnete Oszillatorengruppe *f* ~ parachute Ringfallschirm *m* ~-pendulum tachometer Ringpendeltachometer *m* ~ periphery Ringumfang *m* ~-piston flow meter Ringkolbenflüssigkeitszähler *m* ~ pivot Ringzapfen *m* ~ plain Wallebene *f* (des Mondes) ~ plate Armkreuz *n* ~ pole shoe (or piece) Ringpolschuh *m* ~-porous ringporig ~ rail movement Wagenführung *f* ~-roll mill Pendelmühle *f*, Roulette *n* ~ scaler rückgekoppelte Zählschaltung *f* ~ screw Doppelschraube *f*, Schraubenring *m* ~ seal Ringverschmelzung *f*
ring-shaped ringförmig ~ area Kreisring *m* ~ channel Ringtrog *m* ~ gunpowder Ringpulver *n* ~ horn Ringtrichter *m* ~ induction furnace Induktionsofen *m* mit ringförmigem Herd, rinnenförmiger Induktionsofen *m* ~ precision wiping resistor Ringfeinschleifwiderstand *m* ~ stiffener Aussteifungsring *m*
ring, ~-shearing apparatus Kreisringapparat *m*, Ringscherapparat *m* ~ sight Kreiskimme *f*, Ringvisier *n* ~ single earth (or ground) electrode Ringstirnelektrode *f* ~ slot Ringspalt *m* ~ spanner Schlüsselring *m* ~ spanner box end wrench Ringschlüssel *m* ~ spindle system Ringspinnsystem *n* ~ spring Ringfeder *f* ~ sticking Festkleben *n* der Kolbenringe ~ stopper (Ankerkette) Ringstopper *m* ~ target Ringscheibe *f* ~ thread gauge Gewindelehrring *m* ~ time Abklingzeit *f* (rdr) ~ torch Ringbrenner *m* ~ transformer Ringübertrager *m* ~ traveller Ringläufer *m* ~ trumpet Ringtrichter *m* ~ twisting Ringzwirnen *n* ~-type gas burner Ringgasbrenner *m* ~-type joint Ringfuge *f*, Ringverbindung *f* ~-type orifice plate Ringkammernormblende *f* ~-type shim (or washer) Beilegescheibe *f* ~-type transformer Ringtransformator *m* ~ valve Ringventil *n* ~ vortex distribution Ringwirbelverteilung *f* ~ winding Ringwicklung *f* ~ wire Ader *f* zum Stöpselhals, b-Ader *f*, (plug) Stöpselringzuführung *f*
ringer Brechstange *f*, Rufstromdynamo *m*, Ruf(strom)maschine *f*, Wecker *m* ~ test Prüfung *f* der Rufsätze
ringing Anrufen *n*, Läuten *n*, Ruf *m*, Rufen *n*, (of a tube) Selbsttönen *n*, Wecken *n* ~ changeover switch Weckerumschalter *m* ~ circuit Weckstromkreis *m* ~ code Rufschlüssel *m* ~ connection Rufschaltung *f* ~ converter Rufumsetzer *m* ~ current Ruf-, Weck-strom *m* ~ current indicator Rufstromanzeiger *m* ~ cycle Rufphase *f*, Zeitraum *m* der Rufstromsendung ~ device Läutewerk *n* ~ dynamo Rufstromdynamo *m* ~ failure Rufstörung *f* ~ frequency Rufstromfrequenz *f* ~ inductor Läutinduktor *m* ~ key Ruf-schalter *m*, -taste *f* ~ lead Rufstromzuführung *f* ~ loop Weckstromkreis *m* ~ machine Ruf(strom)maschine *f* ~ noise Klingfähigkeit *f* ~ oscillator Eichmarkenoszillator *m*

~ periodicity Rufperiode *f* ~ pilot lamp Rufüberwachungslampe *f* ~ plug Verbindungsstöpsel *m* ~ position Rufstellung *f* ~ relay Anruf-, Läute-, Ruf(anschalt)-relais *n* (through-) ~ relay Durchrufrelais *n* ~ reversing key Rufstromumkehrtaste *f* ~ set Niederfrequenzrufsatz *m* ~ sound helle Klangfarbe *f* ~ test Klangprobe *f* ~ time Schwingzeit *f* (rdr) ~ tin Feinzinn *n* ~ tone Freiton *m*, Freizeichen *n* ~-trip relay Trennrelais *n* ~ vibrator Polwechsler *m*, Rufstromanzeiger *m*
rinse, to ~ abschwemmen, abspulen, abwaschen, auswaschen, schweifen, schwemmen, spülen, wässern to ~ casks faßschwanken to ~ out ausspülen, herausspülen
rinsing Abspülen *n*, Spülen *n*, Spülwasser *n* ~ apparatus for toilet Abortspülapparat *m* ~ buddle Schlämmfaß *n* ~ device Spülteller *m* ~ line Spülleitung *f* ~ medium Brausetrübe *f*, Spülmittel *n* ~ paddle Spülhaspel *f* ~ plant Spülanlage *f*, Spülerei *f* ~ roller Wasserwalre *f* ~ room Spülhalle *f* ~ screen Abbrausesieb *n*, Brausesieb *n*
rip, to ~ aufreißen, bereißen, nachreißen, schlitzen to ~ off herausschlagen to ~ open aufschlitzen, aufschneiden, auftrennen to ~ up auftrennen
rip, ~-bottom tank Leerlaßbehälter *m* ~ cord Abreiß-, Aufzug-, Reiß-, Zug-leine *f* ~ cord on parachute Abreißschnur *f* ~-cord handle Abreißknopf *m* ~-cord spring Abreißfeder *f* ~ panel Reißbahn *f* ~-saw Kerb-, Spalt-säge *f* ~-stop rißhemmend ~ tooth Hobelzahn *m* ~ wire Reißdraht *m*
riparian uferanliegend ~ lands Ufergelände *n*
ripe reif
ripeness Reife *f* ~ of the viscose Reifezustand *m* der Viskose
ripening Reifung *f* ~ substance Reifungskörper *m*
ripidolite Chlorit *m*
ripper Nachreißer *m*
ripping Nachreißen *n* (des Gesteins), Reißen *n*, Schlitzen *n*, Trennen *n*; spaltend, trennend ~ bar Brechstange *f*, Kistenöffner *m* ~ chisel Brech-, Stemm-eisen *n*, Hobel-, Stech-beitel *m* ~ hammer Aufreißhammer *m* ~ iron Naht-, Werg-haken *m* ~ machine Schrämmaschine *f* ~ panel Aufreißwand *f* ~ saw Brett-, Spalt-säge *f* ~ toggle Reißknebel *m*
ripple, to ~ (water) kräuseln, (flax) reffen, (flax) riffeln
ripple Kabbelung *f*, kleine Welle *f*, Kräuselung *f*, Pulsation *f*, Rippe *f*, Welligkeit *f*; pulsierend, wellig ~ cloth Kräuselstoff *m* ~ component Welligkeitskomponente *f* ~ current pulsierender oder welliger Strom *m*, Wellenstrom *m* ~ filter Wellenstromfilter *m* ~ filter choke Netzdrossel *f* ~-finish Kräusellack *m* ~ frequency (commutator) Kommutierungsfrequenz *f*, Reston *n*, Welligkeits-frequenz *f*, -reston *n* ~ mark Wellenfurche *f* (geol.) ~ potential Brummspannung *f*, wellige Gleichspannung *f* ~ ratio Brummspannungsverhältnis *n* ~ voltage Brummspannung *f*, pulsierende Spannung *f*, wellige Gleichspannung *f*
rippled gekräuselt, wellenförmig, wellig ~ condition Welligkeit *f* ~ line Wellenlinie *f*

**rippling** Kräuselung f, Rieseln n, Riffelbildung f, (Flachs) Riffeln n, Wellung f ~ **saw** Baumsäge f ~ **sea** Kabbelsee f

**riprap** Bruchsteinschütterung f, (bottom paving) Sohlenverkleidung f, Stein-schlag m, -schüttung f ~ **foundation** Gründung f auf Steinschüttung ~ **ped slope** Steinböschung f

**rise, to** ~ abheben, anschwellen, ansteigen, anwachsen, aufgehen, aufstehen, aufsteigen, emporkommen, entspringen, richten, schwellen, steigen, zunehmen **to** ~ **in the back** aufhauen (min.) **to** ~ **in resonance** hinaufpendeln **to** ~ **by steps** staffeln **to** ~ **up** aufhissen, emporsteigen **to** ~ **vertically** senkrecht steigen

**rise** Anschwellen n (des Wassers), Ansteigen n, Ansteigung f, Anstieg m, Anwachsen n, Aufbrechen n, Aufbruch m, Aufgang m, Aufhauen n, (naval arch.) Aufkimmung f, Aufschlag m, Aufsteigen n, Bodenerhebung f, Bogenhöhe f, Emporsteigen n, Erhebung f, Erhöhung f, Hochgehen n, Höhe f, Höhenausdehnung f, Pfeilhöhe f, Steigen n, Steigerung f, Steigung f, Zunahme f

**rise,** ~ **of arch** Pfeilhöhe f des Bogens ~ **of an arch** Pfeil m ~ **of current** Stromzunahme f ~ **of a curve** Pfeil m ~ **of deflection** Pfeilhöhe f des Bogens ~ **off ground** Bodenstart m ~ **of miter (or pointing) sill** Drempelvorsprung m ~ **in price** Preisaufschlag m ~ **in prices** Hausse f ~ **of a propeller** Aufkimmung f einer Luft- oder Schiffsschraube ~ **in temperature** Temperaturerhöhung f, -steigerung f ~ **of temperature** Temperaturanstieg m ~ **of the tide** Tidenhub m ~ **of the transverse pulse** steigende Flanke f des Querimpulses ~ **of value** Wertsteigerung f ~ **of the water** Steigen n des Wassers ~ **off water** Wasserstart m

**rise,** ~-**and-fall-rest** in der Höhe verstellbarer Support m ~ **drift** Über-brechen n, -hauen n ~ **time** (of pulse) Anstiegszeit f, Einschwingzeit f ~-**time correction** Laufzeitentzerrung f ~-**time degeneration** Anstiegszeit-Verschlechterung f ~-**time distortion** Laufzeitverzerrung f ~-**to-span ratio of arch** Stich m

**riser** (Gießerei) Anguß m, Aufhängseil n, Aufsatz m, die Bauhöhe vergrößerndes Zwischenstück n, (of a staircase) Futterstufe f, Gußzapfen m, Hebesteigteil m, Hubsteigstück n, Setzstufe f, Steigeleitung f, Steiger m, Steigetrichter m, Steigrohr n, (Fallschirm) Tragegurt m ~ **gate** Steiger m, Steigetrichter m ~ **pipe** (in drainage well) Einhänger m

**rising** Aufsteigen n, Erhöhung f, Steigen n, Steigung f; ansteigend, aufsteigend ~ **of a floor (timber)** (ship building) Aufkimmung f einer Bodenwaage ~ **on the left** linkssteigend ~ **on the right** rechtssteigend

**rising,** ~ **angle** Einfallswinkel m ~ **cloud** Aufgleitwolke f ~ **drop indicator** Steigtropfenzeiger m ~ **floor** Hebebühne f ~ **force** Steigkraft f ~ **gate** Steiger m, Steigtrichter m ~ **generation** Nachwuchs m ~ **ground** Erhebung f ~ **gust** Steigbö f ~ **head** Gußzapfen m ~ **main** Feuerlöschsteigerohr n, Steigeröhre f, Steigleitung f, Steigrohr n ~-**main hose connection** Feuerlöschstutzen m ~ **pipe** Aufsteigrohr n ~ **scaffold bridge** Auflauf m, Fahrbrücke f, Laufbrücke f ~ **slope** aufsteigendes Gelände n ~ **solution**

erstes Pickelbad n ~ **stem** steigende Spindel f ~ **stone drift** ins Gestein getriebene schwebende Strecke f ~ **tide** Tidenstieg m ~ **warm front** Aufgleitfront m ~ **wheel** feststehendes Handrad n

**risings** Buchtenweger m

**risk, to** ~ gefährden, aufs Spiel setzen, wagen

**risk** Gefahr f, Risiko n, Wagnis n ~ **of breakage** Bruchgefahr f ~ **of shock** Berührungsgefahr f (electr.)

**risky** bedenklich, gefährlich, gewagt

**rival** Konkurrent m

**rivalry** Konkurrenz f

**rive, to** ~ aufspalten

**rivelling** (Farbe) Runzeln n

**river** Fluß m, Wasserlauf m ~ **bank** Flußufer n ~ **basin** Stromgebiet n ~ **bed** Flußbett n ~ **block** Flußsperre f ~ **boat** Kahn m ~ **cable** Flußkabel n ~ **capture** Flußanzapfung f ~ **course** Flußlauf m ~ **crossing** Fluß-kreuzung f, -übergang m ~ **crossing of cables** Flußüberführung f ~-**crossing equipment** Übersetzmittel n ~ **discharge** Abflußmenge f ~ **flats** Flußniederung f ~ **flow** Wasserführung f ~ **gold** Waschgold n ~ **line** Flußlinie f ~ **material** Flußtrübe f ~ **meander** Flußschleife f ~ **mouth** Strommündung f ~ **police** Stromwache f ~ **pollution** Flußverunreinigung f ~ **port** Binnen-, Fluß-hafen m ~ **power station** Laufwasserkraftwerk n ~ **reconnaissance** Flußerkundung f ~-**run power plant** Flußkraftwerk n ~ **sand** Flußsand m ~ **side toe of a dike** Deichfuß m ~ **valley** Flußniederung f ~ **water** Flußwasser n, Oberwasser n ~ **weir** Flußwehr f ~ **widening** Flußausbauchung f ~ **width** Flußbreite f ~-**width measuring instrument** Flußbreitenmesser m

**rivers board** Flußamt n

**rivet, to** ~ befestigen, einnieten, nieten, vernieten **to** ~ **on** annieten

**rivet** Niet m, Niete f, Nietnagel m ~ **with buttonhead** Halbrundniet n ~ **with circular head** Niet m mit Halbrundkopf ~ **under shear stress** Scherniet f

**rivet,** ~ **allowance** Nietabzug m ~ **anvil** Nietamboß m ~ **arrangement** Nietbild n ~ **bats** Nieteneisen pl ~ **edge spacing** Nietrandabstand m ~ **efficiency** Nietkraft f ~ **forge** Nietfeuer n, Nietglühofen m, -wärmofen m ~ **furnace** Nietglühofen m ~ **gun** Nietkanone f, pneumatische Nietmaschine f ~ **head** Nietkopf m ~ **headers** Nietenkopfzieher pl ~ **hole** Nietloch n ~ **inserting** Niet-einsatz m, -zuführung f ~ **joint** Nietung f ~-**(ed) joint** Nietverbindung f ~ **not transmitting a force** Heftniet n ~ **nut** Annietmutter f ~ **pattern** Nietmuster n ~ **pin** Nietzapfen m ~ **pitch** Niet-abstand m, -teilung f ~ **pliers** Nietenzange f ~ **pointer** Nietkopf m ~ **remover** Nietenlöser m ~ **row spacing** Nietreihenabstand m ~ **set** Nieten-döpper m, -kopfaufsetzeisen n, -setzer m, Schellhammer m ~ **(ing) set** Schelleisen n ~ **setter** Nietenzieher m ~ **shank** Nietschaft m ~ **spacing** Nietabstand m ~ **stamp** Nietpfaffe m ~ **steel** Nieteisen n ~ **stem** Nietschaft m ~ **tongs** Niet(en)hälter m ~ **transmitting a force** Kraftniet n ~ **wire** Nietendraht m

**rivetability** Nietbarkeit f

**riveted** genietet ~ **assembly** Niet *n* ~ **back-strap shovel (or spade)** Spaten *m* mit genieteter Dülle ~ **casing** Nietröhre *f* ~ **chain** Kette *f* mit eingenieteten Bolzen, Stiftkette *f* ~ **drive chain** Bolzenkette *f*, Nietbolzenkette *f*, Stahlbolzenkette *f* ~ **girder** zusammengenieteter Träger *m* ~ **joint** Niet *n*, Nietnaht *f* ~ **mill chain** Stahlbolzenkette *f* ~ **seam** Niet *n*, Nietnaht *f* ~ **tank** genieteter Tank *m* ~ **tube** genietete Blechverrohrung *f*

**riveter** Nieteneinschläger *m*, Nieter *m*, Nietmaschine *f*

**riveting** Niet-arbeit *f*, -naht *f*, Nietung *f*, Vernietung *f* ~ **in groups** verjüngte Nietung *f*

**riveting,** ~ **block** Nietkolben *m* ~ **fixture** Ausnietvorrichtung *f* ~ **gang** Nietkolonne *f* ~ **hammer** Niethammer *m* ~ **jig** (Ausnietgroßvorrichtung) Aug *f*, Ausnietgroßvorrichtung *f* ~ **machine** Nietmaschine *f*, Nietstanze *f*, Nist *f* (Nietstanze) ~ **press** Nietpresse *f* ~ **pressure** Nietdruck *m* ~ **punch** Nietmeißel *m* ~ **rig** (Ausnietgroßvorrichtung) Aug *f*, Ausnietgroßvorrichtung *f* ~ **roll for link pins** Nietrolle *f* für Kettenbolzen ~ **set** Döpper *m*, Niet-pfaffe *m*, -stempel *m* ~ **snap** Döpper *m*, Schellhammer *m* ~ **socket** Nietbüchse *f* ~ **stock** Niet-bank *f*, -platte *f* ~ **tongs** Niet-kluppe *f*, -zange *f* ~ **tool** Nietwerkzeug *n*

**riving knife** Spalt-klinge *f*, -messer *n*

**rivnut** Nietmutter *f*

**rivulet** Bach *m*

**r-meter** Röntgenmeter *n*

**roach** Hinterwelle *f*

**road, to** ~ **test** einfahren

**road** Fahrstraße *f*, Förderstrecke *f* (min.), Landstraße *f*, Straße *f*, Strecke *f*, Weg *m* ~ **through portal of mine** Ausfahrweg *m*

**road,** ~ **approach** An- und Abmarschweg, Zufuhrstraße *f* ~ **ballast** Straßenschotter *m* ~ **barricade** Wegsperre *f* ~ **bearer** Streckträger *m* ~ **bed** Bahnkörper *m*, Bettung *f*, Gleis-, Steinbettung *f* ~ **bed foundation** Bahnplanum *n* ~ **behavior** Verkehrsdisziplin *f* ~ **bend** Wegbiegung *f* ~ **block** Straßen- und Wagensperre *f*, Wegsperre *f* ~ **bound** straßengebunden ~ **broken up by traffic** ausgefahrene Straße *f* ~ **building** Straßenbau *m*, Wegebau *m* ~ **camouflage** Weghängemaske *f* ~ **center** Straßenknotenpunkt *m* ~ **condition** Wegverhältnis *pl* ~ **crossing** Straßenüberführung *f*, Wegebiegung *f* ~ **discipline** Straßendisziplin *f* ~ **ditch** Straßengraben *m* ~ **fork** Straßengabel *f* ~ **grader** Planiermaschine *f* ~ **grader and tamper** Bodenverdichter *m* ~ **harrow** Straßenbauegge *f* ~ **holding capacity** Straßenlage *f* ~ **hole** Schlagloch *n* ~ **intersection** Straßenknotenpunkt *m* ~ **jam** Straßenverstopfung *f* ~ **junction** Spinne *f*, Straßen-einmündung *f*, -spinne *f*, Wegespinne *f* ~ **laborer** Straßenarbeiter *m* ~ **limitation** Straßenbegrenzung *f* ~ **lugs** Straßen-greifer *m*, -sporen *pl* ~ **maintenance** Straßenunterhaltung *f* ~ **maker** Wegebauer *m* ~ **making material** Straßenbaumaterial *n* ~ **map** Straßen-, Strecken-, Weg-karte *f* ~ **marker** Wegmarke *f* ~ **metal** Straßen-beschotterung *f*, -schotter *m* ~ **metal plant** Schotteranlage *f* ~**-metal preparing machine** Schottermaschine *f* ~ **metalling** Wegebeschotterung *f* ~ **network** Straßennetz *n*,

Wegenetz *n* ~ **oil** Straßenbauöl *n* ~ **overpass** Straßenüberführung *f* ~ **planer** Straßenhobel *f* ~ **plow** Pflug *m* für Wegebauzwecke ~ **protection** Straßenschutz *m* ~ **reinforcement** Straßenbetonierung *f* ~ **resistance** Fahrwiderstand *m* ~ **ripper** Straßenbrecher *m* ~ **roller** Straßenwalze *f* ~ **sander equipment** Sandstreuanlage *f* ~ **screen** Weghängemaske *f* ~ **shoulder** Straßenrücken *m* ~ **side** Straßen-, Weg-rand *m* ~ **sign** Wegweiser *m* ~**-skid quality** Griffigkeit *f* ~ **stead** Reede *f* ~ **stone** Pflasterstein *m* ~ **surface** Straßen-befestigung *f*, -belag *m*, -fläche *f* ~**-surface friction** Bodenreibung *f* ~ **system** Straßennetz *n* ~ **tanker work** Tankwagenbetrieb *m* ~ **testdriver** Einfahrer *m* ~ **test-vehicle** Einfahrfahrzeug *n* ~ **traction** Bodenhaftung *f* ~**-traffic studs** Straßenverkehrsnägel *pl* ~ **transportation** Straßenverkehr *m* ~ **turn** Straßenbiegung *f*

**roadway** Damm *m* der Straße, Fahr-bahn *f*, -damm *m*, -straße *f*, -weg *m*, Förderbahn *f*, (with ties) Schwellenbahn *f* ~ **over a bridge** Fahrbahn *f*

**roadway,** ~ **beam (or girder)** Fahrbahnträger *m* ~ **plate replacing three main griders** Dreierplatte *f* ~ **winch** Streckenhaspel *f*

**road work** Straßenarbeit *f*

**roads** (navy) Reese *f*

**roadster** offener Zweisitzer *m*

**roam, to** ~ streifen, umherschweifen, wandern

**roamer** (map) Planzeiger *m*

**roaming electron** freies Elektron *n*

**roan** graurötlich

**roar, to** ~ brausen, brüllen, rauschen

**roar** Gebrüll *n*

**roaring** Brausen *n* ~ **flame** brausende oder nichtleuchtende Flamme *f* ~ **noise caused by blast** Blasegeräusch *n*

**roast** Röstung *f*; abrösten, abschwellen, ausglühen, backen, brennen, rösten, zubrennen ~ **heap** Rösthaufen *m* ~**-re-acting process** Röstreaktionsarbeit *f* ~**-reduction process** Röstreduktionsarbeit *f*

**roasted** geröstet, rösch ~ **blende** Röstblende *f* ~ **ore** Garerz *n* ~**-ore dust** Röststaub *m* ~ **product** Röst-erzeugnis *n*, -produkt *n* ~ **pyrites** Kiesabbrand *m* ~ **sulphur ore** Schwefelkiesabbrand *m*

**roaster** Röstofen *m* ~ **copper slag** Schwarzkupferschlacke *f*

**roasting** (preliminary) Abröstung *f*, Brennen *n*, Röste *f*, Rösten *n*, Röstprozeß *m*, Röstung *f* ~ **of matte (or of regulus)** Steinmetzrösten *n*

**roasting,** ~ **apparatus** Röst-apparat *m*, -vorrichtung *f* ~ **bed** Röstbett *n* ~ **blast furnace** Röstschachtofen *m* ~ **chamber** Röstkammer *f* ~ **charge** Röste *f*, Röst-gut *n*, -posten *m* ~ **cup** Glühschale *f* ~ **device** Röstvorrichtung *f* ~ **furnace** Brenn-, Röst-ofen *m* ~ **hearth** Röstherd *m* ~ **kiln** Röstofen *m* ~ **operation** Röst-arbeit *f*, -vorgang *m* ~ **oven** Bratrohr *n* ~ **pan** Bratpfanne *f* ~ **plant** Röst-anlage *f*, -einrichtung *f*, -hütte *f* ~ **practice** Röstbetrieb *m* ~ **process** Brennprozeß *m*, Röst-arbeit *f*, -prozeß *m*, -verfahren *n* ~**-reaction method** Röstreaktionsverfahren *n* ~ **residue** Abbrand *m*, Röstrückstand *m* ~ **smelting** Röstschmelzen *n* ~ **stall** Röststadel *m* ~ **test** Ansiedescherben *m* ~ **time** Röstbetriebsdauer *f*

**robot** Automat *m*, automatische Vorrichtung *f*, Roboter *m*; automatisch ~ **pilot** Selbststeuergerät *n*
**robust** stämmig, stark, unempfindlich, widerstandsfähig ~ **construction** unverwüstliche Ausführung *f*
**robustness** Unempfindlichkeit *f*
**roche moutonnée** Rundhöcker *m*
**Rochelle,** ~ **salt** Kaliumnatriumtartrat *n*, Rochelle-, Seignette-salz *n* ~-**salt crystal** Seignettesalzkristall *m*
**roching of alum** Wachsmachen *n* des raffinierten Alauns
**Röchling-Rodenhauser furnace** (reservoir-type induction) Röchling-Rodenhauserofen *m*
**rock, to** ~ (Rotornabe an der Welle) anschlagen, hin- und herbewegen, hin- und herschwingen, hin und her schwingen, schaukeln, schwingen, pendeln, wackeln, wiegen **to** ~ **a well** ein Bohrloch *n* schaukeln **to** ~ **the well** den Schacht *m* mit Gebirg (mit Abraum) ausfüllen
**rock** Fels *m*, Felsblock *m*, Felsen *m*, Gestein *n*, Gesteinsmasse *f*, Klippe *f*, Naturstein *m*, Stein *m* ~ **alum** Bergalaun *m* ~ **arrangement** Gebirgslagerung *f* ~ **bed** Flöz *n* ~ **bit** Steinmeißel *m* ~-**blasting explosive** Gelatit *n*, Gesteinsprengmittel *n* ~ **borer** Gesteinsbohrer *m* ~ **breaker** Steinbrecher *m* ~ **burst** Gebirgsschlag *m* ~ **butter** Bergbutter *f* ~ **chisel** Kuttermeißel *m* ~ **core bit** Rollenkrone *f* ~ **crossing a lode** Quergestein *n* ~ **crusher** Steinbrecher *m* ~-**crushing mill** Gesteinsmühle *f* ~ **crystal** Bergkristall *m*, Kristallstein *m* ~ **debris** Felsgeröll *n*, (from landslide) Gehängeschutt *m* ~ **debris in arid regions** aride Schüttwannen *pl* ~ **decay** Gesteinverwitterung *f* ~ **disintegration** Lockerung *f* des Gesteinszusammenhaltes ~ **drill** Bohrknarre *f*, -maschine *f*, Gestein(s)bohrer *m*, Preßlufthammer *m*, Senker *m*, Stein-bohrer *m*, -bohrmaschine *f*, Stoßbohrmaschine *f* ~-**drill sharpening machine** Gesteinsbohrerschärfmaschine *f* ~-**drill upsetting machine** Gesteinsbohrerstauchmaschine *f* ~ **drilling** Gesteinsbohrung *f* ~-**drilling plant** Gesteinsbohranlage *f* ~ **dust** Steinmehl *n* ~-**faced rubbles** rohes (felsenartiges) Bruchsteinmauerwerk *n* ~ **faults** Steilhänge *pl* ~-**fill dam** Steinschüttdamm *m* ~ **filling** Bruchstein-bettung *f*, -füllung *f*, geschüttete Bruchsteine *pl*, Steinschüttung *f* ~ **flint** Bergkiesel *m* ~ **flour** Mehlsand *m*, Steinmehl *n* ~ **formation** Gesteinsbildung *f* ~ **material** Felsgestein *n* ~ **milk** Bergmilch *f* ~ **niche** Balme *f* ~ **oil** Naphta *n f* ~ **perforator** Gesteinsbohrmaschine *f* ~ **pile shoe** Pfahlschuh *m* für Felsrammung ~ **pressure** Lage(r)druck *m* ~ **salt** Berg-, Stein-salz *n* ~ **seam** Felsenader *f* ~ **shaft** hin- und hergehende Welle *f* ~ **shelter** Balme *f* ~ **slide** Felsschlipf *m*, Mure *f* ~ **soil** Felsboden *m* ~ **steadiness** Bildstabilität *f* (film) ~ **stratification** Gebirgsschichtung *f* ~ **vein** Felsenader *f* ~ **weathering** Gesteinsverwitterung *f* ~ **wood** Holzasbest *m*
**rocker** Mitnehmer *m*, Rollkufe *f*, (Kran) Schwenge *f*, Schwunghebel *m*, Wiege *f*, Wiegekufe *f*, Wippe *f* ~ **for brushes** (radial type) Bürstenbrücke *f* ~ **of a smith's bellows** Blasebalgschwengel *m*
**rocker-arm** Gleitstein *m* einer Kulisse, Kipphebel *m*, Kulisse *f*, Schwinge *f*, Spannwirbel *m*,

Steuerhebel *m*, Unterbrecherhebel *m*, Ventilkipphebel *m* ~ **bearing** Kipphebellagerung *f* ~ **bracket** Kipphebelbock *m*, Ventilkipphebelachse *f* ~ **cup** Kipphebelkugelpfanne *f* ~ **drive** Schwinghebelantrieb *m* ~ **shaft** Kipphebelachse *f*, -welle *f* ~ **way** Gleitklotzbahn *f*
**rocker,** ~ **bearing** Kipp-, Schwinghebel-lager *n* ~ **bearings** Schwinghebellagerung *f* ~ **box** Kipphebelgehäuse *n* ~-**box horn** am Zylinderkopf angegossenes Kipphebelgehäuse *n* ~ **conveyer** Schaukelförderer *m* ~ **gear** Kipphebelsteuerung *f* ~ **grease** Kipphebelfett *n* ~ **joint** Wälzgelenk *n* ~ **shaft** Kipphebelachse *f*, Schaukelwelle *f*, Schwinghebelwelle *f* ~-**shaft bearing bracket** Kipphebelbock *m* ~-**shaft bearing pedestal** Kipphebellagerbock *m* ~-**starter and battery-ignition unit** Pendel-Lichtanlaßbatteriezünder *m* ~ **support box** Kipphebelgehäuse *n* ~-**type switch** Wipp(en)schalter *m*
**rocket** Blasebalgschwengel *m*, Leucht-kugel *f*, -rakete *f*, Rakete *f*, Zielvorrichtung *f* ~ **with coiled pipes as fuel tanks** Rohrschlangengerät *n*
**rocket,** ~ **aircraft** Rakenetflugzeug *n* ~-**assisted takeoff** Raketenstart *m* ~ **attachment** Raketensatz *m* ~ **base** Raketenabschußbasis *f* ~ **belt** Raketengürtel *m* ~ **bomb** Bombenrakete *f* ~ **booster** Raketentriebwerk *n* ~-**case paper** Raketenhülsenpapier *n* ~ **composition** Raketensatz *m*, Treibsatz *m* ~-**device** R-Gerät *n* ~-**driven aeroplane** Raketenflugzeug *n* ~-**driven propeller** Raketenschraube *f* ~ **ejection seat** Raketenschleudersitz *m* ~ **engine** Raketen-antrieb *m*, -motor *m* ~ **equipment** R-Gerät *n* ~ **flight** Raketenflug *m* ~-**jet nozzle** Raketendüse *f* ~ **launcher** Raketen-abschußvorrichtung *f*, -werfer *m* ~ **motor** chemischer Kraftstoffmotor *m* ~-**motor test panel** Triebwerkspult *n* ~ **paper** Raketenhülsenpapier *n* ~ **plane** Raketenflugzeug *n* ~-**powered** raketengetrieben ~-**propelled** mit Raketenantrieb *m* ~-**propelling charge** Raketentreibsatz *m* ~ **propulsion** Raketen-verfahren *n*, -vortrieb *m* ~-**propulsion unit** R-Gerät *n* ~-**starting assist** Raketenstarthilfe *f*
**rocketry** Raketentechnik *f*
**rocking** Hin- und Herbewegung *f*, Schaukeln *n*, Schwankung *f*, Schwingung *f*, Schaukel . . ., Schwing . . ., Wipp . . ., schwankend, schwingend ~ **arm** Schwinge *f* ~ **beam** Waagebalken *m* ~ **conveyor** Schaukeltransporteur *m* ~ **curve** Schwankungskurve *f* ~ **cylinder** schwingender Zylinder *m* ~ **device** Wiege *f* ~ **lever** hin- und hergehender Hebel, Schwenkhebel *m*, Schwinghebel *m*, Schwunghebel *m* ~ **motion** Hin- und Herschwanken *n*, Schaukelbewegung *f*, Schlängeln *n* ~ **movement** hin- und hergehende Bewegung *f* ~ **path** Rollbahn *f* ~ **pier** schwingender Pfeiler *m* ~ **position** Kippstellung *f* ~ **shaft** Kipp-, Schwenk-welle *f* ~ **suspension** pendelnde Aufhängung *f* ~-**through dish bath** Schaukelkurvette *f* ~ **tree** Ladenprügel *m*, Prügel *m*, Schwingbaum *m* ~ **turntable** Drehringwippe *f* ~ **the well** mit Abraum ausfüllen
**rocklike** fels(en)artig
**Rockwell,** ~ **hardness** Rockwellhärte *f* ~-**hardness number** Rockwellhärtezahl *f* ~-**hardness test** Rockwellhärteprüfung *f*

rocky felsig ~ **bottom** Felsgrund *m* ~ **ground** Felsboden *m* ~ **reef** Felsenriff *n*
**rod, to** ~ einstampfen, mit Stangen *pl* versehen
**rod** Auslösestift *m*, Bügel *m*, Meßlatte *f*, Nadel *f*, Pegel *m*, Querhaupt *n*, Rund-stab *m*, -stange *f*, Rute f, Stab *m*, (anat) Stäbchen *n*, Stange *f*, (cleaning) Stock *m*, Strebe *f*, Treib-, Verbindungs-, Zug-stange *f*, Zain *m* ~ **to belt shifter for reversing** Verbindungsstange *f* zum Riemenausrücken ~ **for placing explosives** Sprengstange *f* ~ **of recuperation piston** Vorholerkolbenstange *f* ~ **for sealing** Dichtungsstange *f*
**rod,** ~ **antenna** Stab-antenne *f*, -strahler *m*, Stielstrahler *m* ~ **assembly** Schubstange *f*, Stabanordnung *f* ~ **barrier** Einlegeschranke *f* ~ **boring** Gestängebohren *n* ~ **cart** Gestängewagen *m* ~ **chisel** Stielmeißel *m* ~ **chuck** Stangenspannfutter *n* ~ **control** Stangensteuerung *f* ~ **copper** Stangenkupfer *n* ~ **core** Stabkern *m* ~ **couplings** Gestängebefestigung *f* ~ **cover** Stoßstangenkammerdeckel *m* ~ **covering a joint** Deckleiste *f*, Fugenleiste *f* ~ **cutter** Gestängeschneidezeug *n*, Stab(eisen)schere *f* ~ **drawing** Stabziehen *n* ~ **drill** Stangenbohrer *m* ~ **elevator** Gestängeaufzug *m* ~ **end** Stangenkopf *m* ~ **eye** Stabkopf *m* ~ **feeding attachment** Stangenzuführung *f* ~ **guide** Gestänge-führung *f*, -steuerung *f* ~ **heater** Heizstab *m* ~ **insulator** Stabisolator *m*, Stützisolator *m* ~ **iron** Stangeneisen *n*, Zaineisen *n* ~ **joint** Gestängestoß *m* ~ **lattice** Stabgitter *n* ~ **loaded for pressure column** auf Druck belasteter Stab *m* ~ **mill** Drahtstraße *f*, Drahtwalzwerk *n*, Stabmühle *f* ~-**mill rolling train** Drahtstraße *f* ~ **milling** Draht-walzen *n* -walzung *f* ~-**milling plant** Drahtwalzerei *f* ~ **mirror** Stabreflektor *m* ~-**mold attachment** Stab-riester *m*, -streichblech *n* ~ **notch** Schieberkerbe *f* ~ **packing** Stangenpackung *f*, Stopfbüchsenpackung *f* ~ **packing gland** Stopfbüchsenring *m* ~ **pass** Rundeisenkaliber *n* ~ **reel** Drahthaspel *f* ~ **reflector** Stabreflektorwand *f* ~ **rolling** Rundeisenwalzung *f* ~-**rolling mill** Rundeisenwalzwerk *n* ~ **screen** Streifenschirm *m* ~ **shaft** Maschinenschacht *m* ~-**shaped** stabförmig, stangenartig ~ **spring** Stangenfeder *f* ~ **steel** Stangenstahl *m* ~ **stock** Stangenmaterial *m* ~ **straightening machine** Stangenrundrichtmaschine *f* ~ **strainer** Spannschloß *n* ~ **support** Hängedraht *m* ~ **swivel** Förderstuhl *m*, Gestängewirbel *m* ~ **tap** Gestängefalldorn *m* ~ **threaded at both ends** Stange *f* mit Gewinde zu zwei Enden ~ **wrench** Gestängeschlüssel *m*
**rodding** Stellgestänge *n*
**rodless** stangenlos
**rodlet** Stäbchen *n*, kleine Stange *f*
**rods** (of nerve endings) Augenzäpfchen *pl*, Rund-eisen *n*, -stahl *m*, Stangenzug *m*, Zaineisen *n* (main) ~ Gestänge *n* ~ **with shoulder** Wulstgestänge *n*
**roentgen** (unit) Röntgen *n*, Röntgeneinheit *f* ~ **ray** Röntgenstrahl *m* ~-**ray installation** Röntgeneinrichtung *f* ~ **rays caused by collision and checking** Röntgenbremsstrahl *m* ~ **tube** Röntgenröhre *f*
**roentgenization** Röntgenverfärbung *f*
**roentgenography** Röntgenaufnahmetechnik *f*
**roentgenology** Röntgenologie *f*

**roentgenometer** Röntgenmesser *m*
**roentgenoscope** Durchleuchtungsgerät *n*
**roentgenotherapy** Röntgenbestrahlung *f*
**roestone** Rogenstein *m*
**rogersite** Rogersit *m*
**roll, to** ~ abwalzen, drehen, fahren, fließen, fortwälzen, kreisen, kugeln, rollen, schlingern, verwalzen, walzen, wälzen, wellen, wickeln, sich wiegen **to** ~ **along runway** anrollen **to** ~ **around** umwälzen **to** ~ **back** aufrollen, nullen, wiederholen (data proc.), zurückdrängen **to** ~ **back and forth** hin- und herrollen **to** ~ **cylindrical(ly)** rundwalzen **to** ~ **double** dublieren **to** ~ **down** abwalzen, (Blech oder Band auf geringere Dicke) auswalzen, herunterwalzen **to** ~ **on edge** (in rolling) stauchen **to** ~ **out** auswalzen, strecken **to** ~ **to final gauge** fertigwalzen **to** ~ **flat** flachwalzen **to** ~ **the form** Farbe *f* auf die Form auftragen, Schwärze *f* auftragen **to** ~ **by guide** walzen aus Führung *f* **to** ~ **by hand** walzen aus freier Hand *f* **to** ~ **in** einwalzen **to** ~ **into** verwalzen **to** ~ **in long lengths** walzen in langen Zügen **to** ~ **off** abrollen **to** ~ **on** aufwalzen **to** ~ **over** umrollen, wenden **to** ~ **and run on the same track** spuren **to** ~ **to a stop** ausrollen **to** ~ **up** aufrollen, wickeln
**roll** Auflaufhaspel *m*, Ballen *m* (Stoff), Krängung *f*, Liste *f*, (textiles) Locke *f*, Rolle *f*, Rollen *n*, Rundleiste *f*, Schlingern *n*, Schriftstück *n*, Walze *f*, Walzen *n*, Welle *f*, Wickel *m*, Wulst *m* ~ **of clay** Lettennudel *f* ~ **for cold-rolling** Kaltwalze *f* ~ **for hot rolling** Warmwalze *f* ~ **to the left** Rolle *f* links ~ **for rails** Schienenwalze *f* ~ **to the right** Rolle *f* rechts ~ **of wadding** Watterolle *f*
**roll,** ~-**away ironer** Bügelmaschine *f* ~ **axis** Längs-, Quer-achse *f* ~ **back** Zurück-rollen *n*, -werfen *n* ~-**back point** Wiederholpunkt *m* ~-**back routine** Wiederholprogramm *n* ~ **bar** Grundwerkschiene *f*, Holländer -messer *m*, -schiene *f* ~ **bars** (paper) Blätter *pl* ~ **body** Walzen-ballen *m*, -körper *m*, Walzkörper *m* ~ **call** Namen-aufruf *m*, -verlesung *f* ~ **change gears** Wälzwechselräder *pl* ~ **changing crane** Walzenwechselkran *m* ~ **clearance** Walzenspalte *f* ~ **coater** Aufwalzvorrichtung *f* ~ **coating machine** Walzenbeschichtungsmaschine *f* ~ **core** Rollenkern *m* ~ **coverer** Pergamentaufzieherin *f* ~ **crusher (or crushing mill)** Walzenmühle *f* ~ **cumulus** Rollkumulus *m* ~ **cutout switch** Rolldämpferschalter *m* ~ **damper** Rollenfeuchtmaschine *f* ~ **designer** Walzenkalibreur *m* ~ **designing** Kalibrieren *n*, Kalibrierung *f*, Walzenkalibrierung *f* ~-**end-yaw factor** Rollgiermoment *n* ~ **engraver** Druckwalzengraveur *m* ~ **engraving attachment** Walzengraviervorrichtung *f* ~-**fed offset printpress** (rotary) Offsettrollenrotations-Druckmaschine *f* ~-**fed photogravure rotary attachment** Tiefdruck-Rollenrotationsmaschine *f* ~-**feed attachment** Walzenschubapparat *m* ~ **film** Rollfilm *m* ~-**film camera** Filmkamera *f* ~ **film magazine** Rollfilmkassette *f* ~ **finish** Prägepolieren *n* ~-**forming machine** Wellmaschine *f* ~ **gap** Walzen-spalte *f*, Walz-spalt *m* ~ **grinder** Walzenriffler *m* ~ **groove** Walzkaliber *n* ~ **holder lock** Verriegelungshebel *m* des Spulenhalters ~ **housing** Walzenständer *m*

~ **housing of pilgrim mill** Pilgergerüst *n* ~**-jaw crusher** Walzenmühle *f*, Walzwerk *n* ~ **joint** Rundfalz *m* ~ **lathe** Walzendrehbank *f* ~ **mandrel** Walzdorn *m* ~ **mill** Walzenmischer *m* ~ **motion** Abwälzbewegung *f* ~**-off** (in frequency response) Abfall *m* ~**-out table** Rolltisch *m*

**roll-over,** ~ **device** ausfahrbare Wendevorrichtung *f* ~ **frame** Wenderahmen *m* ~ **jolter** Rüttelformmaschine *f* mit ausfahrbarer Wendevorrichtung ~ **mechanism** Umrolleinrichtung *f*, Wendelmechanismus *m* ~ **plate** Umrollplatte *f* ~ **ring** Wende-rahmen *m*, -ring *m* ~ **squeezer with swing-out table** Wendeplattenpreßformmaschine mit ausschwenkbarem Tisch ~ **table** (of a mold machine) ausfahrbare Wendeplatte *f*, Wendetisch *m* ~**-table jolter** Wendeplattenformmaschine *f* mit eingebautem Rüttler ~**-table molding machine** Wendeplattenformmaschine *f* mit ausfahrbarer Wendeplatte ~ **table press molding machine** Preßformmaschine *f* mit ausfahrbarer Wendeplatte ~**-type molding machine** Wendeformmaschine *f*

**roll,** ~ **pass** Walz(en)kaliber *n* ~**-picker needle** Flockfangnadel *f* ~ **piercing mill** Schrägewalzwerk *n* ~ **piercing process** Schrägwalzverfahren *n* ~ **polish** Prägepolieren *n* ~ **press** Walzenpresse *f* ~ **rate** Rollgeschwindigkeit *f* (g/m) ~ **recorder** Schlingeranzeiger *m* ~ **roofing** Dachpappe *f* ~ **scale** Walzen-schlacke *f*, -sinter *m*, Walz-sinter *m*, -zunder *m* ~ **scale car** Sinterwagen *f* ~**-shearing knives** Rollenschermesser *n* ~ **sheet iron** Rollenblech *n* ~ **shutter** Rolltuch *n* ~**-shutter darkslide** Jalousiekassette *f* ~ **shutters** Rollgitter *n* ~ **spindle** Walzenspindel *f* ~ **stand** Walzgerüst *n* ~**-stand housing** Gerüstständer *m* ~ **stands arranged in separate lines** gestaffelte Walzgerüstanordnung *f* ~ **straightener** Rollenrichtmaschine *f* ~ **strength** Walzenfestigkeit *f* ~ **sulfur** Stangenschwefel *m* ~**-surface velocity** Walzenumfangsgeschwindigkeit *f* ~ **table** Rollgang *m* ~**-table foreman** Rollgangführer *m* ~ **tape measure** Rollbandmaß *n* ~ **tire** Walzenbandage *f* ~ **top** Rollverschluß *m* ~**-top filter element** Wickelfiltereinsatz *m* ~ **train** Walzenstrecke *f* ~ **turner** Walzen-drechsler *m*, -dreher *m* ~**-turning shop** Walzendreherei *f* ~**-type condenser** Wickelkondensator *m* ~**-type dryer** Bandtrockner *m* ~**-type paper filter** Wickelpapierfilter *m* ~ **vanes** Drahtsegel *pl* ~ **welding** Rollenschweißung *f* ~**- welding equipment** Rollenschweißmaschine *f*

**rollable** walzbar

**rolled** gewalzt ~ **angle** Winkeleisen *n* ~ **beam** Walzträger *m* ~ **brass** Walzmessing *n* ~ **condition** Walzzustand *m* ~ **girder** Walzbalken *m*, Walzträger *m* ~ **glass** gezogenes Glas *n* ~ **gold** Golddoublé *n*, Walzgold *n* ~ **iron** Walzeisen *n* ~**-iron beam** Walzeisenträger *m* ~ **iron transporting crane** Walzeisenförderkran *m* ~ **laminated tube** gewickeltes Rohr *n* ~ **lead** Walzblei *n* ~ **lead wire** Walzbleidraht *m* ~ **metal** Walzmetall *n* ~**-on flange** angewalzter Flansch *m*, Walzflansch *m* ~**-out** abgewickelt ~ **plate** Walzblech *n* ~ **powder** Plattenpulver *n* ~ **product** Walzprodukt *n* ~ **products** Walzeisen *n* ~ **rounds** Walzrundmaterial *n* ~ **rubber thread** gewalzter Kautschukfaden *m* ~ **section** Walz-

profil *n* ~ **section for guide and check rails** Zwangsschienenwinkel *m* ~ **sections** Formeisen *n* ~ **shapes** gewalzte Profile *pl* ~ **sheet** Streckstahl *m* ~ **sheet metal** Walzblech *n* ~ **plits** lisierte Spalte *f* ~ **spring eye** gerolltes Federauge *n* ~ **steel** Walz(en)stahl *m* ~**-steel channel** U-Eisen *n* ~ **steel joist** gewalzter Träger *m* ~ **steel member** Fassoneisen *n* ~ **steel rod** Walzstange *f* ~ **steel section** Walzprofil *n* ~ **stock** Walzgut *n*, Walzmaterial *n* ~ **thread** gerolltes oder gewalztes Gewinde *n* ~ **timber** Windbruch *m*, windbrüchiges Holz *n* ~ **tin** Walzzinn *n* ~ **tube** Walzrohr *n* ~ **wire** Walzdraht *m*

**roller** Druckwalze *f*, Fördermann *m*, Führungs-, Garn-, Gleit-rolle *f*, Kabelführungsrolle *f*, Läufer *m*, Lauf-rolle *f*, -walze *f*, Rädchen *n*, Rolle *f*, Rollenelektrode *f*, Roller *m*, Rollklotz *m*, Trommel *f*, Vorderwalzer *m*, Wagenstößer *m*, Walze *f*, Walzendreher *m*, Walzer *m*, Welle *f*, Wickel *m*, Zylinder *n* ~ **of the baby press unit** Vorpreßwalze *f* ~ **with flywheel effect** Schwungmassenrolle *f* ~ **for gripper operating lever** Greiferrolle *f* ~ **of a pulley** Kranrolle *f* ~ **with spiral twigging** Spiralwalze *f*

**roller,** ~ **assembly** Walzenkranz *m* ~ **bar** Abrollstange *f*, Walzenmesser *n* ~ **beam** Kettenbaum *m* ~ **bearing** Führungsrollenlager *n*, Querlager *n*, Rollager *n*, Rollen-lager *n*, -lagerung *f*, Walzenlager *n*, Wälzlager *n*, Zylinderrollenlager *n* ~ **bearing with shoulders** Schulterrollenlager *n* ~**-bearing axlebox** Rollenachsenlager *n* ~**-bearing drop hanger** Rollenhängelager *n* ~**-bearing pillow block** Rollenstehlager *n*, Stehrollenlager *n* ~**-bearing steel** Rollenlagerstahl *m* ~ **bed** Rollgang *m* ~ **bit** Rollmeißel *m* ~ **blind** Roll-laden *m*, -jalousie *f*, -vorhang *m*, -tuch *n* (photo) ~**-blind dark slide** Schieberkassette *f* ~**-blind instantaneous shutter** Schlitzmomentverschluß *m* ~**-blind shutter** Rouleauverschluß *m* ~ **block** Rollenblock *m* ~ **box** Rollenkasten *m* ~ **box tool in turret** Absatzdrehwerkzeug *n* ~ **bridge** Rollbrücke *f* ~ **buckle** Walzenschnalle *f* ~ **bumper** Abfangwalze *f* ~ **cage** Rollring *m*, (of bearing) Walzenkäfig *m* ~ **card** Walzen-karde *f*, -krempel *f* ~ **carriage arm** Walzenstuhl *m* ~**-carriage optical apparatus** Rollenkranz *m* ~ **casing** Laufwerkgehäuse *n* ~ **chain** Gallsche Kette *f*, Rollenkette *f* ~ **chain drive** Rollenkettenbetrieb *m* ~**-chain sprocket** Rollenkettenrad *n* ~**-chasing-type dry pan** Läuferwerk *n* ~ **chuck** Rollenfutter *n* ~ **clearance** Rollenspiel *n* ~ **clutch** Freilaufkupplung *f*, Rollenkupplung *f* ~ **coating** Aufwalzen *n*, Walzenbezug *m* ~**-conveyer jack** Rollenbahnstütze *f* ~**-conveying installation** Rollenförderanlage *f* ~ **conveyor** Rollenband *n*, Rollgang *m* (Schwerkraftförderer) ~ **crusher** Brechwalze *f* ~ **dam** Walzenwehr *n* ~ **delivery bed** Ablaufrollgang *m* ~ **dies** Roll-, Walzen-stempel *m* ~ **discharge** Walzenabnahme *f* ~ **drier** Walzentrockner *m* ~ **ejector** Rollenausstoß *m* ~ **ends** Walzenzapfen *pl* ~ **expander** Dichtmaschine *f* ~ **fading** Seegangsschwundeffekt *m* ~ **feed** Walzenvorschub *m* ~ **feeder** Walzenaufgabevorrichtung *f* ~ **film gate** Rollenfenster *n* ~ **frame** Walzenstuhl *m* ~ **gate** Rollschütze *f*, Walze *f*, Walzenwehr *n* ~ **gauge** Einstellehre *f*

für Auftragswalzen ~ **gear** Rollenlaufwerk *n*
**~-gig** Rollrauhmaschine *f* ~ **gin** Walzenegre-
ni(e)rmaschine *f* ~ **grate** Walzenrost *m* ~ **grease
box** Rollenschmierbuchse *f* ~ **grinding machine**
Walzenschleifmaschine *f* ~ **grinding mill** Wal-
zenreibmaschine *f* ~ **guide** Rollschlitten *m* ~
**jaw** Rollbacke *f* ~ **key clip** Rollentasterkluppe
*f* ~ **lathe** Trommeldrehbank *f* ~ **lever** Rollen-
hebel *m*, -kurbel *f* **~-lever steering** Wälzhebelsteuerung *f* ~ **map case** Kartenroller *m* ~ **marks**
Walzeneindrücke *pl* ~ **mill** Mahlmaschine *f*,
Quetschwerk *n*, Walzen-mahlwerk *n*, -mühle *f*,
-straße *f*, -stuhl *m*, Walz-maschine *f*, -werk *n*,
-straße *f* ~ **pan** Läuferwerk *n* ~ **path** Flur *m*,
Flurbalken *m* **~-pawl** Klinke *f* mit Rolle, Rollenklinke *f*, **~-pin** Rollenbolzen *m* ~ **pitch** Rol-
lenteilung *f* ~ **plate-bending machine** Walzen-
blechbiegemaschine *f* ~ **press** Walzenaufzieh-
presse *f* ~ **pressure** Druckgebung *f* ~ **printing**
Walzendruck *m* ~ **printing color** Walzendruck-
farbe *f* **~-processed dried milk** Walzenpulver *n*
~ **product** Walzerzeugnis *n* ~ **profile** Tonnenlager *n* ~ **race** Faß-, Rollen-lager *n* ~ **ratchet
mechanism** Rollenschaltwerk *n* ~ **seat** Roll-
pfanne *f* ~ **sets** Rollensatz *m* ~ **setting worm**
Einstellschnecke *f* ~ **shaft** (for cam) Rollen-
stößel *m* **~-shaped** walzig ~ **shaving holder**
Formrollengegenhalter *m* (b. Drehen m. Form-
stahl) ~ **shell** Walzenmantel *m* ~ **shrinkage**
Walzenschwund *m* **~'s side** Einsteck-, Ein-
stich-seite *f* ~ **sluice** weir Walzenschutzwehr
*n* ~ **spiral** Rollenbahnspirale *f* ~ **spot welding**
Rollenschrittverfahren *n* (Schweißen) ~
**squeeze** Rollwalze *f* ~ **stand** Zylinderstanze *f*
~ **steady** Rollen-gegenführung *f*, -setzstock *m*
~ **striking clutch** Rollenschlagkupplung *f* ~
**table** Walzentisch *m* **~-table sandblast machine**
Rollentischsandstrahlgebläse *n*, Sandstrahlgebläse *n* mit Rollentisch ~ **thrust bearing** Rollenlager *n* ~ **tooth** Rollenzahn *m* ~ **track** Rollbahn *f* ~ **track with blowtube** Farbwerkschiene
*f* mit Blasrohr ~ **train** Rollenbahn *f* **~-train
frame** Wange *f* der Rollenbahn ~ **transporter**
Transportrollgang *m* ~ **traverse** Laufrollen-
traverse *f* ~ **traversing** Rollenführung *f* ~
**turner** Rollengegenlager *n* **~-type conveyer**
Rollbahn *f* **~-type heating furnace** Rollofen *m*
~ **vat** Rollfaß *n* ~ **V-rest** Rollenlünette *f* ~
**washing device** Walzenwascheinrichtung *f* ~
**way** Rollschiene *f* ~ **weir** Walzenwehr *n* ~
**wind** Trommelwinde *f* ~ **winder** Aufroller *m* ~
**wire guide** Rollendrahtführer *m*
**rollers,** ~ **for a bridge** Brückewalzen *pl* ~ **for
painters** Malerwalzen *pl* ~ **for rollerjournals**
Zylinderrollen *pl* für Zylinderrollenlager ~ **of
type bed** Rollenwagen *m*
**rolling** Abrollung *f*, (petroleum) Auswalzen *n*,
Einwalzen *n*, Rollen *n*, (movement) Schlinger-
bewegung *f*, (of a ship) Schlingern *n*, Strecken
*n*, Transversaldrehung *f*, Walzen *n*, Walzung *f*;
rollend ~ **by hand** Freihandwalzen *n* ~ **and
pitching motion** Schlinger- und Stampfbewegung *f* ~ **of sectional iron** Profilwalzung *f* ~ **of
sections** Profileisenwalzen *n* ~ **of the selvedges**
Einrollen *n* der Leisten ~ **and slabbing mill**
Brammen- und Platinenwalzwerk *n* ~ **and yawing moment** Schieberollmoment *n*
**rolling,** ~ **action** Abrollung *f* ~ **angle of an air-**

**craft** (or of a ship) Schlingerwinkel *m* ~ **ball
banding gear** Rollballfahrgestell *n* ~ **band**
Wälzband *n* ~ **barrel** Putzfaß *n*, Putztrommel
*f*, Rollfaß *n*, Rommel *f*, Rummelfaß *n*, Scheuer-faß *n*, -trommel *f* ~ **barrier** Rolltorschranke
*f* ~ **bobbin** Abrollspule *f* ~ **burr** Walznaht *f* ~
**circle** Wälzkreis *m* ~ **colter** Scheiben-sech *n*,
-vorschneider *m* ~ **cone** Wälzkegel *m* ~ **contact**
Umfangsreibung *f* ~ **contact lever** Wälzhebel *m*
~ **crack** Walzriß *m* ~ **crusher** Walzenbrecher *m*
~ **curve** Walzbahn *f* ~ **cutter type bits** Rollen-
meißel *m* **~-damping** Rolldämpfung *f* ~ **defect**
Walzfehler *m* ~ **die** Einwalzstempel *m*, Rollbacke *f* ~ **disc** Wälzscheibe *f* ~ **distance** Rollstrecke *f* ~ **draft** Walzdruck *m* ~ **edge** (on
rolled material) Kaliberrand *m*, Walzkante *f*,
Walzrand *m* ~ **energy** Walzarbeit *f* ~ **equipment** Walzarmaturen *pl* ~ **fabrics without
doubling** einfachbreites Wickeln *n* ~ **face** Walzfläche *f* ~ **fin** Walznaht *f* ~ **fixture** Rollbock *m*
~ **friction** gleitende, rollende, oder wälzende
Reibung *f*, Rollwiderstand *m*, Wälzreibung *f* ~
**friction at the rails** Schienenreibung *f* ~ **friction coefficient** spezifischer Fahrwiderstand *m*
~ **gate** Rolltorschranke *f*, Walzenwehr *n* ~
**hand** Umwälzer *m* ~ **ingot** Walzbarren *m* ~
**instability** Queraufrichtungsstabilität *f* **~-key
clutch** Drehkeilkupplung *f* ~ **ladder** Rolleiter *f*
~ **load** Arbeitsbelastung *f* ~ **lock gate** Rollponton *n*, Schiebespule *f* ~ **machine** Dubliermeß-
und Wickelmaschine *f*, Walzmaschine *f* ~ **mark**
Walzriefe *f* ~ **material** Walzmaterial *n* ~ **metal**
Walzmetall *n*
**rolling-mill** Streckwerk *n*, Walzenmühle *f*, Walz-
hütte *f*, -straße *f*, -strecke *f*, -werk *n* ~ **for
annular shapes** Ringwalzwerk *n* ~ **for circular
shapes** Rundwalzwerk *n* ~ **for corrugated tubes**
Wellrohrwalzwerk *n* ~ **for the mint** Münzstreckwerk *n* ~ **for sharpening wires** Anspitz-
walzwerk *n* für Draht ~ **for smoothing tubes**
Röhrenglättwalzwerk *n*
**rolling-mill,** ~ **bearings** Walzwerk(s)lager *n* ~
**construction** Walzwerks-bau *m*, -technik *f* ~
**crane** Walzwerkskran *m* ~ **engineer** Walzwer-
ker *m* ~ **equipment** Walzenpark *m*, Walzwerks-
einrichtung *f* ~ **gear transmissions** Walzwerks-
getriebe *n* ~ **motor** Walzenzugmotor *m* ~
**operation** Walzbetrieb *m*, Walzwerks-betrieb
*m*, -technik *f* ~ **path drive** Walzenstraßenantrieb
*m* ~ **products** Walzwerks- und Hüttenprodukte
*pl* ~ **rolls** Walzwerkswalzen *pl* ~ **shears** Walz-
werksschere *f* ~ **technique** Walzwerkstechnik *f*
~ **train** Walzenstraße *f*
**rolling,** ~ **mills** Walzwerkbetrieb *m* ~ **moment**
Roll-, Quer-moment *n* ~ **moment due to slip**
Schieberollmoment *n* ~ **momentum** Drehmoment *n* ~ **motion** Rollbewegung *f* ~ **mount**
Gleitlafette *f* ~ **operation** Walzvorgang *m* ~
**-over device** Umrollvorrichtung *f* ~ **pin** Rollholz
*n* ~ **plan** Walzprogramm *n* ~ **plane** Walzebene
*f* ~ **platform** Walztisch *m* ~ **point** Walzpunkt *m*
~ **press** Kupferdruckpresse *f*, Mangel *f*, Roll-
stanze *f*, Satiniermaschine *f*, Walzenpresse *f*
~ **pressure** Walzdruck *m* ~ **print** Abdruck *m*
durch Abwälzen ~ **process** Walzvorgang *m* ~
**property** Walzbarkeit *f* ~ **resistance** Fahrtwiderstand *m*, rollende Reibung *f*, Rollen-
widerstand *m* ~ **resistance at the rails** Schienen-

reibung $f$ ~ **scale** Walzzunder $m$ ~ **schedule** Walz-plan $m$, -programm $n$ ~ **screen** Drehsieb $n$ ~ **shapes** Profilwalzen $n$ ~ **shop** Walzwerk-halle $f$ ~ **skin** Walzhaut $f$ ~ **slab** H-Barren $m$, Walzplatte $f$ ~ **speed** Walzgeschwindigkeit $f$ ~ **stability** Rollstabilität $f$ ~ **stand** Walzgerüst $n$ ~ **stock** Betriebsmittel $pl$, rollendes Material $n$, Rollmaterial $n$ (r.r.) ~**-stock clearance gauge** Umgrenzungslinie $f$ für Eisenbahnfahrzeuge ~ **surface** Lauffläche $f$ ~**-table-type sandblast machine** Rollbahntischsandstrahlgebläse $n$ ~ **tape measure** Bandmaß $n$, Rollmaß $n$ ~ **test diagram** Abrolldiagramm $n$ ~ **thread cutter** Gewindewalze $f$ ~ **tolerance** Walztoleranz $f$ ~ **total tabulator** Summentabulator $m$ ~ **train** (for light sections) Feineisenstraße $f$, Walzen-straße $f$, Walz-straße $f$, -strecke ~ **tripod** Schlingerstativ $n$ ~ **truck** Rollenwagen $m$ ~ **undercarriage** Radfahrgestell $n$ ~**-up arrangement** Aufwicklung $f$ ~**-up bracket** Wickelbock $m$ ~**-up device** Rollvorrichtung $f$ ~ **vertical-lift gate sluice valve** Rollschütz $n$ ~ **width** Bandbreite $f$ (bei Walzen) ~ **window blind (or curtain)** Fensterrolladen $m$ ~ **work** Walzarbeit $f$, Walzdruck $m$
**rolls** Streckwerk $n$ ~ **for merchant iron** Grobeisenwalzwerk $n$
**roman character (or type)** Antiqua $f$
**Roman,** ~ **candle** römische Kerze $f$ ~ **cement** Patentzement $m$ ~ **surface** römische Fläche $f$
**ronde** Ronde $f$
**rondle bar** Rohrschiene $f$
**roof, to** ~ bedachen, eindecken, überdachen, überdecken
**roof** Dach $n$, Schicht $f$, hängende Schicht $f$ ~ **over building berth** überdachte Helling $f$ ~ **over** bewegungen $pl$ ~ **ore** Firstenerz $n$ ~ **pads** Dach- **a pulpit** Kanzeldach $n$
~**roof,** ~ **antenna** Dachantenne $f$ ~ **arch** Deckengewölbe $n$, Gewölbe $n$ ~ **batten** Dachlatte $f$ ~ **beam** Deckenträger $m$, Pfette $f$ ~ **beams** Dachgebälk $n$ ~ **bent** Dachbinder $m$ ~ **boarding** Dachbretter $pl$ Dachschalung $f$ ~ **bracket** Dachstütze $f$ (electr.) ~ **brick** Gewölbestein $m$ ~ **control** Gebirgskontrolle $f$ ~**-end post (or stand)** Abspanndachständer $m$ ~ **glazing** Glaseindeckung $f$ ~ **(top) heliport** Hubschrauber-Dach-Flugplatz $m$ ~ **joists** Dachunterzug $m$ ~ **lamp** (Auto) Deckenlampe $f$ ~ **laths** Dachlattung $f$ ~ **membrane** Dachhaut $f$ ~ **movements** Gebirgsbewegungen $pl$ ~ **ore** Firstenerz ~ **pads** Dachpolsterung $f$ ~ **platform** (erected on roofs for workmen) Laufbrett $n$ ~ **point** Firstpunkt $m$ ~ **pole** Dachständer $m$ ~ **pressure** Gebirgsdruck $m$ ~ **principle** Dachrahmen $f$ ~ **protecting shoe** Dachschuh $m$ ~ **ring** Gewölbe $n$ ~ **rock** Dachgebirge $n$ ~ **saturant** Dachbedeckungsimpregniermittel $m$ ~ **seat** Verdecksitz $m$ ~**-shaped antenna** dachförmige Antenne $f$ ~ **sheating** Dachhaut $f$ ~ **sheet of the outside of the firebox** Stehkesseldecke $f$ ~ **sill** Dachunterzug $m$ ~ **standard** Dach(abspann)gestänge $n$, Dachständer $m$ ~ **strengthening** Gewölbespannung $f$ ~ **support** Dachstützpunkt $m$ (electr.) ~ **timber** Dachgebälk $n$ ~**-top helicopter airport** Hubschrauber-Dach-Flugplatz $m$ ~**-top transfer chain** dachförmige Transportkette $f$ ~ **trap door** Aussteigeluke $f$ ~**-tree party** Richtfest $n$ ~ **truss**

Dach-binder $m$, -gebinde $n$ ~ **truss (or framing)** Dachstuhl $m$ ~ **truss bearing** Dachbinderauflager $n$ ~ **tubes** Dachrohre $pl$, Gewölbebogenröhre $f$ ~ **valley** Dachkehle $f$ ~ **ventilator** Dachventilator $m$
**roofer** Dachdecker $m$
**roofing** Abdachung $f$ (Gelände), Bedachung $f$, Dachbau $m$, Dachbelag $m$, Dachdeckerarbeit $f$, Dachhaut $f$, Dachung $f$, Eindeckung $f$, Sparrenwerk $n$ ~ **asphalt** Pech $n$ für Dachwerk $n$ ~ **felt** Dachpappe $f$ ~ **material** Bedachungs-, Dachdeckungs-, Deckungs-material $n$ ~ **pan** (or tile) Dachpfanne $f$ ~ **pasteboard** Dachpappe $f$ ~ **sheets** Bedachungsbleche $pl$ ~ **slate** Dach-, Tafel-schiefer $m$ ~ **tile** Flachziegel $m$
**roofless** unbedeckt, ungedeckt
**rooflike** dachartig
**room** Kammer $f$, Raum $m$, Saal $m$, Stube $f$ ~ **for experiments** Versuchsraum $m$ ~ **to move** Bewegungsfreiheit $f$ ~ **of operations** Betriebssaal $m$
**room,** ~ **acoustics** Raumakustik $f$ ~**-and-pillar system** Kammer- und Pfeilerbau $m$ ~ **heating** Raumheizung $f$ ~ **noise** Saalgeräusch $n$ ~ **position** Platz $m$ ~ **temperature** Raumtemperatur $f$, Raumwärme $f$ (normal), Zimmertemperatur $f$
**roomy** geräumig
**root, to** ~ **out** ausrotten
**root** Ansatz $m$, Fuß $m$, Grund $m$, Wurzel $f$ ~ **of the breakwater** Landanschluß $m$ der Mole ~ **of gear tooth** Zahnwurzel $f$ ~ **of a groin** Buhnenwurzel $f$, Landanschluß $m$ der Buhne ~ **of the notch** Kerb-grund $m$, -scheitel $m$ ~ **of the path** Bahnwurzel $f$ ~ **of propeller blade** Schraubenblattwurzel $f$ ~ **of spline** Keilfuß $m$ ~ **of thread** Gewindefuß $m$ ~ **of tooth** Zahnfuß $m$ ~ **of unity** Einheitswurzel $f$ ~ **of the weld** Schweißwurzel $f$
**root,** ~ **angle scale** Kegelwinkelgradskala $f$ ~ **bead** Wurzelraupe $f$ ~ **blight** Wurzelbrand $m$ ~ **chord** Wurzeltiefe $f$ ~ **clearance** Grundspiel $n$ ~ **crack** Wurzelriß $m$ ~**-crop rot** Mietenfäule $f$ ~ **cutter** Wurzelschneider $m$ ~ **diameter** (of gearing) Fuß-, Kern-durchmesser $m$ ~ **diameter of screw** Kerndurchmesser $m$ der Schraube oder des Schraubengewindes ~ **element of a ring** Wurzelgröße $f$ eines Ringes ~ **end** Wurzelende $n$ ~ **face** Wurzel $f$ **(welding)** ~ **fiber** Wurzelfaser $f$ ~ **form** Grundform $f$ ~ **gap** Wurzel-abstand $m$, -kerbe $f$ ~ **hole** Wurzel-gang $m$, -loch $n$ ~ **layer** Wurzellage(-raupe) $f$ ~ **line** Fußkreis $m$ ~ **loci** (servo) Wurzelortskurven $pl$ ~ **locus plot** Wurzelortskurve $f$ ~ **mean square (RMS)** Effektivwert $m$, quadratischer Mittelwert $m$ ~**-mean-square current** effektiver Strom $m$, Effektivstrom $m$ ~**-mean-square value** Effektivwert $m$, quadratischer Mittelwert $m$ ~**-mean-square voltage** effektive Spannung $f$, Effektivspannung $f$ ~ **pulper** Wurzel-quetsche $f$, -quetschmaschine $f$ ~ **radius** (of gear) tooth fillet Fußabrundung $f$ ~ **rib** Einleitungsrippe $f$ ~ **rot** Wurzelbrand $m$ ~ **section** Grunddurchschnitt $m$ (Querdurchschnitt) des Flügels ~ **sign** Wurzelzeichen $n$ ~**-sum-square value** quadratischer Mittelwert $m$ ~ **thread diameter** Gewindeflankendurchmesser $m$
**rooted** eingewurzelt
**rootlet tube** Wurzelröhrchen $n$

**Root's,** ~ **blower** Rootsgebläse *n* ~ **supercharging blower** Kapselgebläse *n* ~ **type compressor** Roots-Gebläse *n*, Roots-Lader *m*
**rope, to** ~ aufschießen, zusammenschnüren, Fäden ziehen ~ **to** ~ **down** abseilen
**rope** Leine *f*, Kabel *n*, Strang *m*, Strick *m*, Tau *n*, Trosse *f* ~**-acidifying apparatus** Strangsäureeinrichtung *f* ~**-and-cable lashing** Leinen- und Drahtbund *m* ~ **balancing sheave** Seilausgleichrolle *f* ~ **block** Seilflaschenzug *m*, Taukloben *m* ~ **bridge** Taubrücke *f* ~ **clamp** Seilklemme *f* ~ **coil** Wirbelleine *f* ~ **coupling** Seilkupplung *f* ~ **drilling** Seilbohren *n* (min.) ~ **drive** Seil(an)trieb *m* ~ **driving** Seilbetrieb *m* ~ **drop hammer** Seilfallhammer *m* ~ **drum** Seiltrommel *f* ~ **end** Tauende *n* ~ **end safety socket** Sicherheitshülse *f* ~ **eye** Seilschleife *f* ~ **fender** Fendertau *n* (navy), Taukranz *m* ~ **fork drive** Gabelantrieb *m* (Seilbahn) ~ **grab** Seilfänger *m*, Seilfanghaken *m* ~ **grease** Seilschmiere *f* ~ **grommet** Führungsbandschutz *m* ~ **groove** Seilrille *f* ~ **guide** Seilführung *f*, Seilkante *f* ~ **handle** Strickgriff *m* ~ **hoist** Seilflaschenzug *m* ~ **knife** Hakenseilmesser *n* ~ **ladder** Strickleiter *f* ~ **lashing** Leinenbund *m* ~ **loop** Seilschleife *f* ~ **machine** Verseilmaschine *f* ~ **made of strands** Litzenseil *n* ~ **maker** Seiler *m* ~**-maker's stove** Teerküche *f* ~ **making** Seilerei *f* ~ **path** Seillauf *m* ~ **piling device** Rüsselstrangeinleger *m* ~ **post** Seilpost *f* ~ **pulley** Seilrolle *f*, Seilscheibe *f* ~**-pulley frame** Seilscheibengerüst *n* ~ **relief** Seilentlastung *f* ~ **ring** Seilring *m* ~**-scouring machine** Strangwaschmaschine *f* ~ **screw socket** Seilhülsenbohrschraube *f* ~ **setter** Riemenaufleger *m* ~ **sheave** Seilrolle *f*, Zugseiltragrolle *f* ~ **sling** Seilschlaufe *f* ~ **slippage** Seilschlupf *m* ~ **socket** Seilhülse *f*, Seilwirbel *m* ~ **spear** Seilspeer *m* ~ **speed** Seilgeschwindigkeit *f* ~**-spinning machine** Seilschlagmaschine *f* ~ **splice** Seileinband *m*, Seilpleißung *f* ~ **splicing** Seilpleißung *f* ~ **-squeezing apparatus** Ausquetschapparat *m*, Strang (warp), Strangausquetschapparat *m* ~ **start** Zugstart *m* ~ **stopper** Taustopper *m* ~ **tensiometer** Seilzugmesser *m* ~ **tension** Seilspannung *f* ~ **tensioning** Seilunterspannung *f* ~ **thimble** Seilhülse *f* ~**-traction weighers** Seilzugwaagen *pl* ~ **transmission** Seiltrieb *m* ~ **tread** Seillauf *m* ~**-trolley transporter** Verladebrücke *f* mit Seilzuglaufkatze ~ **walk** Seilerbahn *f* ~ **washer** Clapot *m* ~**-washing machine** Strangwaschmaschine *f* ~**-way** Drahtseilbahn *f*, Seilbahn *f* ~**-way bucket** Hängebahnkübel *m* ~**-way car** Seilbahnwagen *m* ~**-way crane** Seilbahnkran *m* ~ **winch** Seilwinde *f* ~ **wire** Seildraht ~ **work** Seilfabrik *f* ~ **yarn** Kabelgarn *n* ~**-yarn knot** Kabelgarnsteg *m*
**roper** Seiler *m*
**ropes** Tauwerk *n*
**ropy** fadenziehend, klebrig ~ **lava** Fladen- oder Stricklava *f*
**rosary,** ~ **pliers** Rosenkranzzange *f* ~ **trompe** Kettengebläse *n*, Paternostergebläse *n*
**rose** Beschlag *m*, Brause *f*, kreisförmige Skala *f*, Manschette *f*, Saugkorb *m* ~ **bit** Kranzkopf *m*, Krauskopf *m*, Senkfräser *m*, Senker *m*, Versenkbohrer *m*, Versenker *m* ~ **chucking reamer** Maschinengrundreibahle *f* ~ **color** Rosenrot *n*, Röte *f* ~ **copper** Scheibenkupfer *n* ~ **copper**

**disk** Garscheibe *f* ~ **cut** Rosettenschliff *m* ~ **engine** Guillochiermaschine *f* ~ **quartz** Rosenquarz *m* ~ **reamer** Grundreibahle *f* ~ Hilfsrippe *f* ~ **steel** Rosenstahl *m* ~ **top** Kronenaufsatz *m* ~ **window** Rosettenfenster *n* ~ **wood** Cypernholz *n*, Jacarandaholz *n*, Rosenholz *n*
**roselite** Roselith *m*
**rosette** Nest *n*, Wandrosette *f* ~ **copper** Gar-, Rosetten-, Scheiben-kupfer *n* ~ **plate** Röllchenplatte *f* ~**-shaped path** (of electrons) Rosettenbahn *f*
**rosin** Geigenharz *n*, Harz *n*, Kolophon(ium) *n* ~ **connection** schlechte oder kalte Lötstelle *f* ~**-core solder** Kolophoniumzinn *n* ~ **ester** Harzester *f* ~ **grease** Harzfett *n* ~ **joint** kalte Lötstelle *f* ~ **oil** Harzöl
**Rossby chart** Rossbysches Diagramm *n*
**rot, to** ~ faulen, spakig werden, verfaulen, verderben, verrotten, verwesen
**rot** Fäule *f*, Fäulnis *f* ~ **proof** fäulnis-beständig, -hindernd, -widrig
**rotameter** Rotadurchflußmesser *m*
**rotary** Drehtisch *m*, dreh . . ., drehbar, drehend, rotierend, umlaufend ~ **actuator** rotierender Betätiger *m* ~ **air filter** Umlauffilter *m* ~ **air pump** Ringluftpumpe *f* ~ **alternating axis** Drehspiegelachsen *pl* ~ **attachment** Rundtisch *m*
**rotary-attachment,** ~ **base** Rundtischunterteil *n* ~ **binder** Hebel *m* zum Festklemmen des Rundtisches ~ **feed clutch** Kupplung *f* für den Selbstgang des Rundtisches ~ **feed rod** Antriebswelle *f* für den Rundtischvorschub ~ **handwheel** Handrad *n* zum Vorschub des Rundtisches (von Hand)
**rotary,** ~ **attenuator** Drehstreifenabschwächer *m* ~ **axial pump** umlaufende Achsialpumpe ~ **ball squeezer** Luppenmühle *f* ~ **beacon** Drehfeuer *n* ~ **bending** Verwindung *f* ~ **beam** Drehrichtstrahler *m* ~ **beating mill** Quetschmangel *f* ~ **bits** Rotary-Bohrmeißel *m* ~ **blow** Drallschlag *m* ~ **blow roller coupling** Drehschlagrollenkupplung *f* ~ **blow motion** Drehschlagbewegung *f* ~ **blower** Kreisel-gebläse *n*, -vorverdichter *m*, rotierender Bläser *m*, Umlaufgebläse *n* ~**-blower-type supercharger** Abgasturbinengebläse *n*, Kreisel-gebläse *n*, -vorverdichter *m* ~**-body-type(gas) producer** Gaserzeuger *m* mit drehbarem Schacht ~ **boiler** Drehkocher *m* ~ **boring** (earth) Drehbohren *n* ~ **boring clamp** Drehbündel *n* ~ **breaker** Kreiselbrecher *m*, Rundbrecher *m* ~ **bridge crane** Brückendrehkran *m* ~ **button** Drehknopf *m* (tape rec.) ~ **calciner** Drehofen *m* ~ **calcining kiln** Drehtrommelröstofen *m* ~ **carbon brush holder** rotierender Kohlenhalter *m* ~ **cement kiln** Zement-drehofen *m*, -rotierofen *m* ~ **cleaning brush** Drehbürste *f* ~ **coil movement** Drehspulmeßwerk *n* ~ **combustion chamber** drehbarer oder rotierender Verbrennungsraum *m* ~ **compressor** Drehkolbengebläse *n*, Kreiselverdichter *m* ~ **converter** Dreh-, Einanker-umformer *m*, Umformeranlage *f*, Umformersatz *m*, umlaufender Umformer *m* ~ **converter with tube regulator** röhrengeregelter Umformer *m* ~ **crane** Drehkran *m* ~ **creasing and glueing machine** Rill- und Anleimmaschine *f* ~ **cross bits** Rotary-Kreuzmeißel *m* ~ **crusher** Kreiselbrecher *m* ~ **current** Drehstrom *m* ~**-current continuous-current**

converter Drehstromgleichstromeinankerumformer *m* ~-current continuous-current dynamotor Drehstromgleichstromumformer *m* ~-cycle drilling machine Rundtaktbohrmaschine *f* ~ diaphragm Revolverblende *f* ~ die collars Fangglocken *pl* ~ discharger rotierende Funkenstrecke *f* ~ disc Laufplatte *f* ~ disc valve flacher Drehschieber *m* ~ distributor multiple-spark photography) Auslösepropeller *m*, umlaufender Verteiler *m* ~ doctor rotierender Schaber *m* ~ drier Drehrohrofen *m*, Trockentrommel *f*, Trommeltrockner *m* ~ drill Drehbohrer *m* ~ drill rig with diamond-crowned points rotierende Bohrmaschine *pl* mit Diamanten gespitzt ~ drilling Drehbohren *n* ~ drilling bit Drehbohrer *m* ~ drive Umlaufantrieb *m* ~ drum Drehtrommel *f* ~ drum filter Trommeldrehfilter *m* ~ drum strainer Rotationsfilter *n* ~ drying kiln Trockentrommel *f* ~ dump-type separator Kreiselwipper *m* ~ engine Umlaufmotor *m* ~ exchange Vermittlung *f* mit Durchwähler ~ exhaust valve Auspuffdrehschieber *m* ~ feed Drehvorschub *m* ~ field Drehfeld *n*, umlaufendes Feld *n* ~ field converter Drehfeldumformer *m* ~ filter Drehfilter *n* ~ flicker shutter rotierende Umlaufverschlußblende *f* ~-floor sandblast room Putzhaus *n* mit Drehboden ~ force Drehkraft *f* ~ furnace Dreh(herd)ofen *m*, Trommelofen *m* ~ -furnace method of volatilizing ores Wälzverfahren *n* ~ grate generator Drehrostgaserzeuger *m* ~-grate-type gas producer Gaserzeuger *m* mit drehbarem Rost ~-grate-type producer Generator *m* mit drehbarem Rost ~-grid generator Drehrostgaserzeuger *m* ~ head Drehkopf *m* ~ helipotentiometer Umdrehungshelipot *n* ~ hoe Bodenfräse *f* ~ hose Rotarybohrschlauch *m* ~ hunting Wahl *f* in einer einzigen Ebene ~ hunting connector Mehrfach-, Sammel-leitungswähler *m* ~ impulse Drehstoß *m* ~ induction system Abgasturbinenvorverdichter *m*, kreisendes Induktionssystem, Umlaufeinlaßverfahren *n* ~ intaglio printing Rotationstiefdruck *m* ~ interrupter umlaufender Unterbrecher *m* ~ interrupter contacts Ruhekontakt *m* des Drehmagneten ~ inversion axis Drehinversionsachsen *pl* ~ jar Rutschschere *f* ~ kiln Drehofen *m*, (cylindrical) Drehrohrofen *m*, Drehtrommel *f*, rotierender Ofen *m* ~ knob Drehknopf *m* ~ light cut-off rotierende Umlaufverschlußblende *f* ~ limit switch Schaltwalze *f* ~ line switch Drehvorwähler *m* ~ machine Rotationsmaschine *f* ~ machine ink Farbe *f* für Rotationsdruck ~ magnet Drehmagnet *m* ~ (electro) magnet Drehelektromagnet *m* ~ magneto Umlaufmagnetzünder *m* ~ milling machine Drehtischfräsmaschine *f* ~ mixer Rotationsmischmaschine *f* ~ mixing table Drehteller *m* einer Zuteilvorrichtung ~ motion Dreh-, Rad-, Rotations-bewegung *f* ~ mounted grinding tool Schaftschleifscheibe *f* ~ off-normal contacts Wellenkontakt *m* ~ oscillation Drehschwingung *f* ~ pan mixer Eirich-Gegenstromschnellmischer *m* ~ pendulum movement Drehpendelbewegung *f* ~ pendulum tachometer (or speedometer) Drehpendeltachometer *n* ~ phase changer Drehphasenschieber *m* ~ phase converter Blindleistungsmaschine *f* ~ phase shifter

Drehphasenschieber *m* ~ photogravure Tiefdruck *m* ~ piercing mill Hohlwalzwerk *n* ~ pin Drehstift *m* ~ piston Drehkolben *m* ~ piston compressor Rollkolbenverdichter *m* ~-piston driven motor Drehkolben-Antriebsmotor *m* ~ -piston meter Ringkolbenzähler *m* ~ piston pump Drehkolbenpumpe *f* ~ planer Rotationshobelmaschine *f* ~ platen Karussel *n* ~ plow Bodenfräse *f* ~ pneumatic drill Preßluftbohrmaschine *f* ~ power Drehungs-, Rotations-vermögen *n* ~ press for high speed printing Rotationsschnelldruckpresse *f* ~ press for illustration printing Illustrationsrotationsmaschine *f* ~ pressure filter Druckdrehfilter *n* ~ printing Rotationsdruck *m* ~ printing press Rotationsdruckmaschine *f* ~ puddling furnace Drehpuddel-, Puddeldreh-ofen *m* ~ pump Drehkolben-, Flügel-, Kreisel-, Rotary-, Rotations-pumpe *f*, rotierende Pumpe *f*, Schrauben-, Umlauf-, Verdränger-pumpe *f* ~ pushbutton switch Drehtastenschalter *m* ~ radial pump umlaufende Radialpumpe *f* ~ reflection Drehspiegelung *f* (cryst.) ~ release solenoid drehbare Auslöse-Magnetspule *f* ~ repeater umlaufende Übertragung *f* ~ results Drehergebnis *n* ~ rig Drehapparat *m* ~ sampling switch Meßstellen-Umschalter *m* ~ scanning and switch unit (rdr) Drehverteiler *m* ~ screen (or sieve) Drehsieb *n*, Siebtrommel *f* ~ search on one level Wahl *f* auf einer Ebene ~ selector Drehwelle *f* ~ (pre)selector Drehvorwähler *m* ~ selector bank radial angeordnetes Kontaktfeld *n* ~ shaping device Rundhobelvorrichtung *f* ~ shears Kreis-, Rotations-schere *f* ~ shield rotierendes Schutzschild *n* ~ shutter Umlauf-verschluß *m*, -blende *f* ~ slide valve Drehschieber *m* ~ slide-valve steering Drehschiebersteuerung *f* ~ slide-valve switch Drehschieberweiche *f* (Rohrpost) ~ slide valve vacuum pump Drehschiebervakuumpumpe *f* ~ sliding vane type compressor Drehschieberpumpe *f* ~ snap switch Drehnockenschalter *m* ~ snowplow Schnee-pflug *m* , -schleuder *f* ~ spark umlaufender Funke(n) *m* ~ spark gap Abreißfunkenstrecke *f*, rotierende oder umlaufende Funkenstrecke *f* ~ spool for control valve Drehschieber *m* zum Steuergerät ~-spool valve Drehschieberventil *n* ~ stabilizer (motion-picture projector) Ausgleichschwingscheibe *f*, Ölschwungmasse *f*, rotierender Dämpfer, Schwungmassenrolle *f* ~ static spark gap Umlauffunkenzieher *m* ~ step Drehschritt *m* ~ strainer rotierender Knotenfänger *m* ~ substation Umformerwerk *n* ~ substitutes Rotary-Übergänge *pl* ~ suction cell filter Saugzellen-Drehfilter *m* ~ swaging machine Rotationsgesenkdrückmaschine *f* ~ sweeper Aufrauhbürste *f* ~ switch Drehschalter *m*, Drehschaltung *f*, Drehungswähler *m*, Kollektor *m*, umlaufender Schalter *m*, Wähler *m* mit einziger Bewegungsrichtung ~ system Drehbohren *n*, Drehwählersystem *m* ~ table Drehscheibe *f*, Drehtisch *m*

rotary-table, ~ blast cabinet Blashaus *n* mit Drehtisch ~ mill Drehtischanlage *f* ~ sandblast cabinet Sandstrahlgebläse *n* mit Drehtisch ~ sandblast machine Drehtisch-gebläse *n*, -sandstrahlapparat *m*, Rotationstischsandstrahlgebläse *n* ~ sandblast room Sandstrahlgebläse-

haus *n* mit Drehtisch ~ **sandblast unit** Dreh-
tischfunker *m* ~ **sandblaster** Drehtischwirbel-
strahler *m*
**rotary,** ~ **taper-taps** Fangdorne *pl* ~ **telescope**
Rundblickfernrohr *n* ~ **tiller for gardens** Gar-
tenfräse *f* ~ **tin printing press** Blechdruckrota-
tionsmaschine *f* ~**-tower crane** Turmdrehkran
*m* ~ **transmissions** Drehdurchführungen *pl* ~
**-type boiler** Drehkessel *m* ~**-type compressor**
Umlaufverdichter *m* ~**-type engine** Sternum-
laufmotor *m* ~ **valve** Rotaryhahn *m* ~ **valve
drill** Ventildrehbohrer *m* ~ **vane pump** Drehschieberpumpe *f* ~ **viscosimeter** Rotations-
viskosimeter *m* ~**-wing aircraft** Drehflügelflugzeug *n*, Drehflügler *m* ~ **wiper motor** Umlauf-
wischermotor *m*
**rotatable** umdrehbar ~ **coaxial-coil transformer**
Klapptransformator *m* ~ **coil** drehbarer Rahmen *m*, Drehrahmen *m*
**rotate, to** ~ antreiben, drehen, kreisen, laufen,
rotieren, rundlaufen, schwenken, sich drehen,
umdrehen, umkreisen, umlaufen, umwälzen,
verdrehen
**rotate register** zyklische Vertauschung *f* (info
proc.)
**rotating** Drehen *n*; drehbar, drehend, rotierend
~ **and reciprocating parts** bewegliche Teile
*pl*
**rotating,** ~ **airfoil** Drehflügel *m* ~ **anemometer**
Drehungswindmesser *m* ~ **anode X-ray tube**
Drehanodenröntgenröhre *f* ~ **arc** Wälzbogen
*m* ~ **arm** Dreharm *m* ~ **armature relay** Dreh-
ankerrelais *n* ~ **band** (of projectile) Führungsband *n*, Führungs-, Geschoß-ring *m*, (ammunition, gun barrel) Geschoßführungsband *n* ~
**band of projectiles** Führung *f* ~**-bar fatigue test**
Dauerbiegeversuch *m* mit Stabumdrehung ~
**-bar-impact fatigue test** Dauerschlagbiegeversuch *m* mit Stabumdrehung ~ **barrel** Drehtrommel *f* ~**-base mount** Sockellafette *f* ~
**beacon** Drehfeuer *n* (aviat.) ~ **beacon light**
Drehscheinwerfer *m* ~**-beam test** Umlaufbiege-
probe *f*, -prüfung *f* ~**-beam testing machine**
Drehstabdauerfestigkeits-, Umlaufbiege-prüf-
maschine *f* ~ **bending test** Umlaufbiegeversuch
*m* ~ **block for pouncing** Tourmaschine *f* ~ **brush**
umlaufende Bürste *f* ~ **buddle** Drehherd *m* ~
**cabin** Drehstand *m* ~ **casing** Umlaufgehäuse *n*
~ **coil** drehbare Spule *f*, Dreh-, Rotor-, Suchspule *f* ~**-coil variometer** Dreh-(spul)variometer
spule *f* ~ **combustion engine** Kreiskolbenmotor *m*
~ **combustion head** Oberteil *m* des Verbrennungs-
raumes ~ **compass card** drehende Kompaß- oder
-Kursrose *f* ~ **component** Drehkomponente *f* ~
**cradle** Schwenkkran *m* ~ **crane** Baudrehkran *m*
~ **crystal** Dreh-, Schwenk-kristall *m* ~ **cylindrical kiln** Drehrohofen *m* ~ **deck radius** Dreh-
radius *m* (Bagger) ~ **device** Umlaufgerät *n* ~
**direction finder** umlaufender Peiler *m* ~**-disk
shutter** Drehscheibenverschluß *m* ~ **drum** Drehtrommel *f* ~ **electrical slipring** rotierender elektrischer Schleifring *m* ~ **field** Drehfeld *n*, rotierendes oder umlaufendes Feld *n* ~**-field deflection** Drehfeldablenkung *f* ~**-field instrument**
Drehfeldinstrument *n* ~ **field magnet** Sekundär-
anker *m* ~ **field motor** Drehfeldantrieb *m* ~
**-field oscillation** Drehfeldschwingung *f* ~ **field
pole** Drehpol *m* ~ **firing post** Walzenblende *f* ~

**frame** Drehrahmen *m* ~**-frame directionfinding
principle** Drehrahmenpeilverfahren *n* ~ **furnace**
Rotierer *m*, Rotierofen *m* ~ **gun mounting** Drehkranz *m* ~ **gun ring** Maschinengewehrdreh-
kranz *m* ~ **jack cam** Drehen *n* der Platinen-
excenter ~ **joint** Drehkupplung *f* ~ **key**
Schwenktaster *m* ~ **kiln** Rotierer *m*, Rotierofen *m*, Trockentrommel *f* ~ **lock** Schleusen-
trommel *f* (Rohrpost) ~ **loop antenna** drehbare Rahmen *m* ~ **loop antenna** drehbare Rahmenantenne *f*, Drehrahmenantenne *f* ~ **loop beacon**
Drehrahmenpeilfunksender *m*, umlaufende
Richtfunkbake *f* ~ **loop direction-finding
antenna** Drehrahmenpeiler *m* ~ **mechanism**
(lens clock) Drehwerk *n* ~ **mirror instrument**
Drehspiegelinstrument *n* ~ **motion** Schwenkbewegung *f* ~ **mount** Pivotlafette *f* ~ **phase
advancer** rotierender Phasenschieber *m* ~
**piston compressor** Drehkolbenverdichter *m* ~
**piston pump** Sogpumpe *f* ~ **plate with fixed
studs** Stiftscheibe *f* ~**-plate condenser** Dreh-
(platten)kondensator *m* ~ **puddling machine**
Puddeldrehofen *m* ~ **rack** schwenkbarer Rahmen *m* ~ **ring** Schwungring *m* ~ **shaft** Schwinge
*f* ~ **shear** umlaufende Schere *f* ~ **shutter** Revolverblende *f* ~ **stabilizer** rotierender Dämpfer *m*
~ **stage** Drehtisch *m* ~ **table** drehender Kehr-
herd *m* ~ **table with adjustable scraper** Abstreif-
teller *m* ~**-table press** Drehtischpresse *f* ~
**track** Drehkranz *m* ~ **tub** Drehbottich *m* ~
**valve** Umlaufventil *n*, Walzenschieber *m* ~
**water bomb** Rotationswasserbombe *f* ~ **wedge**
Drehkeil *m* (opt.) ~**-wheel anemograph** Windradanemograf *m* ~ **wind vane** Windradanemograf *m* ~ **wing** Drehflügel *m* ~**-wing aircraft**
Drehflügelflugzeug *n*, Drehflügler *m*
**rotation** Antrieb *m*, Drehbewegung *f*, Drehung *f*,
Einstelldrehung *f*, Kreisbewegung *f*, Raddrehung *f*, (of a vector) Rotation *f*, Turnus *m*, Umdrehung *f*, Umlauf *m*, Umschwung *m*, Verdrehung *f*, (of a vector) Wirbel *m* without ~
rotationsfrei
**rotation,** ~ **of axes** Achsendrehung *f*, Drehung *f*
des Achsenkreuzes ~ **of the camera** Schwenkung
*f* der Kamera ~ **of crank** Kurbeldrehung *f* ~ **of
the earth** Erddrehung *f*, (upon its axis) Erd-
umdrehung *f* ~ **of plane of polarization** Drehung *f* der Polarisationsebene ~ **of projectile**
Drehung *f* des Geschosses (**direction of**) ~
**propeller** Luftschraubendrehrichtung *f* ~ **of
rotor** Läuferumdrehung *f*
**rotation,** ~ **anemometer** Rotationsanemometer
*n* ~ **box** Gehäuse *n* ~ **coma** Drehungskoma *n*
~ **cooling** Umlaufkühlung *f* ~ **cutter** Schnittrichtung *f* ~ **diagram** Drehdiagramm *n* ~**-free**
drehungsfrei ~ **frequency** Umlaufschwingungs-
zahl *f* ~ **gear** Drehwerk *n* ~ **plane** Kreisebene *f*
~ **range** Umlaufsbegrenzung *f* (Probenwechsler) ~ **roller path** Rollkranz *m* ~ **spark-gap** rotierende Funkenstrecke *f* ~ **spectrograph** Dreh-
spektrogramm *n* ~ **speed** Aufrichtgeschwindig-
keit *f* ~ **vibration band** Rotationsschwingungs-
band *n* ~ **vibration spectrum** Rotationsschwin-
gungsspektrum *n*
**rotational** kreisend, selbst umdrehend, turnus-
mäßig ~ **band spectra** Rotationsbandenspektra
*pl* ~ **deviation** Ausschwenkung *f* aus der Mittel-
lage ~ **distortion** Bildzerdrehung *f* ~ **field** Wir-

belfeld *n* ~ **flattening** Rotationsabplattung *f* ~ **frequency** (of electrons) Umlauffrequenz *f* ~ **inertia** Beharrungsmoment *m*, Drehwucht *f*, Massenmoment *n* ~ **invariance** Drehinvarianz *f* ~ **model** Spinmodell *n* ~ **motion** Rotationsbewegung *f* ~ **partition function** Rotations-Zustandssumme *f* ~ **quantum number** Rotationsquantenzahl *f* ~ **reversibility** reversible Magnetisierungsrichtung *f* ~ **speed** Drehzahl *f*, Radumlaufgeschwindigkeit *f*, Umlaufgeschwindigkeit *f* ~ **switching** Spinrichtungsänderung *f* ~ **threshold curve** Drehsinnschwellenkurve *f* ~ **wave** Schubwelle *f* (acoust.)

**rotationally symmetric motion** rotationssymmetrische Bewegung *f*

**rotative** rotierend, sich drehend ~ **moment** Schwungmoment *n* ~ **speed** Drehgeschwindigkeit *f*, Tourenzahl *f*

**rotator** Drehtisch *m* (opt.), (quantum theory) Drillachse *f*, (ship's log) Flügel *m* des Patentlogs, Massenpunkt *m*, Rotor *m*

**rotatory** dreh . . ., kreisend ~ **dispersion** Drehungsdispersion *f* ~ **flow** Umlaufbewegung *f* ~ **motion** Umdrehungsbewegung *f* ~ **polarization** Drehpolarisation *f*, optische Drehung ~ **rotation** Rotationspolarisation *f* ~ **tide** Drehtide *f*

**rotodyne** Rotodyne *m*

**rotogravure** Kupfer-, Zylinder-tiefdruck *m* ~ **press** Tiefdruckrollenmaschine *f*

**roton** Roton *n*

**rotor** Anker *m*, Dreh-flügel *m*, -körper *m*, Flügelrad *n*, Läufer *m*, Laufrad *n*, Rotor *m*, Schaufelrad *n*, Tragschraube *f*, Unterbrecher *m*, Verteilerrotor *m* ~ **with slip rings** Schleifringläufer *m*

**rotor**, ~ **armature** Lautsprecheranker *m* ~ **blade** (power plant) Läuferschaufel *f*, Laufschaufel *f*, (Hubschr.) Rotorblatt *n* ~ **blade retention** Blattanschluß *m* ~ **blading** Laufradbeschaufelung *f* ~ **body** Ankerkörper *m* ~ **brake** Rotorbremse *f* ~ **cage impeller** Trommelkreiselrührer *m* ~ **coil** drehbare Spule *f*, Drehspule *f* ~ **cooling jet** Läuferkühlung *f* ~ **craft** Drehflügelflugzeug *n*, Hubschrauber *m* ~ **current** Rotorstrom *m* ~ **disc** Läufer-, Rotor-scheibe *f* ~ **drive shaft** Rotorwelle *f* ~ **entrance** Laufradeintritt *m* ~ **gimbal** Rotorkreuzgelenkring *m* ~ **hub** Rotor-kopf *m*, -nabe *f* ~ **inlet** Laufradeintritt *m* ~ **lobe** Rotorflügel *m* ~ **periphery power** Radumfangsleistung *f* ~ **plane** Drehflügel-, Rotor-flugzeug *n* ~ **plate** Rotor-paket *n*, -platte *f* ~ **pole shoe** Läuferpolschuh *m* ~ **pylon** Rotorbock *m* ~ **rheostat** Läufer-, Regel-anlasser *m* ~ **runner** Laufzeug *n* ~ **segment** Rotorsegment *n* ~ **shaft** Läuferwelle *f* ~ **sleeve** Läuferbuchse *f* ~ **speed** Rotorumlaufgeschwindigkeit *f* ~ **starter** Läuferanlasser *m* ~ **terminal** Läuferklemme *f* ~ **thrust** Rotorschub *m* ~ **tilt** Rotorneigung *f* ~ **transmission** Rotorgetriebe *n* ~ **voltage** Läuferspannung *f*

**rotorable grip** rotierende Sperre *f*

**rotosyne**, ~ **converter set** Rotosynumformer *m* ~ **power line** Rotosynanlage *f*

**rotten** angefault, faul, faulbrüchig, morsch(ig), stockig, verfault ~ **hole** Fäule *f* ~ **lime** Fauläscher *m* ~ **stone** Tripel *n*, Tripelerde *f*

**rottenness** Faulbrüchigkeit *f*, Fäule *f*, Fäulnis *f*, Morschheit *f*

**rotting** Faulen *n*, Fäulnis *f*, Vermoderung *f*; angefault

**rotund** rund

**rotunda** Rundbau *m*

**rouge** (polishing powder) Englischrot *n*, Polierrot *n*, Rouge *f*, Schminkrot *n* ~ **for polishing** Schleifrot *n*

**rough, to** ~ rauhschleifen, schruppen, strecken, (work) vorbearbeiten, vorstrecken, vorwalzen **to** ~**-bore** vorbohren **to** ~**-cast** berappen **to** ~**-cut** zurichten **to** ~ **down** auswalzen, herunterwalzen, rauhschleifen, strecken, vorstrecken, vorwalzen **to** ~ **down the blooms** die Luppen *pl* auswalzen **to** ~**-draw** skizzieren **to** ~**-drill** vorbohren **to** ~**-face** planschruppen **to** ~**-file** vorfeilen **to** ~**-flat** flachwalzen **to** ~**-grind** schroten **to** ~**-hammer** grob hämmern **to** ~**-machine** grobbearbeiten **to** ~**-out** aushauen, vorfräsen **to** ~**-plane** schroppen, schruppen, vorhobeln **to** ~**-plane timber** Holz *n* abhobeln **to** ~**-ream** vorfräsen, vorreiben **to** ~ **with sand and water** matt schleifen **to** ~**-turn the axle** die Achse vordrehen **to** ~**-work** herausarbeiten

**rough** Rauheit *f*, Rohzustand *m*, (Gleitschutz) Stollen *m*, Unebenheit *f*; derb, grob, groberdig, rauh, roh, schroff, uneben, unfertig, (cloth) ungewalkt ~ **analysis** Rohanalyse *f* ~**-and-ready** grob gearbeitet, unfertig ~**-and-ready rule** Faustregel *f* ~ **average** annähernder Durchschnitt *m* ~ **balance** rohes Gleichgewicht *n* ~ **bloom** gezängte Luppe *f* ~ **calculation** Faustformel *f*, Überschlagsrechnung *f* ~ **cast** Anwurf *m*, Berapp *m*, Bewurf *m*, Rapputz *m*, Spritzbewurf *m*, Rohguß *m* ~**-cast plate glass** Spiegelrohglas *n* ~ **chisel** Grobbeitel *m* ~ **coal** Förderkohle *f* ~ **coat** Rauhputz *m* ~**-cored die** Rohholzziehstein *m* ~ **covering** Deckenzeug *n* ~**-cutting tool** Schruppstahl *m* ~ **drilling** Vorbohren *n* ~ **drop forging** Gesenkschmiede-Rohling *m* ~ **engine (operation)** rauher Gang *m* des Motors *m* ~ **estimate** Voranschlag *m* ~ **file** Gewichts-, Pack-, Schrupp-, Stroh-feile *f* ~ **finish** Schruppfläche *f* ~ **finishing** Grobschlichten *n* ~ **floor** Greifhaufen *m* ~ **flotation cells** Grobflotator *m* ~ **forging** Schmiederohrteil *n* ~**formula** Überschlag(s)formel *f* ~ **formulation** Näherungsansatz *m* ~ **grading** Rohplanum *n* ~ **grinder** Vorschleifer *m* ~ **grinding** Vorschleifen *n* ~**-ground** rohabgeschliffen ~ **hewn** roh behauen ~ **indication** Anhalt *m* ~ **landing** Bumslandung *f* ~ **lignite** Rohbraunkohle *f* ~**-machined shaft** vorgearbeitete Welle *f* ~ **machining** Vorarbeiten *n* auf Rohmaß ~ **map sketch** Kroki *n* ~ **matt** extra matt ~ **melt** Rauhschmelze *f* ~**-milled** vorgefräst ~ **milling** Schruppfräsen *n* ~ **milling attachment** Vorfräseinrichtung *f* ~ **milling attachment slide screw** Schlittenspindel *f* für Vorfräseinrichtung ~ **milling attachment table screw** Tischspindel *f* für Vorfräseinrichtung ~ **planer** Abrichtmaschinenarbeiter *m* ~ **plaster** Berapp *m*, Bewurf *m*, Spritzbewurf *m* ~ **polishing** Vorpolieren *n* ~**-pressed block** Vorpreßling *m* ~**-pressed wheel** vorgepreßtes Rad *n* ~ **punching** Vordornen *n* ~ **reamer** Vorreibahle *f* ~ **regulator** Grobregler *m* ~ **rolled** vorgewalzt ~ **rolls** Stachel-

walzen *pl* (Zackenwalzen) ~ **rule** Faustregel *f*
~ **running** unruhiger Wagenlauf *m* ~ **scale**
Grobteilung *f* ~ **sea** stürmische See *f* ~ **separat-**
**ing** (of ores) rohes Sortieren *n* der Erze ~ **side**
**of the belt** Fleischseite *f* des Riemens ~ **size**
Rohmaß *n* ~ **sketch** Faustskizze *f* ~ **slate plates**
Schieferrohtafeln *pl* ~ **smalls** Fördergrus *m* ~
**spot** (on glassware) Taststrecke *f* ~ **stamping**
Schmiederohling *m* ~ **stoneware** Tonsteingut *n*
~ **string** Stützbalken *m* ~ **survey** flüchtige Vermessung *f* (geol.) ~**-tanned** unzugerichtet ~
**tuning** grobe Abstimmung *f* ~**-vacuum pump**
Vorvakuumpumpe *f* ~ **walling of a furnace**
Ofenstock *m*
**roughed out** ausgeschruppt
**roughed,** ~ **bloom** gestreckter Walzblock *m* ~
**gears** vorgeschruppte Räder *pl* ~ **machined** vor-
geschroppt
**roughen, to** ~ aufrauhen, einhacken, einkerben,
rauhen, rauhwerden, schrubben, schruppen
**roughened** aufgerauht ~ **surface** (on glass)
Brandfleck(en) *m*
**roughening** Aufrauhen *n*, Einkerben *n*, Rauhen
*n*, Rauhwerden *n* ~ **machine** Anreibemaschine *f*
~ **plate** Stockerplatte *f*
**rougher** Schruppwerkzeug *n*, Streckgerüst *n*,
Vorgerüst *n*, Vorschmiedegesenk *n*
**roughing** Rauhen *n*, Schruppen *n*, Vorkaliberwalzen *n*, Vorwalzen *n* ~ **and smoothing of glass**
Glasveredelung *f*
**roughing,** ~ **block** Grobzug *m* ~ **cut** Schrupp-
schnitt *m* ~ **cutter** Schruppfräser *m*, Vorfräser
*m*, Vorschneidfräser *m* ~ **cylinder** Luppen-
walze *f* ~ **device for hoop iron** Bandeisen-
schrappvorrichtung *f*, Schrappvorrichtung *f*
für Bandeisen ~ **diamond pass** Rautenvorwalz-
kaliber *n* ~**-down of piles** Vorhämmern *n* der
Pakete ~**-down mill** Grobwalzwerk *n* ~ **flat**
Flachwalzen *n* ~ **flotation** Grobflotation *f* ~
**hackle** Abzugshechel *f* ~ **hob** Schruppabwälz-
fräser *m* ~ **lathe** Schruppdrehbank *f* ~ **machine**
Schroppmaschine *f* ~ **mill** Schmirgelrad *n*, Vorstraße *f*, Vorwalzwerk *n* ~**-out** Ausschruppen *n*
~**-out chisel** Grobmeißel *m* ~**-out machine** Vor-
fräsmaschine *f* ~**-out pass** Vorstreckkaliber *n* ~
**pass** Schnellvorwalzkaliber *n*, Streckkaliber *n*,
Vorstich *m* ~ **reamer** Schruppreibahle *f*, Vorreibahle *f* ~ **roll** Schnellvorstreckwalze *f*,
Streckwalze *f*, Vorbereitungskaliber *n*, Vorkaliberwalze *f*, Vorstreckwalze *f*, Zängwalze *f*
~ **rolls** Schweißwalzen *pl* ~ **separatory cell**
Grobflotator *m* ~ **speed** Drehzahlen *pl* der
Schruppreihe ~ **speed range** Schlichtdrehzahl-
reihe *f* ~ **stand** Block-, Streck-, Vor-gerüst *n* ~
**steel** Schruppmeißel *m* ~ **strand of rolls** Vorwalzstrecke *f* ~ **surface** Schruppoberfläche *f* ~
**tool** Schroppmesser *n*, Schruppstahl *m* ~ **tool-**
**holder** Vorschneidstahlhalter *m* ~ **train** (of
rolls) Vorstraße *f* ~ **work** Schrupparbeit *f*
**roughly** in groben Zügen, überschlägig ~ **made**
**in general** ausgeschruppt
**roughness** Glanzlosigkeit *f*, Rauheit *f*, Rauhig-
keit *f*, Unebenheit *f* ~ **of the pipe wall** Rauhig-
keit *f* der Rohrwand
**roughness,** ~ **gauge** Rauhigkeitsmesser *m* ~
**meter** Feintaster *m* ~**-smoothness-meter** Rauhigkeitsmesser *m*
**roughometer** Rauhtester *m*

**roulette of an elipse** Rollkurve *f* der Ellipse
**round, to** ~ abrunden, ausrunden, runden, umbiegen **to** ~ **off** abkanten **to** ~ **off (or out)** abrunden **to** ~ **off edges** Ecken *pl* abrunden **to** ~
**-off end** ankuppen **to** ~ **out** abfangen (Flugzg.
b. d. Landung), ausbeulen
**round** Bucht *f*, (of ammunition) Geschoß *n*,
Runde *f*, Schuß *m*; kreisförmig, rund, voll-
tönend ~ **the loop** über die Schleife hinweg ~ **of pattern** (weaving)
Wiederholung *f*
**round,** ~ **angle** Vollwinkel *m* ~ **arch** Rundbogen
*m* ~ **bar** Rundstab *m*, Stabstahl *m* ~**-bar spiral**
**spring** Rundfeder *f* ~ **belt** Rundriemen *m* ~
**belt pliers** Riemenlochzange *f* ~ **billet** Knüppel
*m* (Rundbarren), Rohrblock *m*, Rundblock *m*,
Rundknüppel *m* ~ **block rest** Rundsupport *m*
~ **bloom** Rundblock *m* ~**-bottom flask** Rund-
kolben *m* ~**-bottom flask with ring neck** Rund-
kolben *m* mit umgelegtem Rand ~**-bottom**
**rotary dump car** Rundkippwagen *m* ~ **bracing**
**and reinforcing irons** Rundeisen *n* ~ **brackets**
runde Klammern *pl* ~ **broach** Polierahle *f* ~
**buddle** Kegelherd *m* (min.), Rundherd *m*, ~
**cable** (switchboard) Rundkabel *n*, Rundseil *n*
~**-chamber kiln** Ringofen *m* ~ **chart** Registrier-
scheibe *f* ~**-chart instrument** Kreisblattinstru-
ment *n* ~**-coil measuring instrument** Rundspul-
instrument *n* ~ **container** Rund-behälter *m*,
-dose *f* ~ **core** Rundkern *m* ~**-corner beveling**
**machine** Eckenrundstoßmaschine *f* (print.) ~
**corner cutter** Abrundmaschine *f* (print.) ~
**corner punching machine** Eckenrundstanz-
maschine *f* ~ **cornering machine** Eckenabrunde-
maschine *f* ~ **down-draft kiln** Kammerringofen
*m* ~**-edge joint file** Scharnierplatzfeile *f* ~**-edged**
**hole** Bördelloch *n* ~ **edges** gerundete Kanten *pl*
~ **elastic** Gummikordel *f* ~**-ended feather** Feder
*f* in Nut mit runden Enden, rundstirnige Keilnutenfeder *f* ~**-ended sunk key** Einlegekeil *m*
mit runden Enden ~**-ended taper-sunk key**
Einlegekeil *m* ~**-eye punch** runder Durchtreiber
*m* ~ **file** Rundfeile *f* ~ **fire pit** Rundschacht *m*
~ **flame** Rundflamme *f* ~ **flint work** abgerun-
detes Geröllmauerwerk *n* ~ **gravel** Rollkies *m*
~ **groove** Rundkaliber *f* ~ **groove milling**
**machine** Rundnutenfräser *m* ~**-hand** Rund-
schrift *f*
**roundhead** (of screws) Halbrund-, Molen-, Rundkopf *m* ~ **buttress dam** Sperre *f* mit rundköpfi-
gen Strebepfeilern ~ **counter-sunk rivet** Linsen-
senkniete *f* ~ **counter-sunk screw** Linsensenk-
schraube *f* ~ **grooved pin** Halbrundkerbnagel
*m* ~ **rivet** Halbrund-, Rundkopf-niete *f* ~
**screw** Halbrundschraube *f*, Rundkopfschraube
*f* ~ **wood screw** Rundkopfholzschraube *f*
**roundheaded** oben rund (arch.), rundköpfig ~
**notched nail** Halbrundkerbnagel *m* ~ **rivet**
rundköpfiges Niet *n* ~ **screw** Halbrundschraube *f*
**round,** ~ **hilltop** Kuppe *f* ~ **hole screen** Rund-
lochsieb *n* ~**-house** Lokomotivschuppen *m*,
Rundhaus *n*, Turm *m* ~ **ingots** Rundgüsse *pl*
~ **iron** Rund-eisen *n*, -stahl *m* ~ **iron bar** Rundeisen *n* ~ **iron strap** Rundeisenbügel *m* ~ **key**
Rundkeil *m* ~ **kiln** Ringofen *m* ~**-link chain**
Rundgliederkette *f* ~ **log** Rundklotz *m* ~ **mold-**
**ing** Rundbogen *m* ~**-nose chisel** Hohlmeißel *m*

**~-nose iron** Hohlkehleneisen *n* **~-nose(d) pliers**
Rundzange *f* **~-nose tool** Hohlschneidestahl *m*
für runde Nuten, Rundstahl *m* **~-nosed clamps**
Flügelklemme *f* **~-nosed cutting tool** rundge-
schliffener Drehstahl *m* **~-nosed plane** Schrupp-
hobel *m* **~-nosed tool** Hohlkehlenstahl *m* **~
notch** Rundkerb *m* **~ nut** Rundmutter *f* **~-off
error** Rundungsfehler *m* **~ pass** Rundkaliber *n*
**~ pile** runder Pfahl *m* **~ point** Linsenkuppe *f*
**~-point setscrew** Druckschraube *f* mit Linsenkuppe **~ puncheon** Ringdorn *m* **~ robin** Rund-
schreiben *n* **~ rolling machine** Umrollmaschine
*f* **~ rope** Rundseil *n* **~-rubber ring** Rundgummi-
ring *m* **~ section** Rundprofil *n* **~ sleeker for
shot molds** Glättplatte *f* **~ slot milling machine**
Rundnutenfräser *m* **~ socket** Rundmuffe *f* **~
specimen head** Klöppel *m* (Prüfmaschine) **~
steel** Rundstahl *m* **~ steel bar** Rundeisen *n*,
Rundstahlstange *f* **~ stock** Rundmaterial *n*
**~-stock straightening machine** Rundricht-
maschine *f* **~ table with graduated circle** Rundtisch *m* mit Teilkreis **~ taper sunk key** Einlegekeil *m* mit abgerundeten Ecken **~ thread** Rundgewinde *n* **~-threaded** rundgängig **~ timber**
Ganzholz *n*, Knüppelrost *m*, Rundholz *n* **~ tip**
Linsenkuppe *f* **~ trip** Rundgang *m* **~ trip echo**
Mehrfachecho *n* **~-tube drip cooler** Beriese-
lungskühler *m* mit Rundrohrflächen **~-type
construction** Rundbauart *f* **~ water collector**
Ringsammler *m* **~ wire** Runddraht *m* **~ wire
armoring** Runddrahtbewehrung *f* **~-wire windings** Runddrahtwindungen *pl* **~-working** rund-
arbeitend

**rounded** gerundet **~ back** gewölbter Rücken *m* **~
cheek** abgerundete Kurbelwange *f* **~ circumferential groove** Gewindeübergangstaille *f* **~
corner** Ecke *f* der Schleuseneinfahrt **~ grain**
Kügelchen *n* **~ hammer** Kehlhammer *m* **~ head**
gewölbter Boden *m* **~ material** Schottermasse *f*
**~ off** abgerundet, rundlich **~ tile** Rundziegel *m*
**~-wave keying** Weichtastung *f*

**rounding** Bogen *m*, Rundung *f*; sich rundend,
rundlich **~ advance** Abrundungsvorschuß *m* **~
error** Rundungsfehler *m* (info proc.) **~-off** (a
curve) Abflachung *f*, Ausrundung *f*, Wälzung *f*
**~-off circle** Abrundung *f* **~ tools** Abrund-stahl
*m*, -werkzeug *n*

**roundness** Rundung *f*, Verrundung *f*

**rounds** Rund-profil *n*, -stab *m*, -stahl *m* **~
counter** Schlußzähler *m* **~ per minute** Schuß *m*
pro Minute

**rouse, to ~** aufrütteln, aufwecken

**route, to ~** dirigieren, (over a line) leiten, einen
Leitweg geben, schicken, steuern, auf den Weg
befördern, zuleiten

**route** den Arbeitsgang *m* festlegen, Fahrt *f*, Kurs
*m*, (traffic) Leitweg *m*, Linien-führung *f*, -zug
*m*, Reiseweg *m*, Schiffahrtsweg *m*, Strecke *f*,
Weg *m* **on ~** unterwegs **~ of approach** Anmarschweg *m*, Annäherungsweg *m* **~ of flight**
Flugstrecke *f* **~ of line** Linienführung *f* **~ for
subscriber's main cable** Hauptkanal *m*

**route, ~ angle** Kurswinkel *m* **~ avigation** Strek-
kennavigation *f* **~ beacon** Kursfeuer *n*, Orien-
tierungsfeuer *n*, Strecken-feuer *n*, -licht *n* **~
calculator** Kursschieber *m* **~ card** Laufkarte *f*
**~ chart** Kursskizze *f* **~ compass** Reisekompaß
*m* **~ curvature** Bahnkrümmung *f* **~ dialling**

Wegwahl *f* (teleph) **~ guide** Streckenhandbuch
*n* **~ lighting** Streckenbeleuchtung *f* **~ pattern**
Streckennetz *n* **~ selection** Trassenwahl *f*,
Trassierung *f* **~ setting relay** Fahrstraßensteller
*m* **~ survey** Streckenführung *f*

**routed** gefräst **~ beam** ausgeschweifter Balken *m*
**~ section** ausgeschweiftes Profil *n*

**router** Vorlageschneidemaschine *f*

**routes of communication** Verbindungen *pl*

**routine** Geschäftsgang *m*, gewohnheitsmäßige
Übung *f*, Maschinenprogramm *n* (info proc.),
Routine *f*; gewohnheitsmäßig, mechanisch,
normal, regelmäßig, vorschriftsmäßig **by ~**
handwerksmäßig

**routine, ~ acknowledgment** betriebliche Bestätigung *f* **~ analysis** Massenanalyse *f* **~ call** gewöhnliches Ferngespräch *n* **~ check** Routine-
überprüfung *f* **~ determination** Reihenbestim-
mung *f* **~ expenditure** Eigenbefugnis *f* (im
Budget) **~ library** Programmbibliothek *f* **~
maintenance** regelmäßige Unterhaltung *f* **~
method** Massenanalyse *f* **~ microscope** Arbeits-
mikroskop *n* **~ repair work** laufende Instand-
setzungsarbeit *f*, Pflege *f* **~ replacement** regelmäßiger Austausch *m* **~ test** regelmäßige Messung *f* oder Prüfung *f* **~ work** laufende Arbeiten *pl*

**routing** (as of beams) Aussteifung *f*, Fräsarbeit *f*
mit Führung des Werkstückes von Hand,
Leiten *n*, Leitweg *m* (rdo), Schablonenschnei-
den *n*, (through selectors) Steuerung *f* **~ for
cutting tool** Ausschneiden *n* mit Führungsgerät
für das Schneidewerkzeug **~ and trimming
machine** Rautingfräser *m*

**routing, ~ bulletin** Leitübersicht *f* **~ chart** Instradierungsplan *m*, Leitwegplan *m* **~ code
sender** Laufnummerngeber *m* **~ cutter** Oberfrä-
ser *m* **~ desk** Leitstelle *f* **~ guide** Leitwegver-
zeichnis *n* **~ head** Ausfräskopf *m* **~ instructions**
Beförderungsvermerk *m* **~ line** Leitwegzeile *f*
**~ plan** Instradierungsplan *m*, Leitübersicht *f* **~
plane** Grund-, Nut-hobel *m* **~ sheet** Arbeitsfolgenplan *m*

**routining, ~ supervision** laufende Überwachung *f*

**rove, to ~** vorspinnen

**rover** Vorspinner *m*, Vorspinnmaschine *f*

**roving** (textiles) Lunte *f*, Vordergespinst *n*, Vorgespinst *n*, Vorspinnen *n* **~ frame** Vorspinnmaschine *f* **~ head** Kammgarn-Vorspinnma-
schine *f* **~ reel** Vorgespinstspule *f* **~ wheel**
Meßrolle *f*

**rovings** Kardenband *n*

**row, to ~** aneinanderreihen, rudern

**row** Baufluchtlinie *f*, Flucht *f*, (of seats) Rang *m*,
Reihe *f*, Zeile *f* **in a (one) ~** in einer Reihe **in ~**
flüchtend (Rohranordnung)

**row, ~ of blocks** Blockreihe *f* **~ cylinders** Zylinderreihe *f* **~ of feed holes** Führungslochreihe *f*
**~ of keys** Schalter-, Tasten-reihe *f* **~ of lights**
Feuerreihe *f* (d. Befeuerung) **~ of mirrors** (in
scanner) Spiegelkranz *m* **~ of nozzles** Düsenreihe *f* **~ of pegs** Pfahlreihe *f* **~ of piles** Pfahlwand *f* **~ of rivets** Nietnaht *f*, Nietreihe *f* **~ of
scavenging ports** Spülschlitzreihe *f* **~ of stakes**
Pfahlreihe *f* **~ of tires** Düsenreihe *f* **~ of
tuyeres** Formreihe *f*

**row, ~ boat** Kahn *m*, Ruderboot *n* **~ house**
Reihenhaus *n* **~ lock** Dolle *f*, Riemen-, Ruder-

klampe f ~ **locking** Fahrstraßenfestlegung f
(r.r.) ~ **matrix** Zeilenmatrix f
**rows, ~ and columns** Zeilen pl und Spalten pl ~
**of teeths displaced relatively to each other** gegeneinander versetzte Zahnreihen pl
**royal post** Königszapfen m
**royalties** Tantiemen pl
**royalty** Abgabe f, Benutzungsgebühr f **on a ~
basis** auf Lizenzbasis f ~ **fee** Lizenzgebühr f ~
**tenth** Grundzehnter m
**r.p.m.** U.p.M. ~ **of compressor** Verdichterdrehzahl f
**rub, to** ~ abschleifen, frottieren, polieren, reiben,
schaben, scheuern, schleifen, streichen **to ~
down** verreiben **to ~ in** einreiben **to ~ off** abreiben **to ~ on** anreiben **to ~ open** aufreiben
**to ~ the pitch out of a cask** ein Faß n ausreiben
**to ~ with pumice** abbimsen
**rubbed, ~ with emery** abgeschmirgelt ~ **finish
of furniture** Möbelmattierung f ~ **lightly along
the surface** gestrichen (streichen) ~
**rubber** Abziehstein m, Glätteisen n, Grobfeile f,
Gummi m, Gummielastikum n, Kautschuk m,
Poliertuch n, Raspel f, Reib-tafel f, -zeug n,
Richt-, Streich-eisen n, (Buchbinderei) Rückeneisen n, Schmirgelpapier n
**rubber, ~ apron** Gummitransportband n ~ **back
velvet mats** Vorlagematten pl ~ **balloon** Gummiballon m ~ **band** Gummi-band n, -ring m ~
**base** Gummiunterlage f ~ **belt** Gummi(treib)
riemen m ~-**belt conveyer** Gummibandtransporteur m ~ **belting** Gummiriemen m ~ **blanket**
Gummituch n ~ **block** Gummi-klischee n,
-polster n ~ **boat** Gummi-, Schlauch-boot n ~
**boots** Gummi-schuhe pl, -stiefel pl ~ **brakelining** Gummibremsbelag m ~ **buffer** Gummibuffer m, -puffer m ~-**buffer-mounted** auf Gummipuffern pl montiert ~-**buffer mounting** elastischer Einbau m mittels Gummipuffer ~
**buffers for typewriter** Schreibmaschinenschalldämpfer m ~ **bulb** Gummidruckball m ~ **bung**
Gummistopfen m ~ **calender** Gummikalander
m ~ **canvas hose (or sleeve)** Gummileinenmuffe f ~ **carrying disc** Gummimitnehmer m ~
**caterpillar** Gummiraupe f ~ **cement** Gummilösung f ~ **club** Gumminüppel m ~ **coat**
Gummimantel m ~ **coated multicore line (or
cable)** Gummiaderleitung f ~ **coating** Gummierung f, Gummi-isolierung f, -überzug m ~ **coating sheet** Gummibahn f ~ **collar** Gummimanschette f ~ **connection** Gummimuffe f ~ **cord**
Gummiseil n ~ **cord deal** Gummischnurring m
~ **cord springing** Gummilitzenfederung f ~
-**cord start** Gummiseilstart m (aviat.) ~ **core**
Gummieinlage f ~ **coupling member** (Kupplung) Gummireifen m ~ **cover** Gummihaut f
~ **cover for cylinders** Walzenbezug m ~-**covered**
gummiisoliert ~-**covered distributor roller**
Gummiverreibwalze m ~-**covered lead-in** gummiisolierte Leitungseinführung f ~-**covered
wire** Gummi-ader f, -draht m, mit Gummi isolierter Draht m ~ **covering** Gummibelag m ~
**cushion** Gummipolster m ~-**cushion helve
hammers** Aufwurfhammer pl mit Gummipuffer
~ **cushioned** gummigelagert ~ **cushioning pad**
Gummipolsterunterlage f ~ **diaphragm** Gummimembrane f ~ **dinghy** Schlauchboot n ~-**driven
model** Gummimotormodell n ~ **earpiece**

**Gummimuschel** f ~ **eraser** Radiergummi m ~
**excentric** Nitschelstanze f ~ **eyepiece** Gummimuschel f ~ **fabric clutch** Gummigewebekupplung f ~ **ferrule** Gummitülle f ~ **filling valve**
Gummikonus m des Füllventils ~ **film** Gummischicht f ~ **fixing hook** End-, Gummi-haken m
~ **flooring** Gummibodenbelag m ~ **foam**
Kautschukschaum m ~ **galoshes** Gummiüberschuhe pl ~ **gas mask** Gummimaske f ~ **gasket**
Gummi-dichtung f, -lasche f ~ **gear** Nitschelzeug n ~ **gloves** Gummihandschuhe f ~ **goods**
Gummiwaren pl ~ **grommet** Dichtgummi n,
Gummiring m (für el. Kabel) ~ **hand roller**
Gummihandwalze f ~ **heel** Gummiabsatz m ~
**hose** Gummi-muffe f, -schlauch m ~-**hose
connection** Gummischlauchmuffe f **(India)** ~
Lederharz n ~ **industry** Gummiindustrie f ~
**insert** Gummieinsatzstück n ~-**insulated** gummiisoliert ~-**insulated cable** Gummikabel n ~
-**insulated leader** gummiisolierte Leitung f ~
-**insulated wire** Gummikabel n ~ **insulating
material** Gummiisolationsbedarf m ~ **insulation** Gummi-isolation f, -isolierung f ~ **insulator**
Gummiisolator m ~ **jacket** Gummihaut f ~
**jacket cords** Gummischlauchleitung f ~ **joint**
Gummiabdichtung f ~ **knife** Kautschuk-Bandsäge f ~ **latex** gummiartiges Harz n, Gummimilchsaft m ~-**lined canvas hose** Schlauch m mit
Gummieinlage ~ **lining** Gummi-auskleidung f,
-zwischenlage f ~ **make-ready material** Zurichtungsmaterial n ~ **mallet** Gummihammer m ~
**mat** Gummi-decke f, -matte f ~-**metal bumper**
Schwingmetallpuffer m ~ **motor** Gummimotor
m ~ **mouthpiece** Gummimundstück n ~ **overshoes** Gummiüberschuhe pl ~ **packer** Kautschukpacker m ~ **packing** Gummi-dichtung f,
-einlage f, -packung f ~ **packing gasket** Abdichtungsgummierung f ~ **packing ring** Abdichtungsgummierung f ~ **packing washer** Manschette f ~ **pad**
Gummi-kissen n, -klotz m ~ **patch** Gummiflecken m, -pflaster n ~ **plate** Gummiplatte f
~ **plug** Gummistopfen m ~ **printing cork** Gummidruckkork m ~ **printing plate** Gummidruckklischee n ~ **product** Gummierzeugnis n ~
-**(tire) repair equipment** Gummiinstandsetzungsmittel n ~-**ring gasket (or packing)** Gummiringdichtung f ~ **roller printing** Gummiwalzendruck m ~ **scraper** Gummi-kratzer m, -schaber
m ~ **seal** Gummidichtung f ~ **shackleblock**
Gummifederlager n ~ **shock absorber** Gummi
-abfederung f, -federer m, -federung f, Gummistoß-dämpfer m, -fänger m ~ **shock-absorber
strut** Gummifederbein n ~ **sleeve** Gummi-ärmel
m, -manschette f, -muffe f, Schlauchmuffe f ~
**socket** Gummitülle f ~ **solution** Gummilösung f
~ **sponge** Kautschukschaum m ~ **sponge
grinding disk** Schwammgummischleifteller m ~
**sponge washer** Schwammgummiunterlage f ~
-**spring shackle** Gummifederlage f ~ **stamp**
Gummistempel m ~ **stamp box** Stempelkissendose f ~ **stamp compositor** Stempelsetzer m ~
**stamp ink** Stempeltinte f ~ **standard** Gumminorm f ~ **stereo** Gummiklischee n ~ **stopper**
Gummi-pfropf m, -pfropfen m, -stopfen m,
Kautschukstopfen m ~ **strip** Gummifaden m
~-**studded tire** Gummigleitschutzreifen m ~
**substitute** Gummiersatzstoff m ~ **surrogate**

Gummiersatzstoff *m* ~ **suspension** Gummifederung *f* ~ **tape** Paraband *n* ~ **tire** Gummireifen *m* ~**-tired** gummibandagiert ~**-tired pad roll (or pressure)** Gummiandruckrolle *f* (film) ~**-tired track roller** Laufrolle *f* mit Gummibandage ~**-tired wheel** Gummirad *n* ~ **tires** Gummibereifung *f* ~ **torsion suspension** Gummitorsionsfederung *f* ~ **track** Gummi-gleiskette *f*, -kette *f*, -raupe *f* ~ **tread** Gummiüberzug *m*, Laufgummi *m* ~ **tread of a pad roller** Laufbahn *f* einer Rolle mit Gummiüberzug, Rollenlaufbahn *f* mit Gummiüberzug ~ **truncheon** Gummiknüppel *m* ~ **tube** Gummi-rohr *n*, -schlauch *m* ~**-tube connection** Schlauchverbindung *f* ~**-tube wiring** Gummischlauchleitung *f* ~ **tubing** Gummischlauch *m*, Kautschukschlauch *m* ~ **varnish** Gummilack *m* ~ **V-belt** Gummikeilriemen *m* ~ **washer** Gummi-buffer *m*, -unterlage *f* ~**-woven goods** Gummiweb-und -wirkwaren *pl*
**rubber(ed) tape** Gummiband *n*
**rubberize, to** ~ gummieren
**rubberized** gummiert ~ **make ready material** gummierter Zurichtungsstoff *m* ~ **material** Gummistoff *m*
**rubberizing** Gummieren *n*, Gummierung *f*
**rubberoid sheathing** Ruberoidbedeckung *f*
**rubbers** Gummischuhe *pl*
**rubbing** Abreibung *f*, Abrieb *m*, (surfaces) Aufeinanderarbeiten *n*, Friktion *f*, Polieren *n*, Reiberdruck *m* (print), Reibung *f* ~ **on the condenser** Würgeln *n* (Nitscheln) ~ **of labyrinth seals** Kämmen *n* der Turbinendichtungen ~ **of the skin** Hautreibung *f*
**rubbing,** ~ **action** (surfaces) Reibungswirkung *f* ~ **contact** Streichkontakt *m* ~ **corrosion** Reiboxydation *f* ~**-down liquid** (Lack) Schleifflüssigkeit *f* ~ **face** (jamb of recess) Anschlagleiste *f* ~ **implement** Frottierartikel *m* ~ **leather** Nitschelleder *n*, Würgelzeug *n* ~ **motion** Nitschelwerk *n* ~**-off** (of a coating) Abfärben *n* ~ **parts** Arbeitsfläche *f* ~ **pile** Scheuer-, Streich-, Stütz-pfahl *m* ~ **speed** Gleitgeschwindigkeit *f* ~ **strip** Randleiste *f* ~ **stroke** Nitschelhub *m* ~ **surface** Reibfläche *f* ~ **surfaces** aufeinander arbeitende Flächen *pl* ~ **varnish** Schleiflack *m*
**rubbish** Abfall *m*, Abraum *m*, Kaff *n*, Kehricht *m*, Müll *m*, Schutt *m*, taubes Gestein *n*
**rubble** Abraum *m*, Bruchsteine *pl*, Bruchsteinschüttung *f*, grober Kies *m*, Rollstücke *pl*, Schotter *m*, Schüttstein *m*, Stein-schutt *m*, -schüttung *f* ~ **bed** Bruchstein-füllung *f*, -unterlage *f*, Schüttsteinunterlage *f*, Steinschüttung *f*, Unterlage *f* ~ **drain** Sickergraben *m* ~ **filter** Kiesfilter *n* ~ **floor** Estrich *m*, gepflasterter Fußboden *m* ~ **masonry** Bruchsteinmauerwerk *n* ~ **mound** geschütteter Körper *m* ~ **ore** Mulm *m* ~ **pile** Trümmerhaufen *m* ~**-slope** Schutthalde *f* ~ **stone** Bruchstein *m*, Geröll *n* ~ **walling** Füllmauer *f*
**rubbly culm coke** Feinkoks *m*
**rubidium** Rubidium *n* ~ **alum** Rubidiumalaun *m*
**rubout signal** Irrungszeichen *n*
**rubproof paint** Binderfarbe *f*
**rubric** Gattung *f*, Rubrik *f*
**ruby** Rubin *m* ~ **blende** Rotgültig *n* ~ **rod** Rubinstab *m*
**ruche** Rüsche *f*

**rudder** Ruder *n*, (assembly) Seitenleitwerk *n*, (pedal) Seitenruder *n*, Steuer *n*, Steuerruder *n* **to put on** ~ Seitensteuer *n* geben
**rudder,** ~**-actuating mechanism** Ruderantriebsmaschine *f* ~ **angle** Ruder-lage *f*, -stellung *f*, Seitenruderwinkel *m* ~ **axis of rotation** Ruderdrehachse *f* ~ **bag cord** Steuersackleine *f* ~ **balance** Seitenruderausgleich *m* ~ **bar** Seitenruderfußhebel *m*, Seitensteuer *n*, Seitensteuerfußhebel *m*, -hebel *m*, -pedal *n* ~ **bracing** Ruderverspannung *f* ~ **brake** Ruderbremse *f* ~ **connection** Ruderanschluß *m* ~ **control stand** Seitensteuerstand *m* ~ **controls** Seitensteuerung *f* ~ **drive** Rudergetriebe *n* ~ **enlargement** Leitwerkaufstockung *f* ~ **helmsman** Seitensteuermann *m* ~ **hinge** Seitenruderfußhebel *m* ~ **hinge beam** Seitenruderscharnierträger *m* ~ **pedal** Seiten-ruderhebel *m*, -steuer *n*, -steuerpedal *n* ~ **pit** Rudergrube *f* ~**-position indicator** Ruderlagezeiger *m* ~ **post** Hinter-, Ruder-steven *m* ~**-post bracket** Ruderbock *m* ~**-setting indicator** Ruderlagezeiger *m* ~ **speed** Ruderlaufgeschwindigkeit *f* ~ **station** Seitensteuerstand *m* ~ **surface** Steuerfläche *f* ~ **torque** Seitenruderdrehung *f* ~ **trim tab** Seitenruderhebel *m* ~ **trimming** Seitenrudertrimmung *f* ~ **tube** Ruderachsrohr *n* ~ **vibration** Seitenruderschütteln *n*
**rudders** Leitwerk *n*
**ruddle** Rötel *m*
**rudiment element** Anfangsgrund *m*
**rudimentary** grundsätzlich
**rue oil** Rautenöl *n*
**rug** Decke *f*, Teppich *m*
**rugged** kompakt, stabil, uneben, unempfindlich
**ruggedness** Holprigkeit *f*, unebene Stelle *f*, Unebenheit *f*
**ruggerized construction** verstärkte Konstruktion *f*
**rugosity** Unebenheit *f*
**Ruhmkorff coil** Funken-geber *m*, -induktor *m*, Induktorium *n*, Ruhmkorffsches Induktorium *n*
**ruin, to** ~ hinrichten, verderben, zerstören
**ruin** Ruine *f*, Trümmerhaufen *m*
**rule, to** ~ herrschen, linieren, schalten
**rule** Gesetz *n*, Grundsatz *m*, Kolumnenmaß *n* (print), Maß-, Meß-stab *m*, Norm *f*, Normalfall *n*, Regel *f*, Relaisscheit *n*, Richt-linie *f*, -maß *n*, -scheit *n*, -schnur *f*, Schmiege *f*, Vorschrift *f*, Winkel-eisen *n*, -maß *n* ~ **of hurricane** Orkanregel *f* ~ **for the prevention of accidents** Unfallverhütungsvorschrift *f* ~ **of proportion (or three)** Regeldetri *f* ~ **of thumb** Faust-formel *f*, -regel *f*, Fingerprobe *f*
**rule,** ~ **addition** Zusammenzählregel *f* ~ **bender** Linienbiegeapparat *m* ~ **bending apparatus** Linienbiegeapparat *m* ~ **border** Linien-einfassung *f*, -rand *m* ~ **cutter** Linienschneide-apparat *m* ~ **drawing frame** Linienziehbank *f* ~**-of-thumb method** Faustregelmethode *f* ~ **paper** Muster-, Tupf-papier *n* ~ **principle** Leitsatz *m* ~ **setting** Linienmanier *f* ~ **slide** Liniengießblock *m*
**ruled,** ~ **box** Linienkasten *m* ~ **glass** Linienglas *n* ~ **grating** Strichgitter *n* ~ **line** vorgedruckte Linie *f* ~ **paper** Koordinatenpapier *n*, Linien-papier *n*, -netz *n*, liniertes Papier *n* ~ **plate**

Linienraster *m* ~ **shooting board** Linienhobel
*m* ~ **surface** Regelfläche *f*
**ruler** Lineal *n*, Linien-reißer *m*, -zieher *m*, Richtscheit *n*
**rules** Linienmaterial *f* ~ **of the air** Luftfahrt-
regeln *pl*, Luftverkehrsregeln *pl* ~ **to be observed
in operating** (ship, plane, plant, etc.) Betriebsvorschriften *pl* ~ **of procedure** Verfahrensordnung *f* ~ **of the (water) road** Seestraßen-
ordnung *f*
**ruling** Linieren *n*, Netzteilung *f*; vorherrschend ~
(of diffraction grating) Furche *f* ~ **device** Linien-
zieher *m* ~ **distance** Linienabstand *m* ~ **engine**
Teilmaschine *f* ~ **groove of a grating** Gitterstrich *m* ~ **grooves** (of diffraction grating)
Gitterfurchen *pl* ~ **ink** Linienfarbe *f* ~ **machine**
Liniermaschine *f* ~ **pen** Reißfeder *f* ~ **technique**
Gitterschneidetechnik *f* ~ **work** Liniersatz *m*
**rumble, to** ~ poltern, rattern, rumpeln
**rumble** Poliertrommel *f* ~ **barrel** Scheuerfaß *n*
~ **seat** Klappsitz *m*
**run, to** ~ (von Anstrich) ablaufen, (eines Arbeits-
vorgangs) ablaufen, antreiben, arbeiten, (color)
auslaufen,betreiben, in Betrieb sein, fahren,
fließen, funktionieren, (wires) führen, laufen,
rennen, rinnen, wirken, zerfließen, zulaufen
**run, to** ~ **aground** auf den Grund *m* laufen, stranden **to** ~ **away** abwetzen, ausrücken, davon-
laufen, durchbrennen, durchgehen **to** ~ **back**
zurücklaufen **to** ~ **by (or from) a battery** aus
einer Batterie *f* betreiben **to** ~ **a cable** auslegen
(ein Kabel) **to** ~ **out a cable** ein Kabel *n* ausrollen **to** ~ **in the clamps with lead** die Klammern *pl* mit Blei vergießen **to** ~ **copal** Kopal
schmelzen **to** ~ **down** (spring) ablaufen,
(Motor) auslaufen, auspumpen, erschöpfen,
(storage cells) überentladen **to** ~ **empty** leerlaufen **to** ~ **end on to shore** gerade auf den
Strand *m* laufen **to** ~ **up an engine** einen Motor
warmlaufen lassen **to** ~ **fast** voreilen, vorgehen
**to** ~ **free** leer laufen **to** ~ **without fuel** freilaufen
**to** ~ **in full width** breitlaufen **to** ~ **hot**
heißlaufen, warmlaufen **to** ~ **idle** leer gehen **to**
~ **in** einlaufen, einlaufen lassen, vergießen, zulaufen **to** ~ **into** eintreiben **to** ~ **out with leaders**
hinterführen **to** ~ **light** freilaufen, leer gehen
**to** ~ **off** abfließen, ablaufen, ausscheren **to** ~
**off (or down) quickly** schnell ablaufen **to** ~ **on**
zulaufen **to** ~ **out** ausgießen, auslaufen, ausrechnen **to** ~ **over** überfahren **to** ~ **(a wire)
overhead** oberirdisch verlegen **to** ~ **parallel**
parallel laufen **to** ~ **for shelter** unterlaufen **to** ~
**short** ausgehen **to** ~ **slow** nacheilen, nachgeben
**to** ~ **up to speed** auf Drehzahl *f* kommen **to** ~
**through** durchfahren, durchlaufen **to** ~ **together** zusammenlaufen **to** ~ **out of true** verziehen
**to** ~ **up** anlaufen, auflaufen, steigen, warmlaufen **to** ~ **up a wall** eine Mauer *f* aufhöhen **to** ~ **before the wind** lenzen **to** ~ **a wire**
einen Draht *m* führen, verlegen
**run** Anflug *m*, (in series of parallel experiments)
Ansatz *m*, Auflage *f*, Bahn *f*, Durchlauf *m*
(data proc.), Fabrikation *f*, Feldeslänge *f*, (of
a furnace) Gang *m*, Lauf *m*, Meßreihe *f*, Rinnsal *n*, Schacht *m*, Schweißdraht *f*, Verlaufen *n*,
Versuch *m* (rolling) ~ Rollstrecke *f* **in the long**
~ auf die Dauer
**run,** ~ **of fabric** Wareneingang *m* ~ **of the fiber**

(or grain) Faserverlauf *m* ~ **of measurements**
Meßserie *f* ~ **of mine coal** Förder-, Misch-,
Roh-kohle *f*, Gasförderkohle *f*, grubenfeuchte
Kohle *f* ~ **of operation** Arbeitsgang *m* ~ **of the
paper web** Papierlauf *m* ~ **of a pipe** Verlauf *m*
einer Rohrleitung ~ **of a smelter (blast)
furnace** Gang *m* eines Schmelzofens ~ **of spot
welds** Schweißpunktreihe *f*
**runaway** Durchgehen *n* ~ **governor** Geschwindigkeitsregler *m* ~ **speed** Durchgehdrehzahl *f*
**run,** ~-**down battery** erschöpfte Batterie *f*
~-**down drum (or tank)** Destillatsammelgefäß *n*
~ **grain** rinnende Narben *pl* ~-**of-the-oven coke**
Großkoks *m*, Hochofenkoks *m* ~ **oil** Ablauföl
*n*
**rung** (ladder) Sprosse *f*, Stufe *f*
**runnable** befahrbar
**runner** Abstichrinne *f*, Aufschiebeschlaufe *f*,
Brücke *f*, Drillschar *f*, Einguß *m*, Einguß-kanal
*m*, -lauf *m*, -trichter *m*, (textiles) Fliege *f*,
Flügelrad *n*, Gieß-kopf *m*, -trichter *m*, -zapfen
*m*, Gußpfeife *f*, Hebeklaue *f*, laufender Block
*m*, Läufer *m*, (turbine) Lauf-rad *n*, -ring *m*,
-rinne *f*, -rolle *f*, Maschinist *m*, Rolle *f*, Schaufelrad *n*, Schieber *m*, Schlaufe *f*, Schnellwalze *f*,
Ski *m*, (in foundry) Tangentialeingußkanal *m*,
tangentialer Zulauf *m*, Trichter-einlauf *m*,
-lauf *m*, -zulauf *m*, Walze *f*, Zeilenzähler *m*
(print), Zugbegleiter *m*, Zulauf *m* **(edge)** ~
Läufer *m* ~ **of the high pressure section** Hoch-
druckläufer *m* ~ **for inking rollers** Auftragrolle
*f* ~ **for mixing** Misch-gang *m*, -koller *m*
**runner,** ~ **basin** Eingußsumpf *m*, Gießtümpel *m*
~ **blade** Laufschaufel *f* ~ **brick** Kanalstein *m*
~ **cup** Gießtrichterschale *f* ~ **frame** Läufer-
zarge *f* ~ **gate** Brücke *f*, Einguß-kanal *m*, -lauf
*m*, -trichter *m*, Gußtrichter *m* mit tangentialem
Zulauf ~ **head** Gußtrichter *m* ~ **opening** Gieß-
loch *n* ~ **planter** Einsteckpflanzmaschine *f*,
Schuhmaisdrill *m* ~ **post of message center**
Läuferposten *m* ~ **ring** Laufring *m* ~ **stick** Eingußstock *m*, Gießlochzapfen *m*, Stäbchen *n* ~
**stone** Läuferstein *m* ~ **waste** (in pouring) Knochen *m*
**runners** Knochen *m*, Spundbohlen *pl*
**running** Betrieb *m*, Destillat *n*, Einguß *m*, Einlassen *n*, Inseln *pl* im Film, Lauf *m*, Laufen *n*,
Laufkraft *f*; aufeinanderfolgend, fahrend,
fließend, laufend, ununterbrochen, zirkulierend
~ **in ball bearings** kugelgelagert ~ **of cable(s)**
Kabelführung *f* **to be in** ~ **condition** in Schuß
sein ~ **in** einlaufend ~ **of a motor** Motorbetrieb
*m* **in** ~ **order** betriebsfähig ~ **in the same direction** gleichläufig ~ **of wires** Leitungsführung *f*
**running,** ~ **axle** Laufachse *f* ~ **balance** dynamische Auswuchtung *f* ~ **block** laufender Block
*m* ~ **board** Lauf-, Tritt-brett *n* ~ **cable** Lauf-
kabel *n* ~ **capacitor** Betriebskondensator *m* ~
**center** (tail) drehbare Reitstockspitze *f*, (of
lathe) umlaufende Spitze *f* ~ **cetane rating**
Cetanzahlbestimmung *f* im laufenden Motor
(Zündverzugsverfahren) ~ **charge** Betriebskosten *pl* ~ **clearance** Drehsitzspiel *n* ~ **cost**
laufende Unkosten *pl* ~ **costs** laufende Ausgaben *pl* ~ **down** Ablaufen *n*, Auspumpen *n*, Erschöpfung *f* ~-**down of a battery** Erschöpfung *f*
der Batterie ~ **edge of rail** Fahrkante *f* der
Schiene, Schieneninnenkante *f* ~ **equipment**

Betriebsausrüstung *f* ~ **expenses** laufende Betriebsunkosten *pl* ~ **features** Laufeigenschaften *pl* ~ **fire** Lauffeuer *n* ~ **fit** Drehsitz *n*, Gangeinstellung *f* (aviat.), Laufanpassung *f* (aviat.), Laufsitz *m* ~ **friction** Laufwiderstand *m* ~ **gear** Laufwerk *n* ~ **hand** fließende Schrift *f* ~ **head** Kolumnentitel *m* ~ **hot** Warmlaufen *n* ~ **hour** Betriebsstunde *f* (Motor)

**running-in** Einlassen *n*, Einlauf *m*, (Motor) Einlaufen *n*, (tools, casing) Einlaufvorgang *m* ~ **oil** Einschleiföl *n* ~ **period of engine** Einfahrzeit *f* des Motors ~ **surface finish** durch Einlaufen entstandene Oberflächenbeschaffenheit *f* ~ **test** Laufprobe *f*, Probelauf *m* ~ **test log sheet** Einlaufprotokoll *n* ~ **time** Einlaufzeit *f*

**running,** ~ **knot** Schiebe-, Schleif-knoten *m* ~ **light** Positionslaterne *f*, Rampenlicht *n* ~ **magneto** Betriebsmagnet *m* ~ **noise** Laufgeräusch *n* ~ **nut** Laufmutter *f* ~-**off single editions** Auflagendruck *m* (print.) ~-**out fire** Frisch-feuer *n*, -herd *m* ~-**out spring** Vorholfeder *f* ~-**out test** Auslaufversuch *m* ~ **position** Fahrstellung *f* ~ **program** betriebsfertiges oder laufendes Programm *n* ~ **pulley** Kreuzrolle *f* ~ **range** Laufbereich *m* (aut. contr) ~ **reserve** Gangreserve *f* ~ **rigging** Kreuztaurolle *f* ~ **rock** schwimmendes Gebirge *n* ~ **sand** Schwimmsand *m*, wandernder Sand *m* ~ **schedule** Fahrzeit *f* ~ **shot** Fahraufnahme *f*, Verfolgungsaufnahme *f* ~ **slag** Laufschlacke *f* ~ **smoothly** stoßfrei laufend ~ **speed** Fahr-, Umlauf-geschwindigkeit *f* ~ **speed of passage** Durchlaufgeschwindigkeit *f* ~ **steel** unruhiger Stahl *m* ~ **stone** Mahlstein *m* ~ **surface of rail** Schienenlauffläche *f* ~ **take-off** Rollstart *m* ~ **temperature** Betriebswärmegrad *m* ~ **term** Laufterm *m* ~ **test** Laufprobe *f* ~ **through** Durchlauf *m* (of a film) ~ **time** Laufzeit *f* ~ **title** Kolumnentitel *m* ~ **tread of rail** Schienenlauffläche *f* ~-**up bolt** Armbolzen *m* ~-**up brake** Auflaufbremse *f* ~-**up pressure** (Anhänger) Auflaufdruck *m* ~ **variable** Laufzahl *f*; laufend veränderlich ~ **wave** Wanderwelle *f*

**runoff** Abfluß *m*, Ablauf *m*, Abschlackung *f*, Abstich *m*, (slag) Abwerfen *n* ~ **and evaporation relations** Wasserhaushalt *m* ~ **gutter** Ablaufrinne *f*

**runout,** ~ **with live rolls** Abfuhrrollgang *m* ~ **of thread** Gewindeauslauf *m* ~ **table** Formkastenwagen *m*, Rolltisch *m*

**run-over** Überlauf *m*

**runway** Ablauf-bahn *f*, -schurre *f*, Abstichkanal *m*, Anlaufbahn *f*, Fahrbahn *f*, Landebahn *f*, Laufbahn *f*, Lauffeld *n*, Piste *f* (aviat.), Schurre *f*, Startbahn *f*, Start- und Landebahn *f* ~ **of crane** Kranfahrbahn *f*

**runway,** ~ **alignment beacon** Pistenrichtungsfeuer *pl* ~ **barrier** Fang-haken *m*, -vorrichtung *f* ~ **ducting and cabling** Röhrenzuführung *f* und Verkabelung *f* (einer Start- und Landebahn) ~ **edge** Pistenrand *m* ~ **lighting** Pistenbefeuerung *f*, Start- und Landebahnbefeuerung *f* ~ **lights** Pistenfeuer *pl* ~ **marker** Startbahnmarkierung *f* ~ **rail** Laufschiene *f* ~ **slope** Pistenneigung *f* ~ **threshold lighting** Schwellenbefeuerung *f* ~ **turn-off** Pistenausgang *m*

**rupture, to** ~ abreißen, (von chem. Verbindung) aufbrechen, brechen, reißen, zerbrechen, zerreißen

**rupture** Abbruch *m*, Bruch *m*, (of insulation) Durchbruch *m*, Unterbrechung *f*, Verbruch *m*, Zerreißen *n*, Zubruchgehen *n* ~ **disc** Zerreißscheibe *f* ~ **(frangible) disc** Zerreißventil *n* ~ **limit** Bruchgrenze *f* ~ **line** Bruchlinie *f* ~ **modulus** Bruchmodul *m* ~ **pieces** Bruchstücke *n* ~ **resistance** Trennwiderstand *m* ~ **strength** Reißfestigkeit *f*

**rupturing** Abschalt-, Schalter-leistung *f* ~ **capacity** (of a switch) Unterbrechungsvermögen *n* ~ **strength** Durchschlagsfestigkeit *f* ~ **voltage** Durchschlagsspannung *f*

**rural** ländlich, landwirtschaftlich ~ **automatic exchange** kleines Wahlamt *n* ~ **cultivation** Landespflege *f* ~ **line** Farmerleitung *f* ~ **party line** Gemeinschaftsanschlußleitung *f* für Landteilnehmer ~ **plant** Landzentrale *f* ~ **power station** Überlandzentrale *f* ~ **subscriber line** Überlandleitung *f* ~ **telephone network** Netzgruppe *f* ~ **telephone plant** Landfernsprechnetz *n* für Selbstanschlußbetrieb

**ruse** Trick *m*

**rush, to** ~ anschießen, beschleunigen, dahineilen, dahinrasen, erstürmen, rauschen, sausen, stürzen, treiben **to** ~ **back** zurückstürzen **to** ~ **on the fiber** schnell aufziehen **to** ~ **forward** vorstürmen **to** ~ **out** ausstürzen **to** ~ **together** zusammenprallen

**rush** Andrang *m*, Anlauf *m*, Ansturm *m*, Binse *f*, Eile *f*, Saus *m*, (of traffic) Verkehrsandrang *m*, Zudrang *m* ~ **of current** Anschwellen *n* des Stromes ~ **by groups (or sections)** Gruppensprung *m*

**rush,** ~ **basket** Binsenkorb *m* ~ **hour** Hauptverkehrszeit *f*, Stoßzeit *f* ~ **hours** Stunde *f* des stärksten Verkehrsandranges ~ **print** Schnellkopie *f* (film)

**rushing** Gebrause *n*

**russet toning** Röteltönung *f*

**rust, to** ~ rosten, verrosten **to** ~ **in** einrosten **to** ~ **into** festrosten **to** ~ **off** abrosten

**rust** Rost *m* ~ **cement** Eisen-oxydkitt *m*, -zement *m* ~ **color** Rostfarbe *f* ~ -**colored** rostfarbig ~ -**free** nichtrostend, rost-frei, -rein ~ **inhibiting compound** Rostverhütungsmittel *n* ~-**preventative** Rostschutzmittel *n* ~-**preventing** korrosionshindernd, rost-hindernd, -schützend, -verhütend ~-**preventing quality** rostschützende Wirkung *f* ~ **preventive** Rostverhütungsmittel *n* ~-**preventive oil** Rostverhütungsöl *n*

**rustproof** Korrosionsbeständigkeit *f*; korrosionsfest, -sicher, nichtrostend, rost-beständig, -fest, -frei, -sicher, (steel) rostlos, schwerrostend ~ **dope** Rostschutzlack *m* ~ **paint** Rostschutzfarbe *f* ~ **steel** nichtrostender Stahl *m*

**rustproofing** Rostsicherheit *f* ~ **grease** Rost- und Korrosions-Schutzfett *n* ~ **process** Rost-behandlung *f*, -verfahren *n* ~-**treatment** Korrosionsbehandlung *f*

**rust,** ~ **protection** Korrosions-, Rost-schutz *m* ~ **removal** Rostentfernung *f* ~ **remover** Entrostungsmittel *n* ~ **removing agent** Rostlösemittel *n* ~ **resisting** korrosionsbeständig, rostbeständig, rostfest ~-**resisting paint** Rostschutzfarbe *f* ~-**resisting property** Korrosions-, Rost-beständigkeit *f* ~ **solvent** Rostlösungsmittel *n* ~ **spot** Rostfleck *m* ~-**stained** rostfleckig

**rustable** leicht rostend

**rusted** verrostet
**rusticated** mit dem Hammer *m* roh bearbeitet
**rustiness** Rostigkeit *f*
**rusting** Korrosion *f*, Rosten *n*, Rostung *f*, Verrosten *n*, Verrostung *f* ~ **of the pipe** Anrostung *f* des Rohres ~ **agent** Korrosionsbildner *m*
**rustle, to** ~ (silk) knistern, krachen, rasseln, (in a tube) rauschen
**rustle** Prasseln *n*
**rustless** korrosionsfrei, rostbeständig, schwer-

rostend ~ **steel** rostbeständiger, rostfreier oder witterungsbeständiger Stahl *m*
**rustlessness** Rostbeständigkeit *f*
**rusty** rostig, verrostet **to become** ~ rosten
**rut** Fahrrinne *f*, Furche *f*, Rast *f*, Wagenspur *f*
**ruthenic acid** Rutheniumsäure *f*
**ruthenium** Ruthenium *n* ~ **monoxide** Rutheniumoxydul *n*
**rutilite** Rutilit *m*
**R wire** b-Ader *f*

# S

saber Säbel *m*
sabotage Sabotage *f*, vorsätzliche Beschädigung *f*
sabulous griesig
saccharate Saccharat *n*
saccharic acid Zuckersäure *f*
sacchariferous zuckerhaltig
saccharification of wood Holzverzuckerung *f*
saccharify, to ~ verzuckern
saccharimeter Polarisationsapparat *m*
saccharimetry Saccharimetrie *f*, Zuckerbestim-/
mung *f*
saccharin Saccharin *n*
saccharometer Zucker(gehalts)messer *m*
sack, to ~ sacken
sack Beutel *m*, Sack *m*, Tüte *f* ~-beating plant
Sackausklopfanlage *f* ~ cell Beutelelement *n*
~ cloth Sack-leinen *n*, -leinwand *f* ~ elevator
Sackstapler *m* ~-filling device Absackvorrich-
tung *f* ~ gross weigher Bruttoabsackwaage *f* ~
net weigher Nettoabsackwaage *f* ~-shaking
plant Ausschüttanlage *f* für Säcke, Sackaus-
schüttelanlage *f* ~-ticking weaving mill Sack-
leinenweberei *f*
Sack's (universal three-high) rolling mill Sack-
sches Walzwerk *n*
sacked eingesackt
sacking Sackleinen *n*, Sackleinwand *f* ~ elevator
Einsackelevator *m* ~ mechanism Absackvor-
richtung *f*
sadden, to ~ abdunkeln, gedämpft färben
saddening Irisdruck *m* ~ agent Abdunklungsmittel *n* ~ liquor Dunklungsbad *n*
saddle, to ~ satteln
saddle Auflager *n*, Bergjoch *n*, Bettschlitten *m*,
Buchrücken *m*, (low part of anticline) Einsenkung *f* der Scheitellinie, Flözsattel *m*,
Kabel-block *m*, -sattel *m*, Konsol-, Langdreh-,
Quer-schlitten *m*, Querholz *n*, Sattel *m*, Sattel-
stütze *f*, Stuhlplatte *f* (r. r.), Support *m*, Sup-
port-schlitten *m*, -unterschlitten *m*, Türschwelle *f*, Unterlage *f*, Unterschlitten *m*
(engin.), Werkzeugschlitten *m* ~ of a bed
Sattel *m* eines Flözes ~ of microscope Mikroskopbügel *m* ~ of pole Stangenkappe *f*
saddle, ~ apron Support-räderplatte *f*, -schloß-
platte *f* ~-backed coping sattelförmige Deckplatte *f* ~ bag Pack-, Sattel-tasche *f* ~ bar
Ballen *m* ~ bearing Schwengellager *n* ~-bind-
ing lever Unterschlittenklemmhebel *m* (engin.)
~ blanket Satteldecke *f*, Woilach *m* ~ bottom
Sattelboden *m* ~-bottomed self-emptying truck
Sattelbodenselbstentlader *m* ~ bracket Bügel *m*
~ clamp bolt Schlittenfeststellschraube *f*
~-clamp lever Hebel *m* zum Feststellen des
Konsolschlittens ~ cloth Sattel-, Unterleg-
decke *f* ~ elevating screw Konsolschlitten-
höhenverstellung *f* ~ escape Flutauslaß *m*
(hydraul.), Freiarche *f* ~ flange gebogener
Flansch *m*, Sattelflansch *m* ~ frame Sattelge-
stell *n* ~ grate Sattelrost *m* ~ horse Reit-,
Sattel-pferd *n* ~ joint übergreifende Falzverbindung *f* ~ key Hohlkeil *m* ~ landing gear

Sattelfahrwerk *n* ~ limber Sattelprotze *f* ~ pad
Sattelkissen *n*, Trachtenkissen *n* ~ piece
Montagebügel *m* ~ pin Sattelhalter *m* ~ pipe
Abzweigrohr *n* ~ plate (locomotive boiler)
Gürtelplatte *f* ~ point Sattelpunkt *m* ~ roof
Satteldach *n* ~-shaped sattelförmig ~ surface
Sattelfläche *f* ~ tank Satteltank *m* ~ trappings
Sattelbekleidung *f* ~ traverse Bettschlittenweg
*m* ~ tree Sattel-baum *m*, -stock *m* ~ trough
Spindelschlitten *m* ~ wire an der Stangenspitze
geführte Leitung *f*
saddler Sattler *m*
saddler's knife Halbmondmesser *n*
safe, to ~-guard sichern
safe Geldschrank *m*, Panzerschrank *m*, Safe *n*,
Sicherheit *f*, Stahlschrank *m*, Tresor *m*; gesichert, kreditwürdig, sicher (playing) ~ unter
der gefährlichen Grenze *f* bleibend ~ to operate
betriebssicher
safe, ~ bearing capacity Sicherheitsbelastung *f*
~ charge zulässige Belastung *f* ~-conduct Ge-
leitschein *m* ~ delivery (of a message) Durchdringen *m* ~ filament current zuverlässiger
Heizstrom *m* ~-fire switch Sicherheitsschalter *m*
~ guard Geleit *n* Schutz, Sicherung *f* ~-guard-
ing against breakage Bruchsicherung *f* ~-guard-
ing by shearing pin Abscherstiftsicherung *f*
~-keeping sichere Aufbewahrung *f* ~ life
sichere Lebensdauer *f* ~ light Farblampe *f* ~
limit Sicherheitsgrenze *f* ~ load Belastung *f*,
spezifische zulässige Belastung *f*, Tragfähigkeit *f*, Zuladung *f*, zulässige Beanspruchung *f*
~-load factor Lastvielfaches *n* ~ operating
temperature betriebszulässige Temperatur *f* ~
plates Geldschrankplatten *pl* ~ range of stress
Arbeitsfestigkeit *f*, Ermüdungsgrenze *f* ~
strain zulässige Beanspruchung *f* ~ stress
Sicherheitsbeanspruchung *f*, zulässige Beanspruchung *f* ~ value zulässiger Wert *m* ~
working stress zulässige Beanspruchung *f*
safety Betriebssicherheit *f*, Gewahrsam *n*,
Sicherheit *f*, Sicherung *f* (mech.) ~ against
buckling Knicksicherheit *f* ~ against explosion
accidents Explosionssicherheit *f* ~ against fracture Bruchsicherheit *f* ~ of fuse while in barrel
Rohrsicherheit *f* des Zünders ~ against high-
potential breakdown hochspannungssicher
safety, ~ apparatus for cages Fallschutzvorrich-
tung *f* ~ apparatus of cages Fangvorrichtung *f*
~ apparatus for winding engines Förderma-
schinensicherheitsapparat *m* ~ appliance Si-
cherheits-apparat *m*, -vorrichtung *f* ~ base
Azetatträger *m* (film) ~ beam Lichtgitter *n* ~
belt Anschnallgurt *m*, Rettungsgürtel *m*,
Sicherheitsgurt *m* ~ block Sicherheits-flaschen-
zugkette *f*, -riegel *m*, Sperrvorrichtung *f* ~
block provisions Blocksicherung *f* ~ board
Gestänge-, Sicherheits-bühne *f* ~ bolt Schieber
*m*, Sicherheitsbolzen *m*, Sicherungsbolzen *m* ~
brake of lift Fangzeug *n* ~ brake release Öl-
drucklöseeinrichtung *f* ~ bunging apparatus
Sicherheitsspundapparat *m* ~ buoy Rettungs-

boje *f*, -ring *m* ~ **cable** Sicherungsseil *n* ~ **cam** Sicherheitsdreieck *n* ~ **cap** Sicherungskappe *f* ~ **capstan** Sicherheitssternrad *n* ~ **catch** Abzugssicherung *f*, Arretierklinke *f*, Fang-korb *m*, -riegel *m*, Sicherungs-hebel *m*, -klinke *f*, -knaggen *m*, -riegel *m* ~ **catch against reversing** Rückdrehsicherung *f* ~ **catch on ripping line** Reißklinke *f* ~ **catches** Fangvorrichtung *f* ~ **chain** Sicherheitskette *f* ~ **cheque paper** Scheck-sicherheitspapier *n* ~ **cinch** Obergurt *m* ~ **clevis** Sicherheitsschäkel *n* ~ **clip** Sicherungs-gabel *f*, -vorstecker *m* ~ **clutch** Überlastungskupplung *f* ~ **clutch of lift** Fangzeug *n* ~ **(pet)cock** (in gasoline engine) Brandhahn *m* ~ **code** Sicher-heitsvorschriften *pl* ~ **cofferdam** Hilfsfange-damm *m* ~ **connector** Sicherheitseinrückung *f* ~ **contact for safe** Geldschranksicherung *f* ~ **containment** Sicherheitsbehälter *m* ~ **coupling** Notkupplung *f* ~ **crab** Sicherheitskabelwinde *f* ~ **cutout** Abschmelz-, Schmelz-sicherung *f* ~ **device** Schutz(erdungs)vorrichtung *f*, Sicher-heits-gerät *n*, -vorrichtung *f*, Überwachungs-gerät *n*, Unfallverhütungsvorrichtung *f* ~ **dike** Schlafdeich *m* ~ **disconnecter** Notausrücker *m* ~ **earthing** Sicherheitserde *f* ~ **elevator** Sicher-heitsförderstuhl *m* ~ **explosive** Sicherheits-sprengstoff *m* ~ **factor** Sicherheits-faktor *m*, -maß *n*, -ziffer *f* ~ **fence** (Straße) Sicherheits-leitplanke *f* ~ **film** unentflammbarer oder un-verbrennbarer Film *m*, Schutzschicht *f* ~ **floor** Gestängebühne *f* ~ **fork** Sicherungsgabel *f* ~ **friction adjusting shaft** Sicherheitsfriktions-mutter *f* ~ **fuel** Sicherheits-brennstoff *m*, -kraftstoff *m* ~ **fuse** Abschmelz-draht *m*, -sicherung *f*, -streifen *m*, Leitfeuer *n*, Schmelz-sicherung *f*, Sicherheits-zünder *m*, -zündschnur *f*, Sicherung *f*, Zeitschnur *f* ~ **fuse for mines** Sicherheitsbrandlunte *f* für Bergwerke ~ **(collecting) gap** Schutzfunkenstrecke *f* ~ **(spark) gap** Sicherheitsfunkenstrecke *f* ~ **gear** Auffang-, Sicherungs-stange *f* ~ **glass** Plexiglas *n*, Sicherheitsglas *n*, Splitterfreiglas *n* ~ **goggles** Schutzbrille *f* ~-**grip spring** (pistol) Schließfeder *f* ~ **gripper** Sicherheitsbügel *m* ~ **hoist** Sicherheitsaufzug *m* ~ **hook** Sicherheits-, Sicherungs-haken *m* ~ **inspector** Sicherheits-inspektor *m* ~ **interlock** Sicherheitsverriege-lung *f* ~ **island** Verkehrsinsel *f* ~ **joint** Sicher-heitsmuffe *f* ~ **joints** Sicherheitsgestängever-binder *m* ~ **keep flange** Sicherungslasche *f* ~ **ladder** Schiffs-, Sicherheits-, Steige-leiter *f* ~ **lamp** Davysche Lampe *f*, Gruben-, Sicherheits-, Sturm-, Wetter-lampe *f* ~ **lever** Sicherheits-, Sicherungs-hebel *m* ~ **light** Sicherheitslampe *f* ~ **limit** Sollbruchstelle *f* ~-**limit switch** Sicher-heitsgrenzschalter *m* ~ **link** Sicherheitsabreiß-vorrichtung *f* ~ **lock** Sicherheitsschloß *n* ~ **lock gates** Schutzschleuse *f* ~ **loop** Sicherheits-schlinge *f* ~ **lug** Sicherungs(klemm)stück *n* ~ **magazine** (of film projector) Feuerschutztrom-mel *f* ~ **margin** Sicherheitsfaktor *m* ~ **match** Sicherheits-streichholz *n*, -zündholz *n* ~ **material** Sicherheitsmaterial *n* ~ **measure** Sicherheitsmaßregel *f* ~ **mining explosive** Wetter-sprengmittel *n*, -sprengstoff *m* ~ **nut** Sicherungsmutter *f* ~ **order** Vorsichts-maß-nahme *f*, -maßregel *f* ~ **paint** Sicherheitsfarbe *f* ~ **pawl** (Kohlenförderer) Sicherheitsanschlag

*m* ~ **pillar** Sicherheitspfeiler *m* ~ **pin** Arretier-stift *m*, Fertigbolzen *m*, Sicherheitsnadel *f* ~ **piston** Sicherheitskolben *m* ~ **platform** (oil drilling) Gestänge-, Sicherheits-bühne *f* ~ **plug** Sicherungsstöpsel *m* ~ **plug socket** Schutzkon-taktdose *f* ~ **precaution** Sicherheitsmaßnahme *f* ~ **rail** Leit-, Schutz-, Sicherheits-schiene *f* ~ **razor** Sicherheitsrasierapparat *m* ~ **regulator** Schnellschluß-, Sicherheits-regler *m* ~ **release stop** Sicherheitsauslösung *f* ~ **rod** Sicherheits-stange *f* ~ **roller** Schutzwalze *f* ~ **rope** Not-, Zug-leine *f* ~ **rules** Sicherheitsvorschriften *pl* ~ **screw** Rettungsschraube *f*, Sicherheitsver-schraubung *f*, Sicherungsschraube *f* ~ **shoulder** Sicherheitsverstärkung *f* ~ **siding** Fanggleis *n* ~ **signal** Sicherheitszeichen *n* ~ **sliding bolt** Sicherheitsschubriegel *m* ~ **sling** Abseilschlinge *f* ~ **slot** Sicherungsraster *m* ~ **spring** Zugfeder-schraube *f* ~-**spring catch** Zugfederhaken *m* ~ **spring tension screw** Zugfederspannschraube *f* ~ **starting crank** Sicherheitsandrehkurbel *f* ~ **stirrup** Sicherungsbügel *m* ~ **stop** Sicherungs-bolzen *m* ~ **stop cable** Abfang-, Grenz-seil *n*, Fangschlaufe *f*, Grenzstropp *n* ~ **strap** Sicher-heits-, Steh-gurt *m* ~ **switch** Endlagenschalter *m*, Sicherheitsschalter *m*, Sicherungsweiche *f* ~ **tank** Sicherheitsbehälter *m* ~ **tap chucks** Sicherheitsfutter *pl* ~ **thread** Sicherungsfaden *m* ~ **toe angle** Bordwinkel *m*, Fußleiste *f* ~ **trailer coupling** Sicherheitsanhängekupplung *f* ~ **train** Hilfzug *m* ~ **tread** Gleitschutzplatte *f* ~ **tread plate** gleitsicheres Trittbrett *n* ~ **valve** Ablaßventil *n*, Entlüfter *m*, Schnarrventil *n*, Sicherheitsventil *n*, (for sodium permanganate) Z-Entlüfter *m* ~ **valve complete** (**press**) Über-druckventil *n* ~-**valve weight** Sicherheitsventil-belastung *f* ~ **wire** Sicherheitsdraht *m*, Siche-rungsvorstecker *m* ~ **zone** gedeckter Raum *m*, Sicherheitsabstand *m*

**safflorite** Arsenikkobalt *m*, Eisenkobalt, Eisen-kobalt-erz *n*, -kies *m*, Safflorit *m*, Schlacken-kobalt *m*

**sag, to** ~ abfallen, absacken, durch-biegen, -hängen, einbiegen, nachsacken, sacken, sen-ken, sich senken, sinken

**sag** Absacken *n*, Absackung *f*, Biegung *f*, Durch-biegung *f*, -hang *m*, -sacken *n*, -senkung *f*, Läufer *m*, (of cables) Leitungsdurchgang *m*, (of lines) Leitungshang *m*, Nachlassen *n*, Senkung *f* ~ **of a belt** Riemendurchhang *m*

**sag,** ~ **gauge** Winkelhaken *m* für Durchgangs-prüfung ~ **magnification** Vergrößerung *f* des Durchganges ~ **wedge** Sackungskeil *m*

**saggar** (**sagger**) Brennkapsel *f*, Muffel *f* ~ **clay** Kapselton *m*

**sagging** Bodensenkung *f*, Durch-federung *f*, -hängen *n*, Einbiegung *f*, Sackung *f*, Ver-sackung *f* ~ **of the beds on hillsides** Hakenwurf *m*

**sagittal,** ~ **focal line** sagittale Brennlinie *f* ~ **plane** zweiter Hauptschnitt (opt.) ~ **ray** Strah-lenpfeil *m*

**sagittate** (**or sagittiform**) pfeilförmig

**sail, to** ~ segeln **to** ~ **about** (**or around**) um-schiffen **to** ~ **in close order** geschlossen fahren **to** ~ **before the wind** lenzen **to** ~ **off the wind** (**or with flowing sheets**) raumschots segeln

**sail** (of windmill) Flügel *m*, Segel *n* ~ **cloth** Segeltuchstoff *m* ~ **flying** Segelfliegen *n* ~ **loft** Segelkoje *f* ~**-plane** Luftsegler *m*, Segelflugzeug *n* ~**-planing** Segelfliegen *n*

**sailing** Schiffahrt *f*, Segeln *n* ~ **barge** (**or boat**) Segelschiff *n* ~ **chart** Segelkarte *f* ~ **directions** Segelanweisungen *pl* ~ **flight** Segelflug *m* ~ **line** Schiffahrtsrinne *f* ~ **permit** Schiffahrtserlaubnis *f* ~ **race** Segelregatta *f*, Wettsegeln *n* ~ **surface** (Zettelrohrpost) Treibfläche *f* ~ **vessel** Segelschiff *n*

**sailor** Matrose *m*, Schiffer *m*, Seemann *m*

**sake, for the** ~ **of** um ... willen **for the** ~ **of completeness** zur Vervollständigung *f*

**sal,** ~ **ammoniac** Ammoniumchlorid *n*, Chlorammonium *n*, Salmiak *m*, Salmiaksalz *n* ~**-ammoniac cell** Salmiakelement *n*

**salable** absatzfähig, gangbar, gängig, lieferbar **to be** ~ Absatz *m* finden

**salamander** Bodensau *f*, Gebäudetrockner *m*, Härtling *m*, Ofenwolf *m*

**salary** Gehalt *n*

**sale** Abnahme *f*, Absatz *m*, Absetzen *n*, Umsatz *m*, Verkauf *m*, Vertrieb *m* **for** ~ verkäuflich, zu verkaufen ~ **by agent** Kommissionsverkauf *m*, Verkauf *m* durch Kommissionar ~ **by auction** Verkauf *m* an den Meistbietenden ~ **in blank** Blankogeschäft *n* ~ **upon sealed tenders** Submissionsvergebung *f*

**sales,** ~ **account** Verkaufsabrechnung *f* ~ **agreement** Verkaufsvertrag *m* ~ **engineer** Verkaufsingenieur *m* ~ **executive** Verkäufer *m* ~ **expense** Vertriebskosten *pl* ~ **manager** Verkaufsdirektor *m*

**salesman** Agent *m*, Verkäufer *m*, Vertreter *m*

**salicylate** Salizylpräparat *n*

**salience** Ausbuchtung *f*, ausspringender Winkel *m*, Einbuchtung *f*, Frontvorsprung *m*, Herausragen *n*, Knie *n*, Kniestellung *f*, vorspringendes Stelle *f*, vorspringender Winkel *m*, Vorsprung *f*

**salient** (poles) ausgeprägt, heraus-ragend, -springend, -stehend, hervorspringend, überstehend, vorspringend **to be** ~ hervorragen, vorspringen

**salient,** ~ **angle** ausspringender Winkel *m* ~ **pole** ausgeprägter Pol *m* ~**-pole generator** Generator *m* mit an den Polkanten erweitertem Luftspalt ~**-pole machine** Schenkelpolmaschine *f* ~ **spark position** vorgezogene Funkenlage *f*

**saliferous** salzführend, salzhaltig

**salifiable base** salzfähige Base *f*

**saline** Salzlösung *f*; salzartig, salzig ~ **coefficient** Achsenkoeffizient *m* ~ **deposit** Abraumsalz *n* ~ **dome** Salzdom *m* ~ **encrustation** Salzhaut *f* ~ **solution** Kochsalzlösung *f* ~ **spring** Salzquelle *f* ~ **water** salziges Wasser *n*

**salinelle** Salse *f*

**saliniy** Salz-gehalt *m*, -haltigkeit *f*

**salinometer** Salzmesser *m*

**saliva** Speichel *m*

**salmon pink** lachsrot

**salpeter** Salpeter *m*

**salse** Salse *f*

**salt, to** ~ **out** aussalzen

**salt** Salz *n* (common) Siedesalz *n* ~ **bath** Salzbad *n* ~**-bath casehardening** Salzbadeinsatzhärtung *f* ~**-bath hardening** (**or tempering**) **furnace** Salzbadhärteofen *m* ~**-bearing** salzhaltig ~ **boring** Solbohrloch *n* ~ **bridge** Zwischenflüssigkeit *f* ~

**cake** Natriumsulfatkurchen *m* ~ **content** Salzgehalt *m*, Salzhaltigkeit *f* ~ **dome** Salzdom *m* ~ **dressing** Salzbeize *f* ~ **gauge** Solwaage *f* ~ **glaze** Salzglasur *f* ~**-grained sludge** salzbreiartige Masse *f* ~ **hopper** Salztrichter *m* ~ **immersion tester** Salzbadtester *m* ~ **interceptor** Salzabschneider *m* ~ (**velocity**) **method** Salzungsmethode *f* ~ **mine** Saline *f*, Salzbergwerk *n* ~ **pan** Salzsiedepfanne *f*, Salzsiederei *f*, Verdunstungsbassin *n* ~ **screen** Fluoreszenzschirm *m* ~ **solution** Salzlösung *f* ~**-spray test** Salzsprüh(nebel)versuch *m* ~ **spring** Solquelle *f* ~ **test** Salzprüfversuch *m* ~ **vapor** Salzdampf *m* ~ **water** Galle *f*, Salzwasser *n*, Sole *f* ~**-water elevator** Solheber *m* ~ **well** Solbrunnen *m*, Solequelle *f* ~ **works** Saline *f*, Salz-fabrik *f*, -siederei *f*

**saltatory** sprunghaft

**salted** (goods) eingemacht, gepökelt, gesalzen

**salting,** ~ **the hides** Einsalzen *n* der Felle ~**-out coefficient** Aussalzungskoeffizient *m*

**saltpeter** Kalisalpeter *m* ~ **blasting powder** Sprengsalpeter *m* ~ **sweepings** Gayerde *f*, Gaysalpeter *m*

**salts of heavy metals** Schwermetallsalze *pl*

**saltworks** Saline *f*

**salty** salzig

**salutary lighting apparatus** Lichtheilapparat *m*

**salvage** Bergen *n*, Bergung *f*, Bergungsgut *n*, Rettung *f*, Wiedergewinnung *f*, Wiederverwertung *f* ~ **of scrap materials** Altmaterialsammlung *f*

**salvage,** ~**-collecting equipment** Berg(e)gerät *n* ~**-collecting service** Bergedienst *m* ~ **column** Bergungskolonne *f* ~ **crane** Abschleppkran *m* ~ **dump** Sammelstelle *f* ~ **operations on disabled ships** Bergungen *pl* (naut.) ~ **party** Bergungskommando *n* ~ **plant** Bergungsanlage *f* ~ **ship** Hebeschiff *n* ~ **value** Schrottwert *m*, Wiederverwendungswert *m* ~ **vessel** Bergungsschiff *n* ~ **work** Aufräumungsarbeiten *pl*, Bergungsarbeiten *pl* (nav.) ~ **yard** Altmateriallager *n*

**salvaged stock** aufgearbeiteter Bestand *m*

**salve** Salbe *f*

**salvo** Lage *f*, Reihenabwurf *m*, Salve *f* ~ **bombing** Massenabwurf *m* ~ **fire** geschützweises Feuer *n*, Gruppenfolge *f*, Lagenfeuer *n*, Salvenschuß *m* ~ **release** Schüttwurf *m* ~**-release equipment** Dauerkontaktschaltgerät *n* ~ **switch** Salvenschalter *m*

**salvy** salbig

**samarium** Samarium *n*

**samarskite** Samarskit *m*, Uranotantal *m*, Yttroilmenit *m*

**same** nämlich **the** ~ **as above** wie oben **in the** ~ **direction** gleichsinnig **to the** ~ **effect** gleichlautend **of the** ~ **height** (**or length or magnitude**) gleich groß **of the** ~ **name** gleichnamig **of the** ~ **sign** von gleichem Vorzeichen *n* **of the** ~ **tint** gleichfarbig **of the** ~ **type** Artgleichen *pl* (Komponenten im selben Zustand)

**sameness** Gleichheit *f*

**sample, to** ~ ausprobieren, bemustern, eine Probe *f* nehmen, probieren, prüfen

**sample** Exemplar *n*, Fassonstück *n*, Formstück *n*, Lehre *f*, Muster *n*, Probe *f*, Probe-gut *m*, -körper *m*, -länge *f*, -stück *n*, Versuchsstück *n*,

Warenprobe *f* ~ **to be drawn** Ziehprobe *f* ~ **of fabric** Arbeitsmuster *n* ~ **in kind** Substanzmuster *n* ~ **of metal** Werkprobe *f* ~ **of molten metal** Chargenprobe *f* ~ **of precipitation** Niederschlagsprobe *f* ~ **of tin** Zinnprobe *f* ~ **of no value** Muster *n* ohne Wert

**sample, ~-bag clips** Musterbeutelklammern *pl* ~ **book** Musterbuch *n* ~ **card** Musterkarte *f* ~ **cock** Probehahn *m* ~ **drawn** Ziehprobe *f* ~ **glass** Schauglas *n* ~ **material** Probe-gut *n*, **-stoff** *m* ~ **pan** Probenschälchen *n* ~ **print** Druckleistungsprobe *f* ~ **representing a working face** Förderwagenprobe *f* ~ **sheet** Musterbogen *m*, Papiermusterbogen *m* ~ **spoon** Probelöffel *m* ~ **taking** Probeentnahme *f* ~**-taking cylinder** Stanzrohr *n*

**sampled** abgetastet (TV)

**sampler** Probenehmer *m*

**samples of the aggregate** Gesteinsproben *pl*

**sampling** Bemusterung *f*, Muster-kollektion *f*, -sammlung *f*, -stück *n*, -ziehung *f*, Nehmen *n* von Proben, **(at random)** Probe-entnahme *f*, -nahme *f*, -nehmen *n*, -ziehen *n* ~ **of molasses** Melasseprobenahme *f*

**sampling, ~ action** periodische Einstellung *f* oder Korrektur *f* ~ **circuit** Diskriminator *m* ~ **cock** Prob(ier)hahn *m* ~ **device** Sonde *f* ~ **inspection** Musterprüfung *f* ~ **pipe** Entnahme-pipette *f* ~ **tube** Entnahmerohr *n* ~ **valve** Probe-, Prüf-hahn *m* ~ **vial** Probefläschchen *n*

**Sampson post** Schwengelbock *m*

**sanction, to** ~ billigen, genehmigen

**sanction** Bestätigung *f*, Genehmigung *f*

**sand, to** ~ besanden to ~ **the commutator** den Kollektor *m* abschmirgeln

**sand (coarse)** Grieß *m*, Sand *m* ~ **for casting purposes** Form- und Kernsand *m* für Guß-zwecke ~ **for cleaning platinum** Platinsand *m* ~ **for use in glazes** Glasursand *m*

**sand, ~-aerating apparatus (or aerator)** Sand-auflockerungsmaschine *f* ~ **agglutinant** Sand-bindemittel *n*

**sandbag** Sandsack *m* ~ **line** Sandsackaufhänge-leine *f*, Sandsackleine *f* ~ **loop** Sandsackaufhängöse *f*

**sand-bagged** mit Sandsäcken *pl* befestigt

**sandbank** Sandbank *f*

**sand, ~ bar** Barre *f*, längliche Sandbank *f* ~ **bath** Sandbad *n* ~ **beach** sandiger Strand *m* ~**-bearing** sandführend ~ **bellows** Sandstrahl-gebläse *n*

**sandblast, to** ~ sandstrahlen, mit Sandstrahl *m* abblasen

**sandblast** Sandstrahl *m*, Sandstrahlgebläse *n*, Sandgebläse *n* ~ **action** Sandblas-, Sandstrahl-wirkung *f* ~ **barrel** Trommelgebläse *n* ~ **barrel mill** Drehtrommelsandstrahlgebläse *n*, Sand-strahlgebläse *n* mit umlaufender Trommel, Sandstrahlputztrommel *f* ~ **blower** Sandstrahl-gebläse *n* ~ **booth** Strahlhaus *n* ~ **cabinet** Blas-haus *n*, Sandstrahl-blasgehäuse *n*, -gebläsehaus *n* ~ **cleaning** Putzen *n* mit Sandstrahlgebläse, Sandstrahlreinigung *f* ~ **cleaning room** Sand-strahlputzerei *f* ~ **effect** Sandstrahlwirkung *f* ~ **equipment** Sandstrahlgebläseanlage *f* ~ **finish** mit dem Sandstrahl *m* behandelt ~ **in-stallation (founding)** Sandstrahlgebläseanlage *f* ~ **machine** Putzmaschine *f*, Sandstrahlma-

schine *f*, Trommelgebläse *n* ~ **(tank) machine** Sandstrahlapparat *m* ~ **machine with stationary nozzle** Sandstrahlgebläse *n* mit feststehender Düse ~ **mill** Sandstrahlzwergtrommel *f* ~ **nozzle** Blas-, Sandblas-, Sandstrahl-düse *f*, Sandstrahlgebläsedüse *f* ~ **pressure system** Sandstrahldrucksystem *n* ~ **rolling barrel** Scheuertrommel *f* mit Sandstrahlgebläse ~ **room** Putzhaus *n*, Sandblashaus *n*, Sandstrahl-gebläsehaus *n* ~ **room with downdraft ventilation** Putzhaus *n* mit Staubabsaugung nach unten ~ **room with overhead trolley** Putzhaus *n* mit Hängebahn ~ **room with track arrangement** Putzhaus *n* mit Gleisanlage ~ **sand** Putzsand *m* ~ **sprocket-table machine** Sprossentischsand-strahlgebläse *n* ~ **suction system** Sandstrahl-saugsystem *n* ~**-tank cabinet** Zwergsandstrahl-gebläse *n* ~ **tank cabinet** Kleinstrahlgebläse *n* ~ **tank machine with automatic return of sand** Sandstrahlgebläse *n* mit Sandkreislauf ~ **tube** Sandstrahlrohr *n* ~ **tumbling barrel** Scheuer-trommel *f* mit Sandstrahlgebläse ~ **unit** Sand-strahlgebläse *n*

**sandblasted** mit dem Sandstrahlgebläse *n* ge-reinigt

**sandblasting** Sandbestrahlung *f*, Sandstrahl-be-handlung *f*, -blasen *n*, -putzen *n*, -reinigung *f* ~ **equipment** Sandstrahlgebläse *n* ~ **drum** Sand-strahltrommel *f* ~ **installation (or practice)** Sandstrahlbläserei *f* ~ **sand** Gebläsesand *m*, Spritzsand *m* ~ **shop** Sandstrahlerei *f*

**sandbox** Sandkasten *n*

**sandcast** in Sand *m* gießen, Sand(form)guß *m*

**sand, ~-casting** Sandgießen *n*, Sandguß *m* ~**-casting plant** Sandgießerei *f* ~ **castings in gunmetal** Formgußerzeugnisse *pl* aus Rotguß ~ **catcher** Grießabscheider *m* ~ **(-mixing) chamber** Sandraum *m* ~ **channel** Sandreiße *f* ~ **circulation** Sandkreislauf *m*, Sandumlauf *m* ~ **cleaning** Sandreinigung *f* ~ **compacting** Sand-verdichtung *f* ~ **content** Sandführung *f* ~**-con-trol valve** Sandregulierhahn *m* ~ **conveyer** Sandförderer *m*, Sandzubringer *m* ~**-conveying** sandführend ~ **core** Sandkern *m* ~ **core blast machine** Sandkernblasmaschine *f* ~**-delivery pipe** Sandleitung *f* ~ **density** Sanddichte *f* ~ **dike (blasting)** Sandreiße *f* *f* ~**-discharge pipe** Sandauslaufrohr *n* ~ **disintegrator** Sand-schleuder *f*, -mühle *f* ~ **disintegrator with siev-ing arrangement** Sandsiebschleudermaschine *f* ~**-drying oven** Sandtrockenofen *m* ~ **dust** Sand-staub *m* ~ **elevator** Sandbecherwerk *n* ~**-faced brick** Sandformziegel *m* ~ **feed** Sandzufuhr *f*, Sandzuführung *f* ~**-feed valve** Sandregulier-hahn *m* ~ **filling machine** Sandzuleitungsma-schine *f* ~ **flag** Sandsteinplatte *f* ~ **flotation** Sandflotation *f* ~ **flow** Sandfluß *m* ~ **frame (in core making)** Füllrahmen *m*, Sandrahmen *m* ~ **glass** Sanduhr *f*, Stundenglas *n* ~ **grain** Sand-korn *n* ~ **gutter** Sandrinne *f* ~ **hill** Düne *f* ~ **hole** Sandloch *n* ~ **hose** Sandschlauch *m* ~ **jet** Sandstrahl *m* ~ **ladle** Sandkelle *f* ~**-lime brick** Kalksandstein *m* ~ **line** Schlammseil *n* ~ **line spool** Löffelseiltrommel *f* ~ **load** Ersatzlast *f* aus Sand ~**-loading test** Sandbelastungsprüfung *f* ~ **mill** Sandmühle *f* ~ **mixer** Sandmisch-maschine *f* ~**-mixing chamber** Sandmischkessel *m* ~ **model** Sandkasten *m* ~ **mold** Sandform *f*,

(of molding) verlorene Form *f* ~**-mold cast iron** Sandguß *m* ~ **molding** Masseformerei *f* ~ **nozzle** Sanddüse *f* ~ **packing** Sandverdichtung *f* **sandpaper, to** ~ abreiben mit Sandpapier *n*, abschmirgeln, schmirgeln
**sandpaper** Glaspapier *n*, Sandpapier *n*
**sand,** ~ **passage** Sanddurchgang *m* ~ **pile** Sandhaufen *m* ~ **pillar** Staubsäule *f* ~ **pin** Formernagel *m* (aviat.), Formerstift *m* ~ **pipe** geologische Orgel *f*, Streurohr *n* ~ **pit** Sandgrube *f* ~ **pocket** Sandeinschluß *m*, Sandtasche *f* ~ **preparation** Sandaufbereitung *f* ~**-preparing plant** Sandaufbereitungsanlage *f* ~ **pressing** Sandpressen *n*, Sandverdichtung *f* ~ **property** Sandbeschaffenheit *f* ~ **pump** Kolbensandpumpe *f*, Ventilbohrer *m* ~ **pump shaft** Löffelwelle *f* ~**-pump spool** Löffeltrommel *f* ~ **pumping** Löffeln *n* ~ **quality** Sandbeschaffenheit *f*, Sandsorte *f*, ~**-recovery conveyer** Rückförderband *n* für Sand ~ **reel** Schlammkran *m*, Schöpfhaspel *f*, Winde *f* ~**-retaining flange (or rib)** Sandleiste *f* ~ **scaffolding** Sandgerüst *n* ~ **scattering** Sanden *n* ~**-scattering equipment** Sandstreuanlage *f* ~ **scraper** Sandschaber *m* ~ **screen** Sandsieb *n* ~ **screw conveyer** Sandförderschnecke *f* ~ **seal** Sandtasse *f* ~ **separation** Sandreinigung *f*, Sandsichtung *f* ~**-shaking sieve** Sandschüttelsieb *n* ~ **sheave** Schlammrolle *f* ~ **sifter** Sandsieb *n* ~ **sifting** Sandreinigung *f* ~ **skin** Gußkruste *f* ~ **slinger** Sandschleuderformmaschine *f*, Sandslinger *m* ~ **soap** Putzstein *m*, Sandseife *f* ~ **spout** Sandhose *f* **sandstone** Sandstein *m* ~ **block** Sandsteinquader *m*
**sand,** ~ **stopping** Schlackenwehr *n* ~**-storage hopper** Sandvorratsbehälter *m* ~ **streaks** Sandschmitzen *pl* ~ **structure** Sandgerüst *n* ~ **supply** Sandvorrat *m*, Sandzuführung *f* ~ **tank** Sandkasten *m* ~ **test** Prüfung *f* durch Besandung ~ **thrower** Sandschleuder *f* ~ **trap** Sandfang *m*, Sandfänger *m* ~**-valve control mechanism** Sandventilsteuerung *f* ~ **washery (or washing plant)** Sandwäsche *f*
**sandweld, to** ~ sandschweißen
**sand-wheel lugs** Sandradsporen *pl*
**sandwich, to** ~ dazwischenschieben, einlegen, plattieren, auf beiden Seiten *pl* mit anderen Schichten bedecken
**sandwich** Zwischenlage *f* ~ **construction** Schichtenkonstruktion *f* ~ **film** doppelbeschichteter Film *m*, Doppelschichtfilm *m*, mehrfachbeschichteter Film *m* ~ **girder** Verbundbalken *m* ~ **monocoque construction** Sandwich-Schalenbauweise *f* ~ **spacing** Schichtabstand *m* ~ **windings** Transformations-Scheibenwicklung *f* ~ **wound relay** Relais *n* mit Scheibenwicklung
**sandwiched** geschichtet ~ **metal** doppelseitig plattiertes Blech *n*
**sandalwood** Sandelholz *n*
**sander** Sandpapierschleifmaschine *f*
**sanding** Saugstreuung *f* ~ **attachment** Schleifeinrichtung *f* ~ **up** Versandung *f*
**sandy** sandführend, sandig ~ **bottom** Sandgrund *m* ~ **soil** Sandboden *m*
**sanforized** sanforisiert
**sanforizing (nonshrinking process)** Sanforisieren *n*
**sanidine** Sanidin *m*

**sanitary** gesundheitlich, hygienisch, sanitär ~ **pumps** Bad- und Klosettpumpen *pl*
**sanitation** Gesundheitswesen *n*, Sanität *f*, Sanitätspflege *f*
**sap, to** ~ unterminieren **to** ~ **the foundation** den Grund *m* untergraben
**sap** Saft *m*, Sappe *f*, Splint *m*, Splintholz *n*
**saphead** Grabenkopf *m*, Sackgrube *f*, Sappenkopf *m*
**sapless** splintfrei
**saponifiable** verseifbar
**saponification** Seifenbildung *f*, Verseifung *f* ~ **agent** Verseifungsmittel *n* ~ **factor** Verseifungsgrad *m* ~ **number** Verseifungszahl *f*
**saponify, to** ~ verseifen
**saponifying** ablaugend ~ **medium** Ablaugmittel *n*
**saponine** Saponin *n*
**saponite** Seifenstein *m*
**sapper** Sappeur *m*
**sapphire** Saphir *m* ~ **quartz** Saphirquarz *m*
**sapping tools** Minierwerkzeug *n*
**sappy** saftreich ~ **wood** weiches, saftreiches Holz *n*
**sapropel** (putrid mud) Faulschlamm *m*, Sapropel *n*
**sapwood** Jungholz *n*, Splint *m*, Splintholz *n* ~ **tree** Splintbaum *m*
**sarcolite** Sarkolith *m*
**sardonyx** Sardonyx *m*
**sartorite** Bleiarsenglanz *m*
**sash** (window sash) Flügelrahmen *m*, Glas-, Schiebe-fenster *n*, Zarge *f* ~ **bar** Fensterstab *m* ~ **bolt** Fensterwirbel *m*, Schubriegel *m* ~ **frame** Fenster-einfassung *f*, -gestell *n*, -stock *m*, -zarge *f* ~ **gate** Schütz *n*, Schutzfalle *f* ~ **gate valve of floodgate** Schleusenschieber *m* ~ **line** Gewichtsschnur *f* an Schiebefenstern ~ **lock** Schützenschleuse *f* ~ **pulley** Schiebefensterrolle *f* ~ **saw** Fein-, Mühl-, Schlitz-säge *f* ~ **sliding valve of floodgate** Schleusenschieber *m* ~ **sluice** Schützenschleuse *f* ~ **square** Fensterfach *n* ~ **strap** Fensterriemen *m* ~ **tool** Fensterrahmenspindel *f* ~ **window** Aufzieh-, Fall-, Schiebe-fenster *n*, in der Höhe verschiebbares Fenster *n* ~ **wing** Fensterflügelrahmen *m*
**satchel** Aktenmappe *f*
**sateen** englisches Leder *n*, glänzender Futterstoff *m*
**satellite** Begleiter *m*, Nebenlinie *f* (astron., phys.), Satellit *m*, Trabant *m*, kleines Zahnrad *n* des Planetengetriebes ~ **airfield** Ausweichflugplatz *m* ~ **carrier** Planetenträger *m*, Satellitenträger *m*, Zapfenstern *m* ~ **community center** Trabantenstadt *f* ~ **effect** Begleiterscheinung *f* ~ **exchange** Hilfsamt *n*, Teilamt (teleph.) ~ **lines** Begleitlinien *pl* ~ **pulse** Nebenimpuls *m* ~ **switching networks** Teilamtsnetzwerke *pl* ~ **transmitter** Hilfssender *m*
**satiate, to** ~ sättigen
**satiation** Sättigung *f*
**satin, to** ~ glätten, satinieren
**satin** Atlas *m*, Satin *m* ~ **de laine** Wollsatin *m* ~**-finished** mattiert ~ **spar** Atlasspat *m*, Faserkalk *m*
**satisfactory** befriedigend, einwandfrei, genügend, zufriedenstellend
**satisfy, to** ~ befriedigen, entsprechen (math.), erfüllen (math.), genügen, gerecht werden, sättigen **to** ~ **a relation** eine Beziehung *f* erfüllen

**saturable** sättigungsfähig ~ **core reactor** sättigungsfähiger Drosselkern *m* ~ **reactor** Sättigungsdrossel *f*, sättigungsfähige Drosselspule *f*, Steuerdrossel *f* ~**-reactor type compass** Erdinduktionskompaß *m*
**saturant** Imprägnier- oder Sättigungsmittel *n*
**saturate, to** ~ (acids) abstumpfen, anschwängern, durch-nässen, -tränken, sättigen, saturieren, (with water) schlemmen, tränken
**saturated** gesättigt, getränkt, satt, vollgesogen ~ **activity** Sättigungsaktivität *f* ~ **adiabatic lapse rate** gesättigte adiabatische Maßveränderung *f*, gesättigter adiabatischer Zustand *m* der Luft ~ **hydrocarbons** gesättigte Kohlenwasserstoffe *pl* ~ **steam** gesättigter Dampf, Sattdampf *m* ~ **vapor** Sattdampf *m*
**saturating device** Sättiger *m*
**saturation** Durch-feuchtung *f*, -nässung *f*, -setzung *f*, -tränkung *f*, Sättigung *f*, Saturation *f*, Tränkung *f* ~ **activity** Sättigungsaktivität *f* ~ **bend** Sättigungsknie *n* ~ **capacity** Sättigungskapazität *f* ~ **characteristic** (curve) Sättigungskennlinie *f* ~ **current** Sättigungsstrom *m* ~ **deficit** Sättigungsunterschuß *m* ~ **density** Sättigungsdichte *f* ~ **line** Linie *f* der Sättigung, Sättigungslinie *f* ~ **point** Sättigungs-grenze *f*, -moment *m*, -punkt *m* ~ **potential** Saugspannung *f* ~ **pressure** Sättigungsdruck *m* ~ **strength jamming** sättigungsstarke Störung *f* ~ **tank** Saturationspfanne *f* ~ **value** Grenz-, Sättigungs-wert *m* ~ **vessel** Saturateuer *m* ~ **voltage** Sättigungsspannung *f*
**saturator** Sättiger *m*, Sättigungsapparat *m*, Saturateuer *m*
**saturn** Saturn *m* ~ **ring** Saturnring *m*
**sauce** Sauce, Tunke *f* ~ **pan** Kochtopf *m*
**saucer** Napf *m*, Schale *f*, Untertasse *f* ~ **bosh** zylindrischer Kohlensack *m* ~**-shaped** schalenförmig ~**-shaped furnace bottom** Herdmulde *f*
**sausage** Wurst *f* ~ **ballon** Beobachtungs-, Fesselballon *m* ~ **filler** Wurstfüllmaschine *f* ~ **machine** Wurstfüllmaschine *f*
**save, to** ~ bergen, bewahren, ersparen, retten, sparen
**save-all** Saftabscheider *m*, Sparvorrichtung *f*, Tropf-behälter *m*, -schale *f*
**saving** Einsparung *f*, Ersparnis *f* ~ **of time** Zeitersparnis *f* ~ **in weight** Gewichtsersparnis *f* ~ **of weight** Gewichtseinsparung *f*
**saving, (water-)** ~ **chamber** Sparkammer *f*
**saw, to** ~ sägen, zersägen **to** ~ **off** absägen **to** ~ **out** aussägen **to** ~ **square** ablängen **to** ~ **timber** Holz *n* schneiden
**saw** Säge *f* ~ **for loaf sugar** Brotesäge *f* ~ **for sugar slabs** Plattensäge *f*
**saw,** ~ **blade** Sägeblatt *n* ~**-blade holder** Sägeangel *f* ~ **bow** Sägebügel *m* ~ **cut** Sägeschnitt *m*
**sawdust** Säge-mehl *n*, -späne *m pl*, -staub *m* ~ **filter** Sägemehlfilter *m* ~ **waste** Sägeabfall *m*
**saw,** ~ **edge** Säge *f* ~ **file** Säge(*n*)feile *f* ~ **frame** Gatterrahmen *m*, Sägebogen *m*, Sägegatter *n*, Weife *f* ~ **grindstone** Sägeschleifstein *m* ~**-horse-ramp obstacle** Rampensperre *f* ~ **log** Sägeblock *m*
**sawmill** Säge-gatter *n*, -mühle *f*, -werk *n* ~ **with one saw** Blockgatter *n* ~ **waste** Sägespäne *m*
**saw,** ~ **nick** Sägeschnitt *m* ~**-pit frame** Unterlage *f* zum Brettschneiden ~ **set** Schränkeisen *n*

~**-set pliers** Sägeschränkzange *f*, Schränkzange *f* ~ **steel** Sägestahl *m* ~ **tooth** Sägezahn *m* ~ **tooth with wide set** geschränkter Sägezahn *m*
**saw-tooth,** ~ **antenna** Sägezahnantenne *f* ~ **current** Sägezahnstrom *m* ~ **curve** Sägezahnkurve *f* ~ **generator** Sägezahn-erzeuger *m*, -generator *m* ~ **oscillation** Intermittenz-, Kipp-, Sägezahn-schwingung *f* ~ **oscillations** Kippschwingungen *pl* ~ **oscillator** Kimmschwingschaltung *f*, Kippgerät *n*, Kipp(schwing)schaltung *f* ~ **producer** Sägezahnerzeuger *m* ~ **roof** Sägedach *n* ~ **shape** Sägezahnform *f* ~ **shaped** sägezahnartig ~**-shaped sweep current** sägezahnförmiger Kippstrom *m* ~**-shaped sweep voltage** sägezahnförmige Kippspannung *f*, Sägezahnspannung *f*, Stromsägezahn *m* ~ **wave** Intermittenzschwingung *f*, Relaxationsschwingung *f*, Sägezahnwelle *f*
**saw,** ~**-toothed** sägezahnartig ~ **web** Säge(n)blatt *n* ~ **whetstone** Sägeschleifstein *m*
**sawing** Sägen *n* ~ **jack** Sägeblock *m*
**sawlike** sägeartig
**S-bend** gekröpftes Rohr *n*
**scabbard** Futteral *n*, Trägervorrichtung *f*
**scabbing** Schülpenbildung *f*
**scabs** Mattschweiße *f*
**scaffold, to** ~ berüsten, mit einem Gerüst *n* versehen oder stützen
**scaffold** (of a blast furnace) Ansatz *m*, Arbeitsbühne *f*, -gerüst *n*, Baugerüst *n*, Feststauen *n*, Gerüst *n*, Gestell *n*, Hänge-gerüst *n*, -gewölbe *n*, Leergerüst *n*, Lehrgerüst *n* ~ **controller** Beleuchter *m*, (studio) Oberbeleuchter *m* ~ **lamp** Soffittenlampe *f* ~ **light** Oberlicht *n*, Soffittenlampe *f* ~ **lighting** senkrechte Beleuchtung *f* ~ **material** Rüstmaterial *n* ~ **pole** Richtstange *f* ~
**sashing** Gerüstzurrung *f*
**scaffolding** Baugerüst *n*, Bühne *f*, Errichtung *f* eines Gerüstes, Feststauen *n*, Hängen *n*, (in a blast furnace) Hängenbleiben *n*, Rüstung *f* ~ **of girders** Balkengerüst *n*
**scaffolding,** ~ **imp** Lantenne *f* ~ **pole** Gerüststange *f*, Lantenne *f*, Rüstbaum *m*
**scalar** skalar ~ **product** inneres Produkt *n*, skalares Produkt *n* ~ **property** Skalar *n* ~ **quantity** Skalar *n*, Skalargröße *f*
**scald, to** ~ abbrühen, abdämpfen, ausbrühen, brühen
**scald** Brandwunde *f*, Verbrennung *f*
**scalding** (bleaching) Nachbeuchen *n* ~ **chamber** Brühraum *m* ~ **juice** Brühsaft *m* ~ **tub** Abbrühkessel *m*
**scale, to** ~ abblättern, abklopfen, ablösen, abschälen, abtragen, ausglühen, bemessen, den Kesselstein *m* entfernen, mit einer Teilung *f* versehen, verschlacken, wiegen **to** ~ **the boiler** den Kessel *m* ausklopfen **to** ~ **down** maßstäblich verkleinern **to** ~ **off** abblättern (aircraft maneuver), abklopfen, abschuppen, abzundern, schilfern, schuppen **to** ~ **off a length** eine Strecke *f* abtragen
**scale** Abschilferung *f*, Blättchen *n*, Brandkruste *f*, Glühspan *m*, Gradeinteilung *f*, (casting) Gußhaut *f*, Hammerschlag *m*, Kesselstein *m*, Maß *n*, Maßstab *m*, Meßstab *m*, Pegellatte *f*, Schale *f* (of a balance), Schuppe *f*, Sinter *m*, Sinterschlacke *f* Skala *f*, Splitter *m*, Staffelung *f*, Steinansatz *m*, Stufenleiter *f*, Teilung *f*, Ton-

leiter *f*, Verhältnismaßstab *m*, Waage *f*, Walzhaut *f* (aviat.), Wiegen *n*, Zumeßwaage *f*, (oxide) Zunder *m*; maßstabgetreu, meßbar sein **not to** ~ wilder Maßstab *m*

**scale,** ~ **of alburnum** Ukeleischuppe *f* ~ **of degrees** Gradteilung *f* ~ **of eight** achtfacher Untersetzer *m* ~ **of eight circuit** Achterzählrohr *n* ~ **for evaluation** Beurteilungsmaßstab *m* ~ **of forces** Kräftemaßstab *m* ~ **of hardness** Härteskala *f* ~ **of height** Höhenmaßstab *m* ~ **of image** Bildmaßstab *m* ~ **for ink mixing** Farbmischskala *f* ~ **of latitude** Breitenskala *f* ~ **of length** Längenmaßstab *m* ~ **of the (plotting) machine** Maschinenmaßstab *m* ~ **of ordinate** Ordinatenmaßstab *m* ~ **of photograph** Abbildungsmaßstab *m*, Bildmaßstab *m* ~ **of picture** Bildmaßstab *m* ~ **of protection** Schutz-ausmaß *n*, -skala *f* ~ **of quantities** Flächen-, Größenskala *f* ~ **of reduction** Verjüngungsmaßstab *m* ~ **of reproduction** Abbildungsmaßstab *m* ~ **of the sounding tube** Lotröhrenmaßstab *m* ~ **of sparks** Funkenskala *f* ~ **of surfaces** Flächen-, Größen-skala *f* ~ **of the survey** Aufnahmemaßstab *m* ~ **of ten circuit** Dekadenuntersetzer *m* ~ **of time** Zeitmaßstab *m* ~ **of turbulence** (wind tunnels) Ausdehnung *f* der Turbulenzelemente, Turbulenzgrad *m* ~ **of two circuit** binärer Zähler *m* ~ **from water** Wasserstein *m* ~ **with zero at end** einseitiger Ausschlag *m*

**scale,** ~ **balance** Waagebalken *m* ~ **base paper** Millimeterrohpapier *n* ~ **beam** Waagebalken *m* ~ **board** Furnierplatte *f* ~ **board for back of small looking glasses** Hinterlage *f* ~ **chain** Schuppenkette *f* ~ **coating** Glimmhaut *f* ~**-correction card** Skalenberichtigungstabelle *f* ~ **deflection** Skalenausschlag *m* ~ **deposit** Wassersteinsatz *m* ~ **disk** Skalenscheibe *f* ~ **division** Skalen-intervall *n*, -teil *m*, -teilung *f*, Teilstrich *m* ~ **drawing** Meßbild *n* ~ **effect** Größeneinfluß *m*, Maßstab(s)einfluß *m* ~ **efflorescences** Zünderausblutungen *pl* ~ **factor** Skalenfaktor *m* (info proc.) ~**-free annealing** Zunderfreiglühen *n* ~ **graduation** Skalenteil *m* ~ **height** Skalenhöhe *f* ~ **interval** Skalen-, Teilungs-intervall *m* ~ **law** Skalenverlauf *m* ~ **magnifying glass** Meßlupe *f* ~ **maker** Waagenmacher *m* ~ **mark** Skalenteilstrich *m* ~ **micrometer** Mikrometermaß *n* ~ **microscope** Skalenmikroskop *m* ~ **model** maßstäbliches Modell *n* ~ **number** Maßstabzahl *f* ~ **pan** Waagschale *f* ~ **pit** Sintergrube *f* ~ **plate** Teilplatte *f* ~ **range** Skalenbereich *m* ~ **reading** Skalenablesung *f* ~**-resistant** zunderbeständig ~ **retaining frame** Skalenrahmen *m* ~ **rule** Maß-stab *m*, -stock *m* ~ **sector** Skalenteil *m* ~ **size** Maßstabgröße *f* ~ **spacing** Teilstrichabstand *m* ~ **span** Skalenlänge *f* ~ **stair** gerade Treppe *f* ~ **transfer car for weighing up charges** Gattierunglaufgewichtswaage *f* ~ **tube** Skalenrohr *n* ~ **wax** Schuppenparaffin *m*

**scaled** mit einer Skala *f* versehen ~ **up** vergrößert

**scaled,** ~ **cam setting motion** Exzenterskaleneinstellung *f* ~ **detector** geeichter Detektor *m* ~ **down** in verkleinertem Maßstab *m*, verkleinert ~**-down version** Ausführung *f* im kleinen Maßstab ~ **form** Modellausführung *f* ~ **travelling grate** Schuppenwanderrost *m*

**scaleless** schuppenfrei

**scalene** ungleichseitig (geom.)

**scalenous triangle** ungleichseitiges Dreieck *n*

**scaler** Absperrmittel *n*, (color) Einlaßmittel *n*, elektronisches Auslöse- und Zählgerät *n*, Impulszähler *m* mit Schaltvorrichtung, Schaber *m* ~ **and ratemeter** Strahlungsmeßgerät *n*

**scales** Wägevorrichtung *f*

**scaling** Ablätterung *f*, Blättchenbildung *f*, Glühspanbildung *f*, Verzunderung *f*, Zunderung *f* (paint) ~ **chipper** Abklopfer *m*, Abklopfhammer *m* ~ **couple** Frequenzhalbierschaltung *f* ~**-down vacuum tube** untersetzender Verstärker *m* ~ **equipment** Eskaladiergerüst *n* ~ **hammer** Abklopfer *m*, Abklopfhammer *m*, Pickhammer *m* ~**-off** Entzunderung *f* ~**-off of rail** Abblätterung *f* der Schiene ~**-resistant** verzunderungsbeständig ~ **test** Ausbrandversuch *m* ~ **tool** Stahl *m* zum Abdrehen, zum Abstechen der Kruste

**scallop, to** ~ (textiles) auskerben, auszacken, am Rande zackig ausschneiden, zacken

**scallop** Zacke *f*, Zacken *m*

**scalloped** zinnenförmig

**scalloping** (of electron beam) Periodizität *f* ~ **frame** Ausboge-, Auszack-maschine *f*

**scalp, to** ~ häuten

**scalpel** kleines Messer *n*, Seziermesser *n*, Skalpell *n*

**scalping** Oberflächen-hautentfernung *f*, -zerspannung *f*

**scaly** schalig, schuppig ~ **foliated gypsum** schaumiger Gips *m* ~ **fracture** Schieferbruch *m* ~ **gypsum** Schneegips *m* ~ **red hematite** Eisenmann *m* ~ **red iron ore** Eisenschaum *m*

**scan, to** ~ (scanning disk) abfühlen, absuchen, abtasten, bestreichen, sorgfältig untersuchen, zerlegen

**scan** Ablenkungswechsel *m*, Absuchen *n*, Abtastung *f* (TV, rdr), Bildauflösung *f*, Blick-, Sicht-weite *f* (TV, rdr), Skandierung *f* (metr.) ~ **area** Rasterplatte *f* ~ **frequency** (= lines × frame frequency) Abtastfrequenz *f* ~ **generator** Ablenkgerät *n* ~ **generator unit** Abtasterablenkgerät *n* ~ **hole** Rasterblende *f* ~ **pattern** Rasterplatte *f* ~ **period** Regenerierungsintervall *m* ~ **selector switch** Abtastwahlschalter *m*

**scandium** Skandium *n*

**scanned** abgetastet ~ **area** Raster *m* (TV)

**scanner** Abgreifer *m*, Abtaster *m*, Fernsehabtaster *m*, Fühlhebel *m*, Taster *m*, Zerleger *m*, Zerlegungsvorrichtung *f* ~ **amplifier** Abtastverstärker *m*

**scanning** Abtastblendung *f*, Abtasten *n*, Abtastung *f*, Bild-abtastung *f*, -punktverteilung *f*, -raster *m*, -synthese *f*, -wiedergabe *f*, -zerlegung *f*, -zusammensetzung *f*, Bildfeldzerlegung *f* (rdo, TV), Tasten *n* **(line)** ~ Zeilenabtastung *f* ~ **by deflection of the image** Abtastung *f* durch die Ablenkung des Bildes ~ **with an electron (or light) pencil** Strahlabtastung *f* ~ **by illumination** Durchleuchtungsabtastung *f* ~ **with a moving diaphragm** Abtastung *f* mit bewegter Rasterblende ~ **of the sweep function** Abtasten *n* der Kippfunktion (electron.) ~ **of a tuning range** Abstimmbereichüberstreichung *f*

**scanning,** ~ **aperture** Abtastblende *f* ~ **apparatus** Quirl *m* ~ **beam** Abtaststrahl *m* ~ **beam of light** Abtastlichtstrahl *m* ~ **belt** Abtastband *n* ~ **brush** Lichtstrahlabtaster *m* ~ **circuit** Ablenk-, Abtast-schaltung *f* ~ **condenser** Fühler-kondensator *m* ~ **device** Bildauffänger *m*, Rasterblende *f* ~ **device at the receiving end** Bildpunktverteilung *f*, Bildzusammensetzvorrichtung *f* ~ **device at the transmitter end** Bildabtastvorrichtung *f* ~ **device at the transmitting end** Abtasteinrichtung *f*, Bildpunktabtaster *m*, Bildzerleger *m*, Bildzerlegungsvorrichtung *f* ~ **diaphragm** Abtastblende *f* ~ **disk** Abtastscheibe *f* (TV) ~ **drum** Blenden-, Loch-, Tonabnehmer-trommel *f* ~ **electron beam** abtastender Elektronenstrahl *m* ~ **element** Bildelement *n*, Bildpunkt *m*, Raster-element *n*, -punkt *m* ~ **field** Abtastfeld *n*, Bildfeld *n*, Bildraster *m*, Raster *m* ~ **frequency** Rasterfrequenz *f* ~ **gate** Abtastfenster *m* (film) ~ **hole** Abtast-, Zerleger-, Zusammensetz-blende *f* ~ **impulse** Rasterstoß *m* ~ **interference** Rasterstörung *f* ~ **key** Bildpunktabtaster *m* ~ **lead** Sondenführung *f* ~ **lens** Abtastoptik *f* ~ **light** Tonbelichtungsstelle *f* ~ **line** Abtast-, Bild-zeile *f* ~ **linearity** Abtastlinearität *f* (TV) ~ **loss** Abtastverlust *m* ~ **motion** Hinlauf *m* ~ **needle with spherical point** (sound engraved film) Kugelspitzenfühlnadel *f* ~ **pattern** Raster *m*, Rasterfläche *f* ~-**pattern image** Rasterbild *n* ~ **pencil** Lichtstrahlabtaster *m* ~ **pencil of light** Abtastlichtstrahl *m* ~ **pitch** Zeilenabstand *m* ~ **point** Belichtungsrolle *f* ~ **positive slides** Positivabtastung *f* ~ **rate** Abtastgeschwindigkeit *f*, Bildwechselzahl *f* ~ **ring** Umschaltkette *f* (electron.) ~ **separation** Zeilenabstand *m* ~ **sequence** Abtastfolge *f* ~ **slit** Abtastschlitz *m* (film) ~ **slot** Abtastspalt *m* ~ **speed** Abtastgeschwindigkeit *f* ~ **spot** abtastender Lichtfleck *m*, Abtastfleck *m* (rdo, rdr), Elektronenfleck *m* ~ **spotlight** wandernder Lichtstrahl *m* ~ **stage** Bildablenkstufe *f* ~ **strip** Abtastzeile *f*, Zeile *f* ~ **surface** Fernsehraster *m* ~ **switch** Bildpunktabtaster *m* ~ **traverse** Abtastvorschub *m* ~ **unit** Beobachtungsgerät *n* ~ **yoke** Ablenkjoch *n*

**scansion** Bild-punktverteilung *f*, -synthese *f*, -wiedergabe *f*, -zerlegung *f*, -zusammensetzung *f*, Hinlauf *m*, Zeilenhinlauf *m* ~ **and flyback** Hin- und Rückführung *f* (TV) ~ **of image** Bilderzeugung *f*

**scant** kärglich

**scantling** Abfallstück *n*, kleines Eckstück, Übergangsstück *n*

**scanty** dürftig, knapp, nicht ausreichend, notdürftig, spärlich

**scar** Narbe *f*, Schramme *f* ~ **seam** Narbe *f*

**scarce** knapp, rar, selten, spärlich vorhanden ~ **goods** Mangelware *f*

**scarcity** Knappheit *f*, Mangel *m*, Seltenheit *f*

**scarf, to** ~ abschärfen, abschrägen, anschärfen, (rails) überblatten, zusammenlaschen **to** ~ **timbers** Balken *pl* bündig überschneiden **to** ~-**weld** überlappt schweißen

**scarf** Blattung *f*, Lasch *m*, Laschung *f*, Scherbe *f*, Überlappung *f*, zugeschärfter Rand *m* ~ **cloud** Schleierwolke *f* ~ **joint** Blatt-fuge *f*, -verbindung *f*, Falzverbindung *f*, schräger oder verblatteter Stoß *m*, Verblattung *f* ~-**milling machine** Ausschärffräsmaschine *f* ~ **mount** (gun) Drehkranz *m*, Drehring *m* ~-**planing machine** Ausschärfhobelmaschine *f* ~ **weld** überlappte Schweißung *f*

**scarfed** angeschärft ~-**built beam** verschränkter Balken *m* ~ **end of rail** abgeschrägtes Schienenende *n* ~ **joint** Verbindung *f* mit angeschärften Enden

**scarfer** Zuschärfmaschine *f*

**scarfing** Flammstrahlen *n* (metall.), Verblattung *f* ~ **angle** Abschrägungswinkel *m*

**scarifier** Messeregge *f*, Reißpflug *m*, Scharriereisen *n*, Skarifikator *m*, Straßenaufreißer *m*

**scarlet** scharlachrot

**scarp** Abhang *m*, Erdwall-Innenwand *f*, steile Böschung *f*, Terrainstufe *f*

**scarped** abgeböscht, abschüssig, steil

**scarred** benarbt

**scars** (in furnace) Ansätze *pl*

**scarving** Verblattung *f*

**scatter, to** ~ ausbreiten, ausstreuen, flackern, spratzen, sprühen, streuen, versprengen, verstürzen, verteilen, zerstreuen **to** ~ **around** umherspritzen, umherstreuen **to** ~ **with fire** (artil.) abstreuen

**scatter** Streu..., Streubereich *m*, Streuung *f* ~ **of direction** Richtungszerstreuung *f*

**scatter,** ~ **curve** Streuungskurve *f* ~ **fading** Streuschwund *m* ~ **propagation** Streustrahlausbreitung *f* ~ **unsharpness** Unschärfe *f* durch Streuung

**scattered** diffus, vereinzelt, verstreut, zerstreut ~ ~ **at random** willkürlich verteilt

**scattered,** ~ **light** Streulicht *n* ~ **radiation** Streustrahlung *f* ~ **shots** vereinzelte Schüsse *pl*

**scatterer** Dibbelmaschine *f* (agr.)

**scattering** Auflösung *f*, Ausblendung *f* (of rays), Breitsäemaschine *f* (agr.), Spratzen *n*, (of powder) Streuung *f*, Zerstäubung *f*, Zerstreuung *f*, (of electrons) Zerstreuungswirkung *f*, streuend ~ **by a dislocation line** Streuung *f* durch eine Versetzungslinie ~ **of electromagnetic waves** Streuung *f* der elektromagnetischen Wellen ~ **by imperfections** Streuung *f* an Gitterstörungen ~ **by the internal boundaries** Streuung *f* in den inneren Grenzen ~ **of a light source** Streuung *f* einer Lichtquelle ~ **of the plotted points** Streuung *f* der aufgetragenen Meßpunkte ~ **by static imperfections** Streuung *f* an statischen Fehlstellen

**scattering,** ~ **angle** Streuwinkel *m* ~ **area ratio** Streuungsquerschnitt *m* ~ **cell** Streuzelle *f* ~ **cross section** Streuquerschnitt *m* ~ **direction** Streurichtung *f* ~ **effect** Streuung *f* ~ **factor** (of X rays) Streufaktor *m* ~ **layer** Streuschicht *f* ~ **mean free path** mittlere Streuweglänge *f* ~ **power** Streuvermögen *n* ~ **term** Streuglied *n* ~ **value** Streufaktor *m*

**scavenge, to** ~ (gases) ausfegen, durchspülen, entionisieren, reinigen, säubern, spülen

**scavenge pump** Ölabsug-, Rücklauf-, Spülölpumpe *f*

**scavenged sprue** abgeschnittener Anguß *m*

**scavenger** Spülmotor *m*, Straßenreinigungsmaschine *f* ~ **flotation machine** Nachschäumer *m* ~ **flow (or jet)** Spülfluß *m* ~ **pump** Ölsaugpumpe *f*

**scavenging** Abführsystem *n*, Ausstoßen *n* (gases), Entionisierung *f*, Spülen *n*, Spülung *f* ~ **and charging period** Spül- und Ladeperiode *f* ~ **of the clearance space** Totraumspülung *f* ~ **of residual gases** Restgasausspülung *f* ~ **of two-**

**stroke-cycle engines** Nachladung *f* von Zweitaktmaschinen ~ **of the working cylinder** Ausspülung *f* des Arbeitszylinders
**scavenging,** ~ **air** Spülluft *f* ~ **air blower** Spülgebläse *n* ~ **air channel** Spülluftkanal *m* ~ **air pump** Kolbenluftpumpe *f* ~ **air valve** Spülluftklappe *f* ~ **area** Atmungsraum *m* (Zündkerze) ~ **blower** Spülluftgebläse *n* ~ **duct** Spülleitung *f* ~ **efficiency** Spülwirkungsgrad *m* ~ **engine** Spülmotor *m* ~ **gas** Spülgas *n* ~ **period** Spülperiode *f* ~ **port** Einlaß-, Spül-schlitz *m* ~ **pressure** Spüldruck *m* ~ **pump** Rückförder-, Spül-pumpe *f* ~ **service** Fäkalienabfuhr *f* ~ **stroke** Ausspülhub *n* ~ **valve** Spülventil *n*
**sceleton forming** Rahmenverfomung *f*
**scelometric hardness** Ritzhärte *f*
**scenario** Drehbuch *n*, Handlung *f*, Manuskript *n*
**scene** (of action) Schauplatz *m*, Szene *f* ~ **of accident** Unfallstelle *f*
**scenery** Aufmachung *f*, Landschaft *f*, Szenerie *f*
**scenic** (or light) **effects** Effektszenen *pl*
**scenographical scale** perspektivischer Maßstab *m*
**scent** Duft *m*, Geruch *m*, Spur *f*, Witterung *f*, Wohlgeruch *m* ~ **extract** Duftauszug *m* ~ **trail** Riechspur *f*
**scents** Gerüche *pl*
**schapbachite** Wismut(h)bleierz *n*, Wismutsilber *n*
**schedular** tabellarisch
**schedule, to** ~ eintragen (in Liste), planen
**schedule** Aufstellung *f*, Beilage *f*, (train) Fahrplan *m*, Liste *f*, Plan *m*, Schema *n*, Stundenplan *m*, Tabelle *f*, Verzeichnis *n* ~ **of fits** Passungsliste *f* ~ **of hours** Personalkurve *f* ~ **of individual costs** Einzelkostenaufstellung *f* ~ **of limits** Passungsliste *f* ~ **of operation** Arbeitsschema *n* ~ **of periodic tests** Maßplan *m*, Plan *m* für die regelmäßigen Messungen
**schedule forecast** Terminvoraussage *f*
**scheduled** (flight) flugplanmäßig, planmäßig ~ **air service** flugplanmäßiger Luftverkehr *m* ~ **maintenance** regelmäßig durchgeführte Instandsetzung *f* ~ **speed of aircraft** flugplanmäßige Geschwindigkeit *f* ~ **stop** planmäßiger Halt *m* ~ **train** fahrplanmäßiger Zug *m*
**scheeletine** Scheelblei-erz *n*, -spat *m*
**scheelite** Scheelerz *n*, Scheelit *m*, Schellspat *m*, Schwerstein *m*, Tungstein *m*
**Scheibe cell** Scheibeküvette *f*
**Scheimpflug's principle** Scheimpflugsches Prinzip *n*
**schematic** schematisch, wesenhaft ~ **diagram** (construction) Aufbauplan *m*, schematisches Schaltaderbild *n* ~ **drawing** schematischer Entwurf *m*, Umriß *m* ~ **illustration** Blockzeichnung *f*
**schematize, to** ~ schematisieren
**scheme, to** ~ austüfteln, entwerfen, planen
**scheme** Abriß *m*, Entwurf *m*, Plan *m*, Planung *f*, Projekt *n*, Schema *n*, System *n*, Übersichtsskizze *f* ~ **of crossings** Kreuzungsfolge *f* ~ **of forces** Kraftschema *n*
**schiller spar** Bastit *m*, Schillerspat *m*
**schillerization** Schillern *n*
**schist** Schiefer *m* ~ **oil** Schieferöl *n*
**schistose** schieferig
**schistosity** Schieferung *f*
**schistous** schieferartig, schief(e)rig ~ **lignite** Braunkohlenschiefer *m*

**schlieren** Glasschlieren *n*, Schlieren *pl*, Schlierenoptik *f* ~ **photograph** Schlierenaufnahme *f* ~ **photography process** Schlierenverfahren *n*
**Schlippe's salt** Schlippe'sches Salz *n*
**Schoen mill** Scheibenradwalzwerk *n*
**schoepite** Schöpit *m*
**scholar** Gelehrter *m* ~ **ship** Freistelle *f* für Studium, Stipendium *n*
**scholarly** gelehrt, wissenschaftlich
**school** Schule *f* ~ **of commerce** Handelshochschule *f* ~ **of mines** Bergakademie *f* ~ **and training plane** Schul- und Übungsflugzeug *n*
**school,** ~ **bench** Schulbank *f* ~ **flying** Schulflug *m* ~ **form** Schulbank *f* ~ **plane** Schulflugzeug *n* ~ **requisites** Schulbedarf *m*
**schooner** Schoner *m*
**Schopper ring testing machine** Schoppermaschine *f*
**Schottky,** ~ **effect** Funkeleffekt *m*, Schottky-Effekt *m*, Schroteffekt *m* ~ **line** Schottkysche Gerade *f* ~ **noise** ungeschwächter Schroteffekt *m*
**Schreiner calender** Seidenglanzkalender *m*
**schroeckingerite** Dakeit *m*, Schröckingerit *m*
**Schuko-type coupling piece** Schukokupplung *f*
**Schwartz-type propeller blade** Schwartz'sches Luftschraubenblatt *n*
**science** Forschung *f*, Lehre *f*, Wissenschaft *f*, Wissensgebiet *n* ~ **of color** Farbenlehre *f* ~ **of communications** Nachrichtentechnik *f* ~ **of metals** Metallkunde *f* ~ **of mineral deposits** Lagerstättenkunde *f* ~ **of roentgen rays** (or X-rays) Röntgen-kunde *f*, -lehre *f* ~ **of transport** Verkehrstechnik *f*
**scientific** wissenschaftlich ~ **article** (or paper) Fachaufsatz *m* ~ **researcher** Forscher *m*
**scientist** Gelehrter *m*, Naturwissenschaftler *m*, Wissenschaftler *m*
**scintillant** Szintillationssubstanz *f*
**scintillate, to** ~ flimmern, flittern, (when excited by rays) fluoreszieren
**scintillating** funkelnd, glänzend, schimmernd, strahlend, szintillierend
**scintillation** Flimmern *n*, Funkeleffekt *m*, Funkeln *n*, Funken-sprühen *n*, -werfer *n*, Lichtblitz *m*, Scintillation *f*, Sprühen *n*, Szintillation *f* ~ **counter** Szintillationszähler *m* ~ **detector** Szintillationszähler *m* ~ **layer** Szintillationsschicht *f* ~ **well crystal** Szintillations-Bohrloch-Kristall *m*
**scintillometer** Flimmermesser *m*, Scintillometer *n*
**scintiscanner** Szintigraf *m*
**scissor, to** ~ schneiden, zuschneiden
**scissor,** ~ **arm** oberer Arm *m* der Mitnehmerschere ~ (helicopter) **link** unterer Arm *m* der Mitnehmerschere
**scissors** Schere *f* ~ **blade** Scheren-blatt *n*, -messer *n* ~ **grinder** Scherenschleifer *m* ~-**pattern binoculars** Handscherenfernrohr *n* ~-**shaped collector** Scherenstromabnehmer *m* ~ **shutter** Scherenbelnde *f* ~ **spring** Scherenfeder *f* ~ **telescope** Scherenfernrohr *n* ~-**type arc lamp** Bogenlampe *f* mit breitwinklig gestellten Kohlen, Bogenlampe *f* mit weitwinkligen Kohlenstiften
**sclera** Lederhaut *f*
**sclerometer** Härte-messer *m*, -prüfer *m*, -prüfgerät *n*, Ritzhärteprüfer *m*, Sklerometer *n*

**scleronomic system** skleronomes System *n*
**scleroscope** Skleroskop *n* ~ **hardness** Kugelfall-, Skleroskop-härte *f* ~**-hardness test** Kugelfallprobe *f*
**sclerotic lamp** Sklerolampe *f*
**sclerotium disease** Mietenfäule *f*
**sconce** Wandleuchter *m*
**scoop, to** ~ schöpfen **to** ~ **out** aus-höhlen, -löffeln
**scoop** Aufnehmer *m*, Hutze *f*, Löffelschaufel *f*, Rührflügel *m*, Schaufel *f*, Schaufeln *n*, Schippe *f*, Schippen *n*, Schöpfen *n*, Schöpf-gefäß *n*, -kelle *f*, Schub *m*, Stoß *m* ~**-shaped speculum** Löffelspiegel *m* ~ **shovel** Exkavator *m* ~ **wheel** Schöpfrad *n*, Wurfrad *n*
**scooping** Ausschöpfen *n* ~ **basin** Schöpfbecken *n*
**scope** Anwendungsbereich *m*, Ausdehnung *f*, Bereich *m*, Bildschirm *m* (rdr), Feld *n*, Geltungsbereich *m* (Norm), Rahmen *m*, Raum *m*, Sphäre *f*, Spielraum *m*, Umfang *m*, Weite *f*, Ziel *n* ~ **of an agreement** sachliches Vertragsgebiet *n* ~ **of application** Aufgabenbereich *m* ~ **of business** Geschäftskreis *m* ~ **of invention** Erfindungsgeltungsbereich *m* ~ **of an invention** Geltungsbereich *m* einer Erfindung ~ **of patent** Erfindungsgeltungsbereich *m*
**scorch, to** ~ anschmoren, anvulkanisieren, dörren, rösten, sengen, verschmoren, versengen
**scorch retarder** Vorvulkanisationsverzögerer *m*
**scorching** Schmoren *n* ~ **together of contacts** Kontaktzusammenschmoren *n*, Zusammenschmoren *n*
**score, to** ~ ausbrennen, einschneiden, kratzen, (Walzenzapfen) riefen
**score** Bewertung *f*, Einschnitt *m*, Kerbe *f*, Kerbholz *n*, Punkt-wert *m*, -zahl *f*, Ritz *m*, Ritze *f*, Satz *m* von zwanzig Stück ~ **board** Resultattafel *f* ~ **book** Anschreibeblock *m* ~**-cut principle** Durchschnittsystem *n*
**scored,** ~ **cylinder** gefurchter Zylinder *m* ~ **pulley** Keilriemenscheibe *f* ~ **scoring** Riefenbildung *f*
**scoria** Braschen *pl* (metall.), Gesteinsschlacke *f*, Schlacke *f*, Skorie *f*
**scoriaceous** schlackenähnlich, schlackenartig, schlackig ~ **surface** schaumige Oberfläche *f*
**scorification** Schlackenbildung *f*, Verschlackung *f*
**scorified lava** Schlackenlavat, schlackige Lava *f*
**scorifier** Ansiedescherben *m*, Scherbe *f*, Schlaken-scherbe *f*, -scherben *m*
**scorify, to** ~ schlacken, verschlacken
**scoring** Abnutzung *f*, Einschnitt *m* (geol.), (of barrel) Grat *m*, Kerbe *f*, Narbe *f* (Oberflächenfehler), Zerfressen *n* ~ **and creasing machine** Rill- und Ritzmaschine *f* ~ **apparatus** Rillapparat *m* ~ **machine** Ritzmaschine *f* ~ **stage** Synchronisieratelier *n*
**scorious** verschlackt
**scorodite** Arseniksinter *m*
**scorper** Grab-, Messer-stichel *m*
**scotch, to** ~ **a wheel** ein Rad *n* hemmen
**scotch** Brems-keil *m*, -klotz *m*, Hemmschuh *m*
**Scotch,** ~ **mist** dichter Nebel *m* ~ **pine** Föhre *f* ~ **tape** Klebeband *n*
**scotopia** Dunkel-adaptation *f*, -anpassung *f*
**scotopic vision** Nachtsehen *n*
**Scott system** Scottsche Schaltung *f*

**scour, to** ~ abbeizen, abputzen, abscheuern, beizen, (a wire) blank machen (Draht), dekapieren, (wool) entfetten, entzundern, fressen, putzen, reinigen, schaben, scheuern, schrubben, schruppen **to** ~ **the silk** die Seide entschälen
**scour** Abrieb *m*, Abschliff *m*, Auskolkung *f*, Ausschwemmen *n*, Furche *f*, Kolkbildung *f*, Reinigung *f*, Säuberung *f*, Unter-spülung *f*, -waschung *f* ~ **gate** Spülschütze *f* ~ **prevention** Kolkabwehr *f* ~ **valve** Spülventil *n*
**scoured** gekocht, (silk) sacht ~ **silk** entschälte Seide *f*
**scourer** Getreidereinigungsmaschine *f*, Reinigungs-apparat *m*, -mittel *n*, Spüler *m*
**scouring** Ausfressung *f*, Entfetten *n*, Kolk *m*, Scheuern *n*, Schruppen *n* ~ **of a sluice** Spülen *n* einer Schleuse
**scouring,** ~ **barrel** Scheuer-faß *n*, -trommel *f* ~ **basin** Spülbecken *n* ~ **box** Putzkasten *m* ~ **cloth** Putzlappen *m* ~**-cloth weaving mill** Putzlappenweberei *f* ~ **liquor** Schmutzlauge *f* ~ **powder** Scheuerpulver *n* ~ **properties** Waschvermögen *n* ~ **sand** Fege-, Reinigungs-sand *m* ~ **sluice** Ablaßschütz *n*, Spülbecken *n* ~ **solution** Flotte *f* ~ **water** Fleckwasser *n*
**scout, to** ~ aufklären, auskundschaften, ausspähen, erkunden
**scout** Aufklärer *m*, Erkunder *m*, Kundschafter *m*, Pfadfinder *m*, Späher *m* ~ **vehicle** Erkundungsfahrzeug *n* ~ **vessel** Aufklärungsfahrzeug *n*
**scouting** Aufklärungsdienst *m*, Aufsuchung *f*, Kundschaft *f*, Schürfung *f* ~ **mission** Aufklärungsauftrag *m* ~ **plane** Aufklärungs-, Erkundungs-, Pfadfinder-flugzeug *n*
**scow** Kipprahm *m*
**scram** Schnellschluß *m* ~ **button** Notschalter *m* ~ **rod** Schnellstopstab *m* ~ **switch** Notschalter *m*
**scramble, to** ~ verwürfeln
**scramble** Alarmstart *m*
**scrambled,** ~ **speech** verschlüsselte oder verwürfelte Sprache *f* ~ **take-off** Durchstarten *n* ~ **telephony** Geheimtelefonie *f*
**scrambler** Entfritter *m*, Misch-, Verwürfelungs-vorrichtung *f*, Verwürfler *m* ~ **phone exchange** Kreuzvermittlung *f* ~ **circuit** Bandspalter *m*
**scrambling** Verwürfelung *f*
**scrap, to** ~ abwracken, ausrangieren, als Ausschuß ausscheiden, auswerfen, verschrotten
**scrap** Abfall *m*, Abfälle *pl*, Altmaterial *n*, Ausschuß *m*, (image) Bildausschnitt *m*, Brosamen *pl*, Bruch *m*, Entfall *m*, Schnipsel *n*, Schnipselchen *n*, Schrot *m*, Schrott *m*, Schrottentfall *m*, Stoffabfall *m* ~ **from sorting rags for rag-paper stock** Schrenz *m*
**scrap,** ~ **billets** Knüppelschrott *m* ~ **blade** Quermesser *m* ~ **brass** Altmessing *n* ~ **breaker crane** Fallwerkskran *m* ~**-briquetting press** Schrottpaketierpresse *f* ~ **cutter** Schrottschere *f* ~ **dealer** Schrotthändler *m* ~ **drive** Altstoffsammlung *f* ~**-handling magnet crane** Schrottmagnetkran *m* ~ **iron** Abfall-eisen *n*, -haufen *m*, Alteisen *n*, angearbeitetes Eisen *n*, Brucheisen *n*, Eisen-abfall *m*, -schrott *m*, Gußbruch *m*, Schrott *m* ~ **lead** Bleiabfall *m* ~ **leather** Lederabfälle *pl*, Leimleder *n* ~ **list** Ausschußliste *f* ~

**market** Schrottmarkt *m* ~ **melting** Schrott-schmelze *f* ~ **metal** Altmetall *m*, Bruchmetall *n*, Schrott *m*, Umschmelzmetall *n* ~ **pile** Schrott-haufen *m* ~ **rubber** Altgummi *n* ~-**shearing machine** Schrottschere *f* ~ **silver** Bruchsilber *n* ~ **smelting** Schrottverhüttung *f* ~ **steel** Ab-fallstahl *m* ~-**stock yard** Schrottlagerplatz *m* ~-**to-finish product ratio** Ausschußziffer *f* im Verhältnis zum Ausbringen ~ **value** Schrott-wert *m* ~ **view** Teilansicht *f* (aviat.) ~ **wire** Drahtabfall *m* ~ **yard** Schrott-lager *n*, -platz *m* ~-**yard crane** Schrottlagerkran *m*

**scrape, to** ~ abkratzen, aufrakeln, (den Kessel) ausschaben, bekratzen, reiben, schaben, scharren, scheuern, schürfen **to** ~ **the bearing** das Lager einschaben, nachschaben **to** ~ **and clean brightly** (Draht) blank machen **to** ~ **loose** los-wuchten **to** ~ **off** abkratzen, abschaben, ab-streichen **to** ~ **off hair** abhaaren **to** ~ **out** aus-kratzen, ausradieren, herauskratzen, losbre-chen, loswuchten **to** ~ **up the roll** die Walze ab-streichen (print.)

**scrape** Kratzen *n*, Kratzer *m*, Kratzstelle *f*, Rührrücke *f*, Schürfung *f* ~ **resin** Scharrharz *n*

**scraped to fit each other** aufeinander eingeschabt

**scraped,** ~ **cork** abgekratzter Kork *m* ~ **leather** gestrichenes Leder *n* ~ **pulp** Schabstoff *m* ~-**surface oil chilling machine** Ölkratzkühler *m*

**scraper** (of a conveyer) Abschaber *m*, Abstrei-cher *m*, Abstreich-eisen *n*, -messer *n*, Aus-streichmesser *n*, Kratze *f*, Kratzeisen *n*, Kratzer *m*, Löffelräumer *m* (min.), Planierpflug *m*, Pflug *m*, Räumlöffel *m*, Schabeisen *n*, Schaber *m*, Schab(e)messer *n*, Schare *f*, Scharrer *m*, Schlichteisen *n*, Schrapper *m*, Schrotmeißel *m*, Streich-eisen *n*, -messer *n*, Ziehklinge *f* ~ **of a conveyer** Abkratzer *m*

**scraper,** ~ **chain conveyer** Kratzkettenförderer *m* ~ **conveyer** Schiebetransporteur *m* ~ **disk** Kratzerscheibe *f* ~ **flight conveyer** Kratzband *n*, Kratzer-kettenförderer *m*, -transporteur *m*, Verladeschiebe *f* ~ **knife** Schabeisen *n* ~ **motion control** Rakelsteuerung *f* ~ **plane** Schabhobel *pl* ~ **plate** Kratzerblech *n* ~ **ring** Abstreifring *m*, Ölabstreifring *m*

**scrapers for ground rails** Auskratzer *m* für Rillen-schienen

**scraping** Abfall *m*, Kratzen *n*, Schaben *n*, Späne *pl* ~ **iron** Kratze *f*, Kratzeisen *n* ~ **knife** Ab-streifmesser *m*, Schab(e)messer *n* ~ **machine** Abgratmaschine *f* ~-**out** Losbrechen *n* ~ **tool** Abschabe-, Räum-werkzeug *n*

**scrapings** Schabsel *pl* ~ **of liquation** Seigerkrätze *f* ~ **ejector** Abfallausstoßer *m*

**scrapped,** ~ **parts** ausgeworfene Teile *pl* ~ **vessel** Abwrackschiff *n*

**scrapper blade** Schaberklinge *f*

**scrapping** Schrott-geben *n*, -zugabe *f*, Ver-schrottung *f* ~ **floor** Schrottbühne *f* ~-**off** Korrosion *f*

**scraps of wire** Drahtabfall *m*

**scratch, to** ~ aufkratzen, (with scriber) einreißen, einritzen, einschneiden, (loud speaker) klirren, kratzen, ritzen, scharren, schürfen **to** ~-**mark** markieren **to** ~ **off** abkratzen, abritzen **to** ~ **out** auskratzen **to** ~ **up** aufkratzen

**scratch** Abschürfung *f*, Kratzen *n*, Kratzer *m*, Riß *m*, Ritz *m*, Ritze *f*, Ritzen *n*, Schleifriefe *f*,

Schramme *f*, Schrammen *n* ~ **awl** Markier-, Reiß-nadel *f* ~ **filter** Geräuschfilter *n*, Krach-filter *n* (phono) ~ **gauge** Streichmaß *n* ~ **hardness** Ritzhärte *f* ~-**hardness number** Ritz-härtezahl *f* ~-**hardness tester** Ritzhärteprüfer *m* ~ **marking** Streichmaß *n* ~ **method** Ritzver-fahren *n* ~ **noise** Schrammenrauschen *n* ~-**pad** Notizblock *m* ~ **test** Rißhärteprobe *f*, Ritz-härteverfahren *n*, Ritzprobe *f*

**scratched** geritzt, zerkratzt ~ **line** Markierungs-strich *m* ~ **slit** Ritzspalt *m*

**scratching** Kratzen *n* ~ **machine** Ritzmaschine *f* ~ **noise** Kratzgeräusch *n* ~ **stone** Abritzstein *m* ~ **test** Kratzprobe *f*

**scratchy noise** kratzendes Geräusch *n*

**scrawl, to** ~ abschmieren, kritzeln

**screed** Abgleich-, Abstreif-bohle *f*

**screeding** Strichschicht *f*

**screen, to** ~ abschirmen, aussieben, blenden, durchsieben, klassieren, maskieren, schirmen, sichten, sieben, verblenden, vergittern, (by smoke) verneblen **to** ~ **lights** abblenden **to** ~ **off** abfangen **to** ~ **out** abschirmen, absieben, aussieben **to** ~ **sand** Sand *m* durchwerfen

**screen** Ablesefenster *n*, Abschirmung *f*, Bild-feld *n*, -schirm *m*, Draht-gitter *n*, -netz *n*, Durchwurf *m*, Filter *n*, Gitter *n*, Kanzelle *f*, Leinwand *f*, Maske *f*, Maskierung *f*, Projek-tionswand *f*, Punktraster *m*, Raster *m*, Raster-platte *f*, Roentgenschirm *m*, Rost *m*, Schirm *m*, Schleier *m*, Schutz *m*, Schutzwand *f*, Sieb *n*, Wand *f*, Windschutzscheibe *f*, Wurfgitter *n* ~ **of the scanning tube** Leuchtschirm *m* der Ab-taströhre

**screen,** ~ **analysis** Siebanalyse *f* ~ **aperture** Ma-schenweite *f* ~ **area** Siebfläche *f* ~ **basket** Korb-sieb *n* ~ **beater mill** Siebschlagmühle *f* ~ **border** Schirmrand *m* ~ **brightness** Beleuchtungsstärke *f* (CRT) ~ **burning** Schirmeinbrennung *f* (CRT) ~ **by-pass capacitor** Schirmgitterblock *m* ~ **can** Abschirmtopf *m* (rdr) ~ **classifier** Siebplan-sichter *m* ~ **cloth** Siebgewebe *n* ~ **credit** Vor-spann *m* (film) ~ **cylinder** Abschirmzylinder *m* ~ **discharge ball mill** Siebkugelmühle *f* ~ **dissipation** Schirmgitterverlustleistung *f* ~ **dropping resistor** Schirmgitterwiderstand *m* ~ **edge** Schirmrand *m* ~ **efficiency** Schirmwir-kungsgrad *m* ~ **factor** Filterfaktor *m* ~ **filter** Spülspaltfilter *m* ~ **filter element** Siebfilterein-satz *m* ~ **grid** Anodenschutznetz *n*, (of tetrode) Beschleunigungsgitter *n*, Schirmgitter *n* (rdo), Schutzgitter *n*, (Entaschung) Trenngitter *n*

**screen-grid,** ~ **bias** Schirmgitterpotential *n* ~ **current** Schirmgittergleichstrom *m* ~ **dissipa-tion loss** Schirmgitterverlustleistung *f* ~ **output tube** Schirmgitterendröhre *f* ~ **power loss** Schirmgitterverlustleistung *f* ~ **tetrode** Schirm-gitter-, Vierpolschirm-röhre *f* ~ **tube** Raum-ladegitter-, Schirmgitter-, Vierpolschirm-röhre *f* ~ **tube with plate-current rectification** Schirm-gitterrichtverstärker *m* ~ **vacuum tube** Doppel-gitterröhre *f* ~ **valve** Schirmgitter-, Vierpol-schirm-röhre *f* ~ **voltage** Schirmgitterspannung *f*

**screen,** ~ **holder** Rasterkassette *f* ~ **hole** Sieb-öffnung *f* ~ **line distance** Rasterabstand *m* ~ **material** Leuchtschirmmaterial *n* ~ **measure-ment** Schirmmessung *f* ~ **meshing** Schirmgeflecht n ~ negative Rasternegativ n ~ opening Sieböffnung f ~ pattern photographed with outside camera Außenaufnahme f ~ pipe Filterrohr n ~ plate Rasterelement n ~ play Drehbuch n ~ printing Schablonen-, Sieb-druck n ~ recess Schutzgitternische f ~ riddle Vibrationssieb n ~ scan microscope Rastermikroskop n ~ segment Siebsegment n ~ separation Siebsichtung f ~ sizing Korn-, Sieb-klassierung f ~ trough Siebtrog m ~ voltage (or volts) Schirmgitterspannung f ~ wall Schirmwand f, Schutzmauer f ~-wall counter abgeschirmtes Zählrohr n ~ width Schirmbreite f ~ wire Maschendraht m ~ work Rasterarbeiten pl

screenage Abschirmung f

screened abgeschirmt, funkentstört, geschirmt, gesiebt, verdeckt ~-air intake valve abgeschirmtes Einlaßventil n ~ antenna abgeschirmte Antenne f, Gegengewichtsantenne f ~ cabin abgeschirmter Raum m ~ coal gesiebte Kohle f, Gruskohle f ~ conductor Schirmleiter m ~-conductor cable induktionsfreies Kabel n, Kabel n mit abgeschirmten Leitern ~ diaphragm (of microphone) Schirmmembrane f ~ down-lead abgeschirmte Niederführung f (rdo) ~ grid geschirmtes Gitter n, Schirmgitter n ~-grid tube Röhre f mit geschirmtem Gitter ~ grid valve Schutzgitterröhre f ~ ignition abgeschirmte Zündung f ~ magneto abgeschirmter Magnet m ~ ore Scheideerz n ~ oven coke Brechkoks m ~ pair abgeschirmte Doppelader f ~ pentode Hochfrequenzpentode f ~ plug vollentstörte Kerze f ~ producer coal Generatorstückkohle f ~ room abgeschirmter Raum m ~ spark plug abgeschirmte Zündkerze f ~ transformer abgeschirmter Transformator m

screening Abblendung f (photo), Abschirmen n, Abschirmung f, Abschützung f, Absieben n, Absiebung f, Funkenentstörung f, Klassierung f, Kornklassierung f, Panzerung f, Projizierung f, Rastern n, Schirmung f, Sichtung f, Sieberei f, Siebklassierung f, Siebsichtung f, Siebung f, Störabschirmung f, Tarnung f, Überprüfung f, Verfilmung f, Verschleierung f ~ of grain Korngrößentrennung f

screening, ~ agent Nebelstoff m ~ apparatus Verdunk(e)lungseinrichtung f ~ box Abschirmgehäuse n (rdr), Abschirmung f, Kasten m mit Schirmwänden, Schutzschirm m ~ cage Faradaykäfig m ~ can Abschirm-becken n, -büchse f ~ cap Entstörkappe f ~ capacity Siebleistung f ~ device Scheide-anlage f, -vorrichtung f, Sturzsieb n ~ drum Siebtrommel f ~ effect Abschirmungs-effekt m, -wirkung f, Raumladungseffekt m, Schirmwirkung f ~ effect of rail and cable-sheath currents Induktionsminderung f durch Schienenstrom und Kabelmantelstrom ~ frame Siebkranz m ~ harness Kabelgeschirr n für Störabschirmung ~ house Siebanlage f ~ machine Siebmaschine f ~ means Ausblendmittel n ~ metal braid Entstörgeflecht n ~ plant Separations-, Sieb-anlage f ~ plate Abschirm-, Sieb-blech n ~ refuse Siebabfall m ~ sample Siebprobe f ~ shield Abschirmung f ~ sleeve Abschirmschlauch m, Entstörmuffe f (electr.) ~ sphere Schirmeffektkugel f ~ stack (Kühlmaschine) Siebpaket n

~ test Siebprobe f ~ tube Blendenrohr n ~ wall Schirmwand f

screenings Ausgesiebtes n, Durchfall m, Feinkohle f, Koksgrus m, Schrenzpackpapier n, Siebdurchfall m, Siebrückstand m (von Pigmenten)

screens for securing the carbon rods Kohlenführung f aus Speckstein

screw, to ~ schrauben, verschrauben, zusammenschrauben to ~ down (nut) anziehen, einschrauben, (nut) festziehen, senken to ~-fasten festklampen to ~ into einschrauben to ~ off abschrauben, losschrauben to ~ on andrehen, anschrauben, aufschrauben, einschrauben, (up or tightly) festschrauben to ~ open aufdrehen to ~ in place einschrauben to ~ tight zuschrauben to ~ up anziehen, aufziehen

screw Bolzen m, Schraube f, (without nut) Schraube f ohne Mutter (endless) ~ Schnecke f (lead) ~ Spindel f to rub the pitch out of the ~ bung Fässer pl ausreiben

screw, ~ for adjusting horizon Horizontierschraube f ~ for anchorage in masonry Steinschraube f ~ for bearing cover Lagerdeckelschraube f ~ to clamp saddle to crossrail Supportfeststellbolzen m ~ for elevating crossrail Spindel f zum Heben des Querbalkens f ~ with head Kopfschraube f ~ with hexagonal recessed hole Innensechskantschraube f ~ for lowering lever Schraube f für Herablaßhebel ~ with pivot pin Mutter f mit Pinnenzapfen ~ for rimless mounting Glasbrillenklemmschraube f ~ with a triangular thread scharfgängige Schraube f ~ with undercut neck Halsschraube f ~ for wooden casing clamp Bündelschraube f

screw, ~ action Schraubwirkung f ~ anchor Schraubenklemme f ~ anchor Schraubenbrunnen m ~ arbor Werkzeugspindel f mit Gewindefutter, Gewindehandstück n ~ auger Schnecken-, Schrauben-, Spiral-bohrer m ~ axis Schraubenachse f ~ base Schraubenfuß m, Schraubsockel m ~ bit Schlangenbohrer m ~ blast (or blowing) machine Spiralgebläse n ~ body Schraubenkern m ~ bolt Schraubbolzen m ~ box Schrauben-büchse f, -docke f ~ brake Spindelbremse f ~ bushing Schraubenhülse f ~ cap Gewindestopfen m, Schraub-deckel m, -verschluß m, Schraubenkopf m, Überwurfmutter f, Verschluß-kappe f, -stück n, Verschraubung f ~ cap for filling hole Fülllochschraube f ~ channel Schneckenkanal m ~ chase Schraubenrahmen m ~ chuck Schraubenfutter n ~ clamp Schrauben-klemme f, -zwinge f, Schraubknecht m ~-clamp-shaped adjusting gauge schraubzwingenähnliche Einstellehre f ~ clip Schraubenklemme f ~ collar Ansatzfläche f des Objektivgewindes f, Ringmutter f ~ collar ring Überwurfring m ~ connection Schraubverbindung f ~ control (of recording cutter) Spindelführung f ~ conveyer Förderschnecke f, Schnecke f, Schneckenförderer m, Schraubentransporteur m, Transportanlage f mit Schraubengewinde ~-conveyer trough Schneckenförderrinne f ~ coupling Schraubenkupplung f, Verschraubung f ~ cover Schraubdeckel m ~ cutter Schraubenschneidemaschine f

**screw-cutting** Gewindeschneiden *n* ~ **attachment** Gewindeschneideinrichtung *f* ~ **die** Gewindeschneide-backe *f*, -backen *m* ~ **gear** Gewindeschneideeinrichtung ~ **gearbox** Gewindekasten *m* ~ **lathe** Gewindeschneidedrehbank *f*, Schrauben-drehbank *f*, -schneidbank *f* ~ **machine** Schraubenschneidemaschine *f* ~ **oil** Gewindeschneideöl *n* ~ **tool** Gewindeschneidestahl *m*
**screw,** ~ **diameter** Schneckendurchmesser *m* ~ **die** Gewinde-backe *f*, -eisen *n*, -schneideeisen *n*, Kluppenbacke *f*, Schneidbacken *m*, Schneideisen *n* ~ **die stock** Schneidkluppe *f* ~ **dislocation** Schraubenversetzung *f* ~ **displacement** Schraubung *f* ~ **dog** Schraubenführungklaue *f* ~ **dolly** Nietwinde *f* ~-**down gear** Anstellvorrichtung *f* (Walzwerk) ~-**down mechanism** Anstellmechanismus *m* ~ **drier** Schneckentrockner *m* ~ **drive** Bewegungs-, Transport-spindel *f*
**screwdriver** Schrauben-dreher *m*, -zieher *m* ~ **blade** Schraubenzieherklinge *f* ~ **capacity** Schraubleistung *f*
**screw,** ~ **elevator** Förder-rohr *n*, -röhre *f* ~ **extractor** Gewindezieher *m*, Schraubenausdreher *m* ~ **extruder** Schneckenpresse *f* ~ **eye** Ösenschraube *f* ~ **fan** Flügelmutter *f* mit Stoßarmen, Schraubenlüfter *m* ~ **fastening** Klemmverbindung *f* ~ **feeding** Schneckenaufgabe *f* ~ **gauge** Schraubenlehre *f* ~ **gear** Hebewinde *f* ~(-**reversing**) **gear** Schraubensteuerung *f* ~-**geared pulley** Schraubenflaschenzug *m* ~ **gearing** Schnecken-, Schrauben-gewinde *n* ~ **head** Schraubenkopf *m* ~ **hole** Schraubenloch *n* ~ **hook** Hankenschraube *f*, Schraubstollen *m* ~ **jack** Förderwerk *n*, Hebeschraube *f*, Schraubenblock *m*, -heber *m*, Spindel *f*, Stellschraube *f*, Wagenwinde *f*, Winde *f* ~ **joint** Gewindekupplung *f*, Schraubenverbindung *f* ~ **key** Schraubenschlüssel *m* ~ **lag** Grundbolzen *m* ~ **line** Schraubenlinie *f* ~-**locking device** Schraubensicherung *f* ~ **machine** Fassondrehbank *f* ~ **man** Walzensteller *m* ~ **microscope** Schraubenmikroskop *n* ~ **mooring** Schraubenanker *m* ~ **motion** Schraubung *f* ~ **neck** Schraubenführungshals *m* ~ **nipple** Gewindestutzen *m* ~ **nut** Mutterschraube *f* ~-**on lug** Anschraublappen *m* ~-**on surface** Anschraubfläche *f* ~-**on thread** Anschraubgewinde *n* ~ **pile** Schraubenpfahl *m* ~ **pinchcock** Schraubenquetschhahn *m* ~ (**ed**) **pipe joint** (**or coupling**) Rohrverschraubung *f* ~ **pitch** Abstand *m* (je) zweier Schraubengänge, Gewindehöhe *f* ~-**pitch gauge** Gewinde-kaliber *n*, -schablone *f*, -steigungslehre *f* ~ **plate** Gewinde-eisen *n*, -schneider *m*, Gewindeschneidebacke *f*, -backen *m*, -eisen *n*, -kluppe *f*, -werkzeug *n*, Kluppe *f*, Schneidkluppe *f*, Schraubeneisen *n* ~-**plate die** Kluppenbacken *m* ~ **plate set** Gewindeschneidgarnitur ~-**plate tap** Schneideisengewindebohrer *m* ~ **plug** Gewindestopfen *m*, -stöpsel *m*, Verschluß-pfropfen *m*, -schraube *f* ~ (**ed**) **plug** Schraubstollen *m* ~-**plug header** Rücklaufbüchse *f* mit Gewindezapfen ~-**plug washer** Verschlußschraubendichtung *f* ~ **point** (of a bit) Einziehschraube *f* ~ **post** Schraubpfahl *m* ~ **press** Prägepresse *f*, Preßschnecke *f*, Schraubenpresse *f*, Spindelpresse *f*, Stoßwerk *n* ~ **propeller** Flügelrad *n* ~ **pump** Schraubenpumpe *f* ~-**punching press**

**Schraubenstanze** *f* ~-**push starter** Schubschraubtreibanlasser *m* ~ **rack** Schraubenzahnstange *f* ~ **reed and pin** Mutter *f* mit Pinnenzapfen ~ **ring** Augenbolzen *m*, Gewindeöse *f* ~ **riveting machine** Schraubennietmaschine *f* ~ **rotation** Schraubung *f* ~ **seat of objective amount** Ansatzfläche *f* des Objektivgewindes ~ **shaft** Schraubenschaft *m* ~ **shoulder** Ansatzfläche *f* des Objektivgewindes ~ **slot milling machine** Schraubenschlitzfräsmaschine *f* ~-**slotting cutter** Schrauben(kopf)schlitzfräser *m* ~ **socket** Gewindebuchse *f* ~-**socket wrench** Schrauben- und Steckschlüssel *m* ~ **spanner** Schraubenschlüssel *m* ~ **spar coupling** Holmverschraubung *f* ~ **speed** Schneckendrehzahl *f* ~ **spike** Gewindedorn *m*, Schwellenschraube *f* ~ **spindle** Leitspindel *f* ~-**spindle steering gear** Schraubenspindelsteuerung *f* ~ **stay bolt** Schraubenstehbolzen *m* ~ **steel** Schraubeneisen *n*, -material *n* ~ **stirrer** Schraubenrührwerk *n* ~ **stock** Gewinde-, Schneid-kluppe *f*, Schrauben-kluppe *f*, -material *n* ~ **stop cap** (**or socket**) Schraubenanschlagsockel *m* ~ **take-up** (Förderband) Schraubenspannvorrichtung *f* ~ **tap** Gewinde-, Schrauben-bohrer *m* ~ **terminal** Befestigungsklemme *f*, Schraubklemme *f* ~ **thread** Bolzengewinde *n* (auf Drehbänken), Schraubengewinde *n*, Schraubengang *m*, -windung *f*
**screw-thread,** ~ **comparator** Gewindekomparator *m* ~ **gauge** Gewindelehre *f* ~ **measuring** Gewindemessung *f* ~ **micrometer calipers** Gewinde-Schraublehre *f* ~ **runout** Gewindeauslauf *m* ~ **standards** Gewindenormen *pl*
**screw,** ~ **threading die** Schraubengewindeschneideisen *n* ~ **tieplate** Schraubenspannplatte *f* ~ **trimmer** Schraubentrimmer ~ **trunk** Schraubenbrunnen *m* ~-**tubes** Lochpfeifen *pl*
**screw-type,** ~ **compressor** Schraubengebläse *n* ~ **conduit fitting** Kabelverschraubung *f* ~ **core** Schraubkern *m* ~ **fan** Schraubenlüfter *m* ~ **lock** Schraubenverschluß *m* ~ **obstacle picket** Hindernisschraubpfahl *m* ~ **spindle** Feinzustellung *f* ~ **valve** Schraubventil *n*
**screw,** ~ **union** Einschraubstutzen *m* ~ **well's grating** Brunnengräting *f* ~ **wheel** Flügel-, Schrauben-, Schnecken-rad *n* ~ **wrench** Schraubenschlüssel *m*
**screwed** geschraubt, mit Gewinde *n* versehen ~ **down** zusammenklappbar ~ **in** eingeschraubt ~ **on** angeschraubt ~ **and shrunk on** warm aufgeschraubt und geschrumpft ~ **with spring action** elastisch verschraubt
**screwed,** ~ **blades** aufgeschraubte Messer *pl* ~ **bonnet shut-off valve** Kopfstück-Absperrventil *n* ~ **bush** Gewindebuchse *f* ~ **bushing** Gewindebuchse *f* ~ **cap** Schraubkappe *f* ~ **dowel** Schraubendübel *m* ~ **ends** verschraubte Enden *pl* ~ **filler cap** Einfüllverschraubung *f* ~ **flange** Gewindeflansch *m* ~ **gland** Verschraubungsstopfbüchse *f* ~ **hinge pin** Scharnierschraube *f* ~ **hose connection** Schlauchverschraubung *f* ~ **jack spur** Anschlagsporn *m* ~ **joint** Verschraubung *f* ~ **male** Einschraubzapfen *m* ~ **nut** Schraubenmutter *f* ~-**on cover** Schraubdeckel *m* ~ **pin** Gewindestift *m* ~ **pipe connection** Anschlußgewinde *n* für Röhren ~ **pipe joint** Schraubverbindung *f* ~ **pipeline** Gewinderohr-

leitung *f* ~ **plug** Mundlochdeckel *m*, Schraubendübel *m* ~ **reducing nipple** Zwischennippel *n* ~
**rivet** Scharnierschraube *f* ~ **rod** Gewinde-
stange *f* ~ **sleeve (or socket)** Gewindemuffe *f* ~
**spigot** Führungszapfen *m* mit Gewindeansatz
~ **spindle** Gewindespindel *f* ~ **steel ferrule**
Schraubennippel *m* ~ **stop pin** Schaftschraube *f*
~ **test bar** Probestab *m* mit Gewindekopf ~
**tube** Gewinderohr *n*
**screwing** Gewindeschneiden *n* (mit dem Schneideeisen), Verschraubung *f* ~ **attachment** Gewindeschneideeinrichtung *f* ~ **die** Schrauben-
gesenk *n* ~ **down** Senken *n* ~ **equipment** Gewindeschneideeinrichtung *f* ~ **plate** Gewinde-
schneidplatte *f* ~ **tap** Schneidgewindebohrer *m*
**scribble, to** ~ abschmieren, kritzeln, schrubbeln
**scribbler card** Grobkrempel *m*, Schrubbelmaschine *f*, Vorkrempel *m*
**scribe, to** ~ anreißen, anzeichnen mit Reiß-
nadel, vorreißen
**scribed line (or mark)** eingeritzte Marke *f*
**scriber** Anreißnadel *f*, Markiernadel *f*, Metallbeschrifter *m*, Reißnadel *f*, Stahlgriffel *m*
**scribing,** ~ **awl** Reißspitze *f* ~ **block** Flächen-
lehre *f*, Reißstock *m*, Winkelstreichmaß *n* ~
**iron** Markierungseisen *n* ~ **needle** Markiernadel *f* ~ **point** Reißnadel *f*
**scrimp rollers** Ausstreichwalzen *pl*
**script** Manuskript *n*, Schreibschrift *f*
**scroll** Schnecke *f*, Schnörkel *m*, Spirale *f* ~ **case**
Spirale *f* ~ **chuck** Dreibackenfutter *n*, selbstzentrierendes Spannfutter *n* ~ **saw** Hubsäge *f*
(Decoupiersäge)
**scroop, to** ~ knirschen, kratzen
**scrooping agent** Aviviermittel *n*
**scrub, to** ~ abtreiben, berieseln, scheuern,
schrubben, schruppen, waschen **to** ~ **out**
herauswaschen
**scrubbed solvent** gereinigtes Lösungsmittel *n*
**scrubber** Berieselungsturm *m*, Naßreiniger *m*,
Rieseler *m*, Rieselturm *m*, Skrubber (gas) *m*,
Wäscher *m*
**scrubbing** Berieselung *f*, Schruppen *n* ~ **process**
Wasch-prozeß *m*, -vorgang *m* ~ **tower** Berieselungsturm *m*, Rieselturm *m*, Wäscher
*m*
**scrutinize, to** ~ bis ins einzelne erforschen, eingehend prüfen, untersuchen
**scrutinizing** genaue Untersuchung *f*
**scrutiny** Nachforschung *f*
**scud, to** ~ lenzen
**scud** Bö *f*, Eilen *n*, Laufen *n*, Lenzen *n*, Wolken-
fetzen *pl*
**scuff** Abnutzung *f*
**scuffing** Fressen *n* ~ **tendency** Freßneigung *f*
(metall.)
**sculling** Wriggeln *n*
**sculpture, to** ~ aushauen, ausmeißeln, gravieren,
schnitzen, skulpturieren
**sculpture** Bildhauerei *f*, Plastik *f*, Skulptur *f* ~
**scum, to** ~ abschäumen, abstreichen, schlacken
**scum** (Ab)Schaum *m*, Abstrich *m*, Abzug *m*,
Ausschlag *m*, Schlacke *f* ~ **cake** Schlamm-
kuchen *m* ~ **juice** Schlammsaft *m* ~ **mixture**
Schaumgemisch *n* ~ **pile** Schlammberg *m* ~
**plant** Abzugswerke *pl* ~ **riser** Schaum-kopf *m*,
-trichter *m*, (in casting) Schlackenkopf *m* ~
**thickener** Schlammeindicker *m* ~ **washing**

(sugar mfg.) Schlammabsüßen *n* ~ **works** Abzugswerke *pl*
**scummer** Abstreichlöffel *m*
**scumming pan** Abschäum-pfanne, *f*, -wanne *f*
**scupper** Überlauf-sammelschale *f*, -stutzen *m*
**scuppers** Speigatten *pl*
**S-curve** Haltepunktskurve *f*
**scutch, to** ~ entwirren
**scutch** Flachs-messer *n*, -schwingmaschine *f*,
Schwingwerg *n*, Ziegelhammer *m*
**scutched hemp** Strähnenhampf *m*
**scutching machine** Schwingmaschine *f*
**scuttle, to** ~ anbohren, durch Anbohren versenken, versenken
**scuttle** Bütte *f*, Dachluke *f*, Kiepe *f*, Korb *m*,
Luke *f*, Niedergang *n*, Spritzbrett *n*, Stirnwand
*f* ~ **air bleed** Luftöffnung *f*
**scythe** Hippe *f*, Schleifsense *f*, Sense *f* ~ **anvil**
Dengelamboß *m* ~ **hammer** Dengelhammer *m*
~-**sharpening appliance** Dengelgerät *n* ~ **stick**
Sensenbaum *m*
**S-distortion** S-Verzeichnung *f*
**sea** Meer *n*, See *f* **to be ready to go to** ~ seeklar
sein ~ **of clouds** Wolkenmeer *n* **above** ~ **level**
über dem Meeresspiegel *m* **below** ~ **level** unter
dem Meeresspiegel *m* ~ **of light** Licht(er)meer
*n* ~ **of secondary importance** Nebenmeer *n*
**sea,** ~ **air** Seeluft *f* ~ **anchor** See-, Treib-, Wasseranker *m* (aviat.) ~ **annual** Seehandbuch *n* ~
**area** Meeresgebiet *n* ~ **bed** Meeresgrund *m*
~ **board** Küstenstrich *m* ~ **breeze** Seebrise *f* ~
**channel** Seekanal *m* ~ **chart** hydrografische
Karte *f*, Paßkarte *f*, Seekarte *f* ~ **clutter** See-
gangs-echo *n*, -reflex *m* (rdr) ~ **coal** Steinkohle
*f* ~ **coast** Meeres-, See-küste *f*, Strandlinie *f* ~
**cock** Boden-, Bord-ventil *n*, Seehahn *m* ~ **gauge**
Tiefenmesser *m*
**seagoing,** ~ **aircraft** Hochseeluftfahrzeug *n* ~
**plane** Überseeflugzeug *n*
**sea,** ~ **inlet** Meeresarm *m* ~ **level** Meeres-höhe *f*,
-niveau *n*, -spiegel *m*, (mean) Normalnull *n*,
Seehöhe *f*, Wasserstand *m* ~ **level in storms**
Sturmflut *f*, Wasserstand *m* von Sturmfluten
~-**level altitude** Meeresspiegelhöhe *f* ~-**level
canal** Seekanal *m* ~-**level equivalent of power**
ideelle Bodenleistung *f* ~-**level-rated engine**
Bodenmotor *m* ~ **lights** Seefeuer *n* ~ **lock** Seeschleuse *f* ~ **mark** Seezeichen *n* ~ **meteorological service** Ozeanwetterdienst *m* ~ **mine** See-
mine *f* ~ **plane** Flugboot *n*, Wasserflugzeug *n* ~
**port** Hafenstadt *f*, Seehafen *m* ~ **quake** Seebeben *n* ~ **return(s)** Seegangreflexe *pl*, See-
rückleitung *f*, Seezeichen *pl* (rdr) ~ **sand**
Meeres-, See-, Silber-sand *m* ~ **shore** Meeres-
küste *f* ~ **side town** Küstenstadt *f*, Seestadt *f* ~
**voyage** Seefahrt *f* ~ **wall** Damm *m*, Hafen-
damm *m*, Strandmauer *f*, wellenbrechendes
Uferschutzwerk *n* ~ **watch** Seewache *f* ~
**water** Meer-, See-wasser *n* ~-**water cooling**
Seewasserkühlung *f* ~ **wave** Meereswelle *f* ~
**wax** Meerwachs *n* ~ **wind** Seewind *m* ~ **wing**
Schwimmerstummel *m* ~ **wings** Gleitflossen-
stabilisator *m*, Seitenflossenkippsicherung *f* ~
**worthiness** Seefähigkeit *f*, Seetüchtigkeit *f*
~-**worthy** seefest, seetüchtig ~-**worthy packing**
seemäßige Verpackung *f*, Seeverpackung *f*
**seal, to** ~ abdichten, abschließen, absperren,
dichten, (glass) einschmelzen, isolieren, plom-

bieren, verdichten, verkitten, verschließen, versperren, zumachen, zuschmelzen **to ~ in** einschließen, (a wire or lead in a tube) einschmelzen **to ~ legally** gerichtlich versiegeln **to ~ off** dicht abschließen

**seal** Abschluß *m*, Abschluß-schild *n*, -vorrichtung *f*, Dichtung *f*, (spring-water) Dichtungsabschluß *m*, Einschmelzstelle *f*, Kapsel *f*, Plombe *f*, Siegel *n*, Versager *m*, Verschluß *m*, Verschlußstück *n* **~ box** druckdichte Durchführung *f* **~ connection to dipole** Sperrtopf *m* **~ drain** Leckablaßleitung *f* **~ housing** Schlußring *m* **~ retainer** Dichtungshalter *m* **~ tank** Fallwasserkasten *m* **~ wire and lead** Plombendraht *m* und Bleiplombe *f*

**sealant** Isoliergrund *m*

**sealed** abgedichtet, abgeschlossen, dicht, isoliert, luftdicht verschlossen, vakuumdicht **~ assembly rolling process** Walzschweißverfahren *n* **~ bellows** Aneroid *n* **~ cable end** Kabelstumpf *m* **~ chamber terminal** Kabelendverschluß *m* **~ combustion chambers** geschlossene Brennräume *pl* **~ contact joints** Berührungsdichtungen *pl* **~ end of a line** isoliertes Ende *n* einer Leitung **~-in wire** Einschmelzdraht *m* **~-off tube** abgeschmolzene Röhre *f* a **~ offer** ein versiegeltes Angebot *n* **~ rectifier** pumpenloses Ventil *n* **~ tender** versiegeltes Angebot *n* **~ tube** Einschlußröhre *f* **~-tube furnace** Schießofen *m*

**sealing** Abdichten *n*, Abdichtung *f*, Dichten *n*, Einschmelzung *f* (electr.), Verschließen *n* **~ a spring** Dichten *n* einer Quelle

**sealing, ~ band** Dichtlinie *f* **~ bushing** Abdichtbuchse *f* **~ chamber** Vergußkammer *f* **~ composition** (for joint boxes) Füllmasse *f* **~ compound** Kabelvergußmasse *f*, Vergußmasse *f* **~ cone** (E-Ofen) Tauchglocke *f* **~ cord** Dichtungsstrick *m* **~ cover** Dichtungsdeckel *m* **~ device** Plombiervorrichtung *f* **~ diaphragm** Abschlußmembran *f* **~ disc** Dichtscheibe *f* **~ end** Endverschluß *m* **~ end of cable** Ausgießen *n* des Kabels **~ eyes** für Versiegeln vorgesehene Löcher *pl* **~ felt** Abdichtfilz *m* **~ ferrule** Dichtungskegel *m* **~ filler** Vergußmasse *f* **~ flange** Verschlußflansch *m* **~ gap** Dichtungsspalte *f* **~ head** Verschließkopf *m* **~-in** Befestigung *f* im Mauerwerk **~-in contact** Selbstspeisekontakt *m* **~-in machine** Einschmelzmaschine *f* **~-lacquer** Isoliergrund *m* **~ ledge** Dichtungsleiste *f* **~ liquid** Absperrflüssigkeit *f*, Sperrflüssigkeit *f* **~ machine** Kapsel-, Zuschmelz-maschine *f* **~ material** Dichtmasse *f* **~ medium** Abdichtmasse *f* **~ member** Manschette *f* **~ nipple** Dichtnippel *n* **~-off burner** Abschmelzbrenner *m* **~ plate** (gasket) Dichtplatte *f*, Verschlußblech *n* **~ point** Dichtungsstelle *f* **~ ram** Verschlußstempel *m* **~ range** Verschließbereich *m* (electr.) **~ rims** Dichtränder *pl* **~ ring** Dicht-, Abdichtungs-, Dicht-ring *m*, **~ ring washer** Dichtungsring *m* **~ rope** Dichtungsstrick *m* **~ run** Kappnaht *f* **~ screw** Dichtungsschraube *f* **~ sleeve** Dichtmanschette *f* **~ strip** Dichtgummi *n*, Dichtungsschnur *f* **~ surface** Dichtfläche *f* **~ tape** Dichtungsstreifen *m* **~ thread** Plombenfaden *m* **~ tip** Verschlußspitze *f* **~ unit** Verschließeinrichtung *f* **~ varnish** Dichtungs-, Plombier-lack *m* **~ voltage** Verschließspannung *f* **~ washer** Dichtring *f* **~ wax** Siegel-

lack *m* **~ weld** abdichtende Naht *f* **~ yoke** Dichtbrücke *f*

**seam, to ~** abbinden, nähen, säumen, sicken, Metall *n* durch Schweißnaht verbinden **to ~-weld** nahtschweißen

**seam** Ader *f*, Bart *m*, Bruchstelle *f*, Falz *m*, Faser *f*, Flöz *n* (geol.), Fuge *f*, Gang *m* (min.), Grat *m*, Lötstelle *f*, Naht *f*, Nahtkante *f*, Ritz *m*, Saum *m*, Schicht *f*, (reinforcing) Sieke *f*, Spalt *m*, Sprung *m*, Trum *n* **~ of the plates** Naht *f* der Platten

**seam, ~ hammer** Seckenhammer *m* **~ roller** Sickenwalze *f* **~ size** Nahtstärke *f* **~ welder** Nahtschweißmaschine *f* **~ welding** Nahtschweißung *f* **~ welding roller** Kontaktrolle *f*

**seaming, ~ die** Falzgesenk *n* **~ lace** Abschlußborte *f* **~ machine** Verschließmaschine *f* **~-on** Auffalzen *n*

**seamless** nahtlos **~ drawn tube** nahtlos gezogenes Rohr *n* **~ lining** fugenlose Ausmauerung *f* **~ pipe** nahtlos gezogenes Rohr *n* **~ sheathing** nahtloser Mantel *m* **~ sockets** gezogene Hülsen *pl* **~ steel spigot and faucet tubes** nahtlose Stahlmuffenröhre *pl* **~ steel tube** nahtloses Stahlrohr *n* **~ tube** nahtloses Rohr *n*

**sear** Abzugs-stange *f*, -stollen *m* **~ bar** Auslösestange *f* **~ nose** Abzugsstollen *m* **~-release lever** Druckstück *n* **~ spring** Zugfeder *f*

**search, to ~** abtasten, aufsuchen, durchsuchen, forschen, schürfen, suchen, untersuchen **to ~ the grounds** das Gelände absuchen

**search** Absuchen *n*, Aufsuchen *n*, Erkundigung *f*, Nachforschung *f*, Prüfung *f*, Schürfen *f* **in ~ of** auf der Suche *f* (nach)

**search, ~ coil** Auffangspule *f*, Forschungsspule *f*, Kopplungsschleife *f*, Maßspirale *f* (rdr), Meßspule *f*, Prüfspule *f*, Suchspule *f* **~ electrode** Sonde *f* **~ installation** Suchanlage *f* **~ lamp** Suchscheinwerfer *m*

**searchlight** Lampe *f*, Scheinwerfer *m*, Suchlicht *n* **~ beam** Lichtkegel *m* **~ control pillar** Richtungswerfer *m* **~ density** Scheinwerferdichte *f* **~ equipment** Scheinwerfergerät *n* **~ lane** Scheinwerferstraße *f* **~ mechanic** Lampenwart *m* **~ plane** Scheinwerferflugzeug *n* **~ platform** Scheinwerferstand *m* **~ section (or squad)** Beleuchtungstrupp *m* **~ station** Scheinwerferstellung *f* **~ trailer** Beleuchtungs-anhänger *m*, -anlage *m* **~ wagon** Beleuchtungswagen *m*

**search, ~ lighting** Zielanleuchtung *f* durch Radar **~ plane** Suchebene *f* **~ radar** Suchradar *m* **~-radar tracking** Flugzielverfolgung *f* (mit Suchradar) **~ receiver** Suchempfänger *m* **~ stopper** Suchstoppimpuls *m* **~ track(ing) radar** Such- und Verfolgungsradar *n*

**searcher** Schürfer *m*, Sucher *m*, (blacksmith) Wirkeisen *h*

**searching** Schurf *m* **~ gate** Suchtor *n* **~ instrument** Suchgerät *n* **~ iron** Brenneisen *n* **~ sector** Beobachtungsraum *m*, Suchsektor *m* **~ time** Suchzeit *f*

**season, to ~** ablagern lassen, (wood) ablagern, auswintern, auswittern, (wood) trocknen, würzen **to (super-) ~** altern **to ~ in summer** (tile) durchsommern **to ~ in winter** (tile) durchwintern, wintern **to ~ wood** Holz *n* austrocknen lassen **to ~ wood artificially** Holz *n* dörren

season Jahreszeit *f*, Saison *f* ~ **crack** Altersriß *m*, Alterungsriß *m* ~ **ticket** Zeitkarte *f*
**seasonable** saisonabhängig, saisonbedingt
**seasonal,** ~ **inflow (or influence)** Saisoneinfluß *m* ~ **load curve** Belastungsgebirge *n* ~ **road** Sommerweg *m* ~ **tariff** Saisontarif *m* ~ **variations** jahreszeitliche Änderungen *pl*, jährliche Schwankungen *pl*
**seasoned** (wood) abgelagert, erfahren, gelagert, getrocknet ~ **paper** abgelagertes (reifes) Papier *n*
**seasoning** Ablagerung *f*, Austrocknen *n*, Auswitterung *f*, (a gauge block) Entspannung *f*, Trocknung *f*, Würze *f* (**super**) ~ Altern *n*, Alterung *f* ~ **of timber** Ablagern *n* des Holzes ~ **kiln** Trockenofen *m*
**seat, to** ~ auf-setzen, -legen, sitzen
**seat** Auflage *f*, Auflagefläche *f*, Auflager *n*, Fundament *m*, Paß-fläche *f*, -sitz *m*, Pfanne *f*, Platz *m*, Sattelsitz *m*, Schienen-Stuhllager *n* (r. r.), Sitz *m*, Sitz-brett *n*, -fläche *f*, (of valve) Teller *m* ~ **of generation** Entstehungsort *m* ~ **for loose poises** Gewichtsschale *f* ~ **of magnetic field (or flux or induction)** magnetische Belegung *f* ~ **of piston** Sitz *m* des Kolbens ~ **of production** Entstehungsort *m*
**seat,** ~ **adjustable in flight** im Fluge verstellbarer Sitz *m* ~ **adjustment** Sitzverstellung *f* ~ **attachment** Sitzvorrichtung *f* ~ **bracket** Sitzstütze *f* ~ **bracket assembly** Sitzabstützung *f*, Sitzaufhängung *f* ~ **cushion** Sitzpolster *n* ~ **frame** Sitzrahmen *m* ~ **key** Sitzkeil *m* ~ **pack parachute** Sitzkissenfallschirm *m* ~ **pan** Sitzwanne *f* ~ **raising handle** Sitzhöherverstellhebel *m* ~ **reservation system** Platzbuchungsanlage *f* ~ **stone** Gesenk *n* ~**-type parachute** Sitzkissenfallschirm *m*
**seating** Lagerung *f*, Sitzfläche *f* ~ **of projectile** Geschoßführung *f*
**seating,** ~ **accommodation** Sitzgelegenheit *f* ~ **area** Auflagerfläche *f* ~ **arrangement** Sitzanordnung *f* ~ **capacity** Fassungsraum *m*, Sitzzahl *f* ~ **shoulder (or ledge or rim)** Auflageleiste *f* ~ **(or attachment) stud** Haltezapfen *m* (Stift *m* zur Aufnahme des X-Tubus)
**sebaceous** talgartig
**sebacic** fettsauer ~ **acid** Fettsäure *f*, Sebacinsäure *f*, Talgsäure *f*
**secant** Schnittlinie *f*, Sekans *m*, Sekante *f*
**secede, to** ~ abfallen
**seclusion** Abgeschlossenheit *f*
**secohm** Henry *n*
**second** Beiton *m* (acoust.), das Folgende *n*, das Nächste *n*, Sekunde *f*, das Untergeordnete *n*, Waren *pl* zweiter Güte, zweit(e, er, es) zweiter Gang *m* (Auto), zweite Güte *f* oder Qualität *f* **per** ~ sekundlich ~ **in command** Stellvertreter *m* ~ **to the power** hoch zwei (z. B. 10²)
**second,** ~ **accelerator** Beschleunigungsanode *f*, zweite Anode *f* ~ **annealing** zweite Glühung *f* ~ **anode** Beschleunigungsanode *f*, zweite Anode *f* ~ **boiling-off bath** Repassierbad *n* (Seide) ~ **channel frequency** Spiegelfrequenz *f* ~ **cheek stone** Anfangsstein *m* ~**-class seed** Nachzuchtsamen *m* ~ **coat** (on wall) Aufzug *m* ~**-cut file** Halbschlichtfeile *f* ~ **detector** Audion *n*, Gleichrichterstufe *f*, zweiter Detektor *m*, Zwischenfrequenzgleichrichter *m* ~

**etching** Mittelätzung *f* ~ **growth** Nachwuchs *m* ~ **gun** Beschleunigungsanode *f*, zweite Anode *f*
**second-hand** gebraucht, indirekt, sekundär erworben, übernommen, aus zweiter Hand ~ **bookseller** Antiquar *m* ~ **bookshop** Antiquariat *n* ~ **car** Gebrauchtwagen *m* ~ **stock** aufgearbeiteter Bestand *m*
**second,** ~ **harmonic** zweite Harmonische *f* ~ **lens** Hauptsammellinse *f* (CRT) ~ **line finder** zweiter Anrufsucher *m* ~ **mashing** Aufmaischen *n* ~ **member of an equation** rechte Seite *f* einer Gleichung ~ **mold of vellum** Dünnquetsche *f* ~**-motion shaft** Gegenwelle *f* ~ **operations** Zweitarbeitsgänge *pl* ~ **power** Quadrat *n* ~ **preselector** zweiter Vorwähler *m* ~ **press proof** Nachsehbogen *m* ~ **product** Nachprodukt *n* ~ **quantum number** azimutale Quantenzahl *f* ~ **rate** zweitklassig ~ **state creep** sekundäres Kriechen *n* ~ **time around echo** Überreichweitenecho *n* (rdr) ~**-trace echoes** Sekundärechos *pl* ~**-washpot compartment** Durchführkessel *m* (metall.) ~ **wire** Meßader *f* ~ **working** Widerdruck *m*
**secondarily** sekundär
**secondary** Sekundärseite *f*; indirekt, nachfolgend, sekundär, untergeordnet, zusätzlich **of** ~ **importance** nebensächlich, zweitrangig
**secondary,** ~ **action** Begleiterscheinung *f* ~ **air** Bei-, Misch-, Neben-, Zusatz-, Zweit-luft *f* ~**-air blower** Zweitluftgebläse *n* ~ **air fan** Zweitluftventilator *m* ~ **air mixer** Mischflosse *f* ~ **alternating current** Sekundärwechselstrom *m* ~ **aluminum** Umschmelzaluminium *n* ~ **amplifier** Sekundärverstärker *m* ~ **arm of a canal** Zweigkanal *m* ~ **axis** Nebenachse *f* ~ **battery** Ladungssäule *f*, Neben-, Sammler-, Sekundär-batterie *f* ~ **beam** Nebenbalken *m* ~ **beams** Nebenzipfel *pl* (rdo) ~ **bond** Nebenbindung *f* ~ **burst** Sekundärdurchbruch *m* ~ **cell** Sammlerzelle *f*, Sekundär-batterie *f*, -element *n* ~ **cellulose acetate** Cellit *n* ~ **charge** Zweitpatrone *f* ~ **circuit** Nebenstromkreis *m*, Sekundär(strom)kreis *m*, Zweitkreis *m* ~ **circuit reception** Zwischenkreisempfang *m* ~ **clock** Nebenuhr *f* ~ **clock of an electric-clock installation** Nebenuhr *f* einer elektrischen Uhrenanlage ~ **coil** Ausgangswicklung *f*, Oberspannungswicklung *f*, Sekundärspule *f* ~ **cold front** sekundäre Kaltfront *f* ~ **color** Mittelfarbe *f* ~ **column** Nebenkolonne *f* ~ **combustion chamber** Nachbrennkammer *f* ~ **compression** Nachverdichtung *f* ~ **condition** Nebenbedingung *f* ~ **conductivity** strukturempfindliche Leitfähigkeit *f* ~ **constituent of mixture** Nebengemengteil *m* ~ **consumer** Nebenverbraucher *m* ~ **control surface** sekundäre Ruderfläche *f* ~ **cooler** Schlußkühler *m* ~ **creep** sekundäres Kriechen *n* ~ **crusher** zweiter Steinklopfer *m* ~ **crystallization** sekundäre Kristallisation *f* ~ **current** Sekundärstrom *m* ~ **cyclone** Rand-, Teil-tief *n* ~ **deflection** Nebenausschlag *m* ~ **depression** Rand-, Teil-tief *n* ~ **diagonal wiring** Feldverspannung *f* ~ **dike** Schlickfänger *m* ~ **discharge** Nebenentladung *f* ~ **distribution** Unterverteilung *f* ~ **drive** Abtrieb *m* ~ **ear** Nebenmaximum *n* ~ **effect** Nach-, Nebenwirkung *f* ~**-effect phenomenon** Nachwirkungserscheinung *f* ~ **effort** Nebenangriff *m* ~ elec-

**trode** Nebenelektrode f ~ **electron** Sekundärelektron n ~ **electron multiplier** Sekundärelektronenverstärker m ~ **emission** Sekundär-emission f, -strahlung f
**secondary-emission, ~ amplification** Sekundäremissionverstärkung f ~ **current** Sekundäremissionsstrom m ~ **factor** Sekundäremissionsfaktor m ~ **multiplier** Sekundäremissionsvervielfacher m ~ **noise** Sekundäremissions-Rauschen n ~ **rate** Sekundäremissionsausbeute f
**secondary, ~ emitting** sekundäremittierend ~ **exchange** Knotenamt n ~ **face** Nebenfläche f ~ **flow** Nebenströmung f ~ **fluorescence** Sekundärfluoreszenz f ~ **frequency** Ausweichwelle f, Neben-, Zweit-frequenz f ~ **gear** Nebenzahnrad n ~ **gearbox** Sekundärgetriebe n ~ **guide** Nebenleitung f ~ **hardness** Anlaß-, Sekundärhärte f ~ **ingredient** Nebenbestandteil m ~ **interference radiator** Störsekundärstrahler m ~ **line** Lokalbahn f, Nebenlinie f, Sekundärleitung f ~**-line switch** Mischwähler m ~ **line switch** zweiter Anrufsucher m oder Vorwähler m ~ **lobe** Nebenmaximum n ~ **low** sekundärer Tiefdruck m, Teiltiefdruck m (aviat.) ~ **meaning** Verkehrsgeltung f ~ **member** sekundäres Glied n ~ **metal** Altmetall m ~ **natural radionuclides** sekundäre natürliche Radionuklide pl ~ **objective** Nebenabsicht f ~ **operation** Nebenoperation f ~ **phenomenon** Begleiterscheinung f ~ **pipe** (in an ingot) sekundärer Lunker m ~ **plane** zweiter Hauptschnitt ~ **potential** Sekundärspannung f ~ **process** Nebenbetrieb m ~ **proof** Nebenbeweis m ~ **protection tube** äußeres Schutzrohr n ~ **radar** Sekundärradar n ~ **radiation** Nebenmaximum n, Seitenstrahlung f, Sekundärstrahlung f ~ **radiator** Sekundärstrahler m ~ **radio station** Ersatzfunkstelle f ~ **railway** Sekundärbahn f ~ **reaction** Folge-, Neben-reaktion f ~ **reception** Sekundär-, Zweikreis-empfang m ~ **repeater station** Nebenverstärkeramt n ~ **resistance** Sekundärwiderstand m ~ **road** Nebenstraße f ~ **rock** Flözgebirge n ~ **searchlight** Nebenscheinwerfer m ~ **squall front** Böenfront f ~ **stall** erneuter Strömungsabriß m ~ **storage** Außenspeicher m, externer Speicher m, Fremdspeicherung f ~ **stress** Neben-, Sekundär-spannung f ~ **structure** sekundärer Aufbau m, sekundäre Struktur f, Sekundärstruktur f ~ **target** Ausweich-, Neben-ziel n ~ **tone association** Phonismus m ~ **tones** Nebentöne pl ~ **union** Nebenbindung f ~ **valency** Nebenvalenz f ~ **voltage** Sekundärspannung f ~ **winding** Ausgangs-, Oberspannungs-wicklung f, Sekundär-seite f, -wicklung f, Zweitwicklung f ~ **wire** Hochspannungsdraht m
**seconds** Blech n 2. Güte ~ **control** Sekundenkontrolle f ~ **counter** Sekundenzähler m
**secrecy** Geheimhaltung f, Geheimnis n ~ **relay** Geheimschaltungsrelais n
**secret** Chefsache f, Geheimnis n; geheim, geschlüsselt ~ **attachment for teleprinters** Geheimzusatz m ~ **ballot** geheime Abstimmung f ~ **controlling counter** Geheimprüfzähler m ~ **item** geheimer Gegenstand m ~ **language** geheime Sprache f ~ **lock** geheime Tastenverriegelung f (typewriter) ~ **order** Geheimver-

fügung f ~ **text** verschlüsselter Text m ~ **writing** Geheimschrift f
**secretary** Schriftführer m, Sekretär m, Sekretärin f
**secrete, to** ~ absondern, ausscheiden, verheimlichen
**secretion** Absonderung f
**sectile** schneidbar
**section, to** ~ ab-, ein-, unter-teilen, querzerteilen
**section** Abbaufeld n, Absatz m, Abschnitt m, Abteilung f, Aussparung f (aviat.), Bruchteil m, Durchschnitt m, Fasson f, Feld n, Gegend f, Glied n, Gruppe f, Halbzug m, Kapitel n, Kettenglied n erster Art, (of line) Linienabschnitt m, Profil n, Querschnitt m, Schnitt m, Schnittfläche f, Schrank m. Segment n, Sektion f, Stockwerk n (geol.), Strecke f, Streckung f, Teil m, Wandstärke f **in** ~**s** gruppenweise
**section, ~ of the body** Rumpfquerschnitt m ~ **of circuit** Leitungsabschnitt m ~ **of a city** Stadtteil m ~ **of core** Kerndurchschnitt m ~ **of earthworks** Massenprofil n ~ **of float** Schwimmerausschnitt m ~ **of groove** Kaliberteil m ~ **of line** Leitungsabschnitt m ~ **of a line** Teilstrecke f ~ **of multiple** Gruppenschaltung f der Verbindungsleitungen ~ **of a network** Kettenleitermasche f ~ **of pole** Gestängestück n ~ **of the populace** Volksschicht f ~ **on which principal stress occurs** Hauptschnitt m ~ **of a recurrent structure** Teilvierpol m ~ **of rim** Kranzquerschnitt m ~ **of shaped steel** Formeisenprofil n ~ **of skew gear on transverse section** Tangentialprofil n des Schraubrades ~ **of a sphere** Kugelausschnitt m ~ **of structure** Bauglied n ~ **of a town** Stadtteil m ~ **of a valley** Taleinschnitt m (geol.) ~ **of vibration** Schwingungsabschnitt m ~ **of worm gear on transverse section** Stirnprofil n des Schraubrades
**section, ~ block** Durchgangsblockwerk n ~ **block and blocking** (traction) Blockstrecke f ~ **blocking** Streckenblock m (r. r.) ~ **bunging apparatus** Kolonnenspundapparat m ~ **chamber** Teilkammerkessel m ~ **circuit (switch) breaker** Streckenschalter m ~ **cutter for grain examination** Farinatom n ~ **elevation on center line** Längsschnitt m auf der Mittellinie m ~ **groove** Formkaliber n ~ **insulator** Streckentrenner m ~ **(al)iron** Fasson-, Form-, Profil-eisen n ~**-iron cutter (or shears)** Formeisenschere f ~ **lifter** Schnittfänger m ~ **line** Reviergrenze f ~**-lined** schraffiert ~ **lineman** Leitungsaufseher m ~ **mill** Formprofilwalzwerk n, Profil-(eisen)walzwerk n ~ **modulus** (metal testing) Widerstandsmoment n ~ **point** Festpunkt m, Pupinisierungsfestpunkt m, (loading) Spulenfestpunkt m ~**-press** Schliffpresse f ~ **reduction in area** Querschnittverformung f ~ **rolling** Profilwalzen n ~ **rolling mill** Formeisenwalzwerk n ~ **shape** Profil n ~ **steel** Fasson-, Formeisen n, Formstahl m ~ **(al) steel** Profileisen n ~ **switch** Kuppelschalter m ~ **wire** Fasson-, Form-draht m
**sectional, ~ area** Querschnitt m, Querschnitts-, Schnitt-fläche f ~ **area of cut** Spannquerschnitt m ~ **area of spar** Holmquerschnittsfläche f ~ **beam** Teilbaum m ~ **boiler** Teilkammerkessel m ~ **chamber** Teil-, Wasser-kammer f ~ **change**

Wandstärkenübergang *m* ~ **chart** Sektionskarte *f* ~ **cutout of wing** Flügelausschnitt *m* **~-cylinder selfactor** Mehrzylinderselfaktor *m* ~ **density** Durchschnittsdichte *f* ~ **drawing** Schnitt *m* ~ **electrolytic copper tubes** Elektrolytprofilkupferrohre *pl* ~ **elevation** Ansicht *f* in Aufriß und Schnitt ~ **form** Formprofil *n* ~ **forming ratio** Querschnittverformung *f* ~ **girder** Profileisenträger *m* ~ **groove** Formkaliber *n* ~ **header boiler** Teilkammerkessel *m* ~ **mast** Steckmast *m* ~ **moment of inertia** Schnittträgheitsmoment *n* ~ **part** Fassonteil *m* ~ **plane** Riß *m*, Schnitt-ebene *f*, -fläche *f* ~ **screw** Gliederschnecke *f* ~ **shape** Querschnitt *m* ~ **side elevation** Seitenschnitt *m* ~ **steel** Formprofileisen *n*, Profilstahl *m* ~ **steel plate** Profilstahllamelle *f* ~ **strandrubber** Gummiprofilschnüre *pl* ~ **strip** Profilstreifen *m* ~ **system** Baukastenprinzip *n* ~ **toll switching** Auftrenntechnik *f* (aut. teleph.) ~ **vertical antenna** mehrfach unterteilte Vertikalantenne *f* ~ **view** Ansicht *f* im Schnitt, Schnittansicht *f* ~ **wire** Fasson-, Form-draht *m*

**sectionalize, to** ~ **a fault** eine Störung *f* eingrenzen

**sectionalized,** ~ **antenna** unterteilte Vertikalantenne *f* ~ **radiator** Elementkühler *m*

**sectionalization** Eingrenzung *f* ~ **of a trouble** Fehlereingrenzung *f*

**sectioning** Aufteilung *f*, Profilierung *f*, Streckentrennung *f*

**sectionize, to** ~ **in Abschnitte** *pl* zerlegen

**sector** Abschnitt *m*, Ausschnitt *m*, Gelände-abschnitt *m*, -teil *m*, (testing-machine) Kraftsklaenbogen *m*, Kreisbogen *m*, Peilabschnitt *m*, Proportionalzirkel *m*, Schneise *f*, Segment *n*, Sektor *m*, Streckenabschnitt *m*, Streifen *m* ~ **of circle** Kreisausschnitt *m* ~ **of dial** Winkelbereich *m* ~ **of a sphere** Äugelsektor *m* ~ **of upper river course** Oberlaufstrecke *f*

**sector,** ~ **alignment indicator** Schnellwertgeber *m* ~ **angle** Zentriwinkel *m* ~ **arc** Kalottenbogen *m* ~ **boundary** Abschnittsgrenze *f*, Naht *f* ~ **cone** Sektormembran *f* ~ **control panel** Abschnittskontrolltafel *f* ~ **diaphragm** Sektormembran *f* ~ **display** Sektordarstellung *f* (rdr) ~ **exchange** Knotenamt *n* ~ **gate** Fächertor *n* ~ **instantaneous shutter** Sektormomentverschluß *m* ~ **light** Sektorfeuer *n* ~ **map** Streifenkarte *f* ~ **scan(ning)** Sektorabtastung *f* (rdr) ~ **shaft** (Auto) Segmentwelle *f* ~**-shaped basin** sektorförmiges Becken *n* ~ **turntable** Drehweiche *f*

**sectorial cone** (loudspeaker) Sektorenmembran *f*

**sectors, by** ~ abschnittweise

**secular** hundertjährig, säkular, weltlich ~ **acceleration** säkulare Beschleunigung *f* ~ **determinant** Säkulardeterminante *f* ~ **equation** Säkulargleichung *f* ~ **fluctuation** Säkularschwankung *f* ~ **retardation** Säkularverzögerung *f* ~ **trend** Säkularschwankung *f*

**secularize, to** ~ einziehen, säkularisiern

**securable** feststellbar

**secure, to** ~ anhalten, arretieren, befestigen, bewirken, bremsen, einfügen, einlegen, einsetzen, fangen, fest-halten, -machen, -stellen, -ziehen, schützen, sichern, sperren **to** ~ **anchors for sea** die Anker seefest zurren **to** ~ **by bolts** fest-

schrauben **to** ~ **to a buoy** an der Boje *f* festmachen **to** ~ **fore and aft** vorn und hinten festmachen **to** ~ **a patent for** . . . ein Patent *n* erwirken für . . . **to** ~ **to a pier** an einen Pier *m* (oder Kai) gehen **to** ~ **in position** festlegen

**secure** fest, sicher ~ **from shear** (or **from sliding**) schubsicher

**secured** gesichert ~ **by wedge** aufgekeilt ~ **creditor** sichergestellter Gläubiger *m*

**securely braked** festgebremst

**securing** Feststellung *f* ~ **bolt** Befestigungsschraube *f* ~ **cord** Befestigungsleine *f* ~ **eye** Zurröse *f* ~ **pad** Anschraubflansch *m* ~ **pin** Gießstift *m* ~ **plate for fan** Halteblech *n* für Ventilatorkopf ~ **ridge** Sicherungssteg *m* ~ **ring** Befestigungsring *m* ~ **srew** Linsenschraube *f* ~ **strap** (or **strip**) Sicherungsstreifen *m* ~ **the tire** Radreifensicherung *f*

**securities** Effekten *pl*

**security** Bürgschaft *f*, Garantie *f*, Sicherheit *f*, Sicherung *f*, Unterpfand *n*, Wertpapier *n* **able to put up** ~ kautionsfähig ~ **against breakage** Bruchsicherung *f* ~ **against buckling** Knicksicherheit *f* ~ **at regular interest** festverzinsliches Wertpapier *n*

**security,** ~ **control** Abwehrangelegenheit *f* ~ **measure** Abwehrmaßnahme *f*

**sedative agent** Beruhigungsmittel *n*

**sedecimal** sedezimal (info proc.)

**sedentary** seßhaft

**sediment, to** ~ **out** absedimentieren, aussedimentieren

**sediment** Ablagerung *f*, Absatz *m*, Ansatz *m*, Bodensatz *m*, Geläger *n*, Grund *m*, Lager *n*, Niederschlag *m*, Satz *m*, Schicht *f*, Schlamm *m*, Schlammabsatz *m*, Sediment *n*, Sinkstoff *m* ~ **beer** Trübbier *n* ~ **bowl** Abscheidungsgefäß *n* ~ **carpet** Sedimentdecke *f* ~ **containing decaying organisms** Faulschlamm *m* ~ **drain screw** Schlammablaßschraube *f* ~ **filter press** Gelägerfilterpresse *f*

**sedimentary,** ~ **material in suspension** Sinkstoff *m* ~ **material transported along stream bed by flowing water** Geschiebe *n* ~ **rocks** Sedimentgestein *n*

**sedimentation** Ablagerung *f*, Absetzen *n*, Absetzung *f*, Schichtenbildung *f*, Schlämmung *f*, Schlämmverfahren *n*, Sedimentation *f*, Sedimentierung *f* ~ **analysis** Schlammanalyse *f* ~ **depth** (Absatzgefäß) Fallhöhe *f* ~**-equilibrium method** (for determination of particle size) Steighöhenmethode *f* ~ **potential** Niederschlagspotential *n* ~ **process** Schlammprozeß *m*, Schlämmvorgang *m*

**sedimenting glass** Spitzbecherglas *n*

**see, to** ~ aufsuchen, einsehen, sehen **to** ~ **at a glance** übersehen

**Seebeck effect** Seebeckscher Effekt *m*, thermoelektrischer Effekt *m*, thermoelektrische Wirkung *f*

**seed, to** ~ besäen, besamen, impfen (kristallisieren), pflanzen, säen

**seed** Blase *f*, Gipsen *m*, Keim *m* (cryst.), Saat *f*, Saatgut *n*, Same(n) *m* ~ **in the glass** Glasblase *f*

**seed,** ~ **beet growing** Samenrübenbau *m* ~ **crystal** Kristallkeim *m* ~ **crystals** Anregekristalle *pl* ~ **dressing** Beizung *f* ~ **drill** Breitsämaschine *f*, Dibbelmaschine *f*, Drillschar *f* ~**-drill**

**plough** Bohrpflug *m* ~ **dusting** Trockenbeizung *f* ~ **runner** Schoß-, Stock-rübe *f* ~ **unit** aktive Einheit *f*

**seeder** Fruchtentkerner *m*, Sä(e)apparat *m*, Sämaschine *f*

**seediness** griesiges Aussehen *n*

**seeding, ~ of the embankment** Schutzpflanzung *f* ~ **center** Keimzentrum *n* ~ **machine** Sämaschine *f* ~ **pencil** Impfstift *m*

**seedless** kernlos, samenlos

**seedling** Kernholz *n* ~ **method** Keimpflanzenmethode *f*

**Seeger, ~ circlip** Seegersicherung *f* ~ **cone** Seeger-Kegel *m* ~ **ring** Seegerring *m*

**seek, to** ~ ausfindig zu machen suchen, streben, suchen

**seeker** Sucher *m* ~ **head** Suchkopf *m* (g/m) ~ **lever** Sucherhebel *m* ~ **parallax** Sucherparallaxe *f* ~ **picture** Sucherbild *n* ~ **toe** Sucherfuß *m*

**seeking switch** Einwegschalter *m*

**seem, to** ~ den Anschein haben, scheinen

**seeming** anscheinend, augenscheinlich, scheinbar

**seep, to** ~ sickern **to** ~ **away** versickern **to** ~ **through** durch-sickern, -strömen

**seep** Ausschwitzen *n*

**seepage** Durchströmung *f*, Sickerung *f*, Unterströmung *f*, Versickerung *f* ~ **face** Sickerfläche *f* ~ **failure** hydraulischer Grundbruch *m* ~ **flow** Filterbewegung *f*, Sickerströmung *f* ~ **line** Sickerlinie *f* ~ **pipe** Sickerleitung *f* ~ **pressure** Strömungsdruck *m* ~ **shaft** Sickerschacht *m* ~ **test** Filterversuch *m*

**seepages of natural gas** Quellen *n* von Erdgas

**seesaw** Schaukeln *n*, Wippen *n* ~ **axis** Schlagachse *f* bei halbstarren Rotorsystemen ~ **circuit** Kathodenbasisverstärker *m* ~ **motion** Schaukelbewegung *f* ~ **oscillations** Kippschwingungen *pl* ~ **pan** Kipp-, Schwung-pfanne *f*

**seethe, to** ~ abbrühen

**S-effect** Oberflächeneffekt *m*

**segment, to** ~ segmentieren, in Abschnitte teilen

**segment** Abschnitt *m*, Ausschnitt *m*, Kolbenring *m*, Lamelle *f*, Zelle *f* ~ **of an ellipse** Ellipsenabschnitt *m* ~ **of the object** Objektausschnitt *m* ~ **of a parabola** Parabelabschnitt *m*

**segment, ~-fed loop** Endkoppelschleife *f* ~ **length** Gliederlänge *f* ~ **pitch** Lamellenteilung *f* ~ **rocker** Segmentschwinge *f* ~ **shift** Segmentumschaltung *f* ~ **shutter** Segment-, Sektoren-verschluß *m* ~ **transmission** Segmentübertragung *f*

**segmental** segmentförmig ~ **arch** Flachbogen *m* ~ **barrage** Segmentwehr *n* ~ **chip** Brockenspan *m* ~ **gate** Segmenttor *n* ~ **piece** Segmentstück *n* ~ **rack** Zahnsegment *n* ~ **valve** Segmentschütz *n*

**segmentally shaped** segmentförmig

**segmentation** Gliederung *f*, Programmunterteilung *f* (data proc), Unterteilung *f*

**segmented** in Segmente *pl* geteilt, (of fiber structures) parzellarisch ~ **branch** Segmentabzweig *m* ~ **combustion chamber** Teilbrennkammer *f* ~ **ring** geteilter Ring *m*, Segmentring *m* ~ **shell** Segmentgranate *f*

**segments of chain** Kettenglieder *pl*

**segregate, to** ~ absondern, aussaigern, ausscheiden, ausseigern, entmischen, saigern, seigern, sondern

**segregated, ~ band** Seigerungsstreifen *m* ~-**band system** Getrenntlageverfahren *n* ~**spot** Seiger-(ungs)stelle *f*

**segregation** Absonderung *f*, Absonderung *f* durch Krystallisation (in castings), Abtrennung *f*, Ausscheidung *f*, Ausseigerung *f*, (of sand and cement) Entmischung *f*, Saigerung *f*, Scheidung *f*, Seigerung *f* ~ **in an ingot** Blockseigerung *f* ~ **line** Seigerungszeile *f*

**seiche** Seiche *f*

**Seignette salt** Seignettesalz *n*

**seine** Zugnetz *n*

**seism** seismische Bewegung *f*

**seismic** seismisch ~ **disturbance** seismische Erschütterung *f* ~ **method** seismografische Untersuchungsmethode *f* ~ **rays** Erdbebenstrahlen *pl* ~ **waves** Erdbebenwellen *pl*

**seismograph** Beben-messer *m*, -zeiger *m*, Erdbeben-anzeiger *m*, -messer *m*, Erschütterungszeiger *m*, Seismograf *m*

**seismological observatory** Erdbebenwarte *f*

**seismology** Erdbebenkunde *f*, Seismologie *f*

**seismometer** Beben-messer *m*, -zeiger *m*, Erdbebenmesser *m*, Seismograf *m*

**seismoscope** Seismoskop *n*

**seizable** greifbar, pfändbar

**seize, to** ~ abfassen, beschlagnahmen, mit Beschlag belegen, Besetztzeichen belegen, einbringen, erfassen, ergreifen, fangen, fassen, fest-binden, -fressen, -setzen, (bearing) festbrennen, sich festfressen (engin.), sich festlaufen, fressen, greifen, (valves) hängen bleiben, kapern, nehmen, pfänden **to** ~ **the bearing** das Lager fressen **to** ~ **a line** eine Leitung *f* belegen

**seized up** (engine) festgefressen

**seizing** Bändsel *n*, Festfressen *n*, Fressen *n*, (of engines) Verschweißerscheinung *f*, Zurrtau *n* ~ **of slides** Fressen *n* der Gleitsteine

**seizing, ~ line** Bändselleine *f* ~ **marks** Freßspuren *pl* ~ **throat** Augbändsel *m*

**seizure** Anfressung *f*, Belegen *n*, Belegung *f*, Beschlagnahme *f*, Besitzergreifung *f*, Einnahme *f*, Fassen *n*, Festnahme *f*, Greifen *n*, Pfändung *f*

**Se-lattice (or structure)** Selengitter *n*

**selbite** Grausilber *n*

**select, to** ~ ausklauben, auslesen, ausnehmen, aussieben, aussuchen, auswählen, seligieren, sieben, suchen, wählen **to** ~ **the site** den Platz *m* auswählen

**select** ausgewählt ~ **code** gewählter Kode *m* (sec. rdr.) ~ **magnet** Stangenmagnet *m* ~ **mode** gewählte Betriebsart *f* (sec. rdr.) ~ **switch-core output** Adressen-Schaltkernsignal *n*

**selectance** Selektivität *f*

**selected** auserlesen, gewählt ~ **area diffraction** Feinbereichsbeugung *f* ~ **(crystal) lattice reflex** definierter (Kristall-) Gitterreflex *m*

**selecting** Suchen *n* ~**and sifting machinery** Auslese- und Sortiermaschine *f* ~ **bar** (cross bar) Einstellschiene *f*, Wählschiene *f* ~ **bar magnet** Stangenmagnet *m* ~ **cam** Wähldaumen *m* ~ **circuit** Selektivkreis *m* ~ **lever** Sucherhebel *m* ~ **magnet** Wählmagnet *m* ~ **mechanism** Wählmechanismus *m*, -organ *n*, -werk *n* ~ **needle** Abfühlnadel *f*, Fühlnadel *f* ~ **path** Einstellweg *m* ~ **pin** Fühlnadel *f* ~ **switch** Einwegschalter *m*

**selection** Abstimmung *f*, Auslese *f*, Auswahl *f*, Belieben *n*, Entwerfen *n*, Klaubarbeit *f*, Klau-

ben *n*, Nummerwahl *f*, Selektion *f*, Trennung *f*, Wahl *f*, Wählen *n*, Wahlvorgang *m* in ~ auszugsweise ~ **of route** Trassenwahl *f*, Trassierung *f*
**selection,** ~ **check** Ansteuerungsprüfung *f* (data proc.) ~ **circuit** Wählstromkreis *m* ~ **principle** Auswahl-prinzip *n*, -regel *f* (in electron transition) ~ **relay** Umschaltsteuerrelais *n* ~ **rule** (in electron transition) Auswahl-prinzip *n*, -regel *f* ~ **system** Adressen-, Ansteuerungs-system *n*
**selective** mit Auswahl und Unterscheidung verfahrend oder erfolgend, selektiv, trennscharf, wählfrei, wahlweise ~ **assembly** Auswahlpassung *f* ~ **call** wahlweiser Anruf *m* ~ **circuit** Selektiv-, Sieb-kreis *m* ~ **control** Anwahlsteuerung *f* ~ **corrosion** örtliche Korrosion *f* ~ **endorsing** wahlweise Indossierung *f* ~ **fading** selektiver Schwund *m*, Selektivschwund *m* ~ **filter** Farbenauszugfilter *m* ~ **hardening** Teilhärtung *f* (eines Werkstückes) ~ **jamming** Selektivstörung *f* ~ **manual control** Steuerorgan *n* mit Wahl der Einstellung von Hand ~ **mechanism** Wählwerk *n* ~ **list control** wahlweise Schreibsteuerung *f* ~ **listing** wahlweiser Einzelgang *m* ~ **localization** selektive Deposition *f* ~ **photoelectric emission** selektiver Fotoeffekt ~ **process** Wahlvorgang *m* ~ **protection** Selektivschutz *m* ~ **ringing** Einzelanruf *m* bei Gemeinschaftsleitungen, selektiver oder wahlweiser Anruf *m*, Wahlanruf *m* ~ **screen** Farbenauszugfilter *m* ~ **sequence electronic calculator** Elektronenrechner *m* mit steuerbarer Rechenfolge ~ **signaling** Wahlanruf *m* ~ **squelch** Schaltung *f* zur Unterdrückung des Rauschens ~ **switch** Tastwahlschalter *m* ~ **system** Siebgebilde *n* ~ **system of gear shifting** Kulissen-, Verschiebe-schaltung *f* ~ **telemetering** Wahlfernmessung *f* ~ **transmission** Selektions-, Selektiv-, Wähl-getriebe *n*
**selectivity** Abstimmschärfe *f*, Auslesefähigkeit *f*, Resonanzschärfe *f*, Selektivität *f*, Trennungsvermögen *n*, -schärfe *f* (rdo) ~ **capacity** Trennwirkung *f* ~ **characteristic** Abstimm-, Selektivitäts-kurve *f* ~ **control** Selektivitätsregelung *f* ~ **curve** Abstimm-, (of a directionfinding receiver) Durchlaß-, Selektivitäts-kurve *f* ~ **discrimination** Trennschärfe *f* eines Filters
**selector** Abstimmeinrichtung *f*, Auslesenknopf *m*, (calling apparatus) Einzelanrufer *m*, Führungsstück *n*, Gruppenwähler *m*, Kulissenplatte *f*, Linienwähler *m*, Schaltgriff *m*, Sucher *m*, Wähler *m*, Wähler mit freier Wahl ~ **of periods** Periodenwähler *m* ~ **with repeater** Wähler *m* mit Übertrager
**selector,** ~ **arm** (wiper) Wählarm *m* ~ **ball** Schaltknopf *m* ~ **bank** radial angeordnetes Kontaktenfeld *n*, Zuteilungsbank *f* ~ **bar** Zuteilungsschiene *f* ~ **calling** Wahlanruf *m* ~ **carrying capacity** Belastung *f* der Wähler ~ **circuit** Selektivkreis *m* ~ **contact** Wählerkontakt *m* ~ **control** Umschalteinrichtung *f* ~ **dial** Wählerscheibe *f* (teleph.) ~ **diaphragm** Bereichsblende *f* ~ **disk** Wählscheibe *f* ~ **gate** Auswahltorimpuls *m* ~ **hunting time** Freiwahlzeit *f* ~ **installation** Linienwähleranlage *f* ~ **key** Ziehkeil *m* ~ **lever** Wählhebel *m* ~ **magnet** Wählmagnet *m* ~ **pin** Abfühlstift *n* ~ **plant** Wähleranlage *f* ~ **plate** Vorstufenlamelle *f*, Zu-

teilungsscheibe *f* ~ **pulse** Wählimpuls *m* ~ **rack** Zuteilungsgestell *n* ~ **release** Zuteilungsauslösung *f* ~-**release trunk** Verbindungsleitung *f* für Wählerauslösung ~ **repeater** Gruppenwähler *m* mit Stromstoßübertrager, Mitlaufwähler *m* ~**rod** Bürsten-arm *m*, -träger *m*, Kontaktarm *m*, -armträger *m* ~ **shaft guide** Führungsschlitz *m* ~ **station** Kommandoplatte *f* (autom. contr.) ~ **stepping magnet** Wählerschaltmagnet *m* ~ **switch** Dreh-, Kreiselüberwachungsschalter *m*, Meßstellenumschalter *m*, Umschalter *m*, Wählschalter *m* ~ **system** Zuteilungssystem *n* ~ **valve** Dreiwegehahn *m*, Umschalt-, Umsteuer-ventil *n*
**selectron** Selectron *n*
**selenate** Selenat *n*
**selenic,** ~ **acid** Selensäure *f* ~ **compass** Selenkompaß *m*
**selenide** Selenid *n*, Selenmetall *n*, Selensalz *n*
**seleniferous** selenhaltig
**selenious** selenig ~ **acid** selenige Säure *f* ~ **cell** Selenzelle *f*
**selenite** Erdglas *n*, Marienglas *n*, Selenit *m*
**selenitic** selentisch
**selenium** Selen *n* ~ **barrier cell** Selensperrschichtfotozelle *f* ~ **barrier layer photovoltaic cell** Selensperrschicht-Fotozelle *f* ~ **blocking-layer photo cell** Selensperrschichtfotozelle *f* ~ **cell** Selen-, Foto-zelle *f* ~ **(tetra)chloride** Selenchlorid *n* ~ **compass** Selenkompaß *m* ~ **compound** Selenverbindung *f* ~ **conductive cell** Selenwiderstandzelle *f* ~ **diethyl** Selendiäthyl *n* ~ **disk** Selenschichtscheibe *f* ~ **lead** Bleiselenid *n* ~ **monochloride** Selenchlorür *n* ~ **mud** Selenschlamm *m* ~ **oxide** Selenoxyd *n* ~-**oxide rectifier** Trockengleichrichter *m* ~ **rectifier** Selengleichrichter *m* ~ **slime** Selenschlamm *m* ~ **sulfide** Schwefelselen *n*, Selensulfid *n* ~ **tetrafluoride** Selentetrafluorid *n*
**selenocyanate** Selenzyanid *n*
**selensulfur** Schwefelselen *n*, Selenschwefel *m*
**self-acting** automatisch, selbsttätig ~ **cock** selbsttätiger Hahn *m* ~ **control** direkte Regelung *f* ~ **controller** direkter Regler *m* ~ **equilibrium brake operated by load itself** Lamellenschraubbremse *f* ~ **instantaneous gear clutch** selbsttätig ausrückende Momentklauenkupplung *f* ~ **mule** Selbstspinner *m* ~ **stream-pipe closing valve** Selbstschlußdampfventil *n*
**self,** ~-**actuating clutch** selbsteinrückende Kupplung *f* ~-**adjoint** selbstadjungiert ~-**adjustable** selbstnachstellbar
**self-adjusting** selbsteinstellend ~ **bearing** Wälzlager *n* ~ **brake gear of a winding machine** Bremsdruckregler *m* einer Fördermaschine ~ **cam** selbsteinstellender Nocken *m* ~**hydraulic tappet** Ventilstößel *m* mit automatischer Spieleinstellung ~ **line finder** selbsttätiger Zeilenanzeiger *m*
**self-adjustment** Selbsteinstellung *f*
**self-alarm receiver** Selbstnot-Empfänger *m*
**self-aligning** selbsteinstellend, selbstfluchtend ~ **ball bearing** Pendelkugellager *n* ~ **bearing** Pendellager *n* ~ **double-row ball bearing** Ringpendellager *n* (zweireihig) ~ **roller bearing** Pendelrollenlager *n*, Tonnenlager *n* ~ **spherical roller** Ringtonnenlager *n* ~ **system** selbstrichtendes System *n*

**self,** ~-**alignment** Selbsteinstellung *f*, selbsttätiges Einstellen *n*, Spurzwang *m* ~-**annealing** Selbsttemperung *f* ~-**aspirating engine** laderloser oder selbstansaugender Motor *m* ~-**balancing** selbstabgleichend ~-**baking** selbstbrennend ~-**bearing** Eigenpeilung *f* ~-**bias** automatische Gittervorspannung *f*, Selbstvorspannung *f* ~-**biased grid** automatisch vorgespanntes Gitter *n* ~-**biased stage** Stufe *f* mit automatischer Vorspannung ~-**biasing of grid** automatische Gittervorspannung *f* ~-**binder** Selbstbinder *m* (Lack) ~-**blocking oscillator** selbstsperrender Schwingungserzeuger *m* ~-**capacitance** (distributed capacity) Eigenkapazität *f*, Selbstkapazität *f*, verteilte Kapazität *f*, Windungskapazität *f* ~-**capacity** Eigenkapazität *f* ~-**capacity of a coil** Eigenkapazität *f* der Spule, Spulenkapazität *f* ~-**centering chuck** selbsttätiges Zentrierfutter *n* ~-**centering three jaw chuck** Dreibackenfutter *n* (zentr. spannend) ~-**charge** Selbstaufladung *f* ~-**checking code** selbstprüfender Kode *m* ~-**cleaning** selbstreinigend ~-**cleaning contact** Wischkontakt *m* ~-**clinkering** Selbstabschlackung *f* ~-**clipping device** Selbstklemme *f* ~-**closing ball valve** Selbstschlußkugel *f* ~-**closing valve** Rohrbruchventil *n* ~-**color** Selbstfarbe *f* ~-**colored** selbstfarbig ~-**commutation** Selbstführung *f* ~-**compensation** selbstausgleichend ~-**computing chart** Fluchtentafel *f*, Nomogramm *m* ~-**consistency** Widerspruchslosigkeit *f* ~-**consistent field** Eigenfeld *n*, eigenes Feld *n* ~-**consistent function** selbstständige Funktion *f*
**self-contained** in sich abgeschlossen, in sich geschlossen, unabhängig ~ **construction** geschlossener Aufbau *m* ~ **drive** (tool machine) Einzelantrieb *m* ~ **instrument** unabhängiges Meßinstrument *n* ~**motor** unabhängiger Motor *m* ~ **unit** unverpackter Artikel *m*, Geräteeinheit *f*
**self,** ~-**cooled** eigenventiliert ~-**correcting code** selbstkorrigierender Kode *m* ~-**damping** Eigendämpfung *f* ~-**defense** Notwehr *f*, Selbstschutz *m* ~-**demagnetizing field** Selbstentmagnetisierungsfeld *n* ~-**destroying shell** Zerleger *m* ~-**destruction device** (Rakete) Selbstzerstörungsanlage *f* ~-**destruction type fuse** Selbstzerlegerzünder *f* ~-**diffusion plasticity** Platzwechselplastizität *f* ~-**discharge** Selbstentladung *f* ~-**discharging centrifugal** selbstentleerende Schleuder *f* ~-**discharging truck** Selbstentladewagen *m* ~-**distribution piston** selbststeuernder Kolben *m* ~-**dumper** Selbst-entlader *m*, -entleerer *m* ~-**electrode** Selbstemissionselektrode *f* ~-**emission electron microscope** Elektronenmikroskopselbstleuchtverfahren *n* ~-**emissive method** (using electrons) Fremdschichtverfahren *n*, Selbst-leuchtverfahren *n*, -strahlungsverfahren *n* ~-**emulsifying oil** selbstemulgierendes Öl *n* ~-**energizing braking** bremsverstärkend ~ **energy** Selbstenergie *f* ~ **evacuating** (Pumpe) selbstsaugend ~-**evaporation** Selbstverdampfung *f* ~-**evident** augenscheinlich ~-**exchange** Selbstaustausch *m* ~-**excitation** Eigenerregung *f*, Selbst-erregung *f*, -schwingen *n* ~-**excitation winding** Rückkopplungswicklung *f*
**self-excited** selbsterregt ~ **field circuit** Eigener-

regung *f* ~ **oscillator** eigenerregter oder selbsterregter Schwingungserzeuger *m* ~ **oscillator transmitter** selbsterregender Röhrensender *m* ~ **oscillator tube (or valve)** selbsterregte Schwingröhre *f* ~ **tube** (in a directdrive circuit) selbsterregte oder selbstgesteuerte Röhre *f* ~ **valve** rückgekoppelte, selbsterregte oder selbstgesteuerte Röhre *f*
**self,** ~ **exciter** Eigenerregermaschine *f* ~-**exciting** selbsterregend ~-**expanding packing** (Kolben) Selbstspanner *m* ~-**feed** Selbstgang *m* (des Vorschubes) ~-**feed attachment (or -feeder)** Selbsteinleger *m*
**self-feeding,** ~ **furnace** Füllschachtfeuerung *f* ~ **furnace with hopper above gate** Schüttfeuerung *f* ~ **rock drill** sich selbsttätig verschiebender Bohrer *m*
**self,** ~-**filling fountain pen** Selbstfüller *m* ~-**flux solder** Lötdraht *m* mit Gußmittel ~-**fluxing** selbstgehend ~-**fluxing ore** selbstgehendes Erz *n* ~-**focusing** Selbstkonzentration *f*; selbstfokussierend ~-**focusing differential arc lamp** Fixpunktdifferentialbogenlampe *f* ~-**focusing shunt arc lamp** Fixpunktnebenschlußbogenlampe *f* ~-**government** Selbstverwaltung *f* ~-**gripping general purpose pliers** Blitzzange *f* ~-**gyro-interaction** Eigenkreiselwirkung *f* ~-**hardening** selbsthärtend ~-**hardening steel** lufthärtender Stahl *m*, Selbsthärter *m* ~-**heating** Selbsterhitzung *f*
**self-heterodyne** Selbstüber-lagerer *m*, -lagerung *f* ~ **receiver** Überlagerungsempfänger *m* mit Selbsterregung *m* ~ **reception** Schwebungsempfang *m* mit Selbsterregung, Selbstüberlagerungsempfang *m*, Überlagerungsempfang *m* mit Selbsterregung
**self,** ~-**heterodyning mixer tube** selbstschwingende Mischröhre *f* ~-**hooped** (gun barrel) autofrettiert, kaltgereckt ~-**hooping** Selbstberingung *f* ~-**igniting lighter** Selbstzünder *m* ~-**ignition** Selbstzündung *f*
**self-illuminating** selbstleuchtend ~ **light modulator** selbstleuchtender Lichtmodulator *m* ~ **method** Selbststrahlungsverfahren *n*
**self,** ~-**imposed assessment** Freistellungsbescheid *m* ~-**inductance** Induktivität *f*, Selbstinduktionskoeffizient *m*, -induktivität *f* ~-**inductance variation** Induktivitätsänderung *f*
**self-induction** Selbst-erregung *f*, -induktion *f* ~ **coil** Selbstinduktionsspule *f* ~ **decade** Selbstinduktionsdekade *f* (rdo) ~ **standard** Selbstinduktionsnormal *n*
**self,** ~-**inductive** selbstinduktiv ~-**instructed carry** selbsttätiger Übertrag *m* ~-**interrupter** Selbstunterbrecher *m* ~-**interrupter contact** Selbstunterbrecherkontakt *m* ~-**interrupting interrupter** Selbstunterbrechungsunterbrecher *m* ~-**interruption** Selbstunterbrechung *f* ~-**intersection** Selbstdurchdringung *f* ~ **inversion** Selbstumkehr *f* ~-**light** Augeneigenlicht *n* ~-**light of retina** Eigenlicht *n* des Auges ~-**limiting chain reaction** selbstbremsende Kettenreaktion *f* ~-**loading gun** Selbstlader *m*
**self-locking** selbst-hemmend, -sperrend, -verriegelnd ~ **gear** selbstsperrendes Getriebe *n* ~ **nut** selbstsichernde Mutter *f* ~ **pipe vise** selbstsichernder Rohrschraubstock *m* ~ **tripod**

Schnappstativ *n* **~ worm gear** selbsthemmendes Schneckengetriebe *n*

**self-lubricating** selbstschmierend **~ bearing bush** Selbstschmierlagerbuchse *f* **~ bush** Kompobuchse *f* **~ bushing** Leerlaufbuchse *f* mit selbsttätiger Ölschmierung *f*

**self-luminous, ~ electron microscope** Elektronenmikroskopselbstleuchtverfahren *n* **~ method** (using electrons) Fremdschichtverfahren *n*, Selbstleuchtverfahren *n* **~ substance** Selbstleuchter *m*

**self, ~-magnetic** eigenmagnetisch **~-maintained discharge** selbständige Entladung *f* **~-maintaining** selbstunterhaltend **~-management** Selbstverwaltung *f* **~ mass** Selbstmasse *f* **~-measuring** selbstmessend **~-motivated spin** Eigendrehbewegung *f*, -impuls *m* (Atomphysik) **~-mutilation** Selbstverstümmelung *f* **~-multiplying chain reaction** selbstvermehrende Kettenreaktion *f* **~-noise** Eigenrauschen *n* **~-oiling** selbsttätige Schmierung *f* **~-oiling bearing** Schmierlager *n* **~-oiling bushing** selbsttätig schmierende Buchse *f* **~-opening die head** selbstöffnender Schneidkopf *m* **~-opening flap** selbsttätige Klappenöffnung *f* **~-operated controller** Regler *m* ohne Hilfsenergie **~-oscillation** Pfeifen *n* der Verstärker, selbständige Schwingungserzeugung *f*, Selbst-erregung *f*, -schwingen *n* **~-oscillatory transmitter** selbsterregender Sender *m* **~-oscillatory tube** selbsterregte oder selbstgesteuerte Röhre *f* **~-oscillatory valve** selbsterregte Schwingröhre *f* **~-packing** selbstdichtend **~-plugging rivet** selbststauchender Niet *m* **~-polymerization** Autopolimerisation *f* **~-positioning switch** Wählschalter *m* **~-potential** Eigenpotential *n* **~-powered** mit eigenem Netzteil *m* **~-powered recorder** Batterie-Tonbandgerät *n* **~-preservation** Selbsterhaltung *f* **~-priming carburetor** Vergaser *m* mit Anlaßeinspritzpumpe *m* **~-priming centrifugal pump** selbsttätig ansaugende Kreiselpumpe *f* **~-priming rotary pump** selbstansaugende Kreiselpumpe *f* **~-propagating** sich von selbst fortpflanzend

**self-propelled** selbstangetrieben **~ car** Triebwagen *m* (r. r.) **~ mount** Motorlafette *f* **~-traveling gear** Eigenfahrwerk *n* **~-type sand slinger** fahrende Sandschleuderformmaschine *f*

**self, ~-propelling** Eigenantrieb *m*, Eigenfortbewegung *f* **~-propelling swinging crane** Drehscheibenkran *m* **~-propulsion** Selbstantrieb *m* **~-protection** Selbstschutz *m* **~-quenched counter tube** selbstlöschendes Zählrohr *n* **~-quenching** Selbstauslöschung *f* **~-quenching oscillator** Sperrschwinger *m* **~-recording** selbst-registrierend, -schreibend **~-recovery** Selbsterholung *f* **~-rectifying tube voltmeter** selbstgleichrichtender Röhrenvoltmesser *m* **~-reducing** selbstreduzierend **~-registering thermometer** Maximal- und Minimalthermometer *n* **~-registering tide gauge** Limnograf *m* **~-regulating** selbstregelnd **~-regulation** Selbstausgleich *m* **~-regulative** selbstregelnd **~-reinforcing of a vibration** Aufschaukeln *n* der Schwingung **~-releasing slag** selbstabfallende Schlacke *f* **~-reliant** selbständig **~-reliant plant** von außen unabhängige Anlage *f* **~ repulsion** Selbstabstoßung *f* **~-restoring drop** Rückstellklappe *f* **~-restoring indicator**

Rückstellklappe *f*, selbsthebende Klappe *f* **~-reversal** Selbst-umkehr *f*, -umkehrung *f* **~-running** controlled timebase method mitgenommene selbständige Kippmethode (TV), selbständige mitgenommene Kippschwingmethode *f* **~-running time bases** selbständige Kippschwingungen *pl* **~-saturating rectifier** Selbstsättigungsgleichrichter *m* **~ scanning** Selbstrücklauf *m* (TV) **~-scattering** Eigenstreuung *f* **~ screening range** Maskierungsreichweite *f* **~-sealed** selbstabdichtend **~-sealed bearing** selbstabdichtendes Lager *n*

**self-sealing** (fuel tank) lecksicher, selbst-abdichtend, -dichtend, -klebend **~ connection** selbstdichtende Kupplung *f* **~ fuel tank** selbstschließender Brennstoffbehälter *m* **~ tank** geschützter Behälter *m*, Sackbehälter *m*, Sicherheitsbehälter *m*, schußsicherer Tank *m*

**self, ~-setting** Selbsteinstellung *f* **~-shadowing** gegenseitige Überlagerung *f* **~-shielded coil** selbstschirmende Spule *f* **~-shielding** selbstabschirmend **~ stabilization** Eigenstabilisierung *f* **~-starter** Selbststarter *m*; selbstangetrieben **~-starting** selbsttätiges Angehen *n*; selbstgehend **~-starting push button** Anlaßdruckknopf *m* **~-steering device** Selbststeuervorrichtung *f* **~-stopping** Selbstsperrung *f* **~-stripping revolving flat card** selbstreinigende Drehscheibenkarde *f* **~-sufficiency** Eigen-, Selbst-versorgung *f*

**self-supporting** freistehend, freitragend, selbsttragend **~ chimney** freistehender Schornstein *m* **~ mast** freitragender Mast *m* **~ rack** freitragendes Gestell *n* **~ radio tower** freitragender Funkmast *m* **~ steel tower** freitragender Stahlturm *m* **~ tower** freistehender oder freitragender Turm *m* **~ ventilated stack** freistehender, ventilierter Kamin *m*

**self, ~-sustained** selbständig, selbstschwingend **~-sustained discharge** selbständige Entladung *f* **~-sustained operation** Selbstanlauf *m* **~-sustained oscillation** selbsterregte Schwingung *f*, Selbsterregung *f* **~-sustaining** von selbst ablaufend **~-sustaining speed** Selbstlaufdrehzahl *f* **~-synchronizing** selbstsynchronisierend **~-tannage** Alleingerbung *f* **~ tapping** selbstschneidend **~-tapping convex fillister head screw** Linsenblechschraube *f* **~-tapping screw** Blech-, Treib-schraube *f*, selbstschneidende Schraube *f* **~-testing feature** Eigenprüfeinrichtung *f* **~-tightening** selbst-dichtend, -spannend **~-timer** Selbstauslöser *m* (photo) **~-toning** selbsttönend **~-ventilation** Eigenbelüftung *f* **~-vulcanizing** selbstvulkanisierend **~-wetting** selbstbenetzend **~-whistles** Überlagerungspfeifen *n* **~-wiping contact** Wischkontakt *m*

**sell, to ~** absetzen, abstoßen, an den Mann bringen, umsetzen, unterbringen, verkaufen **to ~ by auction** verauktionieren, versteigern **to ~ below cost** schleudern **to ~ at a loss** (or **below cost price**) verschleudern

**seller** Verkäufer *m*

**selling by retail** Kleinverkauf *m*

**selling, ~ commission** Verkaufsprovision *f* **~-off** Ausverkauf *m* **~ rate** Briefkurs *m*

**selsyn** Drehtransformator *m*, Geber *m*, Geberanlage *f*, Synchronisierungsmotor **~ receiver** Quittungs-, Ringfeld-empfänger *m* **~ trans-**

**mitter** Ringfeldgeber *m* **~-type regulator** (self-synchronous) Drehregler *m*
**Seltner screen** Seltner-Rost *m*
**selvage (or selvedge)** Borte *f*, Eckplatte *f*, Gewebeleiste *f*, Kante *f*, Salband *n* (geol.), Webekante *f* **~ guide** Leistenführung *f* **~ neutralizing device** Leistenneutralisiervorrichtung *f* **~ printing** Kantendruck *m* **~ printing machine** Bügeldruckmaschine *f* **~-protected belt** Kantenschutz *m* **~ straightener** Leistenaufroller *m*
**selvagee** Kabelgarnstropp *m*
**sematic range** Bedeutungsfeld *n*
**semaphore** Flaggenwinker *m*, optischer Telegraf *m*, Semafor *m*, Signalmast *m* **~ direction indicator** Pendelwinker *m* **~ flag control** Wink(er)flagge *f* **~ message** Winkspruch *m* **~ signal** Armsignal *n*, Flügelsignal *n* **~ station** Semaphorstation *f*
**semi, to ~-finish bore** halbfertig bohren **to ~-sinter** versintern
**semi, ~-amplitude** Halbamplitude *f* **~-annual** halbjährlich **~-anthracite** Halbanthrazit *m*, magere Kohle *f*, Magerkohle *f*, Sandkohle *f* **~-apertural angle** Halböffnungswinkel *m* **~-apochromatic-lens system** Fluoritlinsensystem *n*
**semiautomatic** halbautomatisch **~ controller** halbautomatischer Regler *m* **~ exchange** halbautomatisches oder halbselbsttätiges Fernsprechamt *n*, halbselbsttätige Anlage *f* **~ lathe** Drehhalbautomat *m* **~ operation** Teilautomatisierung *f* **~ plant** halbselbsttätige Anlage *f* **~ rectifier** halbautomatisches Entzerrungsgerät *n* **~ rifle** halbautomatisches Gewehr *n* **~ switch** Gruppenumschalter *m* **~ system** Halbwahlsystem *n* **~ telephone system** Halbwählsystem *n* **~ type sliding-wedge breech mechanism** Schubkurbelverschluß *m*
**semiautomatical** halbselbsttätig
**semi, ~-axis** Halbachse *f* **~-B position** halbautomatischer B-Platz *m* **~-balloon type** Halbballon *m* **~-basic** halbbasisch
**semibituminous, ~ coal** Magerkohle *f*, magere Kohle *f* **~ coal of high rank** backende Sinterkohle *f* **~ coal of medium rank** (short-flaming smokeless) Eßkohle *f* **fat ~ run-of-mine coal** Fettförderkohle *f*
**semi, ~-bleach** Halbbleiche *f* **~-bodied soaps** halbwarme Seifen *pl* **~ boiled nigre** Leimseife *f* **~-bright** halbblank **~-bulkhead** Halbschott *n*
**semicantilever** halbfreitragend **~ low-wing monoplane** halbfreitragender Tiefdecker *m* (aviat.) **~ monoplane** halbfreitragender Eindecker *m* (aviat.) **~ wing** halbfreitragender Flügel *m* (aviat.)
**semicircle** Halbkreis *m*
**semicircular** halbkreisförmig, halbkreisig, halbrund, semizirkular, semizirkular **~ approach** halbkreisförmiger Annäherungsweg *m* **~ arch** Halbkreisbogen *m* **~ contact** Kontakthalbring *m* **~ cutter** Pilzfräser *m* **~ deviation** halbkreisartige Deviation *f* **~ error** Halbkreisfehler *m* **~ groove** Halbkreiseinschnitt *m* **~ head wall** gewölbte Abschlußmauer *f* **~ protractor** Halbkreisgradbogen *m* **~ rail** Vollbogen *m* **~ segment** Binant *m* **~ sling swivel** Halbmondriemenbügel *m*
**semi-coke** Halbkoks *m*, Urkoks *m* **~ from brown coal** Braunkohlenhalbkoks *m* **~ residue of party carbonized lignite** Grudekoks *m*

**semi, ~-coking** Halb(ver)kokung *f* **~-colon** Punktstrich *m*, Strichpunkt *m* **~-conductible** halbleitfähig **~-conducting** halbleitend **~-conducting crystals** Halbleiterkristalle *pl* **~-conduction germanium** halbleitendes Germanium *n* **~-conductor** Halbleiter *m* (electr.) **~-conductor device** Halbleitergerät *n* **~-conductor rectifier** Halbleiterventil *n* **~-conductor technology** Halbleitertechnik *f* **~-continuous** halbkontinuierlich **~-continuous mill** halbkontinuierliches Walzwerk *n* **~-continuous rolling train** halbkontinuierliche Walzstraße *f* **~-controlled mosaic** Bildplanskizze *f* **~-convergent** Halbkonvergente *f* **~-cured** halbvulkanisiert **~-cycle** Halbperiode *f* **~-cylinder** Halbzylinder *m* **~-cylindrical** halbrund **~-darkness** Halbdunkel *n* **~-difference** halbe Differenz *f* **~-direct recovery** halbdirekte Gewinnung *f* **~-diurnal** halbtägig, halbtäglich **~-diurnal tide** Halbtagszeit *f* **~-dry process** Halbtrockenverfahren *n* **~-dull** hellmatt **~elliptic** halbelliptisch **~-elliptical arch**, halbelliptischer Bogen *m* **~-elliptical spring** gesprengte Feder *f* **~-embedded** halbeingebettet **~-empirical** halbempirisch **~-enclosed** halbgeschlossen gegen (zufällige) Berührung geschützt **~-fabricated form** Halbzeugform *f* **~-finish** Halbappretur *f*
**semifinished** vorgefrischt **~ articles** Halbzeug *n* **~ flat** Breiteisen *n*, Platine *f* **~-forgings** vorgearbeitete Schmiedestücke *pl* **~ goods** Vorstoff *m* **~ material** Halbzeug *n* **~ metal products** Metallhalbzeuge *pl* **~ nut** Halbzeugmutter *f* **~ part** Halbprodukt *n* **~ product** Halb-erzeugnis *n*, -fabrikat *n*, Vorerzeugnis *n* **~ products** Halbzeug *n*
**semi, ~-finishing mill** Walzwerk *n* für Halbzeug **~finishing mill train** Halbzeugstraße *f* **~-finite solid coils** einseitig unendliche Vollspulen *pl* **~-fixed** halbeingebettet **~-flat base rim** Halbflach(bett)felge *f* **~-floating axle** halbfliegende Achse *f* **~-floating rear axle** halbschwingende Hinterachse *f* **~-fluid** dickflüssig, halbflüssig, strengflüssig **~-frosted** halbmatt, schwachmattiert **~-fused** halbgeschmolzen **~-gantry** Halbportal *n* **~-gas firing** Halbgasfeurung *f* **~-gloss** Seidenglanz *m* **~-grange of the tide at a port on a day of mean equinoctial springs** Einheitskoeffizient *m* der Tidehöhe eines Hafens **~-graphical procedure** halbzeichnerische Methode *f* **~-hard** halbhart **~-hard rubber** halbharter Gummi *m* **~-holonomic** semi-holonom **~-illuminated** halbhell **~-incandescent lamp** Glimmlampe *f* **~-indirect lighting** halbindirekte Beleuchtung *f*
**semi-infinite, ~ body** (space) Halbraum *m* **~ coil** einseitig unendliche Spule *f* **~ line** quasiunendlich (lange) Leitung *f* **~ solid** einseitig unendlich ausgedehnter Körper *m*
**semi, ~-invariant** Semiinvariante *f* **~-killed steel** halbberuhigter Stahl *m* **~-lateral** halbseitig **~-lens** Halblinse *f* **~-liquid mass** halbflüssige Masse *f* **~-long link** halblanggliedrig **~-machined bolt** halbblanke Schraube *f* **~-machined washer** halbblanke Scheibe *f* **~-manufactured article** Halb-erzeugnis *n*, -fabrikat *n* **~-manufactured goods (or product)** Halbfertigerzeugnis *n* **~-mat** halbmatt **~-mat chromium-plated** halbmatt verchromt **~-mat glaze** Halbmattglanz *m* **~-mat glazing** Halbmattglasur *f*

**semimechanical** halb-automatisch, -selbsttätig ~ **central office** halbselbsttätiges Amt *m* (teleph.) ~ **installation** halbselbsttätige Anlage *f* ~ **position** Zahlengeberplatz *m* ~ **system** Halbwahlsystem *n*

**semi,** ~**-metal** Halbmetall *n* ~**-mobile** verlegefähig ~**-monocoque construction** Halbschalenbauweise *f* (aviat) ~**-monocoque fuselage** Halbschalenrumpf *m*, Rumpf *m* in Halbschalenbauweise ~**-motorized** teilmotorisiert ~**-muffle furnace** Halbmuffelofen *m* ~**-note** Halbton *m* ~**-opaque** halbdeckend ~**-open crystallizer** halbgeschlossene Maische *f* ~**-oscillation** Halbperiode *f* ~**-oval** halbeirund ~**-overhauling** Teilüberholung *f* ~**-parabolic superstructure** Halbparabelfachwerkträger *m* ~**-pasty slag** zähflüssige Schlacke *f* ~**-period** Halbperiode *f*, Halbzeit *f* ~**-permanent** halbdauernd, halbfest ~**-permeable** halbdurchlässig ~**-permeable membrane (or partition)** halbdurchlässige Scheidewand *f* ~**-plane** Halbebene *f* ~**-plastic** halbteigig ~**-plastic press** Halbnaßpresse *f* ~**-porcelain** Halbporzellan *n*

**semiportable** halbbeweglich, halbtragbar ~ **engine** halbstehende Antriebsmaschine *f* ~ **track** halbbewegliches Geleise *n*

**semi,** ~**-portal crane** Halbportalkran *m* ~**-product** Halb-erzeugnis *n*, -fabrikat *n* ~**-product of electric steel** Elektrostahlhalbzeug *n* ~**-product of superrefined steel** Edelstahlhalbzeug *n* ~**-production basis** Halbfließbandfertigung *f* ~**-products** Halbzeug *n* ~**-protected** mit geschützten Wicklungen *pl* ~**-pulp** Rohschliff *m* ~**-purified** vorgefrischt ~**-pyritic process** Halbpyritschmelzen *n* ~**-quantitative** halbquantitativ ~**-recessed** halbversenkt **-recessed switch** halbversenkter Schalter *m* ~**-refined wax** halbraffiniertes Paraffin *n* ~**-remote control** Halbfernsteuerung *f* ~**-rigid** halbstarr ~**-rigid airship** halbstarres Luftschiff *n* ~**-rigid system (or type)** halbstarres System *n* ~**-rotary pump** Flügel-, Würgel-pumpe *f* ~**-selective ringing** Einzelanruf *m* mit verabredeten Zeichen ~**-self-maintained discharge** halbselbständige Entladung *f* ~**-single sideband reception** Restseitenbandempfang *m* ~ **sintered** halbgesintert ~**-skilled** angelernt ~**-skilled operator (or worker)** angelernter Arbeiter *m* ~**-slicing knife** halbschnittiges Messer *n* ~**-solid** festweich ~**-solid phase** Sumpfphase *f* ~**-solid substance** halbfester Stoff *m* ~**-space** Halbraum *m* ~**-span** Halbspannweite *f* ~**-spherical bearing** Kalotte *f* ~**-stall** halbübberzogener Flug *m* ~**-steel** Halbstahl *m*, Mockstahl *m*, schiedbarer Guß *m* ~**-step slope** halbsteile Lagerung *f* ~**-strut** halbstielig ~**-strut-type biplane** Halbstieler *m* ~**-sunk** halbversenkt ~**-sunk switch** halbversenkter Schalter *m* ~**-symmetric** halbsymmetrisch ~**-tone** Halbton *m* ~**-trailer** Sattelschlepper *m* ~**-transparent** halbdurchlässig ~**-trumpet chute** Kegelschnurre *f* ~**-vowel** Halbvokal *m* ~**-vulcanized** halbvulkanisiert ~**-water gas** Dowson-, Halbwasser-, Misch-gas *n* ~**-worsted yarn** Sayet(te)garn *n*

**semolina** Grieß *m* ~ **mill** Grützmühle *f*

**senarmontite** Senarmontit *m*

**send, to** ~ einhängen, geben, remittieren, schikken, senden, übertragen, überweisen **to** ~ **after** nachschicken **to** ~ **again** nochmals senden **to** ~ **on approval** zur Ansicht *f* schicken **to** ~ **back** remittieren **to** ~ **for cash on delivery** nachnehmen **to** ~ **forward** übermitteln **to** ~ **off** absenden, entsenden **to** ~ **out** aussenden **to** ~ **a thing to a person** jemandem etwas zustellen **to** ~ **through** durchleiten

**sender** Absender *m*, drahtloser Sender *m*, Funksender *m*, Geber *m* (telegr.), Handsender *m*, Sender *m*, Sendestelle *f*, Stromstoßsender *m*, Taste *f*, Taster *m*, Zahlengeber *m* (teleph.) ~ **with rotary switch** Drehwählerzahlengeber *m*

**sender,** ~ **armature** Sendeanker *m* ~ **control** Steuersender *m* ~ **end** Sendeseite *f* ~ **finder** Sendesucher *m* ~**-key contact** Tastkontakt *m* ~ **method** Verfahren *n* mit Stromstoßübertragung ~**-receiver converter** Senderempfängerumformer *m* ~ **selection** Vorwahl *f* ~ **selector** Registerwähler *m* ~ **test device** Zahlengeberprüfeinrichtung *f*

**sending** Abgabe *f*, Stromgeben *n* ~ **amplifier** Sendeverstärker *m* ~ **battery** Sendebatterie *f* ~ **brush** Sendebürste *f* ~ **condenser** Sendekondensator *m* ~ **current** abgehender Strom *m*, Sendestrom *m* ~ **direction** Senderichtung *f* ~ **distributor** Sendeverteiler *m* ~ **end** Geberende *n*, Sendeseite *f* ~ **end impedance** Eingangsimpedanz *f* ~ **key** Drucktaster *m*, Handsender *m*, Sendetaste *f*, Sendklavier *n*, Taste *f* ~ **loop** Sendeschleife *f* ~ **point** Absendestelle *f* ~ **power** Sendeleistung *f* ~ **pulses** Impulssenden *n* ~ **station** Funkstelle *f*, Sender *m*, Sende-station *f*, -stelle *f*

**senior** älter ~ **engineer** Betriebsoberingenieur *m* ~ **inspector** Bergrat *m* ~ **official** gehobener Beamter *m*

**seniority** Anziennität *f*, Dienstalter *n*

**sensation** Empfindung *f*, Gefühl *n*, Wahrnehmung *f*

**sensation,** ~ **of brightness** Helligkeits-eindruck *m*, -empfindung *f* ~ **of direction** Ortseindruck *m* ~ **of heaviness** Schwereempfindung *f* ~ **of light** Lichtreiz *m* ~ **of motion** Bewegungseindruck *m* ~ **of movement** kinesthesis Bewegungsempfindung *f* ~ **of saturation** Sättigungseindruck *m* ~ **of temperature** Temperaturempfindung *f*

**sensation,** ~ **increment** Empfindungszunahme *f* ~ **level** Hörschwelle *f* ~ **unit** Lautstärkenstufe *f*

**sense, to** ~ ab-fühlen, -tasten, empfinden, wahrnehmen, wittern

**sense** Sinn *m*, Sinnesempfindung *f*, Vorzeichen *n* **sense,** ~ **of balance** Gleichgewichtssinn *m* ~ **of direction** Ortsgedächtnis *n*, Richtungssinn *m* ~ **of equilibrium (or balance)** Gleichgewichtsgefühl *n* ~ **of feeling** Tastgefühl *n* ~ **of force** Wirkungssinn *m* ~ **of length** Längsrichtung *f* ~ **of light** Helligkeitssinn *m* ~ **of localization** Orientierungssinn *m* ~ **of position** Lagen-, Raum-empfindung *f* ~ **of responsibility** Verantwortungsbewußtsein *n* ~ **of rotation** Drehrichtung *f*, -sinn *m*, Rotations-richtung *f*, -sinn *m*, Umlaufsinn *m* ~ **of sight** Augensehen *n* ~ **of tension** Spannungsgefühl *n* ~ **of touch** Betastung *f*, Tastsinn *m* ~ **of vector** Vektorsinn *m* ~ **of weight** Gewichtsgefühl *n* ~ **of winding** Wicklungssinn *m*

**sense,** ~ **amplifier** Leseverstärker *m* ~ **antenna** Seitenbestimmungsantenne *f* ~ **coil** Lesespule *f* ~ **direction of vector** Vektorsinn *m* ~ **finder** Seitenbestimmer *m*, Zweideutigkeitsaufhebungsgerät *n* ~ **finding** Richtungsbestimmung *f*, Seitenbestimmung *f* (Peiler), Zweideutigkeitsaufhebung *f* ~**-finding switch** Peilseiteschalter *m* ~ **lines** Leseleitungen *pl* ~ **research** Zweideutigkeitsaufhebung *f* ~ **winding** Geber-, Lese-wicklung *f*
**sensed** abgetastet
**sensibility** Empfindlichkeit *f*, Sensibilität *f* ~ **to chrome** Chromempfindlichkeit *f* ~ **to variation (or difference)** Unterschiedsempfindlichkeit *f* ~ **reciprocal** Empfindlichkeitswert *m*
**sensibilization** Empfindlichmachung *f*
**sensible** empfindlich, fühlbar, sensibel, verständig ~ **heat** Eigenwärme *f*, fühlbare Wärme *f* ~ **loudness of sound** Empfindungslautstärke *f*
**sensing** Richtungssinnbestimmung *f*, Zeigerausschlag *m* ~ **cam** Einstellkurve *f* ~ **capacitor** Meßkondensator *m* ~ **coil** Fühlspule *f* ~**-device** Abtasteinrichtung *f* ~ **frame** Abtastgestell *n* ~ **lever** Einstellhebel *m* (teletype) ~ **line** Geberleitung *f*, Meß- oder Steuerleitung *f* ~ **point** Fühlort *m* ~ **unit** Abfühleinrichtung *f*
**sensitive** empfindbar, empfindlich, empfindungsfähig, erregbar, feinstufig, (photo) lichtempfindlich, sensibel
**sensitive,** ~ **to acid** säureempfindlich ~ **to caustic (lye)** laugenempfindlich ~ **to copper** kupferempfindlich ~ **to corrosion** korrosions-, rost-empfindlich ~ **to fire** feuerempfindlich ~ **to hardness** härteempfindlich ~ **to heat** feuer-, hitze-, wärmeempfindlich ~ **to light** lichtempfindlich **not** ~ **to light** lichtunempfindlich ~ **to lime** kalkempfindlich ~ **to radiation** strahlungsempfindlich ~ **to red light** rotempfindlich ~ **to temperature** temperaturempfindlich ~ **to water-spotting** tropfwasserempfindlich
**sensitive,** ~ **altimeter** Feinhöhenmesser *m*, sensitiver Höhenmesser *m* ~ **axis** Empfindlichkeitsachse *f* (gyro) ~ **clay** strukturempfindlicher Ton *m* ~**-coarse altimeter** Fein-Grob-Höhenmesser *m* ~ **contacts** Feinkontakte *pl* ~ **control** Feinstufenregelung *f*, feinstufige Regelung *f* oder Regulierung *f* ~ **element** Fühlglied *n* ~ **element of the fuel tank gauge** Kraftstoffbehältersonde *f* ~ **feed** feinstufiger Vorschub *m* ~ **fuse** hochempfindlicher Zünder *m* **(light-)**~ **layer** empfindliche Schicht *f* ~ **measure** Empfindlichkeitsmaß *n* ~ **operation** feinfühlige Führung *f* ~ **paper** Kopierpapier *n*, lichtempfindliches Papier *n* ~ **position** (of motor or machine) Ansprechstellung *f* ~ **pressure gauge** Feinmeßmanometer *n* ~ **region** empfindlicher Teil *m* **(light-)**~ **surface** empfindliche Schicht *f* ~ **time** Ansprech-, Empfindlichkeits-zeit *f* ~ **tracer** Feintaster *m* ~ **volume** Zählvolumen *n*
**sensitiveness** Empfindlichkeit *f*, Empfindlichkeitsgrad *m*, Erregbarkeit *f*, Feinfühligkeit *f*, Sensibilität *f* ~ **to corrision** Korrosionsempfindlichkeit *f*, Rostempfindlichkeit *f* ~ **to impact** Empfindlichkeit *f* gegen Schlag ~ **of the level** Libellenempfindlichkeit *f* ~ **at all points** allgemeines Empfinden *n* ~ **to position** Lageempfindlichkeit *f*

**sensitivity** Empfindlichkeit *f*, Empfindungsvermögen *n*, (in color change) Indikationsschärfe *f*, Lichtempfindlichkeit *f*, Verstärkungsgrad *m*
**sensitivity,** ~ **of the brake** Feinfühligkeit *f* der Bremse ~ **of deflection** Ablenkempfindlichkeit *f* (electr.) ~ **of discrimination** Unterscheidungsempfindlichkeit *f* ~ **of an echo-suppressor** Empfindlichkeit *f* einer Echosperre ~ **of fuse** Zünderempfindlichkeit *f* ~ **to light** Helligkeitsempfindlichkeit *f* ~ **to overheating** Überhitzungsempfindlichkeit *f* ~ **to percussion** Schlagempfindlichkeit *f* ~ **of the photoelectric cell** Empfindlichkeit *f* der Fotozelle ~ **of the plate** Plattenempfindlichkeit *f* ~ **of response** Ansprechempfindlichkeit *f* ~ **of selection** (in tuning) Feintrennung *f*
**sensitivity,** ~ **constant** Ansprechkonstante *f* ~ **constant of microphone** Mikrofonansprechkonstante *f* ~ **control** Empfindlichkeitsregelung *f* ~ **current** Empfindlichkeitsstrom *m* ~ **decrease** Empfindlichkeitsabnahme *f* (acoust.) ~ **distribution** spektrale Empfindlichkeitsverteilung *f* ~ **drift** Empfindlichkeitsveränderung *f* ~ **factor** Empfindlichkeits-faktor *m*, -ziffer *f* ~ **limit** Empfindlichkeitsgrenze *f* ~ **limits** Ansprechgrenzen *pl* ~ **profile** Empfindlichkeitsprofil *n* ~ **regulator** Empfindlichkeitsregler *m* ~ **speck** Prokeim *m* (photo) ~ **test** Empfindlichkeitsprobe *f*, -prüfung *f* ~ **time control (STC)** Nachechodämpfung *f* (rdr)
**sensitization** Empfindlichmachung *f*, Formierung *f*, Präparieren *n* (photo), Sensibilisierung *f* ~ **of a surface** Oberflächenbeladung *f*
**sensitize, to** ~ (a surface) beladen, empfindlich machen, lichtempfindlich machen, sensibilisieren **to** ~**-chrome** chromieren **to** ~ **a surface with hydrogen** eine Oberfläche *f* mit Wasserstoff beladen
**sensitized** lichtempfindlich ~ **decomposition** sensibilisierte Zerlegung *f* ~ **fluorescence** Sekundärfluoreszenz *f* ~ **material** strahlenempfindliches oder sensibilisiertes Material *n* (photo) ~ **paper** lichtempfindliches Papier *n* ~ **plate** lichtempfindliche Platte *f*
**sensitizer** Reifungskörper *m*, Sensibilisator *m*
**sensitizing** Empfindlichmachen *n*, Sensibilisierung *f* ~ **dye** Sensibilisierungsmittel *n* ~ **switch** Taste *f* für Empfindlichkeitsregelung
**sensitometer** Empfindlichkeitsmesser *m*, Filmsensitometer *m*, Sensitometer *n* ~ **tablet** Stufenplatte *f*
**sensitometric measurement** Empfindlichkeitsmessung *f*
**sensor** Meß-fühler *m*, -geber *m*, -größenumformer *m*, Sensor *m* ~ **transformer** induktiver Geber *m*
**sensory,** ~ **disturbance** Sensibilitätsstörung *f* ~ **equipment** Meßfühlereinrichtung *f*
**sentence, to** ~ verurteilen
**sentence** Entscheidung *f*, Rechtsspruch *m*, (law) Richterspruch *m*, Satz *m*, (law) Strafe *f*, Urteilsspruch *m* ~ **of the court** Gerichtsbeschluß *m*
**sentinel** Posten *m*, Schildwache *f*, Wache *f*
**sentry** Gruppenmarke *f*, Marke *f* (data proc.), Markierung *f*, Posten *m*, Schildwache *f*, Wache *f* ~ **box** Schilderhaus *n*

**separable** abscheidbar, dissoziierbar, lösbar, scheidbar, scheidefähig

**separate, to** ~ abfassen, abreißen, abscheiden, abschneiden, absondern, abspalten, abtreiben, abtrennen, aufbereiten, auseinandergehen, auseinanderweichen, auslesen, ausschalten, ausscheiden, einteilen, einzeln, klassieren, klauben, loslösen, lostrennen, niederfallen, reinigen, scheiden, separieren, sichten, sondern, sorten, sortieren, staffeln (Flugzeuge), teilen, trennen, verschieden zerlegen, zerteilen **to** ~ **into fibers** auffasern **to** ~ **in flakes** ausflocken **to** ~ **as a flocculent (or to** ~ **at the flocculent state)** ausflocken **to** ~ **by fusion** abseigern **to** ~ **into layers** abschichten **to** ~ **out** absondern, aussondern **to** ~ **into parts** entmischen **to** ~ **by sedimentation** durch Absetzen *n* ausscheiden **to** ~ **tar** entleeren

**separate** selbständig, sonder ~ **(d)** getrennt **with** ~ **poles** mit getrennten Polen *pl*

**separate,** ~ **adjustment** Einzelvorstellung *f* ~ **batches** Teilsendungen *pl* ~ **battery** Ortsbatterie *f* ~ **casting method** Einzelgußverfahren *n* ~ **clearing cam** Fangteil *m* eines Hebers ~ **commutation** Fremdführung *f* ~ **development** Sonderentwicklung *f* ~ **drive** Einzel-, Fremdantrieb *m* ~ **excitation** Fremd-erregung *f*, -steuerung *f* ~ **exciter** Fremderregermaschine *f* ~ **filter with lightning arrester** Antennenweiche *f* mit Ableiter ~ **focusing to the eyepieces of a field glass** Einzeltriebeinstellung *f* der Okulare eines Feldstechers ~ **heterodyne** (local oscillator) Fremdüberlagerer *m* ~ **heterodyne receiver** Überlagerungsempfänger *m* mit Fremderregung ~ **heterodyne reception** elektromotorische Schwebung *f* mit besonderem Überlagerer, Schwebungsempfang *m* mit besonderem Überlagerer, Überlagerungsempfang *m* mit Fremderregung ~ **impression** Separatabdruck *m* ~ **leaflet** Druckblatt *n* ~ **letter** Sonderbuchstabe *m* ~ **modulation** Fremdmodulation *f* ~ **operation** Einzelbestimmung *f* (chem.) ~ **plant** Einzelanlage *f* ~ **print** Sonderabdruck *m* ~ **process** Einzelvorgang *m* ~ **scavenging** Fremdspülung *f* ~ **selfexcitation** äußere Mitkopplung *f* ~ **(sanitary) system** Trennverfahren *n* ~ **treatise** Einzeldarstellung *f* ~ **unit** Sondereinheit *f* ~ **ventilation** Luttenbewetterung *f*

**separated** abgeschieden, abgesondert, gerättert, gesiebt, geteilt, getrennt ~ **flow** abgerissene oder nicht anliegende Strömung *f* (aerodyn.)

**separately,** ~ **air-cooled motor** Motor *m* mit Fremdlüftung ~ **cast cylinder** einzelstehender Zylinder *m* ~ **controlled** fremdgesteuert ~ **controlled oscillator** fremdgesteuerter Schwingungserzeuger *m* ~ **cooled** Fremdkühlung *f* ~ **excited** fremderregt, fremdgesteuert ~ **excited alternator** Generator *m* mit Fremderregung ~ **excited dynamo (or generator)** fremderregte Dynamomaschine *f* ~ **excited oscillator** fremderregter oder fremdgesteuerter Schwingungserzeuger *m* ~ **excited oscillator valve** fremderregte Schwingröhre *f* ~ **excited tube** fremderregte Röhre *f* ~ **instructed carry** gesteuerter Übertrag *m* ~ **loading ammunition** Kartuschmunition *f* ~ **mounted solenoid-operated switch** weggebauter Magnetschalter *m* ~ **ventilated** fremdbelüftet

**separating** Sortieren *n* ~ **action** Ausscheidungseffekt *m* ~ **agent** Ausscheidungsmittel *n* ~ **bath** Scheidebad *n* ~ **box** Austragkasten *m* ~ **calorimeter** Abscheidekalorimeter *n* ~ **capacitor** Trennkondensator *m* ~ **cylinder** Trior *m* ~ **dam** Separationswerk *n*, Trennbühne *f* ~ **device** Scheidevorrichtung *f* ~ **edge** Trennkante *f* ~ **effect** Ausscheidungseffekt *m* ~ **factor** Trennfaktor *m* ~ **filter** elektrische Weiche *f*, Frequenzweiche *f*, Trennfilter *m* ~ **flask** Scheidekolben *m* ~ **funnel** Scheidetrichter *m* ~ **glass** Scheidegefäß *n* ~ **line** Scheidelinie *f*, Trennfuge *f*, -linie *f* ~ **medium** Abscheidungsmittel *n* ~ **network** Frequenzweiche *f* ~ **plant** Sicht-(ungs)anlage *f* ~ **prism** Scheideprisma *n* ~ **process** Scheidevorgang *m* ~ **rivet** Abstandniet *m* ~ **screen** Separationstrommel *f*, Setzsieb *n* ~ **sieve** Pulverkornsieb *n* ~ **slide** Trennvorrichtung (Rohrpost) *f* ~ **strength** Trennfestigkeit *f* ~ **sump** Klär-sumpf *m*, -teich *m* ~ **unit** Trenngruppe *f* ~ **valve** Trennventil *n* ~ **web** Trennsteg *m* ~ **zone** Abscheidezone *f*

**separation** Ablösung *f*, Abscheidung *f*, Absonderung *f*, Abspaltung *f*, Abstand *m*, Abteil *n*, Abtrennung *f*, Aufbereitung *f*, Aufspaltung *f*, Auseinanderweichen *n*, Aushalten *n* (der Berge in der Grube), Ausscheidung *f*, Aussonderung *f*, Entmischung *f*, Klassierung *f*, Lösung *f*, Reinigung *f*, Scheiden *n*, Scheidevorgang *m*, Scheidung *f*, Schott *n*, Separation *f*, Sichtung *f*, Sieberei *f*, Sortierung *f*, Spaltung *f*, Staffelung *f* (von Flugzeugen), Trennung *f*

**separation,** ~ **of charge carrier** Ladungsträgertrennung *f* ~ **of charged particles** Ladungsträgertrennung *f* ~ **of constituents** Entmischung *f* des Möllers ~ **of crystals** Kristallausscheidung *f* ~ **of dust** Staubausscheidung *f* ~ **of flow** Ablösung *f* ~ **of graphite** Graphitausscheidung *f* ~ **of iron losses** Trennung *f* der Eisenverluste ~ **of liquid air** Luftzerlegung *f* ~ **of the salt** Ausscheiden *n* des Salzes ~ **of sand from cement** Entmischung *f* ~ **of silver** Silberscheidung *f* ~ **of waste** Abfallausscheidung *f*

**separation,** ~ **box** Separationslutte *f* ~ **column** Trennrohr *n* ~ **energy** Bindungsenergie *f* ~ **factor** Trennungsfaktor *m* ~ **joint** Trennungsfuge *f* ~ **method** Scheideverfahren *n* ~ **negative** negativer Farbauszug *m*, Negativfarbauszug *m* ~ **pay** Wartegeld *n* ~ **plant** Reinigungsfabrik *f*, Trennanlage *f* ~ **point** Ablösepunkt *m*, Grenzschicht(e)ablösungspunkt *m* ~ **precision** Trennschärfe *f* ~ **pressure** Schiebedruck *m* ~ **process** Trennungs-verfahren *n*, -vorgang *m* ~ **processes** Entmischungsvorgänge *pl* ~ **product** Abspaltungsprodukt *n* ~ **trough** Separationslutte *f*

**separative** trennend ~ **element** Trennglied *n* ~ **power** Trennvermögen *n*

**separator** Abscheider *m* (Öl), Ausleser *m*, Distanzhalter *m*, Frequenzweiche *f* (TV), Scheider *m*, Scheidewand *f*, Schleuder *f*, Separator *m*, Sichter *m*, Sortierapparat *m*, Trennstück *n*, Trennsymbol *n* (info proc.), Trennungsapparat *m*, (for video signals) Weiche *f*, Zwischenlage *f* ~ **bolt** Stehbolzen *m* ~ **diaphragm** poröser Scheider *m* ~ **machine** Sichtmaschine *f* ~ **piece** Distanzstück *n* ~ **skimmer** Abscheider *m* ~ **stage** Puffer-, Trenn-stufe *f* ~ **tube** Trennröhre *f* ~ **-type oil purifier** Ölseparator *m*

separatory funnel Scheide-, Schüttel-, Separier-
trichter *m*
sepia Sepia *f* ~ paper Sepiapapier *n*
septal suture Lobenlinie *f*
septate waveguide Hohlleiter *m* mit axialer Scheidewand
septic faulend, septisch ~ poison Fäulnisgift *n*
~ tank Faulbehälter *m*
septum Scheidewand *f* (geol.)
septuple siebenfach
sequel Folge *f*
sequence, to ~ an-, auf-, ein-reihen
sequence Aufeinanderfolge *f*, Folge *f*, Reihe *f*,
Reihenfolge *f*, Serie *f*, Wirkung *f* in ~ der Reihe nach
sequence, ~ of dots Punktfolge *f* ~ of operation
Arbeits-folge *f*, -gang *m* ~ of optic actions Bildfolge *f* ~ of phantom crossings Viererfolge *f* ~
of pictorial actions Bildfolge *f*, (film) Bildreihe
*f*, Reihenbilder *pl* ~ of pictures Reihenbilder *pl*
~ of ratio Schichtfolge *f* ~ of reductions Ab-
nahmefolge *f* ~ of stages (or steps) Stufenfolge
*f* ~ of strata Schichtfolge *f*
sequence, ~ alternator Reihenfolgeschalter *m*
~ calling (or calls) Anmeldung *f* von Reihenge-
sprächen, Reihenanmeldung *f* ~ check Folge *f*,
Kontrolle *f* ~-checking routine Folgeprüfpro-
gramm *n* ~ contact Folgekontakt *m* ~ control
Folgeschaltung *f* ~ control register Befehlszähler *m* (info proc.) ~ illusion Sequenzillusion
*f* ~ list Liste *f* der Gesprächsanmeldungen (eines Teilnehmers) ~ signal pausenfreies Signal
*n* ~ switch Folge-, Steuer-, Stufen-schalter *m*
~ switch cam Folgeschalterkamm *m*, Steuer-
daumen *m*, -nocken *m* ~ unit Kontrolleinheit
*f* ~ valve Folgeventil *n* ~ warning Sperrwarn-
leuchte *f*
sequencer Aufreiher *m*, Zuordner *m*
sequencing Reihenfolge *f*, (of valves) Schaltfol-
ge *f*
sequential aufeinanderfolgend, folgeabhängig,
folgend ~ color system Zeitfolgeverfahren *n*
beim Farbfernsehen ~ computer programmge-
steuerte (digitale) Rechen- oder Datenverar-
beitungs-anlage *f* ~ control Folgesteuerung *f*
(comput.) ~ operation Folgeschaltung *f* ~ scanning Folgeabtastung *f* (TV), unmittelbar aufeinanderfolgende Abtastung *f*, Zeitfolgeverfahren *n* ~ selection Reihenfolgewahl *f*
sequentially nacheinander
serein Tau *m*
serial laufend, reihenweise, serienmäßig in ~
production serienmäßig
serial, ~ access Serieneingabe *f* ~ air-survey
camera Reihen(bild)meßkamera *f* ~ arrangement Reihenanordnung *f* ~ camera Reihen-
(bild)kamera *f* ~-capacity coil Antennenverlängerungsspule *f* ~ construction Serienbau
*m* ~ counter assembly Aufnahmezählwerk *n* ~
development Reihenentwicklung *f* ~ digital
computer Serienrechner *m* ~ examination Rei-
henuntersuchungen *pl* ~ fabrication Serienherstellung *f* ~ film camera Reihenbildkamera *f*
~ hand taps Satzgewindebohrer *pl* ~ manufacture Serienherstellung *f* ~-model modification
Ausbaustufe *f* ~ number Aktenzeichen *n*, Fabrikationsnummer *f*, Gerätnummer *f*, laufende
Nummer *f*, Reihennummer *f*, Seriennummer *f*,

Werknummer *f* ~ number of a call Laufnummer
*f* eines Gesprächs ~ numbering holder Wechsel-
schablone *f* ~ operation Serienbetrieb *m* ~
photogrammetric camera Meßreihenbildner *m*,
Reihenmeßkamera *f* ~ photogrammetry camera
Reihenmeßkamera *f* ~ photograph Reihenbild
*n* ~ plate camera Plattenreihen-(bild)kamera *f*,
-bildner *m* ~ radiography Serienaufnahme *f* ~
section Serienschnitt *m* ~ storage Serienspei-
cherung *f* ~ survey Reihenaufnahme *f* ~ symbol
Aktenzeichen *n* ~ testing Mengenprüfung *f* ~
transfer Serien-transport *m*, -übertragung *f* ~
trials Serienversuche *pl*
seriality reihenweise Anordnung *f*
serially hintereinander ~ connected in Reihe geschaltet
seriatim nacheinander
sericite slate Serizitschiefer *m*
sericose printing Serikosedruck *m*
series Abteilung *f*, (of fabricated products) Baureihe *f*, Folge *f*, Kette *f*, Posten *m*, Reihen-folge
*f*, -schaltung *f*, Serie *f*, Zahlenreihe *f* in ~ hintereinander, in Reihe, der Reihe nach, in Serie liegend, serienmäßig
series, ~ of bearing readings on a direction finder
Serienpeilungen *pl* a ~ of blows (by a pile driver)
Hitze *f*, (eine) Schlagreihe *f* ~ of curves Kurvenzug *m* ~ of elements elektrolytische Spannungsreihe *f* ~ of experiments Versuchsreihe *f*
~ of flashes lasting more than two seconds Blinkfeuer *n* ~ of frequencies Frequenzreihe *f* ~ of a
higher degree Reihe *f* höherer Ordnung ~ of
impulses Stromstoßreihe *f* ~ of loads Lastenzug
*m* ~ of measurements Meßreihe *f* ~ of measures
Meßserie *f* ~ of observations Beobachtungs-
reihe *f* ~ of optic actions Bildfolge *f* ~ of parallel connections Gruppenschaltung *f* ~ of
pictorial actions Bildfolge *f* ~ of pulses Impuls-
folge *f* ~ of readings Meßreihe *f* ~ of shots
Schußserie *f* ~ of size Größenordnung *f* ~ of
soundings Lotungsreihe *f* ~ of strata Schichtfolge *f* ~ of tests (or trials) Versuchsreihe *f* ~
~ of tubes Schachtverrohrung *f*
series, ~-aiding gleichsinnig in Reihe ~ approach
Reihennäherung *f* ~-arc furnace Lichtbogenofen *m* mit der Widerstandserhitzung vereinigt,
Mehrelektrodenofen *m* ~ arc lamp Haupt-
strombogenlampe *f* ~ arm Längs-arm *m*, -zweig
*m* ~ arrangement Reihen-, Serien-schaltung *f*
~ bank (of cylinders) Reihe *f* ~bomb-release
mechanism Reihenabwurf-gerät *n*, -vorrichtung
*f* ~-boring mill Reihenbohrwerk *n* ~ camera
Reihen-bildkamera *f*, -bildner *m*, -kamera *f*,
Serienapparat *m* ~ capacitor Vorschaltkonden-
sator *m* ~ cascade action Reihenkaskadenver-
halten *n* ~ characteristic Reihenschlußverhal-
ten *n* (d. Motors) ~ choke coil Vorschaltdros-
sel *f* (zum Entstörkondensator) ~ circuit Staffelleitung *f* ~ coil Längsspule *f*, Reihenspule *f*
~-connected gleichsinnig in Reihe, hinterein-
andergeschaltet ~-connected spark gap Vor-
funkenstrecke *f* (Vorschalt-Funkenstrecke)
~-connected station Reihenstelle *f* ~ connection
Hintereinanderschaltung *f*, (of resistance) Reihenanordnung *f*, Serienschaltung *f* ~ connection cascading Kaskadenschaltung *f* ~ connections Reihen-bau *m*, -schaltung *f* ~ decay
Kettenzerfall *m* ~ development Serienentwick-

lung f ~ direct-current motor Reihenschluß-
gleichstrommotor m ~ disintegration Kettenumwandlung f ~-drilling machine Reihenbohr-
maschine f ~ drive Gruppenantrieb m ~ dynamo Hauptschlußdynamomaschine f, Hauptstrom-dynamo m, -dynamomaschine f, -ma-
schine f, Reihenschluß(dynamo)maschine f,
Seriendynamomaschine f ~(-wound) dynamo
Reihenschlußdynamo m ~-efficiency diode Se-
rienspardiode f ~ element Längs-, Reihen-
glied n ~ element of a filter Längsglied n eines
Filters ~ element of a network Längsglied n
einer Kette ~ equalizer Längsentzerrer m ~
excitation Reihenschlußerregung f ~ exciter
coil Reihenschlußerregerwicklung f ~ expansion Serienentwicklung f ~ exposure slide Be-
lichtungsreihenschieber m ~-fed seriengespeist
~ feeding Reihen-, Serien-speisung f ~ form
Reihenform f ~ formula Reihenformel f
(math.) ~ impedance Längs-, (line) Reihen-im-
pedanz f ~-impedance element Reihenimpe-
danzglied n ~-impedance-type equalizer Rei-
henimpedanzentzerrer m ~ impedor Reihen-
impedanzglied n ~ inductance Reiheninduktivi-
tät f ~ inverter Speicherwechselrichter m ~
lamp Reihenlampe f ~ limit Seriengrenze f ~
loading Pupinisierung f ~ magnetic coil Hauptspule f ~mesh of a network Längsglied n einer
Kette ~ modulation Reihenröhren-, Vorröhren-
modulation f ~ motor Haupt-schlußmotor m,
-strommotor m, Serienmotor m ~-multiple
connection gemischte Schaltung f ~-multiple
jack Vielfachunterbrechungsklinke f ~-multiple
resonant circuit Resonanz f und Drosselkreis m
in Reihe ~ multiplex telegraph gestaffelter
Mehrfachtelegraf m, Mehrfachtelegraf in Staffelschaltung m, Staffeltelegraf m ~ oil spring-
dashpot Ölkataraktzylinder m ~-opposed gegeneinander in Reihe ~-opposing gegensinnig
in Reihe
series-parallel in Reihenparallelschaltung f ~
arrangement of cells Gruppenschaltung f ~
change by opening the circuits in the network
Reihenparallelschaltung f mit Leistungsunter-
brechung ~ connection Reihenparallelschaltung f ~ control Regelung f durch Reihenparallelschaltung ~ shunt transition Reihenparallelschaltung f mit Widerstandsschaltung ~
switch Serienparallelschalter m ~ transisting
Reihenparallelschaltung f ~ winding (multiplex
wave) Reihenparallelwicklung f
series, ~ photograph flight Reihenbildflug m ~
production Reihen-bau m, -herstellung f, Serienbau m, -herstellung f ~ reactor Saugspule f,
Wellensauger m ~ rectifier circuit Serienschaltung f von Gleichrichtern ~ regulator Haupt-
strom-regelwerk n, -regler m, -werk n ~ resistance Längs-, Reihen-, Serien-, Vor(schalt)-
widerstand m ~ resistance of bulk of semiconductor Bahnwiderstand m ~ resistor Vorwider-
stand m ~ resonance Reihen-, Spannungs-,
Strom-resonanz f ~-resonance circuit Serien-
resonanzkreis m ~ resonant circuit induktive
Reaktanz f, Reihenresonanzkreis m, Resonanzkreis m (Spannungsresonanz), Spannungsre-
sonanzkreis m ~ rheostat Serienwiderstand m
~ self-inductance Serienselbstinduktion f ~
servo Reihensteller m ~-shunt network sym-

metrisches Netzwerk n ~spark gap Vorschalt-
funkenstrecke f ~ switch Serienschalter m ~
system of distribution System n mit Reihen-
schaltung ~ tank Serienbad n ~ transformer
Reihen-bau m, -transformator m, Stromwandler m ~ two-terminal pair network Serienvier-
pol m ~-type supervisory system Gemeinschafts-
verkehr m ~ welding machine Reihen-bau m,
-schlußmotor m, Serien(schluß)motor m ~
winding Reihenwicklung f
series-wound Hauptschluß m, in Serie gewickelt
~ generator Hauptschlußdynamomaschine f,
Hauptstrom-dynamo m, -dynamomaschine f,
Reihenschlußdynamomaschine f, Seriendynamomaschine f ~ motor (starter) Hauptschluß-
motor m, Hauptstrommotor m, Reihen-bau m,
-schlußmotor m, Serien(schluß)motor m
serif Haarstrich m, Querstrich m
serimeter Serimeter m
serious bedenklich, ernst, ernsthaft, schwerwiegend ~ accident schwerer Unfall m
seriously ernstlich ~ injured person Schwerbe-
schädigter m
serpentine Kehrschleife f, Schlangenkurve f,
Serpentine f ~ quarry Serpentinsteinbruch m
serrate, to ~ auszacken, riefen, riffeln, rippen,
verzahnen
serrate gezackt, schartig
serrated gereifelt, gerieft, geriffelt, gerippt, gezahnt, kerbig, sägeartig, sägeartig gezackt,
zackig ~ coin gezähnelte Münze f ~ cross
section gezahnter Querschnitt m ~ edge of the
piston Schlagkrone f des Kolbens ~ hub Kerb-
zahnnabe f ~ jaws gehauene Backen pl ~ joint
verzahnte Teilfuge f ~ knife Zackenmesser n ~
line ausgezackte (ausgezahnte) Linie f, (aus)-
gehackte oder gezackte Linie f ~ pliers Rohrzange f ~ pulse Impuls m mit eingesägtem Gipfel ~ range rings eingesägte Markierungskreise
pl ~ shaft Zahnwelle f ~ slats Zahnleisten pl
~slot Sägeschlitz m ~ wedges (Prüfmaschine)
Beißbacken pl
serration Auszackung f, Kerbverzahnung f, Lappung f, Riffelung f, Riffelverzahnung f, Rif-
felzahn m, Rippe f, sägeförmiger Rand m, Verzahnung f, Zacke f, Zacken m, Zähnelung f ~
hob Kerbverzahnungswälzfräser m
serum Serum n
serve, to ~ ableisten, bedienen, dienen, umhüllen, (sentence) verbüßen, versorgen, vorladen
to ~ with insulating tape mit Isolierband n umwickeln to ~ as a stopgap herhalten
served, to have ~ one's time ausgelernt haben ~
quires fadengehefteter Bogensatz m
service, to ~ bedienen, warten
service Beanspruchung f, Bedienung f, Betrieb
m, Betriebsart f, Dienst m, Dienstleistung f,
Funktion f, Gefälligkeit f, Leistung f, Verkehr
m, Wartung f in ~ im Betrieb "no-hang-up"
~ Sofortverkehr m ~ with grabs Greiferbetrieb m ~ in return Gegenleistung f
service, ~ advice Dienstnotiz f ~ age (Fallschirm) Lebensalter n ~ altitude Arbeitshöhe
f ~ answering jack Dienstabfrageklinke f ~
application Anmeldung f eines Fernsprechanschlusses f (teleph.) ~ apron Abfertigungsvor-
feld n (aviat.) ~ area (in broadcasting) Emp-
fangsgebiet n, Versorgungsgebiet n ~ band

zugeteiltes Frequenzband *n* ~ **board** Bedienungstafel *f* ~ **bonus** Dienstprämie *f* ~ **book** Agende *f* ~ **box** Anschlußkasten *m* ~ **brake** Betriebsbremse *f* ~ **bridge** Bedienungssteg *m*, Betriebsbrücke *f* ~ **bulletin** Betriebs-, Unterrichts-blatt *n* ~ **cable** Verbraucherleitung *f* ~ **call** Dienst-anruf *m*, -gespräch *n* ~**call key** Dienstanrufschalter *m* ~ **catwalk** Bedienungssteg *m* ~ **ceiling** Dienstgipfelhöhe *f*, Gebrauchsgipfelhöhe *f*, praktische Gipfelhöhe *f* ~ **certificate** Dienstzeugnis *n* ~ **circuit** Dienstleitung *f* ~ **condition(s)** Betriebs-beanspruchung *f*, -bedingungen *pl*, -verhältnisse *pl* ~ **connector** Dienstwähler *m* ~ **control** Betriebsüberwachung *f* ~-**control channel** Dienstverkehr *m* ~ **corrosion test** Betriebskorrosionsversuch *m* ~ **deadweight** Betriebseigengewicht *n* ~ **department** Hilfsbetrieb *m* ~ **distortion** Betriebsverzerrung *f* ~ **elevation** Bedienungshöhe *f* ~ **fractures** Betriebsbrüche *pl* ~ **gangway** Bedienungsausgang *m* ~ **hour meter** Betriebsstundenzähler *m* ~ **indications** Dienstvermerke *pl* ~ **instruction** Betriebsvorschrift *f*, Dienstanweisung *f* ~ **instrument set** Abfrageapparat *m* ~ **irregularity** dienstlicher Vorfall *m*, Dienstvorkommnis *n* ~ **jack** Anschalteklinke *f* ~ **jacks** Dienstklinken *pl* ~ **key(ing) circuit** Diensttastenleitungen *pl* ~ **level** Bedienungshöhe *f* ~ **life** Haltbarkeit *f*, Standzeit *f* ~ **line** Abzweig *m*, Hausanschluß *m* ~ **load** Dienstlast *f*, Gebrauchslast *f*, Nutzlast *f*, Zuladung *f* (eines Militärflugzeuges) ~ **manual** Dienstvorschrift *f* ~ **mechanic** Reisemonteur *m* ~ **message** Amtstelegramm *n*, Dienst-spruch *m*, -telegramm *n* ~ **meter** Gesprächszähler *m* ~ **motor vehicle** Dienstkraftfahrzeug *n* ~ **observation** Dienstüberwachung *f* ~-**observing board** Überwachungsschrank *m* ~ **observing desk** Kontrollplatz *m*, Überwachungsplatz *m* ~-**observing summary** Überwachungsblatt *n* ~ **platform** Bedienungsbühne *f* ~ **record** Führungszeugnis *n*, Lebenslaufakte *f*, Leistungsbuch *n* ~ **regulation** Dienstanweisung *f* ~ **rendered in return** Gegendienst *m* ~ **requirement** Betriebserfordernis *n* ~ **room** Dienstraum *m*, (in a lighthouse) Wachtraum *m* ~ **routine** Wartungsprogramm *n* ~ **rules** Dienstvorschrift *f* ~ **section** Betriebsabteilung *f* ~ **side** Bedienungsseite *f* ~ **stage** Betriebsreife *f* ~ **station** Tankstation *f* ~-**station salesman** Tankwart *m* ~ **stress** Betriebsbeanspruchung *f* ~ **switch** Dienstwähler *m* ~ **tank** Betriebs-behälter *m*, -tank *m* ~ **telephone line** Betriebsfernsprechleitung *f* ~ **telephone plant** Betriebsfernsprechanlage *f* ~ **test** praktischer Versuch *m* ~ **test oscillator** Kontroll-, Prüf-sender *m* ~ **toolkit** Bordwerkzeugtasche *f* ~ **traffic** Dienstverkehr *m* ~ **transmission** Leistungsübertragung *f* ~ **walkway** Dienststeg *m*, Fußweg *m* ~ **water** Gebrauchswasser *n* ~ **weight** Betriebsgewicht *n*

**serviceability** Brauchbarkeit *f*, Dauerhaltbarkeit *f*, Leistungsfähigkeit *f*, Wartbarkeit *f*

**serviceable** brauchbar, dienlich, einsatzfähig, leistungsfähig, nützlich, tauglich, wartbar, zweck-dienlich, -entsprechend ~ **condition** (of ammunition) Ladefähigkeit *f* **in** ~ **condition** ladefähig

**services rendered** Dienstleistungen *pl*

**servicing** Bedienung *f*, Behebung *f* der Panne, Instandhaltung *f*, Wartung *f* ~ **area** Wartungsbereich *m* (aviat.) ~ **compound** Pflegemittel *n* ~ **ladder** Bedienungstreppe *f* ~ **platform** Arbeitsbühne *f* ~ **premises** Wartungshallen *pl* ~ **schedule** Wartungsvorschrift *f* ~ **station** Betriebsstelle *f*

**serving** Packung *f*, Überzug *m*, Umhüllung *f*, Umwicklung *f* ~ **of thread** Fadenumschnürung *f* ~ **mallet** (colored) Kleidkeule *f* ~ **thread** Kennfaden *m*

**servo**, ~ **action** Servowirkung *f* ~ **arrangement** Servoeinrichtung *f* ~-**assisted** mit Servoeinrichtung *f* ~ **brake** Servobremse *f* ~ **control** Servo-lenkung *f*, -steuerung *f* ~-**control mechanism** Servomechanismus *m* ~ **coupler** automatisches Abstimmgerät *n* (d. Antenne) ~ **flap** Servoruder *n* ~ **line** Servo-, Steuer-leitung *f* ~ **loop** Regelkreis *m*, Servoschleife *f* ~-**mechanism** Hilfsvorrichtung *f*

**servomotor** Hilfsmotor *m*, Kraftverstärker *m*, Rudergetriebe *n*, Servomotor *m*, Stellmotor *m* ~-**operated controller** Steuerschaltung *f* mit Servomotor ~ **piston** Servomotorkolben *m* ~ **regulator** Servomotorregelung *f*

**servo**, ~-**piston** Kraftverstärkerkolben *m*, öldruckbetätigter Hilfskolben *m*, Servokolben *m*, Steuerschieber *m* ~**rudder (or control surface)** Flettnerseitenruder *n* (aviat.) ~ **shaft** Stellmotorwelle *f* ~ **speed control** Geschwindigkeitsservosteuerung *f* ~**tab** Ausgleichsruder *n*, Flettner(hilfs)ruder *n*, Servoruder *n* ~ **valve** Servo-, Steuer-ventil *n* ~ **valve spool** Steuerschieber *m*, Umsteuerkolben *m*

**sesquicarbonate of soda** anderthalbkohlensaures Natron *n*

**sesquioxide** Sesquioxyd *n*

**sesquiplane** Anderthalbdecker *m*, Eineinhalbdecker *m* (aviat.)

**sesquisilicate** Sesquisilikat *n*

**sesquisulfide of phosphorus** Phosphorsesquisulfid *n*

**sessile** nichtgleitfähig

**session** Sitzung *f*

**set, to** ~ (concrete) abbinden, abschirmen, adjustieren, anhalten, anordnen, anziehen, arretieren, aufspannen, aushärten, binden, bremsen, einbauen, einpassen, (a broken limb) einrichten, einstellen, erhärten, (cement) erstarren, (inserted parts or precious stones) fassen, festhalten, festlegen, feststellen, fest werden, konsolidieren, richten, (a saw) schränken, setzen, spannen, sperren, stellen, verfestigen, (cement) verhärten, (saw teeth) verschränken, zurichten to **be** ~ **up** sich bilden

**set, to** ~ **about** angreifen, in Angriff *m* nehmen **to** ~ **afire** in Brand *m* setzen, in Brand *m* stecken **to** ~ **afloat** flott machen **to** ~ **alight** entflammen **to** ~ **the altimeter** den Höhenmesser *m* auf Null stellen (einstellen) **to** ~ **up an apparatus** einen Apparat *m* zusammenstellen **to** ~ **aside** (marginal portions for sound track on film) aussparen **to** ~ **back** zurückstellen **to** ~ **the blast to work** das Gebläse anlassen **to** ~ **buoys** abbaken (das Fahrwasser) **to** ~ **a certain time** eine Frist *f* setzen **to** ~ **up a connection** eine Verbindung herstellen **to** ~ **a copy** ein Manuskript *n* absetzen **to** ~ **a date for**

a **hearing** (or for **hearing** a case) Termin *m* zur Verhandlung einer Sache anberaumen **to** ~ **the diaphragm** abblenden **to** ~ **down** absetzen (aviat.), zur Landung *f* ansetzen **to** ~ **a fire** Feuer *n* anmachen (anzünden), anstecken, zünden **to** ~ **on fire** anfeuern **to** ~ **free** befreien, betätigen, entbinden, entlassen, freimachen, loslösen **to** ~ **the fuse** den Zünder einstellen, tempieren **to** ~ **going** anregen, inbetriebsetzen **to** ~ **in** einsetzen, eintreten **to** ~ **out the line** die Linie abstecken **to** ~ **up a magnetic field** ein Magnetfeld *n* aufbauen **to** ~ **in motion** anlaufen lassen, anwerfen, ingangsetzen, in Schwung bringen **to** ~ **neutral** neutral einstellen **to** ~ **at normal** normal einstellen, normal schalten, regelrecht schalten **to** ~ **up a number** (on a key set) eine Zahl *f* einstellen **to** ~ **up a number** einstellen **to** ~ **off** ab-drucken, -liegen, -schmieren (slug), -schmutzen (print.), Getter *m*, abschießen, schießen **to** ~ **on** ansetzen **to** ~ **into oscillation** in Schwingung *f* versetzen **to** ~ **out** abstecken, (trip) antreten **to** ~ **a pole** eine Stange *f* setzen **to** ~ **a record** eine Bestleistung *f* aufstellen **to** ~ **right** berichtigen **to** ~ **the sight** Aufsatz *m* stellen **to** ~ **sights** justieren **to** ~ **up** Anlagen *pl* herstellen, ansetzen, aufbauen, aufmontieren, aufpflanzen, aufrichten, aufsetzen, aufstellen, einrichten, erzeugen, montieren, setzen (print.), zusammenbauen **to** ~ **off vaporization of getter** Getter *m* abschießen **to** ~ **to work** in Betrieb *m* setzen **to** ~ **to zero** auf Null *f* einstellen

**set** Anlage *f*, Apparatsatz *m*, Aufbauten *pl* (film), Ausstattung *f*, Batterie *f*, Biegung *f*, Garnitur *f*, Menge *f*, Niet(en)setzkopf *m*, Paket *n*, Rahmen *m*, Satz *m*, Treibrichtung *f*, Verformung *f*

**set,** ~ **at an angle** geneigt um ~ **of arms** Armkreuz *n* ~ **of assembly of code bars** Empfangsschienensatz *m* (telet.) ~ **of blades** Schaufelsatz *m* ~ **of brackets** Konsolausrüstung *f* ~ **of buckets** Schaufelsatz *m* ~ **of cam discs** Kurvenscheibensatz *m* ~ **of cams** Kurvensatz *m* ~ **of characters** Schriftzeichensätze *m* ~ **of coils** Spulensatz *m* ~ **of color plates** Farbsatz *m* ~ **of contact springs** Kontaktfedersatz *m* ~ **of curves** Kurvenschar *f* ~ **of dies** Backenpaar *n* ~ **of drawing instruments** Reißzeug *n* ~ **of equations** Gleichungssystem *n* ~ **of experiments** Versuchsfolge *f* ~ **of filter bags** ein Satz *m* Filterbeutel ~ **of glass plates** Glasplattensatz *m* ~ **of instruments** Besteck *n* ~ **of jaws** Backenpaar *n* ~ **of lenses** Objektivsatz *m* ~ **of lime pits** Äschergang *m* ~ **of lines** Linienschar *f* ~ **of mirrors** (in scanner) Spiegelkranz *m* ~ **of mixing rollers** Mischwalzwerk *n* ~ **of object glasses** Objektivsatz *m* ~ **of piles** Pfählung *f* ~ **of plates** Plattensatz *m* ~ **of points** Weichenstraße *f* ~ **of pulleys** Schraubenzug *m* ~ **of pumps** Pumpensatz *m* ~ **of rolls** Walzenpaar *n* ~ **of sieves** Siebsatz *m* ~ **of springs** Federn-bündel *n*, -paket *n* ~ **of supporting girders** Trägerlager *n* ~ **of templets** Schriftsatz *m* ~ **of testing nozzle** Prüfdüsensatz *m* ~ **of timbers** Geviert *n*, Grubenzimmerung *f* ~ **of tires** Bereifung *f* ~ **of tools** Handwerkszeug *n*, Werkzeug-besteck *n*, -garnitur *f* ~ **of tracks** Schienengeleise *n* ~ **in**

**type** abgesetzt ~ **up for** zugeschnitten auf ~ **of utensils** Besteck *n* ~ **of valves** Röhrenbestückung *f* ~ **of walking pipes** Gelenkrohrkühlung *f* ~ **of warp** Einstelldichte *f* der Kette ~ **of weights** Einsatzgewichte *pl*, (one fitting inside the other) Gewichtssatz *m* ~ **for zero beat** auf Schwebungsfrequenz *f* Null eingestellt, auf Überlagerung *f* der Trägerfrequenz eingestellt

**set** betriebsklar ~ **back** Rückschlag *m*; rückstellbar, versetzt (Gebäude) ~**-back counter** rückstellbarer Einzelzähler *m* ~**-back knob** Rückstellknopf *m* ~ **bar** Riegel *m* ~ **bolt** Kopfbolzen *m*, -schraube *f*, Zugbolzen *m* ~ **c** auf Mitte *f* ~ **collar** Begrenzungsring *m* (Haspel), Stellring *m* ~ **dresser** Bühnenaufsteller *m* ~ **frame** (textiles) Wattenmaschine *f* ~ **hammer** Breithammer *m*, Schellhammer *m*, Setz-hammer *m*, -meißel *m*, -stempel *m*, Wirkmesser *n* ~ **head** Setzkopf *m* ~ **lighting** Aggregatbeleuchtung *f* ~ **lights** Aufnahmelampen *pl* ~ **lining** Studioauskleidung *f* (film) ~ **mark** Einstellmarkierung *f* ~ **noise** Eigengeräusch *n* eines Geräts, Eigenrauschen *n*, Knackstörung *f*, Raumgeräusch *n*, Verstärkerrauschen *n* ~ **nut** Einstellmutter *f*

**setoff** Abdruck *m* (print.), Gegen-posten *m*, -rechnung *f*, (wall) Mauerabsatz ~ **base** vorspringende Grundschicht *f* ~ **button** (piano) Auslöseknopf *m*, Auslöserknopf *m* ~ **demand** Gegenforderung *f* (print.) ~ **device** Abschmutzvorrichtung *f* (print.) ~ **knob** Auslösungsknopf *m* ~ **sheet** Abschmutzbogen *m* (print.)

**set,** ~**-on** Haarseil *n* ~ **paving** (regelmäßiges) Reihenpflaster *n* ~ **pile** Vorschlagpfahl *m* ~ **pin** Haltezapfen *m*, Paßstift *m* ~ **point** Erstarrungspunkt *m*, Fixpunkt *m*, Sollwert *m* (aut. contr.) ~**-point value** Sollwert *m* ~ **pulse** Einstellimpuls *m* ~ **regulator for saw teeth** Schränklehre *f* für Sägezähne

**set-screw** Abdrück-, Ansatz-, Anschlag-, Bedien-, Druck-, Einstell-, Feststell-, Grenz-, Hemm-, Justier-, Klemm-, Kopf-, Maden-, Regulier-, Setz-, Stell-, Stellring-, Stiftschraube *f* ~ **of support** Ständerfußschraube *f* ~ **ring** Spannring *m*

**set,** ~ **square** Reiß-, Schiebe-, Zeichen-dreieck *n*, Winkelmaß *n* ~ **teeth** Einsetzzähne *pl* ~ **term** Frist *f* ~ **tuned** abgeschirmt

**set-up** (Werkzeugmaschinen) Aufbau *m*, Einrichtung *f*, Schema *n*, Schleier *m* (TV) ~ **for differential analyzers** Rechenschaltungen *pl* ~ **comb** Anschlagkamm *m* ~ **disposition** Anordnung *f* ~ **man** Werkzeugmaschineneinrichter *m* ~ **range** Aufsatzherd *m* ~ **weight** Einhängegewicht *n*

**set,** ~**-value control** Festverriegelung *f* ~ **wheel** Stellrad *n* ~ **winding** Einstellwicklung *f* ~ **zero** Nulleinstellung *f*

**sets, in** ~ satzweise

**setter** Einrichter *m*, (tools) Einsteller *m*, Setzmeißel *m*

**setting** (of cement) Abbinden *n*, Abbindung *f*, (radio) Abschirmung *f*, Arretierung *f*, Aufspannung *f*, Bettung *f*, (on points of control) Einpaßverfahren *n*, Einrichtearbeit *f*, Einspannung *f*, Einstelldrehung *f*, Einstellung *f*, (radar) Erdeinstellung *f*, Erhärten *n*, Erstarrung *f*, (jewel) Fassung *f*, Festlegemittel *n*, Justierung *f*,

Regelung *f*, Richten *n*, (saw) Schränkung *f*, Sitz *m*, Stellhemmung *f*, Stellung *f*, (of airscrew-pitch gauge) Uhrstellung *f*, Verfestigung *f*, Verlegen *n*, (of saw teeth) Verschränkung *f*, (by pontoon or floating sheers) Versenken *n*

**setting, ~ and composition** Filmarchitektur *f*, Filmbaukunst *f* **~ into action** Ansatz *m* **~ of the base** Basiseinstellung *f* **~ of beacons** Bebakung *f*, Betonung *f* **~ of blocks** Verlegung *f* der Blöcke **~ of the camera** Schwenkung *f* der Kamera **~ of diaphragm** Abblendung *f*, (action of) Blendeneinstellung *f* **~ of the focal length** Brennweiteneinstellung *f* **~ of the focusing screen** Mattscheibeneinstellung *f* **~ the front cutting edge** Vorderkanteneinstellung *f* **~ of the glue** Anziehen *n* des Leimes **~ of heating current** Heizstromeinstellung *f* **~ of the ignition cable** Zurichten *n* der Zündleitung **~ for impulse** Stellung *f* für Stromschritt **~ of mathematical formulae** mathematischer Satz *m* **~ of the mirror** Spiegelstellung *f* **~ in motion** Inbetriebsetzung *f* **~ in operation** Einrückung *f* **~ of piles** Bocksetzen *n* **~ by spanners** Schlüsseleinstellung *f* **~ for the stitch cam** Einstellung *f* der Senker **~ of the sun** Rüste *f* **~ the tolerances** Toleranzeinstellung *f* **~ by trial and error** systematisches Probieren *n* **~ of the tubbing** Setzen *n* der Cuvelage

**setting, ~ angle** Einstell-, Frei-winkel *m* **~ arm** Hebelübertragung *f* **~-aside margin** Randaussparung *f* **~ back** Zurücklegung *f* **~ bath** Fällbad *n* **~ blow** Setzschlag *m* **~ circle** (Teleskop) Einstellkreis *m* **~ collar** Einstellring *m* **~ condition** (Zement) Abbindeverhältnis *n* **~ (potentiometer) control** Einstellpotentiometer *m* **~ controls** Einstellsteuerungen *pl* **~ cost** Einrichtekosten *pl*, (tools) Einstellkosten *pl* **~ device** Einstellvorrichtung *f*, Stellvorrichtung *f* **~ dial** Sollwertskala *f* **~ dimension** Aufbaumaß *n* **~ disc** Meßscheibe *f* **~ gauge** Einstellehre *f* **~ going** (engine, etc.) Inbetriebsetzung *f* **~ head** Einstellknopf *m* **~ heat** Hydrationswärme *f* **~-in** Einsetzen *n* **~-in of impulse** Impulseinsatz *m* **~-in of oscillations** Schwingungs-anfachung *f*, -einsatz *m* **~ jaw** Schrankbacke *f* **~ knob** Einstellknopf *m* **~ lever** Richthebel *m* **~ magnet** Einstell-, Richt-magnet *m* **~ mark** Einstellmarke *f* **~ means** Sollwertgeber *m* **~ mechanism** Verstelleinrichtung *f* **~ member** Einstellglied *n* **~-off** Klatschdruck *m* **~-out** (surveying) Abstecken *n*, Ausscheiden *n* **~-out buildings** Bauabsteckung *f* **~-out curves** Kurvenabsteckung *f* **~-out machine** Ausstoßmaschine *f* **~ performance** Brechungsgeschwindigkeit *f* **~ period** (Maschinenwesen) Rüstzeit *f* **~ pin** Stellstift *m* **~ plug** Paßdorn *m* **~ point** Stockpunkt *m* **~ quality** (cement) Abbindefähigkeit *f* **~ range** Einstellbereich *m* **~ register** Einstellwerk *n* **~ ring** Stell-mutter *f*, -ring *m* **~ rule** Setzlinie *f* (print.) **~ scale** Einstellskala *f* **~ slicker** Stoßeisen *n* **~ speed** Einstellgeschwindigkeit *f* **~ spindle** Einstellspindel *f* **~ stick** Winkelhaken *m* **~ strength** Bindekraft *f* **~ test** (cement) Erstarrungsprobe *f* **~ time** Abbindezeit *f*, Bindezeit *f*, (f. Werkstück) Spannzeit *f* **~ tools** Stahleinstellen *n* **~ true to the perpendicular** Horizontierung *f*

**setting-up** Aufstellung *f*, Einrichten *n*, Einstellen *n*, Montage *f*, Montierung *f*, Voreinstellung *f* **~ of stresses** Entstehen *n* von Beanspruchungen

**setting-up, ~ apparatus** Aufsatzform *f* **(time of) ~ a call** Ausführungszeit *f* einer Verbindung **~ department** Einrichterei *f*, Einstellerei *f* **~ device** Aufsetzvorrichtung *f* **~ the objective** Zielsetzung *f* **~ time** Aufspann-, Einrichte-, Einstell-, Rüst-zeit *f*

**settings** Aufmachung *f*, Ausstattung *f* **~ (of controller knobs)** Abstimmung *f*

**settle, to ~** (account) abrechnen, absenken, absetzen, absetzen lassen, (affairs) abwickeln, ansiedeln, ausgleichen, begleichen, besiedeln, erledigen, festsetzen, sich festsetzen, feststellen, klären, klassieren, liquidieren, nachsacken, niederlassen, niederschlagen, ordnen, regeln, regulieren, zur Ruhe kommen, saldieren, schlichten, setzen, sich setzen, vermitteln **to ~ an account** die Rechnung bezahlen **to ~ for damage** einen Schaden *m* ersetzen **to ~ a dispute** Streit *m* beilegen **to ~ the dispute** Streit schlichten **to ~ down** absitzen, niederfallen **to ~ out** ausscheiden, heraussetzen, sedimentieren **to ~ into place** sich einnisten **to ~ in water** (im Wasser) absetzen

**settled** ansässig, beständig, festgesetzt, gesetzt, ruhig, ständig **~ curd soap** abgesetzte oder geschliffene Kernseife *f* **~ production** stabilisierte Produktion **~ speed** (Turbinenregler) Beharrungszustand *m*

**settlement** Abfindung *f*, Abkommen *n*, Abschluß *m*, Abschlußrechnung *f*, Absenkung *f*, Abwicklung *f*, Akkord *m*, Anordnung *f*, Ansiedlung *f*, (of accounts) Ausgleich *m*, Begleichung *f*, Besiedlung *f*, Bezahlung *f*, Einigung *f*, Kolonie *f*, Niederlassung *f*, Reg(e)lung *f*, Regulierung *f*, Sackung *f*, Senkung *f*, Setzung *f*, Siedlung *f* **day of ~** Vergleichs-tag *m*, -termin *m* **~ of accounts** Begleichung *f* der Rechnungen **~ of the ground** Senkung *f* des Grundes

**settlement, ~ gauge** Grundpegel *m* **~ range** Siedlungsherd *m* **~ zone** Versinkungszone *f*

**settler** Absetz-behälter *m*, -gefäß *n*, -tank *m*, Absitzbehälter *m*, Ansiedler *m*, Klärbecken *n*, Klärfaß *n*, Kolonist *m*, Verdickungsbehälter *m*, Vorherd *m*

**settling** Absetzen *n*, Absetzung *f*, Erledigung *f*, Klärung *f*, Klassierung *f*, Niederschlagung *f*, Sacken *n*, Sackung *f*, Senkung *f*, Setzenlassen *n*; erledigend **~ area** Absetzfläche *f*, Niederschlagsfläche *f* **~ basin** Absatzbecken *n*, Klärbassin *n*, -becken *n*, -sumpf *m*, -teich *m* **~ bottom** Setzbrett *n* **~ box** Klärkasten *m* **~ cask** Klärfaß *n* **~ chamber** (pumps or dredgers) Ablagerungsbecken *n*, (crushing mill) Absetzbecken *n*, (crushing mill) Absetzkammer *f*, Niederschlagskammer *f* **~ classification** Setzarbeit *f* **~ cone** Klärspitze *f* **~ cylinder** Absitzrohr *n* **~ dish** Satzschale *f* **~ down** Einlaufen *n* (Verschleiß) **~ head** Massel *f*, Saugmassel *f* **~ operation** Setzarbeit *f* **~ pond** Absatzteich *m*, Absetzteich *m*, Klär-sumpf *m*, -teich *m* **~ reservoir** Klärbassin *n* **~ room** (crushing mill) Absetzbecken *n* **~ speed** Absetzgeschwindigkeit *f* **~ sump** Klärtopf *m* **~ tank** Absetzbecken, -behälter *m*, -gefäß *n*, -tank *m*, Absitzbehälter *m*, Klär-bassin *n*, -becken *n*, -behälter

*m*, -bottich *m*, Setz-bett *n*, -maschine *f*, -tank *m*
~ **time** Anlaufzeit *f* ~ **tray (or trough)** Nieder-
schlagswanne *f* ~ **tub** Klär-bassin *n*, -bottich *m*,
Setzfaß *n*, Zusammenlaufbütte *f* ~ **unit (or
trough)** Beruhigungselement *n* ~ **vat** Klärbassin
*n*, Setzbottich *m* ~ **velocity** (falling velocity)
Fälligkeit *f* ~ **volume** Stampfvolumen *n* (von
Pigmenten)
**seven,** ~**-conductor cable terminal** Siebenfach-
kabelklemme *f* ~**-stage** siebenstufig
**sever, to** ~ abtrennen, auftrennen, lostrennen,
scheiden, sondern, trennen **to** ~ **a connection**
eine Verbindung *f* trennen
**severable** abschaltbar
**several** einige, mehrere, verschieden **with** ~
**blades** mehrscharig ~ **stations in a circuit**
mehrere Ämter *pl* in einer Leitung
**severance** Abtrennung *f*, Lösung *f*
**severe** forciert, hart, ohne Nachsicht *f*, scharf,
schwer, stramm, streng ~ **strain** hohe Beanspruchung *f*
**severing** Scheidung *f*, Trennung *f* ~ **relief** Ablösen *n*
**sew, to** ~ (binding) broschieren, heften, nähen
~ **together** zusammennähen **to** ~ **up** zuheften
**sewage** Abwasser *n* ~ **conveyor** Schwemmförde-
rer *m* ~ **disposal** Abwasserwesen *n* ~**-disposal
plant** Abwasserbehandlungsanlage *f* ~ **farm**
Rieselfeld *n* ~ **filter** Klärteich *m* ~ **gas** Faul-,
Klär-gas *n* ~ **powder** Düngpulver *n* ~ **pump**
Kanalisationspumpe *f* ~ **purification by
filtration** Schwemmfilterung *f* ~**-removal enterprise** Abfuhranstalt *f* ~**-treatment plant** Kläranlage *f* ~ **water** Kanalisationsabwässer *pl*
~**-water pump** Kanalwasserpumpe *f*
**sewed** geheftet ~ **quires** fadengehefteter Bogensatz *m*
**sewer** Abflußrohr *n*, Abwasserkanal *m*, (main)
Abzugskanal *m*, Abzugsschleuse *f*, Gassenrinne *f*, Gosse *f*, Hefter *m*, Kanal *m*, Kloake *f*,
Rinnstein *m*, Siel *n*, Straßen-kanal *m*, -rinne *f*
~ **cover** Kanaldeckel *m* ~ **gas** Klär-, Stadt-gas *n*
~ **man** (Abwasser) Kanalarbeiter *m* ~ **pipe**
Abwasserleitung *f*, Kanal(isations)rohr *n* ~
**system** Kanalisation *f*, Sielanlage *f*
**sewerage** Kanalisation *f*, Schwemmkanalisation *f*
**sewing** Heften *n*, Näharbeit *f*, Näherei *f* ~ **kit**
Flick-kasten *m*, -zeug *n*, Nähzeug *n* ~ **machine**
Nähmaschine *f* ~**-machine oil** Nähmaschinenöl
*n* ~**-machine shuttle** Nähmaschinenschiffchen *n*
~ **palm** Nähring *m* ~ **press** Heftlade *f* ~ **thread**
Nähgarn *n*
**sexadecimal,** ~ **notation** hexadezimale Schreibweise *f* ~ **number system** Sedezimalzahlen-
system *n*
**sexagenary (circular) graduation (or scale)** sexagesimale Kreisteilung *f*
**sexagesimal circle graduation** sexagesimale
Kreisteilung *f*
**sexivalent** sechswertig
**sexless connector** Zwitterstecker *m* (rdr)
**sextant** Gradbogen *m*, Sechstelkreis *m*, Sextant
*m*, Winkel-messer *m*, -spiegel *m*
**sexton** Kirchner *m*
**sextuple** sechsfach ~ **pump** Sechsfachpumpe *f* ~
**system** Sechsfachbetrieb *m*
**shabby** abgetragen, schäbig, zerlumpt
**shackle, to** ~ anschäkeln

**shackle** Bügel *m*, Bund *m*, Einspann-kopf *m*,
-vorrichtung *f*, Feder-bund *m*, -kasten *m*,
-lasche *f*, Lasche *f*, Schäkel *n*, Schelle *f*, Spannseil *n* ~ **band** Laschenspannband *n* ~ **bolt**
Gabelbolzen *m*, Gelenkschraube *f* ~ **eye**
Bügelhaken-, Schäkel-auge *n* ~ **hook** Bügel-
haken *m*, -auge *n*, Wirbelhaken *m* ~ **insulator**
Abspannisolator *m* ~ **line** Zuggestänge *n* ~ **pin**
Federlaschenbolzen *m* ~ **spring rigging**
Laschenfedergehänge *n*
**shackling** Schäkeln *n*
**shade, to** ~ (painting) abstufen, beschatten,
nuancieren, schattieren, schirmen, schraffieren,
stricheln **to** ~ **off** abschattieren, abstimmen
(photo)
**shade** Abstufung *f*, dunkelgetönte Farbe *f* oder
Färbung *f*, Farb(en)stufe *f*, Farbton *m*,
Nuance *f*, Schatten *m*, Schattierung *f*, Schirm
*m*, Schlagschatten *m*, Ton *m* ~ **of gray** Graustufe *f* (TV) ~ **to be matched** Mustervorlage *f*
**in the same** ~ tongleich ~ **for side light** Blendschirm *m* für Positionslicht
**shade,** ~ **cards** Farbtonkarten *pl* ~ **mosaic**
Schattenmosaik *n* ~ **temperature** Schatten-
temperatur *f*
**shaded** schraffiert, strichliert ~**-off transition**
verflauter Übergang *m* ~**-pole motor** Spaltpol-
motor *m* ~ **portion of drawing** Schraffur *f* ~
**rule** fettfeine Linie *f*
**shading** Schattierung *f*, (in drawings) Schraffierung *f*, Schraffur *f*, (of colors) Übergang *m*,
Verfärbung *f* ~**-compensation signal** Schatten-
kompensationssignal *n* ~ **condenser** Verdunk-
(e)lungskapazität *f* ~ **control** Rauschpegel-
regelung *f* (TV) ~ **dyestuff** Abtönungsfarbstoff
*m* ~ **generator** Schattengenerator *m* ~ **signal**
Abschattierungsstörsignal *n* ~ **value** Tönung *f*
~**-value signals** Bildhelligkeitssignale *pl* (TV)
~ **values** Helligkeitswerte *pl*
**shadless perforated tape** angelochter Lochstreifen *m* (telet.)
**shadow, to** ~ beschatten
**shadow** (image) dunkelster Bildpunkt *m*,
Schatten *m*, (zone) Schlierenzone *f* ~ **angle**
Schattenwinkel *m* ~**-area** Schattengebiet *n* ~
**bank** Wolkenschatten *m* ~ **border** Schattengrenze *f* ~**-casting jig** schattenwerfender
Schwengel *m* ~**-column instrument** Schatten-
bandinstrument *n* ~ **dyeing** Schattenfärbung *f*
~ **effect** Abschattungsverlust *m* ~ **factory**
Schattenwerk *n* ~ **image method** Schattenbild-
verfahren *n* ~ **images** Schattenbilder *pl* ~
**industry** Notfall-Industrie *f* ~ **mask** Loch-
maske *f* (TV) ~ **mask hole** Maskenloch *n* (TV)
~ **microscope** Schattenmikroskop *n* ~ **painting**
Sichtschutzanstrich *m*, Tarnanstrich *m* ~
**pencil (or pin)** Schattenstift *m* ~ **projection
microscopy** Schattenmikroskopie *f* ~ **range**
Schattenbereich *m* ~ **ratio** Schattenverhältnis *n*
~ **region** Schattenbereich *m*, Schweigezone *f* ~
**scattering** Brechungsstreuung *f* ~ **scratch**
optische Verkratzung *f* (film), Schramme *f* ~
**signal** unechtes Signal *n* ~ **tuning indicator**
Schattenanzeiger *m* ~ **wall** (of a glass tank)
Schirmwand *f*
**shadowgraph** Schatten(an)zeiger *m*
**shadowing** Schrägbedämpfung *f*
**shadows** Schattenpartien *pl*

**shady** schattig

**shaft** Achswelle *f*, Bohrstrang *m*, Bohrung *f*, Brunnen *m*, Dorn *m*, (mortar shell) Flügelschaft *m*, Gelenkrohr *n*, (of hammer) Helm *m*, Kohlenschacht *m*, Nische *f*, (mine) Schacht *m*, Schaft *m*, Schaftstab *m*, Seele *f*, Spindel *f*, Stahlwelle *f*, Stiel *m*, Wagenschere *f*, Walze *f*, Welle *f*, Zahnstange *f* ~ **for blower bar** Spindel *f* für Bläser ~ **with eccentrics forged on** Welle *f* mit angeschmiedeten Hubscheiben ~ **of elevating mechanism** Höhenrichtwelle *f* ~ **with forged flanges** Welle *f* mit angeschmiedeten Flanschen ~ **of a forked thill** Gabel-baum *m*, -rohrschaft *m* ~ **of frame antenna for direction finder** Peilrahmenschaft *f* ~ **with groove (keyway, or slot)** genutete Welle *f* ~ **for handwheel** Führungsspindel *f* ~ **of medium height** Halbhochofen *m* ~ **of mine** Stollenschacht *m* ~ **of rays** Strahlungsbüschel *m* ~ **of the stamp** Stempelstange *f* ~ **of steering arm** Lenkerwelle *f* ~ **for table movement mechanism** Transportwelle *f* ~ **of a winch** Haspelwelle *f*

**shaft,** ~ **alignment** Wellenfixierung *f* ~ **angle** (spiral bevel gear) Achswinkel *m* ~ **bar of a forked thill** Gabelriegel *m* einer Gabeldeichsel ~ **basis system** Einheitswelle *f* ~ **bearing** Achslager *n* ~ **bearing bush** Achslagerbuchse *f* ~ **bender** Gabelbieger *m* ~**-boring machine** Wellenbohrbank *f* ~ **bottom** Schaftboden *m* ~ **bucket conveyance** (Schacht) Gefäßförderung *f* ~ **butt** Wellenstumpf *m* ~ **collar** Wellenbund *m* ~ **cone** Wellenkegel *m* ~ **contact** Wellenkontakt *m* ~ **coupling** Wellenschalter *m* ~**-coupling flange** Kurbelwellenkupplungsflansch *m* ~ **cover** Wellenhülse *f* ~ **covering** Schachtabdeckung *f* ~ **cross section** Schachtscheibe *f* ~ **crucible furnace** Tiegelschachtofen *m* ~ **deflection** Wellendurchbiegung *f* ~ **diameter** Schachtdurchmesser *m* ~ **door** Schachtverschluß *m* ~ **drive (or driving)** Kardanantrieb *m*, Wellenantrieb *m* ~ **end** Wellen-ende *n*, -stumpf *m* ~ **failure** Wellenbruch *m* ~ **fire** Schachtbrand *m* ~ **flange** Wellenflansch *m* ~ **frame** Schachtgeviere *n*, -geviert *n* ~ **furnace** Blau-, Krumm-, Schacht-ofen *n* ~**-furnace melting** Schachtofenschmelzen *n* ~ **gland** Wellendichtung *f* ~ **governor** Achsenregler *m* ~ **hangar** Hängebock *m* ~ **hauling (or hoisting)** Schachtförderung *f* ~ **hole** Hammerauge *n* ~ **horsepower** Wellenpferde-kraft *f*, -stärke *f*, PS *f* (an der Welle) ~ **insert** Buchse *f* ~ **intersection** Achsenschnittpunkt *m* ~ **jack** Schaltspindel *f* ~ **journal** Wellenzapfen *m* ~ **key-seat** Wellennute *f* ~ **key-seater** Wellennutenstoßmaschine *f* ~ **landing** Abzugsbühne *f* ~ **line** Wellenleitung *f* ~ **lining** Schachtausbau *m* ~ **mining** Schachtbetrieb *m* ~ **mouth** Mundloch *n* ~ **nut** (Naben)-Wellenmutter *f* ~ **packing** Wellendichtung *f* ~ **partition** Schachtscheider *m* ~ **passing from end to end** durchgehende Welle *f* ~ **pinion** Wellenritzel *n* ~ **pipe** Schachtrohr *n* ~ **pivot** Wellenende *n* ~ **plate** Gelenkscheibe *f* ~ **portion** Schaftstück *n* ~ **position** Wellenlage *f* ~ **power** Wellenleistung *f* ~ **put down by boring** Bohrschacht *m* ~ **raising** Schachtförderung *f* ~ **revolution counter** Wellendrehzahlzähler *m* ~ **revolution indicator** Wellendrehzahlmesser *m* ~ **rod** Schaftstab *m* ~ **seal** Radialdichtung *f* ~

**sets** Schaftsätze *pl* ~ **setting** Schacht-geviere *n*, -geviert *n* ~ **sheds** Schachthalle *f* ~ **signal hammer** Schachthammer *m* (min.) ~ **signal recorder** Schachtsignalschreiber *m* (min.) ~ **sinking** Schachtabteufen *n* ~ **sinking through a layer (or a stratum)** Durchteufen *n* einer Schicht ~**-sliding mechanism** Steuergetriebe *n* ~ **stave** Schaftstab *m* ~**-straightening machine** Wellenrichtmaschine *f* ~**-straigtening press** Wellenrichtpresse *f* ~ **stub** Wellenstumpf *m* ~ **support** Wellenträger *m* ~ **timbering** Bolzenschrotzimmerung *f* ~ **timbers** Joche *pl* ~ **tower** Schachtgerüst *n* ~ **trace** Hinterstrang *m* ~**-turning lathe** Wellendrehmaschine *f* ~ **wall** Schachtwand *f* ~ **well** Schütznische *f*

**shafting** Wellenleitung *f* ~ **journal friction** Wellenflächenreibung *f*

**shake, to** ausschwenken, erschüttern, rütteln, schlottern, schütteln, schwanken, umschütteln, wackeln, zittern **to** ~ **down (or off)** abschütteln **to** ~ **loose (or off)** losschütteln **to** ~ **out** ausgleichen **to** ~ **out with ether** ausäthern **to** ~ **out a reef** ausreffen **to** ~ **thoroughly** durchschütteln **to** ~ **up** aufrütteln, aufschütteln

**shake** Ablösungsfläche *f*, Erschütterung *f*, Wimmer *m* ~**-free** unverrückbar ~**-key** Trillerklappe *f* ~ **number** Schüttelzahl *f* ~**-out dust** abgeschiedener Staub *m* ~**-proof** erschütterungssicher, schüttelfest, schüttelsicher

**shaker** Schüttel-apparat *m*, -sieb *n* ~ **conveyer trough** Schüttelrinne *f* ~ **loader** Schwingförderrinne *f* ~ **screen** Schüttelsieb *n*

**shaking** Erschütterung *f*, Rütteln *n*, Rütt(e)lung *f*, Schütteln *n*, Umschütteln *n* ~ **apparatus** Rüttelzeug *n*, Schüttelapparat *m*, Sichtezeug *n* ~ **arm** Sichtearm *m* ~ **barrel** Scheuer-faß *n*, -trommel *f* ~ **device** Schüttelvorrichtung *f* ~ **electrode** Schüttelelektrode *f* ~ **feeder** Schüttelaufgabevorrichtung *f*, -speiser *m* ~ **funnel** Schütteltrichter *m* ~ **gear** Schüttelvorrichtung *f* ~ **grate (or grizzly)** Schüttelrost *m* ~ **gutter** Schüttelrinne *f* ~ **machine** Schüttelmaschine *f* ~ **motion** Rüttelbewegung *f* ~ **screen** Planrätter *m*, Rüttelsieb *n*, Schwingsieb *n* ~ **shoot** Schüttelrutsche *f* ~ **sieve** Planrätter *m*, Schüttelsieb *n* ~ **sifter** Schüttelsieb *n* ~ **table** Planrätter *m*, Stoßtisch *m* ~ **trough** Schüttelrinne *f*

**shaky** (crevice) kernrissig, wackelig

**shale** Schiefer *m*, Schieferton *m* ~ **mud** schieferhaltiger Bohrschlamm *m* ~ **oil** Schieferöl *n* ~ **wax** Schieferparaffin *n*

**shaley water** Bergtrübe *f*

**shallow, to** ausbeulen **to** ~ **out a turn** die Kurvenlage verringern (aviat.)

**shallow** flach, niedrig, oberflächlich, seicht, (water) untief ~ **arch** Blendanstrichbogen *m* ~ **boring** flache Bohrung *f* ~ **cut** geringe Spanabnahme *f* ~ **cylindrical valve** niedriges Zylinderventil *n* ~ **depression** flache Mulde *f* ~ **dive** Stechflug *m* (der Bombe) ~ **draft** leichter Tiefgang *m* ~ **drawing** Tiefziehen *n* (metall.) ~ **footing** flaches Einzelfundament *n* ~ **forming** Flachdrücken *n* ~ **foundation** (Gründung) Flachfundation *f* ~ **punch press** Flachstanze *f* ~ **sea** Flachsee *f* ~ **stab-milling cut** Flachspanfräsen *n* ~ **subsidence** Senke *f* ~ **trough** Küvette *f* ~**-water cable** Küstenkabel *n*

**shallowness** Seichtigkeit *f*, Untiefe *f*

sham unecht

shambles Fleischbank *f*

shamfer, to ~ (founding gate) anschneiden

shammy Gemsleder *n*, sämisches Leder *n*

shank (Bohrer) Einsteckende *n*, Gabel *f*, Schaft *m*, Schenkel *m*, Stange *f*, Stiel *m*, Stielansatz *m*, (of casting ladle) Tragschere *f* ~ of anchor Ankerschaft *m* ~ of bit Bohrerzapfen *m* ~ of connecting rod Schubstangenschaft *m* ~ of hook Haken-schaft *m*, -zapfen *m* ~ of the lead Lotspindel *f* ~ of a letter Schriftkegel *m*

shank, ~ board Gelenkstutzpappe *f* ~ cutter Zapfenfräser *m* ~ hob Schaftwalzfräser *m* ~ ladle Gabelpfanne *f*, Gießpfanne *f* mit Handkippvorrichtung *f*, Scherpfanne *f* ~-type cutter Schaftfräser *m* ~-type profile cutter Schaftprofilfräser *m*

shape, to ~ ausgestalten, bilden, fassonieren, formen, fräsen, gestalten, profilieren, umformen, vorstrecken, vorwalzen to ~ course for absetzen to ~ by generating wälzstoßen

shape Bildung *f*, Fasson *f*, Fassonstein *m*, Figur *f*, Form *f*, Format *n*, Gestalt *f*, Gestaltung *f*, Linienführung *f*, Umriß *m*, Verlauf *m*

shape, ~ and appearance Formgestaltung *f* ~ of boiling curve Verlauf *m* der Siedekurve ~ of cam Daumenform *f* ~ of a curve Kurvenform *f* (Verlauf einer Kurve) ~ of face dam Profil *m* ~ of the frame Rahmenform *f* ~ of fuselage Rumpfform *f* ~ of a groove Rillenprofil *n* ~ of a mold Formteil *m* ~ of projectile Geschoßform *f* ~ of surface Oberflächengestalt *f*

shape, ~ correction Formkorrektor *m* ~ dependent parameter potentialformabhängiger Parameter *m* ~ factor Form-faktor *m*, -beiwert *m*, -ziffer *f* ~-independent potentialformunabhängig *f* ~ mill (or rolling mill) Profil(eisen)-walzwerk *n* ~ variations of the (explosives) charge and projectile Gestaltunterschiede *pl* der Ladung

shaped geprägt, gestaltet, profiliert, verformt ~ article abgepaßtes Warenstück *n* ~-beam antenna Antenne *f* mit spezieller Richtcharakteristik ~ blank vorgepreßte Ronde *f* ~ brick Formstein *m* ~ castings Formguß *m* ~ disc scheibenförmig ~ filler pipe for fuel Betankungsstutzen *m* ~ grinding wheel Formschleifscheibe *f* ~-out portion Becherausschnitt *m* ~ part Formstück *m* ~ piece gekumpeltes Stück *n* ~ plate Formblech *n* ~ section Formprofil *n* ~ wire Fasson-, Form-draht *m* ~ wood Profilholz *n*

shapeless formlos, gestaltlos

shaper Feil-, Fräs-maschine *f*

shaping Feilen *n*, Formbildung *f*, Formen *n*, Formgebung *f*, Formung *f*, (Holz) Fräsen *n*, Hobeln *n*, Holzbearbeitung *f*, Profilgestaltung *f*, Schäftung *f*, Umformung *f*, Vorkaliberwalzen *n*, Vorstrecken *n* ~ by hand Handstrich *m* ~ by machine tool with removal of chips (or by machining with removal of material by cutting tools) spanabhebende Formgebung *f*

shaping, ~ machine Fräsmaschine *f*, Kurzhobler *m*, Shapingmaschine *f*, Waagerechtsstoßmaschine *f* ~ machine shaper Stößelhobelmaschine *f* ~ mill Vorwalzwerk *n* ~ operation Fassonierarbeit *f* ~ pass Fertigstich *m*, Formkaliber *n*, Formstich *m*, Vorstich *m* ~ roll

Vorkaliber-, Vorstreck-walze *f* ~ strand Vorstrecke *f*, Vorwalzstrecke *f* ~ tool Fasson-, Form-stahl *m* ~ train (of rolls) Vorstraße *f*

shard cobalt Scherbenkobalt *m*

share, to ~ teilen to ~ in profits am Gewinn *m* beteiligt sein

share Aktie *f*, Anteil *m*, Beitrag *m*, Beteiligung *f*, Los *n*, Quantum *n*, Quote *f*, Teil *m* ~ of capital Kapitaleinlage *f* ~ in a mine Grubenanteil *m*, Kux *m* ~ in profits Dividende *f* ~ of the profits Gewinnanteil *m*

share, ~ capital Aktienkapital *n* ~ tenant Teilpächter *m*

shared anteilig (electr.), gemeinsam

shareholder Aktieninhaber *m*, Aktionär *m*, Gewerke *m*, Teilhaber *m*

shares are at par Aktien *pl* stehen auf pari

sharp (music) Halbton *m*, (music) Kreuz *n*, (music) Obertaste *f*; durchdringend, herb, scharf, schneidend, schneidig, spitz, spitzig, stark ~ angle of shoulder scharfer Ansatz *m* ~ angle of shoulders scharf abgesetzte Stelle *f* (mach.) ~ angle of step scharfer Ansatz *m* ~ angle of steps scharf abgesetzte Stelle *f* (mach.) ~ bank Abschwung *m*, enge Kurve *f* ~ climbing turn hochgezogene Steilkurve *f* (aviat.) ~ color hervorstechende Farbe *f* ~ curve Steilkurve *f* ~ cutoff Eckfrequenz *f* ~ definition Schärfe *f* ~ directional effect scharfe Richtwirkung *f* ~ dot spitzer Punkt *m* ~ edge scharfe Kante *f* ~-edged scharf-begrenzt, -kantig ~-edged gust scharfer Windstoß *m* ~ file Stoßfeile *f* ~ fire Glatt-brand *m*, -feuer *n* ~ focusing Scharfeinstellung *f* ~-freezing Gefrieren *n* bei tiefer Temperatur ~ groove scharfe Kerbe *f* ~ iron Schöreisen *n* ~ leading edge steile Vorderflanke *f* (d. Impulses) ~ nick (or notch) scharfer Kerb *m* ~ picture scharf eingestelltes Bild *n* ~ point Stachel *m* ~ pointed instrument Reißnadel *f* ~ road curve Straßenknie *n* ~ sand scharfer Sand *m* ~ setting Scharfeinstellung *f* ~ slags Abzüge *pl* ~ slit (or slot) scharfer Kerb *m* ~ thread Spitz(en)gewinde *n* ~ tuning Feinabstimmung *f*, feine oder scharfe Abstimmung *f*, Feinregulierung *f* (rdo) ~ turn kurze Kehrtwendung *f*, Steilkurve *f* ~-V thread scharfgängiges Gewinde *n*

sharpen, to ~ abziehen (mit dem Ölstein), anschleifen, schärfen, schleifen, wetzen, zuschärfen to ~ previously vorschärfen to ~ the reflex den Reflex *m* verstärken to ~ a scythe by hammering die Sense dengeln

sharpened zugeschärft

sharpener (pencil) Anspitzer *m* ~ for razor blades Klingenabziehapparat *m*

sharpening (in direction finding) Nullpunktschärfung *f*, Schärfung *f*, Schliff *m* ~ of minonum Enttrübung *f* (rdr)

sharpening, ~ anew Aufhauen *n*, Aufschärfung *f* ~ die Schärfstempel *m* ~ grinding Schärfschliff *m* ~ machine Anschärfmaschine *f*, Anspitzmaschine *f* (für Stangen) ~ machine for bars Stangenanspitzmaschine *f* ~ machine for tubes Rohranspitzmaschine *f* ~ stone Schärf-, Schleif-stein *m* ~ tool Flasche *f*

sharping knife Schab(e)messer *n*

sharpite Sharpit *m*

**sharply, ~ defined** scharf, scharfbegrenzt **~ defined hairline** haarscharfe Trennungslinie *f* **~ defined image** scharfes Bild *n* **~ focused image** scharf eingestelltes Bild *n* **~ peaked impulse** scharf begrenzter Impuls *m* **~ tuned** scharf abgestimmt

**sharpness** Scharfabbildung *f*, Schärfe *f* **~ of center** Mittenschärfe *f* **~ of control** Steuerschärfe *f* **~ of delineation** Schärfenzeichnung *f* **~ of edge** Randschärfe *f* **~ of focus** Strichschärfe *f* **~ of resonance** Abstimmschärfe *f*, Resonanzschärfe *f* **~ of resonance gauge** Peilschärfe *f* **~ of selecting network** Filtersteilheit *f* **~ of selective network** Siebsteilheit *f* **~ of tuning** Abstimm-, Resonanz-schärfe *f* **~ of vision** Sehschärfe *f*

**sharpness limit** Schärfenfeldgrenze *f* (film)

**shatoyant** buntschillernd

**shatter, to ~** platzen, splittern, zerschlagen, zersprengen **~ box** Schlagsieb *n* **~ oscillation** Zerreißfrequenz *f* **~ test** Rüttelprobe *f*, Schlagsiebprobe *f*

**shattering** Zertrümmerung *f* **~ power** Brisanz *f*

**shatterproof** splitter-bindend, -frei **~ glass** splittersicheres Glas *n*

**shave, to ~** bestoßen (print.), schaben, schleifen **to ~ off** abschaben **~ hook** Bleifeile *f*

**shaving** (wood) Hobelspan *m*, Schaben *n* (tape rec.), Span *m* **~ beam** Falzbock *m* **~ exhaust installation** Späneförderanlage *f* **~ knife** Gerbeisen *n*, Schabeisen *n*, Schab(e)messer *n* **~ machine** Drehbankrauter *m*, Schälmaschine *f* **~ suction plant** Späneabsauganlage *f* **~ tool** Schabfräser *m*

**shavings** Holzspäne *m pl*, Schabsel *n*, Schnitzel *pl*, Stoffabfall *m* **~ exhauster** Späneabscheider *m* **~ separator** Spänefänger *m*

**shawl** Schal *m*

**sheaf** Büschel *m*, Garbe *f* **~ of fire** Feuergarbe *f*, Garbe *f*, Geschoßgarbe *f* **~ of fire at point of impact** Einschlagsgarbe *f* **~-binding engine** Garbenselbstbinder *m*

**shear, to ~** abscheren, besäumen, beschneiden, gleiten, schneiden **to ~ fine** feinscheren **to ~ off** abscheren, abschneiden

**shear** Abscherung *f*, Beanspruchung *f*, Querkraft *f*, Schiebkraft *f*, Schub *m*, Vorschub *m* **~ (ing)** Scherung *f* **~ blade** Blechscherenmesser *n*, Scheren-blatt *n*, -messer *n* **~ bolt** Scherbolzen *m* **~ coefficient** Schubzahl *f* **~ cone** Scherungskegel *m* **~ diagram** Querkraftfläche *f* **~ distortion** anisotropische Verzeichnung *f* **~ flow** Scherungsströmung *f* **~ force** Scherkraft *f* **~ fracture** Scherungsbruch *m* **~ intensity** Scherungsstärke *f* **~ knives for textile plants** Scherenmesser *pl* für Textilfabriken **~ ladle** Schergießpfanne *f* **~ layer** Scherungsschicht *f* **~ leg** Pendelstütze *f* **~ legs** Bock *m*, Dreibein *m*, Dreifuß *m*, Rohrlegebock *m* **~ member** Abscherglied **~ mode of vibration** Schubschwingungsart *f* **~ modes** Scherschwingungen *pl* **~ modulus** Drillingsmodulus *m*, Gleitungszahl *f*, Scherungs-modul *m*, -modulus *m*, Schiebungsmodulus *m*, Torsionsmodulus *m* **~ pin** Scherbolzen *m*, Scherstift *m* **~-pin clutch** Abscherkupplung *f* **~ plane** Schiebungsfläche *f* **~ reinforcement** Schubbewehrung *f* **~ relaxation time** Schub-Relaxationszeit *f* **~**

**resistant** schubsteif **~ steel** Gerbstahl *m*, Scherenstahl *m*, Zementgerbstahl *m* **~ stiffness** Scherungssteifigkeit *f* **~ strain** Scherungsanteil *m*, Scherverformung *f* (Scherdehnung), Schubspannung *f* **~ strength** Scher-beanspruchung *f*, -festigkeit *f*, -spannung *f*, Schub-festigkeit *f*, -spannung *f* **~ stress** Scherungskomponente *m*, Schubspannung *f* **~-stress deformation curve** Schubformänderungskurve *f* **~ tension test** Scherzugversuch *m* **~ test** Reibungsversuch *m*, Scherprobe *f*, Scherversuch *m*, Schubversuch *m* **~ trajectories** Scherungslinien *pl* **~ transfer** Scherungsübertragung *f* **~ value** Abscherwert *m* **~ vibration** Quer-, Scher-schwingung *f* **~ viscosity** Schubviskosität *f* (accoust.) **~ wave** Scherungs-, Schub-welle *f* (accoust.) **~ wire** Scherdraht *m* **~ zone** Schubgebiet *n*

**shearer** Glasgießer *m*

**shearing** Abscheren *n*, Abscherung *f*, Scheren *n*, Schur *f*, Verschiebung *f*, Zerschneiden *n* **~ of an image** Bildscherung *f* **~ of the magnetic characteristic** Scherung *f* des Eisens

**shearing, ~ action** Abscherwirkung *f* **~ anvil** Scherbacke *f* **~ arm** (push arm) Schubarm *m* **~ blade** Abschneidemesser *n* **~ block** Abscherblock *m* **~ bolt** Abscherbolzen *m* **~ collar** Abscherring *m* **~ crack** Abscherungs-, Scherriß *m* **~ cross section** Abscherquerschnitt *m* **~ cut** Scherschnitt *m* **~ cylinder bearing** Scherzylinderlagergehäuse *n* **~ deformation** Schiebung *f* **~ demand** Schubbeanspruchung *f* **~ die** Schertempel *m* **~ effect** Abscherwirkung *f*, Scherwirkung *f* **~ force** Querkraft *f*, Scherbeanspruchung *f*, -spannung *f*, Schubkraft *f* **~ knife** Schermesser *n* **~ load** Querkraft *f*, Scherbelastung *f* **~ machine** (textiles) Abstutzmaschine *f*, Maschinenschere *f*, Tuchschermaschine *f* **~ machine for cutting reinforcement of concrete** Betoneisenschere *f* **~ mechanism** Scherzeug *n* **~ modulus of elasticity** Elastizitätsmodul *m* für Schub, Schub-modul *m*, -steife *f* **~ pin** Abscherstift *m* **~ plane** Gleitebene *f* **~ resilience** Scherspannung *f* **~ resistance** Scherfestigkeit *f* **~ strain** Abscherungsbeanspruchung *f*, Beanspruchung *f* auf Abscherung, Gleitung *f*, Scher-beanspruchung *f*, -spannung *f*, spezifische Schiebung *f* **~ strength** Abscher-(ungs)festigkeit *f*, Scherungsfestigkeit *f*, Schubwiderstand *m* **~ stress** Scherkraft *f*, Schubkraft *f*, Schubriegelspannung *f* **~ stress lines** Schublinien *pl* **~ stress-stream diagram** Spannungsschiebungsdiagramm *n* **~ test** Abscher-, Schub-versuch *m* **~-test apparatus** Scherapparat *m* **~ tools** Scherwerkzeuge *pl* **~ viscosity** Scherungsviskosität *f*

**shears** Abschneideschere *f*, Freischnitt *m*, Pappschere *f*, Schere *f* **~ for dividing the wire** Halbierschere *f*

**sheath** Futteral *n*, Hülle *f*, Scheide *f*, Schicht *f*, Schutzhülle *f* **~ (ing)** Armierung *f*, (of a cable) Bewehrung *f*, Mantel *m* **~ converter** Umwandlungselement *n* **~ cutting pliers** (cable) Mantelwirbelstrom *m* **~ stripping head** Abmantelkopf *m* **~ wave** Mantelwelle *f*

**sheathe, to ~** (cable) armieren, bekleiden, beschlagen, bewehren, (ships, etc.) doppeln, einhüllen, planken, umkleiden, ummanteln, verkleiden

**sheathed** armiert, bewehrt, blechumhüllt ~ **electrode** elektrodeblechumhüllte Elektrode *f*, ummantelter Schweißstab *m* ~ **explosive** Mantelsprengstoff *m*

**sheating** Bekleidung *f*, Beplattung *f*, Bohlenbelag *m*, Füllraum *m*, Kantenschutz *m*, Schalung *f*, Schutzmantel *m*, Schutzummantelung *f*, Stanniolüberzug *m*, Stulpwand *f*, Umhüllung *f*, Ummantelung *f*, Verhäutung *f*, Verkleidung *f*, Zimmerung *f* ~ **with copper** Kupfern *n* ~ **wire** Bewehrungsdraht *m*, Schutzdraht *m*

**sheave** Auflaufhaspel *m*, Kettenleitrolle *f*, (guide) Leitrolle *f*, Rillenscheibe *f*, Rolle *f*, Rollkolben *m*, Scheibe *f*, Wirbel *m* ~ **block and tackle** Klobenrolle *f* ~ **blocks** Blöcke *pl* ~ **groove** Scheibenrille *f* ~ **height** Rollenhöhe *f* ~ **man** Haspelversetzer *m* ~ **pulley** Giebelrolle *f*, Umführrolle *f* ~ **ratio** Scheibenübersetzung *f*

**shed, to** ~ schuppen, vergießen **to** ~ **skin** häuten

**shed** Dach *n*, Halle *f*, Isolierglocke *f*, Kaue *f*, Schauer *m*, Schirmdach *n*, Schuppen *m*, Schutzdach *n* ~ **construction** Hallenbau *m* ~ **rods** (of a loom, textiles) Fadenteilstangen *pl* ~ **roof** Halbdach *n*

**shedding of vortices** Wirbelablösung *f*

**sheen** Glanz *m*, Widerschein *m*

**sheep gong** Schalmeiglocke *f*

**sheepsfoot roller** Schaffußwalze *f*

**sheer, to** ~ (a line) ausstraken **to** ~ **off (or away)** abgieren (Schiff), scheren

**sheer** Decksprung *m*, Deckstrak *m* ~ **iron** Strakgewicht *n* ~ **line** Schertau *n*

**sheet, to** ~ **out** auswalzen

**sheet** Blatt *n* (Papier), Blechstreifen *m*, dünnes Blech, (of paper) Bogen *m*, Feinblech *n*, Fläche *f*, Lagergang *m*, Laken *n*, Lamelle *f* einer Gürtung, Mantel *m*, Platte *f*, Scheibe *f*, Streifen *m*, (of metall.) Tafel *f*

**sheet,** ~ **of blotting paper** Löschblatt *n* ~ **of cardboard** Pappbogen *m* ~ **of corrugated iron** Wellblechtafel *f* ~ **of insulating material** Isolierplatte *f* ~ **of metal** Lamelle *f* ~ **of pasteboard** Pappbogen *m* ~ **of plotting drum** Folie *f* ~ **of the quires** Lage *f* des Bogensatzes ~ **of stationery** Briefbogen *m* ~ **of water discharge (or overflowing)** Überfall *m*

**sheet,** ~ **anchor** Hauptanker *m* ~**-and-wire industry** Blech- und Drahtindustrie *f* ~ **antenna** Flächenantenne *f* ~ **baffle** Faltenkörper *m* ~ **bar** Blechstab *m*, Flachknüppel *m*, Platine *f* ~ **bar rolling mill** Platinwalzwerk *n* ~**-bar shears** Platinenschere *f* ~**-bar storage** Platinenlager *n* ~**-bar yard** Knüppellager *n* ~**-bordering machine** Blechbördelmaschine *f* ~ **brass** Blechmessing *n*, -scheiben *pl*, Messingblech *n*, Tafelmessing *n* ~ **brass in rolls** Rolltomback *m* ~**-by-~feeder** schuppenförmiger Bogentransport *m* ~ **calender** Bogenkalender *m* ~ **card** Blätterkarde *f* ~ **copper** Blattkupfer *n*, Kupferblech *n*, Walzkupfer *n* ~ **counter** Bogenzähler *m* ~ **cover** Abdeckblech *n* ~ **cutter** Bogenschneider *m* ~ **delivery** Bogenausleger *m* ~ **doubler** Blechbiegemaschine *f*, Blechdoppler *m* ~ **drying loft** Plattenmansarde *f* ~ **electron** Elektronblech *n* ~**-fed gravure machine** Tiefdruckbogenmaschine *f* ~**-fed offset press** Bogenoffsetmaschine *f* ~**-fed photogravure rotary machine** Tiefdruckrotationsmaschine

*f* für Bogenanlage ~**-fed press** Bogenmaschine *f* ~**-fed rotary letter press** Hochdruckbogenrotationsmaschine *f* ~**-fed rotary machine** Bogenrotationsmaschine *f* ~ **feeding apparatus** Bogenanlegeapparat *m* ~ **filter** Schichtenfilter *m* ~ **fly** Bogenauswerfer *m* ~ **folder** Blechbiegemaschine *f* ~ **furnace** Blech(wärm)ofen *m* ~ **gasket** Scheibendichtung *f* ~ **gelatin** Gelatinefolien *pl* ~ **glass** Tafelglas *n*, Walzenglas *n* ~ **guide** Leitblech *n* (print.), Leitzunge *f* ~ **guide spring** Bogenführungsfeder *f* ~ **heater** Blechwärmer *m* ~ **hook double** Doppelhaken *m* ~ **ingot** Walzbarren *m* ~ **intergrowth** Schichttextur *f*

**sheet-iron** Blech *n*, Blech-eisen *n*, -tafel *f*, (edging) Borte *f* von Blech, Eisenblech *n*, Stahlblech *n* ~ **for deep stamping** Tiefstanzblech *n*

**sheet-iron,** ~ **cover** Blechmantel *m* ~ **diaphragm** Eisenblechmembrane *f* ~ **dust bin** Müllkasten *m* aus Eisenblech ~ **jacketing** Eisenblechummatelung *f* ~ **lining** Blechmantel *m* ~ **mold** Eisenblechschalung *f* ~ **panel** Tafel *f* von Eisenblech ~ **piles** Kanaldiele *f* ~ **pipe** Blechrohr *n*, -röhre *f* ~ **plate** Schwarzblechtafel *f* ~ **shears** Blechschere *f* ~ **shoe** Blechschuh *m*

**sheet,** ~ **jogger** Bogengeradeleger *m* (print.) ~ **lead** Bleiblech *n*, Tafelblei *n*, Walzblei *n* ~ **leveler** Blechrichtmaschine *f* ~ **lightning** Flächenblitz *m* ~ **magnesium** Magnesiumblech *n* ~ **material** Bandmaterial *n*

**sheet-metal** Beplankung *f*, Blech *n*, Blechmaterial *n*, -panzer *m* (furnace), Metall-blech *n*, -platte *f*, Rauch *m*, Walzblech *n* ~ **for forming (or for deep drawing)** Tiefziehblech *n* ~ **for stencils** Schablonenblech *n* ~ **for wall facing** Wandbekleidungsblech *n*

**sheet-metal,** ~ **blade** Blechschaufel *f* ~ **box** Blechkasten *m* ~ **cap** Blechkapsel *f* ~ **case** Blechgehäuse *n* ~ **casing** Blechhülle *f* ~ **clip** Blechklemme *f* ~ **cover** Blech-verkleidung *f*, -dekel *m* ~ **covering** Blechverschalung *f* ~ **cutter** Blechschneidemaschine *f* ~ **cylinder** Blechmantel *m* ~ **deflector** Leitblech *n* ~ **drum** Blechtrommel *f* ~ **goods** Blechwaren *pl* ~ **guide** Leitblech *n* ~ **holder (or bracket)** Halteblech *n* ~ **housing** Blechgehäuse *n* ~ **iron** Blechtafel *f* ~ **manufacture** Blechausführung *f* ~ **plate** Blechtafel *f* ~ **plateholder** Blechkassette *f* ~ **printing** Blechdruck *m* ~ **punch** Blechaushauer *m* ~ **punching** Metallblechstanzen *n* ~ **radiator** Lamellenkühler *m*, Streifenkühler *m* ~ **roof** Blechdach *n* ~ **scrap** Blechschrott *m* ~ **shell** Blechmantel *m* ~ **sleeve** Blechhülse *f* ~ **stamping** Metallblechstanzen *n* ~ **store** Blechlager *n* ~ **strip** Blechstreifen *m* ~ **structure** Blechkonstruktion *f* ~ **tank** Blechbehälter *m* ~**-testing apparatus** Blechprüfapparat *m* ~ **trough** Blechmulde *f* ~ **tube** Blech-rohr *n*, -röhre *f* ~ **varnishing machine** Blechlackiermaschine *f* ~ **work** Klempnerarbeit *f* ~ **worker** Klempner *m*, Spengler *m* ~**-working plant** Blechbearbeitungsanlage *f*

**sheet,** ~ **mica** Glimmer-folie *f*, -platte *f*, -tafel *f* ~ **mill** Blechwalzwerk *n*, Feinblechanlage *f* ~ **mill stand** Blechgerüst *n*, Feinblechgerüst *n* ~**-mill train** Blechstrecke *f* ~ **mills** Feinblechwalzwerke *pl* ~ **pack** Blech-packen *m*, -sturz *m* ~-pack heating furnace Blechsturzwärmofen *m* ~ panel Blechtafel *f* ~-passage Bahndurchlauf *m*, Bogenübergang *m* ~ pile Spundbohle *f* ~ pile bulkhead Fang-, Koffer-damm *n* ~ pile cut-off Spundwandabdichtung *f* ~ pile screen Dichtungsspundwand *f* ~ pile shoe Spundbohlenschuh *m* ~ piles and piling Falzpfahl *m* ~ piling Spund-bohle *f*, -wand *f* ~ piling of reinforced concrete Eisenbetonspundwand *f* ~ plate Blechtafel *f* ~-pole wall Spundwand *f* ~ products Blechwarenerzeugnisse *pl* ~ pusher Bogenschiebeapparat *m* ~ remover Bogenabstreifer *m* ~ resistance Scheibenwiderstand *m* ~ roll Feinblechwalze *f*, Glattwalze *f* ~ rolling Feinblechwalzerei *f* ~-rolling mill Blechwalzwerk *n*, Feinblechstraße *f*, Feinblechwalzwerke *pl* ~-rolling train Blechstraße *f*, Feinblechstraße *f* ~ rubber Blattgummi *m*, Kautschukplatte *f* ~ separator spring Abstreichfeder *f* ~ shears Tafelschere *f* ~ signature Bogenzeichen *n* ~ sizing Leimung *f* im Bogen ~ slow-down Bogenbremse *f* ~ smoother Bogengeradestreifer *m*, Bremsband *n* ~ steadier Bogenstreicher *m* ~ steadier spring Streichfeder *f* ~ steel Stahlblech *n* ~ steel with good punching properties Stanzblech *n* ~ steel casing Stahlblechmantel *m* ~-steel cubicle Stahlblechschrank *m* ~ steel enclosure Blechschutzgehäuse *n* ~ steel wall Zwischenwandstahlblech *n* ~ straightening Blech-richten *n*, -spannen *n* ~-straightening machine Blechspannmaschine *f* ~ support plates Verbreiterung *f* für Rechenstab ~ tension Membranspannung *f* ~ tin Walzzinn *n*, Zinnblech *n* ~ transfer Bogenübergabe *f* ~ transfer mechanism Schwinganlage *f* ~ trimmings Blechstanzen *pl* ~ type Schichtenklasse *f* ~ types Schichtensorten *pl* ~ zinc Zinkblech *n*

**sheeting** Beplankung *f*, Blechverkleidung *f*, (plastics) Folien *pl*, Füllraum *m*, Schalwand *f*, (with plywood) Verschalung *f* ~ board Schalbrett *n* ~ pile Abtreibearbeit *f* in losem Gebirge

**sheetlike, of** ~ form (or nature) flächenförmig

**sheets for electrical industry** Elektrobleche *pl*

**shelf** Balkweger *m*, Brett *n*, Bund *m*, Dünnbrett *n*, Etagere *f*, Fach *n*, Gemeinlade *f*, Gestell *n*, Planke *f*, Regal *n*, Riff *n*, Schelf *m*, Schrankfachbrett *n*, Sims *m*, Untiefe *f*, vorspringende Kante *f* ~ for metal types Schriftregal *n*

**shelf,** ~ drier Trockenschrank *m* ~ furnace Schüttofen *m* ~ life deterioration Lagerungsverschlechterung *f*

**shell, to** ~ abschälen to ~ off abblättern

**shell** Außenhaut *f*, äußere Haut, Blechkörper *m*, Bohrlöffel *m*, Geschoß *n*, Granate *f*, Haut *f*, (of blast furnace) Hemd *n*, Hülle *f*, Kapsel *f*, Klobengehäuse *n* einer Rolle, Lagerstützschale *f*, Mantel-fläche *f*, -gehäuse *n*, Muschel *f*, Muschelschale *f*, Randzone *f*, rundes Gehäuse *n*, Schale *f*, Schild *n*, Schülpe *f*, Wand *f*, Zünderhalter *m* ~ of arc furnace Wanne *f* ~ of the breaker Brechmantel *m* ~ and coil condenser Schlangenkesselverflüssiger *m* ~ of a furnace Rauchschacht *m* ~ of a mold Form-kappe *f*, -kapsel *f* ~ and tube heat exchanger Mantel- und Rohrwärmeaustauscher *m*

**shell,** ~-absorption coefficient Schalenabschirmungskoeffizient *m* ~ aperture Schalenöffnung

*f* ~ auger Hohl-, Löffel-bohrer *m* ~ baffle zylindrisches Leitblech *n* ~ bars Grundmesser *pl* ~ belt Kesselschluß *m* ~ body Schalenrumpf *m* ~ case Granathülse *f*, Hülse *f*, (semifixed) Kartusche *f*, Lagerschalengehäuse *n* ~ casing (ballistics) Ladungshülle *f* ~ cavity Geschoßhöhlung *f* ~ chamber Geschoßkammer *f* ~ circuit Topfkreis *m* ~ construction Schalen-bau *m*, -konstruktion *f* ~ core Mantelkern *m* ~ crater Granat-loch *n*, -trichter *m*, Sprengtrichter *m* ~ distribution Schalenverteilung *f* ~-drawing die Patronenzugstempel *m* ~ drill Aufstecksenker *m* ~ electron Hüllerelektron *n* ~-end mill Aufsteck-, Kopf-fräser *m* ~-end mill with keyway Walzenstirnfräser *m* mit Längskeilnut ~-end mill with slot Walzenstirnfräser *m* mit Querkeilnut ~ end reamer Stangenreibahle *f* ~ filler Granatfüllung *f* ~-firing gun Granaten verfeuernde Kanone *f*, Granatkanone *f* ~ fragment Granatsplitter *m* ~ hook Geschoßheber *m* ~ jacket Geschoßmantel *m* ~-like muschelig ~ lime Muschelkalk *m* ~ lining Mantelfutter *n*, (of converter) Wandauskleidung *f*, (of converter) Wandfutter *n* ~ marl Muschelmergel *m* ~ model Schalenmodell *n* ~ mold Hohlform *f* ~ molding Schalengußverfahren *n* ~ plate Mantelblech *n* ~-plate button hohler Blechknopf *m* ~ plating Haut *f* ~ proof granat-, schuß-sicher ~ reamer Aufsteckreibahle *f*, Hohlreibahle *f* ~ ring Kesselschluß *m* ~ room Granatenlast *f* ~ sand Muschelsand *m* ~-shaped schalenförmig ~ spectrum Hüllenspektrum *n* ~ splinter Granatsplitter *m* ~ theory Schalentheorie *f* ~ transformer Manteltrafo *m* ~-type boiler Zylinderkessel *m* ~-type cooler Mantelkühler *m* ~-type core Schalenkern *m* ~ vacancy Schalenleerstelle *f* ~ wall Geschoßwand *f* ~ wave Mantelwelle *f* ~ zone Randzone *f*

**shellac, to** ~ schellackieren

**shellac** Schalenlack *m*, Schellack *m* ~ substitute Achatschellack *m*

**shelled** schalig

**sheller** Enthülser *m*, Entschäler *m*

**shelter, to** ~ beherbergen, schützen

**shelter** Deckung *f*, Herberge *f*, Obdach *n*, Schauer *m*, Schirm *m*, Schutz *m*, Schutzdach *n*, Schutzraum *m*, Überdeckung *f*, Unterkunft *f*, Unterkunftsraum *m*, Zufluchtshafen *m* ~ trench Deckungsgraben *m* ~ wall Schutzmauer *f*

**sheltered** gedeckt ~ basin Zufluchtshafen *m* ~ harbor geschützter Hafen *m*

**sheltering effect** Abdeckeffekt *m*, Schirmwirkung *f*

**shelving** Fach *n* ~ coast Steilküste *f* ~ dune Steildüne *f* ~ shore abfallendes Ufer *n*

**sherardize, to** ~ sherardisieren, trocken verzinken

**sherardizing** Sherardisieren *n*, Sherardisierung *f* ~ process Sherardisierverfahren *n*

**sherbet** Gefrorenes *n* ~ machine Fruchteiserzeuger *m*

**shield, to** ~ abschirmen, (in radiography) ausbleien, beschirmen, entstören (rdo), (from) schirmen, schützen

**shield** Blende *f*, Funkenentstörer *m*, (electrode) Lichtsteuerelektrode *f*, Lochblende *f*, Panzer *m*,

Schild *n*, Schildklappe *f*, Schirm *m*, Schirmung *f*, Schußbühne *f*, Schutz *m*, Schutzgitter *n*, Steuergitter *n*, Tunnelschild *n*
**shield,** ~**-base** Abschirmungsträger *m* ~ **chamber** Abschirmkammer *f* ~ **coil** Käfigspule *f* ~ **factor** Abschirmfaktor *m*, Schirmfaktor *m* ~ **grid** Schirmelektrode *f* ~ **rings** Abschirmungsringe *pl* ~ **support** Schildstütze *f* ~ **tube** Abschirm-, Schutz-rohr *n* ~ **wire** Schirmdraht *m*
**shielded** gepanzert, (ab)geschirmt ~ **for radio interference** funkentstört
**shielded,** ~ **actinometer** Panzeraktionometer *m* ~ **antenna** abgeschirmte Antenne *f* ~ **arc electrode** Elektrode *f* mit gasumhülltem Lichtbogen ~ **arc welding** Schutzgasschweißung *f* ~ **ball bearing** geschütztes Kugellager *n* ~ **electromagnet** Topfmagnet *m* ~ **galvanometer** Panzergalvanometer *m* ~ **line** abgeschirmte Leitung *f* ~**-metal arc welding** Lichtbogenschweißung *f* unter Schutzgas ~ **nuclide** abgeschirmtes Nuklid *n* ~ **pair** abgeschirmte Doppelader *f* ~ **plug** vollentstörte Kerze *f* ~ **transformer** geschirmter Übertrager *m*, Käfigtransformer *m*, Manteltransformator *m*, Panzertransformator *m* ~ **transmission line** abgeschirmte Leitung *f*
**shielding** Abschirmung *f*, Abschützung *f*, Entstörung *f*, Panzerung *f*, Schirmung *f*, Schutzbedeckung *f*, -bekleidung *f*, Siebung *f* ~ **of cables with wire mesh** Bewehrung *f*
**shielding,** ~ **action** Schirmwirkung *f* ~ **braiding** Schirmgeflecht *n* ~ **can** Abschirmbecher *m* ~ **effect** abschirmende Wirkung *f* ~ **efficiency** Schirmdämpfung *f* ~ **factor** Durchgriff *m*, Schirmfaktor *m* ~ **harness** Entstörgeschirr *n*, Störschutzpackung *f* ~ **radiation** strahlungsabschirmend ~ **sleeve** Abschirmschlauch *m* ~ **surface** Schirmfläche *f* ~ **tube** Entstörschlauch *m*
**shift, to** ~ ausrücken, einrücken, gieken, (a lever) einen Hebel umlegen, schalten, schieben, umladen, (wind) umlaufen, umschießen, umschlagen, (a phase) verdrehen, verlagern, verlegen, verrücken, verschieben, versetzen, verstellen, wandern, wechseln **to** ~ **berth** den Ankerplatz *m* wechseln **to** ~ **the brushes** die Bürsten *pl* verstellen **to** ~ **a cable** ein Kabel *n* verlegen **to** ~ **a gear** ein Getriebe *n* schalten **to** ~ **to exact center** nachzentrieren **to** ~ **by oil pressure** ölhydraulisch verstellen **to** ~ **point of impact** Aufschlagpunkt *m* verlegen
**shift** Arbeits-plan *m*, -schicht *f*, (signal) Figurenwechsel *m*, Schicht *f*, Schieben *n* (data proc.), Schub *m*, Umschaltung *f*, Verlagerung *f*, Verlegen *n*, Verlegung *f*, Verrückung *f*, Verschiebung *f*, Verstellung *f*, Wechsel *m*
**shift,** ~ **of aerodynamic center** Druckpunktwanderung *f* ~ **of an airplane** Flugzeugversetzung *f* ~ **of the fault** Sprungweite *f* ~ **of line position** Zellenverlagerung *f* ~ **in phase** Phasensprung *m* ~ **of spectral lines toward red or longer waves** Rotverschiebung *f* ~ **of vision** Bildsprung *f* ~ **of wind** Windsprung *m*
**shift,** ~ **angle** Fehlwinkel *m* ~ **claw** Schaltgabel *f* ~ **control** Regler *m* für Strahlverschiebung ~ **disc** Umschaltscheibe *f* ~ **dog** Schaltklaue *f* ~ **factor** Verschiebungsfaktor *m* ~ **finger** Schaltfinger *m* ~ **fork** Schaltgabel *f* ~ **forward** Vorverlegung *f* ~ **framing** Phaseneinstellung *f* ~

**gate** Schaltführung *f* ~**-gear transmissions** Schaltgetriebe *n* ~**-head** Rückkopf *m* ~ **interlock** Verriegelungsschieber *m* ~ **key** Umschaltetaste *f*, Wechseltaste *f* ~ **key lever** Umschalthebel *m* ~ **lever** Riemenrück-, Schalt-, Umlenk-hebel *m* ~**-lever dust cover** Schaltdom ~ **lever shaft** Schalthebelwelle *f* ~**-lock** Feststeller *m*, Feststelltaste *f*, Umschaltefeststeller *m* ~**-lock of the keys** Tastenverriegelung *f* ~**-lock key** Feststelltaste *f* ~**-lock keyboard** Tastatur *f* mit Verriegelung ~ **lock mechanism** Feststellung *f* ~ **mechanism** Umschalteinrichtung *f* ~ **operation** Umschaltmaßnahme *f* ~ **period** Schaltperiode *f* ~ **plate** Schaltblech *n* ~ **pulse** Schiebeimpuls *m* ~ **rail** Schaltschiene *f* ~ **rail frame** Führungsrahmen *m* ~ **register** Schieberegister *n*, -speicher *m*, Verschiebungsregister *n* ~ **rocker** Wechselwippe *f* ~ **rod frame** Schaltstangenführung *f* ~ **signal** Wechselzeichen *n* ~ **unit** Verschiebeeinrichtung *f* ~ **work** Schichtarbeit *f*
**shiftable** posaunenartig verschiebbar, verschiebbar ~ **belt conveyor** rückbare Bandanlage *f* ~ **drive terminal** rückbare Antriebsstation *f*
**shifted** verschoben, versetzt ~ **phase** verschobene Phase *f*
**shifter** (belt) Ausrücker *m*, (belt) Ausrückvorrichtung *f*, Einrücker *m*, Schalter *m*, Umleger *m*, Verschiebeeinrichtung *f* (info proc.) ~ **collar** Verschiebemuffe *f* ~ **finger (or fork)** Ausrückergabel *f* ~ **line** Schwenkleitung *f* ~ **pitch-control mechanism** Verstellvorrichtung *f* ~ **rod** Schaltstange *f* ~ **yoke** Ausrückerbügel *m*
**shifting** Ausweichen *n*, Bearbeitung *f*, Einrückung *f*, Fortrückung *f*, Gleitung *f*, (magnetism) Scherung *f*, Schiebung *f*, Umlegung *f*, Umschaltung *f*, Verschieben *n*, Verschiebung *f*, Versetzung *f*, Verstellen *n*, Wandern *n*, Wanderung *f* ~ **of beach sand** Küstenversetzung *f* ~ **of channel** Kanalverschiebung *f* ~ **of forces** Kraftverschiebung *f* ~ **of gravel bars** Wanderung *f* der Kiesbänke ~ **a line** Umlegung der Leitung *f*
**shifting,** ~ **arrangement** Schalteinrichtung *f* ~ **ballast** fliegender Ballast *m*, Flugballast *m* ~ **camshaft** verschiebbare Nockenwelle *f* ~ **clockwise** Ablenkung *f* nach rechts, Rechtsablenkung *f* ~ **counterclockwise** Ablenkung *f* nach links, Linksablenkung *f* ~ **coupling** ausrückbare oder lösbare Kupplung *f*, Schaltkupplung *f* ~ **crane** Versatzkran *m* ~ **device** Ausrückvorrichtung *f* ~ **diagram** (Getriebe) Schaltdiagramm *n* ~ **dune** Wanderdüne *f* ~ **gauge** Reißmaß *n* ~**-jaw clutch coupling** Klauenkupplung *f* ~ **lever** Abschiebehebel *m* ~ **linkage** Schaltgestänge *n* ~ **operation** Wechselbetrieb *m* ~ **pinion** Verschieberritzel *m* ~ **register** Schieberegister *n* ~ **ring** Ansetzerring *f* ~ **river bed** bewegliche Flußsohle *f* ~ **rods** Einsatzstangen *pl* ~ **sand** Flug-, Schwimm-sand *m* ~ **soil** beweglicher Boden *m* ~ **stoppage** Umschaltstillstand *m* ~ **(the) track** Gleisrücken *n* ~ **zero** wandernde Nullinie *f*, wandernder Nullpunkt *m*
**shim** Abstand(s)scheibe *f*, Ausgleichsring *f*, (plate) Beilagefolie *f*, (plate) Beilegescheibe *f*, Blechzwischenlage *f*, (plate) dünne Beilage *f*, Füllstück *n*, Futterholz *n*, Keil *m*, Klemmstück

*n*, Lamelle *f*, Paßring *m*, Paßscheibe *f*, Unterlage *f*, Unterlagplatte *f*, Unterlegscheibe *f* ~ **(plate) liner** Zwischenplatte *f* ~ **plates** Beilagebleche *n pl* ~ **rod** Anpassungs-, Grobregel-stab *m* ~ **separation** Polschuhspaltbreite *f*

**shimmer** Flimmern *n*, Zwischen-linienflimmer *m*, -zeilenflimmer *m*

**shimming** Feldkorrektion *f* durch Einlagen dünner Bleche ~ **procedure** Shimverfahren *n*

**shimmy** Flattern *n* von Schwenkrollen oder von Spornrädern, Schwänzeln *n*, Tanzen *n* ~ **damper** Flatterdämpfer *m*

**shimmying** Flattern *n*

**shims** Ausgleichscheiben *f pl*, (leveling) Ausgleichsscheiben *f pl*

**shine, to** ~ aufleuchten, aufstrahlen, entleuchten, flimmern, leuchten, scheinen, spiegeln, wichsen **to** ~ **through** durch-scheinen, -strahlen **to** ~ **upon** bescheinen

**shine** Lichthof *m*, Politur *f*, Schein *m*, Schiller *m*, (textiles) Spanner *m*

**shiner** Glanzschuß *m*, (textiles) Spannfaden *m*

**shingle, to** ~ recken, zängen

**shingle** Dach-schindel *f*, -ziegel *m*, Grand *m*, Kies *m*, Schindel *f* ~ **bottom** Kieselgrund *m*

**shingler** Zänger *m*

**shingling** Zängearbeit *f*, Zängen *n* ~ **hammer** Zängehammer *m* ~ **hatchets** Hammerbeile *pl* ~ **slag** Schwal *m*, Zängeschlacke *f* ~ **tongs** Scherenzange *f*

**shining** leuchtend ~ **coal** glänzende Kohle *f* ~ **soot** Glanzruß *m*

**ship, to** ~ abtransportieren, befördern, (cargo) laden, schicken, transportieren, verladen, versenden, zusenden **to** ~ **a sea** eine Sturzsee *f* überbekommen

**ship** Schiff *n* ~ **of large tonnage** dickes Schiff *n* ~ **of the line** Linienschiff *n*

**shipboard,** ~ **plane** Bordflugzeug *n*, Schiffsflugzeug *n* ~ **regulations** Schiffsordnung *f*

**ship,** ~-**borne aircraft** Bordflugzeug *n* ~ **bottom** Schiffsboden *m* ~ **broker** Schiffsmakler *m* ~ **brought to** beigedrehtes Schiff *n* ~ **builder** Schiffsbaumeister *m*

**shipbuilding** Schiffbauwesen *n* ~ **bulb angle** Schiffbauwulstprofil *n* ~ **crane** Hellingkran *m* ~ **dock** Baudock *n* ~ **installation** Werftanlage *f* ~ **plant** Hellinganlage *f* ~ **plate** Schiffsblech *n* ~ **port** Bauhafen *m* ~ **yard** Schiffswerft *f*

**ship** ~ **caisson** Schwimmponton *n* ~ **canal** Seekanal *m*, Seeschleuse *f* ~ **charges** Bordgebühr *n* ~ **crane** Werftkran *m* ~ **directionfinding station** Bordpeilstelle *f* ~ **field error** (direction-finding) Bordeffekt *m* ~ **hove to** beigedrehtes Schiff *n* ~ **ladder** Schiffsleiter *f* ~ -**lifting device** Schiffshebewerk *n* ~-**locating radar** Schiffssuchgerät *n* ~ **next ahead** (navy) Vordermann *m* ~ **next astern** Hintermann *m* ~ **owner** Reeder *m*, Schiffsausrüster *m* ~ **plane** Bordflugzeug *n* ~ **radio station** Bord-, Schiffs-funkstelle *f* ~ **stayer** Schildfisch *m* ~-**to-~ communication** Bordverkehr *m* ~ **transmitter** Schiffssender *m* ~ **yard** Bootswerft *f*, Helling *f*, Schiffbauwerft *f*, Schiffswerft *f*, Werft *f*, Werftanlage *f* ~ **winch** Hellingwinde *f* ~ **wreck** Schiffbruch *m* ~ **wright** Schiffszimmermann *m*

**ship's,** ~ **armor** Schiffspanzerung *f* ~ **barometer** Marine-barometer *n*, -luftdruckmesser *m*,

**Schiffsbarometer** *n* ~ **bell** Schiffsglocke *f* ~ **boat** Schiffsboot *n* ~ **certificate of registry** Schiffszertifikat *n* ~ **compass** Bordkompaß *m* ~ **crew** Schiffsbemannung *f*, Seemannschaft *f* ~ **frame** Schiffspanten *pl* ~ **hatch** Schiffsluke *f* ~ **iron** Schiffseisen *n* ~ **journal** Logbuch *n* ~ **lights (or light plant)** Schiffsbeleuchtung *f* ~ **magnetism** Schiffsmagnetismus *m* ~ **manifest** Ladungsverzeichnis *n* ~ **mess** Messe *f* ~ **papers** Schiffspapiere *pl* ~ **position** Besteck *n* ~ **pump** Schiffspumpe *f* ~ **pumps** Bordpumpen *pl* ~ **radio direction finder** Bordpeilstelle *f*, Bordpeiler *m* ~ **radio range finder** Bordfunkpeiler *m* ~ **register** Schiffsliste *f* ~ **routine** der laufende Dienst *m* an Bord ~ **sheathing** Schiffsblech *n* ~ **skin** Außenhaut *f* ~ **winch** Bootswinde *f*

**shipment** Absendung *f*, Fracht *f*, Lieferung *f*, Partie *f*, Sendung *f*, Transport *m*, Verladung *f*, Versand *m*, Wanderung *f*, (of goods) Warensendung *f* ~ **by steamer** Seetransport *m*

**shipped** versandt

**shipper** Verlader *m*, Verschiffer *m* ~ **disc** Schaltscheibe *f* ~ **shaft** Ausrückerwelle *f*

**shipping** Abbeförderung *f*, Schiffahrt *f* ~ **by rail** Bahnversand *m*

**shipping,** ~ **box** Versandkiste *f* ~ **cask** Versandfaß *n* ~ **charges** Transportspesen *pl* ~ **clerk** Expeditions-, Versand-beamter *m* ~ **company** (merchant marine) Reederei *f*, Seetransportgesellschaft *f* ~ **conditions** Verladebedingungen *pl* ~ **dimensions** Umfangmasse *f* bei seemäßiger Verpackung ~ **document** Versandbeleg *m* ~ **flask** Verpackungsflasche *f* ~ **instruction** Versand-anweisung *f*, -vorschrift *f* ~ **manifest** Ladeverzeichnis *n* ~ **note** Schiffszettel *m* ~ **office** Musterungsamt *n* ~ **order** Versandauftrag *m* ~ **plug** Blindkerze *f* für Versand, Blind-, Transport-stopfen *m* ~ **room** Laderaum *m* ~ **slip** Lieferschein *n* ~ **trade** Kauffahrtei *f*

**shirt** Wellenfallenrand *m* (microw.)

**shiver, to** ~ splittern, zerspringen

**shiver** Splinter *m*

**shoal, to** ~ abflachen

**shoal** Bank *f*, Untiefe *f* ~ **water** seichtes Wasser *n*

**shoals** Wattenmeer *n*

**shock, to** ~ erschüttern

**shock** Anprall *m*, Anstoß *m*, Erschütterung *f*, Mandel *f*, Rückstoß *m*, Schlag *m*, Stoß *m*, Stoßfestigkeit *f* ~ **of the waves** Wellenschlag *m*

**shock,** ~-**absorbent** erschütterungsfrei ~ **absorber** Dämpfer *m*, (spring type) Federstoßdämpfer *m*, Puffer *m*, Stoßdämpfer *m*, Stoßdämpfer-kupplung *f*, -vorrichtung *f*, Stoßfang *m*, Stoßfänger *m*

**shock-absorber,** ~ **cord** Abfederungsseil *n* ~ **leg** Feder-bein *n*, -strebe *f* ~ **spring** Stoß(fang)feder *f* ~ **strut** Fahrwerkfederbein *n*

**shock-absorbing** stoßdämpfend ~ **cable** Abfederungskabel *n* ~ **drive** elastischer Antrieb *m* mit Stoßdämpferkupplung ~ **mechanism** Stoßdämpfervorrichtung *f* ~ **mountings** stoßdämpfende Aufhängung *f* oder Lagerung *f* ~ **strut** Federbein *n*

**shock,** ~ **absorption** Abfederung *f*, Erschütterungs-, Schwingungs-, Stoß-dämpfung *f* ~-**and -impact** Stoß *m* und Schlag *m* ~-**and-vibration control mounting** vibrations- und stoßdämpfende Lagerung *f* ~ **angle** Stoßwinkel *m* ~ **bend-**

**ing test** Schlagbiegeprobe *f* ~ **cord** Gummiseil *n*, (in parachute) Zugband *n* ~ **curvature** Stoßfrontkrümmung *f* ~**-dispersing baffle (or beam)** Verteilungsplatte *f* ~ **equalizer with compressed-air-cushion** Stoßausgleicher *m* mit Druckluftbelastung ~ **excitation** Stoßerregung *f* ~ **expansion** Stoßentwicklung *f* ~ **formation** Stoßentstehung *f* ~**-free entry** stoßfreier Eintritt *m* ~ **front** Stoßfront *f* ~ **heating** Schnellerhitzung *f* ~ **landing gear** Stoßfahrgestell *n* ~ **layer** Stoßfrontschicht *f* ~ **load** Schlagbeanspruchung *f*, Stoßbelastung *f*, Stoßkraft *f* ~ **loss** Stoßverlust *m* ~ **motion** Schnittbewegung *f* (acoust.) ~ **mount** stoßdämpfende Halterung *f* ~**-mounted** federnd montiert ~ **output** Stoßleistung *f* ~ **polar** Stoßpolare *f* ~ **potential** Stoßspannung *f* ~ **pressure** Druckstoß *m* ~**-proof** erschütterungsfest, hochspannungssicher, stoßfest ~**-proof plug** berührungssicherer Stecker *m* ~ **propagation** Stoßausbreitung *f* ~ **seed** Anregekristalle *pl* ~ **strength** spezifische Ladung *f* ~ **stress** Schlag-, Stoß-beanspruchung *f* ~ **strut** Federbein *n*, Stoßstrebe *f* ~ **strut mounting trunnion** Federbeinlager *n* ~ **surface** Stoßfront *f* ~ **tester** Stoßprobegerät *n* ~ **tube** Stoßrohr *n* ~ **wave** Knallwelle *f* (acoust.), (ballistics) Kopfwelle *f*, stoßender Luftstrom *m*, Stoßwelle *f*

**shockless** erschütterungsfrei, stoßfrei ~ **jarring machine (or jolter)** stoßfreie Rüttelformmaschine *f*

**shoddy** (wool) Kunstwolle f

**shoe, to** ~ anschuhen, mit einem Schuh *m* versehen **to** ~ **a pile** einen Pfahl *m* beschuhen **to** ~ **a pole** eine Stange *f* anschuhen

**shoe** (bridge construction) Auflagerschuh *m*, (Dichtungslabyrinth) Einsatz *m*, (recoilguide) Gleitschuh *m*, Schuh *m*, (plow) Stelze *f*, Walzbalken *m* (mach.) ~ **of crosshead** Gleitklotz *m* ~ **of an oil press** Fußplatte *f*

**shoe,** ~ **brake** Backenbremse *f* ~ **brush** Schuhbürste *f* ~**-button tube** Knopfröhre *f* ~ **drill** Drillmaschine *f* mit Schuh ~ **heel** Absatz *m* ~ **holder of brake** Bremsklotzhalter *m* ~**-nosed shell with valve** Ventilbohrer *m* ~ **pipe** Ausgußrohr *n*, -röhre *f* ~ **piston** Gleitschuh *m* ~ **scraper** Fußabstreifer *m* ~**-shaped angular fishplate** Schuhwinkellasche *f* ~**-(type) wire holder** Meßdrahthalter *m* (schuhartig)

**shoed pole** angeschuhte Stange *f*

**shoot, to** ~ anschießen, (pictures) aufnehmen, auswerfen, schießen, sprengen, treiben **to** ~ **down** abschießen, herunterholen, niederschiessen **to** ~ **in** einschießen **to** ~ **out** herausschleudern **to** ~ **the edge of a board** ein Brett *n* fügen

**shoot** Eintragfaden *m*, Gleitbahn *f*, Rutsche *f*, Schußfaden *m*, Sprosse *f*, Sprößling *m* ~ **of an arch** Seitenschub *m* eines Bogens

**shootboard** Hobel *m* (print.)

**shooter** Schießarbeiter *m*

**shooting** Bohr- und Schießarbeit *f*, Schießarbeit *f*, Schießen *n*, Sprengarbeit *f*, Torpedieren *n* ~ **angle** Aufnahmewinkel *m* ~ **board** Stoßlade *f* ~ **correction** Schußverbesserung *f* ~ **flame** Stichflamme *f* ~ **flow** schießender Abfluß *m* ~ **gallery** gedeckte Bahn *f* ~ **grounds** Schießplatz *m* ~ **improvement** Schußverbesserung *f* ~ **line** Schießbahn *f* ~ **plane** Abkant-, Abschräg-hobel *n* ~ **position** Schießstellung *f* ~ **range** Schieß-

stand *m*, Schußweite *f* ~ **script** Drehbuch *n* ~ **stick** Keiltreiber *m* ~ **test** Beschußprobe *f*

**shop** Betrieb *m*, Laden *m* ~ **accident** Betriebsunfall *m* ~ **agreement** Betriebsvereinbarung *f* ~ **assembly** Werkstattmontage *f* ~ **coat** Grundanstrich *m* ~ **condition** Betriebsbeanspruchung *f* ~ **conditions** Betriebsverhältnisse *pl* ~ **control instrument** Betriebsüberwachungsgerät *n* ~ **data** Betriebsangaben *pl* ~ **deputy** Vertrauensmann *m* ~ **floor** Hüttenflur *m* ~ **gauge** Arbeitslehre *f* ~ **inspection** Prüfung *f* beim Hersteller ~ **management** Betriebsleitung *f* ~ **news** Werkzeitschrift *f* ~ **overhaul** Grundüberholung *f* ~ **overhead** allgemeine Werkstattkosten *pl* ~ **plug gauge** Arbeitslehrdorn *m* ~ **requirement** Betriebsbeanspruchung *f* ~ **right** Arbeitgeberrecht *n* an Arbeitnehmer-Erfindung, Vorbenutzungsrecht *n* ~ **secret** Betriebsgeheimnis *n* ~ **splice** Werkstoß *m* ~ **talk** Fachsimpelei *f* ~ **term** Betriebsausdruck *m* ~ **test** Prüfung *f* in der Fabrik ~ **tools** Kleinwerkzeuge *pl* ~ **window** Schaufenster *n*

**shopping center** Einkaufsviertel *n*

**shore, to** ~ abspreizen, abstreifen **to** ~ **up** aufstützen, stützen, versteifen

**shore** Gestade *n*, Getriebepfahl *m*, Küste *f*, Schore *f*, Stapelstütze *f*, Strand *m*, Strebe *f*, Ufer *n*, Wall *m* ~ **anchor line** Landankertau *n* ~**-based** landfest (Station) ~**-based radar** Küstenradar *n* ~ **beam** Landschwelle *f*, Uferbalken *m* ~ **bearing** Landpeilung *f* ~**-end of submarine cable** schweres Küstenkabel *n* ~**-end cable** Küstenkabel *n* ~ **fast** Vertauung *f* ~ **horizon** Strandkimme *f* ~ **lighting** Küstenbefeuerung *f* ~ **line** Küsten-, Strand-, Ufer-linie *f* ~ **line effect** Küsteneffekt *m* ~ **mooring** Landverankerung *f* ~ **piling** Endauflager *n* ~ **protection** Uferschutz *m* ~ **side** Landseite *f* ~ **sill** Uferbalken *m* ~ **station** Küsten-, Land-funkstelle *f*

**shored, lying** ~ **up** auf Stapel *m* liegen

**shores** Stützbalken *m*

**shoreward side** Landbord *n*

**shoring** Absteifung *f*, Abstützung *f* (Gebäude), Längsbrett *n*, Stutzen *m*

**short, to** ~ kurzschließen

**short** zu kurz gehendes Geschoß *n* (artil.), Kurzschluß *m* (electr.); abgekürzt, brüchig, kurz, mürbe, spröde, unzureichend ~ **and handy instrument** kurzes handliches Instrument *n* **for a** ~ **time** kurzfristig

**short,** ~ **anneal(ing)** Kurztemperung *f* ~ **balance** Unterwaage *f* ~**-base range-finder** Einstandentfernungsmesser *m* ~**-beam analysis balance** kurzarmige Analysenwage *f* ~ **bill** Wechsel *m* auf kurze Sicht ~ **borer** Meißelbohrer *m* ~ **box spanner** Nußsteckschlüssel *m* ~**-brittle** faul, brüchig, kurzbrüchig ~ **broken** kurzspanend ~ **car** Kurzgondel *f*

**short-circuit, to** ~ kurzschließen

**short-circuit** Ableitung *f*, kurzer Stromkreis *m*, Kurzschließen *n*, Kurzschließung *f*, Kurzschluß *m*, Kurzschlußstromkreis *m*, Schleife *f*, Schleifenberührung *f*, Schluß *m* ~ **to earth chokes and coils** Erdspulen *pl* ~ **to ground** Erdschluß *m*, Erdschlußstrom *m* ~ **between plates** Plattenkurzschluß *m*

short-circuit, ~ admittance Kurzschlußadmittanz
f ~ armature Kurzschlußanker m ~ bottom
Kurzschlußknopf m ~ bow Kurzschlußbug m
~ brake Kurzschlußstrombremse f ~ contact
Kurzschlußkontakt m ~ cover Kurzschluß-,
Unterbrecher-deckel m ~ current Kurzschlußstrom m ~ detèctor Kurzschlußsucher m ~
ground cable Kurzschlußkabel n ~ impedance
Kurzschluß-impedanz f, -widerstand m ~ key
Kurzschlußtaste f ~ line Kurzschlußleitung f ~
member Kurzschlußglied n ~ pipe Kurzschlußleitung f ~-proof kurzschlußsicher ~ protection
Kurzschlußschutz m ~ ratio Leerlaufkurz-
schlußverhältnis n ~ reactance Kurzschluß-
widerstand m ~ reporter Isolationsmeßgerät n
~ rotor motor Kurzschlußläufermotor m ~
section Schlußstück n ~ slide-ring rotor Kurz-
schlußschleifringläufer m ~ spark Kurzschluß-
funke m ~ stable kurzschlußstabil ~ strength
Kurzschlußsicherheit f ~ terminal Körper-
schlußklemme f, Kurzschlußklemme f, Rück-
schlußklemme f ~ (test)ing stand Kurzschluß-
prüfstand m ~ transition Kurzschlußübergang
m ~ voltage Kurzschlußspannung f ~ welding
Kurzschluß-Schweißung f (Blankdraht-Schweis-
sung) ~ wire Körperschluß-, Rückschluß-kabel
n

short-circuited kurzgeschlossen ~ circuit (or ~
line) am Ende kurzgeschlossener Stromkreis m
~ rotor Anker m mit kurzgeschlossener Wicklung, Kurzschluß-läufer m, -wicklung f

short-circuiting, ~ bridge Kurzschluß-brücke f,
-bügel m ~ capacitor Schlußkondensator n ~
contact Kurzschlußschaltstück n ~ device
Kurzschließer m ~ disc (or ring) Kurzschluß-
ring m, -scheibe f ~ push button Kurzschluß-
druckknopf m ~ springs Kurzschlußfedern pl
~ switch Kurzschlußschalter m ~ wire bridge
Kurzschlußdrehbrücke f

short, ~-columnar kurzsäulig, säulig ~ culvert
kurzer Umlauf m ~ cuts Durchschnittssätze pl
(von Kosten) ~-cycle annealing Kurzzeittempe-
rung f, Schnelltemperverfahren n ~ cylindrical
piece Rohrstutzen m ~ distance radio shielding
Nahentstörung f ~-distance receiving Nah-
empfang m ~-distance scatter Nahstreuung f
~-distance traffic Nahverkehr m ~-distance
transporter Nahförderer m ~ drive Kurzantrieb
m ~-duration power output Kurzleistung f ~
-elbowed box spanner Winkelrohrschlüssel m ~
electrode ends Elektrodenreste pl ~ feed pipe
Stutzen m ~ fiber kurzfaserig ~-fibered asbestos
Asbestolith n ~-flaming kurzflammig ~-flaming
coal kurzflammige Kohle f ~-flaming semibituminous coal of selected sizes bestmelierte Eßkohle f of ~ focal distance kurzbrennweitig ~
-focus lens kurzbrennweitige Linse f, kurzbrennweitiges Objektiv n ~-focus objective
kurzbrennweitiges Objektiv n ~ form invoice
Kurzrechnung f ~ gondola Kurzgondel f ~
-haul jet airliner Kurzstrecken-Strahlflugzeug n
~-haul toll circuit Nahverkehrsleitung f ~-haul
toll traffic Nahfernverkehr m (teleph.) ~
-interval ranging Kurzzeitmethode f ~ lengths
Unterlängen pl ~-life (or lived) kurzlebig ~-line
fault Abstandskurzschluß f ~ link Kurzzug-
draht m (zwischen Tasten- und Typenhebel)
~-link cham kurzgliedrige Kette f ~ linked

kurzgliedrig (Ketten) ~ mark Kürzezeichen n
~ measure Untermaß n ~ method of calculation
Kurzrechnungsmethode f ~-necked kurzhalsig
on ~ notice kurzfristig ~ oil fatliquors magere
Fettungsmittel pl ~-oil paint magere Anstrich-
farbe f ~ page Ausgangskolumne f ~ period of
rise kurze Anlaufzeit f ~-period barrage Kurz-
zeitsperre f ~ piece of pipe Rohrstutzen m ~
pipe Ansatzrohr n ~ pitch feingliedrig ~-pitch
winding gesehnte Wicklung f ~-pitched winding
gesehnte oder schrittverkürzte Wicklung f ~
-proceeds Mindererlös n ~ projection Knagge
f ~ pulse Kurzimpuls m ~ pump magerer
Stoff m ~-radius curve stark gekrümmte
Kurve f

short-range nah ~ fading Nahschwund m ~ fading antenna Nahschwundantenne f ~ field Nahfeld n ~ field of antenna Antennennahfeld n ~
focus Naheinstellung f ~ limit Kurzreichweiten-
grenze f ~ radar Nahbereichsradar n

short, ~ residuum konzentrierter Rückstand m
~-run casting mangelhaft ausgelaufener Abguß
m, nicht ausgelaufenes Gußstück n ~-run
production Kleinserienfertigung f ~ shipment
Teilversand m ~ sides of right-angled triangle
Katheten pl ~ splice Kurzpleißung f ~ spring
(of the jack) A-Feder f ~ stack Auspuffstutzen
m ~ staple kürzeres Material n ~ starting run
kurzer Start m ~-stop bath Unterbrechungsbad
n ~-stop landing Landung f mit kurzem Rollweg ~-stroke kurzhubig ~ sweep Kurzabta-
stung f (rdr) (flanged) ~-sweep twin elbow
doppeltes Krümmer-T-Stück n, geschweiftes
T-Stück n ~ taper Kurzkegel m, steiler Kegel m
~-term credit Überbrückungskredit n ~ term
worker Kurzarbeiter m ~-thread milling Kurz-
gewindefräsen n ~ tie.beam Stichbalken m on
~-time basis auf Kurzzeit f ~-time current
Stoßstrom m ~-time duty kurzzeitiger Betrieb
m ~-time observation Kurzzeitbeobachtung f
~-time short-circuit (Hochspannung) Wischer
m ~-time storage kurzzeitige Einlagerung f ~
-time test Kurz(zeit)prüfung f ~-time voltage
Stoßspannung f ~-to-earth Erdschluß m ~
wave kurze Welle f, Kurzwelle f

short-wave kurzwellig ~ adapter Kurzwellenvor-
satzgerät n ~ coil Kurzwellenspule f ~ communication Kurzwellenverbindung f ~ condenser Antennenverkürzungskondensator m, Luft-
drahtverkürzungs-, Verkürzungs-kondensator
m ~ converter Kurzwellen-vorsatzgerät n,
-wandler m ~ (field) range Kurzwellenbereich
m ~ receiver Kurzwellen(verkehrs)empfänger
m ~ reception Kurzwellenempfang m ~ therapeutic apparatus Hochfrequenzheilgerät n ~
transmission Kurzwellenübertragung f ~ transmitter Kurzwellensender m

short, ~ weight Gewichts-abgang m, -manko n,
Minder-, Unter-gewicht n

shortage Ausfall m, Fehlbestand m, Knappheit
f, Mangel m, Manko n ~ of material Stoffmangel m ~ of raw materials Rohstoffnot f ~
of workers Arbeitermangel m

shortage, ~ area Zuschußgebiet n ~ discrepancy
Fehlbestand m shortcoming Mangel m, Unzu-
länglichkeit f

shorten, to ~ abkürzen, abstützen, beschneiden,
kürzen, magern, verkürzen

**shortened,** ~ **diffusion battery** verkürzte Diffusionsbatterie *f* ~ **segment** kleiner Kontakt *m*, verkürztes Segment *n*

**shortening** Kürzungsfette *f*, Stauchung *f*, Verkürzung *f* ~ **of the delay period** Totzeitverkürzung *f* ~ **of ligament** Bandschrumpfung *f*

**shortening,** ~ **agents** Mürbemachemittel *pl* ~ **condenser** (antenna) Antennenverkürzungskondensator *m* ~ **saw** Abkürzsäge *f* für Dauben

**shorthand** Kurzschrift *f*, Stenografie *f*

**shorting,** ~ **to earth** Erdschluß *m*

**shorting,** ~ **bar** Kurzschlußbügel *m* ~ **contact** Schleppkontakt *m* ~ **plug** Kurzschlußstecker *m* ~ **plunger** Kurzschlußschieber *m*

**shortness** Brüchigkeit *f*, Faulbrüchigkeit *f*, Kürze *f*, Sprödigkeit *f*

**shot** Aufnahme *f* (film), (blasting) Bohrschuß *m*, Explosion *f*, Graupe *f*, Körner *pl*, Kugel *f*, Ladung *f*, Schrotkugeln *pl*, Schuß *m*, Schützenschlag *m*; schillernd ~ **from the solid** (blasting) Schuß *m* aus dem Vollen ~ **beyond target** Weitschuß *m*

**shot,** ~ **bag** Schrotbeutel *m* ~ **blast** Strahlsand *m* ~**-blasted** sandgestrahlt ~ **blasting** Kugelstrahlen *n* ~**-colored** schillernd ~ **copper** Kupfergranalien *pl* ~ **drilling** Schrotbohren *n* ~ **effect** Schroteffekt *m*, Störgrieß *m* (phys.) ~ **effect of radio** Schrotrauschen *n* ~ **factory** Schrotfabrik *f* ~ **firing** Minenzündung *f* ~ **gauge** Kaliberlehre *f* ~ **group (or pattern)** Einschlagsgarbe *f*, Scheibentreffbild *n*, Zieldreieck *n* ~ **gun** Schrotgewehr *n* ~ **hardening** Kugelstrahlen *n* ~ **hole** Bohr-, Schieß-, Schuß-, Spreng-loch *n* ~**-hole gauge** Schußlehre *f* ~**-lubrication system** Druckschmierung *f*, Eindruckschmierung *f* ~ **metal** Schrotmetall *n* ~ **noise** Schrot-effekt *m*, -geräusch *n* ~**-noise reduction factor** (electr. tube) Schwächungsfaktor *m* ~**-to-~ variation** Schußgewichtsschwankung *f* ~ **welding** Schnellpunktschweißung *f*, Schußschweißung *f*

**shoulder** (rounded part under bolthead) Abrundung *f*, Absatz *m*, Achsel *f*, Ansatz *m*, Bankett *n*, Blatt *n*, Bund *m*, Kröpfung *f*, Schulter *f*, vorspringende Kante *f*, vorspringender Rand *m*, Vorsprung *m* ~ **of axle collar** Anbau *m* am Achsbund ~ **of bit** Meißelabsatz *m* ~ **on crankshaft** Kurbelwellenbund *m* ~ **of a projectile** Zentrierwulst *f* ~ **of rim** Felgenschulter *f* ~ **of tooth** Zahnfuß *m*

**shoulder,** ~ **belt** Schulterriemen *m* ~ **bevel** Achselschrägung *f* ~ **girder** Tragerand *m* ~ **guard** Abweiser *m* ~ **harness** Schultergurte *pl* ~**-high spaces** achselhoher Ausschluß *m* (print.) ~ **patch** Schulterstück *n* ~ **piece** Achselband *n*, Schulterstück *n* ~ **point** Schulterpunkt *m* ~ **screw** Ansatzschraube *f* ~ **strap** Achselstück *n*, Schulter-gurt *m*, -klappe *f*, Umhängeriemen *m* ~ **stud** Schaftschraube *f*

**shouldered,** ~ **bolt** Bundbolzen *m* ~ **off** abgestuft ~ **pin** Gewindebolzen *m* ~ **rod** abgesetzte Welle *f* ~ **thread chaser** Gewindestahl *m* (als Schulterstahl) ~ **work grinding** Ansatzschleifen *pl*

**shouldering** Anschulterung *f*

**shoulders** Schulterauflauf *m*

**shout, to** ~ brüllen, rufen

**shout** Ruf *m*, Schrei *m*

**shouting** Geschrei *n*

**shove, to** ~ schieben **to** ~ **away** abdrängen **to** ~ **off** abstoßen ~ **off!** ab! **to** ~ **the throttle full open** Drossel *f* ganz öffnen, Vollgas *n* geben

**shove** Schub *m*

**shovel, to** ~ schaufeln **to** ~ **into** einschaufeln **to** ~ **out** ausschaufeln, herausschaufeln

**shovel** Scharre *f*, Schaufel *f*, Schieber *m*, Schüppe *f* ~ **for barreling** Einlaßschaufel *f*

**shovel,** ~ **bucket** Schaufelbecher *m* ~ **conveyor** Schaufelwerk *n* ~ **coverer** Schaufelschützer *m* ~ **furrow opener** Schaufelfurchenöffner *m* ~ **handle** Schaufelstiel *m* ~**-nose tool** Schaufelstahl *m* (single-) ~ **plow** Schaufelpflug *m* ~ **teeth** Greiferschneiden *pl*

**shoveling,** ~ **buckets** Schaufelung *f* ~ **scrap** handzerkleinerter oder schaufelfähiger Schrott *m*

**shoveller** Schaufler *m*

**shoves, by** ~ schubweise

**shoving** Schieben *n*, Schiebung *f*

**show, to** ~ anzeigen, aufweisen, bieten, dartun, ergeben, zeigen **to** ~ **diagrammatically** schematisch darstellen **to** ~ **heavy** (light) mit starken (schwachen) Linien *pl* darstellen **to** ~ **on the surface** unterlaufen **to** ~ **up** hervortreten

**show** Aureole *f*, Ausstellung *f*, Demonstation *f*, Schau *f* ~ **card** Fensterplakat *n* ~ **case** Schaukasten *m* ~**-piece** Schaustück *n* ~ **window** Schaufenster *n*

**shower** Regen-guß *m*, -schauer *m*, Schauer *m* ~ **of cosmic rays** Höhenstrahlen-schauer *m*, -stoß *m*, Raumstrahlenstoß *m* ~ **of sparks** Funken-garbe *f*, -regen *m*

**shower,** ~ **bath** Brausebad *n*, Duschbad *n* ~**-bath truck** Badewagen *m* ~ **head** Brause *f* ~ **radiations** Schauerstrahlen *pl* ~ **unit** Schauerlänge *f*

**showing** Weisung *f*

**showings** Spuren *n*

**shown in full** ausgezogen

**shows** Spuren *n*

**shrapnel** Granat-kartätsche *f*, -schrapnell *n*, Schrapnell *n*

**shred, to** ~ in Streifen *m pl* schneiden, zerreißen, zerschnitzeln

**shred** Fetzen *m*, Lappen *m* ~ **of plate** Blechschnitzel *n*

**shredder** Lumpenzurichter *m*

**shredding machine** Aktenvernichter *m*

**shriek, to** ~ einen schrillen, kreischenden Ton *m* erzeugen, kreischen

**shriek** Schrei *m*

**shrill** gellend, hell, klimpernd, scharf (acoust.) ~ **timbre** helle Klangfarbe *f*

**shrink, to** ~ dekatieren, eingehen, einlaufen, einschreiben, einschrumpfen, (of fibers) einspringen, krimpfen, krumpen, schrumpfen, schwinden, sintern, (dolomite) totbrennen, verschrumpfen, (steel) verziehen, zusammenschrumpfen, -ziehen **to** ~ **back** zurückschrecken **to** ~ **on** aufschrumpfen, aufziehen, warm aufziehen

**shrink** Stich *m*, Zusammenschrumpfen *n* ~ **bar** Längenbegrenzungsstange *f* (Fahrwerk) ~ **bob** Schwindungslot *n* ~ **bobs** Bindesteiger *pl* ~ **cavity** Schwindungshohlraum *m* ~ **fit** Schrumpfsitz *m*, Warmsitz *m* ~**-fitting method** Aufschrumpfmethode *f* ~ **head** Massel *f* ~ **hole** Hunker *m*, Lunker *m*, Lunker-hohlraum *m*, -stelle *f*, schwarze Stelle *f*, Schwind(ungs)hohlraum *m*, Stich *m* ~ **resistance** Krumpffestigkeit *f* ~ **strut** Längenbegrenzer *m* ~ **water** Schwindwasser *n*

**shrinkage** (fabric) Eingang *m*, Einschnürung *f*, Einschrumpfung *f*, (fabric) Einsprung *m*, Einwalken *m*, Entquellen *n*, Kleinerwerden *n*, Kontraktion *f*, Lunker *m*, Lunkerung *f*, Sackmaß *n*, Schrumpfen *n*, Schrumpfung *f*, Schwinden *n*, Schwindmaßverkürzung *f*, Schwindung *f*, Schwund *m*, (phenomenon) Schwunderscheinung *f*, Zusammen-schrumpfen *n*, -ziehung *f* ~ **from drying** Trockenschwindung *f* ~ **in drying** Trockenschwindung *f* ~ **in firing** (ceramics) Brennschwindung *f* ~ **in length** Längenschrumpfung *f*

**shrinkage**, ~ **allowance** Schwindmaß *n*, Schwindmaßzugabe *f* ~ **cavity** Lunkerhohlraum *m*, Schwind-hohlraum *m*, -lunker *m*, Schwindungshohlraum *m*, -loch *n*, -lunker *m* ~ **compensation** Schrumpfungsausgleich *m* ~ **contraction** Schwindmasseverkürzung *f* ~ **crack** Schwind-(ungs)-, Schrumpf-riß *m* ~ **distortion** Schrumpfverformung *f* ~ **equalization** Schrumpfungsausgleich *m* ~ **factor** Schwundzahl *f* ~ **fault** Lunker *m* ~ **gauge** Schwindmaß *n* ~ **hole** Lunker *m* ~ **(contraction) joint** Schwindfuge *f* ~ **limit** Schrumpfgrenze *f* ~ **measure** Schrumpfmaß *n* ~ **strain (stress, or tension)** Schrumpfspannung *f* ~ **test** Schrumpfversuch *m* ~ **value** Schwindmaß *n*

**shrinking** Einengung *f*, Lunkerbildung *f*, Lunkern *n*, (in putty or plastic wood for filling defects) Nachsinken *n*, Schrumpfung ~ **of cloth in width** Breiteneingang *m* der Zeuge ~ **of real shell** Schrumpfen *n* von Schildpatt

**shrinking**, ~ **combination (or connection)** Schrumpfverbindung *f* ~ **head** Saugmassel *f* ~ **hole** Lunker *m* ~ **hollow** Schrumpfhohlraum *m* ~-**on** Warmaufziehen *n* ~ **machine** Dekatiermaschine *f* ~ **thread** Schrumpfgewinde *n* ~ **wall** ausgebauchte Mauer *f*

**shrivel, to** ~ einschrumpfen, schrumpfen

**shriveled** runzelig ~ **finish** Kräusellack *m*

**shrivelling** Faltenwerfen *n*, Runzelbildung *f* (paint.)

**shroud, to** ~ einhüllen

**shroud** Leichentuch *n*, Rad-kranz *m*, -wand *f*, Schwenkseil *n*, Schwungseil *n*, Ummantelung *f*, Want *f* ~ **band** (Turbine) Deckband *n* ~-**laid rope** vierschattiges Tau *n* ~ **line** (parachute) Fangleine *f*, Tragleine *f* ~ **lines** (parachute) Geleine *n*, Leinensystem *n* ~ **ring** (Triebwerk) Bandagierung *f*, Mantelring *m*, (gas turbine) Umhüllungsring *m*, Ummantelungsring *m* ~ **ring of blades** Abschlußring *m*

**shrouded** ummantelt ~ **plug and socket** Kragensteckvorrichtung *f* ~ **propeller** ummantelte Schraube *f*

**shrouding** Umhüllung *f* ~ **ring** Flanschring *m* ~ **tube** Schutzrohr *n* ~ **wire** Versteifungsdraht *m*

**shrouds** (navy) Gut *n*

**shrub** Strauch *m*

**shrubbery** Gebüsch *n*, Gesträuch *n*

**shrunk**, ~ **fit** Aufschrumpfung *f* ~-**on** aufgeschrumpft ~-**on coupling halves** aufgezogene Kupplungshälften *pl* ~-**on gear ring** aufgesetzter Zahnkranz *m* ~-**on hoop** Schrumpfband *n* ~-**on pipe joint** Einsteckklebeverbindung *f*

~-**on rim** Schrumpfwulst *f* ~ **(-on)ring** Schrumpfring *m*

**shudder** Schauer *m*

**shuffle, to** ~ hin- und herschieben, weiterschieben

**shuffle bar** Wimmler *m*

**shuffling** (paper) Aufdrehen *n*

**shun, to** ~ vermeiden

**shunt, to** ~ ableiten, nebenschließen (electr.), Nebenschluß *m* bilden, in den Nebenschluß *m* legen, mit einem Nebenschluß *m* versehen, parallelschalten, rangieren, überbrücken, verschieben **to** ~ **off** abzeigen **to** ~ **onto a siding** ausweichen

**shunt** Ablaß *m*, Ableitung *f*, Abzweig *m*, Nebengleis *n*, -schluß *m*, -stromregelwerk *n*, -widerstand *m*, (circuit) Nebenweg *m*, Parallelweg *m*, Stromableitung *f* **in** ~ **with** in Nebenschluß *m* zu

**shunt**, ~-**admittance-type equalization** Querentzerrung *f* (teleph.) ~-**admittance-type equalizer** Querimpedanzentzerrer *m* ~ **arm** (of network) Querarm, Querzweig *m* ~ **arrangement** Parallelschaltung *f* ~ **branching** (of transmutation products) Verzweigung *f* ~ **capacitor** Querkondensator *m* ~ **capacity** Querkapazität *f* ~ **characteristic** Nebenschlußkreis *m*, Nebenschlußverhalten *n* (d. Motors) ~ **circuit** Nebenschlußleitung *f* ~ **coil** Zweispule *f* ~ **condenser** Nebenschlußkondensator *m* ~ **conductance** Ableitung *f* ~-**conduction** Nebenschlußkommutatormotor *m* ~ **connection** Nebeneinander-, Nebenschluß-, Parallel-schaltung *f* ~ **current** Querstrom *m* ~-**driven vibrator** Zerhacker *m* ~ **dynamo** Nebenschlußdynamomaschine *f* ~ **element** Ableitungsglied *n*, Querglied *n*, Querwiderstand *m* ~ **element of a filter** Querglied *n* eines Filters ~ **excitation** Nebenschlußerregung *f* ~-**excited antenna** Antenne *f* mit Ableitung ~-**fed vertical antenna** parallelgespeiste Vertikalantenne *f* ~ **field** Nebenschlußfeld *n*, Querfeld *n* ~ **field current** Nebenschlußfeldstrom *m* ~ **field rheostat** Nebenschlußwiderstand *m* ~ **generator** Nebenschlußstromerzeuger *m* ~ **impedance** Parallel-, Quer-impedanz *f* ~ **inductance** Querinduktivität *f* ~ **inductor** Querinduktor *m* ~ **line** Querleitung *f* ~ **loading** Querbelastung *f* ~ **magnetic coil** Nebenschlußspule *f* ~ **motor** Abzweigflußmesser *m*, Nebenmotor *m* ~ **path** Parallelweg *m* ~ **rectifier circuit** Parallelschaltung *f* von Gleichrichtern ~-**regulating resistance** Nebenschlußregelwiderstand *m* ~ **regulator** Hauptstromwerk *n*, Nebenschlußregler *m* ~ **relay** Shuntrelais *n* ~ **resistance** Abzweigungs-, Nebenschluß-, Parallel-widerstand *m* ~ **switch** Umgehungsschaltung *f* ~ **system of distribution** System *n* mit Parallelschaltung ~ **terminal** Kurzschlußklemme *f* ~-**type arc lamp** Nebenschlußbogenlampe *f* ~-**type attenuation equalizer (or compensator)** Querentzerrer *m* ~ **winding** Doppelwicklung *f*, gemischte Wickelung *f* ~ **wire** Ableitungsdraht *m*, Abzweigdraht *m* ~ **worm** Nebenschluß *m* ~-**wound** nebenschlußgewickelt ~-**wound electric motor** Nebenschlußmotor *m* ~-**wound generator** Nebenschlußdynamomaschine *f*

**shunted** (across) parallelgeschaltet ~ **across** im Nebenschluß *m* zu ~ **buzzer** Nebenschlußsummer *m* ~ **condenser** Kondensator *m* mit Parallelwiderstand, Maxwell-anordnung *f*, -schaltung *f* ~ **instrument** Instrument *n* mit Nebenwiderstand ~-**telephone method** Parallelohmmethode *f*

**shunter** Kuppler *m*

**shunting** Rangierung *f* ~ **capacitor** Überbrückungskondensator *m* ~ **condenser** Parallel-(impedanz)kondensator *m*, Querkondensator *m*, Überbrückungskondensator *m* ~ **device** Rangieranlage *f* ~-**engine** Verschiebelokomotive *f* ~ **installation** Rangiervorrichtung *f*, Verschiebe-anlage *f*, -vorrichtung *f* ~ **line** Ausweichgeleise *n* ~ **resistor** Ableitungswiderstand *m* ~ **spring** Nebenschlußfeder *f* ~ **station** Verschiebebahnhof *m* ~ **trolley** Verschiebekarren *m* ~ **winch** Rangier-, Verschiebe-winde *f* ~ **yard** Rangiergelände *n*

**shut, to** ~ einrasten, schließen, sperren, verschließen, zumachen **to** ~ **down (or off)** abstellen, außer Betrieb *m* stellen **to** ~ **in** einsperren **to** ~ **off** abschalten (water), abschliessen, absperren, (Schub) drosseln, unterbrechen **to** ~ **up** aufschweißen

**shut** Lötstelle *f*, Schweißstelle *f*

**shutdown** Betriebs-pause *f*, -stillegung *f*, -störung *f*, -unterbrechung *f*, Stillegung *f*, Stillstand *m*, Zeitdauer *f* des Stillsetzens

**shut-in,** ~ **pressure** statischer Druck *m* ~ **well** abgeschlossene Sonde *f*

**shutoff,** ~ **cock** Abstellhahn *m* ~ **device** Absperrorgan *n*, -teil *m*, -vorrichtung *f* ~ **indicating pin** Ausschaltsignalstift *m* ~ **needle valve** Absperrnadelventil *n* ~ **nozzle** Absperrhahn *m* ~ **plug** Absperrpflock *m* ~ **regulator** Anschlagregler *m* ~ **rod** Sicherheitsstab *m* ~ **valve** Absperrventil *n*, Windschieber *m*

**shutter** Abdeckblech *n*, Blende *f*, Deckel *m*, Diaphragm *n*, Gleittafel *f*, Hohlleitersperre *f*, Jalousie *f*, Klappe *f*, Reintonblende *f*, Schalter *m*, Schirm *m*, Schütztafel *f*, Spund *m*, Verlust *m*, Verschluß *m*, Verschluß-blende *f*, -schieber *m*, Zahlschalter *m* **(drop)** ~ **of dark slide** Kassettenschieber *m* ~ **for making multilateral sound track** Vielfachzackenblende *f*

**shutter,** ~ **action** Verschlußbetätigung *f* ~ **annunciator** Gitterschauzeichen *n* ~ **axis** Blendenachse *f* (film) ~ **blade** Abdeckflügel *m* ~ **casing** Verschlußgehäuse *n* ~ **control** (on carburetor) Abdeckvorrichtung *f* ~-**control cord (or line)** Zieh-, Zug-schnur *f* ~ **cover** Blendabdeckung *f* ~ **cutoff frequency** Abdeckfrequenz *f* ( film) ~ **disc** Verschlußscheibe *f* ~ **disc with spiral slot** Abdeckscheibe *f* ~ **efficiency** Lichtwirkungsgrad *m* ~ **frame** Blendenrahmen *m* ~ **housing** Blendengehäuse *n* ~ **leaf** (swinging-out) Verschlußklappe *f* ~ **leaves** Verschlußlamellen *pl* (photo) ~ **lever** Blendenhebel *m* ~ **mask** Abdeckblende *f* ~-**mask method** Abdeckverfahren *n* ~ **movement** Pendelbewegung *f* ~ **openings** Hellsektoren *pl* ~ **period** Schaltperiode *f* ~ **release** Auslösung *f* des Verschlusses, Verschlußauslösung *f* ~ **setter** Verschlußaufzug *m* ~ **setting** Verschlußeinstellung *f* ~ **speed** Verschlußgeschwindigkeit *f* ~ **speed setting ring** Belichtungszeitenring *m* ~ **speedtesting machine** Verschlußzeitenmeßgerät *n* ~ **spring plate** Spannscheibe *f* ~ **tester** Verschlußzeitenmeß-

gerät *n* ~ **trip mechanism** Verschlußauslösevorrichtung *f* (photo) ~ **wind spring** Verschlußaufzugsfeder *f*

**shuttering** Absteifung *f*, Abstützung *f*, Schalung *f*, Sprießung *f*

**shutting** Schließbeschlag *m* ~ **down** Betriebsschluß *m*, Fehlen *n* ~-**off** Abschluß *m*, Abschlußokklusion *f* ~-**off of electric current** elektrische Abstellung *f* ~ **plug** Verschlußstück *n*

**shuttle, to** ~ hin- und herbewegen, hin- und hergehen, pendeln

**shuttle** Doppelventilkegel *m*, Hin- und Herbewegung *f*, hin- und hergehender Körper *m*, Pendelbetrieb *m*, pneumatisches Rohr *n*, Rohrpostbüchse *f*, Schiffchen *n*, (weaving) Schütz *n*, Untergestell *n* (film), Weberschiffchen *n* ~ **armature** (magneto) Doppel-T-Anker *m*, I-Anker *m* ~ **boom** Verschiebeträger *m* ~ **box** Schützenkasten *m* ~ **cam** Doppeldaumen *m* ~ **cam for reciprocating motion** Doppeldaumen *m* für Hin- und Rückgang ~ **cock** Federball *m* ~ **conveyor** Pendelförderer *m* ~ **flight** Weberschiffchenflug *m* ~ **hauler** Pendelförderer *m* ~ **injector** Aufgabeschaukelschieber *m* ~ **list** Pendelliste *f* ~ **mechanism** Schaltwerk *n* ~ **net** Schützennest *n* ~ **peg** Schützenspindel *f* ~ **screen** Schwingsieb *n* ~ **service** Pendelbetrieb *m* ~ **skid** Pendelschlepper *m* ~ **spindle (or tongue)** Schützenspindel *f* ~-**type discharger** Abwurfwagen *m* ~ **valve** Wechselventil *n*

**shuttling** Pendelverkehr *m* ~ **motion** hin- und hergehende Bewegung *f*

**sialography** Sialografie *f*

**Siamese-twin-blade propeller** Luftschraubendoppelflügel *m*

**siamesed** gegabelt ~ **induction pipes** gegabelte Laderohre *pl*

**sibatit** Sibatit *n*

**sibilant sound** Zischlaut *m*

**siccative** Siccativ *n*, Sikkativ *n*, Trocken-mittel *n*, -stoff *m*, Zusatzmittel *n* zur Beschleunigung des Trocknens; schnelltrocknend

**sick** krank, rank ~ **leave** Erholungs-, Genesungs-, Kranken-, Krankheits-urlaub *m* ~ **pay** Krankenlöhnung *f*

**sickle** Hippe *f*, Sichel *f* ~-**shaped** sichelförmig

**side** Flanke *f*, (piece; of angle) Schenkel *m*, Seite *f*, Wand *f* to one ~ einseitig ~ **of a barrage** Hang *m* einer Talsperre (engin.) ~ **and buffer lanterns** Ausschlaglaternen *pl* ~ **of a gallery** Streckenstoß *m* (min.) ~ **of the member** Stabseite *f* ~ **of receiver** Kastenwand *f* **(short)** ~ **of a shaft** Schachtstoß *m* ~ **of ship** Bordwand *f* ~ **and tail boards** Bordwände *pl* ~ **of thread** Gewindeflanke *f*

**side,** ~ **air-lift agitator** Seitenmischlufttührer *m* ~ **antenna** Seitenantenne *f* ~ **appendage** (of a bulb or tube) Ansatz *m* ~ **arm** Ansatz *m* ~ **arm for stand** Unterschenkel *m* für Stativ ~ **armature relay** Seitenankerrelais *n* ~ **armor** Seitenpanzer *m*

**side-band** Seitenband *n*, Übersprechen *n* der Seitenfrequenzen (rdr) ~ **of modulation** Modulationsseitenbänder *pl*

**side-band,** ~ **attenuation** Amplitudenabschwächung *f* ~ **component** Seitenbandkomponente *f* ~ **frequency** Seitenfrequenz *f* ~ **interference**

Seitenbandstörung *f* ~ **oscillation** Neben-
schwingung *f* ~ **splashing** Seitenbandinterferenz *f* ~ **suppression** Seitenbandunterdrückung
*f* ~ **transmission** Seitenbandsenden *n* ~ **transmitter with suppressed carrier** Zweiseitenband-
übertrager *m*
**side,** ~ **bar** Lasche *f*, Seitensteg *m*, Steg *m*,
Trachte *f* ~**-bar pillion** Satteltracht *f* ~ **bearing**
Längslager *n* ~ **binding** Seitenbund *m* ~**-blow
percussion table** Querstoßherd *m* ~**-blowing
converter** Windfrischapparat *m* mit seitlicher
Windeinströmung ~**-blown converter** Konverter
*m* mit seitwärts eintretendem Wind ~ **board**
Bordwand *f* ~ **bonnet** Seitenabdeckung *f* ~ **boys**
Fallreepgasten *pl* ~**-bracket bearing** Flanschlager *n* ~ **branch** Seitenarm *m* ~ **buckling of the
spring** seitliches Ausknicken *n* der Feder ~
**-bump table** Querstoßherd *m* ~**-bay-**~ **valves**
stehende Ventile *pl* ~ **car** Beiwagen *m*, Seiten-
gondel *f*, -wagen *m* ~ **chain** Nebenkette *f* ~
**channel** Gassenrinne *f*, Gosse *f*, Seitengraben *m*
~ **circuit** Stamm-kreis *m*, -leitung *f* ~ **circuit
loading coil** Stammpupinspule *f*, Stammspule *f*
~ **clearance** Regellichtraum *m* an Eisenbahnen
~ **clearance of teeth of a toothed wheel** Spielraum *m* der Zähne eines Zahnrades ~**-clearance
angle** Seitenspiel *n* ~ **clearance angle** Freiwinkel
*m* der Nebenschneide (Drehstahl) ~ **connecting
rod** Seitenkolbenstange *f* ~ **contact** Seitenan-
schluß *m* ~**-contact base** Außenkontaktsockel
*m* ~ **contact rail** seitliche Stromschiene *f* ~**-con-
tact-type base** stiftloser Sockel *m* ~ **cover** Seitenabdeckung *f*, (Motorhaube) Seitenteil *n* ~
**cover plate** Seitendeckel *m* ~ **curtain** Seiten-
vorhang *f* ~**-curtain fade-in** Kulissenaufblen-
dung *f* ~ **curve of roof** Dachwute *f* ~ **cutters**
Seitenschneiderollen *pl* ~ **cutting** Füllgrube *f*,
Seitenentnahme *f* ~**-cutting pliers** Flachzange *f*
mit Seitenschneider ~ **delivery** seitlicher Abzug
*m* ~ **delivery hay rake** Schwadenrechen *m*, seitwärts ablegender Heurechen *m* ~ **direction**
Seitenrichtung *f* ~**-discharge car** Seitenentleerer *m* ~ **(wise) displacement** Seitenversetzung *f*
~ **ditch** Fanggraben *m* ~**-door elevator** Aufzug
*m* mit seitlichem Verschluß *m* ~**-drawing-type
crucible furnace** Überflurtiegelofen *m* ~ **drift**
Flügelort *m* ~**-dump bucket** Kastenhänge *f* mit
Seitenentleerung ~**-dump car** Eisenbahnwagen
*m* mit seitlicher Entladung, Seitenkipper *m* ~
**-dump scow** Klappenprahm *m* mit seitlichem
Laderäumen ~**-dump truck** Holländerwagen *m*
~**-dumping hopper** Seitenentleerer *m* ~ **echo**
Nebenzipfelecho *n* ~ **effect** Nebenwirkung *f* ~
**electrode** Seitenelektrode *f* ~ **elevation** Kreuz-
riß *m*, Seitenansicht *f*, Seiten(auf)riß *m* ~
**emission** Streuemission *f* ~ **entrance** Neben-
eingang *m* ~ **face** Seitenfläche *f* ~**-faced** profiliert ~**-feed firing** Seitenschubfeuerung *f* ~**-
feed stoker** Seitenschubbeschickungsanlage *f*
~ **fillister** Plättbank *f*, Plattenhobel *m* ~ **flange**
Seitenflansch *m* ~ **flap** Seitenklappe *f* ~ **flashing
of lightning** Überspringen *n* ~ **float** Seitenflosse
*f* ~ **force** Seitendruck *m* ~ **frame** Anlegerseiten-
teil *n*, Eckrahmen *m* ~ **frequency** Seiten(band)-
frequenz *f* ~ **friction** Seitenreibung *f* ~ **gallery**
Flügelort *m* ~ **gangway** Seitenlaufgang *m* ~
**gear** Nebenrad *n* ~ **glide** Seitwärtsschleuderung
*f* ~ **guard** Seitenschutzblech *n* ~ **guard rail**

Bügel *m* für Seitenschutz ~ **guard shaft** Lager-
stange *f* ~ **guide (gauge)** Seitenmarke *f* ~ **guide
bearing** Bogenführung *f* ~ **guide clamping lever**
Seitenmarkenhebel *m* ~ **guide rocker arm**
Schalthebel *m* (print.) ~ **guide rollers** Seiten-
führungsrollen *pl* ~ **guiding** (of film strip)
Seitenführung *f* ~ **gutter** Gassenrinne *f*, Seitengraben *m*, -rinne *f* ~ **guy** Seitenanker *m* ~
**handle** Seitengriff *m* ~ **head** Ständer-, Seitensupport *m* ~ **heading** Marginaltitel *m* ~ **iron(s)**
Seiteisen *n* ~ **knurl holder** seitlicher Rändelhalter *m* ~ **lamp** Seitenlampe *f* ~ **lap** Seiten-
überdeckung *f* ~**-lap weld** Seitenschweiße *f* ~
**leakage** Flankenstauung *f*, -streuung *f* ~ **length**
Kantenlänge *f* ~ **light** Ochsenauge *n*, Seiten-
licht *n* ~**-light on fender** Kotflügelleuchte *f* ~
**-light screen** Laternenkasten *m* ~ **lights** Seiten-
leuchten *n* ~ **line** Neben-beruf *m*, -betrieb *m*,
-linie *f* ~ **link** Kettentasche *f*
**side-lobe** Nebenlappen *m* (rdr), Nebenzipfel *m*
(rdr) ~ **echo** Seitenecho *n* (rdr) ~ **intensity**
Seitenzipfelintensität *f* ~ **suppression (SLS)**
Nebenzipfelunterdrückung *f* (sec rdr)
**side,** ~ **lobes** Aufzipfelung *f* ~**-looking airborne
radar** Schrägsichtradar *n* ~**-mark** Anstoß-
marke *f* (print) ~ **member** Seiten-holm *m*,
-träger *m* ~ **mill** Scheibenfräser *m* ~**-milling
cutter** Radial-, Scheiben-, Seiten-fräser *m* ~
**-milling cutter with backed-off teeth** Scheibenfräser *m* mit hinterdrehten Zähnen ~ **note** Randglosse *f*, Seitenanmerkung *f* ~ **opening** Seiten-
öffnung *f* ~ **outrigger** Seitenholm *m* ~ **page
companion** Gassengespan *m* ~ **panel** Seitenblech
*n* ~**-panel end plate** Seitenteil *n* ~ **piece** Seiten-
stück *m* ~ **piling** Aufstürzung *f*, Seitenablagerung *f* ~ **pipe** Nebenrohr *n* ~ **plate** Gleitwand *f*,
Innenflanschlasche *f*, Platine *f*, Seitenblech *n*,
Wange *f* ~ **pole** Seitenstange *f* ~ **press** (open)
Schwanenhalspresse *f* ~ **pressure** indirekter
Druck *m*, (steel mill) indirekter Walzdruck *m*,
Seitendruck *m* ~ **propeller** Seitenschraube *f* ~
**pull** seitlich auftretender Zug *m* ~ **radiator**
Flossenkühler *m* ~ **rail** Rödelbalken *m*, Sicherheitsschiene *f* ~ **rail clamp** Rödelzange *f* ~**-rail
edge aligning device** Kantenvorrichtung *f* für
Rahmenträger ~ **rail lashing** Rödelbund *m*,
Rödelung *f* ~ **rake** Nebenschneide *f*, Span-
winkel *m* ~ **rake and tedder** kombinierter
Schwadenrechen *m* und Wender *m* ~**(-loading)
ramp** Seitenrampe *f* ~ **reaction** Nebenreaktion
*f* ~ **rebate plane** Wandhobel *m* ~ **reflecting
prism** Seitenstrahlprisma *n* ~ **rest** Abstellplatte
*f* ~ **retainer** Seitenscheibe *f* ~ **road** Neben-straße
*f*, -weg *m* ~**-road crossing** Einmündung *f* ~
**rod** Gitterhaltestange *f*, Gittertraverse *f*, Seitenstange *f* ~ **rudder** Seitenruder *n* ~ **screen** Kajü-
tensonnensegel *n*, Seitenwindschutz *m* ~ **seam**
Flankennaht *f* ~**-seaming machine** Längsfalzma-
schine *f* ~ **sheet of firebox shell** Stehkesselseiten-
wand *f* ~ **shield** Seitenschild *m* ~ **sill** Längsträger *m* ~**-skip** Ausbrechen *n* (aviat.) ~ **skimming
attachment** Seitenschäumvorrichtung *f*
**side-slip, to** ~ abgleiten (aviat.), rutschen
**side-slip** Abgleiten *n* (aviat.), Schleudern *n*, seitliches Abrutschen *n* (aviat.), (of an airplane)
seitliches Ausgleiten *n*, Seitengleitflug *m*, Seitwärtsschleuderung *f*, Slip *m* ~ **angle** Schiebe-
winkel *m* ~ **indicator** Seitenabrutschanzeiger *m*

**side,** ~ **slipping** Ausrutschen *n*, Zustand *m* des Abgleitens (aviat.) ~ **slipping danger** Schleudergefahr *f* ~ **slipping a parachute** einen Fallschirm *m* abrutschen lassen ~ **slope** Bankett *n*, Spanwinkel *m* ~**-slope angle** (of a cutting tool) Seitenschliffwinkel *m* ~ **spectrum** Nebenspektrum *n* ~ **stability** einseitige Ruhelage *f* ~ **stick** Anlegesteg *m* (print.), Beschnittsteg *m* ~ **stop** Haltebolzen *m* ~ **stray** Flankenstreuung *f* ~ **stream** Abstrom *m*, Seitendeckung *f* ~ **street** Neben-, Quer-straße *f* ~ **stringer** Seitenstringer *m* ~ **strip** Seitenleiste *f* ~ **strip light** Seitenstreifenfeuer *n* ~ **strip marking** Seitenstreifen *m* ~ **strut** Knickstrebe *f* ~ **supports** Seitenstutzen *pl* ~ **surface** Seitenfläche *f* ~ **switch** Seitenschalter *m* ~ **take-off** Querzug *m* ~ **terminal** Seitenklemme *f* ~ **tester** Probenentnehmer *m* ~ **thrust** Achsenlängs-druck *m*, -schub *m*, Auslenkhärte *f* (phono), Axialschub *m*, Seitendruck *m*, -schub *m*, seitliche Auslenkkraft *f* (acoust.) ~**-thrust effect** seitliche Ablenkung *f* ~**-tip car** Seitenkipper *m* ~**-tip wagon (or tipping dump car)** Seitenentleerer *m*

**side-tone** Mikrofongeräusch *n*, Nebengeräusch *n*, Nebenton *m* (teleph.), Rückhören *n* ~ **attenuation** Rückhördämpfung *f* ~**-circuit** Mithöreinrichtung *f* ~ **device** Mithöreinrichtung *f* ~ **receiver** Mithöreinrichtung *f* ~ **reference equivalent** Rückhörbezugsdämpfung *f* ~ **telephone set** Fernsprechapparat *m* ohne Rückhördämpfung (ohne Geräuschdämpfung)

**side,** ~ **tool** Abflach-, Seiten-stahl *m* ~**-tophantom crosstalk** Mitsprechen *n* ~**-to-phantom far-end cross talk** Gegenmitsprechen *n* (teleph.) ~**-to-~ balance** Übersprechkupplung *f* ~**-to-~ capacity** Viererschleifenkapazität *f* ~**-to-~ crosstalk** Übersprechen *n* ~**-to-~ far-end crosstalk** Gegenübersprechen *n*

**side, to** ~**-track** etwas abbiegen

**side,** ~ **track** Seitengleis *n* ~**-track obstructions** Ausweichen *n* (verlorenen Werkzeugen ausweichen) ~ **tracking** Ablenkung *f*, (petroleum) abweichendes Bohren *n* ~ **trays** Ablegetisch *m* ~ **trimming (or shaving)** Kantenfräsen *n* ~ **tube** Ansatzrohr *n*, Nebenrohr *n*, Seiten-ansatz *m*, -arm *m* ~**-valve engine** seitengesteuerter Motor *m* ~ **vane** Seitenklappe *f* ~ **view** Profil *n*, Seitenansicht *f*, Vorbeiflug *m* ~**-walk** Bürgersteig *m*, Gehbahn *f* ~**-walk manhole** Kabelschacht *m* für Gehbahn ~ **wall** Dockwand *f*, Endwiderlager *n*, Seiten-wand *f*, -wandung *f*, Umfassungsmauer *f*, Wange *f*, Wehrwange *f*, Widerlager *n* ~ **wall of a lock** Kammer-mauer *f*, -wand *f* ~ **wall of a vein** Gangbegrenzung *f* ~**-wall covering** Seitenwandbelag *m* ~ **wall crush test** Mantelstauchprobe *f* ~ **wall sprinkler** Seitenwandsprinkler *m* ~ **wall upright** Seitenwandsäule *f* ~ **wall water boxes** (Kessel) Seitenwangen *pl* ~ **way** Weiche *f* ~ **weld** Flankennaht *f* ~ **wind** Seitenwind *m* ~**-working tool** Seitenwerkzeug *n*

**sidereal,** ~ **day** Sterntag *m* ~ **hour** Sternstunde *f* ~ **time** Sternzeig *f* ~ **year** Polarjahr *n*

**siderical** siderisch

**siderite** Siderit *m*, Spateisenstein *m*, spatiger Eisenstein *m*, Stahlerz *n*, Stahlstein *m*

**siderolite** Mesosiderit *m*

**sideroscope** Sideroskop *n*

**siderugical cement** Eisensportlandzement *m*

**siderurgy** Eisenmetallurgie *f*

**siding** Abstellgeleise *n*, Ausweicheplatz *m*, Ausweich-geleise *n*, -stelle *f*, Nebengeleise *n*, Rangiergleis *n*, Weiche *f* ~ **sidetrack** Seitengang *m* ~ **track** Gleisanschluß *m*

**siege floor** Glasschmelzofenbank *f*

**siegenite** Siegenit *m*

**Siemens,** ~ **arc furnace** Siemens-Lichtbogenofen *m* ~ **(regenerative) furnace** Martinofen *m* ~ **gas** Luftgeneratorgas *n*, Siemens-Gas *n* ~ **gas-reversing valve** Siemens-Gaswechselklappe *f* ~ **independent arc furnace** Siemens-Strahlungsofen *m* ~ **lock and block instrument** Wechselstromblockfeld *n* ~**-Martin furnace** Siemens-Martin Ofen *m* ~**-Martin plant** Martinofenanlage *f* ~**-Martin process** Siemens-Martin-Ofen-Verfahren *n* ~**-Martin steel** Siemens-Flußeisen *n*, Siemens-Flußstahl *m*, Siemens-Martin-Ofen-Stahl *m* ~**-Martin steelworks** Martinstahlhütte *f* ~ **mercury unit** Siemens-Einheit *f* ~ **printer** Ferndrucker *m* ~ **process** Siemens-Prozess *n* ~ **renerative openhearth furnace** Siemens-Regenerativfeuerung *f* ~ **reversing gas valve** Siemens-Gaswechselklappe *f*

**sieve, to** ~ absiebsetzen, durchsetzen **to** ~ **out** aussieben, durchsieben

**sieve** Durch-laß *m*, -wurf *m*, Sieb *n* ~ **with beating arrangement** Klopfsieb *n* ~ **for corning powder** Pulverkornsieb *n* ~ **with a wide mouth** Keubel *m* (min.)

**sieve,** ~ **analysis curve** Siebkurve *f* ~ **bottom** Läuterboden *m* ~ **discharge** Siebabnahme *f* ~ **drum** Siebtrommel *f* ~**-like** siebförmig ~ **netting** Siebgewebe *n* ~ **sheet** Siebblech *n* ~ **shovel** Siebschaufel *f* ~ **socket** Siebstutzen *m* ~ **test** Siebprobe *f* ~ **width** Siebbreite *f* ~**-worm-centrifuge** Siebschneckenzentrifuge *f*

**sieved** gesiebt

**sieving** Absieben *n*, Absiebung *f*, Durchsieben *n*, Sieben *n*

**siferrite** Siferrit *n*

**sift, to** ~ absieben, beuteln, durch-setzen, -sieben, -werfen, sichten, sieben **to** ~ **out** aussieben **to** ~ **sand** Sand *m* durchwerfen

**sift** Sichten *n*

**sifted** gerättet (min.), gesiebt

**sifter** Ausleser *m*, Filter *m*, Klaubhammer *m*, Sieb *n*, Sieb-apparat *m*, -gebilde *n*, -kette *f*, -kettenleiter *m*, -kreis *m*, -maschine *f*, -schaltung *f*, Schüttelsieb *n*, Wellensieb *n* ~ **effect** Sichtwirkung *f* ~ **machine** Sichtmaschine *f*

**sifting** Absieben *n*, Absiebung *f*, Durchsieben *n*, Sandsichtung *f*, Sichtung *f*, Sieben *n*, Sieberei *f*, Siebsichtung *f*, Siebung *f*, Siebwäsche *f* ~ **device** Sichteranlage *f* ~ **drum** Rundsieb *n* ~ **machine** Beutelmaschine *f* ~ **plant** Sicht-, Sieb-anlage *f*

**siftings** Durchfall *m*, Siebfeine *n*, Siebsel *m*

**sight, to** ~ abfluchten, anrichten, anvisieren, anzielen, (einen Gegenstand) visieren, zielen **to** ~ **coast (or land)** Land *n* in Sicht bekommen **to** ~ **in** anrichten **to** ~ **out** abvisieren

**sight** Ansicht *f*, Besteck *n* (astron.), Diopter *n*, Dioptrie *f*, Gesicht *n*, Gesichtssinn *m*, Richtaufsatz *m*, -gerät *n*, Sehen *n*, Sicht *f*, (of a light) Sichten *n*, Zielvorrichtung *f* **(rear)** ~ Visier *n* ~ **(vane)** Seh-schlitz *m*, -spalt *m* **in** ~ **of the ground** mit Bodensicht *f* (aviat.)

**sight, ~ adjustment** Aufsatzstange *f*, Visierstellung *f* **~ angle** Lagewinkel *m* **~ bar** Aufsatzstab *m*, Stangenaufsatz *m* **~-bar arm** Visierstange *f* **~ bowl** Glasbehälter *m* **~ bracket** Aufsatzträger *m*, Visiergestänge *n* **~ carrier** Lattenträger *m* **~ case** Aufsatzgehäuse *n* **~ chamber** Sichtkammer *f* **~ characteristic** Visierkenngröße *f* **~ control** Sichtkontrolle *f* **~ defilade** Sichtdeckung *f* **~ discharge** Auslaufstutzen *m* mit Schauglas **~-discharge attachment** Schlauchentleerungsaufsatz *m* mit Schauglas **~ discharge indicator** Durchflußanzeiger *m* **~ distance** Sichtweite *f* **~ draft** Sichttratte *f* **~-extension bar** Aufsatzstange *f*

**sight-feed, ~ lubrication** Tropfvorrichtung *f* **~ lubricator** Öltropfgefäß *n*, Sichtschmierapparat *m* **~ oil cup** Sichtschmierapparat *m* **~ oiler** Tropf(en)öler *m* **~ regulator** Tropfenregler *m*

**sight, ~ flow indicator** Durchflußrichtungsanzeiger *m* **~ gauge** Außenmanometer *n* **~ glass** Augen-, Schau-glas *n* **~ glass cover** Schaulochdeckel *m* **~-glass spinner** Schauglaspropeller *m* **~ graticule** Abkommen *n*, Absehen *n*, Zielmarke *f* **~ head indicator** Zielgerätekopf *m* **~ height indicator** Visierhöhenmesser *m* **~ hole** (furnaces, etc.) Besichtigungs-fenster *n*, -loch *n* **~ indicator** Aufsatzzeiger *m* **~ leaf** Blättervisier *n*, Visier-klappe *f*, -stange *f* **~ (deflection) leaf** Aufsatzschieber *m* **~ level** Längslibelle *f* **~ line** Betrachtungs-, Blick-richtung *f*, Visierlinie *f* **~ lubricator (or oil feed)** Schau-ölapparat *m* (Schauöler) **~-mount housing** Aufsatzgehäuse *n* **~ notch** Kimme *f* **~ opening** Schauloch *n* **~ peep** Fenster *n* **~ reticle** Abkommen *n* (opt.) **~ reticule** Zielmarke *f* **~ rule** Alhidade *f* **~ setter** Aufsatzeinsteller *m* **~ setting** Visierstellung *f* **~ shield** Visierschild *n* **~ slit** Sichtschlitz *m* **~ socket** Aufsatzbuchse *f* **~ support** Aufsatzträger *m*, Richtaufsatzträger *m* **~ testing chart** Sehprobentafel *f* **~ vane** Diopter *n*, Dioptrie *f*, Kompaßdiopter *n* **~-window frame** Schauglasfassung *f*

**sighting** Zielen *n* **~ aperture** (of compass) Durchblicköffnung *f*, Richtstrich *m* **~ binoculars** Doppelsuchfernrohr *n* **~ collimator** Richtglas *n* für Rundblickfernrohr **~ color** Kennzeichnungsfarbe *f* **~ device** Peilvorrichtung *f*, Visiereinrichtung *f* **~ disk** Visierscheibe *f*, Zeitverschluß *m*, Zielscheibe *f*, Zieltafel *f* **~ distance** Zielgenauigkeit *f* **~ error** Zielfehler *m* **~ instrument** Richtmittel *n* **~ level** Nivellierlibelle *f* **~ line** Schaulinie *f* **~ mark** Einstellmarke *f*, Index *m* bei Maßstäben **~-mark error** Indexfehler *m* **~ mechanism** Richtgerät *n*, Visier-einrichtung *f*, -vorrichtung *f*, Ziel-einrichtung *f*, -gerät *n* **~ mirror** Festlegespiegel *m* **~ peep** Visierlupe *f* **~ pendant** Visierdraht *m* **~ pillar** Signalstein *m* **~ platform** Richtstand *m* (g/m) **~ point** Visierspitze *f*, Zielpunkt *m* **~ a point on the horizon** Anvisieren *n* (nav.) (eines Horizontalpunktes) **~ range** Visierschußweite *f* **~ shot** Einschießen *n* der Länge, Probe-, Richtungs-, Visier-schuß *m* **~ sighting slot** Schauritze *f* **~ telescope** Abwurfsehrohr *n*, Visierfernrohr *n*, Zielfernrohr *n* **~ thread** Kontrollfaden *m* **~ triangle** Dreieckzielen *n*

**sights, (gun) ~** Visiereinrichtung *f* **~ on the horizon** Horizontalbeobachtungen *pl*

**sigma, ~-meson** Sigmameson *n* **~-monogenic function** Sigma-monogene Funktion *f* **~-pile** Sigmareaktor *m*

**sigmoid curve** S-Kurve *f*

**sign, to ~** anmerken, markieren, signieren, unter-schreiben, -zeichnen, zeichnen **to ~ an affidavit** beschwören **to ~ the agreement** mustern **to ~ a bill** einen Wechsel *m* unterschreiben **to ~ per initials** mit den Initialen *pl* abzeichnen **to ~ per procuration** per Prokura *f* zeichnen

**sign** Anzeichen *n*, Aushängeschild *n*, Bezeichnung *f*, Erkennungszeichen *n*, (z. B. für aufgetretene Störung) Kennzeichnung *f*, Mal *n*, Merkmal *n*, Sichtzeichen *n*, Signal *n*, Vorläufer *m*, Wink *m*, Zeichen *n*, Zug *m* **~ (board)** Schild *n* **(plus or minus) ~** Vorzeichen *n* (math.) **~ of cavitation number** Vorzeichen *n* der Kavitationszahl **~ of a charge** Ladungssinn *m* **~ of charge determination** Ladungsvorzeichenbestimmung *f* **~ of equivalence** Kongruenzzeichen *n* **~ of fatigue** Ermüdungsanzeichen *n* **~ of inequality** Ungleichheitszeichen *n* (math.) **~ of integration** Integralzeichen *n* **~ of the time** Zug *m* der Zeit **~ of the zodiac** Himmelszeichen *n*

**sign, ~-board** Wegweiser *m* **~ change** Zeichenwechsel *m* **~ convention** Vorzeichenfestsetzung *f* **~ digit** Vorzeichen-stelle *f*, -ziffer *f* (comput.) **~-off signal** Beendigungszeichen *n* **~ painting** Schriftmalen *n* (Lack) **~ plate** Richtungsschild *n* **~ post** Aushängeschild *n*, Richtposten *m*, Richtungsweise *m* **~ symbol** Begriffszeichen *n* **~-type bar** Zeichentypenhebel *m* **~ writing** Schriftmalen *n*

**signal, to ~** anzeichen, (with lamps) blinken, geben, melden, signalisieren **to ~ with flags and rods** winken

**signal** Anhängegerät *n*, Anschlußbahn *f*, Anzeige *f*, Ausweichvermittlung *f*, Entfritter *m* (teleph.), Lichttag *m*, Melder *m*, Nachricht *f*, (visible) Schauzeichen *n*, Schriftzeichen *n*, Signal *n*, Signalzeichen *n*, Stromschritt *m*, Zeichen *n* **~ for busy trunk lines** Fernbesetztzeichen *n* **~ of distress** Notzeichen *n* **~ to go ahead (or carry on)** Ausführsignal *n* **~ to speed up the operation** Drängelsignal *n*

**signal, ~ ammunition** Erkennungsmunition *f*, Leuchtmunition *f* **~ amplitude** Signalamplitude *f* **~ apparatus for domestic use** Haussignalvorrichtung *f* **~ area** Signal-feld *n*, -platz *m* (aviat.) **~ band** Übertragungsband *n* **~ baton** Zeichenstab *m* **~ book** Signalbuch *n* **~ box** Blockstation *f* **~ call** Anruf *m* **~ carrier frequency** Signalträgerfrequenz *f* **~ cartridge** Leucht-, Signal-patrone *f* **~ center** Fernmelde-, Nachrichten-zentrale *f* **~ check** Sprechprobe *f* (rdo) **~ circuit** Fernmeldeleitung *f*, Nutzstromkreis *m* **~ circuit controller** Signalflügelkontakt *m* **~ code** Signalanordnung *f* **~-code table** Signaltafel *f* **~ coil** (of loud-speaker) Tauchspule *f* **~ combination** Zeichenkombination *f* (teletype) **~ communication** Nachrichtenverbindung *f* **~ communications** Nachrichtenwesen *n* **~ comparator** automatischer Sendemonitor *m* **~ component** Signalteil *m* **~-construction truck** Telegrafenbaukraftwagen *m* **~-controlled synchronization** selbständiges Synchronisieren *n* (TV) **~ converter** Signalwandler *m* **~ cord**

Signalleine *f* ~ **crank** Signalkurbel *f* ~ **curbing** Vorfilter *n* ~ **current** Nutz-, Sprech-, Ton-strom *m* ~**(ing) current** Telegrafierstrom *m* ~ **current coil** Schwingspule *f* ~ **data converter** Meßwertwandler *m* ~ **device** Schlagwerk *n* ~ **distance** Hammingdistanz *f* ~ **distortion** Zeichenverzerrung *f* ~ **distributor** Befehlübermitt(e)lungsapparat *m* ~ **element** kürzester Telegrafierstromstoß *m*, Stromschritt *m*, Telegrafierstromschritt *m* ~**-feeder line** Blockspeiseleitung *f* ~ **field** Netzfeld *n*, Nutzfeld *n* (rdo) ~ **field strength** Signalfeldstärke *f* ~ **fire** Signalfeuer *n* ~ **flag** Signalflagge *f*, Wink(er)flagge *f* ~ **flare** Lichtsignal *n*, Signalbombe *f* ~**-flare projector** Signalwerfer *m* ~ **flasher** Blinker *m* ~ **flow symbol** Signalsymbol *n* (Schaltbild) ~ **forward junction working** Anrufbetrieb *m* ~ **frequency** Modulations-, Signal-, Zeichen-frequenz *f* ~ **front** Zeichenkopf *m* ~ **generator** Meßoszillator *m*, Meßsender *m*, Prüfgenerator *m* ~ **grid** Steuergitter *n* ~ **head** Zeichen-kopf *m*, -stirn *f* ~ **hole** Stanz-, Zeichen-loch *n* ~ **horn** Rufhorn *n*, Signalhorn *n* ~ **indicating position of points** Blocksignal *n* ~ **indicator** Signalrückmelder *m* ~ **indicator pillar** Lichtsignalsäule *f* ~ **inertia drag** Fahnenbildung *f* ~ **installation for bulkhead** Schottensignalanlage *f* ~ **intelligence** Nachrichtenaufklärung *f* ~ **intelligence station** Funkauswertestelle *f* ~ **intensity** Zeichen-intensität *f*, -stärke *f* ~ **interpolation** Zeicheninterpolation *f* ~ **lamp** Blinker *m*, Meldelampe *f*, Morselaterne *f*, Signal-lampe *f*, -laterne *f* ~ **-lamp board** Signallampentafel *f* ~**-lamp communication** Blinkverbindung *f* ~ **level** Nutzlaufstärke *f* ~ **light** Bake *f*, Bakelicht *n*, Feuer *n*, Signallicht *n*, Winker *m* ~ **light indicating direction of turn** Winker *m* ~ **light indicating „not ready"** Unklarlampe *f* ~ **light projector** Zeichengebungsscheinwerfer *m* ~ **limiter** Begrenzungsgerät *n* ~ **line** Fernmelde-, Signalleitung *f* ~ **man** (Bahn-)Wärter *m* (r.r.), Einwinken *m* (aviat.) ~ **mast** Zeichengebungsmast *m* ~ **mean value** Signalmittelwert *m* ~ **message** Signalspruch *m* ~ **meter** Signalstärkemesser *m* ~ **mixer unit** Signalverteilungskasten *m* ~ **mixing** Kursaufschaltung *f* (g/m) ~ **mixing for guide beam** Leitstrahlaufschaltung *f* ~ **mixing motor** Aufschaltmotor *m* (g/m) ~ **mixing ratio** Aufschaltgröße *f*, Aufschaltung *f* (g/m) ~ **modulator** Meßwertumformer *m* ~**-noise ratio** Rauschabstand *m*, Signal-Rauschverhältnis *n* ~ **oscillation** Übertragungsschwingung *f* ~ **panel** Rahmenflagge *f* ~ **pattern** Signalverlauf ~ **pedal** Schienenstromschließer *m* für Signalsteuerung ~ **pin** Signalstift *m* ~ **pistol** Leuchtpistole *f* ~ **plate** Gegenelektrode *f*, (iconoscope) Impulsplatte *f*, Signalplatte *f* ~ **plug** Hinweisstöpsel *m*, Hinweisungsstöpsel *m* ~**-point indicator lamp** Weichenlaterne *f* ~ **port flap** Zeichenklappe *f* ~ **potential** gerichtete Empfangsspannung *f*, Nutzspannung *f*, Zeichenspannung *f* ~ **power** Nutzleistung *f* ~ **reading** Auslesen *n* ~ **receiver** Rufempfänger *m* ~ **red** Signalrot *n* ~ **relay** Signalrelais *n* ~ **reproduction** Zeichenwiedergabe *f* ~ **resistance** Nutzwiderstand *m* ~ **rocket** Leuchtkugel *f*, Signalrakete *f* ~ **rocket with axial staff** Achsstabsignalrakete *f* ~ **scanner** Signalschalter *m* ~ **shaper** Signalformer *m*

~**-shaping amplifier** signalformender Verstärker *m* ~**-shaping network** signalformendes Glied *n* ~ **simulator** Signalgeber *m* ~ **slot circuit** Signalkupplungsstromkreis *m* (r.r.) ~ **spectrum** Mischsignal *n*, Signalverlauf *m* ~ **spread** Zeichenverbreiterung *f* ~ **station** Signalstelle *f* ~ **stick** Zeichenstab *f* ~ **storage** Speicherwirkung *f* ~ **storage tube** Bildspeicherröhre *f* ~ **strength** Lautstärke *f*, Nutzfeldstärke *f*, Telegrafierstromstärke *f*, Zeichenstärke *f* ~ **summing bridge** Addierschaltung *f* ~ **system** Signalwesen *n* ~ **tape** Aufnahmestreifen *m* ~**-to-crosstalk-ratio** Signal-Nebensprechverhältnis *n* ~**-to-noise ratio** Geräuschabstand *m*, Rauschabstand *m*, Signal/Rauschverhältnis *n*, Signalstörverhältnis *n* (rdr.), Störabstand *m* ~ **tracer** Störsuchgerät *n* ~ **tracing** Signalverfolgung *f* ~ **traffic** Blinkrelaisverkehr *m* ~ **transfer** Übersprechen *n* ~ **transmitter** Signalgeber *m* ~ **transmission** Zeichengeber *m*, Zeichenübertragung *f* ~ **truck** Nachrichtenwagen *m* ~ **unit** Telegrafierstromschritt *m* ~ **velocity** Signalgeschwindigkeit *f* ~ **voltage** HF-Gitterwechselspannung *f*, Signalspannung *f* ~ **wave** Arbeitswelle *f*, Betriebswelle *f*, Übertragungsschwingung *f*, Zeichenwelle *f* ~ **waveform** Signalform *f* ~ **whistle** Signalpfeife *f* ~ **winding** Steuerwicklung *f* ~ **windings** Eingangswicklungen *pl* ~ **wire** Klopfgestänge *n* (min.), Signaldraht *m* ~ **working** Anrufbetrieb *m*

**signaling** Markierung *f*, Nachrichten-übermittlung *f*, -übertragung *f*, Sendung *f*, (impulses) Signalgabe *f*, Signalgebung *f*, Signalisierung *f*, Vermarkung *f*, Zeichen-gabe *f*, -gebung *f* ~ **of peak load of traffic** Spitzensignalisierung *f*

**signaling,** ~ **apparatus** Anzeigevorrichtung *f*, Nachrichtengerät *n*, Signalapparat *m* ~ **arrangement** Meldevorrichtung *f* ~ **channel** Sendekanal *m* ~ **circuit** Schwachstromleitung *f* ~ **condenser** Sendekondensator *m* ~ **contact** Signalkontakt *m* ~ **device** Signal-einrichtung *f*, -mittel *n*, -vorrichtung *f*, Stellwerk *n* (r.r.) ~ **disc** Anzeigerdeckung *f*, Kelle *f*, Winkestab *m*, Winker-kelle *f*, -stab *m* ~ **faults** Rufstörungen *pl* ~ **flag** Winkerflagge *f* ~ **frequency** Rufstrom-, Telegrafierfrequenz *f* ~ **impulses** Signalstöße *pl* ~ **installation** Signalanlage *f*, Zeichengebungsanlage *f* ~ **key** Signalkontakt *m* ~ **lamp** Signalscheinwerfer *m* ~ **lens** Signallinse *f* ~ **line** Blinlinie *f* ~ **plant** Lichtsignalanlage *f* ~ **relay** Durchruf-, Ruf-, Sende-relais *n* ~ **service** Blinkbetrieb *m* ~ **set** Niederfrequenzrufsatz *m* ~ **speed** Sende-, Übertragungs-geschwindigkeit *f* ~ **system** Signalanlage *f*, Zeichenübermittlungssystem *n* ~ **telescope** Blinkfernrohr *n* ~ **test** Rufprüfung *f*, Rufversuch *m* ~ **torch** Notsignalfackel *f* ~ **troubles** Rufstörungen *pl* ~ **tube** Signalisierungsröhre *f*

**signalized intersection** geregelte Kreuzung *f*

**signalizing contact** Signalisierkontakt *m*

**signals** Schrift *f* **the ~ run together** die Zeichen *pl* laufen zusammen

**signals,** ~ **equipment** Nachrichtenmittel *n* ~ **officer** Nachrichtenoffizier *m* ~ **section** Nachrichtenabteilung *f* ~ **service** Nachrichtendienst *m*

**signatory to an agreement** Inhaber *m* einer Sprechstelle

**signature** Handzeichnung *f*, Kennungssignal *n*, Namensunterschrift *f*, Unterschrift *f* ~ **of a station** Kennsignal *n*

**signed,** ~ **for** im Auftrag *m* ~ **statement** Protokoll *n*

**signer** Zeichner *m*

**significance** Berücksichtigung *f*, Sinn *m*, Tragweite *f* ~ **of test** Anwendung *f* des Versuches

**significant** ausgeprägt, ausgezeichnet, bedeutend, bezeichnend, groß, sinnvoll ~ **digit** geltende oder wesentliche Zahl *f* oder Ziffer (comput.) ~ **figure** geltende Ziffer *f* ~ **instants** Hauptmomente *pl* ~ **turn** Minderungskurve *f*

**signify, to** ~ bedeuten, bezeichnen

**silence, to** ~ dämpfen, niederkämpfen, zum Schweigen *n* bringen, auf Tonminimum *n* einstellen **to** ~ **the sounder** den Übertragungsklopfer *m* abstellen

**silence** Geräuschlosigkeit *f*, Ruhe *f*, Schweigen *n*, Stille *f* **(position of)** ~ Lautminimum *n*

**silenced** geräuscharm, leise ~ **magazine** leise Kassette *f*

**silencer** Anrufeinrichtung *f* für Übertragungen, Dämpfer *m*, Knalldämpfer *m*, Schalldämpfer *m* ~ **cabinet** Anruferschränkchen *n* ~ **magnet** Anrufermagnet *m*

**silencing** Geräusch-, Schall-dämpfung *f* ~ **amplifier** Regelverstärker *m* ~ **circuit** Geräuschunterdrückungsstromkreis *m*, Grundstromkreis *m*

**silent** geräusch-arm, -los, lautlos, ruhig, still, stillschweigend, tonlos ~ **area** Schweigezone *f* ~ **block** Gummilager *n* ~ **camera** Stummfilmkamera *f* ~ **chain** Hülsenkette *f* ~ **discharge** stille Entladung *f* ~ **file** Glättfeile *f* ~ **film** Stummfilm *m* ~ **partner** stiller Teilhaber *m* ~ **running** ruhiger Gang *m* ~ **third speed** geräuschloser dritter Gang *m* ~ **tooth chain with pin link** Zahnkette *f* mit Gleitgelenk ~ **transmitter** Schweigesenderüberlagerer *m* ~ **tuning** automatische Geräuschbeseitigung *f*, Krachtöter *m*, Leiseabstimmung *f*, leise Abstimmung *f*

**silex** Feuerstein *m*, Kiesel *m*, Kieselerde *f*

**silhouette** Schatten-bild *n*, -riß *m* ~ **target** Figurscheibe *f*

**silica** Kiesel *m*, Kiesel-erde *f*, -säure *f* ~ **black** Sandruß *m* ~ **brick** Silika(t)stein *m*, Silikaziegel *m* ~ **pencil** Silikastab *m* ~ **sand** Silikasand *m*

**silican carbide** Silit *n*

**silicate** Kieselsäureverbindung *f*, Silikat *n* ~ **enamel** Silikatemaille *f* ~ **mixture** Beglasung *f* ~ **slag** Silikatschlacke *f*

**silicated** kieselsauer

**silication** Silizierung *f*

**siliceous** kiesel-artig, -haltig, -sauer, -säurehaltig, silifiziert, siliziumhaltig ~ **calamine** Hemimorphit *m*, Kiesel-galmei *m*, -zinkerz *n*, Zink-glas *n*, -glaserz *n*, -kiesel *m*, -kieselerz *n* ~ **(calcareous) deposit** kieselartiges (kalkiges) Gestein *n* ~ **ferromanganese** Siliziumspiegel *m* ~ **flux** Kieselzuschlag *m* ~ **limestone** Kieselkalk *m* ~ **rocks** Kieselgesteine *pl* ~ **sandstone** Kieselsandstein *m* ~ **sinter** Kiesel-, Quarz-sinter *m*, Sinterquarz *m* ~ **sinter of geyser** Geyserit *m*, Kieseltuff *m* ~ **zinc oxide** Zink-glas *n*, -glaserz *n*

**silicic,** ~ **acid** Kieselsäure *f*, Siliziumdioxyd *n* ~ **-acid-gel container** (gas-testing apparatus) Saug-

steinbehälter *m* ~ **anhydride** Kieselsäureanhydrid *n* ~ **compound** Siliciumverbindung *f*

**silicide** Siliciumverbindung *f*, Silizid *n*

**siliciferous** kieselhaltig

**silicify, to** ~ verkieseln

**silico,** ~ **-fluoric acid** Kieselfluorwasserstoffsäure *f* ~ **-manganese steel** Mangansiliziumstahl *m*, Siliziummanganstahl *m*

**silicon** Silicium *n*, Silizium *n* ~ **alloy** Siliciumlegierung *f* ~ **bromide** Bromsilizium *n*, Siliziumbromid *n* ~ **bronce wire** Siliziumbronzedraht *m* ~ **carbide** Karborund *n*, Silizium-karbid *n*, -kohlenstoff *m* ~ **compound** Siliziumverbindung *f* ~ **content** Siliziumgehalt *m* ~ **dioxide** Siliziumoxyd *n* ~ **dressing plant** Silikaaufbereitungsanlage *f* ~ **iodide** Siliziumjodid *n* ~ **oxycarbide** Siliziumoxykarbid *n* ~ **steel** Siliziumstahl *m* ~ **tetrachloride** Chlorsilizium *n*, Siliziumtetrachlorid *n*

**siliconize, to** ~ silizieren

**siliconizing** Aufsilizieren *n*, Silizierung *f*

**silico-oxalic acid** Siliziumoxalsäure *f*

**silicosis** Silikose *f*

**silico,** ~ **tungstic acid** Kieselwolframsäure *f* ~ **varnish** Flußspatimprägnierungsmasse *f*

**silification** Silizierung *f*, Verkieselung *f*

**silk** Seide *f* ~ **-and-cotton-covered wire** Seidebaumwolldraht *m* ~ **bag** Kartuschbeutel *m* ~ **braid** Seidenzopf *m* ~ **cloth** Seiden-gewebe *n*, -stoff *m* ~ **cotton tree** Wollbaum *m* ~ **-covered brass wire** seidenübersponnener Messingdraht *m* ~ **-covered cable** Seidenkabel *n* ~ **-covered wire** mit Seide isolierter Draht *m* ~ **covering** Seidenbespannung *f* ~ **fabric** Seiden-gewebe *n*, -stoff *m* ~ **filature** Seidenhasplerei *f* ~ **gum** Seidenleim *m* ~ **insulation** Seidenisolation *f* ~ **-like scroop** knirschender Griff *m* ~ **noil** Seidenkämmling *m* ~ **paper** Seidenglanzpapier *n* ~ **parachute** seidener Fallschirm *m*, Seidenfallschirm *m* ~ **scroop** Seidengriff *m* ~ **softener** Lichtdämpfer *m* (film) ~ **spinning** Vorzwirnen *n* ~ **suture** Seidenfadennaht *f* ~ **thread** Seidenfaden *m* ~ **titer** Seidennummer *f* ~ **titration** Seidennummerierung *f* ~ **twist** Seidenzopf *m* ~ **weave** Seidengewebe *n* ~ **weighting** Seidenerschwerung *f* ~ **-winding machine** Seidenspulmaschine *f* ~ **wringing** Ausringen *n* des Seidengarnes

**silking** seidenartiger Glanz *m*

**silky** seidenglänzend, seidig ~ **sheen** seidenartiger Glanz *m*

**sill** Fensterbrett *n*, Fußlatte *f*, Grundbalken *m*, Lagergang *m*, Schwelle *f*, Schellholz *n*, Wehrschwelle *f*, Wehrsohle *f*, Tragebaum *m*, Türschwelle *f* ~ **of floor** Sohlenfalz *m* ~ **of a framework** Bundschwelle *f* ~ **of a pile-driving engine** Rammenschwelle *f* ~ **for sealing base of gates** Drempel *m* ~ **of a seam** Liegendes *n* eines Flözes ~ **of timber** Holzschwelle *f* ~ **of upper sluice head** Oberdrempel *m* ~ **of the weir** Fachbaum *m*

**sillemite** Kieselzinkerz *n*

**sillimanite bricks** Sillimanitsteine *pl*

**sillometer** Sillometer *n*

**silo, to** ~ einmieten

**silo** Bunker *m*, Getreidegrube *f*, Miete *f*, Schachtspeicher *m*, Silo *m*, Zellenspeicher *m* ~ **compartment** Silozelle *f* ~ **effect** Silowirkung *f* ~ **process** Siloverfahren *n* ~ **works** Siloanlage *f*

**siloing of pulp** Schnitzel-einmietung *f*, -einsäuerung *f*

**silt, to ~ up** verlanden, versanden

**silt** Feinkohle *f*, Kohlenschlamm *m*, Schlamm *m*, Schluff *m* ~ **analysis** Schlammanalyse *f* ~ **cement** Schlick *m* ~ **content** Schwebestoff-, Schwemmstoff-führung *f* ~ **stones** Schluffmergel *m* (verfestigt)

**siltation** (with mud) Verschlämmung *f*

**silting** Anschwemmung *f*, Sedimentierung *f* ~ **-up** Aufschlickung *f*, Schlammablagerung *f*, Verschlickung *f*

**silty** schluffig

**silumin** Silumin *n* ~ **alloy** Siliziumaluminium *n*

**silundum** Silundum *n*

**silurian** silurisch (Silur)

**silver, to ~** (a mirror) belegen, (a mirror) foliieren, silbern, übersilbern, versilbern **to ~-coat** versilbern **to ~-plate** übersilbern, versilbern

**silver** Silber *n* ~ **of due alloy** lötiges Silber *n*

**silver, ~ acetate** Silberessigsalz *n* ~ **alloy** Silberlegierung *f* ~ **amalgam** Merkursilber *n*, Silberamalgam *n* ~ **arsenite** Silberarsenit *n* ~ **assay** Silberprobe *f* ~ **bar** Silberbarren *m*, silberner Streifen *m* ~ **bath** Versilberungsbad *f* ~ **bead** Silberkorn *n* ~ **-bearing** silber-führend, -haltig ~ **benzoate** Silberbenzoat *n* ~ **brick** Silbersau *f* ~ **brocade** Silberbrokat *n* ~ **bromide emulsion** Bromsilberemulsion *f* ~ **bromide positive** Bromsilberpositiv *m* ~ **bromide printing** Bromsilberdruck *m* ~ **capping** Silberkappe *f* ~ **center electrode** Silbermittelelektrode *f* ~ **chloride** Kerat *n*, Silberchlorid *n* ~ **chlorobromide** Chlorbromsilber *n* ~ **citrate** Silberzitrat *n* ~ **coating** Feuerversilberung *f*, Silber-belegung *f*, -schicht *f*, Versilbern *n*, Versilberung *f* ~ **-colored** silberfarbig ~ **content** Silbergehalt *m* ~ **crucible** Silbertiegel *m* ~ **currency** Silberwährung *f* ~ **electrum** Silbergold *n* ~ **enamel** Silberlack *m* ~ **extracted from lead ore** Werksilber *n* ~ **extraction** Silbergewinnung *f* ~ **ferricyanide** Ferrizyansilber *n* ~ **film** Silber-belegung *f*, -schicht *f* ~ **fog** dichroitischer oder zweifarbiger Schleier *m* ~ **foil** Silber-blatt *n*, -blech *n*, -papier *n* ~ **-gilt** golden ~ **globule** (of iconoscope mosaic) Rasterkorn *n* ~ **grain** Silberkorn *n* ~ **halide** Halogensilber *n* ~ **ingot** Silber-barren *m*, -sau *f* ~ **iodide** Silberjodid *n* ~ **layer** Silberschicht *f* ~ **-leaching plant** Silberlaugerei *f* ~ **leaf** Blattsilber *n*, Silber-blatt *n*, -schaum *m* ~ **litharge** Silberglätte *f* ~ **luster** Silberschein *m* ~ **metal** Silbermetall *n* ~ **mine** Silbergrube *f* ~ **nitrate** Höllenstein *m*, salpetersaures Silber(oxyd) *n*, Silber-ätzstein *m*, -nitrat *n*, -salpeter *m* ~ **nitride** Stickstoffsilber *n* ~ **nitrite** Silbernitrit *n* ~ **ore** Silbererz *n* ~ **oxysalt** Silberoxydsalz *n* ~ **paper** Silberpapier *n* (thin) ~ **plate** Silberblech *n* ~ **plating** Silber-belegung *f*, -plattierung *f*, Versilbern *n*, Versilberung *f* ~ **-plating solution** Versilberungslösung *f* ~ **plug** Rasterkorn *n* ~ **polish** Silberputz *m* ~ **precipitate** Silberniederschlag *m* ~ **refinery** Silberscheideanstalt *f* ~ **refining** Silber-brennen *n*, -scheidung *f* ~ **-refining hearth** Silberbrennherd *m* ~ **rhodanide** Silberrhodanid *n* ~ **saltmaker** Silberpräparierer *m* ~ **selenide** Selensilber *n* ~ **sheet** Silberblech *n* ~ **smith** Silberarbeiter *m* ~ **solder** Silber-(schlag)lot *n* ~ **solution** Silber-lauge *f*, -lösung *f*

~ **stamping works** Silberprägeanstalt *f* ~ **-style writing** Silberstiftschreibverfahren *n* ~ **subchloride** Silberchlorür *n* ~ **subiodide** Silberjodür *n* ~ **suboxide** Silberoxydul *n* ~ **sulfate** Silbervitriol *m* ~ **sulfide** Schwefelsilber *n*, Silbersulfid *n* ~ **tailings** Silberschlamm *m* ~ **telluride** Tellursilber *n* ~ **tensel** Silberflitter *m* ~ **thaw** Anraum *m*, Duftanhang *m*, Haarfrost *m* ~ **thiosulfate** Silberthiosulfat *n* ~ **thread** Silbergespinst *n* ~ **-tripped contact** Silberschaltstück *n* ~ **ware** Silber-ware *f*, -waren *pl* ~ **white** Kremserweiß *n*; silberweiß ~ **wire** Silberdraht *m* ~ **wire fuse** Silberschmelzsicherung *f* ~ **works** Silberhütte *f*

**silvered, ~ on the front surface** (or on the face) oberflächenversilbert ~ **convex-glass reflector** Glassilberkonvexspiegel *m* ~ **glass mirror** (or **reflector**) Glassilberspiegel *m* ~ **parabolic-glass reflector** Glassilberparabolspiegel *m*

**silvering** Silberbelegung *f*, Spiegel *m* Spiegelbelag *m*, Versilbern *n* ~ **bath** Versilberungsbad *f* ~ **table** (glass mfg.) Belegtisch *m*

**silvery** silber-artig, -glänzend, -hell, silbern, silb(e)rig ~ **gray** silbergrau ~ **luster** Silberglanz *m* ~ **pig iron** Ferrosilizium *n*, Silbereisen *n*

**similar** ähnlich, gleich-artig, -namig ~ **to nature** naturähnlich ~ **pole** gegengesetzter oder gleichnamiger Pol *m*

**similarity** Ähnlichkeit *f*, Ebenbild *n*, Gleichartigkeit *f* ~ **to metal** Metallartigkeit *f*

**similarity, ~ equation for dimensioning airplane models** Froudesche Zahl *f* ~ **principle** Ähnlichkeitstheorie *f* ~ **transformations** Ähnlichkeitstransformationen *pl*

**similarly charged** (or **electrified**) gleichnamig elektrisch

**similitude** Ähnlichkeit *f* ~ **theory** Ähnlichkeitstheorie *f*

**simmer, to ~** sieden, wallen, wellen ~ **gasket** Simmering *m*

**simmrit gasket** Simmritdichtung *f*

**simple** anspruchslos, einfach, einfältig, schlicht ~ **alloy steel** Ternarstahl *m* ~ **-aperture lens** Lochblendenlinse *f* ~ **beam** einfacher Balken *m*, Freiträger *m* ~ **bed lowering lever** einarmiger Hebel *m* ~ **bridge** Einfachbrücke *f* (teleph.) ~ **cable** (one conductor only) Einleiterkabel *n* ~ **controller** Kleinregler *m* ~ **criteria** einfache Deckoperationen *pl* ~ **detector-type receiver** Geradeausempfänger *m* ~ **dipole** Einfachdipol *n* ~ **fixed pulley** feste (am Ort bleibende) Rolle *f* ~ **flap** Wölbungsklappe *f* ~ **glide** Einfachgleitung *f* ~ **-harmonic** einwellig, rein sinusförmig ~ **harmonic motion** rein sinusförmige Bewegung *f* ~ **ignition** Zweifunkenzündung *f* ~ **improvided truss** (building) Bocksprengwerk *n* ~ **lens** Einlinsenobjektiv *n* ~ **-lever cash register** Einzählerhebelkontrollkasse *f* ~ **line grating** Strichgitter *n* ~ **magnifier** einfache Lupe *f* ~ **momentum theory** Schraubenstrahltheorie *f* ~ **movement** einfache Schiebung *f* ~ **parallel winding** eingängige Parallelwicklung *f* ~ **process factor** Stufentrennfaktor *m* ~ **repetend** einzifferige Periode *f* ~ **roof truss** Sparrendach *n* ~ **shear(ing)** einschnitte Abscherung *f* ~ **signal** Einfrequenzsignal *n* ~ **sound source** Allrichtungs-Schallquelle *f* ~ **stationary gas producer** Generator *m* mit fixem Rost ~ **steel** unlegierter Stahl *m* ~ **stress** einfacher Druck *m* ~ **tone** reiner Ton *m* ~

**simplex** einwegig, einfache Verbindung f, Simplexbetrieb m ~ **circuit** Einfach-leitung f, -schaltung f, Simplexleitung f, Wechselverkehrskreis m ~ **needle indicator** Einfachnadelzeiger m ~**-operated telegraph** Einfachtelegraph m ~ **operation** Einfach-betrieb m, -verkehr m, einfacher Verkehr m, Simplex-betrieb m, -verkehr m, Wechselverkehr m ~ **pile** Simplexpfahl m ~ **pipe cutter** Simplexrohrabschneider m ~ **repeater** Einwegeverstärker m ~ **roller chain** Einfachrollenketter m ~ **system** Einfachschaltung f, Simplexverfahren n ~ **telegraphy** Einfachtelegrafie f, einseitige Telegrafie f ~ **working** Einfach-betrieb m, -verkehr m, Simplexverkehr m, Wechselverkehr m

**simplexed**, ~ **circuit** Simultanleitung f ~ **coil** Simultanübertrager m ~ **telegraphy** Unterlagerungstelegrafie f

**simplicity** Einfachheit f
**simplification** Vereinfachung f
**simplify, to** ~ vereinfachen
**simply connected** einfach zusammenhängend
**simulate, to** ~ nachbilden, simulieren, vortäuschen **to** ~ **closely** genau nachbilden
**simulated** vorgetäuscht ~ **echo** simuliertes Echo n ~ **instrument flight** Schein-Instrumentenflug m ~ **jet effect** Strahlnachahmung f ~ **movement** Scheinbewegung f ~ **program** Nachahmungsprogramm n
**simulation** Nachbilden n, Vorschützung f von Gebrechen ~ **of terrain** Geländenachbildung f
**simultaneity,** ~ **concept** Gleichzeitigkeitsbegriff m ~ **factor** Gleichzeitigkeitsfaktor m
**simultaneous** gleichzeitig ~ **broadcast** gemeinsames Programm n ~ **differential equations** partielle Differentialgleichungen pl ~ **equation** Simultangleichung f ~**-movement selector** Mitlaufwähler m ~ **range** gleichzeitige Reichweite f ~ **reception** Simultanempfang m ~ **rising of the punch and lowering of the table** gleichzeitiges Hochgehen n des Stempels und Niedergehen n des Preßtisches ~ **two-way radio communication** Funkdoppelverkehr m
**simultaneousness** Gleichzeitigkeit f
**sine** Sinus m ~ **of angle** Sinuswinkel m
**sine,** ~ **bar rule** Sinuslineal n ~ **component meter** Blindverbrauchszähler m ~ **condition** Sinusbedingung f ~**-cosine card** Funktionspotentiometer m ~**-cosine potentiometer** Sinus-Kosinuskompensator m ~ **current** Sinusstrom m ~ **curve** Sinus-kurve f, -linie f ~ **function** Sinusfunktion f ~ **galvanometer** Sinusbussole f ~ **law** Sinus-gesetz n, -satz m ~ **movement** sinusförmige Bewegung f ~ **oscillation** sinusförmige Schwingung f, Sinusschwingung f ~ **potentiometer** Sinuskompensator m ~ **rule** Sinuslineal n ~ **series** Sinusreihe f ~ **shape** sinusförmig ~ **term** Sinusglied n ~ **voltage** Sinusspannung f ~ **(wave of) voltage** sinusförmige Spannung f, -schwingung f, -welle f **pure** ~ **wave** reine Sinuswelle f ~ **wave of sound** sinusförmige Schallwelle f ~**-wave alternator** Sinuswellenerzeuger m ~**-wave current** sinusförmiger Strom m, Sinusstrom m ~**-wave potential** (distribution) sinusförmige Spannungsverteilung f ~ **wave tone** Dauerton m ~**-wave voltage** Sinusspannung f
**sinew** Sehne f
**sing, to** ~ pfeifen
**singe, to** ~ flambieren, sengen, versengen **to** ~ **the thread** den Faden m brennen
**singeing** Sengerei f ~ **machine** Flamm-, Seng-maschine f ~ **pin** Sengnadel f
**singing** Pfeifen n, Pfeifen n der Verstärker, Schwingungserregung f ~ **arc** singender Lichtbogen m ~ **arc lamp** singende Bogenlampe f ~ **limit** Pfeifgrenze f ~ **margin** Abstand m vom Pfeifpunkt, Pfeif-abstand m, -sicherheit f ~ **path** Rückkupplungsweg m ~ **point** Pfeifgrenze f, Pfeifpunkt m ~ **stability under no-load condition** Leerlaufpfeifsicherheit f ~ **suppressor** Rückkopplungssperre f ~ **valve** Tonröhre f
**single, to** ~ einzeln, vereinzeln **to** ~ **out** aussortieren, heraus-greifen, -suchen
**single** ein, einfach, eingängig, einmalig, einteilig, einzig, ungeteilt, vereinzelt **in** ~ **shear (or with** ~ **cross section)** einschnittig ~ **and twin barrel tanks** Einzel- und Zwillingsbehälter pl
**single,** ~**-acting** einfachwirkend ~**-acting cylinder** einfachwirkender Zylinder m ~**-acting engine** einfachwirkende Maschine f ~**-acting pump** einfachwirkende Pumpe f ~**-action** Einzelvorgang m ~**-action hydraulic cylinder** Einweg-Arbeitszylinder m ~ **action printer** Einzelzeichendrucker m ~**-address code** Einadreßkode m ~**-address instruction** Einadreßbefehl m ~**-angle cutter** einseitiger Winkelfräser m ~**-angle milling cutter** Winkelstirnfräser m ~**-anode rectifier** Einanodenventil n ~ **apron high draft system** Einriemchen-hochverzugsstreckwerk n ~ **arm kneader** einarmiger Kneter m ~ **armature system** Einankersystem n ~**-axle tractor** Einachsschlepper m ~**-axle vehicle** Einachsfahrzeug n ~**-ball** einfach kugelig ~**-band system** Einseitenbandsystem n ~**-banked oar** langer Riemen m ~**-banking** einrudrig ~**-bar** einschenkelig ~**-barreled** einläufig ~**-barreled gun** einläufiges Geschütz n ~**-barreled pointer** Einbalkenzeiger m ~**-bath chrome dyestuff** Einbadchromierfarbstoff m ~**-bath tanning** Einbadgerbung f ~ **battery** Einzelbatterie f ~**-bay** einstielig ~**-bay machine** Einstieler m (aviat.) ~**-beam** einbäumig ~**-beam traveling crane** Einträgerlaufkran m ~**-beam tube** Einstrahlrohr n (rdr) ~**-bearing shaft** einlagerige Welle f ~ **bellows** Handblasebalg m ~ **belt** Einzelriemen m ~ **bevel** einfache Abschrägung f ~ **biplane** einstieliger gestaffelter Doppeldecker m ~ **(double) Blackwall hitch** einfacher (doppelter) Hakenschlag m ~**-blade** einflügelig ~**-blade propeller** Einblattluftschraube f, einflügelige Luftschraube f ~**-blade shutter** Einflügelblende f (film) ~**-blister roof** Einlinsendach n ~ **block** Flaschenzug m, einscheibiger Block m ~ **block furnace** Bauernofen m (metall.) ~ **blow** Einzelschlag m ~**-blow cold upsetting machine** Eindruckkaltpresse f ~**-blow impact value** Bruchfaktor m ~**-branched** einfach verzweigt ~ **breadth of paper hangings** Tapetenbahn f ~ **bridge**

einfache Brücke *f* ~-**bucket excavator** Grief-
bagger *m* ~ **bus bars** Einfachsammelschienen
*pl* ~ **cable** Einfachkabel *n*, (one conductor
only) Einleiterkabel *n*, Einzelkabel *n* ~
**cable in cable network** Masche *f* eines Kabel-
netzes ~ **cage operation** Einkorbbetrieb *m*
~ **capacitance** Einfachkapazität *f* ~ **card**
Kartenblatt *n* ~ **card checking** Einzelkarten-
prüfung *f* ~ **cascade action** einfache Kas-
kadenregelung *f* ~ **cavity mold** Einfachform
*f* ~ **ceiling (or winding)** Einschichtwicklung *f*
~-**chamber brake energizer** Einkammerbrems-
verstärker *m* ~-**chamber drying apparatus** Ein-
kammertrockner *m* ~-**chamber machine**
Einkammerapparat *m* ~-**chamber pneumatic**
**brake valve** Einkammerdruckluftbremsventil
*n* ~-**chamber vacuum filler** Einkammervakuum-
füller *m* ~-**channel** einwegig, Einkanal (teleph.)
~-**channel analyzer** Einkanalanalysator *m* ~-
**channel method** Einkanalverfahren *n* ~-**channel**
**modulation** einfache Modulation *f* ~-**channel**
**pulsing** Einkanaltastung *f* ~-**channel single-**
-**side-band equipment** Einkanal-Einseitenband-
gerät *n* ~-**channel-spectrometer** Einkanalspek-
trometer *m* ~-**channel telegraph** Einfachtele-
graf *m*, einwegiger Telegraf *m* ~-**channel**
**transmission** Einkanalübertragung *f* ~-**chanel-**
ed einläufig ~ **choke** Einfachdrossel *f* ~
**circuit** Einzel-kreis *m*, -leitung *f* ~-**circuit**
**drift tube** Einkreistriftröhre *f* ~-**circuit receiver**
Einkreisempfänger *m*, Einkreiser *m* ~-**circuit**
**receiving set** Primärempfänger *m* ~-**circuit**
**reception** Einkreis-, Primär-, Such-empfang
*m* ~-**circuit transposition** Schleifenkreuzung
*f* ~-**coil** einspulig ~-**coil ringer** Einspulenwecker
*m* ~ **collisions** Einzelstöße *pl* ~-**color dotted-**
-**line recorder** Einfarbpunktschreiber *m* ~-
**color sheet-fed offset press** Einfarbenbogen-
offset *m* ~ **column** einspaltig ~ **column press**
Einständerpresse *f* ~ **commutation** einpolige
Umschaltung *f* ~-**compartment sandblast-tank**
**machine** Einkammersandstrahlapparat *m* ~
**component of deviation** Einzeldeviation *f*
~-**component lense** Einlinsenobjektiv *n* ~-**com-**
**ponent signal** Einfachsignal *n* ~ **conductor**
Einzel-kabel *n*, -leiter *m* ~-**conductor cord**
einadrige Schnur *f* ~-**contact** Einkontakt *m*
~-**contact one-field regulator** Einkontaktein-
feldregler *m* ~-**contact two-field regulator**
Einkontaktzweifeldregler *m* ~ **container** Einzel-
magazin *n* ~-**control tuning** Einknopf-abstimmung *f*, -bedienung *f* ~ **coordinate control**
Aussteuerung *f* in einer Koordinate ~ **copy**
Einzelexemplar *n*, Einzelnummer *f* ~ **cord**
Einschnur(schaltung) *f* ~-**cord operation** Ein-
schnurbetrieb *m* (teleph.) ~-**core** einadrig
~-**core(d) cable** einadriges Kabel *n* ~-**core**
**circuit** Einfachleitung *f* ~-**cored** einkernig
~-**cotton-covered** einfach umsponnen mit
Baumwolle *f* ~-**covered** einfachumsponnen
~-(**double-, triple-**)**covered wire** einfach (doppelt, dreifach) umsponnener Draht *m* ~-**crank**
**compound steam engine** Einkurbelverbund-
dampfmaschine *f* ~-**crank steam engine**
Einkurbeldampfmaschine *f* ~ **crystal** Ein(zel)-
kristall *m* ~-**crystal filament (or growth)**
Einkristallzüchtung *f* ~-**crystal seed** Ein-
kristallkeim *m* ~-**crystal specimens** Ein-

kristallproben *pl* ~-**crystal surfaces** Ein-
kristalloberflächen *pl* ~-**cup insulator** Ein-
fachglockenisolator *m* ~ **current** Einfach-,
Einzel-strom *m* ~ **cut** Einzelschnitt *m* ~-**cut**
**file** einhiebige Feile *f*, (of a file) Schlichthieb
*m* ~ **cutting edge** Einfachschneide *f* ~ **cutting**
**tool** Einzeldrehstahl *m* ~-**cycle state (or**
**condition)** Einwelligkeit *f*
**single-cylinder,** ~ **blower** Einzylindergebläse *n*
~ (**or yankee**) **drier** Einzylindertrockner *m*
~-**exhaust steam engine** Einzylinderauspuff-
dampfmaschine *f* ~ **magneto** Einzylindermag-
netzünder *m* ~ **plunger-type compressor** Ein-
zylindertauchkolbenverdichter *m* ~ **steam engine**
einzylindrige Dampfmaschine *f* ~ **test bed**
Einzylinderprüfstand *m* ~ **test bench** Einzylinderprüfstand *m* ~ **test motor (or unit)**
Einzylindermotor *m* ~ **two-cycle engine** Ein-
zylinderzweitaktmotor *m* ~ **type compressor**
Einzylinderverdichter *m*
**single,** ~ **daylight press** Einetagenpresse *f* ~-
**decade counting unit** einfacher Zehnerunter-
setzer *m* ~ **delivery** Einzelausgang *m* ~-**dial**
**tuning** Einknopf-abstimmung *f*, -bedienung
*f* ~ **diffusion stage** Einzeldiffusionsstufe *f* ~
**diode** Einfachdiode *f* ~-**direction bearing**
einseitig wirkendes Lager *n* ~-**disc clutch** Einscheibenkupplung *f* ~-**disc thrust bearing**
Einscheibendrucklager *m* ~ **disturbance** Ein-
zelstörung *f* ~-**dividing method** Einzelteilverfahren *n* ~-**dome bell** einschaliger Wecker *m*
~ **dotted line recorder** Einfach-Punktschreiber
*m* ~ **drive** Einzelantrieb *m* ~-**drum drive**
Einrollenantrieb *m* ~-**duct** einzügig ~-**duct**
**concrete block** einzügiges Formstück *n*, einzügiges Zementformstück *n* ~-**duct conduit**
einzügige Rohrpost *f*, Kanal *m* aus einem
Vollrohr ~-**edged width sound record (or**
**track)** Einzackenschrift *f* ~-**edged variable-**
-**width sound record (or track)** einspuriger Tonstreifen *m* ~-**enclosure (multipolar switch)**
Einkessel *m* ~ **enclosure switch** Einkessel-
schalter *m* ~-**ended** einendig ~-**ended amplifier**
Seriengegentaktverstärker *m* ~-**ended cable**
**grip** Ziehschlauch *m* ~-**ended push-pull stage**
transformatorlose Gegentaktstufe *f* ~-**end-**
ed **spanner** einfacher Schraubenschlüssel *m*
~-**engined** einmotorig ~ **entrance** Dockschleuse *f* ~-**entry compressor** einseitiger
Kompressor *m* ~ **event** einmaliger Vorgang
*m*, Einzelvorgang *m* ~ **exchange** einfache Vermittlung *f* ~ **exposure** einmalige Bestrahlung
*f* ~-**field regulator** Einfeldregler *m* ~-**field**
**single-contact regulator** Einfeld-Einkontakt-
regler *m* ~-**field single-contact regulator with**
**drooping characteristic curve** Einfeld-Ein-
kontaktregler *m* mit Neigkennlinie ~-**field-**
-**two-element two-contact regulator** Einfeld-
-Zweielement-Zweikontaktregler *m* ~ **filament**
Einzelfaden *m* ~ **filament lamp** Eindrahtlampe
*f* ~-**fillet weld** einseitige Kehlnaht *f* oder
Kehlnahtschweißung *f* ~-**film projector** Ein-
bandprojektor *m* ~ **finishing cutter** Einzel-
fertigschneider *m* ~-**flame** einflammig ~
**flange on tire** Untergriff *m* beim Radreifen
(r.r.) ~-**flash beacon** Einblitzfeuer *n* ~-**flight**
**stairs** einläufige Treppe *f* ~ **float** einfacher
Schwimmer *m*, Einzelschwimmer *m* ~-**float**

plane Einschwimmerflugzeug *n* ~ **fluid cell Element** *n* mit einer Flüssigkeit ~**-fluked anchor** einarmiger Hafenanker *m* ~**-folding magnifier** einfache Einschlaglupe *f* ~ **form cutter** Gewindeformfräser *m* ~ **frame** Einzelständer *m* ~**-frame hammer** Einständerhammer *m* ~**-frame steam hammer** Einständerdampfhammer *m* ~**-frequency tone** Reinton *m* ~**-fuselage plane** Einrumpfflugzeug *n* ~ **gate** (ebbtide gates) Dockschleuse *f* ~ **gearing** Einzelverzahnung *f* ~**-grain structure** konservative Struktur *f* ~**-grid tube (or valve)** Eingitterröhre *f* ~ **gyro compass** Einkreiselkompaß *m* ~**-gyro system** Einkreiselanordnung *f* ~**-handed** eigenhändig ~**-head** einstemplig ~**-head box wrench** Einfachringschlüssel *m* ~**-head wrench** Einfachschraubenschlüssel *m* ~**-hearth furnace** einherdiger Ofen *m* ~**-hole nozzle** Einlochdüse *f* ~ **hood** Halbverdeck *n* ~ **hull** Einzelhülle *f* ~**-indexing process** Einzelteilverfahren *n* ~**-ingot plate** Restplatte *f* ~**-intersection method** Einzelschnittverfahren *n* ~ **isolated wheel load (SIWL)** Einzelradlast *f* (aviat.) ~ **item ejection** Einzelpostenvorschub *m* ~**-jack platform elevator** Einstempelplattformhebebühne *f* ~ **jump** Einzelsprung *m* ~**-knob controlling** Einknopf-abstimmung *f*, -bedienung *f* ~**-layer coil** einlagige Spule *f* ~**-layer film** Einschichtfilm *m* ~**-layer solenoid** einlagiges Solenoid *n* ~ **layer stenter** Planrahmen *m* ~ **leaf springs** einlagige Blattfedern *pl* ~ **leap** Einzelsprung *m* ~**-leg** einschenk(e)lig ~**-leg construction gantry crane** Halbbockkran *m* ~ **lens (or objective)** Einzelobjektiv *n* ~**-lens camera** Einfachkamera *f* ~**-lens magnification** Lupenvergrößerung *f* ~**-lens objective** Einlinsenobjektiv *n* ~**-lens (aerial) survey camera** Einfachreihenmeßkamera *f* ~ **level butt joint** einseitiger V-Stoß *m* ~ **level formula** Einniveauformel *f* ~**-lever contact breaker** Einfachunterbrecher *m* ~**-lever control** Einhebelsteuerung *f* ~**-lever feed control** Einhebelschaltung *f* ~**-lever quick-action control** Einhebelschnellschalteinrichtung *f* ~**-lever slot control** Einhebelschlitzsteuerung *f* ~ **(wire) line** Einzelleitung *f* ~ **line of ducts** Einrohrkanal *m* ~**-line machine** einbahnige Maschine *f* ~**-line mill train (or line roll train)** einachsige Straße *f* ~ **line subscriber** Einzelanschlußteilnehmer *m* ~**-lip and multiple-fluted cutters** Ein- und Mehrschneidefräser *m* ~**-lip cutter** Einschneidefräser *m* ~ **lip cutter grinding attachment** Frässtichelschleifvorrichtung *f* ~**-lip cutters with taper shank** Frässtichel *m* mit Kegelschaft ~ **lipped rifle deep hole drill** Einrillentieflochbohrer *m* ~ **load** Einzellast *f* ~ **loader** Einzellader *m* ~ **lobe cam** Einfachnocken *m* ~ **lock** einfache Kammerschleuse *f* ~**-loop** (servo) einschleifig ~**-loop oscillogram** Füllkurve *f* ~ **magnifier** Einzellupe *f* ~**-meche** Einfachband *n* ~**-mesh filter** eingliedrige Siebkette *f* ~ **metal plate-holder** Metall-Einfachkassette *f* ~ **molar** einmolar ~**-motion selector** Wähler *m* mit einer Bewegungsrichtung ~**-needle compass rose (or card)** Einnadelrose *f* ~**-needle telegraph** Einnadeltelegraf *m* ~**-noded** einknotig ~**-operator punching machine** Einmannlochwerk *n* ~**-operator telegraph** (in-

strument) Einmanntelegraf *m* ~**-operator welding machine** Einzelschweißumformer *m* ~**-oxygen-hose cutting torch** Zweischlauchbrenner *m* ~ **pair of lock gates** Schutzschleuse *f* ~**-part production** Einzelfertigung *f* ~ **particle level** Einteilchenniveau *n* ~**-particle model** Einteilchenmodell *n* ~ **particle wave function** Einteilchenwellenfunktion *f* ~**-passage crusher** Einstufenzerkleinerer *m* ~**-passage grinding** Einstufenmahlung *f* ~ **petticoat insulator** Einzelglocke *f*

**single-phase** einphasig, einwellig ~ **alternating-current** Einphasenwechselstrom *m* ~ **alternating-current motor** Einphasenwechselstrommotor *m* ~ **commutator motor with self-excitation** kompensierter Repulsionsmotor *m* ~ **current** Einphasenstrom *m* ~ **electric railway** Einphasenbahn *f* ~ **electric railway power circuit** Einphasenbahnspeiseleitung *f* ~ **full-wave connection** Einphasenvollwegschaltung *f* ~ **generator** Einphasengenerator *m* ~ **induction motor** Einphaseninduktionsmotor *m* ~ **meter** Einphasenzähler *m* ~ **motor** Einphasenmotor *m* ~ **rectifier** Einphasengleichrichter *m* ~ **starter** Einphasenanlasser *m* ~ **supply** Einphasenzuführung *f* ~ **system** Einphasensystem *n* (soil mech.) ~ **transformer** Einphasentransformator *n*

**single,** ~ **phasing** Einphasenlauf *m* ~ **photograph** Einzelaufnahme *f* ~**-photograph measuring** Einbildmessung *f* ~**-photograph plotting apparatus** Bildkartiergeräte *pl* für Einzelbilder, Einzelbildgerät *n* ~**-pickled sheet** einmal dekapiertes Blech *n* ~**-picture crank and shaft** Eingang-kurbel und -welle *f* (film) ~ **picture movement** Einzelbildwiedergabe *f* (film) ~**-spice furnace** Blasofen *m* ~**-piece production** Einzelanfertigung *f* ~ **pile** einpolig ~**-pillar isolator** Einsäulentrenner *m* ~**-pilot remote control** Eindrahtfernsteuerung *f* ~ **pipe conduit** Vollrohrkanal *m* ~**-piston hand-operated oil pump** Einkolbenhandschmierpumpe *f* ~**-plate clutch** Einscheibenkupplung *f* ~**-plate dark slide** Einzelkassette *f* ~**-plate dry clutch** Einplattentrockenkupplung *f* ~**-plate holder** Einfacheinzelkassette *f* ~**-plug apparatus** Geräteeinzelstecker *m* (electr.) ~**-plunger distributor pump** Einkolbenverteilerpumpe *f* ~**-plunger lubricating pump** Einkolbenschmierpumpe *f* ~ **plunger pump** Einkolbenpumpe *f* ~**-ply** einfach ~**-ply belt** Einfachriemen *m* ~**-point curve-drawing recorder** Einkurvenschreiber *m* ~**-point recorder** Einpunktschreiber *m* ~**-point recording controller** Einkurvenregler *m* ~**-point thread chaser** Spitzstahl *m* ~**-point threading tool** einseitiger Gewindeschneidestahl *m*, einzahniger Gewindestahl *m*, Gewindestahl *m* (als Spitzstahl) ~**-point tool** Spitzstichel *m* ~**-polarity pulse** unipolarer Impuls *m* ~ **pole** Einzelstütze *f* ~**-pole** einpolig ~**-pole line** einfache Stangenlinie *f* ~**-pole receiver** einpoliger Fernhörer *m* ~**-pole switch** einpoliger Schalter *m* ~ **pressing** Einzelpreßling *m* ~**-prism spectrograph** Einprismaspektrograf *m* ~ **process** Einzelvorgang *m* ~ **pulley** Einscheibe *f* ~ **pulley belt drive** Einscheibenriemenantrieb *m* ~ **pulley block** einscheibiger Block *m* ~**-pulley drive** Ein-

scheiben-, Einzelscheiben-antrieb *m* ~-pulse
method Einpulsverfahren *n* (electron) ~-
-purpose automat Sonderzweckautomat *m*
~ purpose computer Einzweckrechner *m*
~-purpose machine Einzweckmaschine *f* ~-
-purpose templets Festschablonen *pl* ~-purpose
unit Einzweckgerät *n* ~ push-button control
Einknopfsteuerung *f* ~ radio shielding Ein-
zelentstörung *f* ~-rail einschienig ~ range
instrument Einzelbereichmeßgerät *n* ~-range
meter Einfachzweckeinfachbereichsinstrument
*n* ~ reaction Einzelvorgang *m* ~ reactor Einfachdrossel *f* ~ rear wheels Einfachbereifung
*f* ~-record instrument Einfachschreiber *m* ~
refining Schwalarbeit *f* ~ refracting einfachbrechend (opt.) ~ region Einzelbezirk *m* ~
relay (intermediate) repeater Einrohrverstärker
*m* ~ release Einzelabwurf *m*, (bombing)
Einzelwurf *m* ~ repetend einziff(e)rige Periode
*f* ~ representation Einzeldarstellung *f* ~-
-revolution machine Eintourenmaschine *f* ~-
-revolution press Eintourenpresse *f* ~-ring-type
furnace Einrinnenofen *m* ~-roll crusher Ein-
walzenbrecher *m* ~-roller clutch Einfach-
kupplung *f* ~-roller wire guide Einrollendraht-
führer *m* ~ roughness Einzelrauhigkeiten *pl*
~ round Einzelschuß *m* ~ routed ship Einzelfahrer *m*
single-row einreihig ~ ball bearing einreihiges
Lager *n*, Rillenkugellager *n* ~ inner race
with taper bore (bearing) Kegelinnenring *m* ~
radial engine einreihiger Sternmotor *m*, Ein-
sternmotor *m* ~ rigid-ball journal bearing
einreihiges Querlager *n* (Kugellager) ~ rigid-
-ball journal bearing with taper clamping
sleeve einreihiges Spannhülsenlager *n* (Kugellager) ~ rigid deep-groove journal bearing
Hochschulterlager *n* ~ riveting einreihige
Nietung *f*
single, ~-sashed window einflügeliges Fenster
*n* ~ scattering (of electrons) Einzelstreuung *f*
~-screw chuck (or clamp) Einschrauben-
klemme *f* ~-screw vise Stufenscheibenantrieb
*m* ~ seam Einfachnaht *f* ~-seat(ed) valve
Einsitzventil *n* ~-seater Einsitzer *m* ~-seating
einsitzig ~-section filter Einzelabschnittfilter
*m* ~-set regulating valve Einsitzregelventil *n*
~ shaft eintrümmiger Schacht *m* ~-shaft
unit system power station Einwellenblock-
kraftwerk *n* ~-shear riveting einschnittige
Nietung *f* ~-shear steel Schweißstahl *m*
~-sheave steel block einrolliger Kloben *m*
~ shed insulator Einzelglocke *f* ~ sheet bend
einfacher Schotsteg *m* ~-ship sweep Einschiff-
gerät *n* ~ shot Einzelschuß *m* ~-shot fire
Einzelfeuer *n* ~-shot multi-vibrator einseitiger
Multivibrator *m* ~-shot weapon Einzellader
*m* ~ shroud knot Want *f* ~ side band Einseiten-
bandbetrieb *m* (teleph.) ~ side-band modulation Einseitenbandmodulation *f* ~ side-band
transmission Einseitenband *n*, Einseitenband-
-senden *n*, -übertragung *f* ~-sided impeller
halboffenes Laufrad *n* ~-sided pattern plate
einfache Modellplatte *f*, Reversiermodellplatte *f*, (in molding) Umschlagplatte *f* ~-sided
(or unilateral) pressure application einseitiger
Anpressungsdruck *m* ~-sided rack einseitiger
Rahmen *m*, einseitiges Gestell *n* ~-signal

Einzeichen *n* ~-silk-covered mit Seide *f* einfach umsponnen ~-skip hoist eintrümmiger
Schrägaufzug *m* ~-sleeve valve engine Ein-
schiebermotor *m* ~-slide bar guides einschienige
Kreuzkopfführung *f* ~-slotted wing einschlitzi-
ger Schlitzflügel *m* ~-sound source Einzel-
schallquelle *f* ~ sound track Einzackenschrift
*f* (acoust.) ~ span Einbereich *n* ~-span stringer
bridge Uferbrücke *f* ~-span tuning Einknopf-
-abstimmung *f*, -bedienung *f* ~ spar Einzel-
sparren *m* ~-spar type Einholmbauart *f* ~
spark gap Einzelfunkenstrecke *f* ~-spark
ignition Einfunkenentzündung *f* ~-sparred
wing einholmiger Flügel *m* ~-spindle automatic Einspindelautomat *m* ~-spindle semi-
-automatic Einspindelhalbautomat *m* ~-spindle
upright shaper Bockfräsmaschine *f* ~ split
form zweiteilige Form *f* ~-spool relay ein-
spuliges Relais *n* ~ spring safety valve ein-
faches federbelastetes Sicherheitsventil *n* ~-
-stage Einzelstufe *f*; einstufig ~-stage blower
Einstufenlader *m* ~-stage compressor einstufiger Kompressor *m* ~-stage grinding mill
Einstufenmahlwerk *n* ~-stage recycle Kreislauf *m* nach Hertz ~ stage resin Resolharz *n*
~-stage supercharger einstufiger Kompressor
*m* ~-stand mill eingerüstiges Walzwerk *n* ~
standard of a hammer einseitiger Hammer-
ständer *m* ~ standard planing machine Einständerhobelmaschine *f* ~ start cutter eingängiges Fräsen *n* ~-station pilot-balloon
spotting Einfachanschnitt *m* ~-station range
finder einstationärer Entfernungsmesser *m*,
Einstandentfernungsmesser *m* ~-step einstufig ~-strand eintrümmig ~-strand chain
einsträngige Kette *f* ~ strength brand einfache
Marke *f* (Farbstoff) ~-stroke bell Einschlag-
wecker *m* ~ stroke force Einzelschlagstärke *f*
~-strut landing gear Einbeinfahrwerk *n* ~-strut
undercarriage Einbeinfahrgestell *n* ~ support
Einzelstütze *f* ~ suspension Einzelaufhängung
*f* ~ sweep einmaliger Vorlauf *m* ~ switchboard
unit Vermittlungskästchen *n* ~ thread eingängig (mech.), eingängiges Gewinde *n*
~-thread screw eingängige Schraube *f* ~-thread
worm eingängige Schnecke *f* ~-throat Venturi
tube Einfachsaugrohr *n* ~-throw crankshaft
einfach gekröpfte Kurbelwelle *f* ~-throw
switch Einregelschalter *m* ~ tile einzügiges
Formstück *m* oder Tonformstück *n* ~-toggle
jaw crusher Einschwingenbrecher *m* ~ tone
Einton *m* ~-tone keying Eintonverfahren *n*
~ track Einzelgeleise *n* ~-track engleisig,
einspurig, eintrümmig ~-track hoist bridge
eintrümmiger Schrägaufzug *m* ~-track railroad Eingeleisebahn *f* ~ (double) track
railroad ein-(zwei)gleisige Eisenbahn *f* ~-
tracked einläufig ~ tree Einspännerwage *f*
~-tree attachment hook Zughaken *m* ~-tube
sweep circuit Einröhrenkippanordnung *f* ~-
-type antenna Einstockantenne *f* ~-type
pressure gauge (or manometer) Einfach-
druckmesser *m* ~ undercoater Einheitsgrund
*m* ~ unit Einzelmaschine *f* ~ V-butt joint
V-Stoß *m* ~ V-butt joint with buck-up V-Stoß
*m* mit Stutzlasche ~ V-groove V-Naht *f* ~
V-weld with root reinforment V-Naht *f* wurzel-
seitig nachgeschweißt ~-valued einwertig

~-valued function einwertige Funktion *f*
~-valve amplifier (or -valve intermediate
repeater) Einrohrverstärker *m* ~-valve receiver Einröhrenempfänger *m* ~ variable-
-density track Gleichtaktsprossenspur *f* ~
view Einzelaufnahme *f* ~-wave rectification
Einweg-, Halbweg-gleichrichtung *f* ~ way
einzügig ~-way duct einzügiger Kanal *m*
~-way multi-spindle drilling machine Einweg-
vielspindel-Bohrmaschine *f* ~-way switch
Einwegumschalter *m* ~-wheel load (SIWL)
Einzelradbelastung *f* ~ winding Einzelwickel
*m* ~-wire einadrig ~-wire circuit Eindraht-,
Einfach-leitung *f* ~-wire feeder Eindrahtspeise-
leitung *f* ~-wire route Einzelleitungslinie *f* ~
wiring (Leitung) Einzelverlegung *f* ~ wooden
plateholder Holzeinfachkassette *f*
singleness Einzelheit *f*
singlet, ~ density Einteilchendichte *f* ~ distribution function Einteilchen-Verteilungsfunktion
*f*
singly, ~ operated drive Einzelantrieb *m* ~
re-entrant armature winding eingängige Ankerwicklung *f* ~ refractive einfachbrechend
singular ausgeprägt, ausgezeichnet, sonderbar
~ event einmaliges Ereignis *n*, einmaliger Vorgang *m* ~ phenomenon einmaliger Vorgang *m*
~ point ausgezeichneter Punkt *m* ~ surfaces
Unstetigkeitsflächen *pl*
singularity Eigenheit *f*, Singularität *f* ~ super-
-position Superposition *f* von Singularitäten
sinistro-propagating surge (or wave) linkslau-
fende Welle *f*
sink, to ~ abflauen, (submarine) absacken,
absenken, absinken, (a shaft) abteufen, eingraben, einsenken, fallen, niedersinken, sacken,
senken, sinken, sinken lassen, teufen, untergehen, versenken, versinken to ~ by driving
einrammen to ~ a hole in stone ein Loch *n* in
Stein einhauen to ~ in einlassen, einsinken,
eintauchen, sich senken to ~ the loops (textiles) kulieren to ~ by piling abtreiben to ~
into a porous stratum in den Boden *m* versickern to ~ the shaft den Schacht *m* abteufen to ~ wells Bohrlöcher *pl* absenken (tiefbohren)
sink Abwaschbecken *n*, Abzugs-kanal *m*,
-schleuse *f*, Ausguß *m*, Ausgußschale *f*, Nebenschluß *m*, (kitchen) Rinnstein *m*, Senke *f*,
Sinkstelle *f*, Umwandler *m*, Umwandlungsvorrichtung *f*, Wirbelsenke *f* (aerodyn.) ~
of a pit Schachtsumpf *m*
sink, ~ bolt Senkbolzen *m* ~ casting Saugguß
*m* ~ float separator Schwimmsinkschneider *m*
~head Wärmhaube *f* ~ hole Abzugsgrube *f*,
Einsturztrichter *m*, Lunker *m*, Senk-grube *f*,
-loch *n* ~ mill Lochwalzwerk *n* ~ trap Senkgrube *f*
sinker Senker *m*, Senkkörper *m* ~ bar Fang-
schwerstange *f* ~ butt Platinenfuß *m* ~ dial
Platinenführungsring *m* ~ lifting bar Platinen-
presse *f* ~ rest ring Platinentragering *m* ~
ring Platinenkranz *m* ~ slot Platinenführungs-
schlitz *m* ~ throat Einschließkehle *f* der Platine
sinking (a shaft) Abteufen *n*, (shaft) Abteufung *f*,
Einsenkung *f*, Löcher *pl* ausfräsen, Sacken *n*,
Senkung *f*, Sinken *n*, Untergehen *n*, Versenken

*n* ~ by the caisson method Senkschachtver-
fahren *n* ~ by piling Abtreibearbeit *f* (mit
senkrechtem Anstecken der Pfähle) ~ of the
subsoil Auftreibung *f* des Bodens
sinking, ~ areas absinkende Gebiete *pl* ~
depth Senktiefe *f* ~ force Sinkkraft *f* ~ fund
Amortisationsfond *m*, Schuldtilgungsfond *m*
~ head Saugmassel *f* ~-in Sacken *n* ~-in the
bore hole Niederbringen *n* des Bohrloches ~
mill Reduzierwalzwerk *n* ~ point Einsinkpunkt
*m* ~ pump Senkpumpe *f* ~ set Senksatz *m* ~
speed sinkende Geschwindigkeit, Sinkgeschwindigkeit *f* ~ trestle Abteufgerüst *n* ~
work Abteufarbeit *f*
sinks Schwergut *n*
sinography Sinografie *f*
sinoide Sinoide *f*
sinter, to ~ aufschmelzen, aussickern, brennen,
erweichen, festbrennen, fritten, rösten, sintern,
verschlacken, versintern to ~ together zusammensintern
sinter Sinter *m* ~ brick Sinterstein *m* ~ cake
Agglomerat *n*, Anhäufung *f*, Ballung *f*, Gesinter *n*, Sinter-erzeugnis *n*, -kuchen *m* ~(ing)
coal Sinterkohle *f* ~-glass filter Planglasfrit-
tenfilter *m* ~ pot Sintertopf *m* ~ slag Schwal
*m*
sintered, ~ corundum Sinterkorund *n* ~ hard
carbide Hartmetall *n* ~ hearth bottom Sinterherd *m* ~ metal bearing bush Sinterlagerbuchse
*f* ~ metal carbide gesintertes Metallkarbid *n*
~ stainless steel filter Sintermetallfilter *n*
sintering Aufschmelzen *n*, Brennen *n*, Erweichung *f*, Fritten *n*, Frittung *f*, Rösten *n*,
Sintern *n*, Sinterung *f*, Verschlackung *f*, Ver-
sinterung *f* ~ furnace Sinterofen *m* ~ heat
Sinterungshitze *f* ~ machine Sinterapparat *m*
~ operation Sinterungsprozeß *m* ~ plant Sinteranlage *f* ~ point Erweichungspunkt *m* ~
process Brennprozeß *m*, Sintervorgang *m*
sinuated ausgeschweift
sinuous geschlungen, gewellt, verschlungen ~
flow Wirbelströmung *f* ~ line Sinus-, Wellen-
-linie *f* ~ motion reine sinusförmige Bewegung
*f* ~ (winding) section of river gekrümmte
Flußstrecke *f*
sinus-multiple disc clutch Sinuslamellenkupp-
lung *f*
sinusoid Sinus-kurve *f*, -linie *f*, -schwingung *f*,
Sinusoide *f*
sinusoidal einwellig, sinusförmig ~ current sinusförmiger Strom *m*, Sinusstrom *m* ~ field
Sinusfeld *n* ~ quantity sinusförmige Wechselgröße *f* ~ voltage Sinusspannung *f* ~ wave
Fundamental-, Sinus-schwingung *f*
sip, to ~ nippen, schlürfen
siphon, to ~ hebern to ~ off abhebern, ablassen
siphon Druckdose *f*, Düker *m*, Faltenbalg *m*,
Flüssigkeitsabschneider *m*, Heber *m*, Heberohr *n*, Kanaldüker *m*, Saugheber *m*, Saugrohr *n*, (of a recorder) Schreibröhrchen *n*,
Siphon *m*, Standrohr *n*, Stechheber *m*, Syphon
*n*, Wasserabschluß *m* ~ for aerated water
Druckheber *m* ~ for carbonated water Druck-
schank *m*
siphon, ~ acidometer Hebersäuremesser *m* ~
barometer Heber-barometer *n*, -luftdruck-
messer *m* ~ diaphragm Membrandose *f* ~

**feeder** Siphonspeiser *m* ~ **gland** Stopfbüchse *f* mit Faltenbalgabdichtung ~ **head** Siphonkapsel *f* ~ **heating** Schleifenheizung *f* ~ **lead tap** automatischer Bleistich *m* ~ **lubricator** Dochtöler *m*, Siphonschmierapparat *m* ~ **pipe** Hebe-, Hosen-rohr *n* ~ **recorder** Heberschreiber *m*, Syphonrekorder *m* ~ **separator** Hebewäsche *f* ~ **tap** Selbststich *m*
**siphoning** Auftreten *n* einer saughebeartigen Wirkung *f* ~ **installation** Heberleitung *f*
**siphuncle** Sipho *m*
**siren** Heul-pfeife *f*, -signal *n*, Sirene *f*
**sirufer**, ~ **(iron-dust) coil** Siruferspule *f* ~ **core** Siruferkern *m*
**sirup** Dicksaft *m*, Sirup *m*, Sirupeinzug *m* ~ **classifying apparatus** Siruptrennvorrichtung *f* ~ **consistency** Sirupkonsistenz *f* ~ **discharging valve** Sirupablaufventil *n* ~ **draft** Sirupzuzug *m* ~ **pump** Siruppumpe *f* ~ **washing** Sirupdecke *f*
**sirutit** Sirutit *n*
**sisal-fiber** Agaven-, Sisal-faser *f*
**sit, to** ~ aufsitzen **to** ~ **close** dicht anliegen **to** ~ **down** setzen, sich setzen
**sit-down strike** Arbeitsverweigerung *f*
**site** Aufstellungsort *m*, Grundstück *n*, Lage *f*, Ort *m*, Platz *m*, Stelle *f* **in the** ~ an Ort und Stelle ~ **of an anchor buoy (or light)** Ankerlage *f* ~ **of break(ing) (or fracture)** Bruchstelle *f* ~ **of a gauge** Lage *f* eines Pegels ~ **of injection** Einspritzpunkt *m* ~ **of a pile** Meilerstätte *f* ~ **of rupture** Bruchstelle *f*
**site,** ~ **assembly** Montagebauverfahren *n* ~ **plan** Lage-, Situations-plan *m* ~**-to-mask clinometer** Deckungswinkelmesser *m*
**sitting** Sitzung *f*
**situ, in** ~ an Ort *m* und Stelle *f*
**situated** angebracht, gelegen ~ **at an infinite distance** in unendlicher Entfernung *f* liegend ~ **inland** binnenländisch
**situation** Lage *f*, Situation *f*, Stelle *f*, Stellung *f*, Umstand *m*, Verhältnis *n*, Zustand *m* ~ **of proximity** Näherung *f* ~ **map** Lagenkarte *f* ~ **report** Lagebericht *m*, Lagenkarte *f*
**six,** ~**-angled** sechswinklig ~ **bladed** sechsflüg(e)lig. ~**-color recorder** Sechsfarbenschreiber *m* ~**-component balance** Sechskomponentenwaage *f* ~**-cylinder in-line engine** stehender Sechszylinderreihenmotor *m* ~**-cylinder-in--line inverted engine** hängender Sechszylinder--Reihenmotor *m* ~**-cylinder radial engine** Sechszylindersternmotor *m* ~**-digit** sechsstellig ~**-engined** sechsmotorig ~**-foot way** Mittelweg *m* zwischen zwei Gleisen ~**-line reeving** sechsfacher Zug *m* ~**-line switchboard** Klappenschrank *m* zu sechs Leitungen ~**-phase rectifier** Sechsphasengleichrichter *m* ~**-point recorder** Sechsfachschreiber *m* ~**-pointed** sechsstrahlig ~**-pole** sechspolig ~**-roller mill** Sechswalzwerk *n* ~**-sided** sechsseitig ~**-speed gearbox** Sechsganggetriebe *m* ~**-spindle automatic lathe** Sechsspindelautomat *m* ~**-spindle bar-type automatic lathe** Sechsspindelstangenautomat *m* ~**-spline shaft** Sechskeilwelle *f* ~**-way plug connection** Sechsfachsteckverbindung *f* ~**-way tool block** Sechsfachstahlhalter *m* ~**-wheel vehicle** Dreiachser *m*

**sixteen, in** ~**s** Sechzehntelformat *n* **one** ~**th degree** Sechzehntelgrad *m*
**sixth** Sechstel *n*
**size, to** ~ klassieren, (paper) planieren, (in polishing) preßglätten, (weaving) schlichten, sondern, sortieren **to** ~ **the warp** die Kette *f* stärken
**size** Abmessung *f*, Ausmaß *n*, Bemessung *f*, Dimension *f*, Feinheit *f* (des Kornes), Format *n*, Größe *f*, Größennummer *f*, Klasse *f*, Leim *m*, Mächtigkeit *f*, Maß *n*, Maßstab *m*, Nummer *f*, Schlichtflüssigkeit *f*, Umfang *m*, Vorlack *m* ~ **of audience in a playhouse** Besetzungsverhältnisse *pl* ~ **of casting mold** Gießformat *n* ~ **of image** Bildformat *n*, Bildgröße *f* ~ **of lump** Stückgröße *f* ~ **of matter** Satzformat *n* ~ **of mesh** Maschenweite *f* ~ **of order** Auftragsumfang *m* ~ **of paper** Bauscht *m* ~ **of photograph** Aufnahmeformat *n* ~ **of picture** Bildformat *n*, Bildgröße *f* ~ **of pipe** Rohrgröße *f* ~ **of pitch** Stichgröße *f* ~ **of pore** Porengröße *f* ~ **of print** Bildformat *n* (photo) ~ **of section** Profilgröße *f* ~ **of thread** Gewindestärke *f* ~ **of type** Schriftgrad *m* ~ **and weight** Größe *f* und Gewicht *n*
**size,** ~ **analysis by sedimentation** Sedimentationsanalyse *f* ~ **boiler** Leimkocher *m* ~ **brushing** Leimaufstrich *m* ~ **color** Leimfarbe *f* ~ **consist analysis** Siebanalyse *f* ~ **cutting** Harzleimverseifung *f* ~ **cylinder** Formatzylinder *m* ~ **distribution** Durchmesser-, Korngrößen-verteilung *f* ~ **factor** Größenfaktor *m* ~**s limit** Korngröße *f* ~ **maker** Schlichtekocher *m* ~ **printing** (Tapeten) Leimdruck *m* ~ **reduction** Zerkleinerung *f* ~ **screening** Rasterweite *f* ~ **test** Leimfestigkeit *f* ~ **tolerance** Maßtoleranz *f*
**sized paper** geleimtes Papier *n*
**sizer, (hand)** ~ Abrichthobelmaschine *f*, Schlichter *m*
**sizes** Kornklassen *pl*
**sizing** animalische Leimung *f*, Dimensionierung *f*, Größenbestimmung *f*, Klassierung *f*, (glue) Klebelack *m*, Leimung *f*, Siebung *f*, Sortierung *f* nach der Größe ~ **of the carded yarn** Feinheitsbezeichnung *f* des Streichwollgarns ~ **of grain** Korngrößentrennung *f* ~ **the warp** Leimen *n* der Kette
**sizing,** ~ **apparatus** Sortierapparat *m* ~ **brush** (Appretur) Schlichtbürste *f* ~ **die Matrize *f* für Feinzug** ~ **drum** Klassiertrommel *f*, Siebtrommel *f* ~ **emulsion** Schlichtemulsion *f* ~ **engine** Leimholländer *m* ~ **machine** (Felgen) Rundstauchmaschine *f* ~ **machine for rolls** Rollenleimmaschine *f* ~ **material** Schlichtmittel *n* ~ **mill** Maßwalzwerk *n* ~ **paste** Schlichtmasse *f* ~ **press** Kalibrierpresse *f* ~ **roll** Maßwalze *f* ~ **screen** Klassiersieb *f* ~ **soap** Schlichtseife *f* ~ **test** Siebanalyse *f* ~ **tooth of a broach** Kalibrierzahn m einer Räumnadel ~ **trommel** Sortiertrommel *f*
**sizz** Zischen *n*
**sizzle, to** ~ knistern, zischen
**sizzle** Knattern *n*, Knistern *n*, Zischen *n*
**skating rink** Eisbahn *f*
**skeet range** Wurftaubenschießstand *m*
**skein** Gebund *n*, Gewinde *n*, Knäuel *n*, Strang *m* ~ **examiner** Fitzenarbeiterin *f* ~**-twisting apparatus** Strängedrillapparat *m*

**skeletal mineral forms** gestrickte Mineralformen *pl*

**skeleton** Fachwerk *n*, Gerippe *n*, Gerippebau *m*, Gerüst *n*, Gerüstbau *m*, Skelett *n* ~ **of a torus** Seelenachse *f* eines Torus

**skeleton,** ~ **agreement** Manteltarif *m* ~ **board** Brettchen *n* ~ **case** Kratte *f* ~ **chase** Sparschließrahmen *m* (print.) ~ **crystal** Kristallskelett *n* ~ **cylinder** Skelettrommel *f* ~ **diagram** Grund-, Prinzip-schaltung *f*, schematisches Schaltaderbild *n* ~ **drums** Windhaspel *f* ~ **form** Eindruckform *f*, Wicklungshalter *m* ~ **key** Dietrich *m*, Nachschlüssel *m* ~ **map** Netzkarte *f* ~ **pier** Gitterwerksbrücke *f* ~ **sketch** schematische Zeichnung *f* ~ **structure** Skelettenausfachung *f* ~ **weight** Montagegewicht *n* ~ **wheel** (of a tractor) Gitterrad *n*

**skeletonize, to** ~ schematisch darstellen

**skelp** Blech-, Röhren-, Rohr-streifen *m*, Streifen *m* ~ **mill** Ziehbank *f* ~ **steel** Schweißstahl *m*

**skelper** Ziehbank *f*

**sketch, to** ~ anreißen, aufnehmen, aufreißen, aufzeichnen, entwerfen, flüchtig entwerfen, skizzieren, zeichnen **to** ~ **a map** kokieren

**sketch** Abbildung *f*, Abriß *m*, Anlage *f*, Aufriß *m*, Aufzeichnung *f*, Beschreibung *f*, Betriebsplan *m*, Entwurf *m*, Entwurfszeichnung *f*, Grundriß *m*, Konzept *n*, Riß *m*, Skizze *f*, Umrißzeichnung *f*, Zeichnung *f* ~ **from bird's-eye view** Skizze *f* aus der Vogelschau ~ **drawn to scale** Maßskizze *f*

**sketch,** ~ **block** Zeichenblock *m* ~ **board** Reißbrett *n*

**sketcher** Reißer *m*

**sketching** Anfertigung *f* der Lageskizze ~ **paper** Krokierpapier *n*

**skew, to** ~ abschrägen

**skew** Kippwinkel *m*, Schiefe *f*, Schrägstellung *f* (der Brücke); aus der Richtung *f*, schiefwinklig, windschief ~ **aileron** schiefer Hilfsflügel *m* (aviat.), seitlich überragendes Querruder *n* ~**back** Auflagerstein *m*, Widerlager *n* ~ **bevel gears** Hyperboloidräder *pl* ~ **factor** Schrägungsfaktor *m* ~ **field** Schiefkörper *m* ~ **gearing** Zahnrädergetriebe *n* mit gekreuzten Wellen ~ **iron** Schiefeisen *n* ~ **joint** Schrägverblattung *f* ~ **lines** windschiefe Gerade *f* ~ **problem** Kipp-Problem *n* ~ **ray** Schrägstrahl *m*, schräger Strahl *m* ~**-resonance characteristic** schiefe Resonanzkurve *f* ~**-rolling** Schrägwalzen *n* ~**-rolling mill** Schrägwalzwerk *n* ~ **spur wheel** Helixgetriebe *n* ~ **surface** windschiefe Fläche *f* ~**-symmetric** (of curves) schief, schiefsymmetrisch ~**-symmetric negative pairs** schiefsymmetrische negative Paare *pl*

**skewed** abgeschrägt ~ **pole tips** abgeschrägte Polschuhränder *pl* ~ **region** gekippter Bezirk *m*

**skewer** Aufsteckspindel *f* ~ **turning machine** Aufsteckspindeldrehbank *f*

**skewing** Schrägstellung *f*

**ski, to** ~ skilaufen

**ski** Kufe *f*, Schi *m*, Ski *m* ~ **airplane** Kufenflugzeug *n* ~ **landing gear** Schneekufenfahrwerk *n* ~ **runner** Schneekufe *f* ~ **undercarriage** Fahrgestell *n* mit Schneekufen, Schneekufenfahrwerk *n*

**skiagraph** Röntgenaufnahme *f*, Röntgenograf *m*

**skiascope** Augen-brechungsmesser *m*, -spiegel *m*

**skiatron** Dunkelschriftröhre *f*

**skid, to** ~ ausgleiten, gleiten, in der Kurve *f* abrutschen, nach außen schieben, rutschen, schleudern **to** ~ **the derrick** einen Bohrturm *m* walzen **to** ~ **a wheel** ein Rad *n* hemmen

**skid** Bergstütze *f*, Gleitschiene *f*, Hemmzeug *n*, Kufe *f*, Ladebock *m*, Ladegestell *n*, Schiene *f*, Schlepper *m* (Walzwesen), Schlitten *m* (aviat.), Schuh *m*, Sporn *m* ~ **of the sailplane** Kufe *f* des Segelflugzeuges

**skid,** ~ **band** Reibband *n* ~ **board** Bremsklotz *m* ~ **fin** Flügelkielflosse *f* ~ **fin antenna** Flossenantenne *f* ~ **frame** Kufenrahmen *m* ~ **hopper** Ladegestell *n* mit Mulde ~ **mark** (road) Bremsspur *f* ~ **marks** Schleuderspur *f* ~ **piece** Kufe *f* ~ **platform** Lade-bock *m*, -gestell *n* ~ **platform fitted with stakes** Ladegestell *n* mit Rungen ~ **runners** Schlittenkufen *pl* ~ **shoe** Hemmschuh *m* ~ **turn** Schiebekurve *f*

**skidding** Ausgleiten *n*, Hängen *n*, Schieben *n*, seitliches Gleiten *n* ~ **on a curve** Schleudern *n* in der Kurve ~ **on the turn** Schieben *n* in der Kurve

**skidding movement without banking** Schieben *n* ohne Schräglage

**skiff** Dingy *m*, Kahn *m*

**skill** Erfahrung *f*, Fertigkeit *f*, Geschick *n*, Geschicklichkeit *f*, Kunst *f*, Übung *f*

**skilled** bewandert, erfahren, geschult, geübt, gewandt, sachkundig ~ **laborer (or worker)** Facharbeiter *m*

**skillet** Tiegel *m*

**skillful** geschickt, kunstfertig

**skim, to** ~ (milk) abrahmen, (slag) abstreichen, abziehen, entschäumen **to** ~ **the milk** die Milch *f* entrahmen **to** ~ **off** abheben, abschäumen, abschöpfen **to** ~ **off slag** abschlakken, Schlacke *f* abziehen

**skim** (residual latex serum obtained on centrifugation of latex) dünne Gummischicht *f* ~ **milk** Magermilch *f*

**skimmed** abgeschäumt ~ **off** abgerahmt

**skimmer** Abhebelöffel *m*, Abschneider *m*, (for removing slag from molten metal) Abstreicheisen *n*, Abstreicher *m*, Abstreichlöffel *m*, Abstreifer *m*, (slag) Krampstock *m*, Kratze *f*, Kratzer *m*, Planier-bagger *m*, -löffel *m*, Pustspan *m*, Schaum-kelle *f*, -löffel *m* ~ **bar** Krampstock *m* ~ **gate** (in casting) Schaumfang *m* ~ **holder** Krätzerbehälter *m*

**skimming** Abschäumung *f*, Abstrich *m*, Gummierung *f*, Schaum-abheben *n*, -entnahme *f* ~ **basin** Gießtümpel *m* mit Schaumumfang ~ **ladle** Schaumkelle *f* ~ **machine** Abschäum--maschine *f* ~ **pan** Abschäum-pfanne *f*, -wanne *f* ~ **plant** primäre Destillationsanlage *f* ~ **spoon** Schaumlöffel *m*

**skimmings** Abhub *m*, (in lead refining) Farbe *f*, Schaum *m*

**skin, to** ~ abisolieren, abstreichen, häuten, hautziehen

**skin** Außenhaut *f* (aviat.), Beplankung *f*, (of a ship) Beplattung *f*, Blutwolle *f*, Fell *n*, (of a

casting) Gußhaut *f*, Haut *f*, Randzone *f*, Schale *f*, Schicht *f*, Schlachtwolle *f*, Schwarte *f*, Überzug *m*, Wand *f* ~ **of aircraft** Oberschale *f* ~ **of fabric** Flügelbespannung *f* ~ **of metal parts** Gußhaut *f*
**skin,** ~ **construction** Hautkonstruktion *f* ~ **coppering process** Hautaufkupferungsverfahren *n* ~ **decontamination agent** Hautentgiftungsmittel *n* ~ **depth** Hauttiefe *f* ~ **dose** Hautdosis *f* ~ **effect** Haut-effekt *m*, -wirkung *f*, Stromverdrängung *f* ~**-effect losses** Hautwirkungsverlust *m* ~**-effect resistance** Oberflächenwiderstand *m* ~**-effect rotor** Stromverdrängungsläufer *m* ~ **friction** Haut-, Mantel-, Oberflächen-, Wand-reibung *f* ~ **pass** Dressierstich *m* ~**-pass edging stand** (Warmwalzen) Egalisiergerüst *n* ~**-pass stand** Dressier-, (Kaltwalzen) Egalisier-gerüst *n* ~ **plate** Dichtungsblech *n*, Zylinderverschluß *m* ~ **resistance** Hautwiderstand *m* ~ **sheet** Hautblech *n* ~ **stressed covering** tragende Außenhaut *f* ~**-type oil radiator** Oberflächenölkühler *m*
**skinned wire** abisolierter Draht *m*
**skinner** Abdecker *m*, Schinder *m*
**skiograph** Skiograf *m*
**skip, to** ~ auslassen, übergreifen, überspringen, verschränken
**skip** Auslassung *f*, Aussetzer *m*, Förder-gefäß *n*, -korbkübel *m*, Füllkorb (min.) *m*, Gefäß *n*, Gichtwagen *m*, Hops *m*, Kübelwagen *m*, Leerbefehl *m* (comput.) ~ **with (or for) concrete** Betonkübel *m*
**skip,** ~ **bars** Sprungschienen *pl* ~ **bridge** Kippwagenaufzug *m* ~ **car** Kipp-gefäß *n*, -kübel *m* ~ **charging** Kippwagen-, Kippgefäß--begichtung *f* ~**-charging gear** Begichtungsanlage *f* mit Kippgefäß, Kippgefäßbegichtungsanlage *f* ~ **distance** Schweigezone *f*, Sprungentfernung *f*, tote Zone *f* ~ **filling** Kippgefäßbegichtung *f* ~ **hoist** Kippaufzug *m*, Kippgefäßaufzug *m*, Kippwagenaufzug *m*, Muldenbevorrichtung *f*, Schrägaufzug *m* ~ **hoist with tipping bucket** Schrägaufzug *m* mit Kippgefäß ~ **instruction** Verweisungsauftrag *m* ~ **jaw** Gleitbacke *f* ~ **keying** Impulsfolgefrequenzteilung *f* ~**-phenomenon** Sprungerscheinung *f* ~ **theory** Gleittheorie *f* ~ **wagon** Kippwagen *m* ~ **winding** Gefäß-, Skip--förderung *f* ~ **zone** Leerbereich *m*, Schlierenzone *f*, Sprungentfernung *f*, stille Zone *f*, tote Zone *f*
**skipped distance** Schweigezone *f*, tote Zone *f*
**skipper** Schiffer *m*; Überspringvorrichtung *f*
**skipping** Verwurf *m* (geol.) ~ **motion** hüpfende Bewegung *f* ~ **teache** Füllbecken *n* (sugar mfg.)
**skips** Ausfälle *pl*
**skirt, to** ~ entlangstreifen, streifen, umspülen, vorbei-führen, -streichen, -streifen **to** ~ **the coast** längs der Küste *f* fahren
**skirt** Rand *m*, Rock *m*, Schutz *m*, unterer Rand *m* ~**board** Seitenleiste *f* ~ **end of a piston** kurbelseitiges Kolbenende *n*, offenes Ende *n* des Kolbens
**skirting** Einfassung *f*, Fußleiste *f*, (board) Scheuerleiste *f*
**skittle ball** Kegelkugel *f*
**skive, to** ~ abschaben

**sklodowskite** Sklodowskit *m*
**skull** Bär *m*, Decke *f*, Hirnschale *f*, Pfannenrest *m*, Schädel *m*, Totenkopf *m* ~ **at the converter mouth** Mündungsbär *m*
**skull,** ~ **cracker** Ballwerk *n*, Fallwerk *n* ~ **guard** Kopfschutz *m*
**skulls** Eisenreste *pl*
**skutterudite** Arsenkobalt *m*, Arsenkobaltkies *m*, Hartkobalt-erz *n*, -kies *m*, Tesseralkies *m*
**sky** Himmel *m*, Luftraum *m*, Rauchkammer *f* ~ **blue** Cölestinblau *n* ~ **conditions** Himmelszustand *m* ~ **filter** Verlauffilter *m* ~ **light** Decken(ober)licht *n*, Deck-fenster *n*, -glas *n*, -licht *n*, einfallendes Licht *n*, Klappfenster *n*, Oberlicht *n*, schräges Fenster *n* ~**light mica** Dachlukenglimmer *m* ~ **light polarization** Himmelslichtpolarisation *f* ~ **light turret** Dachaufsatz *m* ~**line** Horizont *m* ~**scraper** Hochhaus *n*, Turmhaus *n*, Wolkenkratzer *m* ~ **shine** Luftstreuung *f* ~ **symbols** Himmelssymbole *pl* (meteor.) ~ **wave** Himmels-, Höhen-, Luftweg-, Raum-welle *f* ~**way** Brückenstraße *f*, Hochstraße *f* ~**writer** Himmelsschreiber *m*, Rauchschreiber *m* ~ **writing** Rauchschrift *f*
**slab, to** ~ flachwalzen, (in rolling) platinieren
**slab** Decke *f*, Eisenblechbramme *f*, Metallplatte *f*, Platte *f*, (concrete) Plattenbalken *m*, Plattine *f*, (in reinforced concrete) Rippenplatte *f*, Rohbramme *f*, Schalbrett *n*, Schellstück *n*, Schürbel *m*, Schwarte *f*, Steintafel *f*, Sturz *m*, Tafel *f* ~ **of cement** Zementplatte *f* ~ **of guncotton** Körper *m* von Schießbaumwolle ~ **for platforms** Laufbühnenbelag *m* ~ **to be rolled** Walzplatte *f*
**slab,** ~ **bloom** Bramme *f* ~ **coil** Flachspule *f* ~ **continuous over two spans** Zweifeldplatte *f* ~ **copper** Vierkantkupfer *n* ~ **flooring** Plattenbelag *m* ~ **foundation** Flachgründung *f*, Fundamentalplatte *f* ~ **geometry** Plattenanordnung *f* ~**-heating furnace** Brammentiefofen *m* ~ **ingot** Bramme *f* ~ **inking unit** Tischfarbwerk *n* ~ **milling** Flachnutenfräsen *n* ~**-milling machine** Planlangfräsmaschine *f* ~ **pass** Flachstich *m* ~**-piled faggot** flachliegend paketiertes Schweißeisenpaket *n* ~ **piling** flachliegende Paketierung *f* ~ **shears** Brammenschere *f* ~ **stone paving** Plattenbelag *m* ~ **zinc** Plattenzink *n*
**slabbing** Brammen *n*, Flachwalzen *n* ~ **frame** (textiles) Grobspindelbank *f* ~ **mill** Brammen-, Platinen-walzwerk *n* ~ **mill for rolling armor plate** Panzerplattenwalzwerk *n* ~**-mill roll** Flachwalze *f* ~**-mill train** Brammen-, Platinen--straße *f* ~ **roll** Brammenwalze *f*
**slabs** Rohbrammen *pl*
**slack, to** ~ lockern **to** ~ **off** schlaffmachen
**slack** Durch-hang *m*, -senkung *f*, Feinkohle *f*, Grus *m*, Kohlengrus *m*, Lösche *f*, (cable) Slack *m*, Spannungslosigkeit *f*, Steinkohlenklein *n*, Zuschlag *m* zur Leitungsdrahtlänge, (submarine cables; for downward pressure) Zuschlag *m* für Abtrieb, Zuschlag *m* einer Linie; entspannt, (force) locker, lose, schlaff **to be** ~ **Spiel** *n* haben ~ **of chain** Kettendurchgang *m*
**slack,** ~ **adjuster** Gestängenachstellvorrichtung *f* ~ **adjustment** Nachstellvorrichtung *f* für das Spiel ~ **chain** lose Kette *f* ~ **chain switch**

Schlaffkettenschalter *m* ~ **coal** Gruskohle *f*,
Kohlen-grus *m*, -klein *n* ~ **coupling** lose
Kopplung *f* ~ **end of a belt (or a rope)** schlaffes
Trumm *n* ~ **hours** verkehrsschwache Zeit *f* ~
**lime** gelöschter Kalk *m* ~**line** Schlaffseil *n*
~ **puller** Ankerspannschraube *f* ~**-side** Schlaff-
seite *f* ~ **side of a belt** schlaffes Trumm *n* ~
**strand** gezogenes oder loses Trumm *n* ~ **strand**
**of belt** loses Riementrumm *n* ~ **strand of**
**chain** loser Kettenstrang *m* ~ **thread** Längel
*n* ~ **traffic** schwacher Verkehr *m* ~ **water**
Stau-, Still-wasser *n* ~ **wax** Filterkuchen *m*,
(of crude scale) Gatsch *m*, Rohparaffin *n* ~
**wire** Nachlaßdraht *m*
**slacken, to** ~ abschwächen, abspannen, auflockern, ausspannen, entspannen, erschlaffen,
fieren, lockern, los geben, los werden, schlaffmachen, stocken, mehr Tau *m* auslassen,
umdrehen, zurückdrehen **to** ~ **back a screw**
eine Schraube *f* zurückdrehen **to** ~ **a chain**
ausstechen **to** ~ **the fire** kaltgeben **to** ~ **speed**
Fahrt *f* vermindern
**slackening** Abschwächung *f*, Entspannung *f*,
Lockerung *f*, Schlaffwerden *n* ~ **of wire**
Schlaffwerden *n* des Drahtes ~ **motion** abnehmende Bewegung *f*
**slackness** Lagerspiel *n*, Lockerstein *n*, Schlaffheit *f*, Schwäche *f*, Spiel *n* ~ **at work** Arbeits-
bummelei *f*
**slag, to** ~ ausschlacken, schlacken, sintern,
verschlacken, zusammensintern **to** ~ **off**
abschlacken, abziehen, Schlacke *f* abziehen
**slag** Asche *f*, Blachmahl *n*, Braschen *pl*, Dorn
*m*, Gekrätz *n*, Lacht *m*, Schlacke *f*, Skorie *f*
~ **from liquated copper** Kupferdorn *m* ~ **of**
**liquation** Krätzschlacke *f* ~ **from roasting**
Röstschlacke *f*
**slag,** ~ **accumulation** Schlackenansammlung *f*
~ **action** Schlackenangriff *m* ~ **addition**
Schlackenzuschlag *m* ~ **adhering to the con-**
**verter nose** Mündungsbär *m* ~**-bearing** schlak-
kenhaltig ~ **bed** Schwalboden *m* ~ **bottom**
Schwal-, Zerrenn-boden *m* ~**-breaker roller**
Schlackenbrecherwalze *f* ~ **brick** Schlacken-
-stein *m*, -ziegel *m* ~ **bridge** Schlackenschütze
*f* ~ **cake** Topfschlacke *f* ~ **calculation** Schlak-
kenberechnung *f* ~ **car** Schlacken-karre *f*,
-wagen *m* ~ **catcher** Schlacken-kammer *f*,
-kasten *m* ~ **cement** Schlackenzement *m* ~
**clinker hopper** Schlackentrichter *m* ~ **com-**
**position** Schlackenkonstitution *f* ~ **concrete**
Schlackenbeton *m* ~ **cone** Schlackenkuchen *m*
~ **constitution** Schlackenkonstitution *f* ~ **con-**
**trol** Schlackenkontrolle *f* ~ **conveyer** Schlak-
kenband *n* ~**-conveying machinery** Schlacken-
transportanlage *f* ~**-cooling tube** Schlacken-
kühlrohr *n* ~ **corrosion** Verschlackung *f* ~
**cover** Schlackendecke *f* ~ **crust** Schlackenkruste *f* ~ **deposit** Aschenablagerung *f* ~
**discharge** Schlackenabzug *m* ~ **disposal**
Aschenaustragung *f* ~ **dump** Schlackenhalde
*f* ~ **enclosure** Schlackeneinschluß *m* ~ **eye**
Schlackenauge *n* ~ **flow** Schlackenstrahl *m*
~ **forming** Schlackebilden *n*, Schlackenmachen
*n* ~**-forming** schlacke(n)bildend ~**-forming**
**constituent** Schlackenbildner *m* ~**-forming**
**period** Verschlackungsperiode *f* ~**free** schlak-
kenfrei ~ **furnace** Schlacken-herd *m*, -ofen *m*

~ **fusion point** Schlackenschmelzpunkt *m* ~
**gate** Schaumfänger *m* ~ **globule** Schlacken-
korn *n* ~ **granulation** Schlackengranulation
*f* ~ **handling mantrolley transporter** Schlacken-
verladebrücke *f* ~ **handling plant** Schlacken-
verladebrücke *f* ~**-heap conveyer** Schlacken-
haldenbahn *f* ~ **hearth** Schlackenherd *m* ~
**hole** Schlacken-loch *n*, -öffnung *f*, -stich *m*,
-stichloch *n* ~ **hopper** Schmelztrichter *m* ~
**inclusion** Schlackeneinschluß *m* ~ **iron** Schlak-
kenspieß *m* ~ **ladle** Schlacken-kübel *m*,
-pfanne *f* ~ **ladle and car** Schlackenwagen *m* ~
**lead** Krätz-, Schlacken-blei *n* ~ **line** Schlacken-
-stand *m*, -zone *f* ~ **lump** Schlacken-klotz *m*,
-klumpen *m* ~**-making material** Schlackenbildner *m* ~ **nodule** Schlackentröpfchen *n*
~ **nose** Formnase *f* ~ **notch** Abschlacköffnung
*f*, Schlacken-form *f*, -loch *n*, -öffnung *f*,
-stich *m*, -stichloch *n* ~ **occlusion** Schlackeneinschluß *m* ~ **overflow** Schlackenüberlauf *m* ~
**penetration** Verschlackung *f* ~ **pit** Schlackengrube *f* ~ **pocket** Schlacken-kammer *f*, -kasten
*m*, -sammelgefäß *n*, -sammler *m*, -tasche *f* ~
**pot** Schlacken-kübel *m*, -topf *m* ~ **pot cradle**
Schlackenhaubengehänge *n* ~ **process** Sinter-
-prozeß *m*, -vorgang *m* ~ **puddling** Schlacken-
puddeln *n* ~ **pumice** Schaumschlacke *f* ~
**ration** Schlackenziffer *f* ~**-removing plant**
Entschlacker *m* ~ **riser** Schlackentrichter *m*
~ **roasting** Schlackenrösten *n* ~ **screen** Fangrost *m* ~ **seam** Schlackenzeile *f* ~ **separation**
Schlackenabsonderung *f* ~ **skimmer** Schlak-
kenabschneider *m* ~ **smelting** Schlackenarbeit
*f* ~ **spout** Schlackenabstrichrinne *f* ~ **stone**
Schlackenstein *m* ~ **strip** Schlackenstreifen
*m* ~**-tap boiler** Schmelzkammerkessel *m* ~ **tap**
**furnace** Schmelzaschen-, Schmelzkammer-feu-
erung *f* ~ **tapped out** Abstichschlacke *f* ~
**trap (in gating system)** Schlackengraben *m* ~
**treatment** Schlackenbehandlung *f* ~ **trough**
Schlackenrinne *f* ~ **tube** Formnase *f* ~ **tuyère**
Schlackenform *f* ~ **volume** Schlackenmenge *f*
~ **washing (process)** Schwalarbeit *f* ~ **wool**
Schlackenwolle *f* ~ **working** Schlackebilden
*n*, Schlackenmachen *n* ~ **yard** Schlackenplatz
*m*
**slagged out** abgeschlackt
**slagging** Abziehen *n*, Raffinierung *f*, Verschlackung *f* ~ **spout** Abschlackrinne *f*
**slaggy** schlacken-ähnlich, -artig, -reich, schlackig
**slagless** schlackenfrei
**slags dump** Halde *f*
**slake, to** ~ (lime) ablöschen, (lime) erlöschen,
(lime) löschen, verschlacken, zerfallen
**slake-lime pit** Äscher *m*
**slaked lime** angemachter Kalk *m*, gelöschter
Kalk *m*, Kalkschlamm *m*, Sackkalk *m*
**slaking** Ablöschen *n*, Löschung *f* ~ **drum**
Löschtrommel *f* ~ **slag** Zerfallschlacke *f* ~
**test** Löschversuch *m*
**slam, to** ~ zuschlagen
**slam** Knall *m*, Krach *m*
**slant, to** ~ ablöschen, ausschrägen
**slant** Schräge *f* ~ **approach** Schräganflug *m* ~
**bore** Schrägbohrung *f* ~ **distance** Geradeaus-
entfernung *f* ~ **drill** Spiralbohrer *m* (stein-
gängig) ~ **evaporation** Schrägbedämpfung *f*
~ **hole** Schrägbohrung *f* ~ **plane** Schrägebene

*f* ~ **range** Schrägentfernung *f* ~ **range to crossing point (or to mid-point)** Wechselpunktentfernung *f* ~ **range to present position** Meßentfernung *f* ~ **vaporization** Schrägdämpfung *f* ~ **weld** Schrägnaht *f* ~ **wind** Schrägwind *m*
**slanting** abdachig, abschüssig, geneigt ~ **bank** Schrägufer *n* ~ **edge** Schrägkante *f* ~ **gallery** Schrägbohrung *f* ~ **groove** Schrägnute *f* ~ **position** Schräglage *f* ~ **seat valve** Schrägsitzventil *n* ~ **side** Schrägseite *f* ~ **slot** Schrägschlitz *m* ~ **tube** Schrägrohr *n*
**slantwise** schief, schräg
**slap, to** ~ klapsen
**slash** Hieb *m*, Schramme *f* ~ **feed** Spritzspeisung *f*
**slashing weapon** Hiebwaffe *f*
**slashings** (forestry) Schlagabraum *m*
**slat** Hilfsflügel *m*, Keilnut *f*, Latte *f*, Plattenband *n*, Stab *m*, Vorflügel *m* (aviat.) ~ **of reel fly** Haspelholm *m* ~ **conveyor** Plattenband *n* ~ **track** Vorflügelführung *f*
**slate** Schiefer *m*, Schieferstein *m*, Tafel *f* ~ **(board)** Schiefertafel *f* ~ **coal** Blätterkohle *f*, Blattkohle *f* ~ **(-colored)** schieferfarben ~ **drill** steilgängiger Spiralbohrer *m* ~ **gray** schiefergrau ~ **knife (or slaten)** Glatteisen *n* ~ **pencil** Griffel *m* ~ **quarry** Schiefergrube *f* ~ **rock** Schiefergebirge *n* ~ **roof** Schieferdach *n* ~ **spar** Aphrit *n*, Schaumkalk *m* ~ **stab** Schiefertafel *f* ~ **tablet** Schieferplatte *f* ~ **tar** Schieferteer *m*
**slatelike** schieferartig
**slater's** Schieferdecker *m* ~ **anvil** Dachamboß *m* ~ **hammer** Dachdeckerhammer *m* ~ **tools** Dachdeckerausrüstung *f*
**slating** Dachlattung *f* ~ **machine** Glättmaschine *f*
**slaty** schieferartig, schieferig ~ **cleavage** schiefrige Absonderung *f* ~ **coal** Büschelkohle *f* ~ **fracture** Schieferbruch *m* ~ **structure** schieferige Struktur *f*
**slaughter, to** ~ abschlachten, niedermachen, niedermetzeln
**slaughter** Gemetzel *n* ~ **house** Schlachthaus *n*
**slaughtering** Schlachtung *f*
**slave, to** ~ sich abplagen, (gyro) mitnehmen, (gyro) nachführen
**slave** Sklave *m* ~ **clock** Nebenuhr *f* ~ **cylinder** Kraftverstärkerzylinder *m*, Servozylinder *m* ~**-driving system** Antreibesystem *n* ~ **equipment** Tochtergerät *n* ~ **gudgeon pin** blinder Kolbenbolzen *m* ~ **meter** Wiederholungsinstrument *n* ~ **recorder** Zweitschreiber *m* ~ **station** Nebenstelle *f* (rdo nav.) ~ **transmitter** Neben-, Tochter-sender *m* ~ **unit** Kraftverstärker *m*
**slaving** Fesselung *f*, Führung *f*, (gyro) Mitnahme *f*, Stauchung *f*
**sleaker** (striking pin) Recker *m*
**sleazy** fadenscheinig
**sled** Schleife *f*, Schlitten *m* ~ **lister cultivator** Dammkulturhackmaschine *f* auf Schlitten, Reihenkultivator *m* auf Schlitten ~ **plane** Kufenflugzeug *n*
**sledge** Handhammer *m*, Maker *m*, Schlägel *m*, Schlegel *m* ~ **hammer** Boßhammer *m*, Breithammer, Hammer *m* mit Kanten, Schienenhammer *m*, Schlaghammer *m*, Schmiedehammer *m*, Senkhammer *m*, Setzhammer *m*,

Treibfäustel *n*, Vorschlaghammer *m*, Zuschlaghammer *m* ~ **runner** Kufe *f*
**sledging road** Schlittenbahn *f*
**sleek, to** ~ schlichten
**sleeking stick (or tool)** Glättholz *n*, Glättschiene *f*
**sleeper** Brückenbaum *m*, Eisenbahnschwelle *f* (r. r.), Lagerschwelle *f* des Fußbodens, Schwellholz *n*, Unterzug *m*, Weitstab *m* ~ **of a floor** Unterzug *m* eines Fußbodens ~ **of stairs** Treppensohle *f*
**sleeper,** ~ **capping machine** Schwellenkappmaschine *f* ~ **chair** Schwellenstuhl *m* ~ **fishplate** Schwellenlasche *f*
**sleeping,** ~ **bag** Schlafsack *m* ~ **cabin** Schlafkabine *f* ~ **car** Schlafwagen *m* ~ **partner** stiller Gesellschafter *m* ~ **sickness** Schlafkrankheit *f* ~ **top** symmetrischer Kreisel *m*
**sleet** Glatteis *n*, Graupel *f*, Graupeln *pl*, Hagel *m*, Schlosse *f*, Schnee *m* und Regen *m* gemischt ~ **formation** Graupelbildung *f*
**sleeting** Graupelschlag *m* ~ **shower** Graupelschauer *m* ~ **squall** Graupelbö *f* ~ **storm** Hagel-schlag *m*, -sturm *m*
**sleeve, to** ~ mit einer Hülse *f* versehen
**sleeve** Ärmel *m*, Büchse *f*, (motor) Dichtmanschette *f*, Führungskörper *m*, Hülse *f*, Klinkenbuchse *f*, Klobenrolle *f*, Lagerschale *f*, Laufbuchse *f*, Leerlaufbuchse *f*, Manschette *f*, Muffe *f*, Rohrstutzen *m*, Steck(er)buchse *f*, Teiltrommel *f*, (with screws) Trommel *f*, Umkleidung *f*, Verbindungsmuffe *f*, Zylinder *m*, Zylinderlaufbuchse *f* ~ **of commutator** Stromwendermuffe *f* ~ **of control rod** Regelstangenhülse *f* ~ **for grip cocks** Schlauchtülle *f* für Griffhähne ~ **of injection timing device** Spritzverstellermuffe *f* ~ **of jack** Klinkenhülse *f* ~ **for (or of) stuffing box** Stopfbuchsenhülse *f*
**sleeves,** ~ **blade** Hülsenschaufel *f* ~ **brick** Gießlochstein *m*, Lochstein *m* ~ **clutch** Schiebkupplung *f* ~ **conductor** Buchsenleitung *f* ~**-control cord circuits** vereinfachte Schnurschaltungen *pl* ~ **coupling** Hülsen-, Muffen-, (solid) Schalen-kupplung *f* ~ **dipole** Hülsendipol *n* ~ **drive** Scheiberantrieb *m* ~ **gate valve** Muffenabsperrschieber *m* ~ **gear** Zahnrad *n* ~ **guide** Muffenführung *f* ~ **holder** Manschettenhalter *m* ~ **inductor magneto** Magnetzünder *m* mit Umlaufhülse ~ **joint** (twisted) Hülsenbund *m*, Muffenverbindung *f* ~ **joint of tubing** Rohrschellenverbindung *f* ~ **mounting** Hülsenbefestigung *f* ~ **nipple insulator** Hülsennippelisolator *m* ~ **nut** Hülsenmutter *f* ~ **nut for slide adjustment** Mutter *f* zur Stößelverstellung ~ **patch** Ärmelabzeichen *n* ~**-shaped cradle** muffenartige Wiege *f* ~ **socket** Paßeinsatz *m* ~ **target** Schleppsack *m* ~ **tube** Überschiebrohr *n* ~ **twisters** Hebelkluppe *f* ~**-type engine** ventilloser Motor *m* ~ **valve** Buchsenventil *n*, Kolbenschieber *m*, Rohrschieber *m*, Schiebeventil *n*, Zylinderlaufbuchse *f* ~**-valve engine** Schiebermotor *m*
**wire** Ader *f* zum Stöpselkörper, c-Ader *f*, Hülsenleitung *f*, Prüfader *f*, Prüfdraht *m* ~ **(-type) wire holder** Meßdrahthalter *m* (hülsenartig) ~ **yoke assembly** Keilnabenmitnehmer *m*

**sleigh** Schlitten *m* ~ **cradle** Jackenwiege *f* ~ **runners** Schlittenkufen *pl* ~ **trail** Schlittenstutze *f*

**slender** dünn, schlank ~ **beam (or girder)** schlankster Träger *m* ~ **nail** Bodenspieker *m*

**slenderness ratio** Schlankheit *f*, Schlankheitsgrad *m*

**slewability** Schwenkbarkeit *f*

**slewing** Schnellnachführung *f* (gyro) ~ **bracket** (gas, electr.) Gelenkarm *m* ~ **crane** Drehkran *m* ~ **crane with fixed jib** Drehkran *m* mit festem Ausleger ~ **garb crane** Greiferdrehkran *m* ~ **gear (unit)** Drehwerk *n* ~ **jib crane** Auslegerdrehkran *m* ~ **mechanism** Drehwerk *n* ~ **motor** Schnellabtastmotor *m* ~ **operation** Schwenkbetrieb *m* ~ **speed** Drehgeschwindigkeit *f* ~ **track ring** (Kran) Drehkranz *m* ~ **wheel pressure** Drehrollendruck *m*

**slice, to** ~ abschärfen

**slice** Farbeisen *n* (print.), Schaumlatte *f*, Scheibe *f*, Schnitt *m*, Schnitte *f*, Stauvorrichtung (paper mfg.) ~ **cut staples** Schlaufen *n*

**slicer** Schneid-, Schnitzel-maschine *f*, Stauvorrichtung *f* (paper mfg.)

**slicing,** ~ **knife** Schnitzelmesser *n* ~ **lathe** Abstechdrehbank *f* ~ **machine** Schnitzelmaschine *f*

**slick, to** ~ schlichten

**slick,** ~ **of waste metal** Krätzschlich *m* ~ **ground** Moddergrund *m*

**slickenside** Harnisch *m*

**slickensides** Blei-schweif *m*, -spiegel *m*

**slicker** (in molding) Lanzette *f*

**slidable** verschiebbar ~ **lens panel** Linseneinschiebebrett *n* ~ **tension** verschiebbare Spannung *f*

**slide, to** ~ gleiten, rutschen, schieben, schleifen, verschieben **to** ~ **across** verschieben **to** ~ **down** abrutschen, herabgleiten **to** ~ **into each other** ineinanderschieben **to** ~ **off** abgleiten **to** ~ **over** übergleiten

**slide** Aufschiebeschlaufe *f*, Durchziehglas *n*, Erdrutsch *m*, Führung *f*, (carriage) Führungsschlitten *m*, Gleitbacke *f*, Gleitbahn *f*, Gleitfläche *f*, Gleitschiene *f*, (steam engine) Glitscher *m*, (of table or stage) Kugelführung *f*, Läufer *m*, (micros.) Objektglas *n*, Objektträger *m*, (of a slide rule) Rechenschieberläufer *m*, Rutschbahn *f*, Rutsche *f*, Rutschung *f*, Schieber *m*, Schiebevorrichtung *f*, Schleife *f*, Schliffstück *n*, (rail) Schlitten *m*, Stößelführung *f*, Support *m*, Verwerfungskluft *f*, (sight) Visierschieber *m* ~ **of a bank** Rutschung *f* einer Böschung ~ **with projecting teeth** gezahnte Gleitbahn *f* ~ **of a sliding door** Türlaufschiene *f* ~ **of a slope** Rutschung *f* einer Böschung ~ **with two grooves** zweiseitige Gleitbahn *f*

**slide,** ~ **adjustment** Stößelverstellung *f* ~ **(-induction) apparatus** Schlittenapparat *m* ~ **armature brake** Verschiebeankerbremse *f* ~ **-back voltmeter** vergleichendes Voltmeter *n* ~ **bar** Geradführung *f*, Gleitlineal *n*, Gleitschiene *f*, Schieber *m* ~ **-bar bearer** Gleitschienen-bock *m*, -träger *m* ~ **-bar bracket** Gleitbahnträger *m*, Gleitschienen-bock *m*, -träger *m* ~ **base** (Motor) Spannschiene *f* ~ **block** Schiebestein *m* ~ **box** Gleitbüchse *f*, Schieberkasten *m*, Schiebeschachtel *f*, Ventil-

kasten *m* ~ **bridge** Schleifdraht(meß)brücke *f* ~ **buckle** Schiebeschnalle *f* ~ **caliper rule** Schieblehre *f* ~ **carton** Schiebeschachtel *f* ~ **catch** Bajonettverschluß *m* ~ **changing** Diawechsel *m* ~ **changing arrangement** Diawechseleinrichtung *f* ~ **-coil** Schiebespule *f* ~ **contact** Kontaktschieber *m*, Reiter *m* ~ **conveyer** Schlittenführung *f* ~ **distributor** Schleifbahnverteiler *m* ~ **division** Schiebeeinteilung *f* ~ **dog** Anschlagknagge *f* ~ **drilling machine** Schlittenbohrmaschine *f* ~ **empennage** Verschiebeleitwerk *n* ~ **face** Schieberfläche *f* ~ **(-valve) face** Schieberspiegel *m* ~ **-fastener link (or part)** Reißverschlußglied *n* ~ **feed** Gleitzuführung *f* ~ **fit** Schiebesitz *m* ~ **fracture** Gleitbruch *m* ~ **frame** Gleitrahmen *m* ~ **gauge** Schieblehre *f*, Schublehre *f* ~ **guide** Stößelführung *f* ~ **holder** Präparat-, Schlittenhalter *m* ~ **-in chassis amplifier** Kassettenverstärker *m* ~ **-in plateholder** Einlegekassette *f* ~ **-in tray system** Einschubsystem *n* ~ **-in unit (or module)** Einschub(einheit) *f* ~ **lathe** Support-, Zylinder-drehbank *f* ~ **lock** Bajonettverschluß *m*, Gleitriegel *m* ~ **magazine** Diamagazin *n* ~ **main face** Schieberspiegel *f* ~ **mechanism** Gleitvorrichtung *f* ~ **microtome** Schlittenmikrotom *n* ~ **movement** Schlittenbewegung *f* ~ **part** Verschiebekörper *m* ~ **pin** Gleitbolzen *m* ~ **plate** Schieber *m*, Zungenplatte *f* ~ **position** Schieberstellung *f* ~ **-preventing leather device** Ledergleitschutz *m* ~ **projection** Diapositivprojektion *f* ~ **rail** Stellschiene *f* ~ **rails** Spannschienen *pl* ~ **resistance** Schiebwiderstand *m* ~ **resistor** Schiebewiderstand *m* ~ **rest** Drehsupport *m*, Support *m* ~ **retaining screw** Schlittenhalteschraube *f* ~ **rheostat** Schieberwiderstand *m* ~ **ring** Gleitstein *m* ~ **ring packing (or sealing)** Gleitringdichtung *f* ~ **rod** Schieberstange *f* ~ **rod guide** Schieberstangenführung *f* ~ **rule** Gleitlineal *n*, Rechen-schieber *m*, -stab *m* ~ **-rule caliper** Schublehre *f* ~ **scanner** Diaabtastgerät *n* ~ **scanning equipment** Diaübertragungsanlage *f* ~ **screw** Schieberspindel *f* ~ **-screw tuner** verschiebbarer Tauchtrimmer *m* ~ **shacklespring carrier** Abwälzfederbock *m* ~ **shaft (or spindle)** Schieberspindel *m* ~ **spark** Gleitfunken *m* ~ **speed** Gleitgeschwindigkeit *f* ~ **spring** Gleitfeder *f*, Schleiffeder *f* ~ **(guide) spring** Führungsfeder *f* ~ **tappet** Stößel *m* ~ **-top oil cup** Schiebedeckelöler *m* ~ **torque** Rutschmoment *n* ~ **transformer** Schiebetransformator *m* ~ **trombone** Zugposaune *f* ~ **-type spring lock** Bajonettverschluß *m* ~ **valve** Abschluß-, Absperr-schieber *m*, Dampfschieber *m*, Schieber *m*, (gate) Schieberkörper *m*, Schieberventil *n*, Schubventil *n*, Verteilungskammer *f* ~ **valve for the air** Entlüftungsschieber *m* ~ **valve for chimneys** Kaminschieber *m*

**slide-valve,** ~ **air pump** Schieberluftpumpe *f* ~ **damper** Absperrschieber *m* ~ **diagram** Schieberdiagramm *n* ~ **engine** Schiebermotor *m* ~ **face** Schieberauflage *f*, Schieberspiegel *m* ~ **gear** Flach(schieber)steuerung *f*, Muschelschiebersteuerung *f*, Schiebersteuerung *f* ~ **rod** Schiebergestänge *n* ~ **weight** Ventilgegengewicht *n*

**slide,** ~ **wall** Gleitwand *f* ~ **way** Gleitbahn *f* ~ **way grinding machine** Führungsbahnen-Schleifmaschine *f* ~ **ways** Bettführungsbahn *f* ~ **window** Aufziehfenster *n* ~ **wire** (differential) Brückendraht *m* mit Gleitkontakt, (differential) Gleitdraht *m*, Meßdraht *m*, Schleifdraht *m*

**slide-wire,** ~ **bridge** Drahtmeß- (electr.), Gleitdraht-, Schleifdraht-, Widerstands-brücke *f* ~ **measuring bridge** Schleifdrahtmeßbrücke *f* ~ **potentiometer** Spannungsteiler *m* mit Gleitkontakt ~ **resistance** Gleitdrahtwiderstand *m* ~ **resistor** (or **rheostat**) Schleifdrahtwiderstand *m*

**slider** Gleitkontakt *m*, Gleitschieber *m*, Gleitstück *n*, Reiter *m*, Schieber *m*, Schiebering *m*, Schleifer *m* ~ **coil** Spule *f* mit einem Gleitkontakt

**slides** Diarähmchen *pl*, Gleitbügel *m*

**sliding** Erdrutsch *m*, Gleiten *n*, Gleitung *f*; gleitend, verschiebbar ~ **accuracy** Schleifgenauigkeit *f* ~-**and-rotary-type spool** Längsdrehschieber *m* (Steuergerät) ~ **area** Rutschfläche *f* ~ **armature** Schiebeanker *m* ~-**armature starter** Schubankeranlasser *m* ~ **average** gleitender Durchschnitt *m* ~ **ball bearing** Kugelschiebelager *n* ~ **bar** Führungsstange *f*, Geradführung *f*, Gleitstange *f* ~ **barrier** Rollenschranke *f*, Schiebe-barriere *f*, -gatter *n* ~ **bay** (bridge) Rampenstrecke *f* ~-**bay road-bearer of a bridge** Rampenträger *m* ~ **bearing** Gleitlager *n*, Schiebelager *n* ~-**blade isolating switch** Schubtrenner *m* ~ **block** Kulissenstein *m*, Kurbelschwinge *f*, Schwingenstein *m* ~ **block governor** Kulissenregler *m* ~ **block position** Kulissenstellung *f* ~ **bolt** Gleitriegel *m* ~ **(locking) bolt** Schubriegel *m* ~ **bottom** Schieberboden *m* ~ **bow** Bügelschleifkontakt *m* (electr.) ~ **caliper** Schublehre *f* ~ **cam** Gleitnocken *m* ~ **carriage** Schlitten *m* ~ **change gear** Schieberädergetriebe *n* ~ **claw of the clutch** verschiebbare Kupplungsklaue *f* ~ **coil** Schiebe-kern *m*, -spule *f*, verschiebbare Spule *f*, Verschiebespule *f* ~ **contact** Gleit-kontakt *m*, -stück *m*, Läufer *m*, Schiebekontakt *m*, Schleifkontakt *m* ~ **contact bridge** Schleifkontaktbrücke *f* (zum Steuerschalter) ~ **contact cam** (or **finger**) Schleifdaumen *m* ~ **controls** (Mischpult) Flachbahnregler *m* ~ **counterpoise** Entlastungsschiebegewicht *n* ~ **coupling** Schiebkupplung *f* ~-**cowl skirt** verschiebbarer zylindrischer Teil *m* der Haubenverkleidung ~ **cross-over** Schiebeweiche *f* ~ **curve** Schubkurve *f* ~ **cylinder head** Schieberzylinderkopf *m* ~ **cylindrical valve** Ringschieber *m* ~ **damper** Fallschieber *m* ~ **diaphragm** Blendschieber *m*, Schiebeblende *f* ~ **disc** Gleitscheibe *f* ~ **displacement value** Verschiebewert *m* ~ **dog of the clutch** verschiebbare Kupplungslue *f* ~ **door** Bostwickgitter *n*, Schiebetür *f* ~ **drill arm** Bohrschlitten *m* ~ **electrode** verstellbare Elektrode *f* ~ **endwise on a shaft** auf einer Welle *f* frei längsverschiebbar ~ **eyepiece** Steckokular *n* ~ **face** Laufläche *f* ~ **fastener strips** Gleitverschlußbänder *pl* ~ **feather key** Gleitfeder *f* ~ **feed** Leitvorschub *m* ~ **fit** Gleit-passung *f*, -sitz *m*, Schiebesitz *m* ~-**fit shaft** Schiebewelle *f* ~ **free-running** Schiebeleerlauf *m* ~ **friction** gleitende Reibung *f*,

Gleitreibung *f*, rollende Reibung *f*, Schlüpfungsreibung *f*, wälzende Reibung *f* ~ **gate** Gleittafel *f* ~ **gauge** Schublehre *f* ~ **gear** ein- und ausrückbares Zahnrad *n*, Schieberad *n*, Schubgetriebe *n*, verschiebbares Zahnrad *n* ~ **gear clutch** Schieberäderkupplung *f* ~-**gear transmission** Schubradwechselgetriebe *n* ~ **gears** Räderverschiebung *f* ~ **grid** Bostwickgitter *n* ~ **guide** (or **guidance**) Gleitführung *f* ~ **hatch** Schiebeluke *f* ~ **head** Schlitten *m*, Spindelkopfschlitten *m*, Stößel *m* ~ **jaw** Gleitbacke *f*, verschiebbare Backe *f* ~ **joint file** Scharnierfeile *f* ~ **latticework** Bostwickgitter *n* ~ **leaf gate** Gleitschütze *f* ~ **lever** Gleithebel *m* ~-**lever bar** Wanderhebelschiene *f* ~-**lever breechblock** Gleithebelverschluß *m* ~ **lock gate** Stau-, Ziehschütz *n* ~ **magazine lock bar** Riegelschiene *f* ~ **magnet** Magnetschieber *m* ~ **micromanipulator** Mikromanipulator *m* ~ **mount** Gleitlafette *f* ~ **note** gleitender Ton *m* ~ **nut** gleitende (Schrauben-) Mutter *f*, Gleitmutter *f* ~ **objective changer** Objektivschlittenwechsler *m* ~ **packing ring** Gleitsichtring *m* ~ **phenomenon** Rutschvorgang *m* ~ **pin drive** Kurbelschwinge *f* ~-**pinion system** Schubradwechselgetriebe *n* ~ **plane** Gleitungsfläche *f*, Rutschfläche *f* ~ **plate** (Schlepper) Auffahrtblech *n* ~ **platform** Schiebebühne *f* ~ **plug** Schieber *m* ~ **poise** Laufgewicht *n* ~ **position** Verschiebestellung *f* ~ **post** Schiebeleiste *f* ~ **property** Gleitfähigkeit *f* ~ **pulley** Gleitblock *m* ~ **quoin** Keil *m* ~ **rail** Druck-, Roll-schiene *f* ~ **resistance** Gleitwiderstand *m* ~ **rib** Gleitrippe *f* ~ **ring** Schiebering *m* ~ **(contact) ring** Schleifring *m* ~ **rod panel** Schiebedach *n* (beim Auto) ~ **roof** Gleitverdeck *n*, Schiebedach *n* ~ **rotor** Verschiebeanker *m* ~ **sash** Schiebflügel *m* ~ **sash of a sash window** Fensterschieber *m* ~ **saw** Schlittensäge *f* ~ **scale** Gleitskala *f* ~ **seat** Gleitsitz *m* ~ **shackle bearing** Abwälzlager *n* ~ **shaft** Gleit-, Schub-welle *f* ~ **shell** Gleitschale *f* ~ **shift gear** Schubräderschaltgetriebe *n* ~ **shutter** Längsschieber *m*, Schieberverschluß *m* ~ **shuttle** Schleifschütze *m* ~ **sleeve** Abhebezylinder *m*, Ausrückmuffe *f*, Gleitbüchse *f*, Gleitmuffe *f*, Schiebehülse *f*, Schiebemuffe *f*, Schieberohr *n*, verschiebbare Hülse *f* ~ **sleeve valve** Schieberhülse *f* ~ **sluice** Schützenschleuse *f*, Stauschütz *n*, Ziehschütz *n* ~ **socket** Abfallhülse *f* ~ **spool** (Steuergerät) Längsschieber *m* ~ **square** Schub-winkel *m* ~ **square-wedge breechblock** Flachkeilverschluß *m* ~ **stress** Schubriegelspannung *f* ~ **surface** Gleit-, Lauf-, Rutsch-fläche *f* ~ **surface contact plate** Schleifbahnkontaktplatte *f* (zur Gangschaltung) ~ **switch** Schaltschieber *m* ~ **T-bevel** Winkelschmiege *f* ~ **table** Tischschlitten *m* ~ **thwart** Gleitsitz *m* ~ **tongs** Gleitzange *f* ~ **transom of bridge** Rampenlatte *f* ~ **travel** Schubweg *m* ~ **traverse motion** Schubchangierung *f* ~ **tripod** zusammenschiebbares Dreibeinstativ *n* ~ **tube** Einstell-, Schiebe-, Trieb-rohr *n* ~ **valve** Schutzfalle *f*, Stauschütz *n*, Ziehschütz *n* ~ **vane** Nivellierscheibe *f*, Tafel *f* einer Nivellierplatte ~ **wedge-type breechblock** Verschlußkeil *m* ~ **wedge-type breech mechanism** Gleithebel-, Leitwell-verschluß *m* ~ **weight** Läufergewicht *n*, Laufgewicht *n*, verschiebbares Gewicht *n* ~

wheel Schlepprad *n* ~ **window** Schiebefenster *n*
~ **wiper** Schleifer *m*
**slight** gering, leicht, oberflächlich, unbedeutend
~ **audible knock** leichtes gerade noch hörbares
Klopfen *n* (im Verbrennungsmotor) ~ **deformation of the axle** geringe Verbiegung *f* einer
Achse ~ **detonation** schwaches Klopfen *n* (im
Verbrennungsmotor) ~ **taper** schlanker Kegel
*m*
**slightly** obenhin, (cohesive) schwach ~ **crushed**
grob gepreßt, rösch ~ **curved** gradschalig ~
**damaged** leicht beschädigt ~ **damped** schwach
gedämpft ~ **dulled** hellmatt ~ **injured** leicht
verletzt ~ **stained** leicht gefärbt
**slim** schlank ~ **beam (bracket, or girder)** schlankster Träger *m*
**slime** Bohrschmand *m*, Schlamm *m*, Schleim *m*,
Schlich *m*, Schliech *n*, Schmand *n* ~ **of waste
metal** Krätzschlich *m*
**slime,** ~ **blanket** Schlammdecke *f* ~ **formation**
Schlammbildung *f* ~ **jig** Schlammsetzmaschine
*f* ~ **layer** Schlammscheider *m* ~ **pit** Schlammherd *m*, -sumpf *m* ~ **pulp** Schlammtrübe *f* ~
**separator** Schlamm-scheidemaschine *f*, -schei-
der *m* ~ **table** Schlammherd *m* ~ **treatment**
Schlammverarbeitung *f* ~ **yield** Schlammaus-
beute *f*
**sliminess** Schmierigkeit *f*
**slimy** gallartig, gelartig, schlamm-artig, -haltig,
schlammig ~ **ground** Moddergrund *m* ~ **matter**
Schleimstoff *m*
**sling, to** ~ anschlingen, schlenkern, schleudern,
(packs, rifles) umhängen **to** ~ **an engine** einen
Motor *m* an Seilschlingen aufhängen
**sling** Armbinde *f*, Kropfeisen *n*, Länge *f*, Riemen *m*, Schleuder *f*, Schleudern *n*, Schlinge *f*,
Stropp *m*, Teufelsklaue *f* ~ **bolt** Bügelschraube
*f* ~ **chain** Anschlagkette *f* ~ **hangar** Hängebock
*m* ~ **hole** Kropfloch *n* ~ **psychrometer** Schleu-
derpsychrometer *n* ~ **swivel** Riemenbügel *m* ~
**thermometer** Schleuderthermometer *n*
**slinger** Spritzring *m* ~ **ring** (for deicing fluid)
Abspritzring *m*, Schleuderring *m*, Spiritusluft-
schraubenblattenteiser *m*
**slinging action** (in molding) Schleuderwirkung *f*
**slip, to** ~ abgleiten, ausgleiten, fehltreten, gleiten, rutschen, schieben, schleifen, schleudern,
schliefen, schlüpfen **to** ~ **the cable** Kette *f* ab-
schakeln und schlippen **to** ~ **in cables** einstek-
ken **to** ~ **down** sacken **to** ~ **forward** voreilen **to**
~ **into** hineinschieben **to** ~ **(or glide) off** abgleiten, abrutschen **to** ~ **on** aufstecken, aufstreifen **to** ~ **out** herausgleiten **to** ~ **over** über-
schieben, **to** ~ **a rope** ein Tau *n* schlippen, ein
Tau *n* schlüpfen lassen **to** ~ **through** durchschlüpfen **to** ~ **in with others** unterlaufen
**slip** Abfangkeil *m* für Rohre und Gestänge, Be-
zugsmasse *f*, Blättchen *n*, Blatt Papier *n*, Falte
*f*, Fangkeil , Fehltritt *m*, Helling *f*, Luftschrau-
benschlupf *m* (aviat.), Rutschung *f*, (ceramics)
Schlamm *m*, Schleifsel *n*, Schloß *n*, (of propeller, generator or asynchronous motor) Schlupf
*m*, Schlüpfung *f*, Schlüpfungsreibung *f*, Slip *m*,
Streifen *m*, Stürzen *n*, Verschiebung *f* (geol.)
Verwerfung *f* ~ **(ping)** Gleiten *n*, Gleitung *f*
~ **of bar in concrete** Schlupf *m* (einer Eiseneinlage im Beton) *m* ~ **in the coupling** Kupplungs-
schlupf *m* ~ **of fine cardboard** Kartenspan *m* ~

**of propeller** Slip *m* der Schraube ~ **for small
craft** Aufschlepphelling *f* ~ **for sticking on**
(errata slip) Tektur *f*
**slip,** ~ **band** Gleitlinie *f*, Translationslinie *f* ~
**bands** Translationsstreifung *f* ~ **belly tank** Abwurfbehälter *m* ~ **bolt** Riegel *m*, Schubriegel *m*
~ **cap** Schiebedeckel *m* ~ **change gear** Schiebe-
wechselrad *n* ~ **clutch** Schlupfkupplung *f* ~
**coefficient** Gleitungskoeffizient *m*, Schlupfzahl
*f* ~ **conditions** Gleitbedingungen *pl* ~ **cone**
Rutschkegel *m* ~ **coupling** Freilauf-, Leerlauf-,
Schlupf-kupplung *f* ~ **cover** Überzug *m* ~
**difference** Schlüpfdifferenz *f* ~ **dog** Ablauf-
knagge *f* ~ **drawer** Streifen-lade *f*, -schublade *f*
~-**elevator** Abfangkeilheber *m* (Gestänge- oder
Rohrabfangkeilheber) ~ **fit** Schiebesitz *m* ~
**fittings** Gleitausrüstungsteile *pl* ~-**fold** Fälzel *n*
~ **forming** Streckformverfahren *n* ~ **fuel tank**
abwerfbarer Brennstoffbehälter *m*, Abwurfbehälter *m* (aviat.) ~ **galley** Kolumnenschiff *n*
~ **gauge** Endmaß *n*, Fühlerlehre *f*, Parallelendmaß *n* ~ **gear** Schiebezahnrad *n* ~-**in diaphragm** Einsteckblende *f* ~ **joint** biegsame Verbindung *f*, Dehnungsschlitz *m*, Gleitfuge *f*
~-**joint pliers** Blitzrohrzange *f* ~ **joint tubing**
Falzrohr *n* ~ **kiln** Abdampfofen *m* ~ **knot**
Schiffer-, Schleif-knoten ~ **line** Gleitspur *f* ~
**line field** Gleitlinienfeld *n* ~-**line length** Gleit-
linienlänge *f* ~ **line pattern** Gleitlinienbild *n* ~
**lines** Gleitlinienbild *n* ~ **magnifier** Spaltlupe *f*
~ **meter** Schlüpfungsmesser *m* ~ **motions** Gleit-
bewegungen *pl* ~ **multiple** verschränkte Vielfachschaltung *f* ~-**on cable socket** Aufsteck-
kabelschuh *m* ~-**on cap** Aufsteckklappe *f* ~-**on
flange** loser Flansch *m* ~-**on lens segment** Auf-
steckglas *n* ~-**on lever** Aufsteckhebel *m* ~-**on
mount** Vorhängerfassung *f* ~ **pan** Schubfläche *f*
~ **phenomenon** Rutscherscheinung *f* ~ **plane**
Gleitfläche *f*, (cryst.) Schiebungsfläche *f*, Translations-ebene *f*, -fläche *f* (cryst.) ~ **plane in the
body-centered cubic lattice** kubisch raumzentriertes Gitter *n* ~-**plane blocking** Gleitebenen-
blockierung *f* ~ **points** Kreuzungsweiche *f* ~
**proof** Korrekturfahne *f* ~ **regulator** Schlupf-
regler *m* ~-**regulator induction motor** Regulier-
schleifringläufermotor *m* ~ **resistance** Schleif-
widerstand *m* ~ **ring** Gleitring *m*, Schleifer *m*,
Schleifring *m*
**slip-ring,** ~ **armature** Schleifringanker *m* ~ **brush**
Schleifringbürste *f* ~ **end bearing** Schleifring-
lager *m* ~ **engine** Schleifring(anker)motor *m* ~
**motor** Schleifringankermotor *m* ~ **rotor** Schleif-
lederanker *m*, -ringanker *m* ~ **rotor engine**
Schleifringläufermotor *m*
**slip,** ~ **sheet** Schmierbogen *m* ~ **socket** Fangglocke *f*, Keilfänger *m* ~ **speed** Schlupfge-
schwindigkeit *f*, Schraubenstrahlgeschwindig-
keit *f* ~ **stream** Abstrom *m*, Fahrtwind *m*,
Luftschraubennachstrom *m*, Nachstrom *m*,
Nachstrom *m* der Luftschraube, Propeller-
strahl *m*, -wind *m*, Schrauben(nach)strom
*m*
**slipstream,** ~ **characteristic** Abwindverhältnis *n*
~ **deflection** Triebstrahlumlenkung *f* **equalization** Schraubenstrahlausgleich *m* (aviat.). ~
**interference** Strahlbeeinflussung *f* ~ **region**
Strahlfeld *n* ~ **rotation** Strahldrehung *f* ~ **zone**
Strahlfeld *n*

**slip, ~ stub shaft** Keilwellenzapfen *m* **~ surface** Gleitebene *f*, Gleitfläche *f*
**slipless** schlüpflos
**slipway** Aufschleppe *f*, Aufschlepphelling *f*, Gleitbahn *f*, Gleitschiene *f*, Hellinggerüst *n*, (for a vessel) Schiffshelling *f*, Slip *m* **~ crane** Helligkran *m* **~ superstructure** Helliggerüst *n*
**slipways** Aufschleppen *n*, Gleitbalken *pl*
**slip zones** Gleitzonen *pl*
**slippage** Abgleitung *f*, Rutschen *n*, Schlupf *m*, Schlüpfen *n*, Schlüpfung *f*, Translation *f* **~ (of the V belt)** Gleitwiderstand *m* (des Keilriemens) **~ along the wall** Wandgleitung *f*
**slippage, ~ tensor** Gleittensor *m* **~ test** Gleitversuch *m*
**slipped wheel** entkeiltes Rad *n*
**slipper** Bremsklotz *m*, Hemmschuh *m*
**slipper, ~ block** Blattzapfen *m* **~ clutch** Gleit-, Rutsch-kupplung *f* **~ handle** Ankerfallhebel *m* **~-out guide** auslösende Gestängesteuerung *f* **~ piston** Gleitschuhkolben *m* **~ shoe** Blattzapfen *m*
**slipperiness** Schlüpfrigkeit *f*
**slippery** glatt, schlüpfrig **~ ice** Glatteis *n*
**slipping** Ausgleiten *n*, Stürzen *n* **~ of the rails** Laufen *n* der Schienen **~ of a slope** Abbröckelung *f* einer Böschung
**slipping, ~ apparatus** Fallapparat *m* **~ cam** Leerlaufnocke *f* **~ clutch** rutschende Kupplung *f*, Rutschkupplung *f* **~ drive** Gleitantrieb *m*, nachgiebiger Reibantrieb *m* **~ earthwork** rutschender Boden *m* **~ instrument** Fallinstrument *n* **~ plane** Rutschfläche *f* **~ turn** Gleit-drehung *f* (aviat.), -wendung *f*, -widerstand *m*
**slips** Abfangkeile *pl*, (ignition) Aussetzer *m*, Fugen *n*
**slit, to ~** aufschlitzen, ritzen, schlitzen, spalten
**slit** Blende *f*, Blendenausschnitt *m*, Einwurf *m*, Fuge *f*, Gabel *f*, Nut *f*, Nute *f*, Ritz *m*, Ritze *f*, Schlitz *m*, Spalt *m*, (diaphragm) Spaltblende *f*; ritzig **~ of the type bar** Auszahnung *f* am Typenhebel
**slit, ~-and-tongue** Gabelverbindung *f* **~-and-tongue joint** Kerbenfügung *f*, Scherenverbindung *f*, Verbindung *f* durch Schlitzzapfen **~ aperture** Spaltöffnung *f* **~ burner** Schlitz-, Schnitt-brenner *m* **~ collimation** Spaltkollimation *f* **~ cover** Spaltkappe *f* **~ diaphragm** Schlitzblende *f* **~ effect** Schlitzeffekt *m* **~ height** Spalthöhe *f* **~ illumination** Spaltausleuchtung *f* **~ image** Spaltbild *n* **~ jaws** Spaltbacken *pl* **~ lamp** Schlitzlampe *f* **~ lamp microscope** Spaltenmikroskop *n* **~-like nozzle opening** schlitzförmige Düsenöffnung *f* **~ mechanism** Spalteinrichtung *f* **~ nut** Schlitzmutter *f* **~ opening** Schlitzweite *f* **~ pin** Spaltstift *m* **~ source** spaltförmige Quelle *f* **~ stop** Schleifspalt *m* (opt.), Schlitzblende *f* **~ system** Spaltsystem *n* **~ tone** Ausfluß-, Spalt-ton *m* **~ trench** Deckungsgraben *m* **~-type printer** Schlitzkopiermaschine *f* **~-type (rotary-disk) shutter** Schlitzverschluß *m* **~ width** Schlitzbreite *f*
**slitless** spaltfrei
**slitter** Keilhaue *f*, Streifschneidemaschine *f*
**slitting** Schlitzen *n* **~ and cross cutting machine** Quer- und Längsschneider *m* **~ and sheeting** Längs- und Querschnitt *m*
**slitting, ~ arrangement** Längsschneideeinrich-

tung *f* **~ cutter** Schlitzfräser *m* **~ cutter for wood screws** Schlitzfräser *m* für Holzschrauben **~ file** Einstreichfeile *f* **~ knife** Längsschneidemesser *n*, Schlitzmesser *n* **~ plane** Schneidehobel *m* **~ saw** Einstreichsäge *f*
**sliver** Walzsplitter *m* **~ delivery** Kanalausgang *m* **~ doubling attachment** Einrichtung *f* für Bandvereinigung **~ length** Luntenlänge *f* **~ weight** Bandgewicht *f*
**slogan** Schlagwort *n*
**sloop** Schaluppe *f*
**slop, to ~-pad** überklotzen, überpflatschen
**slop** Schlempe *f*, (of a converter) Spucken *n* **~ pad aniline black** Klotzanilinschwarz *n* **~-padded aniline black** Anilinklotzschwarz *n* **~ padding** Pflatschdruck *m* **~ runner** Trebertrockner *m*
**slope, to ~** abböschen, abfallen, abfassen, abschrägen, böschen, flach werden, neigen, senken, sich senken **to ~ away** abdachen **to ~ down** vollständig abfallen, verflachen **to ~ off** abfallen **to ~ steeply** steil abböschen **to ~ upwards** schräg aufwärts verlaufen
**slope** Abdachung *f* (Gelände) Abfall *m*, Abhang *m*, Abhängigkeit *f*, Abschrägung *f*, Anhöhe *f*, Anlauf *m*, Anstieg *m*, Böschung *f*, Böschungsverhältnis *n*, Bremsberg *m* (min.), Entwicklungssteilheit *f*, Filtersteilheit *f*, Gefälle *n*, Gehänge *n*, Geländeneigung *f*, Hang *m*, Neigung *f*, Quergefälle *n*, Rampe *f*, Schiefe *f*, Schräge *f*, Steigung *f*, Steigungsmaß *n*, Talgehänge *n*, Verlauf *m*
**slope, ~ of bank** mittlere Böschungsneigung *f* **~ of curve** Steilheit *f* einer Kurve **~ of a curve** Neigung *f* der Kurve **~ of density** Dichteabfall *m* **~ of a dike** Deich *m* oder Dammböschung *f* **~ of the dune** Dünenböschung *f* **~ of an embankment** Deich *m* oder Dammböschung *f* **~ of the emission characteristic** Steilheit *f* **~ of force lines** Kraftlinienschräglage *f* **~ of the lift curve** Neigung *f* der Auftriebskurve **~ of plateau** Steigung *f* des Geigerbereiches **~ of a river** Stromgefälle *n* **~ of river banks** Uferböschung *f* **~ of a roof** Abdachung *f* eines Daches **~ of surface** Oberflächenneigung *f* **~ of (water) surface** Spiegelgefälle *n* **~ of valve characteristics** Steilheit *f* von Röhrenkennlinien (teleph.)
**slope, ~ angle** Neigungswinkel *m* **~-background** hinterschliffen **~ boring** schräge Bohrung *f* **~ change** Gefälle-, Neigungs-änderung *f* **~ conductance** Steilheit *f*, Transponierungssteilheit *f* **~ conductivity** Anstiegleitfähigkeit *f* **~ contour** Umrißliniensteigung *f* (math.) **~ deflections equations** Elastizitätsgleichungen *pl* **~ deviation** Steigungsabweichung *f* **~ equalizer** Grobentzerrer *m* **~ failure** Böschungs-, Hang-rutschung *f* **~ landing** Landung *f* am Hang (Hubschreuber) **~ line** Böschungslinie *f* **~ polarity** Umpolung *f* **~ soaring** Hangsegelflug *m* **~ spring** Hänge-, Hang-quelle *f* **~ stabilization** Böschungsbefestigung *f* **~ wash** Gehängelehm *m*
**sloped** abgeschrägt, geneigt, schief **~ beam (girder, or post)** Schrägstab *m*
**slopes** Flanken *pl* (von Impulsen)
**sloping** Abböschung *f*; abdachig, abhängig, abschüssig, ansteigend, geneigt, schief, schief-

liegend, schräg, schräg abfallend, mit Steigung *f* verlegt ~ **beach** Rampe *f* ~ **bench** Abwurf-, Schräg-rampe *f* ~ **bottom** geneigter Boden *m* ~ **clamp** Feilkluppe *f* ~ **column** Schrägkolonne *f* ~ **edge** Zuschärfungsfläche *f* ~ **edges** Böschungskante *f* ~ **ground** abfallendes Gelände *n* ~**-hearth-type liquating furnace with a bed of glowing charcoal** Pauschherd *m* ~ **hoop channel** Schlingenkanal *m*, (in rolling) Tieflauf *m* ~ **portion** Steigung *f* ~ **position** Schräglage *f* **terrain** abfallendes Gelände *n* ~ **wall** geböschte Mauer *f*
**slopping of a converter** Konverterauswurf *m*
**slot, to** ~ bestoßen, mit Einschnitten *pl* versehen, einstoßen, kehlen, kerben, nuten, schlitzen, stoßen **to** ~ **a shaft** eine Welle *f* nuten
**slot** Durchgangsöffnung *f*, Einschnitt *m*, Einwurf *m*, Keilnut *f*, Keilrille *f*, Kerb *m*, Kerbe *f*, Langloch *n*, Nut *f*, Nute *f*, Riefe *f*, Schlitz *m*, (for slotted wing) Schlitzklappe *f*, (paper mfg) Spalt *m* des Stoffauflaufs, Spalte *f* ~ **for air inlet** Luftschlitz *m* ~ **and feather** Nut *f* und Feder *f*, Nute *f* und Feder *f* ~ **and keyway milling machine** Schlitz- und Nutenfräsmaschine *f* ~ **of mailbox** Briefeinwurf *m*
**slot,** ~ **armature** Anker *m* mit Nuten ~ **array** (antenna) Schlitzstrahlersystem *n* ~ **bolt** Schlitzschraube *f* ~**-boring machine** Langlochbohrmaschine *f* ~ **burner** Schnittbrenner *m* ~ **covering** Spaltverkleidung *f* ~ **cutter** Langloch-, Nuten-fräser *m* ~ **depth** Schlitztiefe *f* ~ **die** Schlitzform *f* ~ **discontinuity effect** (microwaves) Schlitzstoß *m* ~ **drill** Langbohrer *m* ~**-drilling machine** Nutenbohrmaschine *f* ~ **extender** Verlängerungsstanze *f* ~ **frequency drift** Frequenzverwerfung *f* ~ **gas apparatus** Gasautomat *m* ~ **grinder** Nutenschleifmaschine *f* ~ **guide** Schlitzführung *f* ~ **inclination** Nutenschrägung *f* ~ **inserting** Einsteckschlitz *m* ~ **lantern** Schlitzlaterne *f* ~ **leakage** Spaltstreuung *f* ~ **lining** Nutauskleidung *f* ~**-lip aileron** Schlitzklappenquerruder *n* ~ **meter** Münzzähler *m* ~ **milling** Schlitzfräsen *n* ~ **mixer** Schlitzmischer *m* ~ **mold attachment** durchbrochener Riester *m* ~ **pitch** (of drum winding) Nutenschritt *m*, Nutenteilung *f*, Spulenweite *f* ~ **position** Nutenstellung *f* ~ **radiator** (antenna) Schlitzstrahler *m* ~ **rail** Schlitzschiene *f* ~ **ripple** Ankernutenwellen *pl* (des Gleichstroms), Nutenwelle *f* ~ **ripple frequency** Nuten(wellen)-frequenz *f* ~ **roll** Kulissenrolle *f* ~ **scattering** Nutstreuleitwerk *m* ~ **screw** Schlitzschraube *f* ~ **seam** Schlitznaht *f* ~**-shaped port** schlitzförmige Steueröffnung *f* ~**-siphoning** (jet) Spaltabsaugung *f* ~ **stray** Spaltstreuung *f* ~**-type atomizer** Spaltzerstäuber *m* ~ **velocity** Spaltgeschwindigkeit *f* ~ **wedge** Nutverschlußkeil *m* ~ **weld** Lochschweißung *f*, Schlitz-naht *f*, -schweißung *f* ~ **welding** Dübelschweißung *f* ~ **width** Schlitzbreite *f* ~ **winding** Nutenwicklung *f* ~ **wound** mit Nutenwicklung *f*
**slots** Spalten *pl* (im Filterpaket)
**slotted** genutet, geschlitzt, mit Nuten *pl* versehen ~ **out** ausgeschlitzt
**slotted,** ~ **aileron** Schlitzquerruder *n*, Spaltquerruder *n* ~ **batten** Haftlatte *f* (telegr.) ~ **bolt** geschlitzter Bolzen *m* ~ **cheese head** (cylinder)

geschlitzter Zylinderkopf *m* ~ **cone point set screw** Gewindestift *m* mit Spitze ~ **cup point set screw** Gewindestift *m* mit Kegelkuppe ~ **cylinder** aufgeschnittener Zylinder *m* ~ **dial plate** Skalenschablone *f* ~ **diaphragm** Schlitzblende *f* ~ **disc** Rasten-, Schlitz-scheibe *f* ~ **dog point set screw** Gewindestift *m* mit Zapfen ~ **fillister head screw** Linsenschraube *f* ~ **flap** Spaltklappe *f* ~ **headless screw** Schaftschraube *f* ~ **hole** längliches Loch *n*, Langloch *n*, Schlitz *m* ~ **jaw** Gabel *f* ~ **jaw for brakes** Bremsgabel *f* ~ **lever** Kulissen-, Langloch-hebel *m* ~ **line** Meßleitung *f* ~ **nut** Nuten-, Schlitz-mutter *f* ~ **oil ring** Ölabstreifschlitzring *m* ~ **opening** längliche Öffnung *f* ~ **oval-head screw** Linsensenkschraube *f* ~ **(taper) pin** Kerbstift *m* ~ **pipe** geschlitztes Rohr *n* ~ **plate** Schlitzblech *n* ~ **plug** Schlitzstopfen *m* ~**-ring antenna** Stromschleifenantenne *f* ~ **rocket jet** Ringspaltdüse *f* ~ **round-head screw** Halbrundschraube *f* ~ **sampling vessel** Schlitzgefäß *n* ~ **screw** Schlitz-, Spaltkopf-schraube *f* ~ **section** geschlitzte Leitung *f* ~ **shaft** Schlitzachse *f* ~ **shutter** Blättchen-, Schlitz-verschluß *m* ~ **sieve** Schlitzsieb *n* ~ **spring sleeve** Spannhülse *f* ~ **tube** (rocket) Spaltröhre *f* ~ **tube charge** (rocket) Spaltrohrtreibsatz *m* ~**-type piston** Schlitzmantelkolben *m* ~ **wing flap** Schlitzhilfsflügel *m*
**slotter** Nutstoßmaschine *f*, Stoßmaschine *f* ~ **tool** Nutenstoßwerkzeug *n*
**slotting** Aufspalten *n*, Keilnutenhobeln *n*, Nutenstoßen *n*, Schlitzen *n*, Stoßen *n* ~ **of the knives** Bestoßen *n* der Schnitzelmesser
**slotting, attachment** Stoßapparat *m* ~ **cutter** Schlitzfräser *m* ~ **end mill** Schaftennutenfräser *m* ~ **machine** Nutenstoßmaschine *f*, Stoßmaschine *f* ~ **machine for paring knives (or choppers)** Bestoßmaschine *f* für Schnitzelmesser ~ **tool** Stoß-meißel *m*, -stahl *m*
**slough, to** ~ **thread** Faden *m* abschlagen
**sloughing** Nachrutschen *n* (von Material) ~ **action** Abtragen *n* ~**-off tank** Setzmaschine *f*
**slovenly built wall** fluchtlose Mauer *f*
**slow, to** ~ **down** (Drehzahl) abfallen, auslaufen, langsamer werden, verlangsamen **to** ~ **down the engine** den Gang *m* verlangsamen **to** ~ **up** verringern
**slow** (speed) gering, langsam, sacht, säumig, schwerfällig, träge **to be** ~ zurückbleiben **to go** ~ nachgehen (Uhr) ~ **to develop** entwicklungsträge ~ **to operate relay** anzugsverzögertes Relais *n* ~ **to react** reaktionsträge ~ **to release** langsam auslösend
**slow,** ~**-acting** langsamwirkend ~**-acting relay** Verzögerungsrelais *n* ~**-blowing fuse** träge Sicherung *f* ~**-break switch** Langsamausschalter *m* ~ **burning** Schwelung *f*; schwerbrennbar ~**-burning film** unverbrennbarer Film *m* ~ **burning stove** Dauerbrenner *m* ~ **combustion** schleichende Verbrennung *f* ~**-combustion cooker** Dauerbrandherd *m* ~**-combustion stove** Dauerbrand-, Füll-ofen *m* ~ **(-burning) composition** fauler Satz *m* ~ **disengaging (or dissolving)** langsamlösend ~**-down signal** bedingtes Haltsignal *n* ~ **drift of image** Bildverschiebung *f* ~ **drive** Kriechgang *m* ~**-drying oil** Harttrockenöl *n* ~ **flight** Langsamflug *m* ~

**frequency drift** Frequenzverwerfung *f* ~**-hardening cement** langsam erhärtender Zement *m* ~ **interruptor** Langsamunterbrecher *m* ~ **landing speed** geringe Landegeschwindigkeit *f* ~ **match** Brandlunte, Lunte *f* ~ **motion** Feinbewegung *f*, feine Verstellung *f*, Langsamgang *m*

**slow-motion,** ~ **camera** Zeitdehner *m*, Zeitlupe *f* ~ **dial** Feinstellskala *f* ~ **effect** Zeitlupenverfahren *n* (film) ~ **gear** Zahnradfeinbewegung *f* ~ **method** Zeitlupenverfahren *n* ~ **photography** Zeitlupenaufnahme *f* ~ **picture** Zeitdehneraufnahme *f* ~ **pictures** Zeitlupenaufnahme *f* ~ **screw** Feinbewegungsschraube *f* ~ **starter** Ständeranlasser *m*

**slow,** ~ **moving article** Remittende *f* ~**-moving engine** Langsamläufer *m* ~ **neutron capture** Einfang *m* langsamer Neutronen ~ **neutron fission** Spaltung *f* mit langsamen Neutronen ~ **(fast) operating** langsam (schnell) ansprechend ~**-operating relay** Relais *n* mit Anzugsverzögerung *f* oder mit verzögerter Anziehung *f* ~ **operation** verzögerte Anziehung *f* ~ **reactor** Reaktor *m* mit langsamen Neutronen ~ **release** verzögerte Unterbrechung *f* oder Trennung *f* ~**-release** langsamlösend ~ **releasing** langsamlösend ~**-releasing relay** langsam abfallendes Relais *n*, Relais *n* mit Abfallverzögerung, Verzögerungsrelais *n* ~**-response** langsam ansprechend ~ **roll** langsame Rolle *f* ~**-run rotary valve** Lehrlaufdrehschieber *m* ~ **running** Langsamlauf *m*

**slow-running,** ~ **cutout valve** Leerlaufabstellvorrichtung *f* ~ **engine** langsam laufender Motor *m*, Langsamlauf *m* des Motors ~ **jet** Leerlaufdüse *f* ~ **position** Leerlaufstellung *f* ~ **screw** Leerlaufeinstellschraube *f* ~ **stop** Leerlaufbegrenzung *f*

**slow,** ~**-setting cement** langsam abbindender Zement *m*, Langsambinder *m* ~**-sinking speed** geringe Sinkgeschwindigkeit *f* ~**-speed** langsamlaufend ~**-speed drive** Langsamantrieb *m* ~**-speed interrupter** Langsamunterbrecher *m* ~**-speed jet** Leerlaufbrennstoffdüse *f* ~**-speed well** Leerlaufnapf *m* ~ **storage** langsamer Speicher *m* ~**-sweep** flache Krümmung *f* ~**-sweep television** Industriefernsehen *n* ~ **test** Belastungsversuch *m*, statische Prüfung *f* ~**-to-release relay** langsam abfallendes Relais *n* ~ **train** Bummelzug *m* ~ **welding** Langzeitschweißung *f*

**slowed-down** abgebremst

**slowing-down** Abbremsung *f*, Verlangsamung *f* ~ **brake** Bremsfläche *f*, Nachlaufbremse *f* ~ **control cam** Auslaufsteuerdaumen *m* ~ **density** Brems-, Verzögerungs-dichte *f* ~ **length** Bremslänge *f* ~ **path** Nachlaufweg *m* ~ **power** Bremsvermögen *n* ~ **process** Bremsprozeß *m* ~**-signal** bedingtes Haltesignal *n* ~ **time** Auslaufzeit *f*

**slowly** allmählich ~ **raised** (bei) langsame Steigen *n* ~ **reacting** schwerfühlig

**slowness** Langsamkeit *f*, Trägheit *f*

**sluable** schwenkbar

**slub, to** ~ vorspinnen

**slub** Fluse *f*, Vorgespinst *n* ~ **line** Stichleitung *f*

**slubber** Webereivorbereitungsmaschine *f*

**slubbing** (textiles) Lunte *f*, Vordergespinst *n*, Vorspinnen *n* ~ **frame** Grobfleier *m*, Vorfleier *m* ~ **melange printing machine** Vigourexdruckmaschine *f*

**slude** (refinery) Raffinationsabfall *m*

**sludge** Bohrschmand *m*, Faulschlamm *m*, Kohlenschlamm *m*, Matsch *m*, Schmand *n*, Trübe *f*, wässerige Schneemasse *f* ~ **in the lead chamber of photoelectric cells** Bleikammerschlamm *m*

**sludge,** ~ **activation** Schlammbelebung *f* ~ **analysis** Schlammanalyse *f* ~ **basin** Behälter *m* für Wascheinrichtung ~ **bed** Rieselfeld *n*, Schlamm-beet *n*, -teich *m* ~ **box** Schlammkasten *m* ~ **cock** Schlammhahn *m* ~ **cooler** Laugenkühler *m* ~ **cutting** Bohrschlamm *m* ~ **digestion** Schlammfaulung *f* ~**-digestion tank** Faulraum *m* für Schlamm, Schlammfaulraum *m* ~ **drain** Schlammablaß *m* ~ **draining** Schlammentwässerung *f* ~ **extractor** Schlammfänger *m* ~ **fittings** Entsalzungsanlage *f* ~ **formation** (in fuel) Harzbildung *f* ~ **level indicator** Laugenstandsanzeiger *m* ~ **pit** Schlammkuhle *f* ~ **pocket** Schlammsack *m* ~ **pump** Schlammpumpe *f* ~ **pressure reducer** Laugenentspanner *m* ~ **remover** Schlammräumer *m* ~ **residues** Schlammrückstände *pl* ~ **scraper** (ring) Schmutzabstreifring *m* ~ **trap** Schlammfang *m* ~ **trough** Schmutzfangschale *f* ~ **valve** Schlammventil *n*

**sludger** Bohrlöffel *m*, Schlammlöffel *m*

**sludges** Schlämmen *n*

**sludging value** Säureniederschlagswert *m*

**slue, to** ~ schwenken

**slug** Anpaßkörper *m*, Einheit *f* der Masse, Rohrstück *n* (Rohling), Schnecke *f*, Verzögerer *m*, Würfel *m* ~ **cutter** Zeilenhacker *m* ~ **tuner** Abstimmkörper *m* ~ **tuning** Eisenkernabstimmung *f*

**slugged relay** verzögertes Relais *n*

**slugging machine** Oberfleckstiftmaschine *f*

**sluggish** strengflüssig, träge ~ **in action** träge ansprechend **to be** ~ **in response** langsam ansprechend

**sluggish combustion** schleichende Verbrennung *f*

**sluggishness** Ermüdung *f*, Langsamkeit *f*, (instrument) Trägheit *f*, Zähflüssigkeit *f*

**slugs** Durchschuß *m* (print.)

**sluice** Absperr-glied *n*, -organ *n*, -teil *m*, -vorrichtung *f*, Gerinne *n*, Grundablaß *m*, Rinne *f*, Schleuse *f*, (gate) Schütze *f*, Schützenöffnung *f*, Verschluß *m*, Zugschütz *n* ~ **with circular chamber** Kesselschleuse *f*, Trommelschleuse *f* ~ **on the dock** Dockverschluß *m* ~ **with shutters** Jalousieschütz *n* ~ **with turning doors** Drehtorschleuse *f*

**sluice,** ~ **board** Ablaßschütz *n*, Schütze *f*, Stellfalle *f* ~ **chamber** Schützkammer *f* ~ **dam** Schützenwehr *n* ~ **door** Drehtor *n* ~ **gate** Falle *f*, Schleusentor *n*, Spülschleuse *f* ~ **pipe** (road) Dücker *m*, Durchlaß *m* ~ **stay** Stellfalle *f* ~ **valve** Absperrschieber *m*, Abzugsschieber *m*, Ausblaseventil *n*, Schieberschütz *n* ~ **valve with flat body** Keilflachschieber *m* ~ **valve with open body** Keilovalschieber *m* ~ **valve with oval body** Keilovalschieber *m* ~ **valve with round body** Keilrundschieber *m* ~ **valve with socket ends** Muffenabsperrschieber *m* ~**way** Ablaßschleuse *f*, Durchflußöffnung *f*, Schleuse *f* ~ **weir** Durchlaß-, Schützen-wehr *n*

**sluicing** Spülen *n*

**sluing** Schwenkung *f*; schwenkbar

**slump** Absackung *f*, Einsackung *f*, Senkung *f* ~ **test** Setzprobe *f*
**slung** umgehängt
**slur, to** ~ (founding) ausstreichen, schlichten, schwärzen, verwischen
**slur** Schlichte *f* ~ **page of a bock** erste Seite *f* eines Buches
**slurry, to** ~ schlichten
**slurry** (coal) Abrieb *m*, dünner Mörtel *m*, flüssiger Brei *m*, Kohlenschlamm *m*, Pochschlamm *m*, -schlich *m*, Schlamm *m*, Trübe *f* ~ **coal** Schlammkohle *f* ~ **operation** Sumpfphase *f* ~ **pond** Schlammteich *m* ~ **screen** Schlammsieb *n*
**slush, to** ~ (an engine) mit Konservierungsmitteln behandeln **to** ~ **an engine** den Motor *m* einschmieren
**slush** Schlamm *m*, Schneematsch *m*, Trübe *f* ~ **pit (or pond)** Schlammloch *n* ~ **pump** Spülpumpe *f*
**slushing oil** Konservierungsöl *n*, Rostschutzöl *n*
**small** gering, geringfügig, klein, schmal ~ **in bulk** gedrängt (auf)gebaut, wenig Raum *m* beanspruchend ~ **in diameter** kleinkalibrig
**small,** ~ **angle inelastic collisions** unelastische Kleinwinkelstreuung *f* ~**-angle grain boundary** Kleinwinkelkorngrenze *f* ~**-angle scattering** Kleinwinkelstreuung *f* ~ **area** Weißscher Bezirk *m* ~ **arm** Hand-feuerwaffe *f*, -schußwaffe *f*, -waffe *f* ~ **arms** Schußwaffen *pl* ~ **arms ammunition** Gewehrmunition *f* ~ **ball** Kügelchen ~ **band** Plättchen *n* ~ **bar of metal** Metallstäbchen *n* ~ **base** Böckchen *n* ~ **battery charger** Kleinlader *m* ~ **bead** Tröpfchen *n* ~ **bearing pedestal** Lagerböckchen *n* ~ **bell** Oberglocke *f* ~ **bessemer converter** Kleinbessemerbirne *f* ~ **block** Kleinhebezeug *n* ~ **boat** Kahn *m* (of) ~ **bore** kleinkalibrig ~**-bore barrel** Kleinkaliberlauf *m* ~**-bore machine gun** kleinkalibriges Maschinengewehr *n* ~**-bore target range** Kleinschießplatz *m* ~ **bottle** Fläschchen *n* ~ **bulb** Kügelchen *n* (of) ~ **caliber** kleinkalibrig ~ **calorie** Grammkalorie *f*, kleine Kalorie *f* ~ **can** Kännchen *n* ~**-capacity cable** Kabel *n* mit geringer Adernzahl ~ **carbon products** Kohlekleinerzeugnisse *pl* ~ **cask** Fäßchen *n* ~ **castings** Kleinguß *m* ~ **chain** Kettel *f* ~ **chain for hanging curtain from head of needles** Befestigungskette *f* am Nadelkopf ~ **(air) chamber** Kleinkammer *f* ~ **channel** Priele *f* ~ **channel in tidal flats** Wattpriele *f* ~ **charge** kleine Ladung *f* ~ **chuck** Futterring *m* ~ **circle** kleiner Kreis *m*, Nebenkreis *m* ~ **claim** Bagatelle *f* ~ **coal** Feinkohle *f*, Kohlen-grus *m*, -klein *n* ~ **coin** Scheidemünze *f* ~ **coke** Koksklein *n*, Kokslösche *f* ~ **coke of lignite** Grudekoks *m* ~ **consumer** Kleinverbraucher *m* ~ **control switch** Kleinsteuerschalter *m* ~ **converter** Klein-birne *f*, -umformer *m* ~ **conveyer** Kleinförderanlage *f* ~ **cube** Sparwürfel *m*, Würfelchen *n* ~ **cupola furnace** Kleinkupolofen *m* ~ **cuts** Hüllschnitte *pl* ~ **disc holder** Ventilhalterstück *n* ~ **discharging plunger** Tauchkölbchen *n* ~ **dot** Pünktchen *n* ~ **end** (of piston rod) Kolbenstangenkopf *m* ~ **end of connecting rod** Kolbenbolzenkopf *m* ~ **end of master rod** Kolbenbolzenende *n* des Hauptpleuels ~ **end bush** (piston rod) Kolbenstangenkopfbüchse *f* ~**-end hole** (of connect-

ing rod) Pleuelauge *n* ~ **face** kleines Auge *n* ~ **film** Kleinbildfilm *m* ~ **flange** Kleinflansch *m* ~ **flange connection** Kleinflanschanschluß *m* ~ **flow** Nebenstrom *m* ~ **forceps** Zängchen *n* ~ **forging furnace** Kleinschmiedeofen *m* ~ **fortification** Bunker *m*
**small-gauge** (wire) schwach ~ **line** dünndrähtige Leitung *f* ~ **track** Schmalspur *f* (of) ~ **wire** schwach drähtig ~ **wire** Feindraht *m*; dünndrähtig, dünner Draht *m* ~ **wire cable** dünndrähtiges Kabel *n*
**small,** ~ **generator** Kleinmaschinensatz *m* ~ **gold foil (or leaf)** Goldblättchen *n* ~ **grinding machines** Schleifböcke *pl* ~ **grub srew** Sicherungsgewindestift *m* ~ **hammer** Handfäustel *n* ~ **hand hammer** Fausthammer *m* ~ **handwheel** Handrädchen *n* ~ **hardware** Kleineisen-waren *pl*, -zeug *n* ~ **hill** Holm *m* ~ **holding** landwirtschaftlicher Kleinbetrieb *m* ~ **hole** Grübchen *n* ~ **hook** Häkchen *n* ~ **house** Häuschen *n* ~ **insertion frame** Einlegerähmchen *n* ~ **interval of time** Zeitteilchen *n* ~ **ironware** Kleineisenwaren *pl*, -zeug *n* ~ **jet of the plate type** Düse *f* der Drosselplattenbauart ~ **joint(ing) trowel** Fugenkelle *f* ~ **lace** Nagelschnur *f* ~ **lathe** Drehstuhl *m* ~ **leaf** Blättchen *n* ~ **letter** Minuskel *f* ~ **lifting equipment** Kleinhebezeug *n* ~ **lighting outfit(s)** Kleinbeleuchtung *f* ~ **load** kleine Ladung *f* ~ **mallet** Handfäustel *n* ~ **matter** Kleinigkeit *f* ~ **metering bridge** Kleinmeßbrücke *f* ~ **motorcar** Kleinkraftwagen *m* ~ **nail** Täck *m* ~ **naval craft** Kleinkriegsschiff *n* ~ **nippers** (jeweler's) Federzange *f* ~ **opening** Grübchen *n* ~ **ore** Schlich *m* ~ **package (or parcel)** Päckchen *n* ~ **part** Kleinteil *m*, Teilchen *n* ~ **particle** Partikelchen *n* ~ **photogravure rotary press** Kleintiefdruckrotationsmaschine *f* ~ **pieces of charcoal** Holzkohlenklein *n* ~ **pieces of lead ore** Bleierzfunken *pl* ~ **pile** Kleinstapel *m* ~ **place** Plätzchen *n* ~ **plane** Kleinflugzeug *n* ~ **plant operation** Kleinbetrieb *m* ~ **plate** Plättchen *n* ~ **power electromotor** Kleinelektromotor *m* ~ **power station** Kleinstation *f* ~ **protuberance** (on glassware) Külbelstich *m* ~ **sample** Handmuster *n* ~ **scale** kleines Meßgebiet *n* ~**-scale manufacturing operation** Kleinbetrieb *m* ~**-scale series production** Kleinserienbau *m* ~**-scale test** Handversuch *m* ~**-section mill train** Feineisenstraße *f* ~ **section mills** Feinstraßen *pl* ~ **section rolling mill** Feinstahlwalzwerk *n* ~**-size appliances** Kleingeräte *pl* ~**-size folding machine** Kleinfalzer *m* ~**-size phase shifter (or advancer)** Kleinphasenschieber *m* ~**-size power-factor-correcting capacitor** Kleinphasenschieberkondensator ~**-size tapper** Kleingewindeschneider *m*
**small-sized** kleinstückig ~ **cable** Kabel *n* mit geringer Adernzahl ~ **coke** Klein-, Knabbel-, Perl-koks *m* ~ **jarring machine** Kleinrüttler *m* ~ **scrap** Kleinschlag *m*
**small,** ~ **sluice** Schütz *n* ~ **smoothing iron** Schlichtspan *m* ~ **steel articles** Kleinstahlwaren *pl* ~ **stone** Krebs *m* im Ton ~ **stone filling** Steinschlag *m* ~ **support** Böckchen *n* ~ **tongs** Zänglein *n* ~ **tube boiler** engröhriger Kessel *m* ~ **unit** Kleinanlage *f* ~ **valve** Schütz *n* ~ **valve hammer** Hahn *m* ~ **wire tacks** Drahtzwicken *pl* ~ **wood** Hölzchen *n* ~ **work** Feingewinde *n*

**smaller** minder ~ **amount** Minderzahl *f* ~ **center** Sammelamt *n* ~ **end** schwächeres Ende *n* (der gelochten Hülse) ~ **number** Minderzahl *f*

**smallest** mindest ~ **breadth of the top of an embankment** Mindestkronenbreite *f* ~ **interval** Feinintervall *n* ~ **particle** Grundteilchen *n*

**smallness** Kleinheit *f*, Unbedeutendheit *f*

**smalt** Blaufarbenglas *n*, Schmelzblau *n*

**smaltite** Glanzkobalt *m*, Graupenkobalt *m*, Smaltin *m*, Speisekobalt *m*, Stahlkobalt *m*

**smart, to** ~ brennen

**smart** flink, geschäftskundig, tüchtig ~ **blow with a hammer** leichter Hammerschlag *m*

**smash, to** ~ plattdrücken, quetschen, zerbrechen, zerschlagen, zerschmeißen, zerschmettern, zertrümmern

**smash** Zermalmung *f*

**smashing** Zertrümmerung *f*

**smear, to** ~ bestreichen (mit), einreiben, einschmieren, schmieren, verschmieren, verwischen

**smear** Aufstrich *m*

**smearing** Verschmieren *n*, Verschmierung *f* ~ **effect** Nachzieheffekt *m* (TV)

**smell, to** ~ riechen

**smell** Geruch *m* ~ **of gas** Gasgeruch *m* ~**-tight** geruchdicht

**smelling,** ~ **substance** Geruchstoff *m* ~ **test** Riechprobe *f* ~**-test box** Riechprobe *f*, Riechprobenkasten *m*

**smelt, to** ~ abschmelzen, (blast furnace) blasen, einschmelzen, erschmelzen, schmelzen, verhütten, verschmelzen **to** ~ **down** niederschmelzen

**smelt** Schmelz *m*, Schmelze *f*

**smeltable** schmelzwürdig, verhüttbar, verhüttungsfähig

**smelted** geschmolzen

**smelter** Schmelzarbeiter *m*, Schmelzer *m* ~ **coke** Schmelzkoks *m* ~**-slag pumice** Hüttenbims *m*

**smeltery** Schmelze *f*, Schmelzwerk *n*

**smelting** Abschmelzen, Einschmelzen *n*, Erschmelzung *f*, Schmelze *f*, Schmelzen *n*, Verhütten *n*, Verhüttung *f*, Verschmelzung *f* ~ **of precious metals** Reichschmelzen *n* ~ **of slimes** Schlichtarbeit *f*

**smelting,** ~ **capacity** Schmelzleistung *f* ~ **chamber** Herd *m*, Schmelzraum *m* ~ **finery** Frischhütte *f* ~**-flux electrolysis** schmelzflüssige Elektrolyse *f* ~ **furnace** (melting) Anschmelzherd *m*, Flußofen *m*, Schmelzofen *m* ~ **hearth** Schmelzherd *m* ~ **operation** Schmelzarbeit *f* ~ **ore** Schmelzerz *n* ~ **plant** Hütte *f*, Hüttenwerk *n*, Schmelz-anlage *f*, -hütte *f* ~ **practice** Schmelzführung *f* ~ **process** Schmelz-arbeit *f*, -prozeß *m* ~ **room** Hüttenraum *m* ~ **schedule** Schmelzführung *f* ~ **unit** Umschmelzapparat *m* ~ **work** Hüttenarbeit *f* ~ **works** Schmelzhütte *f* ~**-works requirements** Hüttenbedarf *m* ~ **zone** Schmelzzone *f*

**smith's,** ~ **anvil** Schmiedeamboß *m* ~ **assistant** Zuschläger *m* ~ **coke** Schmiedekoks *m* ~ **fire** Schmiedefeuer *n* ~ **hammer** Schmiedehammer *m* ~ **hearth** Schmiedeherd *m*

**smithing** Schmieden *n* ~ **chisel** Setzeisen *n* ~ **operation** Schmiedevorgang *m* **of** ~ **quality** schmiedbar

**smithsonite** Galmei *m*, Smithsonit *m*, Zinkspat *m*

**smithy** Schmiede *f*, Schmiedewerkstatt *f* ~ **coal** Schmiedekohle *f*

**smoke, to** ~ anschmauchen, dämpfen, rauchen, räuchern **to** ~**-dry** räuchern **to** ~ **the mold** die Gußform *f* anblaken **to** ~ **out** ausdämpfen, ausräuchern **to** ~ **upon** aufdämpfen

**smoke** Dampf *m*, Nebel *m*, (dense) Qualm *m*, Rauch *m*, Smauch *m* ~ **agent** Nebelmittel *n* ~ **ammunition** Nebelmunition *f* ~ **apparatus** Rauchgerät *n* ~ **band** Rauchstreifen *m* ~ **blanket** Nebeldecke *f* ~ **bomb** Nebelbombe *f*, Rauch-ball *m*, -bombe *f* ~ **box** Nebelkasten *m*, Rauch-behälter *m*, -kammer *f*, -kasten *m* ~ **box door** Rauchkammertür *f* ~ **candle** Nebelkerze *f* ~ **canister ejected from projectile on burst** Ausstoßbüchse *f* ~ **cap** Rauchglocke *f* ~ **cartridge** Nebelpatrone *f* ~ **chamber** Flugstaubkammer *f* ~ **cloud** Nebelwolke *f* ~**-colored** rauchfarben ~ **column** Rauchsäule *f* ~ **consumer** Rauchverzehrer *m* ~**-container firing device** Nebelzündmittel *n* ~ **coverage** Nebelausdehnung *f* ~**-cured** rauchgar ~ **curtain** Rauchvorhang *m* ~ **cylinder** Nebelzerstäuber *m* ~ **detector** Rauchgasanzeiger *m* ~ **distress signal** Rauchnotzeichen *n* ~ **drum** Dunst-, Nebel-trommel *f* ~ **dust** Flugstaub *m* ~ **effect** Raucherscheinung *f* ~ **element** Rauch-körper *m*, -ladung *f* ~**-emitting projectile** Rauchspurgeschoß *n* ~ **equipment** Nebel-gerät *n*, -mittel *n* ~ **filament** Rauchfaden *m* ~**-filled chemical cylinder** Nebelabblasgerät *n* ~ **fire** Nebelschießen *n* ~ **float** Rauchdose *f* (mil.), Rauchschwimmer *m* ~ **flue** Rauch-kanal *m*, -rohr *n*, -zug *m* ~**-flue damper** Rauchschieber *m* ~ **gas** Nebelgas *n* ~**-generating message canister** Rauchmeldepatrone *f* ~ **generator** Nebel-apparat *m*, -büchse *f*, -kasten *m*, -topf *m*, -trommel *f*, Raucherzeuger *m* ~ **hand grenade** Nebel(hand)granate *f* ~ **hood** Fuchsanschlußstutzen *m* ~ **hood elbow** Fuchsanschlußkrümmer *m* ~ **house** Schwitzkammer *f* ~**-laying apparatus** Nebel-apparat *m*, -gerät *n* ~**-laying plane** Nebelflugzeug *n* ~ **limit** Rauchgrenze *f* ~ **meter** Rauchdichtemesser *m* ~ **mixture** Rauchsatz *m* ~ **munitions** Nebelmunition *pl* ~ **nozzle** Rauchdüse *f* ~ **nozzle elbow** Fuchsanschlußkrümmer *m* ~ **nucleus** Rauchkern *m* ~ **nuisance** Rauchplage *f* ~ **pot** Nebeltopf *m*, römische Kerze *f* ~ **preventing spraying gun** nebelfreie Farbspritzpistole *f* ~**-producing agent** Nebel-mittel *n*, -stoff *m* ~**-producing unit** Nebelabteilung *f* ~ **projectile** Nebelgeschoß *n*, Rauchgeschoß *n* ~**-puff charge** Kanonenschlag *m* mit Raucherscheinung ~ **screen** künstlicher Nebel, Nebel-decke *f*, -schleier *m*, -wand *f*, Rauch-schleier *m*, -wand *f* ~ **screening** Nebelabblasen *n*, Vernebelung *f* ~**-screening fire** Blendungsschießen *n* ~ **shell** Nebel-, Rauch-geschoß *n*, Rauchmine *f* ~**-shell firing** Nebelschießen *n* ~**-shell mortar** Nebelwerfer *m* ~ **signal** Rauch-signal *n*, -zeichen *n* ~**-signal cartridge** Rauch(melde)patrone *f* ~ **slide valve** Rauch(gas)schieber *m* ~ **spray** Nebelzerstäuber *m* ~**stack** Esse *f*, Kamin *m*, Rauchfangkamin *m*, Schlot *m* ~ **streamer** Rauchfahne *f* ~**-tight** rauchdicht ~ **tracer** Rauchkugel *f* ~ **trail** Rauchfahne *f*, Rauchstreifen *m* ~ **tube** Rauchrohr *n* ~**-writing plane** Rauchschreibflugzeug *n* ~**-writing substance** Rauchschriftmasse *f* ~ **zone** Nebelzone *f*

**smoked,** ~ **glass** Rauchglas *n* ~ **sheet rubber** geräuchertes Gummi *n* in Blättern

**smokeless** rauchfrei, rauchlos, (of powder) rauchschwach ~ **gunpowder** rauchloses Schießpulver *n* ~ **powder** rauchloses oder rauchschwaches Pulver *n*

**smoking** Anräucherung *f*, Verqualmung *f* ~ **compartment** Raucherabteil *n* ~ **fire** Schmauchfeuer *n*

**smoky** rauchig ~ **quartz** Rauchquarz *m* ~ **topaz** Rauchtopas *n*

**smolder, to** ~ (to continue) fortglimmen, schwelen

**smoldering** Schwelen *n*, Schwelung *f* ~ **zone** Schwelzone *f*

**smooth, to** ~ abschleifen, abziehen, ausglätten, ausstreichen, (a way or path) bahnen, beruhigen, blank schleifen, bügeln, ebnen, einebnen, glänzen, glätten, hobeln, nivellieren, planieren, plätten, polieren, schlichten **to** ~ **the cloth** das Tuch *n* ausrecken **to** ~ **down** abflachen, verflachen **to** ~ **the edge of a board** ein Brett *n* fügen **to** ~ **off** abglätten **to** ~ **out** ausstreichen **to** ~ **over** bemänteln

**smooth** blank, glatt, gleich-förmig, -mäßig, homogen, sacht, sanft, schlicht, stoßfrei, weich ~ **air-flow wake (or wash)** glatter Abfluß *m* ~ **approximation** geglättete Näherung *f* ~ **bore** glatte Seele *f*; glattkalibrig ~**-bore barrel** Schrotlauf *m* ~**-bore gun** glattes Geschütz *n* ~ **change of direction without shocks (or change of driving direction)** stoßfreier Fahrtrichtungswechsel *m* ~**-core rotor** Vollpolläufer *m* ~ **cut** Schlicht-hieb *m*, -span *m* ~**-cut file** Schlichtfeile *f* ~ **file** Abziehfeile *f*, Glättfeile *f* ~ **filling** Schlichtfeilen *n* ~ **finish** Glättung *f* ~ **finishing** Glätten *n* ~ **fracture** feinkörnige Bruchfläche *f* ~ **grinding** Blankschleifen *n* ~ **ice** Bareis *n*, Glatteis *n* ~ **idling** ruhiger Leerlauf *m* ~ **leather** abgenarbtes Leder *n* ~ **line** gleichförmige, gleichmäßige oder homogene Leitung *f* ~ **operation** Betriebsannehmlichkeiten *pl* ~ **outline of a body moving in a fluid** Strak *f* ~ **plane** Schlichthebel *m* ~ **planing machine** Abrichthobelmaschine *f* ~ **regulation** gleichförmige Regelung *f* ~ **roller** Ackerwalze *f* ~ **rollers** Glattwalzen *pl* ~ **running** leichter Gang *m*, ruhiger oder stoßfreier Lauf *m* ~ **running of an engine** ruhiger Lauf *m* des Motors ~ **sail** glatter Flügel *m* ~ **sea** Glattwasser *n*, ruhige See *f* ~ **skin** glatte Außenhaut *f* ~ **slide** glatte Gleitbahn *f* ~ **speed regulation** weiche Drehzahlregelung *f* ~ **start of oscillations** weicher Schwingungseinsatz *m* ~ **starting** stoßfreier Anlauf *m* ~ **surface** glatte Oberfläche *f* ~ **surface finish** Oberflächenglätte *f* ~ **test** glatter Probestab *m* ~ **torque** Drehmoment *m* von geringem Ungleichförmigkeitsgrad ~ **working** spielfreies Zusammenwirken *n*

**smoothed** geglättet **to be** ~ flach werden ~ **current** ausgeglichener Strom *m*

**smoother** Spachtel *m*

**smoothing** Abflachung *f*, Glätten *n*, Glatthobeln *n*, (ripples of current, potential etc) Glättung *f*, Schlichtdrehen *n* ~ **of chilled casting** Kokillenschlichte *f*

**smoothing,** ~ **blade** Ausstreichmesser *n* ~ **broach** Glättahle *f* ~ **capacitor** Beruhigungskapazität *f*, Glättungskondensator *m* ~ **choke** Abflachungs-, Glättungs-drossel *f*, Wellensauger *m* ~ **circuit** Beruhigungs-, Reinigungs-kreis *m* ~ **coil** Abflachungsdrossel *f*, Saugspule *f* ~ **compound** Glattmasse *f* ~ **condenser** Abflachungs-, Ausgleichs-, Glättungs-kondensator *m* ~ **device** Spannungsglättung *f* ~ **drift** Schlichtdorn *m* ~ **effect** Glättwirkung *f* ~ **gills** (textiles) Flachshechelmaschine *f*, Glatthechel *m* ~ **iron** Plätteisen *n* ~ **mandrel** Glattdorn *m* ~ **means** Stromreiniger *m* ~ **network** Beruhigungsglied *n* ~**-out** Ausrundung *f* ~ **piece** Planierstück *n* ~ **pipe** Beruhigungsrohr *n* ~ **plane** Glätt-, Schlicht-hobel *m* ~ **press** Glättpresse *f* ~ **reactor** Glättungsdrossel *f* ~ **resistor** Glättungswiderstand *m* ~ **roll** Plättwalze *f* ~**-roller** Glätterwalze *f* ~ **rolls** (paper) Feuchtglätte *f*, Glättwalzwerk *n* ~ **tool** Schlichtstahl *m* ~ **tool for cast iron** Schlichtstahl *m* für Gußeisen ~ **tool for wrought iron (or steel)** Schlichtstahl *m* für Schmiedeeisen oder Stahl ~ **trowel** Spachtel *f* & *m*

**smoothly** stoßfrei ~ **variable** gleitend regelbar

**smoothness** Ebenheit *f*, Geschmeidigkeit *f*, Glätte *f*, Schliff *m* ~ **of action** ruhiger Gang *m*, ruhiger Lauf *m* einer Maschine

**smoothness,** ~ **meter** Glattheitsprüfer *m*, Rauheitsprüfer *m* ~ **tester** Glattheitsprüfer *m*

**smother, to** ~ dämpfen, ersticken, unterdrücken

**smothering** warmes Lagern *n*, Schwelen *n*

**smudge,** ~ **candle** Schwelkerze *f* ~ **fire** Rauchofen *m* ~ **pot** Kokskorb *m* (zum Austrocknen des Baus)

**smudging** Verschmieren *n*

**smudgy** schmierig, unsauber ~ **impression** unsauberer Abdruck *m*

**smut, to** ~ unsauber abziehen

**smut** Faulbrand *m*, Ruß-fleck *m*, -flocke *f*, (of grain crops) Steinbrand *m*

**smutter** Aspirationsreinigungsmaschine *f*

**smutty** schmutzig

**snaffle** Trense *f* ~ **bit** Trensengebiß *n* ~ **rein** Trensenzügel *m*

**snag, to** ~ ausputzen

**snag** Knorren *m* ~ **in wood** Knast *m* im Holz

**snaggy** knästig

**snail** Schnecke *f* ~ **creep** Mauerwerk *n* in Polygonverband ~ **form** Schneckenform *f* ~**-formed** schneckenförmig ~**-formed vaulting** Schneckengewölbe *n* ~**'s pace** Schneckentempo *n*

**snake** Schlange *f* ~**-like response of permanent echo** Schlängel *m* (rdr) ~ **track** Schlangenkette *f*

**snaked** (wire) verdrillt ~ **wire** Drahtsicherung *f*, verdrillter oder zusammengewirbelter Draht *m*

**snaker** Biege-, Verkröpf-gesenk *n*

**snap, to** ~ abbrechen, abreißen (Faden), einrasten, (in, into) einschnappen, schnappen, zersprengen **to** ~ **forward** vorprellen **to** ~ **in(to)** (bolt) einfallen **to** ~ **off** abbiegen, abspringen **to** ~ **out** ausschnappen

**snap** Nietpfaffe *m*, Schnepper *m*, Verschluß *m* ~ **action** Ausgangsstromsprung *m* ~**-action amplifier** Kippverstärker *m* ~**-action contact** Schnappkontakt *m* ~**-action controller** Kippregler *m* ~**-action switch** Schnapp-, Springschalter *m* ~ **angle** Schnappwinkel *m* ~-**button fastener** Druckknopfverschluß *m* ~ **catch** Klappschloß *m* ~ **check** Überwachung *f* durch Stichproben ~ **closure** Schnellverschluß *m* ~ **connection** Schnelltrennstelle *f* ~ **cup** Nietsetzer *m* ~ die Kopfsetzer *m* ~ **dies** Schellhammer *m* ~ **fastener** Druckknopf *m* (Kleidung) ~**flask** abschlagbarer Formkasten *m*, Abschlagkasten *m*, Abziehformkasten *m*, aufklappbarer Formkasten *m*, Klappformkasten *m* ~-**flask unit** Abschlagkasten *m* ~ **gauge** Außengrenzlehre *f*, Grenzlehre *f*, Rachenlehre *f*, Tasterlehre *f* ~ **hammer** Schnellhammer *m* ~ **head** Schnell-, Schließ-knopf *m* ~-**head** die Aufsatzhammer *m*, Schließkopfgesenk *n* ~ **head rivet** Halbrundniet *m* ~ **headed rivet** Schnellkopfniet *m* ~ **hook** Hakensprengring *m*, Karabinerhaken *m* ~ **hook and eye** Karabinerhaken *m* ~ **lock** Klappschloß *n*, Klappverschluß *m* ~ **member** Schnappstück *n* ~ **nut** Schloßmutter *f* ~ **pin** Rastenbolzen *m*, Schnappstift *m* ~ **piston ring** Überstreifring *m* ~ **ring** Schnappring *m*, Seegerring *m*, Seegersicherung *f*, Sprengring *m* ~ **riveting** Schnellkopfnietung *f* ~ **roll** schnelle Rolle *f* ~ **set** Nietstempel *m* ~ **shooter** Momentkamera *f* ~ **shot** Augenblicks-aufnahme *f*, -bild *n*, Moment-aufnahme *f*, -bild *n*, Schnappschuß *m* ~ **switch** Feder-, Nocken-, Schnappschalter *m* ~ **tool** Kopfsetzer *m*, Schelleisen *n* ~ **valve** Quetschhahn *m*

**snapped flint work** roh bearbeitetes Geröllmauerwerk *n*

**snapping-in** Einfallen *n*

**snappy vulcanisates** nervige Vulkanisate *pl*

**snare** Fallbrücke *f*, Falle *f* ~ **drum** kleine Trommel *f*

**snarl, to** ~ knäueln, knurren

**snatch, to** ~ entreißen, schnappen **to** ~ **from** abfangen

**snatch,** ~ **block** Einlegeblock *m*, Führungsrolle *f* (zum Regeln des Drahtdurchhanges), Fußblock *m*, kleine Seilscheibe *f*, Kloben *m* ~**brake** Schnappbremse *f* ~ **operation of the throttle** kurzzeitiges Aufreißen *n* der Drosselklappe ~ **post** Spannrast *f* ~-**post pulley** Führungsrolle *f*

**sneak, to** ~ kriechen, schleichen **to** ~ **away** fortschleichen **to** ~ **upon** heranschleichen

**sneek,** ~ **current** Fremdstrom *m*, Kriechstrom *m*

**sneeze, to** ~ niesen

**sneeze gas** Niesgas *n*

**sniff, to** ~ **at** beriechen

**sniffer** Schnüffler *m* ~ **tube** Schnüfflerröhre *f*

**sniffing valve** Belüftungs-, Schnarch-, Schnüffelventil *n*

**snip, to** ~ schnippeln

**snip** Schnipfel *m*

**snips** Metallschere *f*, Schere *f*

**snore piece of a sinking pump** Senkkorb *m*

**snout** Rüssel *m*, Schnauze *f*

**snow, to** ~ schneien **to** ~ **in (or under)** einschneien

**snow** Schnee *m* ~-**blind** schneeblind ~ **brick** Schneeziegel *m* ~ **chain** Schneekette *f* ~-**covered** verschneit ~ **crystal** Schnee-kristall *m*, -stern *m* ~ **drift** Schnee-schanze *f*, -treiben *n*, -verwehung *f*, -wehe *f* ~ **fall** Schneefall *m* ~ **fence** Ablagerungszaun *m*, Schnee(fang)gitter *f* ~ **field** Firnfeld *n*, Gletscher *m* ~ **flake** Schneeflocke *f* ~ **gauge** Schneemesser *m* ~ **goggles** Schnee-

brille *f* ~ **guard** Schneegitter *n* ~ **landing** Schneelandung *f* ~ **line** Schneegrenze *f* ~ **load** Schnee-belastung *f*, -last *f* ~ **plow** Schneefegewagen *m*, -pflug *m* ~ **propeller** Schneefräse *f* ~ **removal equipment** Schneeräumgerät *n* ~ **roller** Schneeroller *m* ~ **runner** Schneekufe *f* ~ **scraper** Schneeräumer *m* ~ **shoe** Schneereifen *m* ~ **shovel** Schneeschaufel *f* ~ **static** Störung *f* durch Schneepartikeln ~ **storm** Schneesturm *m* ~ **trench** Schneeschanze *f* ~ **water** Schneewasser *n* ~-**white** schneeweiß ~ **(zine) white** Schneeweiß *n* (chem.)

**snowed under** verschneit

**snowy** schneeig ~ **gypsum** Schneegips *m* ~ **weather** Schneewetter *n*

**snub, to** ~ (pressure) ausbauen

**snub pulley** Einschnürrolle *f* (Förderband)

**snubber,** ~ **strut** Anschlagpuffer *m*, Dämpferstrebe *f* (Hubschr.) ~ **valve** Dämpfungsventil *n*

**snuff** (Kerzen) putzen

**snug** Nase *f*; bündig ~ **adhesion** enge Anschmiegung *f* ~ **bolt** Nasenschraube *f* ~ **fit** dichtes Aufeinanderpassen *n*, Festsitz *m*, Paßsitz *m*, saugend passender Gleitsitz *m*

**soak, to** ~ abweichen, aufsaugen, aufweichen, beuchen, dauerglühen, durch-feuchten, -glühen, -nässen, -tränken, -wässern, (brewing) einquellen, einsumpfen, eintauchen, einweichen, einziehen lassen, längere Zeit *f* einer Glühtemperatur aussetzen, nässen, netzen, normalisieren, quellen, tauchen, tränken, weichen, ziehen **to** ~ **in** (into) eindringen, einsickern **to** ~ **in lime water** einkalken **to** ~ **off** losweichen **to** ~ **up** einsaugen

**soak-quench technique** Technik *f* des langsamen Abkühlens und nachfolgenden Abschreckens

**soakage** Feuchtigkeitsaufnahme *f* ~ **pit** Senkloch *n*, Sickergrube *f*

**soakaway** Sickergrube *f*

**soaked** durchnäßt, getränkt ~ **with oil** öldurchtränkt

**soaked,** ~-**in boric acid** borsäuregetränkt ~-**in salt solution** salzgetränkt ~ **weight** Weichgewicht *n*

**soaker** Tiefofen *m* ~ **drum** Reaktionskammer *f* ~ **tube** Reaktionsröhre *f*

**soaking** Durch-nässung *f*, -wärmung *f*, -weichung *f*, Quellung *f*, Tränkung *f* ~ **apparatus** Tränkapparat *m* ~ **furnace** Wärmeofen *m* ~ **hearth** Schweißherd *m* ~-**in period** Einsickerungszeit *f* ~ **liquor** Weichwasser *n* ~ **pit** Ausgleichgrube *f*, Durchweichungsgrube *f*, Tiefofen *m*, Wärmeausgleichgrube *f* ~ **pit crane** Tiefofenkran *m* ~-**pit furnace** Herdtief-, Zellentief-ofen *m* ~ **pitman** Tiefofenmann *m* ~ **time** Ausgleichzeit *f* ~ **tub** Quellbottich *m*, Weichkufe *f* ~ **wheel** Weichhaspel *f* ~ **zone** Heißtemperatur-, Heiz-, Schweiß-, Vollhitze-zone *f*

**soap, to** ~ abseifen, seifen

**soap** Seife *f* ~ **not salted out** Leimseife *f*

**soap,** ~**base** Grünseife *f* ~ **bubble** Seifenblase *f* ~-**bubble method** Seifenblasenmethode *f* ~ **crushing machine** Seifenquetschmaschine *f* ~ **distributor** Seifenspender *m* ~ **flakes** Seifenflocken *pl* ~ **manufacturer** Seifensieder *m* ~ **mill** Piliermaschine *f* ~ **nigre** Leim(sieder)-niederschlag *n* ~ **paste** Seifenleim *m* ~ **powder** Seifenpulver *n* ~-**refining outfit** Pilieranlage *f*

~-shaving **planing machine** Seifenspänehobelmaschine *f* ~-**slab cooling plant** Seifenplattenkühlanlage *f* ~-**solution maker** Laugenbereiter *m* ~ **stock** Federweiß *n* ~ **stone** Piotin *m*, Saponit *m*, Schmerstein *m*, Schneiderkreide *f*, Seifenansatz *m*, Seifenstein *m*, Talkstein *m*, Topfstein *m* ~ **stone packing** Talksteinpackung *f* ~ **suds** Seifen-lauge *f*, -wasser *n* ~ **tincture** Seifenspiritus *m* ~ **waste** Abfallseife *f* ~ **wort emulsion** Seifenwurzelemulsion *f*

**soapy** seifig ~ **water** Seifenwasser *n*

**soar, to** ~ sich in die Luft *f* erheben, schweben, segeln (aviat.), mit einem Segelflugzeug *n* fliegen

**soaring** ~ **demonstration** Segelflugvorführung *f* ~ **flight** Segelflug *m* ~ **glider** Schwingenflieger *m* ~ **over slopes** Hangsegelflug *m* ~ **performance** Schwebeleistung *f* ~ **quality** Schwebe-fähigkeit *f*, -vermögen *n* ~ **site** Segelfluggelände *n*

**social insurance (or security)** Sozialversicherung *f*

**society** Gesellschaft, Verein *m*

**sock** Socke *f*, Windrüssel *m* ~ **pole** Windrüsselstange *f* ~ **stitch** versetzte Naht *f*

**socket** Angelring *m*, (connector) Anschlußbüchse *f*, Ansteckdose *f*, Büchse *f* (electr.), Dose *f*, Dübel *m*, Dülle *f*, (candlestick type) Einstecksockel *m*, Fassung *f*, Flansch *m*, Glocke *f*, Hülse *f*, Klinke *f*, Konushülse *f*, Lager *n*, Lagerstuhl *m*, Lampemfassung *f*, Muffe *f*, Rohransatz *m*, Röhrenfassung *f*, Rohrmuffe *f*, Rohrstutzen *m*, Sockel *m*, Spurzapfenlager *n*, (plug) Steckdose *f*, (plug-in) Steckfassung *f*, Steckhülse *f*, Stutzen *m*, Tülle *f*, Türangelpfanne *f*, Zapfenlager *n*

**socket,** ~ **of the boxes** Muldentasche *f* ~ **for brake circuit** Bremsanschlußdose *f* ~ **for compensation** Ausgleichssteckdose *f* ~ **of the compass needle** Kompaßhütchen *n* ~ **for control current** Steuerstromsteckdose *f* ~ **on dome** Domstutzen *m* ~ **with earth contact** Schukosteckdose *f* ~ **of eye** Augenhöhle *f* ~ **of jack** Klinkenhülse *f* ~ **of a lamp** Klinkenfuß *m* ~ **(end) of a pipe** Hals *m* einer Röhre ~ **of a tube (or valve)** Röhrenfuß *m* with ~ **removed** entsockelt ~ **with shrouded contacts** Kragensteckdose *f* ~ **with valve** Dornfänger *m* ~ **with valve to fish for rope** Klappenseilfänger *m*

**socket,** ~ **adapter** Zwischenstecker *m* (für Röhren) ~ **body** (cast iron) Gußschale *f* eines Steckers ~ **cement** Sockelkitt *m* ~ **chisel** Düllbeitel *m*, Rohrstechbeitel *m* ~ **chuck** Spundfutter *n* ~ **connection** Sockelschaltung *f*, Steckhülsenverbindung *f* ~ **contact** Sitz(um)schalter *m* ~ **control valve** Sockelsteuergerät *n* ~ **coupling** Hülsenverbindung *f* ~ **end** Hülsenende *n* ~ **end of a tube** Muffenende *n* eines Rohres ~ **fitting** Muffenverbindung *f* ~ **head** Kopf *m* ~-**head bolt** Zylinderschraube *f* ~-**head cap screw** Zylinderkopfschraube *f* ~ **head screw** Innensechskant *n*, Kopfschraube *f*, Zylinderschraube *f* ~ **holder** Glockenhalter *m* ~ **jack** Bauwinde *f* ~ **joint** Muffe *f*, Muffenverbindung *f* ~ **line** Büchsenleitung *f* ~ **panel** Buchsenplatte *f* ~ **plate for rubber pad** Federpufferhalter *m* ~-**power unit** Netzanschlußgerät *n*, Netzteil *m* ~ **plug** Hülsenstopfer *m* ~ **pressure pipe** Muffendruckrohr *n* ~ **ram** Glockenramme *f* ~

**screw** Innensechskantschraube *f* ~ **screw wrench** Zweilochmutterschlüssel *m* ~ **stem** Glockenstange *f* ~ **strip** Büchsengruppe *f* (electr.) ~-**switch connection** Steckdosenverbindung *f* ~ **tube** Muffenrohr *n* ~-**turning and -screwing machine** Muffendreh- und Gewindeschneidebank *f* ~ **wrench** Aufsatz-, Aufsteck-, Rohr-, Stangensteck-, Steck-schlüssel *m* ~ **wrench with bar for screw slot** Schlitzschraubenschlüssel *m*

**socketed,** ~ **grip** (movable in socket) Gelenkklemme *f* ~ **pipe** Muffenrohr *n* ~ **pressure pipe** Muffendruckrohr *n* ~ **terminal** (movable in socket) Gelenkklemme *f*

**sockets and plugs** Steckvorrichtungen *pl*

**socketing and flanging machine** Aufweite- und Börder-maschine *f*

**socking brake** Feststellbremse *f*

**socle** Fußgestell *n* ~ **base** Fußstück *n* ~ **wainscoting** Fußsockel *m*

**sod** Grasscholle *f*, Rasen *m*, Sode *f*, Soden *m* ~ **knife** Plaggenpflug *m*, Sodenpflug *m* ~ **squares** (for paving) Rasenziegel *m* ~ **work** Sodendecke *f*

**soda** kohlensaures Natrium, Natron *n*, Soda *f* ~ **alum** Natronalaun *m* ~ **ash** kalzinierte Soda *f* ~ **cellulose works** Natronzellstoffabrik *f* ~ **craft paper** Natronkraftpapier *n* ~ **crystals** Kristallsoda *n* ~ **lime** Natronkalk *m* ~ **liquor** Natronlauge *f* ~ **lye** Natronlauge *f*, Sodalauge *f* ~ **salt** Sodasalz *n* ~ **solution** Sodalösung *f* ~ ~ **water** Brausewasser *n*

**sodded slope** Rasendecke *f*

**soddite** Soddit *m*

**sodium** Natrium *n*, Sodium *n* ~ **acetate** essigsaures Natron *n*, Sodaazetat *n* ~ **alum** Natriumalaun *m* ~ **aluminate** Tonerdenatron *n* ~ **ammonium phosphate** Phosphorsalz *n* ~ **antimoniate** antimonsaures Natrium ~ **arsenate** arsensau(e)res Natron, Natriumarseniat *n* ~ **arsenite** arsenigsau(e)res Natrium *n* ~ **aurochloride** Natriumgoldchlorid *n* ~ **aurous chloride** Goldnatriumchlorür *n* ~ **benzene sulfonate** Natriumbenzolsulfonat *n* ~ **benzoate** benzoesaures Natrium *n* ~ **bicarbonate** doppel(t)-kohlensaures Natron *n* ~ **bisulfate** Natriumbisulfat *n* ~ **bisulfide liquor** Natriumbisulfidlauge *f* ~ **bisulfite solution** Natriumbisulfitlösung *f*, Sulfitlauge *f* ~ **borate** Natriumtetraborat *n* ~ **bromate** bromsaures Natron *n* ~ **bromide** Bromnatrium *n* ~ **carbonate** kohlensaures Natrium *n*, kohlensau(e)res Natron, Natriumkarbonat *n*, Soda *f*, Sodasalz *n* ~ **chlorate** arsensau(e)res Natrium *n* ~ **chloride** Chlornatrium *n* ~ **chloroaurate** Goldsalz *n* ~ **chloroplatinate** Natriumplatinchlorid *n* ~ **chromite** Chromooxydnatron *n* ~ **citrate** zitronensaures Natrium *n* ~ **compound** Natriumverbindung *f* ~ **content** Natriumgehalt *m* ~ **cooled** natriumgekühlt, sodiumgekühlt ~ **cyanamide** Cyanamidnatrium *n* ~ **ferricyanide** Ferrizyannatrium *n*, Ferrozyannatrium *n* ~ **fluosilicate** Kieselfluornatrium *n*, Natriumkieselfluorid *n* ~ **formate** ameisensaures Natron *n*, Natriumformiat *n* ~ **hydride** Natriumwasserstoff *m* ~ **hydroxide** Ätznatron *n*, kaustische Soda *f*, Natrium-hydrat *n*, -hydroxyd *n*, -oxydhydrat *n*, Natron, Natronhydrat *n* ~ **hypo-**

**chlorite** Chlornatron *n*, unterchlorigsaures Natron *n* ~ **hypophosphite** unterphosphoriges Natron *n* ~ **hyposulfate** Natriumsubsulfat *n* ~ **hyposulfite** Antichlor *n*, Fixiernatron *n*, Natriumthiosulfat *n*, unterschwefligsaures Natron *n* ~ **iodide** Jodnatrium *n*, Jodnatron *n* ~ **lamp** Natriumleuchte *f*, selbstleuchtender Lichtmodulator *m* ~ **line** Natriumlinie *f* ~ **line reversal method** Temperaturbestimmung *f* nach dem Verfahren der Natriumlinien-Umkehr ~ ~ **magnate** mangansaures Natrium *n* ~ **nitrate** Chilisalpeter *m*, Natronsalpeter *n*, salpetersaures Natron ~ **nitride** Stickstoffnatrium *n* ~ **oxalate** Natriumoxalat *n*, oxalsaures Natrium *n* ~ **perborate** überborsaures Natron *n* ~ **perchlorate** Natriumperchlorat *n*, übersaures Natrium *n* ~ **periodate** Natriumhyperjodat *n* ~ **permanganate** Z-Stoff *m* ~ **peroxide** Natriumhyperoxyd *n*, -superoxyd *n* ~ **persulfate** überschwefelsaures Natron *n* ~ **phenolate** Phenolnatrium *n* ~ **phosphate** phosphorsaures Natron *n* oder Natrium *n* ~ **phosphide** Phosphornatrium *n* ~ **potassium carbonate** Natriumkaliumkarbonat *n* ~ **potassium tartrate** Natriumkaliumtartrat *n*, Rochellesalz *n* ~ **pyrophosphate** pyrophosphorsaures Natron *n* ~ **salicylate** Natriumsalizylat *n*, salizylsaures Natrium *n* ~ **salt** Natriumsalz *n* ~ **selenite** selenigsaures Natrium ~ **silicate** kieselsaures Natron *n*, Natronwasserglas *n* ~ **stannate** Zinnoxydnatron *n*, zinnsaures Natrium *n*, Zinnsoda *f* ~ **stearate** stearinsaures Natrium *n* ~ **sulfate** Glaubersalz *n* ~ **sulfide** Schwefelnatrium *n* ~ **sulfite** Natriumsulfit *n*, schwefligsaures Natron *n* ~ **sulfocyanate** Natriumsulfozyanid *n*, Rhodanatrium *n* ~ **superoxide** Natriumsuperoxyd *n* ~ **tartrate** weinsau(e)res Natron *n* ~ **tetraborate** Natriumtetraborat *n* ~ **thioantimoniate** Natriumthioantimoniat *n* ~ **thiocyanate** Natriumsulfozyanid *n*, Rhodannatrium *n* ~ **thiosulfate** Fixiernatron *n*, Natriumthiosulfat *n* ~ **titanate** titansaures Natrium ~ **tungstate** Natriumwolframat *n*, wolframsaures Natron *n* ~ **uranate** Urangelb *n* ~ **valerate** Natriumvalerianat *n*, valeriansau(e)res Natrium *n* ~ **vapor** Natriumdampf *m* ~ **wire** Natriumdraht *m* ~ **woodpulp** Natronzellulose *f*

**soffit** innere Wölbfläche *f*, Laibung *f*, Leibung *f*, Sockel *m* ~ **of an arch** Bogenleibung *f*

**soffit,** ~ **board** Schalbrett *n* ~ **lamp** Leuchtröhre *f* ~ **scaffolding** Bogengerüst *n*

**soft, to** ~-**solder** weichlöten

**soft** geschmeidig, leise, locker, mild, mürbe, pastös, sacht, sanft, schlaff, schneidig, ungehärtet, weich ~ **as wax** wachsweich

**soft,** ~-**and fiber board** Weich- und Hartfaserplatte *f* ~-**and high-grade lead** Weich- und Feinblei *n* ~ **asphalt** Weichasphalt *m* ~ **blast** Wind *m* mit geringer Pressung ~-**centered steel** Weichkernstahl *m* ~ **clinker** zähflüssige Schlacke *f* ~ **coal** Braunkohle *f*, milde Kohle *f*, Steinkohle *f*, Weichkohle *f* ~ **component** weiche Komponente *f* ~ **copper** Weichkupfer *n* ~ **copper gasket** Weichkupferring *m* ~-**focus lens** weitwinklige Linse *f*, weichzeichnende Linse *f*, Weichzeichner *m* ~ **hail** Graupel *f* ~ **ice** morsches Eis *n* ~ **impression** weicher Druck *m*

(print.) ~ **India rubber** Weichkautschuk *m* ~ **iron** Weicheisen *n*, weiches Eisen *n*

**soft-iron,** ~ **ammeter** Weicheisenstrommesser *m* ~ **armature** Weicheisenanker *m* ~ **bar** Weicheisenstange *f* ~ **core** Weich-eisenkern *m*, -bleikern *m* ~ **mass** Weicheisenmasse *f* ~ **oscillograph** Dreheisenoszillograf *m* ~ **vane ammeter** Weichbleistrommesser *m*

**soft,** ~ **lead** Frisch-, Weich-blei *n* ~ **light** Langflammenlampe *f* ~ **machinable material** bearbeitbarer weicher Werkstoff *m* ~ **metal** Weichmetall *n* ~ **mud ooze** Moddergrund *m* ~-**nose bullet** Weichspitzengeschoß *n* ~ **packing** Weichpackung *f* ~ **part** Weichteil *m* ~ **pellet** Weichschrot *n* ~ **picture** weiches Bild *n* ~-**pointed bullet** Weichspitzengeschoß *n* ~ **porcelain** Fritt-, Weich-porzellan *n* ~ **quality of a picture** Bild-flachheit *f*, -flauheit *f* ~ **radiation** weiche Strahlung *f* ~ **rays** weiche Strahlen *pl* ~ **resin (or balsam)** Weichharz *n* ~ **rock** mildes Gebirge *n* ~ **roll** Weichwalze *f* ~ **rubber** Weichgummi *n* ~ **rubber gasket** Weichgummidichtung *f* ~ **rubber packing** Weichgummidichtung *f* ~ **rubber strip** Weichgummistreifen *m* ~ **running** leichter Gang *m* ~ **self-bedding metallic sealing** selbstdichtender Weichmetallsitz *m* ~ **shot** Weichschrot *n* ~-**sized paper** schwach geleimtes Papier *n* ~ **soap** Schmierseife *f* ~ **solder** Blei-, Schnell-, Weich-, Weißlot *n*, Zinnlöte *f* ~ **spot** Weichstelle *f* ~ **steel** Flußeisen *n*, -stahl *m*, niedriggekohlter Stahl *m*, weicher Stahl *m*, Weichstahl *m* ~-**steel disc** Flußstahlscheibe *f* ~-**steel scrap** Flußeisenschrott *m* ~-**steel sheet** Flußstahlblech *n* ~-**steel wire** Flußstahldraht *m* ~ **tire** flacher Reifen *m* ~ **valve** weiche Röhre *f* ~ **vein** Loßschicht *f* (geol.) ~ **water** weiches Wasser *n* ~ **water governor** Weichwasserregler *m* ~ **wood** weiches Holz *n*, Weichholz *n* ~ **woodblock paving** Weichholzpflaster *n* ~ **woods** Weichholzarten *pl*

**soften, to** ~ abschwächen, abtönen, anweichen, aufweichen, (steel) ausglühen, enthärten, erweichen, lindern, töten, vorraffinieren, weichen, weichmachen, weichwerden **to** ~ **up** mürbe machen

**softened base bullion** Werkblei *n*

**softener** Ferrosilizium *n*, Weichmacher *m*, Weichmachungsmittel *n*

**softening** Enthärtung *f*, Erweichung *f*, Vorraffination *f*, Weicherwerden *n*, Weichmachen *n* ~ **agents** Mürbemachemittel *pl* ~ **furnace** Vorraffinierofen *m* ~ **material** Milderungsmittel *n* ~ **point** Erweichungspunkt *m* ~-**up process** Abweichung *f*, Schmelzpunkt *m*

**softly shaded transition** verflauter Übergang *m* (TV)

**softness** Mürbigkeit *f*, Sanftheit *f*, Weichheit *f*

**soil, to** ~ abschmutzen (print.), beschmutzen, verunreinigen

**soil** Boden *m*, Erdboden *m*, Erde *f*, Grund *m*, Land *n* ~ **for forest growth** Waldboden *m* ~ **formed by glacial action** Gletscherboden *m* ~ **of medium consistency** schnittfähiger Boden *m*

**soil,** ~ **analysis** Bodenanalyse *f* ~ **bearing** Bodentragfähigkeit *f* ~ **bearing values** Bodenbelastungswerte *pl* ~ **burial test** Bodenverrottungsprobe *f* ~-**cement** Bodenverfestigung *f*

~-cement stabilization Bodenvermörtelung *f* ~ coefficient Bettungsziffer *f* ~ coefficients Bodenkennziffern *pl* ~ colloid Bodenkolloid *n* ~ compaction künstliche Verdichtung *f* von Erdboden ~ conditions Bodenverhältnisse *pl* ~ conductivity meter Bodenleitwertmeßgerät *n* ~ conservation Melioration *f* ~ consolidation Bodenverfestigung *f* ~ constants Bodenkennziffern *pl* ~ corrosion Bodenkorrosion *f* ~ creep Boden-bewegung *f*, -gekriech *n* ~-draining plant Bodenentwässerungsanlage *f* ~ elevation Bodenerhebung *f* ~ exhaustion Bodenerschöpfung *f* ~-failure investigation Grundbruchuntersuchung *f* ~-forming process bodenbildender Vorgang *m* ~ improvement Gütung *f* ~ mechanics Bodenmechanik *f*; bettbildend ~ modulus Bodenkennziffern *pl* ~ moisture Grundfeuchtigkeit *f* ~ movement Boden-fluß *m*, -gekriech *n* ~ nutrient Bodennährstoff *m* ~ pan Schmutzfänger *m* (für Kabelschächte) ~ particles Bodenteilchen *pl* ~ physics Bodenphysik *f* ~ pipe Abtrittschlot *m*, Erdrohr *n* ~ pressure Erddruck *m* ~ pulverizer Bodenfräse *f* ~ reaction Bodenreaktion *f* ~ requirement Bodenbeanspruchung *f* ~ research Bodenuntersuchung *f* ~ sample Bodenprobe *f*, Versuchskörper *m* ~ sampler Entnahmegerät *n* für Bodenproben ~ science Boden-kunde *f*, -lehre *f* ~ segment Bodensegment *n* ~ shifting operations Erdbewegungen *pl* ~ stabilization Bodenverfestigung *f* ~ stack Abfallrohr *n* ~ suction Bodensaugvermögen *n* ~ technology Bodentechnologie *f* ~ temperature Bodentemperatur *f* ~ texture Bodenstruktur *f* ~ trap Abzugsrohrverschluß *m* ~-ventilating device Bodenlüftungsgerät *n* ~ wedge Erdkeil *m*
**soiled** schmutzig
**soils** Erdstoffe *pl*
**sojourn, to** ~ verweilen
**sojourn** Aufenthalt *m*
**sol dispersion** Sollösung *f*
**solar,** ~ activity Sonnentätigkeit *f* ~ activity period Sonnentätigkeitsperiode *f* ~ battery Sonnenbatterie *f* ~ chromosphere Sonnenchromosphäre *f* ~ constant Solarkonstante *f* ~ corona Sonnenhof *m* ~ day Sonnentag *m* ~ elipse Sonnenfinsternis *f* ~ engine Sonnenmotor-Antrieb *m* ~ eruption Sonnenausbruch *m* ~ flare Sonneneruption *f* ~ halo Sonnenring *m* ~ heat latente Wärme *f*, Sonnenwärme *f* ~ observation Sonnenbeobachtung *f* ~ oil Solaröl *n* ~ radiation Sonnen(be)strahlung *f*, Wärmeabführung *f* ~ second Sonnenzeitsekunde *f* ~ tide Wasserziehen *n* der Sonne ~ time Sonnenzeit *f* ~-type reaction Kernreaktion *f* von der auf der Sonne vorkommenden Art
**solarimeter** Sonnen-strahlungsmesser *m*, -wärmemesser *m*
**solarization** Bildumkehrung *f*, Einstrahlung *f*, Insolation *f*, Solarisation *f*, Umkehrerscheinung *f*
**solarometer** Solarometer *n*
**solation** Gel-Sol-Transformation *f*
**sold** vergriffen, verkauft ~ for account of verkauft für Rechnung *f* von ~ by auction meistbietend verkauft ~ free on board verkauft franko Waggon *m* ~ for immediate delivery

verkauft für sofortige Lieferung *f* ~ ex warehouse verkauft ab Lager *n*
**sold,** ~ standing (on the stalk) auf dem Halm *m* verkauft ~ subject to inspection vorbehaltlich der Besichtigung *f* verkauft
**solder, to** ~ anlöten, anschmelzen, löten, verlöten, verschmelzen, weichlöten to ~ with bismuth wismuten to ~ end to end stumpflöten to ~ (in) einlöten to ~ on anlöten, auflöten to ~ tightly ablöten to ~ together zusammenleimen to ~ into a turned groove in eine Ausdrehung *f* einlöten to ~ up zusammenlöten
**solder** Lötmittel *n* ~ beams Lasche *f* der Sprießung ~ line Lot *n* ~-mounted crystal eingelöteter Kristall *m* ~ tag strip Lötösenstreifen *m* ~ union Lötstutzen *m*
**solderable** lötbar
**soldered** eingelötet, gelötet not ~ unverlötet
**soldered,** ~ connection Lötanschluß *m* ~ flange Auflötflansch *m* ~ joint Löt-fuge *f*, -naht *f*, -stelle *f*, -verbindung *f*, Schweißnaht *f*, Wickellötstelle *f* ~ junction Lötstelle *f*, Lötverbindung *f* ~ link Schmelzlotglied *n* ~ member Einlötstück *n* ~ pipe gelötetes Rohr *n* ~ seam Lötnaht *f* ~ stout sprinkler Schmelzlotsprinkler *m*
**solderer** Löter *m*
**soldering** Anlötung *f*, Löterei *f*, Lötung *f*, (soft) Weichlöten *n*, Zinnlöte *f* ~ acid Lotsäure *f* ~ apparatus Lötapparat *m* ~ bit Lötkolben *m* ~ box Lötbüchse *f* ~ copper Lötkolben *m* ~ crucible Schmelztiegel *m* ~ fluid Lötwasser *n* ~ fork Lötgabel *f* ~ installation Lötanlage *f* ~ iron Kolben *m*, Lötkolben *m* ~ jumpers Lötbrücken *pl* ~ lamp Gebläselötlampe *f*, Lötlampe *f* ~ lead Lötzinn *n* ~ lug Fahne *f*, Lötöse *f*, (wire-type) Lötstift *m* ~ lug strip Lötösenleiste *f* ~ material Lötmaterial *n* ~ materials Lötmittel *n* ~ metal Lötmetall *n* ~ method Lötverfahren *n* ~ pan Lötpfanne *f* ~ pin Lötstift *m* ~ powder Lötpulver *n* ~ salt Lötsalz *n* ~ surface Lötfläche *f* (flat-type) ~ tab Lötklemme *f*, Lötöse *f* ~ terminal disc Lötösenscheibe *f* ~ tin Lötzinn *n*
**solderless** lötfrei
**sole, to** ~ besohlen
**sole** Angelring *m*, Etage *f*, Schwelle *f*, Sohle *f*, Türangelpfanne *f*, untere Fläche *f*; einzig ~ of foot Fußblatt *n* ~ of a wall Latsche *f*
**sole,** ~ agent Alleinvertreter *m* ~ bar Bordschwelle *f* ~ cutting machine Sohlenausschneidemaschine *f* ~ distributor Alleinvertreter *m* ~ flue Sohlkanal *m* ~ holder Alleininhaber *m* ~ manufacturer Alleinhersteller *m* ~ owner Alleininhaber *m* ~ plate Grund-, Lager-, Sohlen-platte *f* ~ plate of an oil press Fußplatte *f* ~ plate of wind wheel Windradsockel *m* ~-repairing factory Besohlanstalt *f* ~ right Alleinberechtigung *f*
**solenoid** Binderbetätigungsmagnet *m*, Drahtspule *f*, Hubmagnet *m*, Kolbenmagnet *m*, Magnetspule *f*, Solenoid *n* ~ air valve Magnetluftventil *n* ~ block brake elektromagnetische Backenbremse *f* ~ brake Magnetbremse *f* ~ chronograph Spulenchronograf *m* ~ control valve elektrisches Ventil *n* ~ fuel valve Elektrokraftstoffventil *n* ~ rock drill Solenoidbohrmaschine *f* ~ switch Anlaßregler *m* ~ valve Magnetventil *n*

solenoidal quellenfrei ~ condition Devergenzfreiheit f ~ field quellenfreies Feld n
solfatara stage Solfatarenstadium n
solicit, to ~ auffordern, dringend bitten, nachsuchen, (a transaction) veranlassen
solid Körper m; (line) ausgezogen, derb, fest, gediegen, gleichdick, haltbar, hart, kompakt, körperlich, massiv, stark, undurchschossen (print.), ungeteilt, voll ~ and air dielectric geschichtetes Dielektrikum n ~ of revolution (or rotation) Rotationskörper m, Umdrehungskörper m
solid, ~-amorphous festmorph ~ angle (in spheradian units) Raumwinkel m, (in spheradian units) räumlicher Winkel m ~ angle of a crystal Kristallecke f ~ (or fixed) axle feststehende Achse f ~-back transmitter Mikrofon n mit fester Rückwand, Postmikrofon n ~ bearing einteiliges Lager n ~ bit Vollbohrer m ~ blade Vollschaufel f ~ body fester Körper m, Vollkörper m ~-bolt die Schneideisen n aus einem Stück ~-borne sound Körper-, Trittschall m ~ boundary Festkörperbegrenzung f ~ brick Vollziegel m ~ bullet Vollgeschoß n ~ cable Kabel n mit enger Umwicklung ~ carbon Homogen-, Rein-kohle f ~ cast-steel bit Gußstahlvollbohrer m ~ center width Laufbreite f ~ coil Vollspule f ~ collet Vollpatrone f ~ color Selbstfarbe f ~ coloring Fleckenarbeit f ~ composition enger Satz m (print.) ~ compositor Paketsetzer m, Stücksetzer m (print.) ~ conductor Volleiter m ~ content Geschiebeführung f ~ contents körperlicher Inhalt m ~ contraction feste Schwindung f ~ coupling feste oder ungeteilte Kupplung f ~ crank disc volles Kurbelblatt n ~-cross-section area Vollquerschnittsfläche f ~ crossing Blockherzstück n (r. r.) ~ curve ausgezogene Kurve f ~ cylinder Vollzylinder m ~ deposit Geschiebeablagerung f ~ diamond bit Diamantbohrer m ~ die einteilige Backe f, ungeteiltes Schneideisen n ~-drawn axle gezogene Achse f ~-drawn pipe nahtlos gezogenes Rohr n ~ earthing Nullpunkterdung f ~ end axle Faustachse f ~ expansion thermometer Zeigerthermometer n ~ flier (textiles) voller Flügel m ~ flywheel Massenschwungrad n ~ frame Vollrahmen m ~ friction Festreibung f, Trockenreibung f ~ fuel rocket Feststoffrakete f ~ geometry Raumgeometrie f, Stereometrie f ~ ground gewachsener Boden m ~ incrustation near top of furnace Gichtschwamm m ~ injection Druckeinspritzung f, Strahl-einspritzung f, -zerstäubung f ~-injection engine kompressorloser Motor m ~ jacket Vollmantel m ~ jet geschlossener nicht aufgespaltener Strahl m, geschlossener oder scharfer Strahl m ~ journal bearing Augenlager n ~ line (voll) ausgezogene Linie f ~ lubricant konsistente Fette pl, Starrschmiere f ~ lubricant with high dropping point Starrfott m mit hohem Tropfpunkt ~ masonry weir gemauertes Wehr n ~ master connecting rod einteiliges (ungeteiltes) Hauptpleuel n ~ material obtained gewonnenes Schleudergut n ~ matter Feststoffe pl ~ matter content Feststoffanteile pl ~ matter separation Feststoffabscheidung f ~ measure Festmaß n ~ meter Festmeter m ~ object körperlicher Gegenstand

m ~ opal diffuser disk Vollopalscheibe f ~ phase Bodenkörper m, feste Phase f, Festsubstanz f ~ phase welding Rekristallisationsschweißung f ~ photovoltaic cell Sperrschichtenzelle f ~ portions forming grid meshes Gittersteg m ~ press wheel massive Druckrolle f ~ propellant Festtreibstoff m ~-propellant booster rocket Feststoffrakete f ~-propellant rocket Feststoffrakete f ~ propellant system Feststoffmarschantrieb m ~ residue from evaporation Abdampfrückstand m ~ rib Vollrippe f ~ ring ungeteilter Ring m ~ rivet Vollniete f ~ rock anstehendes Gestein n, festes Gebirge n ~ rocket motor Feststoffraketenmotor m ~ rods massives oder volles Gestänge n ~ rods in gunmetal Vollstangen pl aus Rotguß in ~ rolls walzig ~ rope cable Vollseil n ~-rubber tire Vollgummireifen m ~ scale model (made of solid wood) Holzmodell n ~ section Vollquerschnitt n ~ sector Raumsektor m ~ shades Unitöne pl ~ shaft Vollwelle f ~ shrinkage Schrumpfung f ~-shrinkage strain Schrumpfspannung f ~-skirt-type piston Topfkolben m mit nicht geschlitztem Mantel ~ solution feste Lösung f, Mischkristall n, Verbundwerkstoff m (metall.) ~-solution alloy Mischkristallegierung f ~-solution series Mischkristallreihe f ~ spar Vollholm m ~ sphere Vollkugel f ~ spirit Hartspiritus m ~ state fester Aggregatzustand m (phys.) ~-state body Festkörper m ~-state physics Festkörperphysik f ~ steel squares Flachwinkel pl ~ strut Voll-stiel m, -strebe f ~ styles Uniartikel m ~ surface Festkörperoberfläche f ~ three-dimensional structure Raumgebilde n ~ (rubber) tire Vollreifen m ~ tired vollgummibereift ~ tool Vollstahl m ~ tread Blockstufe f ~-type fabric parachute Gewebeschirm m ~ viscosity innere Reibung f ~ voltaic cells Sperrschichtzelle f ~ wall Massivwand f (with) ~ wall vollwandig (with) ~ web vollwandig ~ wheel Vollrad n, vollwandiges Rad n ~ wire Volldraht m ~ wire solder voller Lötdraht m
solidification Erstarrung f, Festmachen n, Festwerden n, Verdichtung f, Verfestigung f ~ curve Erstarrungskurve f ~ period Erstarrungsinterval n ~ point Erstarrungspunkt m ~ range Erstarrungs-bereich m, -zone f ~ shrinkage Erstarrungsschwindung f ~ temperature Erstarrungstemperatur f
solidified erstarrt to become ~ fest werden ~ melted material Schmelzkuchen m
solidify, to ~ erkalten, erhärten, erstarren, festmachen, konsolidieren, verdichten, verfestigen
solidifying Verfestigung f ~ point Erstarrungs-, (of oil) Kälte-, Stock-punkt m
solidity Ausfüllungsgrad m (der Luftschraubenkreisfläche durch die massiven Schraubenblätter), Derbheit f, Dichte f, Dichtheit f, Dichtigkeit f, Feste f, (tensile) Festigkeit f, Haltbarkeit f, körperlicher Inhalt m, Körperlichkeit f, Standfestigkeit f, Völligkeit f ~ of blades *Blattdichte f
solidity, ~ index Gütezahl f ~ ratio Völligkeitsgrad m
solidly earthed neutral fest geerdeter Nullpunkt m
solids Festsubstanz f ~ pipeline Feststoff-Förderleistung f

**solidus curve** Zustandsdiagramm *n*
**solifluction** (slow creeping of wet soil) Erdflie-
ßen *n*
**solitary** einmalig, vereinzelt **~ wave** Einzelwelle *f*
**sollar stopping place** Bühne *f*
**solo** Solo *n* **~ flight** Allein-, Einzel-, Solo-flug *m*
**~ motorcycle** Einzelkrad *n*
**solodyne, ~ receiver** Einquellempfänger *m*, Einstellskala *f*, Röhrenempfänger *m* ohne Anodenbatterie **~ reception** Solodynempfang *m*
**solstice** Solstitium *n*, Sonnenwende *f*
**solstitial point** Solstitialpunkt *m*
**solubility** Auflösbarkeit *f*, Auflöslichkeit *f*, Auflösung *f*, (of glass) Auslaugbarkeit *f*, Lösbarkeit *f*, Löslichkeit *f*, Lösungskurve *f* **~ coefficient** Löslichkeitszahl *f* **~ curve** Löslich-
keitskurve *f* **~ limit** Auflösbarkeitsgrenze *f* **~
pressure** Löslichkeitsdruck *m* **~ procedure** Lös-
lichkeitsverfahren *n* **~ value** Löslichkeitszahl *f*
**solubilize, to ~** löslich machen
**solubilizer** Löser *m*
**soluble** auflösbar, auflöslich, lösbar, löslich **~ in
alcohol** spritlöslich **~ in one another** ineinander
löslich **~ in fat** fettlöslich
**soluble, ~ combination** lösbare Verbindung *f* **~
cutting oil** wasserlösliches Schneideöl *n* **~ iron
carbide** Härtungskohle *f* **~ matter** Lösliches *n*
**~ oil** Bohröl *n* **~ resin** Harzlösung *f*
**solute** aufgelöster Stoff *m*
**solution** Auflösen *n*, Auflösung *f*, Aufschlemmung *f*, Auswertung *f*, Lösung *f* **~ of alkali**
Alkalilösung *f* **~ of alum and salt** Alaunbrühe *f*
**~ of an equation for n** Auflösung *f* einer Gleichung nach N **~ of gold** (chloride) **in ether**
Goldäther *m* **~ of potash** kaustische Pott-
aschenlauge *f*
**solution, ~ circulation** Laugenumlauf *m* **~
composition** Laugenzusammensetzung *f* **~
concentration** Lösungsstärke *f* **~ feed** Laugen-
zufluß *m* **~ flow** Laugendurchfluß *m* **~ heat-
treatment** Erwärmen *n* und Abschrecken *n* **~
potential** Lösungspotential *n* **~ pressure**
Lösungsdruck *m* **~ theory** Phasenlehre *f* **~
treatment** Homogenisieren *n*, Lösungsglühen *n*,
Vergütungsbehandlung *f* **~ welding** Quell-
schweißen *n*
**solutions required** Reagensbedarf *m*
**solvability** Lösbarkeit *f*
**solvated resin** solvatisiertes Harz *n*
**solvation** Solvatisierung *f*
**Solvay, ~ process** Solvayverfahren *n* **~ soda**
Ammoniaksoda *f*
**solve, to ~** auflösen, lösen **to ~ an equation with
respect to "n"** eine Gleichung *f* nach N auflösen
**solvency** Löse-fähigkeit *f*, (of plastics) -wirkung *f*,
Solvenz *f*
**solvent** Auflösungsmittel *n*, Lösungsmittel *n*,
Solvent *n*; kreditfähig, lösend, solvent, zah-
lungsfähig **~ action** Lösungswirkung *f* **~ benzol**
Lösungsbenzol *n* **~ effect** Lösewirkung *f* **~
fumes** Lösemitteldämpfe *pl* **~ manufacturing
and recovery** Lösungsmittelfabrikation *f* und
Rückgewinnung *f* **~ naphtha** Lösungsbenzol *n*,
Solventnaphtha *n* **~ power** (of telescopes,
photographic, and shortwave telescope) Auflösungsvermögen *n* **~ recovery** Rückgewinnung
*f* der Lösungsmittel **~ soap** Lösungsmittelseife

*f* **~-treating plant** Lösemittelraffinieranlage *f* **~
vapors** Lösungsmitteldämpfe *pl* **~ water** Lö-
sungswasser *n*
**solvents-handling equipment** Lösungsmittelappa-
rat *m*
**solving** Auflösen *n* **~ of a mathematical problem**
Ausführung *f* **~ agent** Löser *m*
**somatic, ~ cells** somatische Zellen *pl* **~ number**
somatische Zahl *f*
**somersault, to ~** sich überschlagen
**somersault** Purzelbaum *m*, Überschlag *m*
**somersaulting of the images** Bild-sturz *m*,
-stürzen *n*, Stürzen *n* der Bilder
**sonant** tongebend **~ consonant** stimmhafter Konsonant *m*
**sonar** Ultraschallunterwasser-Ortungsanlage *f*,
Zielsucher *m* **~ work** Echolotung *f*
**sonde** Sonde *f* (rdo)
**sonic, ~ altimeter** Behm Lot *n*, Echolot *n*,
Relativhöhenmesser *m* **~ appliance** Beschallungsgerät *n* **~ barrier** Schall-grenze *f*, -mauer *f*
**~ boom** Überschallknall *m* **~ delay-line** akustisches Laufzeitglied *n*, akustischer Speicher *m* **~
depth finder** Echolot *n* **~ generator** Schall-
generator *m* **~ impulses** Geräusch *n* **~ pressure**
Schalldruck *m* **~ thrust level control** (rocket)
Schubregelung *f* durch Schallwellen **~ wave**
akustische Welle *f* **~ washer** Schallwascher *m*
**sonometer** Monochord *m*, Schall(stärken)mes-
ser *m*, Sonometer *n*, Tonmesser *m* **~ magnet**
Schallstärkenmessermagnet *m*, Sonometermag-
net *m* **~ spring** Schallstärkenmesser-, Sonometer *f*
**sonorous** weichtönend, wohlklingend **~ figures**
Schwingungskurven *pl* **~ sound** klingende
Sprache *f* **~ tin** Feinzinn *n*
**soot, to ~** anschmauchen, rußen, verrußen
**soot** Ruß *m* **~ arrester** Rußfänger *m* **~ baked
hard** fest gebrannter Ruß *m* **~ blower** Rußbläser *m* **~ catcher** Rußfänger *m* **~ deposit**
Rußansatz *m* **~ flake** Rußflocke *f* **~ formation**
Rußbildung *f* **~ pan** Stuppkasten *m* **~-proof**
rußsicher
**sooting** Verrußung *f*
**sootless flame** nichtrußende Flamme *f*
**sooty** rußig **~ coal** Rußkohle *f*
**sorbite** Sorbit *m*
**sorbitic** sorbitisch **~ structure** sorbitisches Gefüge *n*
**sorbitizing** Vergüten *n*
**sore** wund
**sort, to ~** aushalten, auslesen, aussondern, aussortieren, aussuchen, klassieren, lesen, scheiden,
separieren, sortieren **to ~ a charge** gattieren
**to ~ out** ausklauben, ausscheiden, aussortieren,
klauben, sichten, sondern, sorten, sortieren
**sort** Art *f*, Gattung *f*, Güteklasse *f*, Klasse *f*,
Marke *f*, Qualität *f*, Sorte *f*
**sorted** sortiert
**sorter** Sortierer *m* (comput.)
**sorting** Anreicherung *f*, Aufbereitung *f*, Auslese
*f*, Aussortierung *f*, Gattierung *f*, Klassierung *f*,
Säuberei *f*, Scheidung *f*, Separation *f*, Sichtung
*f*, Sichtwirkung *f*, Siebung *f*, Sonderung *f*,
Sortierung *f*, (of ions in mass spectrometer)
Trennung *f* **~ of charged particles** Ladungsträgertrennung *f* **~ of ores** Erzscheidung *f*
**sorting, ~ belt** Klaube-, Lese-band *n* **~ board**
Klaubtisch *m* **~ disc** Sortierscheibe *f* **~ drum**

Sortierscheibe *f* ~ **house** Scheidehaus *n* ~ **machine** Verlesemaschine *f* ~ **plant** Sichtanlage *f* ~ **room** Sortierraum *m* ~ **table** Klaube-, Sortier-, Verlese-tisch *m* ~ **trough** Sortiermulde *f*
**sorts** Defekte *pl*, Handmatrizen *pl*
**SOS** dringendes Notsignal *n*, Not-zeichen *n*, -anruf *m* ~ **call** Notruf *m* ~ **distress signal** Notmeldung *f*
**sound, to** ~ (the ground) auspeilen, blasen, klingen, loten, messen, peilen, schallen, (the upper air) sondieren, tönen **to** ~ **the alarm** Alarm *m* blasen **to** ~ **booming (or heavy)** dumpf klingen **to** ~ **the ground** den Grund *m* abloten **to** ~ **out** ausforschen **to** ~ **simultaneously** mittönen **to** ~ **through (or above)** übertonen
**sound** Geräusch *n*, Hall *m*, (articulated) Laut *m*, Schall *m*, (device) Sonde *f*, Sprechstrom *m*, Ton *m*; dicht (bei Metallguß), einwandfrei, fehlerfrei, folgerichtig, gesund, kredit-fähig, -würdig, stichhaltig, unversehrt, vertrauenswürdig ~ **of speech** Sprachlaut *m* ~ **on disc attachment** Nadeltonzusatz *m* ~ **on disc method** Nadeltonverfahren *n* ~ **with large number of modes of vibration** Heulton *m* ~ **and picture on disc** Bildschallplatte *f* ~ **on vision** Ton *m* im Bild ~ **on wire-recording system** Stahldrahtaufnahme *f*
**sound,** ~**-absorbent** schallweich ~**-absorbent insulation** Schalldämpfungsmittel *n* ~**-absorbent material** Schallschlucker *m* ~ **absorber** Schalldämpfer *m*, -schlucker *m* ~**-absorbing** schalldämpfend, -schluckend ~**-absorbing lining** schallschluckende Wandbekleidung *f* ~**-absorbing material** schalldämpfender Stoff *m*, Wandstoffbekleidung *f* ~**-absorbing wall draping (or lining)** schallschluckende Wandbekleidung *f* ~ **absorption** Schall-absorption *f*, -abwehr *f*, -tilgung *f*, Schluckwärmung *f*, schalldämmend ~ **absorptivity** Schallschluckung *f* ~ **action** Schallereignis *n* ~ **action accompanying television** Begleiten *n* zum Fernsehbild *n* ~ **adjustment** Tonblendeinstellung *f* ~ **amplification** Schallverstärkung *f* ~ **amplifier** Tonverstärker *m* ~ **analysis** Tonzerlegung *f*, Tonzersetzung *f* ~ **analyzer** Schallanalysegerät *n* ~**-and-flash-ranging** Anschneideverfahren *n* ~**-and flash-ranging slide rule** Schießlineal *n* ~ **animated cartoon** Tontrickfilm *m* ~ **aperture** Schallöffnung *f*, Tonabnahmestelle *f*, Tonspalt *m* ~ **appliance** Beschallungsgerät *n* ~**-attenuating** Schalldämpfend ~ **attenuation** Schall-abwehr *f*, -dämmung *f* ~ **bandwidth** Tonbandbreite *f* ~ **base** Abstrahlbasis *f* (tape rec.) ~ **board** Arche *f*, Kanzeldach *n*, Schallbrett *n*, Resonanzkörper *m* ~ **boarding** Einschubbrett *n* ~ **booth** Tonkabine *f* (film) ~ **box** Schalldose *f*, -kammer *f*, -raum *m*, Sprechkopf *m*, (of organ) Windlade *f* ~ **broadcasting** akustischer Rundfunk *m* ~ **carrier** Tonträger *m* ~**-carrier frequency** Tonträgerfrequenz *f* ~**-carrier wave** Tonträgerwelle *f* ~ **casting** gesunder Guß *m* ~ **cell** Tonzelle *f* ~ **chamber** Schallraum *m* ~ **channel** Tonkanal *m* ~ **clarifier** Tonveredler *m* ~ **collection** Schallempfang *m* ~ **collector** Besprechungsmikrofon *n* ~ **color** Färbung *f* des Schalles ~ **column** Tonsäule *f* ~**-communication device** Schallmittel *n* ~ **concentrator** Schallsammler *m* ~ **conducted through solids**

Körperschaft *m* ~ **conduction** Schallübertragung *f* ~ **conductivity** Schallfähigkeit *f* ~ **conductor** Schalleiter *m* ~ **constitution** gesunde Natur *f* ~**-corrector** Schallregler *m* ~ **cover** Schalldeckel *m* ~ **current** Tonstrom *m* ~ **damper** Dämpfungsmittel *n* ~**-damping** schalldämpfend ~ **deadening** Schalldämmung *f* ~**-deadening material** Schallschlucker *m* ~**-delay computor** Schallverzugsrechner *m* ~ **density** Schalldichte *f* ~**-detection training** Hörausbildung *f* ~ **detector** Abhorchvorrichtung *f*, Abhörgerät *n* ~ **diffraction** Schallbeugung *f* ~ **diffusion** Schallverteilung *f* ~ **discriminator** Tonfrequenzgleichrichter *m* (TV) ~ **distortion** Klangverzerrung *f* ~ **effect** Knallerscheinung *f*, Klangwirkung *f* ~ **effect system** Effektanlage *f* ~**-emission microphone** Abgangsmikrofon *n* ~ **emitter** Schallsender *m* ~ **energy** Schallenergie *f* ~**-energy flux** Schallenergiefluß *m*, Schallfluß *m* ~ **engineer** Schallingenieur *m*, Tonmeister *m* ~ **engineering** Tontechnik *f* ~ **engineering structure** technisch gut durchdachte Bauform *f* ~ **entertainment** Tondarbietung *f* ~ **eraser** Ausradiervorrichtung *f* ~ **fading** Tonüberblendung *f* ~ **field** Schallfeld *n*
**sound-film** Hörfilm *m*, Tonfilm *m* ~ **attachment** Lichtton-gerät *n*, -zusatz *m*, Tonkopf *m* ~**-editing machine** Abhörtisch *m* ~ **feed mechanism** Tonlaufwerk *n* ~ **feed roller** Tontransportrolle *f* ~ **head** Lichtton-gerät *n*, -zusatz *m*, Tonkopf *m*, Tonzusatzgerät *n* ~ **projector** Tonfilmprojektor *m* ~ **transmitter** Tonfilmgeber *m* ~ **used for reproduction** Tonkopie *f*
**sound,** ~ **filter** Klangfilter *m* ~**-flash ranging** Lichtschallmessen *n* ~ **floor** Einschub *m*, Fehlboden *m* ~ **flux** Schallenergiefluß *m* ~ **focus** Schallfokus *m* ~ **frequency** Tonfrequenz *f* ~ **gate** Abnahmestelle *f*, Abnehmerstelle *f*, Belichtungs-, Steuer-, Tonabnahme-, Tonabtast-, Tonbelichtungs-stelle *f*, Tonfenster *n* (film) ~ **generation** Schallerregung *f* ~ **generative** tongebend ~ **generator** Schall-erzeuger *m*, -quelle *f* ~ **hardness** Schallhärte *f* ~ **head** Hörkopf *m*, Lichttonzusatz *m*, Tonabnehmer *m* (film) ~ **head axis** Tonkopfachse *f* ~ **head core** Tonkopfkern *m* ~**-head lens** Tonoptik *f*, Tonwiedergabeoptik *f* ~**-head optic** Tonoptik *f* (film), Tonwiedergabeoptik *f* ~ **hole** Schalloch *n* ~ **image** Schallbild *n* ~**-impressed surface** vertonte Oberfläche *f* ~ **impression** Klangbild *n* ~ **incidence** Schallinzidenz *f* ~ **insulation** Schall-dämpfung *f*, -isolation *f*, -schutz *m* ~ **insulation factor** Dämmzahl *f* (acoust.) ~ **intensity** Schall-intensität *f*, -stärke *f* ~ **interference band** Tonstreifen *m* (TV) ~ **intermediary frequency** Tonzwischenfrequenz *f* ~ **intermediate frequency** Tonzwischenfrequenz *f* ~ **irradiation** Schallbestrahlung *f* ~ **lag** Schallverzug *m* ~**-lag time** Schallverzugszeit *f* ~ **level** Lautstärke *f* ~ **level indicator** Druckpegelmesser *m* ~**-level meter** Geräuschmesser *m*, Schallstärkenmesser *m* ~ **location** akustische Ortung *f*, Horchortung *f*, Schallortung *f* ~ **locator** Abhörgerät *n*, Horcher *m*, Horchgerät *n*, Ortseindruck *m*, Richthörer *m* ~ **magnetic pole** Südpol *m* ~ **measurement** Lautstärkemessung *f* ~**-measuring device** Lautstärkemes-

ser *m*, Schall-messer *m*, -meßeinrichtung *f* ~ **meter** Raumgeräuschmesser *m* ~ **mixer** Tonmischpult *n* ~ **modulation** Tonmodulation *f* ~ **motion picture** Tonfilm *m* ~ **navigation and ranging** Unterwasserschallortungsanlage *f* ~ **overshooting** Überstreuung *f* (acoust.) ~ **panel** Schallwand *f* ~ **particle-velocity** Schallschnelle *f* ~**-path length** Schallweglänge *f* ~ **pattern** Klang-bild *n*, -figur *f* ~ **perception** Schallwahrnehmung *f* ~ **permeability** Schalldurchlässigkeit *f* ~ **pickup** Abnehmerstelle *f*, Schall-aufnehmen *n*, -empfänger *m*, Tonabnahmegerät *n*, Tonabnehmer *m* ~ **pickup microphone** Besprechungsmikrofon *n* ~ **pickup outfit** Besprechungsanlage *f* ~**-pickup point** Tonabtaststelle *f* ~**-picture projector** Tonprojektor *m* ~ **pit** Schallgrube *f* ~ **power** Schalleistung *f* der Sprache ~**-powered telephone** batterieloser Fernsprecher *m* ~ **pressure** (brought on microphone) Beaufschlagung *f*, Schalldruck *m* ~ **pressure increase** Druckstauung *f* (acoust.) ~ **pressure level** Schalldruckpegel *m* ~ **probe** Schallsonde *f* ~**-producing** schallerzeugend ~ **projection** Schallabstrahlung *f* ~ **projection on film** Knallbild *n* ~**-projection opening** Schallaustrittsöffnung *f* ~ **projector** Schalltrichter *m*, Tonprojektor *m* ~**-proof** geräuschsicher, schalldicht, -isoliert, -sicher, -undurchlässig ~**-proof booth (or cabinet)** schalldichte Zelle *f* ~**-proof cell** schalldichte Zelle *f* ~**-proof materials** schalldämpfende Stoffe *pl* ~**-proofed booth (or cabin)** schalldichte Zelle *f* ~**-proofing** Geräuschdämpfung *f*, Schall-schutz *m*, -verkleidung *f* ~**-proofing material** geräuschdämpfendes Material *n* ~**-propagating medium** Schallmedium *n* ~ **propagation** Schallfortpflanzung *f* ~ **quality** Klangfarbe *f* ~ **radar** Schallortung *f* ~ **radiation** Schallabstrahlung *f* ~**-radiation distribution** Schallbestrahlung *f* ~ **range** Tonkreis *m* ~**-ranger** Schallmeßgerät *n* ~ **ranging** akustisches Meßverfahren *n*, Horchortung *f*, Knallmeßverfahren *n*, Schall-meßbeobachtung *f*, -messen *m*, -messung *f*, -richten *n* **sound-ranging**, ~ **altimeter** Behm-Lot *n* (aviat.), Lufttecholot ~ **battery** Schallmeßbatterie *f* ~ **cable** Schallmeßkabel *n* ~ **microphone** Schallmeßmikrofon *n* ~ **net** Schallmeßsystem *n* ~ **station** Schallmeßstelle *f* ~ **system** Schallmeßsystem *n*

**sound,** ~ **ray** Schallstrahl *m* ~**-reading** Höraufnahme *f* ~ **receiver** Schallempfänger *m.* ~ **receiving** Schallaufnehmen *n* ~ **reception** Schallempfang *n* ~ **(or aural) reception** Höraufnahme *f* ~ **record** Tonfolgeschrift *f* ~ **recorder** Schall-anzeiger *m*, -aufzeichner *m*, -schreiber *m*, Ton-aufnehmer *m*, -aufzeichner *m* ~ **recording** Schall-aufnehmen *n*, -aufzeichnung *f*, -auszeichnung *f*, Tonaufnahme *f*, Tonaufzeichnung *f*, Tonstreifen *m* ~**-recording amplifier** Tonaufzeichnungsverstärker *m* ~**-recording camera** Tonkamera *f* ~**-recording drum** Belichtungsrolle *f* ~**-recording steel tape** Lautschriftband *n* ~**-recording stylus** Schallschreiber *m* ~**-recording unit** Tonschreiber *m* ~ **reduction** Schalldämmung *f* ~**-reflecting** schallreflektierend ~ **reflector** schallabstrahlendes Gebilde *n*, Schallscheinwerfer *m* ~ **refraction** Schallbrechung *f* ~ **reinforing means** Schall-

verstärkungsmittel *n* ~ **rejection** Tonunterdrückung *f* (TV) ~ **reproducer** Tonlautsprecher *m* ~ **reproducer head** Tongeberkopf *m* ~ **reproduction** Tonwiedergabe *f* ~**-reproduction quality** Schallwiedergabequalität *f* ~ **rod** Klangstab *m* ~ **scanning** Tonspurabtastung *f* (film) ~**-scanning slit** Tonabtasterspalt *m*, Tonabtast-spalte *f*, -stelle *f* ~ **screen** Tonschirm *m* ~ **sensation** Tonempfindung *f* ~**-sight broadcasting** Tonbildrundfunk *m* ~ **signal** Lärmsignal *n*, Schall-signal *n*, -zeichen *n*, Tonsignal *n* ~**-signal volume** Nutzlautstärke *f* ~ **signals** Gehörsignal *n* ~ **slit** Ton-fenster *n*, -schlitz *m*, -spalt *m* ~ **source** Schallquelle *f* ~ **spectrum** Klanggemisch *m*, Schallspektrum *n*, Tonspektrum *n* ~ **stage** Tonatelier *n* ~ **stimulus** Tonreiz *m* ~ **studio** Besprechungsraum *m* (rdo) ~ **suppression** Schall-dämpfung *f*, -tilgung *f* ~ **system** Takte *pl* des Tones ~ **take-off drum** Tonabnehmertrommel *f*, Tontrommel *f* ~ **tape** Tonband *n* ~ **technician** Schallingenieur *m* ~ **telecast** Hörfilm *m*, Rundfunktonfilm *m* ~ **telecine** (of film) Rundfunktonfilm *m* ~ **telecinematography** Hörfilm *m* ~ **test** Tonprobe *f* ~ **test chamber** Schallzelle *f* ~ **track** Tonfolgeschrift *f*, Tonspalte *f*, Tonstreifen *m* ~**-track film player** Licht-grammofon *n*, -fonograf *m* ~**-track squeezing** Tonstreifenverschmälerung *f* ~**-track support** Tonträger *m* ~ **tracks** (variable area) Amplitudenschrift *f* ~ **transcription** record Schallaufnahmeplatte *f* ~ **transmission** Schalldurchlässigkeit *f* ~ **transmission channel** Schallbündelung *f* ~**-transmission characteristic** Schalldämmzahl *f* ~ **transmittance** Schalldurchlässigkeit *f* ~ **transmitted by air** Luftschall *m* ~ **transmitted by solid bodies** Bodenschall *m* ~ **transmitter** Schall-, Ton-sender *m* ~ **trap** Ton-falle *f*, -sperrkreis *m* (TV) ~ **travel time** Schallaufzeit *f* ~ **truck** Lautsprecherwagen *m*, Tonwagen *m* ~ **tuning** Abstimmung *f* der Tonhöhe ~ **variation** Schallschwingung *f*, Tonschwingung *f*, -variation *f* ~ **velocity** Schallgeschwindigkeit *f* ~ **volume** Klangfülle *f* ~ **wave** Geschwindigkeit *f* der Schallwelle, Schallwelle *f*, Tonwelle *f* **sounded** abgelotet

**sounder** Klopfer *m*, Klopferapparat *m* ~ **key** Klopfertaste *f*, Lotung *f*, Morsetaste *f* **sounding** Anregung *f*, Klingen *n*, Klopfen *n*, Loten *n*, Lotung *f*, Peilung *f*, Tönen *n*; klingend, tönend ~ **the ionosphere** Echolotung *f* der Ionosphäre **sounding,** ~ **apparatus** Lotungsgerät *n*, Tiefenlotapparat *m* ~ **balloon** Ballonsonde *f*, Lot-, Registrier-, Sondier-, Stich-, Versuchs-ballon *m* ~**-balloon station** Drachenwarte *f* (meteor.) ~ **beam** Lotstrahl *m* ~ **board** Resonanzboden *m*, Schall-brett *n*, -deckel *m* ~ **(cask or tun) buoy** Schall-tonne *f*, -tonnenboje *f* ~ **device** Grundtaster *m* ~ **electrode** Sondierelektrode *f* ~ **equipment** Sondervorrichtung *f* ~ **installation** Loteinrichtung *f* ~ **lead** Lot *n*, Peillot..., Wurfblei *n* ~ **line** Lot *n*, Senkschnur *f* ~ **machine** Lotmaschine *f* ~ **method** Peilmethode *f* ~ **pipe** Peilrohr *n* ~ **plate** Klangplatte *f* ~ **pole** Peilstange *f* ~ **position** Lotstand *m* ~ **rod** Grundtaster *m*, Peilstock *m*, Sondiereisen *n*, Visiereisen *n* ~ **test** Sonderversuch *m* ~ **top** Schall-

deckel *m* ~ **tube** Lotröhre *f* ~ **winch** Sonder-
winde *f*
**soundless** klanglos
**soundness** Fehlerlosigkeit *f*, Stichhaltigkeit *f* ~
**test** Güteprüfung *f*
**sour, to** ~ ansäuren, säuren
**sour** sauer **too** ~ übersauer
**sour,** ~ **bark** Sauerlohe *f* ~ **water** Sauerbeize
*f*
**source** Ausgangspunkt *m*, Beginn *m*, Quelle *f*,
Quellgebiet *n*, Ursprung *m* ~ **of acoustic energy**
Schallquelle *f* ~ **of current** Stromquelle *f*,
Zentralspeisung *f* ~ **of danger** Gefahrquelle *f* ~
**of discovery** Fundort *m* ~ **of disturbance**
Störungsquelle *f* ~ **of electromotive force**
Spannungsquelle *f* ~ **of electron gun (or pencil)**
Strahlentstehungsort *m* ~ **of energy** Energie-,
Kraft-quelle *f* ~ **of error** Fehlerquelle *f* ~ **of**
**evaporation** Verdampfungsquelle *f* ~ **of fire**
Feuerquelle *f* ~ **of generation** Entstehungsort
*m* ~ **of heat** Heizquelle *f* ~ **of interference**
Fehlereingrenzung *f* ~ **of light** Lichtquelle *f*
~ **of light exhibiting structural markings** Lichtquelle *f* mit Struktur ~ **of lighting current**
Lichtstromquelle *f* ~ **of noise** Geräuschquelle *f*
~ **of power** Energiegrundlage *f*, Kraftquelle *f*
~ **of power waste** Kraftverlustquelle *f* ~ **of**
**production** Entstehungsort *m* ~ **of radiation**
Strahlungsquelle *f* ~ **of sonic vibrations** Geräuschquelle *f* ~ **of sound** Tonquelle *f* ~ **of**
**suction** Sogversorgung *f* ~ **of supply** Bezugsquelle *f* ~ **of trouble** Fehlereingrenzung *f* ~ **of**
**volatilization** Verdampfungsquelle *f* ~ **of**
**welding current** Schweißstromquelle *f*
**source,** ~ **admittance** (of electr. tube) Quellen-
leitwert *m* ~ **article** Quellenschrift *f* ~ **biblio-**
**graphy** Quellenkunde *f* ~ **book** Quellenschrift *f*
~ **density** Quelldichte *f* ~ **disc** Quellscheibe *f* ~
**distribution** Quellenbelegung *f*, Verteilung *f*
von Strahlenquellen ~ **equivalence** Quellen-
äquivalenz *f* ~ **function** Ergiebigkeit *f* ~ **holder**
Strahlerhalter *m* ~ **impedance** Quellwiderstand *m* ~ **interlock** Spaltraumverriegelung *f* ~
**layer** Quellschicht *f* ~ **material** Rohstoff *m*,
Schrifttum *n* ~ **point** Quellpunkt *m* ~ **program**
Originalprogramm *n* ~ **radiation** Strahlungs-
quellen *pl* ~ **range** Reaktorbetrieb *m* mit Hilfsquelle ~ **region** Ursprungs-gegend *f*, -region *f*
~ **resistance** Innenwiderstand *m* ~ **rock** Mutter-
gestein *n* ~**-sink representation** Quellen-Senken-
Darstellung *f* ~ **strength** Quellstärke *f* ~ **surface**
Quellfläche *f* ~ **technique** Quellentechnik *f* ~
**term** Quellenglied *n* ~**-to-film distance** Quellen-
Filmabstand *m* ~**-type solutions** quellenförmige
Lösungen *pl*
**sources used** Quellennachweis *m*
**sourdine** Dämpfer *m*, Sourdine *f*
**souring,** ~ **machine** Sauermaschine *f* ~ **tank**
Sauerkasten *m*
**sourness** Sauerkeit *f*, Säure *f*
**south** Süden *m*, Südpunkt *m* ~**-east** Südosten *m*
~**-easterly** südostwärts ~ **latitude** südliche
Breite *f* ~**-magnetic** südmagnetisch ~ **pan**
Linkshänder *m* ~ **pedestal** Südlagerung *f* ~ **point**
Südpunkt *m* ~ **pole** Augpunkt *m* ~ **(magnetic)**
**pole** Südpol *m* ~**-ward** südlich, südwärts
~**-west** Südwesten *m* ~**-wester** Südwester *m*
**southern,** ~ **foehn** Südföhn *m* ~ **hemisphere**

Südhalbkugel *f*, südliche Erdhälfte *f*, südliche
Halbkugel *f* ~ **lights** Südlicht *n*
**sovereign territory** Hoheitsgebiet *n*
**sovereignty** Hoheit *f* ~ **of the air** Lufthoheit *f*
**sow, to** ~ aussäen, säen
**sow** (foundry) Abstichgraben *m*, (foundry) Aufläufer *m*, Eisenklumpen *m*, Masselgraben *m*,
Sau *f* ~ **iron** Schaleneisen *n*
**sowing** Aussaat *f* ~ **machine** Säemaschine *f* ~
**time** Bestellzeit *f*
**spa** Badeort *m*
**space, to** ~ einteilen, sperren, Trennstrom *m*
senden, verteilen **to** ~ **apart** auseinanderlegen
**to** ~ **evenly** ausgleichen
**space** Durchschuß *m*, Gelaß *n*, Luftreich *n*,
Lücke *f*, Platz *m*, Pause *f*, Raum *m*, Strecke *f*,
(in telegraphic work) Telegrafierpause *f*,
Zwischenraum *m* **in** ~ räumlich ~ **between the**
**backbands of a book** Rückenfeld *n* ~ **within**
**bulb** Kolbenraum *m* ~ **in the clear** lichter
Raum *m* ~ **of events** Ereignisraum *m* ~ **between**
**the fire bars** Rostfuge *f* ~ **within firing range**
Treffbereich *m* ~ **between (two) lines** Linienbreite *f* ~ **between the links** Laschenteilung *f*
~ **of momentum and energy** Energie-Impuls-
Raum *m* ~ **of numbers** Zahlenraum *m* (math.)
~ **for plugging cable termination** Füllraum *m*
~ **within reach** Fühlraum *m* ~ **of representation**
Darstellungsraum *m* ~ **of states** Zustandsraum
*m* ~ **of states and energy** Energie-Zustandsraum
*m* ~ **of tooth** Zahnlücke *f* ~ **between words**
Wortzwischenraum *m*
**space,** ~ **arrangement** räumliche Anordnung *f* ~
**axis** Raumachse *f* ~ **band box** Keilkasten *m* ~
**band key** Keiltaste *f* ~ **bar** (typewriter) Leer-
taste *f* ~**-bound** raumfest ~ **bracing** Raumverspannung *f* ~ **bushing** Abstandsbuchse *f* ~
**centered** raumzentriert ~ **character** Abstandszeichen *n* ~ **charge** Raumladung *f*
**space-charge,** ~ **accumulation** Raumladungs-
anhäufung *f* ~ **beam spreading** Bündelverbrei-
terung *f* durch Eigenladung ~ **cloud** Raumladungswolke *f* ~ **debunching** Defokussierung
*f* durch Raumladung ~ **density** Raumladungs-
dichte *f* ~ **detector** Raumladungsdetektor *m* ~
**distorted fields** raumladungsverzerrte Felder *pl*
~ **distortion** Raumladungsverzerrung *f* ~
**distribution** Raumladungsverteilung *f* ~ **effect**
Raumladewirkung *f*, Raumladungseffekt *m* ~
**equation** Raumladungsgleichung *f* ~ **field**
Raumladungsfeld *n* ~ **grid** Hilfsgitter *n*, Raum-
lade-gitter *n*, -gitterspannung *f*, -netz *n*, Raum-
ladungs-gitter *n*, -netz *n*, -zerstreuungsgitter *n*,
Sauggitter *n*, Spannungsnetz *n*, Zuggitter *n* ~
**layers** Raumladungsschichten *pl* ~**-limited-**
**current state** Raumladungszustand *m* ~ **region**
Raumladungsgebiet *n* ~ **repulsion** Raum-
ladungsabstoßung *f* ~ **tetrode** Doppelgitter-,
Raumladegitter-, Raumladungsgitter-, Zweigitter-röhre *f* ~ **theory** Raumladungstheorie *f*
~ **wave** Raumladungswelle *f*
**space,** ~ **cone** Rastpolkegel *m* ~ **contact** Arbeitskontakt *m* ~ **cooler** Raumkühler *m* ~
**coordinate** Raumkoordinate *f* ~ **craft** Raum-
fahrzeug *n* ~ **current** Anodenstrom *m*, Elektronenstrom *m*, Emissionsstrom *m*, Entla-
dungsstrom *m*, Raumladestrom *m* ~**-current**
**source** Raumstromquelle *f* (teleph.) ~ **curve**

Raumkurve f ~ **diagram** Raumbild n, Raumgebilde n ~ **disc** Abstandsscheibe f ~ **enclosed** umbauter Raum m ~ **factor** Füllfaktor m, Wicklungsfaktor m ~ **feeling** (in vision) Raumgefühl n ~ **filling** Raumerfüllung f ~ **flight** Raumflug m ~ **focussing** Raumfokussierung f ~ **framework** Raumfachwerk n ~ **freighter** Raumfrachter m ~ **girder in bending** räumlicher Biegungsträger m ~ **grid** Raumgitter n ~ **group** Raumgruppe f ~ **image** Raumbild n ~ **image reversed in depth** tiefenverkehrtes Raumbild n ~ **integral** Raumintegral n ~ **interval** Raumfolge f ~ **key** Zwischenraumtaste f ~ **lattice** Kristallgitter n, Raumgitter n ~**-lattice distance** Netzabstand m ~**-like** raumartig ~ **line** Durchschußlinie f ~ **mark** Raumgitterform f, Raummarke f ~ **multiple** (in crossbar switches) Wegvielfach n (Kreuzschienenschalter) ~ **needle** Raumnadel f ~ **occupied** Platz-, Raum-bedarf m ~ **pattern** Herzcharakteristik f, Peilkurve f, Richt-charakteristik f, -kennlinie f ~ **perception** Raumsinn m ~ **phasing** (antenna) Phasendifferenz f ~ **picture** Zielraumbild n ~ **point** Aufpunkt m, Raumpunkt m ~ **pressure suit** Raumdruckanzug m ~ **probe orbit** Raumsondenbahn f ~ **quad case** Ausschlußkasten m **in** ~ **quadrature** räumlich senkrecht aufeinander ~ **quantization** Raumquantelung f ~ **radiation** Raumstrahlung f, Steilstrahlung f ~ **ray** Raumstrahl m ~ **reflection** Raumspiegelung f ~ **required** (for plant) Grundflächenbedarf m ~ **research** Raumfahrtforschung f ~ **restriction** Raummangel m ~ **rod** Raumlenker m ~ **rotation group** räumliche Rotationsgruppe f ~ **saving** Raumersparnis f ~ **ship** Raumschiff n ~ **signal** Zwischenraumzeichen n (telegr.) ~**-signal pulse** stromlose Zeichenschrift f ~**-stabilized** raumstabilisiert ~ **structure** hinausragender Strukturteil m ~ **telegraphy** Raumtelegrafie f ~ **tensor** Raumtensor m ~**-time axiom** Raum-Zeitkontinuum n ~ **tracking station** Bahnverfolgungsstation f ~ **wave** Höhen-, Luftweg-, Raum-welle f ~ **zone** Sprungzone f

**spaced** unterteilt **to be** ~ **apart (or away)** abliegen ~ **armor** Hohlraumpanzerung f

**spacer** Abstand-halter m, -hülse f, Abstand(s)-ring m, Auflagescheibe f, Ausfüllblock m, Beilage f, Distanz-halter m, -scheibe f, -stück n, Druckplatte f, Käfig m, Kartuschdeckel m, Keil m, Laterne f, Paßring m, Raa f, Sperrtaste f, Spreize f, Unterlegplatte f, Zwischen-ring m, -stück n ~ **bar** Richte-, Verteilungs-eisen n ~ **block** Distanzblock m ~ **bolt** Stehbolzen m ~ **key** Abstandstaste f ~ **sleeve** Abstandsrohr n ~ **tube** Abstandshülse f

**spaces** Ausschluß m (print.)

**spacing** Abstand m, Gedrängtheit f, Lagerung f, Teilung f, Verteilung f, Zwischenstück n ~ **of blades** Schaufelteilung f ~ **between cutoff points** Lochbreite f eines Frequenzsiebes ~ **of fins** Rippenabstand m ~ **of frames** Spantenentfernung f ~ **of ribs** Rippenabstand m ~ **of the rivets** Nietteilung f

**spacing,** ~ **accuracy** Teilgenauigkeit f ~ **bar** Abstandsstange f ~ **battery** Trennbatterie f ~ **cam** Transport-, Vorschub-daumen m ~ **collar** Spindelring m ~ **collars for cutter arbors** Fräsdornringe pl ~ **collet** Spindelring m ~ **contact**

Ruhe-, Trenn-kontakt m ~ **(break) contact** Trennungskontakt m ~ **current** Ruhestrom m, Trennstrom m, Zwischenzeichenstrom m ~ **diagram** Teildiagramm n ~ **drill** Dibbelmaschine f ~ **escapement pawl** Klinke f (telet.) ~ **insulator** Abstandslehre f, Distanzisolator m ~ **jig** Stichmaß n ~ **key** Abstandstaste f ~ **magnet** Fortschubmagnet m, Papierfortschubmagnet m, Vorschubmagnet m ~ **material** Blind-, Füll-material m (print.) ~ **means** Distanzhalter m ~ **pawl** Transporthebel m ~ **piece** Abstandsstück n ~ **position** Trenn-seite f, -stellung f ~ **pulse** Leerimpuls m, Pausenschritt m ~ **ratio** Teilungsverhältnis n ~ **shaft** Vorschubachse (telet.) f ~ **signal** Abstandszeichen n, Blank n ~ **stop** Ruheanschlag m, Rueschiene f ~ **stop lever** Vorschubausschalthebel m ~ **strip** Abstandsleiste f ~ **valve** Trennstromröhre f (telegr.), Zwischenstromröhre f ~ **washer** Abstandscheibe f, Ausgleichsring m ~ **wave** Ruhewelle f, Verstimmungswelle f, Zwischenzeichenwelle f ~ **wedge** Abstandskeil m

**spacings in open deckings** Fugen pl im Bohlenbelag

**spacious** geräumig, räumlich

**spade** Spaten m ~**-handle drive** Steigbügelantrieb m ~**-shaped chisel** Spatmeißel m ~ **terminal** offener Kabelschuh m ~**-type chisel** Schaufelmeißel m

**spall, to** ~ abschülpen, absplittern, klaffen, reißen

**spallation** Absplitterung f ~ **fragment** Spallationsbruchstücke n

**spaller** Erzpocher m

**spalling** Absplittern n (aviat.), Aufspaltung f ~ **test** Spaltprobe f

**spalls** Steinspiller m

**span, to** ~ spannen, überbrücken, umfassen, umspannen, verspannen **to** ~ **a distance** eine Strecke überbrücken

**span** Abspannung f, Bogenweite f, Brückenweite f, Einspannlänge f, Feld n, (dimension) Feldweite f, (of airplane) Flügelbreite f, gesamte Spannweite f, Gewölbebogen m, Längsholm m, Lichtweite f, Mastfeld n, Spanne f, Spannung f, Stützweite f ~ **with anchoring (or span) pole** Abspannstrecke f (electr.) ~ **of arch** Spannweite f des Bogens ~ **of attention** Aufmerksamkeitsumfang m ~ **between bridge** Stützweite f der Brücke ~ **of bridge** Brückenspannweite f ~ **of the elevator** Höhenruderspannweite f ~ **of operation range** Steilheit f ~ **of the stabilizer** Höhenflossenspannweite f ~ **of the tail plane** Höhenflossenspannweite f ~ **of a wing unit** Gesamtspannweite f ~ **with wings folded** Spannweite f mit beigeklappten Flügeln

**span,** ~ **bracket** Abspannmauerbügel m ~ **cable** Abspannseil n ~ **ceiling** Balkendecke f ~**-chord ratio** Flügelstreckung f ~ **drill** Ausfüllung f des Gewölbezwickels ~ **lashing** Spanntau n ~ **length** Spannweite f ~ **loading** Klafterbelastung f ~ **pole** Abspannpfahl m ~ **rope** Abspannseil n, Pardun n, Pardune f ~ **width** Spannweite f ~ **wire** Spanndraht m

**spandrel** Dekorationsplatte f, Gewölbezwickel m, Hohlkehle f ~ **beam** Randträger m ~ **wall** fliegende oder schwebende Mauer f

**spangle** Flitter *m*, Spange *f*
**spangles** Schiefer *m*
**Spanish, ~ burton** Ladetakel *n* **~ fox** (naut.) Fuchsje *f*
**spanker** Besan *m* **~ boom** Besanbaum *m*
**spanking** Haue *f*
**spanner** Bolzenschlüssel *m*, (for nut) Engländer *m*, Mutterschlüssel *m*, Schlüssel *m*, Spannschloß *n* **~ for caps** Kappenschlüssel *m* **~ for contact breaker** Schlüssel *m* für Unterbrecherkontakt **~ for square nut** Vierkantschlüssel *m*
**spanner, ~ board** Schraubenschlüsselbrett *n* **~ clearance** Maulweite *f* des Schraubenschlüssels **~ wrench** englischer Schraubenschlüssel *m*, Vierkantschlüssel *m*
**spanning** Verspannung *f*
**spar** Ausleger *m*, Balken *m*, Dachsparren *m*, Holm *m*, Rundholz *n*, Sparren *m*, Spat *m*, Spiere *f*, Tragholm *m* **~ of corrugated sheet metal** Wellblechholm *m* **~ of metal tubing** Metallrohrholm *m*
**spar, ~ attachment** Holmansatz *m* **~ cross section** Holmquerschnitt *m* **~ coupling** Holmverschraubung *f* **~ depth** Holmhöhe *f* **~ distance** Holmabstand *m* **~ element** Spant *n* **~ fitting** Holmbeschlag *m* **~ flange** Holmgurt *m* **~ gap** Holmabstand *m* **~ gate** Gattertor *n* **~-girder shape** Holmgurtprofil *n* **~-milling machine** Holmfräsemaschine *f* **~ model** Stabmodell *n* **~-section area** Holmquerschnittsfläche *f* **~ trestle** Stangenbock *m* **~ trussed beam** Fachwerkholm *m* **~ varnish** Außenlack *m*, Bootslack *m*, Luftlack *m* **~ width** Holmbreite *f*, -tiefe *f*
**spare, to ~** aussparen, schonen
**spare** Aushelf *m*, Ersatz *m*, Reserveteil *m*; überzählig **~ anchor** Raumanker *m* **~ armature** Ersatzanker *m*, Reserveanker *m* **~ bar** Ersatzhebel *m* **~ barrel** Vorratslauf *m* **~ battery** Ersatz-, Not-batterie *f* **~ branch** Reservestutzen *m* **~ capital** flüssiges Kapital *n* **~ cell** Abschaltzelle *f* **~ circuit** Vorratsleitung *f* **~ coil** Reservewindung *f* **~ conductor (or cable)** Ersatzader *f* **~ contact** Leerlaufkontakt *m* **~-current source** Netzersatzanlage *f* **~ drive** zusätzlicher Hilfsantrieb *m* **~ drum** Reserve-, Vorrats-trommel *f* **~ encoder** Scheibenverschlüßler *m* **~ equipment** Ersatzausrüstung *f* **~ fuse strip** Reservesicherungsstreifen *m* **~ glower** Ersatzglühkörper **~ heater plate** Ersatzheizplatte *f* **~ instrument** Ersatz-, Reserve-apparat *m* **~ knife** Ersatzmesser *m* **~ lamp** Reservelampe *f* **~ lead** (pencil) Ersatzmine *f* **~ leak detector tube** Austauschlecksuchröhre *f* **~ level** unbesetzter Höhenschritt *m* **~ line circuit** Sonderteilnehmer *m* (Tin-Schaltung) **~ member** Ersatzstab *m* **~ part** Austauschstück *n*, Einzelteil *m*, Ersatzbestandteil *m*, -stück *n*, Reserveteil *m*, Zubehörteil *m* **~ part replacement** Ersatzbestükkung *f* **~-parts catalogue** Ersatzteilliste *f* **~-parts depot** Ersatzteillager *n* **~ parts manufacture** Ersatzteilfertigung *f* **~ plate** Abstandsscheibe *f*, Ersatzplatte *f* **~ pump** Reservepumpe *f*, Saug- und Druckpumpe *f* **~ repeater** Vorratsverstärker *m* **~ rim** Extra-, Reserve-felge *f* **~ set** Ersatzanlage *f*, Reservesatz *m* **~ spinner** (textiles) Hilfsspinner *m* **~ tank** Zusatzbehälter

*m* **~ terminal** Leerlaufkontakt *m* **~ terminating set** Vorratsgabelschaltung *f* **~ test box** Ersatzprüfkästchen *n* **~-time engagement** nebenberufliche Beschäftigung *f* **~-time work** Nebenarbeit *f* **~ tire** Ersatzreifen *m* **~-tire equipment** Ersatzbereifung *f* **~-tire lift** Reservereifenaufzug *m* **~ turn** Vorratswindung *f* **~ unit** Ersatzeinheit *f*, Reserve-anlage *f*, -satz *m* **~ warp thread** Ersatzkettengarn *n* **~ wheel** Hilfsrad *n* **~ wheel lock** Ersatzreifensicherung *f* **~ wheel rim** Vorratsfelge *f* **~ wheel support** Reserveradhalter *m* **~ wire** Vorratsader *f*
**spares department** Ersatzteilabteilung *f*
**sparger** Aufgußapparat *m*
**sparging** Durchblasen *n*
**spark, to ~** abbrennen (electr.), feuern, sprühen, überspringen **to ~-over** überschlagen
**spark** Funke(n) *m*, Funken *n* **~ at break** Öffnungsfunke *m* **~ at make (or before contact)** Schließungsfunke *m*
**spark, ~ adjustment** Zündverstellung *f* **~ advance** Frühzündung *f*, Vorzündung *f* **~-advance and -retard mechanism** Zündzeitpunktverstellvorrichtung *f* **~ arrester** (on a locomotive) Funken-fänger *m*, -kammer *m*, -löscher *m* **~ blowout** Funken-anzeiger *m*, -löscher *m* **~ capacitor** Funkenlöschkondensator *m* **~ catcher** (on a locomotive) Funkenfänger *m*, Funkensieb *n* **~ coil** Funken-geber *m*, -induktor *m*, Induktorium *n* **~ collector** Funken-rost *m*, -sieb *n* **~ condenser** Funkensammler *m* **~-condensing chamber** Funkenfänger *m* **~ conductor** Funkenleiter *m* **~ contact breaker (or brush)** Abbrennbürste *f* **~-contact piece** Abbrennkontaktstück *n* **~ control** Zünd-hebel *m*, -regelung *f*, -verstellung *f* **~ counter** Funkenzähler *m* **~ current** Durchschlagsstrom *m* **~ damping** Funkendämpfung *f* **~ discharge** Durchbruchsentladung *f*, Funk(en)entladung *f*, Funken- oder Licht- und Bogenentladung *f* **~ discharge continuum** Funkenkontinuum *n* **~ distance** (of a gap) Schlagweite *f* **~ drawer** Funkenzieher *m* **~ drawing** Funkenziehen *n* **~ escape** Auswurfkamin *m* (Konverter) **~ extinguisher** Funkenlöscher *m* **~-extinguisher magnet** Funkenlöschermagnet *m* **~ extinguishing** Funkenlöschung *f* **~ extinguishing condenser (or capacitor)** Funkenlöschkondensator *m* **~ failure** Zündungsaussetzung *f* **~ frequency** Funkenzahl *f*, Wellenzugfrequenz *f* **~ gap** (rotary) Drehzündbrücke *f*, Elektrodenabstand *m*, (magneto) Funkenabstand *m*, Funkenbrücke *f*, Funkstrecke *f*, Meßfunkenstrecke *f*, Zünderbrücke *f*
**spark-gap, ~ breakdown** Funkenüberschlag *m* **~ chamber** Schlagraum *m* **~ disc** Funkenstrecker *m* **~ face** Funkenstreckenelektrode *f* **~ rotor** Läufer *m* oder Rotor *m* der Funkenstrecke, Rotorfunkenstrecke *f* **~ space** Schlagraum *m* **~ terminal** Funkenstreckenelektrode *f* **~ tube** Nullode *f* **~ type of arrester** Funkenstreckenblitzableiter *m*
**spark, ~ generating spot** funkenbildende Stelle *f* **~ generator** Funkenerzeuger *m* **~ guard** Sprühschutzring *m* **~ handlever tube** Zündspindel *f* **~ igniter** Funkenzünder *m* **~ ignition** Fremdzündung *f* **~-ignition engine** Otto-, Zünder-motor *m* **~ jumping the gap** Abreiß-

funke *m* ~ **killer** Funken-löscher *m*, -schutz *m*
~ **knob** Funkenstreckenelektrode *f* ~ **knock**
Zündfunkenklopfen *n* ~ **lag** Funken-, Zündverzögerung *f* ~ **length** Funkenlänge *f*, Schlagweite *f* ~ **lever** Zünd(er)verstellhebel *m* ~ **line**
Funkenlinie *f* ~ **micrometer** Funkenmikrometer
*n* ~ **note** Funkenton *m* ~-**over** Funkenüberschlag *m* ~-**over path** Überschlagstrecke *f*
~-**over strength** Spannungssicherheit *f* ~ **plate**
(condenser of automobile radio) Funkplatte *f*
~ **plug** Funkkerze *f*, Glühkerze *f*, Kerze *f*,
Zündkerze *f*
**spark-plug,** ~ **adapter** Zündkerzeneinsatz *m* ~
**body** Kerzenkörper *m* ~ **boss** Zündkerzenauge
*n* ~ **bushing** Zündkerzeneinsatz *m* ~ **cleaner**
Kerzenreiniger *m* ~ **connecting nut** Zündkerzen-
anschlußmutter *f* ~ **elbow** Zündkerzenknie-
stück *n* ~ **electrode** Kerzenelektrode *f*, Zünd-
kerzenelektrode *f* ~ **erosion** Zündkerzenab-
brand *m* ~ **gap** Elektrodenabstand *m* der Zündkerze ~ **gauge** Zündkerzenlehre *f* ~ **head**
Kerzenkopf *m* ~ **hole** Zündkerzenbohrung *f* ~
**ignition** Kerzenzündung *f* ~ **insert** Zündkerzen-
einschraubbüchse *f* ~ **interference** Zündkerzen-
interferenzstörung *f* ~ **point** Zündstück *n* ~
**seat** Zündkerzensitz *m* ~ **series** Kerzenreihe *f*
~ **setting tool** Zündkerzeneinstellwerkzeug *n* ~
**shield** Funkenabschirmung *f* ~ **shielding** Zünd-
kerzenabschirmung *f* ~ **spanner (wrench)**
Zündkerzenschlüssel *m* ~ **terminal** Zündkerzen-
klemmschraube *f* ~ **tester** Zündungsprüfer *m*
~ **thread** Kerzengewinde *n* ~ **voltage** Zünd-
kerzenspannung *f* ~ **washer** Zündkerzendich-
tung *f* ~ **wire** Zündkabel *n*
**spark,** ~ **position** Funkenlage *f* ~ **potential** Ein-
satzspannung *f*, Funken-potential *n*, -spannung
*f* ~ **quench** Funkenlöscher *m* ~-**quench circuit**
**values** Funkenlöscherkreiswerte *pl* ~ **quencher**
Funkenlöscher *m* ~-**quenching condenser** Fun-
kenlöscherkondensator *m* ~-**quenching resistor**
Funkenlöschwiderstand *m* ~ **radiation** Funkenlinie *f* ~ **rate** Funkenzahl *f* ~ **(continuous-**
**wave) receiver** Empfänger *m* für gedämpfte
(ungedämpfte) Wellen ~ **reducer** Funken-
schwächer *m* ~ **resistance** Funkenwiderstand
*m* ~ **signal** gedämpftes Zeichen *n* ~ **spacer**
Funkenstrecker *m* ~ **spectrum** Funkenspek-
trum *n* ~ **suppressing gases** funkenverhindernde
Gase *pl* ~-**suppression means on a contact** Kon-
taktenstörung *f* ~ **suppressor** Funkenlöscher-
kasten *m* ~-**suppressor coil** Funkenlöscherspule
*f* ~-**tail formation** Funkenschwanzbildung *f* ~
**telegraphy** Funktelegrafie *f* mit gedämpften
Wellen ~ **test** Funkenprobe *f* ~ **tester** Spannungsprüfer *m* ~ **timing** Zündeinstellung *f*
~-**timing range** Zündverstellbereich *m* ~ **track**
Funkenkanal *m* ~ **transmitter** Funkengeber *m*,
Funkensender *m*, gedämpfter Sender *m* ~
**welding** Lichtbogenschweißung *f*
**sparking** Feuern *n*, Funken *n*, Funken-bildung *f*,
-sprühen *n* ~ **of brushes** Bürstenfeuer *n*
**sparking,** ~ **advance** Verstellen *n* des Zündzeitpunktes ~ **advance handle** Vorzündungsgriff *m*
~-**advance lever** Frühzündungshebel *m* ~ **ball**
Kugelelektrode *f* ~ **distance** Funken-länge *f*,
-schlagweite *f*, -strecke *f*, Schlagweite *f* ~
**frequency** Gruppenfrequenz *f* ~ **instant** Zündzeitpunkt *m* ~-**out** Ausfeuern *n* (beim Schlei-

fen ohne Vorschub auslaufen lassen bis Funken
verschwinden) ~ **paper** (motor) Zündlunte *f* ~
**plug** Funkenstöpsel *m* ~ **power** Zündleistung *f*
~ **printer** Lichtbildtypendrucker *m* ~ **retard**
Nachzündung *f*, Spätzündung *f* ~ **tappet**
Abreißhebel *m*
**sparkle, to** ~ flittern, funkeln, Funken *pl*
sprühen, perlen
**sparkless** funkenfrei, funkenlos (electr.) ~
**breaking** funkenlose Unterbrechung *f*
**sparkling** Funkeleffekt *m*
**sparks emitted** Funkenauswurf *m*
**sparry** spatartig ~ **gypsum** Gipsspat *m*
**spasmodically** ruckweise
**spathic** spat-artig, -förmig, spatig ~ **carbonate**
Carbonspat *m* ~ **chlorite** Chloritspat *m* ~ **iron**
**ore** Flirr *n*, spatiger Eisenstein *m* ~ **ore** Spaterz
*n*
**spathiopyrites** Eisenkobalt-erz *n*, -kies *m*
**spatial** räumlich ~ **average** Raumgrößen-Mittel-
wert *m* ~ **charge density** räumliche Ladungsdichte *f* ~ **distribution** räumliche Verteilung *f*,
Raumverteilung *f* ~ **effect** Raumwirkung *f* ~
**encoder** Scheibenverschlüßler *m* ~ **feeling** (in
vision) Raumgefühl *n* ~ **formula** Raumformel *f*
~ **grid** Raumgitter *n* ~ **mark** Raummarke *f* ~
**model** Raummodell *n* ~ **orientation** räumliche
Orientierung *f* ~ **perception** Raumanschauung
*f* ~ **rod** Raumlenker *m*, räumlicher Lenker *m* ~
**theorem** räumlicher Satz *m* ~ **variables** Raum-
größe *f*
**spatio-temporal** raumzeitlich
**spats** Verkleidung *f* am Fahrgestell
**spatted** verkleidet
**spatter, to** ~ abspritzen, bespritzen, herausschleudern, spratzen, spritzen, verspritzen,
zerstäuben
**spatter** Spritzer *m*
**spattering** Verspritzen *n*, Zerstäubung *f*
**spatula** Rührscheit *n*, Spachtel *m*, Spatel *m* ~
**grinding** Spachtelschliff *m*
**spatular** löffelförmig ~ **strut** Löffelstrebe *f*
**spatulate** spatelförmig
**spavin** Spat *m*

buchstabieren **to** ~ **to** ansprechen
**speaker** Lautsprecher *m*, Redner *m*, Sprecher *m*
~ **cabinet** Lautsprechertruhe *f* ~ **cloth** Laut-
sprecherseide *f* ~ **cone** Lautsprechermembrane
*f* ~'s **lectern** Rednerpult *n* ~ **socket** Laut-
sprecheranschluß *m* (rdo) ~ **wire** Dienstleitung
*f*, Sprechleitung *f*
**speaking** Sprechen *n* ~-**and ringing key** Sprech-
und Rufschlüssel *m* ~ **arc** sprechende Bogenlampe *f*, sprechender Lichtbogen *m* ~ **battery**
Mikrofonbatterie *f*, Sprechbatterie *f* ~ **circuit**
Sprech-kreis *m*, -stromkreis *m*, -weg *m* ~ **clock**
Zeitansagegerät *n*, sprechende Uhr *f* ~-**current**
**supply** Sprechstrom-speisung *f*, -zuführung *f*
~-**current supply tube** Mikrofonspeiseröhre *f* ~
**equipment** Sprecheinrichtung *f* ~ **instrument**
Sprechapparat *m* ~ **key** Sprech-schalter *m*,
-schlüssel *m* ~ **position** Sprechstellung *f* ~
**range** Sprechreichweite *f* ~ **set** Abfragegarnitur
*f* ~ **trumpet** Schallrohr *n* ~ **tube** Sprachrohr *n*,
Sprechschlauch *m*
**spear** Greifdorn *m*, Pricke *f*, Speer *m*, Spieß *m*,
Stange *f* ~ **head** Stoßkeil *m* ~ **pyrites** Sperrkies *m*

**spears** Gestänge *n*
**special** besonder(s), eigen, extra, sonder ~ **aircraft** Spezialflugzeug *n* ~-**alloy tool steel** hochlegierter Werkzeugstahl *m* ~-**analysis steel**
Stahl *m* besonderer Zusammensetzung ~ **body**
Sonderaufbau *m* ~ **case** Sonderfall *m* ~ **cast
iron** Sonder-, Spezial-gußeisen *n* ~ **casting**
Formstück *n* ~ **chart** Sonderkarte *f* ~ **circuit**
Sonderleitung *f* ~ **code selector** Dienst-, Hilfs-
wähler *m* ~ **control position** Störungsüberwachungsplatz *m* ~ **cross connections** Aus-
nahmeverbindungen *pl* ~-**delivery letter** Eil-
brief *m* ~ **design** Sonderausführung *f* ~ **design
features** Besonderheiten *pl* der Konstruktion ~
**development** Sonderentwicklung *f* ~ **dispositions**
besondere Anordnungen *pl* ~ **dividend** Super-
dividende *f* ~-**effects mixer** Trickmischer *m*
~ **end mill** Maschinenstirnreibahle *f* ~ **engineer**
Fachingenieur *m* ~ **equipment** Sonder-aus-
rüstung *f*, -gerät *n* ~ **factors** besondere Einflüsse *pl* ~ **factory** Spezialwerk *n* ~ **feature**
Vorzug *m* ~ **finish** Spezialzurichtung *f* ~ **flap
valve** Froschklappe *f* ~ **foundry coke** Spezial-
gießereikoks *m* ~ **frequency** Sonderfrequenz *f*
~ **grade pig iron** Sonderroheisen *n* ~ **graticules**
Sonderstrichplatten *pl* ~ **headstock** Sonder-
spindelkasten *m* ~ **impression** Separatabdruck
*m* ~ **iron** Spezialeisen *n* ~ **knowledge** Sachkunde
*f* ~ **library** Fachbücherei *f* ~ **locking** bedingter
Verschluß *m* (r. r.) ~ **model** Spezialausführung
*f* ~ **mount** Sondereinstellfassung *f* ~ **nails**
Fassonstifte *pl* ~ **notion** Sonderbegriff *m* ~
**operation** Sonderbetrieb *m* ~ **piece** Fassonstück
*n* ~ **pig** Sonderroheisen *n* ~ **plywood** Blocksperrholz *n* ~-**purpose fixture** Sonderzweck-
vorrichtung *f* ~ **quality tubes** Röhre *f* in Sonderausführung ~ **rims** Schrägschulterfelgen *pl*
~ **rolled-steel bar** Spezialeisen *n* ~ **(ty) rolling
mill** Sonderwalzwerk *n* ~ **rolling-mill section**
Sonderwalzprofil *n* ~ **roof bracket for subscriber's circuits** Dachgesimsstütze *f* ~ **section**
Sonder-, Spezial-profil *n* ~-**section timber**
Profilholz *n* ~ **service** besonderer Dienstzweig
*m*, Sonderdienst *m* ~ **shape** Sonderprofil *n* ~
**slot-rotor type motor** Spezialnutläufermotor *m*
~ **socket-outlet** Spezialsteckvorrichtung *f* ~
**spanner** Fassonschlüssel *m* ~ **steel** Sonder-,
Spezial-stahl *m* ~ **steel for nitriding** Nitrier-
sonderstahl *m* ~ **step** Sondermaßnahme *f* ~
**telephone call** Sondergespräch *n* ~ **temper-
hardening** Sondervergütung *f* ~ **thread** Sonder-
gewinde *n* ~ **tool** Sonderwerkzeug *n* ~ **train**
Sonderzug *m* ~ **trimming machine** Spezialbe-
schneidemaschine *f* ~ **tube** Spezialröhre *f* ~
**type** Sonder-ausführung *f*, -bauart *f* ~-**type
plane** Sonderflugzeug *n* ~ **unit** Sondereinheit *f*
~ **wire** Fasson-, Sonder-draht *m* ~ **write-down**
Sonderabschreibung *f*
**specialist** Facharbeiter *m*, Fachmann *m*,
Sachverständiger *m* ~ **expert** Fachverstän-
diger *m* ~ **foundry of mechanics** Kunstwerker-
hütte *f*
**specialists** Fachleute *pl*
**specialization** Spezialisierung *f*
**specialize, to** ~ spezialisieren
**specialized** spezialisiert ~ **experience** Spezialer-
fahrung *f*
**specially-mounted** in Sonderfassung *f*

**specialty** Besonderheit *f*, Fach *n*, Spezialität *f* ~
**finish** Vered(e)lungslack *m*
**species** Abart *f*, Art *f*, Gattung *f*, Sorte *f*
**specific** arteigen, bestimmt, eigen, eigenartig,
eigentlich, eigentümlich, eindeutig, kennzeichnend ~ **(ally)** spezifisch
**specific,** ~ **capacity** spezifische Kapazität *f* ~
**charge** spezifische Ladung *f*, Verhältnis *n* von
Ladung zu Masse ~ **clearance** bezogenes Spiel
*n* ~ **coding** Speicherfolgekodierung *f* ~ **commodity rates** Spezialfrachtraten *pl* ~ **conductance**
spezifisches Leitvermögen *n* ~ **conductivity**
spezifische Leitfähigkeit *f* ~ **consumption** spezifischer Verbrauch *m* ~ **density** Gewicht *n* ~
**detergent** waschaktiv ~ **efficiency** absoluter
Nutzeffekt *m* ~ **embodiment** Ausführungsform
*f*, (patent) besonderes Kennzeichen *n* ~ **emission** spezifische Emission *f* ~ **energy** spezifische
Kapazität *f* ~ **energy of blow** spezifische Schlagarbeit *f* ~ **fuel consumption** Einheits-betriebs-
stoffverbrauch *m*, -kraftverbrauch *m* ~
**gravity** Eigengewicht *n*, Einheitsgewicht *n*,
spezifisches Gewicht *n*, (of material) Stoffgewicht *n*, Wichte *f* ~ **gravity of acid** Säuredichte
*f* ~ **gravity of the air** Luftwichte *f* ~ **gravity of
separation** Trennwichte *f* ~-**gravity bottle**
Pyknometer *n* ~ **gravity concentration** gravitative Konzentration *f* ~-**gravity spindle** Senkspindel *f* ~ **ground pressure** spezifischer Bodendruck *m* ~ **heat** Artwärme *f*, Eigenwärme *f*,
spezifische Wärme *f*, Wärme-einheit *f*, -größe *f*
~ **humidity** spezifische Feuchtigkeit *f* ~ **inductive capacity** relative Dielektrizitätskonstante *f*,
spezifische induktive Kapazität *f*, (for vacuum)
Verschiebungskonstante *f* ~ **ionization coefficient** differentieller Ionisationskoeffizient *m*
~ **load** Lastdichte *f* ~ **luminous intensity**
Leuchtdichte *f* ~ **period of time** Zeiteinheit *f* ~
**power output** Literleistung *f* ~ **pressure on
ground** Flächeneinheitsdruck *m* ~ **primary
ionization** spezifische Primärionisation *f* ~
**rating** spezifische Leistung *f* ~ **resistance**
spezifischer Widerstand *m* ~ **strength** spezifische Intensität *f* ~ **surface energy of the liquid**
spezifische Oberflächenenergie *f* der Flüssigkeit ~ **symmetry** Eigensymmetrie *f* ~ **type of
construction** Baumuster *n* für bestimmten Verwendungszweck ~ **viscosity** Eigenviskosität
*f*, spezifische Zähigkeit *f* ~ **volume** Eigenvolumen *n* ~ **weight** Gewichtsdichte *f*, Leistungsgewicht *n*, Wichte *f* ~ **weight of air**
Luftwichte *f*
**specification** Angabe *f*, Anweisung *f*, Aufstellung
*f*, Aufzählung *f*, Ausführung *f*, Baubeschrei-
bung *f*, (of a patent) Beschreibung *f*; Bestimmung *f*, eingehende Beschreibung *f*, Einzel-
anführung *f*, -angabe *f*, genaue Bezeichnung,
(Pflichtenblatt) Lastenheft *n*, Liste *f*, Merkblatt
*n*, Pflichten-blatt *n*, -heft *n*, Spezifikation *f*,
Stückliste *f*, Vorschrift *f*, Werkstoffliste *f* ~ **of
engine type** Motormustertafel *f* ~ **of sale** Lieferungsbedingung *f*
**specification,** ~ **sheet** Datenblatt *n* ~ **test** Ab-
nahme-probe *f*, -prüfung *f* ~ **value** Pflichtwert
*m*
**specifications** Anforderungen *pl*, Bestimmungen
*pl*, Normblatt *n*, Oberbegriff *m*, technische Vorschriften *pl* ~ **to be observed in operating** (ship,

plane, plant, etc.) Betriebsvorschriften *pl* ~ **for standard threads** Gewindenormblatt *n*

**specified** vorgeschrieben ~ **dimension** Bestellmaß *n* ~ **maximum permissible deviation** zulässige Regelabweichung *f* ~ **shape** Sollform *f* ~ **timing curve** Sollverstellinie *f*

**specifier** Spezifikationssymbol *n* (info proc.)

**specify, to** ~ bestimmen, genau benennen, spezifizieren, vorschreiben

**specimen** Erstufe *f*, Exemplar *n*, Formstück *n*, Muster *n*, Probe *f*, Probe-objekt *n*, -stab *m*, Prüfling *m* ~ **of ore** Erzprobe *f* ~ **of printing type** Schriftprobe *f* ~ **with semicircular notch** Rundkerbe *f* ~ **for strength test** Bruchprobe *f* ~ **with V-notch** Spitzkerbe *f*

**specimen, ~ air lock** Objektschleuse *f* ~ **book** Musterbuch *n*, Schriftprobe *f* ~ **changer** Probenwechsler *m* ~ **glass** Probenglas *n*, Schauglas *n* ~ **holder** Einspannkopf *m*, Objekthalter *m*, Spannkopf *m* ~ **holder for headed test pieces** Einspannkopf *m* für Stabköpfe ~ **holder for screwed test pieces** Einspannkopf *m* für Gewindeköpfe ~ **region** Objektbereich *m* ~ **room** Präparateraum *m* ~ **safety device** Präparatschutz *m* ~ **sheet** Aushängebogen *m* (print.) ~ **tube** Sammel-, Proben-glas *n*

**speck** faulige oder poröse Stelle *f*, Fleck *m*

**speckle, to** ~ flecken, masern, sprenkeln, tüpfeln

**speckle** Maser *m*

**speckled** fleckig, gemasert, gesprenkelt, getüpfelt, maserig, meliert ~ **with sullage** aschenfleckig ~ **curlwood** Maserholz *n*

**spectacle** Schauspiel *n* ~ **condenser** Brillenglaskondensor *m* ~ **frame** Brillengestell *n* ~ **furnace** Brillenofen *m* ~ **magnifier** Brillenlupe *f*

**spectacles** Brille *f*

**spectator** Schauer *m*, Zuschauer *m*

**spectral, ~ average** Mittelwerte *f* über die Spektralverteilung ~ **centroid** Spektralschwerpunkt *m* ~ **characteristic** spektrale Verteilungscharakteristik *f* ~ **chart** Spektraltafel *f* ~ **color** Spektralfarbe *f* ~ **components** Gemischkomponenten *pl* ~ **composition** spektrale Zusammensetzung *f* (des Lichtes) ~ **distribution** Spektralverteilung *f* ~ **grate ghosts** Gittergeister *pl* ~ **line** Spektrallinie *f* ~ **region** Spektralbereich *m* ~ **representation** Spektraldarstellung *f* ~ **response** (of eye) spektrale Empfindlichkeitsverteilung *f* ~ **response of eye** Augenfarbenempfindlichkeit *f*, Farbenempfindlichkeitsverteilung *f* ~**-response characteristic** spektrale Verteilungscharakteristik *f* ~ **response curve** (of eye cones) Zapfenkurve *f* ~ **selectivity** spektrale Empfindlichkeit *f* ~ **sensitivity of eye** Augenfarbenempfindlichkeit *f* ~ **series** Spektralreihe *f*

**spectrally resolved light** spektral zerlegtes Licht *n*

**spectrobolometer** Spektralbolometer *m*

**spectrogram** Spektrogramm *n*

**spectrograph** Spektralaufnahme *f*, Spektrogramm *n*, -graf *m*

**spectrometer** Spektrometer *n* ~ **absorption** Spektrometeraufnahme *f*

**spectrometric** spektralanalytisch

**spectrometry** Spektrometrie *f*

**spectrophotoelectric sensitivity** spektrale Empfindlichkeit *f*, spektrale Empfindlichkeitsverteilung *f*

**spectrophotoelectrical sensitivity** spektrale lichtelektrische Ausbeutung *f*

**spectrophotometer** Spektralfotometer *n*

**spectroradiometer** Spektroradiometer *n*

**spectroscope** Spektral-apparat *m*, -gerät *n*, Spektroskop *n*

**spectroscopic** spektral, spektralanalytisch

**spectroscopic (ally), ~ analysis** Spektralanalyse *f* ~ **dispersed (or separated) light** spektral zerlegtes Licht *n* ~ **observation** spektroskopische Beobachtung *f*

**spectroscopy** Spectroskopie *f*

**spectrum** Farben-band *n*, -fächer *m*, Spektrum *n*, Wellenband *n* ~ **of electromagnetic waves** Wellenspektrum *n* ~ **without substratum** untergrundfreies Spektrum *n*

**spectrum, ~ analysis** Spektralanalyse *f* ~ **analyzer** Spektralanalysator *m* ~ **line** Spektrallinie *f* ~ **locus** Spektralkurve *f* (TV) ~ **oscillator** Rasteroszillator *m*

**specular** spiegelbildlich, spiegelig, spiegelnd ~ **cast iron** Hartfloß *m* ~ **density** Aussichtsdichte *f* ~ **effect** Spiegelung *f* ~ **forge pig** Puddelspiegel *m* ~ **galena** Bleispiegel *m* ~ **iron mine** Glanzeisenerzgrube *f* ~ **iron ore** Eisenglanz *m*, Glanzeisenerz, Spiegelerz *n* ~ **light** direktes Licht *n* ~ **pig iron** Hartfloß *n* ~ **reflection** Glasflächenspiegelung *f*, Rückspiegelung *f*, spiegelnde Reflektion *f*, spiegelnde Rückstrahlung *f*, Spiegelreflexion *f* ~ **surface** spiegelnde Oberfläche *f* ~ **symmetry** Spiegelsymmetrie *f* ~ **wave** Spiegelwelle *f*

**speculum** Spekulum *n*, Spiegel *m* ~ **image** Spiegelbild *n* ~ **metal** Spiegelmetall *n*

**speech** Anrede *f*, Ansprache *f*, Rede *f*, Sprache *f* ~ **amplifier** Sprachverstärker *m* ~ **(-input) amplifier** Mikrofonverstärker *m* ~ **analysis** Sprachuntersuchung *f* ~ **band** Sprachfrequenzband *n* ~ **band inversion** (secret teleph.) Bandumkehrung *f* ~ **channel** Betriebsgespräch *n* ~ **circuit** Sprechkreis *m* ~ **coil** (of loud-speaker) Schwingspule *f* ~**-frequency range** Frequenzbereich *m* (Sprechbereich), Sprechfrequenzbereich *m* ~ **guarding circuit** Sprachschutz *m* ~**-input equipment** Mikrofonsender *m* ~ **level** Sprachpegel *m* ~ **modulated** besprochen, sprachmoduliert ~**-modulated continuous waves** sprachmodulierte ungedämpfte Wellen *pl* ~**-modulated wave** sprachmodulierte Welle *f* ~ **oscillations** Sprachschwingungen *pl* ~**-plus signaling** Systemwahl *f* ~ **power** Schalleistung *f* der Sprache ~ **quality** Sprachgüte *f* ~ **recorder** Aufnahmevorrichtung *f* ~ **recording** Sprach-, Sprech-aufnahme *f* ~ **rendition** Sprachwidergabe *f* ~ **sidetone** Rückhören *n* für Sprache ~ **study** Sprachuntersuchung *f* ~ **test** Sprechversuch *m* ~ **transformer** Sprechtransformator *m* ~ **transmission** Sprachübertragung *f* ~ **volume** Sprachlautstärke *f* ~ **wave** Sprach-, Sprechschwingung *f* ~ **waves** Sprachwellen *pl*

**speed, to** ~ beschleunigen ~ **a hydroplane through water** ein Wasserflugzeug *n* flächen **to** ~ **up** beschleunigen, sich beeilen, schneller werden

**speed** Drehzahl *f*, Fahrt *f*, (of travel) Fahrtgeschwindigkeit *f*, Gang *m*, Geschwindigkeit *f*, Getriebegang *m*, Laufgeschwindigkeit *f*, Lichtempfindlichkeit *f* (photo), Schnelle *f*, Schnellig-

keit *f*, Tempo *n*, Tourenzahl *f*, Umdrehungs-
zahl *f*, Umlaufzahl *f* **on** ~ auf Touren *f pl*,
Geschwindigkeit *f* oder Drehzahl *f* entspricht
dem gewollten Wert

**speed,** ~ **of absorption** Absorptionsgeschwindig-
keit *f* ~ **of accomodation** Anpassungsgeschwin-
digkeit *f* ~ **of admission** Zulaufgeschwin-
digkeit *f* ~ **of advance** Vorwärtsgeschwindigkeit
*f* ~ **of answer** Verzögerung *f* bei der Beantwortung ~ **of approach** Eintrittsgeschwindig-
keit *f* ~ **of ascent and descent** Aufstieg- und Ab-
stiegsgeschwindigkeit *f* ~ **of autorotation** Ei-
gendrehgeschwindigkeit *f* ~ **of blade** Blattge-
schwindigkeit *f* ~ **of combustion** Verbrennungs-
geschwindigkeit *f* ~ **of compression** Verdich-
tungsgeschwindigkeit *f* ~ **of conversion** Umsetzungsgeschwindigkeit *f* ~ **by dead reckoning**
erkoppelte Fahrt *f* ~ **of declination** Ausweichgeschwindigkeit *f* ~ **of discharge** Entladegeschwindigkeit *f* ~ **of displacement** Verrückungsgeschwindigkeit *f* ~ **of entering velocity** Ein-
trittsgeschwindigkeit *f* ~ **of escape** Abzugs-
geschwindigkeit *f* ~ **of exposure** Aufnahmegeschwindigkeit *f* ~ **of fall** Fallgeschwindigkeit *f*
~ **and feed selector** Drehzahlvorschubwähler *m*
~ **at ground level** Bodengeschwindigkeit *f* ~ **of**
**growth** Anklinggeschwindigkeit *f* ~ **per hour**
Stundengeschwindigkeit *f* ~ **of indication** An-
zeigegeschwindigkeit *f* ~ **of inlet velocity** Ein-
trittsgeschwindigkeit *f* ~ **of the landing run** Ausrollgeschwindigkeit *f* ~ **of a lens (or an objec-**
**tive)** Helligkeit *f* eines Objektivs ~ **at outlet**
Ausströmungsgeschwindigkeit *f* ~ **of plotting**
**maps** Auswertegeschwindigkeit *f*, Kartierungs-
geschwindigkeit *f* ~ **of progression** Fortschreitungsgeschwindigkeit *f* ~ **of propagation** Fortpflanzungsgeschwindigkeit *f* ~ **of response** Ansprechgeschwindigkeit *f* ~ **of rotation** Drehgeschwindigkeit *f*, Rotationsgeschwindigkeit *f*,
Umdrehungsgeschwindigkeit *f* ~ **of rotation**
**while filling the centrifugal** (sugar mfg.), Füll-
drehzahl *f* ~ **of rotation in tailspin** Trudeldrehgeschwindigkeit *f* ~ **of rotations** Drehzahl *f* ~
**of ship** Schiffsgeschwindigkeit *f* ~ **over a**
**straightline course** Geschwindigkeit *f* über
Grundstrecke ~ **of structural change** Gefüge-
umwandlungsgeschwindigkeit *f* ~ **and torque**
**characteristics** Arbeitscharakteristik *f* (electr.)
~ **of transmission** Sendegeschwindigkeit *f* ~ **of**
**travel** Fortschreitungsgeschwindigkeit *f*,
Schweißfolge *f*, Wandergeschwindigkeit
*f* ~ **of travel through rolls** Walzgeschwindigkeit *f* ~ **of a wave** Wellengeschwindig-
keit *f*

**speed,** ~ **adjusting device** Drehzahlverstellvor-
richtung *f* ~ **adjusting motor** Drehzahlverstell-
motor *m* ~ **allowance** Vorhaltwert *m* ~ **boat**
Schnellboot ~ **boost** Stufenladung *f* ~**-box case**
Geschwindigkeitswechselgehäuse *n* ~**-change**
**gear** Gangwechselgetriebe *n*, Geschwindig-
keitswechselrad *n*~**-change lever** Hebel *m* für
den Spindelgeschwindigkeitswechsel ~ **changer**
Drehzahlverstellung *f* ~ **changer range** Bereich
*m* der Drehzahlverstellung ~ **changing** Gang-
schalten *n* ~**-changing lever** Hebel *m* für den
Wechsel der Spindelgeschwindigkeiten ~
**characteristics** Drehzahlverhalten *n* ~ **compen-**
**sator** Drehzahlausgleicher *m* ~ **contactor**

Drehzahlschutz *m* ~ **control** Geschwindigkeits-
regelung *f* ~**-control device** Umlaufregler *m*
~**-control lever** Drehzahlverstellhebel *m* ~
**control relay** Drehzahlsteuerrelais *n* ~ **con-**
**troller (or guard)** Drehzahlwächter *m* ~**-con-**
**trolling driving power** drehzahlregelnde Antriebskraft *f* ~ **counter** Drehzahlmesser *m*,
Fliehpendeldrehzahlmesser *m* ~ **deficiency**
Drehzahlmangel *m* ~ **delivery characteristic**
(pump) Förderleistungskennlinie *f* ~ **diagram**
Geschwindigkeitsschaubild *n* ~ **dressing** Beizmittel *n* ~ **drill** Schnellbohrer *m* ~ **drop (rise)**
Drehzahlabfall *m* (-anstieg) ~ **dynamo** Dreh-
zahldynamo *m* ~ **factor** Verstärkungsfaktor *m*
~ **finder** Geschwindigkeitssucher *m* ~ **fluctua-**
**tions** Geschwindigkeitsschwankungen *pl* ~ **gear**
Geschwindigkeitsgetriebe *n* ~ **governing** Drehzahlregelung *f* ~ **governor** Drehzahlregler *m*,
Geschwindigkeitsregulator *m*, Umlaufregler *m*
~ **governor for cars** Fahrtregler *m* ~**-graph** Geschwindigkeitskurve *f* ~ **increase** Geschwindigkeitssteigerung *f* ~**-increaser drive** Antrieb
*m* mit Übersetzung ins Schnelle, Übersetzungsgetriebe *n* ~ **indicator** Geschwindigkeitsmesser *m*, Tourenzähler *m*, Umdrehungsan-
zeiger *m* ~ **jet** Zusatzdüse *f* ~ **lever** Gang-
hebel *m*, Reglergetriebe *n* ~ **limit** Ge-
schwindigkeitsgrenze *f*, (allowed) zulässige
Geschwindigkeit *f* ~ **limiting device** Geschwin-
digkeitsbegrenzer *m* ~ **load** characteristic
Maschinencharakteristik *f*, Tourenbelastungs-
kennlinie *f* ~ **matching gear** Geschwindigkeits-
gleichhaltungsvorrichtung *f* ~ **measurement**
Geschwindigkeitsmessung *f* ~ **model** Renn-
modell *n* ~ **number** Lichtstärke *f* ~ **plate** Ge-
schwindigkeitsschild *m* ~ **pulley** Stufenscheibe
*f* ~ **range** Drehzahlbereich *n*, Geschwindig-
keitsbereich *m* ~ **range of engine** Drehzahlbereich *n* ~ **ratio** Übersetzungsverhältnis *n* ~
**record** Geschwindigkeitsrekord *m* ~ **reducer**
Geschwindigkeitsminderapparat *m* ~**-reducer**
**drive** Getriebe *n* mit Übersetzung ins Langsame, Untersetzungsgetriebe *n* ~ **reduction**
Drehzahlverminderung *f* ~**-regulating action**
geschwindigkeitsregelnde Wirkung *f* ~**-regulat-**
**ing device** Drehzahlregler *m* ~**-regulating**
**rheostat** Nebenschlußregler *m* ~ **regulation**
Geschwindigkeitsregulierung *f*, Tourenregulie-
rung *f* ~ **regulator** Drehzahlregler *m*, (in shape
of a knob) Tempoknopf *m*, Umlaufregler *m*
~**-regulator lever** Leistungswählhebel *m* ~
**relay** Drehzahlrelais *n* ~ **selection control**
**switch** Steuerschalter *m* zur Gangschaltung ~
**selection dial** Drehzahlwählskala *f* ~ **selection**
**limiter** Schaltbegrenzer *m* (zur Gangschaltung
*f*) ~ **selection system** Gangschaltanlage *f* ~
**selector (preselector)** Gangwähler *m* ~**-selector**
**level** Drehzahlwahlhebel *m* ~ **sense** Ge-
schwindigkeitssinn *m* ~ **switch** Drehzahlschal-
ter *m* ~ **table** Fahrttafel *f* ~ **tele-indicator** Um-
drehungsfernzeiger *m* ~ **test** Schleuderversuch
*m* ~**-time measuring device** Fahrlinienmeß-
gerät *n* ~ **transformer** Radumformer *m* ~ **truck**
Schnellastwagen *m* ~ **variation** Touren-
schwankung *f*, Geschwindigkeitsabstufung *f* ~
**variations** Geschwindigkeitsschwankungen *pl*
~ **way** Drehzahlbereich *n*
**speeding up** Beschleunigung *f*

**speedometer** Drehzahlmesser *m*, Geschwindigkeits-anzeiger *m*, -messer *m*, Meilenzähler *m*, Tachometer *n*, Tachymeter *n*, Tageszähler *m* ~ **assembly** Geschwindigkeitsmesserantrieb *m* ~ **worm** Tachometerschnecke *f*, Tachometertrieb *m*

**speedy** beschleunigt, früh, schleunig ~ **return** Schnellrücklauf *m*

**speiss** Speise *f* ~-**forming constituent** Speisebildner *m*

**spell, to** ~ buchstabieren

**spell** Bann *m*, Zauber *m* ~ **time** Arbeitszeit *f*

**spelling** Orthografie *f*, Rechtschreibung *f*, Schreibweise *f* ~ **system** Buchstabierverfahren *n*

**spelter** Rohzink *n*, Zink *m* & *n* ~ **solder** Hart-, Schlag-lot *n*

**spend, to** ~ aufwenden, ausgeben, verbrauchen, verbringen, verzehren

**spent** erschöpft, verbraucht ~ **air** Abluft *f* ~ **air chamber** Abluftkammer *f* ~ **ball** matte Kugel *f* ~ **catalyst** gebrauchter Katalysator *m* ~ **electrolyte** Elektrolysenendlauge *f* ~ **file** Glättfeile *f* ~ **grain** Treber *f pl* ~-**grains remover** Austrebermaschine *f* ~ **liquor** Ablauge *f*, Natronablauge *f*, schwarze Lauge *f* ~ **lye** Abfall-, Unter-lauge *f* ~ **shale** Schiefertonabfall *m* ~ **tan-bark** ausgelaugte Lohe *f* ~ **yeast** Brannthefe *f*

**sperm**, ~ **oil** Spermazet(i)öl *n* ~ **whale oil** Spermwalöl *n*

**spermaceti** Walrat *m*

**spermatogenesis** Spermatogenese *f*

**spewing** Auswerfen *n*, seitliches Herausquetschen *n* des Nietschaftes zwischen den Blechen

**sphaerosiderite** Kugeleisenstein *m*, Sphärosiderit *m*, strahliger Spateisenstein *m*

**sphalerite** Sphalerit *m*, Strahlblende *f*, Zinkblende *f*, Zinkpecherz *n*

**sphere** atmosphärische Schicht *f*, Bereich *m*, Feld *n*, Kreis *m*, Kugel *f*, Rahmen *m*, Späre *f* ~ **of action** Arbeitsgebiet *n* ~ **of combustion gas** Schwadenkugel *f* ~ **of exclusion** Ausschlußsphäre *f* ~ **of influence** Wirkungskreis *m* ~ **of operation** Wirkungssphäre *f* ~ **of reflection** Reflexionskugel *f*

**sphere**, ~ **gap** Kugel-abstand *m*, -funkenstrecke *f* ~ **gap voltmeter** Kugelspannungsmesser *m* ~ **photometer** Kugelfotometer *n*

**spheric point** Kreispunkt *m*

**spherical** ballig, kugel-artig, -förmig, kugelig, rund, sphärisch ~ **aberration** sphärische Aberration *f* oder Abweichung *f* ~ **area** Kugelfläche *f* ~ **area of indentation** Kugeleindruckfläche *f* ~ **balloon** Kugel-, Rund-ballon *m* ~ **bearing** Kugelgleitlager *n*, Lager *n* mit Kugelbewegung ~ **brass** Kugellagergehäuse *n* ~ **buoy** Kugel-boje *f*, -tonne *f* ~ **bush** Kugellagergehäuse *n* ~ **bushing** Kugelstück *n* ~ **cavity** Kugelhohlraum *m* ~ **coil** Kugelspule *f*, sphärische Spule *f* ~ **combustion chamber** Kugelbrennraum *m* ~ **composed systems** sphärische, zusammengesetzte Systeme *pl* ~ **cone** Kugelkeil *m* ~ **coordinate** Kugelordinate *f* ~ **correction** Richtungsreduktion *f* ~ **corrector** Kugelkorrektor *m*, Quadrantalkorrektor *m* ~ **crystal** Sphärokristall *n* ~ **curve** Raumkurve *f* ~ **cutter** Kugelfräser *m* ~ **densitometer** Kugeldensitometer *n* ~-**end measuring rod** Kugelendmaß *n* ~ **fitting** Kugelleuchte *f* ~ **flask** Kugel-

flasche *f* ~ **float** Kugelschwimmer *m* ~ **flying sphärische** Navigation *f* ~ **function** Kugelfunktion *f* ~ **harmonic analysis** Kugelfunktionsentwicklung *f* ~ **image** sphärische Abbildung *f* ~ **indentation** Kalotte *f*, Kugel-abschnitt *m*, -eindruck *m* ~ **involute** Kugelevolvente *f* ~ **joint** Kugelgelenk *n* ~ **knob** Kugelknopf *m* ~ **lens** Kugellinse *f* ~ **level** Dosenlibelle *f* ~ **mapping** sphärische Abbildung *f* ~ **micrometer** Kugelmikrometer *n* ~ **microphone** Kugelmikrofon *n* ~ **mirror** Kugelspiegel *m*, sphärischer Spiegel *m* ~ **mount** Kugelfassung *f* ~ **mouthpiece** Kugeleinsprache *f* ~ **nipple** (for pipes) Rohrdichtungskegel *m* ~ **part** (of universal joint) Gelenkkugel *f* ~ **point of support** Kugelstützpunkt *m* ~ **polar coordinates** Kugelkoordinaten *pl* ~ **pyranometer** Kugelpyranometer *m* ~ **rear sight** Kugelkimme *f* ~ **receiver** Kugelvorlage *f* ~ **reduction factor** Reduktionsfaktor *m* einer Lichtquelle ~ **reflector** Kugelspiegel *m* ~ **revolving nosepiece** Kugelrevolver *m* ~ **ribbon parachute** Großkreisbänderfallschirm *m* ~ **roller thrust bearing** Scheibentonnenlager *n* ~ **seat** Kugelfassung *f* ~ **seating** kugelig gebildete Auflagerfläche *f*, kugelige Lagerung *f* ~ **sector** Kugelausschnitt *m* ~ **segment** Kugel-abschnitt *m*, -segment *n* ~ **shape** Kugel-form *f*, -gestalt *f* ~ **shell** Kugelschale *f* ~ **shutter** Kugelverschluß *m* ~ **sorce** Kugelstrahler *m* (acoust.) ~ **spark gap** Kugelfunkenstrecke *f* ~ **surface** Kugelfläche *f* ~ **surface displacements** Verschiebung *f* auf Kugelflächen ~ **table** Kugeltisch *m* ~ **tensor** Kugeltensor *m* ~ **thrust bearing** Kugelspurlager *n* ~ **top** Kugelkreisel *m* ~ **top mark** Balltoppzeichen *n* ~ **top molecule** Kugelkreiselmolekül *n* ~ **trench-mortar shell** Kugelmine *f* ~ **triangle** sphärisches Dreieck *n* ~ **trigonometry** sphärische Trigonometrie *f* ~ **turning** Balligdrehen *n* ~ **vault** Helmdecke *f*, Kugelgewölbe *n*, Kuppel *f* ~ **vortex** Kugelwirbel *m* ~ **washer** kugeligballige Unterlegscheibe *f* ~ **wave** Kugel-, Raum-welle *f* ~ **wave horn** Kugelwellentrichter *m* ~ **wedge** Kugeldreieck *n* ~ **zone** Kugelzone *f*

**spherically**, ~ **mounted central** Knüppelmittenschaltung *f* ~ **over-corrected** sphärisch überkorrigiert ~ **seated grip holder** Einspannkopf *m* mit kugelförmigem Widerlager ~ **symmetric** kugelsymmetrisch

**sphericity** Kugelgestalt *f*

**sphero**, ~-**conical** sphärokonisch ~ **crystal** Sphärokristall *n*

**spheroid** Afterkugel *f*, Ellipsoid *n*, Rotationsellipsoid *n*, Sphäroid *n*; keilfach, kugelförmig, sphäroidisch

**spheroidal** kugel-ähnlich, -artig, kugelig ~ **cementite** kugeliger Zementit *m* ~ **concretion** Niere *f* ~ **concretion of marl** Kalksteinniere *f* ~ **core** sphäroidaler Rumpf *m* ~ **function** Sphäroidfunktion *f* ~ **graphite iron** Sphäroguß *m* ~ **jointing** kugelige Absonderung *f* ~ **state** sphäroidaler Zustand *m* ~ **well** Sphäroidpotential *n*

**spheroidization** Zusammenballung *f*

**spheroidize, to** ~ auf kugeligem Zementit glühen, zusammenballen

**spheroidized cementite** körniger Zementit *m*

**spheroidizing** Glühung *f* auf kugeligem Zemen-

tit, Weich-glühen *n*, -glühung *f*, Zusammenballung *f* ~ **property** Ballungsfähigkeit *f*
(cryst.)
**spherometer** Kugelmesser *m*, Sphärometer *n*
**spherosymmetric** kugelsymmetrisch
**spherule** Kügelchen *n*, Tröpfchen *n*
**spherulite** Kugelkörperchen *n*, Nierenstein *m*,
rundes Kristallkörperchen *n*, Sphärolith *m*
**sphygmograph** Pulsmesser *m*
**sphygmo-manometer** Sphygmomanometer *n*
**sphygmometer** Pulsmesser *m*
**spick and span** blitzblank
**spicula** Ähre *f*
**spicule** Spieß *m* (geol.)
**spider** Armkreuz *n* (engin.), (of a wheel) Arm-
stern *m*, Drehkreuz *n*, Klemme *f*, Läuferkörper
*m*, Speichennabe *f*, Speichenstern *m* (mach.),
Speichentriebrad *n*, Spinne *f*, Tragkreuz *n*,
(petroleum) Verteilrohr *n*, (of loud-speaker
coil) Zentrierungsfeder *f* ~ **of loud-speaker** Laut-
sprecherspinne *f*
**spider,** ~ **arm** Arm *m* des Kreuzstückes ~
**bearing** Armkreuzlager *n* ~ **coupling** Federbandkupplung *f* ~ **flange** Spinnenflansch *m* ~
**line** Fadenstrich *m* ~ **lines** Fadenkreuz *n* ~
**patch** Gänsefuß *m*, Hahnenfüße *pl*, Hahnepot
*m* ~**-shaped** radförmig ~ **stand** Dreifuß *m*
(Stativ) ~**-web coil** Korbbodenspule *f*, Spinn-
gewebspule *f*
**spiegel cupola** Spiegeleisenkupolofen *m*
**spiegeleisen** Hartfloß *m*, Spiegel-eisen *n*, -floß *n*
**spigot** Drucklager *n*, glattes Ende *n* (eines Muf-
fenrohres), Wirbel *m*, Zapfen *m*, Zentrier-an-
satz *m*, -bund *m*, (shaft) -zapfen *m* ~ **for gear**
Flanschbolzen *m*
**spigot,** ~**-and-socket** Vorsprung *m* und Rücksprung *m* ~**-and-socket joint** Muffenverbindung
*f* ~ **end** Spitzende *n* ~ **end of a tube** Muffenrohr-
spitzende *n* ~ **(-and-socket) joint** Muffenrohr-
verbindung *f* ~ **mortar projector** Ladungs-
werfer *m* ~ **mounting** Stirnwandbefestigung *f*
(Magnetzünder) ~ **shaft** Mittzapfen *m*
**spike, to** ~ annageln, nageln, spikern
**spike** Ähre *f*, Balken *m* (TV), Dorn *m*, Hakennagel *m*, Impulsspitze *f* (rdr), langer Nagel *m*,
Nagel *m*, Spiker *m*, Spitze *f*, Stift *m*, Zacken *m*
(rdr), Zinken *m* ~ **drawer** Geißfuß *m*, Nagel-
kalue *f* ~ **driver (or hammer)** Nagelhammer *m*
~ **length** Stachellänge *f* ~ **oil** Spiköl *n* ~ **plate**
(tripod) Spornteller *m*
**spiked** dornartig ~ **chain** Bahnkette *f* ~ **elevator**
**apron** Nadeltuch *n* ~ **lattice** Nadeltuch *n* mit
Holzlatten ~ **rollers** Stachelwalzen *pl* (Zackenwalzen)
**spikes** Stachel *pl*
**spiking** Nageln *n* ~ **curb** Nagelhilfskranz *m*
**spiky** stengelig
**spill, to** ~ (acid) ausschütten, gießen, vergießen,
verschütten, verspritzen
**spill** Anpfahl *m*, Grat *m*, Rückstreuungsverlust
*m* ~ **gap** Schutzfunkenstrecke *f* ~**-over** Überlaufen *n* ~**-over echo** bizarres Echo *n* ~**-over**
**position** Überlaufposition *f* ~**-over (or over-**
**flow) valve** Überströmventil *n* ~ **piston** Rück-
lauf-kolben *m*, -nadel *f* ~ **port** Überlaufkanal
*m* ~ **valve** Überströmventil *n*
**spill-way** Entwässerungsanlage *f*, Überfallwehr
*n*, Überlauf *m* ~ **board (dam)** Sturzbett *n* ~

**coefficient** Überfallbeiwert *m* ~ **crest** Überfall-
krone *f* ~ **crest gate** Überlaufverschluß *m* ~
**gate** Überlaufverschluß *m* ~ **weir** Entlastungs-
wehr *n*
**spilled,** ~ **electrolyte** ausgelaufene Säure *f* ~
**liquid** verschüttete Flüssigkeit *f*
**spilling** Verspritzen *n* ~**-over** plötzliches Schwingen *n*, Schwingungsneigung *f*
**spilly** Riß *m* im Leichtmetall, Riß *m* im Stahl
**spin, to** ~ drücken, drücken auf der Drehbank,
kreiseln, schleudern, schnell drehen, trudeln,
umkreisen, wirbeln **to** ~ **out** ausschleudern **to**
~ **over** überspinnen, überwalzen
**spin** Drall *m*, Drehbeschleunigung *f*, Drehen *n*,
Spinnbewegung *f*, Steiltrudeln *n*, Trudeln *n*,
Verdrehung *f* **to put the machine into a** ~ ins
Trudeln *n* bringen ~ **and orbital momentum**
Spin- und Bahnmomente *pl* ~ **of projectile**
Geschoßdrehung *f* ~ **of rope** Seilschlag *m*
**spin,** ~ **alignment** Spinausrichtung *f* ~ **angular**
**moment** Spindrehimpuls *m* ~ **bearing** Lauflager
*n* eines Kreisels ~ **chute** Trudelschirm *m*
~**-dependent force** spinabhängige Kraft *f* ~
**direction** Spinrichtung *f* ~ **distribution** Spin-
verteilung *f* ~ **doublets** Spindublette *pl* ~ **doubl-**
**ing** Spinverdopplung *f* ~**-drier** Wäscheschleu-
der *f* ~ **effect** Spinwirkung *f* ~**-effect nozzle**
Wirbeldüse *f* ~**-eigenstate** Spineigenzustand
*m* ~ **flip** Spinumklappung *f* ~ **flip scattering**
Spinumklappstreuung *f* ~**-magnetic moment**
magnetisches Spinmoment *n* ~**-magnetic reso-**
**nence** magnetische Spinresonenz *f* ~ **matrix**
Spinmatrize *f* ~ **model** Trudelmodell *n* ~**-mo-**
**mentum density** Spinmomentdichte *f* ~ **multi-**
**plet** Spinmultiplette *pl* ~ **orbit coupling** Spin-
bahn-aufspaltung *f*, -kupplung *f*, Spinimpuls-
kopplung *f* ~**-orbit interaction** Spinbahn-
Wechselwirkung *f* ~ **orbit potential** Spinbahn-
potential *n* ~ **partition function** Spinzustands-
summe *f* ~ **phenomenon** Spinerscheinung *f* ~
**quantum number** Impulsquantenzahl *f* ~ **refe-**
**rence axis** Arbeitsstellung *f* der Laufachse *f*
(of gyro) ~ **space** Spinraum *m* ~**-~ interaction**
Spin-Spin-Wechselwirkung *f* ~ **stabilization**
Drallstabilisierung *f*, Stabilisierung *f* durch
Rotation um die Längsachse ~**-stabilized**
drall-stabilisiert ~ **tensor** Drehgeschwindig-
keitstensor *m* ~ **term** Spinglied *n* ~**-up rocket**
Rakete *f* zur Drallerzeugung ~ **zero particle**
Spin-Null-Teilchen *n*
**spindle** Achse *f*, Drehspindel *f*, Druckstange *f*,
Fadenhänger *m*, Pinode *f*, Schraubenspindel *f*,
Schwingschenkel *m*, Spiere *f*, Spill *n*, Spindel *f*,
Welle *f*, Zweieck *n*; zweieckig ~ **for fly frames**
Fleierspindel *f* ~ **for tension discs** Klemmbolzen *m*
**spindle,** ~ **assembly** Schleifspindel *f* ~ **axis on**
**the tail wheel** Spornstrebenachse *f* ~ **band**
Spindelband *n* ~**-base** Spulenfuß *m* ~ **bearing**
Spillenlager *n*, Spindel-lager *n*, -lagerung *f*,
unteres Zapfenlager *n*, Zapfenlager *n* ~ **bear-**
**ings** Werkzeugschieberführungen *pl* ~**-bolt-**
**bearing box** Achsschenkellagerbüchse *f* ~**-bolt**
**nut** Achsschenkelmutter *f* ~ **box** Schleifspindel-
lager *n* ~ **bracket** Stützlager *n* ~ **brake** Spindelbremse *f* ~ **buoy** Spierentonne *f* ~ **capacity**
Wellenleistung *f* ~ **carrier** Ausladung *f*,
Spindelmitnehmer *m* ~ **collar** Spindelhals *m* ~

drawback Spindelrückgang *m* ~ **drawbar cap** Endverschraubung *f* für die Dornanzugstange ~ **drive** Spindelantrieb *m* ~ **driving pulley** Spindelantriebsscheibe *f* ~ **driving shaft** Spindelantriebswelle *f* ~ **drum indexing** Spindeltrommelschaltung *f* ~ **end** Spindel-kopf *m*, -nase *f* ~ **extension** Absperrspindelverlängerung *f* ~ **feed** Drehzahlenbereich *m* ~ **feed cam** Spindelvorschubkurve *f* ~ **flange** Spindelflansch *m* ~ **gauge** Spindelwaage *f* ~ **gearing** Frässpindelgetriebe *n* ~ **guard tube** Spindelrohr *n* ~**-head** Spindelkopf *m* ~**-head bearing** Spindelkopfgleitfläche *f* ~**-head gear mechanism** Spindelstockgetriebe *n* ~ **head longitudinal adjustment** Spindelstockbewegung *f* ~**-hole diameter** Spindelbohrung *f* ~ **lightning protector** Spindelblitzableiter *m* ~ **lower box** unteres Spindellager *n* ~ **nose** Spindel-kopf *m*, -nase *f* ~ **nose collet chuck** Spindelkopfspannzange *f* ~ **nut** Spindelmutter *f* ~ **oil** Spindelöl *n* ~ **pitch** Spindelteilung *f* ~ **point** Meßzapfen *m* ~ **quick-motion turnstile** Drehkreuz *n* zum schnellen Einstellen der Bohrspindel ~ **rail** Spindelbank *f* ~ **resistor** Spindelwiderstand *m* ~ **reversing cam** Spindelumkehrnocke *f* ~ **reversing mechanism** Frässpindelwendegetriebe *n* ~**-shaped** spindelförmig ~ **shuttle** Spindelschütze *m* ~ **sleeve** Pinole *f*, Spindel-büchse *f*, -hülse *f*, -pinole *f*, Spinole *f* ~ **slide** Frässpindelschlitten *m* ~ **slip** Spindelschlupf *m* ~ **slow motion** Handrad *n* zum Feineinstellen der Bohrspindel ~ **socket** Spindelbock *m*

**spindle-speed** Spindeldrehzahl *f* ~ **and feed selector** Drehzahl-Vorschubwähler *m* ~ **indicator** Drehzahlmesser *m* ~ **range** Drehzahlenbereich *m* ~ **selector levers** Schalthebel *pl* für Hauptgetriebe ~ **table** Drehzahlenschaubild *n*

**spindle,** ~ **starting lever** Spindelsteuerungshebel *m* ~ **step** Spindelbuchse *f* ~ **switch** Drehschalter *m*, Drehwähler *m* ~ **transmission** Spindelübertragung *f* ~ **travel** Spindelhub *m* ~**-type elevating machanism** Spindelhubwerk *n* ~**-type universal testing machine** Spindeluniversalprüfmaschine *f* ~ **upper box** oberes Spindellager *n* ~ **wharve** Spindelwirtel *m* ~ **wheel** Spillenrad *f* ~ **whorl** (textile) Zwirnscheibe *f* ~ **yoke** Spindelgabel *f*

**spine** Dorn *m*, Gitterstab *m* (d. Reflektors), Rückgrat *n*, Stachel *m*, Wirbelsäule *f*

**spinel** Magnesium-Titanspinell *n*, Spinell *m* ~ **of iron** Eisenspinell *n* ~ **lattice** Spinellgitter *n*

**spinnability** Spinnbarkeit *f*

**spinnable** spinnbar

**spinner** Abtaster *m*, Nabenhaube *f*, Probellerhaube *f* ~**-ring engine cowling** Ringmotorhaube *f*

**spinneret** Mehrloch-, Spinn-düse *f*

**spinning** Wirbelung *f* ~ **with hot water** Warmnaßspinnerei *f* ~ **upside down** Rückentrudeln *n* (aviat.)

**spinning,** ~ **arrangement** Spinnschema *n* ~ **axis** Trudeldrehachse *f* ~ **bobbin** Spinnspule *f* ~**-bobbin turning machine** Spinnspulendrehbank *f* ~ **bucket process** (synthetic fibers) Spinntopfspinnen *n* ~ **can** Spinntopf *m* ~ **curve** Trudellinie *f* ~ **defect** Spinnfehler *m* ~ **(sector) disc** Sektorenscheibe *f* ~ **draft** Feinspinnverzug *m* ~ **electron** Kreiselektron *n* ~ **emulsion** Spinn-

emulsion *f* ~ **factory** Spinnerei *f* ~ **frame** Feinspinnmaschine *f*, Spinn-maschine *f*, -stuhl *m* ~ **jenny** Jennyspinnmaschine *f*, Wagenspinner *m* ~ **lathe** Drückbank *f* ~ **lubricant** Spinnschmalze *f* ~ **machine** Forcier-, Umspinnmaschine *f* ~**-machine spindle** Spindel *f* für Spinnmaschinen ~ **molecule** Kreismolekül *n* ~ **nose dive** Trudeln *n* mit der Nase nach unten (aviat.) ~ **nozzle** Spinndüse *f* ~ **package** Gespinst *n* ~ **pot** Spinnzentrifuge *f* (Kunstseide) ~**-pot procedure** Topfspinnverfahren *n* ~ **solution** Spinnmasse *f* ~ **top** Bleihandkreisel *m*, Kreiselpendel *n* ~ **tunnel** Trudelwindkanal *m* ~ **yield** Spinnrendement *n*

**spinor** Spinor *m* ~ **calculus** Spinorkalkül *n* ~ **notation** Spinorbezeichnung *f*

**spiral, to** ~ **up** emporschrauben

**spiral** (walk) Schneckengang *m*, Schneckenlinie *f*, Schraubenlinie *f*, Spirale *f*, Spiralgleitflug *m*, Sturzspirale *f* (aviat.), Wendel *f*; schneckenförmig, schraubenförmig, spiralförmig, spiralig ~ **for landing** Spiralgleitflug *m* zur Landung ~ **in space** räumliche Spirale *f*

**spiral,** ~ **agitator** Bandrührer *m* ~ **anchor** Spiralanker *m* ~ **angle** Schraubenwinkel *m* ~ **angle of tooth** Schrägungswinkel *m* der Flankenlinie ~ **axis** Schraubung *f* ~ **band buffing cylinder** Wickelbandwalze *f* ~ **bevel** Spiralkegel *m* ~ **bevel gear** Spiralzahnkegelrad *n* ~**-bevel ring gear** schräg verzahnter Kegelradzahnkranz *m* ~ **bit** Holzspiralbohrer *m* ~**-casing mixed-flow pump** Spiralgehäuse-Schraubenradpumpe *f* ~ **cute** Wende(l)rutsche *f* ~ **coil** Schlangenröhre *f*, Spiralrohrschlange *f* ~ **coil of pipe** Rohrschlange *f* ~ **condenser** Schlangenkühler *m* ~ **condensing pipe (or tube)** Kühlschlange *f* ~ **conveyer** Förderschnecke *f*, Schraubenförderer *m* ~ **countersink** Spiralsenker *m* ~**-course torpedo** Spiralläufer *m* ~ **curve** Spirallinie *f* ~**-cut shell end mill** Aufsteckfräser *m* mit Spiralzähnen ~ **disc** Abtastscheibe *f* ~ **disc scanner** Spirallochscheibe *f* ~ **distortion** Zerdrehung *f* (anisotrope Verzeichnung) ~ **dividing head** Teilkopf *m* zum Spiralfräser ~ **diving** Trudeln *n* (aviat.) ~ **dowel** Spiraldübel *m* ~ **drilling machine** Drehbohrmaschine *f* ~**-eight cable** Doppelsternkabel *n* ~ **elevator** Förderrohr *n*, -röhre *f* ~ **feeder** Spiralaufgabevorrichtung *f* ~ **flute** Spiralnute *f* ~**-fluted** spiral genutet ~**-fluted miller** spiralgenuteter Langlochfräser *m* ~**-fluted reamers** Reihbahlen *pl* mit Spiralzähnen ~ **four** Sternvierer *m* ~ **(ed) four cable** Sternviererkabel *n* ~ **four quad** Sternvierer *m* ~ **gear** Schrauben-rad *n*, -zahngetriebe *n*, Spiral(zahn)rad *n* ~ **gear hob** Schraubenradabwälzfräser *m*, Schraubenwalzfräser *m* ~ **gear lead test** Steigungsprüfgerät *n* für schrägverzahnte Stirnräder ~ **gear wheel** Schraubenzahnrad *n* ~ **gearing** Schraubenradgetriebe *n*, Spiralgetriebe *n*, Spiralverzahnung *f* ~ **gears** Schnecke *f* ~ **glide** Spiralgleitflug *m* ~ **grain** Dreh-wuchs *m*, -wüchsigkeit *f* ~ **guide valve** Schraubenlenkerventil *n* ~ **hose** Spiralschlauch *m* ~ **illusion** Spiraltäuschung *f* ~ **instability** Abrutschinstabilität *f* (aviat.) ~**-jaw clutch** Klauenkupplung *f* mit vier Klauen ~ **key spring** Keildruckfeder *f* ~ **leads** Spiralsteigungen *pl* ~ **line** Berührungs-, Spiral-linie *f*

~-milling attachment Spiralfräseinrichtung *f* ~ orbit spectrometer Spiralbahnspektrometer *m* ~ quad Spiral-, Stern-vierer *m* ~ quad cable Sternviererkabel *n* ~ reinforcement Spiralbewehrung *f* ~-roller conveyer Rollenbahnspirale *f* ~ scanning Spiral-, Wendel-abtastung *f* ~ slot Spiralnute *f* ~ spline shaft Schrägkeilwelle *f* ~ spring Biege-, Schnecken-, Schrauben-, Spiral-, Sprung-, Zug-feder *f* ~-spring equilibrator Spiralfederausgleicher *m* (artil.) ~ springs gewundene Biegungsfedern *pl* ~ suction hose Spiralsaugschlauch *m* ~ target Schneckenscheibe *f* ~ teeth grätenverzahnt ~ tension spring Schraubenzugfeder *f* ~ thread Drall *m* ~ tooth system Spiralverzahnung *f* ~-toothed bevel gear Spiralkegelrad *n* ~ tube Rohrschlange *f*, Schlangen-, Schrauben-rohr *n* ~ tube binder Spiralhülsenwickelmaschine *f* ~ tube piston unit Rohrschlangentriebwerk *n* ~ turning fillet Spiralwendeleiste *f* ~ vault Schneckengewölbe *n* ~ vibrating troughs Wendelschwingrinnen *pl* ~ volute Austrittsspirale *f* ~-weave cable Hochfrequenzlitze *f* ~ wheel Schneckenrad *n* ~ winding fillet Spiralwendeleiste *f* ~ wire Drahtspirale *f*, Spiral-, Wendel-draht *m* ~ worm Spiralschnecke *f*

spirally spiralförmig ~ fluted (or grooved) spiralverzahnt ~ welded spiralgeschweißt ~ welded tube spiralgeschweißtes Rohr *n* ~ wound shaft Spiralwelle *f*

spirals of air stream Strömungsspiralen *pl*
spire (of a screw) Gang *m*, Gewinde *n*, Halm *m*, Schnecke *f*, Turmspitze *f*, Windung *f* einer Wicklung ~ of rope Seilwindung *f*
Spirek shaft (or tile) furnace Spirek-Ofen *m*
spirit Geist *m*, Spiritus *m* ~ of soap Seifenspiritus *m* ~ of verdigris Grünspanessig *m*
spirit, ~ color Spritfarbe *f* ~ compass Flüssigkeitskompaß *m*, Schwimmkompaß *m* ~ gauge Weingeistmesser *m* ~ lacquer Spritlack *m* ~ level Libelle *f*, Nivellier-libelle *f*, -waage *f*, Notvisier *n*, Wasserwaage *f* ~-level clinometer Dosenlibelle *f*, Libellenquadrant *m* ~ refinery Spritraffinerie *f* ~ stain Spritzbeize *f* ~ thermometer Flüssigkeitsthermometer *n* ~ varnish Spirituslack *m*
spirits Spirituosen *pl* ~ of camphor Kampferspiritus *m*
spirometer Atmungsmesser *m*, Pneumonometer *n*
spiro-type air turbine Pfeilradmotor *m*
spit, to ~ spratzen, sprühen, spucken to ~ and mew (like a cat) fauchen to ~ forth entsprühen
spit (of land) Landzunge *f*, Bratspieß *m*, Spieß *m*
spitting Spratzen *n*
spittler Speichel *m*
splash, to ~ spratzen, verspratzen, verspritzen to ~ around herumspritzen, umherspritzen
splash Spratzen *n*, Spritzer *m*, Wassersäule *f* ~ baffle Ablenkplatte *f* ~ board Wellenbrecher *m* ~ lubricant Schleuderöl *n* ~ lubrication Spritzschmierung *f*, Tauch(bad)schmierung *f* ~ plate Ablenkplatte *f* des Schmiersystems ~-proof spritzwassergeschützt ~-proof enclosure Spritzwasserschutz *m* ~ ring Spritzring *m* ~ ring on camshaft Spritzscheibe *f* auf Nockenwelle ~ wall Schallwand *f* (Tankwagen) ~ water Spritzwasser *n*
splasher Radmantel *m*

splashing Angießen *n*, Herausspritzen *n*, Spratzen *n*, Tauchbad *n*, Verspritzen *n*
splatter, to ~ verspratzen
splay, to ~ ausspreizen to ~ out ausschweifen
splay (Fenster)Schmiege *f*, (Leibungs)Schräge *f* ~ of wheel Radsturz *m*
splayed (a pipe mouth) ausgebreitet, ausgeschrägt ~ arch ausgeschrägter Bogen *m* ~ jamb abgeschrägter Pfosten *m* ~ joint abgeschrägte Verbindung *f*, Schrägverblattung *f*
splice, to ~ (cable) ansplissen, aufpfropfen, einspleißen, kleben (tape rec.), schäften, spleißen, stoßen, überlaschen, verlaschen, verspleißen to ~ in einschneiden to ~ the spar den Holm *m* schäften
splice Doppeleinschlag *m*, Lötstelle *f*, Schäftung *f*, Seilverbindung *f*, Spleiß *f*, (cable) Spleißstelle *f*, (cable conductor) Spleißung *f*, Spliß *m*, Splissung *f* ~ angle Deckwinkel *m* ~ bar Spleißstange *f* ~ bar to prevent creeping of the rails Stemmlasche *f* gegen Schienenwandern ~ box Kabelmuffe *f*, Verbindungsmuffe *f* ~ bump Klebestellengeräusch *n* ~ effected gradually over length of member Stufenstoß *m* ~ point Einbindestelle *f* ~ strap Lasche *f* ~ strip Lasche *f* (r.r.)
splicer Kabellöter *m*, Klebepresse *f*, Spleißer *m*, Splisser *m* ~'s tool bag vollständige Gerätetasche *f* für Löter
splicing Endlosmachen *n* (tape rec.), Kleben *n*, Spleißen *n*, Spleißung *f*, Überlaschung *f*, Verspleißung *f*, (textiles) Verstärkung *f* ~ bench Klebetisch *m* ~ block Spleißkloben *m* ~ box Spleißkasten *m* ~ clamp Spleiß-, Verbindungs-klammer *f* ~ ear Verbindungsklemme *f* ~ fid Fid *f* ~ guide Verstärkungsfadenführer *m* ~ needle Spleiß-horn *n*, -nadel *f* ~ pin Spleiß-horn *n*, -nadel *f* ~ sleeve Verbindungsmuffe *f* ~ table Klebetisch *m* ~ tape Klebeband *n* (tape rec.) ~ test Schäftungsversuch *m* ~ tool Spleißgerät *n*, Splisser *m*
spline, to ~ mit Keilnuten *pl* versehen, verkeilen
spline Feder *f*, Feder *f* für Keilnut, Federnute *f*, Keilnut *f* einer Keilverzahnung, Keilzahn *m*, Nut *f*, Nute *f*, Nutung *f*, Rille *f* ~ drive Ritzelantrieb *m* ~ length Keilbahnlänge *f* ~ milling machine Keilachshobler *m* ~ part Keilleiste *f* ~ shaft Keilnabe *f*, Nuten-, Schiebe-, Vielkeil-welle *f* ~-shaft grinding attachment Keilwellenschleifeinrichtung *f* ~ shaft hob Keilwellen(ab)-wälzfräser *m*
splined (externally or internally) kerbverzahnt ~ arbor Keilwelle *f* ~ butt Keilwellenstumpf *m* ~ hub sleeve Keilnabenmitnehmer *m* ~ nut Mutter *f* mit Kerbverzahnung ~ pin Kerbstift *m* ~ shaft Keil-, Riffel-, Segment-welle *f* ~ shaft bush Keilwellenhülse *f* ~ shaft end Keilwellenstumpf *m* ~ shaft pin Keilwellenzapfen *m* ~-shaft starter Schubtriebanlasser *m* ~ trunnion längsverzahnter Drehzapfen *m* ~ yoke Keilnabenmitnehmer *m*, Keilnabenschiebestück *n* ~ yoke-type universal joint Keilnabengelenk *n*
splines Keilnuten *pl*, Keilverzahnung *f*
splining Sternkeilwellen *pl* fräsen
splint, to ~ anschienen, schienen
splint Armschiene *f*, Schiene *f*, Schleiße *f*, Splint *m* ~ bit Splintbohrer *m* ~ coal Schiefer-, Splint-, Splitter-kohle *f*

**splinter, to ~** splittern, zersplittern **to ~ off** absplittern, zerspanen

**splinter** Schiefer *m*, Schleiße *f*, Span *m*, Spleiß *f*, Splitter *m*, Sprengstück *n* **~ bomb** Splitterbombe *f* **~ catch(er)** Splitterfänger *m* **~ density** Splitterdichte *f* **~ forceps** Splitterpinzette *f* **~ -proof** splittersicher **~-proof cover** splittersichere Deckung *f* **~-proof glass** splitterfreies Glas *n* **~ screen** Schild *n* **~ test** Spanversuch *m*

**splintered, ~ pulp** splitterige Papiermasse *f* **~ wound** Splitterverletzung *f*

**splintering** Absplitterung *f* **~ effect** Splitterwirkung *f* **~ fire** Zersplitterung *f*

**splintery** splitt(e)rig **~ fracture** splitterige Bruchfläche *f*

**splints** Beschienung *f*

**split, to ~** aufklaffen, aufreißen, aufspringen, (rays) brechen, klaffen, klieben, platzen, reißen, schlitzen, spalten, spleißen, splittern, springen, spritzen, teilen, zerfallen, zerspalten **to ~ off** abspalten, absplittern, (rays) abzweigen (von Strahlen) **to ~ up** abbauen, aufspalten, aufteilen, spalten, (light beam) zerlegen **to ~ up in shivers** (Holz) sich aufschiefern **to ~ up water** Wasser *n* zersetzen

**split** Ritze *f*, Spalt *m*, Spalte *f*, (point) Verzweigungspunkt *m*; aufgeschnitten, geschlitzt, gespalten, gesprungen, geteilt, rissig, versplintet, zerteilt **~ anode** Schlitzanode *f* **~-anode magnetron** Schlitzanodenmagnetron *n* **~ axle** geteilte Achse *f* **~ battery** geteilte Batterie *f* **~ beam** Strahlspaltung *f* **~ bearing** geteiltes Lager *n* **~ belt pulley** (for fan) Riemenscheibenhälfte *f* **~ billet** Scheit *m* & *n* **~ bushing** Spaltbuchse *f* **~ cable grip** Kabelziehschlauch *m* **~ chuck** Spannpatrone *f* **~ clamp ring** Schellenklemme *f* **~ clinker flags** Spaltplatten *pl* **~ collar** gespaltene Gewindemuffe *f* **~ collet** Keilkegel *m* **~ cone** Schlitzkegel *m* **~ (duplex) connection** Gabelschaltung *f* (Gegensprechgabel) **~ core** geschlitzter Kern *m* **~-core-type transformer** Zangenstromwandler *m* **~ coupling** geteilte Kupplung *f*, Schalenkupplung *f* **~ cover** geschlitzte Kappe *f* **~ dashes** gebrochene Morsestriche *pl* **~ die** geteiltes Schneideisen *n*, Schneideisen *n* mit Schlitz **~ direction finding** Schnittpunktpeilung *f* **~ duct** zweiteiliges Formstück *n* **~ -field-coincidence range finder** Halbbildentfernungsmesser *m* **~-field motor** Motor *m* mit geteiltem Feld **~-field range finder** Schnittbildentfernungsmesser *m* **~ filament** Streufaden *m* **~ filter** Spaltfilter *m* **~ firewood** Scheitholz *n* **~ flap** Spalthilfsflügel *m*, Spaltklappe *f*, Spreizflügelklappe *f* **~ focus** geteilte Abbildung *f*, Zwitterapertur *f* **~ gear** geteiltes Zahnrad *n* **~ glass mold** Preßglasform *f* **~-head rivet** Spaltniet *n* **~ hole** Spaltloch *n* **~ image** geteiltes Bild *n* **~-image rangefinder attachment** Schnittbildeinsatz *m* **~-in piston ring** Kolbenringschlitz *m* **~ landing-gear axle** geteilte Fahrgestellachse *f* **~ lath** Waldlatte *f* **~ level house** terrassiertes Haus *n* **~ lock washer** Federring *m* **~ mold** mehrteilige Form *f* **~-multiples telegraph** Mehrfachtelegraf *m* in Gabelschaltung **~ notation** aufgespaltene Schreibweise *f* **~ nut** Schloßmutter *f* **~ order wire** Dienstleitung *f*, Gilles-Schaltung *f*, Sammeldienstleitung *f*, Sammeldienstleitungsbetrieb *m* **~ pattern** ge-

teiltes, mehrteiliges oder zweiseitiges Modell *n* **~-phase motor** Einphasenmotor *m* **~ pin** Schlitzstift *m*, Splint *m*, Splinte *f* **~-pin extractor** Splintauszieher *m* **~-pin hole** Splintloch *n* **~-pin pliers** Splintzieher *m* **~ pipe** zweiteiliges Rohr *n* **~ plug** Bananenstecker *m*, geschlitzter Stöpsel *m*, **~ point** Mitte *f* oder Scheitel der Differentialspule **~ pole motor** Spaltmotor *m* **~ post** Halbholz *n* **~ pulley** geteilte Riemenscheibe *f* **~ quadruplex** Doppelgegensprechen *n* für Gabelverkehr **~ reel** zweiteilige Spule *f* (film) **~ response** Doppelzeichen *n* (rdr) **~ ring** aufgeschnittener oder geteilter Ring *m*, Spaltring *m* **~-ring mold** mehrteiliges Gesenk *n* **~ riveting** Spaltnietung *f* **~-rudder assembly** geteiltes Seitenleitwerk *n* **~ second** Sekundenbruchteil *m* **~ section of a bearing** Lagerschale *f* **~-skirt piston** Schlitzmantelkolben *m* **~-skirt-type piston** Kolben *m* mit geteiltem Mantel **~ sleep** Spannhülse *f* **~ spoon** (geteilte, aufklappbare) Entnahmesonde *f* **~ stator capacitor** Kondensator *m* mit geteiltem Stator **~ steel key** gespaltener Stahlkeil *m* **~ straw** Lisseret *n* **~ taper sleeve** Spannhülse *f* **~-trail carriage** Spreizlafette *f* **~ transformer** Anzapftransformator *m* **~ tube** Schlitzrohr *n* **~-type axle** geteilte Achse *f* **~-type stabilizer** Verstellflosse *f* **~ undercarriage** achsloses Fahrgestell *n* **~ undercart** Zweibeinfahrwerk *n* **~-up** zerlegbar **~ valve (spring) washer lock** geteiltes Ventilklemmstück *m* **~ washer** geschlitzte Unterlagscheibe *f* **~ wheel** gesprengtes Rad *n* **~ winding** Treppenwicklung *f* **~-wired** mit Schaltschnüren *pl* geschaltet

**splits** Spaltschnitte *pl*

**splitting** Aufspaltung *f*, Brechen *n*, Spaltung *f*, Springen *n* **~ of forces** Kraftzersplitterung *f* **~ of points** Gabelfahrt *f* **~ of rays** Strahlbrechung *f* **~ of rational lines** Aufspaltung *f* der Rotationsbanden

**splitting, ~ button** Trenntaste *f* (teleph.) **~ effect** Sprengwirkung *f* **~ factor** Aufspaltungsfaktor *m* **~ hammer** Keillochhammer *m* **~ key** Abtrenn-, Trenn-schalter *m* **~ knife** Spaltmesser *n* **~ machine** Spaltmaschine *f* (für Leder) **~-off** Abspaltung *f* **~-off acid** Säureabspaltung *f* **~ patterns** Aufspaltungsbilder *pl* **~ phenomena** Aufspaltungserscheinung *f* **~ position** Trennstellung *f* **~ resistance** Spaltfestigkeit *f* **~ test** Spaltversuch *m* **~-up** Aufspalten *n*, Aufspaltung *f*, Zersetzung *f* **~-up line** Ausbiegestelle *f* **~ wedge** Spaltkeil *m*

**splutter, to ~** auswerfen, hervorsprudeln

**splutter prints** Spritzdrucke *pl*

**spluttering** Knallgeräusche *pl* (rdo), Stuckern *n* (des Motors)

**spodium** Tutia *f*

**spoil, to ~** verderben, verschlechtern, verwahrlosen **to ~ by careless work** verpfuschen **to ~ in grinding** vermahlen

**spoil** Bodenaushub *m* **the ~** Baggergut *n* **~ bank** Aufstürzung *f*, Berghalde *f*, Bodenkippe *f*, Seitenablagerung *f* **~ heap** Halde *f* **~ sheet** Ausschußbogen *m*

**spoilage** Ausschuß *m*

**spoiled** verdorben **~ by rain** verregnet **~ casting** Ausschuß *m*, Fehlguß *m*, Wrackguß *m*

**spoiler** Stau *m*, Störklappe *f*, Unterbrecher-klappe *f*, Zusatzflügel *m* ~ **control** Unterbrechersteuerung *f* (aviat,)

**spoiling** Vergällung *f*, Verschlechterung *f* ~ **effect** Wirbelbildung *f* ~ **loss** Wirbelverlust *m*

**spoke, to** ~ **a wheel** ein Rad verspeichen

**spoke** (of wheel) Arm *m*, (of wire wheel) Drahtspeiche *f*, Speiche *f*, Sprosse *f* ~-**and handle-turning machine** Speichen- und Griffdrehbank *f* ~ **fitter** Speicheneinsetzer *m* ~ **flange** Speichenträger *m* ~ **gauge** Speichenmesser *m* ~-**machine** Speichen- und Griffdrehbank *f* ~ **nipple** Speichennippel *m* ~ **shave** Leder-, Speichen-hobel *f*, Ziehklinge *f* ~ **tightener** Speichenspanner *m* ~ **wheel center** Radstern *m* ~ **wire** Speichendraht *m*

**spoking** Radeffekt *n*

**sponge, to** ~ anfeuchten mit dem Schwamm, dekatieren **to** ~ **out** auswischen, (gun) das Rohr auswischen

**sponge** Schwamm *m*, Wischer *m* ~ **charcoal** Schwammkohle *f* ~ **lead** Bleischwamm *m* ~ **rubber** Gummischaum *m*, Schaum-, Schwammgummi *m* ~-**rubber disc** Schwammgummischeibe *f* ~ **staff** Wischstock *m*

**spongelike** schwammig ~ **decay** faulende Rostanfressung *f*

**spongiform quartz** Schwimmkiesel *m*

**sponginess** Lappigkeit *f* ~ **of cast iron** Schwammigkeit *f* des Gußeisens

**sponging establishment** Dekatieranstalt *f*

**spongy** blasig, locker, porös, schwammig ~ **bismuth** Wismuthschwamm *m* ~ **cast iron** schwammiger Guß *m* ~ **copper** Schwammkupfer *n* ~ **lead** Bleischwamm *m* ~ **platinum** Platinschwamm *m*

**sponson** Kaffe *f*, Schwimmerstummel *m*

**sponsor, to** ~ Bestrebung fördern, sich für etwas einsetzen

**sponsored television** Werbefernsehen *n*

**spontaneous** freiwillig, selbst ~ **breakage in a horizontal fracture** Abringeln *n* (of glass) ~ **combustibility** Selbstentzündlichkeit *f* ~ **combustion** Selbst-entzündung *f*, -verbrennung *f*, spontane Verbrennung *f* ~ **decomposition** Selbstzersetzung *f* ~ **discharge** selbständige Entladung *f* ~ **(radioactive) disintegration** spontaner (radioaktiver) Zerfall *m* ~ **gathering** Zusammenlauf *m* ~ **heating** Selbsterhitzung *f* ~ **ignition** Selbstzündung *f* ~ **inflammability** Selbstentzündlichkeit *f* ~ **oscillating** selbständige Schwingungserzeugung *f* ~-**polarization method** Methode der spontanen Polarisation ~ **pre-ignition** Selbstfrühzündung *f* ~ **slaking of lime** Selbstlöschung *f* des Kalks

**spontaneously,** ~ **combustible substance** Hypergole *f* ~ **inflammable** selbstentzündlich

**spool, to** ~ aufspulen, aufwickeln, spulen, umwickeln, wickeln

**spool** Bandspule *f* (tape rec.), Filmkern *m*, Fördertrommel *f*, Haspel *f*, Rolle *f*, (weaving) Schußspule *f*, Spule *f*, Spulen-kasten *m*, -körper *m*, Weife *f* ~ **of thread yarn** Garnrolle *f*

**spool,** ~ **arm** Spulen-arm *m*, -träger *m* ~ **box** Kassette *f* ~ **chain** Haspelkette *f* ~ **disc** Spulenteller *m* ~ **feeder** Spulenaufstecker *m* ~ **flange** (or head) Spulen-flansch *m*, -scheibe *f* ~ **paper** Hülsenpapier *n* ~ **piece** Füllstück *n* ~ **pin** Spul-

**spindel** *f* ~ **shaft** Löffelhaspelwelle *f* ~ **support** Spulenarm *m*, Spulenteller *m* (tape rec.)

**spooler** Spuler *m*

**spooling** Spulen *n*, Spulerei *f* ~ **flange** Trommelseite *f*

**spoon, to** ~ löffeln

**spoon** (torpedo) Führungsschaufel *f*, Löffel *m* ~ **with bowl** (for relining furnace) Flickschaufel *f* ~ **sporn,** ~ **agitator** Löffelrührer *m* ~ **bit** Löffelbohrer *m* ~ **bit carbing chisel** Schaufelbeitelchen *n* ~-**bit gouge** Löffelgutsche *f* ~-**bit parting tool** Löffelgeißfuß *m* ~ **brake** Schuhbremse *f* ~ **die** Löffelgesenk *m* ~ **discharge** Löffelaustragung *f* ~ **rolling mill** Löffelwalze *f* ~ **sample** Löffelprobe *f* ~ **scraper** Löffelschaber *m* ~-**shaped** löffelförmig ~-**shaped prow** Löffelbug *m* ~ **test** Löffelprobe *f* ~ **tool** Löffelwerkzeug *n* ~-**type glass manometer** Glasledermanometer *n*

**spoonful, by the** ~ löffelweise

**sporadic** vereinzelt

**sport** Spiel *n*, Sport *m* ~ **airplane** Sportflugzeug *n* ~ **flying** Flugsport *m*

**sporting,** ~-**cycle lamp** Sportleuchte *f* ~ **flying** Flugsport *m*, Sportflugwesen *n* ~ **powder** Jagdpulver *n* ~ **rifle** Jagdflinte *f*

**sports-pilot** Sportflieger *m*

**spot, to** ~ ankernen, (a surface) anschneiden, aussehen, beflecken, betupfen, flecken, fleckig machen, rasten, sprenkeln, tüpfeln, zentrieren **to** ~ **with beam of light** anleuchten **to** ~-**face** fräsen (von Stirnflächen), plansenken **to** ~ **light** aufhellen **to** ~ **a surface** eine Auflagefläche anschneiden **to** ~-**weld** punktscheißen

**spot** (of arc on cathode) Ansatzpunkt *m* (beam pencil) Brennfleck *m*, Fleck *m*, Flecken *m*, (luminous) Leuchtpunkt *m*, Lichtmarke *f*, Maser *m*, Platz *m*, Stelle *f* **on the** ~ an Ort und Stelle, auf der Stelle ~ **on anode** (of an arc) Anodenansatzpunkt *m* ~ **(of arc) formed on cathode** Kathodenansatzpunkt *m* ~ **of light** Leuchtfleck *m*, leuchtender Punkt *m*, Leuchtscheibe *f* (-scheibchen *n*), Lichtzeiger *m*

**spot,** ~ **analysis** Fleckanalyse *f*, Tüpfelanalyse *f* ~ **arc lamp** Punktlichtlampe *f* ~ **blower** Rußbläser *m* ~ **check** fliegende Kontrolle *f*, Stichprobe *f* ~ **diameter** Fleckdurchmesser *m*, Leuchtfleckdurchmesser *m*, Linienbreite *f*, Zeilenbreite *f* (TV) ~ **distortion** Fleck-verformung *f*, -verzerrung *f*, Lichtfleckverzerrung *f* ~-**face** Auflagefläche *f* rings um ein Loch mit Senker glatt und eben fräsen ~ **facer** Aufstecksenker *m*, Oberflächenfräser *m* ~ **facers** Aufstecknabensenker *m* ~ **facing** Fräsen *n* (von Stirnflächen), Spiegeln *n* (von Auflageflächen für Schraubenköpfe) ~ **facing cutter** Anschneidesenker *m*, Zapfensenker *m* ~ **focus** Fleckschärfe *f* ~ **footing** Einzelfundation *f* ~ **frequency** gerastete Frequenz *f*, Streuwelle *f* (rdr) ~ **glueing** Punktleimverfahren *n* ~ **goods** greifbare Ware *f* ~ **heights** Höhenpunkte *pl* ~ **jamming** (absichtliche) Punktfrequenzstörung *f* ~ **landing** Genauigkeitslandung *f*, Ortslandung *f* ~-**light** Linsen-, Scheinsuch-scheinwerfer *m* ~ **light lens** Sucherglas *n* ~ **light scanner** Lichtstrahlfühler *m* ~ **lighting** Effektbeleuchtung *f* ~-**lighting glow lamp** Punktglimmlampe *f* ~ -**lighting lamp** Punktlichtlampe *f* ~ **method** Tüpfelmethode *f* ~ **noise factor** effektiver

Rauschfaktor *m*, Spektralrauschzahl *f* ~ **overlap** Zeilenüberdeckung *f* ~ **rate** Kassakurs *m* (bei Devisen) ~ **reading** Punktmarke *f* ~ **removing agent** Detachiermittel *n* ~ **sale** Promptverkauf *m* ~ **segregation** Gasblasensteigerung *f* ~-**set** (frequency) gerastet ~ **shape** Punktform *f* ~ **shape anisotropy** Punktformanisotropie *f* ~ **shift** (causing plastic effect) Bildpunktverlagerung *f* (TV) ~-**size distortion** Kugelgestaltsfehler *m* ~ **sugar** greifbarer Zucker *m* ~ **test** Flecken-, Tropfen-probe *f* ~ **transactions** Platzgeschäft *f* ~-**weld** Punktschweißung *f* ~-**weldable** punktschweißbar ~-**welded** punktgeschweißt, punktverschweißt ~-**welder** Punktschweißer *m*, Punktschweißmaschine *f* ~ **welding** Punktschweißen *n*, Punktschweißung *f* (electr.) ~-**welding apparatus** Punktschweißgerät *n* ~-**welding machine** Punktschweißmaschine *f* ~-**welding process** Punktschweißverfahren *n* ~ **wobble** Strahlwobblung *f* ~-**wobbling** Zeilenwobbelung *f* (TV) ~ **zero** Strahlruhelage *f*

**spotless (or stainless)** fleckenfrei
**spotlessness** Makellosigkeit *f*
**spots, ~ of arcing** Schmorstellen *pl* ~ **on film** Inseln *pl* im Film
**spotted** fleckig, gefleckt, gesprenkelt, maserig, scheckig ~ **prints** Pünktchen *pl* im Druck
**spotter** Beobachter *m*, Flecker *m*, Fleckergerät *n*
**spottiness** fleckiges Bild *n*
**spotting** Abhorchen *n* von Flugzeugen ~ **correction** (range) Aufsatzverbesserung *f* ~ **scope on height finder** Beobachterfernrohr *n* ~ **telescope** Beobachtungsrohr *n* ~ **test** Tupfprobe *f* ~ **tool** gebogener Zentrierstahl *m* ~ **top** Fleckerstand *m*
**spotty** gefleckt, soßig ~ **steel** grindiger Stahl *m* ~ **wear** stellenweise Abnutzung *f*
**spout, to** ~ springen, spritzen **to ~ out** herausspritzen
**spout** Ausguß *m*, Ausguß-kasten *m*, -schale *f*, -tülle *f*, Fülltülle *f*, Gußsteinauslauf *m*, Nase *f*, Röhre *f*, Schnabel *m*, Schnauze *f*, Schnauze *f* einer Kanne, Schneppe *f*, Tülle *f*, (of gutter) Wasserluke *f* ~ **of a gutter** Abtraufe *f* der Dachrinne
**spout, ~ adz** Krummdechsel *f* ~ **closing apparatus** Tüllenverschlußapparat *m* ~ **nozzle** Rinnenschnauze *f* ~ **plane** runder Hobel *m* ~ **slag** Rinnenschlacke *f*
**spouter** Sprudelbohrung *f*
**sprag** (automobile) Bergstütze *f*, Fallstütze *f*, Hemmkeil *m* ~ **brake** Knüppelbremse *f* ~ **device** Rücklaufsicherung *f*
**sprain, to** ~ verrenken
**sprain** Verrenkung *f*
**spray, to** ~ abbrausen, aufspritzen (a coat or film) aufstäuben, berieseln, besprengen, besprühen, spritzen, sprühen, streuen, zerstäuben **to ~ around** herumspritzen **to ~ gas from a plane** abregnen **to ~ off (or over)** abspritzen **to ~ on** aufspritzen **to ~ out** herausspritzen **to ~ from a plane** abgießen
**spray** Brause *f*, Düse *f*, (jet) Einspritzdüse *f*, Spritzlackieren *n*, Sprühnebel *m*, (water) Wasserspritzer *m*, Wasserstaub *m*, Zerstäubung *f* ~ (er) Spritze *f* ~-**and-hopper granary** Rieselspeicher *m* ~ **apparatus** Strahlapparat *m*, Zer-

stäubungsapparat *m* ~ **arrester** Schwabbelplatte *f*, Spritzschutzdeckel *m* (d. Batterie) ~ **booth** Lackierkabine *f*, Spritzbox *f* ~ **calorizing** Spritzalitieren *n* (metall) ~ **carburetor** Düsen-, Einspritz-, Spritz-, Zerstäubungs-vergaser *m* ~ **chamber** Diffusionskammer *f* ~ **closure** Spritzverschluß *m* ~ **coat** Spritzauftrag *m*, Spritzgang *m* ~ **color** Spritzfarbe *f* ~ **column** Rieselturm *m* ~ **condenser with horizontal (vertical) pipes (or tubes)** Berieselungsverflüssiger *m* mit liegenden (stehenden) Rohren ~ **cooling** Kühlung *f* durch Zerstäubung ~ **damper** Anfeuchtmaschine *f* ~ **device** Abgießgerät *n* ~ **diffuser** Zerstäubungsdüse *f* ~ **diffuser for wall painting** Anstreichgerät *n* ~ **discharge** Sprühentladung *f* ~ **drying** Zerstäubungstrocknung *f* ~ **dyeing** Spritzfärbung *f* ~ **fluid hose** Schlauch *m* für Flüssigkeit ~ **gun** Spritzpistole *f* ~ **hole cleaner** Spritzlochreiniger *m* ~ **ion** Sprühion *n* ~ **level** Strahlebene *f* ~ **metallizing** Spritzmetallisierung *f* ~ **nozzle** Regner-, Streu-düse *f* ~ **(ing) nozzle** Spritzdüse *f* ~ **nozzle head** Düsenspritzkopf *m* ~-**nozzle tube** (careburetor) Tauchrohr *n* ~-**painting** Farbspritzen *n* ~-**painting apparatus** Farbenspritzapparat *m* ~-**painting plant (or equipment)** Farbspritzanlage *f* ~ **painting stencil** Spritzschablone *f* ~-**painting unit** Farbspritzanlage *f* ~ **pattern** Sprühbild *n* ~ **pipe** Siebrohr *n* ~ **plate** Abdeckscheibe *f* ~ **points** Sprühspitzen *pl* ~ **process** Sprühverfahren *n* ~-**proof (splash proof)** spritzwasserdicht ~-**proofness** Sprühfestigkeit *f* ~ **shroud** Strahlabschirmung *f* ~ **sprinkler** Nebelsprinkler *m* ~ **strip** Spritzwasserleiste *f*, Zerstäubungsstreifen *m* ~ **tank** Flüssigkeitszerstäuber *m* ~ **tower** Berieselungsturm *m*, Rieselturm *m* ~ **tube** Sprührohr *n* ~-**type air cooler** Berieselungsluftkühler *m* ~-**type arc** Sprühregenlichtbogen *m* ~ **unit** Puderapparatbehälter *m* ~-**valve regulation** Spritzventilregulierung *f* ~ **varnishing** Spritzlackieren *n* ~ **washing machine** Traufenwaschmaschine *f* ~ **water** Einspritzwasser *n*, Spritzwasser *n* ~ **water guard** Spritzwasserschutz *m* ~ **water range** Spritzwasserbereich *m*
**sprayed** gespritzt ~ **cathode** bespritzte Kathode *f* ~ **coatings of aluminum (lead or zinc)** aufgespritzte Überzüge *pl* ~ **die casting** Spritzgußteil *m* ~-**on zinc bridge** Zinkspritzbrücke *f* (Durchführungskondensator) ~ **rubber** Zerstäuberlatex *m*
**sprayer** Zerstäuber *m* ~ **containing cover** Zerstäuberdecker *m* ~ **valve** Einspritzventil *n*
**spraying** Berieselung *f*, Versprühen *n*, Zerstäubung *f* ~ **with aluminum** Aufspritzen *n* von Aluminium
**spraying, ~ airplane** Streuflugzeug *n* ~ **apparatus** Nebelapparat *m*, Zerstäubegerät *n* ~ **color** Spritzdruckfarbe *f* ~ **compound** Spritzmasse *f* ~ **damper** Spritzdämpfer *m* ~ **device** Berieselungs-, Bräuse-, Lösch-vorrichtung *f* ~ **drier** Zerstäubungstrockner *m* ~ **equipment for plastics** Kunststoffspritzgerät *n* ~ **fat** Spritzfett *n* ~ **lacquer** Spritzlack *m* ~ **medium** Besprühmittel *n* ~ **method** Beregnungs-, Spritz-verfahren *n* ~ **nozzle** Sprühdüse *f* ~-**over** Verspritzung *f* ~ **pipe** Spritzrohr *n* ~ **plant** Spritzanlage *f* ~ **pressure** Spritzdruck *m* ~ **pump** Spritzpumpe *f*

**spread, to** ~ (pressure) ausbreiten, (a sail) ausreißen, ausspannen, ausstreuen, bestreichen
(mit), breiten, einreißen, erweitern, spreizen,
strecken, streuen, verbreitern, verteilen, zerstreuen
**spread, to** ~ **apart** auseinanderziehen **to** ~
**fanwise** auffächern **to** ~ **and level** einebnen **to**
~ **on** aufstreichen **to** ~ **out** aufspannen, ausbreiten, ausstreichen **to** ~ **out on a line of
bearing** fächerförmig auseinanderziehen **to** ~
**over** aufstreuen, ausbreiten, bestreuen, verstreichen **to** ~ **by rubbing** verreiben **to** ~ **the
trails** die Holme spreizen
**spread** Ausdehnung *f*, Breitenstreuung *f*, Breitung *f*, Streubereich *m*, (of image) Verbreiterung
*f*, Verbreitung *f*; gespreizt ~ **of beam** Strahl-
erweiterung *f* (TV) ~ **of burst** Streugarbe *f* ~ **of
hump** Auseinanderlaufen *n* eines Buckels ~ **of a
light source** Streuung *f* einer Lichtquelle ~ **of
line (or strip)** Linienverbreiterung *f* (TV)
**spread,** ~ **bead** breite Schweißraupe *f* ~ **footing**
liegender Rost *m* ~ **foundation** Flachfundation
*f* (Gründung) ~ **light lens** Streulinse *f*
**spreader** Aufbreitmaschine *f* (textiles), Raa *f*,
Rahe *f*, Spreize *f*, Spreiz-klappe *f*, -körper *m*
~ **bar** Spann-balken *m*, -stange *f*, Spreizstange
*f* ~ **current transformer** Stützerstromwandler *m*
~ **glass** Streuglas *n* ~ **lens** Streuscheibe *f* ~ **plow**
Haldenplaniermaschine *f* ~ **stoker** Wurffeue-
rung *f* ~ **tube** Quetschtube *f* ~**-type travelling
grate** Wurfwanderrost *m*
**spreading** Ausbreitung *f*, Breiten *n*, Streuung *f*;
abstehend ~ **of the blast** Strahlverteilung *f* ~
**of groove** Kaliberbreitung *f* ~ **of weld seams
during welding** Auseinanderklaffen *n* des
Schweißspaltes
**spreading,** ~ **action** Breitungswirkung *f* ~ **brush**
Streichbürste *f* (paper mfg.) ~ **calender** Streich-
kalander *m* ~ **coefficient** Spreizungskoeffizient
*m* ~ **device** Entfaltvorrichtung *f* ~ **effect** Breitenwirkung *f* ~ **frame** Anlegemaschine *f* ~ **knife**
Streichmesser *n* ~ **machine** Aufstreichmaschine
*f*, Streichmaschine *f* ~ **power** (color) Ausgiebigkeit *f* ~ **rate** Ausgiebigkeit *f* ~ **screw** Spreizschraube *f* ~ **spring** Spreizfeder *f* ~ **wedge head**
Spreizstößel *m* ~ **work** Breitungsarbeit *f*
**Sprengler pump** Tropffallpumpe *f*
**sprig** Formerstift *m*, Reis *n*
**spring, to** ~ durchfedern, entspringen, federn,
quellen, reißen, springen **to** ~ **back** zurückfe-
dern **to** ~ **in** einspringen **to** ~ **a leak** leck machen, ein Leck *n* springen, leck werden **to** ~ **off
(or out)** abbersten, ausspringen
**spring** Brunnen *m*, Feder *f*, Feder *f* für Keilnut,
Federung *f*, Formfeder *f*, Frühjahr *n*, (season)
Frühling *m*, Quelle *f*, Spring *f*, Zuhaltung *f* ~
**for bar keys** (accordion) Brummtastenfeder *f*,
Tastenfeder *f* ~ **for conical grip (or grip chuck)**
Spannpatrone *f* ~ **due to damming of underground water** Stauquelle *f* ~ **(skewback) of
extrados (intrados)** Anfang *m* der äußeren (inneren) Bogenfläche ~ **with lefthand helix**
linksgewundene Feder *f* ~ **of low rate** schwache
Feder *f* ~ **for maximum speed governing** End-
regelfeder *f* ~ **and peg-tooth harrow** Acker- und
Federzahnegge *f* ~ **with rolled end** gerolltes
Federende *n* ~ **in tension** gespannte Feder *f* ~
**with upset end** gestauchtes Federende *n*

**spring, a** ~ A-Feder *f* ~ **action** Feder-antrieb *m*,
-wirkung *f*, Federung *f* ~ **activation** Federan-
trieb *m* ~ **adjusting motor** Federverstellmotor
*m* ~ **assembly** Federnpaket *n*, Federsatz *m* ~
**-back** Zurückschnellen *n* ~ **bailer** Federbüchse
*f* ~ **balance** Federwaage *f*, Springfederwaage *f*,
(for the home) Wirtschaftsfederwaage *f* ~
**balance pull** Zugdynamometer *n* ~ **band** Feder-
bund *m* ~**-band dismantling hammer** Feder-
bundabziehhammer *m* ~**-band mounting press**
Federbundaufziehpresse *f* ~ **bank** Feder(n)satz
*m* ~ **bar** federnde Haltevorrichtung *f* ~ **bar box**
Federbüchse *f* ~ **barrel** Feder-haus *n*, -trommel
*f* ~ **base** Abstützbasis *f* ~ **battery** Federpaket *n*
~ **beam** Pranstock *m*, Reitel *m*, Stoßreitel *m* ~
**bearing** federndes Lager *n*, Federlager *n* ~
**bellows** Federbalg *m* ~ **bending machine** Feder-
formmaschine *f* ~ **bit** Federbohrer *m* ~ **blind**
Schnapprouleau *n* ~ **block** Federkolben *m* ~
**board** Sprungbrett *n* ~ **bolt** Federbolzen *m*, Verbindungsstück *n* ~ **bolt lock** Schnäpperschloß *n*
~ **bow compass** Nullzirkel *m* ~**-bow compasses**
Nullenzirkel *m* ~**-bow divider** Teilzirkel *m* ~
**box** Federhaus *n* ~ **bracket** Feder- (vehicle),
-bock *m*, -hand *f*, -tragzapfen *m* ~ **brake** Federbremse *f* ~ **brass** Federmessing *n* ~ **bush** Federbuchse *f* ~ **cage** Federhülse *f* ~ **calipers** Feder-
taster *m*, Taster *m* mit Federspannung ~ **cap**
Federdeckel *m*, Federteller *m*, Gehäusekappe *f*
~ **cap oiler** Federdeckelschmiernippel *m* ~
**capping machine** Federkapselmaschine *f* ~ **carrier** Federbock *m* ~ **carrier arm** (open and closed
types) Federhand *f* ~ **catch** federnde Nase *f*,
Feder-haken *m*, -klinke *f*, -raster *m*, -rost *m*,
Hebelsteuerung *f*, Schnappriegel *m*, Schnappschloß *n* ~ **catcher** Fangfeder *f* ~ **center bolt**
Federbefestigungsstift *m* ~ **center pin** Federstift *m* ~ **chamber of delivery valve** Druckventil-
federraum *m* ~ **characteristic** Federungs-
charakteristik *f* ~ **clamp** Feder-bock *m*,
-büchse *f*, -kapsel *f*, Spannblech *n* ~ **clamp key**
Springschlüssel *m* ~ **clamp plate** Federspann-
platte *f* ~ **clamp ring** Klemmhülse *f* ~**-clamped**
durch eine Feder *f* festgehalten, federn eingespannt ~ **clip** Blattfeder *f*, Feder-bügel *m*,
-bund *m*, -klammer *f*, Haltefeder *f* (Federklammer zum Halten des Objektträgers), Objekthalter *m*, -Klemme *f*, Quetschhahn *m*, Schenkelfeder *f* ~ **clutch** Federkupplung *f* ~ **collar**
Federteller *m* ~ **collet** Spann-büchse *f*, -patrone
*f*, federnder Futterring *m*, federnde Spannpatrone *f* ~ **collet chuck** Spannpatronenfutter
*n* ~ **commutator brush** Stromabnehmerfeder *f*
~ **compasses** Federzirkel *m* ~**-compression
grease cup** Feder(druck)schmierbüchse *f*,
Schmierbüchse *f* mit Federdruck ~ **connection**
Federbefestigung *f* ~ **constant** Elastizitätsmodul
*n*, Federkonstante *f* ~ **contact** Federkontakt *m*,
federnder oder weicher Kontakt *m* ~**-contact
plug** Bananenstecker *m* ~ **contact strip** Feder-
leiste *f* (electr.) ~ **control** Sprungschaltung *f*
~**-controlled switch** Federschalter *m* ~ **cotter**
Schwerspannstift *m* ~ **cotter pin** Federsplint *m*
~ **counterrecoil mechanism** Federvorholer *m* ~
**couple pin** Federverschlußbolzen *m* ~ **coupling-
catch** Kuppelklinke *f*, Springkeil *m* ~ **coupling
key** Kuppelfeder *f* ~ **cover** Feder-gamasche *f*,
-kappe *f*, -verkleidung *f* ~ **cradle** Federwiege *f*

~ cup Federmuffengehäuse n ~ cushion Federkissen n ~-cushioned federgedämpft ~ cutters Federschneide f ~ deflection Federdurchbiegung f, Federweg m ~ detent Federraster m ~ disc Sprengscheibe f ~ dividers Feder-spitzzirkel m, -teilzirkel m, -zirkel m ~ drive federnder Antrieb m, federndes Antriebsrad n ~ -driven clockwork Federuhrwerk n ~ drum Feder-haus n, -trommel f ~ dynamometer Federzugmesser m ~ elongation Federweg m ~ end (side) Federseite f ~ equilibrator Federausgleicher m ~ excursion Federweg m ~ extension Federdehnung f ~ extremity shoe Federschuh m ~ eye Federauge n ~ eye bolt Federstift m ~ eye bushing Federbuchse f ~-fed pond Quellteich m ~ finger (textiles) Preßfinger m ~ flange Spannflansch m ~ flap Federklappe f ~ flapper valve Federklappenventil n ~ fork Federgabel f ~ frame Prellgerüst n ~ free wheeling Federfreilauf m ~ friction clutch Federbandkupplung f ~ gaiter Feder-gamasche f, -schutz m, -schutzhülle f ~ galvanometer Federgalvanometer n ~ governor Feder-kraftregler m, -regulator m ~-grip-control nozzle Zapfhahn m mit Federgriff ~ gripper federnder Greifer m ~ guide Federsäulensatz m, Federteller m ~ hammer Federhammer m ~ hanger Feder-bock m, -bundstütze f, -lasche f, -stuhl m, -wiege f ~ hanger guide Federstützenführung f ~ hanger pin Federbolzen m ~ heating furnace Schlitzofen m ~-hard federhart ~ hinge Federscharnier n ~ holder Stern m (print.) ~ holder attachment Federapparat m ~ hook Federhaken m ~ housing Feder-gehäuse n, (gun) -zylinder m ~ jack Klinke f ~ leaf Federblatt n ~ leaf insert Federblatteinlage f ~ leaf opener Federblattspreizer m ~ leaf retainer Federklammer f ~ leaf sight Federvisier n ~ leaf spreader Federblattspreizer m ~ leg Federbein n ~ length Federlänge f ~ lever Federheber m ~-lid oil cup Federdeckelöler m ~ lifter Federheber m ~ line Beitau n ~ linkage system federndes Verbindungssystem n ~-loaded federbeinbelastet, unter Federspannung f stehend, Federvorspannung f, gefedert ~-loaded brake Federbremse f, Federspeicherbremse f ~-loaded brake cylinder Federspeicherbremszylinder m ~-loaded governor Federregler m ~-loaded idler federnde Leitrolle f ~-loaded nozzle durch Spannfeder schließender Hahn m ~-loaded preselector aperture Vorwahlspringblende f ~-loaded shuttle valve Einfachwechselventil n ~-loaded stirrup gefederter Bügel m ~-loaded strut Schraubenfederbein n ~-loaded valve federbelastetes Ventil n ~ lock (doors) Fallschloß n, (doors) Federschloß n, Schnappschloß n, Schnellschluß m ~ lubrication key Federschmierkeil m ~ lubricator Federöler m, Federschmierbüchse f ~ manometer Feder-druckmesser m, -manometer n ~ mechanism Feder-einrichtung f, -werk n ~ motor Federkraftantrieb m ~ (-wound) motor Federmotor m ~ motor of phonograph Sprechmaschinenlaufwerk n ~ mount Objektschutz m ~ -mounted abgefedert ~-mounted scraper Federabstreifer m ~ needle Hakennadel f ~ operated breaker Federunterbrecher m ~ pad Federauflage f ~ peg Feder f (Bolzen) ~ perch Federsitz m ~ pick loom Federschlagstuhl m ~ pile Fe-

derpaket n ~ pin Federstift m ~ piston ring Überstreifring m ~-pivot seat Federsattel m ~ plate Feder-blatt n, -blattrippe f, -platte f, -scheibe f, -teller m ~ pliers Federzange f ~ plunger Federkolben m ~ pole Bohrhebel m, Federstange f, Prellklotz m, Sprungbaum m ~-pole jacket Federbalkenmantel m ~ pressure Federdruck m ~-pressure gauge Federmanometer n ~-pressure sounding apparatus Federdrucksonde f ~ protecting sleeve Federschutzgamasche f ~ pull Federzug m ~ pulley Federeigenschaft f ~ rate Federkonstante f, Federsteife f ~ reaction Federwirkung f ~ rebound clip Gummipuffer m der Feder ~ recuperator Federvorholer m ~ resetting gear Federspannwerk n ~ rest Federteller m, Klinkenaufleger n, Ladebogen m ~ retainer Federhalter m, Hakensprengring m ~ retainer keys Ventilkeilpaar n ~ rigging gear Federgehänge n ~ ring Feder-, Spreng-ring m ~-ring clutch Spreizringkupplung f ~ ring driver Sprengringzieher m ~ roller bearing Federrollenlager n ~ roller mill Federwalzenmühle f ~ rolls Walzenmühle f ~ saddle Federbügel m ~ safety hook Karabinerhaken m ~ salt Quellsalz n ~ scale Federwaage f ~ seat Feder-büchse f, -hand f, -kapsel f, -sitz m ~-seating washer Federsitzscheibe f ~ separator Federblattspreizer m ~ separator tool Federblattwerkzeug n ~ setting Federeinstellung f ~ shackle Feder-bund m, -gehänge m, -hänger m, -lasche f ~ shackle bolt Federbolzen m ~ shackle bracket Federlaschentragbock m ~ shaft Webstuhlwaage f ~ shock absorber Federstoßfänger m ~ skid Federkufe f ~ sleeve (cylinder) Feder(druck)hülse f, Federtülle f ~ steel Federstahl m ~ steel band Federstahlband n ~ steel bar Federstabstahl m ~-steel cross bands Federbandstahl m ~ steel strip Federstahldrahtgewebe f ~ stiffness Federkonstante f ~ stirrup Federbride f ~ stone Prellstein m ~ stop Federanschlag m ~-stretching handle Federspanngriff m ~ stud Federkopf m ~ support Federträger m ~-supported an einer Feder f angebracht ~-supported bend Federkrümmer m ~ suspension (vehicle) Abfederung f, (of cars, various) Federaufhängung f, Federgehänge n, Federung f ~ suspension of front axle Vorderachsfederung f ~ suspension gear Federgehänge n ~ swivel Federwirbel n ~ tail Springschwanz m ~ temper Federhärte f ~ tension Feder-kraft f, -spannung f ~-tensioned driving plate Dorn m mit federnder Mitnehmerscheibe ~ tensioning screw Federspannschraube f ~ terminal plug Federanschlußklemme f ~-testing apparatus Federprüfapparat m ~-testing machine Federprüfmaschine f ~ thread guide federnder Fadenführer m ~ tide Springflut f ~ timber Frühlingsholz n ~-tooth cultivator Federzahnkultivator m ~-tooth gangs Federzahnsätze pl ~ -tooth harrow Federzahnegge f ~ trap Federfalle f ~ travel Federweg m ~ tray abgefedertes Relaisbrett n ~ trigger Federschnepper m ~ -type starter Federanlasser m ~-type trailer coupling gefederte Anhängerkupplung f ~ U bolts Federbügel m ~ U clamp Federbride f ~ washer Feder-dichtung f, -ring m, -teller m, federnde Unterlagscheibe f ~ water Quellwasser n ~ wedge bolt Zugankerschraube f ~ well

Filter-, Quell-brunnen *m* ~ **winder** Schrauben-federwickelmaschine *f*, Wickelmaschine *f* für Schraubenfedern ~ **winding device** Federwickel-vorrichtung *f* ~ **wire** Feder(stahl)draht *m*, Springfederndraht *m* ~ **wire bow** Federdraht-bügel *m* ~ **wire retaining ring** Sprengring *m* ~ **wire stirrup** Federdrahtbügel *m* ~ **wood** Frühlingsholz *n*
**springer of an arch** Bogenkämpfer *m*
**springiness** Feder-elastizität *f*, -kraft *f*, -wirkung *f*, Federung *f*, Federungsvermögen *n*, Spring-kraft *f*
**springing** Abfederung *f*, Federung *f*, Springen *n* ~ **of the undercarriage** Fahrgestellabfederung *f*
**springing,** ~ **course** Kämpferschicht *f* ~ **stone** Kämpferstein *m* ~ **transom** Kämpfer
**springlike** federartig
**springy** elastisch, federhart, federnd ~ **stage** fe-dernder Tisch *m*
**sprinkle, to** ~ anfeuchten, aufstreuen, benetzen, berieseln, besprengen, bespritzen, bestreuen, einsprengen, einstreuen, sprengen, sprenkeln, sprinklern, sprühen **to** ~ **with fire** abstreuen (artil.)
**sprinkled** gesprenkelt
**sprinkler** Berieselungsapparat *m*, Berieselungs-brause *f*, Brause *f*, Feuerlöschbrause *f*, First-brause *f*, Regner *m*, Sprenger *m*, Sprengwagen *m*, Sprenkler *m*, Spritzapparat *m*, Streuer *m*, Wurfbeschicker *m* ~ **frame** Sprinklerkörper *m* ~ **guard** Sprinklerschutzkorb *m* ~ **installation** Sprinkleranlage *f* ~ **machine** Übergußapparat *m* ~ **method** Beregnungsverfahren *n* ~ **plant** Beregnungsanlage *f* ~ **stopper** Spritzkorken *m* ~ **system** Feuerlöschsystem *n* ~ **truck** Spreng-wagen *m* ~ **valve retaining members** Sprinkler-verschluß *m* ~ **yoke** Sprinklerbügel *m*
**sprinkling** Anflug *m*, Beregnung *f*, Besprengen *n* ~ **apparatus** Aufguß-, Spritz-apparat *m* ~ **bed** Rieselfeld *n* ~ **brush** Netzpinsel *m* ~ **can** Brause *f* ~ **engine** Einsprengmaschine *f* ~ **powder** Auf-streupulver *n* ~ **system** (Kühlturm) Berieselung *f* ~ **system fore fire fighting** Berieselungsanlage *f* ~ **wagon** Wassersprengwagen *m* ~ **wax** Streu-wachs *n*
**sprocket** (Dach) Aufschiebung *f*, Daumenrad *n*, Kettenzahnrad *n*, Perforationswalze *f*, Rolle *f*, Transportrolle *f*, Transportschiene *f*, Zahnrad *n*, Zahntrommel *f* ~ **bit** Stellenbit *n* ~ **chain** Gall'sche Kette *f*, Gelenkkette *f*, Kettenrad-kette *f*, Laschenkette *f* ~ **channel** Stellenspur *f* ~ **cutter** Kettenradfräser *m* ~ **drum** Sprossen-rolle *f*, -trommel *f*, Zackentrommel *f* ~ **gear** Kettenradvorgelege *n* ~ **hob** Kettenradwalz-fräser *m* ~ **hole** Transportloch *n* (teletype), Zähl-loch *n* ~-**hole modulation (or noise)** Perfora-tionsgeräusch *n* (film) ~ **hum** Perforations-geräusch *n* (film) ~ **ring** Kettenkranz *m* ~ **scraper** Kettenradreiniger *m* ~ **shaft** Ketten-welle *f* ~ **table** Sprossentisch *m* ~-**table sand-blast machine** Sandstrahlgebläse *n* mit Sproß-tisch ~ **wheel** (of film feed) Führungsrolle *f*, Kegelrad *n*, Ketten-nuß *f*, -rad *n*, -scheibe *f*, Schaltrolle *f* (film feed), Sprossentrommel *f*, Zacken-rolle *f*, -trommel *f*, Zahnkettenrad *n* ~-**wheel washer** Zahnradscheibe *f*
**sprout, to** ~ auswaschen, guzen, treiben
**sprout** Sprößling *m*

**spruce** Balsamtanne *f*, Fichte *f* ~ **resin** Fichten-harz *n*
**sprue** Einguß *m*, Einguß-kanal *m*, -lauf *m*, -trich-ter *m*, Gießtrichter *m*, Lauf *m*, Spritzer *m*, Trichterlauf *m*, verlorener Kopf *m* (aviat.), ver-lorenes Verbindungsstück *n* ~-**and runner scrap** Trichterschrot *m* & *n* ~ **cutter** Eingußabschnei-der *m* ~ **hole** Eingußloch *n* ~ **lock pin** Einguß-verschlußbolzen *m* ~ **opening** Einguß-loch *n*, -trichter *m*, Gieß-loch *n*, -trichter *m*, Trichter-loch *n*, -mündung *f* ~-**opening gate** Einguß-mündung *f* ~ **slug** Angußkegel *m*
**sprung** gerissen ~ **snap pin** abgefederter Schnapp-stift *m* ~ **tail skid** Federsporn *m*
**spud** Bohr-arm *m*, -schaufel *f*, -spaten *m*
**spudding** Umbohrung *f* (min.) ~ **bit** Schlagmeis-sel *m* ~ **bits** Schlagbohrer *m* ~ **ring** Spudding-scheibe *f* ~ **shoe** Bohrschaufel-, Spudding-schuh *m*
**spun** gesponnen ~ **bearing** Schleudergußlager *n* ~ **casting** Gespinsthülle *f* ~ **concrete** Schleuder-beton *m* ~ **fabric** Gespinst *n* ~ **fiber** Spinnfaser *f* ~ **filament** gereckter Einfaden *m* ~ **glass** Fa-denglas *n*, Glasgespinst *n*, Glasleinen *n* ~ **glass fabrics** Glasgewebe *n* ~-**glass tape wrapping** Glashandabwicklung *f* ~ **gold** Goldgespinst *n* ~-**over wire** besponnener Draht *m* ~ **pipe** Schleudergußrohr *n* ~ **yarn** Gespinst *n*
**spunkt** Zunderschwamm *n*
**spur, to** ~ **on** anspornen
**spur** Abzweig *m*, Ausläufer *m*, Bein *n*, Dorn *m*, (of a pole) Druckstab *n*, Sporn *m*, Stichbahn *f* **on the** ~ **of the moment** aus dem Stegreif
**spur,** ~ **bevel gear** Tellerrad *n* ~ **brake** Sporn-bremse *f* ~ **dike** Buhne *f* ~-**gear(s)** Spurrad *n*, Stirn-getriebe *n*, -rad *n*, -ruder *n*, -zahnrad *n*, Stirn- und Kegelräder *pl*, stirnverzahntes Rad, Zahnkranz *m*, (Stirn) Zahnradgetriebe *n*
**spur-gear,** ~ **cutter** Stirnradfräser *m* ~ **cutting** Stirnradfräsarbeit *f* ~ **cutting machine** Stirnrad-fräsmaschine *f* ~ **generating machine** Stirnrad-abwälzfräsmaschine *f* ~ **grinding machine** Stirnradschleifmaschine *f* ~ **hob** Stirnradwälz-fräser *m* ~ **planing machine** Stirnradhobel-maschine *f* ~ **shaper** Stirnradstoßmaschine *f* ~ **system** Stirnradgetriebe *n* ~ **transmission** Stirn-radübersetzung *f*
**spur,** ~-**geared chain block with planetary-gear system** Flaschenzug *m* mit Stirnradplaneten-getriebe ~ **gearing** Stirnräder-getriebe *n*, -paar *n*, Stirnverzahnung *f* ~ **gearing motor** Stirn-radgetriebemotor *m* ~ **grips** Greifergehänge *n* ~ **insulator** Krückenisolator *m* ~ **jetty** Quer-damm *m* ~ **line** kurze Abzweigleitung *f* ~ **(from main) line** Nebenlinie *f* ~ **pinion** gerades Zahnrad *n*, Stirngetriebe *n*, Stirnradritzel *n* ~ **pinion shaft** Stirnradritzelwelle *f* ~ **rack** Zahn-stange *f* ~ **tooth** Geradezahnung *f* ~ **wheel reduction gearing** Stirnradvorgelege *n* ~ **wheel slewing gear** Stirnrad-Drehwerksgetriebe *n*
**spurious** unecht, untergeschoben ~ **capacities** ungewollte kapazitive Kupplung *f* ~ **capacity** Streukapazität *f* ~ **coupling** Streukopplung *f*, ungewollte kapazitive Kopplung *f*, Verkopp-lung *f*, wilde Kopplung *f* ~ **emission** Neben-wellen *pl* ~ **feedback** Mitkopplung *f* ~ **oscilla-tion** durch Nebenkopplungen hervorgerufene Schwingung *f*, Nebenkopplungsschwingung *f*,

Schwingung *f* durch Nebenkopplung, wilde Schwingung *f* ~ **pattern** Störmuster *n* ~ **periods** Scheinperioden *pl* ~ **pulse** Fehlstoß *m*, Störimpuls *m* ~ **pulse mode** ungewollter Impulsmodus *m* ~ **radiation** Nebenausstrahlung *f*, Seitenstrahlung *f* ~ **response** Unselektivität *f* ~ **signal** Störsignal *n* ~ **wave attenuation** Nebenwellenabschwächung *f*, Nebenwellenleistung *f*

**spurt, to** ~ spratzen, spritzen **to** ~ **out** herausspritzen

**spurt cork** Spritzkorken *m*

**spurting** Spratzen *n* ~ **(or squirting) pipe** Abspritzrohr *n*

**sputter, to** ~ zerstäuben

**sputter** Spratzen *n*

**sputtering** (of metals from filaments) Abtragen *n*, (of a cathode) Zerstäubung *f* ~ **of cathode** Kathodenzerstäubung *f*

**squad** Bedienungsmannschaft *f*, Gruppe *f* **in** ~s gruppenweise

**squadron** Geschwader *n*

**squall** Bö *f*, Windbö *f*, Winddalle *f* ~ **belt** Böenband *n* ~ **cloud** Böenwolke *f* ~ **head** Böenkopf *m* ~ **line** Bö(en)linie *f*

**squally** böig, stoßförmig

**squander, to** ~ verzetteln

**squarable** quadrierbar

**square, to** ~ abgleichen, abschwarten, (lumber) abvieren, begleichen, behauen, beschnieden, (sheets) besäumen, zum Quadrat erheben, quadrieren, saldieren, viereckig oder rechtwinklig machen **to** ~ **an ashlar** steinwinkeln **to** ~ **away** aufbrassen **to** ~ **off** abschneiden, rechtwinklig abstecken **to** ~ **out** ausvieren **to** ~ **wood** Holz *n* abvieren, Holz *n* beschlagen

**square** Anreißwinkel *m*, Aufnahmezapfen *m*, Carreau *n*, Geviert *n*, Mitnehmer *m*, (public) Platz *m*, Quadrat *n*, Quadratzahl *f*, Viereck *n*, Viererlage *f*, Vierkant *m*, (as a tool) Winkel *m*; geviert, quadratisch, recht-eckig, -winklig, viereckig, vierkantig ~ **of the modulus** Quadrat *n* des Betrages ~ **of the orbital angular momentum** Bahndrehimpulsquadrat *n* ~ **for sag regulation** Winkelhaken *m* für Durchhangsreglung (electr.) ~ **with** senkrecht zu

**square,** ~ **bar** Vierkant *m* ~**-bar iron** Vierkanteisen *n* ~ **(steel) bars** Flacheisen *n* ~ **bit** Kronenbohrer *m* ~ **bloom** gewalzter Quadratblock *m* ~ **bolt** Schließnagel ~ **box wrench** Vierkantschraubenschlüssel *m* ~ **bracket** eckige Klammer *f* ~ **broach** Vierkantreibahle *f* ~ **building stone** Quader *m* ~ **butt weld** zweiseitige Vollnaht *f* ~**-cap screw** Bauschraube *f* ~ **cascade** Stufenkaskade *f* ~ **centimeter** Quadratzentimeter *m* ~ **check** Carreau *n* ~ **clincher** Klammerhaken *m*, Zulagklammer *f* ~ **coal scoop** Vierkantkohlenschaufel *f* ~ **coil** quadratische Spule *f* ~ **column** quadratische(r) Pfeiler *m*, Säule *f*, Stütze *f* ~ **course** rechtwinkliger Kurs *m* ~ **decimeter** Quadratdezimeter *m* ~ **deviation** quadratische Abweichung *f* ~ **dimensions** Flächenausdehnung *f* ~ **drop-block breech mechanism** Flachkeilverschluß *m* ~**-edge** scharfkantig ~**-edge joint** Bördelschweißung *f* ~ **electromotiveforce curve** rechteckige Kurvenform *f* der elektromotorischen Kraft ~ **end** Vierkant *m* ~ **file** Vierkantfeile *f* ~ **flag** Quadratflagge *f* ~ **flange** Viereckflansch *m* ~ **(-face)**

**flatter** Schlichthammer *m* ~ **fluctuation** Schwankungsquadrat *n* ~ **foot** Quadratfuß *m* ~ **forging** Viereckigschmieden *n* ~ **foundry shovel** Formerschaufel *f* ~ **frame** Viereckspant *m* ~ **girder** Viereckträger *m* ~ **gouge** Viereisen *n* ~ **grid** Quadratnetz *n* ~ **groove** Quadratkaliber *n* ~ **head** Vierkantkopf *m* ~ **head(ed) bolt** Vierkantkopfschraube *f*, Schraubenbolzen *m* mit Vierkantkopf ~**-headed** vierkantig ~**-headed bolt** viereckiger Bolzen *m* ~**-headed coach bolt (or screw)** Vierkantholzschraube *f* ~**-headed plug** Vierkantstöpsel *m* ~ **headed screw** Schraube *f* mit viereckigem Kopf ~ **hole** Vierkantloch *n* ~ **inch** Quadratzoll *m* ~**-jaw clutch** Klauenkupplung *f* mit drei Klauen ~ **joint** I-Stoß, rechtwinklige Verbindung *f* ~ **key** Quadrat-, Spießkant-, Vierkant-keil *m* ~ **key for quoin** Schlüssel *m* für Schließzeug ~ **key spanner** Vierkanteinsteckschlüssel *m* ~ **kilometer** Quadratkilometer *m* ~ **knee** Winkelknie *n* ~ **lashing** Bockschnürbund *m* ~ **lath** Dachlatte *f* ~ **law** quadratischer Gleichrichter *m*

**square-law,** ~ **of distances** Quadratabstandsgesetz *n* ~ **capacitor** quadratischer Kondensator *m* ~ **characteristic** quadratische Charakteristik *f* ~ **condenser** Kondensator *m* mit wellengerader Kennlinie, Nierenplattenkondensator *m*, quadratischer Kondensator *m* ~ **detection** lineare Gleichrichtung *f* ~ **detector** quadratischer Gleichrichter *m* ~ **instrument** Instrument *n* mit quadratischer Kennlinie ~ **rectifier** quadratischer Detektor *m* oder Gleichrichter *m* ~ **scan** quadratische Abtastung *f* ~ **scanning** quadratische Abtastung *f* nacheinanderfolgender Bildpunkte

**square,** ~ **loop** (antenna) viereckiger Dipolrahmen *m* ~ **measure** Flächenmaß *n* ~ **measurement** Quermaß *n* ~ **mesh screening cloth** Quadratmaschengewebe *n* ~ **meter** Geviert-, Quadrat-meter *m* ~ **mile** Quadratmeile *f* ~ **mining claim** Gebietfeld *n* ~ **navigation chart** Quadratnavigationskarte *f* ~ **net** Quadratnetz *n* ~ **number** Quadratzahl *f* ~ **nut** Vierkantmutter *f* ~**-oval roughing pass** Quadratovalstreckkaliber *n* ~ **pass** Quadratkaliber *n*, Quadrattisch *m* ~ **perforation** Vierkantlochung *f* ~ **pile** Vierkantpfahl *m* ~ **pillar** quadratischer Pfeiler *m*, quadratische Säule *f* oder Stütze *f* ~ **pipe** Vierkantrohr *n* ~ **place coil** quadratische Flachspule *f* ~ **poles** Quadratgestänge *n* ~ **rib** Winkelspant *n* ~**-rigged** mit Rahen *pl* getakelt ~ **roller** Vierkantspule *f* ~ **root** Quadratwurzel *f*, zweite Wurzel *f* ~ **rope** Quadratseil *n* ~ **roughing(-out) pass** Quadratvorkaliber *n* ~ **rule** rechter Zeichenwinkel *m* ~**-ruler** Kantel *n* ~ **screw-mounted flange** quadratischer Anschraubflansch *m* ~ **section** Quadratprofil *n* ~**-section rod** Vierkantenstange *f* ~ **sennit** Vierkantplattung *f* ~ **sets** Abbau *m* mit planmäßiger Geviertzimmerung ~ **shaft** Vierkantschaft *m*, Vierkantwelle *f* ~ **shank of the drawbar** Zugstangenvierkant *n* ~**-shank taper bridge reamer** Kesselreibahle *f* ~ **sight** Rechteckkorn *n* ~ **sluice** Kastenschleuse *f* ~ **socket wrench** Vierkantsteckschlüssel *m* ~ **steel bar** Quadratstahlstange *f*, Vierkantstahl *m* ~ **T washer** Vierkantdoppel-T-Scheibe *f* ~ **thread** flaches oder flachgängiges Gewinde *n*, Flachgewinde *n*, viereckiges Gewinde *n* ~-**thread screw** flachgängige Schraube *f* ~-**threaded** flachgängig ~ **timber** rechteckiger Ausbau *m* mit behauenen Balken, Kantholz *n* ~-**topped electromotive force** elektromotorische Kraft *f* von rechteckiger Kurvenform ~ **tube** Quadratröhrchen *n*, Vierkantrohr *n* ~ **tubing** Quadratröhrchen *n* ~ **turret** Vierkantrevolver(kopf) *m* ~ **turret head** Vierkantrevolversupport *m* ~ **U washer** Vierkant U-Scheibe *f* ~ **upper beam** Vierkantoberwange *f* ~ **washer** Vierkantscheibe *f* ~ **wave** rechteckige Welle *f*, Rechteckwelle *f* ~-**wave calibrator** Rechteckwellen-Eichstufenwähler *m* ~-**wave curve** Rechteckkurve *f* ~-**wave generator** Sprunggenerator *m* ~-**wave impulse** Rechteckimpuls *f* ~-**wave keying** Harttastung *f* ~-**wave pulse** Rechteckschwingung *f* ~-**wave radiator** Rechteckwellengenerator *m* ~-**wave signal** Rechtecksignal *n* ~-**wave voltage** Rechteckspannung *f* ~ **wedge** V-Vierkantmutter *f* ~ **well** Potentialtopf *m* ~ **well potential** Potential *n* mit rechteckigem Topf ~-**winding** quadratische Wicklung *f* ~ **work** schachbrettförmiger Abbau *m* ~ **yard** Quadratyard *n* ~ **yardage** Quadratyardfläche *f*

**squared,** ~ **beam** vollkantiger Balken *m* ~ **flint work** behauenes Geröllmauerwerk' *n* ~ **map** Quadratkarte *f* ~-**nose (tool)** Schrägstahl *m* ~ **paper** gekästeltes oder karriertes Papier *n*, Linienpapier *n* ~-**paper exercise book** Millimeterheft *n* ~ **rubble course** Werksteinmauerwerk *n* in durchgehenden Lagen ~ **scale** quadratische Skala *f* ~ **speed** Geschwindigkeitsquadrat *n* ~ **timber** Eckholz *n*, Vierkantholz *n*

**squareness,** ~ **error** Lochungsfehler *m* (film) ~ **ratio** Quadratsverhältnis *n*

**squares** Quadrat-, Vierkant-eisen *n*

**squaring** Kantung *f* ~ **ax** Zurichtaxt *f* ~ **canvas reticulation** Zeichennetz *n* ~ **circuit** Rechteckkreis *m*

**squash, to** ~ (on turns at full throttle) nach außen schießen, quetschen, zerquetschen,

**squash** Fuß *m*, (lamp) Lampenstempel *m*, Quetschfuß *m*, Quetschfuß *m* einer Röhre, Zermalmung *f* ~ **of a tube (or valve)** Röhrenstempel *m*

**squat, to** ~ hocken

**squat** flach, niedrig, tiefsitzend

**squeak, to** ~ quieken, quietschen

**squeak** Knirschen *n*

**squeal, to** ~ pfeifen, quieken

**squealing** Pfeifen *n* der Verstärker, (of a tube) Selbsttönen *n*

**squeegee, to** ~ absaugen, anreiben (photo), ausquetschen

**squeegee** Gummiquetscher *m*, Rechen *m*, Rollenquetsche *f*

**squeeze, to** ~ abfiltern, abfiltrieren, abkrümmen, anstauchen, drücken, durch-kneten, -krümmen, sich einklemmen, festpressen, klemmen, kneifen, kneten, pressen, quetschen, verdichten, verengen, (metals) zängen, zerquetschen, zusammendrücken **to** ~ **out** ausdrücken, auspressen, ausquetschen **to** ~ **out (or off)** abpressen **to** ~ **through** durchpressen **to** ~ **together** zusammenquetschen

**squeeze** (with a forging hammer) Pressdruck *m*, Pressung *f* ~ (**ing**) Quetschung *f* ~ **bottle**

Spritzflasche *f* ~ -**flask lift machine** Preßformmaschine *f* mit Stiftenabhebung ~ **roll-over pattern-draw machine** Wendelplattenpreßformmaschine *f* mit Modellabhebevorrichtung ~ **rolls** Abquetschwalzen *pl* ~ **section** Quetschleitung *f* ~ **stripper (or stripping machine)** Preßformmaschine *f* mit Abstreifvorrichtung ~ **track** Lichttonstreifen *m* veränderlicher Breite, (sound film) Schnürspur *f* ~ **unit** Abdeckstreifen *m* ~ **unit of three rollers each** Dreiwalzenquetschsystem *n*

**squeezed** abgepreßt

**squeezer** Auspreßmaschine *f*, Preßformmaschine *f*, Quetsche *f* ~ **board** Preßplatte *f* ~ **man** Preßführer *m* ~ **plate** Tischplatte *f* ~ **yoke** Preßhaupt *n*

**squeezing** Klemmen *n*, Kolkung *f*, Verdichtung *f*, Zängearbeit *f*, Zängen *n* ~ **of band** Bandeinengung *f* ~ **the cycle time** Verdichtung *f* der Vorgangfolge ~ **on either side** Pressung *f* ~ **of film** Filmeinschnürung *f*

**squeezing,** ~ **board** Preßklotz *m* ~ **cylinder** Preßzylinder *m* ~ **effect** Abquetscheffekt *m* ~ **hand-lever lift machine** Preßformmaschine *f* mit Handhebelmodellaushebung ~ **machine with flask-lift pins** Preßformmaschine *f* mit Stiftenabhebung ~ **molding machine** Preßformmaschine *f* ~ **operation** Pressung *f* ~-**out** Herausquetschen *n* ~-**out pressure** Ausschubdruck *m* ~ **plate** Preßholm *m* ~ **position** (in molding machines) Preßstellung *f* ~ **roller** Ausquetschwalze *f*, Entwässerungspresse *f* ~ **roller for open width squeezing** Breitabquetschvorrichtung *f* ~ **stroke** Preßhub *m* ~ **table** Preßtisch *m* ~ **valve** Preßventil *n* ~ **yoke** (of molding machines) Preßtraverse *f*

**squegging** Überoszillieren *n* ~ **oscillator** Sperrschwinger *m*, Stoßoszillator *m*

**squelch,** ~ **circuit** Stummabstimmschaltung *f* ~ **control** Krach-, Rausch-sperre *f* (rdo)

**squelching** stille Regelung *f*

**squib** Zündung *f* (min.)

**squint, to** ~ blinzeln, schielen

**squint** (antenna) Winkelfehler *m* ~ **angle** Schielwinkel *m*

**squinting** Schielen *n*

**squirrel-cage** Käfigwicklung *f*, Kurzschlußanker *m* ~ **armature** Eichkatzenanker *m* ~ **induction motor** Käfigankermotor *m* ~ **machine** Käfiganker *m*, Kurzschlußmaschine *f* ~ **magnetron** Käfigmagnetron *n* ~ **motor** Käfigankermotor *m*, Kurzschlußläufermotor *m*, Kurzschlußmotor *m* ~ **rotor** Käfiganker *m*, Käfigläufer *m*, Kurzschlußläufer *m* ~ **winding** Käfigwicklung *f*

**squirt, to** ~ (out) ausspritzen **to** ~ **around** herumspritzen **to** ~ **out** herausspritzen

**squirt** Spritzer *m* ~ **can** Spritzkanne *f*

**squirted** gespritzt ~ **filament** gespritzter Glühfaden *m* ~ **skin** Spritzbewurf *m*

**stab, to** ~ durchstechen, stechen

**stab** Dolchstoß *m*, Stich *m*, Stichwunde *f* ~ **hole** Anstichbohrloch *n*, Entspannungsbohrloch *n* ~ **holes** Entspannungsbohrlöcher *pl*

**stabbing** Durchheftung *f* ~ **awl** Haftahle *f*

**stability** Bestand *m*, Beständigkeit *f*, Dauerhaftigkeit *f*, dynamisches Gleichgewicht *n*, Festigkeit *f*, Gleichgewicht *n*, Haftung *f* am Erdboden, Haltbarkeit *f*, Kippsicherheit *f*, Konstanz *f*, Lagerfestigkeit *f*, Pfeifsicherheit *f*, stabile Luft *f*, Stabilisierung *f*, Stabilität *f*, Standfestigkeit *f*, Standsicherheit *f*, Steifheit *f*. Stetigkeit *f*, Widerstandsfähigkeit *f*, Unverschiebbarkeit *f* ~ **of course** Kursstabilität *f* ~ **of film** Filmregisterhaltigkeit *f* ~ **of frequency** Frequenzkonstanz *f* ~ **against overturning** Standsicherheit *f* gegen Drehen ~ **in rest position** Kippsicherheit *f* der Ruhelage ~ **of shape** Formbeständigkeit *f*, Formfestigkeit *f* ~ **against sliding** Gleitsicherheit *f* ~ **against tilting** Standsicherheit *f* gegen Umkippen ~ **of trajectory** Bahnstetigkeit *f*

**stability,** ~ **area** Stabilitätsgebiet *n* ~ **(lever) arm** Stabilitätsarm *m* ~ **computation** (analysis) Festigkeitsberechnung *f* ~ **condition** Stabilitäts-bedingung *f*, -kriterium *n* ~ **criterion** Stabilitätsbedingung *f* ~ **curve** Stabilitätslinie *f* ~ **determination** Stabilitäts-betrachtung *f*, -untersuchung *f* ~ **diagram** Stabilitäts-diagramm *n*, -schaubild *n* ~ **investigation** Stabilitäts-betrachtung *f*, -untersuchung *f* ~ **margin** Pfeif-abstand *m*, -sicherheit *f* ~ **number** Festigkeitszahl *f* ~ **ratio** Festigkeitszahl *f* ~**-schaltknopf** Stabilitätsschaltknopf *m* ~ **weight** Stabilitätsgewicht *n*

**stabilization** (currency) Aufwertung *f*, Beständigmachen *n*, Konstanthaltung *f*, Stabilisierung *f*, Verfestigung *f* ~ **clause** Wertbeständigkeitsklausel *f*

**stabilize, to** ~ beruhigen, glätten, gleichmachen, konstanthalten, lagefest machen, stabilisieren, stetigen

**stabilized** konstant, stabilisiert ~ **feedback** stabilisierte Rückkopplung *f* ~ **line** feste Linie *f* im Raum ~ **rectifier** stabilisierter Gleichrichter *m*

**stabilizer** abbauverhinderndes Mittel *m*, Führungsmuffe *f*, Gradlaufapparat *m*, Kippsicherung *f*, Leibrett *n*, Schlingerdämpfungsanlage *f*, Schwanzstabilisator *m*, Seitenflossenkippsicherung *f*, Stabilisator *m*, Stabilisieranlage *f*, Stabilisierungsflosse *f* (aviat.) Tastgerät *n*, Vertikalflosse *f* ~ **adjusting wheel** Höhenflossenverstellrad *n* ~ **bar** Stabilisierstange *f* ~ **bar center frame** Mittelrahmen *m* der Stabilisierstange ~ **bar damper** Stabilisierstangendämper *m* ~ **lobe** Stabilisierungswulst *f* ~ **plant** Stabilisationsanlage *f* ~ **setting** Höhenflosseneinstellung *f* ~ **tab** Höhenruder *n* ~ **tower** Stabilisierturm *m* ~ **tube** Spannungsregler-, Stabilisations-, Stabilisier-röhre *f*

**stabilizing** (weight) Beruhigung *f*, Haltbarmachen *n*, Konstanthaltung *f* ~ **feedback** stabilisierende Rückführung *f* ~ **fin** Stabilisierungsflosse *f* ~ **medium** Stützstoff *m* ~ **quartz crystal** Steuerquarz *m* ~ **surface** Dämpfungsfläche *f*, Flosse *f*, Stabilisierungsfläche *f*

**stabilovolt valve** Spannungsstabilisator *m*

**stable, to be** ~ beharren

**stable** Stall *m*; beständig, bodenständig, dauerhaft, fest, feststehend, haltbar, lagefest, stabil, standfest, standhaft, standsicher, stet ~ **in air** luftbeständig ~ **when boiling** kochbeständig ~ **in cement** (Farbstoff) zementecht ~ **in flight** flugstabil ~ **to hard water** härtebeständig ~ **higher alpha values** kreisstabil ~ **to light** lichtbeständig ~ **at one trim attitude** punktstabil ~ **at a red heat** glühbeständig **of** ~ **shape** formbeständig

**stable,** ~ **air** stabile Luft *f* ~ **airfoil** stabile Tragfläche *f* ~ **arc** stehender Lichtbogen *m* ~ **bolt** Bügelschraube *f* ~ **electron group** Elektronenrumpf *m* ~ **equilibrium** stabiles oder stetes Gleichgewicht ~ **gas** konstantes Gas *n* ~ **governor** stabiler Regler *m* ~ **halter** Stalleine *f* ~ **material** totes Material *n* ~ **orbit** stabile Bahn *f* ~ **phase** beständige Phase *f* ~ **reactor period** stabile Reaktorzeitkonstante *f* ~ **state** stabiler oder stationärer Zustand *m*

**stack, to** ~ aufschichten, aufstapeln, häufeln, schichten, stapeln, zusammensetzen **to** ~ **up** aufeinanderhäufen

**stack** Esse *f*, Meiler *m*, Pyramide *f*, (of projection booth) Rauchklappe *f*, Schacht *m*, Stapel *m*, Zug-rohr *n*, -röhre *f* **(smoke)** ~ Schornstein *m* ~ **of the carbonator** Dunstrohr *n* der Scheidepfanne ~ **of emulsions** Emulsionspaket *n* ~ **of plate springs** Blattfederbündel *n* ~ **of plates** Plattenpaket *n* ~ **of pots** Topfstapel *m* ~ **of sheets** Blechpaket *n* ~ **of wood** Holzhaufen *m*

**stack,** ~ **band** Schachtband *n* ~ **brickwork** Schachtmauerwerk *n* ~ **casting** Schachtpanzer *m* ~ **controller** Aufstockregler *m* ~ **cutting** Beschneiden *n* eines Blechstapels *m*, Blechstapel *m* beschneiden ~ **damper** Essenklappe *f* ~ **draft** Essen-, Kamin-, Schornstein-zug *m* ~ **effect** Schornsteinwirkung *f* ~ **gas** Essen-, Gicht-, Schornstein-gas *n* ~ **height** Kaminhöhe *f*, Schornsteinhöhe *f* ~ **lining** Schacht-futter *n*, -mauerung *f* ~ **loss** Kaminverlust *m* ~ **molding** Übereinanderpressen *n* ~ **paint** Schornsteinfarbe *f* für Schiffe ~ **temperature** Essentemperatur *f* ~ **walling** Schachtmauerung *f*

**stacked** aufgestapelt, geschichtet ~ **antenna array** übereinandergestockte Antennenanordnung *f* ~ **measure** Raummaß *n* ~ **wood** Schichtholz *n*

**stacker** Hebetisch *m* (film), Kamerabühne *f*, Stapelvorrichtung *f*, Stapler *m* ~ **truck** Hubstapler *m*

**stacking** (up) Schichtung *f*, Stapelung *f* ~ **crane** Stapelkran *m* ~ **fault** Stapelfehler *m* ~**-fault energy** Stapelfehlerenergie *f* ~ **operators** Stapeloperatoren *pl* ~ **tray** Auffangvorrichtung *f* ~ **truck** Stapelwagen *m* ~ **winch** Staplerwinde *f* ~ **yard** Stapelplatz *m*

**stactometer** Stalagmometer *n*, Tropfenmesser *m*

**stadia** Basismeßlatte *f*, Latte *f*, Meßlatte *f* ~ **line** Meßfaden *m* ~ **lines** Entfernung *f* messende Fäden (Tachymeter) ~ **rod** Basislatte *f* ~ **scale** **(or graduation)** Meßteilung *f*

**stadiometric straightedge** Entfernungslineal *n* (Vermessung)

**stadium** Stadium *n*

**Staedeler condenser** Städelerkühler *m*

**staff** Besetzung *f*, Gefolgschaft *f*, Kader *m* (mil.), Mitarbeiterstab *m*, Personal *n*, Stab *m* ~ **calling** **(system)** Personenrufanlage *f* ~ **costs** Personalkosten *pl* ~ **economies** Personalersparnis *f* ~ **distance** Lattenentfernung *f* ~ **facilities** soziale Einrichtungen *pl* des Betriebes ~ **float** Stangen-, Stock-schwimmer *m* ~ **gauge** Meßlatte *f* ~ **information** Personalunterrichtung *f* ~ **location system** Rufanlage *f* ~ **locator installation** Personensuchanlage *f* ~ **reduction** Abbau *m* ~ **searching system** Rufanlage *f* ~ **section** Lattenabschnitt *m* ~ **slope** Lattenschiefe *f* ~ **suggestion**

Verbesserungsvorschlag *m* (eines Arbeitneh-mers) ~ **support** Rechenstiel *m*
**staffed** besetzt **to be** ~ Personal *n* haben ~ **position** besetzter Platz *m*
**stag** Hirsch *m*
**stage, to** ~ abstufen, berüsten, inszenieren
**stage** Abschnitt *m*, Atelier *n*, Bühne *f*, Entwicklungsstufe *f*, Etappe *f*, Farbläufer *m*, Gestell *n*, Getriebegang *m*, Haltestelle *f*, Höhe *f* des Spiegels, Objektträger *m*, Phase *f*, Reiber *m*, Stand *m*, Ständer *m*, Stativ *n*, Stufe *f*, Teilstrecke *f*, Übersetzungsstufe *f*, Wasserstand *m* ~ **of amplification** Verstärkerstufe *f* ~ **of construction** Bauabschnitt *m* ~ **of degradation** Abbaustufe *f* ~ **of development** Bildungsstufe *f* ~ **of exposure** Belichtungsstufe *f* (photo) ~ **of ionization** Ionisierungsstufe *f* ~ **of manufacture** Verarbeitungsstufe *f* ~ **of motion** Bewegungsstufe *f* ~ **of operation** Arbeitsstufe *f* ~ **of power** Leistungsstufe *f* ~ **of preselection** Vorwahlstufe *f* ~ **of progress** Entwicklungsstadium *n* ~ **of rotation** Drehungswinkel *m* ~ **of selection** Wählstufe *f*
**stage,** ~ **amplifier** Zwischenverstärker *m* ~ **blower** stufiges Gebläse *n* ~ **clamp** Tischfeder *f* ~ **drying** Stufentrocknung *f* ~ **efficiency** Stufenausbeute *f*, -leistung *f* ~ **end** Endstufe *f* ~ **grinding** Stufenmahlen *n* ~ **heating** Stufenerwärmung *f* ~ **lift** Etagensgeheber *m* ~ **lighting** Bühnenbeleuchtung *f* ~**-lighting effects** Bühnenbeleuchtung *f*, Bühnenlichteffekte *pl* ~ **load** Stufenbelastung *f* ~ **micrometer** Objektmikrometer *n* ~ **one** Alarmstufe *f* Eins ~ **power** Stufenleistung *f* ~ **raising nut** Tischmutter *f* ~ **scenery** Ateliersszenen *pl* ~**-selector switch** Stufenschalter *m* ~ **separation** Stufentrennung *f* (mehrstufiger Raketen) ~ **turbine** Stufenturbine *f* ~ **valve** Stufenventil *n*
**staged** abgestuft
**stages, by** ~ absatzweise, etappenweise, sprungweise **in** ~ abschnittsweise, stufenweise ~ **of appeal** Instanzenweg *m* ~ **of finishing (or of processing)** Stadien *pl* der Veredelung ~ **of preselection** Vorwahlstufen *pl*
**stagger, to** ~ staffeln, taumeln, torkeln, (rotationally) verdrehen, versetzen
**stagger** (of wings) Staffelung *f* der Flächen ~ **angle** Staffelungswinkel *m* ~ **bracing** Tiefenverspannung *f* (aviat.) ~ **tuning** versetzte Abstimmung *f* ~ **wire** Gegenseil *n*, Tiefenkreuzdraht *m* ~ **wiring** Tiefenkreuzverspannung *f*
**staggered** gestaffelt, auf Lücke stehen, schachbrettförmig, versetzt, versetzt angeordnet, zickzackförmig ~ **of welding** Zickzackpunktschweißung *f*
**staggered,** ~ **arrangement** Höhenanordnung *f* ~ **array** (antenna) Reuse *f* ~ **biplane** gestaffelter Doppeldecker *m* ~ **cams** versetzte Daumen *m* ~ **circles** versetzte Kreise *pl* ~ **circuits** gestaffelte oder verstimmte Kreise *pl* ~ **driving pinions** gegeneinander vesetzte Zahnscheiben *pl* ~ **filtration** stufenweise Filtration *f* ~ **gear rims** Etagenräder *pl* ~ **holes** (in blasting) versetzte Löcher *pl* ~ **piling** schachbrettförmig eingerammte Pfähle *pl* ~ **radial engine** Doppelsternmotor *m* mit gegeneinander vesetzten Zylindersternen ~ **reflection** Zickzackreflexion *f* ~ **riveting** Versatznietung *f* ~ **rolling train** gestaf-

felte Straße *f*, gestaffelte Walzstraße *f* ~ **row of rivets** versetzte Nietreihe *f* ~ **scanning** alternierende oder springende Zerlegung *f*, Zwischenzeilenverfahren *n* ~ **seam** versetzte Naht *f* ~ **seats** gestaffelte Sitze *pl* ~ **teeth** kreuzverzahnt, Schaftfräser *m* für T-Nuten ~ **tooth cutter** Kreuzzahnscheiben-, Verbundzahn-fräser *m* ~**-tooth gear** Verbundzahnrad *n* ~**-tooth side mill** Verbundzahnscheibenfräser *m*
**staggering** Staffeln, Stufung *f*, Versetzung *f* ~ **of groups** Staffeln *n* von Gruppen ~ **advantage** Versetzungsgewinn *m*
**staging** Gerüst *n*, Landebrücke *f*, Landungssteg *m* ~ **of the compressor** Stufenunterteilung *f* des Kompressors
**stagnant** geschäftslos, leblos, lustlos, stagnierend, still, stillstehend, stockend ~ **condition** Stille *f* ~ **liquid** ruhende Flüssigkeit *f* ~ **water** Stauwasser *n*
**stagnation** Stau *m*, Stillstand *m*, Stockung *f* ~ **of air** Luftstauung *f* ~ **of business** Geschäftsstille *f* ~ **in the market** Absatzstockung *f*
**stagnation,** ~ **line** Staupunktlinie *f* ~ **point** Spaltungspunkt *m* (aerodyn.), Staupunkt *m*, Totlage *f* (aerodyn.), Totpunkt *m* (aerodyn.)
**stain, to** ~ anfressen, aufstreichen, ausbeizen, beflecken, beizen, färben, flecken, verschmutzen **to** ~ **superficially** anfärben
**stain** Beize *f*, Farbe *f*, Farbstoff *m*, Fleck *m*, Flecken *m*, Makel *m* ~ **caused by wringing** Wringfleck *m* ~ **remover** Fleckenentfernungsmittel *n* ~ **resistance** Korrosionswiderstand *m*, Rostwiderstand *m*
**stained** fleckig, gefleckt, gesprenkelt ~ **glass** Farbglas *m* ~ **paper** Marmorpapier *n* ~ **wood** gebeiztes Holz *n*
**staining** Anfraß *m*, Beizen *n*, Einfärbung *f*, Färbung *f* ~ **in toto** Stückfärbung *f* ~ **solution** (microsc.) Farblösung *f*
**stainless** fleckenlos, korrosions-beständig, -fest, -frei, -sicher, makellos, nichtrostend, (steel) rostlos, rostsicher, schwerrostend ~ **property** Korrosionsbeständigkeit *f*, Rostbeständigkeit *f*, Rostsicherheit *f* ~ **steel** korrosionsbeständiger, nichtrostender, rostfreier oder rostsicherer Stahl *m*
**stainlessness** Fleckenlosigkeit *f*
**stains caused by putrefaction** Faulflecke *pl*
**stair** Stufe *f*, Treppenstufe *f* ~**-builder's saw** Gratsäge *f* ~ **horse** Treppenseitenstück *n* ~ **landing** Absatz *m*, Treppenabsatz *m* ~ **stringer** Treppenwange *f* ~ **tread** Treppenstufe *f* ~ **well** Treppenhaus *n*
**staircase** Stiege *f*, Treppe *f*, Treppenhaus *n* ~ **conveyor** Schrägförderer *m* ~ **fitting** Treppenbeschlag *m* ~ **lift** Treppenaufzug *m* ~ **lighting time switch** Treppenlichtzeitschalter *m* ~**-like structure of echelon grating** Gitterblock *m* ~ **waveform** Treppenspannung *f*
**stairs** Treppe *f* ~ **with broken center line (or with landing places)** gebrochene Treppe *f* **in the form of** ~ treppenförmig
**stake, to** ~ (leather mfg.) stollen **to** ~ **off the distance** die Entfernung abgreifen (Karte) **to** ~ **out** abpfählen, abstecken, ausstecken, Linienführung *f* festlegen **to** ~ **out a line** eine Linie abpfählen **to** ~ **out a railway line** eine Bahnlinie *f* abpflöcken

**stake** Absteck-haken *m*, -pfahl *m*, -pflock *m*, -stange *f*, Anteil *m*, Einsatz *m*, Formerstift *m*, Meßmarke *f*, Pfahl *m*, Pflock *m*, Pfosten *m*, Picket *n*, Runge *f*, Spieleinsatz *m*, Stab *m*, Stake *f*, Stange *f*, Wagenrunge *f*, (at r.r. switch) Weichenpfahl *m* ~ **chain** Rungenkette *f* ~ **marked with a number** Nummernpfahl *m* ~ **obstacle** Gittersperre *f*

**staked,** ~ **bearing** festeingebautes Lager *n* ~**-off distance** Stoppstrecke *f*

**staker** Stollblock *m*

**stakes** Bühnenpfähle *pl*, Gestänge *n*

**staking machine** Stollmaschine *f*

**stalactite** Stalaktit *m*, Tropfstein *m*

**stalagmite** Stalagmit *n*

**stalagmometer** Tropfen-messer *m*, -zähler *m*

**stale** abgekämpft, abgestanden, fade, schal, (of air) verbraucht, verjährt ~ **check** alter Scheck *m*

**stalk, to** ~ anpirschen, beschleichen, (sich) heranpirschen

**stalk** Halm *m*, Pfahl *m*, Schaft *m*, Stamm *m*, Stengel *m*, Stiel *m* ~ **cutter** Maisstoppelzerschläger *m*, Stengelschneider *m* ~ **fiber** Stielfaser *f*

**stalked** stengelig

**stall, to** ~ abdrosseln (motor), abrutschen, aufziehen, durchsacken (aviat.), sich festlaufen, durch Hindernis zum Stillstand *m* kommen, hochziehen, überziehen (aviat.) **to** ~ **the airplane** das Flugzeug überziehen **to** ~ **an engine** abwürgen

**stall** Abrutschen *n* (aviat.), Abbaustrecke *f*, Stadel *f*, Stall *m*, Überziehen *n*, überzogener Flug *m* (aviat.) ~ **at a fair** Messestand *m* ~ **without power** Sackflug *m* ohne Kraft

**stall,** ~ **device** Durchsackwarngerät *n* ~ **landing** Sackfluglandung *f*, überzogene Fluglandung *f* ~ **roasting** Röstung *f* in Stadeln, Stadelröstung *f* ~ **test** Abkippmessung *f* ~ **torque** Anzugs-, Kurzschluß-drehmoment *m*

**stalled** bewegungsunfähig ~ **condition** Sackflugzustand *m*

**stalling** (in flight) Geschwindigkeitsverlust *m*, Überziehen *n* ~ **angle** kritischer Anstellwinkel *m*, kritischer Winkel *m*, Höchstauftriebswinkel *m*, Strömungsabreißwinkel *m* ~ **flight** Langsamflug *m*, Sackflug *m* ~ **point** Erstarrungspunkt *m* ~ **speed** Durchsackgeschwindigkeit *f*, kritische Geschwindigkeit *f*, Mindestauftriebsgeschwindigkeit *f* (aviat.), Minimumauftriebsgeschwindigkeit *f*, Sackfluggeschwindigkeit *f* ~ **torque** (polyphase motors) Abfallmoment *n*

**stamina** Ausdauer *f*, Durchhaltevermögen *n*

**Stammer degree** Stammergrad *m*

**stamp, to** ~ abdrucken, abstempeln, anmerken, ausstampfen, eichen, eindrücken, feststampfen, kennzeichnen, markieren, pochen, prägen, pressen, signieren, stampfen, stanzen, stempeln, verpochen **to** ~ **down** einstampfen **to** ~ **in** einstempeln **to** ~ **money** münzen **to** ~ **on** aufstampfen **to** ~ **upon** aufdrücken

**stamp** Abdruck *m*, (postage) Briefmarke *f*, Eindruck *m*, Fallwerk *f*, Klangfarbe *f*, Marke *f*, Mönch *m*, (of various machines) Patrize *f*, (of various machines) Patrone *f*, Pochstempel *m*, Präge *f*, Prägestempel *m*, Schlag *m*, Schlagstempel *m*, Schüsser *m*, Stampfe *f*, Stanze *f*, Stempel

*m* ~ **battery** Pochsatz *m*, Pochstempelreihe *f* ~ **-battery framing** Pochstuhl *m* ~ **bearer** Stempelhalter *m* ~ **capacity** Pochleistung *f* ~ **die** Pochsohle *f*, Poch-, Präge-stempel *m* ~ **duty** Stempel -abgabe *f*, -gebühr *f* ~ **head** Pocheisen *n*, Pochstempelbeschwerer *m* ~ **ink cushion** Stempelkissen *n* ~ **man** Erzpocher *m* ~ **mill** Pochmühle *f*, Pochwerk *n*, Stampfwerk *n* ~**-mill framing** Pochwerkgerüst *n* ~ **milling** Verpochen *n* ~ **moistener** Markenfeuchter *m* ~ **mortar** Pochtrog *m* ~ **(ing) pad** Stempelkissen *n* ~ **printing** Stempeldruck *m* ~ **pulp** Pochtrübe *f* ~ **rock** Pochgestein *n* ~ **screen** Pochblech *n* ~ **shoe** Pocheisen *n*, Pochschuh *m* (min.) ~ **socket** Pocheisen *n*, Pochschuhhülse *f* ~**-system guide** Pochstempelschaft *m* ~ **subbase** Pochunterlage *f* ~ **tax** Stempelsteuer *f*

**stamped** eingeschlagen (Maschinennummer), geprägt, gepreßt (metall) ~ **lamp** geeichte Lampe *f* ~ **metal part** Stanzteil *m* ~ **ore** Pochmehl *n* ~ **plate** gestanzte Bleche *pl* ~ **sheet of paper** Stempelbogen *m* ~ **steel** getanzter Stahl *m* ~ **weight** geeichtes Gewicht *n*

**stampede, to** ~ durchgehen

**stamper** Preßmatrize *f* (phono), Schlagmaschine *f*, Stampfer *m*, Stempel *m*, (in sheet-metal testing) Stößel *m*, Stößer *m* ~ **press** Öllade *f*

**stamping** Abstempeln *n*, Abzeichen *n*, Ausstampfung *f*, Einstampfen *n*, Gepräge *n*, Pochen *n*, Prägen *n*, Prägung *f*, Preßling *m*, Preßstück *n*, Preßteil *m*, Stanze *f*, Stanzen *n*, Stanzstück *n* ~ **calender** Prägekalander *m* ~ **chuck** Abstechfutter *n* ~ **device** Einschlagvorrichtung *f* ~ **down** Feststampfen *n* ~ **die** Prägerstanze *f*, Schlagmatrize *f*, Stanzmatrize *f* ~ **gauge** Stanzmesser *n* ~ **lead** Prägeblei *n* ~ **machine** Formstampfmaschine *f*, Stanzmaschine *f* ~**-machine plant** Stampanlage *f* ~ **mass** Stampfmasse *f* ~ **method** Stanzverfahren *n* ~ **mill** (wet, dry, screen, or grate trough types, for ore) Gitterpochwerk *n*, Stampfwerk *n* ~ **ore** Pocherz *n* ~ **-out press** Ausstanzpresse *f* ~ **press** Aushauschere *f*, Prägewerk *n*, Stanzpresse *f*, Stoßwerk *n* ~ **tool** Prägegesenk *n*, Stanzwerkzeug *n* ~ **trough** Pochladen *m*

**stamps mill** Gitterpochwerk *n*

**stanching,** ~ **blind** Dichtungsjalousie *f* ~ **ring** Dichtungsring *m*

**stanchion** Runge *f*, Säule *f*, Stempel *m*, Wagenrunge *f*

**stancil, to** ~ signieren

**stand, to** ~ aushalten, ausstehen, bestehen, ertragen, stehen, widerstehen **to** ~ **by** bereitstehen **to** ~ **clear** freistehen **to** ~ **for** befürworten, einstehen für **to** ~ **in line** Schlange *f* stehen **to** ~ **off** freistehen **to** ~ **opposed (or opposite)** gegenüberstehen **to** ~ **out** ausstehen (Forderungen), hervorragen **to** ~ **out to sea** die offene See erreichen **to** ~ **toward the anchorage** nach dem Ankerplatz gehen **to** ~ **up** aufstehen **to** ~ **up to** entgegentreten **to** ~ **upon the course** Kurs *m* halten **to** ~ **security** avalieren **to** ~ **still** stillstehen **to** ~ **the test** bewähren

**stand** Bank *f*, (forest) Bestand *m*, Bock *m*, Etagere *f*, Gerüst *n*, Gestell *n*, Objektträger *m*, Säule *f*, Stand *m*, Ständer *m*, Standort *m*, Stativ *n*, Tribüne *f*, Untergestell *n*, Untersatz *m*, Zug *m* ~ **for checking cylindrical pieces on the**

**machine tool** Stativ *n* zum Ausrichten von zylindirschen Körpern auf der Werkzeugmaschine ~ **of rolls** Wälzengerüst *n* ~ **for stitcher** Anrollvorrichtung *f*
**stand-by** Alarmbereitschaft *f*, Beistand *m*, Verstärkung *f*; bereit **at** ~ in Bereitschaft *f*
**stand-by** ~ **area** Warteraum *m* (rdo) ~ **blower** Reservelüfter *m* ~ **boiler** Reservekessel *m* ~ **circuit** Vorratsleitung *f* ~ **current** Ruhestrom *m* ~ **equipment** Hilfs-, Reserve-apparat *m* ~ **fan** Hilfventilator *m* ~ **frequency** Hilfsfrequenz *f*, Wachwelle *f* ~ **generator** Notstromaggregat *n* ~ **light** Hilfsfeuer *n* ~ **losses** (of a power station) Abbrandverlust *m* ~ **operation** Horchempfang *m* ~ **plant** Bereitschaftsanlage *f* ~ **position** Empfangsstellung *f*, Stellung *f* auf Empfang, Wartestellung *f* ~ **receiver** Suchempfänger *m* ~ **stage** Reservestufe *f* ~ **station** Hilfsvoramt *n* ~ **time** (of furnace) Chargierzeit *f* ~ **transmitter** Reservesender *m* ~ **unattended time** Außerbetriebzeit *f* ~ **unit** Hilfsgerät *n* für Notfall, Notaggregat *n* ~ **ventilator** Hilfsventilator *m*
**stand,** ~ **head** Gestellkopf *m* ~ **magnifier** Stativlupe *f* ~-**off** Auslauf *m* ~-**off capacitor** Stützpunktkondensator *m* ~-**off insulator** Abstandisolator *m* ~ **oil** Dicköl *n*, Standöl *n* ~ **pipe** Ausgleichsrohr *n*, (hydraulischer Druck) Standrohr *n*, Steigleitung *f*, Steigrohr *n* ~-**pipe valve** Standrohrventil *n* ~-**still** Ruhestand *m*, Stillstand *m*, Stockung *f* ~-**still corrosion** Stillstandkorrosion *f* ~-**still dimming** Stationärblendung *f* ~ **support** Gehäusefuß *m*
**standard** (used for comparison) Bezugssystem *n*, Eichmaß *n*, Eichnormal *n*, Eichung *f*, Einheit *f*, Fahne *f*, Feingehalt *m*, Flagge *f*, Geländerstütze *f*, Maß *n*, Maßstab *m*, Muster *n*, Niveau *n*, Norm *f*, Normalie *f*, Normalmaß *n*, Pfeiler *m*, Richt-linie *f*, -stange *f*, Standarte *f*, Standpunkt *m*, Stütze *f*, Urmuster *n*, (banking) Valuta *f*, (roll) Walzenständer *m*; anerkannt, geeicht, gewöhnlich, maßgebend, musterhaft, normal, normentsprechend, normgerecht, normig, normrecht, serienmäßig
**standard, of** ~ **gauge** vollspurig (r.r.) ~ **up to** ~ vollwertig ~ **of comparison** Vergleichs-maßstab *m*, -standard *m* ~ **of length** Längenmaß *n* ~ **of life** Lebenshaltung *f* ~ **of measurement** Grundmaß *n* ~ **of mills** Mühlenständer *m* ~ **of reference** Vergleichsmaßstab *m* ~ **of value** Wertmesser *m*
**standard,** ~ **absorber** Normalabsorptionsmittel *n* ~ **accessories** Normalausrüstung *f* ~ **acid** Probesäure *f* ~ **air** Ina-Atmosphäre *f*, Normalluft *f*, reduzierter Luftzustand *m* ~ **air coil** Standardluftspule *f* ~ **air-core coil** Normalluftspule *f* ~ **altimeter** Grobhöhenmesser *m* ~ **artificial cement** (künstlicher) Normalzement *m* ~ **atmosphere** Normal-atmosphäre *f*, -luft *f*, -luftzustand *m* ~ **atmospheric conditions** Normalluftzustand *m* ~ **attachments** Normalvorrichtung *f* ~ **bar** Normalstab *m* ~ **bayonet cap** Normal-Swansockel *m* ~ **boost** Normaldruck *m* ~ **bore** Einheitsbohrung *f* ~ **cable** Standardkabel *n* ~ **cable equivalent** Leitungsäquivalent *n* in Meilen Standardkabel, Standardkabeläquivalent *n* ~ **candle** Dezimalkerze *f* ~ **cap** Normalsockel *m* ~ **capacitor** Standardkapazität *f* ~ **capacity** Normal-, Regel-leistung *f* ~

**caterpillar track** Normal-kette *f*, -raupe *f* ~ **cell** Akku *m*, Normal-, Standard-element *n* ~ **charge** Normalgattierung *f* ~ **chase** Normalschließrahmen *m* (print.) ~-**circuit receiver** Eichkreistelefon *n* ~-**circuit transmitter** Eichkreismikrofon *n* ~ **clearance gauge** Normalprofilschablone *f* ~ **clock** Normaluhr *f* ~ **clutch** Einheitskupplung *f* ~-**color glass** normalisiertes Farbglas *n* ~ **compass** Norm(al)kompaß *m*, Regelkompaß *m* ~ **composition** Normalzusammensetzung *f* ~ **condenser (or capacitor)** Einheitskondensator *m* ~ **condition** Normal-bedingung *f*, -zustand *m* ~ **cord** Einheitsschnur *f* ~ **couple** (spectrum analysis) Ficierungspaar *n* ~ **coupling** normale Muffe *f* ~ **curve** Normenkurve *f* ~ **cutting speed** Normalschnittgeschwindigkeit *f* ~ **density of air** Einheitsluftdichte *f* ~ **design** Normalausführung *f* ~ **deviation** Standardabweichung *f* (mittlere quadratische Abweichung) ~ **distillation test** normalisierte Versuchsdestillation *f* ~ **equipment** Normal-ausrüstung *f*, -zubehör *n*, serienmäßige Ausrüstung *f*, Rüstsatz *m* ~ **female ground taper** Normalmantelschliff *m* ~ **figure** Normungszahl *f* ~ **fit** Normalsitz *m* ~ **footing** Bockbein *n* ~ **frequency** Einheits-, Normal-frequenz *f*, Frequenznormale ~ **fuse** Einheitszunder *m* ~ **gasoline** (for cars) Fahrbenzin *n* ~ **gasoline container** Einheitskanister *m* ~ **gauge** Eichmaß *n*, Maß- und Gewichtsnorm *f*, Normallehre *f*, Normalspur (r.r.), Normspur *f*, Prüflehre *f*, Prüflehrdorn *m*, Prüfmaß *n*, Regelspur *f*, Richtmaß *n*, Spurlehre *f*, Urlehre *f*, Urmaß *n*, Vollspur *f* ~ **(for setting) gauge** Einstellmaß *n*
**standard-gauge,** ~ **rail** Vollbahnschiene *f* ~ **railroad** normalspurige Eisenbahn, Vollbahn *f*, Vollspurbahn *f* ~ **railway** Normalspurbahn *f* ~ **siding** Vollspurgleisanschluß *m*
**standard,** ~ **geopotential foot** geopotentieller Normfuß *m* ~ **gold** Münzfuß *m*, Probe-, Probier-gold *n* ~ **goods** Stapelartikel *m* ~ **grab** Normalkatze *f* ~ **gravity** Normalschwere *f* ~ **grinding wheel size** Schleifscheibengröße *f* ~ **ground apparatus** Normschliffgerät *n* ~ **ground joint** Normalschliff *m* ~ **hole** Einheitsbohrung *f* ~ **horsepower** normale Pferdestärke *f* ~ **instrument** Etalonapparat *m*, (reference) Normalinstrument *n*, Regelinstrument *n* ~ **intensity of light** Vergleichslicht *n* ~ **international atmosphere** internationale Normalatmosphäre *f* (INA) ~ **lamp** Eichlampe *f* ~ **lathe** Normaldrehbank *f* ~ **length** gewöhnliche oder regelrechte Länge *f* ~ **linear measure** Längsnormale *f* ~ **liquid** Normalflüssigkeit *f* ~ **load** Normalbelastung *f* ~-**long-distance telephone cable** Normalfernkabel *n* ~ **magnetic tape** Normalband *n* ~ **make** der üblichen Art ~ **master** Normalstück *n* ~ **measure** Normal-, Probe-, Ur-maß *n* ~ **meridian** Normalmeridian *m* ~ **meter** Urmeter *m* ~ **meter bars** Kontrollnormalmeter *pl* ~ **meter board** Normenzählertafel *f* ~ **microphone** Eich-, Normal-mikrofon *n* ~ **mixture** Normalgattierung ~ **mounting** Normalfassung *f* ~ **muzzle velocity** schußtafelmäßige Anfangsgeschwindigkeit *f* ~-**number series** Normdrehzahlen *pl* ~ **objective thread** übliches Objektivanschraubgewinde *f* ~ **ohm** internationales Ohm *n*, Normalohm *n* ~ **outline** Normal(profil)

figur *f* ~ **output** Regelleistung ~ **paper-size** (Papier) DIN-Format *n* ~ **parallel** Normalparallele *f* ~ **part** Normteil *m* ~ **parts** Massenteile *pl* ~ **performance** Normleistung *f* ~ **picture frame** Normalbildkader *n* ~ **pile** Bordpfahl *m* ~ **pipe** normales Rohr *n* ~ **pitch** Bezugssteigung *f* (eines Propellers), Normalstimmton *m* ~ **plate specimen** (in testing) Normalflachstab *m* ~ **port** Basisstation *f* ~ **poster** Dauerplakat *n* ~ **practice** normale Arbeitsweise *f* ~ **pressure** Normaldruck *m* ~ **pressure altitude** Normaldruckhöhe *f* ~ **price** Einheitspreis *m* ~ **propagation** Normalausbreitung *f* ~ **quality** gleiche Qualität *f* ~ **radiator** Vergleichsstrahler *m* ~ **receiver** Normalfernhörer *m* ~ **reference** Bezugsnormale *f* ~ **reference disc** Parallelelendmaß *n* ~ **reference telephone circuit** Fernsprechvergleichsstromkreis *m*, Vergleichsstromkreis *m* ~ **refraction** Normalbrechung *f* ~ **repeater unit** Einheitsverstärker *m* ~ **resistance** Normalwiderstand *m*, Vergleichswiderstand *m*, Widerstandsatz *m* ~ **reversing mirror** Normalumkehrspiegel *m* ~ **ring gauge** Normallehrring *m* ~ **round test bar** Normalrundstab *m* ~ **sample** Normalprobe *f* ~ **sand** Normensand *m* ~ **scale** Normenmaß *n* ~ **(cross) section** Normalprofil *n* ~ **shade** Stapelnuance *f* ~ **shaft** Einheitswelle *f* ~ **shape** Normalformat *n* ~ **sieve** Einheitssieb *n* ~ **signal generator** Eichgeber *m* (g/m) ~-**signal oscillator** Meßsender *m* ~ **silver** Probesilber *n* ~ **size** gewöhnliche Größe *f*, Normal-format *n*, -größe *f*, Regelgröße *f*, übliche Größe *f* ~ **solution** Normal-flüssigkeit *f*, -lösung *f*, Titerflüssigkeit *f*, Titrierlösung *f*, Vergleichslösung *f* ~ **solvent** Naphthah *n* & *f* ~ **spanner** Normschlüssel *m* ~ **specification** Normen-, Normungs-vorschrift *f* ~ **specifications** Norm *f* ~ **strength of a solution** Titer *m* einer Flüssigkeit ~ **stress** normale Beanspruchung *f* ~ **style** übliche Bauart *f* ~ **tapping machine** Normhammerwerk *n* (acoust.) ~ **target** Einheitsscheibe *f* ~ **television receiver** Einheitsfernsehempfänger *m* ~ **temperatur** Normaltemperatur *f* ~ **test bar** Kontrollstab *m*, Normalprobestab *m* ~ **test sample** Normalversuchsprobe *f* ~ **testing method** normalisierte Prüfmethode *f* ~ **thermometer** Normalthermometer *n* ~ **thread** Einheits-, Normal-gewinde *n* ~ **thread outline** Normalgewindeprofilfigur *f* ~ **thread ring gauge** Normalgewinde-Lehrring *m* ~ **thread tool** normaler Gewindestahl *m* ~ **time** Einheitszeit *f*, die gesetzliche Zeit *f*, Normalzeit *f* ~-**time zone** Einheits(zeit)zone *f* ~ **tin** Probezinn *n* ~ **tire** Normalbereifung *f* ~-**tone horn** Normaltonhorn *m* ~ **track** (sound recording) Normalschrift *f* ~ **trade-marked merchandise** Markenerzeugnis *n* ~ **trajectory** Normalbahn *f*, Sollausgabe *f* ~ **transmitter** Eich-, Normal-mikrofon *n* ~ **truck** Einheitslastkraftwagen *m* ~ **tube** Niveauröhre *f* ~ **type** (construction) Einheitsbauart *f*, Regelbauart *f* ~-**type battery** Verbandanode *f* ~-**type joining member** normaler Stoßriegel *m* ~ **unity** Standardeinheit *f* ~ **(upright)** Standsäule *f* ~ **value** Richtwert *m* ~ **value of deviation** Streuwert *m* ~ **value of focal length of object glasses** Normalbrennweite *f* der Objektive ~ **vinegar** Normalessig *m* ~ **voltage** Nennspannung *f* ~ **wavemeter** Normalwellenmesser *m* ~ **weight** geeichtes Gewicht *n*, Nor-

malgewicht *n*, Probegewicht *n*, übliches Gewicht *n*, Urgewicht *n* ~ **weld** normale Schweißnaht *f* ~ **white plate** Normalweißplatte *f*.

**standardization** Eichung *f*, Einstellung *f*, Normalisierung *f*, Normenaufstellung *f*, Normierung *f*, Normung *f*, Titerstellung *f*, Typenbereinigung *f*, Typisierung *f*, Vereinheitlichung *f* ~ **of frequency characteristic** Normung *f* des Frequenzganges ~ **of frequency-response curve** Frequenzgangnormung *f*

**standardization committee** Ausschuß *m* für Einheiten und Formeln

**standardize, to** ~ eichen, (a solution) einstellen, normalisieren, normen, Norm festsetzen, normieren, schematisieren, vereinheitlichen

**standardized** genormt, nach Norm hergestellt, normalisiert ~ **oil types** Einheitsöle *pl* ~ **position of indicator** Normalstellung *f* eines Anzeigegerätes ~ **production** Einheitserzeugung *f*

**standardizing,** ~ **box** Meßdose *f* ~ **sheet** Normalblatt *n*

**standards** Norm *f*, Ständergerüst *n* ~ **of operation** Betriebsregeln *pl* ~ **committee** Ausschuß *m* für Einheiten und Formeln

**standing** Stellung *f*; stehend ~ **on open circuit** geöffnet ~ **over** überstehend

**standing,** ~ **balance** ruhendes Gleichgewicht *n* ~ **barrage** Notfeuer *n* ~ **bias** feste Vorspannung *f* ~ **crop of wood** Holzbestand *m* ~ **idle** stillgelegt ~-**on-nines carry** Neunerübertragung *m* ~ **platform** Trittpodest *m* ~ **press** Glättpresse *f* ~ **pump** feste Pumpe *f* ~ **room** Stehplatz *m* ~ **still** stillstehend ~ **time of water** Stehzeit *f* vom Wasser ~-**type collar** Stehkragen *m* ~ **value** Zeitwert *m* ~ **valve** Saugventil *n* ~ **vise** Stehknecht *m* ~ **wave** Deckwalze *f*, Interferenzwelle *f*, Knoten *m*, Mantelwelle *f*, stehende Welle *f* ~ **wave indicator** Stehwellenmeßgerät *n* ~-**wave meter** Stehwellenverhältnismesser *m*, Welligkeitsmesser *m* ~-**wave ratio** Welligkeit *f*

**stannate of soda** Präpariersalz *n*

**stannic,** ~ **acid** Zinnsäure *f* ~ **anhydride** Zinnsäureanhydrid *n* ~ **bisulfide** Doppelschwefelzinn *n* ~ **bromide** Zinnbromid *n* ~ **chloride** Stanni-, Zinn-chlorid *n* ~ **compound** Stanni-, Zinnoxyd-verbindung *f* ~ **fluoride** Zinnfluorid *n* ~ **hydroxide** Stanni-, Zinn-hydroxid *n* ~ **iodide** Stanni-, Zinn-jodid *n* ~ **oxalate** Stannioxylat *n* ~ **oxide** Zinn-asche *f*, -kalk *m* ~ **salt** Zinnoxydsalz *n* ~ **sulfide** Musivgold *n*, Zinnsulfid *n* ~ **thiocyanate** Rhodanzinnoxyd *n*

**stanniferous** zinnhaltig

**stannite** Zinnkies *m*

**stannous,** ~ **bromide** Zinnbromür *n* ~ **chloride** Stannochlorid *n*, Zinnchlorür *n*, Zinnoxy(dul)-chlorid *n*, Zinnsalz *n* ~ **compound** Stannoverbindung *f* ~ **hydroxide** Zinnhydroxydul *n*, Zinnoxydhydrat *n*, Zinnoxydulhydrat *n* ~ **iodide** Stannojodid *n*, Zinnjodür *n* ~ **oxalate** oxalsaures Zinnoxydul *n* ~ **oxide** Stannooxyd *n*, Zinnmonoxyd *n*, Zinnoxydul *n* ~ **sulfide** Stannosulfid *n*, Zinnsulfür *n* ~ **thiocynate** Rhodanzinnoxydul *n*

**staple, to** ~ aufschichten, häufeln

**staple** Bügel *m*, Drahtklampe *f*, Drahtöse *f*, Haspe *f*, Kramme *f*, Krampe *f*, (pointed) Krampe *f* mit Stiftspitze, Öse *f*, Stapel *m* ~ **for a bolt** Haken *m* eines Querriegels (Schloss)

**staple,** ~ **color** Stapelfarbe *f* ~ **commodities** Stapelgut *n* ~ **conveyor** Stapelförderer *m* ~ **feeding** Klammerzuführung *f* ~ **fiber** Kunstspinnfaser *f*, Stapelfaser *f*, Zellwolle *f* ~ **goods** gangbarer Artikel *m*, Stapelartikel *m* ~ **knee** Doppelknie *n* ~ **plate** Hakenblatt *n*, Riegelblech *n* ~ **pliers** Drahtreiterklemme *f* ~ **products** Gebrauchsgüter *pl* ~ **shanks** Klammerschenkel *m* ~ **trade** Stapelhandel *m*

**stapling,** ~ **machine** Heftmaschine *f*, Heftmaschine *f* für Drahtheftung ~ **wire** Heftdraht *m*

**star** Stern *m* ~ **of principal sequence** Hauptsequenzstern *m*

**star,** ~ **antenna** Sternantenne *f* ~ **bit** Kreuz-, Stern-meißel *m* ~ **chain** Sternkette *f* ~ **circuit** Stern-glied *n*, -schaltung *f* ~ **cluster** Sternhaufen *m* ~**-connected** sterngeschaltet ~**-connected three-phase system with neutral Y connection** Sternschaltung *f*, Y-Schaltung *f* ~ **connection equivalent** Ersatzsternschaltung *f* ~**-delta** Schaltung *f*, Sterndreieck *n* (electr.) ~**-delta connection** Dreiecksternverbindung *f* ~**-delta starting** Sterndreieckanlauf *m* ~**-delta switch** Sterndreieckschalter *m* ~ **drill** Stein-bohrer *m*, -meißel *m* ~**flight** Sternflug *m* ~**formation by pions** Sternbildung *f* durch Pionen ~ **frame** Sternreifen *m* ~ **gauging** Rohraufmessung *f* ~ **grouping** Sternschaltung *f* ~ **handle** Griffkreuz *n* ~ **indicator** Sternschauzeichen *n* ~ **knob** Kreuzgriff *m*, Sternknopf *m* ~ **map** Sterntafel *f* ~ **network** Sternglied *n* ~ **pattern** Sternmuster *n* ~ **performer** Hauptrollendarsteller *m* ~ **pinion** Sternzahnrad *n* ~ **quad** Sternvierer *m* ~ **role** Hauptrolle *f* ~ **shake** Strahlenriß *m* ~**-shaped** sternförmig ~**-shaped locking handle** Kreuzgriff *m* ~**-shaped shed** Sternhalle *f* ~ **shell** Granatsignal *n*, Leucht-geschoß *n*, -granate *f*, Signalgeschoß *n*, -granate *f* ~**-shell gun** Lichtspucker *m* ~ **signal** Sternsignal *n* ~ **(delta) starter** Sterndreieckanlasser *m* ~ **tensioner** Federkreuz *n* ~ **twisting** Sternverseilung *f* ~**-type impeller** beiderseits offenes Laufrad *n*, Schaufelstern *m* ~ **-type reel-star** Drehstern *m* ~ **voltage** Sternspannung *f* ~ **wheel** Arretierungsscheibe *f*, Rad *n* mit scharfen Zähnen, Rastenscheibe *f*, Schaltstern *m*, Sternrad *n* **(cam and)** ~ **wheel in Geneva movement** Malteserkreuz *n* ~**-wheel idler** Planetenrad *n*

**starboard** Backbordhalsen *n*, Steuerbord *n* **on the** ~ **beam** querab Steuerbord ~ **bow** Steuerbordvoraus *n* ~ **corrector** Steuerbordkorrektor *m* ~ **curve** Rechtskurve *f*, Steuerbordkurve *f* ~ **dolphin** Steuerborddalbe *f* ~ **engine** Steuerbordmaschine *f*, -motor *m* ~ **light** Steuerbord(seiten)licht *n* ~ **motor** Steuerbordmotor *m* ~ **outer engine** der äußere Steuerbordmotor *m* ~ **pale** Steuerborddalbe *f* ~ **signal** Steuerbordzeichen *n*

**starch, to** ~ stärken

**starch** Stärke *f* ~ **dressing** Pappverband *m* ~ **paste** Stärkekleister *m* ~**-potassium iodide solution** ·Jodkaliumstärkelösung *f* ~ **solution** Stärkelösung *f* ~ **sugar** Kartoffelzucker *m*

**starching clay** Stärkeglanz *m*

**starchy** stärkehaltig

**starling of a bridge** Pfeilerhaupt *n*

**starred** gesternt

**starring** Sternen *n*

**starry heaven** Fixsternhimmel *m*

**start, to** ~ abfahren, abfliegen, abgehen, ablaufen, andrehen, anfahren, anfangen, ankurbeln, anlassen, anlaufen, ansetzen, (an engine) anstellen, antreiben, anziehen, anzüden, aufbrechen, ausgehen, beginnen, betätigen, einleiten, einrücken, einsetzen, inbetriebsetzen, in Betrieb setzen, ingangsetzen, in Gang bringen, in Gang setzen, in Schwung bringen, in Wirksamkeit setzen, starten **ability to** ~ (motor) Andrehvermögen **to** ~ **in any position of the crankshaft** in jeder Kurbelstellung *f* anspringen **to** ~ **boiling** ansieden **to** ~ **up boilers** anheizen **to** ~ **bore** anbohren **to** ~ **combustion** in Produktion *f* geraten **to** ~ **construction** auflegen **to** ~ **an engine** andrehen **to** ~ **the engine** den Motor anlassen oder anwerfen **to** ~ **engine by compressed air** anschießen **to** ~ **from** anknüpfen an etwas **to** ~ **on full load** vollbelastet anlaufen **to** ~ **light** leer anlaufen **to** ~ **under load** belastet anlaufen, unter Last *f* anlaufen **to** ~ **without load** leer anlaufen **to** ~ **a mine** eine Grube *f* bauen **to** ~ **without jerk** stoßfrei anlaufen **to** ~ **off** losziehen **to** ~ **and stop** ein- und ausrücken **to** ~ **a train** einen Zug *m* ablassen **to** ~ **up** anregen, anwerfen, inbetriebnehmen

**start** Abfahrt *f*, Ablauf *m*, Abmarsch *m*, Anfang *m*, Anlassung *f*, (of run) Anlauf *m*, Ansatz *m*, Antrieb *m*, Aufbruch *m*, Beginn *m*, Einsetzen *n*, Start *m* ~**(ing)** Anlaufen *n* ~ **of flight** Abflug *m* ~ **of pressure** Druckanfang *m* ~ **with rubber cable** Hangstart *m* ~ **of a top** Anschnitt *m* (Gewindebohrer)

**start,** ~**-and-stop-control-lever** Ein- und Ausschalthebel *m* ~**-and-stop system** Halteschrittverfahren *n* (Anlauf) ~ **magnet** Einrückmagnet *m* ~**(ing) magneto** Anlaßmagnet *m* ~**-of-scale sensitivity** Anfangsempfindlichkeit *f* ~ **relay** Anlaßrelais *n* ~ **signal** Anfangszeichen *n* ~**-stop apparatus** Gestehapparat *m*, Springschreiber *m* ~**-stop button** Ein-und Ausschalterknopf *m* ~**-stop distortion meter** Bezugsverzerrungsmesser *m* ~**-stop distributor** Gehstehverteiler *m* ~**-stop spindle** Gehstehwelle *f* ~**-stop system** Geh-Steh-Prinzip *n* ~**-stop telegraph** Gehstehtelegraf *m* ~**-stop type of telegraph printer** Gehstehapparat *m* ~**-up procedure** Anfahrmethode *f*

**starter** Andrehvorrichtung *f*, (of a machine) Anlasser *m*, Anlaß-schalter *m*, -vorrichtung *f*, Antriebs-ritzel *n*, -führer *m*, Motoranlasser *m* ~ **with speed control** (for electric motors) Anlasser *m* mit Drehzahlregelung ~ **with switching in jerks** Anlasser *m* mit ruckweiser Schaltung (electr.)

**starter,** ~**battery** Starterbatterie *f* ~ **button** Anlaß-knopf *m*, -kontakt *m* ~ **check** Anlaßrückschlagventil *n* ~ **clutch** Andrehklaue *f* ~ **device** Einrückvorrichtung *f*, Anwert-klaue *f* ~ **dog** Anlasser-, Anwert-klaue *f* ~ **Startflagge *f*** ~ **gap** Zündstrecke *f* **gear** Anlaßsteuerung *f* ~ **generator** Dynastart *m* ~ **housing** (or jacketing) Anlasserverkleidung *f* ~ **jaw** Anlasserklaue *f* ~ **level** Anlasserhebel *m* ~ **motor** Anlaßmotor *m* ~ **pawl** Anlasserklinke *f* ~ **pinion** (jet) Anlaß-ritzel *n* ~ **rheostat** Anlaßwiderstand *m* ~ **ring conduit** Anlaßringleitung *f* ~ **safeguard relay** Anlaßsperrschutz *m* ~ **shaft** Anlaßwelle *f* ~

step Anlaßstufe f ~ test bench Anlasserprüf-stand m
starting Anlaß m, Anlassen n, Anlassung f, An-lauf m, Ausgangs . . ., Auslösung f, Hochlauf m (v. Maschinensatz), Inbetriebsetzung f, Ingang-setzung f ~ of control (or regulation) Regulier-einsatz m ~ of current regulator Stromregler-einsatz m ~ with full load Anlauf m unter Vollast ~ without jolt stoßfreier Anlauf m ~ of a mine Grubenabbau m ~ of oscillations (increment of oscillation) Anstoßen n der Schwingungen, Schwingungs-anfachung f, -ein-satz m ~ to run Laufeinschaltung f ~ up Ein-rücken n ~ without shock stoßfreier Anlauf m, stoßfreies Ingangsetzen n
starting, ~ ability of an engine leichtes An-springen n ~ acceleration (of a locomotive) Anfahrbeschleunigung f ~ air Anfahrluft f ~-air bottle Anlaßluftflasche f ~-air compressor Anlaßluftkompressor m ~-air control Anlaß-steuerung f ~-air tank Anlaßluftbehälter m ~-air valve Anlaßventil n ~-and-stopping-lever Ein- und Ausrückhebel m ~ apparatus Anlaß-vorrichtung f, Inbetriebnahme f ~ arrangement Anlaufvorrichtung f (aviat.) ~ borer Anfangs-bohrer m (Steinbrechen) ~ brush Anlaß--kontaktstück n, -stift m ~ button Anlaß-knopf m, -kontakt m ~ cable Anlaßkabel m ~ cam Anlaßnocken m ~ carriage Anlaufgestell n (aviat.), Startwagen m ~ change-over switch Anlaßumschalter m ~ circuit Anlasserkreis m, Anlaßstromkreis m ~ clutch Anlaufkupplung f ~ cock Anlaßhahn m ~ coil Anlaßspule f, Ausgangswicklung f ~ condition Anlaufver-hältnis n ~ connections Anlaßstufen pl ~ control rods Startregulierungsgestänge n ~ crack on the side under tension Anriß m auf der Zugseite ~ cradle Anlaufgestell n (aviat.) ~ crank Andrehkurbel f, Anwerf-, Anwurf-kur-bel f ~-crank bracket Andrehkurbellager-stütze f ~-crank jaw Andrehkurbelklaue f ~-crank jaw pin Anlaßkurbelbolzen m ~ crank shaft Andrehwelle f ~ crew Startmann-schaft f ~ current Anfahr-, Anfangs-strom m ~ current peak Einschaltspitzenstrom m ~ curve of dynamo Angehkurve f der Lichtma-schine ~ deck (for ship planes) Startdeck n ~ device Anfahrvorrichtung f, Anlaß(hilfs)vor-richtung f, Starteinrichtung f ~ distance An-laufstrecke f ~ dog Anlaßklaue f ~ effect An-laufmoment n ~ electrode Zündelektrode f ~ energy Anfahrkraft f ~ equations Ausgangs-gleichungen pl ~ fan Anheizventilator m ~ force Anfahrkraft f ~ friction Anlaufreibung f (metal) ~ fuel injector Anlaßeinspritzanlage f ~ gear Anfahrvorrichtung f (aviat.), Anlaß--gestänge n, -vorrichtung f, erster Gang m, Ingangsetzungsvorrichtung f ~-gear ring on flywheel Anlaßring m am Schwungrad ~ grip Andrehkurbelgriff m ~ groove Einlaufrille f (phono) ~ handle Andrehkurbelgriff m, An-werfkurbel f, (of an electric controller) Hand-hebel m, Manövrierhebel m, Schalthebel m ~-ignition coil Anlaßzündspule f ~ impulse An-laß-, Auslöse-stromstoß m ~ jaw Andreh-kurbel f ~ key Anruftaste f, Startknopf m ~leg Anfangsschenkel f ~ length Anlauflänge f ~ level Anlasser m ~ lever Anfahr-, Anlaß-,

Anlauf- (metal.), Einrück-, Steuerungs-hebel m ~ line Ablauflinie f ~ load Anlaufbelastung f ~ magnet Abgangsmagnet m ~ magnetic switch Einrückmagnetschalter m ~ magneto Anlasser m ~ material Ausgangsmaterial n ~ mechanism Anlaßvorrichtung f ~ (induction) mixture Anlaßgemisch n ~ moment Anfahr-moment n, (driving) Antriebmoment m, An-zugsmoment n ~ operation Anlaßvorgang m ~ output Anlaufleistung f ~ panel Anlassertafel f, Einschaltbrett n ~ performance Abfluglei-stung f, Startleistung f ~ permit Startgenehmi-gung f ~ pipe Anlaßleitung f ~ plunger Anlaß-stempel m ~ point Ablaufpunkt m, Abmarsch-punkt m, Anfangslage f, Anknüpfungspunkt m, Anstoß m, Aufbruchsort m, Ausgangsbasis f, Ausgangspunkt m, Startpunkt m ~ position Anfahrstellung f, Anlaßstellung f, Ausgangs-lage f, Ausgangsstellung f ~ potential Ein-satzspannung f ~ power Anlaufleistung f ~ process (or procedure) Anfahrvorgang m ~ rail Startschiene f ~ ramp Anlauframpe f ~ ratio Anlaufverhältnis n ~ relay Anlaßrelais n ~ resistance Anfahr-, Anlaß-, Anlaufwider-stand m ~ rheostat Anlaßwiderstand m ~ rotating field Anlaufdrehfeld n ~ run Anlauf-strecke f ~ shaft Einrückwelle f ~ sheet Mutter-blech n ~ signal Startkommando n, Start-zeichen n ~ solution Ausgangslösung f ~ speed Anlaßdrehzahl f ~ spring Abschnellfeder f ~ switch Anlaßschalter m, Anwurfkurbel f ~ temperature Anstelltemperatur f ~ time (of a tone) Anklingzeit f, Schaltzeit f ~ torque An-fahrdrehmoment n, Anlauf(dreh)moment n, Anlaßdrehmoment n, Anzugs-drehmoment n, -kraft f, -moment n, Drehmoment n (aviat.), Drehungsmoment n, Torsionsmoment n, Ver-drehungsmoment n ~ train abfahrender Zug m ~ tub Anstellbottich m ~-up connection An-laßschaltung f ~ valve Anlaßventil n, (slide valve) Anfahrschieber m, Start-ventil n, -ver-schluß m ~-velocity error abweichende An-fangsgeschwindigkeit f ~ vessel Anstellbottich m ~ voltage Anlaufspannung f ~ vortex An-fahrwirbel m ~ winding Anlaß-, Ausgangs--wicklung f ~-winding circuit Hilfsphasenkreis m ~ wire Anlaßkabel n ~ wiring (diagram) Anlaßschaltung f ~ work Arbeitsbeginn m
Stassano (independent-arc) furnace Stassano--Ofen m

state, to ~ angeben, aufführen, aussagen, be-haupten, bemerken, bestimmen, erhärten, feststellen, Fehler pl feststellen, konstatieren
state Befund m, Beschaffenheit f, Erscheinungs-form f, Lage f, Land n, Staat m, Stand m, Verfassung f, Zustand m ~ of affairs Sachlage f, Sachverhalt m ~ of aggregation Aggregatzu-stand m, Stoffzustand m ~ of the air Luftver-hältnisse f eins ~ of alert Alarmstufe f eins ~ of being annuled (or not in force, or being void) Un-gültigkeit f ~ of binding Bindungszustand m ~ of charge Ladungszustand m ~ of emergency Ausnahmezustand m, Notstand m ~ of energy Energiestufe f ~ of equilibrium Beharrungszu-stand m, Gleichgewichts-lage f, -zustand m ~ of flow Strömungszustand m ~ of fusion Schmelzfluß m ~ of humidity Feuchtigkeits-zustand m ~ of inertia Beharrungszustand m

~ of insulation Isolationszustand m, ~ of material as supplied Anlieferungszustand m ~ of overvoltage überspannter Zustand m ~ of polarization Polarisationszustand m ~ of quiescence Ruhezustand m ~ of resistance Beharrungszustand m ~ of rest Beharrungszustand m, Ruhe(zu)stand m ~ of siege Belagerungszustand m ~ stress Spannungszustand m ~ supertension überspannter Zustand m ~ suspension Federungszustand m ~ training Ausbildungsstand m
state, ~ continuity Zustandskontinuität f ~ diagram Zustandsschaubild n ~ forest staatlicher Forst m ~ property Staatseigentum n ~room Kammer f ~ secret Staatsgeheimnis n ~ supervision Staatsaufsicht f
stated genannt ~ another way anders ausgedrückt at ~ intervals fristenweise
statement Abrechnungsnachweis m, Angabe f, Ansatz m, Aufstellung f, Aussage f, Darlegung f, Erklärung f, Feststellung f, Liste f, Nachweisung f ~ of account Kontoauszug m ~ of actual strength Stärkenachweis m ~ of charges Anklageschrift f ~ of claim Klageschrift f ~ of contents Inhaltsangabe f ~ of expense Kostenübersicht f ~ of lines Liniennachweis m ~ of local plant Ortsnetzübersicht f ~ of the problem Stellung f der Aufgabe ~ of time Zeitangabe f
statement showing strength Iststärkennachweisung f
statements Angaben pl
statia line Entfernungsmeßfaden m (Vermessung)
static atmosphärischer Störpegel m, atmosphärische Störung f, Funkentstörer m, Gewitterstörung f, luftelektrische Störung(en) pl, Luftstörungen pl, Rauschen n (rdo), Störbefreiung f, Störspannung f; bodenständig, luftelektrisch, f, Störspannung f, bodenständig, luftelektrisch, ruhend, stationär ~(al) statisch ~ airscrew Bremsluftschraube f ~ balance gewichtlicher Ausgleich m, Gleichgewicht n der Ruhelage, statischer Ausgleich m ~ balancer Mittelpunktstransformator m, Spannungsteiler m ~ ball-indentation test Kugeldruckhärteuntersuchung f, Kugel(druck)probe f ~ bearing surface Paßfläche f einer sich nicht drehenden Lagerung ~ bed statisches Medium n ~ bonding connection Erdungsanschluß m ~ bonding wire Erdungskabel n ~ bottom-hole pressure statischer Bodendruck m ~ breakdown condition statische Durchschlagsbedingung f ~ ceiling statische Gipfelhöhe f ~ characteristic (of tube) Kurzschlußkennlinie f, statische Kennlinie f (rdo) ~ charge ruhende Ladung f ~ chart Staudrucktafel f ~ cone-indentation test Kegeldruckprobe f (electro)~ coupling statische Kopplung f ~ crack strength Trennfestigkeit f ~ current changer ruhender Umformer m ~ detector Knackstörpegel m (electr.) ~ discharge statische Entladung f (Reifen) ~ electrocity statische Elektrizität f, Reibungselektrizität f (shield) ~ eliminator Funkentstörer m ~ field elektrostatisches Feld n, Influenzfeld n, ruhendes Feld n ~-free störungsfrei ~ frequency changer ruhender Frequenzwandler m, Umrichter m ~ fre-

quency converter ruhender Frequenzwandler m ~ friction Haftreibung f, ruhende Reibung f ~ governor statischer Regler m ~ hardness test Kugel(druck)probe f ~ head Saugöffnung f, statisches Druckgefälle f, statische Druckhöhe f ~ induction elektrostatische Beeinflussung f ~ inverter Wechselrichter m (g/m) ~ line (parachute) Aufziehleine f, Reißleine f für Fallschirm (zwischen Fallschirm und Flugzeug) ~ line device Zwangsauslösung f ~ load ruhend wirkende Last f, statische Belastung f, statische Last f ~-load fatigue test Dauerversuch m mit ruhender Beanspruchung ~ load test statische Lastprobe f ~ loading Gleichbeanspruchung f ~ loading spring Mittellastfeder f ~ loading stress ruhende Dauerbeanspruchung f ~ luminous sensitivity statische Lichtempfindlichkeit f ~ mutual conductance statische Steilheit f ~ opening Saugöffnung f ~ phase advancer statischer Phasenschieber m ~ plate Durchflußmengenmeßblende f, Stauscheibe f ~-plate manometer Stauscheibendruckmeßrohr n ~ pressure Ruhedruck m, statischer Druck m, statischer Luftdruck m ~-pressure ratio Ruhedruck-verhältnis n, -ziffer f ~ probe statische Sonde f ~ propeller thrust statischer Schraubenzug m ~ radial engine statischer Sternmotor m ~ rating Standgütegrad m ~ soaring statisches Gleitfliegen n, statischer Segelflug m ~ spark gap feststehende Funkenstrecke f ~ stability Aufrichtungsvermögen n, statische Stabilität f ~ stop statische Sperre f ~ storage statischer Speicher m ~ strength check statische Kontrolle f ~ stress Ruhepunktspannung f, statische Beanspruchung f ~ subroutine statisches Unterprogramm n ~ table Staudrucktafel f ~-tensile-test specimen Dauerversuchszugprobe f ~ tension test Zugversuch m mit Schlagbeanspruchung oder mit statischer Beanspruchung ~ test Belastungs-probe f, -versuch m, Dauerversuch m, Standprüfung f, statische Prüfung f, statischer Versuch m ~ test stand Brennstand m (g/m) ~ thrust Standschub m ~-thrust test Feststellung f des stationären Schubs ~ transformer ruhender Transformator m ~(al) transverse field statisches Querfeld n ~ tube Rohr n zur Messung des statischen Druckes, statische Röhre f, Staurohr n ~ turn indicator statischer Wendezeiger m ~ value Ruhewert m ~-vibration test Schüttelfestigkeitsversuch m
statically, ~ balanced control surface statisch balancierte Ruderfläche f ~ cast ruhend gegossen ~ definable (or determinable) statisch bestimmbar ~ determinate statisch bestimmt ~ indeterminate statisch unbestimmt ~ stable statisch stabil ~ stable airplane statisch stabiles Flugzeug n
staticizer Raumverteilung-Zeitfolge-Umformer m, Serienparallelwandler m und Speicher m
staticless störungslos
statics atmosphärische Störungen pl, Baustatik f, Blitz m (film), Empfangsstörungen pl durch atmosphärische Entladung, Filmblitz m, Gleichgewichtslehre f, Statik f
station Amt n, Anhaltspunkt m, Anstalt f, Bahnhof m (r.r.), Beobachtungsort m (photo-

grammetry, trig.), Bezugspunkt *m*, Feuer *n*, Ladeplatz *m*, Posten *m*, Richtpunkt *m*, Sprechstelle *f* (telephone), Station *f*, Stellung *f* ~ **for changing cars** Umsteigbahnhof *m* ~ **of destination** Bestimmungsfunkstelle *f* ~ **of origin** Absendestelle *f*, Ursprungsfunkstelle *f* ~ **in the shaft** Anschlagsohle *f*
**station,** ~ **amplifier** Endverstärker *m* ~ **block** Befehlsstellwerk *n* (r.r.) ~ **capacity** Bahnhofsleistung *f* ~ **control** Kommandogerät *n* ~ **dial** Skala *f* der Sendestationen ~ **indicator** Stationsmelder *m* ~ **line** Teilnehmerleitung *f* ~-**line jack** Anschlußklinke *f* ~ **logbook (or log register)** Betriebsbuch *n* ~ **master** Bahnhofsvorstand *m* ~ **point** Augenpunkt *m* ~ **pointer** Doppeltransporteur *m* ~ **ringer** Apparatwecker *m* ~ **selector** Stationswähler *m* ~ **service (system)** Eigenbedarfsversorgung *f* (Kraftwerk), Eigenversorgung *f* ~ **service unit** Eigenbedarfssatz *m* ~ **signal** Einfahrtsignal *n* ~ **switch** Stationsschutzschalter *m* ~ **tender** Aviso *m* ~ **time** Taktzeit *f* ~-**to-**~ **call** Gespräch *n* zwischen zwei Sprechstellen ~ **voltage** Zentralenspannung *f* ~ **wagon** Kombiwagen *m*
**stationary** bodenständig, fest, feststehend, ortsfest, seßhaft, ständig, stationär, stehend, stillstehend, unbeweglich **to be** ~ stillstehen ~ **and office supplies** Büromaterial *n*
**stationary,** ~-**anode tube** Festanodenröhre *f* ~ **armature with winding** feststehender Anker *m* mit Wicklung ~ **axle** feststehende Achse *f* ~ **base** Sockel *m* ~ **battery** ortsfeste Zelle *f* und Batterie *f* ~ **belt conveyer** ortsfester Bandförderer *m* ~ **block** Festblock *m* ~ **boundary** feste Randfläche *f* ~ **cell** ortsfeste Zelle *f* und Batterie *f* ~ **converter** feststehender Konverter *m* ~ **coupling** starre Kupplung *f* ~ **distortion** stationäre Verzerrung *f* ~ **dive** Senkrechttauchen *n* ~ **drier** Trockner *m* mit ruhendem Gut ~ **engine** feststehende Maschine *f*, Standmotor *m* ~ **ensemble** stationäre Gesamtheit *f* ~ **field cone** Stehfeldtubus *m* ~ **field radiation** Stehfeldbestrahlung *f* ~ **figure** stehende Figur *f* ~ **flow** stationäre Strömung *f* ~ **folder** festformatiger Falzapparat *m* ~ **furnace bottom** Festherd *m* ~ **gantry crane** feststehender Bockkran *m* ~ **grid** stationäre Blende *f* ~ **image** stehendes Impulsbild *n* ~ **installation** feste Verlegung *f* ~ **magneto** Standmagnetzünder *m* ~ **mounting ring** Grundring *m* ~ **nucleus** ruhender Kern *m* ~ **pattern** stehende Figur *f* ~ **plant** feste Anlagen *pl* ~ **point** Umkehrpunkt *m* ~ **point of a curve** Rückkehrpunkt *m* einer Kurve ~ **radiation monitor** Strahlungsüberwachungsanlage *f* ~ **setting up** ortfeste Aufstellung *f* ~ **spark gap** feststehende Funkenstrecke *f* ~ **state** stationärer Zustand *m* ~ **stop** Ruheanschlag *m* ~ **target** feststehendes Ziel *n* ~ **test** Standversuch *m* ~ **thread** Stehfaden *m* ~ **thread guide** feststehender Fadenführer *m* ~ **transfer gantry** feststehender Bockkran *m* ~ **transformer** ruhender Transformator *m* ~ **value** Ruhewert *m* ~ **wave** stehende Welle *f*
**stationer's shop** Papier(waren)handlung *f*
**stationery** Briefpapier *n*, Papierwaren *pl*
**statistic distribution** statische Verteilung *f*
**statistical** statistisch ~ **counter time lag** Zählerverzögerung *f* ~ **data** statistische Angaben *pl*

~ **noise** statistisches Rauschen *n* ~ **weight** Quantumgewicht *n*, thermodynamische Wahrscheinlichkeit *f*
**statistician** Statistiker *m*
**statistics** Statistik *f*
**statometer** Statometer *n*
**stator** Rahmen *m*, (coil) feste Spule *f*, Ständer *m* (electr.), Stator *m* ~ **arrangement** Turbinengitteranordnung *f* ~ **blade** (Verdichter)leitschaufel *f* (des Triebwerks) ~ **blading** Leitradbeschaufelung *f* ~ **bore** Gehäusebohrung *f* ~ **coil** Statorspule *f* ~ **disc** Gitterscheibe *f* ~ **field** Ständerfeld *n* ~ **laminations** Statorbleche *pl* ~ **plates** Statorpaket *n*, Statorplatten *pl* eines Drehkondensators ~ **ring** Leitkranz *m* ~ **stack** Ständerpaket *n* ~ **stage** Leitstufe *f* (Triebwerk)
**statoscope** Aneroidbarometer *n* besonderer Art, Feinhöhenmesser *m*, Höhenmesser *m*, Statoskop *n*, Variometer *m*
**statu, in** ~ **nascendi** nascendie
**statue** Bildsäule *f*, Gesetz *n*
**stature** Gestalt *f*
**status** Stand *m*, Zustand *m*
**statue** Satzung *f* ~ **of limitation** Verjährungsfrist *f* ~ **book** Gesetzbuch *n* ~ **mile** englische Meile *f*, Landmeile *f*
**statutory** gesetzlich ~ **bar** gesetzliches Hindernis *n* ~ **declaration** eidesstattliche Versicherung *f* ~ **meeting** ordentliche Versammlung *f*
**stave, to** ~ **in** einschlagen
**stave** Daube *f*, (barrel) Faßdaube *f*, Korsettenstab *m* (Hochofen), Schove *f*, Sprosse *f* ~-**and headjointing machine** Dauben-und-Bodenfügemaschine *f* ~-**backing and -hollowing machine** Hobel- und Aussparrmaschine *f* für Dauben ~ **crosscut saw** Abkürzsäge *f* für Dauben, Daubenabkürzsäge *f* ~-**dressing machine** Daubenabrichtmaschine *f* ~ **grooving and tonguing machine** Daubenfüge-, Nut- und Federmaschine *f* ~ **holder** Schraubenwinde *f* ~-**jointing machine** Faßdaubenfügemaschine *f* ~ **lining** Daubenauskleidung *f* ~ **pipe** hölzernes Daubenrohr *n* ~-**shortening saw** Daubenabkürzsäge *f* ~ **wheel** Sprossenrad *n*
**staved end** angestauchter Rand *m*
**staving machine** (for casks and barrels) Daubenzurichtmaschine *f*
**stay, to** ~ abspannen, abspreizen, absteifen, anhalten, sich aufhalten, bleiben, stationär bleiben, stützen, verankern, verspannen, verstreben, verweilen **to** ~ **away** ausbleiben
**stay** Anker *m*, Anker-bolzen *m*, -draht *m*, -seil *n*, Aufenthalt *m*, Aufschub *m*, (of proceedings) Aussetzung *f*, Bolzen *m*, Drahtanker *m*, Gestängebelastung *f*, Knagge *f*, Leitstrick *m*, Rast *f*, Spreize *f*, Stag *m*, Steife *f*, Stempel *m*, Strebe *f*, Stütze *f*, Träger *m*, Verbindungsstück *n*, Verspannung *f*, Winkelstutzen *m*, Zuganker *m* ~(**ing**) Verankerung *f* ~ **of proceedings** Inhibierung *f* ~ **for timbering of the roof** Firstenstempel *m*
**stay,** ~ **bar** Bügel *m* ~ **block** Anker-klotz *m*, -pfahl *m* ~ **bolt** Spange *f*, Stand-, Steh-bolzen *m* ~-**bolt drill** Stehbolzenbohrer *m* ~-**bolt driver** Stehbolzenschrauber *m* ~-**bolt tap** Stützbolzenbohrer *m* ~ **clamp** Tragseilklemme *f* ~ **crutch** Druckstab *m* ~ **fixed to pole at**

**ground level** Fußanker *m* ~ **guard** Ankerschutzpfahl *m*, Scheuerpfahl *m* ~ **head** Spreizenkopf *m*, Zugstangenkopf *m* ~ **hook** Ankerhaken *m* ~ **pin** Kettensteg *m* ~ **pole** Abspannstange *f* ~**-poles terminal** in telegraph poles Abspanngestänge *n* ~**-put switch** Apparat *m* ohne Ruhestellung ~ **rod** Ankerpfahl *m* ~ **rods** Abstützgestänge *n* ~ **rope** Ankerdraht *m* ~ **tackle** Fußtau *n* ~ **thimble** Ankerkausche *f* ~ **tightener** Ankerspannschraube *f* ~ **wire** Abspanndraht *m*, Schwungleine *f*, Seilanker *m*
**stayed,** ~ **mast** abgespannter Mast *m* ~ **pole** Abspannmast *m*, verankerte Stange *f* ~ **terminal pole** Endgestänge *n*
**staying** Abspannen *n*, Abspannung *f*, Verspannung *f*, Verstrebung *f* ~ **of a beam** Absteifung *f* eines Balkens ~ **the compressional members** Abstützung *f* der Druckstäbe ~ **members** abstützende Teile *pl*
**stays and guys** Abspannmaterial *n*
**steadfast** beständig
**steadies** Setzstöcke *pl*
**steadiness** Ausdauer *f*, Beständigkeit *f*, Gleichmäßigkeit *f*, Stabilität *f*, Standfestigkeit *f*, Standhaftigkeit *f*, (to shock and vibration) Standsicherheit *f*, Stehen *n*, Stete *f* ~ **of image** Stehen *n* des Bildes ~ **of the wave** Wellenlängenkonstanz *f* ~ **of the wind** Windkonstanz *f*
**steady, to** ~ beruhigen, gleichmachen, gleichmäßig erhalten, gleichmäßig machen, stabilisieren **to** ~ **the plummet** das Lot feststellen
**steady** beständig, drehungsfrei, fest, gleich-bleibend, -förmig, -mäßig, konstant, laufend, ruhend, stabil, standfest, standhaft, ständig, stationär, stet, stetig, stetiger Bildwechsel *m*, stoßfrei ~ **background** ständiger Untergrund *m* ~ **brace** Bogenabzug *m* ~ **conditions** Beharrungszustand *m* ~ **current** gleichförmiger Strom *m*, Gleichstrom *m*, Ruhestrom *m* ~ **current resistance** Ohmscher Widerstand *m* ~ **deflection** gleichmäßiger Ausschlag *m* ~ **discharge** Dauerentladung *f* ~ **feed of film** stetiger Bildwechsel *m* ~ **field** Gleichfeld *n* ~ **flow** plastisches Fließen *n*, ständiger Fluß *m*, stationäre Strömung *f* ~ **holder** Stütze *f* für Werkzeugstangen ~ **illumination** Ruhe-belichtung *f*, -licht *n* ~ **motion of film** stetiger Bildwechsel *m* ~ **pace** Arbeitstempo *m* ~ **parallel flow** gleichförmige Schichtenströmung *f* ~ **pin holder** Quadraturstift *m* ~ **position** Ruhelage *f* ~ **potential** Ruhespannung *f* ~ **(current) resistance** Gleichstromwiderstand *m* ~ **rest** feststehende Brille *f*, Lünette *f*, Setzstock *m* ~ **rests** Setzstöcke *pl* ~ **span** Spanndraht *m* für seitliche Festlegung ~ **state** Dauerzustand *m*, (of current, oscillation, etc.) eingeschwungener Zustand *m*, stationärer Zustand *m* ~**-state behavior** Beharrungsverhalten *n* (aut. contr.) ~**-state charcteristics** Beharrungskennwerte *pl* ~**-state condition** Beharrungszustand *m* ~ **state cross talk** stationäres Übersprechen *n* ~**-state current** eingeschwungener oder stationärer Strom *m* ~**-state deviation** Endabweichung *f* ~**-state gain** Ausgleichsert *m* (aut. contr.), (servo) Übertragungsfaktor *m* ~**-state oscillation** eingeschwungener Zustand *m* ~**-state stability** statische Stabilität *f* ~ **stress component** Mittelspannung *f* ~ **value** Dauer-,

Ruhe-wert *m* ~ **voltage** gleichmäßige Spannung *f* ~ **vorticity motion** stationäre Wirbelbewegung *f* ~ **wind** gleichmäßiger Wind *m* ~ **working of the blast furnace** Gargang *m* des Hochofens
**steadying** Stabilisierung *f* ~ **circuit** Beruhigungs-glied *n*, -kreis *m* ~ **device** Beruhigungsvorrichtung *f* ~ **resistance** Beruhigungswiderstand *m*, Berührungswiderstand *m* ~ **strut** Stützstrebe *f* ~ **tube** Stützrohr *n*
**steal, to** ~ entwenden, stehlen
**steam, to** ~ abdämpfen, ausdämpfen, auslaugen, dämpfen, dekatieren, dunsten, eindampfen **to** ~ **in moist** feucht dämpfen **to** ~ **out** ausdämpfen
**steam** Brodel *m*, Brodem *m*, Dampf *m*, Dämpfe *pl*, Dunst *m*, Wasserdampf *m* ~ **for heating** Heizdampf *m* **to let the** ~ **off** (or out) Dampf *m* ablassen ~ **and water drum** Obertrommel *f* ~ **steam,** ~ **accumulator** Dampfspeicher *m* ~ **admission** (or inlet) Dampfeintritt *m* ~**-admission valve** Dampfeinlaßventil *n* ~**-admitting pipe** Dampf-eingangskanal *m*, -einlaßrohr *n*, Eingangskanal *m* für Dampf ~ **air ejector** Dampfstrahlluftpumpe *f* ~ **apparatus** Dampftopf *m* ~ **atomization** Dampfzerstäuber *m* ~**-atomizer burner** Dampfzerstäubungsbrenner *m* ~ **atomizing** Dampfzerstäubung *f* ~**-atomizing oil burner** Dampfexpansionszerstäuber *m*, Ölbrenner *m* für Dampfzerstäubung ~ **balance hole** Dampfausgleich *m* ~ **barrier** Sperrdampf *m* ~ **black** Dampfschwarz *n* ~ **blast** Dampfgebläse *n* ~ **blower** (or blowing) **engine** Dampfgebläse *n* ~**boat** Dampf-boot *n*, -schiff *n* ~ **boiler** Dampfmaschinenkessel *m*, Kesseldampfmaschine *f* ~ (or generator) **boiler** Dampfkessel *m* ~**-boiler fittings** Dampfkesselarmaturen *pl* ~ **box** Ventilkasten *m* ~**-brake valve** Dampfbremsventil *n* ~ **brewing plant** Dampfsudwerk *n* ~ (or vapor) **bubble** Dampfblase *f* ~ **calorimeter** Kondensationskalorimeter *n* ~ **case** Dampfhemd *n* ~ **chamber** (or accumulator) Dampfsammler *m* ~ **check valve** Dampfrückschlagventil *n* ~ **chemicking** Dampfchloren *n* ~ **chest** Dampfbüchse *f*, Schieberkasten *m*, Ventilkasten *m* ~ **clarifier** Dampfreiniger *m* ~ **clarifier with fixed heating coil** Dampfklärpfanne *f* mit fester Heizspirale ~ **coal** Bunker-, Dampfkessel-, Flamm-, Kessel-kohle *f* ~ **coal slurry** Eßkohlenschlamm *m* ~ **cock** Dampfhahn *m* ~ **coil** Dampf-schlange *f*, -spirale *f*, -spule *f*, Rohrschlange *f* für Dampf ~ **collector** Dampfsammler *m* ~ **conductor** (or lead-to) **pipe** Dampfzuführung *f* ~ **conduit** Dampfleitung *f* ~ **cone of the injector** Ansaugraum *m* der Dampfstrahlpumpe ~ **connection** Dampfanschluß *m* ~ **consumption** Dampfverbrauch *m* ~ **converting valve** Dampfumformventil *n* ~**-cooking installation** Dampfkochanlage *f* ~**-cooled** dampfgekühlt ~ **cooling system** Dampfkühlung *f* ~ **crane** Dampfkran *m* ~ **crane navvy** Dampfkranlöffelbagger *m* ~ **cure** Dampfvulkanisation *f* ~ **cylinder** Dampfzylinder *m* ~**-cylinder oil** Dampfzylinderöl *n* ~ **damping box** Dampffeuchtkasten *m* ~ **dessiccator** (or drier) Dampfentwässerer *m* ~ **diagram** Dampfdruckbild *n*, Druckbild *n* des Dampfes ~ **digester** Dampkochtopf *m* ~ **distillation** Destillation *f* mit Wasserdampf, Wasser-

dampfdestillation *f* ~ **distributor** Dampfver-
teiler *m* ~ **dome** Dampf-dom *m*, -stutzen *m*,
Dom *m* ~ **douche** Dampfsprudler *m* ~ **draff**
**drier** Treberdampftrockenapparat *m* ~ **draff-**
**-drying apparatus** Dampftrebertrockenapparat
*m* ~**-dried** dampfgetrocknet ~ **drive** Dampfan-
trieb *m* ~**-driven aircraft** Dampfflugzeug *n*
~**-driven beet lifter** Dampfrübenheber *m* ~
**driven hoist** dampfgetriebenes Hebewerk *n*,
Hebewerk *n* mit Dampfmaschinenantrieb ~
**drum** Oberkessel *m* ~ **drying apparatus** Dampf-
trockner *m* ~ **education pipe** Dampfableitungsrohr *n* ~ **ejector** Dampfstrahlpumpe *f* ~**-**
**-emulsion number** Dampfemulgierungszahl *f*
~ **engine** Dampf-maschine *f*, -motor *m* ~**-engine**
**drive** Dampfmaschinenantrieb *m* ~**-engine oil**
Dampfmaschinenöl *n* ~ **engine working expansi-**
**vely** Expansionsdampfmaschine *f* ~**-escape pipe**
Dampfabblaserohr *n* ~ **exhaust pipe** Dampf-
austrittsstutzen *m* ~**-exhaust port** Dampfaus-
strömungskanal *m* ~ **extraction** Zwischendampfentnahme *f* ~ **fire engine** Dampf(feuer)-
spritze *f* ~ **fitting** Dampfarmatur *f* ~ **floor clear-**
**er** Dampfesel *m* ~ **friction** Dampfreibung *f* ~
**funnel** Dampftrichter *m* ~ **gate valve** Dampfabsperrschieber *m* ~ **(or pressure) gauge**
Dampfdruckmesser *m*, Dampfdruckzeiger *m*,
Dampfmesser *m* ~**-generating plant** Dampfanlage *f* ~ **generation** Dampf-bildung *f*, -ent-
wicklung *f*, -erzeugung *f* ~ **generator** Dampf-
erzeuger *m* ~ **governor** Dampfregulator *m* ~
**hammer** Dampfhammer *m*, Rammbär *m* ~**-**
**-heated still** Dampfblase *f* ~ **heating** Dampf-
heizung *f* ~**-heating line** Heizdampfleitung *f* ~
**hood** Dampfglocke *f* ~ **injector** Dampf-ein-
blasedüse *f*, -injector *m*, -speisepumpe *f*,
-strahlapparat *m* ~**-inlet port** Dampfeinströ-
mungskanal *m* ~ **installation** Dampfanlage *f*
~ **intensifier** Dampftreibapparat *m* ~ **jacket**
Dampf-hülle *f*, -mantel *m*, Heizmantel *m* ~
**jacketed** mit Dampfmantel *m* versehen ~**jacke-**
**ted kettle** Gefäß *n* mit Dampfmantel ~**-jacketed**
**pan** doppelwandiger Kessel *m* ~ **jet** Dampfstrahl *m* ~**-jet air pump** Dampf-luftejektor *m*,
-strahlluftpumpe *f* ~**-jet blower** Dampfstrahlgebläse *n*, ~**-jet flue-cleaning apparatus** Dampf-
strahlrauchrohrreiniger *m* ~**-jet sandblast**
Dampfsandstrahlgebläse *n* ~ **jet test** Dampf-
strahlprobe *f* ~**-jet valve** Dampfstrahlventil
*n* ~ **laundry** Dampfwäscherei *f* ~ **line** Dampfleitung *f* ~ **lock** Dampfsperre *f* ~ **lubrication**
**apparatus** Dampfschmierapparat *m* ~ **meter**
**(or counter)** Dampfzähler *m* ~ **metering**
Dampfmengenmessung *f* ~**-mixing jet** Dampf-
strahlrührgebläse *n* ~**-motor model** Dampfmo-
tormodell *n* ~ **nozzle** Dampfdüse *f* ~ **operated**
**air ejector** Dampfstrahlluftsauger *m* ~**-outlet**
**pipe** Dampfabzugsrohr *n* ~ **packing** Dampf-
dichtung *f* ~ **pile driver** Dampframme *f* ~
**pile-driving hoist** Dampfraumwinde *f* ~ **pipe**
Dampfrohr *n* ~ **piping** Dampfleitung *f* ~
**piston** Dampfkolben *m* ~**-piston packing**
Dampfkolbenliderung *f* ~ **pitching device**
Dampfentpichmaschine *f* ~ **plant** T-Stoff-
-Anlage *f* (g/m) ~**-plant plug** T-Stoff-Anlagen-
stecker *m* ~ **point** Verdampfungspunkt *m* ~
**port** Dampfkanal *m* ~ **port (or inlet)** Dampfeinlaß *m* ~ **power** Dampf-betrieb *m*, -kraft *m*

~ **power plant** Dampf-kraftanlage *f*, -trieb-
werk *n* ~ **power station** Dampfkraftwerk *n*
~**-powered sawmill** Dampfsäge *f* ~ **pressure**
Dampfspannung *f* ~ **pressure (or compression)**
Dampfdruck *m* ~ **pressure above atmospheric**
Dampfüberdruck *m* ~ **pressure governor**
Dampfdruckregler *m* ~**-pressure intensifier**
Dampfdruckübersetzer *m* ~**-pressure reducing**
**valve** Dampfdruck-minderventil *n*, -regelventil
*n* ~ **processing** Dampfung *f* ~ **pump** Dampf-
pumpe *f* ~ **purifier** Dampfreiniger *m* ~ **raising**
Dampferzeugung *f* ~**-raising coefficient** Verdampfungsziffer *f* ~ **ram** Dampframme *f* ~
**reducer** Dampfdruckverminderer *m* ~ **regulator**
Dampfzulaßventil *n* ~ **retting** Dampfröste *f*
(Flachs) ~ **reversing gear** Dampfumsteuerung
*f* ~ **rig** Dampfkabel *n* ~ **roller** Dampfstraßen-
walze *f*, Straßenwalze *f* ~ **separator** Dampfab-
schneider *m*, Kondens-topf *m*, -wasserableiter
*m* ~ **ship** Dampfer *m*, Dampfschiff *n* ~**ship line**
Schiffahrtslinie *f* ~ **shovel** Bagger *m*, Dampf-
-kranlöffelbagger *m*, -schaufel *f*, Löffelbagger
*m*, Trockenbagger *m* ~**-slide stop valve** Dampfabsperrschieber *m* ~ **sluice valve** Dampfschieber *m* ~ **space** Dampfraum *m* ~ **spraying**
**station** Dampfaussprühstation *f* ~ **stamp**
Pochwerk *n* ~ **(or thermal) station** Dampfzen-
trale *f* ~ **sterilizer** Dampfsterilisator *m* ~ **stop**
**valve** Dampfabsperrventil *n* ~ **strainer** Dampf-
reinigungssieb *m* ~ **superheater** Dampfüber-
hitzer *m* ~ **supply** Dampfzuführung *f* ~**-supply**
**line** Dampfzuleitungsrohr *n* ~ **sweating** Dampf-
schwitze *f* ~ **table** Wärmeschrank *m* (Großküche) ~ **taste** Pasteurisiergeschmack *m* ~
**(or vapor) tension** Dampfspannung *f* ~ **throttle**
Dampfrohr *n*, Dampfzulaßventil *n* ~**-tight**
**(or -proof)** dampfdicht ~ **trace** Begleithei-
zung *f* ~ **tractor** Dampf-traktor *m*, -trecker *m*,
-zugwagen *m* ~ **tramway** Dampfstraßenbahn
*f* ~ **trap** Dampf-klappe *f*, -wassertopf *m*,
Kondens-topf *m*, -wasserableiter *m* ~**-treated**
**asphalt** mit Dampf behandeltes Pech *n* ~
**trough drier** Dampfmuldentrockner *m* ~ **truck**
Dampfwagen *m* ~ **tube** Dampfrohr *n*, Siede-
rohr *n* ~**-tube brush** Siederohrbürste *f* ~ **tube**
**drier** Dampfröhrentrockner *m* ~ **turbine** (exhaust type) Auspuffdampfturbine *f*, (impulse
type) Dampfdruckturbine *f*, Dampfturbine *f*,
(backpressure type) Gegendruckturbine *f* ~
**turbine drive** Dampfturbinenantrieb *m* ~**-tur-**
**bine oil** Dampfturbinenöl *n* ~**-turbine rotor**
Dampfturbinenläufer *m* ~**-turbine test bed**
Dampfturbinenprüfstand *m* ~ **turboblower**
Dampfturbinengebläse *n* ~ **valve** Dampfventil
*n* ~ **veil sealing** Dampfabschluß *m* ~ **velocity**
Dampfgeschwindigkeit *f* ~ **vent** Dampfaustritt
*m* ~ **void** Dampfblase *f* ~ **washing** Dampfdecke
*f* ~**-whistle valve** Dampfpfeifenventil *n* ~
**winch** Dampfwinde *f* ~ **winder** Dampfgöpel *m*
**steamer** Dampfer *m*, (steaming) Dämpfer *m*,
Dampfschiff *n*
**steaming** Eindämpfung *f*, (in water process)
Gasen *n*, Gasperiode *f*, Kaltblasen *n* ~ **appara-**
**tus** Dekatierapparat *m* ~ **calender** Dekatier-
kalender *m* ~ **capacity** Dampfleistung *f*, Ver-
dampfungsleistung *f* ~ **chamber** (steam ager)
Dampfkasten *m* ~ **cone (or dome)** Dampfhaube
~ **economiser** Verdampfungsvorwärmer *m* ~

**machine** Dampfmaschine *f* ~**-out liquor** Ausdampfwasser *n* ~ **period** Dämpfzeit *f*, Gasen *n*, Gasperiode *f*, Kaltblasperiode *f*, Wassergasperiode *f* ~ **plant** Dämpfanlage *f* ~ **radius** Fahrstrecke *f* ~**-vapor method** Dampfstrommethode *f*

**stearate** Stearat *n*

**stearic** stearinsauer ~ **acid** Cetylessigsäure *f*, Stearinsäure *f*, Talgsäure *f* ~ **acid soap** Stearinseife *f*

**stearine** Stearin *n* ~ **cake** Stearinkuchen *m* ~ **pitch** Stearinpech *n* ~ **spew** Stearinausschlag *m*

**stearite** Speckstein *m*, Steadit *m*, Steatit *m* ~ **block** Steatitstein *m* ~ **powder** Specksteinpulver *n*

**steatitic** specksteinartig

**steckling elite seed** Stecklingselitesamen *m*

**steel, to** ~ stählen, verstählen **to**~**-tip** mit einer Stahlspitze *f* versehen

**steel** Eisen *n*, Stahl *m*, Stahlknüppel *m* (made of) ~ stählern ~ **with good high-temperature characteristics** (or for high-temperature service) warmfester Stahl ~ **for making safes or vaults** Tresorstahl *m*

**steel,** ~ **alloy** Stahllegierung *f* ~ **arches** hufeisenförmiger Eisenausbau *m* ~ **area** (in re-inforced concrete section) Eisengehalt *m* ~ **armor-piercing shell** Panzerstahlgranate *f* ~**-armored conduit** Stahlpanzerrohr *n* ~**-backed bearing** Lager *n* mit Stahlstützschale ~**-backed lining** (for bearing) Lagermetallausguß *m* mit Stahlstützschale ~ **ball** Stahl-kugel *f*, -luppe *f* ~**-ball penetrator** Stahlkugelspitze *f* ~ **band** Stahlband *n* ~ **bar** (iron) Stabeisen *n*, Stahl-stab *m*, -stange *f* ~ **barrel** Eisenfaß *n* ~ **bead** Stahlperle *f* ~ **belt-apron conveyer** Blechgurtförderer *m* ~**-belt conveyer** Stahlbandförderer *m* ~ **bit** Stahlkrone *f* ~ **blade** Stahlklinge *f* ~ **blank** Stahlronde *f* ~ **blasting shop** Stahlstrahlerei *f* ~**-block coupling** Klotzkupplung *f* ~**-blue** stahlblau ~ **bolt** Stahlbolzen *m* ~**-bound tote box** Ladekasten *m* mit geschlossenem Eisenrahmen ~ **bristle** Stahlborste *f* ~ **bronze** Stahlbronze *f* ~ **bullet** Stahl-geschoß *n*, -kugel *f* ~**-bushed chain** Stahlbüchsenkette *f* ~**-bushed roller chain** Büchsenkette *f*, Rollenlagerkette *f* ~ **cable** Stahl-draht *m*, -kabel *n*, -seil *n* ~ **case** Stahlhülse *f* ~ **casing** (furnace) Blechpanzer *m* ~ **casing shoe** Stahlschneideschuh *m* ~ **casting** Stahlformguß *m*, Stahlformgußstück *n* ~**-casting foundry** Stahl(form)gießerei *f* ~**-casting production** Stahlformgußerzeugung *f* ~ **castings** Stahlguß *m* ~**-castings plant** Stahlformgießerei *f* ~ **chain** Stahlkette *f* ~ **cheek** Stahlbacke *f* ~ **chimney** Blechschornstein *m* ~ **coil** Stahlwendel *f* ~ **collar** Spannband *n*, Stahlband *n* ~**-colored** stahlfarbig ~ **concrete** Stahlbeton *m* ~ **conduit** Stahlpanzerrohr *n* ~ **construction** Eisenbau *m*, Stahlbau *m* ~**-conversion process** Zementierverfahren *n* ~**-converting furnace** Zementierofen *m* ~ **core** Stahl-kern *m*, -seele *f* ~**-core bullet** Stahlkerngeschoß *n* ~ **coverplate** Stahlblechdeckel *m* ~ **cruciform pile** kreuzförmiger (Profil-) Stahlpfahl *m* (oder Profileisenpfahl) ~ **cutting** spanabhebende Bearbeitung *f* ~ **cutting edge** Stahlschneide *f* ~ **cuttings** Stahlspäne *pl* ~

**cylinder** Stahl-flasche *f*, -zylinder *m* ~ **disc** Stahlscheibe *f* ~ **disc wheel** Stahlscheibenrad *n* ~ **doctor** Stahlrakel *n* ~ **dog** Stahlgußknagge *f* ~ **double ground cone** Doppelschliff *m* aus Stahl ~ **dowel** Stahldübel *m* ~**-drawing works** Zieherei *f* ~ **drill for rock** (or stone) Meißel *m* für Felsbearbeitung ~ **engraving** Stahlstechen *n* ~ **fabric** (im voraus hergestelltes) Bewehrungsnetz *n* ~**-faced wrought-iron armor plate** Kompoundpanzerplatte *f* ~ **flat** (strip of small rectangular section) Bandeisen *n* ~ **forge** Stahlhammer *m* ~ **founder** Stahl-gießer *m*, -werker *m* ~ **foundry** Stahlgießerei *f* ~ **frame** Stahlrahmen *m* ~ **frame construction** Stahlblechrahmenkonstruktion *f* ~ **frame structures** Stahlbauten *pl* ~**-frame super-structure** Eisenhochbau *m* ~ **furnace** Stahlofen *m* ~ **furniture** Stahlmöbel *pl* ~ **gear** Stahlzahnrad *n* ~ **girder** Stahlträger *m* ~ **grade** Stahl-gattung *f*, -güte *f*, -sorte *f* ~**-gray** stahlgrau ~ **grillage** Verankerung *f* (Stahlrost) ~ **grinding balls** Stahlmahlkugeln *pl* ~ **grinding bodies** Stahlmahlkörper *pl* ~ **grit** Stahlkorn *n*, Stahlsand *m* ~ **hammer** Stahlhammer *m* ~**-hardening oil** Härteöl *n* ~ **hawser** Stahltrosse *f* ~ **helmet** Stahlhelm *m* ~ **hoop** Stahlband *n* ~ **hull** (of U-boat) Eisenkörper *m* ~ **impeller** Stahlkolben *m* ~ **industry** Schwerindustrie *f* ~ **ingot** Stahlblock *m* ~ **inlay** Stahleinlage *f* ~ **insert** Eiseneinlage *f* ~ **jacket** Blechmantel *m*, Panzer *m*, Stahlmantel *m* ~ **keys** Stahlkeile *pl* ~ **ladle** Stahl(gieß)pfanne *f* ~**-lattice mast** Eisengittermast *m* ~**-like** stahlähnlich ~ **line** Stahlseil *n* ~ **lining** Stahlfutter *n* ~ **list** Eisenliste *f* ~ **loophole cover plate** Stahlblende *f* ~ **made directly from the ore** Rennstahl *m* ~ **maker** Stahl-erzeuger *m*, -werker *m* ~**making** Stahlbereitung *f*, Stahlherstellung *f*, Stahlschmelzen *n* ~**making furnace** Stahlschmelzofen *m* ~**making plant** Stahlwerks-anlage *f*, -betrieb *m* ~**making process** Stahlgewinnungsverfahren *n* ~**making shop** Stahlwerkshalle *f* ~ **manufacture** Stahl-erzeugung *f*, -fabrikation *f* ~ **measuring tape** Stahlmeßband *n* ~ **melting** Stahlschmelzen *n* ~**melting furnace** Stahlschmelzofen *m* ~**-melting plant** Stahlgießerei *f* ~**-melting process** Stahlschmelzprozeß *m* ~ **member used for erection purposes** Richteeisen *n* ~ **mesh** Stahldrahtgeflecht *n* ~ **metal straightening and rolling machine** Blechwalz- und Richtmaschine *f* ~ **mill** Hütten-, Stahl-werk *n* ~ **mining supports** Stahlgrubenstempel *m* ~ **mold** Stahlgußform *f* ~ **ore** Stahlerz *n* ~ **parts** Stahlbauteile *pl* ~ **pen** Stahlfeder *f* ~ **picket** Haftpflock *m* ~ **pig** Rohstahleisen *n*, Stahleisen *n* ~ **piling** Stahlpfahl *m* ~ **pin** Stahl-bolzen *m*, -stift *m* ~ **pinion** Stahlritzel *n* ~ **pipe** Stahlrohr *n* ~ **plant** Stahlwerk *n* ~ **plate** Flußeisenblech *n*, Stahl-blech *n*, -platte *f*, -teller *m* ~**-plate apron conveyor** Stahlgliederband *m* ~**-plate-engraving relief printing** Stahlstichprägedruck *m* ~ **plate liner** Stahlplattenpanzer *m* ~**-plate lining** Panzer *m*, Panzerung *f*, Stahlblechauskleidung *f*, Stahlplattenpanzerung ~**-plated** verstählt ~ **plating** Verstählung *f* ~ **plow** Stahlpflug *m* ~**-pointed contact tip** Meßhütchen *n* mit Stahlfläche ~ **poured** abgestochene Rohstahlmenge *f* per Schmelze ~ **printing plate** Stahldruckplatte f ~ production Stahlerzeugung f, Stahlfabrikation f ~ projectile Stahlgeschoß n ~ prop Stahlstempel m ~ protecting disc Stahlschutzscheibe f ~ puddling Stahlpuddeln n ~ punch Stahlstempel m ~ quality Stahlgüte f ~ quenched and drawn Stahl m abgeschreckt und angelassen ~ quieted in the last moment in the ingot mold im letzten Augenblick in der Blockform beruhigter Stahl m ~ quill Stahlhohlwelle f ~ rail Stahlschiene f ~ ready for casting unmittelbar vergießbarer Stahl m ~ reciprocating rider Wechselreiter m ~ reed Stahlzunge f ~ refinery Stahlfrischfeuer n ~-reinforced aluminum cable Stahlaluminium-freileitungsseil n ~ reinforcement Armatur f, Eiseneinlage f ~ retainer Bohrerhalter m ~ rider Stahlreiber m ~ rigging stählerne Takelage f ~ rimmer Glättstahl m ~ ring Stahlring m ~ rivet Stahlniet n ~ rod Stahlarm m ~ roller Stahlrolle f ~ roofing tile Stahldachpfanne f ~ rope with moderate twist drallarmes Stahlseil n ~ rubber-lined gummierter Stahl m ~ rule die Stanzform f ~ sample Stahlprobe f ~ scrap Stahl-abfall m, -schrott m ~ scratch brush Stahlkratzbürste f ~ section Stahlprofil n ~ sheet Stahlblechbahn f ~ sheet pile Stahlspundbohle f ~-sheet piling Eisenspundwände pl, Spundstahlwand f, Spundwandprofile pl ~shield Stahlabschirmung f ~ short Stahlkorn m ~ shot Stahlschrot m ~-shot blasting Stahl-kiesstrahl m, -kugelblasen n ~ shot cleaning plant Kugelregenreinigungsanlage f ~ shuttering material Stahlschalungen pl ~ side-bar chain Laschenkette f ~ skeleton for multi-storied buildings Stahlhochbau m ~ skeleton structures Stahlskelettbauten pl ~ sleeve Stahlhülse f ~ socket pipes Stahlmuffenrohre pl ~ spar Stahlholm m ~ spike Stahlnadel f ~ spiral Stahlwendel f ~ spotweld Stahlpunktschweißverbindung f ~ spring Stahlfeder f ~-spring indicator Stahlfederindikator m ~-spring landing gear Stahlfederfahrgestell n ~ square Stahlwinkel m ~ stack Eisenkamin m ~ stake Haftpflock m ~ stamping letters and figures Stahleinschlagstempel m ~ stranded conductor for maximum tension Höchstspannungsstahlleitung f ~ strap (or strip) Stahlband n ~ strapping stretcher with sealer Bandeisenspannapparat m ~ straps Verpackungsbandeisen n ~-studded tire Gleitschutzreifen m ~ supporting structure Stahltragwerk n ~-tank rectifier Ventil n mit Eisengefäß ~-tape armored stahlband-armiert, -bewehrt ~ test Stahlprobe f ~ tie Stahlschwelle f ~ tie rod Stahlanker m ~ tip Stahlspitze f ~-tipped mit einer Stahlspitze f versehen ~ tool Stahlwerkzeug n ~ treadway Stahlspurbahn f ~ tub Innenbehälter m ~ tube Stahlrohr n ~ tube air preheater Stahlrohrlufterhitzer m ~ tube bracket Stahlrohrbock m ~-tube engine mounting Stahlröhrtriebwerksgerüst n ~-tube framework Stahlrohrfachwerk n ~-tube fuselage Stahl-röhrenrumpf m, -rohrgerüst n ~ tubes used for uprights of roof standard Rohrständer m ~ tubing Stahlrohr n ~ tubular pole Stahlrohrmast m ~ turnings Stahlspäne pl ~ vessel Stahlgefäß n ~ wheel Stahlrad n ~ winch Stahlwinde f ~ wire Eisendraht m, Stahl-

-draht m, -seil n ~ wire armor Stahldrahtbeklöppelung f ~-wire block Drahtseilkloben m ~-wire clip Stahlklammer f ~-wire conveyor belts Stahldrahttransportbänder pl ~-wire cup brush Stahldrahttopfbürste f ~ wire insert (or insertion) Stahldrahteinlage f ~-wire rope Stahldrahtseil n ~ wool Stahl -späne pl, -wolle f ~-worker Stahlwerker m ~works Stahl-gießerei f, -hütte f, -werk n ~works practice Stahlwerksbetrieb m ~ worm Stahlschnecke f ~yard Balkenware f, Laufgewichtsbalken m
steeliness Glasigkeit f
steeling Stählung f, Verstählung f
steely stahlartig ~ iron Stahleisen n ~ malt Glasmalz n
steep, to ~ abfischen, anweichen, einquellen, eintauchen, einweichen, netzen, quellen, tauchen, tränken, weichen to ~ in lye laugen
steep abschüssig, jäh, schroff, steil ~-angle bearing Schrägrollenlager n ~-angle taper Steilkegel m ~-angle valve Schrägsitzventil n ~ bank Steilufer n ~ cam Steilnocken m ~ climbing turn steiler gezogener Kurvenflug m ~ coast Steilküste f ~ coast covered with dunes Dünensteilküste f ~ dipping seam steiles Flöz n ~ dune Steildüne f ~ fall steiler Abfall m ~ gliding turn steiler Gleitwendeflug m ~ gradient steiler Abhang m, Steilrampe f ~ measures steile Lagerung f ~ rise (helicopter) steiler Abflug m ~ rocky shore steiles Felsenufer n ~ shore Steilufer n ~ sighting Steilsicht f ~ sighting prism Steilsichtprisma n ~ slope steiler Abhang m ~-spiral starkspiraliger Hochleistungsfräser m ~ spiral drill steilgängiger Spiralbohrer m ~ taper steiler Kegel m ~ thread pitch große Gewindesteigung f ~ turn steiler Wendeflug m, Steilkurve f ~ wave front steile Wellenstirn f oder Wellenfront f
steeped geweicht
steepest descent method Sattelpunktmethode f
steeping Röste f ~ in dye Einfärbung f
steeping, ~ cistern Quellbottich m, Weichkufe f ~ machine (textiles) Entfernungsmaschine f ~ period Weichdauer f ~ trough Quell-, Vormaisch-bottich m, Weichkufe f ~ tub Frischbalge f, Weichfaß n ~ vat Netzkessel m
steeple Kirchturm m
steepness Steilheit f ~ of gradation Emulsionssteilheit f ~ of sides (or slopes) of a curve Flankensteilheit f
steer, to ~ fahren, führen, leiten, lenken, steuern to ~ into (harbor, etc.) einsteuern to ~ toward ansteuern
steerable lenkbar, steuerbar ~ tail skid steuerbarer Sporn m ~ tail wheel lenkbares Spornrad n
steerage Zwischendeck n
steered gesteuert ~ course Kompaß-, Schiffs-, Steuer-kurs m ~ trailer chassis Lenkgestell n
steering (automatic) automatische Steuerung f, Führung f, Lenkung f ~ arm Lenk-gabel f, -stock m, Steuerarm m ~-arm stop Lenkhebelanschlag m ~ axle Leit-, Lenk-achse f ~ box Lenkungsgehäuse n ~ brake Lenkbremse f ~-brake drum Lenkbremstrommel f ~ cam Lenksäulenführung f ~ channel Steuerungskanal m ~ clutch Steuerungskupplung f ~ column Lenksäule f, Steuersäule f (electron.)

~-column bearing Lenkrohrlager *n*, Lenksäulenhalter *m* ~ column seal Lenksäulenabdichtung *f* ~ compass Steuerkompaß *m* ~ connecting rod Lenkschubstange *f* ~ connection Steuerradverbindung *f* ~ control Lenkvorrichtung *f*, Steuergewalt *f* ~-control by homing device Zielsuchsteuerung *f* ~-control current Steuerstrom *m* ~-control equation Steuergleichung *f* ~-control member Lenkgestell *n* ~-control theorem Steuergesetz *n* ~ coupling Kupplungslenkgetriebe *n* ~ device Lenk-, Ruder-einrichtung *f* ~ engine Rudermaschine *f* ~ fin Steuerschwanz *m* (naut., ball.) ~ force Steuerungsantrieb *m* ~ fork Lenkgabel *f* ~ frame Steuerkasten *m* ~ gear Leitwerk *n*, Lenk--einrichtung *f*, -getriebe *n*, -stange *f*, -vorrichtung *f*, Lenkung *f*, Steuer-apparat *m*, -gerät *n* ~-gear arm Lenkstockhebel *m* ~ gear attachment Lenkstockhalterung *f* ~-gear box Lenkgehäuse *n* ~ gear bracket Lenkstockhalterung, *f* ~-gear case Steuergehäuse *n* ~ gear clamp Lenkstockhalter *m* ~-gear housing Lenkgehäuse *n* ~-gear (female) nut Lenkmutter *f* ~ head Achsschenkelträger *m* ~ journal Lenkzapfen *m* ~ knuckle Achsschenkel *m*, Steuer -gelenk *n*, -schenkel *m* ~ knuckle arm Lenk(er)hebel *m*, Lenkschenkel *m*, Spurstangenhebel *m* ~ knuckle bearing Achsschenkellager *n* ~ knuckle carrier Achsschenkelträger *m* ~ knuckle pin Achsschenkelbolzen *m* ~-knuckle pivot Vorderachszapfen *m* ~-knuckle shaft Fingerhebelwelle *f*, Vorderachswelle *f* ~-knuckle type of steering Achsschenkellenkung *f* ~ lever Lenk-, Steuer-hebel *m* ~ lever-shaft Hebelwelle *f* ~ lock Ausschlag *m* der Räder, Lenkungsausschlag *m*, Steuerungsausschlag *m* ~ machine with vane velocity control Laufgeschwindigkeitssteuermaschine *f* ~ mechanism Lenkgetriebe *n*, Steueranlage *f*, Steuerung *f* ~ oar Steuerruder *n* ~ panel Steuerungspult *n* ~ part Steuerteil *m* ~ pillar Lenksäule *f* ~ pivot Lenkzapfen *m* ~ play toter Gang *m* ~ post Steuerleitungsrohr *n* ~ rod Steuerstange *f* ~ (guide) rod Führungsstange *f* ~ screw Steuerschraube *f* ~ sets Steuersätze *pl* ~ shaft Lenksäule *f* ~-shock suspension Lenkstoßfang *m* ~ stop Lenkungsanschlag *m* ~ stop limit Lenkanschlagbegrenzung *f* ~ suspension Lenkgestänge *n* ~ swivel Steuerschenkel *m* ~ system Lenkungsschema *n* ~ table Steuertafel *f* ~ tie rod with jaw ends Steuertraverse *f* mit Gabeln ~ tube Steuerrohr *n* ~ wheel Kraftwagensteuer *n*, Lenk(hand)rad *n*, Lenkungsrad *n*, Steuer *n*, Steuerrad *n* ~ wheel nut Lenkradmutter *f* ~ wheel pullers Lenkradabzieher *pl* ~ wheel shaft (or spindle) Lenkspindel *f* ~ wheel spider Lenkradspeichenkreuz *n* ~ worm Lenk-schenkel *m*, -schnecke *f* ~-worm cam Lenkschraube *f* ~--worm sector Lenksegment *n*

**steersman** Rudergänger *m*

**Steinmetz coefficient** Hysteresiskonstante *f*

**stellar,** ~ **atmosphere** Sternatmosphäre *f* ~ distillation Wasserdampfdestillation *f* ~ dynamics Sterndynamik *f* ~ opacity stellare Opazität *f* ~ radiation Starrolle *f*

**stellate** sternförmig

**stellite** Druckplättchen *n*, Stellit *m*

**stellited** stelliert

**stem, to** ~ aufhalten **to** ~ **the tide** tot segeln **to** ~ **the water** das Wasser anstauen

**stem** Bug *m*, (of a lamp, tube or valve) Fuß *m*, Glashalterung *f*, Querglied *n*, Querwiderstand *m*, Quetschfuß *m*, (of a thermometer) Röhre *f*, Schaft *m*, Seele *f*, Stamm *m*, Stange *f*, Steg *m*, Steven *m*, Stiel *m*, (of valve) Stößel *m*, Vor(der)-steven *m*, (of fuse) Zünderzapfen *m* ~ for casing dogs Backenbremse *f*, Fangkeil *m* ~ of girder Trägersteg *m*

**stem,** ~ **bit** Kolbenbohrer *m* ~ caster Lenkrolle *f* mit Bolzenschaft ~ correction Fadenkorrektion *f*, Thermometerkorrektur *f* ~ equation Stammgleichung *f* ~ guide Spindel-, Stift-führung *f* ~ hook Stengelhaken *m* (zum Aufhängen des Lotes) ~ ossicle Stielglied *n* ~ point Staupunkt *m* ~ press (incandescent lamps) Glasputzen *n* ~ press of the tube Röhrenpreßteller *m* ~ pressure device Staudruckeinrichtung *f* ~ radiation Stielstrahlung *f* ~-winding watch Remontoiruhr *f* ~ wood Stammholz *n*

**stemmer** Bohrstampfer *m*

**stemming** Besatz *m*, Stemmung *f*

**stemple** Stempel *m*

**stemson** Unterlauf *m*

**stench** Gestank *m* ~ trap Geruchverschluß *m*

**stencil, to** ~ schablonieren **to** ~ **on** aufschablonieren

**stencil** Anreißschablone *f*, Malerschablone *f*, Matrize *f*, Patrone *f*, Schablone *f* ~ paper Schablonenpapier *n* ~ varnish Korrekturlack ~ wheel Nummernrad *n*

**stencilling,** ~ **copper** Schablonenkupfer *m* ~ device Schabloniervorrichtung *f* ~ paper Malblatt *n*

**stenopeic spectacles** Schlitz-, Sichel-brille *f*

**stenotype machine** Schnellschreiber *m*

**stenotypist** Stenotypistin *f*

**stenter finish** Heißluftapparatur *f*

**step, to** ~ treten **to** ~ **down** abwärts formieren, abwärts transformieren, herab-, nieder-transformieren (electr.), reduzieren **to** ~ **forward** vorrücken **to** ~ **a mast** einen Mast *m* einspuren **to** ~ **off** abstufen **to** ~ **on** betreten, fortschalten, vorwärtsschalten, weiterschalten **to** ~ **out** austreten **to** ~ **out of** heraustreten **to** ~ **over** übertreten **to** ~ **round the wipers** die Schaltarme *pl* weiterdrehen **to** ~ **in tickets** Gesprächszettel *pl* einordnen **to** ~ **up** aufdrehen, aufwärtstransformieren, aufwärts transformieren, erhöhen, fortschalten, herauf-, hinauf-transformieren (electr.), hochspannen, vorwärtsschalten, weiterschalten **to** ~ **down the voltage** die Spannung herunterdrücken

**step** Aufstieg *m*, Auftritt *m*, Fußsteg *m* (aluminium mfg.), Galerie *f*, Maßnahme *f*, Maßregel *f*, Schritt *m*, Sprosse *f*, (of a mast or beacon) Spurloch *n*, Spurplatte *f*, Strosse *f*, Stufe *f*, Stützlager *n*, Tritt *m*, Wagentritt *m* in ~ in Phase mit ~ of the arc Lichtbogenschritt *m*

**step,** ~ **action** Schrittwirkung *f* ~-and-leaf sight Rahmentreppenvisier *n* ~-back welding Pilgerschrittschweißung *f* ~ bearing Lager *n* einer stehenden Welle, Lagerpfanne *f*, Stützlager *n* (foot) ~ bearing Fußlager *n* ~ block abgestufter Block *m* ~ bracket Aufsteigetritt *m* (am Tankwagen), Auftrittstütze *f*, Tritt-

-brettstütze f, -halter m ~ **brass** Lager n einer stehenden Welle

**step-by-step, ~ action** Stufenwirkung f ~ **automatic telephone system** Schrittschaltselbstanschlußsystem n ~ **calculation** Differenzenrechnung f ~ **call indicator operation** Handbetrieb m mit unmittelbar gesteuertem Nummernanzeiger ~ **excitation** stufenweise Anregung f ~ **magnet** Drehmagnet m ~ **method** Stufenmethode f ~ **resistance** Stufenwiderstand m ~ **seam welding** Rollenschrittverfahren n ~ **selection** Schrittwahl f ~ **selector** Schrittschaltwähler m ~ **selector switch** Schritt-halter m, -wähler m (electr.) ~ **selector system** Schrittwählersystem n ~ **starting** Langsameinschaltung f ~ **switch** Schritt-schalter m, -schaltwerk n ~ **synchronizing** Fernsynchronisierung f, übertragene Synchronisierung f ~ **system** Schrittschaltersystem n ~ **telegraph** Schrittschalttelegraf m

**step, ~ chuck** Stufenfutter n ~ **cone** Stufenscheibe f ~ **control** Schrittregelung f ~ **divider circuit** Frequenzteilerschaltung f ~-**down gear** Reduktionsgetriebe n ~-**down gearing** Untersetzungsgetriebe n, Untersetzung f eines Getriebes ins Kleine ~-**down transformation** Abtransformation f ~-**down transformer** Abwärts-transformator m, -wandler m, Reduktionstransformator m, Reduziertransformator m, spannungserniedrigender Transformator m ~ **fairing** Stufenverkleidung f ~ **faults** Staffelbruch m ~ **formation** Stufenformation f ~ **freezing** absatzweises Frieren n ~ **function** Treppenfunktion f ~ **grate** Etagen-, Stufen-, Treppen-rost m ~-**grate producer** Treppenrostgenerator m ~ **groove** Stufennut f ~ **hanger** Trittbretthalter m ~ **height** Stufenhöhe f ~ **jaw** Stufenspannbacke f ~ **ladder** Stufen-, Tritt-leiter f ~-**landing indicator** Stufenhöhenmesser m ~ **length** Schrittlänge f ~ **light** (on running board) Trittbrettlampe f ~ **milling** Stufenfräsen n ~ **notch** Stufennut f ~ **photometer** Stufenfotometer n ~ **planking** Treppenbelag m ~ **printer** Fensterkopiermaschine f ~ **pulley** Stufenscheibe f ~-**pulley V-belt drive** Stufenscheibenkeilriementrieb m ~ **rail** Trittbrettschiene f ~ **regulator for bulb voltage** Lampenstufenregler m ~ **response** Sprungwiedergabe f ~ **rocket** mehrstufige Rakete f ~ **roll** Stufenwalze f ~ **size** Schrittgröße f ~ **slit** Stufenspalt m ~ **spring** Stufenfeder f ~ **switch** Rasten-, Stufen-schalter m ~ **tablet** Stufenkeil m ~-**time** Schrittzeit f ~-**type grates** Treppenroste pl

**step-up** Aufwärtstransformierung f ~ **in performance** Leistungszunahme f

**step-up, ~ cure** Stufenheizung f ~ **distance** Stufung f ~ **gear** Eilganggetriebe n ~ **gearing** Übersetzungsgetriebe n, Übersetzung f eines Getriebes ins Große ~ **ratio of the signal** Aufschaukelung f des Signals ~ **transformation** Auftransformation f ~ **transformer** Aufspanner m, Aufwärtstransformator m, Aufwärtswandler m (rdo), Spannungsverdoppler m ~

**step, ~ valve** Stufenventil n ~ **voltage regulator** Stufenspannungsregler m ~-**wedge** Stufenkeil m ~ **wheel** Schrittschaltrad n, Stufenrad n ~ **width** Flankenhöhe f, Schrittweite f

**stephanite** Melanglanz m, Rösch-erz n, -gewächs n, Schwarzsilber-erz n, -glanz m, Sprödglaserz n, Stephanit m

**stepless variability** stufenlose Regelung f

**steplike** stufenartig

**stepped** abgestuft, stufenartig, treppenförmig ~ **arrangement** Stufenanordnung f ~ **curve** Treppenkurve f ~ **down** (of a pipe) abgestuft ~-**down formation** nach unten abgestufte Formation f ~ **edges** Aussprung m ~ **gear** Stirnrad n mit Schrägverzahnung, Stufenzahnrad n ~ **hole** abgesetzte Bohrung f ~ **photometric absorption wedge** Stufenkeil m ~ **piston** Stufenkolben m ~ **plunger diameter** Kolbendurchmesserstufung f (Pumpenelement) ~ **protection** Staffelschutz m ~ **pulley drive** Stufenscheibenantrieb m ~ **resistance** abgestufter Widerstand m (electr.) ~ **roller** Stufenrolle f ~ **shaft** abgesetzte Welle f ~ **tool** Stufenstahl m ~-**up combustion** Zusatzverbrennung f ~-**up formation** Höhenstaffelung f, nach oben abgestufte Formation f ~-**up process** Intensivverfahren n ~ **wheel** Stufenrad n ~ **winding** Treppenwicklung f

**stepping** Fortschreitungsbewegung f, Schalten n ~ **circuit** Impulsleitung f ~ **control** Schritt m für Schrittsteuerung f ~ **device** Fortschalt-vorrichtung f, -werk n ~ **down** (of potential, gearing, etc.) Herabsetzung f ~ **electromagnet** Schrittschaltelektromagnet m ~ **line** Impulsleitung f ~ **magnet** Fortschaltmagnet m, Schaltmagnet m ~ **mechanism** Schaltwerk n, Schrittschaltwerk n ~ **motor** Schrittmotor m ~ **pulses** Fortschaltimpulse pl ~ **relay** Fortschaltrelais n, Stromstoßschalter m ~ **switch** Fortschaltrelais n ~ **wheel** Schaltrad n

**steps** Treppe f **by ~** absatzweise, sprungweise, stufenförmig **having ~** stufig

**stepwise** stufenweise ~ **approximation** schrittweise Näherung f ~ **loading** stufenweise Belastung f ~ **reaction** Stufenreaktion f

**steradian** Einheitsflächenwinkel m, räumlicher Winkel m

**stereo** Gußabdruck m, Stereo . . . ~ **acoustics** Raumakustik f ~ **autograph** Stereo-Autograf m ~**camera** Standlinienkammer f ~**chemical** raum-, stereo-chemisch ~**chemistry** Raum-, Stereo-chemie f ~ **comparator** Raumbildmesser m ~**comparator with flicker microscope** Stereokomparator m mit Blinkmikroskop ~ **dressing process** Reliefzurichteverfahren n ~**exposure** Stereoaufnahme f ~**fluoroscopy** Stereodurchleuchtung f ~**gram** Stereogramm n ~**graphic** stereografisch ~ **holder** Stereovorhalter m ~**isomer** Stereoisomer n ~ **isomeric** raumisomer ~ **metal** Stereotypiemetall n ~-**meter** Raumklanbox f (tape rec.) ~**metry** Körperlehre f, Raummessung f ~**microscopy** Mikrostereoskopie f ~-**pair** Zweierbildskizze f ~ **paths** Stereozüge pl

**stereophonic, ~(al)** stereofonisch ~ **effect** Raumwirkung f ~ **hearing** zweiohriges Hören n ~ **image forming** räumliche Schallabbildung f

**stereo, ~ photogrammetric survey** stereofotogrammetrische Aufnahme f ~**photogrammetry** Raumbildmessung f, Stereo(foto)grammetrie f ~**photograph** Stereoaufnahme f ~**photographic plotting machine** Stereoplanigraf m ~**photo-**

**graphy** fotografische Raumdarstellung *f* ~**pho-tomicrography** Mikrostereofotografie *f* ~ **power** Stereovermögen *n* ~**radioscopy** Stereo-durchleuchtung *f* ~**-reflex camera** Spiegelre-flexstereokamera *f*

**stereoscope** Raumglas *n*, Stereoskop *n*

**stereoscopic** plastisch (film), räumlich, stereo-skopisch ~ **complement** Meßbildpaar *n* ~ **contact** (in range finder) Deckung *f* ~ **effect** Bildplastik *f*, (pictures) Plastik *f*, plastische Bildwirkung *f*, Raumeffekt *m* (TV), Raum-wirkung *f*, (of picture) Tiefenwirkung *f* ~ **film** plastischer Film *m* ~ **fusion** Verschmelzung *f* der Halbbilder bei stereoskopischer Betrach-tung ~ **guide** Raumlenker *m* ~ **mark** Meßmarke *f*, Raummarke *f* ~ **measurement** stereoskopi-sches Messen *n* ~ **pair** Zweierbildskizze *f* ~ **photography** Stereofotografie *f* ~ **picture** Raum-bild *n* ~ **picture projection** Raumbildvorfüh-rung *f* ~ **pictures** Raummeßbilder *pl* ~ **plotting machine** Autokartograf *m* ~ **radiograph for locating defects** Tiefenbestimmung *f* ~ **radio-graphy for locating defects** Fehlerscheibchen-tiefenbestimmung *f* ~ **range finder** Raum-bildentfernungsmesser *m*, stereoskopischer Ent-fernungsmesser *m*, Stereotelemeter *n* ~ **sense** Fähigkeit *f* stereoskopisch zu sehen ~ **televi-sion** plastisches Fernsehen *n* ~ **video picture** plastisches Fernsehbild *n* ~ **view** Raumbild *n* ~ **view of range graduation marks** Meßmarken-raumbild *n* ~ **vision** räumliches oder zweiäugi-ges Sehen *n* ~ **vision and measurement** stereo-skopisches Sehen *n* und Messen *n*

**stereoscopically covered surface** stereoskopisch gedeckte Fläche *f*

**stereostop** Stereoblende *f*

**stereotype, to** ~ abklatschen, stereotypieren

**stereotype** Bildstock *m*, Druckstock *m*, Platten-schrift *f*, Stereotyp *m* ~ **metal** Stereotypiemetall *n* ~ **plate** Halbtonklischee *n* ~ **printing** Plat-tendruck *m* ~ **printing machine** Plattendruck-maschine *f* ~ **tempering plant** Stereotypiever-härtungsanlage *f*

**stereotypes** Stereos *pl*

**stereotyping,** ~ **tissues** Stereotypieseidenpapier *n* ~ **workshop** Stereotypieanstalt *f*

**stereotypography** Plattendruck *m*

**steric,** ~ **factor** sterischer Faktor *m* ~ **hindrance** räumliche Behinderung *f*

**sterile** taub, unfruchtbar ~ **mass** taubes Mittel *n* ~ **milk bottle** Sterilmilchflasche *f*

**sterilization** Entkeimung *f*, Entseuchen *n*, Ent-seuchung *f*

**sterilize, to** ~ entkeimen (dauerhaft machen), entseuchen, entwesen, haltbar machen, sterili-sieren

**sterilized** keimfrei

**sterilizer** Sterilisator *m*, Sterilisier-apparat *m*, -maschine *f*

**sterilizing,** ~ **filter** Entkeimungsfilter *n* ~ **sheets** Entkeimungsschichten *pl*

**stern** Achterschiff *n*, (navy) Heck *n* **by the** ~ hinterlastig **to go out** ~ **first** über den Achter-steven *m* auslaufen **to have** ~ **way** Fahrt *f* über den Achtersteven oder achteraus haben

**stern,** ~ **anchor** Heckanker *m* ~ **anchoring point** Heckankerwerk *n* ~ **cap** Heckkappe *f* ~ **com-partment** Hinterkaffe *f* ~ **droop** Knick *m* am

Heck ~ **fast** hintere Landfeste *f* ~ **frame** Heck--ring *m*, -spant *n* ~**-heavy** heck-lastig, -schwer ~ **light** Heck-feuer *n*, -laterne *f*, (tail) Heck-licht *n* ~ **line** Achterleine *f* ~ **mooring point** Heckankerwerk *n*, Heckspitze *f* ~ **post** Ach-tersteven *m*, Anschlagleiste *f* ~ **radiator** Heck-kühler *m* (aviat.) ~ **sheet of boat** hintere Sitz-bank *f*, Hinterteil *m* ~ **stay** Hintersteven *m* (aviat.) ~ **stiffening** Heckversteifung *f* ~ **tip** Heckspitze *f* ~ **tube** (for driving rod) Führungs-rohr *n*, Wellenaustrittsrohr *n* ~ **turret** Heckge-fechtsstand *m* ~**ward** nach achtern ~**ward(s)** nach dem Heck *n* zu ~ **wave** Heckwelle *f* (avi-at.) ~**-weighing device** Apparat *m* zum Messen der Hubkraft am Heck ~**-wheel steamer** Heckraddampfer *m*

**sternutator** Nasen- und Rachenreizstoff *m*

**sterol** Sterin *n*

**stethoscope** Abhörapparat *m*, Hörrohr *n*, Stethoskop *n*

**stevedore** Auslader *m*, Lader *m*, Stauer *m*

**stew, to** ~ abdämpfen, dämpfen, schmoren

**stew pan** Tiegel *m*

**stibium oxide bath** Grauglanzoxydbad *n*

**stibnite** Antomonglanz *m*, Grauspießglanzerz *n*, Schwefelspießglanz *m*, Schwefelspießglanzerz *n*, Spießglanzerz *n*, Stibnit *m*

**stick, to** ~ Anker kleben, ankleben, bleiben, ein-halten, hängen, kleben **to** ~ **fast** festkleben **to** ~ **on bestecken to** ~ **to** anhaften, festhaften, haften

**stick** Ankleben *n*, Büttkrück *m*, Knüppel *m* (aviat.), Latte *f*, Pfahl *m*, Prügel *m*, Spriegel *m*, Stab *m*, Stange *f*, Steuerhebel *m*, Stock *m* ~**-and--latch relay** Stützrelais *n* ~**-and-wire-construc-tion** Gemischtbau *m* (aviat.) ~**-and-wire-fuse-lage** Holzdrahtrumpf *m* ~ **bending machine** Stockbiegemaschine *f* ~ **bombing** Bomben-reihen(ab)wurf *m*, Reihenwurf *m* ~**-candy puller** Zuckerstangenmacher *m* ~**-candy spinner** Zuckerspinner *m* ~ **control** Knüppelsteuerung *f* (aviat.) ~ **coordination** Knüppelverfahren *n* ~ **flute** Stockflöte *f* ~ **force** Knüppel-, Steuer--kraft *f* ~ **gauge** Fühlstab *m* ~ **grip** Knüppel-griff *m* (aviat.) ~ **hand grenade** Stielhandgra-nate *f* ~ **lac** Stangenlack *m* ~ **length** (bomb ballistics) Reihenwurflänge *f* ~ **polish** Stan-genwachs *m* ~ **potash** Stangenkali *n* ~ **right back** Knüppel *m* an den Bauch ~ **time** Aus-rollzeit *f*

**sticker** Festläufer *m* (beim Gießen), (rolling material) Kleber *m*, Klebezettel *m*

**stickiness** Klebrigkeit *f*

**sticking** Festwerden *n*, (of a keeper) Kleben *n* ~ **of the armature** Kleben *n* des Ankers ~ **of contacts** Kontaktkleben *n* ~ **of electrons** Kleben *n* der Elektronen ~ **of needle** Hängen *n* der Nadel ~ **of the safety valve to its seat** Festsitzen *n* des Sicherheitsventils auf dem Sitz ~ **of a valve** Hängenbleiben *n* des Ventils

**sticking,** ~ **temperature** Erweichungspunkt *m* ~ **voltage** Sperrspannung *f* ~ **wax** Wachskitt *m*

**stickler** (colloquial) Paragrafenreiter *m*

**sticks** Prügel-, Rund-holz *n*

**sticky** klebrig, pappig ~ **formation** drückendes Gebirge *n* ~ **limit** (Atterberg) Klebegrenze *f* ~ **shale** (schmieriger) haftender Schieferton *m* ~ **slag** zähflüssige Schlacke *f* ~ **sugar** klumpi-ger Zucker *m* ~ **tape** Klebband *n*

**Stiefel disc-piercing process** Stiefel-Scheiben-walzprozeß *m*

**stiff** steif, starr, stramm ~ **against torsion** drehsteif ~ **colored paper** Natur(kunstdruck)papier *n* ~ **drag brace** Aussteifungsstrebe *f* zur Aufnahme der Kräfte der Blattebene ~**-fissured clay** steifer, geklüfteter Ton *m* ~ **paper binding** Steifbroschur *f* ~ **plastic process** Halbplastischverfahren *n* ~ **pupil** Lichtstarre *f* ~ **upright leather case** Köcher *m*

**stiffen, to** ~ absteifen, aussteifen, verfestigen, versteifen

**stiffened against bending** biegefest

**stiffener** Aussteifung *f*, Imprägnierungsmittel *n*, Verstärkungsstreifen *m*, versteifender Teil *m* ~ **molding machine** Kappenformmaschine *f*

**stiffening** Aussteifung *f*, (shoemaking) Hinterkappe *f*, Sicke *f*, Steifwerden *n*, Verfestigung *f*, Verstärkung *f*, Versteifung *f* ~ **of the web** Stegversteifung *f*

**stiffening,** ~ **angle iron** Versteifungswinkeleisen *n* ~ **beam** Versteifungsbalken *m* ~ **brace** Versteifungsstrebe *f* ~ **calender** Stärkekalander *m* ~ **dope** Spannlack *m* ~ **frame** Auflegerahmen *m* ~ **girder** Versteifungskiel *m* ~ **piece** Gegenlage *f* (teleph.) ~ **plate** Stehblech *n*, Verstärkungsblech *n* ~ **rib** falsche Rippe *f*, Verstärkungsbalken *m*, Versteifungsrippe *f*, Zicke *f* ~ **ring** (Kupplung) Versteifungsring *m* ~ **sheet** Verstärkungsblech *n* ~ **strap** Versteifungsbügel *m* ~ **varnish** Spannlack *m*

**stiffness** Biegefestigkeit *f*, Festigkeit *f*, Starrheit *f*, Steife *f*, Steifheit *f* ~ **of cable** Seilsteifigkeit *f*

**stiffness,** ~ **coefficient** Steifigkeitsziffer *f* ~ **depending on amplitude** amplitudenabhängige Steife *f* ~ **motor** Stutzmotor *m* (gyro) ~ **tester** Steifigkeitsprüfer *m*

**stifle, to** ~ erdrücken, ersticken, unterdrücken

**stifle** Knie *n*

**stigma** Mal *n*

**stigmatic** punktzentrisch

**stigmator** Stigmator *m*

**stilb** Stilb *n*

**stilbene crystal scintillator** Stilben-Kristallszintillator *m*

**still** Abtriebeapparat, Abtreiber *m*, Blase *f*, Destillier-apparat *m*, -blase *f*, -kessel *m*, Erhitzer *m*, ruhendes Bild *n*, Säule *f*; noch, noch immer, regungslos, ruhig, still ~ **air space** Luftinsel *f* ~ **bottom heel** Rückstand *m* ~ **condenser passage** Kühlrohrleitung *f* ~**-film picture** Standbild *n* ~ **picture** ruhendes Bild *n*, Stehbild *n*, unbewegtes Bild *n* ~**-view projection attachment** Dia-Einrichtung *f* ~ **water** Totwasser *n* ~ **wax** Retortenparaffin *n*

**stillage** Gestell *n*

**stilling,** ~ **basin** Beruhigungs-, Brems-kammer *f*, Tos-becken *n*, -kammer *f*, Wasserbecken *n* ~ **sill** (or weir) Gegenschwelle *f*

**stilt** Gerüststange *f* ~ **bit** Stelzenbohrer *m*

**stilts** Stelzen *pl*

**stimulant** Anregungsmittel *n*, Anreiz *m*, Reizmittel *n*

**stimulate, to** ~ anregen, anspornen, antreiben, erregen, stimulieren

**stimulated** angeregt

**stimulating potential** Anregungsspannung *f*

**stimulation** Anregung *f*, Erregung *f*, Reiz *m* ~ **energy** Anregungsenergie *f* ~ **value** Schwellenreiz *m*

**stimulus** Anregung *f*, Betonung *f*, Reiz *m*, Reizmittel *n* ~ **of light** Lichtreiz *m* ~ **error** Reizirrtum *m*

**sting, to** ~ stechen

**sting** Flügelstachel *m* (aviat.), Stachel *m*, Stich *m*

**stink,** ~ **bomb** Stinkbombe *f* ~**stone** Stinkstein *m*

**stipple, to** ~ betupfen, tüpfeln

**stipple graver** Punktierstichel *m*

**stipulate, to** ~ ausbedingen, bedingen, bestimmen, fest -legen, -setzen, vereinbaren **to** ~ **in writing** schriftlich vereinbaren

**stipulated,** ~ **load** vorgeschriebene Belastung *f* ~ **period of wear** (of clothing) Tragezeit *f*

**stipulation** Abmachung *f*, Bedingung *f*, Bestimmung *f*, Festsetzung *f*, Klausel *f*, Übereinkunft *f*, Voraussetzung *f*

**stipulations for acceptance of a manuscript** (or merchandise) Aufnahmebedingungen *pl*

**stir, to** ~ anrühren, bewegen, rühren, stochern, umrühren, verrühren **to** ~ **the fire** das Feuer anschüren oder schüren **to** ~ **in** (or up) aufrühren, einrühren, schüren **to** ~ **to a paste** anteigen **to** ~ **with a spatula** umschaufeln **to** ~ **thoroughly** durchrühren **to** ~ **vigorously** turbinieren **to** ~ **with a whirling motion** quirlen

**stir** Bewegung *f* ~ **test** Rührversuch *m*

**stirrable** aufrührbar

**stirrer** Aufrührer *m*, Maischgitter *n*, Rühr-apparat *m*, -arm *m*, Rührer *m*, Rühr-scheit *n*, -werk *n* ~ **bar** Sticheisen *n* ~ **vessel** Rührgefäß *n*

**stirring** Rührung *f*, Umrühren *n* ~ **apparatus** Rührapparat *m*, Rührwerk *n* ~ **arm** Rührarm *m* ~ **arrangement** Rühreinrichtung *f* ~ **blades working in opposite directions** gegeneinander arbeitendes Rührwerk *n* ~ **effect** (in induction furnace) Badbewegung *f*, Bewegungserscheinung *f* ~ **hole** Schürloch *n* ~ **implement** Rührhaken *m* ~ **machine** Rührmaschine *f* ~ **paddle** Mischflügel *m*, Rühr-flügel *m*, -schaufel *f*, -stab *m* ~ **pole** (or rod) Rührstab *m* ~ **rod** Glasstab *m* ~**time** Rührzeit *f* ~ **wing** Misch-, Rühr-flügel *m*

**stirrup** Bügel *m*, Hakenbügel *m*, Steigbügel *m*, Tragbügel *m*, (of boring apparatus) Wirbelstück *n*, Zange *f* ~ **of a boring apparatus** Kopf *m* des Erdbohrers ~ **of valve** Ventilbügel *m*

**stirrup,** ~ **bolt** Bügelschraube *f* ~ **leather** Steigriemen *m* ~ **pump** Luftschutzbrandspritze *f* ~ **spacing** Bügelteilung *f*

**stitch, to** ~ broschieren, einsteppen, feststecken, heften, ketteln, zuheften **to** ~ **in** einheften **to** ~ **on** anketteln **to** ~ **together** aufheften

**stitch** Masche *f*, Nadelstich *m*, Stich *m* ~ **adjusting** Senkereinstellschraube *f* ~ **cam post** Senkerträger *m* ~**counting** Maschenprüfung *f* ~ **dial** Skalenscheibe *f* für die Senkereinstellung ~ **glass** Maschenzähler *m* ~ **rivet** Heftniet *n* ~ **welding** Punktschweißung *f*

**stitched** geheftet ~ **brake lining** gestepptes Bremsband *n* ~ **pack** gehefteter Block *m* ~**-stuff-and-knitting-machine** Wirk- und Strickmaschine *f* stitcher Hefter *m*, Heftvorrichtung *f* ~ stand Aufroller *m*

stitching Heften *n* ~ apparatus Heftapparat *m* ~ gauze Heftgaze *f* ~ head Heftkopf *m* ~ hook Heftklammer *f* ~ machine Anroll-, Heft-, Stepp-maschine *f* ~ needle Heftnadel *f* ~ test Heftprobe *f* ~ warp Bindekette *f* ~ wire Heftdraht *m*

stitchlike nadelstichartig

Stobie (series-arc) furnace Stobie-Ofen *m*

stochastic stochastisch ~ differentiation stochastische Differentiation *f* ~ process stochastischer Prozeß *m*

stock, to ~ aufbewahren, aufspeichern, lagermäßig führen to ~-pile platzbeschicken

stock Bestand *m*, Grundkapital *n*, Gut *n*, Inventar *n*, Kapital *n*, Lager *n*, Lagerbestand *m*, Material *n*, Schabotte *f*, Schawatte *f*, Schulterstütze *f*, Stock *m*, Stoff *m*, Vorrat *m*, Werkstoff *m* from ~ aus Vorrat in ~ auf Vorrat, vorrätig ~ in hand Warenbestand *m* ~ on hand Iststand *m*, Lagervorrat *m* ~ of plane Hobelgehäuse *n*

stock, ~ account Kapitalkonto *n* ~ anchor Stockanker *m* ~-and-die Wendeeisen *n* ~-and--dies Gewindeschneidzug *m* ~ anvil Schlagstöckchen *n* ~ bin Vorratsbehälter *m* ~ book Bestand(s)buch *n* ~ bottle Vorratsflasche *f* ~ brick Glaskopf *m* (Ziegel) ~ broker Effektenmakler *m* ~ cable Vorratskabel *n* ~ car Serienwagen *m* ~ card index Lagerfachkartei *f* ~ checking and certification Bestandsüberprüfungen *pl* und Bestätigung *f* ~ coke Haldenkoks *m* ~ company Aktiengesellschaft *f* ~ converter Stockkonverter *m* ~ decline Kurssturz *m* ~ decrease Bestandsverminderung *f* ~ density Stoffdichte *f* ~ discharge paste Stammätze *f* ~-distributing gear Schüttvorrichtung *f* ~ emulsion Stammemulsion *f*

stock-exchange Effektenbörse *f* ~ call Börsengespräch *n* ~ call office öffentliche Börsensprechstelle *f* ~ list Kurszettel *m* ~ office Börsenamt *n* ~ report Börsenbericht *m* ~ switch-board Börsenamt *n* ~ telephone action Börsensprechstelle *f*

stock, ~ feed Materialnachschub *m*, Werkstoffvorschub *m* ~-feeding device Materialvorschubvorrichtung *f* ~ ferrule swivel Riemenbügel *m* ~ groove Geschoßanlage *f* ~head Aufsatzstativ *n* ~ holder Aktien-besitzer *m*, -inhaber *m*, Aktionär *m* ~ house Vorratsraum *m* ~ inventory Bestandsnachweisung *f* ~ keeper Verwalter *m* des Vorratslagers ~ length Lieferlänge *f* (Maß) ~ line Beschickungsoberfläche *f* ~ line gauge Gichtanzeiger *m* ~-line indicator (of a furnace) Teufenanzeiger *m* ~ liquor Stammflotte *f* ~ list Lagerliste *f* ~ loss Vorratsverlust *m* ~ mixture Stammansatz *m* ~ movement reports Warenbewegungslisten *pl* ~ movements by grades Bestandsbewegungen *pl* nach Sorten ~ number Anforderzeichen *n*, Fabrikationsnummer *f* ~-pile Bodenablagerung *f*, Bodenkippe *f*, Kippe *f* ~-piling yards for storing pipes Rohrlagerplätze *pl* ~ position Vorratslage *f* ~ preservative Schaftpflegemittel *n* ~ rail Anschlag-, Backen-, Stock-schiene *f* ~ rail of a siding feste Schiene *f* oder Hauptschiene *f* einer Weiche ~ records Bestands-

-aufstellungen *pl*, -unterlagen *pl*, Materialverwaltung *f* ~ removal Materialentfernung *f* ~ report Bestandsnachweisung *f*, Nachweisung *f* ~ resist Stammreserve *f* ~room Lagerraum *m*, Sammelkammer *f* ~ solution Stamm-, Vorrats-lösung *f* ~ statement Bestandsmeldung *f* ~ status report Lagerbestandsbericht *m* ~ stop Anschlagbolzen *m*, Materialanschlag *m*, Werkstoffanschlag *m* ~ taking Bestandaufnahme *f*, Inventaraufnahme *f*, Inventur *f*, Lageraufnahme *f* ~taking verification physische oder effektive Bestandsaufnahme *f* ~ tank Vorratstank *m* ~ thickening Stammverdickung *f* ~ ticker Börsen-, Fern--drucker *m* ~ tub Standgefäß *n* ~-type Lagertyp *m* ~ vat Stammküpe *f* ~yard Lagerplatz *m*, Stapelplatz *m*, Viehhof *m*, Vorratsplatz *m* ~-yard crane Zangenkran *m*

stockade Einzäunung *f*, Pfahlzaun *m*, Staket *n*

stockage Bevorratung *f*, Lagerbestand *m*

stocker Aufräumer *m*

Stockholm tar schwedischer Teer *m*

stocking Beförderung *f* (Schiff), Kniestrumpf *m* ~ cutter Schrupp-, Vorschneid-, Zahnformvor-fräser *m* ~-out Lagerplatzbeschickung *f* ~ tool Breitschneidestahl *m* zum Vorstechen ~ turner Strumpfwender *m*

stocks Effekten *pl* (Bank) ~ of goods in transit durchlaufende Warenbestände *pl* ~ in tanks Tankbestände *pl* ~ in transit schwimmende Bestände *pl*

stocky stockig

Stoddard solvent Lösungsmittel *n*

stoichiometric, ~(al) stöchiometrisch ~ metalloid deficiency stöchiometrischer Metalloidmangel *m*

stoichiometry Elementenmessung *f*

stoke, to ~ schüren, stochen, stochern to ~ up boilers anheizen to ~ the fire das Feuer anschüren oder schüren

stoke, ~ hold Heizerstand *m*, Heizkammer *f* ~hole Heizloch *n*, Schürloch *n*

stoker Heizer *n*, Kohlenbeschickungsvorrichtung *f*, Ofenbrücke *f*, Stocheisen *n*, Stoker *m* ~ with reciprocating grate bars Vorschubrost *m*

stoker, ~ door Schürloch *n* ~ firing selbsttätige Rostfeuerung *f*, Stokerfeuerung *f* ~'s rod Sticheisen *n* ~ scoop Heizerschaufel *f* ~'s shovel Tenderschaufel *f*

Stokes trench mortar Stokes-Gasmörser *m*

stoking Rostbeschickung *f* ~ door Feuer(ungs)tür *f*

stolzite Wolframbleierz *n*

stomach Magen *m* ~ pump Magenpumpe *f*

stomadaeum Schlundrohr *n*

stone, to ~ (cutting edge) abziehen (mit dem Ölstein), (fruit) auskernen, (pavement) besteinen, Bruchsteine trocken verlegen

stone Fels *m*, Felsen *m*, Gestein *n*, Naturstein *m*, Stein *m*, Steinchen *n*, (dry) Trockenmauerwerk *n* ~ with binding materials Steine *pl* mit bindigem Boden ~ for grinding tools Werkzeugschleifstein *m*

stone, ~ age Steinzeit *f* ~ axe Flächenhammer *m* ~ band Bergemittel *n* ~ bit Steinbohrer *m* ~board Steinpappe *f* ~ bolt Steinschraube *f* ~ bottle Kruke *f* ~-breaking hammer Vorschlaghammer *m* für Felsbearbeitung ~ broken to pass a

**specified diameter** Steinschlag *m* von bestimmter Korngröße **~ burnisher** Achatglattmaschine *f* **~ burr mill** Steinschrotmühle *f* **~ chippings** Feinsplitt *m* **~ chips** Splitt *m* **~ coal** Steinkohle *f* **~ console** Kragstein *m* **~ coping** Deckquader *n* **~ corbel** Kragstein *m* **~ cradling** Ausmauerung *f* **~ crusher** Steinbrecher *m* **~ crushing plants** Bergebrechanlagen *pl* **~cutter** Stein-hauer *m*, -metz *m* **~ dike** gepflasterter Deich *m* **~ disc** Steinscheibe *f* **~ drain** Dränage *f* **~ drill** Mauerbohrer *m*, Steinbohrer *m* **~ dust** Gesteinsstaub *m*, Steinmehl *n* **~ dusting** Gesteinsstäubung *f* **~ filled drain** Stein-drainage *f* (-packung *f*) **~ frame saws** Steinsägegatter *n* **~ grainer** Steinkörner *m* **~ grinding** Steinschliff *m* **~ guard** (for lamps) Schutzkorb *m* **~ hammer** Brech-, Stein-hammer *m* **~head** festes Gestein *n* **~ hemp** Steinflachs *m* **~ hewing** Steinbehauerei *f* **~ jar** Kruke *f* **~lifting tongs** Adlerzange *f* **~man** Gesteinshauer *m*, Vorrichtungsarbeiter *m* **~ marking boundary of a mine** Lochstein *m* **~ mason** Stein-hauer *m*, -metz *m* **~mason's hammer** Bossierhammer *m* **~ miner** Gesteinshauer *m* **~ mug** Steinkrug *m* **~ pared on every side** allseitig bearbeiteter Stein *m* **~ pavement** Kopfsteinpflaster *n* **~ paving** Abdeckung *f*, Steinböschung *f* **~ paving of gentle slope** flachgeneigte Steinverkleidung *f* **~ picking** Steinebehauen *n* **~ pine** Arobe *f* **~ pit** Steinbruch *m* **~ plug** Steinkerze *f* **~-powder pulverizing plant** Gesteinstaubmahlanlage *f* **~ radiating oven** Steinstrahlofen *m* **~ runner** Steinläufer *m* **~ setter** Steinsetzer *m* **~ smoothing and planing machine** Steinabricht- und Hobelmaschine *f* **~ spindle** Mühleisen *n* **~ splitter** Keillocher *m* **~-splitting machine** Steinspaltmaschine *f* **~-splitting tools** Steinspaltmesser *n* **~ stopper** Steinstopfen *m* (Säureballons) **~ structure** Quaderbau *n* **~ stud** Eckpfeiler *m* **~ wall** Gesteinswand *f*, Steinmauer *m* **~ware** Stein-gut *n*, -zeug *n*, Tonzeug *n* **~ware bottle** Steinzeugflasche *f* **~ware vat** Steinbottich *m* **~work** Steinarbeit *f* **~working** Steinbearbeitung *f* **~-working machine** Gesteinsbearbeitungsmaschine *f*

**stoned finish** Glättung *f*
**stonelike** steinähnlich
**stoner** (tanning) Abbimser *m*
**stones** Bruchstein *m*
**Stoney gate** Stoneyscher Rollschütz *m*
**stoning machine** (fruit pip or stone extractor) Auskernmaschine *f*, Entkernmaschine *f*
**stony** steinartig, steinig **~ ground** Steingrund *m*
**stool** Bock *m*, Fußschemel *m*, (of an ingot mold) Gespann *n*, (foundry) Gießgespann *n*, Schemel *m*, Untersatz *m*
**stop, to ~** abbrechen abschließen, (engine) absperren, abstellen, anhalten, arretieren, aufhalten, aufhören, ausrücken, (cable) beizeisen, bremsen, einhalten, Einhalt tun, einstellen, enden, fest-halten, -legen, -stellen, grundieren, hintanhalten, innehalten, kaltlegen, sperren, stehen bleiben, stillsetzen, stillstehen, (machine) stocken, stoppen, unterbrechen, versetzen, verstemmen, verstopfen, zukorken, zurückhalten **to ~ the blast** das Gebläse abstellen **to ~ down** (lens aperture) abblenden **to ~ the engine** außer Gang setzen, den Motor *m* abstellen **to ~**

**the flood gates** die Schützen *pl* einstellen **to ~ gradually** auslaufen (Motor) **to ~ with mud** verschlämmen **to ~ out** ausblenden **to ~ at a position** in einer Stellung *f* stehenbleiben **to ~ rolling** ausrollen **to ~ the time** die Zeit abstoppen **to ~ up** dämmen, verstreichen, zustopfen **to ~ the way** (of a ship) abstoppen **to ~ work** die Arbeit einstellen **to ~ the work** den Betrieb *m* einstellen

**stop** Absatz *m*, Anschlag *m* (Widerlager), Arretierung *f*, Aufenthalt *m*, Auflager *n*, Auslösung *f*, Ausschlag *m*, (back) Begrenzungsanschlag *m*, Blendscheibe *f*, Einhalt *m*, (for arresting tool or spring) Fangbügel *m*, Feststellknagge *f*, Feststellung *f*, (in machines) Frosch *m*, Halt *m*, Haltestelle *f*, Hubbegrenzer *m*, Hubbegrenzeranschlag *m*, Klebstift *m*, Pause *f*, Prellbock *m*, (organ) Register *n*, Schwinganschlag *m*, Sperre *f*, Sperr-stift *m*, -stück *n*, Stillstandszeit *f*, Stütz-knagge *f*, -punkt *m*, Unterbrechung *f*

**stop, ~ in counting** Zählstopp *m* **~ down** (aperture of lens) Blendeneinstellung *f* **~ for downfeed** Anschlag *m* für die Vertikalbewegung **~ for governing rack** Regelstangenanschlag *m* **~ for swinging lever** Schwinghebelanschlag *m*

**stop, ~ adjustment operator's side** Anschlagbedienungsseite *f* **~-and-reversing-light-switch** Brems- und Rückfahrlichtschalter *m* **~-and--tail-lights** Brems- und Schlußleuchten *pl* **~-and-waste-cock** Ablaßhahn *m* **~ ball** Arretierkugel *f* **~ band** (Filter) Sperrbereich *m* **~ band attenuation** Sperrdämpfung *f* **~ banging** Mastschlagen *n* **~ bar** Arretier(ungs)stange *f* **~ block** Prell-bock *m*, -klotz *m*, Sperr-klotz *m*, -schuh *m*, Stützwinkel *m* **~ bolt** Grenzriegel *m* **~ button** Abstellknopf, Schnellstopptaste *f* (phono) **~ cable** Stopptau *n* **~ cock** Abschluß-, Abstell-, Block-hahn *m*, Hahn *m*, Küken *n*, Sperrhahn *m* **~cock box** Absperrhahn *m* **~cock grease** Hahnenfett *n* **~-cocking and oil well** stoßweises Springenlassen *n* einer Ölbohrung **~ collar** Anschlagbund *m* **~ control** Abstechvorrichtung *f* **~ control clock** Stoppschaltuhr *f* **~-cylinder press** Haltzylinderpresse *f* **~ device** Haltevorrichtung *f* **~ diaphragm** Sperrmembrane *f* **~ disc** Fallenscheibe *f* **~ disc with spiral slot** Abdeckscheibe *f* **~ dog** Schalt-knagge *f*, -nocke *f*, -nocken *m*, verstellbarer Begrenzugsanschlag *m* **~ down rays** Strahlabblender *m* **~ face** Anschlagfläche *f* **~ fin** Arretiersicke *f* **~ flange** Sperrflansch *m* **~gap** Lückenbüßer *m* **~-gap advertisement** Füllanzeige *f* **~ gauge** Anschlag *m* **~ gear** Sperrwerk *f* **~ groove** Gleitbahn *f* für Aufstellung **~ grooves for sluice** Führungsrinne *f* der Schütze **~ hole** Anschlagloch *n* **~ knob** Arretierknopf *m* **~ lamp** Bremslaterne *f* **~ lamp cable** Stopplichtleitung *f* **~ lever** Anschlag- (print.), Arretier-, Begrenzungs-, Brems-, Festhalte-, Rast-hebel *m* **~ light** Bremslicht *n* **~ light switch** Bremslichtschalter *m* **~ line** Aufhaltelinie *f* **~ link** Ausrückglied *n* **~ location** Raststellung *f* **~ log** Dammbalken *m* **~-log dam** Dammbalkenwehr *f* **~-log grooves (or recesses)** Dammfalze *pl* **~ mechanism** Abstellvorrichtung *f* **~-motion box** Abstellkasten *m* **~-motion device** Zeitraffer *m* **~-motion rails** Abstellgestänge *n* ~ **notch** Sperraste *f* **~-over** Halte-punkt *m*, -stelle *f*, Zwischenlandung *f* **~-over of a seaplane** Zwischenwasserung *f* ~ **pawl** Sperrklinke *f* ~ **pin (instruments)** Anhaltstift *m*, Anschlag-bolzen *m*, -stift *m*, Arretierschraube *f*, einsteckbarer Bolzen *m*, Klebstift *m*, Nase *f* zum Festhalten, Vorsteckstift *m* ~ **plank** Flutschütze *m* ~ **plate** Abschlußplatte *f*, Abstellscheibe *f* ~ **plug** Steckblende *f* ~ **position** Stoppstellung *f* ~ **ring for locking fulcrum** Anschlagring *m* (aviat.) ~ **rod** Platine *f* ~ **screw** Anschlag-, Arretier-, Grenz-schraube *f* ~ **screw for quantity-regulator handle** (accelerator) Anschlagschraube *f* (für Mengenverstellhebel) **~-send signal** Sende-Stoppsignal *n* ~ **signal (traffic)** Haltezeichen *n*, Stoppsignal *n* ~ **sleeve (bush)** Anschlagbuchse *f* ~ **spindle** Ausrückspindel *f* ~ **spring** Festhalte-, Rast-, Riegel-feder *f* ~ **surface** Abstellfläche *f* ~ **switch** Abstellschalter *m* **~-timer** Stopp-Schaltuhr *f* ~ **valve** Abstellventil *n* **~-valve key** Absperrschlüssel *m* für Stoppventil ~ **watch** Chronometer *n*, Sekundenzähler *m*, Stoppuhr *f* **~-watch measurement of time intervals (or values)** Abstoppen *n* von Zeitdauern **~way** Stoppbahn *f* (aviat.) ~ **wire** Anschlagdraht *m* **~work** Feststellung *f* (watch)

**stope, to** ~ **out** stoßweise abbauen **to** ~ **overhand** einen Firstenstoß *m* oder Firstenbau *m* abbauen **to** ~ **underhand** einen Strossenstoß *m* abbauen

**stope** Abbauort *m*, Strosse *f*

**stoping ground** erzführendes Gestein *n*

**stoppage** Anhalt *m*, Anhalten *n*, (factory) Ausfall *m*, Betriebs-pause *f*, -störung *f*, Fehlen *n*, Feststellung *f*, Hemmung *f*, Ladehemmung *f*, Stillegung *f*, Stillstand *m*, Stockung *f*, Unterbrechung *f* ~ **caused by broken cartridge case** Hülsenreißer *m* ~ **of ice** Eisstopfung *f* ~ **in loading mechanism** Ladestörung *f* ~ **of mine work** Einstellung *f* des Betriebes einer Grube

**stopped** abgestoppt, eingestellt **~-down radiation** Strahlenbegrenzung *f* ~ **pipe** gedeckte Pfeife *f* ~ **section** Schutzblockstrecke *f* **~-section circuit** Stromkreis *m* der Schutzblockstrecke *f* ~ **tone** Stopfton *m* ~ **up** verstopft

**stopper, to** ~ absperren, blockieren, pfropfen, sperren, stoppen, stöpseln, verstopfen

**stopper** Absperr-organ *n*, -teil *m*, -vorrichtung *f*, Auffangvorrichtung *f* (print.), Ausrücker *m*, Pfropfen *m*, Sperre *f*, Spund *m*, Stopfen *m*, Stopper *m*, Stoppkette *f*, Stopptau *n*, Stöpsel *m*, Stromreiniger *m*, Verschluß *m*, Verschlußstück *n*, Zeitnehmer *m* **to let go the** ~ seinen Stopper *m* brechen oder abtreiben ~ **of rolling mill** Walzstopfen *m* ~ **with thumb piece** Griffelstopfen *m* ~ **within valve** Ventilstopfen *m*

**stopper,** ~ **bottle** Stöpselflasche *f* ~ **circuit** Drosselkreis *n*, Entkopplungsglied *n*, Sperrfilter *m* ~ **cock** Stöpselhahn *m* ~ **gear** Tätigkeitsunterbrecher *m* ~ **head** Stopfenkopf *m* ~ **ladle** Gießpfanne *f* mit Stopfenausguß, Stopfenpfanne *f* ~ **nozzle** Stopfenausguß *m* ~ **rod** Stopfenstange *f* ~ **screw** Absperrschraube *f*

**stoppered** gespundet ~ **cylinder** Mischzylinder *m* ~ **glass** Stöpselglas *n*

**stopping** Anhalten *n*, Aufhören *n*, Ausrückgang

*m* (print.), Feststellung *f*, Kittmaterial *n*, Still-setzung *f*, -stand *m*, Stopfen *n*, Versatz *m*, Verschluß *m*, Versetzung *f* ~ **of branches** Rinnenabschluß *m* ~ **of the headstock spindle** Hauptspindelstillsetzung *f* ~ **of machine** Druckabstellung *f* ~ **out** Schutzfirnis *n* ~ **of payment** Zahlungssperre *f* ~ **of ventilator** Lüfterstillsetzung *f*

**stopping,** ~ **anode point** Glimmspannung *f* ~ **cam** Auflaufnocken *m* ~ **capacitor** Blockkondensator *m* ~ **condenser** Block-, Sperr-kondensator *m* ~ **contact** Abstellkontakt *m* ~ **cross section** Bremsquerschnitt *m* ~ **device** Abstell-, Anhalte-, Ausschalte-vorrichtung *f* ~ **distance** Brems-strecke *f*, -weg *m* **~-down** (lens aperture) Abblendung *f* **~-down of the crystal image** Kristallbildabblendung *f* **~-down of image of crystal** Abblendung *f* des Kristallbildes ~ **formula** Bremsformel *f* ~ **lens** Verzögerungslinse *f* ~ **lever** Abstellhebel *m* ~ **means** Ausblendmittel *n* ~ **mechanism** Tätigkeitsunterbrecher *m* ~ **oscillation of motor** Auspendeln *n* des Motors ~ **place** Halte-platz *m*, -stelle *f* ~ **place of a pit** Schachtbahn *f* ~ **point** Anhalts-, Halte-punkt *m* ~ **potential** Löschspannung *f* ~ **potential method** Anhaltepotentialmethode *f* ~ **power** (of photographic emulsions) Bremsvermögen *n* ~ **relay** Stopprelais *n* ~ **rod** Abstellstange *f*

**stops,** ~ **on the carriage** Hemmleisten *pl* am Laufbrett (print.) ~ **for adjustment of stroke** Anschläge *pl* zur Hubverstellung

**storability** Lagerfähigkeit *f*

**storage** Ablagerung *f*, Aufbewahrung *f*, Auflagerhalten *n*, Aufspeicherung *f*, Einlagerung *f*, Einräumung *f*, Lagern *n*, Lagerung *f*, Lastenraum *m*, Speicherung *f* ~ **for building materials** Baulager *n* ~ **of water** Wasser-speiser *m*, -speicherung *f*

**storage,** ~ **basin** Speicherbecken *n* ~ **battery** Akkumulator *m*, Akkumulatorbatterie *f*, Lagerbatterie *f*, Sammler *m*, Sammlerbatterie *f*, Sekundärelement *n*, Speicher *m*, Stromsammler *m*

**storage-battery,** ~ **acid** Akkumulatorsäure *f* ~ **cell** Akkumulatorenelement *n*, Akkumulatorzelle *f*, Platteneinsatz *m* ~ **locomotive** Akkumulatorlokomotive *f* ~ **plate** Akkumulatorenplatte *f* ~ **potential** Akkumulatorspannung *f* ~ **traction** Akkumulatorenfahrbetrieb *m* ~ **voltage** Akkumulatorspannung *f*

**storage,** ~ **bin** Sammel-behälter *m*, -bunker *m*, Schachtspeicher *m*, Silo *m*, Vorratstasche *f* ~ **bunker** Sammelbunker *m*, Vorratsbehälter *m* ~ **camera tube** Bildspeicherröhre *f* ~ **capacity** Ladevermögen *n*, Lagerungsfähigkeit *f*, Speicherfähigkeit *f* ~ **case** Aufbewahrungskasten *m* ~ **cathode** Speicherkathode *f* ~ **cell** Bleiakkumulator *m*, Bleisammler *m*, Sammlerzelle *f*, Sekundärbatterie *f* ~ **cellar** Lagerkeller *m* ~ **chain** Speicherkette *f* ~ **charges** Lagergebühren *pl* **~-closet rack** Lagergerüst *n* ~ **condenser** Speicherkondensator *m* ~ **conditions** Lagerungsverhältnisse *pl* ~ **contact spring** vorbereitende Kontaktfeder *f* ~ **crane** Lagerkran *m* ~ **effect** Speicherwirkung *f* ~ **element** Speicherelement *n* ~ **expenses** Lagerkosten *pl* ~ **facility** Lagerungsmöglichkeit *f* ~ **indicator** Inhaltsanzeiger *m* ~ **keyboard** Speichertastatur *f* ~ **layer** Speicherschicht *f* ~ **life** Lagerfähigkeit *f* ~ **location** Speicherzelle *f* (comput.) ~ **loss** Standverlust *m* ~ **magazine** Vorratskassette *f* ~ **pickup** Speicherplatte *f* ~ **place** Lagerplatz *m*, Speicher *m* ~ **pocket** Vorratstasche *f* ~ **pond** Stauteich *m*, Stauweiher *m* ~ **principle** Speicherprinzip *n* ~ **process** Lagerprozeß *m* ~ **property** Lagerfähigkeit *f* ~ **register** Speicherzelle *f* (comput.) ~ **relay** Speicherrelais *n* ~ **reproducibility** Speicherlesbarkeit *f* ~ **room** Sammelkammer *f* ~ **room for explosives** Sprengstoffraum *m* ~ **shed** Lagerschuppen *m* ~ **silo** Lagersilo *m* ~ **space** Lagerraum *m* ~ **stability** Lagerfähigkeit *f* ~ **stack** Speicherpaket *n* ~ **stock** Lagervorrat *m* ~ **surface** Speicheroberfläche *f* ~ **tank** Lager-behälter *m*, -tank *m*, Sammel-behälter *m*, -gefäß *n*, Vorrats-faß *n*, -tank *m* ~ **track** Speicherspur *f* ~ **transmitter** Speichergeber *m* ~ **tube** Speicherröhre *f* ~ **units** Speicherwerke *pl* ~ **vessel** Aufbewahrungsbehälter *m*, Standgefäß *n* ~ **yard** Lager *n*, Lagerplatz *m*

**store, to** ~ ablagern, aufbewahren, aufhäufen, auflagern, lagern, registrieren, schreiben (info proc.), speichern, unterbringen **to** ~ **before using** vorlagern **to** ~ **heat** Wärme *f* aufspeichern **to** ~ **in stock pile** ablagern **to** ~ **up** anlagern

**store** Behälter *m*, Behältnis *n*, Kaufladen *m*, Lager *n*, Magazin *n*, Speicher *m* (info proc.), Verkaufslokal *n*, Vorrat *m* **in** ~ vorrätig ~ **for building materials** Baulager *n*

**store,** ~ **book** Bestandbuch *n* ~ **cellar** Lagerkeller *m* ~ **hand** Lagerarbeiter *m* ~ **house** Depot *n*, Lagerhaus *n*, Niederlage *f*, Warenlager *n* ~**keeper** Hellegatsmann *m*, Lager--aufseher *m*, -halter *m*, -meister *m*, -verwalter *m*, Magazinverwalter *m* ~ **location** Zelle *f* (info proc.) ~ **magazine** Lagerhaus *n* ~**man** Lageraufseher *m* ~**room** Ablage *f*, Hellegat *n*, Lagerraum *m*, Magazin *n*, Vorratsraum *m* ~ **sample** Meßmuster *n* ~ **shed** Materialschuppen *m* ~**ship** Depotschiff *n* ~ **timber** Stapelholz *n* ~ **yard** Vorratsplatz *m*

**stored** gelagert ~ **energy** Vorratsenergie *f* ~**-impulse automatic telephone system** Selbstanschlußsystem *n* mit Stromstoßempfängern ~ **program** Speicherprogramm *n* ~ **program (digital) computer** speicherprogrammierte (digitale) Rechenanlage *f* oder Datenverarbeitungsanlage *f* ~**-up energy** Energieeinhalt *m* ~**-water power station** Speicherkraftwerk *n*

**stores** Proviant *m*, Stoffe *pl* ~ **issues** Magazinentnahmen *pl*

**storing** Aufhäufung *f*, Auflagerhalten *n*, Aufspeicherung *f*, Speicherung *f* ~ **up the calls** Wartefeld *n*

**storing,** ~ **capacity** Platzausnutzung *f*, Speicherfähigkeit *f* ~ **circuit** Speicherorgan *n* ~ **keysender** Speicherzahlengeber *m*

**storm, to** ~ erstürmen

**storm** Orkan *m*, Sturm *m* ~ **area** Sturmfeld *n* ~ **ball** Sturmball *m* ~ **center** Sturmzentrum *n* ~ **cone** Sturmkegel *m* ~ **damage** Sturmschaden *m* ~ **guyed pole** Linienfestpunkt *m* ~ **path** Sturmbahn *f* ~ **sash** Doppelfenster *n* ~ **tide** Sturmflut *f* ~ **warning** Sturmwarnung *f* ~-

-**warning flag** Sturmwarnungsfahne *f* ~**-warning service** Sturmwarnungsdienst *m* ~**-warning signal** Sturmwarnungszeichen *n*

**stormy** stürmisch

**story** Geschoß *n*, Handlung *f*, (of building) Stock *m*, Stockwerk *n* ~ **furnace** Etagenofen *m*

**stout** dick ~ **pole** starke Stange *f*

**stove, to** ~ einbrennen (keramische Farben und Emaillen) **to** ~**-enamel** ofenlackieren **to** ~ **wood** Holz *n* dörren

**stove** Ofen *m*, Vorwärmer *m* ~ **bolt** Herdschraube *f* ~ **casting** Herdguß *m* ~ **castings** Ofenguß *m* ~ **coke** Abfall-, Knabbel-koks *m* ~ **distillate** Heizöldestillat *n* ~**-enamelling** Einbrennlackierung *f* ~ **gloss** Ofenglanz *m* ~ **lacquer paint** Ofenlackfarbe *f* ~ **lid** Herdplatte *f* ~ **plate** Herdplatte *f* ~ **polish** Ofenputzmittel *n* ~ **screw** Ofenschraube *f* ~ **tile** Ofenkachel *f*, Ofenziegel *m* ~ **truck** Schleppe *f* ~ **varnish** Einbrennfirnis *m*

**stoving** Schwefelkastenbleiche *f* ~ **chamber** (Schwefel-)Kammer *f* ~ **enamel** Einbrennlack *m* (Emaillelack) ~ **residue** Einbrennrückstand *m* ~ **temperature** Einbrenntemperatur *f* (Lack)

**stow** stauen, unterbringen, verstauen, vollfüllen **to** ~ **away** einräumen, packen **to** ~ **away the roof** Firste *pl* nachreißen

**stowage** Lagerung *f*, Stauen *n*, Verstauen *n* ~ **chart** Zeichnungssatz *m* ~ **chart for vehicles** Beladeplan *m* ~ **tube** Verpackungsrohr *n*

**stowed away** gestaut

**stower** Aufstapler *m*, Versatzarbeiter *m*

**stowing** Versatz *m* ~ **arrangement** Stauvorrichtung *f* ~ **device for bombs** Tragvorrichtung *f* ~ **elevator** Aufstauelevator *m* ~ **machines** Bergeversatzmaschinen *pl* ~ **rinsing plant** Spülversatzanlage *f*

**strabismometer** Schielmesser *m*

**strabometer** Strabometer *n*

**straddle, to** ~ spreizen

**straddle,** ~ **closure** Spreizverschluß *m* ~ **cutter** Doppelscheibenfräser *m* ~ **gauge** Rieterlehre *f* ~**-legged** breitbeinig ~ **mill** Scheibenfräser *m* für Muttern oder Bolzenköpfe ~ **wrench** Gabelstiftschlüssel *m*

**straddling** rittlings ~ **of axle** Achsschrägung *f* (Kupplung) ~ **of points** Gabelfahrt *f*

**straddling,** ~ **dowel** Spreizdübel *m* ~**position** Grätsche *f*

**straggling** Dispersion *f*, Schwankung *f*, Streuung *f* ~ **parameter** Streuparameter *m*

**straight** (joint) zylindrisches Gewinde *n*; aufrecht, gerade, scheitelrecht ~ **ahead** geradeaus ~ **between two curves** Zwischengerade *f* ~ **as a die** kerzengerade, fadengerade ~ **to the eye** nach Augenmaß gerade **in** ~**-line variation** gleichmäßig fallend oder steigend ~ **on** geradeaus ~ **out** geradeaus ~ **as a rail** baumgerade ~ **as a tree** baumgerade

**straight,** ~**-ahead amplifier** Direktverstärker *m* ~**-ahead reception** Suchempfang *m* ~**-and--circular-knitters** Flachstrickmaschinen *pl* ~ **angle** gestreckter Winkel *m* ~ **(or one-wire) antenna** Linearantenne *f* ~ **approaching line** Näherungsgerade *f* ~ **arm paddle agitator** Balkenrührer *m* ~**-away measurement** Streckenmessung *f* ~ **bead** (welding) Längsnaht *f* ~

**beam** Balkenverhautträger *m* ~ **bending** Geradbiegen *n* ~ **bevel gear** Kegelrad *n* ~ **bit** gerader Bohrmeißel *m* ~ **boiling** Geradekochen *n* (Bodsten) ~**-bore mixing nozzle** Bohrungsmischöse *f* ~ **carbon steel** gewöhnlicher Kohlenstoffstahl *m* ~ **channel** gerades Gerinne *n*, Schußgerinne *n* ~**-circuit receiver** Geradeausempfänger *m* **conductor** gerader Leiter *m* ~ **connection** gerades Verbindungsstück *n* ~**-course chart** gnomonische Karte *f* ~ **crank drive** gerader Schubkurbeltrieb *m* ~ **cut** (saw teeth) Geradschliff *m* (Sägezähne), Geradeschnitt *m* ~**-edge** Abrichtlineal *n*, Abstreich-holz *n*, -lineal *n*, Abstrichlineal *n*, Lineal *n*, Relaisscheit *n*, Richt-latte *f*, -lineal *n*, (drawing) Richtscheit *n* ~**-edge tooth-sharpening machine** Geradschliffmaschine *f* (Säge) ~**-eight engine** Achtzylindermotor *m* ~**-face pulley** Riemenscheibe *f* mit geradem Kranz ~**-face roller** glatte Laufrolle *f* ~**-fiber** schlichte Faser *f* ~**-flat-lap weld** durchlaufende Kehlnaht *f*, Kehlnaht *f* eines überlappten Stoßes, leichte Kehlnaht *f* ~**-flow system** (gas turbines) Geradeflußsystem *n* ~**-flute drill** Bohrer *m* mit geraden Spannuten ~**-fluted** (plain teeth) geradeverzahnt ~**-fluted reamers** Reibahlen *pl* mit geraden Zähnen ~ **forward** einfach, schlicht, unkompliziert ~**-forward operation** Betrieb *m* mit selbsttätiger hörbarer Zeichengebung für die Beamtin des Auskunftsplatzes ~**forward patent application** Patentanmeldung *f* von gewöhnlichem Umfang ~**forward trunking method** Anrufbetrieb *m* ~**forward working** glatte Fabrikation *f* ~ **girder** Balkenverhautträger *m* ~**-glued joint** Leimfuge *f* ~**-grained** längsgefasert ~ **groove** (flute slot) gerade Nut *f* ~ **head** Geradeausspritzkopf *m* ~**-in approach** Geradeaus-Anflug *m* ~ **insulator pin** gerade Stütze *f* ~ **joint** I-Stoß *m* ~**-joint floor** stumpfgefügter Fußboden *m* ~ **joint line** ebene Teilung *f* ~ **knives** Langmesser *pl* ~ **land** Führungsfase *f* (bei Räumnadel) ~ **length** Baufluchtlinie *f* ~ **light** gerade einfallendes Licht *n* ~ **line** Baufluchtlinie *f*, Fluchtlinie *f*, Strahl *m* (geom.), Strahlung *f*; gradlinig ~ **line at infinity** unendlich ferne Gerade *f* ~ **line of weld** durchlaufende Schweißnaht *f* ~ **line connecting two points** Verbindungsgerade *f*

**straight-line,** ~**-capacity condenser** kapazitätsgerader Kondensator *m*, Kreisplattenkondensator *m* ~ **chart** Fluchtentafel *f*, Nomogramm *n* ~ **course** Grundstrecke *f* (geol.) ~ **cutter** Längsschneidemaschine *f* ~ **cutting machine (or guide)** Längsschneidemaschine *f* ~ **detection** lineare Gleichrichtung *f* ~ **detector** geradliniger Gleichrichter *m*, linearer Detektor *m* ~ **flying** Flug *m* in gerader Richtung ~ **frequency capacitor** frequenzgerader Kondensator *m* ~ **frequency condenser** frequenzgerader Kondensator *m*, frequenzgleicher Drehkondensator *m* ~ **lapping** Läppen *n* mit geradliniger Hin- und Herbewegung der Läppwerkzeuge ~ **miller** Längsfräsmaschine *f* ~ **portion of a great circle** Orthodrome *f* ~ **relationship** geradlinige Beziehung *f* ~ **setting telescope** Fluchtfernrohr *n* ~ **spot welding** Reihenpunktschweißung *f* ~ **wave-(length) condenser** wellengerader Kondensator *m* ~ **wave-length**

**Wellenlängenlinear** *n* ~ **wave-length condenser** Nierenplattenkondensator *m*

**straight,** ~**-lined** geradlinig ~ **link** gerades Glied *n* ~ **mineral oil** reines Mineralöl *n* ~ **multiple** gerade Vielfachschaltung *f* ~**narrow forest clearing** Waldschneise *f* ~ **oak timber** Eichenlangholz *n* ~**-on angle shot** Waagerechtaufnahme *f* (film) ~ **oscillator** gerader Oszillator *m* ~**-pane sledge** Vorhammer *m* ~ **parting tool** gerader Geißfuß *m* ~ **peen hammer** Hammer *m* mit gerader Finne ~ **pin** Zylinderstift *m* ~ **pit saw** verjüngte Brettsäge *f* ~ **poker** Löschspieß *m* ~ **portion** Gerade *f* ~ **profile** Streckenprofil *n* ~ **reach** gradliniger Verlauf *m* ~ **reach of river** Flußstrecke *f* ~ **receiver** Geradeausempfänger *m* (rdr) ~**-reinforced** längsbewehrt ~**-run** unmittelbar abdestilliert ~**-run fuel** einfach destilliertes (nicht gekracktes) Benzin *n* ~**run gasoline** klares unverfälschtes Gasolin *n* ~ **scanning** unmittelbar aufeinanderfolgende Abtastung *f* ~ **shaft** gerade Welle *f* ~ **shank** Zylinderschaft *m* ~**-shank drill** Bohrer *m* mit zylindrischem Schaft ~ **shank lathe tool holder** gerader Drehstrahlhalter *m* ~ **shifting line** (magnetism) Scherungsgerade *f* ~**-side rim** Geradseitfelge *f* ~**-side tire** Geradseitreifen *m* ~**-sided** geradflankig ~**-sided crank press** Doppelständerkurbelpresse *f* ~ **signal** richtiges Zeichen *n* ~ **signals** richtige Schrift *f* ~ **slot boring bar** gerade Bohrstange *f* ~ **spar** Balkenholm *m* ~ **spindle** gerade Stütze *f* ~ **stretching** Geradstrecken *n* ~ **T weld** T-Stoß *m* mit durchlaufender Naht ~ **teeth** Geradverzahnung *f* ~ **thread** zylindrisches Gewinde *n* ~**-through air filter** Durchgangsluftfilter *m* ~**-through outlet** freier Austritt *m* ~**-through valve** Durchgangs-, Schrägsitz-ventil *n* ~ **tool** gerader Drehstrahl *m* ~**-toothed bevel gears** Geradverzahnung *f* ~ **tracing** Zeilentasten *n* ~ **track** gerader Strang *m* ~**-tube boiler** Sektionalkessel *m* ~**-way cock** Durchgangshahn *m* ~**-way steam cock** Dampfdurchgangshahn *m* ~**-way valve** Durchgangs-schieber *m*, -ventil *n* ~ **ways** Geradführung *f* ~ **well** gerade senkrechte Bohrung *f* ~ **wiring** waagrechte Gruppierung *f*

**straighten, to** ~ abrichten, ausbeulen, ausrichten, gerade machen, geraderichten, gerade richten, gleichrichten, richten, zurichten **to** ~ **out** aufrichten, ausrichten (fluchten) **to** ~ **a pole** eine Stange *f* ausrichten **to** ~ **whilst hot** warm richten

**straightened,** ~ **border** Randversteifung *f* ~ **brass** poliertes Messing *n*

**straightener** (wind tunnel) Gleichrichter *m*

**straightening** Ausrichten *n*, Ausrichtung *f*, Gerad(e)biegen *n*, Gerade-richten *n*, -strecken *n*, Richten *n* ~ **of wire** Recken *n* des Drahtes ~ **of the wire** Geradelegen *n* der Drähte

**straightening,** ~ **blades** Strömungsgleichrichter *m* ~ **capacity** Richtleistung *f* ~ **device** Richtgerät *n* ~ **press** Richtpresse *f* ~ **roll** Richt-rolle *f*, -walze *f* ~ **rolls** Ausrichtwalzen *pl*, Rollenrichtmaschine *f* ~ **tongs** Richtzange *f* ~ **vane** Ausgleichschaufel *f*

**straightness** Geradlinigkeit *f*

**strain, to** ~ abläutern, abseihen, anspannen, anstrengen, beanspruchen, deformieren, dehnen, durch-geben, -gießen, -lassen, -pressen, -seihen, eindrehen, filtrieren, recken, sieben, (a wire) spannen, überanstrengen, verziehen, verzerren **to ~ back to** abspannen **to ~-harden** kalthärten **to ~ off** (by sieve, etc.) abfiltern, abfiltrieren **to ~ to the utmost limit of the capacity** bis zur Grenze der Leistungsfähigkeit *f* ausnutzen

**strain** Abspann *m*, Anspannung *f*, Anstrengung *f*, Beanspruchung *f*, Belastung *f*, Deformation *f*, Deformierung *f*, Formänderung *f*, Formänderungsarbeit *f*, Kraftäußerung *f*, Materialspannung *f*, Pfeil *m*, Reckspannung *f*, Spannung *f*, Spannung *f* bei der eine bleibende Verformung auftritt, Verformung *f*, Verzerrung *f*, Zerdrückfestigkeit *f*, Zug *m* **free from ~** spannungsfrei **~ at failure** Bruchdehnung *f* **~ of glass** Glasspannung *f* **~ of a mooring cable** Steifwerden *n* einer Kette **~ of pole** Gestängebelastung *f* **~ on poles** Holzbeanspruchung *f* **~ and rotation** Deformation *f* und Drehung *f* **~ of tension** Spannabfall *m* **~ of volume** Dilatation *f* des Volumens

**strain, ~-aging** Stauchalterung *f* **~ amplitude** Verformungsamplitude *f* **~ bolt** Sperrbolzen *m* **~ coefficient** (reciprocal of modulus of elasticity) Dehnungszahl *f* **~ composition** Deformation *f* **~ control** Dehnungsmessung *f* **~ cord** Tragschnur *f* **~ curve** Dehnungslinie *f* **~ disease** Forcierkrankheit *f* (metall.) **~-distribution** Deformationsverteilung *f* **~ ear** Abspann--öhre *f*, -öse *f* **~ energy** Formänderungsarbeit *f* **~ energy method** Arbeitsverfahren *n* **~ figures** Fließfiguren *pl* **~-free glass** spannungs-freies Glas *n*, spannungsloses Glas *n* **~ gauge** Dehnungs-, Spannungs-messer *m* **~ gauge accelerometer** Dehnungsmeßstreifen-Beschleunigungsmesser *m* (gyro) **~ gauge rosette** Dehnungsmeßstreifenrosette *f* **~ hardenability** Kalthärtbarkeit *f* **~ hardening** (forging) Festigung *f*, Kaltverfestigung *f*, Verfestigung *f* durch Recken **~ (or wear) hardness** Kalthärte *f* **~ insulator** Sattel-, Spann -isolator *m* **~ invariants** Verformungsvarianten *pl* **~ magnitude** Dehnmaß *n* **~ mast** Abspannstange *f* **~ measurement** Verzerrungsmaß *n* **~-measuring instrument** Dehnungsmesser *m* **~ path** Verformungsweg *m* **~ pole** Abspannstange *f* **~ rate** Dehnungsgeschwindigkeit *f* **~ relief** Beseitigen *n* von Eigenspannungen (durch Erwärmen), Glühen *n* um die Eigenspannungen zu beseitigen **~-relief anneal (or thermal) treatment** spannungsfreies Glühen *n* **~ relieving notch** spannungsvermindernde Kerbe *f* (Entlastungskerbe) **~ resistance** Knickfestigkeit *f*, Zerbrechungsfestigkeit *f* **~ rod** Führungsstange *f* **~ tensor** Verformungstensor *m*

**strained, ~ sphere** Sphäroid *n* **~ structure** Struktur *f* der Kaltreckung

**strainer** Durchguß *m*, Durchschlag *m*, Filter *m*, Filtereinsatz *m*, Reiniger *m*, Saugkorb *m*, Seihe *f*, Seiher *m*, Sieb *n*, Siebeinsatz *m*, Spanner *m*, Spannklemme *f* **~ bow** Schmutzabscheidergehäuse *n* **~(pouring or grate) core** Siebkern *m* **~ screen** Siebeinsatz *m* **~ texture** Siebgewebe *n*

**straining** Anstrengung *f*, Beanspruchung *f*, Filtern *n*, Filtrierung *f*, Recken *n*, Sieben *n*, Siebung *f*, Verziehen *n*, Verzug *m* **~ apparatus** Läutervorrichtung *f* **~ bag** Seihesack *m* **~**

**beam** Brustriegel *m* **~ clamp** Abspannklemme *f* **~ cross head** Zugtraverse *f* **~ drum** Siebtrommel *f* **~ frame experiment** Zugversuch *m* **~ funnel** Seihetrichter *m* **~ press** Seiherpresse *f* **~ pulley** Spannwinde *f* **~ ring** Spannring *m* **~-sill** Spannschwelle *f* **~ speed** Zerreißgeschwindigkeit *f* **~ tie** Abspannschiene *f*

**strait** Landenge *f* **~(s)** Meerenge *f*

**straiten, to ~** beengen

**straits** Enge *f*, Meeresenge *f*

**strand, to ~** flechten, (together) verdrallen, verdrillen, verlitzen, verseilen

**strand** (of cable or rope) Drahtstrang *n*, (of cable) Ducht *f*, Faser *f*, (testing) Kaliberstecksteckgerüst *n*, (of cable, etc) Kardeel *n*, (of a belt) Riementrum *n*, Seil *n*, Strand *m*, Strang *m*, Trum *n*, Vorkaliberwalzgerüst *n* **~ of cable** Seillitze *f* **~ of chain** Kettentrumm *n* **~ of rolls** Walzenstrecke *f* **~ of rope** Seilstrang *m*

**strand wire** Litze *f*

**stranded** gelitzt, gestrandet, verdrallt, (cable) verseilt **~ aluminum wire** Aluminiumseil *n* **~ cable** Litzenleitung *f* **~ conductor coil** Litzenspule *f* **~ copper** Kupferseil *n* **~ wire** Aderlitzbündel *n*, Drahtlitze *f*, Drahtseil *n*, gelitzter Draht *m*, Litze *f*, Litzendraht *m*, unterteilter Draht *m* **~ wire cable** vielsträhniges Drahtkabel *n* **~ wire coil** Litzenspule *f* **~-wire line** verkabelte Leitung *f* **~ wire rope** geflochtenes Drahtseil *n* **~-wire stay** Drahtanker *m*, Drahtseilanker *m*

**stranding** Verlitzung *f*, (of wires or leads) Verseilung *f* **~ machine** Litzen-, Verlitz-maschine *f* **~ roll** Kaliberstreckwalze *f*, Vorkaliberwalze *f* **~-up machine** Verseilmaschine *f*

**strands of rope (or wire)** Adernbündel *n*

**strange** fremd, merkwürdig, seltsam, sonderbar

**strangeness** Eigentümlichkeit *f*, Fremdheit *f*, Seltsamkeit *f*

**strangle, to ~** erdrosseln

**strangler** Gemischtausgleichvorrichtung *f*, Luftklappe *f* des Vergasers

**strap, to ~** schnallen **to ~ oneself in** sich anschnallen (aviat.) **to ~ to** anschnallen **to ~ together** durch Stege *pl* verbinden

**strap** Band *n*, Bügel *m*, (for steering lock) Fangband *n*, Gurt *m*, Kopfhörerbügel *m*, Länge *f*, Pleuelstangenknopf *m*, Pratze *f*, Riemen *m*, Spannband *n*, Steg *m*, Streifen *m*, Struppe *f*, Verbindungsstück *n*, Ziehband *n* **~ of the limber rail** Gabelriegelband *n*

**strap, ~ adaptor** Schiebeschnalle *f* **~ bolt** Bügelschraube *f* **~ brake** Bandbremse *f* **~ coil** Spiralspule *f* **~ cold shears** Bügelkaltschere *f* **~ freaks** Koppelleitungunterbrechung *f* **~ hinge** Bandscharnier *n* **~ iron** Bandeisen *n* **~ joint** Laschenstoß *m* **~ rail** Flachschiene *f* **~ saw** Bandsäge *f* **~-wrap-machine** (forming press) Steck-Form-Maschine *f*

**strapped** (together) durch einen Bügel *m* verbunden **~ jack spur** Anschnallsporn *m*

**strapping** Blankverdrahtung *f* **~ (of a multiple-cavity magnetron)** Koppelleitung *f* (eines Vielfachmagnetrons) **~ switch** Umhängeschalter *m*

**straps on eccentrics** Bügel *pl* auf Exzenterkörper

**strass** Straß *m*

**strata** Gebirgsschicht *f* **in ~** schichtweise **~ group** Schichtengruppe *f* **~ measuring** Stufenmessung *f* **~ spring** Schichtquelle *f*

**stratagem** List *f*, Trick *m*
**stratameter** Stratameter *n*
**strategic** operativ ~ **map** Generalstabskarte *f*
**strategical** strategisch
**stratification** Blattschichtung *f*, Lagerung *f*
(geol.), Schichtung *f* ~ **of gases** Schichtung *f* der
Abgase ~ **parameter** Schichtungsgröße *f* ~
**plane** Schicht(ungs)fläche *f*
**stratified** geschichtet, schichten-förmig, -weise
~ **body** Schichtkörper *m* ~ **lattice** Schicht(en)-
gitter *n* ~ **rock** Schichtgestein *n* ~ **structure**
Schichtenbau *m*, Schicht-körper *m*, -textur *f*
~ **texture** Schichttextur *f*
**stratiform** schichtenweise
**stratify, to** ~ abschichten, anschichten, aufschichten, einschichten, schichten, stratifizieren
**stratigraphic,** ~ **age** stratigrafisches Alter *n* ~
**hiatus** Schichtlücke *f*
**stratigraphical succession** stratigrafische Folge *f*
**stratigraphy** Stratigrafie *f*
**strato-cumulus cloud** geschichtete Haufenwolke
*f*, Schichtwolke *f*
**stratoplane** Höhenflugzeug *n*
**stratosphere** atmosphärische Schicht *f*, Stick-
stoffsphäre *f*, Stratosphäre *f* ~ **airship** Raum-
luftschiff *n* ~ **engine** Stratosphärenmotor *m* ~
**flight** Stratosphärenflug *m* ~ **plane** Strato-
sphärenflugzeug *n*
**stratum** Ablagerung *f*, Bank *f*, Flöz *n*, Fluh *f*,
Gefüge *n*, Gesteinsschicht *f*, Lage *f*, Lager *n*, Luftschicht *f*, Schicht *f*, Schichtglied *n*, Steinschicht
*f* ~ **of constant temperature** wärmebeständige
Schicht *f* ~ **of warm air** warme Luftschicht *f*
**stratum length** Schichtlänge *f*
**stratus (cloud)** Schichtwolke *f* ~ **cloud** Nebelwolke *f*, Stratuswolke *f*
**straw, to** ~ mit Stroh beflechten
**straw** Stroh *n* ~ **for bedding** Lagerstroh *n*
**straw,** ~ **board** Preßspan *m* ~ **boards** Packpappe
*f* ~ **casings** Strohhülsen *pl* ~ **collector and**
**dump** Strohsammler *m* (mit Abkippvorrich-
tung) ~ **cutter** Hächselbank *f* ~ **file** Strohfeile *f*
~ **knife** Futterklinge *f* ~ **mat** Strohmatte *f* ~
**oil** Waschöl *n* ~**-plait** Strohgeflecht *n* ~ **pulp**
Strohstoff *m* ~**-rope making machine** Strohseil-
spinnmaschine *f* ~ **sheaf** Dachschaube *f*
~**-splitter** Strohspalter *m* ~**-spreader attach-**
**ment** Strohstreuereinrichtung *f*, Strohstreu-
apparat *m*, -vorrichtung *f* ~ **stacker** Strohauf-
schichter *m* ~**-yellow** strohgelb
**strawy grain** strohige Narben *pl*
**stray, to** ~ streuen
**stray** atmosphärische Störung *f*, Streuung *f*, (of
side radiations) Zipfel *m* ~ **capacity** elektrostatische Streuung *f*, kapazitive Nebenkopplung *f*, Streukapazität *f*, (of wiring) verteilte
Kapazität *f* ~ **coupling** Streukopplung *f*, Verkopplung *f*, wilde Kopplung ~ **current** Ab-
leitungsstrom *m*, Erdstrom *m* herumirrender
Strom *m*, Irrstrom *m*, Streustrom *m*, vagabundierender Strom *m* ~ **discharge** Nebenentladung *f* ~ **effect** Einstreuung *f* ~ **field** Stör-,
Streu-feld *n* ~ **flux** Streufluß *m* ~ **inductance**
Streu-induktivität *f*, -kapazität *f* ~ **light** falsches Licht *n*, Streulicht *n* ~ **line** Vorläufer *m* ~
**line pulses** Störimpulse *pl* ~ **losses** Zusatzver-
luste *pl* ~ **neutron** Streuneutron *n* ~ **oscillation**
durch Nebenkopplung hervorgerufene Schwin-

gung *f*, Schwingung *f* durch Nebenkopplung
~ **radiation** flasche Strahlung *f*, Seitenstrahlung *f*, Streustrahlung *f* ~ **rays** flasche Strahlung *f* ~ **shot** Ausreißer *m* ~ **voltage** Streu-
spannung *f*
**strays** atmosphärische Störungen *pl*, Luftstörungen *pl*, Nebengeräusch *n* (rdo), Seetriften *pl*,
Störbefreiung *f*, Störung *f*, Störung *f* der
elektromagnetischen Welle
**streak, to** ~ adern, streifen, stricheln
**streak** Plankengang *m*, Schliere *f*, Streifen *m*, Strich
*m*, Zeile *f* ~ **lightning** Linien- oder Gabelblitz *m*
**streaked** bandstreifig, gemasert, gestreift, maserig, streifig ~ **curlwood** Maserholz *n*
**streakiness** Streifigkeit *f*
**streaking** Maserung *f*
**streaks** (Farben) Notenlinien *pl*, Schlieren *pl*
(photo) ~ **of mist** Nebelschwaden *m*
**streaky** aderig, geädert, schlierig, streifig
**stream, to** ~ abfließen, entströmen, strömen **to**
~ **in** einströmen, zuströmen **to** ~ **paravanes**
Minenabweiser *pl* auslassen **to** ~ **through**
durchströmen
**stream** Bach *m*, Fluß *m*, Gewässer *n*, Strahl *m*,
Strom *m*, Strömung *f*, Wasserlauf *m* **on** ~ in
produktivem Arbeitsgang *m* ~ **of abrasive**
Sandstrahl *m* ~ **of electrons** Elektronenstrom
*m* ~ **of ions** Ionenstrom *m* ~ **of liquid** Flüssigkeitsstrom *m* **the** ~ **of liquid is interrupted** der
Flüssigkeitsstrom reißt ab ~ **of metal** Eisen-
strom *m* ~ **of water** Wasserstrom *m*
**stream,** ~ **function** Stromfunktion *f*, Strömungspotential *n* ~ **jig** Stromsetzmaschine *f*
**streamline, to** ~ stromlinienförmig ausbilden,
windschnittig verkleiden
**streamline** Bahn *f* der Flüssigkeitsteilchen,
Strom-faden *m*, -linie *f* ~ **of discontinuity**
Unstetigkeitsstromlinie *f*
**streamline,** ~ **filter** Spalt(ebenenkristall)filter *m*
~ **flow** plastisches Fließen *n*, Stromlinienfluß
*m* ~ **resistance of a ship** Formwiderstand *m* ~
**sectiontube** Tropfenrohr *n* ~ **shape** Strom-
linienform *f* ~ **wire** Stromliniendraht *m*
**streamlined** aerodynamisch, luftschnittig, profiliert, strom-linienförmig, -liniert, windschlüp-
fig, -schnittig ~ **blocking body** Zeppelinver-
drängungskörper *m* ~ **body** Stromlinienaufbau
*m* ~ **cable** Stromlinienkabel *n* ~ **flow** gleichförmige Strömung *f* ~ **fuselage** stromlinienförmiger Rumpf *m* ~**-section strut** Profilstrebe
*f* ~**-section tube** Profilrohr *n* (aviat.) ~ **shearing**
Abscherfestigkeit *f* ~ **shell** C-Geschoß *n* ~
**tire** Stromlinienreifen *m* ~ **tube** Stromlinien-
rohr *n* ~ **wire** Profildraht *m*
**streamlining of the wheel** Radverkleidung *f*
**stream,** ~ **ore** Seifenerz *n* ~ **sheet** Stromblatt *n*
~ **thread** Stromstrich *m* ~ **tin** Seifenzinn *n*,
Stromzinn *n*, Zinnseife *f* ~ **tube** Stromröhre *f*
~ **ward** wasserwärts ~ **washer** Stromsetzmaschine *f* (hydraul.)
**streamer** Ausläufer *m*, Leuchtfaden *m*, (in-
discharge) Stromfaden *m*, Wimpel *m* ~
**discharge** strahlartige Entladung *f* ~ **onset**
Leuchtfadeneinsatz *m*
**streamers of aurora** Nordlichtdraperie *f*
**streaming,** ~ **calorimeter** Strömungskalorimeter
*n* ~ **flow** strömender Abfluß *m* ~ **potential**
Strömungspotential *n*

street Gasse *f*, Straße *f* ~ **alinement** Straßen-
flucht *f* ~ **barricade** Straßensperre *f* ~ **car**
Straßenbahn *f*, Straßenbahnwagen *m*, Tram-
wagen *m* ~ **car rail** Straßenbahnschiene *f* ~
**car stop** Straßenbahnhaltestelle *f* ~ **cleaning**
Straßenreinigung *f* ~ **cleanser** Straßenkehrmaschine *f* ~ **construction** Straßenbaustelle *f*
~**-drainage inlet** Straßeneinlauf *m* ~ **level**
Straßen-höhe *f*, -niveau *n*, -planum *n* ~
**lighting** Straßenbeleuchtung *f* ~ **macadam**
Straßenschotter *m* ~ **manhole** Kabelschacht *m*
für Fahrbahn ~ **steam tractor** Dampfstraßen-
zugmaschine *f*
**strength** Anreicherungsgrad *m*, (of a unit) Bestand *m*, Bruch-festigkeit *f*, -grenze *f*, Derbheit
*f*, Festigkeit *f*, Gehalt *m*, Kraft *f*, Kräfte *pl*,
Mächtigkeit *f*, Stärke *f*, Stoßfestigkeit *f*,
(materials) Widerstand *m*, Widerstands-fähigkeit *f*, -kraft *f*, Wirkungsgrad *m*, Zähfestigkeit *f*
**strength,** ~ **under alternating torsion stress** Ver-
drehungswechselfestigkeit *f* ~ **of cover** Dek-
kungsstärke *f* ~ **of electric field** elektrische
Feldstärke *f* ~ **of field** Feldintensität *f*, Feldstärke *f* **the** ~ **guarantees a safe working** die
Festigkeit reicht aus ~ **of illumination of a lens**
Lichtstärke *f* eines Objektives ~ **of leaf** Stärke
*f* des Blattes ~ **of materials** Festigkeit *f* der
Stoffe, (science of the) Festigkeitslehre *f* ~ **of**
**poles** Polstärke *f* ~ **of roof** Deckenstärke *f* ~ **of**
**shell** Stärke *f* des Blattes ~ **of solution** Lösungsstärke *f* ~ **of a unit** Kopfstärke *f* ~ **of wood**
Festigkeit *f* des Holzes
**strength,** ~ **coefficient** Festigkeitszahl *f* ~
**factors** Festigkeitswerte *pl* ~ **group** Stärkeklasse *f* ~ **hull** Stärkerumpf *m* ~ **hypothesis**
Festigkeitshypothese *f* ~ **modulus** Festigkeits-
modulus *m* ~ **report** Nachweisung *f* ~ **test**
Festigkeits-prüfung *f*, -untersuchung *f*, Kraft-
probe *f* ~ **tester** Festigkeitsprüfer *m* ~ **-testing**
**machine** Festigkeitsprüfmaschine *f* ~ **-weight**
**ratio** Kraft- und Gewichtsverhältnis *n* ~ **weld**
Festigkeitsschweißung *f*, Festnaht *f*
**strengthen, to** ~ anreichern, aussteifen, befestigen, bekräftigen, bewehren, erstarken,
stärken, verfestigen, verstärken, versteifen
**strengthened** verstärkt
**strengthening** Stärkung *f*, Verfestigung *f*, Verstärkung *f*, Versteifung *f* ~ **rib** Verstärkungs-
rippe *f*
**stress, to** ~ beanspruchen, dehnen, Nachdruck
legen auf **to** ~ **the importance of** nahe legen
**stress** Anspannung *f*, Anstrengung *f*, Beanspruchung *f*, Belastung *f*, Deformation *f*, Dehnung
*f*, Gewicht *n*, Kraft *f*, Kraftäußerung *f*, Spannung *f*, Stärke *f*, Zopf *m*, Zug *m* ~ **in bending**
Knickbiegung *f* ~ **of firing** Beschußbeanspru-
chung *f* ~ **on the masonry** Beanspruchung *f*
(einer Mauer)
**stress,** ~ **analysis** Druckanalyse *f*, Festantennen-
lehre *f*, Festigkeits-lehre *f*, -nachweis *m*, -rech-
nung *f*, -untersuchung *n*, Kräftebestimmung *f*,
statische Berechnung *f* ~ **analyst** Druckanaly-
tiker *m* ~ **anneal** durch Ausglühen *n* spannungsfrei machen ~**-application cycle** Lastwechsel *m* ~ **bi-refringence** Spannungsdoppel-
brechung *f* ~ **calculation** Festigkeitsberechnung *f* ~ **characteristic** Festigkeitskurve *f* ~
**concentration** Druckansammlung *f*, Span-

nungsanhäufung *f* ~**-concentration factor**
Formfaktor *m*, Spannungsanhäufungsbeiwert
*m* ~**-concentration index** Kerbempfindlichkeit *f* ~ **conic** Spannungskegelschnitt *m* ~
**core** (cable) Wickelkeule *f* ~ **cycle** Spannungswechsel *m* ~**-cycle diagram** Wählerschaubild *n*
~**-deformation diagram** Spannungsdehnungsdiagramm *m* ~ **diagram** Kräfteplan *m*, Kraftzug *m* ~ **dispersion capacity** Druckverteilungs-
vermögen *n* ~ **distribution** Druckverlauf *m*,
Druckverteilung *f*, Spannungsverteilung *f* ~
**factor** Sicherheitszahl *f* ~ **increase** Spannungssteigerung *f* ~ **intensity** Spannungsgröße *f* ~
**lamp** Betonungsanzeiger *m* (acoust.) ~ **mark**
Betonungszeichen *n* ~ **mean** Spannungsmittel-
wert *m* ~**-momentum-tensor** Spannungs-Im-
puls-Tensor *m* ~ **pattern** Streifensystem *n* ~
**quadric** quadratische Form *f* des Spannungs-
tensors ~ **raiser** Spannungserhöher *m* ~ **relief**
Spannungsentlastung *f* ~**-relief heat-treatment**
spannungsfreies Glühen *n* ~**-relieving an-
neal(ing)** Entspannungsglühen *n* ~**-relieving**
**lift** Entspannungsverhieb *m* ~ **reversal** Spannungswechsel *m*, Wechselbeanspruchung *f*
~**-strain analysis** statische Kontrolle *f* ~**-strain**
**curve** Dehnungslinie *f* ~**-strain deformation**
**curve** Schubformänderungskurve *f* ~**-strain**
**diagram** Feinmeßdiagramm *n*, Spannungs-
dehnungs-diagramm *n*, -schaubild *n*, Zerreiß-
diagramm *n* ~**-strain ratio** Dehnung-Span-
nungs-Verhältnis *n* ~ **strain relations** Span-
nungs-Verformungs-Beziehungen *pl* ~**-strain**
**tester** Kraftdehnungsmesser *m* ~ **table** (wires)
Belastungstabelle *f* ~ **tensor** Spannungstensor
*m* ~ **tester** Kraftprüfer *m* ~ **trace** Ermüdungs-
spur *f* ~ **trajectory** Spannungstrajektorie *f*
**stressed** (structures) tragend **with** ~ **skin** tragende Außenhaut *f* ~ **skin** mittragende Beplankung *f*, tragende Haut *f* ~**-skin constructed**
glattblechbeplankt ~**-skin construction** Schalen-
konstruktion *f*
**stressing** Beanspruchung *f*
**stretch, to** ~ anspannen, aufspannen, ausdehnen, ausrecken, ausspannen, ausstrecken, ausweiten, ausziehen, dehnen, erstrecken, recken,
spannen, straffen, strecken, weiten, zainen,
ziehen **to** ~ **a belt** einen Riemen *m* spannen **to**
~ **cold** kalt recken **to** ~ **(in a frame)** einspannen **to** ~ **out iron** das Eisen abbreiten **to** ~ **a**
**line** aufstrecken **to** ~ **out of its natural shape**
verstrecken **to** ~ **tightly** festspannen **to** ~ **too**
**far** übergreifen **to** ~ **a wire** aufstrecken
**stretch** Abschnitt *m*, Anspannung *f*, (paper)
Bruchdehnung *f*, elastische Dehnung *f*, Strecke
*f*, Wegstrecke *f*, Zug *m* ~ **(ing)** Dehnung *f* ~ **of**
**a belt** Riemendehnung *f* ~ **of country** Geländestreifen *m*, Streifen *m* ~ **of line** Leitungsstück *n*
**stretch,** ~**-board** Spannbrett *n* ~**-forming** Streckziehen *n* ~**-rolled** gestreckt gewalzt ~ **spinning**
Streckspinnen *n* ~ **test** Reckprobe *f* ~**-wrap**
**forming press** Streckformmaschine *f*
**stretchability** Verstreckbarkeit *f*
**stretched** gespannt, gestreckt ~ **cord** ausgespannte Saite *f* ~ **length** gestreckte Länge *f* ~ **specimen**
gereckte Probe *f* ~ **string** ausgespannte Saite *f*
**stretcher** Bahre *f*, Kranken-bahre *f*, -trage *f*,
Läufer *m*, Reckbank *f*, Spannapparat *m*,
Strecker *m*, Streck-maschine *f*, -mittel *n*, -vor-

richtung *f*, Tragbahre *f*, Trage *f* ~ **bond** Läufer-
verband *m* ~ **course** Läuferlage *f* ~ **footboard**
(in boat) Fußlatte *f* ~ **leveler** Streck-, Tief-
ziehmaschine *f* ~ **lines** Fließfiguren *pl* ~ **mule**
Grobstuhl *m* ~ **rod** Zungen-angriffstange *f*,
-verbindungsstange *f* ~ **strain** Ziehriefe *f*
**stretching** Ausdehnung *f*, Einspannen *n*, Festspannen *n*, Recken *n*, Spannen *n*, Strecken *n*,
Streckung *f*, Streckziehen *n* ~ **block** Spann-
bock *m* ~ **bolt** Spannschloß *n* ~ **course**
Streckerschicht *f* ~ **curve** Längungskurve *f* ~
**device** Spannvorrichtung *f* ~ **effect** Spannwirkung *f* ~ **force** Streckkraft *f* ~ **hammer**
Spannhammer *m* ~ **machine** Spann-, Streck-
maschine *f* ~ **mule** Vorspinnmaschine *f* ~
**power** Reckkraft *f* ~ **rod** Verbindungsstange *f*
~ **rolls** Reckwalzwerk *n* ~ **screw** Spannschloß
*n*, Zugwinde *f* ~ **strain** Zugbeanspruchung *f*
~-**strain limit** Fließgrenze *f* ~ **tensor** Strek-
kungstensor *m* ~ **vibration** Valenzschwingung *f*
~ **wire** Abspanndraht *m*
**strew, to** ~ aufstreuen, bestreuen, einstreuen,
streuen
**strewing arrangement** Streuvorrichtung *f*
**stria** Schramme *f*, Streifen *m*, Streifung *f*,
Strich *m*
**striae** Glasschlieren *pl*, (of glass) Schlieren *pl*
~ **method** Schlierenmethode *f*
**striate, to** ~ riefeln, streifen
**striated** geschichtet, gestreift, streifig ~ **pattern**
Streifenmuster *n* (TV) ~ **rock pavements** ge-
kritzte Geschiebe *n*
**striation** Schichtung *f*, Streifenbildung *f*, Streifung *f* ~ **technique** Schlierentechnik *f* (film)
**strickle, to** ~ abstreichen, schablonieren
**strickle** Abstreichplatte *f*, Schablone *f*, Streichmodel *m*
**strickler** Flachstreicher *m*
**strict** genau, straff, streng
**striction** Striktion *f*
**strictness** Straffheit *f*
**stricture** Verengung *f* ~ **of volume** Dynamikverengung *f* ~ **coil** Striktionsspule *f*
**stride** Doppelschritt *m*, langer Schritt *m*
**strident** klimpernd, scharf, schneidend ~
**timbre** helle Klangfarbe *f*
**striding level** Aufsatzlibelle *f*
**strike, to** ~ anprallen, anschlagen, anschlagen
gegen, (an arc) ansetzen, ansprechen, (an arc)
anzünden, Arbeit einstellen, aufbumsen, (projectile) aufschlagen, auftreffen, berühren,
(Sicherung) durchbrennen, (lightning) einschlagen, hauen, klappen, prägen, rühren,
schlagen, streichen, streiken, treffen **to** ~
**against** anstoßen **to** ~ **the average** havarieren
**to** ~ **back** zurückschlagen **to** ~ **a balance** Bi-
lanz ziehen **to** ~ **the bell** glasen **to** ~ **centers** den
Lehrbogen *m* abschnüren **to** ~ **a flag** eine
Flagge *f* streichen **to** ~ **from** herausschlagen
**to** ~ **the ground** aufprallen **to** ~ **a key** eine
Taste *f* anschlagen oder niederbrechen **to** ~ **a**
**mine** auf eine Mine *f* laufen **to** ~ **off** abschlagen,
abstreichen **to** ~ **out** ausstreichen **to** ~ **a reef**
auf eine Klippe *f* stoßen **to** ~ **off a sheet** einen
Korrekturbogen *m* abziehen **to** ~ **a spark**
einen Funken *m* reißen **to** ~ **out a thing** einen
Strich *m* durch etwas machen **to** ~ **the topmast**
die Stenge *f* fieren

**strike** (molding) Abgleichplatte *f*, Abstreicher *m*,
(molding) Abstreichlatte *f*, Arbeits-einstellung
*f*, -niederlegung *f*, Ausstand *m*, Niederschlag *m*,
(of beds) Richtung *f* der Schichten, (the trend)
Streichen *n*, Streichen *n* einer Schicht (geol.),
Streichen *n* der Schichten (geol.) Streik *m*,
Sud *m* **to be out on** ~ ausständig sein (Arbeiter)
~ **of melis** Melissud *m* ~ **of thick juice** Dick-
saftsud *m*
**strike,** ~ **back** Rückprall *m* ~ **breaker** Streik-
brecher *m* · ~ **direction** Streichrichtung *f* ~
**fault** Längsverwerfung *f* ~ **lever** Klöppelhebel
*m* ~ **measure** Abstreichmaß *n*, Abstrichmaß *n*
~ **note** Schlagton *m*
**striker** Anschläger *m*, Bolzen *m*, Hammerbär *m*,
Mitnehmer *m*, Schlagbolzen *m*, Stößel *m*,
Zuschläger *m* ~ **bar** Anschlagschiene *f*, Mit-
nehmerschiene *f* ~ **clutch** Schaltgabel *f* ~ **fork**
Schaltgabel *f*, Verschiebegabel *f* ~ **nut** Schlag-
bolzenmutter *f* ~ **spring** Schlag(bolzen)feder *f*
~ **spring stop** Gegenlager *n* für Schlagbolzen
**striking** Anschlag *f* (des Weckers), (fuse) Ansprechen *n*, (of electrons) Aufprall *m*, Durchbrennen *n*, Rührung *f*, Zündung *f*; markant,
schlagend, treffend ~ **(of an arc)** Zündung *f*
(eines Lichtbogens) ~ **by surprise** schlagartig
**striking,** ~ **cap** Streichkappe *f* ~ **carbon** Zünd-
sohle *f* ~ **circuit** Schlagstromkreis *m* ~ **deposi-**
**tion potential** Niederschlagspotential *n* ~ **disc-**
**type horn** Aufschlagtellerhorn *n* (electr.) ~
**distance** Luftstrecke *f*, (of a gap) Schlagweite *f*
~ **edge** Kardätsche *f*, Schlag-finne *f*, -schneide
*f* ~ **effect** (typewriter) Anschlagwirkung *f* ~
**energy** Schlagarbeit *f* ~ **energy absorbed** ver-
brauchte Schlagarbeit *f* ~-**extinction potential**
**difference** Zündlöschspannungsdifferenz *f* ~
**face of a hammer** Hammerfinne *f* ~ **force**
Schlagkraft *f* ~ **hammer** Schlagbär *m* ~ **lever**
Schlaghebel *m* ~ **line** (piano) Anschlaglinie *f*
~ **mechanism** Schlagwerk *n* ~ **movement**
schlagende Bewegung *f* ~ **parts** aufeinander
arbeitende Teile *pl* ~ **piece** Prallstück *n* ~
**point** Abschmelzstromstärke *f*, Anschlagstelle
*f*, (fuse) Ansprechstromstärke *f*, Einschlagstelle *f* ~ **potential** Angriffskraft *f*, Zündspannung *f*, Zündungspotential *n* ~ **rod** Zündstab *m* ~ **roller** Einzahnstift *m* (film) ~ **shade**
ausdrucksvolle Färbung *f* ~ **surface** Schlagfläche *f* ~ **velocity** Aufschlag-, Auftreff-, End-,
Schlag-geschwindigkeit *f* ~ **voltage** Zündspannung *f* ~ **wheel** Schlagrad *n*
**string, to** ~ (wires) ausspannen, schlingen,
(wires) verlegen, (wires) ziehen **to** ~ **a pin**
einen Stecknadelkopf *m* aufspießen **to** ~
**pipes together** Rohre *pl* aneinanderreihen **to** ~
**to** anreihen **to** ~ **up** anspinnen **to** ~ **a wire**
einen Draht ziehen
**string** Bindfaden *m*, Faden *m*, Kette *f* (info proc.),
Koppel *f*, Kordel *f*, Litze *f*, Saite *f*, Schnur *f*,
(fingerboard) Spielsaite *f* **to lay the** ~ **pieces on**
**the top of earth piles** ein Pfahlwerk *n* verholmen
**string,** ~ **of beads** Perlschnur *f* ~ **of cartridges**
Patronenreihe *f* ~ **of casing** Bohrkolonne *f*,
Bohrröhren-kolonne *f*, -zug *m*, Futterröhren-
zug *m*, Verrohrung *f* ~ **of casing tubes** Futter-
röhrenstrang *m* ~ **of conductors** Führungs-
röhrenzug *m* ~ **of drill pipes** Gestängezug *m* ~
**of insulators** Isolatorenkette *f* ~ **of ore** Erz-

schnur *f* ~ **of pearls** Perlschnur *f* ~ **of poles** Stangenzug *m* ~ **of tools** Werkzeuggehänge *n*
**string,** ~ **approximation** Saitennäherung *f* ~ **discharge** Schnürenabnahme *f* ~ **drive** Schnurantrieb *m* ~ **electrometer** Saiten-, Schleifenelektrometer *n* ~ **fitter** Saitenaufspanner *m* ~ **galvanometer** Faden-, Saiten-, Seiten-galvanometer *n* ~ **instrument** Streichinstrument *n* ~ **oscillograph** Saiten-, Schleifen-oszillograf *m* ~ **polygon** Seileck *n* ~ **problem** Fadenproblem *n* ~-**proof boiling** Verkochen *n* auf Faden ~-**proof test** Fadenprobe *f* ~-**shadow instrument** Fadenschatteninstrument *n* ~ **tension** Saitenzug *m* ~ **wall** Treppengeländer *n*, Wangenmauer *f*
**stringer** Erz- oder Öader *f*, Gleisbalken *m*, Längs-träger *m*, -versteifung *f*, mittlerer Rumpfholm *m*, Ortbalken *m*, Petroleumvorkommen *n* (min.), Pfette *f*, Streckbalken *m*, Stringer *m*, Tragbalken *m*, (of a chain) Trum *n*, Verlängerung *f* (Gurtförderung), Verschlußstück *n*, Wangenmauer *f* (mach.) ~ **leads** Erztrümmer *pl*, schneidende Gängchen *pl*
**stringers** schneidende Gängchen *pl*
**stringing** Ziehen *n*
**strings** Band *n* ~ **of reads** geschlossene Gruppen *pl* von Lesevorgängen ~ **for removing** Schnürenbespannung *f*
**stringy** zähe ~ **root** Wurzelbrand *m*
**strioscopic method** Schlierenmethode *f*
**strip, to** ~ (teeth) abbrechen, abmontieren, (a mast) abtakeln, abziehen, auseinandernehmen, von der Umhüllung befreien, demontieren, durchziehen, entblößen, streifen, strippen, wegfüllen (min.), wegziehen, zerlegen **to** ~ **bark from wood** abborken **to** ~ **fiber** Fäden *pl* abdecken **to** ~ **off** ablösen, abstreifen, (as roofs) abdecken, putzen **to** ~ **the thread** das Gewinde überdrehen
**strip** Band *n*, Leiste *f*, (for forming tubes) Rohrstreifen *m*, Schiene *f*, Splint *m*, Stoßband *n*, Streifen *m*, Verjüngung *f* ~ **of answering jacks** Abfrageklinkenstreifen *m* ~ **of elastic** Gummizug *m* ~ **of fuses** Sicherungsleiste *f* ~ **of ground** Geländestreifen *m* ~ **with holes** Lochstreifen *m* ~ **of home jacks** Abfrageklinkenstreifen *m* ~ **of indicators** Klappenstreifen *m* ~ **of jacks** Klinkenstreifen *m* ~ **of keys** Tastenstreifen *m* ~ **of lead** Bleistreifen *m* ~ **of paper** Papierstreifen *m* ~ **of successive photographs** Bildstreifen *m* aufeinanderfolgender Aufnahmen ~ **of wood** Holzleiste *f*
**strip,** ~-**and-blind-stall packing** (stowage) Blindortsversatz *m* ~ **attenuator** Bandfernabschwächer *m* ~ **board** Latte *f* ~ **cell** Leistenzelle *f* ~ **chart** Meßstreifen *m* ~-**chart recorder** Linienschreiber *m* ~ **coating** abstreifbarer Überzug *m* ~ **condition** Streifenbedingung *f* ~ **cutter** Streifenschneider *m* ~ **cutting machine** Streifen-abschneidemaschine *f*, -schere *f* ~ **feed** Streifenmagazinzuführung *f* ~ **finder** Streifensucher *m* ~ **flooring** Parkettstäbe *pl* ~ **foundation** Streifenfundament *n* ~ **frequency** Zeilenfrequenz *f* ~ **fuse** Streifensicherung *f* ~ **heater** Heizband *n* ~ **heating member** Streifenheizkörper *m* ~ **helix** Bandwendel *f* ~ **ink chart recorder** Linienschreiber *m* ~ **interval** Streifenabstand *m* ~ **iron** Bandeisen *n* ~ **lamp** Soffitenlampe *f* ~

**light** Leuchtschild *n* ~ **line** Mikrostrip *m*, Streifenleiter *m* ~ **linen** Riementuch *n* ~ **load** Linienbelastung *f*, Streifenlast *f* ~ **locking** Bandverschluß *m* ~ **map** Streifenkarte *f*, Wegskizze *f* ~ **materials** bandförmige Materialien *pl* ~ **mill** Bandstahl-straße *f*, -walzwerk *n* ~ **(rolling) mill** Streifenwalzwerk *n* ~ **(sheet) mill** Bandstahlwalzwerk *n* ~ **mill train** Streifenstraße *f* ~ **milling** Bandwalzerei *f* ~ **mosaic** Bildreihe *f* ~-**mounted set** auswechselbarer Gerätesatz *m* ~ **packing** Rippenversatz *m* ~ **penetrometer** Härtemesser *m* mit Platte ~ **photography** Streifenaufnahme *f* ~ **pit** Tabebaubetrieb *m* ~ **sealing** Streifendichtung *f* ~ **shear** Streifenschere *f* ~ **spoiler** Keulenverformer *m* ~ **(steel) sheet mill** Streifenblechwalzwerk *n* ~ **steel** Bandstahl *m*, Streifenblech *n* ~ **stitcher** Bandheftmaschine *f* ~ **stock** Blechstreifen *m* ~ **stowing** Rippenversatz *m* ~ **survey** Streifenaufnahme *f* ~ **suspension** Bandaufhängung *f* ~ **test** Streifenprobe *f* ~ **theory** Streifentheorie *f* ~ **weld** Laschenstoß *m* ~ **width** Zeilenbreite *f* ~ **winder** Bandwickler ~ **winding machine** Bandwickelmaschine *f*
**stripe, to** ~ streifen
**stripe** Streifen *m* ~ **from the raising-gig** Rauhstreifen *m*
**stripe,** ~ **machine** Ringelmaschine *f* ~ **pattern** Streifenmuster *n* ~ **signal** Randstreifensignal *n* (TV) ~ **signal generator** Randstreifengenerator *m* (TV) ~ **slitting device** Streifentrennvorrichtung *f*
**striped** gestreift, streifig ~ **goods** gestreiftes Zeug *n* ~ **jasper** Bandjaspis *m* ~ **material** gestreiftes Zeug *n* ~ **pattern** Ringelmuster *n* ~ **pedestrian crossing** Zebra-Fußgängerübergang *m*
**striping chain** Ringelkette *f* (textiles)
**striplike,** ~ **coil** flächenhafte Spule *f* ~ **mirror** Lamellenspiegel *m*
**strippable** abstreifbar
**stripped** (off) abgeschält, nackt, streifig ~ **atom** elektronenberaubtes oder nacktes Atom *n* ~ **emulsion stack** trägerfreies Emulsionspaket *n*
**stripper** (ingot) Abstreifer *m*, Abstreifformmaschine *f*, Stripper *m* ~ **crane** Abstreifrahmen *m*, Blockabstreifkran *m*, Stripperkran *m* ~ **gantry** Strippertorkran *m* ~ **guides** Abstreifmeißel *m* ~ **lever** Ausklinkhebel *m* ~ **plate mold** Abstreifform *f* ~ **ram** Stripperkolben *m* ~ **roller** Abstreichwalze *f* ~ **stop** Abstreifanschlag *m* (um das Arbeitsstück vom Steigen mit dem Bohrer zu verhüten) ~ **tongs** Abstreiferzange *f* ~ **well** Sonde *f* mit geringer Produktion
**stripping** Abräumen *n*, Abstreifen *n*, Abziehen *n*, Auseinanderschichten *n* ~ **of alcohol** Entgeistung *f* ~ **with explosives** Abraumsprengung *f* ~ **the forms** Ausschalen *n*
**stripping,** ~ **agent** Abziehmittel *n* ~ **bath** Abziehbad *n* (Entfärbungsbad) ~ **comb** Schabevorrichtung *f* ~ **crane** Abstreifkran *m* ~ **device for rolling mills** Abschiebevorrichtung *f* für Walzwerke ~ **drum** Rektifizierkolonne *f* ~ **edge** Abstreifleiste *f* ~ **fork** Abstreifer *m* ~ **frame** Abstreifrahmen *m* ~ **machine** Enthülsungsmaschine *f* ~ **plate** Abstreif-kamm *m*, -platte *f*, Durchzieh-, Durchzugs-platte *f* ~-**plate** Durchzieh-, Durchzug-formmaschine *f* ~-**plate molding machine** Abstreifformma-

schine *f* ~ **process** Durchziehverfahren *n* ~
**roller** Abzieh-, Putz-walze *f* ~ **room** Abstreif-
raum *m* ~ **shovel** Seilbagger *m* ~ **squeezer**
Preßformmaschine *f* mit Abstreifvorrichtung
~ **table** Abstreiftisch *m* ~ **tongs** Abstreifpin-
zette *f,* Isolationsabziehzange *f* ~ **tower**
Rektifizierkolonne *f* ~ **work** Wegfüllarbeit *f*
**strips** Bandstahl *m,* Streifenmaterial *n*
**strive, to** ~ sich bestreben, streben **to** ~ **for**
zielen auf
**strobe** Markierungsfenster *n,* Schwelle *f* (rdr)
**strobing** Signalauswertung *f* (rdr) ~ **circuit**
Modulationskreis *m* ~ **pulse** Auswerteimpuls *m*
**stroboscope** Stroboskop *n,* stroboskopische
Scheibe *f*
**stroboscopic,** ~ **(al)** stroboskopisch ~ **aberration**
stroboskopischer Fehler *m* ~ **disc (or pattern)**
**wheel** stroboskopische Scheibe *f*
**stroke, to** ~ (magnet) bestreichen, (a string) streichen **to** ~ **over** überstreichen **to** ~ **up** auffeuern
**stroke** Anschlag *m,* (e. g. of a telescopic shock
absorber) Einfederung *f,* Hub *m,* Kolbenhub *m,*
Ruderschlag *m,* Schlag *m,* Stoß *m,* Strich *m,*
(of engine) Takt *m,* Ziehlänge *f,* Zug *m* ~ **of bit**
Meißelschlag *m* ~ **of the carriage** Länge *f* der
Schlittenbewegung ~ **of crank** Kurbelhub *m* ~
**of a flash** Schlag *m* eines Blitzes ~ **of pen** Federzug *m,* -strich *m* ~ **of spindle** Spindelhub *m*
**stroke,** ~ **adjustment device** Hubstellvorrichtung
*f* ~**-arresting device** Hubbegrenzeranschlag *m*
~ **compensator** Stoßausgleicher *m* ~**-connect-**
**ing-rod ratio** Pleuelstangenverhältnis *n* ~ **cure**
Härtezeittest *m* ~ **decrease** Hubverminderung *f*
~ **index** Hublängenskala *f* ~ **limiter** Hubbegrenzung *f* ~ **method** Strichmethode *f* ~ **oar**
Achterriemen *m* ~ **operating device** Hubschalt-
werk *n* ~**-order** Schlagreihenfolge *f* (des
Blitzes) ~ **plate** Hubscheibe *f* ~ **rate change**
**gears** Hubzahlenwechselräder *pl* ~ **ratio** Hubverhältnis *n* ~**-to-bore ratio** Hubverhältnis *n*
zwischen Hub und Bohrung ~ **valve drill**
Ventilstoßbohrer *m* ~ **volume** Hubraum *m,*
Hubvolumen *n* ~ **wheel** Ausstreichrad *n*
**strokes** Anschläge *pl*
**stroking** (a string) Anstreichen *n* ~ **over** Überstreichen *n*
**stromeyerite** Kupfersilberglanz *m,* Silberkupfer-
glanz *m*
**strong** dauerhaft, derb, fest, gewaltig, haltbar,
kräftig, mächtig, massiv, stabil, standfest,
stark ~ **alloy** vergütbare Recklegierung *f* ~ **box**
Panzerschrank *m* ~ **breeze** steife Brise *f* ~
**coupling** feste Kopplung *f* ~ **current** Starkstrom
*m* ~ **down-current** Sturzwind *m* ~**-jack grooving**
**iron** Schroppeisen *n* ~ **liming paste** Anschwöde-
brei *m* ~ **point** Anklammerungs-, Kern-,
Schwer-punkt *m* ~ **wind** starker Wind *m* ~
**wrapping paper** Kaschier *n*
**strontia** Strontianerde *f*
**strontianite** Strontianit *m*
**strontium** Strontium *n* ~ **acetate** Strontium-
azetat *n* ~ **carbide** Strontiumcarbid *n* ~
**chloride** Chlorstrontium *n* ~ **content** Stron-
tiumgehalt *m* ~ **ferricyanide** Strontium-
ferrizyanid *n* ~ **hydrate** Strontianhydrat *n* ~
**hydride** Strontiumwasserstoff *m* ~ **iodide**
Strontiumjodid *n* ~ **nitrate** Strontiansalpeter
*m,* Strontium-nitrat *n,* -salpeter *m* ~ **perborate**

**Strontiumperborat** *n* ~ **salicylate** Strontium-
salz *n* ~ **sulfate** Strontiumsulfat *n* ~ **sulfide**
Strontiumsulfid *n*
**strop, to** ~ (mit dem Ölstein) abziehen, be-
stroppen, (blade) am Streichriemen abziehen
**strop** Strick *m* ~ **insulator** isolierendes Seil *n*
**stropping** Ziehschleifen *n*
**Strowger selector (or switch)** Hebdrehwähler *m,*
Strowgerwähler *m*
**struck** geschlagen ~ **joint pointing** abgeschrägte
Fuge *f*
**structural** aufbauend, baulich, strukturell ~
**abnormality** Gefügenormalität *f* ~ **adhesive**
Montageleim *m* ~ **alloy steels** legierte Bau-
stähle *pl* ~ **alteration** Umbau *m* (Gebäude *n*) ~
**analysis** statische Berechnung *f* ~ **arrangement**
Gefügeanordnung *f* ~ **built-up beam** Fachwerkträger *m* ~ **casting** Bauguß *m* ~ **change**
Gefüge-änderung *f,* -veränderung *f* ~ **com-**
**position** Gefügeaufbau *m* ~ **constituent** Ge-
füge-, Struktur-bestandteil *m* ~ **constitution**
Gefügeaufbau *m* ~ **data** bauliche Angaben *pl*
~ **defect** Konstruktionsfehler *m* ~ **design**
strukturelle Zeichnung *f* ~ **dimensions** Baumaß
*n* ~ **drag** Rumpfwiderstand *m,* schädlicher oder
struktureller Widerstand *m* ~ **element** Bauelement *n* ~ **elements** strukturelle Teile *pl* ~
**engineering** Bautechnik *f* ~ **equation** Bildungsgleichung *f* ~ **failure** organische Schwäche *f*
strukturelles Versagen *n* ~ **fatigue** Struktur-
müdung *f* ~ **features** Aussehen *n* des Gefüges ~
**formula** Formelbild *n,* Wertigkeitsformel *f* ~
**fracture** Bruchgefüge *n* ~ **frame** Baugerüst *n* ~
**girder** Profileisenträger *m* ~ **hollow tile** Hohlblockstein *m* ~ **iron** Baueisen *n,* Flußeisen *n,*
(steel) Profileisen *n* ~**-ironwork contractor**
Eisenbauunternehmer *m* ~ **joint** Gerippunkt *m*
~ **material** Baustoff *m* ~ **member** Bauteil *m,*
Bauwerksteil *n,* tragender oder Kräfte aufnehmender Bauteil *m* ~ **member subject to**
**highest stresses** stark beanspruchter Konstruktionsteil *m* ~ **members** strukturelle Bestandteile
*pl* ~ **mill** Formstahlwalzwerk *n* ~ **part** Bauteil
*n,* Bauwerksteil *n* ~ **part subject to highest**
**stresses** stark beanspruchter Konstruktionsteil
*m* ~ **pivot** Gerippunkt *m* ~ **products sections**
Profileisen *n* ~**-return loss** Rückflußdämpfung
*f* ~ **rolled-steel member** Profileisen *n* ~ **shape**
Formeisen *n,* Konstruktionsform *f* ~ **shape**
**section** Fassoneisen *n* ~ **shapes** Formprofileisen *n,* Konstruktionsprofil *n,* Profileisen *n* ~
**stability** Gefügebeständigkeit *f* ~ **steel** Baustahl
*m,* Fassoneisen *m,* Form-eisen *n,* -stahl *m,*
Konstruktions-eisen *n,* -stahl *m,* Maschinenbaustahl *m,* Stahlkonstruktion *f* ~**-steel rolling**
**mill** Profilstahlwalzwerk *n* ~**-steel shape**
Formeisenprofil *n* ~ **steelwork** Stahlhochbau *m*
~**-steel works** Eisenkonstruktionswerkstätte *f*
~ **strength of pressurized cabin** Überdruck-
festigkeit *f* ~ **tongs** Nietzange *f* ~ **transforma-**
**tion** Gefüge-neubildung *f,* -umwandlung *f* ~
**unit** Bauglied *n* ~ **weight** Bau-, Konstruktions-
gewicht *n* ~ **work below ground level** Tiefbau *m*
~ **wrench** Gabelschlüssel *m*
**structurally dual network** Dualglied *n*
**structure** Aufbau *m,* Bau *m,* Bauart *f,* Bauwerk
*n,* Bildung *f,* Fachwerk *n,* Gebäude *n,* Gebilde
*n,* Gefüge *n,* geologische Beschaffenheit *f,*

Gerippebau *m*, (supporting) Gerüst *n*, Gerüstbau *m*, Gestaltung *f*, Gestänge *n*, Gliederung *f*, Grundgefüge *n*, Konstruktion *f*, (man-made) Kunstbau *m*, Struktur *f* ~ of cold-worked metal Struktur *f* der Kaltreckung ~ of electron shell Schalenbau *m* der Elektronenhülle ~ with entire mass of vibratory system contained in oil körperlose Spule *f* ~ of the filter sheets Schichtenstruktur *f* ~ of fuselage Aufbau *m* des Rumpfes ~ above the ground Hochgerüst *n* (für Tanks) ~ of hair Haargewebe *n* ~ for mooring pier Festmachepfahl *m* ~ for strengthening roofs for standards Sprengwerk *n* any ~ above the surface Hochbau *m* ~ of the transition zone Übergangsgefüge *n*

structure, ~ element Bauglied *n* ~ examinations Gefügeuntersuchungen *pl* ~ factor Strukturfaktor *m* ~ joint Gerippepunkt *m* ~-loading coefficient Wagnerische Kennzahl *f* ~ pivot Gerippepunkt *m* ~ sensitive strukturabhängig, strukturempfindlich

structureless gefügelos, strukturlos

structures Bauten *pl*

struggle, to ~ sich anstrengen, kämpfen, ringen, sich sträuben, streben to ~ against ankämpfen

strum Sauge-korb *m*, -sieb *n*

strut, to ~ abspreizen, absprießen, absteifen, abstützen, aussteifen, ausstreben, verspreizen, versteifen, verstreben

strut Druckstab *m*, Holm *m*, Säule *f*, Spreize *f*, Spreizholz *n*, Steife *f*, Stempel *m*, Stiel *m*, Strebe *f*, Versteifung *f*, Verstrebung *f* ~ (ing) Absteifung *f* ~-attachment fitting Strebenknotenstück *n* ~-attachment ring Strebenanschlußring *m* ~ beam Strebestütze *f* ~-braced verstrebt ~-braced wing verstrebte Zelle *f* ~ bracing Strebenfachwerk *n* ~ camera Spreizenkamera *f* ~ deflection Federbeinweg *m* ~ distribution Strebenteilung *f* ~ fairing Strebenebene *f* (aviat.), Strebenverkleidung *f* ~ fitting Strebenschuh *m* ~ frame Strebenfachwerk *n* ~-frame bridge Sprengbrücke *f* ~ joint Stielknotenpunkt *m* ~ nodal point Stielknotenpunkt *m* ~ palm (of a propeller shaft) Wellenbockbefestigung *f* ~ pile Druckpfahl *m* ~ repartition Strebenteilung *f* ~ section Strebenquerschnitt *m* ~ slope Strebenneigung *f* ~ socket Holmschuh *m* ~ spacing Strebenabstand *m* ~ top Strebenkopf *m* ~-type radiator Strebenkühler *m*

strutted, ~ pole Abspannmast *m*, Stange *f* mit Strebe, verstrebte Stange *f* ~ terminal pole Endgestänge *n* ~ wing verstrebter Flügel *m*

strutting Verspreizung *f*, Verstreben *n*, Verstrebung *f* ~ of poles Stangen *pl* mit Stützen

stub Stutzen *m* ~ antenna Stabantenne *f* ~ axle Vorderachschenkel *m* ~ ball yoke Schweißzapfenmitnehmer *m* ~ bar Kurzbalken *m* ~ boring bar Bohrstange *f* ~ cable Abzweig-, Einführungs-kabel *n* ~ card Abschnittkarte *f* ~ drill Spiralbohrer *m* (kurz eingespannt) ~ exhaust pipe Auspuffstutzen *m* ~ guy Luft-, Überweg-anker *m* ~ head Stangenkopf *m* ~ mast Stumpfmast *m* ~-matched antenna mit Stichleitung angepaßter Dipol *m* ~ plane Gleitflosse *f* ~ pole (wagon) Deichselstumpf *m* ~-reinforced pole angeschuhte Stange *f* ~ reinforcement Stützpfahl *m* ~ shaft Stummelwelle *f* ~ support Stichleitungsträger *m* ~

thread Stumpfgewinde *n* ~ tooth Stumpfzahn *m* ~ tooth cutter Stumpfzahnstahl *m* ~-tooth form Stumpfverzahnung *f* ~-tooth gearing Stumpfverzahnung *f* ~ tube (jet) Leitblech *n* ~ wing Ansatzflügel *m* ~ wing stabilizer Seitenflossenstabilisator *m*

stubbing plow Schälpflug *m*

stubble plow Stoppelpflug *m*

stubborn beharrlich, halsstarrig, hartnäckig, schwer schmelzbar, stockig, zähe

stucco Gips-mörtel *m*, -stuck *m*, Stuck *m*, Stuckmörtel *m*, Verputz *m* ~ work Stuckarbeit *f* ~ works Stuckwaren *pl*

stuck bewegungsunfähig to be ~ festsitzen to get ~ sich festlaufen

stud, to ~ bespicken

stud (piano) Agraffe *f*, Anschlag *m*, Döckchen *n*, Dübel *m*, Knopf *m*, Kontakt *m*, (contact) Kontaktstück *n*, Nase *f*, Pimpel *m*, Ständer *m*, Stehbolzen *m*, Stiel *m*, Stift *m*, Stiftschraube *f*, Stutzen *m*, Warze *f*; stumpf ~ for ball bearing Rollenbolzen *m*

stud, ~ block Schlüssel *m* zum Einziehen von Stiftschrauben ~ bolt (fixed) Gewindestift *m*, Stift-bolzen *m*, -schraube *f* ~ chain Gallkette *f* ~ chuck Zapfenklemme *f* ~ connection Knopfverbindung *f* ~ driver Schlüssel *m* zum Einziehen von Stiftschrauben ~ fastener Drehwirbel *m*, Planenwirbel *m* ~ lathe Stehbolzendrehbank *f* ~-link cable (or chain) Stegkette *f* ~ nut Bolzenmutter *f* ~ pin Kettensteg *m* ~ stave Wagenrunge *f* ~ washer Zapfenscheibe *f*

studded bestiftet, mit Zähnen *pl* versehen ~ disc Zahnscheibe *f* ~-disc discharger Zahnscheibenfunkenstrecke *f* ~ link Kettenglied *n*

studding Bestiften *n*

student Schüler *m*, Student *m*

studio Atelier *n*, Aufnahmeraum *m*, Tonatelier *n* ~ broadcast Studiosendung *f* ~ camera Atelierkamera *f* ~ copy Arbeitskopie *f* ~ light Effektlampe *f* ~ lights Aufnahmelampen *pl* ~ manager Atelierdirektor *m* ~ pick-up Studiosendung *f* ~ print Arbeitskopie *f* ~ shot Innenaufnahme *f* ~ still Werkfoto *n* ~ transmission Studiosendung *f*

studs Metallknöpfe *pl*

studtite Studtit *m*

study, to ~ forschen, lernen, studieren, untersuchen

study Beobachtung *f*, Forschung *f*, Lernen *n*, Studium *n*, Untersuchung *f* ~ of total efficiency Griff-Forschung *f* ~ group Arbeitsgemeinschaft *f*

stuff, to ~ abdichten, ausstopfen, fetten, füllen, polstern, stopfen to ~ hot (Fett) einbrennen

stuff Degras *n*, Feinzeug *n*, Grundstoff *m*, Stoff *m*, taubes Gestein *n*, Zeug *n* ~ chest Ganzzeugkasten *m* ~ grinder Zerfaserer *m*

stuffed (seam) wulstig ~ full vollgepfropft

stuffer hydraulische Strangpresse *f*

stuffing Dichtung *f*, Füllung *f*, Packung *f*, Polster *n* ~ box Stoff-, Stopf-büchse *f* ~ box for casting Bohrrohrstopfbüchse *f* ~ box for hollow rods Gestängestopfbüchse *f* ~ box for water-flush system Spülbohrkopf *m* ~ box bush with external threading Stopfbuchsenverschraubung *f* ~-box cover Stopfbüchsendeckel *m* ~-box guide Stopfbuchsenführung *f* ~-box packing Stopfbüchsen-dichtung *f*, -einsatz *m*

~-box pipe Stopfbüchsenrohr *n* ~-box stud
Stopfbüchsenschraube *f* ~ cone Dichtdeckel *m*
~ drum Schmierwalkfaß *n* ~ material Füllmittel *n*
stull (in den Firstenbauten) Abgestemme *n*,
(min.) Sparren *m*
stulls Kastenzimmerung *f*
stumble, to ~ fehltreten, stolpern, straucheln
stumble Fehltritt *m*
stump, to ~ cloth stumpen
stump Stock *m*, Stubbe *f*, Stubben *m*, Stummel
*m*, Stumpf *m*, Trumm *n*, Wischer *m*
stumpy stockig
stung gestochen
stunt, to ~ verkrüppeln
stunt, ~ flight Kunstflug *m* ~ flying Kunstfliegen *n*, -flug *m*
stupp Stupp *f*
sturdily built robust gebaut
sturdiness Festigkeit *f*
sturdy derb, gedrungen, handfest, klemmig
(min.), kräftig, robust, stabil, standfest, stark,
stetig ~ design Unverwüstlichkeit *f*
style Art *f*, Ausführung *f*, Darstellungsweise *f*,
Fasson *f*, Fassung *f*, Ritzstift *m*, Schick *m*,
Schreib-griffel *m*, -stift *m*, Stichel *m*, Stil *m*,
Zeichenstift *m* ~ of dressing Tracht *f* ~ of
printing Ätzdruck *m*
style number Sortenbezeichnung *f*
stylometer Säulenmesser *m*
stylus Kopfspitze *f* (tape rec.), Meißel *m*, Ritzstift *m*, Schreibfeder *f*, Schreibstift *m*, Stichel *m*,
Zeichenstift *m* ~ collet Schablonenstiftklem-
mung *f* ~ drag Rückstellkraft *f* der Nadel ~
force Auflagedruck *m* (tape rec.) ~-groove
resonance Rille *f* und Nadelresonanz *f* ~-re-
corder and stylus-reproduced sound Nadelton *m*
~ recorder Ritzgerät *n* ~ tip Fühlerschneide *f*,
Nadelspitze *f* ~ trajectory Nadelbewegung *f*
styptic blutstillendes Mittel *n*; blutstillend ~
cotton blutstillende Watte *f*
styrende Styrol
sub, ~ accounts heading Nebenkostenstellen *pl*
~ acetate basisch essigsaures Salz *n* ~ acoustic
speed Unterschallgeschwindigkeit *f* ~ aeration
Subaeration *f* ~ aerial subaerisch ~ aerially in
Luft *f* ~ aqueous cable Unterwasserkabel *n* ~
aqueous rock breaker Unterwasserfelsbrecher *m*
subassembly Montageuntergruppe *f*, sekundäre
Zusammenstellung *f*, Teilmontage *f*, Teilzu-
sammenbau *m* ~ of structural parts Bauteilgruppe *f*
subaudible unterhörfrequent
subaudio infraakustisch, unterhörfrequent ~
frequency Unterhörfrequenz *f* ~ telegraphy Unterlagerungstelegrafie *f* ~ wave Unterschallwelle *f*
sub, ~ base Grundplatte *f* ~-basement Souter-
rain *n* ~ bituminous coal Moorkohle *f*, schwarzer Lignit *m* ~-block Teilblock *m* ~-carrier
Hilfsträger *m*, Zwischenträger *m* ~ channel
Teilkanal *m* ~ chromate basisch-chromsaures
Salz *n* ~ claim Unteranspruch *m* ~ claim of a
patent Patentunteranspruch *m* ~ cloud car
unter die Wolken herabsenkbare Gondel *f* ~
committee Unterausschuß *m* ~ compartment
Unterabteilung *f* ~ contract Untervertrag *m* ~
contracting Vergebung *f*, Vergebung *f* von Auf-
trägen an Unterlieferanten ~ contractor Teil-
unternehmer *m*, Unterlieferant *m* ~-control

unit Unterzentrale *f* ~ cooler Nachkühler *m* ~
cooling Unterkühlung *f* ~ cost head Kosten-
unterstelle *f* ~-critical mass unterkritische
Masse *f* ~ critical state of flow unterkritischer
Strömungszustand *m*
subdivide, to ~ einteilen, unterabteilen, unterteilen
subdivided eingeteilt, unterteilt (finely) ~ core
(fein) unterteilter Kern *m*
subdivision Einteilung *f*, Unterteilung *f* ~ of a
component Teilspule *f* ~ of length Wegabschnitt
*m* ~ of plant Netzunterteilung *f* (electr.)
subdrift contour Aussehen *n* des Gebirges unter
den Anschwemmungen
subdue, to ~ dämpfen, überwältigen
subdued geknickt
suberic acid Korksäure *f*
suberite Suberit *m*
sub, ~ exchange Fernsprechunteramt *n*, Unteramt *n*, Untervermittlungsstelle *f* ~ face Sohl-
fläche *f* ~-floor Untergrund *m* ~ fluvial cable
Flußkabel *n* ~ formant Unterformant *m* ~
frame Hilfsrahmen *m*, Unter-, Zwischenrahmen *m* ~-frequency dialling Unterlagerungs-
fernwahl *f* ~ frequency oscillation subharmonische Unterschwingung *f* ~ glacial unter-
glazial ~ grade Planum *n*, Unterbau *m*, Untergrund *m* ~ grain formation Subkornbildung *f*
~ grain growth Feinkorngrenzenwanderung *f*
~ group Teilgruppe *f*, Untergrund *m*, Untergruppe *f* ~ harmonic Subharmonische *f* ~
harmonic oscillation (or harmonics) subharmonische Unterschwingung *f* ~ head Neben-
titel *m* ~ inspector Einfahrer *m* (min.) ~ insurance Unterversicherung *f*
subjacent darunterliegend, unter-lagert, -liegend
subject, to ~ (to) unterwerfen to ~ to current
anspannen to ~ to gradual application of load
allmählich belasten to ~ to measurement bemessen to ~ to torsional force verwinden to ~
to torsional stress verdrehen
subject Angelegenheit *f*, (matter) Gegenstand *m*,
Objekt *n*, Staatsangehöriger *m*, Stoff *m*, Tatbestand *m*, Thema *n*, Vorwurf *m* (photo) ~ to
ausgesetzt, bedingt ~ to being sold freibleibend
~ to your immediate assent bei umgehender
Zusage *f* ~ to influence bestimmbar ~ to license
arrangements lizenzpflichtig ~ to registration
zulassungspflichtig ~ to stamp duty stempel-
pflichtig ~ to wage tax lohnsteuerpflichtig
subject, ~ index Sachregister *n*, Sachverzeichnis
*n*, Stichwörterverzeichnis *n* ~ matter Bildinhalt *m*, Thema *n*
subjected to thermoplastic treatment warmgepreßt
subjection Abhängigkeit *f*, Unterwerfung *f*
subjective einseitig, subjektiv ~ angular field
subjektives Gesichtsfeld *n* ~ chromatic scale
value empfindungsgerechte oder empfindungs-
gemäße Farbstufe *f* ~ field of view subjektives
Gesichtsfeld *n* ~ grading (chromatic scale)
empfindungs-gemäße Farbstufe *f*, -gerechte
Abstufung *f* ~ loudness subjektive Lautstärke *f*
~ sensation of light Fotopie *f*
sublattice Teil-, Unter-gitter *n*
sub-layer of gravel Kiesunterlageschicht *f*
sublet, to ~ submissionieren, untersubmissio-
nieren, untervermieten

**sublevel** Unter-niveau *n*, -stufe *f*
**sublimable** sublimierbar
**sublimate** Sublimat *n*
**sublimates** Beschläge *pl*
**sublimation** Sublimation *f*, Verflüchtigung *f* (fester Körper) ~ **nuclei** Sublimationskerne *pl*
**sublime, to** ~ sublimieren, treiben
**submachine gun** Maschinenpistole *f*
**submarine** Tauchboot *n*, U-Boot *n*, Unterseeboot *n*; untermeerisch, (flat, bank, shoal) unterseeisch ~ **bell** Unterwasserglocke *f* ~ **cable** Ozeankabel *n*, Seekabel *n*, Überseekabel *n*, unterseeisches Kabel *n*, Unterwasserkabel *n* ~ **chaser** U-Bootjäger *m* ~ **detector** Unterwasserabhörapparat *m* ~ **direction-finding set** Tauchpeilanlage *f* ~ **echo sounder** Atlas-Hochperioden-Echolot *n* ~ **force** Unterseebootsstreitkräfte *pl* ~ **listening device** Unterwasserhorchgerät *n* ~ **oscillator** Wasserschallsender *m* ~ **rescue vessel** U-Bootrettungsschiff *n* ~ **signal** Unterwasser-schallsignal *n*, -zeichen *n* ~ **sound receiver** Unterwasserschallempfänger *m* ~ **sound signal** Unterwasser-schallzeichen *n*, -signal *n* ~ **sound wave** Unterwasserschallwelle *f* **the** ~ **surfaces** das U-Boot taucht auf ~ **telephone cable** Seefernsprechkabel *n*
**submerge, to** ~ eintauchen, tauchen, überschwemmen, untertauchen, versenken to ~ **entirely (or partially)** hineintauchen **to** ~ **a village** ein Dorf überstauen
**submergeable roller gate** versenkbare Walze *f*
**submerged** ersoffen (min.), untergetaucht ~ **arc** verdeckter Lichtbogen *m* ~ **bearing** Unterwasserlager *n* ~ **body** eingetauchter Körper *m* ~ **coil-type cooler** Schlangenkühler *m* ~ **discharge** Ausfluß *m* unter Wasser ~ **float** versenkter Schwimmer *m* ~ **installation** versenkter Einbau *m* ~ **object** eingetauchter Körper *m* ~ **plunger** versenkter Schwimmer *m* ~ **speed** Unterwassergeschwindigkeit *f* ~ **tubbing** schwimmende Cuvelage *f* ~ **tube** Unterwasserrohr *n* ~ **unit weight** Raumgewicht *n* unter Wasser ~ **weir** Grundwehr *n*, unvollkommener Überfall *m*
**submergence** Eintauchung *f*, Senkung *f*, Überschwemmung *f*
**submergible taintor gate dam** Sektorwehr *n*
**submerging** Eintauchen *n*
**submersible** tauchfähig ~ **inspection lamp** Untersäurelampe *f* ~ **roller** Versenkwalze *f*
**submersion** Untertauchung *f* ~-**proof** tauchsicher
**submetallic** metallartig ~ **luster** Halbmetallglanz *m*
**submicrometer** Submikrometer *n*
**sub-miniature,** ~ **channel gyroscope** Subminiaturkreisel *m* ~ **component** Kleinstbauelement *n*
**submission** Ausschreibung *f*, Submission *f*, Unter-ordnung *f*, -werfung *f*, Verdingung *f*
**submissive** demütig, gehorsam
**submit, to** ~ einsenden, unterbreiten, vorlegen **to** ~ **to** herhalten, hinnehmen, sich unterziehen **to** ~ **to an expert** begutachten lassen
**submitting an accounting** Rechnungslegung *f*
**submultiple** Faktor *m* einer Zahl, in einer Zahl ohne Rest aufgehender Teiler *m*, ~ **resonance** ganzzahlige vielfache oder untersynchrone Resonanz *f*, Unterresonanzfrequenz *f*
**submultiplication** Herabsetzung *f*, Teilung *f*, Unterteilung *f*

**subnormal hearing** Schwerhörigkeit *f*
**suboffice** Hilfsamt *n*, Unteramt *n*
**subordinate, to** ~ untergeben, unterordnen, unterstellen **to be** ~ **to** unterstehen, unterstellen
**subordinate** Untergebener *m*; untergeordnet ~ **chain** Nebenkette *f* ~ **group** Untergruppe *f* ~ **patent** Nebenpatent *n* ~ **proposition** Nebensatz *m*
**subordinated** unterlagert
**subordination** Unterordnung *f*, Unterstellung *f*
**suboxide** Oxydul *n*
**subpatent** Neben-, Unter-patent *n*
**subpoena, to** ~ vorladen
**subpoena** Vorladung *f*
**subpressure** Unterdruck *m*
**sub-procurement authorization** Einzelbeschaffungsgenehmigung *f*
**subproduct** Teilprodukt *n*
**subprogram** Teilprogramm *n*
**sub-punch, to** ~ vorstanzen (auf einem kleineren Durchmesser)
**subrefraction** Infrabrechung *f*
**subresonances** Unterresonanzen *pl*
**subrounded** abgekantet
**subroutine** Unterprogramm *n* (info proc.)
**subsample** Teilprobe *f*
**subscribe, to** ~ abonnieren, subskribieren, unterschreiben, eine Unterschrift leisten
**subscribed capital** gezeichnetes Kapital *n*
**subscriber** Abonnent *m*, Inhaber *m* eines Fernsprechanschlusses oder Hauptanschlusses ~ **in arrears** (of payment) Gebührenschuldner *m* (Teilnehmer mit der Gebührenzahlung im Rückstand) ~ **having extension station** Inhaber *m* eines Nebenanschlusses, Nebenanschlußteilnehmer *m* ~ **with several lines** Anschlußteilnehmer *m*, Inhaber *m* mehrerer Anschlüsse, Mehrfachanschlußteilnehmer *m*, Sammelanschlußteilnehmer *m*
**subscriber,** ~'**s account** Teilnehmerrechnung *f* ~'**s annual rental** Jahresgebühr *f* für einen Fernsprechanschluß ~'**s annual rental for supplementary apparatus** Jahresgebühr *f* für Zusatzeinrichtungen ~'**s apparatus fault** Sprechstellenstörung *f* ~'**s box** Fernschaltgerät *n* ~'**s cable** Anschlußkabel *n*, Fernsprechanschlußkabel *n*, Ortskabel *n*, Teilnehmerkabel *n* ~'**s calling equipment** (in an exchange) Anrufeinheit *f*, Anruforgan *n* ~ **clears** Teilnehmer *m* hängt ein ~'**s contract** Teilnehmer-(rechts)verhältnis *n* ~ **dialing** Wählbetrieb *m* (teleph.) ~'**s extension** Nebenstelle *f* ~'**s extension station** (with direct exchange facilities) Nebenstelle *f* mit freier Amtswahl, Teilnehmernebenanschluß *m*, -stelle *f* ~ **installation with extension stations** Nebenstellenanlage *f* ~'**s jack** Teilnehmerklinke *f* ~'**s line** Amtsleitung *f*, Anruf-einheit *f*, -organ *n*, Anschlußleitung *f*, Hauptanschlußleitung *f*, Teilnehmerleitung *f* ~'**s line finder** erster Anrufsucher *m* ~'**s line indicator** Teilnehmeranrufzeichen *n* ~'**s line reserved for incoming (outgoing) calls** Teilnehmerleitung *f* für ankommenden (abgehenden) Verkehr ~'**s loop** Teilnehmerdoppelleitung *f*, -schleife *f* ~'**s main station** Hauptanschluß *m*, Teilnehmerhaupt-anschluß *m*, -stelle *f* ~'**s multiple** Klinkenfeld *n* (Teilnehmerklinken), Teilnehmervielfach(feld) *n* ~ **network** Ortsnetz *n* ~'**s number** Anschluß-

nummer *f*, Rufnummer *f* ~'s **rack** Teilnehmer-anschlußgestell *n* ~'s **set** Teilnehmer-apparat *m*, -sprechstelle *f* ~'s **station** Teilnehmeran-schluß *m* (teleph.), Teilnehmer-leitung *f*, -sprechstelle *f*, -station *f* ~'s **telephone** Teil-nehmersprechstelle *f*

**subscribing firm** Mitgliedsfirma *f*

**subscript** Index *m* (math.), tiefgestellter Index *m*, Tiefzahl *f*

**subscription** Abonnement *n*, Beitrag *m*, Fern-sprechanschluß *m*, Subskription *f* ~ **call** Monats-, Wochen-gespräch *n* ~ **form** Zeich-nungsformular *n* ~ **rate** Abonnementsgebühr *f* ~ **right** Bezugsrecht *n*

**subscripts (and superscripts)** Indices *pl*

**subsection** Unter-abschnitt *m*, -abteilung *f*, -kapitel *n*

**subsector** Unterabschnitt *m*

**subsequence** Teilfolge *f*

**subsequent** anschließend, nachträglich ~ **(ly)** sekundär ~ **adjustment of accounts** nachträg-liche Richtigstellung *f* der Rechnungen ~ **compensation** Nachausgleichung *f*, Nachkom-pensierung *f* ~ **grant** Nachbewilligung *f* ~ **stage** Folgestufe *f* ~ **treatment** Weiter-behand-lung *f*, -verarbeitung *f*

**subsequently,** ~ **added** nachgeschaltet ~ **worked (into)** weiterverarbeitet

**sub,** ~ **series** Teilreihe *f* ~ **servient** untergeordnet ~ **set** Teilmenge *f* ~ **shell** (of electrons) Elek-tronenunterhülle *f*, Unterhülle *f*, (of electrons) Unterschale *f*

**subside, to** ~ absenken, nachlassen, nachsacken, niedergehen, sacken, setzen, sich legen, sich setzen, zusammenbrechen

**subsidence** Absenkung *f*, Absinken *n*, Boden-einsenkung *f*, Sackung *f*, Senkung *f*, Setzmaß *n*, Sichsenken *n* ~ **of the earth** Erdsenkung *f* ~ **of the ground** Senkung *f* des Grundes ~ **of the substructure** Setzen *n* des Zwischenbaues

**subsidence,** ~ **earthquake** Einsturzbeben *n* ~ **ratio** Dämpfungsverhältnis *n*

**subsidiary** Filiale *f*, Tochtergesellschaft *f*; neben, untergeordnet, zusätzlich ~ **circuit** Hilfskreis *m* ~ **company** Tochtergesellschaft *f* ~ **conveyor** Nebenförderer *m* ~ **direction-finder station** Peilnebenstelle *f* ~ **distributing point** Unterver-teilerstelle *f* ~ **flow** Nebenströmung *f* ~ **gallery** Begleitstrecke (min.) ~ **group** Untergruppe *f* ~ **motion** Zusatzbewegung *f* ~ **orientation** Klein-orientierung *f* ~ **quantity** abgeleitete Größe *f* ~ **result** Nebenergebnis *n* ~ **road** Hilfs-, Neben-strecke *f* ~ **role** Nebenrolle *f* ~ **triangu-lation** Kleintriangulation *f* ~ **warehouse** Hilfs-lager *n*

**subsides** Subsidien *pl*

**subsiding** Sacken *n*, Senkung *f* ~ **tank** Absetz-becken *n*

**subsidize, to** ~ subventionieren, unterstützen

**subsidy** Beihilfe *f*, Beisteuer *f*, Subvention *f*

**subsilicate** Subsilikat *n*

**subsist** Abschlaglohn *m*

**subsistence** Auskommen *n*, Erwerbsmittel *n pl*, Lebensmittel *pl* ~ **money** Auslösung *f* ~ **supply** Verpflegungsmittel *n*

**subsoil** Unter-boden *m*, -grund *m* ~ **attachment** Untergrundlockerer *m*

**subsoiler** Untergrundpflug *m*

**subsoiling** Erschütterung *f* des Bodens, Unter-grundlockerung *f* ~ **plow** Untergrundpflug *m* ~ **-water map** Grundwasserkarte *f*

**subsolid** halbfester Körper *m*

**subsonic** Unterschall..., mit Unterschallge-schwindigkeit *f* ~ **blading** Unterschallbeschau-felung *f* ~ **compressible flow** kompressible Unterschallströmung *f* ~ **edge** Unterschall-kante *f* ~ **flow** Unterschallströmung *f* ~ **jet** Unter-schallstrahl *m* ~ **potential flow** Unterschall-Potentialströmung *f* ~ **velocity** Unterschallge-schwindigkeit *f* ~ **wind tunnel** Unterschallkanal *m*

**subspace** invarianter Unterraum *m*

**subspecies** Halbart *f*, Unterart *f*, Unterform *f*

**substage** Beleuchtungsapparat *m* ~ **lamp unit** Beleuchtungsuntersatz *m*

**substance** Bestandteil *m*, Gegenstand *m*, Inhalt *m*, Körper *m*, Masse *f*, Material *n*, Stoff *m*, Substanz *f*, Wesen *n* ~ **chains** Stoffketten *pl* **liquid** ~ Fließkörper *m*

**substandard,** ~ **film** Klein-, Schmal-film *m* ~ **-gauge railroad** Kleinbahn *f* ~ **refraction** unternormale Brechung *f*

**substantial** greifbar, stabil, standfest, stetig, weitgehend

**substantiality** Körperlichkeit *f*

**substantially** halbwegs ~ **synonymous** sinnähnlich

**substantiate, to** ~ begründen, tatsächliche Unterlagen *pl* liefern

**substantiation** (patents) weitere Begründung *f*, Bekräftigung *f*

**substantivity** Substantivität *f*

**substate** Unterzustand *m*

**substation** Hilfsunterwerk *n*, Teilnehmer-an-schluß *m*, -sprechstelle *f*, Unterwerk *n* ~ **of the telephone system** Anschluß *m* an das Fern-sprechnetz

**substation,** ~ **message register** Gebührenan-zeiger *m* ~ **plant** Teilnehmeranlage *f* ~ **switch-board** Nebenstellenumschalter *m* ~ **transformer** Speise-, Stations-transformator *m*

**substitute, to** ~ ausstatten, auswechseln, (a value) einsetzen, ersetzen, substituieren **to** ~ **a spare (ringing) generator** einen anderen Ruf-satz *m* einschalten

**substitute** Aushilfe *f*, Austausch *m*, Austausch-ware *f*, Ersatz *m*, Ersatz-lieferung *f*, -mann *m*, -mittel *n*, Surrogat *n*, Umstellwerkstoff *m*, Wechselstück *n* ~ **action** Ersatzhandlung *f* ~ **circuit** Umwegschaltung *f* ~ **driver** Ersatz-fahrer *m* ~ **form** Ersatzlieferschein *m* ~ **fuel** Ausweichkraftstoff *m* ~ **map** Ersatzkarte *f* ~ **material** Austausch(werk)stoff *m* ~ **part** Er-satzteil *m* ~ **reaction** Austauschreaktion *f* ~ **steel** Austauschstahl *m*

**substitution** Einsetzen *n*, Einsetzung *f*, Sub-stitution *f* ~ **method** Substitutions-methode *f*, -verfahren *n* ~ **process** Substitutionsverfahren *n* ~ **table** Tauschtafel *f*

**substitutional,** ~ **disorder** Fehlordnung *f* durch Fremdatome ~ **induction coil** Nebenschluß-drosselspule *f* ~ **rectifier** Ersatzgleichrichter *m*

**substrate** Substratfläche *f* ~ **heating** Objekthei-zung *f* ~ **surface** Substratoberfläche *f*

**substratosphere** Substratosphäre *f*

**substratum** gewachsener Baugrund *m*, Substrat *n*, Unter-grund *m*, -lage *f*, Unterguß *m* (photo)

**substructure** Fundament *n*, Unterbau *m*, Zwi-

schenbau *m* ~ **for railroads** Eisenbahnunterbau *m*
**subsurface** unter der Oberfläche *f* ~ **draining**
Untergrundabwässerung *f*
**subsynchronous resonance** untersynchrone Resonanz *f* ·
**subtangent** Subtangente *f*
**subtend, to** ~ gegenüberliegen (Seite und Winkel), sich hinziehen unter, unterspannen **to** ~
**an angle** einen Winkel *m* einschalten
**subtense,** ~ **bar** Latte *f*, Meßlatte *f* ~ **line**
Distanzstrich *m*
**subterfuge** Ausflucht *f*, Behelf *m*, Vorwand *m*
**subterranean** unterirdisch ~ **current** Grundwasserstrom *m* ~ **dugout** unterirdischer Unterstand *m* ~ **quarry** Berg(e)mühle *f*, Gesteinsmühle *f* ~ **shelter** unterirdischer Unterstand *m* ~ **steam** geophysische Energie *f* ~ **storage** unterirdische Einlagerung *f*
**subthreshold** unterwertig
**subtilization** Verflüchtigung *f*
**subtle** geschickt, scharfsinnig ~ **medium** dünnes Mittel *n*
**subtotal** Zwischensumme *f* ~ **key** Zwischensummentaste *f*
**subtract, to** ~ abrechnen, abziehen (math.), sub(s)trahieren
**subtraction** Abziehen *n*, Subtraktion *f* ~ **sign** Minuszeichen *n*
**subtractive** subtraktiv ~ **color system** subtraktives Farbensystem *n* ~ **primary** Primärfarben-Subtrahend *m* (TV)
**subtractor cipher** Zahlenwurm *m*
**subtrahend** Subtrahend(us) *m*
**sub-transient three-phase short-circuit current** subtransitorischer Kurzschlußwechselstrom *n*
**subtropical** subtropisch
**suburb** Vorort *m*, Vorstadt *f*
**suburban** vorstädtisch ~ **area** Vorortsgebiet *n* ~ **call** Vorortsgespräch *n* ~ **connection** Vorortsverbindung *f* ~ **junction** Vorortsleitung *f* ~ **position** Vorortsplatz *m* ~ **railway** Vorortbahn *f* ~ **service** Vorortsverkehr *m* ~ **settlement** Stadtrandsiedlung *f* ~ **switchboard** Vorortsschrank *m* ~ **traffic** Vorortsverkehr *m* ~ **train** Vorortzug *m*
**subvariety** Unter-art *f*, -form *f*
**subvolt, to** ~ unterspannen
**subway** Tunnel *m*, Tunnelgang *m*, Untergrundbahn *f* ~ **for heating and electric mains** Heizkanal *m* ~ **crossing** Unterführung *f*
**subzone,** ~ **network** Netzgruppe *f* ~ **toll office** Nebenfernamt *n*
**succeed, to** ~ folgen, nachfolgen, sukzedieren
**succeeding stage** nachgeschaltete Stufe *f*
**succession** Aufeinanderfolge *f*, Folge *f*, Reihe *f*, Reihenfolge *f*, senkrechte Reihe *f* **in** ~ der Reihenfolge nach
**succession,** ~ **of beds** Schichtenfolge *f* ~ **of cranks** Kurbelfolge *f* ~ **of cuttings** Hiebfolge *f* ~ **of impulses** Impuls-, Stromstoß-reihe *f* ~ **of passes** Kaliber-, Stich-folge *f* ~ **of phases** Phasenfolge *f* ~ **of sparks** Funkenfolge *f* ~ **of stages (or steps)** Stufenfolge *f* ~ **of strata** Schichtenfolge *f*
**successive** aufeinanderfolgend, fortlaufend ~ **appeal** (law) Instanzenzug *m* ~ **approximation** Iterationsverfahren *n* ~ **change-over** sukzessive Umstellung *f* ~ **cycles** aufeinanderfolgende

Gänge *pl* ~ **drawing** Weiterziehen *n* ~ **picture** Folgebild *n* ~ **reaction** Stufenreaktion *f*
**successively** nacheinander, der Reihe nach, stufenweise
**successor** Nachfolger *m*
**succinct** bündig
**succinic acid** Bernsteinsäure *f*, Subzinylsäure *f*
**succinite** Goldgranat *m*
**succumb, to** ~ unterliegen
**suck, to** ~ lutschen, nutschen, saugen, ziehen **to** ~ **in** ansaugen **to** ~ **in (or up)** einsaugen **to** ~ **off (or out)** absaugen, aussaugen **to** ~ **through** durchsaugen **to** ~ **up** aufsaugen
**sucked** gezogen ~ **off** abgesaugt
**sucker** Sauger *m* ~ **bar carrier** Saugstangenträger *m* ~ **bar lever** Saugstangenhebel *m* ~ **bar movement** Saugstangenbewegung *f* ~ **bar slide** Saugerschiene *f* ~ **fish** Schildfisch *m* ~ **rod** Pumpenstange *f* ~-**rod hanger (or jack)** Pumpengestängerechen *m* ~-**rod joint** Bohrstangenverbindung *f* ~-**rod steel** Bohrstangenstahl *m* ~ **rods** Sauggestänge *f*
**sucking** Ansaugen *n*, Saugung *f* ~ **off** Absaugung *f*
**sucking,** ~ **action** Saugwirkung *f* ~ **coil** Tauchkernspule *f* ~ **effect** Zug-(Sog)-Wirkung *f* ~ **jet pump** Saugstrahlpumpe *f* ~ **solenoid** Saug-(drossel)spule *f*, Tauchkernspule *f* ~ **table** Saugkasten *m* ~ **wick** Saugdocht *m*
**sucrose** Rohrzucker *m*, Saccharase *f*, Saccharose *f* ~ **octa-acetate** Azetylsaccharose *f* ~-**splitting enzyme** rohzuckerspaltendes Enzym *n*
**suction** Ansaug *m*, Aufsaugung *f*, Einströmung *f*, Saugen *n*, Saugung *f*, Saugwirkung *f*, (of shock wave resulting from an explosion or detonation) Sog *m* ~ **and forcing apparatus** Druckbehälter *m*, Vakuumkessel *m* **to take** ~ saugen (Pumpe)
**suction,** ~ **absorption** Ansaugung *f* ~ **adjusting screw** Ventilschraube *f* ~ **air current** Saugstrom *m* ~ **air piping** (Gasmaschine) Mischluftleitung ~ **apparatus** Nutschapparat *m*, Saugapparat *m*, Sauger *m* ~ **basket** Saugkopf *m* ~ **bell** Ansaugeglocke *f* ~ **bend** Saugkrümmer *m* ~ **blade filter** Saugrohrfilter *n* ~ **blank feed attachment** Teilansaugvorrichtung *f* ~ **bore** Saugbohrung *f* ~ **bottle** Saugflasche *f* ~ **box** Absauger *m*, Saugkasten *m* ~ **brush** Saugbürste *f* ~ **capacity** Ansaugleitung *f* ~ **cell filter** Saugzellenfilter *n* ~ **chamber** Saug-kammer *f*, -raum *m* ~ **cock** Saughahn *m* ~ **control spindle** Bolzen *m* zur Luftabstellung ~ **couch** Sauggutsch *m* ~ **couch roll** Sauggautsche *f* ~ **crucible** Saugtiegel *m* ~ **cup** Saugnäpfchen *m*, Tauchtasse *f* ~ **cupola furnace** Saugkuppelofen *m* ~ **cut-out lever** Anschlaghebel *m* (print.), Griffnocken *pl* ~ **device** Ansaugvorrichtung *f* ~ **dredge** Saug(e)-bagger *m* ~ **drum** Saugwalze *f* ~ **duct** Ansaugekanal *m* ~ **effect** saugender Ventilator *m*, Saugwirkung *f* ~ **end** Saugseite *f* ~ **fan** Sauggebläse *n* ~ **field** Absauge-, Saug-feld *n* ~ **filter** Nutsche *f*, Saugfilter *n* ~ **flange** Saugflansch *m* ~ **flask** Absauge-, Saug-kolben *m* ~ **force** Sogkraft *f* ~ **funnel** Saugtrichter *m* ~ **gas** Generatorgas *n* ~-**gas motor** Sauggasmotor *m* ~-**gas plant** Sauggasanlage *f* ~-**gas producer** gezogener Generator *m* ~ **gas producer** gezogener Gaserzeuger *m*, Unterdruckgaserzeuger *m* ~ **gauge** Saug-, Zug-messer *m* ~ **head**

(gas intake) Gaseinlaß *m*, Unterdruck *m*, Saughöhe *f*, Saugkopf *m* ~ **height** Saughöhe *f* ~ **hole** Saugbohrung *f* ~ **hood** Absauge-, Saughaube *f* ~ **hose** Saugkorb *m*, Saugschlauch *m* ~ **lift** Saughöhe *f* ~ **line** Saugleitung *f* ~ **main** Saugleitung *f* ~ **nose** Saugrüssel *m* ~ **nozzle** Einsaugdüse *f*, (gas intake) Gaseinlaß *m*, Saugdüse *f* ~ **pipe** Ansaugrohr *n*, Saugkopf *m*, (line) Saugleitung *f*, Saug(e)rohr *n* ~**-pipe connection** Saugestutzen *m*, Saugleitungsanschluß *m* ~ **pipe flange** Saugrohrflansch *m* ~ **pipe line** Saugrohrleitung *f* ~ **piping** Einlaßrohrleitung *f*, Saugrohrleitung *f* ~ **pit** Saugschacht *m* ~ **plant** Absauganlage *f* ~ **port** Ansaug-kanal *m*, (Motorzylinder)-öffnung *f*, Saugöffnung *f* ~ **position** Saugstellung *f* ~ **power** Zugwirkung *f* ~ **press roll** Saugpreßwalze *f* ~ **pressure** Saugdruck *m* ~**-pressure recorder** Saugdruckschreiber *m* ~ **(-gas) producer** Sauggaserzeuger *m* ~ **pump** Saugpumpe *f* ~ **receptacle** Saugwindkessel *m* ~ **regulator** Sogregler *m* ~ **resistant** saugfest ~ **rinsing head** Saugspülkopf *m* ~ **rods** Ansaug-, Einlaß-gestänge *n* ~ **rollers** Siebzylinder *m* ~ **scoop** Ansaughutze *f* ~ **screen** Saugsieb *m* (Ölpumpe) ~ **(cross) section** Ansaugquerschnitt *m* ~ **side of wing** Saugseite *f* des Flügels ~ **sieve** (Ölpumpe) Saugsieb *n* ~ **sleeve** Überrohr *n* ~ **slot** Saugschlitz *m* ~ **socket** Saugstutzen *m* (der Pumpe) ~ **strainer** Saugkopf *m*, (Pumpe) Saugkorb *m*, Seiher *m* ~ **strength** Saugfestigkeit *f* ~ **stroke** Ansauge-hub *m*, -takt *m*, Ansaugen *n*, Ansaughub *m*, -periode *f*, saugender Hub *m*, Saughub *m* ~ **stub** Saugstutzen *m* ~ **system** Saugsystem *n* ~**-system sandblast unit** Sandstrahlgebläse *n* nach dem Saugsystem ~ **tap** (Rohrpost) Sauganschluß *m* ~ **transformer** Saugtransformator *m* ~ **tube** Saugrohr *n* ~**-type carburetor** Saugvergaser *m* ~**-type hose sandblast tank** Saugfreistrahlgebläse *n* ~**-type sandblast cabinet** Sandstrahlsaugapparat *m* ~ **valve** Ansaugventil *n*, Saug-hahn *m*, -klappe *f*, -ventil *n* ~**-valve chamber** Ansaugeraum *m* ~**-valve control of fuel pumps** Saugventilregulierung *f* von Brennstoffpumpen ~ **vane** Saugflügel *m* ~ **ventilating fan** Exhaustventilator *m* ~ **ventilator** Luftablaß *m* ~ **wave** Absperrsunk *m*, Sunk *m* ~ **zone** Depressionszone *f*, Saugzone *f*

**suctorial disc** Haftscheibe *f*

**sudden** jäh, plötzlich, sprunghaft **in** ~ **bursts** schlagartig ~ **change** Umschlag *m* ~ **change of direction** scharfkantige Richtungsänderung *f* ~ **change of frequency** Frequenzsprung *m* ~ **change in program velocity** Programmsprung *m* (g/m) ~ **change of weather** Witterungsumschlag *m* ~ **change of wind** Windsprung *m* ~ **decline (or droop)** Sturz *m* ~ **gas absorption** plötzliche Gasaufzehrung *f* ~ **load** Stoßbelastung *f* ~ **(undesired) oscillating** plötzliches Schwingen *f* ~ **roll in the direction of flight** gerissene Rolle *f*

**suddenness** Plötzlichkeit *f*

**sudorific** schweißtreibend

**suds** Absud *m*, Schaum *m*

**sue, to** ~ gerichtlich verfolgen, klagen, verklagen **to** ~ **for a penalty** ausklagen (einer Geldstrafe)

**suède** schwedisches Leder *n*, Wildleder *n* ~ **finish** Feinkräusellack *m*

**suet rendering installation** Talgschmelzanlage *f*

**suffer, to** ~ dulden, erdulden, erleiden, ertragen, leiden, tragen **to** ~ **damage** Schaden *m* erleiden **to** ~ **hits** Treffer *pl* erhalten

**suffice, to** ~ auslangen, genügen, herhalten, reichen

**sufficiency** Auskommen *n*, Hinlänglichkeit *f*, Zulänglichkeit *f*

**sufficient** ausreichend, genug, genügend, hinlänglich **to be** ~ hinreichen

**sufficiently,** ~ **fine** ausreichend fein ~ **roasted ore** garer Rost *m* ~ **safe for investment of trust money** mündelsicher

**suffix** Hauptzeichen *n*

**suffocate, to** ~ ersticken

**suffocating** stickig ~**-vapor discharge** Schwadenabzug *m*

**suffocation** Erstickung *f*

**suffuse, to** ~ unterlaufen

**sugar, to** ~ auskristallisieren

**sugar** Zucker *m* ~ **analysis** Zuckerbestimmung *f* ~ **beet** Zuckerrübe *f* ~**-beet cossette** Zuckerschnitzel *n* ~ **breaker** Knotenbrecher *m* ~ **cane** Zuckerrohr *n* ~**-cane crusher** Zuckerrohrquetsche *f* ~**-cane mill** Zuckerrohr-mühle *f*, -walzwerk *n* ~**-cane mill machinery** Zuckerrohrmaschine *f* ~ **factory** Zuckerfabrik *f* ~ **flask** Polarisationskolben *m* ~**-grinding plant** Zuckervermahlungsanlage *f* ~**-mill scraper** Abstreicheisen *n* ~**-molding machine** Formmaschine *f* für Zucker ~ **press with rotary table** Drehtischzuckerpresse *f* ~ **rasp** Zuckerreibemaschine *f* ~ **refiner** Zuckersieder *m* ~**-refinery** Zuckersiederei *f* ~ **screw conveyer (or scroll)** Zuckertransportschnecke *f* ~ **sifter** Zuckersiebvorrichtung *f* ~ **sirup evaporator** Sirupverdampfer *m* ~ **sirup kettle** Sirupkessel *m* ~**-sorting plant** Zuckersortieranlage *f* ~**-testing apparatus** Zuckeruntersuchungsapparat *m*

**sugaring** Auskristallisieren *n*

**sugary** zuckerhaltig

**suggest, to** ~ anregen, vorschlagen **to** ~ **oneself** sich aufdrängen

**suggested compromise** Vermittlungsvorschlag *m*

**suggestibility** Beeinflußbarkeit *f*

**suggestion** Anregung *f*, Antrag *m* **at the** ~ **of** auf Veranlassung von

**suit, to** ~ angemessen sein, anpassen, eignen, passen

**suit** Anzug *m*, Garnitur *f*, Klage *f* (law), Prozeß *m*, Rechtsfall *m* ~ **for annulment (or cancellation)** Nichtigkeitsklage *f* ~ **for infringement of patent rights** Patentverletzungsklage *f*

**suitability** Anstand *m*, Eignung *f*, Geeignetheit *f*, Schicklichkeit *f*, Zweckmäßigkeit *f* ~ **for parchment manufacture** Pergamentierfähigkeit *f*

**suitable** adaptiert, angebracht, angemessen, angetan, anstellig, brauchbar, empfehlenswert, entsprechend, geeignet, passend, sachgemäß, schicklich, zweckmäßig **to be** ~ sich eignen, passen **to be** ~ **for** sich eignen für ~ **for connection** anschließbar ~ **for crosscountry work (or driving)** geländegängig ~ **for reconditioning** aufbereitbar ~ **for shipment** seemäßig ~ **for smelting** schmelzwürdig ~ **for tracing (or copying)** pausfähig ~ **for tropics** tropenfest ~ **for use in tropical climates** tropenbeständig

**suitcase** Handkoffer *m*

**suite**, ~ **of apartments** Zimmerflucht *f* ~ **of cores** Kernsatz *m*
**suited** tauglich
**sulfanilic acid** Sulfanilsäure *f*
**sulfate, to** ~ sulfatieren
**sulfate** Schwefelsalz *n*, Sulfat *n* ~ **of schwefel-sauer** ~ **of barium** Bariumsulfat *n* ~ **of lead** schwefelsaures Blei *n* ~ **of lime** Kalksulfat *n* ~ **of zinc** schwefelsaures Zink *n*
**sulfate roast** sulfatisierende oder sulfidierende Röstung *f*
**sulfating** Sulfatation *f* ~ **agent** Sulfatisierungs-mittel *n* ~ **(roasting) furnace** Sulfatisierofen *m* ~ **roasting** sulfatisierende oder sulfidierende Röstung *f*
**sulfation** Sulfatierung *f*, Sulfation *f*, Sulfierung *f*, Sulfolyse *f*
**sulfatization** Sulfatierung *f*, Sulfation *f*
**sulfidization** Sulfidierung *f*
**sulfide, to** ~ sulfidieren
**sulfide** Sulfid *n*; sulfidisch ~ **inclusion** Sulfidein-schluß *m* ~ **ore** Schwefelerz *n*, Sulfiderz *n*, sulfidisches Erz *n* ~ **sulfur** Sulfidschwefel *m*
**sulfidic** sulfidisch
**sulfitation** Saturation *f* mit Schwefeldioxyd, Schwefelung *f*
**sulfite liquor (or solution)** Sulfitlauge *f*
**sulfo,** ~ **carbonic acid** Schwefelkohlensäure *f* ~ **compound** Sulfoverbindung *f* ~ **cyanate** Rhodanat *n* ~ **cyanic acid** Rhodanwasserstoff-, Schwefelzyan-, Sulfozyan-, Thidzyan-säure *f* ~ **cyanide** Ferrozyanid *n*, Rhodansalz *n* ~ **cyanogen** Rhodan *n*, Sulfozyan *n* ~ **hydrate** Sulf-hydrat *n*, -hydrid *n*, Sulfohydrat *n*
**sulfolysis** Sulfolyse *f*
**sulfonate, to** ~ sulfieren
**sulfonate** Sulfonat *n*
**sulfonated oil** geschwefeltes Öl *n*
**sulfonation** Sulfierung *f*
**sulfonic acid** Sulfonsäure *f*
**sulfosalicylic acid** Sulfosalizylsäure *f*
**sulfosalt** Sulfosalz *n*
**sulfur, to** ~ schwefeln
**sulfur** Schwefel *m* ~ **bromide** Schwefelbromid *n* ~ **burner** Schwefelzieher *m* ~ **chloride** Chlor-schwefel *m*, Schwefelchlorid *n* ~ **compound** Schwefelverbindung *f* ~ **drop** Entschwefelung *f* ~ **dross** Schwefelschlacke *f* ~ **fluoride** Fluor-schwefel *m* ~ **impression** Schwefelabdruck *m* ~ **moniodide** Schwefeljodür *n* ~ **monobromide** Schwefelbromür *n* ~ **monochloride** Chlor-schwefel *m*, Schwefelchlorür *n* ~ **ore** Schwefel-erz *n* ~ **pit** Schwefelgrube *f* ~ **refinery** Schwefel-hütte *f*, -werk *n* ~ **removal** Entschwefelung *f* ~ **roll** Schwefelstange *f* ~ **salt** Schwefelsalz *n* ~ **test** Schwefelprobe *f* ~ **treatment** Schwefelung *f* ~ **trioxide** Schwefelsäure *f*
**sulfurate, to** ~ (auf)schwefeln
**sulfurated** sulfuriert
**sulfurea** Sulfoharnstoff *m*
**sulfured cossettes** geschwefelte Schnitzel *pl*
**sulfuric,** ~ **acid** Schwefelsäure *f*, Vitriol-öl *n*, -säure *f* **of** ~ **acid** schwefelsauer ~ **acid parting** Schwefelsäurescheidung *f* ~ **acid plant** Schwe-felsäure-anlage *f*, -fabrik *f* ~ **acid treatment** Be-handlung *f* mit Schwefelsäure ~ **anhydride** Schwe-felsäureanhydrid *n* ~ **ether** Schwefeläther *m*
**sulfuring (or sulfurization)** Schwefelung *f*

**sulfurize, to** ~ schwefeln, mit Schwefeldioxyd *n* saturieren
**sulfurizing agent** Schwefelungsmittel *n*
**sulfurous** schwefel-artig, -haltig, schwefelig ~ **acid** schweflige Säure *f*
**sulk, to** ~ bocken
**sulky lister** Sulkydammkulturpflug *m*, Sulky-reihensäe- und Zudeckmaschine *f*
**sully, to** ~ besudeln
**sultry** drückend, schwül
**sum, to** ~ (up) addieren, summieren **to** ~ **up** summen
**sum** Betrag *m*, Inbegriff *m*, Summe *f* ~ **of (all) digits** Quersumme *f* ~ **of the angles** Winkel-summe *f* ~ **of the contract** Verdingungssumme *f* ~ **of momenta** Impulssumme *f* ~ **of money** Geldposten *m* ~ **of squares** Quadratsumme *f*
**sum,** ~ **-and-difference amplifier** Differenzton-faktorverstärker *m* ~ **angles of triangle** Winkel-summe *f* in einem Dreieck ~ **frequency** Summen-menfrequenz *f* ~ **operation** Summenbildung *f* ~ **rule** Permanentsatz *m*, Summen-regel *f*, -satz *m* ~ **subscript** Summenindex *m* ~ **total** Gesamtbetrag *m*
**summarization** Aufzählung *f*
**summarize, to** ~ aufführen, zusammenfassen
**summarized circuit details** Leitungsübersicht *f*
**summarizing article (or report)** zusammenfassen-der Bericht *m*
**summary** Abriß *m*, Angabe *f*, Auszug *m*, ge-drängte Übersicht *f*, (book) Hauptinhalt *m*, Übersicht *f*, Zusammen-fassung *f*, -stellung *f* ~ **of character assessment** Beurteilungsnotiz *f* ~ **of evidence** Tatbestandaufnahme *f* ~ **of traffic statistics** Verkehrszählung *f*
**summary,** ~ **dismissal** fristlose Entlassung *f* ~ **punch** Summenlocher *m* (comput.) ~ **report** Kurzmeldung *f*
**summation** Aufzählung *f*, Summe *f*, Summierung *f* ~ **check** Summen-kontrolle *f*, -probe *f* (info proc.) ~ **convention** Summierungsvorschrift *f* ~ **device** Additionsgerät *n* ~ **equation** Summen-gleichung *f* ~ **formula** Summationsformel *f* ~ **frequency** Summenfrequenz *f* ~ **instrument** Summeninstrument *n* ~ **meter** Summenzähler *m* ~ **method** Summationsverfahren *n* ~ **tone** Summationston *m*, Summenton *m*
**summer** Additionsschaltung *f* (comput.), Ober-, Träger-schwelle *f*, Sommer *m* ~ **lightning** Wetterleuchten *n* ~ **monsoon** Sommermonsun *m* ~ **path** Sommerweg *m* ~ **polder** Sommerpol-der *m* ~ **road** Sommerweg *m* ~ **thunderstrom** Sommergewitter *n* ~ **timber** Sommerholz *n* ~ **timetable** Sommerflugplan *m* ~ **wood** Sommer-holz *n*
**summing amplifier** Summierverstärker *m*
**summit** Ecke *f* (cryst.), Endeck *n*, Gipfel *m*, Gipfel-, Höhe-punkt *m*, Poleck *n* (cryst.), Scheitel *m* ~ **of a dike** Deichkappe *f* ~ **of a mountain** Bergkuppe *f*
**summit,** ~ **pond** Scheitelhaltung *f* ~ **pool** Schei-telhaltung *f*
**summon, to** ~ aufbieten, herbeirufen, vorladen
**summons** Aufforderung *f*, Aufgebot *n*, gericht-liche Vorladung, Vorládung *f*
**sump** Becken *n*, Gesenk *n*, Ölsumpf *m*, Pumpen-sumpf *m*, Sammelbehälter *m*, (for oil) Sammel-napf *m*, Schachtsumpf *m*, Schmierölsammel-

tank *m*, Senkgrube *f*, Sumpf *m*, Tümpel *m*, Wirbelsenke *f* (engin.) ~ **hole** Entwässerungsanlage *f*, Sickeranlage *f* ~ **pit** Ablaßgrube *f* ~ **plug** Sumpfablaß *m* ~ **shaft** Pumpen-, Wasserhaltungs-schacht *m*

**sumpter** Packtier *n*

**sun** Sonne *f* ~'**s altitude** Sonnenhöhe *f* ~**-and-planet gear** Planetenrad-, Umlauf-getriebe *n* ~ **arc** Sonne *f* ~ **beam** Sonnenstrahl *m* ~ **burn** Sonnenbrand *m* ~ **compass** Sonnenkompaß *m* ~ **dog** Nebensonne *f*, Parhelion *n* ~ **down** Sonnenuntergang *m* ~ **glasses** Sonnenschutzbrille *f* ~ **helmet** Tropenhelm *m* ~ **light** Sonnenlicht *n* ~ **pillar** Sonnensäule *f* ~'**s radiation** Sonnenstrahlung *f* ~ **ray** Sonnenstrahl *m* ~ **rise** Sonnenaufgang *m* ~ **rise and sunset colors** Dämmerungsfarben *pl* ~ **set** Sonnenuntergang *m* ~ **set effects** Sonnenuntergangserscheinungen *pl* ~ **shade** Sonnen-schutz *m*, -vorhang *m* ~ **shield** Sonnenblende *f* ~ **shine** Sonnenschein *m* ~ **shine recorder** Sonnenschein-anzeiger *m*, -autograf *m*, -messer *m* ~ **spot** Sonnenfleck *m* ~ **spot cycle** Sonnenfleckperiode *f* ~ **spot number** Sonnenfleckenzahl *f* ~ **spot period** Sonnenfleckperiode *f* ~ **stone** Aventurinfeldspat *m*, Oligoklas *m* ~ **stroke** Sonnenstich *m* ~ **visor** Sonnenblende *f*

**sunder, to** ~ trennen

**sundry** mehrere

**sunk** versenkt, vertieft ~ **by loading** (monolith foundation) Brunnengründung *f*

**sunk,** ~ **caisson** Senkkasten *m* ~ **characters** vertiefte Schrift *f* ~ **key** Versenk *n* ~ **key** Nutenkeil *m* ~ **line** versenktes Geleise *n* ~ **well** Senkbrunnen *m*

**sunken,** ~ **obstacle** versenktes Hindernis *n* ~ **peg (or pin)** Senkstift *m* ~ **road** eingeschnittene Straße *f*, Hohlweg *m* ~ **runway lights** Versenkleuchten *pl*

**sunny** sonnig ~ **weather** sonniges Wetter *n*

**super** Firstendruck *m* ~**-abundant** überreichlich ~ **acidity** Perazidität *f* ~**-annuated** verjährt ~**-annuation** Verjährung *f* ~**-annuation allowance** Alterszulage *f* ~**-audible** überhörfrequent ~**-audible frequency** Überhörfrequenz *f* ~**-audio telegraphy** Überlagerungstelegrafie *f* ~**-calender** Hochglanzkalender *m* ~**-calendered** (Papier) stark satiniert ~**-carburization** Überkohlung *f* ~**-cargo plane** Großfrachtflugzeug *n*

**supercharge, to** ~ aufladen, laden, überverdichten, vorverdichten

**supercharge** Aufladen *n*, Kompression *f*, Zusatzladung *f* ~ **engine** vorverdichtender Motor *m*, Überladerhöhe *f* ~ **gear change** Ladegangwechsel *m* ~ **pressure** Ladedruck *m*

**supercharged** vorverdichtet ~ **aero-engine** Ladermotor *m* ~ **cabin** Überdruckkabine *f* ~ **engine** Auflade-, Gebläse-, Höhen-motor *m*, (für große Nennhöhe) Motor *m* mit Vorverdichtung, überverdichtender Motor *m* ~ **height** Gleichdruckhöhe *f* ~ **ignition** verdichtendes Zündungsgeschirr *n* ~ **motor** Kompressormotor *m*, Motor *m* mit Vorverdichtung

**supercharger** Gebläse *n*, Kompresser *m*, Kompressor *m*, Ladepumpe *f*, Lader *m*, (of engine) Überlader *m*, Überverdichter *m*, Vorverdichter *m*, Zwischengebläse *n* ~ **with high blower ratio**

Höhenlader *m* ~ **with low blower ratio** Bodenlader *m* **eye of the** ~ Eintrittsöffnung *f* des Laders

**supercharger,** ~ **air-flow** Ladeluftdurchsatz *m* ~ **air intercooler** Ladeluftzwischenkühler *m* ~ **assembly** Ladesatz *m* ~ **blast gate control** Verdichterregelung *f* ~ **compression ratio** Ladedruckverhältnis *n* ~ **control lever** Laderbedienhebel *m* ~ **delivery pressure** Laderenddruck *m* ~ **diffuser chamber** Lader-ringkanal *m*, -verteilergehäuse *n* ~ **distribution** Laderringkanal *m* ~ **distribution chamber** Laderverteilergehäuse *n* ~ **drive** Kompressorantrieb *m* ~ **drive gear** Laderantriebpritzel *n* ~ **drive power** Laderantriebsleistung *f* ~ **gear** Ladergang *m*, zweiter Ladergang *m* (volle Aufladung beim Höhenmotor) ~ **housing** Ladergehäuse *n* ~ **impeller** Vorverdichterladerad *n* ~ **intercooler** Ladeluftkühler *m* ~**-pressure limiting control** Ladedruckbegrenzer *m* ~ **ratio** Laderübersetzung *f* für volle Aufladung ~ (pressure) **regulator** Ladedruckregler *m* ~ **rotor impeller** Ladelaufrad *n* ~ **screen** Ladersieb *n* ~ **speed** Laderdrehzahl *f* ~ **spiral** Laderspirale *f* ~**-testing rig** Laderprüfstand *m* ~ **volute** Laderschnecke *f*

**supercharging** Über-, Vor-verdichtung *f*; vorverdichtend ~ **blower** Ladergebläse *n* ~ **equipment** Aufladeeinrichtung *f*

**super,** ~**-chopper** sehr schneller Zerhacker *m* ~**-compressed** überkomprimiert ~**-compression engine** überkomprimierter Motor *m* ~**-conducting** supraleitend ~**-conducting computer device** Supraleitvermögen-Rechenmaschine *f* ~**-conducting transition** Supraleitungsübergang *m* ~**-conduction electron** Supraleitelektron *n* ~**-conductive phase transition** supraleitender Phasenübergang *m* ~**-conductive specimen** supraleitende Probe *f* ~**-conductive transitions** Supraleitungsübergänge *pl* ~**-conductivity** Superleitfähigkeit *f*, Supraleitfähigkeit *f* ~**-conductor** Supraleiter *m* ~**-control tube** Exponentialröhre *f*, Regelpentode *f*, Röhre *f* mit veränderlichem Durchgriff, Röhre *f* mit variabler Steilheit, Röhre *f* mit variablem Verstärkungsfaktor

**super-cool, to** ~ überkalten

**super,** ~**-cooled** überschmolzen ~**-cooled state** Unterkühlung *f* ~**-cooling** Überkühlung *f*, Unterkühlung *f* ~**-cushion tire** Riesenluftreifen *m* ~**-elastic impact** superelastischer Stoß *m* ~**-elevated** überhöht ~ **elevation** (of rails) Überhöhung *f* ~ **elevation under firing-table conditions** Aufsatzwinkel *m* ~**-face** Dachfläche *f* (geol.) ~**-fattening** Überfettung *f*

**superficial** flüchtig, oberflächlich ~ **area of shops** Werkstättenflächenraum *m* ~ **cementation** Einsatzhärtung *f* ~ **current** Oberflächenstrom *m* ~ **deposit** oberfläche Lagerstätte *f* ~ **entropy** Oberflächenentropie *f* ~ **(face) hardening** Oberflächenhärtung *f* ~ **layer** (in sublimation) Anflug *m* ~ **measure** Flächenmaß *n* ~ **pair density** Zweiteilchen-Oberflächendichte *f* ~ **tanning** Angerbung *f*

**superficies** Mantelfläche *f* ~ **area** Raumumfang *m*

**superfine** doppelschlicht, hochfein ~ **file** Feinschlichtfeile *f* ~ **flour** Auszug-, Gabel-mehl *n* ~ **hardening** Weißerde *f*

**superfinish(ing)** Feinstziehschleifen n, Feinst-bearbeitung f ~ **boring mills** Feinstbohrwerke pl ~ **lathe** Genauigkeitsdrehbank f
**super flash** Fotoblitzlampe f
**superfluidity** Suprafluidität f
**superfluous** überflüssig, überzählig, unnütz
**super, ~ fusion** Überschiebungsschmelzung f ~ **gasoline** Extrabenzin m ~ **generation circuit organization** Pendelfrequenzschaltung f
**superheat, to** ~ (beyond saturation) überhitzen
**superheated** überhitzt ~ **slide for locomotives** Heißdampfschieberring m für Lokomotiven ~ **steam** Heißdampf m, Heizdampf m, überhitzter Dampf m
**superheater** Überhitzer m ~ **coil** Überhitzerschlange f ~ **header** Überhitzerkammer f ~ **outlet** Überhitzeraustrittsstutzen m ~ **pipe (or tube)** Überhitzerrohr n
**superheating** Überhitzung f ~ **of the steam** Dampfüberhitzung f ~ **meter** Überhitzungsmesser m
**superheavy nucleus** ultraschwerer Kern m
**superheterodyne** Transponierungsempfänger m ~ **action** Frequenzwandlung f ~ **amplifier** Summendifferenzverstärker m ~ **principle** Überlagerungsprinzip n ~ **receiver** Superheterodyn-, Transponierungs-, Überlagerungs-, Zwischenfrequenz-empfänger m ~ **reception** Superheterodyn-, Überlagerungs-, Zwischenfrequenz-empfang m
**superhigh, ~ draft system** Höchstverzugsstreckwerk n ~ **frequency** Zentimeterbereich m ~ **gear** Schnellgang m ~ **gearshift claw** Schnellgang-schaltklaue f ~ **gear wheel** Schnellgangrad n
**superhoning machine** Feinziehschleifmaschine f
**supericonoscope** (image tube) Bildwandler m, Superikonoskop n
**superimpose, to** ~ aufdrücken, aufsetzen, übereinanderlagern, überlagern
**superimposed** überlagert ~ **box** Aufsetzkasten m ~ **circuit coil** Viererspule f ~ **circuits** Phantomkreise pl ~ **concept** Oberbegriff m ~ **connection** Überlagerungsschaltung f ~ **plant** Vorschaltanlage f ~ **ringing** Vielfachzeichengebung f ~ **short** Simultananschluß m ~ **turbine** Vorschaltturbine f
**superimposing** Mehrfachbelichtung f, Überlagerung f
**superimposition** Über(einander)lagerung f
**superintend, to** ~ bewirtschaften, überwachen
**superintendent** Berginspektor m, Betriebs-leiter m, -oberingenieur m, Fabrikationsleiter m, Fabrikleiter m, Fabrik(s)direktor m, Inspektor m, Oberaufseher m, Werksführer m ~ **(or manager) of construction** Bauleiter m ~ **of shipyard** Werftdirektor m
**superior** Vorgesetzter m; ober, überlegen ~ **alloy steel** Edelstahl m ~ **authority** Oberbehörde f ~ **court** höhere Instanz f ~ **figure** obenstehende Ziffer f ~ **forces** Überzahl f ~ **grade** bessere Sorte f ~ **quality tubes** Rohre pl mit Gütevorschriften ~ **strength** Übermacht f
**superiority** Überlegenheit f, Übermacht f, Vorrang m
**superjacent** übereinanderliegend
**superlative form** Meiststufe f
**superlattice** Überstrukturgitter n ~ **reflections**

Überstrukturreflexe pl
**super-microscopic** übermikroskopisch
**supermultiplet** Supermultiplett n
**supernatant** (liquids) überstehend
**supernumerary call** überzähliger Anruf m
**superordinated** überlagert
**superphantom circuit** Achterkreis m
**superposable** superponierbar
**superpose, to** ~ auflagern, übereinander-lagern, -legen, überlagern
**superposed** übereinanderliegend ~ **on (or with) direct current** gleichstromüberlagert
**superposed, ~ circuit** Simultanverbindung f, überlagerter Strom m, überlagerter Stromkreis m, Vierer(sprech)kreis m ~ **(or abuting) ends** zusammenstoßende Enden pl ~ **field** überlagertes Feld n ~ **loading** Viererpupinisierung f ~ **magnetization** überlagerte Magnetisierung f, Vormagnetisierung f ~ **motion** Zusatzbewegung f ~ **pulse** aufgeprägter Impuls m ~ **ringing** Rufen n mit gleichstromüberlagertem Wechselstrom ~ **ringing current** gleichstromüberlagerter Rufstrom m ~ **sieve** Übersieb n ~ **telegraph** Simultantelegraf m ~ **turbine** Vorschaltturbine f
**super(im)posing** Überlagerung f
**superposition** Aufeinanderschichten n, Superposition f, Über(einander)lagerung f ~ **of one wave upon another** (in a canal) Wellengefüge n ~ **of vibrations** Schwingungsüberlagerung f ~ **approximation** Überlagerungsnäherung f
**super, ~ potential** Überpotential n ~ **power distribution** Stromgroßverteilung f ~ **power station** Großkraftwerk n ~-**press printing blankets** Superpreßdrucktücher pl ~ **pressure** Überdruck m, Überdruckgebiet n ~ **quick fuse** empfindlicher Zünder m ~-**quick impact fuse** empfindlicher Aufschlagzünder m ~ **refined steel plate** Edelstahlblech n ~ **refraction** Superbrechung f
**superregeneration** erhöhte oder übertriebene Dämpfungsverminderung f, (working with quench voltage) Hilfsfrequenzrückkupplung f, (using quench potential) Rückkupplung f mit Hilfsfrequenz, Überrückkupplung f
**superregenerative, ~ circuit** Pendelrückkopplung f, Pendelrückkopplungsschaltung f ~ **receiver** Hilfsfrequenz-empfänger m, -rückkupplungsempfänger m, Pendelrück-koppler m, -kopplungsempfänger m, Superregenerativempfang m, -empfänger m, Überrückkupplungsempfänger m ~ **reception** elektromotorische Rückkopplung f mit Hilfsfrequenz, Rückkupplungsempfang m mit Hilfsfrequenz
**superregenerator** Hilfsfrequenz-empfänger m, -rückkupplungsempfänger m, Pendelrück-koppler m, -kopplungsempfänger m, Superregenerativempfänger m, Überrückkupplungsempfänger m
**super-retroaction** Pendelrückkopplung f
**super-saturate, to** ~ übersättigen
**super, ~-saturated steam** unterkühlter Wasserdampf m ~-**saturation** Übersättigung f
**superseason, to** ~ ablagern lassen
**supersede, to** ~ abschaffen, ersetzen
**supersensitive** hochempfindlich
**supersession** Verdrängung f

**supersonic** oberhalb der Schallgeschwindigkeit liegend, Überschall *m*, Ultraschall *m* ~ **echo sounding** Echolot *n* (mit Ultraschall) ~ **flow** Überschallströmung *f* ~ **frequency** Überhörfrequenz *f*, Über-, Ultra-schallwellenfrequenz *f* ~ **glide** Überschallgleitflugbahn *f* ~ **heterodyne reception** Überlagerungsempfang *m* mit Überhörfrequenz ~ **light relay** Überschallichtzelle *f* ~ **light valve** Überschallicht-, Ultraschall-, Piezoquarz-zelle *f* ~ **nozzle** Überschalldüse *f* ~ **sounding** ultra-akustisches Sondieren *n* ~ **speed** Überschallgeschwindigkeit *f* ~ **stroboscope** Ultraschallstroboskop *n* ~ **transition time** Ultraschall-Laufzeitstrecke *f* ~ **tunnel** Windkanal *m* für Überschallgeschwindigkeit ~ **velocity** Überschallgeschwindigkeit *f* ~ **wave** Über-, Ultraschallwelle *f* ~ **wind tunnel** Überschallwindkanal *m*

**super,** ~ **spectral** überspektral ~ **speed operation** Überdrehen *n* (film) ~ **speed steel** Hochleistungsschnellschnittstahl *m* ~ **structural parts** Aufbauten *pl*

**superstructure** Aufbau *m*, Aufbauten *pl*, Deckaufbau *m*, Oberbau *m*, (of furnace) Oberofen *m*, Torportal *n*, Überbau *m* ~ **of bridge** Brückenoberbau *m*

**supertension** Höchstspannung *f* ~ **cable** Höchstspannungskabel *n*

**superthreshold** (sound) überschwellig ~ **(sound-) intensity level** überschwelliger Schallreiz *m* ~ **visibility** überschwellige Sichtbarkeit *f*

**supervise, to** ~ beaufsichtigen, inspizieren, kontrollieren, überwachen

**supervising,** ~ **authority** Aufsichtsbehörde *f* ~ **desk** Mischpult *n* ~ **panel** Misch-brett *n*, -tafel *f* ~ **specifications** Überprüfwerte *pl*

**supervision** Aufsicht *f*, Aufsichtsbehörde *f*, Beaufsichtigung *f*, Bewachung *f*, Inspektion *f*, Kontrolle *f*, Leitung *f*, Oberaufsicht *f*, Schlußzeichengebung *f*, Überwachung *f* ~ **of answering** Rufüberwachung *f* (teleph.) ~ **of construction** Bauleitung *f* ~ **of drawings** Zeichnungskontrolle *f*

**supervision frequency** Ultraschallfrequenz *f*

**supervisor** Aufseher *m*, Aufsichtführender *m*, Aufsichts-beamter *m*, -beamtin *f*, Inspektor *m*, Kontrolleur *m*, Lehrgangsleiter *m*, Ober-aufseher *m*, -aufsichtsbeamtin *f*, -leiter *m* ~ **'s position** Kontrollplatz *m* ~ **'s section** Aufsichtsabteilung *f*, Platzgruppe *f*

**supervisory,** ~ **air pressure** Überwachungsluftdruck *m* ~ **board** Überwachungstafel *f* ~ **circuit** Schlußzeichenschaltung *f*, Überwachungsstromkreis *m* ~ **cord** Überwachungsschnur *f* ~ **device** Oberkontrolle *f* ~ **equipment** Überwachungseinrichtung *f* ~ **indicator** Schlußzeichen *n* ~ **lamp** Kontroll-, Schluß-, Signal-, Überwachungs-lampe *f* ~ **relay** Kontroll-, Melde-, Schlußzeichen-, Überwachungs-relais *n* ~ **signal** Schlußzeichen *n*, Überwacjimgszeichen *n* ~ **work** Aufsichtsdienst *m*

**supervoltage** Hochspannung *f*

**supinate, to** ~ auswärtsdrehen

**supplant, to** ~ ersetzen, verdrängen

**supple** biegsam, federnd, gelenkig, geschmeidig, weich

**supplement, to** ~ beifügen, ergänzen, nachtragen

**supplement** Anlage *f*, (to a journal) Beiblatt *n*, Beiheft *n*, Beilage *f*, Ergänzung *f*, Ergänzungswinkel *m*, Ersatz *m*, Nachtrag *m*, Supplement *n*, Zugabe *f*, Zusatz *m*

**supplemental** zusätzlich ~ **cost** Überpreis *m* ~ **divider** verlängerter Außenteiler *m* ~ **hopper** Ergänzungstrichter *m*, extra Tank *m* für Getreide

**supplementary** ergänzend, nachträglich, zusätzlich ~ **air** Zusatzluft *f* ~ **allowance** Nachbewilligung *f* ~ **angle** Supplement(är)winkel *m* ~ **apparatus** Zusatz-apparat *m*, -gerät *n* ~ **application** Nachanmeldung *f* ~ **carrier** Zusatzträger *m* ~ **fee** Zuschlaggebühr *f* ~ **feedcock** Zusatzhahn *m* ~ **feed spring** Mehrmengenfeder *f* ~ **feed stop** Mehrmengenanschlag *m* ~ **fire** Zusatzfeuer *n* ~ **firing** Zusatzfeuerung *f* ~ **folder** Nachtragmappe *f* ~ **fount** Ergänzungsschnitt *m* ~ **impedance** Ergänzungsimpedanz *f* ~ **lens** Vorsatz-, Zusatz-linse *f* ~ **magnification** Übervergrößerung *f* ~ **means** Hilfsmittel *n* ~ **motion** Zusatzbewegung *f* ~ **order** Nachbestellung *f*, Nachtragsbefehl *m* ~ **parts** Nachrüstteile *pl* ~ **reduction** Zusatzuntersetzung *f* ~ **regulation** Zusatzbestimmung *f* ~ **resistance** Zusatzwiderstand *m* ~ **rule** Zusatzbestimmung *f* ~ **scavenging** Zusatzspülung *f* ~ **sheet** Deckblatt *n* ~ **slope** Zusatzgefälle *n* ~ **spring** Hilfsfederung *f* ~ **stress** Zusatzspannung *f* ~ **supplies** Ergänzungsbedarf *m* ~ **tank** Zusatztank *m* ~ **test** Zusatzversuch *m* ~ **trial** Zusatzversuch *m*

**suppliant** Petent *m*

**supplied** angegeben, mitgeliefert **to be** ~ **in bulk** in großen Mengen *pl* lieferbar ~ **by continuous current** mit Gleichstrom *m* gespeist

**supplier** Lieferant *m*, Lieferer *m*, Lieferfirma *f*

**supplies** Lebensmittel *pl*, Nachschubgut *n*, Proviant *m*

**supply, to** ~ abliefern, anschaffen, ausrüsten, (a customer) beliefern, beschaffen, liefern, speisen, stellen, versehen, versorgen, zuführen, zuleiten, zuteilen **ability to** ~ Abgabefähigkeit *f* **to** ~ **with oil** mit Öl *n* bespülen **to** ~ **a want** eine Lücke *f* ausfüllen

**supply** Ablieferung *f*, Angebot *n*, Anlieferung *f*, Anschluß *m* an das Netz, Beschaffung *f*, Bestand *m*, Förderung *f*, Lieferung *f*, Mittel *n*, Nachschub *m*, Speisung *f*, Versorgung *f*, (source) Verspannung *f*, Vorrat *m*, Zufluß *m*, Zufuhr *f*, Zuführung *f*, Zulauf *m*

**supply,** ~ **by accumulators** Sammlerspeisung *f* ~ **of coal free to miners** Deputatkohle *f* ~ **of coolant** Kühlölzuführung *f* ~ **of energy** Energiezufuhr *f* ~ **of food** Nachschub *m* von Verpflegung *f* ~ **of labor** Arbeitsangebot *n* ~ **of personnel** Nachwuchs *m* ~ **of power** Energiezufuhr *f* ~ **and return line** Zu- und Rückführung *f*

**supply,** ~ **air chamber** Zuluftkammer *f* ~ **area** Absatzgebiet *n* ~ **base** Ausrüstungsplatz *m*, Lager *n*, Nachschubstützpunkt *m*, Verpflegungsbasis *f* ~ **bin** Vorratsbehälter *m* ~ **boat** Nachschubboot *n* ~ **bridge** Speisebrücke *f* (teleph.) ~ **cable** Verbindungs-, Zuführungskabel *n* ~ **cableway** Drahtseilbahn *f* für Materialien ~ **cart** Dynamokarren *m* ~ **center** Materialstelle *f* ~ **channels** Nachschubwege *pl* ~ **circuit** Speise(strom)kreiş *m*, Zuleitungsstromkreis *m* (electr.) ~ **clerk** Geräteverwalter *m* ~ **column** Nachschubkolonne *f* ~ **connection** Speiseanschluß *m* ~ **container with parachute** Mischlastabwurfbehälter *m* ~ **contract** Lieferungs-kontrakt *m*, -vertrag *m* ~ **current** Speisestrom *m* ~ **depot** Nachschub-lager *n*, -park *m* ~ **-distributing point** Nachschubverteilungsstelle *f* ~ **district** Versorgungsgebiet *n* ~ **drum** Ablaufhaspel *m* ~ **duct** Zulaufrohr *n* ~ **dump** Nachschub-lager *n*, -park *m*, Proviantlager *n* ~ **efficiency** (ratio of actual suction volume to stroke volume) Liefergrad *m* des Saughubes ~ **equipment** Speiseapparatur *f* ~ **fan** Zulüfter *m* ~ **frequency** Netzfrequenz *f* ~ **industry** Lieferindustrie *f* ~ **lead** Speise-, Zuführungs-leitung *f* ~ **line** Leitung *f*, Lichtnetz *n*, Nachschublinie *f*, Zuführung *f*, Zuleitung *f* ~ **-line fluctuation (or ripple)** Netzunruhe *f* ~ **main** Zuführungs-, Zulauf-leitung *f* ~ **mechanism** Ausschüttanlage *f* ~ **meter** Verbrauchsmesser *m* ~ **pipe** Einlaß-, Einlauf-, Zuführungs-rohr *n* ~ **point** Nachschub-ausgabestelle *f*, -umschlagstelle *f* ~ **pointer** Vorratszeiger *m* (Luftpresser) ~ **practice** Nachschubübung *f* ~ **pressure** Vordruck *m* ~ **rectifier** Speisegleichrichter *m* ~ **reel** Transporttrommel *f* (film), Vorwickel-rolle *f*, -trommel *f* ~ **relay** Speiserelais *n* ~ **resistance** Förderwiderstand *m* ~ **restriction** Vordrossel *f* ~ **road** Nachschubweg *m*, Zufuhrstraße *f* ~ **route** Nachschub-straße *f*, -weg *m* ~ **schedule** Lieferumfang *m* ~ **ship** Hilfsbei-, Proviant-, Troß-, Versorgungs-schiff *n* ~ **spool** Transporttrommel *f* (film) ~ **sprocket** Vortransport *m* ~ **station** Abgabestelle *f* ~ **system** Netz *n* ~ **terminals** Netzanschlußkontakt *m* ~ **traffic** Nachschubverkehr *m* ~ **train** Fahrkolonne *f*, Nachschub-kolonne *f*, -zug *m*, Troß *m* ~ **tube** Speiserohr *n* ~ **unit** Speisegerät *n* ~ **units** Nachschubtruppe *f* ~ **valve** Einlaßventil *n*, Zufuhrregler *m* ~ **vehicle** Nachschubfahrzeug *n* ~ **voltage** Ladespannung *f*

**supplying** Versorgen *n* ~ **of water** Wasserversorgung *f* ~ **device** Speisevorrichtung *f*, Zufuhrvorrichtung *f*

**support, to** ~ abspreizen, absteifen, abstützen, auflagern, führen, halten, Halt geben, lagern, stützen, tragen, unter-halten, -stützen, Vorschub leisten

**support** Absteifung *f*, Abstützung *f*, Anhalt *m*, Anschluß *m*, Auflage *f*, Auflagefilter *f*, Auflager *n*, Aufrechterhaltung *f*, Bettung *f*, Durchzug *m*, Etagere *f*, Fuß *m*, Gegenblock *m*, Gestänge *n*, Gestängebelastung *f*, Gestell *n*, Halt *m*, (scaffolding platform) Haltegerüst *n*, Halter *m*, Hilfe *f*, Lager *n*, Lagerung *f*, Säule *f*, Schlitten *m*, Sockel *m*, (for hood or tilt) Spriegel *m*, Spulenträger *m*, Stag *m*, Ständer *m*, Stativ *n*, Strebe *f*, Stütze *f*, Stütz-punkt *m*, -support *m*, Tragfuß *m*, Tragerüst *n*, Unterbau *m*, -lage *f*, -satz *m*, -stützung *f*, Wider-hall *m*, -lager *n* **(tool)** ~ Werkzeugträger *m* **in** ~ Bereitschaft *f* ~ **for footboard** Trittbrettträger *m* ~ **for joint** Gelenkträger *m* ~ **of mines** Grubenausbau *m* ~ **of an objection** Unterstützung *f* eines Einspruches ~ **of a protest** Unterstützung *f* eines Einspruches ~ **for screen** Schirmträger *m* ~ **for a three-bladed air screw** drei-

armiger Flügelhalter *m* für Luftschiffpropeller ~ **of travelling crane** Laufkranständer *m*

**support,** ~ **arm** Führungsarm *m* ~ **bar** Halteschiene *f* ~ **beam** Halteschiene *f* ~ **bearing** Unterstützungslager *n* ~ **bracket** Tragleiste *f* ~ **casting** Haltebock *m* ~ **chock** Unterlegklotz *m* ~ **flange for engine** Motortragflansch *m* ~ **grid with square perforations** Quadratloch-Stützblech *f* ~ **leg** Stützfuß *m* ~ **-load distributing band (jacket or ring)** Tragmantel *m* ~ **pressure** Stützdruck *m* ~ **stock** Stehknecht *m* ~ **strand** Entlastungsseil *n* ~ **structure** Tragkonstuktion *f* ~ **stud** Lagerstutzen *m* ~ **table** Montage *f* ~ **truss** Abstützblock *m* ~ **vise** Bankknecht *m* ~ **yoke** Stützbügel *m*

**supported** aufruhend, gelagert, gestützt **to be** ~ aufliegen ~ **by** angelehnt ~ **flank** angelehnter Flügel *m*

**supporter** Anhänger *m*, Träger *m*, Unterlagscheibe *f*, Verfechter *m*

**supporting** tragend ~ **angle** Tragwinkel *m* ~ **angle piece** Winkelstütze *f* ~ **area** Auflagerfläche *f* ~ **arm** Tragarm *m* ~ **axis** Stehachse *f* ~ **bar** Stützbalken *m* ~ **base** Tragsockel *m* ~ **beam** Schwimmträger *m*, Stütz-, Trag-, Trägerbalken *m* ~ **bearing** Traglager *n* ~ **board** Untersatzbrett *n* ~ **bolt** Haltebolzen *m* ~ **bracket** (Windmühle) Auflageknagge *f*, Tragstutze *f* ~ **bridge** Aufhängebrücke *f* ~ **capacity** Tragfähigkeit *f* ~ **choke coil** Stützdrosselspule *f* ~ **column** Gerüstsäule *f*, Stütz-pfeiler *m*, -säule *f*, Trag(kranz)säule *f* ~ **column of cupola** Kupolofentragsäule *f* ~ **cylinder** Walze *f* ~ **device** Aufnahmevorrichtung *f* ~ **disc** Stütz-, Unterlegscheibe *f* ~ **edge** Auflagerand *m* ~ **film** Folie *f* ~ **flange** Auflageflansch *m* ~ **foils** Trägerfollen *pl* ~ **frame** Gestellwand *m*, Stütz-gerüst *n*, -werk *n*, Trag-gestell *n*, -rahmen *m*, -vorrichtung *f*, Untergestell *n*, Verlagerungsgerüst *n* ~ **framework** Traggerüst *n* ~ **gas** Traggas *n* ~ **girder** Schwimmträger *m* ~ **grate** Traggitter *n*, Tragrost *m* ~ **grid** Traggitter *n* ~ **head** (front gudgeon) Tragkopf *m* ~ **hub** Tragnabe *f* ~ **insulator** Isolatorstütze *f* ~ **journal** Tragzapfen *m* ~ **knife edge** Stützschneide *f* ~ **lever** Stützhebel *m* ~ **mast** Tragarm *m* ~ **member** Tragstab *m* ~ **nave** Tragnabe *f* ~ **pile** Tragpfahl *m* ~ **pillar** Stütz-pfeiler *m*, -säule *f* ~ **pipe** Standrohr *n* ~ **plate** Stützscheibe *f* ~ **pole** Stützstange *f* ~ **post** Stützpfosten *m* ~ **power** Tragkraft *f* ~ **prop** Tragstütze *f* ~ **rail** (harness) Trageschiene *f* ~ **reaction** Auflagerreaktion *f* ~ **ring** Unterlegring *m* ~ **rod** Tragstange *f* ~ **roller pedestal** Stützrollenträger *m* ~ **roller pedestal in box form** kastenartiger Stützrollenträger *m* ~ **schedule** Belegaufstellung *f* ~ **sill** Tragschwelle *f* ~ **socket** Tragsockel *m* ~ **spring** Stützfeder *f* ~ **stay** Tragstütze *f* ~ **strand** Tragseil *n* ~ **strand for overhead cables** Luftkabeltragseil *n* ~ **strip for connecting wires** Anschlußsteg *m* für Verdrahtung ~ **structure** Traggestell *n* ~ **structures** Unterbauten *pl* ~ **strut** Lagerstrebe *f* ~ **stud** Zwischenstück *n* ~ **stud of telescope** Lagerzapfen *m* ~ **surface** Halteflausche *f*, Tragdeck *n*, tragende Fläche *f*, Tragfläche *f* ~ **system** Halterungssystem *n* ~ **table** Auflagetisch *m* ~ **tissue** Traggerüst *n* ~ **trap** Spann-

platte f ~ tube of concentric cable Aufnahme-röhre f ~ wall Stützmauer f ~ wing rib Tragrippe f ~ yoke Stützbügel m
supports Aufsatzstützen pl with ~ at different elevations mit verschieden hohen Fußpunkten pl ~ at the same elevation Fußpunkte pl auf gleicher Höhe
suppose, to ~ annehmen, vermuten to (pre) ~ voraussetzen
supposed receipt Solleinnahme f
supposedly angeblich
supposing angenommen
supposition Annahme f, Vermutung f (pre) ~ Voraussetzung f
supposititious untergeschoben
suppress, to ~ abbremsen, abschaffen, bremsen, (vibrations) dämpfen, Einhalt tun, einziehen, hintanhalten, niederhalten, sperren, unterdrücken to ~ the echo absperren
suppressed gesperrt ~-carrier transmission Senden n mit unterdrückter Trägerwelle ~ carrier wave unterdrückte Trägerwelle f (electr.) ~ frame blanking Halbbildaustastung f ~-frame recording Halbbildverfahren n ~ zero unterdrückter Nullpunkt m
suppressing Abbremsen n ~ network Zerrungskreis m
suppression Abschaffung f, (of interference) Entzerrung f, Geräuschbekämpfung f, Schalltilgung f, Unterdrückung f, Wegfall m ~ of carrier Unterdrückung f der Trägerwelle ~ of feedback Rückkupplungsunterdrückung f ~ of the interferences Störungsunterdrückung f ~ of oscillations Schwingungsabreißen n
suppression, ~ band Sperrbereich m ~ capacitor Entstörungskondensator m ~ circuit Geräusch-unterdrückungsstromkreis m, Grundstromkreis m ~ diode Fangdiode f ~ dust Staubbekämpfung f ~ filter Sperr-filter n, -kreis m ~ range Sperr-, Unterdrückungs-bereich m
suppressor Absperre m, Bremse f, Sperre f, Sperr-gerät n, -glied n, Verriegler m ~ of harmonics Wellensauger m ~ in which magnetization does not proceed at uniform rate Bremsfeldschaltung f
suppressor, ~ choke Entstördrossel f ~ elbow Schutzkrümmer m ~ grid Anodenschutz-, Brems-, Fang-, Stau-gitter n ~-grid keying Bremsgittertastung f ~ grid modulation Bremsgittermodulation f ~ grid tube Schutznetzröhre f ~ impulse Sperrimpuls m ~ plug Entstörstecker m ~ plug for distributor Verteilerentstörstecker m ~ screen Funk(en)entstörer m
supraconductivity Supra-, Über-leitfähigkeit f
supremacy of the air Beherrschung f des Luftraumes
surbased vault Stichbogengewölbe n
surcharge, to ~ überladen
surcharge Aufschlag m, Strafporto n, Zusatzgebühr f ~ on the surface Auflast f
supercharged steam ungesättigter Dampf m
surcingle (harness) Deckengurt m
surd Irrationalzahl f (math.) ~ consonant stimmloser Konsonant m
sure-footed fußsicher
surety Gewähr f, Gewahrsam n to be ~ for gewähren ~ bond Kaution f, Verpflichtungsschein m

suretyship Bürgschafts-, Garantie-leistung f
surf Brandung f, Dünung f, Welle f ~ beats Brandungsschwebungen pl ~ board Wellenreiter m
surface, to ~ abrichten, (excavation) abteufen, aufbringen, flachdrehen (mech.), plandrehen to ~-grind flachschleifen to ~-polish überpolieren
surface Außen-seite f, -wand f, (road) Belag m, Bodenoberfläche f, Faserschicht f, Fläche f, Flächenraum m, (earth) Grundfläche f, Mantel m (math.), Mantelfläche f, Niveau n, Oberfläche f, Parallele f, (ground surface) Planum n, Raummenge f, Tragfläche f to rise to the ~ auftauchen on the ~ zutage ~ with absorbed gas gasbehaftete Oberfläche f ~ of buoyancy Oberflächenauftrieb m ~ of constant slope Böschungsfläche f ~ of cut Schnittfläche f ~ of discontinuity Aufgleitfläche f (meteor.), Diskontinuitätsfläche f, Unstetigkeitsfläche f, Wirbelfläche f ~ of distinct vision Schärfenfläche f ~ of the earth Erdoberfläche f ~ of equivalent drag Widerstandsfläche f ~ of esplanade (or of filling quay level) Kaiplanum n ~ of film Filmfläche f ~ of the focal plane frame Meßrahmenfläche f ~ of fourth order Fläche f vierter Ordnung ~ of fracture Bruchfläche f ~ at given level Niveaufläche f ~ of instability Unstetigkeitsfläche f ~ of lands Führungsfläche f ~ of low work function Oberfläche f mit kleiner Austrittsarbeit ~ of milled work Fräsbild n ~ of misfit Lockerstelle f ~ of the mold Kokillenwand f ~ of no motion stromlose Fläche f ~ of projection Projektionsfläche f ~ of reference Einstellungsfläche f ~ of revolution Rotationsfläche f ~ of rupture Gleitfläche f ~ of section Schnittfläche f ~ of seepage Hangquelle f ~ of separation Grenz-, Trennungs-fläche f ~ of shafting Anschlag m ~ of sharp vision Schärfenfläche f ~ of slide Schlifffläche f ~ of the source Strahlenoberfläche f ~ of spherical indentation Kalottenoberfläche f ~ of subsidence Abgleitfläche f ~ of traveling grate Wanderrostfläche f under(side) ~ of wing Flügelunterseite f
surface, ~ abrasion Oberflächenabnutzung f ~-active kapillaraktiv ~ action Oberflächenwirkung f ~ adhesion Oberflächenhaftung f ~ analyzer Glattheitsprüfer m, Oberflächen-(ab)taster m ~ appearance Oberflächenbild n ~ area Flächeninhalt m ~ attraction Flächenanziehung f ~ band Streckfigur f ~ bands Fließfiguren pl ~ barriers Grenzflächen pl ~ blow (foundry) Kochen n der Form ~ blowhole Außenlunker m ~ blow-off cock Abschabhahn m ~ brightness Flächenhelligkeit f ~ broaching Außenräumen n (Bearbeitungsverfahren) ~ burst Oberflächendetonation f ~ carburizing Oberflächenkohlung f ~ cementation Einsatzhärtung f ~ coat Spachtelschicht f ~ coating Oberflächenüberzug m ~ combustion Oberflächenverbrennung f ~ configuration Oberflächengestalt f ~ constraint Oberflächenspannung f ~ contact flächenhafte Berührung f ~-contact rectifier Flächengleichrichter m ~ content Flächenberührung f ~ cooler Flächenanziehungsberieselungsapparat m, Oberflächenkühler m ~ coverage Bedeckung

*f* ~ **crack** Mantel-, Oberflächen-riß *m* ~ **crazing** (very fine surface cracking) Haarrißformung *f* auf der Oberfläche ~ **cultivator** flacharbeitende Hackmaschine *f* ~ **current** Flächenstrom *m*, Oberflächenströmung *f* ~ **curvature** Flächenkurve *f* ~ **curve** Flächenkurve *f*, Spiegellinie *f* ~ **cutter** Flächenfräser *m* ~ **(and trespass) damage** Flurschaden *m* ~ **damping** Randdämpfung *f* ~ **decarburization** Oberflächenentkohlung *f* ~ **defect** Außenlunker *m*, Oberflächen-fehler *m*, -mangel *m* ~ **digger** Flachgreifer *m* ~ **digging machine** Flachbagger *m* ~ **discharge** Oberflächenentladung *f* ~ **disintegration** Verwitterung *f* ~**-distribution function** Belegungsfunktion *f* ~ **drying** (of lacquers) Antrocknen *n* ~ **duct** bodennaher Kanal *m* ~ **economizer** Oberflächenvorwärmer *m* ~ **effect** (of photocathode) Oberflächeneffekt *m*, Oberflächenwirkung *f*, ~ **element** Flächen-stück *n*, -teilchen *n* ~ **elements** Flächenelemente *pl* ~**-endurance testing machine** Abrieb-, Verschleiß-prüfmaschine *f* ~ **examination** Oberflächenkontrolle *f* ~ **facer** Oberflächenfräser *m* ~ **fall** Bodeneinsenkung *f* ~ **filler** (Lack) Flächenspachtel *f* ~ **film** Oberflächenfilm *m* ~ **finish** Oberflächen-beschaffenheit *f*, -güte *f* ~ **finishing** Oberflächenveredelung *f* ~ **float** Oberflächenschwimmer *m* ~ **foam** (on glassware) Feuerweiß *n* ~ **force** Oberflächenkraft *f* ~ **formation** Deckschicht *f* ~ **friction** Flächendruck *m*, Oberflächenreibung *f* ~ **gauge** Parallelreißer *m* ~ **gloss** Oberflächenglanz *m* ~ **glow** Glimmhaut *f* ~ **glow on the anode** anodische Glimmhaut *f* ~ **glow on cathode** kathodische Glimmhaut *f* ~ **glow lamp** Flächenglimmlampe *f* ~ **grinder (or grinders and rectifiers)** Flächenschleifmaschine *f* ~ **grinding** Flächen-schleifen *n*, -schliff *m*, Planschleifen *n* ~**-grinding machine** Oberflächen-, Plan-schleifmaschine *f* ~ **grub** Erdraupe *f* ~ **hardening** Oberflächenhärtung *f* ~ **hardness** Oberflächenhärte *f* ~ **harmonics** Kugelflächenfunktion *f* ~ **heat of charging** Oberflächenaufladungswärme *f* ~ **heat-transfer coefficient** Oberflächenwärmeübergangsbeiwert *m*, Wärmeübergangsbeiwert *m* ~ **hydrant** Oberflurhydrant *m* ~**-ignition engine** Glühkopfmaschine *f* ~ **imperfection** Oberflächenfehler *m* ~ **influence** Oberflächeneinfluß *m* ~ **interaction** Oberflächenwechselwirkung *f* ~ **interior wiring** Verlegung *f* über Putz ~ **lathe** Scheibendrehbank *f* ~ **lattice** Kreuzgitter *n* ~ **layer** Deck-, Oberflächen-, Rand-schicht *f* ~ **leakage** Kriechen *n*, Oberflächenableitung *f* ~ **leakage current** Kriechstrom *m* ~ **leakage loss** (dielectric) Belagverlust *m* ~**-leakage loss of a condenser** Kondensatorbelagverlust *m* ~**-leakage path** Kriechweg *m* ~ **line** Mantellinie *f* ~ **loading** Flächenanziehungsbelastung *f* ~ **loudspeaker** Flächenlautsprecher *m* ~ **luster** (Lack) Oberflächenglanz *m* ~ **measure(ment)** Flächenmaß *n* ~ **membrane** Oberflächenabdichtungsschicht *f* ~ **milling** Flächenfräsen *n*, Oberflächenfräsen *n* ~ **milling machine** Planfräsmaschine *f* ~ **moisture** grobe Feuchtigkeit *f* ~**-mounted switch** aufgesetzter Schalter *m* ~ **mounting** Aufputzmontage *f* ~**-navigation**

Flächennavigation *f* ~ **noise** Plattengeräusch *n*, Schallplattengeräusch *n* ~ **normal** Flächennormale *f* ~**-observation report** Bodenbeobachtungsmeldung *f* ~ **peat** Faser-, Moos-, Wurzel-torf *m* ~ **phase** Oberflächenphase *f* ~ **plate** Abreiß-, Anreiß-, Richt-platte *f* ~ **preparation** Oberflächenvorbehandlung *f* ~ **pressure** Anpreßdruck *m*, Flächen-druck *m*, -pressung *f*, Manteldruck *m*, Oberflächendruck *m* ~ **printing** (as distinguished from relief and intaglio printing) Flachdruck *m* ~ **printing roller** Reliefwalze *f* ~ **protection** Oberflächenschutz *m* ~ **quality** Oberflächengüte *f* ~ **radiator** Flügel-, Tragdeck-kühler *m* ~ **ray** Bodenstrahl *m* ~ **reflectivity** Oberflächenreflexion *f* ~ **removing** Oberflächenabtragung *f* ~ **residual (internal) stresses** (Oberflächen-) Eigenspannungen *pl* ~ **resistance** Flächen-, Oberflächen-widerstand *m* ~ **resistivity** spezifischer Oberflächenwiderstand *m* ~ **roller** Deckwalze *f* (hydr.) ~ **roughness** Oberflächenrauhigkeit *f* ~ **runoff** oberirdischer Abfluß *m* ~ **scale** Flächenmaßstab *m* ~ **scratch** Oberflächenkratzer *m* ~**-scratching test** Ritzhärteprobe *f* ~ **sharpeners and rectifiers** (emery or girt, for tolls, etc.) Flächenschleifmaschine *f* ~ **sheer stress** Randschubspannung *f* ~ **sizing** Oberflächenleimung *f* ~ **slope** Spiegelgefälle *n* ~ **smoothness** Flächenglätte *f* ~ **socket** Aufputzsteckdose *f* ~ **soil** Bekleidungserde *f* ~ **space-charge** Oberflächenladung *f* ~ **speed** Oberflächen-, Überwasser-geschwindigkeit *f* ~ **spray cooler** Berieselungskondensator *m* ~ **stability (or strength)** Oberflächenfestigkeit *f* ~ **stress** Oberflächenspannung *f* ~ **survey** Vermessung *f* über Tage ~ **table** Abreiß-, Anreiß-platte *f* ~ **tensiometer** Oberflächenspannungs-messer *m*, -prüfer *m* ~ **tension** Oberflächenspannung *f* ~ **traverser** unversenkte Schiebebühne *f* ~ **treatment** Oberflächen-bearbeitung *f*, -behandlung *f*, Vered(e)lung *f* ~ **trench** Obergraben *m* ~**-type instrument** Aufbauinstrument *n* ~ **value** (of pulp or cellulose) Flächenwert *m* ~ **vessel** Überwasserfahrzeug *n* ~ **vibration** Oberflächenrüttelung *f* ~ **voltage** Oberflächenspannung *f* ~ **water** Oberflächenwasser *n*, Tag(es)wasser *n* ~**-water mirror** Oberwasserspiegel *m* ~ **wave** Boden-, Oberflächen-welle *f* ~ **weathering** Verwitterung *f* der Oberfläche ~ **well** Flachbrunnen *m* ~ **winds** Flächenwinde *pl* ~ **wiring** Verlegung *f* auf Putz ~ **(panel) wiring** (of a switchboard) vorder-(rück-)seitige Beschaltung *f* ~ **work** Oberflächenwirkung *f* ~ **working** Tagebau *m* ~ **wrinkling** Runzelbildung *f* ~ **zero** Hypozentrum *n*

**surfacer** Grundierungsmittel *n*, Spachtelmasse *f* **any** ~ **that is sprayed on** Spritzspachtel *f* ~ **applied by drawing** Ziehspachtel *f* ~ **applied by pouring** Gußspachtel *f*

**surfaces** Flächen *pl* ~ **with a boundary** berandete Flächen *pl* ~ **of constant brightness** Flächen *pl* konstanter Helligkeit ~ **of constant Gaussian curvature** Flächen *pl* konstanter Gausscher Krümmung ~ **of constant mean curvature** Flächen *pl* konstanter mittlerer Krümmung ~ **of constant width** Flächen *pl* konstanter Breite ~ **of order higher than the third** Flächen

*pl* höherer als dritter Ordnung **~ of second order** (confocal) Flächen *pl* zweiter Ordnung **~ of the third order** Flächen *pl* dritter Ordnung
**surfacing** Belag *m*, Einebnen *n*, Glattmachung *f* der Oberfläche, Oberflächenbehandlung *f* **~ of a submarine** Auftauchen *n* **~ material** (Straße) Deckeneinbaumaterial *n* **~ pressure** Oberflächendruck *m*
**surfboard effect** Flüssigkeitskeilwirkung *f*
**surfeit** Überschuß *m*
**surficial gravity** Oberflächenschwere *f* (astron.)
**surfused** überschmolzen
**surge, to ~** (back and forth) schwingen **to ~ back and forth** hin und her schwingen, hin- und herschwingen **to ~ out** herausströmen
**surge** allgemeines Ansteigen *n* des Luftdruckes, atmosphärisches Aufwallen *n*, Brandung *f*, (Turbinenmotor) gestörte Verdichterförderung *f*, Pumpen *n*, Saugwelle *f*, Schwebung *f*, Schwingen *n*, Schwingung *f* (electr.), Stoß *m* (electr.), Stoßwelle *f*, Überspannung *f*, Wanderwelle *f*, Welle *f*, zu- und abnehmende Bewegung *f*
**surge, ~ absorber** Überspannungsfunkenstrecke *f* **~ arrester** Überspannungs-funkenstrecke *f*, -schutz *m* **~ capacitor** Stoßkondensator *m* **~ chamber** Ansaugkammer *f* (im Kraftstofftank), Druckraum *m*, Puffergefäß *n* **~ cock** Ausstoßhahn *m* **~ current** Ladestrom *m* **~ gap** Überspannungsfunkenstrecke *f* **~ generator** Impuls-geber *m*, -generator *m*, Wanderwellengenerator *m* **~ guard** (of relay) Pendelsperre *f* **~ impedance** Wellenwiderstand *m* **~ indicator** Wanderwellenmesser *m* **~ limit** Pumpgrenze *f* **~ load** Stoßbelastung *f* **~ peak** Einschaltspitze *f* **~ period** Einbruchperiode *f* **~-power generator** Stoßleistungsgenerator *m* **~-pressure test** Sprungwellenprobe *f* **~-proof** stoßspannungsfest **~ protection** Überspannungs-, Wanderwellen-schutz *m* **~ pump** Membranpumpe *f* **~ recorder** Wanderwellenmesser *m* **~-resisting fuse** träge Sicherung *f* **~ stress** (Transformator) Stoßbeanspruchung *f* **~ suppressor** Ableiter *m* **~ tank** Beruhigungsbehälter *m*, Buffertank *m*, Wasserschloß *n*, Windkessel *m*, Zwischenbehälter *m* **~ voltage** Stoßspannung *f*
**surgical, ~ cotton** Watte *f* **~ dressing** Verband *m* **~ operation** Operation *f* **~ wool** Verbandwatte *f*
**surging** ab- und zunehmende Bewegung *f*, Belastungsschwankung *f* (electr.), Hin- und Herwallen *n*, Pumpen *n*, Pumpen *n* eines Gebläses, Schwebung *f* **~ of the boost control** Schwingen *n* des Ladedruckreglers, Schwingen *n* des Ladedrucks **~ amplitude** Schwebungsamplitude *f*
**surmount, to ~** an Höhe übertreffen, überragen, übersteigen, überwinden
**surmounted** überhöht
**surmounting** Überwindung *f*; überstehend
**surpass, to ~** übertreffen
**surplus** Betriebsüberschüsse *pl*, Überschuß *m*, Zugabe *f*; überschüssig **~ of lime** Kalküberschuß *m* **~ of power** Kraftüberschuß *m* **as ~ to stock** überetatsmäßig
**surplus, ~ gas** Überschußgas *n* **~ heat** Wärmeüberschuß *m* **~ induction** Induktionsüberschuß *m* **~ power** Mehrleistung *f* **~ production**

Überproduktion *f* **~ stock** Plusbestand *m* **~-stock sugar** Übervorratszucker *m*
**surrender, to ~** aufgeben, ausliefern, kapitulieren, sich ergeben, übergeben
**surrender** Aufgabe *f*, Auslieferung *f*, Kapitulation *f*, Übergabe *f*, Unterwerfung *f* **~ of profits** Gewinnabfuhr *f* **~ value** Ablösungswert *m*, Rückkaufswert *m*
**surrogate** Ersatzmittel *n*, Surrogat *n*
**surround, to ~** abschließen, einkreisen, einschließen, umgeben, umkleiden, ummanteln, umringen, umschließen, umspülen **to ~ with a smoke screen** einnebeln **to ~ with walls** ummauern
**surround** Randwulst *f*
**surrounding** Umlagerung *f* **~ lead shielding** Bleidicke *f* **~ noise** Umgebungslärm *m*
**surroundings** Umgebung *f*, Umgegend *f*, Umkreis *m* **~ luminance** Umfeldleuchtdichte *f*
**surtax** Überzoll *m*
**surveillance** Überwachung *f*
**survey, to ~** aufnehmen, (a line) auskunden, besichtigen, durchmustern, messen, überschauen, übersehen, vermessen **to ~ a mine** eine Grube *f* aufnehmen **to ~ by stadia (or by tachymeter)** tachymetrieren
**survey** Aufnahme *f*, Aufnahme *f* des Geländes, Collinscher Hilfspunkt *m*, Gutachten *n*, Orientierung *f*, Überblick *m*, Übersicht *f*, Überwachung *f*, Vermessung *f*, Vermessungsarbeit *f*, zusammenfassender Bericht *m* **~ of cable route** Kabelauskundung *f* **~ and mapping unit** Vermessungs- und Kartenabteilung *f* **~ of parcels of land** Parzellenvermessung *f* **~ by serial photographs** Reihenaufnahme *f* **~ of the trajectory** Bahnvermessung *f*
**survey, ~ book** Abpfählbuch *n* (electr.), Trassierbuch *n* (teleph.) **~ camera** Meßbildkamera *f* **~ course** Repetitorium *n* **~ data** Vermessungsangabe *f* **~ detachment** Vermessungs-abteilung *f*, -trupp *m* **~ diagram** Übersichtszeichnung *f* **~ lens** Übersichtslupe *f* **~ line on surface** Zug *m* über Tage **~ map** Markierungslandkarte *f* **~ measurement** fotogrammetrische Kartierung *f* **~ photography** Übersichtsaufnahme *f* **~ plane** Vermessungsflugzeug *n* **~ point** Bodenpunkt *m* **~ report** zusammenfassender Bericht *m*
**surveyed** graphisch
**surveying** Auskundung *f*, Besichtigung *f*, Feldmessen *n*, Flurvermessung *f*, Meßkunde *f*, Prüfung *f*, Schürfen *n*, Vermessung *f*, Vermessungs-kunde *f*, -wesen *n* **~ apparatus** Aufnahmegerät *n*, Bildfänger *m* **~ camera** Meßbildkamera *f* **~-camera unit** Aufnahmekamera *f* **~ instruments** Feldmeßinstrumente *pl*, Meßgerät *n* **~ the land** Landesvermessung *f* **~ office** Vermessungsanstalt *f* **~ rod** Absteck-pfahl *m*, -pflock *m* **~ sextant** Prismenkreis *m*, Reflexionskreis *m* **~ ship** Vermessungsschiff *n* **~ underground** Markscheiden *n* **~ vessel** Vermessungsfahrzeug *n*
**surveyor** Feldmesser *n*, Geometer *m* **(land) ~** Landmesser *m*, Markscheider *m*
**surveyor's, ~ chain** Feldmesserkette *f*, Meß-band *n*, -kette *f* **~ compass** Stativkompaß *m* **~ flag** Absteckfähnchen *n* **~ level** Fernglaslibelle *f*, Gradbogen *m* **~ rod** Fluchtstab *m*,

Lachterstab *m*, Meßlatte *f* ~ **staff** Latte *f* ~ **steel wire arrows** Markiernägel *pl* ~ **tape** Meßband *n* ~ **transit** Vermessungsinstrument *n* ~ **wheel** Meßrad *n*

**survival, ~ curve** Überlebungskurve *f* ~ **kit** Notvorrat *m*

**survive, to** ~ überdauern

**surviving dependent** Hinterbliebener *m*

**survivor** Hinterbliebener *m*, Überlebender *m*

**susceptance** Blindleitwert *m*, Leitwerk *n*, Leitwert *m*, Rückwirkungsleitwert *m*, Suszeptanz *f*

**susceptibility (to disease)** Anfälligkeit *f*, Aufnahmevermögen *n*, Empfänglichkeit *f*, Magnetisierungskoeffizient *m*, Suszeptibilität *f* ~ **to attack** Angriffsfähigkeit *f* ~ **of being replaced** Ersatzfähigkeit *f* ~ **to stimulus** Reizempfindlichkeit *f*

**susceptibility, ~ binding** Neigung *f* zum Hängenbleiben ~ **corrosion** Korrosionsneigung *f* ~ **flakes** Flockenempfindlichkeit *f* ~ **interference (or noise)** Störanfälligkeit *f* ~ **meter** Suszeptibilitätsmesser *m* ~ **respond** Ansprechvermögen *n* ~ **scratching** Ritzbarkeit *f* ~ **shock** Stoßempfindlichkeit *f* ~ **trouble** Störanfälligkeit *f*

**susceptible** aufnahmefähig, empfindlich ~ **to design** berechnungsfähig ~ **to spotting** fleckenempfindlich ~ **to trouble** störanfällig ~ **to weld-cracking** schweißrißempfindlich

**suspend, to** ~ anschlämmen, aufhängen, aufschlämmen, aussetzen, (in eine Lösung) einhängen, (work) einstellen, hängen, hangen, schwebend halten, suspendieren, unterbrechen **to** ~ **a connection** eine Verbindung *f* aufheben **to** ~ **on gimbals** kardanisch aufhängen **to** ~ **payment** die Zahlungen *pl* einstellen **to** ~ **on springs** federnd aufhängen **to** ~ **work** den Betrieb *m* einstellen

**suspended** aufgehängt, fein verteilt, hängend, in hängender Anordnung, schwebend, (temporarily) zeitweilig eingestellt **to be** ~ freischweben, herabhängen, schweben ~ **in air** freihängend ~ **in points** spitzengelagert

**suspended, ~ body** Schwebekörper *m* ~ **contact box** Hängeanschlußdose *f* ~ **cover ring** Einhängedeckelring *m* (d. Breitstrahlers) ~ **insulator** Baumträgerisolator *m* ~ **isolator** Pendelisolator *m* ~ **lamp** Hängelampe *f* ~ **matter** Schwebestoff *m* ~ **particle** Schwebeteilchen *n* ~ **particle drier** Zerstäubungstrockner *m* ~ **particles** Schwebestoff *m* ~ **piston** freischwebender Kolben *m* ~ **pump** Bohrloch-, Tiefbohrpumpe *f* ~ **railroad** Hängebahn *f* ~ **railway** Schwebebahn *f* ~ **roof** Hängedecke *f* ~ **roof arch** Hängegewölbe *n* ~ **rotating coil** Drehspule *f* (electr.) ~ **substance** Schwebestoff *m* ~ **valve array** hängende Ventile *pl* ~ **wall** Hängewand *f*

**suspender** Aufhänger *m*, Schwebe *f*, Trageband *n* ~ **for aerial cables** Traghaken *m* für Luftkabel ~ **clip** Schnuranschlußklemme *f*

**suspending, ~ liquid** Grundflüssigkeit *f* ~ **wire** Aufhängedraht *m*, Tragdraht *m*, Tragseil *n*

**suspense** Aufschub *m*, Spannung *f* ~ **file** Arbeitsablage *f* ~ **items** Rechnungsabgrenzungsposten *pl*

**suspensing clip** Aufhängelasche *f*

**suspension** Aufhängung *f*, Aufhebung *f*, Aufschiebung *f*, Aufschlemmung *f*, Aufschub *m*, (spring) Federung *f*, feine Verteilung *f* in einer Lösung, Gehänge *n*, Hängen *n*, Lagerung *f*, Schwebe *f*, Suspendierung *f*, Suspension *f* **in** ~ schwebend ~ **with one degree of freedom** Einachslagerung *f* ~ **of diaphragm** Membraneinspannung *f* ~ **in double bows** Doppelbügelaufhängung *f* ~ **of gangway (or walkway)** Laufgangaufhängung *f* ~ **of mine work** Einstellung *f* des Betriebes einer Grube ~ **of the motor** Motoraufhängung *f* ~ **of payment** Zahlungseinstellung *f* ~ **of pendulum** Pendelaufhängung *f* ~ **from position** Dienstenthebung *f* ~ **of the proceedings** Aussetzung *f* des Verfahrens ~ **of the spring** Aufhängung *f* der Feder

**suspension, ~ air** Suspensionsluft *f* ~ **band** Traggurt *m*, Trapez *n* ~ **bar** Hängekorb *m*, Trapez *n* ~ **basket** Hängegondel *f* ~ **belt** Fanggurt *m* ~ **bridge** Hängebrücke *f* ~ **bucket** Hängekübel *m* ~ **cable** Draht-, Hänge-seil *n* ~ **cable of roller train** Drahtseil *n* zur Aufhängung der Rollenbahn ~ **cam** Aufhängenocken *m* ~ **chain** Hängekette *f* ~ **chains** Aufhängeketten *pl* ~ **clamp** Hängeklemme *f*, Seilschelle *f* ~ **clip** Traglasche *f* ~ **eye** Aufhängeöse *f* ~ **fastening** Einhängebefestigung *f* ~ **fiber** Aufhängefaden *m* ~ **frame** Aufhängerahmen *m* (rdr) ~ **gear** Aufhänge-gestell *n*, -vorrichtung *f*, Gehänge *n* ~ **hook** Aufhängehaken *m* ~ **insulator** (single hook) Aufhängeisolator *m*, Hängeisolator *m* ~ **junction box** Hängedose *f* ~ **keel** Versteifungskiel *m* ~ **line** Fang-, Hänge-, Trag-, Trage-leine *f* (aviat.) ~ **link** Aufhängestange *f*, Hängebügel *m*, Pleuelstange *f*, Schubstange *f* ~ **loop** Aufhänge--bügel *m*, -öse *f* ~ **lug** Aufhängelasche *f* ~ **lugs** Aufhängenasen *pl* ~ **member** Hängekörper *m* ~ **method** Schwebemethode *f* ~ **mooring** (of light pontoon bridges) Luftverankerung *f* ~ **patch** Aufhängung *f*, Stoffscheibe *f* ~ **pin** Aufhänge-, Hänge-achse *f* ~ **pin of float** Hängeachse *f* des Schwimmers ~ **pivot** Aufhängezapfen *m* ~ **plug socket connector** Hängesteckdose *f* ~ **power (or property)** Schwebefähigkeit *f* ~ **pulley** Aufhängerolle *f* ~ **rail** Hängeschiene *f* ~ **railroad** Hänge-, Schwebe-bahn *f* ~ **ring** Hängering *m* ~ **rods** Aufhängegestänge *n* ~ **runner** Gehänge *n* ~ **stirrup** Aufhängebügel *m* ~ **strand** Tragseil *n* ~ **switch** Schnurschalter *m* ~ **tower** Tragmast *m* ~ **truss** Hängebanklager *n* ~ **-type mounting** hängende Befestigung *f* ~ **wire** Aufhängedraht *m*, Hänge-, Trag-draht *m* ~ **work** Hängeboden *m*

**suspensions** Aufschlämmungen *pl*

**suspensory** Suspensorium *n* ~ **mounting** Aufhängegestell *n*

**sustain, to** ~ abfangen, tragen, unterstützen

**sustained** ungedämpft ~ **accuracy** Dauergenauigkeit *f* ~ **deviation** bleibende Regelabweichung *f* ~ **earth fault** Dauererdschluß *m* ~ **fire** Dauerfeuer *n* ~ **oscillation** ungedämpfte Schwingung *f* ~ **short-circuit current** Dauerkurzschlußstrom *m* ~ **signals** ungedämpfte Zeichen *pl* (aviat.) ~ **sinusoid** ungedämpfte Sinusschwingung *f* ~ **sound** gehaltener Ton *m* ~ **wave** kontinuierliche oder ungedämpfte Welle *f*

sustainer Marschtriebwerk *n*, (Marsch-)Rake-
tenmotor *m*
sustaining, ~ reaction von selbst ablaufende Reaktion *f* ~ voltage range Vorglimmlichtgebiet *n*
suture Lobenlinie *f* ~ of the twin plane Zwil-
lingsnaht *f* (cryst.) ~ clip Wundklammer *f* ~
line Lobenlinie *f*, Naht *f* ~-needle holder
Nadelhalter *m*
swab, to ~ abdweilen, dweilen, schwabbern
swab Abstrich *m*, Aufwischlappen *m*, Pistonier-
kolben *m*, Schwabbel *m*, Schwabber *m*
swabbing Pistonieren *n*, Schwabberstiel *m*
swadge Birner *m*
swadged nipple Reduziernippel *m*, Übergangs-
nippel *m*
swage, to ~ ankümpeln, einschnüren, einziehen, im Gesenk anspitzen, im Schmiedegesenk
anspitzen
swage Amboßgesenk *n*, Bohrbirne *f*, Gesenk *n*,
Präge *f*, Schmiedegesenk *n*, Stanze *f* ~ for
lowering casing Verrohrungsbirne *f* ~ for
shaping trusses Preßbacke *f*
swage, ~ anvil Gesenkamboß *m* ~ block
Gegenstanze *f*, Gesenkplatte *f*, Lochplatte *f*
~-pressed part Gesenkpreßteil *m* ~ work
Fassonschmieden *n*
swaged, ~ crossbar traction eye gesenkge-
schmiedete Kreuzlappenzugöse *f* ~ head Setzkopf *m* ~ wire verwickelter Draht *m*
swaging Gesenkarbeit *f* ~ die Breitsattel *m* ~
hammer Gesenkhammer *m* ~ machine Anspitzmaschine *f* (für Stangen) ~ mandrels
Stauchbacken *pl* ~ terminal aufgepreßtes
Seilendstück *n*
swallow, to ~ begierig aufnehmen, einsaugen,
schlucken, verschlingen, verschlucken
swamp, to ~ überdecken
swamp Moor *n*, Sumpf *m*, Torfmoor *n* ~ ore
Ortstein *m*
swamped, to be ~ vollschlagen ~ signals durch Störer verdeckte Zeichen *pl* ~ state Untergehen *n*
swamping (of spectral lines) Überstrahlen *n*,
Untergehen *n* ~ of picture signal Untergehen *n*
der Bildspannung ~ of signals Signalüberdek-
kung *f* ~ of video signal Untergehen *n* der
Bildspannung
swamps Polder *m*
swampy sumpfig, versumpft
swan, ~ band Swanband *n* ~ neck Schwanenhals
*m* ~-neck spindle Hakenstütze *f* ~ spectrum
Swanspektrum *n*
swarf breiiger Ölrückstand *m* mit Metallspänen,
Schleifstaub *m* ~ box Spänebehälter *m* ~
disposal Spanabfuhr *f* ~ gutter Spänerinne *f*
swarm Schwarm *m* ~ of particles Teilchenwolke *f*
swarming (of film) Kribbeln *n*, (of film) Rieseln
*n*, Würmerkriechen *n* (film)
swartzite Swartzit *m*
swashplate Taumelscheibe *f* ~ bearings Tau-
melscheibenlager *n* ~ engine Taumelscheibenmotor *m*, Trommelmotor *m* ~ injection pump
Taumelscheibeneinspritzpumpe *f* ~ scissor
lever Taumelscheibenscherenhebel *m*
swatch (of fabric) Länge *f* Tuchmuster *m*
swath, ~ of mist Nebelschwaden *m* ~ board
Schwadenblech *n* ~ level position Schwad-
stellung *f*
swatter Löschwedel *m*

sway, to ~ aufheißen, wackeln
sway brace Feststellpratze *f*, Verschwertung *f*
swaying Querschwingungen *n* ~ of antenna
Antennenschwingung *f*
swear, to ~ beeidigen
sweat, to ~ ausschwitzen, bähen, heißlöten,
schwitzen, weichlöten to ~ out (tin) aussaigern, ausseigern, saigern, seigern
sweat cooling Schwitzkühlung *f*
sweated together feuerverlötet
sweating (of lacquers) Ausschwitzen *n*, Bähung *f*,
Feuerlötung *f*, Schweißung *f*, Schwitzen *n*,
(of tin) Seigern *n*, Weichlöten *n* ~ gutter
Seigergasse *f* ~ heat Abschweißwärme *f*,
Entzunderungswärme *f* ~-out Saigerung *f*
~ stove Schwitzkammer *f* ~ system Antreibesystem *n*
Swedish, ~ charcoal iron schwedisches Holzkohleneisen *n* ~ iron schwedisches Eisen *n* ~
wrench Rollgabelschlüssel *m*
sweep, to ~ abtasten, (fire) bestreichen, bürsten,
fegen, schweifen, übersehen, überstreichen,
wobbeln (rdr) to ~-back pfeilen to ~ a channel
das Fahrwasser räumen to ~ down (wind)
herunterkommen to ~ by fire mit Feuer *n*
bestreichen to ~ with fire abstreuen (artil.) to
~ off abkehren to ~ out ausfegen, bestreichen
to ~ over bestreichen to ~ up kehren
sweep Absuchen *n*, Anschlußbogen *m*, Ausladung *f*, Ausladung *f* einer Radialbohrmaschine (mech.), Bereich *m*, (a line) Durchlauf
*m*, Hinlauf *m*, Kipp *m*, (iron rails) seitliche
Krümmung *f*, Krümmung *f*, Schleife *f*,
Schwenkung *f*, Spur *f* (rdr), Weg *m* eines
drehbaren Armes, Zeitablenkung *f*, Zeitlinie *f*
(rdr) ~ over front axle Kröpfung *f* am Vorderrahmen ~ out of television screen Überstreichen *n* des Leuchtschirmes ~ of wing
Tragflügelpfeilung *f*
sweep, ~ amplifier Ablenkverstärker *m* ~ amplitude Ablenkungsweite *f*, Kippamplitude *f*
~ apparatus Kippgerät *n* ~-back Pfeil-form *f*,
-stellung *f* (aviat.) ~-back angle Pfeilstellungs-
winkel *m* (aviat.) ~-back wing Pfeilflügel *m*,
Pfeilflügelform *f*, Pfeilformflügel *m* ~ circuit
Kipp-anordnung *f*, -gerät *n*, -schaltung *f*,
Strahlkippeinrichtung *f* ~ circuits with high-
-vacuum tubes (hard tubes) Kippschaltungen *pl*
mit Vakuumröhren ~ circuits with two-element
glow lamps Kippschaltungen *pl* mit Zweipol-
glimmlampen ~-coil pair Kippspulenpaar *n*
~ condenser Kippkondensator *m* ~-fly-back
ratio Hinlauf-, Rücklauf-verhältnis *n* ~ forward negative Pfeilform *f* ~ frequency Kippfrequenz *f* ~-frequency oscillograph Frequenz-
kurvenschreiber *m* ~ gauge Übergreiflehre *f*
~ generator Kippgenerator *m*, Wobbelmeß-
sender *m* ~ net Wurfnetz *n* ~ oscillator Ab-
lenkungsoszillator *m* ~ process Kippvorgang
*m* ~ radiation Störstrahlung *f* ~ rake Heu-
schubrechen *m*, Schlepprechen *m* ~ rod
Schiebegestänge *n* ~'s rods Einschiebege-
stänge *n* ~ second-hand ständig umlaufender
Sekundenzeiger *m* ~ trace Zeitlinie *f* ~ transformer Kipptransformator *m* ~ tube Kipp-
röhre *f* ~ unit Ablenkgerät *n*, Zeitablenkungs-
gerät *n* ~ voltage Ablenkungsspannung *f*,
Kippspannung *f*

**sweeper** Aufräumer *m*, Decksaufklarer *m*

**sweeping** Schweifung *f* ~ **with an electron pencil** Strahlabtastung *f* ~ **of a frequency range** Überstreichung *f* eines Frequenzbereiches ~ **with a light pencil** Strahlabtastung *f*

**sweeping,** ~ **coil** Ablenkungsspule *f* ~ **device** (in molding) Schabloniervorrichtung *f* ~ **fire** Breitenfeuer *n*, Schießen *m* mit wechselnder Seitenrichtung ~ **force** (of water) Schleppkraft *f* ~ **machine for gold leaf** Abkehrmaschine *f* für Blattgold, Blattgoldabkehrmaschine *f* ~**-out** Abtastung *f* ~ **yoke** Ablenkungsspule *f*

**sweepings** Fegsel *n*, Goldkrätze *pl*, Schutt *m*, Spillage *f*

**sweet,** ~ **tester** Kondenswasserprüfer *m* ~ **water** Absüßer *pl*, Absüßwasser *n* ~**-water channel** Absüßkanal *m* ~**-water pump** Absüßpumpe *f* ~**-water spindle** Absüßspindel *f*

**sweeten, to** ~ absüßen, versüßen

**sweetened gasoline** angesüßtes Benzin *n*

**sweetening** Entschwefelung *f* ~ **of rubber** Weichwerden *n* des Kautschuks ~ **chemical** Versüßungsmittel *n*

**sweetness** Süßigkeit *f*

**sweets** Füllmasseknoten *m*, Knoten *m* im Zucker, Zuckerknoten *m*

**swell, to** ~ anschweben, anschwellen, anwachsen, aufgehen, aufquellen, blähen, pauschen, quellen, sich ausbauchen, sich bauschen, stauchen, steigen, treiben, wachsen, zunehmen **to** ~ **up** aufblähen, aufquellen, aufschwellen

**swell** Anschwellung *f*, Ausbauchung *f*, Schwall *m*, Seegang *m* ~ **box** Schwellkasten *m* ~ **organ** Echowerk *m*, Schwellwerk *n* ~ **pedal** (organ) Schwellpedal *n* ~**-proof** quellfest

**swelled,** ~ **neck rivet** Niet *m* mit gefrästem Schaftende ~ **up** aufgedunsen

**swelling** Anfluß *m*, Aufbauschung *f*, Aufblähen *n*, Auffederung *f*, Aufquellen *n*, Anschwellung *f*, Anwachsen *n*, Beule *f*, Bund *m*, Geschwulst *f*, Hebung *f*, Quaddel *f*, Quellung *f*, Schwellung *f*, Treiben *n*, Wachsen *n* ~ **of lime** Gedeihen *n* des Kalkes

**swelli·g,** ~ **capacity** Quellvermögen *n* ~ **colloid** Quellungskolloid *n* ~ **guard** Bundhaken *m* ~ **property** Blähungsgrad *m*, Quellbarkeit *f*, Quellungsvermögen *n* ~ **substance** quellbarer Körper *m* ~ **test** Blähprobe *f*

**swept,** ~ **bend** Rohrbug *m* ~ **capacity** Hubraum *m*, Hubvolumen *n*, Zylinderinhalt *m* ~ **forward wing** Flügel *m* mit negativer Pfeilform ~**-frequency oscillator** Wobbelgenerator *m* ~ **gain** zeitabhängige Stärkeregelung *f* ~ **saltpeter** Gaysalpeter *m* ~ **volume** Hubraum *m*, Hubvolumen *n*, Zylinderinhalt *m* ~ **wing** Pfeilflügel *m* (aerodyn.) ~**(-back)wing** gepfeilter Flügel *m*

**swerve** Ausbrechen *n* (aviat.)

**swift** Ablaufkrone *f*, Trömmel *f*; geschwind

**swifter** Fendertau *n*

**swiftness** Schnelle *f*

**swim, to** ~ schwimmen

**swimmer** Schwimmer *m*

**swimming** Schwimmen *n*; flott

**swing, to** ~ ausholen, ausschlagen, hin und her schwingen, pendeln, schaukeln, schlenkern, schwenken, schwingen, schwoien, sich in Angeln drehen **to** ~ **away (or down)** abklappern

**to** ~ **back** zurückpendeln, zurückschwenken **to** ~ **for deviation** Deviation *f* bestimmen **to** ~ **down** herunterklappen **to** ~ **off** abschwenken **to** ~ **on** einschwenken **to** ~ **out** abklappen, ausschwenken, herausschwenken **to** ~ **out (or away)** wegklappen **to** ~ **out of the way** wegschwenken **to** ~ **over** ausschwenken **to** ~ **the propeller** anwerfen, durchdrehen **to** ~ **round** sich drehen **to** ~ **sideways** abschwenken **to** ~ **through** durchschwingen

**swing** Abweichung *f*, Ausladung *f*, Ausschlag *m*, Gestängeschwinge *f*, Kantung *f*, Nachwirkung *f*, Schaukel *f*, Schwenkausschlag *m*, Schwenkbewegung *f*, Schwinganschlag *m*, Schwingen *n*, Schwung *m*, Verkantung *f*, Zeigerausschlag *m* ~ **of amplifier grid** Steuerungsbereich *n* ~ **of a chord** Schwingungsausschlag *m* einer Saite ~ **of frequency** Frequenzabweichung *f* ~ **of a pendulum** Pendelschwingung *f*

**swing,** ~ **arm** Schwingarm *m* ~ **arm bracket** Schwingarm-Lagerbock *m* ~ **arm discharge** Schwenkarmzapfvorrichtung *f* ~**-arm traverse motion** Pendel-, Schaukel-changierung *f* ~**-back** neigbares Hinterteil *n* (photo) ~ **bar** Schwenklineal *n* ~ **base** Wippe *f* ~ **beam** Prügel *m*, Schwingbaum *m* ~ **bearing** Pendellager *n* ~ **bolt** Gelenk-, Klapp-schraube *f* ~ **boom** drehbarer Ausleger *m* ~ **bridge** Drehbrücke *f*, fliegende Brücke *f*, Gierbrücke *f*, Rollbrücke *f* ~ **bucket** Drahtseilkübel *m*, Hängegefäß *n*, Hänge(bahn)kübel *m* ~**-bucket elevator** Schaukelbecherwerk *n* ~ **cam** schwenkbarer Schloßteil *m* ~ **check** Sperrscheibe *f* ~ **check valve** Scharnierventil *n* ~ **clearing cam** schwenkbarer Heber *m* ~ **crane** Drehkran *m*, Wanddrehkran *m* ~ **crushing jaw** schwingende Brechbacke *f* ~ **diameter** Schwingdurchmesser *m* ~**-down** nach unten klappbar ~ **frame** (of a motor-cycle) Schwingrahmen *m* ~ **gate** Drehschranke *f* ~ **gear shifter** Räderschwinge *f* ~**-hammer mill (or -hammer pulverizer)** Hammermühle *f* ~ **joint** Drehanschluß *m*, Gelenkrohr *n*, Gelenkverbindung *f*, Kugelgelenk *n* ~ **lever operation typewriter** Schwinghebelmaschine *f* ~**-out** aufklappbar ~**-out condenser** aufklappbarer Kondensor *m*, Gelenkkondensator *m* ~**-out arm (or arm)** Klappteil *n* ~**-out mandrel** ausschwenkbarer Einlagedorn *m* ~**-out plate** Ausschwingplatte *f* ~**-out table** Schwenktisch *m* ~**-over bed** Drehdurchmesser *m* (über dem Bett) ~**-over gap** Drehdurchmesser *m* (in der Kröpfung) ~**-over lens** Vorschaltlinse *f* ~**-over saddle** Drehdurchmesser *m* (über dem Bettschlitten) ~ **phenomenon** Kipperscheinung *f* ~ **pipe** Schwenk-, Schwing-rohr *n* ~ **plate** Schwingbrett *n* ~ **rest** schwenkbarer Support *m* ~ **roller** Schwingtrommel *f* ~ **saddle** Schwingsattel *m* ~ **sieve** Rätterwäsche *f*, Schwingsieb *n* ~ **stop** Schwenkanschlag *m* ~ **support** (bridge) Kipplager *n* ~ **table** Schwenktisch *m* ~ **tool holder** Schwingstahlhalter *m* ~ **turning diameter** Drehdurchmesser *m* der Drehbank ~**-up** hochklappbar ~ **wheel** Hemmungsrad *n*

**swingable** schwenkbar

**swinging** Ausschlagen *n*, Frequenzschwankung *f*, Pendel *n*, Pendelung *f*, Schwenkung *f*; pendelnd, umlaufend ~ **of antenna** Antennenschwingung *f*

**swinging,** ~ **arm** Schwenkhebel *m* ~ **armature** Wippanker *m* (zum Impulsschütz) ~ **away** wegwendbar ~ **back** (of pendulum) Rückschwingung *f* ~ **base** Ablenkungsanzeiger *m* ~ **boom** Backspiere *f* ~**-bucket elevator** Schaukelbecherwerk *n* ~ **choke** Drossel *f* mit veränderlicher Induktivität, Siebdrossel *f* ~ **crane** Drehlaufkatze *f*, Schwenkkran *m* ~ **device** Schwenkvorrichtung *f* ~ **disc** Schwingscheibe *f* ~ **filament** schwingender Draht *m* ~ **gripper cutout bar** Stange *f* zur Abseilung ~ **gripper lever** Vorgreiferhebel *m* ~ **haystacker** drehbarer Heustapler *m*, Heustapler *m* mit Hochwinde und drehbarer Abladevorrichtung ~ **hoist** Schaukelaufzug *m* ~ **isolator with suspending hook** Pendelisolator *m* ~ **lever** Pendelanker *m* (teleph.), Schwenkarm *m* ~ **lever work** Schwingankerwerk *n* ~ **machine** Schwingmaschine *f* ~**-out** Ausklappen *n*; ausklappbar ~ **platen** Drucktiegel *m* ~ **platform** Schwenkbühne *f* ~ **the propeller** Propeller *m* (mit der Hand) schwingen ~ **roller** Schwingwalze *f* ~ **round of a ship** Wenden *n* eines Schiffes ~ **shoot** Pendelschurre *f* ~ **support** Schwenklager *n* ~ **tool holder** Schwingwerkzeughalter *m* ~**-type trailer coupling** (schwenkbare) Anhängerkupplung *f* ~ **valve** Pendelventil *n*

**swingletree** Waagebalken *m*, Zugscheit *n*

**swipe beam (or counterpoise) of a drawbridge** Zugrute *f* einer Zugbrücke

**swirl,** ~ **angle** Verwindungswinkel *m* ~ **baffles** Dralleinsatz *m* ~ **chamber** Wirbelkammer *f* ~**-nozzle** (Rakete) Dralldüse *f* ~ **plug** Düsenkegel *m* ~ **port** Einlaßdrallkanal *m* ~ **vane** Drallblech *n* ~ **vanes** Dralleinsatz *m*

**swirling,** ~ **of metal** Badbewegung *f* ~ **motion** Durchwirbelung *f*

**swishtailing** den Schwanz *m* von einer Seite zur anderen schwingen (aviat.)

**Swiss,** ~ **bush-type automatic lathe** Langdrehautomat *m* ~ **welding method** Linksschweißung *f*

**swissing** Kalandrieren *n* unter Druck

**switch, to** ~ abschalten (electr.), ausschalten, unterbrechen **to** ~ **on after starting** nach dem Start einschalten **to** ~ **into circuit** einschalten **to** ~ **off ignition** Zündung *f* ausschalten **to** ~ **on** andrehen, anschalten, (to) an die Leitung anlegen, zuschalten **to** ~ **out** abschalten **to** ~ **over** überschalten **to** ~ **in a repeater** Verstärker *m* zünden

**switch** Absperr-glied *n*, -organ *n*, -teil *m*, -vorrichtung *f*, Ausleseknopf *m*, Ausschalter *m*, (closing) Einschalter *m*, Griff *m*, Kommutator *m*, Schalter *m*, Schlüssel *m*, Stromtrenner *m*, (cutoff) Unterbrecher *m*, (selective or automatic) Wähler *m*, Weiche *f*, Zunge *f* ~ **for by-pass light signal** Lichthupenschalter *m* ~ **with counterpoise** Weiche *f* mit Gegengewicht ~ **and crossing ties** Weichenrost *m* ~ **for excitation** Umschalter *m* für Erregung ~ **for generating set** Aggregatschalter *m* für Erregung ~ **for generating set** Aggregatschalter *m* ~ **for reversing light** Rückfahrschalter *m* ~ **with sliding plates** Plattenweiche *f* ~ **with spring contacts** Federschalter *m* ~ **with spring tongs** Federweiche *f*

**switch,** ~ **apparatus** Schaltapparat *m* ~**-arcs** Schaltbögen *pl* ~ **armature** Schalteranker *m* ~ **armature spring** Schalterankerfeder *f* ~ **base** Schaltersockel *m* ~ **base plate** Schaltergrundplatte *f* ~ **blade** Weichenzunge *f* ~**-blade planing machine** Weichenzungenhobelmaschine *f*

**switchboard** Bedienungstafel *f*, Platzgruppe *f*, Schalt-brett *n*, -pult *n*, -schrank *m*, -tafel *f*, -wand *f*, -werk *n*, Schrank *m*, Umschalter *m*, Umschalteschrank *m*, Vermittlungsschrank *m* ~ **for intercommunication sets** Klappenschrank *m* für Reihenanlagen

**switchboard,** ~ **cable** Schrank-, System-kabel *n* ~ **cord** Platzschnur *f* ~ **drop annunciator** (signal) Anrufklappe *f* ~ **frame** Schalttafelgerüst *n* ~ **fuse bases** Schalttafelsicherungssockel *m* ~ **panels** Schalttafelplatten *pl* ~ **position** Abfragestellung *f*, Platz *m*, Vermittlungsplatz *m* ~ **rack** Schalttafelgestell *n* ~ **receiver** Koppeltischempfänger *m* ~ **section** Schrankabteilung *f* ~ **system** Schranksystem *n* ~ **transmitter** Koppeltischgeber *m* ~ **wire** Abfrageschnur *f*

**switch,** ~ **body** Weichenkörper *m* ~ **box** Ausschlußdose *f*, Schalt-dose *f*, -kasten *m*, -schrank *m*, -werk *n*, Stellwerk *n*, Vermittlungskästchen *n*, Weichenbock *m* ~ **breaker** Hebelschalter *m* ~ **cabinet** Schalt-dose *f*, -schrank *m*, -tafel *f*, -werk *n* ~ **capacitance box** Kurbelkondensator *m* ~ **case** Schaltkasten *m* ~ **change** Schaltungsänderung *f* ~ **clerk** Schrankbeamter *m* ~ **clock** Schaltuhr *f* ~ **cock** Schalthahn *m* ~ **column** Schaltsäule *f* ~ **commutator** Umschalter *m* ~ **connection** Bremsschaltung *f* ~ **construction works** Weichenbauanstalt *f* ~ **contact plate** (Scheinwerfer) Schalterkontaktplatte *f* ~**-control room** Eisenbahnstellwerk *n* ~ **correction of traverse** Seitenänderung *f* ~ **cover** (for gear) Schaltdeckel *m* ~ **current coil** Schalterstromspule *f* ~ **desk** Schaltpult *n* ~ **detector** Tastdetektor *m* ~ **element** Schaltmittel *n* ~ **engine** Rangierlokomotive *f* ~ **frame** Schaltergestell *n*, Zuteilungsgestell *n* ~ **fuse** Trennsicherung *f* ~ **gallery** Schaltwarte *f* ~ **gear** Schalt-anlage *f*, -anlagengerät *n*, -schrank *m*, -tafel *f*, -vorrichtung *f*, -werk *n* ~ **gear distributing (junction) box** Schaltdose *f* ~ **gear panel** Schaltfeld *n* ~ **grease** Fett *n* für Ausschalter ~ **handle** Schaltknebel *m* ~ **handle position** Schaltgriffstellung *f* ~ **handle support** Schaltgriffauflage *f* ~ **hook** Hakenumschalter *m* (teleph.) ~ **inductance box** Kurbelinduktivität *f* ~ **interval** Umschaltepause *f* ~ **jack** Feder-, Messer-kontakt *m* ~ **knob** Schaltknopf *m* ~ **lamp** Weichenlaterne *f* ~ **lead** Schaltkabel *n* ~ **lever** Fahrschalthebel *m*, Schalt-hebel *m*, -knebel *m*, Stell-hebel *m*, -vorrichtung *f*, Umlegeschalter *m*, Umleghebel *m*, Umleithebel *m* (r.r.), Weichen-hebel *m*, -stellbock *m* ~ **line** Zweigstellung *f* (r.r.) ~ **machine** Weichenantrieb *m* ~**man** Rangierer *m*, Verschieber *m*, Weichensteller *m* ~ **matrix** Schaltmatrix *f* ~ **method** Flimmerpeilung *f* ~ **motion** Schaltbewegung *f* ~**-off position** Ausschaltstellung *f* ~**-on peak** Einschaltstromspitze *f* ~**-operating mechanism** Weichenstellvorrichtung *f* ~**-over flap** Umschaltklappe *f* ~**-over valve** Umschaltventil *n* ~ **outlet** Schal-

terdose f ~ **panel** Ausschalttafel f ~ **plant**
Schaltanlage f ~ **plate** Grundplatte f der
Weiche ~ **plug** Umschaltstöpsel m ~ **point**
Weiche f, Zungenschiene f (r.r.) ~ **point on
railroad** Herzstück n ~ **position** Schaltstellung f
~**-position indicator** Zungenüberwachungsein-
richtung f ~ **post** Schaltdorn m ~ **push button**
Schalterdruckknopf m ~ **resistance box** Kurbelwiderstand m ~ **room** Apparat-raum m,
-saal m, Schaltraum m, Wähler-raum m, -saal
m ~ **shaft** Schalterachse f ~ **shelf** Wäh-
lerrahmen m ~ **(deflecting) side guard** Ablenk-
führung f ~ **siding** Aufstellgleis n (r.r.) ~ **signal**
Weichensignal n ~ **socket** Hahnfassung f (für
Glühlampe), Schalterfassung f ~ **spring**
Klinke f, Schalterfeder f ~ **stand** Weichen-
back m ~**-stand handle** Weichenstellvorrichtung f ~ **step** Schaltstufe f ~ **system** Schaltschema n, Schaltungsplan m ~ **tank** Schalter-
kessel m ~ **tie** Weichenschwelle f ~ **tongue**
Weichen-schiene f, -zunge f ~ **trench** Annäherungsgraben m ~ **valve** Umschaltventil n ~
**wafer** Schalterplatte f ~ **work** Schalterarbeit f
~ **yard** Freiluftschaltanlage f, Rangierbahnhof m
**switchable** umschaltbar
**switched,** ~ **off (or out)** ausgeschaltet ~ **cardioid** (direction finder) Wechselkardioide f ~
**distribution box** Schalterverteilungskasten m
**switcher** Rangierlokomotive f
**switching** Rangierung f, Schalten n, Schaltung f,
Umschalten n, Umschaltung f, Vermittlung f
~ **action** Fortschaltung f ~ **amplifier** Schleusen-
verstärker m (Regelpult) ~ **arrangement**
Schalteinrichtung f, Schaltungsanordnung f ~
**bar** Schaltstange f ~ **central** Fernsprechvermittlung f ~ **characteristic** Schaltkennlinie f
(Transistor) ~ **coefficient** Oberflächenverluste
pl ~ **condition** Schaltungsbedingung f ~ **contrivance** Schaltvorrichtung f ~ **criterium**
Schaltkennzeichen n ~ **cycle** Schaltspiel n ~
**desk** Schaltertisch m ~ **device** Rangieranlage f,
Schalt-adereinrichtung f, -mittel n ~ **effect**
Schalteffekt m ~ **element** Verknüpfungsglied n
(comput.) ~ **equipment** Schaltadereinrichtung f
~ **frequency** Schalterfrequenz f ~ **gear** Um-
schaltgetriebe n ~ **handle** Arbeitskurbel f ~**-in**
Aufschalten n (teleph.), Einschaltung f ~
**incline** Ablaufberg m ~ **key** Schaltschlüssel m
~ **lever** Schaltwippe f ~ **locomotive** Verschiebelokomotive f ~ **magnet** Schaltmagnet m ~
**magnet circuit** Schaltelektromagnetenstrom-
kreis m ~ **mechanism** Schalt-mechanismus m,
-vorrichtung f ~ **member** Schaltorgan n ~
**network** Schaltkreise pl ~**-off** Abschalten n,
Abschaltung f (electr.), Ausschaltung f ~
Außerbetriebsetzung f ~ **operation** Schaltvorgang m ~ **pad office** Amt n mit künstlichen Ver-
längerungsleitungen ~ **period** Schaltspanne f
~ **point** Abzweigstelle f, Schaltstelle f ~ **principles** Grundsätze pl für die Herstellung der
Verbindungen ~ **process** Schaltvorgang m ~
**pulse** Schaltimpuls m ~ **relay** Schalt-organ n,
-relais n, Umschaltrelais n, Verteilerrelais n
~ **selector repeater** Mitlaufwerk n, (dial
system) Überbrückungsschaltung f ~ **station**
Verteilerwerk n ~ **system** Umschaltungs-
system n ~ **time** Schaltzeit f ~ **track** Rangier-

gleis n, Verschiebegleis n ~ **trunk** Fernplatzrufleitung f, (toll) Vermittlungsleitung f ~
**valve** Schaltröhre f ~ **variable** Schaltvariable f
(comput.) ~ **work** Verschiebedienst m ~ **yard**
Abstell-, Aufstell-, Verschiebe-bahnhof m
**swivel, to** ~ schwenken **to** ~ **back** zurückschwenken **to** ~ **through x degrees** um X Grad
drehen
**swivel** Ankerspannschraube f, Bohrwirbel m,
Dreh-gelenk n, -haken m, -ring m, -teil m,
-wirbel m, Einhängestück n, Kardandrehzapfen
m, Kettenwirbel m, Klammer f, Lyra f, Riemenbügel m, Schenkel m, Spannschloß n, Spül-
(bohr)kopf m, Wirbel m **to get out sweep as far
as** ~ **shackle** Gerät n bis Wirbelschakel ausbringen ~ **of toolholder** Drehteil m des Werk-
zeughalters
**swivel,** ~ **arm** Schwenkarm m ~ **axis** Schwenk-
achse f ~ **axis of index head** Drehachsmitte f
des Teilkopfes ~ **axle** Achsschenkel m ~**(ing)
base** Drehunterteil m ~ **bearing** Drehlager f ~
~ **bearing of drive system** Antriebsschwenk-
lager n ~ **block** Kenterschäkel m ~ **carriage**
Drehtisch m, Schwenkplatte f ~**(ling) carriage**
(of a machine tool) Drehschlitten m ~ **caster**
Schwenkrad n ~ **chair** Drehstuhl m ~ **connection** Gelenkverbindung f, schwenkbare Verbindung f ~ **damper** Drehklappe f, Drehschieber m ~ **doll** Gelenkpuppe f ~ **eye** Gelenköse f, Wirbelauge n ~**-feature** Schwenkbarkeit f ~ **guide bar** Leitlineal n ~ **guide bar
for taper work** Konuslineal n (schwenkbar)
~ **head** Bohrwirbel m, Dreh-support m, -vor-
richtung f ~**-headed tripod** Verfolgungsstativ n
~ **hinge** Drehscharnier n ~**-hoist truck** Kran-
karre f, Straßendrehkran m ~ **hook** drehbarer
Lasthaken m, Haken m mit Drehgelenk,
Wirbelhaken m ~ **joint** Knochengelenk n,
Kugellagerverbindung f, Schwenkverschrau-
bung f ~ **loom** Broschierwebstuhl m ~
**-mounted** schwenkbare Wippe f ~ **neck** Wirbel-
hals m ~ **pan** Kippkessel m ~ **pen** Kurvenziehfeder f ~ **pin** Drehachse f, Schwingbolzen m
~ **plus keystone link** Kabelziehschlauch m
~ **pulley** Wirbelblock m ~ **ring** Gelenköse f
~ **screws** Anziehschrauben pl ~ **shackle** Kettenwirbel m ~ **spindle** Wirbelspindel f ~ **table**
Tischschwenkplatte f ~ **union** Gelenkverbindung f ~ **wheel** schwenkbares Rad n, Schwenkrad n ~ **wrench** Wirbelgabel f
**swivelled** ausgeschwenkt
**swivelling** drehbar, schwenkbar ~ **ball bearings**
Drehkugellager n ~ **bolster** Drehschemel m ~
**boom** Drehausleger m ~ **bracket** Schwenk-
lagerung f ~ **device** Schwenkvorrichtung f
~**-jib** Drehausleger m ~ **link** Drehgelenk n ~
**milling spindle** schwenkbare Frässpindel f ~
**nipple** schwenkbares Rohranschlußstück n ~
**nozzle** schwenkbare Düse f ~ **piece** Drehgelenk n ~ **propeller** Schwenkschraube f
(VTOL aircraft) ~ **soldering nipple** schwenkbares Ringlötstück n ~ **template** Schwenk-
schablone f ~ **tool-holder** schwenkbarer
Stahlhalter m ~ **V-block** Schwenkprisma n
**swollen** aufgedunsen
**swoop,** to ~ **over** überfegen
**sword** Degen m, Klinge f, Säbel m, Schwert n ~
**belt** Degenkoppel f, Koppel n ~ **blade** Degen-

klinge *f* ~ **grip** Degengriff *m* ~ **hilt** Degengefäß *n* ~ **knot** Faustriemen *m*, Portepee *n* ~**like** schwertförmig ~**like part** schwertartiger Teil *m* ~ **point** Degenspitze *f* ~ **pommel** Degenknopf *m* ~**-shaped part** schwertartiger Teil *m* ~ **sheath** Degenscheide *f*

**sworn,** ~ **declaration** eidliche Versicherung *f* ~ **statement** beglaubigte Angabe *f*, eidliche Versicherung *f* ~ **testimony** eidliche Versicherung *f*

**S-wrench** S-Schlüssel *m*

**swung** geschwungen, verkantet **in** ~**-in position** im eingeschwenkten Zustand *m*

**sycamore** Sykomore *f*

**syllable** Silbe *f* ~ **articulation** Silbenverständlichkeit *f* ~ **intelligibility** Silbenverständlichkeit *f*

**syllabus** Lehrplan *m*, Verzeichnis *n* der Vorlesungs- oder Prüfungsgegenstände

**sylvanite** Schrifterz *n*, Sylvanit *m*, Tellursilberblei *n*, Weiß-golderz *n*, -tellur *n*

**sylvite** Hartsalz *n*, Sylvin *n*

**symbol** Betrachtungszeichen *n*, Bezeichnung *f*, Formelzeichen *n*, Kurzzeichen *n*, Sinnbild *n*, Symbol *n*, Zeichen *n* ~ **for earth** Erdungszeichen *n* ~ **of radio-circuit element** Schaltzeichen *n* ~ **for rate of acceleration** Vekt.

**symbol printing** Zeichenschreibung *f*

**symbolic** sinnbildlich ~**(al)** symbolisch ~ **address** Pseudonym *n* (info proc.), symbolische Adresse *f* ~ **coding** symbolische Kodierung *f* ~ **representation** symbolische Darstellung *f*

**symbolization** Versinnbildlichung *f*

**symmetric** spiegelbildlich gleich, spiegelig, symmetrisch ~ **line** Symmetrielinie *f* ~ **wave** Gleichtaktwelle *f*

**symmetrical** gleich-artig, -mäßig, symmetrisch ~ **about an axis** achsensymmetrisch ~ **with respect to rotation** rotationssymmetrisch ~ **about vertical axis** höhensymmetrisch

**symmetrical,** ~ **alternating quantity** symmetrische Wechselgröße *f* ~ **arrangement** Symmetrieschaltung *f* ~ **flight** gleichmäßiges Fliegen *n* ~ **fold** aufrechte Falte *f* ~ **heterostatic circuit** heterostatische Schaltung *f* ~ **line of flange** Flanschsymmetrielinie *f* ~ **profile** spiegelgleiches oder symmetrisches Profil *n*, Streckenprofil *n* ~ **quadripole** symmetrischer Vierpol *m* ~ **section** spiegelgleiches Profil *n* ~ **transducer** symmetrischer Wandler *m* (acoust.) ~ **transformer** symmetrischer Übertrager *m* (electr.) ~ **two-terminal pair network** symmetrischer Vierpol *m*

**symmetrically tilted photographs** gleichmäßig geneigte Aufnahmen *pl*

**symmetrics** Symmetrien *pl*

**symmetrization** Symmetrierung *f*

**symmetrize, to** ~ symmetrieren

**symmetry** Ebenmaß *n*, Gleichmaßanordnung *f*, Spiegeligkeit *f*, Symmetrie *f* **having point (or radial)** ~ punktsymmetrisch ~ **of lattice** Gittersymmetrie *f* ~ **of reflection** Spiegelungssymmetrie *f*

**symmetry,** ~ **circuit** Symmetrieschaltung *f* ~ **operations (of the lattice)** Deckoperation *f*

**Symons disc crusher** Symons-Tellermühle *f*

**sympathetic,** ~ **chord (or string)** mitschwingende Saite *f* ~ **vibration** Mitschwingung *f*, Resonanzschwingung *f*

**sympiesometer** Sympiezometer *n*

**symplectic,** ~ **group** symplektische Gruppe *f* ~ **matrix** symplektische Matrix *f*

**symposium** Zusammenstellung *f*

**symptom** Anzeichen *n*, Begleiterscheinung *f*, Kennzeichen *n*, Krankheitszeichen *n*, Vorzeichen *n*

**synch generator** Impulszentrale *f*, Taktgeber *m* (TV)

**synchro** Drehmelder *m*, Gleichlaufeinrichtung *f*, Synchro *m* ~**-angle** Gleichlaufwinkel *m* ~ **control transformer** extern gesteuerte Gleichlaufeinrichtung *f* ~ **cyclotron** Synchrozyklotron *n* ~ **differential receiver** Differentialsynchroempfänger *m* ~**dyne reception** Synchrodynempfang *m* ~**-magnetic jigsaw** Elektromagnet-Laubsägemaschine *f*

**synchromesh** Synchrongetriebe *n* ~ **(gear)** Gleichlaufgetriebe *f* ~ **transmission** Aphongetriebe *n*

**synchronism** Gleich-gang *m*, -gängigkeit *f*, -lauf *m*, -zeitigkeit *f*, Synchronismus *m* **in** ~ im Rhythmus **to be in** ~ (of phases) übereinstimmen ~ **of the fork** Gleichlauf *m* der Gabel

**synchronization** Gleichlaufschaltung *f*, Mitnahme *f*, Steuerung *f*, Synchronisierung *f* ~ **of the carrier frequency** Synchronisierung *f* des Trägerstroms ~ **of watches** Uhrvergleich *m*

**synchronization,** ~ **control** Gleichlaufregelung *f* ~ **signal** Koinzidenzsignal *n* ~ **stretcher** Pegelstabilisierung *f* ~ **troubles** Gleichlaufstörung *f*

**synchronize, to** ~ abstimmen, eine Frequenz *f* nach einer anderen einstellen, zum Gleichlauf bringen, gleichschalten, gleichstellen, synchronisieren, in Tritt kommen, in Übereinstimmung bringen **to** ~ **watches** die Uhren *pl* vergleichen

**synchronized** gesteuert, gleichlaufend, synchronisiert ~ **by power supply** netzsynchronisiert ~ **from supply mains** netz-, quarz- und fremdsynchronisierbar

**synchronized,** ~ **capacitors** gleichlaufende Kondensatoren *pl* ~ **condenser** gleichlaufender Kondensator *m*, Mehrfach-, Mehrgang-kondensator *m* ~ **elevator** Synchronhöhensteuer *n* ~ **multivibrator** fremdgesteuerter Multivibrator *m* ~ **system** Taktverfahren *n*

**synchronizer** Synchronismusanzeiger *m* ~ **for two guns** Doppelschußgeber *m* ~ **for two guns with electrical coupling** Doppelgeber *m* mit elektrischer Geberkupplung

**synchronizing** mit der Grundfrequenz in Tritt kommen; synchronisierend ~ **of image** Bildsynchronisierung *f*

**synchronizing,** ~ **amplifier** Synchronisierverstärker *m*, Synchronpegel *m* ~ **arm** Synchronisierungsarm *m* ~ **baffle** Abstimmblende *f* ~ **band** Bildsynchronisierlücke *f* ~ **circuit** Synchronisieranlage *f*, Synchronisierungsschaltung *f* ~ **device** Gleichlaufeinrichtung *f*, Synchronisier-einrichtung *f*, -vorrichtung *f* ~ **element** Gleichlaufzeichen *n* ~ **frequency** Steuerfrequenz *f* ~ **gap** Bildsynchronisierlücke *f* ~ **gear** Synchrongetriebe *n*, Synchronisiervorrichtung *f*, Synchronisierungsgetriebe *n* ~ **generator** Taktgeber *m* (TV) ~ **grid** Synchronisiergitter *n* ~ **impulse** (correcting) Gleich-

laufimpuls *m*, Gleichlaufzeichen *pl*, Synchronisier-impuls *m*, -zeichen *n*, Taktimpuls *m* ~
**knob** Synchronisierknopf *m* ~ **leader** Synchro-
nisierband *n* ~ **level** Synchronisierungspegel *m*
~ **mark** Synchronisierzeitmarke *f* ~ **method** Synchronisierverfahren *n* ~ **pilot** syn-
chronisierende Pilotwelle *f* ~ **potential** Syn-
chronisiersignal *n* ~ **pulse** Gleichlaufsig-
nal *n* ~ **pulse generator** Synchronisierim-
pulsgenerator *m* ~**-pulse separation** Impuls-
trennung *f* ~ **separation** Synchronisierimpuls-
-abtrennung *f*, -aussiebung *f* ~ **signal** Gleichlauf-signal *n*, -zeichen *n*, Synchronisier-impuls
*m*, -zeichen *n* ~**-signal generator** Gleichlauf-
generator *m* ~ **voltage** Summerspannung *f*
**synchronometer** Synchronmesser *m*
**synchronoscope** Synchronoskop *n*
**synchronous** gleich-gängig, -gehend, -laufend,
-zeitig, synchron, synchronistisch, zusammenfallend ~ **to crystal** Quarzsynchron *n*
**synchronous, ~ alternator** Synchrongenerator *m*
~ **(electric) clock** Synchronuhr *f* ~ **converter**
Einankerumformer *m* ~ **detector** gleichlaufender Demodulator *m* (TV) ~ **electric clock
system** elektrisches Synchronuhrensystem *n*
~ **fading** synchrone Schwundserscheinung *f* ~
**induction motor** synchronisierter Asynchronmotor *m* ~ **machine** Synchronmaschine *f* ~
**motor** Synchronmotor *m* ~ **operation** Synchron-
lauf *m*, Taktgeberbetrieb *m* ~ **position indicator**
Synchronoskop *n* ~ **revolutions per minute**
synchrone Drehzahl *f* ~ **scanning** Synchron-
abtastung *f* ~ **spark-gap** Synchronfunken-
strecke *f* ~ **speed regulating gear** Gleichlauf-
regeleinrichtung *f* ~ **timer** Gleichlaufzeit-
schalter *m* ~ **vibrator** Synchronzerhacker *m* ~
**voltage** Gleichlaufspannung *f*
**synchro, ~ phasemeter** Synchrophasenmesser *m*
~ **receiver** Synchronhörer *m* ~**scope** Syn-
chroskop *n* ~**-tie** elektrische Welle *f* ~**ton**
Synchroton *n* ~**-transmitter** Synchronsender *m*
**synclinal** gleiche Neigung *f* mit anderen Schichten; muldenförmig ~ **flexure of stratum** Mul-
denfalte *f* ~ **formation** Muldenbildung *f* ~
**valley** Synklinaltal *n*
**syncline** Mulde *f*, Synklinal *n*
**synclinorium** Synklinorium *n*
**syndicate** Konsortium *n*, Syndicat *n*
**syndrome** Syndrom *n*
**synergism** Synergismus *m*
**syngenite** Syngenit *m*
**synonymous** gleich-bedeutend, -sinnig ~**(ly)**
sinnverwandt
**synopsis** Abriß *m*, Montage *f*, Tabelle *f*, Tafel *f*,
Übersicht *f*, Zusammenfassung *f*
**synoptic** ~ **chart** synoptische Karte *f*
~ **meteorology** synoptische Meteorologie *f* ~
**report** synoptische Meldung *f* ~ **table** Über-
sichtstabelle *f* ~ **telegram** Sammeltelegramm *n*
**synoptical sketch** Übersichtsskizze *f*
**synthesis** Aufbau *m*, Synthese *f*, Zusammensetzung *f*
**synthesize, to** ~ künstlich herstellen
**synthetic** Kunst-produkt *n*, -stoff *m*, Preßmasse
*f*; aufbauend, künstlich, synthetisch ~ **cellulose fiber** Zellfaser *f* ~ **compound** Kunstmasse *f*
~ **fiber** Kunst-, Spinn-faser *f* ~ **fuel** synthetischer Brennstoff *m* ~ **gasoline** Kunstbenzin *n*

~ **index** künstlich geschaffener Kennwert *m*
~ **material** Kunststoffbezug *m* ~ **molding
resin** Kunstharzpreßstoff *m* ~ **nitric acid** Luft-
salpetersäure *f* ~ **pig iron** synthetisches Roheisen *n* ~**-plastic material (or plastics)** Kunstharzpreßstoff *m* ~ **product** Kunst-produkt *n*,
-stoff *m* ~ **resin** Edelkunstharz *n*, Kunst-harz
*n*, -stoff *m* ~**-resin cement** Kunstharzleim *m*
~**-resin product** Kunstharzerzeugnis *n* ~ **rubber**
Buna *n*, künstliches Gummi *n* ~ **wool made of
cellulose** Zellwolle *f*
**synthetize, to** ~ synthetisieren
**syntonic** abgestimmt ~ **wireless telegraphy** abgestimmte Funktelegrafie *f*
**syntonization** Abstimmung *f*
**syntonize, to** ~ abstimmen
**syntonized** in Resonanz
**syntonizing, ~ coil** Abstimmspule *f* ~ **inductance**
Abstimminduktanz *f*
**syntony** Abstimmung *f*, Einklang *m*, Resonanz *f*,
Tonbeständigkeit *f*, Tongleichheit *f*, Vorhandensein *n* der Abstimmung *f* **not in** ~ nicht abgestimmt
**syphon, to** ~ **off** absaugen
**syphon** Ansaugheber *m* ~ **barometer** Saug-
heber-Röhrenbarometer *n* ~ **bellows** Feder-
balgdichtung *f* ~ **pipe** Heberohr *n* ~ **recorder**
Heberschreiber *m*
**syringe, to** ~ gauzen
**syringe** Handspritze *f*, Heber *m*, Injektions-
spritze *f*, Spritze *f*
**syringes with radiation protection** Spritzen *n* mit
Strahlenschutz
**syringing** Einspritzung *f*
**syruper** Saftvorfüller *m*
**system** Anordnung *f*, Einrichtung *f*, Formation
*f*, Gebilde *n*, Netz *n*, System *n*, Verfahren *n*
**system, ~ of arrows** Zeigersystem *n* ~ **of asso-
ciative fibers** Assoziationssystem *n* ~ **of
averages** Durchschnittssystem *n* **(coordinate)** ~
**of axes** Achsenkreuz *n* ~ **of baffles** Leitblech-
system *n* ~ **of beds** Gebirgsformation *f* ~ **of
bells** Glockenanlage *f* ~ **of buffer battery**
Pufferbetrieb *m* ~ **of circulation** Umlauf-
system *n* ~ **of coils** Spulensatz *m* ~ **of control**
Steuerungsart *f* ~ **of coordinates** Achsenkreuz
*n*, Koordinaten-achsen *pl*, -kreuz *n*, -system *n*
~ **of crystallization** Kristallsystem *n* ~ **of
current** Stromgattung *f* ~ **of curves** Kurvenschar *f* ~ **of debiting agents** Istsystem *n* ~ **of
duties and preferences** Abgaben- und Präferenz-
regelung *f* ~ **of equations** Gleichungssystem *n*
~ **of fits** Passungssystem *n* ~ **of forces** Kräfte-
system *n* ~ **of forks** Gabeltransportmechanis-
mus *m* ~ **of framework** Fachwerkgebilde *n* ~
**of graphs** Kurvenschar *f* ~ **of heating flues**
Heizkanalsystem *n* ~ **of lenses** Linsen-folge *f*,
-satz *m*, -system *n* ~ **of levers** Hebelverbindun-
gen *pl* ~ **of measuring** Maßsystem *n* ~ **of
meridians** Kartennetz *n* ~ **of non-linear
equation** nichtlineares Gleichungssystem *n* ~
**of ore dressing** Stammbaum *m* ~ **of parallels**
Kartennetz *n* ~ **of particles** Teilchensystem *n*
~ **of piping** Röhrenwerk *n* ~ **of points** Punkt-
system *n* ~ **of the principal axes** Hauptachsen-
kreuz *n* ~ **of progressive gear** Durchzugsschal-
tung *f* ~ **of roads** Wegenetz *n* ~ **of rotor
spiders** Speichensystem *n* ~ **of space axes**

Raumachsenkreuz *n* ~ **of spheres** Kugelpackung *f* ~ **of spherical coordinates** Kreis-
system *n* ~ **of ties** Schwellenrost *m* ~ **of units**
Einheitensystem *n* ~ **of using averages** Durchschnittssystem *n* ~ **of ventilation** Lüftungsart *f*
~ **of waveguides** Hohlleitersystem *n* ~ **of wires**
**and cables** Verteilungsleitungen *pl* ~ **of working**
Abbausystem *n* (min.), Betriebsweise *f*
**system,** ~ **capacity** Netzergiebigkeit *f* ~ **con-**
**stant** Systemkonstante *f* ~ **function** Stamm-
funktion *f* ~ **lead** Strahlengang *m* ~ **line** Netz-
linie *f* ~ **line-ups** Einjustierung *f* ~ **locking**

Systemarretierung *f* ~ **noise** Apparategeräusch
*n* ~ **parameter** Zustandsgröße *f* ~ **performance**
Systemverhalten *n* (servo) ~ **planning** Anlagen-
planung *f* ~**-repeater drawer** Systemverstärker-
lade *f* ~ **trigger** Hauptsteuerimpuls *m* ~
**variable** Zustandsgröße *f* ~ **voltage** Netzspannung *f*
**systematic** planmäßig, rationell, systematisch
~ **reaction** Gesamtverhalten *n*
**systematics** Systematik *f*
**systematology** Systematik *f*
**systogene** Systogen *m*

# T

**T, (flanged)** ~ T-Stück *n*
**tab** Hilfs-klappe *f*, -ruder *n* (aviat.), Lappen *m*, Nase *f* (an Stanzteilen), Öhr *n*, Streifen *m*, Vorsprung *m* **(collar)** ~ Kragenspiegel *m*
**tab,** ~ **angle protractor** Trimmkantenwinkelmesser *m* ~ **index card** Leitkarte *f* ~ **speed** Sammelgang *m* ~ **washer** Sicherungsblech *n* mit Lappen oder Nase
**table, to** ~ verkämmen, verscherben
**table** Auflagetisch *m*, Datentabelle *f*, Liste *f*, Planke *f*, Planscheibe *f*, Tabelle *f*, Tafel *f*, Tisch *m*, Übersicht *f*, (of figures) Zahlentafel *f*
**table,** ~ **of abbreviations** Abbreviaturensatz *m* ~ **of altitudes** Höhentafel *f* ~ **of box type (or box type slotted)** Aufsparnkasten *m* ~ **of clearances and fits** Passungsliste *f* ~ **of contents** Inhalts-angabe *f*, -verzeichnis *n* ~ **of co-ordinates** Koordinatenverzeichnis *n* ~ **of correction** Eichtabelle *f* ~ **of cutting speeds** Schnittgeschwindigkeitsschild *n* ~ **of deviation** Deviationstabelle *f* ~ **of diagram lines** Tabelle *f* der Serienlinien ~ **of dimensions** Maßblatt *n* ~ **of distances** Abstandtafel *f* ~ **of equipment** Ausrüstungsnachweisung *f* ~ **of errors** Fehlertafel *f* ~ **of hardwood** Hartholzplatte *f* ~ **of lighting-up times** Brennkalender *m* (aviat.) ~ **of logarithms** Logarithmentafel *f* ~ **of minimum clearance** Tiefenfeuertafel *f* ~ **of movements** Transportübersicht *f* ~ **of organization strength** Planstärke *f* ~ **of sags** Durchhangtabelle *f* ~ **of specifications** Maßblatt *n* ~ **of values** Wertetabelle *f*
**table,** ~ **adjusting gear** Tischeinstellungsgetriebe *n* ~ **adjustment** Tischverstellung *f* ~ **arm** Tischkonsol *n* ~ **-arm clamping screw** Tischkonsolklemmschraube *f* ~ **bracket** Tischträger *m* ~ **bracket screw** Konsolschlittenspindel *f* ~ **carpet** Tischläufer *m* ~ **carrier (or socket)** Tastbolzenhalter *m* ~ **centerpiece** Tafelaufsatz *m* ~ **centerpin** Tischzapfen *m* ~ **clamp screw** Tischklemmschraube *f* ~ **concentrate** Herdkonzentrat *n* ~ **concentration** Herdarbeit *f* ~ **control-lever** Tischsteuerhebel *m* ~ **drill press** Tischbohrmaschine *f* ~ **-elevating screw** Spindel *f* zur Senkrechtstellung des Tisches ~ **feed** Tischvorschub *f*, Tischvorschub *m* ~
**table-feed,** ~ **handle** Handgriff *m* für die selbsttätige Tischbewegung ~ **lever** Tischvorschubhebel *m* ~ **screw** Längsvorschubspindel *f* ~ **trip block** Ausrückkloben *m*
**table,** ~ **floor** Plattenbelag *m* ~ **frame** Tischbakke *f* ~ **gear rim** Planscheibenzahnkranz *m* ~ **grating (or grid)** Tischrost *m* ~ **guides** Tischgleitbahn *f* ~ **hand vise** Bankkloben *m* ~ **land** Blachfeld *n*, Hochebene *f* ~ **leaf** Tischblatt *n* ~ **locking bolt** Tischklemmschraube *f* ~ **mat** Untersetzer *m* ~ **movement** Tischbewegung *f* ~ **moving across and endwise** Spannplatte *f* mit Kreuzschlitten ~ **moving crosswise and endwise** kreuzbeweglicher Tisch *m* ~ **plate** Tischplatte *f* ~ **power traverse** Tischselbstgang *m* ~ **protector** Plattenblitzableiter *m* ~ **raising nut** Tisch-

mutter *f* ~ **relief** Tischentlastung *f* ~ **rim** Tischzarge *f* ~ **roller** Rollgang *m* ~ **section** Herdabteilung *f* ~ **service** Tafelgeschirr *n* ~ **set** Tischapparat *m*, -empfänger *m* ~ **slide longitudinal feed screw** Tischschlittenwaagerechtbewegung *f* ~ **speed** Planscheibenumlauf *m* ~ **speed-indicator** Tischdrehzahlanzeiger *m* ~ **stand** Tischstativ *n* ~ **stop** Tischanschlag *m*, Tischstillsetzung *f* ~ **sweeper** Tischkehrmaschine *f* ~ **synopsis** Liste *f* ~ **-telephone station** Tischgehäuse *n* ~ **track** Planscheibengleitbahn *f* ~ **travel** Tischhub *m* ~ **-travel control** Klemmgriff *m* zum Ausrücken des Tischselbstganges ~ **traverse** Tischbewegung *f* ~ **treatment** Herdbearbeitung *f* ~ **tripod** Tischständer *m* (film) ~ **work** Herdarbeit *f*
**tables** Register *n*, Tabellen *pl*
**tablet** Pastille *f*, Tablette *f*, Tafel *f* ~ **chart** Schreibplatte *f* ~ **counter** Tablettenzählmaschine *f* ~ **protector** Plattenblitzableiter *m*
**tabling** Herdarbeit *f*
**tabular** tabellarisch, tafelförmig, tafelig ~ **compilation** Tabellenwerk *n* ~ **data** in Zahlentafel zusammengestellte Werte *pl* ~ **deposit** plattenförmige Lagerstätte *f* ~ **form** Tabellenform *f* ~ **inserts** Tabulator-einlagen *pl*, -reiter *pl* ~ **synopsis** Übersichtabelle *f* ~ **work** Tabellensatz *m*
**tabularize, to** ~ tabellarisieren
**tabulate, to** ~ abflachen, aufführen, katalogisieren, listenförmig oder tabellarisch zusammenstellen, tabellarisieren
**tabulated** abgeflacht, abgeschliffen **can be** ~ läßt sich tabellarisch darstellen **in** ~ **form** tabellarisch
**tabulating machine** Tabelliermaschine *f*
**tabulation** Aufstellung *f*, Tabellarisierung *f*, zahlenmäßige Aufstellung *f*
**tabulator** Tabelliermaschine *f*, Tabulator *m* ~ **button** Tabulatorknopf *m* ~ **stop rack** Tabulatorzahnstange *f*
**TACAN (tactical air navigation system),** ~ **beacon** TACAN-Funkfeuer *n* ~ **mode** Betriebsart *f* TACAN
**tach feedback** Drehzahlgeber-Rückführspannung *f*
**tacho,** ~ **alternator** Tachogenerator *m* ~ **generator** Tacho-dynamo *n*, -generator *m* ~ **gram** Tachogramm *n* ~ **graph** Drehzahlschreiber *m*, Geschwindigkeitsschreiber *m*, Tachograf *m*
**tachometer** Drehzahl *m*, Drehzahl-geber *m*, -messer *m*, Ferntachometer *n*, Geschwindigkeitsmesser *m*, Reibungsdrehzahlenmesser *m*, Stichzähler *m*, Streckenmeßtheodolit *m*, Tachometer *n*, Tourenzähler *m*, Umdrehungs-messer *m*, -zähler *m*, Umfangsgeschwindigkeitsmesser *m*, Umlaufzähler *m*, Wirbelstromdrehzahlmesser *m* ~ **drive** Tachometerantrieb *m* ~ **generator** Drehzahlgeber *m* ~ **mounting** Tachometereinbau *m* ~ **recorder** Tachograf *m*
**tachoscope** Drehzahlmesser *m*, Tachoskop *n*
**tachygraph** Geschwindigkeitsmesser *m*

**tachymeter** Tachymeter *m*
**tachymetric** tachymetrisch ~ **survey** tachymetrische Aufnahme *f* ~ **traverse** Tachymeterzug *m*
**tachymetry** Schnellmessung *f*, Tachymetrie *f*
**tack, to** ~ (naut.) drehen, heften, lavieren, nageln, (naut.) sich winden **to** ~**-weld** heftschweißen
**tack** Backbordhalsen *pl*, Kammzwecke *f*, Nagel *m*, Pinne *f*, Stift *m*, Täck *m*, Zwecke *f* ~ **bolt** Verbindungsbolzen *m* ~ **detector** Nagelsucher *m* ~ **rivetting** Heftniet *n* ~ **welding** Heft-nietung *f*, -schweißung *f*, Stiftschweißen *n*
**tacked connection** Haftverbindung *f*
**tacking** Handnaht *f* ~ **effect** Hafteffekt *m* ~ **rivet** Heftniet *n*
**tackle, to** ~ anpacken, in Angriff nehmen **to** ~ **a problem** eine Aufgabe *f* anpacken
**tackle** Flasche *f*, Flaschenzug *m*, Kleingerät *n*, Rollenzug *m*, Takel *n*, Takelung *f*, Taukloben *m*, Tauwerk *n*, Zugwinde *f* ~ **block** Flaschenzug *m* ~ **fall** Taljeläufer *m*, Zugseil *n* ~ **hook** Blockhaken *m* ~ **line** Seilzug *m*
**tacky** klebrig
**tact** Takt *m*
**tactical** taktisch
**tactics** Gefechtslehre *f*, Taktik *f*
**tactile** taktil, tastbar ~ **cell** Tastzelle *f* ~ **corpuscle** Tastkörper *m* ~ **device** Taster *m* ~ **forces** Befühlungskräfte *pl* ~ **indicator** Tastzeichen *n* (telef. und f. Blinde) ~ **lever** Fühlhebel *m* ~ **part** Tastbolzen *m*, Taster *m* ~ **sensitiveness** Fingerspitzengefühl *n*
**tactometer** Taktometer *n*
**taenite** Taenit *m*
**taffeta** Taft *m*
**taffrail log** Heckrelinglog *n*, Patentlog *n*
**tag, to** ~ kennzeichnen
**tag** Anhänger *m*, Anhängezettel *m*, Etikett *n*, Lötklemme *f*, Lötöse *f*, Lötstift *m*, Marke *f*, Markierung *f* (data proc.) ~**-on calculation** Anhängekalkulation *f* ~ **paper** Papier *n* für Anhänger ~ **strip** Löt(ösen)streifen *m* ~ **wire** Etikettendraht *m*
**tagged atom** markiertes Atom *n*
**tagging compound** Markierungsverbindung *f*
**tail, to** ~ befestigen, verbinden
**tail** Ansatz *m*, Ausläufer *m*, (bomb) Bodenschwanz, Endstück *n*, Heck *n*, Schwanz *m*, (Kran) Schwenge *f*, (of a thermometer) Tauchspindel *f*, Unterflügel *m*
**tail, ~ and blinker lamp** Bremsblinkschlußleuchte *f* ~ **of a dressing machine** Überschlag *m* der Sichtmaschine ~ **of a forge hammer** Hammerschwanz *m* ~ **of the fuselage** Rumpfende *n* ~ **and number plate lamp** Schlußkennzeichenleuchte *f*, Schlußnummernleuchte *f* ~ **of projectile** Geschoßzapfen *m* ~ **of the wicket** (weirs) Betätigungsflügel *m*
**tail, ~ assembly** Leitwerk *n* ~ **bands** Schwanzbänder *pl* ~ **base** Heckboden *m* ~ **bay** Balkenfach *n* ~ **bearing** Schwanzlager *n* ~ **board** hintere Klappe *f*, Ladeklappe *f* ~ **boom** Gitterschwanzträger *m* (aviat.), Leitwerksträger *m* ~**-boom strut** Gitterschwanzstrebe *f* ~**-box water seal** Vorlage *f* ~ **buffeting** Leitwerkschütteln *n* ~ **carrier** Schwanzwagen *m* ~ **center** Gegen-, Reitstock-spitze *f* ~ **chain** (of buoy)

Schwanzkette *f* ~ **chute** Bremsfallschirm *m* ~ **cone** (Außentank) Endkegel *m*, (aircraft) Heckkonus *m*, Schubdüsenkegel *m* ~**-controlled** schwanzgesteuert ~ **decalage** Heckschränkung *f* ~**-down landing** Landung *f* mit großem Anstellwinkel ~ **edger** (sawmill) Holzbesäumer *m* ~ **end** Bleirohrkabel *n* zwischen Überführungskasten und Freileitung, hinteres Ende *n*, (of fuselage) Schwanzende *n* ~ **explosion** Heckexplosion *f* ~ **fin** Kielflosse *f*, Leitbrett *n*, Schwanz-, Seiten-flosse *f* ~**-first airplane** Enten-flugmodell *n*, -flugzeug *n* ~ **flange** Abtriebsflansch *m* ~ **flap** Schmutzfänger *m* ~ **float** Schwanzstützschwimmer *m* ~ **flow of a barrage** Abfluß *m* einer Talsperre ~ **flutter** Leitwerkflattern *n* ~ **force** Leitwerkskraft *f* ~ **fuse** Heckzünder *m* ~ **gate** Gegenschütze *m*, (sluice) Niedertor *n*, Untertor *n* ~ **group** Leitwerk *n*, Schwanz-einheit *f*, -leitwerk *n* ~ **group area** Leitwerkflächeninhalt *m* ~ **hammer** Schwanzhammer *m* ~**-heaviness** Schwanzlastigkeit *f* ~**-heavy** hinter-, rumpf-, schwanz-lastig (aviat.) ~ **helve** Schwanzhammer *m* ~ **hook** Greifhaken *m* ~ **hull** Heckrumpf *m*, Heckzelle *f* (ohne Einbauten) ~ **journal** Schwanzzapfen *m* ~ **lamp** Schluß-licht *n*, -leuchte *f* ~**-lamp bracket** Halter *m* für die Schlußlaterne ~ **lamp cable** Schlußlichtleitung *f* ~ **landing** Schwanzlandung *f*
**tail-light** Hecklicht *n*, Schluß-laterne *f*, -leuchte *f*, -licht *n* ~ **chamber** Schlußlichtkammer *f* ~ **margin** unterer Rand *m* (der Buchseite) ~ **window** Schlußlichtfenster *n*
**tail, ~ lighting** Schlußbeleuchtung *f* ~ **lime liquor** letzter Äscher *m* ~ **load** Heck-, Schwanz-last *f* ~**-load moment** Schwanzlastmoment *n* ~ **lock** Ablaufschleuse *f* (hydraul.) ~ **loop** Heckschleife *f* ~ **machine gun** Heckmaschinengewehr *n* ~ **mass** Leitwerkmasse *f* ~ **moment** Leitwerkmoment *n* ~ **nozzle** Heckdüse *f* ~ **parachute** (Heck)bremsschirm *m* ~ **piece** Anhang *m*, Fallenhebel *m*, Finalstock *m*, Schlußleiste *f*, -stück *n*, -vignette *f*, Schwanzstück *n*, Zierleiste *f* (print.) ~ **pin** Reitstockstift *m* ~ **pipe** Abgas-, Strahl-rohr *n* ~ **pipe burning** Nachverbrennung *f* ~ **pipe cone** Strahlrohr-Austrittskegel *m*
**tail-plane** Dämpfungsflosse *f*, Höhenflosse *f*, Höhenleitfläche *f*, Leitwerk *n*, Schwanzfläche *f* ~ **adjusting wheel** Höhenflossenverstellrad *n* ~ **area** Höhenflossenfläche *f*, Schwanzflächeninhalt *m* ~ **stabilizer** Dämpfungs-fläche *f*. -flosse *f* ~ **stabilizer tip** Flossenende *n* ~ **trimming gear** Höhenflossenverstellung *f*
**tail, ~ position of rib** Rippenendstück *n* ~ **pulley** Fußrolle *f* ~ **race** Abfall-, Abzugs-graben *m* ~ **rack** Hinterwand *f* ~**-rod antenna** Heckstabantenne *f* ~ **rope** Hinterseil *n*, offenes Seil *n*
**tail-rotor** Heckrotor *m* ~ **blade** Heckrotorblatt *n* ~ **drive gear** Heckrotorantriebsgetriebe *n* ~ **grip** Heckrotorblattanschluß *m* ~ **hub** Heckrotorkopf *m* ~ **hub bearings** Heckrotorkopflager *pl* ~ **pitch change** Heckrotorblattverstellung *f*
**tail, ~ section** (Außentank) Endkegel *m*, Leitwerk *n*, Rumpfheck *n* ~ **shaft** Kurbelwellenstumpf *m* ~**-shaft bracket** Wellenbock *m* ~ **shield** Kopf-, Schwanz-schild *n*

**tailskid** Heckstütze *f*, Notsporn *m*, Schleifsporn *m*, Schwanz-kufe *f*, -sporn *m*, (aircraft) Sporn *m*, Spornkufe *f* ~ **arrest** Spornrückschlag *m* ~ **control lever** Spornhebel *m* ~ **shoe** Spornschuh *m* ~ **spring** Spornfeder *f* ~ **springing** Spornfederung *f* ~ **stop** Spornrückschlag *m* ~ **towing** Spornschlepp *m*

**tail,** ~ **slide** Abrutschen *n* nach hinten oder über den Schwanz (aviat.), rückwärtiges Abrutschen *n* ~ **spin** Trudeln *n* (aviat.) ~-**spin with power off** Leerlauftrudel *m* ~-**spin with power on** Vollgastrudel *m* ~-**spin curve** Trudellinie *f* ~-**spin speed (or velocity)** Trudelgeschwindigkeit *f* ~ **spindle** Pinole *f*, Reitnagel *m* ~-**spindle lock** Reitnagelklemmhebel *m* ~ **stabilizer chord** Höhenflossentiefe *f* ~-**steering airplane** schwanzgesteuertes Flugzeug *n* ~ **step** Heckstufe *f*

**tailstock** Reitstock *m* ~ **barrel** Reitstock-oberteil *n*, -pinole *f* ~ **base** Reitstock-platte *f*, -untersatz *m*, -unterteil *m* ~ **handwheel** Reitstock-handrad *n* ~ **overhang** Reitstock-Ausladung *f* ~ **pinion** Getriebe *n* zur Verstellung des Reitstockes auf dem Bett ~ **quill** Reitstockpinole *f* ~ **slide** Reitstockgehäuse *n* ~ **(or footstock) spindle** Meßamboß *m* ~ **spindle sleeve** Reitstockpinole *f*

**tail,** ~ **stroke** Fiederungsstrich *m* ~-**strut axis** Spornstrebenachse *f* ~ **surface** Leitwerkfläche *f*, Schwanz(ober)fläche *f* ~-**surfaces** hochliegendes Leitwerk *n* ~-**tank water** Fallwasser *n* ~ **trolley** Schwanzwagen *m* ~-**tube buoy** Schwanztonne *f* ~ **turret** Heckkanzel *f* ~ **type** geschwänzte Schrift *f* ~ **unit** Leitwerk *n* ~ **unit loads** Leitwerk-belastung *f*, -last *f* ~ **unit strut** Leitwerksstrebe *f* ~ **warning radar set** Heckwarnradar *n* ~ **water** Unterwasser *n* ~-**water elevation** Unterwasserspiegel *m* ~-**water port** Unterhafen *m* ~ **wave** Heckwelle *f* ~-**wedge angle** Kantenwinkel *m* ~ **wheel** Heck-, Schwanz-rad *n*, Sporn-rad *n*, -rolle *f* ~-**wheel fairing** Spornradverkleidung *f* ~-**wheel position indicator** Spornradanzeiger *m* ~ **wind** Rückenwind *m*, Wind *m* von hinten

**tailed characteristic** schleichende Kennlinie *f*

**tailing** Abfall *m*, Ausschwingungsversuch *m*, Gatsch *m*, Rückstand *m*, Herschwanken *n*, Schlängeln *n* ~ **motion** Hin- und Herschwanken *n*, Schlängeln *n*

**tailings** Abfall *m*, Abgang *m*, After *m* (min.), Berg *m*, Erzabfälle *pl*, Knoten *m* im Zucker, Zuckerknoten *m* ~ **for mine filling** Berg(e)versatz *m* ~ **launder** Berg(e)gerinne *n*

**tailless** schwanzlos ~ **airplane** Nurflügelflugzeug *n*, schwanzloses Flugzeug *n*

**tailored pattern** zugeschnittenes Feldstärkediagramm *n* (d. Antenne)

**tails** (distilling) Nachlauf *m*

**Taintor,** ~ **gate** Segmenttor *n* ~ **valve** Segmentschütz *n*

**take, to** ~ (current; photo) aufnehmen, dauern, fangen, fassen, hinnehmen, nehmen, zulegen **to** ~ **into account** berücksichtigen, einrechnen **to** ~ **to the air** aufsteigen **to** ~ **along** mitnehmen **to** ~ **apart** abrüsten, auseinandernehmen, demontieren **to** ~ **away** ausbringen, eine Menge abnehmen, entfernen, wegnehmen **to** ~ **up the bearing** das Lager *n* nachstellen **to** ~ **up a cable** ein Kabel *n* aufnehmen **to** ~ **out a clearing**

**cistern** ausschlagen **to** ~ **down a connection** eine Verbindung aufnehmen, trennen **to** ~ **out dents** (in a metal part) ausbeulen **to** ~ **into consideration** bedenken, in Betracht ziehen **to** ~ **down** abreißen, abrüsten, abtragen, auseinandernehmen, notieren **to** ~ **off edge** abkanten **to** ~ **from** entnehmen **to** ~ **off the furniture** das Format *n* abschlagen (print.) **to** ~ **off the gangue** Gestein *n* abtreiben **to** ~ **in hand** übernehmen **to** ~ **in** (a line) einbringen, einnehmen **to** ~ **from the loom** abbäumen **to** ~ **by means of tongs** zängen **to** ~ **out of a mounting frame** abrahmen **to** ~ **off (or down)** abhängen **to** ~ **upon oneself** übernehmen **to** ~ **out** (her)ausnehmen **to** ~ **over** abnehmen **to** ~ **over part of the stock** einen Aktienanteil *m* übernehmen **to** ~ **out a patent** ein Patent *n* nehmen **to** ~ **to pieces** auseinandernehmen **to** ~ **up in plane** mitnehmen **to** ~ **up a position of all-around defense** einigeln **to** ~ **out quantities** (moving dirt or soil) Abtragsmassen *pl* ermitteln **to** ~ **on rapidly (or off)** abführen, abheben, ablassen, ablegen, abnehmen, abschalten, abstellen, abtun, eine Menge abnehmen, vom Boden *m* abnehmen, ausschalten, entfernen, (the printing cylinder) herausrollen, (aircraft) starten **to** ~ **in a reef** einreffen **to** ~ **off rope** Seil *n* abschlagen **to** ~ **out of service** blockieren **to** ~ **off a sheet** einen Korrekturbogen *m* abziehen **to** ~ **up slack** nachspannen **to** ~ **off the slags** abschlacken **to** ~ **up (away, or on)** abnehmen **to** ~ **up wear** ausgleichen **to** ~ **off weight from the axle** die Achse *f* entlasten **to** ~ **off into the wind** gegen den Wind *m* starten

**take, to** ~ **advantage of** ausnutzen, benutzen **to** ~ **aim** anvisieren **to** ~ **an angle** einen Winkel *m* messen **to** ~ **an appeal** (from) Berufung *f* einlegen (gegen) **to** ~ **a bearing** anpeilen, einen Winkel *m* messen, orten, sich orientieren **to** ~ **bearings** Einpeilen *n* eines Richtempfängers auf einen Funksender, peilen, visieren **to** ~ **bearings from** ansteuern **to** ~ **bearing from (or on)** (a radio station) Anpeilen *n* **to** ~ **bearing from a radio transmitter with directional receiver** einpeilen **to** ~ **bearings on** anvisieren **to** ~ **bearings on the stars** Sterne *pl* schießen **to** ~ **bearings on the sun** Sonne *f* schießen **to** ~ **bearings with direction finder** Peilung *f* vornehmen ~ **care!** Obacht! **to** ~ **care of** berücksichtigen, besorgen, betreuen, erledigen, versorgen, wahren **to** ~ **charge of** eine Funktion *f* übernehmen, übernehmen, (something) sich einer Sache *f* annehmen **to** ~ **charge of a function** eine Funktion *f* übernehmen **to** ~ **a course** beschreiten, verlaufen, einen Lehrgang *m* durchmachen **to** ~ **cover** in Deckung *f* gehen **to** ~ **cross bearings** einschneiden **to** ~ **current** Strom *m* abnehmen oder aufnehmen **to** ~ **effect** sich auswirken **to** ~ **an end** ein Ortsgedinge *n* übernehmen **to** ~ **examination** Prüfung *f* machen **to** ~ **an examination** ein Examen *n* machen **to** ~ **exception to** Anstoß *m* nehmen **to** ~ **the finishing cut** nachdrehen **to** ~ **fire** Feuer *n* fangen, zünden **to** ~ **a fix** peilen **to** ~ **a hand in** eingreifen **to** ~ **a heave** verschoben sein (min.), verworfen sein **to** ~ **hold** (in silent tuning) hängen bleiben **to** ~ **hold of** anfassen **to** ~ **ink** die Farbe *f* auf die

Form auftragen **to ~ the lead** die Führung *f* übernehmen **to ~ a lead on** vorhalten **to ~ legal steps** verklagen **to ~ the level** abwägen **to ~ the mean** das Mittel *n* bilden **to ~ an oath** beschwören, einen Eid *m* ablegen, einen Schwur *m* ableisten, schwören **to ~ part** sich beteiligen, teilnehmen **to ~ photographs** Aufnahmen *pl* machen **to ~ place** erfolgen, sich abspielen **to ~ a position** antreten **to ~ possession** antreten **to ~ possession of** beziehen, übernehmen **to ~ precautions** Vorkehrungen *pl* treffen **to ~ a proof** abziehen, einen Druck *m* abziehen **to ~ a radio bearing** einpeilen **to ~ a reading** eine Ablesung *f* nehmen **to ~ a reading on an instrument** die Anzeige *f* eines Meßgeräts ablesen **to ~ roots in** verwurzeln **to ~ a rough proof(copy)** in Fahnen *pl* abziehen **to ~ a set** eine Verformung *f* erleiden **to ~ sights** Höhe *f* nehmen **to ~ soundings** loten (naut.) **to ~ sun's altitude** die Sonnenhöhe *f* messen **to ~ the thrust** den Druck *m* aushalten **to ~ (one's) turn** an die Reihe *f* kommen **to ~ water off (or on)** abwassern
**take-down cell** zerlegbare Küvette *f*
**take-off** *f*, Anlauf *m*, Einsatz *m*, Start *m* (aviat.), Tonunterdrückung *f* (TV) **~ with the wind** Start *m* mit Rückenwind
**take-off, ~-and-wind-up units** Abzugs- und Aufwickelmaschine *f* **~ area** Abflugsektor *m*, Startbereich *m*, Startzone *f* **~ boost** Startdruck *m*, (pressure) Startladedruck *f* **~ climb performance** Startsteigleistung *f* **~ configuration** Startzustand *m* **~ contact** Abhebekontakt *m* **~ conveyors** Abzugsbänder *pl* **~ distance** Abflugstrecke *f*, Startstrecke *f*, Startlänge *f* **~ fee** Startgebühr *f* **~ field** Aufstiegsplatz *m* **~ fuel weight** Kraftstoffgewicht *n* beim Start **~ grossweight** Brutto-Startgewicht *n* **~ hook** Starthaken *m* **~ method** Startart *f* **~ output** Abflugleistung *f* **~ point** Startpunkt *m* **~ power** Startleistung *f* **~ quality** Start-fähigkeit *f*, -vermögen *n* **~ rating** Abflug-einschätzung *f*, -leistung *f*, Startleistung *f* **~ region** Startbereich *m* **~ restriction** Startverbot *n* **~ run** Anlauf *m*, Anlaufstrecke *f*, Start(roll)strecke *f* **~ safety speed** Sicherheitsstartgeschwindigkeit *f* (V2) **~ signal** Abhebesignal *n*, Startkommando *n* **~ speed** Abfluggeschwindigkeit *f* **~ spool** Abnahmespule *f* **~ strip** Abflugbahn *f*, Rollweg *m* **~ surface** Abflugebene *f*, Startfläche *f* **~ thrust** Startschub *m* **~ weight** Startgewicht *n*
**take-over prices** Übernahmepreise *pl*, Übernahmewerte *pl*
**take-up** Riemenspannvorrichtung *f*, Rückspulen *n* (film) **~ of film strip** Auflaufen *n* des Films **~ for spool** Spulenhalter *m*
**take-up, ~ device** Aufwickelvorrichtung *f* (film) **~ elevator** Tankheber *m* (film) **~ end** Aufgabe-, Aufnahme-, Belade-station *f* **~ reel** wickelrolle *f* **~ roller** (knitting) Abzugswalze *f*, Aufwickelapparat *m* **~ spindle** Nachwickelrolle *f* **~ spool** Aufwickelrolle *f* (film) **~ sprocket** Nachwickler *m* **~ sprocket action** Nachtransport *m* **~ washer** Vorsteckscheibe *f*
**taken, not ~** freibleibend **~ off by tanning** abgegerbt **~ on and off** absetzbar
**taker** Gedingenehmer *m* **~ of a bill** Wechselnehmer *m*
**taker, ~-in** Bogenfänger *m* (print.) **~-in grid**

Vorreißerrost *m* **~-in tooth** Vorreißerzahn *m* **~-off** Bogenabnehmer *m* (print.) **~-off flyers** Bogenfänger *m*
**taking** Aufnahme *f* (phot.) **~ up the backlash** (of threads) Ausgleichung *f* des toten Ganges **~ on coal** Abkohlen *n* **~ of current by grid** Gitterstromaufnahme *f* **~ from direction-finding station situated in direction of travel** Längspeilung *f* **~ off the gases of blast furnaces** Gichtgasentziehung *f* **~ of idle current** Blindstromaufnahme *f* **~ in** Einzug *m* **~ of an inventory** Bestandaufnahme *f* **~ a job** Verdingung *f* **~ the levels of points** Einnivellieren *n* von Punkten **~ (a boat) through a lock** Durchschleusung *f* **~ the mean** Mittelwertbildung *f* **~ from one object** Einzelpeilung *f* **~ of power** Leistungsentnahme *f* **~ of reactive current** Blindstromaufnahme *f* **~ of wattless current** Blindstromaufnahme *f*
**taking, ~ attachment** Aufnahmevorsatz *m* **~ back** Zurücknahme *f* **~ bearings** Peilung *f* **~ cover** Deckungnehmen *n* **~ filter** Aufnahmefilter *m* **~ measurements** Maßnehmen *n* **~ objective** Aufnahmeobjektiv *n* **~-out** Außertrittfallen *n* **~-over** Übernahme *f* **~-over detachment** Übernahmekommando *n* **~-over of the plant** Abnahme *f* der Anlage **~-over report** Übernahmebericht *m* **~ place** Zustandekommen *n* **~ possession** Besitznahme *f* **~ (the) readings** Ablesung *f* **~ root** Einwurzeln *n* **~ samples** Nehmen *n* von Proben **~-up** Aufnahme *f*
**Talbot, ~ (tilting) furnace** Talbot-Ofen *m* **~'s law** Talbotscher Satz *m* **~ (continuous-open-hearth) process** Talbot-Verfahren *n*
**talc, ~(um)** Talk *m*, Talkum *n* **~ schist** Talkschiefer *m*
**tale** Bleierzgedinge *n*
**talent** Begabung *f*, Gabe *f*, Talent *n* **~ for technical hobbies** Bastelbegabung *f*
**talented** begabt
**talk, to** Gespräche *pl* wechseln, reden, sprechen **to ~ over** besprechen **to ~ shop** fachsimpeln
**talk** Besprechung *f*, Gerede *n*, Unterredung *f* **~-back** Gegensprechen *n* (rdo) **~-back circuit** Gegensprechschaltung *f* **~-back unit** Kommandosteuerstufe *f* **~-listen button** Sprechtaste *f* **~-listen switch** Hörsprechschalter *m*
**talker** Sprecher *m*, Sprechender *m* **~ echo** Rückhören *n*, Sprecherecho *n*
**talking** Sprechen *n* **~ beacon** akustisches Funkfeuer *n* **~ circuit** Sprech-kreis *m*, -spule *f* **~ condition** Sprechstellung *f* **~ current** Sprechstrom *m* **~ film** Tonfilm *m* **~ key** Sprech-(um)schalter *m*, -schlüssel *m* **~ motion picture** sprechendes Bild *n* **~ path** Sprechweg *m* **~ peak** Sprachspitze *f* **~ position** Sprechstellung *f* **~ range** Sprechverständigung *f* **~ test** Sprechversuch *m*
**tall** groß, hoch **~ building** Hoch-, Turm-haus *n* **~ oil** Tallöl *n*
**tallow, to** einfetten, schmieren, talgen **to ~ leather over a charcoal fire** Leder *n* abflammen
**tallow** Schmiere *f*, Talg *m*, Unschlitt *m* **~ candle** Talgkerze *f* **~-melting plant** Talgschmelzanlage *f* **~ oil** Härteöl *n* **~ wood** Talgbaumholz *n*
**tally** Kennzeichen *n*, Zählstrich *m* **~ lights** Studiosignallampen *pl*
**talon** Kehlleiste *f* (arch.), Schlüsselansatz *m*

**talus** Böschung *f*, Schutthalde *f* ~ **fan** Schutt-kegel *m*

**tamarack** nordamerikanische Lärche *f*, Tama-rakholz *n*

**tambour** Trommel *f* ~ **needle** Tambouriernadel *f*

**tambourine,** ~ **cymbals** Tambourinbecken *n* ~ **jingles** Tambourinbecken *n*

**tame** bebaut, veredelt, zahm

**tamp, to** ~ abrammen, besetzen, einstampfen, festrammen, festtreten, (concrete) rammen, stampfen, unterstopfen, verdämmen, verstem-men **to** ~ **down** aufstampfen **to** ~ **firm** fest-stampfen **to** ~ **full** vollstampfen **to** ~ **a hole** ein Bohrloch *n* besetzen

**tamped** gestampft (Boden) ~ **bottom (or floor)** Stampfboden *m* ~ **lining** Stampffutter *n* ~ **tie** unterstopfte Schwelle *f*

**tamper, to** ~ (with) verfälschen

**tamper** Besetzstempel *m*, Dämmasse *f* (Tamper), Gleisstopfgerät *n*, Schüsser *m*, Stampfer *m*, Stößer *m* ~ **of ties** Schwellenstopfer *m*

**tamper,** ~ **cylinder** Stopferkolben *m* ~ **plate** Stampferplatte *f*

**tamping** Abdämmen *n*, Besatz *m*, Einstampfen *n*, Feststampfen *n*, Stampfen *n*, Unterstopfen *n*, Versetzen *n* ~ **bar** Gerät *n* zum Stopfen des Geleises, Stampfholz *n* ~ **cartridge of clay** Let-tennudel *f* ~ **clay** Stampfmasse *f* ~**-down** Fest-stampfen *n* ~ **iron** Stampfer *m* ~ **machine** (to tamp molds, etc.) Formstampfmaschine *f*, Stampfmaschine *f* ~ **performance** Stopfarbeit *f* ~ **rod** Stopfschläger *m* ~ **roller** Gürtelrad-walze *f*, Stampfwalze *f* ~ **tool** Stopfer *m*, Stößel *m*

**tampion** Mündungspfropfen *m*

**tan, to** ~ beizen, (leather) gerben

**tan,** ~ **pickle** Beizbrühe *f* ~ **pit** Lohgrube *f* ~ **shades** Lederbrauntöne *pl*

**tandem** eines hinter dem anderen, hinterein-ander, hintereinander-angeordnet, -liegend

**tandem,** ~ **aircraft** Tandemflugzeug *n* ~ **area** Bezirks-, Vororts-netz *n* ~ **arrangement** Rei-hen-, Tandem-anordnung *f* ~ **balloons** Ballon-gespann *n* ~ **blower engine** Tandemgebläse *n* ~ **calender** Doppelkalander *m* ~ **central office** Durchgangs-, Knoten-amt *n* ~ **compound en-gine** Reihenverbundmaschine *f* ~ **condenser** Mehrfachkondensator *m* ~ **connection** (of mo-tors) Hintereinanderschaltung *f*, Kaskaden-schaltung *f* ~**-design propeller** Doppelluft-schraube *f* ~ **dialling** Durchgangswahl *f* ~**disc harrow** Doppelscheibenegge *f* ~ **drive** Tandem-antrieb *m* ~ **engine** Reihen-, Tandem-maschine *f* ~ **engines** zwei hintereinanderliegende Moto-ren *pl* ~ **exchange** Durchgangsamt *n* ~ **junction** Verbindungsleitung *f* für Tandembetrieb ~ **load sharing** Lastverteilung *f* bei Doppeltraktion ~ **mill** Tandemwalke *f* ~ **office** Tandemamt *n* ~ **operation** Fernsprechbetrieb *m* mit Durch-gangsvermittlung, Reihenbetrieb *m*, Tandem-betrieb *m*, Überweisungsverkehr *m* ~ **plungers** Tandemkolben *pl* ~ **position** Durchgangsplatz *m* ~ **preselection** doppelte Vorwahl *f* ~ **propel-ler** Tandemschraube *f* ~ **propellers** hinterein-anderliegende Luftschrauben *pl* ~ **seat** Doppel-sitz *m* ~ **selector** Durchgangswähler *m*, II., III., IV. Gruppenwähler *m* ~ **sender** Speicher *m* in Knotenämtern ~ **steam road roller** Dampf-

**tandemwalze** *f* ~ **super-heated steam roller** Heißdampftandemwalze *f* ~ **toll-circuit dial-ing** Fernsteuerung *f* von Zwischenämtern ~**-tube** (push-pull) **circuit organization** Doppel-röhrenschaltung *f* ~**-type powerplant** Hinter-einandertriebwerkanlage *f* ~ **type rollers** Tan-demwalze *f* ~ **wheel undercarriage** Tandem-fahrgestell *n* ~ **winding** Tandemförderung *f*

**tang** Angel *f*, Bohrerangel *f*, Dorn *m*, Griff *m*, Heftzapfen *m*, Lappen *m*, Mitnehmer *m*, Mit-nehmer-lappen *m*, -zapfen *m*, Schaft *m* eines Werkzeuges zur Befestigung im Griff ~ **face** Lappenfläche *f* ~ **slot** Angelnute *f*

**tangent** Berührende *f*, Berührungslinie *f*, Tan-gens *m*, Tangente *f*; berührend **at a** ~ mit plötz-lich geändertem Kurs *m* ~ **to course** (of a ship) Bahntangente *f* ~ **of elevation** Aufsatzwinkel *m* ~ **of flight path** Flugbahntangente *f* ~ **of motion** Bahn-berührende *f*, -tangente *f* ~ **through the point of inflection** Wendetangente *f* ~ **to reversing point** Wendetangente *f* ~ **of slope** Neigungstangente *f*

**tangent,** ~ **arc** Berührungs-, Tangenten-bogen *m* ~**-circle pattern** Doppelkreisdiagramm *n* ~ **compass** Berührungslinie *f* ~ **equation** Tangen-tensatz *m* ~ **friction force** tangentiale Reibung *f* ~ **galvanometer** Tangentenbussole *f* ~ **image** Tangentenbild *n* ~ **key** Tangentialkeil *m* ~ **law** Tangentensatz *m* ~ **line to a curve in space** Tan-gente *f* einer Raumkurve ~ **line to a plane curve** Tangente *f* einer ebenen Kurve ~ **meter** Steil-heitsmesser *m* ~ **plane** Tangentialebene *f* ~ **point** Tangentenpunkt *m* ~ **roadway** gerade Straßenstrecke *f* ~ **section** gerade Strecke *f* (r.r.) ~ **sight** Aufsatz *m*, Stangen-aufsatz *m*, -stab *m*, -visier *n* ~ **sighting mechanism** Stan-genvisiereinrichtung *f* ~ **value** Tangensgetrie-bewert *m*

**tangential** abweichend, berührend, tangential ~ **acceleration** Tangentialbeschleunigung *f* ~ **admission** tangentiale Beaufschlagung *f* ~ **bur-ner** Eckenbrenner *m* ~ **cam** harmonischer Nocken *m* ~ **clamping band** Tangentialspann-band *n* ~ **component** Umfangskomponente *f* ~ **connection** Tangentialanschluß *m* ~ **corner burner** Eckenbrenner *m* ~ **cut** Ader-, Flader-, Sehnen-, Tangential-schnitt *m* ~ **distortion** Bildzerdrehung *f* ~ **feed** Tangentialvorschub *m* ~ **force** Tangentialkraft *f*, Umfang(s)kraft *f* ~ **incidence** tangentialer Einfall *m* (acoust.) ~ **key-way** Tangentkeilnut *m* ~ **lighting** Spitz-lichter *pl* ~ **plane** Berührungsebene *f*, erster Hauptschnitt *m*, Sehnenebene *f* ~ **scavenging** Tangentialspülung *f* ~ **section** Ader-, Flader-, Sehnen-, Tangential-schnitt *m* ~ **stress** Bogen-schub *m*, Horizontalschub *m*, Scher-bean-spruchung *f*, -kraft *f*, -spannung *f*, Schub-be-anspruchung *f*, -kraft *f*, -spannung *f*, Tangen-tial-beanspruchung *f*, -spannung *f*, waagerech-ter Seitenschub *m* ~ **thrust** Bogen-, Horizon-tal-schub *m*, waagerechter Seitenschub *m* ~ **transformation** Berührungstransformation *f* ~ **wheel** Tangentialrad *n* ~ **wind stress** Wind-schubspannung *f*

**tangible** berührbar, fühlbar, greifbar ~ **expend-iture** erfaßbare Ausgaben *pl* ~ **mark (prick mark)** fühlbare Marke *f*

**tangle, to** ~ durchziehen (parachute), verfitzen, verheddern, verwickeln, verwirren

**tank** Ausblasebutte *f*, Bad *n*, Balanziertrog *m*, Barke *f*, Bassin *n*, Becken *n*, Behälter *m*, Bottich *m*, Bütte *f*, Diffuser *m*, F Boot *n*, Gefäß *n*, Kessel *m*, Panzer-kampfwagen *m*, -kraftwagen *m*, -wagen *m*, Sammeltank *m*, Tank *m*, Wanne *f*, Warengleitmulde *f*, Waschbehälter *m*, Wasserstation *f*, Zisterne *f* ~ **for holding liquid** Flüssigkeitsbehälter *m*

**tank,** ~ **aeration** Tankentlüftung *f* ~ **air charging (or pressurizing) line** Tankbelüftungsleitung *f* (g/m) ~ **apparatus** Druckapparat *m* ~ **barrel** Tankmantel *m* (Tankwagen) ~ **block** Wannenstein *m* ~ **body** Panzeraubau *m* ~ **capacity** Kapazität *f* eines Schwingkreises ~ **car** Bassinwagen *m*, Behälterwagen *m*, Eisenbahn-kesselwagen *m*, -tankwagen *m*, Kessel-, Tank-, Zisternen-wagen *m* ~ **carried in the wing** im Flügel aufgehängter Tank *m* ~ **cavity** Pumpensumpf *m* ~ **cession price** Lieferpreis *m* in Tank (im Behälter) ~ **circuit** Abstimm-, Parallelschwing-, Schwingungs-, Stromresonanz-, Topf-kreis *m* ~ **coil** Induktivität *f* ~ **condenser** Speicher-, Vorrats-kondensator *m* ~ **cupola** Kupolofen *m* mit erweitertem Herd, Panzerkuppel *f* ~ **development** (daylight) Dosenentwicklung *f* (photo), Standentwicklung *f* (photo) ~ **discharge valve** Klärbeckenablaßventil *n* ~ **empty float switch** Leer-Warnschalter *m* ~ **farm** Behälter-lager *n*, -park *m* ~ **filler cap** Einfüllstutzen *m*, Tankfüllstutzen *m* ~ **filling hole** Tankeinfüllstutzen *m* ~ **full indicator** (Tankwagen) Füllanzeiger *m* ~ **funnel** Tanktrichter *m* ~ **furnace** (glass mfg.) Wannenofen *m* ~ **gauge** Behälterstandanzeiger *m* ~ **head** Tankboden *m* ~ **heads (or ends)** Behälterböden *pl* ~ **locomotive** Tenderlokomotive *f* ~ **obstacle** Panzerhindernis *n* ~ **oxygen** Bombensauerstoff *m* ~ **plant** Tankanlage *f* ~ **plate** Behälterblech *n* ~ **potential** Badspannung *f* ~ **pressure** Behälterdruck *m* ~ **pressurizing** Tankbelüftung *f* (g/m) ~ **shell** Tankwandung *f* ~ **shutter** Panzerblende *f* ~ **skin** Behälter-behäutung *f*, -haut *f*, -hautblech *n* ~ **standpipe** Tankbelüftungsrohr *n* (g/m) ~ **station** Zapfstelle *f* ~ **steamer** Tankdampfer *m* ~ **supporting slipper** Kesselstuhl *m*, Kesselstützbremsklotz *m* ~ **test** Wasserkanalversuch *m* ~ **trailer** Tankanhänger *m* ~ **trolley** Fahrgestell *n* für Tank ~ **truck** Betriebsstoffwagen *m*, Kraftwagen *m* mit Behälter, Tankentankwagen *m*, Tank(kraft)wagen *m* ~ **truck for powdered coal** Kohlenstaubwagen *m* ~ **truck (draw-off) faucet** Tankwagenzapfhahn *m* ~ **unit** Kraftstoffvorratsgeber *m*, Tanksonde *f* ~ **ventilation** Behälterentlüftung *f* ~ **voltage** Badspannung *f* ~ **wagon** Kesselwagen

**tankage** Tank-inhalt *m*, -vorrat *m* ~ **capacity** Fassungsvermögen *n*

**tankard** Kanne *f*, Krug *m*

**tanker** Tanker *m*, Tank-flugzeug *n*, -schiff *n*, -wagen *m* ~ **transport** Betankungflugzeug *n*

**tanking** Eintauchung *f* in einen Tank oder Behälter

**tannage** Gerbstoff *m*, Gerbung *f*

**tanned** abgebeizt, gegerbt, mit Tannin *n* getränkt ~ **jute** mit Tannin getränkter Flachs *m*, mit

Tannin getränkte Jute *f* ~ **leather** rohgares Leder *n*

**tanner** Gerber *m* ~**'s pit** Äscher *m*

**tannery** Gerberei *f*

**tannic,** ~ **acid** Gerbsäure *f*, Tannin *n* ~ **glue** Gerbleim *m* ~ **print** Tannindruck *m* ~ **resist** Tanninreserve *f* ~ **resist style** Tanninätzartikel *m*

**tannin** Gerbsäure *f*, Gerbstoff *m*, Tannin *n* ~ **bark** Lohe *f*

**tanning** Gerberei *f* ~ **plant** Gerbanlage *f* ~ **substance** Gerbstoff *m* ~ **vat** Lohgrube *f*

**tannometer** Gerbsäuremesser *m*

**tantalate** tantalsaures Salz *n*

**tantalic** tantalsauer ~ **capacitor** Tantal-Elektrolytkondensator *m* ~ **pentafluoride** Tantalpentafluorid *n*

**tantalite** Tantalit *n*

**tantalum** Tantal *n* ~ **ore** Tantalerz *n* ~ **oxide (or pentoxide)** Tantaloxyd *n* ~ **rectifier** Tantalgleichrichter *m*

**tantamount** gleichbedeutend

**T-antenna** T-Antenne *f*

**tap, to** ~ ablassen, abnehmen, abzapfen, abziehen, abzweigen, anbohren, anstecken, anzapfen, aufschließen, (thread) eindrehen, einschneiden, entnehmen (compressor), entwässern, Gewinde *n* bohren, (threads) Gewinde *n* schneiden, klopfen, mithören, schneiden, stecjhen, mit Stöpseln *pl* anschalten (electr.) ziehen **to** ~ **the blast furnace** den Hochofen *m* abstechen **to** ~ ~ **(off) current** abgreifen **to** ~ **iron** (foundry) Eisen *n* abfangen **to** ~ **off** (furnace) anstechen **to** ~ **slag** abschlacken, ausschlacken, Schlacke abziehen **to** ~ **a telegraph line** eine Telegrafenleitung *f* abhören **to** ~ **a wire** abgreifen, abhören

**tap** Abgreifpunkt *m*, Abgriff *m*, Abstich *m*, Absteckhahn *m*, Abzweig *m*, Abzweigpunkt *m*, Abzweigung *f*, Anzapfung *f*, Auslaufventil *n*, Bohrer *m*, Gewindebohrer *m*, (for screw or bolt) Gewinderung *f*, (screw or screwing tap) Gewindeschneidebohrer *m*, Hahn *m*, Niederschraubhahn *m*, Schneidbohrer *m*, Spannungspunkt *m*, Stich *m*, Stromabnehmer *m*, (of battery) Unterteilungspunkt *m*, Verzweigung *f*, Zapfen *m*, Zapfhahn *m*, Zapfstelle *f* ~ **for cutting machine dies** Maschinenbackengewindebohrer *m*

**tap,** ~ **binding** Abzweigbund *m* ~ **bolt** Dübel *m*, Kopfschraube *f*, Stiftschraube *f* ~ **borer** Zapfenbohrer *m* ~ **bottoming** Gewindenachschneider *m* ~ **breakage** Gewindebohrerbruch *m* ~ **catcher** Fangdorn *m* ~ **changer** Anzapfumschalter *m*, Stufenschalter *m* ~**-changing signal** (from tapped transformer) Stufenmeldung *f* ~ **cinder** Dorn *m*, Dörnerschlacke *f*, Puddelschlacke *f*, Schlacke *f* ~ **connection** Zapfstelle *f* ~ **drill** Gewindebohrer *m* ~ **gauge** Gewindebohrerlehre *f* ~ **grease** Hahnfett *n* ~ **handle** Absperrgriff *m* ~ **holder** Bohrfutter *n*, Gewindebohrerhalter *m*

**taphole** Abstich-loch *n*, -öffnung *f*, Gewindebohrung *f*, Stich *m* (metall.), Stich-auge *n*, -loch *n*, Zapfloch *n* ~ **borer** Zapflochbohrer *m* ~ **bush** Zapflochbüchse *f* ~ **displacement** Stichlochversetzung *f* ~ **fitting** Stichlocharmatur *f* ~ **gun** Stichlochstopfmaschine *f* ~ **plug** Stichpfropf *m*

**tap,** **~ ladle** Abstichpfanne *f* **~ lead style resistor** Widerstand *m* mit radialen Anschlüssen **~ line** Stichleitung *f* **~ lock** Zapfenschloß *n* **~ root** Pfahlwurzel *f* **~ spout** Abstich-kanal *m*, -rinne *f* **~ thread** Gewindeschneider *m* **~ water** Leitungswasser *n* **~ weight** Abstichgewicht *n* **~ wrench** Hahn-, Locher-schlüssel *m*, Wende-, Wind-eisen *n*

**tape, to** **~** abisolieren, bewickeln, mit Band *n* bewickeln oder umwickeln, tippen

**tape** Band *n*, Bandmaß *n*, Streifen *m*, Zwirnband *n* **~ antenna** Bandantenne *f* **~ armoring** Bandeisenbewehrung *f*, Bewehrung *f* in Bandform **~ background noise** Bandrauschen *n* **~ backspacing** Bandrückzug *m* **~ bar** Fadenrolle *f* **~ controlled carriage** bandgesteuerter Vorschub *m*, Lochbandvorschub *m* **~ copy** Streifenausfertigung *f* **~ counter** Bandzählwerk *n* **~ data selector** Banddatenwähler *m* **~ deck** Einbaulaufwerk *n* (tape rec.) **~ delay** Bandverzögerung *f* **~ delivery** Bandablage *f* **~ dispenser** Tesafilmabroller *m* **~ drive motor** Tonmotor *m* **~ feed** Bandvorschub *m* (comput.) **~ feed roll detent** Sperrollenhebel *m* **~ gauge** Streifenprüflehre *f* **~ guide** Bandführung *f* **~ helix coaxial cable** Bandwendelkabel *n* **~ hole** Streifenloch *n* **~ identification card** Bandidentifizierungskarte *f* **~ indicator** Bandzähluhr *f* **~ input** Bandeingabe *f* **~ lapping** Bebänderung *f* **~ length indicator** Zählwerk *n* **~ lifting device** Bandabhebevorrichtung *f* **~ loading** Einlegen *n* des Bandes **~ loop** endloses Band *n* **~ measure** Bandmaß *n*, Meßband *n* **~ microphone** Bändchenmikrofon *n* **~ operated printer** bandgesteuerter Drucker *m* **~ perforator** Streifenlocher *m* **~ platform** Streifenbahn *f* **~ plus header** Rücklaufbüchse *f* mit konischem Verschluß **~ printer** Streifen-drucker *m*, -schreiber *m* **~ printing** Streifendruck *m* **~ pulley** Bandrolle *f* **~ punch girl** Streifenlocherin *f* **~ race** Streifenbahn *f* **~ reader** Lochstreifenabtaster *m* **~ record** Bandsatz *m* **~ recorder** Magnetofon *n* **~-recorder tape** Tonband *n* **~ recording** Schallbandaufnahme *f* **~ reel container** Papiertrommel *f* **~ relay center** Lochstreifenvermittlung *f* **~ relay procedure** Lochstreifensendebetrieb *m* **~ relay working** Verkehrsregelung *f* **~ reproducer** Streifendoppler *m* **~ roll** Streifenrolle *f* **~ roll holder** Papierrollenhalter *m*, Rollenhalter *m* **~ roller** Bandwalze *f* **~ run** Bandlauf *m* (tape rec.) **~ shrouding** Bandumwicklung *f* **~ skew** Schräglauf *m* (d. Bandes) **~ splicing equipment** Bandklebevorrichtung *f* **~ spool** Bandspule *f* **~ store** Bandspeicher *m* **~ tension** Bandspannung *f* **~ tensioning motor** Wickelmotor *m* **~ thickness** Streifendurchmesser *m* **~ track** Tonspur *f* **~ transmission** Lochstreifensendung *f*, Streifensendung *f* **~ transmitter** Lochstreifensender *m*, (perforated) Streifengeber *m*, (perforated) Streifensender *m* **~ transport** Magnetbandlaufwerk *n* **~ travel** Bandleserichtung *f* **~-type teleprinter** Streifenschreiber *m* **~ unit** Bandeinheit *f* **~ wheel** Papierrollenträger *m* **~-wound core** bandumwickelter Kern *m*

**taped** umwickelt **~ wire** umwickelter Leitungsdraht *m*

**taper, to** **~** abschrägen, kegeln, konisch machen,

verjüngen, zuspitzen **to ~ away** (allmählich) dünner werden **to ~ to a point** spitz zulaufen **to ~-turn** kegelig drehen

**taper** abnehmender Querschnitt *m* (aviat.), Anzug *m*, (Bricketpresse) Buckelmaß *n*, kegeliger Verlauf *m*, Keilverjüngung *f*, Kerze *f*, Konizität *f*, Rohrstutzen *m*, Steigung *f*, Trichterhorn *n*, Verjüngung *f*, Wachskerze *f*, Widerstandsverteilung *f* **~ of center** Körnerspitzkegel *m* **~ of groove** Kaliberanzug *m* **~ of key** Keilanzug *m* **~ of milling spindle** Frässpindelkegel *m* **~ of the pass** Anzug *m* des Kalibers **~ of wedge** Keilanzug *m*

**taper,** **~ adapter** Konuseinsatz *m* **~ attachment** Kegeldrehvorrichtung *f*, Konus-drehapparat *m*, -leitapparat *m* **~ ball bearing** Konuskugellager *n* **~ bar** Konuslineal *n* **~ bolt** kegeliger Bolzen *m* **~ bore (or hole)** Kegelbohrung *f* **~-bore mounted** fliegend gelagert **~ boring** Kegelbohren *n* **~ bridge reamer** Nietlochreibahle *f* **~ coaxial waveguide** Stufentransformator *m* für konzentrische und Hohlraummischerkreise **~ cone** Verjüngung *f* **~ cut** Kegeldrehschnitt *m* **~ die** Vorschneideisen *n* **~-drill socket** verlängerte Kegelhülse *f* **~ end of crankshaft** Kurbelwellenkegel *m* **~ file** Spitzenfeile *f* **~ fit** Kegelpassung *f* **~ foot roller** Igelwalze *f* **~ gauge** Kegellehre *f* **~ gauging** Kegelmessung *f* **~ gib** Keilleiste *f*, Prismenführung *f* **~ grinding** Konischschleifen *n* **~ guide bar** Kegel-, Konus-schiene *f* **~ key** Anzugskeil *m* **~ (-sunk) key** Treibkeil *m* **~ pin** kegeliger Stift *m*, Keilstift *m*, konischer Stift *m*, (of converter bottom) Nadel *f* **~ pin reamer** Kegel-, Stiftloch-reibahle *f* **~ pin twist drill** kegeliger Spiralbohrer *m* **~ plug** Kegelstopfen *m* **~ pulley** Stufenscheibe *f* **~ ratio** Verjüngungsgrad *m* **~ reamer** Nietlochreibahle *f* **~ ring** Hülse *f* **~ roller** Kegelrolle *f* **~-rolling bearing** Kegelrollenlager *n*, konisches Rollenlager *n*, Konusrollenlager *n*, Schrägrollenlager *n* **~ shank** Kegelschaft *m*, Konus-dorn *m*, -schaft *m* **~ shank drill** Bohrer *m* mit Konusschaft **~-shank reamer** Kegelreibahle *f* **~ sheet pile** Eckbohle *f* **~ sleeve** Einsatzbüchse *f*, -hülse *f*, Kegelhülse *f*, Konus *m*, Reduziereinsatz *m* **~ socket** Konusaufnahme *f* **~ square file** Flachstumpffeile *f* **~ tap** (of a screwthreading device) Vorschneider *m* **~ thread** Kegel-, Rohr-gewinde *n* **~ threads** konisches Gewinde *n* **~ turning** Kegeligdrehen *n*, Konischdrehen *n* **~-turning attachment** Konuslineal *n* **~-turning work** Konischdreharbeit *f*

**tapered** abgeschrägt, hohlflächig, kegelförmig, kegelig, konisch, spitz zulaufend, trichterförmig, verjüngt, zugespitzt

**tapered,** **~ bearing** konisches Lager *n* **~ bore** Würgebohrung *f* **~-bore barrel** konisches Rohr *n* **~ collar** Ringkeil *m* **~ compression piston ring** Minutenring *m* **~ cone** Klemmkonus *m* **~ drum** Kegeltrommel *f* **~ end** verjüngtes Ende *n* **~ face** (of key) Rückenfläche *f* **~ machine handle** Kegelgriff *m* **~ pin** Kegelstift *m* **~ ring** Kegelring *m* **~-roller bearing** Kegelrollen-, Ringkegel-lager *n* **~ roller thrust bearing** Scheibenkegelrollenlager *n* ~ shank arbor Kegeldorn *m* **~ sleeve** Kegel *m* **~-sleeve wedge bearing** Spannhülsenlager *n* **~ socket** Kegel *m*, Konus *m* **~ spigot** Kegelzapfen *m* **~ spindle** konische

Spindel *f* ~ **spur wheel** (Bagger) Kegelstirnrad *n* ~ **square hole** kegeliger Vierkant *m* ~ **tail plane** verjüngte Höhenflosse *f* ~ **tap** Gewindebohrer *m* ~ **thread** konisches Gewinde *n* ~ **thrust ring** konisch ausgebildeter Druckring *m* ~ **top** Kegelkuppe *f* ~**-tube pole** nach oben verjüngte Rohrpost *f* ~ **washer** Schrägscheibe *f* ~ **washout plug** Reinigungsschraube *f* ~ **well** abgeschrägter Potentialtopf *m* ~ **wing** Dreieckflügel *m*, Flügel *m* mit Verjüngung ~ **wings** Tragwerk *n*

**taperer cam** Schrägnocken *m*

**tapering** Verjüngung *f*, Verjüngungsstelle *f*, Zuspitzung *f*; spitzig ~ **in section** Profilverjüngung *f* ~ **of wing** Verjüngung *f* des Flügels

**tapering,** ~ **cable-distributing system** offenes Kabelverteilungssystem *n* ~ **cabling system** offenes Verkabelungssystem *n*, Zweigsystem *n* ~ **scale** Verjüngungsmaßstab *m*

**tapestry** Tapete *f* ~ **work** Tapisserie *f*

**tapia** Piseebau *m*

**taping** Bandumwicklung *f*, Bewicklung *f*, Umlappung *f* mit Band, Umwicklung *f* ~ **machine** Bandwickler *m* ~ **wire** Wickeldraht *m*

**tapped** (wire) abgegriffen, abgelassen, abgestochen, (off) abgezapft, angezapft ~ **blind hole** Grundloch *n* mit Innengewinde, Sackloch *n* mit Gewinde ~ **coil** Abzweigspule *f*, Anzapfspule *f*, angezapfte Spule *f* ~ **condenser** Mehrfachkondensator *m* ~ **hole** Gewinde-bohrung *f*, -loch *n*, Schraubenloch *n* ~ **inductance** Abzweigspule *f* ~ **metal** Stich *m* ~ **rod guide** Ventilstößelführung *f* ~ **transformer** Abzweig-, Anzapf-, Stufen-transformator *m* ~ **winding** angezapfte Wicklung *f*

**tapper** Abgreifer *m*, Dekohärer *m*, Entfritter *m*, Gewindeschneider *m*, Klopfer *m*, Morsetaste *f* ~ **coil** Klopferspule *f* ~ **tap** Maschinenmuttergewindebohrer *m* mit langem Schaft

**tappet** Arm *m*, Daumen *m*, Frosch *m*, Greifer *m*, Hebedaumen *m*, Hebekopf *m*, Hebel *m*, Hebling *m*, Knagge *f*, Nocken *m*, Mitnehmer *m*, Mitnehmerbolzen *m*, Nase *f*, Schieberkerbe *f*, (Ventil) Stößel *m*, Verschlußknagge *f* ~ **of arbor** Wellendaumen *m*

**tappet,** ~ **base** Stößelpfanne *f* ~ **cap** Klaue *f* ~ **chamber** Stoßstangenraum *m* ~ **clamping device** Stößelklemmer *m* ~ **clearance** Spiel *n* der Ventile, Stößelspiel *n* ~ **cover** Stoßstangenraumdeckel *m* ~ **finding groove** Stößelführungsnut *f* ~ **gear** Nockensteuerung *f* ~ **guide** Stößelführung *f* ~ **head** Stößelsteller *m* ~ **hook** Fingerhaken *m* ~ **lever** Schwunghebel *m* ~ **rod** Gleitstößel *m* ~ **rod of eccentric-cam type** Abwälzstößel *m* ~ **rod guiding bush** Ventilstößelführungsbuchse *f* ~ **roller** Anschlagrolle *f* ~ **sleeve** Stößelbuchse *f* ~ **socle** Stößelpfanne *f* ~ **stop** Stößelanschlag *m* ~ **switch** Kippschalter *m*

**tapping** Abgriff *m*, Abhörverfahren *n*, Abstich *m*, Abziehen *n*, Abzweigung *f*, (point) Anzapfstelle *f*, Anzapfung *f* (rdo, rdr), Eindrehung *f*, Gewinde-bohren *n*, -schneiden *n*, Klopfen *n*, Spulenanzapfung *f*, Stich *m*, Stromableitung *f*; intermittierend ~ **of current** Stromübernahme *f* ~ **and straining device** Zapf- und Abseihvorrichtung *f*

**tapping,** ~ **attachment** Bohr-, Gewindebohr-vor-

richtung *f* ~ **bar** Abstechspieß *m*, Laßeisen *n*, Räumeisen *n* ~ **binding** Abzweigbund *m* ~ **circuit** Abhörstromlauf *m* ~ **a circuit** Einschleifen *n* einer Leitung ~ **contact** intermittierender Kontakt *m*, zeitweise Leitungsberührung *f* ~ **device** Anschaltvorrichtung *f* ~ **device attachable set** Anschaltgerät *n* für Morseleitungen ~ **door** Abstichöffnung *f* ~ **funnel** Gußloch *n* ~ **hole** Gußloch *n*, Stich-auge *n*, -loch *n* ~ **knife** Zapfmesser *n* ~ **machine** Gewinde-bohrmaschine *f*, -schneidemaschine *f* ~**-out on signaling lamp** Blinkrelaistasten *n* ~ **plant** Vorschaltanlage *f* ~ **platform** Abstichbühne *f* ~ **point** Abgreifpunkt *m*, Entnahmepunkt *m*, -stelle *f* ~**-point valve** Entnahmeventil *n* ~ **screw** Schneidschraube *f* ~ **selector** Stufenwähler *m* ~ **service** Abhördienst *m* ~ **slag** Abstechen *n* der Schlacke, Abstichschlacke *f* ~ **stage** Entnahmestufe *f* ~ **switch** Regelschalter *m* ~ **technique** Abzweigtechnik *f*

**tappings** Abzweigung *f* von Leitungen, Puddelschlacke *f*

**taps** Abzweigung *f* von Leitungen

**tar, to** ~ beteeren, teeren, verteeren

**tar** Pechharz *n*, Teer *m* ~ **from low-temperature carbonization** Schwelteer *m* ~ **from shale** Schieferteer *m*

**tar,** ~ **acid** Teersäure *f* ~ **ballast** Teerschotter *m* ~ **board** Teerpappe *f* ~ **box** Teerbüchse *f* ~ **brush** Teer-bürste *f*, -quaste *f* ~ **carrying a large amount of dust** stark staubhaltiger Teer *m* ~ **coat** Teerüberzug *m* ~ **coating** Kohlenteeranstrich *m* ~ **condensate** Teerkondensat *n* ~ **constituent** Teerbestandteil *m* ~ **cooler box** (petroleum) Teerkühler *m* ~ **distillery** Teerschwelerei *f* ~**-distilling apparatus** Teerschwelapparat *m* ~ **extraction** Teerabscheidung *f*, Teerwäsche *f* ~**-extraction plant** Teergewinnungsanlage *f* ~ **extractor** Teer(ab)scheider *m* ~**-fired furnace** Teerfeuerung *f* ~ **formation** Teerbildung *f* ~ **fumes** Teer-dampf *m*, -dunst *m* ~ **globule** Teerpröpfchen *n* ~ **incrustation** Teerkruste *f* ~ **mist** Teernebel *m* ~ **mop** Teerquaste *f* ~ **oil** Teeröl *n* ~ **paper** Asphaltpapier *n* ~ **paper (or board)** Dachpappe *f* ~ **pit** Teergrube *f* ~ **residue** Teerrückstand *m* ~ **road** geteerter Weg *m* ~ **scrubber** Teerwäscher *m* ~**-separating sump** Teerscheidegrube *f* ~ **separation** Entteerung *f*, Teerabscheidung *f* ~ **separator** Teer(ab)scheider *m* ~ **settler** Teerabscheider *m* ~ **spraying** Teerung *f* ~**-storage tank** Teergrube *f* ~ **surface** Teerdecke *f* ~ **tab** kleiner Flügel *m* ~ **value** Verteerungszahl *f* ~ **vapor** Teer-dampf *m*, -dunst *m* ~ **washer** Teerwäscher *m*

**tare, to** ~ austarieren, auswägen, auswiegen, tarieren

**tare** Eigen-gewicht *n*, -verbrauch *m*, Leergewicht *n*, Tara *f* ~ **estimation** Schmutzbestimmung *f* ~ **reading** Eigenverbrauchsablesung *f* ~ **room** Prozentstube *f* ~ **ticket** Putzkarte *f* ~ **washer** Probewäsche *f* ~ **weight** Leergewicht *n*

**tared capacitance** Nennkapazität *f*

**target** Angriffsziel *n*, Antikathode *f*, (in a multiplier) Auffanganode *f*, (of cyclotron) Auffänger *m*, Auffänger *m* (phys.), Aufnahme *f*, aufnehmender oder führender Teil *m*, Auftreffplatte f, Kugelobjektiv n, Leuchtschirm m, Meßobjekt n, Mosaikplatte f, Prallplatte f, Reflektor m, Scheibe f, Schicht f, Schießscheibe f, Stellungsunterschied m, Strahlungselektrode f, Weichensignal n (r.r.), Ziel n (rdr), Ziel-kern m, -latte f, -scheibe f

**target,** ~ **acquisition** Zielerfassung f ~ **airplane** Zielflugzeug n ~ **angle** Anodenwinkel m ~ **angular position** Winkelposition f des Ziels ~ **area** Bombenplatz m, Treffläche f, Zielgelände n, -kreis m ~ **beam** Suchstrahl m ~ **board** Justierscheibe f ~ **boat** Zielboot n ~ **break-up** Zielauflösung f ~ **circle** Zielkreis m ~ **course** Zielweg m ~ **course between present and future positions** Auswanderungsstrecke f ~-**course map** Zielwegkarte f ~ **cursor dot** Zielmarkierungsring m des (Bord)Radarbildschirms ~ **data** Zielunterlagen pl ~ **designation** Ziel-ansprache f, -bestimmung f, -bezeichnung f, -darstellung f, -feststellung f ~ **detection** Zielerfassung f, -ortung f ~ **detector** Zielkopf m (electron.) ~ **deuteron** Treffplattendeuteron n ~ **dialling** Zielwahl f (teleph.) ~ **discrimination** Zielauflösung f ~ **displacement** Zielabweichung f ~ **divided into squares** Felderscheibe f ~ **drone** Zielflugkörper m ~ **electrode** Auftreff-, Bildwurf-, Prall-, Rückprall-elektrode f ~ **element** Speicherplattenelement n ~ **fade-out** Zielbildschwinden n, Zielechoschwund m ~ **finder** Zielkopf m ~ **flight** Anflugweg m ~ **frame** Scheibenrahmen m ~ **identification** Zielidentifizierung f ~ **illumination** Zielleuchten n ~-**indication fire** Zielweisungsschießen n ~-**indication searchlight** Zielanweisungsscheinwerfer m ~ **killing** Zielunterdrückung f ~ **lamp** Signallampe f (r.r.) ~ **leg between present and future positions** Auswanderungsstrecke f ~ **level** Planungskennzeichen n ~ **locating set** Zielortungsgerät n (rdr) ~ **location** Zielfeststellung f ~ **marking for bombers** Zielmarkierung f ~-**marking flare** Kaskadenbombe f ~ **missile** Zieldarstellungsflugkörper m ~-**offset distance** Stellungsunterschied m ~ **overlay** Zielspinne f ~ **pickup** Zielauffassung f ~ **plane steered by remote control** ferngelenktes Zielflugzeug n ~ **plate** Auffänger-, Fang-platte f ~-**plate distance** Fokusplattenabstand m ~-**position indicator** Standortszielanzeiger m ~-**practice projectile** Übungsgeschoß n ~ **projector** Lichtpunktsteuergerät n ~ **representation** Zieldarstellung f ~ **resolution** Zielauflösung f ~ **sector** Zielteil m ~-**seeking and DF device** Zielsuch- und Ortungsgerät n ~ **set** Zielausrüstung f ~ **sketch** Zielskizze f ~ **sleeve** Luftziel n ~ **speed** (bomb ballistics) Eigengeschwindigkeit f des Zieles, Zielgeschwindigkeit f ~ **thickness** Auffängerdicke f ~-**towing aircraft** Scheibenschlepper m ~ **tracking** Zielführung f ~ **tracking radar** Zielverfolgungsradar n

**tariff** Gebühr f, Gebührentarif m, Kostenordnung f, Tarif m, Zolltarif m ~ **meter with locking device** Sperrschalttarifzähler m ~ **protection** Zollschutz m ~ **rate of tare** Tarasatz m ~ **regulations** Tarifordnung f ~ **system** Tarifsystem m ~ **treaty** Tarifvertrag m ~ **unit** Gebühreneinheit f, Tarifeinheit f

**tariffs** (airport) Gebühren(ordnung) f

**tarlatan** Baumwollgaze f

**tarless** teerfrei

**tarmac** (Flughafen) Vorfeld n

**tarnish, to** ~ (metall.) abblicken, anlaufen, beschlagen, mattieren, mattwerden

**tarnish** Metallbeschlag m ~ **on glass** Beschlag m ~-**proofness** Anlaufbeständigkeit f

**tarnishable** mattierbar

**tarnished** trübe (glanzlos)

**tarnisher** Mattpunze f

**tarnishing** (of silver) Anlaufen n, Trübung f

**tarpaulin** Decke f, Persenning f, Plache f, Plandecke f, Plane f, Schutzhülle f, Segeltuchplane f, Teerleinwand f, Wagendecke f ~ **for vehicle** Wagenplane f

**tarred** geteert ~ **felt** Teerpappe f ~ **hemp** geteerter Hanf m ~ **pole** geteerte Stange f ~ **road** geteerte Straße f ~ **rope** geteertes Seil n oder Tau n ~ **type** Isolier-, Teer-band n

**tarring** Verteerung f

**tarry, to** ~ säumen

**tarry** teerartig, teerig ~ **matter** Teersubstanz f

**tarrying path** Verweilweg m

**tartar** Weinstein m

**tartaric acid** Wein(stein)säure f

**tartarous** weinsteinhaltig

**tartrate** Tartrat n, weinsau(e)res Salz n

**tasimeter** Tasimeter n

**task** Arbeit f, Arbeitsleistung f, auferlegte Arbeit f, Aufgabe f, Auftrag m ~ **work** Akkordarbeit f

**tassel** Quaste f

**tattered** zerfetzt

**tattoo** Retraite f

**taut** gespannt, prall, straff ~ **end** (of belt) straffes Trumm n ~ **side** (of belt) ziehendes Trumm n ~ **status** Prallheit f ~ **strip suspension** Spannbandaufhängung f (meas. instr.) ~ **tape attachment** Spannbügel-vorrichtung f, -schalter m ~ **wire** Zugdraht m

**tautness** Prallheit f, Spannung f, Straffheit f ~ **warning device** Prallmelder m

**tautomeric transformation** tautomerer Übergang m

**taw, to** ~ abbeizen, weißgerben

**tawed** alaun-, weiß-gar

**tawing liquor** Weißbrühe f

**tawny** braungelb, lohfarben

**tax, to** ~ abschätzen, besteuern, überanstrengen

**tax** Abgabe f, Auflage f, Gebühr f, Steuer f, Taxe f ~ **on increment values** Wertzuwachssteuer f

**tax,** ~ **adjustment** Steuerausgleich m ~-**aided** steuerbegünstigt ~ **balance sheet** Steuerbilanz f ~-**free** gebührenfrei ~ **rate (or municipal rate)** Hebesatz m ~ **remission (or rebate)** Nachlaß m, Steuerermäßigung f ~ **unpaid** unversteuert

**taxable** steuerpflichtig

**taxation** Abschätzung f, Besteuerung f, Schätzung f, Steuerwesen n

**taxes** Steuerwesen n

**taxi, to** ~ gleiten, rollen (aviat.)

**taxi** Mietauto n, Taxi n ~ **channel** Zufahrtrinne f ~ **channel lighting** Zufahrtrinnenbefeuerung f ~ **circuit** Roll-plan m, -route f ~ **holding-position** Rollhalteort m ~ **hook** Feststellhaken m ~ **meter** Fahrpreisanzeiger m, Taxameter n ~ **phone** Münzfernsprecher m ~ **plane** Mietflugzeug n ~ **strip** Zurollweg m (airport)

**taxiway** Rollbahn *f*, Rollweg *m* ~ **fillets** Rollbahnausrundungen *pl* ~ **lighting** Rollbahnbefeuerung *f* ~ **margins** Rollbahnrandzonen *pl* ~ **slope** Rollbahnneigung *f*

**taxiing** (or **taxi-ing**) Rollen *n* ~ **aids** Rollhilfen *pl* ~ **guidance system** Wegweiseranlage *f* für das Rollen ~ **qualities** Roll-eigenschaften *pl*, -verhalten *n*

**T bar** Aufhängevorrichtung *f*, T-Eisen *n* ~ **clamps** Türspanner *f*

**T beam** Plattenbalken *m*, T-Balken *m*, T-Träger *m*

**T bolt** T-Bolzen *m*

**T coating** T-Belag *m*

**teach, to** ~ belehren, lehren, unterrichten

**teacher** Lehrer *m*

**teaching** Unterricht *m*

**teak** Teakholz *n*

**team** Gespann *n*, Mannschaft *f* ~ **competition** Staffelwettbewerb *m* ~ **work** Zusammenarbeit *f*

**tear, to** ~ einreißen, reißen, zerreißen, zerren **to** ~ **apart** auseinanderreißen, lostrennen **to** ~ **asunder** durchreißen **to** ~ **down** abschleifen, (old buildings) auseinanderreißen **to** ~ **off** anreißen, losreißen **to** ~ **off** (or **away**) abreißen **to** ~ **open** aufreißen **to** ~ **out** herausreißen **to** ~ **to pieces** zerfetzen **to** ~ **up** aufreißen

**tear** Einriß *m*, Riß *m* ~ **chip** Reißspan *m* ~**-down** Zerlegung *f* (eines Motors) ~**-down inspection** Befundaufnahme *f* (eines zerlegten Motors) ~**-exciting** tränenerregend ~ **gas** Augenreizstoff *m*, Tränen-gas *n*, -stoff *m* ~**-off cover** (or **cap**) (parachute) Reißlasche *f* ~**-proof** reißfest ~ **resistance** (of rubber) **by slit test** Kerbzähigkeit *f* ~**-shaped** tropfenförmig ~ **strength** Einreißfestigkeit *f*, Reißfestigkeit *f*

**tearing** Einreißen *n*, Zerrung *f* ~ **length** Reißlänge *f* ~ **limit** Zerreißgrenze *f* ~**(-down) method** Zerreißverfahren *n* ~ **resistance** Einreißfestigkeit *f* ~ **strength** Einreißfestigkeit *f*, Zerreißfestigkeit *f* ~**-strength test** Reiß(festigkeits)probe *f* ~ **test** Reißversuch *m* ~**-up thermically** thermisches Aufreißen *n*

**tears of material in fatigue** Ausrisse *pl*

**tease, to** ~ auszupfen, fasern, hecheln, kardieren, rauhen, schüren (Feuer)

**teasel** Karde *f* ~**-raising machine** Rollkardenrauhmaschine *f*

**teaseler** Rauher *m* (Tuch)

**teat drill** Zapfenbohrer *m*

**tecalemiting** Sommerung *f*

**technical** technisch ~ **adviser** Fachberater *m* ~ **assistant** technischer Gehilfe *m* ~ **college** Gewerbeschule *f*, höhere technische Lehranstalt *f* (HTL) ~ **education** Fach(aus)bildung *f* ~ **examination** theoretische Prüfung *f* ~ **expression** Fachausdruck *m* ~ **fair** technische Messe *f* ~ **filter laboratory** filtertechnisches Laboratorium *n* ~ **goods** technische Artikel *pl* ~ **inspection** Waffen- und Geräteuntersuchung *f* ~ **instruction** theoretischer Unterricht *m* ~ **journal** Fachzeitschrift *f* ~ **language** Fachsprache *f* ~ **literature** Fach-literatur *f*, -presse *f* ~ **material** technischer Werkstoff ~ **officer** (or **official**) Ingenieur *m* ~ **paper** Fach-blatt *n*, -zeitschrift *f* ~ **press** Fach-literatur *f*, -presse *f* ~ **proceeding of the survey** Aufnahmetechnik *f* ~ **publication** Fachblatt *n* ~ **school** Fach-,

Gewerbe-schule *f* ~ **schooling** Fachausbildung *f* ~ **staff** technisches Personal *n* ~ **standard** Fachnorm *f* ~ **survey** technische Aufnahme *f* ~ **term** Betriebsausdruck *m*, Fachausdruck *m*, Fachwort *n*, Kunstausdruck *m* ~ **terminology** Fachsprache *f* ~ **test** technischer Wettbewerb *m* ~ **worker** Facharbeiter *m* ~ **world** Fachwelt *f*

**technician** Facharbeiter *m*, Techniker *m* ~ **specialist** Fachbearbeiter *m*

**technics** Technik *f*, (structure) Tektonik *f*

**technique** Arbeitsverfahren *n*, Ausführungsverfahren *n*, Kunstfertigkeit *f*, Technik *f*, Verfahren *n*, Wesen *n*

**technological** technologisch ~ **institute** technische Hochschule *f*

**technologist** Technologe *m*

**technology** Technik *f*, Technologie *f* ~ **of coke** Kokereitechnik *f* ~ **of textiles** Faserstofftechnik *f*

**tectonic features** tektonisches Aussehen *n*

**tedder** Heuwendemaschine *f*, Heuwender *m*, Spannseil *n*

**tedding** Heuwenden *n* ~ **fork** Wendegabel *f*

**tedious** ermüdend, langweilig, langwierig, weitschweifig, zeitraubend

**tee, to** ~ abzweigen, in Brücke schalten (teleph.) **to** ~ **across** in Brücke schalten **to** ~ **in** sich anschalten **to** ~ **together** parallelschalten

**tee** T-Stück *n* ~ **bolt** T-Schraube *f* ~ **branch** Abzweigstück *n* ~ **connection** T-Verbindungsstück *n* ~ **fitting** (Triebwerk) T-Fitting *n* ~ **joint** T-Stoß *m* ~ **slots** T-Nuten *pl* ~ **slotted cross slide** Planschieber *m* mit T-Nuten

**teed across** in Brücke geschaltet

**teeing-off substation** Abzweigungsunterstation *f*

**teem, to** ~ (ingots) abgießen, abstechen, ausgießen, ausschöpfen, gießen, vergießen

**teeming** Gießen *n*, Vergießen *n* ~ **box** Gieß-grube *f*, -platz *m* ~ **ladle** Eispfanne *f* mit Stopfenausguß, Gießpfanne *f*, Gießpfanne *f* mit Stopfenausguß, Pfanne *f*, Stopfenpfanne *f* ~ **lip** Ausgußschnauze *f* ~ **nozzle** Ausflußöffnung *f*, Gieß-loch *n*, -stein *m*, Lochstein *m* ~ **platform** Gießbühne *f* ~ **spout** (of a crucible) Ausgußlippe *f*

**teeth** Bezahnung *f*, Zahnung *f* ~ **closely set** enger Zahnabstand *m* ~ **surfaces** Zahnflanken *pl*

**teething** Bezahnung *f*

**tegmen** Kelchdecke *f*

**telautograph** Handschriftschreiber *m*, Telautograf *m*

**telautographic** telautografisch

**telautography** Schriftfernübertragung *f*, Telautografie *f*

**tele,** ~**ammeter** Stromfernmeßgerät *n* ~**breaker** Fernbrecher *m* ~**camera** Fernkamera *f* ~**cast** Fernseh-sendung *f*, -übertragung *f* ~**casting station** Fernsehsender *m* ~**centric** telezentrisch

**telecine,** ~ **equipment** Filmabtaster *m* ~ **projector** Fernsehbildprojektor *m* ~ **scan** (of film) (Film-)Fernsehabtastung *f* ~ **transmission** Film-abtastung *f*, -übertragung *f*

**telecommunication** Fernmeldeverkehr *m*, Fernmeldewesen *n* ~ **network** (or **system**) Fernmelde-, Nachrichten-netz *n* ~ **service** Fernmeldedienst *m*

**telecommunications** Fernmeldewesen *n*

**telecompass** Fern(seh)kompaß *m*
**telecontrol** Fernbedienung *f*, Fernregelung *f*,
Fernsteuerung *f* ~ **of steering gear** elektrische
Rudermaschine *f* ~ **pulse** Fernwirkimpuls *m*
**telecontrolled,** ~ **aircraft** Fernlenkflugzeug *n* ~
**substation** ferngesteuertes Unterwerk *n* ~
**tachometer** Ferntachometer *m*
**tele,** ~**counter for water meters** Fernzählwerk *n*
für Wassermesser ~**diffusion** Fernausbreitung *f*
~**finder** Telesucher *m* ~**focus cathode** Fern-
fokuskathode *f* ~**gauge** Fernmeßgerät *n* oder
Meßinstrument *n* mit Fernablesung
**telegen effect** Wechselwirkungseffekt *m*
**telegram** Depesche *f*, Drahtnachricht *f*, Drahtung *f*, Kabelbrief *m*, Telegramm *n* ~ **counter**
Telegrammschalter *m*
**telegraph, to** ~ depeschieren, drahten, funken,
telegrafieren **to** ~ **by line grounding** mittels
Erdschluß *m* telegrafieren
**telegraph** Fernschreiber *m*, Telegraf *m* **to-and-
for** ~ Klippklapptelegraf *m* ~ **by-pass set** Tele-
grafenumgehungseinrichtung *f* ~ **cable** Tele-
grafenkabel *n* ~ **channel** Telegrafierweg *m* ~
**clerk** Telegrafenbeamter *m* ~ **code** Telegrafenalfabet *n*, -kode *m*, -schlüssel *m* ~**-construction
office** Telegrafenbauamt *n* ~**-construction tool**
Telegrafenbaugerät *n* ~ **interference (or noise)**
Telegrafiergeräusch *n* ~ **key (or sounder)** Tele-
grafentaste *f* ~ **line** Telegrafenleitung *f* ~**-line
plan** Wegeplan *m* ~ **magnifier** Telegrafenvor-
verstärker *m* ~ **message** Morsespruch *m* ~
**messenger** Telegrafenbote *m*, Telegrammaus-
träger *m* ~ **modulated wave** getastete ungedämpfte Welle *f* ~ **modulator** Telegrafen-
modulator *m* ~ **noise** Telegrafiergeräusch *n*
~ **office** Telegrafenamt *n* ~ **office of origin**
Aufgabetelegrafenanstalt *f* ~ **operator** Fernschreiber *m*, Telegrafist *m* ~ **order-wire working** Summermeldebetrieb *m* ~ **plant** Telegrafenanlage *f* ~ **pole** Telegrafen-leitungsmast *m*,
-stange *f* ~ **post (or pole)** Telegrafenmast *m*
~ **printer** Fernschreiber *m* ~ **receiver** Telegra-
fieempfänger *m* ~ **reception** Telegrafieempfang
*m* ~ **regenerative repeater** entzerrender Tele-
grafieübertrager *m* ~ **repeater (set)** Telegrafen-
übertragung *f*, Übertragung *f* in Telegrafen-
leitungen ~ **route** Telegrafierweg *m* ~ **selector**
Einzelanrufer *m* ~**(ic) signal** Telegrafierzeichen
*n* ~ **signal element** Telegrafieschritt *m* ~ **signal
reception** Telegrafenschreibempfang *m* ~ **speed**
Telegrafiergeschwindigkeit *f* ~ **station** Tele-
grafen-amt *n*, -anstalt *f* ~ **superposed circuit**
Simultantelegrafenleitung *f* ~ **switchboard**
Telegrafenklappenschrank *m* ~ **system** Telegrafenanlage *f* ~**-terminal cable** Telegrafenab-
schlußkabel *n* ~ **transmission potential** Tele-
grafensendespannung *f* ~ **transmitter** Telegrafiesender *m* ~ **wire** Telegrafendraht
*m*
**telegrapher** Telegrafist *m*
**telegraphic** drahtlich, telegrafisch ~ **address**
Drahtanschrift *f*, Telegrammanschrift *f* ~
**alphabet** Telegrafenalfabet *n* ~ **ciphers** Depe-
schenschlüssel *m* ~ **code** Kodeschlüssel *m* ~
**equation** Telegrafengleichung *f* ~ **frequency**
Telegrafierfrequenz *f* ~ **modulator** Telegrafen-
modler *m* ~·**repeater** Relaisübertragung *f* ~
**restitution** telegrafische Wiedergabe *f* ~ **transfer**

telegrafische Überweisung *f* ~ **transmitter**
Telegrafiersender *m*
**telegraphical time-signal distributor** Zeitsignal-
übertrager *m*
**telegraphone** Telegrafon *n* ~ **sound-recording
method** Magnettonverfahren *n*
**telegraphy** Telegrafenwesen *n*, Telegrafie *f* ~
**not to be listened to** unabhörbare Telegrafie *f*
~ **transmitter** Telegrafiesender *m*
**tele,** ~**guided aircraft** ferngesteuertes Flugzeug
*n* ~**hydrobarometer** Wasserstandsfernmelder *m*
~**iconography** Fernbildsendung *f* ~**-indicating**
Fernanzeige ... ~**-lens (or -objective)** Fernobjektiv *n* ~**mechanics** drahtlose Übertragung *f*
elektrischer Energie, mechanische Fernsteuerung *f*, Telemechanik *f* ~**mechanism** Fernsteuerung *f* ~**-meter** Entfernungsmesser *m*, Fernanzeiger *m*, Fernmesser *m*, Fernmeßgerät *n*,
Fernmeßgerät *n* mit Fernablesung, Telemeter *n*
~**metered data** Sendewerte *pl*
**telemetering** Fernmessen *n*, Fernmessung *f* ~
**device (or equipment)** Fernmeßeinrichtung *f* ~
**flight** Meßflug *m* ~ **system** Fernmeßanlage *f*
**telemetric,** ~ **action (by selsyntype motor)** Fernsteuerung *f*, Fernwirkung *f* ~ **instrument**
Wettermeßinstrument *n* mit funkentelegrafischer Fernübertragung ~ **integrator** Fernmeß-
summengeber *m* ~ **law** Entfernungsgesetz *n* ~
**scale** Entfernungsmaßstab *m*
**telemetry** Entfernungsmessen *n*, Fernmeßtech-
nik *f*, Fernmessung *f*, Telemetrie *f* ~ **and remote
control** Fernwirktechnik *f*
**telengiscope** Telengiskop *n* (opt.)
**telephone, to** ~ anklingeln, anrufen, durchspre-
chen, fernsagen, fernsprechen, telefonieren,
zusprechen
**telephone** Fernhörer *m*, Fernsprecher *m*,
(station) Sprechapparat *m*, Telefon *n*, Telefon-
apparat *m* **by** ~ fernmündlich, telefonisch **the**
~ **Telefonsystem** *n* ~ **for towed flight** Schlepp-
flugfernsprecher *m*
**telephone,** ~ **(circuit) amplifier** Leitungsverstärker *m* ~ **amplifying valve** Fernsprechver-
stärkerröhre *f* ~ **area** Sprachbereich *m* ~ **block
system** Raumfolge *f* mit Zugmeldedienst ~
**board** Fernsprechtafel *f* ~ **booth** Fernsprechzelle *f*, Zelle *f* mit öffentlichem Fernsprecher
~ **break-down** Fernsprechstörung *f* ~ **buoy**
Fernsprechboje *f* ~ **cabin(et)** Fernsprechzelle *f*
~ **cable** Fernsprechkabel *n* ~ **call** Anruf *m*,
Ferngespräch *n*, Fernspruch *m* ~ **call box**
Fernsprechkabine *f* ~ **central** Fernsprechvermittlung *f* ~ **central office** Fernsprechvermitt-
lungsstelle *f* ~ **channel** Fernsprech-kanal *m*,
-weg *m*, Sprechweg *m* ~ **channel group** Fern-
sprechbündel *n* ~ **circuit** Fernsprechleitung *f*
~ **code** Fernsprech(er)schlüssel *m* ~ **communication** Drahtnachrichtenverbindung *f*, Gesprächsverbindung *f*, Sprechverbindung *f*,
Vermitteln *n* mit Fernsprechern ~ **com-
munication(s)** Fernsprechverbindung *f* ~ **conduit
wires** Fernsprechrohrdrähte *pl* ~ **connecting plane and towed glider** Schleppflug-
fernsprechanlage *f* ~ **connection** Fernsprechanschluß *m*, Fernsprechvermittlung *f*, Gesprächsverbindung *f*, Sprechverbindung *f*
~**-construction crew** Fernsprechbautrupp *m*
~**-construction truck** Fernsprechbauwagen *m*

~-construction unit Fernsprechbaugerät-
kolonne f ~ conversation Ferngespräch n ~
cord Fernhörerschnur f ~ current Sprechstrom
m ~ cushion Fernhörerkissen n ~ dial Telefon-
wählscheibe f ~ directory Fernsprech-amt-
buch n, -buch n, -teilnehmerverzeichnis n,
Teilnehmerverzeichnis n ~ distribution center
Verteilfernamt n ~ drop Einführungsleitung f
~ earpiece Muschel f des Fernhörers ~ end
cable Fernsprechabschlußkabel n ~ equipment
Fernsprech-anlage f, -gerät n, Telefonanlage f
exchange Amt n, Amtsanschließer m, Fern-
sprech-vermittlung f, -vermittlungsstelle f,
-zentrale f, Vermittlung f, Vermittlungs-amt n,
-stelle f ~-exchange attachment Amtszusatz m
~ extension Haustelefon n, Sprechstelle f,
Telefonanschluß m ~ extension set Fernsprech-
nebenstelle f ~ frequency Fernsprechfrequenz f
~ handset Hand-apparat m, -fernsprecher m,
Mikrotelefon n, Sprechhörer m ~ headgear
receiver Kopfhörer m ~-influence factor Fernsprechformfaktor m ~ installation Fernsprech-,
Telefon-anlage f ~ instrument Fernsprech-
amtapparat m ~-interception post Lauschstelle
f ~-interference factor Fernsprechstörfaktor m
~ intermediate repeater Fernsprechzwischen-
verstärker m, Zwischenverstärker m ~ jacks
Kopfhöranschluß m ~ kiosk öffentliche Fernsprechstelle f ~ lightning protector Fernsprech-
blitzableiter m ~ line Fernsprechleitung f, Linienführung f, Sprechverbindung f ~ line with
protection against high tension hochspan-
nungsgeschützte Fernsprechleitung f ~ line
equalizer Telefonleitungsentzerrer m ~ loop
Fernsprechamtdoppelleitung f ~ message Fernspruch m ~ meter Gesprächszähler m ~ network
Fernleitungs-, Fernsprech-netz n ~ office Fernsprechanstalt f ~ operation Fernsprechbetrieb
m ~ operator Telefonist(in) m, f ~ pick-up
Telefonspule f ~ plant Fern(sprech)anlage f,
(outside) Fernsprechnetz n, Telefonanlage f ~
plug Telefonstecker m ~ pole Leitungsmast m,
Telegrafenstange f ~ receiver Fernhörer m,
Fernsprechempfangsgerät n, Hörer m, Telefon
n, Telefonhörer m ~reception Telefonie-
empfang m ~ (transmission) reference system
Fernsprecheichkreis m ~ relations Sprechbeziehungen pl ~ relay Fernsprechrelais n
telephone-repeater Fernsprechverstärker m, (circuit) Leitungsverstärker m ~ operation Fern-
sprechverstärkerbetrieb m ~ station Fern-
sprechverstärkeramt n ~ tube Fernsprechver-
stärkerröhre f
telephone, ~ responder Anrufbeantworter m ~
ringer Telefonwecker m ~ section Fernsprech-
trupp m ~ service Fernsprechamt-betrieb m,
-dienst m, Sprechbeziehungen pl, Vermitt-
lungsdienst m ~ service with two wires Schleifenleitungsbetrieb m ~ set (or telephone) Fernsprechapparat m ~ short-circuiting contact
Telefonkurzschlußkontakt m ~ signal Tele-
fonie-signal n, -zeichen n ~ station Fernsprech-
amtanschluß m, Fernsprecher m, Fernsprechstelle f ~ switchboard Fernsprech(klappen)-
schrank m ~ switchbord for private-branch
exchange Klappenschrank m für Nebenstellen
~ switch box Telefonanschlußkasten m ~ switch
hook Gabelschalter m ~ switching engineer

Vermittlungstechniker m ~ switching point
Verteilfernamt n ~ system Fernleitungsnetz n,
Fernsprech-netz n, -system n ~ tariff Fern-
sprechtarif m ~ terminal repeater Fernsprech-
endverstärker m ~ traffic Fernsprechverkehr m
~-traffic recorder Fernsprechverkehrsschreiber
m, Verkehrsschreiber m ~-traffic unit Verkehrswert m ~ transcription Fernspruch m ~
transformer Fernsprechübertrager m ~ transmission technique Fernsprechtechnik f ~ transmitter Fernsprechmikrofon n ~ truck Fern-
sprech(kraft)wagen m ~ trunk line Fernleitung
f ~-trunk-line board Fernleitungstafel f ~
trunk zone Fernverkehrs-bereich m, -zone f ~
~-voice recorder Gerät n zum Aufzeichnen von
Gesprächen ~-voltage form factor Fernsprechformfaktor m ~ wire Fernsprechleitung f,
Telefondraht m ~ zone Fernsprechzone f ~
(trunk) zone Taxquadrat n ~-zone center Fern-
sprechzonenhauptpunkt m
telephonic Telefon ...; fernmündlich, telefonisch ~ echo fernmündliches Echo n ~ frequency Sprechfrequenz f ~ relay Fernsprechrelais n ~ transmitter Telefoniesender m (rdo)
telephonically silent generator oberschwingungs-
freier Gleichstromerzeuger m
telephonist Beamtin f, Platzbeamtin f
telephonograph Telefonograf m
telephonometry Fernsprechmeßtechnik f, Lautstärkemessung f
telephony Fernsprechen n, Fernsprechwesen n,
Telefonie f ~ amplification Fernsprechver-
stärkung f ~ transmitter Telefoniesender m
telephoto Fernbild n, Fernrohraufnahme f,
Lichtfunk m ~ attachment Teleansatz m ~ lens
Fernlinse f, Fernobjektiv n, Teleobjektiv n ~
transmission Bildübertragung f
telephotograph Fototelegraf m
telephotographic fernfotografisch, fototelegra-
fisch ~ transmission bildtelegrafische Übertragung f ~ work Teleaufnahme f
telephotography Bildtelegrafie f, Fernaufnahme
f, Fernfotografie f
telephotometry Telefotometrie f
telephoto station (facsimile) Bildstelle f
teleprint, by ~ fernschriftlich ~ connection
Fernschreibvermittlung f ~ exchange Fernschreibvermittlung f ~ gauge Fernschreibgerä-
telehre f ~ lorry (or truck) Fernschreib(kraft)-
wagen m
teleprinter Drucktelegraf m, Empfangsmagnet m
mit Teilkreis, Ferndrucker m, Fernschreiber m,
Fernschreibmaschine f, Typendrucker m,
Typendrucktelegraf m ~ code Fernschreiber-
kode m ~ mechanic Fernschreibmechaniker m
~ network Fernschreibnetz n ~-on-radio
Funkfernschreib ... ~ operator Fernschreiber
m ~ paper tape Papierstreifen m ~ reception
Buchstabenschreibempfang m ~ switching
center Fernschreibvermittlung f ~ tape Aufnahmestreifen m ~ tape switching center Fern-
schreibspeichervermittlung f
tele, ~printing on radio Funkfernschreiben n
~program Fernprogramm n ~psychrometer
Fernfeuchtigkeitsmesser m, Fernpsychrometer
n ~-radium unit Radium-bombe f, -kanone f
~-reception Hellschreiber m ~repeating device
Fernsteuervorrichtung f mit Rückmeldung

**telescope, to** ~ (a coil) eintauchen, ineinander-schieben
**telescope** Fernrohr *n*, Sehrohr *n*, Teleskop *n*
~ **with knee periscope** Winkelsichtfernrohr *n*
~ **of short focal length** kurzbrennweitiges Fern-rohr *n* ~ **with stadia lines** Fernrohr *n* mit Ent-fernung messenden Fäden
**telescope,** ~ **adjustment screw** Fernrohrschlüssel *m* ~ **barrel** Fernrohrbüchse *f* ~ **carrier** Fern-rohrträger *m* ~ **center section** Fernrohrhals *m*
~ **cradle** Fernrohrwiege *f* ~ **head** Fernrohr-kopf *m* ~ **housing** Fernrohrhülse *f* ~ **knob** Triebscheibe *f* zum Fernrohr ~ **mount** Auf-satz *m*, Aufsatzzeiger *m* ~ **seat** Fernrohrlager *n* ~ **shaft** Teleskopwelle *f* ~ **stand** Fernrohr-träger *m* ~ **support** Fernrohr-steg *m*, -träger *m*
~ **tripod** Röhrenstativ *n* ~ **tube** Fernrohr-büchse *f*, -hülse *f*, -körper *m* ~ **tubes** Fernrohr-arme *pl*
**telescoped** ineinanderschiebbar **the parts have been** ~ die Stücke sind durch Ineinander-stecken *n* vereinigt
**telescopic** ausschiebbar, ausziehbar, teleskopisch zusammenschiebbar ~ **altimeter** Visierhöhen-messer *m* ~ **bridge** (Flughafen) Teleskopbrücke *f* ~ **carriage** Stauchlafette *f* ~ **condenser** posaunenartige Leitung *f* ~ **feed shaft** auszieh-bare Antriebswelle *f* ~ **finder** Sucherfernrohr *n*
~ **floodlight** Teleskoplichtmast *m* ~ **jack** Hub-presse *f* ~ **jack retraction line** Hubpressen-rückzugsleitung *f* ~ **landing gear strut (or leg)** Posaunenfahrgestellstrebe *f* ~ **leg** Federstrebe *f* (aviat.) ~ **lens** Triebrohr *n* ~ **magnifier** Fern-rohrlinse *f* ~ **mast** Kurbel-, Teleskop-mast *m*
~ **performance** Fernrohrleistung *f* (photo) ~ **pipe** Degen-, Teleskop-rohr *n* ~ **ram** Hubpresse *f* ~ **screw** Teleskopspindel *f* ~ **shaft** auszieh-bare Gelenkwelle *f* ~ **shock-absorber** Teleskop-federung *f*, -stoßdämpfer *m* ~ **shock course** Federweg *m* ~ **sight** Aufsatzfernrohr *n*, Fern-rohraufsatz *m*, Okulardiopter *m*, Richt-aufsatz *m*, -fernrohr *n*, Visierfernrohr *n*, Zielfern-rohr *n* ~ **sight mount** Zielfernrohrträger *m* ~ **sights** Fernrohrvisier *n* ~ **span wing** Auszieh-flügel *m* ~ **spectacles** Fernrohrbrille *f* ~ **stuffing box** Posaunenstopfbüchse *f* ~ **strut** Auszieh-rohr *n*, (pipe) Posaunenrohr *n*, Tauchrohr *n*, Teleskopbetätigungsstrebe *f*, Triebrohr *n* ~ **support** Teleskopstütze *f* ~ **suspension** Tele-skopfederung *f* ~ **tube** Teleskoprohr *n* ~ **tube stuffing box** Tauchrohrstopfbüchse *f*
**telescoping** Ineinanderschiebung *f*; posaunen-artig verschiebbar ~ **chute** Teleskopschurre *f*
~ **coil** Tauchspule *f* ~**-coil transformer** Tauch-transformator *m* ~ **coils** ineinanderschiebbare Spulen *pl* ~ **means** Posaunenauszug *m*
**tele,** ~**speedometer** Fernfahrtmesser *m* ~**stereo-scope** Telestereoskop *n* ~**studio** Atelier *n*, Auf-nahmeraum *m* ~**-switch** Fernschalter *m*
~**tachometer** Ferndrehzahlmesser *m* ~**thermo-meter** Fernthermometer *n* ~**torium** Atelier *n*, Aufnahmeraum *m*, Tonatelier *n* ~**transmitter** **for differential pressure meters** Ferngeber *m* für Druckdifferenzmesser
**teletype** Fernschreiber *m*, Fernschreibmaschine *f*, Schreibempfänger *m*, Springschreiber *m*, Teletype *m* ~ **apparatus** Ferndrucker *m* ~ **circuit** Fernschreibleitung *f* ~ **control panel**

**(or network)** Fernschreibmeßgestell *n* ~ **keying unit** Fernschreibertastgerät *n* ~ **truck** Fern-schreibwagen *m*
**teletyper** Ferndrucker *m*
**teletypewriter** Fernschreibmaschine *f* ~ **circuit** Fernschreib-leitung *f*, -verbindung *f* ~ **tape** Fernschreiblochstreifen *m*
**teleview, to** ~ fernsehen
**teleview,** ~ **apparatus** Fernsehempfänger *m* ~ **object** Übertragungsgegenstand *m*
**televiewer** Fernsehzuschauer *m*
**televise, to** ~ fernsehen
**televised,** ~ **object** Übertragungsgegenstand *m*
~ **picture** Fernbild *n*
**televising car** Aufnahmewagen *m*, Fernsehauf-nahmewagen *m*
**television** Fernsehen *n* ~ **apparatus** Fernseher *m*
~ **broadcast** Fernsehsendung *f* ~ **cable** Fernseh-kabel *n* ~**-cable transmission** Fernsehkabelüber-tragung *f* ~ **camera** Bildfänger *m*, Fernseher *m*
~ **camera truck** Aufnahmewagen *m* ~ **channel** Fernsehkanal *m* ~ **combined with telephone service** Fernseh-sprechen *n*, -sprechnetz *n* ~ **connection** Fernsehverbindung *f* ~ **dissector tube** Fernsehzerlegerröhre *f* ~ **emitter** Fernseh-sender *m* ~ **frame** Fernsehbildraster *m* ~ **grating** Fernsehraster *m* ~ **image** Fernsehbild *n* ~ **monitoring** Fernsehverkehrslenkung *f* ~ **pattern generator** Fernsehprüfgenerator *m* ~ **pickup** Fernsehaufnahme *f* ~ **picture** Fernbild *n* ~ **projection tubes** Projektionsfernsehröhren *pl* ~ **receiver** Bildempfänger *m*, Fernsehemp-fänger *m* ~ **reception** Fernsehempfang *m* ~ **relay link** Fernsehrelaiskette *f* ~ **reporting van** Fernsehaufnahmewagen *m* ~ **scanning device** Bildabtaster *m* ~ **signal** Fernsehsignal *n* ~**-sound transmitter** FS-Tonsender *m* ~ **speech** Fernsehsprechen *n* ~ **standard** Fernsehnorm *f*
~ **station link** Fernseh-Relaisstrecke *f* ~ **studio** Fernsehaufnahmeraum *m* ~ **synchronisation** Fernsehsynchronisierung *f* ~ **system** Fernseh-einrichtung *f* ~ **telephone** Fernseh-sprecher *m*, -telefon *n* ~ **transmission** Bildfunk *m*, Fern-seh-sendung *f*, -übertragung *f* ~**-transmission pickup tube** Fernsehsenderöhre *f* ~ **transmitter** Fernseher *m*, Fernsehsender *m* ~ **transmitter with electric scanning** Fernsehgeber *m* mit elek-trischer Abtastung ~ **tube** Fernsehröhre *f*
**televisor** Bildabtaster *m*, Fernsehabtaster *m*, Fernseher *m* ~ **tube** Bildschreibröhre *f*
**televoltmeter** Spannungsfernmeßgerät *n*
**telewattmeter** Leistungsfernmeßgerät *n*
**telewriter** Fernschreibemeßgerät *n*, schreibendes Fernanzeigegerät *n*, (apparatus) Teleautograf *m* ~ **service** Teilnehmertelegrafie *f*
**telex** Fernschreiben *n* ~ (teleprint) **exchange system** Fernschreibvermittlungsanlage *f*
**telford base** (Straßenbau) Packlage *f*
**tell, to** ~ angeben, erzählen, sagen **to** ~ **by listening** anhören
**teller mine** Tellermine *f*
**telling** ausschlaggebend, durchschlagend
**telltale** Anzeigevorrichtung *f* mit Rückmelde-einrichtung für Betriebszustände, Kontrolluhr *f*, (selbsttätige) Registriervorrichtung *f*, Wäch-teruhr *f*, Warnzeichen *n* ~ **clock** Kontrolluhr *f*
~ **device** Überwachung *f* ~ **gauge** Axiometer *m*
~ **indicating system** Lichtwarnungssystem *n* ~

**indicator** Anzeigemeßgerät *n* mit Rückmelde-
einrichtung (z. B. bei Einziehfahrwerk) **~ lamp**
Kontroll-, Melde-lampe *f* **~ marking** Unterscheidungsmarkierung *f* **~ means** Schauzeichen
*n* **~ pipe connection** Signalrohranschluß *m* **~**
**spark gap** Kontrollfunkenstrecke *f* **~ system**
Schülerüberwachungsanlage *f* **~ watch** Wächterkontrolluhr *f*
**telluric** tellurisch, tellursauer **~ bismuth** Wismuttellur *n*
**telluride** Tellurid *n*
**telluriferous** tellurführend
**tellurite** Tellurocker *m*
**tellurium** Tellur *n* **~ glance** Tellurglanz *m* **~**
**pre-alloy** Tellurvorlegierung *f*
**tellurous** telluirg
**telpher, ~ fitted with driver's stand** Führerstandlaufkatze *f* **~ line** elektrische Drahtseilbahn *f*,
Telpherbahn *f*
**temper, to ~** (steel) abhärten, abschrecken,
(water content) anfeuchten, anlassen, anmengen, ausglühen, entspannen, härten, kälken,
mäßigen, mildern, mischen, nach-erhitzen,
-glühen, -hitzen, -lassen, temperieren, tempern,
vergüten
**temper** Abschreckung *f*, (of metal) Gare *f*,
Härte *f*, (steel) Härtegrad *m*, Laune *f*, Naturhärte *f*, Stimmung *f* **~ brittleness** Ablaß-,
Anlaß-sprödigkeit *f* **~ carbon** Temperkohle *f*
**~-carbon nodule** Temperkohleknötchen *n* **~**
**color** Ablaß-, Glüh-farbe *f*, Wärmetönung *f*
**~(ing) color** Anlauffarbe *f* **~-effect** Anlaß-
wirkung *f* **~ graphite** Temperkohle *f* **~ harden-**
**ing** Ablaßhärtung *f* **~ heat of iron** Blauwärme *f*
**~ pass mill** Dressierwalzwerk *n* **~ reel** Nachlaßwinde *f* **~ screw** Nachlaß-schraube *f*, -spindel *f*,
Setzschraube *f* **~ structure** Anlaßgefüge *n*
**temperable** härtbar
**temperance** Mäßigkeit *f*
**temperature** Temperatur *f*, Wärme *f*, Wärmegrad
*m*, Warmheit *f*, Witterung *f* **~ of combustion**
pyrometrischer Wärmeeffekt *m*, Verbrennungstemperatur *f* **~ of the cooling water** Kühlwasser-
wärmegrad *m* **~ of the filling (or inflation)**
Füllungswärmegrad *m* **~ of inflation gas** Fül-
lungswärme *f* **~ of the radiator** Kühlerwärme-
grad *m*, -temperatur *f* **~ of reaction** Reaktions-
temperatur *f* **~ at various altitudes** Höhentem-
peratur *f*
**temperature, ~ amplifier** Temperatursignalver-
stärker *m* **~ balance** Temperaturausgleich *m* **~**
**bath** Temperierbad *n* **~ bulb** Temperaturfühler *m* **~ coefficient** Temperatur-koeffizient
*m*, -zahl *f*, Wärme-koeffizient *m*, -zahl *f* **~**
**compensation** Wärmeausgleich *m* **~ concept**
Temperaturbegriff *m* **~ constancy** Temperatur-
beständigkeit *f* **~ control** Temperaturreglung *f*
**~ control equipment** Temperaturregler *m* **~**
**controller** Temperaturregler *m* **~ controller**
**shaft** Temperaturreglerspindel *f* **~ dependence**
Temperaturabhängigkeit *f* **~ dependent** temperaturabhängig **~ difference** Temperaturgefälle
*n*, Wärmegradunterschied *m* **~ distribution**
Temperaturverteilung *f* **~ drawing** Anlaßtemperatur *f* **~ drop** Temperaturgefälle *n* **~**
**eddy** Temperatur-, Wärme-wirbel *m* **~ effect**
Einfluß *m* der Temperatur, Temperatur-,
Wärme-wirkung *f* **~ embrittlement** Warmver-

sprödung *f* **~ emission of electrons** Elektronenverdampfung *f*, glühelektrische Elektronenemission *f*, Glühelektronenemission *f* **~-**
**entropy diagram** Wärmebild *n* **~ feeler** Temperaturfühler *m* **~ field pattern** Temperatur-
feld *n* **~ gauge** Kühlwasseranzeiger *m*, Tempe-
ratur-anzeiger *m*, -messer *m* **~ gauge plug** Tem-
peraturmeßkerze *f* **~ gradient** geothermische
Tiefenstufe *f*, Temperatur-feld *n*, -gefälle *n*,
-gradient *m*, -steigerung *f*, Wärme-gefälle *n*,
-gradient *m* **~ hysteresis** Temperaturhysteresis
*f* **~ increase** Temperatur-erhöhung *f*, -zunahme
*f* **~-indicating equipment** Temperaturregistrier-
anlage *f* **~ indication** Temperaturanzeige *f* **~**
**interlock** (Turbine) Temperaturregelventil *n* **~**
**interval** Temperaturintervall *m* **~ inversion**
Temperaturumkehrung *f* **~ ionization** Wärmeionisation *f* **~ lapse** Temperaturgradient *m*,
Wärmegradverlauf *m* **~ level** Temperatur-
niveau *n* **~ limit** Temperaturgrenze *f* **~ meas-**
**urement** Temperaturmessung *f* **~ observation**
Temperaturbeobachtung *f* **~ probe** Temperaturfühler *m* **~ radiation** Wärmestrahlung *f* **~**
**range** Temperaturbereich *m* **~ recorder** Tempe-
ratur-registrierapparat *m*, -schreiber *m* **~ re-**
**cording chart** Temperaturaufzeichnungsblätter
*pl* **~ reducing set** Temperaturreduzieranlage *f*
**~ regulation** Kühlerreglung *f* **~ regulator**
Temperaturregler *m* **~ response** Temperaturgang *m* **~ responsive** temperaturabhängig
**temperature-rise** (heating) Erwärmung *f*, (gradient) Temperatursteig(er)ung *f* **~ computation**
Wärmeberechnung *f* **~ voltage** maximal zulässige Betriebsspannung *f* **~ within limits**
Grenzerwärmung *f*
**temperature, ~ scale** Temperaturskala *f* **~**
**schedule** Temperaturprogramm *n* **~ selector**
Temperaturwahlschalter *m* **~ sensor** Temperaturfühler *m* **~ stability** Temperaturbeständig-
keit *f* **~ stresses** Temperaturspannungen *pl*
**~ tube** Peilrohr *n* **~ variation** Temperaturschwankung *f* **~ wave** Temperaturwelle *f*
**tempered** gehärtet **~ castings** schmiedbarer Guß
*m* **~ glass** Vulkanglas *n* **~ glass apparatus** Hart-
glasgeräte *pl* **~ specimen (or test) piece** vergütete Probe *f* **~ steel** gehärteter Stahl *m*
**tempering** Abschrecken *n*, Anlassen *n*, Härten *n*,
Rückglühung *f*, Vergüten *n*, Vergütung *f* **~**
**agent** Härte-, Temper-mittel *n* **~ bath** Härte-,
Temperier-bad *n* **~ color** Anlaß-, Tempera-
farbe *f* **~ compound** Härtemittel *n* **~ effect** Ablaßwirkung *f* **~ flame furnace** Härteflammofen
*m* **~ furnace** Ablaß-, Anlaß-, Härte-, Temper-,
Temperier-ofen *m* **~ heat** Anlaßtemperatur *f* **~**
**jacket** Temperiermantel *m* **~ liquid** Abschreckform *f*, Härteflüssigkeit *f* **~ machine** Härtungs-
maschine *f* **~ material** Tempermittel *n* **~ med-**
**ium** Ablaßmittel *n*, Abschreckform *f*, Anlaßmittel *n*, Temperiermedium *n* **~ oil** Ablaßöl *n*,
Talgöl *n* **~ operation** Anlaßvorgang *m* **~**
**period** Anlaßzeit *f* (metall) **~ powder** Härte-
pulver *n* **~ process** Härteverfahren *n* **~ quality**
Vergütbarkeit *f* **~ range** Ablaßstufe *f* **~ tem-**
**perature** Anlaß-, Nachglüh-temperatur *f* **~**
**time** Anlaßdauer *f* **~ water** Lösch-, Tempe-
rier-wasser *n*
**tempest** Sturm *m*, Unwetter *n*
**template, to ~** schablonieren

**template** Anreißschablone *f*, Einstellehrring *m*, Formblatt *n* (print), Formstück *n*, Führungsgerüst *n*, Kragholz *n*, Lehre *f*, Lehrmutter *f*, Mallbrett *n*, Paßlehre *f*, Schablone *f*, Stichmaß *n* ~ **for laying down a bedplate** Fundamentlehre *f*, -schablone *f*

**template, ~ board** Musterbrett *n* ~ **casting** Schablonenguß *m* ~ **follower** Leitschablone *f* ~ **gauge** Gewindelehrmutter *f* ~ **guide** Schablonenführung *f* ~ **machine** Schablonenmaschine *f* ~ **molding** Schablonenformerei *f* ~ **pool** Schablonenspeicher *m* ~ **swiveling device** Schablonenschwenkeinrichtung *f*

**templates, arm support for** ~ Schablonenhalter *m* ~ **in lockmaking** Führungsschnitte *pl*

**temple** Brillenbügel *m*, (weaving) Zeugspanner *m*

**templet** Anpfahl *m*, Dachpfette *f*, Drehbrett *n*, Fußpfahl *m*, Kopierschablone *f*, Zelluloidtafel *f* ~ **ocular head** Revolverokularkopf *m*

**tempo** Tempo *n*, Zeitmaß *n*

**temporal** zeitlich ~ **sense** Zeitsinn *m*

**temporarily** aushilfsweise, kurzzeitig ~ **suspended** zeitweilig eingestellt

**temporary** interimistisch, kurzzeitig, nichtständig, provisorisch, temporär, vorläufig, zeitlich, zeitweilig ~ **antenna** Behelfsantenne *f* ~ **assistance** Aushilfe *f* ~ **bottom** Gleichgewichtsboden *m* ~ **bridge** Knotenverbindung *f* ~ **casing** vorläufige Verrohrung *f* ~ **dam** provisorisches Wehr *n* ~ **emergency lighting** Bedarfsnotbefeuerung *f* (aviat.) ~ **employee** Hilfskraft *f* ~ **end sleeve** Transportverschluß *m* ~ **exchange** fliegendes Fernsprechamt *n*, Notamt *n* ~ **frame** Hilfsjoch *n* (min.), Hilfskranz *m* (min.), (building construction) Notjoch *n* ~ **hand** Aushilfsarbeiter *m* ~ **hardness** (Wasser) Karbonathärte *f* ~ **injunction** einstweilige Verfügung *f* ~ **joint** Knotenverbindung *f* ~ **locking pin** Vorstecker *m*, Vorsteckstift *m* ~ **magnet** temporärer Magnet *m* ~ **overload** Förderspitze *f* ~ **plant** fliegende oder zeitweilige Anlage *f* ~ **raft** Notfloß *n* ~ **railway line** Interimsbahn *f* ~ **ramp** Notrampe *f* ~ **repair** einstweilige Ausbesserung *f* ~ **repairs** Zwischenausbesserung *f* ~ **rivet** Heftniet *n* ~ **service contract** vorübergehender Fernsprechanschluß *m* ~ **stop** Kurzstop *m* (des Bandlaufes) ~ **storage** Zwischenspeicher *m* (comput.), Zwischenspeicherung *f* ~ **structures** Behelfsbauten *pl* ~ **type** Zwischenform *f* ~ **worker** Aushilfsarbeiter *m*

**ten** Zehner *m* ~-**button key set** Zehntastensatz *m* ~ **penny nail** Floßnagel *n* ~-**point preselector** zehnteiliger Vorwähler *m* ~-**point selector** zehnteiliger Wähler *m* ~-**second division** Zehnersekunde *f* ~-**thousands of an inch** hundertstel Millimeter *m* ~-**way telephone station** Fernsprecher *m* für zehn Linien

**tenacious** bruchfest, hartnäckig, zähe, zähfestig

**tenacity** Ausdauer *f*, Fähigkeit *f* des Festhaltens, Festigkeit *f*, Hartnäckigkeit *f*, Reißfestigkeit *f*, Tenazität *f*, Zähfestigkeit *f*, Zähigkeit *f*, Zugfestigkeit *f* ~ **test** Festigkeitsuntersuchung *f*

**tenant** Hausbewohner *m*, Mieter *m*, Pächter *m*

**tend, to** ~ abzielen auf, gerichtet sein, neigen, schwoien (naut.), sich zuneigen, streben, zustreben **to** ~ **to** zielen auf

**tend** Schwojen *n*

**tendency** Neigung *f*, Richtung *f*, Sinn *m*, Streben

*n*, Strömung *f*, Tendenz *f*, Zug *m* ~ **to(ward) corrosion** Korrosionsneigung *f* ~ **to crack at corners** Kantenrissigkeit *f* ~ **to expand** Ausdehnungsdrang *m* ~ **to flow** Fließfähigkeit *f* ~ **to fracture** Bruchneigung *f* ~ **to hunt** Neigung *f* zur Überdrehzahl, Neigung *f* zum Übersteuern **having a** ~ **to the left** linksorientiert ~ **toward oscillating (or spilling over)** Schwingungsneigung *f* ~ **of profiled metal pile to deform** Auslängen *n* **to have a** ~ **to rise** aufstreben ~ **to rotate** Drehstreben *n* ~ **to rust** Korrosionsneigung *f*, Rostneigung *f* ~ **to set itself** Einstellungsbestreben *n* ~ **to sing** Pfeifneigung *f* ~ **to tear at corners** Kantenrissigkeit *f* ~ **to thunderstorm** Gewitterneigung *f* ~ **to tip** Kippbestreben *n* ~ **to work loose** Lockerungsbestreben *n*

**tender, to** ~ darbieten **to** ~ **in evidence** als Beweis *m* vorlegen **to** ~ **the fiber** Faser *f* schwächen

**tender** Anerbieten *n*, Angebot *n*, Hilfsbeischiff *n*, Kostenanschlag *m*, Lieferungsangebot *n*, Schlepper *m*, Submission *f*, Tender *m*, Tendermaschine *f*; innig, mürbe, weich **to make a** ~ eine Submissionsofferte *f* einreichen

**tender, ~ guarantee** Bietungsgarantie *f* ~ **locomotive** Tenderlokomotive *f* ~ **offer** Angebot *n* ~ **water-tank plate** Tenderwasserkastenblech *n* ~ **wheel** Tenderrad *n*

**tenderer** Submittent *m*

**tenderness** Mürbheit *f*, Mürbigkeit *f*, Weichheit *f*

**tendon** Sehne *f*

**tenement** Mietswohnung *f*

**tenet** Lehre *f*

**tennantite** Arsenfahlerz *n*, Kupferblende *f*, Tennantit *m*, Zinkfahlerz *n*

**tennis court** Tennisplatz *m*

**tenon** Dübel *m*, Feder *f*, Vorsprung *m*, Zapfen *m*, Zinke *f* ~-**cutting machine** Zapfenschneidemaschine *f* ~ **saw** Ansatzsäge *f*

**tenoning** Zapfenschneiden *n* ~ **auger** Zapfenbohrer *m* ~ **machine** Zapfen-fertigungsmaschine *f*, -schneidemaschine *f* ~ **machinist** Maschinenzapfenschneider *m* ~ **saw** Ansatzsäge *f* ~ **tenter** Zapfenschneider *m*

**tenor** Beschaffenheit *f*, Gang *m*, Verlauf *m*

**tens, ~ digit** Zehner-satz *m*, -stufe *f* ~ **relay** Zehnerrelais *n*

**tense, to** ~ einspannen, straff werden, stramm machen, sich straffen

**tense** gespannt

**tensibility** Spannbarkeit *f*

**tensible** dehnbar

**tensile** dehn-, spann-, streck-bar ~ **bar** Zugprobekörper *m*, Zugstab *m* ~ **diagram** Spannungsdehnungsdiagramm *n* ~ **fatigue strength** Zugschwellfestigkeit *f* ~ **fatigue test** Zugermüdungsversuch *m* ~ **force** Dehnfestigkeit *f*, Zieh-, Zug-kraft *f* ~ **impact stress** Schlagzerreißbeanspruchung *f* ~-**impact test** Fallzerreißversuch *m* ~ **load** Zug-beanspruchung *f*, -belastung *f* ~ **reinforcing bar** Tragstab *m* ~ **shock test** Schlagzerreißversuch *m* ~ **strain** Dehnungs-, Reck-, Zug-spannung *f*, positive Spannung *f*, Reckbelastung *f*

**tensile-strength** Biegungsfestigkeit *f*, Bruchfestigkeit *f*, -widerstand *m*, Dehnfestigkeit *f*, Reiß-festigkeit *f*, -kraft *f*, Verformungswiderstand *m*, Zerreißfestigkeit *f*, Zugfestigkeit *f*,

Zugfestigkeitswert *m* ~ at knot Knotfestigkeit
*f* ~ test Zugfestigkeitsuntersuchung *f* ~ testing
machine Zugfestigkeitsprüfmaschine *f*
tensile, ~ stress Dehnungsspannung *f*, positive
Spannung *f*, Reckspannung *f*, Zugbeanspruchung *f*, Zugspannung *f* ~ test Reckprobe *f*,
Zerreiß-probe *f*, -versuch *m*, Zugprüfung *f*,
Zugversuch *m* ~ test at elevated temperature
Warmzerreißversuch *m* ~-test machine Prüfmaschine *f* für Zugversuche ~-test rod Zug-
probestab *m* ~-test specimen Zerreißprobe *f*,
Ziehprobe *f* ~-testing machine Zerreiß-, Zugprüfungs-maschine *f*
tensimeter Dampf(druck)messer *m*, Gasmesser
*m*
tensiometer Dehnungs-, Spannungs-, Zugmesser *m*
tension, to ~ (an)spannen, straffen to ~ a spring
eine Feder *f* spannen
tension Anspannung *f*, Anstrengung *f*, Dehnung
*f*, Druck *m*, Federkraft *f*, Materialspannung *f*,
Spannkraft *f*, Spannung *f*, Stromspannung *f*,
Zug *m*, Zugkraft *f* ~ of belt Riemenspannung *f*
~ of the cable Seilspannung *f* ~ and compression
test Zugdruckversuch *m* ~ and compression
testing machine Zugdruckprüfmaschine *f* ~ of
magnetic tape Bandzug *m* ~ of the main Netzspannung *f* ~ of spring Federspannung *f* ~ in
the tie rod Zugkraft *f* im Zugband
tension, ~-adjusting knob Spitzendruckadjustier-
schraube *f* ~ bar Brems-leiste *f*, -stange *f* ~
bar bearing Zahnstangenlager *n* ~ block Spann-
bock *m* ~ brace Zugdiagonale *f* ~ cable Spanndraht *m* ~-cable drum Zugkabeltrommel *f* ~
carriage Spannwagen *m* ~ chain Spannkette *f*
~ characteristic Spannungskennlinie *f* ~
chord Zuggurt *m* ~-compression fatigue testing
machine Zugdruckdauerprüfmaschine *f* ~ cone
Spannkonus *m* ~ cotter pin Spannstift *m* ~
crack Spannungsriß *m* ~ device (for rubber
motor) Aufziehvorrichtung *f* ~ difference
Spannungsunterschied *m* ~ dynamometer Zugmesser *m* ~ face Zugseite *f* ~-field beam Zug-
blechträger *m* ~ flange Zuggurt *m* ~ increase
Spannungssteigerung *f* ~ index of a (tension)
member Spannziffer *f* eines Zugstabes ~ indicator Spannungs-anzeiger *m*, -messer *m* ~ joint
Spannungsverbindung *f* ~ line Windleine *f* ~
load Spannkraft *f*, Zugbelastung *f* ~ lock
Spannschloß *n*, (fuse) Vorstecker *m* ~ loss
Spannungsverlust *m* ~ medium Zugmittel *n* ~
member (Triebwerk) Spannstange *f*, Zugband
*n*, Zugglied *n*, Zugstab *m* ~ pile Zugpfahl *m* ~
pulley Spannrolle *f* ~ rack Zugstange *f* ~
regulator (coil) Spannungsregler *m* ~ release
Entspannung *f* ~-release igniter Zerschneide-
zünder *m* ~ removing device Entspanneinrich-
tung *f* ~ ring (fuse) Stellring *m* ~ rod Brems-
leiste *f*, -stange *f*, Spannbolzen *m*, Zuganker *m*,
Zugband *n* ~ roll Spannwalze *f* ~ roller Spannrolle *f* ~ roller crank Spannrollenarm *m* ~
screw Federspann-, Zugfeder-schraube *f* ~-set-
ting ring of fuse Satzstück *n* ~-setting time ring
of fuse Satzstück *n* ~ shackle Einspannkopf *m*
für Zugversuche ~ side Zugseite *f* ~ spinning
Streckspinnen *n* ~ spring Spann-, Spannungs-,
Zug-feder *f* ~ spring arrestor Fadenwächter-
bügel *m* ~ spring hook Zughaken *m* (Haken an

einer Feder) ~ stop Stellplättchen *n* ~ test
Zerreiß-probe *f*, -versuch *m*, Zugversuch *m* ~
thread Spannfaden *m* ~ voltmeter Spannungs-
anzeiger *m* ~ wrench Drehmomentschrauben-
schlüssel *m*
tensional, ~ force Spannkraft *f* ~ members Zugstäbe *pl*
tensioned cord (or string) ausgespannte Saite *f*
tensioning of valve spring Spannung *f* der Ventil-
feder
tensioning, ~ device Belastungseinrichtung *f* ~
device for light wires Kniehebelklemme *f* ~
eccentric Spannexzenter *m* ~ pulley Spannrolle
*f* ~ segment Spannsegment *n*
tensor absoluter Vektor *m*, Tensor *m* ~ of curvature Krümmungstensor *m* ~ of inertia Träg-
heitstensor *m* ~ of momentum Tensor *f* der
Impulse
tensor, ~ algebra Tensoralgebra *f* ~ components
Tensorkomponenten *pl* ~ invariants Tensor-
invarianten *pl* ~ notation Tensorschreibweise *f*
~ quadric quadratische Form *f* eines Tensors
~ sheet Tensorblatt *n*
tensorial mean square tensorielles mittleres
Quadrat *n*
tent Zelt *n* ~ equipment Zeltausrüstung *f* ~
fittings Zeltbeschläge *pl* ~ peg Hering *m* ~ pin
Zeltpflock *m* ~ pitching Zeltbau *m* ~ pole Zelt-
pfahl *m*, -stange *f* ~ roof Zeltdach *n*
tentative Experiment *n*, Versuch *m*; probierend,
provisorisch, versuchend, vorläufig ~ standard
Richtliniennorm *f*, versuchsweise aufgestellte
Norm *f*, Vornorm *f* ~ standard specification
Normentwurf *m*
tentatively versuchsweise
tenter, to ~ aufspannen, in einen Rahmen *m*
spannen
tenter Spannrahmen *m* ~ drier Spannrahmen-
trockner *m* ~ frame Tuchrahmen *m*
tentering (cloth fabrication) Aufrahmen *m* ~
frame Spannrahmen *m* ~ limit Spannfeld *n*
tenth, ~-normal solution Zehntellösung *f* ~
power Zehnerpotenz *f* ~ value thickness Zehn-
telwertsdicke *f*
tenuity factor Luftverdünnungsfaktor *m*
tenuous dünn, fein
tenure Gehalt *m*, Pacht *f* ~ of office Amtszeit
*f*
tepid lau, lauwarm
terbia Terbinerde *f*
terbium Terbium *n* ~ oxide Terbinerde *f*
term, to ~ bezeichnen, nennen
term (of an equation) Ausdruck *m*, Bedienung *f*,
Bedingung *f*, Begriff *m*, Benennung *f*, Bezeichnung *f*, Frist *f*, Glied *n* (in Gleichung oder
Summe), Niveaustufe *f*, Termin *m*, Zeitdauer *f*
~ of copyright Schutzfrist *f* ~ of delivery Lieferfrist *f*, Lieferungsbedingung *f* ~ in the numerator Glied *n* im Zähler ~ in parentheses Klammerausdruck *m* ~ of a sum Summand *m*
(math.)
term, ~ displacement Gliederversetzung *f* ~
postponement Termverschiebung *f* ~ shift
Termverschiebung *f* ~ station Fristenstelle *f* ~
values Termlagen *pl*
terminable befristet, begrenzbar, bestimmbar,
kündbar
terminal begrenzend

**terminal** Anschlagstelle *f* (electr.), Anschlagstück *n* (electr.), Anschluß *m*, Anschlußklemme *f*, -schraube *f*, Drahthalter *m* (electr.), Drahtklemme *f*, (of an arc) Elektrodenende *n*, (station) Endbahnhof *m*, Ende *f*, (of carrier system) Endeinrichtung *f*, Endpunkt *m*, Endstation *f*, (box) Endverschluß *m*, Klemme *f* (electr.), Klemmschraube *f*, Kontakt *m*, Kontaktstift *m*, Kopfstation *f*, (pole) Polklemme *f*, Spannungspunkt *m*, Stift *m*, Trenndose *f*, Verbindungsklemme *f*, Versandstation *f* ~ **of an element** Pol *m*

**terminal, ~ amplifier** Endverstärker *m* ~ **amplitude** Endamplitude *f* ~ **angle** Endeck *n* ~ **apparatus** Endapparat *m* ~ **bar** Bleileiste *f* ~ **base loading** (of antenna) Fußpunktwiderstand *m* ~ **block** Anschlußklemmleiste *f*, Endverzweiger *m*, Klemmenkasten *m* ~ **board** Klemmbrett *n* ~ **box** Abschlußmuffe *f*, Anschlußkasten *m*, Kabelendverschluß *m*, Klemmenschutzkasten *m*, Kraftspeicher *m* ~ **bracket** Abspannstütze *f* ~ **bushing** Endtülle *f* ~ **cable** Abschluß-, Einführungs-, End-, Fernleitungsend-kabel *n* ~ **charge** Endgebühr *f* ~ **circuit** End(kunst)schaltung *f* ~ **clamp** Leitungs-, Löt-anschlußstück *n* ~ **coil section** Anlauflänge *f* ~ **connection** Klemmenanschluß *m* ~ **connector** Anschlußklemme *f*, Kabelanschlußstück *n* ~ **control area (TMA)** Nahverkehrsbereich *m* (aviat.) ~ **corrosion** Zerfressen *n* der Klemme ~ **cover** Klemmendeckel *m* ~ **current** Klemmenstrom *m* ~ **distribution board** Verteilerklemmbrett *n* ~ **double pin** Abspanndoppelstütze *f* ~ **duplex repeater set** Gegensprechsatz *m* ~ **edge** Polkante *f* ~ **equipment** End-apparat *f*, -satz *m*, Überführungsgerät *n* ~ **exchange** Endanstalt *f* ~ **face of a crystal** Kristallendfläche *f* ~ **flange** Endflansch *m* ~ **head** Anschlußkopf *m* ~ **holder** Klemmenhalter *m* ~ **impedance** Abschlußimpedanz *f*, Endimpedanz *f*, (of antenna) Fußpunktwiderstand *m*, Klemmenwiderstand *m* ~ **insulator** Abspannisolator *m*, Überführungsisolator *m* ~ **line** Stichbahn *f* ~ **loss** Endverluste *pl* ~ **lug** Sockelstift *m* ~ **marking** Klemmenbezeichnung *f* ~ **member** Endglied *n* ~ **moraine** End-, Stirn-moräne *f* ~ **network** Abschlußkunstschaltung *f*, Endkunstschaltung *f* ~ **(or final) nut** Abschlußmutter *f* ~ **office** Endamt *n* ~ **pair** Anschlußklemmenpaar *n* ~ **panel** Klemmenbrett *n* ~ **part** Anschlußteil *m* ~ **pillar** Polbolzen *m* ~ **plate** Arretierungslamelle *f* ~ **point** Endstelle *f* ~ **pole** Abspann-gestänge *n*, -stange *f*, Endmast *m*, Überführungs-gestänge *n*, -säule *f*, -stange *f* ~ **poles** Einführungsstangen *pl* ~ **port** Endhafen *m* ~ **post** Anschlußsäule *f*, (brass) Polbolzen *m* ~ **pressure** Auspuffdruck *m* ~ **punching** Sockelstift *m* ~ **rack** Klemmenleiste *f* ~ **rate** Endgebühr *f* ~ **repeater** Endsatz *m*, Endschaltsatz *m*, Endverstärker *m*, Fernleitungsendverstärker *m*, Schaltsatz *m* ~ **-repeater circuit** Endverstärkerschaltung *f* ~ **resistance** Abschluß- (electron.), End-, Klemmen-widerstand *m* ~ **room** Wähler-raum *m*, -saal *m* ~ **screw** Anschlußbolzenschraube *f* ~ **section** Endstück *n* einer Mole ~ **selector (TeS)** Endwähler *m* (EW) ~ **sleeve** Abschlußhülse *f* ~ **spindle** Abspannstütze *f* ~ **state** Endzustand *m*

~ **station** Endamt *n* ~ **strain insulator** Abspannisolator *m* ~ **strip** Anschlußstreifen *m*, Klemmen-dose *f*, -leiste *f*, -streifen *m*, Klemmleiste *f*, Kontaktbürstenstreifen *m*, Löt(ösen)streifen *m*, Messerleiste *f* ~ **-to-~ continuity** Durchgang *m* von Klemme zu Klemme ~ **tower** Endmast *m* ~ **traffic** Endverkehr *m* ~ **transformer** Abschlußtransformator *m*, Abschlußübertrager *m* ~ **trunk exchange** Überweisungsfernamt *n* ~ **twist** Enddrall *m* ~ **unit** Endsatz *m* ~ **valve** Endröhre *f* ~ **velocity** Freifall-, Grenz-geschwindigkeit *f* ~ **-velocity dive** Sturzflug *m* mit Endgeschwindigkeit ~ **-velocity engine speed** Sturzflugdrehzahl *f* ~ **voltage** Klemmenspannung *f* ~ **VOR (TVOR)** Flugplatzdrehfunkfeuer *n* ~ **wall** Trennschott *m*

**terminals** (binding posts) Klemmer *m*

**terminate, to** ~ abschließen, (a wire) abspannen, aufheben, begrenzen, enden, endigen **to** ~ **an engagement** abkehren **to** ~ **a line in its own impedance** eine Leitung *f* durch ihren Wellenwiderstand abschließen

**terminated** abgeschlossen ~ **on jacks** an Klinken *pl* endigend ~ **by a resistance** durch einen Widerstand *m* abgeschlossen

**terminated line** abgeschlossene Leitung *f*

**terminating** Abschluß *m* ~ **of wire on insulator** Abspannung *f* der Leitungen

**terminating, ~ box** Klemmenendverschluß *m* ~ **capacitor** Endkondensator *m*, Schlußkondensator *m* ~ **capacity** Endkapazität *f* ~ **circuit** endigende Leitung *f* ~ **condenser** Abschluß-, End-, Schluß-kondensator *m* ~ **element** Endelement *n* ~ **inductor** Enddrosselspule *f* ~ **office** Ankunfts-, Bestimmungs-anstalt *f* ~ **resistance** Widerstandabschluß *m* ~ **set** Gabelschaltung *f* ~ **toll center** Ankunfts-, Bestimmungs-anstalt *f*, Überweisungsfernamt *n*

**termination** Abschluß *m*, Abspannung *f*, Ausgang *m*, Begrenzung *f*, Ende *n*, Endung *f*, Kündigung *f*, Schluß *m* ~ **of a cable** Endverschluß *m* ~ **of a filter at mid-series position** Abschluß *m* eines Filters durch ein halbes Längsglied ~ **of service** Auflösung *f* des Dienstverhältnisses ~ **of wires on intermediate (or terminal poles)** Abspannung *f* der Leitungen

**termination charge** Streichungsgebühr *f*

**terminology** Fachausdrücke *pl*, Terminologie *f*

**terminus** Endpunkt *m*, Endstation *f*

**termite** Termite *f*

**terms** Bedingungen *pl*, Bestimmungen *pl*, Festsetzung *f*, Formelausdrücke *pl*, Glieder *pl* ~ **of contract** Submissionsbedingungen *pl* ~ **of delivery** Bezugsbedingungen *pl* ~ **of the first order** Glieder *pl* erster Ordnung ~ **of payment** Zahlungsbedingungen *pl* ~ **of sale** Verkaufsbedingungen *pl*

**ternary** dreizählig, ternär ~ **alloy** Dreistofflegierung *f*, ternäre Legierung *f* ~ **code** Dreieralfabet *n*, Ternärkode *m* ~ **compound** Dreifachverbindung *f* ~ **critical point** kritischer Mischungspunkt *m* ~ **fission** Kernspaltung *f* in drei Bruchstücke ~ **fuel** (Kraftstoff) Dreiergemisch *n* ~ **notation** ternäre Schreibweise *f* ~ **steel** Stahl *m* mit einem Legierungsbestandteil neben Kohlenstoff, Ternärstahl *m* ~ **system** Dreistoffsystem *n*

**terne, ~ plate** Mattblech *n*, Terneblech *n*, verbleites Blech *n*, Weißblech *n*
**terpene hydro-carbon** Terpenkohlenwasserstoff *m*
**terpeneless** terpenfrei
**terpineol** Terpineöl *n*
**terra cotta** Kunsttonwaren *pl*, Terrakotta *f*
**terrace, to** ~ terrassieren
**terrace** Bankett *n*, Berme *f*, Terrasse *f*
**terraced** flach (Dach), terassenförmig ~ **sieve drum** Terrassentrommel *f*
**terracing cut** Terrassenschnitt *m*
**terrain** Gelände *n*, Geländeabschnitt *m*, Grund *m*, Terrain *n* ~ **appreciation** Geländeauswertung *f* ~ **area** Geländeraum *m* ~ **clearance** Bodenfreiheit *f* ~ **(or avoidance) computer clearance** Rechner *m* für Hinderniswarner ~ **conditions** Geländeverhältnisse *pl* ~ **feature** Geländegegenstand *m* ~ **features** Bodengestaltung *f*, Gelände-formen *pl*, -gestaltung *f* ~ **obstacle** Geländehindernis *n* ~ **point** Geländepunkt *m*, Punkt *m* im Gelände ~ **reconnaissance** Geländeerkundung *f* ~ **representation** Geländedarstellung *f* ~ **sector** Geländestreifen *m* ~ **survey** Geländevermessung *f*
**terrazzo** Zementmosaik *n* ~ **grade** Terrazokörnung *f* ~ **strip** Mosaikzierstreifen *m*
**terrestrial** zur Erde *f* gehörig, irdisch, terrestrisch ~ **atmosphere** Erdatmosphäre *f* ~ **equator** Erdäquator *m*, Erdgleicher *m* ~-**magnetic** erdmagnetisch ~-**magnetic activity** Aktivität *f* des Erdmagnetismus, magnetische Aktivität *f* der Erde ~-**magnetic field** Erdmagnetfeld *n*, erdmagnetisches Feld *n* ~-**magnetic pole** magnetischer Pol *m* der Erde ~ **magnetism** Erdmagnetismus *m* ~ **orbit** Erdbahn *f* ~ **photogrammetry** Erdbildmessung *f*, Geländefotogrammetrie *f*, Geofotogrammetrie *f*, terrestrische Fotogrammetrie *f* ~ **photographic camera** Geländekamera *f* ~ **photography** Geländeaufnahme *f* ~ **radiation** Abendthermik *f* ~ **survey** Erdbildaufnahme *f*, terrestrische Aufnahme *f* ~ **telescope** Erdfernrohr *n*
**terre-verte** Veronesererde *f*
**territorial** territorial, Gebiets... ~ **sea** Küstenmeer *n* ~ **section** Geländeabschnitt *m* ~ **waters** Hoheitsgewässer *n* ~ **zone** Territorialzone *f*
**territory** Flur *m*, Gebiet *n*, Land *n*
**tertiary** Tertiär *n* ~ **circuit** Tertiärkreis *m* ~ **creep** tertiäres Kriechen *n* ~ **formation** Tertiärformation *f* ~ **radiation** Tertiärstrahlen *pl* ~ **winding** (of a transformer) Drittwicklung *f*, Tertiärwicklung *f*
**Tesla, ~ coil (or transformer)** Teslatransformator *m*
**tesselated, a ~ type of chiastolite** gewürfelter Hohlspat *m*
**tesselite** Tesselith *m*
**tesseral** tesseral
**tessular** würfelig
**test, to** ~ abnehmen, ausproben, durchprüfen, eichen, erproben, experimentieren, messen, nachprüfen, auf die Probe *f* stellen, proben, probieren, prüfen, untersuchen, versuchen **to** ~ **and check under flight conditions** nachfliegen **to** ~ **for contact** auf Berührung *f* untersuchen **to** ~ **for earth** auf Erdschluß *m* (Berührung, Kurzschluß) prüfen **to** ~ **the engine**

Motor *m* abbremsen **to** ~-**fence** auf Prüfgestell *n* aufbringen **to** ~ **a gun** anschießen **to** ~ **for hardness** Härteprüfung *f* vornehmen **to** ~ **by hydrometer** spindeln **to** ~ **for leaks** auf Dichtigkeit *f* prüfen, auf Dichtheit *f* prüfen **to** ~ **a line for insulation and conductivity** eine Leitung *f* auf Isolation und Leitfähigkeit messen **to** ~ **the line** abfragen (teleph.), abrufen **to** ~ **a shaft for truth** eine Welle *f* auf Rundlauf prüfen **to** ~ **the signaling** (ringing) den Ruf *m* prüfen **to** ~ **by sound** (ranging equipment) auspeilen **to** ~-**splice** aufkreuzen (Adernpaare) **to** ~ **by the spot method** tüpfeln **to** ~ **the valve cone for warpage** Ventilkegel *m* auf Schlag prüfen **to** ~ **for warpage** auf Schlag *m* prüfen
**test** Messung *f*, Probe *f*, Prüfbefund *m*, Prüfmerkmal *n*, Test *m*, Versuch *m* ~ **(ing)** Erprobung *f*, Prüfung *f*, Untersuchung *f* ~ **of material** Stoffprobe *f*
**test, ~ acid** Probesäure *f* ~ **apparatus** Prüfvorrichtung *f* ~-**balancing method** Kapazitätsausgleich *m* oder Kapazitätsausgleichsverfahren *n* durch Adernkreuzung ~ **bar** Probe-stab *m*, -stange *f*, Versuchsstab *m*, Zerreißstab *m* ~-**bar head** Stabkopf *m* ~ **barometer** Prüfungsluftdruckmesser *m* ~ **beaker** Reagensglas *n* ~ **bed** Prüfstand *m* ~ **(ing) bench** Prüfstand *m* **(lens)** ~ **bench** Prüfbank *f* (opt.) ~ **bench running** Probelauf *m* auf dem Prüfstand ~ **board chief's desk** Prüfgestell *n*, Prüfschrank *m*, Prüftisch *m* ~ **boring** Bohrversuch *m* ~ **box** Prüf-kasten *m*, -schrank *m*, Überführungskasten *m*, Untersuchungskasten *m* ~ **brush** C-Kontaktarm *m*, Prüfkontaktarm *m* ~ **buzzer** Prüfscharre *f* ~ **cable** Prüfschnur *f* ~-**cable stub** Prüfstumpf *m* ~ **cage** geschirmter Prüfraum *m* ~ **call** Probe-, Versuchs-verbindung *f* ~-**card** Testbild *n* ~ **case** Kastenuntersuchungsstelle *f*, Prüfschrank *m*, Untersuchungskasten *m*, (cable) Untersuchungs- und Verteilungsstelle *f* ~ **cell** Prüfgefäß *n* ~ **certificate** Prüfschein *m*, Prüfungsattest *n* ~ **chamber** Untersuchungskammer *f* ~ **change-over switch** Prüfumschalter *m* ~ **charge** Versuchsladung *f* ~ **chart** Einstelltafel *f*, Täfelchen *n* ~ **circuit** Kontrollstromkreis *m*, Prüfleitung *f*, Prüfstromkreis *m* ~ **clerk** Prüf(ungs)-, Störungs-beamter *m* ~ **clip** Prüfklemme *f* ~ **cock** Prob(ier)hahn *m*, Prüfhahn *m* ~ **coil** Prüfspule *f* ~ **conditions** Versuchsbedingungen *pl* ~ **connection** Anschlußprüfgerät *n* ~ **connector** Prüf-klemme *f*, -stecker *m*, -wähler *m* ~ **control unit** Spannungsprüfteil *m* ~ **cord** Prüfschnur *f* ~ **cube** Probewürfel *m* ~ **curve** Prüfungsschleife *f* ~ **cylinder** (concrete) Probezylinder *m* ~ **data** erflogene Meßwerte *pl*, Meßwerte *pl* ~ **data board (or table)** Prüfwertetafel *f* ~ **desk** Meßtisch *m*, Prüf-pult *n*, -schrank *m*, -tisch *m*, Untersuchungstisch *m* ~ **drill** Untersuchungsbohrer *m* ~ **dyeing** Probefärbung *f* ~ **equipment** Prüfstandsgerät *n* ~ **evaluation** Versuchsauswertung *f* ~-**fence** Prüfgestell *n* (zur Freilagerprüfung von Lack) ~ **field** Prüffeld *n* ~ **final selector** Prüfwähler *m* ~ **firing** Beschuß *m*, Versuchsschießen *n* ~ **fitter** Prüffeldmonteur *m* ~ **flame** Versuchsflamme *f* ~ **flight** Probefahrt *f*, Erprobungs- *m*, Prüf-, Versuchs-flug *m* ~ **gauge** Eichgerät *n*, Eichinstrument *n* ~ **gears** Prüf-

räder *pl* ~ **glass heater** Reagensglaserhitzer *m*
~ **grid** Prüfgitter *n* ~ **hole** Versuchskanal *m* ~
**hut** Untersuchungshäuschen *n* ~ **instructions**
Prüfungsbescheid *m* ~ **insulator** Trennisolator
*m* ~ **jack** Prüfklinke *f* ~ **jack panel** Meßtisch *m*
~ **jack strip** Meßbuchsenleiste *f* ~ **jar** Probe-
kolben *m* ~ **jet** Prüfstrahl *m* ~ **jig** Prüfgerüst *n*
~ **key** Prüfschalter *m* ~ **(ing) key** Prüftaste *f* ~
**lamp** Testlampe *f* ~ **lead** Meßschnur *f*, Probe-
blei *n*, Prüfleitung *f* ~ **level** Meßpegel *m* ~
**liquid** Probeflüssigkeit *f* ~ **litharge** Probier-
glätte *f* ~ **load** Probe-belastung *f*, -last *f*, Prüflast *f* ~ **log** Versuchs-aufschreibung *f*, -proto-
koll *n* ~ **loop** Meßschleife *f* ~ **lug** Prüfzapfen *m*
~ **man** Meßbeamter *m* ~ **mark** Justiermarke *f*
~ **mast** (antenna) Justiermast *m* ~ **material**
Versuchsmaterial *n* ~ **metal** Probiermetall *n* ~
**method** Prüfmethode *f*, Prüfungsart *f* ~ **(ing)**
**method** Prüfverfahren *n* ~ **model** Prüfling *m*;
Versuchsmodell *n* ~ **modulation** Prüfmodula-
tion *f* ~ **oil supply (or feed) line** Prüfölzuleitung
*f* ~ **oscillator** Meßoszillator *m*, Meßsender *m*,
Prüfgenerator *m*, Prüfsender *m* ~ **panel** Meßfeld *n* ~ **paper** Reagens-, Reaktions-papier *n*
**chemical** ~ **paper impregnated with potassium**
**iodide** Jodkaliumpapier *n* ~ **path** Versuchsbahn
*f* ~ **pattern** Bildmuster *n* (zum Prüfen der Bildgüte), Testbild *n* (TV) ~ **pattern generator**
Bildmustergenerator *m* (TV) ~ **pick** Prüfdraht
*m* ~ **picture** Probeaufnahme *f* ~ **piece** Probe *f*,
Probe-körper *m*, -stab *m*, -stück *n*, Prüfling *m*,
Prüfstück *n*, Untersuchungsobjekt *n*, vergütete
Probe *f*, Versuchsstück *n* ~ **-piece dividing**
**machine** Probestabteilmaschine *f* ~ **pile** Probepfahl *m* ~ **pilot** Ein-, Prüf-, Versuchs-flieger *m*
~ **pit** Schürf-grube *f*, -hoch *n* ~ **plane** Versuchsflugzeug *n* ~ **plant** Prüfstandsanlage *f* ~ **plate**
Prüfschild *n* ~ **plug** Prüfstöpsel *m* ~ **point**
Meßpunkt *m* ~ **pole** Trenn-, Untersuchungsstange *f* ~ **portion** Probenahme *f* ~ **position**
Prüf-, Untersuchungs-platz *m* ~ **potential**
**dispenser** Prüfverteiler *m* ~ **pressure** Probe-,
Prüf-druck *m* ~ **print** Probe-abzug *m*, -kopie *f*
(film) ~ **printing machine** Probendruckma-
schine *f* ~ **probe** Meßsonde *f* ~ **procedure** Prüfverfahren *n* ~ **prod** Prüfspitze *f* ~ **program**
Prüf-, Versuchs-programm *n* ~ **-prong** (autopilot) Meßklemme *f* ~ **propeller** Prüfstands-
luftschraube *f* ~ **pump** Kesselprobierpumpe *f* ~
**rack** Versuchsgestell *n* ~ **rail** Prüfungs-, Ver-
suchs-bahn *f* ~ **reading** Ablesung *f*, Meßergebnis *n* ~ **receiver** Prüfhörer *m* ~ **record** Prüf-
bericht *m*, Testplatte *f* (phono) ~ **(ing) relay**
Prüfrelais *n* ~ **report** Prüf-bericht *m*, -protokoll
*n* ~ **resistance** Prüfwiderstand *m* ~ **result** Meßwert *m*, Prüf-, Prüfungs-ergebnis *n*, Versuchsergebnis *n*, -wert *m* ~ **rig** Prüfgerüst *n* ~ **ring**
Herdrahmen *m* ~ **rings** Prüfringe *pl* ~ **road**
Einfahrstraße *f* ~ **rod** Probestange *f*, Versuchsstab *m* ~ **routine** Prüfprogramm *n* ~ **run**
Durchschaltversuch *m* (g/m), Probelauf *m*
(engin.), Versuchslauf *m* ~ **sample** Prüfling *m*,
Untersuchungsobjekt *n* ~ **scope** Prüf-oszillo-
graf *m*, -oszilloskop *n* ~ **screw plug** Prüfschraube *f* ~ **section** Maß-schnitt *m*, -weg *m* ~
**selector** prüfender Wähler *m*, Prüfwähler *m* ~
**set** Vielfachinstrument *n* ~ **set for steering**
**machine** Heckkoffer *m* ~ **setup** Versuchsaufbau

*m* ~ **shaft** Schurfschacht *m* ~ **sheet** Prüfproto-
koll *n*, Versuchsbericht *m* ~ **shot** kalte Probe *f*
~ **signal** Eichsignal *n*, Prüfungslampe *f* ~
**sleeve** Prüfhülse *f* ~ **slide** Testdia *n* ~ **socket**
Anschlußprüfgerät *n* ~ **specification** Prüfungs-
vorschrift *f* ~ **specimen** Musterstück *n*, Probe-
körper *m*, -stück *n*, Prüfstück *n*, Untersuchungsobjekt *n*, Versuchs-körper *m*, -stück *n* ~ **splice**
Ausdruckkreuzlötstelle *f* ~ **splicing** (method)
Auskreuzen *n* ~ **-splicing method** Adernkreu-
zungsverfahren *n* für Kapazitätsausgleich ~
**stand** Probeblock *m*, Probierstand *m*, (test
floor) Prüfstand *m*, Versuchs-gestell *n*, -träger
*m* ~ **-stand area** Prüfstandbereich *m* ~ **-stand**
**arrangement** Prüfstandaufbau *m* ~ **start** Probestart *m* ~ **station** Erprobungsstelle *f* ~ **switch**
Prüftaste *f* (rdo) ~ **tape** Prüfstreifen *m* ~ **ter-**
**minal box** Prüfklemmenblock *m* ~ **track** Versuchsbahn *f* ~ **trestle** Prüfbock *m* ~ **tube** Probierröhrchen *n*, Reagens-glas *n*, -röhre *f*,
-zylinder *m* ~ **unit** Probierstation *f* ~ **value**
Meß-, Prüf-wert *m* ~ **value storage unit** Meßwertspeicher *m* ~ **vibrating device for bulbs**
Glühlampenschüttelvorrichtung *f* ~ **voltage**
**symbol** Prüfspannungszeichen *n* ~ **voltmeter**
Prüfvoltmeter *n* ~ **-wave generator** Stoßgenerator *m* ~ **weight** Probegewicht *n* ~ **weld**
Probeschweißung *f* ~ **wire** Ader *f* zum Stöpselkörper, c-Ader *f* ~ **wire chief's desk** Prüf-gestell
*n*, -schrank *m*, -tisch *m* ~ **work** Prüfarbeit *f*,
Versuchstätigkeit *f*

**tested** erprobt, geprüft, getestet

**testee** Prüfobjekt *n*

**tester** Meßbeamter *m*, Prüfer *m*, Prüfgerät *n*,
Prüfverrichtung *f* ~ **for calibration circuit**
Eichleitungsprüfer *m*

**testify, to** ~ aussagen, beglaubigen, zeugen

**testimonial** Attest *n*, Zeugnis *n*

**testimony** Aussage *f*, eidliche Versicherung *f*,
Zeugnis

**testing** Ausprobieren *n*, Erprobung *f*, Messung *f*,
Nachprüfung *f*, Probe-machen *n*, -nahme *f*,
-nehmen *n*, Probieren *n*, Prüfen *n*, Prüfung *f*,
Sichtung *f*, Versuch *m* ~ **of hardness** Festigkeits-
probe *f* ~ **of material(s)** Material-, Werkstoffprüfung *f* ~ **of vision** Sehprüfung *f*

**testing, ~ apparatus** Prüfapparat *m*, Prüfgerät *n*,
Untersuchungsapparat *m* ~ **appliance** Prüfvorrichtung *f* ~ **arbor with reticule** Prüfdorn *m*
mit Fadenkreuz ~ **balance** Titrierwaage *f* ~
**bar** Stellungsprüfer *m* ~ **battery** Meß-, Prüf-
batterie *f* ~ **bed** Versuchsstand *m* ~ **button**
Prüfknopf *m* ~ **cart** Meßkarren *m* ~ **circuit**
Meß(strom)kreis *m* ~ **commutator** Meßum-
schalter *m* ~ **current** Meßstrom *m* ~ **-current**
**intensity** Meßstromstärke *f* ~ **department** Erprobungsstelle *f* ~ **device** Prüfeinrichtung *f* ~
**device for current transformer** Stromwand-
prüfeinrichtung *f* ~ **device for receiver sensiti-**
**vity** Empfindlichkeitsprüfer *m* ~ **directions**
Versuchsvorschriften *pl* ~ **engine** Prüfmotor *m*
~ **engineer** Prüfingenieur *m* ~ **fusibility**
Schmelzpunktermittlung *f* ~ **ground** Prüffeld
*n* ~ **installation** Versuchseinrichtung *f* ~ **in-**
**stitute** Prüfanstalt *f* ~ **instrument** Prüfgerät *n* ~
**key** Meßtaste *f* (teleph.) ~ **laboratory** Erprobungsstelle *f*, Prüfungs-, Versuchs-anstalt
*f*

**testing-machine** Prüfmaschine *f* ~ **for bending tests** Prüfmaschine *f* für Biegeversuche ~ **for repetition of torsional stress** Dauerversuchsmaschine *f* für Drehschwingungsbeanspruchung ~ **for torsion test** Prüfmaschine *f* für Drehversuche

**testing,** ~ **mark** Prüfungszeichen *n* ~ **method** Meß-methode *f*, -verfahren *n*, Prüfungsmethode *f* ~ **needle** Probiernadel *f* ~ **office** Meßamt *n*, Untersuchungsamt *n* ~ **officer** Abnahme-, Prüf-beamter *m* ~ **operator** Prüftechnicker *m* ~ **outfit** Prüfeinrichtung *f* ~ **paper** Spurpapier *n* ~ **plant** Versuchsanlage *f* ~ **point** Trenn-, Untersuchungs-stelle *f* ~ **point with remote control** Untersuchungsstelle *f* mit elektrischer Fernsteuerung (teleph.) ~ **position** (operator's) Prüfplatz *m*, Prüfstelle *f* ~ **process** Prüfverfahren *n* ~ **push button** Prüftaste *f* ~ **rig** Prüfstandsbock *m* ~ **room** Prüffeld *n*, Prüfraum *m* ~ **screen** Brennrahmen *m* ~ **section** Untersuchungsabschnitt *m* ~ **set** Prüfeinrichtung *f* ~ **sieve** Prüfsieb *n* ~ **stand** Versuchsstand *m* ~ **station** Abnahme *f*, Versuchsstelle *f* ~ **stop** Prüfstand *m* ~ **technique** Meßtechnik *f* ~ **terminal** Untersuchungsklemme *f* ~ **time** Meßzeit *f* ~ **track** Probefahrbahn *f* ~ **transmitter** Normalgenerator *m* ~ **tube** Niveauröhre *f* ~ **vessel** Probiergefäß *n* ~ **voltage** Prüfspannung *f* ~ **wall** Prüfwand *f* ~ **wire** Prüf-draht *m*, -leitung *f* ~ **work** Versuchsarbeit *f*

**Testor hardness** Testor-Härteprüfer *m*

**tests,** ~ **of the liquid gas vehicle** Dehnungsmessungen *pl* bei Flüssiggastankwagen ~ **by progressive loadings** Stufenversuche *pl* ~ **on selected samples** Stichprobenmessungen *pl*

**tetanthrene** Tetanthren *n*

**tethering links** Fesselbeschläge *pl*

**tetra,** ~ **araban** Tetraaraban *n* ~ **basic** vierbasisch ~ **boric acid** Tetraborsäure *f* ~ **bromoethane** Tetrabromäthan *n* ~ **chloride** Tetrachlorid *n* ~ **chloromethane** Tetrachlorkohlenstoff *m* ~ **decane** Tetradecan *n* ~ **dymite** Markasitglanz *m*, Wismuttellur *n* ~ **ethyl lead** Tetraäthylblei *n* ~ **ethyl lead fuel** Bleibenzin *n*, Bleikraftstoff *m*

**tetrad** Tetrade *f*, vierwertiges Atom *n*, Vierzahl *f*

**tetragonal** vier-eckig, -gliedrig (cryst.), -kantig, quadratisch (cryst.), tetragonal ~ **thread** trapezförmiges Gewinde *n*

**tetrahedral** tetraedrisch, vierflächig ~ **packing of spheres** tetraedrische Kugellagerung *f*

**tetrahedrite** Antimonfahlerz *n*, Fahlerz *n*, Kupferfahlerz *n* ~ **containing mercury** Quecksilberfahlerz *n*

**tetrahedron** Tetraeder *n*, Vierflach *n*

**tetrahexahedron** Pyramidenwürfel *m*

**tetral notation** quaternäre Schreibweise *f*

**tetralinperoxide** Tetralinperoxyd *n*

**tetratomic** vieratomig

**tetratricontadien** Tetratriakontadien *n*

**tetravalence** Vierwertigkeit *f*

**tetravalent** vierwertig

**tetrode** Doppelgitterrähre *f*, Tetrode *f*, Vierelektroden-, Vierpol-, Zweigitter-röhre *f*

**tetroxide** Tetraoxyd *n*

**tetryl** Tetryl *n*

**texrope belt** gewobener Gurt *m*

**text** Wortlaut *m* ~ **in clear** Klartext *m* ~ **book** Handbuch *n*, Lehrbuch *n*, Leitfaden *m*

**textile,** ~ **draftsman** Webereizeichner *m* ~ **drive** Textilantrieb *m* ~ **fabric** Gewebe *m*, Zeug *n* ~ **industry** Faserstoff-gewerbe *n*, -industrie *f* ~ **machine** Textilmaschine *f* ~ **mill** Weberei *f*

**textilose** Textilose *f*

**texture** faseriges Gefüge *n*, Faserung *f*, Gefüge *n*, Gewebe *m*, Gewirk *n*, Struktur *f*, Textur *f*, Zeug *n* ~ **of hair** Haargewebe *n*

**textured paint** plastische Masse *f*

**T-fillet weld(ing)** T-Stoßkehlnahtschweißung *f*

**T girder** T-Träger *m*

**T guide channel** T-förmige Führungsnute *f*

**thalassometer** Gezeitenmesser *m*

**thallic** thalliumsauer ~ **bromide** Thallibromid *n* ~ **chlorate** Thallichlorat *n* ~ **compound** Thalliverbindung *f* ~ **ion** Thalliion *n* ~ **oxide** Thalliumoxyd *n* ~ **salt** Thallisalz *n* ~ **sulfate** Thallisulfat *n*

**thallium** Thallium *n* ~ **alum** Thalliumalaun *n* ~ **chloride counter** Thalliumchloridzähler *m*

**thallofide,** ~ **cell** Thallofidezelle *f* ~ **photoconductive cell** Thalliumsulfidzelle *f*

**thallous,** ~ **bromide** Thalliumbromür *n*, Thallobromid *n* ~ **chloride** Thalliumchlorür *n* ~ **fluoride** Thalliumfluorür *n* ~ **halides** Thalliumhalogenide *pl* ~ **hydroxide** Thallohydroxyd *n* ~ **iodate** Thallojodat *n* ~ **iodine** Thalliumjodür *n* ~ **oxide** Thalliumoxydul *n*

**thalweg** Stromstrich *m* (physiog.), Talweg *m*

**T-handle** Knebelgriff *m* ~ **socket wrench** Stechschlüssel *m*

**thatch** Mauerrohr *n*, Rohrdach *n*

**thatched** strohgedeckt

**thaw, to** ~ auftauen, tauen **to** ~ **off** abtauen **to** ~ **out** austauen

**thaw** Tau *m*, Tauwetter *n*

**thawing** Auftauung *f*, Wiederauftauen *n* ~ **boreholes** Auftauen *n* der Bohrlöcher ~ **point** Taupunkt *m* ~ **weather** Tauwetter *n*

**T-head** **(or -bolt)** Hammer-kopfschraube *f*, -schraube *f* ~**-type engine** Motor *m* mit T-förmigem Verbrennungsraum

**theater** Theater *n*, Schauplatz *m* ~ **of war** Kriegs--gebiet *n*, -schauplatz *m* ~ **fader** Saalregler *m*

**Theisen cleaner (disintegrator, or washer)** Theisen-Wascher *m*

**theme** Aufgabe *f*, Handlung *f*, Kennsignal *n*, Motiv *n*, Sache *f*, Thema *n*

**theodolite** Theodolit *m*, Winkelmesser *m* ~ **with compass** Bussoleninstrument *n* ~ **with stadia lines** Theodolit *m* mit Distanzmeßeinrichtung *f*

**theorem** Lehrsatz *m*, Satz *m* ~ **on circulating motions** Zirkulationssatz *m* ~ **of conservation of areas** Flächensatz *m* ~ **of conservation of momentum** Impulserhaltungssatz *m* ~ **on incidence** Schnittpunktsatz *m* ~ **of momentum** Impulssatz *m* ~ **of the sines** Sinussatz *m* ~ **for vorticity** Wirbelsätze *pl*

**theoretic weight per meter run** rechnungsmäßiges Metergewicht *m*

**theoretical** formal, theoretisch ~ **air requirement** theoretischer Luftbedarf *m* ~ **best gliding angle** theoretisch bester Gleitwinkel *m* ~ **ceiling** Rechnungsgipfelhöhe *f* ~ **cut-off** theoretische Grenzfrequenz *f* ~ **deflection** schußtafelmäßige Seitenverschiebung *f* ~ **mechanics** Festigkeits-

lehre *f* ~ **quantity** Sollmenge *f* ~ **range** schußtafelmäßige Schußweite *f* ~ **rate of revolution** Bezugsdrehzahl *f* ~ **size** Sollmaß *n* ~ **stress-concentration factor** Formziffer *f* ~ **time of flight** (ballistics) schußtafelmäßige Flugzeit *f* ~ **value** Sollwert *m*
**theoretically** begrifflich
**theory** Grundlehre *f*, Lehre *f*, Theorie *f*
**theory,** ~ **of approximations** Berichtigungsverfahren *n* ~ **of beams** Balkentheorie *f* ~ **of the boundary layer** Grenzschichtlehre *f* ~ **of buckling** Stabilitätstheorie *f* ~ **of color** Trägerfrequenz *f* ~ **of compass deviation** Deviationslehre *f* ~ **of concentrated defenses at key points** Schwerpunktgedanke *m* ~ **of continuous action** Nahewirkungstheorie *f* ~ **of elasticity of anisotrope mediums** Elastizitätstheorie *f* anisotroper Medien ~ **of elementary divisor** Elementarteilertheorie *f* ~ **of ellipticity** (Hertz) Abplattungstheorie *f* ~ **of emission** Strahlungslehre *f* ~ **of functions** Funktionentheorie *f* ~ **of interlacing** Bindungslehre *f* ~ **of lift** Aufstiegs-, Auftriebs-theorie *f* ~ **of mesotrons** Mesotrontheorie *f* ~ **of plain bearings** Gleitlagertheorie *f* ~ **of radiation** Strahlungstheorie *f* ~ **of refraction** Beugungstheorie *f* ~ **of relativity** Relativitätstheorie *f*, Standpunktslehre *f* ~ **of sets** Mengenlehre *f* (math.) ~ **of spin** Drehgeschwindigkeit *f* ~ **of stability** Stabilitäts-lehre *f*, -theorie *f* ~ **of strength of materials** Werkstoffestigkeitslehre *f* ~ **of structure** Baustatik *f* ~ **of valences** Basizitätstheorie *f* ~ **of vortices** Wirbellehre *f*
**therapeutic,** ~ **apparatus** Heilgerät *n* ~ **purpose** Heilzweck *m*
**therapy tube** Therapieröhre *f*
**thermactinic radiation** Temperaturstrahlung *f*
**thermal** kalorisch, thermisch
**thermal-agitation** thermische Bewegung *f* ~ **of a tube** Röhrenrauschen *n* ~ **noise** Wärme-geräusch *n*, -rauschen *n* ~ **voltage** Wärmerauschspannung *f*
**thermal,** ~ **ammeter** Hitzbandamperemesser *m*, Hitzdraht-amperemeter *n*, -strommesser *m* ~ **analysis** Thermoanalyse *f* ~ **antecedents** Wärmevergangenheit *f* ~ **balance** Wärmebilanz *f* ~ **barrier** Hitzegrenze *f* (aviat.) ~ **bolt** Reifenüberdrucksicherung *f* ~ **bubble** Thermikblase *f* ~ **bump** Thermitstoß *m* ~ **capacity** Artwärme *f*, spezifische Wärme *f*, Wärmekapazität *f* ~ **change point** Umwandlungspunkt *m* ~ **circle** thermischer Kreisprozeß *m* ~ **coefficient** Wärme-wert *m*, -zahl *f* ~ **condition** Wärmezustand *m* ~ **conduction** Wärmefortleitung *f* ~ **conductivity** Temperaturleit-fähigkeit *f*, -zahl *f*, Wärmeleit-fähigkeit *f*, -vermögen *n*, Wärmeleitungsvermögen *n* ~ **conductivity detector** Wärmeleitfähigkeitszelle *f* ~ **contact** thermische Kopplung *f* ~**-control switchboard** Wärmewarte *f* ~ **convection** Wärmekonvektion *f* ~ **converter** Thermoumformer *m* ~ **crack** Warmriß *m* ~ **cross** Thermokreuzbrücke *f* ~ **curve** Haltepunktskurve *f* ~ **cut-out** thermische Auslösung *f* ~ **cycle** Wärmeübertragungssystem *n* ~ **deformation** Wärmeverformung *f* ~ **delay relay** Thermo-Verzögerungsrelais *n* ~ **detector** Schmelzlotmelder *m* ~ **diffusion** Thermodiffusion *f* ~ **diffusion factor** Thermodiffusionsfak-

tor *m* ~ **diffusivity** Temperaturleit-, Wärmeausbreitungs-vermögen *n* ~ **eddies** Thermikwirbel *pl* ~ **effect** Wärme-tönung *f*, -wirkung *f* ~ **efficiency** Hitzewirkungsgrad *m*, thermischer Wirkungsgrad *m*, Wärme-ausnutzung *f*, -leistung *f*, -wirkungsgrad *m* ~ **effusion** Wärmedifusion *f*, -übertragung *f* ~ **energy** Wärmeenergie *f* ~ **engine** Wärmekraftmaschine *f* ~ **equation of state** thermische Zustandsgleichung *f* ~ **excitation** thermische Anregung *f* ~ **expansion** Wärme(aus)dehnung *f* ~ **expansivity** Wärmeausdehnungsvermögen *n* ~ **explosion** Wärmeexplosion *f* ~ **flow (or flux)** Wärmefluß *m* ~ **gliding** Thermiksegelflug *m* ~ **gradient** geothermische Tiefenstufe *f* ~ **gradiometer** thermischer Neigungsmesser *m* ~ **ice elimination** Warmluftenteisung *f* ~ **inertia** Wärmeträgheit *f* ~ **influence** Wärmewirkung *f* ~ **insulation** Wärmedämmung *f* ~ **ionization** Wärmeionisation *f* ~ **jet** Heißluftstrahl *m* ~**-jet engine** Heißstrahltriebwerk *n* ~ **jet propulsion** Vortrieb *m* durch Heizstrahltriebwerk ~ **lag** thermische Verzögerung *f* ~**-lag switch** Thermoschalter *m* ~ **leakage** thermischer Verlust *m* ~ **loop** Wärmeübertragungssystem *n* ~ **loss** Erwärmungs-, Wärme-verlust *m* ~ **metamorphism** Pyrometamorphose *f* ~ **motion** Wärmebewegung *f* ~ **motion of electrons** Wärmerauschen *n* ~ **neutron** langsames Neutron *n* ~ **noise** Wärmegeräusch *n*, (electr. tube) Widerstandsrauschen *n* ~ **noise generator** Wärmerauschgenerator *m* ~ **overload capacity** thermische Überlastbarkeit *f* ~ **phenomenon** Wärmephänomen *n* ~ **potential** Thermopotential *n* ~ **power** Heiz-, Wärme-kraft *f* ~ **power installation** Wärmekraftanlage *f* ~ **power station** Wärmekraftwerk *n* ~ **process** Wärmevorgang *m* ~ **radiation** Wärmestrahlung *f* ~ **range** thermisches Energiegebiet *n* ~ **ray** Wärmestrahl *m* ~ **reactor** Reaktor *m* mit thermischen Neutronen ~ **receiver** Empfänger *m* mit thermischem Hörer ~ **relief valve** Thermo-Entlastungs- oder Überdruckventil *n* ~ **requirements** Wärmebedarf *m* ~ **response** Temperaturanstiegrate *f* ~ **retardation** Haltezeit *f* ~ **shield** Wärmeabschirmung *f* ~ **shock** plötzliche Temperaturveränderung *f* ~ **soaring** thermischer Segelflug *m* ~ **spectrum** Wärmespektrum *n* ~ **spike** Wärme-Maximum *n* ~ **spring** Mineralbrunnen *m*, Thermalquelle *f* ~ **stability** Wärmebeständigkeit *f* ~ **strain** Wärmespannung *f* ~ **stratification** Temperaturschichtung *f* ~ **stress** Temperaturspannung *f*, Wärmebelastung *f*, -spannung *f* ~ **stress-relief** Spannungsfreiglühen *n* ~ **telephone receiver** Empfänger *m* mit thermischem Hörer ~ **time constant** Kathodenanheizzeit *f* ~ **transmitter** Thermomikrofon *n* ~ **transpiration** thermische Strömung *f*, Wärme-diffusion *f*, -übertragung *f* ~ **treatment** Wärmebehandlung *f* ~ **tuning** thermische Abstimmung *f* ~ **tuning time** thermische Abstimmzeit *f* bei Anwärmen ~ **unit** Kalorie *f*, Wärmeeinheit *f* ~ **upcurrent** thermischer Nutzfaktor *m* ~ **value** Wärmewert *m* ~ **yield** Wärmeausbeute *f*
**thermic** thermisch ~ **flight** Thermikflug *m* ~ **relay** Thermorelais *n* ~ **upwash near ground** Bodenthermik *f*

**thermically controlled contact** thermisch gesteuerter Kontakt *m* (Blinkgeber)
**thermion** Thermion *n*
**thermionic** glühelektrisch, thermionisch ~ **amplifier** Glühkathoden-, Röhren-verstärker *m* ~ **apparatus** Elektronenröhrengerät *n* ~ **arc** thermionischer Lichtbogen *m* ~ **cathode** Glühkathode *f* ~ **constant** Glühemissionskonstante *f* ~ **current** Glühelektronenstrom *m* ~ **discharge** Elektronenentladung *f*, Glühelektronenentladung *f* ~ **electron source** Glühelektronenquelle *f* ~ **emission** Elektronenverdampfung *f*, glühelektrische Elektronenabgabe *f* oder Elektronenemission *f*, Glühelektronen-abgabe *f*, -emission *f* ~ **emission of grid** thermische Gitteremission *f* ~ **emission properties** Glühemissionseigenschaften *pl* ~ **oscillator** Röhrengenerator *m*, -sender *m* ~ **rectifier** Glühkathodengleichrichter *m* ~ **relay** Thermionenrelais *n*, Verstärker-rohr *n*, -röhre *f* ~ **trigger device** Selbstschwingungsvorrichtung *f* ~ **tube** Elektronen(strahl)röhre *f*, Kathodenstrahlröhre *f* ~ **vacuum tube** Elektronenröhre *f*
**thermionic-valve** Detektorenempfänger *m*, Elektronenröhre *f*, Kathodenröhre *f*, Röhrenempfänger *m*, thermionischer Gleichrichter *m* ~ **detector** Röhrendetektor *m* ~ **receiver** Röhrenempfänger *m* ~ **sender** Röhrensender *m* ~ **vacuum tube** Glühkathodenröhre *f*
**thermionic voltmeter** Röhrenvoltmeter *n*
**thermions** Thermionen *pl*
**thermistor** Heißleiter *m*, Thermistor *m*
**thermit welding** aluminiumthermische Schweißung *f*, (pressure) aluminothermisches Schweißverfahren *n*
**thermite** Thermit *n* ~ **bomb** Thermitbombe *f* ~ **charge** Thermitladung *f* ~ **filling** Thermitfüllung *f* ~ **fusion welding** Thermitgießverfahren *n* ~ **incendiary bomb** Thermitbrandbombe *f* ~ **iron** Thermiteisen *n* ~ **process** Thermitverfahren *n* ~ **weld** Thermitschweißung *f* ~ **welding** Thermitschweißung *f* ~ **welding process** aluminothermisches Schweißverfahren *n*, Thermitschweißverfahren *n*
**thermo** thermoelektrisch, Thermo... ~ **ammeter** Thermostrommesser *m* ~**-anelasticity** Thermoanelastizität *f* ~ **barograph** Termobarograf *m* ~ **cautery** Kauterisation *f* (med.), Thermokauterisation *f* ~ **chemical** thermochemisch ~**-chromium pin** Thermochromstift *m* ~ **color** Thermofarbe *f* ~ **converter** Thermoumformer *m*
**thermocouple** thermoelektrisches Element *n*, Thermo-element *n*, -kreuz *n*, -paar *n*, -zelle *f* ~ **ammeter** thermischer Strommesser *m* ~ **carrier** Thermoelementträger *m* ~ **circuit** Thermoelementenkreis *m* ~ **cross** Thermokreuzbrücke *f* ~ **detector** Thermokreuz *n* ~ **element** Thermoelement *n* ~ **galvanometer** Thermogalvanometer *n* ~ **instrument** thermoelektrisches Meßgerät *n*, Thermoumformer-Meßgerät *n* ~ **joint** Anschlußkopf *m* ~ **meter** thermoelektrisches Meßgerät *n* ~ **voltmeter** Thermovoltmeter *n* ~ **wire** Thermoelementendraht *m*
**thermo**, ~ **current** thermoelektrischer Strom *m*, Thermostrom *m* ~ **detector** Thermodetektor *m* ~ **diffusion** Wärme-diffusion *f*, -übertragung *f* ~ **dynamic degeneracy** thermodynamische Ent-

artung *f* ~ **dynamic nuclear rocket** thermische Nuklearrakete *f* ~ **dynamics** Thermodynamik *f*, Wärme-dynamik *f*, -kraftlehre *f*, -lehre *f*, -mechanik *f* ~**-elastic** thermoelastisch ~ **elasticity** Thermoelastizität *f*
**thermo-electric** Wärme-einsaugung *f*, -einstrahlung *f*; elektrokalorisch, thermoelektrisch ~ **couple** Thermoelement *n* ~ **current** thermoelektrischer Strom *m*, Thermostrom *m* ~ **effect** thermoelektrischer Effekt *m*, thermoelektrische Kraftwirkung *f* oder Wirkung *f* ~ **force** Thermokraft *f* ~ **generating set** thermische Generatorgruppe *f*, Wärmekraftmaschinensatz *m* ~ **instrument** thermoelektrisches Meßgerät *n* ~ **inversion** thermoelektrische Umkehrung *f* ~ **pile** Thermosäule *f* ~ **power** Thermo-kraft *f*, -spannung *f* ~ **relay** Thermorelais *n* ~ **thermometer** thermoelektrisches Thermometer *n* ~ **voltage** Thermospannung *f* ~ **wire** Thermoelementendraht *m*
**thermo**, ~ **electrical** elektrokalorisch, wärmeelektrisch ~ **electricity** Thermo-, elektrizität *f* ~ **electromotive** thermoelektromotorisch ~ **electron** Thermoelektron *n* ~ **element** Thermo-kette *f*, -zelle *f* ~ **engine** Warmkraftmaschine *f* ~ **galvanometer** Thermogalvanometer *n* ~ **gauge** Thermomanometer *n* ~ **gen** Wärmestoff *m* ~ **gram** Thermogramm *n* ~ **graph** Schreibthermometer *n*, Thermograf *m*, Wärmegradschreiber *m* ~ **hydrometer** Thermohydrometer *n* ~**-ionic tube** Elektronenröhre *f* ~ **isopleth** Wärmeisoplethe *f* ~ **junction** thermoelektrische Lötstelle *f* ~ **labile** thermolabil ~ **locating** thermische Ortung *f*, Wärmeanpeilung *f* ~ **location** thermische Ortung *f*, Wärmeanpeilung *f*
**thermolysis** Thermolyse *f*
**thermo**, ~**-mechanical transducer** thermomechanischer Wandler *m* ~ **metallurgy** Thermometallurgie *f* ~ **milliammeter** Thermomilliamperemeter *n*
**thermometer** Temperaturmesser *m*, Thermometer *n*, Wärme-gradmesser *m*, -messer *m*, Warmheitsmesser *m* ~ **for heating boilers** Heizungskesselthermometer *n*
**thermometer**, ~ **branch** Thermometerstutzen *m* ~ **bulb** Thermometer-blase *f*, -kugel *f* ~ **column** Thermometer-röhre *f*, -säule *f* ~ **probe** Temperaturfühler *m* ~ **scale** Thermometerskala *f* ~ **screen** Wärmegradmesserhülle *f* ~ **socket** Thermometereinsatz *m* ~ **stem** Thermometer-röhre *f*, -säule *f*, -schaft *m* ~ **tube** Temperaturfühler *m* mit Schutzrohr ~ **well** Thermometer-hülse *f*, -tasche *f*
**thermometric** wärmemessend ~ **column** Thermometerfaden *m* ~ **scale** Temperatur-, Wärmeskala *f*
**thermometrograph** selbstregistrierendes Thermometer *m*
**thermometry** Thermometrie *f*, Wärmemessung *f*
**thermomotive** thermomotorisch
**thermonic** glühelektrisch
**thermo**, ~ **nuclear reaction** thermonukleare Reaktion *f* ~ **phone** Thermofon *n* ~ **phosphorescence** Thermophosphoreszenz *f* ~ **physical** thermophysikalisch ~ **physics** Thermophysik *f* ~ **pile** Thermosäule *f*

**thermoplastic** (substance) härtbare Masse *f* (thermoplastischer Stoff), Thermoplast *n*, thermoplastisch, in der Wärme *f* bildsam ~ **rip cords** Zwillingsleitung *f*

**thermo,** ~ **regulation** Wärmeregulierung *f* ~ **regulator** Temperatur-, Thermo-, Wärmeregler *m* ~ **resistant** hitzebeständig, thermoresistent

**thermos,** ~ **bottle** Thermosflasche *f* ~ **vessel with exhaust jacket** Vakuummantelgefäß *n*

**thermoscope** Thermoskop *n*

**thermo-sensitive** temperaturempfindlich

**thermoset, to** ~ durch Wärme *f* härten, warmhärten

**thermosetting** Hitzehärten *n*; in der Hitze *f* erhärtend, (synthetic resin) durch Wärme *f* härtbar, warmhärtbar ~ **material** in der Hitze *f* aushärtendes Metall *n* ~ **plastics** Duroplast *n*, thermostatoplastischer Werkstoff *m*

**thermosiphon** selbsttätiger Umlauf *m*, Wärmesaugheber *m* ~ **cooling** Thermosyphonkühlung *f*

**thermostable** hitzebeständig

**thermostat** Temperaturregler *m*, Thermostat *m*, Wärme-fühler *m*, -halter *m*, -regler *m* ~ **connection** (from radiator to thermometer) Fühleranschluß *m* ~ **unit** Thermostateinheit *f*

**thermostatic,** ~ **control valve** Kühlerventil *n* ~ **expansion valve** (Tiefkühltruhe) thermostatisches Expansionsventil *n* ~ **regulator** Temperaturregler *m* ~ **switch** Thermoschalter *m*

**thermo,** ~ **statics** Thermostatik *f* ~ **switch** Thermoschalter *m* ~ **telephone** Thermotelefon *n*

**thermotron measuring tube** Thermotronmeßröhre *f*

**thermovalue range** Wärmewertbereich *m*

**thermoviscous number** Thermoviskositätszahl *f*

**thesis** Dissertation *f*, Thema *n*, Thesis *f*

**theta function** Thetafunktion *f*

**thick** (paper) dicht, dick, mächtig, stark, trübe ~ **blotting paper** Löschkarton *m* ~ **cod oil** Brauntran *m* ~ **cover** einseitiger Umschlag *m* ~ **drawing** Grobzug *m* ~ **inwall** Rauhgemäuer *n* ~ **juice** Dicksaft *m* ~ **juice blowup** Dicksaftaufkocher *m* ~-**juice carbonation** Dicksaftsaturation *f* ~-**juice washing** Dicksaftdecke *f* ~ **mash** Dickmaische *f* ~ **metal plate** Grobblech *n* ~ **metal sheets** Grobbleche *pl* ~ **mud** Dickschlamm *m* ~-**ply** dickdrähtig ~ **printing paper** Dickdruckpapier *n* ~ **ripping (or rumpling)** Grobzug *m* ~-**skinned** dickschalig ~ **slag** kurze Schlacke *f* ~ **stranded** dickdrähtig ~ **target** dicke Schicht *f* (Röntgenspektrum), dicke Treffplatte *f* ~ **wall** (of blast furnace) Rauhgemäuer *n* ~-**walled** dickwandig, stark-wandig ~-**walled beaker** Standgefäß *n* ~ **weather** dickes Wetter *n* ~ **wood** einsträhniges Holz *n*

**thicken, to** ~ absetzen, abstreifen, eindicken, gerinnen, sichern, verdichten, verdicken, verstärken **to** ~ **up slag** Schlacke *f* versteifen

**thickener** (flotation) Absetzbehälter *m*, Eindicker *m*, Verdicker *m*, Verdickungsbehälter *m* ~ **overflow** Eindickerüberlauf *m*

**thickening** Absetzen *n*, Absteifung *f*, Eindicken *n*, Eindickung *f*, Festwerden *n*, Verdichtung *f*, Verdickung *f* ~ **of the blades** Verstärkung *f* der Messer ~ **by gravitation** Gleichfälligkeit *f*

**thickening,** ~ **drum for wood pulp** Eindicktrommel *f* für Zellulose ~ **filter** Eindickfilter *n* ~ **substance** Eindickungs-, Verdickungs-mittel *n*

**thicket** Dickicht *n*

**thickness** Dichte *f*, Dicke *f*, Dickte *f*, Grobheit *f*, Mächtigkeit *f*, Stärke *f*, (of plate) Wand *f*, Wandstärke *f*

**thickness,** ~ **of armor** Panzerstärke *f* ~ **at bottom** Bodenstärke *f* ~ **of boundary layer** Grenzschichtdicke *f* ~ **of chip** Spandicke *f* ~ **of coal seam** Kohlenmächtigkeit *f* ~ **of coat** Schichtdicke *f* ~ **of the coil** Wandstärke *f* der Spule ~ **of concrete** Betonstärke *f* ~ **of cover** Deckungsstärke *f* ~ **of cut** Spannstärke *f* ~ **of deposit** Flözmächtigkeit *f* ~ **of fabric** Warenstärke *f* ~ **of film** Schichtdicke *f* ~ **of (crystal) flake** Dicke *f* der Platte (geol.) ~ **for half absorption** Halbwertschicht *f* ~ **of head** Kopfhöhe *f* ~ **of ice** Dicke *f* des Eises ~ **of key** Keilhöhe *f* ~ **of layer** Dicke *f* der Schicht, Schichtdicke *f* ~ **of the lens** Linsendicke *f* ~ **of the lines** Strichstärke *f* ~ **of nut** Mutterhöhe *f* ~ **of overburden** Stärke *f* der Deckschicht (min.) ~ **of paper** Papierblattstärke *f* ~ **of a plate** Blechdicke *f*, -stärke *f* ~ **of profile** Profildicke *f* ~ **of rib web** Rippenstegstärke *f* ~ **of roof** Deckenstärke *f* ~ **of shaving** Spanstärke *f* ~ **of a sheet** Blech-dicke *f*, -stärke *f* ~ **of sheet gauge** Stärke *f* der Blechlehre ~ **of shock layer** Stoßfronttiefe *f* ~ **of spar web** Holmstegstärke *f* ~ **of stratum** Lagerungsdichte *f* ~ **of strut** Stiel-breite *f*, -dicke *f*, Strebendicke *f* ~ **of tooth** Zahnstärke *f* ~ **of veneer** Furnier-dicke *f*, -stärke *f* ~ **of wall** Mauerstärke *f*, Wandstärke *f* (pipe) ~ **of web** Seelenstärke *f* ~ **of weld** Sollnahtdicke *f* ~ **of welded seam** Dicke *f* der Schweißnaht ~ **of wire** Draht-maß *n*, -nummer *f*, -stärke *f*

**thickness,** ~ **board** Hemdbrett *n* ~-**chord ratio** Dickenverhältnis *n* (Profil), relative Profildicke *f* ~ **distribution** Dickenverteilung *f* ~ **feeler** Fühlblech *n* ~-**function curve** Tropfenkurve *f* ~ **gauge** Abstandsmesser *m*, Dickenlehre *f*, -messer *n* ~ **gradient** Dicken-gradient *m*, -stufe *f* ~ **measurement of layers** Schichtdickenmessung *f* ~ **ratio** Sehnendicke-, Sehnenstärke-verhältnis *n* ~ **vibration of a crystal** Kristalldickenschwingung *f* ~ **vibrations** Dickenschwingungen *f* (cryst.)

**thicknessing machine for planks** Bretter-, Dickten-hobelmaschine *f*

**thickset** gestreifter Manchester *m*

**thief,** ~ **pump** Wasserzapfpumpe *f* ~ **rod** Stehstange *f*

**thigh** Schenkel *m* **with one** ~ (or arm) einschenkelig ~ **spring** Schenkelfeder *f*

**thill** Gabeldeichsel *f* ~ **wagon** Gabelwagen *m*

**thimble** Fingerhut *m*, Haube *f*, Herzkausche *f*, Kabelschuh *m*, Kleinkammer *f*, Muffe *f*, Öhrchen *n*, sackartig geschlossenes Rohr *n*, Seilkausche *f*, Zwinge *f* **to put a** ~ **in an eye** eine Kausche *f* in ein Ende splissen

**thimble,** ~ **hook** Haken *m* mit Kausche ~ **ionizing chamber** Fingerhutionisierungskammer *f* ~-**type diffuser** Fingerhutzerstäuber *m*

**thin,** to ~ anschärfen, durchforsten, lichten, vereinzeln, verflüssigen, verjüngen, verschneiden **to** ~ **out** auskeilen, lichten, verdünnen

**thin** dünn, dünnflüssig, geringmächtig, leicht-
flüssig, mager **~ and medium sheet metal working** Fein- und Mittelblechbearbeitung *f* **in ~
scales** dünnschuppig **in ~ sheets** blätterförmig
**thin, ~ bouillon** Kandillen *pl* **~ brazier rivet**
flacher Rundkopf *m* **~ brick** Kanalziegel *m* **~
combustion chamber** flacher Verbrennungsraum *m* **~ film** dünne Folie *f*, Häutchen *n* **~
film memory** Dünnfilmspeicher *m* **~ film
thickness measurement** Dünnfilmdickenmes-
sung *f* **~ film transistor** Dünnfilmtransistor *m*
**~ gauge metal block** Metalldünnklischee *n*
**~-gauge plate** Feinblech *n* **~ gold layer** Gold-
folienschicht *f* **~ gold leaf on silver foil** Quickgold *n* **~-ground stone plate** Gesteinsdünn-
schliff *m* **~ inwall** Kerngemäuer *n* **~ juice**
Dünnsaft *m* **~ juice blow-up tank** Dünnsaft-
aufkocher *m* **~ layer** Blättchen *n* **~ magnetic
film** magnetischer Dünnfilm *m* **~ metal plate**
Feinblech *n* **~ metal target** dünne Metallschicht
*f* **~ mica plate** Glimmerplättchen *n* **~ nut**
flache Mutter *f* **~ orifice plate** Scheibenblende
*f* **~ plate** Feinblech *n* **~ plate orifice** Staurand-
(scheibe *f*) *m* **~ printing paper** Dünndruckpapier *n* **~ profile** dünnes Profil *n* **~ rubber tubing**
dünner Gummischlauch *m* **~ seam** gering-
mächtiger Flöz *m* **~ section** (of rock or metal
for microscopic examination) Dünn-schliff *m*,
-schnitt *m* **~-section castings** dünnwandiger
Guß *m* **~ section wing** dünner Flügel *m* **~
sheet** dünnes Blech *n*, Sturzblech *n* **~-skinned**
dünnschalig **~ slag** dünnflüssige Schlacke *f* **~
straight double plane cutter** Doppelhobeleisen
*n* **~ target** dünne Treffplatte *f* **~-walled** dünn-
wandig **~-walled blast furnace** dünnwandiger
Hochofen *m* **~-walled castings** dünnwandiger
Guß *m* **~ wedges through wooden dowels** Aale
*pl* **~ wire** Feindraht *m*
**think, to ~ up** austüfteln
**thinly liquid** leichtflüssig
**thinned** angeschärft, verdünnt, verjüngt
**thinner** Farbenverdünner *m*, Verdünner *m*, Verdünnungsmittel *n*
**thinness of walls** Dünnwandigkeit *f*
**thinning** Nachstrecken *n*, Verdünnung *f*, Verflüssigung *f* **~ agent** Verdünnungs-, Verflüs-
sigungs-mittel *n*
**thio, ~ acetic acid** Thioessigsäure *f* **~ acid** Thiosäure *f* **~ antimonic acid** Schwefelantimonsäure
*f* **~ carbamine** Thioharnstoff *m* **~ cresol** Thio-
kresol *n* **~ cynate** Rhodanat *n* **~ cyanic acid**
Rhodanwasserstoff-, Schwefelzyan-, Sulfo-
zyan-, Thiozyan-säure *f* **~ cyanogen** Rhodan *n*,
Thiozyan *n* **~ cyanoplatinic acid** Platinirhodan-
wasserstoffsäure *f* **~ cyanoplatinous acid**
Platinorhodanwasserstoffsäure *f*
**thiomic acid** Thiosäure *f*
**thionyl chloride** Thionylchlorid
**thio, ~ phenol** Thiophenol *n* **~ phthene** Thio-
phten *n* **~ salt** Sulfosalz *n* **~ stannic acid** Thiosin-
amin *n* **~ stannic acid** Thiozinnsäure *f* **~
sulfuric acid** Thioschwefelsäure *f*
**third** Drittel *n*; dritt(er, e, es), dritter Gang *m*
(auto) **~ of a sheet** Drittelbogen *m* (print.)
**third, ~ brush dynamo** Dreibürstenmaschine *f*,
Stromregel- und Lichtmaschine *f* **~ carbonation**
dritte Saturation *f* **~-class conductor** Heißleiter
*m*, mit negativem Widerstand behafteter Leiter

*m* **~-class resistance** negativer Widerstand *m* **~
conductor** c-Leitung *f* **~ differential of characteristic** dritte Ableitung *f* der Charakteristik **~
harmonic** dreifache Oberharmonische *f* **~-mo-
tion shaft** dritte Getriebewelle *f* **~ order terms**
Glieder *pl* dritter Ordnung **~ part of a sheet**
Drittelbogen *m* **~ plow attachment** Dreischar-
vorrichtung *f* **~ ply** dritter Ausschnitt *m* **~
points** Drittelpunkte *pl* **~ rail** Stromschiene *f* **~
rail for crane (or trolley) car power supply**
Stromzuführungsschleifleistungsstränge *n* **~
rail gauge** Be- oder Um-grenzung *f* der Stromschiene **~ rail insulator** Fahrschienenisolator *m*
**~-rail system** System *n* der dritten Schiene *f* **~
root** Kubikwurzel *f* **~ space** Drittelspatium *n*
**~ speed** dritter Gang *m* **~ wire** Mittelleiter *m*
**thirty-slot armature** Dreißignutanker *m*
**thistle funnel** Tülle *f*
**thole pin** Dolle *f*
**tholelite dykes** Tholelit-Gänge *pl*
**Thomas, ~ converter** Thomas-birne *f*, -konverter
*m* **~-Gilchrist process** Thomasverfahren *n* **~
low-carbon steel** Thomasflußeisen *n* **~ meal**
Thomasschlackenmehl *n* **~ process** Thomasverfahren *n* **~ slag** Thomasschlacke *f* **~ steel**
Thomas-eisen *n*, -flußeisen *n*, -flußstahl *m*,
-material *n*, -stahl *m* **~ steel plant** Thomasstahlwerk *n*
**Thomson, ~ contact-welding process** Thomson-
schweißverfahren *n* **~ effect** Thomsoneffekt *m*
**~ formula** Thomsonsche Formel *f* **~ meter**
dynamometrischer Motorzähler *m*
**thong** (Leder)Riemen *m*, Streifen *m*
**thoracic somite** Thoraxsegment *n* (geol.)
**thorax** Rumpf *m*
**thoria** (thorium dioxide) Thorerde *f*
**thoriate, to ~** thorieren
**thoriated** thorhaltig, thoriert, thoriumhaltig **~
cathode** thorierte Kathode *f* **~ filament** thorhaltiger (thorierter) Faden *m*, Thoriumfaden *m*
**~ filament valve** Thoriumröhre *f* **~ tungsten
emitter** emittierende Thoriumschicht *f* auf
Wolfram **~ tungsten filament** thorhaltiger oder
thorierter Wolframfaden *m*
**thoric** thorisch
**thorite** Orangit *n*, Thorit *m*
**thorium** Thor *n*, Thorium *n* **~ chloride** Thorium-
chlorid *n* **~ fission** Thoriumspaltung *f* **~ oxide**
Thoroxyd
**thorn** Dorn *m* **~ tongue** Stachel *m*
**thorny** dornig, stachelig
**thorogummite** Nicolayit *m*, Thorogummit *m*
**thorolite** Thorolit *m*
**thoron** Thoriumemanation *f*
**thorough** durchgreifend, eingehend, gänzlich,
gründlich, völlig, weitgehend **~ bend** Durchbiegung *f*
**thoroughfare** Durch-fahrt *f*, -gang *m*, Hauptverkehrsstraße *f*, Verkehrsader *f*
**thoroughly** durchaus, durch und durch **~ reliable**
durchaus zuverlässig
**thousand** tausend
**thousands, ~ of times** Vieltausendfache *n* **~
digit** Tausenderwahlstufe *f* **~ place** Tausender-
stelle *f*
**thousandth** Tausendstel *n*
**thrash** Verschleißkette *f*
**thrashed weight** Abdrusch-, Drusch-gewicht *n*

**thrashing floor** Tenne *f*
**thraulite** Thraulit *m*
**thread, to** ~ einfädeln, Gewinde *n* schneiden, schneiden, umschnüren **to** ~ **into** einschrauben **to** ~ **on** aufreihen, aufschrauben **to** ~ **on(to)** aufstreifen **to** ~ **up the film** Film einlegen
**thread** Draht *m*, Einschraubgewinde *n*, Faden *m*, Faser *f*, (screw) Gang *m*, Gangzahl *f*, Garn *n*, Gespinst *n*, (screw) Gewinde *n*, Gewindegang *m*. Schraubengewinde *n*, Zaser *f*, Zwirn *m* ~ **for draw-in bolt** Anzugsgewinde *n* ~ **of a screw** Gang *m* eines Gewindes ~ **of tool spindle** Werkzeugspindelgewinde *n* ~ **of the weft** Eintragfaden *m*
**thread,** ~ **adjusting core** Gewindeabgleichkern *m* ~ **bacteria** Eisenbakterien *pl* ~**-balancing core** Gewindeabgleichkern *m* ~ **bare** abgetragen, fadenscheinig, kahl ~ **base** Gewindefuß *m* ~ **beam** Fadenstrahl *m* ~ **bolt** Gewindebolzen *m* ~ **bottom** Gewindefuß *m* ~ **brake** Fadenbremse *f* ~**-bulging machine** Gewindedrückmaschine *f* ~ **calipers** Gewindetaster *m* ~ **candy** Fadenkandis *m* ~ **carrier** Fadenhänger *m* ~ **chamfer** Gewindeauslauf *m* ~ **chaser** Gewinde-strähler *m*, -strehler *m* ~ **chasing** Gewindeschneiden *n* mittels Strehler, Gewindestrehlen *n* ~ **chasing attachment** Gewindestrehleinrichtung *f* ~ **contour microscope** Gewindeprofilmikroskop *n* ~ **core diameter** Gewindekerndurchmesser *m* ~ **counter** Fadenzähler *m*, Weberglas *n* ~ **cutter** Gewindestahl *m*
**thread-cutting** Gewindeschneiden *n* ~ **indicator** Gewindeschneideanzeiger *m* ~ **lathe** Gewindedrehbank *f* ~ **machine** Gewinde-herstellungsmaschine *f*, -schneidemaschine *f* ~ **protective cap** Gewindeschutzkappe *f*
**thread,** ~ **diameter** Gewindedurchmesser *m* ~ **electrometer** Fadenelektrometer *n* ~ **feeder** Fadeneinleger *m* ~ **galvanometer** Fadengalvanometer *n* ~ **gauge** Gewinde-kaliber *n*, -lehre *f*, Prüfschraube *f* ~ **gauging wire** Meßdraht *m* ~**-grinding machine** Gewindeschleifmaschine *f* ~ **groove** Gewinde-kanal *m*, -rille *f* ~ **guide** Faden-führer *m*, -leiter *m* ~ **hair** Faden *m* im Okular ~ **interlacing attachment** Fadeneinschußzusatz *m* ~ **interval meter** Fadendistanzmesser *m* ~ **like defect of glass** Glasfaden *m* ~ **hobbing** Gewinde *n* nach dem Wälzverfahren ~ **hobs** Gewindewälzfräser *m* ~ **lapping machine** Gewindelappmaschine *f* ~ **limit** Gewindegrenzmaß *n* ~ **limit roll snap gauges** Grenzgewinderollenlehren *pl* ~ **line** Horizontalfaden *m* ~ **lube** Gewindefett *n* ~ **lug** Gewindeansatz *m* ~ **machine** Fadenheftmaschine *f* ~ **measured in inches** Zollgewinde *n* ~ **milling** Gewindefräsen *n* ~**-milling cutter** Gewinderillenfräser *m* ~**-milling machine** Gewindefräsmaschine *f* ~ **nipple** Gewindenippel *n* ~ **per unit** Gangzahl *f* ~ **picker** Fadenklauber *m* ~ **pitch** Gewindesteigung *f* ~**-pitch diameter** Flankendurchmesser *m* (Gewinde) ~ **plug gauge** Gewindelehrdorn *n* (screw-) ~ **profile** Gewindeprofil *n* ~ **profile projector** Gewindeprojektionsapparat *m* ~ **ray discharge** Fadenladung *f* ~ **representation** Wiedergabe *f* von Rohrgewinden
**thread-rolling** Gewinde-rollen *pl*, -walzen *pl* ~ **die** Gewindewalzbacke *f*, Werkzeug *n* zum

Gewinderollen ~ **dies** Gewindewälzbacken *pl* ~ **head** Gewinderollkopf *m* ~ **machine** Gewindewalzmaschine *f*
**thread,** ~ **setting gauges** Einstellgewindelehren *pl* ~**-shaped** fadenförmig ~ **standards** Gewindenormblatt *n* ~**-strength tester** (textiles) Fadenfestigkeitsprüfer *m* ~**-swing machine** Fadenheftmaschine *f* ~ **system** Gewindesystem *n* ~ **take-up lever** Fadenleiter *m* ~ **template** Spitzenlehre *f* ~**-tension device** Fadenspanner *m* ~ **tolerance** Gewindetoleranz *f* ~ **tool** Gewindestahl *m* ~ **width** Schraubenführungsgangbreite *f*
**threaded** (on) aufgeschraubt, gängig, (rods and sockets) geschraubt, mit Gewinde *n* versehen ~ **angle connection piece** (or **socket**) Einschraubwinkelstutzen *m* ~ **anvil** Amboßschraube *f* ~**-base percussion fuse** Schlagzündschraube *f* ~ **bolt** Schraubenbolzen *m* ~ **bush** Gewindebuchse *f* ~ **bushing** Einschraubstutzen *m*, Gewindebuchse *f* ~ **cap** Deckel *m* mit Gewinde, Endverschraubung *f*, Rohrkappe *f* ~ **collar** ringförmige Mutter *f*, Ringmutter *f* ~ **conduit hub** Gewinderohreinführung *f* ~ **connection** Anschlußgewinde *n* ~ **coupling** Gewindestück *n* ~ **drill-bit nipple** Bohrkronengewindenippel *m* ~ **end of the stud** Einschraubende *n* der Stiftschraube ~ **extension** Gewindefortsatz *m* ~ **eye** Gewindeauge *n* ~ **female chuck** Gewindefutter *n* ~ **flange** Gewindeflansch *m* ~ **hole** Gewindeloch *n* ~ **hose nipple** Schlauchgewindestutzen *m* ~ **joint** Schraubenverbindung *f* ~ **journal** Gewindezapfen *m* ~ **knuckle** Gewindegelenk *n* ~ **nipple** Gewindestück *n* ~ **nose cap** Verschlußschraube *f* ~ **percussion primer** Zündschraube *f* ~ **percussion-primer case** (or **housing**) Zündschraubengehäuse *n* ~ **pin** Gewindestift *m* ~ **pin with tapering end** Gewindestift *m* mit Spitze ~ **pipe** Gewinderohr *n* ~ **pipe connection** Gewindeanschluß *m* ~ **plug** Stopfen *m* ~ **portion of a tap** Gangzahl *f* des Gewindeteiles eines Gewindebohrers ~ **primer** Zünderschraube *f* ~ **ring** Gewindering *m* ~ **rod** Gewindestift *m* ~ **seat** Gewindeeinschraubsitz *m* ~ **setting gauge** Einstellgewindelehre *f* ~ **shouldered pin** Bolzen *m* ~ **sleeve** Gewinde-buchse *f*, -hülse *f*, -muffe *f* ~ **socket** Einschraubstutzen *m*, Mutterstück *n* ~ **spindel** Gewindespindel *f* ~ **test piece** Gewindekopfprobe *f* ~ **tube** Gewinderohr *n* ~ **valve stem** Schraube *f* ohne Ende
**threading** Gewindeschneiden *n*, Umschnürung *f* ~ **film** Film-einfädelung *f*, -einführung *f* ~ **beading and trimming machine** Gewinde-, Drücke-, Sicken- und Beschneidemaschine *f* ~ **comb** Einlegekamm *m* ~ **die** Gewinde-backe *f*, -backen *m*, Gewindeschneide-backe *f*, -backen *m*, -eisen *m*, Kluppe *f*, Schneideeisen *n* ~ **die grinding machine** Schneideisenschleifmaschine *f* ~ **die head** Gewindeschneidekopf *m* ~ **die tapper** Gewinderollwerkzeug *n* ~ **end** Einlaufseite *f* (film) ~ **jaw** Rollbacke *f* ~ **machine** Gewindeschneidemaschine *f* ~ **operation** Gewindeschneidgang *m* ~ **tool** Gewindeschneide-stahl *m*, -werkzeug *n*, Gewinde-schneidwerkzeug *n*, -stahl *m*, Stechstahl *m* mit Rundschneide ~ **tool chaser** Gewinde-schneidzahn *m*, -spitzstahl *m* ~**-tool holder** Gewindezahnhalter *m* ~**-up** Filmeinfädelung *f*

**threads** Verästelung *f* **~ per inch** (pipe) Anzahl *f* der Gänge auf den Zoll, Gänge *pl* auf einen Zoll, Gewindegänge *pl* pro Zoll
**three** drei **in ~ parts** dreiteilig **with ~ steps** dreistufig **of ~ threads** dreidrähtig
**three, ~-address instruction** Dreiadressenbefehl *m* **~-ammeter method** Drei-Amperemeterverfahren *n* **~-bank upright engine** Dreireihenstandmotor *m* **~-bar** dreilamellig **~-bar magneto** dreilamelliger Kurbelinduktor *m* **~-bay biplane** Dreistieler *m* (aviat.) **~-bayed** dreischiffig **~-beam headlight** Dreistrahler *m* **~-beam problem** Dreistrahlproblem *n* **~-bearing crankshaft** dreimal gelagerte Kurbelwelle *f* **~-blade controllable propeller** Dreiblattverstellschraube *f* **~-bladed propeller** dreiflügelige Luftschraube *f*, Dreiflügelschraube *f* **~-body forces** Dreikörperkräfte *pl* **~ button switch** Dreiknopfschalter *m* (electr.) **~-carrier model** Dreiladungsträgermodell *n* **~-case shift** doppelte Umschaltung *f* **~-centered (eliptic) arch** elliptisches Gewölbe *n* **~-channel method** Dreikanalverfahren *n* **~-circuit receiver** Drei-kreisempfänger *m*, -kreiser *m* **~-circuit reception** Tertiärempfang *m* **~-circuit set** Drei-kreisempfänger *m*, -kreiser *m* **~-coil transformer** Symmetrieübertrager *m* (teleph.)
**three-color, ~ camera** Dreifarbenaufnahmekamera *f* **~ equation** dreiglied(e)rige Farbgleichung *f* **~ photography apparatus** Helichromoskop *n* **~ photogravure** Dreifarbentiefdruck *m* **~ printing** Dreifarbendruck *m* **~ process** Dreifarbenverfahren *n* **~ screen** Dreifarbenraster *m* **~ separation** Dreifarbentrennung *f*
**three, ~-colored ribbon** Dreifarbenband *n* **~-column** dreispaltig **~-component alloy** Dreistofflegierung *f* **~-component balance** Dreikomponentenwage *f* **~-component system** Dreistoffsystem *n* **~-condition cable code** trivalenter Kode *m* **~-conductor (lead) cable** dreiadriges Kabel *n* **~-contact system** Dreikontaktsystem *n* **~-control airplane** Dreisteuerflugzeug *n* **~-cornered** drei-eckig, -kantig
**three-cylinder** dreizylinder **~ compressed-air motor** Dreizylinderpreßluftmotor *m* **~ engine** Dreizylindermotor *m*, Drillingsmaschine *f* **~ inverted engine** Dreizylindersternmotor *m*
**three, ~-decade resistance** Dreidekadenstufenwiderstand *m* **~ decimal places** dreistellig **~-digit group** Trigramm *n* **~-digit system** Tausendersystem *n*
**three-dimension** räumliche Fläche *f* **~ direction-finding** Raumpeilung *f* **~ DF station** Raumpeilstation *f* **~ framework** Raumtragwerk *n*
**three-dimensional** dreidimensional, körperlich, räumlich **~ autopilot** Dreiachsensteuerung *f* **~ bracing** Raumverspannung *f* **~ coordinates** räumliches Achsenkreuz *n* **~ cross slide** räumlicher Kreuzschlitten *m* **~ cross-slide system** räumliches Kreuzschlittensystem *n*, Raumschlittensystem *n* **~ design** unregelmäßige Körperformen *pl* **~ flow** räumliche Strömung *f*, Raumströmung *f* (aerodyn.) **~ lattice** Raumgitter *n* **~ object** körperlicher Gegenstand *m* **~ problem** räumliches Problem *n* **~ video picture** plastisches Fernsehbild *n*

**three, ~-draft wire** dreimal gezogener Draht *m* **~-electrode cell** Steuergitterfotozelle *f* **~-electrode tube** Dreipol-, Eingitter-röhre *f*, Triode *f* **~-element control** Dreikomponentenregelung *f* **~-element voltage and current regulator** Dreielementknickregler *m* **~-engine set** Dreimaschinensatz *m* **~-engined dreimotorig **~-engined travelling crane** Dreimotorenlaufkran *m* **~-faced** dreiflächig **~-faced burr** dreikantiger Lochmeißel *m* **~-figure** dreistellig **~-figure exchange** Tausenderamt *n* **~-figure number** dreistellige Zahl *f* **~-figure system** Tausendersystem *n* **~-filament lamp** Dreifadenlampe *f* **~-finger rule** Dreifingerregel *f* **~-flame** dreiflammig **~-flame tube kettle** Dreiflammrohrkessel *m* **~-flamed** dreiflammig **~-floored kiln** Dreihordendarre *f* **~-fold** dreidrähtig, dreifach, dreizählig **~-fold collisions** Dreierstöße *m* **~-furnace process** Triplexverfahren *n* **~-furrow plow** Dreischarpflug *m* **~-gang condenser** Dreifachdrehkondensator *m* **~-gear pump** Drei-räderpumpe *f* **~-girder design** Dreiträgerbauart *f* **~-grid tube** Dreigitterröhre *f* **~-groove drill** dreinutiger Spiralbohrer *m* **~-grooved drill** dreinutiger Spiralbohrer *m* **~-gun turret** Drillingsturm *m* **~-gyro system** Dreikreiselanordnung *f* **~-halves-power law** Emissions-, Entladungs-, Raumladungs-gesetz *n*
**three-high, ~ arrangement** Trioanordnung *f* **~ bar mill** Stabeisentrio *n* **~ bloomer** Blocktrio *n* **~ blooming mill** Blocktrio *n*, Trioblockwalzwerk *n* **~ blooming mill train** Trioblockstraße *f* **~ breaking down rolling mill** Triograbstraße *f* **~ finishing train** Triofertigstraße *f* **~ mill** Trio *n*, Trio-straße *f*, -walzwerk *n* **~ mill arrangement** Triowalzensystem *n* **~ mill stand** Triogerüst *n* **~ plate mill** Blechtrio *n*, Trioblechwalzwerk *n* **~ plate-mill train** Trioblechstraße *f* **~ rail mill** Schienentrio *n* **~ reversing mill** Trioumkehrwalzwerk *n* **~ rolling mill** Doppeltriowalzwerk *n*, Trio *n*, Triowalzwerk *n* **~ rolling mill train** Triostraße *f* **~ rolling stand** Triowalzgerüst *n* **~ rolling train** Triowalzstraße *f* **~ roughing mill** Vorwalztrio *n* **~ roughing mill train** Triovorstraße *f* **~ roughing stand of rolls** Triovorwalzgerüst *n* **~ shape mill** Trioprofileisenwalzwerk *n* **~ shape mill train** Trioprofileisenstraße *f* **~ universal mill** Universaltrio *n* **~ universal mill train** Triouniversalstraße *f* **~ universal rolling mill** Triouniversalwalzwerk *n*
**three, ~-hinged arch** Dreigelenkbogen *m* **~-hole flange** Dreilochflansch *m* **~-jaw chuck** Dreibackenfutter *n* **~-jaw drill chucks** Dreibackenbohrfutter *n* **~-jaw hand operate chuck** Dreibackenfutter *n* mit Handspannung **~-key perforator for cable code** Handlocher *m* für Kabelbetrieb **~-knife trimmer** Dreischneider *m* **~-layer filter** (gas mask) Dreischichteneinsatz *m* **~-leaved twill** dreifädiger Köper *m* **~-leg compasses** Dreispitzzirkel *m* **~-legged** dreischenk(e)lig **~-legged choke coil (or reactor)** Dreischenkeldrossel *f* **~-legged tongs** Dreiarmklemme *f* **~-lens combination (or objective)** dreiteiliges Objektiv *n* **~-limb core** E-Kern *m* **~-lip drill** Dreischneidenbohrer *m* **~-lipped drill** Bohrer *m* mit drei Schneiden *n* **~-lipped twist drills** Spiralsenker *m* **~-mash process** Dreimaischverfahren *n* **~-membered** drei-

glied(e)rig **~-mile limit** Dreiseemeilenzone *f*
**~-momentum** Dreierimpuls *m* **~-motored**
**traveling crane** Dreimotorenlaufkran *m* **~-neck-**
**ed flask** Dreihalskolben *m* **~-needle frame**
(textiles) Dreinadelstuhl *m* **~-nozzle atomizer**
Dreidüsenzerstäuber *m* **~-parameter wave**
**function** dreiparametrige Wellenfunktion *f* **~**
**pass boiler block** Kesselblock *m* mit drei Zügen
**three-phase** dreiphasig **~ A.C.** exciter Dreh-
stromerregermaschine *f* **~ alternating current**
dreiphasiger Wechselstrom *m* **~ alternating**
**motor** Dreiphasenwechselstrommotor *m* **~**
**boost control** Ladedruckregler *m* für drei Regel-
stufen, Ladedruckregler *m* mit drei Stufen **~**
**commutator motor** Drehstromkollektormotor
*m* **~ confluent zone** Drei-Phasen-Kontaktzone
*f* **~ connection** Dreiphasenschaltung *f* (electr.)
**~ current** Drehstrom *m*, Drei(phasen)strom *m*,
Dreiphasen(wechsel)strom *m* **~-current fly-**
**wheel generator** Drehstromschwungradgenera-
tor *m* **~-current generator** Drehstromgenera-
tor *m* **~-current oil transformer** Drehstromöl-
umformer *m* **~-current outputmeter** Drehstrom-
leistungsmesser *m* **~-current plant** Drehstrom-
anlage *f* **~ electricity meter** Drehstromzähler
*m* **~ equilibrium** Dreiphasengleichgewicht *n* **~**
**four-wire system** Drehstromvierleitersystem *n*
**~ full-wave rectifier** Dreiphasen-Vollweg-
gleichrichter *m* **~ furnace** Dreiphasenofen *m* **~**
**furnace current** Drehstromofen *m* **~ generator**
Drehstrom-, Dreiphasen-dynamo *f* **~ induction**
**generator** Asynchrondrehstromgenerator *m* **~**
**inverter** Drehrichter *m* **~ low-voltage plant**
Drehstromniederspannungsanlage *f* **~ mains**
Drehstromnetz *n* **~ motor** Dreiphasenmotor *m*
**~ net-work** Dreiphasennetz *n* **~ network with**
**four wires** Dreiphasenvierleiternetz *n* **~ recti-**
**fier** Dreiphasengleichrichter *m* **~ repulsion**
**motor** Drehstromrepulsionsmotor *m* **~ rever-**
**sing starter** Drehstromreversieranlasser *m* **~**
**sequence indicator** Drehfeldrichtungsanzeiger
*m* **~ series commutator motor** Drehstrom-
reihenschlußmotor *m* **~ series-wound short**
**circuit motor** Drehstromreihenschlußkurz-
schlußmotor *m* **~ squirrel cage motor** Dreh-
stromkurzschlußmotor *m* **~ supply** Drei-
phasenzuführung *f* **~ switch** Drehstromschalter
*m* **~ system** Dreiphasen-, Drehstrom-system *n*
**~ three-wire meter** Drehstromdreileiterzähler *m*
**~ transformer** Drehstromtransformator *m* **~**
**voltage source** Drehstromspannungsquelle *f* **~**
**wattmeter on unbalanced circuit** Drehstrom-
wattmeter *n* bei ungleichbelasteten Stromkrei-
sen **~ welding service** Dreiphasenschweißbe-
trieb *m* **~ winding** Dreiphasenwicklung *f*
**three, ~-phonon process** Dreiphononenprozeß *m*
**~-photon annihilation** Dreiquantenvernichtung
*f* **~-pin central base** Dreistift-Zentralsockel *m*
**~-pin plug** Dreifachstecker *m*, dreipoliger
Normstecker *m*, Drillingsstecker *m* **~-pin**
**safety sockets and plugs** Dreistiftsicherheits-
steckvorrichtungen *pl* **~-piston pressure pump**
doppelte Drillingspreßpumpe *f* **~-piston pump**
Dreikolbenpumpe *f* **~-place number** dreistellige
Zahl *f* **having ~ places** dreisitzig **~-plane**
autopilot Dreiachsensteuerung *f* **~-plunger pres-**
**sure pump** doppelte Drillingspreßpumpe *f* **~-**
**ply** dreischicht, dreischichtig (Sperrholz) **~-ply**

**covered fuselage** Sperrholzrumpf *m* **~-ply twine**
dreischäftiger Bindfaden *m*
**three-point, ~ bearing** Dreipunktlagerung *f* **~**
**(connection)** Dreipunktschaltung *f* **~ contact on**
**the ground** Dreipunktauflage *f* **~ curve drawing**
**recorder** Dreikurvenschreiber *m* **~ jack** dreiteilige Klinke *f* **~ landing** Dreipunktlandung *f*
**~ linkage** Dreipunktaufhängung *f* **~ lock** Drei-
klinkenschloß *n* **~ plug** dreiteiliger Stecker *m*,
dreiteiliger Stöpsel *m* **~ problem** ebener Rückwärtseinschnitt *m*, Rückwärtseinschnitt *m* in
der Ebene **~ problem in space** Rückwärtseinschnitt *m* im Raum **~ resection** Einschneiden *n*
nach drei Punkten **~ space** Dreipunktspatium *n*
**~ spark drawer** Dreispitzenfunkenzieher *m* **~**
**spring suspension** Federung *f* mit Quer- und
Längsfedern **~ suspension** Dreieckaufhängung
*f*, Dreipunktaufhängung *f*
**three, ~-pointed spark gap** Spitzenfunkenstrecke *f* **~-polar (or pole)** dreipolig **~-position**
**key** Hebelschalter *m* mit drei Stellungen
**~-position switch(ing)** Schalten *n* für drei
Stellungen **~ positions relay** Relais *n* mit drei
Stellungen **~-product** Dreigut *n* **~-prong** Drei-
zack *m* **~-purpose tube** Dreifachröhre *f*
**three-quarter, ~ elliptic springs** dreiviertel
elliptische Federn *pl* **~ floating axle** dreiviertel-
fliegende Achse *f* **~ rear view** Dreiviertelrück-
ansicht *f* **~ view** Schrägansicht *f*
**three, ~-rail gravity-roller conveyer** dreigleisige
Rollenbahn *f* **~-rail track** Dreischienengeleise
*n* **~-reel machine** Dreirollenmaschine *f* **~ reel**
**stand** Dreirollendrehstern *m* **~-ring short-**
**circuit armature** Dreinutkurzschlußanker *m*
**~-roll crusher** Dreiwalzenbrecher *m* **~-roll mill**
Dreiwalzenstuhl *m* **~-roll sugar mill** Drei-
walzenzuckermühle *f*
**three-roller, ~ cane mill** Dreiwalzenmühle *f* für
Zuckerrohr **~ compensating device** Drei-
walzenkompensator *m* **~ grinding mill** Drei-
walzenreibemaschine *f* **~ plate-bending ma-**
**chine** Dreiwalzen-blechbiegemaschine *f*, -blech-
rundmaschine *f* **~ refiner for colors** Dreiwalz-
werkfarbenreibemaschine *f*
**three, ~-screw base** Dreifuß *m* (Zwischenstück
zwischen Stativ und Theodolit) **~-seat(ed)**
dreisitzig **~-seater airplane** Dreisitzer *m* (in) **~**
**sections** dreitilig **~-sided** dreiseitig **with ~ spars**
dreiholmig **~-speed backgears** Dreifachvorge-
lege *n* **~-speed transmission** Dreiganggetriebe *n*
**~-square** dreikantig **~ square file** Dreikantfeile
*f* **~-square scraper** dreischneidiger Schaber *m*,
Lagerschaber *m* **~-square tip** Dreikantspitze *f*
**three-stage** dreistufig **~ amplifier** Dreifachver-
stärker *m* **~ automatic boost control** selbsttätige,
dreistufige Ladedruckregelung *f* **~ boost con-**
**trol** Ladedruckregler *m* mit drei Stufen **~**
**heating device** Dreistufenheizung *f* **~ sense**
**amplifier** Dreistufen-Lesesignalverstärker *m*
**three, ~-step cone pully** dreifache Stufenscheibe
*f* **~-step pulley drive** Dreistufenscheiben-
trieb *m* **~-stimulus equation** dreiglied(e)rige
Farbgleichung *f* **~-storied** dreistöckig **~-storied**
**furnace** Dreietagenofen *m* **~-strand cable**
dreischäftiges Kabel *n* **~-strand plated wire**
(armament) drillierter Federdraht *m* **~-strand**
**round rope** dreilitziges Rundseil *n* **~-strand**
**wire rope** dreilitziges Drahtseil *n* **~-stranded**

**rope** dreischäftiges Tau *n* **~-stringed piano** dreihöriges Piano *n* **~-strutter** Dreistieler *m* **~-switch connection** Durchgangsverbindung *f* über drei zwischenstaatliche Leitungen **~-term controller** Dreipunktregler *m* **~-threaded yard** Dreifachgarn *n* **~-throw crankshaft** dreikurbelige Welle *f* **~-throw switch** verschränkte Doppelweiche *f* **~-tone signal** Dreiklangsignal *n* **~ tubes receiver** Dreiröhrenempfänger *m* **~-tuning-bands reception** Eingriffbedienung *f* der drei Abstimmkreise **~ twin window** Fenster *n* mit drei Doppellichtern **~-unit code** Dreieralphabet *n* **~-unit tube** Dreifachröhre *f* **~-valve receiver** Dreiröhrenempfänger *m* **~ valves set** Dreiröhrenempfänger *m* **~-voltmeter method** Dreivoltmeterverfahren *n*

**three-way** Dreiweg... **~ antenna frog** Dreiwegweiche *f* **~ boring and facing mill** Dreiwege-, Bohr- und Plandrehwerk *n* **~ boring mill** Dreiwegebohrwerk *n* **~ (stop)cock** Dreiwegehahn *m* **~ connection** Dreiwegverbindung *f* **~ dumper** Dreiwegkipper *m* **~ flange cock** Dreiwegeflanschenhahn *m* **~ jack** dreiteilige Klinke *f* **~ multiple spindle special drilling machine** Dreiwege-Vielspindelsonderbohrmaschine *f* **~ pipe** Dreiwegestück *n* **~ plug** dreiteiliger Stecker *m* oder Stöpsel *m* **~ series drilling machine** Dreispindelreihenbohrmaschine *f* **~ special drilling machine** Dreiwege-Sonderbohrmaschine *f* **~ switch** Dreiwegeschalter *m* **~ tap** Dreiwegehahn *m* **~ trim switch** Dreiachsentrimmschalter *m* (electr.) **~-type spider** dreiarmiger Getriebestern *m* **~ valve** Dreiwegventil *n*

**three, ~-wheel Diesel motor roller** Dieselmotordreiradwalze *f* **~-wheel loadcarrying truck** dreirädriger Transportwagen *m* **~-wheel superheated steam roller** Heißdampfdreiradwalze *f* **~-wheeled undercarriage** Dreiradfahrgestell *n* **~-winding transformer** Ausgleichstransformator *m* **~ wing bit** Dreiflügelmeißel *m*

**three-wire** dreiadrig **~ automatic telephone system** Dreiselbstanschlußsystem *n* **~ conductor** dreiadriges Kabel *n* **~ cord** dreiadrige Schnur *f* **~ generator** Dreileiterdynamo *f* **~ junction** dreiadrige Verbindungsleitung *f* **~-method** Dreifaden-, Dreinahtmeß-methode *f* **~ supply system** Dreileiternetz *n* **~ system** Ausgleichsstromanzeiger *m*, Drehstromsystem *n*, Dreidrahtsystem *n* **~ system with direct or alternating current** Dreileitersystem *n* **~ three-phase A.C.** Dreileiterdrehstrom *m* **~ trunk** dreiadrige Verbindungsleitung *f*

**thresh, to ~** dreschen, entkörnen

**thresher** Dreschmaschine *f*

**threshing, ~ drum** Dreschtrommel *f* **~ floor** Dreschtenne *f*, Scheune *f*

**threshold** Ansprechwert *m*, Schwelle *f*, (gyro) Schwellwert *m* **~ of acoustic perception** Schallempfindungsschwelle *f* **~ of audibility** Hörschwelle *f*, Schallempfindungsschwelle *f* **below ~** unterschwellig **~ of difference** Unterschiedsschwelle *f* **~ of energy** Energieschwelle *f* **~ of feeling** Schmerz-grenze *f*, schwelle *f* **~ of hearing** Hörschwelle *f* **~ of intensity of a color** Farbschwellenwert *m* **~ of luminescence** Lumineszenzschwelle *f* **~ of perception** Reizschwelle *f* **~ of sensation** Reizschwelle *f* **~ of time-sense** Zeitschwelle *f* **~ of visibility** Sichtbar-

keitsschwelle *f* **~ of vision** Sichtbarkeitsschwelle *f*

**threshold, ~ audiogram** Audiogramm *n* **~ beam** Schwelle *f* **~ bias** Verzögerungsspannung *f* **~ condition** Grenzbedingung *f* **~ contrast bar** Schwellenkontrastbalken *m* **~ curve** Schwellenkurve *f* **~ discriminator** Schwellwertdiskriminator *m* **~ field** Schwellenfeld *n* **~ field curve** Schwellenwertkurve *f* **~ frequency** Schwellenfrequenz *f* **~ law for single ionization** Schwellengesetz *n* für Einfachionisation **~ lighting** Schwellenbefeuerung *f* **~ lights** Schwellenfeuer *pl* **~ limit (or sensitivity)** Schwellenwert *m* **~ relation for streamer discharge** Zündbedingung *f* für Entladung **~ sensivity** Empfindlichkeitsschwellenwert *m* **~ setting** Schwellenfestigung *f* **~-sound intensity** Schallschwellenstärke *f* **~ tube** Schwellenröhre *f* **~ value** Grenzwert *m*, Schwell(en)-reiz *m*, -wert *m* **~ voltage** Ansprechspannung *f*, Schwellenspannung *f* **~ wave length** Grenzwellenlänge *f*

**throat** (welding) Bruchdicke *f*, (carp.) Dünnung *f*, Düsenhals *m* (g/m), (of container or pipe) engste Einlaufstelle *f* Eintrittsöffnung *f*, (of furnace) Gicht *f*, Gurgel *f*, Hals *m*, Kehle *f*, Stärke *f* der Schweißnaht

**throat, ~ of blast furnace** Hochofenkranz *m* **~ of a hook** Hakenkehle *f* **~ of horn** (narrow inlet) Lautsprechertrichtermundstück *n*, (loudspeaker) Trichterhals *m*, Trichtermundstück *n* **~ of loud-speaker horn** Lautsprechertrichterhals *m* **~ of a pipe** Einschnürung *f* eines Rohres **~ of a shaft** Hals *m* einer Welle **~ on threading die** Anschnitt *m* einer Gewindeschneidbacke **~ of the tunnel** Düse *f* des Windkanals **~ of welded seam** Dicke *f* der Schweißnaht

**throat, ~ area** Halsfläche *f* **~ depth** Ausbuchtung *f*, Kehlnahttiefe *f* (of weld), Maultiefe *f* (mech.) **~ gas** Gichtgas *n* **~ halyard** Gaffelfall *n* (harness) Kehlriemen *m* **~ microphone (or mike)** Kehlkopfmikrofon *n* **~ phone** Kehlkopftelefon *n* **~ seam** Kehlnaht *f* **~ seizing** Herzbindsel *n* **~ stopper** Gichtverschluß *m* **~ stopper for blast furnaces** Hochofengichtverschluß *m* **~-stopper winch** Gichtglockenwinde *f* **~ temperature** Gichttemperatur *f*

**throating** Vorsprung *m* mit Wassernase

**throb, to ~** atmen

**throb** Bewegung *f*, Klopfen *n*, Pochen *n*, Pulsation *f*, Stoßton *m*

**throbbing** pulsierend

**throttle, to ~** (motor) abdrosseln, drosseln, erdrosseln **to ~ back** Drossel *f* schließen (drosseln), Gas *n* zurücknehmen **to ~ down** abdrosseln, (gas) absperren **to ~ down the engine** Gas *n* zurücknehmen

**throttle** Ansaugrohr *n*, Drossel *f*, (gas) Gashebel *m* **~ adjusting screw** Leerlaufbegrenzungsschraube *f*, Regelschraube *f* **~ body** Drosselkörper *m* **~ box** Gas-, Motor-hebelbock *m* **~ bush** Drosselbüchse *f* **~ chamber** Drosselkammer *f* **~-clack valve** Drosselabsperrklappe *f*, Gashebel *m* (Kolbenmotor), Leistungshebel *m* **~ control** Drosselsteuerung *f*, Kraftstoffregler *m* **~ control rod** Gasgestänge *n* **~ cover** Drosselüberdeckung *f* **~ crank** Kurbelschwinge *f* **~ effect** (on) Drosselwirkung *f* **~ flange** Meßblende *f* **~ flap** Drosselklappe *f* **~ frame**

Drosselstuhl *m* ~ **frame tenter** Drosselspinner *m* ~ **gate** Kulisse *f* des Gashebels ~ **governing** Drosselregulierung *f* ~ **grip** Drosselgriff *m* ~ **hand-lever tube** Gasspindel *f* des Gashebels ~ **handle** Drosselgriff *m* ~ **lever** Drossel-hebel *m*, -klappenheber *m*, Gas-, Regler-, Leistungshebel *m* ~ **lever for normal flight** Normalgashebel *m* ~ **nozzle** Drosseldüse *f* ~ **opening** Drosseleinlaß *m* ~ **quadrant** Gashebelbock *m*, Gashebelführung *f* ~ **response** Ansprechen *n* (des Motors) auf Verstellung der Drosselklappe ~ **response of motor** Beschleunigungsverhalten *n* des Motors ~ **shaft** Drosselklappenwelle *f*, -spindel *f* ~ **spindle** Drosselspindel *f* ~ **stop** Drosselanschlag *m* ~ **stop horsepower** Motorleistung *f* bei Anschlagstellung der Drossel ~ **stop screw** Leerlaufbegrenzungsschraube *f* ~ **valve** Drossel-klappe *f*, -ventil *n*, Reglerventil *n* ~**-valve spindle** Drosselklappenachse *f* ~ **wide open** mit Vollgas *n*
**throttleable** drosselbar
**throttled** abgedrosselt ~ **passage** eingeengter Kanal *m* ~ **steam** gedrosselter Dampf *m*
**throttling** Abdrosselung *f*, Drosselung *f* ~ **action** Drosselwirkung *f* ~ **calorimeter** Drosselkalorimeter *m* ~ **experiment** Überleitungsversuch *m* ~ **member** Drosselorgan *n* ~ **valve** Sparventil *n*
**through** dadurch, durch, mittels **(cut)** ~ durchgeschaltet (teleph.) ~ **the intermediary of** unter Zwischenschaltung *f* von
**through,** ~**-and-through coal** Förderkohle *f* ~ **bolt** durchgehender Bolzen *m*, Zuganker *m* ~ **bore** Durchgangsloch *n* ~**-call** Durchgangsgespräch *n* ~**-capacitance** Durchgriffskapazität *f* ~ **carved** (textiles) durchbrochen gearbeitet ~ **circuit** Durchgangs(fern)leitung *f*, durchgehende Leitung *f* ~ **clearing** Trennung *f* zu Gunsten des Fernverkehrs ~ **connection** Durchverbindung *f* ~ **connection via ...** Gespräch *n* über ... ~**-dialing** Durchwählen *n* ~**-dialing over toll (or trunk) circuits** Fernwahl *f* über ein Durchgangsamt ~**-dialing over toll (or trunk) circuits with multifrequency impulses** Mehrfrequenzfernwahl *f* ~**-going shaft** durchgehende Welle *f* ~ **hardening** Durchhärtung *f* ~ **hole** Durchgangsbohrung *f*, Durchgangsloch *n* ~ **hole guide pilot** Durchgangslochzapfen *m* ~ **line** Durchgangsleitung *f*, durchgehendes Gleis *n*, durchgehende Leitung *f* ~**-line repeater** (permanently connected in circuit) eingeschalteter Verstärker *m*, (fester) Zwischenverstärker *m* (teleph.) ~ **loading** Durchfracht *f* ~ **message** Übermittlungsspruch *m* ~**-out groove** Ausschaltrille *f* (phono) ~ **perforated tape** durchgelochter Lochstreifen *m* ~**-pin** Durchsteckstift *m* ~ **position** Durchgangsplatz *m*, Durchsprechstellung *f* (teleph.) ~**-position cord pair** Durchgangsverbindungs-Schnurpaar *n* ~**-put capacity** Durchsatzleistung *f* ~**-put of a machine (or unit)** Durchsatz *m* ~**-put rating** Durchgangsleistung *f* ~ **rate** Gesamtgebühr *f* ~ **ringing** Durchrufen *n* ~**-run filter** Durclauffilter *n* ~ **station** Durchgangsbahnhof *m* ~ **switchboard** Durchgangs(fern)schrank *m* ~ **switching board** Ferndurchgangsschrank *m* ~**-switching exchange** Durchgangsanstalt *f* ~ **switching position** Ferndurchgangsplatz *m* ~ **telephone call** Durchgangsfernspruch *m* ~ **toll line**

Durchgangsfernleitung *f* ~ **track** Durchlaufgleis *n* ~ **traffic** Durchgangsverkehr *m* ~ **train** Durchgangszug *m*, durchgehender Zug *m* ~**-type oven** Durchgangsofen *m* ~ **vulcanization** Durchvulkanisation *f* ~**-way valve** Durchgangsventil *n* ~ **wire** durchgehender Draht *m*, Durchgangsdraht *m*
**throw, to** ~ (silk) mulinieren, schleudern, stürzen, (a key) umlegen, veredeln, werfen **to** ~ **away** wegwerfen **to** ~ **back** abweisen, zurückwerfen **to** ~ **off the belt** den Riemen *m* ausrükken **to** ~ **a bridge** eine Brücke *f* schlagen **to** ~ **out the clutch** abkuppeln **to** ~ **down** herunterwerfen, hinwerfen **to** ~ **the driving pulley out of gear** die Antriebsscheibe *f* ausrücken **to** ~ **over the dump** auf Halde *f* stürzen **to** ~ **in gear** (an engine) anwerfen, einkuppeln, in Gang setzen, in Wirksamkeit *f* setzen **to** ~ **out of gear** auslösen, ausrücken **to** ~ **in** anhängen, einklinken, einrücken, einstürzen **to** ~ **into** einblasen **to** ~ **off** kippen **to** ~ **on** einschalten **to** ~ **(a line) onto** an die Leitung *f* anlegen **to** ~ **out** ausklinken, (Getriebe) ausrücken, ausschalten, ausschlagen, ausstoßen, auswerfen, herausschleudern **to** ~ **over** umschalten **to** ~ **overboard** über Bord *n* werfen **to** ~ **a switch** einen Schalter *m* schließen, einen Schalter *m* umlegen **to** ~ **up** aufwerfen, schütten **to** ~ **into vibration** in Schwingung versetzen
**throw** Ausschlag *m*, Galvanometerausschlag *m*, Hub *m*, Hub-höhe *f*, -länge *f*, Kröpfung *f*, Schub *m*, Verwerfung *f*, Vorschub *m*, Wurf *m* ~ **of the fault** Sprunghöhe *f* ~ **of governor** Reglerausschlag *m*
**throw,** ~ **crankshaft** Kurbelwellenkröpfung *f* ~ **key** Schaltschlüssel *m*
**throw-off** Abstellung *f* (electr.), Druckabsteller *m* ~ **(regulating) lever** Abstellhebel *m*
**throw-out** Ausschaltung *f*, Auswerfer *m*, Faltblatt *n*, Fehlfuß *m*, Umschaltklinke *f* ~ **antenna** Wurfantenne *f* ~ **bearing sleeve** Ausrückmuffe *f* ~ **cam** Entkupplungskurve *f* ~ **lever** Ausschaltehebel *m* ~ **mechanism** Schaltstück *n* ~ **sleeve** (of gear clutch) Ausrückbüchse *f* ~ **spiral** Auslauffrille *f* (phono)
**throw-over,** ~ **switch** Umschalter *m* ~ **switch with break** Umschalter *m* mit Unterbrechung ~ **switchboard** Umschalter *m* mit zwei Stellungen
**throw switch** Stellungsschalter *m*
**thrower** Spritzring *m*, Wurfförderer *m*
**throwing** Ausschlagen *n*, (of a gear) Laufeinschaltung *f*, (of a switch) Umlegen *n*, Vered(e)lung *f*, Werfen *n* ~ **of points** Weichenstellung *f*
**throwing,** ~**-in** Einschalten *n* ~**-out** Auswurf *m* ~**-out of gear** Ausrückung *f*, Entkupplung *f* ~**-out of sparks** Funkenwurf *m* ~ **power** Streufähigkeit *f*, Streuung *f* ~ **pull of the cable** Kabelzugentlastung *f* ~ **range** Wurfweite *f* ~ **wheel** Töpferscheibe *f*
**thrown** geworfen, (down) gesunken ~ **in** eingestreut ~ **into** eingeschüttet ~ **off** abgesetzt ~ **up** aufgeworfen
**thrown silk** mulinierte Seide *f*
**throwster** Verarbeiter *m*, Vered(e)lung *f*
**thrummed mat** Serving *f*
**thrust, to** ~ drücken, schieben, stoßen **to** ~ **after** nachstoßen **to** ~ **in** einarbeiten

**thrust** Auflagerdruck *m*, Ausschlag *m*, Axial-druck *m*, axialer Druck *m*, Bodenstoß *m*, Druck *m*, Druckkraft *f*, Erddruck *m*, Längs-druck *m*, Schiebkraft *f*, Schub *m*, Schublei-stung *f*, Stich *m*, Stoß *m*, Vorschub *m*, Vorstoß *m*, Widerlager *n*, Widerlagerdruck *m*, Zug *m* **(propeller)** ~ Schraubenzug *m* (aviat.), Ver-werfung *f* (geol.) ~ **of an arch** Seitenschub *m* eines Bogens ~ **of arch** Druckstoß *m* des Bogens **thrust,** ~ **acceleration** Schubbeschleunigung *f* ~ **augmentation** Schuberhöhung *f* (im Fluge) ~ **augmenter** Schub-vergrößerer *m*, -vermehrer *m*, -(kraft)verstärker *m* ~ **ball bearing** Druck-kugel-, Scheibenrillen-lager *n* ~ **bearing** Axial-(druck)lager *n*, Drucklager *n*, Hochschulter-lager *n*, (concealed-collar) Kammlager *n*, Längslager *n*, Schublager *n*, Spur(zapfen)lager *n*, Tragstützlager *n* ~ **bearing shell** Paßlager-schale *f* ~ **blade switch** Schubtrennschalter *m* ~ **block** Drucklager *n*, Kammlager *n*, Quer-stück *n*, Stützwinkel *m* ~ **block seating** Block-lagersitz *m* ~ **bolt** Druck-bolzen *m*, -schraube *f* ~ **borer** Erdbohrer *m* mit Stoßbewegung, Stoßbohrer *m* ~ **brake** Schubumkehr *f* ~ **cham-ber** Brennkammer *f* ~ **class** Schubklasse *f* (aviat.) ~ **coefficient** Schraubenschubzahl *f*, Schub-beiwert *m*, -belastungsgrad *m* ~ **collar** Druckring *m* ~ **cone** Schubkonus *m* ~ **curve** Schubverlauf *m* ~ **decrease at cutoff** Abschalt-abfall *m* (g/m) ~ **energy** Vortriebsarbeit *f* ~ **fault** (petroleum) abnorme Verwerfung *f* ~ **flange yoke** Druckflansch *m* ~ **frame** Schub-gerüst *n* ~ **frontal area ratio (lb/sq. ft.)** Stirn-flächenschub *m* (kp/m²) ~ **gain** Schubrückge-winn *m* ~ **gauge** Druckkraftmesser *m* ~ **genera-tor** Druckerzeuger *m* ~ **grating** Schubrost *m* ~ **horsepower** effektive Propellerleistung *f*, Stoß-pferdekraft *f*, Zugpferdekraft *f* ~ **journal** Druckzapfen *m* ~ **line** Angriffsrichtung *f* des Luftschraubenschubes, Anliegestrich *m*, Luft-schraubenachse *f* ~ **load** Achsial-, Längs-druck *m*, Schubbelastung *f* ~ **loss** Schubverlust *m* ~**-measuring apparatus** Druckkraftmesser *m* ~**-measuring stand** Schubmeßwaage *f* ~ **motion** Schubbewegung *f* ~ **needle** (jet) Schubdüsen-nadel *f* ~ **nozzle** Rückstoß-, Schub-düse *f* ~ **nozzle fairing (or casing)** Schubdüsenmantel *m* ~ **nut** Druckmutter *f* ~ **pawl** Stoßklinke *f* ~ **peak** Schubspitze *f* ~ **performance** Schublei-stung *f* ~ **piece** Druckstück *n* ~ **pin** Feder-stütze *f*, Gleitbolzen *m* ~ **piston** Schubkolben *m* ~ **plane** Druckebene *f*, Schurf-, Über-schiebungs-fläche *f* ~ **plate** (clutch) Kupplungs-scheibe *f* ~ **point** Schwerpunkt *m* **(propeller)** ~ **power** Schubleistung *f* der Luftschraube ~ **race** Drucklager *n* ~ **recorder** Schubmeßgerät *n* ~ **resistance** Zerknickungsfestigkeit *f* ~ **rever-sal** Schubumkehr *f* ~ **reverser** Schubumkehr-vorrichtung *f* ~ **ring** Schulaufnahmering *m*, Stützring *m* ~ **ring structure** Gerüstring *m* ~ **rod** Schubstange *f* ~ **screw** Preßschraube *f* ~ **shoe** Drucklagerbügel *m* ~ **side** Druckseite *f* (beim Kolben) ~ **spring** (clutch) Schubauf-nahmefeder *f* ~ **strength** Schubstärke *f* ~ **termination** Schubbeendigung *f* ~ **throttling** Schubdrosselung *f* ~ **trunnion projectile** Treib-zapfengeschoß *n* ~ **value** Schubwert *m* ~ **vector** Schubvektor *m* ~ **washer** Druckscheibe

*f*, Stoß-ring *m*, -scheibe *f* ~ **washers** Gegen-druckunterlagsscheiben *pl* ~ **weapon** Stich-waffe *f* ~ **weight ratio** (aircraft) reziproke Schubbelastung *f*, (engin.) reziprokes Schub-gewicht *n*

**thruster** patronenbetätigter Druckgeber *m*
**thrusting** Drücken *n* ~ **power** Schiebkraft *f*
**thrusts, by** ~ schubweise
**thruway branch** Umgehungsstraße *f*
**thulium** Thulium *n*
**thump, to** ~ aufschlagen, prellen
**thump** Daumen *m*, (of a key or contact) Schlag *m*, dumpfer Schlag *m*, Simultangeräusch *n*, (Morse) Telegrafiergeräusch *n* ~ **bolt** Ohren-schraube *f* ~ **head** Lappenkopf *m* ~ **index perforating machine** Registrierloch-Perforier-maschine *f* ~ **jump** Daumensprung *f* ~ **latch** Daumenfalle *f* ~ **lever** Knebel *m* ~ **nut** Daumen-mutter *f*, (sight leaf) Drücker *m*, Flügelmutter *f* ~ **plate** Daumenplatte *f* ~ **rule** Daumenregel *f*, Dreifingerregel *f* ~ **screw** Daumen-, Druck-, Flügel-, Klemm-, Kordelkopf-, Kordel-, Lap-penschraube *f*, Ohrenschraube *f* ~ **screw sleeve** Stellschraubenhülse *f* ~ **shaft** Daumenwelle *f* ~ **tack** Heftzwecke *f*, Reiß-nagel *m*, -zwecke *f* ~ **wheel** Daumenrad *n*, Einstellknopf *m*
**thumbs,** ~ **width** Daumenbreite *f* ~ **width method of range determination** Daumensprung *m*
**thunder, to** ~ donnern, tosen
**thunder** Donner *m*, Donnerschlag *m* ~ **bolt** Blitzstrahl *m*, Donnerkeil *m* ~ **clap** Donner *m* ~ **clouds** Gewitterwolken *pl* ~ **effect** Donner-effekt *m* ~ **shower** Gewitterregen *m* ~ **squall** Gewitterbö *f* ~ **stone** Belemnit *m*
**thunderstorm** Gewitter *n*, Gewittersturm *m* ~ **DF center** Gewitterpeilzentrale *f* ~ **light** Ge-witterleuchte *f* ~ **sack** Gewittersack *m* ~ **vortex (or whirl)** Gewitterwirbel *m*
**Thury thread** Thuryuhrschraubengewinde *n*
**thwart** Ducht *f* ~**-pole force** Querschiffskraft *f*
**thwartship,** ~ **convex arch of deck** Bucht *f* (a) ~ **iron mass** Querschiffseisenmasse *f* ~ **magnet** Schiffsmagnet *m* (a) ~ **pole** Querschiffspol *m*
**thwartships** querschiffs
**thymic acid** Thymolsäure *f*
**thymol iodide** Thymoljodid *n*
**thymolphthalein** Thymolphthalein *n*
**thymotrol panel** Elektronensteuerung *f*
**thyration** Thyration *f*
**thyratron** Gasentladungsgefäß *n*, gittergesteuerte Gasentladungsröhre *f* oder Ionenröhre *f*, Glühkathodenstromrichter *m*, Ionensteuerrohr *n*, Stromtor *n*, Thyratron *n*, Umformer *m*, Wandstromverstärker *m*
**tick, to** ~ **over** auf dem Flugplatz *m* rollen (aviat.)
**tick impulse** Teilungsstufe *f* (Uhrwerk)
**ticker** intermittierender Kontakt *m*, Schnell-unterbrecher *m*, Taktgeber *m*, Ticker *m* ~ **apparatus** Tickerempfänger *m*
**ticket** Billet *n*, Einlaßkarte *f*, Fahr-karte *f*, -schein *m*, Gesprächsblatt *f*, (toll) Gesprächs-zettel *m*, Karte *f*, Preiszettel *m*, Scheck *m*, Zettel *m* ~ **for incoming call** Ankunftsblatt *n* ~ **for a reserved seat** Platzkarte *f*
**ticket,** ~**-distributing system** Fördereinrichtung *f* ~ **distribution position** Leitstelle *f*, Zettelver-teiler(stelle) *f* ~ **nippers** Karten-locher *m*,

-lochzange *f*, Knipszange *f* ~ **office** Karten-ausgabe *f* ~ **sailing device** Treibfahne *f* (Zettel-rohrpost) ~ **time** Gesprächs-minuten *pl*, -zeit *f* ~ **window** Fahrkartenschalter *m*, Schalter *m*, Zahlschalter *m*

**ticketing** Erzkauf *m*

**ticking** Drell *m*, Drillich *m*, Ticken *n* (Uhrwerk) ~ **frequency** Tickfrequenz *f* (Uhrwerk)

**tickler** (textiles) Decker *m*, Dipper *m*, Mahnzettel *m*, Rückkupplungsspule *f* ~ **coil** (regeneration) Drossel *f*, Rückkupplungsspule *f*, Rückkupp-lungsspule *f* des Schwingaudions, Stromstärke-regler *m* ~ **machine** (textiles) Deckmaschine *f*

**tidal** die Gezeiten *pl* betreffend ~ **air** Luftvor-lagerung *f* ~ **barrage** Flutsperre *f*, Gezeitenstau-damm *m* ~ **basin** Flut-bassin *n*, -becken *n*, -dock *n*, Tidebecken *n* ~ **basin that dries out at low tide** Strandbecken *n* ~ **component** Gezeiten-beschleunigung *f* ~ **computing machine** Ge-zeitenrechenmaschine *f* ~ **corrosion test** Ebbe-flut-Korrosionsversuch *m* ~ **current** Gezeiten-strom *m*, Tideströmung *f* ~ **currents** Gezeiten-strömung *f* ~ **dock** Flutdock *n* ~ **effect** Anstau *m* ~ **flow** Flut *f* ~ **forces** Gezeitenkräfte *pl* ~ **friction** Gezeitenreibung *f* ~ **gravimeter** Ge-zeitengravimeter *n* ~ **harbor** Fluthafen *m* ~ **impulse** Gezeiten-, Tide-bewegung *f* ~ **light** Gezeitenfeuer *n* ~ **mud deposits** Gezeitenabla-gerung *f* ~ **phenomena** Gezeitenerscheinungen *pl* ~ **power plant** Gezeitenkraftwerk *n* ~ **quay** Landungskaje *f* ~ **range** Tidenhub *m* ~ **river** Gezeitenfluß *m* ~ **signal** Wasserstandszeichen *n* ~ **stream** Tideströmung *f* ~ **territory** Ebbe-und Flutgebiet *n* ~ **test stand** Ebbeflut-Prüf-stand *m* ~ **wave** Gezeitenwelle *f*

**tide** Ebbe *f* und Flut *f*, Gezeite(n) *pl*, Tide(n) *pl* ~ **advancing front of the incoming tide** Flutwelle *f* ~-**calculating machine** Gezeitenrechenma-schine *f* ~ **constant** Gezeitenkonstante *f* ~ **gate** Obertor *m* ~ **gauge** Ebbe- und Flutmesser *m*, Flutmesser *m*, Flutuhr *f*, Pegel *m* ~ **head** Flut-grenze *f* ~ **level** (due to storm) Wasserstand *m* von Sturmfluten ~ **lift** Gezeitenfeuer *n*, Tide-hub *m* ~ **lock** Flutschleuse *f* ~ **measurement** Gezeitenvermessung *f* ~ **producing force** ge-zeitenerzeugende Kraft *f* ~ **signal** Gezeiten-signal *n* ~ **table** Gezeitentafel *f*, Wasserstands-tabelle *f* ~ **wave of the incoming tide** Flutwelle *f*

**tidiness** Sauberkeit *f*

**tidy** reinlich, sauber, wohlgeordnet

**tie, to** ~ anstecken, binden, knüpfen, verankern, verknüpfen, zurren **to** ~ **down an aircraft** ein Flugzeug *n* verankern **to** ~ **around** umbinden **to** ~ **to a cable** (or **with cable**) anseilen **to** ~ **in** einbinden **to** ~ **off** abbinden, abschnüren **to** ~ **with ribbon** (or **tape**) bandagieren **to** ~ **with a string** nesteln **to** ~ **to** anbinden, anknüpfen an etwas **to** ~ **together** verbinden, zusammen-schnüren **to** ~ **up** binden, fesseln, festbinden, festklemmen, unnütz belegen, zuschnüren

**tie** Band *n*, Bindedraht *m*, Bindeglied *n*, Boden-schwelle *f*, Bügel *m*, (rail) Bund *m*, Eisenbahn-schwelle *f*, Gleisbalken *m*, gewöhnlicher Draht-bund *n* (teleph.), Haft *f*, Halter *m*, Riegel *m*, Schlinge *f*, Schwelle *f* (r.r.), Verband *m*, Ver-bindungsstück *n* ~ **anchor** Gewölbeanker *m* ~ **band** Fitzband *n* ~ **bar** Anker-bolzen *m*, -stan-ge *f*, Riegel *m*, Spannschloß *n*, Spurstange *f*,

Zuganker *m* ~ **beam** Balkenzug *m*, Dachbin-derbalken *m*, Eckstichbalken *m*, Eisenbahn-schwelle *f*, Gebindesparren *m*, Riegel *m*, Spann-balken *m*, -riegel *m*, Tramen *m*, Verankerung *f*, Verankerungsbalken *m*, Zuganker *m* ~ **beam of a truss frame** Spannriegel *m* eines Hänge-werkes ~ **beam plate** Stoßlasche *f* (des Spann-riegels) ~ **bolt** Anker-, Verbindungs-bolzen *m* ~ **brace** Zugdiagonale *f* ~ **cable** Querverbin-dungskabel *n* ~-**down loop** Ankeröse *f* ~ **en-velope** Schnurbandumschlag *m* ~ **hook** Band-haken *m*, Hakenband *n* ~ **line** Abzweigung *f*, Querverbindung(sleitung) *f*, Stichleitung *f*, (power engine) Verbundleitung *f* ~ **lines** (used in physicalchemical diagrams) Konoden *pl* ~ **lug** (in wiring) Stützpunkt *m* ~-**over credit** Überbrückungskredit *m* ~ **pass** Schwellenkali-ber *n* ~ **piece** Ligatur *f*, Verbindung *f* ~ **piece between uprights** Oberbalken *m* ~ **pile** (pile subjected to tension or pull) Zugpfahl *m* ~ **plate** Anker-, Unterlags-platte *f* ~ **plate rotor** Anker *m* ~ **rod** Anker *m*, Anker-bolzen *m*, -stange *f*, Gestängeverbindung *f*, Haltestrebe *f*, Kuppelstange *f*, Riegel *m*, Spann-schloß *n*, -stange *f*, Spurhalter *m*, Verbindungs-stange *f*, -stück *n*, Zuganker *m*, Zugband *n*, Zugstange *f* ~-**rod material** Zugbandstoff *m* ~ **rods** Lenk-gestänge *n* ~ **section** Schwellenprofil *n* ~ **tamper** Schwellenstampfer *m*

**tie-up**, ~ **for weaving names** Kreuzungsstelle *f*, Monogrammschnürung *f* ~ **basin** Ausweich-stelle *f* ~ **resist** Unterbindereserve *f* ~ **wharf** An-legestelle *f*, Wartestelle *f*

**tie,** ~ **wall** Ankerwall *m* ~ **wire** Bandagendraht *m*, Bindedraht *m*

**tied** (in bundles) gestaucht, unfrei ~ **off** abge-schnürt ~ **up** blockiert, lahmgelegt, stillstehend, in Unordnung *f*

**tier of blocks** Blockreihe *f*

**tier,** ~-**construction platform** Ladegestell *n* mit Etagenaufsatz ~ **frame** (or **stand**) Etagengestell *n* ~-**up** (textiles) Ausschnürer *m*

**tiering,** ~ **attachment** Hubvorrichtung *f* ~ **truck** Hochhubwagen *m*, Hubwagen *m* mit Hebevor-richtung

**tiffany** Flor *m*

**tight** dicht, druckdicht, eng, fest, gespannt, klamm, prall, straff ~ **alignment** Abgleich *m* der Durchlaßkurve, Frequenzkurventrimmung *f* ~-**and-loose-pulley** Antrieb-Los- und Fest-scheibe *f* ~-**binding approximation** Näherung *f* bei fester Bindung ~ **cable** fest verseiltes Kabel *n* ~-**core cable** Kabel *n* mit enger Umwicke-lung ~ **coupling** feste Kopplung *f*, Festkupp-lung *f* ~ **engine** Motor *m* durch verklebtes Öl schwer durchdrehbar ~ **fit** Fest-, Haft-, Treib-sitz *m* ~-**fit screw** Paßschraube *f* ~-**fitting** gut-schließend ~-**fitting baffle** dicht anschließendes, gut passendes Leitblech *n* ~ **framing** enge Um-rahmung *f* ~-**making iron** Dichteisen *n* ~ **pulley** Festscheibe *f* ~ **pulley drive** Festscheibenantrieb *m* ~ **sand** schwer durchlässiger Sand *m* ~ **shooting** Knappschießen *n* ~ **spiral** enge steile Spirale *f* ~-**spun** (rope) festgeschlagen ~ **strand** ziehendes Trumm *n*, Zugseite *f*, Zugtrumm *n* ~ **strand of belt** gespanntes Riementrumm *n* ~ **strand of chain** gezogener Kettenstrang *m* ~-**weld** Dichtnaht *f*

**tighten, to** ~ andrehen, anspannen, anziehen, befestigen, dichten, dicht machen, enger machen, fest anziehen, festschrauben, spannen, straffen, straff spannen, zusammenziehen, zuziehen **to ~ a bolt** festziehen **to ~ a cable** steif holen **to ~ cotter** den Keil *m* anziehen **to ~ a screw** eine Schraube *f* nachziehen **to ~ the screw** Schraube *f* (fest) anziehen **to ~ a spring** eine Feder *f* anspannen **to ~ up** geradelaufen, nachziehen **to ~ in the width** breitstrecken
**tightened** abgedichtet, festgezogen
**tightener** (belt) Riemenspanner *m*, (pulley) Spannrolle *f*, (stay) Spannschloß *n*
**tightening** Anpressung *f*, Anziehen *n*, Befestigung *f* ~ **of a strike of sugar** strammes Verkochen *n* eines Sudes ~ **up the wire** Nachspannen *n* der Leitung
**tightening,** ~ **belt** Spannmutter *f* ~ **device** Streckvorrichtung *f* ~ **disc** Spannscheibe *f* ~ **eyebolt** Ösenspannschraube *f* ~ **handle** Spanngriff *m* ~ **key** Schlüsselkeil *m*, Spannschlüssel *m* ~ **layer** (gas, water) Dichtungsschicht *f* ~ **lever** Spannbügel *m* ~ **nut** Spannmutter *f* ~ **ring** Spannring *m* ~ **screw** Befestigungs-, Spann-, Zuspannschraube *f* ~ **screws** Anziehschraube *pl* ~ **sheet** Abdichtungsblech *n* ~**-spot yield stress** Spannung *f* (Getriebe) ~ **spring** Anzugsfeder *f* ~ **toggle** Aufziehstift *m* ~ **torque** Anzugs-, Festdreh-, Festzieh-moment *n* ~ **wedge** Kreuzkopfkeil *m*, Stellkegel *m*
**tightly,** ~ **coupled** festgekoppelt ~ **joined** fugensicher ~ **sealed** gutschließend ~ **woven** dichtgewebt
**tightness** Abdichtung *f*, Dichte *f*, Dichtheit *f*, Dichtigkeit *f*, Enge *f*, Festigkeit *f* ~ **of money** Geldnot *f*
**tile** (roofing) Dachziegel *m*, Fliese *f*, Formstück *n* (Tonform), Kachel *f*, Steingutformstück *n*, Tonformstück *n*, Ziegel *m*, Ziegelstein *m* ~ **burner** Ziegelbrenner *m* ~ **burning** Ziegel-brand *m*, -brennen *n* ~ **clay** Ziegelton *m* ~ **creasing** Fliesenabdeckung *f* ~ **floor(ing)** Fliesenfußboden *m* ~ **furnace** Schüttofen *m* ~ **hearth** Kachelherd *m* ~ **jacket** (cable) Kachelauskleidung *f* ~ **kiln** Ziegelhütte *f* ~ **layers' work** Fliesenlegerarbeiten *pl* ~ **maker** Ziegelbrenner *m* ~ **ore** Kupferbraun *n*, Kuprit *m*, Ziegelerz *n* ~ **paving** Fliesenpflaster *n*, Plattenwerk *n* ~ **pin** Mauernagel *m* ~ **work** Ziegelmauerwerk *n*
**tiled roof** Ziegeldach *n*
**tiler** Ziegeldecker *m*
**till, to** ~ ackern, (land) bewirtschaften
**till** Grundmoräne *f*
**tillage cutter** Ackerfräser *m*
**tilled soil** Obergrund *m*
**tiller** Bodenfräse *f*, Handgriff *m*, Lenkstange *f*, Ruderpinne *f*
**tilling** Kultur *f*
**tilt, to** ~ ankippen, ecken, kippen, krängen, schwenken, sich neigen, stellen, umlegen, verkanten **to ~ over** überstülpen, umfallen, umkippen **to ~ up** aufbäumen **to ~ up in front** vorn aufkippen (aviat.)
**tilt** Abstrahlwinkel *m*, Bahnneigung *f* (g/m), Impulsabflachung *f*, Kippung *f*, Kippwinkel *m*, Nadirdistanz *f*, Neigung *f* der Antenne (rdr), Neigung *f* der Kamera, Schirmtuch *n*, Wagenplane *f* ~ **of the vertical photograph** Bildneigung *f*

**tilt,** ~ **brace** Hammerband *n* ~ **frame** Hammergebälk *n* ~ **hammer** Aufwerfhammer *m* ~ **indicator** Kippungs-, (camera) Neigungs-anzeiger *m* ~ **meter** Neigungsmesser *m* ~ **mixer** Zeilenverzerrungskompensator *m*
**tiltable** kippbar, schrägstellbar ~ **casting machine** Kippgießwerk *n* ~ **lamp base** kippbare Lampenfassung *f* (Breitstrahler) ~ **lifting beam** kippbare Traverse *f*
**tilted** gekippt, geneigt, schief, schiefwinklig ~ **cylinder mixer** Schrägstrommelwischer *m* ~ **line cross (or cross lines)** liegendes Strichkreuz *n* ~ **photograph** verschwenkte Aufnahme *f* ~ **upwards** herausklappbar ~ **wave front** geneigte Wellenfront *f* oder Wellenstirn *f* ~ **wave head** geneigte Wellenfront *f*
**tilter** Kantvorrichtung *f*, Wipptisch *m*
**tilting** Kantung *f*, Kippen *n*, Schrägstellen *n*, Umlegen *n*, Verkanten *n*; schwenkbar ~ **of strata** Schichtaufrichtung *f*
**tilting,** ~ **bearing** Klapplager *n* ~ **ball** Schnellgewicht *n* ~ **converter** kippbarer Konverter *m* ~ **device** Kantvorrichtung *f*, Kipp-gerät *n*, -vorrichtung *f*, (typewriter) Wippvorrichtung *f* ~ **edge** Kippkante *f* ~ **engine** schwenkbares Triebwerk *n* ~ **equipment** Kippvorrichtung *f* ~ **feed table** schrägstellbarer Anlagetisch *m* ~ **force** Kippkraft *f* ~ **furnace** kippbarer Ofen *m*, Kipp-, Schaukel-ofen *m* ~ **gear** (searchlights etc.) Kippgetriebe *n* ~ **gearing** Kippantrieb *m* ~ **head** Neigekopf *m* ~ **joint** Kippgelenk *n* ~ **ladle** Kipp-Pfanne *f* ~ **level screw** Kippschraube *f* (Nivellierinstrument) ~ **lever** Aufzughebel *m* ~ **lever adapter** Schwenkhebelstutzen *m* ~ **machinery** Kippwerk *n* ~ **manipulator** Kippstuhl *m* ~ **mechanism** Kippmechanismus *m* ~ **milling head** schwenkbarer Fräskopf *m* ~ **milling spindle** schrägstellbare Frässpindel *f* ~ **miter-cutting guide** abklappbarer Gehrungsanschlag *m* ~ **moment** Umsturzmoment *m* ~ **motion** Kipp-, Schaukel-bewegung *f* ~ **oscillation** Intermittenz-, Relaxations-schwingung *f* ~ **oscillations** Kippschwingungen *pl* ~ **plate** Schwenkplatte *f* ~ **platform** Schwenkbühne *f* ~ **position** Auskippstellung *f* ~ **prism** Kipp-Prisma *n* ~ **reflector** Kippspiegel *m* ~ **rods** Kippgestänge *n* ~ **screw** Kippschraube *f* ~ **stage** Drehschwingungsrahmen *m* ~ **table** Hebe-tisch *m*, -trog *m*, Keil-, Kipp-tisch *m*, schräg verstellbarer Tisch *m*, Wipptisch *m* ~ **trap** Kippwassermesser *m* ~ **trestle** Wendebock *m* ~**-type mixer** Kippmischer *m*
**tiltometer** Tiltometer *n*
**timber, to** ~ (a trench) absteifen, verzimmern, zimmern **to ~ wood** Holz *n* abvieren oder beschlagen
**timber** Bauholz *n*, Bergholz *n*, Holz *n*, Nutzholz *n*, Spant *n* ~**-and iron-cased concrete foundations** Mantelgründung *f* ~ **auger** Holzabnehmer *m* (min.) ~ **crib** Balkenstapel *m*, Holzgerippe *n* ~ **economy** Holzbewirtschaftung *f* ~ **facing** Holzverkleidung *f* ~ **fender** Reibholz *n* ~ **fenders** Futterholz *n* ~ **fillets** Futterhölzer *pl* ~ **float** Floßfeder *f* ~ **framing** Holzfachwerk *n* ~ **framework** Holzrahmenwerk *n* ~ **hitch** Zimmermanns-stich *m* oder -stek *m* ~**-lashing cord** Schnürleine *f* ~ **line** Baumgrenze *f* ~ **lining** Holzbekleidung *f* ~ **loader** Holzausschlepper *m*

~ man Stempelsetzer *m* ~ obstacle Balkenverhau *m* ~ pile Holzpfahl *m* ~ planking Holzverschalung *f* ~ platform Schwellwerk *n* ~ ring Holz-, Jahres-ring *m* ~ road block Balkensperre *f* ~ roadway Balkenbahn *f* ~ sheathing Futterholz *n* ~ shelter Holzbunker *m* ~ spandrel Holzbogenfachwerk *n* ~ square tubbing Holzgeviert *n* (min.) ~ stanching frame hölzerne Dichtungsleiste *f* ~ stay lashing Schnurbund *m* ~ store Bretterlager *n* ~ support Bockgesparr *n* ~ tester Zuwachsbohrer *m* ~ tie Holzschwelle *f* ~ upright hölzerne Anschlagleiste *f* ~ water seal hölzerne Dichtungsleiste *f* ~ water seals Holzfutter *n* ~ water stop hölzerne Dichtungsleiste *f* ~ work Gezimmer *n*, Holzverzimmerung *f*, (in blasting) Rahmenvortrieb *m* ~ yard Zimmerplatz *m*

timbered, ~ trench abgesteifte Baugrube *f* ~ wall Bohlwand *f*

timbering Ausbau *m*, Holzverkleidung *f*, Kuvelage *f*, Verschalung *f*, Verzimmerung *f*, Zimmerung *f* ~ of galleries Auszimmerung *f* ~ of mine galleries Verkleidung *f* der Minengänge ~ of a shaft Schachtausbau *m*

timbering joint Anschlußzimmerung *f*

timbre Färbung *f* des Schalles, Klangfarbe *f*, Schallfärbung *f*, Tonfarbe *f* ~-control means Klangfärbemittel *n*

time, to ~ (die Zeit) abstoppen oder bestimmen, einregeln, (ignition) einstellen, einteilen, nach der Zeit *f* abmessen, zeiten, zeitlich bemessen, Zeit *f* messen to ~ the carburetor capacity Vergaserleistung *f* einregeln to ~ the ignition die Zündung *f* einstellen to ~ instruments Geräte *pl* (allgemeine Ausrüstung) einregeln

time Mal *n*, Takt *m*, (signal) Uhrenzeichen *n*, Uhrzeit *f*, Zeit *f* for the ~ being vorläufig at that ~ damals for a long ~ auf die Dauer *f* in modern ~s neuzeitlich in ~ fristgerecht, rechtzeitig in ~ with im Takt *m* mit on ~ beizeiten, pünktlich, zeitig

time, ~ of acceptance Aufgabezeit *f* ~ of appearance Entwicklungsfaktor *m* ~ of arrival Ankunftszeit *f* ~ of ascent Aufstiegzeit *f* (chem.) ~ of beginning of operation Inbetriebnahme *f* ~ of blast Blasezeit *f* ~ of blowing Blase-dauer *f*, -zeit *f* ~ of burning Brennlänge *f* ~ of congestion Gefahrzeit *f* (teleph.) ~ of contact Berührungsdauer *f* ~ taken for cooling off Abkühlzeit *f* ~ of day Tageszeit *f* ~ of decay (of an oscillation) Ausschwing-, Ausschwingungszeit *f* ~ of delivery Lieferzeit *f* ~ of departure Abfahrtzeit *f*, Abflugzeit *f* ~ of departure (or dispatch) Abgangszeit *f* ~ of descent Abfahrtzeit *f* ~ of diffusion Diffusionsdauer *f* ~ of discharge Entladedauer *f*, Entladungszeit *f* ~ for discharging cargo Löschzeit *f* ~ for emptying Entladezeit *f* ~ of etching Ätzdauer *f* ~ of exposure Belichtungs-dauer *f*, -zeit *f*, Bestrahlungszeit *f*, Expositionszeit *f* ~ of fall Fallzeit *f* (bomb) ~ of flow Durchflußzeit *f* ~ of fusing Abschmelzzeit *f* ~ of growth Dauer *f* des Anwachsens ~ of heat Schmelzzeit *f* ~ of heating Erwärmungsdauer *f* ~ taken for heating up Anheizzeit *f* ~ of issuance Ausgabezeit *f* ~ of learning Lehrzeit *f* ~ of liberation Auslösezeit *f* ~ of lighting up (foundry street lights) Anbrennzeit *f* ~ of loan Leihfrist *f* ~ between the

lowest daily temperature and sunrise Wärmedämmerung *f* ~ of melting Schmelzzeit *f* ~ of opening Öffnungszeit *f* ~ of operation Betriebsdauer *f*, Einstellzeit *f* ~ of operation of a relay Ansprechzeit *f* eines Relais ~ of ordering Bestellzeit *f* ~ of origin Aufgabezeit *f*, taktische Uhrzeit *f* ~ of oscillation Perioden-dauer *f*, -zeit *f* ~ between overhaul Zwischenüberholungszeit *f* ~ of overhaul Überholungszeit *f* ~ of payment Verfallzeit *f* ~ of phase transmission Phasenlaufzeit *f* ~ of propagation Betriebslauf-, Latenz-zeit *f* (teleph.) ~ of raise Anhebezeit *f* (Hubzeit) ~ of (flash) ranging measurement Stoppzeit *f* ~ of recovery (or recuperation) Erholungszeit *f* ~ of relaxation Kippzeit *f* ~ of relay Ansprechgeschwindigkeit *f* ~ of release Ausgabezeit *f* ~ for ripening Reifungsdauer *f* ~ of rise Aufstiegszeit *f* (chem.) ~ of rotation Umschwungzeit *f* ~ of rotation in hours Stundenumlaufzeit *f* ~ of running down of the dial Ablaufzeit *f* des Nummernschalters ~ of set Bindezeit *f*, (of gels) Verfestigungszeit *f* ~ of setting up a call Ausführungszeit *f* einer Verbindung *f* ~ of smelting Schmelzzeit *f* ~ of the spark Funkendauer *f* ~ of starting Abfahrtzeit *f* ~ of storage Lagerzeit *f* ~ of throwout (switch) Auslösungszeit *f* ~ of transit (of projectile) Geschoßflugzeit *f*, Laufzeit *f* ~ transition Phasenlaufzeit *f* ~ of vibration Periodenzeit *f*, Schwingungs-dauer *f*, -zeit *f* ~ of warehousing Lagerfrist *f* ~ to warm up Einbrennzeit *f*

time, ~ adjustment Zeiteinstellung *f* ~ advance relay Zeitvorgaberelais *n* ~ allowed for payment Stundungsfrist *f* ~-and-motion expert Arbeitstechniker *m* ~ angle Zeitwinkel *m* ~ announcement Uhrzeitangabe *f* ~ average Zeitmittel *n* ~ axis Zeitachse *f* ~-axis plate Abszissenplatte *f* ~ ball Zeitball *m* ~ bargain Termingeschäft *n* ~ base Ablenkgerät *n*, Ablenkung *f*, Abtastperiode *f* (rdo), Zeitablenkung *f*, Zeitmaßstab *m*

time-base, ~ action Kippvorgang *m* ~ circuit Basis-, Kipp-kreis *m*, Kippspannungsschaltung *f* ~ condenser Kippkapazität *f*, Zeitkreiskondensator *m* ~ deflection amplifier Ablenkverstärker *m* (rdr) ~ device Kippgerät *n* ~ expansion unit (in oscilloscope) Mikroskopzeitbasisgerät *n* ~ frequency Kipp-, Zeitablenk-frequenz *f* ~ generator Zeitbasisgenerator *m* ~ oscillation Intermittenz-, Kipp-schwingung *f* ~ oscillator Kippschwinger *m* ~ period Kipp-, Zeitachsen-periode *f* ~ unit Kippspannungsapparat *m* ~ unlock speed Rückkippgeschwindigkeit *f* ~ voltage Zeitbasisspannung *f*

time, ~ behavior Übertragungsverhalten *n*, Zeitverlauf *m* ~ bomb Bombe *f* mit Verzugszeit ~ calculation Zeitverrechnung *f* ~ card Arbeits-, Stempel-karte *f* ~ characteristic (relays) Zeitstaffelung *f* ~ chart Laufzeitdiagramm *m* ~ charter Befrachtungskontrakt *m* für eine lange Dauer ~ check Telefonometer *n*, Uhrenvergleich *m* ~ check lamp Gesprächsuhr *f* mit Lampensignal ~ circle Annäherungszeitkreis *m* ~ clock Kontrolluhr *f* ~ consolidation curve Zeitsetzungskurve *f* (-linie)

time-constant Zeitkonstante *f* ~ of fall Abfallzeit *f* ~ of resonant amplification Aufschaukel-

zeit *f* ~ **of rise** Anstiegzeitkonstante *f* ~ **of time
delay** Verzögerungszeitkonstante *f*
**time-constant,** ~ **multiplier** Zeitkonstantenver-
vielfacher *m* ~ **setting** Zeitkonstante *f*
**time,** ~ **consuming** langwierig, zeitraubend ~
**control** Zeitprüfung *f* ~ **control pulse** Zeitkontrollimpuls *m* ~**-control unit** Zeitschalter *m*
~**-control wheel** Zeitrad *n* ~ **correction** Zeitver-
besserung *f* ~ **criteria** Zeitkriterien *pl* ~ **cross
section** Zeitquerschnitt *m* ~ **cycle** Spiel *n* ~
**-cycle operation** Übertragungsverhalten *n* ~
**-decision multiplex** Zeitmultiplex-Verfahren *n*
~ **deflection** Zeitablenkung *f*
**time-delay** Verzug *m* ~ **circuit** Verzögerungsschaltung *f*, -kreis *m* ~ **closing** Relais *n* mit
Einschaltverzögerung ~ **connection** Langsam-
schaltung *f* ~ **drop-out** Relais *n* mit Abschalt-
verzögerung *f* ~ **fuse** Zeitverzögerungssiche-
rung *f* ~ **motor** Verzögerungsmotor *m* ~ **relay**
Verzögerungsrelais *n* ~ **switch** Zeitkontakteinrichtung *f*
**time,** ~ **demodulation** Zeitimpulsdemodulation *f*
~ **dependence** Zeitabhängigkeit *f* ~ **dependent**
zeitabhängig ~ **derivative** Zeitableitung *f* ~
**designation** Zeitangabe *f* ~ **difference method**
Zeitunterschiedverfahren *n* ~**-distance-curve
(or graph)** chronische Kurve *f*, Laufzeitkurve *f*
~ **distribution** Zeitverteilung *f* ~ **distribution
system** Zentraluhrenanlage *f* ~**-division multiple
method (or system)** Zeitmultiplexverfahren *n*,
Zeitselektionssystem *n* ~ **domain** Originalbe-
reich *m* ~**-electrical distribution system** Zentraluhrenanlage *f* ~ **element of automatic circuit
breaker** Eigenzeit *f* ~ **element in the development of** zeitliche Entwicklung *f* ~ **error indicator**
Periodenkontrolluhr *f* ~ **exactitude** Zeitge-
nauigkeit *f* ~ **exposure** Zeitaufnahme *f* ~
**expression** Zeitausdruck *m* ~ **extender** Zeitdehner *m* ~ **extension** Aufschub *m*, Fristverlängerung *f*, Nachfrist *f* ~ **factor** Zeitfaktor *m*
~ **fire** Brennzünderschluß *m*, Schießen *n* mit
Doppelzünder ~ **flux** zeitlicher Fluß *m* ~ **fuel
graph** Flugverlaufkurve *f* ~ **function** Zeitfunk-
tion *f* ~ **fuse** Brenn-, Doppel-zünder *m*, Zeit-
sicherung *f*, -zünder *m*, -zündschnur *f* ~**-fuse-
ring** Zündring *m* ~ **group** Zeitgruppe *f* ~ **heading** Zeitgruppe *f* ~ **history** zeitlicher Ablauf *m*
~ **integral** Zeitintegral *n* ~ **interval** Zeit-abstand
*m*, -folge *f*, -intervall *n*, -reserve *f*, Zwischen-
pause *f* ~ **interval between successive photographs** Aufnahmeintervall *n*, Bildfolge *f* ~
**jitter** Flackern *n*, Unsicherheit *f* des Impuls-
einsatzes, Zittern *n* ~ **keeper** Lohnbuchhalter
*m*, Zeit-messer *m*, -nehmer *m* ~ **keeping check**
Kontrollindexmarke *f*
**time-lag** (total period of) Auslösezeit *f*, Nachhinken *n*, Schleppzeit *f*, Trägheit *f*, Verzögerung *f*, Zeit-differenz *f*, -verschiebung *f*, zeitliche Nacheilung *f*, zeitliche Rücktrift *f*, zeitlicher Verzug *m* ~ **of the ignition** Zündverzögerung *f* ~ **due to loading** Ladeverzug *m*, Ladeverzugszeit *f*
**time-lag,** ~ **relay** Steuerverzögerungsrelais *n* ~
**relay (or release)** verzögerte Auslösung *f* ~ **set
for . . . seconds** Ausschalter *m* eingestellt für
. . . Sekunden Auslösezeit ~ **switch** Verzöge-
rungsschalter *m*
**time,** ~**-lapse camera** Zeitraffaufnahmekamera *f*

~**-lapse motion camera** Zeitraffer *m* ~**-lapse
photography** Zeitraffaufnahmeverfahren *n* ~
**-limit protection** verzögerter Schutz *m* ~**-limit
relay** Relais *n* mit Zeitauslösung ~**-limit release**
unabhängig verzögerter Auslöser *m* ~ **mark**
Zeitmark *f* ~ **marker** Kontaktgeber *m*, Zeit-
markierer *m* ~ **marking lamp** Zeitmarkenlampe
*f* ~ **meter** Zeitzähler *m* ~ **metering** Zählung *f*
nach der Zeitdauer ~ **modulation** Zeitimpuls-
modulation *f* ~ **multiplex** Zeitmultiplexsystem
*n* ~**-off** Aussetzen *n* ~**-of-flight** Flugzeit *f* ~**-of-
flight from reference point to future position** Ge-
samtauswanderungszeit *f* ~**-of-flight curve**
Flugzeitkurve *f* ~**-of-flight distribution** Flug-
zeitenverteilung *f* ~**-of-flight indicator** Aufschlagmeldeuhr *f* ~**-of-flight measurements**
Flugzeitmessungen *pl* ~**-of-flight spectrometer**
Laufzeitspektrometer ~**-of-flight technique**
Laufzeitmethode *f* ~**-on** Zeit *f* des Gesprächs-
beginns ~ **pattern control** Zeitplanregelung *f* ~
**per charge** Durchsatzzeit *f* ~ **piece** Zeitgeber *m*
~ **piece with secondhand** Sekundenuhr *f* ~
**policy** Zeitversicherungsschein *m* ~ **pressure
history** zeitlicher Druckverlauf *m* ~**-proportion-
al** zeitproportional ~ **pulse relay** Impulsrelais
*n* ~ **quenching** stufenweises Härten *n* mit unterbrochener Abschreckung *f*, Zeithärtung *f* ~
**-ranging station** Auswertestation *f* ~ **record**
Dauerrekord *m*, Zeitmarkierung *f* ~ **recorder**
Zeitregistrierapparat *m*, Zeitzähler *m* ~ **recorder on film** Meßfilmeinrichtung *f* ~ **recording
apparatus** Zeitmarkengeber *m* ~**-recording
camera** Zeitblinker *m* ~ **reference** Zeitenmaßstab *m* ~ **regulator** Zeitregler *m* ~ **relay** Zeit-
schütz *n* ~**-relay call connection** Zeitrelais-
anrufschaltung *f* ~ **release-assembly** Auslöse-
zeitwerk *n* ~ **required** Zeitaufwand *m* ~ **required
for concrete to set** Bindezeit *f* ~ **required for
damping out of a vibration** Abklingzeit *f* ~
**resolution (or resolving)** Zeitauflösung *f* ~
**reversal invariance** Zeitumkehrinvarianz *f* ~
**-saver** Zeitersparnis *f* ~ **scale** Synchronisierzeitmarke *f*, Zeit-dehnung *f*, -marke *f*, -maß *n*,
Zeitenmaßstab *m* ~ **scale factor** Zeitmaßfaktor
*m* (comput.) ~**-scale instrument** Zeitskalen-
instrument *n* ~ **schedule** Zeiteinteilung *f* ~
**selection band** Adressenwahlspur *f* ~**-selector
transducer** Zeitdiskriminatorkreis *m* ~**-sense**
Zeitsinn *m* ~ **sequence** Zeitfolge *f* ~**-settlement
curve** Zeitsetzungskurve *f* (-diagramm) ~
**-shared amplifier** Zeit-Multiplex-Verstärker *m*
~ **sharing circuit** Zeitteilerschaltung *f* ~ **sheet**
Arbeitsstundenbericht *m* (Wochenzettel) ~
**sheet intercept board** Kontrollkarte *f* ~ **shift**
Zeitverschiebung *f* ~ **shutter** Zeitverschluß *m*
**time-signal** Uhrzeitangabe *f*, Zeitsignal *n*, Zeit-
zeichen *n* ~ **injector** Zeitmarkengeber *m* ~
**station** Zeitzeichenstelle *f* ~ **system** Zeitsignal-
anlage *f*, Zeitzeichenwesen *n* ~ **transmitter,**
Zeitgeber *m*, Zeitsignalgeber *m*, Zeitzeichengeber *m*
**time,** ~ **slope** zeitlicher Verlauf *m* ~ **space** periodische Größe *f*, Zeitabschnitt *m* ~**-spark
system** Taktfunksystem *n* ~ **spiral** Zeitspirale
*f* ~ **stamp** Uhr-, Zeit-stempel *m* ~**-standard for
operation** Vorgabezeit *f* ~ **strength** Zeitfestig-
keit *f* ~ **study** Akkordwesen *n*, Refa-Zeitstudie
*f* ~ **study man** Kalkulator *m*, Refa-Ingenieur *m*

**~ sweep** Zeitablenkung f **~ switch** Schaltuhr f, Zeitschalter m, Zeitschaltwerk n **~ table** Dienstplan m, Fahrplan m, Kursbuch n, Stundenplan m **~ tapper** Kontakt-, Takt-geber m **~ -temperature curve** Zeittemperaturkurve f **~ trace (or mark)** Zeitmarke f **~-train indicator (fuse)** Brennlängenschieber m **~-train travel (recorder)** Einstellzeit f **~-traverse diagram** Zeitwegschaubild n **~-tried** bewährt **~ valve** Taktgeber m **~ varying field** zeitabhängiges Feld n **~ vector** Zeitlinie f **~ yield** Zeitdehnung f **~ zone** Zeitzone f **~ zone meter** Zeitzonenzähler m **~-zone metering** Zeitzonenzähler m

**timed, ~ deflection** taktmäßige Ablenkung **~ ignition** gesteuerte Zündung f **~ oiler** Zeitöler m **~ spark** Steuerfunken m **~ spark discharger** Funkenstrecke f für die Erzeugung ungedämpfter Wellen, Vielfachfeld n für die Erzeugung ungedämpfter Schwingungen **~ spark gap** Knallfunkenstrecke f **~ sparks** gesteuerte Funken pl **~ sweep** taktmäßige Ablenkung f

**timely** angebracht, rechtzeitig, zeitgemäß, zeitlich

**timer** Stoppuhr f, Taktgeber m, Zeitglied n, Zeitmesser m, Zeitnehmer m, Zeitschalter m, Zeitwerk n **~ contact** Zeitkartenkontaktgabe f **~ distributor** (Auto) Zündverteiler m

**times, at ~** verschiedentlich, zeitweilig **in modern ~** neuzeitlich

**timing** Einstellung f, (valve timing) Einstellung f, Feststellung f, Reg(e)lung f, Steuerung f, Synchronisieren n, Zählung f nach der Zeitdauer, Zeitmessung f, Zeitpunkteinstellung f, Zündpunktverstellung f, Zeitzählung f **~ of calls** Festsetzung f, Bestimmung f der Gesprächsdauer (teleph.) **~ of engine** automatische Steuerung f, Steuerung f **~ of ignition** Zündungseinstellung f **~ of the ignition** Zündverstellung f **~ of the valves** Einstellen n der Ventile

**timing, ~ adjustment** Einstellung f der Steuerung **~-adjustment range** Zündverstellbereich m **~ angle** Regulierwinkel m **~ apparatus in fuse** Uhrwerk n **~ box** (Automat) Schaltkasten m **~ chain** Steuerkette f, Zündeinstellungskette f **~ chart** Steuerungsdiagramm n **~ circuit** Zeitglied n **~ constant** Zeitkonstante f **~ curve** (Zündung) Verstellinie f **~ device** Gesprächsuhr f (teleph.), Nockenversteller m **~ diagram** Steuerungsdiagramm n **~ disc** Einstellscheibe f (aviat.) **~ gear** Spritzversteller m, Steuerungsvorrichtung f, Zeitausschalter m **~ gear case** Nockenwellenantriebsgehäuse n **~ generator** Takt-, Zeit-geber m (comput.) **~ impulse** Taktimpuls m **~ interrupter** Zeitkontakteinrichtung f **~ line** Zeitmarke f **~ machine** Zeitregelmaschine f **~ mark** Einstellmarke f **~ motor** Zeitgebermotor m **~ needle** Einstellzeiger m **~ pointer** Einstellzeiger m (zur Einstellung der Steuerzeiten), Zeiger m zur Einstellung der Steuerzeiten **~ pulse generator** Zeitgeber m **~ range** Verstellbereich m **~ register** Gesprächszeitmesser m, Zeitmesser m, Zeitzähler m **~ relay** Zeitrelais n **~ resistance** Verzögerungswiderstand m **~ schedule** Steuerschema n **~ sector** Einstellsektor m **~ shaft** Verteilerwelle f **~ signal recurring every six seconds** Sechserkontakt m **~ system** Zeitmesser m **~ tape** Zeitregulierband n **~ unit** Taktgeber m

**tin, to ~** verzinnen, zinnen **to ~-plate** blechen

**tin** Blechbüchse f, Weißblei n, Zinn n; zinnern **~ acetate** Zinnazetat n **~ alloy** Zinnlegierung f **~ ashes** Zinnasche f **~ assay** Zinnprobe f **~ bar** Zinnbarren m **~ box** Blech-büchse f, -dose f, -emballage f, -kasten m **~-box jointing machine** (clasp joint) Falz-maschine f, -verschlußmaschine f **~-box maker** Blechbüchsenmacher m **~ bromide** Zinnbromid n **~ buddle** Zinnwasche f **~ butter** Zinnbutter f **~ can** Blechbüchse f, -dose f, -hafen m, -kanne f, Weißblechkanne f **~ cap** Zinnkapsel f **~ cashbox** Metallkassette f **~ casting** Zinnguß m **~ chloride** Zinnchlorid n **~ coating** Feuerverzinnen **~ compound** Zinnverbindung f **~-concentrate smelting** Erzarbeit f **~ container** Blechgefäß n **~ containing a great deal of lead** Halbgut n **~ content** Zinngehalt m **~ control plate** Blechmarke f **~ cry** Zinn-geschrei n, -knirschen n, -kreischen n, -schrei m **~ crystal discharge** Zinnsalzätze f **~-cutting and -edging machine** Dosenabschneide- und Bördelmaschine f **~ deposit** Zinnlagerstätte f **~ dioxide** Kassiterit m **~ discharge paste** Zinnätzfarbe f **~ disulfide** Zweifachschwefelzinn n **~ dross** Zinn-abstrich m, -krätze f **~ drum** Weißblechtrommel f **~ dust** Zinnstaub m **~ filing** Zinn-feilicht n, -feilspan m **~ foil** Blattzinn n, Silberpapier n, Spiegelfolie f, Stanniol n, Stanniolstreifen m, Zinnblatt n, Zinnfolie f **~ foil capacitor** Metallfolienkondensator m **~-foil covering** Stanniolüberzug m **~-foil paper** Stanniolpapier n **~ founder** Zinngießer m **~ funnel** Blechtrichter m **~ furnace** Zinnofen m **~ glaze** Zinnglasur f **~-glazed** zinnglasiert **~ glazing** Zinnglasur f **~-lined** mit Blecheinsatz m **~-lined case** mit Blech ausgeschlagene Kiste f **~ lode** Zinnader f, Zinnerzgang m **~-making machine** Blechfaltemaschine f **~ metal printing** Blechdruck m **~ mine** Zinn-bergwerk n, -grube f **~ mordant printing** Zinnbeizendruck m **~ obtained from slag** Schlackenzinn n **~ ore** Zinnerz n **~-ore formation** Zinnerzformation f **~-ore refuse** Zinnafter m **~ output** Zinnproduktion f **~ pest** Zinnpest f **~ phosphide** Phosphorzinn n **~ pipe** Zinnpfeife f, -rohr n, -röhre f **~ placer deposit** Zinnerzseife f **~ plague** Zinnpest f

**tinplate** Blechplatte f, verzinntes Eisenblech n, Weißblech n, weißes Blech n, weißes Eisenblech n, Zinn-blech n, -platte f **~ doubler** Weißblechwalzer m **~ enamel** Blechlack m **~ export** Weißblechausfuhr f **~ folder** Blechfaltemaschine f **~ household ware** Weißblechhaushaltungsgeräte pl **~ ink** Blechdruckfarbe f **~ laquer** Blechlack m **~ manufacturer** Weißblechhersteller m **~ mills** Weißblechwerke pl **~ puncher** Weißblechstanzer m **~ roller** Weißblechwalzer m **~ ware** Weißblechwaren pl **~ works** Weißblechwerke pl

**tin, ~-plated** verzinnt **~-plated container** Blechgefäß n **~ pot** Grobkessel m, Zinnpfanne f **~ powder** Zinnpulver n **~ printing machine** Blechdruckmaschine f **~ printing varnish** Blechdruckfirnis m **~ protochloride** Einfachchlorzinn n **~ pyrites** Zinnkies m **~ refining** Zinnraffination f **~-refining plant** Zinnraffinationsanlage f **~ refuse** Zinn-after m, -gekrätz n, -rückstand m **~ residue** Zinnrückstand m **~ roller** Antriebstrommel f ~ salt Zinnsalz n ~ scrap Weißblech-abfall m ~ scum Zinnabstrich m ~ sheet Zinn-platte f ~ sheeting Blech n ~ skim gates Siebe pl ~ skimming Zinnabstrich m ~ slab Zinnbar-ren m ~ smeltery Zinnhütte f ~-smelting plant Zinnhütte f ~ smith Blechner m, Blechschmied m, Klempner m, Spengler m ~ snips Zinnschere f ~ solder Lötzinn n, Weichlot n, Zinnlot n ~ solution (or spirit) Zinnlösung f ~ stone Kas-siterit n, Zinnstein m ~ strip Weißband n ~ strongly mixed with lead Halbzinn n ~ sulfide Zinnsulfid n ~ tack Tapeziernagel m ~ test Zinnprobe f ~ tetrachloride Tetrachlorzinn n, Zinnbutter f ~ tube Zinn-rohr n, -röhre f ~ vein Zinnader f, Zinnerzgang m ~ vessel with a narrow neck Blechflasche f ~ vessels Zinn-geschirr n ~ vessels and plates Blechgeschirr n ~ ware Blechware f, Weißblechwaren f & pl ~ waste Zinnkrätze f ~-weighing Zinncharge f, Zinnbeschwerung f ~ wire Zinndraht m ~ works Zinnhütte f ~ yield Zinnausbringen n

**tincal** Tinkal m

**tinctorial** färbend ~ power Ergiebigkeit f eines Farbstoffes, Färbekraft f

**tincture** Tinktur f ~ of iodine Jodtinktur f

**tinder** Zunder m, Zunderschwamm n ~ box Feuerzeug n ~ paper Zunderpapier n ~ proof zunderfest

**tindery** zund(e)rig

**tine** Zinke f, Zinken m ~ of a tuning fork Stimm-gabelzinken m

**tinge, to** ~ anstreichen, einen Anstrich m geben, Farbton m, Ton m

**tingling** gellend, klimpernd

**tinker, to** ~ herum-flicken, -pfuschen

**tinker** Kessel-, Pfannen-flicker m

**tinlike** zinnartig

**tinman's pot** Fettkessel m

**tinned** verzinnt ~ iron sheet verzinntes Eisen-blech n ~ wire verzinnter Draht m

**tinner** Blechschmied m ~'s snips Blechschere f

**tinning** Reiblöten n, Verzinnen n, Verzinnung f ~ plant Verzinnerei f

**tinny** (sound) blechern, zinnartig ~ sound ble-cherner Klang m

**tinol** Tinol n

**tinsel** Düppel m (rdr), Flitter m, Lahn m, (cord) Lahnlitze f, Rauschgold n ~ cord Lahnlitzen-schnur f, Litzenschnur f, Schnur f mit Draht-litzenleiter

**tint, to** ~ abtönen, leicht färben, nuancieren, schattieren

**tint** Einkopierraster m, Farbe f, Farbenstufe f, Farbstufe f, Farbton m, hellgetönte Farbe f, Nuance f, Schattierung f, Ton m, Tönung f

**tinted,** ~ cover einseitiger Umschlag m ~ glass Rauchglas n ~ paper Tonpapier n

**tinting,** ~ color Abtönfarbe f ~ strength Färbe-kraft f

**tintometer** Farbenmesser m

**tiny** sehr klein, winzig ~ barbed hook Wider-häkchen n ~ point Pünktchen n

**tip, to** ~ abkippen, ausschütten, betupfen, kan-ten, kippen, umkippen, umlegen, stürzen, zu-spitzen **to** ~ over auf den Kopf m gehen, über-kippen **to** ~ stretch anformen **to** ~ up aufklap-pen, känteln

**tip** Bodenkippe f, Düse f, Kopf m, (of torch)

Mundstück n, (for soldering cutting tools) Plättchen n, Punktschweißelektrode f, Rand m, Schwanz m, Spitze f, (of telephone plug) Steckerspitze f, Trinkgeld n ~ of brush Bürsten-spitze f ~ of plug Stöpselspitze f ~ and sleeve contact Berührung f zwischen Stöpselspitze und Berührungsschaft ~ of tooth Zahnkopf m, Zahnspitze f

**tip,** ~ balance Seitenausgleich m ~ barrow Kipp-karren m ~-change switch Kippumschalter m ~ chord Endtiefe f ~ clearance (gear) Kopfspiel n ~ contact of the tooth Zahnkanteneingriff m ~ cross section Blattspitzenquerschnitt m ~ drag Punkt-, Spitzen-widerstand m ~ engage-ment Kanteneingriff m ~ frame Kipprahmen (g/m) ~ guide Kohlestütze f ~-in Einkleben n ~ loss Rand-, Spitzen-verlust m ~ pan kipp-barer Trog m ~ plate Blattspitzenkappe f ~ radius Außen-, Propeller-radius ~ rock Tan-gentialspiel m ~ screw (scope mount) Kipp-schnecke f ~ size Düsengröße f ~ speed Spitzengeschwindigkeit f ~ stall Flügelende-sackflug m ~ stretching Vorformen n ~ strip Randbogen m ~-up seat Klappsitz m ~ vortex Rand-, Spitzen-wirbel m ~ wire a-Ader f, Ader f zur Stöpselspitze, Stöpselspitzenzuführung f

**tipped,** ~ fill durch Kippen gewonnene Auffül-lung f ~ off abgeschmolzen ~ position Auskipp-stellung f ~ propeller bedeckte Flügelspitze f

**tipper** Bergekipper m, Kipper m, Kippfahrzeug n, Wipper m

**tippet** Dachschaube f

**tipping** Kippen n, Umkippung f ~ of the propel-ler Luftschraubenbeschlag m

**tipping,** ~ angle Kippwinkel m ~ arrangement Sturzanordnung f ~ bracket Abrollbock m ~ bucket Kipp-becher m, -gefäß n, -kübel m, Klappkübel m ~ cart Kippkarren m ~ convert-er kippbarer Konverter m ~ cradle Chargier-vorrichtung f ~ device Kippvorrichtung f, Tipp-vorrichtung f ~ equipment Kippvorrichtung f ~ excavated material Bodenkippen n ~ force Kippkraft f ~ furnace Schaukelofen m ~ gear-ing Kippantrieb m, Kippvorrichtungsgetriebe n ~ gradient Kippneigung f ~ hopper Kipp-mulde f ~ jetty Sturzgerüst n ~ ladle Kippfanne f ~ lever Kipphebel m ~ lorry Kipper m, Kipp-lore f ~ mechanism Kippmechanismus m ~ motion Kippbewegung f ~ platform Kipp-bühne f ~ scale Kippwaage f ~ speed Schluß-drehzahl f ~ stage Kippbühne f, Kipperkatzen-brücke f, Sturzbühne f ~ torch Abschmelz-brenner m ~ trailer Kippanhänger m ~ trough Kippmulde f ~-trough four-bolted wagon Vier-zapfenkipper m ~-type mixer Kippmischer m ~ wagon Kippwagen m

**tipple-mechanism** Kippmechanismus m

**tippler** Kippkasten m, Kreiselwippe f, Wipper m

**tips** Hörner pl

**TIR (total indicator reading)** Gesamtbereich m der Meßuhrablesung

**tire, to** ~ abhetzen, altern, ermatten, ermüden

**tire** Bandage f, Flanschreifen m (of a coupling), Pneu n, Pneumatik m, Radkranz m, Radreifen m, Reif m, Reifen m, Schiene f ~ air-pressure controller Luftdruckkontroller m ~ bands Räderschutzreifen m ~-bending machine Rei-fenbiegemaschine f ~-boring mill Radreifen-

bohrbank *f* ~ **canvas** Reifengewebe *n* ~ **capacity** Reifentragfähigkeit *f* ~ **case** Reifenkoffer *m* ~ **casing** Lauf-decke *f*, -mantel *m*, Reifenhülle *f* ~ **chain** Schneekette *f* ~ **cirumference** Reifenumfang *m* ~ **contact area** Reifenlauffläche *f* ~ **control** Reifenwächter *m* ~-**cord** Reifencord *m* ~ **cover** Decke *f*, Reifenmantel *m* ~ **deflection** Reifeneinsenkung *f* ~ **fastening** Radreifensicherung *f* ~ **flap** Felgenband *n*, Schlauchschoner *m* ~ **gauge** Druckprüfer *m* für Luftreifen, Reifendruckprüfer *m* ~ **grip** Griffigkeit *f* der Reifen, Reifengriffigkeit *f* ~ **indicator lamp** Reifenanzeigelampe *f* ~ **inflating set** Reifenfüllanlage *f* ~ **inflator** Reifenfüller *m* ~ **inflator assembly** Einsteckreifenluftpumpe *f* ~ -**inflator cock** Reifenfüllhahn *m* ~ **inflator cylinder** Reifenfüll-flasche *f*, -ventil *n* ~ **iron** Bandageneisen *n* ~ **lever** Aufzieh-, Montiereisen *n*, Reifenmontierhebel *m* ~ **(rolling) mill** Radreifenwalzwerk *n* ~ **mounting** Reifenmontage *f* ~ **mounting press** Radaufziehpresse *f* ~ **pattern** Reifenprofil *n* ~ **pressure** Reifendruck *m* ~ **pressure gauge** Steckmanometer *n* ~ **pressure indicator lamp** Reifendruckanzeigelampe *f* ~ **protection** Reifenschutz *m* ~ **pump** Reifenpumpe *f* ~ **puncture** Luftschlauchbeschädigung *f* ~ **putty** Reifenkitt *m* ~ **reclaim** Reifenregenerat *n* ~ **removing device** Abwalkmschine *f* ~-**repair kit** Reifenflickzeug *n* ~ **repairing** Reifenvulkanisierung *f* ~ **roll** Felgenwalze *f* ~-**rolling mill** Bandagenwalzwerk *n* ~ -**rolling practice** Bandagenwalzerei *f* ~ **section** Reifenprofil *n* ~ **setting** Bereifung *f* ~ **shoe** Decke *f* ~ **size** Reifengröße *f* ~ **sleeve** Schlauchmanschette *f* ~ **slippage** Reifenschlupf *m* ~ **spreader** Reifenspreizer *m* ~ **steel** Bandageneisen *n* ~ **stock** Reifenmischung *f* ~ **stripping machine** Abwalkmaschine *f* ~ **tool** Montiereisen *n* ~ **track** Reifenspur *f* ~-**tread stock** Laufdeckenmischung *f* ~ **treads** Reifenprofile *pl* ~ **tube** Reifenschlauch *m* ~ **upsetter** Reifenstauchmaschine *f* ~ **valve** Luftventil *n* des Reifens, Schlauchventil *n* ~ **vulcanizing** Reifenvulkanisierung *f* ~ **wear** Reifen-verbrauch *m*, -verschleiß *m* ~ **welding** Radkranzschweißung *f*
**tired** müde, schlaff
**tiredness** Ermüdung *f*, Mattigkeit *f*
**tires** Bereifung *f*
**T iron** Doppel-T-Eisen *n*, T-Eisen *n*, T-Stück *n*
**tissue** Gewebe *n*, Zellgewebe *n*, Zeug *n* ~ **in relief** Reliefgewebe *n*
**tissue,** ~ **belt** Geweberiemen *m* ~ **disease** Gewebezerstörung *f* ~ **dose** Gewebedosis *f* ~ **paper** Gazepapier *n*, Seidenpapier *n* ~ **staining** Stückfärbung *f* ~-**testing machine** Gewebeprüfmaschine *f* ~ **wrapping paper** Packseidenpapier *n*
**titan crane** Titankran *m* ~ **traveling (or truck) pedestal** Laufgerüst *n* des Titankrans
**titanate** Titanat *n*
**titanium** Titan *n* ~ **carbonitride** Kohlenstoffstickstofftitan *n* ~ **chloride** Titanchlorid *n* ~ **compound** Titanverbindung *f* ~ **cyanonitride** Stickstoffzyantitan *n* ~ **dioxide** Titansäure *f* ~ -**dioxide-delustered rayon** titandioxydmattierte Kunstseide *f* ~ **dioxide-rutile** Titandioxyd *n* ~ **glass** Titanglas *n* ~ **halide** Titalhalogen *n* ~ **hybride** Titanhybrid *n* ~ **nitride** Titanstickstoff

*m* ~ **potassium fluoride** Titankaliumfluorid *n* ~ **potassium oxalate** Titankaliumoxalat *n* ~ **salt** Titaniumsalz *n* ~ **vaporizer** Titanverdampfer *m* ~ **white** Titanweiß *n*
**titanosulphuric acid** Titanschwefelsäure *f*
**titanous** titanig
**titer** Titer *m*, Umschlagszahl *f*
**title** Anrecht *n*, Gerechtsame *f*, Kopf *m* einer Zeitung, Titel *m*, Überschrift *f* ~ **to mining claims** Verleihungsurkunde *f* ~ **along the spine** Längsrückentitel *m* ~ **to sue** Aktivlegitimation *f*
**title,** ~ **deed** Besitzurkunde *f* ~ **page** Titelseite *f*
**titled** betitelt
**titrate, to** ~ titern, titrieren **to** ~ **back** zurücktitrieren
**titrating,** ~ **apparatus** Titrierapparat *m* ~ **solution** Titrier-, Meß-flüssigkeit *f*
**titration** Titration *f*, Titrierung *f* ~ **exponent** Titrierexponent *m* ~ **method** Titriermethode *f* ~ **standard** Titer *m* einer Flüssigkeit ~ **value** Umschlagszahl *f* ~ **voltameter** Titrationsvoltmeter *n*
**titrimetric (volumetric)** maßanalytisch, titrimetrisch
**tittle** Lota *n*
**T-joint** Knotenverbindung *f*, T-Stoß *m*
**T-junction** T-Verzweigung *f*
**T-lever** T-förmiger Hebel *m*
**T-matched antenna** angepaßte T-Antenne *f*
**T-mesh** Sternglied *n* ~ **network** Sternglied(er)kette *f*
**T-network** Kettenleiter *m* zweiter Art
**to-and-fro,** ~ **bending tester** Hin- und Herbiegeprüfer *m* ~ **focussing movement** Tiefenfokussierung *f* ~ **motion** Hin- und Herbewegung *f*
**toadstone** Basalt *m*
**tobacco** Tabak *m* ~-**cutting knife** Tabakschneidemesser *n* ~ **dresser** Tabakzurichter *m* ~ -**hoeing attachment** Hackvorrichtung *f* für Tabakkultur ~ **machine wringer** Tabakpresser *m* ~ **manipulator** Tabakzurichter *m* ~ **stripper** Tabakripper *m*
**Tocco hardening** Toccohärtung *f*
**tockens** Fördermarken *pl*
**toe** Zehe *f* **to** ~ **in** gegeneinander stellen ~ **of a dike (or slope)** Deichfuß *m*
**toe,** ~ **failure** Basisbruch *m* ~-**in** Neigung *f* gegeneinander (aviat.), Vorderlauf *m*, Vorspur *f* ~ **method of recording** (underexposure of film) Durchhangverfahren *n* ~ **negative method** Deltaverfahren *n* (photo) ~ **region of film characteristic** Kurvenschwanzstück *n* ~ **wall** Böschungsmauer *f*
**toggle, to** ~ festknebeln, verschränken
**toggle** Anbindekreuz *n*, Gelenk *n*, Kipphebel *m*, Knebel *m*, Schwimmergewicht *n* ~ **action** Froschklemme *f* ~ **assembly** Knebelleinenbund *m* ~ **bolt** Knebel-, Paß-bolzen *m* ~ **clamp** Kniehebelzwinge *f* ~ **collapsible tube press** Kniehebeltubenpresse *f* ~ **drawing press** Exzenterziehpreßmaschine *f* ~ **eccentric** Froschklemme *f* ~ **joint** Gelenkhebel *m*, Knebel-dichtung *f*, -gelenk *n*, Kniegelenk *n* ~-**joint press** Kniegelenk-, Kniehebel-presse *f* ~-**joint riveting machine** Kniehebelniemaschine *f* ~ **lever** Gelenkstück *n*, Kniehebel *m*, Winkelhebel *m* ~-**lever press** Kniehebelpresse *f* ~-**lever tongs**

Kniehebelzange *f* ~ **lock** Kipphebelsperre *f* ~ **loop** Knebelschlaufe *f* ~ **motion** Kniehebelbewegung *f* ~ **plate** Druckplatte *f* ~ **press** Knebelpresse *f* ~ **rope** Knebelleine *f* ~ **switch** Kipp-(hebel)schalter *m*, Kniehebel *m* ~**-type switch** Einbaukippschalter *m* ~ **unit press** Kniehebelpressen *n* ~ **upset** Kniehebelstauchung *f*

**toggles** Kniehebel *m*

**toil** mühselige Arbeit *f*

**toilet** Toilette *f*, Wasserklosett *n* ~ **article** Toilettengegenstand *m* ~ **car** Abortwagen *m* ~ **flush tank** Spülkasten *m* ~ **tank float** Spülkastenschwimmer *m* ~ **vinegar** Toilettenessig *m*

**token** Mal *n*, Merkmal *n*, Papierzeichen *n*

**tolerable** einwandfrei, erträglich, leidlich, mittelmäßig

**tolerance** Abmaß *n*, (of deviation) erlaubte oder gestattete Abweichung *f*, Spiel *n*, Spielraum *m*, Toleranz *f*, Verträglichkeit *f*, Zugabe *f*, Zulaß *m*, zulässige Abweichung *f*, zulässige Maßabweichung *f* ~ **of distortion** Verzerrungstoleranz *f* ~ **on fit** Paß-maß *n*, -toleranz *f* ~ **in size** Maßtoleranz *f*

**tolerance,** ~ **accumulation** Abmaßsummierung *f* ~ **dose** Toleranzdosis *f* ~ **endmeasuring rod with spherical ends** Grenzkugelendmaß *n* ~ **frequency** Frequenzabstand *m* ~ **indicator unit** Toleranzanzeiger *m* ~ **limit** Toleranzgrenze *f* ~**-plug gauge** Grenz-lehrdorn *m*, -lochlehre *f*, Toleranzkaliberdorn *m* ~ **rate** Toleranzdosisrate *f* ~ **space** Toleranzraum *m* ~ **surpassing indicator** Toleranzüberschreitungsanzeiger *m* ~ **unit** Passungseinheit *f* ~ **variation** zulässige Abweichung *f* ~ **zone** Toleranzfeld *n* ~ **zone position** Toleranzlage *f*

**tolerances** Einpaßgrößen *pl* (geom.), (mech. engin.) Plus- und Minusabmaß *n* ~ **across the flats** Schlüsselweiteabmaß *n*

**tolite** Tolit *n*

**toll** Übergangsgebühr *f*, Wegsteuer *f*, Zoll *m*, Zollgeld *n* ~ **answering jack** Fernabfrageklinke *f* ~ **area** Vorortsverkehrsbereich *m* ~ **bridge** Zollbrücke *f* ~ **busy** fernbesetzt ~**-busy condition** Fernbesetztsein *n* ~ **cable** Fernkabel *n* ~**-cable circuit** Fernkabelleitung *f* (mit Sprechstromverstärkern), Fernkabellinie *f* ~**-cable system** Fernkabelnetz *n* ~ **call** Anruf *m* im Fernverkehr, Fernverbindung *f* ~ **call to suburban area** Anruf *m* im Vorortsverkehr, Vorortsverbindung *f* ~ **call pilot lamp** Fernüberwachungslampe *f* ~ **center** (originating or terminating) Endfernamt *n* ~ **center at origin or terminal** äußerstes Fernamt *n* in einer Verbindung ~ **central office** Fernamt *n* ~ **circuit** Fernleitung *f* ~ **circuit with dialing facilities** Nahverkehrsleitung *f* mit Fernwahlbetrieb ~ **collector** Zolleinnehmer *m* ~ **connection** Vorortsverbindung *f* ~ **connector** Fernleitungswähler *m* ~ **directory desk** Fernamtauskunftsstelle *f* ~**-entrance cable** Fernleitungsendkabel *n* ~ **exchange** Fernamt *n*, Fernvermittlung *f* ~ **final selector** Fernleitungswähler *m* ~**-free** gebührenfrei (Straße, Brücke, Tunnel) ~ **information desk** Fernamtsauskunft-(stelle) *f* ~ **intermediate cable** Fernleitungszwischenkabel *n*

**toll-line** Fernleitung *f*, Fernleitungslinie *f* ~ **conversation** Ferngespräch *n* ~ **dialing** Fernwahl *f*,

Selbstwahlfernverkehr *m*, Wählfernsteuerung *f* ~ **jack** Fernleitungsklinke *f*

**toll,** ~ **message** Ferngespräch *n* ~ **network** Fernleitungsnetz *n* ~ **office** Fernamt *n* ~ **operator** Fern-beamtin *f*, -gehilfin *f* ~ **plant** Fernleitungsnetz *n* ~ **position** Fernplatz *m* ~ **preparing connection** (circuit) Fernvorbereitungsschaltung *f* ~**-rate chart** Ferngebührentafel *f* ~ **rate table** Ferngebührentafel *f* ~ **record circuit** Meldeleitung *f* ~ **recording** Gesprächsanmelden *n* ~ **road** zollpflichtige Straße *f* ~ **service** Nahverkehr *m* ~ **service observing** Fernbetriebsüberwachung *f* ~**-service supervision** Fernbetriebsüberwachung *f* ~ **station** Überweisungsamt *n* ~ **switchboard** Fernschrank *m*, Umschalteschrank *m* für den Fernverkehr

**toll-switching** Fernwahl *f* ~ **level** Netzebene *f* ~ **plan** Fernleitungsschaltplan *m*, Schaltplan *m* ~ **position** Fernvermittlungsplatz *m* ~ **trunk** Fernvermittlungsleitung *f*

**toll,** ~ **telephone circuit** Fernleitung *f* ~ **terminal** Teilnehmerleitung *f* für Fernverkehr ~ **test board** Fernprüfschrank *m* ~ **test panel** Klinkenumschalter *m* ~ **ticket** Gesprächsblatt *f* ~ **traffic** Bezirks-, Fern-, Vororts-verkehr *m* ~ **turning lathes** Walzendrehbänke *pl*

**toluene** Toluol *n*

**toluidine** Toluidin *n*

**tombac** Tombak *m*

**tommy** Knebelgriff *m*, Stellstift *m* (mech.) ~ **bar** Drehstift *m*, Knebel *m*, Stecker *m* ~ **bar for lifting case catch** Steckdorn *m* mit Holzgriff ~ **head** Knebelkopf *m* ~ **screw** Knebel(griff)schraube *f*

**tomography** Stratigrafie *f*, Tomografie *f*

**ton, by the** ~ tonnenweise ~ **burden** Tonnentragfähigkeit *f* ~ **control means** Tonblende *f* ~ **kilometer** Tonnenkilometer *m* ~ **measurement** Tonnenmaß *n* ~ **method** Tonnenverfahren *n*

**tonal,** ~ **beat** Tonstoß *m* ~ **center of frequencies** Tonzentrum *n* ~ **frequency** Hörfrequenz *f* ~ **fusion** Tonverschmelzung *f* ~ **generator** Impulsgeber *m*, Tongenerator *m* ~ **pattern** Tonbild *n* ~ **range** Tonumfang *m* ~ **reception** Tonempfang *m* ~ **signal** tönendes Signal *n* ~ **value** Tönungswert *m* ~ **values** Bildtöne *pl*

**tonality** Tonart *f*

**tonalizer** Klang-blende *f*, -färbemittel *n*, -farbenregler *m*, Tonblende *f*, Tonfärbemittel *n*

**Toncan iron** Toncan-Eisen *n*

**tondo** Rundbild *n*

**tone, to** ~ tönen **to** ~ **down** abtönen **to** ~ **up** tonisieren

**tone** Farbe *f*, Klangfarbe *f*, Kolorit *n*, Schattierung *f*, Ton *m*, Tönung *f*, Zeichen *n* ~ **arm** Tonarm *m* ~ **band** Tonstreifen *m* ~ **channel** Tonkanal *m* ~ **color** Färbung *f* des Schalles, Schallfärbung *f*

**tone-control** Aussteuerung *f*, Klang-blende *f*, -farbenregelung *f*, -farbenregler *m*, -färber *m*, -regler *m* ~ **aperture** Helligkeitsskala *f* ~ **choke** (coil) Klangdrossel *f* ~ **means** Klangfärbemittel *n* ~ **means to accentuate (emphasize, or underscore) bass (or lowpitch notes)** Baßanhebung *f* ~ **scale** Tönungsskala *f*

**tone,** ~ **controls** Klangfarbe *f* (tape rec.) ~ **converter** Tonwender *m* ~ **correctness** Tonrichtigkeit *f* ~ **distortion** Frequenzverschiebung *f* ~

**divider** Frequenzteiler *m* ~ **fader** Tonmischer *m* ~ **filter** Tongeber *m*, Tonsieb *n* ~ **generator** NF-Meßsender *m*, Tongenerator *m* ~ **identification** Tonkennung *f* ~**-idle system** Dauerstromverfahren *n* ~ **mixer** Tonmischer *m* ~**-modulated** tonmoduliert ~**-modulated telegraphic transmitter** tonmodulierter Telegrafiersender *m* ~**-modulated wave** tonmodulierte Welle *f* ~ **oscillator** NF-Meßsender *m*, Ton-generator *m*, -summer *m* ~ **perception** Schallwahrnehmung *f* ~ **producer** Tonerreger *m* ~ **quality** Schallfärbung *f*, Tonfarbe *f* ~**-reducer switch** Drosselschalter *m* ~ **rendering** Tonwiedergabe *f* (acoust.) ~**-shading means** Klang-blende *f*, -färbemittel *n*, Tonfärbemittel *n* ~ **signal** Besetztprüfung *f* mit Summerton, Klangsignal *n* ~ **source** Tonquelle *f* ~ **spectrum** Tonbild *n* ~**-sustaining pedal** Tonhaltungspedal *n* ~ **switch** Klangblende *f* ~ **test** Besetztprüfung *f* mit Summerton ~ **tester** Tonprüfer *m* ~ **tuning** Tonabstimmung *f* ~ **value** Tonwert *m* ~ **variator** Tonhöhevergleicher *m*, Tonmesser *m* ~ **wedge** Gradationskeil *m* ~ **wheel** Einstimmungsrad *n*, phonisches Rad *n*, Tonrad *n*, Zahnsirene *f*

**toner** Pigmentfarbstoff *m*, (Lack) Töner *m*

**tones of brightness** Helligkeitsunterschiede *pl*

**tong** Manipulator *m* ~ **arm** Zangenarm *m* ~ **beam** Zangenstempel *m* ~ **riveting machine** Zangennietmaschine *f*

**tongs** Beißzange *f*, Drahtzange *f*, Kluppe *f*, Zange *f* ~ **for burning-in** Einbrennzange *f* ~ **for drill pipe** Gestängezangen *pl*

**tongue** (of a file) Angel *f*, Ausläufer *m*, Dorn *m*, (saw) Knebel *m*, Langbaum *m*, Lasche *f*, Lenkbaum *m*, Ösenblatt *n*, Paßleiste *f*, Ralsianker *m*, Vortrieb *m*, Zunge *f* ~ **of a crossing** Herzstück *n* einer Kreuzung ~ **of fire** Feuergarbe *f* ~ **(or spit) of land separating the haff from the sea** Nehrung *f* ~ **of a relay** Anker *m* eines Relais ~ **of a sliding door** Führungsfeder *f* einer Schiebetür

**tongue,** ~**-and-groove** Nut- und Federfuge *f* ~**-and-groove joint** Anschlitzunge *f*, Spundung *f* ~**-and-groove pass** geschlossenes Stauchkaliber *n* ~ **attachment** Deichseleinrichtungen *pl* ~ **depressor** Zungenspatel *f* ~ **file** Gabelfeile *f* ~ **heel** Zungenwurzel *f* ~**-holding forceps** Zungenhalter *m* ~ **plane** Spundhobel *m* ~ **planing machine** Weichenzungenhobelmaschine *f* ~ **rail** Zungenschiene *f* ~ **roll** Oberwalze *f* ~ **socket** Zungenhülse *f*

**tongued,** ~ **collar** Patrize *f* ~ **wood** gezapftes Holz *n*

**tonguing** Spundung *f* ~ **and grooving machine** Spundmaschine *f*

**tonic,** ~ **sine modulation** Tonüberlagerung *f* ~ **spectrum** Tonspektrum *n* ~ **train signalling** Tonfrequenzsignalisierung *f* ~ **transmitter** Tonsender *m*

**toning,** ~ **and fixing salt** Tonfixiersalz *n* (photo) ~ **bath** Tonbad *n*, Tonfixierbad *n*

**tonnage** Laderaum *m*, Schiffsraum *m*, Tonnengehalt *m*, Tragfähigkeit *f* ~ **of steel tapped per heat** abgestochene Rohstahlmenge *f* per Schmelze

**tonnage,** ~ **deck** Vermessungsdeck *n* ~ **foundry** Fabrikations-, Produktions-gießerei *f* ~ **law**

Vermessungsgesetz *n* ~ **per die** Steinleistung *f* ~ **production** Großerzeugung *f* ~ **scale** Tonnengehaltskala *f*

**tonneau light** Deckenlicht *n*

**tonometer** Tonfrequenzmeßgerät *n*, Tonometer *n*

**tonsit test** Tonprobe *f*

**tonus** Elastizität *f*, Spannkraft *f*, Tonus *m*

**too** ebenfalls ~ **acid** übersauer ~ **coarse grain** Überkorn *n* ~ **heavily loaded** überlastig ~**-heavy** oberlastig ~ **hot** (of a furnace) übergar

**toogle-drawing press** Exzenterzieh-presse *f*, -preßmaschine *f*

**tool, to** bearbeiten, einrichten

**tool** Dreh-bank *f*, -stahl *m*, Hobelkopfschlitten *m*, Meißel *m*, Prägestempel *m*, Stößel *m*, Werkzeug *n*, Werkzeugmaschine *f* ~ **and cutter grinder** Werkzeugschleifmaschine *f* ~ **of fabrication** Fabrikationswerkzeug *n* ~ **for finishing in corners** Stahl *m* zum Aushobeln scharfer Ecken ~ **for noncutting shaping** Werkzeug *n* für spanlose Formung ~ **for observation** Beobachtungsmittel *n* ~ **for rough turning** Schruppmeißel *m* ~ **for sounding mortises** Zapfensonde *f* ~ **for turning up centers** Stahl *m* zum Abdrehen von Spitzen ~ **for working tin and metal** Werkzeug *n* für Blech- und Metallbearbeitung

**tool,** ~ **accessories** Werkzeugzubehör *n* ~ **adjustment** Einstellen *n* der Werkzeuge ~ **angle** Schnittwinkel *n* ~ **arm** Seitensupport *m* ~ **arm slide** Werkzeughalterschlitten *m* ~ **back-out attachments** Stahlrückzugeinrichtung *f* ~ **bag** Werkzeug-beutel *m*, -tasche *f* ~ **bar** Montiereisen *n* ~ **binder** Stahlhälter *m*, Werkzeughalter *m* ~ **bit** Drehstahl *m*, Drehzahn *m* ~ **block** Messerblock *m*, Stahlklaue *f*, Stichelklotz *m* ~ **body** Werkzeugkörper *m* ~ **box** Gerätekasten *m*, Gerätekoffer *m*, Gezähnekasten *m*, Reparaturkasten *m*, Werkzeug-kasten *m*, -kiste *f*, -koffer *m* ~ **cabinet** Werkzeugschrank *m* ~ **car** Bautruppwagen *m* ~ **carriage** Werkzeugschlitten *m* ~ **carrier** Längssupport *m*, Werkzeugaufnehmer *m*, -schlitten *m* ~**-carrying head** Werkzeugsupport *m* ~ **cart** Bautruppwagen *m* ~ **chest** Gezähnekasten *m*, Werkzeug-kiste *f*, -koffer *m*, -schrank *m* ~ **chuck** Werkzeugfutter *n* ~ **collar** Werkzeugbund *m* ~ **crib** Werkzeugausgabe *f* ~ **department** Werkzeug-abteilung *f*, -macherei *f* ~**-dresser** Werkzeug-macher *m*, -schleifer *m* ~ **entrance bowl** Werkzeugführungstrichter *m* ~ **equipment** Werkzeug-ausrüstung *f*, -ausstattung *f* ~ **failure** Werkzeugversager *m* ~ **gate** Schlagbaum *m* ~ **gauge** Gerät-, Werkzeug-lehre *f* ~ **grinder** Schleifmaschine *f* ~**-grinding machine** Werkzeugschleifmaschine *f* ~ **handle** Werkzeuggriff *m* ~ **head** Kopf *m* des Stößels, Stahlhalter *m*, Werkzeugkopf *m* ~ **holder** Stahlhalter *m*, Stichelhaus *n*, Werkzeughalter *m* ~ **holder bit** Einsatzdrehstrahl *m* (für Stahlhalter), Einsatzmesser *n* (zum Einsetzen in Stahlhalter) ~ **holder top slide** Stahlhalter-Obersupport *m* ~ **holding strap** Stahlspannlasche *f* ~ **hook** Bohrhaken *m* ~ **jack** Werkzeugschraubwinde *f* ~ **joint** Außenmuffe *f*, Gestänge-verbinder *m*, -verbindung *f* ~ **joints** Gestängerohrgewinde *n* ~ **kit** Werkzeug-ausrüstung *f*, -ausstattung *f*, -kasten *m* ~

**layout** Einstell-plan *m*, -zeichnung *f*, Einstellung *f* ~ **life** Werkzeugstandzeit *f* ~ **locker** Geräteschrank *m* ~ **maker** Werkzeugmacher *m* ~ **maker's calipers** Federkaliber *n* ~ **maker's flat** Meßscheibe *f* ~-**maker's lathe** Werkzeugdrehbank *f* ~ **manufacture** Werkzeugbau *m* ~ **milling machine** Werkzeugfräsmaschine *f* ~ **operated at standard frequency** Normalfrequenzwerkzeug *n* ~ **parting** Abstechwirkung *f* ~ **position** Werkzeugstellung *f* ~ **post** Längssupport *m*, Stahlhalter *m*, Stichelhaus *n*, Werkzeugschlitten *m* ~-**post grinding** Supportschleifarbeiten *pl* ~ **pusher** Bohraufseher *m* ~ **rack** Halter *m* für Werkzeuge ~ **reconditioning** Werkzeuginstandhaltung *f* ~ **relief** Werkzeugabhebung *f* ~ **rest** Werkzeugsupport *m* ~ **rest holder** Stahlauflageuntersatz *m* ~-**rest swivel** Supportdrehteil *m* ~ **roll** Werkzeug(roll)tasche *f*, zusammenrollbare Werkzeugtasche *f* ~ **room** Werkzeug-ausgabe *f*, -kammer *f*, -lager *n*, -macherei *f*, -magazin *n* ~ **room department** Werkzeugabteilung *f* ~ **room lathe** Leit- und Zugspindeldrehbank *f* ~ **setter** Werkzeugeinrichter *m* ~-**setting** Werkzeugeinstellung *f* ~ **setting gauge** Stahleinstellehre *f* ~-**setting time** Einrichtezeit *f* ~ **setup** Werkzeugeinrichtung *f* ~ **shank** Werkzeugschaft *m* ~ **shed** Remise *f* ~ **shop** Werkzeugmacherei *f* ~ **slide** Stahlhälterschlitten *m*, -support *m*, Stichelschlitten *m*, Supportoberschlitten *m*, Werkzeugschlitten *m* ~ **slide for facing** Plandrehschlitten *m* ~ **slide for turning** Langdrehschlitten *m* ~ **smith** Werkzeug-, Zeug-schmied *m* ~ **spindle** Werkzeugspindel *f* ~ **steel** Werkzeugstahl *m* ~ **steel for the mining industry** Bergbauwerkzeugstahl *m* ~-**storage room** Geräteraum *m* ~ **store** Werkzeugmagazin *n* ~ **swivel** Gerätewirbel *m* ~ **templet** Werkzeugschablone *f* ~ **thrust** Arbeits-, Schneid-druck *m* ~ **tipping** Werkzeugauftragschweißung *f* ~ **wear** Werkzeugverschleiß *m* ~ **withdrawal mechanism** Stahlrückzugeinrichtung *f* ~ **withdrawer** Abziehvorrichtung *f* ~ **wrench** Gerät-, Werkzeug-schlüssel *m*

**tolled** bearbeitet ~ **surface** bearbeitete Fläche *f*
**tooling** Bearbeitung *f*, Einrichten *n* einer Werkzeugmaschine, Werkzeug-ausrichtung *f*, -ausrüstung *f*, -ausstattung *f* ~ **diagram** Einstellplan *m*, -zeichnung *f*, Einstellung *f*, Werkzeuganordnung *f* ~ **layout** Werkzeuganordnung *f* ~ **methods** Bearbeitungsverfahren *n* ~ **quality** Bearbeitbarkeit *f*
**tools** Arbeitsgerät *n*, Gerät *n*, Gerätschaft *f*, Gezäh(e) *n*, Gezähne *n*, Werkgerät *n* ~ **for glueing** Verleimungswerkzeuge *pl*
**tooth, to** ~ auszacken, verzahnen, zähneln, zahnen, mit Zähnen *pl* versehen
**tooth** Absatz *m*, Kamm *m*, (of a grinder) Schneide *f*, Zacke *f*, Zacken *m*, (engin.) Zahn *m*, Zinke *f*, Zinken *m* ~ **of milling cutter** Fräserzahn *m*
**tooth,** ~ **accuracy** Verzahnungsgenauigkeit *f* ~ **bar** Absatzschwelle *f* ~ **brush** Zahnbürste *f* ~ **chain gear** Zahnkettentrieb *m* ~ **chamfer** Zahnabschrägung *f* ~ **chamfering hub** Abkantwälzfräser *m* ~-**chamfering machine for toothed wheels** Zahnabrundfräsmaschine *f* für Zahnräder ~ **construction** Verzahnung *f* ~ **curve** Zahnkurve *f* ~ **cutter** Nachbohrer *m* ~-**cutting**

**machine** Zahnschneidemaschine *f* ~ **depth** Zahntiefe *f* ~ **design** Zahngestaltung *f* ~ **face angle** Zahnbrustwinkel *m* ~ **flank grinding machine** Zahnflankenschleifmaschine *f* ~ **form** Zahnflanke *f* ~ **generation by twin reciprocating tool** Zweistahlabwälzhobel *m* ~ **induction** Zahninduktion *f* ~ **measuring gauge** Zahnmeßlehre *f* ~ **milled free from backlash** spielfreigefräster Zahn *m* ~ **milling cutter** Messerkopf *m*, Verzahnungsfräser *m* ~ **pitch** Zahn-breite *f*, -länge *f*, -teilung *f* ~ **plane** Zahnhobel *m* ~ **pressure** Zahn-anlage *f*, -druck *m* ~ **profile** Zahn(flanken)-form *f* ~ **rest** Handauflage *f* ~ **ripple** Nutenwelle *f* ~ **rolls** Rippenwalzenmühle *f* ~ **rounding machine** Zahnkantenabrundmaschine *f* ~ **sector** Zahn-bogen *m*, -segment *n* ~ **shape** Zahnform *f* ~ **spaces** Zahnlücken *pl* ~ **systems** Verzahnungen *pl* ~ **thickness** Zahndicke *f* ~-**wheel drive** Zahntrieb *m* ~-**wheel synchronizer** Zahnsirene *f*

**toothed** ausgezackt, gekerbt, gezahnt, mit Zähnen *pl* versehen, verzinkt, zackenförmig, zackig, zahnig ~ **attrition (or disc) mill** Zahnscheibenmühle *f* ~ **chisel** Scharnier-, Zahn-eisen *n* ~ **cutter** Zahnschneider *m* ~ **disc** verzahnte Unterlegscheibe *f*, Zahnscheibe *f* ~ **distributor bar** Ablegezahnstange *f* ~ **flywheel rim** Schwungrad *n* mit Anlasserzahnkranz *m* ~ **forceps** chirurgische Pinzette (med.) ~ **forces** Hakenpinzette *f* ~-**gear cutting** Zahnräderschneiden *n* ~ **iron** Zahneisen *n* ~ **lock washer** federnde Zahnscheibe *f* ~ **mica spacer** gezähnte Glimmerscheibe *f* ~ **quadrant** Zahnbogen *m* ~ **rack** gezahnte Stange *f*, Zahnstange *f* ~ **rail** gezahnte Schiene *f* ~ **rim** Zahnkranz *m* ~-**rim turning arrangement** Zahnkranzdrehvorrichtung *f* ~ **rolls** Stachelwalzwerk *n* ~ **sector** Zahnsektor *m* ~ **segment** Zahn-bogen *m*, -segment *n* ~ **spring disc** federnde Zahnscheibe *f* ~ **spring washer** federnde Zahnscheibe *f*, verzahnte und federnde Unterlegscheibe *f* ~ **wheel** Zahnkranz *m* ~-**wheel circuit breaker** Zahnradunterbrecher *m* ~-**wheel gearing** Zahnrädergetriebe *n*, Zahnrad-getriebe *n*, -vorgelege *n* ~-**wheel-tempering machine** Zahnradhärtemaschine *f* ~-**wheel work** Zahnräderwerk *n* ~-**wheels milling machine** Zahnräderbearbeitungsmaschine *f*

**toothing** Bezahnung *f*, Verzahnung *f*, Zahnung *f* ~ **of face (or surface)** Planverzahnung *f* ~ **iron** Zackeisen *n*

**top, to** ~ abschöpfen, köpfen, schopfen **to** ~-**cast** fallend (ab)gießen **to** ~-**pour** fallend ab-, vergießen **to** ~ **a timber** einen Baum zopfen **to** ~ **up** aufdirken, (elements) auffüllen, nachfüllen **to** ~ **up storage cells** sammlernachfüllen
**top** Aufsatz *m*, Dach *n*, Deckel *m*, Firste *f*, Gicht *f*, Gipfel *m*, Höhepunkt *m*, Kappe *f*, (end) Kopf *m*, Kopfende *n*, Krone *f*, Mars *m*, Oberteil *n*, Scheitel *m*, Spitze *f*, Kreisel *m*, (eines Gewindes) Teil *m*, Topf *m*, Verdeck *n*, Verdecksitz *m*; oben, ober . . ., Ober . . .
**top,** ~ **of arch** Bogenscheitel *m* ~ **of cam** Rast *f* ~ **of clouds** Wolkenobergrenze *f* ~ **of crankcase** Kurbelgehäuseoberteil *m* ~ **of a crest** Deichkrone *f*, Wellenkrone *f* ~ **of derrick** Krone *f* des Turmes (Tiefbohranlage) ~ **of a dike** Deich-kappe *f*, -krone *f* ~ **of the drum** Trommeldecke *f* ~ **of Fermi distribution** Fermi-

potential η (in Metallen) ~ of loop Wellenoberkante f ~ of piston Kolbenboden m ~ of pole Mastspitze f ~ of rail (edge) Schienenoberkante f ~ of roadbed Oberbau m (r.r.) ~ of shaft Wellenoberkante f ~ of sleeper Schwellendecke f ~ of a slope Deichkrone f ~ of tank Oberseite f des Tanks ~ of a wave Wellenkamm m

top, ~ air Vorspannluft f ~ anchorage Halsverankerung f ~ antenna Dachantenne f ~ area Deckfläche f ~ back slope Neigungswinkel m einer Schneidekante ~ beam Hainbalken m, Kron-holz n, -schwelle f ~ bend oberes Knie n ~ binding Kopfbindung f (electr.), Oberbund m ~ (side) binding Bindung f im oberen (seitlichen) Drahtlager ~ block Oberflasche f ~-bottom suspension gegenfädige Aufhängung f ~ box Oberkasten m ~ bridge Querhaupt n ~ camber Oberflächenwölbung f ~ cap Anschlußkontakt m an der Spitze des Glasballons, Besohlung f (vulkanisieren); Gitterkappe f, Röhrenkappe f ~ cap of hexode tube Hexodenkappe f ~ (loading) capacity Dachkapazität f, Endkapazität f ~ caps Anodenklappen pl ~ carriage Oberlafette f ~ casting fallender Guß m ~ (dead) center oberer Totpunkt m ~ center-line girder Firstträger m ~ charge (explosives) Aufladung f ~ charging Begichtung f ~-charging gear Schüttvorrichtung f ~ circle Kopfkreis m ~ clamping bar Oberwange f ~ coat Deckanstrich m ~ coat for leather (a lacquer) Egalisierlack m ~ coating Deck-anstrich m, -aufstrich m ~ course Mauer-abdeckung f, -deckplatte f ~ cover Marsbezug m ~ cover end oberer Abschluß m ~ cross-piece Oberriegel m ~ cut Höhenbeschneidung f ~ cutter molding machine Oberfräsmaschine f ~ cylinder casting Zylinderkopf m ~ cylinder cover oberer Zylinderabschluß m ~ dead center höchster oder oberer Totpunkt m ~ deck of the caisson Brücke f des Tors ~ diameter Zopfdruckmesser m (electr.), Zopfstärke f ~ die Patrize f ~ disc Deckelhut m, oberer Teller m (Motoraufhängung) ~ discard abgeschopfter Lunkerkopf m, (of ingots) verlorener Kopf m ~-dressing Kopfdüngung f ~ drive oberer Antrieb m ~-driven mit Antrieb m von oben ~-driven centrifugal Schleuder f mit oberem Antrieb ~ drum Obertrommel f ~ edge Außen-, Ober-kante f ~ edge of guideways Bettoberkante f (mech.) ~ end verlorener Kopf m, (wood) Zopf m, Zopfende n ~ end of an ingot Blockkopf m ~ end of a pipe Lunkerkopf m ~ face (of key) Rückenfläche f ~ fence Gitteraufsatz m ~ fermented obergärig ~ figure obenstehende Ziffer f ~ filling Befüllung f von oben, Obenfüllung f ~ finish Deckappretur f ~ fittings (of blast furnace) Kronenarmatur f ~ fixing bracket Deckenträger m (furnace) ~ flame Gichtflamme f ~ flange Gurtungsblech n, Oberflansch m, Obergurt m (aviat.) ~ flange of truss Binderobergurt m ~ fly Deckelflug m ~ force Oberstempel m ~ gas Gichtgas n ~ girder Oberrahmen m, Oberrahmstück n, oberer Riegel m ~ groove (insulator) Kopfrille f, oberes (seitliches) Drahtlager n ~ gun position Rumpfoberstand m ~-half mold Oberform f ~ header Vorstauchstempel m ~-heavy kopf-, ober-lastig ~ illumination Auflicht n ~ jaw Aufsatzklaue f

~ land (Kolben) Feuersteg m ~ land of a tooth Kopffläche f eines Zahnes ~ layer Oberschicht f ~ layer of a weld Deckraupe f ~ -leaf Hauptfederbett n ~ level obere Grundstrecke f (min.) ~ lever Deckelhebel m ~ light Spitzenfeuer n, Topplicht n ~-light controller Beleuchter m, (studio) Oberbeleuchter m ~ limit of visibility Bewölkungshöhe f ~ line of the tooth Kopflinie f des Zahnes ~-loaded antenna Antenne f mit Dachkapazität, dachbelastete Antenne f ~ mark Toppzeichen n ~ molecule Kreiselmolekül n ~ most oberst ~ most flow line Durchströmungslinie f, Sickerlinie f ~ most indexing position Sperrstellung f ~ motion Kreiselbewegung f ~ opening Gichtöffnung f ~ overhaul Teilüberholung f, (of engine) Überholung f des oberen Teiles eines Motors ~ part of the valve Ventiloberteil n ~ part molding box Oberkasten m ~ performance Gipfelleistung f ~ piece of gun carriage Lafettenaufsatzstück n ~ pipe Scheitelrohr n ~ pivot (trunnion) Halszapfen m ~-plan view Ansicht f von oben ~ plank Firstpfahl m ~ plate Deckplatte f ~ portion of side frame Zylinderlagerdeckel m ~-pour ladle Gießpfanne f mit Ausgußschnauze oder Schnauzenausguß, Pfanne f mit Ausgußschnauze ~-poured fallend gegossen ~ pouring fallender Guß m ~ pressure Dachdruck m (min.), Oberdruck m ~ radius Kopfabrundung f ~ rail of valve (window) Flügelweite f ~ rake Brustwinkel m ~ ram Oberstempel m ~ rib flange Rippenobergurt m ~ roll obere Walze f, Oberwalze f ~ roller with a tilting device hochkippbare Oberwalze f ~ rot Zopffäule f ~ rollers Zylinderbelastung f ~ row obere Reihe f ~ sail halyard Marsfall n ~ secret geheime Kommandosache f ~ section Haubenkörner m, Oberbett n ~-shaft frame Ohrjoch n ~ shed Oberfach n ~ sheet Firstbrett n ~ side Oberseite f; Schanzkleid n ~ side slope Spanwinkel m ~-side view Draufsicht f ~ sill of frame Dekkenstück n ~ slide Ober-schieber m, -schlitten m ~ smoke Gichtrauch m ~ soil Bekleidungserde f, Deckschicht f, Erdkrume f, Mutterboden m, Oberboden m, Obergrund m ~ speed Höchst-fahrt f, -geschwindigkeit f, Spitzen-drehzahl f, -geschwindigkeit f ~ spreader waagerechter Teil m der L-Antenne ~ sprocket obere Führungsrolle f (film) ~ stage Endstufe f (rocket) ~ stairs of a blast furnace Gichttreppe f ~-step landing Antrittsstufe f ~ stratum Deckschicht f ~ surface Deckfläche f ~ surface of wing Saugseite f des Flügels ~ surface covering Oberflächenbelag m ~ swedge hohler Setzhammer m ~ temperature Gichttemperatur f ~ timber line Toppsente f ~ timbering Dach-verkleidung f (min.), -verzimmerung f ~ turret mittlerer oberer Maschinengewehrstand m ~ view Aufsicht f, Grundrißansicht f ~ water Hängendwasser n ~ weight Obergewicht n ~ wing spar Oberflügelholm m

topaz Topas m ~ pebble Topasgeschiebe n

topazolite Goldgranat m

topee Tropenhelm m

topgallant Bram f ~ yard Bramraa f

topic Sache f, Thema n

topogram Topogramm n

topographer Mappeur m, Topograf m

**topographic** topografisch ~ **map** topografische Karte f ~ **section** Kartenbatterie f ~ **survey** Erdvermessung f, topografische Aufnahme f ~ **tachometer** topografischer Entfernungsmesser m

**topographical** topografisch ~ **crest** Böschungskrone f ~ **effects** topografische Einflüsse pl ~ **features** Geomorfologie f, Topografie f ~ **fixing of position** topografische Ortsbestimmung f ~ **mapping** Landesaufnahme f, topologische Abbildung f ~ **particular** Situationsgegenstand m ~ **point** grafischer Punkt m ~ **position finding** topografische Ortsbestimmung f ~ **representation** Geländedarstellung f ~ **survey** Landesaufnahme f, Mappierung f

**topography** Gelände n, Geländekunde f, Oberflächengestaltung f, Ortsbeschreibung f, Topografie f

**topological** topologisch ~ **mapping** stetige bzw. topologische Abbildung f

**topology** Topologie f

**topped** aufgestoppt ~ **crude** Rückstand m der ersten Destillation

**topper-lifter** Köpf- und Rodemaschine f

**topping** Feuerung f für die Toppingsanlage oder Krackanlage, primäre Destillation f, Topping f, Überguß m ~ **in the ground with the topping spade before lifting** Pommritzen n ~ **up** Auffüllen n, (accumulator) Auffüllen n und Laden n

**topping,** ~ **color** Aufsatzfarbe f ~ **device** Köpfvorrichtung f ~ **hoe** Köpfhacke f ~ **lift** Dirk f ~ **spade** Köpfschippe f ~ **turbine** Vorschaltturbine f

**topple, to** ~ (gyro) kippen **to** ~ **backward** nach rückwärts umfallen

**toppled condition of pictures** Bildstromstürzung f

**topsy-turvy** drunter und drüber, in Unordnung f

**torbanite** Torbanit m

**torbernite** Kupferuranglimmer m, Uranglimmer m

**torch** Brenner m, Fackel f, Schweißbrenner m ~ **for lighting blasting fuses** Stichflamme f

**torch,** ~ **battery** Stabbatterie f ~ **form of discharge** Fackelentladung f ~ **handle** Brennergriff m ~ **head** Brennerkopf m ~ **igniter** Fackelzünder m ~ **oil** Torbanit m ~ **welding** Gas-schweißung f, -schweißverfahren n

**torching** Nachzündung f

**tore** Ringwulst f

**toric** torisch, wulstförmig ~ **lens** Punktallinse f ~ **lenses** torische Brillengläser pl ~ **radiant field lens** torische Leuchtfeldlinse f

**torn** zerrissen ~ **tape** geschnittener Lochstreifen m ~ **tape relay installation** nichtautomatische Wiedergabeeinrichtung f (telet.)

**tornado** Tornado m, tropischer Orkan m, Windhose f, Wirbelsturm m, Zyklon m

**toroid** Ringwulst f, Toroid n ~ **core** Ringkern m ~ **doughnut** (Beschleunigermaschine) Ringröhre f

**toroidal** ringförmig, wulstförmig ~ **coil** Ring-, Toroid-spule f ~ **coordinate** Toruskoordinat m ~ **core** Ringkern m ~ **repeating coil** Fernsprechringübertrager m, Ringübertrager m ~ **transformer** Ringübertrager m

**torpedo, to** ~ torpedieren

**torpedo** Alarm-, Spreng-patrone f, Mine f, Torpedo n ~ **boat** Torpedoboot n ~ **crutch** Torpedoaufhängung f ~ **fan** Torpedolüfter m ~ **fuse** Pistole f ~ **gyroscopic angle setting device** Torpedorohrkreiselwinkelrichter m ~ **-launching mechanism** Torpedoabschußvorrichtung f ~ **net** Torpedoschutznetz n ~ **-recovery vessel** Fangboot n ~ **room** Torpedoraum m ~ **tube** Ausstoßrohr n, Torpedorohr n ~ **tube above water** Überwasserrohr n ~ **-tube door** Torpedorohröffnung f ~ **-tube shutter** Torpedorohrverschluß m

**torque** Abfallmoment n, Drall m, Drehkraft f, Drehleistung f, Drehmoment m, Drehung f, Drehungskraft f, Feder f, Raddrehung f, Reaktionskraft f, Torsionskraft f, Verdrehungskraft f **(propeller)** ~ Schraubendrehmoment n (aviat.)

**torque,** ~ **amplifier** Drehmomentverstärker m ~ **arm** Bremshebel m, Dreharm m, Federbeinschere f (Fahrwerk), Torsionsstrebe f ~ **-balance(d)** Beschleunigungsmesser m mit Drehmomentausgleich m ~ **-balance system** Drehmomentkompensator m ~ **balancing tab** Trimmkante f ~ **bar** Bremsblock-Widerlager n ~ **box** Drehmomentkasten m, Drehungskasten m ~ **check gear** Synchrongetriebe m mit Einrichtung zur Beseitigung von Drehmomentstößen ~ **clutch** Rastkupplung f ~ **coefficient** Schraubendrehmoment f ~ **coil** Momentenspule f ~ **compensation** Dreh(moment)ausgleich m ~ **converter** Drehmomentwandler m ~ **converter transmission** Wandlergetriebe m ~ **coupling** Momentkupplung f ~ **cylinder** Drehmomentübertragungszylinder m ~ **drive key** Mitnehmerkeil m ~ **due to gas and inertia forces** Gas- und Massendrehkraft f ~ **dynamometer** Drehmomentwaage f ~ **gauge** Drehmomentdose f ~ **horsepower** Pferdekraft f der Schraubendrehung ~ **imposed on the engine by the slip stream on the cradle frame** durch den Luftschraubennachstrom auf die Pendelwaage des Prüfstandes ausgeübtes Drallmoment n ~ **-indicating wrench** Schraubenschlüssel f mit Drehmomentanzeige ~ **indicator** Drehmoment-meßgerät n, -meßnabe f, Meßnabe f ~ **link** Federbeinschere f ~ **meter** Anzugskraft-, Torsionsmesser m ~ **metering hub** Drehmomentmeßnabe f ~ **-metering spanner** Drehmomentmeßschlüssel m ~ **motor** Drehmomenterzeuger m, Nachsteuermotor m, Schaltmotor m ~ **overload** Rutschkupplung f ~ **overload release clutch** Drehmomentüberlastkupplung f ~ **pickup** Drehmoment-aufnehmer m, -geber m ~ **piston** Drehmomentkolben m ~ **-producing element** Momentenerzeuger m ~ **rating** Drehmomentbemessung f ~ **receiver** Drehmoment-Empfänger m ~ **rod** Schubstange f ~ **r.p.m.** Standdrehzahl f ~ **spider** (Radbremse) Totorsegmentträger m ~ **stand** Probeblock m, Prüfstand m, Untersuchungsstand m für das Schraubendrehmoment ~ **stand truck** Prüfstandwagen m ~ **stay rod** Momentstütze f ~ **synchro** Drehmoment-Drehmelder m ~ **test** Drallprüfung f ~ **transmitter** Drehmomentgeber m ~ **tube** Drehwiderstandsröhre f, Hohlwelle f, Schubrohr n, Verdrehrohr n ~ **-tube aileron control** Verdrehrohrhilfsflügelmechanismus m ~ **-weight ratio** bezogenes Drehmoment m ~ **wrench** Drehmomentschlüssel m

**torquer** Drehmomenterzeuger *m* ~ **magnet** Stellmagnet *n*
**torquing amplifier** Drehmomentsignalverstärker *m* ~ **rate** (gyro) Präzessionsgeschwindigkeit *f*
**torrent** Gießbach *m*, reißender Strom *m*, Wasserstrom *m*, Wildbach *m*
**torrential** reißend, strömend
**torrid** dürr, heiß, verbrannt ~ **zone** heiße Zone *f*
**torsibility** Verdrehungsfähigkeit *f*
**torsimeter** Drehmomentmesser *m*
**torsiograph** Drehschwingungsschreiber *m*, Torsiograf *m*
**torsion** Drehachse *f*, Drehbeanspruchung *f*, Drehen *n*, Drehung *f*, Drillung *f*, Raddrehung *f*, Schiebungsmodulus *m*, Torsion *f*, Verdrehung *f*, Verdrillung *f*, Verwindung *f*, Windung *f*, Zerdrehung *f*, Zusammendrehung *f* ~ **of stream-lines** Windung *f* von Stromlinien
**torsion, ~ alternation testing machine** Torsionswechselprüfmaschine *f* ~ **angle** Drehungs-, Torsions-winkel *m*, Verdrehungs-winkel *m* ~ **balance** Dreh(feder)-, Torsions-wage *f* ~ **bar** Drillstab *m*, Torsionswelle *f* ~ **bar spring** Drehstabfeder *f* ~ **bar springing** Torsionsstabfederung *f* ~ **-bar suspension** Drehstabfederung *f* ~ **circle** Torsionskreis *m* ~ **constant** Torsions-, Verdrehungskonstante *f* ~ **couple** Drillung *f*, Torsionsmoment *n* ~ **endurance test** Dauerverdrehungsversuch *m* ~ **failure** Verdrehungsbruch *m* ~ **-flexure** Torsionsbiegung *f* ~ **head** Torsionskopf *m* ~ **impact** Stoßtorsion *f* ~ **indicator** Torsionsindikator *m*, -messer *m*, Verdrehungsindikator *m* ~ **-magnetometer** Torsionsmagnetometer *n* ~ **meter** Drillschreiber *m*, Torsiograf *m*, Torsionsmesser *m*, Verdrehungsmesser *m* ~ **modulus** Drillungs-, Gleit-, Scherungs-, Torsions-modulus *m* ~ **pendulum** Torsionspendel *n* ~ **rod** (spring) Drehstabfeder *f*, (vehicle) Drillstab *m*, Verdrehstab *m* ~ **spring** Drehungs-, Torsionsfeder *f* ~ **strain gauge** Torsion-Dehnungsmeßstreifen *m* ~ **suspension** Torsionsaufhängung *f* ~ **test** Drehungsversuch *m*, Torsions-probe *f*, -versuch *m*, Verdrehungsversuch *m*, Verwindeprobe *f*, Verwindungsversuch *m* ~ **testing machine** Drehfestigkeitsmaschine *f*, Torsion-(prüf)maschine *f*, Verdrehungs-festigkeitsmaschine *f*, -maschine *f*, -prüfmaschine *f* ~ **torque** Torsionsmoment *n* ~ **tube** Biegerohr *n* ~ **wire** Torsionsfaden *m*
**torsional** dreh . . . ~ **angle of twist** Verdrehwinkel *m* ~ **buckling** Drehknicken *n* ~ **clamp** Torsionsnase *f* (aviat.) ~ **deformation** Drillverformung *f* ~ **-endurance (or fatigue) limit** Drehschwingungsfestigkeit *f* ~ **force** Drall *m*, Dreh-, Drehungs-, Torsions-, Verdrehungs-kraft *f* ~ **frequency** Torsionsschwingungsfrequenz *f* ~ **instability** Drehungsinstabilität *f* ~ **modes** Drehschwingungen *pl* ~ **moment** Dreh-, Drehungs-, Torsions-moment *n*, verdrehendes Moment *n*, Verdrehungsmoment *n* ~ **oscillation** Torsions-, Verdrehungs-schwingung *f* ~ **oscillation resonance** Drehresonanz *f* ~ **quartz crystal** Torsionsquarz *n* ~ **resilience** (intensity of) Drehspannung *f*, Torsionsspannung *f* ~ **resistance** Drehungswiderstand *m*, Verdrehsteifigkeit *f*, Verdrehungswiderstand *m* ~ **resonance frequency** Torsions-Resonanzfrequenz *f* ~ **rigidity** Drill-

steifigkeit *f*, Verdrehsteifigkeit *f*, Verdrehungsfestigkeit *f* ~ **-shear strain** Gleitung *f* ~ **shearing stress** Schubspannung *f* ~ **stiffness** Drillsteifigkeit *f*, Torsionsfestigkeit *f* ~ **strain** Drehbeanspruchung *f*, (intensity of) Drehspannung *f*, Torsionsspannung *f*, Verdrehungsmesser *m* ~ **-strain indicator** Torsionsmesser *m*, Verdrehungsmesser *m* ~ **-strain meter** Verdrehungsmesser *m* ~ **strength** Dreh(ungs)festigkeit *f*, Torsionsfestigkeit *f*, Verdrehsteifigkeit *f*, Verdrehungsfestigkeit *f* ~ **stress** (intensity of) Drehspannung *f*, Dreh(ungs)beanspruchung *f*, Torsions-beanspruchung *f*, -spannung *f*, Verdrehungsbeanspruchung *f* ~ **stress-strain diagram** Spannungsschiebungsdiagramm *n* ~ **vibration** Dreh(ungs)schwingung *f*, Kristalldrehungsschwingung *f*, -drillungsschwingung *f* ~ **vibration of shaft** Drehschwingungen *pl* der Welle ~ **vibration balancer** Drehschwingungsdämpfer *m* ~ **vibration damper** Torsionsdämpfer *m* ~ **vibration damping weight** Drehschwingungsdämpfergewicht *n* ~ **vibration testing machine** Drehschwingungsprüfmaschine *f* ~ **waves** Torsionsschwingungen *pl*
**torsionally mounted** verdreht eingebaut
**torsionless** torsionsfrei
**torsometer** Elastizitätsmesser *m*
**tortuous** gekrümmt, gewunden ~ **path of lightning flash** verschlungene Blitzbahn *f*
**tortuousity** Verwindung *f*
**torus** Drehkörper *m* mit ringförmigem Querschnitt, Ring *m*, Ring-kanal *m* (Kühlturbine), -raum *m*, -wulst *f*, Torus *m*, Wulst *m*
**Tosi flap valve** Tosiklappe *f*
**toss, to** ~ durchschütteln, hin- und herwerfen, prellen, schütteln
**tossing** Rührverfahren *n* beim Schlämmen
**total, to** ~ ausmachen, sich belaufen auf, summieren **to** ~ **up** zusammenrechnen
**total** Aufzählung *f*, Betrag *m*, Endsumme *f*, Ganzes *n*, Gesamte, *n*, (grand) Gesamtsumme *f*, Summe *f*; ganz, gänzlich, gesamt, insgesamt, total, völlig ~ **of employees** Gesamtbelegschaft *f*
**total, ~ absorption** Gesamtabsorption *f* ~ **achievement** restlose Gewinnung *f* ~ **amount of flow** Gesamtdurchsatz *m* ~ **amount of insurance** Gesamtversicherungswert *m* ~ **amplitude of oscillation** Schwingungsbreite *f* ~ **angle** Gesamtwinkel *m* ~ **angle of lead** Gesamtvorhaltwinkel *m* ~ **angular momentum quantum number** Gesamtdrehimpulsquantenzahl *f* ~ **antenna resistance** Gesamtdämpfungswiderstand *m* ~ **area** Gesamt(oder)fläche *f* ~ **area of fuselage bottom** Gesamtrumpffläche *f* ~ **ash** Gesamtasche *f* ~ **attenuation** Dämpfungsmaß *n* ~ **blast-nozzle area** Blasquerschnitt *m* ~ **cable equivalent** Gesamtdämpfung *f* ~ **cable force** Gesamtseilkraft *f* ~ **capacity** Gesamtbelastung *f*, Kapazität *f* eines Schwingkreises ~ **carbon content** Gesamtkohlenstoffgehalt *m* ~ **characteristic** Schwinglinie *f* ~ **charge number** Gesamtladungszahl *f* ~ **chord of tail unit** Gesamtleitwerktiefe *f* ~ **cloudiness** Gesamtbewölkung *f*, Totalbewölkung *f* ~ **connections** Gesamtanschluß *m* ~ **consumption** Gesamtbedarf *m* ~ **correction** Gesamtverbesserung *f* ~ **cost** Gesamtunkosten *pl* ~ **counter** Sammelzähler *m* ~ **cross section measurement** Gesamtwirkungsquerschnittsmessung *f* ~

**cross section of nuclear reactions** Gesamtquerschnitt *m* von Kernprozessen ~ **current** Gesamtstrom *m* ~ **curvature** Gesamtkrümmung *f* ~ **cutting capacity** Gesamtspannquerschnitt *m* ~ **cycle** Summengang *m* ~ **cylinder capacity** Gesamtzylinderinhalt *m* ~ **cylinder displacement** Gesamtzylinderhubraum *f* ~ **declination** Gesamtabweichung *f* ~ **deflection** Gesamtseitenrichtung *f* ~ **departure** Gesamtabweichung *f* ~ **deviation** Gesamt-ablenkung *f*, -deviation *f* ~ **distortion** elektrisches Längenmaß *n*, Gesamtverzerrung *f* ~ **distortion of a telegraph cable** Längenmaß eines Telegrafenkabels ~ **drag** Gesamtwiderstand *m* ~ **duration** Gesamtdauer *f* ~ **economy** Gesamtwirtschaftlichkeit *f* ~ **effect** Gesamtwirkung *f* ~ **efficiency** Gesamtwirkungsgrad *m* ~ **electrode capacitance** Elektrodenkapazität *f* ~ **electromotive force** Ersatzspannung *f* ~ **elevation** Gesamtrhhrerhöhung *f* ~ **emission current** Emissions-, Entladungs-, Raumlade-strom *m* ~ **emission noise** Emissionsrauschstrom *m*, Gesamtemmissionsrauscher *n* ~ **exhaust energy** Auslaß-, Austritts-energie *f* ~ **expansion** Gesamtdehnung *f* ~ **expense** Gesamtaufwand *m*, -unkosten *pl* ~ **extension** Bruch-, Gesamt-dehnung *f* ~ **facilities** Gesamtheit *f* der Einrichtungen ~ **fading** Gesamtfading *n* ~ **field dose** Gesamtfelddosis *f* ~ **flying time** Gesamtflugzeit *f* ~ **flux** gesamte Kraftlinienzahl *f*, Gesamtfluß *m* ~ **force** Gesamtkraft *f* ~ **gauging error (or error of measurement)** Gesamtmeßfehler *m* ~ **ground covered** vollgezogene Flugausdehnung *f* ~ **hangover time** Abfallzeit *f* ~ **hardness** Gesamthärte *f* ~ **head** Förderhöhe *f*, Gesamthöhe *f*, totaler Druckverlust *m* ~ **heat** Gesamt-saugwirkung *f*, -wärme *f*, Gesamtwärmeinhalt *m* ~ **heat of a body** Eigenwärme *f* des Körpers ~ **holding capacity** gesamtes Fassungsvermögen *n* ~-**immersion thermometer** Eintauchthermometer *n* ~ **impact** Gesamtstoß *m* ~ **impulse** Gesamtimpuls *m* ~ **indicator reading (TIR)** Gesamtbereich *m* der Meßuhrablesung ~ **influx** Gesamthäufigkeit *f* ~ **intensity** Gesamthelligkeit *f* ~ **internal work function** Ablösearbeit *f*, Ablösungs-arbeit *f*, -energie *f* des Elektrons, Auslösearbeit *f* ~ **ionozation** Ionisierungsstärke *f* ~ **length** Baulänge *f*, Gesamtlänge *f*, gestreckte Länge *f* ~ **level width** totale Niveaubreite *f* ~ **lift** Gesamtauftrieb *m* ~ **lift of crane hook** Gesamthub *m* ~ **light emitted** Lichtsumme *f* ~ **light flux** Gesamtlichtstrom *m* ~ **load** Gesamt-belastung *f*, -last *f*, Rohlast *f* ~ **load at fracture** Bruchlast *f*, Gesamtbruchlast *f* ~ **load switch** Gesamtlastschalter *m*, Gesamt-dämpfung *f*, -verlust *m*, Totalverlust *m* ~ **losses** Gesamtausfall *m* ~ **manometric height** Förderhöhe *f* ~ **mass number** absolute Massenzahl *f* ~ **mileage odometer** Gesamtkilometerzähler *m* ~ **moisture** Gesamtfeuchtigkeit *f* ~ **moisture contents** Gesamtwassergehalt *m* ~ **moment of resistance** Gesamtwiderstandsmoment *m* ~ **number** Gesamtzahl *f* ~ **oil flow** Gesamtdurchflußmenge *f* ~ **open-loop frequency response** Frequenzgang *m* des Regelkreises ~ **output** Gesamtleistung *f* ~ **permissable transmission equivalent** zulässige Gesamtdämpfung *f*, zulässiges Übertragungsmaß *n* ~ **piston displacement** (Motor) Gesamthub-raum *m* ~ **power output of the engines** Gesamtmaschinenleistung *f* ~ **pressure** Gesamtdruck *m* ~ **printing** Summenschreibung *f* ~ **production figures** Gesamtproduktionswert *m* ~ **radiation** Gesamtstrahlung *f* ~ **ram pressure** Gesamtdruck *m* ~ **range** Gesamtskala *f* ~ **range of controller output** Stellbereich *m* ~ **range of manipulated variable** Stellbereich *m* ~ **receipts** Gesamteinnahme *f* ~ **reduction** Gesamtuntersetzung *f*, Reckgrad *m* ~-**reflection coefficient** Extinktionskoeffizient *m* ~ **refraction** Gesamtbrechung *f* ~ **resistance** Gesamtwiderstand *m*, Wechselwirkungseffekt *m* ~ **resisting effort** Gesamtfahrwiderstand *m* ~ **rolling** Summenübertragung *f* ~ **sale** Gesamtabsatz *m* ~ **sample** Rohprobe *f* (Sammelprobe) ~ **sectional area** Gesamtquerschnittfläche *f* ~ **sensitivity** Gesamtempfindlichkeit *f* ~ **sentence** (law) Gesamtstrafe *f* ~ **situation** Gesamtlage *f* ~ **solids** (refinery) Eindampfrückstand *m*, Gesamttrockensubstanz *f* ~ **soluble matter** Gesamtlösliches *n* ~ **spin** Gesamtspin *m* ~ **spread** Gesamtstreuung *f* ~ **step method** Gesamtschrittverfahren *n* ~ **strain** Gesamtspannung *f* ~ **strength** Gesamtstärke *f* ~ **stress** Summenspannung *f* ~ **suction action** (of a chimney or stack) Gesamtsaugwirkung *f* ~ **telegraph distortion** Gesamtverzerrung *f* ~ **temperature** Betriebstemperatur *f* ~ **transfer** Summenübertragung *f* ~ **transmission** Totaldurchgang *m* (acoust.) ~ **transmission equivalent** Dämpfungsmaß *n*, Gesamtdämpfung *f*, gesamtes Übertragungsmaß *n* ~ **travel** Gesamtweg *m* ~ **travel of crane hook** Gesamthub *m* ~ **travel time** Gesamtfahrzeit *f* ~ **traverse** Gesamtseitenrichtung *f* ~ **tuyére area** Blasquerschnitt *m* ~ **ultimate vacuum** Endtotaldruck *m* ~ **upcurrent (or upward) component** Gesamtaufwind *m* ~ **value** Gesamtwertung *f* ~ **valve-lift period** Gesamtventilöffnungszeit *f* ~ **view** Gesamtansicht *f* ~ **voltage** Klemmenspannung *f* ~ **weight** Gesamtgewicht *n* ~ **white count** Leukozytengesamtzahl *f* ~ **wing area** Gesamtflügelfläche *f*

**totality** Ganzes *n*, Ganzheit *f*, Gesamtheit *f*, Inbegriff *m*

**totalize, to** ~ summieren

**totalizing,** ~ **current transformer** Summenwandler *m* ~ **meter** Sammelzähler *m*

**totally,** ~ **enclosed dynamo** geschlossener Dynamo *m* ~ **overcast** ganz bedeckt ~ **reflecting prism** totalreflektierendes Prisma *n*

**tote box** Ladekasten *m*

**totter, to** ~ schlottern, wackeln, wanken

**tottering** schwankend, wankend

**touch, to** ~ anfassen, angreifen, anrühren, anschlagen, (capital) antasten, berühren, betasten, greifen, (a key of button) niederdrücken, rühren, streifen **to** ~ **and go** aufsetzen und durchstarten **to** ~ **at** (port) anlaufen **to** ~ **down** aufsetzen, landen **to** ~ **a key** eine Taste *f* niederbrechen **to** ~ **up** abziehen (mit dem Ölstein), befeilen, den letzten Schliff *m* geben, nachbessern, retuschieren, überarbeiten

**touch** Anflug *m*, Feilenstrich *m*, Fingerprobe *f*, Fühlung *f*, Gefühl *n*, Griff *m*, Pinselstrich *m*, Tastempfindung *f* **in** ~ **with** angelehnt ~ **(ing)** Berührung *f* ~ **of a key** Anschlag *m* einer Taste

**touch-and-go-landing** Landung f mit anschliessendem Durchstarten

**touchdown** Aufsetz . . ., Aufsetzen n ~ **markings** Aufsetzzonenmarken pl ~ **point** Aufsetzpunkt m ~ **zone** Aufsetzzone f ~ **zone lighting** Aufsetzzonenbefeuerung f

**touch,** ~ **needle** Probier-, Streich-nadel f ~ **rod** Kontaktelektrode f ~ **signals** Berührungszeichen pl ~ **spark** Abreißfunken m ~ **stone** Lydit m, Prüfstein m, schwarzer Kieselschiefer m, Stricheisen n ~ **typewriting** blindes Maschinenschreiben n, blindes Stanzen n ~ **welding** Schweißung f mit Schleppführung

**touchable** berührbar, tastbar

**touching** Berühren n, Streifen n ~ **ground** Aufsetzen n (aviat.) ~ **point** Berührungspunkt m ~**-up** Retusche f, Tuschieren n, Überarbeitung f (photo)

**tough** hart, unnachgiebig, zähe ~ **clinker** zähe Schlacke f ~ **cover** einseitiger Umschlag m ~ **fractures of steel** Brucharten pl des Stahles ~ **grinding** Vorschliff m ~**-pitch** hammergar ~**-pitch copper** hammergares oder zähgepoltes Kupfer n, Zähkupfer n ~**-rubber-sheated cable** Gummischlauchleitung f

**toughen, to** ~ zäh machen, zäh werden **to** ~**-copper** Kupfer n hammergar machen **to** ~ **by poling** zähpolen (metal.)

**toughened iron** Zäheisen n

**toughening** Anlassen n

**toughness** Zähfestigkeit f, Zähigkeit f ~ **of the core** Kernzähigkeit f

**toughness,** ~ **index** Zähigkeitsziffer f ~ **test** Zähigkeits-probe f, -versuch m ~**-test specimen** Zähigkeitsprobe f ~**-testing machine** Zähigkeitsprüfmaschine f

**touring** Wanderfahren n ~ **aircraft** Reiseflugzeug n, Toursitikflugzeug n ~ **car** Reisewagen m, Rundfahrtswagen m ~ **model** Tourenmodell n

**tourmaline** Turmalin n

**tourniquet** Aderpresse f, Arterienkompressorium n, Tourniquet n

**tow, to** ~ ins Schlepptau n nehmen, schleppen, treideln **to** ~ **alongside** längsseits schleppen **to** ~ **astern** achteraus schleppen **to** ~ **away** abschleppen **to** ~ **boats from land** treideln **to** ~ **a glider** ein Segelflugzeug n schleppen **to** ~ **a vessel** ein Fahrzeug n bugsieren

**tow** Packleinwand f, Schleppen n, Schlepp-arbeit f, -seil n, -tau n, -trosse f, Spinnkabel n, Werkgarn n, -tuch n ~ **bar** Schleppstange f ~ **bar pull** Zuggabelkraft f ~ **breaking** Wergreiben n ~ **brush** Bremsfilz m ~ **cable** Schlepptau n ~ **coupling** Anhängerkupplung f ~ **coupling head** Anhängerkupplungskopf m ~ **flannel** Putzlappen m ~ **hook** Schlepphaken m ~**-in** Spreizung f ~ **line** Schleppleine f, Seilzug m, Treil n, Zugleine f ~ **line winch** Verholwinde f ~ **linen** Hede-, -leinwand f ~ **model** Schleppmodell n ~**-path** Leinpfad m, Treidelweg m, Ziehweg m ~ **picker** Wergzupfer m ~ **plane** Schleppflugzeug n ~**-rod** Abschleppstange f ~ **rope** Bugsiertau n, Schlepp-seil n, -tau n, Treidelleine f, Zugtau n ~**-rope with hand hold** Knebelleine f ~ **rope eyelet** Schleppseilöse f ~ **spinning** Hedespinnerei f ~**-target cable brake** Schleppbremse f ~ **truck** Abschleppwagen m ~ **winch** Schleppwinde f

**towage** Schlepplohn m

**toward** auf, gegen ~ **the left bottom corner** schräg nach links unten hin ~ **the right top corner** schräg nach rechts oben hin

**towed** geschleppt **to be** ~ sich schleppen lassen ~ **flight** Schleppflug m ~ **sleeve target** Schleppsack m ~ **start** Schleppstart m ~ **sweep** Kabelfernräumgerät n ~ **take-off** Seilstart m ~ **target** gezogenes Ziel n, Luftsack m, Schleppscheibe f (aviat.)

**towel** Handtuch n, Wischtuch n

**tower, to** ~ auftürmen, sich auftürmen, emporragen **to** ~ **above** überragen

**tower** freitragender Mast m, Gittermast m, Kolonnen-apparat m, -turm m (refinery), Mast m, Schleppstange f, Signalmast m, Turm m ~ **car** Turmwagen m ~ **crane** Turmkran m ~ **drier** senkrechter Trockenofen m ~ **frame** Turmgerüst n ~ **like** turmartig ~ **mill** Turmwindmühle f ~ **scrubber** Berieselungsturm m ~ **silo** Futterturm m ~ **skirt** Grundmauer f, Turmsockel m ~ **slater** Turmdecker m ~ **slewing crane** Turmdrehkran m ~ **telescope** Turmteleskop n ~**-type boiler** (Turmbauart) Turmkessel m ~**-type winder** Turmförderanlage f ~ **washer** Naßreiniger m ~ **winding** Turmförderung f

**towering** burgartig, turmartig ~ **clouds** Wolkenturm m

**towing** Abschleppen n (abiat.), Schleppen n, (navy) Treideln n ~ **by aircraft** Flugzeugschlepp m ~ **of the airship** Einholen n des Luftschiffes ~ **by automobile** Autoschlepp m ~ **with a winch** Windenschlepp m

**towing,** ~ **basin** Schleppbehälter m (aviat.), Schleppkanal m, Wasserflugzeugbassin n ~**-basin test practice** Schleppversuchswesen n ~ **cable** Giertau n, Schleppseil n ~ **coupling** Abschleppkupplung f ~ **detachment detail** Abschleppkommando n ~ **drag** Schleppwiderstand m ~ **drum** (marine cable) Zugtrommel f ~ **equipment** Abschleppgerät n ~ **gear** Schleppgeschirr n ~ **hawser** Schlepptrosse f ~ **line** Treidelleine f ~ **rope** Abschleppseil n ~ **service** Abschleppdienst m ~ **test** Schleppversuch m

**town** Stadt f ~ **center** (Innenstadt) Stadtkern m ~ **drainage** Stadtkanalisierung f ~ **driving** Stadtverkehr m ~ **hall** Rathaus n ~ **network** städtisches Netz n (Elektrizität) ~ **plan** Stadtplan m ~ **planner** Stadtplaner m ~ **planning** Städteentwurf m ~ **tax** Stadtabgabe f ~ **water** Leitungswasser n

**Townsend build-up (or structure)** Ionisierungsspielaufbau m

**toxic** giftig ~ **agent** Giftstoff m ~ **effect** Giftwirkung f ~ **smoke** Giftnebel m ~**-smoke candle** Giftrauchkerze f

**toxicity** Giftigkeit f

**T-piece** (flanged) Kugel-T-Stück n, T-Stück n

**T-plot** T-Diagramm n

**trace, to** ~ abstecken, anreißen, aufnehmen, aufreißen, (curves) auftragen, ausfindigmachen, durchpausen, durchzeichnen, durchziehen, (out) entwerfen, kopieren, Linienführung f festlegen, nachspüren, nachzeichnen, pausen, reißen, schreiben, Spur f verfolgen, trassieren, untersuchen **to** ~ **back** zurückführen **to** ~ **curves** Kurven pl aufnehmen **to** ~ **errors (or**

**faults)** Fehler pl suchen **to ~ in full size** aufschnüren **to ~ with india ink** mit Tusche f ausziehen

**trace** Bindestrang m, Geschirrtau n, Hinlauf m (TV), (of a graph) Linienzug m, Schirmschrift f, Spur f (rdr), Spurpunkt m, Strang m, Strangpresse f, (harness) Tau n, Vorzeichnung f **~ of the course** Kurslinie f **~ of curve** Kurvenverlauf m **~ of rainfall** unmeßbarer Niederschlag m **~ of trajectory** Bahnspur f **~ of wear** Abnutzungs-, Verschleiß-spur f

**trace, ~ chemistry** Mikrochemie f **~ concentration** Spurenkonzentration f **~ element** Spur(en)-element n **~ hook** Zughaken m **~ interval** Zeilenabtastdauer f (TV) **~ loop** Strangschlaufe f **~ rotation system** Zeitachsendrehung f **~ support** (harness) Strangträger m **~ toggle** Brustklappe f **~ tree bar** Vorderbracke f **~ trix** Schleppkurve f

**traceable** nachweisbar, zurückführbar

**traced map (or chart)** Planpause f

**tracer** Fühlfinger m (profile), Fühlstift m, Indikator m, Kenn-faden m, -streifen m, Leitisotop n, Leuchtspur f, Punzen m, Spurensucher m (atom), (bullet) Spurgeschoß n, Stahlstift m, technischer Zeichner m, Vorzeichner m, Zeichenvorrichtung f **~ ammunition** Leuchtspurmunition f **~ bullet** Lichtspur-, Rauchspur-geschoß n **~ composition** Leuchtsatz m **~ compound** Indikatorverbindung f **~ control** Fühlersteuerung f **~ controlled** fühlergesteuert **~ current** Tasterstrom m **~ device** Anreißgerät n **~ element** radioaktiver Indikator m **~ fuse** Leuchtspurzünder m **~ grenade** Rauchbahngranate f **~ head** Fühlerkopf m (Fühlersteuerung) **~ housing** Feintastergehäuse n **~ igniter** Anfeuerungssatz m **~ method** Indikatorenmethode f **~ point** Taststift m **~ pressure** Tasterdruck m **~ projectile** Suchgeschoß n **~-propelled** rauchspurangetrieben **~ rod** Tasterstange f **~ round** Leuchtschuß m **~ shell** Leuchtspur-geschoß n, -granate f, Rauchbahngranate f **~ spindle** Fühler-, Tast-spindel f **~ studies** Indikatoren-, Spuren-untersuchungen pl **~ stylus** Taststift m **~ technique** Indikatormethode f **(colored) ~ thread** Kennfaden m **~ tip** Tasterspitze f **~ travel** Feintasterweg m **~ work** Spürarbeit f

**traces** Spuren pl **in ~** spurenweise **~ of heavier grades** Spuren pl von schwereren Sorten **~ of impurity** Spurenverunreinigung f **~ of natural gas** Erdgasspuren pl

**tracheid** Röhrenzelle f

**trachyte** Trachyt m

**tracing** Anreißschablone f, Aufnehmen n, Durchzeichnung f, Pause f, Suchen n, Zeichnung f **to make a ~** kopieren **(act of) ~** Absteckung f

**tracing, ~ of the circle** Fadkonstruktion f des Kreises **~ of conics by means of pedal points** Fußpunktkonstruktion f der Kegelschnitte **~ of curves** Kurvenaufzeichnung f **~ of an ellipse** Fadenkonstruktion f der Ellipse **~ of the ellipsoid** Fadenkonstruktion f des Ellipsoids **~ of the involutes** Fadenkonstruktion f der Evolventen **~ of the lines of curvature on a ellipsoid** Fadenkonstruktion f der Krümmungslinien auf dem Ellipsoid **~ by means of a string** Fadenkonstruktion f

**tracing, ~ accuracy** Nachfahrgenauigkeit f **~ arm** Zeichenarm m **~ cloth** Paus-leinen n, -leinwand f **~ distortion** Rillenverzerrung f (phono) **~ film** Pausfilm m **~ loss** Abspielverlust m **~ paper** Durchpauspapier n, Pauspapier n **~ point** (of planimeter) Fahrstift m **~ punch** Ziehpunze f **~ routine** Überwachungsprogramm n (info proc.) **~ speed** (oscilloscope) Schreibgeschwindigkeit f **~ tape** Trassierband n

**track, to ~** (blades) Blattlauf m einstellen, nachführen, Spur f halten, treideln **to ~ a target** führen

**track** Auflager n, Bahn f, Bahnspur f, Bettbalken m, Fahrbahn f, Fahrschiene f, Fahrstraße f, Flugweg m, Führungs-bahn f, -rille f (phono), -schiene f, Gefährt n (min.), Geleise n, Gleis n, (for seaplanes) Gleitbahn f, Kurs m, Kurs m über Grund (navig.), Laufbahn f, (bearing) Laufrille f, Schallrille f (phono), Schiene f, Schienenweg m, Speicher-rille f, -spur f, Spur f, Startschiene f, Strecke f, Sturmbahn f, Wegspur f, Zielweg m (Ortung) **(rail) ~** Schienen-bahn f, -strang m **~ of the depression** Depressionsbahn f **~ of external ring of ball bearing** Laufrille f des Kugellageraußenringes **~ of hedge** Heckenpfad m **~ on the ice** Eisbahn f **~ of lightning** Blitzbahn f **~ through portal of mine** Ausfahrweg m **~ of travel** Bewegungsbahn f **~ of the types** Typenbahn f **~ of the undercarriage** Fahrgestellspur f

**track, ~ alignment** Nachlauf m **~ angle** Geleise-, Gleis-, Spur-winkel m **~ beam** Laufschiene f **~ blobs** Bahnspurkleckse pl **~ block** Kettenglied n **~ bolt** Laschenschraube f **~ bolt for fastening fishplate** Laschenbolzen m **~-bolt hole** Laschenloch n **~ chart** Kurskarte f **~ circuit** Geleisestromkreis m, Gleisstromkreis m (mit Ruhestrom) **~ circuit with polarized relays** Gleisstromkreis m für Gleichstrom **~ circuiting for interlocking** Gleisbesetzung f **~-connecting tool (or connective fixture)** Kettenspanner m **~ curvature** Bahnkrümmung f **~ curve** Bahn-, Fahr-kurve f **~ diagram** Gleistafel f **~-diagram interlocking** Gleisbildstellwerk n **~ diagram push-button system** Gleisbildstellwerk n **~-diagrammatic push-button control system** Spurplanstellwerk n **~ displacements** Spurverschiebungen f **~ display** Flugweg-, Zielweg-darstellung f **~ edge** Spurbegrenzung f **~ engraved in film strip** eingeprägte Tonspur f **~ error** Spurfehler m (tape rec.) **~-following** Spurfahren n **~ formation** Bahnspur f **~ gate** Nachlauf-Torimpuls m **~ gauge** Geleisebreite f, Gleisbreite f (r.r.), Spur-lehre f, -maske n **~ gauge of one meter width** Meterspur f (r.r.) **~ gauge narrowing (or tightening)** Spurverengung f **~ groove** Führungsrille f (phono) **~ history** Flugweg-, Zielweg-verlauf m **~ hopper** Einwurftrichter m **~-in** Kameravorschub m **~ initiation** Zielwegbeginn m **~ jack** Zahnstangengewinde n **~ layer** Gleiskettengerät n **~ laying machinery** Gleisbaumaschinen pl **~ laying vehicle** (Gleis-) Kettenfahrzeug n **~ layout** Gleisplan m **~ level** Gleisniveau n **~ lifter** Gleis-heber m, -traghaken m **~ link steering arm** Lenkspurhebel m **~ link (or shoe)** Raupenglied n **~ lock** Gleissperre f **~ locking** (signal gear) Gleisverriegelung f **~ man** Streckenarbeiter m **~ method** Schlierenmethode *f* ~ **noise** Spurverwaschung *f* ~**-out** Zurückfahren *n* der Kamera ~ **packing** Gleisstopfen *n*, Stopfen *n* des Geleises ~ **packing tool** Schwellenstampfer *m* ~ **pin** Kettenbolzen *m* ~ **pitch** Spurverteilung *f* (tape rec.) ~ **point** Spurpunkt *m* ~ **property** Spureigenschaft *f* ~ **put on rear wheels of track-wheel vehicles** Geländekette *f* ~ **relay** Block-, Gleis-relais *n* ~ **resistance** Kriechstromfestigkeit *f* ~ **ring** Laufring *m* ~ **rod** Spurstange *f* ~ **rope** (Drahtseilbahn) Tragseil *n* ~ **runway rail** Laufschiene *f* ~ **scraper** Spurlockerer *m* ~ **section** Gleisabschnitt *m* ~ **sectioning cabin** Blockstelle *f* (r.r.) ~ **set in paving** eingepflastertes Geleise *n* ~ **shield** Kettenabdeckung *f*, Laufwerkpanzerung *f* ~ **shifter** Gleisrückvorrichtung *f* ~ **shoe** Kettenglied *n*, Raupenkettenschuh *m* ~ **shot** Fahraufnahme *f* (film) ~ **spike** Schienennagel *m* ~ **spreader** Kettenspanner *m* ~ **support** (sound on film, disc) Aufzeichnungsträger *m*, Schienenstütze *f* ~**-supporting roller** Stützrolle *f* ~ **switch operation mechanism** Weichenstellen *n* (r.r.) ~ **tamping** Gleisstopfen *n*, Stopfen *n* des Geleises ~ **way** Fahrbahn *f*, Treidelweg *m* ~ **wheel** Laufrad *n* ~ **wheel flat dump car** Flachbodenselbstentlader *m* ~ **width** Geleise-, Tonspur-breite *f*

**trackage for pile driver** Rammbahn *f*

**tracked vehicle** Kettenfahrzeug *n*

**tracker** Abstrakte *f*, Zielverfolgungsgerät *n*

**tracking** Abgleich *m* (rdo), Abtasten *n* (phono), Bahnverfolgung *f*, dynamische Hellsteuerung *f* (TV), Gleichlauf *m*, Kursflug *m* (navig.), Kriechstrom *m*, Kriechwegbildung *f* (electr.) ~ **a target** Zielverfolgung *f*

**tracking,** ~ **apparatus** Bespurungsmaschine *f* (tape rec.) ~ **beam** Spurstrahl *f* (TV) ~ **circuit** Nachlaufschaltung *f* ~ **device** Spurvorrichtung *f* ~ **distortion (or error)** Spurverzerrung *f* ~ **down seat** Fehlereingrenzung *f* ~ **down seat of interference (or trouble)** Eingrenzen *n* oder Eingrenzung *f* von Fehlern ~ **factor** Spurfaktor *m* ~ **force** Auflagedruck *m* (phono), Nadeldruck *m* (phono) ~ **merit** Flugdatenzuverlässigkeit *f* ~ **operations** Blattlaufeinstellung *f* ~ **point** (of planimeter) Fahrstift *m* ~ **radar** Verfolgungsradar *n* ~ **ring** Spurring *m* ~ **rope** Treil *n*, Zugleine *f* ~ **spindle pointer** Spurfühler *m* ~ **spot** Zielmarkierungszeichen *n* ~ **station** Verfolgungsstation *f* (g/m)

**trackless** gleislos, spurlos ~ **trolley bus (or car)** Oberleitungsbus *m*

**tract** Leitungsbahn *f* ~ **oil** Schwerpetroleum *n*

**tractable** behandelbar, gefügig

**traction** Ziehkraft *f*, Ziehung *f*, Zug *m*, Zugkraft *f* ~ **aid** Gleitschutzmittel *pl* ~ **band** Geländekette *f* ~ **battery** Fahrzeugbatterie *f* ~ **bogie** Triebdrehgestell *n* ~ **coefficient** Traktionskoeffizient *m* ~ **current** Fahrstrom *m* ~ **device** Greifvorrichtung *f* ~ **dynamometer** Zugdynamometer *n*, Zugkraftmesser *m*, Zugmesser *m* ~ **engine** Dampftraktor *m*, Lokomobil *n*, Zugmaschine *f* ~ **gear** Geländegang *m* ~ **lever** Zughebel *m* ~ **motor** Fahrmotor *m* ~**-motor blower set** Fahrmotorenlüftersatz *m* ~**-motor isolating contactors** Fahrmotorentrennschütze *m* ~ **power** Durchzug *m* ~ **power supply** Bahnstromversorgung *f* ~ **relief** Zugentlastung *f* ~ **rope** Zugseil

*n* ~ **roller** Zugorgan *n* ~ **station** Zentrale *f* ~ **test** Zugversuch *m*

**tractional resistance** Fahr-, Zug-widerstand *m*

**tractive,** ~ **effort** Zugkraft *f* am Radumfang ~ **effort at starting** Anzugkraft *f* ~ **force** Anfahrkraft *f*, Durchzugskraft *f*, Schleppkraft *f*, (magnet) Tragkraft *f*, Zugspannung *f* ~ **force meter** Zugkraftgeber *m* ~ **power** Durchzugskraft *f*, Zugleistung *f* ~ **quality** Zugleistung *f* ~ **resistance** Fahrtwiderstand *m* ~ **stress** Reibungswiderstand *m*, Schleppspannung *f*

**tractor** Ackerschlepper *m*, Elektrozugkarren *m*, Kraftschlepper *m*, Kraftzug *m*, Lokomobil *n*, Raupe *f*, Schlepper *m*, Traktor *m*, Trecker *m*, Zugkarren *m*, Zugmaschine *f*, Zugwagen *m* ~ **with platform** Traktor *m* mit Ladefläche

**tractor,** ~ **airplane** Zug-Propeller-Flugzeug *n* ~ **airscrew** Zugluftschraube *f* ~ **attachment** Anbaugerät *n* zum Schlepper ~ **binder** Schlepper-, Traktorbinder *m* ~ **brake valve** Zugwagenbremsventil *n* ~ **cleats** Traktorenkeile *pl* (Klampen) ~ **digger** Schlepperroder *m* ~ **disc harrow** Traktorscheibenegge *f* ~ **disc plow** Traktorscheibenpflug *m* ~ **drill** Traktorendrillmaschine *f* ~ **driver** Schlepperführer *m* ~ **engine** Zugschraubenmotor *m* ~ **fuel** Traktorentriebstoff *m* ~ **grub breaker plow** Schlepperbrachenpflug *m*, Traktoraufbrecherpflug *m* ~ **hitch** Schlepperanhänge-, Traktoranhänge-vorrichtung *f* ~ **hitch for corn picker** Traktoranhängevorrichtung *f* für Maispflückmaschine ~ **hitch for grain binder** Traktoranhängevorrichtung *f* für Garbenbinder ~ **hitch for grain drill** Traktoranhängevorrichtung *f* für Getreidedrillmaschine ~ **hitch for harvester thresher** Traktoranhängevorrichtung *f* für kombinierte Mäh- und Dreschmaschine *f* ~ **hitch for manure spreader** Traktoranhängevorrichtung *f* für Düngerstreuer ~ **hitch for mower** Traktoranhängevorrichtung *f* für Grasmäher ~ **hitch for push binder** Traktorhängevorrichtung *f* für Stoßbinder ~ **housing** Traktorengehäuse *n* ~ **lifter** Schlepperheber *m* ~ **lister** Schlepperdammkulturpflug *m* ~ **oil** Traktorentriebstoff *m* ~ **plow** Schlepper-, Traktor-pflug *m* ~ **propeller** Zug-propeller *m*, -schraube *f* ~ **screw** Zugschraube *f* ~ **shovel** Schaufellader *m* ~ **transmission** Schleppertriebwerk *n* ~**-truck for semitrailers** Sattelschlepper *m* ~**-type** Zugtyp *m*

**tractrix** Zuglinie *f*

**trade, to** ~ handeln, Handel *m* treiben

**trade** Beruf *m*, Betrieb *m*, Erwerbstätigkeit *f*, Fach *n*, Geschäft *n*, Geschäftslage *f*, Gewerbe *n*, Gewerbe-art *f*, -zweig *m*, Handel *m*, Handelsverkehr *m*, Handwerk *n*, Kunst *f*, Marktlage *f*, taubes Gestein *n* ~ **on a monetary basis** Geldwirtschaft *f*

**trade,** ~ **accessories** handelsmässiges Zubehör *n* ~ **agreement** Tarifvertrag *m* ~ **association** Fachgruppe *f* ~ **balance** Handelsbilanz *f* ~ **bill** Warenverzeichnis *n* ~ **designation** Handelsbezeichnung *f* ~ **education** Berufserziehungswerk *n* ~ **journal** Fachzeitschrift *f*

**trademark** Devise *f*, Fabrik-marke *f*, -zeichen *n*, Handels-marke *f*, -zeichen *n*, Markenbezeichnung *f*, Schutzmarke *f*, Warenzeichen *n* ~ **counterfeiting** Warenzeichenverfälschung *f* ~ **protection** Musterschutz *m* ~ **rights** Schutzrechte *pl*

**trade,** ~ **name** Deckname *m*, Erkennung *f*, Handels-bezeichnungs *f*, -name *m*, -zeichen *n* ~ **outlook** Konjunktur *f* ~ **school** Fachschule *f* ~ **secret** Betriebsgeheimnis *n* ~ **sign** Bildmarke *f* ~ **tax** Gewerbesteuer *f* ~ **terms** Bezugsbedingungen *pl* ~ **union** Berufsgenossenschaft *f*, Gewerkschaft *f*, Gewerkverein *m* ~ **wind** Passatwind *m* ~-**wind belt** Zone *f* der Passatwinde ~-**wind zone** Passat-gebiet *n*, -zone *f*

**trader** Handelsmann *m*, Händler *m*

**trades,** ~ **directory** Handelsadressbuch *n* ~ **man** Gewerbetreibender *m*, Kaufmann *m*

**trading,** ~ **area** Absatzbereich *m* ~ **company (or firm)** Handelsgesellschaft *f* ~ **fleet** Handelsflotte *f* ~ **house** Kaufhaus *n* ~ **seaport** Handelsseehafen *m* ~ **settlement** Handelsniederlassung

**traffic** Betrieb *m*, Handelsverkehr *m*, Straßenverkehr *m*, Verkehr *m*, Verkehrswesen *n* ~ **having little** ~ verkehrsschwach ~ **for general use** Verkehrserlaubnis *f* ~ **from the opposite direction** Gegenverkehr *m*

**traffic,** ~ **analyst** Betriebsauswerter *m* ~ **area** Verkehrskreis *m* ~ **branch** Verkehrsabteilung *f* ~ **by-law** Verkehrsbestimmungen *pl* ~ **call** Belegungsstunde *f* ~ **carried** geleisteter Verkehr *m* ~ **center** Verkehrspunkt *m* ~ **channel** Leitweg *m* ~ **circuit** Platzrunde *f* (aviat.) ~ **column** Verkehrssäule *f* ~ **control** Verkehrsleitung *f* ~ **curve** Verkehrs-bewegung *f*, -kurve *f* ~ **data** Verkehrswert *m* ~ **density** Verkehrsdichte *f* ~ **department** Betriebsdienst *m* ~ **dispatch** Verkehrsabwicklung *f* ~ **distributor** Anrufverteiler *m* ~ **factor** Betriebsfaktor *m* ~ **fluctuation** Verkehrsänderung *f* ~ **fluctuations** Verkehrsschwankungen *pl* ~ **handling** Verkehrsabwicklung *f* ~ **increase** Verkehrszunahme *f* ~ **intensity** Verkehrsstärke *f* ~ **island** Verkehrsinsel *f* ~ **jam** Straßenverstopfung *f* ~ **light** Verkehrsampel *f* ~ **load** Verkehrs-belastung *f*, -leistung *f*, -stärke *f* ~ **meter** Belegungszähler *m* ~ **noise** Verkehrsgeräusch *n* ~ **observation** Verkehrsbeobachtung *f* ~ **offered** Verkehrsangebot *n* ~ **patrol** Verkehrsstreife *f* ~ **pattern** Verkehrsschema *n*, Platzrunde *f* ~ **pilot** Mehrwegschalter *m* ~ **point** Straßenmarkierungsfarbe *f* ~ **recorder** Gesamtbelegungszähler *m* ~ **regulation** Fahrordnung *f*, -vorschrift *f*, Verkehrs-ordnung *f*, -regelung *f* ~ **regulations** Straßenverkehrsordnung *f* ~ **report** Fahrbericht *m* ~ **requirements** Verkehrsbedürfnis *n* ~ **route** Absatz-, Verkehrs-weg *m* ~ **rule** Verkehrsvorschrift *f* ~ **safety** Verkehrssicherheit *f* ~ **section** Betriebsdienst *m* ~ **sign** Verkehrstafel *f* ~ **sign(al)** Verkehrszeichen *n* ~ **signal** Verkehrssignal *n* ~ **terminal** Verkehrslandeplatz *m* ~ **tunnel** Bahnunterführung *f* ~ **unit** Gesprächs-, Verkehrseinheit *f* ~ **wave** Verkehrswelle *f*

**tragacanth** Tragantgummi *m*

**T-rail** Doppel-T-Schiene *f*

**trail, to** ~ (antenna) abhängen, hinterherschleifen, kleben, nachschleppen, schleifen, verfolgen

**trails** Fahrseil *n*, Lafettenschwanz *m*, Nacheilen *n*, Rücktrift *f* (aviat.), (ballistics) Rücktrieb *m*, Schleppe, Schwanz *m* ~ **of fall** Fallweg *m* (bomb) (line of) ~ Spur *f*

**trail,** ~ **angle** Rücktriftwinkel *m* (aviat.), Spurwinkel *m* ~ **base** Lafettentisch *m* ~ **box** Lafettenkasten *m* ~ **distance** (ballistics) Rücktriftstrecke

*f* ~ **ferry** Gierfähre *f* ~ **ferrying** Gieren *n* ~ **flying bridge** Gierbrücke *f* ~ **handspike** Griffrohr *n*, Lafettenschwanzgriff *m*, Richtbaum *m* ~ **mark** Wegmarkierung *f* ~ **plate** Lafettenplatte *f* ~ **spade** Bremsdorn *m*, Erdsporn *m*, Lafetten(schwanz)sporn *m*, Sporn *n*, Spornblech *n*, Schwanzblech *n* ~-**spade trunnion** Lagerzapfen *m*

**trailer** Anhänger *m*, Anhängewagen *m*, Bespannfahrzeug *n*, Bürstenarm *m*, (of a film) Endstreifen *m*, Wohnwagen *m* ~ **of motor freight car** Lastkraftwagenanhänger *m* ~ **for sailplanes** Segelflugzeugtransportwagen *m*

**trailer,** ~ **air brake** Anhängerbremse *f* ~ **axle** Anhängerachse *f* ~ **block** Ergänzungsblock *m* ~ **brake cylinder** Anhängerbremszylinder *m* ~ **brake linkage** Anhängerbremsgestänge *n* ~ **brake valve** Anhängerbremsventil *n* ~ **card** Folgekarte *f* ~ **coach** Beiwagen *m* ~ **control valve** Anhängersteuerventil *n* ~ **coupling** Anhänger-kupplung *f*, -vorrichtung *f* ~-**dump** Kippanhänger *m* ~ **hitch** Anhängerkupplung *f* ~ **light coupling** Anhängerlichtkupplung *f* ~ **load** Anhängerlast *f* ~ **marker** Anhängezeichen *n* ~ **mount** Lafettenfahrzeug *n* ~ **plug box** Anhängersteckdose *f* ~ **runout** (piece of blank film at end) Filmschwarz *n* ~ **sign** Anhängezeichen *n* ~ **sweeper** Anhängerkehrmaschine *f* ~ **tipping device** Anhängerkippvorrichtung *f* ~ **triangle sign** Anhängerdreieckzeichen *n*

**trailing** freihängend ~ **antenna** Hängeantenne *f* ~ **apron** Schleppsegel *n* ~ **axle** Hinterachse *f* ~ **bogie** Laufdrehgestell *n* ~ **cable** Schleppkabel *n* ~ **contact** Folge-, Schlepp-kontakt *m*

**trailing-edge** Ablaßkante *f*, ablaufende Kante *f*, Ablaufseite *f*, (of an impulse) Abstrich *m*, Achterkante *f* (aviat.), Austrittskante *f*, hintere Kante *f*, hinterer Rand *m*, Hinterkante *f*, Profilhinterkante *f*, Rückkante *f*, Schleppkante *f* ~ **of airfoil** Blatthinterkante *f* ~ **of a propeller blade** austretende Kante *f* eines Propellerflügels ~ **(of pulses)** Hinterflanke *f* (Impuls) ~ **of wing** Endleiste *f* (aviat.), Flügelhinterkante *f*

**trailing-edge,** ~ **angle** Hinterkantenwinkel *m* ~ **flap** hintere Flügelklappe *f*, Landeklappe *f*, Wölbungsklappe *f* ~ **portion of rib** Rippenendstück *n* ~ **rise time** Abfallzeit *f* der Rückflanke ~ **strip** Abschlußleiste *f*

**trailing,** ~ **end** Endband *n* (film) ~ **pointer** Schleppzeiger *m* ~ **pole tip** ablaufende Polkante *f* ~ **shaft** Achsenverlängerung *f* ~ **vortex** Wirbelschleppe *f* ~ **vortices** (air foil) abgehende Wirbelbahnen *pl* ~ **wave front** Endflanke *f* ~ **wheel** Tragrad *n* ~-**wire antenna** freihängende Antenne *f*, Hängedraht-antenne *f*, Kurbel-, (of airplane) Schlepp-antenne *f*, Weich-eiseninstrument *n*

**train, to** ~ abrichten, anlernen, ausbilden, (animals) dressieren, einarbeiten, einweisen, erziehen, schulen, ziehen **to** ~ **on** richten auf

**train** (navy) Depotschiffsverband *m*, Kette *f*, kinematische Kette *f*, Räderkette *f*, Reihe *f*, Schleppe *f*, Zug *m*, Zugvorrichtung *f* **(rolling)** ~ Strecke *f* **(supply)** ~ Kolonne *f*

**train,** ~ **of barges** Schleppzug *m* ~ **of bombs** Bombenreihe *f* ~ **of gears** Räderwerk *n* ~ **of impulses** Impulsreihe *f*, Stromstoß-reihe *f*, -serie *f* ~ **of measurand** Meßwertreihe *f* ~ **of**

**rollers** Rollenbahn *f* ~ **for rolling squeezed balls** Luppenstraße *f* ~ **of rolls** Walz-straße *f*, -strecke *f*, Walzenstraße *f* ~ **of thought** Gedankenfolge *f* ~ **of vehicles** Fahrzeugkolonne *f* ~ **of waves** Wellenzug *m* ~ **of wheels** Räder-, Zähler-werk *n*

**train,** ~ **accident** Zugunfall *m* ~ **blocking** Zugdeckung *f* ~ **call** Zuggespräch *n* ~ **conductor** Zugschaffner *m* ~ **connection** Bahnverbindung *f* ~ **control** (of signals) Signalübertragung *f* auf den Zug ~ **control office** Zugleitung *f* ~ **screw** Zugpersonal *n* ~ **describer** Zugnummernmelder *m* ~ **direction indicator** Zugrichtungsmelder *m* ~ **dispatch service** Zugabfertigungsdienst *m* ~ **dispatcher** Zugdienstleiter *m* ~ **exchange** Zugvermittlungsstelle *f* (teleph.) ~ **ferry** Trajekt *n* ~ **guard** Zugwache *f* ~**-heating plant** Zugvorheizanlage *f*. ~ **indicator** Zuglaufanzeiger *m* ~ **interrupter** Ticker *m* ~ **lighting equipment** Zugbeleuchtungsausrüstung *f* ~ **lighting system** Zugbeleuchtungsanlage *f* ~ **oil** Fischöl *n*, Fischtran *m* ~ **patrol** Zugstreife *f* ~ **progress recording apparatus** Zugzeitschreiber *m* ~ **protection installation** Zugsicherungsanlage *f* ~ **radio** Zugfunk *m* ~ **radio station** Zugbetriebsstelle *f* ~ **service** Zugdienst *m* ~ **staff** Blockstab *m* ~ **telephony** Zugtelefonie *f* ~ **traffic line** Zugverkehrsleitung *f* ~ **vehicle** Troßfahrzeug *n*

**trained** ausgebildet, geschult, vorgebildet ~ **in a profession** sachkundig ~ **mechanic** ausgebildeter oder geschulter Monteur *m* ~ **workers** Fachleute *pl*

**trainee** Anlernling *m*

**trainer** Lehr-, Lern-maschine *f*, Schulmaschine *f* ~ **airplane** Trainingsflugzeug *n*

**training** Ausbildung *f*, Heranbildung *f*, Schulbetrieb *m*, Schulung *f*, Trainieren *n*, Vorbildung *f* ~ **airport** Übungsflugplatz *m* ~ **ammunition** Übungsmunition *f* ~ **apparatus** Ausbildungseinrichtung *f* ~ **area** Übungsplatz *m* ~ **camp** Ausbildungslager *n* ~ **center** Ausbildungsstelle *f* ~ **dike** Leitdamm *m* ~ **director** Lehrgangsleiter *m* ~ **equipment** Ausbildungsgerät *n* ~ **facility** Ausbildungseinrichtung *f* ~ **field** Übungsflugplatz *m* ~ **film** Lehrfilm *m* ~ **flight** Übungsflug *m* ~ **gear** Seitenrichtvorrichtung *f* ~ **institution** Lehreinrichtung *f* ~ **instrument** Ausbildungsgerät *n* ~ **manual** Ausbildungs-, Druck-vorschrift *f* ~ **order** Ausbildungsanweisung *f* ~ **outline** Ausbildungsrichtlinie *f* **(in-factory)** ~ **period** Anlehre *f* ~ **phase** Ausbildungsstufe *f* ~ **plane** Schulflugzeug *n*, Übungsflugzeug *n* ~ **projectile** Übungsgeschoß *n* ~ **regulations** Ausbildungsrichtlinie *f* ~ **schedule** Ausbildungsplan *m* ~ **ship** Ausbildungs-, Schul-schiff *n* ~ **staff** Ausbildungsstab *m* ~ **unit** Übungsverband *m* ~ **vessel** Schulschiff *n* ~ **wall** Leitwand *f*, Leitwerk *n*

**trains run in two sections** Zwillingszüge *pl*

**traject, to** ~ (radiation etc.) hindurchwerfen

**traject** (Eisenbahn)Fähre *f*, Fährschiff *n*, Trajekt *n*, Überfahrt *f*

**trajectory** Bahn *f*, Bahnverlauf *m*, Bandkurve *f*, Bogenbahn *f*, durchstreichende Linie *f*, Fallbahn *f*, -kurve *f*, Flugbahn *f*, (of projectile) Geschoßbahn *f*, Kegelschnittlinie *f*, Schußbahn *f*, Trajektorie *f*, Wegzeitkurve *f*, Wurflinie *f* **(air)** ~ Luftbahn *f* ~ **of the bomb** Wurfbahn *f* der

Fallkurve ~ **of ray** Strahlverlauf *m* ~ **of wind** Windbahn *f*

**trajectory,** ~ **band** Trajektoriengurt *m* ~ **chart** grafische Schuß- oder Schießtafel *f* ~ **computation** Bahnvermessung *f* ~ **diagram** Flugbahnbild *n* ~ **motion** Zugrichtung *f* ~ **oscillation** Bahnschwingung *f* ~ **parabola** Wurfparabel *f* ~ **path track** Bahnrichtung *f* ~ **plane** Bahn-, Flugbahn-ebene *f* ~ **zenith** Scheitel *m* der Flugbahn

**tram** Förderwagen *m*, Laufkarren *m*, Straßenbahn *f* ~**-board** Lauf-bohle *f*, -brett *n* ~**-board bracket** Laufbrettstütze *f* ~ **clearer** Förderaufseher *m* ~ **rail** Falzschiene *f* ~ **road** Förderstrecke *f* ~ **silk** Einschlagseide *f* ~**-way** Schienenweg *m*, Straßenbahn *f*, Trambahn *f* ~**-way rail** Rillenschiene *f*

**trammel** Ellipsen-, Oval-, (beam) Stangen-zirkel *m*, Kesselhaken *m*, Schleppnetz *n*

**trammels or universal dividers** Stangen- oder Universalzirkel *m*

**trammer** Fördermann *m* (min.), Wagenstößer *m*

**trammers** Förderleute *pl*

**tramming** Handförderung *f*

**tramp, to** ~ auf Gelegenheitsladung *f* fahren

**tramp,** ~ **iron separator** Eisenabscheider *m* ~ **steamer** Trampdampfer *m*

**transaction** Geschäft *n*, Marktlage *f*, Transaktion *f* ~ **on account** (exchange) Fixgeschäft *n*

**transactions** Abhandlungen *pl*, Bargeschäft *n*

**transadmittance** Gegenscheinleitwert *m*

**transatlantic,** ~ **plane** Transatlantikflugzeug *n* ~ **receiver** Überseempfänger *m* ~ **trade** Überseehandel *m* ~ **traffic** Atlantikverkehr *m*

**transceiver** Lautsprecherempfänger *m*, Senderempfänger *m*, Sendeempfangs-apparat *m*, -gerät *n*, Sprechhörer *m* ~ **radio set** Empfängersenderkombination *f*

**transcendental** transzendent

**transcident light** durchfallendes Licht *n*

**transconductance** fortschreitende Welle *f*, gegenseitige Leitfähigkeit *f*, Gegenwirkleitwert *m*, Röhrenkonduktanz *f*, Steilheit *f*, Steilheitsröhre *f*, Übergangsleitwert *m* ~ **bridge** Querleitungsbrücke *f*

**transcribe, to** ~ übertragen, umschreiben

**transcriber** Übertragungsgerät *n*, Umschreiber *m*

**transcription** Übertragung *f*, Übertragung *f* eines Textes vom Band, Umschreiben *n* ~ **record** Diktafonplatte *f*

**transcrystalline** transkristallin ~ **crack** quer durch das Kristallkorn verlaufender Riß *m* ~ **rupture** intrakristalliner Bruch *m*

**transcrystallization** Transkristallisation *f*

**transducer** Energieumwandler *m*, Kraftübertrager *m*, Meßwertwandler *m*, Übersetzer *m*, Übertrager *m*, Übertragunssystem *n*, Umformer *m*, Umsetzungsgerät *n*, Umwandler *m*, Umwandlungsvorrichtung *f*, Vierpol *m*, Wandler *m* ~ **dissipation loss** Streuverlust *m* (acoust.), Wärmeableitungsverlust *m* ~ **equation** (quadruple equation) Vierpolgleichung *f* ~ **gain** Wirkverstärkung *f* ~ **loss** Wirkdämpfung *f* ~ **pulse delay** Eingang-Ausgang-Intervall *n*

**transduction** Leistungsumsatz *m*, Umsetzung *f*

**transductor** Regeldrossel *f*, Übertragungssystem *n*, Umsetzungsgerät *n*, vormagnetisierte Spule *f* ~ **reactor** Blindlasttransduktor *m*

**transept** Querschiff *n*
**transfer, to** ~ abdrücken, abgeben, (potential, etc.) abnehmen, abtreten, (money) anweisen, aufstoßen, befördern, transportieren, überführen, überführen (info proc.), übertragen, überweisen, umschalten, umsteigen, verlegen, versetzen **to** ~ **from one account to another** stornieren **to** ~ **to another bed** umbetten **to** ~ **a cable** ein Kabel *n* verlegen **to** ~ **a call** eine Verbindung *f* umlegen **to** ~ **a circuit to a reserve position** eine Leitung *f* auf einen freien Platz legen **to** ~ **property** übereignen
**transfer** Abdruck *m*, Förderung *f*, Fortleitung *f*, Transport *m*, Überführung *f*, Übergang *m*, Überladen *n*, Übertrag *m*, Übertragung *f*, Überweisung *f*, (by pouring) Umguß *m*, Umlegung *f*, Versetzung *f*, Wanderung *f* ~ **(ring)** Verlegung *f*

**transfer,** ~ **of current** Stromübernahme *f* ~ **of energy** Energie-, Sprach-übertragung *f* ~ **of force** Kraftübertragung *f* ~ **by hand** Handübertragung *f* ~ **of liquids** Weiterbeförderung *f* von Flüssigkeiten ~ **of material** Werkstoffwanderung *f* ~ **of possession** Besitzeinweisung *f* ~ **of registry** Registerübertragung *f* ~ **of a telephone connection** Übertragung *f* eines Fernsprechanschlusses
**transfer,** ~ **admittance** Übertragungsleitwert *m* ~ **bead** Randwulst *m* ~ **belt conveyor** Abzugsband *n* ~ **blocking** Blockiersystem *n* ~ **blower** Umfüllventilator *m* ~ **button** Signaltaste *f* ~ **car** Förder-, Transport-, Zubringer-wagen *m* ~ **case** Hauptvorgelege *n* ~ **chain** Förderkette *f*, Transportkette *f* ~ **characteristic** Charakteristik *f* (der Röhre), Übergangsleitwert *m* ~ **charges** Überführungskosten *pl* ~ **check** Transportkontrolle *f*, Übertragungsprüfung *f* ~ **chute** Überleitrutsche *f* ~ **circuit** Anschalte-, Dienst-, Verbindungs-leitung *f* (zwischen zwei Plätzen eines Amtes) ~ **constant** Übertragungskonstante *f* ~ **copying book** Durchschreibebuch *n* ~ **cross section** Übergangswirkungsquerschnitt *m* ~ **deed** Übertragungsurkunde *f* ~ **element** Übertragungsglied *n* ~ **exchange** Überleitungsamt *n* ~ **factor** Übertragungsfaktor *f* ~ **frame** (Rollenkühlbett) Schieber *m* ~ **frequency** Zwischenfrequenz *f* ~ **gantry** Bockkran *m* ~ **grid** Überlaufrost *m* ~ **impedance** Kopplungswiderstand *m* ~ **jack** Umschalte-, Verlege-klinke *f* ~ **joint** Knotenverbindung *f* ~ **key** Zuteilschalter *m* ~ **ladle** Transportpfanne *f* ~ **line** Überweisungsleitung *f* (teleph.) ~ **loss** Kopierverlust *m* (film) ~ **molding** Spritzpreßformung *f* ~ **number** Überführungszahl *f* ~ **operator** Spitzenplatzbeamtin *f* ~ **paper** abziehbares Papier *n*, Bügelpapier *n*, Umdruckpapier *n* ~ **period** Umschlagdauer *f* ~ **picture** Abziehbild *n* ~ **pipe** Übergangsstutzen *m* ~ **plunger cylinder** Preßspritzzylinder *m* ~ **port** Überstromkanal *m* ~ **position** Spitzenplatz *m* ~ **poster** Abziehplakat *n* ~ **printing** Abziehbilderdruck *m* ~ **printing apparatus** Bügeldruckapparat *m* ~ **process** Umdruck *m* (print.) ~ **pump** Überleitungs-, Umfüll-, Zubringer-pumpe *f* ~ **rate** Übertragungsverhältnis *n* ~ **resistance** Übergangswiderstand *m* ~ **roller (or feeder)** Abroller *m* ~ **signal** Umlegungszeichen *n* ~ **star** Überführungsstern *m* ~ **stock** Rollgut *n* ~ **substation** Koppelstelle *f*, Über-

gabestelle *f* ~ **switch** Umschalter *m* ~ **ticket** Verrechnungs-, Überweisungs-scheck *m* ~ **time** Übertragungszeit *f* ~ **tripping** (powerline protection) Freigabesystem *n* ~ **truck** Transportwagen *m* ~ **trunk exchange** Überweisungsfernamt *n* ~**-type (stripping) paper** Abziehpapier *n* (photo) ~ **unit** Aufgabevorrichtung *f*, Übergangszelle *f* ~ **yard** Umladebahnhof *m*
**transferable** übertragbar, verlegefähig ~ **security** übertragbares Wertpapier *n*
**transferee** Erwerber *m*, Indossat(ar) *m*, Zessionär *m*
**transference,** ~ **number** (of solution) Übergangszahl *f* ~ **of oil** Öldurchtritt *m*
**transferor** Übertragender *m*
**transferred charge call** R-Gespräch *n*
**transferrer** Abtrage-, Übertragungs-vorrichtung *f*, Indossant *m*, Indossent *m*, Umdrucker *m*
**transferring** Überschub *m* ~ **device** Aufstoßeinrichtung *f*
**transfiguration** Transfiguration *f*
**transfinite** überendlich
**transfluxor** Transfluxor *m*
**transform, to** ~ transformieren, umbilden, umformen, umgestalten, umsetzen, umspannen, umwandeln, verändern, verwandeln **to** ~ **the current** den Strom *m* umspannen
**transformability** Umwandelbarkeit *f*
**transformable** umwandelbar, verwandelbar
**transformal function** Abbildungsfunktion *f*
**transformation** Abänderung *f*, Änderung *f*, Transformation *f*, Transformierung *f*, Überführung *f*, Umbildung *f*, (of an equation) Umformung *f*, Umsetzung *f*, Umwandelbarkeit *f*, Umwandlung *f*, Verwandlung *f*, Wandlung *f* ~ **of coordinates** Koordinatentransformation *f* ~ **of energy** Energie-umsatz *m*, -umwandlung *f* ~ **of an equation** Umrechnung *f* einer Gleichung ~ **of material into clay** Vertonung *f* ~ **into peat** Vermoorung *f*
**transformation,** ~ **apparatus** Umbildungsgerät *n* ~ **chain** radioaktive Zerfallsreihe *f* ~ **constant** Zerfallskonstante *f* ~ **drawing apparatus** Umzeichengerät *n* ~ **figure** Übersetzungszahl *f* ~ **period** Umwandlungsperiode *f* ~ **point** Umwandlungspunkt *m* ~ **process** Umwandlungsvorgang *m* ~ **range** Umwandlungsbereich *m* ~ **ratio** Nennübersetzung *f*, Übersetzungs-verhältnis *n*, -zahl *f*, Umformungsverhältnis *n*, Umwandlungsverhältnis *n*, Wandlungsverhältnis *f*
**transformed** umgeformt
**transformer** Beförderung *f*, Beträger *m*, Gitterüberträger *m*, Polwechsler *m*, Spannungswandler *m*, Trafo *m*, Trägerformator *m*, Transformator *m*, Turm *m*, Umformer *m*, Umformersatz *m*, Umspanner *m*, Umwandler *m*, Wandler *m* **(regulating)** ~ Auf- und Abspanner *m* (electr.) **(step-down)** ~ Abspanner *m* (electr.) ~ **for automatic regulator** Wandler *m* für Schnellregler *m* ~ **with iron core** Übertrager *m* mit Eisenkern
**transformer** ~ **amplifier** Transformatorverstärker *m* ~ **capacity** Transformator-, Umspannerleistung *f* ~ **circuit** Stamm-, Übertrager-kreis *m* ~ **coil** Übertrager *m* (teleph.) ~ **container** Transformatorgehäuse *n* ~ **core** Transformatorkern *m* ~**-coupled** mit Transformatoren *pl* gekoppelt ~**-coupled amplifier** Übertragerverstärker *m* ~

**coupling** Anpassungsübertragung *f*, induktive Kupplung *f*, Transformatorenkupplung *f* ~ **current** Transformatorenstrom *m* ~ **furnace** Transformatorofen *m* ~ **kiosk** Umspannerzelle *f* ~ **lamination** Transformatorenblech *m* ~ **oil** Transformatorenöl *n* ~**-operated ammeter** Strommesser *m* für Wandleranschluß ~**-operated meter** Zähler *m* für Wanderanschluß ~ **output winding** Transformatorausgangswicklung *f* ~ **rating** Umspannerbelastbarkeit *f* ~ **ratio** Transformatorübersetzung *f*, Übersetzung *f* ~**-repeating amplifier** Verstärker *m* mit magnetischer Kopplung ~ **sheet** Transformatorenblech *m* ~ **shell** Transformatorengehäuse *n* ~ **station** Abspannwerk *n*, Umspannstation *f* ~ **substation** Transformatorenstation *f*, Umspannwerk *n* ~ **sweep unit** Transformatorkippgerät *n* ~ **tank** Transformator-gefäß *n*, -gehäuse *n* ~ **tap** Transformatoranzapfung *f* ~ **time-base** Transformatorkippgerät *n* ~ **truck** Transformatoren-, Verstärkungs-wagen *m* ~ **voltage** Transformatorspannung *f* ~ **winding** Transformatorwicklung *f*

**transformerless output stage** eisenlose Endstufe *f*

**transforming** Entzerrung *f* (photo), Umformung *f* ~ **plant** Umspannungsanlage *f* ~ **section** Transformationsstück *n*

**transfusion** Eingießung *f*

**transgress, to** ~ übergreifen, verletzen

**transgressions** Überschreitung *f*, Übertretung *f*

**transgrid action** Durchgriff *m*

**transiency** Flüchtigkeit *f*

**transient,** ~ **analyzer** Schwingungsmodell *n* ~ **behaviour** Einschwing-, Übertragungs-verhalten *n*, zeitliches Verhalten *n* ~ **component** flüchtige Komponente *f* ~ **creep** Übergangskriechen *n* ~ **current** Anpassungs-, Ausgleichs-, Einschwing-strom *m* ~**-decay current** Nachwirkungsstrom *m* ~ **distortion** Laufzeitverzerrung *f*, Sprungverzerrung *f* (acoust.), Verzerrung *f* durch Ein- und Ausschwingen oder durch Einschwingvorgänge ~ **earth fault** Erdschluß-wischer *m* ~ **effect** Ausgleichsvorgang *m*, Übergangszustand *m* ~ **equilibrium** Übergangsgleichgewicht *n* ~ **imperfections** vorübergehende Störungen *pl* ~ **impulse** flüchtiger Stromstoß *m* ~ **internal voltage** (servo) Hauptfeldspannung *f* ~ **load response** Ausregelzeit *f* ~ **motion** Ausgleichsvorgang *m* (acoust.) ~ **noises** Geräusche *pl* durch Einschwingvorgänge ~ **peak** Einschaltspritze *f* ~ **period** Dauer *f* des Ausgleiches bei Stromschließung, Einschwingzeit *f* ~ **phenomenon** Ausgleichsvorgang *m*, Ein- und Ausschwingen *n*, Einschwingvorgang *m*, flüchtiger Vorgang *m*, Übergangserscheinung *f* ~ **problems** Anlaufvorgänge *pl* ~ **process** Einschaltvorgang *m* ~ **ray** durchfallender Strahl *m* ~ **recovery** Ausgleichsvorgang *m* ~ **response** (servo) Einschwingverhalten *n*, Übergangsfunktion *f*, zeitlicher Verlauf *m* ~ **response curve** Frequenzdurchlaßkurve *f* ~ **response measuring set** Sprungfunktionsmeßgerät *n* ~ **sound** Übergangston *m* ~ **stability** dynamische Stabilität *f* ~ **stage** Übergangszustand *m* ~ **state** Einschwingzustand *m*, vorübergehender Zustand *m* ~ **target** Augenblicksziel *n* ~ **three-phase shortcircuit current** transitorischer Kurzschlußwechselstrom *m* ~ **time** Einschwingzeit *f*

~ **transaction** vorübergehender Vorgang *m* ~ **voltage** Ausgleichspannung *f*, flüchtige Spannung *f* ~ **wave** Wanderwelle *f*

**transientness** Flüchtigkeit *f*, Vergänglichkeit *f*

**transillumination** Durchleuchtung *f* ~ **ball stage** Durchleuchtungskugeltisch *m* ~ **lamp** Durchleuchtungslampe *f*

**transinformation content** Transinformationsgehalt *m*

**transistor** Halbleitertriode *f*, Transistor *m* ~ **circuit** Transistorschaltung *f* ~**-circuit bandwidth** Transistorfrequenzband *n* ~ **metal** Transistorenmetall *n* ~ **pre-amplifier** Transistorvorverstärker *m*

**transistorized** transistorisiert ~ **amplifier** Transistorverstärker *m*

**transit, to** ~ hindurchgehen, passieren, überqueren

**transit** Anschneidetheodolit *m*, Durch-fahren *n*, -fahrt *f*, -fuhr *f*, -gang *m*, Durchgangsverkehr *m*, Nivellierinstrument *n*, Passageinstrument *n*, Richtkreis *m*, Transit *m*, Transport *m*, Umschlag *m*, (additional) Umtelegrafierung *f*, Verpflanzung *f* **in** ~ in Zuführung *f* **to be in** ~ in eins peilen ~ **of armature** Ankerumschlag *m*

**transit,** ~ **administration** Durchgangsverwaltung *f* ~ **angle** Laufwinkel *m*, (of klystron) Wegwinkel *m* ~ **bearing** Deckpeilung *f* ~ **book** Durchgangsbuch *n* (teleph.) ~ **camp** Durchgangslager *n* (mil.) ~ **case** Transportbehälter *m* ~ **charge** Durchgangsgebühr *f* ~ **circuit** Durchgangsleitung *f* ~ **duty** Durchfuhr-, Durchgangs-zoll *m* ~ **exchange** Durchgangszentrale *f* ~ **expenses** Durchfuhrkosten *pl* ~ **goods** Durchgangsgut *n* ~ **instrument** Richtkreis *m* ~ **line** durchgehende Leitung *f* ~ **office** Durchgangsamt *n* ~ **path** Leitbahn *f* ~**-path motion** Leitbahnbewegung *f* ~ **phase angle** Laufzeitwinkel *m* ~ **rate** Durchgangsgebühr *f* ~ **telescope** durchschlagbares Fernrohr *n* ~ **theodolite** Universalinstrument *n*, Theodolit *m* mit durchschlagbarem Fernrohr ~ **ticket** Durchgangsblatt *n* (teleph.)

**transit,** ~**-time** (of armature) Ansprechzeit *f*, Durchlaufzeit *f*, (oscillation) Magnetronleitbahnwelle *f*, Sprungzeit *f*, Übertragzeit *f*, Übertragungszeit *f*, Umschlagszeit *f* ~ **of minority carriers** Laufzeit *f* der Nebenträger

**transit-time,** ~ **compensation** Laufzeitausgleich *m* ~ **compression** Komprimieren *n* der Laufzeit, Laufzeitkomprimierung *f* ~ **correction** Laufzeitentzerrung *f* ~ **damping** Laufzeitdämpfung *f* ~ **distortion** Laufzeitverzerrung *f* ~ **effect** Laufzeit-einfluß *m*, -erscheinung *f* ~ **frequency** Leitkreisfrequenz *f* ~ **frequency of a magnetron** Magnetronleitkreisfrequenz *f* ~ **mode** Laufzeitmodus *m* ~ **oscillation** Leitbahnwelle *f* ~ **oscillations** Laufzeitschwingungen *pl* ~ **tube** Laufzeitröhre *f*, (klystron) Triftröhre *f*

**transit,** ~ **traffic** Durchgangs-, Übergangs-verkehr *m* ~ **trade** Durchgangshandel *m*

**transition** Überführung *f*, Übergang *m* ~ **of electron resulting in emission of radiation** Resonanzsprung *m* ~ **from glow to arc** Übergang *m* von Glimmentladung zu Lichtbogen ~ **with nontransition metals** Übergangs-mit Nicht-Übergangsmetallen *pl* ~ **of radiation heat** Strahlungswärmeübergang *m*

transition, ~ altitude Übergangshöhe *f* (aviat.) ~ anode Kommutierungsanode *f* ~ color Übergangsfarbe *f* ~ curve Übergangs-bogen *m*, -krümmung *f*, -kurve *f* ~ distance Übergangsstrecke *f* ~ effect Übergangseffekt *m* ~ elbow Übergangsbogen *m* ~ energy Übergangsenergie *f* ~ factor Falschanpassungs-, Fehlanpassungsfaktor *m*, Reflexionskoeffizient *m* ~ field Übergangsfeld *n* ~ fit Haftsitz *m*, Passung *f* mit teilweiser Überschneidung der Toleranzfelder ~ flow Übergangsströmung *f* ~ jet (Vergaser) Übergangsdüse *f* ~ layer Übergangsschicht *f* (aviat., opt.) ~ level Übergangsfläche *f* (aviat.) ~ loss Übergangsverlust *m* ~ member Übergangsglied *n* ~ period Übergangszeit *f* ~ piece Übergangsstück *n* ~ point Ablösungspunkt *m*, Sprungpunkt *m*, Übergangs-punkt *m*, -stelle *f*, Umschlagspunkt *m* ~ point of boundary layer Umschlagspunkt *m* (aerodyn.) ~ probability Sprung-, Übergangs-wahrscheinlichkeit *f* ~ range Umschlagsgebiet *n* ~ region Übergangsgebiet *n* ~ relation Übergangsrelation *f* ~ rock Übergangsgebirge *n* ~ screw Durchführungsschraube *f* ~ shore spans Übergangsrampen *pl* (leichte Bauweise) ~ shot Überbrückungsaufnahme *f* (film) ~ stage Übergangs-stadium *n*, -stufe *f*, -zustand *m* ~ steel Übergangsstahl *m* ~ strip Übergangsstreifen *m* ~ substance Übergangsbestandteil *m* ~ temperature Übergangstemperatur *f* ~ time Elektronenlaufzeit *f*, Sprungzeit *f*, Übergangszeit *f* ~-time distortion Laufzeitverzerrung *f* ~ time oscillography Laufzeitoszillografie *f* ~ time stretch Laufzeitstrecke *f* ~-type Übergangs-glied *n*, -muster *n* ~ zone Gleitung *f*, Übergangs-, Umwandlungs-zone *f*

transitional, ~ agreement Übergangsvertrag *m* ~ case Übergangsfall *m* ~ surface Übergangsfläche *f*

transitivity Transitivität *f*

transitory nichtständig, vorübergehend

transitron circuit Transitronschaltung *f*

transitting of the telescope Durchschlagen *n* des Fernrohres

translate, to ~ übersetzen, übertragen, umrechnen, umsetzen

translated übersetzt not ~ unübersetzt

translating, ~ circuit Demodulator *m*, Modulatorschaltung *f* ~ device Übersetzer *m*, Umrechner *m*, Umsetzer *m* ~ place Aufnahmestelle *f* (telegr.) ~ relay Übertragungsrelais *n*

translation fortschreitende Bewegung *f*, Modulation *f*, Parallelverschiebung *f*, Translation *f* (cryst.), Übersetzung *f*, Übertragung *f*, Umrechnung *f*, Umsetzung *f* ~ in the hyperbolic plane Schiebung *f* in der hyperbolischen Ebene ~ of tidal wave Fortpflanzung *f* der Flutwelle

translation, ~ banding Translationsstreifung *f* ~ conversion (automatic telephony) Umrechnung *f* ~ field Umrechner-, Wählerkontakt-feld *n* ~ gliding striae Translationsstreifung *f* ~ plane Translations-ebene *f*, -fläche *f* ~ point Belichtungsrolle *f* (film)

translational weitertragend, Fortschreitungs . . . ~ motor Zusatzrakete *f* ~ movement Fortschreitungsbewegung *f* ~ partition function Translationszustandssumme *f* ~ velocity Längsgeschwindigkeit *f*

translator Übersetzer *m*, Übertrager *m*, Umleiter *m*. Umrechner *m*, Umsetzer *m* (data proc.), Umwerter *m* (teleph.), Zuordner *m* (comput.) ~ case Übertragerkästchen *n* ~ key Translatorenschlüssel *m* ~ system Direktorsystem *n*

translatory translatorisch ~ key Übertragerschalter *m* ~ motion fortschreitende Bewegung *f*, Parallelverschiebung *f* ~ movement Vorwärtsbewegung *f* ~ shift Parallelverschiebung *f* ~ speed Fortbewegungsgeschwindigkeit *f*

transload, to ~ umschlagen

transloading, ~ point Umschlagstelle *f* ~ vehicle Empfangsteilwagen *m*

translocation diastase Translokationsdiastase *f*

translucency Durchsichtigkeit *f*

translucent durch-scheinend, -sichtig, lichtdurchlässig, optisch dünn ~ picture Durchsichtsbild *n* ~ screen Durchprojektionswand *f*

transmarine cable Überseekabel *n*

transmission (message) Abgabe *f*, Aussendung *f* (rdo), Beförderung *f*, Durch-gabe *f*, -laß *m*, Durchlässigkeit *f* (acoust.), Fortpflanzung *f*, (of power) Kraftübertragung *f*, Leitung *f*, Sendung *f*, Transmission *f*, Triebwerk *n*, Triebwerksleitung *f*, Überführung *f*, Übermittlung *f*, Übersetzung *f*, Übertragung *f*, Vermittlung *f*, Verpflanzung *f*, Vorgelege '*n*, (gear) Wechselgetriebe *n*, Weiterbeförderung *f* (radio) ~ Senden *n* (signal) ~ Verständigung *f*

transmission, ~ of communications Nachrichtenübermittlung *f*, -übertragung *f* ~ by double current Doppelstrombetrieb *m* ~ of electrical energy (or power) elektrische Kraftübertragung *f* ~ of feed Vorschubübertragung *f* ~ of heat Wärme-abfuhr *f*, -übertragung *f* ~ of information Benachrichtigung *f* ~ of intelligence Nachrichten-übermittlung *f*, -übertragung *f* ~ of light Durchstrahlung *f* ~ of load Lastübertragung *f* ~ of metering intelligence Meßwertübertragung *f* ~ of morsecode messages Morsen *n* ~ of motion Bewegungsübertragung *f* ~ of orders Befehlsübermittlung *f* ~ of pictures and images Fernseh- und Bildübertragung *f* ~ of radiations Durchstrahlung *f* ~ of readings Fernschreiben *n* ~ by rods Gestängeübertragung *f* ~ of signals Zeichen-gabe *f*, -gebung *f* ~ by simplex current Einfachstrombetrieb *m* ~ of sound Tonsendung *f* ~ of telegrams Telegrammbeförderung *f* ~ of telephone and radio messages Befördern *n* von Fernsprüchen und Funksprüchen ~ of tractive force Schulübertragung *f*

transmission, ~ agent Transmissionsorgan *n*, Übertragungs-mittel *n*, -organ *n* ~ angle Gesamtübertragunswinkel *m* ~ audibility Verständlichkeit *f* ~-balancing network Kunstschaltung *f* ~ band Durchlässigkeitsbereich *m* eines Filters ~ belt Treib-, Übertragungs-riemen *m* ~ brake Getriebebremse *f* ~ brake drum Getriebebremstrommel *f* ~ cable Übertragungsleitung *f* ~ cable of transmission equipment Renkstecker *m* ~ case Getriebeflansch *m* ~ chain Transmissionsgliederkette *f* ~ channel Sendekanal *m*, Übertragungs-kanal *m*, -weg *m* ~ characteristics Durchlaßlinie *f*, Übertragungskenngröße *f* ~ checking camera Fotojustierkamera *f* ~ coefficient Durchdringwahrscheinlichkeit *f*, Durchlässigkeits-, Übertragungs-faktor *m* ~ constant Übertragungskon-

stante *f* ~-**control apparatus** Sendekontroll-
apparat *m* ~ **curve** Übertragungskurve *f* ~
**deception** Täuschung *f* durch Ausstrahlung ~
**delay** Übertragungsverzögerung *f* ~ **device**
Übersetzgerät *n* ~ **diagram** Sendediagramm *n* ~
**diffraction** Durchstrahlungsbeugung *f* ~ **distortion measuring set** Verzerrungsmeßgerät *n* ~
**drive** Transmissionsantrieb *m* ~ **drive shaft**
Getriebewelle *f* ~ **dynamometer** Einschalt(ungs)-
dynamometer *m*, Torsionsindikator *m* ~ **efficiency** Gesamtdämpfung *f*, Gesamtübertra-
gungsmaß *n*, gesamtes Übertragungsmaß *n*,
Übertragungswirksamkeit *f* (teleph.) Übertra-
gungswirkungsgrad *m*, Wirkdämpfung *f* ~-
**efficiency measuring set** Dämpfungsmesser *m*,
Streckendämpfungsmesser *m* ~-**efficiency test**
Streckendämpfungsmessung *f* ~ **equation** Lei-
tungsgleichung *f* ~ **equipment** Sendeeinrich-
tung *f*, Übertragungsamt *n* ~ **equivalent** Be-
zugs-, Leitungs-dämpfung *f*, Übertragungs-
äquivalent *n*, -maß *n* ~ **equivalent in miles of
standard cable** Übertragungsmaß *n* in Meilen
Standardkabel ~ **factor** Durchlässigkeit *f* ~
**filter** durchlässige Kette *f* ~ **filter circuit** Über-
tragungsfilter *n* ~ **frequency** Übertragungs-
frequenz *f* ~-**frequency characteristic** Übertra-
gungs-frequenzlinie *f*, -kurve *f* ~ **frequency
meter** Durchgangswellenmesser *m* ~ **gain** Entdämpfung *f*, Übertragungsgewinn *m* ~ **gear**
Arbeitsrad *n*, Getriebe *n*, Transmissionsvorge-
lege *n*, Vorgelege *n*, Zahnradübersetzung *f* ~
**gear(ing)** Übersetzungsgetriebe *n* ~ **gear case**
Übertragungsradgehäuse *n* ~ **gear ratio** Zahn-
radübersetzungsverhältnis *n* ~ **gearing** Triebwerk *n* ~ **guide** Leitung *f* ~ **housing** Getriebe-
gehäuse *n* ~ **identification** Übermittlungsken-
nung *f* ~ **impairment** Minderung *f* der Übertragungsgüte *f* ~ **intermission** Sendepause *f* ~
**lag** Übertragungsverzögerung *f* (servo) ~ **layout**
Fernsprechübertragungsmaßregeln *f* ~ **level**
Pegel *m*, relativer Pegel *m*, Sprach-höhenlinie *f*,
-niveaulinie *f*, -verlauf *m*, Übertragungs-niveau
*n*, -pegel *m* ~ **level diagram** Pegellinie *f* ~ **level
meter** Pegelmesser *m* ~ **line** Energieleitung *f*,
Feeder *m*, Lecher(draht)system *n*, Triebwerk-
leitung *f*, Übertragungsleitung *f*, Wellenleitung
*f*, Zwischenwelle *f* ~-**line amplifier** Kettenverstärker *m* ~ **line stub** Übertragungsleitungs-
stück *n* ~ **lock** Getriebeschloß *n* ~ **loss** Über-
tragungsverlust *m* ~ **lubricant** Getriebe-fett *n*,
-öl *n*, Schmieröl *n* für Transmission, Transmissionsöl *n* ~ **machinery** Triebwerk *n*, Triebwerks-
anlage *f* ~ **main shaft** Getriebehauptwelle *f* ~
**maintenance work** Netzüberwachung *f* ~
**maximum** Übertragungsmaximum *n* ~ **measure**
Übertragungsmaß *n* ~ **measurement** Durch-
lässigkeits-, Übertragungs-messung *f* ~ **measuring set** Dämpfungszeiger *m*, Pegelmesser *m* ~
**medium** Übertragungsweg *m* ~ **method** (ultramicroscope) Durchstrahlungsverfahren *n* ~
**network** Siebkette *f* ~ **oil** Transmissionsöl *n* ~
**part** Triebwerkteil *m* ~ **path** Übertragungsweg
*m* ~ **performance** (in rating) Übertragungsgüte
*f* ~-**power drive chain** Transmissionstreibkette *f*
~ **range** Durchlaß-bereich *n*, -weite *f*, Durchlässigkeitsbereich *m*, Loch-bereich *m*, -breite *f*,
-breite *f* eines Frequenzsiebes, Frequenzdurchlässigkeit *f*, Sendereichweite *f*, Übertragungs-

bereich *m* ~ **range of band pass** Lochgrenze *f* ~
**rating** Betriebsqualität *f* ~ **ratio** Übersetzungsverhältnis *n* ~ **reference system** Übertragungs-
vergleichssystem *n* ~ **rod** Übertragungsstange *f*
~ **rods** Übertragungsgestänge *n* ~ **rope** Trans-
missions-, Triebwerk-seil *n* ~ **selector** (switch)
Senderwahlschalter *m* ~ **shaft** Getriebe-,
Haupt-, Transmissions-, Triebwerk-welle *f* ~
**shock absorber** Übersetzungsstoßdämpfer *m* ~
**spectrograph** Durchstrahlungspektograf *m* ~
**speed** Übermittlungs-, Übertragungs-geschwindigkeit *f* ~ **standard** Sendenorm *f*, Übertra-
gungs-norm *f*, -normal *n* ~ **step pulley** Ge-
triebestufenscheibe *f* ~ **substation** Netzstation *f*
~ **suspension** Getriebeaufhängung *f* ~ **system**
Drahtrundspruchanlage *f*, Folgezeiger *m*, Über-
tragungs-anlage *f*, -leitung *f*, -system *n*, -weg *m*
~ **target** Durchlaßtreffplatte *f* ~ **technique**
Übertragungstechnik *f* ~ **test** Dämpfungsmessung *f* ~ **theory** Leitungstheorie *f* ~ **time** Laufzeit *f* ~ **tower** Leitungsturm *m* ~ **trouble** mangelhafte Verständigung *f* ~-**type ultramicroscope**
Durchstrahlungsübermikroskop *n* ~ **unit** Über-
tragungs-einheit *f*, -maß *n*

**transmissivity** Durchlässigkeit *f*, (atmosphere)
Durchlässigkeit *f*, (light) Transparenz *f*
**transmit, to** ~ (message) abgeben, aussenden,
befördern, (message) durchgeben, durchlassen,
fortpflanzen, geben, hindurchlassen, senden,
überführen, übermitteln, übersetzen (mech.),
übertragen, überweisen, verpflanzen, verstän-
digen **to** ~ **to a distance** fernanzeigen, fernleiten
**to** ~ **heat** Wärme *f* durchlassen **to** ~ **picture**
bildsenden
**transmit,** ~ **branch** Sendeseite *f* ~-**receive switch
(T/R switch)** Empfängersperrschalter *m* (rdr),
Sende/Empfangs-schalter *m*
**transmittal** Weiterbeförderung *f*
**transmittance** (Rein-)Durchlaßgrad *m*, Durchlässigkeit *f* (acoust.)
**transmittancy** Transparenz *f*
**transmitted** ferngetastet **to be** ~ (current) durchfallen
**transmitted,** ~ **band** Frequenzsieblochbreite *f* ~
**band of frequencies** durchgelassener Frequenzbereich *m*, Lochbreite *f* eines Frequenzsiebes ~
**color** Durchsichtsfarbe *f* ~ **energy** hindurchge-
lassene Energie *f* ~ **frequency band** Übertra-
gungsbereich *m* eines Übertragungssystems ~
**illumination** regrediente Beleuchtung *f* ~ **light**
ausfallendes Licht *n*, durchfallender Strahl *m*
~ **sideband** Hauptseitenband *n* ~ **wave** ausge-
sandte Welle *f*, Sendewelle *f*
**transmitter** Fernsender *m* (beim Messen und
Regeln), Geber *m* (eines Meßgerätes), Sender *m*,
Sendestation *f*, Übermittler *m*, Übertrager *m*
(electr.), Übertragungsmittel *n* ~ **for intermediate and short waves** Grenz- und Kurzwellensender *m* ~ **and receiver** Funkgerät *n* ~ **in
telephone handset** Mikrofon *n*
**transmitter,** ~ **accessories** Senderzubehör *n* ~
**amplifier** Sendeverstärker *m* ~ **antenna** Sendeantenne *f* ~ **apparatus** Gebeeinrichtung *f* ~
**arm** Mikrofon-arm *m*, -träger *m* ~ **battery**
Mikrofon-batterie *f*, -element *n* ~ **blocker cell**
Sendersperröhre *f* ~ **button** Einsatzkapsel *f*,
Mikrofonkapsel *f*, Sprechtaste *f* ~ **circuit**
Mikrofonstromkreis *m* ~ **connection** Sender-

anschluß *m* ~ **construction technique** Senderbau *m* ~ **contact** Sendekontakt *m* ~ **control** Senderaussteuerung *f* ~ **control station** Senderzentrale *f* ~ **cubicle** Sendeschrank *m* ~ **current** Erst-, Mikrofon-strom *m* ~ **current supply** Mikrofonspeisung *f* ~ **cutout** Lauthörknopf *m* ~ **decrement** Senderdekrement *n* ~ **diaphragm** Mikrofonmembran *n* ~ **distributor** Lochstreifensender *m* ~ **frequency** Senderfrequenz *f* ~ **hook-up** Senderanschluß *m* ~ **input** Sendereingang *m* ~ **input polarity** Polarität *f* des Videosignals ~ **inset** Einsatzkapsel *f*, Mikrofon-einsatz *m*, -kapsel *f* ~ **installation** Sendeanlage *f* ~ **mounting frame** Senderahmen *m* ~ **mouth-piece** Mikrofontrichter *m* ~ **noise** Mikrofongeräusch *n* ~ **operator** Sendebeamter *m* ~ **output power** Sendeausgangsleistung *f* ~ **plug** Geberstecker *m* ~- **power monitor** Kontrolldiode *f* ~ **selection panel** Senderwahltafel *f* ~ **set** Senderanlage *f* ~ **standard** Mikrofonnormal *n* ~ **station** Senderanlage *f* ~ **truck** Sende(e)wagen *m* ~ **unit** Sprechkapsel *f* ~ **valve** Senderöhre *f* ~ **voltage** Mikrofonspannung *f*

**transmitting,** ~ **in the clear** Klartextfunken *n* ~ **and receiving attachment** Doppelverkehrszusatz *m* (rdo) ~ **and receiving set** Sende- und Empfangsgerät *n*

**transmitting,** ~ **antenna** Sendeantenne *f* ~ **basis of earth telegraphy** Sendebasis *f* der Erdtelegrafie ~ **cam cylinder (or sleeve)** Sendersteuerbuchse *f* (teletype) ~ **capacity** Übertragungsfähigkeit *f* ~ **characteristics** Durchlaßkennlinie *f* ~ **direction** Durchlaßrichtung *f* ~ **disc** Sendescheibe *f* ~ **distortion** Sendeverzerrung *f* ~ **distributor** Sende-, Übertragungs-verteiler *m* ~ **end** Geberseite *f* (einer Meßeinrichtung) ~ **equipment** Sendeanlage *f* ~ **filter** Sendebandfilter *n* ~ **frequency** Sendefrequenz *f* ~ **key** Sendetaste *f* ~ **lever** Sendehebel *m* ~ **line** Sendeleitung *f* ~ **loop** Sendeschleife *f* ~ **loop loss** Übertragungsverluste *pl* ~ **means** Übertragungsmittel *n* ~ **medium** Übertragungs-medium *n*, -mittel *n* ~ **pentode** Sendepentode *f* ~ **power (or property)** Durchlässigkeit *f* ~ **receiver switch** Sende-Empfangsschalter *m* ~ **relay** Senderelais *n* ~ **resistance** Durchlaßwiderstand *m* ~ **ring** Sendering *m* ~ **segment** Sendesegment *n* ~ **shaft** Senderachse *f* ~ **station** drahtloser Sender *m*, Geber-amt *n*, -station *f*, Sende-amt *n*, -station *f*, -stelle *f*, Senderamt *n*, Senderstation *f*, Vermittlungsstation *f* ~ **stylus** Abtaststift *m* ~ **tape** Sendestreifen *m* ~ **tube** Senderöhre *f* ~ **-type device** Geber *m* (eines Meßgerätes) ~ **-type gauge** Fernmeßinstrument *n*, Meßgerät *m* mit Fernübertragung ~ **valve** Senderöhre *f* ~ **valve cooler** Senderöhrenkühler *m* ~ **wave** Geberwelle *f* (rdo)

**transmutable** verwandelbar

**transmutation** Umwandlung *f* ~ **of radiation heat** Strahlungswärmeübergang *m*

**transmutation,** ~ **constant** Zerfallkonstante *f* ~ **function** Umwandlungsfunktion *f*

**transmuted wood** imprägniertes Holz *n*

**transocean receiver** Überseempfänger *m*

**transoceanic** transozeanisch, überseeisch ~ **air traffic** Ozeanluftverkehr *m* ~ **broadcast** Fernfunk *m* ~ **communication** Übersee-verbindung *f*, -verkehr *m*

**transom** Ausleger *m*, Bodenwrange *f*, Holm *m*, Quer-holz *n*, -riegel *m*, -träger *m*, Riegel *m*, Traverse *f*, Unterzug *m* ~ **of a French casement** Weitstab *m* ~ **of a gin** Riegel *m* eines Hebezeugs

**transom,** ~ **bed** Lenkschemel *m*, Wendeschemel *m* ~ **board** Asbestschieferbrett *n* ~ **boards** Asbestwand *f* ~ **screw** Querholzschraube *f*

**transonic** transsonisch, Transschall … Überschall ~ **flow** schallnahe Strömung *f* ~ **nozzle flow** schallnahe Düsenströmung *f* ~ **region** schallnaher Bereich *m* ~ **speed** Übergang *m* zur Überschallgeschwindigkeit *f* ~ **zone** Übergangsgebiet *n*

**transoral screen** Tonbildwand *f*

**transparence** Durchgriff *m*, Leuchtbild *n*

**transparency** Durch-lässigkeit *f*, -sichtigkeit *f*, Glasbild *n*, Klarheit *f*, Lichtdurchlässigkeit *f*, Schattenverhältnis *n*, Sicht *f*, Sichtigkeit *f*, Transparenz *f*

**transparent** durch-leuchtend, -scheinend, -sichtig, hell, klar, lichtdurchlässig, transparent, wasserhell ~ **to ultraviolet light** ultraviolettdurchlässig

**transparent,** ~ **area** Licht *n* (photo, print.) ~ **cellulose film** Zellglas *n* ~ **color** Lasurfarbe *f* ~ **copy** Durchleuchtungskopie *f* ~ **cover sheet** durchsichtiges Deckblatt *n* ~ **cupola** Kuppel *f* aus durchsichtigem Werkstoff ~ **cut** Dünnschliff *m* ~ **decimal grid** Zielgevierttafel *f* ~ **density** Durchsichtdichte *f* (film) ~ **diapositive** Transparentbild *n* ~ **envelope** Fensterbriefumschlag *m* ~ **finish** Transparentausrüstung *f* ~ **flatting varnish** Lasurschleiflack *m* ~ **index label** Fensterreiter *m* ~ **layer** lasierende Schicht *f*, Lasurschicht *f* ~ **lines** Linienblatt *n* ~ **pane** Lichtscheibe *f* ~ **photocathode** Durchsichtskathode *f* ~ **picture** Diaphanie *f* ~ **plastic material** durchsichtiger Werkstoff *m* ~ **positive** Diapositiv *n* ~ **reflector** Reflexglas *n* ~ **screen** Durchsichtsschirm *m* (TV) ~ **spot** Licht *n* (photo, print.) ~ **window** Lindemann-, Strahlenaustritts-fenster *n* ~ **wings** durchsichtige Tragflächen *pl*

**transpiration manometer** Strömungsmanometer *n*

**transplant, to** ~ verpflanzen, versetzen

**transplantation** (Gewebe) Verpflanzung *f*, Pfropfen *n*, Transplantation *f*, Versetzung *f*

**transponder** Antwortsender *m*, Impulsübertrager *m* (sec. rdr)

**transport, to** ~ befördern, fördern, fortschaffen, transportieren, überführen, übertragen, verfahren, verlegen

**transport** Beförderung *f*, Förderung *f*, Fuhre *f*, Transport *m*, Truppenschiff *n* ~ **of blocks** Beförderung *f* der Blöcke ~ **(ation) by cart** Kartenförderung *f* ~ **of charge** Ladungstransport *m*, Ladungsüberleitung *f* ~ **of energy** Sprachtransport *m* ~ **of momentum** Impulstransport *m*

**transport,** ~ **accessory** Transporteinrichtung *f* ~ **airplane** Transportflugzeug *n* ~ **approximation** Näherung *f* der Transporttheorie ~ **arm** Schaltbolzen *m* ~ **bridge** Transportbrücke *f* ~ **bucket** Transportbecher *m* ~ **car** Förderwagen *m* ~ **case** Transportbehälter *m* ~ **check** Transportkontrolle *f* (comput.) ~ **cross section** Transportquerschnitt *m* ~ **cycle**

Transportrad $n$ ~ **device (or equipment)** Transporteinrichtung $f$ ~ **efficiency** Transportwirkungsgrad $m$ ~ **element** Transportglied $n$ (comput.) ~ **equation** Transportgleichung $f$ ~ **good** Rollgut $n$ ~ **lever** Umsteuerhebel $m$ für vor- und rückwärts ~ **line** Förderleitung $f$ ~ **lock** Sperrvorrichtung $f$ (Kraftheberblock) ~ **mechanism** Laufwerkplatte $f$ (tape rec.) ~ **number** (of ions) Überführungszahl $f$ ~ **pawl** Förderklinke $f$ ~ **phenomena** Übergangsphänomen $n$ ~ **pilot** Verkehrsflugzeugführer $m$ ~ **plane** Transporter $m$ ~ **pool** Fahrbereitschaft $f$ ~ **speed** Transportgeschwindigkeit $f$ ~ **system** Transporteinrichtung $f$ ~ **term** Transportglied $n$ ~ **theorem** Transportsatz $m$ ~ **trailer** Transportanhänger $m$ ~ **truck** Transportkarre $f$ ~ **utensil** Transportgerät $n$ ~ **vessel** Transportgefäß $n$

**transportable** fahrbar, förderbar, transportfähig

**transportation** Abbeförderung $f$, Beförderung $f$, Fortschaffung $f$, Transport $m$, Verkehrswesen $n$, Wanderung $f$ ~ **of excavated material** Bodentransport $m$, Erdtransport $m$ ~ **of freight** Frachtenbeförderung $f$ ~ **in motion toward** Antransport $m$

**transportation,** ~ **box** Transportkasten $m$ ~ **charges (or costs)** Versandkosten $pl$ ~ **device** Fördermittel $n$ ~ **equipment** Förderanlage $f$ ~ **lag** Totzeit $f$ ~ **order** Fahrbefehl $m$ ~ **roller** Tragwalzenroller $m$ ~ **system** Transportsystem $n$ ~ **test** Fahrversuch $m$ ~ **ticket** Fahrkarte $f$

**transporter** Fördereinrichtung $f$, Förderer $m$, Transporteur $m$, Übertragwalze $f$ ~ **crane** Verladebrücke $f$ ~ **travel gear** Brückenfahrwerk $n$

**transpose, to** ~ auf die andere Seite $f$ einer Gleichung schaffen, (wires) kreuzen, transponieren, umordnen, umsetzen, umstellen, umwandeln, versetzen, verstellen, vertauschen **to** ~ **two wires** zwei Leitungen $pl$ kreuzen

**transposed** gestürzt (math.), transponiert ~ **diagrams** versetzte Diagramme $pl$ ~ **line** gekreuzte Leitung $f$ ~ **pair** gekreuzte Fernsprechleitung

**transposer** Umsetzer $m$

**transposing** Kreuzen $n$, Versetzung $f$ ~ **equalization** (cable) Kreuzungsausgleich $m$ ~ **gear** Umwandlungsrad $n$

**transposition** (of wires) Kreuzung $f$, Lagewechsel $m$ (teleph.), Umlagerung $f$, Umsetzung $f$, Umstellung $f$, Verdrillen $n$, Verdrillung $f$, Versetzung $f$, Vertauschung $f$ ~ **of pairs** Platzwechsel $m$, Schleifenkreuzung $f$ ~ **of wires** (line) Drahtkreuzung $f$, Leitungskreuzung $f$

**transposition,** ~ **bracket** Einschiebestütze $f$ ~ **cipher** (signal) Kastenschlüssel $m$ ~ **insulator** Doppel-, Kreuzungs-isolator $m$ ~ **line** Linie $f$ mit gekreuzten Leitungen ~ **point** Abschnitt- (teleph.), Kreuzungs-, Umstellungs-punkt $m$ ~ **pole** Abschnittsgestänge $n$ (teleph.), Kreuzungs-Gestänge $n$, -stange $f$, Platzwechselgestänge $n$ ~ **receiver** Zwischenfrequenzempfänger $m$ ~ **reception** Transponierungsempfang $m$ ~ **section** Kreuzungsabschnitt $m$ ~ **section point** Kreuzungsfestpunkt $m$ ~ **step** Kreuzungsabstand $m$ ~ **system** Kreuzungs--folge $f$, -system $n$, Platzwechselfolge $f$

**transpositions** Platzwechsel $m$

**transrectifier** Gleichrichter $m$ mit Gittersteuerung

**transship, to** ~ überladen, umladen

**transshipment** Überladen $n$, Überladung $f$, Überschiffung $f$, Umladung $f$, Umschiffung $f$ ~ **charges** Umschlagkosten $pl$ ~ **port** Umschlagshafen $m$

**transshipping,** ~ **of goods at the waterside** Uferumschlag $m$ ~ **device** Umladeeinrichtung $f$

**transvection** Überschiebung $f$

**transversal** Extremal-Normale $f$, Quer ..., Standlinienschlitten $m$; diagonal, quer hindurchlaufend, schräg verlaufend, transversal ~ **carriage** Basiswagen $m$ ~ **effect** (of crystal) Quereffekt $m$ ~ **forming** Planformarbeit $f$ ~ **forming attachment** Planformvorrichtung $f$ ~ **lamination** falsche Schieferung $f$ ~ **stay** Windanker $m$ ~ **vibration** Querschwingung $f$

**transversality condition** Transversalitätsbedingung $f$

**transversally ribbed (or flanged)** quergerippt

**transverse, to** ~ durchschneiden

**transverse** Ausleger $m$, Bügel $m$, Querwelle $f$; diagonal, quer, querlaufend, schräg, transversal ~ **to direction hearing** quer zur Pfeilrichtung $f$ geneigt

**transverse,** ~ **aberration** seitliche Abweichung $f$ ~**-advance-clutch** Plan-vor-Kupplung $f$ ~ **axis** Hauptachse $f$, Holmachse $f$, Querachse $f$, Y-Achse $f$ ~ **bar** Quer-bewehrung $f$, -stab $m$ ~ **beam** Biegebalken $m$, Sattelholz $n$ ~ **bend test** Querbiegeversuch $m$ ~**-bending resiliance** Biege-, Biegungs-spannung $f$ ~**-bending strength** Durchbiegungsfestigkeit $f$ ~**-bending test** Biegeversuch $m$ ~**-bending-test machine** Biegeprüfungsmaschine $f$ ~**-bending-test shackle** Einspannkopf $m$ für Biegeversuche ~ **bore** Querbohrung $f$ ~ **bracing** Querverband $m$ ~ **bulkhead** Querschott $m$ ~ **cable** Quertragseil $n$ und -draht $m$ ~ **circuit** Querkreis $m$ ~ **conductivity** Querleitfähigkeit $f$ ~ **contraction** Quer--schrumpfung $f$, -zusammenziehung $f$ ~ **contraction ratio** Querkontraktionszahl $f$ ~ **copying** Plankopierdrehen $n$ ~ **copying attachment** Plankopiervorrichtung $f$ ~ **crack** Querriß $m$ ~ **crest** Querkamm $m$ ~ **cylinder** Planzylinder $m$ ~ **diameter** Querdurchmesser $m$ ~ **direction** Planrichtung $f$, Quere $f$ ~ **elasticity** elastische Durchbiegung $f$ ~ **electric wave** Transversalwelle $f$ ~ **error** Querfehler $m$ ~ **expansion** Querdehnung $f$, Querdehnungsziffer $f$ ~ **failure** Ausknickung $f$ ~**-fatigue test** Biegeermüdungsversuch $m$ ~ **fault** Querverwerfung $f$ ~ **feed** Planvorschub $m$ ~ **feed gear** Planzuggetriebe $n$ ~ **fiber feed** Querfaserspeisung $f$ ~ **field** (magnetic) Querfeld $n$, Transversalfeld $n$ ~ **(magnetic) field loss** Querfeldverlust $m$ ~ **finning** Querberippung $f$ ~ **fins** Querrippen $pl$ ~ **fissure** Querspalte $f$ ~ **flattening test** Querfaltversuch $m$ ~ **flue** Querzug $m$ ~ **flux** Querfluß $m$ ~ **focusing** Querfokussierung $f$ ~ **fold** Querfalte $f$ ~ **folding test** Querfaltversuch $m$ ~ **forward clutch** Planvor-Kupplung $f$ ~ **frame** Rahmenquerspant $n$ ~ **frame of the fuselage** Rumpfspant $m$ ~ **gallery** Quer-

stollen *m* ~ **girder** Binderquer-, Ring-träger *m* ~ **groove** Längsnut *f* ~ **guiding link** Querlenker *m* ~ **hole** Querloch *n* ~ **impact test** Schlagbiegeversuch *m* ~ **inclination** Querneigung *f* ~ **jetty** (at right angles to another) Querdamm *m* ~ **joint** Querfuge *f* ~ **lamina** Holzfüllstück *n* ~ **level** Verkantungslibelle *f* ~ **load** horizontale Last *f*, Querlast *f* ~ **loading** Biegebelastung *f* ~ **longitudinal bearing** Querlängslager *n* ~ **loop** Längsschleife *f* ~ **magnet** Ausgleichungsquermagnet *m* ~ **magnetic field** Quermagnetfeld *n* ~ **magnetic (or magnetron)wave** TM-Welle *f* ~ **metacenter** Breitenmetazentrum *n* ~ **motion** Querbewegung *f* ~ **movement** Plangang *m*, Querbewegung *f* ~ **oscillation** Querschwingung *f* ~ **pitch** Querteilung *f*, Umfangsstellung *f* ~ **plane** Stirnfläche *f* ~ **pole** Querschiffspol *m* ~-**pressure angle** Stirneingriffswinkel *m* ~ **pulse** Querimpuls *m* ~-**read pulse** Querleseimpuls *m* ~ **resilience** Durchbiegungsspannung *f* ~-**return** „Plan-Zurück" ~ **return clutch** Planrücklauf-Kupplung *f* ~ **ring** Querring *m* ~ **ring bracing** Ringverspannung *f* ~ **rotary movement** Planbewegung *f* ~ **scale** Transversalmaßstab *m* ~ **seam** Querriß *m* ~ **seat** Quersitz *m* ~ **section** Quer-profil *n*, -schnitt *m*, Schnitt-ansicht *f*, -darstellung *f* ~ **section of ring** Ringquerschnitt *m* ~ **septum** Markstrahl *m* ~-**sheet pilling** Querspundwand *f* ~ **ship magnet** Querschiffsmagnet *m* ~ **shrinkage** Querschrumpfung *f* (film) ~ **shutter** (blinker) Querschnittsblende *f* ~ **sill** Querschwelle *f* ~ **slope** Querneigung *f* ~ **slot** Quernute *f* ~ **spacing** Querteilung *f* ~ **spar** Querholm *m* ~ **spherical aberration** seitliche sphärische Abweichung *f* ~ **spring** Querfeder *f* ~ **stability** Querstabilität *f* ~ **stiffening** Querversteifung *f* ~ **stop** Plananschlag *m* ~ **strain** Biege-, Biegungs-, Durchbiegungs-spannung *f*, Querdehnung *f*, Querdehnungsziffer *f* ~ **strength** Biege-, Biegungs--festigkeit *f* ~ **stress** Biege-, Biegungs-, Durchbiegungs-spannung *f*, Querzug *m*, Seitenzug *m*, seitlicher Zug *m* ~ **stress-strain diagram** Spannungsdurchbiegungsdiagramm *n* ~ **stripe** Querstreifen *m* ~ **strut** Hilfsachse *f* ~ **support girder** Riegel *m* ~ **suspension** Queraufhängung *f* ~ **taper gib** Querkeilleiste *f* ~ **test** Biegeversuch *m* ~ **testing machine** Biegemaschine *f* ~ **valley** Quertal *n* ~ **vibration** Kristallquerschwingung *f*, Querschwingung *f*, Transversalschwingung *f* ~ **vibrations** Dickenschwingungen *pl* ~ **wave** Querwelle *f*, Transversalwelle *f*
**transversely** im rechten Winkel *m* ~ **corrugated** quergewellt ~ **jointed chain** Kreuzgelenkkette *f* (acoust.) ~ **wave system** Querwellensystem *n* ~ **wire** Quertrag-seil *n* und -draht *m*
**transversing condenser roller** Nitschelwalze *f*
**transverter** Umrichter *m*
**trap, to** ~ (electrons) einfangen, einfassen, einkesseln, fangen
**trap** (separating vessel) Abscheider *m*, Abzugsgrube *f*, Auffangvorrichtung *f*, Fallbrücke *f*, Falle *f*, (petroleum) Flüssigkeitsabscheider *m*, Haltestelle *f* (trans.), Heberohrverschluß *m*, Hinterhalt *m*, Klappe *f*, Sperrer *m*, Sperrkreis *m*, Syphon *n*, Tasche *f*, Verschluß *m*, Wasserschluß *m* ~ **board** Fallklappe *f* ~ **box** Auf-

fangkasten *m* (print.) ~ **cover** Verschlußdeckel *m* ~ **door** Fall-tor *n*, -tür *f*, Klappe *f*, Klapptür *f* ~ **door of a pit** Schachtdeckel *m* ~ **door on roof** Dachausssteigluke *f* ~-**door closer** Klapptürschließer *m* ~-**fork elevator** Fanggabelförderstuhl *f* ~ **system** Schleusensystem *n* ~ **tuff** Basalttuff *m*
**trapeze** Trapez *n* ~ **bar** Hängekorb *m* (aviat.) ~ **superstructure** Trapezfachwerkträger *m*
**trapezium distortion** Schlußsteinverzerrung *f*, Trapez-fehler *m*, -verzeichnung *f*
**trapezohedral** trapezoedrisch
**trapezoid** Trapez *n*, Trapzoid *n* ~ **tool** Trapezstrahl *m*
**trapezoidal** trapezförmig ~ **base** trapezförmige Unterlage *f* ~ **coal** Moorkohle *f* ~ **distortion** Trapezfehler *m* ~ **insert** trapezförmige Einlage *f* ~ **load** Trapezbelastung *f* ~ **pulse** Trapezimpuls *m* ~ **raster** Trapezraster *m* ~ **rule** Trapezregel *f* ~ **thread** Trapezgewinde *n* ~ **washer** Trapezring *m*, ~ **wave** Trapez--schwingung *f*, -spannung *f*, -welle *f*
**trapped** eingefangen, (liquids, air bubbles) eingeschlossen ~ **electrons** angelagerte Elektronen *pl* ~ **fuel** Restkraftstoff *m* ~ **mode** geleitete Schwingungsart *f* ~ **oil** Quetschöl *n*, verdrängte Ölmenge *f* ~ **sand** Sandnest *n* ~ **slag** Schlackeneinschluß *m*
**trapping** Abfangen *n* ~ **spot** Fangstelle *f*
**trash** Abfall *m*, Unrat *m* ~ **catcher** Strohfänger *m* ~ **disposal** Verwendung *f* des Abfalls ~ **rack** Einlaufgitter *n*, (water power, screen) Grobrechen *m*, Schmutzrechen *m* ~-**rack bar** Rechenstab *m* ~-**rack cleaner** Harke *f* zur Reinigung des Rechens
**trass** Duckstein *m*, Traß *m* ~ **concrete** Traßbeton *m* ~ **mortar** Traßmörtel *m*
**Trauzl,** ~ **block** (explosives) Bleiblock *m* ~ **block expansion** Bleiblockausbauchung *f*
**travel, to** ~ befahren, bewegen, fahren, reisen, strömen, wandern **to** ~ **around** umfahren **to** ~ **down** niedergehen **to** ~ **downward** abwärtsstreichen **to** ~ **out** auslenken (g/m) **to** ~ **through** durchstreichen, hindurchlaufen, laufen **to** ~ **through an angle** einen Winkel *m* durchlaufen **to** ~ **up** aufsteigen, auftreiben **to** ~ **upward** emporsteigen
**travel** Arbeitsbewegung *f*, (of a coil diaphragm, instrument needle, or beam) Ausschlag *m*, Bewegung *f*, (Bearbeitung) Durchgang *m*, Hinlauf *m*, Hub *m*, Lauf *m*, Umschlag *m* (mach.), Weg *m*, Zeigerausschlag *m* (aviat.) ~ **of drill spindle** Bohrspindelhub *m* ~ **of hook** Hakenweg *m* ~ **of sand** Sandwanderung *f* ~ **of the slide valve** Schieberhub *m* ~ **of target during dead time** Vorauswanderungsstrecke *f*
**travel,** ~ **allowance** Fahr(t)kostenentschädigung *f*, Reise-spesen *pl*, -zuschuß *m* ~ **compensating spring** Wegausgleichfeder *f* ~ **direction** Fahrrichtung *f* ~ **drive** Fahrtantrieb *m* ~ **expenses** Reisekosten *pl* ~ **limiting spring** Wegfeder *f* ~ **order** Fahr-ausweis *m*, -befehl *m*, Fahrtliste *f* ~ **permit** Fahrausweis *m* ~ **resistance** Wanderwiderstand *m* ~ **shot** Folgeaufnahme *f* ~ **stop** Bewegungssperre *f*, Endanschlag *m*, Spielraumsperre *f* ~ **stroke** Federweg *m* ~ **time** Laufzeit *f* (rdr) ~ **velocity** Wanderungsgeschwindigkeit *f*

**traveler** Ausholring *m* (mech.), Fliege *f*, (stone) Läufer *m*, Reisender *m* ~ **for rivetting shops** Nietkran *m*

**traveling** Bewegung *f*, Wandern *n*, Wanderung *f*; bewegend, laufend, (waves) wandernd ~ **as a conveyer** fahrbar ~ **of ions** Ionenwanderung *f* ~ **on overhead rails** Drehlaufkatze *f*

**traveling,** ~ **apron** Beförderungstuch *n*, Laufschürze *f* ~ **article** Reiseartikel *m* ~-**belt screen** Bandsieb *n* ~ **block** (oil drilling) Hampelmann *m* ~ **blower** Wandergebläse *n* ~ **board** Rollbrett *n* (film) ~ **bracket crane** Konsollaufkran *m* ~ **breaker plate** umlaufende Brechplatte *f* ~ **bridge** fahrbare Brücke *f*, Schiebebrücke *f* ~ **cam** Wandernocken *m* ~ **carriage** Lauf-, Rohr-wagen *m* (artil.) ~ **crab** Laufkatze *f* ~ **crane** Bedienungswagen *m*, Brückenkran *m*, Fahrkran *m*, Kran *m*, Laufkran *m*, Rollkran *m*, Wandlaufkran *m* ~ **crane for emergency gate** Laufkran *m* für Notverschluß ~-**crane girder** Laufkranträger *m* ~ **detector** Detektorsonde *f*, mitlaufender Meßleitungsdetektor *m* ~ **distance** Fahrbereich *n* ~ **feed grippers** Zuführgreifer *m* ~ **feed table** Kettenbandeinleger *m*, Tisch *m* mit automatischer Einlage ~ **field** Wanderfeld *n*, wanderndes Feld *n* ~ **gantry** (for blocks) Portalkran *m* ~ **grate** Wanderrost *m* ~-**grate firing** Wanderrostfeuerung *f* ~-**grate width** Wanderrostbreite *f* ~ **gripper shaft** Welle *f* zum Greifer ~ **hoist** Laufwinde *f* ~ **hopper** Zubringerwagen *m* ~ **ladder** Schiebeleiter *m* ~ **light crane** Beleuchtungsbrücke *f* ~ **masks** bewegliche Masken *pl* (film) ~ **mechanism** Fahrwerk *n* ~ **motion** Fahrbewegung *f* ~ **motor hoist** Motorlaufwinde *f* ~-**overhead-hoist assembly** Katzfahrwerk *n* ~ **paddle mixer** fahrbares Rührwerk *n* ~ **plane wave** fortschreitende ebene Welle *f* ~ **plants** bewegliche Anlagen *pl* ~ **platform** Schiebe-bühne *f*, -brücke *f* ~ **position** Fahrstellung *f* ~ **rail** Unterflanschlaufwerk *n* ~ **ranging pole** Reisefluchtstab *m* ~ **receiver** Reiseempfänger *m* ~ **resonance** laufende Resonanz *f* ~ **screen for measuring discharge of water** Meßschirm *m* ~-**screen washing machine** Siebbandwaschmaschine *f* ~ **shed** Fahrhalle *f* ~ **shot** Fahraufnahme *f* ~ **single-beam crane** Laufkran *m* für einen Hauptträger ~ **sleeve** Gleitmutter *f* ~ **speed** Durchsatz-, Fahr-geschwindigkeit *f* ~ **speed of crab** Katzfahrgeschwindigkeit *f* ~ **steadies** mitgehende Setzstöcke *pl* ~ **stoker feed** Wanderrostbeschickung *f* ~ **stripper crane** Stripperlaufkran *m* ~ **tables** Wandertische *pl* ~ **valve** fahrendes Druckventil *m* ~ **velocity** (electrons, etc.) Fluggeschwindigkeit *f* ~ **wave** fortschreitende Welle *f*, sich ausbreitende Welle *f*, Wanderwelle *f* ~ **wave helix** Wanderwellenschraube *f* ~ **wave maser** Wanderfeldmaser *m* ~ **wave protection** Wanderwellenschutz *m* ~-**wave reflection** Reflexion *f* einer Welle ~ **wave tube** Wanderfeldröhre *f* ~ **wave tube transmitter** Wanderfeldröhrensender *m* ~ **wheel** Laufrad *n* ~-**winch** Lauf-katze *f*, -winde *f*

**traverse, to** ~ ausschwenken, durch-fahren, -fließen, -kreuzen, -messen, -schreiten, kreuzen, querversetzen, sich drehen, schwenken, verfahren, verschränken **to** ~ **and elevate** in Seite *f* und Höhe *f* verstellen **to** ~ **a lever** einen Hebel *m* umlegen

**traverse** quer-liegend, -laufend, kreuzweise, quer, überkreuz

**traverse** Bahnkreuzung *f*, Durchgang *m* (Bearbeitung), Flußkreuzung *f*, Gehänge *n*, Polygon *n*, Polygonzug *m*, Quer-balken *m*, -binde *f*, -deckung *f*, -gang *m*, -haupt *n*, -riegel *m*, -stück *n*, -träger *m*, -vorschub *m*, (Drehbank) Schulterwehr *f*, Straßenkreuzung *f*, Strebe *f*, Streckenzug *m*, Traverse *f*, Traversengleitstück *n*, Vorschub *m*, Zug *m* ~ **bolt** Quer-, Traversen-bolzen *m* ~ **drum** Quertrommel *f* ~ **equalizer** Querausgleichshebel *m* ~ **feed** Querschubeinrichtung *f* ~ **force** Querschiffskraft *f* ~ **glide** gleitendes Querhaupt *n* ~ **guide** Führungsquerhaupt *n* ~ **guide arm** Traverse *f* (artil.) ~ **iron mass** Querschiffseisenmasse *f* ~ **joist** Quer-haupt *n*, -stück *n*, -träger *m* ~ **line** Linienzug *m* ~ **measuring** Polygonmessung *f* ~ **motion** abnehmende Changierung *f* ~ **movement of the cam carriage** Schlittenbewegung *f* ~ **outfit** Polygonausrüstung *f* ~ **ply** Querfurnier *n* ~ **problem** polygoniometrische Aufgabe *f* ~ **sailing** Koppelkurs *m* ~ **sight line** Querprofil *n* ~ **sight profile** Querprofil *n* ~ **speed** Bewegungsgeschwindigkeit *f* ~ **station** Polygonpunkt *m* ~ **support** Querträger *m* ~ **surveys** Polygonierung *f* ~ **table** Streckenzugtafel *f* ~ **table arrangement** Koppeltischanlage *f* ~ **table sender** Koppeltischgeber *m*

**traversed,** ~ **by current** stromdurchflossen ~ **by light** durchdrungen

**traverser** Schiebebühne *f* ~ **truck** Schiebeschlitten *m*

**traversing** Durchlaufen *n* ~ **and luffing excavator plant** Abraumschwenkbagger *m*

**traversing,** ~ **arc** Führungsstück *n* ~ **bed** Richtsohle *f* ~ **clamp** Begrenzungshebel *m* ~ **drive** Seitentrieb *m* ~ **fire** Schießen *n* mit wechselnder Seitenrichtung ~ **gear** Richtmaschine *f*, Schwenkwerk *n*, Seitenricht-antrieb *m*, -maschine *f* ~ **handwheel** Handrad *n* zur Seitenrichtung ~ **handwheel shaft** Seitenrichtschraube *f* ~ **lever** Seitenhebel *m* ~ **lock** Festklemmvorrichtung *f* ~ **mechanism** Seitenricht--maschine *f*, -trieb *m* ~ **microscope** Schlittenmikroskop *n* ~ **pinion** Quertriebrad *n* ~ **platform** Fahrbühne *f*, Lafettenrahmen *m* ~ **saddle** gleitendes Querhaupt *n*, Traversengleitstück *n* ~ **shaft** durchgehende Welle *f* ~ **slide** Führungshülse *f* ~ **stop** Seitenbegrenzer *m* ~ **stop on machine gun** Begrenzungsstift *m* ~ **table** Schiebebrücke *f*, Übergangsscheibe *f* (r.r.) ~ **wheel brake** Laufradbremse *f*

**trawl** Grund-, Schlepp-netz *n*

**trawler** Fischdampfer *m*

**tray** Ablegetisch *m*, Boden *m*, Kübel *m*, Küvette *f*, Mulde *f*, Schale *f*, Schüssel *f*, Servierbrett *n*, Tablett *n*, Teller *m*, Trog *m* ~ **for catching excess oil** Ölauffangblech *n* als Bodenbelag

**tray,** ~ **cap** Bodenkappe *f*, Glocke *f* der Fraktionierböden ~ **cell** Trogelement *n* ~ **development** Rahmen-, Schalen-entwicklung *f* (photo) ~ **downspout** Ablauf *m*, Überlaufrohr *n* ~ **dryer** Hordentrockner *m*, Tellertrockner *m* ~ **elevator** Schaukelelevator *m* ~ **ring** Bodenring *m* ~ **riser** Überlauf *m*, Überlauf *m* der Sprudelplatte ~**-shaped** kübelartig ~ **support** Bodenunterstützung *f*, Plattenträger *m* ~ **weir** Plattenüberlaufswehr *n*

**tread, to** ~ treten **to** ~ **down** niedertreten **to** ~ **upon** betreten

**tread** Auftritt *m*, (dredge) Bodenplatte *f*, Entfernung *f* zwischen den Anlaufrädern (r.r.), (of wheels, rails, etc.) Fahrfläche *f*, Gleitfläche *f* eines Rades, Lauffläche *f*, Laufkranz *m*, Reifenprofil *n*, Sprosse *f*, (of a wheel) Spurkranz *m*, Stufe *f*, Tritt *m*, (of a step) Trittfläche *f* ~ **(of groove)** Rillengrund *m* ~ **of a tripod** Fußwinkel *m* eines Stativs

**tread,** ~ **board** Trittstufe *f* ~ **compound (or stock)** Laufflächenmischung *f* ~ **cracking** Laufflächenrisse *pl* ~ **mine** Tretmine *f* ~ **plate** Laufflächenplatte *f* ~ **roll** Spurkranzwalze *f* ~ **roller** (Gleiskette) Laufrolle *f* ~ **way** Spurtafel *f* ~ **width** (of stairs) Auftritt *m*

**treading,** ~ **contact** Treibkontakt *m* ~ **wheel** Tret-, Tritt-rad *n*

**treadle** Pedal *n*, Tretkurbel *f*, Trittbrett *n* ~ **arrangement** Tretvorrichtung *f* ~ **bar** Sperrschiene *f* ~ **bellows (or blower)** Trittgebläse *n* ~ **drive** Fußbetrieb *m* ~ **hammer** Tritthammer *m* ~ **operation** Fußbetrieb *m* ~ **press** Fußhebelpresse *f* ~ **rod** Kupplungsstange *f* ~ **switch** Fußtrittschalter *m*

**treasure of energy** Energieschatz *m*

**treasury** Finanzministerium *n*, Fiskus *m*, Tresor *m* ~ **bond** Staatsschuldschein *m* ~ **rating** Steuersatz *m* ~ **warrant** Staatskassenanweisung *f*

**treat, to** ~ bearbeiten, behandeln, beschichten, handeln, präparieren, verarbeiten, vergüten, verhandeln, versetzen **to** ~ **with copper** kupfern **to** ~ **by eloxal process** eloxieren **to** ~ **in lime** einkalken **to** ~ **with mordants** ätzen **to** ~ **with pitch** verpichen **to** ~ **subsequently** nachbehandeln **to** ~ **with sulfur** aufschwefeln **to** ~ **with X-rays** röntgenisieren

**treatable** bearbeitbar, behandelbar

**treated** behandelt, raffiniert, zubereitet ~ **boiler feed water** aufbereitetes Kesselspeisewasser *n* ~ **butt** zubereitetes Ende *n*, zubereitetes Stammende *n* oder Stangenende *n* ~ **paper** veredeltes Papier *n* ~ **pole** getränkte oder zubereitete Stange *f*

**treating** Versatz *m*, Versetzung *f* ~ **of optical glass** Glasvergütung *f* ~ **process** Raffinations--prozeß *m*, -verfahren *n*, -vorgang *m*, Raffinierverfahren *n*

**treatise** Abhandlung *f*, Arbeit *f*, Aufsatz *m*

**treatment** Aufbereitung *f*, Aufschluß *m*, (mechanical) Bearbeitung *f*, Behandlung *f*, Heilbehandlung *f* (med.), Verarbeitung *f*, Verfahren *n* ~ **of the boundary** Randbehandlung *f* ~ **of ceilings** Deckenbehandlung *f* ~ **by dense medium** Schwerflüssigkeitssortierung *f* ~ **of optical glass** Linsenvergütung *f* ~ **by ultraviolet radiation** Ultraviolettbestrahlung *f*

**treatment,** ~ **band theory** Bändermodell *n* ~ **cone** Bestrahlungstubus *m* ~ **contrary to regulations** vorschriftswidrige Behandlung *f*

**treaty** Vertrag *m*

**treble, to** ~ verdreifachen

**treble** Diskant *m*, hoher Ton *m*, (frequencies)

Höhen *pl*; dreifach ~ **band** (music) Hochfrequenzbereich *m* ~ **block** Flaschenzug *m* ~**-channel** Dreikanal *m* ~ **control** Höhen-, Hochton-regler *m* (tape rec.) ~ **correction** Höhenentzerrung *f* (rdr) ~ **corrector** Höhenentzerrer *m* (acoust.) ~ **cut** Höhenbeschneidung *f* ~ **draft** dreifache Gabelspannung *f* ~ **lift** Höhenanhebung *f* ~ **loudspeaker** Hochlautsprecher *m*, Hochton-konus *m*, -lautsprecher *m* ~ **range** Hochfrequenzbereich *m*

**trebled** verdreifacht

**trebling** Verdreifachung *f*

**tree** Balken *m*, Baum *m*, Schaft *m*, Welle *f* ~ **anchorage** Baumverankerung *f* ~ **attachment** Baumhaken *m* ~ **entanglement** Baumsperre *f* ~**-guard** Drahthose *f* ~**-lined road (or walk)** Baum-allee *f*, -gang *m* ~ **nail** Dübel *m*, hölzerner Nagel *m*, Holznagel *m*, Pflock *m*, Stuhlnagel *m* ~ **pruner** Baumschere *f* ~ **root** Baumwurzel *f* ~ **screw** Baumschraube *f* ~ **stump** Baumstumpf *m* ~**top** Baumkrone *f* ~ **trimmer** Ausästwerkzeug *n* ~ **trunk** Block *m* ~ **wart** Holzkropf *m*

**treeless** baumlos

**treelike** baumähnlich ~ **crystal** Dendrit *m*

**trellis** Gatter *n*, Geländer *n*, Gitter *n*, Glanzleinwand *f* ~ **for glass roofs** Dachsprosse *f* für Glasdächer, Glasdachsprosse *f* ~ **mast** Gittermast *m* ~ **work** Bindwerk *n*

**trellised gate** Gittertor *n*

**tremble, to** ~ schlottern, schnarren, zittern

**trembler** Gleichstrom-unterbrecher *m*, -wecker *m*, Kontakthammer *m*, Unterbrecher *m* ~ **bell** Wecker *m* mit Selbstunterbrecher ~ **coil** Hammerinduktor *m*, Zündspulenunterbrecher *m* ~ **contact** Selbstunterbrecherkontakt *m*

**trembling** Tanzen *n* des Bildes (film) ~ **bell** Gleichstromwecker *m*

**tremolite** Tremolit *m*

**tremolo effect** Tremolostimmung *f*

**tremor** Beben *n*, Erdstoß *m*, Erschütterung *f*, Zittern *n*

**trench, to** ~ erodieren, Gräben ausheben oder ziehen, einen Kabelgraben herstellen, tief umpflügen, verschrämen, zerschneiden

**trench** Baugrube *f*, Einflußrinne *f*, Gassenrinne *f*, Gosse *f*, Graben *m*, Kabelkanal *m*, Rinne *f*, Sammelgraben *m*, Sappe *f*, Schram *m*, Schramhieb *m*, Schramme *f*, Schürfung *f*, Schützengraben *m*, Straßenrinne *f* ~ **of traverser** Schiebebühnengrube *f*

**trench,** ~ **backfill** Grabenfüllung *f* ~ **board** Grabenholzrost *m* ~ **bottom** Grabensohle *f* ~ **digger** Grabenbagger *m* ~ **ditch** Kabelgraben *m* ~ **excavator** Grabenbagger *m* ~ **floor** Grabensohle *f* ~ **invert** Grabensohle *f* ~**liner** Grabenbagger *m* (mit Eimern) ~ **periscope** Beobachtungsrohr *n*, Deckungsspiegel *m*, Grabenspiegel *m*, Laufgrabenspiegel *m* ~**(ing) plow** Rigolpflug *m* ~ **position** Grabenstellung *f* ~ **profile** Grabenprofil *n* ~ **revetment** Grabenbekleidung *f* ~ **wall** Grabenwand *f* ~ **work** Schanztätigkeit *f*

**trenched** verschrämt

**trenching** Grabenherstellung *f*, Kabelkanal *m* ~ **machine** Graben-aushebemaschine *f*, -bagger *m* ~ **machine of the bucket elevator type** Grabenbagger *m* (mit Eimern) ~ **shovel** Skarpierschaufel *f*

**trend** Entwicklungsrichtung *f*, Gang *m*, Neigung *f*, Neigungstendenz *f*, Orientierung *f*, Strömung *f*, Tendenz *f*, Trend *m*, Verlauf *m*, Zug *m* **~ line** Ausgleichlinie *f*, Kurvenzug *m*, mittlerer Linienzug *m*

**trepan** Bohrmaschine *f* **~ with central water clearing** Bohrmeißel *m* mit zentraler Wasserspülung

**trepanning** Hohlbohren *n*

**trespass, to** **~** übertreten, widerrechtlich betreten

**trespassing, no** **~** (on signs) abgesperrt

**tress of wool** (weaving) Zopf *m*

**trestle** Abstützbock *m*, Bock *m*, Bockkonstruktion *f*, Bühne *f*, Fuchs *m*, Gerüst *n*, Gerüstbock *m*, Gestell *n*, Schragen *m*, Schwelljoch *n*, Stellbock *m*, Tischgestell *n* **~ of a bridge** Brückenbock *m*

**trestle, ~-balance standard** Balanzierblock *m* **~ bay** Bockstrecke *f* **~ bearer** Bockstütze *f* **~ bridge** Bockbrücke *f* **~ footbridge** Beseler-, Bockschnell-steg *m* **~ gear** Bockgerät *n* **~ interval** Stützweite *f* **~ leg** Bockbein *n* **~upright** Jochpfahl *m* **~-work bridge** Gerüstbrücke *f*

**trestles** hölzerne Unterstütze *f*

**triad** Dreier *m*, dreiwertiges Element *n*

**triakisoctahedron** Triakisoktaeder *n*

**trial** Bewährung *f*, Erprobung *f*, Gerichtsverhandlung *f*, Probe *f*, Prüfung *f*, Prüfungsversuch *m*, Verhandlung *f*, Verhör *n*, Verhörung *f*, Versuch *m* **by ~ and error** empirische Lösung *f* **on ~** probeweise **~ and error method** Annäherungsverfahren *n*, empirisches Ermittlungsverfahren *n*, Rechnungsverfahren *n* mit fortschreitenden Näherungswerten **~ and error principle** Erfolgsprinzip *n* **~ and return basis** auf Leihbasis *f*

**trial, ~ assembly** Werkstattmontage *f* **~ bore** Vorbohrung *f* **~ boring** Erdbohrung *f* **~ boring tool** Sondiereisen *n* **~ calculations** Rechnung *f* mit fortschreitenden Näherungswerten **~ cast** Versuchskasten *m* **~ connection** Probeverbindung *f* **~ flight** Einfliegen *n*, Probeflug *m* **~ frame** Versuchsgestell *n* **~ function** Ausgangsfunktion *f* **~ installation** Versuchsanlage *f* **~ needle** Probiernadel *f* **~ order** Probeauftrag *m* **~ piece** Probe-scherbe *f*, -scherben *m* **~ pile** Probepfahl *m* **~ pit** Schürfgrube *f* **~ pumping** Probepumpen *n* **~ rail** Versuchbahn *f* **~ rod** Gareisen *n* **~ run** Probelauf *m*, Versuchsfahrt *f* **~ shipment** Probelieferung *f* **~ shot** Probe-aufnahme *f*, -schuß *m* **~ start set** Probestartaggregat *n* **~ track** Versuchsstrecke *f* **~ trip** Probefahrt *f*

**trials committee** Erprobungsausschuß *m*

**triangle** Dreieck *n*, Gestängekreuz *n*, (drafting) Schiebedreieck *n*, Zeichendreieck *n* **~ in elliptic plane** Dreieck *n* in der elliptischen Ebene **~ of error** Fehlerdreieck *n*, fehlerzeigendes Dreieck *n* **~ of forces** Kräftedreieck *n* **~ of velocities** Geschwindigkeitsdreieck *n*

**triangle, ~ covering a joint** Deckleiste *f* **~ symmetry** Dreiecksymmetrie *f*

**triangular** drei-eckig, -kantig, -seitig, -winklig **~ aperture** Zackenblende *f* **~ bar** Dreikant--stab *m* **~ base** (of gun) Lafettendreieck *n* **~ coil** Dreieckspule *f* **~ corner--piece** Dreikantbeschlag *m* **~ file** Dreikantfeile

**~ filter bag** Spitzbeutel *m* **~ frame** Dreieck--bügel *m*, -rahmen *m* **~ lifting eye** Bügel *m*, Lade-, Last-bügel *m*, Schlaufe *f* **~ mark** Dreiecksmarke *f* **~-mesh gauze** Versatzdrahtgeflecht *n* **~ notches** Dreieckskerben *pl* **~ opening** Dreikantmundstück *n* **~ point** Dreieckspitze *f* **~ prism** dreiseitiges Prisma *n* **~ pulse** Dreiecksimpuls *m* **~ puncher** Schallfilmlochvorrichtung *f* **~ road junction** Straßen-, Wege-dreieck *n* **~ scraper** Dreikantschaber *m* **~ section added to top of triangular dam to provide a horizontal crest** Bekrönungsdreieck *n* **~ shaft** Dreikanthaltespeer *m* **~ sign** (signal) Dreieckzeichen *n* **~ sign indicator lamp** Dreieckzeichenleuchte *f* **~ slices** Dreikantschnitzel *pl* **~ sluice** Dreieckschütze *f* **~ spindle** Dreikantspeer *m* **~ spring** Dreieckfeder *f* **~ stirrup** Dreiecksbügel *m* **~ thread** scharfgängiges Gewinde *n* **~-threaded** scharfgängig **~ timber obstacle** Dreieckbalkensperre *f* **~ transversal swinging arm** Dreieckquerlenker *m* **~ wedge** Dreieckskeil *m* **~ wing** Dreieckflügel *m*

**triangulation** Dreiecks-aufnahme *f*, -messung *f*, Triangulation *f*, Triangulierung *f* **~ from equal angle points** (radial triangulation) Fokalpunktstriangulation *f*, Triangulation *f* aus winkeltreuen Punkten **~ from photographs** Bildtriangulation *f*

**triangulation, ~ balloon** Triangulierungsballon *m* **~ method of surveying** Meßdreieckverfahren *n* **~ net** Triangulationsnetz *n* **~ network** Landesvermessungsnetz *n*, trigonometrisches Netz *n* **~ network not of primary order** Kleintriangulation *f* **~ point** Dreiecks-, Triangulations-punkt *m*, trigonometrischer Punkt *m* **~ station** Dreieck-, Stütz-punkt *m*

**triangulator** (for trigonometric calculations) Dreieckrechner *m*

**triaryl, a ~ azine compound** Azenium *n*

**triassic** triassisch **~ formation** Trias *f*

**triatic** Kettenaufhängung *f*

**triatomic** dreiatomig

**triaxial** dreiachsig

**tribasic** dreibasisch

**triblet** Dorn *m*

**tribometer** Reibungsmesser *m*

**tribrach** Dreifuß *m*, (Zwischenstück zwischen Stativ und Theodolit) horizontierbare Platte *f*

**tribromomethane** Tribrommethan *n*

**tribunal** Altarnische *f*

**tribune** Tribüne *f*

**tributary** zufließend **~ functions** Nebenfunktionen *pl* **~ station** Fernmeldenebenstelle *f* **~ stream** Nebenfluß *m*

**tribute** Abgabe *f*, Hauptgedinge *n*

**tributer** Erzgedingehauer *m*

**trice, to ~** aufholen

**trichlor, ~ acetic acid** Trichloressigsäure *f* **~ ethylene** Trichlorethylen *n* **~-oethylene** Trichloräthylen *n*

**trichord piano** dreichöriges Piano *n*

**tricing line** Aufholer *m*

**trick** Geheimschloß *n*, Kniff *m*, Kunstgriff *m*, List *f*, Trick *m* **~ button** Tricktaste *f* (tape rec.) **~ camera** Trickkamera *f* **~ circuitry** Kunstschaltung *f* **~ effect** Trick *m* **~ exposure**

Trickaufnahme f ~ **picture** Trickbild n ~
**plate** Fräsblech n
**trickle, to** ~ auslaufen, rinnen, sickern, tröpfeln,
tropfenweise rinnen **to** ~ **away** entrieseln **to**
~ **down** abtropfen, herabrieseln, heruntertröpfeln **to let** ~ **down** abrieseln lassen **to** ~
**down drop by drop** aufträufeln **to** ~ **out** aus-,
heraus-sickern **to** ~ **through** durch-sickern,
-sintern
**trickle** Tropfen m ~ **charge** Dauerladung f ~
**charger** Gleichrichter m für Dauerladung,
Kleinlader m ~**-type cooling** Sickerkühlung
f
**trickling** Berieselung f, Sickerung f ~ **plant** Rieselanlage f ~ **water** Rieselwasser n
**triclinic** triklin, triklinisch
**tricolor,** ~ **banded filter** dreifacher Streifenfilter m ~ **cinescope** Dreifarbenröhre f (TV)
~ **reflection density values** Dreifarbenleucht-
dichten pl ~ **tube** Dreifarbenröhre f
**tricyanic acid** Trizyansäure f
**tricycle** Dreirad n, dreirädrig ~ **alighting gear**
Bugradfahrgestell n ~ **landing gear** Drei-
beinfahrwerk n, dreirädriges Fahrwerk n
~ **undercarriage** Bugradfahrgestell n
**tridimensional** dreidimensional
**tridimensionality** Dreidimensionalität f
**triding level** Reiterlibelle f
**tried** erprobt, probiert
**tri-equipartition** Dreiteilung f
**triethylamine** Triäthylamin n
**trieur** Trior m
**trifle** Bagatelle f
**triform optical apparatus** dreifache Leuchte f
**trigatron** gesteuerte Funkenschaltröhre f
**trigger, to** ~ ausklinken, auslösen, entriegeln
**trigger** Abzug m, Anlaßschaltstange f, Auslöse-
impuls m, Auslöser m, Blendeneinstellung f
(Auslösehebel), Drücker m, Hemmzeug n,
Schaltgriff m, Sperrkegel m, Steuerimpuls m
~ **for full-automatic fire** Dauerabzug m
**trigger,** ~ **action** Schaltwirkung f ~**-action
relay** Kipprelais n ~ **arm** Abzugs-hebel m,
-stange f ~**-arm guide** Spannstolle f ~ **bar**
Abzugsschiene f ~ **cable** Abzugskabel n ~
**cam** Drucknase f, Schalt-nocke f, -nocken
m ~ **capacitor** Kippkondensator m ~ **circuit**
Kippkreis m, Röhrenwippe f, Triggerschaltung
~ **circuit tube** Kippschwingkreisröhre f ~
**guard** Abzugsbügel m, Bügel m ~**-guard
screw** Kreuzschraube f ~ **input** Triggersignal
n ~ **magnet** Abzugs-, Auslöse-, Einrück-magnet
m ~ **mechanism** Abziehen n, Abzugsvorrich-
tung f ~ **pin** Abzugsstift m, Bolzen m zum
Abzugsstück ~ **pull** Abzugsgewicht n ~
**pulse** Anstoß-, Auslöse-impuls m ~ **relay**
Schwingrelais n ~ **release** Gewehrabzug m
~ **sear** Abzugs-angel f, -stollen m ~**-sear fork**
Abzugsgabel f ~ **sleeve** Abzugswelle f ~
**spring** Abzugsfeder f ~ **switch with lock**
Drückschalter m mit Feststellvorrichtung
~ **timing pulse** Zeitkontrollimpuls m ~ **tube**
Relaisröhre f
**triggered,** ~ **blocking oscillator** getriggerter
Sperrschwinger m ~ **oscilloscope method** Syn-
chroskopmethode f
**triggering** Ansteuern n, Auslösen n, Tasten n ~
**agency** Auslösungsursache f ~ **electrons** aus-

lösende Elektronen pl ~ **level** Triggerhöhe f,
Triggerpegel m
**trigonal** dreieckig, dreizählig (cryst.) ~ **dodecahedron cuproid (or hemitrisoctahedron)** Tri-
gondodekaeder n ~ **reflector antenna** Dreiecks-
reflektorantenne f
**trigonometric,** ~ **distance measurement** trigono-
metrische Streckenmessung f ~ **function**
Winkelfunktion f ~ **point** Artillerie-, Triangulations-punkt m
**trigonometrical** trigonometrisch ~ **function**
Kreisfunktion f
**trigonometrically fixed point** Dreieckspunkt m
**trigonometry** Dreiecks-berechnung f, -lehre f,
-messung f, Trigonometrie f
**trihedral** Dreikant n ~ **angle** Dreikantwinkel
m, dreiseitiger Winkel m
**trilateral** dreiseitig
**trilinear coordinate** Dreieckskoordinat f
**trilit** Trilit n
**trill, to** ~ trillern
**trilobe screw** Dreinockenschraube f
**trim, to** ~ abgleichen, abgraten, abkanten,
(grinding wheel) abrichten, (off) abstechen,
(a wheel) abziehen, ausputzen, (trees) ausästen,
austrimmen, behauen, besäumen, beschneiden,
bestoßen, einfassen, garnieren, geradeschnei-
den, ins Gleichgewicht n bringen (aviat.),
putzen, repassieren, setzen, stoßen, trimmen,
verputzen, zurichten **to** ~ **in** einlassen, einpassen **to** ~ **up grindstones** Schleifsteine pl
abrichten **to** ~ **off the rough edges** die rohen
Kanten pl abschneiden **to** ~ **roughly** abrauhen
**to** ~ **smooth** glattrichten **to** ~ **trees** Bäume pl
ausästen
**trim** Dekorationsschild n, Schnipsel n, Schnipselchen n, (airship) Steuerlastigkeit f, Trimm
m, (of a vessel) Trimmlage f, Trimmung f
~ **of an airplane** Flugzeuggleichgewicht n ~
**by the bow** buglastig, kopflastig
**trim,** ~ **ballast** Trimmballast m ~ **change**
Trimmänderung f ~ **compensation** Trimmung f ~ **controls** Trimmsteuerung f ~ **gear**
Trimmvorrichtung f ~ **indicator** Lastigkeitswaage f ~ **pattern** Beschneideschablone f ~
**saw** Treck-, Trumm-säge f ~ **size** Schneidmaß
n ~ **strip** Zierleiste f ~ **tab** Ausgleichfläche f,
Bügelblech n, Trimmklappe f, Trimmsegel n
(g/m) ~**-tab position indicator** Trimmungs-
anzeiger m
**trimethyl,** ~ **butane** Trimethylbutan n ~ **butene**
Trimethylbuten n ~ **enetrinitroamine** Hexogen
n ~ **pentane** Trimethylpentan n ~ **phenylallene**
Trimethylphenylallen n
**trimetric** trimetrisch
**trimmed** ausgewogen (aviat.), beschnitten, in
den Bunkern pl gestaut ~ **edge** bestoßene
Kante f ~ **size** beschnittenes Format n
**trimmer** Nopper m, Trimmer m, Trimmerkondensator m, Trimmkondensator m, (carp.)
Wechsel m ~ **of rafters** Sparrenwechsel m
**trimmer,** ~ **capacitor** Trimmerkondensator m
~ **pair** Trimmerplattenpaar n (TV) ~ **punch**
Abgratstempel m
**trimming** Abgratarbeit f, Besatz m, Beschnitt m,
Borte f, Korrektionen pl, Lastigkeitsregelung
f, (cloth) Nachschneiden n, (candle, wick)
Putzen n, Trimmung f ~ **of the fixed-tail**

surface Flossenverstellung f ~ and folding paper Einfaß- und Fälzelpapier n ~ for linen Wäschebesatz m ~ with one or more engines inoperative Gegensteuern n ~ by weights Gewichtstrimmen n

trimming, ~ adjustment Korrekturvorrichtung f ~ aggregate Trimmgerät n ~-back knob Rücktrimmknopf m ~ block Beschneideklotz m ~ capacitor Abgleich-, Nachstimmungs-kondensator m ~ card Tresse f ~ condenser (padding) Abgleichkondensator m, Trimm(er)-kondensator m ~ cutter Abgratfräser m, Besäummaschine f ~ die Abgratmatrize f, Beschneideklotz m ~ expense Besteckungskosten pl ~ file Schrotfeile f ~ glass Beschneideglas n ~ knife Beschneidemesser n ~ knob Trimmknopf m ~ lathe Abstechdrehbank f ~ machine Abfräsmaschine f, Abgratpresse f ~ maker Bortenwirker m, Posamentier n ~ manufacturing Posamentenfabrik f ~ moment Verstellmoment m ~ plan Trimmplan m ~ plane Bestoßhobel m ~ pump Trimmpumpe f ~ press Abkant-, Börtel-presse f ~ ribbon Bordürenband n ~ saw Besäumsäge f ~ shears Abgrat-, Besäum-schere f ~ tab (auxiliary control) kleine Steuerfläche f, Trimm-fläche f, -klappe f, -ruder n (aviat.) ~ tank Trimmtank m ~ testboard Abgleichtisch m ~ tool Abgratwerkzeug n ~ unit Schnippeleinrichtung f

trimmings Abgratschrott m, Garnitur f, Stanzabfälle pl, (on cabinets) Zierleisten pl

trimorphous (or trimorphic) trimorph

trimotor dreimotorig

tringle covering a joint Fugenleiste f

Trinidad lake asphalt Trinidadasphalt m

trinitride Azid n

trinitro-, ~ benzene Trinitrobenzol n ~benzoic acid Trinitrobenzoësäure f ~ phenic acid Bittersäure f, Trinitrophenol n ~phenol Bittersäure f, Pikrinsäure f, Trinitrophenol n ~ resorcinate of lead Bleityphnat n, Bleitrinitroresorzinat n ~ toluene Trinitrotoluol n, Tutol n

trinitrotoluol Füllpulver n

trinominal dreiglied(e)rig

trio mill (or rollers) Drillingswalzwerk n, Triowalzwerk n

triode Dreielektroden-rohr n, -röhre f, Dreipolröhre f, Eingitterröhre f, Triode f ~ gun Triodensystem n

triodes Kristalloden pl

trioxymethylene Paraformaldehyd n

trip, to ~ (relay) ansprechen, ausklinken, (camera) auslösen, ausrücken, (current) ausschalten, einrücken, entriegeln, fehltreten, (circuit breaker) herausspringen, hieven, losreißen, plötzlich auslösen, plötzlich in Gang m setzen, schalten, schleppen, stolpern to ~ out Ausschaltbewegung f auslösen, sich ausschalten (aviat.), plötzlich außer Gang m setzen to ~ over kippen

trip Auslösung f (electr.), Fahrt f, Kohlenzug m (min.), Reise f, Sperrhaken m ~(ping) Auslösung f ~ into a mine Grubenfahrt f

trip, ~ action Instabilität f ~ bar Auslösestange f ~ cam Schalt-knagge f, -nocke f, -nocken m ~ coil Auslösespule f, Ausschaltspule f ~-coil magnet Schaltelektromagnet m ~

dog Antriebsklaue f, Auslöse-anschlag m, -knaggen m, Auslösungsanschlag m, Schalt--knagge f, -nocke f, -nocken m, Sperrklinke f ~-free mechanism Fortschaltungsmechanik f ~-free release Freiauslösung f (electr.) ~ gear Ausklinkmechanismus m, (steam distribution) Ausklinksteuerung f, Bimetallstreifen m, Klinkwerk n ~ hammer Aufwerfhammer. m ~ lever Anschlag-, Auslöse-hebel m ~(ping) lever Einrückhebel m ~ lock Spannabzug m ~ magnet Auslöse-, Einrück-, Lösungs-magnet m ~ mechanism Auslösemechanismus m, Schaltwerk n ~ meter Kilometerzähler m ~-out switch Kippausschalter m ~ pin Anschlagstift m ~ plate Auslöseblech m ~ point Auslösungspunkt m ~ recorder Tageszähler m ~ relay Auslöserelais n ~-releasing catch Auslöseklinke f ~ rod Auslöse- oder Steuerstange f, Bürstenwähler m ~ spark Auslösefunke m ~ spindle Auslösespindel f, Bürstenwähler m ~ spring Auslösungsfeder f ~ stop Betätigungsnocken m ~ switch Betätigungs-, Nocken-schalter m ~ switch timer Zeitauslöserelais n ~ wire (entanglement) Drahtschlinge f, Stolperdraht m ~-wire alarm Alarmschußgerät n ~-wire entanglement Stolperdrahthindernis n ~ worm device Schnekkenfallvorrichtung f

tripartite dreiteilig

tripestone Gekrösestein m

triphase dreiphasig ~ current Dreiphasenstrom m ~ motor Drehspulmotor m

triplane Dreidecker m

triple, to ~ verdreifachen

triple Dreileiter m; dreifach, dreizählig ~ apparatus Dreifachapparat m ~ barrel tank Drillingsbehälter m ~-block chain hoist Triplexflaschenzug m ~ bond(ing) Dreifachbindung f ~ burner Dreibrenner m ~ carbonation dreifache Kohlensäuresaturation f ~-circuit receiver Tertiärempfänger m ~ coil Dreispule f ~ collisions Dreierstöße m ~ core dreiadrig ~-core cable dreiadriges Kabel n, Dreileiterkabel n ~-core cord dreiadrige Schnur f ~ crane Dreifachkran m ~-detector reception Doppelsuperempfang m ~ diode triode Dreifachdiode-Triode f ~-effect evaporator Dreikörperverdampfapparat m ~--electrode tube Dreielektroden-rohr n, -röhre f (rdo) ~ eyepiece revolver dreifacher Okularrevolver m ~ fission Dreifach-Spaltung f ~ flexible Dreifachschnur f ~ fuel mixture Dreiergemisch n ~-gyro arrangement Dreikreiselanordnung f ~ gyro compass Dreikreiselkompaß m ~ gyro master compass Dreikreiselmutterkompaß m ~-harmonic dreifache Oberharmonische f, dreizählige Harmonische f ~ harmonics dreifache Harmonische f, dreifache Oberharmonische f, dreizählige Harmonische f, dritte Harmonische f ~ isomorphism Isotrimorphie f ~ lancet window Drillingsfenster n ~ lead Dreileiter m ~--modulation beacon mehrstrahliges Funkfeuer n ~ nozzle Dreifachdüse f ~-plunger pump Dreikolbenpumpe f ~ point Tripelpunkt m (metall.) ~ pole dreipolig ~ pressure Tripelpunktsdruck m ~ prism dreiteiliges Prisma n ~-pulley drive Dreiriemenscheibenantrieb m

**~ push button** Dreifachdruckknopf *m* **~-rail line** Dreischienengeleise *n* **~ reactor** Dreifachdrossel *f* **~ reflector** Tripelspiegel *m* **~ revolving nosepiece** Objektivrevolver *m* mit drei Objektiven **~-roller machine** Dreiwalzenmaschine *f* **~ root** Dreifachwurzel *f* **~ socket** Kreuzmuffe *f* **~ split** Dreifachaufspaltung *f* **~ stars** Dreifachsterne *pl* **~ thread** dreigängiges Gewinde *n* **~-thread screw** dreigängige Schraube *f* **~ turret** Drillingsturm *m* **~ wire guide** Gabelrollendrahtführer *m*

**tripled** verdreifacht

**tripler** Drilling *m*; dreiadrig **~ distribution function** Drei-Teilchen-Verteilungsfunktion *f*

**triplex, ~ glass** Securit *n* **~ process** Triplexverfahren *n*

**triplicate** dreifache Ausfertigung *f*; verdreifacht **in ~** in dreifacher Ausfertigung *f*

**tripling** Verdreifachung *f*

**triplite** Eisenapatit *n*, Manganpecherz *n*

**triplug** Dreifachstecker *m*

**triply orthogonal systems of surfaces** dreifach orthogonale Flächensysteme *pl*

**tripod** Dreibein *n*, Dreibeinstativ *n*, Dreifuß (Stativ) *n*, Gestell *n*, Ständer *n*, Stativ *n* **~ cable support** Dreibock *m* **~ dolly** Stativwagen *m* **~ drill** Dreifußbohrer *m* **~ extension piece** Dreifußaufsatzstück *n* **~ head** Dreifußkopf *m* **~ headplate** Stativkopf-, Stern-platte *f* **~ landing gear** achsloses Fahrgestell *n* **~ magnifier** Dreifußlupe *f* **~ mast** Dreibeinmast *m* **~ mount** Dreibein-, Dreifuß-lafette *f* **~ ring** Dreifußring *m* **~ screw** Anzugsschraube *f* **~ starter** Dreibein-, Dreifuß-anlasser *m* **~ support** dreibeiniger Bock *m*

**tripole antenna** Tripolantenne *f*

**tripoli earth** Tripel *n*, Tripolerde *f*

**tripper** Abwurfwagen *m*

**tripping** Ankerumlegung *f* (naut.), (anchor) Hieven *n*, (anchor) Losreißen *n*, Schaltung *f* **~ of circuit breaker** Schalterfall *m*

**tripping, ~ bar** Absatzschwelle *f*, Auslösewelle *f* **~ cam** Abstreichdaumen *m*, Ausklinknocken *m* **~ chain** Kippkette *f* **~ characteristic** Auslösekennlinie *f* **~ circuit action** Auslösung *f* **~ coil** Betätigungsspule *f* **~ device** Auslöser *m*, Auslösevorrichtung *f* **~ gear** Auslöser *m* (Fernschreiber), Ausrückvorrichtung *f*, Bimetallstreifen *m* **~ impulse** Kippimpuls *m* **~ line** Aufholer *m* **~ mechanism** Schaltmechanismus *m*, Schnellschluß *m* **~ piece** Spannstück *n* **~ pin** Schaltstift *m* **~ pulse** Auslösesynchronimpuls *m* **~ relay** Auslöse-, Einschalt-, Rufabschalt-, Trenn-relais *n* **~ rope** Schneppertau *n* **~ signal** Gleichlauf-signal *n*, -zeichen *n* **~ solenoid** Arbeitsstromauslöser *m* **~ speed** Schnellschlußdrehzahl *f* **~ spring** Auslösefeder *f* **~ wire** Stolperdraht *m*

**tri-rectangular** dreifachrechtwinklig, dreifach rechtwinklig

**tri-rod coupler** Dreistab-Kopplung *f*

**trisection** Dreiteilung *f*, Teilung *f* in drei gleiche Teile

**trisilicate** Trisilikat *n*

**tri-tet oscillator** Tri-tet-Oszillator *m*

**tritiate** mit Tritium *n* versetzt

**tritium** schwerer Wasserstoff *m* **~ estimation** Tritiumabschätzung *f*

**tritol** Tritol *n*

**triton** Triton *n* **~ bombardment** Tritonbeschuß *m*

**triturable** zerreibbar, zerreiblich

**triturate, to** **~** feinmahlen, verreiben, zerreiben

**trituration** Zerreibung *f*

**triturium** Scheidegefäß *n*

**tritylene** Tritylen *n*

**trivalent** dreiwertig **~ element** dreiwertiges Element *n*

**trivariant** dreifachfrei

**troche** Plätzchen *n*

**trochoidal mass analyzer** Trochoid-Massenspektrograf *m*

**troffer (or trough reflector)** Muldenreflektor *m* (Raumbeleuchtung)

**troilite** Troilit *m*

**trol-e-duct** Stromschienenanschluß *m*

**Trolitul coil form** Trolitul-Spulenkörper *m*

**trolley, to** **~** karren, katzen, katzfahren

**trolley** Förder-hund *m*, -karren *m*, -schacht *m*, Katze *f*, Laufkatze *f*, Rollenstromabnehmer *m* **~ arm** Abnehmerarm *m*, Stromabnehmer *m* **~-arm contact rod** Abnehmerstange *f* **~ base** Federbock *m* für Stromabnehmer **~-beam track** Katzenfahrbahn *f* **~ block** Flaschenzug *m* mit eingebauter Laufkatze **~ boom** Stromabnehmerstange *f* **~ brush** Stromabnehmer *m* **~ bus** Oberleitungsomnibus *n*, Obus *m* **~ cap** Rollenträger *m* **~ capacity** Auslegerarm *m* **~ car** Straßenbahn *f* **~ carriage** (for cranes) Katze *f* **~-conveyer crossing** Hängebahnkreuzung *f* **~ conveyor** fahrbarer Förderer *m* **~ engine** Katzfahrmotor *m* **~ frame** Katzen-gestell *n*, -rahmen *m* **~ frog** Fahrdrahtweiche *f* **~ harp** Rollenträger *m* **~ head** Rollenkopf *m* **~ hoist** fahrbare Kranlaufwinde *f*, Katze *f* oder Laufkatze *f* mit Hebezeug, Laufwinde *f* **~ ladle** Hängebahngieß-, Lauf-pfanne *f* **~ line** Straßenbahn *f*, Straßenbahnlinie *f* **~ locomotive** Fahrdrahtlokomotive *f* **~ motion** Katzenfahrt *f*, Laufkatzenfahrt *f* **~ motor** Katzfahrmotor *m* **~ overhead contact pivoted and with two wheels** schwenkbare Abnehmerstange *f* mit zwei Abnehmerrollen **~ pivot** Stromnehmerdrehzapfen *m* **~ pole** Abnehmerarm *m*, Leitstange *f*, Rollenkontaktstange *f* **~-pole contact rod** Abnehmerstange *f* **~ runway** Katzen-fahrbahn *f*, -gang *m*, Laufkatzen-fahrbahn *f*, -fahrwerk *n*, -gang *m* **~ shoe** Schleifstück *n* **~-span wire** Fahrbahnaufhängung *f* **~ support** isolierter Stromnehmerträger *m* **~ track** Katz(en)bahn *f*, (beam) Laufkatzenfahrbahn *f* **~ travel** Katzenfahrt *f*, Laufkatzenfahrt *f* **~-travel gearing** Katzenfahrwerk *n* **~ travel rail** Katzenschiene *f*, Laufkatzenfahrschiene *f* **~ travel speed** Katzfahrgeschwindigkeit *f* **~ wheel** Katzenlaufrad *n*, Stromabnehmerrolle *f* **~ wire** Fahrdraht *m*, Fahrleitung *f* (electr.), Oberleitung *f* **~-wire insulator** Fahrdrahtisolator *m* **~-wire suspension** Fahrbahnaufhängung *f*

**trombone** ausziehbare Leitung *f*, Posaune *f*, **~ with free reed** (organ) durchschlagende Posaune *f*

**trombone, ~ fashion** posaunenartig verschiebbar **~ feeder** Posaunenabgleich *m* **~ stop** Posaune *f* **~ tuning** Posauneneinstellung *f*

**trommel** Trommel *f*, Trommelsieb *n*
**troop** Eskadron *f*, Schar *f*, Trupp *m* ~ **carrier** Truppen-fahrzeug *n*, -transporter *m* ~-**carrier airplane** Truppentransportflugzeug *n* ~-**carrying glider** Lasten-segelflugzeug *n*, -segler *m* ~ **ship** Truppentransportschiff *n* ~ **train** Truppentransportzug *m*
**troostite** Troostit *n*
**troostitic sorbite** Troostosorbit *m*
**tropadyne** Tropadynempfänger *m*
**tropaeoline paper** Tropäolinpapier *n*
**Tropenas (side-blown) converter** Tropenas-Konverter *m*
**tropic** Wendekreis *m* ~ **proof** tropen-beständig, -sicher ~ **proof quality** Tropenfestigkeit *f*
**tropical** tropisch ~ **air** tropische Luft *f*, tropische Warmluft *f* ~ **air mass** Luft *f* über dem tropischen Meere, tropische Luftmasse *f* ~ **cyclone** tropischer Wirbelsturm *m* ~ **disturbance** tropische Störung *f* ~ **finish** Tropenausführung *f* ~ **gasoline** Tropenkraftstoff *m* ~ **neckcloth** Tropenbinder *m* ~ **radiator** Tropenkühler *m*
**tropicalization** Tropenfestmachen *n*
**tropicalize, to** ~ tropenfest machen
**tropicalized** tropenfest
**tropics** Tropen *pl*
**tropism** Tropie *f*
**tropopause** Tropopause *f*
**troposphere** Troposphäre *f*, Wolkenzone *f*
**tropospheric,** ~ **absorption** atmosphärische Absorption *f* ~ **duct** Troposphärenkanal *m* ~ **mode** troposphärische Schwingungsart *f* ~ **scatter** troposphärische Streuung *f* ~ **wave** troposphärische Welle *f*
**trotyl** Trotyl *n*
**trouble** Arbeit *f*, Belästigung *f*, Beschwerde *f*, (in circuits) Fehler *m*, (motor) Panne *f*, Schererei *f*, Störung *f*, verursachte Mühe *f*, Verwirrung *f* ~ **due to a broken (or open) wire** Störung *f* durch Drahtbruch ~ **in calibration** Eichfehler *m*
**trouble,** ~ **alarm** Störungssignal *n* ~ **bell** Störungsglocke *f* ~ **desk** Störungs-platz *m*, -stelle *f* ~-**free** fehler-, störungs-frei, pannengesichert ~-**free operation** störungsfreier Betrieb *m* ~ **incidence** Störanfälligkeit *f* ~ **indication** Störanzeige *f* ~ **localizer light** Fehler-, Störungs-lampe *f* ~-**location problem** Fehlersuchprogramm *n*, Störsuchaufgabe *f* ~ **man** Störungssucher *m* ~ **position** Störungsüberwachungsplatz *m* ~-**prevention service** Störungsvermeidungsdienst *m* ~-**proof** einwandfrei ~ **record** Verhandlungsschrift *f* ~ **relay** Störungsrelais *n* ~ **shooting** Störungsbeseitigung *f* ~-**shooting device** Prüfgerät *n* ~ **sign** Überhangstörzeichen *n* ~ **table** Störungstafel *f* ~ **tone** Störungssignal *n*
**trough** Abflußrinne *f*, Becken *n*, Behälter *m*, Gerinne *n*, Küvette *f*, Mulde *f*, (of an inclined conveyor) Muldenband *n*, Rinne *f*, Senke *f*, Tiefdruckrinne *f*, Trog *m*, Wanne *f*, Zuführungsrinne *f* ~ **of car body** Karosseriemulde *f* ~ **of the cloud waves** Wolkental *n* ~ **for distribution cable** Verteilungskanal *m* ~ **of the pad** Foulardchassis *n*
**trough,** ~ **band** Muldengurt *m* ~ **battery** Trogbatterie *f* ~ **belt** Muldengurt *m* ~-**charging**

**crane** Muldenbeschickkran *m* ~ **compass** Röhrenbussole *f* ~ **conveyor** Förderrinne *f* ~ **core** Muldenkern *m* ~ **curve** Muldenbiegung *f* ~ **drier** Muldentrockner *m* ~ **fault** Grabenbruch *m*, Versenkung *f* ~ **grate** Muldenrost *m* ~ **horizon** Troghorizont *m* ~ **line** Troglinie *f* (meteor.) ~-**line guide** Führungsrinne *f* ~ **roller** Muldenrolle *f* ~-**shaped** trogförmig ~-**shaped bridge** Trogbrücke *f* ~-**shaped iron** Rinneisen *n* ~-**shaped stratum** Schichtmulde *f* ~ **sheet** Trogblech *n* ~ **strip** Muldengurt *m* ~ **terminal** Trogendverschluß *m* ~ **tip** (wagon) Muldenkipper *m* ~ **tip wagon for mines** Grubenmuldenkipper *m* ~ **washer** Muldenwascher *m*, Rinnenwäsche *f*
**troughability** Muldungsfähigkeit *f*
**troughing** Kabelkanal *m*
**troughlike** wannenförmig ~ **edging guide** (in rolling) Wendetrog *m*
**trousered** mit Hosen *pl* ~ **undercarriage** Fahrgestell *n* mit Hosen
**trowel** Fugkelle *f*, Kelle *f*, Mauerkelle *f*, Spachtel *m* ~ **filler** Spachtelkitt *m*
**troy weight** Goldgewicht *n*, Karatgewicht *n*
**truck** Auflager *n*, Drehgestell *n*, Fahrgestell *n*, Förderwagen *m*, (mine) Grubenwagen *m*, (freight) Güterwagen *m*, Karre *f*, Karren *m*, Kraftwagen *m*, Lastkarre *f*, Lastkraftwagen *m*, Lastwagen *m*, (of railroad car) Laufwagen *m*, (farm) Rollwagen *m*, Schleppwagen *m*, Stoßrad *n* (r.r.), Tausch *m*, (navy) Topp *m*, Untergestell *n*, Wagen *m* ~ **for brush breaker** Karren *m* für Buschrodepflug ~ **for converters** Bodeneinsatzwagen *m* für Konverter ~ **for conveyance of drilling tools** Bohrtransportwagen *m* ~ **to push wagons at railway stations** Wagenaufschieber *m* für Förderwagen
**truck,** ~ **capacity** Lastwagentragkraft *f* ~ **cart** Schleppe *f* ~ **center** Drehzapfenabstand *m* (r.r.) ~ **driver** Kraftfahrer *m* ~ **excavators** Autobagger *m* ~ **head** Ausladespitze *f* ~ **ladle** Gießpfanne *f* mit Wagen, Pfanne *f* mit Wagen, Wagenpfanne *f* ~-**load** Wagenladung *f* ~ **platform** Karrentisch *m* ~ **tank** Tankwagenbehälter *m* ~ **tractor** Lastwagenschlepper *m* ~ **tractor for semitrailers** Sattelschlepper *m* ~ **trailer** Lastanhänger *m* ~-**trailer combination** Lastzug *m* ~ **train** Schleppzug *m* ~ **wheel** Karrenrad *n*
**trucker** Wagenstößer *m*
**trucking,** ~ **business** Rollfuhrunternehmen *n* ~ **charges** Rollgeld *n*
**true, to** ~ ablehren, abrichten, abrunden, (wheel) abziehen, einpassen, einschleifen, nachschleifen, prüfen **to be** ~ zutreffen **to** ~ **a bearing** ein Lager *n* ausrichten **to** ~ **up** ausrichten, auf genaues Maß *n* bringen, nachstellen, Räder *pl* richten
**true** echt, gediegen, genau, getreu, (magnetic needle) rechtsweisend, regelrecht, richtig, schlagfrei, treu, wahr, wirklich **out of** ~ unrund
**true,** ~ **to form** form-gerecht, -getreu ~ **to gauge** kaliber-, lehren-haltig ~ **to life** lebenswahr ~ **to measure** maßhaltig ~ **to nature** naturähnlich ~ **to pattern** modellgetreu ~ **to profile** profilgerecht ~ **to scale** maßstabgerecht ~ **to shape** form-gerecht, -getreu, der Sollform

entsprechend ~ **to size** maß-gerecht, -haltig
~ **to type** mustermäßig
**true,** ~ **absorption** absolute Absorption f ~
**air speed** wahre Fluggeschwindigkeit f oder
Eigengeschwindigkeit f (navig.) ~ **altitude**
Höhe f über NN ~ **bearing** rechtsweisende
Peilung f ~ **coefficient of absorption** wahrer
Absorptionskoeffizient m ~**-color filter** tonrichtiger Filter m (opt.) ~ **color perception**
Farbentüchtigkeit f ~ ~ **copy** gleichlautende
Abschrift f ~ **course** rechtweisender Kurs m
~ **course to steer** rechtweisender Zielkurs m
~ **distance** rechter (verbesserter) Abstand m,
tatsächliche Entfernung f ~ **external form**
äußere Eigenform f ~ **fissure vein** echter
Spaltengang m ~ **heading** rechtweisender
Steuerkurs m ~ **indication** Sollanzeige f ~
**level** wahrer Horizont m ~ **mach number** wahre
Machzahl f ~**-map grid system** Koordinatennetz n ~ **measurement** Maßgenauigkeit f ~
**meridian** wirklicher Meridian m ~ **north**
rechtweisend Nord ~**-phased** richtigphasig ~
**plot** wahre Zeichnung f ~ **power** Wirkleistung
f ~ **radio bearing** beschickte oder verbesserte
Funkseitenpeilung f ~ **rake** Spanfläche f ~
~ **range** tatsächliche Schußweite f ~ **ratio**
reelles Verhältnis n ~ **reproduction** inhaltstreue Abbildung f ~**-running gauge** Rundlauf-
lehre f ~ **scale** reine Skala f, Wirklichkeits-
maßstab m ~ **shape** Sollform f ~ **size** genaues
Maß n ~ **slip** (propeller) wirklicher Schlupf m
~ **specific weight** Eigengewicht n ~ **(air) speed**
wahre Geschwindigkeit f ~ **stress of fracture**
wahre Bruchspannung f ~ **time** wahre Zeit f
~ **ultimate stress** bezogene Spannung f, wahre
Bruchspannung f ~ **vertical deflection** Höhenwinkelvorhalt m
**trued** eingerichtet, paßgerecht, zentriert
**truing** Abdrehen n, Einpassen n, Einschleifen n,
Korrektion f ~ **attachment** Abrichtvorrichtung
f ~ **device** Abdrehapparat m, Abziehvorrichtung f (für Schleifscheibe) ~ **diamond**
Abrichtdiamant f ~ **plane** Abdrehebene f ~
**tool** Abrichter m ~**-up** Ausrichten n, Nachjustierung f ~**-up trunnion** Spannzapfen m
**truly aligned** flucht(ge)recht
**trumped** Eingußtrichter m, Trichter m, Trompete
f ~ **horn** Schneckenhorn n (elektropneumatisch) ~**-shaped** trompetenförmig
**truncate, to** ~ abbrechen, abstumpfen (math.),
runden (comput.)
**truncated** abgestumpft, stumpf ~ **cone** abgestumpfter Kegel m, Kegelstumpf m, stumpfer
Kegel m, Stumpfkegel m ~ **corner** abgeschnittene Ecke f ~ **cylinder** stumpfer Zylinder m
~ **electrode** Stumpfelektrode f ~ **framework
pyramid** Pyramidenflechtwerk n ~ **picture**
abgehacktes Bild n ~ **prism** stumpfes Prisma
n ~ **pyramid** Pyramidenstumpf m ~ **tail end**
Heckboden m
**truncating,** ~ **of threads** Abflachen n oder
Abrunden n von Gewinden ~ **face** Abstump-
fungsfläche f (cryst.)
**truncation** Abflachung f, Abstumpfung f ~
**error** Abbrechfehler m (info proc.), Rundungsfehler m (comput.)
**trundle, to** ~ rollen
**trunk** Baumstamm m, Fernleitung f, Gepäck-

koffer m, Kanal m (comput.), Koffer m, (car)
Kofferraum m, Körper m, Rumpf m, Sammelschiene f (comput.), Schaft m, Stamm m,
Stammleitung f, Stock m, (of polygon) Stumpf
m, Transportkoffer m, Verbindungsleitung f,
Weg m ~ **from concentration switch** Sammeldienstleitung f
**trunk,** ~ **airline** Langstrecken-Fluggesellschaft f
~ **airway** Hauptluftstraße f ~ **answering jack**
Fernabfrageklinke f ~ **barrow** Kastenkarre f
~ **board** Kofferpappe f, Vulkanfiber f ~ **busy**
fernbesetzt ~ **cable** Fernkabel n, Ortsverbin-
dungskabel n ~**(-line) cable** Verbindungs-
(leitungs)kabel n ~**-cable jointer** Fernkabel-
löter m ~ **call** Fernanruf m ~ **call signal** Fernanrufzeichen n ~ **call transfer relay** Um-
legungsrelais n ~ **circuit** Fernamtsschaltung f
~ **communication** Fernleitung f, Fernverbindung f ~ **congestion signal** Anhäufungszeichen
n ~ **connection** Fernanschluß m ~ **connection
cancelling relay** Trennrelais n (TR) ~ **control
center** Lastverteilungszentrale f ~ **cord** Fern-
leitungsschnur f ~ **cost coefficient** Umweg-
faktor m ~ **digit** (P.A.B.X) Amtskennziffer f
~**-directory inquiry** Fernamtsauskunft(stelle)
f ~ **drop** Fernklappe f ~ **engine** Dampfmaschine f mit rohrförmiger Kolbenstange ~
**entering relay** Einsteigerelais n ~ **exchange**
Fernamt n, Fernvermittlung f ~ **exchange
with automatic call distribution** Fernamt n
mit Anrufsuchern ~ **exchange switchboard**
Fernschrank m ~ **final selector** Fernleitungswähler m ~ **group** Leitungsbündel n, Ver-
bindungsleitungsbündel n ~ **group fully
available to outward positions** Gesamtheit f
der abgehenden Verbindungsleitungen ~**-group
selector** (TGS) Amtsgruppenwähler m (AGW)
~ **holding and automatic sequence calls relay**
Halte- und Kettengesprächsrelais n ~**-hunting
switch** zweiter Vorwähler m ~ **inquiry** Fern-
amtsauskunft(stelle) f ~ **jack** Stöpselklinke f
~ **jack lamp** Fernklinkenlampe f ~**-jack panel**
Fernklinkenfeld n ~ **junction** Nahverkehrsleitung f ~**-junction circuit** Fernvermittlungsleitung f, Vorschalteleitung f ~ **junction line**
Fernleitung f
**trunk-line** Amts-, Fern-leitung f, Fernlinie f,
Fernverbindung f, Hauptlinie f, Stamm-kreis
m, -leitung f ~ **bundle** Verbindungsleitungs-
bündel n ~ **construction** Stammleitungsbau m
~ **dialing** Wählerfernsteuerung f (electr.) ~ **feeder** Leitung f (electr.) ~ **finder** zweiter Anrufsucher m
~ **repeaters** Fernleitungsverstärker pl ~ **road**
Fernverkehrsstraße f ~ **telephone** Verbindungsrichtung f
**trunk,** ~ **lines** Sammelleitungen pl ~ **lock** Fallschloß n ~ **loss** Nutzdämpfung f der Fernleitung ~ **network** Fernkabelplatz m ~ **offering
line** Fernleitung f ~**-offering selector** Anbiet-
wähler m ~ **operator** Fernamtsbeamtin f,
Ferngehilfin f, Vermittlungsbeamtin f ~
**order-wire jack** Ferndienstklinke f ~ **pipe line**
Hauptrohrleitung f ~ **piston** Hohlkolben m
~**-piston engine** Hohlkolbenmotor m ~
**position** Fernplatz m ~ **record circuit** Fern-
amtsmeldeleitung f ~ **record position** Meldeplatz m ~ **record section** Fernamtsmelde-
stelle f ~ **riveting machine** Koffernietmaschine

*f* ~**road** Fernverkehrsstraße *f*, Schnellstraße *f* ~ **route** Fernleitungslinie *f*, Hauptfluglinie *f*, Linienzug *m* ~**-route beacon** Hauptflugstreckenfeuer *n* ~ **route traffic** Hauptflugstreckenverkehr *m* (aviat.) ~ **service** A-B-Verkehr *m* ~ **service observing** Fernbetriebsüberwachung *f* ~ **signaling** Fernanruf *m* ~ **signaling working** Anrufbetrieb *m* auf Fernleitungen ~ **stripping knives** Baumschälmesser *n* ~ **subscriber's line** Fernanschluß *m* ~ **supervision** Fernamtsaufsicht *f* ~ **(telephone) switch-board** Fernschrank *m* ~ **switching diagram** Netzgestaltung *f* ~ **switching scheme** Fernleitungsschaltplan *m* ~ **system** Fernleitungsnetz *n* ~ **table** Ferntisch *m* ~ **telephone cable** Fernkabel *n* ~ **telephone circuit** Fernleitung *f* ~ **telephone station** Fernamtsanschluß *m* ~ **test board** Fernprüfschrank *m* ~ **traffic** Vorbereitungsverkehr *m* ~**-type** kofferartig ~**-type piston engine** Tauchkolbenmotor *m* ~ **unit** Kofferapparat *m* ~ **unit for shooting pictures** Kofferapparat *m* für Bildaufnahme ~ **wire** Fernkabel *n* ~ **wood** Stammholz *n* ~ **zone** Fernverkehrsbereich *m* ~**-zone cable** Fernleitungsbezirkskabel *n*

**trunking** Verbindungs-leitungsbetrieb *m*, -verkehr *m* ~ **arrangement** Gruppierung *f* (teleph.) ~ **scheme** Verbindungsaufbau *m* ~ **traffic** Verbindungsleitungsverkehr *m*

**trunks** Verbindungen *pl* zwischen den Ämtern

**trunnel head** Füllöffnung *f*

**trunnion** Achse *f*, Auflager *n*, Dreh-, Kurbel-, Lager-, Pivot-, Schild-, Stirn-, Trag-, (of converter) Wende-zapfen *m*, Zapfen *m* ~ **arms of the compass** Aufhängebügel *m* ~ **bearing** Achsen-, Bolzen-, Pivot-, Schild-(zapfen)-lager *n* ~**-bearing cover** Schildzapfenlagerdeckel *m* ~**-bearing pan** Schildzapfenlagerpfanne *f* ~ **block** Kardanstein *m* ~ **guide** Zapfenlager *n* ~ **mount** Triebwerkaufhängebeschlag *m* ~ **ring** Birnenring *m*, (of converter) Ring *m*, Zapfenring *m* ~ **screw** Schildzapfenschraube *f*, Zapfenschraube *f* ~ **seat** Zapfenlager *n* ~ **tilt** Schildzapfenlage *f* ~ **tip wagon** Zapfenkipper *m*

**truss, to** ~ absteifen, (poles) verstärken

**truss** Absteifbalken *m*, Binder *m*, Bruchband *n*, Fachwerk *n*, Fachwerksträger *m*, Gebinde *n*, Gitterwerk *n*, Gespann *n*, Hallenbinder *m*, Kragstück *n*, Riegel *m*, (hip) Strebe *f*, Stütze *f*, verstrebter Träger *m* ~ **of roofing** Dachbock *m*

**truss,** ~ **bolt** Ankerbolzen *m* ~ **bridge** Fachwerkbrücke *f* ~ **force** Stabkraft *f* ~ **frame** Hängewerk *n* ~ **girder** Fachwerkbalken *m* ~**-guyed pole** Stange *f* mit Verspannung ~ **head screw** Flachrundschraube *f* ~ **hoop** Arbeitsreif *m* ~**-hoop driving machine** (brewing) Anziehmaschinenarbeitsreifen *m*, Arbeitsreifenanziehmaschine *f* ~ **(or frame) joint** Knotenpunkt *m* ~ **pointer** verstrebter Zeiger *m* ~ **post** Hängewerksstrebe *f* ~ **rod** Ankerbolzen *m* ~ **rods** (under trussed beam) Fachwerkglied *n* ~ **spar girder** Fachwerkholm *m* ~ **strut** Unterzugsstrebe *f*

**trussed,** ~ **arch** Fachwerkbogen *m* ~ **beam** unterspannter Balken *m* ~ **frame** Rahmenfachwerk *n*, (car) versteifter Rahmen *m* ~ **girder** armierter Träger *m* ~ **gland** verstärkter Träger

*m* ~ **pole** verspannte Stange *f* (electr.) ~ **resting place** gesprengtes Podest *n*

**trussing** Fachwerk *n*, Steifrahmen *m*, Unterzug *m* ~ **of frame** (motorcar) Unterzug *m* des Rahmens ~ **of the frame** Rahmenunterzug *m* ~ **machine** Antreibmaschine *f* für Faßreifen, Faßreifenauftriebmaschine *f*

**trust** Interessengemeinschaft *f*, Konzern *m*, Trust *m* ~ **account** Anderkonto *n* ~ **company** Treuhandgesellschaft *f* ~ **money** Depositenkasse *f*, Mündelgeld *n*

**trusted agent** Vertrauensmann *m*

**trustee** Bevollmächtigter *m*, Treuhänder *m*

**truth,** ~ **of rotation** Rundlaufen *n* ~ **to scale** Maßstabsgerechtigkeit *f*

**try, to** ~ ausprobieren, experimentieren, proben, probieren, prüfen, versuchen **to** ~ **in court** vor Gericht *n* stellen **to** ~ **the engines** Maschinen *pl* herumschlagen lassen **to** ~ **to get away** fortstreben **to** ~ **on** anproben **to** ~ **out** ausproben, ausprobieren, einschießen, erproben **to** ~ **with the plummet** einloten

**try,** ~ **cock** Probehahn *m*, Zwickel *m* ~ **square** Anschlagwinkel *m*, rechter Winkel *m*

**trying** Ausprobieren *n*, Erprobung *f* ~**-out** Erprobung *f*

**T-screw fitting** T-Verschraubung *f*

**T-section** Kettenglied *n* zweiter Art, T-Profil *n*

**T-shaped,** ~ **antenna** T-Antenne *f* ~ **tube (or pipe)** T-Rohr *n*

**T-slot** Aufspannschlitz *m* ~ **cutter** kreuzverzahnt; Schaftfräser *m* für T-Nuten

**T-square** Kreuzwinkel *m*, Reißschiene *f*

**T-strap** Krückeisen *n*

**T-type,** ~ **antenna** T-Antenne *f* ~ **attenuation network** Entzerrungskette *f* mit Brücken-T-Schaltung ~ **fitting** Winkelanschlußstück *n* ~ **network attenuator** Entzerrungskette *f* mit Brücken-T-Schaltung *f* ~ **section** T-Glied *n* ~ **section attenuator** T-Regler *m*

**tub** Barke *f*, Bottich *m*, Bütte *f*, Faß *n*, Förder--hund *m*, -karren *m*, -wagen *m*, (mine) Grubenwagen *m*, Klärbottich *m*, Kübel *m*, Kufe *f*, Wanne *f*, Zuber *m* ~ **file** Ziehkartei *f* ~ **lift hoist** Kübelaufzug *m* ~**-mold baling press** Topfballenpresse *f* ~**-shaped** wannenförmig ~ **sizing** animalische Leimung *f* ~ **washing** (of ores) Setzwäsche *f*

**tubbing** Cuvelage *f*, Gußringausbau *m*, Kuvelage *f* (min.), Tübbings *pl*, (of shaft) Verrohrung *f* ~ **rings** Aufsatzkränze *pl*

**tube, to** ~ **a hole** ein Bohrloch *n* verrohren

**tube** Hülse *f*, Lichtspritze *f*, Pneumatik *f*, Rohr *n*, Röhrchen *n*, Röhre *f*, Röhrentunnel *m*, Schlauch *m*, Staurohr *n*, Strang *m*, Tube *f*, Tubus *m*, zylindrische Kreuzspule *f* ~ **with bulbs** Kugelrohr *n* **by** ~ durch Rohrpost *f* ~ **for cheeses** Kreuzspulenhülse *f* ~ **of flow** Stromröhre *f* ~ **of flux** Feldröhre *f* ~ **of force** Kraft-linie *f*, -röhre *f* ~ **with ground joint** (electron microscope) Schliffrohr *n* ~ **for lubricating axles** Rohr *n* zum Schmieren der Achsen ~ **of magnetic force** magnetische Kraftröhre *f* ~ **of magnetic induction** magnetische Induktionsröhre *f* ~ **with nine--pin glass-button base** Novalröhre *f* **the** ~ **through which orders are transmitted from**

**bridge to engine room** (on a ship) Befehls-rohr *n* ~ **with oxide-coated filament** Oxyd-kathodenröhre *f* ~ **of the scintillator** Szintil-latorschlauch *m* ~ **with screw cap** Gewindeglas *n*, Glas *n* mit Gewinde ~ **for screwing** Gewin-derohr *n* ~ **for suction air** Saugluftleitung *f* ~ **of steam gauge** Manometerrohr *n*

**tube, ~ adapter** Röhrenzwischenstecker *m* ~ **amplification** Röhrenverstärkung *f* ~ **amplifier** Vakuumröhrenverstärker *m* ~ **aperture** Rohr-öffnung *f* ~ **attachment** Tubusaufsatz *m* ~ **base** Röhrensockel *m* ~ **bend** Rohrbogen *m* ~ **bender** Rohrbiegewerkzeug *n* ~ **bending** Rohrbiegen *n* ~-**bending principle** Rohrbiege-prinzip *n* ~ **bends** Rohrbogen *pl* ~ **bit** Kanonen-bohrer *m* ~ **blank** Hohlkörper *m*, Luppe *f* ~ **body** Rohrkörper *m* ~ **boiler** Röhrenkessel *m* ~ **bracket (or arm)** Klemmschelle *f* ~ **brake** Schlauchbremse *f* ~ **branch** Röhrenindustrie *f* ~ **brush** Feuerrohrbürste *f* ~ **burst** Rohrreißer *m* ~ **carrier** Tubusträger *m* ~ **characteristic** Röhrenwert *m* ~ **checker** Röhrenprüfer *m* ~ **clamp** Rohrschelle *f* ~**cleaner** Kesselstein-abklopfer *m*, Rohrputzturbine *f* ~ **cleaning machine** Rohrreinigungsmaschine *f* ~ **closing** Rohrverschluß *m* ~-**closing machine** Tuben-schließmaschine *f* ~ **coil** Rohrspirale *f* ~ **combustion** Röhrchenentgasung *f* ~ **compen-sator** Rohrausgleicher *m* ~ **complement** Röh-ren-bestückung *f*, -mehraufwand *m* ~ **constant** Röhrenkonstante *f* ~ **containing phosphor** Leuchtstofflampe *f* ~ **contraction** Rohrver-kürzung *f* ~ **control of the fork** Röhrensteu-erung *f* der Gabel ~ **conveyor** Förder-rohr *n*, -röhre *f*, Zuführungsrohr *n* ~ **current** Röhren-strom *m* ~ **cut-off** Röhrensperrpunkt *m* ~ **cutter** Rohrschneider *m* ~ **cutting device** Hülsenabstechmaschine *f* ~ **detector** Kathoden-röhrendetektor *m* ~ **diameter** Hülsenloch-durchmesser *m*, Rohrquerschnitt *m* ~ **dia-phragm** Bleiblende *f*, Diaphragma *n* ~ **di-mension** Rohrmaß *n* ~ **door** Teilkammertür *f* ~ **drawing** Rohrzug *m* ~-**drawing bench** Röhrenziehbank *f* ~ **expander** Aufdornwerk-zeug *n*, Röhrenauswalzvorrichtung *f*, Rohr-erweiterer *m*, Rohrwalzapparat *m*, Siederohr-dichtmaschine *f* ~ **expenditure** Röhrenaufwand *m* ~ **feed** Hülsenzuführung *f* ~ **feeder** Rohrzu-leitung *f* ~ **ferrule** (for condensers) Expansions-büchse *f* ~-**filling machine** Tubenfüllmaschine *f* ~ **filter** Pumpenkorb *m*, Schlauchfilter *n* ~ **firing burner** Tauchrohrbrenner *m* ~ **frame** Gitterrohr *n* ~ **funnel** rohrförmiger Trichter *m*, Trichterrohr *n* ~ **furnace** Röhrenofen *m* ~ **galvanometer** Röhrengalvanometer *n* ~ **gluing machine** Hülsenklebemaschine *f* ~ **gripper** Hülsengreifer *m* ~ **hanger** Rohrauf-hängung *f*, Röhrenhänger *m* ~ **heading ma-chine** Rohreinwalzmaschine *f* ~ **heater** Röh-renofen *m* ~ **heating** Röhrenheizung *f* ~ **hiss** Röhrenrauschen *n* ~ **holder** Röhrenfassung *f* ~ **hole** Rohrloch *n* ~ **illumination** Röhren-beleuchtung *f* ~ **industry** Röhrenindustrie *f* ~ **joint connection** Rohrstoßverbindung *f* ~ **jointing** Rohr-lasche *f*, -schäftung *f* ~ **jointing sleeve** Rohrverbindungshülse *f* ~ **level** Röhren-libelle *f* ~ **line** Kurzschlußleitung *f* zur An-tennenanpassung ~ **line-up** Röhrenbestückung

*f* ~ **making** Röhrenfabrikation *f* ~ **manufacture** Rohrherstellung *f* ~ **melting** Röhrenschmel-zung *f* ~ **mill** Röhrenwerk *n*, Rohrmühle *f*, Rotator *m*, Trommelmühle *f* ~-**mill drying** Röhrentrocknung *f* ~ **mounting** Rohr-halterung *f*, -montierung *f* ~ **noise** Röhrenrauschen *n* ~ **over-loading** Röhrenübersteuerung *f* ~ **papers** Spulenpapier *n* ~ **plate** Rohr-blech *n*, -boden *m*, -wand *f* oder Röhrenblech *n* ~ **pliers** Röhrenzange *f* ~ **plug** Rohrstopfen *m* ~ **plug ram** Rohrstopfstange *f* ~ **pole** Rohr-mast *m* ~ **push bench** Rohrstoßbank *f* ~ **quench** Röhrenlöscher *m* ~ **reactor modulator** Reaktanzröhrenkreis *m* ~ **receiver** Röhren-empfänger *m* ~ **reclaim** Luftschlauchregenerat *n* ~ **recoil** Rohrrücklauf *m* (vacuum-)~ **rectifier** Röhrengleichrichter *m* ~-**regulated** röhrengeregelt ~ **resistance** Röhrenwider-stand *m* ~ **roll** Registerwalze *f* ~ **rolling** Röhrenwalzen *n*, Rohrwalzen *n* ~-**rolling and -flaring equipment** Röhrenwalzgerät *n* ~ **rolling mill** Röhrenwalzwerk *n*, Rohrwalzwerk *n* ~ **rolling train** Röhrenstraße *f* ~ **roughing machine** Schruppmaschine *f* für Röhren ~ **rounds** Rohrrohlinge *pl* ~ **sample boring** Bohr-probenbüchse *f*, Bohrung *f* mit Kernbüchsen ~ **sandpapering machine** Röhrenabschleif-maschine *f* ~ **scraper** Rohrauskratzer *m* ~ **screen** Bildschirm *m* ~-**sealing machine** Tuben-schließmaschine *f* ~ **shaft** Hohlwelle *f* ~-**shaped winding** Röhren-, Zylinder-wicklung *f* ~ **sheet** Rohr-platte *f*, -wand *f* ~ **shield** Röhren-abschirmung *f*, Röhrenschutzgehäuse *n* ~ **shutter** Verschluß *m* für Röntgenröhre ~ **slide** Tubusschlitten *f* ~ **slot** Rohrnut *f* ~ **socket** Röhrenfassung *f*, Rohrhülse *f* ~ **spindle** Schlagrohrstock *f* ~ **spring manometer** Rohr-federdruckmesser *m* ~ **stand** Röhrenstativ *n* ~ **stay** Röhrenstrebe *f* ~ **straightening machine** Rohrrichtmaschine *f* ~ **strip** Röhrenstreifen *m* ~ **support** Rohraufhängung *f*, Röhren-hänger *m*, -träger *m*, Tubusträger *m* ~ **surfaces** Ka-nalflächen *pl* ~ **switch** Rohrweiche *f* ~ **tanging machine** Rohranspitzmaschine *f* ~ **tester** Röh-renprüfer *m*, Röhrenprüfgerät *n* ~ **testing** Brauchbarkeitsprüfung *f* der Vakuumröhren ~-**testing switch** Röhrenprüfschalter *m* (rdo) ~-**time meter** Heizstundenzähler *m* für Röhren ~-**tripping circuit** Röhrenkippschaltung *f* ~ **tunnel** Röhrentunnel *m* ~ **turn** Rohrbogen *m* ~-**type** schlauchförmig ~ **ventilation** Röhren-entlüftung *f* ~ **vise** Rohrschraubstock *m* ~ **voltage** Röhrenspannung *f* ~ **voltage drop** Spannungsabfall *m* (vacuum-)~ **voltmeter** Röhren-spannungsmesser *m*, -voltmeter *n* ~ **volume** Rohrvolumen *n* ~ **wall compartments** (Rohrwände) Schotten *pl* ~ **wattmeter** Röh-renwattmeter *n* ~ **welding** Rohrschweißung *f* ~ **welding machine** Rohrschweißmaschine *f* ~ **well** Röhrenbrunnen *m* ~ **window** Austritts-fenster *n* ~ **works** Röhrenwerk *n* ~-**wrapping machine** Hülsenwickelmaschine *f*

**tubeless tire** schlauchloser Reifen *m*

**tubercular corrosion** grübchenartige Rostan-fressung *f*

**tubing** Pumpenröhre *f*, Rohr *n*, Rohr-anlage *f*, -leitung *f*, -zucker *m*, Röhren-anlage *f*, -leitung *f*, Schachtverrohrung *f*, Schlauch *m*, Stahlrohrleitung *f* ~ **action** Rohrblock *m* ~ **bailer** Verrohrungsbüchse *f* ~ **catcher** Pumpenrohrkrebs *m* ~ **connection** Schlauchstück *n* ~ **disc** Rohrscheibe *f* ~ **hanger** Pumpenrohrhänger *m* ~ **sheet** Schlauchplatte *f* ~ **spider** Pumpenrohranhänger *m*

**tubsized** oberflächengeleimt

**tubular** röhrchenförmig, röhren-artig, -förmig, rohrförmig ~**arch bridge** Rohrbogenbrücke *f* ~ **backbone chassis** Mittelrohrrahmen *m* ~ **bearing** Tonnenlager *n* ~ **boiler** Flammrohr-, Rohrdampf-, Röhren-kessel *m* ~ **bolt** Rohrbolzen *m* ~ **capacitor** Rohrkondensator *m* ~ **chassis** Rohrrahmen *m* ~ **compass** Röhrenbussole *f* ~ **condenser** Wickelblockkondensator *m* ~ **conductor** Hohlleiter *m* ~ **conductor line** Rohrleitungsstück *n* ~ **construction** Rohrbauweise *f* ~ **cooler** Röhrenkühler *m* ~ **core** hohler Kern *m*, Hohlkern *m*, Rohrkern *m* ~ **cross member** Rohrquerträger *m* ~ **drive shaft** Antriebsrohrwelle *f* ~ **fabric** Schlauchware *f* ~ **fabric machine** Schlauchmaschine *f* ~ **fabrics** Hohlgewebe *n* ~ **felt** Rundfilz *m* ~ **frame** Rohr-gerüst *n*, -rahmen *m* ~ **fuse** Röhrensicherung *f* ~ **fuselage** Rohrrumpf *m* ~ **girder** Rohrträger *m*, vollwandiger Träger *m* ~ **goods** Hohlgewebe *n*, Rohrmaterial *n* ~**-grain powder** Röhrchenpulver *n* ~ **grate** Röhrenrost *m* ~ **guide** Führungsrohr *n* ~ **gun carriage** Röhrenlafette *f* ~ **handle** Griffrohr *n* ~ **indicator** Klappe *f* mit Topfmagnet, Mantelklappe *f* ~ **inside-micrometer gauge** Zylinderstichmaß *n* ~ **lamp** Röhren-, Soffitten-, Zweisockel-lampe *f* ~ **lamp wire** Soffittenleitung *f* ~ **level** Röhrenlibelle *f* ~ **magnet** Röhrenmagnet *m* ~ **mast** Röhrenmast *m*, Standrohr *n* ~ **needle** Kanüle *f*, Röhrennadel *f* ~ **nitroglycerin powder** Nitroglyzerinröhrenpulver *n* ~ **pin** Rohrbolzen *n* ~ **plate** Röhrchenplatte *f* ~ **pole** Rohrmast *m*, Rohrständer *m* ~ **powder** Röhrenpulver *n* ~ **product** runder Hohlkörper *m* ~ **propeller shaft** Kardanrohr *n* ~ **radiator** Röhrchenkühler *m*, Röhrenkühler *m* ~ **rivet** Hohl-niet *n*, -niete *f*, Lochniete *f*, Rohrniet *n* ~ **seat** Paßhülse *f* (hülsenförmige Aufsitzfläche) ~ **shaft** hohle Welle *f*, Hohlwelle *f* ~ **shell** Mantelrohr *n* ~ **socket wrench** Steckschlüssel *m* aus Rohr ~ **spar** Röhre(n)holm *m*, Rohrholm *m* ~ **steel frame** Stahlrohrgestell *n* ~**-steel mast (or pole)** Stahlrohrmast *m* ~**-steel pole (or stand)** Stahlrohrständer *m* ~**-steel spar** Stahlrohrholm *m* ~ **stem** röhrenförmiger Schaft *m* ~ **structural members** Bauteile *pl* (aus Rohren) ~ **structures** Rohrtragwerke *pl* ~ **support** Tragrohr *n* ~ **tension rod** Zugstangenrohr *n* ~ **thermograph** Rohrtemperaturschreiber *m*, Rohrtermograf *m*, Rohrwärmegradschreiber *m* ~ **thermometer** Rohrthermometer *n*, Rohrwärmegradmesser *m* ~ **tore** Röhrenwulst *f* ~ **valve** Rohrschütz *m* ~ **valve chipping hammer** Rohrschieber *m* (Meißelhammer) ~ **vaporizer** Röhrenverdampfer *m* ~ **worm conveyor** Förderrohr *n*

**tubulars** Gewindebohrteile *pl*

**tubulate, to** ~ tubulieren

**tubulating machine** Stengelansatzmaschine *f*

**tubulature** Tubus *m*

**tubulure** Rohrstutzen *m*

**tubus** Tubus *m*

**tuck,** ~ **pointing** vorspringende Fuge *f* ~ **position** Fangstellung *f*, Halbeinschließstellung *f* ~ **presser** Musterpresse *f*, (textiles) Preßmaschine *f*

**tucking** Spleißvorrichtung *f* ~ **needle** löffelförmiges Spleißwerkzeug *n*

**Tudor plate** Tudorplatte *f*

**tufaceous limestone** Kalktuff *m*

**tuff** Tuff *m*

**tuft** Baumgruppe *f*, Büschel *m* ~ **of hair** Haarbusch *m*

**tug, to** ~ mitziehen, ins Schlepptau *n* nehmen, zerren, zupfen

**tug** Bindestrang *m*, Geschirrtau *n*, Ruck *m*, Spannseil *n*, Zug *m*, Zugtau *n* ~ **(boat)** Schlepper *m* ~ **boat** Schlepp-dampfboot *n*, -dampfer *m* ~ **rim** gerillte Antriebsscheibe *f*

**tugging** Zerring *f* ~ **device** Schleppeinrichtung *f* ~ **iron** Strickeisen *n* (Montage) ~ **ring** Zugring *m*

**tulip,** ~**-shaped** tulpenförmig ~**-supporting rod** Tulpenhaltestange *f* ~**-type valve** tulpenförmiges Ventil *n* ~ **valve** Tulpenventil *n*

**tulle** Tüll *m* ~ **machine** Tüllmaschine *f*

**tumble, to** ~ (over) durcheinanderfallen, zu Fall *m* bringen, schleudern, sich überschlagen, stürzen, torkeln, trommeln, umfallen **to** ~ **down** abstürzen, herunterstürzen, niederstürzen

**tumble,** ~ **bay** Sturzbett *n* ~ **card** Mehrfachkarte *f* ~ **switch** Schubschalter *m*

**tumbled** gestürzt

**tumbler** Auslösevorrichtung *f*, Nocken *m*, Putz-faß *n*, -trommel *f*, Richtwelle *f*, Rollfaß *n*, Scheuer-faß *n*, -trommel *f*, Schnepper *m*, (cleaning drum) Trommel *f*, Trommeltrockner *m*, Wipptisch *m*, Zahn *m*, Zuhaltung *f* ~ **action** Taumelbewegung *f* ~ **gear** Fallwerk *n*, Wendeherz *n* ~ **lever** Hebelschwinge *f* ~ **movement** Taumelbewegung *f* ~ **shaft** Daumen-, Herz-, Schwingen-welle *f* ~ **spring** (lock) Zuhaltungsfeder *f* ~ **switch** Kipp-, Tumbler-schalter *m* ~ **toe** (lock) Zuhaltungslappen *m* ~ **yoke** Hebelschwinge *f*, Kippbrücke *f* (Magnetschalter)

**tumbling** Kippen *n*, Polierzylinder *m*, Rollfaß *n*, Taumeln *n*, Taumelschwingung *f* ~ **of the images** Bild-sturz *m*, -stürzen *n*, Stürzen *n* der Bilder

**tumbling,** ~ **ball** Schnellgewicht *n* ~ **barrel** Gußputztrommel *f*, Naßputztrommel *f*, Putzfaß *n*, Putztrommel *f*, Rollfaß *n*, Rommel *f*, Rummelfaß *n*, Scheuer-faß *n*, -trommel *f* ~ **barrel with dust exhaust** Putztrommel *f* mit Staubabsaugung ~ **bob** Kipphebel *m* ~ **device for hoop iron** Bandeisenschleudervorrichtung *f* ~ **drier** Taumeltrockner *m* ~ **error** Taumelfehler *m* ~ **hook** Fallhaken *m* ~ **mill** Rollfaßanlage *f*, Trommelmühle *f* ~ **motion** Taumelbewegung *f*

**tumifying water** Quellwasser *n*

**tun** Bottich *m* ~ **casing** Bottichmantel *m* ~ **dish** Trichter *m* ~ **drainer** Bottichseiher *m* ~ **fermentation** Bottichgärung *f* ~ **jacket** Bottichmantel *m* ~ **sediment** (brewing) Bottichgeläger *n*

**tunable** abstimmbar

**tune, to** ~ abgreifen (rdo), abschirmen (rdo), abstimmen, abtönen (telegr.), anpassen, einstellen, rasten, in Resonanz *f* bringen, in Übereinstimmung *f* bringen **to ~ alike** gleichstimmen **to ~ down** herabstimmen **to ~ finely** nachstimmen **to ~ in** abstimmen, anstimmen (electr.), anzapfen, Funkgerät *n* auf einen Sender einstellen **to ~ out** auskoppeln, ausstimmen, entkoppeln **to ~ in a radio** anspannen **to ~ on (or off) a radio** abzapfen **to ~ in a radio transmitter with directional receiver** Funksender *m* mit Richtempfänger einpeilen **to ~ to resonance** auf Resonanz *f* abstimmen **to ~ together (or uniformly)** gleichstimmen **to ~ up** (a motor) einfahren, rennfertig machen **tune** Ton *m*, Tonart *f* **in ~** abgestimmt, in Resonanz *f* befindlich, im Rhythmus *m* **to be out of ~** verstimmt sein **in ~ with** in Resonanz *f*

**tuned** abgestimmt **~ alike** gleich abgestimmt **~ in** eingerastet **~ off resonance** verstimmt **~ to resonance** auf Resonanz *f* abgestimmt

**tuned, ~ amplifier** abgestimmter Verstärker *m*, Resonanzverstärker *m* **~ anode circuit** abgestimmter Anodenkreis *m* **~ antenna** abgestimmte oder gesonderte Antenne *f* **~ buzzer** abgestimmter Summer *m* **~ cavity** abgestimmter Hohlraumresonator *m* **~ circuit** abgestimmter Kreis *m*, Abstimm-, Schwing-kreis *m* **~ dipole** abgestimmter Dipol *m* **~ doublet** abgestimmter Dipol *m* **~-in beacon (or transmitter)** gepeilter Sender *m* **~ plate oscillation** Gitteranodenschwingung *f* **~ radio frequency** Geradeausfrequenz *f* **~ radio-frequency amplifier** Hochfrequenzverstärker *m* **~ radio-frequency receiver** Geradeausempfänger *m* **~-reed indicator** Zungenfrequenzindikator *m* **~-reed rectifier** Pendelgleichrichter *m* **~-reed relay** Resonanz-, Tonfrequenz-, Zungenfrequenz-relais *n* **~ reed selector** Resonanzrelais *n* **~ ringing** abgestimmter Anruf *m*, Wahlanruf *m* mit abgestimmten Einrichtungen **~ state** Anklang *m* **~ system** abgestimmtes System *n* **~ transformer** abgestimmter Transformator *m*, Resonanztransformator *m*

**tuner** Abstimm-apparat *m*, -spule *f*, Vorstimmer *m*

**tung oil** Holz-, Tung-öl *n*

**tungar, ~ rectifier** Edelgasglühkathodengleichrichter *m* **~ tube** Quecksilberdampflampe *f* mit Heizkathode **~-type rectifier** Glühkathodengleichrichter *m* mit Edelgas

**tungstate** Wolframat *n*, Wolfram-salz *n*, -säuresalz *n*

**tungsten** Wolfram *n/m* **~ alloy** Wolframlegierung *f* **~ alloyed steel** wolframlegierter Stahl *m* **~arc lamp** punktförmige Lichtquelle *f*, Punktglimmlampe *f*, Wolframbogenlampe *f* **~ bronze** Safranbronze *f* **~ carbide** Wildiametall *n*, Wolframkarbid *n* **~-carbide milling cutter** Hartmetallfräser *m* **~ carbide tool** Hartmetallwerkzeug *n* **~ contact** Wolframkontakt *m* **~ dichloride** Wolframdichlorid *n* **~ di-iodide** Wolframdijodid *n* **~ disc** Wolframscheibe *f* **~ disulphide** Wolframdisulfid *n* **~ electrode** Wolframelektrode *f* **~ filament** Wolframfaden *m* **~ heater wire** Wolframheizdraht *m* **~ lamp** Metallfadenlampe *f* **~ nitride** Wolframstickstoff *m* **~ ore** Wolframerz *n* **~ oxide** Wolframoxyd *n* **~ preparation** Wolframpräparat *n* **~**

**salt** Wolframsalz *n* **~ silicide** Wolframsilicid *n* **~ steel** wolframlegierter Stahl *m*, Wolframstahl *m* **~ sulfide** Wolframsulfid *n* **~ tetraiodide** Wolframtetrajodid *n* **~ trioxide** Wolframtrioxyd *n* **~ wire** Wolframfaden *m*

**tungsteniferous** wolframhaltig

**tungstic, ~ acid** Wolfram-säure *f*, -trioxyd *n* **~ acid salt** Wolframsäuresalz *n* **~ anhydride** Wolframsäureanhydrid *n* **~ ocher** Wolframocker *m*

**tungstide** Wolframid *n*

**tungstite** Wolframocker *m*

**tuning** Abstimmen *n*, Abstimmung *f*, Anpassung *f*, Rastereinstellung *f*, Trieb *m* **~ of an engine** Einregulieren *n* eines Motors **~ out of feedback** Rückkupplungsunterdrückung *f* **~ of filament circuit** Heizabstimmung *f* **~ in** Anwendung *f* **~ out** Auskopplung *f*, Entkopplung *f* **~ of the television receiver** Abstimmung *f* des Fernsehempfängers **~ in a transmitter station with directional receiver** Einpeilen *n* eines Richtempfängers auf einen Funksender **~ by variable condenser** Abstimmung *f* durch veränderlichen Widerstand

**tuning, ~ antenna** künstliche Antenne *f* **~ apparatus** Abstimmvorrichtung *f*, Richtapparat *m* **~ box for frequency doubler** Verdopplerabstimmkästchen *n* (g/m) **~ bridge** Abstimmbrücke *f* **~ circuit** Abstimmkreis *m* **~ coil** Abstimm-, Kreis-spule *f* **~ condenser** Abstimmkondensator *m* **~ condenser sensibly braked (or detained)** fühlbare Abstimmung *f* **~ control** Abstimmeinrichtung *f*, Frequenzeinstellung *f* **~ core** Abstimmkern *m* **~ crank** Abstimmkurbel *f* **~ device** Abstimmeinrichtung *f* **~ dial** Abstimmskala *f* **~ drift** (due to temperature effects) Abstimmänderung *f*

**tuning-fork** Stimmgabel *f* **~ chronoscope** schreibende Stimmgabel *f* **~ circuit breaker** Stimmgabelunterbrecher *m* **~ control** Stimmgabelsteuerung *f* **~-controlled oscillator** Stimmgabel-generator *m*, -sender *m*, (tube) Stimmgabelröhrengenerator *m* **~ crystal** Stimmgabelkristall *n* **~ gyro** Stimmgabelkreisel *m* **~ interrupter** Stimmgabelunterbrecher *m* **~ oscillator drive** Stimmgabelsteuerung *f* **~ prong (or tine)** Gabelzinken *m*

**tuning, ~ indicator** Abstimm-melder *m*, -zeiger *m*, Abstimmungsanzeiger *m*, (magic eye) Abstimmungskreuz *n*, Amplituden-anzeiger *m*, -glimmröhre *f*, Resonanz-anzeiger *m*, -röhre *f*, Stationsmelder *m* **~-indicator glow tube** Abstimmglimmröhre *f* **~ indicator tube** Abstimmanzeigerröhre *f*, Anzeigeröhre *f* **~ inductance** Abstimmspule *f* **~ key** Stimmschlüssel *m* **~ knob** Abstimmknopf *m*, Frequenzangleicher *m* **~ lamp and choke coil** Syntonisierlampe *f* mit Impedanzspule **~ length** Abstimmlänge *f* **~ lock** Rast *f* (rdo) **~ means** Abstimm-einrichtung *f*, -mittel *m* **~ message** Abstimm-, Stimmungs-spruch *m* **~ meter** Abstimmanzeigeinstrument *n* **~ note** Kennsignal *n* **~ pipe** Stimm-flöte *f*, -pfeife *f* **~ pitch** Stimmton *m* **~ plate** Abgleich-, Abstimm-platte *f* **~ probe** einstellbare Sonde *f* **~ property** Abstimmfähigkeit *f* **~ range** Abstimmbereich *m* **~ ring** Abstimmring *m* **~ scale** Abstimmskala *f*, Einstellskala *f* **~ set** Abstimmungsaggregat *n* **~**

slug Abstimmkern *m* ~ spade Abgleichplatte *f*, Abstimmplatte *f* ~ unit Abstimmungsaggregat *n* ~ wand Abgleich-, Abstimm-platte *f* ~ wave Welle *f* der Abstimmung

T-union T-Verschraubung *f*

T-unit Zeitglied *n*

tunnel, to ~ ausschachten, durchtunneln, untertunneln

tunnel Durchstich *m* (r.r.), Gang *m*, Stollen *m*, Tunnel *m* ~ diode Tunneldiode *f* ~ drier Kanaltrockenapparat *m*, Schachtofentrockner *m*, Trockentunnel *m* (film) ~ drill Bohrer *m* für Streckenbetrieb ~-drying oven Kanaltrockner *m* ~ effect Kellerton *m*, Tunneleffekt *m* ~ furnace (kiln) Stoßofen *m* ~ kiln Kanal-, Tunnel-ofen *m* ~ kiln for annealing Glühtunnelofen *m* ~ mining Stollenbergbau *m* ~ radiator Tunnelkühler *m* ~ radius Kanalradius *m* ~ railway Tunnelbahn *f* ~ timbering Stollenzimmerung *f* ~-type radiator Düsenkühler *m*, einziehbarer Hängekühler *m*, Hänge- oder Bauchkühler *m*, in Luftführung eingeschlossener Kühler *m*

tunneling, ~ company Bergkompanie *f* ~ effect Tunneleffekt *m*

tunoscope magisches Auge *n*

tup Bär *m*, Fallbär *m*, Fallhammer *m*, Rammbär *m* ~ cylinder Bärzylinder *m* ~ weight Fallgewicht *n* (Hammer)

turbid trübe, unklar ~ matters Trübungsstoffe *pl* ~ particles Trübungen *pl* ~ prism Trübglasprisma *n* ~ river water Flußtrübe *f* ~ water Trübe *f*

turbidimeter Trübungsmesser *m*

turbidimetry Trübungsmessung *f*

turbidities Trübstoffe *pl*

turbidity fotografische Unschärfe *f*, Schleier *m*, Trübe *f*, Trübung *f* ~ of the air Lufttrübung *f*

turbidity, ~ coefficient Trübungskoeffizient *m* ~ measuring device Trübungsmeßeinrichtung *f* ~ meter Belichtungsmesser *m*

turbine Kreiselrad *n*, Läufer *m*, (engine) Turbine *f* ~ with rimstraddling buckets Reiterschaufelturbine *f* ~ for use with steam accumulator Speicherturbine *f*

turbine, ~ airplane Turbinenflugzeug *n* ~ bearing casing Turbinenlagerträger *m* ~ blade Laufschaufel *f*, Turbinenschaufel *f* ~ blade filter Turbinenschaufelfilter *n* ~ blower Turbinengebläse *n* ~ break Turbinenunterbrecher *m* ~ bucket Drehschaufel *f*, Schaufel *f* ~ bypass valve Kühlturbinen-Umgehungsventil *n* ~ casing Schaufelzylinder *m*, Turbinengehäuse *n* ~ chamber Turbinenkammer *f* ~ compressor unit Turbinenpumpaggregat *n* ~ cylinder Turbinengehäuse *n* ~ disc (jet) Turbinenläufer *m* ~ drive Turbinenantrieb *m*

turbine-driven, ~ auxiliary Turboaggregat *n* ~ feed water pump Turbospeisepumpe *f* ~ pump Pumpe *f* mit Turbinenantrieb ~ set Turbogeneratorsatz *m* ~ supercharger Abgaslader *m*, Abgasturbolader *m*

turbine, ~ drum Turbinentrommel *f* ~ effluent Turbinenabgas *n* ~ engine Turbomotor *m* ~ exhaust Turbinenabgas *n* ~ exhaust steam Turbinenabdampf *m* ~ gear Turbinenvorgelege *n* ~ generator Turboaggregat *n* ~ generator set Turbogeneratorgruppe *f* ~ house

Maschinenhaus *n* ~ housing Turbinengehäuse *n* ~ impeller mixer Turbinenmischer *m* ~ interrupter Turbinenunterbrecher *m* ~ lock Turbinenverschluß *m* ~ mixer Schaufelradmischer *m* ~ mounting Turbinenlager *n* ~ nozzle Leitschaufel-kranz *m*, -ring *m* ~ nozzle blade Turbinendüsenschaufel *f* ~-nozzle loss factor Turbinendüsenverlustbeiwert *m* ~ oil Turbinenöl *n* ~ pipe line Turbinenrohrleitung *f* ~ piping Turbinenleitungsrohre *pl* ~ pressure Laderdruck *m* ~ pump Kreiselpumpe *f* ~ ring (of guide blade) Führungsschaufelring *m* ~ rotor Turbinenläufer *m* ~ rotor blade Turbinenschaufel *f* ~ running forward Vorwärtsturbine *f* ~ seal Turbinendichtung *f* ~ shaft Turbinenwelle *f* ~ shop Turbinenhalle *f* ~ stator Turbinengehäuse *n* ~ stator blade Turbinenleitschaufel *f* ~ stator guide vanes Turbinenleitgitter *n* ~ steamer Turbinendampfer *m* ~ valve Turbinenschieber *m* ~ vane Drehschaufel *f*, Turbinenschaufel *f* ~ wheel Laufrad *n* ~ wheel disc Turbinenradscheibe *f*

turbo, to ~-charge aufladen (Motor)

turbo Turbinen . . ., Turbo . . ., ~ alternator Turboalternator *m* ~blower Kreiselgebläse *n*, Turbogebläse *n* ~ burner Turbobrenner *m* ~ cannon Turbinengeschütz *n* ~-charger Abgasturbolader *m*, Turbo-gebläse *n*, -lader *m* ~compressor Kreisel-gebläse *n*, -verdichter *m*, Turboverdichter *m* ~ drier Turbotrockner *m* ~ electric drive turboelektrischer Antrieb *m* ~engine Propellerturbinenluftstrahlmotor *m*, Turbinenluftstrahlmotor *m* ~ fan Ventilator *m* ~ gas exhauster Turbogasexhaustor *m* ~ generator Turbinen-dynamo *m*, -satz *m* ~gun Turbinengeschütz *n*

turbo-jet Düse *f*, Strahlturbine *f*, TL-Triebwerk *n*, Turbostrahltriebwerk *n* ~-aircraft Luftfahrzeug *n* mit Turbinenstrahlantrieb ~ engine Turboladermaschine *f*, Turbostrahltriebwerk *n* ~ engine with regeneration Zweikreisturbinenluftstrahlwerk *n* ~ transformer Turbotransformator *m*

turbo-prop Propellerturbine *f* ~ aircraft Luftfahrzeug *n* mit Turbinenpropellerantrieb (PTL) ~ engine Propellerturbomotor *m*, Turbinenpropellermotor *m*

turbo, ~ pump Serien-, Turbinen-pumpe *f* ~ rocket propulsion Turboraketenantrieb *m* ~ set Turbosatz *m* ~ supercharger Abgas-turbinengebläse *n*, -turbolader *m*, Turbo-kompressor *m*, -lader *m* ~ ventilator Kreisellüfter *m*

turbulence Böigkeit *f*, Durchwirbelung *f*, Luftwirbel *m*, Turbulenz *f*, Wirbel *m*, Wirbelbildung *f*, Wirbelung *f* ~ of the air Durchwirbelung *f* der Luft

turbulence, ~ chamber Wirbelkammer *f* ~ nozzle (jet) Dralldüse *f* ~ plane Wirbelebene *f* ~-produced turbulenzerzeugt

turbulent stürmisch, verwirbelt ~ air current Turbulenzströmung *f* ~-chamber engine Wirbelkammermaschine *f* ~ film evaporator Turbulenzschichtverdampfer *m* ~ flow Flachströmung *f*, störende Strömung *f*, Wirbelströmung *f* ~ motion Wirbelbewegung *f* ~ motion of the air Durchwirbelung *f* der Luft ~ movement of gas Gaswirbelbewegung *f* ~ region Strömungsschatten *m* ~ spot Turbulenzflecken *pl* ~

**streams** turbulente Ströme *pl* **~ wedge** Turbulenzkeil *m*

**turf** Grasscholle *f*, Rasen *m*, Torf *m* **~-and--stubble plow** Rasen- und Stoppelpflug *m* **~ cutter** Plaggen-, Soden-pflug *m*, Rasenstecher *m*

**turgid** aufgeblasen

**Turkey,** **~ red** Türkischrot *n* **~ red dyeing** Altrotverfahren *n* **~-red oil** Sulforicinat *n*

**turmeric paper** Kurkumapapier *n*

**turn, to ~** abdrehen, ausschlagen, (sich) drehen, drillen, hinrichten, kehren, kurven, Kurve *f* machen, laufen, marbeln, quirlen, richten, schwenken, (of iron during the blowing) übergehen, umgehen, umsetzen, verwandeln, (cover) wenden, werden, sich winden **to ~ about** umklappen, umwenden, wälzen **to ~ around** umdrehen, umwälzen, schwenken **to ~ aside** ableiten, ablenken, ausweichen **to ~ away (or aside)** abkehren **to ~ back** zurückdrehen **to ~ backward** zurückkurbeln **to ~ and bank** herumlegen (aviat.) **to ~ on the blast** anblasen **to ~ clockwise** rechts drehen **to ~ counter--clockwise** links drehen **to ~ through 90 degrees** um 90 Grad drehen **to ~ down** abdrehen, einbiegen, einziehen, zurückdrehen **to ~ off the echo suppressor** die Echosperre *f* löschen **to ~ on the echo suppressor** die Echosperre *f* zünden **to ~ on edge** aufdrehen, drehen, kanten **to ~ the engine past a compression point** den Motor *m* über einen oberen Totpunkt hinwegdrehen **to ~ over an engine** den Motor *m* durchdrehen **to ~ the heater on** anheizen **to ~ in** (a boat) einschwingen **to ~ into** umschlagen in **to ~ inside out** stülpen, umkrempeln **to ~ letters** blockieren (print.) **to ~ moldy** stocken **to ~ off** abdrehen, abkehren, ablenken, abschalten, (by a valve) absperren, abstellen, ausdrehen, ausschalten, (course) Biegung *f* machen, zudrehen, zurückdrehen **to ~ off (or away)** abwenden **to ~ on** andrehen, anlassen, (tap) aufdrehen (ein Ende), aufmachen, einschalten, zulassen **to ~ out** ausbiegen, ausdrechseln, ausdrehen, ausfallen, herausdrehen **to ~ outward** (evert) herauswenden, winden **to ~ outwards** auswärtsdrehen **to ~ over** aushändigen, kanten, sich überschlagen, umkehren, umwenden **to ~ up the paper** das Papier *n* umschlagen **to ~ into paste** verkleistern **to ~-pike a road** eine Straße *f* beschottern **to ~ upon a pivot** sich auf einem Zapfen *m* drehen **to ~ on a radio** anspannen **to ~ off a repeater** einen Verstärker *m* löschen **to ~ on a repeater** einen Verstärker *m* zünden **to ~ roughly** abschruppen **to ~ round** verdrehen, umwenden **to ~ without rubbing** reibungsfrei abrollen **to ~ with a shovel** umschaufeln **to ~ over on the side** umfallen **to ~ sideways** umwenden **to ~ slightly on edge** ecken **to ~-taper** kegelig drehen **to ~ to** zukehren **to ~ to template** auf Maß *n* drehen **to ~ up** aufdrehen, aufklappen, aufkrempeln, aufstülpen, aufwärtsbiegen, umkanten **to ~ up(ward)** aufbiegen **to ~ upside down** hochkant stellen, stülpen, umkippen, umwenden **to ~ inside the wake** zu kurz drehen **to ~ outside the wake** eine Beule *f* machen

**turn** Aufeinanderfolge *f*, Ausschlag *m*, Drehung *f*, Förderhub *m*, (blank) Kehre *f*, Kehrkurve *f*

(aviat.), Kehrtwendung *f*, Krümmung *f*, Kurvenflug *m*, Querneigung *f*, Reihe *f*, Schicht *f*, Schlängelung *f*, (of a coil) Schraubenwindung *f*, Schwenkung *f*, Tour *f*, Umdrehung *f*, Umlauf *m*, Umschwung *m*, Umwindung *f*, (of a picture) Verdrehung *f*, Wende *f*, Wendeflug *m*, Wendung *f*, Windung *f* **~ about** (theory of ferromagnetism) Umklappen *n* **~ of call** Zeitpunkt *m* zu dem ein Gespräch *n* an der Reihe ist **~ of the crankshaft** Kurbelwellenumdrehung **~ in landing** Wendekurve *f* **~ of the market** Konjunktur *f* **~ to port** Linksdrehung *f* **~ of rope** Seilwindung *f* **~ of a scale** Ausschlag *m* **~ of a spin** Trudelumdrehung *f* **~ of the spring** Federwindung *f* **~ of the tide** Kentern *n* des Stromes **~ of wire** Drahtwicklung *f*

**turn,** **~-and-bank indicator** Wendezeiger *m* **~ area** Windungsfläche *f* **~ bridge** Drehbrücke *f*

**turnbuckle** Spann-anker *m*, -schloß *n*, -schraube *f*, -vorrichtung *f*, -wirbel *m*, Spanner *m*, Vorreiber *m* **~ of a window** Fensterwirbel *m* **~ barrel** Spannschloßmutter *f* **~ eye** Spannschloß-auge *n*, -öse *f* **~ fork** Spannschloßgabel *f* **~ nut** Spannschloßmutter *f* **~ rod** Spannschloßschraube *f* **~ screws** Anziehschrauben *pl* **~ thread** Spannschloßgewinde *n*

**turn,** **~ button** Drehverschluß *m* **~ button of a window** Fensterwirbel *m* **~ control** Richtungsgeber *m* **~ control signal** Vorgabe *f* **~ down collar** Umlegekragen *m* **~ flux** Flußverbindung *f*, Kupplung *f*, Windungsfluß *m* **~ handle** Drehschlüssel *m* **~ indicator** Drehungs-, Kurven-zeiger *m*, Wende-messer *m*, -zeiger *m* **~-key job** gebrauchsfertige Ablieferung *f* **~ meter** Wendegeschwindigkeitsmesser *m*, (of airplane) Wendezeiger *m*

**turn-off,** **~ direction** Abdrehrichtung *f* **~ point** Abrollpunkt *m* (an der Startbahn) **~ time** Abschaltzeit *f*

**turnout** Ausbiegestelle *f* (r.r.), Ausweiche *f*, Ausweich(e)gleis *n*, Ausweichstelle *f*, Linksweiche *f*, Weiche *f* **~ track** Ausweichgleis *n*

**turnover** Güterumschlag *m*, Umsatz *m*, Umschlag *m* **~ board** Aufstampfboden *m*, Leerboden *m* **~ calendar** Umlegekalender *m* **~ capacity** Wendeleistung *f* **~ foot-operated draw-molding machine** Wendeplattenformmaschine *f* mit Fußhebelmodellaushebung **~ frame** Wenderahmen *m* **~ frequency** Übergangsfrequenz *f* (acoust.) **~ gear** Wendegetriebe *n* **~ hand-power draw machine** Wendeplattenformmaschine *f* mit Handhebelmodellaushebung **~ manipulator** Drehgestell *n* **~ molding machine** Formmaschine *f* mit Wendeplatte, Wendeformmaschine *f* **~ molding machine with hand-lever lift** Wendeplattenformmaschine *f* mit Handhebelmodellaushebung **~ molding machine with run-out table** Wendeplattenformmaschine *f* mit Wagen **~ patterndraw machine** Wendeplattenformmaschine *f* mit Abhebevorrichtung **~ plate** Wendeplatte *f* **~ point** Umkehrpunkt *m* **~ power squeezing machine** Wendeplattenpreßformmaschine *f* **~ press molding machine** Preßformmaschine *f* mit Wendeplatte **~ squeezer** Wendeplattenpreßformmaschine *f* **~ table** Wendeplatte *f*, (of a mold machine) Wendetisch *m* **~-table jolter** Rüttelformmaschine *f*

mit Wendetisch, Rüttelwendeformmaschine *f*
**~-table molding machine** Wendeformmaschine *f*, Wendeplattenformmaschine *f* **~ tax**
Umsatzsteuer *f* **~ voltage** Kniespannung *f*
(Transistorkennlinie)
**turn, ~ pike** Zollschranke *f* **~ pinion** Andrehrit-
zel *n* **~ plate** Wendeplatte *f* **~ plow** Wendepflug *m* **~ point** Knickpunkt *m* **~ ratio** Transformatorübersetzung *f* **~ relay** Kurvenrelais *n*
**~ round** Umlenkung *f* **~ round time** Liege-
zeiten *pl* **~ screw** Schraubenzieher *m*, Spannschraube *f* **~ sleeve** Drehhülse *f*
**turnstile** Drehkreuz *n*, Drehkreuz *n* für den Revolverschlitten, Steg-, Sperr-kreuz *n*, Sternrad
*n* **~ antenna** Schmetterlingantenne *f* **~ crane**
Ständerkran *m*
**turntable** Dreh-boden *m*, -kranz *m*, -platte *f*,
-ring *m*, -scheibe *f*, Laufwerk *n*, Plattenteller *m*
(phono), Schallplattenwerk *n*, Schwenktisch
*m*, Wendetisch *m* **~ with flush rails** Drehscheibe *f* mit eingegossener Kreuzspur
**turntable, ~ bolt** Sperrklaue *f* der Drehscheibe
**~ cabinet for sandblast** Blashaus *n* mit Dreh-
boden **~ firemen's ladder** Drehleiter *f* **~ gun
carriage** Drehscheibenlafette *f* **~ lathe** Drehbank *f* mit Drehscheibe **~ plate** Tellerdreh-
scheibe *f* **~ racer** Drehring *m* **~ rumble** Laufgeräusch *n* (phono) **~ sandblast cabinet** Dreh-
bodensandstrahlapparat *m*, Sandstrahlgebläse
*n* mit Drehboden **~ sandblast chamber** Dreh-
tischsandstrahlgebläsehaus *n* **~ sandblast
room** Putzhaus *n* mit Drehscheibe **~ speed**
Plattentellerdrehzahl *f* **~ type** Drehscheiben-
bauart *f*
**turn-wrest plow** Kehrpflug *m*
**turned** abgedreht, gedrechselt **~ bolt** gedrehter
Bolzen *m* **~ character** Fliehgenkopf *m* (print.)
**~-down position** Ausgußlage *f* **~ flanges** aufgepreßte Briden ... *pl* **~ letter** Blockade *f*
(print.) **~ off (or away)** abgewandt **~ piece**
Drehteil *m* **~ up** eingestellt **~ wooden goods**
Holzdrehwaren *pl*
**turner** Drechsler *m*, Dreher *m*
**turner's, ~ chisel** Abdrehstahl *m* **~ cross-
-cutting chisel** Stichbeitel *m* für Drechsler **~
finishing tools** Fertigdrehstahl *m* **~ lathe** Drehbank *f* **~ spring tool** Federstahl *m*
**turnery** Drechslerei *f*
**turning** (in lathe) Abdrehen *n*, Drehen *n*, Drehung *f*, Kanten *n*, Span *m*, Verdrehen *n*,
Wende *f*, Wendung *f*, Windung *f*; Dreh ...,
drehend **~ by air** Luftdrehung *f* **~ along** Lang-
drehen *n* **~ on edge** Drehen *n* **~ the fabric** Umspannen *n* der Ware **~ the flank** Hakenbildung
*f* **~ in flight operation** Abschwung *m* **~ in**
Wendung *f* **~ off** Abdrehen *n* **~ over** Wenden *n*
**~ of the tides** Gezeitenwechsel *m*
**turning, ~ angle** Drehungswinkel *m* **~ arbor of
a hammer shaft** Hammerwelle *f* **~ arc** Wende-
bogen *m* **~ axis** Drehungsachse *f* **~ bar** Wende-
stange *f* **~ basin** Wende-becken *n*, -kreis *m*,
-platz *m* **~ boring and cutting-off bench (or
machine)** Dreh-, Bohr- und Abstechbank *f*
**~ box** Drehladen *m* **~ capacity** Drehbereich *m*
**~ carriage** Bett-, Langdreh-, Längs-schlitten *m*
**~ center** Dreh-mitte *f*, -spitze *f* **~ change-over
switch** Drehumschalter *m* **~ chisel** Drechsler-
beitel *m* **~ circle** Drehkreis *m*, Einschlag *m*,

**Wendekreis** *m* **~ clip** Drehklemme *f* **~ condenser** Drehkondensator *m* **~ cut** Drehschnitt *m*
**~ cylinder** Wendewalze *f* **~ device** Andrehvorrichtung *f* **~ door** Drehtür *f* **~ drum** Wende-
trommel *f* **~ error** Querneigungsfehler *m* **~
finish** Drehbild *n* **~ force** Drehkraft *f* **~ gear**
Drehgetriebe *n*, Durchdrehungsvorrichtung *f*,
Schwenk-, Wende-getriebe *n* **~-gear pinion**
Schwenkritzen *m* **~ gouge** Drehröhre *f* **~ graver**
Dreh-, Grab-stichel *m* **~ groove** Drehriefe *f* **~
gyrometer** Drehungsmesser *m* **~ hook** Kettel-
nadel *f* **~ joint on abutment of a bridge arch**
Kämpfergelenk *n* eines Brückenbogens **~ knife**
Drehstahl *m* **~ lathe** Drehbank *f* **~ length**
Drehlänge *f* **~ lever** Krückel *m*, Wender *m*
**~ limit stop** Einschlagbegrenzung *f* **~ lock**
Drehverschluß *m* **~ lock door** Drehtor *n* **~
machine** Drehwerk *n* **~ mechanism** Drehvorrichtung *f* **~ motor** Wendemotor *m* **~ movement**
Schwenkung *f*, Umgehung *f*, Umgehungsbewegung *f* **~-over device for stored engines**
Durchdrehungsvorrichtung *f* für eingelagerte
Motoren **~ pawl** Drehklinke *f* **~ pin** Drehzapfen *m* **~ piston** umlaufender Kolben *m*
**~ point** Umkehrpunkt *m* (cryst.), Wende-
-marke *f*, -punkt *m* **~ points** Drehweiche *f*
**~ radius** Kurvenradius *m*, Schwenkradius *m*,
(Wagen) Wendefähigkeit *f* **~ round of a ship**
Wenden *n* eines Schiffes **~ sash** Drehflügel *m*
**~ saw** Örtersäge *f* **~ section** Dreherei *f* **~ slides**
Vordrehsupporte *pl* **~ speed** Drehgeschwindigkeit *f* **~ square** Lochwinkel *m*, Schubwinkel *m*,
Tiefenmaß *n* **~ support** Langdrehsupport *m*
**~ table** Drehtisch *m* **~ taper** Drehkonus *m* **~
tendency** Drehbestreben *n* **~ test** Drehversuch
*m* **~ thread of the warp** Polfaden *m* **~ tool**
Abdrehstahl *m*, Dreh-meißel *m*, -messer *n*,
-stahl *m*, Langdrehstahl *m* **~ tools** Drehzeug *n*
**~ unit** Bedienungsgerät *n* **~ wall** Kantwand *f*
**turnings** Drehspäne *pl*, Späne *pl*
**turnip, ~-shaped valve** Verstellpilz *m* **~-tops-
plucking machine** Rübenblätterzerreißmaschine
*f*
**turns, by ~** wechselweise **~ and backs** Seitenver-
änderungen *pl* (aviat.) **~ ratio** Wicklungs-,
Windungs-verhältnis *n*
**turpentine** Terpentin *n* **(spirit of) ~** Terpentinöl
*n* **~ separator** Terpentinabscheider *m* **~ substitute** Testbenzin *n*
**turquoise** Türkis *m*
**turret** Drehbankkopf *m*, drehbarer Panzerturm
*m*, Drehkopf *m*, Kanzel *f*, Revolverkopf *m*,
Sechskantrevolversupport *m*, Turm *m* **~ with
all-around visibility** Vollsichtskanzel *f*
**turret, ~ arrangement** Revolveranordnung *f* **~
axis** Schälachse *f* **~ bushing** Revolverkopf *m*
**~ connection** Turmanschluß *m* **~-deck steamer**
Turmdeckdampfer *m* **~ decker** Turmdeckdampfer *m* **~ face** Revolverkopffläche *f* **~
front** Turmfront *f* **~ frontplate** Turmstirnwand
*f* **~ gun** Turm-geschütz *n*, -kanone *f* **~ hatch**
Dachluke *f*, Turmklappe *f* **~-hatch cover**
Turmlukendeckel *m* **~-head** Revolverkopf *m*,
Werkzeugschlitten *m* **~-head boring mill**
Revolverbohrbank *f* **~ index-disc** Revolver-
kopfschaltscheibe *f* **~ indexing mechanism**
Revolverschaltapparat *m* **~ latch** Turmdeckel
*m* **~ lathe** Revolver(dreh)bank *f* **~-locking**

**clamp** Turmzurrung *f* ~ **mount** Turmlafette *f*
(artil.) ~ **rear** Turmheck *n* ~ **saddle** Revolver-
kopfschlitten *m* ~ **servo motor** Quarztrommel-
motor *m* ~ **shaft** Revolverkopfwelle *f* ~ **side-
-plate** Turmseitenwand *f* ~ **slide** Revolver-
kopfsupport *m*, Revolver-schlitten *m*, -support
*m* ~ **spindle axis** Revolverkopfachse *f* ~ **stop
screw** Revolverhemmschraube *f* ~ **stud** Re-
volverkopf-achse *f*, -zapfen *m* ~ **telescope**
Turmfernrohr *n* ~ **tool hole** Werkzeugloch *n*
~ **traversing mechanism** Turmschwenkwerk *n*
~ **turner** induktiver Kanalwähler *m* (TV) ~
**turning machine** Turmdrehmaschine *f* ~ **work**
Revolverarbeit *f*
**turtleback** gewölbte Decke *f*
**tutor eyepiece** Demonstrationsokular *n*
**tutton salts** komplexe Salze *pl*
**tutty** Ofen-bruch *m*, -schwamm *m*, Tutia *f*,
Zink-ofenbruch *m*, -schwamm *m*
**tuyère** Blaserohr *n*, Düse *f* des Blasebalges,
Düse *f* des Schmiedefeuers, Eßeisen *n*, Form *f*,
Luftdüse *f*, Rohrstutzen *m*, Winddüse *f*, Windform *f* **eye of the** ~ Formauge *n*
**tuyère,** ~ **arc cooler** Formkastengehäuse *n* ~
**arch** Blaseform *f*, Formgemäuer *n*, Wind-
-gewölbe *n*, -schutzkasten *m* ~ **arch cooler**
Kühlkasten *m* ~ **area** Düsenquerschnitt *m* ~
**arrangement** Düsenanordnung *f* ~ **belt** Formen-
zonenring *m* ~ **block** Form-, Kühl-, Wind-
schutz-kasten *m* ~ **bottom** Düsenboden *m* ~
**box** Konverterwindkasten *m*, Windkasten *m*
~ **cap** Düsenstockklappe *f* ~ **connection** Düsenstock *m* ~ **connection pipe** Anschlußstück *n* ~
**controller** Düsenstockregler *m* ~ **cooler housing**
Formnischenkasten *m*, Kapelle *f* ~-**cooling
plate** Windformkühlkasten *m* ~ **gate** Düsen-
absperr-, Wind-schieber *m* ~ **hole** Blaseloch *n*,
Blasöffnung *f*, Formauge *n* ~ **iron** Blasrohr *n*
~ **latch** Düsenstock-, Verschluß-klappe *f* ~
**level** Düsenhöhe *f*, Formebene *f* ~ **line** Wind-
formebene *f* ~ **opening** Blasöffnung *f* ~ **outlet**
Düsenaustritt *m* ~ **pipe** Düsenrohr *n*, Rüssel *m*,
Windrohr *n* ~ **plate** Bodenplatte *f*, Düsenboden *m* ~ **plug** Düsenboden *m*, (in a converter)
Fernenboden *m* ~ **region** Formebene *f* ~ **slag**
Nasenschlacke *f* ~ **stock** Düsenstock *m*, Kniestück *n* ~ **system** Düsensystem *n* ~ **wall** Form-
gewölbe *n* ~ **zone** Formzone *f*
**tuyères** Düsenreihe *f*
**TV tower** Fernsehturm *m*
**tweeter** Hochtonlautsprecher *m*, hoher Ton *m*
~ **loud-speaker** Hochlautsprecher *m*, Hochton-
-konus *m*, -lautsprecher *m*
**tweezers** Federzange *f*, Noppzange *f*, Pinzette *f*,
Zängchen *n*, Zange *f*
**T weld** T-Stoß *m*
**T welding** T-Schweißung *f*
**twelve,** ~-**cell lead battery** zwölfzellige Blei-
batterie *f* ~ **channel group** Vorgruppe *f* ~
**punch** Zwölferlochung *f* ~ **word machine**
Zwölfwortmaschine *f*
**twenty-four-hour service** durchgehender Dienst
*m*, ununterbrochener Dienst *m*
**twibil** Stichaxt *f*
**twice** doppelt, zweifach, zweimal ~ **full size**
doppelte natürliche Größe *f* ~ **repeated** zwei-
malig
**twig** Reis *n*, Zweig *m*

**twilight** Dämmerung *f*, Halbdunkel *n*, Helldunkel *n*, Zwiebelkasten *m*, Zwielicht *n* ~
**after sunset** Abenddämmerung *f* ~ **arch** Däm-
merungsbogen *m* ~**band** Zwielichtzone *f* ~
**effect** Dämmerungseffekt *m* (rdo), Zwielicht *n*
~ **glow** Dämmerungsschein *m* ~ **switch** Däm-
merungsschalter *m*
**twill** Köper *m*
**twilled** geköpert
**twin, to** ~ zu zweien verseilen
**twin** Zwilling *m*, Zwillingskristall *n*; doppel-
adrig, doppelt, zweiadrig ~ **action** Zwillings-
wirkung *f* ~ **air supply** Zweitluftzufuhr *f* ~ **antenna** Antennenpaar *n*, Zwillingsantenne *f* ~-
-**arc light** Zwillingsbogenlampe *f* ~-**band
telephony** Zweibandfernsprechsystem *n* ~
**bearings** Doppelpeilung *f* (rdo) ~ **belt pulley**
Doppelriemenscheibe *f* ~ **blinker** Doppel-
blinkleuchte *f* ~ **block** Zwillingsflaschenzug *m*
~ **cable** doppeladriges Kabel *n*, Zwillingskabel
*n* ~ **calender for glazing paper sheets** Doppel-
bogenkalander *m* ~ **calorimeter** Differential-,
Zwillings-kalorimeter *n* ~ **camera** Doppelkammer *f*, Zweifachkamera *f* ~ **capacitor**
Zwillingskondensator *m* ~ **cathode-ray beam**
Zwillingselektronenstrahl *m* ~ **chamber machine** Zweikammerausführung *f* ~-**chambered
furnace** Zweikammerofen *m* ~ **check** Duo-
kontrolle *f*, Duplikatprüfung *f*, Zwillingskontrolle *f* ~ **choke** Zweifachdrossel *f* ~ **circuit**
Doppelleitung *f* ~ **collecting piece** Zwillings-
sammelstück *n* ~ **condenser** Doppelkonden-
sator *m* ~ **conductor** Doppel-leiter *m*, -leitung *f*
~ **contact wire** Doppelfahrleitung *f* ~ **contacts**
Doppelkontakt *m* ~-**core cable** zweiadriges
Kabel *n*, Zweileiterkabel *n* ~ **crystal** Doppel-
kristall *m*, Kristallzwilling *m*, Zwillinge *pl*,
Zwillingskristall *m* ~ **crystals** Doppelplatten *pl*
**twin-cylinder** Doppelzylinder *m* ~ **drying machine** Zweiwalzentrockner *m* ~ **engine** Zwil-
lingsmaschine *f*, Doppelreihenmaschine *f* ~
**engine with triple bearing** Dreilagerzwillings-
maschine *f* ~ **mixer** Zwillingstrommelmischer
*m*

**twin,** ~ **detonating wires** Doppelsprengkabel *n*
~ **diode** Diodenpaar *n*, Duodiode *f* mit gemeinsamer Kathode ~ **door** Doppeltür *f* ~
**drum** Zwillingstrommel *f* ~-**drum drier** Doppel-
walzentrockner *m* ~ **engine** Zwillingstriebwerk
*n* ~-**engine car (or nacelle)** Zweimaschinen-
gondel *f* ~-**engined** zweimotorig ~-**engined
airplane** zweimotoriges Flugzeug *n* ~ **feeder
cable** Doppelzuleitungskabel *n* ~ **flex** Doppelschnur *f* ~ **float** Doppelschwimmer *m*, Zwei-
schwimmer *m* ~-**float plane** Zweischwimmerflugzeug *n* ~ **floodlight** Doppelreflektorflut-
licht *n* ~ **formation** Zwillingsbildung *f* ~ **fuel
pump** Zwillingsbrennstoffpumpe *f* ~ **gas lever
(or throttle)** Zwillingsgashebel *m* ~ **gates**
Zwillingsschützen *pl* ~ **gliding plane** Zwillings-
gleitungsfläche *f* ~ **grain** Zwillinge *pl* ~-**grid
tube** Doppelgitterröhre *f* ~-**hulled flying boat**
Flugzeug *n* mit zwei Bootskörpern ~-**jack**
Zwillingsklinke *f* ~-**jet** zweistrahlig ~ **lamella**
Zwillingslamelle *f* ~ **lamina** Zwillingslamelle *f*
~ **lamp** Zwillingsleuchte *f* ~ **lead** Verteiler-
kabel *n* (tape rec.) ~-**lead transmission line**
Antennenstegleitung *f* ~ **leader** Doppelader *f*

~-lens reflex camera zweiäugige Spiegelreflexkamera *f* ~ **machine gun** Doppelmaschinengewehr *n* ~ **magazine** Doppelkassette *f* (film) ~-**motor airplane** Zweimotorenflugzeug *n* ~ **optical apparatus** (maritime signals) Doppelleuchte *f* ~ **paradox** Zwillingsparadoxon *n* ~ **pedal** Doppelhebel *m* ~ **photogrammetric camera** Zweifachmeßkamera *f* ~-**piston engine** Doppelkolbenmotor *m* ~ **plates** Doppelplatten *pl* ~ **pneumatic tires** Doppelpneu *m* ~ **power section turbo-prop engine** Zwillingspropellerturbinenmotor *m* ~ **propeller** Zweischraube *f* ~ **pumps** Doppelpumpe *f* ~-**range case** Vorschaltgetriebe *n* ~-**ring-type surface tensiometer** Doppelringoberflächenspannungsmesser *m* ~- -**roller pull** Zwillingsrollenzug *m* ~-**row radial engine** Zweireihensternmotor *m* ~ **rudder** Doppelruder *n*, doppeltes Seitenleitwerk *n* ~ **rudders** Doppelseitensteuer *n*

**twin-screw** Doppelschrauben *pl* ~ **conveyer** Doppelschnecke *f* ~ **elevating mechanism** Doppelschraubenrichtmaschine *f* ~ **model** Zweischraubenmodell *n* ~ **ship** Doppelschrauben-, Zweischrauben-schiff *n*

**twin,** ~ **serial camera** Zweifachreihenbildkamera *f* ~ **sheet piles** Spundbohle *f* mit Nut und Feder ~-**skid chassis (or landing gear)** Zwillingskufengestell *n* ~ **slipping** Zwillingsgleitung *f* ~ **sound locator** Doppelrichtungshörer *m* ~ **spark ignition** Zweifunkenzündung *f* ~ **spark magneto** Zweifunkenmagnetapparat *m* ~- -**station unit** Zweistationanlage *f* ~ **steam engine** Zwillingsdampfmaschine *f* ~ **synchronizing commutator** Doppelsynchronumrichter *m* ~ **synchronizing pulse generator** Doppelimpulsgeber *m* ~-**T network** Doppel-T-Netzwerk *n* ~ **tail** Gabelschwanz *m* ~ **tape transporter** Zwillingstransportorgan *n* ~ **track** Doppelspur *f* ~ **traveling crane** Zwillingslaufkran *m* ~ **triode** Doppeltriode *f* ~ **tube** Doppelröhre *f* ~ **tunnel** Doppeltunnel *n* ~ **turret** (navy) Doppelturm *m* ~-**type collecting connection** Zwillingssammelstutzen *m* ~-**type refrigerator** Zwillingskühlschrank *m* ~-**type screwing connection** Zwillingsschraubstutzen *m* ~ **variable delivery pump** Zweifachregelpumpe *f* ~ **V-belt** Doppelkeilriemen *m* ~ **wheel** Doppelrad *n* ~ **wire** Doppelader *f*

**twine, to** ~ zusammendrehen, zwirnen **to** ~ **around** umwinden

**twine** (binder) Bindegarn *n*, Bindezwirn *m*, Bindfaden *m*, Garn *n*, Kordel *f*, Schnur *f*, Zwirn *m* ~ **basket (or holder)** Bindfadenkorb *m* ~ **knife** (sewing machine) Fadenmesser *n*

**twined** gezwirnt

**twinkle, to** ~ flimmern, funkeln, scintillieren

**twinkling** Flimmern *n*, Funkeln *n*, Scintillation *f*

**twinned** hemitrop

**twinning** Paarigstehen *n* der Zeilen, Verknüpfung *f*, Verzwillingen *n*, Zwillingsbildung *f* ~ **axis** Zwillingsachse *f* ~ **law** Zwillingsgesetz *n* (cryst.) ~ **machine** Vierer-Verseilmaschine *f* ~ **mechanism** Zwillingsbildung *f* ~ **plane** Zwillings-ebene *f*, -fläche *f* (cryst.)

**T wire** a-Ader *f*, Ader *f* zur Stöpselspitze

**twirl, to** ~ quirlen, umdrehen

**twirl** Wirbel *m*

**twist, to** ~ aufdrehen, drehen, drillen, flechten, schlingen, sich schlängeln, umdrehen, verbiegen, verdrallen, verdrehen, (said of wires) verdrillen, verkanten, verseilen, verwerfen, verwinden, verzwicken, wickeln, (together) zusammendrehen, zwirnen **to** ~ **a knob** Knopf *m* drehen **to** ~ **off** abdrehen **to** ~ **together** verdrillen

**twist** Drall *m*, Drehung *f*, Drilling *f*, Torsion *f*, Twist *m*, Verdrillung *f*, Verwindung *f*, Wende *f*, Windung *f* ~(**ing**) Verdrehung *f* ~ **of image** Bildzerdrehung *f* ~ **in opposite direction** Kreuzschlag *m* ~ **in same direction** Längsschlag *m* ~ **per unit length** Verwindung *f* ~ **of wire layers** Lagendrall *m*

**twist,** ~ **angle** Drallwinkel *m* ~ **belt** geflochtener Riemen *m* ~ **bit** Schneckenbohrer *m* ~ **chain** (for wire entanglements) Würgekette *f* ~ **clamp** Würgeklemme *f* ~ **compensator** Drallausgleicher *m* ~ **counter** Drallapparat *m* ~ **direction** Drehungsrichtung *f*

**twist-drill** Schrauben-, Spiral-, Wendel-, Zwick-bohrer *m* ~ **center grinding apparatus** Spiralbohrerspitzenschleifapparat *m* ~ **lifters** Spiralbohrerausheber *m* ~ **spear** Drillbohrspeer *m* ~ **stem** Seele *f* eines Spiralbohrers ~ **tester** Spiralbohrerprüfgerät *n*

**twist,** ~ **drilling knives** Spiralbohrmesser *pl* ~ **effect of slipstream** Drallwirkung *f* ~ **equalizer** Feinentzerrer *m* ~-**free** verdrehungsfrei ~ **grip throttle control** (Motorrad) Gasdrehgriff *m* ~ **gyro** Drehschwingkreisel *m* ~ **iron** Würgeisen *n* ~ **joint** Wickellötstelle *f*, Würgeverbindung *f* ~ **key** Schwenktaster *m* ~-**link chain** Kette *f* mit gedrehten Gliedern, gedrehte Kette *f* ~-**lug mounting** Schränklappenbefestigung *f* ~-**off** Gestänge-ausreißer *m*, -bruch *m* ~ **range** Drehungsbereich *m* ~ **ring** Schwenkring *m* ~ **rope** Torsionsseil *n* ~(**ed**) **spin of electrons** Drall *m* ~ **spindle** Drallspindel *f* ~ **system** Verdrallungsschema *n* ~ **wheel clutch** Drahtuhrkupplung *f*

**twisted,** gedreht, gewunden, gezogen, schraubenförmig, verdrallt, verschlungen, verwunden, windschief ~ **while cold** kalt verwunden ~ **and screened lead** verdrallte und abgeschirmte Zuführung *f*

**twisted,** ~ **auger** Schneckenbohrer *m* ~ **bar** Drilleisen *n*, gedrehter Eisenstab *m* ~ **cable** verseiltes Kabel *n* ~ **concrete cross section steel** Drillwulststahl *m* ~ **cordage** geschlagenes Tauwerk *n* ~ **exponential horn** aufgewundener Exponentialtrichter *m*, (loudspeaker) aufgewundener Trichter *m* ~ **exponential loud-speaker** aufgewundener Exponentialtrichter *m* ~ **grains** verschränkte Körner *pl* ~ **growth** Dreh-wuchs *m*, -wüchsigkeit *f* ~ **hemp** Hanfzopf *m* ~ **horn** (of a loud-speaker) Faltenhorn *n* ~ **joint** (unsoldered) Würgestelle *f* ~ **line** verdrallte Leitung *f* ~ **loop** verdrallte Doppelader *f* (teleph.) ~ **loud-speaker horn** aufgewundenes Lautsprecherhorn *n*, aufgewundener Lautsprechertrichter *m* ~ **pair** Bifilardraht *m*, verdrallte Doppelader *f* (teleph.) ~ **pair lay up** Paarverseilung *f* ~ **rope** geschlagene Leine *f*, Spiralseil *n* ~ **section** (wave guide) Hohlleiterverdrehung *f* ~ **silk** kordonnierte Seide *f* ~ **sleeve joint** Hülsen(würge)bund *m*, Kupferröhrenverbindung $f$ ~ **strip galvanometer** Bändchen-galvanometer $n$ ~ **telephone circuit** verdrallte Fernsprechleitung $f$ ~**-tooth spur gear** Schräg-zahnstirnrad $n$ ~ **wire** verdrillter Draht $m$, ver-seilter Draht $m$

**twister** (in felt manufacture) Anstoßmaschine $f$, Drehsturm $m$, Drehwirbel $m$, Garnzwirner $m$, (textiles) Kettenandreher $m$, Windhose $f$ ~ **crystal** Drehungs-, Drillungs-kristall $m$

**twisters** Drahtzange $f$

**twisting** (in feltmaking) Anstoßen $n$, Draht-gebung $f$, einfacher Drall $m$, Drehen $n$, einfache Drehung $f$, Krümmung $f$, Platz-wechsel $m$, (crane) Schieflaufen $n$, Schraub-wirkung $f$, Verbiegung $f$, Verdrallung $f$, Verdrehen $n$, Verdrillen $n$, Verdrillung $f$, Verseilung $f$, Windung $f$, Zusammendrehung $f$ ~ **of a curve in space** Verwindung $f$ einer Raumkurve ~ **in** (textiles) Anzwirnen $n$ ~ **off** Abdrehen $n$ ~ **of the ropes** Verschlingen $n$ der Seile

**twisting**, ~ **apparatus** (textiles) Drillgerät $n$ ~ **cap** Zwirn $m$, Zwirn-deckel $m$, -wickel $m$ ~ **couple** Drillung $f$ ~ **guide** Drall-büchse $f$, -führung $f$, Drehführung $f$ ~ **machine** Verseil-maschine $f$, Zopfdrehmaschine $f$ ~ **moment** Dreh(ungs)moment $n$, Torsionsmoment $n$, verdrehendes Moment $n$, Verdrehungsmoment $n$ ~ **package** Zwirn-deckel $m$, -wickel $m$ ~ **pliers** Windeisen $n$ ~ **resistance** Drehungs-, Tor-sions-, Verdrehungs-widerstand $m$ ~ **strain** Dreh-, Torsions-spannung $f$ ~ **strength** Dreh-, Drillungs-festigkeit $f$ ~ **stress** Torsions-bean-spruchung $f$, -spannung $f$, Verdrehungsbean-spruchung $f$ ~ **test** Drehungs-, Verdrehungs--versuch $m$, Torsionsversuch $m$, Verwindungs-probe $f$

**twitch, to** ~ reißen, zerren

**twitching** Reißen $n$, Ruck $m$, Zerren $n$

**two, with** ~ **columns** zweistielig, zweispaltig (print.) ~ **and a half plane** Zweieinhalbdecker $m$ **in** ~ **rows** zweireihig **on** ~ **shafts** zweiwellig

**two,** ~ **address instruction** Zweiadressenbefehl $m$ ~ **armature system** Zweiankersystem $n$ ~**-armed** doppelarmig, zweiarmig ~**-armed lever** doppelarmiger oder zweiarmiger Hebel $m$ ~**-arms paddle mixer** Zweischaufelrührer $m$ ~**-axis gyro group** Zweiachsenkreiselgruppe $f$ ~**-axle vehicle** Zweiachser $m$ ~**-band principle** Zweibandverfahren $n$ ~**-bar** Deichsel $f$ für An-hänger ~**-bar support** Deichselstütze $f$ ~**-bath** zweibadig ~**-bay** zweistielig ~**-bay antenna** Zweistockantenne $f$ ~**-bay biplane** zweistieliger Doppeldecker $m$ ~**-beaked anvil** Bankhorn $n$ ~**-bearing crankshaft** zweimal gelagerte Kurbel-welle $f$ ~**-blade mincing knife** Zweischneide-wiegemesser $n$ ~**-blade propeller** Zweiblatt-schraube $f$, zweiflügelige Luftschraube $f$, Zweiflügelluftschraube $f$ ~**-bladed** zweiflügelig ~**-bladed mincing knife** Doppelschneidewiege-messer $n$ ~**-bladed propeller** zweiflügelige Luft-schraube $f$ ~**body collision** Zweierstoß $m$ ~**-body forces** Zweikörperkräfte $pl$ ~**-body problem** Zweikörperproblem $n$ ~**-bolt tube flange** Zweischraubenflansch $m$ ~**-bowl sizing and squeezing machine** Zweiwalzenfoulard $m$ ~**-cabled bucket (or dredger grab)** Zwei-seilgreifer $m$ ~ **car** Schleppwagen $m$ ~-

**-center problem** Zweizenterproblem $n$ ~**-chamber brake cylinder** Zweikammerbrems-zylinder $m$ ~**-chamber vacuum brake cylinder** Zweikammer-Vakuum-Bremszylinder $m$ ~**-channel listening** zweiohriges Hören $n$ ~**-channel type frame** Doppel-U-Eisenrahmen $m$

**two-circuit,** ~ **band-pass filter** Zweikreisband-siebschaltung $f$ ~ **receiver** Zweikreisempfänger $m$, Zweikreiser $m$, Zweipolröhre $f$ ~ **reception** Zweikreisempfang $m$ ~ **set** Zweikreisempfän-ger $m$, Zweikreiser $m$, Zweipolröhre $f$

**two-coil,** ~ **flow** Zweischlangendurchfluß $m$ ~ **indicator** zwischenkelige Klappe $f$ ~ **relay** Relais $n$ mit zwei Wicklungen

**two-color,** ~ **printer** Zweifarbendruckvorrich-tung $f$ ~ **printing** Zweifarbendruck $m$ ~**(ed) ribbon** Zweifarbenband $n$ ~**-writing device** Zweifarbenschreibeinrichtung $f$

**two,** ~**-column apparatus** Zweisäulenapparat $m$ ~**-compartment sandblast tank machine** Zwei-kammersandstrahlapparat $m$ ~**-component al-loy** Zweistofflegierung $f$ ~**-component eutectic** Zweistoffeutektikum $n$ ~**-component lens** (zweiteiliges Objektiv) Zweilinsenobjektiv $n$ ~**-component lens system** zweigliedriges Lin-sensystem $n$ ~**-component system** binäres System $n$, Zweistoffsystem $n$ (of) ~ **components** zweikomponentig ~ **compound needles** asiati-sches Nadelpaar $n$ ~ **compound rests** Doppel-support $m$ ~**-condition cable code** bivalenter Kode $m$ ~**-condition trigger circuit** bistabile Kippschaltung $f$ ~**-conductor cable** zwei-adriges Kabel $n$ ~**-conductor firing wire** Dop-pelsprengkabel $n$ ~**-cone** zweikegelig ~**-connection small-size lubricator** zweistellige Kleinschmierpumpe $f$ ~**-contact one-field regulator** Zweikontakteinfeldregler $m$ ~**-con-tact regulator (or governor)** Zweikontaktregler $m$ ~**-contact single field regulator** Einfeld--Zweikontaktregler $m$ ~**-control system** Zwei-fachsteuerung $f$

**two-cycle** Zweitakter $m$ ~ **aircraft engine** Zwei-taktflugmotor $m$ ~ **engine** Zweitaktmotor $m$ ~ **engine with flat-type crown pistons** Flach-kolbenzweitakter $m$ ~ **principle** Zweitakt-system $n$

**two-cylinder,** ~ **compressor** Zweizylinderver-dichter $m$ ~ **crank shaft** Zweizylinderkurbel-welle $f$ ~ **freon liquefier unit** Zweizylinder--Frigen-Verflüssigeraggregat $m$ ~ **plunger-type compressor** Zweizylindertauchkolbenverdich-ter $m$

**two-digit group** Bigramm $n$

**two-dimensional** eben, zweidimensional ~ **flow** ebene Strömung $f$ ~ **motion** zweidimensionale Bewegung $f$ ~ **problem** ebenes (zweidimensiona-les) Problem $n$ ~ **stress** ebener Spannungszu-stand $m$

**two,** ~**-direction bearing** zweiseitig wirkendes Lager $n$ ~ **directional focussing** Richtungs-doppelfokussierung $f$ ~ **doffer type design** Zweiabnehmerausführung $f$ ~**-draft wire** zwei-mal gezogener Draht $m$ ~**-dynamo operation** Zweimaschinenbetrieb $m$ ~**-edged** zweischnei-dig ~**-electrode tube** Diode $f$, Doppelzweipol-röhre $f$ ~**-electrode vacuum tube** Richtröhre $f$ ~**-electrode valve** Gleichrichterröhre $f$ ~**-element regulator** Zwei-Element-Regler $m$

~-engine drive Zweimotorenantrieb *m* ~-
-engine nacelle Zweimotorengondel *f* ~ equal
forces Kräftepaar *n* ~-event characteristic
function vierdimensionales Eikonal *n* ~-faced
zweiflächig ~-filament bulb (Bilux) Bilux-
-Lampe *f*, Zweifadenlampe *f* ~-flank rolling
gear tester Zweiflankenabrollgerät *n* ~-float
bridge ferry Zweiboot-Brückenfähre *f* ~-floor
zweistöckig ~-floored doppelhordig ~-floored
drying kiln Zweihordendarre *f* ~-floored kiln
Doppeldarre *f* ~-fluid manometer Zweistoff-
manometer *n* ~-flute light spiral single-end mill
with morse taper shank leicht spiralgenuteter
Langlochfräser *m* mit Morsekegel ~-flute light
spiral single-end mill with straight shank leicht
spiralgenuteter Langlochfräser *m* mit Zylinderschaft

twofold diagonal, doppelt, zweifach ~ expansion
steam engine Zweifachexpansionsdampfma-
schine *f* of ~ symmetry zweizahlig

two, ~ fractional scans per frame Zweierverfahren *n* ~ group Bigramm *n* (comp.) ~-gun
turret Doppelturm *m* ~-handed hammer Zuschlaghammer *m* ~-handled zweistielig

two-high, ~ blooming mill Blockduo *n*, Duo-
blockwalzwerk *n* ~ blooming-mill train Duo-
blockwalzstraße *f* ~ cogging mill Duoblockwalz-
werk *n* ~ cogging-mill train Duoblockstraße *f*
~ finishing-mill train Duofertigstraße *f* ~
finishing stand Duofertiggerüst *n* ~ finishing
stands in train Duofertigstrecke *f* ~ intermediate
roll train Duomittelstraße *f* ~ mill Doppelduostraße *f*, Duowalzwerk *n*, Zweiwalzenstraße *f*
~ mill arrangement Duoanordnung *f* ~ mill
train Duowalzstrecke *f* ~ nonreversing mill
Duowalzwerk *n* mit gleichbleibender Dreh-
richtung der Walzen ~ piercing mill Duo-
stopfenwalzwerk *n* ~ plate mill Blechduo *n*,
Duoblechwalzwerk *n* ~ plate- (or sheet-)
rolling train Duoblechstraße *f* ~ reversing
blooming mill Blockreversierduo *n*, Duo-
reversierblockwalzwerk *n* ~ reversing blooming
train Duoreversierblockstraße *f* ~ reversing mill
Duoreversierwalzwerk *n*, Reversier-duo *n*,
-walzwerk *n* ~ reversing-mill train Duorever-
sierstraße *f* ~ reversing plate mill Blechrever-
sierduo *n*, Duoreversierblechwalzwerk *n* ~
reversing plate-rolling train Duoreversierblech-
straße *f*, Reversierduogrobstraße *f* ~ reversing
stand Duoreversiergerüst *n* ~ reversing stand
of rolls for roughing Duoreversierstreckgerüst *n*
~ rolling mill Duo *n*, Duowalzwerk *n* ~ rolling-
-mill stand Duowalzgerüst *n* ~ rolling-mill train
Duostraße *f*, Zweistraße *f* ~ rolling train Duo-
walzstraße *f* ~ rougher Duostreckwalze *f* ~
roughing stands in trains Duovorstrecke *f* ~
sheet-mill train Duofeinblechstraße *f* ~ sheet-
-rolling mill Duofeinblechwalzwerk *n* ~ sizing
mill Duomaßwalzwerk *n* ~ stand Duogerüst *n*
(metal.) ~ stand of rolls Zweiwalzengerüst *n*
~ universal mill Duouniversalwalzwerk *n* ~
universal mill train Duouniversalstraße *f*

two, ~-hole flange Zweilochflansch *m* ~-hole
range Zweilochherd *m* ~-hook for trailer An-
hängerzughaken *m* ~-horse agricultural machine
Zweispännerlandmaschine *f* ~-horse draft (or
pull) zweispanniger Pferdezug *m* ~-hourly
zweistündig ~ hours' duration zweistündig ~

hundreds-selector bay Zweihunderter-Gruppen-
wähler-Gestell *n* ~ hundredweight Doppelzent-
ner *m* ~-image photogrammetry Zweibildmes-
sung *f* ~-jaw chuck Zweibackenfutter *n* ~ jaw
draw-in-type chuck Zweibackenzugfutter *n* ~
jawed collect chuck Zweibackenspannpatrone *f*
~ jet spray gun Zweidüsenspritzapparat *m* ~-
-lattice Flächengitter *n* (cryst.) ~-layer zwei-
lagig ~ layer coiling Zweischichtenwicklung *f*
~-layer film Doppelschichtfilm *m* ~ layer model Doppelschichtmodell *n* ~-leaved door
zweiflügelige Tür *f* ~ legged zweischenklig ~-
-lens camera Doppelkammer *f* ~-lens serial
survey camera Zweifachreihenmeßkammer *f*
~-line letter Majuskel *f*, Titelbuchstabe *m* ~-
-line series scans per frame Zweierverfahren *n*
~-line service Zweileitungsbetrieb *m* (teleph.)
~-lipped end miller zweischneidiger Langlochfräser *m* ~-lobed zweihöckrig ~-magazine
composing machine Doppelmagazinsetzmaschine *f* ~-membered zweigliedrig ~-minute
turn Zweiminutenwendung *f* ~-motion selector
Hebdrehwähler *m* ~-motor drive Zweimotorenantrieb *m* ~-necked zweihalsig ~-noded
oscillation (waves, etc.) zweiknotige Schwingung *f* ~-nozzle atomizer Zweidüsenzerstäuber
*m* ~-pair core Doppelzwilling *m* ~-pair-core
cable zweipaariges Kabel *n* ~ parallel surface
of a cube Grundflächen *pl*

two-part zweiteilig ~ mold zweiteilige Kastenform *f* ~ ring zweiteiliger Ring *m* ~ tariff
Grundgebührtarif *m*

two-phase zweiphasig ~ alternating current Zweiphasenwechselstrom *m* ~ alternating-current
relay Wechselstromphasenrelais *n* ~ alternator
Zweiphasengenerator *m* ~ current Zweiphasen-
strom *m* ~ equilibrium Zweiphasengleichge-
wicht *n* ~ field Zweiphasenraum *m* (metal.) ~
four-wire system Zweiphasenvierleitersystem *n*
~ generator Zweiphasengenerator *m* ~ rectifying valve Netzgleichrichter *m* ~ rotor Zwei-
phasenanker *m* ~ system Zweiphasensystem *n*
~ three-wire system Zweiphasendreileitersystem
*n*

two-piece zweiteilig ~ crankshaft zweiteilige
Kurbelwelle *f* ~ flywheel zweiteiliges Schwungrad *n* ~ master rod geteiltes Hauptpleuel *n*

two, ~-pin base Zweistiftsockel *m* ~-pin plug
Doppel(stift)stecker *m* ~-pivot steering gear
Zweizapfenlenkung *f* ~-place airplane Zwei-
sitzerflugzeug *n* ~-plate holder Doppelkassette
*f* ~-ply doppelt, zweibahnig ~-ply belt Doppelriemen *m*

two-point, ~ boundary value problems Randwertproblem *n* für zwei Punkte ~ characteristic Zweipunktcharakteristik *f* ~ characteristic function Eikonal *n* ~ contact bearing
Zweipunktlager *n* ~ control switch arrangement
Wechselschaltung *f* ~ curve drawing instrument
Zweikurvenschreiber *m* ~ emergency cell
switch Doppelzellenschalter *m* ~ end cell
switch Doppelzellenschalter *m* ~ fixing (or
fastening) Zweipunktbefestigung *f* ~ jack
zweiteilige Klinke *f* ~ landing Flach-, Zwei-
punkt-landung *f* ~ measuring system Zwei-
punktmeßverfahren *n* ~ resection Zweipunkt-
verfahren *n* ~ scanning Zweipunktabtastung *f*
~ socket zweiteilige Büchse *f*

two, **~-pointed** zweipolig **~-pole heater plug** zweipolige Glühkerze *f* **~-pole magnetic system** zweipoliges Magnetsystem *n* **~-pole shunt generator (or dynamo)** zweipolige Nebenschlußmaschine *f* **~-pole switch** doppelpoliger oder zweipoliger Schalter *m*

**two-position** Zweigang *m* **~ action** Zweipunkt-, Zweistellungs-regelung *f* **~ action controller** Zweipunktregler *m* **~ controllable airscrews** Verstellschraube *f* mit zwei Steigungsstellungen, Zweistellungsschraube *f* **~ controllable propeller** Zweigangschraube *f* **~ lever** Zweistellungshebel *m* **~ propeller** verstellbare Flugschraube *f* **~ regulator** zweistufiger Regler *m*, Zweistufenregler *m* **~ six-point switch** sechsstelliger Schalter *m* mit zwei Stellungen **~ switch** Schalter *m* (Umschalter) mit zwei Stellungen **~ wing** Zweistellungsflügel *m*

two, **~-power signal** zweiwertiges Zeichen *n* **~-prism spectrograph** Zweiprismenspektrograf *m* **~-product** Zweigut *n* **~-pronged** zweizackig **~-pronged hoe** Gabelhaue *f* **~-pronged hoes** Doppelhaue *f* **~-pronged thread guide** Zweizackenfadenführer *m* **~-purpose (or two--service) aircraft** Zweizweckflugzeug *n* **~-quantum annihilation** Zweiquantenvernichtung *f* **~-rail gravity conveyer** zweigleisige Rollenbahn *f* **~-rate meter** Doppeltarif-, Vergütungs-zähler *m* **~ ratio gear** Zweiganggetriebe *n* **~-reel rotary printing machine** Zweirollenrotationsdruckmaschine *f* **~-revolution press** Zweitourenmaschine *f* **~-revolution printing press** Zweitourenschnellpresse *f* **~-ribbed nut** Zweirippenmutter *f* **~ right angles** gestreckter Winkel *m* **~-roller corn mill** Zweiwalzenstuhl *m* **~-row corn cultivator** zweireihige Maishackmaschine *f* **~-row eighteen-cylinder radial engine** Achtzehnzylinder-Doppelsternmotor *m* **~-row radial engine** Doppelsternmotor *m* **~-scale pressure gauge** Doppeldruckmesser *m* **~-scale rule** Doppelmaßstab *m* **~ seat** zweisitzig **~-seater** Doppelsitzer *m*, zweisitziges Flugzeug *n* **~ seater airplane** Zweisitzer *m* **~-section filter** zweigliedriges Siebgebilde *n* **~-section network** zweigliedriger Kettenleiter *m* **~-segment magnetron** Zweischlitzmagnetron *n* **~-sheet folder** Doppelbogenfalzmaschine *f* **~-sided mosaic pick-up tube** doppelseitige Bildabtaströhre *f* **~-sidedness** Zweiseitigkeit *f* **~-slot flange** Zweischlitzflansch *m* **~ sounders** Doppelläuter *pl* **~-span** zweischiffig **~-spar** zweiholmig **~-spar type of construction** Zweiholmbauart *f* **~-spar wing** zweiholmiger Flügel *m* **~-spark coil** Zweifunkenspule *f* **~-sparred** zweiholmig

**two-speed,** **~ blower** Zweiganglader *m* **~ drill** Zweiganghandbohrer *m* **~ gear** Doppelübersetzung *f* **~ loader** Schaltlader *m* **~ motor** polumschaltbarer Motor *m* **~ planetary gear** Zweiganggetriebe *m*, Zweiphasenvorverdichter *m* **~ supercharger** Zweiganglader *m*, Zweiphasenvorverdichter *m* **~ two-stage supercharger** Zweistufenzweiganglader *m*

two, **~-spindle lathe** Zweispindelbank *f* **~-split magnetron** Zweischlitzmagnetron *n* **~-spool take-up** Doppelauflauftrommel *f*

**two-stage** zweistufig **~ air compressor** zweistufige Luftpumpe *f* **~ amplifier** Zweifachverstärker *m*, zweistufiger Verstärker *m* **~ blower** Zweistufen-gebläse *n*, -lader *m* **~ cold trap** Doppelkühlfalle *f* **~ compressor** zweistufiger Verdichter *m* **~ control** Regler *m* mit zwei Regelstellungen **~ filter** Doppelfilter *m* **~ magnetic switch** Zweistufenmagnetschalter *m* **~ melting process** Zweimalschmelzerei *f* **~ mixed-pressure steam turbine** Zweidruckdampfturbine *f* **~ reducing valve** Doppeldruckminderventil *n* **~ regulator** Zweistufenregler *m* **~ supercharger** Zweistufenlader *m* **~ transmitter** zweistufiger Sender *m* **~ variable amplifier** zweistufiger Regelverstärker *m*

two, **~-standard hammer** Brückenhammer *m* **~-state-device** Binärspeicherelement *n*, bistabiles Bauelement *n* **~ state process** Zweistufenprozeß *m* **~-station net** Funklinie *f* **~-station pilotballoon spotting** Doppelanschnitt *m* **~-station range finder** Zweistandsentfernungsmesser *m*

**two-step** zweistufig **~ (action) controller** Zweipunktregler *m* **~ drawing** Doppelzug *m* **~ magnetic switch** Doppelmagnetschalter *m* **~ magnetic-type starting switch** Anlaßdoppelmagnetschalter *m* **~ relay** Stufenrelais *n* **~ solenoid-operated switch** in Stufen arbeitender Magnetschalter *m* **~ titration** Zweistufentitration *f*

two, **~-story** zweistöckig **~-story folded dipole antenna** Zweistock-Faltdipol-Antenne *f* **~-story furnace** Zweietagenofen *m* **~-strand** zweitrümmig **~-strand field cable** Feldfernkabel *n* **~-strand twine** zweischäftiger Bindfaden *m* **~-stranded** zweiadrig **~-string oscillograph** Zweischleifenoszillograf *m*

**two-stroke,** **~ cycle** Zweitaktprozeß *m* **~ cycle double-acting engine** doppelwirkender Zweitaktmotor *m* **~ cycle engine** Zweitakter *m*, Zweitaktmotor *m* **~ Diesel engine** Zweitaktdiesel *m* **~ engine** Zweitaktmaschine *f*, Zweitaktmotor *m* **~ internal combustion engine** Zweitaktverbrennungsmotor *m* **~ oil engine** Zweitakt-Dieselmotor *m* **~ unit** Zweitaktmotor *m*

two, **~ sub-carrier system** Zweihilfsträgersystem *n* **~ temperature (snap action) valve** Zweitemperatur(Schnapp-)ventil *n* **~-terminal network** Zweipol *m* **~-terminal-pair network** Vierpol *m*

**two-tone** Preßfehler *m* **~ detector** Doppeltondetektor *m* **~ keying** Zweitonverfahren *n* **~ signal** Zweiklangsignal *n* **~ system** Zweiklanganlage *f* **~ voice-frequency telegraphy** (Wechselstromtelegrafie) Doppelton-WT *f*

two, **~-toned** zweifarbig **~-tooth generating roughers** Zweizahnwälzschrupper *m* **~-turnable type player** Zweitellergerät *n* **~-type graders** Anhängestraßenhobel *m* **~-type scrapers** Anhängeschürfwagen *pl* **~-unit dubber** Zweibandspieler *m* **~-valve amplifier** Zweiröhrenverstärker *m* **~-valve receiver** Zweiröhrenempfänger *m* **~-valve repeater** Doppelrohr-, Zweirohr-verstärker *m* **~-valve two-way intermediate repeater** Zweiröhrenzwischenverstärker *m* **~-valve two-wire repeater** Doppelrohrzwischenverstärker *m* (teleph.) **~-valve two--wire (intermediate) repeater** Zweidrahtdoppelrohrverstärker *m* **~-voltage changeover** Zweispannungumstecker *m* **~-wave property** Zweiwelligkeit *f* (cryst.)

**two-way** doppeltgerichtet, gegenseitig, kreuzweise, wechselseitig ~ **acting valve** Zweiwegeventil *n* ~ **breeches piece** Gabelstück *n* ~ **circles** Verbindungsleitung *f* für Wechselverkehr ~ **circuit** Leitung *f* für wechselseitigen Verkehr ~ **cock** Doppelhahn *m*, Durchgangshahn *m*, Hahn *m* mit zwei Wegen, Zweiweghahn *m* ~ **communication** Wechselverkehr *m*, Zweiwegverbindung *f* ~ **cord** zweiadrige Schnur *f* ~ **intercom** (system) Gegensprechanlage *f* ~ **jack** zweiteilige Klinke *f* ~ **junction** doppeltgerichtete Verbindungsleitung *f* ~ **operation** zweiseitiges Arbeiten *n* ~ **platform** Zweiwegeplattform *f* ~ **plow** Zweiwegepflug *m* ~ **plug** Doppel-stecker *m*, -stöpsel *m* ~ **pump** Zweiwegpumpe *f* ~ **radio** Zweiweg-verbindung *f*, -verkehr *m* ~ **railway line** Doppelbahn *f* ~ **repeater** Zweiwegeverstärker *m* ~ **respiration** Zweiwegeatmung *f* ~ **signals communication** Gabelverkehr *m* ~ **simplex system** Zweikanalsimplex *m* ~ **switch** Schalter *m* mit zwei Stellungen, Zweiwege(um)schalter *m* ~ **tap** Zweiwegehahn *m* ~ **telegraphy** wechselseitige Telegrafie *f* ~ **telephone** Gegensprecher *m* ~ **telephone circuit** gegenseitige Sprechverbindung *f* ~ **telephone conversation** Gegensprechen *n* ~ **telephone conversation** phantom **telephony** Doppelsprechen *n* ~ **telephone line** gegenseitige Sprechverbindung *f* ~ **television** Gegen(fern)sehen *n* ~ **top** Zweiweghahn *m* ~ **traffic** doppeltgerichteter oder doppelseitiger Verkehr *m*, wechselseitiger Verkehr *m* (teleph.) ~ **trunk** gemischt betriebene Ortsverbindungsleitung *f* ~ **trunk circuit** doppeltgerichtete Verbindungsleitung *f* ~ **turntable** Kreuzdrehscheibe *f* ~ **type** Doppelweggleichrichter *m* ~ **valve** Doppelventil *n*, Durchgangs-, Wechsel-ventil *n* ~ **working** (duplex) Arbeiten *n* in beiden Richtungen, Gegensprechen *n*
**two,** ~**-wheel trailer** Einachsanhänger *m* ~**-winding transformer** Übertrager *m* mit zwei Wicklungen ~**-wing door** Doppeltür *f* ~**-wing shutter** Zweiflügelblende *f* (film) ~**-winged rasp** Doppelraspel *m*
**two-wire** doppel-adrig, -drähtig, zweiadrig ~ **amplifier** Zweidrahtzwischenverstärker *m* ~ **antenna** zweidrähtige Antenne *f* ~ **automatic telephone system** Schleifensystem *n* ~ **circuit** Zweidraht-leitung *f*, -schaltung *f* ~ **conductor** Doppelschnur *f*, zweiadriges Kabel *n* ~ **connection** Gabel *f* ~ **core** Doppelader *f* ~**-core cable** zweidrahtiges Kabel *n* ~ **junction** zweiadrige Ortsverbindungsleitung *f* ~ **(four) lead cable** zweiadriges (vieradriges) Bleirohrkabel *n* ~ **line** Doppelleitung *f* ~ **loop circuit** zweidrähtige Leitung *f* ~ **metallic circuit** Doppelleitung *f* ~ **network** Zweileiternetz *n* ~ **operation** Doppelleitungsbetrieb *m*, Zweidrahtbetrieb *m* ~ **repeater** Zweidrahtverstärker *m* ~ **repeater board** Zweiröhrenzwischenverstärkerschrank *m* ~ **route** Doppelleitungslinie *f* ~ **side circuit** Zweidrahtstammleitung *f* ~ **system** Schleifenleitungsbetrieb *m*, Zweidrahtsystem *n* ~ **system with alternating (or direct) current** Zweileitersystem *n* ~ **telephone circuit** Fernsprechamtdoppelleitung *f* ~ **termination** Zweidrahtgabel *f* ~ **trunk** zweiadrige Ortsverbindungsleitung *f* ~ **two-valve intermediate repeater** Zweidraht-

doppelrohrzwischenverstärker *m* ~ **two-way repeater** Zweidrahtzwischenverstärker *m* ~ **winding** bifilare Wicklung *f*
**twyer,** ~ **lip** Formlippe *f* (metal.) ~ **plate** Formzacken *m*
**tye** Drehreep *n*
**tying** Festmachen *n* ~ **in** (textiles) Anschnürung *f* ~ **together** Zusammenschnürung *f*
**tying,** ~ **strap** Anschnallriemen *m* ~ **thread** Bindezwirn *m* ~**-up** Schnürung *f*
**tymp** (Hochofen) Tümpel(stein) *m* ~ **arch** Tümpelgewölbe *n* ~ **brick** Tümpelstein *m* ~ **plate** Tümpeleisen *n*
**tympan** Firstbrett *n* (print.), gespannte Membrane *f*, Ölbogen *m*, Preßdeckel *m*, Trommelschneckenrad *n* ~ **clamp** Aufzugklappe *f* ~ **reel** Spannschiene *f* ~ **shut** Einsteckbogen *m* (print.) ~ **weight** Deckelgewicht *n*
**tympanites** Trommelsucht *f*
**tyndallimetry** Tyndallimetrie *f*
**type, to** ~ schreiben, tippen
**type** Art *f*, Ausführung *f*, Druckletter *f*, Entwurf *m*, Form *f*, Letter *f*, (manufacture) Machart *f*, Marke *f*, Modell *n*, Muster *n*, Schriftcharakter *m*, Setzkasten *m*, Sorte *f*, Typ *m*, Type *f*, Typus *m*, Urbild *n*, Vorbild *n* **of the** ~ **of** förmig
**type,** ~ **acceptance test** Typendruckprüfung *f* ~ **approval** Baumuster-, Muster-zulassung *f* ~**-B waves** gedämpfte Wellen *pl*
**type-bar** Druckorgan *n* (telet.), Typen-druckhebel *m*, -radstange *f*, -schaft *m* ~ **backstop** Typenhebelanschlag *m* ~ **carriage** Typenkästenkorb *m* (telet.) ~ **guide** Typen(hebel)führung *f* ~ **link** Typenhebelzugdraht *m* ~ **mechanism** Typenhebelgetriebe *n* ~ **printer (or) translator** Typendruckhebelübersetzer *m*
**type,** ~ **bars** Typenhebel *m* ~ **basket** (typewriter) Typendruckkorb *m* ~ **bed** Druck-, Schrift-fundament *n*, Satzbett *n* ~ **board** Sachbrett *n* ~ **box** Typenkasten *m* (telet.) ~ **carrier** Buchstabenholer *m* ~ **case** Schriftkasten *m* ~**-casting machine** Typengießmaschine *f* ~ **certficate** Typusbescheinigung *f* ~ **character** Schriftbild *n* ~ **cylinder** Plattenzylinder *m* ~ **description** Sortenbezeichnung *f* ~ **designation** Musterbezeichnung *f* ~ **equation** Gleichungstyp *m* ~ **face template** Schriftschablone *f* ~ **gauge** Lettern-, Zeilen-messer *m* ~**-high gauge** Schrifthöhenmesser *m* ~**-inking recording attachment** Typendruckschreibvorrichtung *f* ~ **lever** Typenhebel *m* ~ **marker** Fächermarker *m* ~ **metal** Lettern-gut *n*, -metall *n*, Schrift-metall *n*, -zeug *n*, Typenmetall *n*, Zeug *n* ~ **model** Ausführungsform *f* ~ **mold** Schriftmutter *f* ~ **pallet** Typenhebel *m* (telet.) ~ **piston** Topfkolben *m* ~ **plate** Bauschild *n* ~ **printer** Drucktelegraf *m*, Typendrucker *m*, Typendruckdrucker *m*, Typendrucktelegraf *m* ~ **printing** Typendruck *m* ~ **printing apparatus** Typendrucker *m* ~ **printing telegraph** Typendruckdrucktelegraf *m* ~ **printing telegraph exchange system** Typendrucktelegrafenvermittlung *f* ~**-pusher** Buchstabenausstoßer *m* ~ **sample** Ausfall-, Qualitäts-muster *n* ~**-script reception** Druckempfang *m* ~ **series** (type line) Typenradreihe *f* ~ **setting and distributing machine** Setz- und

Ablegemaschine *f* ~setting machine Setzmaschine *f* ~ shaft (or shank) Typenschaft *m* ~ specimen Musterexemplar *n* ~ surface Typendruckfläche *f* ~ test Baumusterprüfung *f* ~ Musterprüfung *f*, Typenradteilprüfung *f* ~ wheel Typen(druck)rad *n* ~wheel printing positions Typenradschreibstellen *pl* ~-wheel shaft Typen(druck)radachse *f* ~-wheel translator Typendruckradübersetzer *m* ~ wire rope holder Drahtseilkopf *m*
typewriter Schreibmaschine *f* ~ accessories Schreibmaschinenzubehör *n* ~ keyboard Schreibmaschinentastenwerk *n* ~ oil Schreibmaschinenöl *n* ~ ribbon Farbband *n* ~ ribbon spools Farbbandspulen *pl*
typewriting Maschinen-schreiben *n*, -schrift *f*
typhoon Taifun *m*

typical kennzeichnend, musterhaft ~ solution (colorimetry) Typlösung *f* ~ value Eigenwert *m*
typing Maschinenschreiben *n* ~ perforator Schreiblocher *m* ~ unit base Empfängersockel *m* (telet.)
typist Maschinenschreiber *m*
typization Typennormung *f*
typographic error Schreibfehler *m*
typographical design Satzgestaltung *f*
typography Buchdruck *m*, Buchdruckerei *f*, Typografie *f*
typology Typenlehre *f*, Typologie *f*
typometer Letternmesser *m*
tyre: see tire
tyrolite Kupferschaum *m*
tyrosine Tyrosin *n*
tyrotoxicon Tyrotoxikon *n*
tysonite Tysonit *m*

# U

**U-and-V (packing) rings** Nutringe *pl*
**U-bend** U-Bogen *m*
**ubiquity** Allverbreitung *f*
**U-boat** U-Boot *n*, Unterseeboot *n*
**U-bolt** Bügelbolzen *m*, Bügelschraube *f*, Hakenbügel *m*
**udograph** Regenschreiber *m*
**udometer** Regenmesser *m*
**U-frame** U-Rahmen *m*
**UHF** (ultra high frequency) Dezimeterwellen-(bereich) *m* ~ **link** Dezistrecke *f* ~ **sight unit** Funksehgerät *n* ~ **telemetry chain** Funkfeuermeßanlage *f*
**uintaite** Gilsonit *m*
**U-iron** U-Eisen *n*
**Ulbricht, ~sphere** Ulbrichtsche Kugel *f* ~ **sphere-type photometer** Ulbricht'sche Kugel *f*
**ulexite** Boronatrokalzit *m*
**U-link** Reiter *m*
**ullmanite** Antimonnickelglanz *m*, Nickelantimon-glanz *m*, -kies *m*, Nickelspießglanzerz *n*, Ullmannit *m*
**ulrichite** Ulrichit *m*
**ulterior motive** Nebenzweck *m*
**ultimate** absolut, allerletzt, endgültig ~ **analysis** Elementaranalyse *f* ~ **anode** Sammelanode *f* ~ **bearing resistance** Grenzbelastung *f*, Tragfähigkeit *f* beim Bruch ~ **capacity** Ausbauleistung *f*, Vollausbau *m* ~ **compressive strength** Druckfestigkeit *f* ~ **crushing stress** Endstauchung *f* ~ **damping** Grenzdämpfung *f* ~ **delustering** Nachmattieren *n* ~ **lines** letzte Linien *pl* ~ **load** Bruchlast *f*, Grenzbelastung *f*, Überbelastung *f* ~ **partial pressure** Endpartialdruck *m* ~ **production** Gesamtproduktion *f* ~ **ripening** Nachreife *f* ~ **strength** absolute Festigkeit *f*, Bruch-festigkeit *f*, -spannung *f*, -widerstand *m*, Dehnungsgrenze *f*, Endfestigkeit *f*, Endstabilität *f*, höchste Widerstandskraft *f*, Höchstlast *f*, Prüfbelastung *f*, Reißkraft *f*, Tragfestigkeit *f*, Zerreißfestigkeit *f* ~ **stress** Bruch-belastung *f*, -spannung *f*, Festigkeitszahl *f*, Höchstspannung *f* ~ **stress limit** Bruchgrenze *f* ~ **stress value** Festigkeitswerte *pl* ~ **table** Schlußtafel *f* ~ **tensile strength** Zugfestigkeit *f* ~ **tensile stress** Zerreißbelastung *f* ~ **total pressure** Endtotaldruck *m* ~ **vacuum** Endvakuum *n* ~ **value** Endwert *m*
**ultimately** endlich, schließlich
**ultor** Endanode *f* (TV), Hochspannungsanode *f* ~ **voltage** Anodenspannung *f* (CRT)
**ultra** übermäßig ~ **accelerator** Ultrabeschleuniger *m*
**ultra-audible** überhörfrequent ~ **frequency** Überhörfrequenz *f* ~ **wave** Überschallwelle *f*
**ultra-audion** Einkreisempfänger *m*, rückgekoppeltes Audion *n*, Rückkopplungsaudion *n*, Ultraaudion *n*
**ultra, ~ centrifuge** Ultrazentrifuge *f* ~ **dyne receiver** Ultradynempfänger *m* ~**-filter** Spulenkette *f* ~ **gravity waves** ultraschwere Wellen *pl*

**ultrahigh** ultrahoch ~ **frequency (UHF)** Dezimeterwelle *f*, Dezimeterwellen-Frequenz *f*, Ultrahochfrequenz *f*, ultrahohe Frequenz *f*, Ultrakurzwelle *f* ~**-frequency oscillator** Laufzeitgenerator *m* ~**-frequency receiver** Ultrakurzwellenempfänger *m* ~**-frequency screened** vollentstört für Ultrakurzwellen *pl* ~**-frequency transmitter** Ultrakurzwellensender *m* ~**-frequency tube** Habannröhre *f* ~**-frequency wave** Ultrakurzwelle *f*, ultrakurze Welle *f* ~**-speed airplane** Höchstgeschwindigkeitsflugzeug *n* ~**-speed objective** höchstlichtstarkes Objektiv *n* ~ **vacuum** Hochvakuum *n* ~ **vacuum pressures** Ultrahochvakuum-Drücke *pl* ~ **vacuum range** Ultrahochvakuum-Gebiet *n*
**ultra-ionization** Ultra-Ionisierung *f* ~ **potential** Ultra-Ionisationspotential *n*
**ultra, ~-light** überlicht ~ **lightweight** ultraleicht ~**-long-range search radar** Fernsuchanlage *f* ~ **marine** Ultramarin *n* ~ **micrometer** Übermikrometer *n* ~ **microns** winziges Gewebegebilde *n* ~ **microscope** Ultramikroskop *n* ~ **microscopic** übermikroskopisch, ultramikroskopisch ~ **rapid lens (or objective)** höchstlichtstarkes Objektiv *n* ~ **rays** kosmische Strahlen *pl*
**ultrared** überrot, ultrarot ~ **radiation** Ultrarotstrahlung *f* ~ **ray** ultraroter Strahl *m* ~ **transmitter** Ultrarotdurchlässigkeit *f*
**ultrashort** ~ **wave** extrem kurze Welle *f*, Ultrakurzwelle *f* ~**-wave apparatus** Ultrakurzwellengerät *n* ~**-wave part** Ultrakurzwellenteil *m* ~**-wave preamplifier** Ultrakurzwellenvorverstärker *m* ~**-wave set** Ultrakurzwellenanlage *f* ~**-wave transmission** Ultrakurzwellenübertragung *f* ~ **waves** extrem kurze Wellen *pl*, Meterwellen *pl*
**ultra soft** ultraweich
**ultrasonic** Überschall *m*, Ultraschall *m* ~ **absorption** Ultraschallabsorption *f* ~ **crossgrating** Beugungsgitter *n* (acoust.) ~ **delay line** Ultraschall-Laufzeitglied *n* ~ **detector** Ultraschalldetektor *m* ~ **energy** Ultraschallenergie *f* ~ **frequency** Überhörfrequenz *f* ~ **generator** Ultraschallgenerator *m* ~ **grating constant** Brechungskonstante *f* (acoust.) ~ **sensing (or tracer)** Ultraschallfühler *m* ~ **sounding** Ultraschallotung *f* ~ **stroboscope** Ultraschallstroboskop *n* ~ **switching** Ultraschallschalter *m* ~ **test** Ultraschallprüfung *f* ~ **vibration** Ultraschallschwingung *f* ~ **wave** Ultraschallwelle *f* ~ **wind** Quarzwind *m*
**ultrasonics** Ultraschallehre *f*
**ultrasound** Ultraschall *m* ~ **transit time** Ultraschall-Laufzeitstrecke *f*
**ultraspeed lens (or objective)** hochlichtstarkes Objektiv *n*
**ultrastability** Ultrastabilität *f*
**ultraviolet** überviolett, ultraviolett ~ **lamp** Höchstdrucklampe *f* ~ **light** ultraviolette Strahlen *pl*, UV-Leuchte *f* ~**-radiation measuring instrument** Ultraviolettstrahlungsmesser *m* ~ **rays** ultraviolette Strahlen *pl*, Uviollicht *n*

**ultrawhite region** Ultraweißgebiet *n* (TV)
**umber** Bergbraun *n*, Reh *n*, Umbra *f*
**umbilical** nabelförmig, zentral ~ **cable** Nabelschnur *f* (g/m) ~ **point** Nabelpunkt *m*
**umbilicus** Nabelpunkt *m* (geom.)
**umbo** Schnabel *m*, Wirbel *m*
**umbra** Kernschatten *m*
**umbral lens** Umbralglas *n*
**umbrella** Regenschirm *m*, Schirm *m* ~ **antenna** Schirmantenne *f* ~ **insulator** Pilzisolator *m* ~ **stand** Schirmstativ *n* ~**-type generator** Schirmgenerator *m* ~**-type insulator** Schirmisolator *m*
**Umklapp-resistance** Umklappdiopterwiderstand *m*
**unabating** nicht nachlassend
**unable to dive** tauchunklar
**unabridged** unverkürzt
**unaccomodated eye** akkomodationsloses Auge *n*
**unacted** unangegriffen
**unadjusted** unkorrigiert
**unadulterated** unvermischt, unverschnitten
**unaffected** nicht betroffen, unbeeinflußt ~ **by air** luftbeständig ~ **by buffing** schleifecht ~ **by changes of temperature** temperaturbeständig ~ **by storing** lagerbeständig ~ **by water** wasserecht
**unaffected zone** Schneise *f*
**unaged** ungealtert
**unaided** unbewaffnet ~ **eye** unbewaffnetes Auge *n*
**unaimed** ungezielt
**unallotted** unzugeteilt ~ **number** Reservenummer *f*, unzugeteilte Nummer *f*
**unalloyed** lauter, unlegiert ~ **steel** unlegierter Stahl *m* ~ **tool steel** unlegierter Werkzeugstahl *m*
**unaltered** unverändert
**unambiguity** Eindeutigkeit *f*
**unambigous** eindeutig, unzweideutig
**unamplified** unverstärkt
**unanimated picture** ruhendes oder unbewegtes Bild *n*
**unannealed** ungetempert ~ **castings** ungetemperter Guß *m* ~ **malleable iron** weißes Gußeisen *n*, Temperrohguß *m*
**unanswered** unbeantwortet
**unassignable** unübertragbar
**unassisted** unbedient, unüberwacht ~ **exchange** unbedientes Fernsprechamt *n* ~ **office** unüberwachtes Amt *n* ~ **station** unbemannte Station *f* ~ **terminal station** unbewachte Endstation *f*
**unauthorized** eigenmächtig, unbefugt ~ **person** Unbefugter *m*
**unbacked bituminized felt** nackte Bitumpappe *f*
**unbaffled cylinder** Zylinder *m* ohne Leitbleche
**unbalance** Abgleichfehler *m*, Ausgleichungsfehler *m*, Gleichgewichtsfehler *m*, Nebensprechkopplung *f*, schlechte Abgleichung *f*, schlechte Ausgleichung *f*, Übergewicht *n*, Ungleichheit *f*, Unsymmetrie *f*, (of shafts) Unwucht *f* ~ **current** Nebensprechstrom *m*, Störstrom *m* infolge schlechter Ausgleichung
**unbalanced** nicht ausgeglichen, nicht im Gleichgewicht *n*, unausgeglichen, ungleich, unsymmetrisch ~ **to ground** erdasymmetrisch
**unbalanced,** ~ **end thrust** unausgeglichener Achsialdruck *m* ~ **equation** Ungleichung *f* ~

**impact** Überwucht *f* ~ **section** unkompensierter Abschnitt *m* ~ **wire circuit** unsymmetrischer Kreis *m*
**unbalancing** Gleichgewichtsstörung *f*
**unbend, to** ~ **a rope** ein Tau *n* abstecken
**unbending fold** Gebirgssattel *m*
**unbind, to** ~ abbinden, abheften
**unblanking,** ~ **circuit** Zündkreis *m* (TV) ~ **pulse** Helltastimpuls *m*
**unbleachable** bleichecht
**unbleached** ungebleicht ~ **hemp cord** Bindezwirn *m* ~**-linen cloth** Roh-leinen *n*, Rohleinenstoff *m*
**unblemished** fleckenlos, makellos
**unblocking** Deblockierung *f*, Freigabe *f* ~ **potential** Schaltspannung *f*
**unbolt, to** ~ aufriegeln, losschrauben
**unboosted engine** nicht überladener oder nicht aufgeladener Motor *m*
**unbound** unbefestigt ~ **electron** freies Elektron *n* ~ **levels** unverbundene Niveaus *pl*
**unbounded** schrankenlos, unermeßlich
**unbraced** unverspannt, unverstrebt, verspannungslos
**unbraked** ungebremst
**unbreakable** unbrechbar, unzerbrechlich
**unbridle, to** ~ abzäumen
**unbroken** fortlaufende Reihe *f*; ununterbrochen ~ **line** (in graphs, drawings) ausgezogene Linie *f*
**unbuffered** ungepuffert ~ **solution** ungepufferte Lösung *f*
**unbunching** Entbündeln *n*
**unbung, to** ~ aufspunden
**unbur, to** ~ (wool) kletten
**unburdening** Entlastung *f*
**unburned** unverbrannt ~ **(plastic) clay** Rohton *m*
**unburnt brick** Lehmstein *m*, ungebrannter Ziegel *m*
**unbutton, to** ~ aufknöpfen
**uncage relay** Entriegelungs-, Freigabe-, Löse-, Preisgabe-relais *n*
**uncalcined** ungeröstet
**uncalendered cotton tape** ungeglättetes Baumwollband *n*
**uncalibrated-light** Nichteichstellungs-Anzeigelicht *n*
**uncalled capital** noch nicht eingezahltes Kapital *n*
**uncanned fuel element** nacktes Spaltstoffelement *n*
**uncase, to** ~ entrollen
**uncasizable** überschlag(s)sicher
**uncemented** unverkittet ~ **lens** ungekittete Linse *f*
**uncertain** doppelsinnig, unbestimmt, ungewiß, unsicher ~ **conditions** Unsicherheiten *pl*
**uncertainty** Unsicherheit *f* ~ **in (or of) measurement** Meßunsicherheit *f*
**uncertainty,** ~ **condition** Unschärfebedingung *f* ~ **effects** Unbestimmtheitseffekte *pl* ~ **principle** Unbestimmtheits-, Ungewißheits-prinzip *n* ~ **relation** Unbestimmtheits-, Unschärfe--relation *f*
**uncertified** nicht genehmigt
**unchain, to** ~ losketten
**unchangeable** unveränderlich, unwandelbar
**unchanged** einwellig, unverändert

**uncharged** unelektrisch, ungeladen ~ **engine** Motor *m* ohne Aufladung

**unclad** nichtplattiert

**unclaimed,** ~ **dividend** nicht erhobene Dividende *f* ~ **rivet** Rohniet *n*

**unclamp, to** ~ abspannen, entzurren

**unclasp, to** ~ aufhaken

**unclassified,** ~ **documents** freie Sachen *pl*, Freisachen *pl* ~ **soils** verschiedene Böden *pl*

**unclear** ungeklärt, unklar

**uncleared zero** Funktrübung *f*

**unclinched rivet** noch nicht geschlagenes Niet *n*

**unclutching device** Kupplungsausrückvorrichtung *f*

**uncoagulable** ungerinnbar

**uncoated** unbeschichtet ~ **electrode** Blankelektrode *f* ~ **film base** unbeschichteter Blankfilm *m*

**uncock, to** ~ entspannen

**uncoded** unverschlüsselt ~ **language** offene Sprache *f*

**uncoil, to** ~ abrollen, abspulen, abtrommeln, abwickeln

**uncoiler** Ablauf-, Abroll-haspel *f*, Auseinanderrollmaschine *f*

**uncoiling** Abspulen *n* (film) ~ **spool** Ablaufspule *f* (film)

**uncolored crystals** ungefärbte Kristalle *pl*

**uncombined** frei, ungebunden ~ **carbon** grafitischer Kohlenstoff *m*, ungebundener Kohlenstoff *m* ~ **heat** freie Wärme *f*

**uncomfortable** unbehaglich, unbequem, ungemütlich

**uncompensated** unausgeglichen, unausgelenkt, unkompensiert

**uncompleted call** nicht zustandegekommene Verbindung *f*, verlorengehender Ruf *m*

**uncondensable** unkondensierbar

**uncondensed heavy-current discharge** nichtkondensierte Hochstromentladung *f*

**unconditional** bedingungslos, unbedingt (data proc.) ~ **branch** unbedingter Sprung *m* ~ **transfer** Einschleusung *f*, unbedingter Sprung *m*

**unconfined** (compression) ohne Seitenbehinderung *f* ~**-compression test** Druckversuch *m* mit unbehinderter Seitenausdehnung

**unconfirmed credit** widerruflicher Kredit *m*

**unconformity** Diskordanz *f*, Nichtübereinstimmung *f*, ungleichförmige Lagerung *f*

**uncongested** staubfrei

**uncontrasty picture** weiches Bild *n*

**uncontrollable** unfreiwillig, unkontrollierbar, unübersehbar ~ **spin** unfreiwilliges Trudeln *n* (aviat.)

**uncontrolled** frei, losgelassen, ungesteuert ~ **mosaic** Bildskizze *f* ~ **steam extraction** ungesteuerte Entnahme *f*

**unconventional** außergewöhnlich

**uncooled** ungekühlt

**uncorrected** unkorrigiert ~ **bearing** mißweisender Kurs *m* ~ **color** Farbenreste *pl* ~ **north** mißweisend Nord ~ **radio bearing** unverbesserte oder abgelesene Funkseitenpeilung *f*

**uncouple, to** ~ abkuppeln, auskuppeln, entkuppeln, losbinden

**uncoupling** Auslösung *f*, Ausrücken *n*, Entkupplung *f* ~ **field** Auskoppelfeld *n*

**uncover, to** ~ abdecken, aufdecken **to** ~ **a layer** ein Flöz *n* erschürfen

**uncovered** abgedeckt, blanko, unbespannt ~ **part of a slab (or title)** Freifeld *n*, offenes Feld *n*

**uncrossed** ungekreuzt

**uncrystallizable** unkristallisierbar

**unctuous** fettig, geschmeidig, ölig, salbig,

**unctuousness** Fettigkeit *f*, Öligkeit *f*, Schmierigkeit *f*

**uncultivated** unangebaut **wild** ~ **strip at edge of field** Feldrain *m*

**uncurved** wölbungsfrei

**uncut straight plane** Einfachhobeleisen *n*

**undamaged** unbeschädigt, unversehrt

**undamped** ungedämpft ~ **circuit** Generatorkreis *m* ~ **oscillation** ungedämpfte Schwingung *f*, ungedämpfte Welle *f* ~ **wave** kontinuierliche oder ungedämpfte Welle *f* ~ **wave signals** ungedämpfte Zeichen *pl*

**undecomposed** unzersetzt

**undefinable shade** ausdruckslose Färbung *f*

**undefined** nicht fest umrissen, unbegrenzt, unbenannt, unbestimmt, unerklärt ~ **scale** wilder Maßstab *m*

**undeflected** nicht (aus der Richtung) abgelenkt

**undeformed** undeformiert

**undelustered** nichtmattiert

**under,** ~ **command** manövrierfähig ~ **contract** gebunden, kontraktmäßig ~ **erection** im Bau *m* ~ **test** in der Prüfung *f* begriffen ~ **way in** Fahrt (aircraft)

**underback** Ausschlagbottich *m*

**underbid, to** ~ unterbieten

**underbridge** Unterbrückung *f* (r.r.), Unterführung *f*

**underbrush** Gebüsch *n*, Gesträuch *n*, Gestrüpp *n*, Unterholz *n*

**underburner-type (by-product coke) oven** Unterbrennerofen *m*

**undercar antenna** Unterwagenantenne *f*

**undercarriage** Fahrgestell *n*, Rollwerk *n*, Untergestell *n*, Unterrahmen *m* ~ **with independent legs** Einbeinfahrgestell *n*

**undercarriage,** ~ **bracing** Fahrgestellverspannung *f* ~ **brake** Fahrgestellbremse *f* ~ **door** Fahrgestellklappe *f* ~ **fairing** Fahrgestellverkleidung *f* ~ **half** Fahrgestellhälfte *f* ~ **legs** Fahrgestellbeine *pl* ~ **position indicator** Fahrgestellanzeiger *m* ~ **shock absorber** Landepuffer *m* ~ **socket** Fahrgestellschuh *m* ~ **springing** Fahrgestellfederung *f* ~ **strut** Fahrgestellstrebe *f*, Flugzeugbein *n*

**undercharge, to** ~ zu wenig laden, unterladen (Batterie)

**undercharged mine** Quetsch-ladung *f*, -mine *f*

**undercoat** Grundierungsmittel *n*

**undercoating** Voranstrich *m*

**undercompensate, to** ~ unterkompensieren

**undercompensation** Unterkonoidierung *f*, Unterkorrektur *f*

**undercool, to** ~ überkalten, unterkühlen

**undercooled** unterkühlt ~ **liquid** unterkühlte Flüssigkeit *f*

**undercooling** Unterkühlung *f*

**undercorrection** Unterkorrektur *f*

**undercrossing structure** Unterführung *f*

**undercured** nichtausgehärtet

**undercurrent** Grundströmung *f* der Flut, Unterströmung *f*

**undercut, to** ~ hinterschnitten, unterfahren (min.), unterhöhlen, unterschneiden, unterschnitten, unterschrämen, unterteufen
**undercut** Einbrand-riefe *f*, -kerbe *f*, Fallkerb *n*, Unterschneidung *f*, Unterschnitt *m*, unterschnittener Zahn *m*
**undercutter** Schrämer *m*
**undercutting** Schleuderei *f*, Schlitzarbeit *f* (min.), Unterschrämung *f* **no** ~ kerbfrei ~ **saw** Kollektorsäge *f* ~ **(or recessing) tool** Einstechstahl *m*
**underdamping** Unterdämpfung *f*
**underdeveloped** unterentwickelt
**underdevelopment** Unterentwicklung *f*
**underdough** Bodenteig *m*
**underdrive** Geländegang *m*
**underdriven** mit Antrieb *m* von unten ~ **centrifugal** Schleuder *f* mit unterem Antrieb
**underestimate, to** ~ unterschätzen
**underestimated** zu niedrig veranschlagt
**under-excited** untererregt
**underexpose, to** ~ unterbelichten
**underexposure** Unterbelichtung *f*
**underface** Unterfläche *f*
**underfeed,** ~ **firing** Unterschubfeuerung *f* ~ **furnace** Unterschubfeuerung *f* ~ **grate (or stoker)** Unterschubrost *m*
**underfill** (in rolling) Stoffmangel *m*
**underfire, to** ~ unterfeuern
**underfiring** Nichtgarbrennen *n*
**underfloor,** ~ Diesel-Unterflurdieselmotor *m* ~ **duct** Unterflurleitung *f* ~-**engine bus** Unterflurbus *m*
**underflow** Unterströmung *f*
**underfoot** Untergestell *n* (film)
**underframe** Fahrgestell *n*, Tragwerk *n*, Untergestell *n*
**underfrequency,** ~ **protection** Unterfrequenzschutz *m* ~ **relay** Unterfrequenzrelais *n*
**underfuselage tunnel** Wanne *f*
**underglaze** Unter-farbe *f*, -glasur *f*
**undergo, to** ~ ausgesetzt sein, sich unterziehen **to** ~ **fission** zerfallen
**undergoing tests** in der Erprobung *f*
**undergrade arch crossing** Bahnüberführung *f*
**undergrate,** ~ **blast** Unterwind *m* ~ **blower** Unterwindgebläse *n* ~ **firing** Unterfeuerung *f*
**underground** Untergrund *m*; bodenversenkt, erdverlegt, Tiefbau ..., unter der Erde *f*, unterirdisch, unter Tage **to go** ~ einfahren **to survey** ~ markscheiden
**underground,** ~ **antenna** Bodenantenne *f*, Erdantenne *f* ~ **blackening** Untergrundschwärzung *f* ~ **burst** unterirdische Explosion *f* ~ **cable** Erd-, Land-kabel *n*, versenktes Kabel *n* ~ **cable box** Landkabelmuffe *f* ~ **cable-distributing point** Kabelabzweigpunkt *m* ~ **circuit** unterirdische Leitung *f* ~ **construction engineer** Tiefbauingenieur *m* ~ **conveyance** Förderung *f* unter Tage, unterirdische Förderung *f* ~ **crossing** Wegunterführung *f* ~ **distribution chamber** Kabelkeller *m* ~ **fire** Grubenbrand *m* ~ **furnace** Unterflurofen *m* ~ **garage** Tiefgarage *f* ~ **hauling** Förderung *f* unter Tage, unterirdische Förderung *f* ~ **hydrant** Unterflurhydrant *m* ~ **hydro-electric power station** Kavernenkraftwerk *n* ~ **laying** Erdverlegung *f* ~ **leaching** Laugung *f* in situ

~ **line** Erdkabelleitung *f*, Grubenbahn *f*, unterirdische Linie *f* ~ **machine hall** Kaverne *f* ~ **mine** Tiefbaugrube *f* ~ **mining** Untertagbau *m* ~ **operations** Untertagebetrieb *m* ~ **pipe** Erdleitungsrohr *n*, erdverlegtes Rohr *n* ~ **piping** Erdrohrleitung *f*, Unterflurrohrleitungen *pl* ~ **pit mining** Grubenbetrieb *m* ~ **pole reinforcement** Bodenverstärkung *f* ~ **railroad** Untergrundbahn *f* ~ **repeater station** Unterflur-Verstärkerstelle *f* ~ **shelter** unterirdischer Schutzraum *m* ~ **structures** Tiefbau *m* ~ **tank** Tiefbehälter *m*, versenkter Behälter *m* ~ **tanned leather** lohgrubengegerbtes Leder *n* ~ **transmission path** unterirdischer Leitungsweg *m* ~ **water** Grundwasser *n* ~-**water packing** Grundwasserabdichtung *f* ~ **winning** Abbau *m* unter Tage, Grubenbau *m*, Untertagegewinnung *f* ~ **work** Abbau *m* unter Tage ~ **working** Abbau *m* unter Tage, Arbeit *f* unter Tage, Grubenbau *m*, Untertagebau *m*, Tiefbau *m*
**undergrowth** Unterholz *n*
**underhand,** ~ **puddler** Hilfspuddler ~ **rocker gear** Schwinghebelantrieb *m* mit nicht obenliegender Nockenwelle ~ **stope** Strosse *f*
**underhollow, to** ~ unterhöhlen
**underhung revolving jib** Drehlaufkatze *f*
**underlap** Lückensynchronisierungsverfahren *n*, schlechte Überdeckung *f*
**underlay, to** ~ füttern, unterlegen
**underlay** schräges Flöz *n* (min.), Zurichtebogen *m* (print.) ~ **blotch design** Boden *m* (print.) ~ **feltboard** Fußbodenunterlagspappe *f*
**underlayer** Unterlage *f*, Unterzug *m*
**underlie, to** ~ unterworfen sein
**underlie** Einfallen *n*, Fallen *n*
**underline, to** ~ unterstreichen
**underlinen marking ink** Wäschezeichentinte *f*
**underlining** Unterstreichung *f*
**underload(ing)** Unterbelastung *f*
**underlubrication** nicht ausreichende Schmierung *f*
**underlying** darunterliegend, unterlagert, unterliegend ~ **principle** Grundlage *f* ~ **rock of strata** Untergrund *m*
**undermatching of impedance** Widerstandsunteranpassung *f*
**undermine, to** ~ auskolken, kolken, untergraben, unterminieren
**undermined** angreifbar, unterwaschen
**undermining** Kolkbildung *f*, Unterspülung *f* Unterwaschung *f*, Unterwühlung *f*
**undermodulation** Untersteuerung *f*
**underneath** unten, unterhalb
**underpass** Bahnunterführung *f*, Unterführung *f*, Wegunterführung *f*
**underpin, to** ~ unter-bauen, -mauern
**underpinning** Abfangung *f*, Abstützen *n*, Abstützmaterial *n*, Stützwerk *n*, Unter-bau *m*, -fangung *f*, -lage *f*, -mauern *n*, -mauerung *f* ~ **work** Unterfangarbeiten *pl*
**underplate** (of an oil press) Fußplatte *f*
**underpoled copper** übergares Kupfer *n*
**underpowered motor glider** Segelflugzeug *n* mit Hilfsmotor
**underpressure** Unterdruck *m* ~ **wake** Unterdruck-Nachlauf *m*
**underrate, to** ~ unterschätzen, zu niedrig ansetzen

**underream, to** ~ erweitern, nachbohren, unterschneiden

**underreamer** Nachnahmbohrer *m*, Unterschneider *m*

**underrevving** mit zu niedriger Drehzahl *f*

**underrun, to** ~ klarscheren, unterfahren

**underrunning voltage** Unterspannung *f*

**undersaturate, to** ~ untersättigen

**underscore, to** ~ (sounds or frequency bands) anheben, hervorheben, unterstreichen

**underscour, to** ~ unterspülen

**underscouring** Unter-spülung *f*, -waschung *f*

**undersea** unterseeisch

**undershield** (Chassis) Motorschutzblech *n*, unterer Ölschutz *m*

**undershoot, to** ~ sich verrechnen, (bei einer Landung) zu tief gleiten, nicht weit genug gehen (navy), (Bezugswort) unterschwingen

**undershoot** Einschwung *m*, Unterschuß *m*, Unterschwingen *n*

**undershot** unterschlächtig ~ **water wheel** unterschlächtiges Wasserrad *n*

**undersite** Unter-fläche *f*, -kante *f*, -seite *f* ~ **of fuselage** Rumpfunterseite *f*

**undersize** Innenzugabe *f*, Siebdurchgang *m*, Unter-größe *f*, -korn *n*, -maß *n* ~ **drill** Untermaßsenker *m*

**undersized** Untermaß *n* aufweisend

**undersling, to** ~ unterhängen

**underslung** gekröpft, unterbaut ~ **radiator** Bauch-, Hänge-kühler *m* ~ **spring** unterbaute Feder *f*

**underspeed** mit zu niedriger Drehzahl *f*

**understand, to** ~ auffassen, begreifen, verstehen

**understandable** verständlich

**understanding** Einvernehmen *n*, Einverständnis *n*, Übereinkommen *n*, Vereinbarung *f*, Verständigung *f*, Voraussetzung *f*

**understood signal** Verstandenzeichen *n*

**understructure** Gestell *n*, Untergestell *n*

**undersurface** Unterfläche *f*

**underswing** (of volume indicator) Unteranzeige *f*, Unterschwingen *n*

**undersynchronous** untersynchron

**undertaking** Gesellschaft *f*, Unternehmen *n*, Unternehmung *f*

**undertanned** ungar

**undertone** Unterton *m*

**undertow** Grundsee *f*, Grundströmung *f* der Flut, Sog *m*

**undertripping** (relay) Unterspannungsauslösung *f*

**undervalue, to** ~ unterschätzen

**undervoltage** Spannungsrelais *n* ~ **protection** Unterspannungsschutz *m* ~ **relay** Unterspannungsrelais *n* ~ **release** Unterspannungsauslöser *m* ~ **switch** Unterspannungsschalter *m*

**undervolting** Unterspannung *f*

**underwagon for inclined planes** Unterwagen *m* für Bremsberge

**underwashing** Unterspülung *f*

**underwater, ~ air intake** Luftmast *m* ~ **antenna** Unterwasserantenne *f* ~ **burst** Unterwasserexplosion *f* ~ **cutting and welding detachment** Unterwasserschneidetrupp *m* ~ **cutting burner** Unterwasserschneidbrenner *m*, ~ **cutting torch** Unterwasser-brenner *m*, -schneidbrenner *m* ~ **cutting weld** Schneiden *n* unter Wasser

~ **photograph** Unterwasserlichtbildaufnahme *f* ~ **receiver** Geräuschempfänger *m* ~ **reflection** Unterwasserspiegel *m* ~ **sound transmitter** Unterwasserschallgeber *m* ~ **structure** Untergründung *f*, Unterwassergründung *f* ~ **welder** Unterwasserschneider *m*

**underweight** Gewichtsausfall *m*, Minder-, Unter--gewicht *n*

**underwing refuelling** Unterflügelbetankung *f*

**underwound thread** untergeklemmter Faden *m*

**undesired** unerwünscht ~ **coupling** Verkupplung *f*, wilde Kupplung *f* ~ **oscillating** selbständige Schwingungserzeugung *f* ~ **oscillation** Pfeifen *n*

**undetachable** unlösbar, untrennbar

**undetermined** unbestimmt

**undeveloped** unentwickelt

**undifferentiated dike rocks** ungespaltene Gangsteine *pl*

**undiluted** unverdünnt

**undiminished** unvermindert

**undirected** gleichgerichtet ~ **flow** ungerichtete Strömung *f*

**undiscernible** unkennbar, unmerklich, ununterscheidbar

**undispersed light** unzerlegtes Licht *n*

**undissociated** nichtdissoziiert

**undissolvable** unlösbar

**undissolved sulfates in glass** Galle *f*

**undistorted** form-gerecht, -treu, unverzeichnet, unverzerrt ~ **picture** unverzerrtes Bild *n* ~ **reception** unverzerrter Empfang *m*

**undisturbed** (soil) gewaschen, ruhig, störfrei, unerschüttert, ungestört ~ **soil** Mutterboden *m* ~-**zero output** ungestörtes Nullsignal *n*

**undivided** ungeteilt

**undo, to** ~ aufknüpfen, aufmachen, auseinandernehmen, demontieren, lösen, los-knüpfen, -schnüren, -schrauben, zerlegen **to** ~ **one's belt** sich losschnallen

**undock, to** ~ ausdocken, ein Luftschiff *n* von der Vertäuung losmachen

**undoing** Auflösung *f*

**undoped fuel** Kraftstoff *m* ohne Blei- oder andere Zusätze

**undoubling angle** Entdoublierungsvorrichtung *f*

**undrawn tow** unverstrecktes Kabel *n*

**undressed castings** Rohguß *m*

**undriven rivet** Rohniet *n*, Setzkopf *m*

**unduced-draught blower** Saugzuggebläse *n*

**unductile steel band** undehnbares Stahlband *n*

**unduction noise** induziertes Geräusch *n*

**undue** unangemessen, ungebührlich, unzulässig ~ **strain** unzulässige Beanspruchung *f* ~ **wear** zusätzlicher Verschleiß *m*

**undulate, to** ~ hochziehen und drücken, pendeln, schwingen, sich krümmen, undulieren

**undulated** gewellt, schlangenlinienartig ~ **current** welliger Strom *m*

**undulating** Krümmung *f*; schwingend, undulierend, wellig ~ **current** undulierender Strom *m*, Wellenstrom *m* ~ **flight** Wellenflug *m* ~ **ground** hügeliges Gelände *n* ~ **light** Feuer *m* mit periodisch veränderlicher Lichterscheinung *f* ~ **mechanics** Wellenlehre *f* ~ **quantity** pulsierende Größe *f* ~ **seams** flachwellige Lagerung *f*

**undulation** Riffelbildung *f*, Rippe *f*, Schwankung *f*, Schwingung *f*, Schwingungswelle *f*, Welle

*f*, Wellen-bewegung *f*, -schwingung *f*, Wellung *f* ~ **of ground** Bodenwelle *f*
**undulator** Undulator *m*, Wellenschreiber *m*
**undulatory** schwingend, undulierend, undulös, wellen-artig, -förmig ~ **current** undulierender Strom *m* ~ **line** Wellenlinie *f* ~ **motion** Wellenbewegung *f* ~ **theory** Undulationstheorie *f*
**unduly, to** ~ **strain** unnötigen Belastungen *pl* aussetzen
**undyed** ungefärbt
**unearthed system** ungeerdetes System *n*
**uneconomic(al)** unökonomisch, unrentabel, unwirtschaftlich
**unelastically scattered** unelastisch gestreut
**unelectric** unelektrisch
**unembanked alluvial land** unbedeichte Anlandung *f*
**unemployed** Arbeitsloser *m*, Erwerbsloser *m*; arbeitslos, unbeschäftigt
**unemployment** Arbeitslosigkeit *f*, Erwerbslosigkeit *f* ~ **allowance** Erwerbslosenunterstützung *f* ~ **insurance** Arbeitslosenversicherung *f* ~ **relief** Arbeitslosen-fürsorge *f*, -unterstützung *f*
**unencumbered** unbelastet **to be** ~ freiliegen
**unending** endlos
**unequal** in keinem Verhältnis *n* stehend, ungleich, ungleichförmig, ungleichmäßig ~ **impulse** Ungleichimpuls *m* ~ **sided** ungleichschenklig ~-**sided angle iron** ungleichschenkliges Winkeleisen *n* ~ **T** reduziertes T-Stück *n*, Verengungs-T-Stück *n* ~ **weighing** Unwucht *f*
**unequaled** unvergleichlich
**unequally armed** ungleicharmig
**unequivalve** ungleichschalig
**unequivocal** eindeutig, klar
**unetched** ungeätzt
**uneven** holperig, unausgeglichen, uneben, ungerade, ungleich, ungleichmäßig, ungrad, unstet, unstetig ~ **bottom** unebener Boden *m* ~ **dyeing** Farbunruhe *f* ~ **fracture** irreguläre Bruchfläche *f* ~ **number** ungerade Zahl *f* ~ **page** Rekteseite *f*, Vorderseite *f* ~ **running** (of motor) unruhiger Lauf *m*
**unevenly aged** von ungleichem Alter *n*
**unevenness** Unebenheit *f*, Ungleichheit *f*
**unexamined** ungeprüft
**unexampled** beispiellos
**unexcelled** unübertrefflich, unübertroffen
**unexceptionable earthing** einwandfreie Erdung *f*
**unexcited** unerregt
**unexecuted** unausgeführt
**unexplained** unbekannt ~ **cause** unbekannte Ursache *f*
**unexplored** unerforscht
**unexposed** unbelichtet ~ **negative** Leerbild *n*
**unfailing** unausbleiblich
**unfair** unbillig, ungerecht, unlauter ~ **competition** unlauterer Wettbewerb *m*
**unfashioned part (or piece)** Rohling *m*
**unfasten, to** ~ losmachen
**unfavorable** ungünstig ~ **balance of trade** passive Handelsbilanz *f* ~ **installation of apparatus** ungünstige Anordnung *f* ~ **weather** ungünstiges Wetter *n*
**unfavored** nichtbegünstigt
**unfeasible** unausführbar
**unfeather, to** ~ den Einfallswinkel *m* verändern (gegen den Luftstrom)

**unfertile** unfruchtbar
**unfinished** unbearbeitet, unfertig ~ **part** (of piece) Rohling *m*
**unfired** ungefeuert, ungeheizt ~ **pressure vessel** Stahlbolzenzylinder *m*
**unfit** unfähig, ungeeignet, untauglich ~ **for work** arbeitsunfähig
**unfitness** Unfähigkeit *f*
**unfix, to** ~ abnehmen, losmachen
**unfixed** rollig, unbefestigt ~ **ammunition** getrennte Munition *f*
**unflawed** fehlerfrei
**unfold, to** ~ aufspannen, ausbreiten, auseinanderklappen, ausklappen, darlegen, entfalten
**unfortified** unbefestigt
**unfounded** grundlos
**unfranked** unfrankiert, unfranko
**unfulled** ungewalkt
**unfurl, to** ~ entrollen
**unfused** ungeschmolzen
**ungalvanized** unverzinkt
**ungarbling** Entschlüsselung *f*
**ungasifiable** unvergasbar
**ungear, to** ~ auslösen
**ungeared** getriebelos, unübersetzt ~ **engine** getriebeloser oder nicht untersetzter Motor *m* ~ **motor** direkter Motor *m* ~ **radial engine** getriebeloser Sternmotor *m*
**unglazed** unglasiert
**unglue, to** ~ loskitten
**unground** ungeschliffen
**ungrounded** erdfrei, ungeerdet
**unguarded** unüberwacht
**unhair, to** ~ abhaaren
**unhardened** ungehärtet
**unharness, to** ~ abschirren, ausschirren
**unheated downcomer** kaltliegendes Fallrohr *n*
**unhewn timber** Ganz-, Roh-holz *n*
**unhook, to** ~ abhaken, abhängen, abheften, abspannen, loshaken
**unhurt** schadlos, unverletzt
**uniaxial** einachsig ~ **magnetic anisotropy** einachsige, magnetische Anisotropie *f* ~-**wheel tractor** Einachsschlepper *m*
**unicellular** einzellig
**unicontrol** Einknopfkontrolle *f*
**unicursal** einläufig
**unidimensional** eindimensional
**unidirectional** einseitig, einseitig gerichtet oder wirkend ~ **action** einseitige Richtwirkung *f* ~ **antenna** einseitige (Richt-) Antenne f ~ **conductance** unipolare Leitung *f* ~ **conductivity** unipolare Leitfähigkeit *f* ~ **current** Strom *m* gleicher Richtung ~ **current principle** Gleichstromprinzip *n* ~ **cyclic load** Schellbeanspruchung *f* ~ **direction finding** eindeutige Richtungsanzeige *f* ~ **discharge** aperiodische Entladung *f* ~ **lightning disturbance** einpolige Blitzstörung *f* ~ **motion** Bewegung *f* in nur einer Richtung ~ **pulse** Gleichstromimpuls *m* ~ **ray (or beam) receiver** Strahlempfänger *m* ~ **resistance** Richtwiderstand *m* ~ **transducer** Einrichtungswandler *m* ~ **transmitter** Einstrahlsender *m*, Strahl-sender *m*, -sendestelle *f*
**unification** Vereinheitlichung *f*
**unified** einheitlich, vereinheitlicht ~ **nuclear model** einheitliches Kernmodell *n* ~ **system of filing** Einheitsklassifikation *f*

**unfilar** eindrähtig, einfädig, einfilar ~ **suspension** Einfadenaufhängung *f*, einfädige Aufhängung *f*

**uniflow,** ~ **drier** Gleichstromtrockner *m* ~ **engine** Uniflowmaschine *f* ~ **scavenging** Gleichstromspülung *f* ~ **twin engine** Gleichstromzwillingsmaschine *f* ~**-type compressor** Gleichstromkompressor *m*

**uniform** Uniform *f*; einförmig, einheitlich, gleich, gleichbleibend, gleichförmig, gleichmäßig, stetig, unbeschleunigt ~ **as to grain size** homogen

**uniform,** ~ **acceleration** einheitliche Beschleunigung *f* ~ **change of wind** regelmäßiger Windwechsel *m* ~ **charge** Gleichbelastung *f* ~ **color** homochrom **of** ~ **density** gleichmäßig dicht ~ **discharge temperature** gleichbleibende Austrittstemperatur *f* ~ **field** gleichförmiges Feld *n* ~ **line** homogene Leitung *f* ~ **load** Gleichbelastung *f* ~ **loading** Rechteckbelastung *f* ~ **longitudinal load** Längeneinheitsbelastung *f* ~ **method of analysis** einheitliche Untersuchungsmethode *f* ~ **plane wave** einheitliche ebene Welle *f* ~ **random noise** weißes Rauschen *n* ~ **rifling** gleichbleibender Drall *m* ~ **speed** gleichförmige Geschwindigkeit *f* ~ **stress** gleichförmige Spannung *f* ~ **time** Einheitszeit *f* ~ **velocity** gleichmäßige Geschwindigkeit *f* ~ **waveguide** homogener Hohlleiter *m* ~ **wave-length** Nierenplattenkondensator *m*

**uniformity** Gleichartigkeit *f*, Gleichförmigkeit *f*, Gleichheit *f*, Gleichmäßigkeit *f*, Konstanz *f*, Stetigkeit *f* ~ **in delivery** Gleichmäßigkeit *f* den Ablieferungen *pl* ~ **of reproduction** Herstellungsgleichmäßigkeit *f*

**uniformity,** ~ **error** Lochungsfehler *m* (film) ~ **factor** Gleichmäßigkeit *f* der Beleuchtung *f*, Gleichmäßigkeitsfaktor *m*

**uniformly,** ~ **accelerated motion** gleichförmig beschleunigte Bewegung *f* ~ **continuous** gleichmäßig, stetig ~ **decreasing motion** gleichförmig verzögerte Bewegung *f* ~ **distributed** stetig oder gleichmäßig verteilt ~ **distributed inductance** gleichmäßige oder stetig verteilte Induktivität *f* ~ **distributed load** gleichförmig verteilte Last *f* ~ **distributed loading** gleichförmige Ladung *f* ~ **divided scale** gleichmäßige Teilung *f* ~ **heated** gleichtemperiert, gleichwarm ~ **varying load** gleichmäßig variierende Belastung *f*

**unify, to** ~ vereinheitlichen

**unilateral** einseitig, einseitig wirkend, nicht symmetrisch ~ **antenna** Richtantenne *f* ~ **conductance** (of crystals) asymmetrische Leitfähigkeit *f*, einseitige Leitfähigkeit *f* ~ **conductivity** unipolare Leitfähigkeit *f* ~ **transducer** unilateraler Wandler *m* (acoust.) ~ **variable--area track** Abdeckeinfachzackenspur *f*, Einfachzackenspur *f* ~ **variable-area track made with single-vane shutter** Einfachzackenschift *f* mit Abdeckung *f*

**unilaterally,** ~ **biased** (of grid, magnet) einseitig vorgespannt ~ **open** einseitig offen

**unilateralness** Unsymmetrie *f*

**unilayer** Einfachschicht *f*

**unilines** Einwegleiter *m*

**unimodular** unimodular ~ **condition** unimodulare Bedingung *f*

**unimolecular layer** Monomolekularschicht *f*

**unimpaired** unbeeinträchtigt, ungeschwächt, unvermindert

**unimpeded magnetization** freie Magnetisierung *f*

**unimportant** belanglos, gering, geringfügig, unbedeutend, unwichtig ~ **matter (or item)** Nebensache *f*

**unimflammable** unentzündbar ~ **coal** Sandkohle *f*

**uninformed person** Nichteingeweihter *m*

**uninhabited** unbewohnt

**uninitiated person** Nichteingeweihter *m*

**uninsulated** unisoliert

**unintelligibility** Unverständlichkeit *f*

**unintelligible** unverständlich ~ **crosstalk** unverständliches Nebensprechen *n*

**unintentional interference** unbeabsichtigte Störungen *pl*

**uninterrupted** ununterbrochen ~ **sequence (or series)** fortlaufende Reihe *f*

**uninverted cross talk** verständliches Nebensprechen *n*

**union** Angliederung *f*, (wing-fuselage) Anschluß *m*, Anschlußstück *n*, Arbeitergemeinschaft *f*, Bindung *f*, Gesellschaft *f*, Gewerkschaft *f*, (pipe joint) Rohrverschraubung *f*, Stutzen *m*, Überwurfmutter *f* (mach.), Verband *m*, Verbindung *f*, Verein *m*, Vereinigung *f*, Verschraubung *f* ~ **of sets** Vereinigungsmenge *f* ~ **of shroud ropes** Knebelleinenbund *m*

**union,** ~ **flange** Ansatzflansch *m* ~ **goods** Halbwolle *f* ~ **joint hose** Schlauchverbindung *f* ~ **nipple** Rohrnippel *m* für Verbindung durch Überwurfmutter ~ **nut** Anschluß-, Verschraubungs-mutter *f* ~ **point** Verbindungspunkt *m*, Vereinigungsstelle *f* ~ **reducer** Reduktionsklemme *f* ~ **socket** Verbindungsmuffe *f* ~ **T** Rohrverschraubung *f* mit T-Stück ~ **tailpiece** Verschraubungskonus *m*

**unionized** nichtionisiert, unionisiert

**unipivot,** ~ **bearing** Einspitzenlagerung *f* ~ **moving coil** Drehspule *f* mit einem Lager

**unipolar** (apparatus) einpolig, gleichpolig, homopolar, unipolar ~ **dynamo** Unipolardynamo *m* ~ **peak** Höchstwert *m* ~ **transistor** Unipolartransistor *m*

**unipotential** Äquipotentialkathode *f* ~ **cathode** indirekt geheizte (Glüh)Kathode *f* ~ **lens** (electron microscope) Einzellinse *f*

**unique** eigenartig, eindeutig (math.) einmalig, einzigartig ~ **action** einmaliges Ereignis *n*, einmaliger Vorgang *m* ~ **shape** einheitliche Form *f*

**uniquely determined** klarbestimmt

**uniqueness** Ausschließlichkeit *f*, Eindeutigkeit *f* ~ **theorem** Eindeutigkeitssatz *m* ~ **theory** Eindeutigkeitstheorie *f*

**uniroll mill** Einwalzenstuhl *m*

**uniselector** Einknopfkontrolle *f*, Wähler *m* mit einziger Bewegungsrichtung

**unison** Einklang *m*, Gleichgang *m*, Gleichklang *m*; einstimmig, gleichtönend **to be in** ~ übereinstimmen **in** ~ **with** Rhythmus *m*

**unison,** ~ **impulse** Gleichlaufstromstoß *m* ~ **lever** Nullhebel *m* ~ **signals** Gleichlauf-stöße *pl*, -zeichen *pl*

**unisonant** gleichgehend, gleichtönend

**unit** Abteilung *f*, Anlage *f*, Einer *m* (math.), Einheit *f*, Gerät *n*, Glied *n*, Grundzahl *f*,

Maß n, (of measure) Maßeinheit f, Stromschritt m, Truppe f, Truppen-abteilung f, -teil m, -verband m

**unit,** ~ **of action** Aktionseinheit f ~ **of angle** Winkeleinheit f ~ **of area** Flächen-einheit f, -inhaltseinheit f, Oberflächeneinheit f ~ **of attenuation** Dämpfungsmaß n ~ **of caliber** Kalibereinheit f ~ **of capacity** Arbeitseinheit f ~ **of coinage** Münzeinheit f ~ **of construction** Konstruktionsglied n ~ **of (film) density** Schwärzungseinheit f ~ **of displacement** Verschiebungseinheit f ~ **of dosage** Dosiseinheit f ~ **of energy** Leistungseinheit f ~ **of equipment** Gerätesatz m ~ **of fit** Paßeinheit f ~ **of force** Krafteinheit f, Kraftmaß n ~ **of frequency** Frequenzeinheit f ~ **of hardness** Härteeinheit f ~ **of heat** Wärmemaß n ~ **of length** Längeneinheit f, Meßlängeneinheit f ~ **of . . . lines** Amtseinheit f mit . . . Anschlüssen ~ **of machinery** Satz m ~ **of mass** Maßeinheit f ~ **of measurement** Einheitsmaß n, Meßeinheit f ~ **of power** Kraft-, Leistungs-einheit f ~ **of production** Fertigungsmenge f ~ **of quantity** Mengeneinheit f ~ **of set of machines** Aggregat n ~ **of sound absorption** Schalldämpfung f ~ **of space** Raumeinheit f ~ **of speed** Geschwindigkeitseinheit f ~ **of standard of temperature** Temperaturmaßeinheit f ~ **of structual parts** Bauteilgruppe f ~ **of surface** Flächeneinheit f ~ **of volume** Volume(n)einheit f ~ **of wages** Lohneinheit f ~ **of wave length for X-rays** X-Einheit f ~ **of weight** Gewichtseinheit f, Wichte f ~ **of work** Arbeitseinheit f

**unit,** ~ **area** Einheitenfläche f, Flächen-einheit f, -stück n ~ **area reactance** Reaktanz f pro Flächeneinheit (acoust.) ~ **areas** Flächenelemente pl ~ **assemblage** Geräteeinheit f ~ **assigned to one sector of main line of resistance** Widerstandsgruppe f ~ **automatic exchange** automatische Landzentrale f, kleines Wahlamt n ~ **border** Einheitenrand m ~-**bore system** Einheitsbohrung f ~ **call** Gesprächseinheit f ~ **capacity factor** Ausnutzungs-faktor m, -zahl f ~ **cell** Basis-, Einheits-, Elementar--zelle f, Gittereinheit f ~ **cell vectors** Grundtranslationsvektoren pl ~ **charge** Einheitsladung f, Ladungseinheit f ~-**column design** Blocksäulenbauweise f ~-**connected transformer** Blockumspanner m, Maschinentransformator m ~ **construction** Aufbau m als geschlossene Einheit, Maschinenzusammenbau m ~ **container** Einheitskanister m ~ **control** gemeinsame Regelung f ~ **cord** Einheitsschnur f ~ **cost** Stückpreis m ~ **crystal** Einheitszelle f, Einzelkristall n ~ **cubicle** Einheitsschrank m ~ **(of) current** Stromeinheit f ~ **diagram** Einheitsdiagramm n ~ **duration of signal** Schrittlänge f ~ **element** Einheitselement n ~ **flange** Einheitsflansch m ~ **forks** Drahtgabeln pl, (Vier)-Einheitsgabeln pl ~ **form** Einheits-fläche f, -form f ~ **frame in a series of sequence** Einzelbild n einer Bildreihe ~ **function** Einheitsfunktion f ~ **furniture** An- und Aufbaumöbel pl ~ **fuse alarm lamp** Einzelalarmlampe f ~ **impulse function** Einheitsimpuls m ~ **indicator** Einheitenskala f ~ **inductance** Einheitsinduktivität f ~ **interval** Spulenseitenteilung f, (in windings) Spulenteilung f, Zeichenelement n ~

**interval of the commutator** Lamellenteilung f ~ **interval in a winding** Spulenseitenteilung f ~ **lattice** Einheitsgitter n ~ **length** Einheitenlänge f ~ **lens** (of focusing field) Einzellinse f ~ **light intensity** Einheit f der Lichtintensität f ~ **line** Einheitslinie f ~ **load** Flächeneinheitslast f (aviat.), Ladeeinheit f ~ **nominal register (or roll)** Stammrolle f ~ **number** (Kabelkennung) Verbrauchernummer f ~ **operating factor** Betriebsfaktor m ~ **organ** Multiplexorgel f (acoust.) ~ **path** Wegeeinheit f ~ **place** Einerstelle f ~ **plant** Einzelanlage f ~ **point** Einheits-, Haupt-punkt m ~ **pole** Einheitspol m, Magnetpolstärke f ~-**positive charge** positive elektrische Ladungseinheit f ~ **potential** Potentialeinheit f ~ **power** Leistungsgewicht n ~ **(system) power station** Blockkraftwerk n ~ **price** Einheitspreis m ~ **principle** Einheitsprinzip n ~ **process** Einheitsprozeß m ~ **quantity of electricity** Einheit f der Elektrizitätsmenge ~ **rack** Einheitsgestell n ~ **rack stop** Einheitenwiderstand m, Einheitsquerschnitt m ~ **sensitivity** Einheitsempfindlichkeit f ~ **shaft system** Einheitswelle f ~ **sphere** Einheitskugel f ~ **spin** Einheitsspin m ~ **step** Einheitssprung m ~ **step function** Einheitsstoß m ~-**step response** Sprungübergangsfunktion f ~ **stress** Belastung f, Belastung f der Flächeneinheit, Spannung f, Spannungsgröße f, spezifische Belastung f ~ **surface** Oberflächeneinheit f ~ **system power station** Blockkraftwerk n ~ **(or identify) tensor** Einheitstensor m ~ **test** Prüfung f des Triebwerks als Ganzes, Vollmotorprüfung f ~ **(of) time** Zeiteinheit f ~ **transformation** Einheitstransformation f ~ **value** Einheitswert m ~ **vector** Einheitsvektor m ~ **volume** Einheitsvolumen n, Volumeneinheit f ~ **(of) volume** Raumeinheit f ~ **wave-length constant** Phasenfaktor m, Winkelmaß n ~ **weight** Raumgewicht n (ohne Wasser)

**unitarian** unitär

**unitary** unitär, normiert

**unite, to** ~ binden, einigen, kuppeln, mischen, verbinden, vereinigen, zusammenfügen, zusammentreiben

**united** uniert

**uniting** Zusammenlegung f

**units,** ~ **counting tube** Einer-Zählröhre f ~ **digit** Einer m, Einerstufe f

**unity** Einheit f, Zusammenhalt m ~ **coupling** Kopplung f Eins, Kopplungsfaktor m Eins ~ **gain** Verstärkung f Eins

**univalent** einwertig ~ **radical** einwertiges Radikal n

**univariant** einfachfrei, monovariant, univariant ~ **equilibrium** einfachfreies oder monovariantes Gleichgewicht n

**universal** allseitig, universal, universell ~ **adjustability (or adjustment)** Universaleinstellbarkeit f ~ **amplifier** Allverstärker m ~ **apparatus** Universalgerät n ~ **back rest** Universalsetzstock m ~ **ball joint** Kugelkardan n ~ **bars** Universaleisen n ~ **bevel** Universalschmiege f ~ **bevel protractor** Universalwinkelmesser m ~ **bore gauge** Innen-Drall- und Durchbiegungsmesser m ~ **bridge** Allzweckmeßbrücke f ~ **camera attachment** Universalaufsetzkamera f ~ **chopper for fodder dehydrator** Alleszer-

kleinerer *m* ~ **chuck** Universal(dreibacken)-futter *n* ~ **copy dial** Universalschablonenhalter *m* ~ **cord pair** Einheitsschnurpaar *n* ~ **coupling** kardanisches Gelenk *n*, Kreuzgelenkkupplung *f* ~ **cylindrical grinding machine** Universalrundschleifmaschine *f* ~ **depth gauge** Universaltiefenlehre *f* ~ **diurnal variation** weltzeitlicher Anteil *m* ~ **drill** Universalhandbohrer *m* ~ **drive** Verteilerantrieb *m* ~ **extension** Kreuzgelenkverlängerung *f* ~ **finishing stand** Universalfertiggerüst *n* ~ **flange roll** Universalflanschenwalze *f* ~ **function** universelle Funktion *f* ~ **gauge** Universallehre *f* ~ **grinder (or grinding machine)** Universalschleifmaschine *f* ~ **grip** Universalgreifer *m* ~ **guiding machine** Universalführungsmaschine *f* ~ **head** Universalspindelkopf *m*
**universal-joint** artikuliertes Gelenk *n*, Cardan--Gelenk *n*, Doppelgelenk *n*, Doppelkreuzgelenk *n*, Kardan-gelenk *n*, -kupplung *f*, Kniegelenk *n*, Kreuzgelenk *n*, Kugel-gelenk *n*, -kupplung *f*, Kugelschalen-lager *n*, -lagerung *f*, Rollendoppelgelenk *n*, Rollengelenk *n*, Schubkegel *m*, Universalgelenk *n* ~ **with plates** Laschengelenk *n* ~ **housing** Gelenkgehäuse *n* ~ **shaft** gelenkige Welle *f*, Gelenkwelle *f* ~ **yoke** (Kardangelenk) Mitnehmer *m*
**universal,** ~ **jointed shaft** Kugelgelenkwelle *f* ~ **keyboard** Normaltastenfeld *n*, Universaltastatur *f* ~ **lathe** Stahlwechseldrehbank *f* ~ **lens adapter** Universalobjektivring *m* ~ **machine tool** Vielzweckmaschine *f* ~ **mill** Universalwalzwerk *n* ~ **mill for rolling beams** Universalträgerwalzwerk *n* ~ **mill plate** Breit-, Universal-eisen *n* ~ **mill stand** Universalgerüst *n* ~ **milling attachment** Universalfräskopf *m* ~ **motor** Universalmotor *m* ~ **mounting** vollkardanische Aufhängung *f* ~ **neon-tube tester** Universalglimmlampengerät *n* ~ **network** Allpaßnetzwerk *n* ~ **notching machine** Universalausklinkmaschine *f* ~**-pattern cap** Einheitsmütze *f* ~ **pick** Doppelkeilhaue *f* ~ **pliers** Universalzange *f* ~ **punching and notching machine** Universalloch- und Ausklinkmaschine *f* ~ **radio network** Weltfunknetz *n* ~ **rail anchor** Einheitsklemme *f* ~ **receiver** Allstromempfänger *m* ~ **relay** Einheitsrelais *n* ~ **relieving lathe** Universalhinterdrehbank *f* ~ **rotating stage** Universaldrehtisch *m* ~ **screwdriver** Universaldrehtisch *m* ~ **screw-wrench** Engländer *m*, Universalschlüssel *m* ~ **scroll chuck** Universalspannfutter *n* ~ **seismograph** Universalseismograf *m* ~ **shapes** Universaleisen *n* ~ **shunt** Universalnebenschluß *m* ~ **shunt box** Ayrtonscher Nebenschluß *m* ~ **square** Universalwinkel *m* ~ **surface gauge** Universalreißstock *m* ~ **switch** Generalumschalter *m* ~ **testing machine** Universalprüfmaschine *f* ~ **three axis stage** Universaldrehtisch *m* ~ **three-high mill** Universaltriowalzwerk *n* ~ **time clock** Weltzeituhr *f* ~ **tool and cutter grinder** Universalwerkzeugschleifmaschine *f* ~ **trestle** Universalmontagebock *m* ~ **tup machine** Universalfallwerk *n* ~ **two-high rolling mill** Universalduowalzwerk *n* ~**-type timing impulse regulator** Universalimpulsregler *m* ~ **valve** Universalröhre *f* ~ **wheel dressing and profiling attachment** Universalabziehvorrichtung *f* ~

**wrench** englischer Schraubenschlüssel *m*
**universality** Allgemeinheit *f*
**universe** Weltall *n*, Weltraum *m*
**university** Hochschule *f*, Universität *f* ~ **of commerce** Handelshochschule *f* ~ **graduate** Akademiker *m* ~ **student** Akademiker *m*
**unknown** (quantity) Unbekannte *f*; unbekannt ~ **quantity** unbekannte Größe *f*
**unlace, to** ~ losschnüren
**unlaid paper** Papier *n* ohne Wasserlinien
**unlaminated** unkaschiert
**unlash, to** ~ **a rope** ein Tau *n* abstecken
**unlatch, to** ~ ausklinken, entriegeln, entsperren
**unlawful** gesetzwidrig, ungesetzlich, unrechtmäßig, widerrechtlich
**unlay, to** ~ (a rope) aufdrehen **to** ~ **a rope** ein Tau *n* in Kardeele zerlegen
**unleaded** ungebleit, unverbleit ~ **fuel** unverbleites Benzin *n* ~ **gasoline** Reinbenzin *n*
**unleavened** ungesäuert
**unlevel surface** geneigte Fläche *f*
**unleveled** nichtsöhlig
**unlicensed transmitter** Schwarzsender *m*
**unlignified** unverbolzt
**unlike** abweichend, unähnlich, ungleich, ungleichnamig, verschieden ~ **charges** entgegengesetzte Ladungen *pl* ~ **pole** ungleichnamiger Pol *m*
**unlimited** grenzenlos, unbegrenzt, unbeschränkt, unendlich **of** ~ **stability** unbeschränkt haltbar
**unlimited,** ~ **ceiling** unbegrenzte Bewölkungshöhe *f* ~ **power attorney** Blankovollmacht *f*
**unload, to** ~ abladen, ablasten, ausladen, (a gun or camera) entladen, entlasten, (a ship) löschen, verladen **to** ~ **a kiln** eine Darre *f* abräumen
**unload time** Abspannzeit *f*
**unloaded** unbelastet ~ **circuit** unbelasteter Stromkreis *m* ~ **spring length** ungespannte Federlänge *f*
**unloader** Greifer *m* ~ **relief valve** Überdruckablaseventil *n*
**unloading** Abladen *n*, Abladung *f*, Abspannung *f*, Entladung *f*, Entlastung *f* ~**-and-loading-wharf** Entladestelle *f* ~ **apparatus** Löschapparat *m* ~ **area** Ausladegebiet *n* ~ **bridge** Entladebrücke *f* ~ **machinery** Entladeanlage *f* ~ **plant** Umlade--anlage *f*, -vorrichtung *f* ~ **plant for bulk goods** Ladeanlage *f* für Massengüter ~ **platform** Ausladerampe *f* ~ **point** Auslade-platz *m*, -stelle *f* ~ **rate** Entladeleistung *f* ~ **rod** Entladestock *m* ~ **table** Ausladeübersicht *f* ~ **unit** Ausladeeinheit *f* ~ **wharf** Landungsplatz *m*
**unlock, to** ~ aufschließen, entarretieren, entklammern, entriegeln, entsperren, entzurren, öffnen, eine Röhre *f* öffnen ~ **field** Entriegelungsfeldwicklung *f*
**unlockable** aufschließbar
**unlocked** unverriegelt
**unlocking,** ~ **device** Entriegelung *f* ~ **magnet** Entriegelungsmagnet *m* ~ **mechanism** Auslösevorrichtung *f* ~ **port** Entriegelungsanschluß *m*
**unloop, to** ~ entschlaufen
**unloosen, to** ~ lockern
**unmachinable** unbearbeitbar
**unmachined** unbearbeitet
**unmagnetized** unmagnetisch
**unmanageable** unlenksam
**unmanned** unbemannt

**unmarred** fehlerfrei
**unmatched** ungleich
**unmeltable piece of iron** Fuchs *m*
**unmelted** ungeschmolzen
**unmesh, to** ~ abknebeln, ausrücken
**unmethodical working of a mine** Raubbau *m*
**unmilled** ungewalkt
**unmitigated** nicht abgeschwächt, ungemildert
**unmixed** ungemischt
**unmodified** unmodifiziert ~ **scatter** nichtmodifizierte Streuung *f*
**unmodulated** unmoduliert ~ **density** Ruheschwärzung *f* ~ **groove** leere Rille *f* (phono) ~ **light spot** Raster *m* ~ **lighting** Ruhebelichtung *f*, Ruhelicht *n* ~ **power of the transmitter** Antennenkreisleistung *f* in unmoduliertem Zustand, Telefonie(träger)leistung *f* ~ **screen pattern** Fernsehraster *m* ~ **track** Ruhestreifen *m* ~ **transmission** Ruhetransparenz *f*
**unmonitored control system** rückführungslose Steuerung *f*
**unmounted** unmontiert
**unnail, to** ~ losnageln
**unnamed** unbenannt
**unnavigable** unschiffbar
**unnotched** ungekerbt ~ **(or plain) specimen** Vollstab *m*
**unobjectionable** einwandfrei
**unobstructed** licht ~ **entry** störungsfreier Eintritt *m* ~ **view** freier Ausblick *m*
**unobtainable** unerreichbar ~ **number** unausführbare Verbindung *f*
**unoccupied** leerstehend, untätig ~ **position** Ruhestellung *f*, unbesetzter Platz *m* ~ **time** Leerlaufzeit *f*
**unofficial** inoffiziell, unbestätigt
**unoiled cloth** ungeölter Stoff *m*
**unorthodox** ungewöhnlich
**unpack, to** ~ auspacken
**unpacked** unverpackt
**unpaid** unbeglichen, unbezahlt ~ **charge** rückständige Gebühr *f*
**unpaired electron** Einzelelektron *n*
**unpalatable** ungenießbar
**unparalleled** beispiellos, unvergleichlich
**unpaved** ungepflastert ~ **runway** unbefestigte Piste *f* oder Start- und Landebahn *f*
**unperforated** nicht durchbohrt, durchlöchert oder perforiert
**unpitch, to** ~ entpichen
**unplatinized** unplatiniert
**unplug, to** ~ ausstöpseln
**unplugged** ungestöpselt
**unpolished** matt, ungeschliffen
**unprecedented** beispiellos, noch nicht dagewesen
**unpredictable** nicht im voraus bestimmbar
**unprejudiced** unbefangen, unvoreingenommen, vorurteilsfrei
**unpressurizer** Außendruckbereich *m*
**unprime, to** ~ entschärfen
**unprinted paper of any color** Vakat *n*
**unproclaimed ground** freies Feld *n* (min.)
**unproductive** unproduktiv
**unprofitable** nutzlos, uneinträglich, unvorteilhaft
**unpropitious** unzeitgemäß
**unpropped** strebenlos
**unprotected** ungeschützt

**unpulled waste** (Kunstwolle) ungerissene Abfälle *pl*
**unpunctual** unpünktlich
**unpurified** ungereinigt
**unputrefiable** unverfaulbar
**unqualified** ungeeignet
**unquenched** ungelöscht
**unquiet** unruhig ~ **running** (engine) unruhiger Gang *m*
**unraised** ungerauht
**unravel, to** ~ abfasern, ausfasern, loswickeln
**unraveling machine** Zerfaserer *m*
**unreactive** reaktionsträge, unempfindlich
**unreadable** unleserlich
**unrecognizable** unkennbar, unkenntlich
**unrectified mosaic** Luftbildskizze *f*
**unreduced** unverküpt
**unreeve, to** ~ ausreffen
**unrefined** schmutzig ~ **antimony** Rohantimon *n* ~ **spirit** Rohspiritus *m*
**unregulated bleeding point** ungesteuerte Entnahme *f*
**unrelated** unbezogen ~ **color** unbezogene Farbe *f*
**unreliability** Unzuverlässigkeit *f*
**unresolved** nicht aufgelöst
**unrest** Unruhe *f*
**unrestrained** unbehindert ~ **beam** freitragender Balken *m*
**unrestricted** unbeschränkt ~ **documents** freie Sachen *pl* ~ **flow** freie Strömung *f* ~ **passage area** durchflußfreier Querschnitt *m*, freier Durchflußquerschnitt *m* ~ **rotation** ungehindertes Rundschwenken *n*
**unrib, to** ~ ausrippen
**unrifled** drallfrei, ungezogen
**unrigged** abgetakelt
**unrigging** Abnehmen *n*
**unripened** (viscose) jung
**unrivaled** unübertrefflich
**unrivet, to** ~ abnieten, entnieten, losnieten
**unroasted** ungeröstet
**unroll, to** ~ abrollen, abwickeln, aufrollen
**unrolled veneers** ausgerolltes Furnierholz *n*
**unroof, to** ~ **a house** ein Haus *n* abdecken (das Dach abnehmen)
**unsafe** unsicher ~ **warning light** Warn-lampe *f*, -leuchte *f*
**unsalable** schwer verkäuflich ~ **article** Ladenhüter *m*
**unsalted** ungesalzen
**unsaponifiable** unverseifbar
**unsaturated** ungesättigt ~ **linkage** Lückenbildung *f*
**unscheduled** außerplanmäßig (comput.)
**unscientific** unwissenschaftlich
**unscreened** unabgeschirmt, unsortiert
**unscrew, to** ~ abschrauben, aufschrauben, ausschrauben, auswinden, losdrehen, losschrauben
**unscrewable** abschraubbar
**unseal, to** ~ entplomben, entsiegeln, entsperren, öffnen
**unseat, to** ~ absatteln
**unseaworthy** nichtseefest
**unseparated flow** gesunde Strömung *f*
**unserviceability light** Sperrungsfeuer *n* (aviat.)
**unserviceable** betriebsunklar, unbenutzbar, unbrauchbar, unklar, untauglich

**unsettled** noch nicht zur Ruhe gekommen
**unseverable** untrennbar
**unshackle, to** ~ ausschäckeln
**unshaken** ungebeutelt
**unsharpness** Unschärfe *f*
**unshelled** ungeschält
**unshift, to** ~ zurückbewegen
**unshift** Buchstaben-umschaltung *f*, -wechsel *m*, Rückschub *m* ~-**on-space** selbsttätige Zwischenraumbildung *f*
**unshrinkable** (Stoff) nicht einlaufend
**unshunted** ohne Nebenschluß *m*
**unsifted plaster** grober Gips *m*
**unsightly** unansehnlich
**unsilvered** unversilbert
**unsigned** vorzeichenlos (math.)
**unsinkable** sinksicher
**unsintered** unverschlackt
**unsized** (paper) ungeleimt
**unskilled** ungelernt
**unslaked** ungelöscht ~ **lime** ungelöschter Kalk *m*
**unsmeltable** unverhüttbar
**unsolder, to** ~ ablöten, auflöten, loslöten
**unsoldered joint** Würgestelle *f*
**unsolicited** aufgefordert, unverlangt
**unsolvable** unlösbar
**unsolved** ungelöst, unaufgelöst
**unsound** brüchig, fehlerhaft, mit Fehlerstelle *f* behaftet, undicht, verdorben
**unspaced** undurchschossen
**unspecified** unbestimmt
**unspillable,** ~ **accumulator** Akkumulator *m* mit gelatiniertem Elektrolyt *m* ~ **cell** Trockenelement *n*
**unsplinterable glass** Sicherheitsglas *n*
**unsplit,** ~ **anode** Vollanode *f* ~ **ring** ungeteilter Ring *m*
**unstable** gleitend, haltlos, instabil, kipplig, labil, unbeständig, unstet, unstetig, veränderlich ~ **atmospheric conditions** labile Wetterlage *f* ~ **equilibrium** labiles oder unsicheres Gleichgewicht *n* ~ **governor** labiler Regler *m* ~ **multivibrator** astabiler Multivibrator *m* ~ **oil** zur Zersetzung *f* neigendes Öl *n*
**unstaffed position** unbesetzter Platz *m*
**unstall, to** ~ den Strömungsabriß *m* beenden
**unstayed** strebenlos
**unsteadiness** Haltlosigkeit *f*, Unruhe *f*, Veränderlichkeit *f*, Verwackeln *n* ~ **of image** mangelhaftes Stehen *n* (des Bildes) ~ **of picture** Bildtanzen *n*, Bildzittern *n*
**unsteady** haltlos, instabil, labil, nicht stationär, unbeständig, unruhig, unstabil, unstet, unstetig, veränderlich, wankelhaft, wack(e)lig ~ **arc** flackernder Lichtbogen *m*
**unstick, to** ~ abheben (aviat.), lösen, losmachen ~ **point** Abhebepunkt *m* ~ **speed** Abhebegeschwindigkeit *f* ~ **time** Startzeit *f*
**unstirred** ungeschürt
**unstop, to** ~ ausstöpseln
**unstranded conductor** Volleiter *m*
**unstratified** ungeschichtet ~ **rocks** Massengesteine *pl*
**unstressed** unbelastet ~ **conditions** spannungslos
**unstrippable** nicht abblätternd
**unsuccessful** erfolglos
**unsubsidized** nicht subventioniert

**unsuitable** ungeeignet, unpassend, untauglich, unzweckmäßig, verfehlt
**unsuited** ungeeignet
**unsupported** haltlos, strebenlos ~ **length** Freilänge *f* ~ **length over which buckling occurs** Knicklänge *f* ~ **piston** selbsttragender Kolben *m* ~ **span** Freilänge *f*
**unsurpassable** unübertrefflich
**unsurpassed** unübertroffen
**unsweat, to** ~ auf-, los-löten
**unsymmetric** schief ~-**resonance characteristic** schiefe Resonanzkurve *f*
**unsymmetrical** unsymmetrisch ~ **fold** schiefe Falte *f* ~ **propeller thrust** unsymmetrischer Schraubenzug *m*
**untanned** ungegerbt
**untapped** unabhörbar
**untarnished** blank (metall)
**untarred** ungeteert
**untearable** unzerreißbar
**untempered** unangelassen
**untenable** haltlos, unhaltbar
**untension, to** ~ eine Feder entspannen
**untensioning of a spring** Entspannen *n* einer Feder
**unthrifty** unwirtschaftlich
**untie, to** ~ abschnüren, aufbinden, aufknoten, losbinden, lösen, losknüpfen, loslösen ~ **to** ~ **the form** die Form *f* auflösen **to** ~ **the page cord** die Kolumnenschnur *f* abbinden (print.)
**untied** abgeschnürt
**untight** undicht
**untiled** abgedeckt
**untiltable** kippsicher
**untimely** ungünstig, unpassend, unzeitgemäß, unzeitlich
**untouched** unangetastet, unverritzt
**untreated** unbehandelt, unzubereitet ~ **pole** rohe oder unzubereitete Stange *f* ~ **rubber treads** Rohlaufstreifen *m* ~ **water** Rohwasser *n* ~-**water evaporator** Rohwasserverdampfer *m* ~ **wooden pole** rohe oder unzubereitete Holzstange *f*
**untrue** ungenau, unrund, unwahr
**untuned** nichtabgestimmt, nicht abgestimmt, unabgestimmt, verstimmt ~ **antenna** aperiodische Antenne *f*
**untuning** Verstimmen *n*
**untwine, to** ~ sich aufdrehen
**untwist, to** ~ lösen, sich aufdrehen
**untwist** Drall *m*
**untwisting** Lösung *f*, Rückdrehung *f* ~ **machine** Abwickelmaschine *f*
**untying** Abbinden *n*
**unusable** unverwendbar ~ **fuel supply** tote Treibstoffmenge *f*
**unused** unbenutzt, unverbraucht ~ **side** Blankseite *f* ~ **terminal** Blindklemme *f* ~ **turns of a coil** tote Windungen *pl*
**unusual** außergewöhnlich, ungewöhnlich
**unvariant** nonvariant ~ **equilibrium** nonvariantes Gleichgewicht *n*
**unvarnished** unlackiert
**unvarying** fest
**unveil, to** ~ entschleiern
**unventilated** ungelüftet
**unvitrified** unverglast
**unvulcanized rubber** Rohgummi *m*

**unwanted emission** nicht gewollte Aussendung *f*
**unwarranted** ohne Gewähr *f*, ungerechtfertigt, unverantwortlich
**unwashed** ungewaschen ∼ **slack** ungewaschene Feinkohle *f*
**unwatering** Auspumpen *n*, Trockenlegung *f*
**unweathered** unverwittert
**unweighability** Unwägbarkeit *f*
**unweighable** unwägbar
**unweldable** unschweißbar
**unwieldy** unförmig, unhandlich
**unwilling** agbeneigt, unwillig
**unwind, to** ∼ abwickeln, aufwickeln, loswinden, zasern
**unwinder** Auseinanderrollmaschine *f*
**unwinding** Abspulen *n* (film) ∼ (**or unwrapping**) **mechanism** Abtafel- oder Aufwickelvorrichtung *f* ∼ **shaft** Abrollwelle *f* ∼ **spool** Ablaufspuele *f* (film)
**unwired** ungeschaltet
**unworkable** unbauwürdig, unverhüttbar
**unworked,** ∼ **country** unausgerichtetes Feld *n* (min.) ∼ **piece (or part)** Rohling *m*
**unwound** abgewickelt ∼ **ribbon** abgelaufenes Farbband *n*
**unwrap, to** ∼ aufwickeln, von der Umhüllung *f* befreien
**unwrapping mechanism** Abtafel- oder Aufwickelvorrichtung *f*
**unwrinkle, to** ∼ entrunzeln
**unwrought** unbearbeitet
**unyielding** unbeugsam, unergiebig, unnachgiebig
**unyoke, to** ∼ abjochen
**up** auf, aufwärts, empor, in die Höhe *f* ∼ **an incline** schräg nach oben
**up-and-down** auf und nieder, auf und ab, von oben nach unten gehend ∼ **line** Hin- und Rückleitung *f* ∼ **motion** Auf- und Abwärtsbewegung *f* ∼ **movement** Auf- und Abbewegung *f* ∼ **stroke** Doppelhub *m* ∼ **working** wechselseitiger Betrieb *m*
-**upcast** das Aufgeworfene *n* (min.), Überschiebung *f* (geol.) ∼ **shaft** Wetterschacht *m* ∼ **ventilating shaft** ausziehender Schacht *m* ∼ **ventilation** Aufwärtslüftung *f*, Ausziehstrom *m*
**up-coiler** Rundbiegemaschine *f*
**upcurrent** Aufstrom *m*, Aufwind *m*, Hangaufwind *m* ∼ **of air** augsteigende Strömung *f* ∼ **due to hot air** Thermik *f* ∼ **due to a slope** Hangwind *m*
**update, to** ∼ ergänzen, auf den neuesten Stand *m* bringen
**updraft,** ∼ **carburetor** Aufstromvergaser *m*, Steigstrom-vergaser *m*, -vorrichtung *f*, Vergaser *m* mit aufsteigendem Luftzug ∼ **fire** Oberfeuer *n* ∼-**type furnace** Ofen *m* mit steigender Flamme *f*
**upend, to** ∼ hochkant stellen
**upfold** Sattelbogen *m*
**upgear, to** ∼ Fahrgestell *n* einziehen
**upgrade, to** ∼ aufbereiten
**upgrade** Steigung *f*
**upgrading of coal** Aufbereitung *f* der Kohle
**upheaval** Bodenerhebung *f*
**uphill** ansteigend, bergan, bergauf
**upholster, to** ∼ ausfüttern, polstern

**upholstered** aufgepolstert, bezogen
**upholsterer** Tapezierer *m*
**upholstery** (Möbel) Bezugsstoff *m*, Polsterung *f* ∼ **brush** Polsterbürste *f* ∼-**nail** die Polsternagelstempel *m*
**upkeep** Aufrechterhaltung *f*, Instandhaltung *f*, Unterhalt *m*, Unterhaltung *f*, Wartung *f* ∼ **schedule** Wartungsplan *m*
**upland water** Oberwasser *n*
**uplift** Auftrieb *m*, gehobener Flügel *m* einer Verwerfung (aviat.), Heraushebung *f*, Horst *m*, Unterdruck *m* ∼ **of strata** Aufrichtung *f* der Schichten ∼ **pressure** Unterdruck *m*
**uplock** Einfahrverriegelung *f*
**up-milling process** Gegenlauffräsen *n*
**upon** auf ∼ **request** auf Anfrage *f*
**upper** ober (-er, -e, -es) ∼ **and lower sides of the lode** obere und untere Wandschichten *pl* des Erzganges
**upper,** ∼ **air** Höhenluft ∼-**air chart** Höhenkarte *f* (meteor.) ∼ **band** Oberring *m* ∼-**band spring** Oberringfeder *f* ∼ **beam control lamp** Fernlichtanzeigerleuchte *f* ∼ **bell** Oberglocke *f* ∼ **bend** oberes Knie *n* ∼ **bend of characteristic** oberer Kennlinienknick *m* ∼ **bend of valve characteristic** oberer Knick *m* der Röhrenkennlinie ∼**berth** Oberbett *n* ∼ **blade** Obermesser *n* ∼ **block** feste (am Ort bleibende) Rolle *f*, Oberflasche *f* ∼ **boiler** Oberkessel *m* ∼ **boom** Obergurt *m* ∼ **bosh line** Kohlensack *m* ∼ **brace of fixed frame (or gate)** oberes Querstück *n* des Schützes *m* ∼ **bridge** obere Bedienungsbrücke *f* ∼ **cap** Zünderkappe *f* ∼ **carriage** Oberlafette *f* ∼ **carrying chassis of a crane** Kranoberwagen *m*, Oberwagen *m* ∼ **case** Ziffernfeld *n* ∼ **case for capitals** Kapitalkasten *m* ∼ **channel** Obergerinne *n* ∼ **chord** Druck-, Ober-gurt *m* ∼ **cleaving grain** Oberlager *n* ∼ **coal** Dachkohle *f* ∼ **component of a composite aircraft** Oberflugzeug *n* ∼ **connection contact** Kopfschalterkontakt *m* ∼ **cooling water tank** Kühlwasserhochbehälter *m* ∼ **counterweight** Obergewicht *n* ∼ **course** Oberlauf *m* ∼ **cretaceous sandstone** Quadersandstein *m* ∼ **current** Oberstrom *m* ∼ **cutoff frequency** obere Grenzfrequenz *f* ∼ **cylinder lubricant** (oil) Obenschmieröl *n* ∼ **cylinder lubrication** Kopf-, Oben-schmierung *f* ∼ **deck** Oberdeck *n*, oberes Deck *n*, Peildeck *n* ∼ **deviation** oberes Abmaß *n* ∼ **die** Gesenkoberteil *m*, Patrize *f*, Ober-gesenk *n*, -stempel *m* ∼ **drum** Obertrommel *f*
**upper-edge,** ∼ **of plate** Plattenoberkante *f* ∼ **of probing hole** Sondierloch *n* (obere Kante) ∼ **of separator** Separatorenoberkante *f*
**upper-end,** ∼ **of electrode** Elektrodenkopf *m* ∼ **of the heelpost of a lock gate** Hals *m* der Wendesäule eines Schleusentores ∼ **of lens cone** bildseitige Grundfläche *f*, bildseitiges Stutzenende *n*
**upper,** ∼ **feed sprocket** Vorwickler *m* ∼ **feeder** Oberkanal *m* ∼ **flange** Obergurt *m* ∼ **frog of the hinge** Oberpfanne *f* der Türangel ∼ **gate of double floodgate** Oberfalle *f* ∼ **gate (of the) lock** Oberhaupt *n* ∼ **grade** Oberstufe *f* ∼ **half of pillow block** Stehlageroberschale *f* ∼ **half nut** Mutterschloßoberteil *n* the ∼ **hand** Oberhand *f* ∼**harmonics** Oberharmonische *f* ∼ **heating value** oberer Heizwert *m* ∼ **knife** Obermesser *n*

~ **layer** Deckenschicht *f* ~ **level** höhere Teufe *f* ~ **limit** Obergrenze *f* ~ **limiting filter** Spulenkette *f* ~ **moraine** Obermoräne *f* ~ **new red sandstone** Röt *n* ~ **nozzle** obere Düse *f* ~ **olite** Malm *m* ~ **part** Haubenkörper *m*, Oberteil *n* ~ **part of camshaft casing** Steuerwellengehäuseoberteil *n* ~ **part of gate** Oberflügel *m* ~ **part of water basin** oberes Gebiet *n* ~ **pintle castings** Stützwinkel *m* ~ **plate** Oberplatte *f* ~ **platform** obere Bühne *f* ~ **pool elevation** gestauter Wasserspiegel *m*, Oberwasserspiegel *m*, Wasserstand *m* der oberen Haltung ~ **port** Oberpforte *f* ~ **port sill** Obertrempel *m* ~ **portion of casing drainage well** Aufsatzrohr *n* ~ **prism micrometer head** Triebschraube *f* ~ **rib flange** Rippenobergurt *m* ~ **river** Oberlauf *m* ~ **round head** Leitwand *f* ~ **shell of blast furnace** Ofenkegel *m* ~ **shield** Oberschild *n* ~ **side** Oberseite *f* ~ **side band** oberes Seitenband *n* ~ **side-band position** direkte Lage *f* ~ **slide rest** Obersupport *m* ~ **spherical flux** oberer hemisphärischer Lichtstrom *m* ~ **story** Obergeschoß *n* ~ **stratum** obere Schicht *f* ~ **structure of tank** Panzerkastenoberteil *n* ~ **support for frames** Aufhängebrücke *f* ~ **surface** hängende Schicht *f*, Profilrücken *m* ~ **surface of wing** Flügeloberseite *f*, Saugseite *f* des Flügels ~ **tapes** Oberbänder *pl* ~ **threshold of audibility** obere Hörschwelle *f* ~ **threshold of hearing** obere Hörschwelle *f* ~ **transom** Oberriegel *m* ~ **tuyere** obere Düse *f* ~ **valve** Hubventil *n*, oberes Ventil *n* ~ **water level** Oberwasserspiegel *m* ~ **wind** Höhenwind *m* ~ **wind forecast** Höhenwindvorhersage *f* ~ **wind report** Höhenwindmeldung *f* ~ **wing** Oberflügel *m* ~**-wing span** Spannweite *f* des Oberflügels ~ **wing spar** Oberflügelholm *m*
**up-position,** ~ **lock linkage rod** Sperrhakengestänge *n* ~ **lock switch** Sperrhakenschalter *m*
**upright** Bohrständer *m*, Maschinenständer *m*, Pfeil *m*, Pfeiler *m*, Pfosten *m*, Säule *f*, Ständer *m*, Stütze *f*; aufrecht, hochkant, scheitelrecht, seiger ~ **of frame (or gate)** Vertikale *f* des Rahmens ~ **on upstream side of frame** obere Vertikale *f* des Wehrbocks
**upright,** ~ **conical vault** Trichtergewölbe *n* ~ **converter** stehender Konverter *m* ~ **course** (bricks set on end) Rollschicht *f* ~ **cross** Stehkreuz *n* ~ **cylinder** stehender Zylinder *m* ~ **drill** Rennspindel *f*, Vertikalbohrmaschine *f* ~ **drill press** Senkrechtbohrmaschine *f* ~ **drill stand** Bohrständer *m* ~ **drilling machine** Senkrechtbohrmaschine *f* ~ **fold** stehende Falte *f* ~ **format** Hochformat *n* ~ **lattice** Steiglattentuch *n* ~ **piano** Konzertpiano *n* ~ **picture** Hochaufnahme *f* ~ **position** Hochaufstellung *f* ~ **radiator** Standkühler *m* ~ **screw cutting machine** Ständergewindeschneidmaschine *f* ~ **shell** Wölbung *f* ~ **shot** Hochaufnahme *f* ~ **size** Hochformat *n* (print.) ~**sprinkler** stehender Sprinkler *m* ~ **trussing** Ständerfachwerk *n* ~ **tube of a boiler tube** Siederhals *m*
**uprising** Aufbruch *m*
**up-river sluice chamber of a lock** Flutkammer *f*
**uproot, to** ~ entwurzeln
**uprooting machine** Rodemaschine *f*
**upset, to** ~ anstauchen, aus der Fassung *f* bringen, über den Haufen *m* werfen, kippen, stauchen, umkippen, umschlagen (aviat.),

umstürzen, umwerfen, zusammenstauchen
**upset** Aufhieb *m*, Störung *f*, Um-fallen *n*, -kippen *n*, -stoßen *n*, -werfen *n* ~ **end joint** feste Enddichtung *f* ~ **pass** Stauchkaliber *n* ~ **punch** Börderstempel *m* ~ **welding** Stauchschweißung *f*, Stumpf-schweißen *n*, -schweißung *f*
**upsetting** (end of bar) Anstauchung *f*, Stauchen *n*, Stauchung *f* ~ **bulge** Stauchwulst *f* ~ **clouds** Stauchwolken *pl* ~ **device** Stauchvorrichtung *f*, Vorstauchapparat *m* ~ **die** Preßbacke *f*, Stauchmatrize *f* ~ **effect** Stauchwirkung *f* ~ **factor** Stauchfaktor *m* ~ **machine** Stauchmaschine *f* ~ **press** Röhrenstauchpresse *f*, Stauchstanze *f* ~ **pressure** Stauchdruck *m* ~ **property** Stauchbarkeit *f* ~ **slide** (in welding) Stauchschlitten *m* ~ **temperature** Erweichungstemperatur *f* ~ **test** Stauch-probe *f*, -versuch *m* ~ **tools** Stauchwerkzeuge *pl*
**upshot wood** hochgeschossenes Holz *n*
**upside-down** drunter und drüber, mit der Unterseite *f* nach oben ~ **position** Rückenlage *f*
**upslope wind** Aufwind *m*
**upstairs** oben, im oberen oder im höheren Stockwerk *n*
**upstanders of a windlass** Haspelstützen *pl*
**upstream** (of dam) Bergseite *f*, Oberstrom *m*, Oberwasserseite *f*; flußaufwärts, oberhalb, stromaus, stromaufwärts, wasserseitig ~ **apron** Sohle *f* der oberen Haltung, Vorboden *m* ~ **cofferdam** oberer Fangdamm *m* ~ **discharge** Oberwasser *n* ~ **effect** Stromaufwirkung *f* ~ **face** Oberseite *f* des Wehrkörpers, Wasserseite *f* ~ **floor** Vorboden *m* ~ **limit of tidewater** Flutgrenze *f* ~ **pressure** Eingangsdruck *m* ~ **stepped face** (of dam) abgesteppte Bergseite *f* des Wehrkörpers ~ **toe wall** Herdmauer *f* ~ **valve** Oberklappe *f* ~ **water** Oberhaltung *f*
**upstroke** aufgehender Hub *m*, (of piston) Aufgang *m*, (of piston) Aufstieg *m*, (of an impulse) Aufstrich *m*, Aufwärts-bewegung *f*, -hub *m*, Haarstrich *m* ~ **of the knife** Messerhochgang *m* ~ **twisting** Abzwirnen *n* (über den Kopf)
**uptake** (gas) Aufnahme *f*, (of boilers) Fuchs *m*, Fuchskanal *m*, Hochkanal *m*, Steig-kanal *m*, -leitung *f*, -rohr *n* ~ **factor** Aufnahmefaktor *m* ~ **rate** Aufnahmegeschwindigkeit *f*
**upthrow** Springhöhe *f* ~ **side** hängende Scholle *f* (min.)
**upthrust** Auftrieb *m*, Landschelle *f*
**up-to-date** auf dem Laufenden, modern, neuzeitlich
**up-touching** Nachhilfe *f*
**upturned** umgestülpt ~ **ears** hochgebogener Flansch *m*
**upward** aufwärts, empor, firstenweise ~**-and--downward-motion** Heb- und Senkbewegung *f* ~ **bore-hole** Hochbohrloch *n* ~ **component** Aufkomponente *f* ~ **current of air** aufsteigende Strömung *f* ~ **displacement** bergseitige Verschiebung *f* ~ **draft** Saugwirkung *f* ~ **flow** Aufwärtsströmung *f* ~ **flux** oberer hemisphärischer Lichtstrom *m* ~**-folding wing** Hochklappflügel *m* ~ **force** Auftriebs-, Steig-kraft *f* ~ **modulation** additive Modulation *f* ~ **motion** Aufwärts-, Hausse-bewegung *f* ~ **movement of tension spring** Hochschnappen *n* der Spannfeder ~**-pointing hole** Hoch-

bohrloch *n* ~ **pressure on foundations** Auftrieb *m*, Unterdruck *m* ~ **pull** Saugwirkung *f*, Saugzug *m* ~-**sloping** schrägansteigend ~ **spin** Aufwärtstrudeln *n* ~ **stroke** Aufwärts-gang *m*, -**hub** *m* (Kolben) ~ **tendency** Aufwärtsbewegung *f*, Hausse *f*, steigende Tendenz *f* ~ **view** Sicht *f* nach oben ~ **welding** Schweißung *f* mit Aufwärtsführung *f*

**upwarping** (of strata) Aufwolfung *f*

**upwash** Aufströmung *f*, Aufwind *m*, Frontaufwind *m* ~ **due to heat rising in the evening** Abendthermik *f*

**upwelling** Auftriebwasser *n*

**upwind** Aufwind *m*, Gegenwind *m*, gegen den Wind *m*, windwärts ~ **region** Aufwindzone *f*

**uraconite** Uranocker *m*

**U-rail** Brücken-, Hohl-schiene *f*

**urania green** Uraniagrün *n*

**uranic,** ~ **acid** Uranoxydrot *n*, Uransäure *f* ~ **compound** Uraniverbindung *f* ~ **nitrate** Uraninitrat *n*

**uranides** Transurane *pl*, Uranide *pl*

**uraniferous** uranhaltig

**uraninite** Uraninit *m*

**uranite** Autunit *m*, Pechblende *f*, Uranit *n*

**uranium** Uran ~ **acetate** essigsaures Uranoxyd *n*, Uranazetat *n* ~ **bar** Uranstab *m* ~ **carbide** Urankarbid *n* ~ **compound** Uranverbindung *f* ~ **content** Urangehalt *m* ~ **dioxide resistance** Eisenurdoxwiderstand *m* ~-**graphite lattice** Urangraphitgitter *n* ~ **hydroxide** Uranoxydrat *n* ~ **lead** Uranblei *n* ~ **nitrate** salpetersaures Uranoxyd *n* ~ **nucleus** Urankern *m* ~ **ocher** Uranocker *m* ~ **oxide** Uranoxyd *n* ~ **oxide red** Uranoxydrot *n* ~ **oxychloride** Uranoxychlorid *n* ~ **phosphate** Uranphosphat *n* ~ **pile** Uranreaktor *m* ~ **rays** Uranstrahlen *pl* ~ **red** Uranrot *n* ~ **series** Uranreihe *f* ~ **slug** Uranblock *m* ~ **splitting** (of fission) Uranspaltung *f* ~ **sulfate** Uransulfat *n* ~ **supply** Uranvorkommen *n* ~**toning bath** Urantonbad *n*

**uranmolybdate** Uranmolabdat *n*

**uranochalcite** Urancalzit *m*, Urangrün *n*

**uranocircite** Uranzirkit *m*

**uranolepidite** Vandenbrandeit *m*

**uranophane** Uranophan *n*

**uranopilite** Uranpilit *m*

**uranosouranic** uranïanig

**uranoso-uranic oxide** Uranoxydoxydul *n*

**uranospathite** Uranspathit *m*

**uranospherite** Uranspherit *m*

**uranospinite** Uranspinit *m*

**uranothallite** Uranthallit *m*

**uranothorite** Enolith *m*, Uranothorit *m*

**uranotil** Uranotil *m*

**uranous** uranig ~ **compound** Uranoverbindung *f* ~ **oxide** Uranoxydul *n* ~ **salt** Uranosalz *n*, Uranoxydsalz *n* ~ **uranate** Uranouranat *n*

**uranyl,** ~ **ammonium** Uranoxydchlorid *n* ~ **nitrate** Uranylnitrat *n* ~ **phosphate** Uranylphosphat *n*

**urate** Urat *n* (harnsaures Salz)

**urban** städtisch ~ **area** Stadtgebiet *n* ~ **fog** Stadtnebel *m* ~ **planner** Stadtplaner *m* ~ **rapid transit system** Stadtbahn *f*

**Urdox resistor** Urdox-regler *m*, -widerstand *m*

**urea** Harnstoff *m* ~ **formaldehyde** Harnstoffformaldehyd *n*

**ureometer** Ureometer

**urethrography** Urethrografie *f*

**urgency** Notwendigkeit *f* ~ **message** Dringlichkeitsmeldung *f* ~ **signal** Dringlichkeitszeichen *n*

**urgent** dringend, dringlich ~ **call** Ausnahmegespräch *n*, dringender Anruf *m*, dringendes Gespräch *n* ~ **message** dringendes Telegramm *n*, Dringlichkeitsmeldung *f* ~ **telephone call** dringendes Ferngespräch *n*

**uric acid** Harnsäure *f*

**urn model** Urnenmodell *n*

**usability** Brauchbarkeit *f*, Eignung *f*, Verwendbarkeit *f*, Verwendungsfähigkeit *f*

**usable** anwendbar, brauchbar, gebrauchsfähig, nutzbar, tauglich, verwendbar ~ **copy** Gebrauchsvorlage *f* ~ **length** Nutzlänge *f* ~ **minimum width** Mindestnutzbreite *f* ~ **range** nützliche Strecke *f* ~ **size of the photograph** nutzbare Bildgröße *f*

**usage** Brauch, Gebrauch, Gewohnheit *f*, Verbrauch *m*, Verwendung *f*

**use, to** ~ anwenden, benutzen, brauchen, einsetzen, gebrauchen, verwenden **to** ~ **up** abnutzen, aufarbeiten, aufbrauchen, erschöpfen, verarbeiten, verbrauchen, verfeuern **to** ~ **force** etwas gewaltsam tun ~ **no hooks!** ohne Haken *m* handhaben! **to** ~ **levers** hebeln **to** ~ **violence** etwas gewaltsam tun

**use** Anwendung *f*, Anwendungsmöglichkeit *f*, Bedürfnis *n*, Gebräuchlichkeit *f*, Handhabung *f*, Nutzen *m*, Verbrauch *m*, Verwendung *f*, Verwendungszweck *m*, Verwertung *f* **to be of** ~ dienen, nützen **in** ~ üblich **of** ~ brauchbar, nutzbar ~ **of arms** Waffengebrauch *m* ~ **of phantom circuits** Phantomausnutzung *f*, Viererausnutzung *f* ~ **of two groups of lines on the same route** gerichteter Verkehr *m* ~ **of X-rays** Röntgenstrahlbenutzung *f*

**U-section** Querschnitt *m* mit U-Form, U-Profil *n*, U-Querschnitt *m*

**used** gebraucht **not** ~ ungebraucht ~ **up** abgenutzt, verbraucht

**used,** ~ **air** Abluft *f* ~ **car** Gebrauchtwagen *m* ~ **oil** gebrauchtes Öl *n* ~ **paper** Altpapier *n* ~ **rope** halbgeschlossenes Tau *n* ~ **rubber** Altgummi *n* ~ **sand** Altsand *m* ~ **throughout** durchgängig gebraucht

**useful** brauchbar, dienlich, nutzbar, nützlich, sachdienlich, tauglich, zweckdienlich, zweckentsprechend **to be** ~ taugen

**useful,** ~ **aperture of a lens** nutzbarer Linsendurchmesser *m* ~ **area** Nutzfläche *f* ~ **beam** Nutzstrahl *m* ~ **capacity** Nutzkapazität *f* ~ **circuit** Nutzkreis *m* ~ **conductance** Nutzleitwert *m* ~ **contents** Nutzinhalt *m* ~ **cross section** Nutzquerschnitt *m* ~ **current** Nutzstrom *m* ~ **diameter** lichte Weite *f*, Nutzdurchmesser *m* ~ **effect** Nutz-effekt *m*, -leistung *f*, -wirkung *f* ~ **effiency** Nutzleistung *f* ~ **effort** Nutzeffekt *m*, (of a machine) Nutzwirkung *f* ~ **engine thrust** Motornutzschub *m* ~ **field** Nutzfeld *n* (rdo) ~ **(or effective) field of vision** nutzbares Sehfeld *n* ~ **length** Nutzlänge *f* (der Piste) ~ **life** Lebens-, Nutzungs-dauer *f*, Standzeit *f* ~ **lift** Luftkraft *f*, Nutzauftrieb *m* ~ **load** Gesamtlast *f*, Nutzlast *f*, Nutzgewicht *n*, Zuladung *f* ~ **magnification** Abbildungsmaßstab *m* ~ **mineral deposits** nutzbare Lagerstätte *f* ~ **output** Nutzleistung *f* ~- **performance load** Nutzleistungsbelastung *f* ~ **power** Nutzleistung *f* ~ **reactance** Nutzblindwiderstand *m* ~ **resistance** Nutzdämpfung *f*, Nutzwiderstand *m* ~ **voltage** Nutzspannung *f* ~ **work** effektive oder nutzbare Leistung *f*, Nutz-arbeit *f*, -leistung *f*

**usefulness** Brauchbarkeit *f*, Nutzen *m*, Nützlichkeit *f*, Tauglichkeit *f*, Verwendbarkeit *f*, Verwendungsfähigkeit *f*

**useless** fruchtlos, nutzlos, unbrauchbar, unnütz, untauglich, zwecklos **making** ~ Unbrauchbarmachen *n* ~ **multiplication of measurements** Häufung *f* der Einzelmessungen

**uselessness** Nutzlosigkeit *f*, Unbrauchbarkeit *f*

**user** Teilnehmer *m* (teleph.), Verbraucher *m*, Verwender *m*

**U-shaped** U-eisenförmig, U-förmig ~ **magnetsystem carrier** U-förmiger Systemträger *m* ~ **plate** Kanalblech *n* ~ **valley** Trogtal *n*

**using,** ~ **pick-off gears** Aufstecken *n* der Wechselräder *pl* ~ **up** Aufarbeiten *n*, Verarbeitung *f* ~ **up of filaments** Abtragen *n*

**U.S.standard** amerikanische Norm *f*

**usual** gebräuchlich, häufig, üblich ~ **length** gewöhnliche Länge *f* ~ **practical units** gebräuchliche praktische Einheiten *pl* ~ **standard** gewöhnlicher Gebrauch *m* ~ **terms** übliche Bedingungen *pl*

**usufructuary** Nutznießer *m*

**usurpation of patent rights** Patentanmaßung *f*

**utensil,** ~ **plug** Gerätestecker *m* ~ **socket** Gerätesteckdose *f*

**utensils** Gerät *n*, Gerätschaft *f*, Geschirr *n*

**utilisable** verwertbar

**utilitarian,** ~ **principle** Nützlichkeitsgrundsatz *m* ~ **union** Zweckverband *m*

**utilities** Betriebskraft *f*, Energiebetriebe *pl*, Versorgungsleitungen *pl* ~ **department** Betriebskraftabteilung *f* ~ **flow diagram** Fließschemata *n* für Wasser-, Dampf-, Luft- usw. -versorgung *f*

**utility** Brauchbarkeit *f*, Nutzen *m*, Nützlichkeit *f*, Verwendbarkeit *f* ~ **car** Gebrauchswagen *m* **characteristics** Betriebsmittelcharakteristika *pl* ~ **crockery** Gebrauchsgeschirr *n* ~ **design** Gebrauchsmuster *n* ~ **form** Zweckform *f* ~ **glassware** Glasgebrauchsgegenstand *m* ~ **model** Musterschutz *m* ~-**model patent** Gebrauchsmuster *n* ~ **operating method** ständige Bereitschaft *f* ~ **piping** Betriebsmittelrohrleitungen *pl* ~ **program** Dienstprogramm *n* **spark(ing) performance** Verrußungszündleistung *f* ~ **spot light** Handleuchte *f*

**utilizable** nutzbar, verwertbar

**utilization** Anwendung *f*, Ausnutzung *f*, Benutzung *f*, Gebrauch *m*, Nutzung *f*, Verwendbarkeit *f*, Verwendung *f*, Verwertung *f* ~ **of fuel** Brennstoffausnutzung *f* ~ **of heat** Wärmenutzung *f* ~ **of heat of exhaust gases** Abgaswärmeverwertung *f* ~ **of light** Lichtausnutzung *f* ~ **of opportunities** Ausnutzung *f* günstiger Gelegenheiten ~ **of the slipstream** Strahlausnutzung *f* ~ **of space** Platzausnutzung *f* ~ **of terrain** Geländeausnutzung *f* ~ **of waste** Abfallverwertung *f* ~ **of waste heat** Abdampfverwertung *f* ~ **of the wind** Windausnutzung *f*

**utilization,** ~ **area** Abnutzungsfläche *f* ~ **circuit** Nutzstromkreis *m* ~ **coefficient** Ausnutzungskoeffizient *m* ~ **factor** Ausnutzungsfaktor *m* ~ **period** Verwertungsperiode *f*

**utilize, to** ~ anwenden, ausnutzen, auswerten, benutzen, gebrauchen, nutzbar machen, verwenden, verwerten

**utilized power** Nutzleistung *f*

**utilizing plant** Verwertungsanlage *f*

**U-troughing** Leitungskanal *m*

**utter, to** ~ **low dull sound** dröhnen

**U tube** U-Rohr *n* ~ **sedimentator** Zweischenkelflockungsmesser *m*

**U turn** Wendekurve *f*

**U-type bridge** Trogbrücke *f*

**uvanite** Uvanit *m*

**uviol glass** Uviolflintprisma *n*

**U washer** Vorsteckscheibe *f*

**U.X.A. (unit automatic exchange)** automatische Landzentrale *f*

# V

vacancy (liberation of smoke, etc.) Freiwerden *n*, Leere *f*, (in lattice) Leerstelle *f*, offene Stelle ~ **formation** Leerstellenbildung *f* ~ **mechanism** Leerstellenmechanismus *m* ~ **migration** Leerstellenwanderung *f* ~ **pairs** Leerstellenpaare *pl* ~ **principle** Lückensatz *m*
vacant frei, leer, unbesetzt ~ **anion (or cation)** Leerstelle *f* ~ **contact** Leerlaufkontakt *m* ~ **level** unbesetzter oder freier Höhenschritt *m* ~ **position** Gitterlücke *f*, Leerstelle *f* ~ **post (or situation)** offene Stelle *f* ~ **space (or spot)** Lücke *f* ~ **terminal** freier Kontaktstift *m*, Leer-, Leerlauf-kontakt *m*
vacate, to ~ freimachen (a level)
vaccinate, to ~ einimpfen, impfen
vaccination certificate Impfschein *m*
vaccine glass Lymphglas *n*
vacillation about zero Nulleinspielung *f*
vacuole Vakuole *f*
vacuometer Saug-, Vakuum-messer *m*, Vakuummeter *m*
vacuous luftleer ~ **space** (luft)leerer Raum *m*
vacuscope Vakuskop *n*
vacuum Leere *f*, leerer Raum *m*, Luftleere *f*, luftleerer Raum *m*, Luftverdünnung *f*, Saugluft *f*, Unterdruck *m*, Vakuum *n* ~ **accumulator** Unterdruckspeicher *m* ~ **advance and retard** Unterdruckverstellung *f* ~ **arrester** Luftleerspannungssicherung *f* ~ **bell jar** Vakuumglocke *f* ~ **belt dryer** Vakuumbandtrockner *n* ~ **blower** Absauggebläse *n* ~ **bottle** Isolierflasche *f* ~ **box** Unterdruckdose *f* ~ **brake** Saugluftbremse *f*, Unterdruckbremse *f* ~ **breaker** Rückschlagventil *n* ~ **chamber** Vakuumkammer *f* ~ **cleaner** Staubsauger *m*, Vakuumreiniger *m* ~-**cleaner tube** Staubsaugerschlauch *m* ~ **concrete** entwässerter Beton *m* ~ **connection** (Rohrpost) Sauganschluß *m* ~ **control curve** Unterdruckverstellinie *f* ~ **controller** Unterdruckversteller *m* ~**conveyer tube** Saugluftförderer *m* ~-**crystal** Vakuumquarz *m* ~ **cup** Saugnapf *m* ~ **degassing** Vakuumentgasung *f* ~ **deposition** Vakuumbedämpfung *f* ~ **diffusion pump** Diffusionspumpe *f* ~ **distillation** Vakuumdestillation *f* ~ **draffdrying apparatus** Vakuumtrebertrockenapparat *m* ~ **drier** Vakuumtrockner *m* ~ **drum drier** Vakuumwalzentrockner *m* ~ **drying cabinet** Vakuumtrockenschrank *m* ~ **drying oven** Vakuumtrockenofen *m* ~ **dust cleaner** Staubreiniger *m* ~ **eccentric tumbling drier** Vakuumtaumeltrockner *m* ~ **electron** Vakuumelektron *n* ~ **evaporator** Vakuum-verdampfer *m*, -verdampfungsapparat *m* ~ **feed** Vakuumförderung *f* ~ **fermentation** Unterdruckgärung *f* ~ **filter** Saug-, Vakuum-filter *m* ~ **flask** Isolierflasche *f* ~ **forming** Vakuumverformung *f* ~ **freeze drier** Vakuumtiefkühltrockner *m* ~ **fuel-feed device** Unterdruckbrennstofffförderer *m* ~ **fuel pump** Unterdruckförderer *m* ~ **gauge** Manometer *n*, Unterdruckmesser *m*, Vakuum-messer *m*, -meter *m* ~ **gauge for negative pressure** Unterdruckmeßgerät *n* ~ ge-

nerator Vakuumerzeuger *m* ~ **governor** Unterdruck-regler *m*, -versteller *m* ~ **grating spectrograph** Vakuumgitterspektrograf *m* ~ **grease** Hahn-, Vakuum-fett *n* ~ **interlock** Vakuumschleuse *f* ~ **jacket** Vakuummantel *m* ~ **jet** Dampfstrahlsauger *m* ~ **lamp** Vakuumlampe *f* ~ **leak detector** Leckfinder *m* ~ **lightning arrester (or protector)** Luftleerblitzableiter *m* ~ **lightning protector** Edelgasblitzableiter *m* ~-**line connection** Soganschluß *m* ~ **measurement** Vakuummessung *f* ~ **measuring instruments** Vakuummeßinstrumente *pl* ~ **paddle drier** Vakuumschaufeltrockner *m* ~ **paddle mixer** Vakuumtrockner *m* mit Rührwerk *m*
vacuum-pan Vakuumpfanne *f* ~ **for refined sugar** Raffinadevakuumapparat *m* ~ **supply tank** Einziehkasten *m* ~ **supply tank for runoff** Ablaufkasten *m* ~ **supply tank for thick juice** Dicksafteinziehkasten *m*
vacuum, ~ **photocell** Hochvakuumzelle *f* ~ **pipe** Saug-leitung *f*, -rohr *n* ~ **piping** Unterdruckleitung *f* ~ **piping of diffusion pumps** Vakuumleitung *f* der Diffusionspumpen ~ **plant** Vakuumanlage *f* ~ **pressure** Unterdruck *m* ~ **pressure gauge** Saugdruckmesser *m* ~ **processing** Vakuumtechnik *f* ~ **producing priming device** Selbstansaugvorrichtung *f* ~ **pump** Aussaug-, Absauge-, Sog-pumpe *f* ~ **pump system** Vakuumpumpsatz *m* ~ **rectifier valve** Vakuumgleichrichter *m* ~ **regulator** Vakuumregler *m* ~ **relief valve** Schnarr-, Sicherheitsventil *n* ~ **reservoir** Saugluft-, Unterdruckbehälter *m* ~ **rotary drier** Vakuumtrommeltrockner *m* ~ **servo-brake** Saugluftservobremse *f* ~ **servo-brake cylinder** Saugluft-Servo-Bremszylinder *m* ~ **shelf drier** Vakuumschranktrockner *m* ~ **space** Unterdruckraum *m* ~ **spectrograph** Vakuumspektrograf *m* ~ **stripper** Vakuumausstoßsystem *n* ~ **tank** Unterdruckförderer *m*, Vakuumbehälter *m*, Vakuumkessel *m*, Windkessel *m* ~ **tap** Vakuumhahn *m* ~ **tar** Vakuumteer *m* ~ **thermocouple** Vakuumthermoelement *n* ~ **tight** vakuumdicht ~ **topping** Vakuumdestillation *f*, Vakuum-Topping *n* ~ **tractor brake valve** Vakuumzugwagenbremsventil *n* ~ **trap** Vakuumbehälter *m*
vacuum-tube Vakuumröhre *f* ~ **amplifier** Röhrenverstärker *m* ~ **arrester** Vakuumblitzableiter *m* ~ **characteristic** Kennlinie *f* der Elektronenröhre, Röhren-charakteristik *f*, -kennlinie *f* ~ **constant** charakteristische Größe *f* der Röhre, charakteristische Röhrenkonstante *f* ~ **detector** Röhrendetektor *m* ~ **electrometer** Röhrenelektrometer *n* ~ **generator** Röhrengenerator *m* ~ **modulator** Röhrenmodulator *m* ~ **noise** Verstärker-, Röhrengeräusch *n* (rdo) ~ **oscillator** Röhren-oszillator *m*, -summer *m* ~ **recitifier** Hochvakuum-, Kenotron-gleichrichter *m* ~ **transmitter** Röhrensender *m*, Sender *m* mit Vakuumröhre ~ **voltmeter** Röhrenvoltmeter *n*

vacuum, ~-type bulb luftleere Birne f ~ unit Unterdruckaggregat n ~ valve Hochvakuumröhre f ~ valve receiver Vakuumröhrenempfänger m ~ vessel Saugluftkessel m ~ volatilization Vakuumverflüchtigung f ~ wash Innen-, Saug-wäsche f

vagabond, ~ current Irr-, Streu-strom m, vagabundierender Strom m ~ currents (or streams) Schleichströme pl

vagrant current vagabundierender Strom m

vague unbestimmt, unübersichtlich, verschwommen

vale Tal n

valence atombindende Kraft f, Sättigungswert m, Valenz f, Wert m, Wertigkeit f, Winkelkante f off odd ~ unpaarwertig

valence, ~ band Valenzband n ~ bond Wertigkeitsbindung f ~ electron kernfernes Elektron n, Leuchtelektron n ~ force field Valenzkraftfeld n ~ formula Wertigkeitsformel f ~ shell Valenzschale f ~ stage Valenzstufe f

valency Wertigkeitsstufe f ~ dependence Wertigkeitsabhängigkeit f ~ electron Wertigkeitselektron n ~ stage Wertigkeitsstufe f

valent wertig

valentinite Antimonblüte f, Weißspies-glanz m, -glanzerz n

valeric acid Valeriansäure f

valeryl nitrile Valerylnitril n

valid bündig, gültig, rechtskräftig, richtungweisend, statthaft, stichhaltig to be ~ gelten, gültig sein

validate, to ~ rechtskräftig machen

validation Gültigkeitserklärung f

validity Begründetheit f, Bestehen n, Geltung f, Gültigkeit f, (law), Rechtsgültigkeit f, Stichhaltigkeit f, Wertigkeit f

valley Dachkehle f (arch.), Grundlinie f (der Oberflächenrauheit), Senke f, Tal n ~ board Kehlbrett n (arch.) ~ breeze Talwind m ~ channel Kehlrinne f ~ floor Talboden m ~ formation Talbildung f ~ gravel Talschotter m ~ gutter Kehlrinne f ~ rafter Kehlschifter m, Kehlsparren m ~ sensor Talfühler m ~ terrace Talterasse f ~ through which a glacier moves (or has moved) Gletschertal n ~ tile Kehlziegel m ~ wind Talwind m

valonia Valonia f (Gerbstoff)

valorization Preisregelung f, Valorisation f

valuable Wertgegenstand m; wertvoll ~ work Wertarbeit f

valuables Wertsachen pl

valuation Auswertung f, Beurteilung f, Bewertung f, Schätzung f, Wertbestimmung f, Wertung f ~ adjustment Bewertungsausgleich m ~ factor Bewertungsfaktor m ~ figure Bewertungsziffer f ~ law Bewertungsgesetz n ~ report Bewertungsbericht m

value, to ~ abschätzen, beurteilen, einschätzen, schätzen, taxieren, veranschlagen to ~ tax veranlagen

value Bestand m, Betrag m, Bewertung f, Größe f, Kaufkraft f, (phonetics) Lautwert m, Nützlichkeit f, Rückkaufswert m, Wert m

value, ~ of capacitance Kapazitätswert m ~ of damping (or of attenuation) Dämpfungswert m ~ to be determined Bestimmungsgröße f ~ of

distribution Verteilungswert m ~ of drive Schreibwert m ~ of dynamic pressure Staudruckwert m ~ of elasticity Elastizitätsmaß n ~ of moisture Feuchtigkeitswert m ~ of probability Wahrscheinlichkeitswert m ~ of the respective locus of a center of buoyancy (or lift) Pantokarenenwert m ~ of the self-inductance Selbstinduktionswert m ~ of stability Stabilitätswert m ~ of thrust Schraubenzugzahl f

value, ~-classifying unit Klassiereinschub m ~ items bewertete Gegenstände pl (Posten)

valuer Taxator m

values of momentum Impulswerte pl

valve Absperr-glied n, -organ n, -teil m, -vorrichtung f, Absperr- oder Regelorgan n für eine Strömung, Armatur f, Falle f, Hahn m, Klappe f, Rohr n, Röhre f. Schütze f, Stoßstange f, Ventil n, (sleeve) Ventilbüchse f, Verstärkerlampe f, Zugschutz m ~ of bottle (or cylinder) Flaschenventil n ~ with deflector Schirmventil n ~ of gas container Gasflaschenventil n ~ with ground-in ball-and-socket joint Kugelschliffventil n ~ with renewable seat Ventil n mit auswechselbarem Ventilsitz ~ with shutters Jalousieschütz n ~ at side seitlich angeordnetes Ventil n ~ on the side seitlich angeordnetes Ventil n ~ for superheated gases Heißgasschieber m ~ with variable slopes Exponentialröhre f ~ of a window Fensterflügel m

valve, ~-absorption modulation Modulation f in Röhrenabsorptionsschaltung ~ action Ventilwirkung f ~ adapter Röhrenzwischenstecker m, Zwischenstecker m, Zwischenstecker m für Röhren ~ adjuster Ventileinsteller m ~-adjusting stud Ventilstellschraube f ~ adjustment Ventileinstellung f ~ amplifier Röhrenverstärker m ~ area (valve opening area) Ventilquerschnitt m ~ arm Ventilhebel m ~ arms (or scissors) Füllansatzschere f ~ arrester Kathodenfallableiter m ~ assembly komplettes Ventil n ~ ball Stahlkugel f ~ base Röhrensockel m ~ blade Schütztafel f ~ body Ventil-gehäuse n, -körper m ~ bonnet Ventilhaube f ~ box Ventilgehäuse n ~ buckle Schieberrahmen m ~ bush Schieberbuchse f ~ cage Ventil-käfig m, -korb m, -körper m ~ cap Ventil-aufsatz m, -deckel m, -kappe f, -verschraubung f, Verschlußkappe f ~-cap gasket Ventilverschraubungsdichtung f ~ carbonizing Ventilverpichung f ~ carrier Ventilträger m ~ case Ventil-einsatz m, -sitzbüchse f ~ casing Schiebergehäuse n, Schieberkasten m, ~ cell Ventilzelle f ~ chamber Ventil-gehäuse n, -kammer f ~ characteristic Röhren-charakteristik f, -kennlinie f ~ chest Ventilkammer f ~ clearance Spiel n der Ventile, Ventil-spalt m, -spiel m ~-clearance gauge Ventilabstandsmesser m, Ventilspiellehre f ~ collar Ventilklemmkegel m ~ complement Röhrenbestückung f ~ component Ventilteil m ~ compressor Ventil-kompressor m, -spanner m ~ computation Ventilberechnung f ~ cone Ventilkegel m, Verschlußkappe f ~ connection Klappenstutzen m, Ventilanschluß m ~-control panel Ventiltafel f ~ controlled röhrengesteuert ~ cord gland Ventilleinendurchführung f ~ core Ventileinsatz m ~ cotter Ventilkegelstück n ~ coupling head Ventilkupplungskopf m

~ **cover** Ventilaufsatz *m* ~ **cover gasket** Ventildeckeldichtung *f* ~ **cover plate** Ventilgehäusedeckel *m* ~ **covering** Ventilabdeckung *f* ~ **current** Ventilstrom *m* **(thermionic)** ~ **detector** Audion *n*, Röhrendetektor *m* ~ **diameter** Sitzdurchmesser *m* ~ **disc** Keilabschluß *m*, Ventilteller *m* ~ **drainage** Ventilentwässerung *f* ~ **dropping** Schließen *n* des Ventils ~ **edge** Ventilrand *m* ~ **effect** Richtwiderstand *m*, Ventilzellenwirkung *f* ~ **face** Schieber-fläche *f*, -lappen *m*, Ventilsitzfläche *f* ~ **face width** Ventilsitzbreite *f* ~ **facing** Hartmetallauflage *f* des Ventils ~ **flap** Ventilklappe *f* ~ **fracture** Ventilbruch *m* ~ **frequency meter** Röhrenfrequenzmesser *m* ~ **galvanometer** Röhrengalvanometer *n* ~ **gas tightness** Dichthalten *n* ~ **gear** Steuerung *f*, Steuerungsantrieb *m*, Ventil-antrieb *m*, -gestänge *n*, -steuerung *f*, Ventilsteuerungs-organ *n*, -teil *m* (poppet) ~ **gear air piping** Steuerungsdruckluftleitung *f* ~**-gear housing** Ventilverkleidung *f* ~ **grinder** Ventilschleifer *m* ~ **grinding tool** Ventileinschleifwerkzeug *n* ~ **guard** Fängerglocke *f*, Hubbegrenzung *f* eines Ventils ~ **guide** Ventilführung *f* ~ **handle** Hahn-, Ventil-griff *m* ~ **head** Ventilteller *m* ~ **head of air bottle** Luftflaschenkopf *m* ~ **heterodyne receiver** Röhrenschwebungsempfänger *m* ~ **holder** Fassung *f*, federnde Röhrenfassung *f* ~ **hood** Ventil--haube *f*, -kappe *f*, -klappe *f* ~ **housing** Ventilgehäuse *n*, -kanone *f* ~**-in-head engine** obengesteuerter Motor *m* ~ **installation** Ventileinbau *m* ~ **lag** Ventilverzögerung *f* ~ **land** Steuerschieberkante *f* ~ **lead** Ventilvoreröffnung *f* ~ **lever** Ventilhebel *m* ~ **lift** Hubhöhe *f* des Ventils, Ventil-erhebung *f*, -hub *m* ~**-lift diagram** Ventilerhebungsdiagramm *n* ~**-lift period** Ventilöffnungszeit *f* ~**-lift stop** Hubbegrenzung *f* eines Ventils ~ **lifter** Regulierstößel *m*, Ventil-ausgleich *m*, -heber *m*, -mitnehmer *m*, -stößel *m*, -rolle *f* ~**-lifter adjusting nut** Stößelschraube *f* ~**-lifter guide** Ventilstößelführung *f* ~**-lifting velocity** Ventilöffnungsgeschwindigkeit *f* ~ **linkage** Ventilschere *f* ~ **location** Anordnung *f* der Ventile ~ **manifold** Wechselventilkasten *m* ~ **mechanism** Steuerungsgetriebe *n*, Ventilgestänge *n*, Ventilsteuerung *f* ~ **mounting** Venfileinbau *m* ~ **noise** Mikrofongeräusch *n*, Röhrenrauschen *n* ~ **oil** Ventilöl *n* ~ **opening** Ausschlag *m* des Ventils, Schützöffnung *f* ~ **opening area** Durchströmquerschnitt *m* ~**-opening duration** Ventilöffnungsdauer *f* ~ **operating gear** Ventilsteuerungsvorrichtung *f* ~ **oscillator** Röhrengenerator *m*, -oszillator *m* ~ **outlet** Ventilauslaß *m* ~ **overlap** Überschneidung *f* der Ventilöffnungszeiten, Ventilüberschneidung *f* ~**-packing disc** Ventildichtungsscheibe *f* ~ **pad** Dichtkegel *m* aus Gummi ~ **pet cock** Abflußhahn *m* ~ **petticoat** Ventilschutzhaube *f* ~ piston Saufkolben *m*, Schieberkölbchen *n* ~ **plate** Ventilplatte *f* ~ **plate lever** Plattenhebel *m* ~ **play gauge** Ventileinstellehre *f* ~ **plug** Ventil-gehäuseverschluß *m*, -zapfen *m* ~ **port** Ventil-kanal *m*, -öffnung *f* ~**-position potentiometer** Hubpotentiometer *n* (g/m) ~ **positioner** Ventilstellungsregler *m* ~**-protecting cap** Rohrschutzkappe *f* ~ **pull** Ventilziehen *n* ~ **push**

Stoßstangenkammerdeckel *m* ~ **push-rod** Anhubstange *f* des Ventils, Ventilanhubstange *f* ~ **push-rod ball** Ventilstoßstangenkugelpfanne *f* ~ **push-rod chamber** Ventilstoßstangenkammer *f* ~ **push-rod eye** Ventilstoßstangenauge *n* ~ **push-rod guide** Ventilstoßstangenführung *f* ~**receiver** Röhrenempfänger *m* ~ **reception** Röhrenempfang *m* ~ **recess** Ventiltasche *f* ~ **rectifier** Röhrengleichrichter *m* ~ **reed** Ventilplättchen *n* ~ **refacer** Ventilnachschleifvorrichtung *f*, Ventilsitzfräser *m* ~ **regulator** Regulierventil *n* ~ **rejection test** Röhrenprüfung *f* (im Betrieb) ~ **remover** Ventilherausnehmer *m* ~**-reseating machine** Ventil-Schleifapparat *m* ~ **reseating tool** Ventilsitzfräser *m* ~ **retainer** Ventilanschlag *m* ~ **rim** Ventilrand *m* ~ **ring** Ventilring *m* ~ **rocker** Kipp-, Schwing-, Ventil-hebel *m*, Ventilstößel *m*

**valve-rocker,** ~ **bracket** Kipphebelblock *m*, Schwinghebelblock *m*, Ventilsteuerhebelstütze *f* ~ **bush** Kipphebeldrehpunktbolzen *m*, Schwinghebellagerbuchse *f*, Ventilsteuerhebelbuchse *f* ~ **bushing** Kipphebeldrehpunktbolzen *m*, Schwinghebellagerbuchse *f* ~ **fulcrum pin** Kipphebelzapfen *m*, Schwinghebeldrehpunktbolzen *m*, Ventilsteuerhebelachse *f* ~ **pedestal** Kipphebelblock *m*, Schwinghebelblock *m* ~ **roller** Kipphebelrolle *f*, Schwinghebelrolle *f* ~ **roller pin** Kipphebelrollenbolzen *n*, Schwinghebelrollenbolzen *m* ~ **shaft** Kipphebelachse *f*

**valve,** ~**-rocking lever** Ventilsteuerhebelarm *m* ~ **rod** Schieberschubstange *f*, Schieberstange *f*, Spindel *f*, Ventil-spindel *f*, -stange *f* ~ **roller** Ventilhebelrolle *f* ~ **roller lock** Sicherung *f* für Ventilhebelrolle ~ **rope cutter** Klappseilmesser *n* ~ **rustle** Röhrenrauschen *n* ~ **scavenging** Ventilspülung *f* ~ **scissors** Ventilschere *f* ~ **seal** Ventilabdeckung *f* ~ **sealing surface** Ventildichtungsfläche *f*

**valve-seat** Aufsitzfläche *f*, Sitzfläche *f* des Ventiltellers *f*, Ventilsitz *m*, Ventilsitz-fläche *f*, -ring *m* ~ **of a ball valve** Aufsitzfläche *f* eines Kugelventils ~ **of a flap valve** Aufsitzfläche *f* eines Klappventils

**valve-seat,** ~ **bush** Ventilsitzbuchse *f* ~ **contact** Ventilschluß *m* ~ **grinder** Ventileinschleifer *m* ~ **grinding blade** Ventileinschleifklinge *f* ~ **insert** Ventilsitzring *m* ~ **milling and grinding tools** Ventilsitz-Fräs-und-Schleifwerkzeuge *pl* ~**-ring** Ventilsitzring *m*

**valve,** ~ **seating** Aufsitzen *n* des Ventils, Klappensitzfläche *f*, Ventilsitz *m* ~ **set** Röhrenapparat *m* ~ **setting gauges** Ventileinstellehren *pl* ~ **shaft** Ventilschaft *m* ~ **shutter** Ventilverschluß *m* ~ **slide** Ventilschieber *m* ~ **socket** Fallbüchse *f*, Röhren-fassung *f*, -sockel *m* ~ **spindle** Ventilspindel *f* ~ **spool** Steuer-, Ventil-schieber *f* (hydraul.)

**valve-spring** Ventil-erhebung *f*, -feder *f* ~ **cap** Ventilfeder-gehäuse *n*, -haube *f*, -kappe *f*, -teller *m* ~ **computation** Ventilfederberechnung *f* ~ **cotter** Ventil(feder)keil *m* ~ **housing** Ventilfeder-gehäuse *n*, -haube *f*, -kappe *f* ~ **key** Ventilfederkeil *m* ~ **lifter** Ventilfedermitnehmer *m* ~ **lifting pliers** Ventilfederhebezangen *pl* ~ **remover** Ventilheber *m* ~ **retainer** Federteller *m*, Ventilfederteller *m*

~ **retainer lock** Ventilkeil *m* ~ **seat** Ventilfedersitz *m* ~ **seating** Ventilfederkegelsitz *m* ~ **split collets** Ventilklemmstück *n* ~ **surge** Ventilfederschwingung *f* ~ **testing machine** Ventilfederprüfmaschine *f* ~ **washer** Ventilfederteller *m*

**valve,** ~ **stem** Ventil-schaft *m*, -spindel *f*, -stange *f*, -stößel *m*, Zugstange *f* ~**-stem guide** Ventilschaftführung *f* ~**-stem guide bearing** Ventilführungsbuchse *f* **(the)** ~ **sticks** Ventil *n* hängt ~ **stopper** Ventilstopfen *m* ~ **stroke** Ventilhub *m* ~ **support** Ventilträger *m* ~**-tapped roller pin** Stößelrollenbolzen *m* ~ **tappet** Ventilstößel *m* ~**-tappet guide** Ventilstößelführung *f* ~**-tappet roller** Ventilstößelrolle *f* ~**-tappet roller pin** Ventilstößelrollenbolzen *m* ~ **test run** Luftversuch *m* (g/m) ~**-testing outfit** Röhrenprüfgerät *n* ~ **throat area** Ventildurchströmöffnung *f* ~ **thumbscrew** Ventilflügelschraube *f* ~ **timing** Einstellung *f* der Ventilsteuerzeit, Ventil-steuerung *f*, -steuerzeit *f*, -zündeinstellung *f* ~**-timing diagram** Steuerdiagramm *n* ~ **timing gear** Ventilsteuerungsvorrichtung *f* ~ **transmitter** Röhrensender *m* ~ **travel** Hubweg *m* eines Ventils, Ventilhub *m* ~ **type echo suppressor** stetig arbeitende Echosperre *f* ~**-type lightning arrester** Ventilableiter *m* ~ **voltage drop** Spannungsabfall *m* (in einer Röhre) ~ **voltmeter** Röhrenvoltmeter *m* ~**washer lock** Ventilklemmstück *n* ~ **wattmeter** Röhrenwattmeter *n* ~ **well** Schütznische *f* ~ **wing** Fensterflügelrahmen *m*

**valved line** ventilversehene Leitung *f*

**valveless** ventillos

**valves-on-the-side** stehende Ventile *pl*

**valving** Ventilausrüstung *f* ~ **pressure** Abblasedruck *m*

**van** Gepäck-, Güter-, Möbel-, Pack-wagen *m* ~**-type shutter** Drehklappen-abdeckung *f*, -jalousie *f*

**vanadate** Vanadinsalz *n*

**vanadic,** ~**acid** Vanadinsäure *f* ~ **anhydride** Vanadinsäureanhydrid *n* ~ **sulfate** Vanadinsulfat *n*

**vanadiferous** vandinenthaltend

**vanadinite** Vanadin-bleierz *n*, -bleispat *m*

**vanadium** Vanad *n*, Vanadium *n* ~ **carbide** Vanadinkarbid *n* ~ **chloride** Vanadinchlorid *n* ~ **compound** Vanadinverbindung *f* ~ **mordant** Vanadatbeize *f* ~ **nitride** Stickstoffvanadin *n*, Vanadinstickstoff *m* ~ **pentoxide** Vanadinpentoxyd *n*, -säureanhydrid *n* ~ **salt** Vanadinsalz *n* ~ **steel** Vanadinstahl *m*, Vanadiumstahl *m* ~ **trioxide** Vanadiumtrioxyd *n*

**vanadous** vanadinig ~ **chloride** Vanadiumchlorür *f* ~ **salt** Vanadolsalz *n* ~ **sulfate** Vanadosulfat *n*, Vanadylsulfat *n*

**vandenbrandeite** Vandenbrandeit *m*

**vandendriesscheite** Vandendriesscheit *m*

**Vandyke print** Braunpause *f*

**vane** Abdeckflügel *m*, Dämpferfahne *f*, Fahne *f*, Flansch *m*, Flügel *m*, Propellerflügel *m*, Rippe *f*, Schaufel *f*, Schwingschiene *f* (telet.), Stellglied *n*, Steuerfläche *f* (aviat.), Windflügel *m*, Windmühlenflügel *m*, Windrichtungsanzeiger *m* ~ **of a condenser** Platte *f* eines Kondensators ~ **of a quadrant electrometer** Flügel *m* (fest oder beweglich) eines Quadrantenelekto-

meters ~ **with serrated edge** Zackenblende *f* ~ **with triangular edge** Zackenblende *f* ~ **of vacuum pump** Pumpenlamelle *f*

**vane,** ~ **alignment box** Ruder-abgleichkasten *m*, -abstimmkasten *m* ~ **anemometer** Flügelradwindmesser *m* ~ **angle** Ruderausschlagwinkel *m* ~ **apparatus** Flügelsonde *f* ~ **armature** Flügelanker *m* ~**-armature relay** Flügelankerrelais *n* ~ **attenuator** Fahnen-, Streifen-abschwächer *m* (microw.), Streifenteiler *m* (microw.) ~ **beam** Flügelbalken *m* ~ **chamber** Schaufelkammer *f* ~ **coefficient (or constant)** Ruderkonstante *f* ~ **control** (in fan) Drallregelung *f* ~ **drum** Flügeltrommel *f* ~ **form** Schaufelform *f* ~ **hinge torque** Scharniermoment *n* ~ **linkage** Rudergestänge *n* ~ **magnetron** Fahnenmagnetron *n* ~ **motor** Ruderantrieb *m* (g/m) ~ **piston** Flügelkolben *m* ~ **pocket** Schaufeltasche *f* ~**-position indication** Ruderlagenanzeige *f* ~**-restoring potentiometer** Rückführpotentiometer *n* ~ **ring** Leitschaufelkranz *m* ~ **setting** Flügelstellung *f* ~ **spindle** Flügelstuhl *m*, -welle *f* ~ **support** Ruderhalterung *f* ~ **technique** Lamellentechnik *f* ~**-type fluid meter** Flügelmesser *m* ~**-type pump** Flügelradpumpe *f*, Kapselpumpe *f* ~**-type scavenge oil pump** Flügelradölpumpe *f* ~ **water meter** Flügelradwassermesser *m* ~ **wattmeter** Fähnchenwattmeter *n* ~ **wheel** Fahnenrad *n*, Schaufelrad *n* (of a fly brake) ~ **wheel pump** Schaufelradpumpe *f*

**vaned,** ~ **diffuser** Leitschaufelapparat *m* ~ **plate** Leitapparatverdichter *m*

**vanish, to** ~ verlorengehen, verschwinden

**vanishing** Verschwinden *n* ~ **line** Flucht-gerade *f*, -linie *f*, Verschwindlinie *f*, Verschwindungsgerade *f* ~ **plane** Fluchtebene *f*, Verschwindungsebene *f* ~ **point** Fluchtpunkt *m*, Verschwindungspunkt *m* ~**-point control** Fluchtpunktsteuerung *f* (Entzerrungsgerät)

**vanquish, to** ~ überwältigen

**vantage point** Aussichtspunkt *m*

**V-antenna** V-Antenne *f*

**vapor** Abgase *pl*, Brodel *m*, Brodem *m*, Brüdendampf *m*, Dampf *m*, Dunst *m*, Luftfeuchtigkeit *f*, Nebel *m*, Rauch *m* ~ **barrier** Feuchtigkeitsabdichtung *f* ~ **beam** Dampfstrahl *m* ~ **carryover** Mitreißen *n* von Gas ~ **chamber** Brüdenraum *m* ~ **channel** Zug *m* eines Kohlenmeilers ~ **cloud** Dampfwolke *f* ~ **compression** Brüdenkompression *f* ~ **concentrate pump** Brüdenpumpe *f* ~ **condenser** Brüdenkondensator *m* ~ **cooling** Dampfkühlung *f* ~ **cooling for cathode** Kathodensiedekühlung *f* ~ **escape** Dunstabzug *m* ~**-exhaust system** Dunstabsauganlage *f* ~ **fan** Brüdengebläse *n* ~ **formation** Dunstbildung *f* ~ **head** Dunsthaube *f* ~ **lamp** Höchstdrucklampe *f* ~ **lock** Dampfsack *m*, Gassperre *f*, Unterbrechung *f* des Flüssigkeitsstromes durch Dampfblasenbildung, Verstopfung *f* durch Dampfblasen ~ **locking** Dampfblasenbildung *f* ~ **mist** Dunst-nebel *m*, -wolke *f* ~ **nozzle** Dampfventil *n* ~ **outlet** Dampfaustritt *m* ~ **particle** Tröpfchen *n* ~ **phase** Dampf-form *f*, -phase *f* ~ **pipe** Brüdenrohr *n* ~ **pressure** Dampf-, Verdunstungs-druck *m* ~**-pressure igniter** Dampfdruckzünder *m* ~ **pressure thermometer** Dampfspannungsthermometer *n*

~-proof motor gasdicht gekapselter Motor
m ~ pump Treibmittelpumpe f ~-pump
evaporators Brüdenpumpe f ~ shroud Dampf-
hülle f ~ slide valve Brüdenschieber m ~ space
Gasraum m ~ tension Dampfdruck m ~-tight
(or -proof) dampfdicht ~ trail (engines)
Auspuffschweif m, (engines) Böenschleppe
f, Kondensstreifen m, Wolkenschweif m ~
unit Rauchgerät m ~ velocity Schwadenge-
schwindigkeit f ~ vent condenser Schwaden-
kondensator m

**vaporimeter** Dampfspannungsmesser m
**vaporizability** Verdampfbarkeit f
**vaporizable** verdampfbar, verdampfungsfähig,
verdunstbar, vergasbar
**vaporization** Eindampfen n, Verdampfung f,
Verdunsten n, Vergasung f ~ coil Verdamp-
fungswendel f ~ heat Verdampfungswärme f
~ point Verdampfungspunkt m
**vaporize, to** ~ abdampfen, eindampfen, verdampfen, vergasen, verpuffen, zerstäuben
to ~ upon aufdämpfen
**vaporized** gedämpft, zerstäubt ~ carbon resistance aufgedampfter Kohlewiderstand m ~
refrigerant dampfförmiges Kältemittel n
**vaporizer** Verdampfapparat m, Verdampfer m,
Vergaser m, Zerstäuber m
**vaporizing**, ~boat Verdampfungsschiffchen n
~ carburetor Verdampfungsvergaser m ~
cathodes Aufdampfkathoden pl (electron) ~
operation Aufdämpfen n ~ plate Aufdämpf-
platte f ~ unit Verdampferaggregat n
**vaporous** dampfförmig to be ~ duften ~ envelope
Dunsthülle f ~ envelope (or sheath) Dampf-
hülle f
**var** Var n, Voltampère n ~-hour Var-Stunde f
~-hour meter Blind-leistungszähler m, -ver-
brauchszähler m
**variability** Gestaltungsvermögen n, Schwanken n,
Ungleichförmigkeit f, Variabilität f, Veränderlichkeit f ~ with frequency Frequenzabhängigkeit f ~ of volume Raumveränderlichkeit f
**variable** Größenfaktor m, Kenngröße f, Variable
f, Veränderliche f; abwandelbar, abwechselnd,
bestreichbar, schwankend, unbeständig, ungleich, ungleichmäßig, umlaufend, unterschiedlich, unstet, unstetig, variabel, veränderlich, wandelbar, wechselnd ~ with frequency
frequenzabhängig ~ of state Zustands-größe
f, -variable f, -veränderlich f ~ with time zeitlich veränderlich of ~ area flächenveränderlich
**variable,** ~ address veränderliche Adresse f ~
admission veränderliche Füllung f ~ area
veränderliches Feld n ~ area method Trans-
versalmethode f ~ area optical sound-on-film
Zackenschrift f ~ area (or width) recording
Amplitudenschrift f ~ area recording method
Kurvenlichtmethode f ~ area (or width) sound
track Amplitudenschrift f ~ camber veränderliche Wölbung f ~ camber flap Wölbungsklappe
f ~ camber wing profile (or section) Verstell-
profil n, -querschnitt m ~ capacitor Block-,
Dreh-kondensator m ~ capacity veränderliche
Kapazität f ~ condenser Anodenrückwirkung f,
Drehkondensator m, Drehspannungsgleicher
m, einstellbarer oder regelbarer Kondensator m, veränderliche Kapazität f, variabler
oder veränderlicher Kondensator m ~ con-

nection Wackelkontakt m ~ connector Verzweigungspunkt m ~ control rate Verstell-
geschwindigkeit f ~-crankthrow engine Motor
m mit veränderlichem Hub ~ cycle operation
Asynchronbetrieb m ~-datum boost control
Ladedruckregler m mit veränderlicher Ausgangsstellung der Regeldose ~ delivery pressure pump Regelpreßpumpe f ~ density veränderliche Dichte f ~ density double-squeeze
track Gleichtaktsprossendoppelschnürspur f
~ density method Intensitäts-, Schwärzungs-
verfahren n ~ density recording Dichteschrift
f ~ density recording method Schattierungs-,
Streifenlicht-verfahren n ~ density sound track
Dichteschrift f ~ density squeeze track Gleich-
taktsprossenschnürspur f ~ density tracks
Intensitäts-, Sprossen-schrift f ~ density
wind tunnel Höhenkammer f, Höhenwindkanal
m, Windkanal m für veränderliche Luftdichte
m ~ drive Variator m ~ drive with cone pulleys Kegel(trommel)trieb m ~-drop corn planter
stellbare Maisdibbelmaschine f ~ expansion
veränderliche Ausdehnung f ~ feed attachment
Vorschubwechseleinrichtung f ~ field veränderliches Feld n ~ focus and variable magnification
veränderliche Brennweite f und Größe f ~-
frequency bus Sammelschiene f mit veränderlichen Frequenzen ~-frequency oscillator abstimmbarer Oszillator m ~-gain amplifier
Regelverstärker m ~ gear Wechselgetriebe n
~ gearing Stufenrädergetriebe n ~-head torch
Wechselbrenner m ~-head welding torch
Wechselschweißbrenner m ~-image inverting
system of lenses verstellbares bildumkehrendes
Linsensystem n ~ incidence veränderlicher Anstellwinkel m ~ incidence wing Verstellflügel m
~ inductance veränderliche Induktivität f ~
inductor Induktionsvariometer n ~-intensity
method Schwärzungsmethode f ~-jet nozzle
control Schubdüsenverstellung f ~ lift auf-
triebveränderlich ~ load pulsierende Belastung
f, veränderliche Lasten pl ~ micro-groove
record Füllschriftplatte f
**variable-mu,** ~ amplifier Röhre f mit veränderlichem Durchgriff ~ hexode Exponential-
hexode f ~ high-frequency pentode Exponen-
tial(hochfrequenz)penthode f, Fünfpol-
(schirm)regelröhre f ~ pentode Regelpentode
f ~ screen-grid tube Exponential-schirm-
gitterröhre f, -tetrode f, Schirmgitterröhre
f mit veränderlichem Durchgriff, Vierpol-
(schirm)regelröhre f ~ tetrode Exponential-
schirmgitterröhre f, -tetrode f, Schirmgitterröhre f mit veränderlichem Durchgriff,
Vierpol(schirm)regelröhre f ~ tube Exponentialröhre f, Regelröhre f, Röhre f mit variablem Verstärkungsfaktor oder mit veränderlichem Durchgriff ~ valve Exponentialröhre f, Regelpentode f, Regelröhre f, Röhre
f mit variablem Verstärkungsfaktor oder mit
veränderlichem Durchgriff
**variable,** ~ mutual conductance valve Exponentialröhre f, Regelröhre f, Röhre f mit variablem
Verstärkungsfaktor oder mit veränderlichem
Durchgriff ~-mutual tube (or valve) Regelröhre f ~ number Laufzahl f ~ orifice Katarakt
m (with) ~ pitch mit verstellbarer Steigung f
~-pitch fan Verstellgebläse n ~-pitch propeller

einstellbare Luftschraube *f*, Luftschraube *f* mit verstellbarer Steigung, verstellbare Luftschraube *f*, Verstellschraube *f* **~-point representation** Gleitpunktdarstellung *f* **~ potentiometer** Drehspannungsteiler *m* **~-pressure** Gefällespeicher *m* **~ quality** periodische Größe *f* **~ quantity** Veränderliche *f* (math.) **~-ratio gear** stufenloses Getriebe *n* **~ reactance amplifier** Reaktanzverstärker *m* **~ reluctance** veränderlicher magnetischer Widerstand *m* **~-reluctance microphone** elektromagnetisches Mikrofon *n* **~ resistance** Regel-, Regulier-, Überschaltwiderstand *m*, regelbarer, veränderlicher oder verstellbarer Widerstand *m* **~ resistance for running** Einstellungswiderstand *m* für Fahren **~ resistor** Dreh(gleit)widerstand *m*, Potentiometer *n*, Regelwiderstand *m* **~ restrictor** variabler Durchflußbegrenzer *m* **~ rheostat** Heizregler *m* **~ selectivity** veränderliche Selektivität *f* **~ span wing** Ausziehflügel *m* **~ spark-gap** veränderbare Funkenstrecke *f* **~ speed axle-driven generator** Achsgenerator *m* für veränderliche Drehzahl *f* **~ speed direct-current motor** Gleichstromregelmotor *m* **~ speed drive** Regelantrieb *m* **~ speed gear** sufenloses Regelgetriebe *n* **~ speed gearing** Regelgetriebe *n* **~ speed governor** Verstellregler *m* **~ speed modulation** Zeilensteuerung *f*, Zeitmodulation *f* **~-speed motor** Motor *m* mit Drehzahlregelung, Regelmotor *m* **~-speed scan** Geschwindigkeitsabtastung *f* **~-speed scanning** Abtastung *f* mit veränderlicher Geschwindigkeit *f*, Zeitmodulation *f* **~ stress component** Spannungsausschlag *m* **~ surface** Teleskopflügel *m* **~ temporary duty** kurzzeitiger Betrieb *m* mit veränderlicher Belastung **~ transformer** regelbarer Übertrager *m* **~ tuning capacitor** Drehkondensator *m* **~-voltage regulator** Spannungsregler *m* **~ volume pump** Pumpe *f* mit veränderlicher Förderleistung **~ wave length** bestreichbarer Wellenbereich *m* **~-width sound recording** Zackenschrift *f* **~-wing airplane** Flugzeug *n* mit verstellbaren Flügeln

**variance** (of a system) Freiheitsgradzahl *f*, Unstimmigkeit *f* **~ (difference)** Spielraum *m*

**variant** Abart *f*

**variation** Abart *f*, Abmaß *n* (aviat.), Abstufung *f*, Abwechslung *f*, Abweichung *f*, Änderung *f*, (compass) Deviation *f*, (of a curve) Gang *m*, (compass) Mißweisung *f*, Schwanken *n*, Schwankung *f*, Spielart *f*, Umänderung *f*, Unterschied *m*, Variation *f*, Veränderung *f*, Verlauf *m*, Wechsel *m*

**variation,** **~ of amplitude** Amplitudenänderung *f* **~ in apparent bearing** Peilstrahl-schwankung *f*, -wanderung *f* **~ of armature voltage** Ankerspannungsverstellung *f* **~ of audio voltage** Sprechwechselspannung *f* **~ of boiling point** Siedepunktverlauf *m* **~ in capacity** Kapazitätsänderung *f* **~ in combustion-cutoff point** Brennschlußstreuung *f* (g/m) **~ of compass** Tagesänderung *f* **~ of compass diurnal** Änderungen *pl* des Kompasses **~ of cross-section** Querschnittverlauf *m* **~ of density** Dichteverteilung *f* **~ in diameter** Durchmesserabweichung *f* **~ in dimension** Maßtoleranz *f* **~s in the earth's magnetic field** Änderungen *pl* des magnetischen Erd-

feldes **~ of exposure** Lichtwechsel *m* **~ of the feed pressure** Förderdruckveränderung *f* **~ of field intensity** Feldstärke-änderungen *pl*, -schwankungen *pl* **~ of frequency** Frequenzschwankung *f* **~ with frequency** Änderung *f* mit der Frequenz **~ in intensity** Intensitätsverlauf *m*, -schwankung *f* **~ of keyway width from standard** Abmaß *n* der Nutbreite (mech. engin.) **~ in longitude** Längenabweichung *f* **~ in path angle** Bahnwinkeländerung *f* **~ of play** Spielschwankung *f* **~ of pressure** Druckschwankung *f* **~ in program** Programmstreuung *f* **~ in slope** Steigungsänderung *f* **~ of the speed** Geschwindigkeitsänderung *f* **~ in temperature** Temperaturänderung *f* **~ of temperature** Temperatur-, Wärme-gang *m* **~ of wall thickness** Wandstärkeverlauf *m* **~ in weight per unit area** Flächengewichtsschwankung *f*

**variation,** **~ angle** Ortsmißweisung *f* (nav.) **~ compass** Peilkompaß *m* **~ measurement** Variationsmessung *f* (aviat.) **~ method** Variationsmethode *f* **~ problem** Variationsproblem *n*

**variational,** **~ condition** Variationsbedingung *f* **~ ionization potential** Ionisierungspotential *n* durch Variationsmethode **~ principle** Variationsprinzip *n*

**varicolored** bunt, verschiedenfarbig

**varied** mannig-fach, -faltig, verschieden, verschiedenartig

**variegated** vielfarbig **~ sandstone** Buntsandstein *m*

**variegation** geflammtes Muster *n*

**variety** Abart *f*, Abwechslung *f*, Art *f*, Auswahl *f*, Mannigfaltigkeit *f*, Verschiedenheit *f*, Vielfältigkeit *f* **~ of applications** Verwendungsbereich *m* **~ of heavy spar** Leuchtspat *m* **~ of lead** Bleiart *f* **~ of uses** Verwendungsmöglichkeit *f* **in a ~ of ways** auf verschiedene Arten

**variety trial** Sortenversuch *m*

**varindor** Varindor *m*

**vario-coupler** Vario-koppler *m*, -kuppler *m*, veränderliche Kupplung *f*, veränderliche Kupplungsspule *f*

**variode regulator** Variodenregler *m*

**variometer** Dreh-, Schiebe-drossel *f*, variable Induktionsspule *f*, Variometer *m* **~ coil** Variometerspule *f* **~ compass** Variometerkompaß *m* **~ rotor** drehbare Variometerspule *f* **~ stator** feste Variometerspule *f*

**various** vielerlei, vielfach, verschieden, verschiedenartig **~ forms of steel** Austenit *m* **~ grain sizes** gemischtkörnig **~-purpose milling machine** Mehrzweckfräsmaschine *f* **~ types of dams** verschiedene Sperren *pl*

**varistor** Varistor *m*, spannungsabhängiger Widerstand *m*

**varitone tablets** Färbetabletten *pl*

**Varley,** **~ loop test** Erdfehlerschleifenmessung *f* nach Varley **~'s loop test** Varleyschleife *f* (teleph.)

**varmeter** Blindleistungsmesser *m*

**varnish, to** **~** anfrischen, anstreichen, firnissen, lackieren, streichen, überglasen

**varnish** Anstrichstoff *m*, Beglasung *f*, Firnis *m*, Glasur *f*, Lack *m*, Lacküberzug *n*, Politur,

Zapon(lack) *m* ~ **for iron goods** (Lack) Eisenfirnis *m* ~ **(or enamel) for molds** Modellack *m* ~ **for outdoor use** Außenlack *m* ~ **for screens** Ofenschirmlack *m* ~ **for sheet-metal tin plate** Blechlack *m*

**varnish,** ~ **base** Lackunterlage *f* ~ **boiling** Lacksud *n* ~ **coat** Lackanstrich *m*, Lacküberzug *m* ~ **color** Lackfarbe *f* ~ **content** Firnisgehalt *m* ~ **filler** Lackspachtel *f* ~ **made with goldanlegeoil** Goldanlegeöllack *m* ~**-making plant** Firnisanlage *f* ~ **paint** Firnisfarbe *f* ~ **remover** Lackabbeizmittel *n* ~ **shop** Lackiererei *f*, Lackiererwerkstatt *f* ~ **thinner** Lackverdünner *m*

**varnishable** lackierfähig

**varnished** lackiert ~ **cambric** Öl-leinen *n*, -stoff *m* ~ **cambric wire** Lackpapierdraht *m* ~ **cloth** Öltuch *n* ~ **conductor** Lackader *f* ~ **fabric** Lackgewebe *n* ~ **packing paper** gefirnistes Packpapier *n* ~ **paper** gefirnistes Papier *n* ~ **paper board** Elektrolackpappe *f* ~ **rush** Lackrohr *n* ~ **ware** Lackware *f* ~ **wire** Lack(ader)draht *m* ~ **wiring** Lackleitung *f*

**varnishing,** ~ **booth** Lackierkabine *f* ~ **coat** Lackauftrag *m* ~ **machine** Lackauftragmaschine *f* ~ **plant** Lackieranstalt *f*

**varved clay** Bänderton *m*, dünngeschichteter Ton *m*, Warwenton *m*

**vary, to** ~ abändern, abarten, abwandeln, abwechseln, abweichen, (up or down) nach oben oder unten abweichen, schwanken, sich ändern, sich verändern, umändern, variieren, verändern, verschieden sein, wandeln, wechseln **to** ~ **colors** schillern **to** ~ **at the. . .power** sich mit der. . .Potenz ändern

**varying** Wechseln *n*; veränderlich, wechselnd ~ **luster** Schillerglanz *m* ~ **pressure** gleitender Druck *m* ~ **stress** wechselnde Beanspruchung *f* ~ **with time** zeitabhängig

**vascular tissue** Holzfaser *f*

**vasomotor receptacle furnace** Gefäßreflex *m*

**vast** ausgedehnt, riesig, weitgehend

**vat** Ausblasebütte *f*, (dyeing) Back *m*, Barke *f*, Bottich *m*, Bütte *f*, Diffuseur *m*, Faß *n*, Klärbottich *m*, Kübel *m*, Küpe *f*, Trog *m*, Wanne *f*, Warengleitmulde *f*, Waschbehälter *m* ~ **with excess of alkali (sharp vat)** überschärfte Küpe *f*

**vatman** Büttgeselle *m*, Eintaucher *m*, Former *m*, Schöpfer *m* ~**'s shake** Schütteln *n*

**vatting property** Verküpbarkeit *f*

**vault, to** ~ einwölben, wölben **to** ~ **over** überwölben

**vault** Gewölbe *n*, Keller *m*, Wölbung *f* ~ **of cloud** Gewitter-, Wolken-kragen *m* ~ **of the sky** Himmelswölbung *f* ~ **support** Gewölbeträger *m*

**vaulted** gewölbt ~ **ceiling** Gewölbe *n* ~ **cell** Gewölbefach *n* ~ **passage** Halbkuppel *f* ~ **recess** gewölbte Rückwand *f*

**vaulting** Gewölbe *n*, gewölbtes Mauerwerk *n*, Unterkellerung *f*, Wölbung *f* ~ **pillar** Gewölbepfeiler *m* ~ **ragstone** Wölbpläner *m* ~ **ruler** Wölbrichtscheit *n*

**V bar control** Vertikalbalkenregler *m* (TV)

**V-beam** V-Bündel *m*, V-Strahl *m*

**V-belt** Dreikant-, Keil-riemen *m* ~ **drive** Keilriemen-antrieb *m*, -übersetzung *f* ~ **pulley** Keilriemen-ritzel *n*, -scheibe *f* ~ **pulley half** Keilriemenscheibenhälfte *f*

**V-block** Auflegebock *m*, Bohrprisma *n*

**V-bottom surface** gekielte Fläche *f*

**vector, to** ~ mit Radar *n* führen

**vector** (radius) Leitstrahl *m*, Vektor *m* (aerodyn., math.), vektorielle Größe *f*, Zeiger *m* (math.) ~ **calculus** Vektorrechnung *f* ~ **density** Vektordichte *f* ~ **diagram** Vektordiagramm *n*, Vektorzeichnung *f*, Zeigerdiagramm *n* ~ **equation** Vektorgleichung *f* ~ **field** Vektorfeld *n* ~ **group** Schaltgruppe *f* ~ **line** Vektorlinie *f* ~ **meter** Vektormesser *m* ~ **point** geometrischer Ort *m* ~ **potential function** Vektorpotentialfunktion *f* ~ **power** Leistungs-vektor *m*, -zeiger *m* ~ **quantity** Vektor *m*, Vektorgröße *f* ~ **radiant** Strahlungsvektor *m* ~ **representation** vektorielle Darstellung *f* ~**-response index** Vektorindikation *f* ~ **sheet** Vektorblatt *n* ~ **sum** Vektorsumme *f* ~ **tube** Vektorröhre *f*

**vectored thrust engine** Schwenkdüsentriebwerk *n*

**vectorial** vektoriell ~ **angle** Vektorwinkel *m* ~ **product** Vektorenprodukt *n* ~ **recorder** vektorieller Schreiber *m*

**vectoring computer** Einlenkrechengerät *m*

**vee, to** ~ **out** auskreuzen

**vee** V-förmig ~**-and-flat-way** Prismen- und Flachführung *f* ~**-belt drive agitator** Antriebsrad *m* für Strudelrad ~**-coupling** Keilkupplung *f* ~**-engine** V-Motor *m* ~**-groove** prismatische oder schwalbenschwanzförmige Nut *f* ~**-notch** Kimme *f* ~**-pulley** Riemenscheibe *f* ~**-shaped** prismen-, V-förmig ~**-shaped mark** Pfeilmarke *f* ~**-way** Prismaoberkante *f* ~**-way of the tailstock** Reitstockprisma *n*

**veer, to** ~ (of wind) rechts drehen, umschwingen, umspringen (meteor.), wenden **to** ~ **a cable** die Kette ausstecken **to** ~ **off the course** ausbrechen ~ **to the left** Ablenkung *f* nach links **to** ~ **off** fieren **to** ~ **around suddenly** umschlagen

**veering** Ausschießen *n* des Windes, Kursschwankung *f*, Rechtsablenkung *f*

**vees (or V-ways)** Führungsprismen *pl*, Prismenführung *f*

**vegetable** pflanzlich ~ **butter** Pflanzenbutter *f* ~ **charcoal** Pflanzenkohle *f* ~ **color** Holzfarbstoff *m* ~ **fat** Pflanzenfett *m* ~ **fiber** Pflanzenfaser *f* ~ **glue** Harzleim *m* ~ **mold** gewachsener Boden *m* ~ **oil** Pflanzenöl *n* ~ **parchment** Pergamentersatz *m* ~ **size** Harz-leim *m*, -leimung *f* ~ **soil** Humus *m* ~ **wax** Pflanzenwachs *n*

**vegetation** Bodenbewachsung *f*, Pflanzenwuchs *m* ~ **moisture** Bodenbewachsungs-, Vegetations-wasser *n*

**vehemence** Heftigkeit *f*

**vehicle** Beförderungsmittel *n*, Bindemittel *n* Fuhre *f*, Fuhrwerk *n*, Gefährt *n*, Träger *m*, Verkehrsmittel *n*, Wagen *m* ~ **for drugs** Beibringungsmittel *n* ~ **of a mixture** Grundbestandteil *m* eines Gemisches

**vehicle,** ~ **mounted on wheels** Räderfahrzeug *n* ~**-registration certificate** Kraftfahrzeugschein *m* ~ **suitable for tropical climate** Tropenfahrzeug *n*

**vehicular radio interference suppression capacitor** Autoentstörungskondensator *m*

**veil, to** ~ verschleiern

**veil** Schleier *m*

**veiled** trübe

**veiling** Verschleierung f ~ **on the margin** Randschleier m

**vein, to** ~ adern, marmorieren, masern

**vein** Ader f, Bergader f, Berggang m, gangartige Erzausscheidung f, Streifen m, Trumm n ~ **of rock** Gesteinsgang m

**vein,** ~ **accompaniments** Ganggefolgschaft f **(the)** ~ **crops** das Flöz streicht zu Tage aus ~ **following bedding planes** Lagergang m ~ **infilling** Gangfüllung f ~ **ore** Gangerz n ~ **stuff** Bergart f, Ganggestein n ~ **tin** Bergzinn n ~ **walls** Lettenbesteg m

**veined** geädert, (of wood) geflammt, gemasert, maserig, streifig ~ **structure** Geäder n ~ **wood** geadertes Holz n

**veining** Maserung f ~ **gouge** Kanneliergutsche f

**veiny** aderig

**vellum** Kalbspergament n, Pergament n, Velin n

**velocity** Anströmungsgeschwindigkeit f (electron.), Geschwindigkeit f, Schnelle f, Schnelligkeit f **(initial)** ~ Eigengeschwindigkeit f (phys.)

**velocity,** ~ **of advance** Wandergeschwindigkeit f ~ **of agitation** ungeordnete Geschwindigkeit f ~ **of air flow** Luftdurchsatzgeschwindigkeit f ~ **of approach** Anflußgeschwindigkeit f ~ **in blower stream** Anblasgeschwindigkeit f, (wind tunnels) Anströmgeschwindigkeit f ~ **of conversion** Umsetzungsgeschwindigkeit f ~ **of deflection** Ausweichgeschwindigkeit f ~ **of departure** Abgangsgeschwindigkeit f ~ **of descent** Sinkgeschwindigkeit f ~ **of draining (or efflux)** Abflußgeschwindigkeit f ~ **of electrons** Elektronengeschwindigkeit f ~ **of emission** Emissionsgeschwindigkeit f ~ **of flame propagation** Brenngeschwindigkeit f ~ **of flow** (of oil) Leistungs-, Strömungs-geschwindigkeit f ~ **of flow from initial direction** Anströmgeschwindigkeit f ~ **of impact** Aufschlaggeschwindigkeit f ~ **of an ion** Wanderungsgeschwindigkeit f eines Ions ~ **of light** Lichtgeschwindigkeit f ~ **at main cutoff signal** Hauptkommandogeschwindigkeit f (g/m) ~ **of motion** Bewegungsgeschwindigkeit f ~ **of a periodic wave** Fortpflanzungsgeschwindigkeit f einer periodischen Welle ~ **of progression** Laufgeschwindigkeit f einer Welle ~ **of propagation** (of a wave) Ausbreitungs-, Fortpflanzungs-, Wander-geschwindigkeit f ~ **of propagation of sound** Fortpflanzungsgeschwindigkeit f des Schalles ~ **of propagation of wave front** Geschwindigkeit f der Wellenfront (Wellenstirn) ~ **of rebound (or ricochet)** (ballistics) Abprallgeschwindigkeit f ~ **of rotation** Rotationsgeschwindigkeit f ~ **of scanning** Abtastgeschwindigkeit f ~ **of seismic waves** Geschwindigkeit f der seismischen Wellen ~ **of side slipping** Abrutschgeschwindigkeit f ~ **of slip stream** Abflußgeschwindigkeit f ~ **of sound** Schall-, Tongeschwindigkeit f ~ **in trajectory** Bahngeschwindigkeit f (g/m) ~ **of translation of an undulation (or wave)** Fortschrittsgeschwindigkeit f einer Welle ~ **of transport** Transportgeschwindigkeit f ~ **of wake (or wash)** Abflußgeschwindigkeit f ~ **of waste flow** Abflußgeschwindigkeit f ~ **of wave** Gruppengeschwindigkeit f einer Welle ~ **of wave propagation** Fortpflanzungsgeschwindigkeit f der Wellen

**velocity,** ~ **amplitude** Bewegungsamplitude f ~ **analysis** Geschwindigkeitsanalyse f ~ **antiresonance** Geschwindigkeitsgegenresonanz f (acoust.) ~ **change caused by dislocation** Geschwindigkeitsänderung f infolge von Versetzung ~ **correction** Geschwindigkeitskorrektur f ~ **correlation function** Geschwindigkeitseinflußfunktion f ~ **curve** Geschwindigkeitskurve f ~ **dependence** Geschwindigkeitsabhängigkeit f ~ **diffusion (or equalization)** Geschwindigkeitsausgleich m ~ **distribution** Geschwindigkeitsverteilung f ~ **distribution in a duct** Geschwindigkeitsprofil n ~ **factor** Verkürzungsfaktor m ~ **field** Geschwindigkeitsfeld n ~ **focussing** geschwindigkeitsfokussierend ~ **gradient** Geschwindigkeitsgradient m ~ **head** Geschwindigkeitshöhe f, Staudruck m **~-head speed indicator** Staudruckfahrtmesser m (aviat) **~-head tachometer** Staudrucktachometer r ~ **increase** Geschwindigkeits-zunahme f, -zuwachs m ~ **increment** Zusatzgeschwindigkeit f ~ **jump** Abgangsfehler m, Geschwindigkeitssprung m ~ **law** Geschwindigkeitsgesetz n ~ **level** Geschwindigkeitspegel m (acoust) ~ **limit** Grenzgeschwindigkeit f ~ **memory** Geschwindigkeitsspeicher m ~ **microphone** Bändchenmikrofon n, Bewegungs-empfänger m, -mikrofon n, Druckgradienten-empfänger m, -mikrofon m, Geschwindigkeitsmikrofon n, Schallgeschwindigkeitsmikrofon n, Schallschnellempfänger m **~-modulated electron beam (or pencil)** geschwindigkeitsgesteuerter Elektronenstrahl m, **~-modulated oscillator** geschwindigkeitsmodulierter Oszillator m, **~-modulated tube** Laufzeitrohr n (electron.) ~ **modulating effect** Laufzeiteffekt m ~ **modulation** Geschwindigkeits-, Laufzeiten-, Linien-, Zeilen-steuerung f, Spannungsmodulation f, Zeitmodulation f ~ **modulation tube** Durchlaufzeitröhre f ~ **modulation valve** Laufzeit-, Reflexions-röhre f ~ **pick-up** Geschwindigkeitstonabnehmer m ~ **potential** Geschwindigkeitspotential f ~ **profile** Geschwindigkeitsprofil n ~ **range curve** Geschwindigkeit-Weglinie f ~ **rate** Verkürzungsfaktor m ~ **ratio** Geschwindigkeitsaufteilung f ~ **reduction** Geschwindigkeitsverminderung f ~ **resolution** Geschwindigkeitsauflösung f ~ **resonance** Phasenresonanz f ~ **selector** Geschwindigkeitsselektor m ~ **sensitive** geschwindigkeitsempfindlich ~ **sorting** Geschwindigkeitssortierung f ~ **space** Geschwindigkeitsraum m ~ **spectrograph (or analyzer)** Geschwindigkeitsspektrograf n ~ **stage** Geschwindigkeitsstufe f ~ **staging** Geschwindigkeitsabstufung f ~ **step** Geschwindigkeitsstufe f ~ **vector** Geschwindigkeitsvektor m ~ **vector diagram** Geschwindigkeitsdreieck n

**velograph** Schnellpause f

**veluring** Nachtouren n

**velvet** Sammetstoff m, Samt m, samtartiger Stoff m ~ **brown** Sandbraun n ~ **copper ore** Kupfersamterz n ~ **printing** Samtdruck m ~ **trap** Samtdichtung f

**vending machine** Verkaufsautomat m

**vendor** Lieferfirma f

**veneer, to** ~ furnieren, verkleiden **to** ~ **on both sides** gegenfurnieren

**veneer** Deckblatt n, Deckspan m, Dünnschnitt m, Furnier-blatt n, -holz n, Verputz m ~ **assembling machine** Furnierzusammensetzmaschine f ~ **clamper** Furnierpresser m ~ **construction** Verblendbau m ~**-cutting machine** Furnierschneidemaschine f ~ **frame saw** Furnierrahmensäge f ~ **knife** Furniermesser n ~ **layer** Furnierblatt n ~ **(cutting) machine** Spanschneidemaschine f ~ **press** Furnierpresse f ~ **saw** Furniersäge f ~**-saw frame** Furnierschneidemaschine f ~ **sheets** Furnierplatten pl ~ **spoke shaves** Furnierschabhobel pl

**veneerer** Furnierschreiner m

**veneering** (of wood) Plattierung f ~ **circular saw** Furnierkreissäge f ~ **saw** Furniersäge f

**Venetian,** ~ **blind** Fensterjalousie f, Jalousie f, Klappladen m, Schalterladen m ~**-blindtype shutter** Jalousieschütz n ~ **red** Zementrot n

**Venice turpentine** Lärchenterpentin n

**venom** Gift n

**venomous** giftig

**vent, to** ~ auflockern, durchlüften, entlüften, ventilieren

**vent** Abzug m, Abzugsöffnung f, Auslaßröhre f, Austritt m, Belüftung f, Druckausgleichsöffnung f, Dunstloch n, Düse f, Entlüfterstutzen m, Entlüftung f, Entlüftungsöffnung f, Luftabzug m, Lüfter m, Mund m, Rauchklappe f, (charcoal) Raum m, Raumloch n, Scheitelöffnung f, Schlitz m, Schlot m, Steiger m, Steigetrichter m, Ventil n, Ventilationskamin m, Zündkanal m ~ **in the crater** Krateröffnung f

**vent,** ~ **branch** Entlüftungsstutzen m ~ **bush** Zündlochstollen m ~ **cap** Entlüftungsklappe f ~ **collar** (parachute) Lüftungsansatz m, Stoffkragen m ~ **covering** Schlitzabdeckung f ~ **filter** Entlüftungsfilter n ~ **flange** Flansch m für Entlüftungsleitung ~ **float valve** Entlüftungs-, Schwimmer-ventil n ~ **hole** Brandloch n (of blasting cap), Entlüftungsloch n, Luftloch n, Lüftungsschlitz m, Windfang m, Windpfeife f, Zugloch n ~ **line** Entlüftungsleitung f, Leckleitungsrohr n ~ **manifold** Belüftungsverteiler m, Entlüftungssammler m ~ **outlet** Entgasungsstutzen m ~ **pipe** Abdampfrohr n, Abzugsrohr n, Abzugsröhre n, Dunstabzugsrohr n, Windpfeife f, Zug-rohr n, -röhre f ~ **plug** Abdampfstöpsel m, Entlüfterstutzen m ~ Luftnadel f, (in molding) Luftspieß m, (in core forming) Stoßnadel f ~ **screw** Entlüfterschraube f ~ **stack** (of arc lamp) Abzugs-rohr n, -röhre f ~ **stack of arc lamp** Bogenlampenabzugsrohr n ~ **system** (Be)Lüftungsanlage f ~ **tube** Zug-rohr n, -röhre f ~ **valve** Atmungsventil n, Entlüfter m, Entweichungsventil n, Lüftungsventil n **(air)** ~ **valve** Entlüftungsventil n ~ **valve thrust control** Schubregelung f durch Abblasventile (g/m)

**ventilate, to** ~ auslüften, belüften, bewettern, durchlüften, entlüften, lüften, ventilieren **to** ~ **with a fire kibble** kesseln (min.)

**ventilated** gelüftet ~ **commutator** kollektorbelüftet ~ **radiator** Radiatorkühlung f ~ **ribbed surface** mit belüftetem Rippengehäuse n

**ventilating,** ~ **air** Atemluft f ~ **aperture** Ventilationsklappe f ~ **blade** Ventilationsflügel m ~ **channels** Abzüge pl (electr.) ~ **chimney** Dunstkamin m ~ **cover** Schachtabdeckung f mit

Entlüftungsschlitzen ~ **current** Wetterstrom m ~ **duct** Entlüftungsrohr n, Kühlschlitz m, Luftkanal m ~ **exhausts** Ablaufstutzen pl ~ **fan** Wetterrad n ~ **outlets** Abluftstutzen pl ~ **pipe** Dunstrohr n ~ **plate** Lüftungsblech n ~ **shaft** Entlüftungs-kanal m, -schlot m, -schornstein m ~ **slot** Belüftungsspalt m ~ **stubs** Abluftstutzen pl ~ **system** Belüftungsanlage f, Lüftungsanlage f ~ **tube** Lutte f

**ventilation** Belüftung f, Bewetterung f, Entlüftung f, Lüftung f, Lüftungsanlage f, Luftwechsel m, Ventilation f, Wetterhaltung f (min.), Wetterung f ~ **of commutator** Kollektorbelüftung f ~ **of the core** Kernlüftung f ~ **of mines** Bewetterung f von Grubenbauten, Grubenbewetterung f ~ **from outside** fremdbelüftet ~ **of passenger cabin** Fluggastraumlüftung f

**ventilation,** ~ **appliance** Lüftungsvorrichtung f ~ **cap** Saughutze f ~ **dam** Wetterdamm m (min.) ~ **flap** Lüftungsflügel m ~ **hole** (Luftloch) Atmungsöffnung f ~ **hood** Abzugshaube f ~ **openings** Lüftungsöffnung f ~ **pipe** Lüftungsrohr n ~ **plant** Lüftungsanlage f ~ **plug-in unit** Lüftereinschub m ~ **slot** Entlüftungsschlitz m ~ **system (or arrangement)** Entlüftungseinrichtung f ~ **top** Ventilationsaufsatz m

**ventilator** Bläser m, Entlüfter m, Entlüftungsschlitz m, Fächer m, Flügelgebläse n, Gebläse n, Lüfter m, Lüftergebläse n, Rauchabzugskanal m, Ventilator m, Wetterrad m, Windzug m ~ **of the superstructure** Aufbaulüfter m

**ventilator,** ~ **bolt** Lüfterschraube f ~ **chimney** Dunstschlot m ~ **hub** Lüfternabe f ~ **impeller** Entlüftungslaufrad n ~ **indicator lamp** Lüfteranzeigelampe f ~ **separator** (grain) Gebläseseparator m ~ **shroud** Lüftertunnel m ~ **thrust** Pressung f ~ **tube** Dunstschlot m

**venting** Entlüftung f, Gasabfuhr f ~ **by piercing the rammed sand** (in molding) Luftstechen n

**venting,** ~ **pressure** Entlüftungsdruck m ~ **property** Durchlässigkeit f, Luftdurchlässigkeit f ~ **purpose** Entlüftungszweck m ~ **quality** Luftdurchlässigkeit f ~ **screw** Entlüftungsschraube f ~ **valve** Belüftungsventil n ~ **wire** Luftspieß m, Spieß m

**ventless delay-action cap** gasloser Zeitzünder m

**ventral,** ~ **fin** Kielflosse f ~ **gun mount** Bodenlaffette f ~ **tank** (aircraft) Bauchbehälter m ~ **turret** Bodenkanzel f

**ventriculography** Ventrikulografie f

**venture** Risiko n

**Venturi** Glühhals m ~ **head** Ofenkopf m (g/m) ~ **meter** Einsatzmesser m, Venturimesser m ~**-pitot tube** Saug- und Staurohr m, Venturi-Pitotrohr n ~ **section** Venturi-Einsatz m ~**-shaped** venturiförmig ~ **tube** Düse f, Lufttrichter m, Saug-düse f, -rohr n, Venture-düse f, -rohr n

**veratric acid** Veratrumsäure f

**veratrine sulfate** Veratrinsulfat n

**Verdan system** Verdanverfahren n

**Verdet's constant** Rotationsvermögen n

**verdict** Ausspruch m, Entscheidung f, Gutachten n, Rechtsspruch m, Urteil n ~ **of guilty** Schuldigerklärung f

**verdigris** Grünspan m, Kupfer-grün n, -rost m

verge, on the ~ of oscillating Schwingungsneigung *f*

vergency Vergenz *f* ~ of rays Strahlvergenz *f*

verifiable kontrollierbar

verification Befund *m*, Beglaubigung *f*, Eichung *f*, Feststellung *f*, Kontrolle *f*, Nachprüfung *f*, Prüfung *f*

verifier (Lochkarte) Kartenprüfer *m*, Prüfer *m*, Verifiziergerät *n* ~ operator Lochprüferin *f*

verify, to ~ beglaubigen, beurkunden, beweisen, erhärten, feststellen, konstatieren, nachprüfen, prüfen, überprüfen to ~ a busy report Besetztsein *f* prüfen to ~ the fuses Sicherungen *pl* prüfen

verifying Nachmessung *f*

vermillion Zinn-farbe *f*, -oberfarbe *f*, -oberrot *n* ~ substitute Zinnoberersatz *m*

vermin Schädling *m*, Ungeziefer *n* ~ destroyer Ungeziefervertilgungsmittel *n*

vernal equinox Frühlings-äquinoktium *n*, -nachtgleiche *f*

vernalize, to ~ jarowisieren

vernier Feineinsteller *m*, Feinstellschraube *f*, Feinsteller *m*, Feinstellvorrichtung *f*, Gradteiler *m*, Nonius *m*, Noniuseinrichtung *f*, Steuerrakete *f*, Transporteur, Vernier *n* ~ adjustment Feineinstellung *f* ~ caliper Schublehre *f* ~ caliper with locating arbor Schieblehre *f* mit Aufnahmedorn ~ capacitor Nachstimmungskondensator *m* ~ condenser Vernierkondensator *m* ~ control Fein-abstimmung *f*, -regelung *f* ~ control condenser Feineinstellkondensator *m* ~ coupling Kreuzkupplung *f* ~ depth gauge Tiefenlehre *f* mit Nonius ~ dial Feinstellskala *f* ~ division Noniusteilstrich *m* ~ gauge Schieb-, Schub-lehre *f* ~ height gauge Höhenschieblehre *f* ~ lock plate Vielnuten-Sperritzel *m* ~ magnifier Noniuslupe *f* ~ reading Noniusablesung *f* ~ scale Nonius-einteilung *f*, -skala *f* ~ scale mark Noniusstrich *m* ~ tube Noniusröhre *f* ~ tuning Feineinstellung *f*, Frequenzangleich *m*, Vernier *n* ~ tuning condenser Feineinstellkondensator *m*

Vernon-Harcourt lamp Pentanlampe *f*

versatile vielseitig

versatility Anpassungsfähigkeit *f*, Vielseitigkeit *f*, Wandelbarkeit *f* ~ of service vielseitige Verwendbarkeit *f*

version Ausführung *f*, Ausführungsart *f*

versorial force Direktionskraft *f*

vertebra Wirbel *m*

vertebrate waveguide Gliederhohlleiter *m*

vertex Gipfelpunkt *m*, Scheitel *m*, Scheitel-höhe *f*, -punkt *m*, Spitze *f* ~ of lens Linsenscheitel *m* ~ of the triangle Scheitel *m* des Dreiecks

vertex, ~ angle Scheitelwinkel *m* ~ depth Pfeilhöhe *f* ~ plate (antenna) Scheitelplatte *f* ~ power Scheitelbrechwert *m* ~ refraction Scheitelbrechwert *m* (opt.) ~ refractionometer Scheitelbrechwertmesser *m* ~ tangent Scheiteltangente *f*

vertical Lotlinie *f*, Senkrechte *f*, Vertikale *f*; aufrecht, im Lot, lotrecht, scheitelrecht, seiger, senkrecht stehend, vertikal ~ and rotary switch Drehwähler *m* ~ and rotary selector Hebdrehwähler *m*

vertical, ~ acceleration Vertikalbeschleunigung *f* ~ accuracy Höhengenauigkeit *f* ~ adjustability

Absenkbarkeit *f* ~ adjustment Hoch-, Senkrecht-, Tief-verstellung *f* ~ adjustment screw Höhenstellschraube *f* ~ aerial photograph Senkrechtluftaufnahme *f* ~ anemometer Vertikalanemometer *n* ~ angle entgegengesetzter Winkel *m*, Scheitel-, Vertikal-winkel *m* ~ angular acceleration Höhenbeschleunigung *f* ~ angular velocity Höhengeschwindigkeit *f* ~ antenna Stabstrahler *m* ~ axis Achsensenkrechte *f*, Hauptachse *f* (of a rhombic crystal), senkrechte Achse *f* Stehachse *f*, Vertikalachse *f*, Z-Achse *f* ~ axis of a chariot Schlittenachse (telegr.) ~ axis of plane Hochachse *f* ~ bank Steilkurve *f*, vertikale Querneigung *f* (aviat.) ~-bar oscillator Vertikalbalkengenerator *m* (TV) ~ beams of cofferdam Dammfalze *pl* ~ bearing Fuß-, Steh-lager *n* ~ beds Kopfgebirge *n* ~ blanking Zeilen-austastung *f*, -unterdrückung *f* ~ blanking impulse Bildaustastzeichen *n* ~ bore Senkrechtbohrung *f* ~ boring mill Karusseldrehbank *f*, Vertikalbohrmaschine *f* ~ brace Versteifungsschiene *f*, Vertikalverband *m* ~ bucket elevator Becherwerk *n* für senkrechte Förderung ~ capacity (of a testing machine) Prüfhöhe *f* ~ capstan Welle *f* ~ capstan shaft Achse *f* ~ capstan winch Spill *n* ~ cattle guard Scheuerpfahl *m* ~ center-line-wire Senkrechtfaden *m* ~ center section Meridianschnitt *m* ~ central shaft stehende Zentralwelle *f* ~-chamber oven Vertikalkammerofen *m* ~ circle Höhen-, Vertikal-kreis *m* ~ clearance Durchfahrthöhe *f*, lichte Höhe *f* ~(-flue) coke oven vertikaler Koksofen *m* ~ collimation Höhenkollimation *f* ~ commutator Stirnkollektor *m* ~ component lotrechte Komponente *f*, Vertikalkomponente *f* ~ component of earth's magnetic field Vertikalintensität *f* ~ component of speed Vertikalgeschwindigkeit *f* ~ component of velocity Sinkgeschwindigkeit *f* ~ control Höhen(ver)messung *f*, Nivellierung *f* ~ creel Aufsteckgatter *n* ~ crossrail screw Spindel *f* für die Querbalkenhöhenstellung ~ crushing mill Kollermühle *f* mit aufrechten Steinen *f* ~ crystallizer stehender Kristallisator *m* ~ cutters Steilhauer *m* ~ definition Vertikalauflösung *f* ~ deflection Ablenkungswechsel *m*, Höhenunterschied *m*, Vertikalablenkung *f* ~-deflection cycle Ablenkungswechsel *m* ~ deflection electrodes Vertikalablenkelektroden *pl* ~ depth Scheiteltiefe *f* ~ deviation Höhenabweichung *f* ~ diaphragm line Zielstrich *m* ~ difference Höhendifferenz *f* ~ diffusion Vertikalausbreitung *f* ~ diffusion of sound Schallvertikale *f* ~ dimension Höhenmaß *n* ~ direction Lotrichtung *f*, senkrechte Richtung *f*, Vertikalrichtung *f* ~ dispersion Höhenstreuung *f* ~ displacement Höhenverstellung *f* ~ distance between ship bottom and channel bed Flottwassertiefe *f* ~ distribution Höhenverteilung *f* ~ dive Kopfsturz *m*, Sturzflug *m* ~ double-anemometer Senkrechtdippelanemometer *m* ~ drilling machine Senkrecht-, Vertikal-bohrmaschine *f* ~ drive Senkrechtantrieb *m* ~ driving shaft Vertikalantriebswelle *f*, -welle *f* ~ driving-shaft gear Antriebsrad *n* zur Vertikaleinstellung ~-drop machine Fallwerk *n* ~ dynamic balance Vertikalkraftwaage *f* ~ edge mill Kollermühle *f* mit aufrechten Steinen ~ elevator Steilaufzug *m* ~

engine Reihenmotor *m*, stehender Motor *m*
~ erection pickup Lotgeber *m* ~-erection reversing switch (autopilot) Horizontaufrichtung
*f* ~ factor Vertikalkomponente *f* ~ feed Senkrechtvorschub *m* ~ feed lever Zustellhebel
*m* ~ feed pinion Vertikalvorschubzahnrad *n*
~ feed rod Vertikalvorschubspindel *f* ~ feed
trip blocks feststehender Anschlag *m* für die
Auslösung der selbsttätigen Vertikalbewegung
~ feed wheels Handräder *pl* für die Vertikaleinstellung ~ figure of eight senkrecht stehende
Loopingacht *f* ~ fin Kiel-, Seiten-flosse *f* ~-
fin adjustment Seitenflossenverstellung *f* ~
fine adjustment Höhenfeinverstellung *f* ~ fire
Wurffeuer *n* ~ flue Vertikalzug *m* ~-flue oven
Horizontal-, Vertikal-kammerofen *m* ~-flued
regenerative oven Verbundofen *m* ~ focussing
Vertikalfokussierung *f* ~ force Senkrechtkraft
*f*, Vertikalintensität *f* ~ force scale Vertikalkraftwaage *f* ~ frame member Seitenprofil *n*
~ frequency Teilbild-, Vertikal-frequenz *f* ~
girder Ständer *m* ~ guidance Vertikalnavigation *f* ~ guide Vertikalführung *f* ~ gust Verti-
kalbö *f* ~ gust bump Steigbö(e) *f* ~ gust recorder Steigböenanzeiger *m* ~ gyro Vertikal-
kreisel *m* ~ gyro control Vertikalkreiselführung
*f* ~ head Senkrecht-abzug *m*, -spritzkopf *m*
~-heating furnace Herdtiefofen *m* ~ hold (control) vertikaler Bildfang *m* ~ hole Senkrechtbohrung *n* ~ hunting Tanzeffekt *m* ~ illumination Auflicht *n* ~ illuminator Opak-, Vertikal-
illuminator *m* ~ index bubble Höhenindexlibelle *f* ~ index head Vertikalteilapparat *m*
~ index level Höhenindexlibelle *f* ~ induction
Senkrechtinduktion *f* ~ ingot-heating furnace
Tiefofen *m* ~ intensity Senkrechtstärke *f*,
Stärke *f* der lotrechten Komponente ~ internal
grinding machine Senkrechtinnenschleif-
maschine *f* ~ interruptor contact Ruhekontakt
*m* des Hebmagneten ~ interval Höhenabstand
*m*, vertikaler Zwischenraum *m* ~ jump of a gun
(ballistics) Abgangsfehler *m* ~ kiln Vertikal-
ofen *m* ~* lathe Karusseldrehbank *f* ~-lift
bridge Hebebrücke *f* ~-lift gate Hubtor *n* ~
lift valve Zugschutz *m* ~-light Zenitlicht *n* ~
lime kiln Kalkschachtofen *m* ~ line Lot *n*,
Lotrechte *f*, Lotriß *m*, Scheitellinie *f* ~ lobe line
Steuersackleine *f* ~ lode seigerer Gang *m* ~
loop senkrechte Schleife *f* ~ magazine Vertikal-
magazin *n* ~ magnet Hebemagnet *m* ~ magnetic force senkrechte magnetische Kraft *f* ~
milling machine Senkrecht-, Universal-, Verti-
kal-fräsmaschine *f* ~ motion Freifallbewegung
*f* ~-motion screw Höhenstellschraube *f* ~
navigation Vertikalnavigation *f* ~ obstacle
clearance vertikale Hindernisfreiheit *f* ~ off-
normal contacts Kopfkontakte *pl* ~ optimeter Optimeter *m* mit senkrechtem Ständer
~ parallax Höhen-, Vertikal-parallaxe *f* ~-
parallax slide Vertikalparallaxenschlitten *m*
~ pattern Vertikaldiagramm *n* ~ photograph
Nadiraufnahme *f*, Senkrecht-aufnahme *f*,
-bild *n*, Vertikalaufnahme *f* ~ picture Hochaufnahme *f* ~ pipe Standrohr *n* ~ piston pump
stehende Kolbenpumpe *f* ~-pit-type furnace
senkrechter Schachtofen *m* ~ plan Aufriß *m*
~ plane Senkrecht-, Vertikal-ebene *f* ~-plane
directional pattern Vertikaldiagramm *m* ~

plane-milling machine Senkrechtlangfräsma-
schine *f* ~ plate Stehblech *n* ~ pointer Höhen-
richtmann *m* ~ pointing correction Höhenverbesserung *f* ~ position Senkrechtstellung *f*,
Vertikallage *f* ~ power rod Steuerwelle *f* für den
Senkrechtvorschub ~ pressure (or stress) on
the center line Meridianspannung *f* ~ pressure
gradient senkrechte Luftdrucksteigung *f* ~
projection Aufriß *m*, Projektion *f* auf die
Senkrechtebene, Vertikalprojektion *f* ~ projection of conveyer length Förderhöhe *f* ~
pull-down broaching machine Senkrechträummaschine *f* mit Abwärtszug ~ pull-up broaching
machine Senkrechträummaschine *f* mit Auf-
wärtszug ~ pump stehende Pumpe *f* ~ punching and shearing machine Senkrechtstanze- und
Schermaschine *f* ~ radiator Standkühler *m* ~
range Steighöhe *f* ~ recording Tiefenschrift *f*
~ resolution Senkrecht-, Vertikal-auflösung *f*
~ retort stehende Retorte *f* ~ reversement
senkrechte umgekehrte Kurve *f* ~-ring type
roller mill Walzenringmühle *f* ~ roll Vertikal-
walze *f* ~ rolls Kopfwalzensystem *n* ~ row
senkrechte Reihe *f* ~ row index Spaltenindex
*m* ~ row matrix Spaltenmatrix *f* ~ scansion
Vertikalhinlauf *m* ~ scrubber Standkasten *m*
~ section Aufriß *m*, Profil *n*, Profillinie *f*,
Seigerriß *m*, senkrechter Schnitt *m*, Vertikalschnitt *m* ~ seismograph Vertikalseismograf
*m* ~ separation Höhenstaffelung *f* (aviat.) ~
shaft Königswelle *f*, seigerer Schacht *m*, Senk-
rechtwelle *f*, stehende Welle *f* ~ shift Höhenverstellung *f* ~ shot Hochaufnahme *f* ~ side
Amtseite *f* (des Hauptverteilers), senkrechter
Schenkel *m* ~ sighting (or view) telescope
Vertikalsichtfernrohr *n* ~ size control Teil-
bildhöhenregler *m* ~ sliding breechblock Fallblockverschluß *m* ~ slot Senkrechtschlitz *m* ~
slotting machine Senkrechtstoßmaschine *f* ~
solid angle Polecke *f* ~ speed Senkrecht-
geschwindigkeit *f* ~-speed indicator Variometer *m* ~ spin tunnel senkrechter Windkanal
*m* ~ spindle grinder Senkrechtschleifmaschine
*f* ~ spindle head Senkrechtfräskopf *m* ~ spread
Höhenstreuung *f* ~ spread of sound Schallvertikale *f* ~ stabilizer Kiel-, Seiten-flosse *f* ~
step Heb-, Höhen-schritt *m* ~ step of selector
Wählerhebeschritt *m* ~ stepping-down movement Zeilenschaltung *f* ~ stepping magnet
Hubmagnet *m* ~ strip camera Senkrechtreihen-
kamera *f* ~ sweep Bildabtastung *f* (TV), vertikale Abtastung *f* ~ swivel mount setting scale
Einstellteilung *f* für Senkrechtschwenklager ~
synchronization Vertikalsynchronisierung *f* ~
synchronizing cycle Vertikalwechsel *m* ~ synchronizing impulse Bildgleichlaufzeichen *n* ~
synchronizing pulse Vertikalwechsel *m* ~ tail
area senkrechte Schwanzoberfläche *f* ~ tail
fin Kielflosse *f* ~ tail surface senkrechte
Schwanzoberfläche *f* ~ tail surface load
Seitenleitwerkskraft *f* ~ take-off (VTO)
Senkrechtstart *m* ~ take-off and landing
(VTOL) aircraft Lotrechtstarter *m*, Senkrechtstartflugzeug *n* ~ temperature gradient senkrechter Temperaturgradient *m* ~ thermal air
current Senkrechtströmung *f* der Luft ~ thread
Vertikalfaden *m* ~ tracking scope Höhenrichtfernrohr *n* ~ trapezoidal sluice Rollkeilschütz *n*

~ **travel** Senkrechtbewegung *f* ~ **trussing** Ständerfachwerk *n* ~**-tube boiler** Steilrohrkessel *m* ~**-tube evaporator** Steilrohrverdampfer *m* ~ **turn** senkrechte Kurve oder Wendung *f* ~ **turret lathe** Karusselrevolverdrehbank *f* ~ **type** stehende Bauart *f* ~**-type carburetor** Steigstrom-vergaser *m*, -vorrichtung *f* ~**-type support** Senkrechthalter *m* ~ **upward motion** Hebbewegung *f*, Heben *n* ~ **valve** stehendes Ventil *n* ~ **vein** Seigergang *m* ~ **visual angle** Höhenwinkel *m* ~ **wall duct** Hochführungskanal *m*, -schacht *m* ~ **water scrubber** wasserberieselter Koksskrubber *m* ~ **weld** senkrechte Schweißung *f*, stehende Naht *f* ~ **wind component** vertikale Windgeschwindigkeit *f* ~ **wind currents** vertikale Luftströmungen *pl* ~ **wind tunnel for spinning tests** Trudelwindkanal *m* ~ **wire** Vertikalfaden *m* ~ **wire antenna** lineare Antenne *f*, Vertikalantenne *f*

**verticality** Senkrechtstellung *f*

**vertically** in der Höhe ~ **adjustable** hochstellbar ~ **adjustable bottom** Hebeboden *m* ~ **adjustable seat** in der Höhe verstellbarer Sitz *m* ~ **cast** stehend gegossen ~ **downward** nadirwärts ~ **polarized wave** senkrecht polarisierte Welle *f* ~ **split casing** Topfgehäuse *n* ~ **striped** langgestreift

**very** gar, sehr ~ **active** rege ~ **adhesive** festhaftend ~ **big** ganz groß ~ **busy** vielbeschäftigt ~ **fine sand** Mehl-, Staub-sand *m*, Steinmehl *n* ~ **fine sulfur** Ventilatorschwefel *m* ~ **fine writing** Perlschrift *f* ~ **firm** mauerfest ~ **granular** kornreich ~ **heavy** überschwer ~**-high frequency (VHF)** Meterwellen(bereich) *m*, UKW-(Bereich) *m*, Ultrakurzwellen(bereich) *m* ~**-high-frequency reception** UKW-Empfang *m* ~ **low-carbon steel** Fluß-eisen *n*, -stahl *m* ~ **low frequency (VLF)** Längstwellen(bereich) *m*, Myriameterwellen(bereich) *m* ~ **profitable wood** hochgeschoßtes Holz *n* ~ **pure clay** Letten *m* ~ **quickly** sehr schnell, mit wenigen Handgriffen ~ **sharp** haarscharf ~ **short waves** Meterwellen *f* ~ **small print** Augenpulver *n* ~ **soft picture** flaches Bild *n* ~ **solid** nagelfest ~ **thick film of rust** Flugrost *m*

**vesicant** ätzender Kampfstoff *m*, Blasenzieher *m*; blasenziehend, hautätzend ~ **agent** ätzender oder blasenziehender Kampfstoff *m*, Hautgift *n* ~ **gas** ätzender Kampfstoff *m*

**vesicants** Ätzstoffe *pl*

**vesicle** Bläschen *n*

**vesicular** aufgebläht, blasig

**vessel** Becken *n*, Behälter *m*, Birne *f*, Bottich *m*, Fahrzeug *n*, Gefäß *n*, Geschirr *n*, Küvette *f*, Schiff *n*, Wasserfahrzeug *n* ~ **is towed ein** Schiff wird geschleppt ~ **of a tube (or of a valve)** Röhrenkolben *m* ~ **for water heating** Wasserblase *f*

**vessel, ~ holder** Gefäßhalter *m* ~ **index** (list of empty liquid containers) Gebindekartei *f*

**vestibule** Flur *m*, Vorhalle *f*

**vestigal sideband** Restseitenband *n*

**vestige** Überbleibsel *n*

**V.F. (voice-frequency)** Tonfrequenz *f* ~ **currents** Sprechströme *pl* ~ **ringer** Tonfrequenzrufsatz *m* ~ **ringing set** Tonfrequenzrufsatz *m*

**V fork** Wegeinmündung *f*

**V formation** Keil *m*, Keilform *f*

**V-groove, to** ~ auskreuzen, (welding) ausvauen

**V-groove** Keilspundung *f*

**VHF** Meterwellenbereich *m* ~ **direction finder** UKW-Peilanlage *f* ~ **homer** UKW-Peiler *m* ~ **omnidirectional radio range (VOR)** UKW-Drehfunkfeuer *n* ~ **radio telephony** UKW-Sprechfunk *m*

**via, ~ administration** Durchgangsverwaltung *f* ~ **center** Durchgangszentrale *f* ~ **circuit** Durchgangsleitung *f* ~ **traffic** Durchgangsverkehr *m*

**viable** lebensfähig

**viaduct** Brücke *f*, Straßenbrücke *f*, Talüberführung *f*, Transitstromkreis *m*, Überführung *f*, Viadukt *m*, Wegüberführung *f*

**vial** Fläschchen *n*, Phiole *f*

**vibrate, to** ~ beben, erschüttern, erzittern, pendeln, prellen, rütteln, schnarren, schütteln, schwingen, vibrieren, zittern **able to** ~ schwingfähig **to** ~ **to a harmonic** in einer Harmonischen *f* schwingen **to** ~ **thoroughly** durchschütteln

**vibrated** eingerüttelt ~ **concrete** geschüttelter Beton *m*

**vibrating** schwingend, vibrierend ~ **bell** Gleichstromwecker *m*, Wecker *m* mit Selbstunterbrechung ~ **break** Hammerunterbrecher *m* ~ **capacitor** Schwingungskondensator *m* ~ **chord** schwingende Saite *f* ~ **chute** Vibrationsrinne *f* ~ **circuit** Vibrationskreis *m* ~ **condenser** Schwingkondensator *m* ~ **contact** schwingender Kontakt *m* oder Unterbrecher *m* ~ **contactor** Zerhacker *m* ~ **conveyor** Schwingförderer *m* ~ **electrode** Zitterelektrode *f* ~ **governor (or regulator)** Zitterregler *m* ~ **head** Rüttelkopf *m* ~ **mill** Schwingmühle *f* ~ **mirror** Kippspiegel *m*, schwingender Spiegel *m*, Schwingspiegel *m* ~**-mirror scanner** Schwingspiegelabtaster *m* ~ **plate** Schwingungsteller *m* ~ **plate compactor** Platten-vibrator, -rüttler *m*, Rüttelplatte *f* ~ **point** Rüttelnadel *f* ~ **rectifier** Pendel-gleichrichter *m*, -umformer *m*

**vibrating-reed** schwingende Blattfeder *f* oder Zunge *f* ~ **break** Zungenunterbrecher *m* ~ **frequency meter** Zungenfrequenzmesser *m* ~ **gyro** Schwingzungen-Kreisel *m* ~ **instrument** Vibrationsmeßgerät *n*, Zungenfrequenzmesser *m* ~ **rectifier** Nadelschalter *m*, Schwingkontaktgleichrichter *m* ~ **tachometer** Resonanztachometer *n* ~ **transmitter** Stimmgabelsender *m*

**vibrating, ~ relay** Tonfrequenz-, Vibrations-, Zungenfrequenz-relais *n* ~ **screen** Schlag-, Schüttel-, Vibrations-, Zitter-sieb *n* ~ **shock** Rüttelstoß *m* ~ **strength** Schüttelfestigkeit *f* ~ **string** schwingende Saite *f* ~ **string accelerometer** (gyro) Saitenbeschleunigungsmesser *m*, Vibrationsdraht-Beschleunigungsmesser *m* ~ **string extensometers** Dehnungsmesser *m* mit schwingenden Saiten ~ **table** Vibriertisch *m* ~ **troughs** Schwingrinnen *pl* ~ **wire** schwingender Draht *m* ~**-wire interrupter** Saitenunterbrecher *m*

**vibration** Beben *n*, Eigenschwingung *f*, Erschütterung *f*, Rütt(e)lung *f*, Schwingung *f*, Schwung *m*, Vibration *f*, Vibrieren *n*, Zitterung *f* ~ **of air (or gas)** Luftschwingung *f* ~ **of antenna** Antennenschwingung *f* ~ **of contacts** prellende Kontaktöffnung *f* ~ **of form** Rütteln *n* ~ **of mold** Schalungsrütteln *n* ~ **of a molecule** Mole-

külschwingung *f* ~ **of the package cradle** Rahmenschwingung *f* ~ **of soil** Erschütterung *f* des Bodens
**vibration,** ~ **absorber** Schwingungsaufnehmer *m* ~ **absorption** Schwingungsdämpfung *f* ~ **clamp** Flatterbock *m* ~ **contacts** Erschütterungskontakt *m* ~ **dampening connector** Schwingmetall *n* (aviat.) ~ **damper** Schwingungs-, Ton-dämpfer *m* ~ **damping properties** Dämpfung *f* ~ **direction** Schwingungsrichtung *f* ~ **energy** Schwingungsenergie *f* ~ **failure** Schwingungsbruch *m* ~-**free mounting** schwingungsfreies System *n* ~ **frequency** Schwingungs-, Schwing-zahl *f* ~ **frequency meter** Schwingungsmesser *m* ~ **fuse** Erschüttzünder *m* ~ **galvanometer** Resonanz-, Saiten-, Vibrations-galvanometer *n* ~ **machine** Schütteltisch *m* ~-**measuring apparatus** Vibrograf *m* ~ **measuring instrument** Schwingungsmeßgerät *n* ~ **meter** Erschütterungsmesser *m* ~ **method** Schüttelverfahren *n*, Schwebungsmethode *f*, Schwingungsverfahren *n* ~ **mill** Vibrator *m* ~ **motion infinitely adjustable** stufenlos einstellbare Querreibung *f* ~ **mount** Gummimetallelement *n*, Schwingmetall *n* ~ **nodal point (or node point)** Schwingungsknoten *m* ~ **period** Schwingzeit *f* ~ **pickup** Schwingungs-abtaster *m*, -aufnehmer *m* ~ **process** Rüttelverfahren *n* ~-**proof** erschütterungsfest ~-**proof construction** erschütterungsfreie oder rüttelfeste Bauart *f* ~ **quantum** Schwingungsquant *n* ~ **relay** Gulstadrelais *n* ~ **remaining undamped** Restschwingung *f* ~ **research** Schwingungsforschung *f* ~ **resistance** Vibrationsfestigkeit *f* ~ **resonance** Schüttelresonanz *f* ~-**rotation spectra** Rotationsschwingungsspektren *pl* ~ **sander** Vibrationsschleifmaschine *f* ~ **sifter** Vibrationssieb *n* ~**(al) spectrum** Schwingungsspektrum *n* ~ **strain pickup** Gebergerät *n* zur Messung von Schwingungsbeanspruchungen ~ **strength** Schwingungsfestigkeit *f* ~ **stress** Schwingungsbeanspruchung *f* ~ **superposition** Schwingungsüberlagerung *f* ~ **table** Schütteltisch *m* ~ **test** Schüttel-versuch, -prüfung *f*, Schwingungsversuch *m* ~ **test stand** Rüttelprüfstand *m*
**vibrational** oszillatorisch ~ **amplitude** Schwingungsamplitude *f* ~ **analysis** Schwingungsanalyse *f* ~ **behavior** Schwingungsverhalten *n* ~ **energy** Vibrationsenergie *f* ~ **mode** Eigenschwingung *f* ~ **quantum number** Oszillationsquantumszahl *f* ~ **spectrum of a lattice** Schwingungsspektrum *n* eines Kristalls ~ **state** Schwingungszustand *m* ~ **threshold** Vibrationsschwelle *f*
**vibrationless** erschütterungsfrei, schwingungsfrei
**vibrations** Schwingungserscheinungen *pl* ~ **during free decay** Ausschwingversuch *m*
**vibrator** Resonator *m*, Rüttelapparat *m*, Schwinger *m*, Schwingungsprüfmaschine *f*, Sender *m*, Summer *m*, Unterbrecher *m*, Vibrator *m*, Wechsel(gleich)richter *m*, Zerhacker *m*, Zunge *f* ~-**and-distributor-roller** Reib- und Hebwalze *f* ~-**and-distributor-roller stock** Reib- und Hebwalzenspindel *f* ~ **blade** Pendelfeder *f* ~ **converter** Zerhackerumformer *m* ~ **ignition** Summerzündgerät *n* ~ **oscillograph** Schleifen-

oszillograf *m* ~ **rectifier** Schüttgleichrichter *m* ~ **reed** Pendelfeder *f* ~ **roller** Hebwalze *f* ~-**separator** Vibrosichter *m* (zum Entstäuben) ~ **shaft** Heberwelle *f* ~ **spring** Pendelfeder *f* ~ **unit** Wechselsatz *m*
**vibratory** schwingungsfähig, stoßweise, vibrierend ~ **electrode** Zitterelektrode *f* ~ **feeder** Schüttelzuführer *m* ~ **grate** Rüttelherd *m* ~ **inverter** Pendelwechselrichter *m* ~ **mill** Schwingungsmühle *f* ~ **movement** Schwingbewegung *f* ~ **power** Schwungkraft *f* ~ **shock load** stoßweise Belastung *f* ~ **spring** Schwingfeder *f* ~ **stress** Schlagbeanspruchung *f* ~ **test** Dauerversuch *m* ~ **testing machine** Schwingungsprüfmaschine *f* ~ **voltage (or tension)** Schüttelspannung *f*
**vibrato-tactile device** Vibrationstastgerät *n*
**vibrograph** Schwingungsaufzeichner *m*, Vibrograf *m*, Wellenschreiber *m*
**vibrometer** Schwingungsmeßgerät *n* (acoust.)
**vibromotive force** schwingungserzeugende Kraft *f*
**vibropile** Vibropfahl *m*
**vibroplex key** Morsetaste *f* mit selbsttätiger Punktgebung
**vibroscope** Schwingungsmesser *m*
**vicarious element** Vertauschungselement *n*
**vice** Spannblech *n* ~ **dog** Schraubenstockklaue *f* ~ **jaw** Schraubhacke *f*
**vicinity** Gegend *f*, Nachbarschaft *f*, Nähe *f*, Nahfeld *n*, Umgegend *f* ~ **of antenna** Antennennahfeld *n*
**video** Bild . . ., Fernseh . . ., Video . . . ~ **amplifier** Bildverstärker *m* ~ **automatic gain control** automatische Video-Verstärkung *f* ~ **bus** Aufnahmewagen *m*, Fernsehaufnahmewagen *m* ~ **camera** Fernseher *m* ~ **carrier** Bildträger *m* ~ **channel** Videokanal *m* ~ **consolette** Bildmischpult *n* ~ **converter** Videokonvertor *m* ~ **current** Bildstrom *m* ~ **detection** Bildgleichrichtung *f* ~ **detector** Bildgleichrichter *m* ~ **distributor** Bildpunktverteiler *m* ~ **engineer** Bildingenieur *m* ~ **frequency** Bild-, Bildmodulations-, Bildpunkt-frequenz *f* ~-**gain control** Intensitätsbegrenzer *m* ~ **impulses** Bildschwingungen *pl* ~ **insertion** Video-Einblendung *f* ~ **mapping** Kartenvideoeinblendung *f* ~ **mixer** Bild-mischer *m*, -mischpult *n* ~ **output** Ausgangsleistung *f* des Bildgleichrichters ~ **pickup camera** Bildfänger *m* ~ **signal** Bild(punkt)-, Fernseh-, Video-signal *n* ~ **signal with blanking** ausgetastetes Bildsignal *n* ~ **signal cable** Videokabel *n* ~ **signal patching panel** Bildsignal-Wähler *m* ~ **signal wave potential** Bildwellenspannung *f* ~ **switch** Bildpunktabtaster *m* ~ **tape recording** Magnetbildverfahren *n* ~ **telephone** Fernseh-sprecher *m*, -telefon *n* ~ **transmitter** Bildsender *m*
**Vienna polishing chalk** Wiener Kalk *m*
**Vierendeel,** ~ **construction** Rahmenkonstruktion *f* ~ **truss** Pfostenfachwerk *n*
**vietinghofite** Vietinghofit *m*
**view, to** ~ ansehen, besichtigen, betrachten, blicken, sehen
**view** Anschauung *f*, Ansehen *n*, Ansicht *f*, Ausblick *m*, Aussicht *f*, Betrachtung *f*, Bildkreis *m* (photo), Blick *m*, Schnittzeichnung *f*, Sicht *f*, Veranschaulichung *f*, Zeichnung *f* ~ **from above** Höhenschau *f* ~ **from headrace** Ober-

wasserblick *m* ~ **through ocular** Okularein-
blick *m* **in** ~ **of** in Anbetracht, in Hinsicht auf
~ **to the sides** Sicht *f* nach der Seite
**view,** ~ **finder** Aufsichtssucher *m*, Bildsucher
*m* (photo), Sichtattrappe *f*, Visier *n*, Visier-einrichtung *f*, -vorrichtung *f* ~**-finder grid** Bild-
suchergitter *n* ~**-finder image** Sucherbild *n*
~ **looking aft** Ansicht *f* von hinten ~ **looking
forward** Ansicht *f* von vorn ~ **meter** Bildmesser
*m* (photo) ~**-point** Veranschaulichung *f* ~
**room** Feinmeßraum *m*
**viewing** Betrachtung *f* ~ **of slides** Vorführung *f*
von Dias *f*
**viewing,** ~ **angle** Betrachtungs-, Seh-, Sicht-
winkel *m* ~ **apparatus** Betrachtungsapparat *m*
~ **apparatus for negatives** Negativbetrach-
tungsgerät *n* ~ **box** Betrachtungseinrichtung *f*
~ **distance** Betrachtungsabstand *m* ~ **filter**
Blaufilter *n* (photo) ~ **hole** Schauloch *n* ~ **hood**
Tubus *m* ~ **lens** Beobachtungsobjektiv *n*,
Schaulinse *f* ~ **lens frame** Lupenrahmen *m* ~
**mirror** Spiegelreflektor *m* ~ **objective** Beobachtungsobjektiv *n* ~ **screen** Negativschaukasten
*m* ~ **slit** Augenschlitz *m* ~ **stereoscope** Stereo-
betrachter *m* ~ **system** optisches Überwa-
chungssystem *n* ~ **tube** Bildschreibröhre *f*,
Sehrohr *n* ~ **visor for finder** Einblicktubus *m*
für Sucher
**vigilance** Aufmerksamkeit *f*, Wachsamkeit *f*
**vignette, to** ~ vignettieren
**vignette** Letter *f*, Vignette *f*
**vignetter** Vignettierapparat *m*
**vignetting** Vignettierung *f* ~ **of cone of light**
Lichtbündelvignettierung *f*, Vignettieren *n* des
Lichtbündels ~ **effect** Vignettierwirkung *f* ~
**mask** Vignettiermaske *f*
**Vignoles' rail** Vignolschiene *f*
**vigor** Frische *f*, Heftigkeit *f*, Stärke *f*
**vigorous** frisch, heftig, kräftig, rüstig, stark,
wuchtig
**village** Dorf *n*, Ort *m*, Ortschaft *f*
**villous tuft** Zottenbüschel *n*
**vinasse** Schlempe *f*
**vinculum** Kopfstrich *m* (math.)
**vinegar** Essig *m* ~ **naphtha** Äthylacetat *n* ~**-spot
test** Essigtropfenprobe *f* ~ **tester** Essigprüfer *m*
**vineyard** Weingarten *m* ~ **cultivator** Weinbergpflug *m* ~ **plow** Weinbergpflug *m*
**vinometer** Weinmesser *m*
**vinous** weinartig ~ **fermentation** geistige Gärung *f*
**vintager's machine** Winzermaschine *f*
**vinyl,** ~ **plastic tape** Polyzinband *m* ~ **resin**
Vinylharz *n*
**violation** Gesetzübertretung *f*, Übertretung *f*,
Verletzung *f*, Zuwiderhandlung *f* ~ **of contract**
Vertragsbruch *m* ~ **of duty** Pflichtverletzung *f*
**violence** Heftigkeit *f*, Stärke *f*
**violent** heftig, stark, stürmisch ~ **agitation** Aufwallen *n* ~ **bubbling** Aufschäumen *n*, Schäumen *n* ~ **fluctuation** Sprunghaftigkeit *f* ~ **gust**
schwere Bö *f* ~ **jump** Sprunghaftigkeit *f* ~
**proof** Gewaltprobe *f* ~ **striking** Anprall *m*
**V.I.R. (vulcynised india rubber),** ~ **cable** Gummikabel *n* ~ **wire** Gummidraht *m*
**virgin** neu, rein, ungebraucht, unvermischt ~
**cork** Naturkork *m* ~ **curve** Gesetz *n* der Erst-
belastung *n*, jungfräuliche Kurve *f* ~ **curve of
pressure-void ratio diagram** Hauptast *m* des

Druckporenzifferdiagrammes ~ **field** unver-
ritztes Feld *n* ~ **metal** Frischmetall *n*, Hüttenaluminium *m*, Neumetall *n* ~ **neutron flux**
jungfräulicher Neutronenfluß *m* ~ **neutrons**
jungfräuliche Neutronen *pl* ~ **test** Erstprüfung *f*
**virginal groove** unmodulierte Rille *f* (phono)
**Virginia red cedar** Weichzeder *f*
**virial** Virial *n* (kinetische Größe) ~ **coefficient**
Virialkoeffizient *m* ~ **theorem** Virialsatz *m*
**viridine** Viridin *n*
**virtual** faktisch, tatsächlich, virtuell ~ **anode**
virtuelle Anode *f* ~ **cathode** scheinbare oder
virtuelle Kathode *f* ~ **focus** Zerstreuungspunkt
*m* ~ **height** effektive Höhe *f* (der ionisierten
Schicht), scheinbare Höhe *f*, Reflexionshöhe *f*
~ **image** virtuelles Bild *n* ~ **level** quasistatisches
Niveau *n* ~ **mass** scheinbare Masse *f* ~ **model**
virtuelles Modell *n* ~ **particle** virtuelles Teilchen *n* ~ **quantum** virtueller Quant *m* ~ **source**
Bildquelle *f* ~ **value** Effektivwert *m*, quadratischer Mittelwert *m* ~ **work** virtuelle Arbeit *f*
**viscid** dickflüssig, klebrig
**viscidity** Dick-, Zäh-flüssigkeit *f*
**visco-elastic** viskoelastisch
**viscoelasticity** Viskoelastizität *f*
**viscometer** Dichte-, Dichtigkeits-messer *m*
**viscose** Viskose *f*; viskos, zähflüssig ~ **foil
printing press** Zellglasdruckmaschine *f* ~
**spangle** Viskoseflitter *m*
**viscosimeter** Flüssigkeitsgradmesser *m*, Konsistenzmesser *m*, Viskometer *n*, Viskosimeter *n*,
Zähigkeitsmesser *m*
**viscosity** (of gels, starch) Ausgiebigkeit *f*,
Flüssigkeitsgrad *m*, Klebrigkeit *f*, Konsistenz *f*,
Reibung *f*, Viskosität *f*, Zähegrad *m*, Zähfluß
*m*, Zähflüssigkeit *f*, Zähigkeit *f*, Zähigkeits-
grad *m* ~ **breaking** Herabminderung *f* der
Viskosität ~ **concept** Zähigkeitsbegriff *m* ~
**factor** Zähigkeitsfaktor *m* ~ **friction** innere
Reibung *f* ~ **index** Viskositäts-, Zähigkeits-
index *m* ~ **manometer** Molekulardruckmano-
meter *n* ~ **measurement** Viskositätsmessung *f*
~ **ratio** Viskositätssatz *m* ~ **temperature
coefficient** Viskositätstemperaturkoeffizient *m*
**viscothermal** viskothermisch
**viscous** dickflüssig, klebrig, (flow) plastisch,
schwerflüssig, strengflüssig, viskos, zähe, zähflüssig, zähig **to become** ~ (Öl) starr werden
**viscous,** ~ **damper** Flüssigkeitsdämpfer *m* ~
**fermentation** Schleimgärung *f* ~ **flow** plastisches
Fließen *n* ~**-impingement-type air filter** Prall-
filter *m* für Luft mit Ölbenetzung ~ **loss** Zähig-
keitsverlust *m* (acoust.) ~ **products** zähflüssige
Medien *pl* ~ **slag** zähflüssige Schlacke *f* ~
**stress** Reibungsspannung *f*
**viscousness** Dickflüssigkeit *f*
**vise** Aufspannblock *m*, Klemmer *m*, Kloben *m*,
Maschinenschraubstock *m*, Schraubstock *m*
(bench), Spannstock *m*, Zange *f*, Zwinge *f* ~
**clamp** Spannkluppe *f* ~ **clamps** Reifkloben *m*
~ **coupling** Schraubkupplung *f* ~ **dog** Schraub-
stockklaue *f* ~ **jaw** Reifkloben *m* ~ **jaws**
Futterplatten *pl*, Schraubstockbacken *pl*
**visibility** Gesichtsfeld *n*, Klarheit *f*, Sicht *f*,
Sichtbarkeit *f*, Sichtbereich *m*, Sichtigkeit *f*,
Wahrnehmbarkeit *f* ~ **curves** Eichreizkurven *pl*
(electron.) ~ **indicator** Sichtanzeiger *m* ~ **meter**
Sichtbarkeitsmesser *m*, Sichtmesser *m*, Sicht-

meßgerät *n* ~ **navigation** Sichtnavigation *f* ~ **range** Sichtbereich *m*

**visible** absehbar, augenscheinlich, mit dem Auge wahrnehmbar, scheinbar, sichtbar, wahrnehmbar ~ **brickwork** Ziegelrohbau *m* ~ **horizon** Kimm *m*, natürlicher Horizont *m* ~ **light** das Sichtbare *n* ~ **measuring bowl** sichtbares Meßgefäß *n* ~ **oil gauge** Ölsichtkontrolle *f* ~ **pump** sichtbarmessende Pumpe *f* ~ **range** Reichweite *f* des Lichts ~ **signal** sichtbares Signal *n* oder Zeichen *n*

**vision** Gesichtssinn *m*, Sehen *n*, Sehkraft *f*, Sichten *n* ~ **on sound** Bild *n* im Ton

**vision,** ~ **bandwidth** Bildbandbreite *f* ~ **carrier** Bildträger *m* ~ **channel** Bildkanal *m* ~ **frequency** Bildfrequenz *f* ~ **IF amplifier** Bildzeilenfrequenzverstärker *m* ~ **normal** normalsichtig ~ **signal** Bildsignal *n* ~ **transmitter** Bildsender *m*

**visit, to** ~ bereisen **to** ~ **the works** Betriebsbesichtigung *f*

**visor** Schirm *m*, Visier *n*

**vista shot** Fernaufnahme *f*

**visual** Seh ..., Gesichts ...; sichtbar, visuell ~ **acuity** Gesichts-, Seh-schärfe *f* ~ **aids** optische Hilfen *pl* ~ **aiming device** optisches Zielgerät *n* ~ **alarm** optische Alarmvorrichtung *f* ~ **angle** Sehwinkel *m* ~ **apparatus indicator** Sichtgerät *n* (rdo) ~ **approach** Sichtanflug *m* ~ **approach slope indicator system** Gleitwinkelbefeuerung *f* (aviat.) ~ **aural range** optisch-akustischer Leitstrahlsender *m*, Vierkursfunkfeuer *n* mit Sicht- und Höranzeige ~ **axis** Gesichtsachse *f* ~ **bearing** optische Peilung *f* ~ **broadcasting** Fernsehrundfunk *m* ~ **busy lamp** Besetzt-, Freimelde-lampe *f* ~ **busy signal** optische Besetztprüfung *f* ~ **carrier** Bildträger *m* (TV) ~ **communication** optischer Fernmeldeverkehr *m*, Zeichenverbindung *f* ~ **communication by flag (or hand) signals** Winkverbindung *f* ~ **conditions** Sichtbedingungen *pl* ~ **cone** Sehzäpfchen *n* ~ **contact** Sichtberührung *f* ~ **control** Sichtsteuerung *f* ~ **control facility** Beobachtungseinrichtung *f* ~ **D.F. receiver** Sichtpeilempfänger *m* ~ **direction finder** Sichtpeiler *m* ~ **direction finder** Sichtpeiler *m* ~ **direction finding** optische Peilung *f* (aviat.) ~ **distance** Sehweite *f* ~ **education** Anschauungsunterricht *m* ~ **education material** Anschauungsmittel *n* ~ **engaged lamp** Besetztlampe *f* ~ **engaged signal** optisches Besetztzeichen *n* ~ **engaged test with key control** optische Besetztprüfung *f* mit Druckknopf ~ **excitation** Sehreiz *m* ~ **faculty** Sehvermögen *n* ~ **field angle** Gesichtsfeldwinkel *m* ~ **flashing beacon** Kennungsfeuer *n* ~ **flight rules** Sichtflugregeln *pl* ~ **frequency** Bildfrequenz *f* ~ **function** Sehfunktion *f* ~ **ground aids** optische Bodenhilfen *pl* ~ **hallucination** Falschsehen *n* ~ **impression** Gesichts-, Lichteindruck *m* ~ **indicator** Ausgangsspannungsmesser *m*, Leuchtschild *n*, Sichtanzeiger *m* ~ **indicator tube** Anzeigeröhre *f* ~ **inspection** Inaugenscheinnahme *f* ~ **line of a telescope** Gesichtsfeldlinie *f* eines Fernrohres ~ **means of communication** optische Nachrichtenmittel *pl* ~ **movement** Blickbewegung *f* ~ **observation** Betrachtung *f*, subjektive Beobachtung *f* ~

**observation method** optische Ortung *f* ~ **paging system** Personensuchanlage *f* ~ **photometer** Lichtsinnprüfer *m*, subjektives Fotometer *n* ~ **power** Sehkraft *f* ~ **presentation** optische Einblendung *f* ~ **purple** Sehpurpur *m* ~ **range** Fernsicht *f*, Sehweite *f*, Sichtweite *f* ~ **ray** Gesichts-, Seh-strahl *m* ~ **reading instrument** Anzeigevorrichtung *f* ~ **reconnaissance** Augenaufklärung *f*, -erkundung *f* ~ **reference to ground (or terrain)** Erdsicht *f* ~ **requirements** verlangtes Sehvermögen *n* ~ **sensation** Helligkeit *f* (als psychologische Größe) ~ **sighting mechanism** optisches Zielgerät *n* ~ **sign** Sehzeichen *n* ~ **signal** (magnetic) Schau-, Sehzeichen ~-**signal battery** Schauzeichenbatterie *f* ~ **signal device** optische Signalvorrichtung *f* ~ **signaling** optische Telegrafie *f* ~ **stimulus** Sehreiz *m* ~ **substance** Sehstoff *m* ~ **system** Beobachtungs-, Betrachtungs-system *n* ~ **threshold** Sehschwelle *f* ~-**transmitter power** Bildsendeleistung *f* ~ **tuning** Abstimmanzeigerröhre *f* ~-**tuning indicator** Abstimmanzeiger *m*, magisches Auge *n* ~ **video signal** Sichtanzeige *f*

**visualization** Sichtbarmachung *f*

**visualize, to** ~ bloßlegen, sichtbarmachen

**vital** edel ~ **goal** lebenswichtiges Ziel *n* ~ **part** lebenswichtiger Teil *m* ~ **point** lebenswichtiges Ziel *n*

**vitiate, to** ~ entkräftigen, in der Brauchbarkeit beeinträchtigen, schädlich machen, verderben, verseuchen, verunreinigen

**vitiation** Fälschung *f*, Verschlechterung *f*

**vitrain** Glanzkohle *f*

**vitreosity** Glasartigkeit *f*, Glasigkeit *f*

**vitreous** glasähnlich, glasartig, gläsern, glasig ~ **arsenic trioxide** Arsenikglas *n* ~ **body** Glaskörper *m* ~ **clayware** Tonzeug *n* ~ **copper** Kupferglas *n* ~ **electricity** Glaselektrizität *f* ~ **enamel** Glasemaille *f* ~-**enamelling** Feueremaillierung *f* ~ **humor** Glaskörperflüssigkeit *f* ~ **inhomogeneity** Knoten *m* ~ **lava** Glaslava *f* ~ **sand** Glassand *m* ~ **state** Glaszustand *m*

**vitrifiable** verglasbar ~ **color** Schmelzfarbe *f* ~ **pigment** Schmelzfarbe *f*

**vitrification** Glasfluß *m*, Sintern *n*, Sinterung *f*, Überglasung *f*, Verglasung *f* ~ **of clay** Klinkerung *f*

**vitrifications** Glasschmelzwaren *pl*

**vitrified** glasiert, verglast ~ **bond** keramische Bindung *f* ~ **brick** Klinker *m* ~ **coating** Schmelzüberzug *m* ~ **malt** Glasmalz *n*

**vitrifier** Glasbildner *m*

**vitrify, to** ~ (ceramics) dicht brennen, glasieren, sintern, überglasen, verglasen

**vitriol** Vitriol *m* & *n* ~ **ore** Vitriolerz *n*

**vitriolic,** ~ **acid** Vitriol-öl *n*, -säure *f* ~ **ore** Vitriolerz *n*

**vivianite** Blaueisen-erde *f*, -erz *n*, -spat *m*, Blauerz- *n*, -spat *n*, -stein *m*, Eisen-blau *n*, -spat *m*, Vivianit *m*

**vivid** stürmisch

**V joint** V-Stoß *m*

**V-jointed** (mit) Keilspundung *f*

**VLF (very low frequency)** niedriger Langwellenbereich *m*

**V notch** scharfer Kerb *m*, Scharfkerb *m*, V-Kerbe *f*

**vocabulary** Wortschatz *m*

**vocal** mündlich, stimmlich, tönend ~ **call sign** Sprechrufzeichen *n* ~ **compass** Stimmumfang *m* ~ **consonant** stimmhafter Konsonant *m* ~ **cords** Stimmbänder *pl* ~ **horn** Flügelhorn *n* ~ **recording** Sprachaufnahme *f* ~ **register** Stimmregister *n*, -umfang *m*

**vocational** Berufs . . ., beruflich ~ **disease** Berufskrankheit *f* ~ **education** Berufserziehung *f* ~ **guidance** Berufsberatung *f* ~ **school** Berufsschule *f* ~ **test** Berufseignungsprüfung *f* ~ **training** Berufs-, Fort-bildung *f*

**vodas** Rückkupplungssperre *f*

**voder** Sprachsimulator *m*, synthetischer Sprecher *m*

**voice** Sprache *f*, Stimme *f* ~**-actuated** besprochen ~**-actuated modulator** besprochener Modulator *m*, Sprachmodulator *m* ~ **coil** Schwing-, Tauch-spule *f* ~ **control** Beeinflussung *f* durch die Sprache *f*, Besprechung *f* (einer Röhre), Sprachbeeinflussung *f*, Tauchspule *f* ~**-controlled device** Sprachsteuerung *f* ~ **current** Sprechstrom *m* ~**-ear measurement** Messung *f* durch Hörvergleich mit Sprache ~**-ear test** Sprech-Hör-Versuch *m*

**voice-frequency** Sprachfrequenz *f* ~ **dialing** Tonfrequenzfernwahl *f* ~ **key sending** Tonfrequenzwahl *f* ~ **ringer** Tonfrequenzrufsatz *m* ~ **ringing** Tonfrequenzruf *m* ~ **ringing set** Tonfrequenzrufsatz *m* ~ **selective signaling system** Einrichtung *f* für Tonfrequenzfernwahl ~ **signaling** Tonfrequenzruf *m*, Tonfrequenzzeichengebung *f* ~ **signaling current** tonfrequenter Rufstrom *m* ~ **signaling test** Tonfrequenzrufstrommessung *f* ~ **telegraphy** Tonfrequenz-, Wechselstrom-telegrafie *f* ~ **telephony** Niederfrequenzfernsprechen *n*

**voice,** ~**-impressed modulator** besprochener Modulator *m* ~**-modulated** sprachmoduliert ~**-modulated telephony transmitter** sprachgeschalteter Telefoniesender *m* ~**-operated device** sprachbetätigtes Gerät *n*, Sprachsteuerung *f* ~ **operation demonstrator (Voder)** Sprachsimulator *m* ~ **paging system** Personenrufanlage *f* ~ **power** Sprachenergie *f* ~ **radio** Sprechfunk *m* ~**-radio communication** Funksprechübermittlung *f* ~ **recording** Sprach-, Sprech-aufnahme *f* ~ **reproduction** Sprachwiedergabe *f* ~**-signal current** Tonstrom *m* ~ **test(ing)** Sprachmessung *f*, Sprechprüfung *f* ~**-test result** mit Sprache gemessener Wert *m* ~ **transmission by radio** Funksprechen *n*

**vioced consonant** stimmhafter Konsonant *m*

**voicer** Stimmer *m*

**voicing** Intonation *f*

**void, to** ~ ungültig machen

**void** Leere *f*, Leerstelle *f*, Lücke *f*, Luftleere *f*, Lunker *m*, Pore *f*; leer, luftleer, nichtig, unausgefüllt, ungültig ~ **coefficient** Aussparungsfaktor *m* ~ **content** Porengehalt *m* ~ **patent** verfallenes Patent *n* ~ **ratio** (volume of voids to volume of solids) Porenziffer *f*, Porenzahl *f*

**voile** Voile *f*, Voilestoff *m*

**voix céleste** schwebende Stimme *f*

**volatile** ätherisch, flüchtig, leichtflüchtig, verdampfbar, verdampfungsfähig **not** ~ schwerflüchtig ~ **constituent** flüchtiger Bestandteil *m* ~ **fission** flüchtige Spaltprodukte *pl* ~ **ingred-**

**ient(s) of a mixture** flüchtiger Anteil *m* einer Mischung ~ **memory** leistungsabhängiger Speicher *m* (comput.) ~ **spirit** flüchtiger Anteil *m* einer Mischung ~ **storage** leistungsabhängiger Speicher *m*

**volatility** Flüchtigkeit *f*, Verdampfbarkeit *f*, Verdampfungsfähigkeit *f*, Verflüchtigungsfähigkeit *f* ~ **number** Flüchtigkeitszahl *f* ~ **product** Verflüchtigungsprodukt *n* ~ **test** Verflüchtigungsprobe *f* ~ **test of the palm of the hand** Handprobe *f*

**volatilizable** verdunstbar

**volatilization** Abdampfen *n*, Abdampfung *f*, Entgasung *f*, Flüchtigmachung *f*, Verdampfung *f*, Verdunsten *n*, Verflüchtigung *f* ~ **loss** Verdampfungsverlust *m*, Verflüchtigungsverlust *m* ~ **roasting** Verdampfungsröstung *f* ~ **temperature** Verdampfungstemperatur *f*

**volatilize, to** ~ abdämpfen, verdampfen, verdunsten, verflüchtigen

**volatilizer** Verflüchtiger *m*

**volatilizing,** ~ **roasting** verflüchtigendes Rösten *n* ~ **tube** Riechrohr *n*

**volcanic** vulkanisch ~ **glass** Glaslava *f* ~ **rock** Eruptivgestein *n* ~ **scoria** vulkanische Schlacke *f* ~ **tremors** vulkanisches Beben *n* ~ **tuff** Traß *m*, vulkanisches Tuffgestein *n* ~ **vent** Vulkanschlot *m*

**volcanite** Schwefelselen *n*

**volcano** Vulkan *m*

**volplane** Gleitflug *m*

**volt** Volt *n*

**volt-ammeter** Stromspannungsmesser *m*, Voltamperemeter *m* ~**-ampere** Voltampere *n* ~**-hour meter** Scheinleistungszähler *m* ~ **meter** Scheinleistungsmesser *m* ~ **reactive** Var *n* (Einheit der Blindleistung)

**volt,** ~**-ampere** Voltampere *n* ~**-ampere characteristic** Stromspannungskennlinie *f* ~**-ampere reactive hour** Blindwattstunde *f* ~**-amperes diagram** statische Charakteristik *f* ~**-and-ammeter** Strom- und Spannungsmesser *m* ~**-hour meter** Voltstundenzähler *m*

**voltmeter** Gewichtsspannungsmesser *m*, Gewichtsstrommesser *m*, Spannungsmesser *m*, Spannungszeiger *m*, Voltmesser *m*, Voltmeter *n* ~ **change-over switch** Spannungsumschalter *m* ~ **multiplier** Voltmeter *n* mit Hilfsnebenwiderständen ~ **phase select switch** Voltmeterumschalter *m* ~ **plug** Stecker *m* für Voltmeter ~ **resistor** Voltmeterwiderstand *m* ~ **switch** Voltmesserumschalter *m*, Voltmeter-schalter *m*, -umschalter *m*

**volt,** ~**-milliampere meter** Volt-Milliamperemeter *n* ~**-second** Voltsekunde *f*

**Volta,** ~ **cell** Voltasches Element *n* ~ **effect** Voltaeffekt *m* ~ **electromotive force** Voltpotential *n* ~ **potential** Volta-potential *n*, -spannung *f*

**voltage** elektrische Spannung *f*, Lichtbogenspannung *f*, Stromspannung *f*, Voltspannung *f*, Voltzahl *f* ~ **across the circuit** Kreisspannung *f* ~ **and current regulator** Knickregler *m* ~ **of filament battery** Fadenspannung *f* ~ **of the lighting circuit** Lichtspannung *f* ~ **between lines of a polyphase system** Leiterspannung *f* ~ **against neutral (or star) point** Spannung *f* gegen den Sternpunkt ~ **due to nervous action**

Nervenaktionsspannung *f* ~ **to neutral** Leiter-erd-, Leitersternpunkt-spannung *f* ~ **on outside lines of three-wire system** Außenleiterspannung *f* ~ **to ground** Spannung *f* gegen Erde ~ **at starting speed** (Zündapparat) Anlaßschlagweite *f* ~ **at the terminals** Klemmspannung *f*

**voltage,** ~ **adapter** Spannungsumschalter *m* ~ **amplification** Spannungsverstärkung *f* ~-**amplification factor** Spannungsverstärkungsfaktor *m* ~ **amplifier** Heiz-draht *m*, -faden *m*, Spannungsverstärker *m* ~ **amplifying tube** Spannungsverstärkerröhre *f* ~ **amplitude** Spannungsamplitude *f* ~ **arrester** Überspannungsableiter *m* ~ **attenuation** Spannungsdämpfung *f* ~ **breakdown** Isolationdurchschlag *m*, Zusammenbrechen *n* der Spannung ~ **bypassing** Spannungsverschiebung *f* ~ **carbon discharge gap** Kohlenblitzableiter *m* ~ **characteristic** Spannungsverlauf *m* ~ **circuit** Spannungskreis *m* ~ **clipper** Abflacher *m* (telegr.) ~ **coefficient** Spannungskoeffizient *m* ~ **coil** Spannungsspule *f* ~ **collapse** Spannungszusammenbruch *m* ~ **component** Spannungskomponent *m* ~ **control** Spannungsregelung *f* ~-**control relay** Spannungsreglerrelais *n* ~ **controlling dynamo** spannungsregulierende Lichtmaschine *f* ~ **curve** Stromspannungskurve *f* ~ **cutout** Spannungssicherung *f* ~ **degree** Spannungsstufe *f* ~ **detector** Spannungssucher *m* ~ **difference** Spannungsdifferenz *f* ~-**discharge cap** Blitzableiter *m*, Spannungsbegrenzer *m* ~ **distribution** Spannungsverteilung *f* ~ **divider** Spannungs(ver)teiler *m*, Widerstandskette *f* ~-**divider circuit** Spannungsteilerschaltung *f* ~ **doubler** Spannungs-verdoppler *m*, -schaltung *f* ~ **drop** Spannungsabfall *m*, -abnahme *f*, -gefälle *n*, Voltverlust *m* ~ **factor** Verstärkungsfaktor *m* ~ **feedback** Spannungsgegenkopplung *f* ~ **fluctuation** Spannungsschwankung *f* ~ **fuse** Spannungssicherung *f* ~ **gain** Spannungsverstärkung *f* ~ **gradient** Spannungsgradient *m* ~ **grading electrode** Spannungsteilergitter *n* ~ **increase** Spannungserhöhung *f* ~ **indicator** Spannungszeiger *m* ~ **jump** Spannungssprung *m* ~ **leap** Spannungssprung *m* ~ **level** Spannungspegel *m* ~-**limiting device** Spannungsbegrenzer *m* ~ **magnification** Verstärkungs-verhältnis *n*, -ziffer *f* ~ **modulation** Spannungsmodulation *f* ~ **network** Spannungsnetz *n* ~ **nodal point** Spannungsknotenpunkt *m* ~ **node** Spannungsknoten *m* ~ **nullifying** Voltvernichtung *f* ~ **optical** spannungsoptisch ~ **overstress (or overload)** Spannungsüberbeanspruchung *f* ~ **peak** Spannungsspitze *f* ~-**peak indicator** Spannungsspitzenanzeiger *m* ~-**peak limiter** Spannungsspitzennivellierer *m* ~ **pickoff** Spannungsabgriff *m* ~ **power amplifier** Spannungsverstärker *m* ~ **proof test** Spannungsprüfung *f* ~ **pulsation** Spannungspulsation *f* ~ **pulse** Spannungsstoß *m* ~ **rating of a capacitor** Kondensatornennspannung *f* ~ **ratio** Spannungs-übersetzung *f*, -verhältnis *n* ~ **recorder** Spannungsschreiber *m* ~ **reduction** Spannungsrückgang *m* ~-**regulating valve** Stabilivolt *n* ~ **regulation** Spannungs-regulierung *f*, -stabilisierung *f* ~ **regulator** Baretter *m*, Reglerschalter

*m*, Spannungs-regler *m*, -regulator *m* ~ **relay** Spannungsrelais *n* ~ **response of an exciter** Erregungsgeschwindigkeit *f* ~ **rise** Spannungserhöhung *f* ~ **select switch** Spannungsumschalter *m* ~-**selecting switch** Spannungswahlschalter *m* ~ **sensitive resistance** spannungsabhängiger Widerstand *m* ~ **sensitivity** Spannungsempfindlichkeit *f* ~ **source** Spannungsquelle *f* ~ **stability** Spannungskonstanz *f* ~ **stabilization** Spannungsstabilisierung *f* ~ **stabilizer** Spannungsgleichschalter *m*, Stabilivolt *n* ~ **stabilizing** spannungsregelnd ~-**stabilizing circuit** Spannungsgleichhaltungskreis *m* ~ **stabilizing tube** Glimmstreckenstabilisator *m*, Stabilisator *m* ~ **standard** Spannungsnormal *n* ~ **standardizer** Spannungsstabilisator *m* ~ **standing wave ratio (VSWR)** Stehwellen-indikator *m*, -verhältnis *n* ~ **step** Spannungssprung *m* ~ **step-up** Spannungserhöhung *f* ~ **step-up means** Spannungserhöher *m* ~ **step-up transformer** Spannungserhöhungstransformator *m* ~ **strength** Spannungssicherheit *f* ~ **surge** Spannungswelle *f* ~ **telephone-influence factor** Fernsprechformfaktor *m* der Spannung ~ **test** Prüfung *f* der Speisestromspannungen ~ **test box** Spannungsprüfkasten *m* ~ **test terminal** Spannungsprüfklemme *f* ~ **transformer** Spannungstransformator *m* ~ **trebling** Spannungsverdreifachung *f* ~-**type telemeter** spannungsgekoppeltes Telemeter *n* ~ **variation** Spannungsschwankung *f* ~ **variation of mains** Netzspannungsschwankung *f* ~ **vector** Spannungszeiger *m* ~ **wave** Spannungswelle *f*

**voltaic** galvanisch, voltaisch ~ **arc** Lichtbogen *m* ~ **cell** galvanisches Element *n*, galvanische Kette *f*, Primärelement *n* ~ **couple** galvanisches Element *n*, galvanische Kette *f*, Voltapaar *n* ~ **pile** voltaische Säule *f*

**volume** Band *n*, Buch *n*, Dicke *f*, Größe *f*, Inhalt *m*, körperlicher Inhalt *m*, Lautstärke *f*, Menge *f*, Raum *m*, Raum-inhalt *m*, -maß *n*, -menge *f*, -teil *m*, Umfang *m*, Volumen *n*

**volume,** ~ **of air drawn in** angesaugte Luftmenge *f* ~ **of the bell jar** Glockenvolumen *n* ~ **of blast** Windmenge *f* ~ **of coke substance** Koksvolumen *n* ~ **of ebb** Ebbewassermenge *f* ~ **of the flood** Flutwassermenge *f* ~ **of reception** Empfangslautstärke *f* ~ **of the reservoir of a barrage** Stauinhalt *m* einer Talsperre ~ **of sea packing** Kollimaß *n* ~ **of sound** Lautstärke *f*, Tonvolumen *n* ~ **of speech** Lautstärke *f* der Sprache ~ **of traffic** Verkehrsumfang *m* ~ **of voids** Porenvolumen *n* ~ **of water** Wasser-inhalt *m*, -menge *f* ~ **of water discharging on ebb tide** Ebbewassermenge *f* ~ **of water entering on the flood tide** Flutwassermenge *f* ~ **of winding** Wicklungsraum *m*

**volume,** ~ **adjustment** Lautstärkeregelung *f* ~ **brightness** Raumhelligkeit *f* ~ **calculation** Kubizierung *f* ~ **comparison** Sprech-Hör-Versuch *m* ~ **compression** Verdichtsteife *f* ~ **constancy test** Raumbeständigkeitsprüfung *f* ~ **contraction** Dynamikverringerung *f*, Raumverminderung *f*, Volumenkontraktion *f*

**volume-control** (marking on knob) Lautstärke *f*, Lautstärkeregelung *f*, Lautstärkeregler *m*, Mengenregelung *f*, Verstärkungsregler *m*, Volumregler *m* ~ **circuit** Schwundausgleich *m* ~

**potential** Schwundregelspannung $f$ ~ **system** Schwundausgleich $m$

**volume,** ~ **current** Geschwindigkeitsamplitude $f$, Schallschnelle $f$ ~ **density of impurity centers** Störstellendichte $f$ ~ **dilatometer** Volumendilatometer $n$ ~ **discount** Mengenrabatt $m$ ~ **energy** Volumenenergie $f$ ~ **equivalent** Bezugsdämpfung $f$ (eines Übertragungssystems), Nutzdämpfung $f$ ~ **expansion** Dynamiksteigerung $f$ ~ **flutter** Volumenschwankung $f$ ~ (of a substance) **forced away** verdrängtes Volumen $n$ ~ **gauge** Volumenmesser $m$ ~ **governor** Mengenregler $m$ ~ **implant** Volumenimplantation $f$ ~ **indicator** (sound recording) Aussteuerungsanzeiger $m$, Lautstärkeanzeiger $m$, Lautzeiger $m$, Prallanzeiger $m$, Tonmesser $m$, Volumenzeiger $m$ ~ **integral** Volumintegral $n$ ~ **ionization** Raumionisation $f$ ~ **lifetime** Trägerlebensdauer $f$ ~ **limiter** Lautstärkebegrenzer $m$ ~ **loss** Bezugsdämpfung $f$ ~ **loss per unit of length** Bezugsdämpfung $f$ je Längeneinheit ~ **meter** Lautstärkemesser $m$ ~-**percentage scale** Volumprozentskala $f$

**volume-range** Lautstärke-bereich $m$, -umfang $m$ ~ **characteristic** Dynamiklinie $f$ ~ **compression** Dynamikpressung $f$ ~ **control means** Dynamikregler $m$ ~ **limiter** Dynamikbegrenzer $m$ ~ **ratio** Lautstärkeverhältnis $n$, Volumenverhältnis $n$

**volume,** ~ **recombination** Volumenrekombination $f$ ~ **recombination rate** Volumenrekombinationshäufigkeit $f$ ~ **rectification** Volumengleichrichtung $f$ ~ **regulation** Pegelung $f$ ~ **regulator** Lautstärkeregler $m$ ~ **relation** Raum-, Volum-verhältnis $n$ ~ **resistance** Durchgangswiderstand $m$ ~ **resistivity** spezifischer Widerstand $m$ ~ **source** Raumquelle $f$ ~ **strain** Volumdilatation $f$ ~ **unit** (**vu**) Volumeinheit $f$ (acoust.) ~ **unit meter** Druckpegelmesser $m$ ~ **velocity** Schallfluß $m$ (acoust.) ~ **voltameter** Volumenvoltameter $n$ ~ **weight** Volumgewicht $n$

**volumenometer** Mengenmesser $m$

**volumeter** Durchflußmengenmesser $m$

**volumetric** volumetrisch ~ **analysis** Maßanalyse $f$, volumetrische Analyse $f$ ~ **apparatus** Titrierapparat $m$ ~ **capacity** Fassungsvermögen $n$ ~ **density** Raumdichte $f$ ~ **determination of unsaturation** Enometrie $f$ ~ **efficiency** Füllungsgrad $m$, volumetrischer Wirkungsgrad $m$, Völligkeitsgrad $m$ (engin.) ~ **increase** Volumenvergrößerung $f$ ~ **liquid** Mengenmeßgerät $n$ für Flüssigkeiten ~ **meter** Volumenzähler $m$ ~ **modulus of elasticity** Elastizitätsmodul $n$ für Druck ~ **output** Mengenleistung $f$ ~ **part** Maßteil $m$ ~ **pipette** Vollpipette $f$ ~ **solution** Maßlösung $f$ ~ **strain** Quetschung $f$ ~ **weight** Raumgewicht $n$

**voluminous** massig, stark, umfangreich, voluminös

**volute** Ausströmraum $m$, Kegel $m$, Ladereintrittsspirale $f$, Rollenschnecke $f$, Schnecke $f$, Spirale $f$, Spiralgehäuse $n$ ~ **buffer spring** Wickelfeder $f$ ~ **casing** Gehäuse $n$ der Ladereintrittsspirale $f$ ~ **chamber** Auslaufraum $m$, Diffusor $m$ ~ **collecting chamber** Sammlerspirale $f$ ~ **compasses** Spiralenzirkel $m$ ~ **spiral spring** Kegelfeder $f$

**vortex** Luftwirbel $m$, Strudel $m$, Turbulenz $f$, Vortexring $m$, Walze $f$, Wirbel $m$ ~ **center (or**

**core)** Wirbelkern $m$ ~ **chamber** Wirbelkammer $f$ ~ **distribution** Wirbelbelegung $f$ ~ **element** Wirbelelement $n$ ~ **filament** Wirbelfaden $m$ ~ **generation** Wirbelentstehung $f$ ~ **line** Wirbellinie $f$ ~ **line stretching** Wirbellinienstreckung $f$ ~ **motion** Wirbelbewegung $f$ ~ **motion of gas** Gaswirbelbewegung $f$ ~ **nucleus** Wirbelkern $m$ ~ **path** Wirbelstraße $f$ ~ **period** Wirbelzeit $f$ ~ **ring** Wirbelring $m$ ~ **sheet** Wirbel-band $n$, -blatt $n$, -schicht $f$ ~ **source** Wirbelquelle $f$ (aerodyn.) ~ **street** Wirbelschleppe $f$ ~ **stretching mechanism** Wirbeldehnungsmechanismus $m$ ~ **theorem** Wirbelsatz $m$ ~ **trail** Wirbelschleppe $f$ ~ **train** Wirbelzopf $m$ ~ **tube** Wirbelröhre $f$ ~-**type flow** Wirbelströmung $f$ ~ **whistle** Wirbelpfeife $f$

**vortical** kreisend

**vorticity** Verwirbelung $f$, Wirbelbildung $f$, Wirbeligkeit $f$, Wirbelung $f$ ~ **average theorem** Wirbelmittelwertsatz $m$ ~ **effect** Wirbeleinfluß $m$ ~ **equation** Wirbelgleichung $f$ ~ **formula** Wirbelformel $f$ ~ **measure** Wirbelmaß $n$ ~ **number** Wirbelmaß $n$ ~ **tensor** Wirbeltensor $m$ ~ **theorem** Wirbelsatz $m$ ~ **transport theory** Wirbeltransporttheorie $f$ ~ **vector** Wirbelvektor $m$

**vote of confidence** Vertrauensfrage $f$

**voucher** Anweisung $f$, Beleg $m$, Rechnungsbeleg $m$, Unterlage $f$

**vouchsafe, to** ~ gewährleisten

**voussoir** Wölbkeil $m$

**vowel** Selbstlaut $m$, Vokal $m$ ~ **articulation** Selbstlautdeutlichkeit $f$

**V-position** Keilstellung $f$

**V-rope drive** Keilriemenantrieb $m$

**V-shape** Gabelausführung $f$, Keilform $f$; (engine) V-förmig

**V-shaped** wiegenartig ~ **bearing** wiegenartiges Zapfenlager $n$ ~ **notch** Scharfkerb $m$, V-förmige Kerbe $f$ ~ **planing bottom** (seaplane hull) Wellenbindenform $f$

**V slot** V-Verkantmutter $f$

**V stay** Doppelanker $m$, V-Anker $m$

**V strut** V-Strebe $f$

**V tail** Schmetterlingsleitwerk $n$, V-Leitwerk $n$

**V thread** scharfes oder scharfgängiges Gewinde $n$, Spitz(en)gewinde $n$ ~ **tool** Spitzgewindestahl $m$

**V threaded** scharfgängig

**VTOL aircraft** Lotrechtstarter $m$, Senkrechtstartflugzeug $n$

**V tube** Schenkelrohr $n$

**V turn** Wegeknie $n$

**VTVM (vacuum tube voltmeter)** Röhrenvoltmeter $m$

**V-twin engine** Zweizylinder-V-Motor $m$

**V-type,** ~ **engine** V-förmig, Fochermotor $m$, Gabelmotor $m$, Pfeilmotor $m$, V-Motor $m$ ~ **radiator** Spitzkühler $m$ ~ **strut** V-Stiel $m$, V-Strebe $f$

**vug** Druse $f$

**vuggy** Drusen $pl$ enthaltend

**vulcan gear** Vulkangetriebe $n$

**vulcanicity** Vulkanismus $m$

**vulcanite** Hartgummi $m$, Vulkanit $m$ ~ **sheathing** Hartgummiüberzug $m$ ~ **stage** Harttisch $m$ (opt.)

**vulcanizate** Vulkanisat *n*
**vulcanization** Vulkanisation *f*, Vulkanisierung *f*
**vulcanize, to** ~ schwefeln, vulkanisieren
**vulcanized** vulkanisiert ~ **asbestos** Vulkanasbest *m* ~ **caoutchouc** vulkanisierter Kautschuk *m* ~ **fiber** Vulkan-faser *f*, -fiber *f* ~-**fiber base paper** Vulkanfiberrohpapier *n* ~ **fiber grinding wheels** Vulkanfiberschleifscheiben *pl* ~ **india rubber cable** Gummikabel *n* ~ **india rubber wire** Gummidraht *m* ~-**rubber cable** Gummiader *f* ~-**rubber wire** Gummidraht *m*

**vulcanizer** Vulkanisierapparat *m*
**vulcanizing** Vulkanisierung *f* ~ **agent (or ingredient)** Vulkanisationsagens *m* ~ **heater (or pan)** Vulkanisationskessel *m* ~ **pan** Vulkanisierkessel *m* ~ **plant** Vulkanisiereinrichtung *f* ~ **press** Autoklavpresse *f* ~ **styles** Vulkanisationsartikel *pl*
**vulnerability** Verletzbarkeit *f*
**vulnerable** verletzbar, verletzlich, verwundbar
**V way** Prismenführung *f*
**V weight** V-Nettogewicht *n*
**V welding with binder** Keilschweißen *n*

# W

**wable, to** ~ torkeln, wackeln

**wad, to** ~ wattieren

**wad** Filzpfropfen *m*, Ladepfropfen *m*, Wad *n*, Wattepfropfen *m* ~ **punch** Henkellocheisen *n*

**wadding** Bausch *m*, Ladepfropfen *m*, Watte *f* ~ **of cellulose** Zellulosewatte *f* ~ **of glass** Glaswatte *f* ~ **manufacturing machine** Watteherstellungsmaschine *f* ~ **pick** Füllschuß *m*

**wade, to** ~ waten **to** ~ **through** durchwaten

**wafer, to** ~ befestigen, verrippen, zukleben

**wafer** Oblate *f*, Platte *f*, Schaltebene *f*, Schaltersegment *n*, Siegelmarke *f*, Waffel *f* ~ **headed riveting** Waffelnietung *f* ~ **loudspeaker** Flachlautsprecher *m* ~ **socket** Flachsockel *m* ~ **switch** Flachbahnregler *m*, Mischpult *f*

**wafered** gerippt, gewaffelt

**waffle** Waffel *f* ~ **sheet metal** Waffelblech *n*

**waft flag** Notflagge *f*

**wag, to** ~ hin- und herbewegen, wedeln

**wages** Arbeitslohn *m*, Besoldung *f*, Dienstlohn *m*, Erwerb *m*, Gehalt *n*, Lohn *m*, Löhnung *f*, Lohnwesen *n* ~ **for piecework** Stücklohn *m*, Stücklohnsatz *m*, Stückzeit *f*

**wagging frequency** Wedelschwingung *f*

**Wagner,** ~ **earth** Wagner-Hilfszweig *m* ~ **ground** Wagnersche Erdung *f*

**wagon** Förderwagen *m*, Fuhre *f*, Fuhrwerk *n*, Kutsche *f*, Lastwagen *m*, Wagen *m*, Waggon *m* ~ **with shafts** Gabelwagen *m*

**wagon,** ~ **boiler** Kofferkessel *m* ~ **cession price** Lieferungspreis *m* in Wagen ~ **checker** Wagenkontrolleur *m* ~ **circulation** Wagenumlauf *m* ~ **elevator** Wagenelevator *m* ~ **ferry** Wagenfähre *f* ~ **fitting** Wagenbeschlag *m* ~ **grease** Wagenfett *n* ~ **hoist** Wagenaufzug *m* ~ **jack** Wagenheber *m* ~ **load** Wagenladung *f* ~ **loader** Wagenlader *m* ~ **park** Fuhr-, Wagenpark *m* ~ **pushing device** Wagenstoßvorrichtung *f* ~ **radio set** Karrenstation *f* ~ **road runner** Zugbegleiter *m* (min.) ~ **shaft** Deichsel *f* ~ **spring** Trag-, Wagen-feder *f* ~ **stop** Wagenhalteplatz *m* ~ **tipple** Wagenkipper *m* ~ **tippler** Füllwagen *m* ~ **tongue** Wagendeichsel *f* ~ **way** Förderbahn *f*

**wailer** Klauber *m*, Klaubjunge *m*

**wailings** Klaubeberge *pl*

**wainscot, to** ~ austäfeln, verkleiden, vertäfeln

**wainscot** Getäfel *n*, Holz-, Wand-verkleidung *f*, Paneel *n*, Täfelung *f*, Tafelwerk *n*

**wainscoted socle** Fußsockel *m*

**wainscoting** Getäfel *n*, Täfelung *f*, Täfelwerk *n*, Vertäfelung *f*, Wand-bekleidung *f*, -täfelung *f*

**waist** eingezogenes Mittelstück *n*, Einschnürung *f*, Mitteldeck *n*, Prüfstabeinschnürung *f*, Taille *f* ~ **anchor** Hauptanker *m* ~ **band** Bauchgurt *m* ~ **sheet** Sattelplatte *f*

**wait, to** ~ **upon** aufwarten

**waiting** Aufwartung *f*, Bedienung *f*, Wartezeit *f* ~ **list** Warteliste *f* ~ **period** Karenzzeit *f* ~ **position** Wartestellung *f* ~ **room** Warte-raum *m*, -saal *m* ~ **signal** Wartezeichen *n* ~ **time** Wartezeit *f*

**waive, to** ~ aufgeben, (rights) verzichten

**waiver** Verzichterklärung *f*

**wake, to** ~ wecken **to** ~ **up** aufwachen, aufwecken

**wake** Abstrom *m*, Luftschraubenstrahl *m*, Nachlauf *m*, Nachstrom *m*, Nachströmung *f*, Schraubennachstrom *m*, Sog *m*, Wirbelschleppe *f* (aerodyn.) ~ **of a ship** Schiffsspur *f*

**wake,** ~ **drag** Schleppleistung *f* ~ **effect** Wirbelschleppeneffekt *m* ~ **grain** Schraubenbrunnen *m* ~ **light** Hecklicht *n* ~ **rake** Meßharke *f* ~ **survey** Nachlaufmessung *f* ~ **water** Kielwasser *n*

**wale** (of a ship) Bergholz *n*

**wales** (shipbuilding) Krummholz *n*

**waling** Gurtholz *n*, Stützholm *m* ~ **cap** (Längsträger) Holm *n*

**walings** (in ports) Zangen *pl*

**walk, to** ~ gehen, wandeln **to** ~ **through** durchschreiten

**walk** Gang *m*, Weg *m* ~ **in cooler** begehbare Kühlzelle *f*

**walkable deck** begehbares Deck *n*

**walked** zurückgelegt

**Walker wheel** Walker-Tischmaschine *f*

**walkie-talkie** Feldfunksprecher *m*, Kleinfunkapparat *m*, Kleingerät *n*, Langwellenkleinfunkapparat *m*

**walking,** ~ **beam** Balancier *m*, Bohrschwengel *m*, mehrarmiger Hebel *m* ~ **beam conveyor** Fördereinrichtung *f*, Hubbalken *m* ~ **beam furnace bottom** Schwingbalkenherd *m* ~ **beam link** Schwenkbalken-Verbindungsstange *f* ~ **beam post** Bohrdock *m* ~ **beam saddle** Schwengelsattel *m* ~ **beet puller** Rübenheber *m* mit Handführung ~ **cultivator** Hackmaschine *f* mit Handführung, Handhackmaschine *f* ~ **gang plow** mehrschariger Pflug *m* ohne Sitz ~ **jib crane** Velozipedkran *m* ~ **legs** Schreitwerk *n* ~ **lister** Handdammkulturpflug *m*, Handlister *m* ~ **pace** Schrittgeschwindigkeit *f* ~ **pipe** Wasserschwinge *f* ~ **plow** Handpflug *m* (Stell- oder Schwingpflug, Gangpflug)

**walkway** Bedienungssteg *m*, Hubsteg *m*, Kranbühne *f*, (airship) Laufbrücke *f*, Laufgang *m*, Steg *m*, Verbindungssteg *m* ~ **for operating bridges** Gang *m* zum Operieren der Tore ~ **girder** Laufgangträger *m*

**wall, to** ~ **in the form of stairs** eine Mauer *f* abtreppen **to** ~ **off** ummanteln **to** ~ **up** ausmauern, vermauern, zumauern **to** ~ **up (or in)** einmauern

**wall** Abbaustoß *m*, Aufhauen *n*, Einfassung *f*, (masonry) Mauer *f*, Schallplattensteg *m*, Scheidewand *f*, (hoof) Trachte *f*, Wall *m*, Wand *f*, Wandung *f*

**wall,** ~ **of the basket** Korbwand *f* ~ **of compass bowl** Kesselwand *f* ~ **of a container** Behälterwandung *f* ~ **with one side worked fair** einhäuptige oder einseitig abgeglichene Mauer *f* ~ **of pipe** Rohrwand *f* ~ **of planks** Bohlwand *f* ~ **of the rivet hole** Lochleibung *f* oder Loch-

wand(ung) *f* des Niets ~ **of a shaft** Schachtstoß
*m* ~ **of signals** Signalwand *f* ~ **of timbers**
Bohlwand *f*
**wall,** ~ **absorption** Wandabsorption *f* ~ **anchor**
**plate** Wandankerplatte *f* ~ **attachment** Wand-
befestigung *f* ~ **base** Mauerfuß *m* ~ **batten**
Mauerleiste *f* ~ **beam** Randträger *m*
**wall-bracket** Mauer-bügel *m*, -stütze *f*, Säulenhängelager *n*, Wand-arm *m*, -bock *m*, -gestell
*n*, -konsole *f*, -stütze *f*, Winkel -arm *m*, -konsole
*f* ~ **bearing** Mauerlager *n* ~ **crane** Konsolkran
*m* ~ **hanger** Wandarmlager *n*
**wall,** ~ **bushing insert** Durchsteck-Wanddurch-
führung *f* ~ **channel** Mauerdurchführung *f*,
Mauerkanal *m* (electr.) ~ **charge density** Wand-
ladungsdichte *f* ~ **chest** Wandschrank *m* ~
**chisel** Steinbohrer *m* ~ **clamp** Maueranker *m*,
Rohrschelle *f*, Stichanker *m* ~ **clearance** Wand-
abstand *m* ~ **coating** Innenkolben-, Wand-
-belag *m* ~ **coercive force** Wandkoerzitivkraft
*f* ~ **coil** Wandkühlsystem *n* ~**column** Wandpfeiler *m* ~ **coping** Mauerdach *n* ~ **covering**
Wandbedeckung *f* ~ **crane** Wandkran *m* ~
**current** Wandstrom *m* ~ **curvature** Wandkrüm-
mung *f* ~ **diagram** Wandtafel *f* ~ **displace-**
**ment** Wandverschiebung *f* ~ **displacement**
**braking** Wandverschiebungsbremsung *f* ~
**drill** Wandbohrmaschine *f* ~ **effect** Wandeinfluß *m* ~ **entrance** Durchführung *f* ~ **entrance**
**insulator** Durchführungsisolator *m* ~ **evapora-**
**tor** Wandverdampfer *m* ~ **face** Mauerfeld *n* ~
**facing** Wandbekleidung *f* ~ **fixture** Wand-be-
festigung *f*, -anordnung *f* ~ **flag** Wandfliese *f*
~ **friction** Wand(Mantel-)reibung *f* ~ **hand basin**
Wandwaschbecken *n* ~ **heater** Wandheizkör-
per *m* ~ **hook** Fang-, Glücks-, Rohr-haken *m*,
Rohrschelle *f*, Wandhaken *m* ~ **indicator** Wand-
standanzeiger *m* ~ **influence** Wandeinfluß *m* ~
**insulator** Mauerisolator *m* ~ **lead-in insulator**
Wanddurchführungsisolator *m* ~ **line** Mauer-
flucht *f* ~ **map** Wandkarte *f* ~ **motion** Wandverschiebung *f* ~ **mounted bracket lamps**
Wandarmleuchten *pl* ~ **mounting** Wandmon-
tage *f* ~ **nozzle** Wandsauger *m* ~ **panel heating**
Wandflächenheizung *f* ~ **paper** Firnis-, Tape-
ten(roh)-papier *n* ~ **paper embosser** Tapeten-
präger *m* ~ **paper embossing varnish** Tapeten-
prägelack *m* ~ **pattern switchboard** Wandum-
schalte-, Wandvermittlungs-schrank *m* ~
**piece** Joch *n* ~ **plaster** Putzgips *m*, Verputz *m*
~ **plate** Jochholz *n*, Mauerlatte *f*, Vorlegeplatte
*f*, Wandplatte *f* ~ **plug** Steckdose *f*, Wandstecker *m* ~ **plug with movable contact pins**
Wandstecker *m* mit beweglichen Kontaktstiften
~ **poster** Maueranschlag *m* ~ **pressure** Wand-
druck *m* ~ **primer** Wandgrundierung *f* ~ **pump**
Wandpumpe *f* ~ **radial drill** Wandradialbohr-
maschine *f* ~ **scraper** drehender Unterschneider *m* ~ **screw** Steinschraube *f* ~ **siphoning**
Wandabsaugung *f* ~ **slabs and flooring flags**
Wand- und Bodenbelagplatten *pl* ~ **socket** An-
schluß-, Ansteck-, Steck-dose *f* ~ **switchboard**
Wandschalttafel *f* ~ **telephone set** Fernsprechapparat *m* mit festem Mirofon, Wandapparat
*m* ~ **telephone station** Wandfernsprecher *m* ~
**temperature** Wandungs-temperatur *f*, -wärme
*f* ~ **thickness** Wanddicke *f* ~ **tile** Kachel *f* ~
**tile making machine** Wandplattenherstellungs-

maschine *f* ~ **tube** Durchführungsrohr *n*
(electr.) ~ **tube insulator** Durchführungsisolator *m* ~ **ventilator** Wandlüfter *m* ~ **winch**
Wandwinde *f* ~ **winches** Wandwinden *pl*
**waller's hammer** Abputzhammer *m*
**walling** Ausbau *m*, Mauerung *f* ~ **in mines**
Grubenmauerung *f* ~ **of a shaft** Schachtausbau *m*
**walling,** ~ **hammer** Abputzhammer *m* ~-**off**
Verschließen *n*
**walls,** ~ **of borehole** Bohrlochwand *f* ~ **of**
**interior cooling** (Kompressor) Außenwandun-
gen *pl* ~ **of slot** Seitenwände *pl* des Schlitzes
**Walloon,** ~ **iron** Walloneneisen *n* ~ **process**
Wallonenarbeit *f*, Zweimalschmelzerei *f*
**walpurgite** Walpurgit *m*
**wander, to** ~ abirren, wandeln, wandern **to** ~**off**
**course** sich verfranzen (aviat.)
**wander mark** (of telescope) Einstellmarke *f*
**wandering** Irre *f*, wandernder Nullpunkt *m* ~
**direction of arrival** schwankende Einfallsrichtung *f* ~ **lead** lose Zuführung *f* ~ **stroke**
Hubverlegung *f*
**wane, to** ~ abnehmen
**wane** Baumkante *f*
**waning** ausgehend
**want, to** ~ begehren, brauchen
**want** Armut *f*, Bedarf *m*, Bedürfnis *n*, Mangel
*m*, Not *f* **for** ~ **of** mangels ~ **of money** Geldnot
*f* ~ **of place** Platzmangel *m*
**wanted** gesucht, gewollt, verlangt ~ **sub-**
**scriber** angerufener oder verlangter Teilnehmer *m*
**war** Krieg *m* ~ **equipment** Kriegsausrüstung *f*
**warble, to** ~ trillern, wirbeln, wobbeln (zur
Verschlüsselung)
**warble** Wobbeln *n* ~ **note** Trillernote *f*, Wechsel-
note *f*, Wechselton *m* ~ **produced by warbler-**
**-type heterodyne oscillator** Heulton *m* ~ **sound**
Wechsel-note *f*, -ton *m*
**warbled carrier current** zeitlich veränderlicher
Trägerstrom *m*
**warbler** Wobbler *m* ~ **oscillator** Schwebungssummer *m*
**warbling condensor** Heulkondensator *m*
**ward, to** ~ **off** abwehren
**warding file** Schlüsselfeile *f*
**Ward-Leonard,** ~ **hoist** Leonardschaltung *f* ~
**regulation** Leonardsatz *m* ~ **system** Ward-
Leonard-Antrieb *m*
**wardrobe** Kleiderschrank *m*, Schrank *m*
**ware** Ware *f*
**warehouse** Ablage *f*, Depot *n*, Kaufhaus *n*, Lager
*n*, Lagerhaus *n*, Magazin *n*, Niederlage *f*,
Schuppen *m*, Speicher *m*, Stapel *m*, Vorratsraum *m*, Warenlager *n* ~ **charges** Lagerspesen
*pl* ~ **crane** Lagerkran *m* ~ **man** Warenhaus-
wärter *m* ~ **receipt** Lagerschein *m*
**warehousing** Einlagerung *f*
**warm, to** ~ erwärmen, wärmen **to** ~ **up** anwärmen, aufwärmen, erwärmen **to** ~**up the engine**
warm laufen lassen
**warm to touch** handwarm
**warm** warm ~ **air** Heißluft *f*, Warmluft *f* ~
**air current** Thermik *f* ~ **air mass** warme
Luftmasse *f* ~ **coiled** warmgeformt ~ **front**
Warmluftfront *f* (**with**) ~-**moist temperature**
feuchtwarm ~ **part of first stage** Warmast *m*

der ersten Stufe (chem.) ~ **pressing** Warmpreß-
arbeit f ~ **seating** Warmwasserraumschwitze
f ~ **sector** der warme Sektor m ~ **separator**
Warmabscheider m ~ **storage test** Warm-
lagertest m ~-up Aufwärmen n ~-up **time** Anheizzeit f ~ **water retting** Warmwasserrotte
f ~ **water retting of lax** Dampfrotte f des
Flachses
**warming** Erwärmung f, Wärmung f ~ **cupboard**
Wärmeschrank m ~ **plate** Wärmplatte f ~-up
**shot** Anwärmeschuß m ~-up **time of an engine**
zum Anwärmen des Motors erforderliche Zeit f
**warmth** Wärme f
**warning** Achtung f, Alarm m, Androhung f,
Ankündigung f, Anzeichen n, Anzeige f,
Aufkündigung f, Kündigung f, Vormeldung f,
Warnung f ~ **and danger signal** Warnsignal
n ~ **of danger** Gefahrenmeldung f
**warning,** ~ **apparatus** Ankündigungsvorrichtung
f ~ **board** Warnungstafel f ~ **circuit** Vormelde-
stromkreis m ~ **device** Warnanlage f ~ **gauge**
Durchsackwarngerät n, Füllungsbegrenzer m,
~ **horn** Warnhorn n ~ **indicator** Warnanlage
f, Warnungsanzeiger m ~ **lamp** Warnlampe f
~ **light** Signallampe f, Warn-, Warnungs-licht n
~ **light strip** Warnleuchtschild n ~ **notice**
Warnmeldung f ~ **order** Vorbefehl m ~ **plate**
Warnungsschild n ~ **pointer zero (or end)**
**position** Warnzeigerruhestellung f ~ **pressure**
**pointer** Warndruckzeiger m ~ **relay** Störungsrelais n ~ **ring** Warnungsring m ~ **sensing**
**line** Warnsteuerleitung f ~ **shot** Warnungs-
schuß m ~ **sign** Warnungszeichen n ~ **signal**
Ankündigungssignal n, Vorsignal n, Warnungs-
signal n ~ **system** Meldenetz n
**warp, to** ~ anzetteln, (weaving) anscheren,
biegen, krümmen, krumm machen oder werden, mitziehen, sich krümmen, (textiles)
schweifen, verbiegen, (textiles) verholen, verziehen, verwerfen, sich werfen, zetteln, ziehen
(naut.)
**warp** Formänderung f, Kettbaum m, Kette f
(im Gewebe), Krimpe f, Trosse f, Verholtau
n, (weaving) Zettel m ~ **and filling** Kette f
und Schuß m
**warp,** ~ **beam** Garn-, Ketten-baum m ~ **beam**
**hydroextractor** Kettbaumschleuder f ~ **beam**
**sizing machine** Kettbaumschlichtmaschine f ~
**chain** Warpkette f ~ **control** Verwindungshebel
m ~ **drawing-in** Ketteneinziehen n ~ **dressing**
Kettenschlichten n ~ **drying machine** Ketten-
trockenapparat m ~ **fabric** (textiles) Kettenware f ~ **loom** Kettenstuhl m ~ **print** Kettdruck
m ~ **protector** Schußwächter m ~ **shiner** Kett-
spanner m ~ **sizer** Kettenleimer m ~ **spool**
Zettelkötzer m ~ **thread** Ketten-, Schuß-faden
m
**warpage** Verziehung f, Werfen n
**warped** gekrümmt, (wood) geworfen, schief, verzogen, windschief
**warper** Zettler m ~'s **bobbin** Zettelspule f
**warping** Krummwerden n, Schmeißer n, Verbiegung f, Verdrehung f, Verformung f, Verwerfung f, Verwindung f, Verziehen n, Verzug m, Werfen n, Zettlerei f ~ **of the gate** Verbiegung f des Tors ~ **of the model** Modellverbiegung f
**warping,** ~ **block** Scherblock m ~ **cable** Ver-

windungs-kabel n, -seil n ~ **cylinder** Schertrommel f ~ **drum** Spillkopf m ~ **end of winch**
Außentrommel f ~ **frame** Kettenscher-
maschine f ~ **mill** (textiles) Kettenscherma-
schine f, Schertrommel f ~ **lever** Verwindungshebel m ~ **machine** Zettelmaschine f ~ **pipe**
Verholklüse f ~ **wing** verwindbarer Flügel m
~ **wire** Verwindungsseil n ~ **wires** Drehdrähte
pl (aerodyn.), Drehkabel pl (aviat.)
**warrant, to** ~ garantieren, gewähren, gewährleisten
**warrant** Erlaß m, Ermächtigung f, Garantie f,
Gerichtsbefehl m, Lagerschein m, Vollziehungsbefehl m ~ **of arrest** Haftbefehl m
**warranted** gesetzlich geschützt
**warranty** Garantie f, Gewähr f, Gewähr-
leistung f, Sicherheit f
**Warren,** ~ **truss** Gitter-balken m, -träger m ~
**type bracing** Querspannung f ~-**type bridge**
**girder** (Brücke) pfostenloses Fachwerk n ~-
-**type rib** Gitterrippe f
**wash, to** ~ abdecken, (ore) abschwemmen,
(ore) abschlämmen, (ore) abläutern, abtreiben,
abwaschen, (ore) aufbereiten, aufschwemmen,
auswaschen, berieseln, decken, elutrieren,
laugen, läutern, reinigen, schlämmen, schwemmen, spülen, verwaschen, waschen (photo),
wässern **to** ~ **away** auskolken, fortspülen,
kolken, wegspülen **to** ~ **a cask** Faß m schlupfen
**to** ~ **coal** Kohle f waschen **to** ~ **off** abspülen,
abspritzen, abwaschen **to** ~ **out** auslaugen,
ausspülen, auswaschen, heraus-spülen, -wa-
schen, (a magnetic record) löschen, verwaschen
**to** ~ **with a spray** abspritzen **to** ~ **by** squiring
ausspritzen **to** ~ **thoroughly** durchwaschen
**wash** Abfluß m der Strömung (aerodyn.), Abstrom m, Decksirup m, glatter Abfluß m, der
Strömung, glatter Strömungsabfluß m (aviat.),
Lauge f, Nachstrom m, Schlichte f, Sog m,
Wäsche f ~ **of an oar** Ruderblatt n
**wash,** ~ **basin** Abwaschbecken n ~ **board**
Setzbord m ~ **board tension** (coning machine)
Gitterbremse f ~ **boring** Spülbohrung f ~ **bottle**
Auswaschflasche f ~(**ing) bottle** Waschflasche
f ~ **box** Berieselungshütte f ~ **column** Beriese-
lungs-, Riesel-turm m, Wäscher m ~ **gilding**
Vergoldung f auf Bronze ~ **heat** Abschweiß-,
Entzunderungs-wärme f ~ **house** Badekaue f
~-**in** positive Verwindung f (aviat.) ~ **leather**
Gemsleder n, samisches Leder n ~ **liquor**
Deckflüssigkeit f, Deckkläre f, Deckklarsel
n ~ **oil** Waschöl n ~ **ore** Flutwerk n, Wascherz
n ~-**out** negative Verwindung f (eines Flügels)
~-**out of thread** Gewindeauslauf m ~-**out**
**magnet** Löschmagnet m ~ **plate** Schlingerplatte
f ~ **point** Spülsonde f ~ **primer** Haftgrundmit-
tel n ~-**proof** waschbeständig ~ **rack** Wasch-
platz m (Autos, Flugzeuge) ~ **sirup** Deckablauf m, Decksirup m ~ **solution** Wasch-
lösung f ~ **tower** Berieselungsturm m, Rieselturm m, Skrubber m, Wäscher m ~ **trough**
Setzfaß n ~ **tub (or basin)** Waschzuber m
~-**up blade** Messingrakel n ~ **water** Kiel-,
Schwemm-, Spül-, Wasch-wasser n ~ **water**
**containing mud (or slime or sludge)** Schlamm-
wasser n
**washability** (of ores) Aufbereitbarkeit f ~ **curve**
Verwaschungskurve f

**washable** abwaschbar, (ores) aufbereitbar, auswaschbar, waschecht ~ **silk** Waschseide *f*
**washed** gespült ~ **out** abgeschlämmt
**washed,** ~ **drawing** getuschte Zeichnung *f* ~ **electrode** dünngetauchte Elektrode *f* ~ **graphite** gereinigter Grafit *m* ~ **ore slime** Schlammschlich *m* ~-**out ailerons** Querruder *pl* mit verjüngten Enden ~-**out depression** Kolk *m* ~-**out relief** Auswaschrelief *n* (photo) ~ **surface** Wurmlöcher *pl*
**washer** Abdeckscheibe *f*, Abdichtungsring *m*, Berieselungsturm *m*, Bezug *m*, Dichtungsscheibe *f*, Erz-Kohlenwäscher *m*, Fächerscheibe *f*, Gummiabdichtung *f*, Halbholländer *m*, Lagerdichtung *f*, Reiniger *m*, Reinigungsapparat *m*, Rieselturm *m*, Ring *m*, Scheibe *f*, Unterlagsplatte *f*, Unterlagscheibe *f*, Wasch-apparat *m*, -maschine *f*, Zwischenlagscheibe *f*
**washer,** ~ **for bearings** Kissen *n*, Mutterblech *n* ~ **for the removal of carbon dioxide** Kohlensäureabscheider *m*
**washer,** ~ **(thermo)couple** als Unterlegscheibe ausgebildetes Thermoelement *n* ~ **foreman** Wäschemeister *m* ~ **leather** Dichtungsleder *n* ~ **plant** Auslaugerei *f* ~ **remover** Abknöpfer *m*
**washing** (ore) Abläuterung *f*, Anspülung *f*, (ores) Aufbereitung *f*, Auswaschen *n*, Decke *f*, Decken *n*, Läuterung *f*, Naßaufbereitung *f*, Naßwäsche *f*, Reinigung *f*, Schlämmen *n*, Schlämmung *f*, Verwaschen *n*, Waschen *n*, Waschung *f*, Wässern *n*, Wässerung *f* (photo) ~ **of gold** Goldwäsche *f* ~ **with a jet** Strahlwäsche *f* ~ **of ores** Erzschlämmen *n* ~ **with pure-sugar solution** Klärseldecke *f* ~ **and scrubbing** Läuterung *f*
**washing,** ~ **accomodation** Wascheinrichtung *f* ~ **agent** Waschmittel *n* ~ **apparatus** Schlämmvorrichtung *f*, Waschapparat *m* ~ **away** Unterwaschung *f* ~ **basin** Waschbecken *n* ~ **bench** Auswaschtisch *m* ~ **benches for workmen** Arbeiterreihenwaschanlage *f* ~ **bottle** Läuterspritzflasche *f* ~ **chamber** Setzbett *n* ~ **cylinder** Läutertrommel *f* ~ **device** (sugar) Deckvorrichtung *f*, Schlämmvorrichtung *f* ~ **drum** Waschtrommel *f* ~ **engine** Halb-, Wasch-holländer *m* ~ **fluid** Waschflüssigkeit *f* ~ **liquid** Waschflüssigkeit *f* ~ **liquid bath** Laugenbad *n* ~ **liquid picked** Laugenverschleppung *f* ~ **liquid spray** Laugenspritzung *f* ~ **machine** Waschmaschine *f* ~ **machine with beetles** Pantsch-, Prätsch-maschine *f* ~ **machine in full width with three rollers (or with triple rolling system)** Breitwaschmaschine *f* mit Dreiwalzensystem ~ **pipe** Spülleitung *f* ~ **plant** Aufbereitungs-, Reinigungs-anlage *f*, Schlämmerei *f*, Spül-, Wasch-, Wäsche-anlage *f* ~ **process** Flotationsverfahren *n*, Schlämmverfahren *n*, Waschprozeß *m*, Waschvorgang *m* ~ **rack** Wässerungsgestell *n* (photo) ~ **refuse** Wäscheabgang *n* ~ **rolls** Auspreßwalzen *pl* ~ **soda** Waschsoda *n* ~ **solution** Abwaschmittel *n* ~ **train** Waschbatterie *f* ~ **trommel** Waschtrommel *f* ~ **trough** Waschbottich *m*, Waschtrog *m* ~ **tub** (gas) Waschaufsatz *m*, Waschbütte *f* ~-**up machine** Aufwaschmaschine *f* ~ **vat** Waschbottich *m* ~ **water** Waschwasser *n*
**wasp-shaped glass tank** Taillenwanne *f*

**waste, to** ~ schwinden, vergeuden, verschwenden, verzetteln
**waste** Abbrand *m*, Abfall *m*, Abraum *m*, Altmaterial *n*, Ausschuß *m*, Bruch *m*, (paper mfg.) Fabrikationsabfälle *pl*, Entfall *m*, Gekrätz *n*, Makulatur *f*, Schrott *m*, Schund *m*, Schutt *m*, Stoffabfall *m*, Überlauf *m*, Vergeudung *f*, Verlust *m*, Versatzberg *m*, Verschwendung *f*, Zehrung *f* ~ **of coal** Kohlenverlust *m* ~ **of fuel** Brennstoffverschwendung *f* ~ **in mining** Abbauverlust *m* ~ **of oil** Ölvergeudung *f* ~ **of space** Raumverschwendung *f* ~ **of timber** Verschnitt *m* ~ **of time** Zeitvergeudung *f*
**waste,** ~ **air tube** Abluftleitung *f* ~-**and-test valve** Prüf- und Entleeruhgsventil *n* ~ **bagasse** Bagasse *f* ~ **bin** Abfall-eimer *m*, -kasten *m* ~ **casting** Fehlgußstück *n* ~ **chamber** Abfallraum *m* ~ **coal** Abfallkohle *f* ~ **coke-oven gas** Koksofenabgas *n* ~ **copper** Kupferabfall *m* ~ **cotton** Putzbaumwolle *f*, Twist *m* ~ **current** Abfallstrom *m* ~ **current (or energy)** Abfallenergie *f* ~ **cutter** Abfallschneider *m* ~ **disposal** Beseitigung *f* radioaktiver Abfälle (atom.) ~ **dump** Halde *f* ~ **end** Abfallende *n* ~ **energy** Abfallenergie *f* ~ **fat from hide-glue manufacture** Leimfett *n* ~ **flue-gas drier** Abgastrockner *m*
**waste-gas** Abgas *n*, Abzugsgas *n*, (from blast furnace) Gichtgas *n* ~-**collecting channel** Abgassammelkanal *m* ~ **escape tube** Abgasaustrittsrohr *n* ~ **flue** Rauch-kanal *m*, -zug *m* ~ **flue (or duct)** Abgaskanal *m* ~ **purifying plant** Abgasreinigungsanlage *f* ~ **utilization** Abgasverwertung *f*
**waste,** ~ **gases** Feuergase *pl* ~ **gate** Abgasregelklappe *f*, Freiarche *f*, Nebenauslaß *m* ins Freie für zum Betrieb der Turbine nicht benötigtes Abgas (Abgasturbolader) ~ **head** verlorener Kopf *m* ~ **heap** Halde *f*
**waste-heat** abgehende Hitze *f*, Abhitze *f*, Abwärme *f*, Wärmeabfuhr *f* ~ **boiler** Abgas-, Abhitz-kessel *m* ~ **economy** Abhitzeverwertung *f* ~ **flue** Abhitzekanal *m* ~ **loss** Abwärmeverlust *m* ~ **recovering** Abhitzerückgewinnung *f* ~ **recovery** Wärmerückgewinnung *f* ~ **recuperator** Abhitzeverwertungsanlage *f*, Abwärmeverwerter *m* ~ **utilization** Abwärmeverwertung *f*
**waste,** ~ **instruction** Leerbefehl *m* (data proc.) ~ **liquor** Ablauge *f*, Natronlauge *f* ~ **liquors** schwarze Lauge *f* ~ **material** Schrottentfall *m* ~ **material (or product)** Abfallstoff *m* ~ **metal** (dross, sprue sweepings, scrapings, shavings, etc.) Krätze *f*, Metallgekrätz *n* ~ **oakum** Abwerg *n* ~ **oil** Aböl *n* ~ **oil regeneration plant** Ölablaßpfropfen *m* ~ **ore** Erztrübe *f* ~ **paper** Altpapier *n*, Ausschußpapier *n*, Makulatur *f*, Papierabfall *m* ~ **paper cleaning plant** Altpapierreinigungsanlage *f* ~ **paper cutting plant** Altpapierschneideanlage *f* ~ **pickle liquor** Abfallbeize *f* ~ **pickling water** Beizabwasser *n* ~ **pipe** Abflußrohr *n*, Abgangsrohr *n*, Ablaufleitung *f*, Abzugs-rohr *n*, -röhre *f*, Ausflußrohr *n*, Ausguß-röhre *f*, -rohr *n*, Dachröhre *f* ~ **pocket** Bergetasche *f* ~ **product** Abfallerzeugnis *n*, Abfallprodukt *n* ~ **recovery** Abfallaufbereitung *f* ~ **rubber** Altgummi *m* ~ **sheet** Makulaturbogen *m* ~ **sorter** Abfallsortierer *m* ~ **space** schädlicher Raum *m* ~ **steam** Ab-

dampf *m* ~ **steam heating** Abdampfheizung *f* ~ **steam pipe** Dampfableitungsrohr *n* ~ **steel** Stahlabfall *m* ~ **tip** Halde *f* ~ **top gas of a cupola furnace** Kupolofengichtgas *n* ~ **tow** Abwerg *n* ~ **tube** Abzugs-rohr *n*, -röhre *f* ~ **utilization** Abfälleverwertung *f* ~ **utilization plant** Abfallanlage *f* ~ **utilizing plant** Abfallverwertungsanlage *f* ~ **valve** Abdampfventil *n* ~ **washing** Abfallwäscherei *f* ~ **water** abgängiges Wasser *n*, wilde Flut *f* ~ **water clarification** Abwasserklärung *f* ~ **water-purifying plant** Abwasserreinigungsanlage *f* ~ **weir** Fluder *m* ~ **well** Schwind-, Senk-grube *f* ~ **wool** Twist *m*

**wasted, to be** ~ entfallen ~ **energy** Arbeitsverbrauch *m* im Leergang, Leerlaufarbeit *f*

**wasteful** unökonomisch ~ **resistance** Verlustwiderstand *m*

**waster** (in tinplate manufacture) Schrottblech *n*, (in casting) Schrottstück *n* ~ **casting** Fehlgußstück *n* ~ **(metal) reject** Ausschußstück *n* ~ **treatment** Abwasseraufbereitung *f*

**wasting**, ~ **of energy** Energieverschwendung *f* ~ **of time** Aufwand *m* an Arbeit, Zeitverschwendung *f*

**watch, to** ~ aufpassen, beobachten, bewachen, überwachen

**watch** Bewachung *f*, Uhr *f*, Wache *f*, Wacht *f*, Zeitmesser *m* ~ **buoy** Verholtonne *f* ~ **case** Uhrgehäuse *n* ~ **case receiver** Dosenfernhörer *m* ~ **case telephone** Dosenfernhörer *m* ~ **frequency** Wachfrequenz *f* ~ **glass** Uhrglas *n* ~ **maker** Uhrmacher *m* ~ **maker's magnifier** Uhrmacherlupe *f* ~ **maker's tool** Uhrmacherswerkzeug *n* ~ **man** Bahnwärter *m*, Wächter *m* ~ **man's clock** Kontrolluhr *f* ~ **man's control advertiser** Wächterkontrollmelder *m* ~-**man's control clock** Wächterkontrolluhr *f* ~ **man's control system** Wächterkontrollanlage *f* ~ **men's time detector** Wächteruhr *f* ~ **panel (or slide)** Uhrhalter *m* ~ **pivot** Uhrzapfen *m* ~ **receiver** Dosenfernhörer *m* ~ **spring** Uhrfeder *f* ~ **test** Uhrprüfung *f* ~ **tower** Wachtturm *m*, Warte *f*

**watching** Überwachung *f*

**water, to** ~ ablöschen, befeuchten, begießen, berieseln, besprengen, bewässern, (coke) löschen, moirieren, nässen, tränken, verwässern, wässern

**water** Wasser *n* ~ **of combustion** Verbrennungswasser *n* ~ **of condensation** Kondensations-, Kondens-, Schweiß-wasser *n* ~ **of constitution** Konstitutionswasser *n* ~ **of contraction** Schwindungswasser *n* ~ **of crystallization** Kristallwasser *n* ~ **for fire fighting** Löschwasser *n* ~ **in gaseous state** gasförmiges Wasser *n* ~ **for general use** Gebrauchwasser *n* ~ **of hydration** Hydratwasser *n* ~ **for industrial use** Betriebswasser *n* ~ **for radiator** Kühlwasser *n* ~ **and sediment** Gehalt *m* an Wasser und Absatz ~ **of vegetation** Bodenbewachsungs-, Vegetations-wasser *n*

**water,** ~ **absorbent** wasserbindend ~ **absorbing** wasseraufsaugend ~ **absorbing capacity** Wasserabsorptionsvermögen *n* ~ **absorption capacity** Wasserkapazität *f* ~ **absorptive capacity** Wasserbindvermögen *n* ~ **aerodrome** Wasserflugplatz *m* ~ **air interface** Wasser-Luft-Trennungsfläche *f* ~ **alarm** Wasserstandsmel-

der *m* ~ **anchorage** Wasserverankerung *f* ~ **appliance** Wasserarmatur *f* ~ **atomizer** Zerstäuber *m* ~ **attracting** wasseranziehend ~ **balance** Wasser-aufzug *m*, -waage *f* ~ **ballast** Wasserballast *m* ~ **basin** Wasserbecken *n* ~ **bath** Wasserbad *n* ~ **bath desuperheater** Wasserbad-Dampfumformer *m* ~ **beam pump** Wasserstrahlpumpe *f* ~ **bearing** wasserführend; Wasserlager *n* ~ **bearing stratum** wasserführende Schicht *f* ~ **blast** Wasserstrahlgebläse *n* ~ **board** Setzbord *m* ~ **boiler reactor** Siedewasserreaktor *m* ~ **bore** Füllschwall *m* ~ **borne** flott ~ **bottom gas producer** Wasserabschlußgaserzeuger *m* ~ **box** Wasserkasten *m* ~ **brake** Wasserbremse *f* ~ **breaking-in** Wassereinbruch *m* ~ **bucket** Wassereimer *m* ~ **budget** Wasserhaushalt *m* ~ **burst** Wasserdurchbruch *m* ~ **calender** Wasserkalender *m* ~ **calorimeter** Mischungskalorimeter *n* ~ **chamber** Wasserkammer *f* ~ **channel** Wasserrinne *f* ~ **circulating pump** Kühlwasserpumpe *f* ~ **circulation** Wasser-kreislauf *m*, -umlauf *m* ~ **circulation jacket surrounding the electrodes** Elektrodenwasserkühlmantel *m* ~ **clarifier** Wasserreinigungsanlage *f* ~ **clearing plant** Wasserkläranlage *f* ~ **closet** Abortanlage *f*, Abtritt *m* mit Wasserschluß ~ **closet with flat wash down** Flachspülklosett *n* ~ **closet with radial flush system** Tiefspülklosett *n* ~ **coat** Wasserhaut *f* ~ **cock** Wasserhahn *m* ~ **color** Feuchtwasserfarbe *f* ~ **column pressure** Wassersäulendruck *m* ~ **column pressure gauge** Wassersäulendruckmesser *m* ~ **company** Wassergenossenschaft *f* ~ **compartment** Wasserkasten *m* ~ **conduit** Wasserleitung *f* ~ **connecting piece** Wasseranschlußstutzen *m* ~ **connection** Wasseranschluß *m* ~ **conservation** Wasserhaushalt *m* ~ **consumption** Wasserverbrauch *m* ~ **contamination** Wasserverunreinigung *f* ~ **content** Wassergehalt *m* ~ **conveying by wind** Wasserförderung *f* mit Windbetrieb ~-**cooled** wassergekühlt ~-**cooled electrode collar** wassergekühltes Elektrodenanschlußstück *n* ~-**cooled tube** wassergekühlte Röhre *f*, Wasserkühlröhre *f* ~ **cooler** Trinkwasser-, Wasser-kühler *m* ~ **cooling** Wasserkühlung *f* ~ **cooling installation** Wasserkühlanlage *f* ~ **cooling system** Kühlsystem *n*, Wasserkühlung *f* ~ **cooling tower** Wasserkühlturm *m* ~ **course** Fleet *n*, Gewässer *n*, Spülkanal *m*, Wasserlauf *m* ~ **craft** Wasserfahrzeug *n* ~ **cushion** Beruhigungskammer *f*, Tosbecken *n*, Toskammer *f*, Wasser-kissen *n*, -polster *n* ~ **cure** Heißwasservulkanisation *f* ~ **curtain** Wasservorhang *m* ~ **decoction apparatus** Abkochgerät *n* ~ **demanganesing** Wassermanganung *f* ~ **depth gauge** Pegel *m* ~ **discharge** Wasserabfluß *m* ~ **discharge vent** Wasserablaßhahn *m* ~ **disoxydation** Wasserentsäuerung *f* ~ **dispersed varnish** Wasserlack *m* ~ **distributing point** Wasserverteilungsstelle *f* ~ **distribution** Wasserausgleich *m* ~ **distribution plant** Wasserverteilungsanlage *f* ~ **distributor** Wasserführung *f* ~ **diviner** Rutengänger *m* ~ **drain** Wasser-ableiter *m*, -ableitung *f*, -leckleitung *f* ~ **drain pipe** Wasserablaßrohr *n* ~ **drain valve** Wasserablaßventil *n* ~ **draw-off cock** Wasserablaßhahn *m* ~ **drip** Unterschneidung *f* ~ **driven centrifugal** Schleuder *f* mit

Wasserantrieb ~ **drop** Wassertropfen *m* ~ **droplet** Wassertröpfchen *n* ~ **drum** Unterkessel *m* ~ **duct** Wasserführung *f* ~ **ejector** Wasserablaß *m* ~ **electrolysis** Hydrolyse *f* ~ **eliminating plant for tar** Teerentwässerungsanlage *f* ~ **evaporator** Wassereindampfgerät *n* ~ **fall** Wasserfall *m* ~ **fascine** Senkmaschine *f*, Sinklage *f* ~ **faucet** Wasser-hahn *m*, -kran *m* ~ **feed** Wasserzufluß *m* ~ **feed heater** Vorwärmer *m* ~ **feeding piston** Wasserzuführungskolben *m* ~ **film** Wasserhülle *f* ~ **finder** Rutengänger *m* ~ **finish** Feuchtglätte *f* ~ **fitting** Wasserarmatur *f* ~ **flood** Wasserüberschwemmung *f* ~ **flow detector** Strömungswächter *m* ~ **flow pyrheliometer** Durchfluß-Pyrheliometer *m* ~ **fluid brake** Wasserbremse *f* ~ **flush boring poles** Spülgerät *n* ~ **flush system** Spülverfahren *m*, **foam** Wasserschaum *m* ~ **freight** Wasserfracht *f*
**water-gas** (karburiertes) Wassergas *n* ~ **drive** Gasantrieb *m* ~ **equilibrium** Wassergasgleichgewicht *n* ~ **generator** Wassererzeuger *m* ~ **press welding** Wassergas-Preßschweißung *f* ~ **producer** Wassergas-erzeuger *m*, -generator *m* ~ **tar** Wassergasteer *m*
**water,** ~ **gate** (of a mill or pond) Fluttor *n* ~ **gathering ground** Sammelgebiet *n*
**water-gauge (w.g.)** hydraulischer Druckmesser *m*, Peil *m*, Wasser-messer *m*, -säule *f*, -stand *m*, Wasserstands-anzeiger *m*, -rohr *n*, -zeiger *m* ~ **cock** Wasserstandshahn *m* ~ **cock (or tap)** Hahn *m* am Wasserstandsanzeiger ~ **glass** Wasserstand(s)-anzeiger *m*, -glas *n* ~ **level** Pegelhöhe *f* ~ **pocket** Wasserstandsstutzen *m* ~ **preserving tube** Wasserstandsschutzglas *n* ~ **tap** Wasserstandshahn *m*
**water,** ~ **gig** Verstreichrauhmaschine *f* ~ **gilding** Wasservergoldung *f* ~ **glass** Wasserglas *n* ~ **guard** Wasser-fänger *m*, -schild *m* ~**hammer** Druckstoß *m*, Wasserstoß *m*, Widderstoß *m* ~ **hammering** Wasserschlag *m* ~ **hardening** Abschrecken *n* in Wasser, Wasser-härtung *f*, -vergüten *n* ~ **hardening steel** Wasserhärtungsstahl *m* ~ **hose** W.asser-schlange *f*, -schlauch *m* ~ **imbibent** wasserbindend ~ **incrustation** Kesselstein *m* ~ **injection** Wassereinspritzung *f* ~ **inlet** Wasser-einlaß *m*, -eintritt *m* ~ **inlet of radiator** Einlaßstutzen *m* des Kühlers, Kühlereinlaßstutzen *m* ~ **inlet pipe** Wassereinlaßrohr *n* ~ **irrigation** Wasserberieselung *f* ~ **jacket** Kühl-, Kühlwasser-, Wasser-mantel *m* ~ **jacket furnace** Schachtofen *m* ~ **jacket plug** Kernlochstopfen *m* ~ **jacket-type producer** Wassermantelgenerator *m* ~ **jacketed furnace** Wassermantelofen *m*
**water-jet** Wasserstrahl *m* ~ **aspirator** Wasserluftpumpe *f* ~ **(vacuum) injector** Wasserstrahlpumpe *f* ~ **installation** Spülvorrichtung *f* ~ **nozzle** Wasserstrahldüse *f* ~ **pipe** Spühlrohr *n* ~ **(vacuum) pump** Wasserstrahlpumpe *f* ~ **scrubbing** Wasserberieselung *f* ~ **type lightning arrester** Wasserstrahlerder *m*
**water,** ~ **kennel** Gefluder *n* ~ **landing** Wasserlandung *f*.
**water-level** Grundwasserspiegel *m*, Grundstrecke *f* (min.), Libelle *f*, Pegelstand *m*, Stau *m*, Wasser-stand *m*, -spiegel *m*, Wasserstandslinie *f* (geol.), Wasserwaage *f* ~ **in well** Tauchtiefe *f*

**water-level,** ~ **gauge** Pegellatte *f*, Wasserstandsmesser *m* ~ **indicator** Wasserstandsanzeiger *m* ~ **overflow** Wasserüberlauf *m* ~ **regulator** Wasserstandsregler *m* ~ **tele-indicator** Wasserstandsfernmeldeapparat *m* ~ **transmitter** Wasserstandsmelder *m*
**water,** ~ **line** Wasser-linie *f*, -spiegel *m*, -stand *m* ~ **line model** Wasserlinienmodell *n* ~ **load** Wasser-Endlast *f* ~ **loads** Belastung *f* durch Wasserkräfte ~ **lock** Flüssigkeitsheber *m* ~ **logged** voll Wasser *n* ~ **logging** mit Wasser *n* durchtränken ~ **loop** Ausbrechen *n* auf dem Wasser ~ **main** Hauptrohr *n* der Wasserleitung ~ **maintenance machine** Wasserhaltungsmaschine *f* ~ **mark** Spiegelpfahl *m*, (on beach) Wasserlinie *f*, Wasserstandsmarke *f* ~ **mark post** Pegel *m*, Peil *m* ~ **marked paper** Wasserzeichenpapier *n* ~ **measuring vane** Wassermesserflügel *n* ~ **meter** Wasser-messer *m*, -uhr *f* ~ **meter with mechanism and indicator operating (dry) in air** Trockenläufer *m* ~ **milling** Wasserwalke *f* ~ **monitor** Wasserüberwachungsgerät *n* ~ **nozzle** Wasserhahn *m* ~ **obstacle** Wassersperre *f* ~ **outlet** (to water pump) Austrittsstutzen *m*, Wasser-auslaß *m*, -austritt *m*, -durchlaßeinrichtung *f* ~ **outlet of radiator** Kühlerausflußstutzen *m* ~ **outlet valve** Wasseraustrittsventil *n* ~ **overflow** Wasserüberlauf *m* ~ **particle** Wassertröpfchen *n*, -tropfen *m* ~ **passage** Wasserführung *f* ~ **permeability tester** Wasserdurchlässigkeitsprüfer *m* ~ **phantom** Wasserphantom *n* ~ **pigment colors** Wasserdeckfarben *pl* ~ **pigment finish** Wasserdeckfarbe *f* ~ **pipe** Wasser(leitungs)rohr *n* ~ **pipes** Wasserleitung *f* ~ **pocket** Wasser-ecke *f*, -sack *m* ~ **point** Schöpfstelle *f* ~ **polishing tool** Schlichtstahl *m* mit Seitenmesser ~ **post** Wasserstandsmarke *f* ~ **pot** Wassertopf *m* ~ **power** Wasserkraft *f* ~ **power cane mill** Rohrmühle *f* mit Wasserkraftantrieb ~**power plant** Wasserkraftanlage *f* ~ **pressure** Wasserdruck *m* ~ **pressures balanced** ausgespiegelte Wasserstände *pl*
**water-proof, to** ~ imprägnieren, wasserdicht machen
**water-proof** wasser-beständig -dicht, -fest, -undurchlässig, -unlöslich ~ **finish** wasserdichte Appretur *f* ~ **glue** Marineleim *m* ~ **material** wasserdichtes Zeug *n* ~ **paint** wasserabhaltender Anstrich *m* ~ **paper** Guttaperchapapier *n* ~ **paper roof** Pappdach *n* ~ **tissue** wasserdichtes Zeug *n*, wasserdichter Webstoff *m*
**water-proofing** Abdichtung *f*, Dichtung *f* ~ **agent** Wasserdichtmachungsmittel *n* ~ **asphalt** Pech *n* für wasserdichte Anlagen ~ **solution** Fluatlösung *f*
**water-proofness** Wasserundurchlässigkeit *f*
**water-pump** (diluting pump) Wasserpumpe *f* ~ **drive** Wasserpumpenantrieb *m* ~ **impeller** Wasserpumpenflügelrad *n* ~ **nut pliers** Wasserpumpenzange *f* ~ **shaft** Wasserpumpenwelle *f* ~ **thurst bush** Wasserpumpenstopfbüchse *f* ~ **windmill** Wasserschopfwindmühle *f*
**water,** ~ **pumping by wind** Wasserförderung *f* mit Windbetrieb ~ **pumping machine** Wasserhebungsmaschine *f* ~ **pumping packing gland** Wasserpumpenpackung *f* ~ **purification** Was-

serreinigung *f* ~ **purification plant** Wasserreinigungsanlage *f* ~ **purification unit** Trink-
wasserbereiter *m* ~ **purifier** Wasser-reinigungs-
apparat *m*, -vorlage *f* ~ **purifying** Wasseraufbereitung *f* ~ **purifying plant** Wasserkläranlage *f* ~ **purity meter** elektrischer Salzge-
haltsmesser *m* ~ **pyrometer** Wasserpyrometer *n*
~ **quantity-recording apparatus** Wassermengen-
registrierapparat *m* ~ **quenching** Abschrecken *n*
in Wasser ~ **radiator** Wasserkühler *m* ~
**raising machine** Wasserhebungsmaschine *f* ~
**rate test** (turbine) Dampfverbrauchsversuch *m*
~ **ratio** Wasserverhältnis *n* ~ **receiver around
the exhaust valve** Wassermantel *m* um das
Auspuffventil ~ **receptacle** Wassertopf *m* ~
**recooling system** Wasserrückkühlung *f* ~ **recovery** Ballastgewinnung *f*
**water-repellent** wasserabstoßend ~ **concrete**
Sperrbeton *m* ~ **effect** Wasserabperleffekt *m*
~ **mortar course** Sperrmörtelschicht *f* ~ **plastering** Sperrputz *m*
**water,** ~ **replacement** Wasserersatz *m* ~ **requirement** Wasserbedarf *m* ~ **reservoir** Wasserbecken *n* ~ **residue** Wasser-rest *m*, -rückstand
*m* ~ **resistance** Flüssigkeitswiderstand *m*
~**-resistant** wasserbeständig ~ **retaining** wasser-
haltend ~**-retted flax** Wasserflachs *m* ~ **retting**
Wasserröste *pl* ~ **rights** Wasserrecht *n* ~ **ring
air pump** Wasserringluftpumpe *f* ~ **rising**
hochsteigende Wasser *pl* ~ **room** Wasserraum
*m* ~ **rudder** Wasser(seiten)ruder *n* ~ **sample**
Wasserprobe *f* ~**-saturated** wassersatt ~**-saving**
Wasserersparnis *f* ~**-scooping machine** Was-
serschöpfmaschine *f* ~ **screen** Abschreck-,
(Kohlenstaubfeuerung) Granulier-, Kühl-rost
*m* ~ **screening plant** Wassergroßreinigungsanla-
ge *f* ~ **screw** Wasserschnecke *f* ~ **seal** hydraulische Dichtung *f*, Sicherheitsvorlage *f*, Wasser-
-abschluß *m*, -schluß *m*, -verschluß *m*, -vorlage *f*,
wasserdichter Abschluß *m*, (in gas producer)
Wassertasse *f* ~**-seal weld** Azetylensicherheits-
vorlage *f* ~**-sealed gas producer** Wasserabschlußgaserzeuger *m* ~**-sealed valve** Wasser-
abschlußventil *n* ~ **sensitive** wasserempfindlich
~ **separator** Dephlegmator *m*, Entwässerer *m*,
Entwässerungsvorrichtung *f*, Wasserabschei-
der *m* ~ **shed** Einzugsgebiet *n*, (boundary)
Wasserscheide *f* ~ **shock** Wasserschlag *m* ~
**shortage** Wasserklemme *f* ~ **shut-off** Was-
sersperrung *f* ~**-side** wasserseitig ~**-side
leg** wasserseitige Stütze *f* ~ **sluice valve**
Wasserschieber *m* ~ **snake** Wasserschlange *f*
~ **softener** Wasserenthärtungsanlage *f* ~
**softening** Wasser-enthärtung *f*, -enteisenung *f*
~ **softening plant** Wasserenteisenungsanlage *f*
~**-soluble** wasserlöslich ~ **space** Wasserraum *m*
~ **spout** Wasser-hose *f*, -leitungshahn *m* ~
**spray** Wasser-sprühregen *m*, -staub *m* ~
**spraying** Wasser-berieselung *f*, -rieselung *f* ~
**stain** (Holz) Wasserbeize *f* ~ **sterilizing bag**
Wasserreinigungssack *m* ~ **still** Wasserdestil-
lierapparat *m* ~ **stone** Wasserabziehstein *m* ~
**stop wall** Wehrmauer *f* ~ **stored for fire fighting** Brandvorrat *m* ~ **streak** Wasserstreifen
*m* ~ **string** Bohr-, Wassersperr-kolonne *f*
**water-supply** Wasserspeisung *f*, -versorgung *f*,
-vorrat *m*, -zulauf *m*, -zufluß *m* ~ **and pumping
station** Wasserhebewerk *n* ~ **from springs**

Hochquellenleitung *f* ~ **pipe** Wasserzuleitungs-
rohr *n* ~ **plant** Wasserversorgungsstelle *f* ~
**tank** Wasserversorgungstank *m*
**water,** ~ **surface** Wasser-fläche *f*, -spiegel *m*
**below** ~ **surface** unter Wasser *f* ~ **switch** Wasserschalter *m* ~ **system** Wasserleitung *f* ~ **table**
Grundwasser *n*, Rollengerüst *n* ~ **take-off**
Wasserabflug *m* ~ **tank** Wasserbecken *n*, -be-
hälter *m*, -kasten *m*, -kessel *m*, -tank *m* ~ **tap**
Wasser-hahn *m*, -leitungshahn *m* ~ **temperature**
Wassertemperatur *f* ~ **temperature gauge** Fernthermometer *n* ~ **tempering** Wasservergüten *n*
(metall.) ~ **test** Wasserprobe *f* ~ **thermometer**
Kühlwasserthermometer *n* ~ **throttle** Drossel-
schraube *f* ~ **thrower** Wasserabweisring *m*
**water-tight** wasserdicht, wasserfest ~ **bulk-head**
wasserdichtes Schott *n* ~ **compartment** wasserdichtes Abteil *n* ~ **curtain** Dichtungsjalousie *f*
~ **joint** wasserdichte Verbindung *f* ~ **protection**
Druckwasserschutz *m* ~ **ring (or seal)** Dichtungsring *m*
**water tightness** Dichtheit *f*, Dichtigkeit *f*, Ver-
schlußzustand *m*, Wasserdichtigkeit *f*
**water,** ~**-to-carbide gas generator** Überschwem-
mungsentwickler *m*, Wasserzuflußentwickler *m*
~ **tower** Wasserturm *m* ~ **transportation**
Wassertransport *m* ~ **trap** Wassersack *m* ~
**treating plant** Wasseraufbereitungsanlage *f* ~
**treatment** Wasseraufbereitung *f* ~ **trough**
Wasserrinne *f* ~ **trunks** Wasserballasthosen *pl*
~ **tube boiler** Siederohrkessel *m*, Wasser-röh-
renkessel *m*, -rohrkessel *m* ~ **tube boiler with
two headers** Zweikammerwasserrohrkessel *m*
~ **tube cooler** Wasserkühler *m* ~ **tube radiator**
Wasserröhrchenkühler *m* ~ **tube-type cooler**
Wasserröhrenkühler *m* ~ **turbine** Wasserturbine *f* ~ **turbine oil** Wasserturbinenöl *n* ~
**used in brake** Bremswasser *n* ~**-up-stream of
point in question** Oberwasser *n* ~ **vacuum pump**
Wasserluftpumpe *f* ~ **valve cover** Wasserventil-
deckel *m* ~ **vapor** Wasserdampf *m* ~ **vapor
addition** Wasserdampfzusatz *m* ~ **vapor band**
Wasserdampfbad *n* ~ **vat** Wassergefäß *n* ~
**wagon** Sprengwagen *m* ~**-walled** Kühlrohre
*pl* ~ **way (bridge)** Durchflußprofil *n*, Wasser-
-gang *m*, -straße *f*, -weg *m* ~ **way of a bridge**
Flutraum *m* einer Brücke ~ **way construction**
(for navigation) Verkehrswasserbau *m* ~ **way
marking** Fahrwasserbezeichnung *f* ~ **wheel**
Wasserrad *n* ~ **wheel generator** Wasserkraft-
generator *m* ~ **wheel paddle** Wasserradschaufel
*f* ~ **white** wasserhell ~ **willow** Flechtweide *f* ~-
-**works** Wasser-anlage *f*, -versorgung *f* ~**-works
plant** Wassertriebwerk *n*
**watered** bespült, geflammt, (of fabrics) flam-
micht ~ **paper** Metallpapier *n*
**watering** Ablöschen *n*, Berieselung *f*, Bewässerung *f*, Löschung *f*, Tränkung *f* ~ **apparatus**
Wasserspritze *f* ~ **bridle** Wassertrense *f* ~ **can**
Gießkanne *f* ~ **car** Löschwagen *m*, Wassersprengwagen *m* ~ **device** Tränkanlage *f* ~
**ditch** Berieselungsgraben *m* ~ **equipment**
Sprengvorrichtung *f* ~ **house** Löschturm *m* ~
**place** Tränke *f* ~ **plant** Regenanlage *f*
**waterless gas holder** Scheibengasbehälter *m*
**waters** Gewässer *n*, Seegebiet *n*
**watery** dünnflüssig, wasserartig, wässerig
**watt** Watt *n* ~**s of battery current** (~**s of current**

dissipation, or ~s of plate current) Anodenbelastung *f*
**watt,** ~ **component** Wattkomponente *f*, Wirk-
komponente *f*, Wirkstrom *m* ~ **consumption**
Wattverbrauch *m* ~ **current** Arbeits-, Verlust-
-strom *m*
**watt-hour** Wattstunde *f* ~ **capacity** Leitungskapazität *f* ~ **capacity of a storage cell** Kapazität
*f* oder Leistungskapazität *f* eines Sammlers ~
**consumption** Wattstundenverbrauch *m* ~ **meter**
Induktions-, Wattstunden-, Wirkleistungs-zäh-
ler *m*, Wirkleistungsverbrauchszähler *m*
**wattage** Wattleistung *f*, Wattverbrauch *m*,
Wattzahl *f*, Wirkleistung *f*
**wattless** blind, energielos, leistungslos, ohne
Energieverbrauch *m*, stromlos, wattlos ~
**component** Blindkomponente *f*, wattlose Komponente *f* ~ **component of the current** Blindstromkomponente *f* ~ **component of electric
values** Blindwert *m* elektrischer Größen ~
**component of electromotive force** Blindspannungskomponente *f* ~ **component of voltage**
Blindspannungskomponente *f* ~ **component
meter** Blindverbrauchszähler *m* ~ **current**
Blindstrom *m*, Querstrom *m*, wattloser Strom
*m* ~ **current meter** Blindstrommesser *m* ~
**power** Blindleistung *f*
**wattlework** Flecht-werk *n*, -zaun *m*, Packwerk *n*
**wattmeter** Energiemesser *m*, Leistungs-messer
*m*, -zeiger *m*, Watt-messer *m*, -meter *n*, Wirk-
leistungsmesser *m* ~ **method** Leistungsmethode
*f*
**wave, to** ~ flattern, schwingen, wellen
**wave** Becken *n*, Guilloche *f* (print.), Schwingung
*f*, Welle *f*, Wellenkamm *m* ~ **for daytime transmission** Tageswelle *f* ~ **of medium to high frequency** Grenzwelle *f* ~ **of modulation** Modula-
tionswelle *f* ~ **for nighttime transmission**
Nachtwelle *f* ~ **of sound** Schallwelle *f* ~ **on
outer surface of coaxial cable** Mantelwelle *f*
~ **of translation** Übertragungswelle *f*
**wave,** ~ **action** Wellen-schlag *m*, -stoß *m* ~
**amplitude** Wellenausschlag *m* ~ **antenna**
Wellenantenne *f* ~ **attenuation constant**
Wellendämpfungskonstante *f* ~ **band** Wellen-
band *n*, -bereich *m* ~ **band filter** Bandfilter *n*,
Filtersiebkette *f* ~ **band indicator** Wellenbe-
reichmelder *m* ~ **band switch** Wellenbereich-
schalter *m* ~ **breaker** Wellenbrecher *m* ~
**caused by a falling stage** Senkungswelle *f* ~
**caused by a rising stage** Hebungswelle *f* ~
**change switch** Wellenbereichsschalter *m* ~
**(length) changing switch** Verstimmungsschalter
*m* ~ **clutter** Seegangechos *pl* (rdr), Wellenrefle-
xion *f* ~ **collector** Empfangsantenne *f* ~ **constant**
Wellenkonstante *f* ~ **converter** Modusumfor-
mer *m* ~ **crest** Scheitel *m* einer Welle, Wellen-
berg *m* ~ **detector** Wellen-anzeiger *m*, -detek-
tor *m* ~ **diffraction** Wellenbrechung *f* ~ **director circuit** Direktor *m* ~ **disturbance** Wellen-
störung *f* ~ **drag** (of supersonic flight) Wellenwiderstand *m* ~ **duct** Leitschicht *f* (rdr) ~
**equation** Wellengleichung *f* ~ **excitation**
Schwingungsanfachung *f* ~ **exciter** Wellen-
erreger *m* ~ **extent** Wellenausschlag *m* ~
**face** Wellenstirn *f* ~ **field** Wellenfeld *n* ~
**filament** Wellenstrahl *m* ~ **form** Kurven-,
Wellen-form *f* ~ **form converter** Wellenform-

wandler *m* ~ **form distortion** Amplituden-,
Klirr-verzerrung *f*, nichtlineare oder ungradlinige Verzerrung *f* ~ **form response** Wellen-
form-Charakteristik *f* ~ **frequency** Wellen-
frequenz *f* ~ **frequency control** Wellenmesser
*m* ~ **front** (surge front) Anstiegflanke *f*, Wellen-
-front *f*, -kopf *m*, -stirn *f* ~ **front of sweep
voltage** Kippflanke *f* ~ **front angle** Wellenfrontwinkel *m* ~ **function** Wellenfunktion *f* ~
**generation** Schwingungsanfachung *f*, Wellen-
anregung *f* ~ **generation by sheltering effect** Wel-
lenanfachung *f* durch Leerwirbel ~ **generator**
Wellen-erzeuger *m*, -generator *m*, ~ **group**
Wellengruppe *f*
**wave-guide** Hohlleiter *m*, Wellen-führung *f*,
-leiter *m*, -leitung *f* ~ **circuit** Hohlraum-
mischerkreis *m* ~ **cutoff frequency** Hohlleiter-
grenzfrequenz *f* ~ **elbow** Hohlleiterknie *n* ~
**feed** Hohlleiterzuführung *f* ~ **gasket** Wellen-
leiterdichtung *f* ~ **modes** Hohlleiterwellen *pl*
~ **plunger** Hohlleiterabstimmkolben *m* ~ **shim**
Flanschzwischenlage *f*, Hohlleiterkontaktblech
*n* ~ **shutter** Hohlleiterblende *f* ~ **slotted line**
Hohlleitermeßleitung *f* ~ **stub** Stichleitung *f* ~
**switch** Hohlleiterschalter *m*, Hohlleiterum-
schalter *m* ~ **tuner** einstellbarer Hohlleiter-
transformator *m* ~ **twist** Hohlleiterverdrehung
*f*
**wave,** ~ **head** Wellenstirn *f* ~ **impedance** Wellenleitwert *m*
**wave-length** Wellenlänge *f* ~ **in air** Wellenlänge
*f* in Luft ~ **changing switch** Wellenumschalter *m*
~ **constant** Phasenmaß *n*, Wellenlängenkonstante *f*, Winkelmaß *n*, Winkelmaß *n* je Längeneinheit, Winkelkonstante *f* ~ **constant per
section** Kettenwinkelmaß *n* je Glied ~ **meter**
Wellenmesser *m* ~ **prolongation** Wellenver-
längerung *f* ~ **shortening** Wellenverkürzung *f*
~ **switching** Wellenumschaltung *f*
**wavelike** wellen-artig, -förmig
**wave,** ~ **line** Wellenlinie *f* ~ **line recorder**
Wellenlinienschreiber *m* ~ **loop** Wellenbauch *m*
(rdr) ~ **mechanical** wellenmechanisch ~
**mechanics** Wellenmechanik *f* ~ **meter** Fre-
quenz-kontrollgerät *n*, -messer *m*, -prüfgerat *n*,
Wellenmesser *m* ~ **modulation** Wellenmode-
lung *f* ~ **mold** (paper) Velinform *f* ~ **molding**
Sprungleisten *pl* ~ **motion** Wellenbewegung *f*
~ **normal** Wellenfrontnormale *f* ~ **number**
Wellenzahl *f* ~ **packet** Wellen-bündel *n*, -paket
*n* ~ **packets** Bündel *pl* ~ **passage button**
Welleneinstellknopf *m* ~ **profile** Wellenform *f*
~ **propagation** Wellen-ausbreitung *f*, -fort-
pflanzung *f* ~ **quenching** Wellenauslöschung *f*
~ **range** Wellenbereich *m*, Wendungsfeld *n*
~ **range switch** Wellen(um)schalter *m* ~
**range switching** Wellenbereichumschaltung *f*
~ **receiver** Wellenempfänger *m* ~ **reduction**
Wellenvernichtung *f* ~ **reflection** Wellenrück-
strahlung *f* ~ **retarding means** Phasenver-
schiebungsplättchen *n* ~ **screen** Wellensieb *n* ~
**shape** Kurvenform *f*, Wellen-form *f*, -gestalt *f*
~ **shape analysis** Frequenzanalyse *f* einer
Wellenform ~ **signals** Wellenbetrieb *m* ~
**superposed on direct current** gleichstromüberlagerte Welle *f* ~ **surface** Wellenfläche *f* ~
**switch range** Wellenschalterbereich *m* ~ **tail**
Wellenschwanz *m* ~ **tilt** Wellenfrontwinkel *m* ~

train Wellenzug *m* ~ train frequency Gruppen-,
Wellenzug-frequenz *f* ~ trap Beruhigungs-
becken *n*, Sperr-kreis *m*, -topf *m*, Wellen-sau-
ger *m*, -saugkreis *m*, -schlucker *m* ~ trough
Tal *n* einer Welle, Wellental *n* ~-type separator
Wellenseparator *m* ~ variable Wellenverän-
derliche *f* ~ velocity Phasen-, Wellen-ge-
schwindigkeit *f* ~ voltage Wellenspannung *f* ~
winding Wellenwicklung *f* ~ winding attachment Kreuzspulzusatz *m* ~ window Skala *f*
(rdr)
waved geflammt
waver, to ~ schwanken, wanken, zaudern
wavering schwankend, unstet ~ of signal beam
Trübung *f*
waviness Welligkeit *f*
waving Wellenbildung *f*
wavy gebuchtet, gewellt, wellen-artig, -förmig,
wellig ~ distortion by wave action Wellenstrahl *m* ~ fibered growth Wellenfaserigkeit *f*
~ groundmass Grundgewebe *n* ~ slip-lines
Welligkeit *f* der Gleitlinien
wax, to ~ mit Wachs *n* überziehen, wachsen
wax Bienenwachs *n*, Paraffin *n*, Wachs *n* ~
for bleacher Paraffin für Wascher ~ for candle
manufacturers Paraffin *n* für Kerzenfabrikan-
ten ~ for irons Paraffin *n* für Bügeleisen ~ for
matches Paraffin *n* für Streichhölzer oder für
Zündhölzchen
wax, ~ block Joly-Fotometer *n* ~ cake Wachs-
boden *m* ~ candle Wachsfirnislicht *n* ~ cement
Wachskitt *m* ~ chandler Wachszieher *m* ~
cloth for flooring Wachstuch *n* für Fußbodenbelag ~ coat Wachsüberzug *m* ~ coated paper
Wachsschichtpapier *n* ~ composition Wachsfir-
nismasse *f* ~ crayon Wachsstift *m* ~ disc
Wachsplatte *f* ~ distillate paraffinhaltiges
Destillat *n* ~ finish Wachsappretur *f* ~ impression Wachsabdruck *m*, Wachsdruck *m* ~
layer Wachsüberzug *m* ~ master Wachsplatte
*f* ~ match Wachsstreichhölzchen *n* ~ melting
boiler Wachsausschmelzkessel *m* ~ melting pot
Wachsschmelzkessel *m* ~ mold Wachsform *f*
~ ointment Wachssalbe *f* ~ paper Butterver-
packungs-, Wachs-papier *n* ~ paper extracting
process Wachspapierverfahren *n* ~ pattern
Wachsmodell *n* ~ recording Wachsaufnahme
*f* ~ resin resist Wachsfirnisharzreserve *f* ~
seal Lacksiegel *n* ~ sprayer and atomizer
Bohnerwachszerstäuber *m* ~ stain Wachsbeize
*f* ~ tailings paraffinhaltiger Rückstand *m* ~
taper Wachsstock *m* ~ varnish Wachsfirnis *m*
~ wick Wachsfaden *m*
waxed gewachst ~ cotton-covered wire Wachsdraht *m* ~ tracing paper Wachspauspapier *n*
waxing Polieren *n*, Wachsen *n* ~ and waining
of noise Atmen *n* des Störgeräusches, Stör-
geräuschatmen *n* ~ paper Wachsrohpapier *n*
~ period Einregelungszeit *f*
waxlike wachs-ähnlich, -artig
waxy wächsern, wachshaltig, zähflüssig
way Art *f*, Art und Weise *f*, Bahn *f*, (of a ship)
Fahrt *f*, Führungsbahn *f*, Gang *m*, Gebrauch
*m*, Lauf *m*, Weg *m*, Weise *f* on the ~ unterwegs,
~ of cock Hahnöffnung *f*, Hahnweg *m* ~ of
craftsmen Handwerksbrauch *m* by ~ of
example beispielsweise ~ of a railway Fahrbahn *f* ~ out (right or left) Ausweg *m* ~ of

reading Ablesemöglichkeit *f*
waybill Begleitschein *m*, Frachtbrief *m*, Passagierliste *f*
way, ~ board Hängebank *f*, Lettenkluft *f* (min.)
~ circuit Gemeinschaftsleitung *f*, Omnibusleitung *f* ~ cleaner Schienenputzer *m* ~ leave
Wegerecht *n* ~ leave charges Anerkennungs-,
Benutzungs-gebühr *f* ~ shaft Gegenlenker-
schaft *m* ~ side signal Streckensignal *n* ~
slide Schmierplanke *f* ~ station Zwischen-amt
*n*, -bahnhof *m* ~ there Hinweg *m*
ways (tool machines) Führungen *pl* an Werkzeugmaschinen, (shipyard) Gleitbahn *f*, (shipbuilding) Stapelblöcke *pl* ~ (of the bed) (Drehbank) Bettführung(en) *f*
weaf Gewirk *n*
weak (mixture) arm, kraftlos, matt, schwach,
schwächlich, verdünnt, weich ~ absorber
schwach absorbierendes Mittel *n* ~ acid
schwarze Säure *f* ~ ammonia liquor (or water)
Gaswasser *n* ~ burst Fehlzerspringer *m* ~
charge Ausbläser *m* ~ coupling lose Kopplung
*f* ~ current schwacher Strom *m*, Schwachstrom
*m* ~ current cable Schwachstromkabel *n* ~
current cable fittings Schwachstromkabel-
garnituren *pl* ~ gas Schwachgas *n* ~ gas
mixture gasarmes Gemisch *n* ~ image flaues
Bild *n* ~ mixture mageres Gemisch *n*, Spargemisch *n* ~ mixture cruising conditions Reiseflug *n* mit Spargemisch ~ picture weiches Bild
*n* ~ point schwacher Punkt *m* ~ spot Weich-
fleckigkeit *f* (metall.) ~ spring diagram
Schwachfederdiagramm *n*
weaken, to ~ abbauen, abschwächen, entkräftigen, erschlaffen, nachlassen, schwächen, verarmen, verschwächen to ~ the coloring power
koupieren
weakening Abschwächung *f*, räumliche Dämpfung *f*, Schwächung *f*, Verschwächung *f* ~ of
the color Verwaschen *n* der Farbe ~ of the
field Feldschwächung *f* ~ of tensile strength
Reißentfestigung *f*
weakening apparatus (or device) Schwächungsvorrichtung *f*
weakly, ~ damped schwach gedämpft ~ sized paper schwach geleimtes Papier *n*
weakness Schwäche *f*, Schwachheit *f* ~ of sight
Sehschwäche *f*
Weald clay Wealdenton *m*
wealth Reichtum *m*, Wohlstand *m* ~ under
ground Bodenschätze *pl*
weapon Waffe *f*
wear, to ~ abnutzen, abschleifen, fressen, scheuern, tragen, verschleißen to ~ away verschleißen to ~ down ermatten to ~ down to the utmost
bis auf das äußerste Maß ausnutzen to ~ out
by friction ausscheuern to ~-harden kalthärten
to ~ off abscheuern to ~ off (or away) abnutzen to ~ out abnutzen, abscheuern, ausschleifen, (Lager) auslaufen, erschöpfen, sich
abnutzen, verschleißen, zerreißen to ~ by
rubbing abreiben
wear Abnutzung *f* durch Gebrauch, Abrieb *m*,
Abschliff *m*, Verschleiß *m* ~ of the bearings
Abnutzung *f* der Lager ~ of cylinder Unrundwerden *n* des Zylinders ~ at the edges Kanten-
verschleiß *m* ~ of rolls Walzenverschleiß *m* ~
on swage Gesenkverschleiß *m* ~ and tear Ab-

nutzung *f*, (film) Absatz *m*, Filmabsatz *m*, Tragewert *m*
**wear, ~ characteristics** Verschleißverhalten *n* ~ **hardening** Stauchhärtung *f* ~ **plate** (hydr. pump) Gleitring *m* ~ **resistance** Verschleißwiderstand *m* ~ **resistant** abriebbeständig, verschleißfest ~ **test** Verschleißprüfung *f* ~ **testing machine** Abnutzungsmaschine *f*
**wearability** Abnutzbarkeit *f*, Abnutzungsbeständigkeit *f*, Verschleißfestigkeit *f*
**wearer** (mask) Träger *m* ~ **of glasses** Brillenträger *m*
**wearing** abnützend, verschleißend ~ **action** Schleißwirkung *f* ~ **capacity** Abnutzbarkeit *f* ~ **face** Schleiffläche *f* ~ **neck** Verschleißzapfen *m* ~**-out wire drawing** Strapazierzug *m* von Drähten ~ **plate** Reibungsplatte *f*, Schleißblech *n*, Verschleißplatte *f* ~ **property** Verschleißblech *n* ~ **shoe (or side bar)** Schleifsteg *m* ~ **surface** Arbeitsfläche *f* ~ **test** Abnutzungsprüfung *f*, Reibungsprobe *f*
**wearisome** zeitraubend ~ **experiment** mühevoller Versuch *m*
**weather, to** ~ auswintern, auswittern, verwittern **to** ~ **a storm** bewettern
**weather** Wetter *n*, Witterung *f* ~ **analysis** Wetteranalyse *f* ~ **board** Dachschindel *f*, Setzbord *n* (naut.) ~ **bridge bulkhead** Brückenfrontschott *n* ~ **bureau** Wetter-amt *n*, -warte *f* ~ **change** Wetteränderung *f* ~ **chart** Wetterkarte *f* ~ **cock** Wetterhahn *m* ~ **cock stabilization** Pfeilstabilisierung *f* ~ **code** Wetterschlüssel *m* ~ **condition** Witterungsverhältnis *n* ~ **conditions** Wetter-lage *f*, -verhältnisse *pl* ~ **contact** Wetterberührung *f* ~ **echo** Wetterecho *n* (rdr) ~ **exposure** Ausgesetztsein *n* ~ **exposure test** Bewetterungs-probe *f*, -prüfung *f*, Bewitterungsprobe *f* ~ **factor** Wetterelement *n* ~ **forecast** Wetter-prognose *f*, -voraussage *f*, -vorhersage *f* ~ **groove** Unterschneidung *f* ~ **house** Wetterhäuschen *n* ~ **information** Wetternachrichten *pl* ~ **leakage** Wetternebenschluß *m* ~ **limit** Wetterscheide *f* ~ **lore** Wetterkunde *f* ~ **map** Wetterkarte *f* ~ **maxim** Bauernregel *f*, Volkswetterregel *f*, Wetter-regel *f*, -spruch *m* ~ **message** Wettermeldung *f* ~ **minimum** Minimumwetterzustand *m* ~ **molding** Kranzleiste *f*
**weathermometer** Verwitterungsmesser *m*
**weather, ~ observer** Wetterbeobachter *m* ~ **outlook** Tendenz *f* ~**-proof** wetter-beständig, -fest ~**-proof paint** wetterfeste Farbe *f* ~ **protection** Wetterschutz *m* ~ **reconnaissance** Wettererkundung *f* ~ **report** Wetter-bericht *m*, -meldung *f* ~ **research** Wetter-beobachtung *f*, -forschung *f* ~ **resistance** Bewitterungsverhalten *n* ~ **resistant** wetterbeständig ~ **resisting** witterungsbeständig ~ **satellite** Wettersatellit *m* ~ **section** Wetterzug *m* ~ **service** Warnungsdienst *m*, Wetterdienst *m* ~ **service radio station** Wettersendestelle *f* ~ **shooting** Wetterschießen *n* ~ **side** Luv *n*, Luvseite *f*, Windseite *f* ~ **signal** Wetter-signal *n*, -zeichen *n* ~ **stability** Wetterbeständigkeit *f* ~ **station** Funkwetterwarte *f*, Wetter-stelle *f*, -warte *f* ~ **strip** Zugluftabschließer *m* ~ **stripping** Fensterdichter *m* ~ **symbol** Wettersymbol *n* ~ **tight construction** wetterdichte Konstruktion *f* ~ **vane**

Wetterfahne *f*, Windanzeiger *m*, Windfahne *f* ~ **vaning effect** Windfahneneffekt *m* ~ **warnings** Sturmwarnung *f*, Unwetterwarnung *f*
**weathered pointing** abgeschrägte Fuge *f*
**weathering** Auswitterung *f*, Bewitterung *f*, Lüften *n*, Verwetterung *f*, Verwitterung *f* ~ **of a molding** Abwässerung *f* eines Simses
**weathering, ~ factors** Atmosphärilien *pl* ~ **test** Bewetterungs-probe *f*, -prüfung *f*, Bewitterungsprüfung *f*, -versuch *m*, Witterungsversuch *m*
**weave, to** ~ klöppeln, pendeln, weben **to** ~ **a plot** anzetteln
**weave** Zwischenlinien-, Zwischenzeilen-flimmer *m* (TV)
**weaver** Weber *m* ~**'s glass** Weberglas *n* ~**'s knot** Weberknoten *m* ~**'s nippers (or tweezers)** Nopp-, Weber-zange *f*
**weaving** Weberei *f* ~ **accessories** Webereigeschirr *n* ~ **cone** Webkone *f* ~ **factory** Webwarenfabrik *f* ~ **frame** Wirkmaschine *f* ~ **loom fitter** Webstuhlsetzer *m* ~ **mill** Weberei *f* ~ **product** Webereierzeugnis *n*
**web** Arm *m*, Aussteifung *f*, Band *n*, Borte *f*, Fußrippe *f*, Gespinst *n*, Gewebe *n*, Gewirk *n*, Hals *m*, Kern *m* des Spiralbohrers, Netz(werk) *n*, Papierbahn *f*, Radscheibe *f*, Sägeblatt *n*, Schenkel *m*, Schneide *f*, Steg *m*, Stiel *m*, Versteifung *f*
**web, ~ of cloth** Stoffbahn *f* ~ **of fabric** Stoffbahn *f* ~ **of girder** Trägersteg *m* ~ **of graticule** Faden *m* im Okular ~ **of paper** Papier-bahn *f*, -rolle *f*, -strang *m* ~ **of railroads** Schienennetz *n* ~ **of rib** Rippensteg *m* ~ **of spar** Holmsteg *m* ~ **of a wheel** Radscheibe *f*
**web, ~ brake** Papierbremse *f* ~ **break detector** Papierreißsicherung *f* ~ **breaking** Papierreißen *n* ~ **breaking roller** Abreißwalze *f* ~ **breaks** Risse *pl* in der Papierbahn ~ **calendered paper** Rollensatinage *f* ~ **(or sheeting) drier** Bahnentrockner *m* ~ **fed rotary letter** Buchdruckrollenrotation *f* ~ **feeding speed** Einziehgeschwindigkeit *f* ~ **flange** Flanschpartie *f* ~ **frame** Rahmenspant *m* ~ **gear** Stegrad *n* ~ **girder** Stegträger *m*, Stehblech *n*, vollwandiger Träger *m* ~ **glass** Weberglas *n* ~ **holder** (textiles) Einschließkamm *m* ~ **member** (of truss) Füllstab *m*, Füllungsstab *m* ~ **plate** Fußrippe *f*, Stehblech *n* ~ **plate joint** Stegblechstoß *m* ~ **punching** Steglochung *f* ~ **reinforcement** Schubbewehrung *f* ~ **rib** Stegrippe *f* ~ **rivet** Halsniet *m* ~ **roll** Scheibenwalze *f* ~ **rounding** Kurbelwagenausrundung *f* ~ **saw** Furniersäge *f* ~ **squeezing machine** Florquetsche *f* ~ **strap fabric** Riementuch *n* ~ **thickness** Stegstärke *f* ~ **thickness control** Bahndickenregelung *f* ~ **type impeller** halb offenes Laufrad *n*
**webbed** verrippt ~ **wheel** Stegrad *n*
**webbing** Gurtband *n*
**wedge, to** ~ einkeilen, festkeilen, füttern, keilen, klemmen, sich festklemmen, unterlegen, verkeilen, zusammenkeilen **to** ~ **in** einzwängen **to** ~ **on** aufkeilen **to** ~ **out** auskeilen, sich auskeilen **to** ~ **over** verstemmen **to** ~ **up** festkeilen
**wedge** Backe *f*, Bergeisen *n*, Froschklemme *f*, Füllkeil *m*, Fütterung *f*, Hochdruckkeil *m* (meteor.), Holzkeil *m*, Knagge *f*, Spitzkeil *m*, spitzwinkliger Keil *m*, Unterlage *f*, Unterleg-

keil *m*, Zwickel *m*, Zwinger *m* ~ **for fastening helve to the hammer** Helmkeil *m* ~ **of iron** Fimmel *m* ~ **with steps** Stufenkeil *m* ~ **of wood** Sperrholz *n*

**wedge,** ~ **actuated collet chuck** Keilspannfutter *n* ~ **angle** Keilwinkel *m* ~ **belt** Keilbolzen *m* ~ **bolt** Backenschraube *f*, Fixierschraube *f* Nachstellkeil *m* ~ **brick** Keil-stein *m*, -ziegel *m* ~ **coupling** Verbindungskeil *m* ~ **crown (lock or key)** Klemmkrone *f* ~ **error** Keilfehler *m* ~ **filter** Graukeil *m* (opt.), Keil-, Verlauf-filter *n* ~ **formation** Winkel *m* ~-**formed part** Keilstück *n* ~ **friction gear** Keilrädergetriebe *n* ~ **friction wheel** Keilrad *n* ~ **gate valve** Absperrschieber *m* ~ **gib** Keil *m* ~ **grip** Backe *f*, Einspann-hacke *f*, -klaue *f* ~ **grip with serrated grooves** Beißkeil *m* ~ **hammer** Keilhammer *m* ~ **hole** Spalt-Keilloch *n* ~ **lock** Keilschloß *n* ~ **notched friction wheel** Keilnutenreibrad *n* ~ **pass-gap-setting device** Keilanstellung *f* ~ **photometer** Keilfotometer *n* ~ **rail anchor** Keilklemme *f* ~ **ring closure (or fastening)** Keilringverschluß *m* ~ **shape** Keilform *f*

**wedge-shaped** keil-ähnlich, -artig, -förmig ~ **air space** Luftkeil *m* ~ **bar wound armature** Keilstabanker *m* ~ **gate** (founding) Keiltrichter *m* ~ **guard** Keilschutz *m* ~ **lightning arrester** Schneidenblitzableiter *m*

**wedge,** ~ **sluice valve** Keilschieber *m* ~ **spectrograph** Graukeilspektrograf *m* ~ **stirrup coupling** Keilbügelkupplung *f* ~ **strips** Ausrichtkeile *pl* ~ **surface** Keilfläche *f* ~-**type breechblock** Keil *m*, Keilverschluß *m* ~-**type flat slide valve** Keilflachschieber *m* ~-**type printing frame** Keilkopierrahmen *m* ~-**type screw** Keilschraube *f* ~ **valve** Keilschieber *m* ~ **zero** Keilnull

**wedged** festgekeilt, (joint) gekeilt ~ **floor** verkeilte Dielung *f*

**wedges for lifting stones** Steinwolf *m*

**wedging** Festklemmen *n*, Keilung *f*, Klemmen *n* ~ **away** Auskeilung *f* ~ **effect** Schmierkeilwirkung *f* ~ **of a tubbing** Verkeilung *f*

**weed, to** ~ jäten

**weed** Unkraut *n* ~ **attachment** Krautschneidevorrichtung *f*, Unkrautschneidebalken *m* ~ **(or brush) bar** Kraut- oder Buschschneidebalken *m* ~ **catcher** Strohfänger *m* ~ **control** Unkrautbekämpfung *f* ~ **hook** Unkrauthaken *m* ~ **killer** Unkrautvertilgungsmittel *n*

**weeder** Unkrautjäter *m*

**weep,** ~ **drain** Sickerdrainage *f* ~ **hole** Entwässerungsrohr *n*

**weeper** (bridge building) Abzugsloch *n*

**weeping core** tropfender Kern *m*

**weft** Einschlagfaden *m*, Einschuß *m*, Eintrag *m*, Gewirk *n*, Schuß-faden *m*, -garn *n*, Zeug *n* ~ **bobbin** Schußspule *f* ~ **counter** Schußzähler *m* ~ **distributor** Spulenverteiler *m* ~ **fork** (textiles) Abstellgabel *f*, Einschlaggabel *f* ~ **fork grates** (textiles) Abstellgabelgitter *n* ~ **pile** Florschuß *m* ~ **silk** Einschlagseide *f* ~ **winding** Einschlaggarnspulen *n*

**weftless** (fabrics) einschlaglos

weigh, to ~ abwägen, anwiegen, (anchor) einwinden, erwägen, wägen, wiegen to ~ again nachwiegen to ~ anchor Anker *m* lichten to ~ in einwägen, einwiegen to ~ off abwägen

to ~ out auswägen, ausswiegen, verwiegen to ~ wrongly sich verwiegen

**weigh** Wiegen *n* ~ **bin** Wägetasche *f* ~ **board** Lettenkluft *f* ~ **bridge** Brückenwaage *f*, Wiegebrücke *f* ~ **lever** Wäghebel *m*

**weighable** wägbar

**weighed** gewogen ~ **object (or sample)** Einwage *f* ~ **off** ausgewogen

**weigher** Wäger *m*, Wiegemeister *m*

**weighing** Wägen *n*, Wiegen *n* ~ **accuracy** Wiegegenauigkeit *f* ~ **apparatus** Wägapparat *m* ~ **appliance** Wägevorrichtung *f* ~ **device** Wiegevorrichtung *f* ~ **device working to fire tolerances** eichfähige Wiegevorrichtung *f* ~ **dish** Waagschale *f* ~ **equipment** Wiegeeinrichtung *f* ~ **feeders** Waagespeiser *m* (Selbstaufleger) ~ **filter** Ohrfilter *n* ~ **glass** Wägeglas *n* ~ **hopper** Wiegebehälter *m* ~ **lever** Hebelwaage *f* ~ **lever with movable jockey** Laufgewichtswaage *f* ~ **machine** Waage *f* ~ **machine for cars** Fuhrwerkwaage *f* ~ **platform** Waagebühne *f* ~ **room** Waagezimmer *n* ~ **scale** Waageschale *f* ~ **tube** Wägeröhrchen *n*

**weight, to** ~ belasten, beschweren, bewerten, (Seide) chargieren, wiegen

**weight** Gewicht *n*, Last *f*, Schwere *f*, Schwereempfindung *f*, Senkung *f* ~ **of the air** Luftgewicht *n* ~ **and balance clearance** Lademanifest *n* ~ **of balloon gas** Füllungsgewicht *n* ~ **of charge** Schmelzgewicht *n* ~ **of an error** Fehlereinflußzahl *f* (math.), Fehlergewicht *n* ~ **in flying order** Fluggewicht *n* ~ **of flywheel** Schwungkranzgewicht *n* ~ **on front axles** Vorderachsdruck *m* ~ **of fuel** Brennstoffgewicht *n* ~ **of the guided missile at combustion cutoff termination** Brennschlußgewicht *n* ~ **per horsepower** Gewicht *n* je Pferdestärke, Leistungsgewicht *n* ~ **of the hull** Hüllengewicht *n* ~ **of the lead** Lotgewicht *n* ~ **of lifting (or inflation) gas** Füllgasgewicht *n*, Traggasgewicht *n* ~ **per liter** Litergewicht *n* ~ **of load** Ladegewicht *n* (unit) ~ **of material** Stoffgewicht *n* ~ **of material to be supplied** Liefergewicht *n* ~ **of missile** Aggregatgewicht *n* ~ **of object** Einwaage *f* ~ **of oil** Brennstoffgewicht *n* ~ **of oil carried** Schmierstoffgewicht *n* ~ **of the power plant dry** Triebwerk-leergewicht *n*, -trockengewicht *n* ~ **of projectile** Geschoßgewicht *n* ~ **of propellant** Treibstoffgewicht *n* ~ **in quantity** Einwaage *f* ~ **in running order** Betriebsgewicht *n* ~ **of sample** Einwaage *f* ~ **and space** Gewicht *n* und Raumbedarf *m* ~ **of standard air** Normalluftgewicht *n* ~ **of the stretched leather** Streckgewicht *n* ~ **on tail skid (or wheel)** Sporndruck *m* ~ **at take-off** Startgewicht *n* ~ **of unit** Volumgewicht *n* ~ **per unit area** Flächengewicht *n*, Massenbelegung *f* ~ **per unit area gauge** Flächengewichtsmeßanlage *f* ~ **per unit of volume** Raumgewicht *n* ~ **by volume** Volumengewicht *n* ~ **on the wheel** Radgewicht *n*

**weight,** ~ **accumulator** Gewichtssammler *m* ~ **ball** Kugelgewicht *n* ~ **batching** Gewichtsdosierung *f* ~ **box** Gewichtskasten *m* ~ **brake** Gewichtsbremse *f* ~ **compensating lenses** Gewichtsausgleichsgläser *pl* ~ **density** spezifisches Gewicht *n* ~ **distribution balance** Gewichtsverteilung *f* ~ **driven clock** Uhrwerk *n*

mit Gewichtsantrieb *m* ~ **empty** Leergewicht *n*
~ **equalizer** Gewichtsausgleicher *m* ~ **factor**
Gewichtszahl *f* ~ **governor** Gewichtsregler *m*
~ **(of water) head** Fallhöhe *f* ~ **holder** Gewichtsträger *m* ~ **lever system** Gewichtshebeleinrichtung *f* ~ **lifting capacity** Hublast *f* ~
~ **loaded** Betriebsgewicht *n* ~ **loaded friction**
gewichtsabhängige Friktion *f* ~ **notice** Gewichtsnoten *pl* ~ **operated feed** Gewichtsvorschub *m* ~ **percentage of balloon gas** Füllungsgewichtprozent *n* ~ **percentage scale** Gewichtsprozentskala *f* ~ **piece** Gewichtsstück *n* ~
**rate of flow** Durchsatz *m* (electr.), Durchsatz
*m* eines strömenden Mittels ~ **ratio** Gewicht/
Massen-Verhältnis *n* ~ **saving construction**
Leichtbauweise *f* ~ **stand** Bodengewicht *n* ~
**stand hook** Gewichtsträgerhaken *m* ~ **ton** Gewichtstonne *f* ~ **velocity of flow** Mengenstrom
*m*, sekundliche Fördermenge *f* oder Durchflußmenge *f* ~ **voltameter** Gewichts-, Massen-
-Voltameter *n*
**weighted** gewichtsbelastet ~ **antenna** Eierantenne
*f* ~ **index** bewertete Kennziffer *f* ~ **pendulum
testing machine** Prüfmaschine *f* mit Pendelwaage ~ **spring** mit einem Gewicht versehene
Feder *f* ~ **tension roller** Belastungsrolle *f*
**weighting** (of silt) Beschwerung *f*, Bewertung *f*,
Gewicht *n*, Voreinstellung *f* ~ **with iron** Eisenbeschwerung *f* ~ **of leather** Lederbeschwerung
*f* ~ **with sugar** Zuckerbeschwerung *f* ~ **with tin
phosphate** Zinnphosphatbeschwerung *f*
**weighting**, ~ **agent** Beschwerungsmittel *n* ~
**factor** Einflußzahl *f* ~ **filler** Beschwerungsmittel *n* ~ **function** Einfluß-, Gewichts-funktion
*f* ~ **network** Bewertungsfilter *n* ~ **size** Beschwerungs-appretur *f*, -masse *f*
**weightless** ohne Gewicht *n*, leicht, schwerelos
**weightlessness** Gewichtslosigkeit *f*, Leichtheit *f*,
Schwerelosigkeit *f*
**weights loaded** Abgangsgewichte *pl*
**weighty** schwer
**weir** Damm *m*, Fangbuhne *f*, Überfall *m*, Überlauf *m*, Wehr *n*, Wehrdamm *m*, Wehrkörper *m*
~ **with a free wall** vollkommenes Wehr *n*
**weir**, ~ **box** Staukasten *m*, Überlauf *m* ~
**building** Wehrbau *m* ~ **coefficient** Überfallbeiwert *m* ~ **dam plate** Stirnplatte *f* des Überlaufs ~ **plant** Wehranlage *f*
**welch plug** bombierter Verschlußdeckel *m*
**weld, to** ~ abfassen (metall.), ausbrennen, auskreuzen, einbrennen, einschließen, einschweißen, schweißen, verschweißen, versetzen **to**
~ **down-hand** abwärtsschweißen **to** ~-**harden**
schweißhärten **to** ~ **on** anschweißen, aufschweißen **to** ~ **short and thick** aufstauchen **to** ~
**together** verschweißen, zusammenschweißen
**to** ~ **from the top down, from the bottom up**
an der stehenden Wand schweißen **to** ~ **up**
zusammenschweißen **to** ~ **up-hand** aufwärtsschweißen
**weld** Brennerdüse *f*, Naht *f*, Schweißarbeit *f*,
Schweiße *f*, Schweißkuppe *f*, Schweißstelle *f*,
Schweißung *f*, Spleißstelle *f*, V-Stoß *m*,
(gelber Farbstoff) Wau *m* ~ **bead** Schweißraupe *f* ~ **cracking** Schweißrissigkeit *f* ~
**deposit** Ausschweißung *f* ~ **dimension** Nahtstärke *f* ~-**in flange** Einschweißstutzen *m* ~
**iron** Schweiß-eisen *n*, -schmiedeeisen *n* ~

**metal** eingeschweißtes Material *n*, Schweißgut
*n* ~ **metal recovery** Schweißgutausbeute *f* ~
**nugget** Schweiß-linse *f*, -punktkern *m* ~-**on terminal** Lötanschlußstück *n* ~ **recorder** Schweißungsaufzeichner *m* ~ **reinforcement** Schweißüberhöhung *f* ~ **size** Bruchquerschnitt *m* ~
**stalk** Waustengel *m* ~ **steel** Schweißstahl *m* ~
**timer** Schweißzeitbegrenzer *m*
**weldability** Schweißbarkeit *f*
**weldable** schweißbar ~ **metal** Schweißmetall
*n* ~ **steel** schweißbarer Stahl *m*
**weldableness** Schweißbarkeit *f*
**welded** angeschweißt, aufgeschweißt, geschweißt
~ **area** Schweißstelle *f* ~ **design** Schweißkonstruktion *f* ~ **dome body** geschweißter
Dommantel *m* ~ **framing** geschweißte Rahmenkonstruktion *f* ~ **gasoline tank** geschweißter Benzintank *m* ~ **joint** Schweiß-gutverbindung *f*, -naht *f*, -stelle *f*, -verbindung *f* ~
**neck flange** geschweißter Flansch *m* ~ **overlap
joint** überlappte Schweißstelle *f* ~ **rail joint**
Schweißnahtstoß *m* ~ **steel-tube fuselage**
rumpfgeschweißtes Stahlrohrgerüst *n* ~ **steel
tubing** geschweißtes Stahlrohr *n* ~ **structures**
Schweißkonstruktion *f* ~ **tubular structure**
geschweißte Rohrkonstruktion *f* ~ **vessel test
plates** Prüfbleche *pl* für geschweißte Behälter
**welder** Schweißer *m*, Schweiß-maschine *f*, -vorrichtung *f* ~-**'s outfit** Schweißerausrüstung *f*
**welding** Aufschweißen *n*, Kunstschweißung *f*,
Schweißarbeit *f*, Schweißen *n*, Schweißung *f*,
Verschweißung *f* ~ **of preheated cast iron** Gußwarmschweißung *f* ~ **without preheating** Kaltschweißung *f* ~ **with pre- and postheating**
Warmschweißung *f* ~ **by sparks** Abbrennschweißung *f*
**welding**, ~ **apparatus** Schweiß-apparat *m*, -gerät
*n* ~ **arc** Schweiß(licht)bogen *m* ~ **arc voltage**
Schweißspannung *f* ~ **area** Schweißstelle *f*
~ **bay** Schweiß-box *f*, -platz *m* ~ **bead** Schweißtropfen *m* ~ **burner** Schweißbrenner *m* ~
**cable** Schweißkabel *n* ~ **capacity** Schweißleistung *f* ~ **carbon electrodes** Schweißkohlen *pl* ~ **characteristics** Schweißeigenschaften *pl* ~ **cinder** Schweißschlacke *f* ~
**circuit** Schweißstromkreis *m* ~ **collars** Vorschweißbunde *pl* ~ **compound** Metallmischung
*f* zum Schweißen, Schweißmittel *n* ~ **converter**
Schweißumformer *m* ~ **current** Schweißstrom
*m* ~ **current regulator** Schweißstromregler *m*
~ **data** Schweißbedingungen *pl* ~ **defect**
Schweißfehler *m* ~ **die** Schweißbacke *f* ~
**dribble** Schweißtropfen *m* ~ **dynamo** Schweißgenerator *m* ~ **electrode** Schweiß-elektrode
*f*, -stab *m* ~ **electrode covering press** Schweißelektrodenumhüllungspresse *f* ~ **end** Anschweißende *n*, Schweißende *n* ~ **equipment**
Schweiß-anlage *f*, -ausrüstung *f*, -einrichtung *f*
~ **fire** Schweißfeuer *n* ~ **flame** Schweißflamme
*f* ~ **flux** Flußmittel *n*, Schweiß-mittel *n*, -pulver
*n* ~ **furnace** Schweißofen *m* ~ **generator**
Schweißgenerator *m* ~ **glove** Schweißhandschuh *m* ~ **goggles** Schweißbrille *f* ~ **gun**
Punktschweiß-bügel *m*, -zange *f*, Schweißpistole *f* ~ **handle** Pistole *f*, Schweiß-kolben *m*,
-pistole *f* ~ **head** Schweißkopf *m* ~ **heat**
Schweiß-flut *f*, -glut *f*, -hitze *f*, -temperatur *f*,
-wärme *f* ~ **helmet** Schweißhelm *m* ~ **interrup-**

ter Schweißschalter *m* ~ **jig** Schweißvorrichtung *f* ~ **joint** Schweißfuge *f* ~ **line** Schweißstraße *f* ~ **machine** Schweiß-apparat *m*, -maschine *f* ~ **material** Schweißgut *n* ~ **nozzle tip** Schweißdüsenmundstück *n* ~ **operation** Schweißerei *f* ~ **operator** Schweißer *m* ~ **outfit** Schweißausrüstung *f* ~ **period** Punktfolge *f*, Schweißzeit *f* ~ **plant** Schweißerei *f* ~ **plate** Schweißplatte *f* ~ **point** Anschweißstelle *f*, Schweißpunkt *m* ~ **position** Schweißstelle *f* ~ **powder** Schweißpulver *n* ~ **practice** Schweißerei *f* ~ **process** Schweiß-prozeß *m*, -verfahren *n* ~ **property** Schweißgüte *f* ~ **puddle** Schweißbad *n* ~ **quality** Schweißgüte *f* ~ **resistance (or resistor)** Schweißwiderstand *m* ~ **rod** Schweiß-draht *m*, -stab *m* ~ **sand** Schweißsand *m* ~ **seam** Schweiß-fuge *f*, -naht *f* ~ **sequence** Schweißfolge *f* ~ **set** Schweißsatz *m* ~ **set generator** Schweißumformer *m* ~ **shop** Schweißerei *f*, Schweißwerkstatt *f* ~ **soldering terminal** Schweißlötöse *f* ~ **spot** Schweißpunkt *m* ~ **steel** schweißbarer Stahl *m*. Schweißstahl *m* ~ **technique** Schweißtechnik *f* ~ **temperature** Schweißtemperatur *f* ~ **test** Schweißversuch *m* ~ **time** Schweißzeit *f* ~ **timer** Schweißtakter *m*, Schweißzeitgeber *m* ~ **tip** Schweißdüse *f* ~ **tongs** Schweißzange *f* ~ **tool** Schweißwerkzeug *n* ~ **torch** Schweißbrenner *m* ~ **training shop** Schweißlehrwerkstatt *f* ~ **transformer** Schweiß-transformator *m*, -umspanner *m* ~ **upset** Schweißdruck *m* ~ **unit** Schweißaggregat *n* ~ **wheel** Schweißrad *n* ~ **wheel shaft** Schweißradwelle *f* ~ **wire** Schweißdraht *m* ~ **yoke** Schweißzapfenmitnehmer *m* ~ **yoke type universal joint** Schweißzapfengelenk *n*

**weldless** nahtlos ~ **links** nahtlose Hebebügel *pl* ~ **pipe** nahtlos gezogenes Rohr *n* ~ **tire** nahtloser (Eisenbahn)radreifen *m* ~ **tube** nahtloses Rohr *n*

**well, to** ~ **up** aufquellen **to** ~ **up on** beschlagen

**well** Absturzschacht *m*, Behälter *m*, Bohrung *f*, Brunnen *m*, Gestell *n*, Grube *f*, Laderaum *m*, Quelle *f*, Schacht *m*, Tauchrohr *n*; gut, recht, passend ~ **with perforated pipe casing** Abbessinierbrunnen *m* ~ **of thermometer** druckfestes Schutzrohr *n*

**well,** ~ **base rim** Tiefbettfolge *f* ~ **boring drill** Brunnenbohrer *m* ~ **boring plant** Brunnenbohranlage *f* ~ **burnt plaster** fetter Gips *m* ~ **casing** Bohrlochfutterrohr *n* ~ **chamber** Brunnenstube *f* ~ **constructed machine** durchkonstruierte Maschine *f* ~ **crystal** Bohrlochkristall *n* ~ **curbing** Brunnenabschluß *m* ~ **depth** Schacht-, Topf-tiefe *f* ~ **depth parameter** Potentialtopftiefe *f* ~ **dimension** Bohrlochabmessung *f* ~ **dish** Kippschale *f* (photo) ~**drill** Seil-bohren *n*, -bohrung *f* ~ **lining** Brunnenschalung *f* ~ **obtained by driving pipe into waterbearing stratum** Rammbrunnen *m* ~ **ordering theorem** Wohlordnungssatz *m* (math.) ~ **outlet** Brunnenauslauf *m* ~ **pit** Brunnen-, Schacht-feld *n* ~ **point** Brunnen *m* (soil mech.), Filterbrunnen *m* ~ **points** Filterbrunnenaggregat *n*, Rammfilterbrunnen *m* ~ **room** Sodraum *m* ~ **seated valve** gut aufsitzendes Ventil *n* ~ **shaft** Brunnenschacht *m*, Quellfassung *f* ~ **shooting** Bohrlochschießen *n* ~

**sinker** Brunnen-ausschalter *m*, -macher *m* ~ **sinking enterprise** Tiefbohrgeschäft *n* ~ **tube** Fassungsrohr *n* ~**-type counter** Zähler *m* mit Probenkanal ~**-type ionization chamber** Schachtionisationskammer *f* ~**-type scintillation counter** Szintillationszähler *m* mit Bohrlochkristall ~ **ventilated** gut gelüftet ~ **wagon** Tieflader *m*, Tiefladewagen *m* ~ **wall** Schachtwand *f* ~ **water** Brunnenwasser *n* ~ **width** Schachtweite *f*

**Welsbach,** ~ **burner** Auerbrenner *m* ~ **light** Auerlicht *n* ~ **mantle** Auerstrumpf *m*, Glühkörper *m*

**Welsh process** Waleser Verfahren *n*

**welt** Einfassung *f*, Falz *m*, Kante *f*, Lasche *f*, Leiste *f*, Rahmen *m*, Vorschuh *m* ~ **and rand attaching machine** Rahmen- und Lederheftmaschine *f* ~ **position** Unterlegestellung *f*

**welting leather** Rahmenleder *n*

**west** Westen *m*, Westpunkt *m* ~**ward** westlich, westwärts ~ **winds** Westwinde *pl*

**westerly** westlich

**western** westlich ~ **amplitude** Abendweite *f*

**Weston cell** Westonelement *n*

**wet, to** ~ anfeuchten, annässen, befeuchten, begießen, benetzen, durchfeuchten, durchnässen, einsumpfen, feuchten, nässen, netzen **to** ~ **back** aufwalken **to** ~**out** vornetzen

**wet** feucht, naß, auf nassem Wege, niederschlagsreich ~ **and dry bulb psychrometer** Aspirationspsychrometer *n* ~ **and dry bulb temperature** Feuchtigkeitstemperatur *f* ~ **and dry bulb thermometer** Feuchtigkeitsmesser *m*

**wet,** ~ **abrasion effect** Naßschauerwirkung *f* ~ **adiabatic lapse rate** feuchter adiabatischer Zustand *m* der Luft ~ **air cleaner (or filter)** Naßluftfilter *n* ~ **air washer** Naßluftreiniger *m* ~ **battery (or cell)** Naßelement *f* ~**-beaten (pulp)** schmierig gemahlen ~ **beaten pulp** fettig gemahlener Stoff *m* ~ **bottom boiler** Schmelzkammerkessel *m* ~ **bottom gas producer** Gaserzeuger *m* ohne Rost, rostloser Gasererzeuger *m* ~ **bottom producer** Abstichgenerator *m* ~ **bulb temperature** Feuchttemperatur *f* (Klimaanlage), Naßwärmegrad *m* ~ **bulb thermometer** feuchtes (Kugel)thermometer *n*, Naßglasthermometer *n*, Naßthermometer *n*, Thermometer *n* mit feucht gehaltener Kugel ~ **cell** nasses Element *n* ~ **cleaner** Naßreiniger *m* ~ **cleaning** Naßaufbereitung *f*, Naßreinigung *f*, Naßwäsche *f* ~ **cleaning process** nasses Putzverfahren *n* ~ **cobber** Naßseparator *m* ~ **compression machine** Naßkompressionsmaschine *f* ~ **crushing** Naßmahlen *n*, Naßpochen *n*, Naßzerkleinerung *f* ~ **cuts** Naßaushub *m* ~ **day** Regentag *m* ~ **de-ashers** Naßentäscher *pl* ~ **developing method** Naßentwicklungsverfahren *n* ~ **decatizing machine** Naßdekatiermaschine *f* ~ **dial water meter** Naßläufer *m* ~ **decatizing machine** Naßdekatiermaschine *f* ~ **dial water meter** Naßläufer *m* ~ **dock** Flutbecken *n*, Hafendock *n* ~ **drawing machine** Naßzug *m* ~ **elongation** Naßdehnung *f* ~ **end** nasse Partie *f* (paper mfg.) ~ **fat-liquoring** Naßfettung *f* ~ **felt** Kartonfilz *m*, Naßpreßfilz *m* ~ **filter** Naßfilter *m* ~ **finishing process** Naßbehandlung *f* ~ **flashover voltage** Überschlagsspannung *f* (bei nassem Isolator) ~ **fog** nässender Nebel *m* ~

**gas** nasses Erdgas *n* ~ **gas meter** nasser Gasmesser *m*, nasse Gasuhr *f* ~ **grained** naßlevantiert ~ **grinding** Naßmahlen *n*, Naßschliff *m*, Naßschroten *n* ~ **grinding process** Naßmahlverfahren *n* ~ **grindstone** Naßschleifstein *m* ~ **lattice** Eintauchgitter *n* ~ **lead clarification** Bleiessigerklärung *f* ~ **liner** vom Kühlmittel umspülte Zylinderlaufbüchse *f* ~ **machine** Entwässerungsmaschine *f*, Längsbauten *pl*, Siebtischentwässerungsmaschine *f* ~ **machine for scraped pulp** Schabstoffmaschine *f* ~ **mechanical analysis** Schlammanalyse *f* ~ **meter** Naßzähler *m* ~ **mill concentration** naßmechanische Aufbereitung *f* ~ **offset printing** Naßoffsetdruck *m* ~**paint** frisch gestrichen ~ **pan for grinding and tempering brick material** Tonkneter *m* ~ **pan grinding** Naßmahlgang *m* ~ **pan mill** Kneter *m*, Knetmaschine *f* ~ **part** Langsiebpartie *f* der Papiermaschine ~ **pipe (or sprinkler) installation** Sprinkler-Naßanlage *f* ~ **pipe (alarm) valve** Naßalarmventil *n* ~ **practice** Naßbetrieb *m* ~ **precipitation** Naßniederschlagung *f* ~ **press roll** Naßpreßwalze *f* ~ **process** Naßverfahren *n* ~ **processing** Naßappretur *f* ~ **pudding** Fettpudeln *n* ~ **pulp** Naßschnitzel *n* ~ **purification** Naßreinigung *f* ~ **purifier** Naßreiniger *m* ~ **rolling barrel** Naßtrommel *f* ~ **rot** Naßfäule *f* ~ **scrubbing** Naßreinigung *f* ~ **seed dressing** Naßbeizung *f* ~ **separation** nasse Scheidung *f* ~ **separation plant** Naßabscheidungsanlage *f* ~ **separator** Naßscheider *m* ~ **shrinking** Wasserkrumpe *f* ~ **slag** dünnflüssige Schlacke *f* ~ **stage** Naßpartie *f* ~ **stamp mill** Naßpochwerk *n* ~ **start** Anlaßüberflutung *f* ~ **steam** Naßdampf *m*, nasser Dampf *m* ~ **storage battery** Naßbatterie *f* ~ **strength** Naßfestigkeit *f* ~ **sump** feuchter Ölsumpf *m* ~ **tenacity** Zugfestigkeit *f* im Feuchten ~ **test (of insulator)** Beregnungsversuch *m*, Regenversuch *m* ~ **tool grinder** Naßschleifständer *m* ~ **tool grinding machine** Naßschleifmaschine *f* ~ **tumbler** Naßputztrommel *f*, Naßtrommel *f* ~ **tumbling** nasses Putzverfahren *n*, Naßtrommeln *n*, Naßscheuern *n* ~ **turning attachment** Kühlwassereinrichtung *f* für Drehbänke ~ **turning equipment** Naßdreheinrichtung *f* ~ **turning installation** Naßdreheinrichtung *f* ~**-type clinker remover** Druckwasserentaschung *f*, Spülentaschung *f* ~**-type drum separator** Naßtrommelscheider *m* ~ **vapor** Naßdampf *m*, nasser Dampf *m* ~ **washer** Naßreiniger *m* ~ **washing** Naßreinigung *f* ~ **work** Wasserarbeit *f*

**wetness** Feuchtigkeit *f*, Nässe *f*, (paper mfg.) Schmierigkeit *f*

**wettability** Benetzbarkeit *f*

**wettable** benetzbar

**wetted** benetzt, bespült ~ **air filter** Naßluftfilter *n* ~**area** benetzte Fläche *f* ~ **perimeter** benetzter Umfang *m* des Querschnitts ~ **surface** benetzte oder feuchte Oberfläche *f* ~ **type surface air cooler** Naßluftkühler *m*

**wetting** Benetzung *f* ~ **of the clay** Toneinsumpfen *n*

**wetting,** ~ **agent** Netzmitteln *n* ~ **angle** Randwinkel *m* ~ **bench** Papierfeuchte *f* ~ **board** Feuchtbrett *n* ~ **effect** Netzwirkung *f* ~ **liquid** benetzende Feuchtigkeit *f* ~ **machine** Feucht-

apparat *m* ~ **machine for yarns** Befeuchtungsmaschine *f* für Garne

**wetting-out,** ~ **agent applied by the cold method** Kaltnetzer *m* ~ **bath** Netzbad *n* ~ **capacity (or property)** Netzvermögen *n* ~ **effect** Netzkraft *f* ~ **figure** Netzzahl *f* ~ **liquor** Netzflotte *f* ~ **property** Netzfähigkeit *f*

**wetting,** ~ **room** Feuchtkammer *f* ~ **screen** Naßsieb *n* ~ **table** Feuchttisch *m*

**w. g. (water gauge)** WS (Wassersäule)

**whale** Wal *m* ~ **boat** Rettungsboot *n* ~ **oil** Walfischtran *m*

**whaler** Walfänger *m*

**whaling** Walfang *m*

**wharf** Anlandevorrichtung *f*, Ausladungsplatz *m*, Bollwerk *n*, Flußdamm *m*, Kai *m*, Ladeplatz *m*, Landebrücke *f*, Landungssteg *m*, Pier *m*, Quaianlage *f* ~ **for fuel-oil bunkering** Tankanlage *f*

**wharf,** ~ **crane** Hafenkran *m* ~ **revolving crane** Hafendrehkran *m*

**wharfage** Ausladungsgebühr *f*, Kaigebühr *f*

**wharfinger** Kaimeister *m*

**wheat** Weizen *m* ~ **drill** Weizendrillmaschine *f* ~ **moistener** Getreidenetzapparat *m* ~ **scourer** Getreideschälmaschine *f* ~ **screen** Weizensieb *n*

**wheel, to** ~ drehen, fahren, rollen, wälzen **to** ~ **the ground** Erde *f* abkarren

**wheel** Blockscheibe *f*, Drehung *f*, Kreisbewegung *f*, Laufrad *n*, Rad *n*, Rädchen *n*, Scheibe *f*, Schieberad *n*, Walze *f* ~ **and axle-set grinder** Achssatzschleifmaschine *f* ~ **and disc device** Reibradantrieb *m* ~ **for front support** Lenkrolle *f* ~ **with spokes** Speichenrad *n* ~ **with stepped teeth** Zahnrad *n* mit Stufenzähnen ~ **in steps** Stufenrad *m* ~ **with tire cast on** Rad *n* mit angegossenem Reifen

**wheel,** ~ **arbor** (Schleifmaschine) Scheibenachse *f* ~**axle** Radachse *f* ~ **balancing device** Laufradschwinge *f* ~ **barometer** Zeigerbarometer *n* ~ **barrow** Flurfördermittel *n*, Handkarren *m*, Karre *f*, Karren *m*, Laufkarren *m*, Schiebkarre *f*, Schub-karre *f*, -karren *m*, Transportwagen *m* ~ **barrow handles** Karrenschenkel *pl* ~ **barrower** Wagenstößer *m* ~ **base** Achsen-abstand *m*, -entfernung *f*, -stand *m*, Achsstand *m*, Rad(ab)stand *m*, Spurweite *f* ~ **bearing** Radlager *n* ~ **bearing cap** Achslagerdeckel *m* ~ **bearing spindle bushing** Achslagerspindelbüchse *f* ~ **blade** Radschaufel *f* ~ **body** Radkörper *m* ~ **bolt** Radbolzen *m* ~ **box** Nabenbüchse *f* ~ **brace** Rad-abzieher *m*, -mutternschlüssel *m*, -stütze *f* ~ **brake** Fahr-, Radbremse *f* ~ **brakes with shoes** Balkenbremse *f* ~ **braking cylinder** Radbremszylinder *m* ~ **brush** Bürstenscheide *f* ~ **cap** Staubschutzkappe *f* ~ **carriage** Schleifsupport *m* ~ **casing** Radverkleidung *f* ~ **center** Radkörper *m* ~ **center disc** Radscheibe *f* ~ **centering cams** Radzentrierführung *f* ~ **chair** Rollstuhl *m* ~ **chock** Bremsklotz *m*, Rad-keil *m*, -klöse *f* ~ **control** Radsteuerung *f*, Handradsteuer *n* ~ **controls** Handradsteuerung *f* (aviat.) ~ **cylinder** Radzylinder *m* ~ **deflection** Radausschlag *m* ~ **diameter** Raddurchmesser *m* ~ **disc** Radscheibe *f*, Verblendscheibe *f* ~ **drag** Hemmschuh *m*, Hemmzeug *n* ~ **dresser** Abrunder *m*, Abziehvorrichtung *f* (für Schleifscheibe), Abziehwerkzeug *n*, Steinabrunder *m* ~ **drive** Radantrieb *m* ~ **drive key** Bremsscheiben-Mitnehmerkeil *m* ~ **driving pulley** Schleifscheibenantriebsscheibe *f* ~ **elevator** Heberad *n* ~ **escapement** Radschaltung *f* ~ **fairing** Radverkleidung *f* ~ **felloe** Grundfelge *f* ~ **fit** Nabensitz *m* ~ **flange** Felgenhorn *n*, Radflansch *m* ~ **fork** Lenker *m*, Radgabel *f* ~ **fork assembly** Radgabel *f* ~ **fork bearing** Radgabellager *n* ~ **frame** Fahrgestell *m* ~ **glass cutter** Rollglasschneider *m* ~**(grinding) guard** Schleifscheibenschutzhaube *f* ~ **guide** Rödelbalken *m* ~ **harness** Sielengeschirr *n* ~ **horse** Stangenpferd *n* ~ **house** Radkasten *m* (auto) ~ **hub** Rad-nabe *f*, -körper *m* ~ **hub cap** Radnabenkappe *f* ~ **hub drive shaft** Radnabenantriebszapfen *m* ~ **illusion** Radphänomen *n* ~ **landing** Radlandung *f* ~ **load** Radbelastung *f*, Radlast *f* ~ **load weighing machine** Raddruckwaage *f* ~ **lock** Radsperre *f* ~ **molding machine** Räderformmaschine *f* ~ **motion** Radbewegung *f* ~ **motor** Achsentriebmotor *m* ~ **mount** Räderlafette *f* ~ **nave** Radnabe *f* ~ **nut** Radbefestigungsmutter *f*, Radmutter *f* ~ **operated means** Radtaster *m* ~ **pressure** Raddruck *m* ~ **puller** Radabzieher *m* ~ **quill** Schleifscheibenhülse *f* ~ **remover** Radabzieher *m* ~**retaining nut** Radstellmutter *f* ~ **rim** Bindereifen *m*, Radfelge *f* ~ **rim joint** Felgenstoß *m* ~ **rolling mill** Radwalzwerk *n* ~ **rotation** Radbewegung *f* ~ **rut** Wagengeleise *n* ~ **scraper** Radabstreicher *m* ~ **seat** Achsenkopf *m*, Nabensitz *m* ~ **set** Radsatz *m* ~**-shaped** radförmig ~ **shimmy** Radflattern *n* ~ **side cover** Laufradverkleidung *f* ~ **slide** Schleifspindelschlitten *m* ~ **spats** Radverkleidung *f* ~ **spider** Radstern *m* ~ **spin** Durchdrehen *n* der Räder, Gleiten *n* des Laufrades ~ **spindle** Achsschwingschenkel *m* ~ **(grinding-) spindle** Schleifscheibenspindel *f* ~ **spoke** Radspeiche *f* ~ **stand** Schleifbock *m* ~ **stand revolving plate** Schleifbockdrehplatte *f* ~ **stand slide** Schleifbockschlitten *m* ~ **stud** Rad-bolzen *m*, -stift *m* ~ **tank** fahrbarer Tank *m* ~ **testing machine** Laufrädermaschine *f* ~ **tie bolt** Felgenschraube *f* ~ **tire** Radbandage *f* ~ **tooth** Radzahn *m* ~ **track** Radspur *f*, Spurweite *f*, Spurweite *f* des Fahrgestells, Wagenspur *f* ~ **tracked vehicle** Räderraupenfahrzeug *n* ~ **tractor** Radschlepper *m* ~ **trolley** Stromabnehmerrolle *f* ~ **turning lathe** Raddrehbank *f* ~**-type of milling cutter** Frässcheibe *f* ~**-type landing gear** Landfahrwerk *n* ~ **undercarriage** Radfahrgestell *n* ~ **web rolling mill** Radscheibenwalzwerk *n* ~ **weight** Nabengewicht *n*, Radgewichtscheibe *f* ~ **weights** Druckräder *pl* ~ **well** Fahrwerkschacht *m*, Radkammer *f* ~ **work with conical gearing** Kegelradwerk *n* ~ **wright** Radmacher *m*, Stellmacher *m*, Wagner *m* ~ **wright's compass** Speichenmesser *m* ~ **wright's machine** Wagnereimaschine *f* ~ **wright's shop** Stellmacherei *f*

**wheeled** fahrbar, mit Rädern *pl* versehen ~ **cable drum carriage** Auf- und Abwindhaspel *f* ~ **gun carriage** Radlafette *f* ~ **litter carrier** Räderbahre *f* ~ **plow** fahrbarer Pflug *m* ~ **prime mover** Radzugmaschine *f* ~ **stand** Rolltisch *m*

**wheeler** Radlandung *f* ~**'s chisel** Kantbeitel *m*

**wheeling** Schwenkung *f* ~ **by cart** Karrenförderung *f*

**wheels are chocked** Räder *pl* durch Bremsklötze fest ~ **with brakes** bremsbare Räder *pl* ~ **with independent legs** Einbeinfahrgestell *n* ~ **up landing** Bauchlandung *f*

**whelps** Kälber *pl* (mach.)

**whet, to** ~ (mit dem Ölstein) abziehen, ausschleifen, schärfen, schleifen, wetzen

**whet** Schärfen *n*, Schleifen *n*, Wetzen *n* ~ **steel** Wetzstahl *m* ~ **stone** Abziehstein *m*, Schleifstein *m*, Sensenstreicher *m*, Speckstein *m*, Wetzstein *m*

**whetting** Schleifung *f* ~ **machine** Abzieh-, Schärf-maschine *f*

**whining** Winseln *n*

**whip** Klappläufer *f*, Peitsche *f*, Wippe *f* ~ **antenna** Peitschenantenne *f*

**whipper** Schaumschläger *m*, Wipper *m*

**whipping** Besatz *m*, Bewicklung *f*, Umspinnung *f*, Umwicklung *f* ~ **cable** Eisendrahtbeflechtung *f* ~ **crane** Wippkran *m*

**whirl, to** ~ sich drehen, (Welle) flattern, schleudern, turbinieren, wirbeln **to** ~ **high** hochwirbeln

**whirl** Wirbel *m* ~ **pool** Strudel *m*, Walze *f*, Wirbel *m*, Wirbelstrom *m* ~ **wind** Windhose *f*, Wirbelwind *m*

**whirler shoe** Wirbelschuh *m*

**whirling** Flattern *n*, Wirbelströmung *f*, Wirbelung *f*; wirbelnd ~ **of combustion gases** Rotation *f* der Verbrennungsluft

**whirling,** ~ **arm** Rundlauf *m* ~ **chamber** Wirbelkammer *f* ~ **currents** Wirbelströme *pl* ~ **effect** Wirbelwirkung *f* ~ **line** Wirbelleine *f* ~ **motion** Wirbelbewegung *f* ~ **speed** Schwirrgeschwindigkeit *f* ~ **test** Schleuderprobe *f* ~ **test stand** Schleuderprüfstand *m*

**whisk** Fügebank *f*

**whisker** Kontaktdraht *m*, Metallspitze *f*

**whiskers** Geschwindigkeitsschwankungen *pl*

**whispering pump** Flüsterpumpe *f*

**whistle, to** ~ pfeifen, rauschen, sausen **to** ~ **in intermediate frequency** einpfeifen

**whistle** Pfeife *f*, Pfeifen *n*, Pfiff *m* ~ **with double tone** Zweiklangpfeife *f* ~ **signal** Pfeifsignal *m* ~ **valve** Pfeifventil *n*

**whistler** Steigetrichter *m*

**whistling** Pfeifen *n* ~ **buoy** Heul-boje *f*, -tonne *f* ~ **kettle** Flötenkessel *m* ~ **margin** Pfeif-abstand *m*, -sicherheit *f*

**white** hell, licht, weiß **in the** ~ ohne Oberflächenschutz *m*, roh, ungestrichen

**white** Füll-holz *n*, -weiß *n* (print.), Weiß *n*, Helle *f*, Vakant *n* ~ **annealing** zweite Glühung *f* ~ **antimony** Weißpies-glanz *m*, -glanzerz *n* ~ **area** Weißfläche *f* ~ **arsenic** Arsenblüte *f*, Fliegenstein *m*, Giftstein *m*, weißer Arsenik *m* ~ **arsenical pyrite** Silberkies *m* ~ **beech** Weißbuche *f* ~ **brass** Weißmessing *n* ~ **cap** Schaumkrone *f* ~ **cast iron** Weißguß *m* ~ **castings** ungetemperter Guß *m* ~ **compression** Weißpegelkompression *f* (TV) ~ **copper ore** Weißkupfererz *n* ~ **discharge** Weißsätze *pl* ~ **factice (or substitute)** weißer Gummiersatz *m* ~ **film** Schimmel *m* ~ **finish** Weißzurichtung *f* ~ **finishing** Weißappretur *f* ~ **fir** Weißtanne *f* ~ **flake** Schieferweiß *n* ~ **fracture** weißbrüchig ~ **frost** Anraum *m*, Duftanhang *m*, Haarfrost *m* ~ **glow** Blauglut *f* ~ **gold** Weißgold n ~ **ground** Weißboden m ~ **ground wood** Weißschliff m ~ **grub** Engerling m ~ **heart malleable iron** weißer Temperguß m, Weißkernguß m ~ **heat** Glühhitze f, weiße Glut f, Weiß-glühhitze f, -glut f, Weißhitze f ~ **hot** weißglühend, weißwarm ~ **hot ingots** glühende Stahlblöcke pl ~ **(malleable cast) iron** Weißeisen n ~ **iron pyrites** Markasit m ~ **key** Untertaste f ~ **lead** basisches Bleikarbonat n, Bleiweiß n, Deckweiß n, kohlensau(e)res Bleioxyd n ~ **lead ore** Weißbleierz n ~ **lead paint** Bleiweißfarbe f ~ **lead slurry** wässriger Bleiweißbrei m ~ **level** Weißpegel m (TV), Weißwert m ~ **level control** Weißautomatik f ~ **light** weißes Licht n ~ **lime** Fettkalk m, Weiß-äscher m, -kalk m ~ **lime slag** weiße Kalkschlacke f ~ **line** Durchschuß m ~**liquor** Läuterbeize f (Türkisschrotfärberei) ~ **malleable cast iron** Weißguß m ~ **material** Albedo n ~ **metal** Büchsenmetall n, (copper) Konzentrationsstein m, Lagermetall n, Weiß-eisen n, -metall n, weißer Kupferstein m ~ **metal bearing** Weißmetallager n ~ **nickel ore** Weißnickel-erz n, -kies m ~ **noise** weißes Rauschen n ~ **oil** farbloses Paraffinöl n ~ **page** Leerseite f ~ **paste** Erregerpaste f ~ **peak** Maximum n an Weiß (TV) ~ **pig iron** Floßeisen n, weißes Roheisen n, Weißstahl m ~ **pigment** Weißfarbe f ~ **pine** Tannenholz n ~ **poplar** Alle f ~ **pot** Tontiegel m ~ **print** Weißpause f ~ **products** Weißprodukte pl ~ **radiation** Bremsstrahlung f ~ **resin** Burgunderpech n, Galipot n, trockenes Fichtenharz n ~ **resist** Weißpapp n ~ **rot** Weißfäule f ~ **rust** Zinkrost m ~ **saturation** Weißpegelsättigung f (TV) ~ **slag** Einschmelz- (electrometal), Fertig-, Reduktions-schlacke f ~ **slag meltdown** Schmelzen n ohne Oxydation ~ **smith** Weißblechhersteller m ~ **spirit** Terpentinölersatz m, Testbenzin n ~ **structure** Albedo n ~ **sugar massecuite** Weißzuckerfüllmasse f ~ **sugar sifter** Weißzuckersichter m ~ **tint** Weiß-nuancierung f, -tönung f ~ **tissue** Albedo n ~ **vitriol** Kupferrauch m, weißer Vitriol m, Zinkvitriol n

**whiten, to** ~ blanchieren, bleichen, tünchen, weißen

**whitener** Weißkocher m

**whitening** (von Fellen) Ausschlichten n, Bleichen n, Falzen n, (von Zucker) Terrieren n, Tünchen n, Verzinnen n, Weißen n ~ **blade** Blanchierstahl m

**whitewash, to** ~ tünchen, weißen

**whitewash** geschlämmte Kreide f, Kalkmilch f, Kalktünche f, Tünche f, Weiß-kalk m, -tünche f

**whiting** Grund m, Kalktünche f, Schlämmkreide f

**whitish** weißlich

**whittle, to** ~ schnitzeln, schnitzen

**whittle** Schnitzer m

**whitworth,** ~ **fine thread** Whitworthfeingewinde n ~ **thread** Whitworthgewinde n

**whiz, to** ~ pfeifen, sausen, schleudern, schwirren, zentrifugieren **to** ~ **out** ausschwirren

**who-are-you locking lever** Wer-da-Sperrhebel m

**whole** Ganzes n, Ganzheit f; ganz, gänzlich, ungeschürt, ungeteilt, völlig **as a** ~ insgesamt **the** ~ Gesamte n, Gesamtheit f

**whole,** ~ **current ammeter** direkt angeschalteter Strommesser m ~ **current meter** Zähler m für direkten Anschluß ~ **depth of tooth** Zahngesamthöhe f ~ **effect** Gesamtleistung f ~ **line** (in graphs, etc.) ausgezogene Linie f ~ **multiple** ganzes Vielfaches n ~ **number** ganze Zahl f ~ **step** ganze Note f (acoust.), ganzer Ton m ~ **stuff** Ganzzeug n ~ **timber** volles Holz n ~ **tone** Ganzton m (acoust.)

**wholesale** in Bausch und Bogen, engros, Großverkauf m ~ **articles** Massenartikel pl ~ **dealer** Großhändler m ~ **industry** Großgewerbe n ~ **manufacture** Großgewerbe n ~ **price** Engrospreis m, Großhandelspreis m ~ **quota of raw materials** Globalkontingent n ~ **trade** Großhandel m

**wholesaler** Grossist m

**whorl** Umgang m, Wirbel m, (textiles) Wirtel m, (textiles) Würfel m

**wick** Docht m ~ **for miner's lamps** Zündband n für Wetterlampen

**wick,** ~ **carburetor** Dochtverdampfer m ~ **feed system** Dochtschmierung f ~ **holder** Dochtträger m ~ **lubricator** Dochtöler m, Dochtschmierapparat m, Dochtschmierung f ~ **oiler** Dochtöler m ~ **trimming gauge** Dochtlehre f ~ **wire** Dochtspieß m ~ **yarn** Dochtbaumwolle f, Dochtgarn n, Lichtgarn n

**wicker** Korbweide f, Weidenzweige pl ~ **basket** Kiepe f, Weidenkorb m ~ **goods** Flechtwaren pl ~**work** Weidengeflecht n, Zaine f

**wicket** Gegenschütze m, Hub-schütz n, -schütztafel f, Stauflügel m, Zwischenrahmen m ~ **of a lock gate** Klinket n eines Schleusentores ~ **door** Drosseltür f

**wide** ausgedehnt, breit, mächtig, stark, weit

**wide-angle,** ~ **aerial camera** Weitwinkelkamera f ~ **cathode** Weitwinkelkathode f ~ **field glass** Weitwinkelfeldstecher m ~ **lens** Weitwinkelobjektiv n ~ **lens camera** Weitwinkelkamera f ~ **magnifier** Weitwinkellupe f ~ **photograph** Weitwinkelaufnahme f ~**picture** Weitwinkelbild n

**wide-angled** weitwinklig ~ **picture** weitwinklige Aufnahme f

**wideband,** ~ **amplifier** Breitband n, Breitbandverstärker m ~ **antenna** Breitbandantenne f ~ **cable** Breitbandkabel n ~ **dipole** Breitbandverstärker m ~**improvement** Gewinn m am Rausch-Signal-Abstand ~ **oscillograph** Breitbandoszilograf m ~ **property** Breitbandeigenschaft f ~ **ratio** Frequenzbandausbeute f ~ **receiver** Breitbandempfänger m ~ **set** Breitbandgerät n

**wide,** ~ **beam headlight** Breit-, Weit-strahler m ~ **bottom flange rail** Breitfußschiene f ~ **field of uses** vielseitige Einsatzmöglichkeit f

**wide-flanged** breitflanschig ~ **beam** Breitflanschträger m ~ **steel I-beam** Differdingerträger m, Peinerträger m ~ **structural steel I-beam** Greyträger m

**wide,** ~ **front** breite Front f ~ **gauge** Breitspur f ~**meshed** grobmaschig, weitmaschig ~ **meshed grid** weitmaschiges Gitter n ~ **mouthed** weithalsig ~ **mouthed bottle (or jar)** weithalsige Flasche f ~ **necked** weithalsig ~ **open** weitgeöffnet ~ **open cone** weitgeöffneter Konus m ~ **open V-rearsight** Hohlvisierung f ~ **range**

governer Regler *m* mit großem Regelbereich
~ **range instrument transformer** Großbereichs-
wandler *m* ~ **range print** Großumfangkopie *f*
~ **range speaker** Breitstrahler *m*
**widespread** ausgedehnt, weitverbreitet
**wide,** ~ **strip** Breitband *n* ~ **strip mill** Breitbandwalzwerk *n* ~**-track undercarriage** breitspuriges
Fahrgestell *n* ~**-tracked** weitspurig ~ **tread
lister** Weitspurdammkulturpflug *m*
**widely,** ~ **circulated** weitverbreitet ~ **divergent**
sehr verschieden ~ **extended** weitverzweigt ~
**spaced** weitausladend
**widen, to** ~ auseinandertreiben, ausweiten,
breiten, erweitern, nachbohren, sich ausbauchen, verbreitern, weiten **to** ~ **by blasting**
nachschießen
**widener** Nachreißer *m*
**widening** Ausbauchung *f*, Erweiterung *f*, Verbreiterung *f*, Weitung *f* ~ **of a lode** Ausbauchung *f* eines Ganges ~ **out** Breitauffahren *n*,
Nachreißen *n* (des Gesteins)
**Widia,** ~ **cutting metal** Widia-Schneidmetall *n*
~ **cutting tip** Widia-Plättchen *n*
**width** Bogenweite *f* (arch.), Breite *f*, Mächtigkeit *f*, Umfang *m*, Weite *f*
**width,** ~ **of band pass** Durchlaßbreite *f* ~ **of
base of rail** Fußbreite *f* der Schiene ~ **over
bearers** Breite *f* über Stützen ~ **of bed** (Drehbank) Bettbreite *f* ~ **at bottom** Sohlenbreite *f*
~ **across corners** Eckenmaß *n* ~ **of cut** Schnittweite *f* ~ **of distribution** Verteilungsbreite *f* (of
pressure) ~ **of field** Bildfeldgröße *f* ~ **of flange**
(in T beams) Plattenbreite *f* ~ **over flats of
hexagonal nut** Schlüsselweite *f* ~ **of framing**
Gestellweite *f* ~ **of front cutting edge** Vorder-
kantenbreite *f* ~ **of fuselage** Rumpfbreite *f* ~
**of gear ring** Zahnkranzbreite *f* ~ **of groove** Kali-
berbreite *f* ~ **at half maximum intensity**
Halbwertbreite *f* (opt.) ~ **at half transmission**
Halbwertsbreite *f* ~ **of illustration picture** Bildweite *f* ~ **between jaws** Rachenweite *f* ~ **of kerf**
Schnittbreite *f* einer Säge ~ **of keyway** Nutbreite
*f* ~ **of layer** Schichtbreite *f* ~ **of line** Linienbreite *f* ~ **of material** Materialbreite *f* ~ **of members** Breite *f* der Stäbe ~ **of milling cutter**
Fräserstärke *f* ~ **of navigable passage under a
bridge** Durchfahrtbreite *f* unter einer Brücke
~ **of opening** Öffnungsweite *f* ~ **of passage**
Durchgangsöffnung *f* ~ **of port opened** Er-
öffnungsweite *f* ~ **of propeller blade** Blatt-
breite *f* ~ **between rails** Spurmaß *n* ~ **of scoop**
Löffelbreite *f* ~ **of shadow** Schattenbreite *f*
~ **of slab** Plattenbreite *f* ~ **of strip** Streifenbreite *f* ~ **of strip photographed** Aufnahme-
breite *f* ~ **of strut** Stieltief *f* ~ **of suction connection** Sauganschlußweite *f* ~ **of summit**
Kammbreite *f* ~ **between support** Stutzweite *f*
~ **of the supports** Stuhlungsbreite *f* ~ **of the
thread** Gangbreite *f* der Schraube ~ **of tooth**
Zahnbreite *f* ~ **of tooth space** Zahnlückenbreite
*f* ~ **of top** Kammbreite *f* ~ **of transition interval**
Flankensteilheit *f* ~ **of undercarriage** Spurweite
*f* des Fahrgestells ~ **of the vault** Gewölbespann-
weite *f* ~ **at water level** Wasserspiegelbreite *f*
~ **of web** Wangenbreite *f* ~ **of the weir at the
base** Grundmauer *f* ~ **of wire** Siebbreite *f*
**width,** ~ **adjusting spindle** Breitstellspindel *f* ~
**control** Bildbreiteregelung *f* ~ **gauge** Breiten-

maß *n* ~ **indicating lamp** Begrenzungslampe *f*
**Wien bridge** Kapazitätsbrücke *f*
**Wigner effect** Wigner-Effekt *m*
**wigwag, to** ~ sich auf- und abbewegen, wedeln
**wigwagging** Winkerdienst *m* ~ **signal** Winkerzeichen *n*
**wild** stürmisch, wild, wüst
**wildcat** Schurfbohrung *f* (petroleum) ~ **well**
Aufschlußbohrung *f*
**wild,** ~ **gasoline** sehr flüchtiges Benzin *n* ~ **shot**
Ausreißer *m* ~ **steel** beunruhigter Stahl *m* ~
**traverse motion** wilde Changierung *f* ~ **well**
wildes Bohrloch *n*
**wildness** Aufwallen *n*, Aufwallung *f*
**Wilfley,** ~ **concentrator** Wilfley-Herd *m* ~ **table**
Wilfley-Herd *m*
**willemite** Kieselzinkerz *n*
**Williams tube** (Williams-)Speicherröhre *f*
**will-o'-the-wisp** Irrlicht *n*
**willow, to** ~ krempeln, reißen, wolfen
**willow** Haderndrescher *m* (paper mfg.), Lumpenwolf *m* (textile mfg.), Siebtrommel *f* ~
**bark** Weidenrinde *f* ~ **weave** Weidengeflecht *n*
**willower** Öffner *m*, Reißwolf *m*, Wolfbrecher *m*
**willowing,** ~ **drum** Reißtrommel *f* ~ **machine**
Reißwolf *m* (Zerreißmaschine)
**Wilson,** ~ **chamber method** Nebelspurmethode *f*
~ **cloud chamber photograph** Wilsonkammer-
aufnahme *f* ~ **cloud-track method** Nebelspurmethode *f*, Wilsonnebelspurmethode *f* ~
**electrometer** Kippelelektrometer *n*
**wimble** (offene) Schappe *f*, Zapfenbohrer *m*
**wince pit** blinder Schacht *m*
**winch** Baumwolle *f*, Bockwinde *f*, Drehstock *m*,
Erdwinde *f*, Förderwerk *n*, Haspel *f*, Hebe-
haspel *f*, Knüppel *m*, Kurbel *f*, Räderwinde *f*,
Spulwinde *f*, Winde *f*, Windwerk *n* ~ **of the
dragline for pulling in the ropes** Einziehwinde *f*
des Schleppschaufelbaggers ~ **for drawing in
rope** Seilchenwinde *f* ~ **for the grab** Greiferwinde *f* ~ **for manual operation** Winde *f* zur
Handoperierung
**winch,** ~ **beck** Haspelkufe *f* ~ **box** Windekasten
*m* ~ **condenser** Hilfskondensator *m* ~ **drum**
Hubtrommel *f* ~ **dying machine** Haspelfärbe-
maschine *f* ~ **frame** Windenrahmen *f* ~
**handle** Handkurbel *f* ~ **house** Windenhaus *m* ~
**housing** Windengehäuse *n* ~ **jack** Winde *f* ~
**launching** Windenstart *m* ~ **lever** Windenknüp-
pel *m* ~ **mechanism** Windwerkanlage *f* ~ **motor**
Windtriebmotor *m* ~ **operating-engine** Auf-
zugsmaschine *f* ~ **room** Windenhaus *n* ~ **rope**
Haspelseil *n* ~ **shed** Windenhaus *n* ~ **truck**
Windewagen *m* ~ **tube for anchor rope** Ducht-
rohr *n* ~ **vat** Haspelkufe *f* ~ **wagon** Windewagen *m* ~ **washer** Haspelklapot *m*
**wind, to** ~ aufspulen, aufwickeln, bewickeln,
docken, sich schlängeln, schlingen, spulen,
umwickeln, weifen, wickeln, winden, sich
winden **to** ~ **out the antenna** Antenne *f* auskurbeln **to** ~ **around** umschlingen, umwinden
**to** ~ **in** Antenne *f* einziehen **to** ~ **off** abspulen,
(a coil) abwickeln **to** ~ **on** anblasen, umwickeln
**to** ~ **the rubber motor** Gummimotor *m* aufziehen **to** ~ **up** abwickeln, aufhaspeln, aufwinden,
(die Nummernscheibe) aufziehen, hochziehen
**wind** Atem *m*, Förderzug *m*, Verwehung *f*,
Wind *m* **(into) the** ~ gegen den Wind *m* **with**

the ~ mit Rückenwind *m* ~ **by day** Tagwind *m* ~ **at high altitudes** Höhenwind *m* ~ **off** Ende *n* des Blasens

**wind,** ~ **angle** Windwinkel *m* ~ **antenna** Windfühler *m* ~ **arrow** Windpfeil *m* ~ **arrow head** Windpfeilspitze *f* ~ **arrow tail** Windpfeilende *n* ~ **belt** Windmantel *m* ~ **board** Bordbrett *n*, Windbrett *n* ~ **box** Windkasten *m*, Windmantel *m* ~ **brace** Schwungleine *f*, Windstrebe *f* ~ **bracing** Querverspannung *f*, Wind-träger *m*, -versteifung *f* ~ **break** Windbruch *m* ~ **breaker** Windjacke *f*, Windverband *m* ~ **cap** Windschutzkappe *f* ~ **carved pebble** Windkanter *m* ~ **catching area** Windangriffsfläche *f* (antenna) ~ **channel** Windkanal *m* ~ **chart** Windkarte *f* ~ **chest** Luft-, Wind-kasten *m*, Windlade *f* ~ **cloud** Windwolke *f* ~ **compass card** Windrosendarstellung *f* ~ **component** Windkomponente *f* ~ **component indicator** Wetterspinne *f* ~ **conditions** Windverhältnisse *pl* ~ **cone** Windkegel *m*, Windsack *m* ~ **correction** Ausschalten *n* des Windes ~ **correction angle** Windkorrigierungswinkel *m* ~ **course** Windbahn *f* ~ **cracked** windrissig ~ **deposited** äolisch ~ **direction** Wind-kurs *m*, -richtung *f* ~ **direction angle** Windrosenwinkel *m* ~ **direction computer** Windfahrtrechner *m* ~ **direction indicator** Windrichtungsanzeiger *m* ~ **drift** Windabtrift *f* ~ **driven electric power station** Windelektrizitätswerk *n* ~ **driven generator** Fahrtwindgenerator *m* ~ **factors** Windverhältnisse *pl* ~ **fall** Windbruch *m*, windbrüchiges Holz *n* ~ **find** Windbestimmung *f* ~ **force** Windstärke *f* ~ **furnace** Blas-, Zug-ofen *m* ~ **gate** Hauptwetterstrecke *f*, Wetterloch *n*, Wettertür *f* (min.) ~ **gauge** Anemomesser *m*, Anemometer *n*, Wind(stärken)messer *m* ~ **governor controlling gear** Windrosensteuerung *f* ~ **gradient** Windgradient *m* ~ **guard** Windschutz *m* ~ **gun** Windbüchse *f* ~ **increase** Windverstärkung *f* ~ **indicator** Windanzeiger *m*, Windfahne *f*, Windrichtungsanzeiger *m* ~ **influence** Windeinfluß *m* ~ **instrument** Blasinstrument *n* ~ **intensity** Windstärke *f* ~ **jacket** Überziehjacke *f* ~ **jet** Windstrahl *m* ~ **kanter** Windkanter *m* (geol.) ~ **laid deposits** Windablagerung *f*

**wind-lass** Ankerspill *n*, Auflaufhaspel *m*, Bratspill *n*, Förderkorbkranwinde *f*, Haspel *m*, Haspelwelle *f*, Pumpspill *n*, Räderwinde *f*, Spill *n*, (toothed) Winde *f*, Zahnradwinde *f* ~ **drilling** Bohrhaspel *f* ~ **driver** Haspelwärter *m* ~ **housing** Windengehäuse *n* ~ **room** Windenhaus *n* ~ **rope** Haspelseil *n* ~ **shed** Windenhaus *n*

**wind-lasser** Haspeler *m*

**wind,** ~ **layer** Windschicht *f*, Windstufe *f* ~ **load** Windbelastung *f* ~ **loading** Windbelastung *f* ~ **measurement** Windmessung *f* ~ **measuring device (or apparatus)** Windmeßvorrichtung *f*

**wind-mill, to** ~ (Propeller) sich im Fahrtwind *m* drehen

**wind-mill** Windmühle *f*, Windschraube *f* ~ **for drainage** Paltrock-, Polder-mühle *f* ~ **driven fuel pump** Windschraubenbrennstoffpumpe *f* ~ **driven generator** durch Windschraube angetriebener Generator *m* ~ **generator** Windkraftgenerator *m* ~ **motor** Windmotor *m*

**wind-milling** Antrieb *m* der Luftschraube durch den Fahrtwind, durch den Luftstrom *m* angetrieben (of propeller), Schleppdrehzahl *f* (aviat.) ~ **drag** Schleppwiderstand *m*, Widerstand *m* der Luftschraube eines ausgefallenen Motors, die vom Fahrtwind angetrieben wird ~ **propeller** frei umlaufende, vom Fahrtwind angetriebene Luftschraube *f* ~ **torque** Propellerschleppdrehmoment *n*

**wind,** ~ **moment** Windmoment *n* ~ **movement** Windbewegung *f* ~ **~-off and wind-up reel** Abwickel- und Aufwickelhaspel *m* ~ **pipe** Windzuführungsrohr *n* ~ **polished rocks** Windschliffe *pl* ~ **pressure** Luftdruck *m*, Windbelastung *f*, Winddruck *m* ~ **pressure formula** Winddruckformel *f* ~ **pressure gauge** Winddruckmesser *m* ~ **reference number** Windziffer *f* ~ **reservoir** Luftkasten *m* ~ **resistance** Windwiderstand *m* ~ **resistant** windfest ~ **resolver** Windfunktionsdrehmelder *m* ~ **rose** Windrose *f* ~ **sail** Windsegel *n* ~ **scale** Windskala *f*

**wind-screen** Wind-scheibe *f*, -schirm *m*, -schutz *m*, -schutzscheibe *f* ~ **of a blast furnace** Gichtschirm *m* ~ **adjuster** Fenstersteller *m* ~ **defroster** Scheibenentfroster *m* ~ **defrosting** Scheibenentfrostung *f* ~ **demister** Scheibenkläranlage *f* ~ **wiper arm with rubber blade** Wischerarm *m* mit Wischgummi ~ **wiper shaft** Wischerachse *f*

**wind,** ~ **shaft** Flügel-Kronenwelle *f* ~ **shaken** windrissig ~ **shear** Windscherung *f*

**wind-shield** Schutzscheibe *f*, Windschutzscheibe *f* ~ **bearing** Windschildlager *n* ~ **cleaner** Schutzscheibenwischer *m* ~ **de-icer** Windschutzscheibenenteiser *m* ~ **wiper** Scheibenwischer *m*, Windschutzscheibenwischer *m*

**wind,** ~ **shift** Windveränderung *f* ~ **sifting machine** Windsichtmaschine *f* ~ **sock** Wetterfahne *f*, Windsack *m* ~ **speed** Windgeschwindigkeit *f* ~ **~-speed reckoner** Windfahrtrechner *m* ~ **~-speed transmitter** Windgeber *m* ~ **stacker** Gebläse *n*, Strohelevator *m* ~ **star** Windstern *m* ~ **stress** Spannung *f* durch Windbelastung ~ **surface** Windangriffsfläche *f* ~ **T** Wind-T *n* (aviat.) ~ **test** Windmessung *f* ~ **testing apparatus** Windmeßgerät *n* ~ **thresh** Pfeifen *n* des Windes ~**-tipped** windschief ~**-tipped stereocomparator** windschiefer Stereokomperator *m* ~ **triangle** Winddreieck *f* ~ **trunk** (of an organ) Kondukte *f* ~ **truss** Windträger *m*

**wind-tunnel** Kanal *m*, Strömungskanal *m*, Wind-kanal *m*, -tunnel *m* ~ **of closed circuit-type** Ringkanal *m* ~ **balance** aerodynamische Waage *f*, Windkanalwaage *f* ~ **bell (or funnel)** Windkanaldüse *f* ~ **investigation** Windkanaluntersuchung *f* ~ **measurement** Messung *f* im Windkanal ~ **measurements** Windkanalmessung *f* ~ **model** Windkanalmodell *n* ~ **return flow** Windkanalrückleitung *f* ~ **test** Windkanalversuch *m* ~ **throat** Meßstrecke *f* eines Windkanals

**wind,** ~ **unit** Windeingabegerät *n* ~ **vane** Wetterfahne *f*, Windrichtungsmesser *m* ~**-vane anemograph** Windradanemograf *m* ~**-vane sight** Windausgleichsvorrichtung *f*, Windfahnenkorn *n* ~ **velocity** Windgeschwindigkeit *f*, Windstärke *f* ~**-velocity indicator** Anemotachometer *n*, Windgeschwindigkeitsmesser *m*

~ **volume** Luft-, Wind-menge f ~ **wall on the mouth of a blast furnace** Gichtmantel m, Gichtmauer f

**wind-ward** luvwärts, windseitig, windwärts **to be to** ~ am Wind liegen ~ **bank** windseitiges Ufer n ~ **eddy** Luvwirbel m ~ **side** Luv n, Luvseite f

**wind,** ~ **wave** Luftwegwelle f, Luftwoge f, Windwoge f ~**-wave front sight** Windfahnenkorn n ~ **wheel** Windrad n ~ **wheel anemograph** Windradanemograf m ~ **wheel drum** Windradzylinder m

**windage** Spaltreibung f, Spielraum m (electr.) ~ **correction** Windverbesserung f ~ **loss** Luftreibungs-, Ventilations-verlust m ~ **power loss** Ventilationsverlustleistung f

**winded** aufgewickelt werden, sich wickeln oder winden, sich werfen

**winder** Haspeler m, Wendelstufe f ~ **brake** Seiltrommelbremse f ~**'s knife** Winkelmesser n

**winding** Aufspulen n, Bewicklung f, Gewinde n, Haspeln n, Magnetspule f, Spulerei f (coils), Umwickeln n, Umwicklung f, Wickeln n, Wicklung f, Winden n, Windung f (room); gewunden, schlangenförmig ~ **with cotton rope** Seilbewicklung f ~ **of reversing pole** Wendepolwicklung f ~ **of the shutter** Verschlußaufzug m ~ **from tower-type headgear** Turmförderung f ~ **of wire** Drahtwindung f

**winding,** ~ **barrel** Fördertrommel f ~ **bobbin** Aufwickelspule f ~ **cable** Förder-, Schacht-seil n ~ **capacitance** Wicklungskapazität f ~ **coil** Relaisspule f (electr.) ~ **condition** Aufwindungsverhältnis n ~ **cone for dyeing** konische Färbespule f ~ **construction** Wickelaufbau m ~ **core** Wickelkern m ~ **cycle** Aufzugsbewegung f ~ **direction** Wickelsinn m ~ **drum** Auflaufhaspel f, Bobine f, Fördertrommel f, Wickel-rolle f, -trommel f ~ **engine** Förderhaspelmaschine f, Hebezeug n ~ **face** Wickelstirnseite f ~ **form** Wicklungshalter m (electr.), Wicklungsschablone f ~ **gear** Fördereinrichtung f ~ **head** Wickelkopf m ~ **insulation** Windungsisolation f ~ **knob** Aufzugsknopf m ~ **level** Fördersohle f (min.) ~ **machine** Spulmaschine f ~ **machine for bobbins** Bobinenspulmaschine f ~ **machine for textile industry** Garnwindemaschine f für die Textilindustrie ~ **mandrel** Wickeldorn m ~ **material** Wicklungsbaustoff m ~ **mechanism** Aufzugsmechanismus m ~ **mechanism of the shovel** Löffelbaggerwindwerk n ~ **motor** Fördermotor m, Windantriebsmotor m ~ **nut** Spannmutter f ~**-off** Abwicklung f ~**-on bobbin** Aufrollspule f ~**-on cam** Aufwindefortschaltkurve f ~**-on speed** Aufwindedrehzahl f ~ **pin** Wickeldorn m ~ **pit** Göpelschacht m ~ **pitch** Wicklungsschritt m ~ **plane** Windungsebene f ~ **process** Spulvorgang m ~ **retainer** Wicklungshalter m ~ **rope** Förder-, Schacht-seil n, Windetau n ~ **shaft** Windewelle f ~ **sheet** Wickelblatt n ~ **sleeve** Aufwickelhülse f ~ **slide** Wendelrutsche f ~ **space** Wicklungsraum m ~ **speed** Fördergeschwindigkeit f (min.) ~ **spool** Aufwickelspule f ~ **stairs** Wendeltreppe f ~ **stem** Aufziehachse f ~ **stock** Vorderbock m ~ **support** Wicklungstabelle f (electr.) ~ **table** Wicklungstabelle f ~ **tackle** Flaschenzug m, Schwertakel n, Talje f ~ **tackle block** Gienblock m ~ **tape** Wickelband n ~ **terminal** Wicklungsklamme f ~ **tower** Förderturm m ~ **turnspit** Bratspießdrehvorrichtung f ~ **unit** Spindelzahl f ~**-up** Aufziehen n ~**-up machine** Aufbäummaschine f ~**-up mechanism** Aufziehwerk f ~ **width** Wickelbreite f ~ **width reducing lock** Wickelbreitenverkürzungsschloß n ~ **winch** Aufziehwinde ~ **wing** verwindbarer Flügel m ~ **wire** Wickeldraht m

**windless** unbewettert

**window** Düppel m, (in envelope of a cell, iconoscope etc.) Fenster n, Fensterbriefumschlag m, Folie f, Störfolie f (rdr) ~ **of the measuring chamber** Meßkammerfenster n ~ **of a meter (or instrument)** Guckfenster n ~ **of a photocell** Fenster n einer Fotozelle

**window,** ~ **bar** Fenster-sprosse f, -stange f ~ **bench** Fenster-bank f, -schwelle f ~ **burst** Düppelstoß m (rdr) ~ **case** Fenster-einfassung f, -gestell n, -stock m, -zarge f ~ **catch** Fensterriegel m ~ **cloud** Düppelstraße f (rdr) ~ **corridor** Düppelschneise f, verdüppelte Zone f (rdr) ~ **crank** Fensterkurbel f ~ **deception** Düppeltäuschung f ~ **de-icer** Fensterenteiser m ~ **dial** Fensterskala f ~ **envelope** Fensterumschlag m ~ **fastening** Windeisen n ~ **frame** Fensterrahmen m ~ **glass** Fensterglas n ~ **head** Fenster-schluß m, -sturz m, -träger m, -überlage f ~ **holder** Schauglashalter m ~ **jamming** Folienstörung f ~ **ledge** Fenster-brüstung f, -schwelle f ~ **lifter** Fensterheber m ~ **louvres** Fensterklappen pl ~ **noise** Foliengeräusch n (electr.) ~ **opening to the outside** nach außen aufschlagendes Fenster n ~ **pane** Fenster-fach n, -scheibe f, Glasfenster n ~ **post** Fensterpfosten m ~ **rabbet** Fensteranschlag m ~ **railing** Brüstung f ~ **rider** Fensterreiter m ~ **ring** Fensterring m ~ **sash** Fenster-füllung f, -rahmen m ~ **seat** Fensterbank f ~ **service** (in offices) Schalterdienst m ~ **shield** Sichtscheibe f ~ **shutter** Fensterladen m ~ **sill** Einfassung f, Fenster-bank f, -brüstung f, -gesims m, -schwelle f, -sohle f, Lattenbrett n ~ **square** Fensterfach n ~ **strap** Fensterriemen m ~ **surface** Fensterfläche f ~ **transparencies** Fensterbilder pl ~ **valve** Drehflügel m ~ **weight per unit area** Massenbelegung f des Fensters ~ **wiper** Schutzscheibenwischer m

**windowless** fensterlos ~ **X-ray tube** fensterlose Röntgenröhre f

**winds aloft report** Bericht m über die Bewegung der höheren Winde

**windy** windig

**wine** Wein m ~ **cellar** Weinkeller m ~ **cellarage machine** Weinkellereimaschine f ~ **press** Kelter f ~ **racker** Weinabzieher m ~ **storage** Weinlagerung f ~ **testing apparatus** Weinuntersuchungsgerät n ~ **tunner** Weinabzieher m

**wing, to** ~ beflügeln, beschleunigen, mit Flügeln, Tragflächen usw. versehen

**wing** Anode f einer Elektronenröhre, Ansatz m, Ende n, Flügel m, Geschwader n (aviat.), Kotflügel m, Schenkel m, (of a bird) Schwinge f, (aviat.) Tragdeck n, Tragfläche f, Türflügel m

**wing,** ~ **of bridge** Brückennock m ~ **of the compasses** Zirkelbügel m ~ **and rolleron** Flügel m

und Rolldämpfer *m* (g/m) ~ **of a saddle** Bett-
schlittenflügel *m* ~ **with two different dihedral
angles** Knickflügel *m* ~ **in two sections** zweiteiliger Flügel *m* ~ **of a window** Fensterflügel *m* ~
**of working** Flözflügel *m*
**wing,** ~ **aileron** Flügelendquerruder *n*, Hilfsflügel *m* an der Flügelspitze ~ **altitude** Flügelanstellung *f* ~ **area** Flügel-fläche *f*, -flächen-
inhalt *m*, Tragflächeninhalt *m* ~ **arrangement**
Flügelanschluß *m* ~ **assembly** Tragwerk *n* ~
**attachment** Flügel-, Holm-anschluß *m* ~
**beam** Flügelholm *m* ~ **bit** Backenmeißel *m* ~
**bolt** Flügelschraube *f* ~ **booster flap** Hilfsklap-
pe *f* ~ **bow** Tragflächenkrümmung *f* (aviat.) ~
**bracing** Flügel-verspannung *f*, -verstrebung *f* ~
**burner** Flachbrenner *m* ~ **calipers** Bogenzirkel
*m* ~ **cam** Nadelsenker *m* ~ **camber (or curvature)** Flächenwölbung *f* ~ **cannon** Flügelkanone
*f* ~ **cell** Tragflächenzelle *f* ~ **center section**
Flügelmittelstück *n* ~ **center section built
integral with the fuselage** Flügelmittelstück *n*
fest am Rumpf ~ **chord** Flügel-sehne *f*, -tiefe
*f*, Profilsehne *f*, Tragflächensehne *f* ~ **compass**
Flügel-, Tragdeck-, Tragflügel-kompaß *m*
*m* ~ **contour** Flügel-grundriß *m*, -umriß *m* ~
**control** Flügelsteuerung *f* ~ **controlled** flügel-
gesteuert ~ **covering** Bespannung *f*, Flügel-
-behäutung *f*, -beplankung *f*, -bespannung *f* ~
**depth** Flügelbauart *f*, Tragflächentiefe *f* ~
**distortion** Flügelverdrehung *f* ~ **drag** Flügel-
widerstand *m* ~ **duct** Flügelkanal *m* ~ **edge**
Randkappe *f* ~ **electrode** (of a magnetron)
Seitenelektrode *f* ~ **fabric** Außenhaut *f* (aviat.),
Flügelhaut *f* ~ **failure** Flügelbruch *m* ~ **fan**
Flügelgebläse *n* ~ **fastening** Tragflügelbefesti-
gung *f* ~ **fillet** Anschlußstück *n*, Flügelüber-
gang *m*, Übergangsstück *n* ~ **fittings** Flügel-
beschläge *pl* ~ **fixing** Flügelbefestigung *f* ~
**flap** Flügelklappe *f*, Hilfsflügel *m*, Hilfsflügel-
klappe *f*, Querruder *n*, Verwindungsklappe *f*
~ **flap extended speed** Geschwindigkeit *f* bei
ausgefahrenen Flügelklappen ~ **flare** Flügel-
-endfackel *f*, -leuchtpatrone *f*, Lande-fackel
*f*, -leuchte *f* an der Flügelspitze ~ **float** Flügel-
-schwimmer *m*, -stützschwimmer *m*, Gleitflosse
*f*, Hilfsschwimmer *m*, Seitenschwimmer *m*,
Stützschwimmer *m* ~ **float stump** Flossenstummel *m* ~ **flutter** Flattern *n* der Tragflügel,
Flügel-flattern *n*, -schwingung *f* ~**fold position**
Flügelfaltstellung *f* ~ **form** Flügelgrundriß *m* ~
**former** Randbogen *m* ~ **frame** Fensterflügelrahmen *m* ~ **gap** Flächenabstand *m* (aviat.),
Flügelabstand *m* (aviat.), Tragflächenabstand
*m* ~ **gasoline tank** Flügelbehälter *m* ~ **ground
support** Flügelkufe *f* ~**-guided** mit Richtungsflügel *m* versehen ~ **half** Halbflügel *m* ~
**hand grip (or handle)** Flügelhandgriff *m* ~
**headed bolt** Flügelschraube *f* ~ **heaviness**
Querlastigkeit *f* (aviat.) ~ **heavy** flügel-,
quer-lastig ~ **hillers** (agr. mach.) Flügelhäufler
*m* ~ **incidence** Flügelanstellung *f* ~ **junction**
Flügelanschluß *m* ~ **leading edge** Flügelvor-
derteil *m* ~ **load** Flächenanziehungsbelastung *f*
~ **loading** Flächenanziehungsbelastung *f*, vom
Flügel *m* getragene Last *f*, Tragflächenbela-
stung *f* ~ **loss** (uplift of airplane) Flügel-
spitzenauftriebsverlust *m* ~ **momentum** Flügel-
moment *m* ~ **nacelle** Flügelgondel *f* ~ **nut** Flü-

gel-mutter *f*, -schraube *f* ~ **overhang** Flügel-
überhang *m* ~ **panel** Flügelfeld *n* ~ **passage**
bulkhead Wallgangschott *n* ~ **pivot** Flügelan-
lenkung *f* ~ **plan** Flügel-grundriß *m*, -umriß *m*
~ **planking** Flügel-beplankung *f*, -bespannung
*f* ~ **plate** Flügelendscheibe *f* ~ **power** Flächenleistung *f* ~ **profile** Flügel-form *f*, -profil *n*,
-querschnitt *m*, -schnitt *m* ~ **profile thickness**
Flügelprofildicke *f* ~ **pump** Würgelpumpe *f* ~
**pylon** Außenspannturm *m* ~ **radiator** Flügel-
flächenkühler *m*, Tragdeckkühler *m*, Trag-
flächenkühler *m* ~ **rail** (of a crossing) Horn-
schiene *f* ~ **rake** Flügelspitzenumriß *m* ~ **rib**
Flügel-rippe *f*, -spant *m* (aviat.), Rippe *f* ~ **rib
former** Formrippe *f*, Nasenleiste *f* (aviat.),
Stirnleiste *f* ~ **roll** Flügelrollmoment *m* ~ **root**
Flügel-anschluß *m*, -ansatz *m*, -wurzel *f* ~
**root fairing** Rumpfflügelübergang *m* ~ **rotor**
Rollflügel *m* ~ **row** Flügelbiegung *f* ~ **screw**
Daumenflügelschraube *f* ~ **section** Flügel-
(quer)schnitt *m* ~ **sections** Flügeleinteilung *f*
~ **shape** Flügelform *f* ~ **skeleton** Flügelgerippe
*n* ~ **skid** Flügelkufe *f* ~ **skin** Beplankung *f*,
Flügelbehäutung *f*, Tragflächenhaut *f* ~ **slot**
Flügel-nute *f*, -schlitz *m*, -spalt *m*, -spaltflügel
*m* ~ **socket** Flügeltülle *f* ~ **span** Flügelspann-
weite *f* ~ **spar** Flügel-holm *m*, -sparren *m*, Trag-
flächenholm *m*, Tragflügelholm *m* ~ **spar box**
Holmschuh *m* ~ **spar connector** Holmanschluß-
belag *m* ~ **spread** Spannweite *f* ~ **strip** Randbogen *m* ~ **structure** Flügelaufbau *m*, Trag-
flügelgerippe *n* ~ **strut** Flügel-stiel *m*, -strebe
*f* ~ **stub** Flügel-ansatz *m*, -stummel *m* ~
**support** Flügelbefestigung *f* ~ **support strut**
Abfangestrebe *f* ~ **surface** Flügelfläche *f* ~
**tank** Flügelstützschwimmer *m* ~ **taper** Flügel-
-verjüngung *f*, -zuspitzung *f* ~ **thickness**
Flügel-, Profil-dicke *f* ~ **tip** Flügel-ende *n*,
-spitze *f* ~ **tip clearance** Flügelendfreiheit *f* ~
**tip float** Flügelendschwimmer *m* ~ **tip vortex**
Randwirbel *m* ~ **top** Schlitzaufsatz *m* (chem.)
~ **trailing edge** Flügelhinterteil *m* ~ **truck**
Flächenwagen *m* ~ **truss** Flügelfachwerk *n*
~ **truss structure** Flugwerk *n*, Tragwerk *n*,
Tragzelle *f* ~ **twist** Flügelverwindung *f* ~
**twist jig** Prüflehre *f* für Flügelverwindung ~
**twisting** Flügelverdrehung *f* ~ **unit** Tragwerk *n*
~ **vortex** Flügelspitzenwirbel *m* ~ **walk** Tret-
matte *f* ~ **walls** Flügelmauer *f* ~ **warping**
Flächenverwindung *f*
**winged** flügelig ~ **cross** Flügelkreuz *n* ~ **head**
Flügelkopf *m* ~ **nut** Flügelmutter *f* ~ **pole**
Flügelstange *f* ~ **screw** Flügelschraube *f* ~
**tap** Flügelhahn *m* ~ **wedge** Flügelkeil *m*
**winning** Gewinnung *f* ~ **of a seam (or vein)**
Abbau *m* eines Flözes
**winning,** ~ **expenses** Gewinnungskosten *pl*
~ **face** Abbaustelle *f* ~ **headway** Abbauförder-
strecke *f*
**winter** Durchbalken *m* (print.) ~ **equipment**
Winterausrüstung *f* ~ **jet** Winterdüse *f* ~
**monsoon** Wintermonsun *m* ~ **schedule** Winter-
flugplan *m* ~ **solstice** Wintersonnenwende *f* ~
**thunderstorm** Wintergewitter *n*
**winteriting** Winterisierung *f*
**winze** Blindschacht *m*
**wipe, to** ~ abblenden, abstreifen, bestreichen,
entlangstreifen, schleifen, wischen **to** ~ **with**

**chamois skin** abledern **to ~ off** abtupfen, abwischen, auswischen **to ~ out** ausrotten, verwischen

**wipe** Tricküberblendung f **~ resistance** Wischbeständigkeit f **~ spark** Unterbrechungsfunke m

**wiped joint** Lötstelle f, (cable) Lötwulst m

**wiper** Abgreifer m, Abstreicher m, Abstreifer m, Bürste f, Hebedaumen m, Kohlenbürste f, Kontaktarm m, (rotary) Kontaktbürste f, Rakel n, Schaltaderarm m, Schleif-arm m, -feder f, Schmierkamm m, Tastarm m, Wischer m~ **of arbor** Wellendaumen m

**wiper, ~ arm** (of a switch) Abstreifarm m, Schleifarm m, Wischhebel m **~ assembly** Kontaktarmsatz m **~ bar** Wischstab m **~ blade** Scheibenwischerschiene f, Wischer-blatt n, -schneide f **~ blade holder** Wischerblattfassung f **~ contact** Wischkontakt m **~ lubrication** Abstreifölung f **~ set** Kontaktarmsatz m **~ shaft** Bürstenarmspindel f, -träger m, -welle f, Kontaktarmträger m **~ tongs** Abstreiferzange f

**wiping, ~ cloth** Lötlappen m **~ efficiency** Wischleistung f **~ frequency** Löschfrequenz f **~ operation** Gleitvorgang m **~ roller cloth** Wickelwalzenstoff m **~ solvent** Abwischlösemittel n **~ speed** Wischgeschwindigkeit f

**wire, to** anschließen, aufrollen, bedrahten, beschalten, Draht m legen, schalten, telegrafieren, umwickeln, verdrahten (electr.) **to ~ up** bedrahten, mit Stöpseln anschalten

**wire** Ader f, Draht m, Drahtschloß n, Faden m, Faden m im Okular, Rundeisen n in Drahtform, Stahldraht m, Telegramm n **with two ~s** zweidrahtig **~ of a cable** Kabellader m **~ for indoors installation** Zimmerleitungsdraht m **~ for nuts** Mutterdraht m

**wire, ~ annealing** Drahtglühen n **~ antenna** Drahtantenne f **~ arm** Drahthebel m **~ arm with wiping rubber** Drahthebel m mit Wischgummi **~armoring** Drahtbewehrung f, Profildrahtbewehrung f **~ articles** Drahtwaren pl **~ bar** Drahtbarren m, I-Barren m, Knüppel m **~ bar copper** Leitungskupfer n **~ basket** Drahtkorb m **~ bending apparatus** Drahtbiegegerät n **~ billet** Knopf m oder Knüppel m für Draht **~ binding** Drahtverbindung f **~ binding pliers** Drahtbindezange f **~ bound box** Drahtspankiste f (Drahtbundkiste) **~ bow net** Drahtreuse f **~-braced** drahtverspannt **~-braced low-wing monoplane** verspannter Tiefdecker m **~-braced wing** verspannte Zelle f **~ bracing** Drahtverspannung f **~ bracing wing to body** Rumpfanschlußseil n **~ braiding machine** Drahtflechtmaschine f **~ brazer** Drahtlöter m **~ broadcast** Leitungs-übertragung f, -rundfunk m **~ broadcasting** Draht(rund)funk m **~ brush** Drahtbürste f **~ brush-finished** gebürstet **~ bundle** Drahtbündel n **~ cable** Drahtseil n **~ cable obstacle** Drahtseilhindernis n **~ cage** Drahtschutzkappe f **~ carrier system** Drahtfunksystem n **~ carrying current** stromdurchflossene Leitung f **~ cart** Kabelkarren m **~ center** Drahtseele f **~ channel** Drahtkanal m **~ chief's desk** Prüftisch m **~ clamp** Drahtklammer f **~ clamp lever** Drahtklemmhebel m **~ clamp sheath** Seilklemmhülse f **~ cloth**

**Draht-gaze** f, -geflecht n, -gewebe n, endloses Sieb n, Metall-gewebe n, -tuch n, Siebband n **~ cloth for the paper machine** Papiermaschinensieb n **~ coating apparatus** Drahtumspulgerät n, Umspulapparat m (Draht) **~ coil** Draht-spule f, -spirale f **~ coil handling crane** Drahtbündeltransportkran m **~ coiling** Drahtwicklung f **~ communication** Drahtverbindung f **~ compensator** Spannwerk n **~ connection** Drahtverbindung f **~ connector** Drahtklemme f, Leitungsverbinder m **~ core** Draht-einlage f, -kern m, -seele f, unterteilter Eisenkern m **~ core coil** Drahtkernspule f **~ cramp** Drahtklammer f **~ cross bracing** Diagonalverspannung f **~ crossing** Drahtkreuzung f **~ cushion** Drahtkissen n **~ cutter** Draht(ab)schneider m, Drahtschere f, Scherenzange f **~ cutters** Drahtzange f **~ distribution** Rundfunkvermittlung f **~ drawer** Drahtzieher m, Zieher m **~-drawing** Drahtziehen n **~ drawing in exhaust** Drosselung f im Auslaß **~ drawing in intake** Drosselung f im Einlaß **~ drawing bench** Drahtziehbank f **~-drawing department** Drahtzieherei f **~ drawing die** Durchzugmatrize f **~ drawing effect** Engpaß m, Einschnürung f **~-drawing machine** Drahtziehmaschine f **~-drawing plant** Drahtzieherei f **~ dresser** Drahtzurichter m **~ edge** Grat m an der Schneide nach dem Anschleifen **~ enameling machine** Drahtemailiermaschine f **~ enclosure** Drahteinzäunung f **~ entanglement** Draht-hindernis n, -verwicklung f **~ eyelet** Drahtöse f **~ feed** Drahteinlauf m **~ feed lever** Drahtvorschubhebel m **~ feeding** Drahtzuführung f **~ feeding roller** Drahtzuführungswalze f **~ fence** Draht-einzäunung f, -zaun m **~ fencing** Drahtgeflechtzaun m, Zaungitter n **~ ferrule** Drahtzwinge f **~ filter** Drahtfilter n **~ foot mat (or scraper)** Drahtfußmatte f **~ frame** Drahtgestell n **~ frame view finder** Ikonometer n **~ framing** Drahtumwicklung f **~ fuse** Drahtsicherung f **~ galvanization** Drahtverzinkung f **~ gauge** Draht-kaliber n, -lehre f, -messer m, -stärke f, -stärketabelle f, -tabelle f **~ gauge plate** Bohrerlehre f **~ gauge table** Draht(stärken)tabelle f **~ gauging device** Drahtmeßvorrichtung f

**wire-gauze** Draht-gaze f, -gewebe n, -netzelektrode f, -schale f **~ with flat asbestos center** Drahtgewebe n mit eingepreßter Asbestschicht (in der Mitte) **~ with hemispherical asbestos center** Asbestdrahtnetz n mit kugelförmiger Vertiefung **~ of a safety lamp** Drahtzylinder m einer Sicherheitslampe

**wire-gauze, ~ cap** Siebkappe f **~ electrode** Netzelektrode f

**wire, ~ glass** Drahtglas n **~ goods** Drahtwaren pl **~ grating** Drahtgitter n **~ grid** Drahtgitter n **~ grip** Draht-spanner m, -ziehstrumpf m, Kniehebelklemme f, Ziehstrumpf m **~ guard** Drahtschutzkorb m, Führungsrohr n **~ guide frame** Drahtführerrahmen m **~ guided** drahtgelenkt **~ guided telephony** leitungsgerichtete Hochfrequenztelefonie f **~ guides** Drahtführer m, Harfe f (Drahtwalze) **~ guiding lock** Drahtführungsschloß n **~ guy** Drahtseilanker m **~ hand brush** Drahthandbürste f **~ hand tacks** Drahthandtäckse pl **~ hanger** Meßdrahthalter m (galgenartig) **~ hardening plant** Drahthärterei

*f* ~ **haulage rope** Drahtzugseil *n* ~ **head** Draht-, Nachrichten-kopf *m* ~ **head casing** Drahtkopfgehäuse *n* ~ **hawser** Drahttrosse *f* ~ **helix** Drahtspirale *f* ~ **hurdle** Drahtborde *f* ~ **identification** Kabelkennung *f* ~ **impulse transmitter** Stoßdrahtgeber *m* ~ **ingot** Drahtbarren *m* ~ **inlet (or insertion)** Drahteinlage *f* ~ **kiln floor** Drahthorde *f* ~ **lacquering machine** Drahtlackiermaschine *f* ~ **lashing** Drahtbund *m* ~ **lattice maker** Drahtgitterflechter *m* ~ **laying on poles** Stangenbau *m* ~ **lead-in point** Kabelaufführungspunkt *m* ~ **leads** Schlagkasten *m* (Schlingenkanal) ~ **line** Bildzeile *f*, Draht-leitung *f*, -seil *n*, -zug *m*, Kabelleitung *f* ~ **line clamp** Drahtseilklammer *f* ~ **loop** Draht-schleife *f*, -schlinge *f*, Schleife *f* ~ **man** Bauarbeiter *m*, Einrichter *m*, Telegrafenarbeiter *m* ~ **man's tent** Löterzelt *n* ~ **manifold** Kabelführungs-rohr *n*, -geschirr *n*, Nabenrohr *n* ~ **mark** Siebmarke *f*, Markierung *f*
**wire-mesh** Drahtgaze *f*, Maschen-draht *m*, -geflecht *n*, -netz *n* ~ **fence** Maschendrahtzaun *m* ~ **mat** Drahtmatte *f* ~ **road** Drahtstraße *f* ~ **screen** Maschendrahtabdeckung *f* ~ **sieve** Drahtsieb *n*
**wire,** ~ **meshing** Drahtgewebe *n* ~ **mill** Drahtwalzwerk *n*, Drahtzieherei *f* ~ **milling** Drahtwalzen *n* ~ **mounting (or holder)** Meßdrahthalter *m* ~ **nail** Drahtstift *m* ~ **nails** Drahtnägel *pl* ~ **net** Drahtnachrichtennetz *n*, Drahtnetz *n* ~ **net system** Leitungsnetz *m* ~ **netting** Draht-geflecht *n*, -gewebe *n*, -maschengitter *n*, -tuch *n*, Eisendrahtgewebe *n*, Maschendraht *m* ~ **netting insertion** Drahtgewebeeinlage *f* ~ **netting machine** Drahtflechtmaschine *f* ~ **netting sieve** Drahtsieb *n* ~ **netting tapping** Abzweigung *f* einer Leitung ~ **network** Drahttuch *n*, Eisendrahtgewebe *n* ~ **noose für corked bottles** Drahtschleife *f* für verkorkte Flaschen ~ **number** Kabel-kennung *f*, -nummer *f* ~ **peeling machine** Drahtschälmaschine *f* ~ **penetrometer** Drahthärtemesser *m* ~ **pin** Drahtstift *m* ~ **plant** Leitungsanlage *f* ~ **plug** Kabelstecker *m* ~ **polishing device for string wires** Musikdrahtpoliereinrichtung *f* ~ **probe for measuring flame temperature** Drahtsonde *f* zur Flammentemperaturmessung ~ **product** Drahterzeugnis *n* ~ **program distribution** Drahtfunk *m* ~ **pull** Drahtseilzug *m*, Drahtzug *m* ~ **reel** Drahthaspel *f*, Winde *f* zum Abnehmen von Drähten ~ **reinforced glass** Drahtglas *n* ~ **reinforcement** Drahteinlage *f* ~ **relay** Drahtbruchrelais *n* ~ **release** Drahtauslösung *f* (Verschluß) ~ **release lever** Drahtauslösehebel *m* ~ **riddle** Luftspieß *m*
**wire-rod** Draht-eisen *n*, -haspel *f*, -knüppel *m*, -stab *m*, -stange *f*, Runddraht *m*, Walzdraht *m* ~ **finishing mill** Drahtfertigstraße *f* ~ **guide** Drahtführung *f* ~ **mill train** Drahtstrecke *f* ~ **milling** Drahtwalzen *n* ~ **pass** Drahtkaliber *n* ~ **rolling mill** Drahtzieherei *f* ~ **rolling train** Drahtstraße *f*
**wire,** ~ **roll** Drahtwalze *f* (entanglement) ~ **roller** Drahtwalzer *m* ~ **rolling mill** Drahtstraße *f* ~ **rolling mills** Drahtwalzwerk *n*
**wire-rope** dralloses Kabel *n*, Draht-seil *n*, -trosse *f*, Stahltrosse *f* ~ **with hemp center (or core)** Drahtseil *n* mit Hanfseele

**wire-rope,** ~ **block** Drahtseilflaschenzug *m* ~ **cableway** Kabelkran *m* ~ **chopper** Seilreißer *m* ~ **clamp** Drahtseilklemme *f* ~ **compound** Drahtseilfett *n* ~ **core** Seele *f* des Drahtseiles ~ **drive** Drahtseiltrieb *m* ~ **manufacture** Kabelherstellung *f* ~ **press** Metallstrangpresse *f* ~ **pulley** Drahtseilblock *m* ~ **sheave** Drahtseil-rolle *f*, -scheibe *f* ~ **sling** Drahtseilschlaufe *f* ~ **stay** Drahtseilanker *m* ~ **straightening machine** Seilrichtmaschine *f*
**wire,** ~ **ropeway** Drahtseilbahn *f* ~ **screen** Drahtschirm *m*, -sieb *n*, Siebgaze *f* (aus Draht) ~ **screen basket** Drahtsiebkorb *m* ~ **sharpening machine** Drahtanspitzmaschine *f* ~ **shears** Drahtschere *f* ~ **sheating** Drahtbewehrung *f* ~ **shield** Drahtschutz *m* ~ **shoe** schuhartiger Meßdrahthalter *m* ~ **side of paper** Siebseite *f* ~ **sieve** Drahtsieb *n* ~ **signal** Leitungssignal *n* ~ **size** Drahtkaliber *n* ~ **skimmer** Abisolierer *m*, Entisolierer *m* ~ **sling** Drahtschlinge *f* ~ **snap** Schnappfeder *f* ~ **snare** Drahtschlinge *f* ~ **spiral** Drahtspirale *f* ~ **spoke wheel** Drahtspeichenrad *n* ~ **spool** Drahtspule *f* ~ **spring** Draht-bogen *m*, -feder *f* ~ **spring relay** Federdrahtrelais *n* ~ **stain gauge** Dehnungsmeßstreifen *m* ~ **staple** Drahtheftklammer *f* ~ **staple pusher** Klammerschieber *m* ~ **stay** Drahtseilanker *m* ~ **stitch** Heftklammer *f* ~ **stitched quires** drahtgehefteter Bogensatz *m* ~ **stitching** Drahtheftung *f* ~ **stitching apparatus** Drahtheftapparat *m* ~ **straightener** Drahtrichtapparat *m* ~ **straightening and cutting machine** Richtmaschine *f* ~ **strainer** Drahtsieb *n* ~ **strand** Ankerseil *n*, Draht-, Seil-litze *f* ~ **strand core** Drahtlitzenseele *f* ~ **stranding machine** Drahtlitzenmaschine *f* ~ **strap (or bow)** Drahtbügel *m* ~ **stress table** Drahtzugtabelle *f* ~ **stretcher** Drahtspanner *m*, Spannvorrichtung *f* ~ **string** Drahtsaite *f* ~ **strippers** Abstreif-, Abisolier-zange *f* ~ **supports** Drahtträger *m* ~ **suspension** Drahtaufhängung *f* ~ **system** Drahtnachrichtennetz *n*, Leitungssystem *n*, Signalleitungsnetz *n* ~ **table** Siebtisch *m* ~ **tape** Drahtband *n* ~ **tapping** Abhören *n*, Mithöreinrichtung *f* ~ **tapping clamp** Lauschzange *f* ~ **telegraphy** Drahttelegrafie *f* ~ **telephony** Draht-fernsprechen *n*, -telefonie *f* ~ **temperature** Drahttemperatur *f* ~ **tension table** Spannungs-tabelle *f*, -tafel *f* ~ **thermocouple** Drahtthermoelement *n* ~ **thread guide** (textiles) Sauschwanz *m* ~ **tightener** Drahtspannschloß *n* ~ **tinning plant** Drahtverzinnerei *f* ~ **tissue belt** Drahtgurtband *n* ~ **transmission of pictures** Bildtelegrafie *f* ~ **transport** Drahtmitnehmer *m* ~ **trench** Kabelgraben *m* ~ **turn** Drahtschleife *f* ~ **twisting apparatus** Draht-torsionsgerät *n*, -verwindungsgerät *n* ~ **ware** Drahtwaren *pl* ~ **wave telegraphy** Drahtwellentelegrafie *f* ~ **wheel** Drahtspeichenrad *n* ~ **winding** Drahtschleife *f* ~ **winding apparatus** Drahtumspulgerät *n* ~ **wiper arm** Drahtwischhebel *m*
**wire-wound,** ~ **armature** Anker *m* mit Knäuelwicklung *m* ~ **resistor** Drahtwiderstand *m* (drahtgewickelt) ~ **trimmer** Drahttrimmer *m* ~ **variable resistor** Drahtdrehwiderstand *m*
**wire,** ~ **wrap connection** Wickeldrahtverbindung *f* ~ **wrapped** mit Draht *m* umwickelt ~ **wrapped gun** Drahtrohr *n* ~ **zincification** Drahtverzinkung *f*

**wired on tire** Reifen *m* mit Stahldrahtbefestigung
**wired,** ~ **glass** Drahtglas *n* ~ **message** Kabelnachricht *f* ~ **photograph** Funkbild *n* ~ **phototelegraphy** Bildtelegrafie *f* ~ **picture** Bildfunk *m* ~ **radio** Drahtfunk *m*, Drahtwellentelegrafie *f*, Hochfrequenzverbindung *f* (über Leitungen), Leitungsübertragung *f* ~ **radio system** Hochfrequenzdrahtsystem *n* ~ **wave telegraphy** leitungsgerichtete Trägerwellentelegrafie *f* ~ **wireless** Drahtfunk *m* ~ **wireless telegraphy** Wellentelegrafie *f* über Leitungen
**wireless** Rundfunk *m*; drahtlos ~ **amateur** Funk-, Radio-amateur *m* ~ **apparatus** Funk-apparat *m*, -gerät *n* ~ **beam** Funkbüschel *n* ~ **communication** drahtlose Verbindung *f*, Funkverkehr *m*, Funkwesen *n* ~ **compass** Funkkompaß *m* ~ **controlled** ferngelenkt ~ **direction finder** Funkpeilgerät *n*, Peilgerät *n* ~ **direction-finding** drahtlose Richtungsbestimmung *f* ~ **direction-finding transmitter** Peilfunksender *m* ~ **equipment** Funkausrüstung *f*, Funkeinrichtungen *pl*, Funkentelegrafieausrüstung *f*, Fundfunkgerät *n* ~ **exhibition** Funkausstellung *f* ~ **installation** Radioanlage *f* ~ **interception service** Horchfunk *m* ~ **message** Funkspruch *m*, Radiogramm *n* ~ **meteorological service** Funkwetterdienst *m* ~ **officer** Funkbeamter *m*, Funkoffizier *m* ~ **operator** Funker *m*, Funktelegrafiegast *m* ~ **picture telegraph** Bildrundfunkempfänger *m*, Funkbildgerät *n* ~ **plant** Funkanlage *f* ~ **receiver** Funkempfänger *m*, Empfänger *m*, Empfangsgerät *n* ~ **reception** Funk-, Radioempfang *m* ~ **remote control** Funkfernführung *f* ~ **set** Funkapparat *m*, Funkgerät *n* ~ **set-back panel** Radiorückwand *f* ~ **silence** Funkstille *f* ~ **station** Funkerstand *m*, Funkstation *f*, Funkstelle *f* ~ **telegram** Funkmeldung *f*, Funktelegraf *m*, Tastverkehr *m* ~ **telegraphy** drahtlose Telegrafie, Funktelegrafie *f*, Radiotelegrafie *f*, Wellentelegrafie *f* ~ **telegraphy link** Funkwiederholer *m* ~ **telegraphy screening** Funkentstörer *m* ~ **telegraphy security** Funküberwachung *f* ~ **telephone** Funktelefon *m* ~ **telephone receiver** Empfangsfernhörer *m* ~ **telephony** Funkfernsprechen *n*, Funktelefonie *f*, Radiotelefonie *f* ~ **transmission** Funken *n* ~ **transmission of image currents** drahtlose Übertragung *f* der Bildströme ~ **transmitter** Funksender. *m* ~ **transmitter receiver** Funksenderempfänger *m* (rdr) ~ **wave** Radiowelle *f* ~ **waves** Hertzsche Wellen *pl* ~ **weather service** Funk-Wetterdienst *m*
**wires for vehicles** Fahrzeugleitungen *pl*
**wiring** Anschluß *m* an das Netz, Beschaltung *f*, Draht-einlegen *n*, -leitung *f*, -verbindung *f*, Leitung *f*, Leitungen *pl*, Leitungs-führung *f*, -netz *n*, Schaltung *f*, Verdrahtung *f*, Verkabelung *f*, Verspannung *f*, Zündleitungen *pl* ~ **in diagonal pairs** Diagonalgruppierung *f*
**wiring,** ~ **arrays** Schaltmatrix *f* ~ **change** Umlötung *f*, Umschaltung *f* ~ **clip** Drahtverschluß *m* ~ **diagram** Bedrahtungsplan *m*, Installationsplan *m*, Leitungsnetzschema *n*, Schaltaderbild *n*, -bild *n*, -plan *m*, -schema *n*, -skizze *f*, Schaltungs-schema *n*, -zeichnung *f*, Stromlaufzeichnung *f*, Verdrahtungsplan *m* ~ **harness** Kabelbaum *m* ~ **layout** Bauschaltbild *n* ~ **machine** Drahteinlegmaschine *f*, Drahthefter

*m* ~ **mat** Drahtgitter *n* ~ **plan** Beschaltungsplan *m*, Schaltschema *n* ~ **plate** Drahtschluß *m*, Schaltungsplatte *f* ~ **scheme** Beschaltungsplan *m*, Leitungsnetzschema *n*, Schaltplan *m* ~ **specification sheet** Schaltliste *f* ~ **switch** Installationsschalter *m*
**wishbone** (Auto) Radquerlenker *m* ~ **links** (Geländewagen) Querschwingen *n*
**wisp of straw for tying sheaves** Garbenband *n*
**withdraw, to** ~ abbrechen, sich absetzen, abziehen, aufholen, ausheben, ausweichen, einziehen, entziehen, herausziehen, stornieren, wegziehen, zurückziehen, sich zurückziehen **to** ~ **horizontally** waagerecht herausziehen **to** ~ **a plug** aufheben, unterbrechen **to** ~ **from service** aus dem Betrieb ziehen
**withdrawable** ausfahrbar
**withdrawal** Abhebung *f*, Ableitung *f*, Absetzbewegung *f*, Abzug *m*, Aufnahme *f*, Ausheben *n*, Aushub *m*, Austritt *m*, Ausziehen *n*, Entziehung *f*, Herausziehen *n*, Rückzug *m*, Zurückziehung *f* ~ **from circulation** Außerkurssetzung *f* ~ **of slag** Schlackenziehen *n* ~ **of telephone service** Fernsprechsperre *f*
**withdrawal,** ~ **cam** Abzugskurve *f* ~ **door** Auszugstür *f* ~ **finger** Kupplungsausrückgabel *f* ~ **form** Entnahmeschein *m* ~ **lever** Rückzughebel *m* ~ **opening** Aushebeöffnung *f* ~ **plate** Empfangsplatte *f* ~ **resistance** Rückzugswiderstand *m* ~ **roller** Abzugsrolle *f* ~ **sleeve** Abziehhülse *f*
**withdrawing,** ~ **door** Ausziehöffnung *f* ~ **movement to new position** Ausweichbewegung *f* ~ **opening** Ausziehöffnung *f*
**withdrawn** abgeschnitten, abgezogen, isoliert ~ **plug** Leerstecker *m*
**wither, to** ~ abschwelken
**withered** geschwelkt ~ **leaf** welkes Blatt *n*
**witherite** Witherit *m*
**withers** Widerrist *m*
**withhold, to** ~ einbehalten, vorbehalten, vorenthalten **to** ~ **delivery** Lieferung *f* zurückhalten
**within** binnen, innen, innerhalb ~ **the limits (or range)** im Bereich von ~ **the time** innerhalb der Frist *f*
**within,** ~ **reasonable distance** in erreichbarer Nähe *f* ~ **specified time** fristgemäß ~ **striking distance** in erreichbarer Nähe *f* ~ **time limit** fristgerecht
**without** außen, außerhalb, ohne ~ **accessories** ohne Zubehör *n* ~ **after-effect** nachwirkungsfrei ~ **angular displacement between parts** ungeschränkt ~ **assuming any obligations** ohne Gewähr *f* ~ **compressor** kompressorlos ~ **contrast** flach, flau ~ **delay** anstandslos ~ **disturbance** störungslos ~ **drainage** abflußlos ~ **electrodes** elektrodenlos ~ **flywheel** schwingradlos ~ **groove** angußlos ~ **hesitation** anstandslos ~ **interruption** lückenlos ~ **intervals** pausenlos ~ **key-way** nutenlos ~ **knots** astrein ~ **light** lichtlos ~ **limit** unbegrenzt ~ **means** mittellos ~ **a motive** unmotiviert ~ **a net** netzlos ~ **obligation** unverbindlich ~ **prejudice** unbeschadet ~ **residue** restlos ~ **result** ergebnislos ~ **shearing** bruchlos ~ **slowing down** nicht nachlaufend ~ **soldering** lötlos ~ **spatter** reflexionsfrei ~ **spools** hülsenlos ~ **springs** ungefedert ~ **sunshine** sonnenlos ~ **surcharge** zuschlagfrei ~ **surfacing**

ohne Belag *m* ~ **tension** spannungslos ~ **time limit** befristet, fristlos ~ **a trace** spurlos ~ **undulations** wellenfrei ~ **water** wasserleer
**withstand, to** ~ aushalten, standhalten, widerstehen
**withworth,** ~ **leads** Gewindegangzahlen *pl* ~ **screw** Withworth-Gewinde *n*
**witness point** Vergleichsziel *n*
**wittichenite** Kupferwismuterz *n*, Wismutkupfererz *n*
**wobble, to** ~ flattern, (wheels) schwanken, taumeln, torkeln, wobblen
**wobble** Schlag (phono, rdo), (eccentric motion) Schlag *m* einer Welle, Schwanken *n*, Taumel *m*, Taumelbewegung *f* (exzentrische Bewegung des Bohrgestänges), Taumeln *n*, Wackeln *n*, Wobbeln *n*, ~ **frequency** Taumelfrequenz *f* ~ **modulation** Wobbeln *n* ~ **plate** Taumelscheibe *f* ~ **plate engine** Taumelscheibenmotor *m* ~ **pump** Taumelscheiben-, Wobbel-pumpe *f* ~ **type notch saw** Wanknutsäge *f*
**wobbled carrier current** zeitlich veränderlicher Trägerstrom *m*
**wobbler** Kleeblatt *n*, (mach) Kuppelzapfen *m*, (mach) Kupplungszapfen *m*, Taumelscheibe *f*, Treffer *m*, Verschleißzapfen *m*, Wobbler *m* (electr.), Zapfenhohlkehle *f* (mach)
**wobbling** Taumelsendung *f*, Wobbeln *n* ~ **condensor** Heulkondensator *m* ~ **frequency** Heultonfrequenz *f*, überlagerte niederfrequentierte Schwingung *f*, Wobbelfrequenz *f*
**wobbulated** gewobbelt (rdr)
**wobbulation** Frequenzschwankung *f*
**Wofatite filter** Wofatitfilter *n*
**Wöhler,** ~ **diagramm** Wöhlerschaubild *n* ~ **method** Wöhler-Verfahren *n*
**wolf** Wolf *m* ~ **furnace** Wolfsofen *m*
**wolfram blue** Wolframblau *n*
**wolframic acid** Wolframsäure *f*
**wolframite** Wolfram *n* & *m*, Wolframit *m*
**Wollaston,** ~ **prism** Wollastonprisma *n* ~ **wire** Haardraht *m*, Wollastondraht *m*
**wollastonite** Kieselkalkspat *m*, Wollastonit *m*
**Woltmann's sailwheel** Woltmannscher Flügel *m*
**womp** intensiver Lichtfleck *m*
**wood** Baunutzholz *n*, Holz *n*, Wald *m*, Waldung *f* ~ **of acacia** Akazienholz *n* **the** ~ **is casting** das Holz wirft sich ~ **with crooked fibers** widerwüchsiges Holz *n* ~ **having fissures** eisklüftiges Holz *n* **the** ~ **is warping** das Holz wirft sich
**wood,** ~ **alcohol** Holzgeist *m*, Methylalkohol *m* ~ **barrel** Holzfaß *n* ~ **beam** Holzbalken *m* ~ **(-boring) beetle** Holzbohrkäfer *m* ~ **bit** Holzbohrer *m*
**wood-block** Holz-block *m*, -klotz *m*, -schnitt *m*, Pflasterklotz *m* ~ **brake** Holzklotzbremse *f* ~ **paving** Holz-pfeiler *m*, -pflaster *m* ~ **printing** Holztafeldruck *m* ~ **tire** Holzblockreifen *m*
**wood,** ~ **board** Maschinenholzkarton *m* ~ **board flap** Brettertür *f* ~ **border** Holzleiste *f* ~**-boring** Holzbohrarbeit *f* ~**-boring beetle** Bohrkäfer *m* ~**-boring drill** Holzbohrer *m* ~ **box** Holzkasten *m* ~ **bulkhead** Holzquerspant *n* ~ **burytype** Metalldruckverfahren *n* ~ **capping** hölzernes Paßstück *n* ~ **carving** Holzbildhauerei *f* ~ **carving knife** Holzschraubenmesser *n* ~ **casing** Holzmantel *m* ~ **cellulose** Holz-stoff *m*, -zellulose *f* ~ **cement** Kaltleim *m* ~ **center** Holzsteg

*m* ~ **chip board** Holzspanplatte *f* ~ **chips** Holzspäne *pl* ~ **chisel** Holzmeißel *m* ~ **chopper** Gerbholzraspelmaschine *f* ~ **cleaver's ax** Schrotaxt *f* **(natural)** ~ **color** Holzfarbe *f* ~ **construction** Holzbauweise *f* ~ **consuming gas generator** Holzgasanlage *f* ~ **copper** Holzkupfererz *n* ~ **cover** Mantelholz *n* ~ **covering** Holzbelag *m* ~ **creosote** Holzkohlenkreosot *n* ~ **culvert** Holzrinne *f* ~ **cut** Holzschnitt *m* ~ **cutter** Holzhauer *m* ~ **cutting** Holzschlag *m* ~ **distilling apparatus** Holzvergaser *m* ~ **dust** Holzmehl *n* ~ **dye** Holzfarbstoff *m* ~ **engraver** Holzschneider ~ **engraver's art** Holzschneidekunst *f* ~ **engraving** gravierter Holzstock *m*, Holzschnitt *m*, Holzstecherkunst *f* ~ **felling** Holzabrieb *m*, Holzschlag *m* ~ **fiber** Holzfaser *f*, Holzwolle *f*, Zellulose *f*, ~ **fiber building** Holzfaserplatte *f* ~ **fiber hardboard** Holzfaser-Hartplatte *f* ~ **fiber sheet** Hartfaserplatte *f* ~ **filling** Holzfüllstück *n* ~ **flawy on one side** einwüchsiges Holz *n* ~ **flour** Holzmehl *n* ~ **fore planes** Langhobel *pl* ~ **frame** Holzbalken *m* ~ **gas** Holztreibgas *n* ~ **gas engine** Holzgasmotor *m*, Treibgasmotor *m* ~ **gas generator** Holzkohlengasgenerator *m* ~ **gas producer** Holzgasgenerator *m* ~ **granted free of charge** Abgabeholz *n*, Deputatholz *n* ~ **grinder** Holzschleifer *m* ~ **heavier than water** Sinkholz *n* ~ **hurdle** Holzborde *f* ~ **impregnation** Holzimprägnierung *f* ~ **insulation** Holzisolierung *f* ~ **lagging** Belagbrettchen *n* ~ **lamina** Holzlage *f* ~ **land** Holzboden *m*, Wald *m*, Waldgelände *n* ~ **lath** Holz-latte *f*, -leiste *f* ~ **lattice mast** Holzgittermast *m* ~ **layer** Holzlage ~ **man** Holzhauer *m* ~ **man's hammer** Waldhammer *m* ~ **material** Leichtbauplatte *f* ~ **milling cutter** Holzfräser *m* ~ **milling cutter knife** Holzfräsermesser *n* ~ **molder** Leistenmacher *m* ~ **molding** Holzleiste *f*, Holzsims *m* ~ **oil** Holzöl *m* ~ **oil finishing varnish** Holzölüberzugslack *m* ~ **oil rubbing varnish** Holzölschleiflack *m* ~ **packing** Holzbeilage *f* ~ **packing (or filling) piece** Futterholz *n* ~ **panel** Holztafel *f* ~ **paneling** Holztäfelung *f* ~ **pattern** Holzmodell *n* ~ **pattern maker** Modell-schreiner *m*, -tischler *m* ~ **peeling machine** Furnierschälmaschine *f*, Holzschälmaschine *f* ~ **pile** Holz-pfahl *m*, -stoß *m* ~ **piling** Holzverschalung *f* zum Absteifen eines Kabelgrabens, Längsbrett *n* ~ **pipe (of organ)** Holzpfeife *f* ~ **pitch** Holzpech *n* ~ **planing machine** Holzhobelmaschine *f* ~ **powder** Holzmehl *n* ~ **preservation oil** Impägnierungsöl *n* für Holz ~ **preservative** Holzkonservierungsmittel *n* ~ **primer** Holzgrundierfarbe *f* ~ **printing press** Holzbedrukkungsmaschine *f* ~ **pulley** Holzriemenscheibe *f*
**wood-pulp** Holzstoff *m*, Zellstoff *m*, Zellulose *f* ~ **board** Holz-karton *m*, -pappe *f* ~ **making machine** Zelluloseherstellungsmaschine *f* ~ **paper** Holz-, Zellulose-papier *n* ~ **works** Zellulosefabrik *f*
**wood,** ~ **rasp** Holzraspel ~ **resin** Baum-, Holzharz *n* ~ **revetment** Holzverkleidung *f* ~ **saw** Baum-, Holz-, Wald-säge *f* ~ **saw blade** Spannsägeblatt *n* ~ **scaffold** Holzgerüst *n* ~ **scraps (or waste wood)** Abfallholz *n* ~ **screw** Holzschraube *f* ~ **screwknife** Holzschraubenmesser *n* ~ **screw thread** Holzschraubengewinde *n* ~ **sculptor** Holzschnitzer *m* ~ **separator** Holzschneider *m*

~ **separator in storage battery** Holzzwischen-
lage *f* ~ **shaving basket** Spankorb *m* ~ **shavings**
Holzwolle *f* ~ **shoring** Holzverschalung *f* zum
Absteifen eines Kabelgrabens, Längsbrett *n* ~
**slat conveyer** Holzplattenbandförderer *m* ~
**spirit oil** Holzgeistöl *n* ~ **spoked wheel** Holz-
speichenrad *n* ~ **stack** Holzstoß *m* ~ **stain**
Holz-beize *f*, -farbe *f* ~ **steel construction** Holz-
eisenkonstruktion *f* ~ **stereotype** Inkunabel *f* ~
**stock** Holzschnitt *m* ~ **stores** Holzlager *n* ~
**structure** Holzstruktur *f* ~ **sugar** Holzzucker *m*
~ **tank** Holzkasten *m* ~ **tannin** Holzgerbstoff *m*
~ **tar** Holzkohlenteer *m*, Holzteer *m* ~ **thimble**
Holzkausche *f* ~ **timbering sunk in a shaft** Senk-
zimmerung *f* ~ **tin** Holzzinn *n* ~ **tooth bevel**
**gear** Kegelrad *n* mit Holzverzahnung ~ **trans-**
**port** Holzabfuhr *f* ~ **trimmer** Bestoßmaschine *f*
~ **trough** Holzrinne *f* ~ **turning chisel** Holz-
handdrehstahl *m* ~ **turning lathe** Holzdreh-
bank *f* ~ **type** Holzbuchstabe *m* ~ **varnish**
Holzlack *m* ~ **waste** Holzabfälle *pl* ~ **waste**
**panel** Holzspannplatte *f* ~ **wind instrument**
Holzblasinstrument *n* ~ **window blind** Holzja-
lousie *f* ~ **wool** Holzwolle *f* ~ **wool filter** Holz-
wollefilter *m* ~ **work** Holzarbeit *f*, Holzwerk *n*
~ **working**. Holzbearbeitung *f* ~ **working**
**machine** Holzbearbeitungsmaschine *f* ~ **working**
**shop** Holzbearbeitungswerkstatt *f* ~ **working**
**tool** Holzbearbeitungswerkzeug *n* ~ **worm**
Bohr-, Holz-wurm *m*
**wooded** bewaldet ~ **area** Waldgelände *n*
**wooden** hölzern ~ **adjustable-pitch propeller** einstellbare Holzschraube *f* ~ **aircraft** Holzflug-
zeug *n* ~ **air duct** Holzlutte *f* ~ **airscrew with**
**metal-sheathed leading edge** Holzschraube *f* mit
metallgeschützten Kanten ~ **apron** Sohlenverkleidung *f* von Holz ~ **baffle** Holzgehäuse *n* ~
**base** Holzfuß *m* ~ **bay** Holzjoch *n* ~ **beading**
**material** Holzstabgewebe *n* ~ **beam floor** Holz-
balkendecke *f* ~ **block** (for polishing glassware)
Bügelholz *n*, Holzblock *m* ~ **block flooring**
Holzklotzpflaster *n* ~ **block paving** Holzklotzpflaster *n* ~ **board** Holzbrettchen *n* ~ **box** Holz-
kiste *f* ~ **box-mine** Holzmine *f* ~ **brake shoe**
Bremsholz *n* ~ **building** Holzbau *m* ~ **cabinet**
Holzkasten *m* ~ **cage** Gitterkorb *m* ~ **card lag**
Kardenbrettchen *n* ~ **case** Holzetui *n* ~ **center**
Holzdorn *m* ~ **checker** Hordenwascher *m* ~
**clamp** Hirnleiste *f* ~ **culvert** Drumme *f* ~ **disc**
Holzscheibe *f* ~ **dowel** Holzdübel *m* ~ **flange**
Holzkranz *m* ~ **float** Holzschwimmer *m* ~ **foil**
Holzfolie *f* ~ **form** hölzerne Schale *f* ~ **frame**
Holzgerüst *n*, Holzrahmen *m*, Vorderrahmen
*m* ~ **framework** hölzerne Unterzüge *pl* (beim
Stapeln von Stangen) ~ **framework erected to**
**support poles during renewals** Stützbock *m* (für
den Einbau von Stangenfüßen) ~ **framework**
**fuselage** Fachwerkrumpf *m* ~ **funnel** Holztrichter *m* ~ **furniture** Holzstege *f* (print) ~
**fuselage** Holzrumpf *m* ~ **girder** Holzträger *m* ~
**grates** Lattenrost *m* ~ **grating** Holzrost *m* ~
**grid** Holzgitter *n* ~ **gauge** Holzlehre *f* ~ **hood**
Holzhaube *f* ~ **hub** Holznabe *f* ~ **knife fork and**
**spoon** Holzbesteck *n* ~ **lattice** Holzgitter *n* ~
**ledge** Holzleiste *f* ~ **lid** Holzdeckel *m* ~ **lining**
Holzbelag *m*, hölzerne Vertonung *f* ~ **mast**
Holzmast *m* ~ **mitering gate** hölzernes Stemmtor *n* ~ **monocoque fuselage** Holzschalenrumpf

*m* ~ **panel** Holzquerspant *n* ~ **partition** Holz-
wand *f* ~ **parts** Holzbestandteile *pl* ~ **peg** Holz-
pfosten *m*, -ständer *m* ~ **pin** Holzdübel *m*, hölzerner Hagel *m*, Pflock *m* ~ **planking** hölzerner
Verschlag *m* ~ **plate** Holzplatte *f* ~ **plateholder**
Holzkassette *f* ~ **plug** Holzdeckel *m* ~ **pole**
Holzstange *f* ~ **projectile** Holzgeschoß *n* ~
**propeller** hölzerne Luftschraube *f*, Holz(luft)-
schraube *f* ~ **rack** Holz-gestell *m*, -verschlag *m*
~ **reglet** Holzleiste *f* ~ **reinforcement** Stützpfahl
*m* ~ **rib** Holzrippe *f* ~ **rim sieve** Holzrandsieb *n*
~ **ring** Holzkranz *m* ~ **shanks** Holzgelenke *pl*
~ **shed** Bretterschuppen *m* ~ **shell body** Holzwickelrumpf *m* ~ **shield** Holzschutz *m* ~ **shoe**
Holzschuh *m* ~ **slat** Holzstab *m* ~ **spar** Holz-
holm *m* ~ **stake** Holz-pfosten *m*, -ständer *m* ~
**stay (or strut)** Esel *m* ~ **sump box** Holzkasten *m*
zum Sumpfen ~ **support** Grubenholzausbau *m*,
Holzträger *m* ~ **tank** Holzwanne *f* ~ **target**
Holzscheibe *f* ~ **tower** Holzturm *m* ~ **tripod**
Holzstativ *n* ~ **tube** Holzrohr *n* ~ **type** Holz-
schrift *f* ~ **utensils** Holzgeräte *pl* ~ **waterproof-**
**ing** hölzerne Verspundung *f* ~ **wedge** Holz-
birne *f* ~ **wheel** Holzrand *n* ~ **wheel truck** Rollwagen *m* mit Holzrädern ~ **winch** Holzhaspel *f*
~ **window jamb** Fensterpfosten *m* ~ **wing** Holz-
flügel *m*
**Woodruff,** ~ **key** Scheiben-feder *f*, -keil *m* ~
**keyseat cutter** Schlitzfräser *m* für Scheibenfeder-
nuten ~ **key-shaped** scheibenkeilartig ~ **key slot**
**(or groove)** Scheibenkeilnut *f*
**Wood's alloy** Woodmetall *n*, Woodsches Metall
*n*
**woody** hölzern, holzförmig, holzig ~ **brown cut**
'Lignit *m* ~ **fiber** Längenfaser *f* im Holz ~
**fracture** Holzbruch *m*
**woof** Einschlag *m*, Einschuß *m*, (cloth) Schuß *m*,
(in cloth) Schußgarn *n*
**woofer** Konusmembran *f*, Tiefenkonus *m*, (Lautsprecher), Tieftoneinheit *f*, Tieftonlautsprecher
*m*
**wool** Wolle *f* ~ **blanket** Wolldecke *f* ~ **bleacher**
Wollbleicher *m* ~ **card** Wollkratze *f* ~ **dressing**
**machine** Wollaufbereitungsmaschine *f* ~ **felt**
Wollfilz *m* ~ **grease extracting plant** Wollfettgewinnungsanlage *f* ~ **imitation staple fiber**
Kräuselfaser *f* ~ **oil** Schmalzöl *n* ~ **picking** Entkletten *n* der Wolle ~ **picking machine** Wollzupfmaschine *f* ~ **scouring** Entschweißung *f* der
Wolle ~ **sorting** Lesen *n* der Wolle ~ **supplier**
Wollaufleger *m* ~ **utilizing plant** Wolleverwer-
tungsanlage *f* ~ **waste** Wollabfall *m* ~ **wax**
**alcohol** Wollwachsalkohol *m* ~ **yarn grease**
Wollgarnfett *n* ~ **yolk** Wollschweiß *m*
**woolen** Streichgarn *n*, Wollstoff *m* ~ **bolting cloth**
Wollbeuteltuch *n* ~ **cloth** Wollzeug *n* ~ **felt**
**screen** Wollfilzschirm *m* ~ **goods** Streichgarn-
gewebe *n* ~ **industry** Wollindustrie *f* ~ **satin**
Wollatlas *m*
**Woolff's bottle** Woolff'sche Flasche *f*
**wooliness** übermäßiger Nachhall *m* (acoust)
**wooly** weich, wollartig, wollig
**wootz steel** Wootzstahl *m*
**word, to** ~ (Spruch) abfassen, aufsetzen; 
(greases) walken
**word** Befehl *m*, Nachricht *f*, Zahlen-, Zeichen-
gruppe *f* (comput) ~ **in parenthesis** Einschaltung *f*

**word, ~ counter** Wortzwischenraum *m* **~ decoding** Wortentschlüsselung *f* **~ driver** Worttreiber *m* **~ equipment** Wortanlage *f* **~ field limit** Wortfeldgrenze *f* **~ intelligibility** Wortverständlichkeit *f* **~ line** Wortleitung *f* **~ mark** Wortmarke *f* **~ organized memory** Speicher *m* mit Wortadresse **~ selection system** Wortaufrufsystem *n* **~ time** Wortlaufzeit *f* (comput) **~ trade mark** Wortzeichen *n*

**wording** Abfassung *f*, (of document) Fassung *f*, Wortlaut *m*

**work, to ~** abbauen, arbeiten, aufgehen, ausbeuten, bauen, bearbeiten, bedienen, behandeln, betreiben, in Betrieb sein, bilden, funktionieren, im Gang sein, hantieren, hereingewinnen, verarbeiten, verfahren, verformen, wirken **to (re) ~** aufarbeiten

**work, to ~ against** entgegen-arbeiten, -wirken **to ~ cold** (blast furnace) kalt gehen **to ~ by contract** im Akkord arbeiten **to ~ correctly** einwandfrei arbeiten **to ~ to the dip** abhauen **to ~ down** herabarbeiten, herunterfrischen, (liquids) nach unten durchsickern **to ~ hard** schuften **to ~ highly** hochtreiben **to ~ a hoist** Fördermaschine *f* bedienen **to ~ in(to)** einarbeiten **to ~ loose** lockern **to ~ with a material** einen Werkstoff *m* verarbeiten **to ~ a mine** eine Grube *f* abbauen, einschlagen **to ~ in niello** niellieren **to ~ off** abarbeiten, abdrucken **to ~ onto** abarbeiten **to ~ in open circuit** mit Arbeitsstrom *m* betreiben **to ~ with open exhaust** mit Auspuff *m* arbeiten **to ~ in opposition** gegenarbeiten **to ~ out** ausarbeiten, ausstoßen, herausarbeiten, sich auswirken **to ~ over** umarbeiten **to ~ in paste** pappen **to ~ at Pattinson's method** pattinsonieren **to ~ by piece** im Akkord arbeiten **to ~ at red heat** warmarbeiten **to ~ out a shift** Schichtmachen *n* (min) **to ~ on short circuit** im Kurzschluß *m* arbeiten **to ~ out the task** die Schicht verfahren **to ~ through** durcharbeiten **to ~ together** mitarbeiten **to ~ unsteadily** (engine) unruhig arbeiten **to ~ up** (barograph) aufarbeiten, aufbereiten, verbrauchen **to ~ for volume** auf Menge *f* arbeiten

**work** Arbeit *f*, Kraft *f*, Lauf *m*, Leistung *f*, Merkmal *n*, Werk *n*, Werkstück *n* **~ (ing)** Betrieb *m*

**work, ~ of braking** Brems-arbeit *f*, -energie *f* **~ of cohesion** Kohäsionsarbeit *f* **~ of compression** Verdichtungsarbeit *f* **~ by contract** Verdingarbeit *f* **~ by the day** Arbeit *f* in Schichtlohn, Tagewerk *n* **~ of defence** Schutzwerk *n* **~ of deformation** Formänderungsarbeit *f* **~ of elastic strain cushioning action** Federungsarbeit *f* **~ is interrupted** die Arbeit des Werkes ist unterbrochen **~ in phase opposition** Gegentaktarbeiten *m* **~ of protection** Schutzwerk *n* **~ of throttling** Drosselungsarbeit *f* **~ of upsetting** Staucharbeit *f* **~ (ing) of a vein** Flözabbau *m* (min) **~ to be welded** Schweißgut *n*

**work, ~ bar** (textiles) Scheuerblech *n* **~ bench** Werkbank *f* **~ bench hand** Werkbankschreiner *m* **~ center** Arbeitsmilieu *n* **~ chute** Werkstückrutsche *f* **~ clothing** Arbeits-anzug *m*, -kittel *m*, -kleidung *f* **~ connection** Wirkverbindung *f* **~ cycle** Ablauf *m* **~ day** Arbeitstag *m* **~ done** Leistung *f* **~ done in deformation** Verformungsarbeit *f* **~ done under full pressure** Volldruck-

**arbeit** *f* **~ driving arm and pin** Mitnehmer *m* für das Arbeitsstück **~ driving dog** Mitnehmerherze *m* für die Arbeitsstücke **~ due to friction** Reibungsarbeit *f* **~ fixture setting angles for spiral grinding work** Einstellwinkel *pl* des Werkstückträgers beim Spiralschleifen **~ function** Austrittsarbeit *f* (electr.) **~ function of electrons** Elektronenaustrittsarbeit *f* **~ hardening** Verfestigung *f* **~ hardening curve** Verfestigungskurve *f* **~ head** Aufspann-kopf *m*, -spindelstock *m* **~ hindering** leistungshemmend **~ holding arbor** Aufspanndorn *m* **~ holding device** Spannteil *n* **~ holding equipment** Werkstückaufnahme *f* **~ holding spindle** Aufspannspindel *f* **~ input** hineingesteckte Arbeitsleistung *f* **~ lead** Werkblei *n* **~ locating fixture** Spannvorrichtung *f*

**workman** Arbeiter *m*, Arbeitskraft *f*, Gewerke *m*, Handwerker *m* **~-like** fachgemäß, handwerksmäßig **~ ship** Ausführung *f*, Bearbeitung *f*, Bearbeitungsgüte *f*, Kunstfertigkeit *f*, Werkstattarbeit *f* **~'s side of a double reverberatory furnace** hintere Seite *f* eines Doppelflammofens

**work, ~ mounting** Werkstückaufspannung *f* **~ number** Werknummer *f* **~ order** Werkstattauftrag *m* **~ performed by dyes during flocculation** Ausflockbarkeit *f* der Farbstoffe **~ performed during expansion** Ausdehnungsarbeit *f* **~ piece** Arbeitsstück *n*, Werkstück *n* **~ piece holder** Werkstückauflage *f* **~ plan** Arbeitsplan *m* **~ planner** Arbeitsvorbereiter *m* **~ progress slip** Laufzettel *m* **~'s representative** Werkvertreter *m* **~ (ing) roll** Arbeitswalze *f* **~ sheet** Kalkulationszettel *m* **~ shift** Arbeitsmilieu *n*

**workshop** Betrieb *m*, Betriebswerkstatt *f*, Fertigungshalle *f*, Werft *f*, Werkstatt *f*, Werkstätte *f*, Werkstelle *f* **~ for industrial art** kunstgewerbliche Werkstätte *f* **~ for numbering** Numerieranstalt *f* **~ for stucco** Stuckwerkstatt *f*

**workshop hall** Werk-statt *f*, -halle *f*

**work, ~ slide** Werkstückschlitten *m* **~ softening** Verformungsentfestigung *f* **~ space** Nutzraum *m* **~ spindle** Werkstoffspindel *f* **~ squad** Arbeitstrupp *m* **~ stoppage** Betriebsumstellung *f* **~ stroke** Explosionstakt *m* **~ study** Arbeitsstudie *f* **~ table** Arbeitstisch *m* **~ toughening** Hochtrainieren *n* **~ tray** Werkstückkasten *m*

**workability** Bearbeitbarkeit *f*, Bearbeitungsgrad *f*, Formbarkeit *f*, Verarbeitbarkeit *f*, Verarbeitungsfähigkeit *f*

**workable** abbauwürdig, baufähig, bauwürdig, bearbeitbar, behandelbar, betriebs-fähig, -mässig, verarbeitbar, verformbar **~ in cold state** kalt verformbar **~ deposits** bauwürdiges Vorkommen *n* **~ ore field** abbauwürdiges Erzvorkommen *n*

**worked** bearbeitet, durchgearbeitet **~ by hand** handbetätigt **~ and other tests for greases** Walkpenetration *f* **~ out** abgebaut **~ by the syndicate** ausgenutzt von der Gesellschaft **~ up** aufgearbeitet

**worker** Arbeiter, Arbeitskraft *f*, Halbholländer *m* (paper mgf)

**workers** Belegschaft *f*

**working** Abbau *m* (min), Arbeiten *n*, Ausbeutung *f*, Baufeld *n*, Bearbeitung *f*, (of a machine) Gang *m*, Geschäftsgang *m*, Recken *n*, (of slag) Schlackenarbeit *f*, Verarbeitung *f*, Verformung *f*; betrieblich, im Betrieb **to be ~** im Gang sein **~ of a bed** Flözbau *m* **~ ba cart** Karrenförderung *f* **~ on dry compression** Überhitzungsverfahren *n* **~ of a furnace** Gang *m* eines Ofens **~ the heat** Schmelzführung *f* **~ of heat** Ofenführung *f* **~ without load** leergehend, leerlaufend **~ of lodes** Gangbergbau *m* **~ of mines** Bergbau *m* **~ by overhand stopes** Leistenbau *m* **~ of a patent** Benutzung *f* eines Patentes **~ in roof** Überbrechen *n* (min) **~ of a seam** Abbau *m* eines Flözes **~ of the switches** Umlegen *n* der Weichen **~ of trenches** Schrämarbeit *f* **~ of a vein** Abbau *m* eines Flözes

**working** Abbau *m* (min), Arbeiten, Be- und Verarbeitung *f*, Fermentieren *n*, Gärung *f* **~ aisle** Arbeitslaufgang *m* **~ angle** Ausladung *f* **~ aperture** Blendenöffnung *f* (film) **~ area lighting fixtures** Werkplatzleuchten *pl* **~ back pressure** Betriebsgegendruck *m* **~ barge** Boot *n* **~ barrel** Kolbenrohr *n*, Pumpengehäuse *n*, Tiefpumpenzylinder *m* **~ barrel pump** Kolbenrohrpumpe *f* **~ barrel valve** Kolbenrohr **~ batch** (of furnace) Nutzeinsatz *m* **~ blade** (Kompressor) Arbeitsschieber *m*, Flügelblende *f* (film) **~ body (or surface) of piston** Kolbenfläche *f* **~ cam** Arbeitsnocke *f* **~ capacity** Arbeits-leistung *f*, -vermögen *n* **~ capital** Betriebskapital *n* **~ chamber** Arbeitskammer *f* **~ characteristic** Arbeitskennlinie *f* **~ circuit** betriebsfähige Leitung *f* **~ clearance** Betriebsspiel *n* (der Maschinenteile) **~ coil** Betätigungsspule *f* **~ condition** Arbeitsbedingung *f*, betriebsfähiger Zustand *m*, Betriebszustand *m* **~ condition(s)** Betriebsbedingungen *pl* **~ contents** Nutzinhalt *m* **~ control** Gangkontrolle *f* **~ copy** Gebrauchskopie *f* **~ costs** Bedienungskosten *pl*, Betriebskosten *pl* **~ current** Arbeits-, Betriebs-strom *m* **~ curve** Arbeitskennlinie *f* **~ cycle** Arbeits-gang *m*, -takt *m*, Spiel *n* **~ cylinder** Arbeitszylinder *m* **~ data** Betriebsdaten *pl* **~ day** Arbeits-, Werk-tag *m* **~ depth** Angriffshöhe *f*, Eingriffsstrecke *f*, Meßtiefe *f* **~ depth of tooth** wirksame Zahnhöhe *f*, Zahnangriffstiefe *f* **~ distance** Arbeitsabstand *m* **~ door** Arbeitstür *f*, (of a pudding furnace) Schummelloch *n* **~ drawing** Konstruktions-, Werk(statt)-zeichnung *f* **~ dress** Arbeitszeug *n* **~ due** (Patent) Ausübung *f* fällig **~ edge** Hochkante *f* **~ efficiency** Nutzeffekt *m* **~ environment** Arbeitsmilieu *n* **~ equation** Fehlergleichung *f* **~ face** Angriffsstelle *f*, Stollenort *m* **~ face of the tooth of a gear** Eingriffsfläche *f* eines Zahnradzahnes **~ field** Abbaufeld *n* **~ floor** Arbeits-ebene *f*, -flur *m* **~ fluid** Arbeitsflüssigkeit *f*, (vapor pump) Treibmittel *n* **~ frame** Schütztafelrahmen *m* **~ gauge** Arbeitslehre *f* **~ gear** Bewegungs-mechanismus *m*, -vorrichtung *f*, Laufwerk *n* **~ hours** Arbeitsstunden *pl*, Betriebszeit *f*, Betriebsstunden *pl* **~ impulse transmitted on voltage substantially above normal** Hochtastung *f* **~ instructions** Bedienungsvorschrift *f* **~ instrument** Betriebsapparat *m* **~ life** lebensdauer *f* **~ limit gauge** Grenzarbeitslehre *f* **~ line** Belastungskurve *f* **~ load** Arbeits-belastung *f*, -last *f*, Betriebsbelastung *f*, Drucklast *f*, Gebrauchslast *f*, Nutzlast *f*, Schwungmasse *f* **~ losses** Betriebsverluste *pl* **~ magnet** Arbeitselektromagnet *m* **~ material** Arbeitsunterlage *f*, Betriebsstoff *m* **~ medium**

Arbeitsstoff *m* **~ method** Abbau-, Arbeits-verfahren *n* **~ methods** Arbeitsablauf *m* **~ model** Arbeitsmodell *n*, Gebrauchsmuster *n* **~ motion** Schnittbewegung *f* **~ operation** Arbeitsgang *m* **~ order** Betriebs-fähigkeit *f*, -zustand *m* **in (good) ~ order** betriebsfähig **~ part** Triebwerksteil *m*, Werkstück *n* **~ parts** Arbeitsteile *pl* **~ path** Schnittweg *m* **~ piston** Arbeit(s)kolben *m* **~ pit** Göpelschacht *m* **~ place** Abbauort *m*, Arbeitsstelle *f* **~ plan** Werkzeichnung *f* **~ plane** Arbeitsfläche *f*, Meßebene *f* **~ platform** Arbeitsbühne *f* **~ play** Ablauf *m* (des Wagenspieles) **~ point** Angriffs-, Ruhe-punkt *m*, Schaltstufe *f* **~ point on characteristic curve of tube** Arbeitspunkt *m* **~ point of the valve characteristic** Arbeitslage *f* auf der Röhrenkennlinie **~ points** Fahrstufen *pl* **~ position** Arbeitsstellung *f* **~ power** Wirkungsvermögen *n* **~ pressure** Ab-blase-, Arbeits-, Betriebs-, Kessel-druck *m* **~ pressure gauge** Arbeitsmanometer *n* **~ procedure** Arbeitsgang *m* **~ process** Arbeits-gang *m*, -prozeß *m* (in engine), -verfahren *n* **~ promoting** leistungsfördernd **~ Q** Lastgüte *f* **~ qualities** Betriebseigenschaften *pl* **~ radius** Ausladung *f* **~ range** Betriebsbereich *m* von Motoren **~ records** Betriebsunterlagen *pl* **~ reference system** Arbeitseichkreis *m* **~ resistance** Arbeitswiderstand *f* **~ revolution** den Arbeitshub enthaltende Umdrehung *f* des Motors **~ roll** Arbeitswalze *f* **~ rollers drive gear** Knetwalzenantriebsrad *n* **~ scale** Arbeitsskala *f* **~ schedule** Arbeitsgang *m* **~ scheme** Arbeitsplan *m* **~ scope** Arbeitsbereich *m* **~ sequence** Arbeitsablauf *m* **~ service** Lebensdauer *f* **~ set** Betriebsapparat *m* **~ shaft** Zugstange *f* **~ space** (beam tube) Anfachraum *m*, Arbeitsraum *m* **~ speed** Arbeitsgeschwindigkeit *f*, (commercial) Betriebsgeschwindigkeit *f* **~ stage** Arbeitsgang *m* **~ standard** Arbeitseichkreis *m* **~ station** Arbeitsstation *f* **~ stem** Zugstange *f* **~ step** Arbeitsgang *m* **~ storage** Arbeitsspeicher *m* **~ strength** Arbeitsfestigkeit *f* **~ stress** Gebrauchs-, Nutz-spannung *f* **~ stroke** Arbeits-hub *m*, -takt *m* **~ surface** Angriffs-, Arbeits-fläche *f*, Ballenoberfläche *f*, (of cylinders) Gleitfläche *f* **~ surface of cylinder** Zylinderlauffläche *f* **~ table** Betriebstisch *m* **~ tanks** Pufferbehälter *m* **~ temperature** Betriebswärmegrad *m* **~ tension** Belastungsspannung *f*, Betriebsdruck *m* **~ time** Arbeitszeit *f*, Betriebsdauer *f* **~ tolerance** Fertigungstoleranz *f* **~ traverse** Arbeitsweg *m* **~ up** Aufarbeiten *n* **~ valve** Arbeitsventil *n*, Bedienungshahn *m* **~ voltage** Nutzspannung *f* **~ wave** Arbeitswelle *f* **~ wedge** Arbeitsquarzkeil *m* **~ width** nutzbare Breite *f*, Nutzbreite *f*

**works** Fabrik *f*, Werk *n*, Werkanlage *f* **~ for berthing** Anlandevorrichtung *f* **~ for protecting the strand** Seeuferbau *f*

**works, ~ certificate** Gütezeugnis *n* **~ chemist** Betriebschemiker *m* **~ council law** Betriebsverfassungsgesetz *n* **~ engineer** Betriebsingenieur *m* **~ inspector** Fertigungsprüfer *m* **~ management** Betriebsdirektion *f* **~ manager** Werksleiter *m* **~ number** Baunummer *f* **~ public-address system** Betriebslautsprecheranlage *f* **~ railway** Industrie-, Werk-bahn *f* **~ refractometer** Kesselrefraktometer *m* **~ standards** Fabriknormen *pl* **~ super-intendent** Betriebsdirektor *m* **~ test**

Betriebswerksprüfung *f*
**world** Welt *f* ~ **acceleration vector** Weltbeschleu-
nigungsvektor *m* ~ **air traffic** Weltluftverkehr
*m* ~ **commerce** Welthandel *m* ~ **invariant** welt-
invariant ~ **market** Weltmarkt *m* ~ **market
price** Weltmarktpreis *m* ~ **news association**
Weltnachrichtenverein *m* ~ **radio network**
Weltfunknetz *n* ~ **record** Welthöchstleistung *f*,
Weltrekord *m* ~ **tensor form** vierdimensionale
Tensorform *f* ~ **tensor representation** Weltten-
sor-Darstellung *f* ~ **trade and industry** Welt-
wirtschaft *f* ~ **traffic** Weltverkehr *m* ~ **velocity
field** Weltgeschwindigkeitsfeld *n* ~ **velocity
gradient** Weltgeschwindigkeitsgradient *m* ~
**vorticity tensor** Weltwirbeltensor *m* ~ **wide**
weltumspannend
**worm, to** ~ trensen
**worm** Gewinde *n*, Kühlschlange *f*, Raupe *f*,
Rohrschlange *f*, Schlange *f*, Schlangenrohr *n*,
Schnecke *f*, Schneckengewinde *n*, Schraube *f*,
Schraube *f* ohne Ende, Schrauben-gewinde *n*,
-nut *f*, -zahngetriebe *n*, Trense *f*, (gear) Winde *f*,
(endless screw) Wurm *m* ~ **and sector steering
mechanism** Schneckenlenkung *f* ~ **and sector
steering system** Segment-, Zahnkranz-lenkung *f*
~ **and wheel drive** Schneckentrieb *m*
**worm,** ~ **bearing** Lenkstocklager *n* ~ **blank**
Schneckenkörper *m* ~ **conveyer** Förderschnecke
*f*, Schneckenförderer *m*, Transportschnecke *f*
~ **cutter** Schneckenfräser *m* ~ **drive** Buchse *f*,
Schnecken(rad)getriebe *n*, Spindeltrieb *m* ~
**drive shaft** Schneckenantriebswelle *f* ~ **drive
starter** Schraubtriebanlasser *m* ~ **driven
quadrant** Schneckenquadrant *m* ~ **eaten** wurm-
stichig ~ **'s eye view** Ansicht *f* von unten, Frosch-
perspektive *f*
**worm-gear** Getriebschnecke *f*, Schnecken-ge-
triebe *n*, -rad *n*, -vorgelege *n*, Schrauben-ge-
triebe *n*, -rad *n*, Spindeltrieb *m*, Wurmgetriebe
*n*, Wurmrad *n* ~ **drive** Schnecken(an)trieb *m* ~
**hob** Schneckenradwälzfräser *m* ~ **housing**
Schneckengehäuse *n* ~ **parts** Schneckentrieb-
teile *pl* ~ **rear-axle drive** Schneckenhinterachs-
antrieb *m* ~ **reducers** Schneckenuntersetzungs-
getriebe *n* ~ **shaft** Schneckenradwelle *f* ~
**spindle** Gewindespindel *f* ~ **(steering)** Schnek-
kenlenkung *f* ~ **tooth** Schneckenradzahn *m*
**worm,** ~ **geared crane ladle** Krangießpfanne *f* mit
Schneckenradkippvorrichtung ~ **geared winch**
Schneckenwinde *f* ~ **gearing** Schnecken-getrie-
be *n*, -radgetriebe *n*, -trieb *m* ~ **generating hob**
Schneckenwälzfräser *m* ~ **grinding machine**
Schneckenschleifmaschine *f* ~ **housing** Schnek-
kenlager *n* ~ **piston principle** Schraubenkolben-
prinzip *n* ~ **reducer** Schneckenuntersetzungs-
getriebe *n* ~ **reduction gear** Schneckenreduzier-
getriebe *n* ~ **screw** Fänger *m*, Fanggerät *n*,
Schneckenschraube *f* ~ **shaft** Schnecken-achse
*f*, -welle *f* ~ **shaft spur gear** Schneckenwellen-
stirnrad *n* ~ **-shaped** schneckenförmig ~ **thread**
Schneckengewinde *n* ~ **thread milling cutter**
Schneckenfräser *m*
**worm-wheel** Flügel-, Schrauben-, Schnecken-,
Wurm-rad *n* ~ **of expansion** Expansions-
schnecke *f*
**worm-wheel,** ~ **axle** Schneckenradachse *f* ~
**cutter** Schneckenradfräser *m* ~ **cutting machine**
Schneckenradschneidemaschine *f* ~ **drive**

**Schneckentrieb** *m* ~ **generating** Schneckenrad-
erzeugung *f* ~ **generating machine** Schnecken-
radabwälzfräsmaschine *f* ~ **hob(ber)** Schneckenradfräser *m* ~ **lifting** (screw) Schneckenradhub-
schraube *f* ~ **roller bearing** Schraubenrad *m*
mit Lagerbüchse ~ **segment** Schneckenrad-
segment *n*
**worming** Trense *f* ~ **pair** Adernpaar *n* (zum Ausfüllen der Lücken der Kabelseele), Trensen-
adernpaar *n*, -doppelader *f*
**wormy surface** Wurmlöcher *pl* (min)
**worn** abgegriffen, abgetragen, schäbig, verschlissen ~ **contact** ausgefressener Kontakt *m* ~
**cylinder** unrunder Zylinder *m* ~ **hollow** ausgelaufen ~ **-out** abgenutzt, abgejagt, verbraucht,
verschlissen ~ **-out bearing** ausgelaufenes Gleitlager *n* ~ **-out die** aufgezogener Stein *m* ~ **-out
type** abgenutzte Schrift *f*
**worsted (yarn)** Kammgarn *n*
**wort** Würze *f* ~ **cooler** Würzekühler *m* ~ **drawing
apparatus** Ablaufapparat *m* für Würze, Würze-
ablaufapparat *m* ~ **masher** Maischapparat *m*
**worth** Wert *m* ~ **mining** abbauwürdig
**worthiness of being worked** Bauwürdigkeit *f* (min)
**worthless** geringhaltig, nichtswürdig, wertlos
**wound** gewickelt, gewunden ~ **on bobbins** gespult
~ **in duplicate** bifilargewickelt ~ **on mica** auf
Glimmer aufgewickelt ~ **off** abgewickelt ~ **to a
resistance of an ohm** auf ein Ohm gewickelt ~
**by shooting** anschießen ~ **in spools** gespult ~ **up**
(spring) aufgezogen
**wound,** ~ **capacitor of fixed value** Wickelblock-
kondensator *m* ~ **coil** Doppelwendel *n* ~ **filter
element** Wickelfiltereinsatz *m* ~ **hook** Wundhaken *m* ~ **rotor induction motor** Schleifringläufermotor *m*
**woven,** ~ **and knitted goods** Wirk- und Strickwaren *pl* ~ **glass inlay** Glasgewebeeinlage *f* ~
**goods** Webwaren *pl*, Wirkwaren *pl* ~ **hair** Haartuch *n* ~ **tissue from silk** Wirkstoff *m* aus Seide
~ **wire** gewobener Draht *m* ~ **wire ware** Drahtgewebe *n* ~ **wood fabric** Holzdrahtgewebe *n*
**wow and flutter** Gleichlaufschwankungen *pl*
(phono)
**wows** Geschwindigkeitsschwankungen *pl* (electr.)
**wrap, to** ~ bewickeln, (package) einschlagen,
umhüllen, umlappen, umlegen, umschlagen,
wickeln, winden **to** ~ **around** herumschlagen, umbinden, umschlingen, umwickeln **to** ~ **with cloth**
bombagieren **to** ~ **the joint** die Lötstelle umwickeln **to** ~ **with ribbon (or tape)** bandagieren
**to** ~ **round** bewickeln **to** ~ **up** aufwickeln, einpacken **to** ~ **up (or in)** einwickeln **to** ~ **upon**
aufwickeln
**wrap** (single filaments) Aufläufer *m*, Wickler *m*
~ **-around type** rundgewickelter Typ *m* ~ **width**
Streifenbreite *f*
**wrapped** eingehüllt, paketiert ~ **cable** bespon-
nenes Kabel *n* ~ **capacitor of fixed value**
Wickelblockkondensator *m* ~ **tube** gewickeltes
Rohr *n* ~ **wire** Manteldraht *m*
**wrapper** (newspaper) Kreuzband *n*, Umschlag *m*
~ **glueing machine** Streifbandumklebeautomat
*m*
**wrapping** Bewehrung *f*, Bewicklung *f*, Hülle *f*,
Packung *f*, Umhüllung *f*, Umlappung *f*, Umwicklung *f* ~ **of jute** Jutewicklung *f*
~ **machine** Einhänge-, Umschnürungs-maschine

*f* ~ **machinery** Einschlagmaschine *f* ~ **paper** Einschlag-, Einwickel-, Emballage-, Umschlagpapier *f* ~ **test** Wickelprobe *f*
**wreck, to** ~ abtragen, abwracken, zertrümmern
**wreck** Wrack *n* ~ **or other floating things in sea** Seetriften *pl* ~ **buoy** Wrackboje *f*, Wracktonne *f* ~ **signal** Wracksignal *n*
**wreckage** Trümmer *pl*
**wrecked** gescheitert, zerstört **to be** ~ scheitern
**wrecker** Abschleppwagen *m*, Bergungsschiff *n* ~ **salvage truck** Abschleppfahrzeug *n*
**wrecking** Abbruch *m*, Abwracken *n*, Abwrackung *f* ~ **bar** Brecheisen *n*, Brechstange *f*, Kistenöffner *m* ~ **car** Autoschlepp *m* ~ **company** Abbruchsunternehmen *n* ~ **section** Abschleppkommando *n* ~ **truck** Abschleppkraftwagen *m*
**wrench, to** ~ reißen, verdrehen, ziehen **to** ~ **out** auswinden
**wrench** Bolzenschlüssel *m*, Dorn *m*, Mutterschlüssel, Schlüssel *m*, Schraubenschlüssel *m*, (of kettle drum) Stimmschlüssel *m*, Verstauchung *f* ~ **with rim** Schlüsselkranz *m* ~ **for socket head cap srews** Imbusschlüssel *m*
**wrench,** ~ **board** Bohrbank *f* ~ **circle** Drehschlüsselkranz *m* ~ **hammer** Schlagschlüssel *m*
**wrenches** Windeisen *n*
**wrenching clamp** Renkung *f*, Schlüsselkopf *m*
**wrest,** ~ **pin** Stimmschlüssel *m* (acoust.), Stimmwirbel *m* ~ **plank** (piano) Stimmstock *m*
**wrestle, to** ~ ringen
**wretched** elend, kümmerlich, trübe
**wriggle, to** ~ schlängend hin- und herbewegen, sich schlängeln, sich winden
**wriggle instability** Torsionsunbeständigkeit *f*
**wring, to** ~ auswringen, drücken **to** ~ **out** wringen
**wringer** Auswindemaschine *f*, Wringer *m*, Wringmaschine *f*
**wringing** Drücken *n*, Pressen *n*, Wringen *n* ~ **distance** Ansprengdistanz *f* ~ **fit** Schiebesitz *m* ~ **layer** Ansprengschicht *f* ~ **pin** Wringholz *n* ~ **pole** Wring-eisen *n*, -pfahl *m* ~ **post** Wringpfahl *m* ~ **stick** Wring-holz *n*, -stock *m*
**wrinkle, to** ~ knittern, (sich) kräuseln, runzeln, zusammenschrumpfen
**wrinkle** Falte *f*, Furche *f*, Kniff *m*, Runzel *f* ~ **finish** Kräusel-, Runzel-, Schrumpf-lack *m* ~ **proofness** Bauschelastizität *f* ~ **varnish** Runzellack *m* ~ **washer** gewellte federnde Unterlegscheibe *f*, Riffelscheibe *f*
**wrinkled** faltig, runzelig
**wrinkling** Faltenbildung *f*, Faltung *f*, Kräuseln *n*, (color) Runzeln *n*, Verbeulung *f* ~ **of the paper** Rumpeln *n* des Papiers, Wellig-liegen *n*, -werden *n* ~ **limit** Beulgrenze *f*
**wrist** Handgelenk *n* ~ **compass** Armbandkompaß *m* ~ **pin** Anlenk-, Kolben-bolzen *m*, Zapfen *m* ~ **pin bearing** Kolbenbolzenlager *n* ~ **plate** Gelenkplatte *f* ~ **strap** Gelenkriemen *m* ~ **watch** Armbanduhr *f* ~ **watch strap** Uhrarmband *n*
**writ** Gerichtsbefehl *m*, Vorladung *f* ~ **of execution** Vollziehungsbefehl *m*, Zwangsvollstreckungsbefehl *m* ~ **of summons** Prozeßladung *f*
**write, to** ~ abfassen, schreiben **to** ~ **down** an-,

auf-, nieder-schreiben **to** ~ **in** einschreiben **to** ~ **up a message** ein Telegramm *n* aufnehmen **to** ~ **off** abbuchen, gutbringen **to** ~ **off as a loss** als Verlust *m* abschreiben **to** ~ **on** (something) beschriften **to** ~ **out** schreiben (data proc) **to** ~ **shorthand** stenografieren **to** ~ **up** abschreiben
**write,** ~ **head** Schreibkopf *m* ~**-off** Tilgung *f* ~**-off for depreciations** Abschreibung *f* ~ **pulse** Schreibimpuls *m* ~ **transient** Einschwingen *n* des Schreibimpulses ~ **winding** Schreibewicklung *f*
**writer** Farbschreiber *m*, Schreiber, Schreibsteller *m*, Urheber *m*, Verfasser *m*
**writing** Schreiben *n*, Schrift *f*; brieflich, schriftlich ~ **bar** Schreibstange *f* ~ **case** Briefmappe *f* ~ **desk** Schreibpult *n* ~ **head** Schreibkopf *m* ~ **lever** Schreibgestänge *n* ~ **materials** Schreibzeug *n* ~ **mechanism** Schreibapparat *m* ~**-out of a report** Ausfertigung *f* ~ **pad** Schreibblock *m*, Schreibunterlage *f* ~ **panel** Schreibfläche *f* ~ **shelf** Schreibpult *n* ~ **slate** Schreibtafel *f* ~ **speed** Schreibgeschwindigkeit *f* ~ **surface** Schreibfläche *f* ~ **table** Schreibtisch *m* ~ **tablet** Schreibtafel *f* ~ **timer** Schreibtaktgeber *m* ~**-up** Abschreiben *n*
**written** schriftlich ~ **evidence** Beweismaterial *n* ~ **guide** schriftliche Anleitung *f* ~ **off** abgesetzt, abgeschrieben ~**-out program** abgeschriebenes Programm *n* ~ **proceedings** schriftliches Verfahren *n* ~ **record** Niederschrift *f* ~ **reply** Gegenschrift *f*
**wrong** falsch, unrichtig ~ **connection** falsche Verbindung *f*, Verschaltung *f* ~ **connection indicating relay** Fehlbedienungsrelais *n* ~ **manipulation** Fehlgriff *m* ~ **number call** Falschanruf *m* ~ **reaction** Fehlreaktion *f* ~ **setting** Fehleinstellung *f* ~ **way-round insertion** falsches Einstecken *n* (Stecker) ~ **work** Fehlarbeit *f*
**wrought** gehämmert, geschmiedet ~ **alloy** Knet-, Reck-legierung *f* ~ **aluminum** geknetetes Aluminium *n*
**wrought-iron** Gerbeisen *n*, geschmiedetes Eisen *n*, Luppeneisen *n*, schmiedbares Eisen *n*, Schmiedeeisen *n*, (for rolling) Schweiß-eisen *n*, -metall *n*, -stahl *n*; schmiedeeisern ~ **bar** Schweißeisenstab *m* ~ **four-pointed knife rest** schmiedeeiserner Vierspitz *m* ~ **lip union** Lippenkupplung *f* ~ **pipe** schmiedeeisernes Rohr *n* ~ **plate** Schweißeisenblech *n* ~ **scrap** Schmiedeeisenschrott *m*
**wrought,** ~ **nail** Schmiedenagel *m* ~ **steel** Gerbstahl *m*, schmiedbares Eisen *n*, Schmiedeeisen *n*, Schmiedestahl *m*, Schweißstahl *m*, warmverformter Flußstahl *m* ~ **steel pole** Stahlrohrmast *m* ~ **steel shoe** schmiedeeiserner Pfahlschuh *m*
**wryness** Schiefheit *f*
**W-shape** W-Form *f*
**W-type engine** Fächermotor *m*
**Wulf electrometer** Bifilarelektrometer *n*
**wulfenite** Gelbbleierz *n*, Molybdänbleispat *m*, Wulfenit *m*
**Wunderlich valve** Wunderlich-Röhre *f*
**wurtzite structure** Wurtzit-struktur *f*, -typ *m*

# X

xanthate Xanthat *n*, Xahnthogenat *n*
xanthein Xanthein *n*
xanthene Xanthen *n*
xanthic oxide Xanthogenoxyd *n*
xanthite Xanthit *m*
xanthogallolic acid Xanthogallosäure *f*
xanthogen Xanthogen *n*
xanthogenate Xanthogenat *n*
xanthogenic acid Xanthogensäure *f*
xanthone coloring matter Xanthonfarbstoff *m*
xanthosiderite Gelbeisenstein *m*
x-axis Abzissenachse *f*, X-Achse *f*, Zeitachse *f*
x-brace kreuzförmige Versteifung *f*
X-cut- X-geschnitten
xenolite Xenolith *m*
xenomorphic allotriomorph
xenon Xenon *n* ~ bubble chamber Xenonblasen-
kammer *f*
xenotime Xenotim *m*, Ytterspat *m*
X operation X-Betrieb *m*
X particles schwere Elektronen *pl*
X plates Horizontalablenkplatten *pl*
X punch X-Lochung *f*
X radiation Röntgenstrahlung *f*
X-ray, to ~ röntgen
X-ray Röntgen-licht *n*, -strahl *m*, X-Strahl *m* ~
absorption Röntgenabsorptionskontinua *pl* ~
absorption spectra Röntgenabsorptionsspek-
tren *pl* ~ analysis Röntgen(spektral)analyse *f*,
Untersuchung *f* mittels Röntgenstrahlen *pl* ~
apparatus Röntgenapparat *m* ~ beam Röntgen-
strahlenbündel *n* ~-bremsstrahlung Röntgen-
bremsstrahlung *f* ~ coverage Röntgenstrahlen-
feld *n* ~ crystal density Röntgenstrahlenkristall-
dichte *f* ~ diffraction exposure Röntgenbeu-
gungsaufnahme *f* ~ diffraction pattern Röntgen-
beugungsbild *n* ~ diffraction picture Röntgen-
beugungsaufnahme *f* ~ diffractometer Rönt-
genbeugungsgerät *n* ~ dosimeter Röntgen-dosi-
meter *n*, -strahlenmesser *m* ~ equipment Rönt-
genanlage *f* ~ examination Röntgen-prüfung *f*,
-untersuchung *f* ~ flash interferences Röntgen-
blitzinterferenzen *pl* ~ inspection radiografische
Untersuchung *f* der Metalle ~ intensimeter

Röntgenstrahlenmesser *m* ~ iontoquantimeter
Röntgenstrahlenmesser *m* ~ levels Röntgen-
terme *pl* ~ metallography Röntgenmetallografie *f* ~ outfit Röntgeneinrichtung *f* ~ output
Strahlenausbeute *f* ~ photograph Durchleuch-
tungsbild *n*, Röntgen-aufnahme *f*, -bild *n*,
Röntgenograf *m* ~ photon Röntgenstrahlpho-
ton *n* ~ picture Röntgen-aufnahme *f*, -bild *n* ~
port Strahlenaustritt *m* ~ processing trailer
Röntgenschirmbildauswerteanhänger *m* ~ protective glass Röntgenschutzglas *n* ~ radiometer
(or roentgenometer) Röntgenstrahlenmesser *m*
~ spectrogram Röntgenstrahlenspektrogramm
*n* ~ spectrograph Röntgenspektograf *m* ~
spectrometer Röntgenspektrometer *n* ~ spectroscopy Röntgenstrahlenspektroskopie *f* ~
spectrum Röntgenspektrum *n* the ~ spectrum
spektrale Zusammensetzung *f* des Röntgen-
lichts ~ technique Röntgentechnik *f* ~ therapy
applicator Bestrahlungstubus *m* ~ treatment
Röntgenbehandlung *f* ~ tube Röntgenröhre
*f*
X's atmosphärische Störungen *pl*, Betriebsstörung *f*, Interferenz *f*
X-shaped X-geschnitten
X subzone center zweiter Zonenhauptort *m*
X-type engine X-Formmotor *m*, X-Motor *m*
xylene Xylol *n* ~ hydrochloride Xylidinchlor-
hydrat *n*
xylic acid Xylylsäure *f*
xylidine Xylidin *n* ~ hydrochloride Xylidin-
chlorhydrat *n*
xylographer Xylograf *m*
xylography Holzdruck *m*, Xylografie *f*
xylohydroquinone Xylohydrochinon *n*
xylol Xylol *n*
xylolite goods Xylolithwaren *pl*
xylolith (plastic from sawdust and sorel cement)
Steinholz *n*
xylometer Xylometer *n*
xylonic acid Xylonsäure *f*
xylonite Xylonit *m*
xyloplastic Xyloplastik *f*
xylose Holzzucker *m*, Xylose *f*

# Y

yacht Jacht *f*, Segelboot *n*
yangona acid Yangonsäure *f*
**Yankee paper machine** Selbstabnahmemaschine *f*
**Y-antenna** Delta-Antenne *f*
yard, to ~ lagern to lie in ~ during construction auf Stapel liegen
yard Elle *f*, Rahe *f*, Yard *n*, Werkgelände *n* ~ **arm** Rahnock *f* ~ **belt** Platzladeband *n* ~ **crane** Lagerplatzkran *m* ~ **foreman** Lagerplatzmeister *m* ~ **indicator** Zählwerk *n* ~ **lage** Metergedinge *n* ~ **paving** Geländepflasterung *f* ~ **radio** Rangierfunkgerät *n* ~ **stick** Maßstock *m*, Vergleichsmaßstab *m*, Yardmaß *n* ~ **storage tanks** Lagertanks *pl* im Gelände ~ **transporter crane** Lagerplatzbrücke *f* ~ **travelling crane** Hoflaufkran *m*
yardage Flächenleistung *f*
yarn Faden *m*, Flachs *m*, Garn *m*, Zwirn *m* ~ **beam** Garn-, Ketten-baum *m* ~ **brake** Fadenbremse *f* ~-**brushing machine** Garnbürstmaschine *f* ~ **carrier** Garnausgeber *m* ~-**changing attachment** Garnwechseleinrichtung *f* ~ **counter** Fadenprüfer *m* ~ **finishing** Garnappretur *f* ~ **goods** Gespinst *n* ~-**guiding groove** Fadenführungsnut *f* ~ **mender** Garnanknüpferin *f* ~ **package** Spulenfilz *m* ~ **packing** Garndichtung *f* ~ **reel** Garnhaspel *f* ~-**reel core** Garnrollenkern *m* ~ **sizing** Garnschlichtung *f* ~ **steamer** Garnverdämpfer *m* ~-**strength tester** Garnfestigkeitsprüfer *m* ~ **taker-off** Garnaufhänger *m* ~-**tentering machine** Garnstreckmaschine *f* ~ **winding machine** Garnhaspelmaschine *f* ~ **worming** Garntrense *f*
yarning chisels Strickeisen *pl*
yaw, to ~ abtreiben, gieren (naut.), scheren, wenden
yaw Gierung *f*, Kursabweichung *f* ~ **of an airplane** Flugzeugversetzung *f* ~ **and roll gyroscope** Vertikant *m*
yaw, ~ angle Windwinkel *m* ~ axis Hoch-, Gier-achse *f* ~ condition Gierzustand *m* ~ fin Seitenruder *n* ~ line Bugseitenleine *f* ~ meter Gierschlag-, Gierungs-messer *m*, Winkelsonde *f* ~ override switch Gierdämpferschalter *m* ~-roll gyro Gier/Roll-kreisel *m* ~-roll instability Abrutschinstabilität *f* ~ trim control Seitenruder-Trimmschalter *m* ~ zero pulse Gierwinkel-Nullimpuls *m*
yawed wing schiebender Flügel *m*
yawing Gieren *n*, Gierschwingung *f*, Kursschwankung *f*, Seitenbewegung *f* ~ angle Gierwinkel *m* ~ attitude Gierzustand *m* ~ conditions Gierbedingungen *pl*, Schiebeflugzustände *pl* ~ gauge designed by Anschütz Anschütz-Punkter *m* ~ moment Gier-, Kurs-, Scher-, Seiten-, Wende-moment *n* (aviat.) ~ motion in yaw Gierbewegung *f* ~ a recording gauge Punkter *m*
yawl Jolle *f*
y-axis Querachse *f*, Y-Achse *f*
Yaxley type switch Paketschalter *m*
Y chute Hosenschurre *f*

**Y-component** Hochwertkompenente *f*
**Y-connected** sterngeschaltet ~ **three-phase system with star connection** Y-Schaltung *f*
**Y connection** Sternschaltung *f*
**Y-coordinate number** Hochwert *m*
**Y-cut** Y-geschnitten
year Jahrgang *m* of last ~ vorjährig ~ of construction Baujahr *n* ~ of manufacture Fertigungs-, Herstellungs-jahr *n*
year, ~ book Jahrbuch *n* ~ counter Jahreszähler *m*
yearly jährlich ~ settlement Jahresabrechnung *f*
yeast Gest *f*, Hefe *f* ~-cultivating apparatus Hefeaufziehvorrichtung *f* ~ fermenter Hefekulturapparat *m* ~ nutrient Nährhefe *f* ~-raising apparatus Aufziehapparat *m* für Hefe ~ room Heferaum *m* ~-rousing apparatus Hefeaufziehvorrichtung *f* ~ stopper Hefepfropfen *m* ~ tank Hefebehälter *m* ~ tub Bränke *f*
yellow gelb ~ arsenic sulfide Auripigment *n*, gelbes Arsensulfid *n* ~ brass Gelbkupfer *n*, Messing *n*, Neumessing *n* ~ cadmium Kadmiumgelb *m* ~-copper ore Kupferkies *m* ~ earth Gelberde *f* ~ filter Gelbfilter *m* ~ grass-tree gum Akaroidharz *n* ~ heat Gelbglut *f* ~ lead ore Gelbbleierz *n* ~ lead oxide gelbes Präzipat *n* ~ metal Kompositionsmetall *n*, Neumessing *n* ~ mica Goldglimmer *m* ~ ocher Eisen-gilbe *f*, -glanzerz *n*, Gilbe *f* ~ oxychloride of lead englisches Gelb *n* ~ pine Gelbtannenholz *n* ~ precipitate gelbes Präzipitat *n* ~ prussiate potash gelbes Blutlaugensalz *n* ~ screen Gelb-filter *n*, -scheibe *f* ~ solution Gelbblättrigkeit *f* ~-spot disease Gelbfleckenkrankheit *f* ~ trefoil Goldklee *m* ~ tungsten bronze Wolframgelb *n* ~ uranium oxide Urangelb *n*
yellowish gelblich, gelbstichig ~ tinge Gelbstich *m*
yenite Yenit *m*
yew Eibenholz *n*, Taxusbaum *m*
**Y-fitting** Hosenstück *n*, Zweifachverteiler *m*
**Y-grouping** Sternschaltung *f*
yield, to ~ abgeben, anfallen, ausbringen, einbringen, einräumen, enthalten, ergeben, erzielen, hervorbringen, liefern, nachgeben, nachlassen, tragen, weichen to ~ an income (or a revenue) sich rentieren to ~ integers in ganze Zahlen *pl* eingehen
yield Anfall *m*, Ausbeute *f*, Ausbeutegrad *m*, Ausbringen *n*, Ausgiebigkeit *f*, Empfindlichkeit *f*, Entfall *m*, Ergebnis *n*, Ertrag *m*, Fertigung *f*, Förderung *f*, Gehalt *n*, Gewinnung *f*, Produktion *f*, Tragfähigkeit *f*
yield, ~ of crystal sugar Kristallzuckerausbeute *f* ~ of depreciation Abnutzungssatz *m* ~ of ores Metallgehalt *m* ~ of pulp Zellstoffausbeute *f* ~ of radiation Strahlungsausbeute *f* ~ by weight Gewichtsausbringen *n*
yield, ~ bending moment Fließmoment *n* ~ condition Fließbedingung *f* ~ criterion Ableitung *f* ~ deformation (or distortion) Fließverzug *m* ~ force Fließkraft *f* ~ hinge Fließgelenk *n* ~ load Fließbelastung *f*, Nachgebedruck *m* (aviat.),

Weichdruck *m* ~ **locus** Fließspannungsort *m* ~
**phenomenon** Streckgrenzenerscheinung *f*
**yield-point** Ausweich-, Bruch-punkt *m*, (according to application) Federungsgrenze *f*, Fließgrenze *f*, -punkt *m*, Nachgebepunkt *m*, Streckgrenze *f* ~ **under pressure** Quetschgrenze *f* ~ **of torsional shear** Dreh-, Torsions-, Verdrehungsgrenze *f*
**yield-point,** ~ **phenomenon** Streckgrenzeüberhöhung *f* ~ **ratio** Streckgrenzverhältnis *n*
**yield,** ~ **rectangle** Fließrechteck *n* ~ **resistance** Fließwiderstand *m* ~ **strain** Fließdehnung *f* ~ **strength** Streckgrenze *f* ~ **strength at elevated temperature** Warmstreckgrenze *f* ~ **stress** Fließ-, Streck-spannung *f* ~ **value** (of paints) Ausgiebigkeit *f*, (of paints) Ausgiebigkeitswert *m*, Fließ-grenze *f*, -punkt *m*, (of paints) Meßwert *m* für Verlauf
**yielding** Übergabe *f*; federnd, fließend, gehaltvoll, nachgebend, (rock) schneidig, weich ~ **drive** nachgebender Reibantrieb *m* oder Gleitantrieb *m* ~ **stop mechanism for control rod** nachgiebiger Regelstangenanschlag *m* ~ **voltage regulator** nachgiebiger Spannungsregler *m*
**yieldingness** Ergiebigkeit *f*, Nachgiebigkeit *f*
**yields** Ausbeutemengen *pl*
**ylem** Urplasma *n*
**Y-line of map grid** senkrechte Gitterlinie *f*
**Y-needle** Y-Nadel *f*
**yohimboa acid** Yohimboasäure *f*
**yoke** Ablenkjoch *n* (CRT), Bügel *m*, Exzenterrolle *f* (film), Gabel-gelenk *n*, -kopf *m*, Joch *n*, Kreuzkopf *m*, Magnetjoch *n*, Querhaupt *n* ~ **to frame** Rahmenaufsatz *m* ~ **of a magnet** Magnetjoch *n* ~ **of a relay** Joch *n* eines Relais
**yoke,** ~ **arbor** Jochwelle *f* ~ **armature** Jochanker *m* (Zündspule) ~ **bearing** Jochlager *n* ~ **core** Jochkern *m* ~ **driver** Mitnehmerkreuz *n* ~ **end** Gabelkopf *m* ~ **frame (or earth)** Gehäusemasse

*f* (Lichtmaschine) ~ **ring** Jochring *m* ~ **test** Bügelversuch *m*
**yolk** Eigelb *n* ~-**producing plant** Wollfettgewinnungsanlage *f*
**young** jung ~ **forest plantation** Schonung *f*, Waldschonung *f* ~ **fustic** Fisettholz *n* ~ **mountain ranges** Ketten *pl* junger Gebirge
**Young's modulus** Elastizitäts-modul *m*, -zahl *f*, -ziffer *f*
**youth model** Kindermodell *n*
**yperite** Yperit *n*
**Y-piece** (flanged) Gabel-, Hosen-rohr *n*
**Y-pipe** Hosenrohr *n*, Y-Rohr *n*
**Y-plates** Vertikalablenkplatten *pl*
**Y-punch** Zwölferlochung *f*
**Y-shaped** Y-förmig ~ **branches** halbschräge Abzweige *f* ~ **fittings** schräge Abzweige *f*
**Y-splice** Abzweigspleißstelle *f*
**ytterbia** Ytterbin *n*, Ytterbinerde *f*
**ytterbite** Ytterbit *n*
**ytterbium** Ytterbium *n* ~ **carbonate** Ytterbiumkarbonat *n*
**yttria** Ytter-erde *f*, -oxyd *n*
**yttrialite (or yttriferous)** ytterhaltig
**yttrious salt** Yttersalz *n*
**yttrium** Yttrium *n* ~ **acetate** Yttriumazetat *n* ~ **carbide** Yttriumkarbid *n* ~ **compound** Yttriumverbindung *f* ~ **oxide** Ytter-erde *f*, -oxyd *n* ~ **phosphate** Ytterspat *n*
**yttrocerite** Ytterflußspat *m*, Yttrozerit *m*
**yttrocrasite** Yttrokrasit *m*
**yttrofluorite** Yttrofluorit *m*
**yttroilmenite** Yttriolmenit *m*
**yttrotantalite** Yttrotantalit *m*
**yttrotitanite** Yttrotitanit *m*
**Y-tube** Y-Rohr *n*
**yucca** Palmlilie(nholz) *n*
**Yukawa particle** Yucon *n*
**Y-voltage** Sternspannung *f*

# Z

zantewood Fisettholz n
Zap-flap Zap-Klappe f
zapon, to ~ zaponieren
zapon, ~ thinner Zaponverdünnung f ~ varnish
Zaponlack m
zaratite Zaratit m
Z-axis, ~ modulation Helligkeitsmodulation f,
Z-Modulation f
Z-bar Z-Eisen n
zener diode Zener-Diode f
zenith Zenit(h) m ~ of trajectory Gipfelpunkt m
zenith, ~ angle Zenitwinkel m ~ distance Schei-
telabstand m, Zenit-distanz f, -entfernung f ~
ocular (or eyepiece) Zenitokular n ~ photograph
Zenitaufnahme f ~ plumbing Zenitlotung f ~
point Projektions-, Zenit-punkt m
zenithal line Scheitellinie f
zeolite Zeolit m
zephyr Zephir m
zero, to ~ nullsetzen to set at ~ in die Nullstellung f bringen to set on ~ auf Null f stellen
zero Grundzahl f, Null f, Nullpunkt m, Null-
stelle f ~ of operating range Einsatzpunkt m ~
at the scale end seitlicher Nullpunkt m
zero, ~-access storage zugriffzeitfreie Speicherung f (info proc.) ~ access-store Schnellspeicher m (info proc.) ~-address instruction Null-
adressenbefehl m ~ adjuster Nullkorrektur f,
Nullsteller m ~ adjusting Nullpunkteinstellvor-
richtung f ~-adjusting lever Einstell-, Null-hebel
m ~ adjustment Nullpunkteinstellung f ~ angle
of lift Auftriebnullpunkt m ~ axis Nullachse f,
Nullinie f ~ balance Nullage f ~ balancing
Nullkontrolle f ~ beat Schwebungslücke f,
Schwebungsnull f ~-beat frequency Schwebungsfrequenz f Null ~-beat reception Empfang
m mit schwingendem Audion im Schwebungsnull, Homodynempfang m ~ bias Nullpoten-
tial n, Nullvorspannung f ~ branch line Null-
zweiglinie f ~ calorie Regnaultsche Kalorie f
~-carrier Nullamplitude f ~ clearing Bereini-
gung f des Minimums, Minimumbereinigung f,
Minimumschärfen n ~ clearing cam Funkbeschickungskurvenscheibe f ~ clearing frame
Enttrübungsrahmen m ~ clip Nullenklammer f
~ condition Nullförderung f (hydraul.) ~-cross-
ing Nulldurchgang m ~ crossover Nulldurchgang m ~ current Nullstrom m, Strompause f ~
current indicator Nullstromanzeiger m ~ cutout
Nullausschalter m ~ delivery Nullförderung f
~ deviation Nullpunktabweichung f ~-dimen-
sional nulldimensional ~ direction Nullpunkts-
richtung f ~ dispersion Nulldispersion f ~
displacement Nullpunktverschiebung f ~ divisor
Nullteiler m ~ drift Nullpunktswanderung f ~
edge Nullkante f ~ effect Nulleffekt m ~-energy
level Nullenergieniveau n ~ error Abweichung
f des Nullpunktes, Nullpunktabweichung f ~
field emission feldlose Emission f ~ fluctuations
Nullpunktschwankungen pl ~ frequency Frequenz f Null ~-frequency component Gleichstromkomponente f ~-frequency current Gleichstrom m ~ frequency spacing Nullfrequenzab-

stand m ~ fuel weight Leertankgewicht n ~
graduation (or mark) Nullstrich m ~ graph
Zerograf m ~ heeling position Krängungsnull-
lage f ~ hour Angriffs-, Null-zeit f ~-incidence
Nullanstellung f ~ level Nullpunkt m ~ level of
radiation Strahlungsnullpegel m ~-level sensitivity Empfindlichkeit f beim relativen Pegel
Null, Nullpunktempfindlichkeit f ~ lift angle
Nullaufstiegswinkel m ~-lift angle of attack
Nullauftriebswinkel m ~-lift line Nullauftriebslinie f ~ line Grundrichtung f, Nullinie f ~ line
of the vernier Noniusnullstrich m ~ line
displacement Nullinienverlagerung f ~-line
system Kantensystem n ~ load Nullast f ~-loss
circuit verlustlose Leitung f ~-magnetostrictive
composition Nullmagnetostriktion f ~ mark
Null-marke f, -strich m ~ meridian Anfangs-
-längenkreis m, -meridian m, Null-meridian m,
-methode f, -mittagskreis m, -punktmethode f
~ method Nullmethode f ~ moment Nullmo-
ment n ~ output Nullsignal n ~ passage Nulldurchgang m ~-phase modulation Nullphasenmodulation f ~ phase setting Nullphase f ~
photocurrent Foto(zellen)strom m Null ~ place
Nullsteller m ~ point Grundrichtungs-, Null-
-punkt m, Nullpunktfehler m der Teilkreise pl
~-point clearing Nullpunktschärfung f ~-point
director Gerätnullpunkt m ~-point motion Null-
punktsunruhe f ~ position Normalstellung f,
Nullage f, Nullstellung f ~ position of space
rods Lenkernullstelle f ~ position error Null-
stellungsfehler m ~-position indicator Null-
stellungsanzeiger m ~ potential Null-potential
n, -spannung f ~ power brechkraftlose Wirkung
f ~ power-factor characteristic Belastungscha-
rakteristik f für eine Blindlast ~-power reactor
Nullenenergiereaktor m ~ prediction position
Nullstellung f des Vorhaltspiegels ~ pressure
Nullbezugsdruck m, Nulldruck m ~ range Null-
reichweite f ~ reader Universal-Fluganzeige-
gerät n ~ reading Nullablesung f ~ rest mass
Nullruhemasse f ~-return nullkernig ~ rule
Knotensatz m ~-scattering Streuung f ohne
effektive Ablenkung ~ sequence component
Nullkomponente f ~ sequence field impedance
Nullimpedanz f ~ sequence homo-coordinate
Nullkoordinate f ~ setting Eichphasenschieber
m, (plotting apparatus) Grundstellung f, Null-
linienverlagerung f, Nulleinstellknopf m, Nullstellung f ~ sharpening Minimumschärfen n
(direction finder) ~ sharpening frame Enttrübungsrahmen m ~ shift Nullpunktänderung
f ~ signal direct anode current Anodenruhestrom m ~-signal directionfinding method Minimumpeilung f ~ spring Nullfeder f ~ state Nullzustand m ~ steadystate output verschwindende
Statik f ~-suppression Entnullen n, Nullunter-
drückung f ~ terminal Nullklemme f ~-thrust
pitch (Propeller) Nullschubsteigung f ~ time
Nullzeit f ~ time reference Nullbezugszeit f
~-torque pitch Steigung f bei Drehmoment Null
~ trim Nulltrimmausschlag m ~ value Null-
wert m ~ variation Nullpunktabweichung f ~

**visibility** keine Sicht *f* **of ~ visibility** unsichtig **~ voltage** Nullspannung *f*, Nullung *f* (electr.) **~ watt speed** Nullwattdrehzahl *f* **~ wing fuel weight** (zulässiges) Fluggewicht *n* mit leeren Flügeltanks **~ yaw** Gierwinkel *m*, Null, ohne Gieren *n*
**zeroing** Nullstellung *f* **~ error** Nullpunktfehler *m*
**zeronize, to ~** auf Null stellen
**zeronizing knob (head or button)** Nulleinstellknopf *m*
**zigzag, to ~** im Zickzack *n* laufen oder verlaufen, sich zickzackförmig bewegen **~ antenna** Sägezahnantenne *f* **~ barrage** Zickzacksperre *f* **~ connection** Zickzackschaltung *f* **~ lightening** Zickzackblitz *m* **~ line** Zickzacklinie *f* **~ motion** Zickzackbewegung *f* **~ reflections** Mehrfachreflektionen *pl* **~ resistance** Zickzackwiderstand *m* **~ ribbing** Zickzackverrippung *f* **~ riveting** Zickzacknietung *f* **~ seam** Schlangennaht *f* **~ spiral filament** Wendelzickzack *m* **~ target** Zickzackscheibe *f*
**zinc, to ~-plate** verzinken
**zinc** Zink *n* **~ acetate** Zinkazetat *n* **~ alloy** Zinklegierung *f* **~-aluminum alloy** Zinkaluminiumlegierung *f* **~ ash** Zink-asche *f*, -kalk *m* **~-bearing** zinkführend **~ blende** Zinkblende *f* **~ bloom** Zinkblüte *f* **~ box** Fällkasten *m* **~ brand** Raffinatzink *n* **~ bromide** Zinkbromid *n* **~ calx** Zinkkalk *m* **~ carbon** Zinkkohle *f* **~ carbonate** kohlensau(e)res Zink *n*, Zinkspat *m* **~ case** Zinkbecher *m* **~ casting** Zinkkitt *m* **~ chloride** Zinkchlorid *n* **~ chloride resist** Zinkchloridreserve *f* **~-chromate primer** Zinkchromatgrundierung *f* **~ coat** Zinkauflage *f* **~-coated wire** verzinkter Draht *m* **~ coating** Verzinken *n*, Zinküberzug *m* **~ compound** Zinkverbindung *f* **~ contained vessel** Zinkbecher *m* **~ container** Zinkbecher *m* **~ content** Zinkgehalt *m* **~ crust** Zinkschaum *m* **~ crust formed during the Parkes process** Reichschaum *m* **~ cup** Zinkbecher *m* **~ cyanide** Zinkzyanid *n*, Zyanzink *m* **~ cylinder** Zinkbecher *m* **~ desilverization** Zinkentsilberung *f* **~-desilverization plant** Zinkentsilberungsanlage *f* **~ die-casting** Zinkspritzguß *m* **~ distillation** Zinkdestillation *f* **~-distilling furnace** Zinkdestillier-, Zinkmuffel-ofen *m* **~ distilling process** Zinkmuffelverfahren *n* **~ drawer** Zinkzieher *m* **~ drawing** Zinkziehen *n* **~ dross** Zinkasche *f*, -gekrätz *n*, -schlicker *m* **~ dust** Zinkmehl *n*, Zinkstaub *m* **~ dust discharge** Zinkstaubätze *f* **~-dust precipitation** Zinkstaubfällung *f* **~-dust purification** Zinkstaubreinigung *f* **~ electro-deposited** galvanischer Zinküberzug *m* **~ etching** Zinkätzung *f* **~-etching solution** Zinkätze *f* **~ ethyl** Zinkäthyl *n* **~ ferrocyanide** Eisenzyanürzinkoxyd *n*, Ferrozyanzink *n*, Zinkferrozyanid *n* **~ filing** Zink-feile *f*, -feilspan *m* **~ filings** Zinkschnitzel *pl* **~ foil** Zinkfolie *f* **~ formate** Zinkformiat *n* **~ founder** Zinkgießer *m* **~ fume** Zinkrauch *m* **~ furnace** Zinkofen *m* **~ granules** Zinkgranalien *pl* **~ gray** Diamant-, Galmei-, Metall-, Zink-grau *n* **~ green** Zinkgrün *n* **~ hydroxide** Zinkhydroxyd *n* **~ iodide** Zinkjodid *n* **~-lined** mit Zinkeinlage *f* **~ metallurgist** Zinkhüttenmann *m* **~ mordant** Zinkbeize *f* **~ mush** Ofengalmei *m* **~ nitrate** salpetersaures Zink(oxyd) *n* **~ ointment** Zinksalbe *f* **~ oleate** Zinkoleat *n* **~ ore** Ofengalmei *n*

**~ ore-roasting plant** Zinkerzrösthütte *f* **~ oxalate** Zinkoxalat *n* **~ oxide** Philosophenwolle *f*, Zinkblumen *pl*, Zinkoxyd *n* **~-oxide producing plants** Zinkweißanlagen *pl* **~ paint** Zinkfarbe *f* **~ perborate** Zinkperborat *n* **~ phosphide** Phosphorzink *n*, Zinkphosphid *n* **~ pipe** Zinkpfeife *f* **~ plate** Zinkblech *n*, Zinkplatte *f* **~ plate stencil** Zinkblechschablone *f* **~ plating** Verzinken *n* **~ pole** Zinkpol *m* **~ powder** Zinkmehl *n* **~ printing** Zinkdruckverfahren *n* **~-printing plate** Zinkdruckplatte *f* **~ refinery** Zinkraffinerie *f* **~ roaster** Zinkröstofen *m* **~ rod** Zinkstab *m* **~ roofing** Zinkbedachung *f* **~ salicylate** Zinksalizylat *n* **~ scum** Zinkschaum *m* **~ shaving** Zinkspan *m* **~ silicate** Zinksilikat *n* **~ sleeve** Zinkbecher *m* (electr.) **~ smelter** Zinkhüttenmann *m* **~ smeltery** Zinkhütte *f* **~ smelting plant** Zinkhütte *f* **~ solution** Zinklösung *f* **~ spar** Galmei *m*, Zinkspat *m* **~ spinel** Automolit *m* **~ stearate** Zinkstearat *n* **~ sulfate** Bergbutter *f*, weißer Vitriol *m*, Zinksulfat *m* **~ sulfide** Zinksulfid *n* **~ sulfite** Zinksulfit *n* **~ tannate** Zinktannat *n* **~ terminal** Zinkpol *m* **~ valerate** Zinkvalerianat *n* **~ vapor** Zinkdampf *m* **~ welding rods** Zinkschweißdrähte *pl* **~ white** Deck-, Zink-weiß *n* **~ white paint** Zinkweißfarbe *f* **~ works** Zinkhütte *f*, Zinkwerk *n*
**zinciferous** zink-führend, -haltig
**zincing fittings** Verzinkungseinrichtung *f*
**zincite** Rotzinkerz *n*, rotes Zinkerz *n*, Zinkit *m*
**zincked sheet iron** verzinktes Eisenblech *n*
**zincky** zinkartig, zinkisch
**zinclike** zinkähnlich
**zinco paper** Zinkopapier *n*
**zincography** Zinkografie *f*
**zincos** Zinkdruckblöcke *pl*
**zinkenite** Bleiantimonerz *n*
**zinnwaldite** Zinnwaldit *m*
**zippeite** Uranblüte *f*, Zippeit *m*
**zipper** Reißverschluß *m*
**zircon** Zirkon *n*
**zirconic acid** Zirkoniumsäure *f*
**zirconica** Zirkonerde *f*
**zirconium** Zirkon *n*, Zirkonium *n* **~ burner** Zirkonbrenner *m* **~ compound** Zirkoniumverbindung *f* **~ filament** Zirkonbrenner *m* **~ fluoride** Zirkonfluorid *n* **~ metal** Zirkoniummetall *n* **~ oxide** Zirkonerde *f* **~ steel** Zirkoniumstahl *m*
**Z-marker** Z-Funkfeuer *n* (aviat.) **~ beacon** Z-Markierungsfunkfeuer *n*
**zodiac** Tierkreis *m*, Zodiakus *m*
**zodiacal light** Tierkreislicht *n*, Zodiakallicht *n*
**zonal** zonenförmig **~ aberration** Flächenaberration *f*, Flächenabweichung *f* (opt.), Zonenfehler *m*, Zwischenfehler *m* **~ mineral** Leitmineral *n*
**zone** Bereich *m*, Bezirk *m*, Distrikt *m*, Fläche *f*, Gebiet *n*, Gebietsteil *m*, Gegend *f*, Gürtel *m*, Schicht *f*, Zone *f*
**zone, ~ of accummulation** Anhäufungszone *f* **~ of action** Regelband *n* **~ of bad weather** Wetterzone *f* **~ of contact** Eingriffs-bereich *n*, -fläche *f* **~ of crystal** Gittergerade *f*, Kristallzone *f* **~ of danger** Gefahrzone *f* **~ of dispersion** Streuungsbereich *m*, Trefferraum *m* **~ of fire** Wirkungsbereich *m* **~ of flame** Flammenzone *f* **~ of flow** Strömungsgebiet *n* **~ of incandescence** Glühzone *f* **~ of intersection** Schnittlinienzone *f* **~ of**

**negative pressure** Depressionszone *f* ~ **of
observation** Beobachtungsstreifen *m* ~ **of operations** Operations-gebiet *n*, -raum *m* ~ **and
overtime registration** Zeitzonenzählung *f* ~ **of
penetration** Eindringungsbereich *m* ~ **of pressure** Druckzone *f* ~ **of reception** Empfangsbereich *m* ~ **of reduction** Reduktionszone *f* ~ **of
regeneration** Umwandlungszone *f* ~ **of resistance** Widerstandszone *f* ~ **of silence** Leerbereich
*m*, Nullzone *f*, Totzone *f* ~ **of wind gusts** Böenband *n* ~ **of zero stress** Nulldruckfläche *f*
**zone,** ~ **adjacent to the weld** Nahtzone *f* ~ **axis**
Zonenachse *f* ~ **center** innerstaatliches Durchgangsamt *n*, Durchgangsfernamt *n*, Verteiler-
fernamt *n*, Zonenmittelpunkt *m*, Zonenhauptort *m* ~ **fire** Flächen-feuer *n*, -schießen *n* ~
**fossel** Leitfossil *n* ~ **leveling** Zonennivellierung
*f* ~ **map** Kartenblatt *n* ~ **metering** Zonenzählung *f* ~ **plate** Ringfigur *f* ~ **position indicator**
Zoneneinweisungszeit *f* ~ **punching** Zonen-
lochung *f* ~ **rate** Zonengebühr *f* ~ **registration**
Zonenzählung *f* ~ **selection** Zonenentnahme *f*
~ **selection common exit** Zonensteuerungsaus-
gang *m* ~ **selector** Zonenwähler *m* ~ **signal**
Quadrantensignal *n*, Zonensignal *n* ~ **system**
Zonensystem *n* ~ **tariff** Zonentarif *m* ~ **television** Streifenabtastung *f* ~ **time** Zonenzeit *f* ~
**vapor** Dampfphase *f*
**zoned separator** Zonenscheider *m*
**zoner** Tarifgerät *n*

**zoning** Voreinstellung *f*, Zonengliederung *f*
**zoom, to** ~ das Flugzeug *n* steil hochreißen,
hochreißen, hochziehen, senkrecht steigen
(aviat.)
**zoom** Chandelle *f*, Hochreißen *n*, schnelles Steigen *n*, Schnellschwenkung *f* (film) ~ **lens**
Gummilinse *f*, Linse *f* für veränderliche Brennweite und Bildgröße, Varioptik *f*
**zooming,** ~ **lens for variable focus and variable
magnification** Linse *f* für veränderliche Brennweite und Bildgröße ~ **up** senkrechtes Hochreißen *n*
**zorgite** Zorgit *m*
**Z-rail** Z-Schiene *f*
**Z-section** Z-Eisen *n*
**Z-time** (Greenwich Mean Time) Z-Zeit *f*
**Z-type of rocket fuel** Z-Stoff *m*
**zublin differential bit** Zahnkugelkopfmeißel *m*
Type Zublin
**zyme** Gärungsstoff *m*
**zymogenic** gärungserregend
**zymology** Gärungslehre *f*, Zymotechnik *f*
**zymometer** Gärungsmesser *m*
**zymoscope** Zymoskop *n*
**zymosimeter** Gärungsmesser *m*
**zymotechnic** zymotechnisch
**zymotechnical** zymotechnisch ~ **analysis** gärungs-
physiologische Analyse *f*
**zymotechnology** Gärungsphysiologie *f*
**zymotic** zymotisch

NOTES — NOTIZEN

NOTES — NOTIZEN